D0860225

ENCYCLOPEDIA OF PHYSICS

ENCYCLOPEDIA OF PHYSICS

SECOND EDITION

Edited by
RITA G. LERNER
American Institute of Physics, New York, New York
GEORGE L. TRIGG
Formerly of American Physical Society, Ridge, New York

VCH PUBLISHERS, INC.
NEW YORK • WEINHEIM • CAMBRIDGE • BASEL

Rita Lerner
Winding Road South
Chauncey
Ardsley, New York 10502

George L. Trigg
275 Beaver Dam Road
Brookhaven, New York 11719

Library of Congress Cataloging-in-Publication Data

Encyclopedia of physics / edited by Rita G. Lerner, George L. Trigg.—
2nd ed.
p. cm.
Includes bibliographical references.
ISBN 0-89573-752-3
1. Physics—Dictionaries. I. Lerner, Rita G. II. Trigg, George L.
QC5.E545 1990
530′.03—dc20 89-21451
CIP

British Library Cataloguing in Publication Data

Encyclopedia of physics.—2nd ed
I. Lerner, Rita G. II. Trigg, George L. (George Lockwood)
530

ISBN 3-527-26954-1

Printed in the United States of America
ISBN 0-89573-752-3 VCH Publishers
ISBN 3-527-26954-1 VCH Verlagsgesellschaft

Printing History:
10 9 8 7 6 5 4 3 2 1

Published jointly by:

VCH Publishers, Inc.
220 East 23rd Street
Suite 909
New York, New York 10010

VCH Verlagsgesellschaft mbH
P.O. Box 10 11 61
D-6940
Federal Republic of Germany

VCH Publishers (UK) Ltd.
8 Wellington Court
Cambridge CB1 1HZ
United Kingdom

CONTENTS

PREFACE TO THE SECOND EDITION

Like the first edition, this Encyclopedia describes physics only as of a particular moment in time. In the ten years since that edition appeared, there have been significant changes in many areas of physics.

Probably the most publicized has been the discovery of materials that become superconductors of electricity at temperatures much higher than previously known. But other developments have been of at least equal importance within physics, if not so striking to the rest of the world. Fractal geometry has been developed to describe irregular shapes and forms, and has been applied to describing physical phenomena; their beauty is apperant to all. Chaotic theory, first explored early in this century, is being used to describe dynamical systems, especially random ones. The study of critical phenomena has been enlarged by the inclusion of dynamic aspects.

Our understanding of the solid state has extended to structures with such names as intermediate-valence compounds and heavy-fermion materials. Scanning tunneling microscopy has led to new insights in surface physics. An established effect, the Hall effect, has been found to have intriguing quantum aspects which have increased its potential for application to measurement standards.

The appearance of Supernova 1987a has excited astronomers and stimulated new theoretical work in astronomy and cosmology. Theoretical physicists have continued their attempts to derive a grand unified theory linking the strong, weak, and electromagnetic interactions and perhaps, through a so-called supersymmetry, to link them to gravitation. And arrangements known as cellular automata have moved computers a small step closer to mimicking some of the brain's operations.

As in the previous edition, articles are arranged in alphabetical order, with some titles inverted in order to group related articles near each other. At the end of most articles, cross-references are given to related articles in this volume. Since it is not possible in a volume of this size to explore any one subject in depth, most of the authors have provided references for further reading; these are usually labeled E, I, or A indicating that the reference is elementary, intermediate, or advanced.

The editors are grateful for the cooperation of many friends and colleagues for their contributions to this edition, and for suggestions and comments. We wish to thank Dr. Martin Grayson of VCH Publishers, Inc., whose interest and enthusiasm in the project have been most helpful, and Barbara Chernow of Chernow Editorial Services for her patience and competence in dealing with manuscripts from more than five hundred authors.

RITA G. LERNER
GEORGE L. TRIGG

MAIN ENTRIES

Absorption Coefficient

F. L. Galeener

The absorption coefficient α measures the spatial decrease in intensity of a propagating beam of waves or particles due to progressive conversion of the beam into different forms of energy or matter. Absorption usually implies the creation of some form of internal energy in the traversed medium, e.g., the production of heat; however, it may also be associated with other inelastic scattering events, such as the ultimate conversion of incident particles into new types, or the change in frequency of waves from their incident values. Removal of intensity from the beam merely by diversion into new directions is called elastic scattering, and this process is not properly included in the absorption coefficient.

The extinction coefficient α_e measures the reduction in beam intensity due to *all* contributing processes and is often represented as a sum $\alpha_e = \alpha + \alpha_s$, where α_s is the coefficient associated with elastic scattering. Additional subdivisions of processes that remove intensity from the beam are possible and sometimes used.

These coefficients appear in the Bouguer or Lambert–Beer law in the form $I(x) = I(0)e^{-\alpha_e x}$, where $I(x)$ is the beam intensity after it has traveled a distance x in the medium, while α_e, α, and α_s have units of inverse length, often written cm^{-1}.

The absorption coefficient α appears frequently in discussions of the optical properties of homogeneous solids, liquids, and gases, where α may be a strong function of the wavelength of the light involved, the temperature, and various sample parameters. The theory of electromagnetic waves relates α to the complex permittivity ϵ and permeability μ of the medium.

See also ELECTROMAGNETIC RADIATION.

BIBLIOGRAPHY

M. Born and E. Wolf, *Principles of Optics*, Chapter 13. Pergamon, New York, 1959. (A)

F. A. Jenkins and H. E. White, *Fundamentals of Optics*, 3d ed., Chapters 22 and 25. McGraw-Hill, New York, 1957. (E)

R. B. Leighton, *Principles of Modern Physics*, Chapter 12. McGraw-Hill, New York, 1959. (I)

Accelerators, Linear

E. A. Knapp

INTRODUCTION

Linear accelerators, or "linacs," constitute a rather restricted class of accelerators that accelerate nuclear or subnuclear particles in a straight line by means of a series of small impulses, each impulse being considerably smaller than the final overall energy gain achieved. Conceptually simple, the linac only recently has been developed to the same extent as other particle-accelerator systems. Efficient, high-powered microwave power sources were required before linac technology could progress. In general, the individual impulses involved in acceleration in the linac are produced by an oscillating electric field formed in radio-frequency resonant cavities. The cavities are usually coupled in such a way that the velocity of the accelerated particle traveling through them matches the time of electric field maxima in the chain or string of cavities as the particle energy increases. Figure 1 shows schematically a series of resonant cavities, each excited in the TM_{010} electromagnetic mode and arranged with an electric field distribution on axis suitable for particle acceleration.

Basic Principles and Equations

The linac is a particle accelerator that operates on the principle of phase stability, applicable to a variety of accelerator types. In the case of linacs, the phase stability is conceptually particularly simple. In analogy to riding on the leading edge of a wave, if a particle lags behind the proper or synchronous phase ϕ_s, it receives additional acceleration and "catches up" to the proper point; if ahead on the wave, it falls back toward the proper point. A particle that starts away from the synchronous phase or energy oscillates in phase about the synchronous phase and in energy about the synchronous energy, tracing out an elliptical trajectory in the phase–energy plane. Figure 2 shows a typical stability diagram in energy and phase for a proton linear accelerator. Associated with this phase stability is a radial defocusing of the accelerated particles by the radio-frequency electric fields. In early linacs this defocusing was overcome by electric field distortions produced by grids in the beam aperture. All modern proton and electron linacs use quadrupole magnets or solenoidal magnetic fields for radial beam confinement.

The requirements for high-frequency microwave power in linear accelerators make the generators of this power the major consideration in the system design. In fact, the lack of viable high-frequency rf power sources delayed the development of linacs substantially. A quantity relating acceleration achieved to power dissipated is the shunt impedance (ZT^2), conventionally defined as

$$ZT^2 = \frac{V^2}{P \cdot L} \qquad (1)$$

where V is the maximum energy gain in a length L and P is the rf power dissipated in that length L.

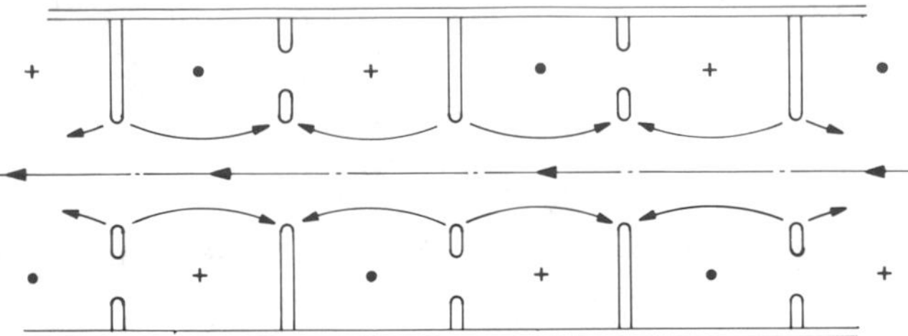

FIG. 1. Series of resonant cavities, each excited in the TM_{010} electromagnetic mode and arranged with an electric field distribution on axis suitable for particle acceleration. Energy coupling from cavity to cavity is via slot in wall.

Typically, ZT^2 varies from about 25 to 100 MΩ/m for typical systems, indicating megawatt power requirements for reasonable cavity lengths and energy gains. Klystrons, superpower triodes, and magnetrons are the only developed microwave sources available in this power-level region.

Linacs are classed as traveling-wave or standing-wave in regard to the behavior of the coupling between the cavities providing the impulses and the rf phase shift per cavity in operation. No fundamental difference in the principles of operation exists between the two classes of accelerators. In the case of traveling-wave linacs, the coupling between adjacent cavities results in a phase shift that is not 180° but some intermediate value, a very strong function of driving frequency. At the end of a chain of cavities, the power left over is coupled out into a load resistor and no power reflection to the input occurs. Cavity lengths and phase shifts are adjusted to match the crest of the electric field wave passing down the chain of cavities to the particle as it also passes through the cavity chain.

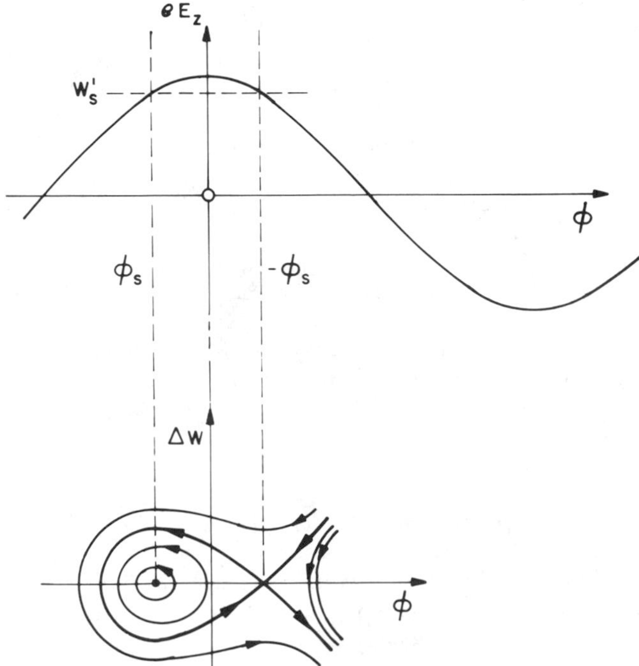

FIG. 2. Energy and phase relations for proton linear accelerator; W is the particle energy, ΔW the deviation from synchronous energy, ϕ the particle phase, ϕ_s the synchronous phase.

In the case of the standing-wave accelerator structure, the coupling between adjacent cavities is very strong. There is a strong reflection of energy from each end of the cavity chain, and the electric fields in adjacent high-Q cavities are constrained, in the limit of zero loss conditions, either to oscillate in phase, to oscillate 180° out of phase, or to be unexcited. This constraint restricts the accelerator designer in his choice of cavity lengths, but provides simplification in control that dramatically improves system design for certain applications. Almost all early electron linacs were of the traveling-wave variety; the side-coupled electron linacs and all proton linacs are of a standing-wave design. In the case of the side-coupled linac structure, an unexcited cavity is moved off axis in order to improve efficiency while retaining some major stability features of the cavity mode, which has alternate cavities unexcited at resonance.

ELECTRON LINEAR ACCELERATORS

Development work on early electron linear accelerators was done at Massachusetts Institute of Technology and Stanford in the late 1930s. Both standing-wave (MIT) and traveling-wave (Stanford) designs were built, but in the 1950s it became apparent that the traveling-wave design was in many ways superior to the resonant-cavity design. The Stanford two-mile linear electron accelerator (SLAC) is the largest electron linac built and has achieved 20 GeV of energy with an average current of 50 μA. This machine has proved to be very reliable and has had an extremely productive physics history since its completion in 1964. Recently a new standing-wave accelerator structure, the side-coupled system, has been applied to short, lower-energy linacs for industrial and medical use. Somewhat more efficient than the older traveling-wave systems, this structure has allowed major simplifications in accelerator system design that have reduced costs of modern linac systems substantially and made electron linear accelerators widely available for modern radiotherapy treatment of cancer.

PROTON LINEAR ACCELERATORS

The first proton linear accelerator was built at the University of California at Berkeley and first operated in 1946. This accelerator was a "drift-tube" accelerator, built as a single, long, resonant tank with drift tubes hanging inside an outer copper shell. The drift tubes produce a field distribution similar to that produced by a chain of cavities with zero phase shift between them, as described previously. The accelerated particle is effectively "hidden" from the cavity electric fields while inside the drift tube during field reversal. This is a standing-wave accelerator. The largest existing proton accelerator is the LAMPF (Los Alamos Meson Physics Facility) proton linac at the Los Alamos Scientific Laboratory of the University of California. It is capable of accelerating a 1-mA (average current) proton beam to a final energy of 800 MeV. This accelerator consists of a drift-tube linac with quadrupole-magnet focusing and operates in a resonant mode modified by post couplers to an energy of 100

MeV, followed by a side-coupled accelerator structure to accelerate the protons from 100 to the 800-MeV final energy. Simultaneous acceleration of both H^+ and H^- beams to full energy is accomplished in this accelerator. Proton linear accelerators are used as injectors for many large synchrotrons in high-energy physics establishments throughout the world.

HEAVY-ION LINACS

The drift-tube linear accelerator is also used to accelerate heavy ions to high energy. The hilacs at Berkeley and Yale, built around 1952, were the first heavy-ion linear accelerators. The upgraded hilac at LBL is still producing excellent particle beams for research and serves as an injector for the Bevalac project, a synchrotron for heavy-ion acceleration. In the case of heavy-ion acceleration, very low velocities are necessary at the beginning of the accelerator resonator, requiring strong magnetic field focusing and a relatively low frequency of excitation for the system. The LBL superhilac has achieved acceleration of ions to an energy of 8.5 MeV/amu.

Recently a new heavy-ion accelerator utilizing a variety of accelerator cavity schemes and/or modes was completed at Darmstadt, West Germany. Called the Unilac, it is capable of accelerating up to an energy of 7.5 Mev/amu.

NEW INITIATIVES IN LINAC TECHNOLOGY

Recently extensive research and development work has been directed toward applying superconductivity to the resonators of linear accelerator systems. Superconductivity would allow linacs to be built whose cavities dissipate very little power, yielding highly efficient systems with the capability of continuous operation. Recently, the application of ultrapure niobium material and the judicious choice of cavity shape have led to much improved cavity performance for superconducting linac systems. Accelerating gradients of over 10 MV/m have been achieved in multicavity systems, and single cavity performance of over 20 MV/m has been achieved. A major construction project, CEBAF, is utilizing superconducting cavity linac structures.

The most dramatic new development in proton or heavy-ion linac systems is the Radio Frequency Quadrupole linac, or RFQ. This device uses a quadrupole electric field configuration established in a resonant cavity loaded with four especially shaped vanes to both focus and accelerate a low-velocity ion beam. Due to the nature of the accelerating field configuration and the focusing forces very careful control of the beam-disrupting influences in the system may be maintained. Very superior performance of low-velocity beams has been achieved, and this system is rapidly gaining total acceptance in the low-velocity section of proton and other ion accelerators.

At present, linear accelerator technology is in a state of rapid flux. Applications of the radiations produced by particle accelerators are increasing, and commercial and medical uses of these accelerators are increasing at a rapid rate. Reliable, low-cost sources of radiations for a variety of purposes, from research to commercial production, will be required and are available from linac systems.

See also MICROWAVES AND MICROWAVE CIRCUITRY.

BIBLIOGRAPHY

M. Stanley Livingston and John P. Blewett, *Particle Accelerators*. McGraw-Hill, New York, 1962.

Linear Accelerators (Pierre M. Lapostolle and Albert L. Septier, eds.). North-Holland, Amsterdam, 1970.

Accelerators, Potential-Drop Linear

R. G. Herb and G. M. Klody

INTRODUCTION

Potential-drop accelerators were the first to open up nuclear physics to extensive experimentation. Despite the higher-energy beams that were soon available with the cyclotron, betatron, and later accelerators that provide very high energies through many small-energy increments, potential-drop accelerators have continued to play an important, and often dominant, role in nuclear physics. In addition, potential-drop accelerators have a broad range of other applications in many, diverse fields in industry, research, and medicine.

Potential-drop accelerators consist of a high-voltage terminal supported by an insulating column, a means of generating the high voltage, and an acceleration tube. To reduce their size, many potential-drop accelerators use high-pressure gas for electrical insulation. The electrostatic potential drop between the high-voltage terminal and ground accelerates charged particles to an energy equal to the charge times the terminal voltage ($E = qV$). A variety of ion and electron sources are available to produce the beams of charged particles for acceleration.

The acceleration tube is an insulating assembly, mounted between the high-voltage terminal and ground potential, through which the beam is accelerated. High vacuum inside the tube minimizes beam losses. Mechanical strength is also important for tubes in accelerators insulated with high-pressure gas. The tube design that maximizes voltage-holding capability and voltage stability under these conditions is a laminated series of insulating rings and metal electrodes. The electrodes are connected to a resistive voltage divider from the terminal to ground for a uniform voltage gradient (linear potential drop) along the tube.

Single-stage potential-drop accelerators can accelerate ions or electrons. Two-stage accelerators, called tandems, are for ion acceleration. Tandems accelerate negative ions from ground potential to a positively charged high-voltage terminal, change the ionic charge from negative to positive in the terminal, and accelerate the ions through a second acceleration tube, back to ground potential. For a given terminal voltage, tandems produce higher-energy ions than single-stage accelerators, but usually with less beam current.

Potential-drop accelerators are generally distinguished by the type of generator used to charge the high-voltage terminal. Each high-voltage generator design has advantages for specific applications. For precise measurements of nuclear-energy-level characteristics and many analytical techniques, the electrostatic accelerator is far superior. For certain neutron-induced reactions the Cockcroft–Walton-type accelerator may be most convenient. Needs for intense positive-ion or electron beams of a few MeV energy are most easily met by the Dynamitron.

Modern potential-drop accelerators are very reliable and simple to operate. This has greatly expanded their range of applicability in many fields. They are routinely used for ion implantation in the manufacture of semiconductor devices, nondestructive analysis of materials, surface hardening to reduce corrosion and wear, polymerization, sterilization, pollution monitoring, age determination of samples for archaeology, cosmology, and oceanography, and analysis of biomedical specimens.

Each of the accelerators described below is serving an important need in science and technology, and each is playing an a expanding role.

COCKCROFT-WALTON VOLTAGE MULTIPLIER ACCELERATOR

This accelerator was the first to be used successfully for nuclear transmutation and gained wide recognition when results were published in 1932. It employs a circuit developed by H. Greinacher in 1920, as illustrated in Fig. 1, which utilizes two stacks of series connected capacitors. One capacitor stack is fixed in voltage except for voltage ripple with one terminal connected to ground and the other to the load which, in this case, is an evacuated accelerating tube.

One terminal of the second capacitor stack is connected to a transformer giving peak voltages of $\pm V$, and voltages at all points along the second capacitor stack oscillate over a voltage range of $2V$. The two capacitor stacks are linked by series-connected rectifiers. As voltage on the second stack oscillates, charge is transferred stepwise from ground to the high-voltage terminal. Voltage here is steady except for ripple caused by power drain and stray capacitance. Its value is approximately $2VN$, where N is the number of stages. The power supply furnishes power to an evacuated accelerating tube equipped with an ion source. Usually the tube and ion source are continually pumped and the multisection tube may have one tube section per accelerator stage.

FIG. 2. 850-keV Haefely accelerator which serves to inject pulses of hydrogen ions into a linear accelerator at the Los Alamos National Laboratory.

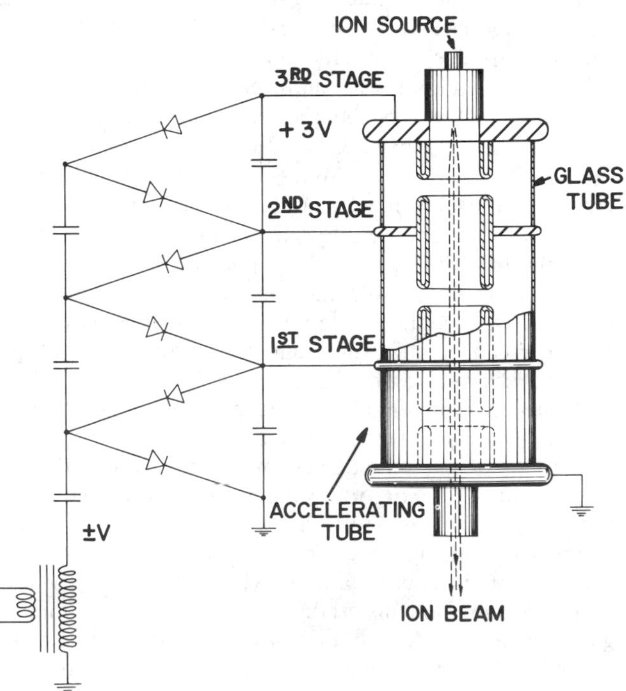

FIG. 1. Schematic drawing of a Cockcroft–Walton accelerator utilizing three stages.

Cockcroft–Walton accelerators operating in open air at 1 million volts are large in size and must be housed in a very large room to avoid voltage flashover. Figure 2 shows an 850-keV accelerator of this type built by Emile Haefely & Co. Ltd. It serves to inject pulses of hydrogen ions into a high-intensity linear accelerator at the Los Alamos National Laboratory for production of intense meson beams.

Size and space requirements increase rapidly as voltage is extended above 1 million volts in open air and the practical upper voltage limit for these accelerators appears to be about 1.5 MV.

G. Reinhold of Emile Haefely & Co. Ltd. has shown that the terminal voltage developed by multipliers, using the circuit of Fig. 1, does not depart greatly from $2VN$ for multipliers going to a few hundred kilovolts. However, above

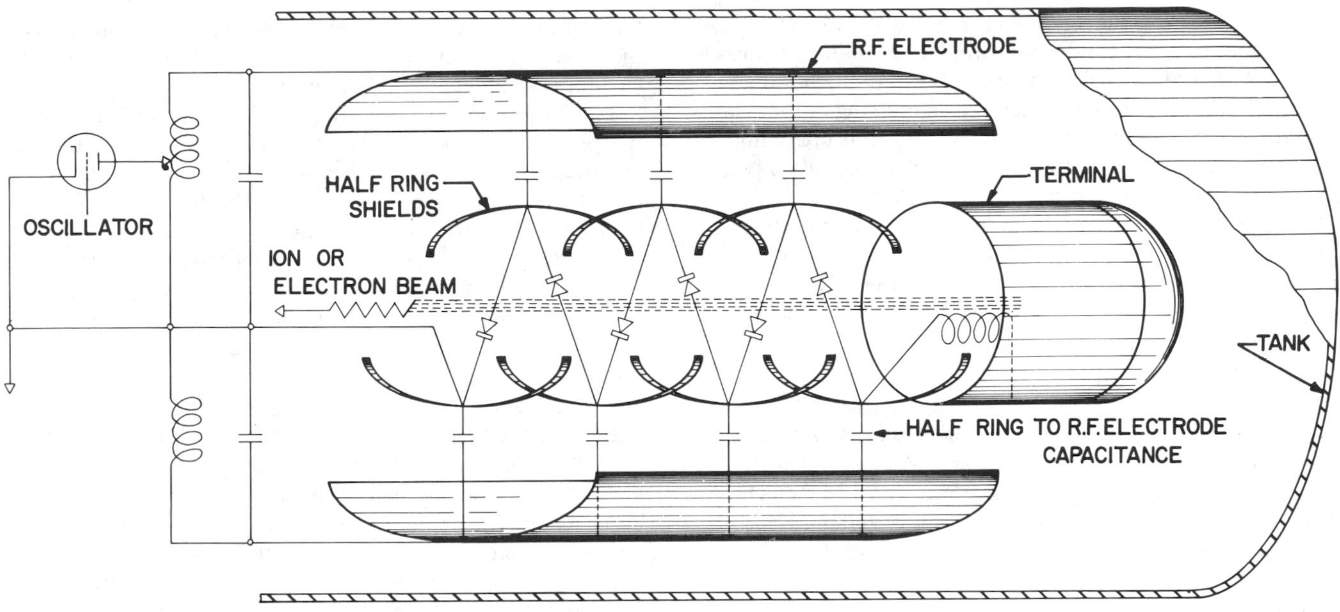

FIG. 3. Schematic diagram of a Dynamitron accelerator in a pressure tank.

about 500 kV, voltages achieved fall substantially below values given by this simple expression because of stray capacitances and other effects. He has developed another circuit called a symmetrical cascade rectifier in which the shortcomings of the simple circuit are largely eliminated. It employs two transformers and two capacitor stacks that oscillate in voltage. Both feed one fixed capacitor stack. Using this symmetric system, Emile Haefely & Co. Ltd. has built an open air rectifier without an accelerating tube going to 2.5 million volts.

The N. V. Philips Gloeilampenfabrieken has also manufactured accelerators with Cockcroft–Walton charging.

Emile Haefely & Co. Ltd. has built accelerators utilizing voltage multipliers housed in tanks containing insulating gases such as SF_6 or a mixture of N_2 and CO_2. These machines have ranged in voltage from about 1 million up to 4 million volts. Nisshin–High Voltage Company also builds SF_6-insulated Cockcroft–Walton accelerators from 500 kV to 3 million volts for their industrial electron processing systems. The use of insulating gases at high pressure permits great savings in the size of equipment for a given voltage.

DYNAMITRON ACCELERATOR

M. R. Cleland invented a cascaded rectifier system termed the Dynamitron in which series connected rectifiers are driven in parallel. The circuit is shown schematically in Fig. 3, and Fig. 4 is a photograph of a 4-million-volt positive ion Dynamitron accelerator.

Rectifiers connected in series between ground and the high-voltage terminal are positioned in two columns on opposite sides of the accelerating tube of the Dynamitron and the high-voltage column is enclosed by half rings that have a smooth exterior surface to inhibit corona and spark discharge. The half-rings serve as capacitor plates coupled capacitively to the large semicylindrical rf electrodes positioned between the walls of the tank and the high-voltage column of the accelerator. The rf electrodes form the tuning capacitance of an *LC* resonant circuit which is driven by a separate power supply through an oscillator tube.

Since ac power is fed in parallel to each of the series-connected rectifiers, the relatively large storage capacitors that are connected between successive stages of other cascaded rectifier systems are not required. Stored energy in the Dynamitron is low and does not differ greatly from that in electrostatic machines. This feature is important since damage due to discharge can be a serious problem in multimillion volt accelerators, especially for discharge in the accelerating tubes.

These accelerators are enclosed in pressure tanks containing high-pressure SF_6 gas. At a pressure of 1 atm, this gas has a dielectric strength about 2.7 times that of air at the

FIG. 4. A 4-MeV Dynamitron positive-ion accelerator with pressure tank rolled away from the high-voltage column.

same pressure. Its dielectric strength rises approximately linearly with pressure up to a few atmospheres and at a pressure of about 7 atm it will sustain fields of about 200 kV/cm.

Radiation Dynamics manufactured many single-stage and double-stage (tandem) positive-ion Dynamitrons operating at terminal potentials up to 4 MV. They are especially advantageous for applications requiring high currents.

A larger proportion of Dynamitrons manufactured have been electron accelerators for industrial applications such as polymerization of plastics and sterilization of disposable medical products. These applications require electron energies up to a few MeV and from 10 to 100 kilowatts of electron beam power.

The most powerful machines built by Radiation Dynamics include a 0.5-MV machine with an electron beam power of 50 kW, a 1-MV machine giving 100 kW of electron beam power, a 1.5-MV machine providing 75 kW, a 3-MV machine providing 150 kW, and a 5-MV machine providing electron beam power of 200 kW.

VAN DE GRAAFF ACCELERATOR

In this accelerator, high voltage is generated by means of an insulating belt which carries charge from ground to the high-voltage terminal. Robert Van de Graaff built the first successful belt-charged high-voltage generator in 1929.

In this device, which is illustrated in Fig. 5, electrical charge is deposited on an insulating, motor-driven belt and is carried into a smooth, well-rounded metal shell which is shown in the figure as a sphere. Here charge is removed from

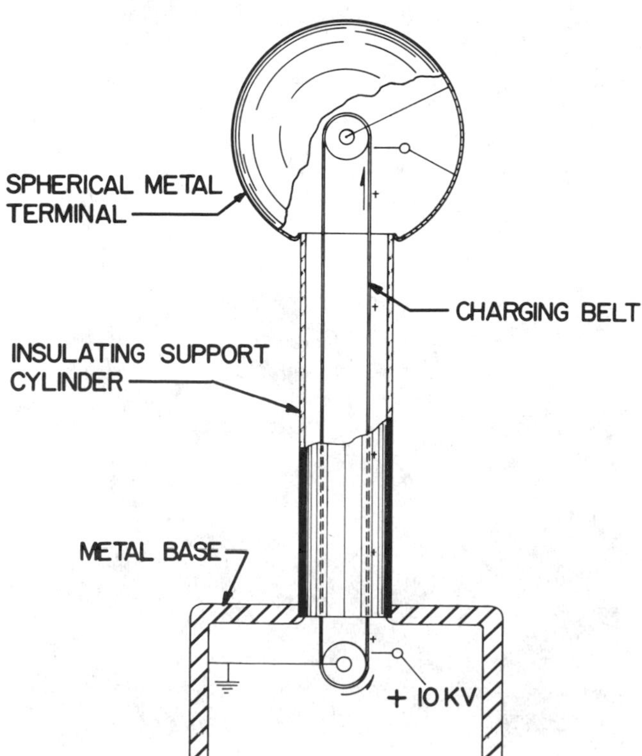

FIG. 5. Schematic drawing of a Van de Graaff generator for operation in atmospheric air.

the belt and passes to the sphere which rises in voltage until the sphere is discharged by a spark or until the charging current is balanced by a load current.

To charge the belt, a corona discharge is maintained between a series of points or a fine wire on one side of the belt and the grounded lower pulley or a well-rounded, grounded, inductor plate on the other side of the belt. If the corona needles are at a positive potential, the belt intercepts positive ions as they move from needle points toward the grounded pulley. Charge is carried into the sphere where it is removed by an array of needle points and passes to the outer surface of the sphere. The generator can provide a high negative voltage if the corona needles are operated at a negative voltage with respect to the grounded pulley.

Belt charging electrodes must be well shielded from the field of the high-voltage terminal and charge must be carried well within the sphere before removal is attempted. Charging current is then completely independent of voltage on the terminal and voltage will rise until limited by corona or sparkover to ground, by leakage current along insulators, or by a load such as ion current through an accelerating tube.

The Van de Graaff belt-type generator was first used to accelerate ions in 1932 at the Department of Terrestrial Magnetism of the Carnegie Institution of Washington, D.C. A machine completed in 1934 at this institution was used extensively for research in nuclear physics and is now on display at the Smithsonian Institute of Washington, D.C.

Open-air belt-type accelerators for 1 million volts or more are large and require a very large enclosure. At Wisconsin this accelerator was adapted to a pressure tank and by 1940 a belt-charged accelerator was operated successfully up to 4.5 million volts.

These machines insulated by high-pressure gas were manufactured from 1946 by the High Voltage Engineering Corporation for accelerating of electrons and for positive ions. In 1988, Vivirad High Voltage Corporation purchased the Accelerator Division of High Voltage Engineering Corporation to build accelerators, for research and materials analysis. High Voltage Engineering–Europa also manufactures accelerators with Van de Graaff belt-charging to about 2 million volts, primarily for ion-implantation applications.

Two-Stage Accelerators

In 1958, the High Voltage Engineering Corporation completed a machine for acceleration of negative ions as illustrated schematically in Fig. 6. Negative hydrogen ions from a source developed for this purpose are accelerated as they pass from ground to the terminal which is at a high positive voltage V. Here they are stripped of both electrons as they pass through a very thin foil or through a small-diameter tube containing adequate gas. The protons are again accelerated as they pass from the terminal to ground and they emerge from the accelerator with an energy of $2Ve$. These machines, which give two stages of acceleration, are called tandems.

Atoms of a large proportion of the elements form stable negative ions. A negative oxygen ion gains an energy of 10 MeV as it goes from ground to the terminal of an accelerator operating at 10 million volts. Here the oxygen atoms may be stripped of all of their eight electrons. The oxygen nuclei

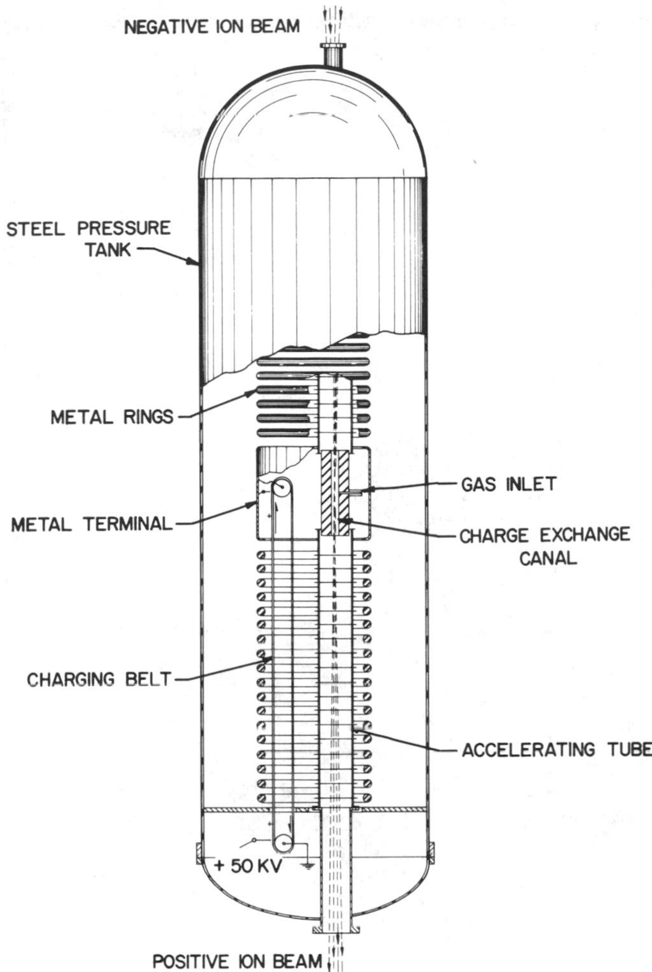

FIG. 6. Schematic drawing of a tandem electrostatic accelerator in a pressure tank.

gain 80 MeV as they pass to ground and they emerge from the machine with an energy of 90 MeV.

The High Voltage Engineering Corporation manufactured a large number of two-stage accelerators, which are in use in laboratories throughout the world. The largest reach terminal voltages of over 16 million volts. Many are used for acceleration of heavy ions for research and materials analysis applications.

Nisshin–High Voltage Company also manufactures tandem accelerators with Van de Graaff belt-charging to 3 million volts for basic research and materials modification and analysis.

TANDETRON ACCELERATORS

In 1980, K. H. Purser began development of the Tandetron, a line of compact tandem accelerators designed to meet the requirements for applications in materials analysis, accelerator mass spectrometry, high-energy implantation, and basic research. An MeV implanter which uses a Tandetron accelerator is shown schematically in Fig. 7.

The high-voltage generator in the Tandetron is a power supply with a parallel-driven cascaded rectifier circuit similar to that of the Dynamitron. Using silicon rectifiers and driven at high frequency (about 50 kHz), the power supply delivers several milliamperes of charging current at up to 3 million volts with high stability and negligible terminal voltage ripple. For accessibility, the high-voltage stack is mounted at right angles to the accelerator column, rather than being built into the column (compare Fig. 3). The accelerator tank contains SF_6 gas at about 500 kPa (5 atm) to insulate the accelerator and the power supply.

Through 1988, General Ionex Corporation manufactured a large number of Tandetron systems, many with specialized accessories that they developed for materials analysis and modification applications. Genus Corporation now manufactures Tandetron systems.

PELLETRON^{TM} ACCELERATORS

Pelletron accelerators use a charging chain, rather than a belt or power supply, to generate high voltage on the accelerator terminal. The chain consists of steel cylinders (pellets) joined by links of solid insulating material such as nylon (Fig. 8). The chain is intrinsically spark-protected (undamaged in test sparks at over 30 million volts). Pellet-charging current is adjustable and highly uniform, so that terminal voltages are easily maintained at very precise values with very little voltage ripple. This is advantageous for many measurements in nuclear physics and is usually required in other applications. Many belt-charged machines have been converted to Pelletron charging.

The pellets are charged inductively, so there is no contact with the charging electrodes (inductors). For a positive terminal voltage, positive charge is induced on the pellets at a motor-driven pulley at ground potential. This charge is removed at a pulley in the terminal, and the pellets are then negatively charged. Thus, the up-going and down-going runs of the chain contribute equally to charging. Reversing the inductor voltage polarities gives a negative terminal voltage.

Pelletron accelerators are manufactured by the National Electrostatics Corporation. Single-stage Pelletrons have either an electron source or a positive-ion source in the terminal. Pelletrons manufactured by this company for industries and laboratories around the world range in voltage from 1 to 25 million volts, including the largest potential-drop accelerator, the 25-million-volt tandem at the Oak Ridge National Laboratory (described below).

Most Pelletrons above 4 million volts are vertically oriented for simplicity of support. The high-voltage column, which supports the terminal, charging chain, and acceleration tubes, is an assembly of standard column modules, each module holding 1 million volts. Most smaller Pelletrons, built for applications requiring only 4 million volts or less, use a simple, inexpensive cast acrylic support column.

Acceleration tubes in Pelletrons have ceramic insulators and titanium electrodes bonded together with a metal for organic-free, ultra-high-vacuum operation. This is important for reliability at high voltage and for contaminant-free vacuum. Earlier tube designs have glass insulators which are bonded to the metal electrodes with organic glues.

Large Pelletrons (8 to 25 million volts) are in operation on five continents. They are used principally for basic research

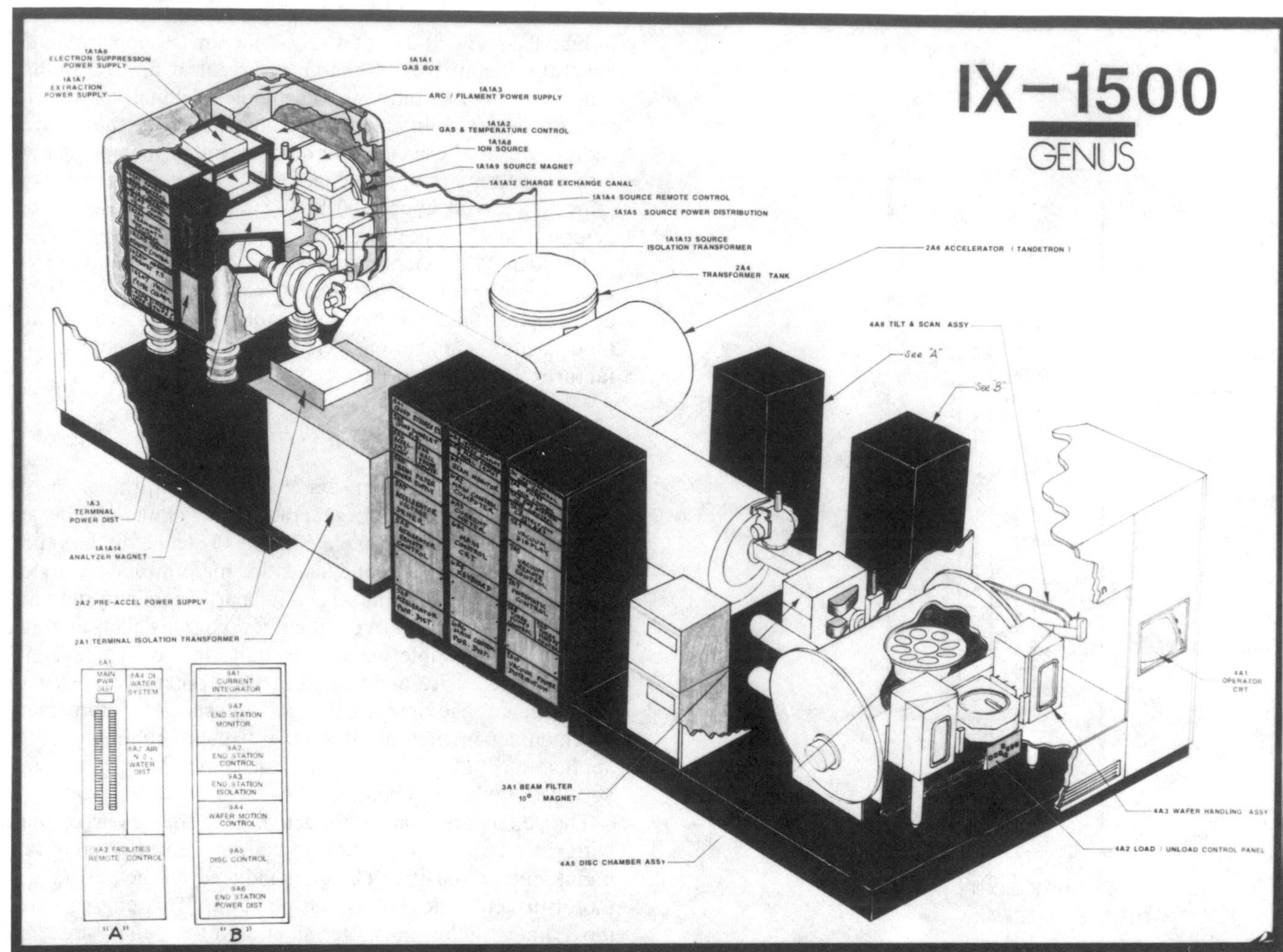

FIG. 7. Schematic diagram of a Tandetron accelerator in an ionimplant system. The high-voltage solid-state multiplier stack is at right angles to the tandem accelerator column, and the high-frequency driver is outside the tank for accessibility.

with heavy ions. Most Pelletrons are smaller, to 5 million volts, and are used primarily for applications such as production ion implantation, industrial materials analysis, accelerator mass spectroscopy, and biomedical analysis.

Folded Tandems

Folded tandems are like straight-through tandems (see Two-Stage Accelerators, above), except that both acceleration tubes are in a single insulating column with a 180° magnet in the terminal to steer the beam from one tube into the other. Figure 9 shows the folded 25-million-volt tandem at the Oak Ridge National Laboratory. The steel tank, which contains about 700 kPa (7 atm) of SF_6 gas for high-voltage insulation, is large (30 m high and 10 m in diameter). The straight-through design, however, would be much taller, because it has insulating columns above and below the high-voltage terminal plus an ion-beam injector above the tank (compare designs in Figs. 6 and 9).

The folded design, first successfully used by Naylor in New Zealand, reduces the costs for the tank, SF_6, and building for very high-voltage tandems. For higher-energy ion beams, Oxford University converted their single-stage accelerator to a folded tandem. General Ionex Corporation manufactured several 660-kilovolt folded tandem systems for materials analysis by Rutherford backscattering. Not only does the folded design give a compact system, but it also locates both the ion source and sample chamber near the system control panel.

With +25 million volts on the terminal, negative ions injected from below the accelerator (see Fig. 9) are accelerated up to the terminal to an energy of 25 MeV, where they are stripped of some or all of their electrons. Stripping produces ions in a range of different charge states, and the energy gained in the second acceleration depends on which charge state is selected. If the magnet in the terminal is set for ions of charge +15, these ions gain 15 × 25 = 375 MeV in the second acceleration for a total energy of 400 MeV. A second stripper, one-third of the way down the second acceleration tube, can increase the charge to +34 to give ions with a final energy of over 700 MeV. The terminal magnet completely filters out all undesired components from the beam before

FIG. 8. Charging chain and pulley.

the second acceleration, so there is no extra beam loading, and ion transmission is straightforward.

The 25-million-volt accelerator column consists of 1-million-volt modules which use the same components as in smaller Pelletrons. Six charging chains generate over 600 μA of current to the terminal. Service platforms inside and outside the column give complete access for maintenance without column disassembly. The computer-based control system uses light link telemetry to transmit over 200 control and monitoring signals to components at high-voltage locations in the accelerator column and terminal.

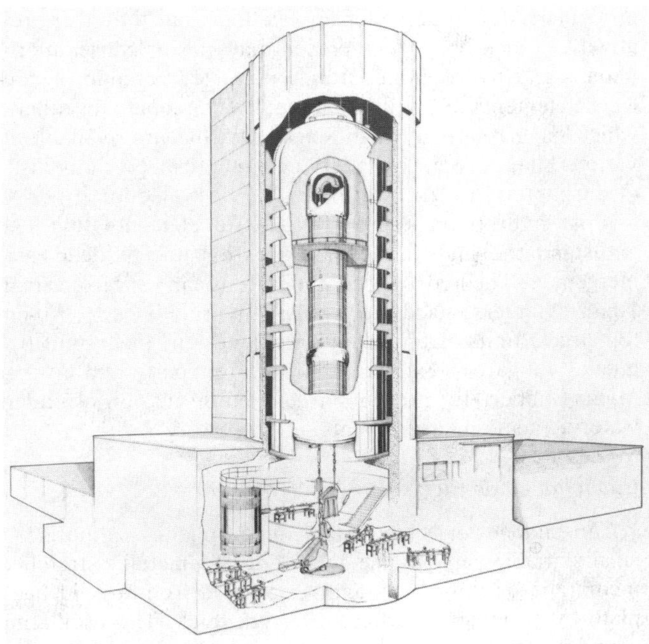

FIG. 9. Drawing of 25-MV folded tandem. Tank is approximately 30 m high and 10 in diameter. Service platforms are shown in use inside and outside the column.

The ion beam injector at Oak Ridge (cylinder in lower left of Fig. 9) operates at −500 kV in air to preaccelerate ions from the negative-ion sources for injection into the accelerator. The 90° magnets below the accelerator provide high resolution of the injected ion mass and high resolution of the accelerated beam energy.

The other large folded tandem is the 20-million-volt Pelletron at the Japanese Atomic Energy Research Institute in Tokai-Mura.

Large Straight-Through Tandems

Argentina's Comisión Nacional de Energía Atómica operates a 20-million-volt Pelletron in their Tandar facility in Buenos Aires.

Daresbury Laboratory in England built a large tandem for scientific research. This accelerator is charged by a Laddertron, which looks like two parallel Pelletron chains with metal bars connecting the pellets of one chain to the adjacent pellets of the other chain.

At Strasbourg, M. Letournel is developing the Vivitron, a design for a very high-voltage belt-charged tandem accelerator. The design uses metal electrodes, called porticos, to modify the electric-field distribution between the accelerator column and the grounded steel tank for higher terminal voltages.

BIBLIOGRAPHY

M. R. Cleland and P. Farrell, "Dynamitrons of the Future," *IEEE Trans. Nucl. Sci.* **NS-12,** 227 (1965).

Large Electrostatic Accelerators (D. Allan Bromley, ed.) [Reprinted from *Nuclear Instruments and Methods* **122** (1974)].

M. S. Livingston and J. P. Blewett, *Particle Accelerators*. McGraw-Hill, New York, 1962.

C. C. Thompson and M. R. Cleland, "Design Equations for Dynamitron Type Power Supplies in the Megavolt Range," *IEEE Trans. Nucl. Sci.* **NS-16,** 124 (1969).

Acoustical Measurements

M. Strasberg

Acoustics is concerned with fluctuations in the value of various mechanical quantities characterizing the state of matter—fluctuations of the pressure or other components of stress, or fluctuations of density, temperature, and the position of individual particles of matter. Primary acoustical measurements determine the magnitude and wave form of the oscillations of one of these quantities at one or more positions in space, whereas secondary measurements characterize the wavelike propagation of these oscilllations through space by determining the speed of propagation, the intensity or rate of propagation of acoustic energy, and the absorption or rate of dissipation of acoustic energy.

Most of the early primary acoustical measurements used mechanical devices that would determine only the magnitude of the oscillatory particle displacement or velocity. The advent of electronic amplifiers led to development of electromechanical transducers which convert the oscillating mechanical quantities into an emf. Nowadays, primary acoustical measurements usually utilize linear electromechanical transducers to generate an oscillating emf that is instantaneously proportional to the oscillating mechanical quantity. The magnitude and wave form of the oscillating mechanical quantity are deduced from measurements of the corresponding characteristics of the emf.

Several textbooks survey acoustical measurements. Wood [1], Stephens and Bate [2], and Meyer and Neuman [3], for example, describe mechanical techniques used before the present electronic age, and the latter two texts also cover contemporary instruments. The textbook *Acoustical measurements* [4] is devoted entirely to the subject, albeit primarily to audible sound. Various ultrasonic measurement techniques are described by Fry and Dunn [5] and are discussed in several chapters in the 18-volume collection edited by Mason [6]. Handbooks published by several manufacturers discuss the use of their instruments for sound measurements [7–9]. Standard procedures and apparatus for performing certain conventional measurements have been published by organizations concerned with standardization [10].

MEASUREMENT OF OSCILLATING PRESSURE

Most present-day acoustic measurements in fluids utilize electromechanical transducers sensing the oscillating pressure. These are called *microphones* when used in gases and *hydrophones* when waterproofed for use in liquids.

Various physical effects are utilized for generating the alternating emf; a comprehensive survey is given in Chapter 8 of Olson [11]. The type chosen for a particular application depends on a compromise among conflicting desirable characteristics, e.g., small size, high sensitivity, constant sensitivity over the frequency range of interest, stability to varying temperature and humidity, and low cost.

Whatever type of transducer is used, the magnitude, wave form, and spectral characteristics of the generated alternating emf are measured with electronic instruments such as voltmeters, oscilloscopes, wave analyzers, and spectrum analyzers. The corresponding characteristics of the oscillating pressure are determined from the measured characteristics of the alternating emf by dividing the electrical magnitudes by a proportionality factor (called a *sensitivity* or *calibration* factor) that is equal to the generated emf per unit sound pressure, including the amplification in the electronic system. Ideally this factor is independent of the magnitude and frequency of the oscillating quantity, but in practical systems there is usually some variation in sensitivity with frequency within the range of interest that must be taken into account.

Carbon Microphone

The carbon microphone, an early pressure-sensing transducer still used in some telephones, depends on the piezoreesistive property of compressed carbon granules. The oscillating sound pressure causes fluctuations in the compression of the granules that result in a fluctuating resistance that is detected electrically as an alternating potential difference developed by a dc current through the granules. The device has a high, albeit unstable, sensitivity; its useful frequency range covers from about 250 to 4000 Hz.

Electromagnetic Microphone

Sometimes called *dynamic* or *moving-coil* microphones, electromagnetic microphones comprise a light coil of wire located in the field of a permanent magnet and attached to a thin diaphragm that vibrates in response to the sound pressure, so that an alternating emf is generated by electromagnetic induction. The vibrating system is designed so that the generated emf is instantaneously proportional to the oscillating pressure. The frequency range of relatively constant sensitivity can extend from perhaps 100 to 10 000 Hz.

Another type of electrodynamic microphone utilizes a limp metallic ribbon instead of the coil. If both sides of the ribbon are exposed to the sound, the ribbon tends to generate an alternating emf proportional to the particle velocity of the gas, and for this reason it is sometimes called a *velocity microphone*, but it is really responding to the gradient of the oscillating pressure.

Piezoelectric Microphones and Hydrophones

The piezoelectric microphones and hydrophones contain piezoelectric elements that develop an emf in response to mechanical stress. In the early days, thin disks cut from natural or synthetic single crystals were used. Rochelle salt crystals were used because of their high sensitivity, but they deteriorated easily. Lithium sulfate crystals were more stable but less sensitive. Disks cut from natural quartz and tourmaline were used if stability and ruggedness were important, particularly for underwater applications, but they were relatively insensitive. Most present-day piezoelectric microphones and hydrophones utilize polarized ceramic piezoelectric elements of barium titanate, lead zirconate, or others, which have relatively high sensitivity and are available in various shapes, e.g., cylinders and spheres as well as disks (see Chapter 3 in Vol. 1A of the Mason collection [6]).

Most hydrophones use piezolectric elements that can withstand the high static pressures existing at deep submergences. The hydrophone dimensions can vary from small 1-mm cylinders, used as "probe" hydrophones, to 10-cm (or larger) units. Depending on the design, the useful frequency range can extend down to a few hertz and up to 1 MHz. Bobber [12] presents a detailed discussion of underwater acoustic measurements.

Capacitance or Electrostatic Microphones

Capacitance or electrostatic microphones comprise a small variable capacitance consisting of a metallic stretched membrane or thin diaphragm, separated from a rigid back plate by an air gap perhaps 10^{-3} cm thick. The oscillating sound pressure results in oscillating deflections of the diaphragm, which change the width of the air gap and the associated capacitance. The variations in capacitance are usually detected by placing a fixed electric charge onto the electrodes of the capacitance through a high-resistance leak from a source of several hundred volts dc, so that variations in capacitance result in a varying potential difference across the capacitor. For very low-frequency measurements, the capacitance can be placed in one arm of an ac bridge so that the capacitance variations result in an amplitude-modulated carrier output from the bridge; alternatively, the varying capacitance can be used to control the frequency generated by an oscillator and thus provide a frequency-modulated carrier.

To eliminate the need for dc polarization, electret capacitance microphones have been developed utilizing polarized electrets holding a permanent charge on the metal electrodes.

Calibration of Microphones and Hydrophones

The ratio of the generated emf to the oscillating pressure generating the emf is called the *sensitivity* of the microphone. The sensitivity is determined by a procedure called *calibration*. Most microphones are calibrated by comparing their generated emf to that of another primary standard microphone subjected to the same sound pressure at various frequencies.

Microphones used as primary standards can be calibrated by one of several absolute methods. The method usually used nowadays is called a *reciprocity calibration*. This procedure involves two sets of measurements, the first being a comparison of the emf generated by the microphone with that of a second microphone when both are responding to the same sound source, and the second being a measurement of the emf of the microphone when the second microphone is itself used as a source (see Chapter 4.2 of Beranek [4]).

Another absolute calibration uses a Rayleigh disk, to be described subsequently, to determine the oscillating-particle velocity. A third method uses an oscillating piston driven by a rotating cam and forming one wall of a small gas cavity; the known volume change is used to calculate the oscillating pressure, taking into account departure from the adiabatic gas law because of heat conduction at the walls of the cavity. Microphones having flat diaphragms can be calibrated by placing an auxiliary electrode close to the diaphragm and applying an alternating potential difference between this electrode and the diaphragm so as to develop a calculable alternating electrostatic pressure on the diaphragm.

Hydrophones are calibrated by methods generally the same as those used for microphones. An absolute method suitable only for hydrophones is to hold the hydrophone in a vibrating container of liquid; the vibration results in an oscillating pressure that can be calculated in terms of the measured acceleration of the container [13].

MEASUREMENT OF DENSITY AND TEMPERATURE OSCILLATIONS

Measurements of fluctuating density are usually done optically, utilizing variations in index of refraction associated with the density variations. The Debye–Sears apparatus is suitable for plane waves of sound, the spatially periodic density variations acting as an optical diffraction grating. The diffraction angles depend on the wavelength of the sound, and the intensity of the diffracted light depends on the magnitude of the oscillating refraction index and density (see Sect. 6.81 of Meyer and Neumann [3]). Optical holographic techniques with laser light sources are now being developed to indicate density oscillations. All these optical methods require relatively large-amplitude sounds and are useful only at high frequencies.

Temperature fluctuations can be observed with small probe wires whose electrical resistance fluctuates with the temperature fluctuations [14].

MEASUREMENTS OF OSCILLATING-PARTICLE MOTION

Direct Visual Observation

The oscillating motion of fluid particles associated with ordinary sound is usually too small to be observed visually, even with a microscope. For example, a sound in air having a frequency of 1000 Hz and an oscillating rms sound pressure of 1 N/m^2 (94 dB re 20 μPa) may be loud enough to cause some damage to hearing, but the oscillatory particle displacement is only about 3.7×10^{-5} cm. However, the oscillatory motion can be observed if the sound is strong enough and the frequency low enough. Photographs through a microscope showing streaks of smoke particles oscillating in an intense sound field are shown on p. 292 of Stephens and Bate [2].

Rayleigh Disk

The Rayleigh disk is an early device used for absolute measurement of the oscillating particle velocity associated with sound in gases. A Rayleigh disk consists of a small thin disk, perhaps 1 cm in diameter, suspended by a fine fiber. The oscillating particle velocity results in a steady torque on the disk, proportional to the mean-square velocity [1–3], tending to turn the disk broadside to the direction of the oscillating velocity. The suspension fiber is used as a balance to indicate the magnitude of the steady torque. Since the torque has been calculated theoretically, the Rayleigh disk permits an absolute measurement of the particle velocity. It is used nowadays mainly to provide an absolute calibration of microphones.

Electromechanical Vibration Pickups

The most common method for measuring the vibration of a solid is to mount an electromechanical transducer on the surface to convert the vibratory motion into an alternating emf. These transducers are called displacement pickups, velocity pickups, or accelerometers, depending on whether they generate emfs instantaneously proportional to the vibratory displacement or to its first or second time derivatives [15,16]. Their sizes range from units smaller than 1 cm and with masses of only a few grams, to units having dimensions of 5 cm or more. Although the larger units are more sensitive, their large mass can perturb the vibration they are intended to measure.

Displacement pickups are used for measuring relatively

low-frequency vibrations covering the frequency range from zero to a few hundred hertz. Velocity pickups are usually electromagnetic and cover the frequency range from perhaps 5 to 1000 Hz. Accelerometers for measuring very low-frequency accelerations utilize active transducers, whereas measurements in the frequency range 10 to 10 000 Hz and higher utilize piezoelectric elements. The sensitivity of vibration pickups can be determined in various ways, including an absolute reciprocity calibration (see Chapter 18 of Harris and Crede [15]).

Variable magnetic reluctance pickups are available which sense the vibratory velocity of a magnetic surface without any part of the pickup being in contact with the surface. Conventional phonograph pickups can be used as low-mass vibration pickups. If the two stereophonic outputs are combined in phase, the pickup is sensitive to vibrations parallel to the axis of the pickup stylus, whereas if combined out of phase, the pickup is sensitive to vibrations perpendicular to the stylus axis. The sensitivity of phonograph pickups can be determined as is done with other vibration pickups, or with a phonograph test record as described in Sec. 10.5 of Olson [11].

Piezoresistive Strain Gauges

The development of semiconductor strain gauges, which are some 50 times more sensitive than the older resistance-wire gauges, has made it possible to measure oscillating strains at the surface of a solid associated with sounds of ordinary magnitude. The fluctuating resistance associated with the oscillating strain can be measured electrically in various ways (see Vaughan [17] or Chapter 17 of Harris and Crede [15]).

Fiber Optics

Fine optical fibers can be used to measure oscillatory strain and displacement. When monochromatic light from a laser is transmitted along an optical fiber, oscillatory strain in the fiber results in an oscillation of the optical path length which in turn causes oscillation of the phase of the transmitted light. Various methods for converting the oscillatory phase into an electrical signal are available. To measure acoustic pressure, the optical fiber is bonded to a diaphram so as to sense the strain in the diaphragm caused by acoustic pressure acting on it (see Chapter 7 of Vol. 16 of Mason [6]).

Optical Measurements of Surface Vibration

Optical interferometry has been used to determine the magnitude of the vibratory displacement of the surface of a solid or liquid. The principle involves splitting a beam of monochromatic light into two beams, one of which is reflected off the vibrating surface and then recombined with the other beam. Motion of the reflecting surface changes the path length and causes the phase of the light beams to change relative to each other, so that the intensity of the recombined beam oscillates in response to vibrations of the reflecting surface. The method can be used to measure the vibration of a small region of a surface [18] as well as to display the distribution of vibration amplitude over a large surface vibrating in a complicated vibration pattern. If a laser is used as the light source, holographic reconstructions of the vibration pattern are possible (see Sec. 6.3.1 of Meyer and Neumann [3]).

Miscellaneous

Other methods for observing and measuring oscillatory displacements are used occasionally. Sensitive flames provided one of the early methods for detecting the oscillating-particle velocity associated with sound in gases (see p. 409B of Wood [1]). Oscillating strains in the interior of a photoelastic solid can be measured using polarized light [19]. The hot wire, involving an electrically heated fine wire whose temperature and resistance fluctuate in response to the cooling effect of a fluctuating flow past the wire, is the conventional device for measuring the fluctuating velocity in turbulent flows in gases; it can also be used to measure the oscillatory particle velocity associated with sound (see p. 440 of Wood [1]). Fluctuating electrochemical potentials detected by a pair of probes inserted in an electrolyte have been used to measure the oscillating-particle velocity associated with sound (see p. 439 of Wood [1]). Finally, mention should be made of the use of sand grains or small ball bearings to indicate when the vertical oscillatory acceleration of a solid surface exceeds $1g$ (see p. 210 of Meyer and Neumann [3]); or of the use of sand to provide a visual display of a complicated pattern of surface vibration (see discussion of Chladni figures on p. 172 of Wood [1]).

SECONDARY MEASUREMENTS

The velocity of propagation of sound can be determined by direct measurement of the transit time of a transient sound, or by observing various phenomena that depend on the wavelength of continuous sounds of constant frequency. These methods are discussed in Chapters V and IX of Herzfeld and Litovitz [20]; also methods for measuring acoustic absorption.

Acoustic Intensity

Measurements of the distribution of acoustic intensity around a sound source have become increasingly commonplace since the advent of digital spectrum analyzers. The intensity is a vector quantity whose component in a specified direction is the acoustic power passing through unit cross section perpendicular to the specified direction, this being equal to the time average of the product of the instantaneous sound pressure and the instantaneous particle velocity component in that direction.

For frequencies below 10 kHz, the intensity can be measured using an intensity probe consisting of two small microphones spaced a small fraction of an acoustic wavelength apart. The intensity can be shown to be proportional to the time average of the instantaneous product of the sum of the sound pressures at the two microphone positions multiplied by the difference between the two pressures. An alternative procedure, especially convenient if a two-channel digital spectrum analyzer is available, is to measure the complex cross-spectral density of the sound pressures at the two-microphone positions; the spectral density of the intensity at any frequency can be shown to be proportional to the imaginary part of the cross-spectral density of the two pressures at that frequency (see Fahy [21]). Although the sum-and-difference method may be used for measurements covering any band of frequencies, the cross-spectral technique is suitable only for narrow bands.

Since the two microphones of intensity probes must be much less than a wavelength apart, they are not suitable for measurements at frequencies above about 10 kHz. For ultrasonic frequencies, radiometers are used which sense the radiation pressure on a small disc in the sound field; the force on the disc is proportional to the acoustic intensity (see Fry and Dunn [5] or Sec. 6.7 of Beyer [22]).

Acoustical Holography

Acoustical holography, as distinct from acoustical imaging, is used to determine the distribution of sound pressure or vibration velocity over the surface of a sound source. The procedure involves measurements of the magnitude and phase of the sound pressure at many positions in the sound field and summation of these pressures with appropriate weighting functions to reconstruct the pressure or velocity distribution over the source surface. Only the radiating or nonevanescent portion of the pressure or velocity distribution can be determined from measurements made several wavelengths removed from the source; but the entire distribution can be reconstructed from measurements in the "near field" (see, e.g., Maynard [23]).

See also ACOUSTICS; PIEZOELECTRIC EFFECT; SOUND, UNDERWATER; TRANSDUCERS; WAVES.

REFERENCES

1. A. B. Wood, *A Textbook of Sound*. G. Bell and Sons Ltd., London, 1957.
2. R. W. B. Stephens and A. E. Bate, *Acoustics and Vibrational Physics*. Edward Arnold Ltd., London, 1966.
3. E. Meyer and E. G. Neumann, *Physical and Applied Acoustics*. Academic Press, New York, 1972.
4. L. Beranek, *Acoustical Measurements*. Acoustical Society of America, New York, 1949.
5. W. J. Fry and F. Dunn, "Ultrasound: Analytical and Experimental Methods in Biological Research," in *Physical Techniques in Biological Research* (W. L. Nastuk, ed.), Vol. 4, Chapter 6. Academic Press, New York, 1962.
6. W. P. Mason, ed., *Physical Acoustics: Principles and Methods*, 18 vols. Academic Press, New York, 1964–1988.
7. A. P. G. Peterson and E. E. Gross, Jr., *Handbook of Noise Measurement*. General Radio Co., Concord, Mass., 1972.
8. J. T. Brock, *Application of B&K Equipment to Acoustic Noise Measurements*. Bruel and Kjaer, Naerum, Denmark, 1971.
9. *Acoustic Handbook*, Application Note 100. Hewlett Packard Co., Palo Alto, Cal.
10. *Catalog of Acoustical Standards*. Acoustical Society of America, New York, 1988.
11. H. F. Olson, *Acoustical Engineering*. Van Nostrand-Reinhold, Princeton, NJ, 1957.
12. R. J. Bobber, *Underwater Electroacoustic Measurements*. U.S. Government Printing Office, Washington, DC, 1970.
13. F. Schloss and M. Strasberg, *J. Acoust. Soc. Am.* **34**, 1958 (1962).
14. J. Hojstrup, K. Rasmussen, and S. E. Larsen, "Dynamic Calibration of Temperature Wires in Still Air," published in *DISA Information*, No. 20, DISA Electronics, Franklin Lakes, N. J., 1975.
15. C. M. Harris and C. F. Crede, eds., *Shock and Vibration Handbook*. McGraw-Hill, New York, 1976.
16. J. T. Brock, *Application of B&K Equipment to Mechanical Vibration and Shock Measurement*. Bruel and Kjaer, Denmark, 1972.
17. J. Vaughan, *Application of B&K Equipment to Strain Measurement*. Bruel and Kjaer, Denmark, 1975.
18. F. J. Eberhart and F. A. Andrews, *J. Acoust. Soc. Am.* **48**, 603 (1970).
19. R. C. Dove and P. H. Adams, *Experimental Stress Analysis and Motion Measurement*. Charles E. Merrill, Columbus, Ohio, 1964.
20. K. F. Herzfeld and T. A. Litovitz, *Absorption and Dispersion of Ultrasonic Waves*. Academic Press, New York, 1959.
21. F. J. Fahy, *Sound Intensity*. Elsevier, New York, 1989.
22. R. T. Beyer, *Nonlinear Acoustics*. U.S. Government Printing Office, Washington, 1974.
23. J. D. Maynard, E. G. Williams, and Y. Lee, *J. Acoust. Soc. Am.* **78**, 1395 (1985).

Acoustics

Robert T. Beyer

Acoustics is the science of sound. What is sound? When you open your mouth and utter speech you are said to produce a sound. Under normal conditions a nearby person with so-called normal hearing says that he hears the sound. In its study of the production and reception of sound and its transmission through material media, acoustics is a branch of physics, though speech and hearing obviously involve biological elements.

Let us examine the physics of what happens when a person speaks. A disturbance is produced in the air in front of the mouth, involving a slight compression of the air or, alternatively, an increase in the air pressure of the order of 0.1 N/m^2. This amount is about one-millionth of the normal atmospheric pressure. Since air is an elastic medium, it does not stay compressed but tends to expand again and hence produces a disturbance in the neighboring air. This in turn passes on the disturbance to the air adjoining it, and the result is a pressure fluctuation that moves through the air in the form of a sound wave. When the wave reaches the ear of an observer, it produces a motion of the eardrum, which in turn moves the little bones in the middle ear, and this movement communicates motion to the hair cells in the cochlea in the inner ear. The ultimate result, by a rather complicated biophysical process that is not yet completely understood, is the hearing of the sound.

While acoustics, as defined above, applies only to audible sound in air, its scope has been gradually expanded until today it encompasses mechanical waves and vibrations in all material media—solids, liquids, and gases, and includes waves of any frequency, as well as aperiodic disturbances, such as shocks and noise. A perusal of the pages of the *Journal of the Acoustical Society of America* will easily demonstrate that this is the case.

It is convenient to divide this discussion of acoustics into three parts: the production, the transmission, and the reception of sound.

PRODUCTION

Any change in stress or pressure leading to a local change in density or displacement from equilibrium in an elastic material medium can serve as a source of sound. We have already mentioned the human vocal mechanism as an example. All sound sources in practical use involve the vibrations of solids, liquids, or gases. Such a vibration is an oscillation of stress or pressure with a definite frequency. The unit of frequency used in acoustics is the hertz (Hz), which is one complete cycle per second. The standard musical instruments are common sources of sound of more or less definite frequency. In sophisticated scientific and technological applications the sound source is called a transducer. Those in standard use are electroacoustic in character, that is, they depend on electrical action to produce mechanical vibrations. An example is the electrodynamic loudspeaker, in which an alternating electric current in a coil of wire placed in a magnetic field produces oscillatory motion in the coil. This motion is communicated to a membrane, whose resulting vibrations are radiated as sound. Another commonly used electroacoustic transducer is based on the piezoelectric effect, according to which certain crystals (e.g., quartz) can be made to vibrate when placed in an oscillating electric field.

Certain ceramic materials, such as barium titanate and lead zirconate, can be polarized by the use of an applied electric field and can thereafter serve as transducers in the same way as that described for quartz. In addition, the change in shape and dimensions of a piece of magnetic material, like nickel, when placed in a magnetic field, can also be used in the construction of transducers. This is known as the magnetostrictive effect. All these types of transducers are useful, not only in the production of sound but also in its reception.

The rapid flow of air over a rough surface or through a nozzle produces sound that can vary over a wide range of frequency and intensity. An example is the aerodynamic sound from a jet engine on an airplane. Sounds of this character are commonly known as noise, one of the chief problems of environmental acoustics in the late twentieth century.

An exciting new field has been the optoacoustic generation of sound by the thermoacoustic effect of absorption of the energy of a pulsed light source (laser beam) by a medium, which thereupon emits acoustic pulses. It is of interest to note that this effect was first observed by Alexander Graham Bell in 1881 (without the use of a laser!).

TRANSMISSION

Sound demands a material medium for its transmission from place to place. As previously mentioned, the propagation takes place by means of wave motion, a good visual example of which is provided by a wave on the surface of water. A sound wave travels with a definite velocity, depending on the type of wave, the physical nature of the medium, and the temperature. Through air at standard room temperature (20°C) sound travels with the velocity 344 m/s, a value that increases as the temperature is raised. Under similar conditions, sound travels through water with a velocity somewhat more than four times as great. The velocity of sound in a highly elastic solid like steel is even greater, being as high as 6000 m/s.

A sound wave represents the transmission of mechanical energy through the medium in which the sound travels. The measure of this transmission is called the intensity of the sound wave. It is defined as the average flow of energy per unit time through a unit area of the surface through which the sound passes (with direction normal to the surface). The strictly scientific unit of sound intensity is the watt per square meter. In practice, however, this unit is replaced by a system in which 10 times the logarithm to the base 10 of the ratio of the given intensity to that corresponding to minimum audibility is taken as the number of decibels (abbreviated dB) represented by the sound in question. Ordinary conversational speech at a distance of 1 m from the speaker has an intensity of about 60 dB, whereas the intensity in the neighborhood of a jet airplane with the engine running can be as high as 140 dB. Sound of intensity in excess of 90 dB at the human ear can be harmful to that delicate and valuable organ of hearing. Sounds of very high intensity are called macrosonic and form the subject of what is termed nonlinear acoustics.

As sound spreads out from a source, its intensity decreases with distance. For a highly localized source from which the sound travels in all directions, the intensity varies inversely as the square of the distance from the source. Sound intensity also falls off with distance traveled through a process known as absorption, a dissipative action in which the energy being transmitted by the sound wave is gradually converted into heat. When a sound wave strikes an obstacle, it undergoes reflection and refraction, following the same laws as those that hold for light waves. The acoustic echo is a well-known phenomenon.

The simplest type of sound wave is the periodic one, in which at any given point in the medium being traversed by the wave the disturbance (e.g., the excess pressure for a wave in air) varies periodically between a minimum and a maximum value and back again a certain number of times a second. As indicated earlier in the case of sound source vibrations, this number is called the frequency of the wave and is measured in hertz. Another important quantity characterizing a periodic sound wave is the wavelength, or the distance between successive points in a wave train at which the disturbance has the same magnitude and is doing the same thing (i.e., is either increasing or decreasing in magnitude). The wavelength is related to the frequency by the simple but fundamental relation that the wavelength is equal to the velocity divided by the frequency. For a given velocity, a long wavelength means a low frequency and vice versa.

RECEPTION

The frequency of a periodic sound wave determines, in a rather complicated way, the pitch at which it is heard. High pitch corresponds to high frequency. Sounds of frequency below 20 Hz are not normally heard by human beings even at very high intensity. These sound waves are called infrasonic. The upper frequency limit of audible sound varies markedly. For young people, this upper limit is at about 20 000 Hz. For older adults, this figure drops to 10 000 Hz or even lower. This phenomenon is known as presbycusis. Frequencies above the audible range are known as ultrasonic; they can be generated and detected into the gigahertz range (10^9 Hz). Ultrasound has many practical applications in such areas as sound signaling, metallurgy, and medicine.

The most important receiver of audible sound in human experience is, of course, the ear, a marvelously sensitive mechanism that can normally detect sound of intensity as low as 10^{-12} W/m^2 and can stand intensity as high as 1 W/m^2 before pain ensues and possible ear damage develops. The normal ear is most sensitive at around 2000 Hz. Loudness, though of course related to intensity in such a way that it increases with intensity, is a subjective quantity connected with the biological response of the listener. The unit of loudness developed by the psychophysicists is the sone, defined as the loudness produced by a tone of 1000 Hz at 40 dB above the minimum audible threshold. A loudness scale in sones has been established by the statistical study of the hearing of a large number of individuals.

The principal artificial sound receiver in common use is the microphone, of which there are many varieties, mainly based on the electroacoustic effects used for transducers in general, as mentioned earlier. Their widespread employment in the audio industry—radio, television, public address systems, sound recording and reproduction, hearing aids, etc.—is well known.

More detailed information about the topics mentioned here involving the application of acoustics to fields like architectural acoustics, ultrasonics, vibration problems, and underwater sound, can be found in the relevant articles in the encyclopedia.

See also ACOUSTICAL MEASUREMENTS; ACOUSTICS, ARCHITECTURAL; ACOUSTICS, LINEAR AND NONLINEAR; ACOUSTICS, PHYSIOLOGICAL; ACOUSTOELECTRIC EFFECT; MUSICAL INSTRUMENTS; NOISE, ACOUSTICAL; SOUND, UNDERWATER; ULTRASONIC BIOPHYSICS; ULTRASONICS; VIBRATIONS, MECHANICAL; WAVES.

BIBLIOGRAPHY

L. L. Beranek, *Acoustics*. Reprint of the 1954 edition, with revisions. Acoustical Society of America, New York, 1986. (I)

F. V. Hunt, *Acoustics*. Reprint of the 1954 edition. Acoustical Society of America, New York, 1982. (I)

L. E. Kinsler, A. R. Frey, A. B. Coppens, and J. V. Sanders, *Fundamentals of Acoustics*, 3rd ed. Wiley, New York, 1982. (I)

James Lighthill, *Waves in Fluids*. Cambridge University Press, Cambridge, 1978. (A)

R. B. Lindsay, ed., *Acoustics: Historical and Philosophical Development*. Dowden, Hutchinson & Ross, Stroudsburg, PA., 1973. (E)

Iain G. Main, *Vibrations and Waves in Physics*. 2nd ed. Cambridge University Press, Cambridge, 1984. (I)

P. M. Morse and K. U. Ingard, *Theoretical Acoustics*. Reprint of the 1968 edition. Princeton University Press, Princeton, 1986. (A)

A. D. Pierce, *Acoustics, An Introduction to its Physical Principles and Applications*. 1981 text reprinted by the Acoustical Society of America, New York, 1989. (I)

J. R. Pierce, *Almost All About Waves*. MIT Press. Cambridge, MA, 1974. (E)

R. W. B. Stephens and A. E. Bate, *Acoustics and Vibrational Physics*, 2nd ed. St. Martin's Press, New York, 1966. (I)

J. W. Strutt (Lord Rayleigh). *The Theory of Sound*. 1877 (reprinted by Dover Publications, New York, 1945). (A)

For information on current progress in acoustical research the *Journal of the Acoustical Society of America* may be consulted. This is published monthly by the American Institute of Physics for the Acoustical Society of America.

Acoustics, Architectural

Thomas D. Northwood

Architectural acoustics may be defined as the science of acoustics applied to the design of buildings. It thus derives from such diverse disciplines as physics, psychology, the arts, architecture, and engineering, the techniques needed to produce acoustical environments acceptable to the building occupants. The objective is twofold: to bring to the occupants the sounds they wish to hear, and to protect them from the sounds they do not wish to hear. The first of these tasks is the more attractive and creative one, but the exercise will be successful only if the second task is handled with equal care.

The profession of architectural acoustics may be said to have been inaugurated by Wallace Clement Sabine (1868–1919), a professor of physics at Harvard University, who in 1895 was assigned the task of curing the acoustical defects of a new lecture theater at the university. In solving that problem he went on to evolve the first quantitative theory of reverberation processes in rooms, and applied it to the solution of problems in many theaters and concert halls, the most famous of which was Symphony Hall in Boston.

For studies of these rooms the instrumentation available to Sabine consisted of a set of organ pipes, a stop watch, and his two ears. He used these with great ingenuity to determine what is now known as the "reverberation time" of a room and, by extension, the "sound absorption coefficients" of typical room surfaces. From his collected data for many rooms he was able to develop the Sabine reverberation theory that is still used by acousticians everywhere.

Another technique first developed by Sabine was the use of two-dimensional models of complex surfaces, utilizing a spark technique and schlieren photography to observe the progression of waves in such models. Today, 80 years later, the same topics are dealt with in more sophisticated ways, but there is scarcely any problem in architectural acoustics on which Sabine did not make a useful contribution.

Most of the sounds of interest in architectural acoustics, such as speech or music, consist of a sequence of transient impulses, and these brief transients constitute the "message." Following the progression of one of these sounds in a room, we can identify three phases: first, the direct transmission of sound through the air from source to listener; then the first reflections from the various room surfaces; and finally, the reverberant field, composed of multiple reflections from the room surfaces.

The acoustical design of a room involves a consideration of these three phases. The direct transmission does not carry much energy very far, but it is an important reference, establishing the location of the source and the timing of the sequence of sounds. The first-order reflections, if they arrive soon enough, provide useful reinforcement of the direct transmission. Strong delayed reflections (echoes) tend to garble the sequence of transient sounds and thus interfere with the perception of speech. Even without forming discrete echoes, the ensemble of multiple reflections forms a persistent reverberation that also can interfere with perceptions of the original sound sequences.

In halls used for speech the design objective is to shape the room surfaces so as to provide short-delay reflections that reinforce the direct sounds and thus extend the range for perception of intelligible speech. For one-way communication the range can be further extended by electroacoustic reinforcement. At the same time the reverberation time must be limited so that it does not blur the sequence of sounds. For musical performances, on the other hand, the hall plays a more active role. A certain optimal amount of reverberation provides a desirable blending and smoothing of musical sounds. In addition, the presence of early reflections, reaching the listeners especially from lateral directions, are essential to the listeners' impression of the space enveloping them and the performers. For the performers an additional requirement is that each should hear his own part in relation to the whole and to what is heard in the hall. To achieve this sort of ambience for every listener and performer and every kind of music is a delicate task, which becomes increasingly difficult as halls and audiences become larger.

The reduction of unwanted or disturbing noises in an enclosure involves the opposite sort of techniques: surfaces are designed to absorb rather than reflect sounds, so that the first-order reflections and the reverberant portions of the sound are made negligible. The direct sound is reduced by interposing partitions or screens between source and listener, or by a partial or complete enclosure around the source. These principles apply with minor variations to all large spaces containing noise sources: factory areas, open-plan offices, restaurants, air terminals, and so on. In many instances the noise may be intrusive speech, for which the annoyance increases rapidly with intelligibility. Reduction of intelligibility can be accomplished either by reducing the level of the transmitted speech or by increasing the level of "background" noise. The latter sometimes takes the form of background music, which is deemed a lesser evil than the intrusive speech. This question of detection of intrusive noises in the presence of acceptable "background" noise is implicit in most noise control problems.

Sound insulation between rooms is mainly a function of the separating wall or floor. In the design of such partition elements it is usual to distinguish between airborne sounds, which travel through the air to reach the partition, and structure–borne sounds, which begin as vibrations in the structure itself. With respect to airborne sounds the most important virtues of a simple partition are that it be impermeable and heavy. For a given total weight, however, it is found more effective to use a multiplicity of relatively independent layers.

The sound transmission loss of a partition increases systematically with frequency in the incident sound, and this must be taken into account in arriving at a representative performance rating for partitions. For sounds such as speech and music (live or by way of radio or television) the customary figure of merit is the sound transmission class (STC), which emphasizes the importance of middle- and high-frequency components.

With respect to structure-borne sounds, such as footsteps and machinery vibrations, the first requirement is structural discontinuity between the vibrating surface and the contiguous structure. A floor, for example, may be composed of a finished floor panel separated from the main structural slab by a soft layer. In the case of machinery it is usual to provide an individually designed mounting tuned to filter out the driving frequency of the machine. Other problems characteristic of modern buildings, especially office buildings, are noise propagation in ventilation ducts and in continuous plenum spaces over suspended acoustical ceilings, and noise produced by plumbing appliances. These are but special cases of the noise mechanisms already described.

See also Noise, Acoustical.

BIBLIOGRAPHY

L. L. Beranek, *Music, Acoustics and Architecture*. Wiley, New York, 1962. (I)

L. Cremer, H. A. Muller, and T. J. Schultz (translator, English edition), *Principles and Applications of Room Acoustics*. Applied Science Publishers, 1982.

Vern O. Knudsen, and Cyril M. Harris, *Acoustical Designing in Architecture*. Wiley, New York, 1950. (E)

Heinrich Kuttruff, *Room Acoustics*. Wiley, New York, 1973. (A)

Anita Lawrence, *Architectural Acoustics*. Elsevier, Amsterdam, 1970. (I)

T. D. Northwood, *Architectural Acoustics, Benchmark Papers in Acoustics, Vol. 10*. Dowden, Hutchinson & Ross, Stroudsburg, PA, 1977.

W. C. Sabine, *Collected Papers on Acoustics*. Dover, New York, 1964. (E)

Acoustics, Linear and Nonlinear

Joshua E. Greenspon

DEFINITIONS AND NOMENCLATURE

Linear acoustics is the study of sounds of relatively small amplitude. Nonlinear acoustics is the study of sounds of relatively large amplitude. The physical phenomena associated with what is called relatively small and what is termed relatively large are, in part, the topics to be discussed in this article. The categories of linear and nonlinear acoustics have to be discussed independently of architectural acoustics, underwater acoustics, physical acoustics, and musical acoustics since linear and nonlinear refers to an amplitude characteristic of the sound whereas the other categories are associated with the acoustic environment and the mechanisms causing the sound. The study of linear acoustics is sometimes referred to as the study of infinitesimally small amplitude waves or just the study of ordinary sound waves.

Nonlinear acoustics is sometimes referred to as finite-amplitude acoustics, high-intensity acoustics, or macrosonics. There are many problems in fluid mechanics which are nonlinear because of the nature of the mathematics involved in their solution. Those physical problems in fluid mechanics which are primarily concerned with the sounds produced are acoustic problems. From a mathematical standpoint the problems involving the classical linear partial differential equation of sound waves (which will be discussed later in this article) are termed linear acoustic problems whereas the ones involving nonlinear partial differential equations (some examples being discussed later) are nonlinear acoustic problems.

PHYSICAL PHENOMENA IN ACOUSTICS

Sound Propagation in Air, Water, and Solids

Many practical problems are associated with the propagation of sound waves in air or water. Sound does not propagate in free space but must have a dense medium to propagate. Thus, for example, when a sound wave is produced by a voice, the air particles in front of the mouth are vibrated, and this vibration, in turn, produces a disturbance in the adjacent air particles, and so on.

If the wave travels in the same direction as the particles are being moved, it is called a longitudinal wave. This same phenomenon occurs whether the medium is air, water, or a solid. If the wave is moving perpendicular to the moving particles, it is called a transverse wave.

The rate at which a sound wave thins out, or attenuates, depends to a large extent on the medium through which it is propagating. For example, sound attenuates more rapidly in air than in water, which is the reason that sonar is used more extensively under water than in air. Conversely, radar (electromagnetic energy) attenuates much less in air than in water, so that it is more useful as a communication tool in air.

Sound waves travel in solid or fluid materials by elastic deformation of the material, which is called an elastic wave. In air (below a frequency of 20 kHz) and in water, a sound wave travels at constant speed without its shape being distorted. In solid material, the velocity of the wave changes, and the disturbance changes shape as it travels. This phenomenon in solids is called dispersion. Air and water are for the most part nondispersive media, whereas most solids are dispersive media.

Reflection, Refraction, Diffraction, Interference, and Scattering

Sound propagates undisturbed in a nondispersive medium until it reaches some obstacle. The obstacle, which can be a density change in the medium or a physical object, distorts the sound wave in various ways. (It is interesting to note that sound and light have many propagation characteristics in common: The phenomena of reflection, refraction, diffraction, interference, and scattering for sound are very similar to the phenomena for light.)

Reflection. When sound impinges on a rigid or elastic obstacle, part of it bounces off the obstacle, a characteristic that is called reflection. The reflection of sound back toward its source is called an echo. Echoes are used in sonar to locate objects under water. Most people have experienced echoes in air by calling out in an empty hall and hearing their words repeated as the sound bounces off the walls.

Refraction and Transmission. Refraction is the change of direction of a wave when it travels from a medium in which it has one velocity to a medium in which it has a different velocity. Refraction of sound occurs in the ocean because the temperature of the water changes with depth, which causes the velocity of sound also to change with depth. For simple ocean models, the layers of water at different temperatures act as though they are layers of different media. The following example explains refraction: Imagine a sound wave that is constant over a plane (i.e., a plane wave) in a given medium and a line drawn perpendicular to this plane (i.e., the normal to the plane) which indicates the travel direction of the wave. When the wave travels to a different medium, the normal bends, thus changing the direction of the sound wave. This normal line is called a *ray*.

When a sound wave impinges on a plate, part of the wave reflects and part goes through the plate. The part that goes through the plate is the transmitted wave. Reflection and transmission are related phenomena that are used extensively to describe the characteristics of sound baffles and absorbers.

Diffraction. Diffraction is associated with the bending of sound waves around or over barriers. A sound wave can often be heard on the other side of a barrier even if the listener cannot see the source of the sound. However, the barrier projects a shadow, called the shadow zone, within which the sound cannot be heard. This phenomenon is similar to that of a light that is blocked by a barrier.

Interference. Interference is the phenomenon that occurs when two sound waves converge. In linear acoustics the sound waves can be superimposed. When this occurs, the waves interfere with each other, and the resultant sound is the sum of the two waves, taking into consideration the magnitude and the phase of each wave.

Scattering. Sound scattering is related closely to reflection and transmission. It is the phenomenon that occurs when a sound wave envelops an obstacle and breaks up, producing a sound pattern around the obstacle. The sound travels off in all directions around the obstacle. The sound that travels back toward the source is called the backscattered sound, and the sound that travels away from the source is known as the forward-scattered field.

Standing Waves, Propagating Waves, and Reverberation

When a sound wave travels freely in a medium without obstacles, it continues to propagate unless it is attentuated by some characteristic of the medium, such as absorption. When sound waves propagate in an enclosed space, they reflect from the walls of the enclosure and travel in a different direction until they hit another wall. In a regular enclosure, such as a rectangular room, the waves reflect back and forth between the sound source and the wall, setting up a constant wave pattern that no longer shows the characteristics of a traveling wave. This wave pattern, called a standing wave, results from the superposition of two traveling waves propagating in opposite directions. The standing wave pattern exists as long as the source continues to emit sound waves. The continuous rebounding of the sound waves causes a reverberant field to be set up in the enclosure. If the walls of the enclosure are absorbent, the reverberant field is decreased. If the sound source stops emitting the waves, the reverberant standing wave field dies out because of the absorptive character of the walls. The time it takes for the reverberant field to decay is sometimes called the time constant of the room.

Sound Radiation

The interaction of a vibrating structure with a medium produces disturbances in the medium that propagate out from the structure. The sound field set up by these propagating disturbances is known as the sound radiation field. Whenever there is a disturbance in a sound medium, the waves propagate out from the disturbance, forming a radiation field.

Coupling and Interaction between Structures and the Surrounding Medium

A structure vibrating in air produces sound waves, which propagate out into the air. If the same vibrating structure is put into a vacuum, no sound is produced. However, whether the vibrating body is in a vacuum or air makes little difference in the vibration patterns, and the reaction of the structure to the medium is small. If the same vibrating body is put into water, the high density of water compared with air produces marked changes in the vibration and consequent radiation from the structure. The water, or any heavy liquid, produces two main effects on the structure. The first is an added mass effect, and the second is a damping effect known as radiation damping. The same type of phenomenon also occurs in air, but to a much smaller degree unless the body is traveling at high speed. The coupling phenomenon in air at these speeds is associated with flutter.

Deterministic (Single-Frequency) Versus Random Linear Acoustics

When the vibrations are not single frequency but are random, new concepts must be introduced. Instead of dealing with ordinary parameters such as pressure and velocity, it is necessary to use statistical concepts such as autocorrelation and cross-correlation of pressure in the time domain and auto- and cross-spectrum of pressure in the frequency domain. Frequency is a continuous variable in random systems, as opposed to a discrete variable in single-frequency systems. In some acoustic problems there is randomness in both space and time. Thus statistical concepts have to be applied to both time and spatial variables.

SOME OF THE PRACTICAL PROBLEMS IN LINEAR AND NONLINEAR ACOUSTICS

The majority of problems in architectural and musical acoustics involve small-amplitude sounds and therefore come under the category of linear acoustics. In fact, as will be seen later, all sounds which are below the threshold of pain* are well within the linear acoustic region. Most problems in submarine and surface-ship sound radiation which involve interaction of a vibrating structure with water are also linear acoustic problems since only very small motions of both the structure and the medium in contact with it, are involved.

The problem of propagation of explosive waves is a large-amplitude, nonlinear acoustic problem. The transition from subsonic to supersonic flow and the associated production of shock waves is also a problem in nonlinear acoustics.

REFERENCES TO FUNDAMENTAL DEVELOPMENTS IN LINEAR AND NONLINEAR ACOUSTICS

There are a number of books on acoustics which contain various degrees of mathematical sophistication. The only presentations on acoustics that the writer would classify as elementary are the accounts given in standard physics texts such as Duff *et al.* [1] and Sears and Zemansky [2]. There are many of these texts, and they usually give good elementary discussions of acoustics. The reader can almost pick any college physics text at random, and it will usually give a reasonably good presentation of elementary acoustics. Of the books which are devoted entirely to acoustics, the writer would rate as number one the treatise of Lord Rayleigh [3]. The writer would classify Rayleigh's two volumes as advanced. The reader who is reasonably familiar with elementary differential equations would do well to go through such books as Kinsler and Frey [4], Beranek [5], and Morse [6] before going to Rayleigh's work. Two very fine books of recent vintage which the writer would classify as advanced are the one written by Skudrzyk [7] and the treatise by Morse and Ingard [8]. The most recent reference is Beyer's excellent work on nonlinear acoustics [13].

* The threshold of pain is the intensity at which an average person starts to feel pain in his ears and at which permanent damage to the ears could result if exposure to the sound is sustained.

The most complete mathematical treatment to date on linear and nonlinear wave motion is contained in a very recent book by G. B. Whitham [9] which is devoted entirely to this subject. A most easily readable account of the mathematics associated with both linear and nonlinear acoustics is given by R. B. Lindsay [10]. In this book Lindsay devotes an entire chapter to discussing sound waves in fluids, and he considers both small- and large-amplitude motions in a general way. Short but readily readable accounts of finite-amplitude waves in acoustics and their comparison to linear waves are given in both Rayleigh's [3] and Lamb's [11] texts on sound. Finally, the *Journal of the Acoustical Society of America,* which is a monthly technical publication published by the American Institute of Physics, has two of its sections devoted entirely to work in general linear acoustics and nonlinear acoustics or macrosonics. Finally, a review of the important aspects of linear acoustics is contained in a recent article [14].

BASIC GOVERNING EQUATIONS OF ACOUSTICS

The fundamental concepts of both linear and nonlinear acoustics can be covered for the one-dimensional case in a simple but general way. From these notions the reader will be able to grasp a physical feeling concerning sounds produced in both the linear and nonlinear acoustic regimes. The extension to two and three dimensions is much more mathematically complicated, but involves no new physical concepts. For this discussion the writer will follow a combination of the presentation on plane sound waves by Kinsler and Frey [4] and Lamb [11]. Let

- x = equilibrium coordinate of a given particle of the medium from some given origin (see Fig. 1);
- ξ = particle displacement along the x axis from the equilibrium position;
- u = velocity of the particle = $\partial\xi/\partial t$;
- ρ = instantaneous density at any point;
- ρ_0 = equilibrium density of the medium;
- S = condensation at any point, which is defined as $S=(\rho-\rho_0)/\rho_0$ so that $\rho=\rho_0(1+S)$;
- $\bar{p}$ = instantaneous pressure at any point;
- p_0 = equilibrium pressure in the medium;
- p = excess pressure (which is the acoustic pressure) at any point, thus $p=\bar{p}-p_0$;
- C = velocity of propagation of the wave.

It will be assumed that the wave is plane, i.e., all particles on the plane at x have the same displacement, and that this displacement is a function only of space coordinate x and time t. We employ three basic concepts to derive three independent equations and then we will combine these equations into a single equation. The first of these concepts is the conservation of mass. We apply the principle of conservation of mass to a cross-sectional area $\bar{S}$ of the undisturbed fluid contained between planes positioned at x and $x+dx$. The mass of this undisturbed fluid is $\rho_0\bar{S}dx$. Assume that upon passage of a sound wave the plane at x is displaced a distance ξ to the right (see Fig. 1) and that the plane at $x+dx$ is displaced a distance $\xi+d\xi$ $[d\xi=(\partial\xi/\partial x)\,dx]$. The volume enclosed is therefore changed to $\bar{S}dx(1+\partial\xi/\partial x)$ and its mass is $\rho\bar{S}dx(1+\partial\xi/\partial x)$. Equating the original mass to the new mass (i.e., employing the conservation of mass) gives

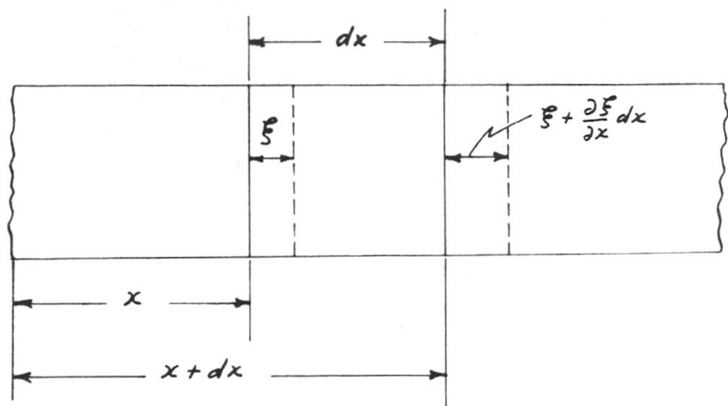

FIG. 1. Propagation of plane one-dimensional sound wave.

$$\rho\bar{S}dx\left(1+\frac{\partial\xi}{\partial x}\right)=\rho_0\bar{S}dx; \tag{1}$$

noting that $\rho=\rho_0(1+S)$ we obtain

$$(1+S)\left(1+\frac{\partial\xi}{\partial x}\right)=1. \tag{2}$$

If both S and $\partial\xi/\partial x$ are *small* (i.e., $\ll 1$), then we can neglect the product of S and $\partial\xi/\partial x$ and Eq. (2) reduces to

$$S=-\frac{\partial\xi}{\partial x}. \tag{3}$$

This equation is known as the equation of continuity.

The second basic concept that shall be employed is Newton's equation of motion of the element. The resultant pressures on the two faces of the volume element $\bar{S}dx$ will be slightly different from each other producing a net force which will accelerate the element. The external force acting on each face is equal to the product of the pressure and the area of the face. The net force acting upon $\bar{S}dx$ in the positive x direction is

$$dF_x=\left[p-\left(p+\frac{\partial p}{\partial x}dx\right)\right]\bar{S}=-\frac{\partial p}{\partial x}dx\bar{S}. \tag{4}$$

This net force is equal, by Newton's second law of motion, to the product of the element's mass and its acceleration, thus

$$-\frac{\partial p}{\partial x}=\rho_0\frac{\partial^2\xi}{\partial t^2}. \tag{5}$$

One other relation is necessary to combine the equation of continuity (2) or (3) with the equation of motion (5), and this is a relation between the pressure and the density. If we assume that the process is isothermal, then the temperature does not change during the passage of the sound wave and the pressure density relation follows Boyle's law, i.e.,

$$\frac{\bar{p}}{p_0}=\frac{\rho}{\rho_0}. \tag{6}$$

However, in ordinary sound waves the condensation S changes sign so frequently and the temperature, consequently, rises and falls so rapidly, that there is no time for transfer of heat between adjacent portions of the fluid. The flow of heat has hardly gone from one element to another before its direction is reversed; therefore the conditions are close to being adiabatic, i.e., no heat transfer occurring, and the pressure density relation becomes

$$\frac{\bar{p}}{p_0} = \left(\frac{\rho}{\rho_0}\right)^{\gamma}, \tag{7}$$

where γ is the adiabatic constant having a value of about 1.4 for air.

For large-amplitude adiabatic waves we combine (2), (5), and (7) as follows:

From before we had, by definition

$$\rho = \rho_0(1+S); \tag{8}$$

Eq. (2) gives

$$(1+S) = \frac{1}{1+\partial\xi/\partial x}.$$

Thus

$$\rho = \frac{\rho_0}{1+\partial\xi/\partial x}; \tag{9}$$

Eq. (7) gives

$$\frac{\bar{p}}{p_0} = \left(\frac{\rho}{\rho_0}\right)^{\gamma} \tag{10}$$

or

$$\frac{\bar{p}}{p_0} = \left(\frac{\rho_0}{\rho_0(1+\partial\xi/\partial x)}\right)^{\gamma} = \left(\frac{1}{1+\partial\xi/\partial x}\right)^{\gamma}, \tag{11}$$

but

$$\bar{p} = p_0 + p. \tag{12}$$

So

$$\bar{p} = \frac{p_0}{(1+\partial\xi/\partial x)^{\gamma}}. \tag{13}$$

So

$$\frac{\partial p}{\partial x} = \frac{\partial \bar{p}}{\partial x} = -p_0\gamma\left(1+\frac{\partial\xi}{\partial x}\right)^{-\gamma-1}\frac{\partial^2\xi}{\partial x^2}. \tag{14}$$

Thus (5) gives

$$\frac{p_0\gamma}{\rho_0}\,\frac{\partial^2\xi/\partial x^2}{(1+\partial\xi/\partial x)^{\gamma+1}} = \frac{\partial^2\xi}{\partial t^2}. \tag{15}$$

Let

$$C^2 = p_0\gamma/\rho_0. \tag{16}$$

Then the final adiabatic equation of motion for large amplitude nonlinear plane waves is

$$C^2\frac{\partial^2\xi/\partial x^2}{(1+\partial\xi/\partial x)^{\gamma+1}} = \frac{\partial^2\xi}{\partial t^2}. \tag{17}$$

If the process is isothermal, $\gamma = 1$.

For very small motions the condensation S is very small. When we employ this assumption and eq. (3) the equation of motion (17) reduces to the linear equation of sound waves

$$C^2\frac{\partial^2\xi}{\partial x^2} = \frac{\partial^2\xi}{\partial t^2}. \tag{18}$$

The general solution of Eq. (18) can be written in the form

$$\xi = f_1(Ct-x) + f_2(Ct+x). \tag{19}$$

This is easily verified by substituting (19) into (18) and performing the indicated differentiations. The solution (19) states that small displacements propagate with velocity C without change of shape. If we neglect any losses in the medium, a sound disturbance which is started at a given point will propagate with the velocity C without being distorted as it propagates. This is not true of large-amplitude waves which satisfy Eq. (17). To see this consider the transitional region between small- and large-amplitude waves in a long straight tube in which a piston (at $x=0$) is made to move in some arbitrary manner described by

$$\xi = f(t). \tag{20}$$

If we expand the denominator of Eq. (17) and neglect the terms greater than the second derivative of ξ, we obtain the following approximate equation for the transition region:

$$\frac{\partial^2\xi}{\partial t^2} = C^2\frac{\partial^2\xi}{\partial x^2} - (\gamma+1)C^2\frac{\partial\xi}{\partial x}\frac{\partial^2\xi}{\partial x^2}. \tag{21}$$

Table I Sound Pressure Level (in dB relative to 0.0002 dynes/cm²)[a] as a Function of Condensation

Condensation S	Sound Pressure Level (SPL)
10^{-10}	−3
10^{-9}	17
10^{-8}	37
10^{-7}	57
10^{-6}	77
10^{-5}	97
10^{-4}	117
10^{-3}	137
10^{-2}	157
10^{-1}	177
0.2	183
0.3	186
0.4	189
0.5	191
0.6	193
0.7	194
0.8	195
0.9	196
1.0	197
2.0	203
3	207
4	209
5	211
6	213
7	214
8	215
9	216
10	217
12	219
14	220
16	221
18	222
20	223

[a] 1 dyn/cm² is called a microbar, written μbar

By means of a procedure adopted by Airy [11] an approximate solution can be constructed. First we note that the solution of (18) is

$$\xi = f(t - x/C). \tag{22}$$

Substituting this value of ξ in the last term on the right-hand side of (21) we obtain

$$\frac{\partial^2 \xi}{\partial t^2} = C^2 \frac{\partial^2 \xi}{\partial x^2} - \frac{1}{2}(\gamma + 1)\frac{\partial}{\partial x}\left\{f'\left(t - \frac{x}{C}\right)\right\}^2. \tag{23}$$

The solution of this equation is

$$\xi = f\left(t - \frac{x}{C}\right) + \frac{\gamma + 1}{4C^2}x\left\{f'\left(t - \frac{x}{C}\right)\right\}^2. \tag{24}$$

Now assume that the motion of the piston at one end of the tube is simple harmonic, i.e.,

$$f(t) = a \cos \omega t. \tag{25}$$

Formula (24) then gives

$$\xi = a \cos \omega\left(t - \frac{x}{C}\right) + \frac{(\gamma + 1)\omega^2 a^2}{8C^2}x\left\{1 - \cos 2\omega\left(t - \frac{x}{C}\right)\right\}. \tag{26}$$

It is thus seen that the displacement of any particle is no longer simple harmonic at $x>0$ but is distorted from the original wave shape.

SOUND INTENSITY AND PRESSURE IN LINEAR AND NONLINEAR ACOUSTICS—SINE WAVES (SIREN-TYPE SOUNDS)

In order to give the reader a physical feeling of the relative sounds coming from various processes in both linear and nonlinear acoustics let us first compute the intensity of sine waves that might come from any physical source such as a siren or transducer of some sort. The sound intensity is defined as the time average of the power flow per unit area (or rate at which energy is transmitted per unit area [6]) as follows:

$$I = \frac{1}{T}\int_0^T p\dot{\xi}dt \tag{27}$$

where I is the intensity, p is the excess pressure, and $\dot{\xi}$ is the velocity of the medium particle. The value of T is taken as some arbitrary value. In the case of sine waves we will take T to be a number of periods of the wave. In the last section it was found that the pressure p was (for adiabatic processes)

$$\bar{p} = p_0(\rho/\rho_0)^\gamma, \tag{28}$$

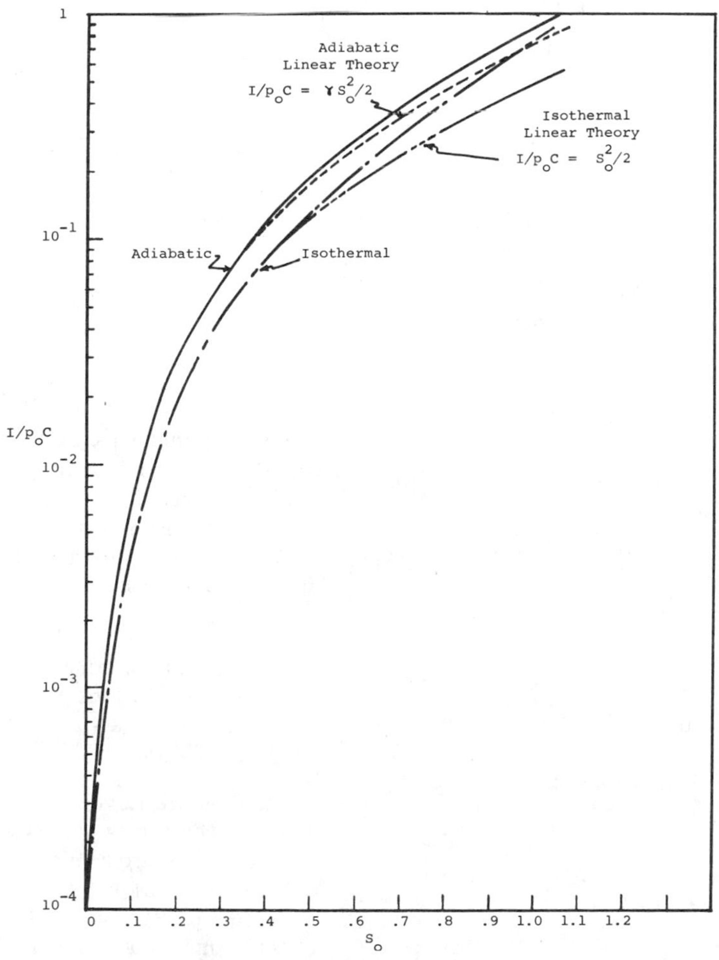

FIG. 2. Intensity as a function of condensation for sine waves.

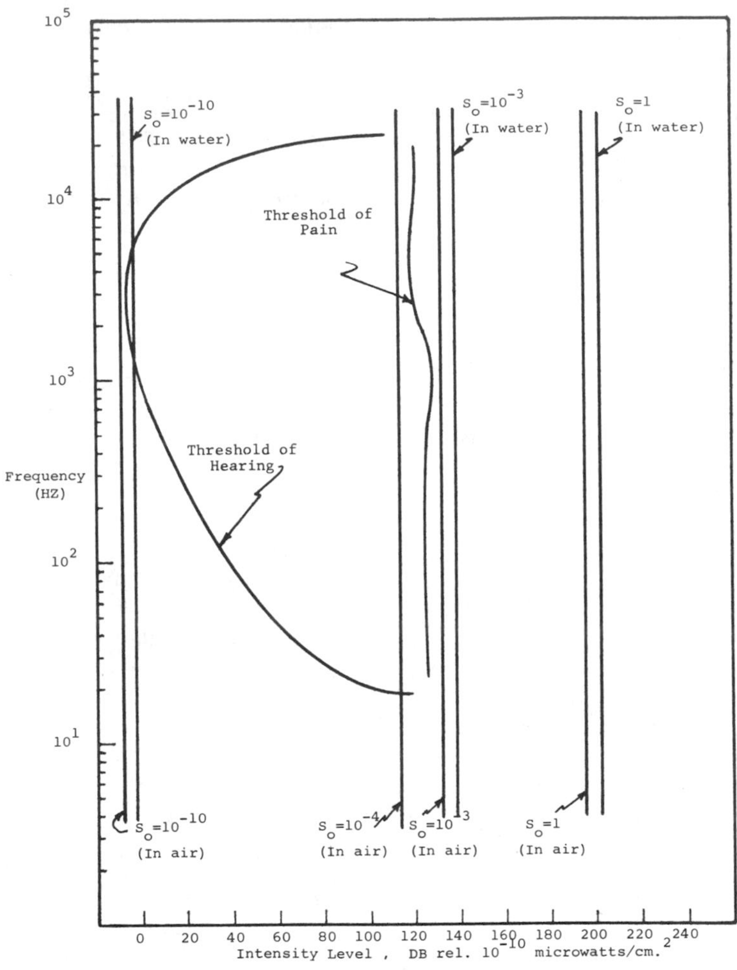

FIG. 3. Intensity of sine waves as a function of frequency.

(for an isothermal process $\gamma=1$) and that the density ρ was connected to the condensation S by the relation

$$S=\rho/\rho_0-1. \tag{29}$$

It has been found [11] that the velocity $\dot{\xi}$ has the following values for a general nonlinear wave:

$$\dot{\xi}=\pm C\log(1+S) \qquad \text{(for an isothermal process);} \tag{30}$$

$$\dot{\xi}=\pm\frac{2C}{\gamma-1}\left[1-(1+S)^{(\gamma-1)/2}\right] \quad \text{(for an adiabatic process).} \tag{31}$$

For both processes it is easily verified that for the linear case ($S \ll 1$)

$$\dot{\xi}=\pm CS. \tag{32}$$

For sine waves of small amplitude

$$I=\frac{1}{T}\int_0^T p_0\gamma CS^2\,dt, \tag{33}$$

so

$$\frac{I}{p_0C}=\gamma\frac{1}{T}\int_0^T S^2\,dt,$$

$$S=S_0\cos\omega t,$$

therefore

$$\frac{I}{p_0C}=\gamma\frac{S_0^2}{2} \qquad \text{(for the adiabatic case),} \tag{34}$$

$$=S_0^2/2 \qquad \text{(for the isothermal case).}$$

For the nonlinear case we just substitute (28), (29), (30), or (31) into (27) and integrate numerically. The result is given in Fig. 2 along with the linear approximation. It is seen that the linear theory holds for values of S_0 less than 0.4.

The sound pressure generated for either the linear or nonlinear adiabatic case is

$$p=p_0(1+S)^{\gamma}-p_0.$$

Assuming that the ambient pressure is sea level pressure (i.e., 1 atmosphere) the pressure level in dB relative to 0.0002 dyn/cm^2, which is a standard measure for pressure levels in air, is as follows:

Sound Pressure Level $= 20\log_{10}p+74$ dB relative to 0.0002 dyn/cm^2 where p is expressed in dynes/cm^2. Table I gives the sound pressure levels for the adiabatic case as a function of the condensation S.

In order to get a physical feeling of the order of magnitude of the sound as a function of condensation S_0, Fig. 3 illustrates the intensity as a function of frequency showing the threshold of hearing and the threshold of pain [6]. The vertical lines, which are not frequency dependent, illustrate the intensity values for various values of condensation amplitude S_0. The intensity level is defined as follows:

$$\text{Intensity Level} = 10 \log_{10}(10^9 I) = 90 + 10 \log_{10} I, \qquad (35)$$

where I is expressed in ergs per square centimeter per second. Thus the intensity level is in dB relative to 10^{-10} microwatts per square centimeter per second. Note that at 3000 Hz (i.e., 3000 cyles per second) the threshold of hearing is about -6 dB relative to 10^{-10} microwatts/cm^2. This corresponds to a sine wave condensation of about 10^{-10} in air and would even correspond to a much smaller S_0 in water. This means that an average person could hear a sound in air at a frequency of 3000 Hz if the condensation amplitude was as small as 10^{-10}. The threshold of pain, which Fig. 3 shows is almost frequency independent, corresponds to condensations of the order of 5×10^{-4} in air. Thus the entire auditory area (i.e., from threshold of hearing to threshold of pain) is well within the region of linear acoustics. Comparing Table I with Fig. 3 it is seen that the sound pressure level corresponding to a condensation of 10^{-10}, which corresponds to the intensity at threshold of hearing, is about -3 dB relative to 0.0002 μbar and that the sound pressure level at threshold of pain is of the order of 125 dB relative to 0.0002 μbar.

LINEARITY AND NONLINEARITY AS RELATED TO ELASTICITY OF THE MEDIUM

In any physical system which involves forces and displacements it is the usual understanding that if the force is a linear function of displacement, then the system is linear and when the force becomes a nonlinear function of displacement, the system becomes nonlinear. The same concept holds true in acoustics. In order to see this use Eqs. (28), (29), (30), (31) and write the pressure in terms of velocity as follows:

$$\frac{\bar{p}}{p_0} = \left(1 + \frac{\gamma - 1}{2C}\,\dot{\xi}\right)^{2\gamma/\gamma - 1} \quad \text{(Adiabatic Case)};$$
$$\frac{\bar{p}}{p_0} = e^{\dot{\xi}/C} \quad \text{(Isothermal Case)}. \qquad (36)$$

For sine waves $\xi = \xi_0 \sin \omega t$, so

$$\left(\frac{\bar{p}}{p_0}\right)_{\max} = \left(1 + \frac{\gamma - 1}{2}\frac{\omega\xi_0}{C}\right)^{2\gamma/\gamma - 1} \quad \text{(Adiabatic Case)},$$
$$\left(\frac{\bar{p}}{p_0}\right)_{\max} = e^{\omega\xi_0/c} \quad \text{(Isothermal Case)}, \qquad (37)$$

but $p = \bar{p} - p_0$.

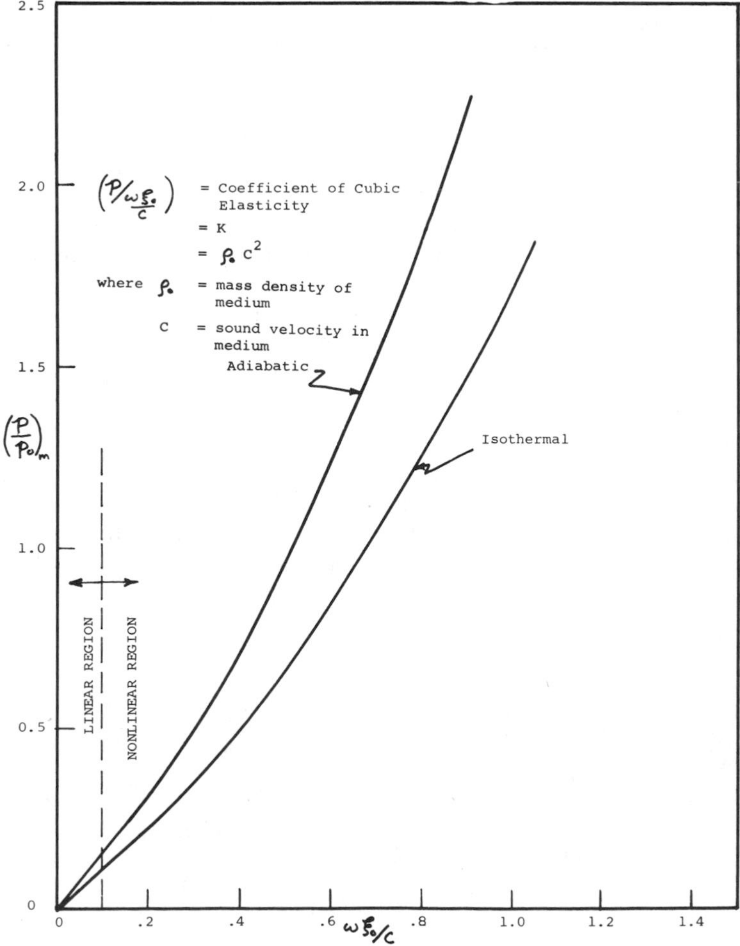

FIG. 4. Sound pressure as a function of displacement for sine waves.

Table II Pressure and Condensation for Blast Waves

$\bar{R}$	P_1/P_0	dB relative to 0.0002 μbar P_1	S_1	P_2/P_0	dB relative to 0.0002 μbar P_2	S_2
0.10	67.9	231	6.0	585	249	32.2
0.2	20.4	220	3.9	146	237	17.1
0.3	7.3	211	2.6	37.7	225	9.0
0.4	3.5	205	1.7	15.3	218	5.1
0.5	2.1	200	1.1	9.4	213	3.2
0.6	1.4	197	0.8	6.1	210	2.1
0.8	0.77	192	0.5	2.6	202	1.1
1	0.51	188	0.33	1.3	196	0.66
2	0.16	178	0.11	0.36	185	0.22
3	0.089	173	0.063	0.19	180	0.12
4	0.062	170	0.044	0.13	176	0.087
5	0.047	167	0.033	0.095	174	0.066
6	0.037	165	0.027	0.077	172	0.053
8	0.026	162	0.019	0.054	169	0.039
10	0.020	160	0.014	0.040	166	0.028
20	0.0087	153	0.0062	0.018	159	0.012
30	0.0054	149	0.0039	0.011	155	0.0077
40	0.0039	146	0.0028	0.0079	152	0.0056
50	0.0030	144	0.0022	0.0061	150	0.0043
60	0.0025	142	0.0018	0.0050	148	0.0035
80	0.0018	139	0.0010	0.0036	145	0.0021
100	0.0014	137	0.00082	0.0028	143	0.0016
500	0.00024	122	0.00017	0.00049	128	0.00033
1000	0.00012	116	0.000082	0.00023	121	0.00017

In Fig. 4 the dimensionless pressure ratio, p/p_0, is plotted as a function of the dimensionless deformation parameter $\omega\xi_0/C$. It is seen in Fig. 4 that when the displacements are small the pressure is a linear function of displacement. As the displacements become larger the medium becomes stiffer and a small change in displacement gives a proportionally larger increase in pressure. The nonlinear region starts at $\omega\xi_0/C \approx 0.1$. This corresponds to an $S \approx 0.1$, which is a much better criterion for linearity than Fig. 2.

EXPLOSIVE WAVES

Consider next the sounds generated by explosive-type waves. Unless the charge weight of the explosive is exceptionally small, the sounds generated by the explosion are in the nonlinear acoustic region at the point of the explosion, but as they die out away from the explosion, they become smaller and enter the linear region. In his book *Explosions in Air* [12], Baker gives the necessary information needed to estimate the sounds obtained from the explosions. The waves produced by the explosions are shock waves and have steep wave fronts. Therefore they have to be analyzed in a different manner from the relatively simple way that has been explained previously in this article. However, the results for shock waves will be given here in order to give the reader a feeling of the sounds produced by these waves as a function of the explosive weight, the distance from the explosion, and the type of explosive. Table II gives the pressure levels and condensation values for explosive waves.

In Table II $\bar{R} = Rp_0^{1/3}/E^{1/3}$, where R is the distance from the explosion, p_0 is the ambient pressure, and E is the total energy in the explosive charge. P_1, S_1 correspond to the pressure and condensation in the incident blast wave (i.e., the wave coming directly from the explosion to the point at which the pressure is being measured) and P_2, S_2 correspond to the pressure and condensation in the reflected wave. The reflected wave parameters are determined from the assumption that the reflection occurs from a rigid wall.

Table III contains the energy characteristics for several of the more important explosives [12]. In Table III the symbol # represents a pound of force (the weight) while lb_m represents a pound of mass. In order to obtain the energy for a given weight of explosive we use the following relation:

$$E = (E/M)(W/g).$$

Thus the energy contained in 1000 # of TNT is

$$E = 18.13 \times \frac{1000}{386} = 47 \times 10^6 \text{\# in.}$$

The value of $\bar{R}$ at 1 mile from an explosion of 1000 # of TNT would be

$$\bar{R} = 5280 \times 12 \times \left(\frac{14.7}{47 \times 10^6}\right)^{1/3} = 430$$

Examination of Table II indicates that for $\bar{R} = 430$ the incident blast pressure is of the order of 125 dB relative to 0.0002 μbar, and the reflected pressure of the order of 130 dB relative to 0.0002 μbar, both being around the threshold of pain for hearing. However, since the blast wave acts only for a very short time, the pain in the ear will undoubtedly not have time to develop for such an explosion.

Table III Characteristics of Explosives

Explosive	Specific Energy E/M (in./lb_m)
Pentolite	20.50×10^6
TNT	18.13×10^6
RDX	21.5×10^6

The values of condensation S contained in Tables I and II compare very favorably for sound pressure levels less than 200 dB relative to 0.0002 μbar, i.e., for S values less than 1. For the larger S values the characteristics of the shock wave front enter the problem and there is no longer any correlation between the value contained in the two tables.

See also ACOUSTICS, ARCHITECTURAL; FLUID PHYSICS; NONLINEAR WAVE PROPAGATION; SHOCK WAVES AND DETONATIONS; SOUND, UNDERWATER.

REFERENCES

1. A. W. Duff (ed.), *Physics for Students of Science and Engineering*. Blakiston Co., Philadelphia, PA, 1937. (E)
2. F. W. Sears and M. W. Zemansky, *College Physics*. Addison-Wesley, Reading, MA, 1960. (E)
3. Lord Rayleigh, *The Theory of Sound*. Dover, New York, 1945. (A)
4. L. E. Kinsler and A. R. Frey, *Fundamentals of Acoustics*. Wiley, New York, 1962. (I)
5. L. L. Beranek, *Acoustics*. McGraw-Hill, New York, 1954. (I)
6. P. M. Morse, *Vibration and Sound*. McGraw-Hill, New York, 1948. (I)
7. E. Skudrzyk, *The Foundations of Acoustics*. Springer Verlag, Wein, 1971. (A)
8. P. M. Morse and K. U. Ingard, *Theoretical Acoustics*. McGraw-Hill, New York, 1968. (A)
9. G. B. Whitham, *Linear and Nonlinear Waves*. Wiley, New York, 1974. (A)
10. R. B. Lindsay, *Mechanical Radiation*. McGraw-Hill, New York, 1960. (I)
11. H. Lamb, *The Dynamical Theory of Sound*. Dover, New York, 1960. (I)
12. W. E. Baker, *Exlplosions in Air*. University of Texas Press, Austin Texas, 1973. (I)
13. R. T. Beyer, *Nonlinear Acoustics*. Naval Sea Systems Command, 1974. (I)
14. J. E. Greenspon "Acoustics, Linear", *Encyclopedia of Physical Science and Technology, Vol. I,* p. 135. Academic Press, San Diego, California 1987.

Acoustics, Musical *see* Musical Instruments

Acoustics, Physiological

Juergen Tonndorf

Physiological acoustics, an expression coined after Helmholtz's *Physiological Optics* (1856), concerns itself with analytical assessments of the reception of sound by the ear and of the further processing of the signals thus received at the various levels of the central auditory nervous system.

This work requires close cooperation between physiologists and anatomists. Formal analyses are made possible by inputs from fluid mechanics (inner-ear dynamics); biochemistry (chemical events underlying the responses of the sense organ); systems analysis, including electrical network theory, and advanced statistics (electrical responses in both the organ and the central nervous system, equivalent network analysis); and many others.

The following brief description of the current state of the art in physiological acoustics must include some anatomical remarks.

The ear is traditionally divided into three parts: the outer, middle, and inner ears (Fig. 1). Outer and middle ears help shape the acoustic signal for optimal reception by the inner ear and its receptor cells.

The outer ear consists of the pinna and ear canal. The ear canal, a continuation of the funnel-shaped pinna, is an open tube terminated at its inner end by the tympanic membrane. The latter seals off the middle ear from the outside. The middle ear, an air-filled cavity (with a volume of approximately 2 cm^3), can be aerated via the Eustachian tube, a connection with the upper pharynx. The tympanic membrane, a thin elastic structure, vibrates in response to sound. It is connected by a mechanical transmission chain, consisting of a series of three small leverlike ossicles, to one of the two "windows" of the inner ear, the oval window. The other one, the round window, looks likewise into the middle ear, but it is simply closed over by a membrane. The inner ear is deeply hidden in hard dense bone. Like that of all vertebrates, it is a fluid-filled cavity, very complex in shape, whence its classical name, otic labyrinth. In addition to the auditory receptor organ, the inner ear houses five other organs that have to do with spatial orientation and maintenance of equilibrium, the so-called vestibular system. (The "vestibule" is part of the inner ear.)

The functions of outer and middle ears are (a) protection of the inner ear (foremost by virtue of their position, which shields the inner ear against physical insults; then there are middle-ear nonlinearities occurring at high amplitudes; and finally there are the two small middle-ear muscles; their reflex contraction on sound exposure attenuates middle-ear transmission); (b) optimization of transmission of acoustic energy into the inner ear (the impedance of the fluid-filled

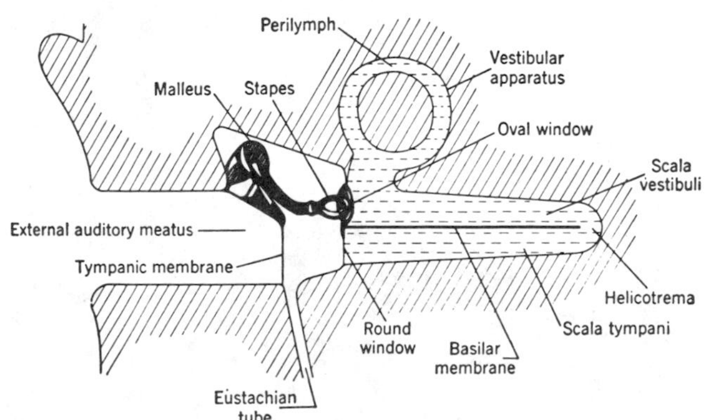

FIG. 1. Highly schematic outline of the ear (from von Békésy and Rosenblith, 1951).

inner ear is much higher than that of the air on the outside; thus, impedance matching is needed; this task is accomplished by a series of mechanical transformers that are completely integrated with one another, involving both the outer and middle ears).

In addition to the route just described, i. e., via tympanic membrane–ossicular chain–oval window (so-called air conduction), mechanoacoustic energy may also be brought into the ear when the bones of the head are set in vibration by contact with a vibrating object. This bone-conduction mode plays an important role in the clinical diagnosis of hearing disorders.

The auditory receptor organ is located in the lower, or cochlear portion of the inner ear. (The vestibular system occupies the upper portion.) The cochlea is a long (35 mm) but narrow bony chamber, coiled up $2\frac{1}{2}$ times like a snail shell (which is what cochlea means in Latin). It is subdivided by a number of membranes into a system of three compartments ("scalae") (Fig. 2). The receptor organ, the organ of Corti, lies in the middle scala and stretches over the whole length of the cochlea. It consists of about 16 000 to 20 000 receptor cells, the hair cells, distributed in a characteristic four-row pattern and held in place by a supporting-cell structure. Each hair cell carries a tuft of 80–100 sensory hairs on its top surface. Sensory cells of this kind are also found in other receptor organs; all of them are stimulated by a mechanical, sideways deflection of their hairs. In the cochlea, the necessary mechanism is provided by the connection of the sensory hairs with a membrane that covers the organ of Corti from side to side for its full length. This tectorial membrane (Fig. 2) executes a sliding ("shearing") movement across the organ that leads to the deflection of the sensory hairs. This constitutes the ultimate mechanical input to the hair cells; it is the final one in a series of interlinked mechanical events that are elicited when mechanoacoustic energy enters the cochlea, usually via the oval window. Such signals set up a series of displacements of the cochlear membranes, including the basilar membrane, on which the organ of Corti is situated (Fig. 2). These displacements progress along the basilar membrane in the manner of traveling waves (Fig. 3), invariably in the direction from the cochlear base to its apex. In this respect, the cochlea acts like a mechanical delay line. For a given sine-wave input, the traveling-wave mechanism creates a displacement maximum at a distinct, frequency-dependent place along the membrane ("place principle"). For high frequencies, the maxima are formed near the cochlear base, and as frequency goes lower this place moves toward the apex in a systematic manner (Fig. 4). Therefore, given sine-wave signals stimulate only limited regions along the basilar membrane and hence distinct

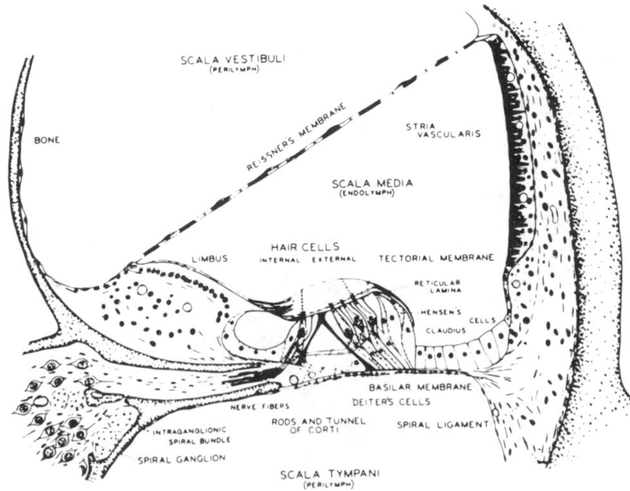

FIG. 2. Cross section of scala media (guinea pig) (from Davis *et al.*, 1953).

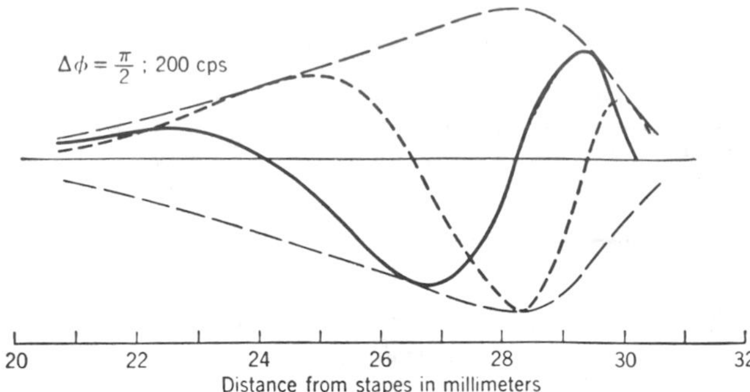

FIG. 3. Schematic outline of cochlear traveling wave at two instances, 90° apart in phase. Waves are moving from base to apex, i. e., left to right in the figure; amplitudes overstated (from von Békésy and Rosenblith, 1951).

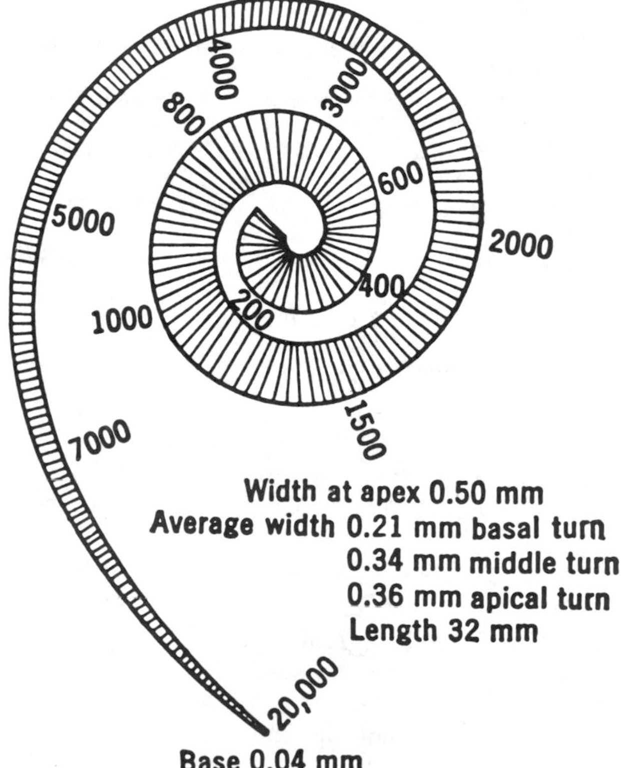

FIG. 4. Outline of the human basilar membrane. Note its width increasing with distance. Places of frequency maxima as indicated (from Stuhlmann, 1943).

groups of hair cells. This tonotopic relation is maintained throughout the entire central auditory system. It enables the latter to process frequencies, by substituting place for frequency, up to approximately 20 kHz, while single fibers of the auditory nerve are capable of responding to frequencies not higher than 3–4 kHz.

At its bottom end, each hair cell is supplied by a fiber (or fibers) of the cochlear (sensory) nerve. There are some 25 000 to 30 000 individual fibers, and their pattern of distribution to the hair cells is complex but systematic. The mechanoacoustic signals received by the hair cells elicit—in a multiple-step operation that bridges the hair-cell–nerve junction—nerve-action potentials (bursts of negative electrical impulses) of essentially the same kind as those observed in fibers of all other nerves. Their energy source is chemoelectric and inherent to the nervous system. These potentials represent a signal code particularly suited for neural transmission and processing.

After entering the part of the brain known as the brain stem, the auditory fibers run in well-defined tracts that go sequentially from one central station ("nucleus") to the next higher one (Fig. 5). This chain of nuclei finally terminates in the auditory portion of the cortex located in the temporal lobe of the brain. In each nucleus, the signal carried by the

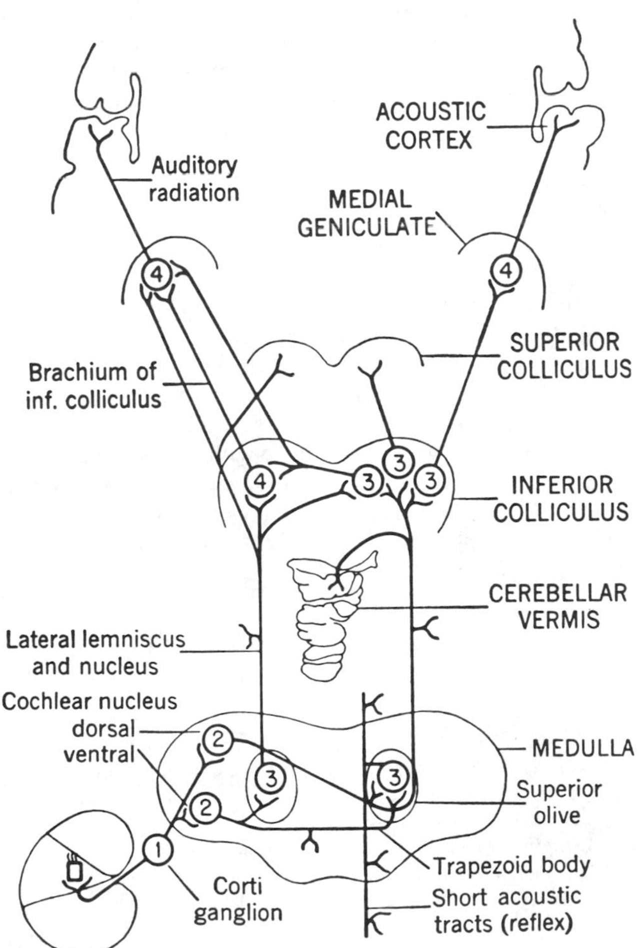

FIG. 5. Schematic outline of the afferent auditory system (from Davis, 1951).

incoming fibers is switched onto a new set of outgoing fibers. The underlying networks are structurally very intricate, allowing for complex signal processing. Most, but by no means all, fibers in each tract cross over to the opposite side, so that the left brain receives primarily signals from the right ear, and vice versa. The left auditory cortex appears to handle mainly signals for which analytical processing is of importance (e. g., speech signals), while the right one is primarily concerned with signals of emotional importance (e. g., music). In the region where the two tracts first cross over, signals received by the two ears are brought together in a set of special nuclei (superior olive; Fig. 5) initiating neural processing that concerns spatial hearing. These ascending ("afferent") fiber tracts are paralleled by a similar system of descending ("efferent") tracts that appear to exert a central (feedback) control upon the input at various levels, mainly at the hair-cell–nerve junction.

See also FLUID PHYSICS; NETWORK THEORY, ANALYSIS AND SYNTHESIS; STATISTICS.

BIBLIOGRAPHY

G. von Békésy, *Experiments in Hearing*. McGraw-Hill, New York, 1960.

G. von Békésy and W. Rosenblith, in *Experimental Psychology*, (S. S. Stevens, ed.), Chapter 27. Wiley, New York, 1951.

P. Dallos, *The Auditory Periphery*. Academic, New York, 1974.

H. Davis, in *Experimental Psychology* (S. S. Stevens, ed.) Chapter 28. Wiley, New York, 1951.

O. Stuhlmann, Jr., *Introduction to Biophysics*, Wiley, New York, 1943.

Acoustoelectric Effect

Esther M. Conwell

The acoustoelectric effect is the appearance of a dc electric field when an acoustic wave propagates in a medium containing mobile charges. It was first named and discussed, theoretically, by R. H. Parmenter in 1953. As pointed out by G. Weinreich, who first detected it experimentally, it is an example of the general phenomenon of "wave–particle drag," of which the operation of a linear accelerator and the motion of driftwood toward a beach are other examples. Although first proposed for metals, it is only significant in semiconductors. It can be quite strong in piezoelectric semiconductors, such as CdS, ZnO and GaAs, but is also seen in nonpiezoelectric semiconductors. In the piezoelectric case, the wave and mobile charges interact through the electric field arising from the strain associated with the wave. This interaction is significant when the electric field is longitudinal, i.e., parallel to the wave propagation direction, and in what follows we assume that the type of wave (i.e., longitudinal or shear) and sample orientation have been chosen to make this the case. In nonpiezoelectric media the interaction is through the shift of carrier energy induced by the strain, the so-called deformation potential. At the frequencies ordinarily used, $\lesssim$1 GHz, the latter interaction is much smaller than that in piezoelectric materials. In either case, however, because of the interaction the passage of an acoustic wave through the medium causes a periodic spatial variation of the potential energy of the charge carriers. If the mean free path l of the carriers is small compared to the acoustic wavelength λ, as is the case for most of the usual (ultrasonic) frequency range, this results in a bunching of the carriers in the potential-energy troughs. Since the wave is propagating, it drags the bunches along with it. This is the origin of the acoustoelectric field and clearly also causes attenuation of the wave. The effect is stronger, the stronger the bunching. Thus it is very weak in ordinary metals, where space-charge effects prevent appreciable bunching. It is enhanced in those nonpiezoelectric semiconductors where space-charge effects are minimized by either (1) the presence of both positively and negatively charged carriers or (2) the existence of different groups of carriers (many-valley band structure) whose energy is affected differently by the strain so that they bunch in different phases of the wave. Being sensitive to the number of free carriers, acoustoelectric interaction has proven to be a powerful tool in nondestructive testing of semiconductors, testing that does not even require contacts.

The foregoing discussion suggests that if an external electric field were applied to the sample to give the carriers a drift velocity, v_d, greater than the wave velocity, v_s, the carriers should drag the wave, i.e., the wave should be amplified. This conjecture was verified experimentally on CdS samples at frequencies of 15 and 45 MHz by Hutson, McFee, and White in 1962. They found acoustic gain for shear waves at fields greater than 700 V/cm, at which field v_d equals the shear wave velocity. For not too large acoustic wave amplitudes it was possible to account quite well for the size of the gain and its variation with frequency, etc., with a linear phenomenological theory taking into account the currents and space charge produced by the piezoelectric fields that accompany the acoustic wave. The gain, called *acoustoelectric gain,* is found to be low at low frequencies (less than the conductivity relaxation frequency σ/ϵ), where the carriers can redistribute themselves quickly enough to essentially cancel out the piezoelectric field. It peaks at the frequency for which the acoustic wavelength is of the order of the Debye length, where the bunching is optimum. In a fairly strong piezoelectric like CdS acoustoelectric gains as high as 40 dB/cm have been found, leading to consideration of this effect for practical use as an amplifier (*acoustoelectric amplifier*). This type of amplification has been found particularly useful for amplification of surface acoustic waves (SAWs). Because the amplitude of a SAW decays exponentially with distance below a free surface, the surface acts as a waveguide for such a wave. SAWs at microwave frequencies are easily introduced into a piezoelectric material such as $LiNbO_3$ by coupling in microwaves through a suitable transducer. A similar transducer can reconvert the SAWs into microwaves. The SAW velocity being smaller by a factor of 10^5 than electromagnetic wave velocity, a short length of a SAW-propagating material is useful as a delay line and for various types of signal processing. For maximum utility the losses of the SAWs in the guide are conveniently overcome by incorporating acoustoelectric gain. If the piezoelectric material is insulating, this may be accomplished by providing a conducting layer, e.g., Si, in contact with it. Alternatively, a conducting piezoelectric material, e.g., GaAs, may be used to support both the SAW and the electrons drifting in the electric field.

The theory and effects considered so far are linear in the sound amplitude provided it is small, i.e., small enough to bunch only a small fraction of the carriers. At large sound amplitude new effects appear. When two acoustic waves are present, the interaction of the piezoelectric field of one with the bunched carriers of the other may result in the generation of the difference or sum frequency of the two. As a special case of this, for a single large wave the interaction of its piezoelectric field with its own carrier bunches results in the generation of a dc current, called the *acoustoelectric current,* flowing in a direction opposite to the usual or Ohmic current, and may result in the generation of the second harmonic.

When an outside acoustic wave is not introduced, application to a highly conducting sample of CdS or ZnO, for example, of a field high enough to make $v_d > v_s$ causes a large amplification of the thermal equilibrium acoustic waves or flux present in the sample. This gives rise to unusual behavior, the exact nature of which depends on the details of sample inhomogeneity and contacts. One possibility is that immediately after the high field is applied a dc current will flow of the expected magnitude for the Ohmic resistance, i.e., the resistance displayed for $v_d < v_s$, but in a short time the current will drop to a much smaller value and remain there. The smaller value is due to the opposing acoustoelectric current arising from the flux amplification. Another frequently seen possibility is the onset of strong current oscillations with a period equal to the length of the sample divided by v_s. These oscillations are due to the creation close to the cathode of a domain or narrow region of high acoustic flux density, which typically moves down the sample with velocity v_s. When it exits at the anode, the current rises to its Ohmic value, strong flux generation begins again at the cathode, and the process is repeated.

The above discussion has been couched in terms appropriate for $l \ll \lambda$, since this is the case for most of the experimental situations studied. However, both linear and nonlinear processes have been studied with a microscopic theory that does not make this restriction. In this theory acoustic gain, for example, may be thought of as due to an excess of stimulated phonon emission by the carriers over absorption.

See also PIEZOELECTRIC EFFECT; SEMICONDUCTORS.

BIBLIOGRAPHY

N. G. Einspruch, "Ultrasonic Effects in Semiconductors," in *Solid State Physics* (F. Seitz and D. Turnbull, eds.), Vol. 17, p. 217. Academic Press, New York, 1965.

J. H. McFee, "Transmission and Amplification of Acoustic Waves in Piezoelectric Semiconductors," in *Physical Acoustics* (W. Mason, ed.), Vol. IV, part A, p. 1. Academic Press, New York, 1966.

H. N. Spector "Interaction of Acoustic Waves and Conduction Electrons," in *Solid State Physics* (F. Seitz and D. Turnbull, eds.) Vol. 19, p. 291. Academic Press, New York, 1966.

R. Bray, "A Perspective on Acoustoelectric Instabilities," *IBM J. Res. Devel.* **13**, 487 (1969). See also other articles in this volume, pp. 494–510.

N. I. Meyer and M. H. Jorgensen, "Acoustoelectric Effects in Piezoelectric Semiconductors with Main Emphasis on CdS and ZnO," in *Festkörper Probleme X* (O. Madelung, ed.), p. 21. Vieweg, Braunschweig, Germany, 1970.

E. M. Conwell and A. K. Ganguly, "Mixing of Acoustic Waves in Piezoelectric Semiconductors," *Phys. Rev.* **B4**, 2535 (1971).

Adsorption

J. G. Dash

The surfaces of all solids and liquids adsorb foreign molecules from the atmosphere. The adsorbed molecules change most of the chemical and physical properties of the underlying substrate: Adhesion, catalysis, corrosion, fracture, lubrication, and wear are affected by the topmost molecular layers on a surface. An understanding of these changes involves the study of the films themselves.

The forces of attraction that cause adsorption are the relatively weak and long-range interactions that exist between all atoms, molecules, and macroscopic objects. These "dispersion" forces are due to the attractions between the electric dipole moments induced in each body by the fluctuating fields of their neighbors. Dispersion forces between neutral atoms and molecules are also known as "van der Waals forces," and they are responsible for the condensation of vapors to liquid or solid phases at sufficiently low temperature. Temperature plays a similar role in adsorption, lower temperatures favoring the adsorbed state and decreasing the vapor pressure of the surface film.

The states of surface films depend on their thickness, temperature, and composition, as well as the structure, uniformity, and constitution of the substrate. As a result of this interplay films display a great variety of distinctive regimes. Monolayer films exhibit two-dimensional analogs of the familiar vapor, liquid, and solid phases of bulk matter, as well as others that have no three-dimensional equivalents. In thicker films some layers may behave with separate thermodynamic individuality while the remainder is diffuse. On irregular surfaces films possess a distribution of properties, different regions reacting to the local surface environment. The various regimes are experimentally studied under controlled laboratory conditions. They can be examined by one or more of a wide variety of experimental techniques, including calorimetry, nuclear resonance, electron-x-ray, and neutron diffraction, molecular scattering, and Mössbauer spectroscopy. Development and application of suitable methods for film studies have proceeded rapidly since the 1960s. The measurements have provided the impetus for considerably greater theoretical interest.

Theoretical attention is focused on the special film-substrate combinations and their experimental parameters, but also on more general questions involving the physics of two-dimensional systems. The physical properties of lower-dimensional systems have interested theorists for many years, since it had been believed for many years that in one- and two-dimensional matter there can be no perfectly ordered states or structures; in other words, two-dimensional crystals, magnets, superconductors, and superfluids cannot exist (except at absolute zero). The reason for this sweeping conclusion is the special richness of low-lying modes of excitation in lower-dimensional phase space, excitations which cause a breakdown of the long-range order at all finite temperatures. However, more recent considerations have shown that there are at least two forms of long-range order, and that only one of them is destroyed in two dimensions while the second can persist.

The new ideas were first outlined by Kosterlitz and Thouless in 1972, and then developed in considerable detail by many others. The theory shows that long-range structural or "topological" order can exist in two-dimensional systems over a finite temperature range: for example, a 2D crystalline lattice should have a finite rigidity up to a definite melting temperature, where its static shear modulus vanishes. The development of complete liquidlike disorder may take place via two successive continuous transitions: first, the unbinding and growth of pairs of dislocations, and then, the unbinding and growth of disclinations. In the interval between the two transitions the structure retains a short-range "hexatic" orientational order similar to that seen in layered liquid crystal films. However, the theory allows that continuous melting may be preempted by a first-order transition, depending on the relative magnitudes of the shear modulus and the defect energies. This may explain why many adsorbed monolayer films undergo first-order melting; up to the present, no unambiguous continuous melting transitions have been observed.

In the following sections several recognized regimes are described, taken in approximate order of increasing surface density.

1. Two-dimensional gases. When a uniform surface is sparsely covered, the adsorbed molecules can act like a two-dimensional gas. Its essential characteristics are adhesion to the substrate, which at sufficiently low temperatures leads to a "freezing out" of higher states of motion normal to the surface, and low surface density, which is equivalent to weak or negligible interactions between the molecules. Beyond these essentials the gas regimes embrace a wide range of properties depending on the atomic structure of the surface, i.e., its "texture." Surfaces composed of relatively small and densely packed atoms are relatively smooth to the adsorbed molecules, which are then free to move about the surface. Examples of very smooth surfaces are the dense (111) planes of close-packed metals and the basal plane of graphite. Low-density helium films on graphite exhibit heat capacities resembling classical two-dimensional gases at temperatures of 4 K and above, thus indicating that the atoms have considerable thermal mobility on the surface and are not strongly affected by interactions with each other. Films on more strongly textured surfaces are less mobile, and the atoms spend longer times in definite adsorption sites before evaporating to the three-dimensional gas phase or hopping to a neighboring site. This restricted mobility can be described in terms of a narrower band structure of the atoms, according to a two-dimensional band model analogous to the conventional theory of band structures of electrons in metals. If the energy bands have narrow widths and are separated by wide gaps, the atoms are relatively immobile; the film is "localized." At the other extreme the band structure is equivalent to a two-dimensional mobile gas, and a continuous range of intermediate mobility is possible. Theoretical band structures of helium on graphite yield wide bands and comparable width gaps, consistent with the thermal measurements. Heavier gases have quite narrow bands, and fragmentary experimental results are consistent with these indications of restricted mobility.

2. Imperfect gases and phase condensation. Interactions between the adsorbed atoms are more important at low temperatures and high surface densities. The effects of the interactions begin as small corrections to the equation of state of the low-density "ideal" film (see point 1). These changes are analogous to the deviations occurring in three-dimensional (3D) "imperfect" gases, conventionally described by a series expansion in the density, the so-called virial expansion. The second virial coefficients of several film–substrate combinations have been obtained experimentally by analysis of calorimetric data or vapor-pressure isotherms, and they are in fair to excellent agreement with theoretical coefficients calculated from empirical free-atom pair potentials. Phase condensation in monolayers to two-dimensional (2D) liquid or solid phases occurs in many monolayer systems when cooled to low temperatures. Critical temperatures are somewhat less than half of the critical temperatures of the bulk phases for most of the noble gases and other simple molecular gases. The gas–liquid critical point is especially interesting, since it belongs to the universality class of the two-dimensional Ising model, in which the critical exponent β is predicted to be equal to $\frac{1}{8}$. Measurements on several monolayer films yield values in good agreement with the theory.

3. Two-dimensional solids. Several monolayer systems have two-dimensional solid phases at high density and low temperature. The exemplary systems are ^3He and ^4He on basal-plane graphite, both of which exhibit classical signatures such as heat capacities varying as T^2, the two-dimensional analog of the well-known Debye T^3 law. The magnitudes of their coefficients indicate that the monolayer solids have elastic constants approximately equal to those of two-dimensional solid helium, but with some significant shifts due to substrate structure.

4. Registered phases. In some films the substrate structure imposes a regularity on the atomic arrangement in the monolayer. The simplest cases are denoted as 1×1 structures, in which adsorbed atoms have the same density and symmetry as substrate atoms. Various arrangements, classified as $n \times m$ structures, are results of the competition between substrate–atom and atom–atom interactions. Low-energy-electron-diffraction studies have disclosed structures as complex as 7×15. The number of distinct structures that could conceivably occur in a macroscopic film is extremely large, but energy considerations reduce the actual number of phases to a very few. The possible phase transitions between registered phases of different symmetries and between registered and nonregistered phases are very interesting for their relevance to theoretical models. Particular interest is attached to the order–disorder transition of the registered phases of ^3He and ^4He on graphite, which exhibit strong heat capacity peaks shaped like the power-law singularities of the appropriate theoretical model. Other registered phases of special interest include Kr and H_2 on graphite, which have triangular arrangements with lattice spacings $\sqrt{3}$ greater than the graphite mesh; Xe on Pd and Ni facets, which have subtle superlattice structures; and polar molecular gases such as Br_2 and $FeCl_2$, in which orientational ordering appears to take place.

5. Multilayer films. Films of several layers thickness display a variety of habits. In some examples the one or two layers closest to the substrate behave as relatively distinct two-dimensional solids while a topmost third layer is effectively a two-dimensional gas. In other films there may be as many as five distinct layers evidenced by the stepwise character of the vapor pressure during their formation. For many systems there is a transition between layer formation and cluster growth, generally occurring at a few layers thickness. The instability of layer formation *vis à vis* cluster formation is an important phenomenon in all types of films. It can occur in many systems involving strongly cohesive metal films as well as those of weakly bound van der Waals molecules. In strongly adsorbed solid films the substrate attraction tends to produce strained layers next to the substrate, which prevents the formation of very thick uniform deposits. This effect belongs to a complex of *wetting* phenomena, which are of fundamental interest and practical importance. Adsorbed multilayer films are also valuable as test systems for the study of phenomena that can occur on typical surfaces of bulk solid materials. Recent examples are surface roughening and surface melting, which have been observed in multilayer noble gas and light molecular films on graphite and magnesium oxide.

See also CATALYSIS; ISING MODEL; ORDER–DISORDER PHENOMENA; PHASE TRANSITIONS; SURFACES AND INTERFACES; THIN FILMS.

BIBLIOGRAPHY

W. A. Steele, *The Interaction of Gases with Solid Surfaces*. Pergamon Press, Oxford, 1974. (A)

J. G. Dash, *Films on Solid Surfaces*. Academic Press, New York, 1975. (A)

J. G. Dash, "Between Two and Three Dimensions," *Physics Today* **38,** 26 (1985). (E)

K. Strandburg, "Two Dimensional Melting," *Rev. Mod. Phys.* **60,** 161 (1988). (A)

S. Dietrich, in *Phase Transitions and Critical Phenomena, Vol. 12,* C. Domb and J. Lebowitz, Eds., Academic Press, London, 1988. (A)

M. den Nijs, in *Phase Transitions and Critical Phenomena, Vol. 12,* C. Domb and J. Lebowitz, Eds., Academic Press, London (1988). (A)

Aerosols

Franklin S. Harris, Jr.

An aerosol is a suspension of liquid, solid, or mixed particles in a gas, usually air. The size of the particles ranges from about 10^{-9} m, just larger than molecules, to a radius of about 25 μm, as in cloud droplets and dusts with short-time stability due to gravitational settling. Examples are hazes, mists, fogs, clouds, smokes, and dusts, as well as living bacteria, viruses, and molds. Aerosols are important in atmospheric electricity, cloud formation, precipitation processes, atmospheric chemistry, air pollution, visibility, radiation transfer, and hence climate.

The smallest particles, called Aitken nuclei, are from molecular sizes up to about 0.05 μm in radius. They vary in concentration from a few particles per cubic centimeter over the South Pole Plateau to 300 in clean continental air, up to hundreds of thousands in a polluted city or downwind from a combustion source. The condensation nuclei serve as centers upon which cloud and fog droplets form. The large droplets, as in fogs and clouds, ordinarily range from 1 to 25 μm with a concentration of 20–500/cm^3 and a liquid water content up to 1 g/m^3.

The tropospheric aerosol sources are (1) inorganic gas-to-particle conversion, primarily SO_2, NH_3, NO_x, both natural and man-made; (2) mineral dust, primarily from arid zones and deserts; (3) sea salt; and (4) organic matter of apparently complex but still unidentified origins, but thought to have heavy contributions from plant-derived terpene compounds, forest fires, and oxygenated hydrocarbons. The residence time in the lower troposphere is about 3–6 days for the particles. Above the ocean the maritime aerosol is found only at the lower altitudes; at higher altitudes there is continental aerosol such as the Sahara Desert dust over the North Atlantic Ocean. The wind (aeolian) transport of aerosols is sometimes as far as 10 000 km. By using elemental tracers regional pollution aerosols of both North America and Europe have been followed for several thousand kilometers downwind. Dusts from the central and eastern Asian deserts have been carried to Hawaii and the Marshall Islands in the Pacific Ocean. The continental atmospheric aerosol has approximately 60% of the total mass water-soluble material and 25–30% organic matter; 25% of the material is volatile at temperatures below 150°C. The hazes over remote areas may be due to photochemical transformation of terpenes from vegetation. Recent work has shown the importance of sulfate regionally distributed sources. Clean air background in remote areas of the earth is about 10 μm/m^3. The standard mass loading established by the U.S. Environmental Protection Agency, not to be exceeded appreciably, is 75 μm/m^3. The mass loading sometimes reaches to about 2000 μg/m^3. The number concentration varies from 10^2 to 10^7/cm^3. In the stratosphere the total number above about 0.01 μm is about 10 particles/m^3, primarily sulfates. There is a maximum at an altitude of about 20 km, the amount depending on the length of time since a major volcanic eruption.

Often a simple function has been found to represent the natural aerosol from 0.1 to 20 μm when in equilibrium, the Junge power law, $dn/d \log r = Cr^{-b}$, where dn is the number of particles in a logarithmic size interval, C a constant, r the radius, and b usually a value of about 3. No simple size model can represent the wide variety of sources and complex interactions in the atmosphere. With differing sources, often a log normal or modified gamma distribution can be used for each, or a combination of distributions with one for each source, such as one for small particles from combustion and one for larger dust particles. The particle sizes important for health through retention in the human body are those retained in the breathing system after passing through the nose, in the range 4.5 μm down to 0.25 μm radius. The particle size distribution in the atmosphere is affected by the type of source, the changes due to gas-to-particle conversion, condensation, and removal through aggregation, precipitation formation and washout, and gravitational settling.

Experiments have shown that at 75–95% relative humidity (RH) 0.3–0.9 of the submicron aerosol mass can be liquid water, and that even at 50% RH 0.1–0.2 may be liquid water. Actual dry maritime aerosol particles collected over the Atlantic Ocean may increase in volume from 5 to 15 times when the RH is increased to 96%. In air pollution from combustion sources the particles are originally small or are formed by gas-to-particle conversion, and in the Los Angeles, California, basin, 50–80% of the submicron aerosol mass is volatile at 220°C, with primarily sulfate, nitrate, noncarbonate carbon, and liquid water.

The optical behavior of the particles is determined by the complex refractive index (which includes the wavelength-dependent real and absorption parts), the shape, the size relative to the wavelength of the radiation, and the size distribution. For particles small compared to the wavelength, the scattering intensity is proportional to the inverse fourth power of the wavelength. For spherical particles (many particles are not), in the range of the radiation wavelength size, the Lorenz–Mie theory must be used in which complicated functions describe the polarization parameters as a function of scattering angle, refractive index, and size distribution. The maximum scattering per unit volume for visible light is for particles 0.5 μm in radius. The amount of solar energy absorbed is comparable in amount with the absorption by atmospheric gases. The absorption part of the refractive index is the critical parameter in determining whether such particles on a world-wide basis will tend to cause cooling or warming of the earth and hence climatic change. For a variety of purposes aerosols are often produced by using a gas under pressure to disperse liquids or solids into the atmosphere. One of the propellants commonly used is a group of chlorofluoromethanes which are chemically quite stable and nontoxic. Currently, however, there is serious investigation of the possible accumulation in the stratosphere and by complex processes reducing the ozone, letting more solar ultraviolet reach the earth's surface.

See also Atmospheric Physics.

BIBLIOGRAPHY

G. Bouesbet and G. Brehan, eds. *Optical Particle Sizing*. Plenum, New York, 1988. (A)

Ardash Deepak, ed. *Atmospheric Aerosols, Their Formation, Optical Properties and Effects*. Deepak, Hampton, VA, 1982. (A)

S. K. Friedlander, *Smokes, Dust, and Hazes, Fundamentals of Aerosol Behavior*. Wiley, New York, 1977. (I)

Peter V. Hobbs and M. P. McCormick, eds. *Aerosols and Climate*. Wiley, 1988. (A)

Kenneth Pye, *Aeolian Dust and Dust Deposits*. Academic Press, New York, 1987. (I)

Parker C. Reist, *Introduction to Aerosol Science*. Macmillan, New York, 1984. (E)

Allotropy and Polymorphism

F. J. DiSalvo

The equilibrium crystal structure of some solids changes when the external conditions, such as pressure or temperature, are varied. In addition, the structure of some compounds depends upon the preparation conditions. One of these structures may be the thermodynamically stable structure, while the remainder are metastable phases. This phenomenon is called allotropy when it occurs in an element, and polymorphism when it occurs in a compound. Three common examples of allotropy are presented below.

SULFUR

Solid sulfur consists of nearly flat S_8 molecular rings that are stacked on top of one another. When heated to 95°C the molecules change orientation, forming a differently stacked structure. Sulfur at room temperature is called rhombic sulfur, and above 95°C, monoclinic sulfur (after the shapes of their respective crystallographic unit cells). Monoclinic sulfur melts at 120°C.

Sulfur can also exist in an amorphous form. When liquid sulfur is heated to several hundred degrees centigrade, most of the S_8 molecules break open and join with others to form long sulfur chains. If this liquid is rapidly cooled to room temperature, the chains remain intact and are randomly packed together to form a rubbery solid. At room temperature, amorphous sulfur will very slowly change back into rhombic sulfur. Rhombic sulfur is the stable, or equilibrium, form of sulfur at room temperature. By other preparation methods a number of other metastable forms of sulfur can be obtained at room temperature. Consequently, sulfur has a large number of allotropes; however, rhombic and monoclinic sulfur are the only equilibrium forms (in their respective temperature ranges of stability *and* at atmospheric pressure).

IRON

At room temperature, iron has a body-centered cubic (bcc) structure; the unit cell is shown in Fig. 1a. (The structure can be visualized by imagining space to be filled with closely packed cubes. At each corner, where eight cubes come together, place an iron atom and then put another in the center of each cube.) When iron is heated to 910°C its structure changes to face-centered cubic (fcc); a unit cell is shown in Fig. 1b. (In this structure an iron atom is placed at each cube corner and one iron atom on each face of the cube, where two cubes touch.) Iron changes back to the bcc structure at 1390°C and melts at 1536°C.

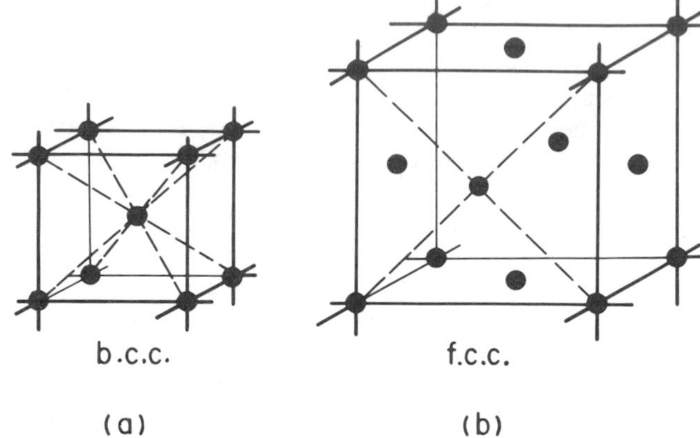

FIG. 1.(a) A unit cell of body-centered cubic (bcc) iron. Iron atoms are spheres that touch along the dotted lines (body center to cell edge). (b) A unit cell of face-centered cubic (fcc) iron. The cubic unit cell is larger than the bcc cell because the iron atoms now touch from the face center to the cell edge.

The allotropy of iron is very important for the production of steels. Carbon is moderately soluble in fcc iron, the carbon atoms occupying some of the holes between the iron atoms in this structure (at the center in Fig. 1b). However, the solubility of carbon is much lower in bcc iron. If iron containing several weight percent of carbon is cooled from 1100°C (fcc phase) to room temperature, the carbon not soluble in bcc iron forms a compound, Fe_3C. The Fe_3C exists in small plate-like regions dispersed in bcc iron. Fe_3C is called cementite, since it makes the iron much stronger. Iron prepared in this manner is called carbon steel.

CARBON

Carbon exists in two structural forms: graphite and diamond. Graphite has a layered structure with weak interlayer bonding. Because of the weak interlayer bonds the layers slide easily over each other. Consequently graphite is used as a lubricant and in pencil lead. Graphite becomes diamond under high pressure (greater than 10 000 atm). With Ni as a catalyst, small diamonds can be manufactured at 1200°C and high pressure. These diamonds are used in making grinding wheels and cutting tools, since diamond is a very hard material. Recently, diamond films up to millimeters thick have been prepared in the laboratory by a low-pressure plasma deposition technique using methane as a source gas.

CONCLUSION

Obviously the properties of some materials are quite affected by a change in their structure, and a knowledge of the allotropic or polymorphic forms of materials is important to the development of many technologies.

See also CRYSTAL SYMMETRY; METALLURGY.

BIBLIOGRAPHY

B. Meyer, *Elemental Sulfur*. Wiley (Interscience), New York, 1965. (I)

A. L. Ruoff, *Introduction to Materials Science*. Prentice-Hall, Englewood Cliffs, NJ, 1972. (I)

W. J. Moore, *Physical Chemistry*, 3d ed. Prentice-Hall, Englewood Cliffs, NJ 1963. (A)

Alloys

P. L. Leath

An alloy is a macroscopically homogeneous mixture of metals or, as in the case of carbon steel, a metallic mixture of metals and nonmetals. Most, but not all, alloys are metallic (for one exception, indium antimonide is a semiconductor). Most pairs of metals are miscible (i.e., form *binary alloys*) at some concentrations, although there are many notable exceptions (e.g., indium is insoluble in gallium). Since there are 70 elemental metals, the subject of alloys is immense, and an enormous variety of electronic and other physical properties is possible. The subject has now expanded even further with recent interest in ternary (three-component) alloys, tertiary (four-component) alloys, etc.

Alloys are often classified into ordered (or stoichiometric) and disordered alloys. The ordered alloys have the symmetry of a Bravais lattice with a multiatomic unit cell. Their structure is specified by giving the location of each atom in the unit cell. Some alloys exist essentially only as ordered alloys over the corresponding very narrow ranges of composition necessary for stoichiometry; these alloys are called *intermetallic compounds*. Other alloys (such as β-brass) have ordered phases at stoichiometric concentrations when the temperature is below a phase transition temperature but are disordered otherwise. (That is, they undergo an order–disorder phase transition.) More recently *quasiperiodic* alloys or *quasicrystals* (most notably Al_4Mn) have been discovered which display sharp diffraction peaks that form three-dimensional icosohedral patterns with 5-fold symmetry axes and which thus are not Bravais lattices and do not have the translational symmetry of crystals but do have point symmetries.

Disordered alloys, also called *solid solutions,* occur usually over appreciable ranges of composition. Most common are *substitutional* alloys, where the various types of atoms randomly occupy the normal sites of a lattice. But there are also *interstitial* alloys (such as carbon in γ-iron), where the solute atoms are small enough to occupy randomly the interstices between the normal lattice sites of the host metal; and recently there has been interest in *amorphous* alloys, where the atoms are not on sites of a regular lattice but are randomly placed, as in a liquid or glass. The occupations of the sites in a disordered alloy by the atomic types may be purely random but generally there is some degree of short-range order; that is, the species occupation of a particular site may be dependent on the occupation of the neighboring sites (e.g., in the disordered phases of brass, the copper atoms are more likely to have zinc than copper nearest neighbors).

Certain principal variables that qualitatively give the alloy structures and phases were pointed out in the classic work of Hume-Rothery and Jones. It is, however, only rarely possible to use these few variables to predict detailed behavior of alloy phases. Clearly the relative sizes of the atoms constitute a vital factor in alloys because the volume-dependent potentials in the cohesive energy are an order of magnitude larger than the interatomic rearrangement potentials. This size factor is especially important in interstitial alloys and certain intermetallic compounds (e.g., interstitial alloys are generally not formed when the ratio of the radii of the solute atoms to those of the host atoms exceeds about 0.6). Electrochemical differences are such that generally we find only intermetallic compounds or very restricted ranges of solubility for elements widely separated in the electrochemical series. Particularly interesting are the interstitial alloys of hydrogen in metals.

When size and electrochemical factors allow solid solutions, the alloy structure can in some cases be directly related to the electron density or electron-to-atom ratio. According to the Hume-Rothery rules, in nearly free-electron alloys the stable crystal structure at a particular electron-to-atom ratio will be that which minimizes the energies of the electrons in the crystal potential; thus the position of the Fermi surface relative to the Brillouin zone faces is an essential ingredient. These rules seem to work qualitatively well for the *d*-band transition metal alloys (especially copper, silver, and gold alloys), but they fail for the more nearly free-electron alkali metal alloys because of electrochemical differences. Clearly the *d* bands play an important role.

Only recently have basic calculational methods been developed to predict the physical behavior of alloys accurately from first principles. The pseudopotential, orthogonalized plane wave (OPW), augmented plane wave (APW), and Korringa–Kohn–Rostoker (KKR) Green's function methods of calculating electronic energy-band structure of metals beginning with the atomic potentials (the atomic potentials in alloys look very much the same as those atomic potentials do in the respective pure metals) are now in many cases being directly applied somewhat successfully in the calculation of such physical properties of alloys as energy-band structure, crystal structure, lattice vibration spectra, electrical and thermal resistivity, and magnetic and superconducting properties. In those cases where the potentials of the constituent atoms do not vary greatly the average potential may be used; this is called the *virtual crystal* (or rigid band, or common band) *approximation*. In the cases of strong disorder, when the potentials differ greatly, the *average t-matrix approximation* (ATA) and the *coherent potential approximation* (CPA), which are capable of producing the separate energy bands for each atomic species, are used. Although these calculations have been somewhat successful, such effects as charge transfer between atoms and atomic cluster effects are often important but are not included in the simple approximations just mentioned. A fine review of the experimental electronic properties of alloys is given by Sellmyer (1978).

Disordered alloys are dramatically different from ordered alloys and pure metals in their electrical resistance at low temperatures. In ordered metals there is a striking decrease in resistance with decreasing temperature that is absent in disordered alloys. For example, in very pure disordered brass at liquid-helium temperatures the electrical resistance is about half its room-temperature value, in contrast to drops by factors of about 10^{-4} in comparable ordered metals. This phenomenon is caused by electronic scattering off of the disorder or of those regions where the periodicity is destroyed.

Finally, the physical properties of alloys are often greatly affected by heat and mechanical treatment, which may introduce or eliminate such defects as vacancies, dislocations, or grain boundaries. For example, wrought alloys, which have been hot or cold worked and hence are generally very anisotropic and fibrous in contrast to cast alloys, which are generally crystalline, are generally more ductile. And there are *shape memory alloys* (notably Ti-Ni) which under certain treatment will return to an original shape. The effect of such defects is only understood qualitatively, although progress is rapidly being made.

See also ELECTRON ENERGY STATES IN SOLIDS AND LIQUIDS; METALS.

BIBLIOGRAPHY

G. Alefeld and J. Volkl, eds. *Hydrogen in Metals,* Vols. I & II. Springer-Verlag, Berlin, 1978.

R. Banks, *Shape Memory Effects in Alloys.* Plenum, New York, 1975.

C. S. Barrett and T. B. Massalski, *Structure of Metals,* 3d ed. McGraw-Hill, New York, 1966. (I)

R. J. Elliott, J. A. Krumhansl, and P. L. Leath, "The Theory and Properties of Randomly Disordered Crystals and Related Physical Systems," *Rev. Mod. Phys.* **46,** 465–543 (1974). (A)

M. Hansen and K. Anderko, *Constitution of Binary Alloys,* 2nd ed. (1958); R. P. Elliott, 1st suppl. (1965); F. A. Shunk, 2nd suppl. (1969). McGraw-Hill, New York. (A compendium of data on specific alloys.)

V. Heine and D. Weaire, "Pseudopotential Theory of Cohesion and Structure," in *Solid State Physics* (F. Seitz, D. Turnbull, and H. Ehrenreich, eds.), Vol. 24, pp. 249–463. Academic, New York, 1970. (A)

J. Janssen, M. Fallon, and L. Delacy, *Strength of Metals and Alloys* (P. Haasen, ed.). Pergamon, London, 1979.

N. F. Mott and H. Jones, *The Theory and Properties of Metals and Allloys.* Oxford, London and New York, 1936. (E)

P. S. Rudman, J. Stringer, and R. I. Jaffee, eds., *Phase Stability in Metals and Alloys.* McGraw-Hill, New York, 1967. (I)

D. J. Sellmyer, "Electronic Structure of Metallic Compounds and Alloys," in *Solid State Physics* (H. Ehrenreich, F. Seitz, and D. Turnbull, eds.), Vol. 33, pp. 83–248. Academic, New York, 1978.

P. J. Steinhardt and S. Ostlund, *The Physics of Quasicrystals.* World Scientific, Singapore, 1987.

Alpha Decay

Irshad Ahmad

Soon after the discovery of radioactivity by Becquerel in 1896, it was established that three types of radiations are emitted by radioactive substances. The most easily absorbed radiations were named alpha (α) rays. In 1909 Rutherford and Royds obtained a direct experimental proof that α particles are doubly ionized helium atoms. Since α particles are electrically charged they are deflected in electric and magnetic fields and produce intense ionization in matter. The thickness of material required to stop an α particle is called its range and it depends on the kinetic energy of the α particle and on the nature of the stopping medium. The range of an α particle with the typical energy of 6.0 MeV is ~5 cm in normal air and ~0.05 mm in aluminum.

At present more than 400 α-emitting nuclides are known and most of these are produced artificially. The kinetic energies of α particles range from 1.83 MeV for ^{144}Nd to 11.65 MeV for ^{212m}Po; these energies correspond to α particle velocities of (1–3) × 10^9 cm/s. The measured half-lives of known α emitters vary from 3.0 × 10^{-7} s for ^{212}Po decay to 2.1 × 10^{15} years for ^{144}Nd decay. Normally α decay occurs from the ground state of the parent nucleus and several groups of α particles (each group contains monoenergetic α particles) are emitted leaving the daughter nucleus in its ground state or in an excited state. In a few cases α particles are also emitted from an excited state of the parent nucleus. These α particles have kinetic energies of 9–12 Me V and are called long-range α particles. Examples of such α emitters are ^{212}Po and ^{214}Po.

Alpha decay has recently been used to characterize newly produced transactinide elements. By following the decay chain down to a known nuclide, it has been possible to determine the atomic number of a new element. This procedure has been used to identify elements 108 and 109.

For a nucleus to be unstable toward α decay, its mass must be greater than the sum of the masses of the daughter nucleus and the α particle. If we write the α decay of a nucleus with mass number A and atomic number Z as

$$^{A}Z \rightarrow {}^{(A-4)}(Z-2) + {}^{4}_{2}\text{He} \tag{1}$$

then the α decay energy, also called Q value, is given by

$$Q = (M_A - M_{(A-4)} - M_{\text{He}})c^2. \tag{2}$$

In the above equation, M represents the atomic mass and c is the velocity of light. Q values have been calculated from known atomic masses. Calculations show that Q values are positive for all β-stable nuclei with $A > \sim 150$. Although such nuclei are thus unstable with regard to α emission, in many cases the half-life is too long for the α decay to have been detected. Experimentally α radioactivity has been detected in most translead and some rare-earth nuclei.

In order to conserve linear momentum, the decay energy Q is divided between the α particle and the daughter nucleus in inverse proportion to their masses. The energy imparted to the daughter nucleus is called recoil energy. The α particle energy E_α and the recoil energy E_R are given by the equations

$$E_\alpha = (M_{A-4}/M_A)Q \quad \text{and} \quad E_R = (M_\alpha/M_A)Q. \tag{3}$$

The laws of conservation of angular momentum and of parity (even or odd character of the state wave function) plus the fact that the α particle has no intrinsic spin and even parity lead to simple selection rules for α decay. The orbital angular momentum L of the emitted α particle is restricted to integral values between the sum and the difference of the total spins of the initial and final nuclear states. If the parent and the daughter nuclear states have the same parity, only even values of L are permitted; if their parities are opposite, only odd L values are allowed.

The energies of α particles and the intensities of α groups are measured with gas ionization counters, solid-state detectors, or magnetic spectrographs. At present Au-Si surface-barrier detectors are widely used in spectroscopic measurements. These silicon detectors, under the best conditions, have resolutions [full width at half-maximum

(FWHM) of the α peak] of 10.0 keV and efficiencies of ~30%. Magnetic spectrographs, on the other hand, have low transmission (~0.1%) but can achieve resolution (FWHM) of less than 3.0 keV for 6.0-MeV α particles. Very thin, essentially massless, sources are used in these measurements. The absolute energies of the α particles emitted by a few nuclides have been measured with high precision by Rytz using a magnetic spectrograph; energies of other α groups are measured relative to these standards. Because of the monoenergetic character of α groups plus the fact that their energies and intensities can be measured with high precision, α particle spectroscopy has been extensively used in nuclear-structure studies of heavy elements.

The systematic relationship between the α decay half-life and the decay energy was first discovered by Geiger and Nuttall in 1911. According to a modified version of their rule, when the logarithm of the α decay half-life is plotted against the inverse square root of Q, a straight line is obtained for each element; i.e.,

$$\log t_{1/2} = A/Q^{1/2} + B, \qquad (4)$$

where A and B are constants and depend on the atomic number Z. This relationship applies only to α transitions between ground states of even–even nuclei. Because of the strong dependence of the α decay rate on the Q value, only states up to a few hundred kilovolts excitation are measurably populated.

The mechanism of α decay and the Geiger–Nuttall rule were first explained by Gamow and independently by Gurney and Condon in 1927. In the potential-energy diagram (see Fig. 1) the maximum occurs at R, where R is equal to the sum of the radii of the α particle and the residual nucleus. At distances $r<R$ the potential is attractive because of the short-range nuclear force and for $r>R$ electrostatic repulsion gives a positive potential which decreases with increasing r according to Coulomb's law. The typical energy of an α particle within the nucleus, with respect to the zero of energy at $r=\infty$, is 6.0 MeV and the height of the potential barrier at R (called the Coulomb barrier) for heavy elements is ~20 MeV. Since the kinetic energy of the α particle is less than the barrier height, according to classical mechanics the α particle can never leave the nucleus. However, the wave nature of matter permits a 6.0-MeV α particle occasionally, on the nuclear time scale, to "tunnel" through the 20-MeV potential barrier. Using a simplified shape for the potential and assuming that the α particle pre-exists as a clustered entity inside the nucleus and is constantly impinging on the barrier, Gamow derived an expression for the α decay rate which explains the observed exponential dependence of transition rates on Q values.

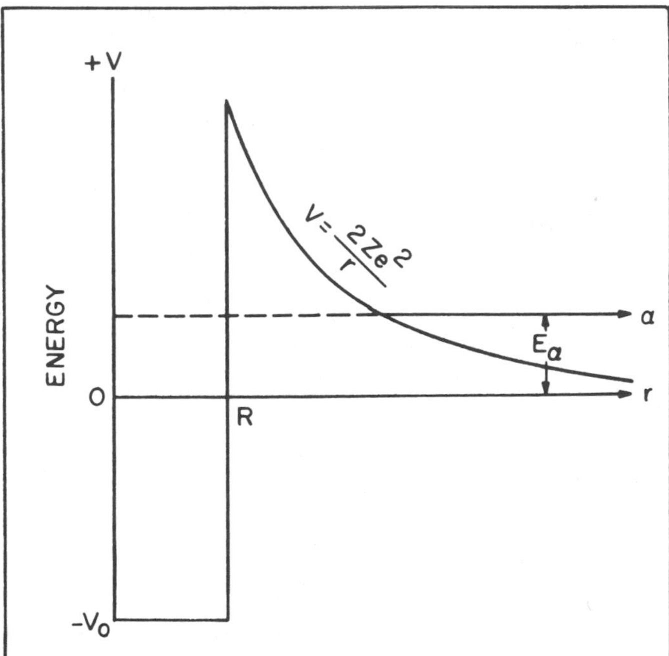

FIG. 1. Schematic representation of the potential energy of an α particle in the vicinity of a heavy nucleus. The potential energy is plotted against the distance r between the centers of the α particle and the residual nucleus.

The measured partial half-lives of α groups in the decay of odd-mass and odd–odd nuclei and α transitions to the excited states of even–even nuclei are found to be longer than the partial half-life $(t_{1/2})_{e-e}$ of an α group of the same energy emitted in the decay between ground states of even–even nuclei. The relative retardation of the former decay is called its hindrance factor and its reciprocal gives the relative reduced α transition probability. The values of $(t_{1/2})_{e-e}$ are calculated either by Eq. (4) or by some other theory. In most recent publications the values of $(t_{1/2})_{e-e}$ are computed with the one-body α decay theory of Preston. In this theory, as in Gamow's, the α particle is assumed to pre-exist inside the nucleus and is ejected with no orbital angular momentum. Radius parameters of even–even nuclei are obtained by normalizing to the measured transition rates between their ground states and the radius parameters of odd-mass nuclei are determined by interpolation between the values of adjacent even–even nuclei.

Hindrance factors for α transitions of odd-mass nuclei vary from unity to several thousands and these yield significant information on the parent and daughter states involved in the decay. Alpha transitions of odd-mass and odd–odd nuclei with hindrance factors of 1–4 are called favored transitions; in these decays the parent and the daughter states have similar wave functions. Since the α transitions are only mildly inhibited by angular momentum changes L, high hindrance factors give a clear indication that the α particle does not exist as a clustered entity all the time in the corresponding nucleus. Instead, the α particle is formed from four nucleons (two protons and two neutrons) at the nuclear surface at the time of its ejection. The probability for the formation of an α particle from four nucleons can be calculated theoretically. Alpha-decay rates for spherical nuclei in the lead region and spheroidal actinide nuclei have been calculated by Mang and Rasmussen and these reproduce the general trend in the observed α decay rates. Although in most cases the calculated and measured rates agree within a factor of 2, there are several unfavored transitions for which the calculations and measurements differ by a factor of ~10. Despite these deficiences, these calculations are extremely useful in nuclear structure studies.

See also NUCLEAR PROPERTIES; RADIOACTIVITY.

BIBLIOGRAPHY

R. D. Evans, in *McGraw-Hill Encyclopedia of Science and Technology*, Vol. 1, p. 305. McGraw-Hill, New York, 1971. (E)

I. Perlman and J. O. Rasmussen, in *Handbuch der Physik*, Vol. 42, p. 109. Springer-Verlag, Berlin, 1957. (I)

J. O. Rasmussen, in *Alpha-, Beta-, and Gamma-Ray Spectroscopy* (K. Siegbahn, ed.), Vol. 1, p. 701. North-Holland, Amsterdam, 1965. (A)

Ampere's Law

Louis T. Klauder, Jr.

The name Ampere's law has been applied to several of the formulas that give magnetic effects of time-independent electric currents. (Formulas given here assume rationalized mks units.)

The equation

$$\oint_C \mathbf{H}\cdot d\mathbf{l} = \mathbf{I} \tag{1}$$

relating the magnetic intensity along a closed curve C and the current I linking C is commonly referred to as Ampere's law or Ampere's circuital law. This nomenclature has become popular in recent years because it associates a useful elementary formula with the most important of the original investigators. Historically, this law was discovered by Gauss with help from the theorem of Ampere stating that the field produced by a magnetic shell is the same as that due to a current flowing around the boundary of the shell. The modern form distinguishing between B and H was first given by Maxwell.

A few authors apply the term Ampere's law to both the integral relationship (1) and the corresponding differential equation

$$\nabla \times \mathbf{H} = \mathbf{j} \tag{2}$$

where $\mathbf{j}$ is the electric current density. (For the generalization to cases in which fields are time dependent, see MAXWELL'S EQUATIONS.)

A number of authors apply the term Ampere's law to the formula for the force exerted by a current element $I_2 d\mathbf{l}_2$ on another current element $I_1 d\mathbf{l}_1$:

$$\begin{aligned} d\mathbf{F}_{12} &= \frac{\mu_0}{4\pi}\frac{I_1 I_2}{r_{12}^3} d\mathbf{l}_1 \times (d\mathbf{l}_2 \times \mathbf{r}_{12}) \\ &= \frac{\mu_0}{4\pi}\frac{i_1 I_2}{r_{12}^3}[(d\mathbf{l}_1\cdot\mathbf{r}_{12})d\mathbf{l}_2 - (d\mathbf{l}_1\cdot d\mathbf{l}_2)\mathbf{r}_{12}] \end{aligned} \tag{3}$$

where $\mu_0 = 4\pi \times 10^{-7}$ is the permeability of free space and $\mathbf{r}_{12}$ is the vector from $d\mathbf{l}_1$ to $d\mathbf{l}_2$. This formula is the basis for the SI unit of electric current referred to as the absolute ampere.

The existence of the interaction between electric currents was discovered by Ampere in 1820, and he subsequently carried out a remarkable program of experiments and analysis that led him to a formula related to Eq. (3). The reason for the difference is itself interesting. Ampere shared the general view that electrostatic and gravitational forces were cases of action at a distance and assumed that the same was true of the force between currents. Thus, in the interest of conservation of momentum, he assumed that the force between two current elements would have to be directed along the line between them. Accordingly, his result lacked the first term in the second line of Eq. (3) but included another term directed along $\mathbf{r}_{12}$ and causing the entire expression to conform to his experimental result that the force exerted by a closed electric circuit on a current element is perpendicular to the current element. When applied to complete circuits, the formula deduced by Ampere gives the same results as Eq. (3). It can be shown that Eq. (3) is consistent with conservation of momentum as long as the momentum of the electromagnetic field is taken into account.

Following a suggestion by Heaviside, some authors have applied the term Ampere's law to the related formula

$$d\mathbf{F} = I\,d\mathbf{l} \times \mathbf{B}, \tag{4}$$

giving the force exerted by the magnetic field $\mathbf{B}$ on the current element $I\,d\mathbf{l}$.

Finally, a number of authors apply the term Ampere's law to the formula

$$d\mathbf{H}_1 = \frac{1}{4\pi}\frac{I_2}{r_{12}^3} d\mathbf{l}_2 \times \mathbf{r}_{12}, \tag{5}$$

giving the contribution of a current element $I_2 d\mathbf{l}_2$ to the magnetic intensity at location 1. However, this equation is a little more frequently referred to as the Biot–Savart law. By their experiments, Biot and Savart established the r^{-1} dependence of the force on a magnetic pole due to current in a long straight wire, and Biot credited Laplace with having inferred from their result that the field contribution from a current element must have the r^{-2} behavior exhibited in formula (5).

See also MAXWELL'S EQUATIONS; ELECTRODYNAMICS, CLASSICAL; ELECTROMAGNETS.

BIBLIOGRAPHY

For physical explanations of the formulas in this article see any college physics text.

For a discussion of Ampere's work from the point of view of the history of ideas, see the article on Ampere in the *Dictionary of Scientific Biography* (C. C. Gillespie, ed.). Scribners, New York, 1970.

Historical references for Eqs. (1) and (2) are C. F. Gauss's article "Allgemeine Theorie des Erdmagnetismus" in *Carl Friederich Gauss, Werke*, Vol. 5, pp. 170, 171; (Göttingen, 1867) and J. C. Maxwell's article "On Faraday's Lines of Force" in *Trans. Camb. Phil. Soc.* **10,** 27 (1856) [reprinted in Vol. 1 of *The Scientific Papers of J. C. Maxwell* (Cambridge, 1890)].

Ampere's counterpart to Eq. (3) is discussed in E. T. Whittaker's *A History of the Theories of Aether and Electricity*, 2nd ed., Vol. 1, pp. 85–87, (London, 1951; reprinted by Harper, New York, 1960) and in J. C. Maxwell's *A Treatise on Electricity and Magnetism*, 3d ed., Vol. 2, pp. 163–174 (Oxford, 1892). Translations of most of Ampere's papers are available in R. A. R. Tricker's *Early Electrodynamics: The First Law of Circulation* (Pergamon, New York, 1965).

For a demonstration that formula (3) does not violate conservation of momentum when the role of the electromagnetic field is included, see the article by L. Page and N. E. Adams in *Am. J. Phys.* **13,** 141 (1945). For an extended treatment see F. Rohrlich, *Classical Charged Particles* (Addison-Wesley, Reading, Mass., 1965).

Anelasticity

A. S. Nowick

The term anelasticity, although once used loosely to refer to nonelastic behavior, was given a more specific meaning by C. Zener in 1946; this meaning has since been widely adopted. Anelasticity is a generalization of Hooke's law of elasticity, which allows for time-dependent effects. Hooke's law may be stated as $\epsilon = J\sigma$, where ϵ is strain, σ is stress, and J is the compliance constant. In anelasticity, the instantaneous response and single-valuedness inherent in Hooke's law are discarded. However, two restrictions are retained: (a) *linearity*, in the sense that doubling the stress doubles the strain at each instant of time; and (b) a *unique equilibrium relationship*, which means that to every value of stress there corresponds a unique value of strain that is attained if sufficient time is allowed. The simplest relation between stress and strain and their time derivatives that obeys these conditions is

$$J_R\sigma + \tau J_U\dot{\sigma} = \epsilon + \tau\dot{\epsilon}, \tag{1}$$

involving three constants τ, J_R, and J_U. Any material that obeys Eq. (1) is called a *standard anelastic solid*. Equation (1) can be solved under conditions of constant stress, say σ_0 (which constitutes a "creep" experiment), to give

$$\frac{\epsilon(t)}{\sigma_0} = J_U + (J_R - J_U)\left[1 - \exp\left(\frac{-t}{\tau}\right)\right]. \tag{2}$$

From this equation the meaning of the constants becomes clear: J_U, called the unrelaxed compliance, corresponds to the instantaneous response, $\epsilon(0)/\sigma_0$, at $t=0$; J_R, the relaxed compliance, is ϵ/σ_0 as $t\to\infty$; τ is the relaxation time (at constant stress). Figure 1 shows this creep behavior, as well as the time-dependent recovery, or "elastic aftereffect," which takes place after the stress is removed. Equation (1) can also be solved under conditions of constant strain to obtain an exponentially decreasing stress, describing a "stress-relaxation experiment." The most important manifestation of anelasticity, however, occurs in the dynamical case, where stress and strain are both periodic, with the strain lagging behind the stress by a phase angle ϕ. Then we can express a complex compliance by $J^* = \epsilon/\sigma \equiv J_1 - iJ_2$ where J_1 is the real part, which is in phase with the applied stress, and J_2 the imaginary part, which lags σ by $\pi/2$. Using Eq. (1), we can express J_1 and J_2 as functions of the angular frequency ω:

$$J_1(\omega) = J_U + (J_R - J_U)/(1+\omega^2\tau^2), \tag{3}$$

$$J_2(\omega) = (J_R - J_U)\omega\tau/(1+\omega^2\tau^2). \tag{4}$$

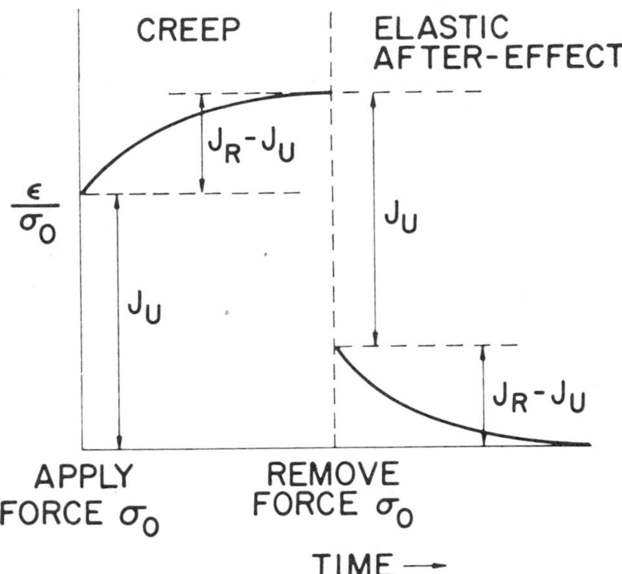

FIG. 1. Behavior of the standard anelastic solid upon application of a static stress σ_0 (creep), and upon the subsequent release of this stress (elastic aftereffect).

Equations (3) and (4) are the celebrated Debye equations. The function $J_2(\omega)$ when plotted versus $\log(\omega\tau)$ gives a symmetrical peak centered about $\log(\omega\tau)=0$ (i.e., $\omega\tau=1$), which is called a Debye peak. Figure 2 shows the variation of both J_1 and J_2 with the variable $\log(\omega\tau)$. The phase angle ϕ by which ϵ lags behind σ is given by $\tan\phi = J_2/J_1$. This quantity, which also takes the form of a Debye peak, is often called the internal friction, since it is a measure of the energy dissipated per cycle.

Since the Debye peak depends on the variable $\omega\tau$, it can be traced out either by varying the frequency, ω, or by changing τ. Although a continuous variation of the vibration frequency is sometimes possible, the usual experimental methods make it preferable to work at one frequency. It is

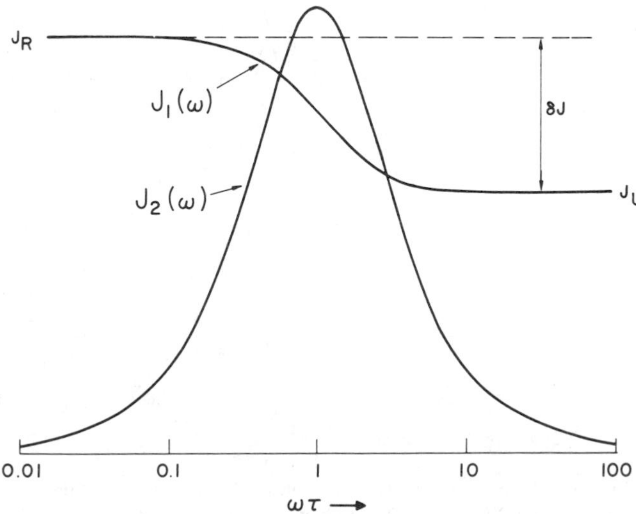

FIG. 2. Dependence of the dynamical functions $J_1(\omega)$ and $J_2(\omega)$ on $\omega\tau$ for the standard anelastic solid.

therefore quite valuable to have a method for tracing out a Debye peak by varying τ while keeping ω constant. This is possible when τ is controlled by a thermally activated process involving an Arrhenius-type relation

$$\tau = \tau_0 \exp(Q/kT) \tag{5}$$

where Q is the activation energy for the process, τ_0 is the preexponential constant, k is Boltzmann's constant, and T is the absolute temperature. In fact, it then turns out that a plot of $\tan\phi$ (or J_2) versus T^{-1} gives a symmetrical peak like that in Fig. 2 except for a change in scale factor. If the peak is then obtained at two or more different frequencies, the activation energy is readily obtained from the shift of the peak location with frequency. Figure 3 shows an example of this type of plot, which is most important in studying anelastic phenomena.

Many phenomena in solids are describable in terms of the equations of the standard anelastic solid. Often, however, this simple model is insufficient to describe the behavior of a material. For example, the internal friction may show a superposition of two or more Debye peaks instead of a single one, or in other cases a single peak may be obtained that is broader than a Debye peak. To treat such cases, it is necessary to introduce a spectrum of relaxation times in place of the single relaxation time of the standard anelastic solid. Thus, instead of a single exponential in the creep function of Eq. (2) there may be a summation of terms with different τ values and weighting factors. Correspondingly, Eq. (4) becomes a sum of Debye peaks. Such behavior is called a discrete relaxation spectrum. In more complex cases there may be a continuous variation in τ, a continuous spectrum, described by an appropriate distribution function. In either of these situations, the display of data in the form of a plot of internal friction ($\tan\phi$) versus T^{-1} is widely used and interpreted.

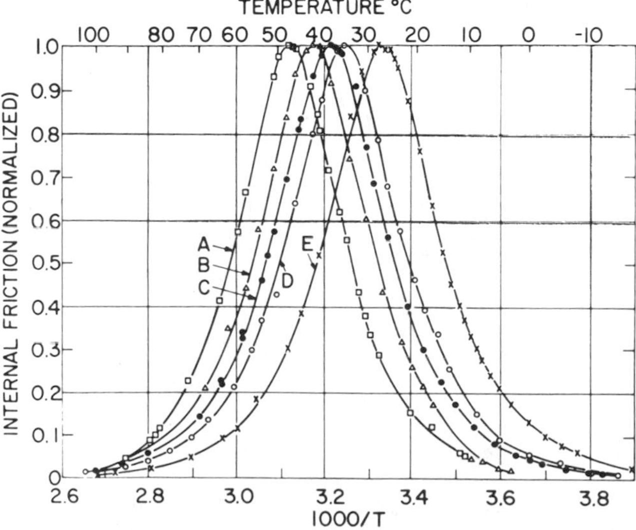

FIG. 3. A series of internal friction peaks for an Fe–C alloy as a function of $1/T$ for five different frequencies: A, 2.1; B, 1.17; C, 0.86; D, 0.63; and E, 0.27 Hz. From C. Wert and C. Zener, Phys. Rev **76**, 1169 (1949).

The physical origins of anelasticity are very varied and encompass almost all aspects of solid-state physics. In crystalline materials, anelastic behavior can result from any of the following mechanisms:

1. Point-defect relaxations: redistribution of point defects (whose symmetry is lower than that of the crystal) into sites that become preferential in the presence of a stress field.
2. Dislocation relaxations: motion of dislocation segments, present either from growth or from plastic deformation, in a variety of ways with the aid of jogs, kinks, and impurity atoms on the dislocation lines.
3. Grain-boundary relaxation: viscous sliding of one grain over another in a polycrystalline material.
4. Phonon relaxation: change in the frequency distribution of phonons (lattice vibrations) due to stress.
5. Magnetic relaxations: magnetoelastic coupling via magnetostriction of a ferromagnetic material, giving rise to a number of different relaxation processes.
6. Electronic relaxations: change in the energetics of the electronic configuration produced by stress leading to a redistribution of both free and bound electrons in various materials.

Glasses (including amorphous alloys) are also capable of showing a variety of relaxations, many of which are similar in origin to those found in crystalline materials.

In amorphous polymers a major relaxation is associated with the glass transition and attributed to large-scale rearrangements of the main polymer chain. Secondary relaxations, at lower temperatures, are due to side groups that are capable of independent hindered rotations.

See also ELASTICITY; RELAXATION PHENOMENA; RHEOLOGY.

BIBLIOGRAPHY

W. Benoit and G. Gremaud, eds., "Internal Friction and Ultrasonic Attenuation in Solids." *J. Phys. (Paris)* **42,** Colloque No. 5 (1981).

R. De Batist, *Internal Friction of Structural Defects in Crystalline Solids*. North-Holland, Amsterdam, 1972.

R. De Batist and J. Van Humbeeck, eds., "Internal Friction and Ultrasonic Attenuation in Solids." *J. Phys. (Paris)* **48,** Colloque C8 (1987).

J. D. Ferry, *Viscoelastic Properties of Polymers*. Wiley, New York, 1961.

A. V. Granato, G. Mozurkewich, and C. A. Wert, eds., "Internal Friction and Ultrasonic Attenuation in Solids." *J. Phys. (Paris)* **46,** Colloque C10 (1985).

R. R. Hasiguti and N. Mikoshiba, eds., *Internal Friction and Ultrasonic Attenuation in Solids*. Univ. Tokyo Press, Tokyo, 1977.

D. Lenz and K. Lücke, eds., *Internal Friction and Ultrasonic Attenuation in Crystalline Solids*, Vols. I and II. Springer, Berlin and New York, 1975.

W. P. Mason and R. N. Thurston, eds., *Physical Acoustics*, Vols. 1–18. Academic Press, New York, 1964–1988.

N. G. McCrum, B. E. Read, and G. Williams, *Anelastic and Dielectric Effects in Polymeric Solids*. Wiley, New York, 1967.

A. S. Nowick and B. S. Berry, *Anelastic Relaxation in Crystalline Solids*. Academic Press, New York, 1972.
R. Truell, C. Elbaum, and B. B. Chick, *Ultrasonic Methods in Solid State Physics*. Academic Press, New York, 1969.
C. Zener, *Elasticity and Anelasticity of Metals*. Univ. of Chicago Press, Chicago, 1948.

Angular Correlation of Nuclear Radiation

Rolf M. Steffen

The probability of emission of a particle or a quantum by a decaying nucleus depends, in general, on the angle between the nuclear spin axis **I** and the direction of emission **k**. In most cases (e.g., ordinary radioactive sources) the total radiation is isotropic, because the nuclear spin axes are randomly oriented in space. An anisotropic intensity distribution of the radiation is only observed if it is emitted from an ensemble of nuclei that is *not* randomly oriented, i.e., in which the spin axes of the decaying nuclei show some preferred direction in space. Such an ensemble is called an *oriented ensemble*.

Oriented ensembles of nuclei can be prepared, e.g., by placing a radioactive sample at a very low temperature in strong magnetic or electrostatic gradient fields, thereby polarizing or aligning the nuclei by virtue of the interaction of the magnetic and electric moments of the nuclei with the external fields. The angular distribution of the radiation emitted by such an oriented source is then, in general, anisotropic with respect to the direction of the applied fields.

Another method of preparing an oriented subensemble of nuclei is based on selecting only those nuclei whose spin axes happen to be in a preferred direction. Nuclear reactions or decay processes that lead to the formation of nuclei in a particular excited state of spin I_1 (the "intermediate" state) can be used in such a selection process.

Many nuclei decay through the *successive emissions of two radiations* R_1 and R_2 via a short-lived (lifetime $\tau \lesssim 10^{-9}$ s) intermediate nuclear state of spin I_1. Some examples of such cascade decays are depicted in Fig. 1A. The observation of R_1 in a fixed direction $\mathbf{k}_1$ selects from the originally random ensemble of nuclei with spin I_0 a subensemble of nuclei in the intermediate state with spin I_1. This subensemble has, in general, a preferred direction of the spin axes $\mathbf{I}_1$ with respect to the observation direction $\mathbf{k}_1$ of R_1, because the radiation emission probability depends on the angle between $\mathbf{k}_1$ and $\mathbf{I}_0$. Since the second radiation R_2 is now emitted from this *oriented* subensemble of spin I_1, the intensity of R_2, observed in a direction $\mathbf{k}_2$, depends, in general, on the angle θ between $\mathbf{k}_1$ and $\mathbf{k}_2$; i.e., the second radiation R_2 has an anisotropic angular distribution with respect to the direction $\mathbf{k}_1$ in which R_1 has been observed. The angular distribution of R_2 with respect to $\mathbf{k}_1$ (or of R_1 with respect to $\mathbf{k}_2$) is called the *angular correlation* of the radiations R_1 and R_2.

If only the propagation directions (no polarization phenomena) of the two radiations are measured, the *directional* correlation is observed. If the linear or circular polarization

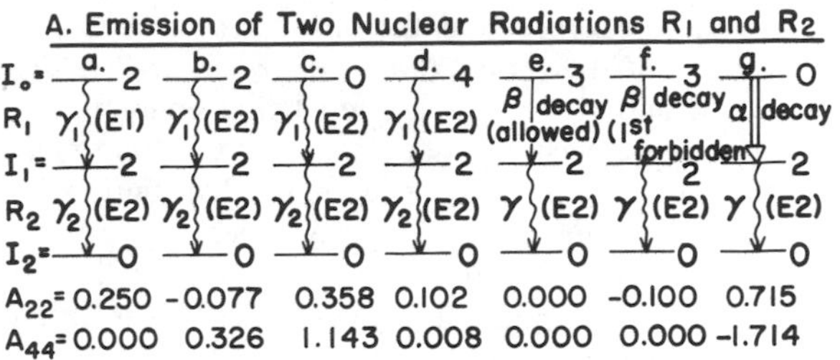

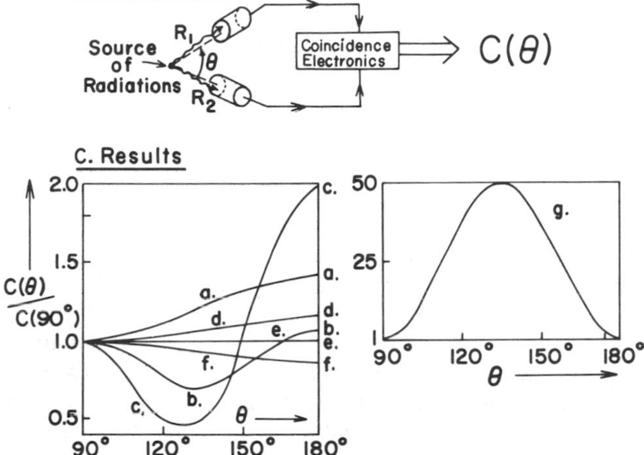

FIG. 1. Examples of directional correlations. (A) Typical gamma–gamma radiation cascades [(a)–(d)], beta–gamma radiation cascades [(e)–(f)], and an alpha–gamma cascade (g). Below each cascade are given the values of the directional correlation coefficients. (B) Experimental arrangements for measurement of directional correlations. (C) Directional correlations of the radiation cascades (a)–(g).

of one or of both of the radiations is measured, a *polarization–directional* correlation or a *polarization–polarization* correlation, respectively, is observed. The term angular correlation comprises all three cases.

The observation of an angular correlation requires a coincidence experiment, i.e., the two radiations R_1 and R_2 must be recorded, each in one of two detectors that respond only if R_1 and R_2 strike the detectors simultaneously (actually within a very short time interval $\tau_0 \approx 10^{-9}$–10^{-8} s) in order to maximize the probability that the observed radiations R_1 and R_2 are emitted from the same nucleus. A directional correlation experiment consists thus simply of measuring the coincidence rate $C(\theta)$ of R_1 and R_2 as a function of the angle θ between the axes of the two detectors (Fig. 1B).

The relative probability $W(\theta)\,d\Omega$ that R_2 is emitted into the solid angle $d\Omega$ at an angle θ with respect to the propagation direction $\mathbf{k}_1$ of R_1 is characterized by the angular correlation function $W(\theta)$. For an *ordinary directional correlation*, $W(\theta)$ can be expressed in the general form

$$W(\theta) = 1 + A_{22}P_2(\cos\theta) + A_{44}P_4(\cos\theta). \quad (1)$$

The angular functions $P_i(\cos\theta)$ are Legendre polynomials, i.e., $P_2(\cos\theta) = (3\cos^2\theta - 1)/2$ and $P_4(\cos\theta) = (35\cos^4\theta - 30\cos^2\theta + 3)/8$. The directional correlation coefficients A_{ii}

($i=2,4$) can be expressed as the product, $A_{ii}=A_i(R_1,I_1;I_0)\cdot A_i(R_2,I_1;I_2)$, of two directional *distribution coefficients* $A_i(R_1,I_1;I_0)$ and $A_i(R_2,I_1;I_2)$, each being characteristic of one of the two emission processes R_1 and R_2 that make up the radiation cascade. The directional distribution coefficients $A_i(R,I;I')$ depend on the properties of the radiation R that is emitted in the transition $I \rightarrow I'$ and on the spins I and I' of the initial and final nuclear states, respectively. In particular, the distribution coefficients depend on the so-called multipolarity L of the emitted radiation R. A 2^L-pole radiation carries away an angular momentum of $L\hbar$ with respect to the center of the emitting nuclei. The directional distribution coefficients, however, do not depend on the reflection symmetry of the emitted radiation. For emission of gamma radiation, e.g., the directional distribution coefficients do not distinguish between electric 2^L-pole (EL) and magnetic 2^L-pole (ML) radiation. The observation of the linear polarization of the gamma radiation is required to distinguish between EL and ML radiation.

The directional distribution coefficients $A_i(\gamma,I;I')$ for gamma transitions do not depend on the energy of the gamma transitions. For alpha-particle emission the $A_i(\alpha;I,I')$ depend on the energy of the alpha particles only if two (or more) alpha-particle waves of different L interfere with each other. In beta emission two particles are emitted simultaneously, an electron (or positron) and an antineutrino (or neutrino), of which only the electron (or positron) is, in general, observed in an angular correlation observation. The electrons (or positrons) have a continuous energy spectrum up to a maximum energy E_0 and the directional distribution coefficients for these electrons (or positrons) depend on the energy of the observed particle.

Theoretical expressions for the directional distribution coefficients (and for polarization distribution coefficients) are available for all types of radiations and for all cases of interest. For details see Refs. 1–4. Four illustrative examples of various multipole gamma–gamma directional correlations are shown in Figs. 1A and 1C, (a)–(d). Beta-gamma directional correlations involving so-called allowed beta transitions are isotropic (e), first-forbidden beta–gamma directional correlations (f) are, in general, nonisotropic. Alpha–gamma directional correlations can show very large anisotropies (g).

Angular correlations are, in general, observed with the initial nuclear state of spin I_0, which emits R_1, randomly oriented (ordinary angular correlation). If R_1 itself is emitted from an oriented state, e.g., from a state produced in a nuclear reaction, the angular correlation from an oriented state (ACO) or the directional correlation from an oriented state (DCO) is observed.

An equivalent situation prevails in triple angular correlations where the angular correlation of the radiations R_1 and R_2 is observed with respect to the observation direction $\mathbf{k}_0$ of a preceding radiation R_0 that is emitted from a random ensemble I_{00} resulting in an oriented nuclear ensemble I_0 from which the radiation R_1 is emitted.

Directional correlations of two successively emitted gamma radiations R_1 and R_2 *emitted by an oriented nuclear ensemble* I_0 that is axially symmetric with respect to a direction $\mathbf{k}_0$ are characterized by a correlation function of the general form

$$W(\theta_1,\theta_2;\varphi) = \sum_{i,k,l} B_l(I_0)A_l^{ki}(R_1,I_0,I_1) \times A_{ll}(R_2,I_1,I_2)H_{ikl}(\theta_1,\theta_2,\varphi) \qquad (i,k,l) = 0,\ 2,\ 4 \tag{2}$$

where θ_1 and θ_2 are the polar angles, with respect to the orientation axis $\mathbf{k}_0$, of the directions $\mathbf{k}_1$ and $\mathbf{k}_2$ in which the radiations R_1 and R_2, respectively, are observed and the azimuthal angle φ is the angle between the planes determined by $\mathbf{k}_0\mathbf{k}_2$ and $\mathbf{k}_0\mathbf{k}_1$. The parameter $B_l(I_0)$ describes the state of orientation of the nuclear ensemble I_0 and $A_l^{ki}(R_1,I_0,I_1)$ is a generalized directional distribution coefficient. Expressions for the latter and for the angular function $H_{ikl}(\theta_1,\theta_2,\varphi)$ can be found in Ref. 5. DCO measurements are particularly useful in assigning spins to nuclear states that are produced in nuclear reactions and in exploring the multipole character of gamma radiations between such states.

In many experimental situations the time t elapsed between the formation of the intermediate oriented state I_1 by the radiation R_1 and the time moment of emission of the second radiation R_2 is long enough ($\sim 10^{-9}-10^{-6}$ s) to cause an appreciable change of the orientation of the nuclear en-

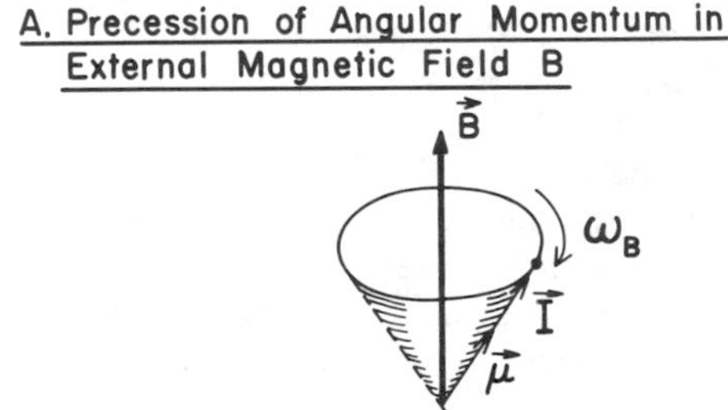

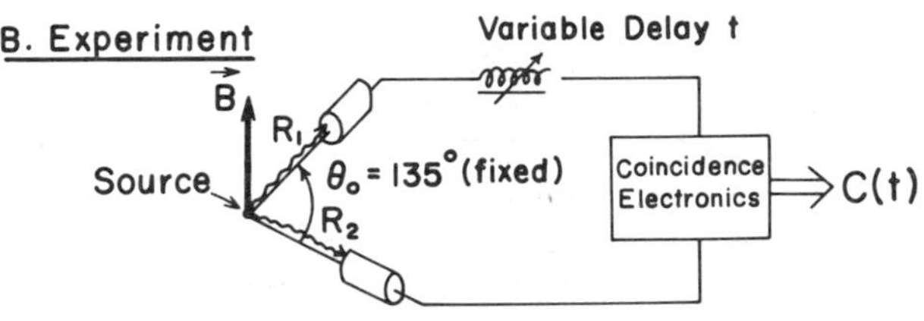

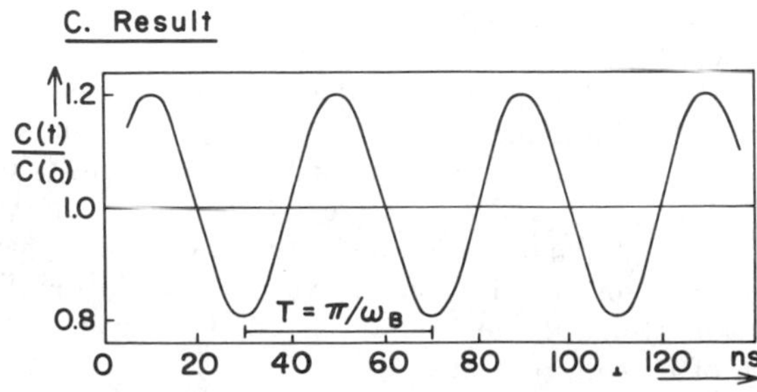

FIG. 2. Extranuclear perturbations by an external magnetic field. (A) Precession of spin about a magnetic field **B** (Larmor precession). (B) Delayed-coincidence observation. (C) Periodic variation with time t of the relative coincidence rate $C(t)/C(0)$ reflecting the spin precession in the intermediate nuclear state.

semble through the interactions of the electromagnetic nuclear moments (magnetic-dipole moment μ, electric-quadrupole moment Q) of the individual nuclei with external fields. In such cases the angular correlation can be influenced by the external fields and a perturbed angular correlation (PAC) is observed (see Refs. 6–8).

A strong external (or internal atomic) magnetic field B, e.g., causes a precession of the magnetic moment μ of the nucleus in the intermediate state about the direction of B as axis (Fig. 2A) with a frequency ω_B that is proportional to B (Larmor precession). The angular distribution pattern of R_2 is then rotated about the direction of B by an angle $\Delta\theta = \omega_B t$. By observation of the angular shift $\Delta\theta$ of the angular correlation pattern, ω_B can be determined and thus either μ or B can be measured. For larger values of $\omega_B\tau$ (i.e., $\omega_B\tau \gtrsim 1$) and if $\tau \gg \tau_0$ the precession of the angular distribution pattern of R_2, i.e., the precession of nuclei in the intermediate state, can be directly observed by measuring the coincidence rate $C(t)$ in two fixed detectors as a function of the time during which the intermediate nuclear state is exposed to the magnetic field B. In practice this is done through delaying (electronically) the detector signal caused by R_1 by a time t before it reaches the coincidence circuit (Fig. 2B). The oscillating behavior of the observed coincidence rate as a function of t represents the spin precession of the nuclei in the intermediate state I_1 (Fig. 2C).

Angular correlation observations are a very important tool in nuclear spectroscopy for the determination of angular momenta and electromagnetic moments of excited nuclear states and for precise measurements of the multipolarities of nuclear radiations. Perturbed angular correlation experiments are also used to explore the magnetic and electric field gradients at the site of nuclei and thus can be applied to atomic, solid-state, and liquid-state problems.

See also ALPHA DECAY; BETA DECAY; GAMMA DECAY; MULTIPOLE FIELDS; NUCLEAR POLARIZATION; POLARIZATION.

REFERENCES

1. H. Frauenfelder and R. M. Steffen, "Angular Correlations," in *Alpha, Beta and Gamma Ray Spectroscopy* (K. Siegbahn, ed.), pp. 997–1198. North-Holland, Amsterdam, 1965. (I)
2. R. M. Steffen and K. Alder, "Angular Distributions and Correlations of Gamma Radiation: Theoretical Basis," in *The Electromagnetic Interaction in Nuclear Spectroscopy* (W. D. Hamilton, ed.), pp. 505–581. North-Holland, Amsterdam, 1975. (A)
3. S. Devons and L. J. B. Goldfarb, in *Handbuch der Physik* (S. Flügge, ed.), Vol. 42, p. 362. Springer-Verlag, Berlin and New York, 1957. (A)
4. A. J. Ferguson, *Angular Correlation Methods in Gamma Ray Spectroscopy*. North-Holland, Amsterdam, 1965. (A)
5. K. S. Krane, R. M. Steffen, and R. M. Wheeler, "Directional Correlations of Gamma Radiations Emitted from Nuclear States Oriented by Nuclear Reactions or Cryogenic Methods," *Atomic and Nuclear Data Tables, Vol. 11*, pp. 351–405. Academic Press, New York, 1975. (A)
6. R. M. Steffen, "Extranuclear Effects on Angular Correlations of Nuclear Radiation," *Adv. Phys.* (*Phil. Mag. Suppl.*), **4,** 293–362 (1955). (E)
7. R. M. Steffen and H. Frauenfelder, "The Influence of Extranuclear Perturbations in Angular Correlations," in *Perturbed Angular Correlations* (E. Karlsson, E. Mathias, and U. Siegbahn, eds.), pp. 1–89. North-Holland, Amsterdam, 1964. (A)
8. R. M. Steffen and K. Alder, "Extranuclear Perturbations and Angular Distributions and Correlations," in *The Electromagnetic Interaction in Nuclear Spectroscopy* (W. D. Hamilton, ed.), pp. 583–643. North-Holland, Amsterdam, 1975. (A)

Angular Momentum *see* Rotation and Angular Momentum in Classical Mechanics

Antennas *see* Transmission Lines and Antennas

Antimatter

Gary Steigman

All quantum theories consistent with the special theory of relativity and the requirement of causality require that particles exist in pairs. Particles and their antiparticles have the same masses and lifetimes; electrically charged particles (e.g., electron, proton) have antiparticles (e.g., positron, antiproton) with equal but opposite electric charges; some electrically neutral particles (e.g., photon) are their own antiparticles (self-conjugate). Following the discovery of the positron (Anderson 1933) there was a hiatus of some 22 years before the antiproton was produced and detected at an accelerator (Chamberlain et al. 1955). Subsequent accelerator experiments have provided strong confirmation that all particles do, indeed, exist in pairs and, further, that particles carry certain quantum numbers (baryon number, lepton number, etc.) which *seem* to be conserved in all reactions. If these conservation "laws," inferred from experimental data, are exact, then matter is restricted to appear (creation) or disappear (annihilation) only as particle–antiparticle pairs. This apparent symmetry (about which, more later) in the laws of physics has stimulated serious speculation on the antimatter content of the Universe (Goldhaber 1956) and the possible astrophysical consequences of macroscopic amounts of antimatter (Burbidge and Hoyle 1956). In approaching the issue of the matter–antimatter symmetry (or, asymmetry) of the Universe, it is valuable to distinguish between two distinct questions: Is the Universe symmetric? Must the Universe be symmetric? The first question will be considered before a "modern" (a la "Grand Unified Theory") answer is given to the second question. For further details and references see Steigman (1976).

SEARCHING FOR ANTIMATTER IN THE UNIVERSE

Antimatter is, in principle, trivially easy to detect. You place your sample in a detector (the most rudimentary device will suffice) and, if the detector disappears (annihilates), the sample was made of antimatter. Unfortunately, only the solar system and the cosmic rays provide a sample of the Universe which may be subjected to such a direct test. Lunar landings and the Venus probes establish that the Moon and Venus are made of ordinary matter. The absence of annihilation gamma rays when the solar wind sweeps through the solar system establishes that the Solar System consists only of (ordinary) matter; were any of the planets made of antimatter, annihilation of the solar wind particles which strike their surfaces would have made them the strongest gamma ray sources in the sky.

Cosmic rays provide the only direct sample of extrasolar system material in the Universe. The cosmic rays, perhaps the debris of exploding stars (supernovae) or the accelerated nuclei of interstellar gas, bring information about the material in our Galaxy. As the cosmic rays traverse the Galaxy they collide with interstellar gas nuclei, occasionally producing (secondary) antiprotons. Therefore, antiprotons in the cosmic rays do not provide an unambiguous signal for the presence of "primary" sources of antimatter (e.g., antistars) in the Galaxy. In contrast, virtually no antihelium (antialpha) nuclei would be present as secondaries in the cosmic rays. The discovery of even one antialpha particle in the cosmic rays would provide compelling evidence for the existence of antimatter in the Galaxy. None has ever been found. In contrast, antiprotons have been observed in the cosmic rays (Golden *et al.* 1979; Bogomolov *et al.* 1979). However, the upper limits to the $\bar{p}$ flux at low energies (Ahlen *et al.* 1988) are completely consistent with a secondary origin; these latest results are in conflict with—and cast doubt on the reality of—an earlier claim (Buffington *et al.* 1981) of a positive detection.

Astronomy is an observational science, often relying on the interpretation of indirect evidence. When matter and antimatter meet, they annihilate producing, among the debris, gamma rays of energy from several tens to several hundred MeV. The annihilation gamma rays can provide an indirect probe for antimatter in the Universe. However, since gamma rays may be produced by other astrophysical processes (synchrotron radiation, Compton scattering, etc.), they are not an unambiguous signal; the best approach is to use the observed gamma ray flux to place upper limits on annihilation and, hence, on the amount of mixed matter and antimatter at various astrophysical sites (Steigman 1976).

Gamma ray observations of the Galaxy limit the antimatter fraction in the interstellar gas to less than 10^{-15}. Such a small limit is not surprising once it is realized that the lifetime—against annihilation—of an antiparticle in the interstellar gas is less than 300 years (the Galaxy is at least 10 billion years old). Our galaxy is clearly made entirely of ordinary matter. What of other galaxies?

Clusters of galaxies are the largest astrophysical entities which can be probed by gamma rays for a possible antimatter component. Many rich clusters shine in the x-ray part of the spectrum; the x-rays are the thermal bremsstrahlung emission from a hot intracluster gas. From the absence of gamma rays, less than one part in 10^5 of that hot gas could be made of antimatter. Thus, the observational data shows no evidence for astrophysically interesting amounts of antimatter in the Universe. If antimatter were present, it would have to be separated from ordinary matter on scales at least as large as clusters of galaxies ($\sim 10^{15} M_{\odot} \sim 10^{48}$gm; $\sim 10^{72}$ protons). Could matter and antimatter be separated in the Universe on such large scales?

SYMMETRIC COSMOLOGIES

In the context of the hot big bang model, particle–antiparticle pairs were present in great abundance during the early ($\lesssim 10$ μs), hot ($T \gtrsim$ few hundred MeV) epochs in the evolution of the Universe. As the Universe expanded and cooled, however, these pairs annihilated. In a completely symmetric, perfectly mixed Universe the annihilation is so efficient that less than one nucleon–antinucleon pair remains for each 10^{18} microwave background photons in the Universe; this is less matter than is present in our Galaxy alone! Either the Universe at these early times ($\lesssim 10$ μs) was asymmetric or, possibly, matter and antimatter were separated. However, although the Universe was very dense, it was also very small and only ~1 g of nucleons (antinucleons) could have been separated by any causal process (none is known which would effect such a separation). Inevitably, then, astrophysicists have led the way to the conclusion that the early Universe ($\lesssim 10$ μs) was asymmetric.

BARYON ASYMMETRY AND GUT

Inspired by the (then recent) discovery of *CP* violation in the $K^0 - \bar{K}^0$ system, Sakharov (1967) outlined the recipe for generating a baryon asymmetry in an initially (matter–antimatter) symmetric Universe. First, there must be interactions which violate conservation of baryon number (matter–antimatter symmetry is a "broken" symmetry). Next, he noted a technical—but crucial—requirement that *CP* conservation be violated (e.g., the branching ratios for decays into certain channels for particles and antiparticles differ; note that the total lifetimes are still required to be equal). Finally, Sakharov (1967) pointed out that these *B*- and *CP*-violating processes must occur "out of equilibrium." The offspring of the marriage of particle physics (Grand Unified Theories) and cosmology (the expanding, hot big bang model) is endowed with the requisite properties (Yoshimura 1978; Dimopoulos and Susskind 1978; Ellis, Gaillard and Nanopoulos 1979; Toussaint *et al.* 1979; Weinberg 1979). Baryon number- and *CP*-violating interactions occurring very early ($\lesssim 10^{-35}$ s) in the evolution of the Universe would have, due to the expansion and cooling of the Universe, dropped out of equilibrium and left behind a small matter–antimatter asymmetry (the net baryon number is $\sim 10^{-10}$–10^{-9} of the photon number). This tiny relic, the legacy of the earliest epochs in the evolution of the Universe, is responsible for the presently observed, matter–antimatter asymmetric Universe.

See also COSMOLOGY; ELEMENTARY PARTICLES; POSITRON.

BIBLIOGRAPHY

S. P. Ahlen *et al.*, *Phys. Rev. Lett.* **61,** 145 (1988).

C. D. Anderson, *Phys. Rev.* **43,** 491; **44,** 406 (1933).

E. A. Bogomolov *et al.*, *Proc. of the 16th Int. Cosmic Ray Conf.* **1,** 330 (1979).

A. Buffington, S. M. Schindler, and C. R. Pennypacker, *Astrophys. J.* **248,** 1179 (1981).

G. R. Burbidge and F. Hoyle, *Nuovo Cimento* **1,** 558 (1956).
O. Chamberlain, E. Segré, C. Wiegand, and T. Ypsilantis, *Phys. Rev.* **100,** 947 (1955).
S. Dimopoulos and L. Susskind, *Phys. Rev.* **D18,** 4500 (1978).
J. Ellis, M. K. Gaillard, and D. V. Nanopoulos, *Phys. Lett.* **B80,** 360 (1979).
R. L. Golden *et al., Phys. Rev. Lett.* **43,** 1196 (1979).
M. Goldhaber, *Science* **124,** 218 (1956).
A. Sakharov, *JETP Lett.* **5,** 24 (1967).
G. Steigman, *Ann. Rev. Astron. Astrophys.* **14,** 339 (1976).
D. Toussaint, S. Treiman, F. Wilczek, and A. Zee, *Phys. Rev.* **D19,** 1036 (1979).
S. Weinberg, *Phys. Rev. Lett.* **42,** 850 (1979).
M. Yoshimura, *Phys. Rev. Lett.* **41,** 381 (1978).

Arcs and Sparks

Uwe H. Bauder

GENERAL PROPERTIES

The electric arc is characterized by high current densities, low potential differences between the electrodes, and small differences compared to other discharges between the temperatures of the different particle species present in the column. In equilibrium temperatures exceeding 50 000 K have been reached [1]. Arcs can be operated over a wide pressure range: from several millibars gas pressure to 1000 bars and in all gaseous media. The initiation of an arc discharge may be achieved either by contacting the electrodes or by preionizing the gas in the discharge channel by breakdown processes following Paschen's law [2]. The unstationary discharge preceding a sustained arc discharge is the spark; besides duration time the main difference between arcs and sparks relates to electrode effects. Most stationary arc discharges operate with a combined thermal and field emission of electrons at a hot cathode [3], whereas duration times of sparks are too small to allow for a substantial increase of the bulk temperature of electrodes.

Depending on the external conditions under which the arc is operated one distinguishes between electrode-stabilized (short) arcs, flow-stabilized arcs (longitudinal or swirl flow), wall-stabilized arcs (operated in a segmented cascaded tube if high power levels have to be achieved), high pressure arcs, and vacuum arcs.

THE HIGH-PRESSURE ARC COLUMN

Arcs operated at atmospheric pressure or above are high-pressure arcs. Once established by one of the ignition processes mentioned above, the arc column is sustained by ionizing processes whose energy is supplied by the electric field generated by the power supply. Free electrons coming from the cathode are accelerated in the cathode fall region and in the column during a free path length between two collisions. High-pressure arc columns are characterized by mean free path lengths which are much smaller than the geometrical dimensions of the arc column. Due to their small mass, electrons reach higher velocities in the field than ions; mainly collisions of electrons with molecules, atoms, and ions lead to further ionization. Ionization is possible from the ground state as well as starting from excited states. Depending on the gas species as well as on the plasma density and temperature, photoionization processes may also play a role. Carriers are lost by recombination processes. In some application plasmas such as SF_6 plasmas, electron attachment with the formation of negative ions may also reduce the number density of free electrons in the column. The large difference between the masses of electrons and heavy particles frequently leads to thermal nonequilibrium in the plasma. It takes approximately 10^3–10^4 elastic collisions of electrons with atoms or ions until thermalization is achieved. Due to this fact local thermodynamic equilibrium (LTE) is only achieved in high-temperature arc plasmas with high collision rates (elevated pressures); in other arc columns partial local thermodynamic equilibrium (PLTE) prevails. In the case of PLTE only the higher excited states are populated in equilibrium with the electron temperature, the ground state being largely overpopulated.

Considering the differential energy balance of the cylindrical arc column without convection losses [4] as it is realized experimentally in the cascade arc [5],

$$\sigma E^2 + \frac{1}{r}\frac{d}{dr}\left(r\kappa\frac{dT}{dr}\right) - e + a = 0, \qquad (1)$$

it can be noted that the energy input to the volume element of the plasma column is mainly due to Ohmic heating (σ being the electrical conductivity and E the electric field strength). Absorption processes may also play a role; the quantity a is the volumetric absorption coefficient integrated over all frequencies. The energy is lost from the column by radiation (e = volumetric emission coefficient) and by thermal conduction. The thermal conductivity κ is composed of the contact conductivities of all species; in the temperature regions of dissociation of molecules and that of ionization a large contribution to κ may be due to the radial transport of dissociation and ionization energies. Due to this transport, temperature profiles of arc discharges operating in molecular gases exhibit a "shoulder" as shown in Fig. 1 [6] for the case of an atmospheric pressure nitrogen arc operated in a cascade channel of 5 mm diameter. By contrast, the $T(r)$ profiles of noble gas cascade arc columns at elevated pressures have a flat central portion and very steep gradients at the wall (Fig. 2 [7]). The energy balance of the central portion of such arcs is largely radiation dominated. It has been shown [8] that up to 80% of the energy of a high-pressure argon arc may be lost by radiation; this property of the column is used in high-energy radiation source applications.

The field-strength/current (E–I) characteristic of the high-pressure arc column exhibits a falling portion at small currents and a rising part in the higher current range. Stable arc operation at small currents (point C in Fig. 3 [9]) therefore requires an Ohmic resistance in series with the arc; thus the resulting total characteristic can be made to increase at the current of interest and stable operation is achieved. Introducing the heat flux potential $S = \int\kappa\, dT$ into the transport

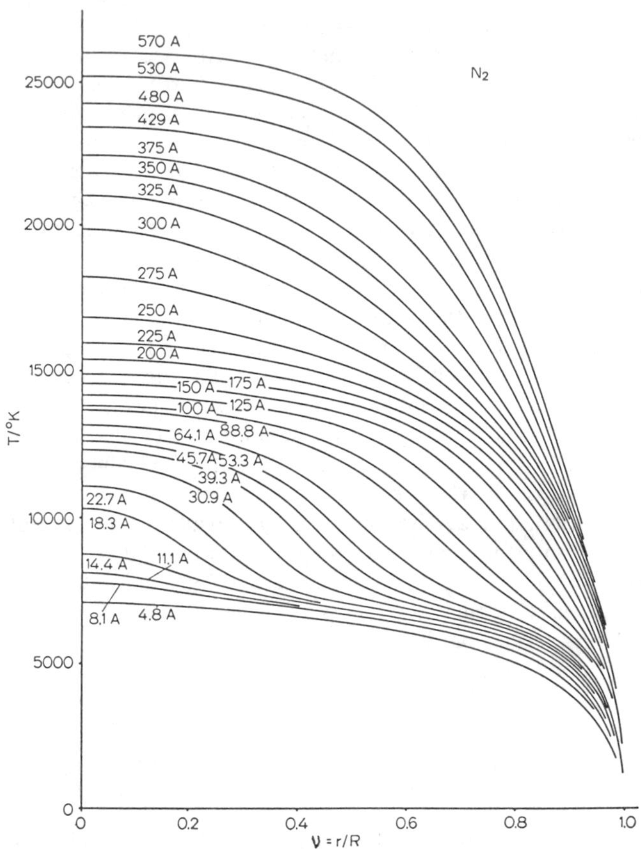

FIG. 1. Temperature distributions in a 5-mm nitrogen cascade arc (from [6], with permission).

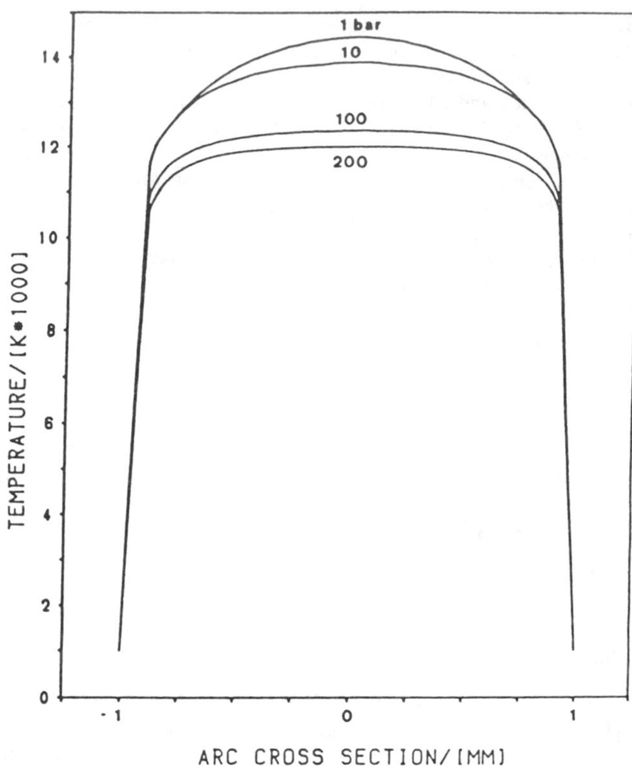

FIG. 2. Temperature distributions in an argon arc at elevated pressures. Arc current 60 A [7].

function $\sigma(T) = \sigma(S)$ and neglecting radiation Eq. (1) may be rewritten

$$\sigma E^2 + \frac{1}{r}\frac{\partial}{\partial r}\left(r\frac{\partial S}{\partial r}\right) = 0. \qquad (2)$$

Defining the Ohmic heating per unit arc length as $L = IE$ and normalizing the arc radius $v = r/R$ (R being the total radius) Eq. (2) yields

$$\frac{\sigma L}{\pi\int_0^1 \sigma d(v^2)} + \frac{1}{v}\frac{\partial}{\partial v}\left(v\frac{\partial S}{\partial v}\right) = 0. \qquad (3)$$

Since Eq. (3) is independent of the total radius R of the column, equal values of L lead to identical $T(r)$-distributions. This independence on radius may be used to predict electrical arc data (E, I) for arcs of different diameters (similarity laws).

Due to its importance for applications the interaction of flow fields and magnetic fields with the arc column has been studied extensively [10–12]. By solving the MHD conservation laws with the appropriate boundary conditions, predictions of arc motion, gas flow in the arc and arc stability may be obtained. The energy Eq. (1) has to be used in its

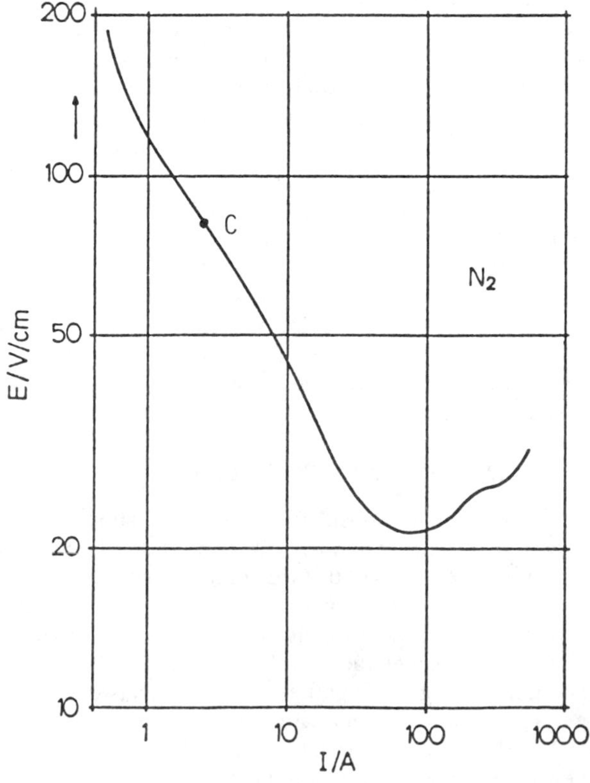

FIG. 3. Electrical arc characteristic (from [9], with permission).

more general form:

$$\rho \frac{dh}{dt} = \rho \frac{\partial h}{\partial t} + \rho \mathbf{v}\cdot\nabla h = \sigma E^2 + \nabla\cdot\kappa\nabla T - e + a, \quad (4)$$

where $\mathbf{v}$ is the velocity of the gas, ρ is its mass density, and h the enthalpy. In addition, the momentum and the mass balance equations (5) and (6) have to be considered:

$$\rho \frac{d\mathbf{v}}{dt} = \rho \frac{\partial \mathbf{v}}{\partial t} + \rho \mathbf{v}\cdot\nabla\mathbf{v} = \mathbf{j} \times \mathbf{B} - \nabla p + \nabla\cdot\tau, \quad (5)$$

$$\frac{\partial \rho}{\partial t} + \nabla\cdot(\rho\mathbf{v}) = 0 \quad (6)$$

with τ being the friction tensor.

VACUUM ARCS

This arc discharge is initiated in an evacuated environment. Since plasma can only be formed if ionizable gas is present, the vacuum arc has to generate its own gaseous environment. Vaporization of the electrodes, mainly that of the cathode material, provides the necessary atoms. Vacuum arcs therefore operate typically in a metal vapor atmosphere. Since the crater formation process which is necessary for the evaporation at the cathode ends after a certain crater size has been reached [13], vacuum arcs are not stationary in nature. The cathode spot only remains at a given crater for times shorter than 50 ns; thus a single (undivided) short vacuum arc is very similar in nature to a spark.

If magnetic fields are present, the vacuum arc moves against the $\mathbf{j} \times \mathbf{B}$ direction; this "retrograde motion" is due to cathode jet phenomena. At higher currents (above 50–100 A) the single vacuum arc splits up; a multitude of cathode or anode spots may coexist in such arcs.

APPLICATIONS

Arc discharges are used in a host of scientific and technical applications. Due to the temperature range which is achievable in the column ($T \leqq 50\,000$K) different ionization stages of the gas atoms are reached. Fundamental data such as collision cross sections, transition probabilities, index of refraction, and spectral line broadening parameters can be determined from the investigation of arc plasmas.

Technical applications also make use of the high plasma temperatures which lead to an excellent electrical conductivity, to high radiative power losses, to high enthalpies in the gas—to name only the most important properties. The high electrical conductivity of an arc column, together with the short recovery time of dielectric strength after current zero, is used in AC-circuit breakers. Very high voltage levels can be handled with gas filled gas breakers (SF_6) where the interaction of flow fields and magnetic fields with the arc column plays an important role. Vacuum circuit breakers are much more compact, however, their voltage handling capability is reduced. At present, voltages of 50–60 kV can be handled. The high specific radiation of noble gases and rare earths is used in radiation source applications, while the high enthalpies reached in arc heaters allow for special chemical reactions and for surface treatment processes, such as spark etching and plasma coating.

See also Corona Discharge; Ionization; Lightning; Photoionization.

BIBLIOGRAPHY

H. Maecker and W. Finkelnburg, *Elektrische Bögen und thermisches Plasma, Handbuch der Physik,* Bd. XXII. Springer-Verlag, Berlin, 1957.

J. M. Lafferty, ed., *Vacuum Arcs, Theory and Application.* Wiley, New York, 1980.

H. Raether, *Electron Avalanches and Breakdown in Gases.* Butterworths, London, 1964.

K. Günther and R. Radtke, *Electric Properties of Weakly Nonideal Plasmas.* Birkhaeuser, Basel, 1984.

K. Ragaller, ed., *Current Interruption in High-Voltage Networks.* Plenum, New York, 1978.

REFERENCES

1. F. Burhorn, H. Maecker, and Th. Peters, *Z. Phys.* **131,** 28 (1951).
2. F. Paschen, *Ann. Phys. (Leipzig)* **37,** 69 (1889).
3. G. Burkhard, Ph.D Thesis, Technische Hochschule Ilmenau (1971).
4. W. Elenbaas, *Physica* **3,** 947 (1936).
5. H. Maecker and S. Steinberger, *Z. Angew. Phys.* **23,** 456 (1967).
6. E. Schade, *Z. Phys.* **233,** 53 (1970).
7. H. Poisel, F. J. Landers, P. Höß, and U. H. Bauder, *IEEE Trans. Plasma Sci.* **PS-14,** 306 (1986).
8. U. H. Bauder and P. Schreiber, *Proc. IEEE* **59,** 633 (1971).
9. U. Plantikow, *Z. Phys.* **237,** 388 (1970).
10. H. Maecker and H. G. Stäblein, *IEEE Trans. Plasma Sci.* **PS-14,** 291 (1986).
11. N. Sebald, *Proc. XIIth ICPIG,* North Holland/American Elsevier Eindhoven, p. 187 (1975).
12. J. Blass and U. H. Bauder, *Proc. IIW Asian Pacific Regional Welding Congress,* p. 528. Hobart, Australia, 1988.
13. J. Prock, *IEEE Trans. Plasma Sci.* **PS-14,** 482 (1986).

Astronomy, Neutrino*

John N. Bahcall

How does the Sun shine? Why do stars explode? Does the neutrino have a mass? Can a neutrino change its lepton number in flight? These are some of the questions that motivate the study of neutrino astronomy.

A neutrino is a weakly interacting particle that travels at essentially the speed of light and has an intrinsic angular momentum of $\frac{1}{2}$ unit ($\hbar/2$). Neutrinos are produced on Earth by natural radioactivity, by nuclear reactors, and by high-energy accelerators. In the Sun, neutrinos are produced by weak interactions that occur during nuclear fusion. There are three known types of neutrinos, each associated with a massive lepton that experiences weak, electromagnetic, and gravitational forces, but not strong interactions. The known leptons are electrons, muons, and taus (in increasing order of their rest masses).

* Adapted, with permission of the publisher, from the introduction to *Neutrino Astrophysics* (Cambridge University Press, 1989) by John N. Bahcall.

Neutrino astronomy is interesting for the same reason it is difficult. Because neutrinos only interact weakly with matter, they can reach us from otherwise inaccessible regions where photons, the traditional messengers of astronomy, are trapped. Hence, with neutrinos we can look inside stars and examine directly energetic physical processes that occur only in stellar interiors. We can study the interior of the Sun or the core of a collapsing star as it produces a supernova.

Large detectors, typically hundreds or thousands of tons of material, are required to observe astronomical neutrinos. These detectors must be placed deep underground to avoid confusing the rare astronomical neutrino events with the background interactions caused by cosmic rays and their secondary particles, which are relatively common near the surface of the Earth.

For two decades, the only operating solar neutrino experiment (using ^{37}Cl as a detector) yielded results in conflict with the most accurate theoretical calculations. This conflict between theory and observation, which has recently been confirmed by a new experiment, is known as the solar neutrino problem.

More is known about the Sun than about any other star and the calculations of neutrino emission from the solar interior can be done with relatively high precision. Hence, the solar neutrino discrepancy has puzzled (and worried) astronomers who want to use neutrino observations to try to understand better how the Sun and other stars shine. The solar neutrino problem could be a clue to something new under the Sun.

Neutrinos from the Sun and from supernovae provide particle beams for probing the weak interactions on energy or time scales that cannot be achieved with traditional laboratory experiments. Since neutrinos from the Sun and from supernovae travel astronomical distances before they reach the Earth, experiments performed with these particle beams are sensitive to weak-interaction phenomena that require long path lengths in order for slow weak-interaction effects to have time to occur. The effects of tiny neutrino masses ($\gtrsim 10^{-6}$ eV), unmeasurable in the laboratory, can be studied with solar neutrinos.

The solar neutrino problem can be stated simply. Both the theoretical and the observational results are expressed in terms of the solar neutrino unit, SNU, which is the product of a characteristic calculated solar neutrino flux (units: cm^{-2} s^{-1}) times a theoretical cross section for neutrino absorption (unit: cm^2). A SNU has, therefore, the units of events per target atom per second and is chosen for convenience equal to 10^{-36} s^{-1}.

The predicted rate for capturing solar neutrinos in a ^{37}Cl target is (Bahcall and Ulrich, 1988; Bahcall, 1989)

$$\text{Predicted rate} = (7.9 \pm 2.6)\ \text{SNU}, \tag{1}$$

where the indicated uncertainty represents the total theoretical range including three-standard-deviation (3σ) uncertainties for measured input parameters and a full range of estimated values for input parameters that must be calculated theoretically. The rate observed by R. Davis, Jr. (1986) and his associates in a chlorine radiochemical detector is

$$\text{Observed rate} = (2.1 \pm 0.9)\ \text{SNU}, \tag{2}$$

where the error is again a 3σ uncertainty.

The disagreement between the predicted and the observed rates constitutes the solar neutrino problem. There is no generally accepted solution to the problem although a number of interesting possibilities have been proposed.

The discrepancy between calculation and observation has recently been confirmed by an independent technique using the Japanese detector of neutrino–electron scattering, Kamiokande II (Hirata et al. 1989). The preliminary Kamiokande II result is

$$\frac{\phi_{\text{observed}}}{\phi_{\text{predicted}}} = 0.39 \pm 0.09, \tag{3}$$

where the neutrino flux, ϕ, is from the rare ^{8}B solar neutrinos discussed below (see Table 1) and the quoted error is the 1σ uncertainty. There is also a small systematic uncertainty.

The predictions used in Eqs. (1) and (2) are valid for the combined standard model, that is, the standard model of electroweak theory (of Glashow, Weinberg, and Salam) and the standard solar model.

The central question for solar neutrino research is easily stated. Is the solar neutrino problem caused by unknown properties of neutrinos or by a lack of understanding of the interior of the Sun? In other words, is this a case of new physics or faulty astrophysics? Experiments to be performed in the next decade will provide the answer to this question.

From the astronomical point of view, solar neutrino experiments test in a direct and rigorous way the theories of nuclear energy generation in stellar interiors and of stellar evolution. These tests are independent of many of the uncertainties that complicate the comparison of the theory with observations of stellar surfaces. For example, convection and turbulence are generally believed to be important near stellar surfaces but unimportant in the solar interior.

The Sun shines by converting protons into α particles. The overall reaction can be represented symbolically by the relation

$$4p \rightarrow \alpha + 2e^{+} + 2\nu_e + 25\ \text{MeV}. \tag{4}$$

Protons are converted to α particles, positrons, and neutrinos, with a release of about 25 MeV of thermal energy for every four protons burned. Each conversion of four protons to an α particle is known as a termination of the chain of energy-generating reactions that accomplishes the nuclear fusion. The thermal energy that is supplied by nuclear fusion ultimately emerges from the surface of the Sun as sunlight. About 600 million tons of hydrogen are burned every second to supply the solar luminosity. Nuclear physicists have worked for half a century to determine the details of this transformation.

The main nuclear burning reactions in the Sun are shown in Table 1, which represents the energy-generating *pp* chain. This table also indicates the relative frequency with which each reaction occurs in the standard solar model.

Table I The *pp* chain in the Sun. The average number of *pp* neutrinos produced per termination in the Sun is 1.85. For all other neutrino sources, the average number of neutrinos produced per termination is equal to the termination percentage/100.

Reaction	Number	Termination[a] (%)	ν energy (MeV)
$p + p \rightarrow {}^2H + e^+ + \nu_e$	1*a*	100	≤ 0.420
or			
$p + e^- + p \rightarrow {}^2H + \nu_e$	1*b* (*pep*)	0.4	1.442
${}^2H + p \rightarrow {}^3He + \gamma$	2	100	
${}^3He + {}^3He \rightarrow \alpha + 2p$	3	85	
or			
${}^3He + {}^4He \rightarrow {}^7Be + \gamma$	4	15	
${}^7Be + e^- \rightarrow {}^7Li + \nu_e$	5	15	(90%) 0.861 (10%) 0.383
${}^7Li + p \rightarrow 2\alpha$	6	15	
or			
${}^7Be + p \rightarrow {}^8B + \gamma$	7	0.02	
${}^8B \rightarrow {}^8Be^* + e^+ + \nu_e$	8	0.02	<15
${}^8Be^* \rightarrow 2\alpha$	9	0.02	
or			
${}^3He + p \rightarrow {}^4He + e^+ + \nu_e$	10 (He*p*)	0.00002	≤ 18.77

[a] The termination percentage is the fraction of terminations of the *pp* chain, $4p \rightarrow \alpha + 2e^+ + 2\nu_e$, in which each reaction occurs. The results are averaged over the model of the current Sun. Since in essentially all terminations at least one *pp* neutrino is produced and in a few terminations one *pp* and one *pep* neutrino are created, the total of *pp* and *pep* terminations exceeds 100%.

The fundamental reaction in the solar energy-generating process is the proton–proton (*pp*) reaction, which produces the great majority of solar neutrinos; however, these *pp* neutrinos have energies below the detection thresholds for the ${}^{37}Cl$ and Kamiokande II experiments. Experiments with ${}^{71}Ga$ that are in progress in the Soviet Union and in Europe are sensitive to neutrinos from the *pp* reaction.

Most of the predicted capture rate in the ${}^{37}Cl$ experiment comes from the rare termination in which ${}^7Be$ captures a proton to form radioactive ${}^8B$ (reaction 7). The ${}^8B$ decays to unstable ${}^8Be$, ultimately producing two α particles, a positron, and a neutrino. The neutrinos from ${}^8B$ decay have a maximum energy of less than 15 MeV. Although the reactions involving ${}^8B$ occur only once in every 5000 terminations of the *pp* chain, the total calculated event rates for the ${}^{37}Cl$ and Kamiokande II experiments are dominated by this rare mode.

The neutrino spectrum predicted by the standard model is shown in Figure 1, where contributions from both line and continuum sources are included. For Kamiokande II, only the ${}^8B$ and He*p* neutrinos (reaction 10) have enough energy to produce recoil electrons above the dominant backgrounds.

The solar neutrino fluxes calculated from the standard solar model are shown in Table 2. The uncertainties in the calculated neutrino fluxes are also shown in Table 2.

The beautiful ${}^{37}Cl$ experiment of Davis and his collaborators was for two decades the only operating solar neutrino detector. The reaction that is used for the detection of the neutrinos is

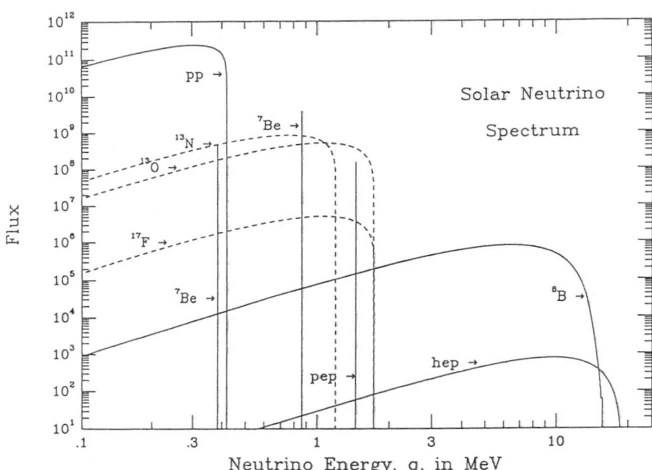

FIG. 1. Solar neutrino spectrum. This figure shows the energy spectrum of neutrinos predicted by the standard solar model. The neutrino fluxes from continuum sources (like *pp* and ${}^8B$) are given in the units of number per cm^2 per second per MeV at one astronomical unit. The line fluxes (*pep* and ${}^7Be$) are given in number per cm^2 per sec. The spectra from the *pp* chain are drawn with solid lines; the CNO spectra are drawn with dotted lines.

$$\nu_e + {}^{37}Cl \rightarrow e^- + {}^{37}Ar, \quad (5)$$

which has a threshold energy of 0.8 MeV. The target is a tank containing 10^5 gallons of C_2Cl_4 (perchloroethylene, a cleaning fluid), deep in the Homestake Gold Mine in Lead, South Dakota. The underground location is necessary in order to avoid background events from cosmic rays. Every few months, for about 20 years, Davis and his collaborators extracted a small sample of ${}^{37}Ar$, typically of order 15 atoms, out of the total of more than 10^{30} atoms in the tank. The ${}^{37}Ar$ produced in the tank is separated chemically from the C_2Cl_4, purified, and counted in low-background proportional counters. The typical background counting rate for the counters corresponds to about one radiactive decay of an ${}^{37}Ar$ nucleus a month! Experiments have been performed to show that ${}^{37}Ar$ produced in the tank is extracted with more than 90% efficiency.

Figure 2 shows all the experimental data that have been published by Davis and his colleagues, B. Cleveland and K. Rowley.

Table II Calculated Solar Neutrino Fluxes

Source	Flux (10^{10} cm^{-2} s^{-1})
pp	6.0 (1 ± 0.02)
pep	0.014 (1 ± 0.05)
He*p*	8×10^{-7}
${}^7Be$	0.47 (1 ± 0.15)
${}^8B$	5.8×10^{-4} (1 ± 0.37)
${}^{13}N$	0.06 (1 ± 0.50)
${}^{15}O$	0.05 (1 ± 0.58)
${}^{17}F$	5.2×10^{-4} (1 ± 0.46)

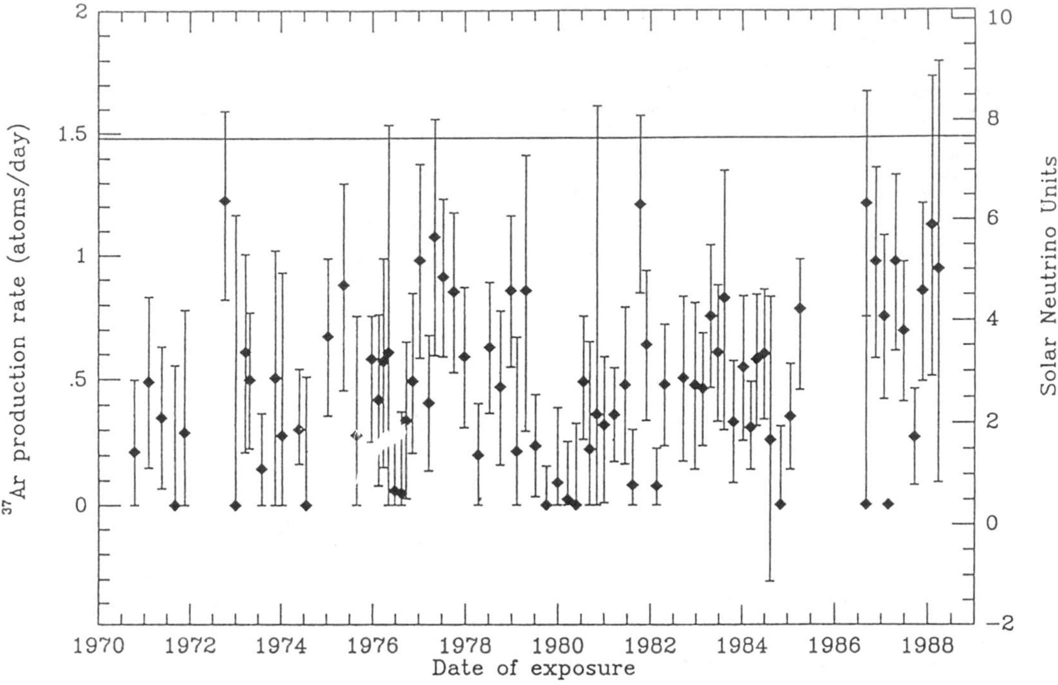

FIG. 2. ^{37}Cl observations. The observed rate in the ^{37}Cl solar neutrino experiment (from observations by Davis, Cleveland, and Rowley). The line at 7.9 SNU across the top of the figure represents the prediction of the standard model.

The Kamiokande II experiment, which is located in the Japanese Alps, detects Cherenkov light emitted by electrons that are scattered in the forward direction by solar neutrinos. The reaction by which the neutrinos are observed is

$$\nu + e \rightarrow \nu' + e', \tag{6}$$

where the primes on the outgoing particle symbols indicate that the momentum and energy of each particle can be changed by the scattering interactions. This is the first of several solar neutrino experiments that are planned or are in progress in which ν–e scattering will be studied. For the higher-energy neutrinos (> 5 MeV, i.e., ^{8}B and Hep neutrinos only) that can be observed by this process using available techniques, the scattering provides additional information not available with a radiochemical detector. Neutrino–electron scattering experiments furnish information about the incident neutrino energy spectrum (from measurements of the recoil energies of the scattered electrons), determine the direction from which the neutrinos arrive, and record the precise time of each event.

The preliminary results from the Kamiokande II detector yield a ^{8}B neutrino flux that is approximately 0.45 of the standard model flux, about 3σ away from zero and from the standard model value. This result applies for recoil electrons with a minimum total energy of 9.3 MeV. A significant forward peaking of the recoil electrons is observed along the direction of the Earth–Sun axis. This result is of great importance since all of the previous observational results on solar neutrinos came from a single ^{37}Cl experiment.

A geochemical experiment has been developed over the past several years at Los Alamos National Laboratory, in which neutrinos are detected via their absorption by molybdenum atoms shielded from atmospheric phenomena by the covering of a deep mine. The absorption of neutrinos produces an unstable but long-lived isotope (4 million years lifetime) of technetium that would not be present in a steady state situation in which there were no high-energy solar neutrinos. The method uses the neutrino capture reaction

$$\nu_e + {}^{98}\text{Mo} \rightarrow {}^{98}\text{Tc}^* + e^-. \tag{7}$$

This experiment will test the constancy of the flux of ^{8}B neutrinos over the past several million years.

Standard ideas about the time scale for solar evolution (estimated to be ~ 10^{10} years) imply that the time-averaged flux of ^{8}B neutrinos measured with the ^{98}Mo experiment will be the same to within 1% (much less than the experimental uncertainties) as the contemporary flux determined from the ^{37}Cl and Kamiokande II detectors. The largest uncertainty

Table III Predicted Capture Rates for a ^{37}Cl Detector

Neutrino source	Capture rate (SNU)
pp	0.0
pep	0.2
He*p*	0.03
^{7}Be	1.1
^{8}B	6.1
^{13}N	0.1
^{15}O	0.3
^{17}F	0.003
Total	7.9 SNU

in the predicted rate, about a factor of 2, comes from the neutrino absorption cross section.

Two radiochemical solar neutrino experiments using ^{71}Ga are under way, one by a primarily Western European collaboration [with U.S. and Israeli participation (see Kirsten 1986)] (GALLEX) and the second by a group working in the Soviet Union (with U.S. participation). The GALLEX collaboration plans to use 30 tons of gallium in an aqueous solution; the detector will be located in the Gran Sasso Laboratory in Italy. The Soviet experiment will use about 60 tons of gallium metal as a detector in a solar neutrino laboratory constructed underneath a high mountain in the Baksan Valley in the Caucasus Mountains of the Soviet Union. The amount of detector material used in each of these experiments is impressive considering that, at the time the experimental techniques were developed, the total world production of gallium was only 10 tons per year!

The gallium experiments can furnish unique and fundamental information about nuclear processes in the solar interior and about neutrino propagation. The neutrino absorption reaction is

$$\nu_e + {}^{71}\text{Ga} \rightarrow e^- + {}^{71}\text{Ge}. \quad (8)$$

The germanium atoms are removed chemically from the gallium and the radioactive decays of ^{71}Ge are measured in small proportional counters. The threshold for absorption of neutrinos by ^{71}Ga is 0.233 MeV, which is well below the maximum energy of the *pp* neutrinos. No other solar neutrino experiment has a demonstrated capability to detect the low-energy neutrinos from the basic *pp* reaction (reaction 1a of Table 1).

The gallium experiments may indicate which class of solution is correct for the solar neutrino problem, faulty astrophysics or new physics. Most nonstandard solar models predict event rates that are not very different from the standard solar model. The minimum rate that is consistent with the assumption that nuclear fusion currently balances the solar luminosity is about 60% of the standard-model value, provided that no physics beyond the standard electroweak model affects the propagation of the neutrinos. Some explanations of the solar neutrino problem that involve particle physics imply that the event rate in the gallium experiments will be much less than the standard-model prediction, perhaps no more than 10% of the standard value. The Soviet experimentalists have 60 tons of gallium available and hope to begin taking data some time in 1989; the GALLEX experiment is expected to begin in 1990.

Two powerful new detectors are being developed as next-generation experiments. One is a proposed 1-kiloton heavy water experiment (D_2O), to be placed in an INCO nickel mine near Sudbury, Ontario (Canada). The other is a 3-kiloton detector of liquid argon, to be placed in the Gran Sasso underground laboratory in central Italy. The deuterium experiment is a collaboration between Canadian, American, and British scientists; the argon detector is primarily an Italian experiment with (so far) limited American participation. The deuterium detector measures the energy and direction of recoil electrons by observing their Cherenkov light with photomultipliers. The argon detector forms a three-dimensional electronic image of the positions of recoil electrons produced by neutrino absorption or scattering, by drifting the electrons in a homogeneous electric field.

These experiments are sensitive to ^8B and He*p* neutrinos, but the other solar neutrinos will be below the energy thresholds that are set at several MeV in order to avoid numerous lower-energy background events.

Both the argon and the deuterium experiments utilize multipurpose detectors, which can study solar neutrinos by neutrino capture and by electron–neutrino scattering. Neutrino absorption is sensitive only to solar ν_e whose type (flavor) is unchanged in transit to the Earth. For electron–neutrino scattering, the cross section for ν_μ or ν_τ is about one-seventh the cross section for ν_e. The solar origin of the events can be tested using the angular distribution (with respect to the Earth–Sun direction) of the electrons scattered by the neutrinos.

Absorption reactions make possible the measurement of individual neutrino energies, E_ν, using the relation $E_\nu = E_e + \text{constant}$, where E_e is the energy of the electron that is produced and the constant is equal to the difference of initial and final nuclear masses. The measurement of individual energies of neutrinos will constitute a test of the predicted shape of the energy spectrum.

The deuterium experiment may include a detection mode that is equally sensitive to all three types of neutrinos, ν_e, ν_μ, and ν_τ. In this neutral-current mode, deuterium nuclei are disintegrated into their constituent neutrons and protons without changing the charge of the nucleons. The measurement of the neutral-current disintegration of deuterium can provide a determination of the total flux of solar neutrinos; radiochemical experiments (with ^{37}Cl, ^{71}Ga, and ^{98}Mo detectors) detect only the component of the flux that is in the form of electron neutrinos. Neutrino–electron scattering is relatively insensitive to all types except ν_e.

Several detectors are being developed to observe the basic (low-energy) *pp* neutrinos by electronic methods, but it is not possible at this time to say which methods will provide practical detectors.

There are two classes of "solutions" to the solar neutrino problem, those that modify the astrophysical description of the Sun in order to obtain a nonstandard solar model and those that hypothesize properties of neutrinos in order to obtain a nonstandard theory of electroweak interactions.

Nonstandard solar models are constructed by changing something, either the physics or the input data, from our current best estimate to something that is less plausible. Some nonstandard models invoke enhanced element diffusion, convective instabilities, or a solar interior greatly deficient in heavy elements.

One suggested solution has been made that might solve simultaneously the famous "dark matter" problem in astronomy and the solar neutrino problem. In this suggestion, weakly interacting massive particles, WIMPs, are produced in the early universe in an abundance that accounts for the unseen matter in large astronomical systems. The properties of the so-far undiscovered WIMPs could be just such that they redistribute some of the heat in the solar interior, smoothing out the temperature gradient, and reducing the flux of ^8B neutrinos to be consistent with the experimental limit.

In the second class of solutions, the flux of neutrinos in the interior of the Sun is predicted correctly by the standard solar model, but these neutrinos are transformed from ν_e to another type of neutrino on their way to the terrestrial detector. For nearly all proposed solutions, this transformation requires a nonzero neutrino mass.

A massive neutrino may have a magnetic moment. If this moment were as large as 10^{-10} to 10^{-11} μ_B (Bohr magnetons), it is possible that the familiar left-handed ν_e could have its spin flipped to a right-handed ν_e in passing through magnetic fields of a few thousand gauss in the outer part of the Sun. Right-handed neutrinos do not have normal weak interactions and would therefore not be detected. This solution was proposed many years ago and has been revived recently by Soviet scientists to account for an apparent anticorrelation of the neutrino capture rate in the ^{37}Cl experiment with the sunspot cycle. The greatest change, a strong decrease in the observed capture rate, occurred near the time of the onset of solar cycle 21 in 1979 and 1980. The large magnetic moment needed for this solution, while barely consistent with existing laboratory limits of the order of 10^{-10} μ_B, is much larger than is obtained in calculations with conventional electroweak models. An alternative explanation of the apparent anticorrelation of SNUs with spots is that it is due to a rather improbable coincidence. Observations of the neutrino capture rate with the ^{37}Cl experiment in 1990 and 1991 will help to resolve this question.

No conclusive evidence exists that any of the neutrinos have a nonzero mass. Direct experiments and interpretations of the observations of neutrinos from SN1987A have yielded, with certainty, only upper limits on the masses of about 20 eV. There are also arguments based upon cosmology that suggest that none of the neutrinos has a mass greater than 80 eV.

The standard model of electroweak interactions is normally presented in a form in which neutrinos are massless. However, in most extensions of the model, neutrinos have a mass.

The physics community was electrified in 1985 when an elegant solution for the solar neutrino problem was proposed that is consistent with expectations from GUT theories of neutrino mass. According to this solution, a ν_e created in the solar interior is almost completely converted into ν_μ or ν_τ as the neutrino passes through the Sun. This conversion reflects the enhancement by the matter in the Sun of the probability that a neutrino of an electron type oscillates into a neutrino of a different type; it is usually referred to as the Mikheyev–Smirnov–Wolfenstein (MSW) effect in honor of its discoverers.

In order for the MSW effect to occur, the flavor eigenstates ν_e, ν_μ, and ν_τ must be different from the mass eigenstates. The flavor eigenstates are created in weak decays and have weak interactions with their associated charged leptons (electron, muon, and tau) that can be written in a simple (diagonal) form. The mass eigenstates, which have diagonal mass matrices, are the states in which neutrinos propagate in a vacuum. The mass eigenstates are often denoted by ν_1, ν_2, and ν_3. For a simplified description in terms of two eigenstates, the relation between flavor and mass eigenstates in vacuum is described by a single mixing angle θ_V, where $\tan \theta_V$ is the relative amplitude of ν_1 and ν_2 in the ν_e wave function ($\nu_e = \cos \theta_V \nu_1 + \sin \theta_V \nu_2$).

The large suppression of the ^8B flux observed in the ^{37}Cl experiment can be explained if the mixing angle $\theta \gtrsim 0.6°$. The mass of the heavier neutrino can lie between 10^{-4} and 10^{-2} eV.

The MSW effect is an attractive solution to the solar neutrino problem that can be tested by carrying out additional experiments on solar neutrinos.

Future experiments can discriminate between solutions to the solar neutrino problem based on nonstandard solar models and those based on new neutrino physics. In nonstandard solar models, the shape of the neutrino spectrum from individual neutrino sources, such as the ^8B spectrum, will be the same as in Figure 1. The total flux from a given source can be altered by astrophysical considerations but the shape is fixed by nuclear physics. On the other hand, the transformation of ν_e to ν_μ or ν_τ due to the MSW effect would be energy dependent and therefore the spectrum would be distorted. Nearly all proposed nonstandard solar models predict the same flux as the standard model for the predominant ν_e from the *pp* reaction. In contrast, the MSW effect can, for certain parameters, decrease significantly the flux of ν_e *pp* neutrinos.

In most oscillation scenarios, the missing ν_e arrive at Earth as ν_μ or ν_τ. Detectors based on neutrino–electron scattering or pure neutral-current interactions should be able to detect the missing neutrinos.

On February 6, 1987, two water Cherenkov detectors (Kamiokande II in Japan and IMB in the United States) made a historic observation of neutrinos from the explosion of a supernova in the Large Magellanic Cloud. Two liquid scintillation detectors (one in the Soviet Union and one in the Mont Blanc Tunnel) with less mass than the Kamiokande II and IMB detectors may also have observed some neutrinos. A total of 19 neutrinos were detected by Kamiokande II and IMB, most of them by the absorption reaction

$$\bar{\nu}_e + p \rightarrow n + e^+. \tag{9}$$

Because of the distance of the supernova (~150,000 light years), the observations did not provide sufficient detail to test precise predictions of theoretical models of stellar collapse. However, the experimental results did confirm the basic ideas of the theory of supernova explosions developed over the preceding 50 years. About $10^{53.5}$ ergs was emitted in all forms of neutrinos over a period of time of several seconds; the characteristic temperature of the observed $\bar{\nu}_e$s was about 4 MeV. The observations also set stringent limits on some characteristics of electron neutrinos, including a proof that the mass is less than 20 eV and the charge is less than 10^{-17} times the charge on the electron. In addition, the observations provided strong evidence that the neutrinos propagated along the same geodesics from the Large Magellanic Cloud as did the photons emitted by the supernova.

Stellar collapses with detectable neutrino emission are expected to occur only rarely and sporadically in the Galaxy, perhaps once every 10 or 100 years. Observations of neutrinos from galactic stellar collapses can provide crucial information about the astrophysics of stellar collapses and about the properties of neutrinos of different types. A number of neutrino detectors are being developed which will have the capability of observing well neutrinos from our own and from nearby galaxies.

BIBLIOGRAPHY

J. N. Bahcall, *Neutrino Astrophysics*. Cambridge University Press, New York, 1989. Systematic treatment of the subject. Contains both elementary and advanced treatments.

J. N. Bahcall and R. K. Ulrich, *Rev. Mod. Phys.* **60,** 297 (1988). State-of-the-art calculations of solar models, neutrino fluxes, and helioseismological frequencies. (A)

I. R. Barabanov, *et al.*, in *Solar Neutrinos and Neutrino Astronomy,* M. L. Cherry, W. A. Fowler, and K. Lande, eds., p. 175. American Institute of Physics, New York, 1985. A clear description of the Soviet experiment. (A)

H. A. Bethe, *Phys. Rev. Lett.* **56,** 1305 (1986). A beautiful physical explanation of the MSW effect that ignited widespread interest in matter oscillators among western scientists. (A)

H. A. Bethe and G. Brown, *Sci. Am.* **252,** 60 (1985). A clear physical description of how a supernova explodes, presented in beautiful and entertaining prose. What every physicist wants to know and what every astronomer should know. (E)

F. A. Boehm and P. Vogel, *Physics of Massive Neutrinos*. Cambridge University Press, Cambridge, 1987. An excellent summary of neutrino properties with special emphasis on experiments that provide information about masses and charge conjugation. (A)

D. D. Clayton, *Principles of Stellar Evolution and Nucleosynthesis*. University of Chicago Press, Chicago, 1983. The standard textbook from which students (and professors) have learned the fundamentals of the subject for almost two decades. This edition contains a new preface which lists many of the modern references. (I)

G. A. Cowan and W. C. Haxton, *Science* **216,** 51 (1982). Classical paper proposing the ^{98}Mo experiment and discussing with great insight both the experimental and theoretical aspects. (A)

J. P. Cox and R. T. Guili, *Principles of Stellar Structure*. 2 vols. Gordon and Breach, New York, 1968. Comprehensive summary of the theory. (A)

R. Davis, Jr., *et al.*, in *Solar Neutrinos and Neutrino Astronomy,* M. L. Cherry, W. A. Fowler, and K. Lande, eds., Vol. 1, p. 1. American Institute of Physics, New York, 1978. Classic description of the experiment by the master. (A)

R. Davis, Jr., in *Proceedings of Seventh Workshop on Grand Unification, ICOBAN'86, Toyama, Japan,* J. Arafune, ed., p. 237. World Scientific, Singapore, 1987. An excellent summary from a contemporary viewpoint. (A)

A. S. Eddington, *The Internal Constitution of the Stars*. Cambridge University Press, Cambridge, 1926. A beautifully written summary of the early theory of stellar evolution. Chapter 8 contains a fascinating account of the first gropings toward understanding of the source of stellar energy generation. (E)

G. T. Ewan, *et al.*, in *Sudbury Neutrino Observatory Proposal* (Sudbury Neutrino Observatory Collaboration: Queen's University at Kingston) Pub. No. SNO-87-12, 1987. Detailed and convincing discussion of solar neutrino experiments and theoretical issues. (A)

K. Hirata, *et al., Phys. Rev. Lett.* **58,** 1490 (1987); R. M. Bionta, *et al., Phys. Rev. Lett.* **58,** 1494 (1987). Historic detection of supernova neutrinos by two water Cherenkov detectors. (A)

K. Hirata *et al.* Institute for Cosmic Ray Research, University of Tokyo, preprint (1989).

T. Kirsten, in *Massive Neutrinos in Astrophysics and in Particle Physics,* O. Fackler and J. Tran Thanh Van, eds., p. 119. Editions Frontières, Paris, 1986. The GALLEX experiment described authoritatively by the senior spkoesman. (A)

S. P. Mikheyev and A. Yu. Smirnov, *Sov. J. Nucl. Phys.* **42,** 913 (1986). *Nuovo Cimento* **9C,** 17 (1986); *Sov. Phys. JETP* **64,** 4 (1986); in *Proceedings of 12th Intl. Conf. on Neutrino Physics and Astrophysics,* T. Kitagaki and H. Yuta, eds., p. 177. World Scientific, Singapore, 1986. Epochal papers, exciting to read. Mikheyev and Smirnov obtained by numerical integration the principal results for matter oscillations in the Sun and presented them succinctly, together with a clear physical explanation. (A)

B. Pontecorvo, *Sov. JETP* **26,** 984 (1968); V. Gribov and B. Pontecorvo, *Phys. Lett.* **B28,** 493 (1969). The original papers suggesting that neturino flavor oscillations explain the solar neutrino problem. Founded a subject. Revolutionary ideas presented with clarity and brevity. (A)

C. Rolfs and W. Rodney, *Cauldrons in the Cosmos*. University of Chicago Press, Chicago, 1988. A valuable review of nuclear astrophysics with special emphasis on laboratory experiments that determine the cross sections. (A)

S. P. Rosen and J. M. Gelb, *Phys. Rev. D* **34,** 969 (1986). One of the earliest numerical explorations and interpretations of MSW solutions. Emphasized the importance of the energy spectrum as a diagnostic test. (A)

M. Schwarzschild, *Structure and Evolution of the Stars*. Princeton University Press, Princeton, NJ, 1958. Classical description of the theory of stellar evolution with emphasis on physical understanding. The clearest book ever written on the subject. (I)

L. Wolfenstein, *Phys. Rev. D,* **17,** 2369 (1978); *Phys. Rev. D* **20,** 2634 (1979). Presented the fundamental equations for neutrino propagation in matter, the basis for the MSW effect. It took seven years for the physics community to recognize the significance of Wolfenstein's brilliant insight. (A)

G. I. Zatsepin, *Sov. JETP* **8,** 205 (1968). The original paper pointing out that antineutrino detectors could set limits on masses of neutrinos produced in supernova. Simple and beautiful. (A)

Astronomy, Optical

David M. Zipoy

SCOPE

For the purposes of this article, optical astronomy will include observations of celestial objects using their ultraviolet, visible, or infrared radiation. The purpose of the observations is to provide information that can be used in explaining the composition, structure, formation, and evolution of these objects using a few physical laws.

Virtually all celestial objects can be studied by means of optical observations. The list includes the solar system, stars, star clusters, galaxies, clusters of galaxies, and the universe as a whole.

TELESCOPES

All large telescopes use mirrors rather than lenses because mirrors are much easier to mount in such a way that they will not distort and degrade the images as the telescope points in different directions. Telescopes are usually used with a small convex "secondary" mirror placed in the converging beam from the "primary" just before it comes to a focus. The beam is reflected by the secondary through a hole in the center of the primary and comes to a focus just behind the primary's mount. This is the "Cassegrain" focus and is a much more convenient place to mount instruments than the "prime" focus of the primary. The convex secondary magnifies the image, that is, it increases the effective focal length of the telescope by typically a factor of 3 or more. This design provides a long focal length in a compact size.

Telescopes are specified by the diameter of their primary. Larger telescopes have greater light-gathering power and therefore can see fainter objects than smaller ones. Some new designs are being built that get around the fact that large massive mirrors are hard to mount rigidly. One way is to make the mirrors thinner and lighter. They are much more flexible and so the design of the mount is more difficult. Another way is to use many small telescopes and superimpose their images to make one bright image. The University of Arizona's Multiple Mirror Telescope (MMT) consists of four 72-in. telescopes on a common mount which form a common image. A third way is to use a segmented mirror. The ideal shape for a single large telescope mirror is approximately a parabola. In this technique the mirror is made in smaller segments that have the surface shape appropriate to their location in the mirror. The new 10-m (400-in.) Keck Telescope on Mauna Kea, Hawaii, is of this type.

INSTRUMENTS

Optical instruments range from simple cameras to complex computer-controlled spectrometers. Cameras normally just consist of a box to hold a photographic plate at the telescope's focus. On the other hand, the Schmidt Camera is an especially designed telescope that is used to take high-quality wide-angle photos. The famous "Palomar Sky Survey" used the 48-in. Schmidt on Mt. Palomar to photograph the entire northern sky on 935 pairs of 14-in. square plates (6 × 6 degrees each).

When studying individual objects one normally wants information about its spectrum: its intensity versus wavelength. The filter photometer is a low-resolution device for doing this. It consists of a detector, be it photographic or photoelectric, and filters that only pass one band of colors each. It turns out that the ratio of the intensity of a star in red light to that in blue light is a good measure of the star's temperature, and each pair of photographs in the Palomar survey consists of a red and a blue photo; comparing the same star on the two photos immediately gives its temperature.

In order to get more detailed information about an object, its spectrum has to be measured at much higher wavelength resolution. A diffraction-grating spectrometer is normally used for these measurements. With such a spectrum from a star one can measure the star's temperature, composition, and surface gravity. From this information and a liberal amount of theory one can deduce the entire structure of the star right down to its center. Another important use is as a speedometer. An object's relative speed toward or away from you can be found by measuring the Doppler shift of its entire spectrum; the amount of shift is proportional to the speed. Among other things this is how the cosmological red-shift is measured.

Due to the noisy nature of infrared detectors a completely different type of spectrometer has been developed for that spectral region, the Fourier-transform spectrometer. It is much more complicated than a grating instrument but it produces a spectrum after extensive computations. There are many other more special purpose instruments such as Fabry–Perot interferometers, spectroheliographs, occultation photometers, speckle interferometers, etc.

DETECTORS

Photographic plates were the first recording detectors and they are still being used. The spatial resolution of a plate is 10–20 μm and so a 4 × 5-in. plate contains over 10 million independent measurements. Their main problem is that they are not very sensitive to light (low efficiency), and it is very difficult to obtain accurate intensities from them.

The other main class of detectors are the photoelectric devices, that is, anything that converts incident light to an electrical signal. The photomultiplier tube is an ultraviolet and visible light detector. It uses the photoelectric effect to convert an incident photon into an electron which is then amplified a million-fold within the tube. Its output current is accurately proportional to the incident light intensity. Its main deficiencies are that it is only about 10% efficient in converting photons to electrons (still 100 times better than a plate), and it is only a single detector as opposed to over 10 million on a plate. They are used extensively in filter photometry.

Infrared detectors are generally semiconductor devices that use an "internal" photoelectric effect. These detectors can have over 80% efficiency but they are fairly noisy in most cases unless they are cooled down to under 4 K. The advent of these detectors opened up the field of infrared astronomy to the point where it is now a major branch of observational astronomy.

A TV camera is essentially a photoelectric version of a photographic plate; however, ordinary TV cameras are very noisy. The development of the charge-coupled device (CCD) camera changed all of this. It is a semiconductor TV camera that is sensitive from the near ultraviolet to the near infrared and that has an efficiency of up to 80%. Their main disadvantages are that they are a bit noisy; it takes a dozen or so photons absorbed in a single picture element (pixel) to be reliably detected over the noise level, and their picture area is small compared to a plate; a typical CCD has a quarter million pixels in a 500 × 500 array. They are getting bigger. They are the detector of choice for many if not most optical observations.

OBSERVATORIES

There are many small and large observatories in this country and throughout the world. They are owned or operated by research organizations, colleges, universities, and nations. The United States has three large national observatories: one is located on Kitt Peak near Tucson, Arizona, another is located on Cerro Tololo near La Serena, Chile, and the third, the National Solar Observatory, is located on Sacramento Peak near Sunspot, New Mexico.

See also COSMOLOGY; GALAXIES; MILKY WAY; SOLAR SYSTEM; SUN.

BIBLIOGRAPHY

M. A. Seeds, *Horizons: Exploring the Universe*. Belmont, Calif., Wadsworth, 1987. (E)

J. H. Robinson, *Astronomy Data Book*. Wiley, New York, 1972. *Astronomy* (Astromedia Corp., Milwaukee, 1972). (E)

B. V. Barlow, *The Astronomical Telescope*. Wykeham Publ. Ltd., 1975. (I)

Telescopes, G. P. Kuiper and B. M. Middlehurst, ed. University of Chicago Press, Chicago, 1966. (I)

Sky and Telescope (Sky Publication Corp., Cambridge, Mass. 19■■)

C. R. Kitchin, *Astrophysical Techniques*. Adam Hilger Ltd., 1984. (A)

W. H. Steel, *Interferometry*, 2nd Ed. Cambridge University Press, Cambridge, 1983. (A)

Astronomy, Radio

A. A. Penzias and B. E. Turner

Radio astronomy differs from the other observational branches of astronomy in that the incident energy is coherently amplified. Unlike film, bolometers, or particle counters, the phase of the incident wave is preserved in the process. The terminals of the receiver are located at the focal point of the *antenna*, where the field in the aperture is coherently added. The most generally used antenna and the one which most nearly resembles its optical counterpart is the parabolic reflector (Fig. 1). The angular distribution of the antenna response, the "antenna pattern," has an angular width at half-intensity of $\sim\lambda/a$ radians, where a is the width of the antenna aperture.

Mechanical limitations on antenna size and precision limit the resolution (minimum beamwidth) of the largest single antennas to about 10 arc seconds. To obtain higher resolution, an array of antennas is connected together to form an *interferometer* (Fig. 2).

When the signals obtained by a pair of antennas are multiplied together, the product is proportional to the cosine of the difference between the phases of the received signals. The angular dependence of the response of such a system to a distant source of monochromatic radiation will have a one-dimensional periodicity in the direction which is coplanar with the line of sight between the two antennas. The angular frequency of the response is given by the projected distance, in wavelengths, between the antennas as viewed from the source.

By observing the source with a number of differently spaced antenna pairs one can obtain enough angular frequency components to construct a Fourier synthesis of the angular distribution of the source intensity. These different spacings can be obtained by use of a number of antennas with different spacing, by moving one or more of the antennas between observations, and by making use of the earth's rotation to change the projected spacing. Two-dimensional information can be obtained by arranging the antenna array in the form of a cross or "Y" (Fig. 2) or making use of the change in orientation of the array axis with respect to the source caused by the rotation of the earth. The resulting intensity map of the source extending over the entire area

FIG. 1. The 15-m submillimeter telescope of the Swedish–European Southern Observatory consortium, at an elevation of 2350 m at La Silla, Chile. The surface accuracy (deviation from a perfect paraboloid) is 35 μm rms, allowing operation to wavelengths as short as 550 μm (Onsala Space Observatory photo).

FIG. 2. The Very Large Array, consisting of 27 antennas, each 25 m in diameter, in a Y configuration. Shown is the most compact configuration (arm length 0.6 km). At its shortest wavelength, 1.3 cm, and largest configuration (arm length 21 km), the VLA provides a spatial resolution of 0.08 arcsec (NRAO photo).

covered by the relatively broad beams of the individual antennas can have the resolution equivalent to that of a single antenna whose aperture encompasses the entire array; hence the name, *aperture synthesis*.

RADIOMETRY

The intensity scale used in radio astronomy is antenna temperature. Consider a warm uniform opaque cloud extending over an angular area much larger than the antenna beam. The power radiated per unit area by this cloud is given by the usual blackbody formula

$$B=\frac{4\pi h\nu^3}{c^2(e^{h\nu/kT}-1)}\,\Delta\nu,$$

and, in the limit $h\nu/kT \ll 1$, the power radiated per unit solid angle becomes the familiar Rayleigh–Jeans formula

$$b=\frac{2kT}{\lambda^2}\,\Delta\nu.$$

At some distance R from the cloud we put an antenna with an aperture of width a pointed toward it. The antenna will receive energy only from that portion of the cloud intercepted by its beam, i.e., $\sim(\lambda/a)^2R^2$. Furthermore, we must multiply by the solid angle subtended by the antenna at the cloud surface, $\sim(a/R)^2$. Thus the power incident upon the antenna terminals in one plane of polarization is

$$\frac{kT\Delta\nu}{\lambda^2}\times\frac{\lambda^2R^2}{a^2}\times\frac{a^2}{R^2}=kT\Delta\nu.$$

This relation leads to the definition of antenna temperature as the power per unit bandwidth received at the antenna terminal divided by Boltzmann's constant. Note, therefore, that antenna temperature is an intensity which only corresponds to a thermodynamic temperature in the Rayleigh–Jeans limit.

Radio astronomy observations are simply the measurements of antenna temperature as a function of angle and frequency. To relate the observed quantity to the property of the emitting medium we introduce the concept of equivalent brightness temperature of an astronomical object. This is the physical temperature of a perfect absorber with the same angular dimensions as the object which would produce the observed antenna temperature. To obtain the equivalent brightness temperature from the antenna temperature we must take the response pattern of the antenna into account. If the object is small compared to the antenna beam, we must determine its angular size by other means. Failing that, we can only assign a flux (incident power density per unit bandwidth per unit frequency interval) to the object by dividing the observed antenna temperature by Boltzmann's constant and the effective area of the antenna, and multiplying the result by 2. (The multiplication by 2 reflects the fact that a coherent receiver responds to only one polarization of the incident field.)

Antenna temperature is measured by means of a radiometer, which in its simplest form consists of an amplifier, frequency-selective filter, rectifier–detector, and an integration and voltage-measurement circuit. Incident power levels from astronomical sources upon even the largest antennas are extremely small and must be measured in the presence of the generally much larger noise power of the radiometer itself.

Only the portion of this power contained within the particular frequency interval selected is detected; it is thus useful to characterize the radiometer's sensitivity by a quantity proportional to its noise power per unit frequency interval, its *equivalent noise temperature*, T_N, defined as the temperature of a perfect absorber placed at the input which delivers the same output noise power for an equivalent noiseless amplifier.

The fluctuation in output due to receiver noise is expressed in terms of the equivalent change in noise power at its input in units of antenna temperature by the radiometer formula

$$\Delta T_{\mathrm{RMS}}=T_N/\sqrt{B\tau},$$

where B is the bandwidth and τ is the postdetection integration time. (A heuristic understanding of the origin of this relation may be obtained by thinking of $B\tau$ as the number of independent measurements one may make of the noise temperature within the integration time.)

Practical radiometry normally employs comparisons in making measurements. The most common such comparisons are made between: the antenna and a reference termination; two positions in the sky; or two different radiometer frequencies. A common method of periodic comparison used is called synchronous detection. The output of the receiver is inverted in synchronism with the switching of the input from the signal to reference condition. Thus, over a number of cycles the integration accumulates a voltage proportional to the difference between the two inputs. The great virtue of this arrangement can be illustrated by considering the effect of a small change in gain ΔG. In the unswitched case this would cause a change in output of ΔG times the total system temperature, whereas in the switched case the fluctuation in output is ΔG times the difference of two temperatures, which can be made as small as desired by proper selection of the reference temperature. The penalty paid is a decrease in the signal-to-noise ratio by approximately a factor of 2. An array of such radiometers, usually sharing a common amplifier and with their filters spaced adjacently in frequency, is the most generally used system for line studies, the multichannel line receiver. An alternative to the multichannel radiometer is the autocorrelation receiver, which uses the fact that the autocorrelation function of the time variation of the input signal is the Fourier transform of the power spectrum in frequency.

It is useful to compare the sensitivity of radiometers with that of incoherent devices as detectors of astronomical radiation. The radiometer contains active elements which amplify both the incident power as well as its own noise to levels much higher than the noise associated with its detector. Thus, it is the noise in the amplifier which limits the sensitivity of the device. Since this noise is proportional to bandwidth, whereas that of the incoherent detector is not, a meaningful comparison of the two types of devices can only be made when the bandwidth of the observation is specified. We may relate the noise equivalent power, NEP, of an incoherent detector to the minimum detectable increment in antenna temperature:

$$\Delta T_{\mathrm{RMS}} = \mathrm{NEP}/kB\sqrt{\tau}.$$

ASTROPHYSICAL SOURCES OF RADIO EMISSION

Free–free emission, often called "thermal emission" for historical reasons, arises from the interaction (acceleration) between unbound charged particles in an ionized gas (HII region). It has a characteristic spectrum whose brightness temperature decreases roughly as the inverse square of the frequency above a certain "turnover" frequency, below which it is equal to the temperature of the ionized gas. The turnover frequency marks where the gas becomes opaque to radiation and ranges from 100 MHz for large, diffuse HII regions to as high as 30 GHz for small, dense ones. Several hundred HII regions are known in the Milky Way; they serve as signposts for massive hot stars since the gas is ionized by uv photons at wavelengths shorter than 912 Å from the central stars. The balance between uv heating rate and cooling by emission from trace elements (O,N) endows most HII regions with a temperature of 10,000 K. Superimposed upon the broadband spectrum of the ionized gas are *recombination lines* caused by electronic transitions in the constituent atoms (mostly H, He) as given by the Rydberg formula ($n \sim 50$ to 500). The line-to-continuum intensity ratio is a sensitive indicator of physical conditions in HII regions.

Synchrotron emission is generated by high-energy electrons moving in a magnetic field. It is called "nonthermal emission" because its intensity is related not to the temperature of the emitter but to the strength of the field and the number and energy distribution of the electrons. It produces much of the radiation from supernova remnants, radio galaxies, and quasars. Radio galaxies and quasars comprise extragalactic radio astronomy. Their emission falls into two categories: the extended structure (which is transparent) and the compact structure (which is opaque to its own radiation). The extended emission is typically associated with galaxies, but in many cases with quasars with no visible optical extent. Most compact sources are identified with quasars or with active galactic nuclei. In less powerful radio galaxies, the radio emission is often confined to the region of optical emission (about 30,000 light-years in size), but in more powerful radio galaxies the emission comes from two well-separated regions hundreds of thousands of light-years across. Figure 3 shows a radio image of the powerful radio galaxy 3C175, made with the VLA at 6 cm wavelength at a spatial resolution of 0.35 arcsec. The unresolved bright spot near the center is a compact source coincident with a quasar at a redshift of 0.77. The long, narrow, one-sided radio jet is typical of powerful double-lobed (extended) sources. There are prominent hot spots in both lobes suggesting they have both been recently supplied with relativistic particles despite the appearance of only a single jet.

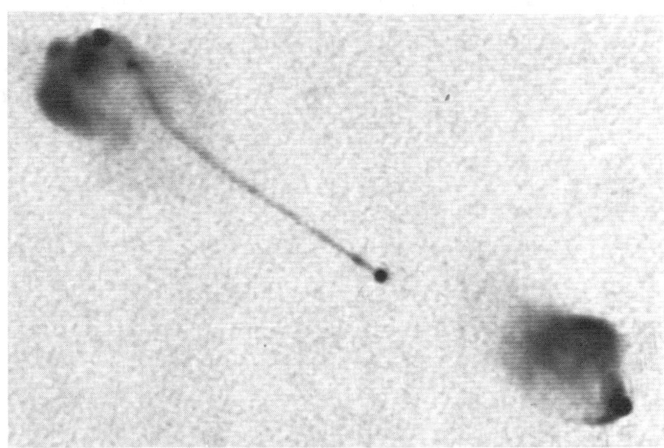

FIG. 3. The radio galaxy 3C175, observed with the VLA at a resolution of 0.35 arcsec at 6 cm wavelength (photo courtesy A. Bridle).

Our own galaxy radiates largely by synchrotron emission. An all-sky map of the galaxy's emission at 408 MHz is shown in Fig. 4; it was made at a single resolution of 0.85 degrees using three of the world's largest parabolic reflectors. The galactic center serves as the center of symmetry for the entire sky, with the low galactic latitude intensity dropping away steeply on each side out to longitudes ~60 and 280 degrees. Large-scale prominences, rising from low galactic latitudes and extending nearly to the poles in both hemispheres, distort the symmetry of the high-latitude emission. These loops and spurs trace large corresponding features in the magnetic field of the galaxy.

Line emission provides a powerful method for the study of neutral (and some ionized) interstellar matter. Hydrogen, the most abundant element, is studied in our own galaxy, in external galaxies, and on cosmological scales by means of its ground-state magnetic hyperfine transition at 1420 MHz (21 cm wavelength). Together with optical redshift data, HI data has established the existence of large-scale "voids" over cosmological distances, regions that are underpopulated at least in matter concentrated in normal galaxies. Cur-

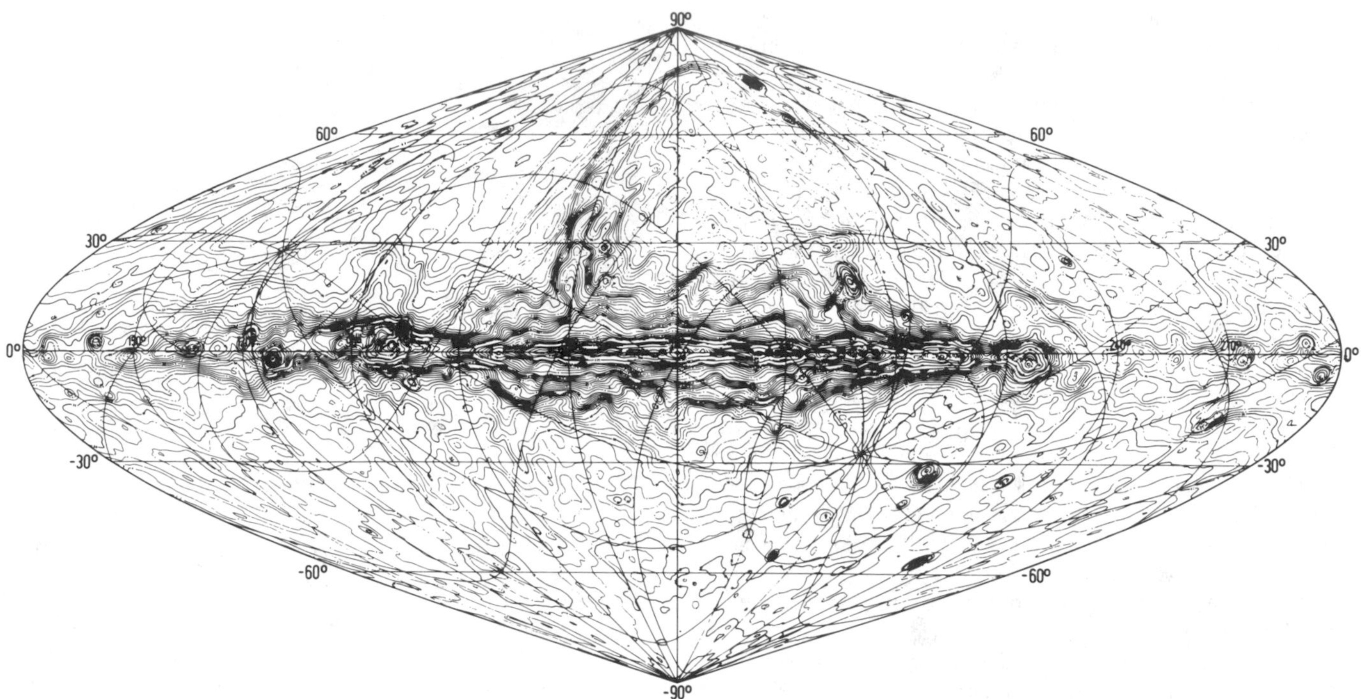

FIG. 4. An all-sky map of the Milky Way at 408 MHz (photo courtesy G. Haslam).

rent cosmological HI research is aimed at determining whether significant amounts of baryonic matter, capable of contributing importantly to the mass density distribution and therefore to the question whether the universe is closed or open, may reside in noncollapsed intergalactic clouds of HI.

HI studies of clusters of galaxies have shown gas deficiencies in cluster spiral galaxies—evidence that gas has been swept from at least their outer regions by passage through the center of the cluster. Several mechanisms may be at work (galaxy collisions, tidal interactions, ram pressure sweeping by the intracluster medium, evaporation) and may remove up to 90% of the initial HI mass. Reduced star formation rates are evident.

HI studies of noncluster galaxies, especially nearby ones for which good spatial resolution exists, serve to relate the structure of the interstellar medium with that of the stellar populations. Clearly defined HI disks correspond to optical disks but extend much further out, allowing galaxy mass determinations. Spiral arms seen in HI often reveal important differences from their optical counterparts that relate to star formation mechanisms. Strong warps are often seen in the outer regions of HI disks, which in many cases appear to be self-maintaining rather than resulting from perturbations such as tidal interactions. HI studies of velocity distributions in galaxies show in the case of spirals that the dynamical masses greatly exceed the total mass (by factors of 2 to 3) seen within the optical radius. HI studies also show that low-luminosity galaxies contain the same large mass-to-light ratio as luminous galaxies do, and that some types also contain dark halos. Since low-luminosity galaxies are by far the most numerous, these HI results are important to the question of the "missing mass" needed to close the universe.

Line emission is also observed from a large number of interstellar molecules, which reside in the denser regions of the interstellar medium. Together with studies of HI in the Milky Way, these lines have delineated the spiral structure and have established the existence of several thousand giant, massive molecular clouds (10^5 solar masses or greater) as well as many more smaller ones. These "GMCs" are the most massive entities in our galaxy and have been established as the sites of massive star formation. In other galaxies as well as our own, the GMCs are now known to trace the spiral arms. Molecular CO has now been observed at redshifts up to 0.16, and promises to become a cosmological tool as important as HI.

Some 85 interstellar molecular species have now been identified in dense molecular clouds, giving rise to the new science of *astrochemistry*. The interstellar chemistry is carbon rich, as on earth, but produces many exotic species not found on earth, as well as many familiar in the laboratory. These molecules play a major role in how stars form from the interstellar gas.

Blackbody radiation is characteristic of solar system objects. At long radio wavelengths the brightness temperature of the *sun* is very large, $\gtrsim 10^6$ K, because the observed emission comes from the corona and material ejected during solar flares. This ionized matter becomes essentially transparent at shorter wavelengths, and the millimeter wavelength brightness temperature of the sun is essentially that of the photosphere, ~6000 K. Conversely, in the case of a *planet*, the long wavelengths are able to penetrate un-ionized atmosphere better than shorter wavelengths. Thus the long-wavelength brightness temperature is that of the surface, whereas at shorter wavelengths the brightness temperature corresponds more closely to that of the cooler atmosphere.

Blackbody radiation fills the entire sky and is the remnant

of the big-bang fireball expanded and cooled to 3 K at the present epoch. It is unique in that it has the same brightness temperature, 3 K, at all wavelengths. The radio spectrum of "empty" sky is a superposition of galactic radiation, unresolved distant radio sources, and the cosmic background radiation.

See also ASTROPHYSICS; BLACKBODY RADIATION; GALAXIES; INTERFEROMETERS AND INTERFEROMETRY; PHOTOSPHERE; RADIOMETRY; SUN; SYNCHROTRON RADIATION; TRANSMISSION LINES AND ANTENNAS.

BIBLIOGRAPHY

J. D. Kraus, *Radio Astronomy*. Cygnus-Quasar Books, 1986. An intermediate level book emphasizing technical aspects, with advanced references.

G. L. Verschuur and K. I. Kellermann (eds.), *Galactic and Extragalactic Radio Astronomy*. Springer-Verlag, New York, 1988. A graduate-level text covering 15 major areas of radio astrophysical research.

Astronomy, X-Ray

Paul Gorenstein and Wallace Tucker

INTRODUCTION

X-ray astronomy involves photons of cosmic origin in the energy band 0.2 to 30 keV. The low-energy limit is determined by the opacity of the interstellar medium, the higher limit by the falling spectra of sources. Observations take place above the absorption of the Earth's atmosphere. X-ray production in a cosmic setting is associated with thermal radiation from plasmas with temperatures from 10^6 to 10^8 K and synchrotron radiation from highly energetic electrons in a magnetic field. Solar x-ray emission was detected in 1948 and a rocket flight in 1962 made the first positive detection of sources outside the solar system.[1] A series of spinning satellites, bearing collimated proportional counters, beginning with UHURU in 1970 and continuing through HEAO-1 in 1977, surveyed the sky and cataloged about 800 sources. Their celestial distribution shows two components: one distributed along the plane of the galaxy and another having the isotropy characteristic of extragalactic objects. In addition, there is an intense isotropic background which may or may not be fully explained as unresolved extragalactic sources.

Beginning in the late 1970s, pointed studies with grazing incidence imaging telescopes, spectroscopic, and large-area photometric-timing detectors were carried out aboard satellites developed by the United States (Einstein Observatory), West Europe (EXOSAT), and Japan (Tenma and Ginga). The number of sources detected by the Einstein Observatory exceeds 7000. Optical identifications have revealed that x-ray emission is detectable from a variety of objects. In the following sections, the sources are discussed by categories.

GALACTIC X-RAY SOURCES

The bright galactic sources are associated with the final phases of stellar evolution. They are identified with remnants of supernova explosions or binary systems containing a compact object such as a neutron star or a black hole. They are among the most luminous objects in the galaxy, radiating 10^{36} to 10^{38} ergs/s in the x-ray band. In comparison, the sun, an average star, radiates 4×10^{33} ergs/s, principally at optical wavelengths. A much lower level of x-ray emission, 10^{28}–10^{32} ergs/s, has been detected from many relatively nearby stars. The x-ray emission from these objects is produced by processes that are analogous to solar coronal and flare activity or involve mass transfer on a scale that is less energetic than in the binaries containing a neutron star or a black hole.

Compact X-ray Binaries

All of the bright luminous galactic x-ray sources that have not been identified with supernova remnants fall into a class designated as "compact x-ray binaries." The essential features of these sources are (i) a 1–10-keV luminosity in the range 10^{36}–10^{38} ergs/s; (ii) membership in binary systems, as evidenced by eclipses or periodic variations on a time scale of days in the radiation of the compact object or its companion star; (iii) a spectrum similar to that produced by radiation from a hot gas having a temperature in the range 50–500 million degrees; (iv) fast, and in some cases periodic, variations or quasi-periodic oscillations on a time scale of seconds or less.

The model consists of matter lost from the primary star in a close binary system accreting onto a neutron star or black hole companion. Gravitational potential energy heats a gas which radiates the input energy as x-rays. If the secondary has a mass M_x and a radius R, the gravitational energy released per gram would be on the order of GM_x/R. For a mass accretion rate $\dot{m}$, the energy released per second is

$$L \sim GM_x\dot{m}/R \sim 10^{41}(M_x/M_\odot)\,(R_\odot/R)\dot{m} \text{ ergs/s} \qquad (1)$$

where $M_\odot$ and $R_\odot$ are the mass and radius of the sun and $\dot{m}$ is the accretion rate in units of solar masses per year.

For both neutron stars and black holes $R_\odot/R \sim 10^5$; we have, therefore,

$$L \sim 10^{46}\dot{m} \text{ ergs/s} \qquad (2)$$

Equation (2) shows that mass accretion rates in the range 10^{-10}–10^{-8} solar masses per year can produce x-ray luminosities in the range 10^{36}–10^{38} ergs/s.

For comparable accretion rates, the expected luminosity for white dwarf companion stars ($R/R_\odot \sim 10^{-2}$) is three orders of magnitude smaller.

Compact binary systems also differ with respect to the nature of the noncompact primary star. In young binary systems the primary is a giant blue star that has a mass more than 10 times that of the sun and whose age is less than 10 million years. They are found in regions of active star formation, such as the galactic spiral arms. In old or low-mass x-ray binaries, the noncompact star is often less massive than the sun. These systems have existed for at least a hundred million years and show no preference for spiral arms.

The difference between young and old binaries manifests itself in other ways. Young neutron stars have intense magnetic fields which modify the accretion flow and produce an asymmetric radiation pattern that, when coupled with the rotation of the neutron star, appears to a distant observer as a series of pulses with eclipses and Doppler variations from the binary motion. For old x-ray binaries, two physical effects change the nature of their radiation. First, the accreting plasma transfers angular momentum to the neutron star, causing it to spin faster over time. Second, the magnetic field of the neutron star has weakened to the point where it can no longer effectively channel the accreting plasma. Consequently, these sources are characterized, not by stable periodic pulses on a time scale of seconds, but by quasi-periodic oscillations down to a scale of milliseconds.

This latter behavior is similar to that observed from an accreting black hole. Strong magnetic fields cannot exist in the vicinity of black holes and the period of the last stable orbit around a black hole having the mass of a few solar masses is on the order of milliseconds. The similarity in the radiation patterns of accreting black holes and old neutron stars has made it impossible to identify black holes conclusively on the basis of x-ray data alone. In general, successful black hole candidates meet two requirements: (1) they are luminous x-ray sources that exhibit large, rapid and sometimes quasi-periodic fluctuations on a time scale of milliseconds; (2) optical observations of the primary star indicate that it has an invisible companion with a mass greater than $3M_{\odot}$, the theoretical upper limit for the mass of a neutron star. To date, three systems have been discovered which qualify: Cygnus X-1, LMC X-3, and A0620-00. The best estimates of the masses of the black holes in these three systems are about 10 solar masses.

Supernova Remnants

Forty-four sources in our galaxy have been identified with supernova remnants (SNR). Nearly all are characterized by a fragmentary shell with a diameter that gets larger and a spectrum that gets softer with age. The evolution of luminosity is more complex. Twenty SNR show evidence for the existence of a neutron star. Four of these are fast pulsars that are losing rotational energy at a rate sufficient to explain their luminosity. Figures 1 and 2 show x-ray images of two relatively young SNR: Tycho, and the Crab Nebula which contains a rotationally powered pulsar. Their x-ray spectra are shown in Figure 3. Tycho contains x-ray lines of highly ionized silicon, sulfur, calcium, and argon, indicative of radiation from a hot plasma ($\sim 10^7$ K) with enriched elemental abundances expected to be associated with supernova ejecta. The spectrum of the Crab Nebula is featureless, and the radiation is partly polarized. Both results are indicative of synchrotron radiation. The rapidly rotating neutron star is seen directly as a point source when the radiation is beamed in our direction. In addition, particles accelerated by the neutron star lose energy by synchrotron radiation in the nebula. The evolution of a supernova remnant can be related to our current understanding of supernova explosions which are believed to fall into two basic categories. One is the disruption of a star which had previously collapsed to a

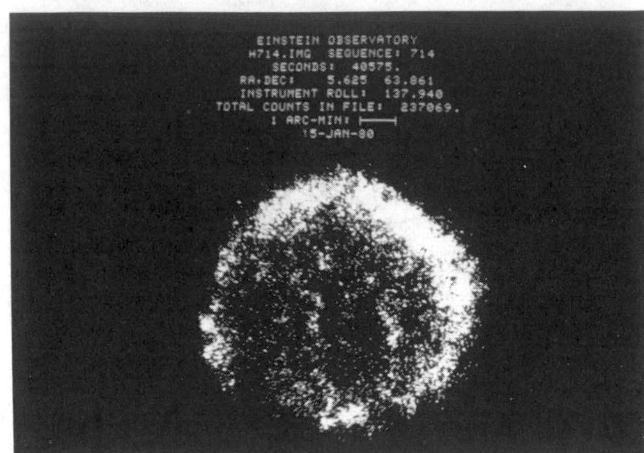

FIG. 1. X-ray image of the supernova remnant Tycho (SN 1572) taken with the Einstein Observatory. There is no evidence for a neutron star.

5000-km radius white dwarf, a phase in which electrons are highly degenerate. In the model, the white dwarf material, mostly helium and carbon, explodes as it fuses to form heavier elements, mostly radioactive nickel-56 which decays by beta emission to cobalt and then iron. Approximately 1 MeV per neutron of binding energy is released for a total of 10^{51} ergs which appears mostly as kinetic energy of the ejecta. Subsequent radioactive decay of ^{56}Ni is believed to be associated with Type I supernova whose visible radiation has a half-life of 55 days and is characterized by the absence of hydrogen lines.

The second category of explosion is believed to occur in a massive star when its core collapses to form a neutron star plus neutrinos, or a black hole. In the process, more than 10^{52} ergs of gravitational potential energy is liberated and transferred, in part, to the star's envelope, resulting in ki-

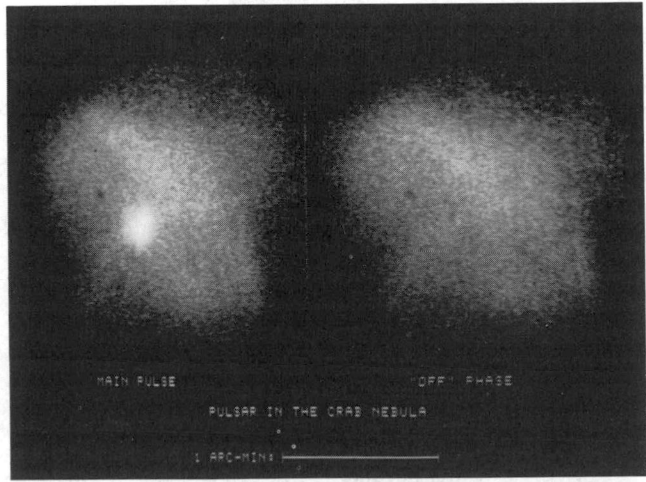

FIG. 2. X-ray images of the Crab Nebula (SN 1054) showing phases of the 33-ms period when radiation directly from the pulsar is, and is not, beamed towards the Earth. The Crab Nebula is unusual among supernova remnants in that there is no shell produced by supernova ejecta interacting with interstellar material.

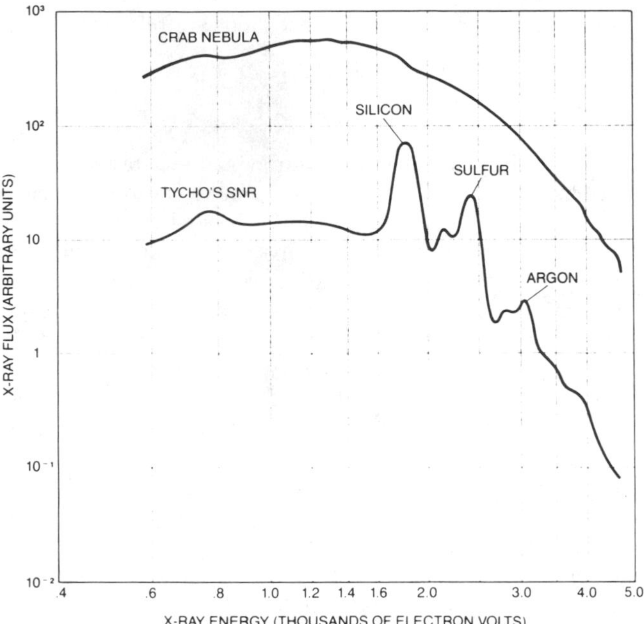

FIG. 3. X-ray spectra of Tycho and the Crab. Strong elemental lines in Tycho's spectrum are indicative of thermal radiation from a hot plasma containing enriched material released from the exploded star. The absence of lines in the Crab's spectrum and the presence of polarization is consistent with synchrotron radiation.

netic energy of ejecta. Fusion to ^{56}Ni apparently takes place in the shock-heated ejecta though on a smaller scale than in the first type of explosion. This is the current model for a Type II supernova explosion whose optical spectrum contains hydrogen lines and where the light curve is less luminous and more erratic in its decay than Type I. Given that the presupernova star was probably rotating, if a neutron star (rather than a black hole) is formed in the collapse of the core, it follows that it will be rapidly rotating from conservation of angular momentum. Hence, the SNR would have considerable energy stored in the form of rotational kinetic energy of a neutron star as well as in translational kinetic energy of the fast ejecta (2000–10 000 km/s). A slow transformation of energy from these reservoirs into radiation by thermal and nonthermal processes results in a discrete x-ray source for as long as 10^5 years. The shell-like appearance of SNR is a result of the propagation of the ejecta into the interstellar medium. Its form is affected by homogeneties preexisting in the interstellar medium and by winds from the presupernova star being overtaken by the ejecta as well as by possible asymmetries in the explosion itself.

In February 1987, a Type II supernova occurred in the Large Magellanic Cloud, a member of the local group of galaxies. This was the closest supernova in over 400 years. X-ray emission was first detected 6 months later, presumably because the ejecta became more transparent. The >50-keV component was explainable as degraded ~1-MeV gamma rays from decaying ^{56}Co. However, the intensity of softer x-rays was more persistent, suggesting that a neutron star is present and acting as an additional source through one of several possible mechanisms.

Stellar Coronae

In the past decade it has become apparent, primarily as a result of discoveries made with the Einstein x-ray observatory, that most stars emit x-rays at some level. This radiation reveals that the very low-density outer layers of a star's atmosphere are typically very hot, with a temperature of a million degrees Kelvin or higher. Evidently a small fraction $\sim 10^{-2}$–10^{-5} of the star's total energy output is channeled into the production of a hot corona around the star. In the hotter stars, the outer layers of the atmosphere are expanding away from the star in a vigorous stellar wind. In stars with surface temperatures below about 7000 K, such as the sun, the corona is probably heated by the interplay between internal convection, rotation, and magnetic fields. Presumably convection and rotation drive a dynamo which generates a magnetic field with twisted magnetic flux tubes that extend into the corona from subsurface layers, as appears to be the case in the solar corona. The resistive dissipation of the free magnetic energy in these twisted magnetic flux tubes provides the heat source for the corona. The closely coupled fields of stellar x-ray emission and dynamo theories are currently in a stage of rapid development. They should receive a significant boost from the next generation of solar and x-ray telescopes.

EXTRAGALACTIC SOURCES

Active Galactic Nuclei

Well over 500 x-ray sources have been identified with quasars, Seyfert galaxies, radio galaxies, and BL Lacertac objects, known collectively as active galactic nuclei (AGN). The x-ray luminosity of AGN is an appreciable fraction of their total radiative output.

There are at least two and, not unlikely, three components in the x-ray spectrum of a quasar. They apparently conspire to mimic a simple power law, $dn/dE \sim E^{-1.7}$ photons/cm^2 s, in the 2–10-keV band. According to models, the primary component is a nonthermal mechanism involving relativistic particle beams. Radio-emitting quasars have higher x-ray luminosity. In several cases, the x-ray emission has been observed to vary significantly on time scales down to 20 min. This variability limits the diameter of the x-ray emitting region to about 10^{14} cm, consistent with the conversion of gravitational potential energy into radiation at about 10 times the Schwarzchild radius for a 10^8 solar mass black hole. Further out, the infalling matter is probably contained in an accretion disk with a temperature of several hundred thousand degrees. This disk may be the source of a second component, soft x-rays produced thermally. A possible third x-ray component is connected with the nonthermal mechanism responsible for the intense infrared luminosity of quasars.

All galactic nuclei presumably contain massive black holes. Larger black holes and/or larger supplies of gas result in larger luminosity. In its simplest form this model suggests that the typical galactic nucleus was once a quasar and that the typical quasar will eventually become the nucleus of a normal galaxy when all the matter susceptible to accretion is consumed. This process might take a few million years, or a few hundredths of a percent of the age of a galaxy, consistent with the observed ratio of quasars to normal galaxies.

Clusters of Galaxies

Clusters of galaxies are detected as x-ray sources extending over a region with a diameter of a million light years. The emission is thermal with temperatures in the range (1–10) × 10^7 K, central densities of ~10^{-3}/cm^3, and relative elemental abundances, based mainly upon the measurement of x-ray lines from Fe ions with one and two electrons, that are between one-quarter and one-half of the solar abundance (Fe/H ~3 × 10^{-4}). Because the gas is in hydrostatic equilibrium in the cluster's gravitational potential, the morphology of the x-ray source reflects the distribution of mass in the cluster. Hence, the underlying mass of subcluster components and extended dark halos around individual galaxies can be studied by measuring the distribution of gas density and temperature. Figure 4 is an example of a double cluster with substructure. X-ray studies have revealed a dark halo exceeding 2.5 × 10^{13} solar masses within a radius of 200 kpc around M 87 plus massive halos around other elliptical galaxies. Despite some uncertainties in the gas temperature at large distances from the center, x-ray measurements provide independent support for the idea that clusters contain sufficient dark matter to be bound gravitationally, a conclusion that was reached in the late 1930s from studies of the dynamics of the member galaxies.

In many cases where the cluster contains a massive central galaxy a pronounced increase in density and a decrease in the temperature is observed at closer distances. Some investigators interpret this as indicating that substantial quantities of gas are flowing toward the center. Gas is apparently cooling as it flows, and disappearing along the way, perhaps forming numerous low-mass stars that are relatively invisible.

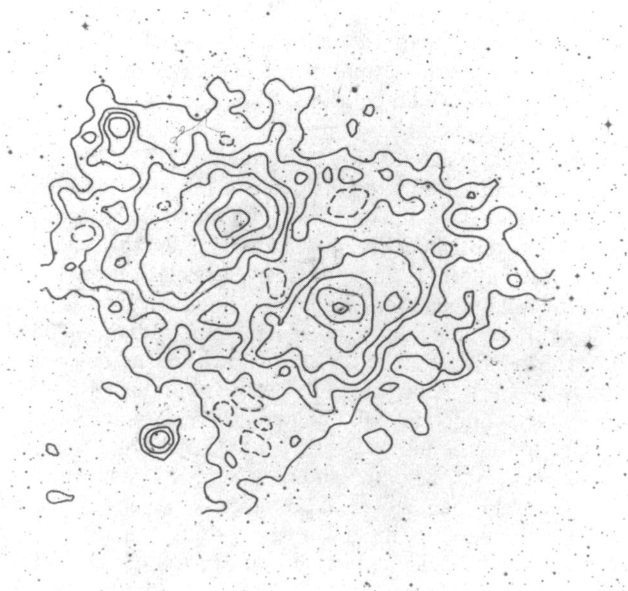

FIG. 4. X-ray contour image of a double cluster of galaxies superimposed upon the optical field.

The X-ray Background Radiation

One of the most striking features of the x-ray sky is the existence of a strong background whose isotropy indicates that it originates at distances of cosmological interest. The origin of the x-ray background has been discussed in terms of two basic models: (1) radiation from intergalactic matter, such as bremsstrahlung radiation from a hot intergalactic medium; or (2) the summation of discrete sources. The hot intergalactic medium model can explain the spectrum of the x-ray background, between 3 and 100 keV, which matches that of optically thin thermal bremsstrahlung from a gas with a temperature $kT = 40$ keV. However, the amount of matter needed to produce the radiation is very large (20–50% of the density required to close the universe) and an inordinate amount of energy (about 100 times the energy produced by all the quasars over an assumed lifetime of 10^8 years) is needed to heat the amount of intergalactic gas to the required temperature.

The case for the discrete source model is based upon the numerous active galactic nuclei that have been detected as strong x-ray emitters, plus some assumptions on the evolution of the luminosity function, to explain the intensity of the background. Limits on the small-scale fluctuations in the background indicate that the number of discrete sources would have to exceed 5 × 10^3/degrees^2, and that the average x-ray luminosity of the sources is less than 10^{45} ergs/s. A major liability of this model is that the spectra of active galactic nuclei in the 2–10-keV band is not consistent with the observed spectrum of the x-ray background.

It has been suggested that either a new class of sources is needed to explain the x-ray background or that active galactic nuclei have passed through an earlier, still unobserved phase during which they produced x-radiation which when redshifted to the present epoch results in the correct intensity and spectrum. Indeed, this is not precluded given the uncertainties in the spectra of active galactic nuclei. At present, it is impossible to determine whether any of these alternatives will work.

CONCLUDING REMARKS

The schedule of x-ray astronomy missions for the 1990s involves focusing telescopes with substantially larger collecting area and better angular resolution plus more powerful spectrometers. The Soviet Union as well as the United States, Western Europe, and Japan are each planning to lead a major program. However, the increasing cost and complexity of these missions has resulted in the need for much more international collaboration. These missions are expected to detect the most distant extragalactic objects in the universe and may resolve the question of the origin of the background.

See also BLACK HOLES; GALAXIES; INTERSTELLAR MEDIUM; NEUTRON STARS; PULSARS; QUASARS; SYNCHROTRON RADIATION

REFERENCES

1. R. Giacconi, H. Gursky, F. Paolini, and B. Rossi, *Phys. Rev. Lett.* **9,** 439 (1962).

BIBLIOGRAPHY

M. Begelman, R. Blandford, and M. Rees, *Rev. Mod. Phys.* **56,** 255 (1984).

E. Boldt, *Phys. Rept.* **146,** 215 (1987).

M. Elvis, "Current Issues in the X-ray Properties of Active Galactic Nuclei," *Publ. Astron. Soc. Pacific* **98,** 148 (1986).

R. Giacconi, H. Gursky, F. Paolini, and B. Rossi, *Phys. Rev. Lett.* **9,** 439 (1962).

J. McClintock, "X-ray Properties of Galactic Black Holes," *Adv. Space Res.* **8.** No. 2–3, (2) 191 COSPAR (1988).

C. Sarazin, *Rev. Mod. Phys.* **58,** 1 (1986).

V. Trimble and L. Woltjer, *Science* **234,** 155 (1986).

W. Tucker and R. Giacconi, *The X-ray Universe*. Harvard University Press, Cambridge, 1985.

IAU Colloquium 115, *High Resolution X-ray Spectroscopy of Cosmic Plasmas*, P. Gorenstein and M. Zombeck eds. Cambridge University Press, Cambridge.

Astrophysics

L. H. Aller

Astrophysics generally refers to the application of physics to astronomy. Classical celestial mechanics, positional astronomy, stellar motions, and usually galactic structures are excepted. Virtually all branches of modern physics have played crucial roles in the development of astronomy. Historically, the initial impetus came from development of the quantum theory and interpretation of line and continuous spectra in terms of outer electronic structures of atoms and molecules and temperature and density of the ambient medium. Spectroscopy remains one of the most important fields of astrophysics since diagnostics of stellar and nebular plasmas depends on the spectra which they emit. Originally, spectroscopic studies were confined to the optical region, approximately 3000–10,000 Å (300–1000 nm), although a few investigations could be carried out in the near infrared through "windows" in the Earth's atmosphere. Next the radio-frequency region (millimeter to meter range) was explored, and finally the development of spacecraft and rockets permitted investigation of the ultraviolet ($\lambda < 3000$ Å), the x-ray and the γ-ray region, and the far infrared. The thermal radiation of the sun and stars falls mostly in the "visible" range.

Hence it is not surprising that most of the optical-region spectroscopic observations could be interpreted in terms of thermal radiation from hot gases, sometimes near thermodynamic equilibrium in stellar atmospheres and sometimes deviating very far therefrom in the interstellar medium.

Radio frequency data showed that much of the observed emission could not be interpreted in terms of thermal radiation. Radiation from certain supernova remnants and various sporadic solar phenomena is often strongly polarized and seems to indicate synchrotron radiation from relativistic electrons accelerated in the presence of large-scale magnetic fields. Perhaps the most striking conceptual development in mid-20th century astrophysics was the recognition of the omnipresence of large-scale magnetic fields in astrophysical sources and the role they played in accelerating charged particles to high energies.

Magnetic fields of the order of 3000 G had been discovered from the inverse Zeeman effect on atomic spectral lines observed in sunspots by George Ellery Hale shortly after the turn of the century. With the refinement of observational techniques, the presence of a general solar magnetic field was established as well as small, intensely magnetized areas in the neighborhood of sunspots. H. W. Babcock discovered magnetic fields ranging in strength up to 30,000 G in stars whose spectra often showed strong lines of lanthanides but which otherwise seemed to be normal, hydrogen-burning main sequence stars. Recent work has shown that white dwarf stars (which have no nuclear energy sources and obey the degenerate gas law) can have fields of 5×10^6 G!

The sun is the only star which can be studied in detail. Its well-established sunspot cycle is basically a magnetic cycle. Plasma clouds can be accelerated in the magnetized regions associated with sunspots; particles can be accelerated to cosmic-ray energies there. Although objects such as the Crab Nebula and other old supernovae are believed to produce cosmic rays, only in the sun can we actually observe, at relatively close hand, the generation of low-energy cosmic rays.

Large-scale magnetic fields exist in interstellar space and presumably dominate vast clouds of ionized gas in radio galaxies and near the centers of relatively "normal" galaxies. An outstanding problem of astrophysics is how rapidly moving plasma clouds can produce magnetic fields in low-density regions and how the polar and toroidal or sunspot fields are generated in the sun. Strong magnetic fields are believed to be associated with pulsars or neutron stars and may play a role in accelerating charged particles to cosmic-ray energies.

In our own solar system, both the Earth and Jupiter have strong magnetic fields which trap charged particles in magnetospheres around them. The origin of this field is believed to lie in dynamo action in liquid metallic cores, iron in the case of the Earth and hydrogen in the case of Jupiter. In stars such as the sun, some regenerative or dynamo action must also exist, although there is strong indication that the magnetic cycle gradually weakens as the star ages. The strong magnetic fields of white dwarfs and presumably also neutron stars probably come from fields that are enhanced as the star contracts when nuclear energy sources are exhausted.

Developments in nuclear physics and an understanding of the properties of radiation and matter at high temperatures and densities provided the background for an interpretation of a vast amount of astronomical data in terms of theories of stellar structure and evolution. Although already in the 19th century it was demonstrated by elementary arguments that the stars must have very high central temperatures if they were gaseous throughout, real progress in understanding their structures and evolution required advances in nuclear physics.

Painstaking observational efforts showed that stellar luminosities, masses, and diameters were not correlated with one another in a haphazard fashion. Most stars belonged to what is called the *main sequence* wherein the luminosity and surface temperatures are correlated with the mass: the more massive the star, the higher its luminosity and surface temperature. Other stars (giants and supergiants) deviated from this correlation in the sense that for a given mass their diameters were larger and their luminosities were often higher than for main sequence stars. There also existed a group of very dense, faint stars—called white dwarfs—which were recognized very early as objects near the ends of their lives.

Studies of clouds of dust and gas in our own and in other galaxies and the relationship of bright, hot stars to these disorganized clouds of material showed that stars must be forming more or less continually out of this interstellar material. Condensations of the normally tenuous dust and gas are accentuated by gravitational action to form protostars which eventually shine by thermonuclear reactions, converting hydrogen into helium in their cores. Such stars remain on the main sequence until all of the hydrogen in the central regions has been converted into helium. The evolution of the star gradually becomes very complex. It leaves the main sequence, becomes a giant or supergiant, and eventually converts helium into carbon and heavier elements. Massive stars may undergo explosive evolution as supernovae. For most stars, however, the outer layers simply escape into space with much of the material returning to the interstellar medium while the hydrogen-depleted core settles down as a white dwarf.

Filling in the details of the life history of a star requires extensive input from many fields of physics and chemistry. Consider first the nuclear reactions. The sun shines by the proton–proton reaction, the first step of which requires bashing two protons together until they form a deuteron with the emission of a positive electron and neutrino. This process has never been observed in the laboratory. Evaluation of the rate requires an accurate theory of nuclear reactions, as the cross section is too low to be observed experimentally. A vast amount of spectroscopic data supports the hypothesis that chemical elements are built in stars by a sequence of complex processes. Up to the iron group of elements, the formation of progressively heavier nuclei continues to deliver energy through exothermic reactions. Yet heavier elements such as Ga, Ba, the lanthanides, etc., can only be found by the addition of neutrons to nuclei of the iron group. Neutrons can be produced slowly in the orderly evolution of a giant star or rapidly in the catastrophic explosion of a supernova. The Burbidges, Fowler, and Hoyle outlined the process of element building in stars; details and refinements have been offered by numerous investigations. Hence studies of stellar chemical composition throw many sidelights on problems of stellar structure and evolution.

Carbon and nitrogen are built in stars up to about 5–8 solar masses. Heavier elements are made in more massive stars, particularly in supernovae. SN 1987A provided an important confirmation of much theoretical work on supernovae, in particular, the creation of large amounts of ^{56}Co which decays to ^{56}Fe, and the production of a large flux of neutrinos.

Construction of stellar models also requires knowledge of the relevant equations of state, of the opacity of the material as a function of temperature and density, and of hydrodynamics. Large portions of stellar interiors are in convective equilibrium. Energy is transported outward by mass motions of material. The kinematics of the solar atmosphere suggests that this motion often tends to be disorderly or turbulent rather than strictly coherent as in well-established convective motions. The theory of convection in a compressible gas in a gravitational field with a strong density and temperature gradient has not yet been developed with sufficient accuracy.

Throughout most of the volume of typical stars the density is so low that the appropriate equation of state is virtually that of a perfect gas. In white dwarf stars the degenerate gas law holds and at very high densities the gas becomes relativistically degenerate. Degenerate electron gas was recognized in the stars before it was identified on the Earth!

The upper limit of the mass of a degenerate star is well defined. If more mass is added, the electrostatic forces between particles can no longer prevail against gravitational forces. Electrons are driven into the nuclei of atoms and a neutron star results. Pulsars are believed to be rapidly rotating neutron stars—core remnants of supernovae. The state of matter in neutron stars has been the subject of much recent speculation.

A very large, cold mass could not settle down even as a neutron star but would evolve into a black hole from which no radiation could ever escape and within which all material would be crushed. A great deal of effort has been expended in searches for black holes. They are most likely to be found in binary systems where more or less normal stars have invisible companions of very large mass.

Neutrino astronomy has emerged as an important topic. The SN 1987A neutrino data indicated that the mass of this elusive particle is less than 23 eV, the most probable value being zero. The theoretical solar neutrino flux greatly exceeds the observed value providing one of the gravest problems in modern astrophysics.

Like other mechanical systems, stars have free periods of oscillations. At certain stages in their post-main-sequence evolution, i.e., after the hydrogen in the core is exhausted, a star may be driven to oscillate in its fundamental period or even in its overtones. The best known examples of such pulsating stars are the Cepheid variables, so-called after the prototype Delta Cephei, and the so-called RR Lyrae stars, which have periods less than a day and are often found in globular clusters. Although radial modes of pulsation appear to produce the most easily observable effects, nonradial modes may also exist. Even running or standing waves (such as the 5-min oscillation in the sun) may occur.

Gaseous nebulae and the interstellar medium offer engaging problems in plasma diagnostics, hydrodynamics, and complex organic chemistry. Hot plasmas ranging in gas kinetic temperature from 6000 K to several hundred degrees occur in shells around dying stars (planetary nebulae), in envelopes of exploding stars (novae and especially slow novae), behind the shock fronts of supernovae, in the active cores of certain galaxies (such as the so-called Seyfert galaxies), in many radio galaxies, and in the mysterious quasars which are believed by many to be regions of high energy release in cores of unusually active galaxies. The emission spectra of such regions are characterized by so-called forbidden lines of abundant familiar elements such as N, O, Ne, S, Ar in various ionization stages and permitted lines of gases such as hydrogen and helium. Target areas for collisional excitation of the metastable levels whence the forbidden lines arise have been calculated for most of the lighter ions but are urgently needed for the several observed ionization stages of ubiquitous iron, which often serves as a good plasma diagnostic for very high-temperature regions.

The cool dense clouds of the interstellar medium supply especially interesting challenges. From such clouds it is believed that stars and solar systems are formed eventually. Here are found unexpected organic molecules including formic acid, methanol, and ethanol. Spectral lines of some of these molecules show complex maser actions, including even refrigeration effects. Within our own solar system it is believed that comets may contain similar materials left over from the formation of our solar system. Evidence is given by the omnipresent 2.7 K radiation and by the abundance ratios of primordial helium, deuterium, and lithium to hydrogen. If the expansion hypothesis is accepted, the age of the universe can be estimated from the Hubble constant (Doppler shift per megaparsec) and also from isotope ratios of certain natural radioactive elements such as Th/Pb, U/Pb, and Re/Os, while constraints can be set by the ages of star clusters. The value of the Hubble constant is uncertain. Estimates from 50 to 100 km/s per megaparsec have been vigorously championed by opposing camps. Recent time delay measurements of transient events in a double quasar suggest a value of 60 km/s.

A closely related problem is the question of an open versus a closed universe. Will the universe go on expanding forever, or will it collapse back upon itself? Current thinking requires that the value of the cosmological constant, Ω, be exactly 1, but the amount of detectable matter in the known universe does not supply a sufficient density for this to happen. Some of the missing mass may be provided by baryons (ordinary matter) but much may be due to hypothetical weakly interacting particles (WIMPS) whose presence is felt by their gravitational effects only.

See also BLACK HOLES; COSMOLOGY; INTERSTELLAR MEDIUM; NEUTRON STARS; NUCLEOSYNTHESIS; PULSARS; QUASARS; STELLAR ENERGY SOURCES AND EVOLUTION; SUN; UNIVERSE.

BIBLIOGRAPHY

The best references on astrophysics are current journals, review volumes, and reports of symposia.

Journals

Astrophysical Journal, published by the University of Chicago Press, Chicago, Ill.

Astrophysics and Space Science, published by Reidel Publishers, Dordrecht, The Netherlands.

Astronomy and Astrophysics, published by Springer-Verlag, Berlin.

Monthly Notices of the Royal Astronomical Society, published by Blackwell Scientific Publications, Oxford, England.

Review Volumes

Annual Review of Astronomy and Astrophysics. Annual Reviews Inc., Palo Alto, Calif.

Stars and Stellar Systems, 8 vols. University of Chicago Press, Chicago, Ill., 1960–1975.

Reports of Symposia

International Astronomical Union Symposia Proceedings, published by Reidel Publishers, Dordrecht, The Netherlands. Examples: pertaining to the *Sun*: Vols. 35,43,56,68,71,86; pertaining to *galactic structure, galaxies, quasars,* etc.: Vols. 38,44,58,60, 63,64,69,74,79,84,92,97,104,106,108,116,117,119,124,126,127, 130; pertaining to *planets and meteors*: Vols. 33,40,47,48,62,65, 89,90; pertaining to *gaseous nebulae and the interstellar medium*: Vols. 34,39,46,52,76,87,103,120,131; pertaining to *stars, stellar spectra, and stellar evolution*: Vols. 42,50,52,54,55,59, 66,67,70,72,75,83,95,98,99,101,105,113,115,122,123,125.

Atmospheric Physics

John N. Howard

Atmospheric physics is the discipline concerned with the physical processes that occur in the atmosphere, the gaseous or vaporous shell that surrounds the earth. When early man first sought to understand his physical environment, his greatest awe surely concerned the sun, moon, planets, and stars (which became the basis of the science of astronomy or astrophysics), and next he wondered at the winds and storms, rain and snow, clouds, lightning, and thunder that collectively made up his weather. The Greek root *atmos*, meaning air or vapor, originally derived from the word for winds.

METEOROLOGY

The earliest meteorological instruments were the wind vane and the rain gauge, both of which date to antiquity. For all practical purposes, however, scientific meteorology began early in the seventeenth century with the Italian invention of the air thermoscope (by Santorio Santorre of Padua in 1612 and by Galileo at about the same time). Soon thereafter came the barometers of Torricelli (1644), Pascal (1646), and Hooke (1668); the sealed-in-glass thermometer (1641); and the temperature scales of Rømer (1702), Fahrenheit (1708), Celsius (1742), and Réaumur (1731). Other instruments were devised to measure atmospheric humidity, duration of sunshine, cloud height, and motion.

The first impetus to meteorological theory was given by the puzzling behavior of the barometer in relation to the weather, but it remained for Vilhelm Bjerknes, a physics professor at Bergen, Norway, around 1900 to formulate weather systems mathematically in terms of high- and low-pressure areas, warm fronts, and cold fronts. In England, Napier Shaw began scientific research on atmospheric processes in 1885 at the Cavendish Laboratory. By 1918 weather forecasting had become a science, much as we know it today.

CLIMATOLOGY

At a given location, the weather is considered to be the day-to-day variation of temperature, windiness, storms, and so on, as well as the seasonal changes from summer to winter. The prevailing weather at a given location is called its climate, and climate varies with geographical latitude, terrain features, and altitude.

ATMOSPHERIC CIRCULATION

Two major processes drive atmospheric weather systems: the rotation of the earth and uneven heating by solar radiation. In addition, the $23\frac{1}{2}°$ tilt of the earth's axis to the plane of its orbit and earth's elliptical orbit about the sun add winter and summer seasonal changes to the solar radiation received. These radiation processes are further altered by variable cloud cover, which reflects radiation that would otherwise warm the surface, and by the heat absorbed in the vaporizing of water or released by the condensation of water vapor into rain or snow. The earth's surface receives more solar radiation in the equatorial regions than at higher latitudes. Air warmed near the surface rises and creates a convective circulation pattern in the upper atmosphere from the equator toward the poles. However, because the earth (and its atmosphere) is also rotating, the rate of eastward rotation is a function of latitude, being about 0.7 of the equatorial velocity at latitude 45° (and 0.5 at latitude 60°). This motion superposes a shift to the right on the wind pattern in the Northern Hemisphere (and to the left in the Southern Hemisphere). The general circulation of the atmosphere is further modified by uneven local heating and by differences in heat absorption of oceans and land. There are five general regions of prevailing winds in the Northern Hemisphere (and, in reversed direction, in the Southern Hemisphere). Near the equator the surface air is heated by the direct rays of the sun and rises. This creates a region of equatorial calms, or doldrums. Between the equator and latitude 30°N is a region of prevailing northeast trade winds. Between latitude 30° and 60° is a region of prevailing westerly winds. Between these two regions are the so-called horse latitudes, another region of calm, where the air is sinking and warmed adiabatically. Above latitude 60° is a region of polar easterlies. At the surface, the moving air forms into whirling masses of high-pressure cells, called highs, or anticyclones (which circulate clockwise in the Northern Hemisphere); and low-pressure cells, called lows or cyclones (which circulate counterclockwise). Highs generally bring fair weather and lows unsettled weather.

Although weather phenomena are atmospheric processes, the field of physical meteorology is now developed to the point that it is considered separate from atmospheric physics, which is ordinarily restricted to those phenomena other than weather.

ATMOSPHERIC COMPOSITION

The earth's atmosphere is the remainder of the primordial gaseous matter, parts of which cooled down to form the earth and sea. The natural atmosphere consists of about 21% oxygen, 78% nitrogen, 1% argon and other gases, and a slight trace of carbon dioxide. Another variable constituent is water vapor, which may be present up to about 3%. Living animals breathe in oxygen and use it to oxidize their food, breathing out carbon dioxide. Plant life containing chlorophyll consumes carbon dioxide in the presence of sunlight to build plant tissues and return oxygen to the atmosphere. In this way the biosphere maintains a balance between oxygen and carbon dioxide. Cosmologists such as Lloyd Berkner and Harold Urey have projected the composition of the atmosphere from earliest prehistory (before the water vapor condensed into the oceans and before life was possible).

ATMOSPHERIC STRUCTURE AND TEMPERATURE

Atmospheric composition is essentially constant up to 90 km, above which the molecular weight of air begins to vary because of molecular dissociation. Pressure and density decrease approximately exponentially with height. Water vapor decreases more rapidly with height than the permanent constituents.

Temperature variation with height is the most pronounced feature of the atmosphere, and layers and shells are described most easily by temperature. Usually temperature decreases with increasing height in the lowermost several kilometers; any increase of temperature with height in this region is called an inversion. The lapse rate is defined as $-dq/dz$, the rate of decrease with height z of any atmospheric variable q.

Except for the first 2 km above the earth's surface, the normal lapse rate is about 6.5°C km^{-1} in the troposphere, which extends from the surface (about 300 K) to the tropopause (at about 220 K). The tropopause is the atmospheric surface at which either the temperature decrease stops abruptly or the lapse rate becomes very low. The height of the tropopause varies with latitude, season, and weather; in general, it is lowest (8–10 km) in arctic regions in winter and highest (16–18 km) in tropical and equatorial regions. Above the tropopause, temperatures usually increase with height, slowly at first, then more rapidly up to about 50 km, where the average temperature is again close to the freezing point (0°C). The region between the stratopause and the tropopause is called the stratosphere. Above the stratosphere there is another region, called the mesosphere, in which temperature again decreases with height. This region ends at the mesopause, at 80–90 km, where the lowest temperatures of the atmosphere are found (about −90°C). Above the mesopause is another region, called the thermosphere, whose upper limit is undefined and in which temperature again increases with height up to about 200 km. Above this level the temperature varies widely according to the degree of solar activity; it is about 600°C when the sun is quiet and possibly 2000°C during sunspot maxima. The exosphere (about 550 km) is the region where the mean free path is so great that particles escape from the atmosphere. In this region the temperature can no longer be defined in the usual way.

ATMOSPHERIC DENSITY

Except for highly variable aerosol and water-vapor content, the lower atmosphere behaves as a homogeneous mixture of the constituents indicated in Table I, and its pressure, density, and temperature obey the gas laws: The density is directly proportional to the pressure (Boyle's law); inversely proportional to the absolute temperature (Charles's law); and directly proportional to the sum obtained by adding to-

Table I Mass of the atmosphere and its constituents

Substance	Volume percentage dry air	Molecular weight	Total mass ($kg \times 10^{11}$)
Total atmosphere			51 300 000
Dry air	100.00	28.97[a]	51 170 000
Nitrogen	78.09	28.02	38 648 000
Oxygen	20.95	32.00	11 841 000
Argon	0.93	39.88	655 100
Water vapor		18.02	130 000
Carbon dioxide	0.03	44.00	23 320
Neon	0.0018	20.0	636
Krypton	0.0001	82.9	146
Helium	0.00053	4.00	37
Ozone		48.00	30
Xenon	0.000008	130.2	18
Hydrogen	0.00005	2.02	2

[a] Effective.

gether the molecular weights of the constituents, each multiplied by the ratio of its partial pressure to the total pressure (Avogadro's law). The average atmospheric pressure at sea level is about 101.3 kPa and the density 1.2 kg m^{-3}. Both decrease exponentially with height, and by several earth radii the density can be said to be that of interplanetary space. Drag effects due to atmospheric density can generally be neglected for rockets and spacecraft more than 200 km above the surface; satellites in such orbits have lifetimes of several years.

ATMOSPHERIC RADIATION

The sun is a huge gaseous sphere (about 1.39×10^6 km in diameter) whose temperature at the radiation surface is about 6000 K. The radiant solar energy incident on the earth outside the atmosphere is about 1390 W m^{-2} (with an annual variation of $\pm 3.5\%$ because of the ellipticity of the earth's orbit). If the earth had no atmosphere, but were a rotating sphere with an overall reflectivity similar to that of the moon, it would absorb this solar radiation until it had warmed up to an average temperature of just over 200 K, for then it would be radiating blackbody radiation to the 4π sphere of free space at about the same rate that radiation is received from the sun.

The atmosphere provides a warm thermal blanket around the earth, because it is largely transparent to the incoming 6000-K blackbody radiation, which is mostly at wavelengths in the visible and near infrared, but the atmospheric constituents ozone, carbon dioxide, and water vapor have strong absorption bands that trap a large fraction of the longer-wavelength infrared radiation emitted by the earth. As a result, the surface temperature of the earth in the temperate regions warms to about 300 K.

Some of the oxygen molecules of the lower atmosphere diffuse upward but are then dissociated into atomic oxygen by incident solar ultraviolet radiation. Atomic oxygen strongly absorbs all solar radiation shorter than 1850 Å. There is a region about 30 km above the earth's surface where the upward-diffusing molecular oxygen and the downward-diffusing atomic oxygen react to form a tenuous layer of ozone. The ozone molecule strongly absorbs all radiation below about 3150 Å, thereby screening out solar ultraviolet radiation that would otherwise be lethal to life at the surface. The absorption of radiation by ozone at 30 km and by atomic oxygen and nitrogen higher in the atmosphere is the mechanism responsible for the thermal structure of the atmosphere.

ATMOSPHERIC OPTICS

The most important optical effects of the atmosphere are the screening out of lethal ultraviolet solar radiation and the raising of the radiation temperature of the earth by trapping infrared thermal radiation. In addition there are a wide variety of curious optical effects that are interesting physics, but not of practical importance or dangerous to life: such phenomena as mirages, rainbows, halos, sun dogs, glories, and heiligenschein. These are caused by temperature inversions, scattering by aerosols, fog, or ice crystals, and so on. At one time it took an alert and skilled observer to see such phenomena; today many of these effects can be seen by airplane travelers. M. Minneart and W. J. Humphreys treat the physics of these phenomena.

THE IONOSPHERE

The bombardment of the atmosphere by ultraviolet radiation and energetic particles from the sun causes dissociation and ionization of atmospheric constituents into a region of the upper atmosphere that is called the ionosphere. The ionosphere itself is classified into regions at different altitudes, each due to differing atmospheric constituents. Above 150 km is the F region, due chiefly to ionized atomic nitrogen and atomic oxygen (N^+ and O^+). The E region occurs between 85 km and 150 km, and is largely due to O_2^+ and NO^+. A weaker D region, between 50 and 85 km, is largely due to hydrated ions. These ionized layers can strongly affect radio communication, reflecting the wavelengths shorter than about 30 MHz. The ionosphere has seasonal and diurnal variations, is strongly latitude dependent, and is also a function of the amount of sunspot activity. The airglow and aurora result from emission by these ionized constituents of the upper atmosphere.

THE MAGNETOSPHERE

In addition to its atmosphere, the earth is surrounded by a magnetic field. The main field originates inside the earth. The magnetic axis is inclined at an angle of about 11° to the axis of rotation. The field strength at the surface is about 0.6 G in the polar regions and 0.3 G in the equatorial region. Fields originating outside the earth are much more variable and are weaker; in general, they last a few days or less. These fields are caused by electric-current systems in the lower ionosphere. During magnetic storms these fields may have fluctuations of 5% of the main field in auroral zones. At altitudes of several earth radii interactions between the main field and corpuscular radiation from the sun (the solar wind) become more important. Magnetic disturbances can interrupt communications and affect the behavior of spacecraft; they are also linked to the phenomenon of auroras in the high-altitude ionosphere.

See also AURORA; IONOSPHERE; MAGNETOSPHERE; METEOROLOGY.

BIBLIOGRAPHY

W. J. Humphreys, *Physics of the Air*, 2nd ed. McGraw-Hill, New York, 1929. (Reprinted by Dover, New York, 1964.)

W. E. K. Middleton, *The History of the Barometer*. Johns Hopkins Press, Baltimore, Md., 1964.

W. E. K. Middleton, *A History of the Theories of Rain*. Franklin Watts, New York, 1966.

W. E. K. Middleton, *The Invention of the Meteorological Instruments*. Johns Hopkins Press, Baltimore, Md., 1969.

M. Minneart, *Light and Colour in the Open Air*. G. Bell and Sons, London, 1940. (Reprinted by Dover, New York, 1954.)

A. S. Jursa, ed., *Handbook of Geophysics and the Space Environment*. National Technical Information Service, Springfield, Virginia, 1985.

Atomic Spectroscopy

Allen Lurio† and Anthony F. Starace

INTRODUCTION

The progressively more detailed understanding of the emission and absorption spectra of atoms has led from the Bohr theory of the hydrogen atom (1913), to the discovery of electron spin by Uhlenbeck and Goudsmit (1925), to the Pauli exclusion principle (1926), and ultimately to the nonrelativistic quantum-mechanical Schrödinger equation (1926).

From these developments we find the following principles apply to the interpretation of all atomic spectra.

1. An atomic system can exist only in discrete stationary states corresponding to a well-defined energy E (within the Heisenberg uncertainty limit $\Delta E \Delta t \sim \hbar$, where Δt is the lifetime of the state). Transitions between these states, including the emission and absorption of radiation, require the complete transfer of an amount of energy equal to the difference in energy between these states.
2. The frequency of the emitted or absorbed radiation in going from state 2 to state 1 is given by $\omega = (E_2 - E_1)/\hbar$ (ω negative is absorption).
3. Each stationary state has associated with it a definite quantized angular momentum **J** and a definite parity (defined later). The projection of the angular momentum on any chosen direction in space is quantized with allowed values $m_J = J, J-1, \ldots, -J$.

Each atom has its own unique spectrum. The interpretation of this spectrum has led to the classification of many of the stationary-state energy levels of neutral and several-times-ionized atoms. These results are tabulated in a classic three volume NBS publication by Charlotte Moore. We shall attempt here to give a simplified treatment of the physical basis of this classification.

† Deceased

ELECTRONIC CONFIGURATIONS

The specification of the stationary states of an N-electron atom is given by the solution of the time-independent Schrödinger equation

$$H\Psi = E\Psi, \tag{1}$$

where the dominant terms contributing to the Hamiltonian H are

$$H = \sum_{i=1}^{N} \frac{p_i^2}{2m} - \sum_{i=1}^{N} \frac{Ze^2}{r_i} + \sum_{i>j=1}^{N} \frac{e^2}{r_{ij}}. \tag{2}$$

The first term is the kinetic energy of the electrons, and the other terms are the potential energies of Coulomb interaction of the electrons with the nucleus and with each other. A good starting approximation to H is the "central-field approximation" in which one replaces the second and third terms of Eq. (2) by $\Sigma_{i=1}^{N} V(r_i)$, where $V(r_i)$ is the spherically symmetric average potential seen by the ith electron due to all other electrons. Equation (1) is now solvable with a wave function which is the product of N single-electron wave functions. The Schrödinger equation for each of these single-electron wave functions is like that for hydrogen (see Burke) except that the hydrogenic potential-energy term e^2/r is replaced by $V(r)$. Consequently, the same set of quantum numbers n, l, s, m_l, m_s used to describe the hydrogenic electron apply here. These are respectively, the principal quantum number n, the orbital and spin angular momentum quantum numbers, and their projections on the quantization axis.

Complete specification of an N-electron state requires N sets of these quantum numbers with the Pauli restriction that no two sets can be identical. For a given n and l if all possible m_l and m_s states are occupied, we have a closed subshell; if all $2n^2$ states ($l = 0, 1, \ldots, n-1$) are filled, we have a closed shell. Closed shells and subshells have exactly spherically symmetric charge distributions and zero net angular momentum.

The standard notation for designating the individual electron configurations or orbitals (n, l values) which are combined to form an N-electron product state is shown in Table I. To illustrate, the ground-state configuration of sodium ($Z = 11$) is written $(1s)^2(2s)^2(2p)^6 3s$, where the superscripts indicate the number of electrons of a given n,l type.

SPECTROSCOPIC NOTATION

To completely specify an atomic state, besides listing as above the individual electron orbitals, we must also specify the coupling of the angular momenta for all unfilled subshells. The closed shells couple to give zero angular momentum.

Table I Designation of electron states

	$l=0$ s	$l=1$ p	$l=2$ d	$l=3$ f	$l=4$ g
$n=1$	$1s$				
$n=2$	$2s$	$2p$			
$n=3$	$3s$	$3p$	$3d$		
$n=4$	$4s$	$4p$	$4d$	$4f$	
$n=5$	$5s$	$5p$	$5d$	$5f$	$5g$

The Hamiltonian of Eq. (2) commutes with $\mathbf{L} = \Sigma_{i=1}^{N} \mathbf{l}_i$, the total orbital angular momentum, and with $\mathbf{S} = \Sigma_{i=1}^{N} \mathbf{s}_i$, the total spin angular momentum, and thus with $\mathbf{J} = \mathbf{L} + \mathbf{S}$, the total electronic angular momentum of the atom. Within a given configuration, therefore, one may take linear combinations of products of the central-field-approximation orbitals to form states having exact values of the quantum numbers L^2, S^2, J^2, and M_J. All these states are degenerate in energy in the central-field approximation. When we use perturbation theory to take into account the difference between the central-field-approximation potential energy and the potential-energy terms of Eq. (2) we find that states with different L and S have significantly different energies. This treatment, which works well for many atoms (especially the light ones), is called the *LS* or Russell–Saunders coupling scheme. In the *LS* coupling scheme we couple vectorially all open-shell electron spins to obtain a number of different S values and couple all the open shell L_i to obtain a number of different L values. In early atomic spectroscopy the *vector model* was used to visualize these different coupling schemes. White gives an extensive discussion of this model. The number of permitted L and S values is discussed in detail by Condon and Shortley. L and S are now coupled vectorially to form the total angular momentum $\mathbf{J} = \mathbf{L} + \mathbf{S}$. J takes values from $|L + S|$ to $|L - S|$.

At this point we add the spin–orbit interaction $H_{so} = \xi \mathbf{L} \cdot \mathbf{S}$ (see FINE AND HYPERFINE SPECTRA AND INTERACTIONS) which removes the degeneracy between the different J values of the same L and S. The spectroscopic notation to designate these *LS* coupled states is $^{2S+1}L_{2J+1}$ where, similar to Table I, for $L = 0, 1, 2, 3$ we use the capital letters S, P, D, F, etc. To illustrate, if $S = 1$, $L = 2$, and $J = 1$ we would write $^3D_1$. The levels of a given J have $2J + 1$ magnetic sublevels which in the presence of an external magnetic field split apart (see ZEEMAN AND STARK EFFECTS).

TRANSITION RATES

Any isolated atom in an excited state will decay spontaneously to lower energy states emitting radiation and ultimately ending in the ground state. Also, in the presence of an external radiation field of the proper frequency, an atom can absorb radiation and make a transition to an excited state.

We will discuss only allowed electric dipole transitions. Classically the time-averaged power radiated per unit solid angle in direction $\mathbf{n}$ by a harmonically varying charge distribution $\rho(\mathbf{r})$ with an electric dipole moment $\mathbf{P} = \int_0^\infty \mathbf{r}\rho(\mathbf{r})\, dv$ is (in Gaussian units)

$$\frac{dw}{dt} = \frac{ck^4}{8\pi} \,|\, nx(nxP) \,|^2,$$

which for a linearly oscillating dipole moment parallel to the z axis reduces to $dw/dt = (ck^4/8\pi)\,|\,P\,|^2 \sin^2\theta$, where $k = \omega/c$ and θ is the angle between the z axis and the direction of radiation $\mathbf{n}$. For a charge distribution rotating about the z axis, $dw/dt = (ck^4/8\pi)\,|\,P\,|^2(1 + \cos^2\theta)$. We make the connection with the quantum-mechanical description of the atomic system by $P = 2P_{ij}$, where $P_{ij} = \int \bar{\psi}_i \mathbf{P} \Psi_j dv$, and by setting the harmonic frequency ω equal to $\omega = (E_i - E_j)/\hbar$, where i and j refer to the initial and final states of the transition.

SELECTION RULES

Electric dipole transitions are possible only between certain atomic energy levels. Rules which tell us which transitions are allowed are called selection rules. We will consider only electric dipole transitions, i.e., "allowed" transitions. If electric dipole transitions are forbidden, transitions can still occur by other radiation processes such as higher-order multipole radiation (see Garstang) but these transitions are much weaker.

The parity operation $P\mathbf{r}_i = -\mathbf{r}_i$ commutes with H (i.e., with the complete atomic Hamiltonian, not just our approximation) and for a product wave function yields $P\Psi(\mathbf{r}_i) = (-1)^{\Sigma l_i}\Psi(\mathbf{r}_i)$. If Σl_i is (even/odd) we say the state is an (even/odd) parity state. In spectroscopic notation an upper right-hand superscript "o" is sometimes used to indicate explicitly states having odd parity. Electronic dipole transitions *only* take place between states of *different* parity so that for the common case of a one-electron jump, $\Delta l = \pm 1$. No condition on Δn is required.

By arguments similar to those given to explain hydrogenic selection rules, we also find

$\Delta S = 0$ (no spin change), $\quad \Delta J = 0, \pm 1$ ($0 \to 0$ forbidden),

$\Delta L = 0, \pm 1, \quad \Delta M_J = 0, \pm 1.$

Radiation from a $\Delta M_J = 0$ transition is linearly polarized ($\sin^2\theta$ dependence); radiation from a $\Delta M_J = \pm 1$ transition is circularly polarized [$(1 + \cos^2\theta)$ dependence] when viewed along the axis of space quantization.

The strength of a transition depends on the magnitude of the radial part of the P_{ij} integral, which is quite sensitive to the approximations used in finding ψ and is difficult to calculate accurately.

EXAMPLES OF SIMPLE SPECTRA

1. Alkali-like Spectra

Alkalis and ions with a single electron outside of closed subshells have energy levels and spectra very similar to hydrogen because the core electrons, not taking any role in optical transitions, act principally to screen the nuclear charge. The larger the n of the valence electron, the larger its orbit and so the more complete is the screening. The energy level diagram of sodium is shown in Fig. 1. The configurations responsible for these levels are:

$(1s)^2(2s)^2(2p)^6 ns\ ^2S_{1/2}, \quad n = 3, 4, \ldots,$

$(1s)^2(2s)^2(2p)^6 np\ ^2P_{1/2, 3/2}, \quad n = 3, 4 \ldots,$

$(1s)^2(2s)^2(2p)^6 nd\ ^2D_{3/2, 5/2}, \quad n = 3, 4, \ldots.$

Early spectroscopists called the $nd\ ^2D - 3p\ ^2P$ transitions the diffuse series because in low resolution the three allowed lines, $^2D_{5/2} - {}^2P_{3/2}$ and $^2D_{3/2} - {}^2P_{3/2, 1/2}$, appeared as a blend.

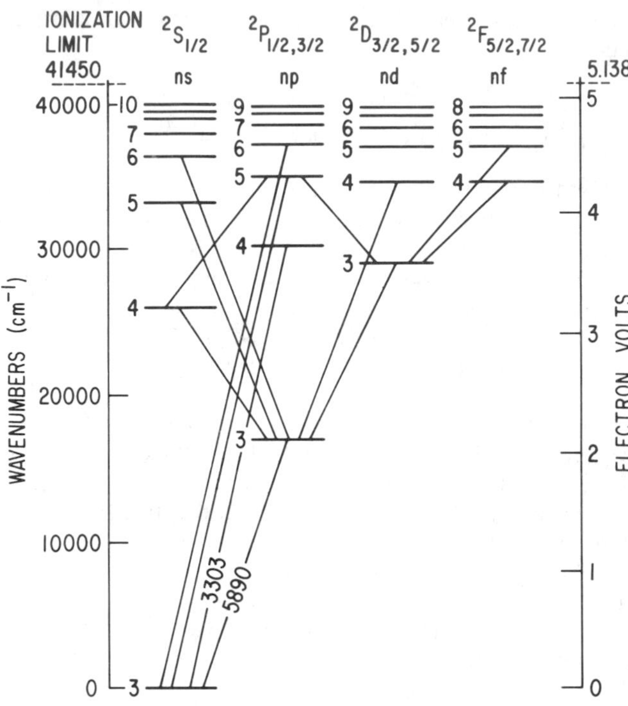

FIG. 1. Term diagram of sodium.

2. *Two-Electron Spectra*

The alkaline-earth elements Be, Mg, Ca, Sr, and Ba are representative of atoms with two electrons outside of closed shells. We shall consider Be, whose term diagram is shown in Fig. 2. The low-lying excited states arise from excitation of one of the ground state $(2s)^2$ electrons. The low-lying excited configurations are

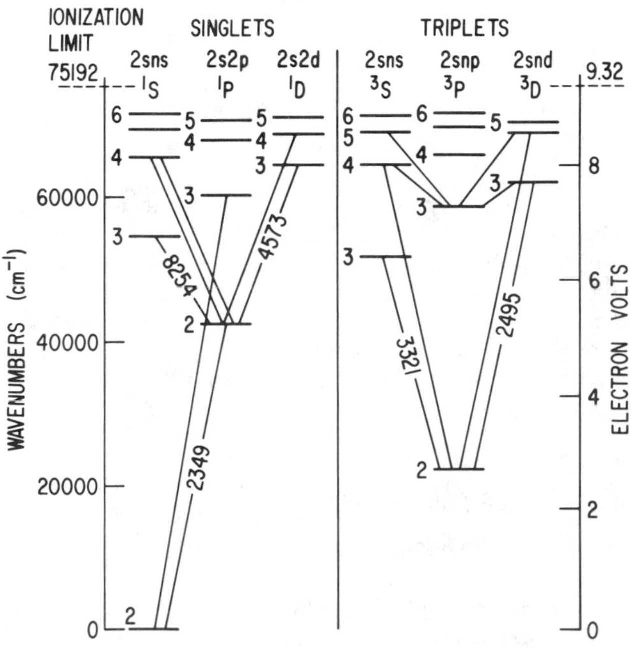

FIG. 2. Term diagram of beryllium.

$$(1s)^2 2s\ ns\ {}^1S_0\ {}^3S_1, \qquad n = 3,4, \ldots,$$

$$(1s)^2 2s\ np\ {}^1P_0\ {}^3P_{2,1,0}, \qquad n = 2,3,4,$$

$$(1s)^2 2s\ nd\ {}^1D_0\ {}^3D_{3,2,1}, \qquad n = 3,4,5 \ldots.$$

In each configuration the $s = \frac{1}{2}$ spins of the two unpaired electrons are coupled to form the resultant total spin $S = 0$, 1. The L value is that of the excited electron. To illustrate: for the D states, S and L are combined to form the resultant J as follows: (for the singlet) $L = 2$, $S = 0$, $J = 2$; (for the triplets) $L = 2$, $S = 1$, $J = L + S, L + S - 1, L - S = 3, 2, 1$. The $\Delta S = 0$ selection rule prohibits singlet to triplet transitions. The allowed ^3D to ^3P triplet transitions are shown in Fig. 3. The same methods can be applied to more complicated spectra (see White and Kuhn).

QUANTUM DEFECT THEORY

A key feature of the attractive Coulomb field in which atomic electrons move is that it supports an infinite number

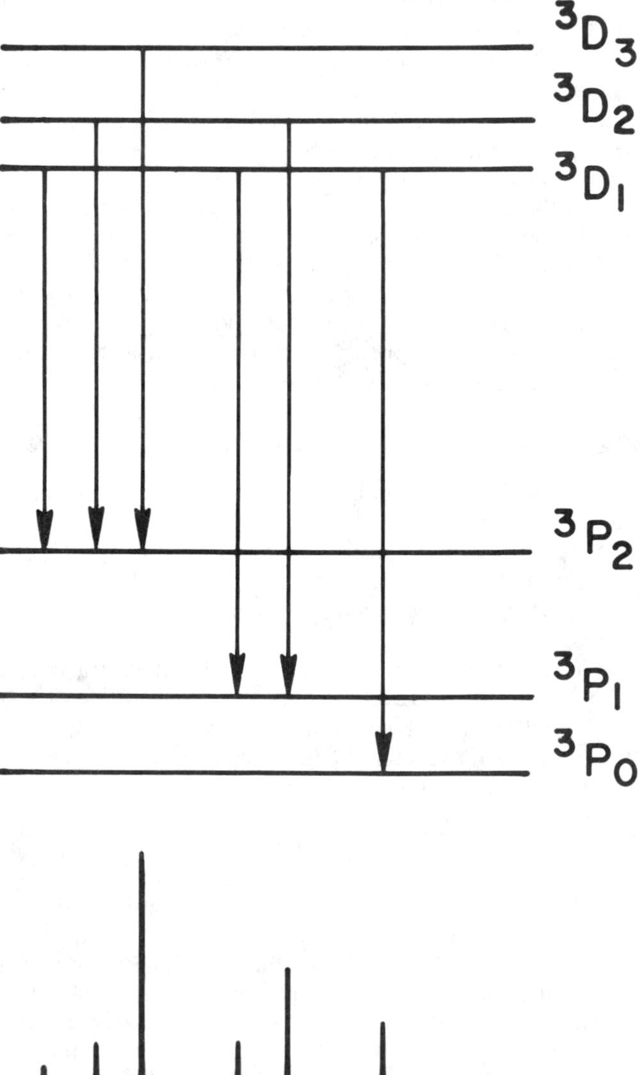

FIG. 3. Allowed ^3D–^3P transitions and relative line strengths.

of bound states which converge in energy to a particular ionization threshold. These states may be grouped in series. In simple cases each series may be identified by the term level of the ion to whose threshold the series converges, the orbital angular momentum of the excited electron, and the coupling of the electron to the atomic core. In general, the various series of states may interact with each other, thereby complicating the analysis in the very region where the number of levels is becoming infinite.

The quantum defect theory (QDT) is a method of using the analytically known properties of excited electrons moving in a pure Coulomb field to describe such atomic spectra in terms of a few parameters. These parameters may be determined either from experimental data or from *ab initio* theoretical calculations. In addition, they are usually nearly independent of energy in the threshold energy region (i.e., within a few electron volts of the atomic ionization threshold). Thus the determination of these parameters at any *single* energy suffices to predict the *variation with energy* of numerous atomic properties in the threshold energy region such as total and partial photoionization cross sections, photoelectron angular distributions, discrete line strengths, autoionization profiles, etc. These properties are often very strongly energy dependent and difficult to measure or to calculate by other methods. Yet all these phenomena, according to the QDT, depend on only a few essential parameters which represent the proper interface between theory and experiment.

The QDT assumes that the configuration space for an excited atomic electron can be divided into two regions: an *inner region*, $0 \leqq r \leqq r_0$, where electron correlations are strong and difficult to treat, and an *outer region*, $r_0 \leqq r \leqq \infty$, where the electron–ion interaction potential is assumed to be purely Coulombic and where the form of the electron wave function is known analytically. The boundary radius r_0 between the two regions is typically of the order of the atomic radius.

Consider the simple problem of an excited electron of angular momentum l in an alkali atom: the electron sees a Coulomb field for $r \geqq r_0$, where r_0 is roughly the ionic radius. We measure the energy ϵ of the excited electron relative to the ionization threshold as $\epsilon = -0.5\nu^{-2}$, where the parameter ν is our measure of energy. The Schrödinger equation for $r \geqq r_0$ has two solutions, one regular and one irregular for small values of r:

$$f(\nu,r) \sim r^{l+1} \quad \text{as } r \rightarrow 0, \tag{3a}$$

$$g(\nu,r) \sim r^{-l} \quad \text{as } r \rightarrow 0. \tag{3b}$$

A general solution of the Schrödinger equation for $r \geqq r_0$ is a linear combination of $f(\nu,r)$ and $g(\nu,r)$ with coefficients to be determined by application of boundary conditions at infinity and at r_0. This general solution may be written as

$$\psi(\nu,r) = N_\nu\{f(\nu,r) \cos \pi\mu - g(\nu,r) \sin \pi\mu\} \quad \text{for } r \geqq r_0, \tag{4}$$

where N_ν is a normalization factor which is determined by the behavior of $\psi(\nu,r)$ at large r. μ, on the other hand, is the relative phase with which the regular and irregular solutions are superimposed. Its value is determined by the behavior of $\psi(\nu,r)$ in the core region, $0 \leqq r \leqq r_0$, where the effective potential is non-Coulombic: i.e., μ has that value which allows the analytically determined $\psi(\nu,r)$ given by Eq. (4) for $r \geqq r_0$ to be joined smoothly at $r = r_0$ onto the numerically determined portion of $\psi(\nu,r)$ that obtains in the inner core region, $0 \leqq r \leqq r_0$.

Alternatively, μ may be determined semiempirically from atomic spectral data on energy levels, as we show here. Consider the asymptotic behavior of $\psi(\nu,r)$ in the case of excited electron energies below threshold, i.e., $\psi(\nu,r)$ must tend toward zero. The asymptotic forms of the regular and irregular Coulomb functions are

$$f(\nu,r) \rightarrow u(\nu,r) \sin \pi\nu - v(\nu,r) \exp i\pi\nu \quad \text{as } r \rightarrow \infty, \tag{5a}$$

$$g(\nu,r) \rightarrow -u(\nu,r) \cos \pi\nu + v(\nu,r) \exp i\pi(\nu + \tfrac{1}{2}) \quad \text{as } r \rightarrow \infty, \tag{5b}$$

where $u(\nu,r)$ is an exponentially increasing function of r and $v(\nu,r)$ is an exponentially decreasing function of r. Substituting Eq. (5) in Eq. (4) gives

$$\psi(\nu,r) \rightarrow N_\nu\{u(\nu,r) \sin \pi(\nu + \mu) - v(\nu,r) \exp i\pi(\nu + \mu)\} \quad \text{as } r \rightarrow \infty. \tag{6}$$

In order that $\psi(\nu,r)$ tend toward zero at large values of r, the coefficient of $u(\nu,r)$ must be zero; i.e., $\sin \pi(\nu + \mu) = 0$ or $\nu + \mu = n$, where n is an integer. Substituting $\nu = n - \mu$ in the expression for the electron's energy gives (in atomic units)

$$\epsilon_n = -\frac{1}{2\nu^2} = -\frac{1}{2(n - \mu)^2}. \tag{7}$$

μ is thus the quantum defect of spectroscopy and may be determined directly from Rydberg energy level data for the alkalis.

It is well known empirically in atomic spectroscopy that the quantum defect μ is a nearly constant function of energy near the ionization threshold. Theoretically, μ *should* be only a slowly varying function of energy since it is determined from the wave function in the inner core region, where the electron's large instantaneous kinetic energy makes it insensitive to the relatively small energy differences between the energy levels near threshold. Knowledge of the parameter μ therefore enables one to predict the energies of a whole series of atomic energy levels according to Eq. (7), thereby illustrating the ability of QDT to describe large amounts of atomic spectral information in a very compact way.

QDT may also be used to understand the variation with energy of the intensities of atomic spectral levels. From Eq. (4) we can see directly that for small $r \geqq r_0$, the wavefunction $\psi(\nu,r)$ depends on energy mainly through the normalization factor N_ν since μ is weakly energy dependent and so are $f(\nu,r)$ and $g(\nu,r)$ at small r [cf. Eq. (3)]. The normalization factor N_ν is determined by the asymptotic behavior of $\psi(\nu,r)$ and may be very energy dependent. The point of this discussion thus is that at small radii $\psi(\nu,r)/N_\nu$ is likely to be quite insensitive to energy in the threshold energy region. Yet it is in this small r region that spectral transitions between the ground and the excited states occur. We conclude that the intensities of these transitions along a series of lines converging to the ionization threshold should depend on the energies of the excited levels in proportion to N_ν^2. QDT shows that N_ν^2 is proportional to $\nu^{-3} = (n - \mu)^{-3}$ for discrete (i.e., negative) electron energies. This implies that multiplication of the measured intensity of the nth level by $(n - \mu)^3$ will produce a renormalized intensity which is only a slowly varying function of n. Hence, accurate measurements of only a few level intensities allows one to determine this slowly varying function and therefore to predict the intensities of all other levels in the series.

The QDT may also be used to describe atomic spectra more complicated than those of the alkali metals as well as to relate an atom's discrete spectrum to collision processes occurring at energies above the atom's ionization threshold. These topics, however, are beyond the scope of this article. The interested reader is referred to review articles on QDT by Fano and by Seaton.

See also Bohr Theory of Atomic Structure; Hamiltonian Function; Rotation and Angular Momentum; Schrödinger Equation; Zeeman and Stark Effects.

BIBLIOGRAPHY

E. U. Condon and G. H. Shortley, *Theory of Atomic Spectra*. Cambridge University Press, Cambridge, 1953. (A)

U. Fano, "Unified Treatment of Perturbed Series, Continuous Spectra, and Collisions," *J. Opt. Soc. Am.* **65**, 979 (1975).

R. H. Garstang, "Forbidden Transitions," in *Atomic and Molecular Processes*. Academic Press, New York, 1962.

G. Herzberg, *Atomic Spectra and Atomic Structure*, 2nd ed. Dover, New York, 1944. (E)

H. G. Kuhn, *Atomic Spectra*. Academic Press, New York and London, 1962. (E)

C. E. Moore, *Atomic Energy Levels*, Circular 467, Vols. I, II, and III. National Bureau of Standards, Washington, DC, 1949, 1952, 1958.

M. J. Seaton, "Quantum Defect Theory," *Rept. Prog. Phys.* **46**, 167 (1983).

B. W. Shore and D. H. Menzel, *Principles of Atomic Spectra*. Wiley, New York, 1968. (A)

J. C. Slater, *Quantum Theory of Atomic Structure*, Vols. I and II. McGraw-Hill, New York and London, 1960. (Vol. I, I; Vol. II, A)

I. I. Sobelman, *Introduction to the Theory of Atomic Spectra*. Pergamon Press, Oxford, 1972. (I)

H. E. White, *Introduction to Atomic Spectra*. McGraw-Hill, New York and London, 1934. (E)

Atomic Structure Calculations—Electronic Correlation

R. K. Nesbet

If electrons did not interact with each other, the electronic wave function of an atom could be a single Slater determinant, an antisymmetrized product of one-electron "orbital" wave functions. Hartree–Fock equations [1] are the variational equations for the occupied orbital functions in a Slater determinant, or in a simple linear combination of such functions when the atomic state in question is not invariant under rotation. These integrodifferential equations are similar in form to the one-electron Schrödinger equation, but contain a nonlocal "self-consistent-field" potential, giving the spatially averaged effect of Coulomb and exchange interactions between electrons. This is an independent-particle model, which neglects correlations between the locations of different electrons. Since the electronic Hamiltonian operator contains the two-electron Coulomb interaction, the electronic energy depends on such correlations and cannot be computed exactly in the Hartree–Fock approximation [2].

The expression "electronic correlation" refers generally to corrections to the Hartree–Fock approximation. For atoms other than the closed-shell rare gases, rotational averaging is usually imposed on the effective one-electron potential, resulting in one-electron "correlation" effects such as spin or rotational symmetry breaking [3]. Otherwise the dominant electronic correlation effect is an electron–pair interaction that causes relaxation of the wave function in relative coordinates in response to the electronic Coulomb repulsion.

One-electron mean-value properties computed in the closed-shell Hartree–Fock approximation are subject only to second-order correlation corrections [2,3]. The practical effect of this is that the total electronic density distribution is well described for many purposes in the Hartree–Fock approximation, but reliable and consistent theoretical results for atomic properties sensitive to open-shell structure or to the response to external perturbations require a quantitative treatment of electronic correlation. Such properties include hyperfine structure, polarizabilities, oscillator strengths, and electron scattering cross sections [4].

While some physical properties are described more or less accurately in the Hartree–Fock approximation, others are not described at all. For example, the van der Waals or dispersion potential energy of two spatially separated electronic systems is simply the long-range limit of the correlation energy of spatially separated electrons [5]. The polarization potential that dominates low-energy electron scattering is the same effect. In the usual Hartree–Fock approximation, the magnetic hyperfine structure constant is zero by symmetry in the ground states of nitrogen and phosphorus. The observed values differ from zero because of a combination of spin symmetry-breaking and Coulombic correlation effects [4,6].

Theoretical methods for the computation of correlation effects are usually based on a preliminary Hartree–Fock calculation. This defines a reference function [7] or "vacuum state" for formal perturbation theory [8]. Because of the dominant effect of the nuclear Coulomb attraction in an atom, a shell model is appropriate for the electrons. Occupied orbitals for a particular state in the Hartree–Fock approximation are labeled by quantum numbers n, l, m_l, m_s (or n, j, m) appropriate to the one-electron Schrödinger equation for a central potential. The conventional open-shell Hartree–Fock equations [1] are spherically averaged so that for each value of orbit angular momentum l there is a set of orthonormal radial functions $R_{nl}(r)$, not dependent on the axial quantum numbers m_l and m_s. A "configuration" is defined by a set of occupation numbers $d_{nl} \leq 2(l + 1)$ which assign occupied orbital functions to subshells of given n, l but arbitrary m_l, m_s, subject to

$$m_l = -l, -l + 1, \ldots, l; \qquad m_s = -\tfrac{1}{2}, \tfrac{1}{2}. \qquad (1)$$

The conventional Hartree–Fock theory is formulated in terms of eigenfunctions of total orbital angular momentum and spin constructed from the Slater determinants of a specified configuration.

The set of radial functions R_{nl} for occupied orbitals in a reference configuration can be extended to a complete orthonormal set. This generates a complete basis for the atomic N-electron wave function as a hierarchy of virtual excitations, defined by substitution of "unoccupied" orbitals (from the extended set) for orbitals occupied in a particular configuration. An n-electron virtual excitation is defined by n such substitutions [7].

Since the electronic Hamiltonian contains only one- and two-electron operators, only one- and two-electron virtual excitations contribute to the first-order wave function of perturbation theory. The closed-shell Hartree–Fock approximation causes one-electron matrix elements with the reference state to vanish [2], but one-electron virtual excitations cannot be neglected for open-shell states. The total energy or correlation energy can be expressed exactly in terms of the coefficients of all one- and two-electron virtual excitations of the reference state [7]. Valid estimates of these coefficients can be made either by formal perturbation theory [8], or by approximate solution of the matrix eigenvalue problem defined in the basis of all virtual excitations of all orders $n \leq N$, for N electrons. The latter method, superposition of configurations or "configuration interaction," has been widely applied and highly developed in its computational and data-handling aspects [9]. The direct use of relative coordinates for pairs of electrons is not computationally feasible for more than two or three electrons, although it has given very accurate results for two-electron atoms and ions [10].

Because of the great complexity of such calculations for virtual excitations with $n > 2$, several methods have been introduced for approximate incorporation of terms of higher order. These methods either treat electron pairs as uncoupled from each other [11] or modify the variational equations for such separated pairs to allow for the higher-order virtual excitations implied by a cluster expansion of the N-electron wave function [12]. This level of approximation appears to be the most natural extension of theory beyond Hartree–Fock to include electronic pair correlation. In the multiconfiguration Hartree–Fock method, configuration interaction is incorporated into an iterative variational calculation. A numerical version of this method has been used for accurate calculations of atomic oscillator strengths and photoionization cross sections [13].

These methods have been tested by numerous calculations of total atomic energies or of excitation energies [4], primarily for atoms in the first third of the periodic table. Their most important application, however, has been to the calculation of physical atomic properties that are difficult or impossible to measure experimentally. Theoretical calculations have helped to resolve discrepancies between conflicting experimental data on oscillator strengths [4], and have provided values of polarizabilities, particularly for atomic excited states, that have not been measured.

Hyperfine-structure calculations of high accuracy have helped to establish the fact that electronic correlation affects the three tensorially distinct magnetic hyperfine-structure interaction operators differently, so that three independent parameters must be used to fit experimental data [14]. Theoretical calculations of the electric field gradient at a nucleus, which cannot be measured directly, have made it possible to obtain accurate nuclear quadrupole moments from measured quadrupole hyperfine coupling constants [4].

Low-energy electron scattering by neutral atoms is dominated by the electric dipole polarization potential, essentially a correlation effect. Theoretical calculations by methods that can describe this effect quantitatively have been carried out for electron scattering by hydrogen, helium, alkali metals, and several other atoms [15]. These theoretical results have helped to elucidate observed structural features (resonance and threshold structures) and to establish absolute values of cross sections.

See also ATOMIC STRUCTURE CALCULATIONS—ONE-ELECTRON MODELS; ATOMIC STRUCTURE CALCULATIONS, RELATIVISTIC ATOMS; FINE AND HYPERFINE SPECTRA AND INTERACTIONS.

REFERENCES

1. D. H. Hartree, *The Calculation of Atomic Structures*. Wiley, New York, 1957. C. Froese-Fischer, *The Hartree–Fock Method for Atoms*. Wiley-Interscience, New York, 1977.
2. C. Møller and M. S. Plesset, *Phys. Rev.* **46,** 618 (1934); L. Brillouin, *Les Champs "Self-Consistent" de Hartree et de Fock*. Hermann et Cie, Paris, 1934.
3. R. K. Nesbet, *Proc. Roy. Soc.* (*London*) *A* **230,** 312 (1955).
4. A. Hibbert, *Rept. Prog. Phys.* **38,** 1217 (1975).
5. F. London, *Z. Phys.* **63,** 245 (1930).
6. N. C. Dutta, C. Matsubara, R. T. Pu, and T. P. Das, *Phys. Rev. Lett.* **21,** 1139 (1968); *Phys. Rev.* **177,** 33 (1969).
7. R. K. Nesbet, *Adv. Chem. Phys.* **9,** 321 (1965).
8. H. P. Kelly, *Adv. Chem. Phys.* **14,** 129 (1969); in *Atomic Physics 2*, G. K. Woodgate and P. G. H. Sandars, eds., p. 227. Plenum Press, New York, 1971.
9. I. Shavitt, in *Methods of Electronic Structure Theory*, H. F. Schaefer III, ed., p. 189. Plenum Press, New York, 1977.
10. C. L. Pekeris, *Phys. Rev.* **115,** 1216 (1959); **126,** 1470 (1962); K. Frankowski and C. L. Pekeris, *Phys. Rev.* **146,** 46 (1966).
11. O. Sinanoglu, *Adv. Chem. Phys.* **6,** 315 (1964); **14,** 237 (1969); R. K. Nesbet, *Adv. Chem. Phys.* **14,** 1 (1969).
12. J. Cizek, *Adv. Chem. Phys.* **14,** 35 (1969); J. Cizek and J. Paldus, *Int. J. Quantum Chem.* **5,** 359 (1971); W. Meyer, *J. Chem. Phys.* **58,** 1017 (1973); W. Kutzelnigg, in *Methods of Electronic Structure Theory*, H. F. Schaefer III, ed., p. 127. Plenum Press, New York, 1977.
13. C. Froese-Fischer, *Comput. Phys. Commun.* **14,** 145 (1978); H. P. Saha and C. Froese Fischer, *Phys. Rev. A* **35,** 5240 (1987); H. P. Saha, C. Froese Fischer, and P. W. Langhoff, *Phys. Rev. A* **38,** 1279 (1988).
14. J. D. Lyons, R. T. Pu, and T. P. Das, *Phys. Rev.* **178,** 103 (1969);

R. K. Nesbet, *Phys. Rev. A* **2**, 661 (1970); J. D. Lyons and T. P. Das, *Phys. Rev. A* **2**, 2250 (1970).

15. B. L. Moiseiwitsch, *Rep. Prog. Phys.* **40**, 843 (1977); R. K. Nesbet, *Variational Methods in Electron-Atom Scattering Theory*. Plenum Press, New York, 1980.

Atomic Structure Calculations—One-Electron Models

Frank Herman

ONE-ELECTRON THEORY OF MANY-ELECTRON SYSTEMS

The one-electron theory of atoms, molecules, and solids [1,2] has enjoyed wide success in many branches of physics and chemistry. This theory is based on the idea that the exact wave function for a many-electron system can be represented with acceptable accuracy by an approximate many-electron wave function constructed from one-electron wave functions or spin-orbitals. According to this approach, emphasis shifts from a consideration of the many-electron wave function—which is very complex because it depends on the coordinates of all the electrons—to one-electron wave functions.

One-electron wave functions are much simpler to deal with because they depend only on the spatial and spin coordinates of single electrons. The ground state of the many-electron system can then be described in terms of the occupied spin-orbitals, and excitations in terms of transitions between occupied and unoccupied spin-orbitals.

The one-electron theory provides a physically appealing description of the electronic structure and properties of many-electron systems such as complex atoms. The theory provides the basis of atomic spectroscopy [3] and many other applications. In addition to its formal and conceptual aspects, the theory provides a convenient and systematic framework for carrying out detailed numerical calculations for many-electron atoms [4–6]. In such calculations, it is usually necessary to introduce many simplifying assumptions, both physical and mathematical, in order to make progress. For some applications, where one is primarily interested in gross electronic properties, it may be sufficient to adopt very crude phenomenological models [7]. In other instances, where the effects being considered are more subtle, it may be necessary to employ highly sophisticated theoretical models [8,9] even though their study might involve extensive numerical computation. Fortunately, large-capacity high-speed electronic digital computers are widely available, as are atomic structure computer codes [5,10]. Extensive tabulations of the results of atomic structure calculations are also readily available [5,11].

An important characteristic of the subject of atomic structure calculations is the wide spectrum of available methods and points of view. In large measure, this diversity arises from the fact that different approximations are most advantageous for different applications. In this introductory sketch we can hardly do justice to the many important techniques and concepts that have been developed over the years for dealing with the electronic structure and properties of complex atoms. We will attempt to provide a general perspective, as well as a representative list of references where the interested reader can find more detailed information.

INDEPENDENT PARTICLE MODELS: HARTREE AND HARTREE–FOCK METHODS

In orbital theories of many-electron systems, one derives a set of one-electron wave equations by using the variational method. The form of these equations is determined by the manner in which the exact many-electron wave function Φ is represented by one-electron wave functions or spin orbitals ϕ_q, where q denotes all the relevant quantum numbers. By using the variational method, one guarantees that the solutions of the one-electron wave equations, the ϕ_q, are the best possible solutions consistent with the assumed form of Φ.

In the Hartree approximation, Φ is represented by a simple product of spin-orbitals ϕ_q, each factor corresponding to a different occupied state q. In the Hartree–Fock approximation, Φ is represented by an antisymmetrized product of the ϕ_q, also known as a determinantal wave function. Both of these treatments lead to an independent particle model in which each of the electrons moves independently of all the others in a time-averaged potential field produced by all the other electrons and the nucleus. Because the field acting on any one electron is obtained by averaging over the motions of all the others, spatial correlations in the motions of pairs of electrons produced by their instantaneous Coulomb repulsion are neglected.

In the Hartree–Fock approximation, Φ is represented by a determinantal wave function in order to take into account the fact that a many-electron wave function must be an antisymmetric function of the electron coordinates, a feature closely related to the requirement that electrons satisfy the Fermi–Dirac statistics and the Pauli exclusion principle. In contrast to the Hartree approach, the use of determinantal wave functions in the Hartree–Fock method leads to additional terms in the one-electron wave equations, the exchange terms, which tend to keep electrons of like spin out of each other's way, as required by the Pauli exclusion principle. Thus, the Hartree–Fock approximation includes a certain type of spatial correlation between like-spin electrons. This is often called statistical correlation because it arises from the Fermi–Dirac statistics.

It is common practice to describe all spatial correlation over and above statistical correlation (exchange) simply as correlation. Thus, it can be said that the Hartree approximation neglects both exchange and correlation, while the Hartree–Fock approximation includes exchange but neglects correlation. In many applications the agreement between theory and experiment can be improved considerably by including exchange only. Thus, Hartree–Fock atomic structure calculations are usually much favored over Hartree calculations. However, for the determination of delicate electronic properties, such as transition probabilities, the neglect of correlation can have serious consequences, and so it is desirable to include correlation as well as exchange. For detailed discussions of correlation effects, and extensive references, see Ref. [8] as well as the accompanying article by Nesbet [9].

SELF-CONSISTENT ITERATION AND THE CENTRAL FIELD APPROXIMATION

Since the potential terms that appear in the Hartree and Hartree–Fock one-electron wave equations depend on the wave functions which are the solutions of these equations, it is necessary to solve these equations by iterative techniques. One begins by choosing a starting set of wave functions and inserting these into the potential terms. Solving the wave equations, one obtains a final set of wave functions for this cycle. The next cycle begins by using a new set of starting wave functions constructed from the initial and final wave functions of the preceding cycle. The process continues until the initial and final wave functions in a given cycle agree with one another to some specified accuracy. In this way we obtain a self-consistent solution in the sense that the wave functions are generated by wave equations whose potential terms are determined by these very wave functions.

The self-consistent field equations for an atomic system (such as the Hartree–Fock equations) can be simplified considerably by taking advantage of the spherical symmetry of the atomic field. One introduces the central field approximation by considering only the spherically averaged component of the atomic field. The complete three-dimensional wave equation can then be separated into an angular wave equation and a radial wave equation by writing the spin-orbital as the product of a radial function $R(r)$, an angular function $Y(\theta,\phi)$, and a spin function. The angular functions are the well-known spherical harmonics $Y_{lm}(\theta,\phi)$, where l and m are the azimuthal and magnetic quantum numbers characteristic of a central field. The radial functions, $R_{nls}(r)$, in general depend on n, l, and s, where n and s are the principal and spin quantum numbers. The $R_{nls}(r)$ can be determined by solving the radial wave equations numerically [4–6] or by the expansion method [6,12]. In the expansion method, the $R_{nls}(r)$ and all the potential terms are expanded in terms of a suitably chosen set of analytic functions, and the wave equations are solved by matrix methods.

In the case of closed shell atoms, it is possible to represent Φ by a single determinantal wave function. For the more general case of open shell atoms, it is usually necessary to represent Φ by a linear combination of determinantal wave functions, each corresponding to a different assignment of the one-electron quantum numbers ($q = n,l,m,s$) compatible with the assumed overall symmetry of the many-electron atom. This leads to interactions over and above the central field approximation, and to the multiplet structure which is of such great importance in atomic spectroscopy [3].

FREE-ELECTRON EXCHANGE APPROXIMATION: HARTREE–FOCK–SLATER EQUATIONS

It is possible to simplify the Hartree–Fock equations and obtain nearly the same level of accuracy by replacing the rather complex (nonlocal) exchange terms by a simple (local) expression derived from the theory of the free-electron gas, $V_{\text{exch}}(r)$, which will be defined shortly. When this so-called free-electron exchange approximation is employed, the one-electron wave equations become known as the Hartree–Fock–Slater (HFS) equations [1,5,13].

If we confine ourselves to atomic systems containing equal numbers of electrons with spin up and spin down (balanced spins), the radial HFS equations take the form

$$\left(-\frac{d^2}{dr^2} - \frac{2Z}{r} + V_{\text{Coul}}(r) + V_{\text{exch}}(r) + \frac{l(l+1)}{r^2}\right) P_{nl}(r) = E_{nl}P_{nl}(r), \quad (1)$$

where the energy eigenvalues are denoted by E_{nl} and the radial eigenfunctions by $P_{nl}(r) = rR_{nl}(r)$. We will measure distances in Bohr units (1 Bohr = 0.529 Å) and energies in Rydberg units (1 Ry = 13.6 eV). The first term on the LHS is the kinetic energy operator, the second term is the nuclear Coulomb energy (Z = nuclear charge), $V_{\text{Coul}}(r)$ is the spherically averaged electronic Coulomb energy, $V_{\text{exch}}(r)$ is the free-electron exchange energy, and the last term is the centrifugal energy.

The radial wave functions $P_{nl}(r)$ must vanish at the origin and at infinity. They have $n-l-1$ nodes between the origin and infinity. The azimuthal quantum number l ranges from 0 to $n - 1$, and the magnetic quantum number m from $-l$ to $+l$. With the radial wave functions normalized, $\int[P_{nl}(r)]^2dr = 1$, the spherically averaged electronic charge density $\rho(r)$ and Coulomb energy $V_{\text{Coul}}(r)$ are

$$\rho(r) = -(4\pi r^2)^{-1} \sum_{nl} \omega_{nl}[P_{nl}(r)]^2 \quad (2)$$

and

$$V_{\text{Coul}}(r) = -(8\pi/r) \int_0^r \rho(t)t^2dt - 8\pi \int_r^\infty \rho(t)t\,dt, \quad (3)$$

where ω_{nl} is the occupation number for the orbital nl (both spins). In the special case of a closed shell, $\omega_{nl} = 2(2l + 1)$. The total number of electrons in the atom is $N = \sum_{(nl)}\omega_{nl}$, and the ionicity is $Z - N$. The free-electron exchange energy is given by

$$V_{\text{exch}}(r) = -6\alpha[(3/8\pi)|\rho(r)|]^{1/3}, \quad (4)$$

where α is an exchange parameter whose value is $\frac{2}{3}$ for a free-electron gas. This exchange approximation is also known as the $X\alpha$ exchange approximation (X for exchange) [1,2]. It is possible to optimize the value of α for any given atom, and when this is done it is found that the optimum value of α is a function of Z. In an early study, the Hartree–Fock–Slater equations (with α = 1) were solved for all normal neutral atoms in the periodic table, and the calculated energy eigenvalues E_{nl} and radial eigenfunctions $P_{nl}(r)$ were discussed and tabulated [5].

In the Hartree–Fock theory, the energy eigenvalues $E_{n\ell}$ represent one-electron ionization energies (Koopmans' theorem). This is not the case for the Hartree–Fock–Slater theory, where ionization and excitation energies are determined by the so-called transition state method [13]. The HFS equations are widely used, not only in atomic structure calculations, but also in molecular and solid-state calculations [2,13,14].

IMPROVED TREATMENTS OF EXCHANGE AND CORRELATION EFFECTS: QUANTUM-CHEMICAL METHODS

Correlation effects must be taken into account in order to obtain many-electron wave functions that are accurate enough to yield reliable values for such delicate physical quantities as oscillator strengths, polarizabilities, hyperfine coupling coefficients, and electron scattering potentials [8,9].

The most straightforward approach for including correlation effects is multiconfiguration Hartree–Fock theory, also known as the configuration interaction (CI) method. According to this approach, the many-electron wave function Φ is expanded as a linear combination of determinantal wave functions, each representing a different electronic configuration. In practice, it may be necessary to include a large number of configurations in order to obtain a highly accurate description of correlation. This approach is very popular with quantum chemists, but is limited to systems with relatively few electrons because of the considerable computational effort involved. In some cases it is possible to make substantial improvements on independent particle models by carrying out limited configuration interaction calculations. Other advanced methods for dealing with correlation effects are based on many-body perturbation theory [8,9,15].

IMPROVED TREATMENTS OF EXCHANGE AND CORRELATION EFFECTS: DENSITY FUNCTIONAL METHODS

There is inevitably a trade-off between computational effort and physical rigor. One can avoid time-consuming calculations and at the same time achieve reasonably accurate descriptions of many-electron systems by turning to local density functional methods [16–20]. These methods lead to equations similar to the HFS equations, but with the free-electron exchange term replaced by an exchange-correlation term which is derived from a consideration of exchange and correlation effects in a free-electron gas [17,18]. In this way one retains the simplicity of the Hartree–Fock–Slater equations and at the same time introduces an effective potential for exchange and correlation which has a deeper theoretical basis than the $X\alpha$ potential.

The essential advantage of local density functional methods is that the effective exchange-correlation potential depends only on the (local) value of the charge density (and possibly its spatial gradients), and so is quite easy to evaluate, particularly for complex many-electron systems. Special functionals known as local spin density functionals are available for dealing with systems containing unbalanced spins [18]. Density functional methods have long been popular with solid-state physicists. They are now receiving increasing attention from atomic, molecular, and surface physicists because they can be used to solve a wide variety of electronic structure problems with relatively modest computational effort [19,20].

In the density functional approach, the wave equations analogous to the HFS equations are called the Kohn–Sham equations. The electronic structures obtained by solving the Kohn–Sham equations using any one of several exchange-correlation potentials [17,18] are generally more accurate than corresponding solutions of the Hartree–Fock–Slater equations based on optimized $X\alpha$ potentials, but less accurate than extended configuration interaction solutions [21]. Once new density functional approximations have been tested out on atomic systems, they can usually be applied to more complex many-electron systems. The same cannot be said for quantum-chemical methods, which often prove intractable in applications to surfaces, liquids, and crystals.

A serious shortcoming of local density functional theory is the incomplete cancellation of self-interaction effects [21–23]. Such effects, reflecting the interaction of an electron with itself, are spurious and should not appear in the exchange-correlation potential. In the case of atoms, it is possible to minimize or eliminate these effects by ad hoc methods. These schemes do improve the solutions, but they introduce undesirable features such as different potentials acting on different electrons [22] or nonlocal potentials [23]. Unfortunately, these techniques are not particularly well suited for molecular or solid-state problems.

Most generalized attempts to go beyond local density functional theory are based on ideas derived from many-body perturbation theory. The screened-exchange plus Coulomb hole approximation [17] is a case in point. This not only eliminates the self-interaction problem, but also treats exchange and correlation effects more adequately than can be done by local density approximations alone. The power of this approach has recently been demonstrated in studies of the electronic energy band structure of semiconductors [24].

RELATIVISTIC EFFECTS

In dealing with heavy atoms, where the innermost electrons orbit the nuclei with velocities approaching the speed of light, it is desirable in some applications and essential in others to take relativistic effects into account. The more accurate relativistic atomic structure calculations are based on the one-electron Dirac equation (as adapted to many-electron systems). The Dirac equation describes three effects that are relativistic in origin: the mass–velocity correction, the spin-orbit coupling interaction, and the Darwin fluctuation or s-shift. The Darwin term arises from the time average of relativistic oscillatory electron motion (Zitterbewegung) in the Coulomb field of the nucleus. This term shifts the energies of s electrons, which are the only ones having finite probability of being at the nucleus.

Relativistic effects tend to contract the orbits of the innermost electrons and increase their binding energies. Because all the electrons are coupled together through the self-consistent field, even the outermost electrons are affected (slightly) by relativistic effects. Particularly for the inner electrons of heavy atoms, the agreement between theoretical electron binding energies and experimental x-ray levels is significantly improved when relativistic effects are taken into account. Spin-orbit splittings are readily measured and play an important role in various forms of electronic and optical spectroscopy.

Relativistic calculations [25–28] are more complicated than their nonrelativistic counterparts because there are two angular momentum states (spin up and spin down) for every nonrelativistic state, and these are coupled to one another through the spin-orbit interaction. Accordingly, it is necessary to solve pairs of coupled wave equations instead of single wave equations for each orbital (*nl*). Recently, simplified treatments of relativistic effects in atoms have appeared [29].

Most relativistic atomic structure computer programs are based on the Dirac equation analogue of nonrelativistic Hartree, Hartree–Fock, or Hartree–Fock–Slater atomic codes. Thus, exchange is neglected, or treated within the spirit of the Hartree–Fock method, or approximated by the free-electron or statistical exchange model. In the most sophisticated relativistic atomic calculations, nuclear size effects are taken into account, as are two-electron relativistic interactions (such as spin-other-orbit coupling). By now, relativistic calculations have been carried out for all atoms in the periodic table, as well as for hypothetical superheavy elements. For additional details and extensive references, see Refs. [25–27] as well as the companion article by Waber [28].

PSEUDOPOTENTIALS

In high-energy applications such as x-ray spectroscopy, it is essential to treat the innermost (core) electrons as well as the outermost (valence) electrons. But in low-energy physics and most branches of chemistry, one is primarily interested in the ground and lower excited states of the outermost electrons. To the extent that core and valence electrons are separable, it is possible to construct an effective potential which represents the effect of the core electrons on the valence electrons. This effective potential is usually called a pseudopotential and can be obtained either empirically or from a first-principles analysis [30,31]. Once the core electrons are eliminated from direct consideration through the introduction of a pseudopotential, atomic structure calculations become substantially simpler and less expensive: one then has to deal only with the valence electrons explicitly. This feature of pseudopotentials becomes particularly important in studies of heavier atoms, where the number of core electrons is very large.

Considerable theoretical effort has been devoted to the development of first-principles pseudopotentials which can be used to carry out rigorous atomic structure calculations with negligible loss of accuracy (relative to all-electrons calculations) [30,31]. As the number of electrons in the system becomes larger, the need for simplifying assumptions and approximations becomes more and more acute. Because of the computational (and conceptual) simplifications afforded by pseudopotentials, these continue to be used extensively in atomic, molecular, and solid-state calculations.

To conclude, the subject of atomic structure calculations continues to be an active field of research, particularly in the realm of applications. The latest original contributions appear in such publications as *Physical Review A* and *B*, *Journal of Chemical Physics, Journal of Physics B* and *C*, and *Physica Scripta,* among others.

See also ATOMIC STRUCTURE CALCULATIONS—ELECTRONIC CORRELATION; ATOMIC STRUCTURE CALCULATIONS, RELATIVISTIC ATOMS; FINE AND HYPERFINE SPECTRA AND INTERACTIONS; HIGH-FIELD ATOMIC STATES.

REFERENCES

1. J. C. Slater, *Quantum Theory of Atomic Structure* (McGraw-Hill Book Co., New York, 1960), Vols. 1 and 2.
2. J. C. Slater, *Quantum Theory of Molecules and Solids* (McGraw-Hill Book Co., New York, 1963, 1965, 1967, 1974), Vols. 1 to 4.
3. I. I. Sobelman, *Introduction to the Theory of Atomic Spectra* (Pergamon Press, Oxford, 1972); R. D. Cowan, *The Theory of Atomic Structure and Spectra* (University of California Press, Berkeley, 1981); B. R. Judd, *Rept. Prog. Phys.* **48,** 907 (1985).
4. D. R. Hartree, *The Calculation of Atomic Structures* (Wiley, New York, 1957). For a personal account of Hartree's work in atomic physics, see B. S. Jeffreys, *Comments Atom. Molec. Phys.* **20,** 189 (1987).
5. F. Herman and S. Skillman, *Atomic Structure Calculations* (Prentice-Hall, Englewood Cliffs, NJ, 1963).
6. C. Froese-Fischer, *The Hartree–Fock Method for Atoms* (Wiley-Interscience, New York, 1977); B. C. Webster, M. J. Jamieson, and R. F. Stewart, *Adv. Atom. Mol. Phys.* **14,** 87 (1978); M. Cohen and R. P. McEachran, *Adv. Atom. Mol. Phys.* **16,** 2 (1980).
7. A. E. S. Green and G. J. Kutcher, *Intern. J. Quantum Chem.* **10S,** 135 (1976); A. Hibbert, *Adv. Atom. Mol. Phys.* **18,** 309 (1982).
8. A. Hibbert, *Rept. Prog. Phys.* **38,** 1217 (1975); H. F. Schaeffer III, *Ann. Rev. Phys. Chem.* **27,** 261 (1976); W. A. Goddard III and L. B. Harding, *Ann. Rev. Phys. Chem.* **29,** 363 (1978); *Many-Body Theory of Atomic Systems: Proceedings of the 46th Nobel Symposium,* I. Lindgren and S. Lundqvist, eds., *Phys. Scripta,* **21** (1980), Issues 3 and 4.
9. R. K. Nesbet, "Atomic Structure Calculations—Electronic Correlation" (this volume).
10. Atomic structure programs are available through *Computer Physics Communications*, Department of Applied Mathematics, Queen's University, Belfast, Northern Ireland, and *Quantum Chemistry Program Exchange (QCPE),* Department of Chemistry, Indiana University, Bloomington, Indiana. Moreover, well-documented copies of public domain computer codes may sometimes be obtained from active workers in the field upon request.
11. S. Fraga, J. Karwowski, and K. M. S. Saxena, *Handbook of Atomic Data* (Elsevier, Amsterdam, 1976). Many tabulations appear in *Atomic Data and Nuclear Data Tables,* for example: E. Clementi and C. Roetti, *Atom. Data Nucl. Data* **14,** 177 (1974).
12. C. C. J. Roothaan and P. S. Bagus, in *Methods in Computational Physics,* B. Alder, S. Fernbach and M. Rotenberg, eds. (Academic Press, New York, 1963), Vol. 2, p. 1. For a concise discussion, see D. Potter, *Computational Physics* (Wiley, New York, 1973), Chap. 7.
13. J. C. Slater, *Phys. Rev.* **81,** 385 (1951); *Adv. Quantum Chem.* **6,** 1 (1972); see also Ref. 2, particularly Vol. 4.
14. K. H. Johnson, *Adv. Quantum Chem.* **7,** 143 (1973); *Ann. Rev. Phys. Chem.* **26,** 39 (1975); D. A. Case, *ibid,* **34,** 151 (1983).
15. I. Lindgren and J. Morrison, *Atomic Many-Body Theory,* 2nd ed. (Springer-Verlag, Berlin, 1986).
16. P. Hohenberg and W. Kohn, *Phys. Rev.* **136,** B864 (1963); W. Kohn and L. J. Sham, *Phys. Rev.* **140,** A1133 (1964).

17. L. Hedin and S. Lundqvist, *Solid State Phys.* **23,** 1 (1969).
18. L. Hedin and B. I. Lundqvist, *J. Phys. C* **4,** 2064 (1971); U. von Barth and L. Hedin, *J. Phys. C* **5,** 1629 (1972); O. Gunnarsson and B. I. Lundqvist, *Phys. Rev. B* **13,** 4274 (1976); D. M. Ceperley and B. J. Alder, *Phys. Rev. Lett.* **45,** 566 (1980).
19. S. Lundqvist and N. H. March, eds., *Theory of the Inhomogeneous Electron Gas* (Plenum Press, New York, 1983); R. M. Dreizler and J. da Providencia, eds., *Density Functional Methods in Physics* (Plenum Press, New York, 1985).
20. M. Schluter and L. J. Sham, *Phys. Today,* Feb. 1982, p. 36; R. G. Parr, *Ann. Rev. Phys. Chem.* **34,** 631 (1983); J. Callaway and N. H. March, *Solid State Phys.* **38,** 136 (1984).
21. O. Gunnarsson and R. O. Jones, *Phys. Rev. B* **31,** 7588 (1985).
22. J. P. Perdew and A. Zunger, *Phys. Rev. B* **23,** 5048 (1981).
23. O. Gunnarsson and R. O. Jones, *Phys. Scripta* **21,** 394 (1980); D. C. Langreth and M. J. Mehl, *Phys. Rev. B* **28,** 1809 (1983).
24. M. S. Hybertsen and S. G. Louie, *Phys. Rev.* **34,** 5390 (1986); **35,** 5585, 5602 (1987).
25. I. P. Grant, *Adv. Phys.* **19,** 747 (1970); I. P. Grant and H. M. Quiney, *Adv. Atom. Mol. Phys.* **23,** 37 (1988).
26. For a discussion of advanced relativistic computer codes, see K. G. Dyall, I. P. Grant, C. T. Johnson, F. A. Parpia, and E. P. Plummer, *Computer Phys. Commun.* **55** (1989).
27. G. L. Malli, ed., *Relativistic Effects in Atoms, Molecules, and Solids* (Plenum Press, New York, 1983); W. C. Ermler, R. B. Ross, and P. A. Christensen, *Adv. Quantum Chem.* **19,** 139 (1988).
28. J. T. Waber, "Atomic Structure Calculations, Relativistic Atoms" (this volume).
29. R. D. Cowan and D. C. Griffin, *J. Opt. Soc. Am.* **66,** 1010 (1976); J. H. Wood and A. M. Boring, *Phys. Rev. B* **18,** 2701 (1978).
30. M. L. Cohen, *Phys. Today,* July 1979, p. 40, and references cited.
31. G. B. Bachelet, D. R. Hamann, and M. Schluter, *Phys. Rev. B* **26,** 4199 (1982), and references cited.

Atomic Structure Calculations, Relativistic Atoms

James T. Waber

Relativistic atomic calculations have become available for all of the atoms of the periodic table in the last decade and have extended our understanding in many areas of physics and chemistry. Some of these will be reviewed briefly below.

Apparently, the first such calculation [1] was made for the $_{29}Cu^{+1}$ ion in 1940. It passed relatively unnoticed until Mayers [2] did $_{80}Hg$ in 1957. These were formidable tasks done without the aid of modern computers. Only a limited number of atomic calculations of any kind existed before 1963 when Herman and Skillman published their book on *Atomic Structure Calculations* [3]. Within a few years [4–7] a large number of nonrelativistic calculations started to appear with increasing array of complications. With these in hand, the importance of the various improvements could be assessed. Excellent agreement with experiment for energy values was demonstrated in 1975.

Several differences between the nonrelativistic Hartree–Fock–Slater (discussed elsewhere by Herman [8]) and the relativistic Dirac–Slater wave functions [9] are important. Instead of a single orbital $P(\mathbf{r})$ for each electron with the quantum numbers (nlm), one writes

$$\Psi_k(\mathbf{r}) = \frac{1}{r}\begin{Bmatrix} i^l F(r)\Omega_{jlm}(\theta, \varphi) \\ i^{l'} G(r)\Omega_{jl'm}(\theta, \varphi) \end{Bmatrix} \qquad (1)$$

as the orbital for the kth electron, where $F(r)$ stands for the major radial component and $G(r)$ for the minor. Instead of one spherical harmonic or associated Legendre polynomial $Y_l^m(\theta, \varphi)$ to represent the angular dependence, there are several. The angular dependence for a given component is defined as

$$\Omega_{lj\mu} = A\{B(lj)Y_l^{\mu-\sigma}(\theta, \varphi)\left[\begin{smallmatrix}0\\1\end{smallmatrix}\right] + C(lj)Y_l^{\mu+\sigma}(\theta, \varphi)\left[\begin{smallmatrix}1\\0\end{smallmatrix}\right]\}, \qquad (2)$$

where A is a normalizing constant, $B(lj)$ and $C(lj)$ are coefficients dependent on the angular momenta, and the two functions $Y_l^{\mu\pm\sigma}$ are the same type of spherical harmonic as occurs in the nonrelativistic case. Finally $\left[\begin{smallmatrix}0\\1\end{smallmatrix}\right]$ and $\left[\begin{smallmatrix}1\\0\end{smallmatrix}\right]$ represent the Pauli spin matrices which represent up and down spin, respectively. Because the angular momentum l' differs from l by the quantity a (which may be either plus or minus 1), the angular dependence of the major and minor components depend either on l or l' or vice versa. Hence one involves an odd and the other an even power of cosθ.

The angular momentum j has the definition $j = l + \frac{1}{2}a$. Another aspect of a will be taken up below, and the μ values differ by 1 since $|\sigma| = \frac{1}{2}$. Concerning the two $Y_l^{\mu\pm\sigma}$ functions in the formula for $\Omega_{jl\mu}$ the quantity μ is the resolved component of j. For these reasons $\Psi(\mathbf{r})$ is a four-component wave function. Details of the formulation are given by Grant [10] but his notation differs slightly from these formulas. The reader is cautioned that phase factors may differ in the various representations which have been published.

A simple table (Table I) will indicate these quantum numbers as well as Dirac quantum number κ, and how many electrons can occupy a complete subshell. Thus, for example, the p shell is divided into two subshells; one is called $2p_{1/2}$ and the other $2p_{3/2}$. Each contains $2j+1$ electrons. This is one additional way in which the relativistic and nonrelativistic wave functions and atomic calculations differ.

The phenomenon just mentioned is called spin–orbit splitting and the energy separation between the two subshells varies roughly as Z^4, where Z is the atomic number. This fact will indicate why relativistic effects have only limited importance when dealing with the electrons of atoms and ions with low Z values but become very important for heavy elements.

Some of the details of making a self-consistent-field calculation have been indicated in this volume [8]. For greater

Table I Quantum numbers for relativistic atoms

		Symbol	s	p	d	f	g
	a	j \ l	0	1	2	3	4
Number of electrons	−1	$l-\frac{1}{2}$	0	2	4	6	8
	+1	$l+\frac{1}{2}$	2	4	6	8	10
Sum			2	6	10	14	18
Kappa value	−1		0	1	2	3	4
	+1		−1	−2	−3	−4	−5

detail, the reader is referred to recent papers [11–13]. The probability of finding the ith electron in a spherical shell of radius r reduces to the simple formula

$$\rho_i(r) = [F_i(r)]^2 + [G_i(r)]^2 \qquad (3)$$

after integration over the angles θ and φ. The total charge density $\rho(r)$ is obtained by summing this over all the occupied orbitals. Slater's approximation [14] to exchange potential is related to the diameter of a Fermi hole, and hence is given by

$$V_{\mathrm{ex}}(r) = \alpha_s \left[\frac{3}{8\pi} \sum \rho_i(r) \right]^{1/3} \qquad (4)$$

where α_s is an adjustable constant in the range of $\frac{2}{3}$ to 1 and the summation runs over all of the occupied orbitals. Mann [15] has discussed the slightly more complicated equations which arise when exchange is handled in the Hartree–Fock manner.

A convenient way [7] to write the radial equations is as a matrix for a given set of quantum numbers:

$$\frac{d}{dr}\begin{pmatrix} F(r) \\ G(r) \end{pmatrix} = \begin{pmatrix} \dfrac{-\kappa}{r} & \dfrac{(V - E_0 - W)}{r} \\ \dfrac{-(V + E_0 - W)}{r} & \dfrac{\kappa}{r} \end{pmatrix} \begin{pmatrix} F(r) \\ G(r) \end{pmatrix}, \qquad (5)$$

so that the derivative of F is equal to the sum of radial factors times both F and G—a similar situation applies for the minor component G. That is, the two radial functions are coupled together by first-order differential equations.

In Eq. (5), V is the Coulomb potential Z/r, E_0 is the rest mass of the electron, and W is the energy eigenvalue $E(nlj)$. The relation of the quantum number κ to the more familiar ones, l and j, is listed above.

By differentiating G with respect to r, and substituting, one can get from (5) a second-order differential equation in the major component. This is similar to the Schrödinger equation but contains additional components. One of these additional terms gives rise to the spin–orbit splitting. By assuming that such terms are only a small perturbation, one can estimate the relativistic effects with single-component nonrelativistic wave functions. This was the traditional approach found in textbooks before the relativistic calculations became available.

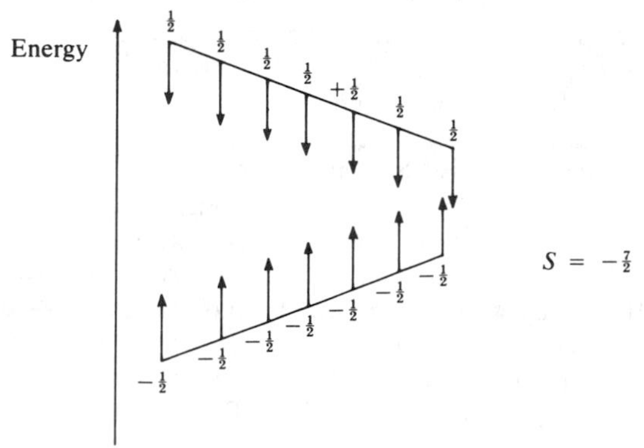

FIG. 1. Schematic of energy involved in adding electrons in LS coupling.

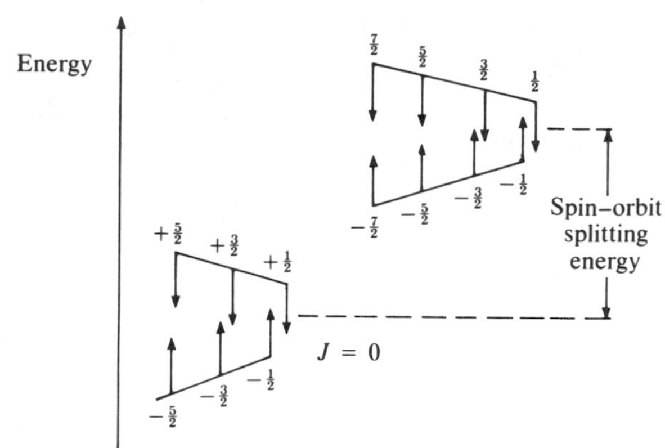

FIG. 2. Energy sketch of adding electrons with various spins in jj coupling.

Relativistic atomic calculations are important for treating phenomena which involve the interaction between the nucleus and electrons. References to typical areas are: Mössbauer spectroscopy [12], hyperfine interaction [13], and beta capture [17].

Another area where relativistic effects are important is in connection with the valence electrons of heavy elements. This can be illustrated by the two diagrams which contrast the "order" of filling electrons in the lowest-energy state of a given shell nl. In a nonrelativistic treatment of the electrons in a rare earth (lanthanide) or actinide element, one first arranges the spin of the f electrons parallel according to LS or Russell–Saunders coupling scheme (Fig. 1). Progressively one spin is occupied after another and the energy increases due to electron–electron interaction. With seven electrons, the resultant is $J = \frac{7}{2}$, corresponding to the seven unpaired spins. Completion of the shell by adding the other $2l+1$ electrons brings the total spin to $J = 0$.

In Fig. 2, there are the two subshells with six and eight electrons separated by a large spin–orbit splitting. According to the modified Hund's rule [18], the first three electrons go in a parallel arrangement hence $J = -\frac{5}{2} - \frac{3}{2} - \frac{1}{2} = \frac{9}{2}$. However, the next three are opposite in sign so that the sum of the six individual J values, namely J, becomes zero. The seventh electron would lead to a J value of $-\frac{7}{2}$ just as was found for LS coupling. In fact, in many cases, the result found by jj coupling (which is characteristic of a relativistic treatment) is the same as one would obtain by LS coupling; in others, the two results differ. For example, for six electrons, LS coupling would give -3 for J, where as the better answer is 0.

Another area which involves the outer electrons is the anomalous behavior of the angular distribution of photoelectrons. The jj coupling between the bound and the emergent electron in the continuum state leads to observed dependence and explains the occurrence of the Cooper minimum [19] in the asymmetry coefficient β as well as its energy dependence. In contrast, LS coupling gives a constant value [20,21] for β.

The accuracy of the relativistic atomic calculations is indicated by the calculated theoretical binding energy of the $1s_{1/2}$ electron [22]:

Energy eigenvalue	−142.929 keV
Magnetic contribution	+0.715
Retardation effect	−0.041
Vacuum fluctuation	+0.457
Vacuum polarization	−0.155
Total binding energy (theoretical)	141.953 (±0.053) keV
Experimental value [23]	141.963 (±0.013) keV

The agreement is very good with the experimental value of Dittner *et al.* [21] and with the independent theoretical value (also based on a Dirac–Fock calculation) of Freedman *et al.* [23].

This listing also serves to indicate the magnitude of other relativistic effects which cannot be discussed fully here. Both the magnetic and the retardation corrections are involved in the Breit interaction—a coupling of two electrons by means of a virtual photon. The vacuum fluctuation is also called Zitterbewegung. Polarization of the vacuum results from the strong Coulomb field of the nucleus. Its estimation is discussed by Pyykkö [24].

Most of these calculations have assumed that the field experienced is spherically symmetric. While this is guaranteed by Unsold's theorem for a closed shell, it is not true for open shells and a more complex treatment is required for dealing with even relatively simple atoms. The reader is referred to the article in this volume by Nesbet [25].

The other important use of such relativistic wave functions is in the construction of a reasonable molecular or crystal potential when heavy elements are concerned. Two types of relativistic molecular calculations have been done. The "discrete variational method" was used in connection with a linear combination of numerical (relativistic) orbitals by Rosén and Ellis [26–28]. At nearly the same time, Yang and his collaborators developed a relativistic version of the scattered wave formalism [29–32]. A review article by Pyykkö [24] covers a number of aspects of relativistic quantum chemistry. The earliest relativistic band calculations were made by Loucks [33, 34] who employed the augmented plane wave (APW) method. Sommers [35] developed a relativistic Kohn–Korringa–Rostoker method and Soven [37] a relativistic orthogonalized plane wave. More recently, Koelling and Freeman have published a series of relativistic APW calculations [38–41].

With the ready availability of even multiconfigurational atomic Dirac–Fock MCDF programs of Desclaux *et al.* [42] and Grant *et al.* [43], progress has been made in the last 10 years on the fronts of molecular and atomic structure calculations. Despite this fact, experimentalists continue to use a collection of approximation and *Ansatz* corrections to NR calculations to bring about improved comparisons with their data. These corrections are of questionable validity and may pertain to only portions of the periodic table. The MCDF calculations are "exact" and now with current computers are easy to perform.

While the word "exact" has been used, some difficulties remain; for example, the question of how to treat the interaction between two electrons is not resolved.

In the comprehensive review in 1970 by Grant [44] of relativistic atomic calculations, two problems had not emerged, even though Brown and Ravenhall [45] had pointed out that the simple Dirac–Coulomb equation did not yield normalizable solutions. The success of obtaining good numerical eigenfunctions had diverted attention from such fundamental problems.

When finite basis sets were employed, unrealistically negative solutions were obtained and the "Brown–Ravenhall disease" and "variational collapse" became of significance. One cause [46] of the latter was that the same basis set was used to expand the major and minor components. The reader is referred to the studies of Kunzelnigg [47] and Goldman [48]. Grant has made a number of comments [49] on these types of problem in recent conference volumes. These difficulties were attributable to the fact that the single-electron relativistic Hamiltonian is not bounded from below, i.e., there exists a negative energy continuum which has features in common with positive energy Rydberg states, and spurious roots with $E \sim -2c^2$ were obtained [50]. The recommended procedure is to project out properly behaved solutions and constrain the solutions to pertain only to a fixed number of electrons, as was effectively done with numerical orbitals [51].

The use of basis sets was motivated by interest in following the NR procedures of Roothan [52] for molecules. Goldman and Dalgarno implemented a variational procedure for two-electron atoms which avoided spurious roots, variational collapse, and continuum dissolution. Results with $Z \leq 10$ were published [53]. It should be mentioned that the proper nodal behavior is complex. That is, in representing the orbital by such basis functions, the minor component has one more node than the major [54], i.e., the significance increases rapidly with Z.

Concerning the importance of utilizing relativistic orbitals, Torboem, Fricke, and Rosén [55] discuss isomer shifts for low-lying states of IIa and IIb elements and compare nonrelativistic NF, DF, and MCDF results. The inadequacy using the contact term $|\psi_{ns}|^2$ rather than the wave functions integrated over the range of the nuclear charge distribution is illustrated. They note that $np_{1/2}$ orbitals contribute very little to the overall electronics charge density in this region. However, they note that the $ns(n-1)d$ states make an important contribution in the multiconfigurational treatment.

Kim, Huang, Cheng and Desclaux [56] emphasize that care should be taken when comparing DF calculations and experimental spin-orbit splittings ${}^2P_{1/2}$–${}^2P_{3/2}$. They presented results for two negative ions, B^- and F^-. Problems arose because ${}^2P_{1/2}$ and ${}^2P_{3/2}$ orbitals do not converge to the same nonrelativistic results and can yield spurious nonrelativistic contributions when one attempts to obtain good agreement with experimental data.

Pitzer and colleagues [57] have investigated the use of an effective potential obtained from DF calculations for molecular problems.

While retardation interaction between two electrons is not very large when treating an atom [54], it will become sig-

Table II Some typical relativistic atomic calculations

Name of type	Author	Type of exchange	Maximum Z calculated	Year	Footnote
Hartree–Fock	Cohen	None	92	1951	a
Dirac–Slater numerical	Carlson *et al.*	Statistical	126	1966	b
	Liberman *et al.*	Statistical	126	1966	c,q,p
	Schofield	Statistical	104	1974	d
	Rosén, Lindgren	Modified Slater	95	1968	e
	Band *et al.*	Statistical	95	1977	f
	Huang *et al.*	Statistical	106	1965	g
Analytic expansion	Fraga *et al.*	Fock integrals	102	1965	h
	Kim		40	1967	i
	Kagawa		50	1975	j
Dirac–Fock	Mann, Waber	Fock integrals plus	131	1970	k
		Breit and correlation corrections	126	1970	l
Multiconfiguration Dirac–Fock	Desclaux	Fock integrals plus	120	1973	m
	Desclaux	Breit and	126	1976	n
	Fricke, Soff	correlation corrections	173	1977	o

[a] S. Cohen, *Phys. Rev.* **118,** 489 (1960).
[b] C. C. Lu, T. A. Carlson, F. B. Malik, T. C. Tucker, and C. W. Nestor, Jr., *At. Data* **3,** 1 (1971); (erratum) **14,** 89 (1974); **2,** 63 (1970).
[c] D. Liberman, J. T. Waber, and D. T. Cromer, *Phys. Rev.* **137,** A27 (1965).
[d] J. H. Schofield, *At. Nucl. Data Tab.* **14,** 121 (1974); *Phys. Rev.* **9,** 1041 (1974).
[e] A. Rosén and I. Lindgren, *Phys. Rev.* **176,** 114 (1968); see also I. Lindgren and A. Rosén, *Atom. Phys.* **4,** 93 (1974).
[f] I. M. Band, M. A. Listengarten, M. B. Trzhaskovskaya, and V. I. Fomichev. *Leningrad. Inst. Idernov. Fiz. Report* **298** (April 1977).
[g] K. H. Huang, M. Aoyagi, M. H. Chen, B. Craseman, and H. Mark, *At. Data* **18,** 243 (1976).
[h] S. Fraga and K. M. S. Saxena. *At. Data* **3,** 323 (1971); **4,** 255 (1972); **4,** 269 (1972); **5,** 467 (1973).
[i] Y. K. Kim, *Phys. Rev.* **154,** 17 (1967); (erratum) **159,** 190 (1967).
[j] T. Kagawa, *Phys. Rev. A* **12,** 2245 (1975).
[k] J. B. Mann and J. T. Waber, *J. Chem. Phys.* **53,** 2397 (1970).
[l] J. B. Mann, *J. Chem. Phys.* **51,** 841 (1969).
[m] J. P. Desclaux, *At. Data* **2,** 311 (1973).
[n] J. P. Desclaux, *At. Data* **18,** 243 (1976).
[o] B. Fricke and G. Soff, *At. Nucl. Data Tab.* **19,** 83 (1977).
[p] D. A. Liberman, *Phys. Rev. B* **2,** 244 (1970).
[q] W. Kohn and L. T. Sham. *Phys. Rev.* **140,** A1133 (1965).

nificant in treating molecules because of larger interelectronic distances. The necessity of treating atomic cases with nonorthogonal orbitals also has not been evaluated for molecular problems.

Zangwill and Liberman [58] have made some improvements of the Liberman–Waber–Cromer program to take in account time-dependent optical response.

Relativistic continuum orbitals have been discussed in the texts by Berestetskii, Lifschitz, and Pitaerskii [59] and Rose [60]. A computer code for such orbitals has been accepted by Computer Physics Communications [61] by Perger and a study involving both discrete and positive energy continuum orbitals expressed in an *ab initio* basis set obtained from the Dirac equation is near completion.

One way of summarizing this article on atomic calculations is to indicate the typical references to various kinds of relativistic atomic calculations which are available as well as some differences (Table II). Most of these are available as numerical tabulations. The results of Fraga, Saxena, etc. are as coefficients of algebraic basis functions. The maximum atomic number studied for each type is indicated.

Pyykkö [62] organized an exhaustive bibliography of several hundred references covering the period 1916 to 1985. The reader is directed to it for locating the literature on various relativistic studies. The recent paper by Grant and Quiney [63] is an excellent review of the theoretical and computational situation.

See also Atomic Structure Calculations—One-Electron Models.

REFERENCES

1. A. O. Williams, *Phys. Rev.* **58,** 723 (1970).
2. D. F. Mayers, *Proc. Roy. Soc. (Lond.) A* **241,** 93 (1957). J. P. Desclaux, D. F. Mayers, and F. O'Brien, *Phys. Rev. B* **4,** 631 (1971).
3. F. Herman and S. Skillman, *Atomic Structure Calculations* (Prentice-Hall, Englewood Cliffs, N.J., 1963).
4. M. A. Coulthard, *Proc. Roy. Soc. (Lond.)* **91,** 44 (1967); **91,** 421 (1967).
5. F. C. Smith and W. R. Johnson, *Phys. Rev.* **160,** 136 (1967).
6. C. C. Lu, T. A. Carlson, F. B. Malik, T. C. Tucker, and C. W. Nestor, Jr., *At. Data* **3,** 1 (1971); (erratum) **14,** 89 (1974); **2,** 63 (1970).
7. D. Liberman, J. T. Waber, and D. T. Cromer, *Phys. Rev.* **137,** A27 (1965).
8. F. Herman, this Encyclopedia.
9. V. M. Burke and I. P. Grant, *Proc. Phys. Soc. (Lond.)* **90,** 297 (1967).
10. I. P. Grant, *Adv. Phys.* **19,** 747 (1970). A slightly different notation was used in earlier papers [*Proc. Roy. Soc. A* **262,** 555 (1961); *Proc. Phys. Soc. (Lond.)* **86,** 523 (1965)].
11. D. Liberman, J. T. Waber, and D. T. Cromer, *Comp. Phys. Commun.* **2,** 107 (1971).
12. B. Fricke and J. T. Waber, *Actinide Rev.* **1,** 433 (1971); *Theo. Chim. Acta* **21,** 235 (1971); *Phys. Rev., B* **5,** 3445 (1972).
13. J.-P. Desclaux, *Int. J. Quantum Chem.* **6,** 25 (1972); J.-P. Desclaux and N. Bessis, *Phys. Rev. A* **2,** 1623 (1970).
14. J. C. Slater, *Phys. Rev.* **81,** 385 (1951).
15. J. B. Mann, *J. Chem. Phys.* **51,** 841 (1969).
16. J. B. Mann and J. T. Waber. *At. Data* **5,** 201 (1973); *J. Chem. Phys.* **53,** 2397 (1970).
17. T. A. Carlson, C. W. Nestor, Jr., F. B. Malik, and T. C. Tucker, *Nucl. Phys. A* **135,** 57 (1969).
18. T. E. H. Walker and J. T. Waber, *Phys. Rev. A* **7,** 1218 (1973).
19. T. E. H. Walker and J. T. Waber, *Phys. Rev. Lett.* **30,** 307 (1973).
20. T. E. H. Walker and J. T. Waber, *J. Phys. B* **6,** 1165 (1973).
21. T. E. H. Walker and J. T. Waber, *J. Phys. B* **7,** 674 (1974).
22. B. Fricke, J.-P. Desclaux, and J. T. Waber, *Phys. Rev. Lett.* **28,** 714 (1972).
23. M. S. Freedman, F. T. Porter, and J. B. Mann, *Phys. Rev. Lett.* **28,** 711 (1972). P. F. Dittmer, CE. Bemis, D. C. Hansley, R. J. Silva, and D. C. Goodmen *Phys. Rev. Lett.* **26,** 1037 (1971).
24. P. Pyykkö, *Adv. Quantum Chem.* (to be published).
25. R. Nesbet, this Encyclopedia.
26. A. Rosén and D. E. Ellis, *J. Chem. Phys.* **62,** 3039 (1975).
27. D. E. Ellis, A. Rosén, and P. F. Walch, *Int. J. Quantum Chem.* **S9,** 351 (1975).
28. P. F. Walch and D. E. Ellis, *J. Chem. Phys.* **65,** 2387 (1976).
29. C. Y. Yang and S. Rabii, *Phys. Rev. A* **12,** 362 (1975).
30. C. Y. Yang, K. H. Johnson, and J. A. Horsley, *Bull. Am. Phys. Soc.* **21,** 382 (1976).
31. C. Y. Yang, *Chem. Phys. Lett.* **41,** 588 (1976).
32. C. Y. Yang and S. Rabii, *J. Chem. Phys.* **78,** 68 (1978).
33. T. Loucks, *Phys. Rev.* **139,** A1333 (1965).
34. T. Loucks, *Augmented Plane Wave Method* (Benjamin, New York, 1966).
35. C. Sommers and H. Amar, *Phys. Rev.* **188,** 1117 (1969).
36. C. Sommers, *J. Phys. C* **3,** 39 (1972).
37. P. Soven, *Phys. Rev.* **137,** A1706 (1965).
38. D. Koelling, *Phys. Rev.* **188,** 1049 (1969).
39. D. Koelling and A. J. Freeman, *Phys. Rev. B* **7,** 4454 (1973).
40. D. Koelling and A. J. Freeman, *Phys. Rev. B* **12,** 5622 (1975).
41. D. Koelling and A. J. Freeman, *Plutonium and Other Actinides,* p. 2911 (H. Blank and R. Lindner, eds.). (North Holland, Amsterdam, 1976).
42. J.-P. Desclaux, *Comp. Phys. Commun.* **9,** 31 (1975).
43. I. P. Grant, *Comp. Phys. Commun.* **21,** 207 (1980).
44. Ian P. Grant, *Adv. Phys.* **19,** 747–811 (1970).
45. G. E. Brown and D. G. Ravenhall, *Proc. R. Soc. London A* **208,** 552 (1951).
46. Y. Ishikawa, R. C. Binning, and K. M. Sand, *Chem. Phys. Lett.* **105,** 189 (1984); **101,** 111 (1983). J. Mark and P. Rozickey, *Chem. Phys. Lett.* **74,** 562 (1980).
47. W. Kunzelnigg, *Int. J. Quantum. Chem.* **25,** 107 (1984).
48. S. P. Goldman, *Phys. Rev. A* **30,** 1219 (1984); **31,** 354 (1985); **37,** 16–30 (1988).
49. I. P. Grant, in *Atom Theory Workshop on Relativistic and QED Effect in Heavy Atoms* (H. Kelly and Y.-Ki Kim, eds., pp. 17–19, 200–203, 299–301, American Institute of Physics, New York, 1985). See also *Phys. Rev. A* **25,** 1230 (1982).
50. See J. Sucher, in *Proceedings of the NATO Advance Study Institute on Relativistic Effects in Atoms, Molecules and Solids* (G. Malli, ed.), Plenum, New York, 1982; also *Proceedings of Argonne Workshop on the Relativistic Theory of Atomic Structure* (H. G. Berry, K. T. Chem, W. K. Johnson, and Y.-Ki Kim, eds.), ANL-80-116, Argonne National Laboratory, Argonne, IL, 1980. See also M. Mittleman, *Phys. Rev. A* **4,** 893 (1971); *A* **15,** 2395 (1972).
51. W. D. Sepp and B. Fricke, in *Atomic Theory Workshop on Relativistic and QED Effects in Heavy Atoms,* AIP Conf. 136 (H. Kelly and Y.-Ki Kim, eds.), pp. 20–25, American Institute of Physics, New York, 1985.
52. C. C. J. Roothan, *Rev. Mod. Phys.* **32,** 179 (1960).
53. S. P. Goldman and A. Dalgarno, *Phys. Rev. Lett.* **57,** 408 (1988).
54. P. J. C. Airts and W. C. Nieuwport, *Chem. Phys. Lett.* **113,** 165 (1985).
55. G. Torboem, B. Fricke, and A. Rosén, *Phys. Rev. A* **31,** 2038–2053 (1985).
56. K. N. Huang, Y.-Ki Kim, K. T. Cheng, and J.-P. Desclaux, *Phys. Rev. Lett.* **48,** 1245 (1982).
57. Y. S. Lee, W. C. Krumler, and K. S. Pitzer, *J. Chem. Phys.* **67,** 5861 (1977); **69,** 976 (1978); **70,** 288–293 (1979); **73,** 360 (1980); **74,** 1162 (1981).
58. A. Zangwill and D. Liberman, *Comp. Phys. Commun.* **37,** 75–82 (1984).
59. V. D. Berestetskii, E. M. Lifschitz, and L. P. Pitaerskii, *Relativistic Quantum Theory,* Vol. 1, pp. 113–115. Pergamon Press, New York, 1971.
60. M. E. Rose, *Relativistic Electron Theory,* p. 82ff, Wiley, New York, 1961.
61. W. Perger, private communication, Michigan Technological University. Accepted by *Jour. Comp. Phys.* (1990).
62. Pekka Pyykkö, in *Relativistic Theory of Atoms and Molecules,* Lecture Notes in Chemistry 41, Springer-Verlag, New York, 1986.
63. Ian Grant and H. M. Quiney, *Adv. At. Mol. Phys.* **23,** 37–86 (1988).

See also ATOMIC STRUCTURE CALCULATIONS—ONE-ELECTRON MODELS.

Atoms

A. P. French

INTRODUCTION

Despite all the discoveries that have been made during the twentieth century in the fields of nuclear and subnuclear physics, the atom remains the most important type of unitary system in our picture of the physical world. This is in part because the electrically neutral, stable atom is the basic building block in the structure of condensed matter as we are most familiar with it. Under more severe conditions, e.g., at the enormously high temperatures and densities characteristic of stellar interiors, the individual atom ceases to be an identifiable unit of the structure. Nevertheless, whenever conditions permit, the atom will establish its existence because of its property of being the system of least energy and greatest stability that can be formed from its constituent particles. This property guarantees the atom a permanent place in our description of nature.

EARLY HISTORY

The birth of the theory that the basic structure of matter is discrete, not continuous, is usually attributed to the ancient Greeks, in particular Democritus (ca. 420 B.C.). (The name "atom" comes directly from the Greek *atomos*—indivisible.) However, it was not until about 1800 that the quantitative study of chemical reactions provided evidence that the behavior of bulk matter might indeed be governed by individual processes on a submicroscopic level. In the latter part of the eighteenth century, A. L. Lavoisier found that the total mass was conserved in chemical reactions, whatever changes in form and appearance took place among the reactants, and J. L. Proust showed that every pure chemical compound contains fixed and constant proportions (by weight) of its constituent elements. (The modern concept of *element* was propounded by Robert Boyle in 1661.) Building on these results, John Dalton in 1808 enunciated a detailed theory of chemical combination, based on the picture that each element is made up of a host of identical atoms and that the formation of a chemical compound from its elements takes place through the formation of "compound atoms" containing a definite (and small) number of atoms of each element.

From the known mass ratios for many different reactions, Dalton was able to suggest values of mass ratios of individual kinds of atoms. However, his scheme led to certain ambiguities and inconsistencies, which were not resolved until it came to be realized that the basic units of a pure element were not necessarily single atoms, but were often compound atoms in the form of diatomic molecules—a hypothesis first put forward by A. Avogadro in 1811, but which met resistance (despite its success in removing internal contradictions from Dalton's theory) because there was no obvious reason why such well-defined compounds of identical atoms should exist.

RELATIVE ATOMIC MASSES

In 1860 an international conference on atomic weights officially adopted the Dalton–Avogadro scheme, and during the succeeding decades a highly accurate tabulation of the relative atomic masses of the elements was built up from the analysis of thousands of different compounds. A natural unit for this scheme of relative atomic masses was the mass of hydrogen, the lightest atom. This was Dalton's choice, but in 1902 the basis was changed to a slightly different value, namely, one sixteenth of the atomic weight of oxygen. (A prime reason for this change was the practical one that oxygen, in contrast to hydrogen, forms stable and tractable compounds with most elements.) However, the discovery of isotopy (that a given element may have atoms of several different characteristic masses) led to a decision, in 1961, to redefine the atomic mass unit as one twelfth of the mass of the particular isotope carbon-12, and to express chemical atomic weights and the masses of individual isotopes as multiples of this unit.

AVOGADRO'S NUMBER AND THE ATOMIC MASS UNIT

The atomic-molecular theory was developed in the absence of any *direct* evidence for the granular structure of matter (although the small irregular movements of microscopic particles in liquid suspension, discovered by the botanist Robert Brown in 1827, were at least suggestive). However, if the theory is assumed to be correct, a definite value must exist for the number of atoms in a given mass of material of known composition. In particular, the *Avogadro number, N,* is defined as the number of atoms in a mass of elementary substance equal in grams to its (relative) atomic weight: The mass of an individual atom is thus equal to its atomic weight (in grams) divided by N, and the atomic mass unit (1 amu) is numerically equal to $1/N$.

One early source of information leading to quantitative estimates of N was the study of the properties of gases. The atomic-molecular kinetic theory of gases led to a picture of a gas as made up of small particles traveling at high speeds (hundreds of meters per second). The slowness of the processes of diffusion and mixing in gases implied, however, that individual molecules travel only very short distances before suffering changes of direction through collisions. If there are n molecules of diameter d per unit volume, the mean free path (which can be related directly to the observed diffusion rate) is of the order of $1/nd^2$. To obtain separately the values of n and d, we can use the fact that the total volume of n molecules is about nd^3, and will represent the volume occupied by these molecules if they are condensed to the almost incompressible liquid phase. Analysis along these lines indicated that N must be of the order of 10^{24}.

Much more precise values of N were obtained later by quite different methods. In 1900, Max Planck inferred a value within 3% of the currently accepted figure as a result of his quantum analysis of the continuous radiation spectrum of hot bodies. In 1916, R. A. Millikan's experimental proof that electric charge exists only in integral multiples of a unit, e, led to a value of N as given by the ratio F/e, where F, the faraday, is the amount of electric charge associated with the transport of 1 gram-equivalent of a substance in electrolysis. Later, in about 1930, J. A. Bearden and others made determinations of interatomic distances in crystals, through the process of diffraction of x rays of known wavelength (measured with a ruled grating); from this the number of atoms in a known mass of crystal could be inferred directly.

The currently accepted value of the Avogadro number is 6.0220×10^{23}, from which follows the result

$$1 \text{ amu} = 1.66056 \times 10^{-24} \text{ g}.$$

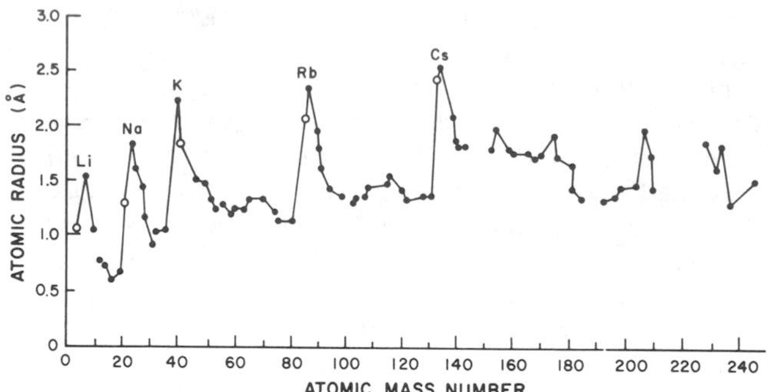

FIG. 1 Atomic radii. Black dots: radius based on experimental analyses of crystals and/or molecules; open circles: radius based on mean-free-path experiments.

MASSES AND SIZES OF INDIVIDUAL ATOMS

The relative atomic weights of the known atoms range from about 1 (hydrogen) to 250 (highly unstable transuranic elements). The corresponding absolute masses range from 1.67×10^{-24} g to about 4.1×10^{-22} g. In this range are about 100 different elements (as characterized by the atomic numbers Z in the periodic table) comprising about 300 distinct atomic species—a given atomic species being characterized by (besides its Z value) its *mass number, A*, the integer closest to its relative atomic weight.

The large range of atomic masses is not accompanied by a correspondingly large or systematic variation in size. Atomic radii all lie between about 0.5 and 2.5 Å (10^{-10} m), with no marked increase from the lightest to the heaviest. One of the simplest sources of this information is a knowledge of the densities of the elements in their solid state; another source is the measurement of cross sections for interatomic collisions. Figure 1 shows how the radii vary with Z; particularly noteworthy are the relatively small radii for the atoms of the noble gases and the especially large radii for the alkali metal atoms that immediately follow them in the periodic table. The general features of Fig. 1 can be understood from the standpoint of the internal structure of atoms (see later).

THE NUCLEAR ATOM

The discovery of the electron by J. J. Thomson in 1897 led quickly to the conclusion that all atoms contain some number of electrons. It followed that atoms, being electrically neutral, must be composed of some combination of electrons and positively charged material. Then, during the period 1911–1913, Ernest Rutherford, with H. Geiger and E. Marsden, carried out the alpha-particle scattering experiments that proved that the positive charge and most of the mass of an atom are concentrated in a minute volume, with a radius of less than 10^{-14} m (i.e., less than 10^{-4} of the outer radius of the atom).

Hard on the heels of Rutherford's discovery came the quantum model of the hydrogen atom, published by Niels Bohr in 1913. Using only the known constants of nature (including Planck's constant h), and without the use of any adjustable parameters, Bohr developed an accurate and quantitative model of the hydrogen atom as a nucleus of charge $+e$ (associated with 99.95% of the mass of the whole atom) with a single electron in orbit around it. For the atom in its ground (lowest-energy) state, Bohr's theory gave a radius of 0.53 Å and an ionization energy of 13.6 eV, in good agreement with observation. Furthermore, and perhaps even more striking, his theory accounted naturally for the systematic series of lines in the visible spectrum (Balmer spectrum) of atomic hydrogen. The basis was, according to the theory, that the possible (quantized) energies of an electron bound to a central charge of magnitude Qe are given by

$$E(n) = -\frac{2\pi^2 me^4 Q^2}{h^2} \times \frac{1}{n^2} \quad (n = 1, 2, 3 \ldots) \tag{1}$$

and that the wavelengths λ of possible spectral lines are defined by the amounts of energy carried away by photons in quantum jumps between levels, using the relations

$$E_{\text{photon}} = E(n_1) - E(n_2)$$

and

$$E_{\text{photon}} = h\nu = hc/\lambda.$$

X RAYS, ATOMIC NUMBER, AND NUCLEAR CHARGE

Bohr's theory had little success in accounting for optical spectra other than those of hydrogen or equally simple systems (e.g., singly ionized helium). However, it was possible to apply the theory to the so-called characteristic x rays emitted by many elements when bombarded with energetic electrons. By 1913 it was established that these x rays are electromagnetic radiations similar to visible light but of much shorter wavelength (typically of the order of 1 Å). In 1913 H. G.-J. Moseley made a systematic study of the characteristic x rays and found that their frequencies ν were just what would be expected, according to the Bohr energy-level formula, for a single electron falling to the lowest ($n = 1$) level from a level with $n = 2$ or $n = 3$. A graph of $\sqrt{\nu}$ against atomic number Z was a pair of straight lines (Fig. 2). Moseley concluded that the characteristic x rays were produced when an atom returned to its ground state after an electron in its lowest possible orbit, close to the nucleus, had been knocked out. He concluded further that the chemical atomic number could be identified with the number of units of positive charge on the nucleus.

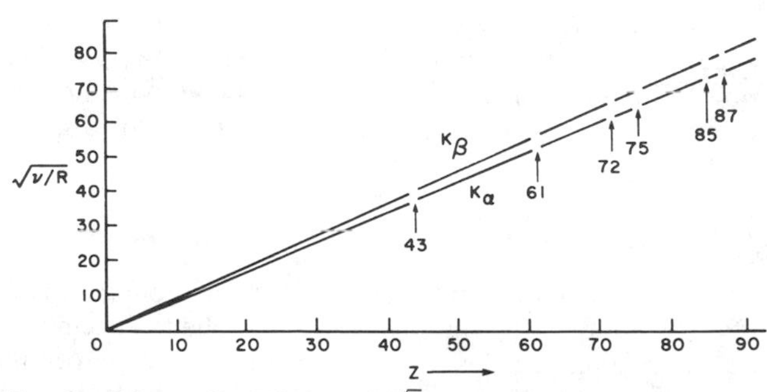

FIG. 2. Linear dependence of $\sqrt{\nu}$ on atomic number for characteristic x rays. The line marked K_α corresponds to the transitions $n = 2$ to $n = 1$; the line marked K_β corresponds to $n = 3$ to $n = 1$. The graphs show gaps at values of Z corresponding to elements not discovered until later. (The unit of measurement for ν is the rydberg, R, equal to $2\pi^2 me^4/h^3$.)

QUANTUM MECHANICS AND ATOMIC STRUCTURE

With the development of wave mechanics by E. Schrödinger in 1926, the picture of an electron in the field of a nucleus was drastically changed. In place of well-defined quantized orbits it became necessary to think in terms of smoothly varying probability distributions that permitted the electron to be found at any distance from the nucleus, or even inside it. The quantized energies of the possible bound states duplicated the results of the Bohr theory, but in every other way the description was quite different.

The spatial state of an individual electron could be characterized with the help of three quantum numbers, n, l, and m, all integers. The value of n, called the *principal quantum number,* defined the electron energy, just as in the Bohr theory, but the identification of a state also required the two quantum numbers l and m that defined the orbital angular momentum {of magnitude $[l(l+1)]^{1/2}h/2\pi$} of the electron and the projection, of magnitude $mh/2\pi$, of this orbital angular momentum along a specified axis. The quantum analysis required $0 \leq l \leq n-1$, and $-l \leq m \leq +l$.

Added to this was a property first recognized by W. Pauli in 1924: states defined by particular values of n, l, and m are in general split into two components of slightly different energy (as manifested, e.g., in the close doublet structure of many spectral lines). This property, subsequently interpreted as the consequence of an intrinsic spin and associated magnetic moment of the electron (G. Uhlenbeck and S. A. Goudsmit, 1925), meant that the full characterization of the quantum state of an electron in an atom required a total of *four* quantum numbers, the last of which simply took on the values $\pm\frac{1}{2}$ to correspond to the two possible quantized projections ($\pm h/4\pi$) of the spin angular momentum. This allows a total of $2(2l+1)$ different quantum states for given values of n and l.

An essential further ingredient had to be supplied, however. This was the *exclusion principle* of W. Pauli (1924), according to which no two electrons can have the same set of quantum numbers. This principle is the fundamental key to the internal structure of atoms, because it means that the electrons in a many-electron atom cannot all congregate in the lowest energy state ($n=1$), but are forced to occupy "shells" of progressively increasing energy. We might expect that a given shell would be defined by a particular value of n, since for a *single* electron in the field of a central charge the energy depends only on n. In a many-electron atom, however, the situation is drastically modified by the partial screening of the nuclear charge by the electrons close to it. The innermost electrons, in states with $n=1$, "see" almost the full nuclear charge Ze, but electrons in states of higher n are exposed to an effective central charge that is less than Ze and depends on l as well as n. Small l, in the quantum-mechanical picture, corresponds to a greater probability for the electron to be near the nucleus and therefore to be exposed to the full nuclear charge. The result of these considerations is to give rise to a fairly well-defined sequence of energy shells, each able to accommodate a certain maximum number of electrons, and corresponding to increasing energy and increasing mean distance from the nucleus. A partial tabulation showing the shell structure up to $Z=86$ is shown in Table I. Completion of a shell leads to a particularly stable and compact structure. The next electron added is relatively weakly bound and can stray quite far from the nucleus.

Table I Electron Shell Structure

Values of (n, l)	Shell capacity	Cumulative total
(1,0)	2	2
(2,0)+(2,1)	8	10
(3,0)+(3,1)	8	18
(3,2)+(4,0)+(4,1)	18	36
(4,2)+(5,0)+(5,1)	18	54
(4,3)+(5,2)+(6,0)+(6,1)	32	86

This theoretical model is in good accord with the empirical features of the periodic table of the elements. The atomic numbers $Z=2$, 10, 18, 36, 54, and 86 are those of the noble gases, He, Ne, A, Kr, Xe, and Rn. The atoms immediately following these ($Z=3$, 11, 19, etc.) are the alkali metals, which have large atomic radii (cf. Fig. 1) and single, weakly bound valence electrons. Many other less prominent features of the atomic sequence can be understood within the framework of this theoretical model.

The reality of the shell structure has been directly demonstrated in a few cases by probing the atomic structure with x rays or electrons. Figure 3 shows some experimental results on the radial variation of electron charge density in argon.

GENERAL STRUCTURE OF A MASSIVE ATOM

To bring together the results discussed thus far, it may be helpful to consider the complete picture of a particular massive atom with many electrons. Let us take the most abundant isotope of tin, with $Z=50$, $A=120$.

The nucleus of this atom has a charge of $+50e$; it contains 50 protons and a number of neutrons equal to $A-Z$, i.e., 70 (but the nuclear structure is not the concern of this article). Closest to the nucleus (on the average) are two electrons in states with $n=1$; these electrons are exposed to almost the full force of the nuclear electric field. According to Eq. (1), with $n=1$, $Q \approx 50$, each electron would be bound with a neg-

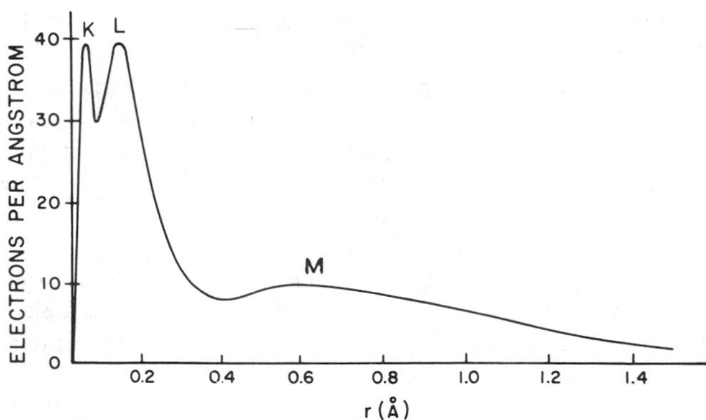

FIG. 3. The radial density distribution for the electron charge cloud in argon atoms, exhibiting the spatial shell structure as inferred from electron scattering experiments. The peaks marked K, L, M correspond to the approximate shell radii for $n = 1$, 2, and 3, respectively. [After L. S. Bartell and L. O. Brockway, *Phys. Rev.* **90,** 833 (1953).]

ative energy equal to almost 2500 times the binding energy of the electron in the hydrogen atom (13.6 eV); this would be of the order of 30 keV. Direct measurements show that it takes x rays with a quantum energy of 29.2 keV to dislodge an electron from this innermost shell. The average distance of these electrons from the nucleus is comparable to the orbit radius calculated from the Bohr theory for $n=1$, $Q=50$; this is about 0.01 Å, or 10^{-12} m—about 150 times the nuclear radius of tin.

Going to the next electron shell (cf. Table I), we have eight electrons for which $n=2$ and for which the nuclear charge is significantly shielded, first by the two innermost electrons; second, by the other electrons in this same shell; and third, by penetrating electrons from shells still farther out. The effective central charge is not easy to estimate in this case, but experiment shows that x rays of about 4 keV can eject electrons from this shell. (Actually, this and subsequent shells have a substructure, based on the involvement of two or more values of the quantum number l, which we shall not go into.) Putting $n=2$ in Eq. (1), and using the observed electron binding energy of about 4 keV, we would infer an effective central charge of about $35e$ and a mean shell radius of about 0.06 Å.

Proceeding in the same way, we find that the third shell (eight electrons, $n=3$) has a critical x-ray absorption energy of about 0.9 keV, corresponding to $Q \approx 24$, $r \approx 0.2$ Å; and the fourth shell (18 electrons) has a critical x-ray absorption energy of about 120 eV, corresponding to $Q \approx 12$, $r \approx 0.7$ Å.

This accounts for 36 out of the 50 electrons in the atom. The remaining 14 belong to the fifth electron shell, which can accommodate up to 18 electrons. The situation in this region of the atom is complicated, but we know that these electrons are the main contributors to the outer parts of the atomic charge cloud, which extends to a radius of about 1.5 Å (see Fig. 1). To remove one of these electrons requires an energy of 7.3 eV (the first ionization potential of tin). Thus we see that the energy-level structure within the Sn atom ranges all the way from weakly bound electrons (less than 10 eV) to very tightly bound electrons (tens of keV).

The chemical and spectroscopic characteristics of an element depend on the details of the quantum states of its outermost electrons. In particular, the chemical valence depends on the extent to which the total number of electrons in an atom represents an excess or a deficit with respect to the more stable configuration of a completed shell. In the case of tin, for example, the total of 50 electrons is four short of completing the fifth shell of Table I. In these terms we can understand why tin is quadrivalent (although it takes a study of finer details to understand in physical terms why it also exhibits divalency). In such matters as these, however, although the properties certainly have their complete basis in electric forces and quantum theory, the theoretical analysis is at best semiempirical.

See also Atomic Spectroscopy; Atomic Structure Calculations—Electronic Correlation; Atomic Structure Calculations—One-Electron Models; Bohr Theory of Atomic Structure; Elements; Isotopes.

BIBLIOGRAPHY

H. A. Boorse and Lloyd Motz, *The World of the Atom*. Basic Books, New York, 1966.

Max Born (tr. J. Dougal), *Atomic Physics*, 8th. rev. ed. Hafner, New York, 1969.

A. P. French and Edwin F. Taylor, *Introduction to Quantum Physics*. Norton, New York, 1978.

H. Haken and H. C. Wolf (tr. W. D. Brewer), *Atomic and Quantum Physics*, 2nd. ed. Springer-Verlag, New York, 1987.

G. P. Harnwell and W. E. Stephens, *Atomic Physics*. McGraw-Hill, New York, 1955.

G. Herzberg, *Atomic Spectra and Atomic Structure*. Dover, New York, 1945.

Alan Holden. *The Nature of Atoms*. Oxford (Clarendon Press), 1971.

H. G. Kuhn, *Atomic Spectra*. Academic, New York, 1962.

F. K. Richtmeyer, E. H. Kennard, and T. Lauritsen, *Introduction to Modern Physics*, 6th ed. McGraw-Hill, New York, 1969.

Auger Effect

Bernd Crasemann

An atom that contains an inner-shell vacancy becomes deexcited through a cascade of transitions that are due to two kinds of competing processes: x-ray emission and radiationless, or Auger, transitions. In either process, the original vacancy is filled by an electron from a higher-energy level. In radiative transitions, the released energy is carried off by a photon; in radiationless transitions, this energy is instead transferred through the Coulomb interaction to another atomic electron, which is ejected. The emitted electron is called a *K-LL* Auger electron, for example, if a *K*-shell vacancy is filled by an *L*-shell electron, and another *L* electron is ejected.

Direct experimental evidence for the existence of radiationless transitions was gained by Pierre Auger through cloud-chamber experiments reported in 1923. X rays traversing the chamber produced photoelectrons; the tracks of these photoelectrons increased if more energetic x rays were used. In addition to the photoelectron tracks, Auger observed numerous short tracks, each of which originated at the same point as a photoelectron track (Fig. 1). The length of the short tracks did not change with x-ray energy, but depended only on the kind of gas that was placed in the cloud chamber. Auger was able to show that the length of many of the short tracks corresponded to the energy that would be released in K-*LL* radiationless transitions. In some cases, Auger found additional tracks, caused by electrons emitted in a second (*L-MM* or *L-MN*) step of the radiationless deexcitation cascade of an atom.

The quantum-mechanical theory of radiationless transitions was formulated by G. Wentzel in 1927. From perturbation theory, the nonrelativistic matrix element for a direct Auger transition is

$$D = \iint \psi^{*}_{n''l''j''}(1)\psi^{*}_{\infty l_A j_A}(2) \left| \frac{e^2}{|\mathbf{r}_{12}|} \right| \psi_{nlj}(1)\psi_{n'l'j'}(2)\, d\tau_1 d\tau_2,$$

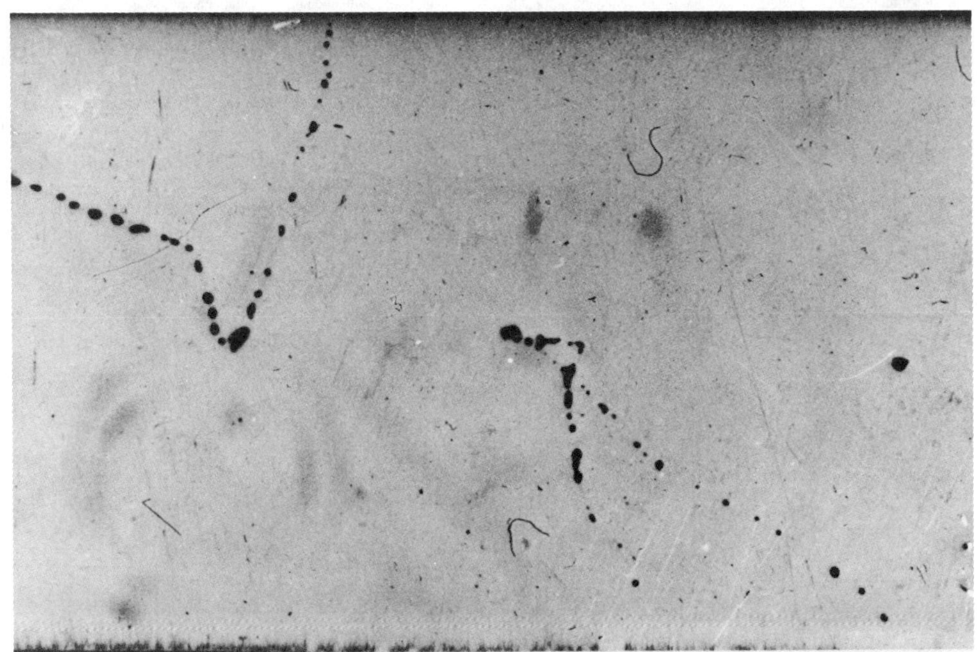

FIG. 1. Photoelectrons and Auger electrons from krypton ionized with 60-keV x rays, photographed in a cloud chamber by Pierre Auger *ca.* 1923. (Courtesy Prof. P. Auger.)

where the quantum numbers n, l, j characterize electrons that are identified schematically in Fig. 2. The state of the continuum (Auger) electron is labeled by $\infty l_A j_A$. In the physically indistinguishable exchange process, described by a matrix element E, the roles of electrons nlj and $n'l'j'$ are interchanged (Fig. 2). The total radiationless transition probability per unit time is

$$w_{fi} = \hbar^{-2} | D - E |^2,$$

if the continuum-electron wave function is normalized so as to correspond to one electron emitted per unit time. The matrix elements D and E can be separated into radial and angular factors. Evaluation of the angular factors depends on a choice of the appropriate angular-momentum coupling scheme. If spin-orbit interaction is neglected, the initial and final two-electron (or two-hole) states can be expressed for different values of the total angular momentum J in the

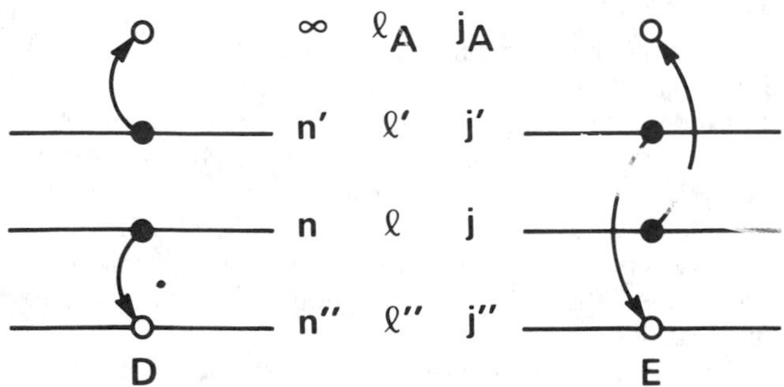

FIG. 2. Energy levels involved in the direct (D) and exchange (E) Auger processes, and notation for the principal, orbital-angular-momentum, and total-angular-momentum quantum numbers that characterize the pertinent electron states.

($LSJM$) representation of Russell–Saunders coupling. For the heavier atoms, inner-shell states are expressed more realistically in *j-j* coupling.

The selection rules governing radiationless transitions require that the total angular momentum and the parity of the final-state system (ion plus emitted electron) must be the same as of the initial-state ion. A large number of different transitions is generally possible from any given initial state; for example, 2784 matrix elements are required to describe the radiationless decay of an L_3 vacancy in a high-Z atom.

The relative probability that a K-shell hole is filled by an Auger process, rather than by x-ray emission, ranges from 0.999 for the lightest elements ($Z \leqslant 5$) to 0.02 for uranium. For vacancies in other shells, the Auger transition rate is generally several orders of magnitude greater than the radiative rate, and thus, essentially determines the lifetime of the vacancy. Auger rates can be very fast. Particularly intense are *Coster–Kronig transitions* that shift a hole to a higher subshell within one major shell. Thus, N_1-$N_{4,5}N_{4,5}$ "super-Coster–Kronig" transitions produce a 4s level width $\Gamma > 60$ eV at $Z \approx 50$, which corresponds to an N_1 vacancy mean life of $< 10^{-17}$ s. The limits of validity of perturbation theory are strained when transition rates of such magnitude are calculated.

The Auger-electron energy is

$$E_{\infty l_A} = E_{n''l''} - E_{nl} - E_{n'l'}.$$

The subscripts pertain to the states indicated in Fig. 2; $E_{n''l''}$ and E_{nl} are (absolute values of) neutral-atom binding energies, while $E_{n'l'}$ is the binding energy of $n'l'$ electron *in the presence of an nl vacancy*. The energy of electrons from solid samples can additionally include a contribution from extra-atomic relaxation, typically of the order of 10 eV.

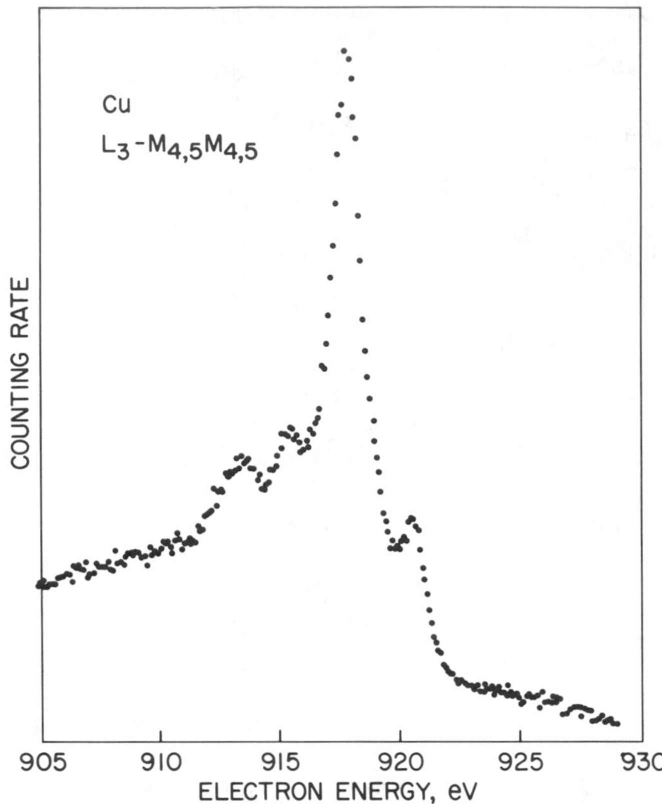

FIG. 3. Energy spectrum of L_3-$M_{4,5}M_{4,5}$ Auger electrons from metallic Cu. The separate peaks result from multiplet splitting due to various couplings of the two final-state holes. (Courtesy Dr. Lo I Yin, NASA Goddard Space Flight Center.)

Measurements of electron spectra from radiationless transitions (Fig. 3) and the theory of Auger processes constitute an active field of research, with relevance to fundamental atomic and solid-state theory as well as to surface physics, chemistry, and materials science.

New impetus has been given to Auger spectrometry in recent years with the advent of tunable synchrotron radiation. Selective photoexcitation of atomic subshells near threshold leads to Auger spectra that reveal details of electron rearrangement process, including correlation effects which produce a multifaceted many-electron response to inner-shell ionization. Important new insights into atomic structure and dynamics are being generated through such studies.

See also ELECTRON ENERGY STATES IN SOLIDS AND LIQUIDS.

BIBLIOGRAPHY

P. Auger, "The Auger Effect," *Surf. Sci.* **48,** 1 (1975). (E) A first-hand account of the discovery of radiationless transitions.

W. Bambynek, B. Crasemann, R. W. Fink, H.-U. Freund, H. Mark, C. D. Swift, R. E. Price, and P. Venugopala Rao, "X-Ray Fluorescence Yields, Auger, and Coster–Kronig Transition Probabilities," *Rev. Mod. Phys.* **44,** 716 (1972). (I) A comprehensive review of the subject.

E. H. S. Burhop and W. N. Asaad, "The Auger Effect," *Adv. At. Mol. Phys.* **8,** 164 (1972). (I) A review of theory and experiment.

B. Crasemann (ed.), *Atomic Inner-Shell Processes*. Academic Press, New York, 1975. (A) This treatise deals with radiative and radiationless transitions, higher-order processes, electron spectrometry, and related topics.

W. Mehlhorn, "Auger-Electron Spectrometry of Core Levels of Atoms," in *Atomic Inner-Shell Physics*, B. Crasemann, ed., Plenum Press, New York, 1985. (A) The theory of Auger transitions and recent experimental results are reviewed in this monograph.

K. S. Sevier, *Low-Energy Electron Spectrometry*. Wiley, New York, 1972. (A) A comprehensive monograph.

Aurora

E. C. Zipf

An aurora is an often spectacular optical phenomenon that occurs at altitudes above 95 km in the polar regions of the north and south hemispheres. Auroral displays occur most frequently in relatively narrow doughnut-shaped regions which encircle the geomagnetic poles. The Earth rotates underneath these patterns which are fixed with respect to the sun, resulting in a characteristic diurnal vibration in the morphology and temporal behavior of aurora at a particular geographic location. In addition to these geometrical effects, the physical thickness and diameter of the auroral ovals varies considerably with the degree of geomagnetic activity. Enhancements in the magnitude of the solar wind due to solar flares or to a general increase in particle flow from the sun during the normal solar cycle result in the equatorward expansion of the auroral zone which is also accompanied by an increase in the frequency of auroral displays and in their average intensity. The instantaneous position of the auroral zone is determined primarily by the precipitation of electrons with energies in the range 1–20 keV which are guided into the polar regions from the plasma sheet by the Earth's magnetic field lines. As these charged particles descend into the denser regions of the atmosphere, they lose their energy in inelastic collisions with atmospheric atoms and molecules. Most of this energy is consumed in ionizing, dissociating, and heating the atmosphere in the altitude range 95–300 km. Less than 5% of the energy of the primary particles is used to produce the visible radiation for which the aurora is so noteworthy. The plasma densities created by the precipitating electrons are comparable in magnitude to those produced by solar radiation in the normal ionosphere. When auroral activity is unusually intense, still larger ionospheric plasmas are created that cause interruptions in global radio communications (polar blackout).

The most common auroral form is the arc or band which is striking because of its extreme length (hundreds or thousands of kilometers) compared with its width (typically 1–10 km). Auroras are frequently classified in terms of their internal structure: homogeneous or rayed; their apparent motion: active or quiet; their brightness: on a scale from I through IV corresponding to an intensity variation from the brightness of the Milky Way to that of the full moon, respectively; and their visual color: the ordinary green or whitish aurora is designated Type C, while the dramatic veil auroras that fill the entire sky with red light are classified as Type-A aurora.

Three other types of auroral forms deserve special mention. The first is the evening hydrogen arc which is produced as the result of proton and electron precipitation and appears in the form of a broad diffuse arc equatorward of the main portion of the auroral oval during the evening hours. The second is the Polar Cap Absorption event (PCA) which is produced by very energetic solar cosmic rays (1–100 MeV) that enter the atmosphere over the entire polar cap down to a geomagnetic latitude of 60° where the cosmic ray cutoff limits further penetration. PCA events are associated with major flare activity on the sun and give rise to a bright uniform glow over the poles that is often accompanied by a complete radio blackout in the polar region. The third unusual auroral form is the midlatitude red arc (M-arc) which is a subvisual, broad arc elongated in the geomagnetic east–west direction and located generally between geomagnetic latitudes 41° and 60°. M-arcs are approximately 600 km wide in north–south extent and are found at altitudes above 300 km. Their east–west extent is for thousands of kilometers, possibly circling the entire globe. The light emitted from these arcs is essentially monochromatic consisting of two atomic lines with wavelengths of 630.0 and 636.6 nm emitted by metastable oxygen atoms in the ^2D state. This unusual spectrum can be contrasted with the variety of features found in normal auroral radiation. These include many molecular bands emitted by N_2, N_2^+, O_2, and O_2^+ as well as more than 100 spectral lines radiated by atomic O and N and their ions.

Our knowledge of auroral morphology and spectroscopy has been enhanced significantly by recent sounding rocket and satellite experiments. The auroral spectrum has been measured quantitatively from the x-ray region (0.1 nm) to the deep infrared (100 μm), and excitation models for the principal emission features have been developed. Satellites with vacuum ultraviolet imagers have also discovered a new type of auroral form: the theta aurora. Theta arcs are aligned in north–south direction and bisect the auroral oval giving it the appearance of the Greek letter "θ" when viewed from afar.

Perhaps the most dramatic auroral phenomenon is the substorm or breakup event. This highly dynamic display develops from a quiet arc and is characterized by intense swirls, surges, and eddies of multicolored light that expand poleward from the original position of the arc and, in a matter of minutes, cover most of the sky. These substorms will frequently last 20 min and are accompanied by disturbances in the local magnetic field that indicate the presence of large currents flowing in the auroral ionosphere. The field-aligned electric currents, which flow into the auroral zone from the magnetosphere and are the source of the magnetic effects associated with aurora, are called Birkeland currents. When these generally vertical currents enter the auroral ionosphere, they change direction and begin flowing horizontally in an east–west sense at altitudes near 100 km until they ultimately find their way along other field lines back to the plasma sheet. The horizontal portion of the current, which is often as much as 1000 km long, is called the auroral electrojet, and it carries currents as high as 10^7 A on some occasions. There is some evidence that the auroral substorms are triggered by plasma instabilities in the electrojet that develop and grow dramatically when the current density in the electrojet exceeds a threshold value. A variety of plasma waves are also generated in auroras. These include Alfvén waves as well as electrostatic drift waves. The excitation of these waves provides a mechanism for heating electrons locally to very high temperatures. As these energetic electrons cool, they produce vibrationally excited molecules that can modify the ion and neutral chemistry of the aurora.

Large electric fields (~50 mV/m) are occasionally observed along the borders of auroral arcs. These fields produce kinetically energetic ions and contribute to the thermal economy of the upper atmosphere through Joule heating. Some auroras also generate substantial x-ray fluxes with energies of 20 keV or above. These x rays penetrate down into the stratosphere where they contribute to the formation of nitric oxide. This mechanism is one example of how energy deposited high in the atmosphere (>90 km) can be coupled into the stratosphere where it can affect the polar ozone budget and may influence long-term climate patterns.

See also ARCS AND SPARKS; ATMOSPHERIC PHYSICS: IONOSPHERE; MAGNETOSPHERE; SOLAR WIND.

BIBLIOGRAPHY

Joseph W. Chamberlain, *Physics of the Auroras and Airglow*. Academic Press, New York, 1961. A still useful comprehensive text.

A. Vallance Jones, *Auroras*. D. Reidel, Holland, 1974. A comprehensive contemporary text on the physics of auroras.

B. M. McCormac (ed.), *The Radiating Atmosphere*. D. Reidel, Holland, 1971. Emphasis on auroral morphology, particle precipitation, and the aurora as a visual manifestation of large-scale magnetospheric processes.

A. Omholt, *The Optical Aurora*. Springer-Verlag, New York, 1971. Text emphasizes the emission spectroscopy of aurora.

Balmer Formula

B. G. Segal†

When an electric discharge (a spark) is passed through a tube containing hydrogen gas at a few millimeters pressure, it dissociates some of the H_2 molecules into H atoms and the tube is observed to glow with a red light. If the radiation from the discharge tube is passed through a slit, dispersed by a prism, detected by means of a photographic plate, and then analyzed, it is found that there are only four wavelengths emitted in the visible region and many more in the near ultraviolet. These emitted wavelengths, called "lines," since they are images of the slit, appear at different positions on the photographic plate. The four lines in the visible region, arranged in order of decreasing intensity, are observed to have the wavelengths (λ) shown in Table I. These four wavelengths were measured by the Swedish physicist A. J. Ångstrom during the second half of the nineteenth century; in addition, several lines in the near ultraviolet were measured by Vogel and by Huggins.

Johann Jacob Balmer, a Swiss physicist, was intrigued by the four discrete lines in the visible region whose wavelengths had been very carefully measured by Ångstrom. He studied these four values and looked for a pattern, for some relation between them. In 1885 Balmer showed that the wavelengths of the first four lines of the emission spectrum of the H atom are related by the equation

$$\lambda = 3645.6\left(\frac{n^2}{n^2-4}\right)$$

where $n=3$, 4, 5, and 6, and the result is in angstroms. On the basis of this equation, Balmer predicted that there should be a fifth line, whose wavelength he obtained by setting $n=7$. He did not know that Vogel and Huggins had measured wavelengths in the near ultraviolet, but on being informed of their measurements, he showed that the wavelengths of the first nine lines of the H atom spectrum all agreed, within experimental error, with his formula, by setting n at all integral values between 3 and 11.

The original Balmer formula is now more usually given as an expression for the wave number, $1/\lambda$, and is

$$\frac{1}{\lambda} = \frac{10^8}{3645.6}\left(\frac{n^2-2^2}{n^2}\right) = \frac{4\times10^5}{3.6456}\left(\frac{1}{2^2}-\frac{1}{n^2}\right)\ \mathrm{cm}^{-1}$$

or

$$\frac{1}{\lambda} = \mathcal{R}\left(\frac{1}{2^2}-\frac{1}{n^2}\right)$$

where $\mathcal{R}$ is called the Rydberg constant after the Swedish spectroscopist, J. R. Rydberg. The Rydberg constant is one of the most accurately known physical constants and has the value $109\ 677.576 \pm 0.012\ \mathrm{cm}^{-1}$.

With the more sensitive emission spectrometers available since the time of Balmer's calculations, other spectral regions have been investigated and many more lines have been detected. All of the observed spectral lines for the H atom fit the relation

$$\frac{1}{\lambda} = \mathcal{R}\left(\frac{1}{n_L^2}-\frac{1}{n_H^2}\right)$$

where n_L and n_H are both integers, with n_H having a value higher than that of n_L. The set of lines for a given value of n_L constitutes a "series" named after its discoverer. Within each series there is a striking pattern: as the wavelength decreases, the spacing between adjacent lines also decreases and the intensity decreases as well. The principal series in the spectrum of atomic hydrogen are shown in Table II. Each series converges to a limit: as $n_H \to \infty$, $1/\lambda \to \mathcal{R}/n_L^2$.

Balmer's work showed that the emission spectrum of hydrogen has an underlying unifying simplicity, but the reason why the atomic spectrum of H (and of all other atoms) should consist of a discrete set of lines instead of being a continuum, as predicted by classical mechanics, remained a great puzzle until the pioneering theoretical work of Niels Bohr in 1913. By introducing the concept of quantization of the energies allowed for the hydrogen atom, Bohr was able to derive an

Table I

Name	Color	λ (nm)
H_α	Red	656.28
H_β	Blue-green	486.13
H_γ	Blue-violet	434.05
H_δ	Violet	410.17

Table II

Series	Spectral region	n_L	n_H
Lyman	Far ultraviolet	1	2,3,4,. . .
Balmer	Visible and near ultraviolet	2	3,4,5,. . .
Paschen	Infrared	3	4,5,6,. . .
Brackett	Infrared	4	5,6,7,. . .
Pfund	Infrared	5	6,7,8,. . .

† Deceased

expression for the Rydberg constant in terms of fundamental constants:

$$\mathcal{R}=2\pi^2\mu e^4/ch^3$$

where e is the charge on the electron; c the speed of light; h Planck's constant; μ the reduced mass of the H atom, $m_em_p/(m_e+m_p)$; m_e the mass of the electron; and m_p the mass of the proton. The calculated value of the Rydberg constant is 109,678 cm^{-1} and the agreement between the theoretical and experimental values was such a triumph for the Bohr theory that it hastened the acceptance of the revolutionary concepts of the quantum theory.

BIBLIOGRAPHY

J. J. Balmer, *Ann. Phys. Chem.* **25,** 80 (1885); translated in *The World of the Atom* (H. Boorse and L. Motz, eds.), Vol. I, p. 365. Basic Books, New York, 1966.

Gerhard Herzberg, *Atomic Spectra and Atomic Structure,* Chapter 1. Dover, New York, 1944.

Gerald W. King, *Spectroscopy and Molecular Structure,* Chapters 1 and 2. Holt, Rinehart & Winston, New York, 1964.

Baryons

David G. Hitlin

All known elementary (that is, subatomic) particles may be classified into four families: *gauge bosons, leptons, mesons,* and *baryons*. The gauge bosons are the carriers of the forces which govern the structure of matter; the leptons, mesons, and baryons, names derived from the Greek for light, medium, and heavy, comprise both stable and unstable varities, some of which are the constituents of atoms and nuclei, while others are produced in high-energy collisions at accelerators or in cosmic ray interactions. The mesons and baryons are further grouped together as *hadrons,* denoting that they interact through the nuclear or strong force. The original mass-derived nomenclature has long since been rendered obsolete, as leptons and mesons more massive than many baryons have been discovered; the basic distinction is now one of internal structure.

Leptons, which are fermions, having half-integral spin, are thought to be truly elementary; that is, they have no known substructure. Mesons, which are bosons, having integral spin, and baryons, which are fermions, have a substructure; they are composed of particular combinations of quarks. It is thus the quarks which are the truly elementary constituents. Mesons are made up of particular combinations of a quark and an antiquark; baryons are composed of three quarks (their antiparticles, antibaryons, are composed of three antiquarks). There are three known *generations* of quarks: (*up, down*), (*charm, strange*), and (*top, bottom*), a total of six *flavors* of quarks. The question of whether there are additional quark generations is under active investigation; a measurement which could settle this question is possible within a few years. All possible combinations of quarks of the same and different flavors can produce a baryon.

The lightest baryon is the proton, the nucleus of the hydrogen atom, which has a mass of 938 MeV. It has traditionally been thought to be stable, but grand unified theories (GUTs) of elementary particles, which seek to find a common origin of the known forces, predict that protons, in fact, decay. Such predictions set off a flurry of experimental activity, which to date has yielded no evidence for proton decay; the current lifetime is known to be greater than 3×10^{32} years, which rules out the simplest GUTs. The next heaviest baryon is the neutron, 1.293 MeV heavier than the proton, discovered by Chadwick in 1932. A free neutron decays via the weak interaction to a proton, an electron, and a neutrino with a lifetime of 896 s. The atomic properties and the bulk of the nuclear properties of matter can be accounted for in detail by the strong and electroweak interactions of protons, neutrons, electrons, and neutrinos. The similarities of the proton and neutron led to the concept of *isospin,* the proton and neutron being viewed as two different states of a single object, the *nucleon*. With the discovery of more massive baryons in cosmic rays by Rochester and Butler and later at accelerators, this concept was eventually generalized, leading to the current picture of the quark substructure of hadrons.

These more massive baryons, the Λ^0, $\Sigma^{\pm}$, Σ^0, Ξ^-, and Ξ^0, are identified by the characteristic "V" signatures they produce upon their decay. They all have spin $\frac{1}{2}$, as do the proton and neutron, and their predominant decays, $\Lambda^0\rightarrow p\pi^-$, $\Sigma^{\pm}\rightarrow n\pi^{\pm}$, $\Sigma^0\rightarrow\Lambda^0\gamma$, and $\Xi^{\;,0}\rightarrow\Lambda^0\pi^{\;,0}$, with subsequent Λ^0 decay, all result in production of the baryon ground state, a proton or a neutron, accompanied by π mesons, the lightest mesons. These heavier baryons, produced in high-energy strong interactions, decay with lifetimes of the order of 10^{-10} s, which are characteristic of weak processes, instead of lifetimes of $\sim10^{-23}$ s which are characteristic of strong interactions. The solution to this puzzle was provided by Gell-Mann and Nishijima, who postulated the existence of a new quantum number, *strangeness,* possessed by the heavier baryons but not by the proton and neutron. The strangeness quantum number is conserved in the strong interaction production process, but not in the weak decay process. The observed decay chains also provide for yet another quantum number, *baryon number,* which is conserved both in production and decay. It is the postulated violation of baryon number, a common feature of GUTs, which has engendered the intense interest in searches for proton decay, as the decay of the lowest known baryon state would manifestly be evidence for nonconservation of baryon number.

There is another class of baryons, which decay with lifetimes characteristic of strong interactions, called the *baryon resonances*. The first of these, the $\Delta(1232)$, was originally called the "3,3" resonance, signifying that it simultaneously possessed the novel properties of having both spin and isospin of $\frac{3}{2}$. There are thus four charge states of the $\Delta(1238)$: $++$, $+$, 0, and $-$. This first baryon resonance has been joined by a host of others. The lowest-mass baryon resonances are bound states of light quarks with nonzero orbital angular momentum. There are also the Y^*, baryon resonances containing strange quarks.

The modern picture of the structure of baryons is the quark model. All known baryons, whether they are stable(?), namely, the proton, or decay via the weak interactions (n, $\Lambda^0, \Sigma^\pm, \Xi^{-,0}$, . . .), the electromagnetic interaction (the Σ^0), or the strong interaction [$\Delta(1232)$, $N(1520)$, $\Lambda(1405)$] are composed of specific combinations of three quarks, arranged in *multiplets*. This regularity was first identified by Gell-Mann and Ne'eman, using techniques which involved the eight generators of the Lie group $SU(3)$, and was called "the eightfold way." This approach was very successful in organizing the then-known baryon states, allowing the accurate calculation of the masses and decay rates of baryon states. It received striking confirmation with the discovery of the Ω^- baryon by Samios and co-workers. This unusual particle, possessing three units of strangeness and existing in only a single (negative) charge state, had been predicted to exist as the tenth member of the spin-$\frac{3}{2}$ baryon multiplet.

The eightfold way predicted baryon multiplets with up to 27 members, while only smaller groupings, with 8 or 10 members, have been identified. The answer to this puzzle was implicit in the quark model, devised independently by Gell-Mann and Zweig. By postulating that the baryons are composed of three quarks with noninteger electric charge [initially the up (u), down (d), and strange (s) quark, but now extended to include the charmed (c), bottom (b) and the (undiscovered) top (t) quark], only those multiplets with 1, 8, and 10 members, which have actually been found experimentally, appear. The larger multiplets, which contain baryonic states with "exotic" combinations of quantum numbers, do not appear in the quark model, and none have been identified in the more than 25 years since the genesis of the idea. The u, c, and t quarks have charge $+\frac{2}{3}$ (in units of the electron charge, e), while the d, s, and b quarks have charge $-\frac{1}{3}$. Thus baryons composed of three quarks can have charges as large as $+2e$ or as small as $-e$, in accord with observation. Antibaryons, composed of three antiquarks, can have charge states ranging from $-2e$ to $+e$. The proton is composed of (uud) quarks, the neutron of (udd), the Λ^0 of (uds) quarks, and so forth. The Ω^-, whose prediction and discovery firmly established the role of $SU(3)$ symmetry in the structure of hadrons, is composed of (sss) quarks, and thus has strangeness -3 and charge $-e$.

There are several known examples of families of baryons which include the heavier quark species. The first of these to be discovered is the lightest of them, the $\Lambda_c{}^+$, which is composed of a (udc) quark combination and which decays via the weak interactions. Several other members of the lowest *charmed baryon* multiplet are also firmly established. No firm evidence has as yet appeared for baryons containing b quarks. The rich spectroscopy of baryons containing heavier quarks has just begun to be explored; much interesting experimental work remains.

While the success of the quark model in explaining the properties of the known mesons and baryons has firmly established the "reality" of quarks, no evidence for free quarks has yet been found. The signature would be striking; a particle with charge $\pm\frac{2}{3}$ or $\pm\frac{1}{3}$. Searches in cosmic rays, in seawater, and in rocks, both terrestrial and from the moon, have yielded no evidence for free quarks. An explanation is to be found in quantum chromodynamics (QCD), which successfully describes the strong interaction as a gauge theory of quarks and *gluons*. The detailed properties of the QCD interaction cause quarks to be bound tightly into hadrons, and do not allow a single quark to be removed.

The culmination of the quark model of baryons is found in its extension to $SU(6)$ symmetry, which describes baryon states in terms of products of an $SU(3)$ portion (the quark content) and an $SU(2)$ portion (the spin degrees of freedom). Those combinations of quark species and spin alignment which produce states which are totally antisymmetric under the interchange of quarks, that is, those combinations which are *fermions*, are identified with physical baryon states. The $SU(6)$ model incorporates the successes of the simple quark model, but in addition, allows the successful prediction of more detailed properties of baryons, such as their magnetic moments.

See also ELEMENTARY PARTICLES IN PHYSICS; HADRONS, HYPERONS; QUARKS; SU(3) AND HIGHER SYMMETRIES.

Beams, Atomic and Molecular

Howard A. Shugart

An atomic or molecular beam is a narrow, collision-free stream of electrically neutral atoms or molecules. In experimental practice a high-vacuum system provides the collision-free environment for the beam, while apertures shape its cross section. Since 1911 when it was demonstrated that a stream of atoms remained collimated in a sufficiently good vacuum, beam techniques have contributed to basic understanding in physics, chemistry, and engineering. This success in elucidating fundamental properties of nature results because each atom or molecule in the beam is essentially isolated from the others in the beam, from residual background gas, and from the containment apparatus. Such a "free" state is ideal for studying the properties of an individual particle, and its interactions with other particles, with electric and magnetic fields, or with a surface.

A beam apparatus usually consists of the following components: (1) vacuum enclosure, (2) beam source(s), (3) interaction region(s), and (4) beam detector(s). The *vacuum enclosure* provides the collisionless environment for the beam. In a gas the average distance an atom or molecule will travel between collisions (mean free path) is inversely proportional to the pressure. Whereas this distance is about 4×10^{-7} m in standard atmospheric air, it increases to about 300 m if the pressure is reduced to 10^{-6} mm Hg, a typical maximum operating pressure in beam experiments. Vacuum techniques currently achieve pressures in the range 10^{-8} to 10^{-10} mm Hg so that scattering is negligible in most applications. The *beam source* shown in Fig. 1 consists of an "oven" having a small orifice from which the gas of atoms or molecules emerge. The oven may be the end of a tube for gases or high-vapor-pressure liquids, a heated tantalum or ceramic crucible for metals requiring high-temperature evaporation, a molten spot on a refractory surface heated by electron bombardment, a microwave discharge for dissociating gaseous molecules or for exciting metastable states,

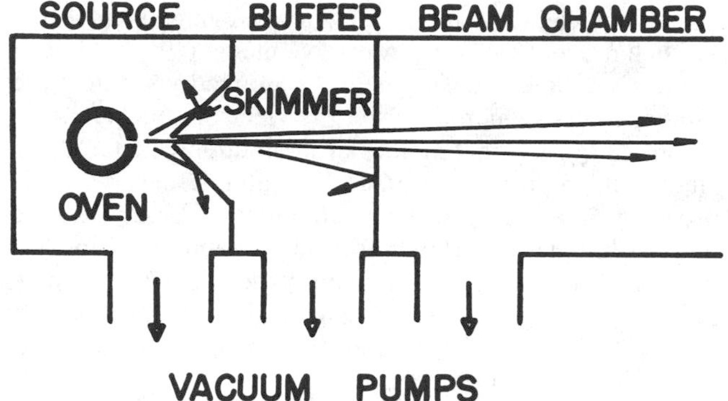

FIG. 1. Diagram of an atomic or molecular beam source. After emerging from a small orifice in the oven, atoms or molecules pass through regions of successively lower pressure until the beam is collision-free.

or one of the numerous other devices which evaporate and/or excite a substance. At low oven temperatures (mean free path≥exit slit dimensions) particles effuse from the oven with a modified Maxwell–Boltzmann velocity distribution

$$N(v)dv \approx (v^3/\alpha^4)\exp(-v^2/\alpha^2)dv,$$

where $N(v)dv$ is the number of atoms with velocity between v and $v+dv$, dv is a small increment in velocity, $\alpha=(2kT/m)^{1/2}$ is the most probable velocity, k is Boltzmann's constant, m is the molecular mass of the particles, and T is the absolute temperature. Depending on the values of m and T, the most probable velocity can range from 100 to 10,000 m/s. At much higher pressure in the oven the mean free path decreases until collisions between atoms become important and gaseous expansion occurs beyond the nozzle. Under these supersonic jet conditions the beam is much more intense, travels faster, and has a smaller velocity spread than the Maxwell–Boltzmann beam from a low-pressure effusion oven at the same temperature. Because of the larger gas flow from a jet, a well-designed skimmer and buffer stages of differential pumping are needed to isolate the high source-chamber pressure from the low beam-chamber pressure.

Detection of weak beams requires methods which are simultaneously sensitive, efficient, and selective of particular atoms or molecules. (a) The earliest detectors condensed the beam on a cooled surface. The optical density and pattern of the deposit gave both intensity and spatial distribution information. Deposits have also been assayed by neutron activation or by direct counting of nuclear decay when the beam contained a radioactive isotope. (b) For beams consisting of alkali atoms (sodium, potassium rubidium, and cesium) or others with low ionization potential, the detector consists of a hot filament of high work function (tungsten) that produces ions which are counted or measured as a current. (c) A universal detector uses ionization of a beam by electron bombardment with subsequent mass analysis for discrimination against background gas ions. Although such systems have an efficiency of only 10^{-4} to 10^{-5}, they excel in fast response, and are capable of selecting a particular mass species in the beam. (d) Single beam atoms or molecules are selectively detected by fluorescence using monochromatic laser light. This detection method resolves not only different types of atoms or molecules but also the particular quantum state of the particle. (e) Other beam detectors include radiometers, pressure manometers (Pirani gauge), thermopiles, bolometers, and space-charge-limited diodes.

Figure 2 shows the basic elements of a beam apparatus for magnetic resonance detection in atoms, molecules, or clusters of atoms. After preparation in the source chamber, the beam passes through a polarization region A, a resonance region C, and an analysis region B before encountering a detector D. The polarization and analysis regions contain an inhomogeneous magnetic field ($\partial H/\partial z \neq 0$) which deflects the particles by virtue of the force on their effective magnetic moment $F_z=\mu_{\text{eff}}(\partial H/\partial z)$, where $\mu_{\text{eff}}=-\partial W/\partial H$ is the negative slope of energy W versus magnetic field H. This force produces a deflection which spatially separates atoms with different magnetic moments. In the C region an oscillating or rotating radio-frequency magnetic field of frequency ν induces transitions between levels of energy E_2 and E_1 when the resonant condition $h\nu=E_2-E_1$ is satisfied (h is Planck's constant). In the B region spatial deflection once again discriminates levels and the detector senses any change of magnetic moment and hence of path induced by the application of the radio-frequency field.

For example, suppose in a magnetic field H an atom has two energy levels, $W=-\mu H$ and $+\mu H$, where the effective magnetic moments are $+\mu$ and $-\mu$, respectively. If the gradients, $\partial H/\partial z$, in the A and B regions are parallel, a typical trajectory of the system remaining in the state $\mu_{\text{eff}}=-\mu$ is shown by (1) in Fig. 2. If a transition to the other state with $\mu_{\text{eff}}=+\mu$ is induced by radio-frequency field ($\nu=2\mu H/h$, amplitude, polarization, and transit time must be correctly related), the atom will now follow path (2) and reach the detector. Thus an increase in beam flux at the detector is an indication that a change of state occurred in the C region. By suitable arrangement of beam stops and of directions and amplitudes of H_A, H_B, H_C, $\partial H_A/\partial z$, and $\partial H_B/\partial z$, a variety of experimental arrangements permit flexible measurements of various transitions in atoms and molecules.

Enormous versatility results when the A and/or B deflecting magnets are replaced by intense light from tunable single-mode lasers. This monochromatic light may deplete or fill

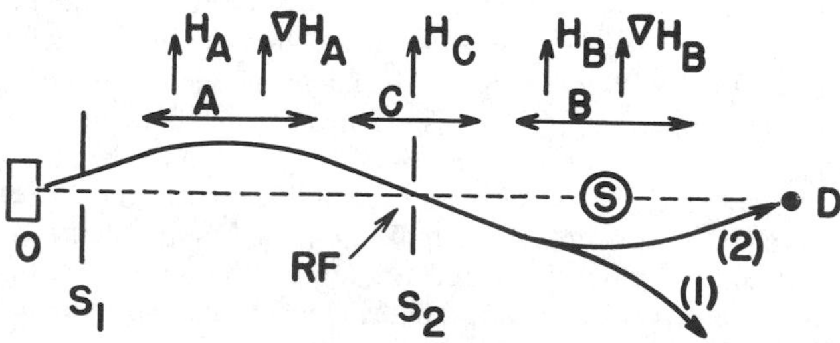

FIG. 2. Schematic of an atomic-beam, magnetic resonance apparatus. An atom leaving the oven O and having an effective magnetic moment $-\mu$ follows path (1) through the inhomogeneous magnetic deflecting fields in regions A and B. If a transition to a state with $+\mu$ is induced by a radio-frequency field (rf) in region C, the atom follows path (2) and reaches the detector D. Stop S intercepts fast atoms which experience small deflections, and apertures S_1 and S_2 collimate the beam.

(pump) a single energy level in the beam, act as a shutter for time-of-flight or beam modulation, detect a single state through fluorescence, and produce short-lived excited electronic states for study.

The history of atomic beams reveals a succession of important accomplishments. By 1911 vacuum pumps and technology had improved to the point that Dunoyer was able to produce the first collision-free collimated beam of sodium atoms. Beginning in 1919 Otto Stern and collaborators commenced a series of experiments which demonstrated the power, simplicity, and directness of the "molecular-ray method." Using macroscopic apparatus to study isolated neutral atoms or molecules, they experimentally confirmed several fundamental assumptions of contemporary theory. These early experiments (1920), for example, established the Maxwell–Boltzmann velocity distribution predicted by kinetic theory. Using deflection of the silver atom by the force on its magnetic moment in an inhomogeneous magnetic field, Stern and Gerlach (1924) demonstrated the validity of the space quantization of angular momentum and established the electron spin as 1/2. This remarkable experiment supported a basic hypothesis of quantum mechanics and led to studies of electromagnetic properties of other atoms, molecules, and nuclei. In 1929 Knauer and Stern observed the de Broglie wave character of neutral matter [λ(wavelength) = h(Planck's constant)/p(momentum)] by diffracting a helium beam from a cleaved NaCl crystal surface. Frisch and Stern (1933) and Estermann and Stern (1933) measured the proton magnetic moment to be about 2.5 nuclear magnetons rather than the 1 nuclear magneton expected if it were a simple Dirac particle. In the same year Frisch and Stern observed the radiation-reaction deflection of a beam atom when a photon of light is emitted or absorbed.

High-precision radio-frequency spectroscopy began in 1937 when I. I. Rabi combined the magnetic resonance proposal (Gorter, 1936) with the state selection and analysis capability of an atomic beam apparatus. This development permitted very accurate measurement of differences between energy levels of a system. The frequency ν at which transitions occur between states of energy E_2 and E_1 is given by the resonance condition $h\nu = E_2 - E_1$, mentioned previously. For transitions between states of long lifetime, the frequency width, $\Delta\nu$, of the resonance is determined by the time, Δt, that the system interacts with the radio-frequency field, according to the Heisenberg uncertainty principle $\Delta\nu\Delta t \gtrsim (4\pi)^{-1}$. Subjecting the beam to a longer interaction region produces narrower resonances for which the peak can be more accurately determined. Using radio-frequency methods Kellogg, Rabi, Ramsey, and Zacharias (1939) established a nonzero quadrupole moment for the deuteron, a discovery which indicated the tensor character of strong forces between nuclear particles. Kusch and Foley (1949) found the anomalous electron magnetic moment. Lamb and Retherford (1947) observed the finite Lamb shift in the atomic hydrogen fine structure. Subsequent refinements in quantum electrodynamic theory have explained both these effects to the accuracy of the experimental measurements. During the 1950s and 1960s a large effort went into the atomic beam study of nuclear spins (I) and of magnetic-dipole (μ) and electric-quadrupole (Q) moments of both stable and radioactive nuclei. This information proved valuable in testing the shell model (1949) and collective model (1953) of nuclei. At the same time, measurements of atomic hyperfine structure constants, which describe the electron–nucleus interaction, and of internal molecular interaction constants contributed to improving the theory of atomic and molecular structure. The most widespread practical application of beam techniques has been in frequency standards employing the cesium "clock" (1952). Presently the hyperfine structure frequency of ^{133}Cs in zero magnetic field is defined to be exactly 9192.631770 MHz, and its resonant detection in an atomic beam provides the current operational standard of time and frequency. Although frequency comparisons are now commonly made to 13 decimal places, it is anticipated that measurements to a few parts in 10^{15} will be soon possible. The ammonia maser (Gordon, Zeiger, and Townes, 1954) and the hydrogen maser (Kleppner, Crampton, and Ramsey, 1960) employ beam techniques for quantum state selection and function as secondary frequency standards.

During the 1970s beam techniques in conjunction with lasers produced significant spectroscopic and chemical reaction data. The decade witnessed developments in isotope separation, in frequency standards, in sources of spin-polarized particles for nuclear accelerators, and in neutral beam injectors for fueling fusion plasmas.

The slowing or cooling of atomic beams with lasers was accomplished in the decade of the 1980s and led to trapping of single neutral atoms in a small volume of space. Velocities corresponding to equivalent temperatures in the few microkelvin range have been obtained. In earlier experiments charged particles in electromagnetic traps were cooled with laser light. Laser cooling occurs when the interaction light is tuned to the "red" or low-frequency side of a resonant absorption transition of the beam atoms. Atoms heading into the laser beam encounter Doppler-shifted radiation nearer the peak of the absorption curve and have a higher probability of interacting than those atoms traveling in the same direction as the laser light, since the latter encounter light Doppler shifted away from the absorption peak. Each time an atom absorbs a photon head-on, its forward momentum is decreased slightly. The subsequent reemission of a photon takes place isotropically (averaged over many absorption–emission cycles) and does not change the average momentum of the atom. After many thousand such absorption–emission cycles the atom's velocity can be reduced to a few centimeters per second. The resulting slow particles have a very small second-order Doppler shift and are amenable to the highest precision spectroscopy and to implementing atomic frequency standards.

In recognition of important advances utilizing atomic or molecular beam methods, Nobel Prizes have been awarded to several scientists: Otto Stern (1943) "for his contribution to the development of the molecular-ray method and for his discovery of the magnetic moment of the proton"; Isidor I. Rabi (1944) "for his application of the resonance method to the measurement of the magnetic properties of atomic nuclei"; Polykarp Kusch (1955) "for his precision determination of the magnetic moment of the electron"; Willis E. Lamb, Jr., (1955) "for his discoveries concerning the fine structure of the hydrogen spectrum"; Charles H. Townes,

Nikolai G. Basov, and Alexander M. Prochorov (1964) "for fundamental work in the field of quantum electronics, which has led to the construction of oscillators and amplifiers based on the maser–laser principle"; and Dudley R. Herschbach, Yuan T. Lee, and John C. Polanyi (1986) "for their contributions concerning the dynamics of chemical elementary processes."

See also CLOCKS, ATOMIC AND MOLECULAR; FINE AND HYPERFINE SPECTRA AND INTERACTIONS; KINETIC THEORY; KINETICS, CHEMICAL; MAGNETIC MOMENTS; MASERS; MICROWAVE SPECTROSCOPY; OPTICAL PUMPING; VACUUMS AND VACUUM TECHNOLOGY; ZEEMAN AND STARK EFFECTS.

BIBLIOGRAPHY

W. J. Childs, "Hyperfine and Zeeman Studies of Metastable Atomic States by Atomic-Beam, Magnetic-Resonance," in *Case Studies in Atomic Physics*, Vol. 3, No. 4. North-Holland, Amsterdam, 1973.

T. C. English and J. C. Zorn, "Molecular Beam Spectroscopy," in *Methods of Experimental Physics*, 2nd ed., Vol. 3. Academic Press, New York, 1973.

M. A. D. Fluendy and K. P. Lawley, *Chemical Applications of Molecular Beam Scattering*. Chapman and Hall, London, 1973.

P. Kusch and V. W. Hughes, in *Handbuch der Physik*, Vol. 37/1. Springer-Verlag, Berlin, 1959.

W. A. Nierenberg, "Nuclear Spins and Static Moments of Radioactive Isotopes," *Ann. Rev. Nucl. Sci.* **7**, 349 (1957).

Laser-cooled and trapped atoms/Workshop on Spectroscopic Applications of Slow Atomic Beams, National Bureau of Standards, Gaithersburg, MD; W. D. Phillips, ed. Oxford; New York, Pergamon Press, 1984. Series title: *Progress in Quantum Electronics* V. 8, No. 3/4.

N. F. Ramsey, *Molecular Beams*. Oxford University Press, London, 1963 (reprinted).

K. F. Smith, *Molecular Beams*. Methuen, London, 1955.

Beta Decay

D. E. Alburger

In a chart of the elements the naturally occurring atoms lie along the so-called valley of nuclear stability. If a nucleus, either existing naturally or produced artificially, has too many or too few neutrons, it will have an excess energy with respect to neighboring nuclei in the valley. Such a nucleus is said to be radioactive and it rids itself of the excess energy by undergoing a spontaneous transformation in which either alpha particles or beta rays are generally emitted. Beta decay is the process in which the radioactive nucleus changes into a neighboring element differing in atomic number, or nuclear electrical charge, by +1 or −1, but with no change in atomic weight. (There is actually a small decrease in the mass because of the energy released.) The relationship is referred to as a mother-to-daughter decay.

Before their nature was understood, beta rays were so called because their power of penetrating through matter was intermediate as compared with alpha rays (now called alpha particles and known to be helium nuclei), which penetrated through very little matter, and gamma rays (now known to be electromagnetic radiation or high-energy photons), which penetrated through large thicknesses of matter. It was shown that beta rays were the most strongly deflected by a magnetic field and that the sign of their electric charge was negative. From their measured charge and their ratio of charge to mass it was demonstrated that beta rays, designated by β^-, were not rays at all but were identical to ordinary electrons. In spite of this the term beta ray is still commonly used.

The simple picture might have been that beta decay occurs when an electron residing inside the nucleus comes out, with a resulting change of +1 in the charge on the residual nucleus. It is now known, however, that nuclei are composed only of protons and neutrons. Furthermore, it has been shown by calorimetric methods that only about 40% of the energy that is given off by a radioactive beta-decaying sample appears as beta-ray energy, and it has been further demonstrated that an individual beta ray can emerge with any energy from nearly zero up to the maximum possible allowed by the mother–daughter mass difference. This continuous intensity distribution has a maximum at somewhat less than half the maximum energy. Another feature is that the beta ray has a spin of $\frac{1}{2}$ unit, whereas the nuclear states connected by beta decay always differ from each other by either zero or an integral number of units of angular momentum.

The puzzle as to where the major portion of the beta-decay energy was going and why angular momentum was apparently not being conserved was solved by Pauli, who postulated that another particle is emitted simultaneously but is undetected in the beta-decay process. Not knowing its properties, he called this particle the neutrino, or little neutron (as discussed in the following, it is actually an antineutrino $\bar{\nu}$ that accompanies β^- emission). The neutrino was presumed to have a spin of $\frac{1}{2}$ unit so as to conserve the total angular momentum of the system and to take away, on the average, the missing 60% of the decay energy. After the neutrino had been shown experimentally to have no measurable charge or mass and an extremely high penetrating power, it was eventually detected by means of a large and complex apparatus.

Beta decay, as described theoretically by Fermi and as now understood, occurs when one of the neutrons in a radioactive nucleus transforms into a proton with the creation and simultaneous emission of a negative beta ray (moving with a speed approaching the velocity of light) and an antineutrino (moving at the velocity of light). In analogy to the atomic case, where a photon or light wave does not exist before an excited atomic electron makes an orbital transition, thereby creating the photon, the beta ray and antineutrino do not exist prior to the beta-decay event. Thus the beta ray and antineutrino emerge with their spins oppositely directed and the sum of their energies is exactly equal to the difference in the mother–daughter energies. They can share the energy in all possible divisions, resulting in the bell-shaped distribution of intensity versus energy for the beta rays.

The discovery of the positron (or positively charged beta ray) in 1932 confirmed Dirac's theory, according to which the electron can exist in an antimatter form. The first observation of a positron, or β^+, decay came about through the formation of the isotope ^{13}N, a positron activity produced in a nuclear reaction in which elemental boron was exposed to a radioactive source emitting alpha particles. It was not until particle accelerators were invented that the means became available to easily produce neutron-deficient radioactive nuclei on the other side of the valley of stability. These nuclei are the positron emitters that, on decaying, change the atomic number of the nucleus by -1. The theory requires that positrons be accompanied by the simultaneous emission of a neutrino and that all the same requirements for conservation of energy and angular momentum as in β^- decay be fulfilled.

An alternative process to positron–neutrino emission, one that also changes the atomic number by -1, is the capture of an orbital atomic electron by the nucleus (electron capture, EC) in the course of the internal transformation of a proton into a neutron. In this case the orbital electron disappears but a neutrino is still emitted. We now include in the term beta decay all three of the processes described thus far: $\beta^- - \bar{\nu}$, $\beta^+ - \nu$, and EC$-\nu$.

When a β^- particle is emitted it is found experimentally that its spin vector points backward with respect to the direction of motion, a situation referred to as negative helicity; β^+ particles, on the other hand, emerge with positive helicity. This has led to the conclusion that in the weak beta-decay interaction occurring in the nucleus, parity (see below) is not conserved, since otherwise we would expect β^- and β^+ emissions each to result in particles of both positive and negative helicity.

As with all types of radioactivity, the most striking feature of beta decay is the exponential rate of decay, characterized by a half-life that can vary all the way from 0.009 sec for ^{13}O to 6×10^{14} yr for ^{115}In. The free neutron, which is the simplest beta-ray emitter of all, decays into a proton and has a half-life of 12 min. For a given radioactivity the decay constant λ, a number inversely proportional to the half-life $t_{1/2}$, is a precisely fixed quantity that, except in a few very unusual cases of electron-capture decay, cannot be altered at all by laboratory measuring conditions. The value for the half-life is fixed mainly by the energy available for the decay (the higher the energy, the shorter the half-life) and by the relationship between the properties of the mother and daughter states. In Fermi's theory and in later theoretical work by Gamow and Teller, selection rules were developed according to which a beta decay can take place most easily if the spins of the mother and daughter nuclei do not differ by more than one unit, and if the parities are the same. Parity is a symmetry property of a nucleus and is always described as either even (+) or odd (−). Beta-ray transitions of this type are called *allowed*. If the spin difference is greater than one unit and/or the parity changes, the transition is called *forbidden*, which is really a misnomer and should be taken to mean inhibited. There are various degrees of forbiddenness, depending on the amount of spin change: first-forbidden, second-forbidden, etc., and for each higher degree of forbiddenness, other things being equal, the half-life is longer by a factor of ~1000.

Beta decay can take place not only to the lowest-energy or ground state of the daughter nucleus but to excited states as well. Each such decay is characterized by a branching ratio and a consequent partial half-life, or the half-life that the mother nucleus would have if only that particular decay mode were to take place. Evidently the observed half-life results from the composite effect of all possible partial half-lives, or in other words, the overall decay constant λ is the sum of the decay constants of the individual branches, $\lambda_1 + \lambda_2 + \cdots$.

All beta decays, whether they are branches in the decay of a given radioactive nucleus or from radioisotopes in different regions of the table of elements, can be reduced to a common denominator by the use of a function f, which is found in tables and effectively removes the energy and atomic-number dependence of the decay. The product $ft_{1/2}$ is called the comparative half-life and is expressed in seconds. Since this number can vary from ~10^3 to ~10^{14} sec for most radioisotopes, it is more convenient to discuss the decay in terms of the $\log_{10} ft_{1/2}$.

The study of beta decay is made partly to learn about the properties of nuclear states. Thus it is known that the log ft value depends on the degree of forbiddenness, which in turn places restrictions on the spin and parity changes in the decay. The shape of the beta-ray intensity versus energy distribution can also be measurably different in forbidden decay, and the departure from the "allowed" shape can sometimes be predicted theoretically. Measurements of the shapes of beta-ray spectra can best be carried out with several types of magnetic spectrometer. In order to make the shape analysis easier to see, the bell-shaped intensity distribution can be converted, by use of the Fermi function F, into a plot, the so-called Fermi–Kurie plot, which for allowed beta-ray spectra is a linearly decreasing function of energy. Not only can departures from the allowed shape be seen more clearly in such a plot, but the extrapolation of the plot to the point of zero intensity is an accurate measure of the maximum energy of the beta rays. The total decay energy obtained from the end point of a ground-state decay, or for a beta decay to an excited state with the subsequent gamma-ray energy added, is essentially a measure of the mother–daughter mass difference.

Although the general features of beta decay are fairly well understood, there are many important details that are continuing to occupy the attention of both theorists and experimentalists. For example, attempts to predict accurately the log ft values for beta decay are being made on the basis of theories such as the shell model of the nucleus, but these calculations are thus far only partially successful. Intensive efforts are still being made to understand the exact nature of the weak interaction responsible for beta decay, and experiments of increasingly greater precision are being carried out to aid in these studies.

See also NEUTRINO; PARITY; RADIOACTIVITY; WEAK INTERACTIONS.

BIBLIOGRAPHY

E. J. Konopinski and M. E. Rose, in *Alpha-, Beta- and Gamma-Ray Spectroscopy* (Kai Siegbahn, ed.), Vol. 2, Chap. 23. North-Holland, Amsterdam, 1965. (A)

O. Kofoed-Hansen and C. J. Christensen, in *Handbuch der Physik* (S. Flügge, ed.), Vol. 41/2. Springer-Verlag, Berlin and New York, 1962. (I)
J. M. Soper, in *Encyclopedic Dictionary of Physics*, Vol. 1, p. 394. Pergamon, New York, 1961. (E)
R. D. Evans, in *McGraw-Hill Encyclopedia of Science and Technology*, Vol. 11, p. 286. McGraw-Hill, New York, 1971. (E)
Encyclopaedia Britannica, Vol. 1, p. 1028. Encyclopaedia Britannica, Chicago, 1973. (E)
Van Nostrand's Scientific Encyclopedia, 4th ed., p. 198. Van Nostrand, Princeton, 1968. (E)

Betatron

Donald W. Kerst

The name betatron is applied to a device for accelerating charged particles by continuously increasing the magnetic flux within the particle's orbit. Since the method seemed especially suitable for the acceleration of electrons, the word beta was incorporated in the name of the accelerator. The electron starts executing an orbit in the magnetic field while a time-varying flux within the orbit provides a voltage gain (see FARADAY'S LAW) for each revolution which the electron makes in its orbit. The action is the same as that in an electrical transformer for which the secondary voltage is proportional to the number of turns of wire.

To enable the electron to circulate around the enclosed flux approximately a million times a typical electron must closely follow a path in a well-evacuated vessel. For betatrons operating with a time-varying magnetic field of 180 Hz a vacuum better than 10^{-5} Torr is required, and a scheme of stably confining the particle to its lengthy orbit within the vacuum vessel is provided by shaped magnetic fields which produce so-called betatron or orbital oscillations.

In betatrons used for practical applications there are typically 100 eV of energy gained per turn of the electron around its orbit and in terms of the magnetic bending field at the orbit the particle's energy is $300Br$ MeV, where B is the magnetic field in teslas and r is the orbit radius in meters.

There were numerous attempts to achieve induction acceleration of electrons prior to the first successful achievement of acceleration at the University of Illinois. These attempts have been described in some detail.* The necessary theory for injecting particles into a time-varying magnetic field indicated that if electrons were injected from an internal gun with a speed sufficient to avoid serious scattering on background gas molecules, and if the magnetic field varied rapidly enough in time, the electron orbit executed a transient oscillation of decreasing amplitude about an equilibrium orbit. The decreasing amplitude allowed the electron to avoid striking the injection structure thereby allowing the beam to be captured† for its long journey.

* D. W. Kerst, *Nature* **157**, 90 (1946).
† D. W. Kerst and R. Serber, *Phys. Rev.* **60**, 54 (1941); D. W. Kerst, *Phys. Rev.* **60**, 47 (1941).

The axial and vertical focusing forces which cause the electrons to oscillate around the equilibrium orbit are produced by shaping the guide field so that it decreases less rapidly in magnitude than $1/r$. Since the centripetal force required to hold the particle in a circular orbit varies as $1/r$, the magnetic force from the shaped field will be too large to hold the electron at a radius larger than the equilibrium radius and too small to hold it in a circle smaller than the equilibrium radius. The particles will then oscillate across the equilibrium orbit with an ever decreasing amplitude as the magnetic restoring forces continue to rise.

For axial focusing, or oscillation about the equilibrium orbit, the decrease of magnetic field with increasing radius provides a curvature of the lines of force or a bulging of the field which supplies a little focusing force directed back to the plane of the orbit. This same slightly bulging field was also used in the cyclotron to provide axial focusing of particles.

A further condition on the magnetic field is necessary if the orbit is to be prevented from continually spiralling into smaller radius as the guide field rises. The particles can be held at a large fixed radius so they enclose a maximum flux change by supplying extra flux to the center of the orbit. This can be seen from the equations for the equilibrium orbit position. We have the momentum $p = eBr$, while the electrical field is $d\phi/dt$ divided by $2\pi r$, provided that the particle is following in the orbit of radius r. The rate of gain in momentum is then

$$\frac{dp}{dt} = \frac{e}{2\pi r}\frac{d\phi}{dt}$$

Integrating this we find that the final momentum is

$$p = \frac{e}{2\pi r}(\phi_2 - \phi_1).$$

This gives the condition on the change in flux

$$\phi_2 - \phi_1 = 2\pi r^2 B,$$

and thus the change in flux linkage during acceleration should be twice that which would occur if the magnetic field were uniform within the orbit. The extra flux thus provides the momentum to keep the orbit at large radius as the bending field increases. It is evident that this central flux could be biased in a reverse direction at the outset so that the iron transformer core providing this flux change could be operated between a negative value close to saturation and a positive value again close to saturation. This enables the iron core to be efficiently used.

In addition, the application of alternating-gradient focusing, in place of the simple shaped field previously described, can allow dc guide magnets to contain both the initial low-energy orbit and the high-energy orbit without a great change in orbital radius. Then the rising central accelerating flux can bring continuously injected electrons from the low-energy orbit to the high-energy orbit in a stream lasting almost as long as the rising flux. This so-called fixed-field alternating-gradient betatron would thus not have the single accelerated pulse for every magnetic field cycle but rather it would provide a high-powered stream of electrons for perhaps 25% of each cycle.

Typical medical betatrons between 25 and 35 MeV generate x rays having an intensity of about 100 R/min at 1 m. They can have the accelerated electron beam extracted for therapeutic purposes giving similar dosage rates for similar areas of treatment with about 2 MeV of their energy being absorbed per centimeter of penetration into tissue. X rays from a 25-MeV electron beam have the property of being the most penetrating in iron for industrial radiography. Photons of higher energy are more strongly absorbed because of increasing electron pair production, while photons of less energy are more strongly absorbed because of Compton scattering.

For nuclear physics research the photons and the electrons have been useful, since with energies of several million electron volts they can photodisintegrate the nucleus. Radiation from the largest betatron for 320 MeV can produce mesons.

The time-average currents of pulses from a medical betatron is very small and approaches only a microampere. However, the circulating electron beam current in the chamber is about 3 A. Numerous attempts to raise this circulating current have included the use of plasma effects to neutralize the space-charge limitation due to beam charge; and subsequently the addition of a toroidal magnetic field parallel to the orbit in the so-called modified betatron. This greatly increases the space charge limit into the 1000-A region if high-energy injection is used and if the limitations from instabilities can be surmounted. Alternating gradient focussing in addition to the longitudinal field has been successful in achieving 1000 A circulating at 20 MeV with injection filling assisted by plasma effects. The control of instabilities is a major problem with the intense currents.

BIBLIOGRAPHY

A. Arnold, P. Bailey, and J. S. Laughlin, *Neurology* **4,** 165 (1953). (I)

H. Gartner, *Strahlentherapie* **96,** 201 (1955); **96,** 378 (1955). (I)

D. Greene, *Br. J. Radiol.* **34,** 129 (1961). (I)

D. Harder, G. Harigel, and K. Schultze, *Strahlentherapie* **115,** 1 (1961). (I)

H. E. Johns, E. K. Darby, R. N. M. Harlam, L. Katz, and E. L. Harrington, *Am. J. Roentgenol. Rad. Ther.* **62,** 257 (1949). (I)

D. W. Kerst, *Rev. Sci. Instru.* **13,** 387 (1942). (I)

D. W. Kerst, *Handbuch der Physik,* Vol. XLIV. Springer-Verlag, Berlin, 1959.

C. A. Kapetanakos, P. S. Sprangle, D. P. Chernin, S. J. Marsh, and I. Haber, *Phys. Fluids* **26,** 1034 (1983).

Ishizuka, Leslie, Mandelbaum, Fisher, and Rostoker, *IEEE Trans. Nucl. Sci.* **NS-32,** 2727–2729 (1985).

D. Lamarque, *J. Radiol. Electrol.* **32,** 491 (1951). (I)

J. S. Laughlin, *Nucleonics* **8,** 5 (1951). (I)

J. S. Laughlin, J. W. Beattie, J. E. Lindsay, and R. A. Harvey, *Am. J. Roentgenol.* **65,** 787 (1951). (I)

J. S. Laughlin, J. Ovadia, J. W. Beattie, J. W. Henderson, W. J. Harvey, and L. L. Haas, *Radiology* **60,** 165 (1953). (I)

N. Maikoff and M. Sempert, *Acta Radiol. Suppl.* **313,** 95 (1972). (I)

B. Markus, *Strahlentherapie* **97,** 376 (1955). (I)

M. Sempert, *Radiology* **74,** 105 (1960). (I)

F. Wachsmann, *Nippon Acta Radiol.* **23,** 375 (1963). (I)

R. Wideroe, *Arch. Elektrotech.* **21,** 387 (1928). (I)

R. Wideroe, *Strahlentherapie* **35,** 266 (1956). (A)

Bethe–Salpeter Equation

Franz Gross

The Bethe–Salpeter equation [1] is a relativistic two-particle equation of a particular form. If $M(p,p';P)$ is the two-body scattering amplitude (or, more generally the four-point function) with all four external particles off-shell, where P is the total four-momenta of the system (assumed to be conserved), and p and p' are the relative four-momenta of the final and initial particle pair (see Fig. 1), then the Bethe–Salpeter (BS) equation for M is

$$M(p,p';P) = V(p,p';P) - \int \frac{d^4k}{(2\pi)^4} V(p,k;P)G(k,P)M(k,p';P), \quad (1)$$

where V is the kernel of the equation and G is the full two-particle propagator

$$G(k,P) = D(\tfrac{1}{2}P+k)D(\tfrac{1}{2}P-k) \quad (2)$$

with $D(k_1)$ being the complete one-particle propagator, including all self-energy contributions. The two essential features which define Eq. (1) as a BS equation are (i) the integration over all four components of the relative momentum of the internal particle pair and (ii) the linear form.

Consider a renormalizable quantum-field theory described by a local Lagrangian with a meaningful perturbation expansion. If V is chosen to be the infinite sum of all Feynman diagrams [2] for the four-point function that cannot be split apart into disconnected diagrams by cutting two lines, then the BS equation gives the exact result for the full four-point function. In this case the equation can be viewed as producing and summing all diagrams which have a two-particle cut (are two-particle reducible) by combining diagrams which have no such cut. However, because the infinite sum of two-particle irreducible graphs is probably as difficult to calculate as the amplitude M itself, and since the kernel V exists order by order in perturbation theory, the kernel is usually approximated by the first few terms of its perturbation expansion. In theories where boson exchange is believed to describe the important physics, such as photon exchange in atomic physics, gluon exchange in perturbative quantum chromodynamics (QCD), and meson (in particular pion) exchange in low-energy nuclear physics, V is often approximated by the lowest-order one-boson exchange diagram. In this case, the BS equation can be viewed as summing the ladder diagrams, as illustrated in Fig. 2. While its solution in this case does indeed correspond to the exact sum of this infinite series when the series converges, and is an analytic continuation of this sum for coupling constants

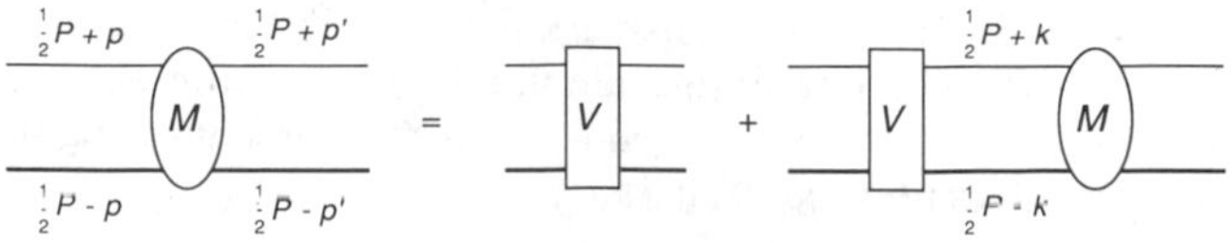

FIG. 1. Bethe–Salpeter equation in diagrammatic form, with momenta labeled. M is the four-point function and V is the kernel.

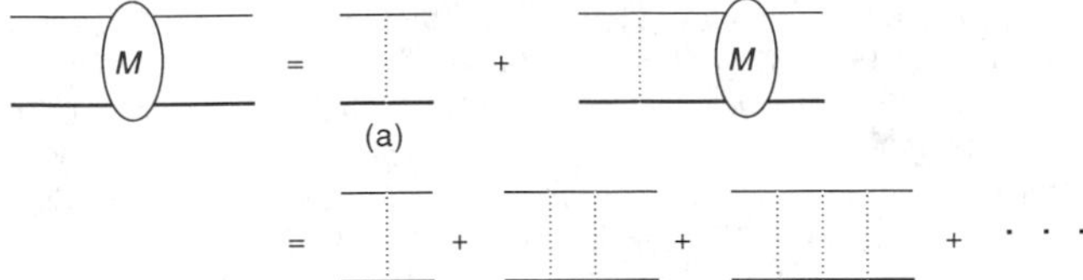

FIG. 2. The ladder approximation to the BS equation. The iteration of the one-boson exchange kernel (a) gives the infinite ladder sum.

too large for the series itself to exist, it represents, of course, only an infinitesimally small subclass of all the perturbation diagrams needed for the exact solution. Care must be taken in reading the literature, because some investigators give the impression that the ladder approximation to the BS equation is an exact solution to the problem. While this is not the case, the ladder approximation to the BS equation is nevertheless a very useful tool for the study of relativistic theories because it sums an infinite class of diagrams exactly.

In general, the BS equation is useful whenever it is essential to evaluate such a sum. The study of bound states, perhaps one of the most important questions in contemporary physics, is one such case. A bound state produces a pole in the scattering matrix in the $s=(p_1+p_2)^2$ channel. If the bound state is truly composite, no such pole exists in any Feynman diagram (or any finite sum); it can only be generated by an infinite sum. The same observations apply to the description of elastic scattering near threshold; an exact treatment of elastic unitarity requires an infinite number of diagrams.

In a broader context, the BS equation has played an important role in the development of other relativistic two-particle equations, and in the study of the reduction of such equations, in appropriate limits, to a one-particle equation, or to a nonrelativistic Schrödinger equation with an effective potential. To maintain relativistic invariance, it is not necessary to keep the full four-dimensional integration d^4k. A number of equations, sometimes referred to as quasipotential equations, have been introduced which reduce the internal integration to three dimensions by restricting the fourth component k_0 of the relative momenta four-vector in a covariant way. These equations all have an integration volume of the form

$$d^4k\delta(f(k_0,\bar{k})), \tag{3}$$

where the function f is a scalar and the δ function therefore eliminates k_0 in a covariant fashion. Well-known choices for f are (i) $f=k\cdot P$, which in the rest system fixes [3] $k_0=0$, and (ii) $f=m_2{}^2-(P/2-k)^2$, which restricts particle two to its mass shell [4]. While these equations are new equations with different properties, they are occasionally referred to in the literature as "Bethe–Salpeter" or "Bethe–Salpeter-like" equations, and in this sense the term "Bethe–Salpeter equation" is sometimes used to refer (incorrectly) to any relativistic quantum-mechanical equation.

The BS equation has been used in ladder approximation, justified in some cases by the validity of perturbation theory, for the study of a variety of systems. One technique which has proved useful in obtaining solutions is to rotate the k_0 integration contour from the real to the imaginary axis [5], which converts the relativistic four-dimensional space with indefinite metric into a space with a positive definite metric, permitting standard techniques to be used. Exact analytic solutions for the case of spinless particles interacting through the exchange of a massless scalar particle have been obtained [5,6] by exploiting the symmetry of such a system. The equation has been very useful in obtaining exact asymptotic results for bound systems in QCD [7], and in classifying and studying the small-distance and scaling behavior of current-correlation functions, as measured in deep inelastic electron scattering [8,9]. Also it has enjoyed some modest success in developing our understanding of Feynman graph models for the asymptotic behavior of scattering processes at fixed momentum transfer [10]. In addition, in the study of electromagnetically bound systems, the equation has been useful in the calculation of relativistic corrections to binding energies and magnetic moments [11,12]. The equation has been applied to the description of nucleon–nucleon scattering, where it has been solved numerically [13].

For example, Feynman diagram models of deep inelastic electron scattering can be studied by applying the equation to the calculation of the Compton scattering amplitude for virtual photons with spacelike momentum q scattering in the forward direction from a particle of mass m. The kinematical region is one where the total four-momenta is zero, $p_1{}^2-p_1'^2-p^2=m^2$, and $p_2{}^2=p_2'^2=q^2<0$ (spacelike). Recall that the scaling observed in electroproduction experiments shows, apart from kinematical factors, that the total cross section, which is related to the absorptive part of this forward Compton amplitude and is in general a function of the two scalar variables q^2 and $2p\cdot q$, becomes a function of the scaling variable $\omega=-q^2/2p\cdot q$ at large q^2. In most field theory models studied, the scaling law is found [8] to be violated by powers of q^2. In Yang–Mills theories, the scaling law is violated [9] by powers of $\ln q^2$. The operator product expansion can also be developed from the Bethe–Salpeter equation. This approach has led to an approximate description of the scaling violations.

In a somewhat different context, the Bethe–Salpeter equation and the closely related multiperipheral model were used to study Regge behavior with Feynman diagram models. Here the important region is fixed $t=(p_1+p_2)^2$ and large $s=(p_1-p_1')^2$ and fixed external masses. The BS equation provides a way to sum the infinite classes of contributions which must be summed in such an approach. In ϕ^3 theory, Regge pole behavior s^α, where α is the Regge trajectory, emerges in certain situations [10].

Unfortunately, the ladder approximation to the BS equation has some serious limitations. In situations where the physics could be expected to be dominated by the ladder diagrams of Fig. 2, it is likely that the crossed ladders (see Fig. 3) will also be important. In some theories (ϕ^3 for example) it can be shown [14] that when one of the particle masses (say m_2) becomes very large, the crossed ladders continue to give contributions of the same order in $1/m_2$ as the ladders, and are required in order to obtain the correct one-body equation for the lighter particle (m_1) in the limit

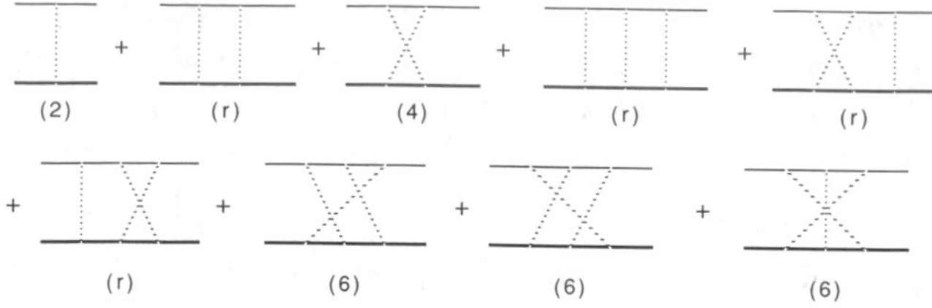

FIG. 3. Ladder and crossed ladder diagrams up to sixth order. Diagrams labeled (2), (4), and (6) are second-, fourth-, and sixth-order irreducible contributions to the kernel. Those labeled (r) are reducible, and are generated by iterating the kernel.

as m_2 approaches infinity (the one-body limit). A similar situation obtains for peripheral interactions at high energy; the eikonal limit cannot be obtained without consideration of crossed ladders [15]. Hence the BS equation will give good nonrelativistic and one-body limits only if its kernel includes irreducible crossed ladder diagrams to all orders. Ironically, the simpler quasipotential equations referred to above do not have this problem. The ladder approximation to these equations already gives the correct one-body and nonrelativistic limits, and this, together with the ease with which they can be solved and interpreted (arising from their three-dimensional nature), makes them better suited for many applications [16].

Recently, the BS equation has been used to study quark models of hadrons, and the dynamical breaking of chiral symmetry [17]. In these applications, a phenomenological kernel is used.

A complete review of this subject prior to 1969 has been given by Nakanishi [18].

See also FEYNMAN DIAGRAMS; QUANTUM FIELD THEORY; QUANTUM MECHANICS.

REFERENCES

1. Y. Nambu, *Prog. Theoret. Phys.* (*Kyoto*) **5,** 614 (1950); E. E. Salpeter and H. A. Bethe, *Phys. Rev.* **84,** 1232 (1951); M. Gell-Mann and F. E. Low, *Phys. Rev.* **84,** 350 (1951).
2. See the text by J. Bjorken and S. Drell, *Relativistic Quantum Mechanics,* Vol. 1, and *Relativistic Quantum Fields,* Vol. 2. McGraw-Hill, New York, 1965.
3. R. Blankenbecler and R. Sugar, *Phys. Rev.* **142,** 1051 (1966); I. T. Todorov, *Phys. Rev. D* **3,** 2331 (1971).
4. F. Gross, *Phys. Rev.* **186,** 1448 (1969).
5. G. C. Wick, *Phys. Rev.* **96,** 1124 (1954).
6. R. E. Cutkosky, *Phys. Rev.* **96,** 1135 (1954).
7. G. R. Farrar and D. R. Jackson, *Phys. Rev. Lett.* **43,** 246 (1979); S. J. Brodsky and G. P. Lepage, *Phys. Rev. D* **22,** 2157 (1980).
8. A. Mueller, *Phys. Rev. D* **9,** 963 (1974).
9. D. Gross and F. Wilczek, *Phys. Rev. Lett.* **30,** 1343 (1973); H. Politzer, *Phys. Rev. Lett.* **30,** 1346 (1973).
10. M. Baker and I. J. Muzinich, *Phys. Rev.* **132,** 2291 (1963); B. W. Lee and R. F. Sawyer, *Phys. Rev.* **127,** 2266 (1962); D. Amati, A. Stanghellini, and S. Fubini, *Nuovo Cimento* **26,** 896 (1962); D. Z. Freedman and J.-M. Wang, *Phys. Rev.* **153,** 1596 (1967).
11. T. Fulton and R. Karplus, *Phys. Rev.* **93,** 1109 (1954); T. Fulton and P. C. Martin, *Phys. Rev.* **95,** 811 (1954).
12. S. J. Brodsky and J. R. Primack, *Ann. Phys.* (*N. Y.*) **52,** 315 (1969).
13. J. Fleischer and J. A. Tjon, *Nucl. Phys.* **B 84,** 375 (1975); *Phys. Rev. D* **15,** 2537 (1977); **21,** 87 (1980).
14. S. Deser, *Phys. Rev.* **99,** 325 (1955); F. Gross, *Phys. Rev. C* **26,** 2203 (1982).
15. M. Levy and J. Sucher, *Phys. Rev.* **182,** 1852 (1969); H. D. I. Abarbanel and C. Itzykson, *Phys. Rev. Lett.* **23,** 53 (1969); S. J. Wallace and J. A. McNeil, *Phys. Rev. D* **16,** 3565 (1977).
16. See, for example, G. P. Lepage, *Phys. Rev. A* **16,** 863 (1977); G. T. Bodwin, D. R. Yennie, and M. A. Gregorio, *Phys. Rev. Lett.* **41,** 1088 (1978); G. T. Bodwin and D. R. Yennie, *Phys. Rep.* **43,** 267 (1978).
17. See, for example, A. LeYaouanc, L. Oliver, S. Ono, O. Pene, and J.-C. Raynal, *Phys. Rev. D* **31,** 137 (1985).
18. N. Nakanishi, *Prog. Theoret. Phys.* (*Kyoto*) **43,** 1 (1969).

Binding Energy

N. B. Gove

The energy required to separate an atom of atomic number Z and mass number A into Z atoms of hydrogen (^{1}H) and N neutrons (^1n), where $A = Z + N$, is called the binding energy, or BE, for that atom. The BE is thus also the energy equivalent of the mass difference between the hydrogen atoms plus neutrons and the atom:

$$\mathrm{BE}(A,Z) = Zm(^1\mathrm{H}) + Nm(^1n) - m(A,Z)$$

(nuclear and atomic ground states are assumed here). Since the hydrogen mass, $m(^1\mathrm{H})$, and the neutron mass, $m(^1n)$, are by now well known, the BE can be computed if the atomic mass $m(A,Z)$ is known. The atomic mass is known for about 1700 (A,Z)-species through combinations of experimental data, for example, mass spectrometer doublets, reaction Q values, radioactive decay energies. Table I shows a few binding energies (BE) and binding energy per nucleon (BE/A) based on a least-squares adjustment of experimental results.

The binding energy per nucleon shows some small peaks at magic numbers but generally rises to a maximum of 8.79 at $A \approx 60$ and drops gradually to 8.6 at $A \approx 100$ and 7.6 for the heavy nuclei. Thus, about 1 MeV/nucleon is released in fission of heavy nuclei.

Table I

Nucleus	BE (MeV)	BE/A
^1n	0	0
^{1}H	0	0
^{2}H	2.224574 ± 0.000006	1.112287
^{3}H	8.481855 ± 0.000013	2.827285
^{4}H	5.66 ± 0.38	1.4
^{4}He	28.295875 ± 0.000026	7.074
^{12}C	92.16239 ± 0.00014	7.680
^{14}C	105.28522 ± 0.00014	7.520
^{16}O	127.62022 ± 0.00015	7.976
^{40}Ca	342.0549 ± 0.0013	8.551
^{58}Fe	509.9508 ± 0.0016	8.792
^{62}Ni	545.2650 ± 0.0016	8.795
^{100}Ru	861.9338 ± 0.0024	8.619
^{200}Hg	1581.1215 ± 0.005	7.906
^{235}U	1783.881 ± 0.003	7.591

Two related terms are proton binding energy, $B(p)$, and neutron binding energy, $B(n)$. The former is the net energy required to remove a proton; the latter is the energy required to remove a neutron. (It is generally not possible to extract a proton with zero velocity. The term "net energy" means here the minimum of externally supplied energy that must be converted into internal energy in a proton extraction process.) Thus:

$$B(p) = m(A-1, Z-1) + m(^1\text{H}) - m(A,Z)$$

$$B(n) = m(A-1, Z) + m(^1n) - m(A,Z)$$

For deuterium, the binding energy, the proton binding energy, and the neutron binding energy are the same.

More generally, the binding energy of any bound system is the energy equivalent of the difference between the mass of the bound system and the sum of the masses of its free constituents. The expression above for BE(A,Z) treats hydrogen atoms and neutrons as the constituents. Other definitions of constituents are possible. The binding energy of a "bare" nucleus (i.e., fully ionized, no orbital electrons) would differ from the above because of electron energies:

$$\text{BE}_{\text{bare}}(A,Z) = Zm(p) + Nm(^1n) - m_{\text{bare}}(A,Z),$$

where $m(p)$ is the proton mass.

See also NUCLEAR PROPERTIES; NUCLEAR STRUCTURE; NUCLEON.

BIBLIOGRAPHY

D. Halliday, *Introductory Nuclear Physics*, 2nd ed., pp. 261–263. Wiley, New York, 1960.

A. H. Wapstra and N. B. Gove, "The 1971 Atomic Mass Evaluation," *Nuclear Data Tables* **9**, 267 (1971).

A. H. Wapstra and G. Audi, "The 1983 Atomic Mass Evaluation," *Nuclear Physics* **A431**, 1 (1985).

Biophysics

R. B. Setlow

Biophysics may be defined as the application of physical techniques and ideas to biological problems. Like any field which borders several disciplines, it is difficult to determine precisely where it begins and ends. On the one hand, it overlaps molecular biology and physical biochemistry, and, on the other hand, cell biology and physiology. It blossomed as a separate discipline in the 1950s with the rapid influx of physicists, and their techniques, into biology. The early biophysicists, the natural scientists of their day, also worked at the interface between biology and quantitative science. Such early biophysicists were, for example, Galvani, who started the field of electrophysiology; Mayer, who was first to enunciate the law of conservation of energy; and D'Arcy Thompson, who in his momentous book *Growth and Form* beautifully laid out the application of dimensional analysis to biological problems, such as why all animals, including fleas and horses, can jump to approximately the same height.

At the present time biophysics—what biophysicists do—can be divided into molecular, cellular, and organismal biophysics. We concentrate in this article on the first two. However, it is worth emphasizing that advances in instrumentation designed to elucidate molecular structures have had important applications in medical physics. General references will be found at the end.

The most important molecules in cells are DNA, RNA, and protein. They are large polymers of subunits called deoxyribonucleotides, ribonucleotides, and amino acids, respectively. The genetic information of cells is in their genes which are represented by sequences in DNA of four possible deoxyribonucleotide subunits containing adenine (A), guanine (G), thymine (T), or cytosine (C). The sequence of these four so-called bases is transcribed by the cellular machinery into RNA which, in turn, is translated with the aid of a number of complex cellular molecules, structures, and organelles to give the sequence of amino acids in proteins. Thus, although the major information of cells is carried by DNA, the day-to-day reactions are carried out by RNA and protein. If we are to understand these reactions and the others catalyzed by enzymes—proteins with specific catalytic activities—it is important that we be able to relate structure of the macromolecules to their function. All of them are large; for example, a typical protein molecular weight may be 6×10^4 (~600 amino acids), whereas DNA may range from 10^6 for a small virus to 10^9 or greater for bacteria. Most DNA molecules are found to be double-stranded helical structures (see below) in which the helices are held together by complementary base pairs, A pairing with T and G pairing with C. A DNA of mass 10^9 daltons would contain ~1.5 M base pairs. Since the two DNA strands are complementary to one another, specification of the sequence of one strand sets the sequence of the second. This redundancy in the information carried by DNA is essential if DNA damaged by environmental, physical, or chemical agents, or by endogenous chemical reactions is to be repaired efficiently so that the genetic information is relatively stable. The haploid DNA content of the human cell is ~3 B base pairs and if all this material contained in the 23 chromosomes were laid end to end, it would extend for ~1 m.

The physical techniques for handling and analyzing large molecules are different from those used to analyze atoms and nuclei. For example, DNA is a long very thin molecule (diameter ~ 2 nm) and is readily subject to shear degradation by hydrodynamic forces. Thus, it took many years to realize that DNA was such a large molecule and to develop techniques appropriate to handling and carrying out reproducible measurements on it. Large molecules may be considered to have an overall shape (the tertiary structure), the folding of the polymer subunits of which they are composed (the secondary structure), and the order of the subunits themselves (the primary structure). The classical way to analyze the tertiary structure is by measurements of sedimentation in centrifugal fields and by the rate of diffusion. However, by the conventional techniques the latter measurement is laborious and almost impossible for large molecules and because the diffusion constant is small and the experiment times are long, there is a very low signal-to-noise ratio. A solution to the problem is to use laser scattering and measure, in times of the order of minutes, the frequency shift of the laser light resulting from molecular Brownian motion. Here is a perfect example of the application of a new physical technique to solve a biological problem.

The local conformation of these long-chain polymers is also important. For example, DNA is a double helical molecule (see below), but its conformation could be linear, circular, or twisted circular. Such possibilities are not as esoteric as they may seem because the replication of a linear molecule often proceeds by way of a circular one. Moreover, a twisted molecule can be caused to open to a circular molecule by a single nick in one of the two chains. Such a change is easy to detect as an alteration in the frictional coefficient of the molecule, observed by sedimentation or electrophoresis, and hence such twisted circular molecules are often used to measure the deleterious physical effects of radiation or environmental chemicals on DNA. The local conformation of molecules may also be assessed by their ultraviolet absorption (molecules with stacked units tend to absorb less) or by circular dichroism, and the interaction between neighboring groups can be measured by nuclear magnetic resonance by observing the fine structure that results from such interactions. These estimates of local conformation are important because to understand how macromolecules work it is necessary to know both inter- and intramolecular interactions among subgroups. For example, it is desirable to discover how the structure of hemoglobin facilitates its interaction with oxygen, and how subtle changes in such structure affect this interaction, as in the hemoglobin of individuals with sickle-cell anemia. Such structural information with a high degree of resolution is necessary to understand the enormous specificity of the various interactions between large molecules and small (such as enzymes and substrates), and between large and large (such as antibodies and antigens).

X-ray diffraction techniques have been used to determine the detailed three-dimensional structure of a large number of proteins and to demonstrate that DNA is normally a double-stranded helical molecule. In the latter case, however, the x-ray diffraction data were not sufficient, and chemical information was necessary to infer that the molecule was helical with complementary base pairs. The application of x-ray diffraction techniques to macromolecules was more than just an extension of the ideas for small molecules. Not only is it more difficult to crystallize large ones, but new schemes for determining experimentally the phases of the diffracted waves had to be developed in order to analyze the structures. The advent of dedicated synchrotron light sources with high-intensity beam lines for biological research has not only speeded up the rate of data collection from protein crystals by orders of magnitude but, by reducing radiation damage to the crystals, improved resolution. The ability to change wavelength makes it possible to phase the reflections by virtue of the anomalous dispersion of heavy-metal atoms. The use of insertion devices, wigglers and undulators, results in intensity increases suitable to record a Laue diffraction pattern in times of the order of milliseconds, making it possible to investigate dynamic changes in macromolecular conformation in muscle contractility and in the interactions between enzymes and substrates or between controlling proteins and DNA. A big challenge to the effective use of synchrotron radiation is the design and construction of two-dimensional detectors capable of recording data at about 10^8 counts/min.

A potentially important spin-off of synchrotron technology is the possibility of performing coronary angiography with little risk by imaging the heart with monochromatic radiation at the K edge of a contrast agent and subtracting from this image the absorption by background radiation off the K edge. The resulting greatly increased contrast is sufficient to enable the reconstruction of the coronary arteries after the injection of the contrast agent into a peripheral vein.

A complementary technique to x-ray diffraction is neutron diffraction. The technique is especially useful because hydrogen has a large negative scattering power and scatters strongly whereas it cannot be visualized by x-ray diffraction because of its very small scattering cross section. Moreover, by substituting deuterium for hydrogen, the magnitude and sign of the scattering power is changed making it possible to examine a structure in which the scattering has been changed without a change in conformation. The labeling of parts of molecular and cellular organelles with deuterium permits one to detect their positions and conformation readily. The visualization of macromolecules has been enhanced by the use of scanning transmission electron microscopy in which the object is scanned by a small, 2.5-Å diameter, electron beam. The electrons transmitted and scattered by the object are collected quantitatively and permit the estimation of the scattering cross section and hence, with high precision, the atomic mass of the object or a portion of the object from molecular weights of 20 000 on up. It is possible to detect clusters of heavy-metal atoms bound covalently to specific macromolecular binding sites so as to locate these sites on proteins or on nucleic acids within < 8 Å. The techniques discussed above are only a small part of the extensive ones applied to studies of macromolecules. Table I lists these techniques.

In addition to the large number of techniques used to determine molecular structures, a number of others are used to alter the characteristics of a molecule, its structure and its function, to gain insight into the way such molecules work. For example, there are a number of so-called relaxation methods in which the temperature or the pH in a molecular or cellular system is changed suddenly. The rate at which the system approaches a new steady state is then measured by one of the techniques of Table I and, from this rate, information is gained about the stability of the new and the old molecular conformations. This is one approach used in attempts to determine how proteins—polyamino acids—manage to fold to a unique functional structure rather than just a random nonfunctional one.

A second group of techniques involves measurements of fluorescence and phosphorescence of molecular or cellular systems. From the excitation and the emission spectra, both the absorbers and the emitting subunits may be identified. Measurements of the lifetimes, the yields, and the fluorescence depolarization make it possible to estimate the proximity of the absorber and the emitter, and to infer any special aspects of the local environment such as hydrophobic or hydrophilic characteristics. Such measurements also give in-

Table I Physical techniques used for studying macromolecules

Technique	Information obtained
Osmotic pressure	M
Diffusion	(L, V, H_2O)
Sedimentation	
equilibrium	M, ρ
velocity	(L, V, H_2O), M_{DNA}
Viscosity	(L, H_2O)
viscoelastometry	M_{DNA}
Birefringence	
electric	(L, H_2O)
flow	(L, H_2O)
Dispersion	
dielectric	(L, H_2O)
optical rotatory	S, H
Light scattering	M, anisotropy, diffusion constant
Light absorption	Composition, S
Dichroism	
linear	O, C
circular	S, H
magnetic circular	C, anisotropy
Luminescence	Composition, S, O, local environment, energy transfer
Electrophoresis	(charge, L, V, H_2O), M
X-ray	
scattering	(L, V), M, O
diffraction	C, O, atomic coordinates
Neutron	
scattering	(L, V), O
diffraction	C, O, atomic coordinates
Electron microscopy	L, V, M
Electron paramagnetic resonance	Free radicals, reaction mechanisms
Nuclear magnetic resonance	O, S, C, H_2O
Irradiation	
ionizing	Sensitive volumes and their location in cells (radiobiology)
nonionizing	Sensitive components in cells, damage to DNA, photosynthesis (photobiology)

M: molecular weight; L: shape; V: volume; H_2O: bound H_2O; ρ: buoyant density; S: stacking of subunits; H: helical content; O: orientation of subunits; C: conformation.

formation about excited states, and by inference reaction mechanisms of interest in photobiology. Fluorescent depolarization of reporter groups (fluorescent subunits) has been used to measure the fluidity of membranes.

The effects of ionizations or excitations on biological systems are also a part of biophysics. Studies of ionizing radiation are necessary for quantitative evaluation of its hazards, and they also may yield target volumes of molecules, which can lead to some of the same types of information given by other techniques of Table I. However, this particular technique is unique in that it can be used not only with impure preparations, but also with complexes of molecules, or even whole cells. Photobiologists not only investigate basic mechanisms in photosynthesis, but also the killing and mutation of cells by ultraviolet radiation of different wavelengths. Such action spectra were one of the first lines of experimental evidence that nucleic acids are the important component of the genetic material. They have also led to the discovery of cellular repair mechanisms that serve to ameliorate most of the deleterious changes made in cellular DNA by radiation and chemicals. The examples given above are some of the clearest ones showing the impact of physical techniques and physical methods of thought on the study of biological systems.

Acknowledgments

This work is supported by the Office of Health and Environmental Research of the U.S. Department of Energy.

See also CRYSTALLOGRAPHY, X-RAY; DIFFRACTION; FLUCTUATION PHENOMENA; LUMINESCENCE; NEUTRON DIFFRACTION AND SCATTERING; NUCLEAR MAGNETIC RESONANCE; OPTICAL ACTIVITY; POLYMERS; RADIOLOGICAL PHYSICS; RELAXATION PHENOMENA; SEDIMENTATION AND CENTRIFUGATION.

BIBLIOGRAPHY

Advances in Biological and Medical Physics **1** (1948) to present. (E)
Annual Review of Biophysics and Bioengineering **1** (1972) to **13** (1984). (I,A)
Annual Review of Biophysics and Biophysical Chemistry **14** (1985) to present. (I,A)
Biophysical Journal **1** (1960) to present. (A)
Progress in Biophysics (and Molecular Biology) **1** (1950) to present. (E)
Quarterly Review of Biophysics **1** (1968) to present. (I,A)

Black Holes

Stephen W. Hawking and Werner Israel

Gravity is by far the weakest interaction known to physics: The gravitational attraction between two electrons is completely dominated by the electrical repulsion between them, which is about 10^{43} times stronger. Despite their weakness, gravitational forces are much more a matter of everyday experience than electrical forces. The reason is that gravity makes up for its extreme weakness by having a property that no other interaction has, the property of *universality* or *equivalence*, which means that gravity affects the trajectories of all freely moving particles in the same way, no matter what their internal constitution. This characteristic has been verified to an extremely high accuracy both in laboratory experiments and by laser ranging on the moon. Mathematically, the property of universality or equivalence is expressed by saying that gravity couples to the energy-momentum tensor of matter. Together with the fact that quantum mechanics seems to require that the energy density be nonnegative (otherwise the vacuum could spontaneously decay into infinite numbers of positive- and negative-energy particles), this implies that gravity is always attractive. The gravitational fields of all the particles in a large body like the earth therefore add up to produce a significant field at the surface. For the earth this field is still weak compared to the electrical and exclusion-principle forces between atoms, so that the density of the terrestrial material is not much higher than it would be in the absence of gravity. For more massive bodies, however, gravity becomes more and more important, until it can dominate all other forces and give rise to catastrophic collapse inward.

The best-understood way in which such collapse can occur is for stars that are more than a few times the mass of the sun. A star is thought to be formed by condensation out of a cloud of interstellar gas. At first it would contract under its own gravity. As it did so, the temperature at the center would rise until it became high enough to start the thermonuclear reaction that converts hydrogen into helium. The heat generated by this process would create enough thermal pressure to prevent the star from contracting any further. The star would spend the next thousand million years or so in a quasi-stationary state burning hydrogen into helium and radiating the heat into space. Eventually, however, the nuclear fuel would all be used up and the star would begin to contract again. If the star were less than about one and a half times the mass of the sun, the collapse could be halted by degeneracy pressure of electrons or neutrons. In the first case the resultant body is called a white dwarf and has a radius of a few thousand kilometers, whereas in the second case it is called a neutron star and has a radius of the order of 10 km. For stars of more than a couple of solar masses, however, there is no final equilibrium state. Some stars may manage to reduce their masses to below this limit by throwing off material, but it seems virtually certain that this will not occur in all cases. It seems that such stars must continue to shrink until their density becomes infinite, creating what is called a singularity of space-time, a place where the notion of space-time as a continuum or manifold breaks down, as do all the laws of physics because they are formulated on a space-time background.

As the star contracts, the escape velocity from the surface will rise. In a spherical collapse, the escape velocity will become equal to the speed of light when the star reaches the Schwarzschild radius $r_s = 2GM/c^2$ (about $3 \times M/M_\odot$ km). After this, any further light emitted from the star cannot escape to infinity, but is dragged inward by the intense gravitational field. The region of space-time from which light cannot escape to infinity is called a *black hole*. Its boundary is called the *event horizon* and is formed by the wavefront of light that just fails to escape to infinity but remains hovering at the Schwarzschild radius. It acts as a sort of one-way membrane: objects can fall into the black hole through the event horizon but nothing can come out of the black hole because the event horizon (as viewed by a local observer in free fall) is moving outward at the speed of light and, according to special relativity, nothing can travel faster than light.

The collapse of a rotating, aspherical star is expected to be qualitatively similar, leading again to formation of a space-time singularity enclosed within a black hole. According to theorems of Penrose and Hawking, a singularity inevitably forms whenever outgoing light waves can no longer expand because of the growing force of gravity. (Essentially the only assumption required is that the energy density of the contracting material remains positive and larger than the sum of the three principal pressures.) That such a singularity will be surrounded by an event horizon which permanently screens it from outside view is a widely accepted, though still unproven, hypothesis, known as "cosmic censorship," whose plausibility is based chiefly on a persistent failure to find persuasive counterexamples. If this is true, then in the wake of the collapse, the horizon and the exterior gravitational field will settle, like a newly formed soap bubble, into the simplest configuration compatible with the external constraints. The end state, known as a Kerr–Newman black hole, is accordingly characterized by just three parameters: mass, angular momentum, and (in principle) charge. Other characteristics originally anchored in the stellar precursor are precipitated into the hole or radiated to infinity. In the words of John Wheeler, "a black hole has no hair."

The "stationary limit" of a black hole is the surface which marks out the innermost set of points at which a physical particle can be at rest with respect to a distant stationary observer while remaining subluminal (i.e., not exceeding the speed of light with respect to a local observer in free fall). It generally lies outside the horizon, but coincides with it for a spherical (nonspinning) black hole. In the intervening layer, called the ergosphere, subluminal orbiting observers, able to communicate with the outside, can still exist but are now forced into partial corotation with the hole. On the horizon itself only a single subluminal observer can exist at each point. He orbits with the speed of light, and his angular speed (the same at all latitudes on the horizon) is called the angular velocity of the hole.

The negative gravitational potential is so large in the ergosphere that a weight can in principle be lowered into this region in such a way as to recover more than its rest-mass energy as work; the extra energy is derived from the rotational energy of the hole. In a realistic astrophysical setting, a black hole's rotational energy can be tapped through hydromagnetic interactions with an ambient plasma. In such interactions, the event horizon behaves much like the surface of a conductor. (For instance, electric field lines threading the hole must cross the horizon orthogonally in the frame of a corotating observer.) Thus a black hole spinning in a uniform paraxial magnetic field anchored in a surrounding accretion disk will behave like a unipolar inductor. Processes of this kind may be responsible for jets and other forms of activity observed in active galactic nuclei.

Astronomical evidence for the existence of black holes is not yet overwhelming, but is steadily growing. A handful of binary x-ray sources in our own galaxy and in the Large Magellanic Cloud are thought (on the basis of more-or-less trustworthy lower bounds for the mass of the unseen component) to contain a black hole of a few solar masses in orbit around a somewhat more massive normal star (see Astronomy, X-Ray). On a larger scale, the motion of gas and stars in the central regions of our own and other galaxies, and the detection of rapidly variable γ- or x-ray emission from "hot spots" in these regions, point to the presence of central condensations of a million or so solar masses, which may be black holes. The most plausible source for the highly luminous quasars is accretion onto black holes of up to a billion solar masses.

The event horizon, the boundary of a black hole, has the property that its surface area always increases when more matter or radiation falls into the black hole. If two black holes collide and merge to form a single black hole, the area of the event horizon around the final black hole is greater than the sum of the areas of the two original black holes. There is thus a similarity between the area of an event horizon and the concept of entropy in thermodynamics, with the law of the area increase being the analog of the second law of thermodynamics. There is also a black-hole analog of the first law of thermodynamics, which relates the change in energy (or mass) of a system to the change in entropy (or area). In the thermodynamic case the constant of proportionality is called the temperature. In the black-hole case it turns out to be a quantity called the *surface gravity*, which is a (redshifted) measure of the strength of the gravitational field at the horizon, equal to GM/r_s^2 for an uncharged spherical black hole. An analog of the third law states that the surface gravity cannot be reduced to zero in a finite (advanced) time by injecting material with bounded, positive density and bounded stresses into the hole. (The surface gravity of an uncharged black hole becomes zero when its angular momentum reaches GM^2/c, the maximum possible for a black hole of mass M.)

Although the area of the event horizon and the surface gravity of a black hole have a strong resemblance to entropy and temperature, respectively, it is not possible, in a purely classical theory, to regard some multiples of the area and the surface gravity as the actual entropy and temperature of the black hole. For according to classical theory, a black hole can absorb radiation but cannot emit anything. It therefore could not be in equilibrium with blackbody radiation at any nonzero temperature. It was, nevertheless, suggested by Bekenstein that the entropy S of a black hole is a universal multiple of its area A. This idea was vindicated when Hawking, taking quantum theory into account, proved in 1974 that one must, after all, assign a finite temperature T to a black hole, because it emits particles and radiation by a quantum tunneling process. The Bekenstein–Hawking relations are

$$S = \frac{1}{4}\frac{kc^3}{G\hbar}A, \qquad T = \frac{1}{2\pi}\frac{\hbar}{kc}\kappa,$$

where κ is the surface gravity. The temperature of a spherical black hole of mass M is $10^{-7}(M/M_\odot)^{-1}$ K.

A black hole radiating into empty space will slowly evaporate. Its area, and therefore its entropy, will decrease. However, this decrease is more than compensated by the entropy of the emergent radiation. In the quantum domain the second laws of black-hole dynamics and thermodynamics are subsumed and sublimated into the statement that the total entropy of a black hole and its surroundings can never decrease.

One way in which the quantum-mechanical emission process can be understood is as follows. The uncertainty principle implies that "empty" space is filled with virtual particle–antiparticle pairs that come into existence together at some point of space-time, move apart, and then come together again and annihilate each other. They are called "virtual" because unlike "real" particles they cannot be observed by a particle detector but their indirect effects can be measured in cases such as the Lamb shift in the spectrum of atomic hydrogen and the Casimir effect between two parallel conducting plates. When a black hole is present, one member of a virtual pair may fall into the hole, leaving the other without a partner with which to annihilate. The forsaken particle or antiparticle may follow its mate into the black hole but it may also escape to infinity where it will appear to be radiation emitted by the black hole. An alternative way of looking at the process is to regard the member of the pair that falls into the hole (say an antiparticle) as a particle that is traveling backward in time. It would come out of the black hole and then, when it reached the point where the particle–antiparticle pair first appeared, it would be scattered by the gravitational field into a particle traveling forward in time. In effect, the particle would have quantum mechanically tunneled through the potential barrier represented by the event horizon, a barrier that it would be impossible for it to surmount classically.

The reason this tunneling process gives a thermal spectrum for the emitted particles is that these particles come from the region inside the event horizon, which someone outside the black hole cannot observe directly. According to the "no hair" property, all such a person could measure about this region would be the three quantities mass, angular momentum, and charge. It implies that a priori probabilities to emit all different configurations of particles with the same total energy, angular momentum, and charge are the same. A black hole could indeed emit a television set or a belly dancer but the number of configurations represented by these possibilities is very small. The overwhelming probability is that the radiation will be very nearly thermal because this represents by far the largest number of possible configurations.

That the particles come from a region about which an external observer has only a very limited knowledge means that the emission from black holes has an extra degree of unpredictability or uncertainty over and above that normally associated with quantum mechanics. In classical physics we can predict with certainty both the position and velocity of a particle. In ordinary quantum mechanics we can make definite predictions of either the position or the velocity or some combination of the two. In this case of particles emitted from a black hole, however, we cannot definitely predict any observable quantity. All we can do is deduce the probabilities that particles will be emitted in certain modes. Thus there is associated with black holes an additional degree of randomness that seems to arise from their containing singularities of space-time at which the laws of physics break down.

According to some recent speculations, a conceivable way in which this extra randomness might arise is through loss of information channeled through "wormholes" into microscopic regions ("baby universes"), which form by quantum fluctuations of the gravitational field near the singularity, and which subsequently detach themselves from our universe. But here we trespass beyond the borders of present knowledge and understanding.

See also ASTROPHYSICS; SPACE-TIME.

BIBLIOGRAPHY

Popular

S. W. Hawking, *A Brief History of Time: From the Big Bang to Black Holes*. Bantam, New York, 1987.

J. P. Luminet, *Black Holes*. Cambridge University Press, Cambridge, 1990.

Elementary

W. Israel, *Sci. Prog.* (*Oxford*) **68**, 333–363 (1983).

R. D. Blandford, "Astrophysical Black Holes," in *300 Years of Gravitation* (S. W. Hawking and W. Israel, ed.), pp. 277–329. Cambridge University Press, Cambridge, 1987.

Advanced

K. S. Thorne, R. H. Price, and D. A. MacDonald, *Black Holes: The Membrane Paradigm*. Yale University Press, New Haven, CT, 1986.

I. D. Novikov and V. P. Frolov, *Physics of Black Holes*. Kluwer, Boston, 1989.

Blackbody Radiation

R. E. Bedford

Every object in the universe continuously emits and absorbs energy in the form of electromagnetic radiation. This kind of energy transport differs from the processes of conduction and convection in requiring neither the presence of a medium nor the transport of màtter. The radiated energy originates from the internal energy associated with atomic and molecular motion and the accompanying accelerations of electrical charges within the object. When the atoms or molecules are only loosely coupled, as in gases at ordinary temperatures and pressures, they can radiate semi-independently with the emitted energy appearing in discrete spectral lines or bands characteristic of the internal energy-level structures. In opaque solids and liquids, on the other hand, the atoms and molecules interact so strongly with one another that they cannot radiate independently and the emitted energy becomes entirely determined by the object's temperature and surface structure. For this reason it is called *thermal radiation*. Thermal radiation is distributed continuously over the entire electromagnetic spectrum but conventionally is taken to lie within the spectral range from roughly 0.2 to 500 μm. When the object's temperature is between 10 and 5000 K, the radiation is concentrated in the infrared, the region commonly associated with the production of heat.

For an object to maintain thermal equilibrium its rates of emission and absorption of energy must be equal. This equality leads to *Kirchhoff's Law* (established in 1859) that, if $M_\lambda(T)$ (variously called spectral radiant emittance, spectral self-exitance, or spectral emissive power) is the power emitted per unit area per unit of wavelength at the wavelength λ, and $\alpha_\lambda(T)$ (spectral absorptance) is the fraction of the incident power per unit area per unit of wavelength that is absorbed, then for every particular wavelength, all objects in thermal equilibrium at the same absolute temperature T have the same ratio of $M_\lambda(T)/\alpha_\lambda(T)$. It follows that an object having the maximum possible absorptance $[\alpha_\lambda(T) = 1]$ also has the maximum possible spectral radiance emittance $[M_\lambda^b(T)]$. Such a perfect absorber and perfect emitter is called a *blackbody* and the electromagnetic energy it emits is called *blackbody radiation*. No true blackbody exists, but a small hole through which radiation escapes from an isothermal enclosure is an excellent approximation. Any real substance is characterized by a spectral emissivity $\epsilon_\lambda(T)$ $[0 < \epsilon_\lambda(T) < 1]$ expressing its spectral radiant emittance as some fraction of that of a blackbody at the same temperature, $M_\lambda(T) = \epsilon_\lambda(T) M_\lambda^b(T)$. With this, Kirchhoff's Law becomes $\epsilon_\lambda(T) = \alpha_\lambda(T)$. This is an important result—a good emitter of thermal radiation is also a good absorber. It follows also that a good reflector is a poor emitter because, for opaque substances, $\rho_\lambda(T) + \alpha_\lambda(T) = 1$ [where $\rho_\lambda(T)$ is the spectral reflectance], and so $\rho_\lambda(T) = 1 - \epsilon_\lambda(T)$. The incident radiation is usually absorbed in a thin surface layer whose properties then determine the value of $\epsilon_\lambda(T)$. As a rule of thumb, smooth or polished surfaces have lower emissivities than rough, grainy, or oxidized surfaces; metals have lower values than nonmetals. In general $\epsilon_\lambda(T)$ varies only slowly with temperature, but rather more with wavelength. For a large number of materials, however, $\epsilon_\lambda(T)$ is approximately constant over a fairly wide wavelength range. These are called *graybodies*.

Materials are also characterized by a total emissivity $\epsilon(T)$ relating the total radiant emittance $M(T)$ (total power emitted per unit area) to that of a blackbody $M^b(T)$ at the same temperature by $M(T) = \epsilon(T)M^b(T)$. $\epsilon(T)$ is an integrated average of weighted values of $\epsilon_\lambda(T)$ using the blackbody spectral radiant emittance as the weighting function. Typical values of these emissivities for a few solids are given in Table I.

The theoretical foundations of blackbody radiation have played an important role in the history of physics. In 1879 Josef Stefan deduced the empirical relation that the total radiant emittance varies directly with the fourth power of the absolute temperature, i.e.,

$$M^b(T) = \sigma T^4, \tag{1}$$

where σ is a constant. Ludwig Boltzmann established the

Table I. Representative Values of Emissivities of Solids

Material	$\epsilon_\lambda(T)$ ($\lambda = 0.65 \times 10^{-4}$ cm)	$\epsilon(T)$ ($T = 300$ K)
Aluminum foil	0.15	0.02
Copper, polished	0.1	0.03
Copper, oxidized	0.8	0.5
Iron, polished	0.4	0.08
Iron, oxidized	0.9	0.8
Carbon	0.9	0.8
Red brick	0.7	0.9
Concrete	0.6	0.94
Soot	0.97	0.95
Flat black paint	0.98	0.94
Flat white paint	0.2	0.87

theoretical proof of Eq. (1) (the Stefan–Boltzmann law) in 1884 from an analysis of the thermodynamics of the pressure exerted by radiation. In searching for a specific relation between the spectral radiant emittance of a blackbody and temperature, Wilhelm Wien showed (1893), on the basis of classical thermodynamics, that $M_\lambda^b(T)$ must have the form

$$M_\lambda^b(T) = c_1\lambda^{-5}f(\lambda T) \tag{2}$$

where c_1 is a constant and $f(\lambda T)$ is some function of the product λT only. Although thermodynamics could not provide the specific form of $f(\lambda T)$, Wien was able to prove that the product $\lambda_m T$ (where λ_m is the wavelength at which $M_\lambda^b(T)$ has a maximum value) is a universal constant having the value 0.2898 cm K. It follows from this *displacement law* that (a) the maximum of $M_\lambda^b(T)$ and, correspondingly, most of the radiated energy shift to lower wavelengths as the temperature increases; (b) the value of $M_{\lambda_m}^b(T)$ increases with the fifth power of the temperature. In 1896 Wien suggested that the function $f(\lambda T)$ should be set equal to $e^{-c_2/\lambda T}$ (where c_2 is a constant), but this was found to depart increasingly from experimental observations as λT increased (a difference of 1% when λT = 0.3124 cm K). For these studies Wien received the Nobel Prize in 1911. In 1900 Lord Rayleigh deduced from classical statistical mechanics that $f(\lambda T) = \lambda T/c_2$. Although a good approximation for large values of λT, this result is clearly wrong for small λ because it predicts the emission of an infinite amount of energy.

The dilemma was finally resolved in 1900 by Max Planck who introduced the revolutionary idea that energy can be emitted (or absorbed) only in discrete amounts that are integral multiples of a fundamental *quantum* of energy hc/λ (where c is the speed of electromagnetic radiation in vacuum and h is a universal constant known as *Planck's* constant). With this assumption Planck derived the correct formula

$$M_\lambda^b(T) = 2\pi hc^2\lambda^{-5}(e^{hc/\lambda kT} - 1)^{-1}, \tag{3}$$

where k is the Boltzmann constant. Equation (3) may be written in equivalent forms in terms of frequency (ν) or numbers of photons (n) instead of wavelength (λ). This formula satisfies the demands of Eq. (2) and also, by equating to zero the first derivative with respect to λ, gives $\lambda_m T = hc/4.9651k$, which is Wien's displacement law. When $\lambda T \gg hc/k$, Eq. (3) reduces to Wien's approximation, and when $\lambda T \ll hc/k$ it reduces to Rayleigh's. Integration of Eq. (3) over all wavelengths leads to the Stefan–Boltzmann Law

$$M^b(T) = \frac{2\pi^5k^4}{15h^3c^2}T^4 \tag{4}$$

which relates σ to the other universal physical constants. For this spectacular success, which marked the beginnings of the quantum theory, Planck received the Nobel Prize in 1918. Equation (3) was subsequently derived by others (notably Einstein and Bose) from completely different lines of reasoning.

The Planck distribution [Eq. (3)] is illustrated in Fig. 1 for some representative temperatures, for one of which the Wien and Rayleigh approximations are also shown. The variation of λ_m with temperature is also indicated. Both Eqs. (3) and (4) are valid only when the blackbody radiates into a vacuum; otherwise $M_\lambda^b(T)$ and $M^b(T)$ must be multiplied by n^2, where n is the refractive index of the surrounding medium and λ is the wavelength in vacuum. For most gases (including air), n is very near to unity so the factor n^2 may be neglected.

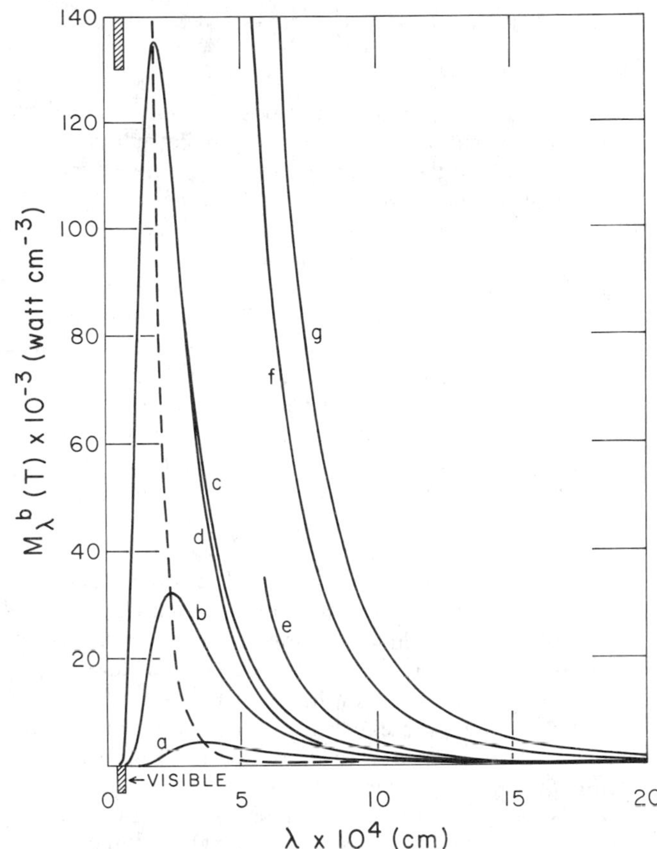

FIG. 1. Spectral radiant emittance of a blackbody: (a) 800 K, (b) 1200 K, (c) 1600 K, (d) 1600 K (Wien), (e) 1600 K (Rayleigh), (f) 6000 K, (g) 10,000 K. The variation of λ_m with temperature is shown by the dashed curve.

Blackbody radiation is important in such diverse areas of physics as metrology, where it underlies standards of radiometry and radiation temperature measurements, and cosmology, where it is a powerful probe for elucidating the early history of the universe.

See also HEAT.

BIBLIOGRAPHY

Blackbody radiation is discussed in most textbooks that deal with heat, heat transfer, infrared radiation and detection, or radiometry. Examples of good, detailed treatments are:

G. R. Noakes, *A Text-Book of Heat*, 3rd ed., pp. 368–409. Macmillan, London, 1953. (E)

R. Siegel and J. R. Howell, *Thermal Radiation Heat Transfer*. McGraw-Hill, New York, 1972. (I)

F. Reif, *Fundamentals of Statistical and Thermal Physics*, pp. 373–388. McGraw-Hill, New York, 1965. (I)

H. P. Baltes and F. K. Kneubühl, "Thermal Radiation in Finite Cavities," *Helvetica Phys. Acta* **45**, 481–529 (1972). (A)

Bohr Theory of Atomic Structure

Edward A. Burke†

The idea that all matter is composed of basic indivisible parts was first suggested by the Greek philosopher Democritus (who coined the word atom in the fourth century) and subsequently alluded to in medieval writings. Early in the 19th century, John Dalton proposed an atomic theory of matter. This theory assumes that (1) all matter is composed of atoms, (2) all atoms of a given element have a distinctive mass, and (3) in chemical reactions whole atoms, never fractions of atoms, are involved. The theory proved invaluable in understanding chemical reactions. Later on in the 19th century, the works of Rudolf Clausius, James Clerk Maxwell, and Josef Loschmidt on the kinetic theory of gases, climaxing the work done by Daniel Bernoulli in the middle of the 18th century, firmly established the concept of an atom as the fundamental constituent of matter which differentiated one element from another.

However, it soon became apparent that the atom has a structure. In particular, the electrolysis research performed by Michael Faraday indicated the existence of electrically charged constituent parts, and in 1891 G. Johnstone Stoney proposed the name electron for the basic unit of charge in an atom. In 1895, the experiments of Jean-Baptiste Perrin showed that the cathode rays emitted from electrical discharge through gases consisted of negative charges, and in 1897 Joseph J. Thomson performed his now famous experiments on these cathode rays, passing them through combined electric and magnetic fields. This established the ratio of charge to mass, e/m, of this electrical constituent of atoms. The concept of an electron as a negatively charged fundamental constituent of an atom was further reinforced through an analysis of the Zeeman effect, in which spectral lines were observed to broaden in a strong magnetic field. The magnitude of the electron charge was determined by Robert A. Millikan in his classical oil drop experiments begun in 1906.

Since atoms are normally uncharged electrically, the discovery that constituent parts were negatively charged (and by inference that an equal amount of positive charge was in the atom) required a theory of atomic structure to accommodate charged particles, such that the total atomic charge would be zero. By assuming that an atom consists of an equal number of positive and negative ions that obey a force law that is slightly different from an inverse square law, James H. Jeans in 1901 was able to calculate some of the observed features of atomic spectra. However, this model could not account for the observed periodicity of the chemical properties of the elements. In 1904, Thomson proposed an atomic structure consisting of a uniformly and positively charged sphere embedded with negatively charged electrons. The total charge was taken to be zero and the electrons confined to positions on rings concentrically located about the center of the sphere. For equilibrium, the numbers of electrons on each ring for a given atom were assumed identical. Thomson calculated the distribution of electrons on the rings for various elements. This model proved effective in describing the periodicity of the elements, but was not very useful in explaining the spectral lines that were observed in atomic spectra. In the same issue of the *Philosophical Magazine* in which Thomson proposed his model, H. Nagaoka, relying on earlier (1890) calculations of Maxwell on the rings of Saturn, first proposed a nuclear atom with extranuclear electrons. He recognized the difficulties associated with such a model in that the electrons could not remain stationary for otherwise they would be drawn into the nucleus by the Coulomb force. Although mechanically stable if they moved about the nucleus, they should radiate energy by virtue of their accelerated motion then slow down, and ultimately collapse. These objections were left unanswered. Radiation was accounted for by the vibrations of the electrons within their orbits. This model was unable to predict more than a few of the observed spectral lines. More importantly, this model did not gain general acceptance because of the inherent instability of the proposed physical system. In order to improve Thomson's model so that spectroscopic phenomena could also be explained, Lord Rayleigh (1906) assumed that the number of electrons could be infinite. This proposal preceded calculations published by Thomson on the basis of an analysis of several different experimental observations that the number of electrons (referred to by Thomson as corpuscles) was the same order as the atomic weight. In 1911 Barkla more accurately ascertained that the number of electrons was more nearly half the atomic weight, except for the lightest elements. And so Rayleigh's modification of Thomson's atomic model was found to be untenable. Also, in 1906, G. A. Schott proposed other modifications of Thomson's model to account for more details of atomic spectra. At this time the accepted model of atomic structure was that due to Thomson. In this model, radiation, as observed in spectral lines, was accounted for by the mechanism of electronic vibration.

However, the experiments of Hans Geiger and Ernest Marsden begun in 1909 at the suggestion of Ernest Rutherford, and the subsequent analysis of the results by Rutherford, first published in 1911, indicated that an atom must consist of a small but massive and positively charged central nucleus surrounded by extranuclear electrons. Following the discovery of radioactivity by Antoine Henri Becquerel in 1896, Rutherford had identified α rays as ionized helium atoms moving very rapidly. Geiger and Marsden observed the scattering of α rays through thin metallic foils. The surprising result of these experiments was an unexpectedly large amount of backscattering. In the words of Rutherford: "It was almost as incredible as if you fired a 15-inch shell at a piece of tissue paper and it came back and hit you." It was not possible with the Thomson model to predict any significant backscattering, even if one assumed multiple (from many atoms) scattering. On the other hand, Rutherford predicted, and Geiger and Marsden confirmed, that the scattering pattern observed could be completely understood by assuming a nuclear model for atoms. Rutherford was able to conclude that the dimension of a nucleus is of the order of 10^{-13} cm. It can be deduced from kinetic theory that an atom has dimensions of the order of 10^{-8} cm. Hence an atom is mostly empty space. In 1906 Thomson's experiments indicated most probably that the smallest known atom, hydrogen, contained just a single electron.

† Deceased.

In 1913, Niels Bohr (a student of Rutherford), in a paper entitled, "On the Constitution of Atoms and Molecules," which appeared in Vol. 26 of the *Philosophical Magazine*, first proposed his atomic theory. This theory is known as the Bohr theory of atomic structure. The principal success of the theory was in explaining the spectra of atomic hydrogen.

If we assume a circular orbit of an electron about a stationary nucleus, the total energy of the system W is given by

$$W = mv^2 - eQ/r, \tag{1}$$

where m is the electron's mass, v is its speed, e and Q are, respectively, the magnitudes of the electron and nuclear charge, and r is the radius of the orbit. Following Bohr, cgs units are employed. In order for the electron to remain in a circular orbit, the centripetal force must be supplied by the Coulomb force between the electron and the nucleus, and hence

$$\frac{mv^2}{r} = \frac{eQ}{r^2}. \tag{2}$$

Combining Eqs. (1) and (2) we obtain

$$W = -\tfrac{1}{2}eQ/r = -\tfrac{1}{2}mv^2 \tag{3}$$

so that the energy of the bound system is less than 0, whereas for the free electron, $r=\infty$, the energy is taken to be zero. If the frequency of orbital motion is f, then

$$v = 2\pi rf. \tag{4}$$

By eliminating r and v from the preceding relations, we obtain

$$f = \frac{\sqrt{2}}{\pi}\frac{(-W)^{3/2}}{eQ\sqrt{m}}. \tag{5}$$

In describing blackbody radiation, Max Planck (1900) assumed that radiation must be emitted in separate emissions (quanta) of $h\nu$, where h is Planck's constant and ν is the frequency of the radiation. (This concept of quantized radiation was successfully employed later by Albert Einstein in 1905 in describing the photoelectric effect.) Following his predecessors in relating radiation frequency to the mechanical frequencies of the electrons, Bohr assumed that the radiation frequency emitted by an atom when an electron goes from an unbound to a bound state would be the average of the mechanical (in this case the orbital) frequencies of the initial and final state. Since the orbital frequency of the unbound electron is zero and its bound frequency is f, then the relationship between f and the radiation frequency ν is

$$\nu = f/2 \tag{6}$$

and, following Planck's hypothesis, the energy $\mathcal{W}$ radiated would be

$$\mathcal{W} = nh\nu = nh(f/2), \tag{7}$$

where n is an integer. Combining Eq. (7) with Eq. (5), noting that $\mathcal{W} = -W$, we obtain

$$W = -\frac{2\pi^2e^2Q^2m}{n^2h^2} \tag{8}$$

and

$$f = \frac{4\pi^2me^2Q^2}{n^3h^3}, \tag{9}$$

with

$$r = \frac{n^2h^2}{4\pi^2meQ}. \tag{10}$$

We note that the most stable state, the state with the lowest energy, occurs for $n=1$. For the hydrogen atom $Q=e$, and if we use the experimental values known to Bohr of

$$e = 4.7\times10^{-10}\ \text{statcoulombs},$$

$$e/m = 5.31\times10^{17}\ \text{statcoul/q},$$

and

$$h = 6.5\times10^{-27}\ \text{erg s},$$

we get, to the accuracy of these data,

$$W = -13\ \text{eV},\ f = 6.2\times10^{15}\ \text{Hz},\ r = 0.55\times10^{-8}\ \text{cm}.$$

These values are at least the same order of magnitude as the ionization potential of hydrogen, the observed optical frequencies, and atomic dimensions.

Larger integer values of n correspond to higher-energy stationary states of the system. We note, from Eq. (8), (9), and (10) that, as $n\to\infty$, W and $f\to0$, and $r\to\infty$. We approach the unbound state. All the energy of the atom is emitted as radiation when the electron proceeds from an initial unbound state to the bound state, and from Eq. (8) this positive amount of energy is

$$W_n = \frac{2\pi^2e^4m}{n^2h^2} \tag{11}$$

for the hydrogen atom. The amount of energy emitted by the passing of the system from a state corresponding to $n=n_1$ to one corresponding to $n=n_2<n_1$ is, consequently,

$$W_{n_2} - W_{n_1} = \frac{2\pi^2e^4m}{h^2}\left(\frac{1}{n_2^2} - \frac{1}{n_1^2}\right). \tag{12}$$

If the frequency of the radiation is ν, then according to Planck this energy must be $h\nu$ so that

$$W_{n_2} - W_{n_1} = h\nu \tag{13}$$

and hence

$$\nu = \frac{2\pi^2me^4}{h^3}\left(\frac{1}{n_2{}^2} - \frac{1}{n_1{}^2}\right). \tag{14}$$

By selecting a value of n_2 and letting n_1 take on all integral values larger than n_2, we obtain a formula for a sequence of spectral lines. In particular, by selecting $n_2=2$, this formula gives results of the series discovered by Johann Balmer in 1884. If we rewrite Eq. (14) as

$$\nu = Rc\left(\frac{1}{2^2} - \frac{1}{k^2}\right), \tag{15}$$

we then have the same form as written by Balmer. R is the Rydberg constant, which, by the Bohr theory, is

$$R = \frac{2\pi^2me^4}{ch^3}. \tag{16}$$

(More properly we should replace m by μ, the reduced mass of the electron–nucleus system. In 1932 this refinement led to the discovery of deuterium.) Putting $n_2=3$ we obtain the series first observed by Friederich Paschen in 1908. For $n_2=1$ we get the Lyman series. Theodore Lyman discovered the first term in this ultraviolet region in 1906 but did not attribute this to hydrogen until after the publication of Bohr's theory. For $n_2=4$ and $n_2=5$ we have series in the infrared. These series, referred to as the Brackett and Pfund series, were predicted by Bohr's theory and were subsequently observed in the 1920s.

Thus the Bohr theory proved spectacularly successful in accounting for known spectral series and predicting future series of hydrogen spectra. With the Bohr theory it also becomes possible to explain the greater number of spectral lines observed from interstellar space. The additional lines can be considered to arise from transitions from very high values of n_1 in Eq. (14). This, in turn, implies orbits of very large radii which could only be possible in the rare atmosphere of outer space.

One can show (as was done by Bohr) that the choice of $\mathcal{W}=nhf/2$ in Eq. (7) is consistent with the known Balmer series expansion and leads to a correspondence of radiation frequency with orbital frequency for transitions between two adjacent highly excited states.

We can state Eq. (7) in an alternative way. We see from Eq. (3) that $\mathcal{W}=-W=T$, where T is the kinetic energy of the electron and for circular motion $T=mr^2f^22\pi^2$, hence

$$2\pi^2mr^2f^2=nhf/2. \tag{17}$$

But the angular momentum p of the system is $2\pi mr^2f$. From this we obtain

$$p=n\hbar \tag{18}$$

where $\hbar=h/2\pi$. The importance of angular momentum had been indicated earlier (1912) by J. W. Nicholson. The condition represented by Eq. (18) is referred to as the quantization of orbital angular momentum. Implicitly, the condition includes all orbitals, i.e., stationary states of the system, for which the angular momentum is an integral multiple of $\hbar$ and precludes all other possibilities. Furthermore, the existence of stationary states implies, by definition, that the atom cannot radiate when in any of its allowed states. Finally, all transitions between stationary states result in emission or absorption of radiation in quanta of energy equal to $h\nu$ with ν the frequency of the radiation. Transitions to lower-energy states result in emission while transitions to higher-energy states are required for absorption.

One can derive the Bohr theory of atomic structure starting with the quantization of angular momentum, Eq. (18), and proceed from Eqs. (1) through (5) above.

The quantization of the angular momentum condition may be thought of as a necessary result of the wave nature of matter, as proposed by Louis de Broglie in 1924. In this theory, we ascribe a wavelength λ to a particle with linear momentum P by $\lambda=h/P$. For circular orbitals in stationary states, we would thus require that the allowed orbits would be integral multiples of the particle wavelength λ. Thus we would require

$$2\pi r=n\lambda=nh/P. \tag{19}$$

Angular momentum p is related to linear momentum P in this case by $p=rP$. Hence Eq. (19) reduces to Eq. (18), which represents the quantization of angular momentum.

That the electron can remain in stationary states can be understood somewhat from quantum mechanics. Quantum mechanically, the electron is spread out in space and if the electrostatic attraction of the nucleus were to confine it within a small region near the nucleus, the indeterminancy principle would yield a great uncertainty in momentum. But this, in turn, would mean a higher expected energy, thereby enabling the electron to move away from the nucleus. As a result, there is an electrical equilibrium allowing for the possibility of stationary states.

Despite the success of the Bohr model, it fails in several important ways, viz.,

1. It does not account for the intensities of the observed spectral lines.
2. Despite some moderate success in describing the spectra of alkali-like atoms, the model is insufficient for atoms of more than one electron.
3. It is unable to explain chemical binding.
4. It is unable to explain fine details observed in the hydrogen spectra.

Even with its failures, the Bohr theory represents a significant advance in thought toward understanding atomic structure. It represents a model which bridges the gap between classical and quantum mechanics. Some of the more important, permanent contributions to our understanding of atomic structure that come from the Bohr theory of atomic structure are

1. the existence of stationary atomic states,
2. energy and momentum quantization,
3. the explanation of emission and absorption of radiation as transitions between stationary states.

See also ATOMS; ATOMIC SPECTROSCOPY; ATOMIC STRUCTURE CALCULATIONS—ONE-ELECTRON MODELS; BALMER FORMULA.

BIBLIOGRAPHY

For an historical survey see Robert M. Besançon (ed.), *The Encyclopedia of Physics*. Reinhold, New York, 1966; and Kenneth W. Ford, *Basic Physics*. Ginn/Blaisdell, Waltham, Mass., 1968.

For a collection of early publications see G. K. T. Conn and H. D. Turner, *The Evolution of the Nuclear Atom*. American Elsevier, New York, 1965.

For an elementary discussion see Richard T. Weidner and Robert L. Sells, *Elementary Modern Physics*. Allyn and Bacon, Boston, 1965.

Bose–Einstein Statistics

O. W. Greenberg

Identical atomic, nuclear, and subnuclear quantum-mechanical particles and systems of particles are indistinguishable. Experimental evidence confirms the relativistic quantum field theory prediction that the wave functions that describe several identical particles with integral spin must be symmetric under joint permutations of the coordinates and spins associated with the identical particles. This property of the wave functions leads to a counting, or "statistics" (called *Bose–Einstein statistics*), of the probability that a given set of identical integral-spin particles will be distributed in the accessible energy levels in a given way. Bose–Einstein statistics leads to characteristic statistical-mechanical (partition, etc.) and thermodynamic (energy distribution, etc.) functions. Composite systems obey Bose–Einstein statistics unless they contain an odd number of constituents that obey Fermi–Dirac statistics; in this case they obey Fermi–Dirac statistics.

Among elementary particles that obey Bose–Einstein statistics are the photon, the pions, the kaons, and mesons generally. Among composite systems that obey Bose–Einstein statistics are the hydrogen atom, the deuteron, the ^{4}He atom, and even atomic number nuclei generally. Particles that obey Bose–Einstein statistics are called *bosons*.

Bose derived the Planck frequency distribution for black-body radiation by treating the photon as a massless particle that obeys Bose–Einstein statistics. Bose–Einstein statistics produces a tendency for identical particles to cluster in the same quantum state. For example, consider two particles, 1 and 2, each of which can be in either of two states, A or B. If these particles are classical distinguishable particles that obey Maxwell–Boltzmann statistics, then at high temperatures the following four configurations each have probability $\frac{1}{4}$: 1 and 2 in A, 1 in A and 2 in B, 1 in B and 2 in A, and 1 and 2 in B. If the particles 1 and 2 are identical and obey Bose–Einstein statistics, then there are three possible symmetric wave functions: $A(1)A(2)$, $2^{-1/2}[A(1)B(2)+B(1)A(2)]$, and $B(1)B(2)$, and the probabilities are $\frac{1}{3}$ for the state corresponding to each wave function. Thus the particles that obey Bose–Einstein statistics are in the same state (either both in A or both in B) two thirds of the time, while the particles that obey Maxwell–Boltzmann statistics are in the same state only one half of the time. Identical particles that obey Fermi–Dirac statistics are never in the same state (Pauli exclusion principle). This clustering tendency, which increases with the number of particles, acts in the absence of forces between the particles; however, forces between the particles can modify this tendency. The clustering tendency plays an important role in processes, such as superradiant states and the action of the laser and maser, that involve coherent electromagnetic radiation.

Einstein pointed out an extreme case of such clustering: A gas of identical particles obeying Bose–Einstein statistics will undergo a phase transition to a state in which a finite fraction of the particles occupies the lowest-energy quantum state when the temperature of the gas is brought below a critical temperature. This fraction goes to unity when the temperature goes to absolute zero. Liquid ^{4}He, which obeys Bose–Einstein statistics, undergoes such a transition (the λ transition) when it is cooled below 2.18 K. The detailed properties of the transition and of the superfluid state below the λ point depend on the Bose–Einstein statistics obeyed by ^{4}He and on the spectrum of excitations in the ^{4}He liquid. The λ transition does not occur in ^{3}He, which obeys Fermi–Dirac statistics.

See also FERMI–DIRAC STATISTICS; MAXWELL–BOLTZMANN STATISTICS; QUANTUM STATISTICAL MECHANICS; STATISTICAL MECHANICS.

BIBLIOGRAPHY

R. P. Feynman, R. B. Leighton, and M. Sands, *The Feynman Lectures in Physics*, Vol III. Addison-Wesley, Reading, Mass., 1965. (E)

K. Huang, *Statistical Mechanics*, 2nd ed. Wiley, New York, 1987. (A)

L. D. Landau and E. M. Lifshitz, *Statistical Physics*. Addison-Wesley, Reading, Mass., 1958. (A)

J. E. Mayer and M. G. Mayer, *Statistical Mechanics*. Wiley, New York, 1940. (A)

R. K. Pathria, *Statistical Mechanics*. Pergamon, Oxford, 1985. (A)

F. Reif, *Fundamentals of Statistical and Thermal Physics*. McGraw-Hill, New York, 1965. (I)

E. Schrödinger, *Statistical Thermodynamics*. Cambridge, London and New York, 1962. (A)

Boundary Layers

C. W. Van Atta

Boundary layers are a ubiquitous feature of natural flow systems and of fluid flows in scientific and technological applications. For most boundary layers of geophysical or technological interest, the Reynolds number of the layer $R = U\delta/\nu$, where U is the mean velocity, δ the layer thickness, and ν the kinematic viscosity, is sufficiently large that the flow is fully turbulent. Geophysical cases include the atmospheric surface layer, oceanic surface and benthic layers, western inertial boundary currents like the Gulf Stream and Kuroshio, and turbidity currents in submarine canyons and alluvial fans. Technological examples include external boundary layers on aircraft, ships, land vehicles, and buildings, and internal boundary layers in pipes, ducts, turbomachinery, shock tubes, wind tunnels, chimneys, and sewer outfalls. Biological examples include flow in the boundary layers on swimming fish (it has been suggested that the unusual speed of some fish may be due to reduction of the viscous boundary layer drag by compliant boundaries or secretion of drag-reducing additives), and blood flow in large arteries, for which the flow properties in the boundary layer may be important for determining the formation and degree of clotting.

Boundary layers are formed whenever a flowing fluid encounters a solid boundary or free surface, or is subject to other dynamically constraining forces over some restricted domain of the flow. The thickness of the layer is generally small compared with a characteristic dimension in the main flow direction. In some boundary layer problems, the external "free stream" conditions far from the boundary and conditions on the boundary are given, and we are interested in the variation of the flow variables (e.g., velocity, temperature, salinity, species concentration) as a function of distance from and along the boundary, or in the flux of momentum, energy, or other transferable property across the layer and the consequent shear stress on the boundary, rate of heat transfer between fluid and boundary, rate of ablation or evaporation of the boundary, etc.

In a viscous boundary layer mean pressure gradients produce a flux of mean vorticity across the fluid–solid boundary. When the Reynolds number of the flow is large enough, certain spatial derivatives of flow variables with respect to the principal flow direction along the boundary are small compared with derivatives in the transverse direction across the boundary layer, leading to justifiable neglect of certain terms in the Navier–Stokes equations for the fluid motion compared with other terms, an approximation first introduced by Ludwig Prandtl. When the Reynolds number of the flow is small enough for the motion to remain laminar, this leads to a well-posed theoretical problem whose solution is obtained by solving the boundary layer momentum and energy equations along with appropriate boundary conditions. When the flow is turbulent, fluctuating vorticity associated with coherent flow structures is produced by local, highly intermittent, flow instabilities. The result is much greater momentum and scalar transport, and hence substantially increased drag and heat transfer compared with laminar flows. The Reynolds-averaged Navier–Stokes equations do not lead to a closed set of equations for the statistical quantities describing the turbulent motion, and no theoretical solutions can be found without invoking further assumptions about correlations between fluctuating variables. These assumptions have taken many forms, from simple models for turbulent stresses patterned after molecular diffusion to elaborate modeling of higher-order correlations occurring in the hierarchy of statistical equations associated with the "closure" problem of turbulence theory. Some types of modeling have been found to yield computation schemes that are satisfactory for engineering calculations and for simulation of certain aspects of geophysical boundary layers.

Direct numerical simulations of the unaveraged Navier–Stokes equations for the case of a turbulent boundary layer on a smooth flow surface accurately predict the mean flow structure and turbulence statistics, and the computed unaveraged instantaneous flow exhibits many of the coherent flow structures observed in experiments.

In addition to the viscosity, thermal conductivity, and other physical properties of the fluid, the parameters of importance in describing the characteristics of any given boundary layer flow depend on the scale of the flow involved and on the various dominant physical mechanisms operating in the flow. The characteristic length scales of all of the technological applications mentioned earlier are sufficiently small that the effect of the rotation of the earth, and hence the Coriolis parameter f, can be neglected, but this parameter is of dominant influence in the planetary boundary layer. The thickness of a nonrotating turbulent boundary layer scales roughly like $\delta \sim (\nu/U)^{0.2}x^{0.8}$, where x is the distance along the boundary. In contrast, for the rotationally influenced atmospheric surface layer or upper mixed layer in the ocean, the thickness of the "Ekman layer" is independent of x and ν and is proportional to $\tau^{1/2}/f$, where τ is the surface skin friction drag. In the nonrotating case the unopposed tendency of the turbulent eddies to mix momentum downward increases the vertical extent of the boundary layer with downstream distance. In the rotating case the vertical growth of the layer is apparently limited by the spontaneous formation of inertial oscillations when the layer depth exceeds a certain value. These instability waves then radiate energy away from the boundary layer, removing part of the source that would otherwise be available for vertical diffusion and growth. In the ocean, the stratification of the underlying thermocline plays an important role in determining the depth of the turbulent boundary layer. The dimensionless parameter describing the degree of stability of a stratified boundary layer is the Richardson number,

$$R_i = -g\frac{d\rho}{dz}\bigg/\rho\left(\frac{dU}{dz}\right)^2$$

where ρ is the fluid density and z the vertical coordinate. The structure of the layer, as well as the heat, momentum, and density differences transported by it, is a sensitive function of R_i. R_i is approximately proportional to z/L, where z is the vertical height in the boundary layer and L is the Monin–Obukhov length, $L = u_*^3/kN$, where $u_* = (\tau/\rho)^{1/2}$, k is von Kármán's constant, and N is the net flux of buoyancy across the interface, proportional to the heat flux for the atmospheric boundary layer. Experimental results are then cast in the form of universal functions for dimensionless turbulent and mean quantities as functions of R_i or z/L. The Monin–Obukhov similarity theory has been successful in describing available data for the surface layer, and numerical models have developed to a state wherein some of the universal functions can be predicted for comparison with measurements.

The Coriolis forces are not important in the near surface layers ($z \leq 100$ m or so) of the atmosphere and ocean, but become increasingly important with increasing height or depth. The balance of Coriolis forces and turbulent stresses outside the inner surface layer produces the turbulent Ekman boundary layer, in which the direction of the mean velocity vector changes continuously throughout the depth of the layer from its external geostrophic direction to that of the surface stress at the boundary. The net fluid transport in the Ekman boundary layer is responsible for the wind-driven ocean circulation patterns of the oceans through the Sverdrup relation $\tau_N = -\text{curl}\ \tau/(\partial f/\partial y)$, which states that the total northward fluid transport is equal to the inverse of the northward gradient of the Coriolis parameter times the curl of the wind stress on the ocean surface. The invalidity of this relation at a western boundary is related to the formation of inertial western boundary layers in the oceanic circulation. A westward-directed current having small relative vorticity tends to intensify into a thin jet as it is forced toward high latitudes by the eastern continental margins. The thickness of these major features, such as the Gulf Stream and the Kuroshio, is proportional to $(\bar{T}/\bar{f})^{1/2}$, where $\bar{T}$ is the total volume transport of the jet and $\bar{f}$ is a Coriolis parameter characteristic of the entire jet.

The relatively thin boundary layers formed on the vertical boundaries of turbulent convection cells for high Rayleigh numbers are thought to be important in convection in stars, in the earth's molten core, and in the earth's crust. The crustal flow is highly viscous and nonturbulent. At the vertical boundaries of the mantle convection cells, relatively thin, hot ascending jets drive spreading centers at oceanic ridges and descending jets form subduction trenches at opposite margins of the global tectonic plate system.

The experimental and theoretical evidence indicates that boundary layers are the preferred natural modes for the transport of momentum, heat, and other forms of energy on the macroscopic level.

See also HYDRODYNAMICS; RHEOLOGY; TURBULENCE; VISCOSITY; VORTICES.

BIBLIOGRAPHY

G. K. Batchelor, *An Introduction to Fluid Mechanics*. Cambridge, London and New York, 1967. (E)

T. Cebeci and A. M. O. Smith, *Analysis of Turbulent Boundary Layers*. Academic, New York, 1974. (A)

E. A. Eichelbrenner, "Three-Dimensional Boundary Layers," *Annu. Rev. Fluid Mech.* **5**, 339–360 (1973). (A)

H. Greenspan, *The Theory of Rotating Fluids*. Cambridge, London and New York, 1968. (A)

J. O. Hinze, *Turbulence*. McGraw-Hill, New York, 1959. (I)

J. Kim, P. Moin, and R. D. Moser, "Turbulence Statistics in Fully Developed Channel Flow at Low Reynolds Number," *J. Fluid Mech.* **177**, 133–166 (1987).

J. L. Lumley, "Drag Reduction by Additives," *Annu. Rev. Fluid Mech.* **1**, 367–384 (1969). (I)

A. S. Monin and A. M. Yaglom, *Statistical Fluid Mechanics: Mechanics of Turbulence*, Vols. 1 and 2. MIT Press, Cambridge, Mass., 1971, 1975. (I–A)

E. Reshotko, "Boundary-Layer Stability and Transition," *Annu. Rev. Fluid Mech.* **8**, 311–349 (1976). (A)

W. C. Reynolds, "Computation of Turbulent Flows," *Annu. Rev. Fluid Mech.* **8**, 183–208 (1976). (A)

A. R. Robinson, "Boundary Layers in Ocean Circulation Models," *Annu. Rev. Fluid Mech.* **2**, 293–312 (1970). (A)

L. Rosenhead, *Laminar Boundary Layers*. Oxford (Clarendon Press), London and New York, 1963. (I–A)

H. Schlichting, *Boundary Layer Theory*, 6th ed. McGraw-Hill, New York, 1968. (I)

E. A. Spiegel, "Convection in Stars: I. Basic Boussinesq Convection," *Annu. Rev. Astron. Astrophys.* **9**, 323–352 (1971); "II. Special Effects," **10**, 261–304 (1972). (A)

H. M. Stommel, *The Gulf Stream*, 2nd ed. Univ. of California Press, Berkeley, 1965. (I)

I. Tani, "Boundary-Layer Transition," *Annu. Rev. Fluid Mech.* **1**, 169–196 (1969). (I–A)

A. A. Townsend, *The Structure of Turbulent Shear Flow*, 2nd ed. Cambridge, London and New York, 1976. (I–A)

D. L. Turcotte and E. R. Oxburgh, "Mantle Convection and the New Global Tectonics." *Annu. Rev. Fluid Mech.* **4**, 33–68 (1972). (I–A)

J. S. Turner, *Buoyancy Effects in Fluids*. Cambridge, London and New York, 1973. (I–A)

W. W. Willmarth, "Structure of Turbulence in Boundary Layers," *Adv. Appl. Mech.* **15**, 159–254 (1975). (A)

Bremsstrahlung

Leonard C. Maximon

One of the straightforward consequences of the classical theory of electrodynamics is that a charged particle, when accelerated, radiates energy in the form of an electromagnetic wave. The large radio broadcasting antennas are a familiar example of this phenomenon on a macroscopic scale. At the other end of the spectrum is an equally familiar example, used for medical diagnosis and therapy: the tubes in which an electron beam is accelerated by a high potential, focused, and directed at a target, with the emission of x rays. It was in describing the continuous spectrum of these x rays that the term bremsstrahlung (literally, braking radiation) was first coined by Sommerfeld in 1909. Here, however, the force acting on the electron is that of the screened nuclear charge, so that the acceleration takes place over atomic distances, necessitating a quantum-mechanical treatment. This was first given by Sommerfeld in 1931, for nonrelativistic electrons ($v/c \ll 1$), for the idealized problem in which the atom is represented by a pure Coulomb potential; the screening of the nucleus by the atomic electrons is neglected. With this simplification he was able to give the intensity, angular distribution, and polarization of the emitted x rays in a closed, analytic form. A fully relativistic calculation of the bremsstrahlung process, in which the electron wave functions are solutions of the Dirac equation, first appeared in 1934, given independently by Racah, by Sauter, and by Bethe and Heitler, the last two names generally being associated with the cross section for this process. These calculations require, for their validity, that the nuclear charge be small and the electron velocities high—specifically $2\pi(Z/137)/(v/c) \ll 1$—in that they include the interaction between electron and nucleus only to first order (the Born approximation). Of greatest significance, both for analysis of experiments and for an understanding of the details of the process, is, however, the work of Bethe (1934) in which, for the experimentally important case of high electron energies (greater than a few MeV), he obtained the bremsstrahlung spectrum (the differential cross section integrated over the angles of the final particles) in a very simple form as an integral over the atomic form factor, accounting in this way for the screening of the nucleus by the atomic electrons. This analysis not only permitted comparison with experimental measurements, but also brought out very clearly the significant theoretical details of the process. Since then, and particularly in the last 25 years, when high-energy electron accelerators have become a most important tool for physics research, the increasing accuracy of experiments has provided an important incentive for the calculation of corrections to these theoretical expressions. Nonetheless, the articles just mentioned have remained an essential basis for most of the theoretical work that has been done on this subject.

There are a variety of reasons why the bremsstrahlung process occupies such an important place in physics. They are related to the fundamentals of the theory, to experiment, and to applications—to the nature of the process itself, in that bremsstrahlung is in fact present in every actual scattering of a charged particle: Since any detection apparatus has perforce a finite energy resolution, particles that have scattered elastically cannot be distinguished from those that have radiated a small amount of energy in the form of bremsstrahlung. This fact has important consequences for the analysis of almost all experiments using electron scattering as a probe to investigate nuclear structure, since the bremsstrahlung that accompanies the scattering (the radiative tail) must be subtracted in order to extract the purely nuclear information. Within the framework of the theoretical formalism, bremsstrahlung is very closely related to a number of other fundamental quantum-electrodynamical processes: inverse bremsstrahlung, pair production, radiative capture, and the photoelectric effect. The amplitude for each of these processes is closely related to that for bremsstrahlung; they differ formally only in that the particles are described by either outgoing or ingoing waves, positive or negative energy states, continuum or bound wave functions.

Thus far we have spoken only of the radiation from electrons. Experimentally they have been the most important source of bremsstrahlung, the cross section for the emission of bremsstrahlung from nonrelativistic charged particles being inversely proportional to the square of the mass of the particle. As the velocity of the particle approaches that of light, however, its mass becomes of decreasing importance. The radiation from heavier particles is therefore not easily detected unless the particles have relativistic velocities. With the advent of high-energy accelerators and the production of beams of heavier particles, the bremsstrahlung produced by the scattering of muons, pions, and protons has also been studied. In particular, the bremsstrahlung produced in proton–proton collisions has been studied extensively, both theoretically and experimentally, as a possible means of gaining greater understanding of the nucleon–nucleon force than is obtained solely from the analysis of elastic scattering experiments.

Another extension of the original work on bremsstrahlung concerns the variety of possible scatterers of the incident particle. In the aforementioned theoretical considerations the atomic electrons served only to screen the Coulomb potential of the nucleus, the recoil momentum being taken up by the atom as a whole, which remains in its ground state. However, the bremsstrahlung process may also take place in the field of the atomic electrons; they then absorb the recoil momentum and are either ejected or raised to an excited state. This effect was first calculated by Wheeler and Lamb in 1939. Subsequent improvements of their calculation have appeared, some quite recently, but in all theoretical work on this topic, both the screening of the scatterer by the other atomic electrons and its binding to the nucleus is neglected; the atomic electrons are treated as free electrons at rest. The closely related process of pair production in the field of the electron (triplet production) has also been the subject of recent experimental and theoretical research.

The field that serves to scatter the incident particle may, however, be magnetic rather than electric. The process is then known as magnetic bremsstrahlung (also referred to as cyclotron radiation when the emission is from nonrelativistic electrons, or as synchrotron radiation when it is from relativistic electrons). It was predicted on the basis of classical electromagnetic theory at the end of the nineteenth century, but the first observation was not made until 1947, and a detailed theoretical calculation of its properties (angular distribution, polarization, total energy loss) was first given by Schwinger in 1949. Since then, and particularly in the last few years, synchrotron radiation has become an increasingly important research tool. As a source of photons for research in other fields of physics, synchrotron radiation complements the more traditional bremsstrahlung, which is produced by scattering electrons from high-energy accelerators in targets placed in the beam. The latter has served for over 20 years as the primary source of photons for photonuclear research in the energy range 1 MeV to 1 GeV. Synchrotron radiation now provides an important source of photons for research in solid-state physics in the ultraviolet and soft x-ray regions and covers the energy range 5 eV to 50 keV (1–10 000 Å).

Most recently, the bremsstrahlung process has become the subject of numerous studies concerned with the design of very high-energy (TeV) electron–positron linear colliders. In order to obtain the luminosity necessary for high-energy physics experiments, the beams in such colliders must have very small transverse dimensions ($\sim 10^{-4}$–10^{-8} cm) and the high density of charged particles provides strong electromagnetic fields as viewed by the particles of the opposite beam. The particles in each beam thus emit intense synchrotron radiation, termed *beamstrahlung*, in the collision region. These studies aim at providing a quantitative description of beam–beam interactions in linear colliders and investigating the new physics they make possible.

The foregoing discussion of bremsstrahlung has referred primarily to laboratory-produced phenomena. However, some of the most interesting research of the last few years has been in a number of branches of astrophysics, where bremsstrahlung and synchrotron radiation occur naturally, and as essential mechanisms in the physical processes. In solar physics, cyclotron and synchrotron radiation are largely responsible for the radio and optical emission from flares, while the emission of x rays and gamma rays is due mainly to bremsstrahlung arising in electron–electron and electron–ion (hydrogen and helium) collisions. [We note here the analogy to laboratory plasmas, for the operation of which both bremsstrahlung and cyclotron radiation are crucial processes. Their measurement serves as an important diagnostic tool; from it can be determined the plasma density, the electron temperature, the impurities (ions other than hydrogen isotopes), and the presence of high-energy particles in the plasma.] Farther away, but within the galaxy, synchrotron radiation is the primary source of radio waves, and both the continuum gamma-ray sources and the diffuse soft x-ray background are due to bremsstrahlung. Finally, there are the very important extragalactic sources of radiation. The x-ray emission from clusters of galaxies is believed to be due largely to bremsstrahlung, while the radio waves from the largest known energy source, the quasars, are due to synchrotron radiation.

See also ELECTRODYNAMICS, CLASSICAL; QUANTUM ELECTRODYNAMICS; SYNCHROTRON RADIATION.

BIBLIOGRAPHY

Classical Radiation

J. D. Jackson, *Classical Electrodynamics*, 2nd ed. Wiley, New York, 1975. (I)

Quantum-Mechanical Bremsstrahlung

V. B. Berestetskii, E. M. Lifshitz, and L. P. Pitaevskii, *Relativistic Quantum Theory*, Vol. 4 of *Course of Theoretical Physics*, Part I. Pergamon, New York, 1971.

H. A. Bethe and E. E. Salpeter, "Quantum Mechanics of One- and Two-Electron Systems," *Handbuch der Physik* (S. Flügge, ed.), Vol. 35, *Atoms I*. Springer-Verlag, Berlin and New York, 1957.

G. Diambrini Palazzi, "High-Energy Bremsstrahlung and Electron Pair Production in Thin Crystals," *Rev. Mod. Phys.* **40,** 611 (1968).

W. Heitler, *The Quantum Theory of Radiation*, 3rd ed. Oxford, London and New York, 1954.

H. W. Koch and J. W. Motz, "Bremsstrahlung Cross-Section Formulas and Related Data," *Rev. Mod. Phys.* **31,** 920 (1959).

J. W. Motz, Haakon A. Olsen, and H. W. Koch, "Pair Production by Photons," *Rev. Mod. Phys.* **41,** 581 (1969).

Haakon A. Olsen, *Applications of Quantum Electrodynamics* (Springer Tracts in Modern Physics, Vol. 44). Springer-Verlag, Berlin and New York, 1968.

A. Sommerfeld, *Atombau und Spektrallinien*, 3rd ed., Vol. II. F. Vieweg, Braunschweig, 1960.

Synchrotron Radiation

G. N. Kulipanov and A. N. Skrinskii, "Utilization of synchrotron radiation: current status and prospects," Usp. Fiz. Nauk **122,** 369 (1977). English translation Sov. Phys. Usp. **20** (7), 559 (1977).

C. Kunz, ed. *Synchrotron Radiation, Techniques and Applications* (Springer Topics in Current Physics, Vol. 10). Springer-Verlag, Berlin, Heidelberg, New York, 1979.

E. M. Rowe and J. H. Weaver, "The Uses of Synchrotron Radiation," *Sci. Am.* **236,** 32 (1977). (I)

H. Winick and A. Bienenstock, "Synchrotron Radiation Research," *Annu. Rev. Nucl. Part. Sci.* **28,** 33 (1978).

H. Winick and S. Doniach, eds., *Synchrotron Radiation Research*. Plenum Press, New York, 1980.

Plasma Physics

G. Bekefi, *Radiation Processes in Plasmas*. Wiley, New York, 1966.

Beamstrahlung

M. Bell and J. S. Bell, "Quantum Beamstrahlung," *Particle Accelerators* **22,** 301 (1988); "End Effects in Quantum Beamstrahlung," *Particle Accelerators* **24,** 1 (1988).

R. Blankenbecler and S. D. Drell, "Quantum Treatment of Beamstrahlung," *Phys. Rev.* **D36,** 277 (1987); "Quantum Beamstrahlung from Ribbon Pulses," *Phys. Rev.* **D37,** 3308 (1988); "Quantum Beamstrahlung: Prospects for a Photon–Photon Collider," *Phys. Rev. Lett.* **61,** 2324 (1988).

P. Chen, "An Introduction to Beamstrahlung and Disruption," in *Frontiers of Particle Beams*, M. Month and S. Turner, eds. Lecture Notes in Physics, Vol. 296, pp. 495–532. Springer-Verlag, Berlin, Heidelberg, New York, 1988; "Review of Linear Collider Beam–Beam Interaction," SLAC-PUB-4823 (Jan. 1989), SLAC, Stanford, CA, also contributed to U.S. Particle Accelerator School, Batavia, IL, July 20–August 14, 1987; "Disruption, Beamstrahlung, and Beamstrahlung Pair Creation," SLAC-PUB-4822 (Dec. 1988), SLAC, Stanford, CA, also contributed paper at the DPF Summer Study Snowmass '88, High Energy Physics in the 1990's, Snowmass, Col., June 27–July 15, 1988.

M. Jacob and T. T. Wu, "Quantum Approach to Beamstrahlung," *Phys. Lett.* **B197,** 253 (1987); "Quantum Calculation of Beamstrahlung," *Nucl. Phys.* **B303,** 373 (1988), **B303,** 389 (1988).

B. Richter, "Very High Energy Colliders," *IEEE Trans. Nucl. Sci.* **NS-32,** 3828 (1985).

P. B. Wilson, "Future e^+e^- Linear Colliders and Beam–Beam Effects," SLAC-PUB-3985 (May 1986) SLAC, Stanford, CA.

Solar and Astrophysics

T. Bai, "Studies on solar hard x-rays and gamma-rays; Compton backscatter, anisotropy, polarization, and evidence for two phases of acceleration," Ph.D. thesis, Univ. of Maryland, Goddard Space Flight Center Report X-660-77-85, April 1977. See also references in this work.

R. Giacconi and H. Gursky, eds., *X-ray Astronomy* (Astrophysics and Space Science Library, Vol. 43). Reidel, Dordrecht, 1974.

V. L. Ginzburg, *Elementary Processes for Cosmic Ray Astrophysics*. Gordon & Breach, New York, 1969.

V. L. Ginzburg and S. I. Syrovatskii, *The Origin of Cosmic Rays*. Pergamon, New York, 1964.

F. B. McDonald and C. E. Fichtel, eds., *High-Energy Particles and Quanta in Astrophysics*. MIT Press, Cambridge, Mass., 1974.

A. G. Pacholczyk, *Radio Astrophysics*. W. H. Freeman, San Francisco, 1970.

G. L. Verschuur and K. I. Kellermann, eds., *Galactic and Extra-Galactic Radio Astronomy*. Springer-Verlag, Berlin and New York, 1974.

T. C. Weeks, *High-Energy Astrophysics*. Chapman & Hall, London, 1969. (I)

R. J. Weymann, T. L. Swihart, R. E. Williams, W. J. Cocke, A. G. Pacholczyk, and J. E. Felton, *Lecture Notes on Introductory Theoretical Astrophysics*. Pachart, Tucson, Arizona, 1976.

Brillouin Scattering

James L. Hunt

Brillouin scattering in a pure fluid is the scattering of light by transitory and localized thermally driven fluctuations in the density of the scattering medium. The effect was first predicted by Brillouin, later by Mandel'shtam (in the U.S.S.R. it is known as "Mandel'shtam–Brillouin scattering"), and was first observed by Gross. The history of the subject before the advent of the laser is reviewed by Rank [1].

The thermally driven density fluctuations in a fluid are of two types: the adiabatic (pressure) fluctuations and the isobaric (entropy) fluctuations. The former are collective modes of motion of the molecules and propagate in the fluid with the velocity of sound, and, thus, light scattered from them experiences a Doppler shift. The latter do not propagate and so do not alter the frequency of the scattered light; they correspond to a diffusional motion of the molecules.

The propagating fluctuations may be viewed as a set of Debye waves travelling in all directions with speed v in the fluid (see Fig. 1). The choice of a direction IX for the input beam of light and an observing direction XO selects two sets of plane waves which satisfy the law of reflection. The components of the sound speed of these two sets in the direction of observation are $\pm v \sin(\theta/2)$ which yield equal and opposite Doppler shifts (Δf) whose magnitude relative to the frequency (f) of the incident light is

$$\left|\frac{\Delta f}{f}\right| = 2n\frac{v}{c}\sin\frac{\theta}{2}, \qquad (1)$$

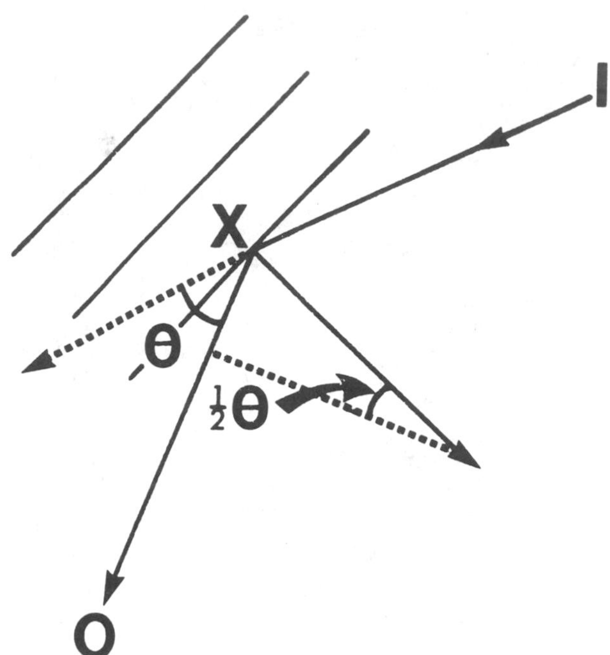

FIG. 1. Reflection of light from plane sound waves.

In Eq. (1), n is the index of refraction and c is the velocity of light in vacuum.

The expected "Brillouin spectrum," i.e., the distribution of scattered light intensity as a function of frequency, should thus consist of three components: one at frequency f (scattered from the nonpropagating fluctuations), one at $f+|\Delta f|$, and one at $f-|\Delta f|$. The whole spectrum is sometimes called the Rayleigh–Brillouin triplet. In all of the foregoing, it is assumed that the light is polarized with the electric vector perpendicular to the plane of IXO and that the scattered light is also observed in this polarization.

The recording of Brillouin spectra requires special techniques since the frequency shifts, Δf, are small ($\Delta f/f = 10^{-6}$), and high-resolution optics are required, e.g., Fabry–Perot etalons. The use of the laser as a light source has resulted in greatly increased activity in Brillouin scattering research. Polarized Brillouin spectra have been observed in almost all isotropic media, e.g., organic and inorganic liquids, amorphous solids, crystals, and gases. An example of the Rayleigh–Brillouin triplet is shown for benzene in Fig. 2.

The ratio of the intensity, I_c, of the central component to the intensity, $2I_B$, of the components shifted in frequency is called the "Landau–Placzek ratio":

$$\frac{I_c}{2I_B} = \frac{c_p - c_v}{c_v} = \gamma - 1, \tag{2}$$

where c_p and c_v are the specific heats of the fluid at, respectively, constant pressure and constant volume, and $c_p = \gamma c_v$. The formula is only approximately correct and many refinements to it have been made for various media.

Even if the incident light were strictly monochromatic and the spectrum were infinitely well resolved, the shifted components in the spectrum would have a finite width due to the damping of the sound waves in the medium. Thus the width of the shifted components is related to the viscosity of the fluid. The width of the unshifted component is related to the thermal diffusivity.

Another broad non-Lorentzian component of the scattered spectrum with its central frequency at the unshifted position is called the "Mountain mode." As the temperature of the scattering sample is lowered, it narrows and appears as a contribution to the Rayleigh line. This contribution arises because of the dispersion of the shear and bulk viscosities and the resulting distribution of relaxation times.

A different spectrum results when the incident light is polarized with the electric vector perpendicular to the plane IXO but observations are made on light polarized in the plane. This depolarized light is in part due to the reorientation of anisotropic molecules; this scattering is not usually called "Brillouin scattering."

Brillouin scattering has also been observed in binary mixtures of fluids. In this case light scattering may also occur as a result of concentration fluctuations; thus information on diffusion coefficients and reaction rates may be obtained. A further type of scattering is "stimulated Brillouin scattering." In this type of scattering the intensity of the scattered light increases nonlinearly with the dimensions of the scattering volume. The effect is observed when very intense sources of light, such as the pulsed ruby laser, are used. In this case, the electric field of the light is sufficiently high to alter (through electrostriction) the nature of the sound waves. Thus the incident and scattered light now alters the motion of the medium whereas in normal Brillouin scattering the motion of the medium is completely described by the Debye waves.

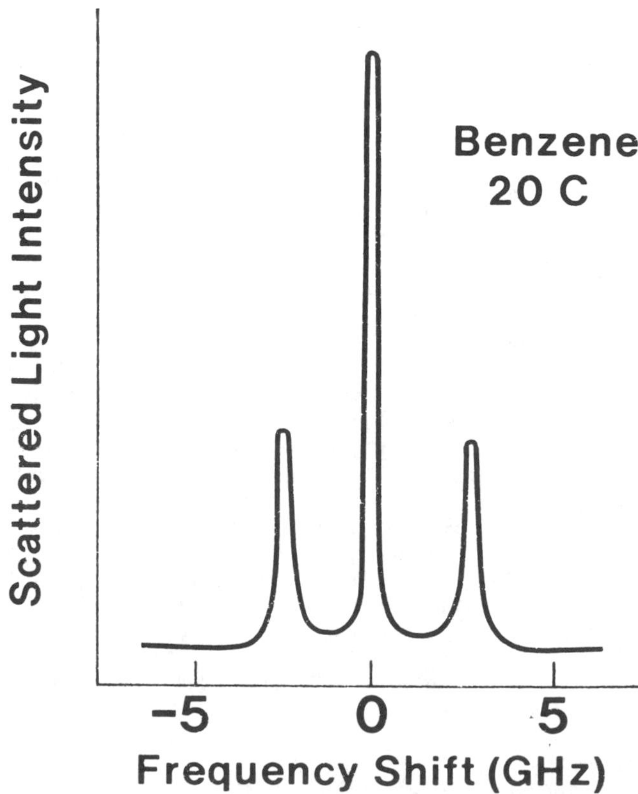

FIG. 2. The Rayleigh–Brillouin triplet for benzene at 20°C.

Brillouin scattering is a widely used analytical tool for investigating properties of matter at the molecular level (e.g., relaxations) and bulk properties (e.g., sound velocity and moduli).

A thorough discussion and bibliography of Brillouin scattering up to 1967 can be found in the book by Fabelinskii [2]. More recent reviews are those of Patterson and Carroll [3] dealing with fluids and Dil [4] on condensed matter including solids.

See also LIGHT SCATTERING.

REFERENCES

1. D. H. Rank, *J. Acoust. Soc. Am.* **49,** 937 (1971).
2. I. L. Fabelinskii, *Molecular Scattering of Light.* Plenum Press, New York, 1968.
3. G. D. Patterson and P. J. Carroll, *J. Phys. Chem.* **89,** 1344 (1985).
4. J. G. Dil, *Rept. Prog. Phys.* **45,** 285 (1982).

Brownian Motion

Frank A. Blood, Jr.

Brownian motion is a random motion of microscopic particles suspended in a fluid. The Scottish botanist Robert Brown, although not the first to see this phenomena, observed it in 1827 for many different types of materials—particles within various pollen grains, powdered coal, glass, and metals—suspended in liquids [1]. Smoke particles in air also show this random motion. A satisfactory explanation of these observations was given in 1906 by Einstein (and separately by Smoluchowski) and constituted the most direct proof up to that time of the existence of atoms. The small random motions occur because atoms bombard the smoke particle in a haphazard manner, sometimes applying more force on one side than another. This random variation in the force causes the particle to move in a random fashion, sometimes called a random walk [2]. An analogy is that of BBs being shot from all directions toward a billiard ball. Slight fluctuations in the number of BBs hitting one side will cause the billiard ball to move.

Einstein derived a formula [3–5] which can be compared with observation. Suppose you observe one particle and note how far it is from where it started after 1 min, 2 min, etc. Then you make the same measurements for many particles and take the average distance travelled after 1 min, 2 min, etc. Einstein's formula says that the average distance travelled is proportional to the square root of the time. That is, a particle, on the average, is 1.414 times farther from the origin after 2 min than it was after 1 min. Furthermore, the average distance travelled depends on the temperature of the fluid the particle is in. The reason for this is that the atoms move faster at higher temperatures and therefore kick the particle harder and farther. The average distance moved also becomes smaller as the size of the particle becomes larger, and, in fact, the Brownian motion is not even observable for large smoke particles.

Another method of observing Brownian motion is to look at the angular movements of a very small mirror suspended from fine quartz fibers. An accurate determination of Boltzmann's constant can be made from such observations [6–7].

An analogous phenomena, Johnson noise [8–10], occurs in electrical circuits. The density of electrons in a wire will fluctuate, again because of random thermal motions. Too many electrons (or too few) in a particular region will produce an electric field and therefore a voltage. The average size of the random fluctuation in voltage in a resistor depends on the resistance and, again, on the temperature. As an example, a 400 000-Ω resistor at room temperature will have an average spurious voltage fluctuation of about 10 μV when fluctuations faster than 20 000 Hz are discounted.

See also FLUCTUATION PHENOMENA.

REFERENCES

1. *Dictionary of Scientific Biography*, Vol. II, p. 516. Scribners, New York, 1970.
2. R. P. Feynman, R. B. Leighton, and M. Sands, *The Feynman Lectures on Physics*, Vol. I, Sec. 6, p. 5. Addison-Wesley, Reading, Mass., 1963. (E)
3. F. Reif, *Fundamentals of Statistical and Thermal Physics*, pp. 560–564. McGraw-Hill, New York, 1965. (I)
4. G. Wannier, *Statistical Physics*, pp. 475–478. Wiley, New York, 1966.
5. A. Sommerfeld, *Thermodynamics and Statistical Mechanics*, pp. 180–183. Academic Press, New York, 1965. (I)
6. R. P. Feynman *et al.*, in Ref. 1, Sec. 41, p. 2.
7. A. Sommerfeld, in Ref. 4, pp. 486–489.
8. R. P. Feynman *et al.*, in Ref. 1, Sec. 41, p. 2.
9. G. Wannier, in Ref. 3, pp. 486–489.
10. A. L. King, *Thermophysics*, pp. 212–214. Freeman, San Francisco, 1962. (I)

Bubble Chambers *see* Cloud and Bubble Chambers

Calorimetry

Edgar F. Westrum, Jr.

A calorimeter is a device for determining energetic quantities typically by measurement of temperature changes produced upon input of electrical work to a system or by heat released by some reaction or process. Typical examples are:

Heat capacities, defined by $C_x = (dq/dT)_x$. Here dq is the heat involved (e.g., as laser radiation, the electrical supplied work, etc.) and x represents the constraints under which the experiment is performed such as constant pressure, constant volume, constant magnetic field, etc.

Enthalpy of phase transitions (e.g., solid/solid, fusion, vaporization, sublimation) or enthalpy increments associated with reactions, mixing, solution processes, or with transitions involving order/disorder, magnetic, ferroelectric, electronic (e.g., Schottky contributions associated with electronic energy levels) phenomena, etc.

Transport properties of materials such as the thermal diffusivity.

Such experimental quantities provide precise characterization of the energetic states of matter. At very low temperatures, the heat capacity assumes a primary role in such matters. At increasingly higher temperatures—but still cryogenic—one enters a thermal chaos in the complexity of the energy spectrum of the thermal properties, such that heat-capacity studies in this region are of considerably less interest to physicists than are those at extremely low temperature. On the other hand, the region above 10 K to—and above—ambient temperatures is of great interest to chemical thermodynamicists concerned with thermophysical properties at or above ambient temperatures. The region from 30 to 300 K is the region over which most of the entropy and the enthalpy are developed. Since the energy interpretation becomes largely statistical in nature, the entropy—particularly the entropy increments associated with the various transformations of matter—often become a more relevant parameter or criterion than does the heat capacity from which they were derived. Hence, this temperature region also assumes a crucial role in the critical evaluation of ambient-temperature chemical thermodynamic values for science (physics, chemistry, materials science, biology, geology, mineralogy, and petrology).

Thus, entropy S_x in a heat-capacity or gradual transformation region provides a definitive thermodynamic criterion:

$$S_x = \int_0^T \frac{C_p(T)dT}{T} + S_0$$

in which upon evaluation of the entropy at zero Kelvin (S_0), practical "absolute" values of the entropy can be ascertained at higher temperatures for thermodynamic utilization.

For an isothermal phase transition,

$$\Delta_{\text{trs}} S = \Delta_{\text{trs}} H / T_{\text{trs}}.$$

Such thermophysical and thermochemical data are invaluable in establishing thermodynamic properties which in turn are needed in the design of physical apparatus or of technological chemical operations. The design of calorimetric apparatus ranges over many types since a great variety of substances, changes of state, and thermodynamic processes may be studied. In physics, quantum effects are often studied with low-temperature heat-capacity cryostats employing solid or liquid nitrogen, liquid hydrogen, liquid helium-4, or liquid helium-3 as refrigerants.

A typical aneroid cryostat is depicted in Figure 1. The sample under investigation, usually a massive, crystalline, or even a finely divided specimen, is contained in a calorimeter (**D**), sealed so that a few kPa of helium may be added to ensure thermal diffusivity between sample, heater, and thermometer. An electrical "heater" (i.e., a coil of resistance wire through which electrical work may be provided) and a capsule-type resistance thermometer (**E**) are often inserted in an axial entrant well with a film of grease to provide adequate thermal contact. A surrounding jacket or shield surrounds the calorimeter to reduce the heat exchange with the surroundings to a negligible value. This is achieved by controlling the temperature of the shield (**C**) or the principal portions of it dynamically to follow the temperature of the calorimeter within the ability of thermocouples (**F**) to monitor the difference. Separate channels of an analog microprocessor control the adiabatic shield. To enhance the quality of this control two other channels of such control monitor the temperature of the surrounding guard shield (**B**) as well as that of the ring (**A**) to "temper" the temperature of the bundle of electrical lead wires—a main source of thermal leak at low temperatures.

Usually the determination of the temperature increment (10^{-6} to 10 K) is the limiting factor in the accuracy of the heat-capacity measurement as the energy increment can be readily ascertained. Temperature determination involves not only the measurement of the resistance of the thermometer, but the stability and calibration of the thermometer as well. Moreover, the establishment of the temperature scale is relevant.

A series of certified heat-capacity standards established by the (U.S.) Calorimetry Conference have enabled calorimetrists to test the reliability of their calorimetric systems.

Because thermometry often limits the accuracy of the calorimetric results, it should be noted that any temperature-dependent property such as electrical resistance, magnetic susceptibility, vapor pressure, or *PVT* relations of gases can be used as the thermometric property. Electrical resistance thermometers, capsule-type with a fine helically wound coil

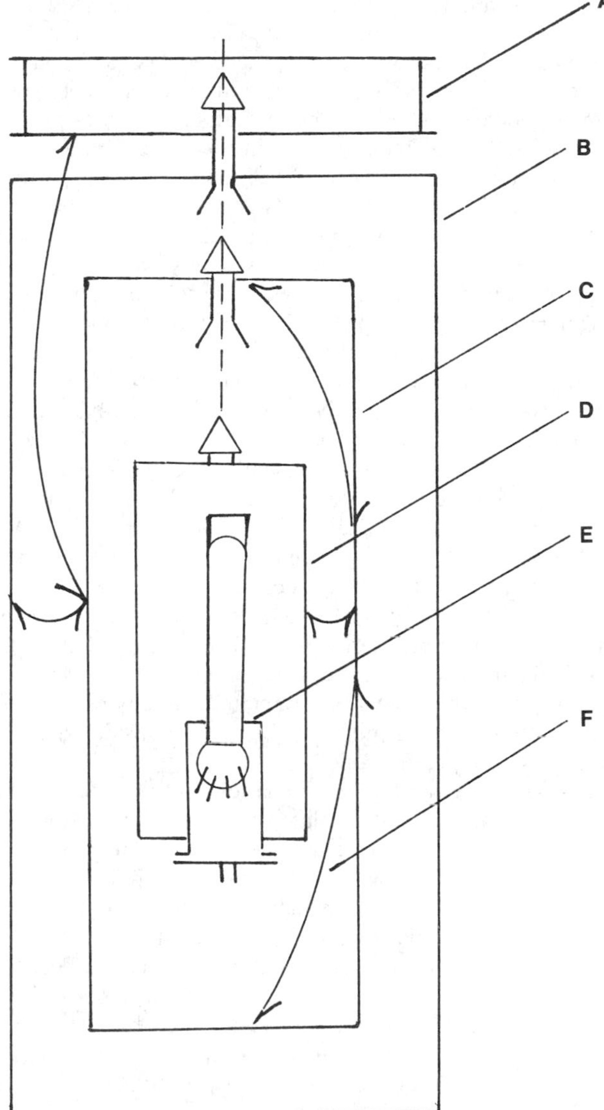

FIG. 1. Schematic cryogenic, adiabatic, heat-capacity cryostat (a superambient thermostat would be similar except in the provision of additional heated and/or floating guard shields). **A,** ring for tempering lead bundle; **B,** guard shield (heated); **C,** adiabatic shield (three separate controlled temperature regions); **D,** calorimeter containing sample; **E,** resistance thermometer (within "heater" sleeve); **F,** one of five differential thermals (multijunction thermocouples).

of wire wound in a strain-free and noninductive configuration on a notched mica cross-like frame, are commonly employed in cryogenic calorimetry. These are typically sealed within a platinum capsule sealed with a glass head to permit entry of the leads and to provide for the sealing of a small amount of helium gas to assist thermal diffusion. Thermal contact with calorimeter vessel or the "heater" sleeve is achieved with vacuum grease or low-melting solder. The sensitivity of platinum resistance thermometers diminishes rapidly below about 20 K and they become too insensitive for accurate thermometry. The use of an iron–rhodium thermometric alloy avoids this problem to below the temperatures of liquid helium (4 K). At sufficiently low temperatures semiconducting resistors with negative temperature coefficients of resistance—even ordinary carbon resistors—are often used. Encapsulated doped germanium resistance thermometers provide a more accurate thermometer which may be cycled over the temperature range without effect.

More exotic thermometers are used below 0.1 K including magnetic moment and γ- and x-ray emission. (Such thermometers are described in Ref. [2].)

The utilization of thermocouples and thermals in ascertaining temperature differentials in the control of adiabatic shields and in relating the temperature of the calorimeter to a more remotely located (more massive) accurate temperature sensor is almost universal.

Although modes of calorimeter operation other than the adiabatic exist—or the quasiadiabatic where a slight offset in the control temperature relative to that of the calorimeter provides effective cancellation of the energy supplied by thermometric current—adiabatic working does permit measurements *at equilibrium*. Isoperibol operation—with a shield at a more or less fixed temperature—does have some convenience at very low temperatures.

At higher temperatures, calorimetry by the method of mixtures or drop calorimetry is often useful. Inverse temperature drop calorimetry has certain advantages in eliminating slow equilibrium and hysteresis effects.

Because of the availability of computerized commercial units requiring only milligram size samples, differential scanning calorimetry (dsc) is often useful above 100 K.

By computerized automation in calorimetric operation and data acquisition even equilibrium, adiabatic measurements do not require much more operator endeavor than does dsc.

A recent comprehensive monograph on thermophysical calorimetric techniques [1] updates the earlier treatise [2] published by The International Union of Pure and Applied Chemistry as well as the excellent discussion of cryogenic and thermometric techniques at extremely low temperature [3]. A more general treatise is also available [4].

See also HEAT CAPACITY; MAGNETIC COOLING; TEMPERATURE; THERMOMETRY.

REFERENCES

1. C. Y. Ho (ed.) *Specific Heat of Solids, Cindas Data Series on Materials Properties, Volume I—II.* Hemisphere, New York, 1988. (I)
2. J. P. McCullough and D. W. Scott, *Experimental Thermodynamics, Vol. I: Calorimetry of Non-reacting Systems.* Butterworth, London, 1968. (A)
3. O. V. Lounasmaa, *Experimental Principles and Methods below 1 K.* Academic, London, 1974. (A)
4. W. Hemminger and G. Hohne, *Calorimetry—Fundamentals and Practice.* VCH Publishers, New York, 1984. (Also available in a German edition.) (E)

Capillary Flow

J. C. Melrose

The term capillary flow refers to the motion of two or more fluid phases within the interstices of a finely porous solid. Viscous flows of this type are encountered in a wide variety of contexts in natural science and technology. The transport of water in soils and the recovery of oil and gas from reservoir rocks are examples of particular importance. Other well-known examples are found in such diverse fields as paper and textile manufacturing, food processing, and powder metallurgy.

The driving forces responsible for capillary flow involve the fluid pressure differences associated with the highly curved interfaces between the several fluid phases (menisci). For most cases in which capillary flow is of significance, the average diameter of the pores will be of the order of 10^{-5}–10^{-1} mm. Consequently, the effect of gravity on the curvature of the interfaces is either negligible or can be accounted for by small corrections. Also, the flows in question will usually be sufficiently slow that hydrostatic conditions are approached in the vicinity of the interfacial regions separating any pair of fluid phases.

In these circumstances, the pressure differences that give rise to the driving forces for capillary flow are determined, at least approximately, by the classical hydrostatic principles applicable to fluid interfaces. The first of these is the Young–Laplace equation, $P_c = \gamma J$. Here, P_c is the static pressure difference (capillary pressure) between a pair of contiguous bulk fluids, γ is the corresponding fluid/fluid interfacial tension, and J is the (constant) mean curvature of the interface separating the two phases. A rigorous thermodynamic analysis of the Young–Laplace equation, as developed initially by Gibbs, indicates that the dependence of γ on curvature is of significance only when the reciprocal of the curvature, J^{-1}, approaches molecular dimensions.

For nonaxisymmetric interfaces the precise description of a surface of constant mean curvature involves the solution of a nonlinear partial differential equation of the elliptic type. However, in the simple case of a very small cylindrical capillary tube with uniform wetting characteristics, such a surface is spherical in shape, and the mathematical difficulties arising in the general case are avoided.

A second principle is the Young equation, $\gamma_{12} = \gamma_{13} + \gamma_{23} \cos \theta$. Here, subscript 1 refers to the solid phase, while subscripts 2 and 3 refer to the two fluid phases in contact with the solid. The Young equation thus interrelates the properties (tensions) of three interfaces that meet at a three-phase contact line. Included in this relationship is the geometrical parameter characterizing the confluence of phase boundaries (the contact angle, θ). This principle and the Young–Laplace equation represent the line and surface analogs of the condition for hydrostatic equilibrium applicable to bulk fluid phases. Whereas the Young–Laplace equation corresponds to a differential equation, the contact angle (as defined by the Young equation), together with the configuration of the solid surface comprising an individual pore, provides the boundary condition required for its solution.

Interfacial properties and the physical principles that determine static interfacial configurations are thus of central importance in capillary flow. The dynamical process by which one fluid displaces another in a cylindrical tube of small diameter is a simple and well-known example of such a flow. In this special case the curvature of the meniscus separating the two fluids under static conditions is given by twice the ratio of the cosine of the contact angle (assumed to be uniform) to the radius of the tube. Under dynamic conditions, however, the flows that occur in the vicinity of the moving fluid interface are quite complex, and it is observed that the so-called dynamic contact angle is generally rate dependent. When this complicating factor is ignored, the analysis of the capillary flow in a small cylindrical tube is straightforward, yielding the well-known Washburn equation.

More generally, typical porous solids possess very complex systems of interconnecting pores, any one of which has a varying cross-sectional area available for flow. Regular or irregular packings of spherical solid particles provide examples of such complex pore systems. The pores in this more general type of porous solid constitute channels that are alternately convergent and divergent. Consequently, the capillary pressure and the meniscus curvature, for a particular static configuration of the meniscus, no longer depend simply on the cosine of the contact angle. This more complex dependence on contact angle is frequently overlooked in analyzing capillary flow behavior in real porous solids.

The structural feature associated with pore channels which are alternately convergent and divergent also plays an important role in determining which configurations of the fluid/fluid interfaces are actually stable under hydrostatic conditions. A third physical principle, also governing static capillary phenomena, must therefore be considered in the case of typical porous solids.

This principle is a stability condition that must be satisfied by the various individual fluid/fluid interfaces within the pores. It is found that the quasi-static motion of an interface through a given pore involves interfacial configurations that are alternately stable and unstable. A rather sudden hydrodynamic event (Haines jump or rheon) will occur when the configuration of an interface within a particular pore becomes unstable. Displacement within those pores that are occupied by interfaces is therefore initiated by a slow flow process, which is then followed by a rapid burst of flow. The process is completed when the interface achieves a stable configuration in a neighboring pore.

Sudden flow, or rheons, resulting from unstable fluid/fluid interfacial configurations, are thus a microscopic feature of the highly irregular displacement front characterizing displacement processes in typical porous solids. Also associated with such hydrodynamic events is the trapping of segments of the displaced fluid in certain pores or in certain portions of typical pores. The interfacial configurations and the relative volume of the displaced phase that is trapped depend on the degree to which the pores are interconnected. These factors are also dependent on the prevailing curvature (convex or concave with respect to the displaced phase) of the fluid interfaces, i.e., on whether the displacement is the so-called drainage or imbibition type of process.

It should be noted in this connection that the distinction between drainage and imbibition is difficult to make in the case of intermediate wetting conditions, i.e., when the contact angle is more than about 40° or less than about 140°. Under strongly wetting conditions, however, the term drainage unambiguously refers to the displacement of the wetting phase (contact angle less than 40°) by the nonwetting phase, while imbibition refers to the reverse process. If these processes are each carried out under quasi-static conditions, two different relationships between the interfacial curvature (proportional to the static pressure difference between the two bulk fluid phases) and the relative volume of one of the fluids are defined. These relationships therefore characterize the hysteresis associated with the two different quasi-static displacement paths. This type of hysteresis is different from and should be distinguished from the hysteresis that is frequently observed in the value of the contact angle itself.

The term capillary flow often refers to a displacement process of the imbibition type, since external driving forces are required in drainage processes. Consequently, the interfacial curvature is concave with respect to the displaced phase. Thus, it is the nonwetting phase that is subjected to the fluid-phase entrapment associated with this type of capillary flow. It is found that the relative volume of trapped fluid varies from about 10% to as much as 50%, depending on the structure of the porous solid.

In many applications, capillary flow processes of the drainage type are also encountered. Often, one of the fluid phases is a gas, while the other is a volatile liquid. In this case, adsorption at the gas/solid interface of those components of the liquid which are volatile must be taken into account. First, it is found that the adsorption is usually sufficient to establish a zero value of the contact angle. Second, the total volume or mass of the wetting liquid retained within the solid phase at a particular value of the capillary pressure depends on the thickness of the adsorbed (wetting) film, as well as on the principles of static capillary phenomena already considered.

This extension of the physical principles involved in capillary flow clearly requires a consideration of diffusional transport in the gas phase, as well as viscous flow in the liquid phase. The classical hydrostatic principles applicable to fluid interfaces must be supplemented by the Gibbs principle of chemical equilibrium between the gas and liquid phases. This principle is expressed by the well-known Kelvin equation (even though not originally so conceived).

See also HYDRODYNAMICS; SURFACE TENSION.

BIBLIOGRAPHY

F. P. Buff, "The Theory of Capillarity," in *Handbuch der Physik*, S. Flugge, ed., Vol. 10, pp. 281–304. Springer-Verlag, Berlin and New York, 1960.

F. A. L. Dullien, *Porous Media*. Academic Press, New York, 1979.

E. B. Dussan V., "On the Spreading of Liquids on Solid Surfaces: Static and Dynamic Contact Lines," *Ann. Rev. Fluid Mech.* **11**, 371–400 (1979).

D. H. Everett, "Pore Systems and their Characteristics," in *Characterization of Porous Solids*, K. K. Unger, J. Rouquerol, and K. S. W. Sing, eds., pp. 1–22. Elsevier, Amsterdam, 1988.

R. Finn, *Equilibrium Capillary Surfaces*. Springer-Verlag, New York, 1986.

P. G. De Gennes, "Wetting: Statics and Dynamics," *Rev. Mod. Phys.* **57**, 827–863 (1985).

J. C. Melrose, "Thermodynamics of Surface Phenomena," *Pure and Applied Chemistry* **22**, 273–286 (1970).

J. R. Philip, "Flow in Porous Media," *Ann. Rev. Fluid Mech.* **2**, 177–204 (1970).

G. F. Teletzke, H. T. Davis, and L. E. Scriven, "How Liquids Spread on Solids," *Chem. Eng. Comm.* **55**, 41–81 (1987).

R. A. Wooding and H. J. Morel-Seytoux, "Multiphase Fluid Flow through Porous Media," *Ann. Rev. Fluid Mech.* **8**, 233–274 (1976).

Carnot Cycle

*Mark W. Zemansky**

A heat engine, such as an internal combustion engine or a steam engine, is a device for converting heat into work. To this end, a substance, such as a mixture of fuel and air or a mixture of water and steam, called the *working substance*, is caused to undergo a series of processes, called a *cycle*, in which the working substance returns periodically to its initial state. During a part of a heat engine cycle, some heat Q_H is absorbed from a chemically reacting mixture of fuel and air, known as the *hot reservoir*, and during another part of the cycle, a smaller amount of heat Q_C is rejected to a *cold reservoir*. The engine is therefore said to operate between these two reservoirs. The difference between the heat Q_H and the smaller quantity Q_C is converted to work W, so that

$$W = Q_H - Q_C$$

The work W is the output of the engine and the heat Q_H is the input, so that the efficiency E is

$$E = \frac{W}{Q_H} = \frac{Q_H - Q_C}{Q_H} = 1 - \frac{Q_C}{Q_H}.$$

Since it is a fact of experience that some heat is always rejected to the cooler reservoir, the efficiency of an actual engine is never 100%. If we assume that we have at our disposal two reservoirs at given temperatures, it is important to answer the following questions: (1) What is the maximum efficiency that can be achieved by an engine operating between these two reservoirs? (2) What are the characteristics of such an engine? (3) Of what effect is the nature of the substance undergoing the cycle? The importance of these questions was recognized by Nicolas Léonard Sadi Carnot, a brilliant young French engineer, who in the year 1824, before the first law of thermodynamics was firmly established, described, in a paper entitled "Reflexions sur la puissance motrice du feu," an ideal engine operating in a particularly simple cycle known today as the *Carnot cycle*.

A Carnot cycle is a set of processes that can be performed by any thermodynamic system whatever, whether chemical, electrical, magnetic, or other. The working substance is imagined first to be in thermal equilibrium with a cold reservoir at the temperature T_C. Four processes are then performed, in the following order:

1. A reversible adiabatic process is performed in such a direction that the temperature rises to that of the hotter reservoir, T_H.

* Deceased.

2. The working substance is maintained in contact with the reservoir at T_H and a reversible isothermal process is performed in such a direction and to such an extent that heat Q_H is absorbed from the reservoir.

3. A reversible adiabatic process is performed in a direction opposite to (1) until the temperature drops to that of the cooler reservoir, T_C.

4. The working substance is maintained in contact with the reservoir at T_C and a reversible isothermal process is performed in a direction opposite to (2) until the working substance is in its initial state. During this process, heat Q_C is rejected to the cold reservoir.

An engine operating in a Carnot cycle is called a *Carnot engine*. A Carnot engine operates between two reservoirs in a particularly simple way. All the heat that is absorbed is absorbed at a constant high temperature, namely, that of the hot reservoir. Also, all the heat that is rejected is rejected at a constant lower temperature, that of the cold reservoir. Since all four processes are reversible, the Carnot cycle is a reversible cycle. The expression "Carnot engine," therefore, means "a reversible engine operating between *only* two reservoirs."

It was shown by Kelvin that an absolute temperature scale could be defined in terms of the operation of a Carnot engine. If T_H and T_C are the Kelvin temperatures of the hot and cold reservoirs, respectively, then Kelvin showed that

$$\frac{Q_C}{Q_H}=\frac{T_C}{T_H},$$

and the Carnot efficiency becomes

$$E(\text{Carnot})=1-\frac{T_C}{T_H}.$$

For a Carnot engine to have an efficiency of 100% T_C must be zero. Since nature does not provide us with a reservoir at absolute zero, a heat engine with 100% efficiency is a practical impossibility.

A temperature–entropy diagram is particularly suited to display the characteristics of a Carnot cycle. The two reversible adiabatic processes are vertical lines, and the two reversible isothermal processes are horizontal lines lying between the two vertical lines, so that the Carnot cycle is represented by a rectangle, as shown in Fig. 1. This is true regardless of the nature of the system.

In describing and explaining the behavior of this ideal engine, Carnot made use of three terms: *feu, chaleur,* and *calorique*. By feu he meant fire or flame, and when the word is so translated no misconceptions arise. Carnot gave, however, no definitions for chaleur and calorique, but in a footnote stated that they had the same meaning. If both of these words are translated as heat, then Carnot's reasoning is contrary to the first law of thermodynamics. There is, however, some evidence that, in spite of the unfortunate footnote, Carnot did not mean the same thing by chaleur and calorique. Carnot used chaleur when referring to heat in general, but when referring to the motive power of heat that is brought about when heat enters at high temperature and leaves at low temperature, he used the expression "chute de calorique," never "chute de chaleur." It is the opinion of a few scientists that Carnot had in the back of his mind the concept of entropy, for which he reserved the term calorique. This seems incredible, and yet it is a remarkable circumstance that, if the expression chute de calorique is translated "fall of entropy," many of the objections to Carnot's work raised by Kelvin, Clapeyron, Clausius, and others are no longer valid. In spite of possible mistranslations, Kelvin recognized the importance of Carnot's ideas and put them in the form in which they appear today.

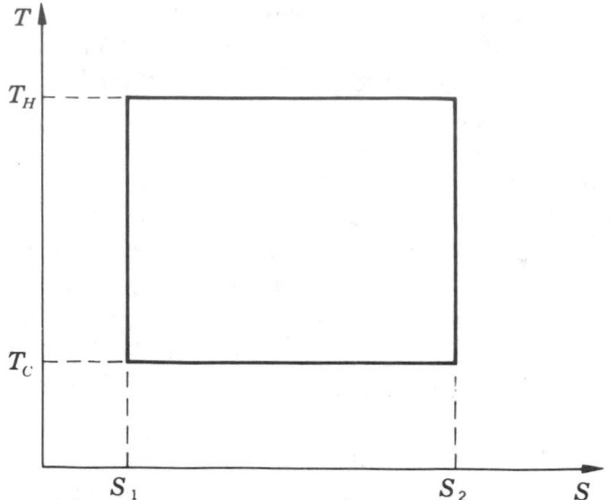

FIG. 1. A Carnot cycle of any system when represented on a *T–S* diagram is a rectangle.

See also THERMODYNAMICS, EQUILIBRIUM; HEAT ENGINES.

Catalysis

Gerhard Ertl

If a chemical reaction is favored thermodynamically, but proceeds too slowly, its rate may be enhanced by the addition of a catalyst. The term "catalysis" (derived from the Greek and meaning "break down") was introduced in 1835 by the Swedish chemist J. Berzelius and received its final definition in 1895 by W. Ostwald: A catalyst is a substance which affects the rate of a chemical reaction without appearing in its end products. Although in principle this phenomenon can refer to either an increase or diminution in the rate, one in practice refers to the agent which accelerates a reaction as a *catalyst* and to an agent which decreases its rate as an *inhibitor*. In short, this effect is caused by the ability of the catalyst to form intermediate bonds with the molecules involved in the reaction, such that an alternate reaction path with an enhanced overall rate is offered. Let us consider a general chemical reaction of the type $A + B \rightarrow AB$ (1), whereby the back-reaction ($AB \rightarrow A + B$) may be neglected for the moment; then a catalyst C may form an intermediate species $A + C \rightarrow AC$ (2) reacting further to $AC + B \rightarrow AB + C$ (3). The reaction will be catalyzed by C if the sum of the rates (2) and (3) exceeds that of the direct reaction (1).

The efficiency of a catalyst is characterized by its *activity* and *selectivity*. The activity may be identified with the number of product molecules formed per "active site" of the catalyst per second ("turnover frequency"). In general from a given set of interacting molecules, the reactants, not only a single but several reactions may occur, yielding different reaction products. Selectivity is then defined as the ratio of the rate of the desired reaction over the sum of the rates of all possible reactions.

Catalysis is a very widespread phenomenon which is of greatest importance for living organisms as well as technical processes: The molecules catalyzing the very complex chemical reactions in biological systems are called enzymes and are unique with respect to their extreme selectivity. Solid catalysts form, on the other hand, the basis of chemical and petroleum industries as well as of processes for controlling atmospheric pollution.

Nonbiological catalysis is usually classified into *homogeneous* and *heterogeneous* catalysis. In the former the molecules involved in the reaction, including the catalyst, are within a single (gaseous or liquid) phase, while in the latter case the catalyst is in a different state of aggregation (often solid) from the reacting species (gas or liquid), and the catalytic action occurs at the interface between these phases. In the following, a few examples will serve to illustrate the basic principles underlying the various forms of catalysis.

HOMOGENEOUS CATALYSIS

A simple example for a reaction sequence as outlined above is offered by the reaction $2SO_2 + O_2 \rightarrow 2SO_3$, which is catalyzed by nitric oxide, NO, through the intermediate steps

$$O_2 + 2NO \rightarrow 2NO_2,$$

$$NO_2 + SO_2 \rightarrow SO_3 + NO.$$

This is the basis of an old and important technical process for the formation of sulfuric acid.

NO as a catalyst may, on the other hand, have fatal consequences in the destruction of ozone (O_3) in the stratosphere, which reaction may be schematically formulated as

$$NO + O_3 \rightarrow NO_2 + O_2,$$

$$O_3 + h\nu \rightarrow O_2 + O,$$

$$NO_2 + O \rightarrow NO + O_2.$$

Obviously, the catalyst NO is not consumed in the overall reaction $2O_3 + h\nu \rightarrow 3O_2$, whose rate is, however, considerably enhanced by the presence of the atmospheric pollutant.

Acid–base catalysis is the most important form of homogeneous catalysis in liquid phase. The reaction

$$N_2CHCOOC_2H_5 + H_2O \rightarrow HOCH_2COOC_2H_5 + N_2,$$

is, for example, catalyzed by hydrogen ions through the intermediate step

$$N_2CHCOOC_2H_5 + H^+ \rightleftharpoons {}^+N_2\text{–}CH_2COOC_2H_5,$$

whereby the product decomposes readily into

$$^+N_2\text{–}CH_2COOC_2H_5 + H_2O \rightarrow HOCH_2COOC_2H_5 + N_2 + H^+,$$

and the catalyst (H^+) is formed back. The overall rate will in this case obviously be determined by the *p*H (= H^+ concentration) or, more generally, by the strength of the acid available for the hydrogen ion transfer process.

A variety of technical processes is based on homogeneous catalysis by transition metal compounds, such as the Wacker process for converting ethylene (C_2H_4) to acetyldehyde (CH_3CHO) in which the species $PdCl_4^{2-}$ plays the role of the catalytic agent. Separation of the catalyst from the product after completion of the reaction is, however, a general problem in processes of this type occurring in homogeneous phase.

HETEROGENEOUS CATALYSIS

The just-mentioned difficulty is almost nonexistent if, e.g., the catalyst is in the form of a solid which is in contact with the reacting species from the gaseous phase. The formation of chemical bonds between these species and the *surface* of the solid (chemisorption) is obviously of fundamental importance for this phenomenon of heterogeneous catalysis. The development of surface physical methods in recent years was of tremendous impact toward elucidation of the underlying elementary steps.

The activity of a given amount of a heterogeneous catalyst will depend on its surface area, and hence usually finely dispersed materials with a high specific surface area (typically up to 100 m^2/g) are applied. This may be achieved by depositing the catalytically active material (e.g., a noble metal) as small particles on a high surface area support such as alumina (Al_2O_3) or silica (SiO_2), by forming small catalyst particles during the chemical pretreatment (e.g., reduction) which may be stabilized by structural "promoters," or by using substances with a large *internal* surface area such as the important class of zeolites (crystalline aluminosilicates).

In general, the surface of a heterogeneous catalyst will be structurally and chemically nonuniform which renders this field so complex: The catalyst particle will expose various crystal planes exhibiting, in addition, defects such as steps and kinks. Moreover, the chemical composition of the surface may differ considerably from the (nominal) bulk composition; purposely added "electronic" promoters such as alkali atoms may enhance its specific activity, while the latter may be sensitively suppressed by other species acting as "poisons."

Access to the atomic processes underlying heterogeneous catalysis can be sought via the "surface science" approach: Well-defined single crystal surfaces under ultrahigh vacuum conditions serve as model systems for which the interaction with gaseous particles can be studied in detail. The "gap" to "real" catalysis can, on the other hand, be bridged by analyzing the surface properties of the practical catalysts and by systematically varying the properties of the model system (e.g., defect or impurity concentrations).

The reaction

$$2CO + O_2 \rightarrow 2CO_2$$

is readily catalyzed by the platinum group metals and is of practical relevance in the control of car exhaust gases. Its mechanism is basically quite simple and may be formulated as

$$CO + * \rightleftharpoons CO_{ad}$$

$$O_2 + 2* \rightarrow 2O_{ad}$$

$$CO_{ad} + O_{ad} \rightarrow CO_2 + 2*;$$

* denotes schematically a bare site on the catalyst surface to which either a CO molecule or an O atom is chemisorbed. The adsorbed oxygen atoms are formed through dissociation of O_2 molecules which process proceeds with high efficiency on the platinum surface, but which, on the other hand, is suppressed by a too high concentration (coverage) of chemisorbed CO. That is why under steady-state conditions the temperature has to be high enough in order to drive part of the adsorbed CO back into the gas phase (desorption).

The synthesis of ammonia (NH_3) from nitrogen (N_2) and hydrogen (H_2) is a large-scale industrial process (Haber–Bosch) which forms, inter alia, the basis for the production of fertilizers. This reaction proceeds with a promoted iron catalyst. Its mechanism, i.e., the sequence of elementary steps, could recently be elucidated following the surface science strategy, whereby the dissociative chemisorption of nitrogen, viz., $N_2 + 2* \rightarrow 2N_{ad}$, was identified as rate-limiting. In addition, based on the kinetic parameters derived from low-pressure studies with single crystal surfaces, even the yields under industrial conditions could be successfully modelled theoretically.

Progress in this area is rapid, and the complex area of heterogeneous catalysis is currently developing from a "black art" to a physical science.

ENZYME CATALYSIS

Biological reactions are catalyzed by macromolecules which belong to the class of proteins and which are called enzymes. This word is derived from the Greek "yeast," a material which played a key role in the development of this branch of science: E. Buchner showed in 1897 that a cell-free extract of yeast is able to catalyze the conversion of sugar into alcohol and carbon dioxide. The classification into "homogeneous" and "heterogeneous" catalysis breaks down in this case: The enzyme molecules are in solution in the form of colloidal dispersions, while interaction with the reacting molecules can, on the other hand, be considered as bond formation at the "surface" of such a macromolecule.

The key for understanding the catalytic properties of enzymes, and in particular of their fascinating selectivities towards specific reactions, lies in their complex structures. In 1894, E. Fischer compared an enzyme molecule with a lock which fits only with a single key, namely, the "correct" reacting molecule—a view which has essentially been confirmed by application of modern x-ray crystallography. Since the basic types of reactions of nonbiological catalysis are also present with enzyme catalysis, several formal analogies between both fields may be found. However, the high structural complexity of biological molecules introduces also cooperative phenomena and novel qualitative aspects.

See also ADSORPTION; KINETICS, CHEMICAL; SURFACES AND INTERFACES.

BIBLIOGRAPHY

Advances in Catalysis. Academic Press, New York, 19XX.

J. R. Anderson and M. Boudart, eds. *Catalysis—Science and Technology*. Springer-Verlag, Berlin, 19XX, 8 Volumes.

M. L. Bender and L. J. Brubaker, *Catalysis and Enzyme Action*. McGraw-Hill, New York, 1973.

G. C. Bond, *Heterogeneous Catalysis—Principles and Applications*. Clarendon Press, Oxford, 1974.

M. Boudart and G. Djéga-Mariadasson, *Kinetics of Heterogeneous Catalytic Reactions*. Princeton University Press, Princeton, NJ, 1984.

Catastrophe Theory

Gérard Toulouse and Francis J. Wright

The theory of catastrophes is one product of the study of singularities. For a long time, mathematics has been concerned with the study of regular objects (functions, mappings, differential equations, etc.) and theorems have been so phrased as to eliminate the singularities. But actually, it quite often appears that the singularities contain a great deal of qualitative information, and a strong trend has developed in mathematics to make a systematic study of them. A general approach consists in building a classification of archetypical singularities in order of increasing complexity. That is the content of Thom's classification of elementary catastrophes.

Elementary catastrophes are singularities associated with the bifurcation of extrema of smooth real functions $f\colon \mathbf{R}^n \rightarrow \mathbf{R}$. As such, they are also singularities of differentiable mappings $Df\colon \mathbf{R}^n \rightarrow \mathbf{R}^n$, and singularities of differential equations of gradient type, $dx_i/dt = X_i$, with $X_i = \partial f/\partial x_i$. Within the theory of differential equations (also called the theory of dynamical systems), gradient systems are a simple restricted class (e.g., the dynamical systems, often used in physics, with limit cycles or strange attractors are clearly outside the gradient class). The term *catastrophe* is sometimes applied outside the elementary framework to describe a breakdown of stability of any equilibrium causing a system to *jump* into another state. But usually *catastrophe* is used to mean *elementary catastrophe* as defined above, and it will be mainly used that way here.

Before giving specific details about the nature and classification of elementary catastrophes, it is appropriate to discuss where such a mathematical theory may be relevant in physics. Very many physical laws are expressed as variational principles: in optics (Fermat's principle), mechanics (principle of least action), thermodynamics (maximal entropy or minimal free energy), etc. In all these cases, the physical state (or trajectory) is obtained as an extremum of a real-valued function. When a bifurcation of extrema occurs, a sudden qualitative change of the physical state may follow; such sudden changes happen in optics (light caustics), mechanics (instabilities), hydrodynamics (shock waves), thermodynamics (phase transitions), etc. Catastrophe theory may be used as a unifying language in these various domains, and as a way of treating general rather than special cases.

DEFINITIONS AND THEOREM

What follows is a presentation of the concepts that are necessary for a first approach and are immediately accessible to a nonmathematician. Many mathematical expositions, at various levels of abstraction, may be found in the literature, with the help of the references in the Bibliography.

Let $f(a_1, \ldots, a_i, \ldots, a_r; x_1, \ldots, x_j, \ldots, x_n)$ be a smooth real-valued function of $r+n$ real variables:

$$f: R^r \times R^n \to R.$$

The variables a_i $(i=1, \ldots, r)$ will be called *control* parameters; the variables x_j $(j=1, \ldots, n)$ will be called *state* variables. The function f is then seen as an r-parameter family of functions of n variables.

Consider then the stationary values (local extrema) of f with respect to variation of the state variables x_j. The set of points in $R^r \times R^n$ for which f is stationary,

$$\frac{\partial f}{\partial x_j} = 0 \qquad (j=1, \ldots, n),$$

is called M.

At this stage in the definitions, a simple example facilitates understanding and the introduction of other concepts. Consider a one-parameter family of functions of one variable $f(a; x)$. Figure 1 shows, on a graph of x versus f, for a range of values of x, three typical curves corresponding to three values of the parameter a. The upper curve, parametrized by a_1, shows two extrema (one minimum of abscissa $\bar{x}_1$, one maximum of abscissa $\bar{x}_1'$); the lower curve, parametrized by a_3, has no extremum; the intermediate curve, parametrized by a_2, has an inflection point at the abscissa value $\bar{x}_2$. It is then useful, and this is done in Fig. 2, to draw the set M (as defined above) as a graph of a versus x: for each value of a are plotted the values $\bar{x}$ of x that make f stationary. In this example, M is seen to be a curve, whose projection on the parameter axis has a singularity at a_2 (vertical slope) corresponding to the coalescence of two extrema (as visible in Fig. 1). On the control-parameter axis there is one point a_2, called a catastrophe or bifurcation point, separating two domains ($a>a_2$ and $a<a_2$) where the shapes of f are qualitatively different, with different numbers of extrema.

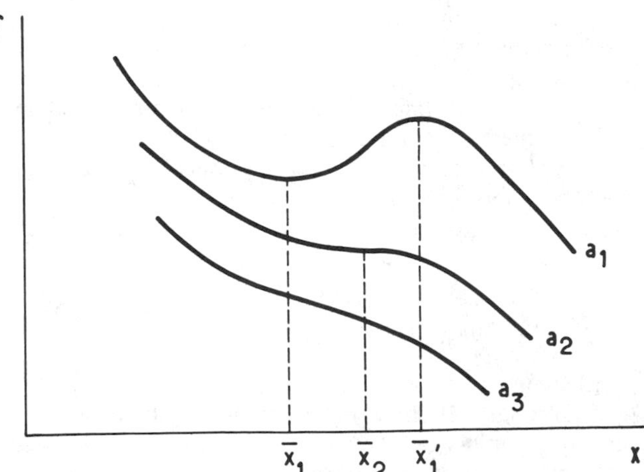

FIG. 1. Three representative curves of a one-parameter family of functions of one variable $f(a; x)$; a is the parameter, x the variable. The stationary values of f have abscissas $\bar{x}$.

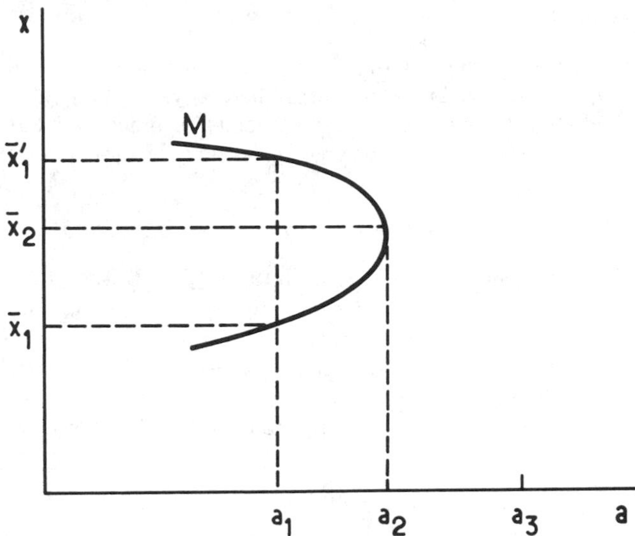

FIG. 2. The fold catastrophe. For a one-parameter family of functions of one variable $f(a; x)$, the set M of values of the variable making f stationary is plotted as a function of the parameter a. The fold singularity occurs at point $(a_2, \bar{x}_2)$ where the slope is vertical.

More generally, instead of a one-dimensional control-parameter axis, there will be an r-dimensional control-parameter space Σ; the set M will be a manifold; the points of Σ where the projection of M on Σ is singular (vertical slope) will constitute a set, called the catastrophe or bifurcation set.

Obviously, it is possible to imagine functions f that have very complicated singularities. Clarification occurs if we decide to look for the singularities that occur generically (generic means typical). This concept is not unfamiliar to physicists: Terms that are not forbidden by symmetry are (generically) nonzero, but they may become (accidentally) zero if some parameter is varied (such is the distinction between systematic and accidental degeneracies of levels in quantum mechanics; cf. also the discussion on the order of a transition in the Landau theory of phase transitions). A precise definition of genericity is clearly crucial for the mathematical proofs (see the mathematical references in the Bibliography). With a hypothesis of genericity for f, it is then possible to give a classification of the singularities, as follows.

If f is generic, then M is an r-dimensional manifold and the singularities of the projection of M onto the control-parameter space Σ are equivalent to a number of types called elementary catastrophes. The number of elementary catastrophes depends only on r (not on n) and is finite for $r\leq 5$, infinite for $r\geq 6$.

THE ELEMENTARY CATASTROPHES

For $r = 1$, there is only one elementary catastrophe (illustrated in Fig. 2) called, rather naturally, the fold catastrophe. In the vicinity of a fold point the function f has, generally after a smooth change of coordinates, the canonical local form

$$x_1^3 + ax_1$$

plus a diagonal quadratic form in the x_j ($j = 2, \ldots, n$).

For $r=2$, there are fold curves (imagine Fig. 2 continued smoothly with another horizontal axis) and another elementary catastrophe called the cusp catastrophe: The catastrophe set (in the control-parameter space Σ) is made up of fold arcs joining in cusps (Fig. 3). The canonical form of f near a cusp catastrophe is

$$x_1^4 + a_2x_1^2 + a_1x_1 + \cdots.$$

For $r=3$, three new elementary catastrophes appear; their names, together with their associated canonical local forms, follow:

1. the swallowtail,
$$x_1^5 + a_3x_1^3 + a_2x_1^2 + a_1x_1 + \cdots;$$
2. the elliptic umbilic,
$$x_1^2x_2 - x_2^3 + a_3(x_1^2 + x_2^2) + a_2x_2 + a_1x_1 + \cdots;$$
3. the hyperbolic umbilic,
$$x_1^2x_2 + x_2^3 + a_3(x_1^2 + x_2^2) + a_2x_2 + a_1x_1 + \cdots.$$

Sketches of the catastrophe sets (in space Σ) are shown in Fig. 4. Note that for the two umbilics, the canonical forms necessarily involve two state variables; canonical forms involving two or more state variables cannot in general be uniquely defined, and slightly different choices are sometimes used.

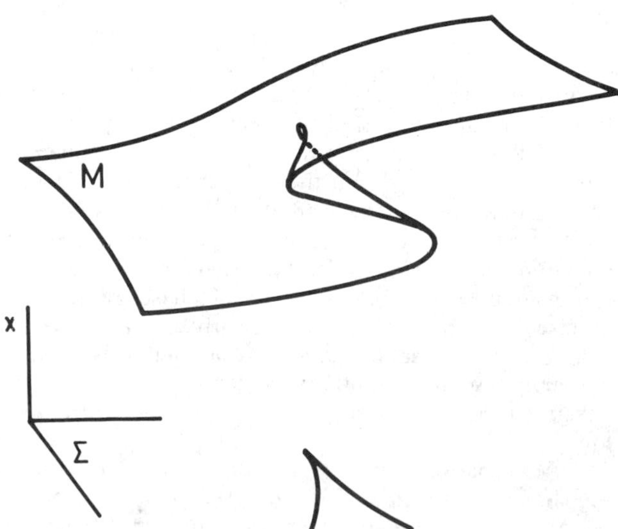

FIG. 3. The cusp catastrophe. The set M is a surface; its projection on the control-parameter plane Σ is singular along the catastrophe set, which exhibits a characteristic cusp shape.

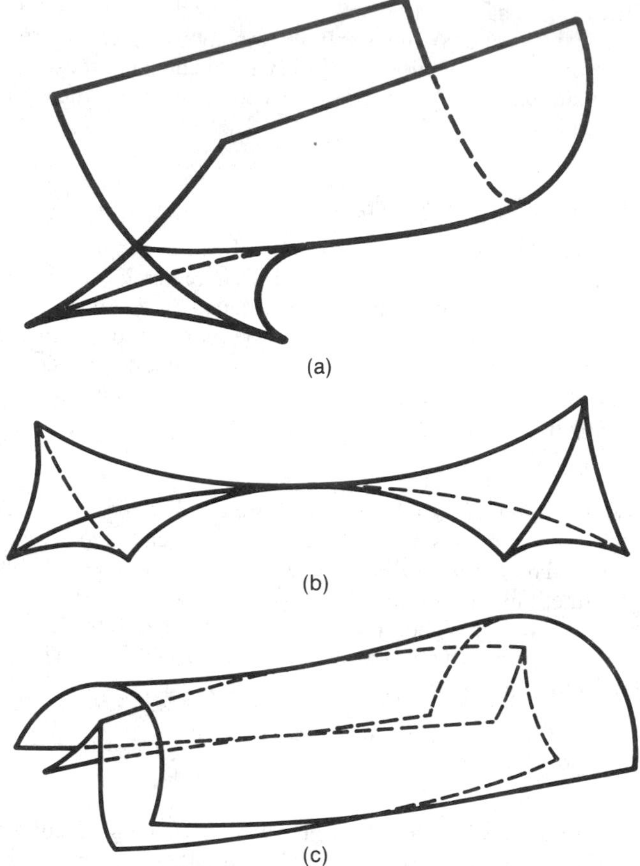

FIG. 4. Sketches of the (a) swallowtail, (b) elliptic umbilic, and (c) hyperbolic umbilic catastrophe sets in three-dimensional control space.

For $r = 4$, two new elementary catastrophes appear: the butterfly and the parabolic umbilic. In many applications, the control parameters will be space and time parameters and therefore $r = 4$. If we add up all the elementary catastrophes that can appear when $r = 4$, we get the famous total of seven elementary catastrophes.

For $r = 5$, four new catastrophes appear. For $r \geq 6$, the classification becomes infinite because there are equivalence classes depending on continuous parameters called *modal parameters*. The simplest such class involving two state variables is the *uni-modal* catastrophe called (by Arnol'd) X_9, which is important for example in optical systems that are nearly rotation-symmetric about their optical axis. Part of X_9 contains two cusp catastrophes and is therefore sometimes called the *double cusp*.

APPLICATIONS IN PHYSICS

Elementary catastrophe theory (ECT) has been applied to a wide range of topics, including most kinds of wave propagation, stability of mechanical structures, stability of ships and aircraft, laser action, singularities of phonon spectra, fluid flow, electric and magnetic phenomena, various phase transitions, nuclear and high-energy physics, semiclassical quantum mechanics, stellar evolution, relativity, geology, meteorology and climatology. Here are sketches of two application areas.

The canonical form for a cusp catastrophe is identical to the expression used to describe an ordinary critical point in the Landau theory of phase transitions. The meanings of x_1, a_1, a_2, and f are now (respectively) order parameter, field coupled to the order parameter, temperature measured from the critical temperature, and free energy density. Similarly, the canonical form for a butterfly catastrophe has its analogy in the expression used to describe tricritical points. However, symmetry considerations are of paramount importance in phase-transition problems and usual symmetry restrictions eliminate many of the elementary catastrophes listed above. Also, the predictions that can be obtained from such expressions for the free energy density ignore spatial fluctuations, which play a dominant role near the critical point. To include these effects, we have to add gradient terms to the free energy density and to sum over contributions due to the spatial fluctuations of the order parameter. This program is very remote from catastrophe theory.

Improvements on the results of simple catastrophe models have been achieved, but only at the expense of some *ad hoc* additions, such as coordinate changes outside those allowed by ECT, or use of a fractal lattice of dimension 3.09. However, the very successful renormalization group technique may be viewed as a catastrophe-theoretic model. A contribution of catastrophe theory to the field of phase transitions might be in a rational vocabulary for higher-order critical points.

A very neat application of catastrophe theory has been made in the theory of caustics. Caustics are envelopes of geometrical rays of a wave field. When the wavelength becomes small, the geometrical approximation of the wave theory becomes appropriate and most of the intensity of the wave field is concentrated on the caustics, which become the dominant features of the wave pattern. In this semigeometrical limit, not only does ECT produce a useful classification of caustics, but also of the wavefields—called *diffraction catastrophes*—that surround the caustics.

The geometrical rays are obtained via a variational principle (this is in essence the content of Fermat's principle), the main control parameters are space coordinates of points where the wave amplitude is to be measured, and the state variables are space coordinates of the intersections of the rays with some arbitrary wave surface. Typically, in a three-dimensional geometry, the number of control parameters is three and, from the classification above, it is possible to predict that the caustic pattern will generically consist of fold surfaces joining along cusp lines meeting in points which may be swallowtails, elliptic umbilics, or hyperbolic umbilics.

The wavefield around a caustic described by the catastrophe with canonical form f is modelled by a canonical diffraction integral of the form

$$\int e^{if(a;x)}\,dx.$$

These integrals have been evaluated numerically for all the caustics that can occur generically in three dimensions and used to produce simulated diffraction catastrophes, which agree spectacularly well with observations.

On a caustic, the amplitude A of the wave diverges as a power law of the wavelength λ:

$$A\sim\left(\frac{1}{\lambda}\right)^{\sigma}, \qquad \lambda\to 0.$$

By scaling out the wavelength from the physical diffraction integral to produce the above canonical form, the value of the index σ is obtained from the canonical form of the elementary catastrophe associated with a given caustic:

$$\begin{aligned} \text{fold},\quad & \sigma=\tfrac{1}{6}\,; \\ \text{cusp},\quad & \sigma=\tfrac{1}{4}; \\ \text{swallowtail},\quad & \sigma=\tfrac{3}{10}; \\ \text{elliptic or hyperbolic umbilics},\quad & \sigma=\tfrac{1}{3}. \end{aligned}$$

By a technique of averaging over caustics, Berry has been able to successfully predict the "twinkling" statistics of starlight, using a second set of scaling indices describing the wavelength dependence of the spatial size of caustics. The amplitude and size indices are quantitative results, which can be measured by varying the wavelength λ.

PERSPECTIVES AND REFERENCES

Catastrophe theory is now sufficiently mature that there are a number of articles reviewing its applications in the physical sciences; that by Stewart is a good place to start. The classic book by René Thom gives marvellous insight into the motivation for this branch of mathematics, although it was E. C. Zeeman who coined the name "catastrophe theory" while developing many of its applications, some of which appear in his book of selected papers. The Russian mathematician V. I. Arnol'd vastly extended Thom's original classification of elementary catastrophes, using a complex rather than real setting, and contributed significantly to the more general version known as "singularity theory." Applications in wave theory in general have been particularly successful, and the background is set out in the review by Berry and Upstill; the paper by Nye is a more recent example of the beauty of diffraction catastrophes and that by Dangelmayr and Güttinger discusses the inverse problem.

ECT grew partly out of the bifurcation theory of solutions of differential equations, and recently has fed back into that area as "imperfect" bifurcation theory. This considers the effects of perturbations on an idealized model, and the perturbation parameters play an analogous role to the control parameters in ECT. Thus some of the relative computational ease of ECT can be brought to bear on "conventional" bifurcation theory, and this has been particularly successful for including symmetry and symmetry-breaking effects, as described in the two volumes by Golubitsky *et al.* For example, it has been applied to "mode jumping" in the buckling of plates and to the way that fluid flows can display highly symmetric cell structures (e.g., planar Bénard convection and cylindrical Taylor–Couette flow). The range of physical systems to which ECT can be applied directly is important but limited, and the influence that it has had back on its origins—dynamical systems and bifurcation theory, which have broader applicability—may yet prove to be its greatest contribution.

See also PHASE TRANSITIONS.

BIBLIOGRAPHY

V. I. Arnol'd, *Catastrophe Theory*. Springer-Verlag, New York, 1984.

V. I. Arnol'd, *Singularity Theory: Selected Papers*. Cambridge University Press, London, 1981.

M. V. Berry and C. Upstill, "Catastrophe Optics: Morphologies of Caustics and Their Diffraction Patterns," *Prog. Opt.* **18,** 257–346 (1980).

G. Dangelmayr and W. Güttinger, "Topological Approach to Remote Sensing," *Geophys. J. R. Astr. Soc.* **71,** 79–126 (1982).

M. Golubitsky and D. G. Schaeffer, *Singularities and Groups in Bifurcation Theory, Vol. I.* Springer-Verlag, New York, 1985.

M. Golubitsky, I. N. Stewart, and D. G. Schaeffer, *Singularities and Groups in Bifurcation Theory, Vol. II.* Springer-Verlag, New York, 1988.

J. F. Nye, "The Catastrophe Optics of Liquid Drop Lenses," *Proc. R. Soc. Lond.* **A403,** 1–26 (1986).

T. Poston and I. N. Stewart, *Catastrophe Theory and its Applications*. Pitman, London, 1978.

I. N. Stewart, "Catastrophe Theory in Physics," *Rept. Prog. Phys.* **45,** 185–221 (1982).

R. Thom, *Structural Stability and Morphogenesis*. Benjamin, Reading, Mass., 1975.

E. C. Zeeman, *Catastrophe Theory: Selected Papers, 1972–1977*. Addison-Wesley, Reading, Mass., 1977.

E. C. Zeeman, "Bifurcation and Catastrophe Theory," *Contemp. Math.* **9,** 207–272 (1982).

Cellular Automata

Tommaso Toffoli

Cellular automata are abstract dynamical systems that play in discrete mathematics a role comparable to that played in the mathematics of the continuum by partial differential equations. In terms of structure as well as applications, they are the computer scientist's counterpart to the physicist's concept of a "field" governed by "field equations." It is not surprising that they have been reinvented innumerable times under different names and within different disciplines. The canonical attribution is to Ulam and von Neumann (ca. 1950).

In a cellular automaton, space is represented by a uniform array. To each site of the array, or *cell* (whence the name "cellular"), there is associated a state variable ranging over a finite set—typically just a few bits worth of data. Time advances in discrete steps, and the dynamics is given by an explicit *rule*—say, a lookup table—through which at every step each cell determines its new state from the current state of its neighbors. Thus, the system's laws are *local* (no action-at-a-distance) and *uniform* (the same rule applies to all sites); in this respect, they reflect fundamental aspects of physics. Moreover, they are *finitary*: even though one may be dealing with an indefinitely extended array (in which case the state space has the cardinality of the continuum, just as with differential equations), the evolution over a finite time of a finite portion of the system can be computed *exactly* by finite means (unlike with differential equations).

A more formal characterization of cellular automata may be given in terms of topological dynamics on the Cantor set.

In spite of their structural simplicity, cellular automata can support complex behavior. They supply useful models for many investigations in natural sciences, combinatorial mathematics, and computer science. One of their most natural and successful applications is the modeling of macroscopic physical phenomena (e.g., fluid dynamics, flow through porous media, phase separation) by reduction to extremely simple fine-grained mechanisms. Indeed, with current progress in computer technology, reductionistic models of this kind—which are in any case conceptually very attractive—are also becoming computationally accessible.

As a simple example, consider a two-dimensional array where each cell can be in one of two states—black or white—and can see as a neighborhood the entire 3×3 window centered on and including the cell itself. At each step, let the new state of the cell follow the *majority* of its neighbors, i.e., become *white* if at least five of the nine neighbors are currently *white*, and *black* otherwise. If we start the array in a random state, as in Fig. 1a, with this rule the cellular automaton will converge in a few steps to the state of Fig. 1b; no further change will take place.

With a slight modification of the voting rules ("win" if you have 4, 6, 7, 8, or 9 votes, "lose" with 0, 1, 2, 3, or 5), the *annealing* of the black and white phases will proceed indefinitely, yielding ever larger and more rounded domains, as shown in Fig. 1c after a few hundred steps. This represents a rudimentary model of phase separation and surface tension.

For more accurate models, a good approach is to devise rules that capture, albeit in a stylized way, other essential aspects of the underlying physics—such as conservation of molecular species, energy, momentum, etc. (In the above phase-separation example, one form of potential energy would be represented by the length of the interface—i.e., the number of black–white edges.) By further demanding microscopic reversibility one ensures the conservation of fine-grained entropy—and thus obtains a model that obeys the second principle of thermodynamics.

Recent developments in the theory and practice of cellular automata have made it straightforward to build such "detailed balance" constraints into a rule, as well as custom-tailor many other desirable properties.

Figure 2 shows a model of "smokestack emission." The black specks undergo Brownian motion in a uniformly drifting fluid. The underlying cellular-automaton rule is a very simple kind of *lattice gas*, in which particles move at constant speed in one of four directions and collide elastically. Similar lattice gases support full-fledged hydrodynamics that, in appropriate macroscopic limits, reproduce the Navier–Stokes equations.

Having proven to be natural tools for investigations in statistical mechanics, cellular automata are now beginning to be used in an exploratory role in quantum-mechanical and relativistic modeling.

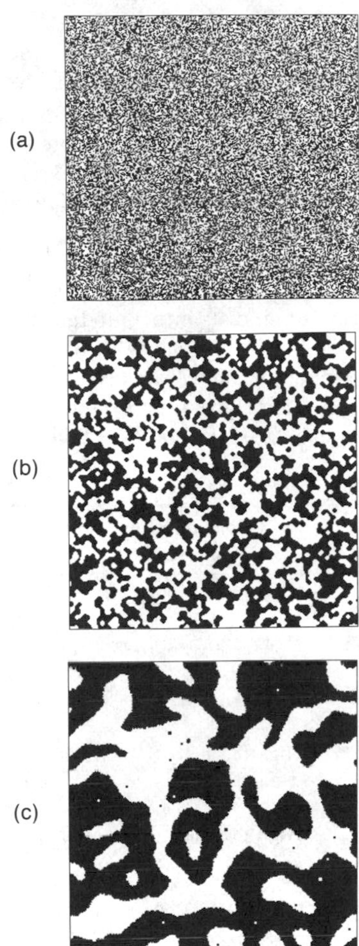

FIG. 1. From a random mix of black and white cells (a), in a few steps the majority rule produces clearly separated phases (b), with a variant of the majority rule, annealing of the two-phase system proceeds indefinitely (c).

The cellular-automaton scheme of computation is unique in the way it lends itself to extremely efficient concrete implementations.

We have seen that volume elements of a physical system to be modeled are directly mapped, in a one-to-one fashion, into volume elements of a cellular automaton. Conversely, when the time comes to concretely realize and "run" such a model, one can directly map volume elements of the cellular automaton into volume elements of an appropriate physical computer—a *cellular automata machine*. Such a massively parallel architecture is *indefinitely scalable*, i.e., the number of cells in the array can be made as large as desired without having to slow down the rate of updating of individual cells.

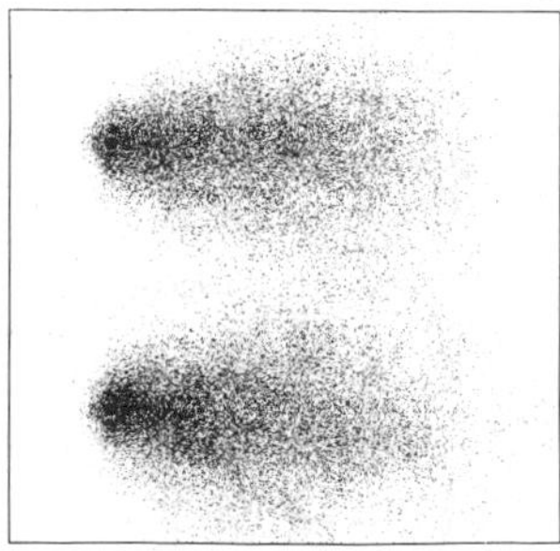

FIG. 2. A lattice-gas model of smokestack emission.

The Turing-machine paradigm characterizes, according to Church's thesis, *what* can in principle be computed by physical means. A stronger version of this thesis is gaining credence today, namely, that the cellular-automaton paradigm characterizes *how efficiently* anything can in principle be computed by physical means.

Thus, cellular automata are relevant not only when one asks how much physical simulation one can get out of computers, but also how much computing power one can get out of physics.

BIBLIOGRAPHY

Arthur Burks, *Essays on Cellular Automata*. University of Illinois Press, Chicago, 1970.

Tommaso Toffoli, "Cellular automata as an alternative to (rather than an approximation of) differential equations in modeling physics," *Physica* **10D,** 117–127 (1984).

Stephen Wolfram (ed.), *Theory and Applications of Cellular Automata*. World Scientific, Singapore, 1986.

Tommaso Toffoli and Norman Margolus, *Cellular Automata Machines—A new environment for modeling*. MIT Press, Cambridge, 1987.

Tommaso Toffoli and Norman Margolus, "Invertible Cellular Automata: A Review," *Physica* **D44** (1990).

Gary Doolen *et al.* (ed.), *Lattice Gas Methods for Partial Differential Equations*. Addison-Wesley, Reading, MA, 1990.

Center-of-Mass System

Philippe Eberhard and Morris Pripstein

For an ensemble of bodies the center-of-mass system is that inertial system of reference axes in which the vectorial sum of the momenta of the bodies is zero:

$$\sum_i \mathbf{p}_i^* = 0 \qquad (1)$$

where $\mathbf{p}_i^*$ is the momentum of the ith body. It is a system in which the center of mass of the ensemble is at rest but, because of the definition [Eq. (1)], it is often referred to as the center-of-momentum system. It is used because the kinematics are generally simpler.

Let m_i be the mass of the ith body and $\mathbf{r}_i$ be its position in any inertial rest frame. The position of the center of mass is defined nonrelativistically by the vector

$$\mathbf{R} = \frac{\sum_i m_i \mathbf{r}_i}{M} \qquad (2)$$

where

$$M = \sum_i m_i \qquad (3)$$

The position $\mathbf{r}_k{}^*$ and the velocity $\mathbf{v}_k{}^*$ of the kth body in the center-of-mass system are defined as

$$\mathbf{r}_k{}^* = \mathbf{r}_k - \mathbf{R} \tag{4}$$

and

$$\mathbf{v}_k{}^* = \mathbf{v}_k - \mathbf{V} \tag{5}$$

where $\mathbf{v}_k$ and $\mathbf{V}$ are the derivatives of $\mathbf{r}_k$ and $\mathbf{R}$, respectively, with respect to time.

Whenever relativistic corrections cannot be neglected, Eq. (5) is not valid because a Lorentz transformation has to be applied instead to ensure that Eq. (1) is satisfied. The momentum $\mathbf{p}_k{}^*$ and energy $e_k{}^*$ of the kth body in the center-of-mass system can always be defined from the momentum $\mathbf{p}_k$ and energy e_k of the body in the initial reference frame by the equations

$$\mathbf{p}_k{}^* = \mathbf{p}_k - \left[\frac{\gamma e_k}{c} - \frac{\gamma^2}{\gamma+1}(\boldsymbol{\beta}\cdot\mathbf{p}_k)\right]\boldsymbol{\beta}, \tag{6}$$

$$e_k{}^* = \gamma e_k - \gamma(\boldsymbol{\beta}\cdot\mathbf{p}_k)c \tag{7}$$

where c is the velocity of light, and

$$e_k = \frac{m_k c^2}{[1-(\mathbf{v}_k/c)^2]^{1/2}}, \tag{8}$$

$$\boldsymbol{\beta} = \frac{\sum_i \mathbf{p}_i c}{\sum_i e_i}, \tag{9}$$

$$\gamma = \frac{\sum_i e_i}{E^*}, \tag{10}$$

and

$$E^* = [(\sum_i e_i)^2 - (\sum_i \mathbf{p}_i)^2 c^2]^{1/2}. \tag{11}$$

As in any Lorentz transformation, the component of the momentum $\mathbf{p}_k$ that is orthogonal to $\boldsymbol{\beta}$ is conserved while that which is parallel to $\boldsymbol{\beta}$, and the energy e_k, are modified. The position of the center of mass in the center-of-mass system is now defined by Eqs. (2) and (3) but with m_i and M replaced by $e_i{}^*$ and E^*, respectively.

Because of Eq. (1), the number of parameters necessary to describe the ensemble in the center-of-mass system can in general be easily reduced, and the description of the motion of the individual bodies is made simpler. For example, in the center-of-mass system of two stars revolving around each other, the trajectory of each one is an ellipse. In particle physics, the scattering of two particles can be described by only one parameter, the scattering angle θ^*, because the momenta of the two particles of the ensemble are always opposite to one another in the initial and in the final states.

See also KINEMATICS AND KINETICS; DYNAMICS, ANALYTICAL.

Ceramics

Lionel M. Levinson

The conventional understanding of ceramics encompasses the production and use of clays and silicate materials for pottery, bricks, and refractories. Modern ceramics also include a wide variety of inorganic nonmetallic materials with precisely controlled properties which play critical roles in the function of numerous technological devices. The need for materials with closely specified characteristics has transformed the ancient ceramic art into a modern-day branch of materials science and engineering.[1] We will briefly list some of the modern technological ceramics and then examine more closely two particular ceramic systems of interest.

In Table 1 we present a selection of technical ceramic ma-

Table I A List of Some Modern Technical Ceramic Materials

Material	Outstanding Property	Use
Alumina, Al_2O_3	High-temperature strength Low electrical loss	Furnace ware, thermocouples IC multilayer package; microcircuit substrates
Beryllia, BeO	High thermal conductivity	Electronic component heat sink
Aluminum nitride, AlN	High thermal conductivity and low thermal coefficient of expansion	Si IC substrates
Magnesia, MgO	High-temperature resistivity	Heater element insulation
Ferrite, MFe_2O_4 (M = Mn, Fe, Ni, Zn, Cu, Mg, etc.)	Ferrimagnetic insulator	Computer memories, high-frequency transformers
Garnet, $R_3Fe_5O_{12}$ (R = rare earth)	Low microwave loss Ferrimagnetic insulator	Microwave devices
Zinc oxide, ZnO	Nonlinear conductor	Overvoltage transient suppression
Insulating barium titanate, $BaTiO_3$	High dielectric constant	Capacitors
Conducting barium titanate, $BaTiO_3$	Positive temperature coefficient of resistivity	Relay, sensor
Lead zirconate titanate (PZT), $Pb(Zr, Ti)O_3$	Piezoelectric	Ultrasonic transducers
	High-temperature superconductivity	
$YBa_2Cu_3O_{7-x}$	T_c = 92° K	Magnetic shielding,
$Bi_2Sr_2Ca_2Cu_3O_{8-x}$	T_c = 110°K	Levitation, Magnetic field detection *plus*
$Tl_2Ca_2Ba_2Cu_3O_{8-x}$	T_c = 125°K	others pending

terials along with some special features and uses of the ceramic. It should be emphasized that the table makes no pretence at completeness and that the use of a particular material in a given application will usually depend on a spectrum of material properties in addition to the particular feature listed. Of the various polycrystalline ceramics in Table 1 it is clear that the utility of some lies in the ease of fabrication and low cost inherent in ceramic processing. For example, single-crystal alumina could provide superior microcircuit substrates and the use of single-crystal lead zirconate titanate would result in the fabrication of better ultrasonic transducers. There are, however, some ceramic devices in which the ceramic has properties essentially not feasible in the single-crystal analog of the ceramic. In this case the ceramic microstructure produces the new effect. We shall examine two such examples: (1) the high-frequency ferrite and (2) the ZnO varistor.

MANGANESE ZINC FERRITE

The double oxide spinel $MO{\cdot}Fe_2O_3$, where M can be (among others) Mn, Zn, or ferrous iron, is a magnetic ceramic material which combines the properties of high magnetic permeability with high electrical resistivity. The latter property results in low eddy-current loss so that the material can be used at high (tens of kilohertz) frequencies. A typical application of such ferrites is in the tuned circuit used for frequency-synthesis Touch-Tone® telephones.[2] The ferrite must have high quality factor Q, low sensitivity to variations in the ambient temperature, and good stability with time. Since Q is the ratio of energy stored ($\propto$ permeability μ) to energy dissipated ($\propto$ conductivity σ), the highest possible value of μ/σ is desired. To maximize μ the $Mn{:}Zn{:}Fe^{2+}{:}Fe^{3+}$ ratio is chosen so that the magnetic anisotropy and magnetostriction are very small. Ideal properties require close control of the purity of the raw materials and processing techniques such as milling and firing atmosphere. A typical processing flowchart is shown in Fig. 1.

The necessary presence of both Fe^{2+} and Fe^{3+} in manganese zinc ferrite produces an increase in the ferrite conductivity by allowing easy electron transfer from Fe^{2+} to Fe^{3+} sites. This in turn produces an unwanted high-frequency loss by eddy-current effects. This loss, while unavoidable in single-crystal material, can be controlled in fine-grain ceramics by the addition of controlled impurities. CaO and SiO_2 have been found[3] to segregate to the grain boundaries to form a low-conductivity phase, thereby substantially decreasing eddy-current losses. Figure 2 gives the effect of small amounts of CaO on μ and σ. Since μ decreases only slightly with CaO content while σ drops precipitously, a few tenths mol% CaO addition substantially improves the performance of the ceramic ferrite.

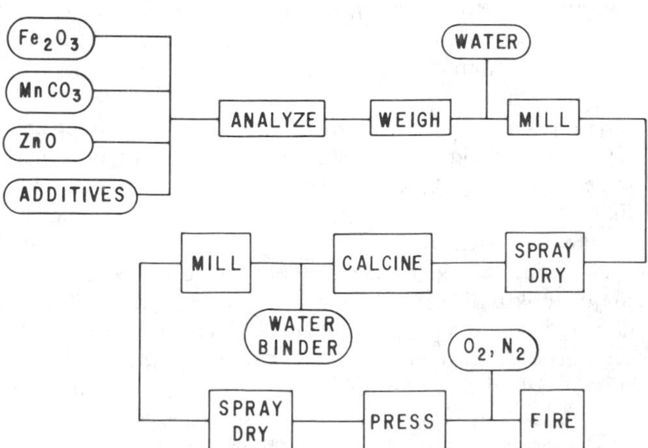

FIG. 1. Typical flowchart for processing of manganese zinc ferrite.

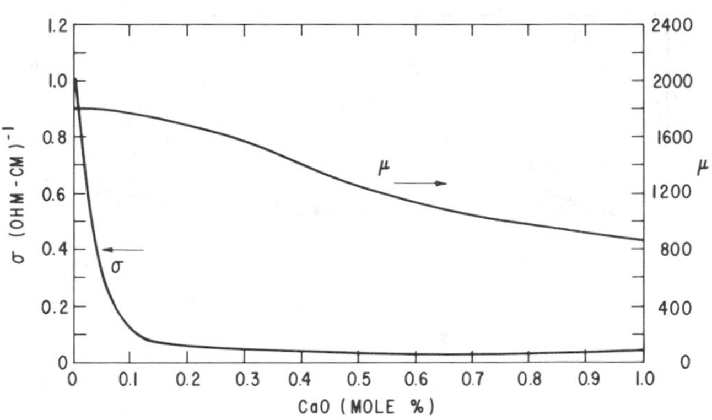

FIG. 2. Conductivity σ and permeability μ as a function of CaO addition to manganese zinc ferrite.

ZnO VARISTORS

Zinc oxide is an ohmic semiconductor. However, it has been found[4] that ZnO ceramics containing a few mol% of other oxide additives (for example, Bi_2O_3, Co_2O_3, MnO_2, and Sb_2O_3) exhibit a marked electrical nonlinearity in their current-voltage characteristic. Figure 3 gives a typical cur-

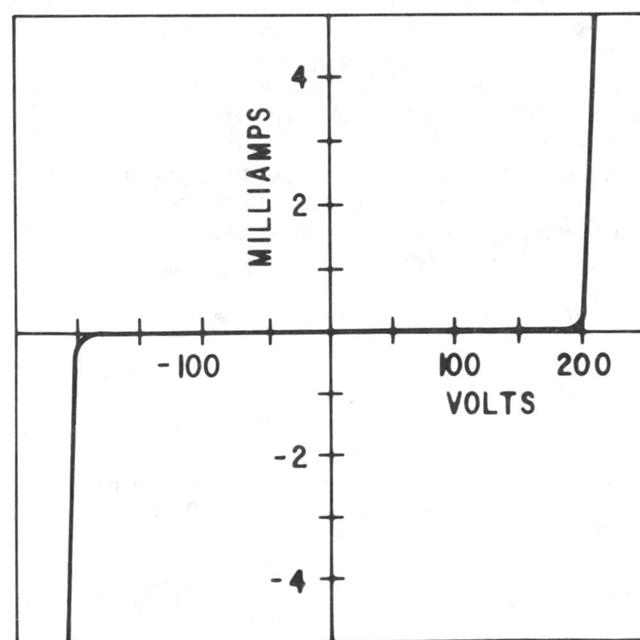

FIG. 3. Current-voltage characteristic of a ZnO varistor.

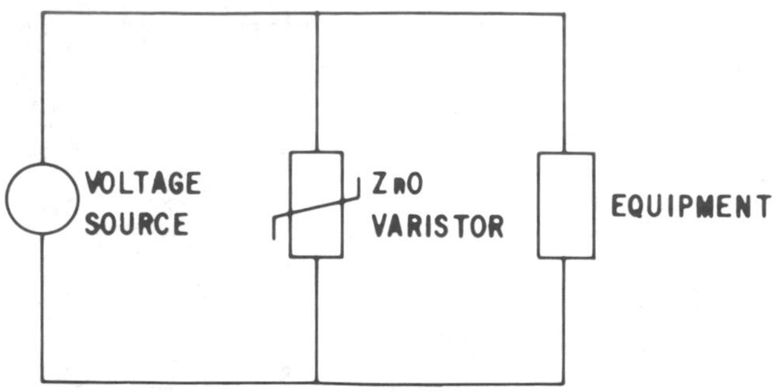

FIG. 4. Use of ZnO varistor to protect electrical equipment.

rent-voltage characteristic of one of these devices. At voltages below a critical value the device acts as an insulator; above that voltage it acts as a conductor. In the vicinity of the turn-on voltage a 5% increase in voltage will typically cause the current to increase by an order of magnitude!

Figure 4 indicates how the ZnO varistor is used to protect electrical and electronic equipment from overvoltage transients due to switching surges or lightning strokes. The varistor is connected in parallel with the load and its turn-on voltage chosen so that it is insulating in normal use. The device conducts when a high-voltage surge in excess of the turn-on voltage appears on the line. This shunts the current and prevents the surge from reaching the protected equipment.

The origin of the varistor effect lies in the ceramic microstructure of this device.[5,6] The ZnO grains are conducting, but during sintering they become surrounded with an adsorbed layer of aliovalent dopants (e.g., Bi) which segregate to the grain boundary region. The dopants are derived from oxides specifically added to produce the nonlinear electrical behavior in the ZnO ceramic. The dopant segregation results in the creation of electrical depletion layers at the grain boundaries which inhibit electron flow at low applied voltages. When increasing voltage is applied to the varistor, very high fields appear across the thin insulating depletion layers since the interior of the ZnO grains is conducting and will not sustain a voltage. At a critical applied voltage[6] (about 3 volts per grain boundary) it becomes possible for electrons to transfer through the depletion layers from one conducting ZnO grain to a neighboring grain, thereby giving rise to the observed high nonlinearity in the electrical characteristics.

See also CONDUCTION; MAGNETIC MATERIALS.

REFERENCES

1. W. D. Kingery, H. K. Bowen, and D. R. Uhlmann, *Introduction to Ceramics*. Wiley, New York, 1976.
2. F. J. Schnettler, "Microstructure and Processing of Ferrites," in *Physics of Electronic Ceramics*, Part B, p. 833, L. L. Hench and D. B. Dove (eds.). Marcel Dekker, New York, 1972.
3. T. Akashi, *Trans. Jpn. Inst. Metals* 2, 171 (1961).
4. M. Matsuoka, *Jpn. J. Appl. Phys.* 10, 736 (1971).
5. L. M. Levinson and H. R. Philipp, *J. Appl. Phys.* 46, 1332 (1975).
6. L. M. Levinson and H. R. Philipp, *American Ceramic Society Bulletin 65*, 639 (1986).

Čerenkov Counters *see* Scintillation and Čerenkov Radiation

Čerenkov Radiation

Michael Danos

In 1934 P. A. Čerenkov reported that light is emitted by relativistic particles traversing a nonscintillating liquid; he also described the properties of this radiation. (Čerenkov radiation had been observed, but not understood, by Marie Curie around the turn of the century.) Soon thereafter the nature of this effect was completely explained by I. M. Frank and I. E. Tamm as being the electrodynamic analog to the Mach waves of hydrodynamics: A charged particle traveling with velocity v through a nonconducting medium (or close to such a medium) having an index of refraction $n(\nu)$ emits electromagnetic radiation at all frequencies ν for which the phase velocity c/n is less than v. In principle, a neutral particle having a magnetic moment (e.g., a neutron) also will emit Čerenkov radiation. Because of the smallness of the magnetic moment, this radiation will be negligibly small. The radiation is emitted at one angle, the Čerenkov angle $\theta = \arccos(c/nv)$, with respect to the momentum vector of the particle. It is 100% linearly polarized, the electric vector being coplanar with the particle momentum vector. The emitted power is given by $dE/dz = (4\pi^2e^2/c^2)\int \nu\, d\nu\, [1-(c/nv)^2]$ where the integration is limited to those frequencies for which the Čerenkov condition $(c/nv)<1$ is fulfilled. Typically, relativistic electrons in water radiate about 250 visible light photons per centimeter of travel. The radiated energy integrated over all frequencies is rather small, since even for $v \approx c$ the Čerenkov condition is fulfilled only over a limited frequency interval; it is less than 1 keV/cm (i.e., its contribution to the stopping power is negligible).

On the other hand, the Čerenkov condition is fulfilled by the vacuum itself for all frequencies for particles that travel at $v>c$, i.e., for tachyons. This case was treated in terms of classical electrodynamics in prerelativity times (1892) by A. Sommerfeld, who thus in effect anticipated the Čerenkov effect. He found that a charged tachyon would radiate an energy equal to its rest mass while traveling a distance of the order of its classical radius, i.e., $d \approx e^2/mc^2$. This result appears to eliminate the possibility that charged tachyons may exist in nature.

The principal applications of the Čerenkov effect are in detectors for high-energy particle physics. We utilize here the existence of a threshold velocity: Only charged particles that fulfill the Čerenkov condition emit radiation, which then is detected, e.g., by photomultipliers. Such counters can be used either to select particles above a certain energy (e.g., the Čerenkov threshold energy for protons in air of normal density is about 55 GeV), or to discriminate against the heavier particles in a beam of given momentum containing a mixture of particles. Alternatively, Čerenkov counters have been built that respond to particles of a particular velocity by detecting only light emitted at a predetermined angle with respect to the beam direction. Such counters can be used to select particles of a given rest mass from a fixed-momentum beam.

See also ELECTRODYNAMICS, CLASSICAL; TACHYONS.

Channeling

J. A. Golovchenko

When beams of fast charged particles, such as those obtained from ion accelerators or radioactive decay, are directed into single crystals, it is experimentally observed that various reaction probabilities depend dramatically on the relative angular orientation of the target and incident beam. For example, consider an angularly well-collimated 1-MeV proton beam impinging on a thin silicon crystal. For angles of incidence $\lesssim 10$ mrad to the $\langle 110 \rangle$ axial direction, the measured Rutherford backscattering yield from collisions with silicon atoms is reduced to ~3% of the value found at larger angles of incidence where the results are similar to those obtained for an amorphous silicon target. For other positive particles, similar reductions in yield of all processes requiring small impact parameters (e.g., nuclear reactions, K-shell ionization) are observed near low-index axial or planar directions.

The explanation of the foregoing and other closely related phenomena relies on the concept of channeling. By examining the possible classical trajectories of positive particles in a crystal, we find that there exists a class of motions that have a remarkable stability against approaching target atoms at small impact parameters. Particles executing these motions are called *channeled*. To illustrate the most important aspects of channeling, we need only consider the basic influence that a closely packed atomic string (row) or plane of atoms has in determining these trajectories.

Figure 1 shows the string case. For a particle approaching the string at a small angle, a series of large-impact-parameter, small-angle scattering events leads to an overall repulsion from the vicinity of the string. As the angle of incidence is increased, the distance of closest approach to the string, r_{min}, decreases along with the number of individual atomic deflections that make up the string collision. Finally, at a sufficiently large incident angle, only a few atomic collisions are involved with such large scattering angles that the essential stability of trajectories, determined collectively by many atoms on the string, is lost. Up until this limit is reached, the particle is channeled.

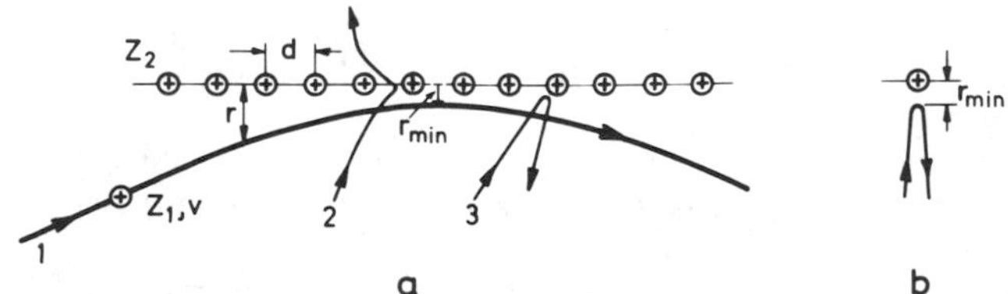

FIG. 1. Single-string collision. (a) Longitudinal view. Trajectory 1 is channeled, 2 and 3 are not. (b) The string is perpendicular to the plane of the paper here.

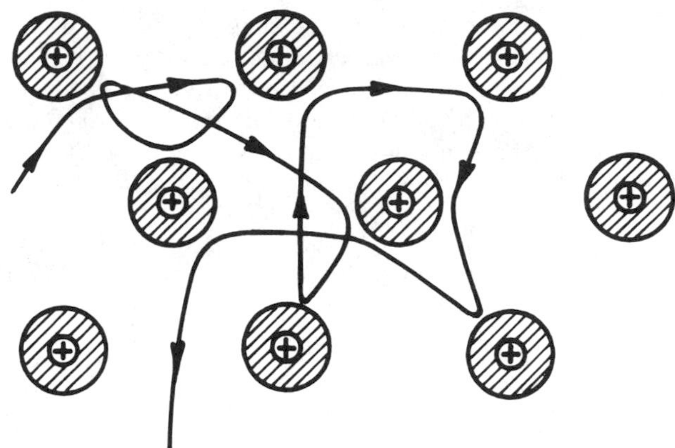

FIG. 2. Motion in a transverse plane showing many-string collisions. Strings are directed into the plane of paper, and only the motion perpendicular to the strings is shown.

Since the crystal is made up of many parallel strings along the axial direction, the overall penetration of the channeled particle is determined by a succession of many string collisions. This is illustrated in Fig. 2, where a typical channeling trajectory is viewed in a plane transverse to the string direction. Note that although it is possible for a trajectory to be trapped in the "channel" formed by several adjacent strings of atoms, in general an axially channeled particle can wander quite freely from string to string in the motion transverse to the strings.

A similar description applies to the case of planar channeling. Here, for sufficiently small angles of approach, a positive particle is gently repelled from the vicinity of the plane of atoms and again small-impact-parameter events are suppressed. In this case, the extended trajectory will be confined between two parallel atomic planes and an oscillatory motion results.

The preceding physical description may be extended theoretically to provide further insight into the channeling process and qualitative estimates of various experimental quantities. The following discussion is generally confined to the string case, although analogous results may be deduced for the planar case, where channeling effects are somewhat less pronounced.

Channeling can be described within classical mechanics. This is, at first, surprising, since the quantity $2Z_1Z_2e^2/\hbar v$ is small compared to unity in many cases of interest. Z_1 and Z_2 are the atomic numbers of the projectile and target atoms, respectively; v is the particle velocity; and $\hbar$ is Planck's constant divided by 2π. In this situation, a classical description of an *isolated* atomic collision in terms of impact-dependent scattering angles is not possible, and a quantum-mechanical approach is required. However, to determine the applicability of a classical description for channeled motion, an analysis of an entire string collision is required. Such a procedure shows that the spread of suitably constructed wave packets obeying classical mechanics is small enough to render all the classical variables meaningful for heavy channeled particles (e.g., protons), even for $v \to c$. Indeed, channeling is basically a classical concept.

The application of classical mechanics to the crystal-penetration problem shows that channeling may be associated with an approximately conserved quantity, the transverse energy $E_\perp$. For an axially channeled particle of mass M and energy $E(=\frac{1}{2}Mv^2)$ with an instantaneous angle and distance from the ith string ψ_r and $\mathbf{r}-\mathbf{r}_i$, $E_\perp$ is given by

$$E_\perp = \frac{1}{2} mv^2\psi_r^2 + \sum_i U(|\mathbf{r}-\mathbf{r}_i|). \tag{1}$$

The continuum potential $U(r)$ is given by

$$U(r) = \frac{1}{d}\int_{-\infty}^{\infty} V[(r^2+Z^2)^{1/2}]\, dZ. \tag{2}$$

Here d is the distance between atoms on the string, which is taken to be in the Z direction, and $V(R)$ is the screened Coulomb-potential energy of interaction between the projectile and target atom at separation R. The first term in Eq. (1) is the kinetic energy of the particle motion projected onto a plane perpendicular to the strings, while the second term is an average potential energy obtained by smearing the atomic potential V along the string direction. The stability of channeled motion referred to earlier corresponds to conservation of $E_\perp$ with the velocity parallel to the strings being nearly unchanged during a string collision.

Using simple analytical estimates for V and examining limiting particle trajectories, we find that conservation of $E_\perp$[i.e., governed motion in the transverse Hamilton, Eq. (1)], breaks down for $E_\perp \gtrsim 2Z_1Z_2e^2/d$. Putting this restriction in Eq. (1) leads to a characteristic limiting angle of approach for axial channeling of

$$\psi_r\,(\max) = \psi_1 \equiv \left(\frac{2Z_1Z_2e^2}{Ed}\right)^{1/2}. \tag{3}$$

This is a central result of the theory. It applies at energies E so high that $\psi_1 < a/d$. Here, a is the Thomas–Fermi screening length $[a = 0.885a_0/(Z_1^{2/3}+Z_2^{2/3})^{1/2}$ where a_0 is the Bohr radius], which measures the range of $V(R)$ for fast collisions. At lower energies, ψ_1 overestimates $\psi_r(\max)$, which now depends in more detail on the form of $V(R)$.

Upon entering the crystal, each particle in a well-collimated beam of energy E will acquire a value of $E_\perp$ [see Eq. (1)] given by $E_\perp = E\psi^2 + U(r_0)$, where ψ is the angle between the beam and the axis directions and r_0 is the point of entry in the transverse plane. The resulting distribution in $E_\perp$ is strongly peaked about $E\psi^2$ since $U(r)$ is quite small over most of the transverse space. Thus as ψ approaches ψ_1, the distribution in $E_\perp$ is weighted toward the highest values of $E_\perp$, and the yield of small-impact-parameter interactions increases. For $\psi = 0$, only particles entering the crystal within an area $\sim\pi\rho^2$, with ρ^2 the host-atom mean-square thermal-vibration amplitude perpendicular to the strings, will be able to interact strongly with target atoms. The channeling dip measured relative to random penetration ($\psi \gg \psi_1$) will be of order $\chi_{\min} = \pi\rho^2/$(transverse unit cell area). Most of the particles, i.e., the fraction $1-\chi_{\min}$ ($\approx 0.95-0.99$ for strong axes) will be well channeled away from the vibrating string atoms.

In transmission experiments through thin crystals, this component of the beam can exit from the back of the crystal within ψ_1 of the axis, the random multiple scattering of such particles being greatly reduced. It is also found that the energy loss due to electronic excitation is reduced by up to about a factor of two for channeled particles. This reduction arises mainly from the low density of target electrons along the path of low-$E_\perp$ trajectories. Coupled with the large reduction in energy lost to recoiling atoms, this also explains the dramatic increase found in the range of channeled particles in thick targets.

To account for effects observed in thick targets, the foregoing theory must be extended to include dechanneling (i.e., those processes that cause $E_\perp$ to change during penetration). For example, multiple scattering on target electrons and variations in the continuum potential resulting from thermal vibrations cause $E_\perp$ to fluctuate, and the statistical accumulation of these fluctuations generally leads to an increase in the nonchanneled component of the beam with increasing depth. For the example discussed earlier, 1-MeV protons→Si ⟨110⟩, the depth of penetration at which half the protons have dechanneled is ~15 μm at room temperature.

A few applications of the channeling effect deserve special attention. The first is the lattice location of impurity atoms in crystals. If the impurity atoms are in an atomic position of high symmetry, it will normally be found that a characteristic fraction of these atoms lies in a given set of strings or planes and therefore will not interact with channeled particles. By studying this fraction along different crystallographic directions, the impurity site can often be deduced simply. In these studies, the energy transfer in a large Rutherford scattering or reaction products from a nuclear reaction are commonly used as signals from the channeled-particle–impurity interaction. This technique has been extensively used for locating impurities and evaluating lattice disorder introduced by ion implantation.

It is also possible to measure directly the lifetime of short-lived (10^{-16}–10^{-18} s) compound nuclei by using channeling. Here a nucleus sitting in a lattice site is excited by the absorption of a particle from an external (unchanneled) beam. The particle momentum absorbed by the nucleus causes it to recoil from the lattice site. If the nucleus decays by emission of a positively charged particle, a study of the angular distribution of these decay products ejected from the target, near axial or planar directions, indicates the extent to which the nucleus has recoiled into the channel. Knowing the recoil velocity (from kinematics), we can then deduce the nuclear lifetime. This is called the "blocking lifetime" technique because of the blocking action the strings have on preventing charged-particle emission from the target along the axial direction in the zero-lifetime case.

Finally, to indicate the scope of the channeling concept, we note that channeling effects have been observed for particles with energies ranging from the kilo-electron-volt to the giga-electron-volt region. Negative particles are also subject to the conditions of governed motion, but because of the attractive nature of the particle–string interaction, the persistence of these motions can be greatly impaired. Furthermore, for the case of light particles (e.g., electrons and positrons) quantum-mechanical considerations may need to be included (particularly for $E \lesssim 1$ MeV) for an overall description of the crystal penetration. To the extent that these considerations are required, this subject lies outside the scope of the channeling concept.

See also SCATTERING THEORY.

BIBLIOGRAPHY

D. S. Gemmell, *Rev. Mod. Phys.* **46,** 129 (1973); a review article covering virtually all areas of channeling. (A)

W. Gibson, *Ann. Rev. Nucl. Sci.* **25,** 465 (1975). The application of the "blocking lifetime" technique to compound nuclei is reviewed. (A)

J. Lindhard, *Mat.-Fys. Medd. Danske Vid. Selsk.* **34,** No. 14 (1965). This is the most in-depth theoretical treatment of channeling. (A)

J. W. Mayer, L. Eriksson, and J. A. Davies, *Ion Implantation in Semiconductors*. Academic, New York, 1970. The role of channeling in ion implantation, the measurement of lattice damage, and lattice-location studies are discussed with the stress on applications in semiconductor technology. (A)

Chaos

Edward Ott

The word chaos, as used in this article, describes a type of behavior of dynamical systems in which the time evolution can be very complicated. (A definition of chaos is given later.) In particular, chaotic systems, although strictly deterministic, nevertheless display many apparent attributes which are reminiscent of randomness. Indeed when viewing the time evolution of such systems one often has the feeling that a statistical description is called for. Surprisingly such complicated chaotic dynamics can be present even in deceptively simple systems.

The existence of chaotic dynamics has been discussed in the mathematical literature for many decades starting with Poincaré at the turn of this century, with subsequent important early contributions made by the mathematicians Birkhoff, Cartwright, Littlewood, Levinson, Smale, and Kolmogorov and his students, among others. Nevertheless, it is only comparatively recently that the wide-ranging impact of chaos in physics has been recognized. Specific examples where chaotic dynamics arise are celestial mechanics, convection in fluids, oscillations in lasers, heating of plasmas by electromagnetic waves, the determination of the limits to weather forecasting, stirred chemical reactor systems, and an ever-growing host of other examples.

In what follows we discuss some of the concepts, definitions and results basic to the field of chaotic dynamics.

DYNAMICAL SYSTEMS

This is a system of equations that allows one to predict the future given the past. One example is an autonomous system of first-order ordinary differential equations in time, $d\mathbf{x}(t)/dt = \mathbf{F}(\mathbf{x})$, where $\mathbf{x}(t)$ is a D-dimensional vector and $\mathbf{F}$ is a D-dimensional vector function of $\mathbf{x}$. Given $\mathbf{x}(0)$, the differential equation determines $\mathbf{x}(t)$ for $t \geq 0$. Another example is a *map* which is an equation of the form $\mathbf{x}_{t+1} = \mathbf{G}(\mathbf{x}_t)$ where here the "time" t is discrete and integer-valued. Thus given $\mathbf{x}_0$, the map gives $\mathbf{x}_1$, which when inserted in $\mathbf{G}$ then gives $\mathbf{x}_2$, and so on. The importance of maps derives from the fact that continuous time systems such as ordinary differential equations can often be reduced to maps by the Poincaré surface-of-section technique: pick some (D-1)-dimensional surface (the surface of section) in the D-dimensional phase space of a continuous-time dynamical system, and consider the orbits of the system and where they pierce the surface of section. Since the location $\mathbf{x}_n$ of the nth piercing uniquely determines the location of the $(n + 1)$th piercing $\mathbf{x}_{n+1}$ (via integration of the ordinary differential equations), there is, in principal, a functional relation $\mathbf{x}_{n+1} = \mathbf{G}(\mathbf{x}_n)$ relating $\mathbf{x}_{n+1}$ to $\mathbf{x}_n$. That is, there is a map.

CONSERVATIVE AND NONCONSERVATIVE SYSTEMS

For the purposes of the following discussion it is useful to distinguish between "nonconservative" dynamical systems and "conservative" dynamical systems. By a conservative system we mean one under which volumes in phase space are preserved as time evolves (or else, if they are not preserved, they can be made to be preserved by a smooth change of variables). Thus, if we consider all the points on the surface of a subset of phase space as initial conditions and evolve these initial conditions in time, then the surface continuously contorts its shape, but the total volume enclosed by it does not change. Hamiltonian systems are conservative in this sense. The most important difference between conservative and nonconservative dynamical systems is that the latter typically have attractors, while the former do not.

ATTRACTOR

If one considers a system and its phase space, then the initial conditions in some region B may be asymptotic as $t \to \infty$ to some smaller set A contained in B. In such a case A is said to be an attractor. For example, the phase-space variables specifying the state of a damped harmonic oscillator, $d^2x/dt^2 + v\, dx/dt + \omega^2 x = 0$, are the position x and velocity $v = dx/dt$. Any initial condition eventually comes to rest at the point $(x, v) = (0, 0)$, and this is the attractor for the system. Thus, here the attractor is very simple, a single point, which is a set of dimension $d = 0$. It is often the case, however, that for chaotic systems the attracting set can be geometrically much more complicated. In fact, it can have fractal geometry with a dimension d which is not an integer. In such cases the attractor is often called a strange attractor.

DIMENSION

There are various ways to define the dimension of a set. Perhaps the simplest is the box-counting dimension (also called the capacity dimension) which is given by the formula

$$d = \lim_{\epsilon \to 0} \frac{\ln N(\epsilon)}{\ln (1/\epsilon)}, \qquad (1)$$

where we imagine the set to be covered by small D-dimensional cubes of edge length ϵ (D denotes the Euclidian dimension of the phase space) and $N(\epsilon)$ is the minimum number of such cubes needed to cover the set. For example, if the set were a single point, then $N(\epsilon) = 1$ independent of ϵ, and Eq. (1) yields $d = 0$, as it should. If the set is a smooth curve, then $N(\epsilon) \sim l/\epsilon$, where l is the length of the curve, and Eq. (1) yields $d = 1$, again as it should. A more interesting example is the case of a Cantor set: Take the interval on the real line from 0 to 1; divide it in thirds; discard the middle third; take the two remaining thirds; divide each of them in thirds; discard the two middle thirds from these; and continue this process. In the limit that the process is applied an infinite number of times the remaining set is a Cantor set. This set is uncountable, and application of Eq. (1) shows that its box-counting dimension is $d = (ln\ 2)/(ln\ 3)$. Thus d is a number between 0 and 1 and the set is fractal. This type of geometric structure is typical of strange attractors.

CHAOS

The dynamics on an attractor for a nonconservative dynamical system is said to be chaotic if typical orbits on the attractor display sensitivity to initial conditions. That is, consider an orbit $\mathbf{x}(t)$ evolving from an initial condition $\mathbf{x}(0)$ in the basin B of a chaotic attractor A (the basin of an attractor is the set of all initial conditions leading to that attractor). Now give the initial condition an infinitesimal perturbation $\delta\mathbf{x}(0)$, and consider the orbits $\mathbf{x}(t)$ and $\mathbf{x}(t) + \delta\mathbf{x}(t)$ which, respectively, evolve from the initial conditions $\mathbf{x}(0)$ and $\mathbf{x}(0) + \delta\mathbf{x}(0)$. If, for typical choices of $\delta\mathbf{x}(0)$, the distance between the two orbits grows exponentially with time, $|\delta\mathbf{x}(t)| \sim \exp(ht)$ with $h > 0$, then we say that the dynamics on A is chaotic. The quantity h defined by

$$h = \lim_{t \to \infty} \frac{1}{t}\, ln \frac{|\delta\mathbf{x}(t)|}{|\delta\mathbf{x}(0)|}$$

is called the Lyapunov exponent. For Hamiltonian systems there are no attractors but chaos is said to be present if typical orbits yield $h > 0$. The extreme (i.e., exponential) sensitivity to initial conditions displayed by chaotic systems has the practical importance that small errors (such as computer round-off) eventually make it impossible to obtain the exact longtime behavior of the system. This was originally pointed out in the context of weather prediction in the seminal 1961 paper of Edward Lorenz. The condition $h > 0$ gives what is perhaps the most common definition of chaos. This definition, however, is not universally accepted. For example, other attributes of chaos which might be taken as its defining property are the infinity of unstable periodic orbits embedded in chaotic regions of phase space, and the property of positive topological entropy. (The topological entropy is a quantitative measure of how rapidly the number of distinct system orbits one can discern under finite resolution grows with the length of observation time.) These latter properties are more relevant when discussing unstable chaotic sets.

UNSTABLE CHAOTIC SETS

Attractors refer to sets which "attract" orbits and hence determine typical long-term behavior. It is also possible to have sets in phase space on which the dynamics can be exceedingly complicated, but which are not attracting. In such cases orbits placed exactly on the set stay there forever, but typical neighboring orbits eventually leave the neighborhood of the set, never to return. One indication of the possibility of complex behavior on such nonattracting (unstable) sets is the presence on the set of an infinite number of unstable periodic orbits whose number increases exponentially with their period, as well as the presence of the uncountable number of nonperiodic orbits. Even though nonattracting, unstable chaotic sets can have important observable macroscopic consequences. Three such consequences are the phenomena of chaotic transients, fractal basin boundaries, and chaotic scattering.

CHAOTIC TRANSIENTS

In chaotic transients one observes that typical initial conditions initially behave in an apparently chaotic manner for a possibly long time, but, after a while, then rapidly move off to some other region of phase space, perhaps asymptotically approaching a nonchaotic attractor. The length of such chaotic transients depends sensitively on initial conditions and exhibits a characteristic Poisson distribution for randomly chosen initial conditions.

FRACTAL BASIN BOUNDARIES

Basin boundaries arise in dissipative dynamical systems when two, or more, attractors are present. In such situations each attractor has a basin of initial conditions which lead asymptotically to that attractor. The basin boundaries are the sets which separate different basins. It is very common for basin boundaries to contain unstable chaotic sets. In such cases the basin boundaries can have very complicated fractal structure. Because of this complicated very fine-scaled structure, fractal basin boundaries can pose an impediment to predicting long-term behavior. In particular, if an initial condition is specified with only finite precision, it may be very difficult *a priori* to determine in which basin it lies if the boundaries are fractal.

CHAOTIC SCATTERING

In the classical dynamics potential scattering problem one considers a Hamiltonian $H = p^2/2m + V(\mathbf{r})$, where the potential V approaches zero for large $|\mathbf{r}|$. One then asks how outgoing orbits at large $|\mathbf{r}|$ depend on incoming orbits. For example, one might plot scattering angle as a function of impact parameter. In typical cases, such functions can have exceedingly complex behavior, where the function is singular on a fractal (uncountable) set of impact parameter values. This type of behavior is indicative of the presence of an unstable chaotic set in the dynamics.

ROUTES TO CHAOS

It is a common procedure in experiments to examine the observed behavior as some parameter of the system is varied. One can then attempt to observe transitions between regions of parameter space where qualitatively different properties occur (e.g., phase transitions and critical phenomena in condensed matter physics). In nonlinear dynamics, particular attention attaches to the study of transitions to chaos in which one observes nonchaotic behavior (e.g., periodic motion) for some range of the parameter, but then observes a chaotic attractor as the parameter is varied. The question then is *how* does the chaotic attractor come into being as the parameter is varied. Generally, it is found that chaotic attractors come about in a limited number of often-observed characteristic ways. These include period doubling cascades, crises, intermittency, and quasiperiodic transitions. In the following we discuss the first two of these routes to chaos.

PERIOD DOUBLING

In some range of the parameter, we might have time periodic behavior of a relevant dynamical variable, $x(t) = x(t + T)$, where T is the period. As the parameter, call it p, increases through a value p_1, the period of $x(t)$ can double to $2T$ in the following way. For $p < p_1$, the periodic signal has a peak (maximum) which repeats every period T. For $p > p_1$, the signal bifurcates so that peaks separated by T are now unequal but repeat after every $2T$. (For $p \to p_1$ from above, the difference between the two adjacent maxima approaches zero.) This is called a period-doubling bifurcation. It is often observed that period doublings occur in cascades. That is, one finds that as p is increased, there is a succession of period doublings at $p_1 < p_2 < p_3 < \cdots < p_\infty$, where p_n accumulates geometrically on some finite value p_∞. The rate of geometric accumulation is universal and has been determined by the renormalization group technique by Feigenbaum. He obtains $\lim_{n\to\infty}(p_n - p_{n+1})/(p_{n+1} - p_n) = 4.669201. . . .$ For $p > p_\infty$ there is typically attracting chaotic behavior.

CRISES

Another type of transition to a chaotic attractor is the crisis. Basically, what happens in this case is that the unstable chaotic set responsible for a chaotic transient becomes stable as the parameter p is increased through a critical crisis value p_c. When it becomes stable, the chaotic set formerly responsible for the chaotic transient becomes a chaotic attractor. For parameter values in the transient range, $p < p_c$, there is typically a characteristic dependence of the mean duration of chaotic transients on p. Namely, $\tau \sim (p_c - p)^{-\gamma}$, where the critical exponent γ can be obtained from a knowledge of the instability properties of certain unstable periodic orbits on the chaotic set. This dependence of τ on p makes clear the nature of the transition: τ increases to infinity as p approaches p_c from below, thus converting the transient to long-term time-asymptotic behavior.

KAM SURFACES

Dynamics in conservative systems, and, in particular, Hamiltonian systems, can differ qualitatively from that in nonconservative systems. In particular, Hamiltonian systems are characterized by the absence of attractors and by the typical occurrence of KAM (for Kolmogorov, Arnol'd, and Moser) surfaces. KAM surfaces are N-dimensional toroidal surfaces in the $2N$-dimensional phase space of coordinates and momenta $(q_1, q_2, . . . , q_N; p_1, p_2, . . . , p_N)$ on which orbits execute N-frequency quasiperiodic motion. (In such quasiperiodic motion the orbit winds around each of the N possible angular paths on the torus with N frequencies which are incommensurate.) For completely integrable systems, such surfaces permeate all of the phase space. Small perturbations from perfect integrability typically lead to chaotic motions, but only in a relatively small volume of the phase space, the remainder of the phase-space volume (Lebesgue measure) being occupied by KAM surfaces. In such cases the KAM surfaces lie on a Cantor set of positive volume; in particular, an arbitrarily small neighborhood of a KAM surface will typically contain chaotic orbits. As the perturbation from exact integrability is increased, the relative volume of chaos increases and can eventually occupy an order-one fraction of the phase-space volume.

QUANTUM CHAOS

According to the correspondence principle, there is a limit where classical behavior as described by Hamilton's equations becomes similar, in some suitable sense, to quantum behavior as described by the appropriate wave equation. Formally, one can take this limit to be $h \to 0$, where h is Planck's constant; alternatively, one can look at successively higher energy levels, etc. Such limits are referred to as "semiclassical." It has been found that the semiclassical limit can be highly nontrivial when the classical problem is chaotic. The study of how quantum systems, whose classical counterparts are chaotic, behave in the semiclassical limit has been called quantum chaos. More generally, these considerations also apply to elliptic partial differential equations that are physically unrelated to quantum considerations. For example, the same questions arise in relating classical acoustic waves to their corresponding ray equations. Among recent results in quantum chaos is a prediction relating the chaos in the classical problem to the statistics of energy-level spacings in the semiclassical quantum regime. Other notable work has concerned Anderson localization phenomena and microwave ionization of atoms in high-energy states.

THE FUTURE

Much remains to be done in applying chaotic dynamics to specific physical systems. In addition, many important fundamental questions in chaotic dynamics remain unanswered. Some of these are the following: characterization of universal properties of dynamics in the chaotic regime; the interaction of spatial patterns and temporal chaos; the understanding of long chaotic transients in higher-dimensional dynamical systems; bifurcations involving chaotic behavior in higher-dimensional dynamical systems; how to extract information from experimental data obtained from chaotic processes; how to find the minimum dimensionality of a dynamical system needed to describe given chaotic data; how best to characterize and determine the properties of fractal sets arising in chaotic dynamics; how to use small external perturbations to control chaotic processes; the development of new computer methods for the study of chaos; the behavior of random (or noisy) dynamical systems. This is but a partial list. What seems certain, however, is that research in chaotic dynamics will be both important and exciting for a long time to come.

See also DYNAMIC CRITICAL PHENOMENA; FRACTALS.

Charge-Density Waves

P. B. Littlewood

A charge-density wave (CDW) in a solid refers to a weak periodic perturbation of the valence-band electronic charge density, accompanied by a periodic lattice distortion (PLD) of the crystal, sometimes also called a Peierls distortion. The appellation CDW is usually reserved for small perturbations in the charge density of a material which would be otherwise a metal, in distinction to periodic structural distortions in insulators (e.g., $NaNO_2$) where the driving mechanism is different.

The physical reason for the distortion is most easily pictured in a model of a one-dimensional chain of atoms; the Fermi energy corresponds to partial filling of the band up to the Fermi wave vector k_F (see Fig. 1). If now a small lattice distortion is introduced with a wavelength $2\pi/Q$, the band structure will have a gap at a wave vector $Q/2$ because of scattering of electronic Bloch waves by the new periodicity in the potential. States of wave vector $|k| < \frac{1}{2}Q$ have their energy lowered, while those of larger momenta have their energy raised; consequently the single-particle energy will be a minimum when $Q = 2k_F$, so that the gap lies exactly at the Fermi level. Because the electronic energy is lowered by the presence of the gap, the undistorted metallic state is unstable to the periodically deformed CDW.

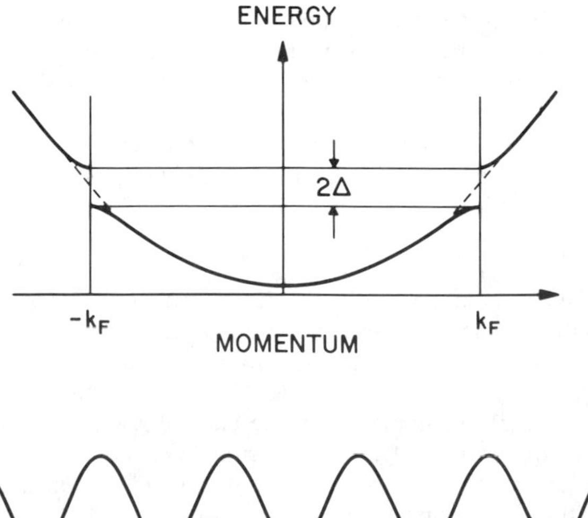

FIG. 1. Electronic energy spectrum (top) in the presence of a periodic lattice distortion (bottom) with wavelength π/k_F. The dashed curve is the metallic spectrum in the absence of the CDW. The periodic CDW is also shown.

In a three-dimensional material the new gaps introduced by the distortion will lie along planes in momentum space (the new Brillouin zone boundaries); a significant lowering of the total energy will require that much of the Fermi surface lies in the new gaps, which in turn requires that the metallic Fermi surface must have flat pieces translated by the CDW wave vector **Q**. This "nesting" condition explains why CDW are prevalent in materials which are electronically low dimensional, typically having either chain-like structures [e.g., $NbSe_3$, $(TaSe_4)_2I$, and some organic conductors of which tetrathiafulvalene-tetracyanoquinodimethane (TTF–TCNQ) is the prototype] or layer structures (e.g., $NbSe_2$, $TaSe_2$). If a gap is opened over the whole Fermi surface, the CDW will be semiconducting; in many cases the Fermi surface is not completely gapped, and the material is then a semimetal.

A closely related concept is that of the spin-density wave (SDW). A SDW can be regarded as a combination of two CDW, one for each spin polarization, arranged 180° out of phase; thus the spin density is periodic while the charge density is unchanged from that of the metal. The driving force for the transition is the exchange interaction between electrons rather than the direct Coulomb interaction (mediated by the lattice) in the CDW case.

In the simple model picture given above, the period of the density wave is related to the "nesting" wave vector, and may not be related to the period of the underlying lattice. In such a case the CDW is termed incommensurate. Also commonly observed is a commensurate CDW, where the period of the CDW is a multiple of the lattice constant. A further complexity is that the lattice symmetry may be such that different directions of the CDW wave vector $\mathbf{Q}_i$ may be related by symmetry. In this case interactions between CDW in different directions may give rise to a single-**Q** CDW, or a multiple-**Q** CDW where all periods coexist. Different combinations of these situations can exist even in the same material at different temperatures. A well-studied example is the layer compound 2*H*-$TaSe_2$, where the individual layers have triangular symmetry, and there are three equivalent CDW under 120° rotation. The low-temperature state is commensurate, with a periodicity of three times the lattice constant. With increasing temperature, there is a phase transition where two of the three **Q** vectors become incommensurate (the "striped" phase); at a higher temperature all three become incommensurate and equivalent before the final phase transition to the normal metallic phase when the CDW amplitude vanishes.

Such complex situations are not uncommon, and are best understood within a phenomenological Ginzburg–Landau expansion of the free energy in terms of the CDW order parameter. For a CDW, one might use as an order parameter the lattice distortion itself, or the periodic components of the charge density; the latter is conventional, and one may write the charge density as $\rho(\mathbf{r}) = \rho_0(\mathbf{r})[1 + \alpha(\mathbf{r})]$. Here ρ_0 is the charge density in the normal state, and $\alpha(\mathbf{r}) = \mathrm{Re}\,\Psi(\mathbf{r})$ the real order parameter. In the case of a simple incommensurate CDW we shall have $\Psi(\mathbf{r}) = \Psi_0 \exp[i(\mathbf{Q}\cdot\mathbf{r} + \phi)]$.

The value of ϕ measures the position of the CDW relative to the underlying lattice; clearly if **Q** is incommensurate, the value of ϕ is arbitrary, if constant. More generally, there will be some interaction with the lattice which will attempt to "lock" the **Q** vector to some nearby commensurate value, say $\mathbf{Q}_c$. In this case, we shall obtain local regions of the CDW which are commensurate separated by periodic jumps in phase ("discommensurations"); the phase ϕ is now a periodic function of position. The situation is illustrated in Figure 2; it bears close similarity with models for atoms adsorbed on surfaces (e.g., Kr on graphite) as well as dislocations in solids.

If the interaction with the underlying lattice is unimportant, then the freedom in choosing ϕ suggests that an incommensurate CDW may be free to slide and conduct a current. Such an idea was proposed by Fröhlich in 1954 as a possible mechanism for superconductivity. In fact, incommensurate CDW are always pinned by impurities (via local fluctuations in the phase ϕ); furthermore a sliding CDW dissipates energy quite rapidly so this is not a plausible mechanism for superconductivity. In recent years, a number of sliding CDW have been discovered, principally among inorganic linear chain compounds (e.g., $NbSe_3$, $K_{0.3}MoO_3$) but also reported in TTF–TCNQ, and possibly also a sliding SDW in tetramethyltetraselenafulvalinium nitrate, $(TMTSF)_2NO_3$.

In all of these materials a finite electric field (the threshold field E_T, typically ~100 mV/cm) must be applied for the CDW to break free from impurities and begin to slide. In the pinned state at an electric field $E < E_T$, the CDW behaves like a collection of damped harmonic oscillators, with a distribution of oscillator frequencies reflecting the disorder induced by the pinning. As in all disordered systems, the pinned configuration is not unique, leading to hysteresis in the electrical polarization of the pinned CDW.

In the sliding state there is a net current transported by the CDW; in many systems, this is found not to be uniform in time, but has both a nearly periodic component ("narrow band noise," NBN) and a low-frequency broadband noise with an approximately $1/f$ power spectrum. The frequency of the NBN varies proportionately to the average velocity **v** of the CDW, and matches the "washboard" frequency $\omega_0 = \mathbf{Q}\cdot\mathbf{v}$ of the CDW. The appearance of this internal frequency naturally reflects the periodicity of the CDW interacting with underlying pinning sites.

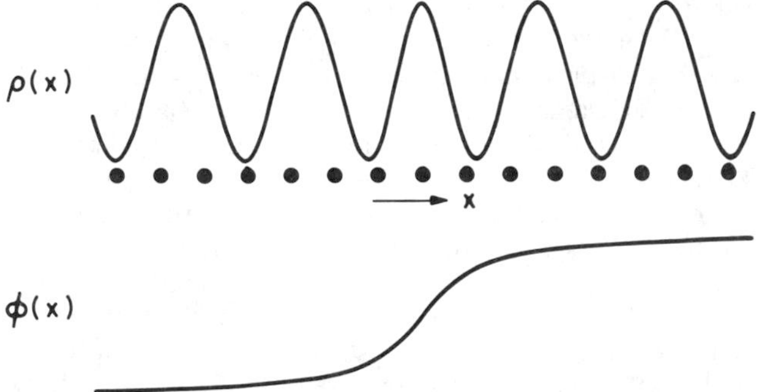

FIG. 2. A discommensuration in a CDW whose wavelength is just less than three lattice constants. At the top is the CDW charge density, and the bottom curve shows the spatial dependence of the CDW phase.

If, in addition to a DC electric field, a fairly large AC component of the electric field is introduced there develop interference features in the response whenever the washboard frequency is a harmonic or subharmonic of the driving frequency. In extreme cases the CDW velocity can mode-lock to the AC driving frequency ω so that $\omega_0 = (p/q)\omega$, with p and q integers; at the same time, the broadband noise vanishes. This appears as plateaus in the current–voltage characteristics, often called Shapiro steps after a similar phenomenon observed in flux flow in type II superconductors. A remarkable phenomenon has been discovered under repetitive pulsed driving, whereby the phase of the transient oscillations at the washboard frequency becomes entrained to the *end* of the pulse. Such experiments, and the models to explain them, have given rise to ideas concerning dynamical selection of states which may have general application to the dynamics of driven nonlinear systems.

In semiconducting CDW at low temperatures, additional phenomena are observed, especially an abrupt rise in conductivity of many orders of magnitude at a second threshold field higher than E_T. This behavior may reflect the Coulomb self-interaction of the CDW, unscreened at low temperatures, and producing a rigid CDW; thus we have the possibility of true superconductivity as originally envisioned by Fröhlich, although only at the absolute zero of temperature.

BIBLIOGRAPHY

Charge Density Waves in Solids, L. P. Gor'kov and G. Grüner, eds. *Modern Problems in Condensed Matter Sciences*. North-Holland, Amsterdam, 1989. (A)

G. Grüner, "The Dynamics of Charge-Density Waves," *Rev. Mod. Phys.* **60,** 1129 (1988). (A)

Low-Dimensional Conductors and Superconductors, D. Jérome and L. G. Caron, eds. NATO ASI Series B: volume 155, Plenum, New York, 1987. (I,A)

K. Bechgaard and D. Jérome, "Organic Superconductors," *Sci. Am.* **247(1),** 52, 1987.

Charged-Particle Optics

J. D. Lawson

Beams of electrons, protons, or heavy ions have many applications in basic science, technology, and industry. The formation of these beams and their transportation and focusing are accomplished with the aid of the principles of charged-particle optics. This is a very wide field, with many specialized branches, relevant to such varied devices as electron microscopes, spectrometers, particle accelerators, cathode ray tubes, beam probe analysis and isotope separators. Nevertheless, despite much specialized technical elaboration, all depend on relatively few, rather general, and basically simple principles.

In the broadest sense, charged-particle optics is concerned with the motion of particles in external electric and magnetic fields. Once these fields are specified, the trajectory of a particle is uniquely determined in terms of its charge q, rest mass m_0, and the three components of velocity at some given point. The equation of motion has the simple form

$$\mathbf{F} = \frac{d\mathbf{p}}{dt} = q(\mathbf{E} + \mathbf{v} \times \mathbf{B}), \tag{1}$$

where the momentum $\mathbf{p}$ is equal to $\gamma m_0 \mathbf{v}$, and γ is the relativistic factor $(1 - v^2/c^2)^{-1/2}$. The essential physics is contained in this equation. In order to find appropriate and convenient solutions of practical value, concepts originally developed in light optics, such as "focusing," "dispersion," and "aberrations," are introduced. At the most fundamental level the curvature of particle trajectories in electric and magnetic fields may be compared with the curvature of light rays in a nonuniform medium. Starting with equivalence between Fermat's principle in optics and the principle of least action in mechanics, a suitable refractive index for charged particle optics may be defined. Although formally elegant, this method is not the most convenient for practical purposes; it is simpler to apply the laws of mechanics more directly. Most applications are concerned with an ensemble of particles in the form of a beam. At any point along the beam these particles are moving in roughly the same direction and have a small spread in energy.

Lenses and prisms in charged-particle optics consist of suitable localized configurations of electric and magnetic fields, and the design of electrodes, coils, and iron magnets to produce fields of appropriate shape for different applications is a highly developed art. In a "perfect" system, the image of a planar object is sharp, and geometrically similar to the object. This idealization is, in charged-particle optics, characterized by the paraxial ray equation, which specifies the particle motion in terms of the electrostatic potential and magnetic field on the axis of the beam. (The axis may be the natural symmetry axis, or, more generally, the trajectory of a particular particle.) It is assumed that particle trajectories (or "rays") make a small angle with the axis, and that the deflecting fields can be expressed as first-order expansions of the fields on the axis.

The detailed form of this equation, of which several derivations may be found in the standard texts, depends on the symmetry and geometrical form of the axis. For axial symmetry the assumptions imply that field components directed along the axis are independent of the distance from the axis, and radial components are proportional to it. The nonrelativistic form of the paraxial equation is then

$$r'' + \frac{\phi' r'}{2\phi} + \left(\frac{\phi''}{4\phi} - \frac{qB_z^2}{8m_0\phi}\right) r + \frac{q\Psi_0^2}{8\pi^2 m_0 \phi} \frac{1}{r^3} = 0, \tag{2}$$

where primes denote d/dz, ϕ is the electrostatic potential on the axis such that $-q\phi$ is equal to the kinetic energy of the particle at that point, and B is the magnetic field. The quantity Ψ_0 will be explained after consideration of the subsidiary angular equation

$$\theta - \theta_0 = -\int_0^z \frac{q}{2m_0 v}\left(B_z - \frac{\Psi_0}{\pi r^2}\right) dz, \tag{3}$$

which, together with Eq. (2), is necessary to specify both r and θ when a magnetic field is present. The magnetic field introduces a feature absent in light optics: the image can be *rotated* as well as scaled in size from the object. It will be seen from the form of Eq. (3) that Ψ_0 is related to the initial angular velocity of a particle about the axis; if, at some radius r_0, $d\theta/dz = 0$, then Ψ_0 is the magnetic flux through a circle of radius r_0 where $d\theta/dz = 0$. Alternatively, $\Psi_0 = 2\pi P_\theta/q$, where $P_\theta = p_\theta + qAr$, the conserved canonical angular momentum of the particle about the axis.

Since ϕ and B are known functions of z and r, trajectories can be calculated. By following a trajectory with initial conditions $r = 1$, $r' = 0$ through the lens, both the focal length and image rotation (for magnetic lenses) may readily be found. For a "thin" lens, in which the radial position of the particle remains essentially constant during its passage through the lens, the focal lengths of electrostatic and magnetic lenses in nonrelativistic approximation are, respectively,

$$\frac{1}{f_e} = \frac{3}{16}\left(\frac{\phi_1}{\phi_2}\right)^{1/4} \int \left(\frac{\phi'}{\phi}\right)^2 dz, \tag{4}$$

$$\frac{1}{f_m} = \int \left(\frac{qB_z}{2m_0 c}\right)^2 dz. \tag{5}$$

For an electrostatic lens ϕ_1 and ϕ_2 represent potentials at entry and exit. Two features of these lenses are evident from the form of these expressions. First, f is always positive, so that all lenses are focusing, and second, the focusing strength depends on the *square* of the fields. These are "second-order" lenses, as a detailed physical consideration of just how they work will reveal.

The paraxial Eq. (2) may be put into linear form by introducing the Larmor transformation, which decouples the radial and angular motion. By observing the orbits in a reference frame which rotates about the axis with frequency $\Omega_L = -\frac{1}{2}qB_z/m_0$, the force on the particles appears purely radial, and it becomes possible to specify the motion in rotating rectangular coordinates, with equations of the form

$$\begin{aligned} x'' + \alpha_x(z)x' + \kappa_x(z)x &= 0, \\ y'' + \alpha_y(z)y' + \kappa_y(z)y &= 0, \end{aligned} \tag{6}$$

where $\alpha_x = \alpha_y$ and $\kappa_x = \kappa_y$. These are now linear, and from this fact many useful relations can be established.

Although introduced here in a particular context, equations of the same form apply in all paraxial situations when there is no coupling between motion in the x and y directions. They describe oscillations about orbits in cyclic particle accelerators, for example, where $B_z = 0$. On the other hand, the focusing is different in the x and y planes, so that the coefficients are not equal. If the accelerating field is negligible or absent (as in a beam transport system), then $\alpha = 0$. When κ is independent of z, the orbits are evidently sinusoidal or exponential in shape.

A useful property which follows from this linear form is that of matrix representation. The values of x and x' at any point z_2 are related to those at an "upstream" point z_1 by the equation

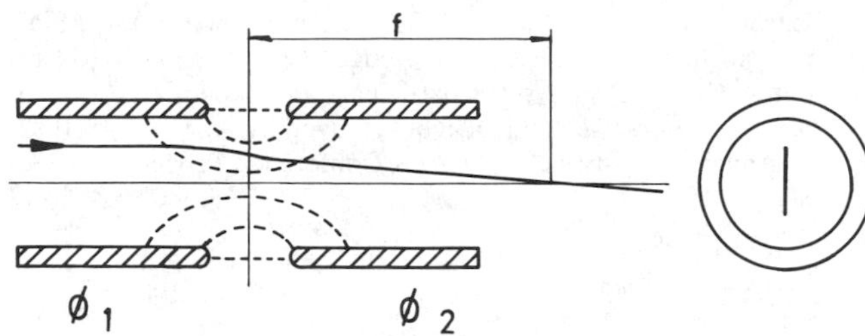

FIG. 1. Some commonly used elements in electron and ion optics. Solid and dashed lines denote orbits and fields, respectively. (a) Electrostatic lens (across which there is a difference of potential $\phi_2-\phi_1$). Hollow disks or cones can be used in place of cylinders. A cylinder with two gaps of opposite polarity, so that the total difference of potential across the pair is zero, is known as an "einzel lens."

$$\begin{pmatrix} x_2 \\ x_2' \end{pmatrix} = M \begin{pmatrix} x_1 \\ x_1' \end{pmatrix}, \tag{7}$$

where M is a 2×2 matrix. Furthermore, when $\alpha=0$, $|M|=1$. Once the appropriate matrices for various elements are known, groups of particles may readily be traced through the system. Particles lying on an ellipse in x-x' space remain on an ellipse. The shape and orientation of the ellipse may change, but its area remains inversely proportional to the momentum of the particles. This fact forms the basis of the emittance concept, which leads to useful design procedures for systems of lenses and accelerator magnet lattices.

The lenses discussed earlier possess axial symmetry. An important lens which does not have this symmetry is the quadrupole, in which the radial and circumferential fields vary as $\cos\theta$ and $\sin\theta$, respectively. (Both types are illustrated in Fig. 1.) In the two orthogonal symmetry planes the field directions are such that there is focusing in one plane and equal defocusing in the other. For a magnetic quadrupole, with field gradient $\partial B_\theta/\partial x$, the paraxial equation in the two planes is

$$\begin{aligned} x'' + \kappa_x x &= 0, \\ y'' - \kappa_y y &= 0, \end{aligned} \tag{8}$$

where $\kappa = q(\partial B_\theta/\partial x)/\gamma m_0 v$. The abandonment of axial symmetry permits first-order focusing in one plane at the expense of defocusing in the other. If, however, two quadrupoles are arranged as a pair with opposite polarity, a net focusing effect is produced. Focusing of this type is much stronger than that obtainable by lenses with axial symmetry. Known as alternating gradient focusing, it forms the basis of the focusing in large particle accelerators where economy of power and size is essential.

In Eq. (2) the coefficients are functions of particle energy, so that energy spread in the beam results in the formation of "blurred" images, a phenomenon known by analogy with light optics as "chromatic aberration." Likewise, if the optical axis is curved, particles with identical initial conditions but different energies follow different trajectories, and a unique "axis" can only be defined with reference to a particle of particular energy. This property, sometimes made use of and sometimes an embarrassment, is known as "dispersion." It may be handled analytically by using 3×3 matrices, with a third row for $\Delta p/p$, the fractional excess of momentum.

Many of the basic ideas of *linear* or paraxial optics have now been introduced. Once the appropriate matrix elements for various devices have been determined, straightforward design procedures, making use of simple computer programs, can be developed. When nonlinear features are introduced, the elegant simplicity disappears; so also does the generality. Aberration problems encountered in different fields tend to be special and different, and need to be considered individually. Small nonlinearities in the focusing systems of accelerators and storage rings, especially those which couple the motion in the two planes, again introduce a wide range of new phenomena. Sophisticated techniques for dealing with these aberrations and nonlinearities have been developed. Recent methods employing Lie algebra may be noted. These are now used particularly in the design of very large accelerators and storage rings.

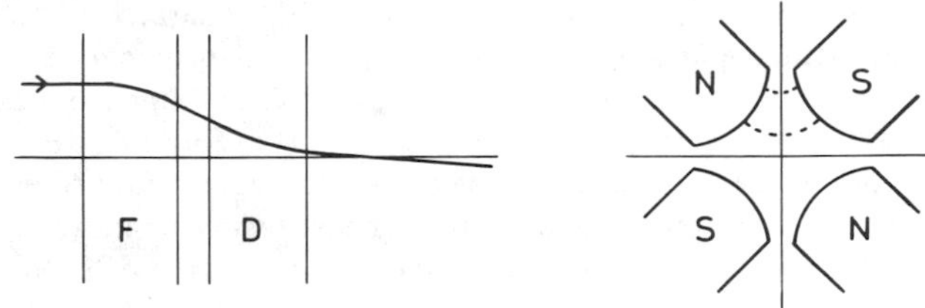

FIG. 1. (c) Magnetic quadrupole pair, in which defocusing is followed by focusing. Pole pieces approximate the ideal hyperbolic shape which gives rise to a pure quadrupole field in the absence of edge effects and saturation.

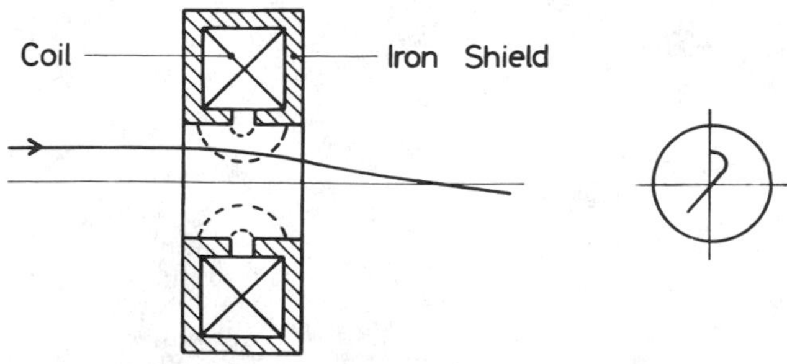

FIG. 1. (b) Magnetic lens. The coil is clad in an iron shield containing a gap. Radial and azimuthal components of velocity interact with azimuthal and radial components of field, and this produces a rotation of the image.

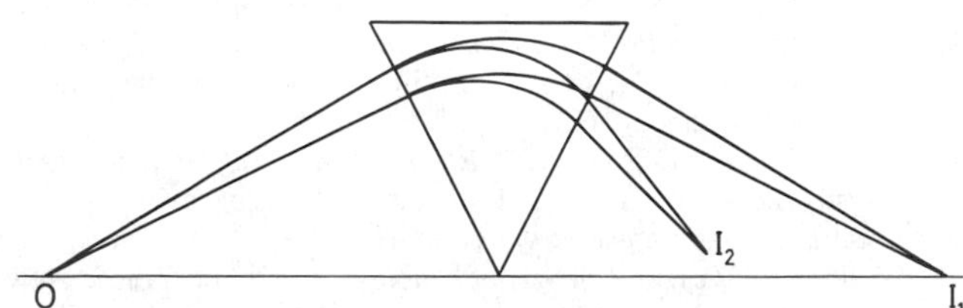

FIG. 1. (d) Magnetic prism, showing bending, focusing, and dispersion. Object, apex, and image are collinear for a magnet with uniform field, normal incidence and exit, and "hard" edges.

So far, attention has been confined to the characteristics of individual particle orbits. Many practical problems are concerned with ensembles of orbits, arising perhaps from a Maxwellian energy distribution of electrons or ions in a cathode or source plasma. The subject can be developed further to investigate these collective properties, using the ideas of statistical mechanics, and particularly the theorem of Liouville. According to this theorem the density of noninteracting particles in phase space remains invariant, and this property is made use of in defining the beam emittance, $\pi\epsilon$. It is often possible to decouple the longitudinal and transverse motion, and assign an emittance to each. Then, by dividing the transverse momentum by $\beta\gamma m_0 c$, the emittance is defined as $1/\pi$ times the projected area on the x-dx/dz or y-dy/dz planes. The meaning of the area when the density of points tapers gradually to zero requires further consideration, and the rms value of the distribution is often used. A more general quantity is the normalized emittance, equal to $\beta\gamma\epsilon$, and the rms value of this quantity is invariant in a linear focusing system even in the presence of acceleration along the axis. The emittance (un-normalized) corresponds to the Helmholtz–Lagrange invariant in light optics, and means physically that at a waist or position of maximum radius the angular spread of the particles in the beam multiplied by the beam diameter remains constant. This may also be thought of as a "gas law" for a beam considered as a drifting gas; the temperature multiplied by the two-dimensional volume remains constant. In intense beams, effects arising from self-fields and scattering may become significant. Here again, however, we leave the realm of optics and begin to impinge on the discipline of plasma physics. Indeed, the study of charged-particle beams in full depth and breadth embraces the three separate disciplines of charged-particle optics, statistical mechanics, and plasma physics.

See also CHARGED-PARTICLE SPECTROSCOPY; ELECTRON AND ION BEAMS, INTENSE.

BIBLIOGRAPHY

P. Dahl, *Introduction to Electron and Ion Optics*. Academic Press, New York, 1973. Clear introduction to basic principles. (E)

P. Grivet and A. Septier, *Electron Optics*, 2nd ed. Pergamon Press, Oxford, 1972. Basic electron optics, with application to a wide range of practical devices. Good bibliography. (I)

S. Humphries, Jr. *Principles of Charged Particle Acceleration*. John Wiley, New York, 1983. Broad coverage, with good introduction to basic principles. (I)

J. D. Lawson, *The Physics of Charged Particle Beams*. Clarendon Press, Oxford, 1977. Synoptic view of basic physical ideas. (I)

J. J. Livingood, *Principles of Cyclic Particle Accelerators*. Van Nostrand, Princeton, N. J., 1969. Rather old, but clear elementary treatment of theory. (E)

P. Sturrock, *Static and Dynamic Electron Optics*. Cambridge University Press, Cambridge, England, 1955. Elegant but formal treatment making use of variational methods. (A)

H. Wollnik, *Optics of Charged Particles*. Academic Press, New York, 1987. Comprehensive theoretical treatment. (I)

The advanced theory of particle accelerators may be found in the proceedings of a number of Accelerator Schools in the USA and Europe from 1982 onwards. These are published in the American Institute of Physics Conference Series, or as CERN reports, CERN, Geneva. In particular, an account of the techniques making use of Lie algebra is given by A. J. Dragt (1982) in American Institute of Physics Conference Proceedings No. 87, p. 145.

Charged-Particle Spectroscopy

Gerard M. Crawley

The study of the properties of nuclear quantum states by bombarding nuclei with charged-particle beams and detecting the emitted charged particles is called charged-particle spectroscopy. Information sought by charged-particle spectroscopy includes the mass (energy) of the nuclear states and their spin (J), parity (π), and isospin (T), the ultimate goal being to obtain the complete wave functions of individual nuclear states.

A typical experimental arrangement is shown in Fig. 1. A beam of charged particles from an accelerator impinges on a "target" in an evacuated chamber. The particles in the beam that pass through the foil without interacting are collected in an insulated Faraday cup, the charge thereby collected giving a measure of the number of beam particles. The relatively few particles that actually interact with nuclei in the target can transfer energy, angular momentum, and nucleons between the beam and the target nucleus. The emitted particles are collected by a particle detector that subtends a solid angle $d\Omega$ at a mean scattering angle θ with respect to the incident beam. The probability of particle emission into solid angle $d\Omega$ is termed the *differential cross section*, $\sigma(\theta)$, which is measured in units of square centimeters per steradian or, commonly, millibarns per steradian (mb/sr), where 1 mb = 10^{-27} cm^2. An *angular distribution* refers to the measurement of $\sigma(\theta)$ at a number of different scattering angles θ.

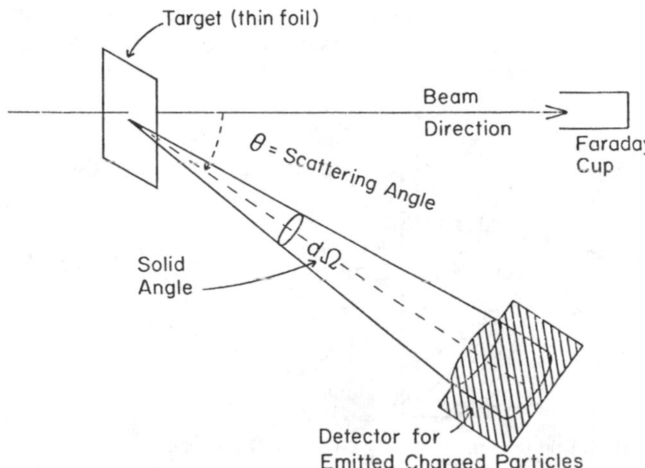

FIG. 1. Schematic diagram of standard experimental arrangement for charged-particle spectroscopy.

The plot of the number of emitted particles of a particular kind as a function of their energy is called a *spectrum*, by analogy with an optical spectrum, in which intensity is plotted as a function of wavelength or frequency. An example of a spectrum of inelastically scattered protons from ^{40}Ca using a 35-MeV incident beam is shown in Fig. 2. The spectrum consists of a number of peaks, each denoting the excitation of a particular nuclear state. As increasing amounts of energy are transferred to the nucleus, states of higher excitation energy in the nucleus are produced and the emitted particles have lower energies. The spectrum shown has an experimental energy resolution of 4.5 keV, which is the full width at half-maximum height of the peaks from ^{40}Ca(p,p'). This spectrum illustrates the need for good energy resolution in order to distinguish peaks from the isotope under study from contaminants in the thin foil targets and, more particularly, to resolve close-lying states.

Excellent energy resolution can also be obtained at lower energies in resonance reactions where compound-nucleus states are studied. This kind of experiment has been pioneered at Duke University where widths of nuclear states as small as 130 eV have been measured. (See, for example, reference 2.)

Thus the mass (energy) of a particular nuclear state can be obtained from the energy spectrum. The spin and parity (J^π) are obtained by comparing the measured angular distribution with theoretical predictions. In the case of resonance reactions, the shape of the yield versus bombarding energy curve also indicates the J^π of the state.

The experimental challenges of charged-particle spectroscopy also include the preparation of incident beams of particles with accurately known energies and with small energy spread, the preparation of clean uniform isotopic targets, and particularly the use of detector systems with both good energy resolution and the ability to discriminate between different particle types emitted from the target. Each of these aspects is discussed in what follows.

ACCELERATORS AND BEAM PREPARATION

Since charged particles must have sufficient energy to overcome the Coulomb repulsion of the positively charged target nuclei, special particle accelerators have been constructed to produce the charged-particle beams. The accelerators most commonly used today are Van de Graaff accelerators (including the tandem type), linear accelerators, and cyclotrons, each having certain advantages as well as limitations. For example, tandem Van de Graaff accelerators produce beams with good energy resolution whose energy can be readily varied, making them particularly suitable for studies of resonance reactions. However, the maximum energy of beams of particles with charge Q from even the largest of such accelerators is presently less than about 25(1 + Q) MeV.

Modern cyclotrons with azimuthally varying magnetic fields can readily accelerate protons to much higher energies (up to a few hundred MeV) and the beam quality from these machines is now comparable to that obtained from tandem Van de Graaffs. Using dispersion-matching techniques, in

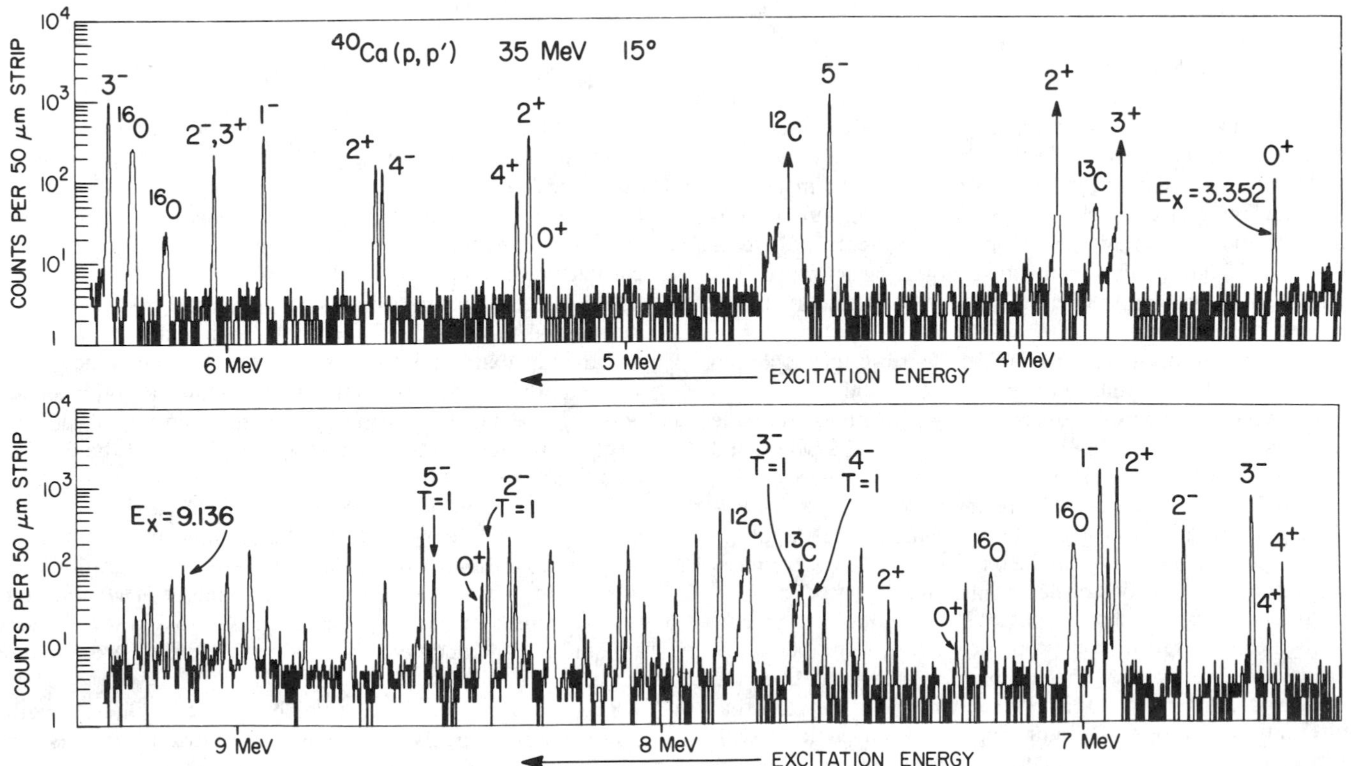

FIG. 2. ^{40}Ca(p,p')^{40}Ca spectrum taken at $\theta_{lab} = 31.2°$ for $E_p = 34.78$ MeV (from [3]).

which beam energy spread is canceled by the dispersion of a magnetic spectrograph, final-state energy resolution of 1 part in 10^4 in energy has been obtained. Because of their higher energy capability, cyclotrons are generally used for direct reaction studies where the beam energy remains constant during the experiment.

Linear accelerators (linacs) have been used to produce large-intensity beams of very high-energy protons. The beam quality from linacs is generally poorer than that from cyclotrons or Van de Graaff accelerators. However, since the beam intensity is high, excellent energy resolution can be obtained even at high energy by selecting a small fraction of the beam and using dispersion matching. For example, an energy resolution of better than 50 keV has been obtained at the 800-MeV proton linac (LAMPF) at Los Alamos National Laboratory.

Superconducting technology has had a significant impact on accelerator design. For example, superconducting radio-frequency cavities have been used to provide post acceleration at several Van de Graaff laboratories, the first being at Argonne National Laboratory in 1978. Superconducting technology has also been applied to cyclotron magnets allowing much higher field strengths at comparatively modest cost. This in turn has allowed the acceleration of more massive charged particles (heavy ions). New heavy ion cyclotrons have recently begun operating at Michigan State University, Texas A&M University, the Chalk River Laboratories, Canada, and at GANIL at Caen, France. Except for the last mentioned, all the other cyclotrons use superconducting magnets. The development of electron cyclotron resonance (ECR) ion sources, which produce intense beams of highly charged ions, has also contributed significantly to the more extensive use of heavy ions for charged particle spectroscopy. Another very recent development in accelerator technology has been the production of beams with very small intrinsic energy spread which are stored in a ring of magnetic elements. The beams are "cooled" by repeated interaction with electron beams having very small energy spread. Such a system was demonstrated in 1988 at Indiana University and similar systems are being built in Heidelberg, West Germany and Uppsala, Sweden. The use of these very high-quality beams for charged particle spectroscopy holds the promise of dramatic improvements in energy resolution.

Beam preparation systems consist of a series of magnetic-quadrupole lenses and dipoles to focus and switch the charged-particle beams to different target positions. A dipole magnet can also provide energy dispersion of the beam, and combined with a slit system can improve the energy resolution of the incident beam. More sophisticated systems produce dispersed as well as focused beams at the target, so that dispersion matching can be used to provide good energy resolution without loss of beam intensity.

At low energies, below about 20 MeV, collimator systems are generally used to constrain geometrically the beam position and direction. At higher energies, scattering from the collimator slits makes this method unsatisfactory, so that alternative methods of monitoring the beam position with retractable detectors (scintillators or wire chambers) are usually employed.

SCATTERING CHAMBERS AND BEAM MONITORING

Since charged-particle beams lose energy and scatter when passing through air, the beam lines and experimental apparatus through which the beam or reaction products pass must be evacuated. Oil diffusion pumps, which were once used extensively in vacuum systems, have largely been replaced by oil-free pumping systems to eliminate carbon and silicon impurities from the system since these can contaminate targets. Modern systems generally employ cryogenic pumping either with a molecular sieve adsorber at liquid nitrogen temperatures or with surfaces held at liquid helium temperatures. Turbo-molecular pumps, carbon vane pumps, titanium sublimation pumps, and ion pumps are also used to obtain clear vacua.

Scattering chambers, which house the target and often the charged-particle detectors, come in many shapes and sizes. The principal design criteria for a general-purpose chamber are (1) accurate setting of the geometry of the experiment, particularly the laboratory scattering angle, (2) the availability of multiple detector systems that can be moved independently; (3) the ability to use a number of targets without breaking vacuum; and (4) the accurate recording of the total charge of the beam that passes through the target in an insulated Faraday cup.

Beam current, target thickness, and beam alignment can be measured by one or more monitor detectors, which are placed at some fixed angles and which can record a prolific reaction such as elastic scattering. The measurement of cross section at a series of angles relative to the number of counts recorded by the monitor provides an accurate measure of relative cross section.

TARGETS

The success of many charged-particle reaction experiments depends on the preparation of suitable targets. Most targets are in the form of a thin foil of the isotope of interest, although in some cases compounds containing the isotope are more convenient. In some cases thin backing foils of carbon or various plastics like Formvar are used as a substrate for the isotopic foil. Gas targets are also used and differential pumping is sometimes employed to enable the entrance or exit window of the gas cell to be eliminated.

The choice of target thickness is usually a compromise between the small counting rate from a thin target and the large energy loss and consequent degradation of the energy resolution from a thick target.

The elimination of unwanted impurities, particularly carbon and oxygen, can, however, be an important element in target preparation. This often involves the preparation and transfer of targets to the scattering chamber *in vacuo*, and the availability of clean high-vacuum systems. In some cases a liquid-nitrogen-cooled shroud may be used around the target.

The use of cooled beams which repeatedly interact with a target poses special problems since ultrathin targets are required. Gas jet and powder targets are under development for these applications.

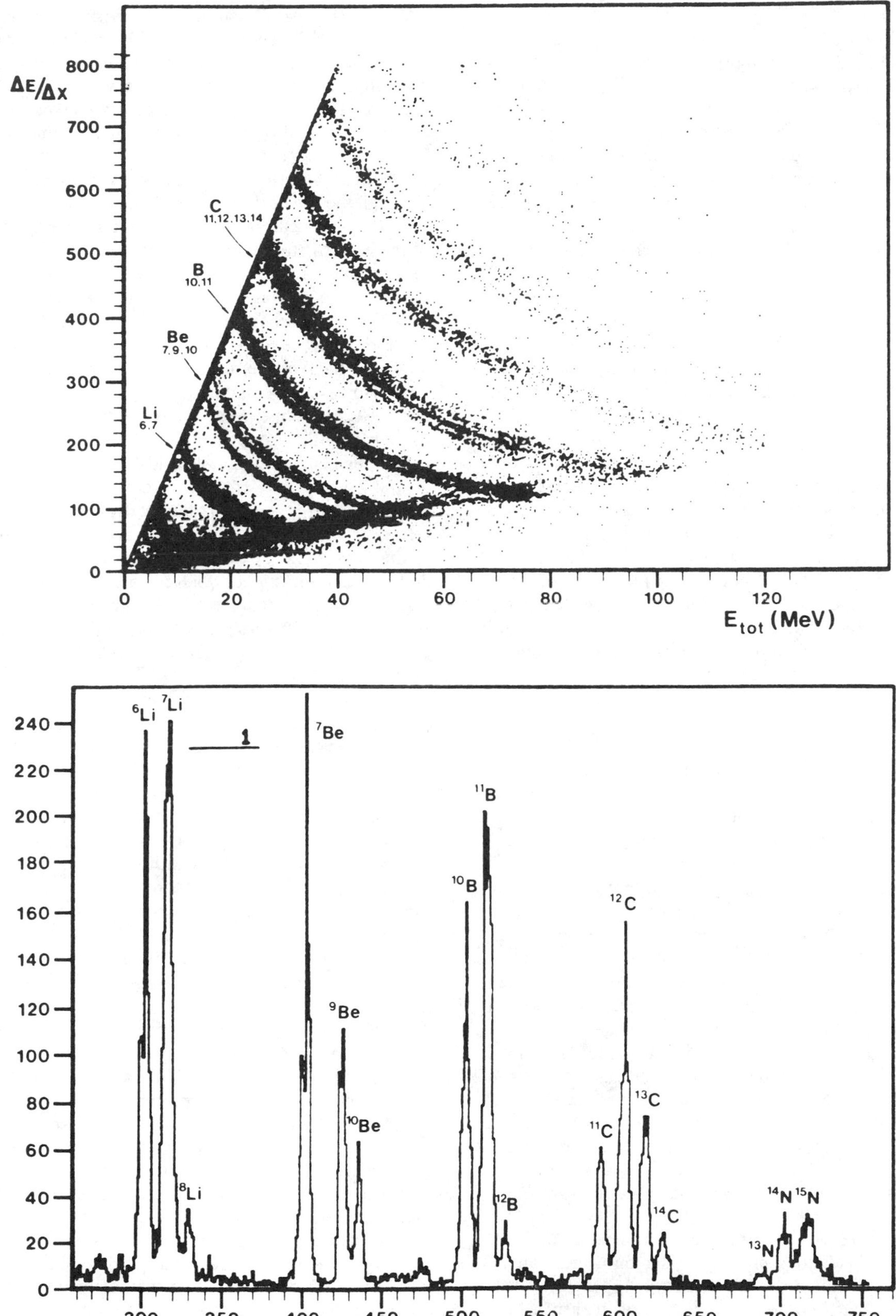

FIG. 3. The two-dimensional *dE*, *E* plot shows the typical result for charge resolution in an ionization chamber. The reaction is 84 MeV/u ^{12}C + Au. If the particle energy is above 3 MeV/u, the resolution is good enough to separate the isotopes of these light ions. This is seen in the lower part of the figure, where the projection on the particle identification (charge) axis is shown (from reference 4).

DETECTORS

Perhaps the single most important factor in a charged-particle reaction experiment is the detector for the emitted charged particles. Just as accelerator technology has produced higher-energy and better-quality beams, recent detector technology has led to much improved energy resolution and better particle discrimination for charged particles. The use of heavy-ion beams has placed even more demands on detectors and has helped to drive the improvement of the technology. There are four general types of particle detectors in current use: (1) scintillation counters, (2) gas counters, (3) solid-state detectors of silicon or germanium, and (4) magnetic spectrometers plus a particle detector in the focal plane.

Scintillation counters consist of inorganic crystals such as NaI, CsI, BaF_2, or CaF_2 as well as many kinds of plastic scintillators, bonded to a photomultiplier tube (PMT). Bismuth germanate (BGO), which has even higher stopping power, has also been used as a charged particle detector. These detectors can be sufficiently thick to stop high-energy charged particles and can also be used when a large solid angle is required. However, the energy resolution obtainable with these detectors is not very good (e.g., about 2% for 30-MeV protons) and they are quite sensitive to gamma rays and neutrons, which are often present as unwanted background in charged-particle experiments. In cases where timing information is needed, a plastic scintillator mounted on a fast photomultiplier tube can be used. The pulses produced in plastic scintillators have rapid rise times, and timing resolution better than 0.5×10^{-9} s can readily be obtained.

A gas counter consists of a gas volume containing an electric field. The primary electrons, produced by the passage of a charged particle through the detector, drift to the anode, often a thin wire, to produce an electrical signal. If the field is low enough, the size of the electrical signal is proportional to the energy loss of the charged particle (ionization and proportional counters). A typical gas mixture used is argon (90%) plus methane (10%). Gas ionization chambers are particularly useful as detectors for heavy ions or low energy light ions because of the small mass of material traversed by the particle. Other useful features are the large areas which can be produced, the lack of radiation damage and the fact that they can be made position sensitive. Gas detectors come in many configurations and have many applications. One of the most common applications is their use as a thin front energy loss detector in a multielement telescope used for particle identification. This is illustrated in Figure 3 where charge and even isotope identification is obtained using a gas-filled ionization chamber. Another common use is as a position sensitive detector in the focal plane of a magnetic spectrograph.

Solid-state ionization chambers made from reverse-biased diodes of semiconducting crystals of silicon or germanium are also very common detectors for charged particles. These detectors have far better energy resolution than gas ionization chambers, since the energy required to produce an ion pair is only about 3 eV in Si compared to about 30 eV in a gas. Thus about 10 times more ion pairs are produced in a silicon detector, with consequently better statistical accuracy in the energy determination. An energy resolution of about 0.1% can now be obtained for such devices.

Solid-state detectors are limited in size because of the difficulty of producing large pure crystals of silicon or germanium. However, thicker detectors can be produced by drifting lithium through the material to compensate exactly for impurities. These lithium-drifted detectors [Si(Li) or Ge(Li)] can be produced in thicknesses up to about 7 mm. Through the use of thin entrance and exit windows, all solid-state detectors can be stacked to produce a thicker amount of material to stop higher-energy particles. Germanium detectors have greater stopping power per unit thickness but have the disadvantage that they must be operated at the temperature of liquid nitrogen. Silicon detectors can also be made position sensitive by evaporating a resistive coating along the back of the detector and taking signals from both ends. However, the maximum length of such devices is only about 10 cm and their position resolution is at best about 1% of their length.

One further advantage of solid-state detectors is their use in multielement systems to differentiate between different particles emitted from a reaction. In a typical two-detector telescope, the thin front detector gives a signal proportional to the rate of energy loss (dE/dx) of the traversing particle, and the second, thicker detector measures the total energy loss (E). (More accurately, the sum of the energy signals from the two detectors gives the total energy loss.) Since the rate of energy loss (dE/dx) of a charged particle is approximately proportional to MZ^2/E, the product $(dE/dx)\cdot E$ is proportional to MZ^2 and so can be used to differentiate particles of different mass and charge. This is the basis for particle identification using a series of solid-state detectors where the logic is carried out with electronic hardware, or more commonly in a computer.

A similar arrangement of a pair of scintillation detectors connected to a single PMT is called a phoswich detector. Either a CaF_2 and plastic scintillator or two plastic scintillators such as NE102 (decay time 2.5 ns) and NE115 (decay time 225 ns) are used. The total light pulse is sampled with both a short and a long gate to distinguish the signals from the fast and slow scintillators. Much larger area two-element detector systems are possible using phoswich scintillators than can be obtained with solid state Si or Ge detectors.

Various kinds of magnetic spectrometers are used to detect the charged products following nuclear reactions. Most modern types include dipole, quadrupole, and higher-multipole elements, and focus particles both radially and vertically (double focusing) to obtain a large solid angle. Some new spectrographs, e.g., Big Karl at Jülich and the K600 at Indiana University allow variation of the dispersion in the focal plane. The radius of curvature gives a measure of the momentum and therefore the energy of the emitted charged particle, and is normally obtained by measuring the position of the particle in the focal plane of the spectrometer. Originally the preferred device to measure position was a nuclear emulsion. This is still the most precise technique but the inconvenience of emulsions, their lack of discrimination of particle type, and, particularly, the improvement in position sensitive gas and solid-state detectors has almost eliminated the use of nuclear emulsions. The two main types of position

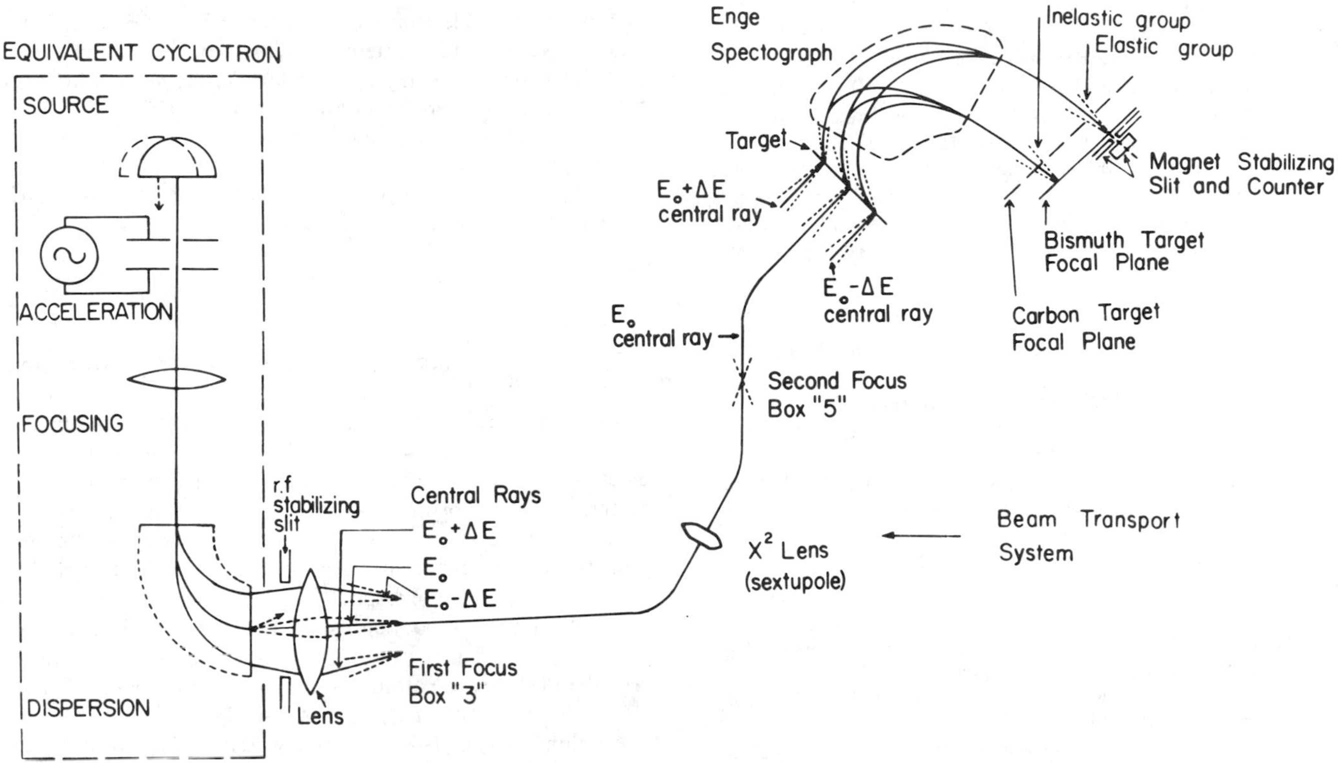

FIG. 4. "Equivalent circuit" of cyclotron and high-resolution beam line showing dispersion matching. The correlated energy dispersion of the beam on target is canceled by the dispersion of the spectrograph (from [1]).

sensitive gas detectors are multiwire and single wire proportional counters. Multiwire proportional counters (MWPC) have a position resolution which depends on the wire spacing but is typically ≤0.5 mm. They also have high count rate capability. Single-wire proportional counters (SWPC) use various methods to measure the position of the incident particle, including charge division of the signals from either end of the counter or measurement of the transit time of the signal along a delay line. By such methods a position resolution of between 0.1 and 0.2 mm can be obtained.

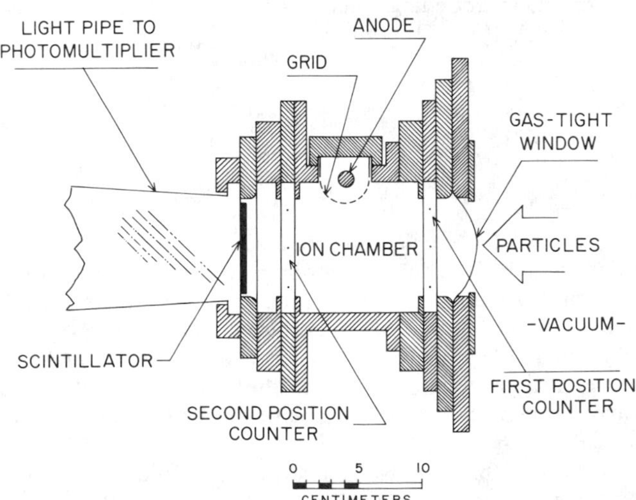

FIG. 5. Spectrometer focal-plane detector system. Gas is usually 200 Torr propane (from *Lawrence Berkeley Lab. Ann. Rep.* **1975**, p. 355).

Apart from large solid angles, another advantage of magnetic spectrometers is that unwanted reaction products with high count rates can often be "bent off" the detector. Very good final-state energy resolution can also be obtained with a magnetic spectrometer, as was illustrated in Fig. 2. This spectrum was measured using a dispersion-matching technique where the energy dispersion of the incident beam at the target position is canceled by the dispersion of the spectrometer, so that only the uncorrelated energy spread gives a contribution to the energy resolution (Fig. 4).

By using multiple detector elements in the focal plane of a spectrometer, particle identification can be carried out very reliably. A typical setup would consist of a series of gas proportional counters to give position and energy-loss information followed by either a solid-state detector or a plastic scintillator to record total energy or time of flight. Such combinations allow very small cross sections to be measured in the presence of prolific backgrounds of other particles. An example of a focal-plane detector used at Berkeley is shown in Fig. 5.

See also ANGULAR CORRELATION OF NUCLEAR RADIATION; CYCLOTRON; NUCLEAR REACTIONS; NUCLEAR SCATTERING; RADIATION DETECTION; SEMICONDUCTOR RADIATION DETECTORS.

BIBLIOGRAPHY

Experimental Aspects and Resonance Reactions

J. Cerny, ed., *Nuclear Spectroscopy and Reactions,* Vols. 1–4. Academic, New York, 1974–1975. (Updates and expands upon the material in the following book.) (I)

W. W. Buechner, "The Measurement of the Spectra of Charged Nuclear Particles"; H. T. Richards, "Charged Particle Reactions"; N. S. Wall, "Charged-Particle Detectors," in *Nuclear Spectroscopy,* Part A (F. Ajzenberg-Selove, ed.). Academic, New York, 1960.

Experimental Methods

B. L. Cohen, *Concepts of Nuclear Physics,* Chapter 9. McGraw-Hill, New York, 1971. (I)

H. Enge, *Introduction to Nuclear Physics,* Chapters 7 and 12; Section 3 of Chapter 13. Addison-Wesley, Reading, Mass., 1966. (I)

Instrumentation for Heavy Ion Nuclear Research, Vol. 7 of Nuclear Science Research Conference Series, D. Shapira, ed. Harwood Academic Publishers, New York, 1985.

REFERENCES

1. H. G. Blosser, G. M. Crawley, R. deForest, E. Kashy, and B. H. Wildenthal, *Nucl. Instr. Methods* **91,** 61 (1971).
2. G. A. Keyworth, G. C. Kyker, E. G. Bilpuch, and H. W. Newson, *Nucl. Phys.* **89,** 590 (1966).
3. J. A. Nolen and R. J. Gleitsmann, *Phys. Rev.* **C11,** 1159 (1975).
4. H. Sann, *Instrumentation for Heavy Ion Nuclear Research,* Vol. 7, p. 27, D. Shapira, ed. Harwood Academic Publishers, 1985.

Chemical Bonding

Henry A. Bent

INTRODUCTION

That chemical combination is essentially an electrical phenomenon Faraday reasoned must be so from the fact that passage of an electrical current between inert electrodes immersed in aqueous solutions produces at the electrodes chemical decomposition ("electro-lysis"). Mathematical development of Faraday's views had to await, however, the introduction into physical theory of the electron, electron spin, the nuclear atom, the wave equation, and the exclusion principle. Chemical theory, meanwhile, developed along purely phenomenological lines, graphically illustrated in the structural theory of organic chemistry and the doctrine of coordination. In recent decades a union of physical and chemical theory has been achieved through the use of the principle of indistinguishability and the creation of localized molecular orbitals that correspond closely to the graphic formulas of classical structural theory.

PHYSICAL MODELS OF THE CHEMICAL BOND

The concept of a chemical *bond* arose from the use in chemistry of graphic formulas to illustrate the rule that carbon atoms in stable compounds, such as CH_4, CH_2Cl_2, CO_2, and HCN, are "tetravalent"—hydrogen and chlorine (in, e.g., HCl) being taken as "univalent"; oxygen (as in H_2O) "divalent"; nitrogen (as in NH_3) "trivalent":

```
  H          H
  |          |
H-C-H      H-C-Cl      O=C=O      H-C≡N
  |          |
  H          Cl

  1          2           3          4
```

For half a century no connection existed between the "bonds" or "valence strokes" in graphic formulas such as **1–4** and physical theory. Then, soon after Moseley's determination of atomic numbers and, thereby, the number of electrons in compounds (usually an even number, for compounds of the nontransition metals), Lewis suggested that in such graphic formulas as **1–4**, (i) each valence stroke represents, improbable as it might seem, *two* "valence-shell" electrons and that, correspondingly, (ii) the symbols for the elements, H, C, N, O, represent the corresponding atoms' positively charged "kernels" or "atomic cores" [after Rutherford and Bohr, the atoms' nuclei and inner-shell (here two *K*-shell or 1*s*) electrons]; e.g., in **1–4**: H^+, C^{4+}, N^{5+}, O^{6+}.

A valence stroke (–) may be viewed as the union of a pair of oppositely directed Faraday lines of force (in the organic chemist's convention: ←→) that stretch between a valence-shell electron pair and the two adjacent atomic cores that that pair helps to bond together. In accordance with the principle of local electrical neutrality and Gauss's law (suitably modified, numerically), the number of lines of force that terminate on an atomic core generally equals that core's positive charge.

In, for example, HCN, the nitrogen atom's kernel, N^{5+}, has, according to Lewis's theory, like C^{4+}, four valence-shell electron pairs, only three of which, however, are simultaneously in the valence shell of another atom (carbon) and shown in the molecule's graphic formula by conventional, straight, bonding valence strokes. The fourth pair is "unshared." It may be shown as

H–C≡N: (**5**) H–C≡N⊃ (**6**)

Like lines of force, valence strokes never cross each other. Valence-shell electrons obey a principle of spatial exclusion: Two but no more than two electrons may be in the same place at the same time. About each spatially coincident electron pair is a Fermi hole, a van der Waals-like domain of diameter approximately equal to the electrons' de Broglie wavelengths, into which other valence-shell electrons cannot easily penetrate.

Because of the decrease in a particle's kinetic energy with increasing de Broglie wavelength, the van der Waals-like domains of electron pairs tend to expand to fill the available space. In a full axiomatization of structural theory, there is associated with valence strokes, as with Faraday's lines of force, a stereochemically significant property of mutual repulsion.

STEREOCHEMICAL MODELS OF CHEMICAL BONDING

Together, the exclusion principle and the kinetic-energy operator in the molecular Hamiltonian, reinforced by Coulombic repulsions, cause the four electron pairs about the carbon atom of methane (CH_4) and its derivatives [such as CH_2Cl_2 (**2**)] to spread out tetrahedrally in three dimensions 7 rather than to remain crowded together in one plane. [Graphic formula **2**, for example, implies, contrary to observation, that there should be *two* isomers of dichloromethane: a *cis* isomer, depicted, and a *trans* isomer.] Similarly, ammonia is pyramidal **8** not planar, and water is bent **9** not linear.

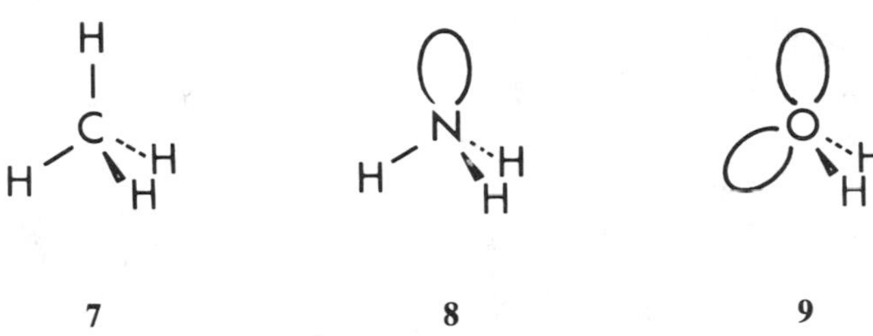

In thought, methane can be transformed into ammonia via an *al*chemical movement of the atomic core of, for example, the upper hydrogen atom in **7**, a bare proton, together with its attached line of force, to the central, heavy-atom nucleus, thereby converting C^{4+} to N^{5+} ("proton capture") and changing the electron pair of the C–H bond of methane into the unshared valence-shell electron pair of ammonia (electron "capture" by the heavy-atom's core). Ammonia and water are similarly related. The three molecules, **7**, **8**, **9**, have similar bond angles (HCH 109.5°, HNH 107°, HCH 104.5°) and, by inference, similar electronic structures. They are said to be "isoelectronic." Likewise, H–C≡N (hydrogen cyanide) and H–C≡C–H (acetylene), both linear molecules, are isoelectronic with N≡N (N_2). Another isoelectronic family of molecules is $H_3C–CH_3$ (ethane), $H_3C–OH$ (methyl alcohol), $H_2N–NH_2$ (hydrazine), HO–OH (hydrogen peroxide), and F–F (fluorine).

The tetrahedral model of directional chemical affinities applies chiefly to C, N, O, F, and other electronegative, "octet-rule"-satisfying atoms (ones that in chemical combination have eight, or four pairs, of valence-shell electrons) in so-called "covalent" compounds (ones that contain no large-atomic-core, electropositive, metallic-like elements).

The tetrahedral model accounts for the nonexistence of quadruply bonded C_2 and, more importantly, for most of the facts of organic stereochemistry, particularly the stereochemistry of many displacement reactions (as described in the last section) and the planar and linear geometries about double and triple bonds, Fig. 1.

Single, double, and triple bonds may be viewed, Fig. 1, as the sharing, respectively, of a corner, an edge, and a face of two tetrahedral polyhedra of electron pairs.

Similar terminology is used to describe the structures of Born-type "ion compounds." Calcium oxide, for example,

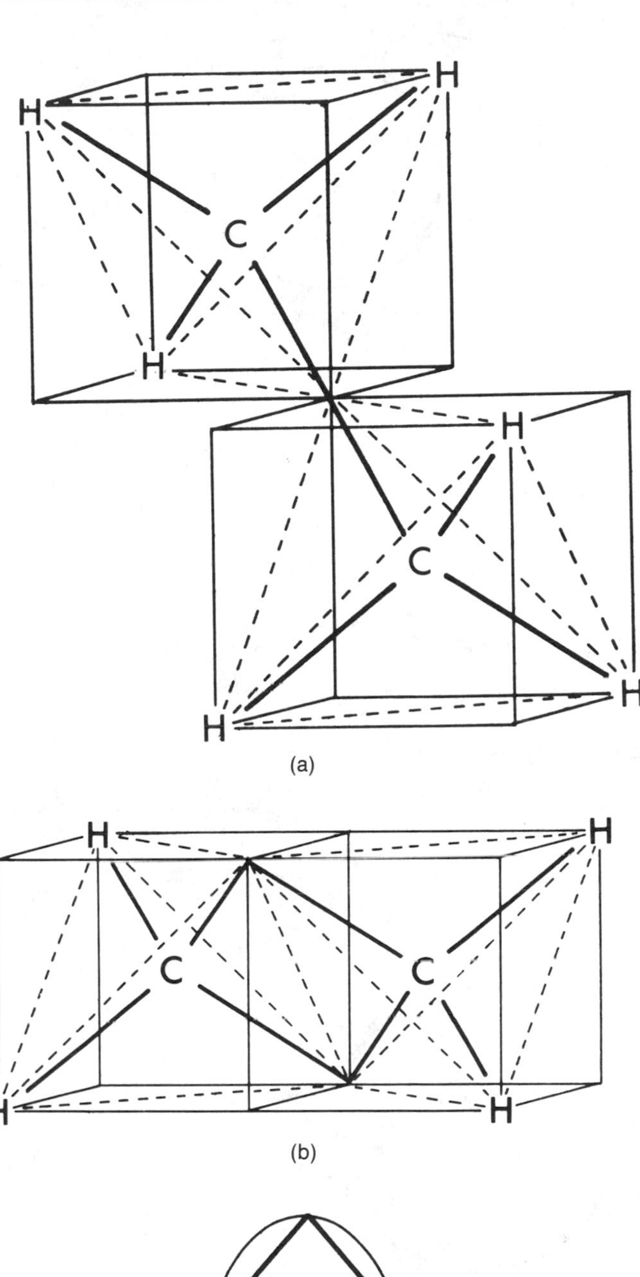

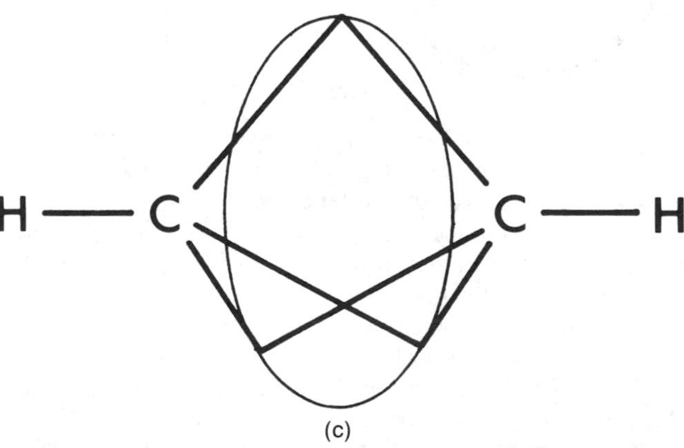

FIG. 1. Sharing of, respectively, a corner, an edge, and a face by a pair of tetrahedra of electron pairs about two atomic cores, C^{4+}, in (a) ethane ($H_3C–CH_3$), (b) ethylene ($H_2C=CH_2$), and (c) acetylene (HC≡CH). Heavy lines represent valence-shell electron pairs. The two carbon atoms of a double bond (b) and their four substituents (H) lie in a plane. The carbon atoms of a triple bond (c) and their substituents lie on a straight line.

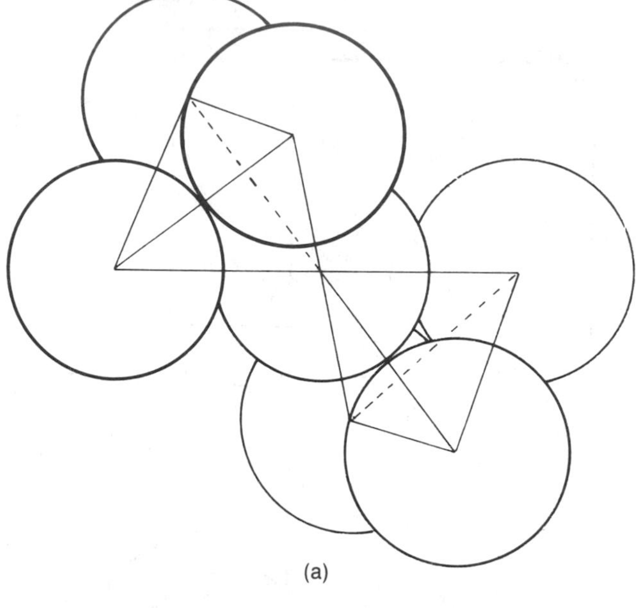

(a)

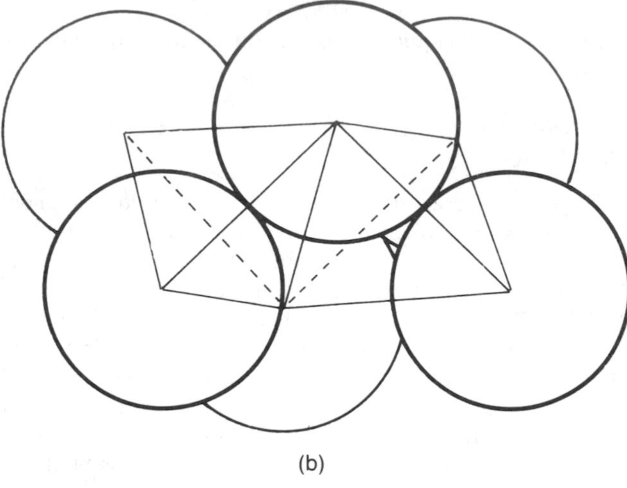

(b)

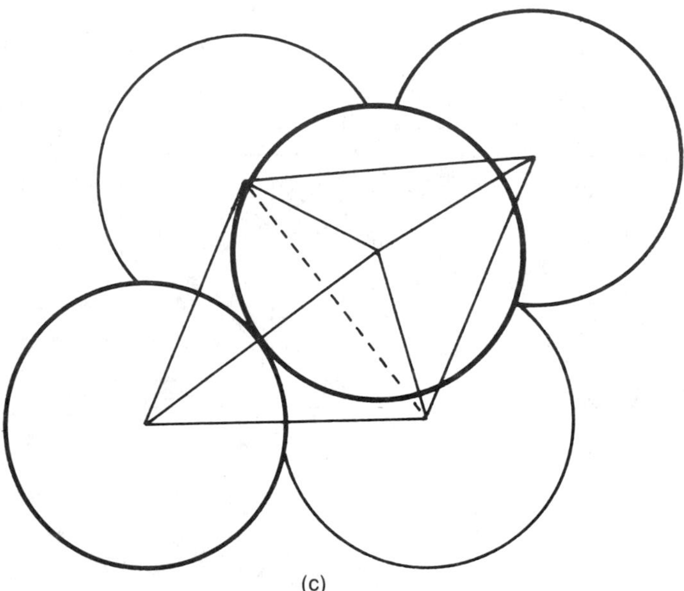

(c)

FIG. 2. Sharing of (a) a corner, (b) an edge, and (c) a face by a pair of tetrahedra of anions (e.g., O^{2-}) about two cations (e.g., Si^{4+}, not shown), after Pauling. There exists an isomorphism with Fig. 1: anions correspond to electron pairs, cations to atomic cores. Edge and face sharing tends to destabilize structures, owing to enhanced cation–cation Coulombic repulsions.

may be viewed as Ca^{2+} cations (a cation being the atomic core of a metallic, large-core element) surrounded by a coordinated polyhedron of (''valence-shell'') O^{2-} anions. When such compounds are ''anion deficient'' (number of anions per cation less than the cation's usual coordination number), the polyhedra of anions about the cations generally share corners, edges, or faces, Fig. 2.

With the exception of (possibly) He and Ne, all uncombined atoms are ''electron deficient'': they contain fewer than the maximum number of valence-rule-allowed electrons. They are, so to speak, ''coordinately unsaturated.'' Particularly electron deficient are atoms to the left of carbon in the periodic table. To utilize fully the low-potential-energy space for electrons about atomic cores of elements from Groups I, II, and III of the Periodic Table, electrons, or pairs of electrons, must often be shared by more than two atomic cores.

B_2H_6, which is isoelectronic with C_2H_4 (**10**), has the electronic structure depicted in **11**.

10 **11**

The boron–boron bond in ''electron deficient'' B_2H_6 may be described variously as: a ''protonated double bond''; as two B–H–B ''three-center bonds''; as two electron pairs simultaneously in the valence shells of three atoms; as two ''bridging'' hydride ions (H^-); or as two tetrahedral BH_4^- ions sharing an edge.

Atoms with large atomic cores may contain in their valence shells more than eight electrons. Whereas oxygen forms only one mono-oxygen fluoride, OF_2, sulfur forms, in addition to SF_2, SF_4 (**12**) and coordinatively saturated, extremely inert SF_6 (**13**).

12 **13**

Preferred geometries for coordination of five and six anions, or electron pairs, are the capped tetrahedron or trigonal bipyramid (**12**) and the octahedron (**13**).

To achieve stability, highly electron-deficient, large-core elements—the metals—may adopt simultaneously the bonding strategies of boron and sulfur. Potassium hydride, for example, a saltlike substance iso-structural with potassium and sodium chloride, may be viewed as a Born-type, ion compound, K^+H^-, in which each potassium atomic core is surrounded by six protonated electron pairs (H^- ions), each of which is in the valence shell of six potassium atoms, forming thereby "seven-center" bonds.

Isoelectronic with K^+H^- is the structure $Ca^{2+}E_2^{2-}$, a model similar to one first proposed by Thompson for the face-centered-cubic modification of calcium metal. E_2^{2-} stands for the deprotonated hydride ion, i.e., in the present instance, for a six-center electron-pair bond.

Historically, bonds have been termed covalent, ionic, or metallic. Those three bond types correspond to electrons shared between or among, respectively, solely small atomic cores (radius <0.05 nm), small and large atomic cores, and exclusively large atomic cores.

ORBITAL MODELS OF CHEMICAL BONDING

Mathematical theories of chemical bonding require a quantitative expression for the wavelike behavior of electrons. Usually used is Schrödinger's equation. Approximations to molecular wave functions for use in Schrödinger's equation generally are constructed from atom-centered, hydrogen-like orbitals.

To mimic the directional properties of chemical bonding, linear combinations of the spectroscopic, nondirectional, doughnut-shaped p orbitals of the hydrogen-atom problem were first used to produce bond orbitals p_x, p_y, p_z that point along the coordinate axes. Formation of additional linear combinations with the s orbital of the same principal quantum number (a process called "hybridization") yields a set of tetrahedrally directed orbitals.

To produce from component orbitals a wave function that does not distinguish between indistinguishable electrons and that does satisfy the exclusion principle requires antisymmetrization with respect to electron labels. Antisymmetrization yields a sum of orbital products that, for closed-shell molecules, may be written as a determinant, each column of which refers to a different orbital.

Since a determinant is unchanged when one column is added to or subtracted from another column, the component orbitals of a determinantal wave function may be described in several ways. In conventional molecular-orbital descriptions of a molecule, the component orbitals extend over the entire molecule. From linear combinations of those orbitals can be created localized molecular orbitals that occupy nearly mutually exclusive domains.

Localized orbitals correspond closely to Lewis's bonding electron pairs, unshared valence-shell pairs, and inner-shell electrons. They constitute a bridge between quantum-mechanical theories of chemical bonding and Lewis's electronic interpretation of classical structural theory, when taken with the discovery of the "classically nondescribable two-valuedness of the electron" called "spin" and the rule that two spin-opposed electrons may share the same spatial orbital.

Calculated energies of many-electron systems are always improved, however, if different spatial orbitals are used for electrons of different spins. Molecular oxygen, for example, is in fact paramagnetic in its ground state. It has five valence-shell electrons of one spin, seven of the other. One may picture the five-membered spin set as having a configuration of maximum probability (CMP) that is triple-bond-like (two tetrahedra sharing a face), with the seven-membered spin set having a single-bond-like CMP (two tetrahedra sharing a corner). Electrons of opposite spin are, thus, somewhat spatially anticoincident. In the bonding region are placed altogether $3+1=4$ electrons, as in a conventional double bond.

THE VIRIAL THEOREM AND CHEMICAL BONDING

As a result of electron–electron repulsions, individual electron, or electron-pair, orbitals cannot be rigorously defined for many-electron systems. Applicable, however, to all systems (for which the Born–Oppenheimer approximation is valid) is the virial theorem.

The virial theorem states that the average kinetic and potential energies, $\bar{T}$ and $\bar{V}$, of, for example, a diatomic molecule are related to the molecule's total energy E and internuclear separation R according to the relations

$$\bar{T}=-E-R\frac{dE}{dR};\quad \bar{V}=2E+R\frac{dE}{dR}.$$

Analysis of the exact E-vs-R curve for the hydrogen molecule ion, H_2^+, reveals that as R approaches from above that value for which E is a minimum, the decrease in E that produces the minimum, and chemical bonding, arises from a decrease in potential energy, owing to contraction of the wave function into the internuclear, bonding region; concomitantly, the kinetic energy rises. As R decreases still further, the fall of potential energy owing to a continued enhancement of nuclear–electron attraction is offset by its rise owing to nuclear–nuclear repulsion, which causes the potential energy, and total energy, to approach infinity as R approaches zero.

When E is a minimum, $dE/dR=0$ and $E=\frac{1}{2}\bar{V}$. In polyatomic, many-electron systems $\bar{V}$ receives contributions from nuclear–electron attractions, nuclear–nuclear repulsions, and electron–electron repulsions:

$$E_{\min}=\tfrac{1}{2}\bar{V}=\tfrac{1}{2}(\bar{V}_{ne}+\bar{V}_{nn}+\bar{V}_{ee}).$$

Only the first term, V_{ne}, is negative (with respect to infinite separation of the parts) and contributes to chemical

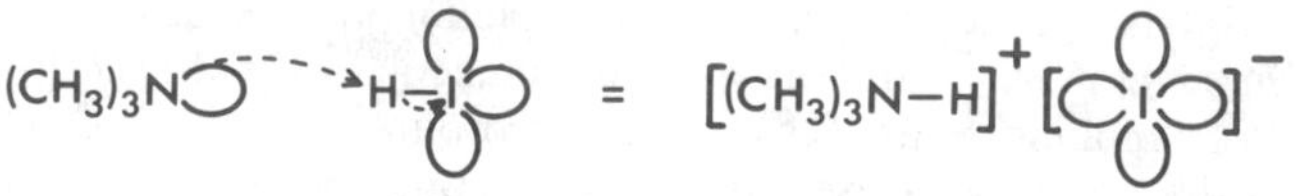

$(CH_3)_3N \cdots H{-}I = [(CH_3)_3N{-}H]^+ [I]^-$

FIG. 3. Graphic representation of diversions of lines of force in the creation of a chemical bond between nitrogen and hydrogen and the simultaneous annihilation of a chemical bond between hydrogen and iodine in the reaction of the base trimethyl amine, $(CH_3)_3N$, with the acid hydrogen iodide, HI. Valence strokes, straight and curved, represent, respectively, localized, doubly-occupied bonding and nonbonding (or unshared) valence-shell molecular orbitals. Symbols H, C, N, I stand for the atomic cores H^+, C^{4+}, N^{5+}, I^{7+}. Formed in the first step of the proton transfer is a hydrogen-bond, ·N–H–I, in which the bond angle NHI is 180°.

bonding. Its central role in theories of chemical bonding is reflected in such words and phrases as coordination, valence saturation, the octet rule, inner-shell electrons, atom-centered orbitals, and spatial pairing of electrons (in, especially, bonding regions).

The role of nuclear–nuclear repulsion term, V_{zz}, in chemical bonding is reflected in the rule that atomic cores (or cations) with large charges tend not to share electrons (or anions) with each other. F–F, HO–F, and HO–OH are thermodynamically highly reactive molecules. The structure of nitrous oxide is NNO (not NON), of nitrosyl fluoride FNO (not NOF), of stable cyanate ion NCO^- (not CNO^-, explosive fulminate).

The role of the electron–electron repulsion term, V_{ee}, is reflected in the use of different orbitals for electrons of different spin.

INTERMOLECULAR FORCES AND REACTION MECHANISMS

Nuclear–nuclear repulsion between the proton of a hydrogen atom and a highly charged atomic core of an electronegative atom to which it is chemically bonded, as, e.g., in hydrogen iodide, H–I, produces departures from local electrical neutrality, dipole moments, and a positive patch, "electrophilic center," or "acidic site" on the molecule's surface that, via stray feeler lines of force, may interact with a negative patch, "nucleophilic center," or "basic site" on an adjacent molecule or ion, usually an unshared valence-shell electron-pair, as, e.g., in the trimethyl derivative of ammonia, $(CH_3)_3N$. Formed thereby is an intermolecular "hydrogen bond," the first step in a proton-transfer, "Bronsted acid-base reaction." Fig. 3.

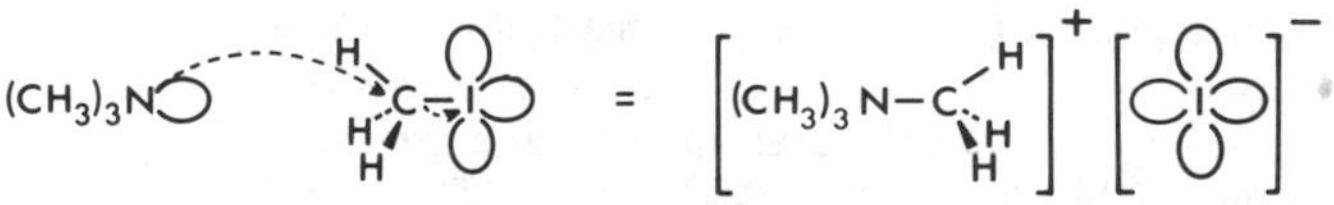

$(CH_3)_3N \cdots H_3C{-}I = [(CH_3)_3N{-}CH_3]^+ [I]^-$

FIG. 4. Reaction of $(CH_3)_3N$ with methyl iodide, CH_3I (rather than, as in Fig. 3, hydrogen iodide, HI). Formed in the first step of the transfer of the methyl cation CH_3^+ from I to N is a "charge-transfer complex" or "face-centered bond," N–C–I, in which the bond angle NCI is 180°. As the carbon core, C^{4+}, leaves I^{7+} and approaches the unshared pair about N^{5+}, it passes through the plane defined by its three hydrogen atoms. The methyl group undergoes, it is said, "Walden inversion." It turns inside out, like an umbrella: $H_3C{-} \rightarrow {-}CH_3$.

$(CH_3)_3N$ interacts similarly with the electrophilic C^{4+} center of methyl iodide, $H_3C{-}I$, through "backside attack." The base $(CH_3)_3N$ approaches with its nucleophilic, unshared electron pair that face of the tetrahedron of substituents surrounding the acidic, C^{4+} cation opposite to the eventual "leaving group," I^-. Formed thereby is an intermolecular "charge-transfer complex," the first step in a methyl-group transfer via a "Walden inversion" at carbon, Fig. 4.

Caveats concerning the notation used in Figs. 3 and 4 have been cogently expressed by Michael Faraday (*Experimental Researches in Electricity*, 14th Series, 1838, par. 1684 [Added remarks appear in brackets]).

> "The terms *free charges* and *dissimulated electricity* convey erroneous notions if they are meant to imply any differences as to the mode or kind of action. The charge upon an insulator [e.g., immobile $(CH_3)_4N^+$] in the middle of a room [or ionic crystal] is [with respect to lines of force] in the same relation to the walls of that room [and surrounding ions, I^-] as the charge upon the inner coating of a Leyden jar [or the carbon kernel of CH_3I] is to the outer coating [the surrounding "ions": H^- and I^-] of the same jar [or molecule]. The one is not more *free* or *dissimulated* [or, with respect to lines of force, more *detached*] than the other; and when sometimes we make electricity appear [as, e.g., in the formation of $(CH_3)_3N^+I^-$] where it was not evident before, as upon the outside of a charged jar [or molecule] when, after insulating it [$(CH_3)_3I$] [by placing it in solution], we touch the inner coating [with $(CH_3)_3N$:] it is only because we divert more or less of the inductive force from one direction into another [from C to N rather than to I; from I to H rather than to C; etc.]."

In a full graphical representation of a reaction mechanism, the path traced by the curly arrows employed in Figs. 3 and 4 would not be left open. The diverted lines of force would form a closed, Gauss circuit. Therein lies the major role of the solvent in many chemical reactions.

See also MOLECULAR STRUCTURE CALCULATIONS; MOLECULES.

BIBLIOGRAPHY

H. A. Bent, "Isoelectronic Systems," *J. Chem. Ed.* **43,** 170 (1966) (E); "The Tetrahedral Atom," *Chemistry* **39,** 8 (1966) (E); **40,** 8 (1967) (E); "Ion-Packing Models of Covalent Compounds," *J. Chem. Ed.* **45,** 768 (1968). (E-I) Written for high school and college students and teachers.

H. A. Bent, *Isoelectronic Molecules in Molecular Structure and Energetics: Chemical Bonding Models,* Vol. 1, pp. 17–50 (Joel S. Liebman and Arthur Greenberg, eds.). VCH Publishers, New York, 1986. (I)

H. A. Bent, The Isoelectronic Principle and the Periodic Table, in *Molecular Structure and Energetics: From Atoms to Polymers: Isoelectronic Analogies,* pp. ix–xii (Joel S. Liebman and Arthur Greenberg, eds.). VCH Publishers, New York, 1988. (E)

C. A. Coulson, *Valence.* Oxford, London, 1961. (I) A readable, relatively nonmathematical introduction to quantum-mechanical theories of valence by a major contributor to the field.

C. Edmiston and K. Ruedenberg, "Localized Atomic Molecular Orbitals," *Rev. Mod. Phys.* **35,** 457 (1963). (A) A landmark paper that has stimulated much research on the creation of quantum-mechanically based, chemically interpretable, transferable molecular orbitals.

R. J. Gillespie, *Molecular Geometry*. Van Nostrand Reinhold, New York, 1972. (E-I) A nonmathematical account of numerous applications of the author's "valence-shell-electron-pair-repulsion" model for allowing in structural theory for the effects of electron correlation between electrons of parallel spin.

G. N. Lewis, *Valence and the Structure of Atoms and Molecules*. Dover, New York, 1966 (reprint). (E) A lively, personal, historical, and still-provocative account of the introduction and uses of the concept of the electron-pair bond. A deservedly ever-popular classic.

E. H. Lieb, "The Stability of Matter," *Rev. Mod. Phys.* **48,** 553 (1976). (A) An explanation in terms of the exclusion principle and electrostatic screening of the fundamental paradox of classical physics as to why matter, which is held together by Coulombic forces, neither implodes nor explodes.

J. W. Linnett, *The Electronic Structure of Molecules: A New Approach*. Wiley, New York, 1964. (E-I) A largely nonmathematical account of the author's "doublet-quartet" model for allowing in structural theory for the effects of electron correlation between electrons of opposite spin.

L. Pauling, *The Nature of the Chemical Bond*. Cornell University Press, Ithaca, New York, 1960. (I) The author has introduced more concepts into structural theory than any other living scientist. Although his theory of resonance has, in many minds, been superseded (though not the classification of molecules based upon it), this classic remains a mine of interesting ideas and information.

J. C. Slater, *Quantum Theory of Molecules and Solids*, Vol. I, *Electronic Structure of Molecules*. McGraw-Hill, New York, 1963. (A) A lucid introduction to the mathematical theory of chemical bonding by an early and long-active contributor to the field. Includes an expanded account of the author's classic discussion of applications of the virial theorem to H_2^+.

Chemical Kinetics *see* Kinetics, Chemical

Chemiluminescence

A. Paul Schaap and Richard S. Handley

In the 1880s Eilhard Wiedemann was investigating various phenomena which resulted in the emission of light. He was the first to use the term "Chemiluminescenz" for chemical reactions which produced light. In 1888, he wrote: *"Das bei chemischen Processen auftretende Leuchten würde Chemiluminescenz. . . ."* Chemiluminescence is produced as a result of the generation of electronically excited products of a chemical reaction which subsequently emit photons. Such reactions are relatively uncommon as most exothermic chemical processes release energy as heat. The initial step is termed *chemiexcitation*, and involves conversion of chemical energy into electronic excitation energy with an efficiency denoted as Φ_{CE}. If the excited state species is fluorescent, the process is *direct* chemiluminescence and occurs with an overall efficiency (Φ_{CL}) that is a product of the efficiencies for the two steps: $\Phi_{CL} = \Phi_{CE} \times \Phi_F$. If the initial excited state species transfers energy to a molecule which then emits light, the process is *indirect* chemiluminescence and occurs with an efficiency that is a product of the efficiencies for the three steps (Fig. 1).

The thermochemical requirement for chemiluminescence is that the total energy available from the reaction, i.e. the enthalpy of activation, $\Delta H^\ddagger$, and the enthalpy of reaction, ΔH_R, be at least as great as the energy of the lowest excited state of one of the products: $\Delta H^\ddagger - \Delta H_R \geq E_{ex}$. The most widely studied type of chemiluminescent reaction which meets this criterion is the decomposition of 1,2-dioxetanes to produce two carbonyl-containing products. In general there is only sufficient energy available to produce one of the products in the excited state.

The thermal stability of 1,2-dioxetanes varies widely, with half-lives at room temperature ranging from minutes for most simple dioxetanes prepared in the laboratory to several years for dioxetanes with bulky polycyclic alkyl groups such as adamantylideneadamantane dioxetane shown below in Eq. (1). In addition, these dioxetanes are inefficient producers of chemiluminescence generating predominantly triplet state products so the yield of direct chemiluminescence is low.

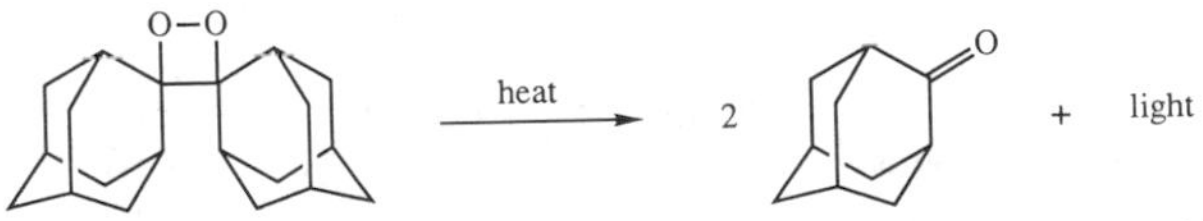

O–O ... heat → 2 ... + light (1)

In contrast, model dioxetanes containing electron-rich aromatic groups are markedly less stable and can produce direct chemiluminescence with high efficiencies. This type of dioxetane more closely resembles the behavior of intermediates in biological processes such as the familiar firefly bioluminescence. These properties have been utilized to design dioxetanes which are stable indefinitely at room temperature but which can be "triggered" to undergo efficient chemiluminescent decomposition on demand. Removal

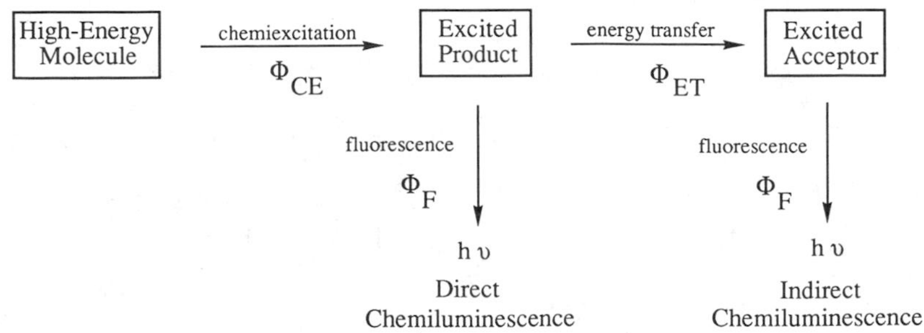

FIG. 1. Direct and indirect chemiluminescence.

of a protecting group from the dioxetanes shown below in Eq. (2) converts them to the unstable aryloxide form which rapidly decomposes with emission of light.

O–O OCH$_3$ — activating agent → O–O OCH$_3$

stable OX — unstable O$^-$

X = Si(CH$_3$)$_2$t-Bu
X = PO$_3$Na$_2$

chemiexcitation

O OCH$_3$

light ← fluorescence

O$^-$
electronically excited product

(2)

The activating agent may be a simple chemical reagent or an enzyme. In the case of the t-butyldimethylsiloxy-substituted dioxetane shown above, reaction of the dioxetane with fluoride ion in DMSO produces brilliant bluish chemiluminescence with an efficiency of 25%. The phosphate-substituted dioxetane is triggered in aqueous solution by alkaline phosphatase and can be used in extremely sensitive assays with a detection limit of 10^{-21} moles of enzyme. These types of dioxetanes will be widely used for medical and research applications.

See also LUMINESCENCE.

BIBLIOGRAPHY

A. K. Campbell, *Chemiluminescence: Principles and Applications in Biology and Medicine*. Ellis Horwood Ltd., Chichester (England), 1988.

K. D. Gundermann and F. McCapra, *Chemiluminescence in Organic Chemistry*. Springer-Verlag, New York, 1987.

A. P. Schaap, *Photochem. Photobiol.*, Volume 47S, 50S, 1988.

R. Schreiner, M. E. Testen, B. Z. Shakashiri, G. E. Dirreen, and L. G. Williams, "Chemiluminescence," Chap. 2 in *Chemical Demonstrations* (B. Z. Shakashiri, ed.). University of Wisconsin Press, Madison, Wisconsin, 1983.

T. Wilson, "Chemiluminescence in the Liquid Phase: Thermal Cleavage of Dioxetanes," Chap. 7 in *International Review of Science, Physical Chemistry, Series Two,* Volume 9 (D. R. Herschbach, ed.). Butterworth, Boston, 1976.

K. Van Dyke, Ed., *Bioluminescence and Chemiluminescence: Instruments and Applications,* Volumes I and II. CRC Press, Inc., Boca Raton, Florida, 1985.

Circuits, Integrated

Sol Triebwasser

The invention of the transistor in 1948 opened the way to the rapid development of a series of increasingly sophisticated and useful device structures. Within 10 years, transistors had become the basis of a major industry which has revolutionized the way we live. Integrated circuit chips, consisting of millions of devices for applications in computer memories, have been fabricated, as well as chips that perform the basic functions of today's sophisticated personal computers.

At the other end of the spectrum, we have seen the displacement of the slide rule by powerful hand-held calculators and remarkably accurate, inexpensive timepieces.

Silicon integrated circuits use active devices of two basic types, the field-effect transistor (FET) and the bipolar transistor.

The FET operates on a simple and easily understood principle: application of a voltage to a capacitively coupled electrode creates an electric field which alters the number of charge carriers in a semiconductor, thus modulating its conductivity. The electrode has been realized as a *p-n* junction (junction FET, J-FET), as a Schottky barrier (metal–silicon FET, MESFET), or as a plate separated from the semiconductor by an insulating layer (insulated gate FET, IGFET). This last structure is usually a metal plate insulated from the surface of the semiconductor by silicon dioxide (metal–oxide–semiconductor FET, MOSFET). The MOSFET, or, more commonly, the MOS device, is of two forms, the *n* channel and *p* channel, distinguished by the nature of the current-carrying species, electrons and holes respectively. In addition, the metal gate has been replaced by heavily doped (for enhanced conductivity) polycrystalline silicon or, in some cases, alloys of Si with certain metals. Today's integrated circuit industry is dominated by the Complementary Metal Oxide Semiconductor (CMOS) technology in which the basic circuit configuration consists of an *n*-channel MOS device in series with a *p*-channel device. This circuit has virtues of being simple to design, but, more important, it draws virtually no current except while being switched. Hence, it represents a very low-power technology, a dominant consideration as the number of circuits being fabricated in very small areas continues to increase almost geometrically with time. Figure 1 shows a cross section of an advanced CMOS device configuration.

The bipolar transistor, unlike the FET, does not rely on capacitive coupling. For an *n-p-n* transistor, it consists of a thin *p* region between two *n* regions. These regions correspond to the emitter, base, and collector of the transistor. In operation, the emitter–base junction is forward-biased such that electrons flow from the emitter region into the base. The base–collector junction is usually reverse-biased,

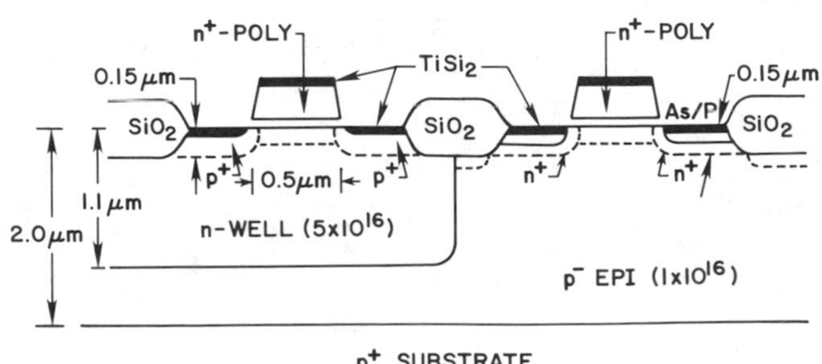

FIG. 1. Cross section of complementary MOS device structures.

that is, the n region is positive with respect to the base p region. The collector junction therefore collects the electrons that do not recombine in the base region. Current and voltage gain between base input and collector output can be obtained by using the base electrode as the input.

A p-n-p transistor operates similarly except that the emitter injects holes into the base region. Again, because electrons are more mobile than holes, n-p-n transistors with the same dimensions perform at higher frequencies than p-n-p transistors. Figure 2 shows a cross section of a modern n-p-n bipolar transistor.

Until about 1959, all semiconductor circuits were fabricated with discrete bipolar devices such as those just discussed. These were used, much like vacuum tubes, in circuits that also contained other, passive components such as diodes (simple p-n junctions), resistors, and capacitors. The introduction of transistors made it possible to operate the circuits at much lower voltage and power levels and offered significant improvements in reliability. Integrated electronics as we know it today had its start in 1958–59, when government-sponsored and industrial research sparked the development of semiconductor integrated circuits. Whereas germanium had dominated the technology in the 1950s, silicon now wins out, since it grows a natural oxide that serves as an insulator, as a passivation layer, and, most importantly, as a diffusion mask.

Integrated circuits are fabricated by a series of processes that form interconnected active and passive circuit components on a single piece of monocrystalline material, usually a thin silicon wafer, perhaps 125 mm in diameter and $\frac{1}{3}$ mm thick. Millions of devices can be fabricated and interconnected simultaneously. It is this batch fabrication capability that makes integrated circuits so inexpensive. The following briefly describes the principal steps in fabricating the typical integrated device shown in Figs. 1 and 2.

The first step is the production of single-crystal material, most commonly by the Czochralski method. Wafers are cut from single crystals, and lapped and polished to a mirror finish. An epitaxial film, that is, a film whose lattice structure is an extension of the substrate crystal, is grown on the surface to a thickness of several micrometers. This film is generally made to have conductivity opposite to that of the substrate, so that a p-n junction occurs at the interface; this is desirable because it provides an isolating junction. In other applications, the film has the same conductivity type as the substrate but a different doping level. An oxide layer (SiO_2), several thousand Ångstroms thick, is grown on the Si surface by heating the wafer in an oxidizing atmosphere. The oxide layer is then coated with a film of photosensitive material called photoresist. The resist has the characteristics that it can be polymerized by radiation; originally optical radiation, but, by today, electrons, ions and x-ray radiation are being used as well. The resist is patterned either through a mask or by directed electron or ion beams. The wafer is then rinsed in a developer, which removes the nonpolymerized unexposed film. After the remaining resist is baked, the wafer is etched. Where the resist has been removed, the etchant leaves openings in the oxide layer. The wafer is now ready for ion implantation or diffusion. With the oxide layer acting as a mask, materials such as phosphorus or boron are ion implanted or diffused (mostly implanted today) through the openings in the oxide layer. Such processes of oxidation, photolithography and implantation are repeated to create the multiple-layer structures shown in Figs. 1 and 2 and to form resistors and diodes, isolation regions, and, finally, several layers of metal conductors to complete the highly functional integrated circuit chips that are on the market. Today's advanced chips require 15 or more lithographic steps and hundreds of additional processing steps.

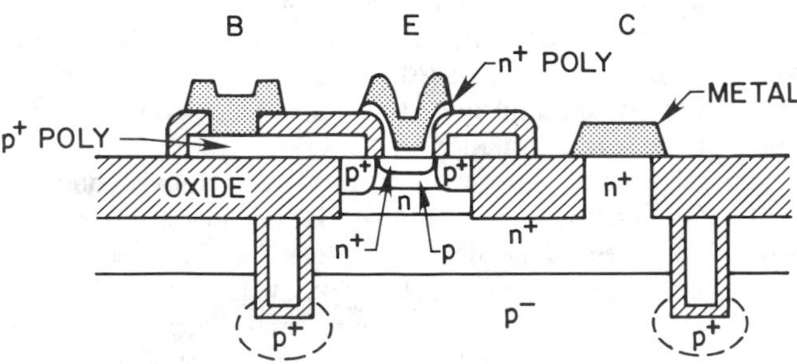

FIG. 2. Cross section of a bipolar transistor. Minimum dimensions are similar to those shown in Fig. 1.

FIG. 3. Photograph of a 4×10^6-bit integrated circuit dynamic random access memory chip. This chip was fabricated using 0.7 μ minimum dimension.

Once processed, the wafers are cut into chips which may measure 10 mm or more on a side. A chip may contain millions of components interconnected to form an electronic function such as a microprocessor, or millions of bits of an electronic memory. Figure 3 shows a chip containing 4×10^6 bits of memory and all the peripheral circuitry required to address the memory locations, write and read the information in this random access memory (RAM). This chip is less than 80 mm^2 in area and contains enough storage capacity to store 400 pages of double-spaced typewritten pages. The speed of operation is such that all of this information may be read out or transferred in $\frac{1}{4}$ s.

The chips are attached to a protective package which provides the interconnecting leads for signal and power external connections. Chips can be attached to packages with wire bonds, beam leads, beam tape, or solder pads.

At present, the minimum line width and line spacings used by industry are about 0.5 μ, a dimension that is taxing the capability of optical systems. Innovative approaches will be required to meet the lithographic challenges of the decade of the '90s.

Integrated circuits have revolutionized computers, television, radio, and electronic products, and have made possible lower costs, higher reliability, savings in size, weight, power, and better performance and have made a major impact on how we live.

See also SEMICONDUCTORS, CRYSTALLINE; TRANSISTORS.

BIBLIOGRAPHY

S. M. Sze, *Semiconductor Devices, Physics and Technology*. Wiley, New York, 1985.

D. A. Hodges and H. G. Jackson, *Analysis and Design of Integrated Circuits*, 2nd ed. McGraw-Hill, New York, 1988.

C. Mead and L. Conway, *Introduction to VLSI Systems*. Addison-Wesley, Reading, MA, 1980.

I. Brodie and J. J. Murray, *The Physics of Microfabrication*. Plenum, New York, 1982.

S. K. Ghandi, *VLSI Fabrication Principles*. Wiley, New York, 1983.

Clocks, Atomic and Molecular

James A. Barnes and John J. Bollinger

Almost any clock consists of three main parts: (1) a pendulum or other nearly periodic device, which determines the rate of the clock; (2) a counting mechanism, which accumulates the number of cycles of the periodic phenomenon; and (3) a display mechanism to indicate the accumulated count (i.e., time).

An atomic clock makes use of an atomic resonance to control the periodic phenomenon. Similarly, a resonance in a molecule could be used to control the periodic phenomenon. The atomic or molecular clock is a very good clock because these resonances are determined by the atom's properties rather than by the man-made dimensions of an artifact; they are among the most stable and accurately measured phenomena known to man. Since all clocks are just devices that count and display the total of a series of periodic events (such as swings of a pendulum, passages of the sun overhead, or oscillations of an atom), the accuracy of the clock depends directly on the stability and accuracy of the periodic phenomenon used to establish the rate of the clock.

The most accurate clocks today make use of a microwave resonance in the ground state of cesium. In fact, the unit of time, the second, is defined in terms of this microwave resonance in cesium, and the national standards of time and frequency for the United States and other countries are cesium clocks. Time (and frequency) can be measured with the smallest uncertainty of any physical quantity. Current estimates of possible errors in various national standards laboratories are of the order of a few parts in 10^{14}. This can be expressed by saying that independent cesium clocks can maintain synchronism with one another to better than one-millionth of a second after one year's operation. This is more than 100 000 times more predictable than the earth's rotation on its axis.

The operation of a cesium atomic clock depends on the observation of a particular resonance in cesium atoms. The atoms are not radioactive, and radioactive decay processes play no part in the scheme. Neutral atoms boiled off from a quantity of liquid cesium are allowed to escape through narrow holes in a small oven and form a beam, which traverses an evacuated chamber. To prevent the cesium atoms from colliding with air molecules and being scattered out of the beam, a good vacuum must be maintained in the chamber.

After passing through the strong, inhomogeneous magnetic field of a Stern–Gerlach magnet, the beam of atoms is separated into two beams with opposite magnetic polarizations. In many cesium-beam devices one of the polarized beams is absorbed in graphite and is of no further interest, while the other continues down the chamber.

Farther down the chamber is another strong, inhomogeneous magnetic field nearly identical to the first. At the end of the chamber there is a detector (which is sensitive to cesium atoms) placed in just such a position as to detect only those atoms that somehow change their polarization while traveling between the two magnetic field regions. Thus, the detector would not detect cesium atoms unless something happened to the atoms between the two strong magnetic field regions to change their polarity.

What happens is that the atoms are exposed to microwave radiation at a frequency of about 9 GHz. If this frequency is adjusted very precisely to the proper resonance frequency of cesium (9 192 631 770 Hz), the magnetic polarization of the atoms reverses, and the beam is deflected by the second magnetic field toward the detector. The detector indicates the presence of cesium atoms by means of an electric current. In actual operation the frequency of the microwave signal is controlled electronically to maximize the detector current, so the resonance condition of the microwave signal with the cesium atoms is ensured. A clock is obtained by counting the cycles of the microwave radiation.

Other kinds of atomic and molecular clocks use similar principles to extract frequency information from the atoms or molecules. The first atomic or molecular clock ever developed (completed in 1949 by Harold Lyons of the U.S. National Bureau of Standards) used the absorption of a microwave signal in ammonia to control the frequency, while hydrogen maser clocks use the stimulated emission of microwave radiation, and rubidium gas cell clocks use absorption of microwave radiation.

Several manufacturers produce atomic clocks commercially. The most accurate commercial devices are based on a resonance technique using a beam of cesium atoms like the various national standards. Somewhat less expensive atomic clocks are based on rubidium vapor. There is a trade-off between cost and stability or accuracy. Most clocks in use today are based on resonances in atoms rather than molecules. At present there are a few tens of thousands of atomic clocks in routine use in many areas. For example, atomic clocks are used to control Loran-C, Omega, and GPS navigational systems, to do very long baseline interferometry and measure continental drift, to control network television signals, and to define an internationally accepted time-of-day system that is the time reference for most of the world.

Recent experimental advances and techniques are improving the accuracy of atomic clocks. The spectral width or possible frequency fluctuations of the cesium resonance used in the cesium atomic clock is approximately equal to the inverse of the transit time of a cesium atom down the beam. More narrow spectral widths have been obtained on atomic resonances with ions stored in traps. Ion traps use the force of electric and magnetic fields on ions (which have a net charge) to confine the ions to a small region in a good vaccum. Confinement times of several hours can be obtained routinely. Two types of traps have been used in the development of new atomic clocks: the rf or Paul trap, and the Penning trap. The rf or Paul trap uses spatially inhomogeneous rf fields to confine the ions, much as an rf quadrupole mass filter works in a mass spectrometer. The Penning trap uses static magnetic and electric fields to confine the ions. In general there is a trade-off between the number of trapped ions which gives good stability and the accuracy in a stored-ion clock. One of the first atomic clocks using ion storage is based on a microwave resonance in the mercury ion with about one million ions stored in an rf trap.

Lasers are starting to be incorporated into present atomic clocks with anticipated improvements in performance. The Stern–Gerlach magnets used in conventional cesium atomic clocks are being replaced with diode lasers. Through the technique of optical pumping, the polarization of the laser light is used to give the cesium atoms a magnetic polarization. The diode lasers are also used at the end of the beam line to detect the polarization state of the cesium atoms. The diode lasers enable all of the atoms in the beam to be used in measuring the atomic resonance.

Laser cooling or the use of radiation pressure from lasers to slow atoms or ions may provide further improvements for atomic clocks. According to Einstein's special theory of relativity, moving clocks tick slower than clocks at rest. Laser cooling enables both the spread in the atomic velocities and the mean atomic velocity to be reduced. This could improve the accuracy of the cesium clock. With the use of laser cooling, stored-ion frequency standards based on microwave resonances in ions are expected to have accuracies and stabilities better than one part in 10^{15}.

In general, atomic clocks are based on resonances in atoms at microwave frequencies, that is, frequencies less than 100 GHz. This is because of the technical difficulty of counting the cycles of higher frequencies. In addition, there is a lack of readily available, stable, narrow-band sources at high frequencies, especially at infrared and optical frequencies. Current research is making progress on both these problems, and in the future atomic clocks may be based on optical resonances in atoms or ions. An advantage of optical resonances is that the ratio of the optical frequency to the spectral width of the resonance can be very high. In a stored-ion clock, this means a good stability can be obtained with only a single ion in the trap. The projected accuracy of an atomic clock based on a single ion in an rf trap is on the order of one part in 10^{18}.

See also BEAMS, ATOMIC AND MOLECULAR; LASERS; MASERS; OPTICAL PUMPING; TIME.

BIBLIOGRAPHY

J. Jespersen and J. Fitz-Randolph, *From Sundials to Atomic Clocks*. Dover, New York, 1982.

N. F. Ramsey, "History of Atomic Clocks," *J. Res. Natl. Bur. Stand.* **88**(5), 301–320 (1983).

N. F. Ramsey, "Precise Measurement of Time," *Am. Sci.* **76**(1), 42–49 (1988).

J. Vanier and C. Audoin, *The Quantum Physics of Atomic Frequency Standards, Vol. I, II*. Adam Hilger, Bristol and Philadelphia, 1989.

D. J. Wineland, "Trapped Ions, Laser Cooling, and Better Clocks," *Science* **226**, 395–400 (1984).

Frequency Standards and Metrology. Proc. 4th Symposium, Ancona, Italy. (A. DeMarchi, ed.). Springer-Verlag, Berlin, Heidelberg, 1989.

Cloud and Bubble Chambers

R. P. Shutt

In order to detect nuclear interactions, for instance in radioactivity, cosmic radiation, or at high-energy accelerators, one can use the trails of ions and of more energetic "knock-on" electrons produced by charged particles colliding with atoms when passing through matter. In cloud chambers, first built by C. T. R. Wilson in 1912, tracks consisting of drops of liquid are formed along these trails, whereas in bubble chambers, invented by D. A. Glaser in 1952, the tracks consist of bubbles in a liquid. Both techniques have many important discoveries to their credit, but have been almost completely displaced by electronic methods. Electronic detectors are now able to distinguish between passages of two charged particles in space as well as the older techniques, in addition to their superior ability to distinguish in time between successively passing particles. Among the last major applications of bubble chambers were studies of neutrino interactions. Neutrinos do not leave ion trails and have very small nuclear interaction cross sections, so that intense, partially collimated neutrino beams could be passed through large bubble chambers without producing much background radiation, but producing weak-interaction events in the dense liquid, which acted as a target as well as a particle-detector medium.

In cloud chambers a gas containing a saturated vapor is expanded adiabatically, lowering the temperature, so that the vapor becomes supersaturated. The liquid surface tension prevents spontaneous drop formation without the presence of some kind of condensation nuclei. The electrostatic field of ions opposes the effect of the surface tension and permits drops to form and grow beyond a critical radius if the vapor is sufficiently supersaturated. Surface tension also prevents spontaneous bubble formation in a liquid in a superheated state, which in a bubble chamber is accomplished by letting the liquid expand elastically. But here the ionization energy due to a passing particle is very quickly converted to molecular kinetic energy, thus producing "heat spikes," locally raising the vapor pressure, which now can overcome the contracting effect of the surface tension, enabling bubbles to grow beyond a critical radius. In gases ions diffuse, but do so sufficiently slowly that cloud chamber expansions can be triggered for interaction events of special interest by means of electronic detectors. This is not possible for bubble chambers because the heat spikes responsible for bubble formation diffuse in about 10^{-12} s.

Chamber vessels usually must withstand pressure differences of less than 1 atm for most cloud chambers, up to 25 atm for bubble chambers. Excellent temperature control must be provided to obtain uniform drop or bubble size and to minimize convection currents, which produce track distortions. Expansions are provided by adequately sealed diaphragms, bellows, or pistons, or by valves letting gas pass through heat regenerators. Volume expansions range from $<1\%$ for liquid hydrogen to 30% for air saturated with water vapor and must take only 10–20 ms to reduce convective distortions and to allow cloud chamber expansions to be triggered. Recompression must be equally fast in order to restore temperature equilibrium as soon as possible, so that the "dead time" during which a chamber is not ready for further expansions can be short. Dead times as short as 20 ms have been achieved with small bubble chambers, several seconds with small cloud chambers, but have lasted up to 30 min in high-pressure cloud chambers operated with argon at 300 atm. Time resolution between successive particles is of the order of 1 ms, sensitive time somewhat longer.

A special kind of continuously sensitive cloud chamber, the diffusion chamber, was built by A. Langsdorf in 1939, and for several years before the advent of bubble chambers it was used for particle physics research, mostly filled with about 20 atm of hydrogen. Here a temperature gradient is impressed on the gas volume, usually by cooling the bottom of the pressure vessel and heating the top, where a source of vapor of a liquid is also provided. As the vapor diffuses downward, it becomes supersaturated at some level, again resulting in track formation along charged-particle trails. Such chambers are relatively simple, since no expansion system is needed, but track-sensitive regions are only up to 8 cm high, and intense radiation can flood them with background tracks. Nevertheless, a chamber simply cooled with dry ice and filled with air or argon at atmospheric pressure, and with methyl alcohol as a vapor supply, serves well for demonstration of cosmic rays or other radiation and can be used to explore background near extended radiation sources. Just as in expansion cloud chambers, an electrostatic field is needed to sweep out unwanted ions.

During the short time while tracks are present in a chamber, they must be illuminated and photographed through suitable glass or quartz windows. Large bubble chambers now employ relatively small spherically curved windows, so that tracks are photographed through a layer of the operating liquid. Illumination is usually provided by xenon in a tube discharged at energies of 20–300 J during less than or about 1 ms. Windows and camera lenses must be of excellent optical quality, often also accommodating wide angles. The depth of the chamber volume, up to 3 m, requires large depth of focus and therefore small aperture as well as small magnification. Photography is thus diffraction limited. Of course, the film grain size, contrast, and speed are also of great importance. Usually, 35- to 70-mm films have been employed. To avoid distortion, the film must be held consistently flat. Illumination at about 90° to the lens axis has been sufficient for liquids with high refractive index, resulting in "dark-field illumination." The low index of, for instance, hydrogen requires light scattered at small angles, $\sim 1°$, and large chambers universally employ a specially developed Scotchlite, which reflects light from a source placed around the camera lens back into the latter, resulting in bright-field illumination. Three or more stereo views are usually photographed for redundancy and accuracy for all track orientations.

Expansion cloud chambers were most frequently used with argon and a water–propyl-alcohol mixture requiring expansion ratios as low as 1.08. In bubble chambers various fluids have been in use, which are operated at one half to two thirds their critical pressure. Propane, operated near 20 atm and 60°C, has a density of 0.43 g/cm^3 and a radiation length of 108 cm, and contains more hydrogen per unit volume than liquid hydrogen. It is therefore suited for studies of interactions with individual protons (whose recoil tracks can also act as neutron detectors) as targets and of neutral interaction products decaying into γ rays, which have a reasonably high probability for conversion into electron pairs in propane chambers of sufficient size. The latter advantage can be reinforced by mixing propane with 30% by volume of methyl iodide, which results in a radiation length of 12 cm. Xenon has a radiation length of only 3.5 cm but, of course, contains no hydrogen and is extremely expensive. Freons, with density of 1.5 and radiation length of 10 cm, are very suitable for bubble chamber operation but contain no hydrogen. When electron-pair conversion is not needed and the large number of interactions with the heavier nuclei that occur in the mentioned liquids cannot be tolerated, pure liquid hydrogen (or deuterium, for interactions with the extra neutron) is used. The density of liquid hydrogen is 0.06, its operating pressure only 5 atm, but its operating temperature is 25–27 K, requiring cryogenic techniques. Window and flange seals are usually made with indium–silver alloy wire often pushed against seal surfaces by means of helium-inflated bellows. For insulation, chambers are almost surrounded by multilayers of aluminum-coated Mylar in a vacuum chamber with windows for photography. Temperature control is provided by hydrogen- or helium-operated refrigerators.

Solid plates can be mounted in the chambers to aid in particle identification—for instance, to distinguish muons from electrons or from pions—but this can create optical problems and promote turbulence. Cryogenic chambers can also be used with liquid-neon–hydrogen mixtures. Pure neon has a radiation length of 25 cm, so that its admixture results in increased pair conversion. "Track-sensitive targets" are relatively small transparent chambers with flexible walls, filled with liquid hydrogen or deuterium. When one is inserted into a neon–hydrogen-filled chamber, interactions can be produced in pure hydrogen or deuterium, and pair conversion can be observed outside the target. Finally, helium has also been a liquid of interest in K^--capture experiments, since the helium nucleus has zero spin. It is operated at 3.2 K and about 1 atm.

Most chambers have been positioned in magnetic fields, which cause a curvature in the tracks that is proportional to the field strength and inversely proportional to the momentum of a charged particle. Superconducting magnets employing copper-stabilized niobium-titanium alloy, cooled to liquid-helium temperatures, have been used, providing fields up to almost 4 Tesla. These magnets require very much less electric power than conventional electromagnets but require an additional cryogenic installation. Large cryogenic bubble chamber installations were found at the Fermi National Accelerator Laboratory, Brookhaven and Argonne National Laboratories, and CERN, the European Organization for Nuclear Research, where 15-ft, 7-ft, 12-ft, and 3.7-m chambers, respectively, were located, the largest of which contained about 30 m^3 of liquid. Installation costs have ranged between $7 and $20 million, and around-the-clock operation required 4–6 persons per shift. Another very large chamber, with a conventional magnet, was located at Serpukhov, USSR.

At particle accelerators, chambers can be exposed to beams of protons, antiprotons, electrons, or charged pi and K mesons, selected by arrays of magnets and electrostatic or radio-frequency beam separators from a spray of particles emerging from a target exposed to the primary circulating beam. Exposures to γ rays, neutrons, and, particularly, neutrinos are made possible by means of different arrangements. Exposures can amount to millions of pictures per experiment and can be distributed to many research groups at various locations where the photographs are scanned for events of interest and analyzed.

Locations of bubbles on the stereo views of event tracks are measured accurately with respect to images of fiducial marks located in the chamber. Angles between tracks and their curvatures can then be determined. Bubbles (or drops) along the tracks or gaps between bubbles can also be counted. The bubble density is inversely proportional to the square of the particle velocity, thus allowing determination of particle masses at not too relativistic energies. Energies of particles stopping in the chamber liquid can be measured from their range. The geometry and particle momenta in an event are fitted to hypotheses that are consistent with conservation of energy, momentum, and all relevant quantum numbers. Goodness of fit is tested with well-established statistical methods, such as the χ^2 test, and best fits are selected for inclusion in final results. Occasionally, a single event has proved the existence of a long-lived new particle or of a new quantum number. Usually large statistics are required, for instance, to show the existence of a short-lived resonant state occurring in strong interactions. For meaningful results, event selection must be as definite as possible, and therefore uncertainties in the tracks and their measurements must be as small as possible. Overall errors in bubble position, due to chamber turbulence, photography, and measuring, can be made as small as 50 μm, but multiple Coulomb scattering on atoms produces spurious curvatures, especially in the heavier liquids. Great care is required in evaluation of all possible errors in cloud and bubble chamber experiments.

High-statistics experiments involving hundreds of thousands of events have only been possible with the help of automatic measuring machines, guided by digital computers, and of elaborate pattern-recognition-type programs for event reconstruction, recognition, and compilation.

An example of an automatic measuring machine has been the flying-image digitizer at Brookhaven National Laboratory, where rotating mirrors sweep a projected image over small light-sensitive elements that move in a direction perpendicular to the motion of the image. By proper digitization and synchronization of motions, a picture can here be measured in a few seconds. Previous examples were the spiral reader, where a fine slit spirals over a photograph, producing light signals when passing over bubble images; and the flying-spot digitizer, where a small light spot, 10–15 μm in size, scans pictures. Events that do not result in satisfactory fits can often be saved by injecting into the analysis procedure additional information derived by human inspection.

See also RADIATION DETECTION.

BIBLIOGRAPHY

J. G. Wilson, *The Principles of Cloud-Chamber Technique*. Cambridge, London and New York, 1951.

C. M. York, "Cloud Chambers," in *Handbuch der Physik* (S. Flügge, ed.) Vol. 45, 260. Springer-Verlag, Berlin and New York, 1958.

D. A. Glaser, in *Handbuch der Physik* (S. Flügge, ed.), Vol. 45, 314. Springer-Verlag, Berlin and New York, 1958.

D. V. Bugg, "The Bubble Chamber," *Progr. Nucl. Phys.* **7,** 1 (1959).

H. Bradner, "Bubble Chambers," *Ann. Rev. Nucl. Sci.* **10,** 109 (1960).

R. P. Shutt, ed., *Bubble and Spark Chambers, Principles and Use*. Academic Press, New York, 1967.

Yu. A. Aleksandrov, G. S. Voronov, V. M. Gorbunkov, N. B. Delone, and Yu. I. Nechayev, in *Bubble Chamers* (W. R. Frisken, ed.), Indiana Univ. Press, Bloomington, IN, 1967.

Colliding Beams *see* Positron-Electron Colliding Beams; Hadron Colliders at High Energy

Collisions, Atomic and Molecular

Arnold Russek

INTRODUCTION

One of the fundamental tools available to explore the submicroscopic domain of quantum physics is the performance and analysis of collision experiments. The present body of knowledge of the structure of, and the interactions between, atomic and molecular systems stems in large measure from the interpretation of the results of scattering experiments. Indeed, it was the analysis by Sir Ernest Rutherford in 1911 of the α-particle scattering experiments of Geiger and Marsden that established the structure of the atom as a small, massive, positively charged nucleus about which the light, negatively charged electrons move. Scattering by a point Coulomb field is to this day referred to as Rutherford scattering.

Experimental research in atomic collisions continued on a very modest scale throughout the 1930s. Theoretically, however, even with the fragmentary evidence then available, L. Landau, C. Zener, and E. C. G. Stückelberg in 1932 outlined the conceptual understanding of molecular processes in collisional excitation. In the same year, W. Weisel and O. Beeck, using the newly created molecular model of F. Hund and R. S. Mullikan, fully described the role of a particular orbital, the $4f\sigma$, in excitation processes. Forgotten, the role of the $4f\sigma$ was rediscovered in 1965.

In recent years, as a result of massive technological advances in vacuum systems, electronics, and particle detectors, a flowering of experimental research has taken place. Beginning in the late 1950s with the pioneering experiments of E. Everhart and his collaborators at the University of Connecticut and of N. V. Fedorenko and his collaborators at the Ioffe Physical-Technical Institute in Leningrad, the analysis of moderate-energy collisions (in the energy range of 1–100 keV) between atoms has provided a significant increase in the understanding of the energy states of diatomic molecules, of the limitations and breakdown of adiabatic behavior, and of the nonadiabatic couplings responsible for excitation processes. Although the experiments are carried out in a much higher kinetic energy range than that at which most chemical reactions take place, the knowledge gleaned from these experiments and their interpretations bears directly on that subject. In this connection, the important parameter is the ratio of the relative translational velocity v_r to the orbital velocity v_o of the electron involved. If v_r/v_o is very much less than unity, a molecular description is in order. The orbital velocity of an electron in a hydrogen atom, which is essentially characteristic of all outer-shell electrons, is equal to the velocity of light divided by 137, or roughly 2×10^6 m/s. For a hydrogen atom of kinetic energy 25 keV, the ratio $v_r/v_o = 1$, while an argon atom must have a kinetic energy of 1 MeV before this ratio is equal to unity for its outer-shell electrons, and of 18 MeV for inner-shell electrons. At higher energies, up to several MeV, collisional excitation of inner-shell electrons has been explored. Much of the theoretical understanding of the molecular processes observed in the experiments from the early 1960s on is due to the interpretations by D. R. Bates, U. Fano, W. Lichten, and M. Barat.

In the analysis of any collision experiment, the confrontation between theory and experiment is to be found in the cross section for some particular process (e.g., the excitation of a given electronic state). The cross section is a quantity with the dimensions of an area that can be both experimentally determined and calculated from the theory being tested. It is the probability that a scattering event for the process will occur when one projectile is incident per unit area and one target is located somewhere with that unit area. A typical order of magnitude for an atomic cross section is ~ 1 Å^2, or 10^{-16} cm^2. In mathematical terminology, the cross section σ_i for the ith process is defined by the relationship

$$N_i = (N_P/A)n_t Al\sigma_i = N_P n_t l\sigma_i \tag{1}$$

where N_i is the number of events that give rise to the process under consideration, N_P the number of projectiles, n_t the number of targets per unit volume, A the cross-sectional area of the beam of incident projectiles, and l the length of the path of the beam in the target gas. Thus Al is the volume of target gas intercepted by the beam and $n_t Al$ the total number of targets intercepted by the beam. Experimentally, it is important that the product of target density and path length in the target be kept sufficiently small so that the probability of making any collision at all (not just the ith process) is very much less than unity. This ensures the "single-collision regime" wherein the likelihood of making more than one collision is negligible. This is usually accomplished by reducing the target gas density in successive measurements and verifying that the number of scattering events N_i varies linearly with target density. Typical pressures are 10^{-4} Torr. Information can be gained from observation of the incident projectiles, scattered targets, and photons and electrons generated in the collision. Basically, collision experiments fall into one of two general categories. Total cross-section measurements lump together all collisions that give rise to a given process at a given energy, independent of impact parameter. They are dominated by small-angle scatterings that come from large-impact-parameter collisions.

On the other hand, differential cross-section measurements, $d\sigma_i/d\omega$, single out only those collisions that give rise to the process under consideration and, at the same time, result in scattering into a small angular region at a prescribed angle of scattering. The differential cross section is the number of events per unit solid angle per target for each incident particle per unit area.

EXPERIMENTAL METHODS

Atomic and molecular collision experiments require a source and accelerator to produce a collimated beam of the desired projectile at the desired energy, a scattering chamber containing the target gas, and a detector able to detect the projectile and/or recoiling target and/or secondary particles with known efficiency. Projectiles can be atoms or molecules or their ions; targets can only be atoms or molecules. Figure 1 schematically illustrates a typical (perhaps it should be called ideal) experimental arrangement. Ions are extracted from an ion source and electrostatically accelerated to the desired energy. Removal of unwanted ions coming out of the ion source (and there will always be some) is accomplished by passing the ion beam through a magnetic field, which separates components with different values of charge-to-mass ratio Q/M. If neutral projectiles (which cannot, of course, be electrostatically accelerated) are to be used, these must be produced by passing the ion beam, after acceleration, through a neutralization chamber containing a gas for which there is substantial probability of electron capture by the ion. Very little energy is exchanged in this process, so

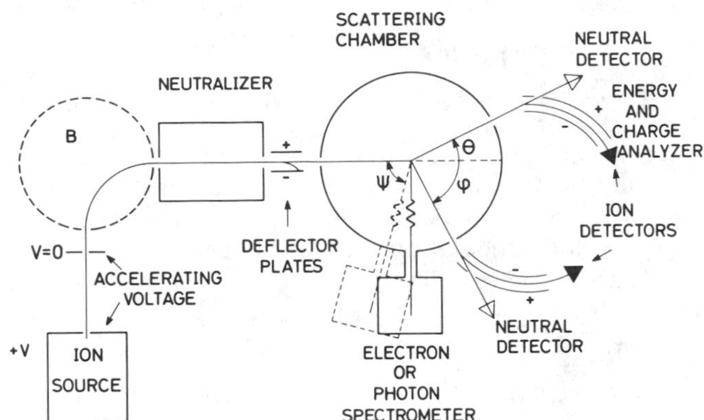

FIG. 1. Principal components of an ideal collision experiment. The vacuum system, gas-handling system, and electronics for counting detected particles are not shown. As depicted here, the apparatus is set up for neutral projectiles. The neutralizer and deflector plates are not used if the projectiles are ions. In this illustration, scattered projectiles, recoiling targets, and secondary electrons or photons are all detected and analyzed. Commonly, however, only one or two of these are encompassed in a collision experiment.

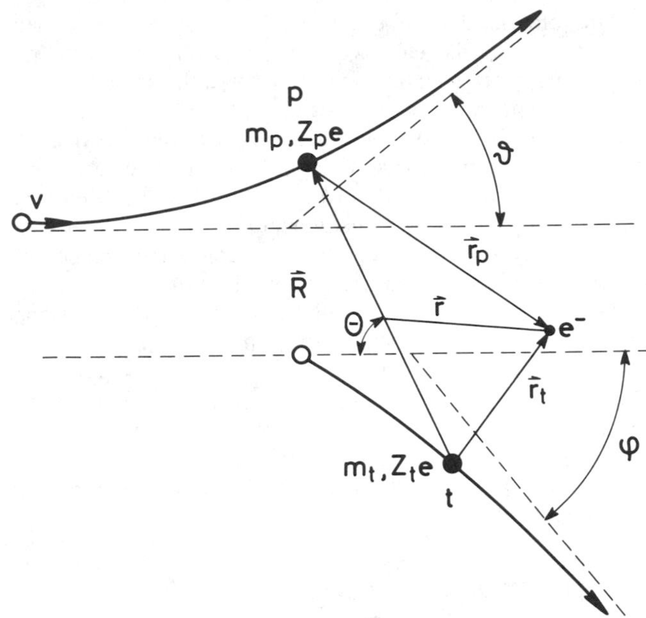

FIG. 2. Collision trajectory in the laboratory frame of reference. **R** is the vector separation between target and projectile; **r** locates an electron in the time-varying quasimolecule formed during the collision process. In general, of course, there will be many electrons. The angles θ and ϕ represent the angle of scattering of the incident projectile and the angle made by the recoiling target with the incident-beam direction, respectively. Mathematically, the quasimolecule is quantum mechanically described in terms of the non-Newtonian rotating coordinate system, called the *body-fixed system*, for which **R** is the polar axis. The change in the angle Θ with time generates the rotational coupling term in Eq. (14).

that a neutral atom at the energy of the accelerated ion passes on to the scattering chamber. Ions that have not been neutralized must be deflected out of the beam by means of deflection plates.

Since the target is a gas or vapor, it must be confined in a scattering chamber with small apertures to allow the incoming beam to enter and the scattered particles to leave. The same, of course, is true for the neutralization chamber. These apertures are generally integrated into the collimation systems that define the respective beams. The target gas (and neutralizer gas, if used) will inevitably leak out through the apertures, so that the regions between the components must all be kept evacuated by constant pumping of these regions. At the same time, target and neutralizer gases must be constantly replenished. In a dynamic situation such as this, the pressure of the target gas (from which its density is determined) must be measured within the scattering chamber itself. Before detection, the energy and charge state can be determined by electrostatic means if an ion, or by time-of-flight means if a neutral. In the latter case, the beam must be pulsed, in order to establish a zero time. The experimental arrangement in Fig. 1 shows a detector for scattered projectile and one for recoiling target. Since atoms in collision behave somewhat like billiard balls, the motion will take place in a plane. It is not too difficult to detect both scattered and recoil atoms *in coincidence* (more precisely, with a slight, but exactly known, time difference, due to the different velocities of projectile and recoiling target).

As illustrated in Fig. 2, the collision is described by the angle θ through which the projectile is scattered and the angle ϕ made by the recoiling target with the incident-beam direction. These angles, as well as the final energies E_p and E_t of projectile and target, can all be determined in a coincidence experiment. Since the incident-projectile energy E_0 is set by the accelerator voltage, and the initial energy of the target can safely be taken equal to zero (thermal energies are ~1/40 eV), the inelastic energy transferred from translational kinetic energy to internal energy (i.e., the excitation energy) can be determined simply by conservation of energy and conservation of momentum. Letting Q denote the excitation energy, we have

$$E_0 = E_p + E_t + Q,$$
$$(2M_pE_0)^{1/2} = (2M_pE_p)^{1/2}\cos\theta + (2M_tE_t)^{1/2}\cos\phi,$$
$$0 = (2M_pE_p)^{1/2}\sin\theta - (2M_tE_t)^{1/2}\sin\phi. \qquad (2)$$

With five final quantities, Q, θ, ϕ, E_p, and E_t, and with three relations existing between these five, it is clear that measurement of any two automatically determines the others. Thus, the inelastic energy Q can be determined just by measuring the angles θ and ϕ, or alternatively, just by measuring E_p and θ. A coincidence measurement is not essential to determine the inelastic energy.

Ideally, the experiment should measure the differential cross section for a collision with specified final states of both projectile and target. Most often, however, it is understandably necessary to settle for less. The situation is made more difficult by the fact that, except for special cases of long-lived states, transit times through the scattering region are much longer than electronic rearrangement times in the atoms themselves. These rearrangements, generally termed deexcitation processes, occur by photon emission or electron ejection. The latter is called Auger emission, or autoionization, and involves an energy exchange between two electrons in the atoms. One drops down to fill a vacancy in a lower energy state, giving up its energy to the second, which then has enough energy to escape, and does so with a characteristic energy. As a consequence of these rearrangements, it is often the case that the states of excitation produced by the collision must be inferred, rather than measured directly. For this reason, many experiments also detect and energy-analyze electrons or photons coming from the collision region. Often, the angular distributions of these secondary particles are measured. As an example, collisional excitation of an inner-shell electron in an atom leaves a vacancy in that inner shell immediately after the collision. Before the atom leaves the scattering region, either an Auger electron or an x ray will be emitted with an energy characteristic of the shell excited. Much can be learned about the process if the angular distribution of Auger electrons or x rays is determined.

THEORETICAL METHODS

Theoretically, the cross sections are most often discussed within the framework of one or the other of two very important approximations which can only be briefly described here.

The Born Approximation

The Born approximation is generally applied to high-energy collisions. It is based on the physical assumption that the collision takes place so rapidly that there is little time for the systems to adjust to the changing internuclear separation. The wave function describing the state of the overall system is therefore expanded in terms of the states of the isolated projectile and target. The Hamiltonian H for the system is written in the form

$$H = T_R + H_p + H_t + V_{pt} = H_0 + V_{pt}. \tag{3}$$

Here, $\mathbf{R}$ is the internuclear separation vector, T_R represents the kinetic energy of translational motion, H_p and H_t are the Hamiltonians describing the isolated projectile and target, and V_{pt} is the sum of all potential energy terms between pairs of components one of which is in the projectile and one in the target. The wave function for the system is expanded in terms of the eigenstates of H_0, which are of the form $\phi_p\phi_t \times \exp(i\mathbf{K}\cdot\mathbf{R})$, with energy $\epsilon_p + \epsilon_t + \hbar^2K^2/2m$, where ϕ_p and ϕ_t are the internal states of projectile and target, m is the reduced mass, and $\hbar^2K^2/2m$ is the translational kinetic energy in the center-of-mass reference frame. The differential cross section for scattering $\mathbf{K}_i \rightarrow \mathbf{K}_f$ together with electronic transition $\phi_p \rightarrow \phi_{p'}$ and $\phi_t \rightarrow \phi_{t'}$ can be equally well derived from either time-dependent or time-independent perturbation theory (where the final state is in the continuum). The differential cross section is given by the expression

$$\frac{d\sigma}{d\omega} = \left(\frac{m}{2\pi\hbar^2}\right)^2 \frac{K_f}{K_i} |M(p,t \rightarrow p',t')|^2 \delta(E_i - E_f) \tag{4}$$

where the δ function ensures that energy is conserved between initial and final states of the overall system. The matrix element M is

$$M(p,t \rightarrow p',t') = \int d^3\mathbf{R}\, e^{i(\mathbf{K}_i - \mathbf{K}_f)\cdot\mathbf{R}} \times \int d^3\mathbf{r}\, \phi_{p'}{}^* \phi_{t'}{}^* V_{pt} \phi_p \phi_t \tag{5}$$

where $\mathbf{r}$ stands for all electronic coordinates.

A variant of the above procedure that is sometimes used treats the nuclear motion classically, with well-defined trajectory as a function of time, $\mathbf{R}(t)$, yet continues to expand the electronic states in terms of those of the isolated systems H_p and H_t.

The Molecular Approximation

The molecular approximation (also called the "adiabatic" or "quasi-adiabatic" approximation) is generally applied to slow collisions. It is based on the physical assumption that the collision process evolves sufficiently slowly in time that the electrons are able not only to adjust continuously to the changing internuclear separation, but also to take energy from, and give energy to, the kinetic energy of nuclear motion. In this way, the electronic motion acts as a potential for the nuclear motion which supplements the repulsive internuclear Coulomb potential. The adiabatic electronic state is defined as the eigenfunction of the electronic Hamiltonian, $\mathcal{H}$:

$$\mathcal{H} = \sum_i \left(\frac{p_i^2}{2m_e} - \frac{Z_p e^2}{r_{ip}} - \frac{Z_t e^2}{r_{it}} + \sum_{j<i} \frac{e^2}{r_{ij}} \right) \tag{6}$$

where r_{ip} and r_{it} are the respective distances of the ith electron to projectile and target nuclei, and m_e is the electronic mass. $\mathcal{H}$ includes the kinetic energy of each electron along with its potential energy of attraction to both nuclei and, finally, the interelectron repulsive term between each pair of electrons. Thus, $\mathcal{H}$ contains the internuclear separation $\mathbf{R}$ as a parameter; it is implicitly contained in r_{ip} and r_{it}. As a consequence, both the eigenstate and eigenenergy of $\mathcal{H}$ depend on $\mathbf{R}$ as a parameter:

$$\mathcal{H}\Phi_n(\mathbf{r};\mathbf{R}) = \mathcal{E}_n(R)\Phi_n(\mathbf{r};\mathbf{R}) \tag{7}$$

where again $\mathbf{r}$ stands for all electronic coordinates.

In the adiabatic approximation, the electronic energy $\mathcal{E}(R)$ is one of the potentials influencing the nuclear motion, so that the wave function χ describing the nuclear motion is a solution to the eigenvalue equation:

$$[-(\hbar^2/2m)\nabla_R{}^2 + Z_p Z_t e^2/R + \mathcal{E}(R)]\chi(\mathbf{R}) = E\chi(\mathbf{R}). \tag{8}$$

The wave function Ψ describing the overall system in the adiabatic approximation is the product of the two functions:

$$\Psi(\mathbf{r};\mathbf{R}) = \Phi(\mathbf{r};\mathbf{R})\chi(\mathbf{R}). \tag{9}$$

It is not an exact eigenfunction of the total Hamiltonian H for the overall collision system. To see this, H, given by (3), is regrouped as

$$H = T_R + Z_p Z_t e^2/R + \mathcal{H}. \tag{10}$$

From (7) and (8) it follows that

$$H\Psi = (T_R + Z_p Z_t e^2/R + \mathcal{H})\Phi(\mathbf{r};\mathbf{R})\chi(\mathbf{R})$$
$$= E\Psi - (\hbar^2/2m)[2(\nabla_R\Phi)\cdot(\nabla_R\chi) + \chi\nabla_R{}^2\Phi]. \tag{11}$$

If the last two terms on the right-hand side of (11) are neglected, then Ψ is indeed an eigenfunction of H, and this is the adiabatic approximation. These terms are, in fact, quite negligible for vibrational states of diatomic molecules and for collisions in the eV energy range. However, for increasing collision energy, the second term on the right-hand side of (11) becomes increasingly important, since $\nabla_R \chi / m$ is essentially the relative velocity of collisions. This gradient term couples different adiabatic states and is responsible for deviations from adiabatic behavior and, therefore, for electronic excitation. In order more easily to describe these couplings mathematically, several variants of the adiabatic states defined by (7) have been introduced. These include "diabatic" states and "traveling orbitals," two subject areas of recent activity. Adiabatic electronic states are calculated at fixed nuclear positions, a built-in contradiction for a theory describing a collision process. Traveling orbitals allow for the nuclear motion by incorporating traveling wave factors $\exp^{iv_N r}$ into the electronic states, where the velocity v_N is that of the nucleus about which the basis function is centered. This subject has been reviewed by Delos in 1981.

Several formulations of diabatic states have been advanced, all of which have in common an attempt to maintain for each diabatic state a well-defined topological "character" that remains invariant throughout the collision. This subject was reviewed by Russek and Furlan in 1989. By contrast, adiabatic states of the same symmetry rapidly interchange their characters in near-degeneracy situations, and the adiabatic energy levels do not intersect. This avoided crossing phenomenon was first elucidated by von Neumann and Wigner and is known as the "noncrossing rule."

The characters of diabatic states can be formulated in terms of topologically defined projection operators which select out specified nodal structures. The diabatic states are eigenfunctions of a diabatic Hamiltonian, $\mathcal{H}_D$:

$$\mathcal{H}_D = P\mathcal{H}P + Q\mathcal{H}Q. \tag{12a}$$

It is not hard to show that a pseudosymmetry operator $\pi = P - Q$ commutes with $\mathcal{H}_D$, so that crossings of the diabatic energy levels do not constitute a violation of the noncrossing rule. The actual electronic Hamiltonian, $\mathcal{H}$, is the sum of $\mathcal{H}_D$ and an interaction term $\mathcal{H}'$, which includes the remaining terms of $\mathcal{H}$ not included in $\mathcal{H}_D$:

$$\mathcal{H} = \mathcal{H}_D + P\mathcal{H}Q + Q\mathcal{H}P = \mathcal{H}_D + \mathcal{H}'. \tag{12b}$$

Writing the state describing the collision system in the form

$$\Phi = \sum_{n=1} c_n(t)\phi_n(\mathbf{r};\mathbf{R}(t)) \exp(-i \int^t \mathcal{E}_n d\tau / h), \tag{13}$$

the time-dependent Schroedinger equation yields a set of coupled first-order differential equations for the coefficients:

$$\dot{c}_m = -\sum_n [i\langle\phi_m|\mathcal{H}'|\phi_n\rangle + \langle\phi_m|\dot{\phi}_n\rangle] c_n \exp[i \int^t (\mathcal{E}_m - \mathcal{E}_n) d\tau/\hbar]. \tag{14}$$

Except at very small angles of scattering, the nuclear motion can be treated in terms of a classical trajectory $\mathbf{R}(t)$ because of the very small wavelength (when compared with atomic dimensions) associated with this motion. The quantum description must, however, be retained for electronic behavior.

Denoting by v_R the radial component of internuclear velocity and by $\dot{\Theta}$ the angular rotation of the line joining the two nuclei, we can show that

$$\dot{\phi}_n = v_R \frac{\partial \phi_n}{\partial R} + \frac{\dot{\Theta}}{2\hbar}(L_+ - L_-)\phi_n \tag{15}$$

where L_+ and L_- are the raising and lowering operators for the component of angular momentum along the internuclear axis. The first term on the right-hand side of (14) is called radial coupling and leaves unaltered the component of angular momentum along the internuclear axis, whereas the second term, called rotational coupling, changes this quantity by $\pm\hbar$.

Figure 3 illustrates an example of energies of a set of diabatic molecular orbitals plotted as functions of internuclear distance. It was used by U. Fano and W. Lichten to analyze inner-shell excitation produced in collisions of argon atoms

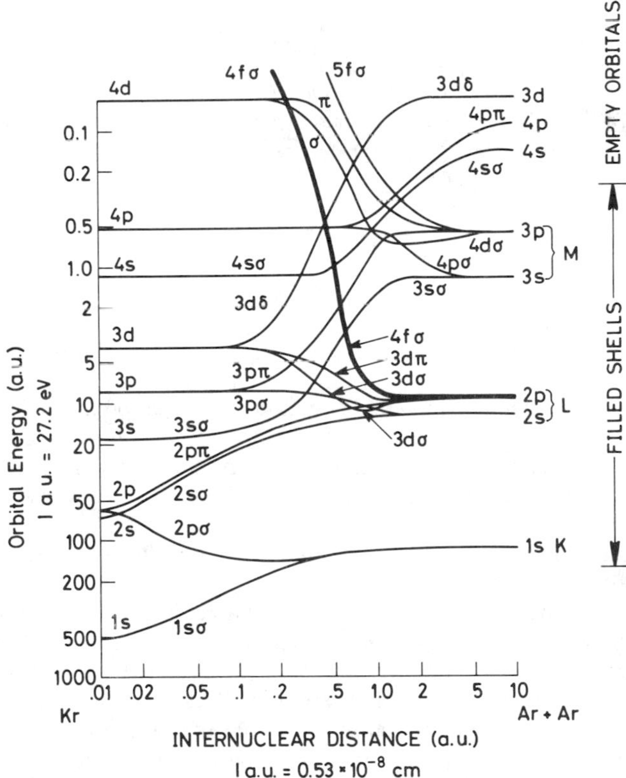

FIG. 3. This molecular orbital diagram, taken from the work of U. Fano and W. Lichten, has been used to explain the detailed behavior of inner-shell excitations in argon–argon collisions. Energies of various orbitals are plotted as functions of internuclear separation R. To use this diagram, start from the right-hand side ($R = \infty$) and proceed to the left down to the minimum internuclear separation for the collision under consideration, and then back out to $R = \infty$. The filled orbitals are indicated on the right-hand side. The two electrons in the $4f\sigma$ orbital, which rises rapidly as internuclear separation decreases, are excited into normally vacant states provided that the minimum internuclear separation is less than approximately 0.5 a.u. (0.26×10^{-8} cm), where the first level crossing with unoccupied orbitals (e.g., $3s\sigma$) occurs.

(or ions) on argon atoms, which have filled *K*, *L*, and *M* shells. For example, *L*-shell excitation can be explained as follows: The $4f\sigma$ orbital, which contains two electrons, rises rapidly as *R* decreases below 0.5 a.u., crossing many vacant orbitals. These two electrons have a substantial probability of being coupled to one of the orbitals crossed, leaving one or two vacancies in the $4f\sigma$ orbital, which separates, after the collision, into a $2p$ vacancy in the argon *L*-shell.

See also ATOMS; AUGER EFFECT; BEAMS, ATOMIC AND MOLECULAR; SCATTERING THEORY.

BIBLIOGRAPHY

P. W. Atkins, *Molecular Quantum Mechanics*, Parts I, II, and III. Oxford (Clarendon Press), London and New York, 1970. (E)

Bernd Crasemann, ed., *Atomic Inner-Shell Processes*, Vols. I and II. Academic Press, New York, 1975. (A)

J. B. Delos, *Rev. Mod. Phys.* **53**, 287 (1981).

Quentin C. Kessel and Bent Fastrup, "The Production of Inner Shell Vacancies in Heavy Ion–Atom Collisions," *Case Studies in Atomic Physics*, Vol. 3, No. 3, pp. 137–213. North-Holland, Amsterdam, 1973. (I)

A. Russek and R. J. Furlan, *Phys. Rev.* **39**, 5034 (1989).

James C. Slater, *Quantum Theory of Molecules and Solids*, Vol. I. McGraw-Hill, New York, 1963. (I)

Ta-You Wu and Takashi Ohmura, *Quantum Theory of Scattering*. Prentice-Hall, Englewood Cliffs, NJ, 1962. (I)

Color Centers

Paul W. Levy

Normally transparent crystals and glasses often appear colored because they contain color centers. More specifically, their colored appearance is attributable to absorption bands associated with defects, impurities, or more complex imperfections. Many different types of color centers are formed when the charge or valence state of these imperfections is changed. The defects or imperfections may be incorporated into the lattice in one charge state and not produce observable absorption bands. Subsequently, if the charge state is altered, most often by exposure to ionizing radiation, absorption bands are formed. Alternatively, color centers are often formed by impurities that were introduced into the lattice in their normal valence state. Familiar examples of solids that appear colored because they contain color centers include:

Solids colored by radiation—e.g., crystals and glasses colored by exposure to nuclear radiation, to x rays, to high-energy particles, or to sunlight for a long period of time (solarization). In fact, almost all transparent solids will develop color centers when exposed to any radiation that will create electron–hole ionization pairs in the interior of the solid.

Colored minerals—e.g., ruby is Al_2O_3 with Cr impurity; smoky quartz usually contains Al and alkalis and the formation of the color requires exposure to ionizing radiation. The many colored fluorites are due to the presence of different rare-earth ions and other impurities.

Colored glasses—e.g., "stained" glass windows are made of different glasses with a variety of impurities.

COLOR-CENTER PROPERTIES

Almost all of the properties of color centers are included in recent books [1–4] and may be illustrated by describing F centers in alkali halides. Historically, this was the first center investigated in any detail and has been studied more extensively than any other center. All of the alkali halides, such as NaCl, KCl, and LiF, become colored when exposed to sufficiently large doses of x-rays, nuclear radiation, or any other radiation producing ionization. The most commonly occurring bands, induced in a typical alkali halide by x-ray irradiation at liquid-nitrogen temperature, are shown in Fig. 1. In many alkali halides the most prominent band is in the visible. Much of the original research on these bands was done in Germany, and they were labeled F bands after the German word for color, *Farbe*. Likewise, the center associated with the F band was labeled the F center. The nature of the F center was established by studying additively colored crystals that were prepared by exposing pristine samples to alkali-metal vapor. For example, KCl crystals that have been heated in potassium vapor and rapidly cooled to room temperature contain a prominent absorption band that is identical to the largest band induced by irradiation. During exposure to the potassium vapor potassium atoms attach themselves to the KCl lattice. To maintain the lattice-ion and defect concentrations appropriate for the crystal and vapor temperature, each potassium ion incorporated into the lattice surface is accompanied by the introduction of a Cl^- ion vacancy into the lattice. To maintain charge neutrality, the electron released from each potassium atom, as it becomes a lattice ion, is transported into the crystal (it "wanders

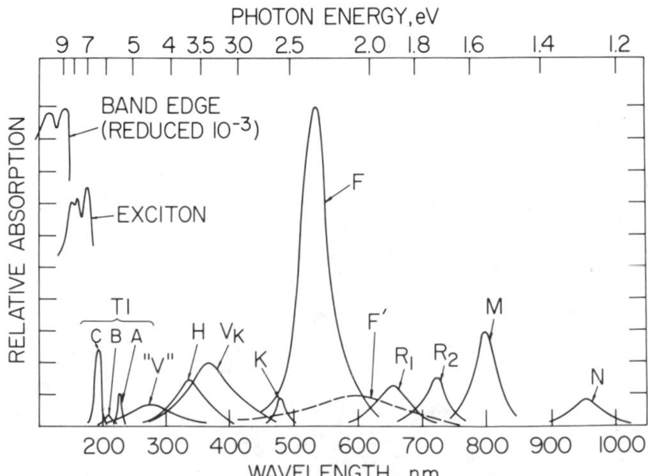

FIG. 1. Principal color-center absorption bands in a "typical" alkali halide. The energies and intensities of bands other than the exciton bands have been chosen to approximate the spectra observed after x-ray irradiation near liquid-nitrogen temperature, roughly 88 K. Exciton bands are observed only near 0 K.

about'' in the conduction band) until it is trapped by a Cl^- ion vacancy.

The defect (i.e., the configuration) formed when a single electron is trapped on an isolated monovalent negative-ion vacancy—such as a Cl^- vacancy in KCl—is called an F center. It, and most other color centers, contain a ground state and at least one excited state. Associated with the F center is the F-center optical absorption band which is due to a transition between the ground state and the next highest electronic state. In most crystals the next highest state lies in the band ''gap'' a few tenths of an electron volt below the conduction band. The lifetime of the excited state is usually quite short and radiative transitions from the upper to the lower state are often observed as luminescence. Also, and increasingly so with increasing temperature, the center can return to the ground state by nonradiative transitions, or the electron in the upper level, i.e., in the excited state, can thermally untrap into the conduction band.

Additional features of F centers and other defect-related color centers are illustrated by measurements made on additively colored crystals. If a crystal is cooled rapidly to room temperature after exposure to a hot alkali-vapor environment and then cleaved into thin sections, it is found that the F-center concentration is a maximum at the surface and diminishes toward the interior. A uniform distribution of F centers is obtained by heating an additively colored crystal for a comparatively longer time period in a vacuum. This clearly illustrates the introduction of F centers at the surface and the relatively rapid diffusion of centers in the crystal. Additively colored crystals that are slowly cooled, in contrast to the rapidly cooled ones just described, exhibit additional absorption bands that occur on the long-wavelength side of the F band and with one-tenth, or less, of its intensity. One of these bands is called the M band and it has been identified with a center formed by two adjacent F centers along ⟨110⟩ directions. Its properties were established by optical absorption measurements on strained and unstrained crystals that utilized light polarized in different directions with respect to the crystal axes. In a similar way other bands have been attributed to R centers, which consist of three adjacent F centers in an equilateral-triangle arrangement with the sides along ⟨110⟩ directions. This arrangement is energetically favored over a linear three F-center configuration along a single ⟨110⟩ direction. The appearance of multidefect M and R centers in slowly cooled additively colored crystals is attributable to the coagulation of F centers by diffusion processes. These multiple-defect centers form only at temperatures high enough to ensure F-center mobility but low enough to prevent the complete dissociation of previously formed multiple defects. Pure additively colored alkali halides contain very weak absorption bands on the short-wavelength side of the F band. These are called L bands and usually are observed only at liquid-nitrogen or lower temperatures. It has been shown that they are coupled in some manner to the F center. However, they are not well understood.

Additional electronic properties of F centers can be demonstrated by exposing additively colored crystals to light that is absorbed by the F band (termed F light). As the exposure to light increases, the F band diminishes and an additional broad band, lying principally to the long-wavelength side of the F band, appears. This broad band is called the F′ band (F-prime band) and is formed by the capture of a second electron by existing F centers. During exposure to F light a fraction of the F center absorbs photons, which are excited to an excited state lying slightly below the conduction band. The lifetime of this state is sufficiently long to allow the electron to be excited thermally into the conduction band. Once in the conduction band the electron is mobile and ''wanders about'' until it is captured by a negative-ion vacancy to recreate an F center, captured by an F center to become an F′ center, or removed from the conduction band in some other way such as by electron–hole recombination. The F′-center formation process may be reversed by exposing the crystal to light that is absorbed by the F′ center or by raising the crystal temperature. Both processes release electrons from the F′ center into the conduction band, which leads to the reestablishment of F centers. Bleaching experiments such as these have provided much of the existing information on the nature of electron-trap centers and their interaction with mobile charges in the crystal. Bleaching measurements also have been very useful in studies of the electrical antimorphs of the electron centers, namely, the (electronic) hole centers.

The properties of the alkali-halide centers described in this article, and other more commonly occurring centers, are included in Table I.

The electronic nature of F, M, and R centers, or more specifically the fact that they are formed by trapping of electrons on defects, is readily understood for additively colored crystals. As described earlier, the electron released by each metal atom as it is incorporated into the surface of the lattice is trapped on the negative-ion vacancy concomitantly formed to maintain electronic neutrality. These and other considerations suggested that there should be centers that could be described as the electronic and ionic antimorphs of the electron centers, i.e., centers formed by trapping (electronic) holes on positive-ion vacancies. Historically, answers to this and related questions were more readily obtained by studies on crystals colored by exposure to radiation, for reasons that are difficult to describe briefly, than by investigating additively colored crystals prepared by heating them in halogen vapor. In the latter case we would expect the halogen atom to join the lattice and release an electronic hole (capture an electron), which in turn is trapped (by the release of an electron) on a positive-ion vacancy. The resulting configuration (i.e., a hole trapped on an isolated positive-ion vacancy) is often referred to as the V center. However, it has not been established that the simple V center described here actually exists. As described later, the hole centers that have been identified are usually more complicated.

To understand the formation of color centers by radiation it is essential to describe the more important processes that occur in solids during exposure to energetic radiation, such as ultraviolet and visible light, x rays, gamma rays, electrons, alpha particles, neutrons, and high-energy particles. Most important is the formation of electron–hole ionization pairs, which are often referred to simply as ion pairs. Electron–hole ionization pairs are created whenever the incident ra-

Table I Principal alkali-halide color-center absorption bands[a]

Classical[b] notation	Defect associated with given band, properties of band, etc.
α	Isolated negative lattice-ion vacancy
F	One electron on a single isolated negative lattice-ion vacancy
F′	Two electrons on a single isolated monovalent negative-ion vacancy
K	Transition(s) to level(s) above the F-band absorption level
$L_{1,2,3}$	Very weak absorption associated with F centers; origin not known
M	Two adjacent F centers in ⟨110⟩ directions
N	Four adjacent F centers: arrangement not certain
$R_{1,2}$	Three adjacent F centers in "triangle" arrangement
M^+	One electron on two adjacent negative lattice-ion vacancies (ionized M center)
F_A	F center adjacent to a single substitutional, usually alkali-ion, impurity
F_B	F center adjacent to two substitutional, usually alkali-ion, impurities
F_A(I)	F_A center that relaxes to a "single-well" configuration after excitation by photon absorption
F_A(II)	F_A center that relaxes to a "double-well" configuration after excitation by photon absorption
M_A^+, M_B^+	M^+ center adjacent to single (M_A^+) or two (M_B^+) substitutional impurities
Z_1	One electron on coupled substitutional divalent impurity and adjacent positive-ion vacancy
β	Isolated positive lattice-ion vacancy
V (or V_1)	The postulated antimorph of the F center, consisting of one hole on a single isolated positive lattice-ion vacancy (or an adjacent negative ion); it is not certain that this center has been observed
$V_{1,2...}$	Bands in V region on the short-wavelength side of the F center; often assumed to be produced by hole centers; may be identical with "identified" hole centers
V_K	One hole trapped on a pair of adjacent negative-ion lattice ions to produce a single ionized two-atom molecule, e.g., Cl_2^-; aptly called the "self-trapped hole"
I	Isolated ionized lattice atom in an interstitial site
H	One hole trapped on a pair of negative lattice atoms centered on a single lattice-atom site, i.e., one hole, one lattice atom, and one interstitial; often described as a "split interstitial molecular ion on a single lattice atom site," e.g., a Cl_2^- molecular ion on a Cl^- site
H_A	H center adjacent to a substitutional positive-ion impurity
V_A	V center adjacent to a substitutional negative-ion impurity such as Br^- or OH^- on a Cl^- site
Exciton bands	Transitions between the "Bohr-atom" like levels of bound electron–hole pairs, i.e., excitons. Since exciton stability decreases rapidly with temperature stable, or nontransient, exciton bands are observed only near zero degrees Kelvin
Colloid or "X" bands	Bands arising from the combined absorption and scattering, called extinction, of colloid particles; usually consisting of regions, in the lattice, containing 5–15 or more metallic ions and few, if any, negative ions

[a] Unless otherwise stated, the absorption bands given in this table refer to ground-state to first-excited-state transitions.

[b] Originally the letter designations, such as F in F band, referred to specific absorption bands. More recently, it has been common to use the letter designation to refer to the center responsible for a given absorption band. This often causes confusion, e.g., there is a K band but not a K center. This multiple usage, and the use of other systems for designating absorption bands and/or defects that are often similarly ambiguous, require the reader of color-center literature to pay particular attention to nomenclature.

diation is sufficiently energetic to excite lattice electrons, i.e., valence-band electrons, into the conduction band. Once formed, the electron member of the electron–hole ionization pair can "wander about" independently of the concomitantly formed hole. Almost all of the ionization charges enter into recombination. Usually ionization holes are trapped on defects, impurities, or other lattice imperfections more rapidly than ionization electrons are trapped on "opposite-sign" lattice defects. Hole centers usually have large cross sections for electron–hole recombination processes. Thus recombination accounts for a very large fraction, usually more than 99%, of all of the ionization pairs produced during irradiation. The charges that escape recombination during irradiation remain trapped (e.g., to form various electron and hole centers) when the irradiation is terminated.

The ionization-producing radiation, other than photons, is always a charged particle and interacts with (excites) the lattice electrons through a Coulombic (charge–charge) interaction. As a result, almost all outer-shell electrons on atoms along the path of the incident charged particle are excited to some extent. Those that receive a small amount of excitation lose excitation energy by contributing phonons to the lattice, i.e., heating the lattice. Electrons that receive sufficient excitation, usually an amount equal to or greater than the band "gap," become ionization electrons; the ionization electron and concomitantly formed hole constitute an electron–hole ion pair. To create one ion pair, the average energy given up by the lattice traversing charged particle is very roughly twice the band-gap energy in low- (average) Z materials and increases to roughly three times the gap energy in higher- (average) Z materials.

To retain electrical neutrality the total number of trapped electrons and holes must remain very nearly equal. However, large local charge imbalance may exist in certain circumstances. For example, when samples are subjected to low-energy electron bombardment from one direction, the trapped charges are unequally distributed and discharges can occur between regions of opposite charge. Also, excess charge can discharge out of the target. The resulting tree-like discharge patterns are known as Lichtenberg figures.

However, for radiation-induced color-center formation the total number of electrons trapped in electron-type color centers is approximately equal to the number of holes trapped in hole centers.

The nature of the more simple identified hole-trapping centers is more complex than the nature of the antimorph electron centers. To achieve local charge neutrality the missing positive lattice-ion charge must be compensated by a trapped hole. To a large extent the electronic nature of hole centers can be understood if it is remembered that the electronic hole is a convenient fiction and that holes are in reality "missing electrons." A hole trapped on a positive-ion vacancy is shared by, i.e., moves between, each of the adjacent negative ions. Thus in the portion of the crystal containing a hole on an isolated positive-ion vacancy there would be local charge neutrality. However, as mentioned earlier it is not certain that this conceptually simple center actually exists. A discussion of reasons why it does not exist or has not been observed is too long to include in this article. The more commonly occurring and established centers formed by trapped holes are described in what follows.

Among the various kinds of V centers that definitely do occur in alkali halides, the V_K center has been studied most extensively. This center assumes an unexpected and surprising configuration. It can enter into a number of color-center formation processes and is comparable to the F center in controlling radiation-induced color-center formation at all temperatures. It is stable, or nearly so, at temperatures below roughly 100 K. At temperatures up to the 550-K region it is stable long enough to enter into a number of reactions, as described below. The V_K center is aptly described and referred to as a "self-trapped hole." The center is formed when an ionization-created hole attaches itself to a pair of adjacent halogen atoms to create a halogen negative ion. For example, in KCl, the V_K centers are Cl_2^- ions oriented in the six equivalent ⟨110⟩ directions. The V_K centers are unique: The available evidence indicates that they are formed in the initially unperturbed or defect-free regions of the lattice. In contrast, other centers are formed at one or more existing lattice defects. The processes controlling the formation of these centers, i.e., the factors determining exactly where in the crystal a V_K center will be formed, or if it is formed in a mobile state where it will localize, are not understood in detail.

The properties of self-trapped holes have been very extensively studied—especially in alkali halides—by optical absorption using polarized and unpolarized light, electron-spin resonance, luminescence, etc. The characteristics of this center are illustrated by the numerous processes that take place at different temperatures. At temperatures appreciably less than 100 K, for most alkali halides, the V_K centers are stable and oriented in ⟨110⟩ directions. At higher temperatures, specific to each type of crystal, the defect axis may change from the initial ⟨110⟩ to any one of the ⟨110⟩ directions at an angle of 60° to the original; reorientation to the ⟨110⟩ directions at 90° from the original direction apparently does not occur. At even higher temperatures, where the orientation occurs very rapidly, the defect becomes quite mobile and usually enters into several different kinds of interactions. It is likely to form any number of other hole centers, such as H or V_A centers. Also it can interact with electron centers to undergo electron–hole center neutralization or recombination. During recombination, light (i.e., luminescence) may be emitted, or energy may be released as heat or phonon emission and, as described below, provide the energy that results in the formation of a negative ion vacancy and a negative ion interstitial, the process called *ionization damage*.

A second type of commonly occurring hole center is the H center. It should be carefully contrasted with the V_K center. The V_K center is a negative molecular ion, such as Cl_2^-, formed by hole trapping in the normal or unperturbed lattice. The V_K center, i.e., the Cl_2^- ion, occupies the volume normally containing two Cl^- lattice ions. In contrast, the H center can be regarded as a similar Cl_2^- molecular ion formed by hole trapping on an interstitial negative atom and an associated lattice atom. In its stable configuration the H center is a negative ion pair, e.g., Cl_2^- ion occupying the volume normally containing a single Cl^- lattice ion. The lattice surrounding the H and V_K centers—in fact the lattice surrounding practically all centers—distorts, or relaxes, to accommodate the atoms (ions) in the center.

An additional commonly occurring hole center is formed when H centers couple with other lattice defects, such as substitutional impurities, to form centers usually labeled H_A centers. Often the A atom is identified, e.g., an H center coupled to a Li atom is labeled H_{Li}.

Next, the *exciton* bands will be described. These bands are stable, i.e., have lifetimes long enough to permit their being easily observed, only near 0 K. They normally occur at wavelengths near, but longer than, the normal band "gap" valence-band to conduction-band transitions (see Fig. 1). To understand exciton bands, recall first that electron–hole ionization pairs formed at high temperatures, e.g., room temperature, appear to dissociate immediately and the charges move independently of each other. At low temperatures, e.g., near 0 K, some of the ionization pairs remain coupled. More precisely, some (ionization) electrons remain coupled to ionized atoms. The coupled pair has many of the properties of a "Bohr-like" atom. The observed energies of exciton absorption bands are, in fact, in reasonable agreement with the Bohr atom model. Near 0 K excitons appear to have little tendency to dissociate but there is considerable evidence they are quite mobile. They can be observed well after being formed and are localized or trapped on lattice defects such as V_K centers. There is increasing evidence that the dominant process for exciton formation is the trapping of mobile electrons by trapped holes, particularly electron trapping by V_K centers. The fact that ionization damage (see below) is observed in NaCl at 250–350°C, and that excitons are essential for this process, indicates that excitons have appreciable lifetimes—in terms of lattice vibration frequencies, $\sim 10^{13}$/s—at these temperatures.

Exciton bands observed in crystals at low temperatures disappear from the absorption spectra as the crystal temperature is raised. This disappearance can be attributed to the dissociation of the excitons with the individual charges entering into the processes usually associated with ionization. Alternatively, they could have become mobile and interacted with lattice defects, surface states, etc. Such interactions produce luminescence or, alternatively, result in the release of an amount of energy, as phonons or heat, slightly less than the band-gap energy. Also, as described below, the recombination energy can contribute to defect formation.

CENTERS FORMED BY OPTICALLY ABSORBING IMPURITIES

A large variety of light-absorbing impurities may be incorporated into alkali halides, oxides, and other crystals. The more common examples include:

Alkali impurities in alkali halides, e.g., Na in KCl, KBr, etc.
Alkaline earths in alkali halides, e.g., Ca and Sr in KCl, LiF, etc.
Metals such as Pb, Cu, Tl.
Molecular ions like SO_4^-, CN^-, OH^-
Nonmetals, including S^{2-}, O^{2-}.

In fact, a very large fraction of the conceivable ions can be incorporated in the alkali halides and simple oxides.

Crystals prepared with specific impurities are described by the *colon notation*. Thus NaCl with lead impurity is specified as NaCl:Pb. This is ambiguous if the Pb was introduced as a compound, e.g., $PbCl_2$. In terms of this example, it has become increasingly common to specify NaCl with lead introduced as a metal by the notation NaCl:Pb, and when the lead is incorporated as $PbCl_2$ by the notation NaCl:$PbCl_2$.

In the development of color-center physics, measurements on alkali halides containing thallium substituted for the alkali metals played a pivotal role. The Tl^+ ion produces three easily observed absorption bands, usually called A, B, and C bands. These three bands occur on the long-wavelength side of the valence-band to conduction-band transition. There is appreciable evidence for a fourth band, the D band, which lies very close to or on the band-gap absorption. The properties of the A, B, and C bands have been deduced theoretically by considering how the electronic energy levels of the free Tl^+ ions are modified by embedding the ion in the host alkali-halide crystal. The A and C bands are attributable to transitions that are allowed in the free Tl^+ atom. The B band is associated with a normally forbidden electronic transition in which the forbiddance is removed by interaction with lattice phonon modes. Inasmuch as this interaction increases with increasing temperature, the B-band absorption increases with increasing temperature. This is a unique property, observed only with these and similar atomic impurity bands. The origin of the D band is not unequivocally established. It has been attributed to transitions between the ground state and levels above the A-, B-, and C-band levels and, alternatively, to transitions from the ground state to the conduction band. A large fraction of the properties of the impurity-related absorption bands are very similar to the defect-related bands. Both types are described well by the configuration coordinate formalism described below.

Many transparent but colored minerals and various kinds of colored glasses are examples of materials colored by the presence of specific light-absorbing atoms. Ruby is Al_2O_3 (corundum) containing Cr^{3+} atoms substitutionally replacing Al^{3+} atoms. Emerald is $Be_3Al_2Si_6O_{18}$ (beryl) containing Cr^{3+} in Al sites. Aquamarine is beryl containing Fe^{2+} and Fe^{3+} impurities in Al sites. Citrine is quartz given a yellow color by the presence of Fe^{3+} ions, most likely in Si^{4+} sites. Examples of colored glasses include the common blue cobalt glass that obviously includes Co, and pink-tinted spectacle lenses, made from didymium glass, which contains a mixture of rare-earth elements. The ubiquitous beer bottle's brown color is due to Fe^{3+} ions, and the light green of wine bottles is ascribable to Fe^{2+} ions.

Almost all colored crystals, in contrast to host crystals that do not contain other than negligible quantities of impurities, exhibit absorption bands in the visible. A large fraction of these bands can be attributed to transitions between atomic energy levels in the impurities, which are often split by crystal-field effects. Ruby and emerald are in this category. The absorption bands in another large group, which includes aquamarine, are due to molecular-orbital and/or charge-transfer transitions. A small number of crystals are colored because they possess valence-band to conduction-band transitions. Cinnabar (HgS) is a good example in this category. Finally, there are a number of crystals that appear colored because they contain defect-related color centers of the type described earlier for the alkali halides, i.e., they contain color centers formed by charge trapping on lattice defects. Smoky quartz and some, but not all, colored fluorites are in this group.

To reduce the existing confusion, it is necessary to mention three types of crystals that exhibit color not attributable to any of the processes mentioned so far, i.e., not caused by defect and impurity color centers. First, halite, which is the natural form of NaCl and is usually called rock salt, occasionally exhibits an intense color ranging from very light blue to a deep blue-black purple. Often the color is layered along ⟨110⟩ planes. These blue crystals exhibit a *de facto* absorption band in the red that is attributable to a combination of absorption and scattering from colloid Na metal particles. This blue coloring can be induced by exposing NaCl crystals to large doses of ionizing radiation, particularly above room temperature, and/or exposing crystals to room light after irradiation. Second, crystals can appear colored if they contain numerous microinclusions of a second mineral, e.g., the color in blue quartz is attributed to microscopic rutile (TiO_2) inclusions. Third, crystals that are composites of regularly arranged microcrystals exhibit colors because they diffract light. The classic example is opal. This mineral consists of nearly equal-size spheres of a hydrated silica located on a uniformly spaced three-dimensional lattice.

OPTICAL PROPERTIES OF COLOR CENTERS AND THE CONFIGURATION COORDINATE DIAGRAM

A large fraction of the optical properties of color centers can be described by utilizing a configuration-coordinate diagram such as Fig. 2. Such diagrams describe the electron and vibrational, or thermal energy, levels of simple electron or hole centers. To a moderately good approximation such centers can be described by quantum-mechanical calculations based on the Born–Oppenheimer approximation. This procedure separates the Schrödinger equation solution into

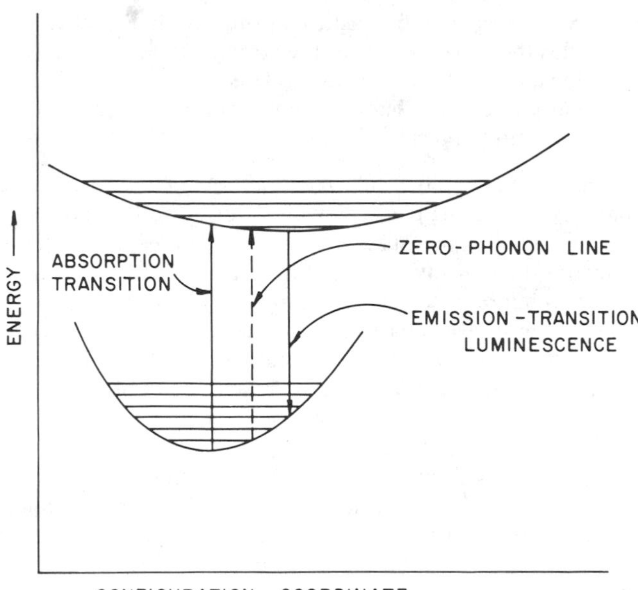

FIG. 2. A typical color-center configuration-coordinate diagram. Many quantitative optical properties of color centers can be deduced from such diagrams. Typical absorption, luminescence, and a "zero-phonon" absorption transition are shown.

an electronic part and a vibrational part. The solution of the vibrational part is obtained by using a "spring-constant" potential which leads to harmonic-oscillator wave-function solutions. The solution of the electronic part, for the lowest two electronic states, is represented by the curved envelopes, properly called "adiabatic potential-energy curves." Superimposed on the electronic states are the vibrational or phonon states, shown as closely spaced lines. Usually the lowest electronic level is an *s* state and the next highest level a *p* state. At temperatures approaching 0 K the system can be described by the *s*-state envelope, the *p*-state envelope, and the lowest-lying phonon levels in each state. At low temperatures the absorption transitions will originate from the lowest-lying levels at the minimum in the ground state and terminate on the phonon levels of the excited-state envelope which lie directly above in the configuration-coordinate diagram. Such transitions are indicated schematically in Fig. 2 as the absorption transition. The resulting absorption band will be quite narrow, since all of the absorbing transitions have nearly the same energy. At higher temperatures the transitions originate and terminate from wider distributions of phonon states, and as the temperature increases the absorption bands become wider. Optical transitions from upper to lower electronic states result in the emission of light, i.e., luminescence. Similarly, the luminescent emission is narrow at low temperatures and increases with increasing temperature. This approach to the optical properties of color centers provides a good description of absorption, luminescence, and many other properties of color centers. The phonon contribution is well approximated by a Boltzmann distribution for all temperatures except those near 0 K. Near 0 K accurate calculations require the inclusion of discrete phonon levels and the zero-point energy. Also, for most calculations the electronic envelope can be regarded as parabolic; in the pure harmonic-oscillator approximation it is precisely parabolic. Calculations made with this model lead to the following expressions for the temperature dependence of the absorption band peak E_0 and the full width of the band at half maximum U_0 (usually both E_0 and U are given in electron volts):

$$E_0(T) = C_1 - C_2 \coth(\hbar\omega/2kT), \quad (1)$$

$$U^2(T) = 8 \ln 2(\hbar\omega)^2 S \coth(\hbar\omega/2kT), \quad (2)$$

where T is the temperature in kelvins, C_1 and C_2 are constants, $\omega/2\pi$ is the effective vibration frequency in s^{-1}, $\hbar = h/2\pi$ where h is the Planck constant, k is the Boltzmann constant, and S is the Huang–Rhys factor. The last-mentioned factor is difficult to explain briefly [5]. It is a coupling constant that is a measure of the relative horizontal displacement of the adiabatic potential energy curves along the configuration-coordinate axis and is approximately equal to the mean vibrational quantum number in the transition.

The optical absorption of a crystal containing color centers can be calculated by considering it as an assemblage of a large number of (harmonic oscillator) absorbing centers. The result is known as *Smakula's formula*, which relates the measured optical absorption coefficient α (in cm^{-1}) to the concentration of color centers N:

$$Nf = 0.87 \times 10^{17} \frac{n}{(n^2 + 2)^2} U\alpha_m, \quad (3)$$

where f is the oscillator strength, which must be determined independently for each color center and is approximately 0.9 for most alkali-halide centers; n is the index of refraction of the host crystal; U is the full width at half maximum of the absorption band (in eV), and α_m is the absorption coefficient (in cm^{-1}) at the peak of the band. This formula is based on the assumption that the absorption band is precisely Gaussian shaped. The numerical coefficient changes by roughly 10% for other common shapes. Precisely Gaussian-shaped bands are observed in many cases and in crystalline materials represent cases where the lower adiabatic potential-energy curve for the centers can be approximated by parabolas and simple harmonic oscillator wave functions are obtained. In many cases the actual potential cannot be represented by simple parabolic curves and the observed absorption and emission bands are not well described by Gaussian-shaped bands; in fact, the observed bands may be quite asymmetric. In glasses and amorphous solids the absorption (luminescence) bands may be Gaussian shaped because they represent a superposition of absorption (emission) by centers randomly influenced by their surroundings, e.g., the randomness superimposed on centers in glasses by the random atomic separation and band angles.

The usefulness of the configuration-coordinate diagram approach and the behavior of many color centers at low temperatures is illustrated by zero-phonon line phenomena. A good example is the R-center absorption in LiF at liquid-helium temperatures, i.e., near 4 K. The absorption spectrum consists of a number of very sharp lines superimposed on a broad underlying continuum. The continuum resembles the spectrum observed at higher temperatures but is slightly narrower and shifted in accord with the peak-energy and bandwidth formulas given earlier. The sharp lines are not observed at higher temperatures. They arise from transitions between individual low-lying phonon states in the upper and lower levels. The most prominent sharp-line transition is between the lowest-lying states, and since these do not include lattice vibration, i.e., phonon modes (the vibrational quantum number is zero), it is called the "zero-phonon line." This transition is illustrated in Fig. 2. It is often observed when the other sharp lines are obscured.

RADIATION-INDUCED COLOR-CENTER FORMATION

Most studies on color centers were undertaken to determine the properties of the different types of centers. More specifically, they were designed to identify the atomic and electronic configuration of each center and to determine its properties. A less emphasized but equally interesting aspect of color-center physics, one that has numerous applications, is the study of radiation-induced color-center formation. Curves of color-center concentration versus irradiation time or dose depend on a large number of physical parameters. The more important are:

1. The nature of the material being irradiated. Some aspects of the color-center formation kinetics are common to most materials. Other aspects depend on the material type. There are features peculiar to each of the following: alkali halides, oxides, mixed ionic and covalent crystals (such as nitrates and bromates), covalent crystals, semiconductors, glasses, glass-ceramics, and organic glasses.
2. Sample temperature.
3. The type of radiation producing the coloring, which includes infrared, visible, ultraviolet, x-ray, and gamma-ray *photons* and the various charged particles such as electrons or beta rays, protons, alpha particles, fission fragments, and the entire list of charged high-energy particles. Low-energy photons can create ionization by optical absorption processes such as band "edge" transitions. Higher-energy photons efficiently create ionization almost entirely by the formation of charged recoils resulting from the photoelectric, Compton, and pair-production processes. Energetic neutral particles, such as reactor neutrons, produce copious ionization and radiation damage, by creating charged lattice atom recoils from elastic or inelastic (nuclear event) collisions.
4. Most importantly, the incident particle energy, the total energy, i.e., *dose*, imparted to the material and the rate of energy deposition, i.e., the *dose rate*. Often the irradiation conditions are given in terms of the total number of particles incident on the sample, i.e., the *fluence*, or the irradiation time and the *flux*, or particles per unit time (fluence = integration of flux over irradiation time). Color-center formation often depends on details of the irradiation conditions such as continuous versus pulse irradiation, pulse length and interval between pulses, etc.
5. The presence of impurities and defects in the lattice. Impurities control the coloring of many materials.
6. Strain state and/or plastic deformation. In the alkali halides the coloring curves are strongly influenced by dislocations and dislocation-related defects. Plastically deformed alkali halides color much more readily than unstrained ones.

Until recently most color-center formation studies were made by irradiating a crystal with x rays, or other radiation, and then making measurements at a later time. If the crystal temperature was low enough, e.g., liquid-nitrogen temperature, to suppress all temperature-dependent decay occurring during and after irradiation, most of the color-center versus dose or irradiation-time curves are linear or monotonic increasing and are true representations of the color-center formation kinetics. However, at higher temperatures and particularly at room temperature, the center concentrations change both during and after irradiation and the measured coloring curves do not accurately describe the color-center growth kinetics. In many crystals, particularly the alkali halides, the coloring curves contain three distinct stages. The initial part, stage I, increases monotonically at a continuously decreasing rate until a low- or zero-slope plateau is reached. The plateau region is called stage II. The irradiation time included in stage II varies from close to zero to very large values. The following region, stage III, when it is observed, usually is concave upward and continues until the irradiation is terminated. At very high doses some materials exhibit growth beyond stage III which flattens out to a plateau, or occasionally reaches a maximum and then decreases. Sometimes this last region is referred to as stage IV.

First- and second-stage coloring curves obtained by making measurements at some controlled time after irradiation and at a fixed temperature (e.g., at room temperature) are crudely approximated by an expression consisting of the sum of a saturating exponential and a linear component. However, the stage-I and -II color-center versus irradiation-time curves *obtained by making measurements during irradiation* [6, 7] are very accurately described by the expression

$$a(t) = \sum_{i=1}^{n} A_i(1 - e^{-a_i t}) + \alpha_L t. \tag{4}$$

The number of saturating exponential terms, n, usually does not exceed four. Also, the possibility that this expression is merely a good empirical curve-fitting formula is almost certainly eliminated by the occurrence, in different materials, of stage-I curves containing one exponential component or curves with two or three unambiguously resolvable exponential components.

An understanding of the various coloring curves that are obtained during irradiation and the changes occurring after irradiation can be obtained by considering the coloring of alkali halides (LiF is often an exception). While Eq. (4) can be regarded as an empirical equation, it can be readily derived from relatively simple considerations. The primary coloring processes is the conversion of traps, often called precursors, into color centers by charge capture. For example, vacancies are precursors in the sense that they can be converted to F centers by electron capture. Usually, prior to irradiation a crystal will contain a given concentration of precursors for each type of color center. Also, it is likely that more than one type of precursor may produce the same color center. For example, in the alkali halides F-center precursors may exist as (1) isolated defects, (2) vacancies coupled to one or more different kinds of impurities, (3) coupled vacancies with trapped charges, such as M or R centers, that separate during irradiation and become F centers or vacancies, and (4) additional vacancies may be introduced by radiation-damage processes. In other words, during irradiation the precursor concentration may be increased (or in rare cases decreased) by radiation-induced processes. Let N_0 be the preirradiation concentration of traps—that are converted to color centers by capturing ionization-produced electrons—K the rate that traps are introduced during irradiation (the assumption of a linear defect formation rate appears to apply to almost all material at low total doses [6,7]), f the fraction of empty traps converted to color centers per unit time, ϕ a measure of the radiation-induced ionization electron concentration (usually dose rate), and n the color center concentration at time t. During irradiation at time t the number of empty traps is $N_0 + Kt - n$ and the rate these empty traps are converted to color centers is

$$dn/dt = \phi f(N_0 + Kt - n) \tag{5}$$

which has the solution, with $n = 0$ at $t = 0$,

$$n = (N_0 - K/\phi f)\,(1 - e^{-\phi f t}) + Kt. \tag{6}$$

If the same center is formed by independent processes, e.g., from two different types of precursors, the growth is described by the superposition of a number of equations like (6), i.e., Eq. (4).

For long, or very long, irradiations this simple treatment must often be modified to include other processes such as electron–hole recombination. In other words, a color center may be converted back into a precursor by recombination with opposite-sign charges; e.g., during irradiation F centers can be converted to vacancies by recombination with holes.

Measurements made after irradiations are terminated show several different features. Such measurements are particularly meaningful if they are made with apparatus for studying color-center formation during irradiation [6,7]. First, in most materials, when the irradiation is terminated, the measured color-center concentrations undergo abrupt changes. In most crystals and glasses the coloring decreases rapidly. In others, one or more centers increase and in some they increase initially and then decrease. Often the changes occurring after irradiation can be resolved unambiguously into decreasing exponential components or into combinations of increasing saturating exponential and decreasing exponential components. Second, at room temperature the radiation-induced color centers usually do not completely disappear after irradiation. The coloring remaining a week or longer after irradiation ranges from a few percent to 98–99% of that present at the end of irradiation. In most materials the decrease is in the 20–60% range. Third, the removal of color after irradiation is almost always increased by raising the sample temperature. The changes occurring after irradiation indicate that one or more parameters in Eq. (6) include both growth and decay process [7]. Also, a large fraction of the radiation-induced coloring often can be removed by exposure to strong visible and/or ultraviolet light.

The formation of precursors that become color centers during irradiation, as mentioned above, is usually attributable to either the *displacement damage* [1] or the *ionization damage* [6–8] process. The latter is particularly important in some alkali halides. *Displacement damage* occurs when an incident particle makes an elastic (elastic event) or occasionally an inelastic (nuclear event) collision with a lattice atom that transfers sufficient energy, usually about 25 eV, to the struck atom to move it from its normal lattice position to a nearby interstitial position. Often the newly created nearby interstitial undergoes immediate recombination with the vacancy at its original lattice position. If the conditions are "right," e.g., if the interstitial is sufficiently removed, a stable vacancy–interstitial pair is formed. The vacancy and/or the interstitial can capture ionization-induced charges to become color centers. The incident bombarding particle, that initiates the process, is usually a neutral or charged particle. It must be emphasized that gamma rays also produce appreciable displacement damage; but by a two-step process. The first step is the formation of a recoil electron by photoelectric, Compton, or pair-formation processes. Once formed, the recoil electron gives up energy—primarily by ion-pair formation—as it moves through material. If sufficiently energetic, it can transfer enough energy to a lattice atom, by making an elastic collision, to create a vacancy–interstitial pair. Once it has degraded below the threshold energy, which is primarily a function of the initial recoil energy and lattice atom mass, it cannot transfer enough energy to displace atoms. Yet, this is a surprisingly efficient process. Very roughly, 1 out of 10^4–10^5 incident 1-MeV gamma rays will create a single displacement damage pair in most low-Z materials. In contrast, only 10–100 incident 1-MeV electrons are required to create a single displacement damage pair in the same material.

The importance of the *ionization damage* process has become widely accepted only in the last decade [6–8]. This process utilizes the energy released as a trapped exciton, i.e., a trapped hole–electron "Bohr-atom" like ion pair, undergoes recombination. Although this process is not completely understood, it is best described as it is presumed to occur in a typical alkali halide, namely, NaCl. The first step is the formation of a V_K center (a Cl_2^- ion in the space normally occupied by two adjacent Cl^- lattice ions) by the "self-trapping" of an ionization-produced hole. This step is accompanied by the relaxation, or distortion, of the nearest and next nearest lattice atoms. The V_K center is stable, or decays very slowly, at temperatures around 80 K and its lifetime decreases as the lattice temperature increases. In NaCl at 250–350°C it is long enough for the ionization damage process to proceed. The next step is the capture of an ionization electron by the V_K center to form a trapped exciton. It is likely that additional rearrangement of the atoms in and surrounding the V_K center occurs as the exciton is formed. The lifetime of the trapped-exciton V_K-center complex is quite short. It is likely that the next step, the recombination step, can occur in two distinct ways. First, it can decay radiatively by the emission of luminescence, i.e., the emission of a photon. In this case all of the energy available in the exciton—

and perhaps some contained in lattice distortion—is emitted as luminescence; or, a large part of the available energy is emitted as light and the rest nonradiatively as lattice phonons, i.e., heat. This sequence will restore the lattice to its original unaltered condition. Second, the decay step can occur by the available energy being transferred to one of the Cl atoms, in the Cl_2^- ion, in a way that provides enough energy for the atom to move to an interstitial position. The probability of this occurring is increased by the possibility that the Cl atom movement is assisted by the formation of a focused series of successive Cl atom displacements, in favored crystallographic directions, to produce an interstitial Cl^- ion four, five, or more Cl^- separations from the site of the original V_K center. Both theoretical and experimental evidence support the existence of such focused collision sequences. This second decay process creates a Cl^- vacancy and an interstitial Cl at a sufficiently distant site to create a stable vacancy–interstitial pair. The interstitial Cl atom is usually neutral, the valence electron having been trapped on the (or an equivalent) Cl^- vacancy to convert it to an F center. It is correct to say that the *ionization damage* process creates (well-separated, uncorrelated) F-center Cl^0 interstitial pairs.

The average energy required to produce a single F-center interstitial pair by the ionization damage process is extremely low compared to the energy required to produce the same defect by displacement damage. As mentioned above, the average energy required to produce an electron–hole ion pair ranges from twice the band-gap energy, in low-Z materials, to three times the gap energy in high-Z materials. Only a fraction of the ion pairs formed by incident radiation produce stable ionization damage: a large fraction of the electron–hole ion pairs formed undergo recombination without forming excitons, a fraction of the excitons radiatively recombine, a fraction of the vacancy–interstitial pairs immediately recombine, and other processes occur that compete with the ionization-damage process. An average of 400–500 eV is required to produce a single vacancy–interstitial pair in NaCl by ionization damage. In NaCl this process is 4,000 to 10,000 times more efficient than displacement damage. If the figure of merit for ionization damage in alkali halides is 4,000, it is roughly 2,000 for alkaline earth halides and very roughly 1 for fused silica. A technically important consequence of ionization damage is included in the Applications section.

COLOR-CENTER APPLICATIONS

Color-center research, because it was the first aspect of "defect solid-state physics" studied in detail, played a pivotal role in the development of a number of applications of solid-state concepts. Examples include luminescence, e.g., fluorescent lights; video display tubes; semiconductors; lasers; etc. Yet, only a few applications based directly on defect color centers can be cited. One application, which is commercially viable and directly utilizes defect color centers—in contrast to substitutional impurities or to molecules—is the *color-center laser* [10,11]. Color-center lasers are based on the same principles that apply to many other laser active materials, particularly dye lasers. These principles are described in laser articles. Quite a number of different color centers, in a variety of host crystals, have been shown to be laser active. These centers and the corresponding host crystals include (but are not limited to) the following: M^+, LiF; M^+, KF; M_A^+, KCl:Na; M_A^+, KCl:Li; $M^+:O^{2-}$, NaCl:O_2; F_B(II), KCl:Na; F_A(II), KCl:Li and F^+, CaO. Most color-center lasers operate in the infrared, particularly in the 1–3-μ region, some are useful in the visible, a few operate in the ultraviolet, and at least one in the vacuum ultraviolet.

Two other direct applications involving color centers are photochromic (spectacle) glasses that darken in strong ultraviolet, such as sunlight, and radiation dosimeters, especially for high total dose applications, that are based on the radiation-induced color-center formation kinetics (coloring) described above.

The development of optical devices that operate in intense light or in non-negligible x-ray and/or nuclear radiation fields, particularly if they include light-transmitting elements such as lenses and/or windows, must always include steps to demonstrate that they will not be rendered inoperable by radiation damage; more specifically, if they are susceptible to non-negligible radiation-induced color-center formation [6]. The number of devices subject to this kind of radiation damage is surprisingly large. Included are satellites and space probes (cosmic rays and Van Allen belt radiation), optical viewers for reactors and hot cells, fiber optics (some fibers are exposed to radioactivity from soil and the seabed), and particle detectors for high-energy physics research—especially in colliding beam experiments where the detectors must be located close to the intersecting accelerator beams.

A most surprising application of color centers has emerged in the search for a permanent way to dispose safely of radioactive waste [7,11]. About three-fourths of the initial disposal effort assumed repositories would be located in the worldwide natural rock salt formations, occurring 1–3 km below the surface and often more than 1 km thick, that cover much of Europe and the southwest United States. Salt deposits were considered resistant to radiation damage. However, recent studies show that both natural and synthetic rock salt is particularly susceptible to radiation induced F-center and colloid center, i.e., X-band, formation because of the ionization damage process. Since one Cl vacancy and interstitial atom pair are formed for each 400–500 eV (not MeV) deposited in the salt by radiation from the radioactive waste, the damage formation rate is very high. Curves of F-center concentration versus dose increase monotonically to a plateau that "levels off" at 10^6–10^7 rad. As the F centers reach saturation a sodium metal colloid absorption band appears that is accurately described by classical nucleation and growth curves with induction period and power law growth. The colloid particle nucleation is related to dislocations; the induction period is reduced markedly by straining samples prior to irradiation. Colloid particles grow by the diffusion of F centers and Cl^- vacancies to particle surfaces. Because of the high defect formation rate, typical waste doses of 10^{10} rad will convert 1–10% of the irradiated salt to colloid in about 400 years and 10^{11} rad will convert 10–50%. When salt irradiated to 10^{10} rad is dissolved in brine, pH values of 12 to 14 are obtained. Such solutions are highly corrosive and make it difficult to find acceptable corrosion-resistant materials to contain the waste.

This color-center and colloid particle formation process is only partially understood [7,11]. The colloid formation rate is low at 90°C, increases to a maximum at 150–170°C, and decreases to a negligible level at 250–350°C. The colloid formation rate *increases* as the dose rate decreases. Measurements made during irradiation show that decreasing (increasing) the dose rate, after an initial irradiation, decreases (increases) the F-center concentration and increases (decreases) the colloid formation rate. After irradiation, large changes occur in the F-center and colloid band intensities that depend primarily on irradiation temperature.

This is one of many examples of research pursued for purely basic reasons, in this case the study of ionization damage, that played an important role in a practical and pressing national and international problem, the disposal of radioactive waste. The studies provided information on the levels of damage expected in natural salt and explained why they were so high; effects that had not been included in the original plans for radioactive waste repositories in salt.

ADDITIONAL INFORMATION ON COLOR CENTERS

The information on color centers given in this article is a very much abbreviated overview of the subject. Many aspects of color-center physics have been mentioned but not described and others are not included. Examples are: electron spin resonance (ESR); electromechanical and magneto-optical properties; luminescence, which includes fluorescence, phosphorescence, and thermoluminescence; ionic, electronic, and photoconductivity; diffusion of defects producing color centers; thermal annealing; dielectric properties; and strain-induced reorientation. Only trivial color-center theory is included. The bibliography includes books that contain advanced level information on almost all aspects of color-center physics.

See also Crystal Defects; Electron Energy States in Solids and Liquids; Electron–Hole Droplets in Semiconductors; Excitons; Lasers; Luminescence; Radiation Damage in Solids; Radiation Interaction with Matter; Thermoluminescence.

BIBLIOGRAPHY

1. J. H. Crawford, Jr. and L. M. Slifkin (eds.), *Point Defects in Solids,* Vols. I, II, III. Plenum, New York. 1972.
2. W. B. Fowler (ed.), *Physics of Color Centers.* Academic, New York, 1968.
3. J. J. Markham, *F Centers in Alkali Halides.* Academic, New York, 1966.
4. J. H. Schulman and W. D. Compton, *Color Centers in Solids.* Macmillan, New York, 1962.
5. A. M. Stoneham, *Theory of Defects in Solids.* Clarendon Press, Oxford, London, and New York, 1975.
6. P. W. Levy, *Radiation Effects in Optical Materials, SPIE Proc. 541* (P. W. Levy, ed.), pp. 2–24. SPIE, Bellingham, WA, 1985.
7. P. W. Levy, *Defect and Impurity Center in Ionic Crystals—I, Optical and Magnetic Properties* (P. W. M. Jacobs, ed.), *J. Phys. Chem. Solids:* Special Topics Issue, in press.
8. F. C. Brown and N. Itoh (eds.), *Recombination Induced Defect Formation in Crystals. Semiconductors and Insulators:* Special Topics Issue 5, 1983.
9. L. F. Mollenauer, *The Laser Handbook* (M. Stich and M. Bass, eds.), Chap III. North-Holland, Amsterdam, 1985.
10. L. F. Mollenauer, *Topics in Applied Physics,* Vol. 59 (L. F. Mollenauer and J. C. White, eds.), Springer-Verlag, Berlin, 1987.
11. P. W. Levy, J. M. Loman, K. J. Swyler, and R. W. Klaffky, *The Technology of High-Level Nuclear Waste Disposal, 1* (P. L. Hofmann, ed.), pp. 136–167. Technical Information Center, U.S. Dept. of Energy, Oak Ridge, TN, 1981.

Combustion and Flames

D. J. Seery

Combustion is usually understood to mean any self-sustaining exothermic process involving the reaction of a fuel with an oxidizer. When visible light is also produced, the term flame is often applied. The type of fuel and the conditions under which the fuel and oxidizer air are brought together are determined by the intended application of the combustion process, which may include domestic and industrial heating, electric power generation, or transportation. Other manifestations of combustion include such areas as incineration, explosives, rocket propulsion, unwanted fires, and some illuminating devices. The manner in which the heat release from combustion is transformed into useful kinetic energy will differ with end use, for example, in gas turbines, boilers, piston engines, or other applications.

Combustion processes have been the major source of energy for our civilization. While this situation is likely to continue for the foreseeable future, we must also contend with the polluting exhausts from these combustion systems (even as we extend their use to new areas such as incineration of hazardous wastes). In addition, the infrared-active combustion products CO_2 and NO_2 may be producing significant "greenhouse" warming of the earth's surface. This ominous possibility must be included with the pollution of the lower atmosphere as major deleterious effects of combustion. Resolving these problems of combustion exhaust products requires a quantitative understanding of the details of combustion. Because many of the present-day combustion problems often relate to species in very low concentrations (ca. <100 ppm), control strategies require specific information on chemical reactions and species not normally considered when only bulk effects are being investigated.

Both chemical and physical processes are combined in combustion and either may dominate in different applications. For example, most practical combustion involves diffusion flames: the reactants are initially unmixed and the progress of the reaction is dominated by the rate of mixing. For liquid fuels, the mixing is also dependent on atomization and vaporization which may vary greatly. In most combustion applications the overall progress of the reaction is determined by a combination of mass transport, heat transport, and chemical reaction rates.

One of the simplest cases of combustion occurs when a gaseous mixture of fuel and oxidizer is placed in an adiabatic vessel and is heated by the chemical reaction. If the temperature of the gases rises to a value, T, which is higher than the wall temperature, T_0, then the rate of heat loss by conduction to the walls, Q_L, is proportional to the difference between them:

$$Q_L = hS(T - T_0),$$

where h is the heat transfer coefficient and S is the surface area of the vessel. The rate of heat production, Q_G, is proportional to the reactant concentrations and the reaction rate which has an exponential dependence on temperature

$$Q_G = f(N)H_0Ae^{-E/RT},$$

where H_0 is the heat of reaction, $f(N)$ is an appropriate function of concentration, and $Ae^{-E/RT}$ is the kinetic rate constant (KINETICS, CHEMICAL). The energy conservation equation can be written

$$\rho C_v \frac{dT}{dt} = f(N)H_0Ae^{-E/RT} - hS(T - T_0),$$

where the left-hand side represents the energy accumulation in the gases with density ρ and heat capacity C_v.

For an idealized system, plots of these heat-rate terms are shown in Fig. 1. The rate of heat loss is a straight line passing through T_0. The rates of heat production form a family of curves related to the concentrations (pressures) of the reactants. If the reactants are admitted into the container at concentrations corresponding to curve A, the heat gain exceeds the heat loss and reaction will heat the system to temperature T_1. Here the curves cross and a stable situation is attained where reaction proceeds without self-acceleration. If the gas is heated by an additional source to the second intersection T_2, the reaction autoaccelerates and ignition occurs. If the reactant concentrations correspond to curve C, the mixture immediately autoaccelerates to ignition since $Q_G > Q_L$. For the conditions of curve B, the heat-loss line forms a tangent to the heat-gain curve, the intersecting points T_1 and T_2 coinciding at T_3 for these conditions. The reaction system will heat slowly to T_3, then rapid acceleration will occur. Curve B represents the minimum conditions for self-ignition. Such an energy balance helps to explain the critical limits observed for flame phenomena. For example, for mixtures of fuels with air or oxygen there are limits of composition for which flame propagation can occur and outside of which no self-sustaining flame can be initiated. It appears that for many flames these composition limits correspond to some minimum final flame temperature achieved in the combustion. For methane (CH_4) this temperature is about 1400 K. Outside the composition limits ("flammability limits"), the flame is not hot enough to be self-sustaining.

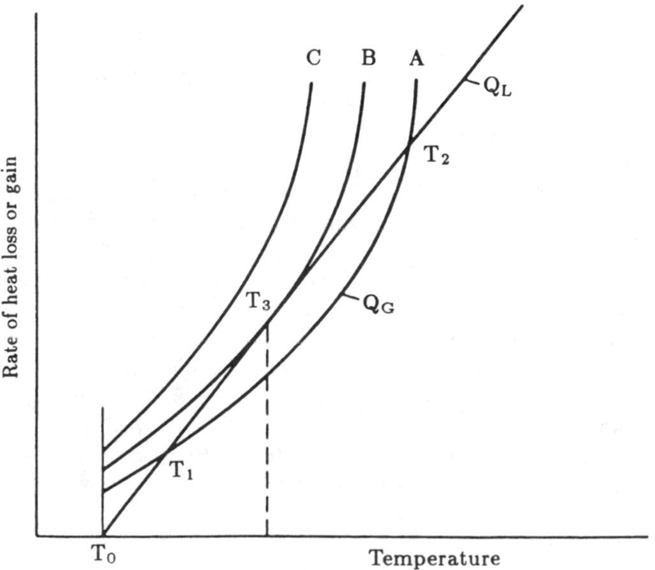

FIG. 1. Heat balance in a closed vessel with exothermic reaction.

At the elementary level, the chemistry of oxidation reactions and the transport properties of the gaseous species control ignition, heat release rate, and the rate at which a flame moves through a combustible mixture. Probably the simplest arrangement for studying a propagating flame, either experimentally or by computer modeling, is a premixed flat flame often operated at low pressures (~5 kPa). The low pressure in effect lowers the concentration of reactants thus slowing down the overall combustion process and stretching out the reaction zone. In this configuration the temperature profile and the species concentration profiles can be used to delineate the entire combustion process, including chemical reactions, molecular diffusion, and thermal conduction. The most popular measurement techniques for species concentration include optical spectrometry *in situ* or sample extraction by a probe followed by chemical analysis using gas chromatography or mass spectrometry. Optical spectrometry has the advantage of being nonintrusive and can be highly accurate but is generally limited to only a few of the important species. Extractive probe sampling has the advantage of using analytical techniques that can measure all of the stable species and, in the case of mass spectrometry, even atoms and radicals. An example of the concentration profiles for major species in a methane (17.9%)–oxygen (31.6%)–argon (50.5%) flame burning at 4.67 kPa (35 Torr) is shown in Fig. 2. The flame is slightly rich; that is, it contains excess fuel or, stated differently, the oxygen concentration is inadequate to burn all the fuel to CO_2 and H_2O. These data were obtained using a mass spectrometer coupled to a molecular beam sampling system.

An examination of the profiles reveals the fairly smooth decline of the reactants CH_4 and O_2 and the rise of the products H_2O and CO_2. The intermediate species CO and H_2 are also produced in quantity and continue downstream relatively unchanged because of the lack of O_2. These data can be conceptually divided into (a) a preheat zone, for the first 2–3 mm above the burner, during which the reactants are heated and critical concentrations of atoms and radicals accumulate either from reaction or by diffusion; (b) a reaction zone, from 3–7 mm above the burner, where rapid reaction takes place and the steepest temperature and concentration gradients are found, and (c) an equilibration zone, greater than 7 mm above the burner, where the concentrations of species approach thermodynamic equilibrium.

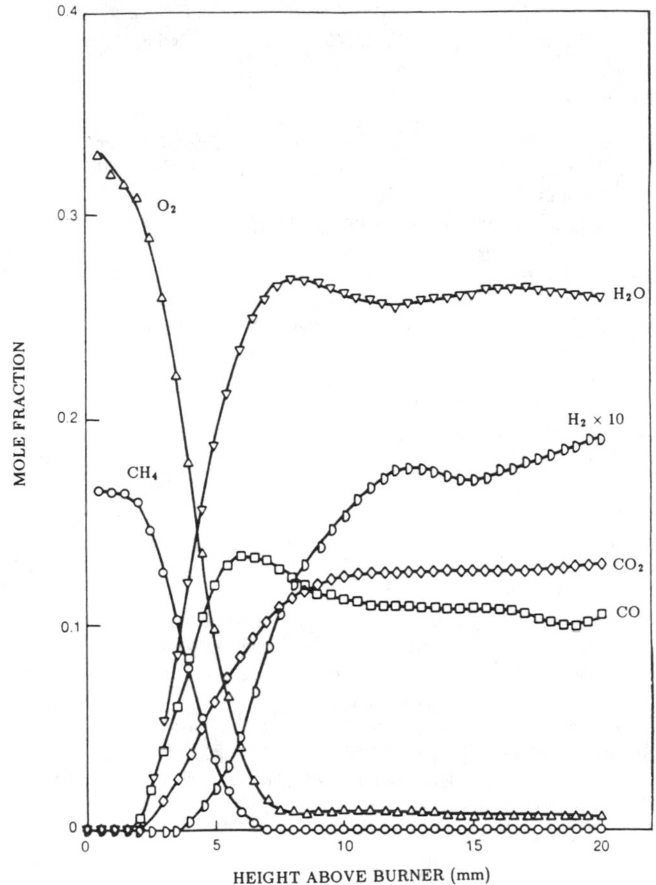

FIG. 2. Major species concentration profiles—methane flame.

Detailed profiles, such as these, form the main testing ground for models of the reaction mechanisms of combustion chemistry. At this writing the analytical models can reasonably predict the major species profiles in simple flames. Detailed calculations of minor species profiles are more challenging and are the subject of active research.

For the case of nonpremixed fuel and oxidizer, the combustion process is largely controlled by mixing rates. This class of flames contains most practical applications including the combustion of gaseous, solid, and liquid fuels. For confined, slow parallel flows, if the fuel is flowing upward through a central tube and the oxidizer is flowing upward through a concentric tube, then a laminar diffusion flame can be stabilized at the end of the tube where the reactants begin mixing. In these flames the mixing is by molecular diffusion and the reactants are all consumed at the interface. A simplified diagram of the radial concentration profiles at some distance above the burner tubes is shown in Fig. 3.

In many practical systems the fuel is introduced as a spray of droplets, and the burning is rapid, the flows are fast, and mixing is associated with aerodynamic turbulence (see TURBULENCE). In these cases the combustion process is dominated by the aerodynamics, except in the case of extreme turbulence where mixing may become so rapid that chemical reaction kinetics limit the combustion rates.

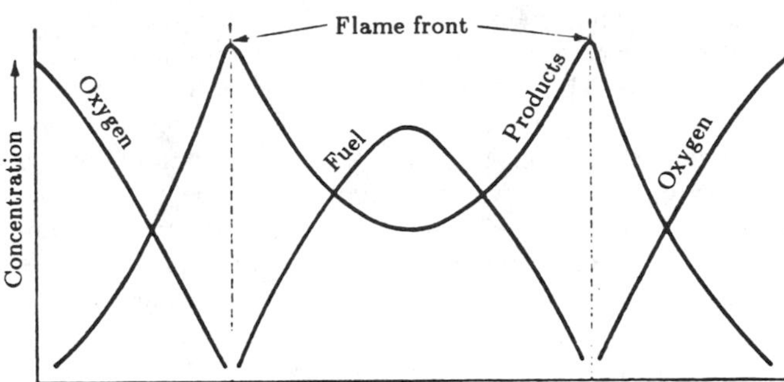

FIG. 3. Concentration profiles through a laminar diffusion flame.

Use of liquid and solid fuels further complicates combustion systems in that rates of vaporization, heat and mass transport, thermal decomposition, or surface reactions may assume key roles in the control and/or the description of the combustion process.

See also HIGH TEMPERATURE.

BIBLIOGRAPHY

A. G. Gaydon and H. G. Wolfhard, *Flames: Their Structure, Radiation and Temperature,* 4th ed. Wiley, New York, 1979. (E)

B. Lewis and G. von Elbe, *Combustion, Flames and Explosion of Gases,* 3rd ed. Academic Press, New York, 1987. (I)

I. Glassman, *Combustion,* 2nd ed. Academic Press, New York, 1987. (I)

R. A. Strehlow, *Combustion Fundamentals.* McGraw-Hill, New York, 1984. (I)

F. A. Williams, *Combustion Theory,* 2nd ed. Benjamin/Cummings, Menlo Park, 1985. (A)

Complementarity

Jeffrey Bub

The concept of complementarity forms the core of Niels Bohr's version of the Copenhagen interpretation of quantum mechanics, the standard or orthodox interpretation of the theory. Early versions of the quantum theory emphasized a wave-particle duality: microsystems seem to manifest wave aspects under certain experimental conditions and particle aspects under other conditions. The uncertainty principle, derivable from an analysis of a wave packet as a representation of a relatively localized system and as a theorem in quantum mechanics, may be regarded as a formal expression of this duality. It says, in effect, that there are certain quantities, like position and momentum, that necessarily satisfy a reciprocal relationship with respect to the precision with which these quantities can be specified in any possible experimental arrangement. More precisely, if a measuring instrument fixes the position q of a system to within a certain latitude Δq, then the momentum p can be fixed only to within a latitude Δp, where Δp is proportional to $1/\Delta q$ (the proportionality constant being a multiple of Planck's constant). Now, this reciprocal relation between position and momentum might reflect a necessary disturbance of position values in any momentum measurement, and conversely, as required by the existence of a quantum of action that is non-negligible at the microlevel, or it might concern the very definition of these quantities. On Bohr's view, it follows from the finite value of the quantum of action that the conditions for the precise applicability of the concept "position" exclude the conditions for the precise applicability of the concept "momentum." He terms the relationship between quantities like position and momentum "complementary," since precise values for both quantities are required for a complete specification of the classical state of the system.

For Bohr, the wave and particle pictures of the early quantum theory are complementary, in the sense that they involve different aspects of a single physical system revealed under mutually incompatible experimental conditions. These experimental conditions must be described classically, i.e., on Bohr's view macrosystems *qua* measuring instruments are necessarily characterized in terms of the concepts of classical physics (like position and momentum). It follows that an interaction between a microsystem and a measuring instrument cannot be sharply separated from the undisturbed or independent behavior of the microsystem itself, because the classical description of the functioning of the measuring instrument means that the unavoidably finite action of the microsystem on the measuring instrument cannot be controlled more accurately than is compatible with the uncertainty principle. For this reason, the interaction between a microsystem and a macroscopic measuring instrument forms an integrable part of the behavior of the microsystem as revealed in the interaction. Bohr used the term "phenomenon" to refer to an observation obtained under a complete specification of classically described experimental conditions, including an account of the whole experimental arrangement. Such a quantum phenomenon cannot be "subdivided" by a further experimental arrangement, because every such attempt would yield a new phenomenon incompatible with the original phenomenon. In this sense, a quantum phenomenon exhibits a feature of "wholeness" or "individuality." The concept of complementarity characterizes phenomena which require mutually exclusive experimental conditions, such as the conditions for space-time coordination and the conditions for the applicability of dynamical conservation laws, both of which are required for the complete specification of a classical mechanical state.

Bohr regarded the concept of complementarity as a generalization of causality appropriate for the description of systems whose undisturbed behavior cannot be sharply separated from the interactions with the measuring instruments that reveal this behavior. The deterministic description of classical physics characterizes the behavior of systems for which the quantum of action can be neglected in measurement interactions. The statistical character of the quantum-mechanical description reflects the fact that the behavior of microsystems can only be revealed in complementary phenomena and is to that extent irreducible. The principle of complementarity provides an interpretation of quantum mechanics, insofar as it explains why no possible measurement can fix canonically conjugate quantities like position and momentum with greater accuracy than that given by the uncertainty principle.

See also History of Physics; Philosophy of Physics; Quantum Mechanics; Uncertainty Principle.

BIBLIOGRAPHY

D. Bohm, "On Bohr's Views Concerning the Quantum Theory," in *Quantum Theory and Beyond* (T. Bastin, ed.). Cambridge, London and New York, 1971. (I)

N. Bohr, (a) *Atomic Theory and the Description of Nature*. Cambridge, London and New York, 1934. (I)

(b) *Atomic Physics and Human Knowledge*. Wiley, New York, 1958. (I)

(c) *Essays 1958–1962 on Atomic Physics and Human Knowledge*. Wiley (Interscience), New York, 1963. (I)

P. K. Feyerabend, (a) "Complementarity," *Aristotelian Soc. Suppl.* **32,** 75–104, (1958). (I)

(b) "Niels Bohr's Interpretation of the Quantum Theory," in *Current Issues in the Philosophy of Science* (H. Feigl and G. Maxwell, eds.). Holt, Rinehart & Winston, New York, 1961. (E)

(c) "On a Recent Critique of Complementarity," Parts I and II, *Philos. Sci.* **35,** 309–331 (1968); **36,** 82–105 (1969). (A)

H. J. Folse, *The Philosophy of Niels Bohr*. North-Holland, Amsterdam, 1985.

W. Heisenberg, "Quantum Theory and Its Interpretation," in *Niels Bohr* (S. Rozental, ed.). North-Holland, Amsterdam, 1967. (E)

M. Jammer, *The Conceptual Development of Quantum Mechanics*. McGraw-Hill, New York, 1966. (I)

A. Petersen, *Quantum Physics and the Philosophical Tradition*. M.I.T. Press, Cambridge, Mass., 1968. (E)

E. Scheibe, *The Logical Analysis of Quantum Mechanics*. Pergamon, New York, 1973. (A)

Compton Effect

Karl Berkelman

The elastic scattering of a photon by an electron is called the Compton effect when the interaction can be considered as the collision of two otherwise free particles. This is true when the energy $h\nu$ of the photon is comparable to or higher than the rest energy m_ec^2 of the electron (0.51 MeV).

The scattering of electromagnetic radiation of lower quantum energies by electrons was explained by J. J. Thomson in terms of the classical oscillatory motion of the electron in the incident electromagnetic wave field and the consequent reradiation of the absorbed energy. In this picture the scattered radiation has the same frequency as that of the motion of the electron, which in turn is equal to the frequency of the incident wave.

However, in 1923 A. H. Compton observed x rays scattered at various angles in thin targets of light elements and noted that the scattered wave lengths were longer than the incident wavelength and increasing with scattering angle. This was in fact one of the decisive early demonstrations of the quantum theory. For if we treat the x rays as particles (called photons) having energy $h\nu$ and momentum $h\nu/c$, and apply energy and momentum conservation to the photon–electron collision (using relativistic kinematics), we can solve for the final photon energy:

$$h\nu' = \frac{m_ec^2}{1+\cos\theta + m_ec^2/h\nu},$$

which is reduced from the incident energy by the energy of electron recoil. The corresponding wavelength shift

$$\lambda' - \lambda = (h/m_ec)(1-\cos\theta)$$

reproduces Compton's observed dependence on scattering angle. The quantity h/m_ec has been named the "Compton wavelength of the electron." Note that in the low-energy limit, in which $h\nu \ll m_ec^2$ (or $\lambda = c/\nu \gg h/m_ec$ in terms of wavelength), the classical Thomson relation between incident and scattered frequencies (or wavelengths) holds. Compton's experiment is a beautiful example of the wave–particle duality in quantum mechanics. The particle nature of electromagnetic radiation is demonstrated by wavelength measurements.

In one of the earliest applications of relativistic quantum mechanics, Klein and Nishina calculated in 1928 the differential cross section (i.e., the probability of scattering through an angle θ into unit solid angle, in a target of unit thickness containing one electron per unit volume) for Compton scattering:

$$\frac{d\sigma}{d\Omega} = \frac{r_e^2}{2}\left(\frac{\nu'}{\nu}\right)^2\left[\left(\frac{\nu}{\nu'}+\frac{\nu'}{\nu}-\sin^2\theta\right)\pm\left(\frac{\nu}{\nu'}-\frac{\nu'}{\nu}\right)\cos\theta\right].$$

Here $r_e = e^2/m_ec^2 = 2.82\times10^{-13}$ cm, and is called the classical radius of the electron. The ± on the last term apply if the initial photons are right or left circularly polarized, respectively. The differential cross section for scattering of unpolarized photons (plotted in Fig. 1) is obtained by omitting the ± term. At low energies $h\nu$ it becomes the classical Thomson cross section $\frac{1}{2}r_e^2(1+\cos^2\theta)$, while at higher energies the scattering is strongly peaked forward. For plane-polarized photons the differential cross section is

$$\frac{d\sigma}{d\Omega} = \frac{r_e^2}{2}\left(\frac{\nu'}{\nu}\right)^2\left(\frac{\nu}{\nu'}+\frac{\nu'}{\nu}-2\sin^2\theta\cos^2\phi\right)$$

where ϕ is the angle between the plane of the incident electric vector and the scattering plane. The yield is a maximum when the photon scatters in the plane normal to the polarization plane. The total cross section, integrated over all angles, decreases from the classical Thomson value, $(8\pi/3)r_e^2 = 0.84\times10^{-24}$ cm², monotonically as the photon energy is increased. The probability that a photon will survive without interaction after passing through x cm of material is $e^{-\mu x}$, where the attenuation coefficient μ is the product of the total cross section and the number of electrons per unit volume of material. In Fig. 2 is plotted the total attenuation coefficient for aluminum and lead, including the effects of the photoelectric effect, which dominates at low energies $h\nu$, and of electron–positron pair production, important at high energies. Compton scattering is the dominant photon attenuation process only in materials of low atomic number Z and for energies $h\nu$ near 1 MeV. Since the scattering probability per atom is proportional to number of electrons Z, it becomes overshadowed by photoelectric absorption, proportional to Z^5, and pair production, proportional to Z^2, as Z becomes large.

The elastic scattering of photons from charged particles other than the electron is often called Compton scattering also. The kinematic relations between incident and scattered photon energies and scattering angle are changed only by substituting the appropriate particle mass for the electron mass. For photons of low energy the proton Compton scattering cross section is again the Thomson cross section, re-

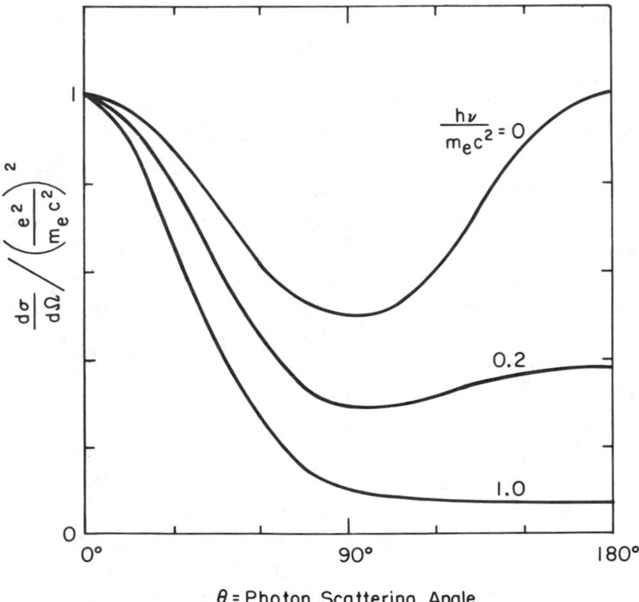

FIG. 1. The differential cross section for Compton scattering of unpolarized photons, expressed in terms of $(e^2/m_ec^2)^2 = 7.95\times10^{-26}$ cm², plotted as a function of the photon scattering angle for several values of the incident photon energy $h\nu$ divided by the electron rest energy $m_ec^2 = 0.51$ MeV.

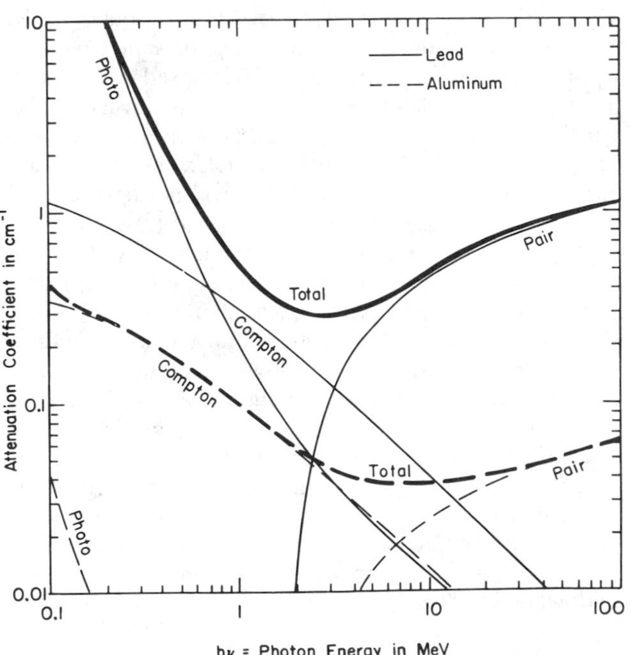

FIG. 2. The attenuation coefficient μ for a typical light and heavy element, plotted as a function of incident photon energy $h\nu$.

duced, however, by the square of the electron–proton mass ratio. At energies above the pion photoproduction threshold (about 150 MeV) proton Compton scattering occurs mainly through the excitation and radiative decay of the various nucleon resonance states observed in pion–proton scattering and pion photoproduction. Above 1 GeV, proton Compton scattering becomes diffractive and is understood in terms of intermediate states involving vector mesons, such as the ρ^0, ω, and ϕ.

See also SCATTERING THEORY.

BIBLIOGRAPHY

An elementary discussion of the Compton effect is given in every introductory modern physics text; see, for example, H. Semat and J. R. Albright, *Introduction to Atomic and Nuclear Physics*, 5th ed., pp. 143–149. Holt, Rinehart & Winston, New York, 1972. (E)

R. D. Evans, *The Atomic Nucleus*, Chapter 23. McGraw-Hill, New York, 1955. (I)

The relativistic quantum-mechanical description of the Compton effect is discussed in many advanced texts; see, for example, J. D. Bjorken and S. D. Drell, *Relativistic Quantum Mechanics*, pp. 127–132. McGraw-Hill, New York, 1964. (A)

R. D. Evans, "Compton Effect," in *Encyclopedia of Physics* (S. Flügge, ed.), Vol. 34, pp. 234–298. Springer-Verlag, Berlin and New York 1958. (A)

Conduction

S. Kirkpatrick and C. J. Lobb

Conduction processes in condensed matter consist of the transport of heat, electric charge, mass, or some combination of the three, in response to an imposed temperature gradient, electric field, or density gradient. (Convection, distinguished among conduction processes by the fact that macroscopic motion of all the atoms or molecules occurs, also contributes to transport in fluids, but will not be discussed here.) Conductivities vary more widely among materials than do other physical properties. While the densities of the liquids and solids which form at room temperature vary over roughly one decade, electrical conductivities vary by 20 decades or more from metals to insulators, and may be infinite in superconductors at sufficiently low temperatures. At cryogenic temperatures, thermal conductivities also exhibit a range of many decades, with metals offering the highest thermal conductivities, and insulators the lowest.

The usefulness of specific materials is very often due to their high or low conductivities. Discovery of new materials with extreme conductivities frequently opens novel applications, e.g., superconducting magnets for producing very large magnetic fields. But here we shall be concerned with conduction, especially at low temperatures where the largest variations are observed, as a probe of the microscopic particlelike excitations which carry the heat, charge, or mass currents. These particles are most commonly phonons (long-wave length mechanical vibrations) or electrons, but ions may contribute (*see* ELECTROCHEMISTRY) and Cooper pairs carry the lossless current in superconductors (*see* SUPERCONDUCTIVITY THEORY).

A semiclassical description, first employed in the kinetic theory of gases, is convenient for discussing conduction by electrons or ions. A mobility μ is defined as the mean carrier drift velocity induced by a unit applied force. For electrons,

$$\mu \equiv \langle v \rangle / e\mathscr{E} = \tau / m^*,$$

where τ is a relaxation time (10^{-14} s is typical for electrons in metals at room temperature), and m^* the effective mass of an electron moving in the attractive potential of the ions in the material. The conductivity σ is obtained from the associated electric current $j^{(e)}$ by

$$\sigma \equiv j^{(e)}/\mathscr{E} = ne^2\tau/m,$$

with n being the density of carriers and e being the electronic charge. Diffusion processes, in which a particle current $j^{(d)}$ is induced by a density gradient, are characterized by a diffusion coefficient D:

$$j^{(d)} = -D\nabla n.$$

The two types of process are intimately connected. The Einstein relation,

$$\mu = eD/k_B T,$$

holds for any particles which obey classical Maxwell–Boltzmann statistics, for example, electrons and holes in semiconductors, or ions in electrolytes.

The thermal conductivity κ is defined from the heat current j^q which flows in response to a temperature gradient:

$$j^{(q)} \equiv -\kappa \nabla T.$$

A useful expression for κ is

$$\kappa = \tfrac{1}{3} C_V v^2 \tau,$$

where C_V is the specific heat at constant volume, and v is the instantaneous velocity of the particle. For electrons, v is the Fermi velocity; for phonons, the speed of sound is appropriate in this expression.

Both n and τ can vary widely with temperature and from one material to another. In insulators, n vanishes as the temperature tends to zero. If E is the excitation energy required to create a free carrier, then n is proportional to $\exp(-E/k_B T)$. "Arrhenius plots" of $\ln(\sigma)$ against $1/T$ yield straight lines for most semiconductors, and provide a means of determining E. Heat is transported principally by the phonons in insulators. These scatter only from defects or sample boundaries at low temperatures, so in this limit $\kappa \propto C_V \propto T^3$ is observed.

In metals, n and v are not sensitive to temperature, but τ can be. Phonon scattering alone gives $\tau \propto T^{-5}$, while scattering from impurities gives a constant contribution to τ. As a result, in a nonsuperconducting metal σ tends to a constant at low temperatures while $\kappa \propto T$.

In noncrystalline solids, such as insulating or semiconducting glasses, localized electronic and vibrational excitations introduce novel transport properties. The dominant electrical conduction mechanism at low temperatures may involve "hopping" of carriers between spatially separated long-lived states, assisted by the absorption of phonons. The characteristic distance and phonon energy involved in hopping change in a complicated way with decreasing temperature. The result, first explained by N. F. Mott, is

$$\ln \sigma \propto T^{-1/4}.$$

Localized vibrations give rise to a phonon scattering time $\propto T^{-1}$ in glasses. As a result, κ is proportional to T^2 at the lowest temperatures.

One additional idea should be mentioned which is not included in the classical theory of electronic conduction. At cryogenic temperatures, electrons maintain their quantum-mechanical phase coherence over longer and longer distances as inelastic scattering becomes less frequent. The resulting quantum-mechanical interference causes a number of interesting effects. One of these effects is localization: the backscattering rate is effectively enhanced due to interference, leading to a reduced σ as the temperature is lowered. Another is the occurrence of quantum conductance fluctuations: when the magnetic field applied to a sample is changed, the phases of the wave functions are shifted, altering the amount of interference. This leads to noisy, seemingly random variations in σ as the field is varied, except that the noise reproduces itself as the magnetic field is swept back and forth.

See also DIFFUSION; ELECTROCHEMISTRY; INSULATORS; METAL–INSULATOR TRANSITIONS; METALS; PHOTOCONDUCTIVITY; SEMICONDUCTORS, CRYSTALLINE; SUPERCONDUCTIVITY THEORY; TRANSPORT PROPERTIES; TUNNELING.

BIBLIOGRAPHY

A good short discussion of transport is given by C. Kittel and H. Kroemer in *Thermal Physics*, 2nd ed., pp 397–406, W. H. Freeman and Company, New York, 1980. A recent review on localization is Gerd Bergman, "Weak Localization in Thin Films, a Time of Flight Experiment with Conduction Electrons," *Phys. Rep.* **107**, 58 (1984). Quantum conductance fluctuations are reviewed in Sean Washburn and Richard A. Webb, "Aharonov–Bohm Effect in Normal Metal Quantum Coherence and Transport," *Adv. Phys.* **35**, 375 (1986).

Conservation Laws

Lincoln Wolfenstein

A conservation law equates the value of a physical quantity in the initial state to its subsequent values, in particular to its value in the final state, for some process. The great importance of conservation laws is that they provide significant constraints on complicated processes for which a detailed mathematical description may be practically impossible.

The simplest form of a conservation law involves scalar quantities. The conservation of the sum of the kinetic energies and potential energies of a set of interacting particles provides an extremely useful integral of the motion in classical mechanics. The basic discovery leading to modern chemistry in the 1800s was the discovery of the conservation of mass, equating the total mass of the reagents to that of the products for a chemical reaction. The theory of special relativity requires a modification of these two conservation principles because of the possible interconversion of mass and energy. Although this modification is quantitatively insignificant for chemical reactions, it is of essential importance in nuclear and elementary-particle reactions. The law appropriate for these reactions is the *conservation of the relativistic energy*, sometimes called mass-energy:

$$\sum_i (T_i + m_i c^2) = \sum_f (T_f + m_f c^2), \qquad (1)$$

where T_i and m_i are the kinetic energy and rest mass of an initial particle and T_f and m_f are for a final particle.

As a simple application, consider the photodisintegration of the deuteron into a neutron and a proton,

$$\gamma + d \rightarrow n + p,$$

where Eq. (1) states (for a deuteron at rest) that

$$E_\gamma + M_d c^2 = M_n c^2 + M_p c^2 + T_n + T_p. \qquad (2)$$

The minimum γ-ray energy E_γ for this reaction is 2.3 MeV, which corresponds to the fact that the mass of the neutron plus that of the proton is greater than that of the deuteron. Note that the potential energy between the neutron and proton and their kinetic energy inside the deuteron are *not* included in this equation because the deuteron is considered a single particle. An alternative approach is to consider the deuteron a composite system bound by nuclear forces. In this approach we can write

$$M_d c^2 = M_n c^2 + M_p c^2 + \langle V_{np} \rangle + \langle T_n \rangle + \langle T_p \rangle, \qquad (3)$$

where $\langle V_{np} \rangle$ is the average value of the potential energy associated with the nuclear force inside the deuteron and $\langle T_n \rangle$, $\langle T_p \rangle$ are average values of the kinetic energies inside the deuteron.

In elementary-particle physics there are much more dramatic examples of the conversion of mass into energy. In the annihilation of a positron (e^+) by an electron (e^-), the total mass is converted into the energy of gamma rays:

$$e^+ + e^- \rightarrow \gamma + \gamma,$$

so that for annihiliation at rest each γ-ray is emitted with the energy $m_e c^2 = 0.51$ MeV.

Every elementary particle, and therefore every system of particles, can be characterized by a set of numbers, often referred to as quantum numbers. The most familiar of these are charge Q, baryon number B, lepton numbers L_e and L_μ, and strangeness S (or hypercharge Y). For a set of particles the value of one of these numbers is the sum of the values for the members of the set. In any physical process each of these numbers is conserved except that S (or Y) is not conserved for the weak interaction. Since antiparticles can combine with particles to produce photons (which have a zero value for each of these numbers), these quantum numbers for an antiparticle must be the negative of their values for a particle.

The most familiar of these rules is the *conservation of electric charge Q*. The constituents of ordinary matter, proton, neutron, and electron, have electric charges Q of $+1$, 0, and -1. The conservation law states that the algebraic sum of the charges of all the particles does not change during a reaction. A very intriguing aspect of charge conservation is that it is the electric charge that determines the strength of the interaction of the particle with the electromagnetic field. In terms of Q this strength is given by Qe where $e = 1.5 \times 10^{-19}$ C. This suggests a fundamental relation between a conserved quantity and a particular form of interaction. This relation is embodied in the concept of gauge invariance in field theory.

The reason that only a small fraction (<1%) of the mass is converted into energy in nuclear reactions is the conservation of the number of nucleons (neutrons plus protons) in interactions involving normal matter. The more general form of this law is the *conservation of baryon number B*; this law states that the number of baryons minus the number of antibaryons is conserved in any reactions. A baryon is an elementary particle with $B = 1$, such as the neutron or proton or the strange baryons Λ, Σ, Ξ, and Ω. An antibaryon is the antiparticle of one of these, such as the antineutron or antiproton. According to the quark model the elementary constituents of baryons are quarks with $B = \frac{1}{3}$ and a baryon is made of three quarks. As an example, consider a possible annihilation of an antiproton ($\bar{p}$) by a deuteron

$$\bar{p} + d \rightarrow \bar{\Lambda} + \Sigma^+ + n + \pi^- .$$

The total baryon number B on both sides is $+1$. Note that $\bar{\Lambda}$ and $\bar{p}$ are antibaryons, the deuteron has baryon number $B = 2$, and the π^- is a meson with $B = 0$. It is the conservation of baryon number that is responsible for the stability of normal matter.

In ordinary reactions involving normal matter the number of electrons is also conserved. The conservation of the number of electrons plus the number of nucleons may be considered as the basis for the conservation of mass in chemistry. Analogous to the conservation of baryon number there exists a conservation of electron number L_e. In electrodynamic processes this law states that electrons cannot be created or destroyed except as e^+–e^- pairs. When weak interactions are included, however, this law must be generalized to include the electron-type neutrino ν_e and antineutrino $\bar{\nu}_e$. Then ν_e and e^- have $L_e = +1$; $\bar{\nu}_e$ and e^+ have $L_e = -1$. In nuclear beta decay, for example, either an e^-–$\bar{\nu}_e$ or e^+–ν_e pair is emitted. A similar law is the *conservation of muon number* L_μ, for which the set $(e^- e^+ \nu_e \bar{\nu}_e)$ is replaced by $(\mu^- \mu^+ \nu_\mu \bar{\nu}_\mu)$. The law is also believed to hold for the τ lepton giving the conservation of L_τ although direct evidence for the corresponding neutrino ν_τ has not been obtained. The conservation of $L_e + L_\mu + L_\tau$ is called the *conservation of lepton number*. If we label the lepton number L, the conservation of $B - L$ may be called *fermion number conservation*.

No exceptions to the conservation laws for Q, B, L_e, L_μ, or L_τ are known, and so it is reasonable to consider them exact. However, many extensions of the standard model of weak interactions, particularly grand unified theories, allow for nonconservation of baryon number. This has led to an extensive unsuccessful search for proton and neutron decay. Lower limits on τ_p/(BR) of the order 10^{32} years exist for a number of possible decay modes ($e^+\pi^0$, $\mu^+\pi^0$, e^+K^0) where τ_p is the proton lifetime and (BR) is the branching ratio to these modes.

Tests of the conservation of lepton number are much less precise. One test looks for the violation of the separate lepton numbers L_e, L_μ with conservation of the total lepton number L by searching for decays in which μ converts to e. The best limit is on the branching ratio for $\mu^- \rightarrow e^- e^+ e^-$ which is less than 10^{-13}. The best test for the conservation of total lepton number L is the search for neutrinoless double beta decay, a process in which a nucleus emits two electrons but no neutrinos which would indicate a charge of two units in lepton number ($\Delta L = 2$). Experiments provide a lower limit of a few times 10^{23} years for the half-life of ^{76}Ge for this decay mode. If $\Delta L = 2$ is not exactly forbidden, it is expected that neutrinos would acquire a nonzero mass of the Majorana type. If neutrinos have a mass, it is possible to probe for the violation of the separate lepton numbers L_e, L_μ, L_τ by a search for neutrino oscillations in which, for example, ν_μ transforms to ν_e. No convincing evidence for a nonzero neutrino mass exists at present.

The fact that K mesons and Λ and Σ baryons were produced in strong interactions but decayed only weakly led to their designation as strange particles. In the strong interactions these are produced in pairs, typically K^+ or K^0 mesons together with Λ or Σ. This leads to the law of *conservation of strangeness*. Each particle is assigned an integer, analogous to charge, called strangeness S: K^+ and K^0 have strangeness $S = +1$, Λ and Σ have $S = -1$, while "normal" particles like nucleons and pions have $S = 0$. The sum of the strangeness numbers is conserved; thus in a reaction such as $\pi^- + p \rightarrow K^0 + \Lambda$ the sum equals zero at the beginning and at the end. Strangeness is conserved in both the strong and electromagnetic interactions but not in weak interactions. As a consequence, strange particles decay into normal particles by means of the weak interaction. This is an example of a conservation law that is not exact but holds only for a class of interactions. Another example is isospin (q.v.) conservation that holds only for the strong interactions but not for electromagnetic or weak. Instead of strangeness S we often use hypercharge Y defined as $S + B$, where B is the baryon number, which is separately conserved.

Each conserved quantity discussed so far is *additive* in the sense that its value for a set of particles equals the sum of the values for the individual particles. In contrast, parity (q.v.) is *multiplicative*: the parity of a set of particles is the product of the orbital parity times the intrinsic parities of each of the particles. The parity is plus or minus one, depending on the behavior of the quantum-mechanical state under inversion of the spatial coordinates. For a pair of particles the orbital parity is $(-1)^l$ where l is the relative orbital angular momentum in units of $\hbar$. For a composite system the intrinsic parity may simply represent the orbital parity of its components; for example, the parity of the ground state of ^{7}Li is negative because the valence neutron is in a p state

($l = 1$). On the other hand, for elementary particles that can be created or destroyed, the intrinsic parity is a basic characteristic of the particle. The intrinsic parity of the antiparticle of a fermion is opposite to that of the particle. Mesons may have either positive or negative intrinsic parity; the lightest mesons (π and K) have been found to have negative parity. This is understood in the quark model because π and K consist of a quark and an antiquark in an $l = 0$ state. Similarly the ground state of positronium has odd parity.

Parity conservation does not hold for weak-interaction processes, such as beta decay. In a weak decay of a particle like the neutron, which has a well-defined parity, the final state is a mixture of both even and odd parities. This shows up in observations sensitive to interference between the final states of even and odd parity. Such interference effects show up as nonzero expectation values of pseudoscalar quantities. For example, in beta decay the electrons emerge predominantly left-handed, corresponding to a nonzero expectation value of $\boldsymbol{\sigma}\cdot\mathbf{p}$, where $\boldsymbol{\sigma}$ is the spin vector and $\mathbf{p}$ the momentum vector of the electron. Since $\boldsymbol{\sigma}$ is an axial vector and $\mathbf{p}$ is a polar vector, $\boldsymbol{\sigma}\cdot\mathbf{p}$ is a pseudoscalar. The parity nonconservation in the weak interactions is not a small effect but appears to be as large as possible.

It is believed that parity conservation holds for the strong and electromagnetic interactions. However, processes governed by these interactions may demonstrate a small amount of parity nonconservation as a result of the perturbation by the weak interactions. This has been demonstrated by observations of admixtures of electric dipole (odd parity) with magnetic dipole (even parity) in certain nuclear gamma-ray transitions (^{181}Ta, ^{175}Lu). The admixture results in a small net circular polarization of the emitted gamma ray.

For every system there exists a system related to it by changing every particle into its antiparticle. For some systems the resulting system is identical to the original system except possibly for a multiplicative factor C equal to plus or minus one. For example, the positronium atom made up of a positron and an electron has the value $C = +1$ for the $^1S_0$ state and $C = -1$ for the $^3S_1$ state. Some particles are also their own antiparticles with definite C values: for examples, $C = -1$ for a photon and $C = +1$ for a neutral pion π^0. C is called the *charge conjugation* (or the particle–antiparticle conjugation) *quantum number*. It is believed that the *conservation of C* holds for strong and electromagnetic interactions, although the evidence is much less precise than for parity conservation. For example, the state of n photons has $C = (-1)^n$, so that the $^1S_0$ state of positronium decays into an even number of photons but the $^3S_1$ state must decay into an odd number.

The weak interactions that do not conserve parity P also do not conserve charge conjugation C. For example, the decays $\pi^+ \to \mu^+\nu_\mu$ and $\pi^- \to \mu^-\bar{\nu}_\mu$ are related by C, but in these decays the μ^+ is emitted with left-handed polarization and the μ^- with right. The $V - A$ theory developed in 1957 yielded the result that both C and P are not conserved in weak interactions but that the product CP is conserved. The failure of CP conservation was discovered in 1964 from a study of K^0 decays. The antiparticle of the K^0 meson ($S = 1$) is $\bar{K}^0$ ($S = -1$). Since strangeness is not exactly conserved the states with definite mass and lifetimes are linear combinations of K^0 and $\bar{K}^0$, called K_s and K_L. If CP is conserved, then K_s and K_L should be CP eigenstates: $K_s = (K^0 - \bar{K}^0)/\sqrt{2}$ ($CP = +1$) and $K_L = (K^0 + \bar{K}^0)/\sqrt{2}$ ($CP = -1$). We find, in accordance with this expectation, that the major decay modes of K_s are $\pi^+\pi^-$ and $2\pi^0$ ($C = +1$, $CP = +1$), whereas an important decay mode of K_L is $3\pi^0$ ($C = +1$, $CP = -1$). These decays are consistent with CP conservation and also demonstrate nonconservation of C. However, it was discovered that K_L has as a rare decay mode the 2π state with $CP = +1$. Thus the conservation of CP is not an exact law, although it appears to be much better than conservation of C or P separately.

The interaction responsible for the nonconservation of CP has not been identified. Within the standard Weinberg–Salam gauge theory of weak interactions it is possible to incorporate CP violation by means of a mechanism suggested by Kobayashi and Maskawa. Further experiments are needed to show whether this is the correct explanation. It follows from the *CPT* theorem (q.v.) that the interaction responsible for CP noninvariance must also violate time-reversal invariance. The most sensitive search for a failure of time reversal invariance is an experiment attempting to find an electric dipole moment of the neutron.

Certain conservation laws relate vector quantities. The *conservation of momentum* requires that the vector sum of the momenta of a set of interacting particles remain constant during any process. In the theory of special relativity, momentum and energy form a four-vector so that the *conservation of the energy-momentum four-vector* defines one covariant law. The *conservation of angular momentum* (q.v.) requires that the vector sum of the angular momenta of a set of particles interacting by means of a rotationally invariant force law be conserved. In quantum mechanics it is in general impossible to specify all three components of the angular momentum of a system since the different components are represented by noncommuting operators. Thus effectively in quantum mechanics we use the conservation of the square of the angular momentum J^2 and one component J_z. In calculating the angular momentum of a set of particles it is necessary to add vectorially the orbital angular momenta plus the spin angular momenta of all the particles.

Conservation laws are closely related to *invariance principles* (q.v.). In quantum mechanics invariance principles are stated in terms of a unitary transform U, which leaves the Hamiltonian invariant:

$$UHU^{-1} = H. \tag{4}$$

If U is a continuous transformation, it is convenient to consider an infinitesimal form of U

$$U = 1 + i\epsilon F,$$

where ϵ is an infinitesimal parameter and F is a Hermitian operator. From Eq. (4) to order ϵ it is required that

$$FH - HF = 0. \tag{5}$$

The fact that F commutes with H guarantees that it is conserved, since it is H that governs the time development of a quantum-mechanical system. The eigenvalues of the operator F thus are conserved quantities. In the case of a dis-

Table I Summary of Conserved Quantities

Quantity	Symbol	Type	Interactions for which it is not conserved	Associated invariance principle
Energy	E	Additive	None	Time translation
Momentum	$\mathbf{p}$	Additive vector	None	Space translation
Angular momentum	$\mathbf{J}$	Additive vector	None	Space rotation
Parity	P	Multiplicative	Weak	Space inversion
Charge conjugation	C	Multiplicative	Weak	Charge conjugation
CP	CP	Multiplicative	Weak(?)	CP; time reversal
Charge	Q	Additive	None	Gauge invariance
Baryon number	B	Additive	None	$U(1)$
Electron number	L_e	Additive	None	$U(1)$
Muon number	L_μ	Additive	None	$U(1)$
Strangeness	S	Additive	Weak	$U(1)$
Hypercharge	Y	Additive	Weak	$U(1)$
Isospin	$\mathbf{I}$	Additive vector	Electromagnetic; weak	$SU(2)$

crete transformation such as space reflection, we can define an operator $\mathcal{P}$, which is both unitary and Hermitian, so that $\mathcal{P}$ plays the role of both U in Eq. (4) and F in Eq. (5). The eigenvalue of $\mathcal{P}$ is the conserved quantity called parity P. It is possible to define a kind of gauge invariance [often referred to as a $U(1)$ group] to be associated with each of the conserved additive numbers Q, B, L_e, L_μ, L_τ, and S. However, although in the case of the original gauge invariance associated with Q there is a profound relation between this invariance and the law of electromagnetic interactions, the significance of the other gauge invariances is not clear.

Table 1 contains a list of conserved quantities. For each there is indicated whether it is additive or multiplicative, the range of validity, and the invariance principle with which it is associated.

See also CPT THEOREM; DYNAMICS, ANALYTICAL; INVARIANCE PRINCIPLES; ISOSPIN; KINEMATICS AND KINETICS; MOMENTUM; PARITY.

BIBLIOGRAPHY

Classical Conservation Laws

G. Holton and S. G. Brush, *Introduction to Concepts and Theories in Physical Science,* Part E. Addison-Wesley, Reading, Mass., 1973. (E)

L. D. Landau and E. M. Lifschitz, *Mechanics*. Pergamon, New York, 1960. (I)

Conservation Laws in Quantum Physics

L. I. Schiff, *Quantum Mechanics,* 4th ed., Chapter 7. McGraw-Hill, New York, 1968. (A)

R. P. Feynman, R. B. Leighton, and M. Sands, *The Feynman Lectures on Physics,* Vol. III, Chapters 17, 18, and 20. Addison-Wesley, Reading, Mass., 1965. (I)

Conservation Laws in Elementary-Particle Physics

R. P. Feynman, *The Character of Physical Law,* Chapters 3 and 4. MIT Press, Cambridge, Mass., 1965. (E)

H. Frauenfelder and E. M. Henley, *Subatomic Physics,* Part III. Prentice-Hall, Englewood Cliffs, NJ, 1974. (I)

W. M. Gibson and R. Pollard, *Symmetry Principles in Elementary Particle Physics*. Cambridge University Press, Cambridge, England, 1976. (A)

D. H. Perkins, *Introduction to High Energy Physics*. Addison-Wesley, Reading, Mass., 1982. (I)

Constants, Fundamental

B. N. Taylor

INTRODUCTION

This article touches upon four main topics: (1) The motivation for "the romance of the next decimal place," or why the fundamental physical constants are important and why their determination to ever greater levels of accuracy can have a profound effect on physics; (2) how a self-consistent set of "best values" of the fundamental constants is obtained, with emphasis on the 1986 least-squares adjustment of the constants (the most recent comprehensive study carried out); (3) new developments in the fundamental constants field since the 1986 adjustment was completed and their impact on the recommended set of best values resulting from that adjustment; and (4) future trends—where the field is heading over the next 5 to 10 years.

ORIGINS

Three distinct "sources" of fundamental constants may be identified. The first is physical theory (and its application to the real world), such as Maxwell's theory of electromagnetism, Einstein's theories of relativity, quantum mechanics, and quantum electrodynamics (QED). The speed of light in vacuum, c, the Planck constant, h, the fine-structure constant, α, and the Rydberg constant for infinite mass, R_∞, fall within this category.

The second is the fundamental particles such as the electron, proton, and neutron. The elementary charge, e, the rest masses of the electron and proton, m_e and m_p, and the magnetic moment of the proton in units of the nuclear magneton, μ_p/μ_N, where $\mu_N = h/4\pi m_p$, are all examples of quantities that characterize a basic property of nature's elementary building blocks.

The third is conversion factors. Although not true fundamental constants like those in the first two categories, these quantities nevertheless play a critical role in the fundamental-constants field, since knowledge of their values is essential to many fundamental-constant determinations. Examples in this category include the local value of the gravitational acceleration, g, which relates mass to force; the ratio of the old x-ray unit of length or kilo-x-unit to the angstrom, Å; and the ratio of various so-called as-maintained electrical units to their International System of Units (SI) definitions, for example, Ω_{NIST}/Ω, where Ω_{NIST} is the National Institute of Standards and Technology (NIST) ohm defined in terms of the mean resistance of a particular group of wire-wound standard resistors, and Ω is the SI ohm.

It should also be noted that $\alpha = \mu_0 ce^2/2h$ and $R_\infty = m_e c\alpha^2/2h$ are examples of constants that are actually combinations of other quantities, but are considered fundamental constants in their own right since the combination always appears in theoretical equations in the same way ($\mu_0 = 4\pi \times 10^{-7}$ N/A^2 exactly is the permeability of vacuum); and that e and m_e are examples of constants used as basic measurement units, for example, the mass of the muon is expressed as (approximately) 207 electron masses or $207m_e$.

IMPORTANCE

There are at least four reasons why the fundamental physical constants play a critical role in science and technology, and thus must be known to as high an accuracy as possible. First, accurate values of the constants are required for the critical comparison of theory with experiment; and it is only such comparisons that enable our understanding of the physical world to advance. A closely related idea is that by comparing the numerical values of the same fundamental constants obtained from experiments in the different fields of physics, the self-consistency of the basic theories of physics themselves can be tested.

Second, determining the fundamental constants to ever greater levels of accuracy fosters the development of state-of-the-art measurement methods that may have wide application. Determining the next decimal place is never trivial and usually requires an entirely new measurement technology. An example is the determination of the Avogadro constant, N_A, which necessitated the development of techniques to measure the lattice spacing of pure, single crystals of silicon in meters to an accuracy of a few tenths of a part per million (ppm). The end result has been the extension of our length scale to the picometer range.

Third, the fundamental constants are the obvious key to the development of a reproducible and invariant system of measurement units, a major goal of measurement science or metrology. If our measurement system could be based on fundamental constants, there would be no need for artifacts such as the platinum–iridium cylinder kept in a vault at the International Bureau of Weights and Measures (BIPM), Sèvres, France, which defines the kilogram.

Finally, values of the fundamental constants are required for computations and measurements throughout science and technology—for example, the calculation of the excited states of atoms and molecules important in the fields of air pollution and nuclear fusion.

As an example of the significant role improved measurements of fundamental constants can play in increasing our understanding of physical theory, we consider the determination of $2e/h$ using the ac Josephson effect in weakly coupled superconductors, which was first reported in 1967. When a Josephson tunnel junction, for example, two thin films of lead separated by a 1-nm-thick thermally grown oxide layer, is irradiated with microwave radiation of frequency f, its current vs. voltage curve exhibits current steps at quantized Josephson voltage U_J. The voltage of the nth step, $U_J(n)$, where n is an integer, is related to f by

$$U_J(n) = nf(2e/h)^{-1}. \tag{1}$$

A measurement of f and U_J thus yields $2e/h$. The value obtained in 1968 using x-band microwaves (8–12 GHz) was

$$2e/h = 4.835\,976(12) \times 10^{14}\ \text{Hz/V}_{\text{NIST}}\ (2.4\ \text{ppm}), \tag{2}$$

where V_{NIST} is the unit of voltage maintained at the National Institute of Standards and Technology, which at the time was based on the mean emf of a group of electrochemical standard cells. [In Eq. (2), the first number in parentheses is the one-standard-deviation uncertainty in the last digits of the quoted value; the second is its fractional equivalent in ppm.] The difference between this value when expressed in SI units and the previous best value, that resulting from the 1963 least-squares adjustment of the fundamental constants, was (35 ± 10) ppm. As we shall see, the cause of the discrepancy was the use in the 1963 adjustment of a value of α derived from the early 1950s measurement of the deuterium fine-structure splitting, which subsequently turned out to be incorrect.

Because QED, which describes the interaction between leptons (electron, muon, tauon) and electromagnetic radiation, is one of the most precise and important theories of modern physics, it is essential to see how well it can withstand experimental tests. The Josephson $2e/h$ determination is therefore significant mainly because a reliable indirect value of α independent of quantum electrodynamic theory can be derived from it, and this value can in turn be used to compare QED theory and experiment critically and unambiguously. This is in marked contrast to the situation that existed prior to 1967 when no such value was available and tests of QED were mainly checks of internal consistency.

The equation relating α to $2e/h$ may be written as

$$\alpha = \left[\frac{4R_\infty(\Omega_{NIST}/\Omega)\gamma_p'(\text{low})_{NIST}}{c(\mu_p'/\mu_B)(2e/h)_{NIST}}\right]^{1/2}, \tag{3}$$

where as before Ω_{NIST}/Ω is the ratio of the NIST as-maintained ohm to the SI ohm; μ_p'/μ_B is the magnetic moment of the proton in units of the Bohr magneton, $\mu_B = h/4\pi m_e$ (throughout, the prime means for protons in a spherical sample of pure H_2O at 25°C); and $\gamma_p'(\text{low})_{NIST}$ is the gyromagnetic ratio of the proton obtained by the so-called low- (magnetic) field method measured in terms of the NIST as-maintained unit of current $A_{NIST} = V_{NIST}/\Omega_{NIST}$. (It should be noted that the subscript NIST may be replaced by the more general lab; and that $\gamma'_p = \omega'_p/B = 4\pi\mu'_p/h$, where ω'_p is the proton nuclear magnetic resonance or spin-flip angular frequency in the magnetic field B. In the low-field method, B is established via a precision solenoid of known dimensions carrying a current known in terms of A_{lab}.) The value of α derived from Eq. (3) in 1968 using the best data available was

$$1/\alpha = 137.036\,08(26)\ (1.9\ \text{ppm}). \tag{4}$$

Included among the quantities that require an accurate value of α for comparing theory and experiment are the g-factors of the electron and muon; the energy levels in hydrogen-like atoms, especially the $n = 2$ Lamb shift ($2^2S_{1/2} - 2^2P_{1/2}$ splitting); and the ground-state hyperfine splitting in hydrogen, muonium (an electron bound to a positive muon or μ^+e^- atom), and positronium (a bound electron-positron pair or e^+e^- atom). Of particular interest in the late 1960s was the hydrogen hyperfine splitting (hfs). This quantity, which is essentially the energy difference between a hydrogen atom in which the electron and proton spins are aligned and one in which they are in opposite directions, can be measured to the extraordinary accuracy of 1 part in 10^{12} using the hydrogen maser. In contrast, the theoretical QED equation for the hydrogen hfs, which involves only well-known constants and α, is limited to an accuracy of a few ppm because of the difficulty in calculating some of the terms in the equation from theory.

The most uncertain term is $\delta_N^{(2)}$, the proton polarizability contribution, which arises from the various excited states or internal structure of the proton. In 1968, theoretical calculations predicted $\delta_N^{(2)} = (0 \pm 5)$ ppm. This was in conflict with what was implied by the value of α accepted at that time, which, as noted earlier, was derived from a measurement of the fine-structure splitting in deuterium. When this value of α was used to calculate a theoretical value of the hydrogen hfs, when this result was compared to the hydrogen-maser value, and when their difference was assumed to arise solely from the existence of a polarizability term, it was found that $\delta_N^{(2)} = (43 \pm 9)$ ppm. This meant that the probability for $\delta_N^{(2)}$ to be as small as predicted by direct calculation was only 1 in 20,000, an obvious inconsistency. In contrast, when the value of α derived from the Josephson-effect measurement of $2e/h$ [Eq. (4)] was used in place of the deuterium value, it was found that $\delta_N^{(2)} = 2.5 \pm 4.0$ ppm, in keeping with the theoretical calculations. Hence the Josephson-effect value of α removed a discrepancy that during the 1960s was termed one of the major unsolved problems of QED. This example further illustrates the overall unity of science—a low-temperature solid-state physics experiment has provided information about the excited states of the proton—as well as how precise measurements of the fundamental constants can illuminate apparent inconsistencies in our physical description of nature.

Another example of a critical test of QED, but one that is still under active experimental and theoretical investigation in the early 1990s, involves the quantum Hall effect (QHE) and the g-factor of the electron, $g_e = 2(1 + a_e) = 2\mu_e/\mu_B$, where a_e is the electron magnetic moment anomaly and μ_e is the magnetic moment of the electron. The deviation of g_e from 2, that is, the finite size of a_e ($\sim 1.16 \times 10^{-3}$), arises from the electron's virtual emission and absorption of photons and the polarization of the vacuum with electron–positron pairs—so-called radiative corrections.

An experimental value for the electron g-factor, g_e(expt), with a relative uncertainty currently believed to be about 4×10^{-12}, has been obtained from a direct experimental determination of the anomaly, a_e(expt), based on measurements on a single electron stored in a Penning trap and having an uncertainty believed to be about 0.004 ppm. The theoretical expression for the g-factor obtained from QED is $g_e(\text{theor}) = 2[1 + a_e(\text{theor})]$ where

$$a_e(\text{theor}) = C_1(\alpha/\pi) + C_2(\alpha/\pi)^2 + C_3(\alpha/\pi)^3 + C_4(\alpha/\pi)^4 + \ldots + \delta a_e. \tag{5}$$

C_1 is exactly known whereas C_2 and δa_e have negligible uncertainties. The coefficients C_3 and C_4 are currently believed to have uncertainties (arising from numerical integrations) such that if α were exactly known, then a_e(theor) would have an uncertainty of 0.007 ppm and thus g_e(theor) a relative uncertainty of 7×10^{-12}. To test QED by comparing g_e(theor) with g_e(expt) to the full accuracies of a_e(expt) and a_e(theor) requires, therefore, a value of α having an uncertainty significantly less than the 0.008 ppm uncertainty of the value of α obtained by equating a_e(expt) with Eq. (5), say 0.002–0.003 ppm. Such a value is not yet available but one with only 10 times this uncertainty has already been obtained from the QHE.

The QHE is characteristic of certain high-mobility semiconductor devices of standard Hall-bar geometry when in an applied magnetic field of order 10 T and cooled to about 1 K. Hence, like the Josephson effect, it is a low-temperature solid-state physics phenomenon. For a fixed current I through a particular type of QHE device, the curve of Hall voltage, U_H, vs. magnetic field displays regions of constant Hall voltage termed Hall plateaus. The Hall resistance of the ith plateau, $R_H(i)$, defined as the quotient of the Hall voltage of the ith plateau, $U_H(i)$, to the current I, is quantized and given by

$$R_H(i) = U_H(i)/I = (h/e^2)i^{-1}. \tag{6}$$

(We consider only the integral QHE where i is an integer.) Since

$$\alpha = \mu_0 ce^2/2h = \mu_0 c/[2iR_H(i)], \tag{7}$$

and $c = 299\,792\,458$ m/s exactly as a result of the 1983 redefinition of the meter in terms of the speed of light, a measurement of the quantized Hall resistance, $R_H = iR_H(i)$, in SI units with a given uncertainty will yield a value of the fine-structure constant α with the same uncertainty. In practice, $R_H(i)$ is measured in terms of Ω_{lab} so that the conversion factor Ω_{lab}/Ω must be determined in a separate experiment using an apparatus known as a calculable capacitor. Nevertheless, the QHE has yielded a value of α with an uncertainty of 0.024 ppm. Using this value to compare g_e(expt) with g_e(theor), one finds agreement within statistically acceptable limits and thus confirmation of QED to the unprecedented level of about 5 parts in 10^{11}.

RELATIONSHIPS

Equation (3), Eq. (5) together with a_e(expt), and Eq. (7) all give a value of α. However, because in practice a value of Ω_{lab}/Ω is required to obtain α from Eq. (7) as well as from Eq. (3), the two values will not be completely independent. Moreover, $2e/h$ (in SI units) can be obtained from the relation

$$2e/h = \left[\frac{16R_\infty(m_p/m_e)N_A}{\mu_0 c^2 M_p \alpha}\right]^{1/2}, \qquad (8)$$

where M_p is the molar mass of the proton; and γ_p'(low) from the equation

$$\gamma_p'(\text{low})_{lab} = \frac{F_{lab}(K_{A\text{-}lab})^2(\mu_p'/\mu_B)(m_p/m_e)}{M_p}, \qquad (9)$$

where F_{lab} is the Faraday constant measured in terms of a particular laboratory's as-maintained ampere, A_{lab}, and $K_{A\text{-}lab} = A_{lab}/A = (V_{lab}/\Omega_{lab})(V/\Omega)^{-1} = (V_{lab}/V)(\Omega_{lab}/\Omega)^{-1} = K_{V\text{-}lab}/K_{\Omega\text{-}lab}$. Further, the Faraday constant is related to the Avogadro constant by $F = N_A e$, and $K_{A\text{-}lab}$ may be obtained from direct measurements using a so-called ampere force balance as well as from the equation

$$K_{A\text{-}lab} = \gamma_p'(\text{low})_{lab}/\gamma_p'(\text{high})_{lab}, \qquad (10)$$

where γ_p'(high) is the gyromagnetic ratio of the proton as obtained by the so-called high- (magnetic) field method. (The magnetic field B is determined by measuring the force it exerts on a conductor of known dimensions carrying a current known in terms of A_{lab}.)

It should now be clear that complex relationships can exist among groups of fundamental constants and conversion factors, and that a particular constant may be obtained either by direct measurement or indirectly by appropriately combining other directly measured constants. If the direct and indirect values have similar accuracy, both must be considered in order to arrive at a best value for the quantity. However, each of the various routes that can be followed to a particular constant, both direct and indirect, provides a somewhat different numerical value. The best way to handle this situation is by the mathematical technique known as least squares.

LEAST SQUARES

The least-squares method furnishes a well-defined procedure for calculating best "compromise" values of the constants from all of the available data. It takes into account all possible routes to a particular constant and yields a single value for each constant by weighting the different routes according to their relative uncertainties. The weights themselves are obtained from the *a priori* uncertainties assigned to the individual direct measurements that constitute the original set of data.

Least-squares adjustments (or studies) of the fundamental physical constants provide a self-consistent set of best values for use by science and technology in any given epoch. More important, however, is that the critical data review that must accompany such studies forces the reviewer to reassign uncertainties to individual experiments on the same basis, thus making possible the ready identification of discrepancies among different measurements and calculations. This identification, in turn, can stimulate new experimental and theoretical work.

Least-squares studies of the constants were pioneered in the late 1920s by R. T. Birge and continued by others, notably J. W. M. DuMond and E. R. Cohen. The three most recent reviews (*see* Bibliography) were those of Taylor, Parker, and Langenberg in 1969, based on their Josephson-effect determination of $2e/h$; of Cohen and Taylor in 1973, carried out under the auspices of CODATA (Committee on Data for Science and Technology); and of Cohen and Taylor in 1986 which was also carried out under CODATA auspices. The last is the most recent adjustment available and the recommended set of constants resulting from it are those officially adopted by CODATA for international use; values of selected constants are given in Table I.

THE 1986 ADJUSTMENT

In general, the data entering a least-squares adjustment of the constants are divided into two groups: (1) the more precise data or *auxiliary constants* and (2) the *stochastic input data*. The auxiliary constants are either defined quantities such as μ_0 and c with no uncertainty, or constants such as R_∞ and M_p with assigned uncertainties sufficiently small in comparison with the uncertainties assigned the stochastic input data with which they are associated in the adjustment that they can be taken as exact (i.e., their values are not subject to adjustment in contrast to the stochastic data). The auxiliary constants used in the 1986 adjustment, none of which had an assigned uncertainty greater than 0.02 ppm, included μ_0, c, R_∞, M_p, m_p/m_e, g_e, the g-factor of the muon g_μ, and the magnetic moment ratios μ_e/μ_p and μ_p/μ_B.

Also included in the auxiliary constant category were the representations of the volt, ohm, and ampere in terms of which all electrical-unit-dependent stochastic data were expressed: $V_{76\text{-}BI}$, Ω_{BI85}, and $A_{BI85} = V_{76\text{-}BI}/\Omega_{BI85}$, respectively. Here $V_{76\text{-}BI} = [E(2e/h)^{-1}]$ V is the representation of the volt (i.e., "practical unit" of voltage) realized by the Josephson effect at the International Bureau of Weights and Measures (BIPM) with E = 483 594 GHz/V exactly. This latter value was suggested by the Consultative Committee on Electricity of the International Committee of Weights and Measures in 1972 and was to be used by all national standards laboratories that based their national representation of the volt on the Josephson effect. The "practical unit" of resistance Ω_{BI85} is the BIPM representation of the ohm based on the mean resistance of a particular group of wire-wound resistors at BIPM as they existed on 1 January 1985.

To carry out a least-squares adjustment, a subset of M constants is chosen in terms of which all of the stochastic input data can be individually and independently expressed, if necessary, with the aid of the auxiliary constants. It is actually the M constants constituting this subset, termed the adjustable constants or unknowns, that are directly subject

Table I Recommended values of selected physical constants from the 1986 CODATA least-squares adjustment by Cohen and Taylor. (The digits in parentheses are the one-standard-deviation uncertainty in the last digits of the given value.)

Quantity	Symbol	Value	Units	Relative uncertainty (ppm)
Speed of light in vacuum	c	299 792 458	$m\ s^{-1}$	(exact)
Permeability of vacuum	μ_0	$4\pi \times 10^{-7}$	$N\ A^{-2}$	
		$= 12.566\ 370\ 614\ldots$	$10^{-7}\ N\ A^{-2}$	(exact)
Permittivity of vacuum, $1/\mu_0 c^2$	ϵ_0	$8.854\ 187\ 817\ldots$	$10^{-12}\ F\ m^{-1}$	(exact)
Newtonian constant of gravitation	G	6.672 59(85)	$10^{-11}\ m^3\ kg^{-1}\ s^{-2}$	128
Planck constant	h	6.626 075 5(40)	$10^{-34}\ J\ s$	0.60
$h/2\pi$	$\hbar$	1.054 572 66(63)	$10^{-34}\ J\ s$	0.60
Elementary charge	e	1.602 177 33(49)	$10^{-19}\ C$	0.30
Magnetic flux quantum, $h/2e$	Φ_0	2.067 834 61(61)	$10^{-15}\ Wb$	0.30
Electron mass	m_e	9.109 389 7(54)	$10^{-31}\ kg$	0.59
Proton mass	m_p	1.672 623 1(10)	$10^{-27}\ kg$	0.59
Proton–electron mass ratio	m_p/m_e	1 836.152 701(37)		0.020
Fine-structure constant, $\mu_0 c e^2/2h$	α	7.297 353 08(33)	10^{-3}	0.045
Inverse fine-structure constant	α^{-1}	137.035 989 5(61)		0.045
Rydberg constant, $m_e c \alpha^2/2h$	R_∞	10 973 731.534(13)	m^{-1}	0.0012
Avogadro constant	N_A, L	6.022 136 7(36)	$10^{23}\ mol^{-1}$	0.59
Faraday constant, $N_A e$	F	96 485.309(29)	$C\ mol^{-1}$	0.30
Molar gas constant	R	8.314 510(70)	$J\ mol^{-1}\ K^{-1}$	8.4
Boltzmann constant, R/N_A	k	1.380 658(12)	$10^{-23}\ J\ K^{-1}$	8.5
Stefan–Boltzmann constant, $(\pi^2/60)k^4/\hbar^3 c^2$	σ	5.670 51(19)	$10^{-8}\ W\ m^{-2}\ K^{-4}$	34
Josephson frequency-voltage quotient	$2e/h$	4.835 976 7(14)	$10^{14}\ Hz\ V^{-1}$	0.30
Quantized Hall resistance, $h/e^2 = \mu_0 c/2\alpha$	R_H	25 812.805 6(12)	Ω	0.045
Bohr magneton, $e\hbar/2m_e$	μ_B	9.274 015 4(31)	$10^{-24}\ J\ T^{-1}$	0.34
Nuclear magneton, $e\hbar/2m_p$	μ_N	5.050 786 6(17)	$10^{-27}\ J\ T^{-1}$	0.34
Bohr radius, $\alpha/4\pi R_\infty$	a_0	0.529 177 249(24)	$10^{-10}\ m$	0.045
Compton wavelength, $h/m_e c$	λ_c	2.426 310 58(22)	$10^{-12}\ m$	0.089
Electron magnetic moment	μ_e	928.477 01(31)	$10^{-26}\ J\ T^{-1}$	0.34
in Bohr magnetons	μ_e/μ_B	1.001 159 652 193(10)		1×10^{-5}
Electron magnetic moment anomaly, $\mu_e/\mu_B - 1$	a_e	1.159 652 193(10)	10^{-3}	0.0086
Electron g factor, $2(1 + a_e)$	g_e	2.002 319 304 386(20)		1×10^{-5}
Muon-electron mass ratio	m_μ/m_e	206.768 262(30)		0.15
Muon magnetic moment anomaly, $[\mu_\mu/(e\hbar/2m_\mu)] - 1$	a_μ	1.165 923 0(84)	10^{-3}	7.2
Muon g factor, $2(1 + a_\mu)$	g_μ	2.002 331 846(17)		0.0084
Muon–proton magnetic moment ratio	μ_μ/μ_p	3.183 345 47(47)		0.15
Proton magnetic moment	μ_p	1.410 607 61(47)	$10^{-26}\ J\ T^{-1}$	0.34
in Bohr magnetons	μ_p/μ_B	1.521 032 202(15)	10^{-3}	0.010
in nuclear magnetons	μ_p/μ_N	2.792 847 386(63)		0.023
Proton gyromagnetic ratio	γ_p	26 752.212 8(81)	$10^4\ s^{-1}\ T^{-1}$	0.30
	$\gamma_p/2\pi$	42.577 469(13)	$MHz\ T^{-1}$	0.30
Molar volume (ideal gas), RT/p $T = 273.15$ K, $p = 101\ 325$ Pa	V_m	0.022 414 10(19)	$m^3\ mol^{-1}$	8.4
		Non-SI units used with SI		
Electron volt, $(e/C)J = \{e\}J$	eV	1.602 177 33(49)	$10^{-19}\ J$	0.30
(unified) atomic mass unit, $1\ u = m_u = \frac{1}{12}m(^{12}C)$	u	1.660 540 2(10)	$10^{-27}\ kg$	0.59

to adjustment. In the 1986 effort there were $M = 5$ unknowns and these were taken to be:

1. the inverse of the fine-structure constant α^{-1};
2. $K_V = V_{76\text{-BI}}/V$, the ratio of the BIPM volt representation based on the Josephson effect and the adopted value of E to the SI volt, which implies that $2e/h = E/K_V$;
3. $K_\Omega = \Omega_{BI85}/\Omega$, the ratio of the BIPM ohm representation as it existed on 1 January 1985 to the SI ohm;
4. $d_{220}(Si)$, the (2,2,0) lattice spacing of a perfect crystal of pure silicon at 22.5°C and in vacuum; and

5. μ_μ/μ_p, the ratio of the magnetic moment of the muon to that of the proton.

An equation which relates a stochastic input datum to the M adjustable constants (with the aid of the auxiliary constants if necessary) is known as an *observational equation*. Thus, if there are N items of stochastic input data, there are N observational equations. However, the observational equations for data items of the same type are of the same form. The weight, w_i, of each stochastic datum and hence of each observational equation is related to the datum's *a priori* assigned uncertainty, s_i, by $w_i = 1/s_i^2$. Consequently, if one value of a datum for a particular quantity has half the uncertainty of another, it carries four times as much weight in the adjustment. Taking into account the weight assigned each observational equation and with the aid of a computer, the least-squares adjusted values of the M unknowns may be readily obtained. "Best" values in the least-squares sense for these M adjustable constants, with their variances and covariances, are thus the immediate output of the adjustment. However, optimal values in the least-squares sense for most other constants of interest not directly subject to adjustment may then be calculated from the M adjustable constants with the aid of the auxiliary constants. (But this does not apply to the auxiliary constants themselves, since for the purpose of the least-squares adjustment they are assumed to be exactly known.)

In contrast to the auxiliary constants of the 1986 adjustment for which no uncertainty exceeded 0.02 ppm, the uncertainties assigned the 38 items of stochastic input data considered in the 1986 adjustment were in the range 0.065 to 9.7 ppm. The 38 items were of 12 distinct types with the number of items of each type ranging from 1 to 6. Hence, there were $N = 38$ observational equations of 12 distinct types. The 38 items of stochastic data may be succinctly summarized as follows:

1. K_Ω determined from calculable capacitor ohm realizations (5 values with uncertainties in the range 0.11–0.36 ppm);
2. $K_A = A_{BI85}/A = K_V/K_\Omega$, where $A_{BI85} = V_{76\text{-}BI}/\Omega_{BI85}$, determined using ampere force balances (6 values, 4.1–9.7 ppm uncertainty range);
3. K_V determined using volt force balances (2 values, 2.4 and 0.60 ppm uncertainties);
4. F, the Faraday constant determined by coulometry and expressed in terms of A_{BI85} (1 value, 1.33 ppm uncertainty);
5. γ_p' (low) expressed in terms of A_{BI85} (6 values, 0.24–3.25 ppm uncertainty range);
6. γ_p' (high) expressed in terms of A_{BI85} (4 values, 1.0–5.4 ppm uncertainty range);
7. d_{220}(Si) determined by combined x-ray and optical interferometry (2 values, 0.10 and 0.23 ppm uncertainty);
8. V_m(Si), the molar mass of pure silicon at 22.5°C and in vacuum determined from the molecular weight of silicon and its density (1 value, 1.15 ppm uncertainty);
9. R_H determined from the quantum Hall effect and expressed in terms of Ω_{BI85} (6 values, 0.12–0.22 ppm uncertainty range);
10. α^{-1} determined from a_e(expt) and a_e(theor), and from the fine structure of atomic helium (2 values, 0.065 and 0.60 ppm uncertainties);
11. μ_μ/μ_p obtained from measurements related to ν(Muhfs), the ground-state hyperfine splitting of muonium, and from a resonance experiment (2 values, 0.36 and 0.53 ppm uncertainties); and
12. ν(Muhfs) [1 value, 0.14 ppm uncertainty taking into account the uncertainty in the theoretical expression for ν(Muhfs)].

Because new results which can influence a least-squares adjustment of the constants are reported continually, it is always difficult to choose an optimal time at which to carry out a new adjustment and to revise the recommended values of the constants. In the case of the 1986 adjustment, all data available up to 1 January 1986 were considered for inclusion, with the recognition that any additional changes to the 1973 recommended values that might result by taking into account more recent data would be much less than the changes resulting from the data available prior to that date. Indeed, the improved measurements of R_H and Ω_{lab}/Ω and the improved measurement of a_e(expt) and calculation of a_e(theor) which, as mentioned earlier (*see* Importance section), have yielded values of α with uncertainties of about 0.024 and 0.008 ppm, respectively, were not available by 1 January 1986. Their impact and that of other more recent results on the 1986 recommended values will be discussed later in this article.

The following is an example of a typical observational equation in the 1986 adjustment, that used to relate each of the five measured values of the proton gyromagnetic ratio determined by the high- (magnetic) field method (Type 6 data) to the five adjustable constants:

$$(\alpha^{-1})^{-2}(K_V)^{-2}(K_\Omega)^1(d_{220})^0(\mu_\mu/\mu_p)^0 = F_6\gamma_p'(\text{high}), \quad (11)$$

where F_6 is the combination of auxiliary constants $4R_\infty/[c(\mu_p'/\mu_B)E]$. Note that each adjustable constant is raised to some power and that γ_p'(high) actually depends on only three of the five adjustable constants. An example of how a quantity not directly subject to adjustment is obtained from the adjustable constants with the aid of the auxiliary constants is the Bohr magneton, $\mu_B = eh/4\pi m_e$. It can be shown that

$$\mu_B = [2\pi\mu_0 R_\infty E]^{-1}(\alpha^{-1})^{-3}(K_V)^1, \quad (12)$$

where the quantities in the brackets are auxiliary constants. The uncertainty of μ_B is calculated from the variances and covariances of α^{-1} and K_V, which are a byproduct of the adjustment.

Because the weight of each stochastic datum included in a least-squares adjustment of the constants is taken as $w_i = 1/s_i^2$, critical analysis of the input data in order to decide what *a priori* uncertainty s_i should be assigned each measurement is the main problem in successfully carrying out an adjustment. The uncertainty problem is made especially difficult because in most experiments sufficient data are taken to reduce the random or statistical uncertainty to negligible amounts, and the final uncertainty assigned the experiment is determined primarily from estimates of the systematic uncertainties. These frequently arise from effects that the experimenter knows little about, and hence their estimation is often subjective and only derivable from educated guesses.

Another problem area is deciding how to handle "discrepant" data, that is, measurements that differ from each other by statistically significant amounts in comparison with their assigned uncertainties s_i. Such data cannot be included in an adjustment uncritically because the inconsistencies imply either incorrect uncertainty assignments or the presence of unknown measurement errors.

When confronted with such a situation, the constants reviewer can in general either include the inconsistent data, but only after expanding (increasing) their s_i so that they are no longer discrepant; or decide, on as sound a theoretical and experimental basis as possible, which of the inconsistent data are least reliable, and discard them. These two approaches clearly imply that different reviewers might treat the same data differently, thereby obtaining a somewhat different set of best values.

In their 1973 adjustment, Cohen and Taylor not only expanded the uncertainties of certain like items of stochastic data in order to obtain better internal consistency among them, but also eventually discarded 4 of the 31 items originally considered for inclusion in the adjustment because they were found to be inconsistent with the remainder of the data. These included two values of the Faraday constant F. However, as an example of the pitfalls of discarding data, it is instructive to compare the 1969 adjustment of Taylor *et al.* with the 1973 adjustment of Cohen and Taylor.

The most critical problem in the 1969 adjustment was the internal inconsistency among the five values of μ_p'/μ_N then available, and the inconsistency between two of these values and the one available measurement of the Faraday constant. After much thought and analysis, Taylor and colleagues decided to discard the two "high" values of μ_p'/μ_N and to retain the remaining three and F.

The most difficult problem in the 1973 adjustment also had to do with μ_p'/μ_N and F. In the intervening four years, however, two new, very accurate (sub-ppm) μ_p'/μ_N determinations were completed which showed that the high values of μ_p'/μ_N discarded in 1969 were more nearly correct than the three retained low values, and that it was the two values of F that were probably in error and should be discarded. This shift in outlook accounts for the large changes in the recommended values for certain constants as given in the 1969 and 1973 adjustments. These changes are readily apparent in column 5 of Table II which compares the values of selected constants resulting from the 1973 adjustment with their counterparts from the 1969 adjustment. As an additional instructional aid, the latter are compared in column 7 to their counterparts resulting from the 1963 adjustment of Cohen and DuMond. As previously discussed, the changes between the 1963 and 1969 values were primarily due to the change in α resulting from the measurement of $2e/h$ using the Josephson effect.

Motivated by the all-too-frequent occurrence of large changes in the recommended values of many constants from one adjustment to the next, new algorithms for handling discrepant data in least-squares adjustments of the constants were developed during the years between the 1973 and 1986 adjustments. These new algorithms, which added increased objectivity to expanding *a priori* uncertainties and discarding inconsistent data, were used extensively by Cohen and Taylor in carrying out the 1986 adjustment. Indeed, with their aid, the initial group of 38 items of stochastic input data was reduced to 22 items by deleting those that were either highly inconsistent with the remaining data or had assigned uncertainties so large that they carried negligible weight.

The 1986 adjustment represents a major advance over its 1973 counterpart; the uncertainties of the recommended values have been reduced by roughly an order of magnitude due to the enormous advances made throughout the precision measurement–fundamental constants field between 1973 and 1986. This can be seen by comparing columns 2 and 3 of Table II. It is also clear from column 3 that unexpectedly large changes have occurred in the 1973 recommended values of a number of these constants (i.e., a change which is large relative to the uncertainty assigned the 1973 value). These changes are a direct consequence of the 7.8 ppm decrease from 1973 to 1986 in the quantity K_V and the high correlation between K_V and the calculated values of e, h, m_e, N_A, and F. Since $2e/h = E/K_V$, the 1986 value of K_V also implies that the value of $2e/h$ adopted by the Consultative Committee on Electricity in 1972, which was believed to be consistent with the SI value and which most national standards laboratories adopted to define and maintain their

Table II Comparison of recommended values of selected physical constants resulting from different least-squares adjustments[a]

Quantity	Uncertainty of 1986 value	Difference 1986–1973 and uncertainty of 1973 value (ppm)		Difference 1973–1969 and uncertainty of 1969 value (ppm)		Difference 1969–1963 and uncertainty of 1963 value (ppm)	
α^{-1}	0.045	−0.37	0.82	+0.15	1.5	−20	4.4
e	0.30	−7.4	2.9	−1.6	4.4	+57	12
h	0.60	−15.2	5.4	−3.0	7.6	+91	24
m_e	0.59	−15.8	5.1	−2.6	6.0	+52	14
N_A	0.59	+15.2	5.1	−21	6.6	−58	15
m_p/m_e	0.020	+0.64	0.38	+23	6.2	+7.2	5.5
F	0.30	+7.8	2.8	−22	5.5	−3.1	5.2
$2e/h$	0.30	+7.8	2.6	+1.7	3.3	−36	9.7

[a] From the 1986 and 1973 least-squares adjustments of Cohen and Taylor; the 1969 adjustment of Taylor, Parker, and Langenberg; and the 1963 adjustment of Cohen and DuMond.

national representation of the volt, is actually 7.59 ppm smaller than the SI value. (As will be discussed shortly, this unsatisfactory situation was rectified starting on 1 January 1990, as was the similar unsatisfactory situation with national representations of the ohm based on wire-wound resistors.)

The large change in K_V and hence in many other quantities between 1973 and 1986 would have been avoided if the two determinations of F which seemed to be discrepant with the remaining data had not been deleted in the 1973 adjustment. In retrospect, the disagreement was comparatively mild. In view of this further example of the pitfalls of discarding data, it is important to recognize that there are no similar disagreements in the 1986 adjustment; the measurements which were deleted were so discrepant that they obviously could not be correct, or of such low weight that if retained the adjusted values of the five unknowns would change negligibly. Thus, it is unlikely that any alternative evaluation of the data considered in the 1986 least-squares adjustment could lead to significant changes in the 1986 recommended values. Moreover, the quality of the 1986 data, its redundancy, and the use of the new least-squares algorithms to analyze it would seem to preclude future changes in the 1986 recommended values relative to their uncertainties comparable to the changes which occurred in the 1973 values. But of course, only future least-squares adjustments, such as the next one scheduled by CODATA for completion in 1993, can prove the correctness of these statements.

In summary, Table II clearly shows that our knowledge of the numerical values of the fundamental constants continually advances as new measurements become available; that the constants are so interrelated that a significant shift in the value of one will usually give rise to large shifts in the values of others; and that no set of recommended constants such as is given in Table I should be taken as final.

To further reinforce this last point, Figure 1 shows the variation of the recommended value of the electron rest mass resulting from several different adjustments carried out since 1950. Clearly, significant changes from one adjustment to another are not uncommon.

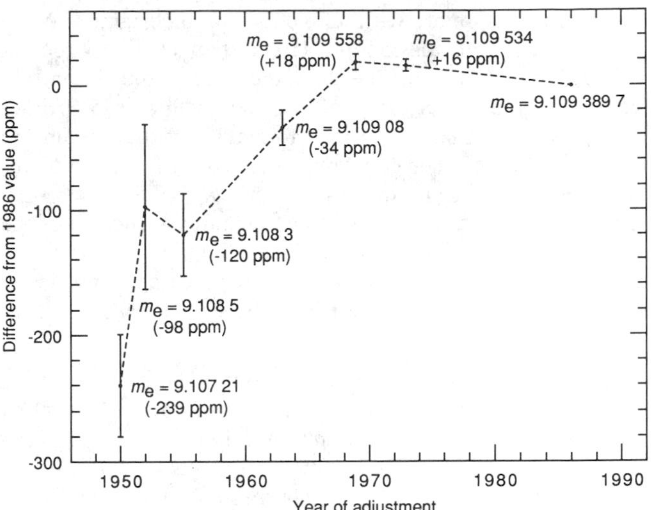

FIG. 1. Variation of the recommended value of the electron rest mass resulting from least-squares adjustments carried out in the years indicated. The quantity shown is the ppm difference between the various adjusted values of m_e and the 1986 value. (The differences are the numbers in parentheses.) The actual resulting values of m_e (in units of 10^{-31} kg) are also given. The error bars are the plus and minus one-standard-deviation uncertainties for each value. (The 0.59-ppm uncertainty of the 1986 value is so small that it can only be shown as a dot.)

RECENT DEVELOPMENTS

Since the completion and publication of the 1986 least-squares adjustment in late 1986, a number of new results have been reported which have implications of varying importance for the 1986 recommended values. For example, improved measurements of R_∞, in particular, that reported in 1989 which used Doppler-free two-photon laser spectroscopy to determine the transition frequencies from the $2S$ to the Rydberg nD states (n = 8, 10, 12), have yielded a value of R_∞ which is 0.0033 ppm larger than the 1986 recommended value and has an uncertainty about three times smaller. While the change is nearly three times the 0.0012-ppm uncertainty assigned the 1986 value, it would have negligible impact on the other recommended values since its magnitude is so small.

A measurement of the gas constant, R, reported in 1988 would have a more profound effect. Determined from measurements of the speed of sound in argon using a hollow spherical acoustic resonator of known volume found by weighing the mercury required to fill it, the new value of R is 4.7 ppm smaller than the 1986 value and has a five times smaller uncertainty—1.7 ppm vs. 8.4 ppm. Although the 4.7-ppm change is only 1.8 times the uncertainty of the 1986 value and thus of limited significance, the reduction of the uncertainty by a factor of 5 is highly significant. This is not only because of the obvious improvement in our knowledge of R itself, but because there would be a comparable reduction in uncertainty in *all* of the 1986 recommended values of the constants that are derived from R. These include the Boltzmann constant, k, the molar volume of an ideal gas, V_m, and the Stefan–Boltzmann constant, σ.

New measurements reported in 1988 of $2e/h$ in SI units, that is, of the ratio K_V [recall $K_V = E(2e/h)^{-1}$ where E = 483 594 GHz/V exactly], would also have a profound effect on the 1986 recommended values. This is especially true of the value for $2e/h$ obtained from a realization of the SI watt. Based on equating electrical and mechanical power using a moving-coil force balance, that is 0.48 ppm larger than the 1986 recommended value and has an uncertainty nearly four times smaller—0.077 ppm vs. 0.30 ppm. Again, although the 0.48-ppm change is only 1.6 times the uncertainty of the 1986 value and hence not of great consequence, the reduction of the uncertainty by a factor of 3.9 is highly important. This is because a large number of the 1986 recommended values strongly depend upon the value of K_V. For example, the uncertainties of e, h, m_e, N_A, and F would all be reduced by a comparable factor.

The improved values of α resulting from the improved measurements of R_H and Ω_{lab}/Ω and the improved measurement of a_e(expt) and calculation of a_e(theor) as discussed earlier are also of great import, especially the value of α from the electron anomaly with its 0.008-ppm uncertainty. While this value is only 0.015 ppm larger than the 1986 recommended value of α or one-third the latter's 0.045-ppm uncertainty, the uncertainty of the new value is 5.6 times smaller. While this would lead to similar reductions in the uncertainties of other constants such as the Bohr radius, $a_0 = \alpha/4\pi R_\infty$, and Compton wavelength, $\lambda_C = h/m_e c = \alpha a_0$, it would also drastically influence the entire structure of the 1986 least-squares adjustment since it implies that α can be taken as an auxiliary constant. Indeed, a traditional least-squares adjustment may not even be necessary in light of the following:

1. Realizations of the SI ohm are now so closely tied to measurements of R_H carried out in terms of laboratory representations of the ohm that they may be combined together and treated as measurements of R_H in ohms. Since $R_H = \mu_0 c\alpha^{-1}/2$, such measurements can be expressed in terms of α^{-1}.
2. Measurements of R_H carried out in terms of laboratory representations of the ohm have advanced to the point where the ohm representation Ω_{BI85} can be replaced with an ohm representation Ω_{Ref} defined by $\Omega_{Ref} = R_H/25\,812.8$ exactly, which can be readily realized in any laboratory. This allows the variable K_Ω to be replaced by α^{-1} since $R_H = \mu_0 c\alpha^{-1}/2$.
3. A discrepancy between the two values of $d_{220}(\mathrm{Si})$ considered for use in the 1986 adjustment has now been resolved experimentally so that $d_{220}(\mathrm{Si})$ and $V_m(\mathrm{Si})$ may be combined and the measured quantity taken as N_A. This allows the variable $d_{220}(\mathrm{Si})$ to be eliminated and, moreover, N_A can be expressed in terms of α^{-1} and K_V.
4. A new stochastic input datum is now available from watt realization experiments as discussed above. The measured quantity may be taken as K_W, the ratio of W_{Ref} to the SI watt W, where $W_{Ref} = V_{76\text{-}BI}^2/\Omega_{Ref}$.
5. In certain cases it is more appropriate to combine $\gamma_p'(\mathrm{low})$ and $\gamma_p'(\mathrm{high})$ measurements from the same laboratory to obtain a single value of γ_p' in SI units rather than to reexpress them individually in terms of $A_{Ref} = V_{76\text{-}BI}/\Omega_{Ref}$.

The net result of these changes is that the five adjustable constants or unknowns of the 1986 adjustment may be reduced to just three: α^{-1}, K_V, and μ_μ/μ_p. There are still 12 distinct types of stochastic input data but some are different from 1986. They are (1) R_H (in ohms); (2) K_V; (3) K_A ($K_A = A_{Ref}/A$); (4) K_W ($K_W = W_{Ref}/W$); (5) F; (6) $\gamma_p'(\mathrm{low})$; (7) $\gamma_p'(\mathrm{high})$; (8) γ_p'; (9) N_A; (10) α^{-1}; (11) μ_μ/μ_p; and (12) $\nu(\mathrm{Muhfs})$. The observational equation for $\gamma_p'(\mathrm{high})$ in this new scheme is

$$(\alpha^{-1})^{-1}K_V^{-2}(\mu_\mu/\mu_p)^0 = F_7\gamma_p'(\mathrm{high}), \qquad (13)$$

with $F_7 = 8R_\infty R[\mu_0 c^2(\mu_p'/\mu_B)E]^{-1}$, where $R = 25\,812.8\ \Omega$ exactly. If α is now taken as an auxiliary constant, this new least-squares adjustment scheme reduces to (a) determining K_V from a single weighted mean of individual values of K_V obtained from stochastic input data items K_V, K_A, K_W, F, $\gamma_p'(\mathrm{high})$, γ_p', and N_A; and (b) determining μ_μ/μ_p from a simple weighted mean of μ_μ/μ_p values obtained from stochastic input data items μ_μ/μ_p and $\nu(\mathrm{Muhfs})$. Thus the complex, coupled, five-variable least-squares adjustment of 1986 will have been reduced to two simple, independent, single-variable weighted averages. For the first time in several decades, the need for a complete least-squares adjustment would be unnecessary.

The final recent development to be mentioned is of considerable practical importance because it deals with the need of commerce, industry, and science for long-term repeatability and worldwide uniformity of voltage and resistance measurements. Because this need often exceeds the accuracy with which the SI units for such measurements, the volt and ohm, can be readily realized, by international agreement, a new representation of the SI volt based on the Josephson effect to replace that of 1972, and a representation of the SI ohm based on the quantum Hall effect to replace artifact resistance standards, came into effect worldwide starting on 1 January 1990. Following the 1988 recommendations of the International Committee of Weights and Measures (CIPM) and its Consultative Committee on Electricity (CCE), the same conventional values for $2e/h$ and h/e^2 were adopted by the national standards laboratories for deriving a standard reference voltage and standard reference resistance from the Josephson and quantum Hall effects based on the relations given earlier:

$$U_J(n) = nf(2e/h)^{-1}, \qquad (1)$$

$$R_H(i) = (h/e^2)i^{-1}. \qquad (6)$$

The internationally adopted conventional values were obtained by CCE working groups from their analyses of the data available up to 15 June 1988. Denoting $2e/h$ and h/e^2 in the above equations by the symbols K_J and R_K following the CIPM and CCE, these conventional values are

$$K_{J\text{-}90} = 483\,597.9\ \mathrm{GHz/V}, \qquad (14)$$

$$R_{K\text{-}90} = 25\,812.807\ \Omega, \qquad (15)$$

exactly, where the subscript 90 indicates that these values came into effect starting 1 January 1990 and not before. K_J and R_K are termed the Josephson and von Klitzing constants, respectively, after the discoverers of the two effects. In actual fact, the CIPM and CCE view K_J and R_K as phenomenological constants characteristic of the Josephson and quantum Hall effects as described by Eqs. (1) and (6); they are not intended to represent the combination of constants $2e/h$ and h/e^2. However, for the purpose of including data from measurements of fundamental constants in the derivation of their recommended values of K_J and R_K, the CCE working groups assumed that $2e/h = K_J$ and $h/e^2 = R_K$. This data included the watt realization value of $2e/h$ with its 0.077-ppm uncertainty and the a_e value of α with its 0.008-ppm uncertainty discussed above. However, the CIPM and CCE took a very conservative approach in analyzing the data and were only willing to state that an ideal representation of the volt based on the Josephson effect and $K_{J\text{-}90}$ is expected to be consistent with the SI volt to within an assigned relative one-standard-deviation uncertainty of 0.4 ppm. The similar figure for an ideal representation of the ohm based on the quantum Hall effect and $R_{K\text{-}90}$ is given as 0.2 ppm. The CIPM and CCE's conservatism is perhaps justified in view of the history of recommended values of the constants as exemplified by Table II and Figure 1, and because the new conventional value of K_J exceeds the CCE's 1972 value by 8.065 ppm even though at the time the latter was adopted it was believed to be consistent with the SI value to within an uncertainty of 0.5 ppm.

TRENDS

The fundamental-constants field is ever evolving; new experiments are undertaken with each major new advance in measurement science. Especially great progress was made immediately after World War II as a direct result of the technological advances made during the war in the fields of microwaves and electronics. During the 1970s and 1980s, the laser and computer in particular have made possible significant improvements in the determination of a number of constants, for example, c, R_∞, N_A, γ_p'(low), and $2e/h$.

While it is always difficult to predict what will happen in this field in the future, a number of experiments currently under way can provide clues. For example, a new experiment to determine both μ_μ/μ_p and ν(Muhfs) is being undertaken which, together with improvements in the QED-dependent theoretical expression for ν(Muhfs), could yield a value of α with an uncertainty of about 0.01 ppm. While this value would depend on QED, comparing it with the value of α derived from a_e(expt) and a_e(theor) would still provide a highly useful check on the internal consistency of QED.

A somewhat related experiment is aimed at determining the g-factor of the muon, $g_\mu = 2(1 + a_\mu)$, with significantly improved accuracy by directly measuring the muon magnetic moment anomaly a_μ with the aid of a new muon storage ring. Coupled with improvements in the QED theoretical expression for a_μ and the completion of associated experiments needed to improve the calculation of the so-called hadronic contribution to a_μ, a comparison of a_μ(expt) and a_μ(theor) will test predictions of the "Standard Model" of particle physics which describes the electromagnetic, weak, and strong forces through which the elementary particles interact. A reliable, QED-independent value of α is also required for this comparison.

Another area of considerable promise involves measurements of the mass ratios of the same or different atoms in various ionization states by measuring the cyclotron resonance frequencies of single ions stored in a Penning trap. Accuracies (in the ratios) approaching a few parts in 10^{12} may be feasible and would lead to several orders of magnitude improvement in the atomic masses of atoms as well as a new means for measuring atomic binding energies. The method might even allow the determination of the Lamb shift in the ground state of completely ionized heavy atoms such as uranium. As noted earlier, comparing experimental and theoretical values of the Lamb shift is yet another way to test QED critically.

Replacing the artifact kilogram is the final area of active research to be touched upon in this section. However, it may well be the most significant because of its practical implications. The kilogram is the only base unit in the SI still defined by a material artifact. The current definition of the kilogram is:

> the kilogram is the unit of mass; it is equal to the mass of the international prototype of the kilogram.

The platinum–iridium prototype was established by the Convention du Mètre which was signed on 20 May 1875 and is rarely used for fear of damage. Indeed, a series of international comparisons of national kilogram standards begun in 1988 constitute only the third use of the prototype.

One approach to replacing the artifact kilogram is to base it on the mass of a fundamental particle such as the proton or electron, for example, on n_e, the number of free electrons, at rest, in one kilogram, defined to be numerically equal to $1/m_e = N_A/M_e$, where M_e is the molar mass of the electron. The formal definition might read:

> the kilogram is the unit of mass; it is equal to the mass of n_e free electrons at rest [n_e to be given].

n_e could be obtained by determining M_e from mass ratio measurements as discussed above involving the electron and ^{12}C ions and N_A from measurements of the lattice spacing, molecular weight, and density of a single crystal of silicon. An alternate and perhaps more promising route to n_e is to take advantage of the following relation:

$$n_e = \{1/m_e\} = \{\mu_0 c^2 (2e/h)^2 / 16 R_\infty \alpha^{-1}\}, \qquad (16)$$

where { } indicates numerical value only and all quantities are in SI units.

It is generally agreed that any replacement for the kilogram should have an uncertainty associated with it no greater than about 1 part in 10^8. The uncertainty of R_∞ is already at the few parts in 10^{10} level and only slight improvement in a_e(expt) and a_e(theor) is required to reduce the uncertainty in α obtained from their combination to a few parts in 10^9. Thus, since μ_0 and c are defined constants, the key issue is determining $2e/h$ with an uncertainty of about 5 parts in 10^9. Although difficult, it does not seem totally out of the question. Indeed, an improved version now under construction of the experiment mentioned in the previous section which yielded a value of $2e/h$ with an uncertainty of less than 8 parts in 10^8 by equating electrical and mechanical power (i.e., realization of the SI watt) just might possibly achieve the desired accuracy. It is quite conceivable that by the turn of the century, the unit of mass in the SI will be based on fundamental physical constants of nature.

"NUMEROLOGY" AND TIME VARIATIONS

No article on the fundamental physical constants would be complete without at least a brief discussion of why the constants have the values they do, and whether they are in fact really constant. For example, why does the ratio of the rest mass of the proton to that of the electron happen to be 1836.15 . . . ? Why does α^{-1} happen to equal 137.035 . . . ? Why not other, perhaps simpler, values? Three rather different points of view have evolved about the constants, especially their dimensionless combinations.

One viewpoint is that the numbers are not at all arbitrary but can be calculated from some as yet unknown basic theory (or theories) in much the same way that the g-factors of the electron and muon can be calculated from QED, or even in the way that the ratio of the circumference of a circle to its diameter (i.e., the constant π) need not be experimentally determined but can be calculated to arbitrary accuracy from mathematics. For example, in 1969 Wyler derived an expression for α involving only simple integers and π based on the volumes of certain bounded spaces associated with the invariance group of a relativistic quantum wave equation, and giving a value for α in excellent agreement with the recommended value at the time. However, the physical basis for the derivation was not at all clear, and it is safe to say that there does not yet exist a physically meaningful "derivation" of an accurate value for a fundamental constant, or dimensionless ratio or combination of constants.

The second viewpoint is not aimed at explaining the values of the physical constants but simply notes that if they were terribly different from their currently observed values, we would not be here to measure them—life as we know it on earth, and perhaps the existence of the universe itself, depends on complex physical processes that require the constants to have their observed values. This is the idea behind the so-called weak anthropic principle. The strong anthropic principle goes even further by speculating that the fundamental constants and laws of physics *must* be such that life and the universe as we know it can exist.

An extreme statement of the third point of view is that the fundamental constants change with time as the universe evolves and therefore the currently observed values are not particularly significant. This idea had its origins in 1937 when Dirac noted that the ratio of the electric to gravitational forces between the electron and proton is of the order 10^{40}, very nearly equal to the age of the universe expressed in an appropriate atomic time unit, for example, the time required for light to travel a distance equal to the classical electron radius. He speculated that either the equality is accidental or it indicates a causal relationship between electromagnetism, gravitation, and cosmology. If such a relationship is in fact true, then since the universe is continually aging, the equality would remain unchanged only if one or more of the fundamental constants entering the two ratios changed, for example, the Newtonian gravitational constant G or elementary charge e.

These speculations have stimulated a considerable amount of theoretical and experimental work over the last 50 years, the latter often involving astronomical or geophysical observations. As a result, stringent limits (albeit frequently model dependent) have been placed on the possible time dependence of a number of physical constants. For example, it has been shown that h changes less than 2 parts in 10^{12} per year, that G changes less than 3 parts in 10^{11} per year, and that α changes less than 5 parts in 10^{18} per year. Clearly, for a long time to come, the changes in the values of the fundamental constants caused by man's ineptitude (see Table II and Fig. 1) will dominate any changes that might be caused by Nature's laws!

CONCLUSION

The following quotation by a famous physicist of the early decades of this century seems to be a quite fitting conclusion to this article:

> Why should one wish to make measurements with ever increasing precision?
> Because the whole history of physics proves that a new discovery is quite likely to be found lurking in the next decimal place.
>
> F. K. Richtmyer, 1931

See also JOSEPHSON EFFECTS; QUANTUM HALL EFFECT; SYMBOLS, UNITS AND NOMENCLATURE.

BIBLIOGRAPHY

E. R. Cohen and J. W. M. DuMond, "Our Knowledge of the Fundamental Constants of Physics and Chemistry in 1965," *Rev. Mod. Phys.* **37,** 537 (1965). (I-A)

E. R. Cohen and B. N. Taylor, "The 1973 Least-Squares Adjustment of the Fundamental Constants," *J. Phys. Chem. Ref. Data* **2,** 663 (1973). (I-A)

E. R. Cohen and B. N. Taylor, "The 1986 Adjustment of the Fundamental Physical Constants," *Rev. Mod. Phys.* **59,** 1121 (1987). (I-A)

P. H. Cutler and A. A. Lucas, eds., *Quantum Metrology and Fundamental Constants,* NATO ASI Series, Series B: Physics, Vol. 98. Plenum Press, New York, 1983. (E-I)

A. Ferro-Milone, P. Giacomo, and S. Leschiutta, eds., *Metrology and Fundamental Constants,* Proceedings of the International School of Physics "Enrico Fermi," Course LXVIII. North-Holland, Amsterdam, 1980. (E-I)

D. N. Langenberg and B. N. Taylor, eds., *Precision Measurement and Fundamental Constants,* National Bureau of Standards Special Publication 343. U.S. Government Printing Office, Washington, DC, 1971. (I-A)

B. W. Petley, *The Fundamental Physical Constants and the Frontier of Measurement.* Adam Hilger Ltd., Bristol, 1988. (E-I)

B. N. Taylor, W. H. Parker, and D. N. Langenberg, *The Fundamental Constants and Quantum Electrodynamics.* Academic, New York, 1969. Also published in *Rev. Mod. Phys.* **41,** 375 (1969). (I-A)

B. N. Taylor and W. D. Phillips, eds., *Precision Measurement and Fundamental Constants II,* National Bureau of Standards Special Publication 617. U.S. Government Printing Office, Washington, DC, 1984. (I-A)

B. N. Taylor and T. J. Witt, "New International Electrical Reference Standards Based on the Josephson and Quantum Hall Effects," *Metrologia* **26,** 47 (1989). (I-A)

Coriolis Acceleration

Robert H. March

The apparent or "fictitious" acceleration attributed to a body in motion with respect to a rotating coordinate system (e.g., the frame of reference fixed with respect to the earth's surface) is known as the Coriolis acceleration. It is fictitious in the sense that if the same motion is described in a nonrotating (inertial Newtonian) reference frame, no such acceleration is present. It is merely an apparent deviation from inertial motion due to the choice of a noninertial reference frame.

Coriolis acceleration is experienced *only* by bodies with nonzero velocity in the rotating frame, unlike *centrifugal* acceleration, which is independent of the state of motion or rest.

If the body has linear velocity **v** with respect to a frame rotating with angular velocity **ω**, the Coriolis acceleration **a** is given by the pseudovector expression

$$\mathbf{a} = -2\boldsymbol{\omega}\times\mathbf{v}.$$

In certain cases, the origin of Coriolis effects is intuitively obvious if the motion is described in an inertial frame. The simplest examples are those where the motion brings the object closer to or farther from the axis of rotation, changing the linear speed of motion due to the frame's rotation. For example, a falling body released from rest will be "deflected" to the east of a vertical line, because its point of release is moving faster than points that are below it and hence closer to the axis of rotation. By a similar argument, a projectile in the northern hemisphere will be deflected to the right, and one in the southern hemisphere to the left. At the equator, or for a trajectory that is due east or west, **a** is vertical and hence affects the range rather than the direction of the projectile.

Terrestial applications of Coriolis acceleration include corrections to long-range artillery tables and the analysis of the precession of Foucault's pendulum. It is responsible for establishing the sense of rotation of large-scale weather systems, so that the wind pattern of a major storm is always "cyclonic" (counterclockwise) in the northern hemisphere and anticyclonic (clockwise) in the southern.

The phenomenon is named after Gaspard Gustave de Coriolis (1792–1843), who in 1835 published the first complete analysis of motion in rotating coordinate systems. Some Coriolis effects, however, were known much earlier. The eastward deflection of a falling body, for example, was cited by Robert Hooke and Isaac Newton as a possible experimental proof of the earth's rotation. In general relativity, certain terms that appear in the metric tensor of rotating coordinate systems are known as Coriolis terms, because for velocities much less than the speed of light and in weak gravitational fields they reduce to the classical expression for Coriolis acceleration.

See also DYNAMICS, ANALYTICAL.

BIBLIOGRAPHY

Keith R. Symon, *Mechanics*, 3rd ed., Chap. 7. Addison-Wesley, Reading, Mass., 1971. (I)

Herbert Goldstein, *Classical Mechanics*, 2nd ed., pp. 135ff. Addison-Wesley, Reading, Mass., 1980. (A)

Corona Discharge

Ernesto Barreto

A protruding convex region of radius a in an electric field E concentrates the electric flux by inducing excess charge on its surface. Thus an isolated region of high field intensity is produced when the gap d, over which a voltage difference exists, is much larger than the convex radius of curvature considered (e.g., when $d \gg a$). In a gas, at sufficiently high voltage, this condition produces local ionization that does not propagate across the whole gap d because the Laplacian field fades away in the low field region ($r \gg a$) that represents most of the space being considered. Instead, weak luminosity may be observed that can be either diffuse or attached to the stressed surface in one or more bright spots that move or become localized. This is a corona discharge. It might require dark-adapted eyes or even photon multiplication to detect its luminosity onset. As the voltage is increased across the gap d, the region of luminosity grows and eventually produces thin filaments that are called streamers. Positive streamers from a thin needle in air are particularly conspicuous and can be emitted at different angles to the axis of symmetry faster than the eye can resolve sequential events. They are responsible for the name corona. On the mast of a ship, sailors observe bunches of streamers and call them St. Elmo's fire. When streamers cross a discharge gap they lead to filamentary sparks. A pressure decrease also causes the luminous corona region to expand because ionization is a function of E/n, with n the neutral number density. However, at a critical pressure (in air ~ 10 mm Hg) coronas become impossible because of the onset of a new, diffusion-dominated, glow discharge. This is characterized by a weakly ionized glowing plasma ($T_e > T_n$) that fills most of the region between electrodes and can follow the contours of a discharge tube (e.g., a neon sign). In two dimensions, coronas occur when a wire or thin rod is surrounded by a much larger cylinder or placed parallel to a plane surface. In three dimensions they occur whenever an object protrudes into a uniform field configuration. Examples are an asperity in a metal surface, a needle facing a plane electrode, a tree high above others, or an airplane. Subsistence of a corona depends on the ability of the electrodes to remove the charge that reaches them. If isolated or nonconducting objects are involved, charge of polarity opposite to that originally on their surface will reach them. The discharge stops but a net charge transfer has taken place. This may charge or neutralize the objects involved. For instance, sharp wires are placed on the tips of airplane wings to promote local ionization that maintains the aircraft close to the potential of the region where it flies. Conversely, a nonconducting surface is purposely charged by coronas in electrophotography. Whenever a charged dust, powder, or spray is produced, induced coronas on the walls of the containing surface provide free charge that limits the space-charge density. These induced coronas indicate the onset of a dangerous condition in an explosive atmosphere and are detrimental for the purposes of electrostatic spraying in painting or agricultural fumigation. Coronas from liquid surfaces are associated with the production of filaments, charged aerosols, and surface oscillations.

For the sake of concreteness, consider only unipolar metal electrodes at atmospheric pressure. Avalanching electrons in the corona region are accelerated between collisions in the field direction (eE/m_e) but collide primarily and at high frequency with neutral molecules ($\nu_{en} \sim 10^{12}\ \mathrm{s}^{-1}$). Very rapidly, electrons obtain an equilibrium velocity distribution ($nt \sim 10^{15}\ \mathrm{m}^{-3}$ s; or 4×10^{-11} s at $n = 2.5 \times 10^{25}\ \mathrm{m}^{-3}$). The electron average energy is limited by inelastic collisions and is, therefore, always significantly smaller than the thresholds for optical excitation and ionization of the gas particles. Ionization is produced only by a small number of electrons in the high-energy tail of a primarily isotropic electron velocity distribution with an average speed, c_e, that is much larger than the electron drift velocity, v_d, in the field direction: $c_e \gg v_d(E)$. Consequently, for ionization, it makes little difference whether the electrons drift into a positive electrode or away from a negative one. By contrast, the positive ions stay practically in place during the time it takes electrons to get into equilibrium with the electric field. Their space charge plays a dominant but very different role in positive and negative coronas.

Near a negative point, positive ions left by avalanches enhance the electric field at the electrode surface but reduce the extent of a high-field region from which free electrons have just been expelled at speeds thousands of times the drift velocity of ions (~100 m/s). If the electrons do not attach in the low-field drift region, positive ions will eventually tunnel or sputter electrons from the metal cathode and the discharge goes into a spark. If the electrons attach, they become slow negative ions. These interact with the positive ions and the discharge stops until the negative ions can disperse far into the low-field region. A Trichel pulse is formed. For instance, negative coronas in pure nonattaching gases such as H_2 or N_2 (with outgassed, baked electrodes) are not stable and in pure Ar there are no coronas at all. However, the addition of 0.1% O_2 to any of these gases produces Trichel pulses. In dry air, for point radii above 20 μm, a series of equally spaced Trichel pulses are produced at their onset (~1.0 kHz, 1.5 ns rise time, milliamperes peak current and ~50 ns decay time). An increase in voltage increases their frequency but does not change their shape; this depends on the gas used. Each pulse maximum has been identified with the cathode region of a miniature glow discharge. This glow discharge is fully developed with Crooke's dark space, negative glow, Faraday dark space, and a positive column. This last one starts only about 100 μm from the surface and fades away in another 100 μm in the shape of a luminous, sharply divergent cone or brush. Cathode damage confirms interaction with the electrons in the metal. Trichel pulses start at frequencies of kHz, increase to MHz, and disappear to make the current steady (at about 120 μA) with only a small jump in the current–voltage characteristic. At the high-frequency limit, pulses might become irregularly timed because two or more cathode spots may act simultaneously.

Near a positive electrode, positive ions decrease the geometrical field intensification and stop the discharge. This will start again when the positive ions disperse into the low-field region, or when a free electron is produced by photoionization in a nearby region that is, nevertheless, sufficiently far not to be affected by the positive space charge. A burst pulse is produced. These have 10^6–10^7 ions, durations of 10–100 μs, and average currents of 10^{-12}–10^{-10} A. Bursts traveling along a wire in an easily photoionizable mixture of gases were fundamental to the development of Geiger counters. If the burst has 10^8–10^9 ions the localized positive space charge facing the drift region provides a local field sufficient to ionize. Filamentary preonset streamers (~20 μm in diameter) are produced that travel a few millimeters at speeds of 10^5–10^6 m/s (larger than the electron drift velocity) into the low-field region.

The existence of a drift region, where ions of the same polarity as the stressed electrode can accumulate, is an intrinsic part of coronas of either polarity. If this region is not provided (small gaps, nonattaching gases), coronas do not exist. However, it must be emphasized that coronas depend on localized fields due to a geometrical configuration and that it is not possible to specify onset conditions equivalent to those in a uniform field. Characteristic times and lengths in the ionization region are nanoseconds and micrometers, while in the drift region they become milliseconds and meters. In air, coronas start at about 10 kV if a rounded rod 0.1 cm in radius is used in a 1.0-cm point-to-plane gap. However, if a rod 0.02 cm in radius is used in the same gap, a corona starts at about 5.0 kV. Atmospheric pressure, humidity, and polarity affect the actual onset value. In all cases, the electric field near the stressed electrode is many times the magnitude of the breakdown value in a uniform field. An increase in voltage causes ions in the drift region to disperse faster and the corona pulses (Trichel, or preonset streamers) to become closer. Eventually steady current and luminosity are established around the stressed electrode. The current density distribution in a point-to-plane drift region is given by $j(\theta) = j(0) \cos^m \theta$; θ is the angle from the axis of symmetry. For positive coronas $m = 4.82$, for negative ones $m = 4.65$. $j(\theta) = 0$ for $\theta > 60$ degrees. This distribution is known as Warburg's $\cos^5\theta$ law. The average corona current was calculated in 1914 by Townsend to be proportional to $V_{gap}(V_{gap} - V_{onset})$.

As the current increases, the corona luminous region also grows and the field becomes less divergent. Eventually, a breakdown streamer about 50 μm in diameter will be emitted into the low-field region ($r \gg a$) at speeds between 10^4 and 10^6 m/s depending on a, d, and the gas composition. In uniform fields, streamers propagate simultaneously in both directions when an avalanche is purposely started in midgap and electron amplification reaches about 10^8 in a distance of 3–5 mm. Negative streamers (anode directed) are timed to be faster than the positive ones. In all cases, streamers have speeds larger than the corresponding electron drift velocity in the given electric field and may exhibit changes in direction that do not follow the Laplacian field configuration. This shows that they are driven by local space-charge accumulation that provides the field which pushes them into unionized gas. In molecular gases such as air (usually "laboratory air": 79% N_2, 21% O_2) the electrons in a streamer have a small average energy (2–4 eV) fixed by a high peak in the cross section of N_2 due to a vibrational excitation peak at 2.5 eV. Only molecular radiation is observed and indicates that the gas is not heated. The electron densities are 10^{18}–10^{21} m^{-3}. These numbers suggest that although the electrons are collision dominated ($\nu_{ee} = 1.41\nu_{ei} \ll \nu_{en}$), the degree of ionization (n_e/n) has reached the stage of multiple Coulomb interactions and that single-particle effects are less important than collective effects between them ($\nu_{ee} \geqq m_e\nu_{en}/m$). The streamer channel behaves like a weakly ionized plasma that screens external fields in sheaths of the order of the Debye length, $\lambda = 69(T_e/n_e)^{1/2}$ m, that can form in times given by the plasma frequency $\omega/2\pi = 8.97 n_e^{1/2}$ s^{-1}. For the range

$10^{18} < n_e < 10^{21}$ m^{-3} and assuming $T_e = 1.0$ eV one obtains $7.4 < \lambda < 0.2$ μm and $110 > 2\pi/\omega > 3.52$ ps. This picture of a streamer as a fairly conducting filament where net charge is limited to very small distances contrasts with an earlier, alternative, model where a positive streamer has a head with net positive space charge that results because of the difference in mobility between electrons and ions but without multiple interactions or Coulomb shielding. Positive streamers can be injected through a small aperture into a uniform field and propagate at field strengths between 4×10^5 and 5×10^5 V/m for distance at least, but probably exceeding, 1 m. When streamers cross a discharge gap, the current either significantly decreases or saturates before a spark is produced.

The lower part of a thunderstorm is usually negatively charged. Inside this cloud, corona discharges from ice or water initiate lightning at heights of 6–8 km above sea level. Ionization inside the cloud persists from tens to hundreds of milliseconds before a luminous channel that exhibits atomic radiation (H_α) emerges into the air. It is not understood how this channel develops from initial coronas. The emerging negative discharge travels in steps photographed to last ~1 μs and to travel tens of meters. Each new section of this stepped leader may choose a new direction of travel thus indicating strong space-charge accumulations. A very long pause between steps (~50 μs) seems to require some interaction with the cloud in order to maintain a conducting channel (similar to restrikes in long laboratory discharges). When the stepped leader approaches ground, it induces an upward moving positive discharge (misnamed a "streamer") from the grounded region of maximum induced stress such as a lightning rod. This upward-moving discharge meets the stepped leader and a very steep fronted wave, the return stroke, is produced. This travels to the cloud along the initial discharge channel at speeds of the order of that of light (~c/3), and with some attenuation with height (typically ~30 kA peak in ~1 μs producing a fully ionized plasma, T ~ 3×10^4 K). The return stroke reaches the cloud in about 100 μs. After ten to hundreds of milliseconds, a new luminous wave, the dart leader (~50 m long), emerges again from the cloud but travels continuously, in the same channel, to ground at speeds of the order of 10^6 m/s. It produces another return stroke. Continuous dart leader propagation suggests that the plasma made by the previous return stroke is still conducting. The sequence leader return stroke may repeat several times (with a world average of 3.5) and constitutes a single cloud-to-ground flash. Lightning may also start first as a positive discharge induced directly on a tall structure by the negative lower portion of a thunderstorm. Tall buildings, rockets, long wires rapidly ejected into the air (triggered lightning), and, as we all know, even kites are suitable objects. When the upper positive section of a storm deviates from vertical (a tilted dipole), a positive leader may be formed by the cloud. This does not step and produces a single return stroke with a long period of continued current.

All stages of lightning associated with its extension into the cloud or in air are associated with incipient corona streamers that then change to a higher-current type of discharge. Within the cloud this includes initiation and discharge extension between strokes; in the air, the propagation of positive and negative leaders. The differences with polarity may just reflect the difference between coronas. However, the physics is poorly understood and the temptation to extrapolate from laboratory discharges has to be very carefully evaluated.

See also Arcs and Sparks; Electrophotography; Lightning.

BIBLIOGRAPHY

M. A. Aronov, E. S. Koletchitsky, V. P. Larionov, V. F. Minein, and J. G. Sergeev, *High Frequency Electrical Discharges in Air*. Energia, Moscow, 1969.

L. G. H. Huxley and R. W. Crompton, *The Diffusion and Drift of Electrons in Gases*. Wiley, New York, 1974.

E. E. Kunhardt and L. H. Luessen, eds. *Electrical Breakdown and Discharge in Gases,* NATO ASI Series Vols. 89a and 89b. Plenum Press, New York, 1983.

L. B. Loeb, *Electrical Coronas: Their Basic Physical Mechanisms*. University of California Press, Berkeley, 1965.

G. L. Rogoff, Special Issue on Applications of Weakly Ionized Plasmas, *IEEE Trans. Plasma Science*. (Scheduled to appear April 1990).

R. S. Sigmond, Corona Discharges, in *Electrical Breakdown of Gases* (J. M. Meek and J. D. Craggs, eds.). Wiley, New York, 1978; also, *J. Electrostatics* **18,** 249–272 (1986).

M. A. Uman, *The Lightning Discharge*. Academic, New York, 1987.

Cosmic Rays—Astrophysical Effects

Stirling A. Colgate

Cosmic rays are energetic particles incident upon earth from sources that include the sun, the galaxy, and other galaxies. The most important particles are nuclei, but an electron flux of a few percent is also present. In addition to high-energy gamma rays and secondary products, neutrinos and antiprotons are also now considered as cosmic rays. The extraordinary pulse of neutrinos from the neutron star collapse initiating the supernova event 1987a in the Large Magellanic Cloud should also be considered as cosmic rays. The characteristics of cosmic rays that are significant astrophysically are (1) the energy spectrum; (2) the composition or number frequency of various nuclei or positron–electron ratio; and (3) the angular distribution, commonly measured as the degree of anisotropy.

The energy spectrum describes the most surprising feature of cosmic rays, namely, the extraordinary energy of individual particles and the large total energy flux. The kinetic energy of individual particles is measured variously in eV (electron volts) per nucleon, MeV, GeV, and TeV. When cosmic rays were first discovered in the early 1920s as residual ionization in an ion chamber and later as pulses in a Geiger counter and tracks in a cloud chamber, the implied energy necessary to penetrate the earth's atmosphere because of ionization loss alone was necessarily greater than a GeV, i.e., the rest-mass of the proton. The flux was like rain, of the order of 1 cm^{-2} s^{-1}. Figure 1 shows a composite

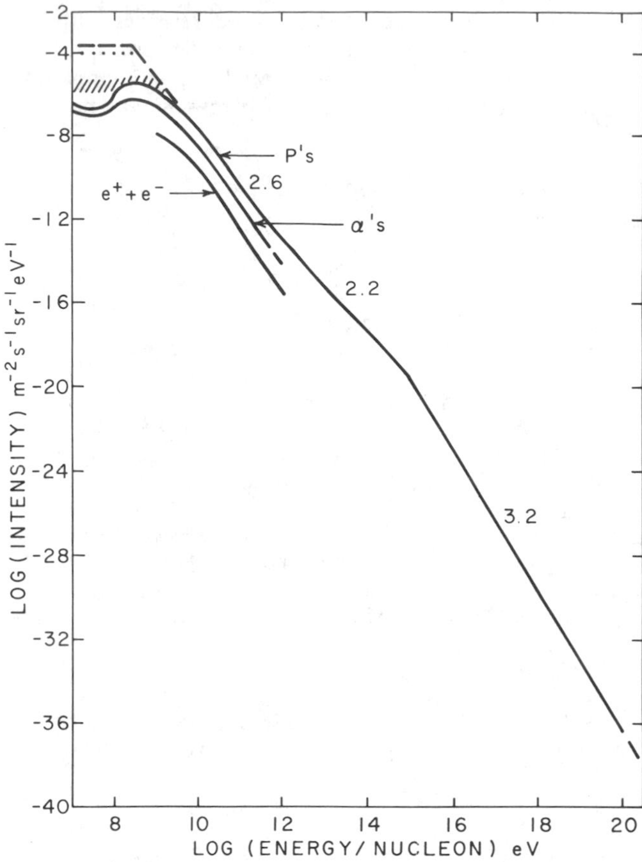

FIG. 1. The differential energy spectra of cosmic rays as log intensity versus log energy is shown. The numbers attached to the curve are the negative exponent of the corresponding power law that approximates the measurements. Above 10^{15} eV, the primary particles may be nuclei rather than protons (p's) as indicated at lower energy. The low-energy spectra of helium nuclei (α's) are shown and the curves for the heavier nuclei are similar but of lower intensity as discussed in the text. The electrons (e^+,e^-) have similar spectra. The shaded curve is the estimated galactic flux as interpreted (with some questions) from satellite measurements. The dashed curve is the flux required for interstellar cloud heating and the dotted is that required for pressure support of the galactic disk.

of the presently most likely spectrum of nonsolar cosmic rays. The most striking feature is the extension of a nearly constant power law over a range of 10^{11} in energy and then also the extraordinary upper energy limit of 10^{20} eV. A particle of this energy is only slightly deflected by the magnetic field of the whole galaxy and so compels the conclusion that some particles must reach us from the universe outside our galaxy. Hence the term galactic cosmic rays is too parochial. Finally, the exclusion of the solar flux of cosmic rays allows one to consider cosmic rays as a unified subject where the measurements have been exceedingly numerous and self-consistent.

Solar cosmic rays, on the other hand, have been recognized as correlated with solar activity and contribute a highly variable composition and spectrum which is almost exclusively less than 1 GeV and is mostly in the region of 10 to 100 MeV. The lack of consistency of solar-cosmic-ray measurements has made interpretation so far almost impossible.

Galactic cosmic rays exclusive of high-energy gamma rays are constant in time, in all measured quantities, and the total flux has evidently not changed significantly in several hundred million (10^8) years as measured by fossil tracks in lunar rocks.

The bulk of the energy and, hence, pressure of and heating by cosmic rays that give rise to the principal astrophysical effects of cosmic rays on our galaxy is contained in the low-energy part of the spectrum, $\lesssim$1 GeV. In Fig. 1 the flattening of the spectrum below 1 GeV is variously attributed to the properties of the source(s), energy loss in propagation, and partial exclusion from the solar-system magnetosphere of cosmic rays originating from elsewhere in the galaxy. (A small solar modulation is superimposed due to solar activity modification of the solar magnetosphere.) The actual galactic flux at low energy ($E\lesssim$300 MeV) is not known accurately although several secondary indicators are considered. The satellite measurements out to 3 A.U. would indicate a galactic flux within the shaded area of Fig. 1. This is less than the flux required for the heating of the interstellar medium by low-energy cosmic rays, which would require a flux indicated by the dashed line. Finally, the support of the gas of the galactic disk by cosmic-ray pressure would require an intermediate flux indicated by the dotted curve. The heating and pressure support of the galactic gas by cosmic rays are each major theories of galactic structure. The fact that both these lines lie above the extrapolation of the satellite measurements is recently taken as an indication that these major theories of galactic structure may have to be revised. The expected degree of ionization of the interstellar gas from cosmic-ray heating is also not observed. On the other hand, two theories of the origin of cosmic rays—the shock ejection by supernova and the quasicontinuous pulsar acceleration—both lead to spectra at least as steep as that required to meet galactic structure theories. What is least known or perhaps controversial in the theoretical sequence of events leading to a prediction of the energy spectrum is the adiabatic energy loss from the expansion of the nebula associated with any given energetic event that accelerates cosmic rays. (The cosmic rays blow a bubble and this takes all the cosmic-ray energy.) Only if cosmic rays are accelerated fairly uniformly and continuously throughout the whole galaxy can the difficulty of adiabatic energy loss be totally avoided; on the other hand, if the interstellar medium is as inhomogeneous as present measurements would indicate with minimum densities less than 10^{-2} of the mean, then single-origin events like supernovae or pulsars can allow cosmic rays to escape into low-density regions where presumably the magnetic field is consistently low and the necessary confinement by magnetic fields for adiabatic loss less likely. Such low-density regions naturally interconnect and have been described as "tunnels" in the interstellar medium. The modification of the spectra of the low-energy cosmic rays as they propagate in such an inhomogeneous region is unknown so that the source spectrum of low-energy cosmic rays below 1 GeV, although of major importance to galactic understanding, must be accepted as uncertain. A recent series of satellite measurements of the 100-MeV gamma-ray distribution in space has been interpreted as due to the creation of π^0 mesons by cosmic rays of energy $\simeq$1 GeV colliding with protons of the interstellar medium and then these mesons subsequently decaying to two gamma rays. If these cosmic rays are indeed distributed with constant flux within the galaxy, then the distribution of gamma rays is consistent with the lower flux that conflicts with the two theories of cosmic-ray heating and cosmic-ray pressure support of the galactic gas. On the other hand, the natural tendency for cosmic rays to fill the low-density holes or tunnels in the interstellar medium leaves these major conclusions uncertain. Finally, the magnetic confinement postulated as necessarily causing adiabatic loss may not be adequate as has been found in the solar corona and in the laboratory so that supernovae or pulsars may inject cosmic rays into the galaxy without the problem of adiabatic cooling.

As one considers higher energy, $10^{10} \leq E \leq 10^{12}$ eV, the measurements are still definitive, but the theoretical impact on galactic phenomena is less; instead the measurements lead to information concerning the properties of the source of cosmic rays. There are two measurements of dominant concern: (1) the near independence of the energy spectra in eV per nucleon regardless of the wide range of nuclear species observed, and (2) the incredible isotropy of the flux (i.e., independence of arrival direction) at energies great enough to be unaffected by terrestrial and solar-system magnetic fields.

In general the energy spectrum of each nuclear species when measured in energy per nucleon, i.e., total energy of the nucleus divided by atomic number, appears to be very closely the same, independent of whether we measure protons, helium nuclei, or carbon through uranium. Of course one can measure helium at higher energy (<[or up to] 10^{12} eV) than the less abundant heavier nuclei, but in general wherever measurements can be made, the most striking feature is that it appears as if the acceleration mechanism is not selectively sensitive to nuclear charge. It now appears that the spectrum of Fe up to 10^{12} eV/amu is flatter, i.e., harder by $\cong$10% in the exponent, sufficiently to indicate a significant enrichment of the Fe in the composition above 10^{14} eV per iron nucleus (2×10^{12} eV per nucleon). In addition there is small but recognizable modulation of the composition correlated with the first ionization potential of the respective element as if injection was influenced by ionization state. However, the general conclusion holds that acceleration and propagation are remarkably insensitive to charge state and because of it, strongly argues against the most outstandingly attractive theory of the origin of cosmic rays proposed by Fermi. In Fermi's theory ionized nuclei, ions, "thermalize" with interstellar clouds by scattering off of the collective and presumed concentrated magnetic field of these matter condensations (100 times the average density) called interstellar clouds. These clouds can be thought of as superparticles and the approach to thermodynamic equilibrium requires that ultimately all particles have the same energy. Since the mass of these clouds is so large (10^2–10^3 solar masses) compared to a single particle, the kinetic energy of the whole cloud even at modest intercloud velocities, $\simeq 10^5$–10^6 cm s^{-1}, is essentially infinite compared to individual particles so that thermalization implies a general heating of the particles by the clouds. For this heating of the particles to occur requires relativistic velocities, $E \gtrsim 10^9$ eV, of the particles so that ionization losses are less than the gain by scattering. Hence particles must be "injected" into the scattering "accelerator." Since the ionization losses are proportional to Z^2, where Z is the nuclear charge, it is almost impossible to imagine an "injection" mechanism into such an exceedingly charge- and energy-sensitive accelerator that would not reflect major charge dependence. In addition, the cosmic rays measured in the energy region 10–300 MeV that might be thought of as currently being "injected" are also nuclear-charge independent in energy spectra and would certainly produce a charge-dependent high-energy spectrum by such a scattering accelerator. Finally, as was pointed out earlier, the total energy in the low-energy (currently-being-injected) particles is likely to be at least as large as, if not larger than, the cloud energies and hence the cosmic-ray pressure will move the clouds rather than vice versa. It should be pointed out that all proposed acceleration mechanisms must meet the essential criteria of (1) enough total energy input into the galaxy $\cong$ several $\times 10^{40}$ ergs s^{-1}, i.e., ten million suns' luminosity; (2) near charge independence of injection and acceleration, and (3) production of a power-law spectrum over a large range in energy ($\cong 10^{11}$). So much energy input should have a large effect on the galactic environment. Charge independence and near-solar composition implies a near-universal acceleration process throughout the galaxy; and finally, less restrictive is the power-law acceleration. A power-law spectrum of the accelerated particles merely defines an accelerator with fractional losses proportional to fractional gain or $dE/E \propto -dN/N$ or $\log E = -\alpha \log N$ or $N \propto E^{-\alpha}$. Thus we should "see" such a lossy and uniformly distributed accelerator, but so far it has not been identified.

As one considers energies above 10^{12} eV per amu, then the technique of measurement becomes a significant limitation. Below 10^{12} eV, measurements of total energy, ionization loss, and hence nuclear charge and mass can be made by stopping the particle within the detector, but above this energy the earth's atmosphere becomes the only feasible detector and then only the "shower" particles of the nuclear cascade in the atmosphere can be sampled and the primary energy is inferred. Because of this relatively complicated sequence of events, there exists a small but modest doubt concerning the reality of the spectrum changes for $E \geq 10^{12}$ eV per amu. The flattening of the spectrum for $10^{13} \leq E \leq 10^{15}$ eV is now fairly well accepted as well as the steepening of the slope for $E \geq 10^{15}$ eV. Very little firm information exists concerning the nuclear composition, but what little there is suggests that protons are still the primary constituent of cosmic rays at these extreme energies. If they are, then the flattening of the slope cannot be explained by a change in composition. Similarly it is hard to suggest a more effective confinement within the galaxy for increasing energy and so the flattening for $E \geq 10^{13}$ eV probably reflects a flattening from the galactic source(s) spectrum. If this is so, then the subsequent steepening of the slope above 10^{15} eV makes good sense as a progressively faster leakage of still higher-energy cosmic rays from the galaxy. The Larmor orbit of a proton at 10^{15} eV approaches the dimension of an interstellar cloud and so the diffusion time of a particle from a galactic source out of the galaxy due to scattering from a distribution of cloud sizes decreases linearly with increasing energy producing the steeper slope above 10^{15} eV. Since cosmic rays can diffuse into as well as out of the galaxy, one would expect a progressively increasing contribution from extragalactic sources at higher energies. Indeed at 10^{19} eV a proton makes less than a Larmor orbit within our galaxy and so the degree of isotropy becomes a key measure of probable extragalactic contribution. It would now appear that if most galaxies are similar to ours, then the intergalactic flux contributed from all galaxies is comparable to our own flux at $\simeq 3\times10^{18}$ eV. An alternative view is that all higher-energy cosmic rays, perhaps $E>10^{13}$ eV, are due to extragalactic sources and they then diffuse into our galaxy. A problem with this interpretation is the extreme isotropy observed up to 10^{16} to 10^{17} eV and then the requirement of explaining two source mechanisms rather than one with the arbitrary changeover at 10^{13} eV. Below this energy the necessary total energy needed to fill, and the energy density within, the intergalactic medium become improbably large and so galactic source(s) are required. Finally at the extreme energy $E\gtrsim10^{18}$ eV additional energy loss mechanisms must be considered due to the formation of $e^{\pm}$'s at $E>10^{18}$ eV and π^0's at $E>10^{20}$ eV by the interaction of the cosmic rays with the primordial blackbody radiation quanta of $h\nu\simeq10^{-3}$ eV. Measurements at the "Fly's Eye" CR shower detector indicate that the spectrum has a near cutoff at $E\gtrsim10^{20}$ eV as expected by theory and possibly with a slight bump due to energy-diffusion loss by multiple collision events with the cosmic blackbody photons. If all cosmic rays of this energy were extragalactic in origin, then the attenuation due to the $e^{\pm}$'s and π^0's occurs fortuitously just at that energy where the flatter slope should emerge above the galactic diffusion-loss steepening. Then the preponderance of evidence is that somehow the incredible range of particle energy from a few tens of MeV to 10^{20} eV is mostly created within our as well as other galaxies.

The nuclear composition of cosmic rays gives us a chance to sample matter that obviously comes to us from distant parts of our galaxy—if not the universe. The relative ratio of nuclear species—in particular with respect to the primary constituent, protons (i.e., hydrogen)—is remarkably similar to the inferred element distribution both in the solar system as well as in the galaxy as a whole. Whatever the nuclear-synthesis conditions may be, they seem to be sufficiently universal that the cosmic-ray source(s) as well as the average interstellar medium seem to have been subject at some time to similar nuclear conditions. This partial agreement may be fortuitous or fundamental, but let us first look at the departures of the cosmic-ray composition from the elemental universal abundances.

(1) Cosmic rays are enriched in heavier nuclei by a factor roughly proportional to nuclear charge. Hence iron is 26 times more abundant than expected, and the few nuclei observed in the lead to uranium group are correspondingly enriched.

(2) The light element group, Li, Be, B, are enriched roughly five orders of magnitude above universal abundance. On the other hand, the abundance of nuclei of odd–even charge approximately reflects the ratios found in nature for the group carbon to iron. The extreme abundance of the Li, Be, B group is entirely expected, even to the observed isotopic ratios, as a result of a secondary effect—namely, the partial spallation of the more abundant heavier nuclei, carbon to iron, colliding with interstellar matter—primarily hydrogen. Indeed, it is this spallation that is interpreted as giving a mean galactic age of cosmic rays of several million years on the assumption that cosmic rays are confined to the galactic disk. There is one nuclear species, ^{10}Be, that acts as a radioactive clock compared to the stable nucleus ^{9}Be, giving a mean age of $\gtrsim 2\times10^7$ years, which is significantly larger. The spallation age corresponds to the traversal of a thickness of matter by the cosmic rays which is significantly less than a spallation mean free path ($\frac{1}{3}$ to $\frac{1}{10}$) and is of course chosen so that the observed abundant nuclei like ^{56}Fe would not be completely spalled before reaching the earth. This is called the "leaky box" model. A necessary corollary of this assumed mean age is that cosmic rays must therefore escape from the galaxy at a relatively young age before undergoing complete spallation. The great age of the ^{10}Be nuclei implies that cosmic rays must spend a long time in a low-density region like a galactic "halo." The escape from, and/or confinement in, the galaxy is determined by the diffusion properties of the interstellar medium. This requirement for partial confinement implies the presence of a quasirandom magnetic field in the galaxy of $\simeq 2\times10^{-6}$ G whose energy density is only slightly less than the cosmic rays that it confines. This is why gravity acting across the plane of the galactic disk is invoked to help confine the cosmic rays.

The spallation age of 3×10^6 years and the volume of the galactic disk and cosmic-ray flux dictate that the source(s) of cosmic rays must produce $\simeq 3\times10^{40}$ ergs s^{-1} of particles with $E\gtrsim10^9$ eV, or the equivalent of $10^7\times$ the solar luminosity. This is so large that only supernovae seem capable of producing this energy. Supernovae are also the probable site of most of the nucleosynthesis that produces the more abundant heavy elements of the interstellar medium so that the partial agreement between the two compositions is an attractive feature for interpreting supernovae as the source of both cosmic rays and the nuclei, $Z>2$, of the interstellar matter. On the other hand, another interpretation of cosmic-ray composition is possible. We assume that cosmic rays at the earth are composed of two components, the one recent in origin ($\leq 10^5$ years) and a second presumed to be the equilibrium composition and flux resulting from perfect confinement in the galaxy and attenuated by collisions alone. The resulting protons are therefore primarily a spallation product and since they have a smaller cross section for destruction than their parent nuclei, they accumulate in number. The advantage of this model is that the proton lifetime is increased to 10^8 years and the energy requirement of the source(s) is correspondingly reduced.

The second model is perhaps most sensitive to test by measurements of the isotropy of cosmic rays relative to galactic coordinates as a function of nuclear charge. One would expect a difference in the degree of isotropy between recent cosmic rays from a nearby source (heavy nuclei) and the final long-life degraded product (protons). Such a measurement has not yet been made.

Instead a major effort has been spent upon measuring the isotropy of the total flux of cosmic rays, mostly protons. Figure 2 shows a summary of these measurements for energies great enough ($E>10^{11}$ eV) such that the solar-system

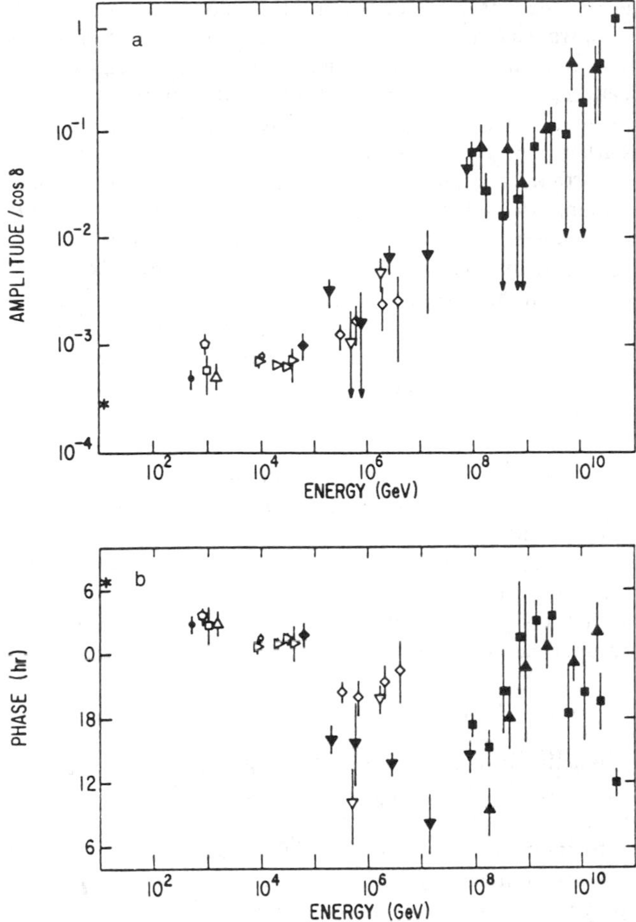

FIG. 2. Cosmic ray anisotropy. The panels show measurements of the equatorial projected amplitude and phase of the first harmonic of count rate versus energy (δ being the latitude of the experiment), compiled by J. Linsley, Proc. 18th International Cosmic Ray Conf. (Bangalore) **12,** 135 (1983).

magnetic field does not significantly affect the result. The first and most striking feature is the incredible isotropy of cosmic rays—as small as 10^{-3} up to 10^{16} eV. This implies that cosmic rays must either undergo substantial diffusion from the source to the earth or, on the other hand, be produced exceedingly isotropically. Since arguments of total energy and lifetime of the ultraheavy cosmic rays rule out an extragalactic origin at least for the low-energy and very heavy cosmic rays, we are confronted with the nonspherical nature of galactic structure, on the one hand, and the extreme isotropy of cosmic rays, on the other, and so are forced to the interpretation of at least partial confinement by diffusion. Furthermore, if supernovae are the source of cosmic rays, the relatively random occurrence of such events within the galaxy and the relatively large number within the cosmic-ray age (10^4–10^6 events) offers no conflict with the observed isotropy. Finally at very high energy we observe an increasing anisotropy above 10^{17} eV. The magnitude of this effect is still so small within the limitations of statistics that one cannot be absolutely sure of its existence in any one measurement, but the recent internal consistency of the measurements at the several large air-shower arrays gives a strong assurance that some anisotropy exists at ultrahigh energy, $\geq 10^{19}$ eV. An increasing anistropy above 10^{15} eV is expected theoretically due to progressively greater leakage from the galaxy with greater energy as pointed out before. Concomitant with this leakage out is the possibility of leakage into the galaxy from other galaxies or sources. We therefore expect both an isotropic and a highly anisotropic component at ultrahigh energies. The irony of this is that if all galaxies produce the full spectrum of cosmic rays like our own, then the extragalactic and galactic fluxes should be comparable at the point in energy where the galactic source can easily escape so that the sought-for effects may still be small in a region of most difficult measurements.

The electron component of cosmic rays is most likely the result of the cosmic rays making nuclear collisions with the interstellar medium. The spectrum is not yet well enough measured for $E \geq 10^{11}$ eV to test conclusively the theory: namely, do electrons produce the radio synchrotron radiation in the galaxy at a rate that is self-consistent with a measured electron lifetime (determined by the high-energy part of the electron spectrum) and total flux and inferred mean magnetic field? Most of this argument is indeed consistent but the modification of the high-energy spectrum due to a decreased lifetime has yet to be confirmed. Ultrahigh-energy gamma rays with energies incredibly up to 10^{17} eV have been identified with extragalactic sources—Cygnus X-3 and Hercules X-1—with excellent positional identification and with good timing in phase relative to the known x-ray emission characteristics. In order to produce these prodigious gamma-ray fluxes requires still greater energies in accelerated particles. If these accelerated "other" particles should freely escape, they could produce sufficient particle fluxes to explain high-energy cosmic rays in the galaxy. The source mechanism, particle composition, and number of sources in the galaxy necessary to explain the homogeneity of low-energy cosmic rays as interpreted from π^0-induced gamma rays all become problems.

Cosmic rays are a major probe for the astrophysical understanding of our galaxy. They are the only sample of matter that reaches us from the rest of the galaxy outside the solar system. They also may originate in the most energetically extreme events of our galaxy, namely supernovae, and are thus closely related to the significant tests of our theoretical understanding of our galaxy and therefore our universe.

See also COSMIC RAYS—SOLAR SYSTEM EFFECTS; GALAXIES; INTERSTELLAR MEDIUM; NUCLEOSYNTHESIS.

BIBLIOGRAPHY

The most comprehensive treatment of cosmic rays in recent years is still the authoritative book, V. L. Ginzburg and S. I. Syrovatskii, *The Origin of Cosmic Rays*. Pergamon Press, New York, 1964. The current scientific changes are reviewed in the "Rapporteur" papers of each International Cosmic Ray Conference (every two years) and the details appear in the published papers. A recent summary of these papers is given in P. Sokolsky, *Introduction to Ultrahigh Energy Cosmic Ray Physics*. Addison-Wesley, Reading, MA, 1989.

Cosmic Rays—Solar System Effects

Joseph V. Hollweg

INTRODUCTION

Cosmic rays were discovered in the early decades of this century by investigations of the electrical conductivity of air. These investigations indicated that the air was being ionized by a residual source of radiation in addition to the radioactive sources known at that time. The residual radiation was found to increase with height above the earth's surface, and the existence of an extraterrestrial radiation source was surmised. Subsequent observations of the *latitude effect* and *east-west effect* (see below) showed that this radiation source consists of high-energy positively charged particles impinging on the earth's atmosphere from space.

The term cosmic rays refers to these high-energy charged particles, but it now includes particles not only near the earth, but in the solar atmosphere, in interplanetary space, in our galaxy, and perhaps throughout the universe as well (*see* Cosmic Rays—Astrophysical Effects). Cosmic rays in the solar system are divided into two groups, depending on whether they originate at the sun (solar cosmic rays) or impinge on the solar system from interstellar space (galactic cosmic rays); high-energy particles that originate elsewhere, such as the earth's Van Allen radiation belts, in the earth's magnetosphere during geomagnetic storms, in planetary magnetospheres, or at shock waves propagating in interplanetary space, are usually not denoted cosmic rays, although there is frequently considerable overlap. However, as will be seen below, energetic particle detectors on spacecraft have revealed that shock waves do energize charged particles, and it is believed that strong shock waves may be the energy source for the solar and galactic cosmic rays. Charged particles in the solar system are found in an extremely wide energy range from several electron volts up to about 10^{20} eV, and the designation cosmic rays is reserved for the high-energy particles; the lower limit of this interval is ill-defined, but it is generally a few hundred keV for protons and a few tens of keV for electrons.

The study of cosmic rays in the solar system touches on a wide variety of areas. Solar cosmic rays carry information about the solar atmosphere and about the particle-acceleration processes that take place in solar flares; the particle-acceleration processes in flares may in turn be similar to acceleration processes occurring in the earth's magnetosphere during geomagnetic storms or in other astrophysical situations. Solar and galactic cosmic rays are detected only after they have propagated through the interplanetary medium (the solar wind) and in some cases through the earth's magnetosphere and atmosphere: the resulting modulation of the cosmic rays can yield information about physical conditions in these regions, while conversely it is necessary to take the modulation into account if the properties of the cosmic rays at their sources are to be deduced from observations from earth or from space vehicles. An understanding of the modulation and acceleration processes in the solar system, where direct observations can be made, can provide valuable clues about physical processes in other astrophysical situations, where direct observations cannot be obtained; the observed acceleration of charged particles by shock waves is an important example. Solar cosmic rays play a role in a variety of geophysical phenomena, such as modifications of ionospheric structure, modifications of the atmospheric ozone layer, and possibly influences on the weather. The interactions of high-energy cosmic rays with the earth's atmosphere have been used to study fundamental problems in high-energy nuclear physics. The radioactive isotopes produced by interactions with the atmosphere have been used in a variety of terrestrial dating methods (e.g., dating of archaeological objects using cosmic-ray-produced ^{14}C), and in tracing and dating such geophysical effects as oceanic sedimentations and the deposition of glacial and polar ice. Isotopes and radiation damage produced by cosmic-ray bombardment have been used to study the histories of meteorites and lunar samples. Finally, cosmic rays are an important factor in evaluating the radiation doses to human passengers in space flight or in atmospheric flight at military altitudes.

The following paragraphs summarize some features of solar and galactic cosmic rays in the solar system, and their interactions with the earth's atmosphere, magnetosphere, and interplanetary medium.

GALACTIC COSMIC RAYS

Galactic cosmic rays are composed of protons (86%), helium nuclei (12.7%), and heavier nuclei (1.3%); there are also about 1% electrons.

The integral energy spectra of the nuclei (i.e., the flux of particles having energies exceeding E), denoted $j(>E)$, obey a power law for $E \gtrsim 10$ GeV/nucleon:

$$j(>E) = KE^{-\delta},$$

where K and δ are positive constants. The spectra roll off below 10 GeV/nucleon, however: the rolloff is due to diffusion and adiabatic energy losses in the solar wind. The flux levels below about 10 GeV/nucleon vary during the 11-year sunspot cycle, roughly in inverse correlation with the sunspot number; this effect is presumably due to the reduced diffusion and enhanced energy loss in the solar wind at times of enhanced sunspot number, but solar wind variations with the sunspot cycle are not well understood at present.

SOLAR COSMIC RAYS

The most energetic solar cosmic rays are emitted nearly impulsively by solar flares, whereas lower-energy particles may be emitted nearly continuously from solar active regions, from magnetic clumps at solar supergranulation cells, or from ephemeral active regions (i.e., small but intense bipolar magnetic regions). Lingering cosmic-ray emission for several hours or days following a flare may indicate either particle storage at the sun or continued particle acceleration following the flare, perhaps in association with shock waves produced by gradual changes in the magnetic field topology; the latter seems more likely since losses in the solar atmosphere are too severe to permit appreciable storage. It is also interesting to note that there is increasing evidence that some parts of the solar corona are heated by many small impulsive events, called microflares or even nanoflares; these events may give rise to a quasi-continuous emission of low-energy cosmic rays. Flare-produced cosmic rays enter the solar wind via solar magnetic field lines that open into interplanetary space, but they often exhibit a broad distribution in heliographic longitude, implying nearly isotropic diffusion near the sun, particle transport via complicated solar magnetic field geometries, or particle acceleration in regions far removed from the flare site (perhaps via flare-associated shock waves), but a definitive explanation is not available. The time profiles of flare-produced cosmic rays observed near earth generally show a rapid rise (a few hours or less) of the particle flux followed by a slow decay (several hours to several days), with the particles frequently arriving nearly isotropically during the decay phase; these data imply diffusive propagation from the sun to the earth, but other events sometimes show anisotropic arrival, implying nearly scatter-free propagation. Flare-produced cosmic rays observed near the earth come preferably from the western solar hemisphere, because the interplanetary magnetic field lines are spiral shaped in the sense that they connect the earth to the western solar hemisphere. The composition of flare-produced cosmic-ray nuclei roughly resembles the photospheric abundances, but large variations (in energy and time) in the ratios between protons and helium or heavier nuclei indicate fractionation in the flare region and/or variable particle-acceleration processes. Fractionation is known to be part of the story, because the abundances are observed to correlate to some extent with the ease with which the various elements are ionized. The energy spectra of solar cosmic rays fall off steeply with energy, and there are very few solar cosmic rays above 10 GeV/nucleon.

ATMOSPHERIC INTERACTIONS

Earth-based detectors almost never observe a primary cosmic ray before it is destroyed by interactions with the earth's atmosphere. Primary particles with energies exceeding a few hundred MeV/nucleon interact with atmospheric nuclei to produce neutral pions, charged pions, and nucleons (protons or neutrons). The neutral pions decay into γ rays, which in turn produce electron–positron pairs in a cascading process called an extensive air shower. The charged pions decay into muons. The nucleons interact further with the atmosphere to produce neutron–proton cascades. In all three cases only the secondary products are detected near the ground. The production of secondary products is sensitive to atmospheric temperature and pressure.

MAGNETOSPHERIC INTERACTIONS

Cosmic-ray paths are bent by the earth's magnetic field, with the result that cosmic-ray particles with energies below a certain cutoff can be deflected to such an extent that they are unable to penetrate into the earth's magnetic field. The magnetic field thereby gives rise to the *east-west effect,* which states that positively charged particles tend to arrive preferentially from the west because the cutoff energy for westward-arriving particles is lower than that for eastward-arriving particles (e.g., the cutoff energy at the geomagnetic equator is 10 GeV for westward-arriving protons and 60 GeV for eastward-arriving protons), and to the *latitude effect,* which states that the flux of cosmic-ray particles is a minimum at the geomagnetic equator because the cutoff energy is maximum at the equator (e.g., the cutoff energy for vertically arriving protons is 15 GeV at the equator and 2.7 GeV at 50° geomagnetic latitude). Cosmic-ray motion in the earth's magnetic field is complicated by the highly asymmetric structure of the magnetosphere and by the possibility of diffusive entry of particles into the magnetospheric tail.

SOLAR WIND INTERACTIONS

The frequently observed nearly isotropic arrival of solar cosmic rays at the earth implies that their motion is randomized by frequent scattering interactions with magnetic irregularities in the solar wind. The nature of the irregularities is still not adequately identified, but they include magnetohydrodynamic waves and discontinuities (especially Alfvén waves of solar origin), "magnetic bubbles" ejected into the solar wind by solar flares and eruptive solar prominences, and high-speed solar wind streams. The motion of both solar and cosmic rays in the solar wind is described in terms of diffusion along the interplanetary magnetic field, convection by the magnetic irregularities as they move outward from the sun, adiabatic energy losses resulting from interactions with the irregularities as they expand with the solar wind flow (analogous to the adiabatic cooling of an expanding gas), and an azimuthal drift associated with the corotation of interplanetary magnetic field lines with the sun. The diffusion-convection description is used to interpret the time profiles of flare-produced cosmic rays and the 11-year solar-cycle modulation of galactic cosmic rays. It is also used to explain *Forbush decreases*: these are large-scale decreases, lasting for a few days to several weeks, in the cosmic-ray flux, which are understood to result from changes of diffusion and convection associated with interplanetary shock waves, high-speed solar wind streams, or magnetic bubbles. The diffusion-convection description can also explain the radial gradient of the galactic cosmic-ray intensity observed by the Pioneer 10 and 11 and Voyager 1 and 2 missions, the diurnal variation in cosmic-ray intensity at earth, which is understood to result from the anisotropy induced by the corotation of the interplanetary magnetic field, and the semi-diurnal variation at the earth, which may imply the existence of strong solar wind variations with heliographic latitude.

But there are still lacunae in our understanding. The observed radial gradient of the galactic cosmic-ray intensity can only be explained with special assumptions about the nature of the magnetic irregularities, or if there is a significant flux of cosmic rays from high heliographic latitudes toward the solar equatorial plane. There is at present insufficient data on either of these points, but current research is placing increasing emphasis on the latitudinal drifts as being an essential feature in the propagation of galactic cosmic rays. The latitudinal drifts are associated with gradients in the magnetic field strength, the curvature of the interplanetary magnetic field lines, and cross-field gradients of the number of charged particles; the latter is called the magnetization drift. There are also problems in our understanding of how the diffusion coefficients are related to the properties of the magnetic irregularities. The theory seems to work well for particles of high rigidity, but low-rigidity particles are observed to be scattered more weakly than theory predicts; and the diffusion of cosmic rays in the directions across the magnetic field seems to be weaker than predicted by theory. The correct mathematical description of cosmic-ray diffusion is a difficult problem in nonlinear plasma physics, which is still in need of a definitive solution, and a full solution of the cosmic-ray propagation equations is still in need of more information about the physical state of the solar wind and the magnetic irregularities, especially at high heliographic latitudes which have not yet been explored by spacecraft.

SHOCK WAVE INTERACTIONS

A significant advance in recent years has been an understanding of how charged particles can be accelerated to high energy by shock waves. Shocks are abundant in the solar system: "bow shocks" are produced when the solar wind is deflected by planetary magnetic fields, shocks are produced by solar flares, and when fast solar wind streams overtake slower streams. In all cases, it has been observed that some solar wind particles attain high energies near the shocks. The mechanism by which this happens is called "diffusive shock acceleration." Magnetic irregularities in the solar wind are convected through a shock, and slowed down as the flow passes through the shock. They can then be thought of as converging. The charged particles are scattered by the irregularities, and diffuse back and forth across the shock, gaining energy as they repeatedly interact with the converging irregularities. Moreover, the energized particles can then produce more irregularities via plasma instabilities, and these irregularities too scatter the particles. Mathematical models which describe this process have been developed, and they explain most aspects of the data. This process may be able to account for the cosmic-ray production in solar flares, and even in the galaxy, particularly in association with supernovae which are believed to generate strong shock waves.

Shock acceleration may also account for the "anomalous component" of cosmic rays. This component is thought to originate from interstellar neutral particles which become ionized via collisions with the solar wind particles or by solar photons (photoionization). Then subsequent acceleration to cosmic ray energies may occur at the termination shock of the solar wind, which is presumed to occur when the solar wind is slowed down by the interstellar gas. The termination shock has not yet been detected by spacecraft, and its properties and effects are currently speculative.

See also COSMIC RAYS—ASTROPHYSICAL EFFECTS; MAGNETOSPHERE; SOLAR WIND.

BIBLIOGRAPHY

O. C. Allkofer, *Introduction to Cosmic Radiation*. Verlag Karl Thiemig, Munich, 1975. (I)

L. A. Fisk, "The Interactions of Energetic Particles with the Solar Wind" in *Solar System Plasma Physics: A Twentieth Anniversary Overview*, C. F. Kennel, L. J. Lanzerotti, and E. N. Parker, eds. North-Holland, Amsterdam, 1979. (A)

L. A. Fisk, "Solar and Galactic Cosmic Rays" in *Solar-Terrestrial Physics: Principles and Theoretical Foundations*, R. L. Carovillano and J. M. Forbes, eds. D. Reidel, Dordrecht, 1983. (A)

M. A. Forman and G. M. Webb, "Acceleration of Energetic Particles" in *Collisionless Shocks in the Heliosphere: A Tutorial Review*, R. G. Stone and B. Tsurutani, eds. American Geophysical Union Monograph 34, Washington, D.C., 1985. (A)

J. R. Jokipii, "Propagation of Cosmic Rays in the Solar Wind," *Rev. Geophys. Space Phys* **9**, 27 (1971). (A)

M. A. Pomerantz, *Cosmic Rays*. Van Nostrand-Reinhold, New York, 1971. (E)

M. A. Pomerantz and S. P. Duggal, "The Sun and Cosmic Rays," *Rev. Geophys. Space Phys.* **12**, 343 (1974). (I)

K. Sakurai, *Physics of Solar Cosmic Rays*. University of Tokyo Press, Tokyo, 1974. (I)

G. Wibberenz, "Interplanetary Magnetic Fields and the Propagation of Cosmic Rays," *J. Geophys.* **40**, 667 (1974). (I)

Cosmic Strings*

John Preskill

Cosmic strings are hypothetical line defects that arise in some relativistic quantum field theories as a consequence of the spontaneous breakdown of gauge or global symmetries. These defects are one-dimensional objects with a width of order m^{-1} and an energy per unit length of order m^2 (in units with $\hbar = c = 1$), where m is a characteristic mass scale of the symmetry breakdown. No cosmic strings are expected in the "standard model" that describes the electroweak and strong interactions at energies of order a few hundred GeV and below. Thus, if cosmic strings can actually occur in nature, they must be associated with very short-distance physics that has not yet been directly explored in accelerator experiments.

Although there is at present no direct observational evidence indicating that cosmic strings exist, it has been proposed that cosmic strings play a significant role in the physics of the very early universe. This proposal is the origin of the use of the adjective "cosmic" as applied to these objects. In current usage, the term "cosmic string" distinguishes the composite strings that arise in relativistic field theories from fundamental strings (such as the superstring) that are truly elementary dynamical entities.

That strings may arise in relativistic quantum field theory was emphasized in 1973 by Nielsen and Olesen, though the mathematics underlying this suggestion had been worked out in 1957 by Abrikosov and applied by him to the theory of superconductivity. These strings were considered in the context of cosmology in 1976 by Kibble, who observed that strings could have been produced in a phase transition in the very early universe. That such cosmic strings might have stimulated the formation of galaxies and other types of large-

* This work supported in part by the U.S. Department of Energy under Contract No. DE-AC0381-ER40050.

scale structure in the universe was proposed in 1980 by Zeldovich, and, in a more specific form, by Vilenkin.

Cosmic strings may be classified into two general types, gauge strings and global strings, according to the nature of the spontaneously broken symmetry with which they are associated. In fact, one-dimensional defects that are analogous to cosmic strings occur in some condensed-matter systems, and for each type there is a corresponding condensed-matter analog that serves as a prototypical example. The concept of a gauge string is nicely illustrated by a magnetic flux tube in a type II superconductor. Magnetic fields are expelled from a superconductor by the Meissner effect, but a type II superconductor may be penetrated by filaments that carry the quantum of magnetic flux. These filaments have a characteristic finite thickness that is known as the penetration depth of the superconductor. In this example, the spontaneously broken symmetry associated with the string is the $U(1)$ gauge symmetry of electrodynamics; this symmetry is spontaneously broken in a superconductor, and the photon acquires a mass m via the Higgs mechanism, where m^{-1} is the penetration depth.

Analogously, one can construct a relativistic quantum field theory in which the $U(1)$ gauge symmetry of electrodynamics is spontaneously broken in the vacuum state. Strings appear in such a theory as stable solutions to the classical field equations. Similar string solutions occur in other theories with various patterns of gauge symmetry breaking. In general, a field theory contains cosmic strings whenever there is a conserved magnetic flux that is carried only by gauge fields that have acquired mass via the Higgs mechanism; then the conserved flux becomes confined by a generalized Meissner effect to a tube of finite width. Thus, the existence of gauge strings is predicted by some "grand unified" theories, in which the gauge group of the standard model is embedded in a larger group that undergoes the Higgs mechanism at a large mass scale.

The concept of a global string is illustrated by a vortex line in superfluid helium. In the superfluid, there is a conserved particle number, and the associated $U(1)$ global symmetry is spontaneously broken by Bose condensation. The vortex is a thin tube of normal fluid that is trapped by the surrounding superfluid. A characteristic property that distinguishes gauge and global strings is that global strings have a long-range interaction mediated by the Goldstone bosons to which they couple, while gauge strings typically have no long-range interaction other than gravity.

Cosmic strings have no ends; they either form closed loops or extend to infinity. (Actually, other topologies are also possible, depending on the details of the associated pattern of symmetry breakdown. But strings that form closed loops seem to be of the greatest cosmological interest.) The tension in a string is equal to its energy per unit length, so that its motion is highly relativistic; transverse oscillations propagate along the string with velocity c, and a closed loop of length L executes a periodic motion with a period comparable to L/c.

When two strings cross one another, they are apt to break at their point of intersection and change partners before rejoining, a process known as intercommuting. A typical closed loop may intersect itself at some point during its motion; then intercommuting may cause the loop to fission into two "offspring" loops. An oscillating closed loop that does not self-intersect can decay by emission of gravitational radiation.

It is not known whether the actual pattern of symmetry breakdown that governs particle physics at very short distances is such that cosmic strings are expected to exist in nature. But if so, it is likely that a phase transition occurred in the very early universe in which strings were copiously produced. Symmetries that are spontaneously broken at low temperature tend to be restored when the temperature exceeds a critical value T_c of order m, where m is the symmetry-breaking mass scale. As the rapidly expanding universe cools down through such a critical temperature, defects inevitably freeze in.

The cosmic strings thus formed comprise a complicated network of vibrating open strings and oscillating closed loops. As the network evolves, strings frequently cross and intercommute. It is believed that the strings eventually attain a self-similar equilibrium configuration that is only weakly dependent on the details of the structure of the network when it was initially formed.

The string network generates inhomogeneous perturbations in the primordial distribution of matter in the universe. Although initially small, these perturbations grow because of gravitational instabilities. They may thus account for the origin of galaxies, clusters of galaxies, and other large-scale inhomogeneities that we observe in the universe today. Because the dynamics of the string network is quite complicated, it is not easy to extract from this scenario detailed predictions about large-scale structure. But one can infer that the primordial mass-density perturbations have the important property of being independent of length scale, as a consequence of the self-similarity of the equilibrium string network. The predicted scale independence is in qualitative agreement with some observed features of large-scale structure. One can also infer that, if cosmic strings did play a significant role in the origin of galaxies and other large structures, then they must have an energy per unit length of order $(10^{16}\ \text{GeV})^2$, in units with $\hbar = c = 1$. Such strings, with a width of order 10^{-30} cm, would be a relic of the first 10^{-38} s after the big bang, and would be associated with particle physics at an energy far beyond the reach of foreseeable accelerator experiments.

Even though precise predictions concerning large-scale structure are not easily extracted from the cosmic string scenario for galaxy formation, this scenario may nevertheless be subject to observational confirmation or refutation. Decaying loops of cosmic string generate a stochastic background of gravitational radiation that should soon be detectable in pulsar-timing measurements. Also, strings might be detected as gravitational lenses, or through the characteristic way that they distort the cosmic microwave background radiation.

In some field theory models, cosmic strings have the property that electric charge carriers are bound to the core of the string, and the string therefore behaves like a superconducting wire. (This was pointed out by Witten in 1985.) Such superconducting strings are capable of carrying very large currents and can be enormously powerful sources of electromagnetic radiation. If superconducting strings occur in nature, the characteristic radiation that they emit might be detected by radio telescopes.

For now, the proposal that cosmic strings stimulated the formation of galaxies and other large-scale structures must be regarded as highly speculative. But this speculation may be confirmed or refuted. Confirmation would establish a truly remarkable connection between particle physics at very short-distance scales and cosmology at very long-distance scales.

See also COSMOLOGY; GALAXIES; GAUGE THEORIES; SYMMETRY BREAKING, SPONTANEOUS.

BIBLIOGRAPHY

R. H. Brandenberger, "Inflation and Cosmic Strings: Two Mechanisms for Producing Structure in the Universe," *Int. J. Mod. Phys.* **A2**, 77–131 (1987).

N. D. Mermin, "The Topological Theory of Defects in Ordered Media," *Rev. Mod. Phys.* **51**, 591–648 (1979).

J. Preskill, "Vortices and Monopoles," in *Architecture of Fundamental Interactions at Short Distance* (P. Ramond and R. Stora, ed.), pp. 235–338. North-Holland, Amsterdam, 1987.

A. Vilenkin, "Cosmic Strings and Domain Walls," *Phys. Rep.* **121**, 263–315 (1985).

A Vilenkin, "Cosmic Strings," *Sci. Am.* **257** (No. 6), 94–102 (1987).

E. Witten, "Superconducting Cosmic Strings," *Nucl. Phys.* **B249**, 557–592 (1985).

Cosmology

P. J. E. Peebles

HOMOGENEITY, HUBBLE'S LAW, AND THE EXPANSION OF THE UNIVERSE

In the standard and observationally successful cosmology the universe is homogeneous and isotropic in the large-scale average, which means the universe appears much the same, apart from local fluctuations, viewed from any galaxy. Also, space is uniformly expanding, which means that well-separated bits of matter are moving apart, the mean recession velocity at separation **r** being

$$\mathbf{v} = H\mathbf{r}, \qquad h = 100h \text{ km s}^{-1}\text{ Mpc}^{-1}. \tag{1}$$

This is Hubble's law. The functional form is dictated by homogeneity. The length unit is 1 Mpc = 3.086 × 10^{24} cm. The value of Hubble's constant, H, is uncertain because it is hard to fix absolute distances of galaxies; the dimensionless factor h is thought to be in the range $0.5 < h < 1$.

The cosmological redshift, z, of a distant object is defined by the ratio of the observed wavelength λ_o to emitted wavelength λ_e of spectral features in the object:

$$1 + z = \frac{\lambda_o}{\lambda_e} = \frac{a_o}{a_e} \tag{2}$$

In the relativistic theory, this ratio also is equal to the factor by which the universe has expanded between now and the epoch of emission of the radiation, as indicated in the last part of the equation. It is standard practice to label epochs in the early universe by the redshift factor. The present distance to an object on our light cone and at very high redshift is on the order of the Hubble length, $cH^{-1} = 3000h^{-1}$ Mpc.

The first large-scale survey of the galaxy distribution was the Lick catalog based on visual counts of galaxy angular positions by C. D. Shane and C. A. Wirtanen, completed in 1967. Recently, improved automated large-scale counts have been obtained by groups at the Institute of Astronomy in Cambridge and the Royal Observatory in Edinburgh. All these surveys extend to depths ~ 10% of the Hubble length.

The surveys are consistent with large-scale homogeneity and a very clumpy small-scale distribution. One measure of this is the reduced two-point correlation function, $\xi(r)$, defined by the joint probability of finding a galaxy in each of the volume elements dV_1 and dV_2 at separation r,

$$dP = n^2 dV_1 dV_1[1 + \xi(r)], \tag{3}$$

where n is the mean space number density. At $r < 10h^{-1}$ Mpc, $\xi(r)$ is close to a power law, $\xi \propto r^{-1.77}$. This with similar power law behavior of the higher-order correlation functions indicates that the small-scale galaxy distribution approximates a nested clustering hierarchy, or fractal, with dimension $D = 1.23$. The hierarchy terminates on the small-scale end at about the size of an individual galaxy.

Redshifts are much more powerful tracers of structure than angular positions alone because the redshift reduces the ambiguity in distance. However, redshifts require much more observing time so redshift surveys still are relatively small. They have revealed in the small-scale galaxy distribution a curious pattern, that has been variously called frothy, bubbly, and cellular, in which rather linear structures surround empty regions, or voids.

The coherence length in the galaxy distribution, where $\xi = 1$, is $hr_o \sim 5$ Mpc. Fluctuations on scales larger than this are relatively small. For example, the rms fluctuation in galaxy counts within a sphere of diameter $100h^{-1}$ Mpc ~ 3% of the Hubble length is about 20% of the mean. One can pick out apparently coherent structures in these fluctuations, the largest well-accepted ones being ~ $50h^{-1}$ Mpc across.

Large-scale fluctuations in the mass distribution are indicated by the gravitational production of departures from the pure Hubble flow of Eq. (1). In linear perturbation theory, the peculiar velocity, or departure from Hubble flow, caused by a spherical density fluctuation $\delta\rho/\rho$ extending over radius R is

$$v \sim \frac{1}{3}\frac{\delta\rho}{\rho} HR. \tag{4}$$

The dipole ($\propto \cos\theta$) anisotropy of the microwave background radiation discussed below is most reasonably interpreted as the result of the peculiar motion of our neighborhood at about 600 km s^{-1}. The nearby galaxies at distances ~ $10h^{-1}$ Mpc are moving at about the same peculiar velocity. There are indications that the coherence length of this peculiar velocity field may be as large as ~ $50h^{-1}$ Mpc, the local motion roughly converging on the "Great Attractor" in the direction of Centaurus. With $R = 50h^{-1}$ Mpc and the above-mentioned $\delta\rho/\rho \sim 0.2$, Eq. (4) gives $v \sim 300$ km s^{-1}, on the order of what is observed.

Most measures of structure on still larger scales yield only upper bounds. Apart from the dipole term, the microwave background is isotropic to better than one part in 10^4 on angular scales down to $\sim$ 10 arc min. Since this radiation is not now appreciably interacting with matter, the isotropy says nothing directly about the present matter distribution, but it does show that the initial entropy density and the cosmological redshift in different parts of space must have been very close to uniform. The extragalactic x-ray background comes from some combination of clusters of galaxies, exploding galaxies, quasars, and hot intercluster gas. This background is isotropic to better than 5% on scales greater than a few degrees. Thus the surface density of these sources integrated to the Hubble length must be isotropic to like accuracy.

In some versions of the inflation scenario mentioned below, we observe at the Hubble distance part of an island outside of which the universe is very different. However, there is no known way to substantiate this. To the limits of our observations, the universe is very close to isotropic on large scales.

THE COSMIC BACKGROUND RADIATION

The microwave–submillimeter background radiation (also called the primeval fireball, relict, or cosmic background radiation, or CBR for short) has spectrum close to blackbody at temperature $T = 2.76 \pm 0.03$ K. There is no natural way to produce a blackbody spectrum in the universe as it is now, for the universe is optically thin (radio sources are observed at redshifts $z > 1$ without appreciable attenuation), so this radiation must have been thermalized when the universe was denser and hotter than it is now. Isotropic and homogeneous expansion of the universe automatically preserves the thermal character of the radiation, with no need for a thermalizing agent: expansion simply lowers the temperature.

We have two candidate theories for the epoch of origin of this radiation. It is possible that dust made the universe optically thick at redshift $z \sim 100$, and that enough starlight was present to be thermalized by the dust to produce the CBR. However, the demands on the dust and the energy production by stars at $z \sim 100$ are so tight that most workers do not consider this picture very likely. Also, we have evidence for the other possibility, that the CBR originated at extremely high redshift.

The evidence comes from the production of light elements. Knowing the present CBR temperature, and assuming entropy was nearly conserved back to high redshifts, we can trace the thermal history of the universe back in time to temperatures high enough to have driven thermonuclear reactions. Computations of the nuclear reactions, under the assumptions (1) the early universe is accurately homogeneous and isotropic, (2) the conventional laws of physics apply, and (3) the universe is not filled with a sea of degenerate neutrinos, predict that most matter comes out of this "hot big bang" as hydrogen, with about 20% by mass helium, and significant amounts of deuterium, ^{3}He, and ^{7}Li. The computation depends on the mean baryon number density, n_b. It is remarkable that a choice of n_b consistent with what is observed to be present in and around galaxies yields computed abundances concordant with what is seen in old stars. This is evidence of reality of the hot big bang, though a fuller test should be possible with improved precision in the measurement of abundances in the oldest stars.

THE RELATIVISTIC COSMOLOGICAL MODEL

The relativistic cosmological model involves the following parameters:

(1) The present rate of expansion, H [Eq. (1)]. This fixes the scales of length, time and mass.

(2) The mean mass density and pressure, ρ and p. By homogeneity, ρ and p are functions of the world time t kept by any observer at rest relative to the local matter distribution. The order of magnitude of ρ is set by the combination H^2/G, where G is Newton's constant. In the standard model, the prefactor is on the order of unity, and $\rho \sim 1$ proton per cubic meter, which is within an order of magnitude of what is observed.

(3) The ages of the oldest objects. In classical general relativity theory, the expansion of the universe traces back to a singularity at a finite time in the past. The world time since the singularity is $\sim H^{-1} \sim 10^{10}$ years, the prefactor again depending on the cosmological parameters. The singularity presumably reflects a failure of the classical theory, but since no object we could hope to date could have survived from preclassical epochs the ages of the oldest objects must be less than the model age. A test of this is limited by the uncertainties in H and the evolutionary ages of stars and the elements; we only know that predicted and observed ages agree to a factor of about 3.

(4) The rate of change of the expansion rate. This is measured by the "acceleration parameter," q:

$$\frac{dH(t)}{dt} = -(q + 1)H(t)^2. \qquad (5)$$

This fixes the first-order relativistic corrections to angular sizes and counts of galaxies as a function of redshift, and so in principle is measurable. Until recently, it has not been possible to correct the measurements for systematic errors: distant galaxies are seen at an earlier world time, when evolution may have made the mean properties of galaxies significantly different from now. The situation has been improving with advances in infrared detectors, where evolutionary corrections are smaller, and in the understanding of constraints on evolution from the spectra of galaxies and their luminosity distributions, so a significant test may be within reach.

(5) The cosmological constant, Λ. This acts as constant contributions ρ_Λ and p_Λ to the net mass density and pressure, with $\rho_\Lambda = -p_\Lambda$, a Lorentz invariant arrangement. (Here and below units are chosen so the velocity of light is unity.) Einstein introduced this term in his gravitational field equations so as to allow a static homogeneous universe with non-negative pres-

sure in ordinary matter. When the expansion of the universe was discovered he proposed that Λ be abandoned; opinion since then has been mixed. A Λ term might be expected from zero-point energy in a Lorentz-invariant theory, but a "reasonable" value would be much larger than would be acceptable to cosmology. Such a large effective Λ is present during inflation in the scenario discussed below.

(6) Space curvature, $R(t)^{-2}$. If $R^{-2} > 0$, space sections at fixed cosmic time are curved and closed like a sphere, with radius of curvature $R(t)$. The volume of space is finite [and proportional to $R(t)^3$] even though there is not a spatial boundary. If $R^{-2} < 0$, space curvature resembles that of a saddle, and there is not direct closure of space. If R^{-2} is negligibly small, the model is said to be cosmologically flat (though space-time is curved).

The gravitational field equations yield the relations

$$(1 - \Omega)H^2 = \Lambda/3 - R^{-2},$$
$$H^2[\Omega(1 + 3p/\rho) - 2q] = 2\Lambda/3, \qquad (6)$$

where the density parameter is

$$\Omega = 8\pi G\rho/3H^2. \qquad (7)$$

The factor $1 + 3p/\rho$ ranges from 1 for nonrelativistic matter to 2 for electromagnetic radiation or zero-mass particles, an unimportant spread compared to the observational uncertainties in the other parameters.

In the Einstein–de Sitter model, Λ and R^{-2} both are negligibly small, so Eq. (6) says $\Omega = 1$. This is a particularly attractive case because the universe presents us with no fixed characteristic scales. (In all other observationally possible cases, the model defines a characteristic world time that must coincidentally be the same order of magnitude as the epoch of life on earth.) However, direct measurements of ρ based on galaxy masses derived from the dynamics of systems of galaxies fall short of the Einstein–de Sitter prediction [Eq. (7) with $\Omega = 1$] by a factor ~ 5. Opinion is divided on whether this is because the dynamical studies miss mass not in systems of galaxies (the "biasing" hypothesis) or whether it might mean the model fails.

In the more general two-parameter set of models with $\Lambda = 0$, it is easy to categorize the time histories. Here Eq. (6) may be rewritten as

$$\Omega H^2 - H^2 = R^{-2}. \qquad (8)$$

The mean mass density varies with time as $\rho \propto R(t)^{-3}$ (if $p \ll \rho$; if pressure is large ρ varies still more rapidly with R because of pdV work). Therefore Eq. (7) says ΩH^2 is decreasing more rapidly than $R(t)^{-2}$. Thus in a closed model, where $R^{-2} > 0$, Eq. (8) indicates there must come a time when $\Omega H^2 = R^{-2}$ and H therefore vanishes, which means the universe stops expanding. That is, a closed model with $\Lambda = 0$ expands from a "big bang" that has singular space curvature (in the classical theory) through the present epoch to a point of maximum expansion, after which it contracts back to a singular "hot crunch."

An open model expands forever after the big bang, the deceleration parameter q approaching zero, which means the gravitational slowing of the expansion becomes negligible. In the Einstein–de Sitter case, $R^{-2} = 0$, expansion also continues forever but with $q = \frac{1}{2}$. Unlimited expansion has been characterized as a heat death, because stars and the elements eventually exhaust the energy supply needed to power observers (though there would be isolated bursts of action for an exceedingly long time from relativistic gravitational collapse of galaxies and clusters of galaxies). The debate on whether the universe ends with a heat death or a hot crunch has the virtue that it concentrates our attention, but better physics is likely to come out of attempts to understand the past history of the universe, because that may have left clues.

DARK MASS

Masses of galaxies and systems of galaxies can be estimated from dynamics, through the gravitational field needed to account for observed motions. In compact stable-looking systems, one assumes gravity is strong enough to bind the system against its tendency to fly apart; in extended systems, one assumes gravity is the cause of perturbations from Hubble flow, as in Eq. (4).

In the bright central parts of galaxies, the dynamical mass estimates are consistent with what is seen to be present in the stars. In the outer parts of galaxies, the dynamical mass is significantly greater than what is seen: if our understanding of gravity physics is adequate, many bright galaxies have massive dark halos with density that varies with radius, r, as $\rho \propto r^{-2}$. It is interesting that the constant of proportionality has rather a hard upper bound, reflected in the observation that the circular velocity in a galaxy seldom is much in excess of 300 km s^{-1}.

The discrepancy between what is seen and what is inferred from dynamics increases with increasing size of the system, reaching a ratio of inferred to seen ~ 10 in rich clusters. As noted above, if the Einstein–de Sitter model were valid the discrepancy would increase by another order of magnitude for the universe as a whole.

What is the nature of the dark mass? The most conservative guess is starlike objects, "brown dwarfs," with low luminosities because their masses are below the limit $\sim 0.1\ M_\odot$ for hydrogen burning. Most of the stars in our neighborhood have masses close to this limit. There could not be a large number of less massive brown dwarfs, because we do not have a large local mass problem, but it is easy to imagine that the ratio of brown dwarfs to visible stars increases with increasing distance from the center of the galaxy, matching the observed increase of mass per unit of starlight. The best hope for probing this is improved measures of the frequency distribution of star masses at $M > 0.1 M_\odot$ as a function of position in the galaxy.

The dark mass need not be baryonic. Among exotic dark mass candidates, massive neutrinos are the most conservative guess, because neutrinos are known to exist. Thermal production of neutrinos in the hot big bang leaves a present residue of about 100 cm^{-3} per neutrino type. If one type had mass ~ 30 eV, the mass density would be about right for the Einstein–de Sitter model, and there would be the pleasant coincidence that the wanted density of neutrinos in galaxy halos would just saturate the phase space bound.

Modern particle physics suggests a host of other exotic candidates, and the possibility of their detection in the halo of our galaxy, through the occasional particle that scatters in a cold detector as it passes through the earth. It would be hard to overstate the impact on cosmology of a detection of exotic halo particles, but even if the dark mass is not exotic the identification of its nature is a great discovery waiting to be made.

THE PHYSICS OF THE VERY EARLY UNIVERSE: INFLATION

In the classical cosmological models, the expansion of the universe traces back to a singularity. This means our theory is physically incomplete; worse, it means we have no way to predict the large-scale structure of the universe. A particular problem is the particle horizon: in the classical models, material now inside our light cone back to the singularity was outside our light cone in the past. The curious consequence is that we can see distant galaxies that at the time we see them were not in causal connection with each other subsequent to the singularity. How then can we understand why well-separated parts of space look so similar to each other, much less understand the small systematic differences that we call large-scale structures? The problem is compounded by the fact that the cosmological models are gravitationally unstable: inhomogeneities in mass density tend to grow with time. Thus if causal connection were established in the past, how could it happen that this promotes the observed homogeneity rather than clustering?

An elegant possible solution is the inflation scenario pioneered by A. Guth, A. Linde, P. Steinhardt, and others. It is based on the idea mentioned above that quantum physics encourages us to think space may contain something approximating the effect of a cosmological constant, Λ. The present value of this effective Λ must be small, but there may have been a time in the very early universe when it was large. In a patch of space-time where this term dominated the effect of matter and curvature, the first of Eqs. (6) gives $H \sim (\Lambda/3)^{1/2}$, which is to say the patch is expanding at a large and nearly constant rate. Where this persisted for a long time, the expansion would make the density of any remaining matter negligibly small, which would eliminate inhomogeneities, and it would make the volume of the patch very large. This "inflationary" epoch of nearly constant expansion would have to terminate in a phase transition in which the large energy density associated with the Λ-like term is converted into the entropy we see as the CBR, and the entropy would have to make the baryons out of which we are made.

A patch that contains observers like us would have to last a long time, because we required a long time to appear. A long-lived patch has to be big, so we might not expect to see the gradients across it. Thus the universe could appear homogeneous to us, even though it truly is chaotic.

It is to be emphasized that there is no very strong reason to believe any of this really happened; this is a scenario rather than a theory. Nonetheless the idea has been influential because it is the only route by which we have been able to see how we might have made *ab initio* computations of the origin of the observed large-scale structure of the universe.

There is one generic testable prediction. Primeval inhomogeneities are supposed to be reduced by the large expansion during inflation, which made the length scale over which any quantity varies much greater than the Hubble length. The same surely would apply also to $R(t)$. Thus we would predict that the first of Eqs. (6) reduces to

$$(1 - \Omega)H^2 = \Lambda/3.$$

This is testable, though the observational uncertainties are still somewhat too large. If in addition the present value of Λ is negligibly small, $\Omega = 1$. As noted above, the status of this prediction is controversial.

COSMOGONY

The origin of planets and stars generally is taken to be a problem for astronomy and astrophysics, while galaxies and the structure in the galaxy distribution is left to cosmology. The division is historical and not necessarily rational.

What are the main effects involved in the formation of galaxies and clusters of galaxies? The list generally is thought to include gravity, because that is what holds these objects together; dissipative gas dynamics, because that certainly is needed to produce the stars that dominate the mass in the central parts of bright galaxies; and perhaps also explosions, because the effect of supernovae is very visible in the interstellar medium. Light pressure ("mock gravity") from early generations of stars may also play a role. Primeval magnetic fields could induce galaxy formation by inhomogeneous magnetic stress and could also account for the observations of cosmic magnetic fields. Cosmic strings, that act like magnetic flux tubes in a superconductor, could enter through gravity or the pressure generated by annihilation. Which played the central roles in galaxy formation is a matter of considerable debate in the cosmology community. A few general aspects of the considerations are worth listing.

A partial answer to the problem almost certainly is to be found in the gravitational instability of the relativistic cosmological models. That means a mass concentration such as a cluster of galaxies has to be growing with time; that is, the large-scale structure in the galaxy distribution is growing by gravitational instability.

For a more complete answer, we need the character of the primeval density fluctuations out of which the gravitational clustering of mass grew. A popular possibility is the Zeldovich spectrum, in which the space curvature caused by the mass fluctuations varies as the logarithm of the length scale, so we get convergence by cutting off the spectrum at some exceedingly small and large scales. This has the virtue of simplicity (we need not specify the values of the cutoffs) and is predicted in some versions of the inflation scenario. Numerical solutions for the growth of structure out of this spectrum do seem to resemble the observed "bubbly" appearance of the galaxy distribution. The most serious challenge is that this spectrum may be unable to account for the largest known coherent structures.

Did the same instability operating at an earlier epoch produce compact groups and clusters of galaxies, and galaxies themselves? One might see evidence for it in the fact that the mean value of the mass density at distance r from a bright galaxy, averaged over a fair sample of galaxies, varies as r^{-2}. This applies within a galaxy, at scales 1 kpc $< r <$ 30 kpc, and it applies with about the same constant of proportionality to the mean mass density in the pattern of clustering around a galaxy, for another factor $\sim$ 30 in radius. This scaling property might mean that galaxies and clusters formed by scaled versions of the same process, which would be gravitational instability.

Scaling cannot be the whole story, however, for a galaxy is a distinct and coherent mass distribution, while the small-scale galaxy distribution resembles a fractal. Perhaps gas dynamics redistributed mass on the scale of galaxies; perhaps galaxies did not form by gravitational instability. Essential hints surely will come from the recent progress in ability to observe galaxies at high redshifts by more efficient detectors, larger telescopes, and space-based observations. All this promises to give us an increasingly better view of galaxies as they were in the distant past. The details of what these young galaxies are like will be followed with great interest.

See also GALAXIES; HUBBLE EFFECT; UNIVERSE; ASTRONOMY, X-RAY

BIBLIOGRAPHY

For a longer account see J. Silk's *The Big Bang*. Freeman, New York, 1989.

For more detailed accounts see P. J. E. Peebles, *Physical Cosmology*, Princeton University Press, Princeton, NJ, 1971; S. Weinberg, *Gravitation and Cosmology: Principles and Applications of the General Theory of Relativity*, Wiley, New York, 1972; and P. J. E. Peebles, *The Large-Scale Structure of the Universe*, Princeton University Press, Princeton, NJ, 1980.

Recent research can be traced through the following: for the cosmological tests see the references in P. J. E. Peebles, *Publ. Astron. Soc. Pacific* **100,** 670, 1988; for large-scale velocity fields, see *Large-Scale Motions in the Universe*, V. C. Rubin and G. V. Coyne, eds., Princeton University Press, Princeton, NJ, 1989; for element production in the hot big bang see J. Yang *et al., Astrophys. J.* **281,** 493, 1984; for the dark mass problem see *Dark Matter in the Universe*, J. Kormendy and G. R. Knapp, eds., Reidel, Dordrecht, 1987; papers on the inflationary scenario are collected in *Inflationary Cosmology*, L. F. Abbott and So-Young Pi, eds., World Scientific, Singapore, 1986; for recent progress in theories of galaxy formation see *Large Scale Structures of the Universe*, J. Audouze *et al.*, eds., Reidel, Dordrecht, 1988.

Counting Tubes

Georges Charpak

The term counting tubes is applied to instruments detecting particles, charged or neutral, such as electrons, protons, gammas, and neutrons. Originally the most common shape was a tube with a central wire and an appropriate gas filling. Now the shape is much more diversified and the structure quite different. Since 1968 important properties of multiwire structures have been brought to light. The counters are used not only for counting, but more for localizing the position of the radiations. This has extended the application to new domains in nuclear physics, biophysics, astrophysics, and medicine. The considerable reduction in the cost of the electronics has played a great role in the evolution of the counters. The correlation between the time of detection of a signal and the ionization localization has led to an important new class of detectors, the drift chambers, which permit the construction of detecting surfaces of several square meters, with localization accuracies of better than 100 μm.

SOME BASIC PHENOMENA IN COUNTING TUBES

All counting tubes are filled with gases. Liquefied noble gases are used in very rare cases. The particles to be detected and counted liberate free ions in the gas, either by direct interaction or by extracting charged particles from the walls of the tube.

An electric field, applied in the gas, drifts the ions to different electrodes, according to their sign. The ions are either collected without multiplication (ionization counter mode) or various types of multiplication may occur. In the proportional counter, each electron gives rise to a single, well-localized avalanche in the intense field surrounding a thin wire. In the Geiger–Müller counter each avalanche is followed by a series of avalanches propagating along the wire. In the limited streamer mode a succession of avalanches along the electric field lines gives rise to a much stronger current pulse than in the proportional mode. It has the advantage, over the preceding mode, of a very localized dead region along the wire with the same advantage of an almost constant pulse height but much faster. It is now used in very large detectors, in high-energy physics, because of the simplification it brings to the electronics. In the spark counter, the avalanches produce a conductive channel of avalanches bridging the anode–cathode gap, with a subsequent spark. The propagation is mediated by ultraviolet photons. Figure 1 illustrates the different modes of operation and Fig. 2 shows the current amplification factor in a cylindrical counter operating in the different modes.

Among the many factors influencing the choice of the filling gases are the following.

1. Ability to operate in the desired mode: Although all gases are suitable for ionization chambers, those working in stable conditions in a multiplicative mode are rare.
   Noble gases mixed with CO_2, or various organics such as alcohols, isobutane, and methane are often chosen. In these gases electrons liberated by the radiations are not captured by the atoms and move freely under the influence of electric fields.
   The additives to noble gases are chosen for their action on the electron drift velocity or their ability to control the mean free path of the ultraviolet radiations emitted by the atoms excited by electronic collisions.
2. Interaction properties with the radiations to be de-

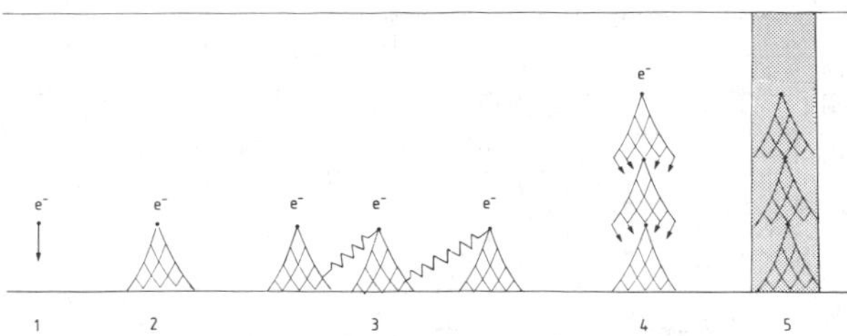

FIG. 1. The different mechanisms of electron multiplication around a wire. (1) Ionization chamber. The electrons are attracted by the anode wire. No charge gain. (2) Proportional counter. The field close to the wire is more intense, and by inelastic collisions each electron gives rise to an avalanche. Maximum charge gain $\sim 10^6$. (3) The ultraviolet light emitted by the excited atoms left behind by an avalanche produces secondary electrons in the close vicinity of the avalanche. This gives rise to a succession of nearby avalanches along the wire. Propagation speed 200 ns/cm: maximum charge gain $\sim 10^{10}$. (4) Limited streamer mode. A succession of avalanches propagates along the electric field lines but does not bridge the gap. (5) A conductive channel is produced after the first avalanches and a spark short-circuits the electrodes.

tected: For low-energy (up to 30 keV) x rays xenon is the best filling gas. Above these energies, the gaseous efficiency drops so drastically that the efficiency is dominated by the electrons extracted from the walls. For slow neutrons, ^{3}He or BF_3 are the best suited, since they have a considerable cross section for slow-neutron absorption.

For charged particles, argon is the main component because of its low cost.

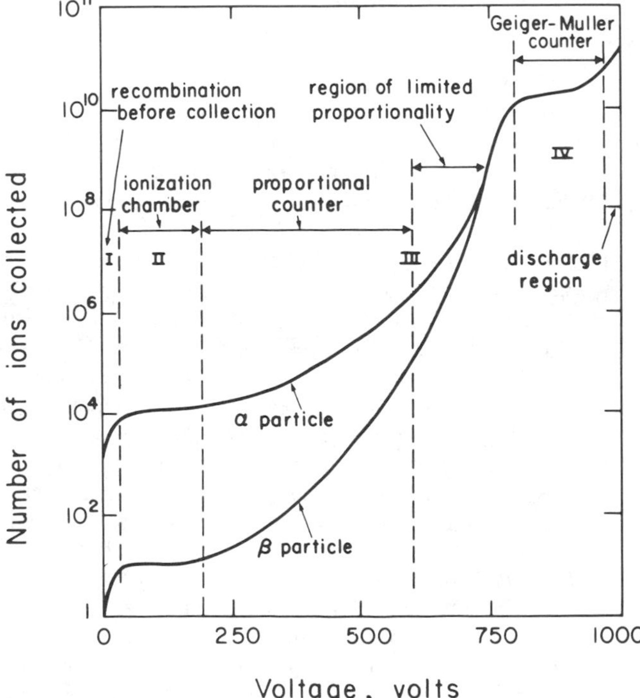

FIG. 2. Charges collected in a cylindrical counter as a function of applied voltage.

THE DIFFERENT GASEOUS DETECTORS

Ionization Counters

Any electrode shape is suitable for a counting tube that is to be used as an ionization counter.

In intense radiation fields the average collected current can be measured with a high accuracy. It is the ideal instrument for monitoring slowly varying radiation intensities. An example of this application is the measurement of the density of the products flowing in a pipeline, which can easily be determined, to accuracies of better than 10^{-3}, from the average current collected in a high-pressure ionization chamber detecting the 0.66-MeV γ rays from a ^{131}Cs source on the opposite side of the pipe.

For heavily ionizing particles, such as α particles or fission fragments, the charge pulses collected from every single track are detectable. The progress in the cost and noise of amplifiers progressively pushes down the minimum detectable energy lost in the gas. The limit is around 1 keV for very expensive amplifiers and around 100 keV for very cheap amplifiers, whereas in proportional counters one single electron liberated in the gas can easily be detected. This is why detectors with gaseous amplification have a wider range of applications.

Proportional Wire Counters

Proportional counters usually consist of a single thin wire in the axis of a cylinder. The wire is at a positive potential with respect to the cylinder. The electric field, as a function of the distance r to the axis, is equal to $2q/r$ where q is the charge per unit length. Close to the wire, which is usually 10–50 μm in diameter, the electric field is so high that the free electrons acquire, between two successive collisions, a sufficient energy to ionize the gas atoms and liberate a new electron. The mean free path at atmospheric pressure for an ionizing collision is close to 1 μm. Since the maximum gain is around 10^6, a maximum length of about 20 μm is required for this amplification.

The majority of the electrons, which are produced in the last few microns, are collected in times less than 10^{-10} s and give rise to an undetectable pulse: it is too fast and too small; its size is proportional to the potential drop experienced by the electrons in the last few microns close to the wire.

The useful pulse is produced by the motion of the positive ions away from the wire. They have to undergo the total potential drop between cathode and anode, and despite their large mass, the field is so intense that the risetime of the induced pulse can be of the order of 20 ns. For a point charge the pulse rises like $\log(1 + t/t_0)$ where t_0 is close to 2 ns for most common counter parameters.

The important properties are listed below.

1. *Energy response*. Since the mechanism of pulse formation is close to a wire, all the electrons freed in the gaseous volume give rise to the same pulse amplitude, and the charge induced in the cathode is proportional to the number of electrons and hence to the energy lost in the gas.

2. *Time resolution*. When an electron is freed in the counter it will produce no pulse as long as it has not reached the anode. The delay in its response is thus a function of its distance to the axis. For particles irradiating a counter uniformly, this results in a time jitter that is equal at most to the maximum drift time of electrons in the volume of the cylinder. It is of the order of 1 μs for a counter 1 cm in diameter.
3. *Dead time*. The electric field between the positive-ion sheath left behind by an avalanche and the anode is reduced. For high amplification factors the anode may be locally dead for times as long as 100 μs, but over a region extending over only 0.2 mm. High counting rates, of the order of 10^4 counts s^{-1} mm^{-1} of wire, are tolerated by proportional counters.

MULTIWIRE PROPORTIONAL CHAMBERS

These consist of a layer of thin anode wires, of diameter d, separated by equal distances s, sandwiched between two cathode planes at a distance L. The usual diameter ranges from 3 to 100 μm; the wire distance s is from 0.5 to a few millimeters (Fig. 3 shows an example of construction).

The charge per unit length is

$$q = \frac{V_0}{2(\ln \sinh \pi L/s - \ln \sinh \pi d/2s)}$$

where V_0 is the cathode–anode potential difference. The field varies like $1/r$ close to the wire and is uniform at distances larger than $s/2$. Since all the multiplication processes occur within about 20 μm from the wire, the amplification processes around the wire are exactly the same as in a single-wire cylindrical counter.

Two essential properties characterize such a structure:

1. The electric field is sharply separated into two regions: Far from the wires the field is nearly uniform, whereas close to the wire it is similar to the radial field produced in a cylindrical tube with a central wire (Fig. 4). When electrons are freed in the gas they drift in the uniform field and then experience inelastic collisions close to the wire, as in a cylindrical counter.
2. On a wire that is collecting the electrons from an avalanche, the pulse-producing mechanism is the same as in a cylindrical counter, the positive-ion sheath moving away in the intense field inducing the fast negative pulses. But the essential point is that it induces, at the same time, positive charges on the neighboring wires. Despite the strong capacitive coupling between wires, this ensures the localization of the detecting wire.

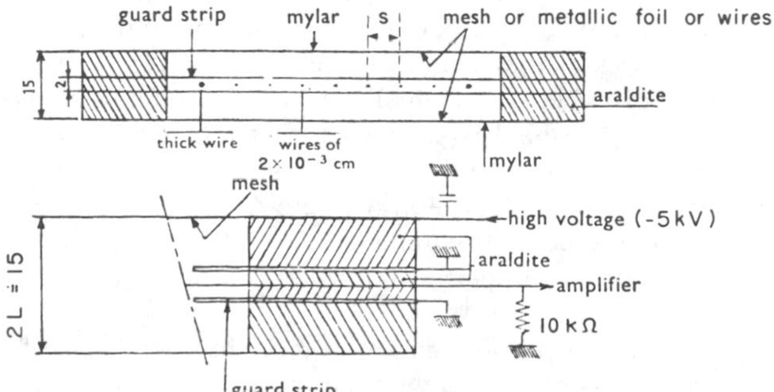

FIG. 3. Example of a multiwire proportional counter. Anode wires 20 μm in diameter, regularly spaced at distances of 2 mm between two cathode planes. Gas filling: argon + isobutane + methylal. High voltage 5 kV.

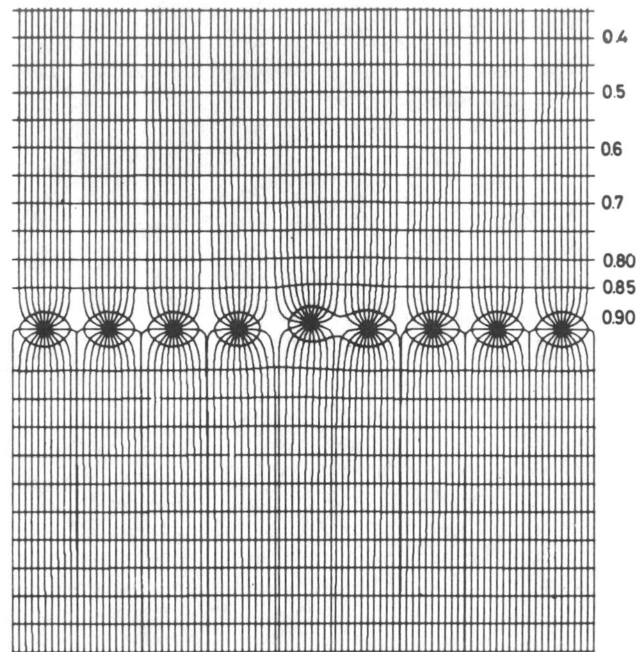

FIG. 4. Equipotentials in a multiwire chamber. Potential 1 on the anode wires, 0 in outer cathodes. $s = 1$ mm, $L = 8$ mm. Two regions of field: around the wire the field is cylindrical; in most of the volume it is uniform.

While large surfaces can be constructed, it is easy to localize with precision the position of the trajectory of the ionizing particle by detecting the wire carrying the pulse. The most remarkable properties of such structures, explaining their expanding use since 1968, are the following:

1. *Time resolution*. The time resolution is determined by the maximum distance between a trajectory and the nearest wire. With 2-mm wire spacing it is 25 ns, considerably better than in the cylindrical tube where the freed electrons had to drift in low fields over 1 cm.
2. *Dead time*. After the detection of an avalanche, the dead region is localized to about 0.2 mm along a wire and lasts for tens of microseconds. Since every wire acts as an independent detector, large counting rates can be reached; 10^7 counts/s over surfaces of 100 cm^2 are possible.
3. *Two-dimensional localization and imaging properties*. The motion of the positive ions, responsible for this localization of the detecting wire, induces positive pulses on the cathode. If the cathode is made of wires or strips, orthogonal or at an angle to the wire, the determination of the centroid of the induced pulses yields the position of the avalanche along the wire with

an accuracy of the order of 0.1 mm. This is an essential property for the two-dimensional localization of neutral radiations such as x rays or neutrons. Since the secondary ionizing radiation that they produce is usually of a very small range and cannot cross several chambers, the coordinates in a plane have to be obtained from a single gap. Several methods have been designed to obtain the two-dimensional readout: delay line methods, current division methods, etc.

4. *Examples of applications.* Chambers several meters in length, with a total of 20 000–100 000 wires, are common in high-energy physics, where they are used to localize trajectories. The imaging of slow neutrons or x-ray distributions with proportional chambers has growing applications in medicine and biology.

DRIFT CHAMBERS

Detecting Drift Spaces

Detection volumes can be attached to wire chambers. They are used to extend the useful volume in case of gaseous detection, or they can be filled with various solid converters for an efficient conversion of γ rays or slow or fast neutrons. The electrons are transferred to the multiwire chamber by appropriate electric fields.

Localization Drift Chambers

The interval of time separating the production of ionization in a gas and the detection of a pulse at a wire is strictly correlated to the distance between the trajectory of the ionizing event and the wire. If the drift velocity of electrons is known, the measurement of time gives the trajectory position.

A typical drift velocity is 5×10^6 cm/s in argon–isobutane mixtures at fields of 1000 V/cm. A great variety of drift-chamber shapes are possible; an example of a widely used chamber is given in Fig. 5. The sense wires are spaced by 5 cm. The cathode planes, 6 mm apart, are made of wires or strips at a linearly increasing potential, producing an electric field parallel to the chamber plane.

The accuracy in position determination can reach 50 μm. Detectors of several square meters are common. It is the cheapest accurate detector so far built.

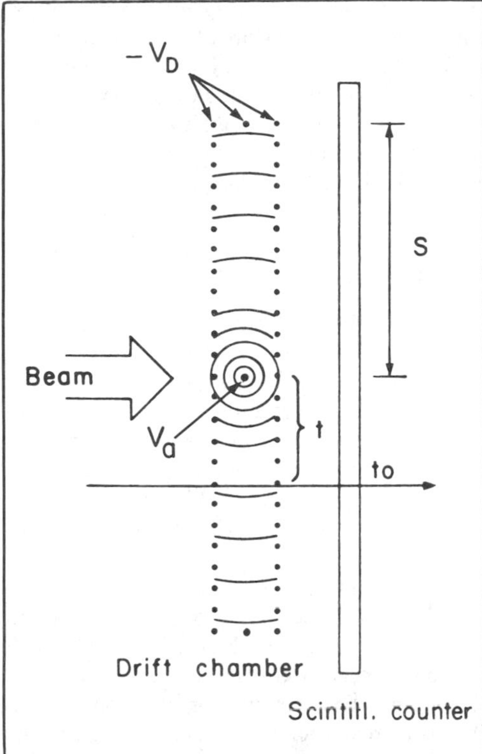

FIG. 5. Example of a drift chamber. The cathode is made of wires or strips at potentials growing linearly from 0 to $-V_D$. The drift field V_D/s is about 1000 V/cm. The electrons are detected on the anode wire at potential V_a. The length varies from 1 cm to 100 cm. The time t is a linear function of position. The zero time is given by a scintillator; the accuracy is better than 100 μm.

GEIGER–MÜLLER COUNTERS

Geiger–Müller counters are constructed like proportional counters. When the field around a wire goes beyond some limit, the positive ions produced close to the wire are so dense that they produce a local electrostatic shielding that will prevent further development of avalanches on the same spot. On the sides of the avalanche, the electric field is high, photoelectrons produced by the ultraviolet photons emitted by the excited atoms will give rise to rapidly developing new avalanches, and the discharge will spread along the wire at a velocity depending on many factors but close to 10^7 cm/s. The pulse shape is governed by this propagation mechanism. It consists of a constant current as the discharge propagates to both sides of the wire; it then drops by a factor of two when one end has been reached, the total pulse width being a function of the position of the initial avalanche. The total charge, however, is constant and easy to detect, since it is much greater than in a proportional counter.

Time Resolution

Time resolution is the same as in proportional counters; the time jitter is governed by the same mechanism, the time taken by the free electrons to reach the anode wire.

Dead Time

After a discharge the electric field between the ion sheath and the wire is reduced all along the wire, and the chamber is paralyzed for times of the order of 100 μs. For this reason Geiger counters have lost much of their interest, since it is now extremely easy and cheap to amplify the pulses from proportional counters to the same level without paralyzing the counter for such long times. Geiger counters can be used in multiwire structures. The propagation of the discharge along the wires permits extremely simple methods of localization, although these methods are much less accurate than in the proportional mode.

Limited Streamer Counters

A new mode of operation of wire counters has been discovered which is adopted for many large size detectors used in high-energy physics. With a proper choice of gases it is possible to have a succession of avalanches propagating along the field lines and stopping before reaching the cathode. This gives rise to large pulses of current, easy to handle by the electronics. The dead region is limited to a small region of the wire near the avalanches, while in a Geiger counter all the wire is shielded by the slow moving positive ions.

PROPORTIONAL COUNTERS WITH PARALLEL-PLANE ELECTRODES

Electron avalanches between parallel electrodes can be used for the detection of the electrons liberated by charged particles. It is more difficult than with wires to reach large amplification. The limiting factor is the occurrence of sparks at some points because of inhomogeneities in the electric field, or because of some contamination by α emitters, which are often present in most gases. With wires the decrease of the electric field far from the wire is unfavorable for spark propagation.

A great extension in the use of the amplification between parallel electrodes can be foreseen with the introduction of multistep structures (Fig. 6). It has been shown that with electrodes made of grids transparent to electrons it is possible, by having a transfer gap for electrons between successive gaps, to reach amplifications even larger than with wire counters permitting for instance the detection and localization of single electrons liberated by the photoionization in the gap by ultraviolet photons.

MULTISTEP AVALANCHE CHAMBERS

The electrons liberated in a conversion gap by ionizing events drift in an electric field to a preamplification gap made of two wire meshes where they are multiplied by a factor 10^4 by an adequate electric field and a fraction of them is transferred to a region of lower field and then to a second amplification gap where the gain can be 10^3. Taking into account all the transfer losses gains of 10^6 are easy to achieve. An intermediate grid permits the gating, between the transfer regions 1 and 2, within times of few tens of nanoseconds, of the swarm of drifting electrons. This detector is ideal for single VUV photon detection and localization.

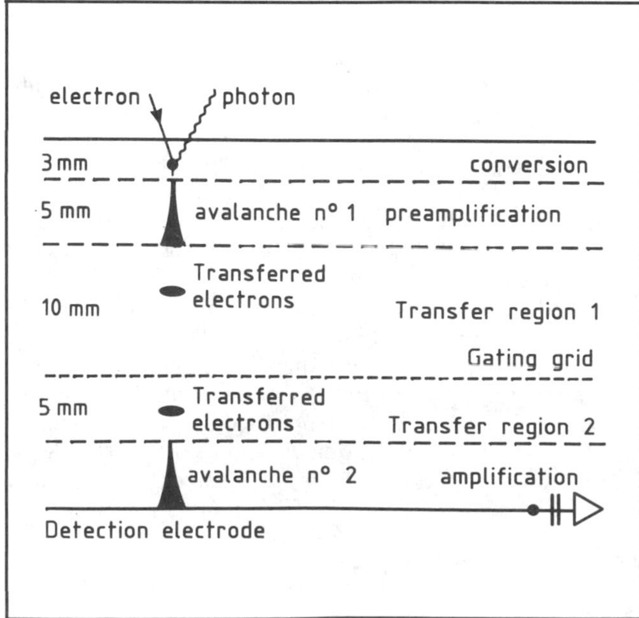

FIG. 6. Example of a multistep structure.

COUNTERS SENSITIVE TO UV OR VUV PHOTONS

An important class of new gaseous detectors has been introduced with the use of vapors which can be ionized by vacuum ultraviolet photons. For instance TMAE (tetrakis dimethylamine ethylene) can be ionized by radiations of a wavelength shorter than 230 nm, thus permitting, by introducing these vapors in wire counters or multistep parallel-plate counters, detection and imaging of VUV photons. The discovery of heavy scintillators, like BaF_2, emitting radiations at wavelength short enough to be detected by the gaseous detectors, permits one in some cases to replace photomultipliers or photodiodes and opens the way for a new wide range of applications.

SCINTILLATING PROPORTIONAL COUNTERS

The scintillating proportional counter relies on a mechanism of gaseous amplification that is quite different from the one in operation in the preceding counters.

If electrons are drifted in some gases (such as pure xenon) in appropriate electric fields, they excite light-emitting atomic levels without ionizing the gas. The photons emitted by these atomic levels are detected by photomultipliers. The advantage is a better energy resolution. Typically, a full width of 8.5% is achieved with x rays of 6 keV, as against 14% with a proportional counter.

See also RADIATION DETECTION.

BIBLIOGRAPHY

G. Charpak, R. Bouclier, T. Bressani, J. Favier, and C. Zupančič, *Nucl. Inst. Methods* **62,** 235 (1968).

G. Charpak, D. Rahm, and H. Steiner, *Nucl. Instr. Methods* **80,** 13 (1970).

G. Charpak and F. Sauli, *Phys. Lett.* **78B,** 523 (1978).

P. Rice Evans, *Spark, Streamer, Proportional and Drift Chambers.* Richelieu Press, London, 1974. (Contains a detailed description of the basic processes and the operation modes of most gaseous detectors in use prior to 1974.)

T. Ferbel, *Experimental Techniques in High Energy Physics.* Addison-Wesley, Reading, Mass., 1987.

CPT Theorem

Raymond W. Hayward

The *CPT* theorem is one of the most far-reaching theorems in quantum field theory. This theorem states that any local

Lagrangian theory which is invariant under *proper* Lorentz transformations is also invariant with respect to the combined operations of charge conjugation, C, space inversion, P, and time reversal, T, taken in any order, although the theory may not be separately invariant under the individual operations C, P, or T. One important corollary of this theorem is that if a physical process is not invariant under one of the operations C, P, or T, it is not invariant under at least one of the remaining two processes. A second corollary is that if a physical process is invariant under one of the operations C, P, or T, it is invariant under the product of the other two.

The validity of the *CPT* theorem is based on the following assumptions:

(a) The theory is invariant under proper orthochronous Lorentz transformations. No specific assumptions concerning invariance under space and time reflections need be made.
(b) All interaction densities are local and constructed out of field operators and derivatives of field operators of at most a finite order.
(c) The normal spin-statistics relations are obeyed. Interaction densities are properly symmetrized or antisymmetrized according to whether bosons or fermions are involved.
(d) All interaction densities are Hermitian.

A proper homogeneous Lorentz transformation from one coordinate frame, S, to another, S', is given by

$$x_\mu{}' = \Lambda_{\mu\nu} x_\nu,$$

where $x_\mu = (x_1, x_2, x_3, x_4) \equiv (x, y, z, ict)$ and summation over a repeated index is implied. The $\Lambda_{\mu\nu}$ are constants independent of the space-time coordinates such that the relativistic interval remains constant, i.e.,

$$x_\mu{}' x_\mu{}' = x_\mu x_\mu.$$

The invariance requires that

$$\Lambda_{\mu\nu}\Lambda_{\mu\lambda} = \delta_{\nu\lambda}.$$

The *usual* criteria for preserving the reality conditions of the space-time coordinates are that Λ_{ij}, with $i,j = 1,2,3$, and Λ_{44} are real while Λ_{4i} and Λ_{i4} are imaginary.

This transformation is continuous, being obtained from a series of infinitesimal transformations from unity, and is characterized by det $\Lambda = 1$. This is an orthochronous transformation having $\Lambda_{44} \geq 1$, i.e., it does not change the direction of time. The proper homogeneous Lorentz transformations form a six-parameter continuous group containing the identity operator.

Such a formalism can be used to describe the proper rotations of a physical system and "boosts" to a different velocity of the physical system with respect to the space-time coordinates. The requirement of equivalence between different Lorentz frames leads directly to the laws of conservation of energy, momentum, and angular momentum.

In addition there are the improper Lorentz transformations that preserve the relativistic invariance of the theory but are disjoint and not continuously connected, unlike the proper Lorentz transformations. These improper transformations do not contain the identity operator. The first of these improper transformations is that of space inversion in which $x_i \rightarrow -x_i$ while $x_4 \rightarrow x_4$. Here det $\Lambda = -1$ while $\Lambda_{44} \geq 1$. The second is that of time inversion in which $x_4 \rightarrow -x_4$ while $x_i \rightarrow x_i$. Here det $\Lambda = -1$ while $\Lambda_{44} \leq -1$. A third improper transformation is that of space-time inversion where $x_\mu \rightarrow -x_\mu$, which is a combination of the previous two improper transformations. However, the subset corresponding to space-time inversion may equivalently be obtained by a finite complex *proper* Lorentz transformation in which all of the $\Lambda_{\mu\nu}$ are real, thus not maintaining the reality of the space-time coordinates except in a particular case. This transformation corresponds to a real rotation through an angle π in the $x_1 x_2$ plane followed by a real "rotation" through an angle π in the $x_3 x_4$ plane.

In a relativistic theory, time inversion is not equivalent to an operation where every velocity is replaced by the opposite velocity so that the position of a particle at t becomes the same as it was, without time inversion, at $-t$. This latter operation may properly be called velocity reversal or motion reversal, but in quantum field theory it is usually, perhaps illogically, called time reversal or Wigner time reversal.

Time inversion is actually the product of time reversal, T, and charge conjugation, C, an operation where all particles are changed into their antiparticles. Charge conjugation is a relativistic quantum-mechanical concept only and, like space and time inversion, is a discontinuous operation.

In quantum mechanics the interacting elementary particles are described by a wave function $\Phi(\mathbf{r},t)$ which is itself a function of the space-time coordinates. In a quantum field theory the functions $\Phi(\mathbf{r},t)$ are field operators. The one-particle expectation value of these field operators are the wave functions. We denote by $\Phi^{(P)}(\mathbf{r},t)$ the field operator in the space reflected system which is obtained by a unitary parity transformation P:

$$\Phi^{(P)}(\mathbf{r},t) = P\Phi(\mathbf{r},t)P^{-1} = O_P\Phi(-\mathbf{r},t).$$

The quantity O_P is the product of a unitary matrix operator operating on the individual spin components (if any) of $\Phi(-\mathbf{r},t)$ and a phase factor η_P. The eigenvalue η_P is the parity of the field. When $\eta_P = 1$ the field $\Phi(\mathbf{r},t)$ is said to have even parity and when $\eta_P = -1$ the field is said to have odd parity.

For a meaningful and satisfactory formulation of time reversal and charge conjugation, it is essential to consider the field operators taking into account their ordering and their commutation relations appropriate to Bose statistics for integer spins and Fermi statistics for half-integer spins. All bilinear covariants of field operators are assumed to be properly symmetrized or antisymmetrized according to whether they obey Bose or Fermi statistics, respectively.

The field operator in the time-reversed system is denoted by $\Phi^{(T)}(\mathbf{r},t)$ and is obtained by the antiunitary time-reversal operator, T, the antiunitary feature arising from the reversal of the ordering of all operators:

$$\Phi^{(T)}(\mathbf{r},t) = T\Phi(\mathbf{r},t)T^{-1} = O_T\Phi^*(\mathbf{r},-t).$$

The quantity O_T is the product of a unitary matrix operator operating on the individual spin components (if any) of $\Phi^*(\mathbf{r},-t)$ and a phase factor η_T.

Table I Transformation of intrinsic fields under space inversion, time reversal, charge conjugation, and strong reflection

Intrinsic Field	$\Phi(\mathbf{r},t)$	P	T	C	R
Scalar	$\varphi(\mathbf{r},t)$	$\varphi(-\mathbf{r},t)$	$\varphi^*(\mathbf{r},-t)$	$\varphi^*(\mathbf{r},t)$	$\varphi(-\mathbf{r},-t)$
Dirac spinor	$\psi(\mathbf{r},t)$	$\eta_P\gamma_4\psi(-\mathbf{r},t)$	$\eta_T\gamma_3\gamma_1\psi^*(\mathbf{r},-t)$	$\eta_C\gamma_2\psi^*(\mathbf{r},t)$	$\gamma_5\psi(-\mathbf{r},-t)$
Polar four-vector	$A_\mu(\mathbf{r},t)$	$(-1)^{1+\delta_{\mu4}}A_\mu(-\mathbf{r},t)$	$(-1)^{1+\delta_{\mu4}}A_\mu^*(\mathbf{r},-t)$	$-A_\mu^*(\mathbf{r},t)$	$-A_\mu(-\mathbf{r},-t)$
Lorentz six-vector	$\mathbf{E}(\mathbf{r},t)$	$-\mathbf{E}(-\mathbf{r},t)$	$\mathbf{E}^*(\mathbf{r},-t)$	$-\mathbf{E}^*(\mathbf{r},t)$	$\mathbf{E}(-\mathbf{r},-t)$
	$\mathbf{B}(\mathbf{r},t)$	$\mathbf{B}(-\mathbf{r},t)$	$-\mathbf{B}^*(\mathbf{r},-t)$	$-\mathbf{B}^*(\mathbf{r},t)$	$\mathbf{B}(-\mathbf{r},-t)$
Axial four-vector	$C_\mu(\mathbf{r},t)$	$(-1)^{\delta_{\mu4}}C_\mu(-\mathbf{r},t)$	$(-1)^{1+\delta_{\mu4}}C_\mu^*(\mathbf{r},-t)$	$C_\mu^*(\mathbf{r},t)$	$-C_\mu(-\mathbf{r},-t)$
Pseudoscalar	$\pi(\mathbf{r},t)$	$-\pi(-\mathbf{r},t)$	$-\pi^*(\mathbf{r},-t)$	$\pi^*(\mathbf{r},t)$	$\pi(-\mathbf{r},-t)$

Likewise, the field operator in the charge-conjugate system is denoted by $\Phi^{(C)}(\mathbf{r},t)$ and is obtained by application of the unitary charge-conjugation operator:

$$\Phi^{(C)}(\mathbf{r},t)=C\Phi(\mathbf{r},t)C^{-1}=O_C\Phi^*(\mathbf{r},t).$$

The quantity O_C is the product of a unitary matrix operator operating on the individual spin components (if any) of $\Phi^*(\mathbf{r},t)$ and a phase factor η_C.

The unitary matrix operators in O_P, O_T, and O_C are determined in a way such that the Lagrangian density and the derivable energy-momentum tensor involving the wave functions or field operators $\Phi(\mathbf{r},t)$ are form-invariant under the discrete transformations P, T, and C. The exact form of the matrix operators depends, of course, on the particular representation employed.

The phases η_P, η_T, η_C for those particles that can be created or destroyed can be found experimentally for bosons from the properties of their mutual interactions assuming conservation of P, C, and T in these interactions. Since fermions always appear in bilinear combinations the phases always occur as $|\eta_P|^2=|\eta_T|^2=|\eta_C|^2=1$ and are indeterminate. Conventional usage assigns $\eta_P=1$, i.e., positive intrinsic parity for fermions. Any fermion state of total angular momentum j can be specified by Φ_{jls}, where $\mathbf{j}=\mathbf{l}+\mathbf{s}$, and has total parity $(-1)^l$ by convention. Similar arguments define a "time parity" and a "charge parity" for each quantum-mechanical state.

The field operator in the system that is rotated first in the x_1x_2 plane followed by a rotation in the x_3x_4 plane is obtained by application of the antiunitary operation of strong reflection, R, which is the nomenclature for this particular proper rotation followed by a reversal of the order of all operators:

$$\Phi^{(R)}(\mathbf{r},t)=R\Phi(\mathbf{r},t)R^{-1}=O_R\Phi(-\mathbf{r},-t).$$

The matrix operator O_R is given by

$$O_R=\exp(iS_{12}\pi+iS_{34}\pi),$$

where S_{12} and S_{34} are the spin operators, corresponding to the particular representation of $\Phi(\mathbf{r},t)$, for a rotation in the x_1x_2 and x_3x_4 planes, respectively.

The transformation under the operations P,C,T, and R for a number of fields with distinct Lorentz transformation properties are listed in Table I. We have employed a Cartesian representation for the polar-vector, Lorentz six-vector, and axial-vector fields and the Dirac–Pauli representation for the spinor field.

Inspection of the table indicates that the operation *CPT* taken in any order is equivalent to the operation of strong reflection. The requirement that any formulation involves a Lagrangian that transforms as a scalar under proper Lorentz transformations ensures the validity of the *CPT* theorem irrespective of its properties under the individual improper transformations of C,P, or T.

One of the better-known experimental applications of the *CPT* theorem occurs in the weak interaction process. In nuclear beta decay it has been observed that the violations of P and C invariance are maximal while the product *CP* is conserved. In all of the weak-interaction phenomena, which include all decays of the elementary particles that were experimentally investigated prior to 1964, except the electromagnetic decays $\pi^0\rightarrow2\gamma$ and $\Sigma^0\rightarrow\Lambda+\gamma$, the violation of C and P invariance and the validity of *CP* and T invariance was fully established. In 1964 experiments showed that the K^0 meson, which should decay only into three π mesons, decayed a small fraction of the time into two π mesons, a process violating *CP* invariance. Vigorous experimental research since that time has confirmed this *CP*-invariance violation as well as a corresponding T-invariance violation but no other process has been observed where *CP* invariance is violated. There have been many theoretical attempts to account for this small *CP*-invariance violation. In 1988, experimental results have agreed with theoretical predictions based on the standard model of weak interactions, considering higher-order processes involving the constituent quark structure of the K^0 and π mesons.

At present only the *CPT* invariance remains unbroken exprementally.

See also BOSE–EINSTEIN STATISTICS; ELEMENTARY PARTICLES IN PHYSICS; FERMI–DIRAC STATISTICS; INVARIANCE PRINCIPLES; LORENTZ TRANSFORMATIONS; MATRICES; OPERATORS; PARITY; QUANTUM FIELD THEORY.

BIBLIOGRAPHY

G. Charpak, R. Bouclier, T. Bressani, J. Favier, and C. Zupancic. *Nucl. Inst. Methods 62,* 235 (1968).

G. Charpak, D. Rahm, and H. Steiner, *Nucl. Instr. Methods* **80,** 13 (1970).

P. Rice Evans, *Spark, Streamer, Proportional and Drift Chambers*. Richelieu Press, London, 1974. (Contains a detailed description of the basic processes and the operation modes of most gaseous detectors in use prior to 1974.)

I. Ferbel, *Experimental Techniques in High Energy Physics*, Addison–Wesley, Cambridge, Mass., 1987.

G. Lüders, "Proof of the *TCP* Theorem," Ann. Phys. **2,** 1 (1957). (I)

W. Pauli, "Exclusion principle, Lorentz group, and reflection of space-time and charge," in *Neils Bohr and the Development of Physics* (W. Pauli, ed.). Pergamon Press, London, 1955. (I)

J. J. Sakurai, *Invariance Principles and Elementary Particles*. Princeton University Press, Princeton, N.J., 1964. (I)

R. F. Streater and A. S. Wightman, *PCT, Spin and Statistics, and All That*. Benjamin, New York, 1964. (A)

Critical Points

J. D. Litster

If the pressure on a fluid, such as water, is increased, the boiling temperature will also increase. At the same time, the density difference between the liquid and vapor phases becomes smaller, and vanishes continuously as the liquid–gas critical point (at temperature T_c and pressure P_c) is approached. This is commonly called a second-order phase transition, because a material property known as the order parameter (here, the density difference between the liquid and vapor phases) vanishes continuously. Near the critical point anomalous behavior (divergence) is observed in the range over which order parameter fluctuations are correlated as well as in a number of thermodynamic derivatives, such as the compressibility, and the specific heat. Nearly all materials near a second-order phase transition have been found to show behavior similar to that of a pure fluid near its critical point. This is because properties near a critical point are determined primarily by the correlation length for fluctuations in the order parameter; when this length exceeds the range of the interactions responsible for the phase transition, the behavior becomes insensitive to the detailed nature of the interactions. With the order parameter given in parentheses, some examples of critical points are: ferromagnets (magnetization), antiferromagnets (sublattice magnetization), separating binary fluid mixtures (concentration of components in the two phases), ferroelectrics (spontaneous polarization), and superconductors and superfluids (condensate wave function). While there are quantitative differences in the critical behavior near these different second-order phase transitions, it is the similarity that is most striking.

The classical models for these phase transitions all predicted the same behavior in the vicinity of the critical point as was given by Landau's 1937 theory [1] of phase transitions. The anomalous quantities diverged according to powers of $T - T_c$ (e.g., the compressibility diverged as $[T - T_c]^{-1}$) and the specific heat was predicted to have a discontinuity rather than to diverge. Some materials, such as ferroelectrics and superconductors, were found to follow the classical theory closely, but most showed power-law divergences with different exponents than predicted as well as specific heats that diverged approximately logarithmically.

The Landau model failed because it does not consider properly the divergent order-parameter fluctuations near the critical point. The renormalization group (RG) method [2] can be used for statistical mechanical calculations in the presence of strong fluctuations, a discovery for which Kenneth Wilson was awarded the Nobel Prize in Physics in 1982. As a result of this breakthrough, we have now a quantitative understanding of the role of critical fluctuations and an appreciation of the importance of the effects of spatial dimensionality and symmetry of the order parameter on critical behavior. Critical points fall into different "universality classes," which should show identical behavior, according to the symmetry, and hence the number of degrees of freedom, of the order parameter [3]. The effect of thermally induced fluctuations of the order parameter diminishes as the spatial dimension increases. There is an upper "marginal" dimension, which is four for most critical points, where the classical Landau theory becomes valid. There is also a lower marginal dimension, two for most critical points; below this dimension the effect of the fluctuations is so strong the phase change does not occur. (Rare gases adsorbed on the surface of crystals are a realizable example of a two-dimensional system.)

The RG methods have been used to explain theoretically a wide variety of second-order phase transitions. This incudes multicritical points where two or more lines of critical points join. An example is a tricritical point, where three lines of critical points meet and the transition becomes first order; this commonly occurs when two order parameters compete in a system. When order parameters are of different universality classes, RG methods can be used to calculate the "crossover" behavior as one dominates the other, and also to calculate the interplay between two- and three-dimensional physics in a wide variety of materials. For example, RG theories of multicritical phenomena in anisotropic magnets have been applied to a detailed model for the growth of two- and three-dimensional liquid crystal phases [4]. Thus the experimental elucidation of the properties of fluids near critical points and their theoretical explanation, which were among the most challenging problems in physics during the 1960s, have enabled us to understand in detail such diverse physical systems as rare earth magnets, thin film substrates on solid substrates, lamellar (high T_c) superconductors, liquid crystals, and biological membranes.

See also Ferroelectricity; Ferromagnetism; Fluctuation Phenomena; Helium, Liquid; Liquid Crystals; Phase Transitions; Renormalization; Superconductivity Theory.

REFERENCES

1. L. D. Landau, "On the Theory of Phase Transitions" (English translation), in *The Collected Papers of L. D. Landau* (D. Ter Haar, ed.). Gordon & Breach, New York, 1965. (E)
2. K. G. Wilson, *Phys. Rev.* **B4,** 3174, 3184 (1971). (A) M. E. Fisher, *Rev. Mod. Phys.* **46,** 597 (1974). (I)
3. See the extensive discussions of various theoretical approaches in the multivolume series *Phase Transitions and Critical Phenomena* (C. Domb and M. S. Green, eds.). Academic, New York, 1972, 1974. (A)
4. J. D. Brock, R. J. Birgeneau, J. D. Litster, and A. Aharony, "Liquids, Crystals, and Liquid Crystals," *Physics Today* (July 1989), and references therein. (E)

Cryogenics

A. C. Anderson

The field of cryogenics involves the production and application of low temperatures and is generally associated with the use of various liquefied gases. It is a broad field. In the industrial sector cryogenics includes the preparation and storage of food and biological materials, the production and delivery of liquid oxygen to medical and metallurgical facilities, the transport and storage of liquefied natural gas, and the liquefaction of hydrogen for the space program. One current area of developmental work is directed toward the application of low-temperature techniques to the generation, transmission, storage, and utilization of electrical energy. The metals employed have small or negligible electrical resistance and hence provide a considerable reduction in power loss.

In experimental physics, cryogenic techniques are employed either because a phenomenon of interest occurs only at low temperature, or because a physical property being studied becomes less complex and more amenable to theoretical interpretation. Phenomena that occur only at low temperatures include superconductivity, the superfluid phases of liquid ^{3}He and liquid ^{4}He, and the effects of the motion of atoms and molecules that persists after thermal motion has been suppressed through a reduction in temperature. Since these phenomena can be understood only in terms of quantum mechanics, cryogenics provides the opportunity for a rigorous test of the modern theory of quantum mechanics applied to condensed matter.

A well-documented example of the use of cryogenics to isolate and simplify physical properties, and hence facilitate theoretical interpretation, has been the study of nonmetallic crystals. Thermal expansion, thermal conductivity, and specific heat can be described in terms of collective atomic vibrations or lattice waves called phonons. At low temperatures those phonons that are of greatest importance have the velocity of a sound wave and a wavelength large compared to the atomic spacing in the solid. The phonon behavior is then rather independent of the detailed atomic arrangement and bonding, and the phonon may be treated as a classical sound wave. This permits the development of a simple theory that still contains quantization, and that may be tested quantitatively against experimental data. Historically this procedure has been of great importance in the development of our knowledge of the physics of solids. Cryogenics continues in this manner to provide the physicist with a highly effective experimental tool.

In measuring low temperatures, the absolute or Kelvin scale of temperature is generally used. On this scale the freezing and boiling points of water are approximately 273 and 373 K, respectively. To appreciate the Kelvin scale and in fact to understand the importance of cryogenics to the physicist, it is necessary to recognize that it is often the ratio of two temperatures, not the difference, that controls a physical or chemical process. Therefore the temperature interval 0.001–0.01 K may contain as much interesting physics as the much "larger" interval from 100 to 1000 K.

The calibration of a low-temperature thermometer is based on a variety of phase transitions, such as the temperature of the triple point of hydrogen or the equilibrium vapor pressure of liquid helium. The temperatures of the phase transitions have been estimated through much exacting experimental work; the accuracy of the determinations is constantly being improved. A thermometer generally utilizes an electrical or magnetic property as the thermometric parameter. Examples are the electrical resistivity of metals and semiconductors or the weak magnetic susceptibility of certain salts and metals. Each type of thermometer has been carefully studied to be certain that impurities, strain, and other variables do not influence the calibration. A variety of thermometers, some precalibrated, are available commercially for use down to 0.01 K.

Providing a research laboratory with appropriate refrigeration facilities has been greatly simplified in recent years. Liquid nitrogen at 77 K and liquid helium at 4 K are available from commercial sources or may be liquefied on site using commercial liquefiers. Gaseous ^{4}He is extracted from natural gas wells in several countries and therefore is sufficiently abundant, at present, that many laboratories vent the evaporated ^{4}He to the atmosphere. Temperatures between 4 K and 300 K are obtained by partial thermal isolation of the experimental samples from the liquid cryogen. Temperatures below 4 K are produced with liquid ^{4}He by rapidly pumping away the evaporating gas. Since helium, unlike other liquids, does not solidify during this process, a liquid bath at a temperature of about 1 K can be produced. Below 2 K, liquid ^{4}He is a superfluid having an extremely large effective thermal conductivity. As a result, the helium bath has the desirable property of being nearly isothermal. On the other hand, a superfluid film creeps up the walls of the cryostat, a process that is often the most important contribution to the liquid loss rate.

The lighter isotope of helium, ^{3}He, has become available as a by-product in the manufacture of nuclear devices. As a liquid, ^{3}He has a higher vapor pressure than ^{4}He and is not a superfluid above 0.003 K. The liquid may be pumped to 0.25 K. The price of ^{3}He dictates that it be used in a closed system. The evaporated gas is either collected for reuse or returned to the cryostat to form a closed-cycle, continuously operating refrigerator. Pumped ^{4}He at 1–2 K is required to absorb the heat as the ^{3}He gas is reliquefied.

A more important closed ^{3}He cycle has recently been developed in which the liquid ^{3}He is dissolved in a small stationary bath of liquid ^{4}He. This process absorbs heat and therefore provides refrigeration. Commercial ^{3}He–^{4}He dilution refrigerators are available that provide temperatures down to 0.01 K continuously. Dilution refrigerators are simple to operate and may be constructed in a variety of configurations to match the experimental problem.

Refrigeration may be provided for temperatures below 0.01 K but not, at present, in a continuous fashion. The ^{3}He–^{4}He dilution process may be used to roughly 0.002 K by not returning the ^{3}He gas to the cryostat. Alternatively, pure ^{3}He can be solidified by the application of a pressure of about 2.9×10^6 Pa (29 atm). Since the freezing characteristics of pure ^{3}He below 0.3 K are such that it cools with the application of pressure, compression and solidification of liquid ^{3}He can provide temperatures to about 0.002 K. Still lower temperatures can be obtained by the use of slightly paramagnetic materials. Application of a large magnetic field causes such a material to warm. The evolved heat may be absorbed by a dilution refrigerator. Removal of the magnetic field then causes the paramagnetic material to cool below the temperature of the refrigerator. The successful application of this technique of adiabatic demagnetization to provide temperatures below 0.001 K is still in its infancy.

It is to be emphasized that the temperatures mentioned in the foregoing refer to the sample or device that is to be cooled. The temperature of the refrigerator or refrigerating material may be much lower, since the transport of heat to the refrigerator becomes an increasingly severe problem with decreasing temperature. It is therefore important to minimize any heat influx to the sample. This is done by providing several stages of stepwise reduction in temperature between 77 and 0.02 K to reduce heat transport by thermal conduction. In addition, and especially at temperatures below 1 K, heat influx from vibration and radiation must also be eliminated. For radiation this includes the electromagnetic spectrum from light waves to radio waves, and many low-temperature laboratories are therefore constructed inside copper-shielded rooms.

See also HEAT CAPACITY; HEAT TRANSFER; HELIUM, LIQUID; MAGNETIC COOLING; PARAMAGNETISM; TEMPERATURE; THERMAL EXPANSION.

BIBLIOGRAPHY

Bulletin of the International Institute of Refrigeration. (A)

E. M. Codlin, *Cryogenics and Refrigeration* (A Bibliographical Guide). MacDonald, London, 1968; Part 2, IFI/Plenum, New York, 1970.

F. E. Hoare, L. C. Jackson, and N. Kurti, *Experimental Cryophysics*. Butterworth, London, 1961. (A)

O. V. Lounasmaa, *Experimental Principles and Methods Below 1 K*. Academic, New York, 1974. (A)

K. Mendelssohn. *The Quest for Absolute Zero*. McGraw-Hill, New York, 1977. (E)

H. M. Rosenberg, *Low Temperature Solid State Physics*. Oxford, London and New York, 1965. (A)

M. W. Zemansky, *Temperature Very Low and Very High*. Van Nostrand, Princeton, NJ, 1981. (E)

Crystal and Ligand Fields

C. J. Ballhausen

The basic idea of crystal field theory is that the electrons located upon a transition-metal ion which is embedded in an ionic lattice are subjected to an electric field originating from the surrounding electric charges. The "lattice" can also be taken to consist of the groups or ions directly associated with the metal ion, called the ligands. For instance in the CrF_6^{-3} complex the $(3d)^3$ electrons, taken to be located upon the Cr^{+3} center, experience an electric intramolecular field of octahedral symmetry coming from the six F^- ligands. In $Ni(H_2O)_6^{++}$ the $(3d)^8$ electrons located on the Ni^{++} ion are subjected to a field originating from the six water ligands. Historically this notion was first used by Van Vleck in the early 1930s to explain the paramagnetism of the complexes of the rare earths, $(4f)^n$, and the $(3d)^n$ transition series.

It was known that for a free ion in vacuum the magnetic moment would have a contribution both from the total orbital angular momentum L and from the total spin angular momentum S of the electrons. For complexes of the $(3d)$ series, in contrast to those of the $(4f)$ series, it was, however, found that the orbital angular-momentum contribution was quenched. Van Vleck explained this by pointing out that the degeneracy of the orbital angular momentum would be lifted by the crystal field—an intramolecular Stark effect occurs. In the Cr^{+++} ion, for instance, the atomic ground state is a ^4F state. This has sevenfold orbital degeneracy. An octahedral crystal field will now split this state into three levels, being respectively three- three-, and onefold degenerate (Fig. 1). The ground state is onefold degenerate, and separated from the first excited state by several thousand wave numbers. It is therefore only the ground state which is populated at room temperature, and since an orbital contribution to the magnetic susceptibility requires degeneracy, there will be none such in octahedral Cr^{+3} complexes.

On the other hand, for the complexes of the $(4f)^n$ series, the f electrons are so well "shielded" from the electric charges of the ligands that they experience only a small crystal field. The degeneracies are indeed broken up, but the splittings are small compared to kT. Consequently it is only at low temperatures that deviations from free-ion behavior are observed.

The splittings of the atomic levels due to the crystal field had been worked out by Hans Bethe in 1929, using point-group symmetry. Applying Bethe's theory, Van Vleck and his co-workers then performed quantitative calculations of the magnetic susceptibilities of transition-metal complexes. Using only the symmetry of the complex—octahedral, tetrahedral, or a distorted arrangement of the ligands—and

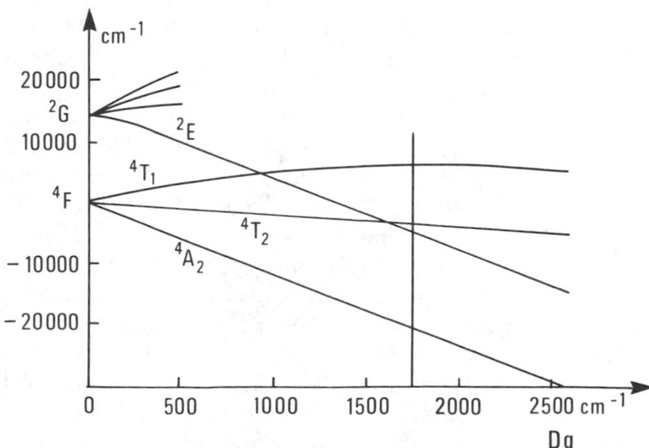

FIG. 1. Partial orbital splitting diagram for $(3d)^3$ in an octahedral field. For $Cr(H_2O)_6^{+++}$, $Dq = 1740\ cm^{-1}$ with the first spin-allowed band $^4A_2 \rightarrow {}^4T_2$ placed at 17 400 cm^{-1}. The $^4A_2 \rightarrow {}^2E$ lines are found at approximately 15 000 cm^{-1}. The second spin-allowed band $^4A_2 \rightarrow {}^4T_1$ is placed at 24 500 cm^{-1}.

with the "strength" of the crystal field as a parameter, they succeeded in giving a complete explanation of the experimental features.

The idea of an operative "crystal field" was, however, not accepted at the time by the chemists. The valence bond treatment of the complexes, as taught by Linus Pauling in his famous book *The Nature of the Chemical Bond,* had a strong grip on the chemical community. It was therefore a major breakthrough when Van Vleck showed in 1935 that there are no contradictions between the two theories: The "crystal field splitting" is indeed the unavoidable consequence of the chemical bonding between the ligands and the metal ion. Taking this into account, the name "crystal field" has been supplanted by the name "ligand field" to acknowledge the role played by the "covalent" ligand–metal bond.

An important consequence of the theory which was not appreciated until much later is the impossibility of calculating the crystal field splittings using an electrostatic approach as originally done by Van Vleck. These splittings can only be calculated by doing a full molecular orbital calculation, including all the many important "exchange integrals." Such a calculation was first performed in 1956 by Tanabe and Sugano. The difficulties of calculating the splittings from first principles are, however, not of any practical importance, since the strength of the crystal field, measured in units of Dq, is always extracted from experiments. Dq is thus a semiempirical parameter analogous with the β parameter in Hückel theory.

In addition to accounting for the magnetic features, crystal or ligand field theory can also explain the often beautiful colors of the transition-metal complexes as being due to transitions from the ground state to the excited atomic levels as split by the crystal field. The first absorption band which was identified is a transition in octahedral Cr^{+++} complexes. This was done by Finkelstein and Van Vleck in 1940. The atomic configuration of $(3d)^3$ gives rise to (among others) ^4F and ^2G states, the latter being placed some 15 000 cm^{-1} above ^4F (Fig. 1). When a cubic crystal field is added, the ^4F splits as indicated, while ^2G gives rise to (among others) a ^2E state. In the diagram, the levels are calculated as a function of Dq, and with a value of Dq = 1820 cm^{-1} Finkelstein and Van Vleck could place the $^4A_2 \rightarrow {}^2E$ transition at 14.900 cm^{-1}.

The weakness of the intensity of this transition is explained by its being both spin- and parity-forbidden. The transitions which take place between the split components of (for example) ^4F in $(3d)^3$ are also parity-forbidden, taking place inside the $(3d)^3$ configuration, but being spin-allowed, they have a somewhat larger intensity. The bands are broad and structureless. The identification of such bands had, however, to wait until after World War II—the first such identification being made by Ilse and Hartmann for $Ti(H_2O)_6{}^{+++}$ in 1951. Extensive correlation diagrams which enable one to identify the excited states of all $(3d)^n$ complexes, have been published by many authors.

The crystal or ligand field theory is exceptional among semiempirical electronic theories in that it permits one to perform identifications of excited states solely from a knowledge of the symmetry of the complex ion. However, the lower the point-group symmetry of the complex, the more splitting parameters have to be introduced. Only in octahedral and tetrahedral symmetries is there only one one-electron parameter, Dq. Fortunately, the Dq parameter is rather insensitive to lower fields, and a gross identification is rarely obscured by the presence of such fields. The order of magnitude of Dq is 1000 cm^{-1} for doubly ionized metal ions of the first transition $(3d)^n$ series. For triply ionized ions, it is approximately 2000 cm^{-1}.

The aim of the so-called "angular overlap models" is to establish a correlation between the orbital splittings caused by the lower symmetries and chemical behavior. Unfortunately, the experimental observations rarely permit an unambiguous assignment of the splitting parameters, and ingenious "chemical" assumptions have to be introduced. The inherent postulate that all "effects" are solely related to the one-electron parameters is also a necessity. Indeed, overrefinements of the crystal and ligand field theories obscure their main merits: the deep understanding they have given us of the ground and excited states of many electronic systems.

See also CHEMICAL BONDING; GROUP THEORY IN PHYSICS; MAGNETIC MOMENTS.

BIBLIOGRAPHY

C. J. Ballhausen, *Introduction to Ligand Field Theory*. McGraw-Hill, New York, 1962.

C. J. Ballhausen, *Molecular Electronic Structures of Transition Metal Complexes*. McGraw-Hill, New York, 1979.

F. A. Cotton, *Chemical Applications of Group Theory*. Wiley, New York, 1979.

T. M. Dunn, D. S. McClure, and R. G. Pearson, *Crystal Field Theory*. Harper & Row, New York, 1965.

J. P. Fackler (ed.), *Symmetry in Chemical Theory*. Dowden, Hutchinson & Ross, Allentown, PA, 1972.

B. N. Figgis, *Introduction to Ligand Fields*. Wiley, New York, 1966.

J. S. Griffith, *The Theory of Transition-Metal Ions*. Cambridge University Press, Cambridge, 1961.

Crystal Binding

J. C. Phillips

Solids are often classified according to the nature of their binding. Four broad types of binding are covalent, ionic, metallic, and molecular.

The simplest of these is ionic binding, which arises when there is substantial charge transfer from one or more cations to one or more anions. As an example, consider NaCl (table salt). In the crystal the electronic configuration is Na^+Cl^-. The electrostatic interaction energy of point charges centered on the ions is called the Madelung energy; in NaCl this energy is about −9 eV per atom pair. The cohesive energy of NaCl is −8 eV per atom pair; about + 1 eV per atom pair arises from closed-shell repulsion (exclusion principle). Many naturally occurring minerals have ionic binding with complex crystal structures, e. g., mica $[KAl_3Si_3O_{10}(OH)_2]$. Ions are rather incompressible, so that ionic radii empirically determined from interatomic distances in simple ionic crystals can be used to rationalize complex ionic structures.

Metallic binding ideally occurs when the valence electrons of metallic (cationic) atoms such as Na or Al are weakly bound to the ion cores. In the solid these electrons then form a nearly free electron gas, with a cohesive energy of about 1 eV per valence electron. In this case there is no classical electrostatic interaction (as in ionic crystals) that can account for the observed cohesion. Instead, quantum-mechanical exchange forces (between electrons of parallel spin) or dynamical correlation (avoidance) between electrons of antiparallel spin takes place; these interaction energies are large in metals just because the electrons are spread out away from the ionic centers.

Three kinds of metals should be distinguished: "normal" metals, with s and p valence electrons only; transition metals (with d electrons); and rare-earth metals (unfilled f shell). Only the s and p valence electrons can become free and contribute fully to cohesion, but the d electrons contribute substantially and the f contribution in the rare earths is small. Because there are in some elements more d electrons than s and p electrons in normal metals, the elements with the largest cohesive energies are those in the middle of the transition series. The $3d$, $4d$, and $5d$ electrons contribute with increasing effectiveness to cohesion. Tungsten, in the middle of the $5d$ series, has the greatest cohesion of any element (9 eV per atom).

Covalent bonding describes sharing of valence electrons in pairs, as in classical chemical theories. Atoms with $N=4$ to $N=7$ s–p valence electrons often form covalent bonds, and in the elemental crystals the coordination configurations are such that each atom tends to have $8-N$ nearest neighbors. The classical covalent crystals are those with $N=4$ (tetrahedral bonding with sp^3 hybrids, in chemical terms), including diamond and Si. With the exception of elements from the first period, covalent binding energies are about $(8-N)$ eV per atom.

Molecular binding arises because of van der Waals interactions between, e. g., benzene molecules in solid benzene. It is weak and is typically of the order of 0.1 eV per molecule or less.

In most crystals the binding is best described as a mixture of ionic, covalent, and metallic interactions. We can distinguish between ionic and metallic binding on the one hand and covalent binding on the other because covalent crystals (such as grey Sn, with the diamond structure) are about 20% less dense than their metallic morphotropes (white Sn, in this example). When there is more than one kind of atom in the crystal there will always be some ionic contribution to the binding. For example, there are many tetrahedrally coordinated compounds with eight valence electrons per atom pair, such as III–V and II–VI compounds; here the bonding is predominantly covalent (tetrahedral coordination) but it contains an ionic component, which chemists associate with the heat of formation of the compound from the elements. An example of mixed metallic–ionic bonding is provided by transition-metal carbides (e. g., WC), which contain solids with the highest known melting points.

See also CHEMICAL BONDING; INTERATOMIC AND INTERMOLECULAR FORCES; METALS; RARE EARTHS; TRANSITION ELEMENTS.

BIBLIOGRAPHY

A. Navrotsky and J. C. Phillips, "Ionicity and Phase Transitions," *Phys. Rev.* **11B**, 1583 (1975). (A)

J. C. Phillips, "Covalent-Ionic and Covalent-Metallic Transitions," *Phys. Rev. Lett.* **27**, 1196 (1971) (A)

J. C. Phillips, "Chemical Bonds in Solids," in *Treatise on Solid State Chemistry* (N. B. Hannay, ed.). Plenum, New York, 1974. (I)

E. P. Wigner and F. Seitz, "Qualitative Analysis of the Cohesion in Metals," *Solid State Phys*, **1**, 97 (1955). (E)

Crystal Defects

George D. Watkins

A crystalline solid with no defects would consist of a perfectly periodic three-dimensional array of unit cells, each containing the identical arrangement of atoms. Such an idealized crystal, however, cannot exist. In a real crystal, there are always structural imperfections.

If a perfect crystal could exist, it would be very strong mechanically. Aside from this desirable feature, however, it might be rather dull. Indeed, many of the important properties of crystalline solids—electrical, optical, mechanical—are determined not so much by the properties of the perfect crystal, but rather by its imperfections: The semiconductor industry is based on the ability to dissolve trace amounts of chemical impurities in silicon or germanium to control the conductivity level and type. The optical absorption and luminescence of solids is controlled primarily by impurities, and other simple lattice imperfections. The ductility and strength of structural materials are determined by imperfections and, as a result, can be modified and controlled by the addition of impurities, heat treatment, cold work, etc.

In many respects, therefore, it is often instructive to consider the host crystal as simply the matrix (or solvent) for the defects, which, in turn, play the dominant role in determining the physical properties of the material. The basic structural defects can be classified as follows:

1. Point defects. These include *lattice vacancies* (lattice sites which are missing an atom), *interstitials* (extra atoms not on regular lattice sites), and *foreign atoms* (chemical impurities).
2. Line defects, such as *dislocations*.
3. Surface defects, such as *grain boundaries* and *stacking faults*.

LATTICE VACANCIES AND INTERSTITIALS

In any real crystal, there must always be a finite number of lattice vacancies and interstitials (see Fig. 1). This can be seen simply as follows: The external surface of a crystal provides an inexhaustible source of vacancies (the empty space beyond) and interstitial atoms (the last atomic layer.) The equilibrium vacancy and interstitial concentrations in the interior of the crystal therefore represent the finite solubility of each of these point defects in the lattice matrix

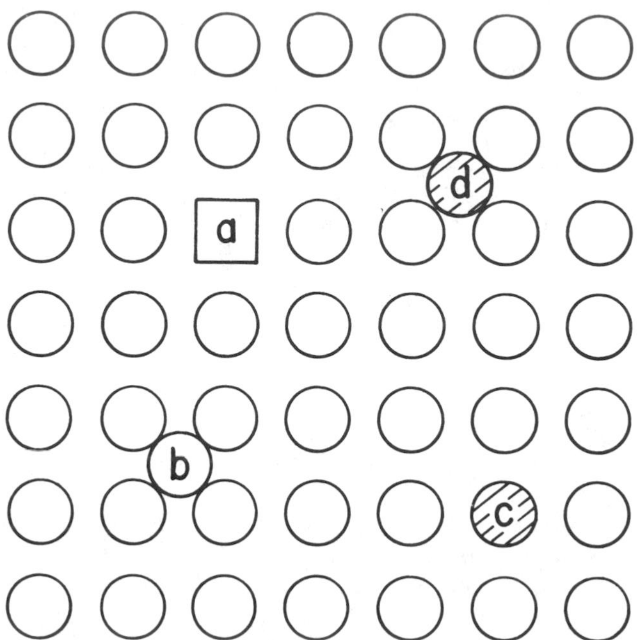

FIG. 1. Simple point defects: (a) vacancy; (b) interstitial; (c) substitutional impurity; (d) interstitial impurity.

(solvent) as they dissolve in from the surface. For a monatomic metal, for instance, simple thermodynamic arguments lead to the expression for the equilibrium solubility concentration of each species,

$$c_i \sim \exp(-E_i/kT), \tag{1}$$

where E_i is the energy required to transfer each defect into the crystal (its formation energy). At any finite temperature T, there is therefore always a finite stable concentration, which increases exponentially as the temperature is raised. For most materials, E_I, the energy to form an interstitial, is greater than E_V, that for the vacancy, and the vacancy tends to be the dominant defect. Typically, near the melting point, a crystal may have a vacancy concentration of 0.01–0.1%.

In insulators and compound materials (more than one atom species per unit cell), the thermodynamics can be somewhat more complicated in that defects can carry charge, and chemical reactions with crystal components in the gas phase must also be considered (e.g., an oxide in an oxygen atmosphere). However, a proper treatment still leads to equilibrium concentrations of the form of Eq. (1), but where E_i may require a more general definition.

The presence of vacancies and interstitial atoms in a crystal provides a mechanism by which mass transport (diffusion) can occur in the lattice. The vacancy provides a missing site into which a neighboring atom can jump. When the atom jumps, the vacancy has moved, now occupying the original site of the neighbor. Similarly, the interstitial may move by jumping to a nearby empty interstitial site, or by pushing a neighbor into an adjacent interstitial site and taking its place. Both defects therefore stir up the lattice atoms as they randomly jump from one lattice position to another. The jump rate for a defect will have a temperature dependence of the form

$$\nu_i \sim \exp(-U_i/kT), \tag{2}$$

where U_i represents an energy barrier that the atoms must overcome in order to accomplish the jump. The contribution of a defect to atomic diffusion in the crystal is therefore proportional to its jump rate (2) times the number of the defects (1) leading to a temperature dependence for the diffusion coefficient:

$$D \sim \exp[-(E_i + U_i)/kT]. \tag{3}$$

In nonmetals, vacancies and interstitials may also be charged. In this case, defect motion produces electrical (*ionic*) conductivity which is also proportional to (3). This is often the dominant source of electrical conductivity in ionic-compound insulators such as alkali halides and ceramic oxides.

At temperatures well below the melting temperature, the *equilibrium* concentration of vacancies and interstitials may be small, because of the exponential temperature dependence, Eq. (1). However, vacancies and interstitials are often present greatly in excess of this amount as a result of (a) quenching from a high-temperature equilibrium, (b) mechanical distortion of the crystal, and (c) radiation damage, where the incoming particle displaces atoms. In these cases, return to equilibrium (annealing) may be extremely slow, being governed by the migration rate (2) which is also exponentially dependent on temperature.

The presence of vacancies and interstitials in semiconductors can alter the *electronic* properties of the material in that they serve to trap and scatter free electrons and holes. Electronic transitions of the trapped charges may also give rise to optical absorption and luminescence bands in semiconductors and insulators.

FOREIGN ATOMS

Chemical impurities may dissolve in a crystal as extra atoms (interstitial) or by replacing normal lattice atoms (substitutional). Even the purest crystals inevitably contain trace amounts of impurities. In addition, impurities may be introduced intentionally by addition during crystal growth, by diffusion into the crystal from the surface at elevated temperatures, or by ion bombardment, where a high-energy ion of the impurity is driven in through the surface of the material. Transmutation doping has also been used, where a nuclear reaction converts a host atom into an impurity. Like vacancies and interstitials, impurities may trap and donate electrons or holes and alter the electronic and optical properties of the crystal. In addition, they may introduce some of their characteristic atomic optical absorption or luminescent spectra to the crystal. Their elastic interactions with dislocations are important in determining the strength of materials.

DISLOCATIONS

A perfect crystal should be very strong, the critical shear stress necessary to slide one perfect atom plane past another being comparable to the shear modulus G of the crystal. Contrary to this, however, it is found experimentally that

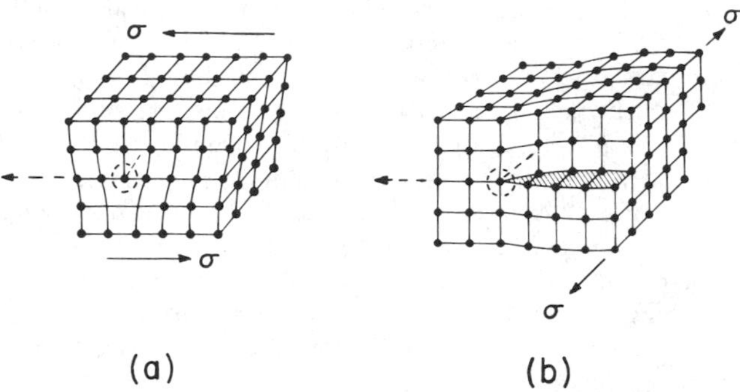

FIG. 2. (a) An *edge* dislocation; (b) a *screw* dislocation (after B. Henderson, *Defects in Crystalline Solids*. Edward Arnold, London, 1972). Applied shear stress σ causes the upper half of the crystal to slip one atom spacing as the dislocation line moves through the crystal to the left.

the purer and more "perfect" the crystal is made, the softer it usually tends to become. The critical shear stress in a well-annealed single crystal may be only 10^{-5}–$10^{-4}G$. The reason for this is the presence of *dislocations*.

Two basic types of dislocations are illustrated in Fig. 2. In Fig. 2a, the *edge* dislocation can be viewed as having been formed by slicing the crystal with a knife partway in from the top and inserting an extra half plane of atoms in the cut. In Fig. 2b, the *screw* dislocation could be produced by slicing part way in from the right and shearing the crystal parallel to the slice by one atomic spacing. For both, the imperfection is the line which terminates the cut, the surrounding material being a strained but otherwise perfect crystal. As shown in the figure, applied shear stress σ allows the upper part of the crystal to slip one atomic spacing as the dislocation moves to the left. A reduction in the critical shear stress arises because now only a single line of atoms is required to move at a time, rather than the whole plane of atoms together.

Dislocations are not thermodynamically stable, the energy to form one in a bulk crystal being too great. However, they are almost always present in crystals as a result of accidents in growth. Typically, densities for an as-grown crystal may be $\sim 10^{12}/m^2$. The rapid and chaotic manner in which atoms fall into place during growth makes the formation of lattice misfits of this type likely. Once formed on a small scale, they can propagate as the crystal grows. Also vacancies and interstitials, present in high concentrations during growth from the melt, can nucleate dislocation loops as they condense out upon cooling.

Dislocation motion can be impeded by the presence of point defects, precipitates, or other dislocations in or near its slip plane. Interaction with point defects and precipitates is an important mechanism for the strengthening of constructional metals such as steel, where controlled additions of impurities and subsequent heat treatments to disperse the impurities or allow precipitation are used to tie up the dislocations to the desired degree. *Work hardening* is another mechanism for strengthening materials. Here dislocation movement under stress generates more dislocations, ultimately leading to a massive tangle of "pinned" dislocations.

SURFACE DEFECTS

The interface between two regions of different crystal orientation is called a *grain boundary*. These can originate in growth as two separately growing regions merge. Grain boundaries can also form by regrouping of dislocations during anneal. *Stacking faults* are the result of an error in layer growth. A stacking fault for a crystal normally growing in the layer order *ABCABCABC* might be *ABCBCABC*.

See also COLOR CENTERS; CRYSTAL GROWTH; DIFFUSION; ELECTRON ENERGY STATES IN SOLIDS AND LIQUIDS; LUMINESCENCE.

BIBLIOGRAPHY

A. H. Cottrell, *Dislocations and Plastic Flow in Crystals*. Oxford Univ. Press, Oxford, 1953. (I–A)

J. H. Crawford, Jr. and L. M. Slifkin, eds., *Point Defects in Solids*, Vols. 1–3. Plenum Press, New York, 1972. (I)

B. Henderson, *Defects in Crystalline Solids*. Edward Arnold, London, 1972. (I)

D. Hull, *Introduction to Dislocations*, 2nd ed. Pergamon Press, Oxford, 1975. (E–I)

F. A. Kröger, *The Chemistry of Imperfect Crystals*, 2nd ed., Vols. 1–3. North-Holland, Amsterdam, 1973. (A)

F. Seitz, in *Imperfections in Nearly Perfect Crystals*, Chap. 1 (W. Shockley, J. H. Hollomon, R. Maurer, and F. Seitz, eds.). Wiley, New York, 1952. (E, I)

Crystal Growth

R. L. Parker

Among the many techniques for growing crystals, the following may be mentioned as representative examples: growth from the melt (solidification), growth from the vapor phase, growth from solution (aqueous, molten salts, other solvents), growth by precipitation from solid solution, growth by chemical (vapor) transport reaction, epitaxial growth from vapor or solution, electrocrystallization, and growth from the solid phase (strain annealing). It is useful, in classifying the various growth processes, to consider the thermodynamic phase diagram for the system; it specifies the regions of equilibrium between the desired solid and the mother medium from which the crystal grows, which may be liquid, vapor, solution, or also solid. Thus, in melt growth of a single-component system at constant pressure, the temperature of the liquid is gradually lowered so that a portion of it is at the freezing point; further cooling may then cause nucleation of the new solid phase, which, if properly controlled, may form a seed upon which further precipitation of the liquid can take place in an orderly way, yielding a single crystal.

Probably the most characteristic and fundamental problem in crystal growth is understanding, predicting, and controlling the crystal morphology or form. By morphology is meant not only the external forms of crystals, important and fascinating as these can be (as, for example, in snow crystals), but more generally the complete history of the solid–liquid, solid–vapor, or solid–solid interface morphology during the entire growth process. It is this morphology that determines the microstructure, the distribution of impurities, and the dislocation and imperfection substructure of the crystal, which in turn control the useful properties of the crystalline material. By no means is the prediction of morphology an easy matter. The interface morphology is controlled simultaneously and in varying proportions by bulk transport effects (flow of heat and/or solute) and by interface kinetic effects. Heat flow is clearly important in, for example, the freezing of a pure liquid, since the latent heat must be removed. Solute flow is clearly important in, for example, the precipitation from a solid solution. Interface kinetics, which refers to the structure, on a molecular level, of the surface and to the relative ease or difficulty that a molecule may have in moving from the mother medium to a relatively final position in the crystal surface, may be controlling for particular materials and for particular molecular and crystal structures. Thus, both macroscopic and microscopic theory are involved in understanding crystal growth mechanisms.

Unless a crystal seed is present, the new phase must start by a nucleation process. Nucleation is the formation of very small particles of new phase in the mother medium and may take place when the appropriate thermodynamic variable (temperature, concentration, pressure, etc.) has departed sufficiently from the equilibrium value. It is generally assumed to take place by a random fluctuation process (heterophase fluctuation) analogous to the better-known density fluctuations in a gas; dimers, trimers, and higher species are assumed to form, with the larger sizes having short lifetimes and dilute concentrations. There is found, by an appropriate balancing of the free energy of formation of a particle against its surface energy, to be a critical embryo size for nucleation; particles greater than this size have a tendency to grow, and those below it tend to dissolve. The dependence of the number of these critical nuclei on temperature or on solute concentration is extremely sensitive, so that only a few degrees change in undercooling can cause many orders of magnitude change in nucleation rate. While the preceding discussion is for "three-dimensional" nucleation, "two-dimensional" nucleation of a monolayer disk may take place from the random fluctuation of adsorbed particles on an otherwise plane crystal surface, and provides one kind of interface kinetic mechanism.

Model calculations of nucleation, interface structure, and interface kinetics have been made for many years. Perhaps the best-known model for crystals is the BCF or Burton–Cabrera–Frank model of growth at the emerging step of a screw dislocation at the crystal surface. Modern supercomputers now permit the calculation, using the methods of molecular dynamics, of the actual three-dimensional trajectory of each atom in entire systems of thousands of atoms interacting with each other with a given interatomic potential. As the temperature or average kinetic energy is lowered, both condensation of vapor to liquid or solid, and freezing of liquid to solid, can be observed and studied. Atom-by-atom and layer-by-layer growth, as for example in epitaxy, is thus calculated at intervals of 10^{-12} s, from first principles. This is an improvement over the Monte Carlo technique. In this technique, atoms are deposited at random sites on an initially flat crystal, but their evaporation depends on the number of their first neighbors. Not only interface structure, but growth rates can be calculated. It is possible to confirm and make more precise predictions of surface roughness that were made previously by statistical mechanics (Ising model).

It is often observed that crystal forms are not stable shapes. An example is the quite common dendritic or tree-shaped form, which exhibits periodic branching and sub-branching, as in snow crystals. This results from the unstabilizing point effect of diffusion (of heat or solute) acting on a small protuberance. This effect can be counteracted by the additional surface energy thereby produced, and it can for particular materials be overcome by the size and/or anisotropy of the interface kinetic effect (sluggish kinetics). Morphological stability theory has been applied to this phenomenon, using the transport equations and appropriate boundary conditions, and has been highly successful in such areas as the breakdown of planar to cellular growth of alloy crystals in the presence of undercooling caused by solute rejection, and in such other areas as ice cylinder growth from supercooled water. It seems likely that the ultimate calculation of crystal form will use this macroscopic theory, with boundary conditions determined by computer modeling on the atomic scale.

When the mother medium is a fluid (liquid or gas), fluid flow and convection can greatly influence crystal growth rates and morphology. Although often overlooked in the past, these effects are due to the breakup of relatively long-range diffusion fields of heat or solute. These fields are translated away and are replaced by relatively short-range diffusion fields. The result is that the fluid of bulk concentration or temperature is now quite close to the interface, and is only separated from it by a thin hydrodynamic boundary layer; this will tend to increase growth rates. Thus, modern crystal growth technology (for example, large Czochralski machines pulling rotating dislocation-free silicon crystals from the convecting melt) must concern itself with hydrodynamics. In addition, there are many deep and unexpected effects on interface morphological stability calculations when the fluid flows are included, as, for example, in double-diffusive convection. These simulations also require extensive computer calculations, particularly those in which the diffusive transport and convective transport are strongly coupled, and indeed the two fields, formerly separate, of morphological stability and hydrodynamic stability are now joined.

Crystalline perfection or freedom from defects is not solely due to the growth process, for subsequent processing, annealing, handling, and the like often alter the concentration and types of defects. However, the growth process is certainly the key aspect. Among the defects that may be introduced are point defects (vacancies, interstitials, foreign atoms), line defects (dislocations), planar defects (stacking faults, twins, low-angle boundaries), and three-dimensional defects (striations, cellular growth, voids, inclusions). An example of controlling defects is the use of substantial temperature gradients in directional solidification to remove constitutional supercooling and consequently to prevent cellular growth by morphological instability. Another is to control dislocation density by use of a bottlenecked seed at which dislocations may emerge on the crystal surfaces. Of the many crystal characterization techniques, x-ray tomography is rapidly growing as a major tool for structural defects such as dislocations, including real-time *in situ* observation, using such powerful x-ray sources as coherent synchrotron radiation, during the melt–growth process. Electrically or optically active defects are often detected by measuring appropriate physical properties such as resistivity or absorption.

Finally, while the physicist may be interested in crystals as good specimens in which to measure physical properties or because the mechanisms of growth are complex and fascinating, it cannot be ignored that substantial recent advances in the technology of crystal growth are due to immense demand for production of industrial and consumer products. These demands exist in the electronics, optical, computer and communications industries for such materials as silicon for transistors, integrated circuit chips, random access memory chips and solar cells; GaAs for higher-speed and higher-frequency ICs including microwave and lightwave chips; and metallorganic chemical vapor deposition epitaxial growth of heterostructures for laser diodes. In a typical year's issues of the *Journal of Crystal Growth,* more than 100 different compositions of new or improved crystals may be found.

See also CRYSTAL DEFECTS; CRYSTAL SYMMETRY.

BIBLIOGRAPHY

A. A. Chernov, *Crystal Growth,* Modern Crystallography 3, Springer-Verlag, New York, 1984.

M. E. Glicksman, S. R. Coriell, and G. B. McFadden, "Interaction of Flows with the Crystal-Melt Interface," *Ann. Rev. Fluid Mech.* **18,** 307–335 (1986).

F. Rosenberger, *Fundamentals of Crystal Growth I.* Springer-Verlag, New York, 1979.

D. Elwell and H. J. Scheel, *Crystal Growth from High Temperature Solutions.* Academic Press, New York, 1975.

Journal of Crystal Growth, North-Holland, Amsterdam, 1967–present.

R. L. Parker, "Crystal Growth Mechanisms," in *Solid State Physics* (H. Ehrenreich, F. Seitz, and D. Turnbull, eds.), **25.** Academic Press, New York, pp. 151–299, 1970.

Crystal Symmetry

Martin J. Buerger†

INTRODUCTION

Homogeneity

Crystals are anisotropic homogeneous solids. The homogeneity is on the scale of a cluster of a small number of atoms. It implies that the solid consists of clusters related to each other by direct or opposite isometries, and that the environments of homologous points of all clusters are the same.

† Deceased

Coincidence Operations

Every cluster is related to any other cluster by one of four simple operations: translation, rotation about an axis, reflection across a plane, or inversion through a point; or by acceptable combinations of these four. (The condition for two different operations to form an acceptable combination is that each of the component operations transforms the other into itself.)

In every crystal the clusters are related to each other at least by the operations of translations. Accordingly, any other operations, called *symmetry operations*, must be consistent with translations. It is readily demonstrated that this requires every rotation operation to be through an angle of $2\pi/n$, where $n=1,2,3,4$, or 6.

Each of the operations relating the crystal clusters generates a group. As an aid to describing the operations it is convenient to designate the locus that remains unmoved under the operations of the cyclical group as the *symmetry element* of that group. The symmetry element of a group of rotations is the axis A of the rotation; of a group of reflections is the mirror plane m of the reflection; of an inversion is the point i through which the operation inverts; etc.

Any of the permissible operations that relate a pair of atomic clusters in a particular homogeneous solid must also relate all clusters in the solid. If the solid contains operations beyond the generally required translations, the solid is said to be *symmetrical* with respect to each of the symmetry elements corresponding to the groups of operations relating its atomic clusters.

Crystallographic Groups

Two kinds of symmetries are especially important in crystallography: *point groups* and *space groups*. Point groups are so named because in each group one point remains unmoved under the operations of the group; they contain no translations. Space groups are so named because in each group all three-dimensional space remains invariant under the operations of the group; they contain pure translations and may also contain operations with translation components.

Since all crystals are ideally homogeneous, every crystal must conform to one of the space-group symmetries. Nevertheless it is useful to consider the point groups because many of the properties of crystals are not dependent on its translations. For example, the appearance of a crystal conforms to its point-group symmetry because the eye, even when aided by the microscope, cannot see the tiny translations. Furthermore, the point groups, which are easy to construct, lead to a simple way of enumerating all space groups.

THE CRYSTALLOGRAPHIC POINT GROUPS

Symbolism for Symmetry Elements

The symbols universally used by crystallographers to designate symmetry elements are the Hermann–Mauguin, or *international*, symbols. These are shown in Table I for both point groups and space groups. These symbols replace the 1890 point-groups symbols of Schoenflies, which cannot be suitably extended to space groups. The equivalences are shown in Table II.

Chirality

Any operation or group of operations that relates objects of the same chirality (i.e., objects related by direct isometry, which can be regarded arbitrarily as all right-handed or all left-handed) is said to be of the *first sort*. Any operation or group that relates objects of opposite chirality is said to be of the *second sort*. Pure rotations relate objects of the same chirality; they are called *proper rotations*. Rotoinversions and rotoreflections are called *improper rotations*; they relate a sequence of objects that are alternately right-handed and left-handed. The objects of homogeneous patterns are either all of the same chirality or they are half right-handed and half left-handed.

Monaxial Groups with Proper Rotations

In Hermann–Mauguin notation a proper rotation axis is labeled by the numerical value of its n. Accordingly there are five such crystallographic point groups, namely, 1,2,3,4, and 6, shown in Fig. 1.

Groups with Proper Rotations about Intersecting Axes

It can readily be shown by Euler's construction, or by multiplying together matrices that represent rotations, that

Table I Hermann–Mauguin symbols for the symmetry elements of crystallographic groups

Point-group symbols	Isogonal space-group symbols
1	1
2	$2 \quad 2_1$
3	$3 \quad 3_1 \quad 3_2$
4	$4 \quad 4_1 \quad 4_2 \quad 4_3$
6	$6 \quad 6_1 \quad 6_2 \quad 6_3 \quad 6_4 \quad 6_5$
$\bar{1}$ $(=\tilde{2})$ $=$ i, an inversion center	$\bar{1}=i$
$\bar{2}$ $(=\tilde{1})$ $=$ m, a mirror	Various glide planes†
$\bar{3}$ $(=\tilde{6})$ $=$ $3+i$	$\bar{3}=3+i$
$\bar{4}$ $(=\tilde{4})$ Not decomposable	$\bar{4}$
$\bar{6}$ $(=\tilde{3})$ $=$ $\frac{3}{m}$, i.e., $3 \perp m$	$\bar{6}=\frac{3}{m}$

†Glide planes

Symbol	Translation component
m	0
a	$\frac{1}{2}\underset{\sim}{a}$
b	$\frac{1}{2}\underset{\sim}{b}$
c	$\frac{1}{2}\underset{\sim}{c}$
n	$\frac{1}{2}(\underset{\sim}{a}+\underset{\sim}{b})$ or $\frac{1}{2}(\underset{\sim}{b}+\underset{\sim}{c})$ or $\frac{1}{2}(\underset{\sim}{a}+\underset{\sim}{c})$
d	$\frac{1}{4}(\underset{\sim}{a}+\underset{\sim}{b})$ or $\frac{1}{4}(\underset{\sim}{b}+\underset{\sim}{c})$ or $\frac{1}{4}(\underset{\sim}{a}+\underset{\sim}{c})$ or $\frac{1}{4}(\underset{\sim}{a}+\underset{\sim}{b}+\underset{\sim}{c})$

a rotation through an angle α about an axis A, followed by a rotation through β about B, is equivalent to a rotation through $-\gamma$ about an axis C. In the product notation for combining the elements of a group, this can be expressed as

$$A_\alpha B_\beta = C_{-\gamma}. \tag{1}$$

This implies that

$$A_\alpha B_\beta C_\gamma = 1 \tag{2}$$

where the right-hand side of (2) is the identity operation, indicating no net motion, expressed in Hermann–Mauguin notation. Since

$$\alpha = \frac{2\pi}{n_1}, \quad \beta = \frac{2\pi}{n_2}, \quad \gamma = \frac{2\pi}{n_3} \tag{3}$$

in crystal symmetry, it is obvious that if there are two rotations about intersecting axes, a rotation about a third intersecting axis is implied. (Exceptionally, the third rotation is one of the other two with an opposite rotation.)

Euler's construction allows the angular relations between rotation axes to be obtained from a spherical triangle in which axes A, B, and C occur in a clockwise sense at arc distances $A \wedge B = c$, $B \wedge C = a$, $C \wedge A = b$, and with $A = \alpha/2$, $B = \beta/2$, $C = \gamma/2$. For this spherical triangle, the sum of its angles is

$$\frac{\alpha}{2} + \frac{\beta}{2} + \frac{\gamma}{2} \geq \pi. \tag{4}$$

With the aid of (3) this can be expressed as

$$\frac{1}{n_1} + \frac{1}{n_2} + \frac{1}{n_3} \geq 1. \tag{5}$$

The only distinct solutions of (5) are

$$2,2,n_3 \quad 2,3,3 \quad 2,3,4 \quad 2,3,5.$$

The crystallographic possibilities for these solutions are the following.

4 "dihedral" sets: 2,2,2 2,2,3 2,2,4 2,2,6

1 "tetrahedral" set: 2,3,3

1 "octahedral" set: 2,3,4

The angles a,b, and c between the rotation axes can be found from relations like

$$\cos a = \frac{\cos(\alpha/2) + \cos(\beta/2)\cos(\gamma/2)}{\sin(\beta/2)\sin(\gamma/2)}. \tag{6}$$

The arrangements of these six axial combinations are shown in Fig. 2.

The Axial Frames of Crystallography

The five monaxial point groups and the six sets of polyaxial point groups just demonstrated are the only crystallographic point groups with proper rotations only. These 11 arrangements of axes provide axial frames on whose geometries all other angular relationships in other crystallographic groups are based.

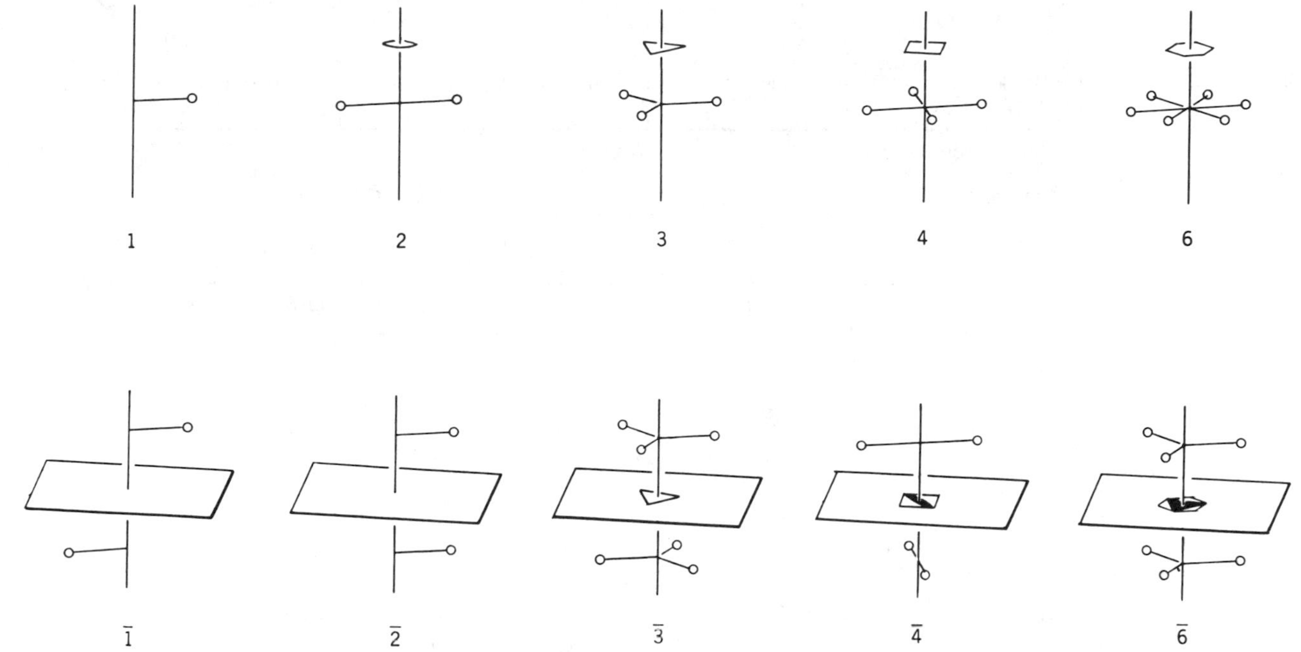

FIG. 1. Top: The symmetry axes of the five crystallographic point groups with proper rotations. Bottom: The symmetry axes of the five crystallographic point groups with improper rotations.

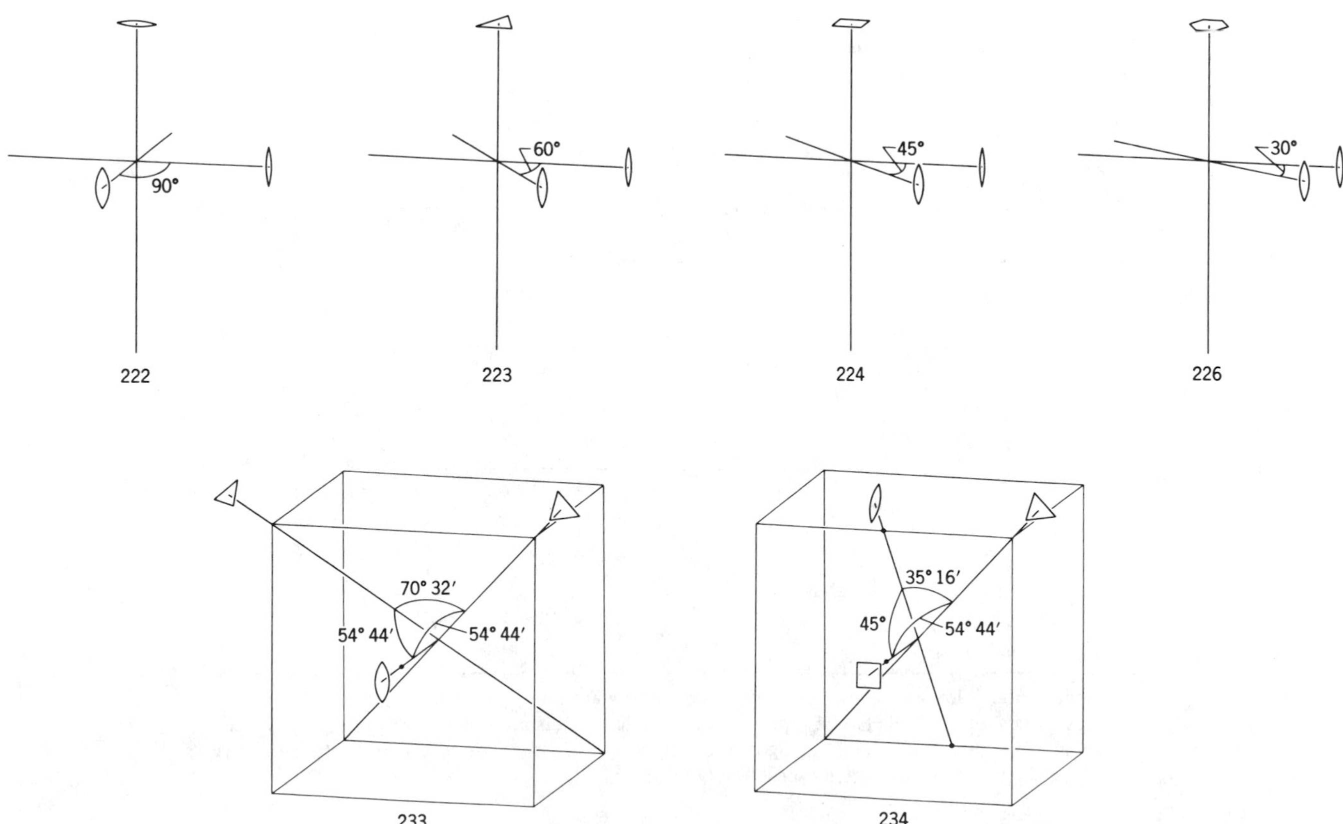

FIG. 2. The six combinations of crystallographic proper rotation axes. [From M. J. Buerger, *Elementary Crystallography* (Wiley, New York, 1963).]

Table II The distribution of point-group symmetries over the crystal systems and their possible lattice types

Point-group symbols		Crystal system		
Schoenflies	Hermann–Mauguin†	Name	Geometry	Bravais-lattice type
C_1	1	Triclinic	$a \neq b \neq c$	P
C_i	$\bar{1}$		$\alpha \neq \beta \neq \gamma$	
C_2	2	Monoclinic§	$a \neq b \neq c$	P, A (or B)
C_s	m		$90° = \alpha = \beta \neq \gamma$	
C_{2h}	$\frac{2}{m}$			
C_3	3	(Trigonal subdivision)		P, R
D_3	32			
C_{3v}	$3m$			
C_{3i}	$\bar{3}$			
D_{3d}	$\bar{3}\frac{2}{m}$ $(\bar{3}m)$			
		Hexagonal	$a = b \neq c$	
			$90° = \alpha = \beta \neq \gamma = 120°$	
C_6	6	(Hexagonal subdivision)		P
D_6	622			
C_{6h}	$\frac{6}{m}$			
C_{6v}	$6mm$			
C_{3h}	$\bar{6}$			
D_{3h}	$\bar{6}m2$ or $\bar{6}2m$			
D_{6h}	$\frac{6}{m}\frac{2}{m}\frac{2}{m}$ $(6/mmm)$			
D_2	222	Orthorhombic	$a \neq b \neq c$	P, I, A (or B or C), F
C_{2v}	$mm2$		$90° = \alpha = \beta = \gamma$	
$D_{2h} \equiv V$	$\frac{2}{m}\frac{2}{m}\frac{2}{m}$ $(2/mmm)$			
C_4	4	Tetragonal	$a = b \neq c$	P, I
D_4	422		$90° = \alpha = \beta = \gamma$	
C_{4h}	$\frac{4}{m}$			
C_{4v}	$4mm$			
S_4	$\bar{4}$			
D_{2d}	$\bar{4}m2$ or $\bar{4}2m$			
D_{4h}	$\frac{4}{m}\frac{2}{m}\frac{2}{m}$ $(4/mmm)$			
T	23	Cubic (isometric)	$a = b = c$	P, I, F
O	432		$90° = \alpha = \beta - \gamma$	
T_h	$\frac{2}{m}\bar{3}$			
T_d	$\bar{4}3m$			
O_h	$\frac{4}{m}\bar{3}\frac{2}{m}$ $(m\bar{3}m)$			

† Hermann–Mauguin symbols in parentheses are abbreviated forms in common use.
§ The geometry and Bravais lattice types for monoclinic crystals are for the first setting, in which the unique axis is c. For the second setting of monoclinic crystals the unique axis is b; the geometry is $a \neq b \neq c$, $90° = \alpha = \gamma \neq \beta$; its lattice types are P, A (or C).

Groups with Improper Rotations

Corresponding to the five monaxial groups with proper rotations are five with improper rotations, symbolized by the rotoinversion axes $\bar{1},\bar{2},\bar{3},\bar{4}$, and $\bar{6}$. As listed in Table I and shown in Fig. 1, all but $\bar{4}$ can also be represented by other symmetry elements.

A proper and an improper rotation axis of the same n can occur along the same line. When these n and $\bar{n}$ axes are written as a fraction $n/\bar{n}$ and reduced, there arise three new groups: $2/\bar{2}\rightarrow 2/m$; $[3/\bar{3}\rightarrow 3/(3+i)\equiv\bar{3}]$; $4/\bar{4}\rightarrow 4/m$; $6/\bar{6}\rightarrow 6/(3/m)\rightarrow 6/m$.

Intersecting axes can be combined as were the three proper rotations, provided either that proper and improper axes coincide, as just noted, or that there are two improper axes and one proper axis. These intersecting combinations generate the other point groups having improper rotations listed in Table II.

The 32 Point Groups

The arrangements of symmetry elements in the 32 point groups are illustrated in Fig. 3. The "point" of the point group is assumed to be at the center of the sphere whose outline is the dotted circle; if there are intersecting symmetry elements, the point of intersection is at the center of the circle. A reflection plane is represented by the line where it intersects the sphere's surface. An n-fold axis is shown as a polygon with n vertices at the point where the axis intersects the sphere's surface. A tiny circle in the center of a diagram indicates that the point group has a center of symmetry.

TRANSLATION GROUPS

Lattices

In three-dimensional space a translation can be represented by its components $\mathbf{t}_1,\mathbf{t}_2$, and $\mathbf{t}_3$ on three noncoplanar vectors. For crystallographic applications, let $\mathbf{t}_1$ be the shortest translation in some direction. A set of points equivalent by translation $\mathbf{t}_1$ constitute a line parallel to $\mathbf{t}_1$. Let $\mathbf{t}_2$ be chosen so that it translates the points of this line to the points of a parallel line that is next adjacent to it in some direction; then let $\mathbf{t}_3$ be chosen so that it translates a point of the plane parallel to $\mathbf{t}_1$ and $\mathbf{t}_2$ to a point in the next parallel plane. When $\mathbf{t}_1$, $\mathbf{t}_2$, and $\mathbf{t}_3$ are chosen in this way they are called *conjugate translations*. There are an infinite number of sets of such translations in a homogeneous solid. Any translation of the solid can then be described by

$$\mathbf{t}_{uvw} = u\mathbf{t}_1 + v\mathbf{t}_2 + w\mathbf{t}_3, \qquad (u,v,w, \text{ integers}). \qquad (7)$$

The translations $\mathbf{t}_{uvw}$ form a group if u,v, and w run over all integers. If the translations of this group operate on an arbitrary point, the resulting collection of points is called a *lattice*. (This is the only correct meaning of "lattice" in crystallography.)

Cells

The parallelepiped, three of whose adjacent edges are conjugate translations, is known as a *primitive cell*. The adjective "primitive" implies that to each cell of this kind there corresponds one point of the lattice. The infinite number of primitive cells of a given crystal all have the same volume. A particular kind of primitive cell, known as the *reduced cell*, is characterized by having the three shortest translations.

Except as noted below, the cell edges of the reduced cell offer a natural coordinate system to which the geometrical features of a crystal can be referred. When a crystal has point-group symmetry other than 1 or $\bar{1}$, however, there are certain cases for which it is mathematically simpler to use as the coordinate system the edges of a more symmetrical cell that contains more than one set of translationally equivalent atoms. These cells are called *multiple cells*, specifically *double*, *triple*, and *quadruple* cells, according as the cell contains 2,3, or 4 sets of translationally equivalent atoms. Such cells have, in addition to a lattice point at the cell origin, one or more additional lattice points at various special positions in the cell. The characteristics of these cells and their designations are given in Table III.

When an appropriate cell has been selected, it is customary to label its three adjacent edges a,b, and c, in such a way as to define a right-handed system of coordinates. The angles between cell axes are designated $\alpha=b\wedge c$, $\beta=c\wedge a$, $\gamma=a\wedge b$.

Crystal Systems

The symmetry of a crystal requires a specialization of its lattice. It turns out that sets of several different point-group symmetries place the same constraints on the lattice, so that, when multiple cells are taken into account, only six kinds of symmetrically different cell edges are required for the coordinate systems of the 32 crystal classes. Crystals that can be referred to the same kind of coordinate system are said to belong to the same *crystal system*. The distribution of the point groups over the crystal systems is given in Table II.

Rational Lines and Directions

An aspect of crystal geometry is said to be rational if it can be expressed in integral numbers of lattice translations. A direction is rational if it has the form

$$\mathbf{r} = u\mathbf{a} + v\mathbf{b} + w\mathbf{c} \qquad (u,v,w, \text{ integers}). \qquad (8)$$

The three integers u,v,w, are known as the *indices of the direction*. Since sets of three integers are used for various indices, to indicate that they refer to a direction they are traditionally written between square brackets. (There is a tendency, especially among metallurgists, to use angular brackets for this purpose.)

Rational Planes

A plane is rational if it contains three lattice points (in which case it contains an infinite number). The translations

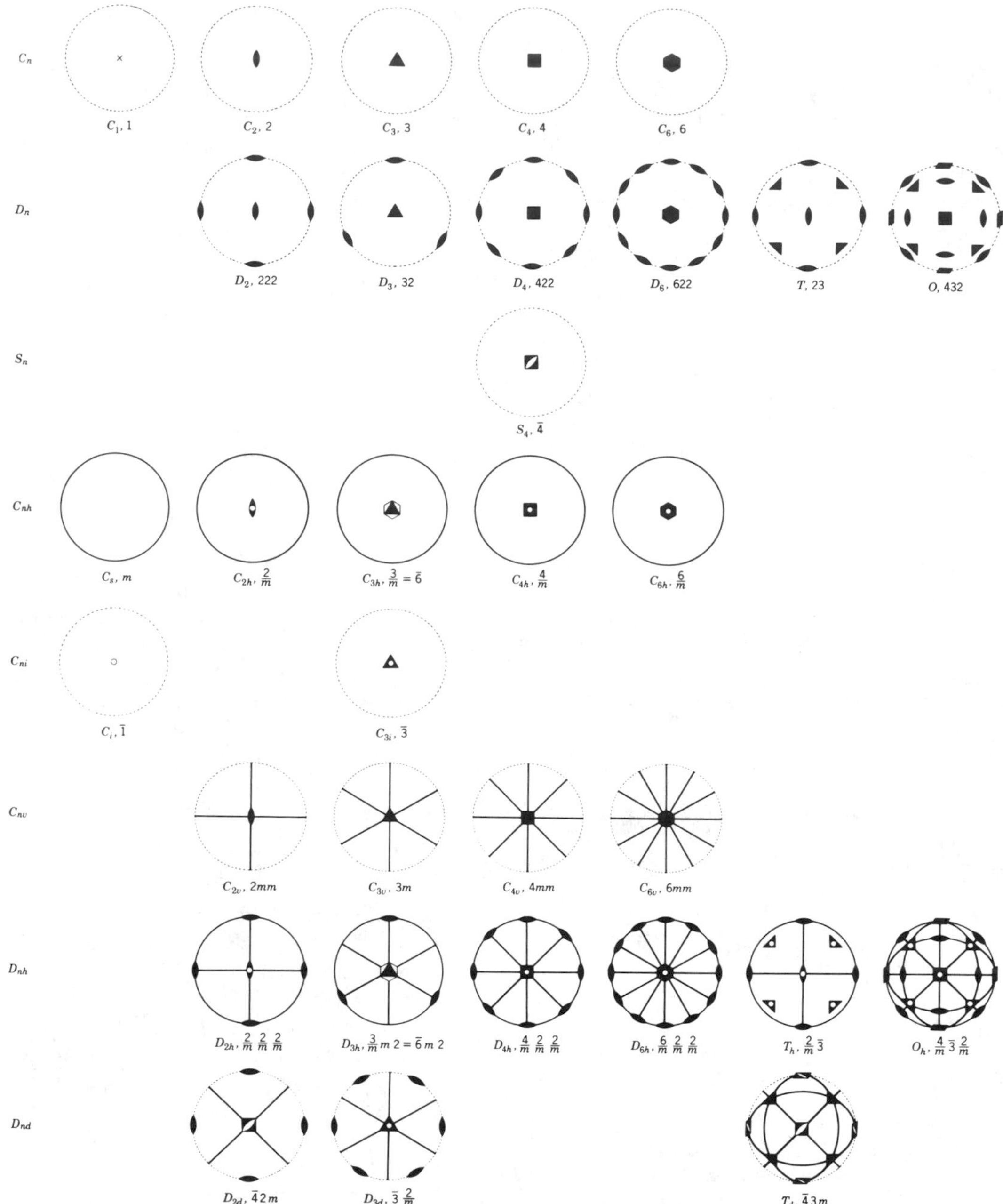

FIG. 3. The arrangements of symmetry elements in the 32 crystallographic point groups. [From M. J. Buerger, *Elementary Crystallography* (Wiley, New York, 1963).]

of the lattice require such a plane to be one of an infinite stack of equally spaced parallel planes whose spacing is designated d. The equation of the nth plane from the origin is

$$hx + ky + lz = n \tag{9}$$

where h,k, and l are integers and x,y, and z are fractions based on a,b, and c, respectively, as units. A rearrangement of (9) for $n = 1$ provides the intercept form for the plane of the stack nearest the origin:

$$\frac{x}{1/h} + \frac{y}{1/k} + \frac{z}{1/l} = 1. \tag{10}$$

This plane has intercepts on the a,b, and c axes of $1/h, 1/k$, and $1/l$, respectively, so the stack cuts these units into h,k, and l parts, respectively. These three integers, placed between parentheses, are the *indices of the stack of planes*. (The designation of rational planes by these indices was proposed by Whewell in 1825 and by Grassman in 1829; the

Table III Characteristics of crystallographic cells

Symbol	Significance of symbol		Coordinates of lattice points in fractions of cell edges a, b, c	Cell multiplicity
P	"Primitive"		000	1
I	"Innenzentriert":	body-centered	000; $\frac{1}{2}\frac{1}{2}\frac{1}{2}$	2
A	"A-centered":	point at center of bc face	000; $0\frac{1}{2}\frac{1}{2}$	2
B	"B-centered":	point at center of ca face	000; $\frac{1}{2}0\frac{1}{2}$	2
C	"C-centered":	point at center of ab face	000; $\frac{1}{2}\frac{1}{2}0$	2
R	"Rhombohedral"	referred to hexagonal axes	000; $\frac{2}{3}\frac{1}{3}\frac{1}{3}$; $\frac{1}{3}\frac{2}{3}\frac{2}{3}$	3
F	"Face-centered"	on A, B, and C faces	000; $0\frac{1}{2}\frac{1}{2}$; $\frac{1}{2}0\frac{1}{2}$; $\frac{1}{2}\frac{1}{2}0$	4

integers became known as "Miller indices" when they were popularized by Miller's 1839 book, *A Treatise on Crystallography*.)

SPACE GROUPS

General Features

Every crystal has a group of translations that are consistent with the symmetry of its point group, specifically one of the Bravais-lattice types of Table II. Its symmetry may include rotational operations with translation components parallel to the axis of rotation, and its reflectional operations may have translation components parallel to the reflection plane, from Table I. The angles of the rotational component of the screws, and the angles between various axes and planes, are not affected by the translations, however; so that the resulting angular geometry of every combination of symmetry elements is *isogonal* with the angular geometry of the corresponding translation-free symmetry elements in some possible crystallographic point group. From the point of view of group theory, each of the 230 space groups is *isomorphous* with one of the 32 point groups. This relation makes it possible to derive every space group by considering, in turn, the combinations of permissible symmetry operations that are isogonal with the symmetry operations of the appropriate point groups.

Groups with Only Operations of the First Sort

For each rotation operation that generates a symmetry element in point-group symmetry, there are several isogonal operations in space-group symmetry with different translation components. The rotation is accompanied by a translation component τ parallel to the axis A such that n translations τ are equal to an integral number p of lattice translations t along the axis:

$$n\tau = pt \qquad (n,p, \text{ integers}). \qquad (11)$$

Thus the space-group operation is characterized by two parameters: n, which is the same as in the isogonal point group; and p, which may be zero. Since these are, in general, helicoidal motions, the symmetry elements are known as *screw axes*. They are designated by the general symbol n_p; the various possibilities are shown in the upper right-hand part of Table I. When $p/n<\frac{1}{2}$, the screw is right-handed; when $p/n>\frac{1}{2}$, it is left-handed; when $p/n=\frac{1}{2}$, it is neutral. The number of threads in the screw is p for $p/n\geq\frac{1}{2}$, but $n-p$ for $p/n\leq\frac{1}{2}$.

A translation $\mathbf{t}'$ normal to the axis A of the screw produces a translation-equivalent operation A' as well as a nonequivalent but equal screw operation about parallel axes B and C located along the perpendicular bisector of AA' and at a distance of t' cot$\alpha/2$ from AA'. These screw operations accumulate along axes B and C, producing screw axes parallel with A. The resulting sets of symmetry elements, repeated by the lattice translations, constitute infinite sets of parallel axes, which have been called "flocks of axes." An interesting way of appreciating the values of the n's for the A,B, and C axes is as follows: If the sphere and its spherical triangle on which the reasoning of (1) through (5) were based is allowed to expand to infinite radius, the three axes may be allowed to become parallel so that the spherical triangle becomes a plane triangle. Then inequalities (4) and (5) become equalities. The solutions for the n's of the three parallel axes and their separations become

$$2,2,[\infty] \quad \text{with} \quad a:b:c=[\infty:\infty]:1,$$

$$2,3,6 \quad \text{with} \quad a:b:c=2:\sqrt{3}:1,$$

$$2,4,4 \quad \text{with} \quad a:b:c=\sqrt{2}:1:1,$$

$$3,3,3 \quad \text{with} \quad a:b:c=1:1:1.$$

The possible space groups isogonal with a particular point group differ from each other by various characteristic translations. Such a set of space groups can be formed by combining the translations of the several Bravais-lattice types permitted to the point group with the different symmetry elements isogonal with each symmetry element of the point group. As a simple example, to derive the space groups isogonal with point group 3, it is seen from Table II that the Bravais-lattice types consistent with this point group are P and R. Table I shows that the space-group symmetry elements isogonal with point group 3 are 3, 3_1, and 3_2. The combination of the two lattice types with the three symmetry elements yields $P3$, $P3_1$, $P3_2$, $R3$, ($R3_1$ and $R3_2$), seen in Fig. 4. These combinations of symbols are, in fact, the Hermann–Mauguin symbols for the space groups. The last two

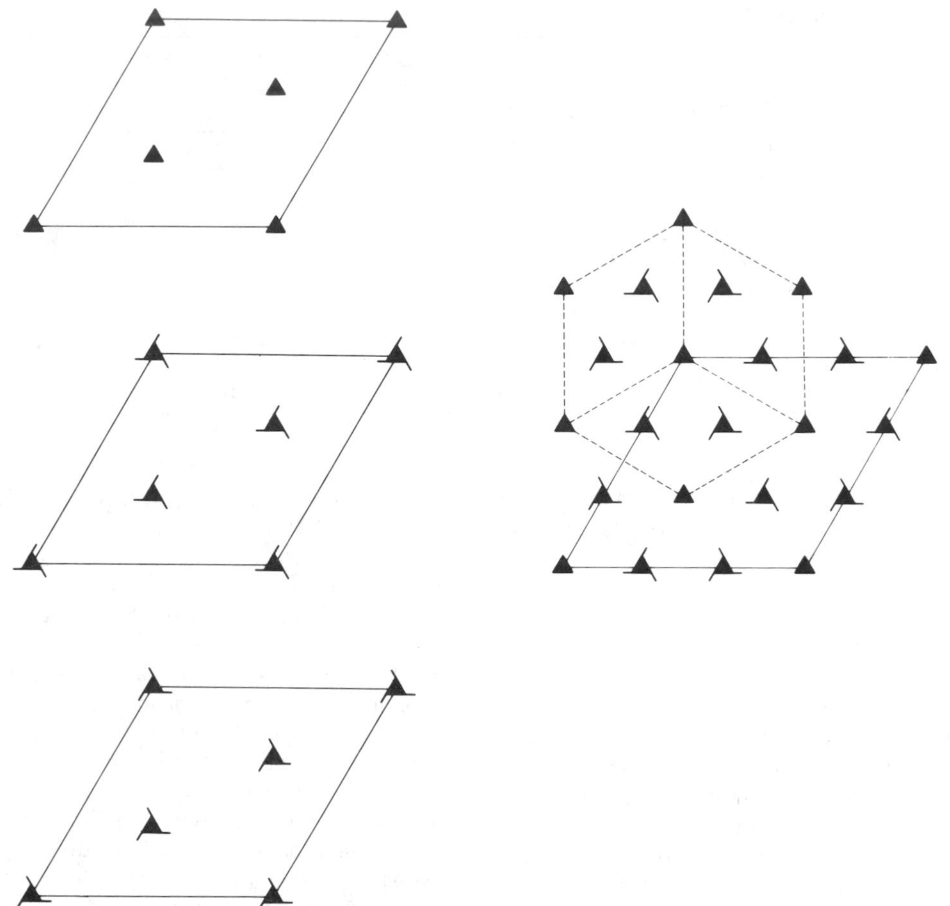

FIG. 4. The arrangement of axes in a cell of each of the space groups isogonal with point group 3. Screw axes are represented by equilateral triangles with propeller-like projections from the three sides. In space group $R3$, the sloping edges of the rhombohedron are shown by broken lines, while the horizontal edges of the corresponding hexagonal triple cell are shown by solid lines. [From M. J. Buerger, *Elementary Crystallography* (Wiley, New York, 1963).]

symbols represent the same space group as $R3$ because, if the operations of any one of the symmetry elements 3, 3_1, or 3_2 is combined with the translations of R, the other two symmetry elements are generated in the B and C positions.

Groups with Operations of the Second Sort

The space-group elements of the second sort are listed in the lower right-hand part of Table I. Only the symmetry elements isogonal with point-group symmetry element m call for special comment. In space groups a reflection may have a translation component in the reflecting plane, since the translation and reflection each transform the other into itself. The resulting *glide reflection* operation generates a *glide plane* as a symmetry element. A sequence of two glide operations is the same as a translation, so the translation component can only be parallel to a lattice translation. The Hermann–Mauguin symbol for a glide plane reveals its translation component; the six possibilities are shown at the lower right in Table I.

Symbolism and Construction of Space Groups

Since a space group differs from the point group with which it is isogonal only in its characteristic translations, its symbol must specify these translations. The symbol always consists of two major parts. The first part is the symbol of the particular Bravais-lattice type that specifies the lattice transformations of the space group. This is one of the several symbols listed in the last column of Table II that are aligned horizontally with the first symbol in column 1 for the isogonal point group. Next there follows a part that has the same general form as the symbol for the correct isogonal point group given in column 2, but with the appropriate isogonal-substituent symbols given in Table I. A space group is completely specified by its symbol and indeed can be constructed from it.

To deduce the geometrical arrangements of the symmetry elements requires a knowledge of the results of combining the various symmetry operations of space groups with the translations of the lattice and with each other. This process is easily carried out with the aid of the *algebra of operations*. After some of the simpler space groups have been derived, the more complicated ones can be deduced from them by a stepwise augmentation.

Equivalent Positions

If a point is placed in a cell, in general the symmetry elements of the space group require the cell to have other symmetrically related points. This symmetrical collection is known as a *set of equivalent positions*, or an *equipoint*. If the location of the original point is unspecialized, the set is said to occupy the *general position* of the space group. Each one of the N operations in the cell generates one of the N points of the general position.

If a point in the general position is shifted until it comes to occupy any one of several specific kinds of symmetry elements (noted later), then a set of s points (characteristic of the kind of symmetry element) coalesce on the symmetry element to form one point. Such a set of equivalent positions is called a *special position*. The number of its points per cell is reduced from the N of the general position to $M = N/s$, an integer known as the *multiplicity* of the equipoint. The coalescence factor s is 2 for inversion centers and mirrors; it is n for a pure n-fold rotation axis and p for a screw axis of p threads. There is no reduction in multiplicity for points on a glide plane or on a screw axis with one thread.

Every space group has a general position. A few space groups have no special positions because they do not have appropriate symmetry elements; for example, $P2_12_12_1$ has only screw axes. Most space groups have several special positions and these may have various multiplicities, depending on the kinds of symmetry elements in the group.

A point placed in the general position has three variable coordinates, x,y, and z. The coordinates of the remaining points of the set can be obtained from this initial one by operating on its coordinates with vectors that can be derived for the individual symmetry elements of the group. Points in special positions have coordinates that are specializations of those of the general position. In general, points on mirrors have two free parameters; those on pure rotation axes and on screws have one free parameter; while points on symmetry centers and on $\bar{n}$ points have no free parameters.

Crystal Structures

The pattern of matter in a crystal consists of sets of integral numbers of atoms in each cell distributed over the available equipoints of its space group. In this way the integral subscripts of a Daltonian compound are accommodated by the integral number of available locations in the cell.

Ideally, the translations of the lattice provide that all regions of the crystal have the same composition. If the conditions of growth, such as temperature, pressure, and composition of the nutrient material, vary with time, however, the composition of the crystal may vary from cell to cell due to the vicarious replacement of one atomic species by another, or of several atomic species by several others, in such a way that the replacing set has certain properties similar to the replaced set.

Finally, we note that a crystal structure is called a "lattice" only by those not acquainted with symmetry theory.

See also GROUP THEORY IN PHYSICS; CRYSTALLOGRAPHY, X-RAY.

BIBLIOGRAPHY

M. J. Buerger, *Introduction to Crystal Geometry*. McGraw-Hill, New York, 1971. (E)

M. J. Buerger, *Elementary Crystallography*. MIT Press, Cambridge, Mass., 1977. (I)

J. J. Burckhardt, *Die Bewegungsgruppen von endlichen Ordnung*. Birkhäuser, Basel, 1947. (I)

E. S. Federow, *Symmetry of Crystals* (transl. by David and Katherine Harker). ACA Monograph 7, 1971. (A)

N. F. M. Henry and K. Lonsdale, *International Tables for X-Ray Crystallography*, Vol. I. Kynoch, Birmingham, 1952. (I)

H. Hilton, *Mathematical Crystallography*. Oxford (Clarendon Press), London and New York, 1903. (I)

I. Janssen, *Crystallographic Groups*. North-Holland, Amsterdam, 1973. (A)

P. Niggli, *Geometrische Kristallographie des Discontinuum*. Borntraeger, Leipzig, 1919. (I)

F. C. Philips, *An Introduction to Crystallography*, 4th ed. Wiley, New York, 1971. (E)

A. V. Shubnikov and V. A. Koptsik, *Symmetry in Science and Art* (David Harker, ed.). Plenum, New York, 1974. (I)

A. Speiser, *Die Theorie der Gruppen von endlichen Ordnung*. Birkhäuser, Basel, 1956. (A)

Crystallography, X-Ray

Martin J. Buerger,† Isabella Karle, and Jerome Karle

INTRODUCTION

Background

Until about the second decade of this century, experimental crystallography was carried out largely by the observation of crystal faces and forms with the aid of the reflecting goniometer and by the study of crystal optics with the aid of the polarizing microscope. Chiefly with these tools the point-group symmetries, crystal systems, and axial ratios of many crystals were established.

The discovery of the diffraction of x rays by crystals afforded the opportunity to obtain much more information concerning the atomic arrangements in matter. Not only is the geometry of the unit cell in a crystal readily determined, but also the locations of all the atoms in crystals of metals, alloys, minerals, salts, organic compounds, natural products, nucleic acids, proteins, and, even, to a certain extent, viruses are determinable. Beams of x rays, electrons, and neutrons are used for diffraction experiments, but x radiation is used primarily because of readily available instrumentation for the individual laboratory, fewer experimental compli-

† Deceased

cations, and well-established techniques. Electron and neutron beams have special purposes. For example, electron beams are suitable for very thin crystalline samples and neutron beams are particularly useful for locating hydrogen atoms. Very intense beams of x rays are available from synchrotron sources and high-intensity neutron beams are available from spallation sources.

Max von Laue's and Sir Lawrence Bragg's descriptions of the diffraction of radiation by three-dimensional gratings are treated in most elementary physics texts. Therefore they are not discussed here.

Space Groups

The morphological study of crystals of different symmetry indicated that they could be referred to seven different sets of crystallographic axes of reference known as triclinic, monoclinic, orthorhombic, tetragonal, trigonal, hexagonal, and cubic. Each of these systems possesses characteristic symmetry. The internal symmetry operations in a crystal may consist of an inversion center, reflection across a plane, 2-, 3-, 4-, or 6-fold rotation, translation, or combinations of these operations called screw axes and glide planes. The seven crystal systems combined with the possible symmetry operators lead to 230 unique space-group types. The characteristics of the 230 space groups are detailed in the International Tables of X-Ray Crystallography, Vol. A.

Space-Group Extinctions

Reflections in a diffraction experiment arise from sets of imaginary parallel planes cutting through a crystal. The planes are indexed with integral numbers that are the reciprocals of their intercepts with the crystallographic axes. The reflected waves are indexed with the integers h, k, and l, the same indices as those of the corresponding planes from which they appear to be reflected. *Extinctions* are defined as certain classes of reflections that are missing in a diffraction pattern owing to the presence of certain symmetry elements, as shown by Paul Niggli in his 1919 book *Geometrische Kristallographie des Discontinuum*. Translation components of glide planes give rise to missing reflections of the type $hk0$, $h0l$, $0kl$, or hhl, and translation components of screw axes cause missing reflections of the type $h00$, $0k0$, $00l$, or $hh0$ (with odd-numbered h, k, or l). Space-group identification is facilitated by the missing reflections.

Friedel's Law

Georges Friedel in 1913 stated that the intensities of reflections from planes $(\bar{h}\bar{k}\bar{l})$ are the same as those from planes (hkl) and consequently it was deemed impossible to determine by means of an x-ray photograph whether a crystal possesses a center of symmetry. However, the statistical distribution of intensities of all the reflections or of certain groups of reflections were shown by A. J. C. Wilson to be different for centric and acentric crystals, generally allowing an unequivocal distinction to be made. Another way of distinguishing centrosymmetric from noncentrosymmetric crystals arises when the wavelength of the incident beam is sufficiently close to an absorption edge of one or more of the atoms present. In that case, Friedel's law is violated by the intensities for noncentrosymmetric crystals, i.e., $I(hkl) \neq (\bar{h}\bar{k}\bar{l})$, but for centrosymmetric crystals, $I(hkl) = I(\bar{h}\bar{k}\bar{l})$.

Reciprocal Lattice

To every crystal lattice there corresponds a related lattice, called the *reciprocal lattice,* that has properties useful in experimental diffraction by crystals. It has the same symmetry as the real space lattice and, for the special cases (orthorhombic, tetragonal, and cubic) when the angles $\alpha = \beta = \gamma = 90°$, the angles of the reciprocal lattice $\alpha^* = \beta^* = \gamma^* = 90°$ and the cell edges are related by $a^* = K/a$, $b^* = K/b$, and $c^* = K/c$, where K can be related to an instrumental constant (see the Precession Method). For the crystal systems with any angle $\neq 90°$, the angle and edge relationships are somewhat more complex.

Application of the Reciprocal Lattice

Bragg's law can be written

$$\sin\theta = \frac{\lambda}{2d(hkl)} = \frac{1/d(hkl)}{2/\lambda}, \tag{1}$$

where 2θ is the angle between the incident beam of wavelength λ and the beam reflected by the planes whose indices are the integers h, k, and l. The distance between such parallel planes is $d(hkl)$.

A geometrical interpretation of the right-hand part of (1), due to Ewald, is illustrated in Fig. 1. The primary x-ray beam is along the direction SO and the origin of the reciprocal lattice is O. Whenever a crystal is so oriented that any reciprocal-lattice point P^*, at the end of a vector of length $1/d$, comes to lie on the surface of a sphere (known as the *Ewald sphere*), Bragg's law is satisfied and a reflected beam

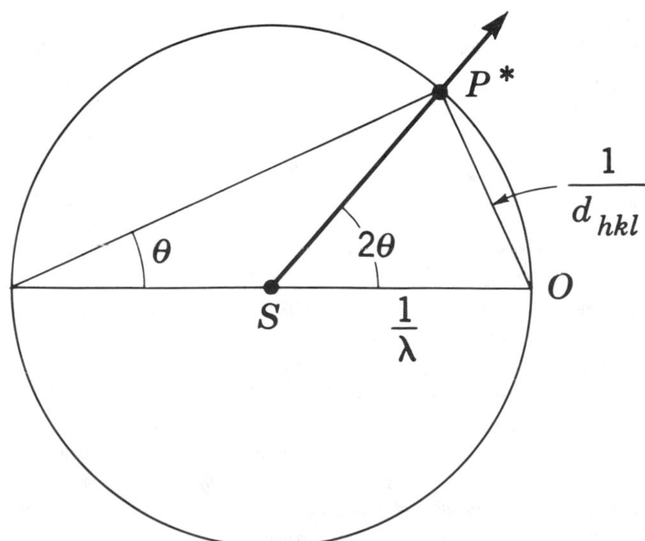

FIG. 1. Ewald's interpretation of Bragg's law with the aid of the reciprocal lattice. [From M. J. Buerger. *Contemporary Crystallography* (McGraw-Hill, New York, 1970).]

develops in the direction SP^* that makes an angle 2θ with the x-ray beam.

The concept of the reciprocal lattice has proven to be of considerable value in the interpretation of the geometrical aspects of diffraction data and the progress of many aspects of diffraction physics.

EXPERIMENTAL METHODS OF X-RAY CRYSTALLOGRAPHY

The Weissenberg Method

If a crystal is rotated about a rational axis, the diffraction directions lie along generators of a nest of Laue cones coaxial with the axis. In the Weissenberg instrument, each cone intersects a coaxial cylindrical film in a circle that, when the film is unrolled, becomes a straight line known as a *layer line*. With the cone containing the incoming x-ray beam counted as the zero cone, the *n*th cone contains reflections with indices *hkn* if the rotation axis is the crystallographic *c* axis. In the Weissenberg method the desired cone is allowed to reach the film through a circular slot in a cylindrical metal screen called the *layer-line screen*. The cylindrical camera can be translated parallel to its axis, and this translation is coupled to the rotation of the crystal about the camera axis (Fig. 2). In this process the diffraction spots are spread out on the two-dimensional surface of the film so that their two variable indices *h* and *k* are determinable. The coordinates *x* and *y* of a spot are readily transformed into indices *h* and *k*.

The photographs taken with a Weissenberg camera have been almost universally recorded by *equi-inclination*, a technique in which the primary beam enters along a generator of the cone whose reflections *hkn* are being photographed. This technique has the advantage that the shapes of rows of the reciprocal lattice as recorded on the Weissenberg film are the same regardless of the value of *n*.

With the introduction of the Weissenberg method in 1924, it became possible for the first time to determine with one instrument not only information about the symmetry of the crystal, but also the indices and intensities of all reflections *hkl*.

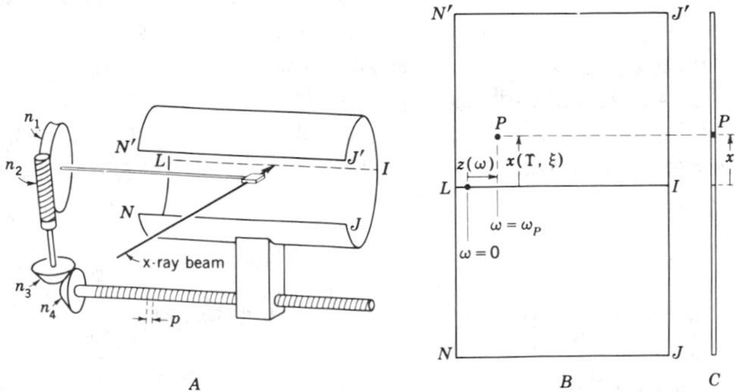

FIG. 2. Diagram of a Weissenberg instrument showing (A) the coupling of the camera translation to the crystal rotation; (B) the resulting position of a reflection on the unrolled film; (C) the position of the corresponding spot along a layer line of a rotating-crystal photograph. [From M. J. Buerger, *X-Ray Crystallography* (Wiley, New York, 1942)].

The Weissenberg method has gained new attention for macromolecular structure analysis. With a film cylinder of larger diameter, a reusable flexible phosphor "plate" for recording the reflections, and automated and computerized scanners for indexing the reflections and reading their intensities from the plate, this method is a viable alternative for collecting the thousands of data for large cells as occur with macromolecules.

The Precession Method

In the precession instrument, a crystal is held in one gimbal and a flat film in another gimbal, as shown in Fig. 3. The unmoved points of these gimbals are aligned along the incoming primary x-ray beam, so that the beam encounters the crystal first. The vertical and horizontal axes of the two gimbals are coupled and a rational axis of the crystal is set parallel with the normal to the film surface. These axes are then inclined at an angle μ to the x-ray beam and caused to precess about it. The nest of Laue cones coaxial with the rational direction of the crystal then also precesses about the x-ray beam. If the rational axis is the crystallographic *c* axis, the reflections of the *n*th cone are *hkn*. The reflections of any chosen cone are allowed to reach the film through an annular opening in a flat layer-line screen that is attached to and precesses with the crystal.

Precession photographs are undistorted pictures of the reciprocal lattice, level by level. Accordingly they not only reveal the symmetry at each level, but the indices of all reflections are obvious by mere inspection. Indeed, indexing for space-group determination is unnecessary since, for example, a glide plane parallel to a level can be detected by simple comparison of the appearances of the zero level and an *n*th level; if a glide plane is present, alternate lines of spots on the zero level are missing in a direction along the translation component of the glide plane. The limitation of the

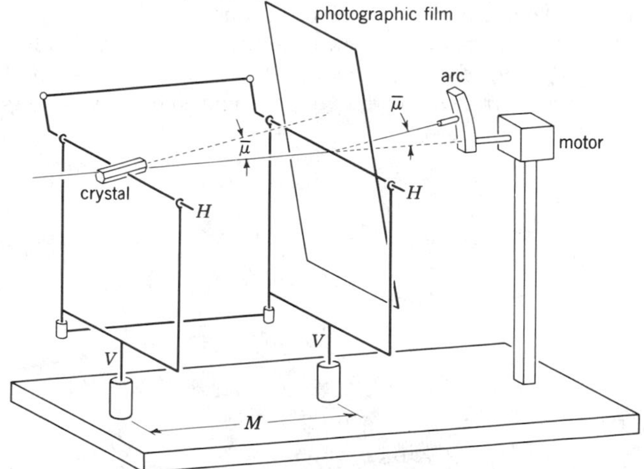

FIG. 3. Diagram of a precession instrument. The motor causes the normal to the photographic film to precess around the direction of the x-ray beam; the coupling of the crystal to the film causes a chosen rational direction of the crystal to precess around the x-ray beam also. [From M. J. Buerger. *Contemporary Crystallography* (McGraw-Hill, New York, 1970)].

method is that only reflection arising at relatively small scattering angles can be recorded on the flat film (as compared with the cylindrical film in the Weissenberg method).

Diffractometer Methods

Several instruments are currently used to record the intensities of x-ray reflections with the aid of photon counters. Two versions have come into common use, one with Weissenberg geometry and the other known as a four-circle diffractometer, shown in Fig. 4. The photons are usually counted by means of proportional counters or scintillation counters.

Diffractometers furnish more precise measurements of the diffraction intensities than the photographic methods and so are preferred for this reason. The operation is automated by a computerized program that sets four angles that are associated with each reflection, scans the diffraction spot, corrects the raw data for various experimental factors, and presents a data record of all the measured reflections in a form suitable for deriving the atomic arrangement in the crystal. The mode of operation is a sequential one in that each reflection is scanned separately.

Other Methods

The Laue, rotating- or oscillating-crystal, and powder methods are among the earliest methods used for recording x-ray diffraction diagrams. Over the years, the facility with which these methods could be used has increased dramatically. This was achieved by the design of more sophisticated instrumentation, the use of high-intensity neutron and x-ray beams, and the advent of convenient advanced computing facilities and associated software packages.

The powder method uses a sample of tiny crystals that are glued or tamped into a pellet or inserted into a thin-walled glass capillary. It usually consists of materials for which larger crystals are not available. Since the powder sample has crystallites oriented in all directions, the diffraction pattern is in the form of concentric circles rather than individual spots. Present technology can analyze a trace of a complex powder pattern (even with overlapped lines) for space-group information and individual *hkl* assignments. Of importance is the least-squares refinement of an approximately known structure by the Rietveld procedure.

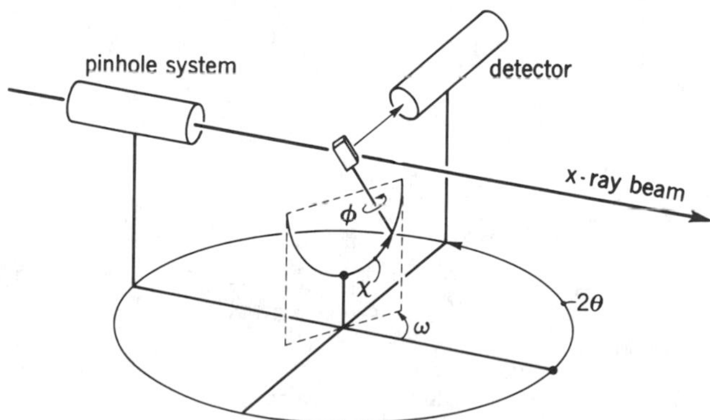

FIG. 4. Diagram of the four-circle diffractometer. [From M. J. Buerger. *Contemporary Crystallography* (McGraw-Hill, New York, 1970).]

In the oscillation method, a variant of the rotating crystal method, the crystal is oscillated through a small-angle range about a particular axis, the x-ray data recorded, the procedure repeated for the next angle range, until the entire pattern for a complete rotation is recorded in segments. In this method, the overlap of spots experienced in the rotation method is avoided. Computerized procedures are used to index the spots and measure their intensities. This method is often used for macromolecules with high-intensity synchrotron radiation that permits very short exposures for each oscillation pattern recorded.

The Laue method uses *white radiation*, that is, radiation consisting of a large range of wavelengths, rather than monochromatic radiation which is used in all other diffraction methods. With white radiation, it is not necessary to continually reorient a crystal by, for example, rotation or precession. A single Laue photograph of a protein crystal, taken at a carefully chosen crystal orientation, with the use of high-intensity synchrotron radiation, can record more than 75% of the useful intensity data in less than a millisecond. This emerging technology has the promise of developing into a method for recording ongoing chemical reactions, for example, between a macromolecule and its substrate.

THE RAW DATA AND THEIR TREATMENT

The Integrated Reflection

It is usual to measure the radiation reflected by a crystal by the *integrated reflection,* which is given by

$$\frac{Q(hkl)\omega}{I_0} = \frac{e^4}{m^2c^4}\lambda^3 N^2 \Delta V L p |F(hkl)|^2. \tag{2}$$

Here $Q(hkl)$ is the energy reflected in the orders *hkl* by a tiny fragment of crystal composed of N cells and volume ΔV, so small that absorption and extinction can be ignored as the crystal is rotated with angular velocity ω in a beam of radiation whose intensity is I_0 and whose wavelength is λ. L and p are the Lorentz and polarization factors, respectively, while e, m, and c are universal constants. The $F(hkl)$ are the amplitudes of the reflected x-ray waves, also called *structure factors*. The left-hand side of (2) has dimensions of power times radians and represents the power of the reflection *hkl* integrated over its angular range. Since the terms before L are fixed for the entire experiment, (2) can be written as

$$Q(hkl)\omega/I_0 = KLp|F(hkl)|^2, \tag{3}$$

where K is a constant for the experimental setup.

Measurement of Intensities

Integrated reflections can be obtained with a single-crystal diffractometer. The reflections can be automatically integrated by counting while the crystal is uniformly rotated through the region of the Bragg reflection; alternatively, the crystal may be held fixed at equally spaced angular intervals for equal times while counting, and the results summed. In either case the background is measured and subtracted.

Correction of Raw Data

To transform the integrated reflections into the $|F(hkl)|^2$, they must be corrected for the Lorentz factor, polarization, absorption, and extinction. Convenient software is available for making the corrections.

FOURIER TRANSFORMATIONS

A crystal structure and its diffraction amplitudes are Fourier mates, so that each can be found if the other is known. The diffraction amplitude $F(hkl)$ is given in terms of the *electron density* $\rho(xyz)$ by

$$F(hkl) = V \int_0^1 \int_0^1 \int_0^1 \rho(xyz) \exp[i2\pi(hx + ky + lz)]\, dx\, dy\, dz. \quad (4)$$

Here V is the volume of the unit cell and x, y, and z are coordinates for the *electron density* expressed in terms of fractions of the cell edges a, b, and c, respectively. The electron density of a crystal cell is the sum of the contributions from a set of N discrete atoms. The scattering power of the jth atom is denoted by f_j, expressed in units of the scattering power of a free electron. The amplitude is then given by

$$F(hkl) = \sum_{j=1}^{N} f_j(hkl) \exp[i2\pi(hx_j + ky_j + lz_j)]T_j. \quad (5)$$

Here x_j, y_j, and z_j are coordinates of the atoms in fractions of the cell edge, and T_j is a factor that represents the effects of the thermal motion.

To recover the *electron density* $\rho(xyz)$ from the $F(hkl)$'s the relation is

$$\rho(xyz) = V^{-1} \sum_h \sum_k \sum_l F(hkl) \exp[-i2\pi(hx + ky + lz)]. \quad (6)$$

Any crystal can be solved for its electron density by making use of (6) if the diffraction amplitudes $F(hkl)$ are fully known. As seen in (4) and (5), however, these are complex quantities, and all that can be determined by experimental methods are the magnitudes $|F(hkl)|^2$. Some progress has been made in deriving phase information rather directly from experiment, by making use of multiple-scattering phenomena. This technique is in its early stages and does not afford the large amount of phase information required for current structure determination. It would appear then that (6) cannot be calculated from the experimental data that are obtained in a standard diffraction experiment. It is quite a remarkable fact, however, that the required phase information is subtly contained in the measured intensities from which, it would seem, only the values of the corresponding $|F(hkl)|^2$ are directly obtainable, i.e., by use of (2). This fact has led to the "direct methods" described below. Other ways to overcome the phase problem will also be noted.

DEDUCTION OF THE CORRECT ARRANGEMENT OF ATOMS

The Patterson Function

In 1934 A. L. Patterson showed that the centrosymmetrical function

$$P(uvw) = V^{-2} \sum_h \sum_k \sum_l |F(hkl)|^2 \cos[2\pi(hu + kv + lw)] \quad (7)$$

has high values at coordinates uvw whenever the vector from the origin to uvw is the same as a vector between two atoms p and q in the crystal cell: i.e., when

$$u = \pm(x_q - x_p), \qquad v = \pm(y_q - y_p), \qquad w = \pm(z_q - z_p), \quad (8)$$

where only all the plus or all the minus signs are taken together. The magnitude of $P(uvw)$ is proportional to $f_p f_q$, so that $P(uvw)$ is large when p and q are both heavy atoms. Thus, the vector distance between heavy atoms is easy to determine, and if the atoms are related by symmetry they can be easily located in the cell.

While the Patterson function provides the interatomic distances in a crystal structure, in practice it does not generally, in itself, solve the structure in a direct way. The function suffers from the fact that if there are n atoms in the crystal cell, there are $n(n-1)$ high values of $P(uvw)$ for $p \neq q$. The clutter due to this overlap increases approximately as n^2, so the usefulness of the function decreases as the number of atoms in the cell increases.

The Patterson function has been particularly useful in applications requiring the location of heavy atoms in structures. This has not only permitted the solution of very small structures, it also has made the solution of macromolecular structures feasible when used in combination with data from the isomorphous replacement and anomalous dispersion techniques. Another worthwhile use of the Patterson function has been to fix the orientation and position of a partially known or sometimes completely known structure. In the case of a partial structure, once its position and orientation are known, there are several procedures for completing the structure.

The Heavy-Atom Method

Crystals that are particularly suited to the application of this method are characterized as consisting of a small number of heavy atoms among many lighter ones. It is often possible to find the locations of the heavy atoms by use of the Patterson function because the interatomic vectors associated with the heavy atoms predominate in this function.

The phases and amplitudes of many of the $F(hkl)$'s of (5) are dominated by the heavy atoms. These phases can then be used as preliminary phases of the coefficients in (6), from which a fair approximation of the electron density can be obtained. This reveals not only the heavy atoms, but some of the light ones as well. From this improved set of atoms an improved set of phases can be computed by (5), which can then be used in another Fourier summation (6). This iteration eventually comes to an end because no phases computed by (5) need to be changed. The heavy-atom method has been responsible for the solution of a great many structures.

Direct Methods in Fourier Space

It is possible to show by use of (5), corrected for the effects of T_j, that because the atoms in a unit cell are discrete and their scattering powers are known to very good approximation, the structure problem is highly overdetermined by the number of independent intensity data obtained from the radiation of an x-ray tube with a Cu target. This applies to the x_j, y_j, z_j coordinates of the atoms as well as the phases. In principle, it should be possible to determine the coordinates directly. It has, however, been found to be much more feasible to determine *phase values* first and then obtain the atomic coordinates from (6).

With strong clues from relations among the $F(hkl)$ that arise from the *non-negativity* of the *electron density* distributions in crystals, phase-determining formulas were obtained by Herbert Hauptman and Jerome Karle in a probabilistic context that has served well in the direct determination of crystal structures containing no heavy atoms. Structures with up to 100 nonhydrogen atoms to be placed can be solved rather readily. With increasing complexity beyond 100 atoms, determinations become more difficult to carry out although a number of the more complex structures have been solved.

The main *phase-determining formulas* are

$$\phi_{hkl} \sim \langle \phi_{h'k'l'} + \phi_{h-h',k-k',l-l'} \rangle_{h'k'l'} \tag{9}$$

and

$$\tan \phi_{hkl} \sim \frac{\sum_{h'k'l'} |E_{h'k'l'} E_{h-h',k-k',l-l'}| \times \sin(\phi_{h'k'l'} + \phi_{h-h',k-k',l-l'})}{\sum_{h'k'l'} |E_{h'k'l'} E_{h-h',k-k',l-l'}| \times \cos(\phi_{h'k'l'} + \phi_{h-h',k-k',l-l'})}, \tag{10}$$

where the E are values of F rescaled in such a way that $\langle |E_{hkl}|^2 \rangle_{hkl} = 1$. Because of the latter property, they are called *normalized structure factors*. The E corresponds to atoms for which the source of scattering is concentrated in a point at the atomic center ("point atoms"). These formulas are used either in a stepwise fashion or with a random set of phase values followed by further refinement with (10). For the stepwise procedure, it is also necessary to have some initial phase values. These are obtained from permitted assignments of some phase values in order to specify the origin in a crystal and the additional use of symbols whose values are determined as the phase-determination proceeds. Other procedures use alternative numerical values for the same phase instead of symbols. The particular phase assignments for a given space group have been tabulated and arise from a theory of phase invariants and structure semi-invariants. The initial phase assignments are used to define the values of additional ones and in a series of steps the phase determination proceeds. The steps are chosen so as to maximize the probability of correctness by use of probability measures associated with (9) and (10). Correct structures are recognized by their good agreement with the data as computed from (5) and by confirming that the bond lengths, bond angles, and connectedness make good chemical sense.

It should be noted that the direct methods of phase determination are particularly effective when applied to structures that contain heavy atoms, because of an enhancement of probability, and when the positions of the heavy atoms are known, development of the remainder of the structure by use of (10) is generally considerably more efficient than the heavy-atom method.

Macromolecular Structure Determination

Heavy atoms have played an indispensable role in the determination of the structures of macromolecules in the crystalline state. The names of the techniques used are *isomorphous replacement* and *anomalous dispersion*. Crystals that have the same unit-cell geometry but differ in chemical composition are called isomorphous. Isomorphous protein crystals are usually comprised of the native protein and the native protein to which some heavy atoms have been added. The information contained in the different intensities of scattering from isomorphous crystals permits the determination of their structures.

If heavy atoms are added to an unsubstituted crystal, the relationship between wave amplitudes is

$$F_R(hkl) + F_Y(hkl) = F_{R+Y}(hkl), \tag{11}$$

where R represents the native substance, Y the added atoms, and $R+Y$ the substituted substance. Analysis of (11) is benefited from the determination of the positions of the atoms Y, for example, from Patterson maps or direct methods which may be applied with enhanced utility to isomorphous replacement data. From this information, $F_Y(hkl)$ may be computed with the use of (5). With knowledge of $F_Y(hkl)$ and the experimentally measured magnitudes in (11), it is possible to derive phase information. If the isomorphous pair of crystals are both centrosymmetric, the phase information will be unique. If the isomorphous pair of crystals are both noncentrosymmetric, the phase information will be obtained with a twofold ambiguity. Many methods have been developed to resolve the ambiguity, e.g., additional substitutions, anomalous dispersion information, probabilistic and algebraic analyses, and a filtering technique.

It has been customary to make several isomorphous replacements when possible in order to compensate for experimental error and lack of precise isomorphy. The method based on several isomorphous replacements is called the *multiple isomorphous replacement* method. It has so far played a predominant role in the progress of protein crystallography.

The second technique, *anomalous dispersion,* is based on the modification of the intensities of scattering as a consequence of absorption processes that take place in some of the atoms that comprise the crystal of interest. The role of the heavy atoms in macromolecules derives from their relatively strong anomalous scattering. The modifications of the scattered intensities provide the information from which the structures can be elucidated.

When atoms are present in a structure that scatter anomalously to a significant extent, the wave amplitude may be written

$$F_\lambda(hkl)=F^n(hkl)+F_\lambda{}^a(hkl), \tag{12}$$

where $F_\lambda(hkl)$ is the amplitude of the scattered wave whose magnitude may be measured by experiment, $F^n(hkl)$ is the wave amplitude that would be obtained if all atoms scattered without any anomalous dispersion, and $F_\lambda{}^a(hkl)$ is the wave amplitude that arises from the anomalous corrections to the normal atomic scattering. In noncentrosymmetric crystals, the Friedel mates are no longer equal when there is significant anomalous scattering so that $F_\lambda(hkl)\neq F_\lambda(\bar{h}\bar{k}\bar{l})$ and a second independent equation follows:

$$F_\lambda(\bar{h}\bar{k}\bar{l})=F^n(\bar{h}\bar{k}\bar{l})+F_\lambda{}^a(\bar{h}\bar{k}\bar{l}). \tag{13}$$

Identification of the anomalous scatterers and their location permits $F_\lambda{}^a(hkl)$ and $F\lambda^a(\bar{h}\bar{k}\bar{l})$ to be computed from (5) in which the atomic scattering factors are replaced by the real and imaginary corrections to the atomic scattering factors. Patterson and direct methods techniques can be used to locate the anomalous scatterers. It was found to be worthwhile to manipulate the right sides of (12) and (13) to express the quantities in somewhat different form. This resulted, with the use of various wavelengths, in providing a useful set of linear simultaneous equations. Phase determination can be made with these equations. If there is only a one-wavelength experiment, the result for noncentrosymmetric crystals will involve a twofold ambiguity. To overcome the ambiguity, it is possible to employ, for example, more wavelengths, isomorphous replacement, probabilistic calculations, and a filtering technique. In Fig. 5 the combination of single isomorphous replacement and one-wavelength anomalous dispersion are illustrated in a manner that emphasizes how their individual ambiguities are mutually resolved.

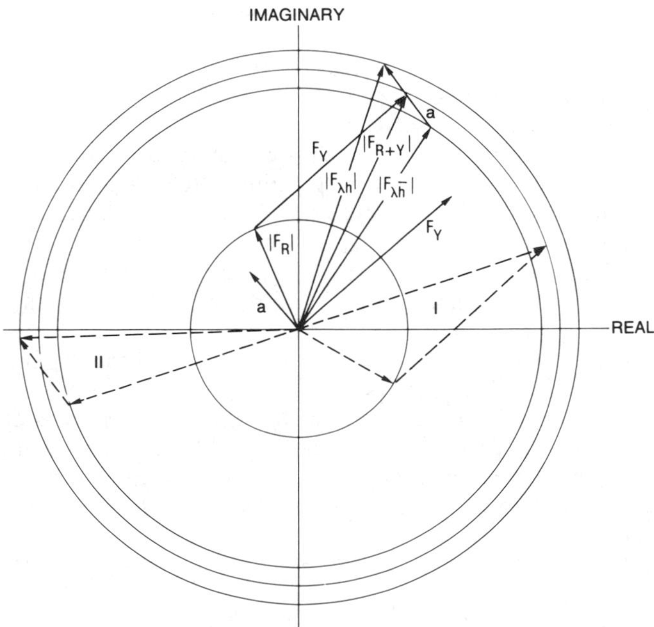

FIG. 5. Diagram illustrating the twofold ambiguity associated with each of single isomorphous replacement and one-wavelength anomalous dispersion and the manner in which the combination of the two techniques resolves the ambiguity. Here we have set $F_{\mathbf{h},R+Y}=0.5(F_{\lambda\bar{\mathbf{h}}}+F^*_{\lambda\bar{\mathbf{h}}})$ to good approximation, $\mathbf{a}=F^a_{\lambda\mathbf{h}}-F^{a*}_{\lambda\bar{\mathbf{h}}}$ and the vector triangles I and II are the incorrect alternatives for the isomorphous replacement and the anomalous dispersion data, respectively [$\mathbf{h}\equiv(hkl)$].

The anomalous dispersion technique is especially facilitated by the use of x-ray radiation from synchrotron sources. Its high-intensity, tunable radiation is well suited to multiple-wavelength anomalous dispersion experiments. Up to the present, anomalous dispersion has been applied to a lesser extent than isomorphous replacement and usually in combination with the latter. The value of each technique is enhanced when used in combination. Anomalous dispersion is beginning to be used more often, however, as the sole method for the investigation of macromolecular structure. It is particularly useful when it appears to not be possible to form isomorphous crystals.

Another approach to macromolecular structure determination arises from the occurrence of symmetry within the unit cell of a crystal in addition to the space-group symmetry. For example, there may be additional rotational symmetry among the atomic arrangements within the unit cell. A methodology for facilitating the analysis of macromolecules based on this so-called noncrystallographic symmetry has been particularly stimulated by the work of Michael G. Rossmann and it has been quite effective in applications to complex viral structures.

REFINEMENT

An approximately solved structure becomes more exactly defined by successive Fourier syntheses, but the final synthesis suffers from errors due to the arbitrary termination of the infinite Fourier series. It also suffers from lack of knowledge of the exact temperature factor. Corrections can be applied by several devices, such as the *difference Fourier synthesis*. This uses coefficients $(F_{obs}-F_{cal})$ for each reflection in (6), where obs and cal signify observed and calculated, respectively. F_{obs} and F_{cal} differ in magnitude. The same phases are used for both. The result of this synthesis is the difference between the observed and calculated electron densities, which shows the magnitudes and directions in which the model must be changed to bring it into conformity with the actual structure.

The progress of refinement is followed by computing at each stage the discrepancy index, a residual summed over the data set,

$$R=\frac{\sum \||F_{obs}|-|F_{cal}\||}{\sum |F_{obs}|}. \tag{14}$$

The value of R is usually between 5 and 10% for an acceptably refined structure, and may be as small as 2% for a very well-refined structure. The acceptable values of R for data from macromolecular structure determinations are generally somewhat higher. The R value is 82.8% for a completely wrong model if centrosymmetrical, and 58.6% if noncentrosymmetrical. A preliminary model that has $R=45\%$ is regarded as promising.

The use of least squares in fitting an approximately correct model to the observed amplitudes was first suggested by E. W. Hughes in 1941 and applied to refinement of the structure of melamine. Least-squares refinement became popular with the rise of electronic digital computers, and currently is the standard way of finishing the details of the structure. Because the structures of most inorganic and the simpler organic structures are overdetermined by the large numbers of their diffraction–amplitude data and because modern computers have large capacities, it is usual not only to refine the three coordinates of each atom, but also to determine the three principal axes of the ellipsoids of thermal motion of each atom and their orientations with respect to the cell axes.

DATA FILES

Compendia of structural information are available on separate computer files for inorganic, organic, and macromolecular substances. An extensive file is also available concerning information obtained by powder diffraction. For details concerning these files, contact International Union of Crystallography, 5 Abbey Square, Chester CH1 2HU, England.

See also CRYSTAL SYMMETRY; FOURIER TRANSFORMS; X-RAY SPECTRA AND X-RAY SPECTROSCOPY.

BIBLIOGRAPHY

F. R. Ahmed, K. Huml, and B. Sedláček, (eds.), *Crystallographic Computing Techniques*. Munksgaard, Copenhagen, 1976. (I)

M. J. Buerger, *X-ray Crystallography*. Wiley, New York, 1942. (E)

M. J. Buerger, *The Precession Method*. Wiley, New York, 1964. (E)

M. J. Buerger, *Contemporary Crystallography*. McGraw-Hill, New York, 1970. (E)

M. A. Carrondo and G. A. Jeffrey (eds.), *Chemical Crystallography with Pulsed Neutrons and Synchrotron X-Rays*. D. Reidel, Dordrecht, 1988. (A)

R. W. James, *The Optical Principles of the Diffraction of X-Rays*. G. Bell and Sons, London, 1948. (A)

T. Hahn (ed.), *International Tables for Crystallography*. D. Reidel, Dordrecht, 1987. (A)

S. R. Hall and T. Ashida (eds.), *Methods and Applications in Crystallographic Computing*. Clarendon Press, Oxford, 1984. (I)

J. Karle, *Angew. Chem.* (English ed.) **25,** 614 (1986). (I)

M. F. C. Ladd and R. A. Palmer (eds.), *Theory and Practice of Direct Methods in Crystallography*. Plenum, New York, 1980. (I)

R. Newnham, *Structure-Property Relations,* Vol. 2 of Crystal Chemistry of Non-Metallic Materials Series. Springer-Verlag, Berlin, 1975. (I)

D. Sayre (ed.), *Computational Crystallography*. Clarendon Press, Oxford, 1982. (A)

A. F. Wells, *Structural Inorganic Chemistry*. Clarendon Press, Oxford, 1984. (I)

H. Wyckoff (ed.), *Diffraction Methods in Biological Macromolecules*. Academic Press, New York, 1985. (A)

Crystals, Liquid *see* Liquid Crystals

Currents in Particle Theory

R. J. Oakes

Currents have played a fundamental role throughout the development of present theoretical ideas about elementary particles. The concept of charges and currents is intimately related to the notion that matter is made up of fundamental constituents and that all the properties of matter are ultimately determined by the nature of these elementary parts and their interactions. For example, the electric charge and electric current densities of matter are superpositions of the individual constituent densities and fluxes weighted by the electric charge carried on each constituent. Although discoveries of new particles and new phenomena have caused drastic revisions in views about which particles are elementary and what laws govern their interactions, the basic concept of charges and currents has remained essentially unchanged.

In nonrelativistic quantum mechanics, where the position of a particle is described by a probability distribution, the electric charge and electric current densities at a point $\mathbf{x}$ at time t due to a particle of charge e are $e\rho(\mathbf{x},t)$ and $e\mathbf{j}(\mathbf{x},t)$ where $\rho(\mathbf{x},t)$ and $\mathbf{j}(\mathbf{x},t)$ are, respectively, the position probability and the probability current density at $\mathbf{x}$ and t. The density $\rho(\mathbf{x},t)$ and current $\mathbf{j}(\mathbf{x},t)$ depend on the Schrödinger wave function $\psi(\mathbf{x},t)$. The local conservation law

$$\frac{\partial \rho}{\partial t}(\mathbf{x},t) + \nabla \cdot \mathbf{j}(\mathbf{x},t) = 0$$

is insured by virtue of $\psi(\mathbf{x},t)$ being a solution to the Schrödinger equation of motion for the particle.

In relativistic quantum field theory, which incorporates the infinite number of degrees of freedom necessary to describe an indefinite number of particles, for each kind of elementary particle there is a canonical field operator that plays a role analogous to the classical canonical coordinates. The Lagrangian describing the entire system of particles is a function of all these fields and their first derivatives. In general, there will be several currents in the theory, corresponding to the charges carried by the particles (e.g., electric charge, isospin, and color). Since these currents are functions of the canonical fields, they can be thought of as analogs of classical (noncanonical) coordinates of the system. The currents and their associated charges have proven to be of enormous practical value in studying the interactions of the elementary particles, since the Euler–Lagrange equations of motion, whose complete solutions would determine all the canonical fields, are exceedingly complex. For each current $J_\mu(\mathbf{x},t)$ there is a charge

$$Q = \int d^3x\, J_0(\mathbf{x},t),$$

which will be a constant of the motion if the current is conserved, thus providing dynamical information about the system. Nonconserved currents have also proven to be very important in the phenomenological studies of elementary-particle interactions.

Currents are essential in describing the interactions among the elementary particles, which at present include at least quarks, leptons, and intermediate vector bosons. While quarks cannot be directly observed, since they carry color charges which are confined to very small regions of space, there is substantial indirect evidence that the observed strongly interacting hadrons, e.g., protons, neutrons, and mesons, are composed of these spin-$\frac{1}{2}$ quark constituents. The leptons, which so far include the electron, the muon, and the tauon, as well as their associated neutrinos, do not participate in strong interactions directly but do have electromagnetic, weak, and, of course, gravitational interactions. The photon (γ), which mediates electromagnetic interactions, and the vector bosons ($W^{\pm}$, Z^0) that mediate the weak interactions have all been directly observed, experimentally. The gluons (g), which mediate the strong interactions, also carry color and, like the quarks, are confined to very small regions of space and cannot, therefore, be directly observed. Nevertheless, there is clear evidence that gluons do exist together with the quarks inside the hadrons, and, indeed, can be emitted or radiated from these hadrons, resulting in observed jets of more hadrons. It is not a settled question, at present, how many quarks, leptons, and intermediate bosons there really are, or what all their properties are, or even if there *must* also be other elementary particles, particularly, elementary scalar bosons.

The electromagnetic interactions of the elementary particles are described by the coupling of the electromagnetic current J_μ to the electromagnetic field A_μ. That is, the interaction Lagrangian is

$$\mathscr{L}_{\mathrm{EM}} = eJ_\mu A^\mu$$

and therefore the quantum of the electromagnetic field, the photon (γ), which is a massless vector boson, mediates the electromagnetic interactions of all the elementary particles with a strength equal to the electric charge of the particle. Very precise experimental tests of quantum electrodynamics, e.g., the Lamb shift, the electron and muon anomalous magnetic moments, and the hyperfine splitting, have confirmed this theory down to very small distances ($\sim 10^{-15}$ cm) or equivalently, to very high energies ($\sim$ 20 GeV).

It has long been known that weak interactions at low energies, e.g., beta decay, muon decay, and muon capture, can phenomenologically be described by an effective coupling of currents to currents. It is now known that this is only an approximation, valid in the low-energy regime, to a theory which unifies these weak interactions with the electromagnetic interaction. In this electroweak theory the weak interactions are mediated by the massive vector bosons ($W^{\pm}$, Z^0) coupled to weak currents, just as the electromagnetic interaction is mediated by the massless vector boson, the photon (γ), being coupled to the electromagnetic current. The weak-interaction Lagrangian then has a form analogous to the electromagnetic interaction; that is,

$$\mathscr{L}_{\mathrm{WK}} = g_C C_\mu W^\mu + g_N N_\mu Z^\mu,$$

where C_μ is a charged current which changes the electric charge by one unit and N_μ is a neutral current which conserves the electric charge. The charged and neutral intermediate vector bosons $W^{\pm}$ and Z^0 are quite massive: approximately, M_W = 80 GeV/c^2 and M_Z = 90 GeV/c^2.

The strong interactions are mediated by massless vector bosons, the gluons (g), which are coupled to the strong currents composed of the particles which carry color: the quarks and the gluons, themselves. These strong interactions are responsible for binding the quarks and gluons together to form the observed hadrons. There are three colors (red, blue, and green) and the unbroken symmetry of the theory among the three colors implies invariance under a symmetry group of transformations, viz., $SU(3)$. As, a consequence, the $3 \times 3 - 1 = 8$ currents corresponding to the possible non-neutral color combinations are conserved, like the electromagnetic current.

It is entirely possible that ultimately all the interactions among the elementary particles can be unified into a grand theory in which the electromagnetic, weak, strong, and even gravitational interactions are all mediated by bosons coupled to currents which are conserved by virtue of the symmetry of the complete theory under a group of transformations. Such theories are called gauge theories, or Yang–Mills theories, and are based on some gauge group of symmetry transformations.

Certainly the electromagnetic interaction is a gauge theory of this type. The conservation of the electromagnetic current and the electric charge is a consequence of invariance of the Lagrangian under a one-dimensional (Abelian) group $U(1)$ of gauge transformations. The weak and electromagnetic interactions have been unified in a generalized (non-Abelian) gauge theory based on the group $SU(2) \otimes U(1)$. The strong interactions are based on the non-Abelian gauge group $SU(3)$, the color symmetry group. The search is now under way for a grand symmetry group, containing as subgroups both the color symmetry group $SU(3)$ and the electroweak symmetry group $SU(2) \otimes U(1)$. If this program succeeds, all the interactions could be understood in terms of intermediate bosons being exchanged between elementary particles in a way prescribed by their gauge-invariant couplings to the elementary-particle currents.

Independent of this ambitious program, currents and their associated charges, whether conserved or not, play an important role in the phenomenology of elementary particles. Measurements of the hadron matrix elements of the electromagnetic and weak currents probe the quark and gluon structure of the hadrons. From elastic electron–nucleon scattering experiments the nucleon form factors of the electromagnetic current have been extracted, while deep inelastic electron-scattering experiments have yielded the proton and neutron structure functions, which naively can be related to constituent quark wave functions. Analogous matrix elements of the weak currents have been measured in neutrino-scattering experiments and, in fact, it was in these neutrino experiments that the existence of *neutral* weak currents in addition to the charged weak currents was first observed. Electron–positron annihilation into hadrons has also been used to probe the matrix elements of the electromagnetic current between the vacuum and various hadronic states yielding many surprising results; particularly the production of new quark–antiquark pairs, which then undergo strong interactions to form the more familiar hadrons.

Currents also play an important role in the study of approximate symmetries of the hadrons. The space integrals of the time components of the currents give charges which close under equal-time commutation to form algebras. The eight vector charges in the algebra associated with the lightest quarks (up, down, and sideways) generate an approximate $SU(3)$ symmetry of strong interactions, which contains the isotopic spin and electromagnetic symmetries as subgroups, and has enjoyed enormous success in classifying the hadrons and relating their properties.

Even though this hadron symmetry is only approximate, and therefore not all the associated currents are exactly conserved, the concept of partial conservation of certain currents has proven very useful. In particular, partial conservation of axial vector currents (PCAC) together with the $SU(2) \otimes SU(2)$ current algebra leads to sum rules, notably the Adler–Weisberger sum rule, which tests the underlying approximate symmetries and how they are broken.

Clearly currents have been exceedingly important in the study of the elementary particles and their interactions and no doubt they will continue to play an essential role in this area of research.

See also Elementary Particles in Physics; Gauge Theories; Quantum Field Theory; $SU(3)$ and Higher Symmetries.

Cyclotron

David L. Judd

The cyclotron is a particle accelerator conceived by Ernest O. Lawrence in 1929 and developed, with his colleagues and students, at the University of California in the 1930s. His goal was to produce beams of high-energy ions, without using high electrostatic voltages, to study their reactions with atomic nuclei. Since then, over 150 cyclotrons of greatly differing sizes, energies, and other properties have been built for a continually growing variety of uses. Three successive generations of cyclotrons ("classic", frequency-modulated, and sector-focused) are described here. In addition, the classic cyclotron is the prototype not only of the later types but also of a wider class of *magnetic resonance accelerators* which include microtrons and various kinds of synchrotrons that accelerate electrons, protons, and heavier ions.

By 1940 "classic" cyclotrons had been built and used at many laboratories throughout the world. Exemplified by the 60-in. cyclotron at Berkeley, their principal components were a dc electromagnet having circular pole pieces with a magnetic gap small compared to their diameter, producing a nearly uniform, axially symmetric magnetic field; two hollow flat D-shaped copper electrodes (dees) placed in the magnet gap, open toward each other, with a small space (dee gap) between them along a diameter; an electric field produced across the dee gap, oscillating at a constant radiofrequency (rf), driven by an external oscillator; a vacuum tank within the magnet gap, enclosing the dees and their supporting structures (dee stems) used as resonant lines; an ion source (electric discharge) between the dees at the center, supplied with a small flow of neutral gas; internal targets to be bombarded at large radii; and a deflection system to bring ion beams outside the accelerator. (See Fig. 1.)

The operation depends on the *cyclotron resonance condition*: a particle of charge e and mass m moving in a circle perpendicular to a magnetic field B circulates with frequency f and angular frequency ω given by $\omega = 2\pi f = eB/m$ (SI units) which is the same for all energies, velocities, and radii for constant B and m. This can be seen from Newton's force law $F = ma$, with $F = eBv$ the centripetal Lorentz force, v the velocity, $a = v^2/r$ the centripetal acceleration, r the radius, and $\omega = v/r$. The kinetic energy at r is $T = \frac{1}{2}mv^2 = \frac{1}{2}(eBr)^2/m$. An ion making a circle around the source may gain energy, if the oscillator has this *cyclotron frequency,* on each crossing of the dee gap. The electric field there will reverse in each half-turn, having the tangential direction to speed up an ion each time the ion "sees" it. Between crossings the ions are shielded, inside a dee, from the electric field. Because of the resonance condition an ion will stay in step; its energy and radius will grow on every turn until it strikes a target or enters a deflector. (The *cyclotron frequency* of a charged particle in a magnetic field is also important in physics of the solid state, plasmas, the ionosphere, and astrophysics.)

It is not hard to get ions started at the center. They are drawn into a dee each time it is negatively charged (for positive ions), start to circulate at once, and tend to bunch at the correct phase relative to the rf within a few turns. However, they will soon drift away from the magnet midplane and strike the inside surface of a dee unless a vertical focusing force directed toward this plane is provided. To produce it the magnetic field strength B is made to decrease slightly with increasing radius, which causes the magnetic field lines to curve as shown in Fig. 2. The Lorentz force, perpendicular to these lines, then has a small component

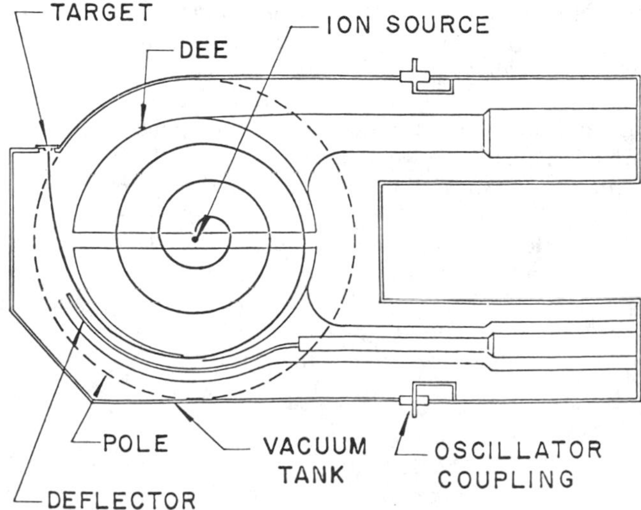

Fig. 1. Some components of a conventional cyclotron.

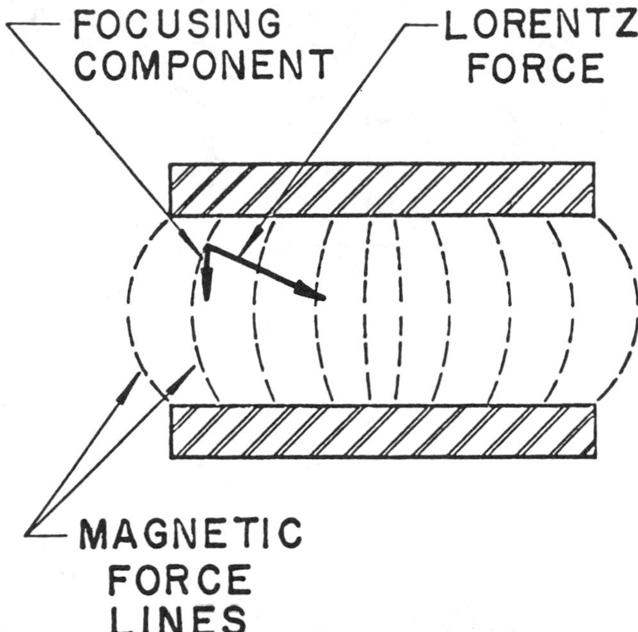

FIG. 2. Curved magnetic field lines showing vertical focusing force.

directed toward the midplane, causing the ions to oscillate slowly up and down across it as they circulate. This small variation of B also serves to define well-centered circular orbits, about which oscillations in radius may occur. Their frequency is determined by the difference between the centripetal acceleration and the inward radial Lorentz force for ions which depart slightly in radius from the proper circle. These vertical and radial motions are called betatron oscillations because they were analyzed in connection with *betatron accelerators* in which the focusing actions are similar. The radial variation of B is described by the *field index* $n=-(r/B)(dB/dr)$; the angular frequencies of the vertical and radial betatron oscillations are, respectively, $\omega_v=\omega n^{1/2}$, $\omega_r=\omega(1-n)^{1/2}$ with ω as above. For $0<n<1$ both frequencies are real, giving stable motions. In cyclotrons n is small, rising from zero at the center to $\leqslant 0.2$ near the outer edge. A resonant effect ($\omega_r/\omega_v=2$) can cause beam loss at $n=0.2$ by transferring energy of radial oscillations, for which there is plenty of room, into vertical oscillations that have limited clearance within the dees.

With radial and vertical motions thus controlled, only an unwanted variation in azimuthal position, or *phase* relative to the rf oscillator, remains. This effect limits the energy of a classic cyclotron, in which there is no *phase stability*. The small decrease of B and (particularly for light ions) the relativistic increase of mass ($m=m_0+T/c^2$, with m_0 the rest mass and c the speed of light) both act to decrease an ion's frequency $\omega=eB/m$ as its energy and radius increase. Its times of gap crossings will lag more and more behind the phase of maximum electric field, and could eventually lead to crossings at times of reversed field, causing energy loss. To reach a high energy in a limited number of turns, before excessive phase lag can accumulate, the largest possible dee voltage is needed. The highest proton energy reached with such a cyclotron was 22 MeV; this required about 500 kV on the dees. Currents of 100 μA were typical, with a maximum of ~1 mA. Deuterons, alpha particles, and heavier ions in higher charge states were also accelerated. Such cyclotrons quickly became the leading tools of nuclear physics research. They were also used to produce radioisotopes for a rapidly growing number of applications in other fields. The first transuranium elements, neptunium and plutonium, were discovered through cyclotron bombardment of heavy targets.

Following World War II new methods were applied which extended the energy and other capabilities of circular magnetic accelerators. The *principle of phase stability* (discovered independently by E. M. McMillan and V. Veksler) led to several new types of accelerators, of which the frequency-modulated cyclotron (synchrocyclotron), electron synchrotron, and proton synchrotron have played important roles in research. They differ from the classic cyclotron by producing particle beams in pulses separated in time (with pulse repetition frequencies ranging from 0.1 to 1000 Hz) rather than as a steady current modulated only at the radiofrequency. In synchrotrons the magnetic field is made to vary periodically at the pulse repetition frequency, but the synchrocyclotron, or FM cyclotron, has a static magnetic field. Its radiofrequency is smoothly lowered during each cycle of acceleration and is followed by that of the ions locked stably to it. Frequency modulation is done with rotating capacitors or vibrating blades. Lawrence's pioneering FM cyclotron at Berkeley accelerated protons to 340 MeV and produced the first man-made pi mesons in 1948. After extensive improvements in 1956 it produced 720-MeV protons, which move at over four-fifths the speed of light, with a mass increase of more than 75%. Deuterons and alpha particles were also accelerated. Protons in the range 600–700 MeV have been produced by similar machines at Dubna, U.S.S.R., and Geneva, Switzerland. These and others producing somewhat lesser energy protons employ magnetic fields on the order of 2 T. Pulse repetition frequencies range from 50 to 1000 Hz and typical currents are 1 μA. Work with these machines greatly enlarged our knowledge of nuclear and meson physics, particularly of pion and muon properties and nucleon–nucleon interactions. Also, large numbers of patients have been treated for certain conditions, particularly pituitary tumors, employing techniques developed at Berkeley.

A different method of overcoming the energy limitation of the cyclotron is based on work by L. H. Thomas in 1938, but the developments needed to put his concept and its later elaborations into practice did not begin until 1949. Because many characteristics of the classic cyclotron (static magnetic field, constant rf frequency, and steady, non-pulsed beam) are retained by the accelerators that have evolved along this line they are called *sector-focused, isochronous, spiral-ridge,* or *azimuthally-varying-field* (*AVF*) *cyclotrons*; the name *fixed-field alternating-gradient* (*FFAG*) *accelerator* has also been used. In these machines the magnetic field (averaged along a full turn of an ion's orbit) *increases* with increasing radius just enough to offset the relativistic mass increase, matching the cyclotron frequency to that of the oscillator at all ion energies. The pole tips are carefully shaped, and correcting fields produced by adjustable currents in pole-face windings called trim coils provide fine tuning so that every turn takes the same time; the orbits are *isochronous*. However, the radially increasing field tends to make the vertical motions unstable. Field line curvature is reversed from that shown in Fig. 2, so the averaged field index n is negative. The focusing force required to achieve net vertical stability by overriding this vertical defocusing effect is obtained by making the orbits noncircular. This is done by introducing *azimuthal variations* in the magnetic field. Threefold symmetry is common on smaller machines, sixfold and eightfold on the largest. In some machines the gradual wave-like variations proposed by Thomas are used, but in others the magnet is divided into *sectors* with field-free spaces between them, in which rf accelerating structures (descendants of the classic dees) are located. In either case, vertical focusing forces result from interaction of the radial component of velocity with the azimuthal component of magnetic field (and from additional interactions for spiralled ridges as in Fig. 4(b) below). The focusing effect is not steady, but varies along an orbit; it becomes a succession of impulses if the field changes at sector edges are steep. Such an impulse, which occurs whenever an ion obliquely crosses a fringing field at a magnet edge in a magnetic spectrometer or other device, is known as *edge focusing*.

To see how edge focusing operates, consider particles of charge q and momentum p entering a magnetic field region with $B_z = -B_0$ at an angle α from the normal to its boundary, as shown in Fig. 3(a); here x, s, and z form a right-hand triad. An ion entering with displacement a_x will encounter "excess" magnetic field (compared with normal entry) along the path Δs, producing excess deflection $\Delta\theta_x < 0$. This deflection represents the focusing effect of a thin converging lens of focal length f_x. The angle is

$$\Delta\theta_x = -\frac{a_x}{f_x} = \frac{\Delta p_x}{p} = \left(\frac{q}{p}\right)\int v_s B_z\, dt$$

$$= \left(\frac{q}{p}\right)\int (-B_0)\, ds = -\left(\frac{qB_0}{p}\right)\Delta s = -\frac{a_x(\tan\alpha)}{r}$$

so that the focal power of the thin lens is $1/f_x = (\tan\alpha)/r$. We have used $\int v_s\, dt = \int ds = \Delta s = a_x \tan\alpha$, and $r = p/qB_0$ = radius of curvature. Similarly, a particle entering above the plane of bend with displacement a_z, as shown in Fig. 3(b), will experience excess deflection $\Delta\theta_z > 0$ that represents the defocusing effect of a thin diverging lens of focal length f_z. The angle is

$$\Delta\theta_z = -\frac{a_z}{f_z} = \frac{\Delta p_z}{p} = \left(\frac{q}{p}\right)\int (-v_s B_x)\, dt$$

$$= -\left(\frac{q}{p}\right)\int B_x\, ds.$$

The field component B_x exists only in the fringing region, and vanishes in the plane $z=0$, so $B_x \simeq z\, \partial B_x/\partial z = z\, \partial B_z/\partial x$ (using curl $\mathbf{B}=0$), so that

$$\Delta\theta_z = -\left(\frac{qz}{p}\right)\int \frac{\partial B_z}{\partial x}(dx\tan\alpha)$$

$$= -\left(\frac{qz\tan\alpha}{p}\right)\int dB_z = +\frac{a_z(\tan\alpha)}{r}$$

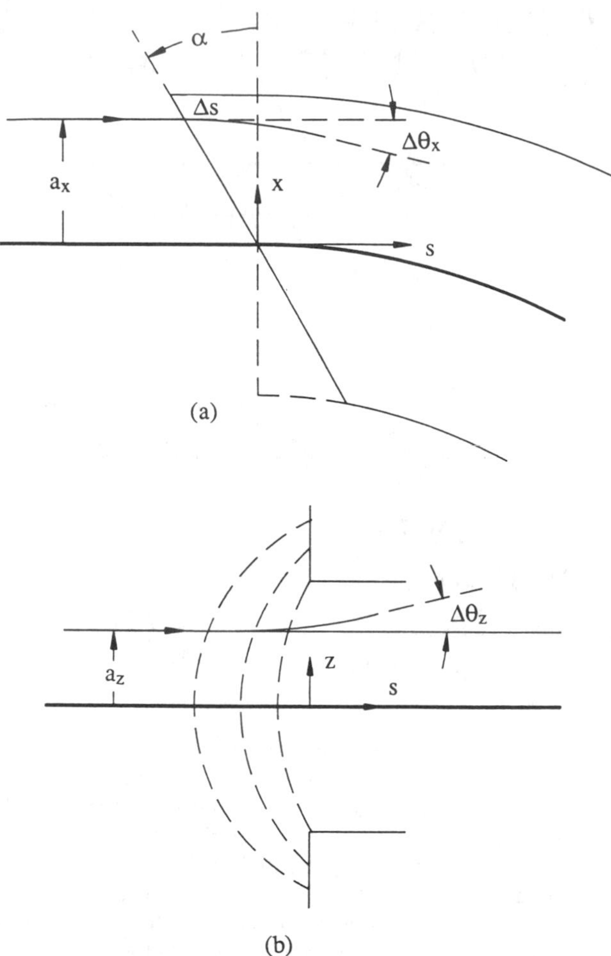

FIG. 3. Orbits experiencing edge focusing from non-normal magnet entry: (a) in bending plane, (b) in vertical plane.

and the focal power of this thin lens is $1/f_z = -(\tan\alpha)/r$. Here we have used $ds = dx\tan\alpha$, and $\int dB_z = -B_0$. These focal effects also occur at non-normal exits. In sector-focused cyclotrons the angles α are on average negative, providing the requisite net vertical focusing.

The first experimental tests of this concept at Berkeley used electron model cyclotrons (1949–1952) to simulate a 150-MeV proton cyclotron. "Thomas shims" to improve the focusing were first inserted into a cyclotron at Los Alamos, New Mexico (~1956). The first isochronous proton cyclotron (12 MeV) was completed at Delft, Netherlands, in 1958. At many laboratories in North America, Europe, the U.S.S.R., and Japan new ideas were explored and programs were started to improve existing machines and to build new ones. In the early 1950s the designs were based on sectors or ridges whose center lines were straight along radii, as proposed by Thomas [Fig. 4(a)]. In the mid-1950s studies at the Midwestern Universities Research Association (MURA) in Michigan and Wisconsin, aimed at developing new types of high-energy accelerators, resulted in the development of spiral-ridge field geometries. It was found that this concept was applicable to cyclotrons, and that by spiralling the ridges or sectors (with larger spiral angle at larger radius) the focusing strength could be greatly increased and higher energies attained [Fig. 4(b)]. [*See* SYNCHROTRON, where alternating-gradient focusing is described.]

The highest proton energies have been attained by the SIN (Zurich, Switzerland, 590 MeV) and TRIUMF (Vancouver, Canada, 520 MeV) machines (Fig. 5). These have been called *meson factories* because of their copious pion production by high-current (100–200-μA) proton beams. The Canadian machine accelerates H^- ions, requiring a low magnetic field (<0.6 T) to avoid stripping an electron by the motional ($\mathbf{v} \times \mathbf{B}$) electric field, and therefore a large orbit radius (~8 m), but beam extraction at any energy is easily accomplished by stripping in a thin foil to H^+ ions which come directly out of the magnet.

In addition to such large, special-purpose machines, many other third-generation cyclotrons of smaller size and energy but greater versatility are now operating, and others are under construction or design. Prototypes for some of these are the ORIC (Oak Ridge, Tennessee) and 88-in. (Berkeley) machines. Among the capabilities of this class are variation in extracted beam energy by changing magnetic field and oscillator frequency, injection and extraction of polarized beams, and acceleration of a wide variety of heavy ions. The further elaboration of unique types of cyclotrons for specialized purposes continues. These include compact, simply controlled cyclotrons for medical treatment, cyclotrons receiving input beams from other accelerators and from newer types of ion sources, multistage cyclotron complexes (sequences of two or three cyclotrons for staged energy gain), intensive development of cyclotrons employing superconducting magnets, and cyclotrons injecting into beam storage and cooling rings.

A recent tabulation of presently active cyclotrons shows about 140 machines in 26 countries. They have many uses in addition to their traditional roles as research tools in nuclear and particle physics. They are used for studies in chemistry, biology, medicine, solid-state physics, and other fields, and for the production of proton-rich radioisotopes for hundreds of applications. The best collected accounts of these diverse accomplishments, activities, and prospects are the 11 volumes mentioned in the last reference in the Bibliography.

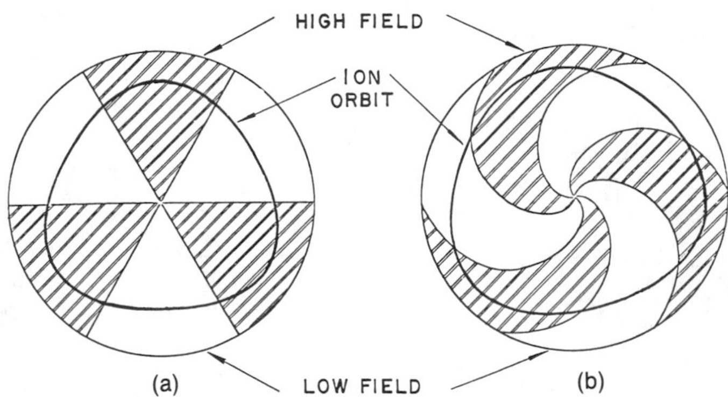

FIG. 4. Sectors of high and low fields; (a) radial, (b) spiral.

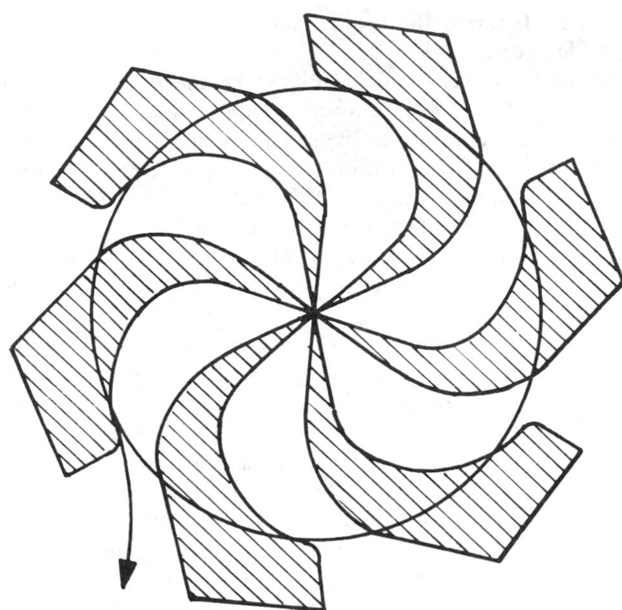

FIG. 5. Separated sectors of TRIUMF H^- cyclotron (Canada).

See also BETATRON; CYCLOTRON RESONANCE; SYNCHROTRON.

BIBLIOGRAPHY

A. A. Kolomensky and A. N. Lebedev, *Theory of Cyclic Accelerators*. Wiley, New York, 1966.

John J. Livingood, *Principles of Cyclic Particle Accelerators*. Van Nostrand, New York, 1961.

M. Stanley Livingston and John P. Blewett, *Particle Accelerators*, McGraw-Hill, New York, 1962.

Edwin M. McMillan, "Particle Accelerators," in *Experimental Nuclear Physics*, Vol. III (E. Segre, ed.). Wiley, New York, 1959.

Proceedings of the Eleventh International Conference on Cyclotrons and their Applications, Tokyo, Japan, 1986 (M. Sekiguchi, Y. Yano, and K. Hatanaka, eds.). Ionics Publishing Co., Tokyo, 1987. [This volume contains references (p. 916) to the Proceedings of the 10 preceding international conferences in this series starting in 1959. Like them, it includes comprehensive discussions of all aspects of cyclotron technology and applications in many fields. It also contains "data sheets" with much detailed information on the cyclotrons of the world.]

Cyclotron Resonance

C. C. Grimes

Cyclotron resonance denotes the resonant absorption of power from an alternating electromagnetic field by charged particles orbiting about a steady magnetic field. The resonance occurs when the frequency of the alternating field is equal to the orbital frequency of the particles. This phenomenon has been extensively employed to study the properties of electrons and holes in solids, and electrons and ions in plasmas. It forms the basic principle utilized in cyclotron particle accelerators.

When a charged particle moves in a magnetic field, it is acted on by the Lorentz force, which continuously rotates the component of the particle's velocity that is perpendicular to the magnetic field. Consequently, the particle moves along a helical orbit with the axis of the helix parallel to the field. The particle's orbital motion is then periodic, having an angular frequency $\omega_c = qB/m^*c$, which is called the cyclotron frequency. Here q/m^* is the charge-to-mass ratio for the particle, B is the magnetic flux density, and c is the velocity of light. If an electromagnetic field having angular frequency equal to ω_c acts on an orbiting particle in a direction that accelerates it, then the particle continually gains energy from the electromagnetic field. The transfer of energy from an electromagnetic field to charged particles is employed in cyclotrons and leads to the resonant absorption of power observed in cyclotron resonance. For a particle to experience a well-defined resonant interaction, it must spend enough time in an orbit to complete a substantial fraction of a revolution about the magnetic field. This condition is expressed by $\omega_c\tau \geq 1$ where τ is the characteristic time between scattering events. The scattering time is limited in solids by the density of impurities, crystal defects, and lattice vibrations, and in plasmas by the density of ions, neutral atoms, and molecules. Examples of typical scattering times, magnetic fields, and frequencies are given below.

Historically, the first observation of cyclotron resonance was in the ionosphere where it was observed that the absorption of radio waves had a peak at a frequency of 1500 kHz. This absorption peak was identified as due to cyclotron resonance of electrons moving in the earth's magnetic field, which has an average flux density of 0.5 G in the ionosphere. Resonance at such a low frequency is possible because the density of scatterers is very low and the electron scattering time is relatively long ($\tau \approx 10^{-5}$ s). Similarly, in gaseous plasmas it is relatively easy to achieve suitable conditions to study cyclotron resonance of electrons or even ions. Since the charge-to-mass ratio of ions is much smaller than for electrons, ion cyclotron resonance occurs at proportionately lower frequencies for the same magnetic field. Cyclotron resonance of electrons and ions in gaseous plasmas constitutes a useful diagnostic tool and provides a means by which energy can be coupled into a plasma to heat it.

Cyclotron resonance has been widely applied to the study of the energy band structure of solids. Its utility lies in the fact that the shapes of the constant-energy surfaces in momentum space (Fermi surfaces) for holes and electrons can be deduced from the variations of the effective masses m^* with magnetic field direction relative to the crystal axes. Typical scattering times in crystals of high purity and perfection cooled to a few degrees Kelvin are $\tau \approx 10^{-11}$ s. In semiconductors m^* is typically $0.1m$ where m is the electron mass. To satisfy the condition $\omega_c\tau \geq 1$, microwave frequencies of 10^{10}–10^{11} Hz are commonly employed in conjunction with magnetic fields of 1–10 kG. In ionic crystals, electrons interact strongly with optical phonons to form polarons, which have frequency-dependent effective masses. Polarons have been studied by cyclotron resonance at frequencies from the microwave region to the infrared region in conjunction with magnetic fields ranging from a few kilogauss to well above 100 kG. Infrared frequencies and high magnetic fields have also been utilized in many recent cyclotron resonance studies on two-dimensional systems such as inversion and accumulation layers at semiconductor–insulator interfaces and electrons in semiconductor heterojunctions and superlattices.

In semimetals and metals the skin effect restricts high-frequency electromagnetic wave penetration to a very thin layer at the surface of a specimen. To achieve cyclotron resonance, the static magnetic field is applied parallel to the specimen surface, so that orbiting electrons and holes near the surface will repeatedly pass through the skin layer and be accelerated by the electromagnetic field for a brief interval during each orbital period. Cyclotron resonance studies of metals and semimetals have yielded a wealth of information on Fermi surface shapes, the density of states at the Fermi surface, the contribution of electron–phonon interactions to the density of states, and the Fermi velocities, as well as some information on the anisotropy of electron scattering times.

See also FERMI SURFACE; RESONANCE PHENOMENA.

BIBLIOGRAPHY

C. Kittel, *Introduction to Solid State Physics*, 6th ed., p. 196. Wiley, New York, 1986.

B. Lax and J. G. Mavroides, in *Solid State Physics* (F. Seitz and D. Turnbull, eds.), Vol. 11, p. 261. Academic Press, New York, 1960. (I,A)

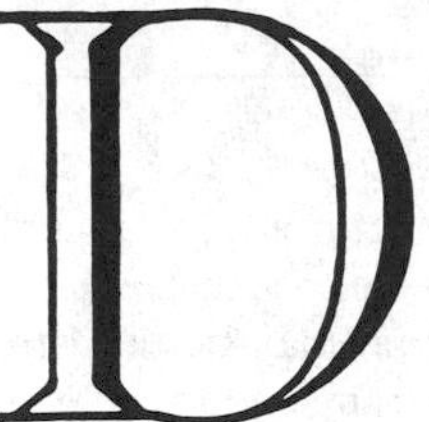

Deformation of Crystalline Materials

R. Bullough and J. R. Matthews

When a crystalline body is subjected to applied forces it will, in general, undergo a change of shape. If the forces are small this shape deformation can be entirely elastic, in which case the atomic structure also suffers a homogeneous deformation (referred to as a homogeneous lattice deformation) that is identical to the shape deformation; removal of the forces after such a deformation results in the body reverting to its original shape. Such a deformation may be termed a simple elastic deformation. On the other hand, if the applied forces are sufficiently large then the formation and movement of extended lattice defects, such as dislocations and twin lamellae, will occur, which ensures that a lattice-invariant or plastic deformation occurs and contributes to the total shape deformation of the body. Removal of the forces now results in the body only partly recovering its original shape. Such a deformation may be termed an elastic–plastic deformation. A complete survey of the physical processes that contribute to the onset and development of the plastic deformation and its detailed contribution to the total shape deformation in, generally, a polycrystalline body is not feasible in this article; it will therefore suffice merely to highlight some of the more important factors involved and refer the reader to the general references for appropriate amplification.

The onset of observable bulk plastic deformation is defined by the *yield strength* of the material, which is simply the applied stress required to produce some specified plastic deformation, usually a few tenths of a percent. This essentially macroscopic concept of yielding implies the *simultaneous sustained* movement of large numbers of dislocations. A single dislocation will glide under an applied (resolved) shear stress, referred to as the Peierls stress for a pure crystal or as the dislocation flow stress, that is much lower than the bulk yield stress and, in fact, if all the preexisting dislocations were simply driven out of the crystal by the applied stress, the total plastic strain would be quite small for annealed materials. It follows that some mechanism for the continuous generation of dislocations is essential and that the macroscopic yield stress is that stress required to generate the dislocations rather than simply to move individual dislocations. Several dislocation sources have been suggested, of which the Frank–Read source is probably the most important, since it clearly demonstrates how certain arrangements of dislocations can interact under the applied stress to generate, purely geometrically, a large number of dislocations and thereby help to define, in dislocation terms, the observed slip lines often associated with plastic deformation.

However, it is certain that the Frank–Read source is not the only athermal source of dislocations, and it is probably not even the most prevalent. Dislocations can be rather easily emitted from various internal stress concentrations, such as surface imperfections and grain boundary irregularities; the precise atomic mechanisms involved have not, as yet, been clearly identified. Since the free energy per atom spacing along a dislocation is about 7 eV, it is quite impossible for thermal fluctuations to generate dislocations spontaneously. Special configurations of mobile dislocations have been identified with observable macroscopic features of the plastic deformation. For example, when the mobile dislocations, which may be causing slip on two or more intersecting crystallographic planes, are confined to a broad band, such a band is referred to as a *deformation band*. If the dislocation motion is confined to a single set of parallel crystallographic planes (single slip) and moreover occurs in a band approximately normal to the slip planes we have a *kink band*. In polycrystalline materials the strain associated with the region of the body where plastic flow begins will often lead to effective softening adjacent to the region, with the result that the plastic zone propagates as a band through the entire crystal. Such deformation bands are often easily visible and are referred to as *Lüders bands*. The plastic deformation of a polycrystalline aggregate involves each grain plastically deforming subject to compatibility with its adjoining grains. Such successful accommodation requires, in general, at least five operative slip systems in each grain to ensure the requisite arbitrary shape changes. Large plastic deformation will often severely distort the grains out of their original shape and tend to rotate the operating slip planes toward the direction of deformation, producing preferred orientations of the grains. In addition to dislocation motion causing the plastic deformation, a further contribution to the lattice-invariant deformation, which is usually conceptually subsumed in the total plastic deformation, can arise by the formation of mechanical twin lamellae. The essential relation between the deformation arising from the dislocation movement and that arising by twinning is perhaps clarified by regarding the twin boundary as a special "surface" dislocation whereby it separates *volumes* of crystal that differ only by a lattice *rotation*, in contrast to the slip dislocation, which is a line that separate regions of different lattice *translation*.

As deformation proceeds, the dislocations multiply and interact with each other. This interaction combined with the overall increase in dislocation density ensures that, in general, the *flow stress* (stress required to maintain plastic deformation) will increase—the material will work-harden. In tensile deformation, however, the load-bearing capacity of the body can reach a maximum, since the effect of the reduction in the cross-sectional area is eventually greater than the increase of flow stress due to work-hardening. At greater loads the specimen will "neck" and finally fracture; the *ductility* of the material is defined by the strain or elongation to fracture.

Although dislocations cannot be formed by thermal fluctuations, temperature will assist their movement. At low temperatures, where the temperature is less than half the absolute melting point, the effect is limited to thermally activated reduction of the flow stress and the enabling of slip on new glide planes, cross slip. At higher temperatures, dislocations can move nonconservatively by a process known as climb involving the absorption and emission of point defects. The freeing of constraints on dislocation motion enables low-energy configuration and annihilation processes to take place, collectively known as recovery. Under these conditions the deformation will continue to increase with time, i.e., the material creeps. On first application of a load at elevated temperature the deformation rate is initially high but continuously decreases as a dislocation substructure is established. If the temperature is high enough, a dynamic balance is eventually established between the deforming dislocation network and its simultaneous recovery, producing a constant creep rate. During this type of creep a stable subgrain structure is frequently observed, with a scale that is inversely proportional to the applied load. Steady-state creep also occurs for small loads and very high temperatures by a purely diffusional process that may involve grain boundaries or dislocations as sources and sinks for the point defects. Under a tensile load, a third stage of creep is observed at large accumulated strains, where the creep rate accelerates. Cavities are generated within the material and failure quickly follows.

Crystalline materials are often used as structural components in the cores of nuclear reactors. In such an environment they will be bombarded by neutrons and other radiation which can strongly affect the response of such materials to any applied forces that are present. Thus, for example, fast neutrons can produce copious numbers of displacement defects (interstitials and vacancies) which at temperatures prevailing in such reactor cores are often very mobile. These displaced atoms can then precipitate into dislocation loops which can harden the material, but under sustained loads displacement damage can produce creep at temperatures well below those expected for conventional creep. The main source of irradiation creep is thought to be the preferred absorption of displaced atoms by dislocations of different orientation, the preference being induced by the applied stress. The resulting dislocation climb may directly produce the creep deformation or the accompanying evolving dislocation substructure may permit dislocation glide. Nuclear fuels undergoing fission damage often exhibit large irradiation creep effects. Materials with anisotropic crystal structures show shape changes without load, known as growth, which are brought about by similar processes.

Engineering structural materials are not usually pure metals or even simple alloys but multicomponent alloys with complex microstructures. A common means of hardening a metal is to add a solute with an atomic size different from the host crystal. These solutes act as obstacles to dislocations increasing the flow stress. Fine precipitates or finely layered eutectic structures also impede dislocation movement, but often at the expense of ductility. Phase transitions within an alloy system may be used to produce hard structures with internal strains, a classic example being the iron–carbon system that may have a wide range of yield strength and ductility according to heat treatment. More recently new classes of alloys have been developed, resistant to creep, for applications at high temperature such as gas turbines. The super-alloys rely on a duplex structure in the alloy with regions of an Ni–Al intermetallic compound alternating with the Ni–Fe matrix. The coherence of the interface between the relatively soft matrix and hard second phase gives great strength with acceptable ductility.

A complete microscopic theory of plastic deformation of crystalline solids does not yet exist. There is a reasonable understanding of the mechanics of single dislocations and very simple dislocation configurations, and a mathematically sophisticated continuum theory of plasticity has been developed; there is, however, no formal bridge between the two. To achieve such a connection we require a mathematical framework for the deformation behavior of a complex dislocation network. Some progress using the continuous distribution theory of dislocations with its associated non-Riemannian representation of the imperfect crystal has been made, but the complete bridge is certainly not yet built to permit a full microscopic understanding of deformation.

See also ANELASTICITY; CRYSTAL DEFECTS; RHEOLOGY.

BIBLIOGRAPHY

A. S. Argon, ed., *Physics of Strength and Plasticity*. MIT Press, Cambridge, Mass., 1969. (I,A)

B. A. Bilby, *Prog. Solid Mech.* **1,** 329 (1960). (A)

R. Bullough, "Microplasticity," in *Surface Effects in Crystal Plasticity*, pp. 5–14. NATO Adv. Study Inst., Sept. 1975, Hohegeiss, Germany. (I)

R. Bullough and J. A. Simmons, "On the Deformation of an Imperfect Solid," in *Physics of Strength and Plasticity*, p. 47. MIT Press, Cambridge, Mass., 1969. (A)

A. H. Cottrell, *Dislocations and Plastic Flow in Crystals*. Oxford (Clarendon Press), London and New York, 1953. (E,I)

R. de Wit, *Solid State Phys.* **10,** 249 (1960). (E)

F. C. Frank and W. T. Read, *Phys. Rev.* **79,** 772 (1950). (I)

J. Friedel, *Dislocations*. Addison-Wesley, Reading, Mass., 1964. (I)

R. Hill, *The Mathematical Theory of Plasticity*. Oxford (Clarendon Press), London and New York, 1950. (A)

J. P. Hirth and J. Lothe, *Theory of Dislocations*. McGraw-Hill, New York, 1968. (I)

U. F. Kocks, A. S. Argon, and M. F. Ashby, *Thermodynamics and Kinetics of Slip*. Pergamon, New York, 1975. (I,A)

K. Kondo, RAAG Memoirs of the Unifying Study of Basic Problems in Engineering and Physical Sciences by means of Geometry **1,** 458 (1955). (A)

E. Kroner, *Arch. Rat. Mech. Anal.* **4,** 273 (1960). (A)

N. H. Loretto, ed., Dislocations and properties of real materials, Proc. Conf. Royal Society, London, Dec. 1984, pub. Inst. of Metals, 1985. (E,I,A)

F. A. McClintock and A. S. Argon, eds., *Mechanical Behaviour of Materials*. Addison-Wesley, Reading Mass., 1965. (E)

W. T. Read, *Dislocations in Crystals*. McGraw-Hill, New York, 1953. (E)

S. Takeuchi and A. S. Argon, *J. Mat. Sci.* **11,** 1542 (1976). (I)

de Haas–van Alphen Effect

M. S. Dresselhaus and G. Dresselhaus

It was in the year 1930 that an oscillatory magnetic field dependence was first observed in the electrical resistance of bismuth by Shubnikov and de Haas and in the magnetization by de Haas and van Alphen. It was not long before Peierls showed how these effects could be understood in principle. Landau had implicitly predicted oscillatory behavior even before the experimental discovery on the basis of his theory of the quantization of the magnetic energy levels normal to the magnetic field direction. Nevertheless these magneto-oscillatory effects remained somewhat of a scientific curiosity for more than 20 years after their first observation. It was only in the 1950s with the observation of magnetic oscillations in many metals other than bismuth and the advent of improved theoretical understanding that it began to be realized that the de Haas–van Alphen effect provided a powerful tool for understanding the electronic structure of metals. Many of the most sensitive experimental techniques were developed by the pioneering work of Shoenberg, who dominated the experimental developments of this field for several decades. The theoretical breakthrough was Onsager's observation that the extremal Fermi-surface cross-sectional area S_0 is related to the de Haas–van Alphen period.

During the 1960s this effect was widely exploited as researchers joined the "band wagon," and with ever-improving experimental and theoretical techniques, an immense amount of detailed information was accumulated about the "Fermiology" of individual metals, including magnetic materials, intermetallic compounds, and alloys. During the 1970s the pace slackened, but new areas emerged in the late 1970s and the 1980s with applications to quantum well structures and to heavy fermion systems.

The de Haas–van Alphen effect refers to a very small magnetic field-dependent oscillatory term in the diamagnetic susceptibility χ. This oscillatory term can be observed under suitable conditions at high magnetic fields B. The de Haas–van Alphen phenomenon is important because the period P of the oscillatory effect is proportional to $1/B$, and P can be very accurately related to extremal cross sections of Fermi surfaces. By varying the direction of the externally applied magnetic field, the Fermi surface of each carrier pocket can be mapped out independently. The de Haas–van Alphen effect can also be used to determine other electronic parameters, such as the cyclotron effective mass and the effective g factor for electrons and holes.

Application of the Bohr–Sommerfield quantization condition to the k-space orbit of electrons perpendicular to the magnetic field leads to the relation for the extremal cross-sectional area

$$S_0 = 2\pi e B_n (n+\gamma)/c\hbar \tag{1}$$

from which the de Haas–van Alphen period is defined as

$$P = \left| \frac{1}{B_n} - \frac{1}{B_{n-1}} \right| = \frac{2\pi e}{c\hbar S_0}, \tag{2}$$

where B_n is a resonant magnetic field corresponding to quantum number n. The observed periods in metals normally are between $\sim 10^{-5}$ and 10^{-9} G^{-1}, corresponding to Fermi surface cross sections ranging from $\sim 10^{13}$ to 10^{17} cm^{-2}. Magneto-oscillatory effects are often reported in terms of the de Haas–van Alphen frequency $\nu = 1/P$.

The significance of this resonant magnetic field can be understood in terms of the energy levels of an electron in a magnetic field. In the effective mass approximation, the magnetic energy levels are given by

$$E_n(k_H) = \frac{\hbar^2 k_H{}^2}{2 m_H{}^2} + \frac{\hbar e B}{m_c{}^* c} \tag{3}$$

in which k_H is the wave vector along the magnetic field **B**, while $m_H{}^*$ is the effective mass component projected along **B**, and $m_c{}^*$ is the cyclotron effective mass corresponding to the motion perpendicular to **B**. For a three-dimensional electron system in a magnetic field, each magnetic subband n has an energy minimum at $k_H = 0$ where the density of the states becomes infinite. According to Eq. (3) the energy of each magnetic subband extremum increases linearly with B. Because of the singularity in the density of the states in a magnetic field, a series of resonant responses is observed, each resonance corresponding to the passage of a magnetic energy subband extremum through the Fermi level:

$$E_F = \frac{\hbar e B_n}{m_c{}^* c} (n + \tfrac{1}{2}) \tag{4}$$

so that the B_n specifies the magnetic field where subband n becomes unoccupied (occupied) with increasing (decreasing) B. From this point of view, for every value of B_n, the density of states at the Fermi level becomes infinite, thereby introducing an oscillatory magnetic-field-dependent term in the density of states. Thus, any observable depending on the density of states also exhibits an oscillatory magnetic field dependence with a period $1/B$. We refer to such oscillations in the electrical conductivity as the Shubnikov–de Haas effect, in honor of its discovery in 1930 by Shubnikov and de Haas. In the intervening time, de Haas–van Alphen oscillations have been found in a large number of observables, including sample temperature (called magnetothermal oscillations), the Hall coefficient, the sound attenuation coefficient, the velocity of sound, the thermoelectric power, and the optical reflectivity.

In order that the de Haas–van Alphen effect (and the magneto-oscillatory phenomena related to this effect) be observable, it is necessary for the electronic motion to be dominated by the magnetic field. This is necessary to exploit the singularity in the magnetic-field-dependent density of states at each magnetic subband extremum. In contrast, there are no singularities in the zero-field density of states for a 3D system; in this case, only a single nonresonant threshold is present at the band extremum. For lower-dimensional systems the functional form of the density of states changes, giving increasing emphasis to the subband extrema as the dimensionality decreases to 2 (quantum wells), to 1 (quantum wires) and finally to 0 (quantum dots).

The magnetic field dominates the electronic motion when B is large enough for a carrier to complete an electron orbit before scattering. This condition is expressed mathematically as

$$\omega_c \tau \gg 1, \tag{5}$$

where the cyclotron frequency ω_c is related to B according to

$$\omega_c = \frac{eB}{m_c^* c} \tag{6}$$

and τ is the mean time between collisions. Thus to observe the de Haas–van Alphen effect, it is necessary to use high magnetic fields to maximize ω_c and to carry out the measurement at low temperatures to minimize the collision probability and thereby maximize τ. From the Lorentz force equation,

$$\dot{\mathbf{p}} = \frac{e}{c} \mathbf{v} \times \mathbf{B}, \tag{7}$$

it follows that the electron orbit in real space has the same shape as in reciprocal space but the real-space orbit is changed by a scale factor (c/eH) and rotated by $\pi/2$.

Since the condition $\omega_c \tau \gg 1$ can be satisfied for many materials, the de Haas–van Alphen and related effects can be used to determine Fermi-surface parameters for a large variety of materials. With ever-improving materials, synthesis techniques, and the improving sensitivity of the magneto-oscillatory measurements, these techniques have been applied to increasing numbers of materials, including new classes of materials. Of particular interest to developments in the 1980s has been the application of magneto-oscillatory techniques to study the Fermi surfaces of heavy fermions and of quantum wells and superlattices in semiconducting and metallic systems.

The detailed interpretation of the de Haas–van Alphen data to yield electronic parameters is made on the basis of the Lifshitz–Kosevich theory, which gives the oscillatory component of the electrodynamic potential as

$$\Omega_{\text{osc}} = 2Vk_B T \left(\frac{eB}{\hbar c}\right)^{3/2} \left(\frac{\partial^2 S}{\partial k_N^2}\right)_{km}^{-1/2} \sum_{j=1}^{\infty} \frac{\exp(-2\pi^2 jk_B T_D/\beta^* B)}{j^{3/2} \sinh(2\pi^2 jk_B T/\beta^* B)} \times \cos\left[\frac{j\hbar e S_m}{eB} - 2\pi j\gamma \pm \frac{\pi}{4}\right] \cos\left[\frac{j\pi g m_c^*}{2m_0}\right], \tag{8}$$

in which $k_H = k_m$ defines the location of the extremal Fermi-surface cross-sectional area S_m, β^* is the effective double Bohr magneton $\beta^* = e\hbar/m_c^* c$, and m_c^* is the cyclotron effective mass $m_c^* = \hbar^2(\partial S/2\pi\partial E)_{E_F}$ on the constant energy surface at E_F and T_D is the Dingle temperature related to the width $\hbar/\tau$ of the Landau levels given by $k_B T_D = \hbar/\pi\tau$. Also, g is the spin-splitting factor, m_s^* is the spin effective mass, and m_0 is the free-electron mass. The upper or lower signs in the phase correspond to a maximum or minimum, respectively, in the external cross-sectional area of the Fermi surface. The parameter γ enters the theory from the quantization condition given in Eq. (1). The magnetic susceptibility is obtained from Eq. (8) through differentiation $\chi = -(\partial^2 \Omega_{\text{osc}}/\partial B^2)_\gamma$. The oscillations in χ arise from the argument of the magnetic-field-dependent cosine function in Eq. (8). Furthermore, the argument of the sinh function gives rise to an exponential temperature dependence of χ, the exponential terms depending also on the parameters m_c^* and T_D. Careful temperature-dependent measurements of χ thus can be analyzed to yield explicit values for m_c^* and T_D. Values of m_c^* can also be determined by other techniques, the most important being cyclotron resonance experiments. When comparisons can be made, values of m_c^* determined by the de Haas–van Alphen effect are in good agreement with m_c^* values obtained by analysis of cyclotron resonance data. However, the effective mass determinations using the de Haas–van Alphen effect are not nearly as accurate as the accurate measurements of Fermi-surface cross-sectional areas.

Introduction of electron spin into the electronic problem results in a spin splitting of the magnetic subbands. In this case, whenever a magnetic subband extremum associated with either spin orientation (parallel or antiparallel to the applied field), crosses the Fermi level, a resonant response will be achieved. Measurement of the de Haas–van Alphen amplitudes as a function of B also can be used to obtain the effective g factor for conduction electrons and holes at the Fermi energy. From analysis of de Haas–van Alphen line shapes, information can be obtained on relaxation times and relaxation processes in the materials under investigation.

"Many-body effects" in an electron gas have been studied on the alkali metals yielding Landau Fermi-liquid parameters for the spin correlation or interaction function. This determination is possible because in Fermi-liquid theory, the effective g factor as measured by a de Haas–van Alphen experiment differs from the corresponding quantity as determined by a microwave electron-spin resonance measurement.

Since about 1980, particular attention has been given to magneto-oscillatory effects in the two-dimensional electron gas, largely related to the discovery of the quantum Hall effect, which is a 2D manifestation of the quantum oscillatory behavior when the magnetic field is normal to the 2D electron gas. Since the zero-field density of states in two dimensions is a constant independent of energy, a δ-function singularity is found for the density of states in a magnetic field for each magnetic subband for the ideal 2D electron gas. Magneto-oscillatory behavior is usually studied in the Hall resistivity ρ_{xy} (appearing as steps in ρ_{xy}), the plateau following each step being associated with a zero in both the transverse magnetoresistance ρ_{xx}, and the transverse magnetoconductivity σ_{xx}; these relations are valid when $\rho_{xy} \gg \rho_{xx}$. Physical realizations of the 2D electron gas are the bound states in semiconductor quantum wells and superlattices and the inversion layer in a MOSFET. In these systems, magneto-oscillatory measurements are used to determine the effective mass of the carriers of the 2D electron gas as the geometry and composition of the quantum wells are varied. Of special interest has been the use of the Shubnikov–de Haas effect to investigate the semiconductor–semimetal transition in InAs/GaSb quantum wells (type II superlattice) as a function of the width of the quantum well.

The de Haas–van Alphen and related magneto-oscillatory effects today represent the most accurate and widely applicable means for the determination of Fermi-surface properties of metals, semimetals, and semiconductors described by a 2D electron gas. Though mature, the field remains active.

See also CYCLOTRON RESONANCE; DIAMAGNETISM AND SUPERCONDUCTIVITY; FERMI SURFACE; HEAVY FERMION MATERIALS.

BIBLIOGRAPHY

N. W. Ashcroft and N. D. Mermin, *Solid State Physics*, p. 264. Holt, Rinehart and Winston, New York, 1976. (E)

W. J. de Haas and P. M. van Alphen, *Leiden Commun.* **208a, 212a** (1930) (of historical interest).

C. Kittel, *Introduction to Solid State Physics*, 6th ed., p. 241. Wiley, New York, 1986. (E)

I. M. Lifshitz and A. M. Kosevich, *Zh. Eksp. Fiz.* **29,** 730 (1955) [English translation: *Sov. Phys. JETP* **2,** 636 (1956)] (a classic paper in the field). (A)

L. Onsager, *Philos. Mag.* **43,** 1006 (1952) (of historical interest).

Proceedings of the 6th International Conference on "Crystal-Field Effects and Heavy Fermion Physics," *J. Magn. Mag. Mater.* **76/77,** 1–42 (1988) (reports on a recent application of de Haas–van Alphen effect in heavy-fermion systems).

D. Shoenberg, *Magnetic Oscillations in Metals.* Cambridge University Press, Cambridge, 1984 (complete book devoted to de Haas–van Alphen effect by one of the pioneers).

D. Shoenberg, *Philos. Trans. R. Soc. London A* **245,** 210 (1952) (a classic paper in the field).

W. Shubnikov and W. J. de Haas, *Leiden Commun.* **207,** 210 (1930) (of historical interest).

Demineralization

M. H. Lietzke

Demineralization is the process of producing water free of dissolved minerals. This may be accomplished either by removing the minerals from the water or by separating the water from the dissolved minerals. The most important methods for demineralizing water on a large scale are distillation and electrodialysis. On a smaller scale ion exchange, hyperfiltration, or the freezing process may be used. Applications of demineralization include the production of potable water from brackish or other mineralized water, such as seawater, and the removal of dissolved salts from water to be used in some industrial processes or in steam boilers, where scaling must be avoided.

The most promising method for producing fresh water from mineralized water on a large scale is distillation. There are several large distillation plants in operation in various parts of the world for recovering fresh water from seawater. Some of these plants utilize heat from electrical power plants and produce over three million gallons ($\sim 10^7$ liters) of fresh water per day. Projected distillation plants with capacities of around 100 million gallons (3.8×10^8 liters) per day would have to be coupled with nuclear reactors to provide the power. On a much smaller scale, solar energy may also be used to provide the heat required for distillation of fresh water from mineralized water. For example, a solar still producing 4000 gallons per day of fresh water from seawater has been constructed on the Greek island of Symi.

Demineralization by electrodialysis is accomplished by passing the stream of water to be demineralized through a cell containing a large number of alternating anion- and cation-selective membranes separated from each other by electrically resistive spacers. The spacers have a central area cut out to permit solution and current to flow between the membranes. At each end of the stack are located electrodes connected to a source of direct current. Two inlet streams and two outlet streams are manifolded to the alternate compartments formed by each pair of membranes and separating gaskets. As flow proceeds through the system the cations in all the cells migrate toward the cathode and the anions migrate toward the anode. In a diluting cell, the cations pass out of the process stream through the cation membrane into the adjacent concentrating cell, while the anions migrate in the opposite direction through the anion-selective membrane into the other adjacent concentrating cell. Because of the high ion selectivity of the membranes, ions are restrained from migrating from the concentrating cells back into the diluting cells. The end result of these ion transfers is that the process stream in the diluting cells is demineralized and the transfer stream in the concentrating cells becomes enriched in electrolyte. Charge balance in the transfer streams is maintained since cations migrate in from one adjacent diluting cell, while anions migrate in from the other. Charge balance at the ends of the stack is maintained by the oxidation–reduction reactions occurring at the electrodes. A number of large-scale electrodialysis plants have been constructed and operated in various parts of the world. Several of these produce over one million gallons (3.8×10^6 liters) of demineralized water per day. Electrodialysis is now considered to be the most economical process available for demineralizing mildly brackish water.

Another promising method for demineralizing brackish water is the process of hyperfiltration or reverse osmosis. The principle of this method can be explained as follows: Suppose we have a membrane through which water can pass but through which salt cannot pass. If pure water is put on one side of the membrane and water containing dissolved minerals on the other side, then water will pass through the membrane from the region of pure water, where its effective concentration is higher, into the water containing the dissolved minerals, where its effective concentration is lower. However, if sufficient pressure is applied to the mineralized water, the flow of water through the membrane may be reversed and water forced out of the salt solution. One of the major problems in designing hyperfiltration plants has been in the development of suitable membranes. Cellulose acetate, which has excellent salt rejection properties, has been most widely used for this purpose. A number of experimental units have been built which produce several thousand gallons ($\sim 10^4$ liters) of fresh water per day from brackish water.

In the ion-exchange process for demineralization, metallic cations are removed by a hydrogen cation exchanger, such as Dowex 50, which is a polystyrene divinyl benzene derivative containing sulfonic acid groups. As the metallic ions are adsorbed an equivalent amount of hydrogen ions is released. Anions are removed by an anion exchanger, such as Dowex 1, which is a polystyrene divinyl benzene derivative containing quaternary amine groups. In this step an equivalent amount of hydroxyl ions is released. The hydrogen and hydroxyl ions set free in the exchange processes then combine to form water. In practice the water to be demineralized may pass through the cation and anion exchangers sequentially, or may pass through a mixed bed of cation and anion exchangers. Carbon dioxide formed from bicarbonates in the exchange process may be removed mechanically by an aerator, degasifier, or vacuum deaerator, or chemically by a strongly basic anion exchanger. The ion-exchange process for demineralization is usually carried out on a much smaller scale than either electrodialysis or distillation.

The principle of the freezing process is that ice frozen from water containing dissolved salts is essentially free of mineral matter. The ice is skimmed off and melted to produce fresh water. A large-scale freezing plant, constructed and operated in Israel in the 1960s and 1970s, has since been shut down.

BIBLIOGRAPHY

K. S. Spiegler, *Principles of Desalination*. Academic Press, New York, 1966. (E)

F. Helfferich, *Ion Exchange*. McGraw-Hill, New York, 1962. (I)

S. Hwang and K. Kammermeyer, *Membranes in Separations*. Wiley, New York, 1975. (E)

Diamagnetism and Superconductivity

Edward A. Boudreaux

The phenomenon of diamagnetism is a consequence of induction, whereby a substance experiences polarization when placed in a magnetic field. In this respect it is similar to the analogous phenomenon of paramagnetism, except that the induced magnetic polarization in diamagnetism is negative and is some hundred to a thousand times smaller than in paramagnetism. Furthermore, diamagnetic substances contain no permanent magnetic dipoles, whereas normal paramagnetic materials do. Also, there is neither a normal temperature dependence nor a magnetic field dependence associated with diamagnetism.

In discussing inductive magnetic effects it is customary to invoke the definition $K = I/H$, where K is the magnetic susceptibility per unit volume of substance, I is the intensity of magnetization of the substance, and H is the strength of the magnetic field. For diamagnetism I is negative and smaller than H, and thus the susceptibility is negative and small (i.e., $\sim 10^{-6}$ emu). The measurement of diamagnetic susceptibilities is attained through the observation of an apparent mass change which the substances exhibit when placed in a sufficiently strong magnetic field. In a diamagnetic material this is observed as a net mass loss resulting from the magnetic force tending to repel the substance from the field.

According to the atomic theory of matter, diamagnetism is a consequence of the orbital motion of electrons, and hence is a universal property of all matter. In fact the majority of substances are observed to be diamagnetic, while, relatively speaking, the net nondiamagnetic behavior of matter is rather rare in nature. Common substances such as wood, plastics, water, sand (SiO_2), nitrogen, CO_2, NaCl, $CaCO_3$, Al_2O_3, etc., and the millions of organic compounds and most biological substances are all diamagnetic. In graphite and certain metals an anomalous large diamagnetism is observed, which is also found to be somewhat temperature and magnetic field dependent. This is called "Landau diamagnetism," after the famous physicist, and results from the large delocalization of the orbital motion of electrons in such systems.

Much of the fundamental theory of diamagnetism and its applications were established by noted scientists such as Langevin, Oxley, Stoner, Pascal, Van Vleck, and others, within the first third of the 20th century. Langevin, for example, is credited with formulating the theory of atomic diamagnetism, while Pascal derived a systematic empirical means for calculating diamagnetic susceptibilities of organic compounds. Van Vleck applied quantum mechanics in extending the theory of diamagnetism from atoms to molecules. Thus he was able to show that in diamagnetic molecules a "temperature-independent paramagnetism" (TIP) must also be inherently present.

The sum total of these effects are outlined in the following flow diagram:

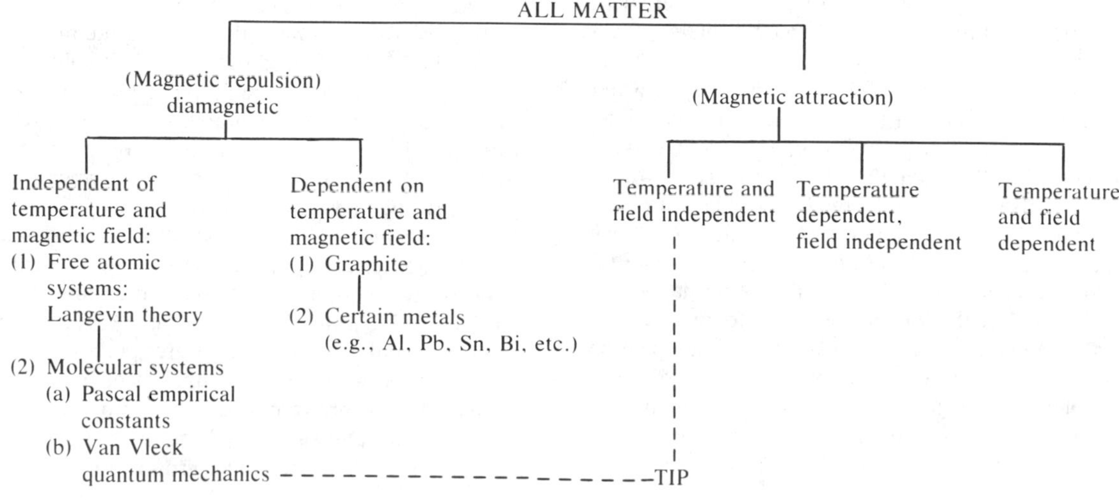

Although the application of static diamagnetic susceptibilities is not currently utilized to any significant extent, diamagnetism still remains an important diagnostic tool to the researcher.

Addendum. According to the general theory of superconductivity, a superconductor generates inherently large diamagnetic supercurrents which persist up to the maximum superconducting critical temperature, T_c. At this point, the material enters into a transition state whereby the magnetic field is completely expelled from it, and the susceptibility goes to zero. That is commonly referred to as the "Meissner effect."

If the superconducting substance were to become perfectly diamagnetic, its susceptibility would be $\chi_m = -1$ emu (or $-\frac{1}{4}\pi$, depending upon the units assigned to the magnetization M). In the case of a high-temperature superconductor such as $YBa_2Cu_3O_7$ ($T_c \approx 90\pi$), the diamagnetic susceptibility ranges from about -1.2×10^{-2} emu/mole for $T \leq 75$ K, up to $\chi_m = 0$ at T_c [1]. Hence, this diamagnetism does have a limited temperature dependence.

Thus far, to this author's knowledge, there has not been reported any quantitative theoretical treatment of these huge diamagnetic susceptibilities. However, recent results obtained from theoretical studies by this author on electronic structure and bonding in extended cluster models of $YBa_2Cu_3O_7$ reveal the presence of antiferromagnetically coupled single electron spins associated with the Cu and O atoms, which are delocalized over large domains of the crystalline lattice [2]. This could be the source of the large diamagnetic currents required to produce the observed superdiamagnetism.

See also PARAMAGNETISM.

REFERENCES

1. S. Daviron *et al.* in *Chemistry of High Temperature Superconductors* (D.L. Nelson, M.S. Whitingham, and T.F. George, eds.), p. 76. ACS Symp. Series 351. American Chemical Society, Washington, D.C., 1987.
2. E. A. Boudreaux and E. Baxter, *Int. J. Quantum Chem.* **23** (1989).

BIBLIOGRAPHY

L. N. Mulay and E. A. Boudreaux, *Theory and Applications of Molecular Diamagnetism.* Wiley-Interscience, New York, 1976.

L. N. Mulay, "Techniques of Magnetic Susceptibility." in *Physical Methods of Chemistry* (A. Weissberger and B. W. Rossiter, eds.). Wiley-Interscience, New York, 1972, Part I, Vol. IV.

L. N. Mulay, *Magnetic Susceptibility* (reprint monograph). Wiley, New York, 1966.

Ya. G. Dorfman, *Diamagnetism and the Chemical Bond.* Edward Arnold, London, 1965.

P. W. Selwood, *Magnetochemistry*, 2nd ed. Interscience, New York, 1956.

J. H. Van Vleck, *The Theory of Electric and Magnetic Susceptibilities.* Oxford, London, 1932.

E. C. Stoner, *Magnetism and Atomic Structure.* Methuen, London, 1926.

Dielectric Properties

Nicholas Bottka

The term *dielectric* applies to the material properties governing the interaction between matter and an electromagnetic field. Induced or permanent electric polarization or magnetization of matter as a function of a static or an alternating electric, magnetic, or electromagnetic field constitutes the *dielectric properties* of the material.

The macroscopic Maxwell equations describe the response of a system to an external electromagnetic field, or probe, described in space and time by the complex field vectors $\mathbf{E}(\mathbf{r}, t)$ and $\mathbf{H}(\mathbf{r}, t)$. The electric displacement $\mathbf{D}(\mathbf{r}, t)$ and the magnetic flux density $\mathbf{B}(\mathbf{r}, t)$ are material-related parameters and, in general, are not proportional to $\mathbf{E}(t)$ and $\mathbf{H}(t)$ at arbitrary times, i.e., the response of the system depends on the past history of the source $\mathbf{E}(t)$. This causality is expressed by the Fourier transform of the Maxwell relation (in cgs electrostatic units)

$$\mathbf{D}(\mathbf{k}, \omega) = \mathbf{E}(\mathbf{k}, \omega) + 4\pi i\mathbf{J}(\mathbf{k}, \omega)/\omega \qquad (1)$$

where $\mathbf{k}$, the wave vector (or propagation direction) of the electromagnetic probe, will have complex components if a wave of angular frequency ω is absorbed by the medium. $\mathbf{J}$ is the current density; it describes all the material properties.

The electromagnetic field can be a time-varying *longitudinal* or *transverse* field; its interaction with a system must conserve both energy and momentum. The electromagnetic wave in optical reflectivity is an example of a transverse probe, coupling directly to the transverse current-density fluctuation of the electrons. Slow neutrons and fast electrons are examples of longitudinal probes, in that they couple directly to elementary excitations in the medium, e.g., density fluctuations in a solid such as phonons and plasmons.

Ideally the external probe is weakly coupled to the material so that the system response can be represented in terms of the properties of the excitations in the absence of the probe. One then speaks of a *linear* response of the system (as opposed to nonlinear interactions, such as discussed under NONLINEAR WAVE PROPAGATION), and Eq. (1) is redefined as

$$\mathbf{D}(\mathbf{k}, \omega) = \mathbf{E}(\mathbf{k}, \omega) + 4\pi i\, \boldsymbol{\sigma}(\mathbf{k}, \omega)\cdot\mathbf{E}(\mathbf{k}, \omega)/\omega$$
$$= \boldsymbol{\epsilon}(\mathbf{k}, \omega)\cdot\mathbf{E}(\mathbf{k}, \omega), \qquad (2)$$

where $\boldsymbol{\epsilon}$ and $\boldsymbol{\sigma}$ are the complex *dielectric function* and *conductivity* of the material, respectively. Analogously one can define a complex magnetic *permeability* $\boldsymbol{\mu}$ relating $\mathbf{B}$ to $\mathbf{H}$. In general $\boldsymbol{\epsilon}$ is a tensor and represents the response of the system to a perturbing field.

For a transverse probe $\mathbf{E} = \mathbf{E}_0 \exp(i\mathbf{k}\cdot\mathbf{r} - i\omega t)$ in an isotropic nonmagnetic medium, the Maxwell relations for the linear case in Eq. (2) yield the dispersion relation

$$\mathbf{k}\cdot\mathbf{k} = \omega^2\boldsymbol{\epsilon}(\mathbf{k}, \omega)/c^2. \qquad (3)$$

When $\boldsymbol{\epsilon}$ is complex, Eq. (3) can be satisfied only by complex $\mathbf{k}$; all solutions correspond to damped waves in a passive medium. The converse is not necessarily true; i.e., even if $\boldsymbol{\epsilon}$ is real there is a solution with complex $\mathbf{k}$ corresponding to an evanescent wave (this wave is damped, but there is no dissipation of energy since ϵ is real). If ϵ and $\mathbf{k}$ are real, Eq. (3) reduces to the *Maxwell relation* $n^2=\epsilon$, relating the refractive index n to the dielectric constant ϵ of the nonabsorbing medium. For optical frequencies, ϵ is usually assumed to be independent of $\mathbf{k}$. The $\mathbf{k}$ dependence of ϵ gives rise to such phenomena as the *anomalous skin effect* in metals and the interaction of light with excitons.

Historically, one of the greatest scientific achievements of our times has been the understanding of the correlation between the experimentally observed *macroscopic* quantities ϵ, μ, and σ and the underlying *microscopic* nature of matter. A class of interactions known as *dipole interactions* contributed substantially to this understanding. For an isotropic atomic system consisting of N elementary dipoles, the macroscopic polarization $\mathbf{P}$ (electric-dipole moment per unit volume) is related to the microscopic average dipole moment $\langle\mathbf{p}\rangle=\langle q\mathbf{r}\rangle$ (in general, represented as a sum of bound electronic, atomic, and permanent dipole moments) by

$$\mathbf{P}=N\langle\mathbf{p}\rangle=N\alpha\mathbf{E}'=(\epsilon-1)\mathbf{E}/4\pi=\chi\mathbf{E}/4\pi, \tag{4}$$

where q is the charge and $\mathbf{r}$ is the displacement from equilibrium of the elementary particle; α is the *polarizability* and χ is the *susceptibility* of the system. $\mathbf{E}'$ is the *local electric-field* strength; in addition to the external field it includes the contribution from induced and permanent dipoles. An expression analogous to Eq. (4) exists for magnetic materials relating the macroscopic magnetization $\mathbf{M}$ to the magnetic susceptibility χ_M and the magnetic field.

For a system of noninteracting oscillating dipoles, the classical model for polarizability yields

$$\epsilon-1=4\pi q^2\sum_s(\omega_s{}^2-\omega^2+i\gamma_s\omega)^{-1}N_s/M_s, \tag{5}$$

where N_s is the number of particles per unit volume of mass M_s in the system, ω_s is the natural resonant frequency, and γ_s is the damping constant of the dipole oscillator. The sum over s indicates the possibility of various dipoles. Figure 1 shows the behavior of the real and imaginary parts of $\epsilon-1$ as a function of frequency in the vicinity of ω_0 for a system of N identical noninteracting oscillators. Strong absorption is indicated by the peak in the imaginary part of ϵ. The frequency dependence of the real part of the dielectric function describes the classical dispersion characteristics of the dielectric medium. Far from ω_0 the real part of ϵ rises with increased frequency; this behavior is called *normal dispersion* in contrast to the *anomalous dispersion* in the half-width region of the spectral line where the real part of ϵ falls with increased frequency.

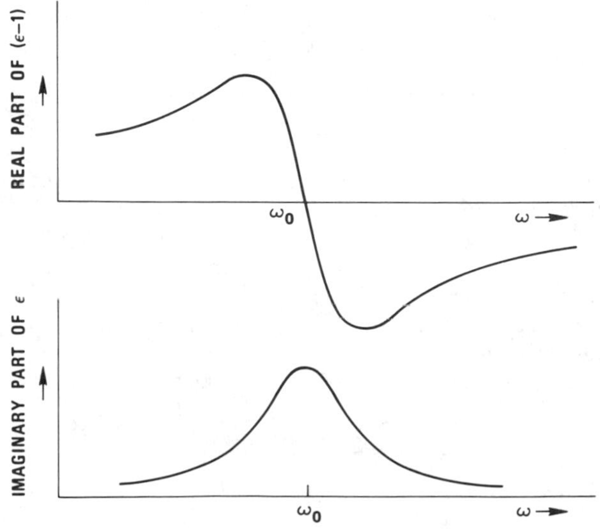

FIG. 1. The real and the imaginary parts of the dielectric function ϵ as a function of frequency near the resonance frequency ω_0.

In the case of electronic polarizability where $\omega_0\simeq10^{15}\ \text{sec}^{-1}$, the strong absorption takes place in the visible region of the spectrum. In the case of ionic polarizability [q and M in Eq. (5) refer to an ion] where $\omega_0\simeq10^{13}\ \text{sec}^{-1}$, resonant absorption takes place in the infrared region.

Although the dispersion formula of classical physics correctly predicts the general appearance of a spectral line, neither the intensity nor the resonance frequency can be derived from such classical considerations. Moreover, the classical picture leads to a number of unexplained physical catastrophes. The more realistic quantum-mechanical calculation leads to the dispersion formula

$$\epsilon-1=(8\pi\hbar^{-1})\sum_{f\neq j}\omega_{jf}|\langle\mu_{jf}\rangle|^2(\omega_{jf}^2-\omega^2+i\gamma\omega)^{-1}(N_j-N_f), \tag{6}$$

where the resonance frequency now corresponds to a transition between two stationary quantum-mechanical energy states E_j and E_f, and $\langle\mu_{jf}\rangle$ is the matrix element of the transition coupling the electromagnetic field to the initial and final quantum-mechanical states j and f; N_j and N_f represent the number of atoms in the lower- and the higher-energy state, respectively.

In addition to the bound electronic and atomic polarizability described above, there are cases where free conduction electrons also contribute to absorption, such as in metals and semiconductors. This occurs at low frequencies, anywhere from microwave to infrared. In such a regime the dielectric function takes the form

$$\epsilon_{\text{free}}=1-4\pi q^2(N/M)/(\omega^2-i\gamma\omega) \tag{7}$$

In conductors and semiconductors, the damping constant γ_s is inversely proportional to the mobility of the free carrier type (which for intrinsic semiconductors can be both electrons and holes). At these frequencies the optical absorption coefficient is proportional to the imaginary part of ϵ, and one speaks of a "λ^2" dependence of the *free-carrier absorption*. For regions of low absorption, the real part of ϵ in Eq. (7) goes through zero at $\omega_p=(4\pi Nq^2/M)^{1/2}$. This is the frequency where the free-electron gas undergoes collective density fluctuations known as *plasma oscillations*. The quanta of energy associated with these oscillations is called the *plasmon*. For most insulators and semiconductors, and many metals, the free-electron model is approximately valid in the frequency range $\omega_v<\omega<\omega_c$, where ω_v is the maximum interband transition of the valence electrons and ω_c is the threshold for interband transitions of the core electrons.

The many excitations discussed above can be put into perspective by considering the absorption spectrum of a hypothetical solid shown in Fig. 2. The strength and the fre-

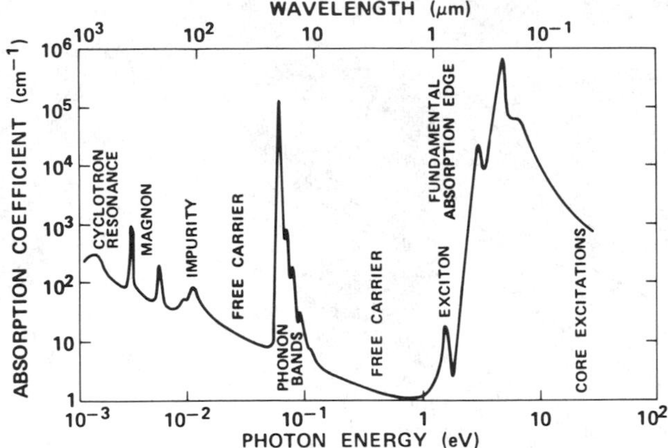

FIG. 2. The absorption spectrum of a hypothetical solid showing the frequency regime where the various excitations give rise to structure in the dielectric function.

quency (or photon energy) of the excitation give an insight into the nature of matter and its dielectric properties.

The dielectric properties of a material, as defined by the real and imaginary parts of the dielectric function $\epsilon(\omega)$ over the frequency range $(0, \infty)$ and as exemplified by some of the absorption processes shown in Fig. 2, constitute a unique "fingerprint" of that material. However, despite their great diversity, the basic dielectric properties of solids are rigorously limited by nature. These limitations take the form of *sum rules* which arise from the requirement of causality mentioned before and reflect the physical laws governing the dynamics of interaction between radiation and matter. These sum rules may be viewed as the ω-space equivalent of the dynamic laws of motion in time-space. They give insight into the mathematical structure of the dielectric function, provide a means of relating different physical properties of the solid without model fits to spectra, and most importantly, serve as self-consistency tests. The best-known sum rule in optics (also known as f sum rule) may be written in a variety of forms useful in optical analysis:

$$\int_0^\infty \omega\epsilon_2(\omega)d\omega = \frac{\pi}{2}\omega_p^2 \tag{8}$$

and

$$\int_0^\infty \omega \, \mathrm{Im}[\epsilon^{-1}(\omega)]d\omega = -\frac{\pi}{2}\omega_p^2. \tag{9}$$

These sum rules involve an infinite frequency interval $(0, \infty)$ and include all absorptive processes. In practice the absorption spectra is only known over limited frequency intervals, i.e., regions corresponding to a single class of absorption processes such as due to electronic or impurity transitions. In order to treat these individual processes, finite-energy sum rules and dispersion relations have been developed. In order for these finite sum rules to be meaningful, the key requirement is that the absorption in question be isolated and not overlap with other absorption spectra. The absorption singularity can be viewed as taking place in a transparent medium with real dielectric function arising from the dispersion of absorptive processes in all other spectral ranges.

In summary, sum rules provide a useful means of testing optical measurements, particularly wide-range composite data, both against theoretical and experimental constraints.

See also DISPERSION THEORY; EXCITONS; MAXWELL'S EQUATIONS; PHONONS; PLASMONS; SUM RULES.

BIBLIOGRAPHY

C. J. F. Böttcher, *Theory of Electric Polarization*. Elsevier, Amsterdam, 1952. (I).

A. R. von Hippel, *Dielectrics and Waves*. Wiley and Sons, New York, 1954. (I).

J. N. Hodgson, *Optical Absorption and Dispersion in Solids*. Chapman and Hall, London, 1970. (A).

D. Pines, *Elementary Excitations in Solids*. Benjamin, Advanced Book Program, Reading, Mass. 1963. (A).

D. Y. Smith, "Dispersion Theory, Sum Rules, and their Application to the Analysis of Optical Data," in *Handbook of Optical Constants of Solids*, Edward D. Palik, ed. Academic Press, New York, 1985.

Diffraction

Myron L. Good

INTRODUCTION

The term diffraction refers to all those examples of the scattering of a wave train wherein a large, often infinite, number of scattered wavelets are involved. Thus we have the term *diffraction grating* for a regular array of ruled parallel lines, each of which generates a scattered wave, in distinction to the interference pattern created by two, or several, slit sources of light.

The passage of a wave train through a small aperture falls into this class, because the emergent wave amplitude may be regarded as the sum of wavelets propagating out from all parts of the aperture.

The phase relation of the scattered wavelets must be taken into account, and it is this that is characteristic of diffraction phenomena, causing distributions of scattered intensity whose angular width is of order λ/d, where λ is the wavelength and d is the relevant transverse dimension of the object. Thus for $d \gg \lambda$, the angular width of the diffraction pattern becomes very small, and if small enough, may often be neglected. This is the realm of geometrical optics, where light is taken to travel in straight lines if the medium is uniform.

Emphasizing as it does the wave nature of light, diffraction encompasses phenomena of great beauty, broad generality, and much practical utility: the diffraction pattern of water waves traversing an opening in a breakwater is the same as that of light transmitted through a narrow slit; the resolving power limit of a lens is to be understood only in terms of its diffraction properties; and the diffraction pattern of a lens,

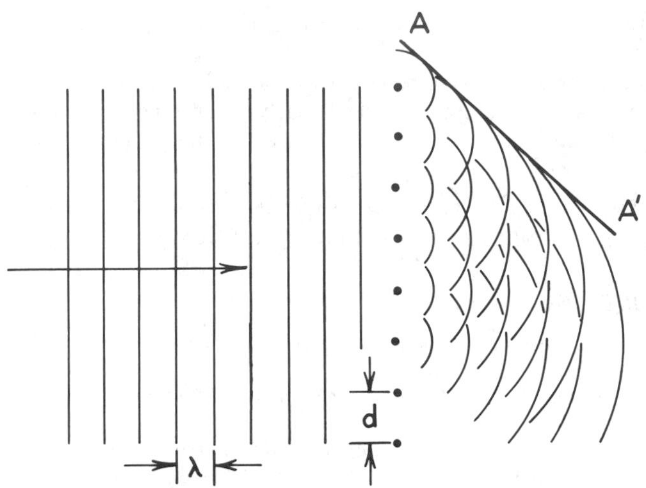

FIG. 1. Diffraction grating.

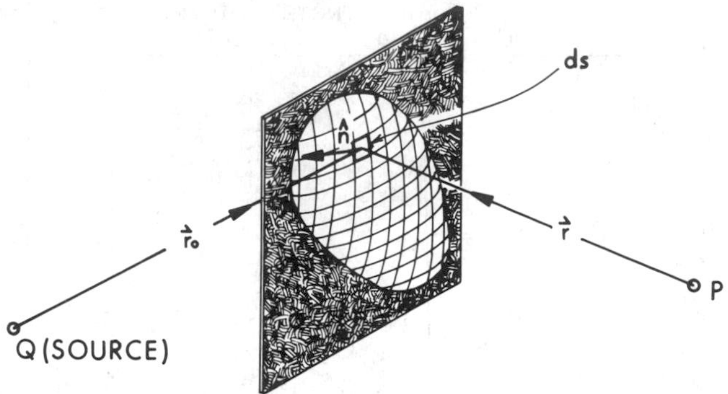

FIG. 2. Nomenclature for the Huyghens–Kirchhoff formulation of diffraction through an aperture.

in turn, is the same as that of a beam of neutrons incident upon a totally absorbing atomic nucleus.

The general problem of diffraction, with the incident wave scattering from an extended scatterer and being itself modified as it progresses through it, is one of forbidding mathematical difficulty. Fortunately, the simplest examples are the most instructive, as well as the most important. Consider the diffraction grating, illustrated in Fig. 1.

DIFFRACTION GRATING

The incident wave is scattered by each of the ruled lines (grooves cut in a glass plate, or molded into a plastic sheet). Scattered wavelets propagate radially outward from each line. Along a line AA' inclined at $\sin\theta=\lambda/d$, all scattered wavelets are in phase. This line constitutes a wave front of a new plane wave, traveling at angle θ with respect to the incident beam, and of the same wavelength and frequency. There are thus sharp maxima in the scattered intensity at angles

$$\theta=\sin^{-1}\left[\frac{n\lambda}{d}\right] \quad (n=0,1,2,\ldots),$$

where n is the number of wavelengths of path difference between wavelets from adjacent rulings ($n=1$ in the example given).

THE HUYGHENS–KIRCHHOFF FORMULATION

In the above, we have implicitly used Huyghens' construction, whereby the new wave front is located by finding the envelope of scattered wavelets. Huyghens' construction has been formally justified by Kirchhoff, who finds, for the amplitude $u(P)$ at a point P caused by a point source at Q diffracting through an aperture s

$$u(P)=\frac{-i}{2\lambda}\iint\frac{e^{i(r+r_0)/\lambda}}{r_0 r} \times[\cos(\hat{n},\mathbf{r})-\cos(\hat{n},\mathbf{r}_0)]ds, \quad r_0,r\gg\lambda, \quad (1)$$

where $\mathbf{r}_0$ and $\mathbf{r}$ are vectors drawn to the surface element ds of the opening from Q and P, respectively, and $\hat{n}$ is a unit vector normal to ds, on the side toward Q (Fig. 2). The source field at the aperture is proportional to $e^{ir_0/\lambda}/r_0$; the coefficient of proportionality has here been taken to be unity, which defines a unit source; $u(P)$ is of course proportional to the square root of the source intensity, but is here written for the unit source.

Equation (1), as written, applies to propagation through an aperture. To describe scattering, as by a diffraction grating, we need only include a factor $S(x,y,\theta)$ inside the integral, giving the scattered amplitude as a function of the position (x,y) and angle of scattering θ. Huyghens' construction is then implicit in Eq. (1) in that as $r\to\infty$, $u(P)$ will be large only when the phase factor $e^{i(r+r_0)/\lambda}$ is the same for all scatterers. The other factors in (1) are extremely slowly varying in comparison to the phase factor, a property which is often useful. For small angles of diffraction the cosine terms (called the obliquity factor) become just $2\cos\alpha$, where α is the angle between QP and the normal to the surface $\hat{n}$.

FRESNEL VERSUS FRAUNHOFER

In applying Eq. (1) we may distinguish two cases. If the observer is infinitely far away, so that the outgoing wave fronts may be treated as plane waves, we have what is called Fraunhofer diffraction; if not, the name of Fresnel is used. The essential distinction is whether the distances AB and $A'B'$ in Fig. 3 are small compared to a wavelength. If so, Fraunhofer diffraction prevails; if not, Fresnel. Fraunhofer diffraction is easier to calculate since only the angle, and not the position, of an emerging ray need be taken into account.

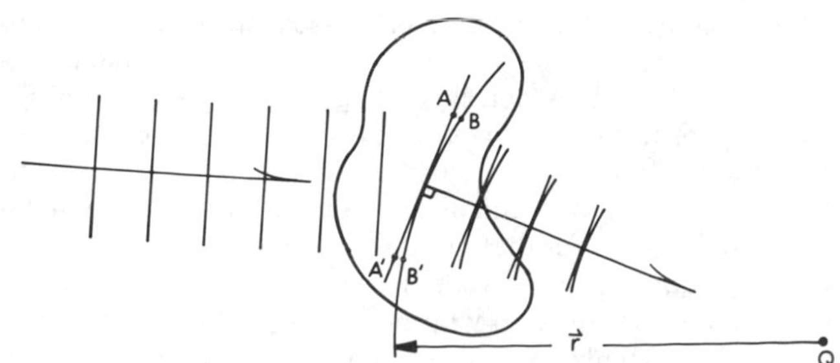

FIG. 3. Fresnel versus Fraunhofer diffraction.

Fraunhofer diffraction may be seen by an observer not at infinity by means of a lens focused on infinity (the human eye, for example). Fresnel diffraction is observed with a diffusing screen or photographic plate, i.e., no lens is needed or desired.

In the Fraunhofer case, Eq. (1) may be simplified by taking the slowly varying factors outside the integral and rewriting the phase factor in terms of the momentum transfer $\hbar\mathbf{q}$ between incident and scattered waves ($\mathbf{q}$ is a unique function of P as $r\to\infty$), and the vector $\mathbf{s}=\hat{\imath}x+\hat{\jmath}y$ which is the two-dimensional position vector of a point (x,y) in the aperture. Equation (1) may then be written as

$$u(\mathbf{q})\sim\iint e^{-i(\mathbf{q}\cdot\mathbf{s})}ds.$$

The Fraunhofer diffracted amplitude is proportional to the two-dimensional Fourier transform of the aperture shape [or of $S(x,y,\theta)=S(\mathbf{S},\mathbf{q})$ taken over the aperture].

This remarkably beautiful and simple result is often useful in the interpretation of data in a variety of fields, ranging from crystallography to high-energy physics.

We are now prepared to examine further simple cases.

CIRCULAR APERTURE

Applying the Huyghens–Kirchhoff formula (1) to a plane wave of wave number k and unit intensity incident on a circular aperture of radius $a=d/2$, integration yields, for the intensity per unit solid angle as $r\to\infty$ (J_1 denotes a Bessel function),

$$\frac{dI}{d\Omega}=r^2|u(\theta)|^2=(\pi a^2)^2\left(\frac{k^2}{4\pi^2}\right)\left[\frac{2J_1(ka\theta)}{(ka\theta)}\right]^2 \qquad (\theta\ll 1), \qquad (2)$$

which has a strong central maximum, a first zero at $\theta=1.22\lambda/d$, and subsequent zeros separated by maxima of rapidly decreasing intensity. At the center, contributions from all radii are in phase; at the first minimum, vector contributions from the entire area combine to form a null vector. The obliquity factor is of no importance, and the problem is greatly simplified by being able to neglect differences in distance from different positions on the aperture, relative to those for a plane wave; i.e., it is of the Fraunhofer type. If the aperture is occupied by a lens of focl length f, the angular diffraction pattern appears in the focal plane of the lens, causing the image of a distant star to form a spot of finite size

$$\delta x\sim f\delta\theta\sim f\lambda/d,$$

which is of the order of, and usually much greater than, the wavelength of the light being used. This is the origin of the limiting resolution of an otherwise ideal lens. It is also the reason why electron microscopes are capable of higher magnification than light microscopes; for an electron beam of a few kilovolts energy, the wavelength is much less than for visible light.

SINGLE-SLIT DIFFRACTION PATTERN; LARGE-ANGLE BEHAVIOR OF DIFFRACTION GRATING

In the case of a diffraction grating made by ruling narrow openings in an otherwise opaque film of metal deposited, say, on glass, the scattering produced by the individual ruling is nothing other than its own diffraction pattern, i.e., that of a narrow slit. Consider such a slit, of width a. Adding the contributions from different segments of this slit gives a one-dimensional integration, which may be visualized by the vector diagrams in Fig. 4. At $\theta=0$ all contributions are equal (Fraunhofer diffraction) and are in phase. As θ increases, a phase difference between contributions from adjacent segments ensues, the vector curls up, and eventually closes into a circle at θ such that a phase difference of 2π exists between the first and the last segment. Thus the first minimum of the diffraction pattern of a single slit is at

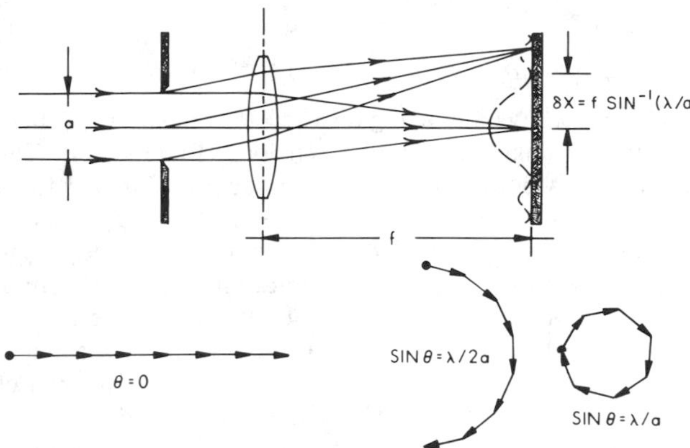

FIG. 4. Diffraction pattern of a narrow slit.

$$\theta=\sin^{-1}(\lambda/a).$$

In a diffraction grating composed of such slits, of width a and spacing d ($a\ll d$), the pattern of regularly occurring sharp maxima [at $\theta=\sin^{-1}(n\lambda/d)$ ($n=0,1,2,\ldots$)] is *modulated* by the diffraction pattern of the individual ruling, which is much wider in angle, since $a\ll d$.

REGULAR THREE-DIMENSIONAL ARRAYS

A crystal is the best example. Here we will not get a diffraction-grating behavior from a series of rows such as AA' (Fig. 5) because that from succeeding rows (such as BB') will be out of phase, and net cancellation will result. Rather we can regard whole planes of atoms (such as AA) as candidates for specular reflection, which at least generates a plane wave. We can then ask, what is the condition for constructive interference between successive parallel planes?

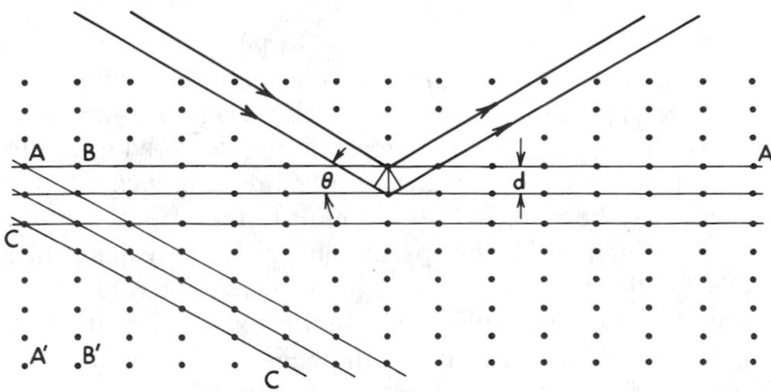

FIG. 5. Diffraction by a crystalline array.

The result is

$$n\lambda = 2d \sin\theta \quad (n=0,1,2,\ldots). \tag{3}$$

Continuous x rays impinging on a crystal will sharply reflect at particular angles in three dimensions (Laue spots). The scattered wavelength is that which satisfies Eq. (3) for a plane which is at angle θ with respect to the beam. For monochromatic x rays, θ is varied until a spot is obtained; measurement of θ then reveals the interatomic spacing d. In a powdered sample, those grains that happen to have the appropriate angle with respect to an incident monochromatic beam will reflect strongly, causing a conical distribution of scattered x rays centered about the unscattered beam. In both cases, many planes can be found (CC, Fig. 5, is a secondary plane) and so the patterns are rich in detail. Their systematic study can be used to determine completely the structure of the unit cell of a crystalline solid. (In the general case, information on intensities and phases is also needed; see Ref. 5.)

Neutrons diffract from crystals as well, in the same way. Neutron diffraction studies start with an intense source of thermal neutrons from a nuclear reactor, from which a monochromatic beam is selected by scattering at a fixed angle from a selected crystal. The monochromatic beam is then scattered from a second crystal (or powder) as with x rays. Different information is obtained from neutron diffraction in that the scattering of the neutrons depends on the atomic number of the different types of atoms in the crystal in a different way from the corresponding dependence for x rays. Neutron diffraction is very useful for locating hydrogen, for example.

Electron diffraction of slow electrons (several eV) from crystals is similar to x-ray and neutron diffraction, except that the wavelength change of the electron on entering the solid must be taken into account. For high-energy electrons (several keV) traversing thin foils, diffraction phenomena of a different character are observed, in which the strong incoherent scattering of the charged electron by the lattice and the basic asymmetry of the crystal caused by the thinness of the foil both play a role.

By use of the above methods, augmented by high-speed computers, the structures of extremely complicated biological molecules have been unravelled; one first forms a crystal, i.e., a regular array of the molecules, and then analyzes its diffraction pattern in detail.

FRESNEL DIFFRACTION

If the observation point P is not at infinity, such that distances such as AB, $A'B'$ in Fig. 3 are not small compared to a wavelength, the phase factor in Eq. (1) no longer varies linearly across the aperture and must be calculated using the actual distances. This in general is more complicated. The integral may be replaced by a sum of terms using a method due to Fresnel: divide the aperture into zones (Fresnel zones) such that the path length $(r+r_0)$ has a definite value along a boundary between two zones, and is exactly a half-wavelength longer along the next such boundary. Since the other factors in the integral are extremely slowly varying, compared to the phase factor, adjacent zones will have almost equal and opposite contributions, and will strongly tend to cancel. The integral becomes a sum, which can often be evaluated by inspection. (The zone boundaries, and the resulting amplitude, depend very much on the point of observation.) For orientation purposes, it is instructive to examine an unrestricted plane wave in this way. The radii r_n of the Fresnel zones are obtained from

$$(r_n^2 + z^2)^{1/2} - z = n\lambda/2 \tag{4}$$

(the observer being a distance z from the chosen wave front), so that $r_n^2 = n\lambda z + n^2\lambda^2/4 \simeq n\lambda z$ (for $z \gg \lambda$). The areas of successive zones are then $\pi\lambda z$, independent of n, and the contributions a_n to the integral may be represented graphically as in Fig. 6. The envelope represents the sum versus n, and diminishes slowly in size because of the obliquity factor and the factor r^{-1} in Eq. (1). The sum, by inspection, is $a_0/2$ and the intensity $a_0^2/4$. Thus the intensity of a plane wave is one-fourth that of the first Fresnel zone.

Consider now a circular aperture. As its radius increases, the intensity at a given point on axis will follow the vector sum in Fig. 6, varying from a_0^2, four times that of a plane wave, to zero and back every time the radius is increased from r_n to r_{n+2}. Since r_n depends on z, a similarly complex behavior is observed for a fixed aperture as z is varied.

As a final (and sharply contrasting) example, consider a circular obstacle. We now have Fig. 6 again, but with the indices relabeled. If the first exposed Fresnel zone is the mth, replace a_0 by a_m, a_1 by a_{m+1}, etc. The sum is then $a_m/2 \simeq a_0/2$ (for m not so large that the slowly varying factors have changed appreciably between 0 and m). Thus a bright spot appears on axis, independent of the size of the object, of the distance z, and even of the wavelength λ, and whose intensity is the same as that of the unobstructed plane wave. This is the celebrated Arago white spot, whose observation is a convincing proof of the wave nature of light and the correctness of Fresnel's formulation of the diffraction problem.

Many other interesting examples may be given. For further details, see Ref. 1.

SHADOW SCATTERING

Whenever a plane wave is incident upon an isolated object, be it light incident on a dust particle or a beam of neutrons on an atomic nucleus, a certain kind of scattering, called shadow scattering or diffraction scattering, is bound to take place, whose origin is rooted in the conservation of probability. The object will itself either absorb or scatter (or, in general, both) the incident wave, through the action of various forces (electromagnetic, nuclear, etc.). This in

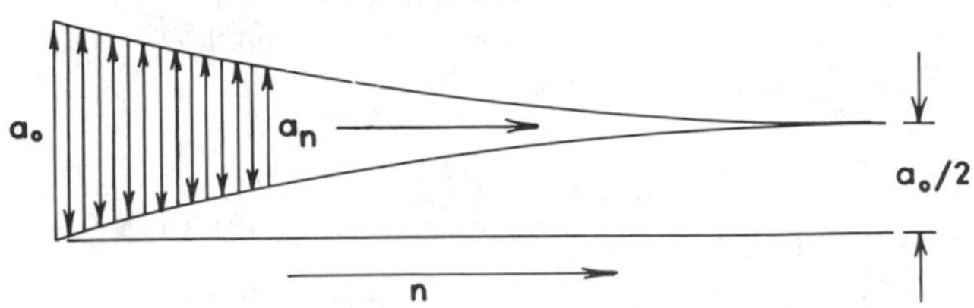

FIG. 6. Vector diagram for Fresnel diffraction.

turn means that the incident flux (or intensity, or probability) must be reduced in the transmitted undeflected wave observed far downstream from the scatterer. In a wave theory, this is brought about by a destructive interference between the incident wave and the forward portion of the wave elastically scattered from the object. The necessary relation is

$$\sigma_{\text{tot}} = 4\pi\lambda\!\!\!^{-}\ \text{Im}\, f(0), \tag{5}$$

which relates the total cross section σ_{tot} to the forward scattering amplitude $f(0)$. In quantum mechanics, this relation, known as the *optical theorem*, is proved from the unitarity condition. In optics, it follows from Maxwell's equations. Since $f(0)$ must be a particular value of a continuous function $f(\theta)$, Eq. (5) implies a strongly forward-peaked elastic-scattering distribution as λ becomes small, or energies high. The origin of this scattering, called shadow scattering, is in the constructive interference, in the near-forward direction, of scattered wavelets from all parts of the extended object. As θ increases to the point where the scattered wavelets from different sides of the object are out of phase ($\theta \sim \lambda/d$), this amplitude falls greatly (to zero in the case of a completely absorbing object), and then rises again to successive maxima of greatly reduced intensity, in an entirely analogous fashion to the diffraction pattern of a circular aperture. In fact the Fraunhofer diffraction patterns of a circular aperture and of a circular obstacle of the same size are identical, as may be seen from the following argument.

BABINET'S PRINCIPLE

Consider a point P in the geometric shadow of an opaque screen with an aperture in it (as $r \rightarrow \infty$, this simply means $\theta \neq 0$). The point P will be illuminated with amplitude a as a result of the diffraction process. Now replace the screen by an opaque obstacle of the same size and shape as the previous aperture. Call the new amplitude at P, a'. Now put back the screen, leaving the obstacle in place. Clearly the amplitude is now zero; but it is also $a + a'$. Hence $a' = -a$, and $I' = I$, where $I = a^2$ and $I' = a'^2$. The diffraction pattern of the obstacle is exactly the same as that of the aperture. This is Babinet's principle.

This seemingly mysterious behavior can be understood as follows. Consider an extended absorbing object as in Fig. 7. The wave front just downstream from the object has a piece removed from it. It may be regarded as the sum of an undisturbed plane wave and a piece of wave front of width d and amplitude -1. This latter looks exactly like the wave front transmitted by an aperture of width d. The undisturbed plane wave propagates on without scattering. The extra piece, of width d, propagates ahead and spreads to the sides, producing a diffraction pattern $|f(\theta)|^2$ identical to that of an aperture of width d. This is the shadow-scattered wave $f(\theta)$, whose forward amplitude $f(0)$ interferes with the incident wave at large distances downstream to produce the diminution in transmitted intensity called for by the total cross section.

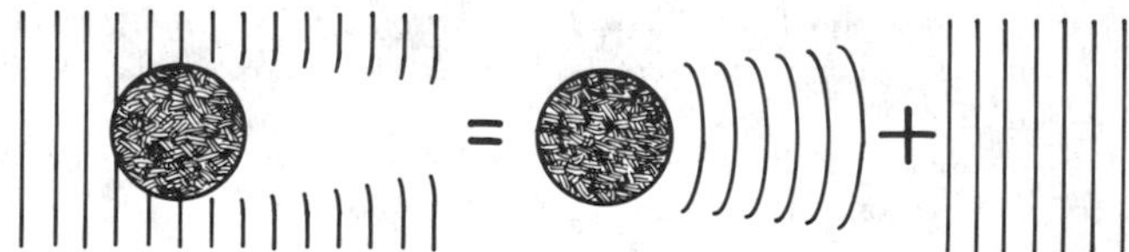

FIG. 7. Diagram illustrating origin of shadow scattering.

For a complete description of the diffraction of elementary particle waves through nuclei, see Ref. 8.

In a discussion as brief as this, it has not been possible to describe many refined applications of diffractive phenomena, among the most prominent of which is the science of holography, described in Ref. 9. For the discussion of these, the reader is referred to other parts of this volume.

Also, polarization effects have not been discussed, but are covered in the cited references.

See also ELECTRON DIFFRACTION; GRATINGS, DIFFRACTION; HOLOGRAPHY; LOW-ENERGY ELECTRON DIFFRACTION (LEED); MICROSCOPY, OPTICAL; NEUTRON DIFFRACTION AND SCATTERING; OPTICS, PHYSICAL; POLARIZED LIGHT; REFLECTION HIGH-ENERGY ELECTRON DIFFRACTION (RHEED); WAVES.

REFERENCES

1. F. Jenkins and H. White, *Fundamentals of Optics*, 4th ed. McGraw-Hill, New York, 1976. (E)
2. M. Born, *Optik; ein Lehrbuch der Elektromagnetischen Lichttheorie*. 2. Unveränderte Aufl. Springer-Verlag, Berlin, New York, 1965; M. Born and E. Wolf, *Principles of Optics*, 5th ed. Pergamon, Oxford, New York, 1975. (I)
3. M. Françon, *Diffraction; Coherence in Optics*. Pergamon, New York, 1966. (I)
4. C. Burnett, J. Hirschberg, and J. Mack, "*Diffraction and Interference*," in *Handbook of Physics*. (E. Condon and H. Odishaw, eds.), pp. (6-77)–(6-108) McGraw-Hill, New York, 1958. (I)
5. W. Zachariasen, *X-Ray Diffraction in Crystals*. Wiley, New York, 1947 (A); G. E. Bacon, *X-Ray and Neutron Diffraction*. Pergamon, Oxford, New York, 1966. (I)
6. D. Hughes, *Neutron Optics*. Interscience, New York, 1954. (E)
7. Z. Pinsker, *Electron Diffraction*. Butterworth, London, 1953. (A)
8. U. Amaldi, M. Jacob, and G. Matthiae, "Diffraction of Hadronic Waves," *Annu. Rev. Nuclear Sci.* **26,** 385–456 (1976). (A)
9. M. Françon, *Holography*, Academic Press, London, New York, 1974. (I)

Diffraction Gratings *see* Gratings, Diffraction

Diffusion

H. B. Huntington

Diffusion is basically the process by which a concentrated quantity is spread out over a wider extent. As applied to matter in three dimensions it is the homogenization that occurs from random (or nearly random) motion in the microcosmos, apart from convective flow.

The basic quantity in diffusion, D, is called the *diffusivity* or the *diffusion constant* and is defined empirically by Fick's first law

$$\mathbf{F}_i = -D\,\nabla c_i \tag{1}$$

where c_i and $\mathbf{F}_i$ are, respectively, the concentration and flux of the ith constituent. (The more basic scientific approach is to recognize that the flux is proportional to the gradient of the chemical potential rather than of the concentration.) Fick's second law is a heat-flow-type equation that results from combining the first law with the continuity equation:

$$\frac{\partial c}{\partial t} = \nabla \cdot D \nabla c. \tag{2}$$

It is applied to determine D from measurements on specimens of simple geometry.

Diffusion occurs in all phases of matter [1]. In gases the kinetic theory analysis shows that D varies as the square of the mean free path. In crystalline solids atoms move by jumps of fixed length related to the lattice constant. As a first approximation the random walk analysis gives for D

$$D = \frac{1}{6} \sum \lambda_i^2 \nu_i, \tag{3}$$

where the summation is over various possible mechanisms, λ_i are the lengths of the respective jumps, and ν_i are the respective frequencies of jumping. Liquids and amorphous solids present an intermediate situation that is less easily modeled but does appear to depend strongly on the "free-volume" character of the structure. In crystalline phases the diffusional jump is almost always associated with the presence of some sort of a structural defect, such as a vacancy or an interstitial. The population of these defects therefore enters as an important factor in Eq. (3). The presence of these defects keeps the jumping process from being completely random and introduces a correlation factor, f, usually less than 1, into the expression for D in Eq. (3).

The study of diffusion in solids [2–11] has both technological and scientific importance. In the field of materials science the term has become synonymous with atom movements. As such it plays a role in all heat treatments: homogenization, aging, grain coarsening, annealing, precipitation, sintering, segregation, tempering, and the like. From the scientific side the study of diffusion provides much basic information on the energies of formation and motion of lattice defects, since diffusion data frequently show an Arrhenius-type dependence on temperature, indicating an activated process for the atom motions. For this reason data on diffusion are almost invariably presented as

$$D(T) = D_0 e^{-Q/RT}, \tag{4}$$

where Q plays the role of activation energy.

The magnitude of Q is directly dependent on the nature of the transport. It is largest for diffusion through the crystalline bulk, smaller for grain boundary diffusions, and least for surface diffusion. This last is quite difficult to determine unequivocally. The best measurements are perhaps those determined by field ion microscopy (FIM). Depending on the crystalline face the diffusivity may be anisotropic even for cubic crystals for which the bulk diffusivity is always isotropic.

The grain-boundary diffusivity is a field which has been attracting ever-increasing attention of late. A complication in its study is the simultaneous presence of bulk diffusion in nearly every situation. For diffusion from the surface in the usual planar geometry three ranges can be distinguished: (A) close to the surface where bulk diffusion predominates, (B) where material first moves out along the grain boundaries and then spreads laterally by bulk diffusion, and (C) which involves diffusion primarily down the grain boundaries. It is the B range which is of most interest in the study of grain-boundary diffusion [12]. The analysis is complicated and has been attacked under various simplifying assumption by several investigators, e.g., Fisher [13], Whipple [14], and Suzuoka [15]. While bulk diffusion gives a concentration falling off as $\exp[-(x^2/4DT)]$ with penetration x, the concentration in the B range appears to be fit best by $\exp(-Ax^{6/5})$.

The techniques for measuring the diffusivity are numerous and depend on a wide variety of material properties. The use of radioisotopes followed by sequential sectioning has on the whole proved to be the most direct, precise, and reliable method. There are, however, several interesting techniques, such as internal friction, nuclear magnetic resonance, and Mössbauer studies, that give atom movement data by determining the jump time of the atoms rather than the overall distances that they travel.

Basic information on the mechanisms for atom motion can be obtained by measurements other than temperature dependence. The influence of pressure on diffusion shows the volume of the activated complex. The effect of impurity enhancement on diffusion can rule out certain mechanisms. Comparative studies of the motion of different radioactive isotopes of the same element can be used to help determine the extent to which a diffusing atom moves alone or as a partner in a cooperative process. Comparison between diffusion measurements and those of ionic conductivity in the appropriate materials can be used to determine, through the application of the Nernst–Einstein equation, the correlation coefficient f involved.

See also KINETIC THEORY; CRYSTAL DEFECTS.

REFERENCES

1. W. Jost, *Diffusion in Solids, Liquids and Gases*. Academic, New York, 1960. (I)
2. A. S. Novick and J. J. Burton, eds., *Diffusion in Solids: Recent Developments*. Academic, New York, 1975. (I)
3. H. I. Aaronson, ed., *Diffusion* (ASM Seminar for 1972). American Society for Metals, Metals Park, Ohio, 1973. (I)
4. L. A. Girifalco, *Atomic Migration in Crystals*. Blaisdell, New York, 1964. (E)
5. P. G. Shewman, *Diffusion in Solids*. McGraw-Hill, New York, 1963. (I)
6. David Lazarus, "Diffusion in Metals," in *Solid State Physics* (F. Seitz, D. Turnbull, and H. Ehrenreich, eds.), Vol. 10 (1960); N. L. Peterson, "Diffusion in Metals," Vol. 22 (1968). Academic, New York. (I)
7. J. R. Manning, *Diffusion Kinetics for Atoms in Crystals*. Van Nostrand, Princeton, NJ, 1968. (A)
8. Y. Adda and J. Philibert, *La Diffusion dans les Solides*. Bibliotheque des Sciences et Techniques Nucleáires, Saclay, 1966. (I)
9. C. P. Flynn, *Point Defects and Diffusion*. Oxford University Press, London and New York, 1972. (A)
10. G. E. Murch and A. S. Nowick, *Diffusion in Crystalline Solids*. Academic Press, New York, 1984. (I)

11. R. J. Borg and G. J. Dienes, *An Introduction to Solid State Diffusion*. Academic Press, New York, 1988. (E)
12. See Ref. 8, Chapter 12.
13. J. C. Fisher, *J. Appl. Phys.* **22,** 74 (1951).
14. R. T. Whipple, *Phil. Mag.* **45,** 1225 (1954).
15. T. Suzuoka, *Trans. Jap. Inst. Met.* **2,** 25 (1961).

Dispersion Theory

Peter A. Carruthers

The original "dispersion relation" of Kramers [1] and Kronig [2] expresses the real part of the frequency-dependent dielectric constant $n(\omega)$ in terms of the absorption coefficient $\alpha(\omega)$ [the attenuation factor is $\exp(-\alpha x)$] according to (P means principal value)

$$n_r(\omega)=1+\frac{cP}{\pi}\int_0^\infty \frac{\alpha(\omega')\,d\omega'}{\omega'^2-\omega^2} \tag{1}$$

where c is the velocity of light. Alternatively, Eq. (1) is a relation between the real and imaginary parts of the complex index of refraction $n(\omega)=n_r(\omega)+i(c/2\omega)\alpha(\omega)$. The structure of Eq. (1) is also a consequence of analyticity in the complex frequency z of the function $n(z)$. This analyticity is in turn related to the physical requirement of causality, i.e., that cause must precede effect. The interesting character of the causal properties of the propagation of light in dispersive media was first clarified by Sommerfeld and Brillouin [3].

The simplest example [4] exhibiting the pertinent analyticity properties is the classical damped harmonic oscillator model of a bound electron subject to an external electric field E of frequency ω. The equation of motion is

$$m(\ddot{x}+\gamma\dot{x}+\omega_0^2x)=-eEe^{-i\omega t}. \tag{2}$$

Here m and e are, respectively, the electron mass and charge, and γ and ω_0 the oscillator damping constant and frequency. The dielectric constant $\epsilon(\omega)$ for a medium of N electrons per unit volume is computed from the relation $P=(\epsilon-1)E/4\pi$ and $P=-Nex$:

$$\epsilon(\omega)-1=\frac{4\pi Ne^2}{m}\frac{1}{\omega_0^2-\omega^2-i\gamma\omega}. \tag{3}$$

The right-hand side of (3) has poles in the lower complex ω plane at $\omega_\pm=-\frac{1}{2}i\gamma\pm(\omega_0^2-\frac{1}{4}\gamma^2)^{1/2}$ for $\omega_0>\gamma/2$ (i.e., "weak" damping). As a consequence of Cauchy's integral theorem we have directly

$$\begin{aligned}\mathrm{Re}(\epsilon(\omega)-1)&=\frac{P}{\pi}\int_{-\infty}^\infty \frac{d\omega'\,\mathrm{Im}\,\epsilon(\omega')}{\omega'-\omega}\\ &=\frac{2}{\pi}P\int_0^\infty \frac{d\omega'\,\omega'\,\mathrm{Im}\,\epsilon(\omega')}{\omega'^2-\omega^2}\end{aligned} \tag{4}$$

since Re ϵ and Im ϵ are even and odd functions of ω. To make connection with (1) we use $\epsilon(\omega)=n^2(\omega)$ and note that for weak absorption, $n\gg n_1$, we have $\mathrm{Re}(\epsilon-1)\cong 2n$ and $\mathrm{Im}\,\epsilon\cong 2n_1$. Actually (1) is an exact consequence of (4), as explained by Nussenzweig [5]. In the foregoing model the positivity of the decay constant γ expresses the causal nature of the process. Clearly the first of Eqs. (4) follows [6] from the less specific assumption that there exists a function $\epsilon(z)$ analytic in the upper half-plane satisfying $\epsilon(z)\to 1$ as $|z|\to\infty$. The functions Re $\epsilon(\omega)$ and Im $\epsilon(\omega)$ are to be interpreted as the boundary values as $z=\omega+i\eta$ approaches the real axis from above, e.g., $\mathrm{Re}\,\epsilon(\omega)=\lim_{\eta\to 0}\mathrm{Re}\,\epsilon(\omega+i\eta)$ with $\eta>0$. The simplification due to the second equality in (4) depends on the symmetry properties of Re ϵ and Im ϵ.

The required analyticity is indeed a consequence of a rather general causality argument, as now shown. Consider the time dependence of a physical quantity $F(t)$ driven by a source $s(t)$ (e.g., electric or magnetic field, current) with a linear response function $R(t-t')$:

$$F(t)=\int_{-\infty}^\infty R(t-t')s(t')\,dt' \tag{5}$$

Causality implies that the behavior of F at time t depends only on the source values at times $t'\leq t$, i.e.

$$R(t)=0, \qquad t<0. \tag{6}$$

In terms of Fourier transforms

$$R(\omega)\equiv\int_{-\infty}^\infty dt\,e^{i\omega t}R(t), \tag{7}$$

Eq. (5) simplifies to $F(\omega)=R(\omega)s(\omega)$, in analogy to the classical oscillator example. Since $R(t)$ vanishes for $t<0$ we can immediately continue (7) into the complex $(z=\omega_1+i\omega_2)$ upper half-plane:

$$R(\omega_1+i\omega_2)=\int_0^\infty dt\,e^{i\omega_1 t-\omega_2 t}R(t), \tag{8}$$

since the integral converges still better than before (convergence assumed!) for $\omega_2>0$.

It is now possible to assert that $R(z)$ is analytic in the upper half-plane (Titchmarsh theorem [7]) under various conditions. As a simple case, assume (7) to be absolutely convergent for $\omega=0$. Then if $R(t)$ is piecewise continuous, (8) is uniformly convergent ($\omega_2>0$), which is adequate to establish the analyticity of (8) in the upper half-plane.

If $R(z)$ vanishes as $|z|\to\infty$ in the upper half-plane, we can write

$$R(\omega+i\eta)=\frac{1}{2\pi i}\int_{-\infty}^\infty \frac{d\omega'\,R(\omega')}{\omega'-\omega-i\eta}, \tag{9}$$

which reduces to

$$\mathrm{Re}\,R(\omega)=\frac{P}{\pi}\int_{-\infty}^\infty \frac{d\omega'\,\mathrm{Im}\,R(\omega')}{\omega'-\omega}, \tag{10}$$

$$\mathrm{Im}\,R(\omega)=-\frac{P}{\pi}\int_{-\infty}^\infty \frac{d\omega'\,\mathrm{Re}\,F(\omega')}{\omega'-\omega}. \tag{11}$$

Such relations are known as Hilbert transforms. They are useful not only in physics but also in electrical engineering.

When $R(z)$ does not vanish as $|z|\to\infty$, a modification of the foregoing argument is needed. Suppose $R(z)$ becomes a constant for large z (as was the case for the dielectric constant). Then the function $[R(z)-R(z_0)]/(z-z_0)$ is analytic as before but vanishes as $1/z$ for large z. Choosing z_0 to be real leads to

$$\text{Re } R(\omega) = \text{Re } R(\omega_0) + \frac{\omega - \omega_0}{\pi} \text{P} \int_{-\infty}^{\infty} \frac{d\omega' \text{ Im } R(\omega')}{(\omega' - \omega)(\omega' - \omega_0)}. \quad (11)$$

Equation (11) is known as a "subtracted" dispersion relation since it results if we formally subtract from (10) the same "equation" evaluated for $\omega = \omega_0$.

The Kramers–Kronig relation may be reexpressed as a connection between the real and imaginary parts of the forward scattering amplitude $f(\omega)$ for light using the Lorentz relation [5]

$$n(\omega) = 1 + \frac{2\pi c^2}{\omega^2} N f(\omega). \quad (11)$$

If we use in addition the optical theorem

$$\text{Im } f(\omega) = \frac{\omega \sigma(\omega)}{4\pi c}, \quad (12)$$

where $\sigma(\omega)$ is the cross section, (1) can be written as

$$\text{Re } f(\omega) = \frac{\omega^2}{2\pi^2 c} \text{P} \int_0^{\infty} \frac{d\omega' \, \sigma(\omega')}{\omega'^2 - \omega^2}. \quad (13)$$

Relations similar to (13) exist for the nonrelativistic and relativistic scattering amplitudes of particles, although considerable care has to be given to the question of subtractions. For example, Eq. (13) fails for scattering off *free* electrons, for which $f(0) \neq 0$ but rather $f(0) = -e^2/mc^2$ (see [5, p. 51]). Dispersion relations also exist for partial-wave amplitudes. These matters are discussed in detail by Goldberger and Watson [8].

The connection between causality and the analyticity of the S matrix has been much studied in scattering theory [6,8,9]. Thus far we have mentioned only one-variable functions, such as the total forward scattering amplitude and the partial-wave amplitude, which obey certain dispersion relations. However, it has been found possible to establish analytic properties not only in the energy variable but also in the momentum transfer variable. In certain nonrelativistic potential scattering problems the full amplitude can be obtained from the associated dispersion relation supplemented by unitarity and plausible assumptions on asymptotic behavior. Efforts to extend this principle to high-energy physics have not succeeded because of the essentially many-body nature of the problem. Nevertheless dispersion relations constitute one of the principal tools of elementary-particle physics and have led to many significant insights. The use of dispersion relations as a dynamical tool is explained in Refs. 8–11.

The expression of the causality condition in nonrelativistic quantum mechanics requires some care since there is no limiting velocity and because it is not possible to construct propagating wave fronts having a sharp edge in time. Various studies of this problem, associated with the names of Schutzer, Tiomno, van Kampen, and Wigner, are discussed in Ref. 5. In relativistic quantum field theory the condition of local commutativity of the field variables [called $\phi(x)$ where $x \equiv (x, y, z, t)$],

$$[\phi(x), \phi(x')] = 0, \qquad x - x' \text{ spacelike}, \quad (14)$$

for spacelike separations gives a precise but operationally obscure definition of causality. The "microscopic causality" expressed in (14) plays the same role as (6) in establishing the analytic properties necessary to write dispersion relations for the scattering amplitude. Since the condition (14) is a very precise geometric constraint it could be concluded that any violation of a dispersion relation might be due to the violation of the local commutativity condition characteristic of quantum field theory. Precise tests of (rigorously valid) dispersion relations for forward pion–nucleon scattering have confirmed [12] rather than called into question the general theoretical structure [including (14)] on which the dispersion relations are based.

See also ABSORPTION COEFFICIENTS; *S*-MATRIX THEORY; SCATTERING THEORY.

REFERENCES

1. H. A. Kramers, *Atti Congr. Int. Fis. Como* **2,** 545 (1927).
2. R. de L. Kronig, *J. Opt. Soc. Am.* **12,** 547 (1926).
3. L. Brillouin, *Wave Propagation and Group Velocity*. Academic Press, New York, 1960.
4. L. Rosenfeld, *Theory of Electrons*. North-Holland, Amsterdam, 1951.
5. H. M. Nussenzweig, *Causality and Dispersion Relations*. Academic Press, New York, 1972.
6. L. D. Landau and E. M. Lifshitz, *Electrodynamics of Continuous Media*, p. 256. Pergamon, Oxford, 1960.
7. E. C. Titchmarsh, *Introduction to the Theory of Fourier Integrals*, 2nd ed., Chapter 5. Oxford (Clarendon Press), London and New York, 1948.
8. M. L. Goldberger and K. M. Watson, *Collision Theory*. Wiley, New York, 1964.
9. R. J. Eden, P. V. Landshoff, D. I. Olive, and J. C. Polkinghorne, *The Analytic S-Matrix*. Cambridge, London and New York, 1966.
10. H. Burkhardt, *Dispersion Relation Dynamics*. North-Holland, Amsterdam, 1969.
11. G. F. Chew, *S-Matrix Theory of Strong Interactions*. Benjamin, New York, 1962.
12. S. J. Lindenbaum, *Particle Physics Interactions at High Energies*, Chapter 5. Oxford, London and New York, 1973.

Doppler Effect

Theo W. Hänsch

A change in the observed frequency of light, sound, and other waves, caused by a relative motion between source and observer is known as the Doppler effect. A familiar example is the change in the pitch of a train whistle as the train approaches and passes. The observed frequency ν' is higher than the source frequency ν if the distance between source and observer is diminishing, and vice versa.

For sound waves that propagate with a characteristic velocity u relative to a medium (air, water), the Doppler shift depends on the velocities of source and listener relative to this medium. The number of waves per second arriving at the observer can be calculated by simply counting the waves emitted per second by the source, and any change per second in the number of waves "in flight" traveling from source to observer. We obtain

$$\nu' = \nu(1 - v_o/u)/(1 - v_s/u) \tag{1}$$

where v_o and v_s are the velocities of observer and source relative to the medium along the direction from source to observer.

The Doppler effect for light is of particular importance for spectroscopy and astronomy. According to the special theory of relativity, the velocity of light has the same value c in all inertial frames. Consequently, the optical Doppler effect, unlike its acoustical counterpart, depends only on the relative velocity v between source and observer. With the help of the familiar Lorentz transformation we calculate a Doppler-shifted frequency

$$\nu' = \nu \frac{1 - v \cos\phi/c}{(1 - v^2/c^2)^{1/2}} \tag{2}$$

where ϕ is the angle between the relative velocity and the line of sight between source and observer. For small velocities, this result is essentially the same as expected classically for a source at rest and a moving observer: a red shift proportional to the velocity if the observer is moving away from the source, and a blue shift if the observer is approaching (*linear* Doppler effect).

For higher velocities, the denominator $(1 - v^2/c^2)^{1/2}$ in Eq. (2) predicts an additional purely relativistic effect: a red shift independent of the direction of the relative velocity between source and observer. This shift persists even if source or observer is moving in transverse direction relative to the line of sight. Though observations of this *transverse* Doppler effect are difficult, they have been carried out in the laboratory with fast-moving atoms, and they give direct evidence for the relativistic "time dilatation" (moving clocks seems to oscillate more slowly).

The linear optical Doppler effect manifests itself much more readily. In astronomy, Doppler shifts of spectral lines are observed which can be used to determine the rotation of planets and stars or to identify binary stars. The red shift of the light from remote galaxies gives proof of the expansion of the universe.

In the laboratory we commonly observe a Doppler *broadening* of spectral lines in gases. The absorption or emission lines appear blurred because atoms or molecules of different velocities contribute with different Doppler shifts. For a gas of temperature T (in kelvins) and mass number M, the relative Doppler width (full width at half maximum) is given by

$$\Delta\nu/\nu = 7.16 \times 10^{-7}(T/M)^{1/2}. \tag{3}$$

For high-resolution spectroscopy and precision measurements it is often important to reduce or eliminate Doppler broadening. This can be accomplished by restricting the range of atomic velocities along the direction of light propagation, as in the transverse observation of a collimated atomic beam. "Doppler-free" spectroscopy of a gas is possible by selectively exciting the absorbing atoms in a narrow velocity interval by a monochromatic saturating laser beam, and by probing the resulting bleaching or "spectral hole burning" with a second, counterpropagating probe laser beam. Resolutions approaching 1 part in 10^{11} have been achieved by this method of saturation spectroscopy or Lamb-dip spectroscopy. Doppler-free spectroscopy without velocity selection is possible by two-photon excitation of a gas with two counterpropagating laser beams, whose linear Doppler shifts cancel. The Mössbauer effect provides a means to eliminate the Doppler shift due to recoil in nuclear spectroscopy.

See also HUBBLE EFFECT; LASER SPECTROSCOPY; MÖSSBAUER EFFECT; RELATIVITY, SPECIAL THEORY; WAVES.

BIBLIOGRAPHY

V. P. Chebotaev and V. S. Letokhov, "Nonlinear Optical Resonances Induced by Laser Radiation," *Prog. Quantum Electron.* **4**(2), 111 (1975). (I–A)

R. W. Ditchburn, *Light*. 1963. (E)

T. P. Gill, *The Doppler Effect*. 1965. (I)

W. C. Michels, "Phase Shifts and the Doppler Effect," *Am. J. Phys.* **24**, 51 (1956). (E–I)

J. J. Snyder and J. L. Hall, "A New Measurement of the Relativistic Doppler Shift," *Lecture Notes in Physics*, Vol. 43 (1975). (I–A)

Dynamic Critical Phenomena

Walter I. Goldburg

The equilibrium properties of systems near the critical point are now quite well understood. We know, for example, that the isothermal compressibility of fluids and the magnetic susceptibility of ferromagnets diverge in the same universal fashion as this point is approached. What is more, the critical exponents that characterize this universal behavior can be calculated, starting with the partition function of statistical mechanics and using the well-justified renormalization group approximations.

Here we are concerned, not with the time-independent properties of critical systems, but (a) with fluctuations about the equilibrium state and (b) with the rate at which the equilibrium state is approached when the system is perturbed. A typical perturbation might be a temperature change, which causes the system to evolve from one equilibrium state to another. Of special interest is the case where the initial state is a "disordered" one, for example, a ferromagnetic material in its paramagnetic state, and the final state is the "ordered" ferromagnetic state. For a fluid in the ordered state, both the liquid and vapor phases are generally present. To study the dynamics of such systems, one might monitor the time dependence of the magnetization or, in fluid systems, the radius or number density of the emerging droplets. In magnetic systems, fluids, and alloys, it is especially interesting to probe the evolving system by scattering electromagnetic radiation or neutrons from it, as discussed below.

RELAXATION OF FLUCTUATIONS ABOUT THE EQUILIBRIUM STATE

Starting from the known interactions between molecules in a system and using the principles of statistical mechanics, one can, in principle, predict the mean square deviation of thermodynamic variables from their equilibrium values, as well as the equilibrium values themselves. However, it is not possible to determine the *rate* at which the fluctuations occur from a knowledge of the interactions alone. Additional assumptions about the dynamical features of the system are required. Near the critical point simple and compelling assumptions about the dynamics can satisfactorily account for the relaxation properties of entire classes of systems. Just as with the equilibrium properties of systems near the critical point, their dynamical properties also display universality and self-similarity.

To illustrate, consider the temporal fluctuations in the density of a fluid $\rho(\mathbf{r},t)$ at the point $\mathbf{r}$ and at the time t. It is actually more useful to focus on spatial Fourier components of this quantity, namely, $\rho(\mathbf{k},t)$. Here $\mathbf{k}$ is the wave number of the fluctuation, its wavelength being $2\pi/k$. Near the critical point, the relaxation rate $\Gamma(k)$ of the long-wavelength fluctuations will go to zero as its critical temperature T_c is approached. Quite generally $\Gamma(k)$ will vary with temperature as $|T - T_c|^\theta$, with the "critical exponent" θ having the same value for entire classes of systems (one-component fluids and fluid mixtures all lie in the same class).

THE DYNAMICS OF PHASE SEPARATION

As an example, consider a fluid, such as CO_2, in a sealed container, its temperature being sufficiently high and its mean density ρ low enough that the system is in the (disordered) vapor phase. Assume, now, that the fluid is abruptly cooled through its coexistence temperature $T_{cx}(\rho)$, so that droplets or domains of liquid form. If ρ differs appreciably from the critical density (or the quench is very shallow), the developing droplets will occupy a fractionally small volume of the container. This type of phase separation is called nucleation. In the special case of systems prepared at the critical density, a quench will give rise to two phases having almost equal volumes. This type of phase separation is very different from nucleation in that there is no surface-energy barrier associated with the creation of the domains of the newly developing phase. This barrier-free phase separation is called spinodal decomposition.

Both nucleation and spinodal decomposition exhibit universal features near the critical point. For example, in nucleation, the emerging phase takes the form of spherical droplets whose mean radius grows as some characteristic power of the time t after the quench. In spinodal decomposition, the two phases form an interconnected structure that has a worm-like appearance in a microscope. Again the mean domain size increases as some characteristic power of time. What is more, critically quenched systems exhibit self-similarity: a micrograph of the structure taken at late times after the quench will look the same as an early time picture, except that it will appear to have been magnified.

The two phases that appear in nucleation (or spinodal decomposition) do not have identical densities or compositions, and hence will scatter neutrons as well as light (if the system is transparent). One can obtain detailed information about the dynamics of phase separation by measuring the scattering cross section to determine the structure factor $S(k,t)$, which is proportional to the mean square modulus of $\rho(k,t)$. In spinodal decomposition the structure factor is found to exhibit a very interesting type of self-similarity that is being actively studied experimentally and theoretically. Spinodal decomposition is observed in diverse types of systems, including fluids, fluid mixtures, polymers, metallic alloys, glasses, liquid crystals, gels, and colloidal systems.

See also Fluctuation Phenomena; Phase Transitions; Renormalization Group; Transport Properties.

BIBLIOGRAPHY

P. C. Hohenberg and B. I. Halperin, *Rev. Mod. Phys.* **49,** 435 (1977).

W. I. Goldburg, in *Light Scattering Near Phase Transitions* (H. Z. Cummins and L. P. Levanyuk, eds.), p. 53. North-Holland, New York, 1983.

J. D. Gunton, M. San Miguel, and P. S. Sahni, *Phase Transitions and Critical Phenomena,* Vol. 8 (C. Domb and J. L. Lebowitz, eds.). Academic, London, 1983.

Dynamics, Analytical

Herbert Goldstein and Michael Tabor

Analytical dynamics is concerned with the classical motion of mass points (called particles), caused either by the forces of mutual interaction or by forces externally imposed on the system. By the adjective "classical" is meant nonquantum, nonrelativistic. The least number of independent coordinates needed to describe the motion of the system of particles is spoken of as *the number of degrees of freedom* of the system. In practice, the field of analytical dynamics is usually restricted to systems where the number of degrees of freedom is small, either because only a few particles are involved or because of constraints on the system (e.g., when the system is a rigid body). Where many independent particles are present, the motion forms the subject of statistical mechanics.

Many of the techniques of analytical dynamics were devised to facilitate solving problems about the motion of material, macroscopic bodies. In present day physics (as distinguished from, say, engineering or celestial mechanics) the interest rather is in the formulations themselves of analytical dynamics, particularly as they may serve as models or springboards for the construction of theories in other areas. What will be summarized here, very briefly and omitting proofs, is the skeleton structure of the various formulations of analytical dynamics. The starting point will be the Newtonian form—the most basic and generally applicable—followed successively by formulations of increasing abstraction and, often, of reduced physical content. Running like a thread throughout all the formulations is the search for constants of the motion, and the connection between the symmetry properties of the system and the existence of meaningful constants of motion.

However, in the end we shall see that most Hamiltonians do not possess sufficient constants of motion to render, in a certain sense, exactly soluble equations of motion. In these situations relatively new results demonstrate that the classical motion can be chaotic.

NEWTONIAN FORMULATION

If $\mathbf{r}_i$ and $\mathbf{p}_i = m_i\dot{\mathbf{r}}_i$ are the position vector and linear momentum, respectively, of the ith particle, then Newton's equations of motion for the system are

$$\frac{d\mathbf{p}_i}{dt} = \mathrm{F}_i^{(e)} + \sum_{j\neq i} \mathbf{F}_{ji}. \tag{1}$$

Here $\mathrm{F}_i^{(e)}$ represents the external force (including constraint forces, if any) acting on the ith particle, and $\mathbf{F}_{ji}$ stands for the interaction force of the jth particle on the ith particle. The equations of motion described by Eq. (1) are of second order and the solutions thus involve two initial conditions for each particle, e.g., of initial position and velocity. First integrals of the equations of motion are constant functions of positions and velocities. Where the first integrals are algebraic expressions of these variables, they are spoken of as *constants of the motion*, or *conserved quantities*. The conditions for the existence of conserved quantities are described in *conservation theorems*.

Depending on the nature of the forces acting on the particles, various conservation theorems can be derived from Newton's equations of motion. For example, if $\mathbf{F}_{ji} = -\mathbf{F}_{ij}$ (the weak form of the law of action and reaction) and the external forces on the particles sum to zero, then the total linear momentum of the system,

$$\mathbf{P} = \sum_i \mathbf{p}_i, \tag{2}$$

is a constant of the motion. When the strong form of the law of action and reaction holds, i.e., $\mathbf{F}_{ji} = -\mathbf{F}_{ij}$ and is along the direction of $\mathbf{r}_i - \mathbf{r}_j$, and the net torque of the external forces about the arbitrary origin vanishes, then the total angular momentum of the system about the given origin,

$$\mathbf{L} = \sum_i \mathbf{r}_i \times \mathbf{p}_i, \tag{3}$$

is a constant of the motion. Conservation of linear momentum can be shown to hold if the system is symmetric under linear translation, i.e., the conditions operating on the system are unaffected if the system is displaced linearly as if it were a rigid body. Similarly, conservation of angular momentum about a point is equivalent to symmetry of rotation about the point. Another conservation theorem can be obtained if all of the forces acting on the ith particle can be obtained as the negative gradient (with respect to the coordinates of the ith particle) of a scalar V, a function of all the coordinates. Under these conditions the *total energy* of the system

$$E = T + V \tag{4}$$

is a constant of the motion, where T is the *kinetic energy*,

$$T = \tfrac{1}{2}\sum_i m_i\,(\mathbf{r}_i)^2, \tag{5}$$

and V is known as the *potential energy*.

LAGRANGIAN FORMULATION

In the presence of constraints, the number of degrees of freedom of a system of k particles is less than the $3k$ position coordinates. If constraints can be expressed by the equations of the form

$$f_i(\mathbf{r}_1, \ldots \mathbf{r}_k, t) = 0, \quad i = 1, \ldots, 3k - n, \tag{6}$$

then the motion of the system can be entirely described by n independent coordinates, which may either be an appropriate subset of the $\mathbf{r}_i$, or new, generalized, coordinates defined by equations of transformation

$$q_j = q_j(\mathbf{r}_1, \ldots \mathbf{r}_k, t), \quad j = 1, \ldots, n. \tag{7}$$

When the constraint equations are of the form of Eq. (6), the constraints (and the system) are said to be *holonomic*. The Lagrangian formulation has its simplest and most useful expression when the mechanical systems are holonomic, although partial extensions to some types of nonholonomic constraints are possible. The constraints defining a rigid body are simple examples of holonomic constraints, and reduce the number of degrees of freedom of a rigid body, no matter how many particles make up the body, to no more than six.

In the Lagrangian formulation it is also assumed that the forces which arise by virtue of the constraint do no work in an infinitesimal displacement of the system coordinates at a given moment (a *virtual displacement*). Almost all constraints in the absence of friction or other dissipative phenomena are workless in this sense, as is also the constraint described by "rolling" even though it requires (rolling) friction. Under these conditions the equations of motion of the system reduce to n independent equations of the form

$$\frac{d}{dt}\frac{\partial T}{\partial \dot{q}_i} - \frac{\partial T}{\partial q_i} = Q_i \equiv \sum_{j=1}^{k} \mathbf{F}_j \cdot \frac{\partial \mathbf{r}}{\partial q_i}, \quad i = 1, \ldots, n, \tag{8}$$

where the Q_i are called *generalized forces*, evaluated in terms of all of the forces on the particles *except* the constraint forces. When the Q_i can be obtained from a generalized potential function $U(q, \dot{q}, t)$ according to the prescription

$$Q_i = -\frac{\partial U}{\partial q_i} + \frac{d}{dt}\frac{\partial U}{\partial \dot{q}_i}, \tag{9}$$

the equations of motion clearly simplify to

$$\frac{d}{dt}\frac{\partial L}{\partial \dot{q}_i} - \frac{\partial L}{\partial q_i} = 0, \quad i = 1, \ldots, n, \tag{10}$$

where L, the Lagrangian function is

$$L = T - U. \tag{11}$$

Most classical interaction forces on a microscopic level, including electromagnetic forces, are derivable from a generalized potential in the form given by Eq. (9). Equations

(10) are customarily called the Lagrangian equations of motion, although Eqs. (8) are often also so designated.

The Lagrangian formulation, as summarized by Eqs. (10), is thus restricted to holonomic systems where the constraints are workless, and where the forces are derivable from a generalized potential. Under such assumptions, an alternative expression of the Lagrangian formulation can be given as a variational principle. The generalized coordinates q_i are considered as the axes in an n-dimensional *configuration space,* in which any given positional configuration of the system corresponds to a point. As the system evolves in time the system point describes a trajectory in configuration space. Between any two fixed points 1 and 2 in configuration space the equations of motion of the system determine a trajectory, and the manner in which the system point traverses it, in such a fashion that

$$\delta \int_1^2 L(q,\dot{q},t)dt = 0; \tag{12}$$

that is to say, the value of the integral in Eq. (12) (known as the *action integral*) evaluated on the actual trajectory has an extremum value relative to all neighboring trajectories with the same end points. Equation (12), *Hamilton's principle,* is a variational principle for which the Euler–Lagrange variational equations are exactly the Lagrange equations of motion of the system, Eqs. (10).

The Lagrangians satisfying Hamilton's principle, and therefore determining the equations of motion, are not uniquely given by the recipe of Eq. (11). Indeed it is clear that if L in any Lagrangian function for which Eq. (12) holds, then an equally suitable Lagrangian is

$$L' = L + \frac{dF(q,t)}{dt}, \tag{13}$$

where F is any (twice differentiable) function of the q's and t. In addition, valid Lagrangians can be found that are not related in this simple manner. Thus, for a single particle moving in one dimension under the influence of a potential V not explicitly dependent on time, then Eq. (11) prescribes the Lagrangian

$$L = \tfrac{1}{2} m\dot{q}^2 - V. \tag{14}$$

As may be easily verified, however, the Lagrangian

$$L = \tfrac{1}{12} m^2\dot{q}^4 + m\dot{q}^2V - V^2 \tag{15}$$

will lead to the same equations of motion, and, consequently, also satisfies Hamilton's principle. The statement of the Lagrangian formulation of mechanics via a variational principle thus leads naturally to extension of the concept of a Lagrangian beyond its initial prescription.

Conservation theorems are easily established in the Lagrangian formulation. If a given coordinate q_i does not appear in L (a so-called *cyclic* and *ignorable* coordinate), then Eqs. (10) show that there exists a first-order constant of the motion,

$$p_i \equiv \frac{\partial L}{\partial \dot{q}_i}. \tag{16}$$

The physical nature of the quantities p_i (known as *generalized* or *conjugate momenta*) depends on what sort of coordinates the q_i are. If q_i is a Cartesian, or translation, coordinate, it can be shown that p_i has the dimensions of a linear momentum; if q_i measures a rotation, p_i is the corresponding angular momentum, etc. Very often, p_i will have no interpretation as a mechanical momentum. The two momentum conservation theorems of Newtonian mechanics can be recovered if q_i is a coordinate measuring either the rigid displacement or rigid rotation, respectively, of the system as a whole.

An additional conserved quantity can be found if L does not contain time explicitly. It can then be shown that a first-order constant of the motion h exists, where

$$h(q,\dot{q}) = \sum_i \dot{q}_i \frac{\partial L}{\partial \dot{q}_i} - L. \tag{17}$$

It very frequently happens that the Lagrangian prescribed by Eq. (11), or otherwise suitably determined, can be written as the sum of three functions, one of which is independent of $\dot{q}_i$, and the other two are homogeneous functions of $\dot{q}_i$ of first and second order, respectively:

$$L = L_0 + L_1 + L_2. \tag{18}$$

While it is not universally required that the Lagrangian conform to this pattern, many of the further developments in classical mechanics are constructed with a view to Lagrangians of the form of Eq. (18). The condition that Eq. (18) holds at least serves to reduce possible ambiguity in the form of the Lagrangian, *vide* Eqs. (14) and (15). If L is of the type described by Eq. (18), application of Euler's theorem on homogeneous functions shows that h, the so-called *Jacobi integral,* is given by

$$h = L_2 - L_0. \tag{19}$$

When further, the Lagrangian is constructed according to Eq. (11) *and* the potential is V independent of $\dot{q}_i$ *and* the defining Eqs. (7) for q_i are not explicit functions of the time, then $L_2 = T$ and $L_0 = -V$, and h is the total energy of the system. Conservation of h then is equivalent to conservation of energy.

To solve problems in analytical dynamics appearing in most engineering situations it is rarely necessary to go beyond the Lagrangian formulation. It has the advantage that the existence of holonomic (and certain types of nonholonomic) constraints can be made visible in the formulation and the forces of constraint obtained by the method of Lagrange multipliers. Finally, extensions can be made to systems with some forms of dissipative forces, through a so-called Rayleigh dissipation function.

HAMILTONIAN FORMULATION

The Hamiltonian formulation is intended for the same restricted class of mechanical systems considered in the Lagrangian formulation, but starts from a radically different viewpoint. It seeks to describe the motion of a holonomic system of n degrees of freedom by $2n$ independent coordinates. Consequently, it must lead to $2n$ first-order equations of motion, which prescribe the trajectory of a system point in the $2n$-dimensional space of the independent variables, known as *phase space*. The $2n$ quantities are divided into two sets, n coordinates q_i and n variables p_i, the conjugate momenta. A function $H(q_i,p_i,t)$, the *Hamiltonian*, is constructed such that the equations of motion are given in terms of the two sets, as

$$\dot{q}_i=\frac{\partial H}{\partial p_i},\quad \dot{p}_i=-\frac{\partial H}{\partial q_i},\quad i=1,2,\ldots n; \tag{20}$$

Eqs. (20) are clearly first-order differential equations in time.

The procedures for the initial selection of the $2n$ variables and the construction of a suitable H are not in principle part of the Hamiltonian formulation. It is usual, however, to start on the basis of some particular generalized coordinates and a corresponding Lagrangian. The conjugate momenta, defined by Eqs. (16), are then selected to play the role of the other half of the independent Hamiltonian coordinates. It can then be shown that Hamiltonian's equations, Eqs. (20), describe the motion if the Hamiltonian is obtained as

$$H(q,p,t)=\sum_i p_i\dot{q}_i-L. \tag{21}$$

In carrying out the prescription implied in Eq. (21) the generalized velocities $\dot{q}_i$ are expressed as functions of q and p by inverting Eqs. (16) so that H is a function only of the Hamiltonian variables and (possibly) the time.

Comparison of Eqs. (17) and (21) shows that h and H have the same values and magnitudes. Functionally (aside from time dependence), however, they are different, for h is a function of the n q_i's and their time derivatives while H is a function of the $2n$ independent variables q_i,p_i. Physically H can be identified as the total energy of the system under the same conditions specified above for h.

Although the Lagrangian and Hamiltonian descriptions of a mechanical system have the same *physical* content their mathematical structure is profoundly different. Velocities transform as *contravariant* variables and the combined set of generalized coordinates and velocities in the Lagrangian description form a $2n$-dimensional manifold known as a *tangent bundle* (TM). The Lagrangian thus provides the mapping from the tangent bundle space to a scalar field, namely, L:TM$\rightarrow R$. By contrast the generalized momenta transform as *covariant* variables and the Hamiltonian phase space of generalized coordinates and momenta is a $2n$-dimensional *symplectic manifold* with a very different geometric structure from the tangent bundle space of the Lagrangian description.

It is important to keep in mind the distinction between the functional bases of the Lagrangian and Hamiltonian formulations, a distinction that is half-hidden by the customary manner of constructing the Hamiltonian from the Lagrangian. In the Lagrangian formulation the conjugate momenta p_i are *defined* by Eqs. (16) as specific functions of q, $\dot{q}$ (and t). As part of the Hamiltonian formulation they are independent variables, whose relationships with q and $\dot{q}$ are *consequences* of the $\dot{q}_i$ half of the equations of motion.

The prescription of Eq. (21) for constructing the Hamiltonian on the basis of the Lagrangian formulation does not always work. If the velocity-dependent part of the Lagrangian is entirely a type L_1, a homogeneous function of the $\dot{q}$'s in first degree, then Eqs. (16) cannot be inverted to find the $\dot{q}$'s as functions of the p's. [For the limiting case in which the $\dot{q}$'s appear only linearly in L, Eqs. (16) do not contain $\dot{q}$ at all!] In effect, then, not all the p's are independent and some of Eqs. (16) act as constraint equations. What one does in this case is not yet clear or universally agreed to. Occasionally (as in relativistic analytical dynamics) one can avoid the problem by seeking an alternative, nonpathologic, Lagrangian which still gives the correct equations of motion.

Hamilton's equation of motion can also be derived from a modified Hamilton's variational principle that says that the equations of motion describe the trajectory in phase space, between fixed end points, which makes the integral

$$\int_1^2\left(H(q,p,t)-\sum_i p_i\dot{q}_i\right)dt \tag{22}$$

an extremum. If the integrand in Eq. (22) is considered as a function of $q,\dot{q},p,\dot{p}$, then the corresponding Euler–Lagrange variational equations are exactly Hamilton's equations of motion. The modified Hamilton's principle contains no physical limitations on the system beyond those applying to Hamilton's principle in the Lagrangian formulation.

Conservation theorems in the Hamiltonian formulation flow directly from the equations of motion. Clearly, from Eqs. (20) the momentum variable p_i is conserved if the conjugate coordinate q_i does not appear in the Hamiltonian. It can be shown easily that a coordinate cyclic in L will also be cyclic in the Hamiltonian constructed from it. From Eqs. (20) it also follows that

$$\frac{dH}{dt}=\sum_i\left(\frac{\partial H}{\partial q_i}\dot{q}_i+\frac{\partial H}{\partial p_i}\dot{p}_i\right)+\frac{\partial H}{\partial t}=\frac{\partial H}{\partial t}. \tag{23}$$

Hence H will be conserved if it is not an explicit function of time. Frequently H is physically the total energy of the system and Eq. (23) then implies a conservation theorem for energy. But it can happen that Eqs. (7) defining q are such that H is conserved when the energy is not conserved, and vice versa, or both H and E can be conserved but may be different constants of the motion.

CANONICAL TRANSFORMATION THEORY

In the Lagrangian formulation, any set of independent generalized coordinates are suitable variables. The Lagrangian has the same significance, $T-U$, in all such sets. In other words, all transformations of configuration space lead to another configuration space with a Lagrangian of the same magnitude, and with the same *form* of the equation of motion. Invariance of this kind does not exist for the Hamiltonian phase space. An arbitrary transformation $(q,p)\mapsto(Q,P)$ will not necessarily lead to variables satisfying equations of motion of the Hamiltonian type. Conditions for transformations leading to Hamilton's equations of motion in the new variables can be expressed in terms of the modified Hamilton's principle. If the transformed set of variables (Q,P) obey equations of motion with a Hamiltonian $K(Q,P,t)$, then in the transformed space the trajectory must make the integral

$$\int_1^2 \left(K - \sum_i P_i \dot{Q}_i \right) dt \tag{24}$$

an extremum. Transformations for which the integrands of (22) and (24), at equivalent points, differ by only the time derivative of an arbitrary point function F are known as *canonical transformations*. It can be shown that if a canonical transformation $(q,p) \mapsto (Q,P)$ does not contain the time explicitly, then $K = H$; otherwise the original and transformed Hamiltonians are related by

$$K = H + \frac{\partial F}{\partial t}. \tag{25}$$

When F is suitably expressed in terms of the old and the new set of variables, it is known as the *generating function* for the canonical transformation; the equations of transformation can be expressed in terms of the partial derivatives of F.

Necessary and sufficient conditions for a canonical transformation can be expressed in a variety of forms; one of the most useful is in terms of *Poisson brackets*. If u and v are two functions of the canonical variables, then the Poisson bracket of u,v with respect to the set (q,p) is the bilinear differential form

$$[u,v]_{q,p} = \sum_i \left(\frac{\partial u}{\partial q_i} \frac{\partial v}{\partial p_i} - \frac{\partial v}{\partial q_i} \frac{\partial u}{\partial p_i} \right). \tag{26}$$

A transformation $(q,p) \mapsto (Q,P)$ is canonical if

$$[Q_i,Q_j]_{q,p} = 0 = [P_i,P_j]_{q,p}, \qquad [Q_i,P_j]_{q,p} = \delta_{ij}. \tag{27}$$

The Poisson brackets appearing in Eq. (27) are called the fundamental Poisson brackets. In words Eq. (27) says that the principal Poisson brackets are invariant under a canonical transformation. It then follows, rather easily, that *all* Poisson brackets are invariant under canonical transformations. Poisson brackets are the classical analogs of the quantum-mechanical commutator, and they share similar algebraic and physical properties.

A more geometric definition of canonical transformations is provided by the use of differential forms. A transformation is canonical if the *differential 2-form* is preserved under the change of variables, namely,

$$\sum_{i=1}^{n} dp_i \wedge dq_i = \sum_{i=1}^{n} dP_i \wedge dQ_i,$$

where $\wedge$ denotes the so-called wedge product.

The solution of the mechanical problem can be looked on as a particular canonical transformation. If the new Hamiltonian is identically zero, the transformed variables are all constants in time. By Eq. (25) such a transformation must necessarily be time dependent. The initial values (q_0,p_0) of the canonical set (q,p) form a suitable set of canonical variables. Hence the transformation $(q_0,p_0) \leftrightarrow (q,p)$ is a canonical time-dependent transformation, for which the equations of transformation give (q,p) as functions of time and the initial values, i.e., the desired solution. In principle, at least, solving the mechanical problem can be reduced to finding the necessary canonical transformation for which $K \equiv 0$. Note that in this view the canonical transformation transforms one system point in phase space (the initial point) to another point in the same space (the system point at time t). This *active* interpretation of a canonical transformation is to be distinguished from the *passive* view in which the transformation relates the coordinates of a given system point in the one phase space to the coordinates in another phase space.

POISSON BRACKET FORMULATION

From the definition of the Poisson bracket, Eq. (26), and Hamilton's equations, Eqs. (20), it follows that the time derivative of any general function $u(q,p,t)$ is

$$\frac{du}{dt} = [u,H] + \frac{\partial u}{\partial t}. \tag{28}$$

Equation (28) is a general form for the equation of motion of quantities related to the mechanical system. With $u = q_i$ or p_i, it reduces to Hamilton's equations of motion. If $u = H$ itself, Eq. (28) leads to Eq. (23). Furthermore, Eq. (28) says that constants of the motion not involving the time explicitly must have vanishing Poisson brackets with H (corresponding to the quantum condition of commuting with H).

The relation between the constants of the motion and the symmetry properties of the system can be derived in an elegant fashion in terms of Poisson bracket behavior. It has been noted previously that the motion of the system point in time from its initial position corresponds to the evolution of a canonical transformation with time as a parameter. There are other canonical transformations depending continuously on some parameter, e.g., spatial rotation. For these transformations one can introduce the concept of an infinitesimal canonical transformation (ICT) corresponding to an infinitesimal change of the parameter from the initial conditions. The change in the value of a function u as the ICT moves the system (on the active viewpoint) from one point to an infinitesimally near point can be shown to be given by

$$\partial u = \epsilon [u,G], \tag{29}$$

where ϵ is the infinitesimal change in the parameter and G is the generating function of the ICT. (It will not come as a surprise to note that $G = H$ for the ICT of motion in time, i.e., where $\epsilon = dt$.) If the system is symmetric under the operation implied by a given ICT, then the Hamiltonian (among other functions of the system) will be unchanged in value by the ICT. Equation (29) then says that the Poisson bracket of H with the generating function G of the given ICT vanishes, which by Eq. (28) immediately shows that G is a constant of the motion, at least where G is not an explicit function of time. More detailed considerations remove even this last restriction, leading to the general theorem: *the constants of motion of a system are the generators of the symmetry transformations of the system, i.e., of ICT's which leave the Hamiltonian invariant.*

The Poisson bracket formulation of the equations of motion is of special consequence in statistical mechanics, where

Eq. (28) leads directly to Liouville's theorem concerning the density of system points in phase space.

COMPLETELY INTEGRABLE HAMILTONIANS AND ACTION-ANGLE VARIABLES

A Hamiltonian system is said to be completely integrable if there exist n (the number of degrees of freedom) single-valued, analytic integrals of the motion $F_1, \ldots, F_n$, with say $F_1=H$, which are in *involution*. This means that all the F_i commute with each other, that is,

$$[F_i,F_j]=0, \qquad i,j=1, \ldots, n.$$

The existence of the n integrals F_i implies that the phase-space trajectories are confined to some n-dimensional manifold in the $2n$-dimensional phase space. It may be shown that this manifold has the topology of an *n-dimensional torus*. This topology imparts a naturally periodic structure to the motion and provides the basis for defining a particularly important set of canonical variables known as the *action-angle* variables.

On an n-dimensional torus one can define n topologically independent closed paths $\mathscr{C}_k$, $k=1, \ldots, n$, which cannot be continuously deformed into each other or shrunk to zero. Using these paths the set of action variables is defined as

$$I_k=\oint_{\mathscr{C}_k} \sum_{i=1}^{n} p_i \, dq_i, \qquad k=1, \ldots, n.$$

The set of conjugate angle variables θ_k are defined via a generating function (found by solving the appropriate Hamilton–Jacobi equation, as discussed below). It is possible to effect a canonical transformation in which the action variables play the role of constant conjugate variables. This results in the transformed Hamiltonian being a function of the action variables only, i.e., $K=K\ (I_i, \ldots, I_n)$ with the consequence that Hamilton's equation take the particularly simple form

$$\dot{I}_k=-\frac{\partial K}{\partial \theta_k}, \qquad \dot{\theta}_k=-\frac{\partial K}{\partial I_k}=\omega_k,$$

where the ω_k denote the set of frequencies associated with a given torus. The equations are trivially integrated leading to the result

$$\theta_k=\omega_k t+\delta_k,$$

where the δ_k are a set of initial phases. The trajectories of completely integrable systems are thus seen to be *multiply periodic* functions of time.

HAMILTON–JACOBI EQUATION AND PERTURBATION THEORY

The observation that the motion of the system can be described by a canonical transformation leads to a reduction of the problem to finding a complete solution of a first-order partial differential equation in $n+1$ variables. Generating functions for finite canonical transformations can be used to provide the equations of transformation if they are functions of a mixture of original and the transformed variables. To find the canonical transformation leading to an identically vanishing K, the generating function, denoted by S, is a function of the old coordinates and the new, constant, momenta. One-half of the equations of transformation are then given by the relations

$$p_i=\frac{\partial S}{\partial q_i}. \tag{30}$$

Equation (25) then says that for $K\equiv 0$, S must satisfy the partial differential equation

$$H\left(q_i, \frac{\partial S}{\partial q_i}, t\right)+\frac{\partial S}{\partial t}=0, \tag{31}$$

known as the Hamilton–Jacobi equation. A complete solution to Eq. (31) will have n nontrivial constants of integration and they, or any n independent functions of them, can serve as the new momenta which we may denote by α_i. The theory of canonical transformations says that the new coordinates, say β_i, are then related to S, known as Hamilton's principal function, by

$$\beta_i=\frac{\partial S(q_i, \alpha_i, t)}{\partial \alpha_i}. \tag{32}$$

Equations (30) and (32), between them, determine (q,p) as functions of the time and $2n$ constants, i.e., solve the problem of the motion in time. A related approach, applicable when H is conserved, is to obtain the solution in terms of the generating function W (Hamilton's characteristic function) of a canonical transformation in which all the new coordinates are cyclic. For conservative systems S in fact is $W-Ht$.

The Hamilton–Jacobi approach has applications to fields outside analytical dynamics. In geometrical optics rays are obtained as normals to the constant-W surfaces. Quantum wave mechanics appears also as an extension of the Hamilton–Jacobi method, in which classical mechanics is to quantum mechanics as geometrical optics is to the wave theory of optics.

One of the main uses of the Hamilton–Jacobi equation is in the formulation of perturbation theory for analytical dynamics. A most important example concerns the case of an integrable Hamiltonian $H_0=H_0(I)$, with associated action-angle variables (I,θ), perturbed by an additional small Hamiltonian term, $H_1=H_1(I,\theta)$. The (I,θ) variables are still canonical but the equations of motion are now

$$\dot{I}_k=-\frac{\partial H_1}{\partial \theta_k}, \qquad \dot{\theta}_k=\omega_k+\frac{\partial H_1}{\partial I_k}.$$

The Hamilton–Jacobi equation can be used to attempt to find a new set of action-angle variables (J,ϕ), where the J are the new actions corresponding to perturbed tori on which the trajectories still execute (modified) multiply periodic motion. In principle, perturbation expansions may be generated to arbitrary order and, for small perturbations, can be useful for calculating the shifts in the frequencies of the motion. Unfortunately, except for special cases, the perturbation expansions generated by the Hamilton–Jacobi formalism are *divergent*. This divergence is more than a mathematical artifact and has extremely important dynamical implications.

THE KAM THEOREM AND THE ONSET OF CHAOS

The problems associated with Hamilton–Jacobi perturbation theory were manifested in celestial mechanics as the problem of *small divisors* encountered in the famous three-body problem which is concerned with, for example, the effect of Jupiter on the orbit of the earth about the sun. The divergences found in this problem suggested the impossibility of making statements about the longtime stability of planetary orbits. An alternative point of view is that these divergences raise serious doubts about the existence of phase-space tori under generic perturbation and hence the fundamental issue of the extent to which phase space is explored by typical trajectories. This fundamental problem was only resolved in the early 1960s by the Kolomogorov–Arnold–Moser (KAM) theorem which proved, for the first time, that for sufficiently small perturbation most tori are preserved. The determination of which tori survive perturbation depends on delicate number-theoretic properties of the frequencies characterizing the unperturbed tori. Despite the preservation of (a large measure of) tori, the perturbed system is fundamentally different from its unperturbed counterpart in that it is *nonintegrable,* that is, it no longer possesses a full complement of constants of the motion. It is for this reason that a global solution to the Hamilton–Jacobi solution no longer exists.

The phase-space structure of a nonintegrable system (such as that created by the generic perturbation of an integrable one) is immensely complicated. The tori which are not preserved break up into sets of alternating stable and unstable orbits—a result which can be proved by a theorem due to Poincaré and Birkhoff. In addition it is possible to prove—and demonstrate by numerical computation—that *chaotic* orbits appear. These orbits are characterized by a great sensitivity to small changes in initial conditions and wander over large regions of accessible phase space in an apparently stochastic (but still deterministic!) manner. Typically, though, such orbits do not explore all of the energy shell, and ergodicity on this shell cannot be claimed. Indeed, it can be proved that a generic Hamiltonian system is neither integrable nor ergodic. Nonetheless it is this appearance of deterministic chaos that provides an understanding of the transition from analytical dynamics to statistical mechanics.

See also Center-of-Mass System; Chaos; Hamiltonian Function; Kinematics and Kinetics; Newton's Laws.

BIBLIOGRAPHY

V. I. Arnold, *Mathematical Methods of Classical Mechanics.* Springer-Verlag, New York, 1978. (A)

V. I. Arnold and A. Avez, *Ergodic Problems of Classical Mechanics.* Benjamin, New York, 1968. (A)

H. Goldstein, *Classical Mechanics,* 2nd ed. Addison-Wesley, Reading, MA, 1980. (I)

L. D. Landau and E. M. Lifshitz, *Mechanics,* 2nd ed. Pergamon Press, Oxford, 1960. (I)

M. Tabor, *Chaos and Integrability in Nonlinear Dynamics: An Introduction.* John Wiley & Sons, New York, 1989. (E–I)

Eigenfunctions

Leonard Eisenbud

Eigenfunctions were introduced initially in relation to the mathematics of function spaces or Hilbert spaces. The concept is a natural generalization of the notion of an eigenvector of a linear operator in a finite-dimensional linear vector space. Eigenfunctions find wide physical applications in classical linear field theories but they are probably best known to physicists for their extensive use in Schrödinger's formulation of quantum mechanics.

In pure mathematics the eigenfunction, or characteristic function, is related to a linear operator defined on a given vector space of functions on some specified domain. The operator is a map or rule of transformation that assigns to each function f in the function space a uniquely defined function, say g, in that space. If we denote the operator by Q and if Q maps f into g, we write $Qf = g$. *If a function h* (but not the null function of the function space) *is mapped by Q into a multiple of itself*, i.e., if $Qh = ch$, where c is a scalar, *h is said to be an eigenfunction of Q belonging to the eigenvalue c*. If to c there belong several independent eigenfunctions, c is said to be degenerate. From the linearity of Q it follows that all linear combinations of the set of independent functions also belong to c. Thus the eigenfunctions of Q belonging to its eigenvalues fill out separate subspaces of the function space.

In the theory of quantum mechanics, states of a physical system may be described by suitably normalized functions defined over the domain of possible *configurations* $\mathscr{C}$ of the system. (This domain requires extension in general to include spin and other nonclassical properties that may be fixed in a particular configuration.) A state function over configuration space is properly normalized if $\int |\psi(\mathscr{C})|^2\, d\mathscr{C} = 1$, where the integration is extended over the range of possible system configurations. These state functions fill out a vector space—the so-called state space of the system. For each observable (e.g., position of a particle, energy, total angular momentum) that can be measured on the physical system, the theory of quantum mechanics defines an associated linear operator on the functions of the state space for the system. For a system consisting of a single particle (to take a simple example) the state space contains all normalizable functions over three-space; the operator for the vector momentum observable $\mathbf{p}$ is given by the theory as $-i\hbar\nabla$ (where $\hbar$ is Planck's constant divided by 2π) and an eigenfunction $\psi(\mathbf{r})$ belonging to the momentum $\mathbf{p}'$ satisfies $-i\hbar\nabla\psi_{p'}(\mathbf{r}) = \mathbf{p}'\psi_{p'}(\mathbf{r})$ so that $\psi_{p'}(\mathbf{r}) = \exp(i\hbar\mathbf{p}'\cdot\mathbf{r})$ (not normalized).

The state functions—or, more accurately, the subspaces of state functions (suitably normalized)—that are eigenfunctions of an operator observable for some particular eigenvalue describe states in which a measurement of the observable is certain to find that eigenvalue. (The observable has no definite value in states not described by eigenfunctions of the associated operator.) If, for example, H is the operator for the observable $\mathscr{E}$—say the energy of the system—then a normalized eigenfunction $\psi_{\mathscr{E}'}$ of H belonging to the eigenvalue $\mathscr{E}'$ (so that $H\psi_{\mathscr{E}'} = \mathscr{E}'\psi_{\mathscr{E}'}$) describes a state of the physical system in which the system *has the energy* $\mathscr{E}'$. The eigenfunction describes the physical system in the following sense: the function $|\psi_{\mathscr{E}'}(\mathscr{C})|^2$ evaluated at a particular configuration $\mathscr{C}$ measures the probability of observing configuration $\mathscr{C}$ in state $\psi_{\mathscr{E}'}$. More generally from a state function $\psi(\mathscr{C})$ the probability of finding observable q to have the value q' on measurement is given by $|\int \chi_{q'}(\mathscr{C})^* \psi(\mathscr{C})d\mathscr{C}|^2$, where $\chi_{q'}(\mathscr{C})$ is the eigenfunction (supposed for simplicity to be nondegenerate) of the operator for observable q with the eigenvalue q'.

See also OPERATORS; QUANTUM MECHANICS.

BIBLIOGRAPHY

Mathematical with Applications to Physics

R. Courant and D. Hilbert, *Methods of Mathematical Physics*. Wiley (Interscience), New York, 1966. (A)

P. M. Morse and H. Feshbach, *Methods of Theoretical Physics*. McGraw-Hill, New York, 1953. (I–A)

F. W. Byron and R. W. Fuller, *Mathematics of Classical and Quantum Physics*. Addison-Wesley, Reading, Mass., 1969. (I)

Quantum Mechanics

J. von Neumann, *Mathematical Foundations of Quantum Mechanics*. Princeton Univ. Press, Princeton, N.J., 1955. (A)

E. Merzbacher, *Quantum Mechanics*. Wiley, New York, 1961. (I)

R. Eisberg and R. Resnick, *Quantum Physics*. Wiley, New York, 1974. (E)

Elasticity

W. E. Bron

Interest in the elastic and inelastic properties of matter dates back to Galileo and has spawned in the interim an overwhelming literature on the mathematical, engineering, and physical aspects of the problem. Much of this body of work ignores the atomic nature of real solids by noting that displacements with wavelengths λ long compared to the interionic spacing a are influenced only weakly by the individual ions and can, accordingly, be treated in the continuum limit. Recently, however, a major interest in the elastic properties of solids has arisen as a result of their applicability to the determination of the interionic forces present in lattice vibrations. It is with the modern goal in mind that this survey is written.

In the long-wavelength limit, $\lambda \gg a$, the equation of motion of a displacement wave can be simply expressed in terms of Hooke's relation between the macroscopic parameters stress and strain. Since most solids are anisotropic, Hooke's law is a tensor relationship of the form

$$\sigma_{ik} = c_{ikjl} \epsilon_{jl} \tag{1a}$$

in which all indices take on the values x, y, z or alternatively 1, 2, and 3, and summation over repeated indices is implied; σ_{ik} are elements of the stress tensor, ϵ_{jl} elements of the strain tensor, and c_{ikjl} are variously called the elastic stiffness constants, the elastic moduli, or the elastic constants. (Their dimension is newtons per square meter.) In order to shorten the cumbersome indexes, a convention exists to transform them to matrix notation as follows:

tensor notation (c)	xx yy zz 11 22 33	yz zy 23 32	xz zx 13 31	xy yx 12 21
matrix notation (C)	1 2 3	4	5	6

such that Hooke's law becomes

$$\sigma_m = C_{mn} \epsilon_n . \tag{1b}$$

This transformation follows from the symmetries

$$c_{ikjl} = c_{kilj} = c_{iklj} = c_{jlik} ,$$

the first three of which are always present; the last follows from the existence of an elastic potential. In the most general case there remain 21 independent elements of the C tensor.

In the matrix notation Eq. (1) explicitly becomes

$$\begin{aligned}
\sigma_1{}^{\mathrm{P}} &= C_{11}\epsilon_1 + C_{12}\epsilon_2 + C_{13}\epsilon_3 + C_{14}\epsilon_4 + C_{15}\epsilon_5 + C_{16}\epsilon_6 ,\\
\sigma_2{}^{\mathrm{P}} &= C_{21}\epsilon_1 + C_{22}\epsilon_2 + C_{23}\epsilon_3 + C_{24}\epsilon_4 + C_{25}\epsilon_5 + C_{26}\epsilon_6 ,\\
\sigma_3{}^{\mathrm{P}} &= C_{31}\epsilon_1 + C_{32}\epsilon_2 + C_{33}\epsilon_3 + C_{34}\epsilon_4 + C_{35}\epsilon_5 + C_{36}\epsilon_6 ,\\
\sigma_4{}^{\mathrm{S}} &= C_{41}\epsilon_1 + C_{42}\epsilon_2 + C_{43}\epsilon_3 + C_{44}\epsilon_4 + C_{45}\epsilon_5 + C_{46}\epsilon_6 ,\\
\sigma_5{}^{\mathrm{S}} &= C_{51}\epsilon_1 + C_{52}\epsilon_2 + C_{53}\epsilon_3 + C_{54}\epsilon_4 + C_{55}\epsilon_5 + C_{56}\epsilon_6 ,\\
\sigma_6{}^{\mathrm{S}} &= C_{61}\epsilon_1 + C_{62}\epsilon_2 + C_{63}\epsilon_3 + C_{64}\epsilon_4 + C_{65}\epsilon_5 + C_{66}\epsilon_6 .
\end{aligned} \tag{2}$$

In Eq. (2) the superscripts P and S refer to pressure (stress) and shear stress, respectively; ϵ_1, ϵ_2, and ϵ_3 refer to simple dilational displacements, whereas ϵ_4, ϵ_5, and ϵ_6 refer to shear displacements.

It follows that

$$C_{mn} = \frac{\partial \sigma_m}{\partial \epsilon_n} . \tag{3}$$

For crystals with cubic lattice symmetry

$$C_{11} = C_{22} = C_{33} , \quad C_{44} = C_{55} = C_{66} , \quad C_{12} = C_{23} = C_{31} ,$$

and

$$\begin{aligned} C_{14} = C_{15} = C_{16} = C_{24} = C_{25} = C_{26} = C_{34} = \\ C_{35} = C_{36} = C_{45} = C_{46} = C_{56} = 0. \end{aligned} \tag{4}$$

Young's modulus Y is defined as the longitudinal stress $\sigma_1{}^{\mathrm{P}}$ per unit strain ϵ_1 when the body is not constrained in the transverse directions, i.e., when ϵ_2 and ϵ_3 are free to change. Accordingly, for a cubic crystal

$$Y_{\mathrm{cub}} = \frac{(C_{11} - C_{12})(C_{11} + 2C_{12})}{C_{11} + C_{12}} .$$

The bulk modulus B, defined in terms of the pressure P and the volume V as

$$B = -V \frac{dV}{dP} ,$$

becomes for a cubic crystal

$$B_{\mathrm{cub}} = (C_{11} + 2C_{12}),$$

and the compressibility $K = 1/B$. The shear modulus (or modulus of rigidity) S becomes

$$S_{\mathrm{cub}} = C_{44} .$$

The corresponding elastic moduli for an isotropic body are

$$Y_{\mathrm{iso}} = \frac{C_{44}(3C_{12} + 2C_{44})}{C_{12} + C_{44}} ,$$

$$B_{\mathrm{iso}} = C_{12} + C_{44} ,$$

$$S_{\mathrm{iso}} = C_{44} .$$

If the forces between lattice ions are central, the elastic constants of isotropic bodies obey the further relations

$$\begin{aligned} C_{12} = C_{66} , \quad \backslash C_{23} = C_{44} , \quad C_{31} = C_{55} , \\ C_{14} = C_{56} , \quad C_{25} = C_{64} , \quad C_{36} = C_{45} . \end{aligned} \tag{5}$$

For cubic crystals this becomes the well-known Cauchy relation; i.e., $C_{12} = C_{44}$.

Interest in the elastic constants in regard to lattice dynamics stems from the fact that the propagation of elastic waves in solids can be written in terms of these quantities. We define the density of the medium as ρ, and u, v, and w as the lattice displacements caused by the elastic wave in a small incremental volume Δx, Δy, and Δz in, respectively, the x, y, and z directions of the volume element. The elastic wave is generated by a stress gradient across the incremental volume. The equation of motion of the medium in, say, the x direction is then

$$\rho \frac{\partial^2 u}{\partial t^2} = \frac{\partial \sigma_1}{\partial x} + \frac{\partial \sigma_6}{\partial y} + \frac{\partial \sigma_5}{\partial z} , \tag{6}$$

and similarly for v and w. From Eq. (3) and the conditions (4) it follows for cubic crystals that

$$\rho \frac{\partial^2 u}{\partial t^2} = C_{11} \frac{\partial \epsilon_1}{\partial x} + C_{12} \left(\frac{\partial \epsilon_2}{\partial x} + \frac{\partial \epsilon_3}{\partial x} \right) + C_{44} \left(\frac{\partial \epsilon_6}{\partial y} + \frac{\partial \epsilon_5}{\partial z} \right) . \tag{7}$$

The elements of the strain tensor are, of course, themselves

Table I Sound velocities

[100] direction	[110] direction	[111] direction
$\rho v_l^2 = C_{11}$	$\rho v_l^2 = \frac{1}{2}(C_{11} + C_{12} + 2C_{44})$	$\rho v_l^2 = \frac{1}{3}(C_{11} + 2C_{12} + 4C_{44})$
$\rho v_t^2 = \rho v_{t'}^2 = C_{44}$	$\rho v_t^2 = C_{44}$	$\rho v_t^2 = \rho v_{t'}^2$
	$\rho v_{t'}^2 = \frac{1}{2}(C_{11} - C_{12})$	$= \frac{1}{3}(C_{11} - C_{12} + C_{44})$

related to the displacements. Specifically

$$\epsilon_1 = \frac{\partial u}{\partial x}, \quad \epsilon_2 = \frac{\partial v}{\partial y}, \quad \epsilon_3 = \frac{\partial w}{\partial z},$$

$$\epsilon_4 = \frac{\partial v}{\partial z} + \frac{\partial w}{\partial y}, \quad \epsilon_5 = \frac{\partial u}{\partial z} + \frac{\partial w}{\partial x}, \quad \epsilon_6 = \frac{\partial u}{\partial y} + \frac{\partial v}{\partial x}.$$

Therefore Eq. (7) can be rewritten as

$$\rho \frac{\partial^2 u}{\partial t^2} = C_{11} \frac{\partial^2 u}{\partial x^2} + C_{44} \left(\frac{\partial^2 u}{\partial y^2} + \frac{\partial^2 u}{\partial x^2} \right) + (C_{12} + C_{44}) \left(\frac{\partial^2 v}{\partial x \partial y} + \frac{\partial^2 w}{\partial x \partial z} \right). \quad (8)$$

Similar equations of motion exist for $\partial^2 v/\partial t^2$ and $\partial^2 w/\partial t^2$.

Solutions to these equations can be written in the form

$$u = u_0 \exp[i(kx - \omega t)], \quad (9)$$

in which k is the wave vector of the wave and ω its angular frequency. The magnitude of the k vector is related to the wavelength by $k = 2\pi/\lambda$, and ω is related to the linear frequency ν by $\omega = 2\pi\nu$. The velocity of propagation is given by $v = \nu\lambda = \omega/k$. Solutions exist for both longitudinal (l) and transverse (t) waves—that is, respectively, waves for which the displacement is in the same direction as the propagation direction and (two) waves in which the displacement is perpendicular to the propagating direction. The velocities depend then on the propagation direction. For waves propagating along the [100], [110], and [111] symmetry directions of cubic crystals the relations determined by substituting Eq. (9) into Eq. (8) and displayed in Table I can be used to determine the propagation (sound) velocity of the displacement waves.

The relations of Table I are, in fact, the ones that permit accurate measurements to be made of the elastic constants. Modern methods of determining the elastic constants proceed by measuring the sound velocity propagating in the body under observation, in various directions and polarizations, until enough relations of the type of Table I are obtained to specify the number of independent elastic constants. In these experiments piezoelectric transducers (such as quartz crystals) are excited to produce ultrasonic pulses whose time of flight across a known length of material yields the desired velocity measurement.

In modern (classical) models for the lattice vibrations of real crystals the ions of the lattice are replaced by nondeformable masses ("cores") and the outer electrons are replaced by deformable "shells." The cores and shells are interconnected by springs that simulate the interionic and polarization forces. In lowest order the springs are considered to be harmonic; i.e., the force exerted by the spring is directly proportional to its change in length. The solutions to the equations of motion of such systems of particles and forces also result in displacement waves whose wave vector and frequency vary over a wider range than those of the waves discussed earlier. The form of these waves is, however, related to that of those of Eq. (9). In fact, in the harmonic limit and in the limit of long wavelength (i.e., small k vectors) the solutions obtained from the lattice of discrete masses and springs reduce to the sound waves in elastic media. This means that the elastic constants, as defined earlier, can be used in part to define the properties of the springs in the dynamical model. This procedure is extensively and successfully performed at this time.

Experimentally the elastic constants are found to depend on temperature. In general they tend to decrease in magnitude with increasing temperature. This behavior results in part from the expansion of the crystal lattice with temperature. Clearly, as the lattice expands the interionic forces change. In addition, however, increasing temperature results in an increase in the average amplitude of vibration of the lattice ions. The combined effects have been shown to be the major source of the observed temperature dependence of the elastic constants, at least in simple solids.

So far the discussion has been limited to small lattice displacement and lattice vibrations in the "harmonic" limit for which the elastic energy stored in the lattice is described by terms up to second order in the strain components. [Within the elastic limit the lattice energy is the integral over the product of the stress and the strain, and hence, according to Eq. (1a), is proportional to the square of the strain components.] Effects resulting from even higher-order products of the strain components are described in elasticity theory by higher-order elastic constants. Third-order elastic constants can be measured through sound-velocity measurements on statically stressed solids, and the relationships between these and even higher-order elastic constants can be derived.

See also ANELASTICITY; DEFORMATION OF CRYSTALLINE MATERIALS; LATTICE DYNAMICS; RHEOLOGY.

BIBLIOGRAPHY

Theory of Elasticity

D. F. Nelson, *Phys. Rev. Lett.* **60**, 608 (1988). (I)

L. D. Landau and E. M. Lifshitz, *Theory of Elasticity*. Addison-Wesley, Reading, Mass., 1959. (A)

H. Leipholz, *Theory of Elasticity*. Nordhoff, Leyden, 1974.

A. E. H. Love, *The Mathematical Theory of Elasticity*, 1st ed. Dover, New York, 1892. (A)

S. P. Timoshenko and J. N. Gordier, *Theory of Elasticity*, 1st ed. McGraw-Hill, New York, 1934. (I)

Application to Solid-State Physics

N. W. Ascroft and N. D. Mermin, *Solid State Physics*. Holt, Rinehart & Winston, New York, 1976. (I)

C. Kittel, *Introduction to Solid State Physics*, 4th ed., Chapter 4. Wiley, New York, 1971. (E)

C. Zener, *Elasticity and Anelasticity of Metals*. Univ. of Chicago Press, Chicago, 1948. (E)

Measurement and Values of Elastic Constants

O. L. Anderson, *Physical Acoustics* (W. P. Mason, ed.), Vol. IIIB, pp. 43–95. Academic, New York, 1965. (E)

R. Truell, C. Elbaum, and B. B. Chick, *Ultrasonic Measurements in Solid State Physics*. Academic, New York, 1969. (E)

Higher-Order Elastic Constants

R. A. Cowley, *Adv. Phys.* **12,** 421 (1963). (I)

R. C. Hollinger and G. R. Barsch, *J. Phys. Chem. Solids* **37,** 845 (1976). (I)

J. M. Ziman, *Electrons and Phonons*, Chapter 3. Oxford (Clarendon Press), London and New York, 1960.

Application to Lattice Dynamics

W. Cochran and R. A. Cowley, *Handb. Phys.* **25/2a,** 59 (1967). (A)

J. A. Reissland, *The Physics of Phonons*. Wiley, New York, 1973. (I)

Electric Charge

Melba Phillips

Electric charge is a property of matter, first observed in the behavior of certain materials we now call dielectrics. Specifically, it was noted in ancient Greece that amber, on being rubbed, attracts bits of straw and other light objects. Hence the name, since the Greek word for amber is *electron*. That electrified bodies repel as well as attract was discovered by Niccolo Cabeo during the first half of the seventeenth century. Nearly a hundred years later Charles-François du Fay demonstrated that there are two kinds of electricity: "Each [body] repels bodies which have contracted an electricity of the same nature as its own, and attracts those whose electricity is of a contrary nature." The two electricities were designated *positive* and *negative* by Benjamin Franklin, a terminology that implies the *conservation of charge*: Charge cannot be created or destroyed, for the algebraic sum of the positive and negative charges in a closed or isolated system does not change in any circumstances. Franklin also established the arbitrary convention as to which is called positive and which negative.

The force between two charges varies with distance in the same way as the gravitational force between two masses, inversely as the square of the distance. This law was first deduced by Joseph Priestley (discoverer of oxygen) from Franklin's observation that there is no electric force *inside* a metal container, and published (1767) in his history of electricity. The empirical discovery of the inverse square law was made by Charles Augustin Coulomb in 1785. (Henry Cavendish anticipated Coulomb's work, but failed to publish.) The mathematical theory of electrical interactions (see ELECTROSTATICS; GAUSS'S LAW) is quite analogous to that of gravitational attraction. The necessary modifications and extensions of the celestial mechanics of Lagrange and Laplace were developed by Poisson in 1812–1813.

In the twentieth century it has become known that charge is intrinsic to the stable components of atoms, electrons and nuclei. This charge is quantized; i.e., it is observed only in multiples of a smallest amount, the electronic charge, usually designated e. In neutral matter the number of elementary charges of one sign is equal to the number of opposite sign. Early in the eighteenth century it was discovered that electricity could be conveyed from one object to another by certain materials, which became known as conductors, and the science of current electricity began. Electric currents result from the motion of charge carriers. In metals the carriers are electrons, each with one elementary charge that is negative in sign. This is essentially because electrons are much more mobile than the massive positive nuclei. In fluids the atomic or molecular charges of both signs may move, and such charge carriers are called ions.

Quantitatively, Coulomb's law for the force F between two point charges q and q' a distance r apart is written

$$F=k(qq'/r^2)$$

where k is a constant that depends on the units. If k is taken to be 1 and dimensionless and both F and r are measured in cgs units, the electrostatic unit of charge (esu), still often used in atomic physics, is defined. In the International System of units (SI) charge is measured in coulombs. The coulomb (C) is derived from the basic SI unit of current, the ampere, and is equal to 1 ampere second (A·s). The standard ampere is determined mechanically from the magnetic force between two current-carrying conductors. In SI units k has a magnitude very nearly equal to 9×10^9; exactly, $k=c^2\times10^{-7}$ where c is the speed of light in meters per second. To three significant figures, the elementary charge $e=1.60\times10^{-19}$ C. Charge is always associated with mass, although some unstable massive particles, the neutron and the neutral mesons and baryons, have no net charge. Neutrinos have no charge, and no mass so far as is presently known.

See also ELECTRON.

BIBLIOGRAPHY

Edmund Whittaker, *A History of the Theories of Aether and Electricity*. Vol. I. Philosophical Library, Philadelphia, 1951.

Richard Becker, *Electromagnetic Fields and Interactions*. Vol. I. Dover, New York, 1982.

J. D. Jackson, *Classical Electrodynamics*. 2nd ed. Wiley, New York, 1975.

Electric Moments

Joseph Macek

The spatial distribution $\rho(x,y,z)$ of a physical quantity is characterized by its moments m_{ijk} given by

$$m_{ijk}=\int \rho(x,y,z)x^iy^jz^k\,d\tau. \quad (1)$$

Under suitable conditions, $\rho(x,y,z)$ can be reconstructed from a knowledge of its moments, and thus ρ or m offer different but equally complete parametrizations of distributions. It is particularly convenient to parametrize electric charge distributions by their moments since the static electric potential $\Phi(\mathbf{r})$ outside of a source relates directly to multipole moments.

The electric potential $\Phi(\mathbf{r})$ of a charge distribution $\rho(\mathbf{r})$ is given by

$$\Phi(\mathbf{r}) = \int \frac{\rho(\mathbf{r}')}{|\mathbf{r}-\mathbf{r}'|}\, d\tau' = \sum_{lm} \frac{2l+1}{4\pi} Y_{lm}(\hat{\mathbf{r}}) \int \frac{r_<^l}{r_>^{l+1}} \rho(\mathbf{r}') Y_{lm}(\hat{\mathbf{r}}')\, d\tau' \tag{2}$$

where $r_<$ is the smaller of the radii r or r' and $r_>$ is the larger. The spherical harmonics $Y_{lm}(\hat{\mathbf{r}})$ are defined in terms of associated Legendre functions $P_l^m(\cos\theta)$ according to

$$Y_{lm}(\theta,\phi) = \left(\frac{2l+1}{4\pi}\frac{(l-m)!}{(l+m)!}\right)^{1/2} P_l^m(\cos\theta)e^{im\phi} \tag{3}$$

where

$$P_l^m(x) = \frac{(-1)^m}{2^l l!}(1-x^2)^{m/2}\frac{d^{l+m}}{dx^{l+m}}(x^2-1)^l. \tag{4}$$

Outside of the charge distribution where $r_> = r$ and $r_< = r'$, Eq. (2) takes the form

$$\Phi(\mathbf{r}) = \sum_{lm} \frac{2l+1}{4\pi} Y_{lm}(\hat{\mathbf{r}}) \frac{q_{lm}}{r^{l+1}}, \tag{5}$$

with

$$q_{lm} = \int \rho(\mathbf{r}') r'^l Y_{lm}(\hat{\mathbf{r}}')\, d\tau'. \tag{6}$$

Equation (6) defines the electric multipole moments of the charge distribution $\rho(\mathbf{r})$ expressed in units of charge $\times(\text{length})^l$. In the physics literature these units are used for the monopole and dipole moments, while higher moments are defined in units of $(\text{length})^l$ by dividing q_{lm} of Eq. (6) by the unit of electric charge $e = 4.8\times10^{-10}$ esu. Since $r^l Y_{lm}(r)$ is a homogeneous polynomial in x,y, and z, the multipole moments q_{lm} equal certain linear combinations of the moments m_{ijk} with $i+j+k=l$. One cannot use the multipole moments to reconstruct $\rho(\mathbf{r})$, however, since there are fewer moments q_{lm} than moments m_{ijk} with $i+j+k=l$; rather the usefulness of the q_{lm}'s rests on Eq. (5).

In essence, Eq. (5) represents the potential outside of a source in terms of the potentials of point multipoles located at the origin. A macroscopic dielectric medium is made up of microscopic constituents whose static electrical properties are characterized by the multipole moments. The potential of the medium is a linear superposition of potentials due to the idealized point multipoles located at a distribution of points in space. The multipole moments thus represent an essential link between microscopic and macroscopic theories of matter. They are also important for the information that is provided about the internal dynamics of the elementary constituents of matter.

The lowest orders of electric moments are of most significance. The zeroth multiple q_{00} just equals the net charge q. The first-order multipoles q_{1m} relate directly to the dipole moment vector $\mathbf{p} = \int \rho(\mathbf{r})\mathbf{r}\, d\tau$ according to

$$q_{10} = (3/4\pi)^{1/2} p_z, \qquad q_{1\pm1} = \mp(3/4\pi)^{1/2}(p_x \mp ip_y), \tag{7}$$

while the multipoles q_{2m} relate to the components of the quadrupole tensor

$$Q_{ij} = \int \rho(\mathbf{r})(x_i x_j - r^2\delta_{ij})\, d\tau,$$

for example,

$$q_{20} = \tfrac{1}{2}(5/4\pi)^{1/2} Q_{33}. \tag{8}$$

It should be noted that the values of the multipole moments depend, in general, upon the arbitrary origin of coordinates; however, the lowest nonvanishing moment does not. For charged ions and atomic nuclei the lowest-order amount is the total charge q. With the origin located at the center of charge the next-order moment, the dipole moment, vanishes by definition. The center of charge need not coincide with the center of mass, but to the best of our knowledge the center of mass and the center of charge of atomic nuclei and charged elementary particles do indeed coincide. Electrically neutral elementary particles, in particular the neutron, exhibit no electric dipole moment despite highly sensitive experiments designed to measure such a moment. The present limit on the electric dipole moment of the neutron is 10^{-23} esu cm.

In addition to a net charge atomic nuclei frequently exhibit nonzero quadrupole moments. The existence of these moments has provided valuable information on nuclear forces; for example, the quadrupole moment of the deuteron, equal to 0.273×10^{-26} cm^2, implies a noncentral character for the force between the constituent proton and neutron.

Atoms and molecules in gases are normally neutral so that $q_{00}=0$; indeed, atoms are normally in isotropic states so that all multipole moments vanish. Application of an external electric field distorts atomic charge distributions by partially separating the positive and negative charges, thereby inducing a nonzero dipole moment. At low values of the electric field the induced dipole moment is proportional to the external electric field. The constant of proportionality is called the polarizability and its value depends on the dynamics of the particular atomic state.

The permanent moments of molecules are defined as the moments of the molecular charge distribution in a hypothetical coordinate frame where the molecules as a whole do not rotate. In such a frame, molecules may exhibit both induced and permanent dipole moments, quadrupole moments, etc. The order of magnitude of molecular dipole moments equals the product of electronic charge 4.8×10^{-10} esu times an atomic diameter 10^{-8} cm, i.e., 4.8×10^{-18} esu cm. The quantity 10^{-18} esu cm is called a debye. Permanent and induced dipole moments are typically measured indirectly by measuring the macroscopic temperature-dependent dielectric constant and applying the Clausius–Mosotti equation to deduce the microscopic dipole moments.

See also ELECTROSTATICS; POLARIZABILITY.

BIBLIOGRAPHY

Amos de Shalit and Herman Feshbach, *Theoretical Nuclear Physics,* Vol. 1, pp. 53–58. Wiley, New York, 19.

J. D. Jackson, *Classical Electrodynamics,* 2nd ed., pp. 136–167. Wiley, New York, 1975.

Norman F. Ramsey, in *Atomic Physics* (Richard Marrus, Michael Prior, and Howard Shugart, eds.), pp. 453–471. Plenum Press, New York, 1977.

Emilio Segrè, *Nuclei and Particles,* pp. 221–244. Benjamin Advanced Book Program, Reading, MA, 1965.

Electrochemical Conversion and Storage

George E. Blomgren

Electrochemical conversion of chemical energy into electrical energy occurs in a Galvanic cell by an oxidation reaction contributing electrons to an external circuit through an electrode called the anode, while a reduction reaction removes electrons from the external circuit through an electrode called the cathode. The active materials which undergo the oxidation and reduction reactions must be in physical (i.e., electronic) contact with the anode and cathode, respectively, so that electronic charge transfer to or from the electrodes can occur. The electrodes must not touch each other, since such contact would cause an internal short circuit, but must each contact a continuous ionically conductive medium such as an electrolyte solution. To ensure that the electrodes do not touch, a porous, insulating material called the separator is often interposed between the electrodes. The separator pores are then filled with electrolyte solution.

An active material which is a sufficiently good electronic conductor can function as its own electrode. Thus, a metal anode often acts as its own electrode and, in the Leclanche dry cell, as the case of the battery as well.

An easily understood example of these definitions is the Daniell cell consisting of zinc anode active material and copper sulfate cathode active material. The anodic and cathodic reactions are

$$Zn \rightarrow Zn^{2+} + 2e^-, \text{ oxidation (anode)},$$

$$CuSO_4 + 2e^- \rightarrow Cu + SO_4^{2-}, \text{ reduction (cathode)}.$$

When a voltage source of opposite polarity to the Galvanic cell can easily reverse the anode and cathode reactions, a storage cell or secondary battery results. A cell made for discharging the active materials only one time is called a primary battery.

Michael Faraday first described the basis of the operation of a Galvanic cell. These statements, now called "Faraday's laws of electrolysis," are: (1) The amount of chemical change produced by electrolysis is proportional to the total amount of electrical charge passed through the cell; (2) the amount of chemical change produced is proportional to the equivalent weight of the substance undergoing chemical change. Thus, the amount of charge in coulombs (C) passed through the cell is proportional to the number of chemical equivalents undergoing chemical change and the constant of proportionality is F or Faraday's constant (equal to 96,484.56 C/equivalent). Under conditions of maximum efficiency, the Gibbs free energy change, ΔG, of the overall chemical reaction of the Galvanic cell is equal to the negative of the electrical work capable of being obtained. The counter electromotive force (EMF) of the device upon which the electrical work is done must be, under these conditions, the reversible or maximum EMF of the cell, E. The electrical work is given by the amount of electricity flowing through the cell (equal by Faraday's laws to nF, where n is the number of equivalents corresponding to the cell reaction) multiplied by the EMF. Thus

$$\Delta G = -nFE. \tag{1}$$

Walther Nernst extended the considerations of Faraday to take into account the dependence of potential on the concentration or activity of reactant and product species. The free energy of each species i (reactant or product) depends on the activity a_i and the standard free energy, G_i^0:

$$G_i = G_i^0 + RT \ln a_i. \tag{2}$$

The reaction free energy depends on the difference between the sum of product free energies and the sum of reactant free energies. With use of (1) and (2) the EMF is given by the Nernst equation as

$$E = E^0 - RT \ln \frac{\prod_i a_i \text{ (products)}}{\prod_i a_i \text{ (reactants)}}. \tag{3}$$

The above discussion shows that the energy obtainable from an ideal Galvanic cell (operating under reversible conditions) is equivalent to the free energy inherently available from the chemical system. The Carnot limitation of a heat engine does not apply to Galvanic cells. The ratio of the free energy change to the total energy change is usually close to unity since the entropy effects are small for most practical cells. However, in spite of theoretical superiority, the difficulty of scale-up and problems of presenting a continuous feed of reactants and removal of products in a Galvanic cell has led to almost universal use of heat engines for large-scale production of electrical energy. Thus, the main application for Galvanic cells is the field of isolated or portable power sources.

Losses of energy in a Galvanic cell consist of voltage losses and Coulombic or active material losses. Active material losses can occur from shedding of active material from the battery plate, by electronic isolation as active materials are surrounded by insulating reaction products, or by parasitic chemical or corrosion reactions which convert active materials into inactive products. Voltage losses are far more complicated in origin, and the study of these losses forms the basis of much present work in electrochemistry. There are three main types of voltage loss or polarization: activation, concentration, and Ohmic. All three depend on the

Table I Properties of Some Presently Manufactured Galvanic Cells

Cell name	Anode active material	Cathode active material	Open circuit voltage	Specific energy (J/kg × 10^{-5})	Energy density (J/m^3 × 10^{-8})	Cell reaction
			Aqueous primary cells			
Leclanche	Zn	MnO_2	1.58	2.4	5.5	$Zn + 2NH_4Cl + 2MnO_2 \rightarrow Zn(NH_3)_2Cl_2 + 2MnO(OH)$
Alkaline MnO_2	Zn	MnO_2	1.55	2.8	7.2	$Zn + 2H_2O + 2MnO_2 \rightarrow Zn(OH)_2 + 2MnO(OH)$
Mercury	Zn	HgO	1.35	4.0	18.0	$Zn + HgO + H_2O \rightarrow Zn(OH)_2 + Hg$
Silver	Zn	Ag_2O	1.60	4.4	18.0	$Zn + Ag_2O + H_2O \rightarrow Zn(OH)_2 + 2Ag$
			Nonaqueous primary cells			
Lithium MnO_2	Li	MnO_2	3.0	7.2	14.4	$Li + MnO_2 \rightarrow Li_xMnO_2$
Lithium carbon fluoride	Li	CF_x	3.1	7.2	14.4	$xLi + CF \rightarrow xLiF + C^x$
Lithium iron sulfide	Li	FeS_2	1.8	5.8	14.7	$4Li + FeS_2 \rightarrow 2Li_2S + Fe$
Lithium sulfur dioxide	Li	SO_2	3.0	10.0	15.8	$2Li + SO_2 \rightarrow Li_2S_2O_4$
Lithium thionyl chloride	Li	$SOCl_2$	3.6	12.8	30.7	$4Li + SOCl_2 \rightarrow 4LiCl + SO_2 + S$
Lithium iodine	Li	I_2 polymer	2.8	8.6	28.0	$2Li + I_2(polymer) \rightarrow 2LiI + polymer$
			Secondary cells			
Lead acid	Pb	PbO_2	2.2	1.6	2.4	$PB + 2H_2SO_4 + PbO_2 \rightleftharpoons 2PbSO_4 + 2H_2O$
Nickel–cadmium	Cd	NiO(OH)	1.35	1.6	3.5	$Cd + 2NiO(OH) + 2H_2O \rightleftharpoons Cd(OH)_2 + 2Ni(OH)_2$

current delivered by the cell in a monotonically increasing way. Ohmic polarization arises from the resistance of all of the elements of the cell and follows Ohm's law for the current dependence. Concentration polarization arises from the creation of concentration gradients in the cell and obeys the Nernst equation. Current dependence enters indirectly through the concentration gradient dependence on the current. Activation polarization is a kinetic factor for the charge transfer reaction and is in general different for each electrode in the functional dependence on current and concentrations and in the kinetic parameters describing the charge transfer reaction. The effect of time of discharge or charge is to increase the voltage loss of the cell. All of these losses contribute to lowering the efficiency of the Galvanic cell.

Many hundreds of couples of active materials have been proposed for use as Galvanic cells. However, very few of these have been manufactured successfully. Table I lists the main types of primary and secondary cells available as articles of commerce, the anode and cathode active materials, the open circuit voltages, the observed energies per unit weight and volume, and the cell reaction on discharge. The Leclanche and alkaline MnO_2 primary cells and the lead acid secondary cell have experienced the most applications by far. The superior energy densities of the silver oxide and mercuric oxide primary cells and the nickel cadmium secondary cell are balanced by their high cost of materials.

In addition, the increasing concerns of the public and governments toward discarding devices which may contain toxic materials such as mercury and cadmium has caused many changes in the marketing and availability of certain batteries. For example, the use of mercury batteries has decreased markedly in nearly all countries, even where laws restricting their use are not in place. In part, this is because the silver-oxide cell and the recently developed lithium cells are more than adequate substitutes. Also, the use of mercury in zinc–MnO_2 cells, acid or alkaline, has been greatly reduced or eliminated in commercial batteries (mercury has traditionally been used on zinc as an additive to improve kinetics and reduce corrosion). The high-energy density and absence of toxic heavy metals have encouraged the gradual introduction of lithium primary batteries, as noted in Table I. The high voltage of unit cells in the cases of manganese dioxide and carbon fluoride cells, however, has made the introduction more difficult since devices must be designed either to accommodate a high voltage or battery compartments must be modified to accommodate fewer unit cells. The lithium–iron–sulfide system can be a direct replacement for zinc cells and has been so used in miniature batteries. The major use of lithium–sulfur dioxide and lithium–thionyl-chloride batteries is for military applications. These batteries are capable of delivering very high currents and have outstanding specific energy and energy density although disposal remains a problem. They are not widely available to consumers for this reason and for reasons of safety. The lithium–iodine battery has been widely used for medical purposes; in fact the widespread use of heart pacemakers relies on the excellent properties of this battery. In spite of its very high energy, however, it is only useful for very small current drains because

Table II Properties of Some Galvanic Cells under Development

Anode active material	Cathode active material	Electrolyte	Specific energy ($J/kg \times 10^{-5}$)	Energy density ($J/m^3 \times 10^{-8}$)
		Primary cells		
Li	MnO_2	Li salt–organic solvent	3.5	7.6
Li	TiS_2	$LiAsF_6$–organic solvent	3.1	6.7
Li	MoS_2	$LiAsF_6$–organic solvent	1.8	5.0
		Secondary cells		
Li (Al alloy)	FeS_2	LiCl–KCl (eutictic melt)	3.6	9.1
Na (liquid)	S (liquid)	β-Alumina (solid)	3.6	—
H_2 (dissolved in metal)	NiO(OH)	$KOH–H_2O$	2.2	7.2
Zn	NiO(OH)	$KOH–H_2O$	2.4	5.4
		Fuel cell		
H_2	O_2	$KOH–H_2O$ or $H_3PO_4–H_2O$	—	—

of its high impedance. For secondary batteries, the use of toxic lead and cadmium continues because there are not sufficiently good alternatives at the present time. In the case of large batteries, most of the lead and cadmium is recycled when new batteries are purchased.

Table II lists some systems which are in advanced stages of development and have been used for various special applications or show promise for greatly improved performance. In recent years, much effort has gone into room-temperature, rechargeable lithium cells with organic solvents, especially for portable power applications for consumers. Although it is still too early to tell if any of these systems will be successful replacements for existing batteries, many of the deficiencies of nickel–cadmium and lead acid batteries have been overcome. Cost, cycle life performance, toxicity and availability of materials, and battery safety must all be established before any of the systems of Table II become viable commercial products. The demands are even more stringent for batteries to be used for traction purposes in electric vehicles. The sodium–sulfur and lithium–iron–sulfide systems are the best hopes at present for the greatly improved specific energy and energy density secondary cells with cost and performance comparable to lead acid cells necessary for vehicle batteries. The hydrogen–oxygen fuel cell provides continuous flow of active materials and continuous (water) product removal which permits power production on an arbitrarily large scale. In spite of many years of effort, the problem of finding a sufficiently cheap or long-life catalyst, which is necessary to lower the electrode polarizations, has prevented full development of an economically viable fuel cell system to date.

See also ELECTROCHEMISTRY; FREE ENERGY.

BIBLIOGRAPHY

D. H. Collins (ed.), *Power Sources*. Pergamon, New York, 1968, et seq. Proceedings of the Biennial International Symposium of Batteries at Brighton, England. Presents state of the art information on electrochemical energy conversion devices and the results of studies on battery related problems. (A)

C. A. Hampel (ed.), *The Encyclopedia of Electrochemistry*. Reinhold, New York, 1964. The entries under "Batteries" and "Fuel Cells" are especially useful as are the entries under specific battery systems. (E)

A. B. Hart and G. J. Womack, *Fuel Cells*. Chapman and Hall, London, 1967. An introduction to and survey of fuel cell systems. (I)

G. W. Heise and N. C. Cahoon (eds.), *The Primary Battery*. Wiley, New York, 1971, Vol. 1; 1976, Vol. 2. Chapter 1, Vol. 1 is a comprehensive history of the development of the science and practice of electrochemical energy conversion. Remainder treats most kinds of primary batteries which have been studied. (A)

R. Jasinski, *High Energy Batteries*. Plenum, New York, 1967. Presents considerations necessary to develop batteries of high specific energy and the progress made to that time in achieving the goal. (A)

Kirk-Othmer Encyclopedia of Chemical Technology, 3rd ed. Wiley, New York, 1964. Entries under "Batteries" survey the field of electrochemical energy conversion. (E)

K. V. Kordesch (ed.), *Batteries*. Marcel Dekker, New York, 1974, Vol. 1. Highly detailed discussions of manufacturing aspects and principles of commercial batteries. (A)

H. A. Liebhafsky and E. J. Cairns, *Fuel Cells and Fuel Batteries*. Wiley, New York, 1968. Very detailed treatment of fuel cell principles and difficulties. (A)

D. Linden, *Handbook of Batteries and Fuel Cells*. McGraw-Hill, New York, 1984. Up-to-date discussion of experimental and production batteries and fuel cells for military as well as consumer and medical applications. (A)

D. Mennie, *IEEE Spectrum,* March, pp. 36–41 (1976). Survey of present and near term future batteries. (E)

Proceedings of the Power Sources Symposium. PSC Publications Committee, Red Bank, NJ, 1946, et seq. Biennial symposium at Atlantic City, New Jersey, on power sources and energy conversion sponsored by several agencies of the U.S. Department of Defense. Presents recent research development results. (A)

Ullmann's Encyclopedia of Industrial Chemistry. 5th ed. VCH Publishers, Deerfield Beach, FL, 1985. Entries under "Batteries" give broad view of batteries and electrochemical energy conversion. (E)

G. W. Vinal, *Primary Batteries*. Wiley, New York, 1950. Emphasis on practical aspects of primary battery technology. (I)

G. W. Vinal, *Storage Batteries,* 4th ed. Wiley, New York, 1955. Emphasis on practical aspects of secondary battery technology. (I)

G. J. Young (ed.), *Fuel Cells*. Reinhold, New York, 1960. Surveys principles and technology of fuel cells. (A)

Electrochemistry

Donald R. Franceschetti and J. Ross Macdonald

Modern electrochemistry is the study of ionic conductors, materials in which ions participate in the flow of electric current, and of interfaces between ionic conductors and materials which conduct current by electron flow, at least in part. It is a truly interdisciplinary field which draws heavily upon many branches of physics and chemistry. Electrochemical phenomena often have analogs in vacuum-tube and semiconductor electronics. Electrochemistry finds numerous applications throughout the natural sciences, the engineering disciplines, and the health-related fields.

IONICS

The study of ionic conductors in themselves has been termed ionics. Ionic conductors are also known as electrolytes. The ionic conductors of interest to electrochemists include (i) liquid solutions of ionic solids (e.g., NaCl in water); (ii) certain covalently bound substances dissolved in polar media, in which ions are formed on solution (e.g., HCl, which is completely dissociated in water, or acetic acid, which is only partially dissociated); (iii) ionic solids containing point defects (e.g., solid AgCl at high temperatures); (iv) ionic solids whose lattice structure allows rapid movement of one subset of ions (e.g., β-Al_zO_3-Na_2O); (v) fused salts (e.g., molten NaCl); and (vi) ionically conducting polymer–salt complexes (e.g., polyethylene oxide–$LiClO_4$).

When an electric field is established in an electrolyte, the migration of positive ions (cations) and negative ions (anions) is observed as a flow of electric current. In defect solid electrolytes, case (iii), the charge carriers are cation and anion interstitials or vacancies (anion vacancies are regarded as positively charged, cation vacancies as negatively charged). Some solid electrolytes are also electronic semiconductors so that the current may also include important contributions from conduction-band electrons and valence-band holes.

The mobility of charge carriers is determined by their interactions with each other and with their environment. Ions in solution often form long-lived aggregates with a characteristic number of solvent molecules. Point defects in solid electrolytes deform the lattice around them. Similarly, the mobility of ions in polymer electrolytes is closely coupled to fluctuations in polymer conformation. A charge carrier is, on the average, surrounded by an "atmosphere" of carriers bearing a net opposite charge. The Debye–Hückel model of dilute electrolytes yields $z_ie_0 \exp(-r/L_D)/\epsilon r$ as the average electrostatic potential at a distance r from a carrier of charge z_ie_0 (e_0 is the proton charge) in a medium of dielectric constant ϵ. The screening or Debye length, L_D, is determined by the concentrations and charges of the carriers.

The flow of current is the result of both carrier drift, in response to the electric field E, and diffusion, which acts to reduce concentration gradients. The current density resulting from charge carriers of species i is given by

$$\mathbf{J}_i = z_ie_0(\mu_i\mathbf{E} - D_i\nabla)c_i,$$

where c_i is the concentration of carriers and μ_i is the carrier mobility, related in dilute electrolytes to the diffusion coefficient D_i by the Einstein relationship, $\mu_i = D_iz_ie_0/kT$. Charge carriers may form pairs, bound by electrostatic or chemical forces, whose members do not readily separate. The equation of continuity for charge carriers of species i,

$$\frac{\partial c_i}{\partial t} = G_i - R_i - (z_ie_0)^{-1}\nabla\cdot\mathbf{J}_i,$$

includes terms for the generation (G_i) and recombination (R_i) of carriers corresponding to the dissociation and formation of bound pairs.

ELECTRODICS

The study of interfaces between ionic conductors and electronic conductors constitutes electrodics, the second major subdivision of electrochemistry. Differences in electrical potential may be determined unambiguously only between electronic conductors; thus electrochemical measurements are usually made on cells with two electrodes. Such cells may be divided into two half-cells, each containing a single electrode, either metal or semiconductor, in contact with an ionic conductor. The half-cell, involving a single interface, is the basic unit studied in electrodics.

The most thoroughly studied electrode material is liquid mercury, for which a clean and atomically smooth interface with electrolyte solutions is readily obtained. Solid metal and semiconductor electrodes are also widely employed, but have not been as extensively characterized. Recently polymer electrodes, which can be formed electrochemically on metal substrates, have become a subject of appreciable interest.

Because the potential difference between the electronic and ionic phases of a half-cell is not measurable, a standard half-cell has been chosen and arbitrarily assigned an electrode potential of zero. This cell, called the standard hydrogen electrode, consists of a platinum electrode in contact with hydrogen ions at 1 atm pressure and an aqueous solution containing hydrogen ions at unit mean activity (see below). The electrode potential of other half-cells is defined as the open-circuit potential of the cell which would be formed with the standard hydrogen electrode.

Electrode potentials provide information about the electrochemical reactions by which charge is transferred between electrode and electrolyte. Simple electrode reactions include (i) ionization of the electrode metal (e.g., $Ag \rightleftarrows Ag^+ + e^-$), (ii) change in the state of an ion (e.g., $Fe^{2+} \rightleftarrows Fe^{3+}$

$+ e^-$), and (iii) ionization of a gas (e.g., $\frac{1}{2}H_2 \rightleftarrows H^+ + e^-$, in the presence of a nonreactive metal). When electrons are removed from a species, the species is said to be oxidized; when electrons are added, the species is said to be reduced. In a half-cell in equilibrium, oxidation and reduction occur at equal rates. Away from equilibrium, the electrode is termed an anode if oxidation predominates over reduction, and a cathode in the opposite case. Most electrode reactions consist of a sequence of chemical and charge transfer steps, some of which may involve short-lived ionic species not present in the bulk electrolyte.

An example of a simple electrochemical cell is the Daniell cell, which consists of zinc and copper electrodes immersed, respectively, in aqueous solutions of $ZnSO_4$ and $CuSO_4$, the two solutions being separated by a membrane which allows the passage of charge but prevents rapid mixing of the two solutions. When electrons are allowed to flow between the electrodes, the zinc electrode dissolves to form Zn^{2+} ions while Cu^{2+} ions are deposited on the copper electrode, in accord with the overall reaction $Zn + Cu^{2+} \rightarrow Zn^{2+} + Cu$. The open-circuit cell potential E is the energy released by the reaction per unit charge transfer between the electrodes and is related to the Gibbs free energy ΔG of the cell reaction by $\Delta G = 2\,FE$, where F is Faraday's constant, 96.485 C mol^{-1}, and two electrons are transferred for each atom of Cu deposited. E is related to the thermodynamic ion activities $a(Cu^{2+})$ and $a(Zn^{2+})$ by Nernst's equation, which in this case becomes

$$E = E^0 - \frac{RT}{2F}\ln\left\{\frac{a(Zn^{2+})}{a(Cu^{2+})}\right\},$$

where E^0 is the cell potential at unit activity and R is the gas constant, 8.314 J mol^{-1} K^{-1}. Activities are quantities related to the ion concentrations (identical at infinite dilution), which take into account ion–ion interactions in the electrolyte.

Nernst's equation is strictly applicable only to systems in thermodynamic equilibrium. In general, the potential of each half-cell is a function of the cell current, which is determined by the slowest step in the electrode reaction sequence. For many half-cells the current is approximately given by the Bulter–Volmer equation, which can be cast in the form

$$i = i_0[\exp(\alpha_{an}\eta F/RT) - \exp(-\alpha_c\eta F/RT)]$$

with

$$\alpha_c + \alpha_{an} = n/\nu.$$

Here i_0 is the exchange current, determined by the rate of the electrode reaction at equilibrium; η is the electrode overpotential, the deviation of the half-cell from its equilibrium value; and n is the number of electrons transferred. The parameters α_{an} and α_c are transfer coefficients for the anodic (oxidation) and cathodic (reduction) processes and ν is the stoichiometric coefficient, the number of times the rate-determining step occurs in the overall half-cell reaction.

Electrodes, or more properly half-cells, are classified as polarizable or nonpolarizable depending on the amount of overpotential required for a fixed ion current flow. Limiting cases, which can be closely approximated in practice, include the perfectly polarizable, or blocking, electrode, one in which no ion current flows regardless of the overpotential, and the perfectly nonpolarizable, or reversible, electrode, one in which the electrode potential retains its equilibrium value regardless of the amount of current flow.

By variation of the potential drop across an electrochemical cell, the rates of the electrode reactions may be altered, and one may even reverse the direction of the net cell reaction. If zinc metal is immersed in $CuSO_4$ solution, copper metal and $ZnSO_4$ are produced spontaneously. By connecting the electrodes of a Daniell cell to a load, useful work may be obtained from the energy of this spontaneous reaction. Cells operated in an energy-producing manner are termed galvanic cells. By rendering the copper electrode sufficiently positive with respect to the zinc, one may effect dissolution of copper and deposition of zinc. The operation of a cell for the production of substances not obtainable spontaneously from the cell materials is termed electrolysis and the cell so operated, an electrolytic cell.

A fundamental theoretical problem in electrodics is the nature of the electrical "double layer," the region of charge separation formed when an electrode is in contact with an ionic conductor. The double layer formed at a metal electrode in an aqueous electrolyte has received particularly intensive study. The traditional Gouy–Chapman–Stern model involves a (usually charged) idealized metal surface, an adjoining plane of chemisorbed water molecules and (often) ions, and a region of increased concentration of cations or anions, depending on the charge on the electrode. The plane of centers of chemisorbed molecules and ions defines the inner Helmholtz plane (ihp), while the plane of closest approach of solvated ions is the outer Helmholtz plane (ohp). The region from the metal surface to the ohp is termed the compact double layer, characterized by an effective dielectric constant which describes the loss in orientational freedom of the adsorbed molecules. The region of space charge beyond the ohp is the diffuse double layer.

More recent studies of the double layer employ a jellium (electron plasma) model for electrons in the metal and a hard-sphere model for both solvent molecules and ions, without making special assumptions about the compact layer. Understanding of double-layer behavior is bound to improve over the next few years as techniques such as scanning tunneling electron microscopy provide a detailed picture of adsorbed layer structure.

APPLICATIONS

Electrochemical methods are employed widely in quantitative and qualitative chemical analysis. Electrolytic methods are the primary industrial means of purifying many metals, of extracting several metals from their ores or salts, and of producing many nonmetallic substances. Electrolytic separation methods make possible the reclamation of valuable materials from industrial waste and reduction in the quantity of pollutants released into the environment. Much industrial research is directed at retarding the corrosion of metals, a phenomenon involving electrochemical reactions at the surface of the metal. The electroplating of metals with thin layers of inert but costly materials is one of a number of electrochemical remedies to this problem.

Electrochemical cells offer an efficient and often portable source of energy. In fuel cells, the energy of a combustion reaction, such as the combination of hydrogen and oxygen to form water, is converted directly to electrical energy, circumventing the thermodynamic restriction on the efficiency of heat engines. The use of solid electrolytes, particularly those whose crystal structure permits rapid ion movements, is a topic of high current interest and offers new possibilities for high-temperature fuel cells and for high-energy-density storage batteries. The properties of semiconductor electrodes are also of interest in their application to photogalvanic energy conversion.

Electrochemical phenomena are also of considerable importance in biology and medicine. The conduction of nerve impulses depends on the current–voltage relationship for sodium-ion transport across the cell membrane. Much of living matter is colloidal, consisting of small (10–10^4 Å) particles suspended in an aqueous solution. Through adsorption of ions, colloid particles acquire a double-layer structure which determines the stability of the suspension. Important recent medical developments include electrochemical control of drug administration across intact human skin and electrically stimulated healing of problem bone fractures.

See also CONDUCTION; CRYSTAL DEFECTS; DIFFUSION; ELECTROCHEMICAL CONVERSION AND STORAGE.

BIBLIOGRAPHY

J. O'M. Bockris and A. K. N. Reddy, *Modern Electrochemistry*. Plenum, New York, 1973. (E)

F. A. Kröger, *The Chemistry of Imperfect Crystals*, 2nd ed., Vol. 3, pp. 131–214. North-Holland, Amsterdam, 1974. (A)

K. J. Vetter, *Electrochemical Kinetics*. Academic Press, New York, 1967. (A)

National Materials Advisory Board, *New Horizons in Electrochemical Science and Technology*. National Academy Press, Washington, 1986. (E)

Electrodynamics, Classical

J. D. Jackson

Electrodynamics, a word used by Ampère in his pioneering researches 150 years ago, may properly be used to encompass all electromagnetic phenomena. There is also a more restricted meaning: electromagnetic fields, charged particles, and their mutual interaction at a microscopic level, excluding in practice, if not in principle, phenomena associated with macroscopic aggregates of matter. *Classical* electrodynamics then consists of the regime where (relativistic) classical mechanics applies for the motion of particles, and the photon nature of electromagnetic fields can be ignored. Its quantum generalization, called quantum electrodynamics, is necessarily employed for phenomena without classical basis (e.g., pair production), as well as where quantum effects are significant.

Separate articles exist on many aspects of macroscopic electromagnetism (e.g., ELECTROMAGNETIC RADIATION, ELECTROSTATICS, MAGNETS AND MAGNETOSTATICS, MICROWAVES AND MICROWAVE CIRCUITRY). The emphasis here is on basic principles and selected results of classical electrodynamics in the restricted sense. The Gaussian system of units and dimensions is used. See the Appendix of Ref. 1 for the connection to the SI or mksa units of practical electricity and magnetism.

MAXWELL EQUATIONS IN VACUUM

The differential equations of electromagnetism in vacuum are the *Maxwell equations*;

$$\nabla\cdot\mathbf{E} = 4\pi\rho, \tag{1a}$$

$$\nabla\times\mathbf{B} - \frac{1}{c}\frac{\partial\mathbf{E}}{\partial t} = \frac{4\pi}{c}\mathbf{J}, \tag{1b}$$

$$\nabla\times\mathbf{E} + \frac{1}{c}\frac{\partial\mathbf{B}}{\partial t} = 0, \tag{1c}$$

$$\nabla\cdot\mathbf{B} = 0. \tag{1d}$$

The quantities ρ and $\mathbf{J}$ are the source densities of charge and current, related because of conservation of charge by the differential continuity equation

$$\frac{\partial\rho}{\partial t} + \nabla\cdot\mathbf{J} = 0. \tag{2}$$

The electromagnetic field quantities $\mathbf{E}$ and $\mathbf{B}$, called respectively the electric field and the magnetic induction, are related to the mechanical force per unit charge according to the *Lorentz force* equation

$$\mathbf{F} = q\left(\mathbf{E} + \frac{\mathbf{v}}{c}\times\mathbf{B}\right) \tag{3}$$

where $\mathbf{F}$ is the force exerted on a point charge q moving with velocity $\mathbf{v}$ in the presence of external fields $\mathbf{E}$ and $\mathbf{B}$. The parameter c that enters the Maxwell equations and the force equation has the dimensions of a speed (length/time). Solution of the Maxwell equations in free space shows the existence of transverse waves propagating with the speed c. It is thus the speed of light and other electromagnetic radiation. The speed of light is now defined to be 299 792 458 m/s (see METROLOGY).

The four Maxwell equations (1a)–(1d) (actually eight scalar equations) are expressions of the experimental laws of Coulomb, Ampère and Maxwell, Faraday, and the absence of magnetic charges, respectively. This can be seen more clearly in integral form—by integration over a finite volume V, bounded by a closed surface S, and use of Gauss's law, for Eqs. (1a) and (1d); by integration over an open surface S_0, bounded by a closed curve C, and use of Stokes's theorem, for Eqs. (1b) and (1c). *Coulomb's law* then reads

$$\oint_S \mathbf{E}\cdot\mathbf{n}\,da = 4\pi\int_V \rho\,d^3r \tag{4}$$

where da is an element of area on S and $\mathbf{n}$ is a unit, outwardly directed normal at da. Equation (4) states that the total electric flux out of the volume V is equal to 4π times the total charge inside. It can be shown that this is a consequence of (a) the inverse-square law of force between point charges, (b) the central nature of that force, and (c) linear superposition. These, plus isotropy of the field of a point charge, are the elements of Coulomb's laws of electrostatics. The corresponding integral of Eq. (1d) has zero on the right-hand side—there are (as far as we presently know) no magnetic charges.

Ampère's (and Maxwell's) *law* has the integral form

$$\oint_C \mathbf{B}\, d\mathbf{l} = \int_{S_0} \left(\frac{4\pi}{c}\mathbf{J} + \frac{1}{c}\frac{\partial \mathbf{E}}{\partial t} \right) \cdot \mathbf{n}\, da \tag{5}$$

where $d\mathbf{l}$ is an element of length tangent to the curve C. The sense of the normal $\mathbf{n}$ to the surface S_0 is determined by the right-hand rule with respect to $d\mathbf{l}$. For static fields the second term on the right is absent. Then Eq. (5) is equivalent to Ampère's laws of forces between current loops. The second term on the right shows that time-varying electric fields produce magnetic fields just as do currents. This term is sometimes called Maxwell's displacement current. It is an essential modification of Ampère's laws for rapidly varying fields.

The analogous integral statement of Eq. (1c), *Faraday's law*, is

$$\oint_C \mathbf{E} \cdot d\mathbf{l} = -\frac{1}{c}\frac{d}{dt}\int_{S_0} \mathbf{B} \cdot \mathbf{n}\, da. \tag{6}$$

The line integral of the electric field around the path or circuit C, called the electromotive force, is proportional to the negative (Lenz's law) time rate of change of magnetic flux through that circuit.

The application of Eqs. (4), (5), and (6) to currents and voltages for conductors and circuit elements is discussed in the articles: NETWORK THEORY, ANALYSIS AND SYNTHESIS, and RESISTANCE AND IMPEDANCE. When the time variations are rapid enough to make the displacement current term in Eq. (5) important, the ideas of lumped circuits fail. The finite speed of propagation and the wave nature of the phenomena must be taken into account: see TRANSMISSION LINES AND ANTENNAS, MICROWAVES AND MICROWAVE CIRCUITY. For explicit discussion of the connection between field and circuit points of view, see Refs. 2 and 3.

CONSERVATION LAWS

The differential continuity equation that follows from the Maxwell equations and expresses conservation of electromagnetic energy is

$$\frac{\partial u}{\partial t} + \nabla \cdot \mathbf{S} + \mathbf{J} \cdot \mathbf{E} = 0 \tag{7}$$

where u, the energy density, and $\mathbf{S}$, called Poynting's vector and representing energy flux (energy per unit area per unit time), are given by the expressions

$$u = \frac{1}{8\pi}(E^2 + B^2), \quad \mathbf{S} = \frac{c}{4\pi}(\mathbf{E} \times \mathbf{B}). \tag{8}$$

The term $\mathbf{J} \cdot \mathbf{E}$ in Eq. (7) represents the rate of work per unit volume being performed on the sources by the electromagnetic fields; it describes the conversion of electrical into mechanical energy of the charged particles that give rise to ρ and $\mathbf{J}$. To see this explicitly, suppose that the current $\mathbf{J}$ is caused by point particles of charges q_j, positions $\mathbf{r}_j(t)$, and velocities $\mathbf{v}_j(t)$. Then the volume integral of $\mathbf{J} \cdot \mathbf{E}$ can be written

$$\int_V \mathbf{J} \cdot \mathbf{E}\, d^3r = \sum_{j(V)} q_j \mathbf{v}_j \cdot \mathbf{E}_j = \sum_{j(V)} \mathbf{v}_j \cdot \mathbf{F}_j. \tag{9}$$

The second expression on the right follows from use of the Lorentz force equation, Eq. (3). Since $\mathbf{v}_j \cdot \mathbf{F}_j$ is the rate of change of mechanical energy of the jth particle, the sum represents the total rate of change of mechanical energy of all the particles within the volume V. The volume integral of Eq. (7) thus expresses conservation of electromagnetic and mechanical energy, the first term being the rate of change of electromagnetic energy, the last that of mechanical energy, and the middle term being the total flux of electromagnetic energy out of the volume.

Conservation of momentum is expressed in integral form by

$$\frac{d}{dt}(\mathbf{P}_m + \mathbf{P}_e)_i = \oint_S \sum_j T_{ij} n_j\, da \tag{10}$$

where the rate of change of mechanical momentum is the volume integral of the Lorentz force density:

$$\frac{d\mathbf{P}_m}{dt} = \int_V \left(\rho \mathbf{E} + \frac{1}{c}\mathbf{J} \times \mathbf{B}\right) d^3r, \tag{11}$$

and the field momentum within the volume V is

$$\mathbf{P}_e = \int_V \frac{1}{4\pi c} \mathbf{E} \times \mathbf{B}\, d^3r = \frac{1}{c^2}\int_V \mathbf{S}\, d^3r. \tag{12}$$

The flow of momentum out of V through the surface S is given by the integral over S of the contraction of the components of the unit normal $\mathbf{n}$ with the *Maxwell stress tensor*:

$$T_{ij} = \frac{1}{4\pi}[E_i E_j + B_i B_j - \tfrac{1}{2}(E^2 + B^2)\delta_{ij}]. \tag{13}$$

The conservation laws for energy and momentum (and angular momentum) show that electromagnetic fields have energy, momentum, and angular momentum, as much as particles do. This seems obvious at the quantum level of photons, but is just as true classically. Although it may seem counterintuitive, electromagnetic angular momentum, $\mathbf{L}_e$, defined by

$$\mathbf{L}_e = \frac{1}{4\pi c}\int_V \mathbf{r} \times (\mathbf{E} \times \mathbf{B})\, d^3r \tag{14}$$

can exist even for static fields and must be included in considerations of conservation of angular momentum for electromechanical systems.

RELATIVISTIC TRANSFORMATIONS OF FIELDS

The equations of classical electrodynamics can be written in a form that exhibits explicitly their covariance under the transformations of *special relativity*. The charge and current densities ρ and $\mathbf{J}$ form the time and space components of a Lorentz four-vector J^μ; the electromagnetic fields are components of an antisymmetric second-rank Lorentz tensor $F^{\mu\nu}$. For the Lorentz-covariant forms of Eqs. (1) and (2) and other details, see Refs. [1, 8, 10, 11]. Of interest are the explicit relations showing how the electric and magnetic fields $(\mathbf{E}',\mathbf{B}')$ in one inertial frame of reference, K', manifest themselves as the fields $(\mathbf{E},\mathbf{B})$ in a frame K moving uniformly with respect to the first. Let the frame K' have a velocity $\mathbf{v}=\boldsymbol{\beta}c$ with respect to the frame K and let $\gamma=(1-\beta^2)^{-1/2}$. Then the relations among the fields are

$$\mathbf{E}=\gamma(\mathbf{E}'-\boldsymbol{\beta}\times\mathbf{B}')-\frac{\gamma^2}{\gamma+1}\boldsymbol{\beta}(\boldsymbol{\beta}\cdot\mathbf{E}') \tag{15a}$$

$$\mathbf{B}=\gamma(\mathbf{B}'+\boldsymbol{\beta}\times\mathbf{E}')-\frac{\gamma^2}{\gamma+1}\boldsymbol{\beta}(\boldsymbol{\beta}\cdot\mathbf{B}'). \tag{15b}$$

The inverse relations can be obtained by interchanging primed and unprimed quantities and reversing the sign of $\boldsymbol{\beta}$. Either set shows that the field components parallel to the velocity are the same in the two frames, but the transverse components of **E** and **B** become mixed. In particular, a purely electric field in K' appears with both electric and magnetic components in K.

The fields of a point charge q in uniform motion in the frame K provide an important illustration. Let the charge be at rest in frame K'. In K, let the coordinates of the charge be $x_1=vt$, $x_2=x_3=0$, and let the fields be observed at a point on the x_2 axis with coordinates $(0,b,0)$. The distance b is the distance of closest approach of the charge to the observer. The purely static Coulombic field in K' has, in the frame K, time-dependent components,

$$E_1=E_1'=-\frac{q\gamma vt}{(b^2+\gamma^2v^2t^2)^{3/2}}, \tag{16a}$$

$$E_2=\gamma E_2'=\frac{\gamma qb}{(b^2+\gamma^2v^2t^2)^{3/2}}, \tag{16b}$$

$$B_3=\gamma\beta E_2'=\beta E_2. \tag{16c}$$

For highly relativistic particles $(\gamma\gg1)$ the transverse components, E_2 and B_3, dominate and are almost equal in magnitude. The fields are highly compressed in time $(\Delta t\approx b/\gamma v)$ and appear like a pulse of electromagnetic radiation with a spectrum of frequencies extending up to $\omega\approx\gamma v/b$. This equivalence between the fields of a relativistic charged particle and a pulse of radiation can be fruitfully exploited to relate charged-particle-induced and photon-induced processes, for example, Compton scattering and bremsstrahlung, or pair production by photons and electrons on nuclei. The basic photon process may be quantum mechanical, but the spectrum of equivalent photons is essentially a classical concept.

EXAMPLES OF CHARGED-PARTICLE MOTION IN EXTERNAL FIELDS

Many phenomena involving particles and fields fall into one of two classes: (a) fields given, particle motion to be determined; and (b) sources (particle motion) given, fields, including radiation, to be found. This division ignores the mutually reactive effects of fields on sources and vice versa. For class (a), the effects of radiation during the motion can often be included in an approximate way—see Ref. 1, Chapter 17. Some examples of charged-particle motion in external fields are given below. Special relativity effects are included; emission of radiation is ignored.

1. *Particle initially at rest at the origin in a constant uniform electric field.* Let the field **E** point in the x direction, the particle's charge and mass be q and m, respectively. At time t its speed and position are

$$v=at/(1+a^2t^2/c^2)^{1/2},\quad x=at^2/[1+(1+a^2t^2/c^2)^{1/2}]$$

where $a=qE/m$ is the initial (nonrelativistic) acceleration. The speed as a function of position is $v=(2ax)^{1/2}\cdot(1+ax/2c^2)^{1/2}/(1+ax/c^2)$. For short times $(t\ll c/a)$, the motion is the familiar nonrelativistic behavior of a particle under constant acceleration. For $t\gtrsim c/a$, the motion becomes relativistic; the particle continues to gain energy at a constant rate but its speed approaches the speed of light and its position is given simply by $x\approx ct$.

2. *Particle in a constant uniform magnetic field.* The particle is initially at the origin, moving with speed v in the y direction. The magnetic field **B** is in the z direction. At time t the positional coordinates of the particle are $x=(v/\omega)(1-\cos\omega t)$, $y=(v/\omega)\sin\omega t$, $z=0$, where ω is the gyration or cyclotron frequency,

$$\omega=qB/\gamma mc=qcB/E. \tag{17}$$

Here $\gamma=(1-v^2/c^2)^{-1/2}$ and $E=\gamma mc^2$ is the total energy of the particle. The particle's trajectory is a circle of radius $R=v/|\omega|=cp/|q|B$. This circular motion in a uniform magnetic field is the basis of operation of the cyclotron. The inverse dependence on γ of the frequency ω is what necessitates modulation of the frequency of the accelerating voltage as the particle gains energy during the acceleration cycle.

3. *Particle drifts in inhomogeneous magnetic fields.* The simple circular motion (or helical, if the particle has a component of velocity parallel to the magnetic field) in a uniform, constant magnetic field is modified if the field varies in strength and/or direction in space. If the variations are slow enough, the short-term motion is circular, but the center of the circle "drifts" slowly in space and perhaps the radius of the orbit changes, too. There are various kinds of drifts, depending on the type of field variation that occurs. These are important for particle motion in the Van Allen radiation belts around the earth and other planets and for particle confinement in plasmas in thermonuclear research. See Chapter 12 of Ref. 1 for some elementary aspects and Ref. 4 for more advanced and detailed discussion. For the planetary particle belts, see Ref. 5.

RADIATION BY CHARGED PARTICLES

A charged particle in uniform motion has only localized "velocity" fields associated with it, fields that, in the inertial frame in which the particle is at rest, are just the inverse-square radial electrostatic field of a point charge. When the particle undergoes acceleration it exhibits additional fields **E** and **B**, equal in magnitude, proportional to the acceleration, transverse to the radius vector from the particle, and falling off with distance only with the first inverse power. These "acceleration" fields represent radiation. There are several features of the radiation by particles that are of interest: total radiated power, angular distribution, and frequency distribution. The limiting examples of nonrelativistic motion and extreme relativistic motion are considered here.

Nonrelativistic Motion

For a nonrelativistic particle of charge q, the instantaneous power radiated per unit solid angle is

$$\frac{dP}{d\Omega} = \frac{q^2}{4\pi c^3} |\dot{\mathbf{v}}|^2 \sin^2\theta \tag{18}$$

where $\dot{\mathbf{v}}$ is the acceleration and θ is the polar angle measured relative to the direction of $\dot{\mathbf{v}}$. The total radiated power is given by Larmor's formula:

$$P = \frac{2}{3}\frac{q^2}{c^3} |\dot{\mathbf{v}}|^2. \tag{19}$$

The frequency distribution of radiated energy is proportional to the square of the Fourier transform of $\dot{\mathbf{v}}(t)$. For *periodic* motion, with period $\tau = 2\pi/\omega_0$, the spectrum consists of discrete lines at $\omega = n\omega_0$, $n = 1,2,3,\ldots$. For *simple harmonic* motion there is, of course, only the fundamental.

Relativistic Motion, Synchrotron Radiation

When the particle's speed is comparable with the speed of light, new features enter. For the same *acceleration,* much more radiation is emitted. For speeds close to c, it is of course difficult to produce large longitudinal accelerations, but transverse accelerations can still be appreciable. The Larmor power formula has the relativistic generalization

$$P = \frac{2q^2}{3c} \gamma^6[(\dot{\boldsymbol{\beta}})^2 - (\dot{\boldsymbol{\beta}} \times \boldsymbol{\beta})^2] \tag{20}$$

where $\boldsymbol{\beta} = \mathbf{v}/c$, $\dot{\boldsymbol{\beta}} = \dot{\mathbf{v}}/c$, and $\gamma = (1-\beta^2)^{-1/2}$. The presence of a high power of γ shows that for extremely relativistic motion ($\gamma \gg 1$) the radiated power can become very large for a given value of $|\dot{\boldsymbol{\beta}}|$.

The angular and frequency distributions are drastically modified for $\gamma \gg 1$. In an inertial frame where the particle is instantaneously at rest, the angular distribution is given by Eq. (15), but the Lorentz transformation to the laboratory causes the radiation to be concentrated almost entirely along the direction of motion, within angles of the order of $\Delta\theta \simeq \gamma^{-1}$. Thus, independent of the details of the acceleration, the particle radiates a narrow "searchlight" beam in its direction of motion. The frequency spectrum, even for simple harmonic motion, contains very many harmonics, up to a maximum frequency of the order of $\omega_{max} = \gamma^3\omega_0$.

A relativistic electron of charge e moving at constant speed βc in a circular orbit of radius R is an idealization of the actual situation in an electron synchrotron or storage ring. The total radiated power per electron is

$$P = \frac{2}{3}\frac{e^2 c}{R^2}(\beta\gamma)^4. \tag{21}$$

With $\omega_0 \simeq 10^7\ \mathrm{s}^{-1}$ and $\gamma \simeq 10^4$, the frequency spectrum is a broad distribution extending up to and beyond $\omega_{max} \simeq 10^{19}\ \mathrm{s}^{-1}$, corresponding to x-ray energies of the order of 10 keV. The radiation, called *synchrotron radiation,* provides an intense, wide-band source of photons for solid-state and biophysical researches. The intensity of the radiation can be enhanced and its properties tailored to specific needs by the insertion of "wiggler" magnets into the storage ring. These magnets have strong bending fields, alternating in polarity in a regular way. The periodicity and electron energy can be chosen to enhance the synchrotron radiation in a desired frequency range. Such amounts of radiation are generated that a long wiggler can serve as the engine for a *free electron laser*.

The unique and well-understood properties of synchrotron radiation (polarization as well as frequency spectrum) permit it to be identified in astrophysical circumstances (radio and infrared emission from planets, radio to x-ray emission from supernovas and pulsars) and aid in establishing the conditions of particle motion there.

OTHER READING

Classical electrodynamics is a vast subject and has applications in every field of science and engineering. Its richness and beauty can only be appreciated by exploring its extensive literature. Some recent basic texts are Purcell [6] at an elementary level, and Griffiths [7] and Schwartz [8] at an intermediate level. More advanced physics texts include Jackson [1], Landau and Lifshitz [9, 10], and Panofsky and Phillips [11]. The history of the development of electrodynamics can be found in Feather [12] and Vol. 1 of Whittaker [13], while the present experimental status is reviewed in the Introduction to [1].

See also Conservation Laws; Electromagnetic Radiation; Electrostatics; Faraday's Law of Electromagnetic Induction; Magnets (Permanent) and Magnetostatics; Maxwell's Equations; Microwaves and Microwave Circuitry; Quantum Electrodynamics; Radiation Belts; Relativity, Special Theory; Synchrotron; Synchrotron Radiation; Transmission Lines and Antennas.

REFERENCES AND SELECTED BIBLIOGRAPHY

1. J. D. Jackson, *Classical Electrodynamics*, 2nd ed. Wiley, New York, 1975. (A)
2. R. B. Adler, L. J. Chu, and R. M. Fano, *Electromagnetic Energy, Transmission and Radiation*. Wiley, New York, 1960. (I)
3. R. M. Fano, L. J. Chu, and R. B. Adler, *Electromagnetic Fields, Energy, and Forces*. Wiley, New York, 1960. (I)
4. P. C. Clemmow and J. P. Dougherty, *Electrodynamics of Particles and Plasmas*. Addison-Wesley, Reading, Mass., 1969. (A)
5. B. Rossi and S. Olbert, *Introduction to the Physics of Space*. McGraw-Hill, New York, 1970. (I–A)
6. E. M. Purcell, *Electricity and Magnetism* (Berkeley Physics Course Vol. 2), 2nd ed. McGraw-Hill, New York, 1985. (E)
7. D. J. Griffiths, *Introduction to Electrodynamics*. 2nd ed. Prentice-Hall, Englewood Cliffs, N.J., 1989. (I)
8. M. Schwartz, *Principles of Electrodynamics*. McGraw-Hill, New York, 1972. (I)
9. L. D. Landau and E. M. Lifshitz, *Electrodynamics of Continuous Media*. Addison-Wesley, Reading, Mass., 1960. (A)
10. L. D. Landau and E. M. Lifshitz, *The Classical Theory of Fields*, 3d rev. English ed. Addison-Wesley, Reading, Mass., 1971. (A)

11. W. K. H. Panofsky and M. Phillips, *Classical Electricity and Magnetism*, 2nd ed. Addison-Wesley, Reading, Mass., 1962. (A)
12. N. Feather, *Electricity and Matter*. University Press, Edinburgh, 1968. (E)
13. E. T. Whittaker, A *History of the Theories of Aether and Electricity*, 2 vols. Nelson, reprinted by American Institute of Physics, New York, 1987. (E–I)

Electroluminescence

M. N. Kabler

Electroluminescence is the efficient generation of light in a nonmetallic solid by an applied electric field. Electroluminescence is "cool" light in the sense that the brightness is far above that characteristic of the temperature alone; thus incandescent light is excluded.

The primary electronic states of nonmetallic solids comprise two bands of allowed states separated by a forbidden gap where only states due to impurity atoms or lattice imperfections can exist. At normal temperatures the higher or conduction band (CB) is empty except for a few mobile electrons, while the lower or valence band (VB) is filled with electrons except for a few vacant states called holes, which are also mobile. Because of interactions with thermal vibrations, any electrons in the CB immediately fall to its low-energy edge and any holes in the VB rise to its high-energy edge. When excess electrons and holes are produced and brought into proximity by the action of an applied electric field, the electrons spontaneously fall into or recombine with the holes. A recombination event releases energy comparable to the band gap, which is dissipated as heat or radiated as electroluminescence.

p-n JUNCTION ELECTROLUMINESCENCE

The most effective structure for producing electroluminescence is the *p-n* junction. Such a junction is illustrated schematically in Fig. 1, which indicates how electron energy varies as a function of distance perpendicular to the junction both with and without an applied electric field. The device is commonly called a light-emitting diode, or LED. Semiconductors selectively doped with impurity atoms can, at normal temperatures, exhibit high conductivities arising either from extra electrons or extra holes. In the former case the extra electrons come from donor states *D* near the CB edge, and the material is termed *n* type; in the latter case holes are created when VB electrons are trapped in acceptor states *A*, and the material is *p* type. When *n*-type and *p*-type regions are made contiguous with each other, a *p-n* junction is formed. Being nearly void of mobile electrons and holes, the interface region acts as an insulator and is the location of a strong electric field. Under a constant forward bias of a few volts, applied through suitable conducting contacts, electrons and holes are swept in opposite directions across the junction, where they can recombine.

The energy of an emitted photon is within the range of transparency of the material, that is, no higher than the band gap. Although many materials have band gaps in the ultraviolet, efficient LEDs operating beyond the yellow-green range are not yet available. This is largely due to difficulties in producing *p-n* junctions in the wider-band-gap materials.

LUMINESCENT PROCESSES

In general, electrons and holes can recombine near atomic impurities as well as in regions of the unperturbed lattice. The color of the luminescence and the relative probability of producing light instead of heat are characteristic of the particular impurities involved. In many semiconductors, luminescent transitions from the bottom of the CB to the top of the VB cannot take place without the intervention of a lattice vibration or imperfection in order to satisfy the law of conservation of momentum. Such luminescent transitions are termed indirect, and the material is said to have an indirect edge. An important example is GaP, with a band gap in the green spectral region near 2.3 eV (540 nm). In other materials with so-called direct edges this transition can occur without restriction; GaAs is an example, its band gap being in the infrared near 1.5 eV. Because of the momentum-conservation requirement, an indirect luminescent transition is relatively unlikely in pure materials and thus electroluminescence is inefficient. However, a crystal with an indirect edge can be doped with impurity atoms which sequentially trap both electrons and holes, thereby providing a radiative recombination path up to several orders of magnitude more efficient. Such a path is included in Fig. 1, where *I* represents the localized impurity state.

LIGHT EMITTING DIODES

Commercial activity has concentrated on developing efficient electroluminescent devices in the visible spectral range. Most materials with band gaps in this range, particularly toward the violet, cannot be made to incorporate *p-n* junctions, and only compounds and alloys of elements from groups IIIA and VA of the periodic table have thus far yielded LEDs of broad utility.

For the green spectral region, N doping of GaP has proved quite effective in mitigating the inefficiency due to the indirect edge while at the same time retaining an emission

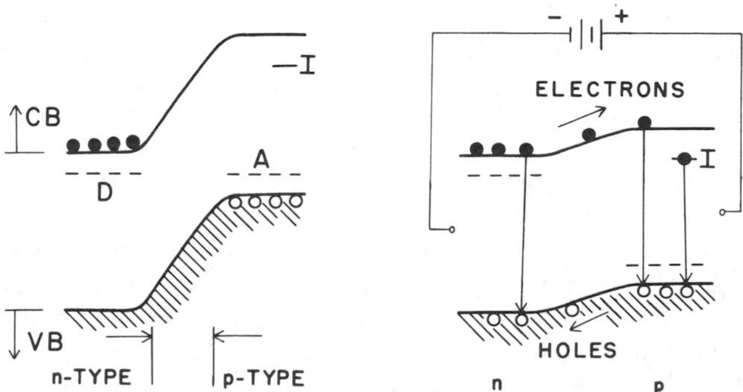

FIG. 1. Schematic representation of electron energies in a *p-n* junction with no bias voltage (left), and with a forward bias in the 1 V range (right). The vertical arrows represent electroluminescent processes, one of which occurs through an impurity level *I*.

wavelength near the band-gap energy. N is isoelectronic with P, for which it substitutes in the lattice. Thus the N normally carries no net charge, and its attractive potential is short range; it traps a conduction electron in a strongly localized orbit and is, for this reason, very effective in providing an interaction through which momentum is conserved during recombination with an approaching hole.

Other dopants, for example, Zn and O incorporated in the lattice as nearest-neighbor pairs, can produce efficient red luminescence in GaP. However, considerable success in the yellow-to-red range has been achieved by N doping of GaAs-GaP alloys. The alloy composition can be chosen to give a particular band-gap energy and emission color. Impurity doping to increase the efficiency of emission is not required for the red and near infrared, since the corresponding alloys have direct edges. The same is true for alloys of AlAs and GaAs.

In many of these *p-n* junction devices the efficiency with which electrical energy is converted to light inside the material can be 10% or greater at room temperature. However, because of geometrical constraints, internal reabsorption, and internal reflection at the semiconductor–air interface, only a small fraction of the photons created actually can emerge into a useful beam. External efficiencies of commercial LEDs generally lie in the 0.1–5% range. The efficiencies of luminescent processes themselves invariably fall at higher temperatures, and thus light output is limited by internal heating due to nonradiative processes. The prominent semiconductors Si and Ge find little utility as LEDs, because their band edges lie in the infrared and are indirect.

LASER DIODES

Electroluminescent lasers can be made by fabricating special *p-n* junction configurations. Since light amplification requires population inversion, the states near the bottom of the conduction band must be more than half filled with electrons and states near the top of the valence band must be more than half empty, i.e., more than half filled with holes. This requires large hole and electron currents into the recombination region and constriction of this region to a small volume in order to maintain high concentrations. An example of an LED which can be driven into laser oscillations is GaAs heavily doped to increase electron and hole concentrations. However, relatively high threshold current densities are required. The most efficient laser diodes are based on the heterojunction, which is a *p-n* junction similar to that of Fig. 1 except that there are two different materials with different band gaps on opposite sides of the junction. This effectively restricts recombination to that side of the junction having the smaller band gap, as well as providing a refractive-index gradient which aids in confining the light.

Heterojunctions are formed by depositing thin layers of the different semiconductors on each other, usually from the liquid phase. When contiguous layers differ in lattice parameter, for example, InAs on GaAs, there occurs near the interface an array of defects or dislocations which act as nonradiative recombination centers and thereby drastically lower the luminescent efficiency. Thus heterojunctions are usually constructed from alloys with nearly identical lattice parameters. A prime example is the AlAs-GaAs system, in which the lattice parameter changes by only 0.14% throughout the entire range of composition. A continuous-wave laser comprising two heterojunctions from AlAs-GaAs alloys is depicted alongside an ordinary sewing needle in Fig. 2. This laser is emitting at a wavelength of 750 nm, and the thickness of the emitting region is only about 500 nm.

HIGH-FIELD ELECTROLUMINESCENCE

It is generally possible to inject electrons or holes from one material into another if a sufficiently high electric field is applied across the interface. Two characteristic examples of electroluminescence originating in this way are the reverse-biased *p-n* junction and certain luminescent insulators containing metallic particles.

If a voltage of polarity opposite that shown in Fig. 1 is applied to a *p-n* junction, the top of the VB on the *p* side can be raised considerably above the bottom of the CB on the *n* side. Electrons from the *p* side can tunnel horizontally through the forbidden gap to the *n* side and into CB states having high kinetic energies. If this kinetic energy is higher than the band-gap energy, the electron can collide with a normal lattice atom and create a new electron–hole pair which can, in turn, recombine radiatively. In practice a considerable amount of energy is lost as heat, and the reverse-biased *p-n* junction is therefore not an efficient electroluminescent source.

The second class of high-field process is exemplified by luminescent ZnS in powder form and doped with Cu_2S. Inside the small ZnS crystallites, the Cu_2S precipitates on lattice imperfections to form submicroscopic, needlelike, electrically conducting particles. The external electric field is strongly intensified near the sharp ends of the needles, where electrons or holes are emitted into the ZnS on alternate half cycles of the applied field. The holes are trapped at impurity atoms, and electrons subsequently arrive to produce recom-

FIG. 2. The minute, star-like luminescence from an AlAs-GaAs heterojunction laser glowing beneath the eye of an ordinary sewing needle. This laser is similar to lasers designed for optical communications. (Photograph courtesy of RCA Laboratories.)

bination luminescence. Cu is the principal luminescent impurity, but many others can play various roles in the process, and most visible colors can be produced. Large electroluminescent panels of this type have been constructed which sandwich the luminescent powder between a metallic base and a transparent conducting electrode such as tin oxide, but their efficiencies and lifetimes are too low to be competitive for general space illumination.

APPLICATIONS

There is now available a wide range of LEDs, which fill an increasingly large fraction of the commercial market for small, low-power sources operating as indicator lights and in symbolic displays. Pocket computers are a conspicuous example. The low voltages required to drive LEDs are compatible with transistor circuitry, a considerable advantage. It is probable that panels of the ZnS type will also gain acceptance for display applications. As the reliability and lifetime of laser diodes improves they will find numerous uses where bright, narrow light beams are required. Optical communications is a particularly promising area, since laser diodes are ideal sources for coupling into fiber-optic waveguides. Their size and operating wavelengths match the diameters and maximum-transmission wavelengths of optical fibers (see Fig. 2), and their output can be modulated at frequencies in the 10^8-Hz range. The less stringent source requirements of many fiber-optic systems can be met by current LEDs. Electroluminescent devices will probably find a role in the interior illumination market, but not before further advances are made in the field of materials science.

See also CRYSTAL DEFECTS; ELECTRON ENERGY STATES IN SOLIDS AND LIQUIDS; LUMINESCENCE; SEMICONDUCTORS, CRYSTALLINE.

BIBLIOGRAPHY

H. Kressel, I. Ladany, M. Ettenberg, and H. Lockwood, "Light Sources," *Physics Today*, p. 38, May 1976.

H. J. Queisser and U. Heim, "Optical Emission from Semiconductors," in *Annual Review of Materials Science* (ed. by R. A. Huggins, R. H. Bube, and R. W. Roberts), Vol. 4, p. 125, 1974.

C. H. Gooch, *Injection Electroluminescent Devices*. Wiley, N.Y., 1973.

R. J. Elliott and A. F. Gibson, *An Introduction to Solid State Physics and Its Applications*. Barnes and Noble, N.Y., 1974.

Electromagnetic Interaction

Toichiro Kinoshita

The electromagnetic interaction is one of the four fundamental interactions known in physics, the other three being the *strong* interaction, which keeps the atomic nucleus together; the *weak* interaction, responsible for spontaneous disintegration (beta decay) of radioactive nuclei; and the *gravitational* interaction, which keeps the stars and galaxies in their orbits. Because of the relative ease with which effects of the electromagnetic interaction can be observed, it is the most thoroughly studied and best understood of the four interactions.

Historically, the structure of the electromagnetic interaction was first determined from the observation of macroscopic electromagnetic phenomena. Its explicit form is given by Maxwell's equations and the Lorentz force acting on the charge, which form the starting point of classical electrodynamics.

Classical electrodynamics fails, however, if it is applied to atomic and subatomic systems; such systems must be treated by quantum mechanics.

Quantization of classical electrodynamics is straightforward if the particles are restricted to nonrelativistic kinematics. In this case, we have only to replace Newton's equation of motion (or, equivalently, Hamilton's equation) by the corresponding Schrödinger equation, and turn the electromagnetic field into a quantized field which means that it can be regarded as a superposition of creation and annihilation operators of photons. (In some applications, the electromagnetic field may be treated as an unquantized field.) The simplest way to introduce the electromagnetic interaction in a quantum-mechanical system is to assume, by invoking the correspondence principle, that it has exactly the same form as in classical electrodynamics. (This does not mean that nothing has changed. In fact, the physics is quite different since the particles now obey the quantum-mechanical equation of motion instead of the classical one.) Such a quantization scheme, if the electron spin is taken into account, is found to be capable of describing practically all properties of atoms and molecules, including emission and absorption of photons.

Quantization of a relativistic particle is considerably more complicated. Straightforward application of the correspondence principle to a relativistic particle leads to the Klein–Gordon equation if the spin of the particle is zero, the Dirac equation if the spin is $\frac{1}{2}$ in units of $\hbar$, etc. However, a satisfactory synthesis of quantum mechanics and relativity requires that not only the electromagnetic field but also the particles be described as quantized fields. The theory thus formulated for the system of interacting electrons and photons is called quantum electrodynamics (QED). This theory can be described by the Lagrangian (or Lagrangian density, to be precise)

$$\mathcal{L}(x) = \mathcal{L}_e(x) + \mathcal{L}_p(x) + \mathcal{L}_I(x) \tag{1}$$

where

$$\mathcal{L}_e(x) = \bar{\psi}(x)\left(i \sum_{\mu=0}^{3} \gamma^{\mu} \frac{\partial}{\partial x^{\mu}} - m\right)\psi(x), \tag{2}$$

$$\mathcal{L}_p(x) = -\frac{1}{4} \sum_{\mu=0}^{3} \sum_{\nu=0}^{3} F_{\mu\nu}(x) F^{\mu\nu}(x), \tag{3}$$

$$\mathcal{L}_I(x) = -\sum_{\mu=0}^{3} j_{\mu}(x) A^{\mu}(x) \tag{4}$$

are the electron, photon, and interaction parts, respectively, of the Lagrangian and

$$F_{\mu\nu}(x) = \frac{\partial A_{\nu}(x)}{\partial x^{\mu}} - \frac{\partial A_{\mu}(x)}{\partial x^{\nu}}, \tag{5}$$

$$j_{\mu}(x) = e\psi(x)\gamma_{\mu}\psi(x). \tag{6}$$

The Greek indices μ and ν take the values 0,1,2,3; 0 for the time axis, and 1,2,3 for the space axis. Four-vectors with lower and upper indices are related to each other by $A_0 = A^0$, $A_i = -A^i$, $i = 1,2,3$. $A_\mu(x)$ is the four-vector potential of the electromagnetic field, γ^μ is the 4×4 Dirac matrix, $\psi(x)$ is a 4×1 matrix representing the "electron field," and $\bar{\psi}(x) = \psi^\dagger(x)\gamma^0$, $\psi^\dagger$ being the Hermitian conjugate of ψ. Finally, e and m are the charge and mass of the electron. (For simplicity we choose the units in which both the velocity of light c and the modified Planck's constant $\hbar$ are equal to one.) The form of the Lagrangian (1) is identical for all observers moving with constant velocity with respect to each other (in other words, invariant under Lorentz transformations). It is also unchanged under another continuous transformation, called the gauge transformation, which is related to the fact that the photon is a massless particle. Finally, it is invariant under discrete transformations C (charge conjugation), P (space reflection), and T (time reversal).

The principle of least action applied to the Lagrangian (1) yields equations of motion for A_μ, ψ, and $\bar{\psi}$. In particular that for A_μ is formally identical (when supplemented by an extra constraint) with the classical Maxwell's equations. The crucial consequence of field quantization is that A_μ is now an operator that creates or annihilates a photon. Similarly, ψ ($\bar{\psi}$) is an operator that annihilates (creates) an electron or creates (annihilates) a positron. The electromagnetic interaction is expressed by $\mathcal{L}_I$ of (4), or equivalently by the interaction energy (i.e., the interaction part of the Hamiltonian density)

$$\mathcal{H}_I(x) = -\mathcal{L}_I(x) = j_\mu(x)A^\mu(x) \tag{7}$$

where the summation over the indices μ is suppressed following the convention.

Thus, we finally have a theory of electromagnetic interaction that satisfies the requirements of both relativity and quantum mechanics. Unfortunately, this theory still has a serious fault: Although the lowest-order predictions of the theory in perturbation expansion (which is essentially an expansion of all quantities in powers of the fine-structure constant $\alpha = e^2/4\pi\hbar c \simeq 1/137$) are in good agreement with experiments, the agreement is destroyed completely if we try to improve it by including higher-order terms that are all infinite. (Actually some of these divergence difficulties are inherited from classical electrodynamics.) It took nearly two decades before it was recognized that these infinities can be eliminated by careful examination (called renormalization) of what is meant by the *observed* mass and charge of an electron.

Renormalized QED as it thus emerged is consistent with all basic principles of physics and is capable of predicting the properties of any process involving electrons and photons as accurately as we might wish. Does this mean that we have finally found the ultimate theory of electromagnetic interaction?

We can answer this question only by performing some critical experiments. There are at least three aspects of the theory that must be examined:

1. How well do higher-order predictions of renormalized QED agree with experiments?
2. Does renormalized QED apply to distances shorter than those already tested?
3. To what extent is the electromagnetic interaction invariant under C, P, T operations?

Some of the most accurate tests of aspect 1 available at present are listed in Table I. Most theoretical values include correction terms of order α and α^2. In case of the magnetic moment anomaly of the electron, which is a small deviation due to the electromagnetic interaction of the magnetic property of the electron from that predicted by the Dirac theory, correction terms of order α^4 have also been included. In all cases, the agreement of theory and experiment is within the uncertainties of experiment and/or theory. Even more stringent tests will be available in the near future.

It is remarkable that the interaction energy (7), which after all is a straightforward adaptation of classical Maxwell theory, has withstood tests of such high precision. Note, for instance, that from relativistic and gauge invariance alone we cannot exclude the possible existence of an additional interaction term

$$\mathcal{H}_I'(x) = k\frac{e}{2m}j_{\mu\nu}(x)F^{\mu\nu}(x),$$

$$j_{\mu\nu}(x) = \frac{i}{2}\bar{\psi}(x)(\gamma_\mu\gamma_\nu - \gamma_\nu\gamma_\mu)\psi(x), \tag{8}$$

Table I High precision tests of quantum electrodynamics

Process	Experiment[a]	Precision in ppm[b]	Theory[c]	Precision in ppm
Hydrogen fine structure[d]	10 969.127(87)[e]	7.9	10 969.034 8(21)	0.2
Hydrogen hyperfine structure[f]	1 420.405 751 766 7(9)	6×10^{-7}	1 420.405 3(45)	3.2
Muonium hyperfine structure[g]	4 463.302 88 (16)	0.036	4 463.302 67(186)	0.42
Helium fine structure[h]	29 616.864(36)	1.2	29 616.834(110)	3.7
Electron magnetic moment anomaly	0.001 159 652 188 4(43)	0.004	0.001 159 652 133(29)	0.025

[a] In units of MHz (10^6 s^{-1}) except for the value of the electron anomaly, which is a pure number.
[b] 1 ppm = 10^{-6}.
[c] The theoretical value is calculated using the value of the fine-structure constant α^{-1} = 137.035 997 9(33) obtained by the quantized Hall effect.
[d] Energy interval between the levels $2P_{3/2}$ and $2P_{1/2}$ of the hydrogen atom.
[e] The quantity enclosed in parentheses represents the uncertainty in the last digits of a numerical value.
[f] Splitting of the ground-state $1S_{1/2}$ level of hydrogen due to the interaction of electron spin and proton spin.
[g] Splitting of the ground-state $1S_{1/2}$ level of muonium (electron-antimuon bound state) due to the spin–spin interaction.
[h] Energy interval between the levels 2^3P_1 and 2^3P_0 of the helium atom.

which would modify the magnetic moment anomaly of the electron by the amount k. Good agreement of theory and experiment implies that the constant k is less than 6×10^{-11}. Actually the presence of such a term would make the theory unrenormalizable. It is reassuring that there is no experimental indication that such a term is present in the Lagrangian $\mathcal{L}$.

The best tests of aspect 2 are provided by electron–positron colliding-beam experiments in which highly accelerated electrons and positrons traveling in opposite directions collide head-on and scatter into large angles,

$$e^+ + e^- \rightarrow e^+ + e^-, \tag{9}$$

or annihilate each other and produce a pair of photons (γ) or muons (μ):

$$\begin{aligned} e^+ + e^- &\rightarrow \gamma + \gamma, \\ &\rightarrow \mu^+ + \mu^-. \end{aligned} \tag{10}$$

Recent measurements of these processes have produced convincing evidence that QED is valid down to lengths of the order of 10^{-17} m, which is nearly two orders of magnitude smaller than the charge radius of the proton.

In spite of the very impressive experimental confirmation of QED, there are reasons to believe that it is not the ultimate theory and that it will have to undergo some changes sooner or later. For instance, the apparent lack of symmetry between the electric and magnetic fields in Maxwell's equations suggests possible existence of a magnetic monopole, the magnetic counterpart of the electric charge. Experimental confirmation of the existence of a magnetic monopole would certainly affect our understanding of electromagnetic interaction in a very profound way.

Another possibility, which has been pursued with a spectacular success, is that the electromagnetic and weak interactions are nothing but different aspects of the same force. One theoretical scheme for such a unification, the Weinberg–Salam model, has recently been given strong experimental support. In a beautiful experiment in which a tiny parity–nonconservation effect was observed in the inelastic scattering of high-energy electrons off deuterium nuclei, the electromagnetic and weak interactions were found to interfere in a manner predicted by this model. Similar (but larger) interference effects have been observed in processes such as (9) and (10) at high energies. Most importantly, W and Z mesons, which are the main ingredients of the Weinberg–Salam model and about 100 times heavier than proton, have been observed experimentally.

We have thus witnessed an end to "naive" QED. At the same time a way was found to incorporate QED into a deeper law of nature. Note that this does not mean that QED has failed at last. In fact the Weinberg–Salam model can be regarded as a natural generalization of QED.

An even grander scheme of unification, including the strong interaction and possibly the gravitational interaction, is purely speculative at present. However, it is not out of the question that it also may receive experimental support in a not-so-distant future.

As for aspect 3, there is no evidence at present that the electromagnetic interaction has components that violate C, P, or T. However, giving precise upper bounds for such violations is not straightforward.

See also ELECTRODYNAMICS, CLASSICAL; GRAVITATION; MAXWELL'S EQUATIONS; QUANTUM ELECTRODYNAMICS; QUANTUM MECHANICS; STRONG INTERACTIONS; WEAK INTERACTIONS.

BIBLIOGRAPHY

Textbooks

L. I. Schiff, *Quantum Mechanics*, 3d ed. McGraw-Hill, New York, 1968.

J. D. Bjorken and S. D. Drell, *Relativistic Quantum Fields*. McGraw-Hill, New York, 1965.

Reviews of QED

B. E. Lautrup, A. Peterman, and E. de Rafael, *Phys. Rep.* **3C,** 193 (1972).

T. Kinoshita and J. Sapirstein, *Atomic Physics 9*, R. S. Van Dyck, Jr. and E. N. Fortson, eds. World Scientific Publishers, Singapore, 1984.

Survey of Fundamental Physical Constants

E. R. Cohen and B. N. Taylor, *Rev. Mod. Phys.* **59,** 1121 (1987).

Test of QED in e^+–e^- Colliding Beams

F. Takasaki, *Nucl. Phys. B* (Proc. Suppl.) **3,** 17 (1988).

Experimental Search for a Magnetic Monopole

P. B. Price *et al., Phys. Rev. Lett.* **35,** 487 (1975).

Review of Unified Theories of Electromagnetic and Weak Interactions

E. S. Abers and B. W. Lee, *Phys. Rep.* **9C,** 1 (1973).

Experimental Evidence for Unification of the Electromagnetic and Weak Interactions

G. Arnison *et al., Phys. Lett.* **122B,** 103 (1983).

C. Y. Prescott *et al., Phys. Lett.* **77B,** 347 (1978).

Electromagnetic Radiation

J. C. Herrera

INTRODUCTION

The most common form of electromagnetic radiation is visible light. But the fact that this type of emission of energy was electrical in nature was hardly suspected until James Clerk Maxwell (1831–1879) conceived the idea of the electromagnetic theory of light in the year 1864. Prior to that time the phenomenon of light had been studied as a mechanical vibration in a medium called the luminiferous ether, while, instead, the electrical interaction between charged bodies and currents was attributed to a force acting through the intervening empty space similar to the force of universal gravitation discovered by Isaac Newton (1642–1727) about 200 years earlier. Though the mathematical description for the electrical force was more complicated than that for the force of gravity, the interaction was still an instantaneous "action at a distance." Maxwell, guided by the unerring physical intuition of Michael Faraday (1791–1867) and his own genius for constructing mathematical models, was able to develop a set of equations that not only accounted for the known electrical phenomena, but in addition predicted that changing electrical charge distributions would result in electromagnetic waves that traveled at the speed of light. The lines of force that Faraday visualized extending between and around electrical bodies were transformed into the electric and magnetic fields of Maxwell's equations. It was Heinrich Hertz (1857–1894) who verified experimentally in the year 1887 that there existed electromagnetic waves other than light that were indeed generated by oscillating currents in electrical circuits. With the advent at the turn of the century of the theory of relativity formulated by Albert Einstein (1879–1955), the ether medium vanished into space, leaving behind electromagnetic fields. Though the interpretation of Maxwell's theory had changed, the efforts of Einstein and H. A. Lorentz (1853–1928) had served to reinforce the fundamental validity of Maxwell's equations and the reality of electromagnetic waves.

Today the phenomenon of electromagnetic radiation covers the entire spectrum of energies, from very high-frequency cosmic gamma rays to low-frequency, long-wavelength radio waves. Radiant heat, ultraviolet light, and x rays are all electromagnetic waves, distinguishable from each other by their frequency or, equivalently, their wavelength.

MAXWELL'S THEORY

Maxwell's equations are four partial differential equations that relate the electric field vector **E** and the magnetic field vector **B** at a particular location in space **x** and at a time t to the electrical charge density per unit volume ρ and the current density per unit area **J** at the same space-time point. Employing the conventional notation of vector analysis, we write them in the classic form

$$\nabla \cdot \mathbf{E} = \frac{\rho}{\epsilon_0}, \tag{1}$$

$$\nabla \times \mathbf{E} = -\frac{\partial \mathbf{B}}{\partial t}, \tag{2}$$

$$\nabla \cdot \mathbf{B} = 0, \tag{3}$$

and

$$\nabla \times \mathbf{B} = \mu_0 \mathbf{J} + \frac{1}{c^2}\frac{\partial \mathbf{E}}{\partial t}. \tag{4}$$

The units used are SI (Système International). We note that the permittivity of free space, ϵ_0, and the permeability of free space, μ_0, are dimensional constants that are related to the speed of light according to the expression

$$c = (\epsilon_0 \mu_0)^{-1/2}. \tag{5}$$

The first of these equations is the differential form of Gauss' law, that is, the equivalent of the statement that the total flux of the electric field vector (the surface integral of **E**) is a direct measure of the total enclosed electric charge. Faraday's law for the induced electric field due to a magnetic field that changes with time is represented by Eq. (2). We observe that for a stationary condition (one that does not vary with time), the partial derivative of the magnetic field with respect to time vanishes ($\partial \mathbf{B}/\partial t = 0$), and then the first two equations are synonymous with Coulomb's law for the electrostatic force between charges with a fixed separation in space. The well-known fact that there are no free magnetic poles, as compared to the existence of free electric charges [Eq. (1)], gives rise to Eq. (3): The divergence of the magnetic field vector is always equal to zero. Maxwell not only expressed Ampère's law (the line integral of the magnetic field along the closed boundary curve of a surface is a measure of the electric current flowing through the surface) by writing Eq. (4), but he also added the last term on the right-hand side (his so-called displacement current). The four equations as they stand are consistent with the conservation of charge ($\nabla \cdot \mathbf{J} + \partial\rho/\partial t = 0$), and when considered together, they predict the radiation of electromagnetic waves.

The operational meaning of Maxwell's equations is as follows: Given *all* the distributions of charges and currents specified by the source functions (ρ,**J**), we can determine the accompanying electromagnetic field properties in space specified by the vector field functions (**E**, **B**). The subsequent step, one that typifies a field theory, is to associate with these fields an energy and a momentum localized in free space. When the field energy within a given spatial volume changes, it is either dissipated as heat, or motion of charges, within the volume, or instead it passes into the surrounding space. This characterization was derived from Maxwell's theory by John H. Poynting (1852–1914) in 1884. Written in vector notation, the energy balance of the field assumes the compact form

$$-\frac{\partial u}{\partial t} = \mathbf{E}\cdot\mathbf{J} + \nabla \cdot \mathbf{S} \tag{6}$$

where the electromagnetic energy per unit volume (in joule per cubic meter) is

$$u = (\epsilon_0/2)(E^2 + c^2B^2), \tag{6a}$$

and the electromagnetic power flux density vector (in watt per square meter) is

$$\mathbf{S} = (1/\mu_0)\mathbf{E} \times \mathbf{B}. \tag{6b}$$

In words, the decrease per second in the field energy density (u) is equal to the power per unit volume delivered to the currents, ($\mathbf{E}\cdot\mathbf{J}$), added to the divergence of the field power per unit area (**S**), the Poynting vector.

PLANE ELECTROMAGNETIC WAVE

The radiation concepts that have just been presented can best be illustrated by considering a plane harmonic wave propagating in free space in the direction of the z axis (x, y, and z form a right-handed coordinate system). The electromagnetic fields that satisfy Maxwell's equations (with $\rho=0$ and $\mathbf{J}=0$) are, in this instance,

$$E_x = E \sin 2\pi(\nu t - z/\lambda), \qquad (7a)$$

$$E_y = E_z = 0, \qquad (7b)$$

$$B_y = (E/c) \sin 2\pi(\nu t - z/\lambda), \qquad (7c)$$

and

$$B_x = B_z = 0. \qquad (7d)$$

This transverse wave is plane polarized in the x direction with the wave front in the (x,y) plane and the wave normal in the z direction. The frequency of vibration (ν) and the wavelength (λ) are simply related to the speed of propagation (c), that is,

$$\nu\lambda = c = (\epsilon_0\mu_0)^{-1/2}. \qquad (8)$$

If we now apply Eqs. (6a) and (6b), we are able to calculate the time average values of the Poynting vector and the electromagnetic energy density. Thus, we obtain

$$\langle S_z \rangle_{av} = \tfrac{1}{2} c \epsilon_0 E^2 \qquad (9)$$

and

$$\langle u \rangle_{av} = \langle S_z \rangle_{av} / c. \qquad (10)$$

We can therefore picture an average transfer, or transmission, of energy taking place in the direction of the wave normal and at a speed equal to that of light.

In Table I we give two examples of the physical magnitudes involved in the propagation of plane electromagnetic waves. The first column of figures is based on an electric field intensity of 1 μV/m, such as might exist at some large distance from a television or radio antenna. The second column of figures has been calculated on the basis of the power arriving at the earth from the sun, that is, corresponding to a solar radiation constant of 20 kcal/m^2 incident every minute (1390 W/m^2). In the last row we have also included the magnitude of the radiation pressure that the plane wave would exert on a totally absorbing material surface. Though the idea of radiation pressure was carefuly discussed by Maxwell in his *Treatise on Electricity and Magnetism*, published in 1873, the experimental corroboration did not occur until about 1900.

RETARDED POTENTIALS

Plane electromagnetic waves, as discussed in the last section, represent an approximation to the observed radiation far away from a localized source. In applying Maxwell's theory to an actual radiating source, it is advantageous to express the two field variables **E** and **B** as functions of the vector and scalar potentials, **A** and ϕ. Thus Eqs. (2) and (3) are satisfied identically if we introduce the two defining relationships

$$\mathbf{B} = \nabla \times \mathbf{A} \qquad (11)$$

and

$$\mathbf{E} = -\nabla\phi - \frac{\partial \mathbf{A}}{\partial t}. \qquad (12)$$

The other Maxwell equations, (1) and (4), containing the source terms, can then be reduced to two similar wave equations

$$\nabla^2 \mathbf{A} - \frac{1}{c^2}\frac{\partial^2 \mathbf{A}}{\partial t^2} = -\mu_0 \mathbf{J} \qquad (13)$$

and

$$\nabla^2 \phi - \frac{1}{c^2}\frac{\partial^2 \phi}{\partial t^2} = -\frac{\rho}{\epsilon_0} \qquad (14)$$

provided that we impose the added stipulation that the potentials satisfy the Lorentz condition

$$\nabla \cdot \mathbf{A} + \frac{1}{c^2}\frac{\partial \phi}{\partial t} = 0. \qquad (15)$$

It should be emphasized at this point in our discussion that Eqs. (11)–(15) are completely equivalent to Maxwell's equations as far as the determination of the electromagnetic fields produced by a given set of charges and currents is concerned. However, this mathematical representation of the theory does facilitate the actual calculation of the fields in many cases; in addition, it plays an important role in the further understanding of the electromagnetic field.

For the case when the electromagnetic radiation is due to localized sources, the physical solutions to the foregoing equations are neatly expressed as the so-called retarded potential functions

$$\mathbf{A}(\mathbf{r},t) = \frac{\mu_0}{4\pi}\int d^3x' \, \frac{\mathbf{J}(x',t^*)}{|\mathbf{r}-\mathbf{r}'|} \qquad (16)$$

and

$$\phi(r,t) = \frac{1}{4\pi\epsilon_0}\int d^3x' \, \frac{\rho(x',t^*)}{|\mathbf{r}-\mathbf{r}'|}. \qquad (17)$$

These integrals over the space distributions of the charge and current densities are evaluated at the time

$$t^* = t - |\mathbf{r}-\mathbf{r}'|/c, \qquad (18)$$

that is, at a time t^*, corresponding to the emission of the radiation, earlier than the observation time t by the time $|\mathbf{r}-\mathbf{r}'|/c$, required for the propagation of the wave from the source point location $\mathbf{r}'$ to the field observation point $\mathbf{r}$.

RADIATION FROM A DIPOLE

A good example of electromagnetic radiation from a localized source is that emitted from a linear electric dipole of moment p (charge multiplied by maximum displacement) oscillating sinusoidally at a frequency ν. The vector potential in the space region far from the dipole, the wave zone, is

$$A_z \simeq \frac{\mu_0 \nu p}{2r} \cos 2\pi\nu\left(t - \frac{r}{c}\right). \qquad (19)$$

Table I Examples of Electromagnetic Plane Waves

		Radio signal	Solar radiation
Electric field	E_x	1×10^{-6} V/m	1025 V/m
Magnetic field	B_y	3.3×10^{-15} T	3.4×10^{-6} T
Poynting vector	$\langle S_z \rangle_{av}$	1.3×10^{-15} W/m^2	1390 W/m^2
Energy density	$\langle u \rangle_{av}$	4.4×10^{-24} J/m^3	4.6×10^{-6} J/m^3
Radiation pressure	$\langle u \rangle_{av}$	4.4×10^{-24} N/m^2	4.6×10^{-6} N/m^2

Here r is the distance from the dipole that is assumed to be located at the origin of the coordinate system and to be oriented along the z direction. The magnetic field vector is transverse to the unit radial vector **n** and is given by

$$\mathbf{B} \simeq (\mathbf{n} \times \mathbf{p}) \frac{\mu_0 \pi \nu^2}{cr} \sin 2\pi\nu\left(t - \frac{r}{c}\right). \tag{20}$$

As expected, the electric field vector is normal to both the radial vector and the magnetic field. Therefore we write it as

$$\mathbf{E} = -c(\mathbf{n} \times \mathbf{B}). \tag{21}$$

The electromagnetic radiation far from an electric dipole can hence be visualized as essentially a plane wave propagating radially outward while the component fields decrease with distance as 1/r. The power associated with such an outward flow is characterized by the Poynting vector, which has the time-average value of

$$\langle S_r \rangle_{\mathrm{av}} = \frac{\pi^2 \mu_0 \nu^4 p^2 \sin^2\theta}{2cr^2}. \tag{22}$$

The angle θ is that between the orientation of the source dipole and the direction of the field observation point with respect to the origin. We notice that, according to Eq. (22), the radiation pattern exhibits a maximum in the plane at right angles to the dipole orientation ($\theta = \pi/2$), while the radiation vanishes for field points along the dipole direction ($\theta = 0$). An integration over all directions yields the total radiated power

$$P = \frac{4\pi^3 \mu_0 \nu^4 p^2}{3c}. \tag{23}$$

We call attention to the dependence of this radiated power on the fourth power of the frequency. It is this characteristic of dipole radiation that basically accounts for the blueness of the sky. Since blue light has a higher frequency than red, it is more effectively reradiated, or scattered, by the bound electrons in the molecules of the atmosphere.

MATERIAL MEDIA

In this brief survey we have presented the basic elements of the theory of electromagnetic radiation in free space. As emphasized by Feynman (see the References), it is this aspect of the great synthesis of Maxwell that is of lasting significance. The exact manner of broadening such a development so as to include radiation through material bodies, which have a crystalline, molecular, and atomic structure, is fundamentally a difficult problem. The usual way of doing this is by introducing some phenomenological parameters such as dielectric constant, permeability, and conductivity. We then speak of the effective electric and magnetic polarizations, the electrical displacement (**D**), and the magnetizing force (**H**). It should be realized, however, that this approach ultimately requires that these macroscopic material parameters be explained by the interaction of the electromagnetic radiation with matter, that is, by the application of electrodynamics and quantum theory.

See also ABSORPTION COEFFICIENTS; ELECTRODYNAMICS, CLASSICAL; LIGHT; MAXWELL'S EQUATIONS.

REFERENCES

1. E. Whittaker, *A History of the Theories of Aether and Electricity, The Classical Theories* (Reprint of 1954 edition). American Institute of Physics, New York, 1987. (I, A)
2. H. H. Skilling, *Fundamentals of Electric Waves*. Wiley, New York, 1948. (E)
3. E. M. Purcell, *Electricity and Electromagnetism*. Berkeley Physics Course, Vol. II, McGraw-Hill, New York, 1963. (E, I)
4. R. S. Elliott, *Electromagnetics*. McGraw-Hill, New York, 1966. (I)
5. R. P. Feynman, R. B. Leighton, and M. Sands, *The Feynman Lectures on Physics, The Electromagnetic Field*, Vol. II. Addison-Wesley, Reading, Mass., 1964. (I, A)
6. J. D. Jackson, *Classical Electrodynamics*. Wiley, New York, 1962. (I, A)

Electromagnets

F. J. Friedlaender

Electromagnets are devices in which magnetic fields are produced by means of current-carrying conductors. Usually the magnetic field in such a device is desired in an air gap or space that is not part of the field-producing structure. Direct-current motors and generators and synchronous motors and generators all require magnetic fields that are produced by either electromagnets or permanent magnets. If time-varying or large fields are required, then the use of permanent magnets is generally ruled out; electromagnets are the only practical means of producing such fields. Lifting magnets, and magnets to produce the fields for (high-gradient) magnetic separators or for MHD generators, are other common applications of electromagnets. Magnetic fields up to over 50 T (500 kG), as required for research purposes, are generally produced by means of electromagnets, and these will be our major concern.

There are, broadly speaking, three classes of electromagnets: (1) those using iron (or a similar ferromagnetic or ferrimagnetic material) in the flux path; (2) those using no iron and having dissipative (normally conducting) coils; (3) those with no iron and nondissipative (i.e., superconducting) coils.

We should also distinguish between pulsed-field or time-varying-field magnets, mostly under classification 2, and the more common steady-field (dc) magnets in all three categories.

Iron in the flux path (category 1) reduces the required coil current greatly, as long as the iron does not saturate (at just over 2.1 T or 21 kG). But iron-clad magnets are useful even at much higher flux densities in the working space (which is always external to the iron). Flux-concentrating means, such as truncated conical iron pole pieces, are commonly used to produce densities of over 10 T (100 kG) in the air gap of iron-clad magnets. As the field is increased in both category 1 and 2 electromagnets, larger currents have to flow in the coils, thus generating increasing thermal losses in the coils requiring adequate cooling to prevent overheating.

There are three major considerations dictating the design of an electromagnet. Of primary concern is the field design, which includes the design of the flux paths for category 1, as well as placement and shapes of coils and current distributions in the coils. Next in importance is thermal design, which also enters into coil design and the provision for adequate cooling, usually by means of water. Somewhat different thermal considerations apply to nondissipative magnets where superconducting coils at cryogenic temperatures are used. Cryogenic fluids (i.e., liquid helium) rather than water are used in such magnets. The third factor is mechanical design: large magnetic stresses occur in high-field electromagnets, and compromises in the design are often necessary in order to achieve a magnet that is mechanically sound.

The current-carrying coils in an electromagnet may have uniform current densities or—to obtain higher efficiencies (i.e., larger fields for the same power)—nonuniform current densities. If the coils have cylindrical symmetry they are called solenoids, but magnets are often designed using rectangular coils. The field H at the center of a field-producing coil can be shown to be related to the applied power P through the equation $H=G(P\lambda/\rho a_1)^{1/2}$ where λ is a winding space factor, ρ the resistivity of the solenoid wire, and a_1 the inside radius of the solenoid. G is a geometrical factor ("G factor," also called *Fabry factor* after Fabry, who first suggested its use).

A major concern of the solenoid designer is the optimization of this G factor by use of an appropriate geometry (i.e., coil dimensions and shape, current distributions, etc.). One such design that has found widespread applications and is named after its inventor, Francis Bitter, uses disks of conductors to form essentially a solenoid with a helical sheet winding. Each disk is a ring with a radial slot. Copper rings are assembled alternately with insulating rings (also with slots), so that a helical conducting path through the copper rings is formed. Axial holes in the disks are provided for water cooling. Gaume modified the Bitter system, in which each disk has the same current density distribution (largest at the inside radius and decreasing with increasing radius), to one in which the thickness of the disk and hence the relative current densities are varied. A Bitter magnet with an inside coil radius of 3.2 cm may require 10 MW to produce 22.5 T (225 kG).

Superconducting magnets can be used to produce fields up to almost 20 T (200 kG). Type II superconductors such as Nb_3Sn with critical fields of over 200 kG are used to make the field coils. Usually it is necessary to stabilize superconducting magnet coils by providing a parallel conducting path of a normal low-resistivity material in good contact with the superconductor. This arrangement allows stable operation of the superconducting magnet and guards against unstable collapse of superconductivity in the entire magnet winding, due to local thermal effects. Superconducting magnets are usually operated at 4.2° K, the temperature provided by the liquid helium that is almost always used as a coolant. Since the coil of a superconducting magnet requires negligible power, superconducting magnets have a considerable economic advantage over other electromagnets for certain applications in which fields of the order of 10 T (100 kG) are required.

In the future, superconductors that operate at liquid-nitrogen or even higher temperatures may become available. Such an event may decrease the cost of the cooling system. But unless superconductors are found that can sustain larger fields and currents, the capabilities of superconducting electromagnets are not likely to change substantially.

In many applications, the only practical method of obtaining the required large fields is by means of pulsed electromagnets. By using a relatively low duty cycle, the high energy required to produce the field can be obtained by means of relatively small average power values. Pulsed magnets provide the only means of obtaining fields over 50 T (500 kG). Usually a bank of capacitors is discharged through the field-producing coil, with suitable electronic control in some recent designs to produce a flat-topped pulse that may last from a fraction of a millisecond to several milliseconds. During the field pulse all the dissipated energy will raise the temperature of the field coil and associated elements, and forced cooling, if any is needed, will depend on the duty cycle (pulse repetition rate) of the magnet. Finally, mechanical considerations also play a substantial role in the design of high-field solenoids. The high fields and large currents give rise to forces which can destroy a solenoid that is not designed properly.

See also MAGNETIC FIELDS, HIGH; MAGNETS (PERMANENT) AND MAGNETOSTATICS.

BIBLIOGRAPHY

D. B. Montgomery, *Solenoid Magnet Design*. Wiley-Interscience, New York, 1969.

H. C. Roters, *Electromagnetic Devices*. Wiley, New York, 1941.

J. Liedl, W. F. Gauster, H. Haslacher, and H. Grössinger "Calculation of the Mechanical Stresses in a High Field Magnet by Means of a Layer Model," *IEEE Trans. Magn.* **MAG-17,** 3256–3258 (1981).

Electron

A. Pais

INTRINSIC PROPERTIES

Ever since its discovery, the electron has been considered an elementary particle, a fundamental building block of matter that cannot be decomposed into more primary constituents. The electron has the following intrinsic properties (the figures in parentheses denote the 1-standard-deviation uncertainty in the last two digits of the main number):

1. *Stability.* The free electron is generally believed to be absolutely stable. Experimental studies of the absolute validity of electric charge conservation have been made by looking for possible disintegrations of the electron. None has

been found. In this way a lower bound of the order of 10^{21} yr for its lifetime was established.

2. *Mass:* $m = 0.511\ 003\ 4(14)\ \text{MeV}/c^2 = 9.109\ 534(47) \times 10^{-28}$ g.
3. *Charge:* $e = 4.803\ 242(14) \times 10^{-10}$ esu.
4. *Spin:* $h/4\pi$ (h is Planck's constant).
5. *Gyromagnetic ratio:* $1.001\ 159\ 656\ 7(35) e/mc$.
6. *Electric-dipole moment.* Not observed; its present experimental upper bound is $3 \times 10^{-24} e$ cm.

DISCOVERY

The discovery of the electron finally settled the question, debated for more than a century, whether there exists a quantum of electricity, a smallest unit of electric charge. From Faraday's law (1833), according to which each gram-atom of any univalent electrolyte carries the same charge, it follows that there is such a unit. To see this, however, we should know that the gram-atom of a pure substance consists of a definite number of identical atoms. This "Loschmidt number" was not determined until the 1860s but even then the notion of an elementary unit of charge seemed to many to be one of terminology rather than of physical reality. For example, in his 1873 treatise on electricity and magnetism Maxwell refers to the electric quantum as "one molecule of electricity. This phrase, gross as it is, and out of harmony with the rest of this treatise ..."; and he adds, "It is extremely improbable that when we come to understand the true nature of electrolysis we shall retain in any form the theory of molecular charges ...." But proponents of the atomistic view of electricity persisted, among them Helmholtz and the Irishman George Johnstone Stoney. The latter has the distinction of having given the first crude estimate of e as early as 1874, and of having baptized this unit with the name electron in 1891. Thus this term was coined prior to the discovery of the quantum of electricity *and* matter that now goes by that name.

The years of its discovery are 1896 and 1897. In his second paper on the Zeeman effect (28 November 1896), Pieter Zeeman recorded that Lorentz "at once kindly informed me" how the motion of an ion [*sic*] in a magnetic field can be determined. Using an oscillator potential model of the atom, Lorentz interpreted the Zeeman effect in terms of the motion in the atom of a particle with $e/mc \sim 10^7$ rad sec^{-1} G^{-1}. This was the first, albeit indirect, hint of the existence of a particle with a novel low mass.

The direct proof of its actual existence was given by J. J. Thomson and was first communicated on 29 April 1897 in a Friday evening discourse at the Royal Institution. His discovery was that cathode rays are electrons. Such rays had been studied for decades, but their constitution was unclear. Some held them to be "molecular torrents," others "aether disturbances." Thomson determined their e/m, noted that his answer was much like Zeeman's, and observed that "the assumption of a state of matter more finely subdivided than the atom is a somewhat startling one." The first fundamental particle had been isolated. The implicit assumption that the e involved is the same as for univalent electrolytic ions was soon verified, especially by Millikan, who measured e with precision and demonstrated its uniqueness.

THE CLASSICAL PRERELATIVISTIC ELECTRON

"We shall ascribe to each electron certain finite dimensions, however small they may be ..., my excuse must be that one can scarcely refrain from doing so if one wishes to have a perfectly definite system of equations...." Thus, with care and caution, Lorentz introduced the classical model of the electron: a charge distribution confined (at rest) to a small sphere of radius r and with the classical mass formula $m = m' + m_e$. Here $m_e = e^2/rc^2$ (up to a number of order unity) is the electromagnetic mass (a concept generally known since 1881) and m', which used to be called the "material mass," is the contribution due to other origins. It was often speculated that $m' = 0$ and hence this particle was ascribed its "classical electron radius": $r = e^2/mc^2 \approx 2.8 \times 10^{-13}$ cm. In fact, in those prerelativistic days it was believed possible to determine experimentally whether or not $m' = 0$ by measuring the electron energy E as a function of velocity v for uniform motion.

For example, if the electron were a rigid sphere (Abraham model), then for $m' = 0$, $E(v)$ is calculable purely as an electromagnetic expression. Progress was not helped by the fact that at first the answer appeared to agree with experiment! For some time this agreement was held as evidence (1) for zero m'; (2) for the rigid model; and (3) against the Lorentz model, in which the finite sphere is contracted in its direction of motion.

THE CLASSICAL RELATIVISTIC ELECTRON

With the advent of Einstein's special theory it became clear that the correct $E(v)$ had to be $E(v) = mc^2/(1 - v^2/c^2)^{1/2}$, regardless of what contributes to m, and therefore that the notion of a determination of m' from velocity measurements was illusory. [Not until about 1915 was this form of $E(v)$ verified experimentally.] The focus now shifted to a new problem: None of the existing models, including the one of Lorentz, gave the Einstein $E(v)$. Indeed in all these models the electron is an open system (i.e., there are unbalanced forces on its boundary), whereas Einstein's answer applies to closed systems only. This discrepancy led to conjectures, starting with Poincaré, about nonelectromagnetic cohesive stresses designed to balance the electromagnetic repulsive stresses. Lorentz was not sure of this: "... perhaps we are wholly on the wrong track when we apply to the parts of the electron our ordinary notion of force."

By about 1925 the focus had changed again. At that time the Russian physicist J. Frenkel wrote, "I hold these riddles ... to be a purely scholastic problem ... the electrons are not only physically but also geometrically indivisible. They do not have any extension in space at all ...." But then there appears still another paradox, since m_e becomes infinite for zero radius. We shall return to this shortly.

ELECTRON SPIN

A critical examination of the anomalous Zeeman effect led Pauli to propose (December 1924) that the doublet structure of alkali spectra is caused by "a two-valuedness, not describable classically," of the quantum properties of the valence electron. This made him assign four quantum numbers (instead of the customary three) to the electron, and in January 1925 he formulated the exclusion principle: In an atom this set of four numbers cannot be the same for any two electrons (a formulation that was much broadened subsequently). In October 1925, Uhlenbeck and Goudsmit suggested that "it is plausible to assign to the electron with its four quantum numbers also four degrees of freedom ... [one of which] ... can be associated with a proper rotation of the electron" and assigned the value $h/4\pi$ to this intrinsic angular momentum or "spin." The quantitative understanding of the fine structure came only after some struggle. The importance of major relativistic effects (the so-called Thomas factor) had to be recognized. Also, a mysterious value e/mc, twice the value expected classically, had to be assigned to the gyromagnetic ratio of the electron. When in May 1927 Pauli succeeded in describing the spinning electron by a two-component "spinor" wave function he still had to incorporate these effects in an ad hoc way. It was an important advance in description but not yet in understanding.

THE DIRAC EQUATION; THE POSITRON

These effects became fully understood when in February 1928 P. A. M. Dirac proposed his relativistic wave equation of the electron, one of the most spectacular advances of modern science. At the same time his equation generated severe new paradoxes.

His equation implies that there are four states for a given momentum **p**. There is a welcome doubling associated with the spin but an apparently paradoxical further doubling due to the fact that the associated energy takes on the two values $\pm(c^2p^2+m^2c^4)^{1/2}$. What is the significance of the inevitable negative-energy states? Speculation arose that these might be associated with the proton, the only other fundamental particle then known. However, "one cannot simply assert that a negative energy electron *is* a proton," Dirac noted. Rather, he argued, we should "assume that there are so many electrons in the world ... that all the states of negative energy are occupied except perhaps a few of small velocity ...." Noting that these holes behave like positively charged particles, "we are therefore led to the assumption that the holes in the distribution of negative energy electrons are protons." This is the earliest version of the "hole theory" (1929).

But this cannot be. Experimentally it would imply that a hydrogen atom would annihilate into photons. Theoretically, as H. Weyl noted, "according to it [the Dirac equation] the mass of the proton should be the same as the mass of the electron ...." The gravity of the situation is illustrated by Weyl's further remark that "the clouds hanging over this part of the subject will roll together to form a new crisis in quantum physics." Once again Dirac found the answer (1931): "a hole, if there were one, would be a new kind of particle, unknown to experimental physics, having the same mass and opposite charge of the electron." This particle, the positron (e^+), was discovered on 2 August 1932 by C. D. Anderson.

THE RELATIVISTIC QUANTUM-MECHANICAL ELECTRON

The revolution of quantum mechanics has erased all pictures of the electron as a tiny sphere. The point model has survived, or (far better stated) the electron is described by a quantized local field. Earlier we noted that for a classical point model $m_e = \lim_{r\to 0} e^2/rc^2$. Also in relativistic field theory the electromagnetic mass (or self-energy) is infinite but the nature of the singularity has changed because of quantum effects. We now have

$$m_e = \lim_{r\to 0}(3\alpha/2\pi)m\ln(h/mcr) + O(e^4)$$

where $\alpha = 1/137$ is the fine-structure constant. Attempts in the 1930s and 1940s to modify the theory such that m_e (and certain other physical quantities) become finite were not successful. However, a major advance has been made since 1947: It is now possible to handle the theory in such a way that in all physical quantities like cross sections or energy levels, only the physical mass m of the electron appears, without hindrance by the infinity in m_e; nor (once again) does the separation into material (or "bare") + electromagnetic mass ever enter. This is the renormalization theory. With most impressive results (e.g., it yields the gyromagnetic ratio given earlier—property 5 in the opening section—to high accuracy) it bypasses the self-energy problem. It does not solve it.

WEAK INTERACTIONS

In December 1933, Fermi postulated a new kind of electron interaction, the weak interaction. Thereby he reduced all of β radioactivity to the occurrence of the fundamental process neutron→proton + electron (e) + "anti-e-neutrino" ($\bar{\nu}_e$). (The latter particle had been hypothesized by Pauli.) Since that time numerous other "weak processes" have been discovered in which an e *and* a $\bar{\nu}_e$ or an e^+ *and* a ν_e (electron neutrino) are created together (see WEAK INTERACTIONS). This has led to the formulation of a new principle, called the conservation of e number: In any reaction the number of $e+\nu_e$ particles minus the number of $e^+ + \bar{\nu}_e$ particles is conserved. This principle ties the mentioned particles together in a "family." In turn, this family is part of a larger family called *leptons* (so named by Møller and Pais). A lepton is a particle that participates in gravitational and weak interactions. It may also participate in electromagnetic interactions (as e and e^+); but it does not (at least not directly) participate in strong interactions (which see). Other known leptons (there may be more) are the muon (μ) and (related to these) another brand of neutrinos called ν_μ.

NEUTRAL CURRENTS; GAUGE THEORIES

A search is under way to unify electromagnetic and weak interactions. In such attempts the electron enters into an interconnected set of currents: (1) the electromagnetic current, coupled to a massless neutral vector field (photons); (2) a "weak charged current," coupled to a massive charged vector boson field (W bosons), which transmits the Fermi-type weak interactions; (3) a "weak neutral current," coupled to a massive neutral vector boson field (Z bosons). (See GAUGE THEORIES; it is not excluded that the set of currents is more complex than sketched here.) This last coupling gives rise to a new class of weak processes in which the electron is *not* accompanied by an e neutrino (but in which e number is conserved). These ideas stimulated a search for such new "neutral-current reactions." Since 1973 there is evidence that these indeed exist.

Theories of this kind are so far speculative (for one thing, W and Z bosons have not as yet been seen), but they appear to be quite promising. They may yet shed new light on fundamental properties of the electron, such as the origin of its mass and the possibility that it carries a tiny electric-dipole moment. It would appear that the days are numbered both for electromagnetism as a force separate from other forces and for the splendid isolation of the electron as a particle unconnected with other particles.

See also BETA DECAY; ELEMENTARY PARTICLES IN PHYSICS; FIELD THEORY, CLASSICAL; GAUGE THEORIES; LEPTONS; QUANTUM ELECTRODYNAMICS; WEAK INTERACTIONS

BIBLIOGRAPHY

E. R. Cohen and B. N. Taylor, *J. Phys. Chem. Ref. Data* **2,** 663 (1973). (Intrinsic properties.)

H. A. Lorentz, *The Theory of Electrons*. Dover, New York, 1952.

A. Pais, "The Early History of the Theory of the Electron: 1897–1947," in *Aspects of Quantum Theory* (A. Salam and E. P. Wigner, eds.), p. 79. Cambridge, London and New York, 1972.

Electron and Ion Beams, Intense

James N. Benford

Theoretical studies of intense relativistic electron beams began in the 1930s with the pioneering works of Bennett on pinched (radially compressed) electron flow and of Alfvén on current limitations in streams of cosmic rays. Practical implementation, however, had to wait until the early 1960s, when J. C. Martin in England succeeded in developing the beginnings of the pulse technology required to deliver the tens of kiloamperes of electron current in the megavolt range that are characteristic of these beams. Since then, vigorous industrial and government laboratory efforts, primarily in the United States, have produced beams with megampere currents at megavolt energies with pulse durations of $\sim 10^{-7}$ s. In general these beams are generated by employing a voltage multiplier source (Marx generator) to pulse-charge an oil or water dielectric line of 1–50 ohms characteristic impedance. The firing of a gas or liquid output switch allows the voltage pulse to be applied to a field-emission cathode from which a burst of electrons is accelerated toward the anode by the potential difference across the gap.

In a diode with no applied axial magnetic field, uniform current flow occurs (planar Child–Langmuir space-charge-limited flow, Fig. 1) until the turning of electrons by the B_θ self-field causes pinching of the beam toward the axis ($F_r \sim v_z B_\theta$). The critical current level for this pinching to occur is obtained by requiring that electrons emitted at the edge of the cathode reach the anode with a velocity parallel to the surface:

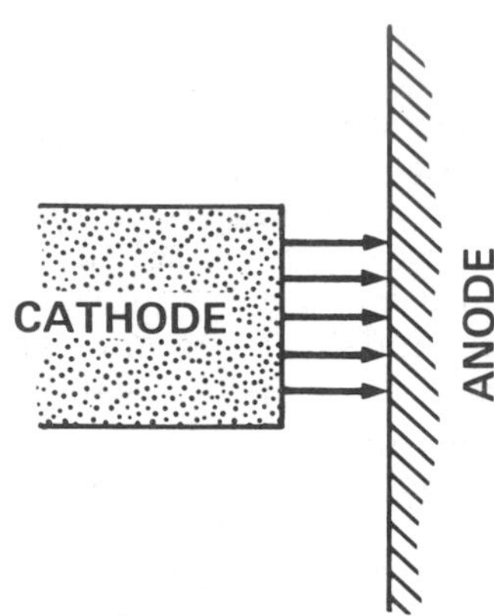

FIG. 1. Flow in unpinched diode.

$$I_c = \frac{2\pi mc\beta\gamma}{\mu_0 e}\frac{R}{d} = 8500\beta\gamma\frac{R}{d}\ \text{A}$$

where R is the cathode radius, d the anode–cathode gap, γ the relativistic factor $\gamma = (1-\beta^2)^{-1/2}$, and $\beta = v/c$. For $I > I_c$, the electrons flow toward the axis, and current follows the "parapotential" flow expression, which is derived from a model in which electrons move along conical equipotentials for which $\mathbf{E} - \mathbf{v} \times \mathbf{B} = 0$ (Fig. 2). The current is given by

$$I = 8500 g\gamma \ln[\gamma + (\gamma^2 - 1)^{1/2}]\ \text{A}$$

where g is a factor that is geometry dependent, equal to R/d for Fig. 2. Many data agree with this expression, which requires a bias current in the diode as a boundary condition. The bias current is in part provided by ions coming from the anode. This anode plasma seems to be restrained from expanding across the diode gap and shorting the generator by the magnetic pressure of the self-field.

When the anode is a thin metal foil effectively transparent to the relativistic beam electrons, the beam will leave the diode region. Propagation will depend on the magnitudes of the repulsive electrostatic force due to the charge of the

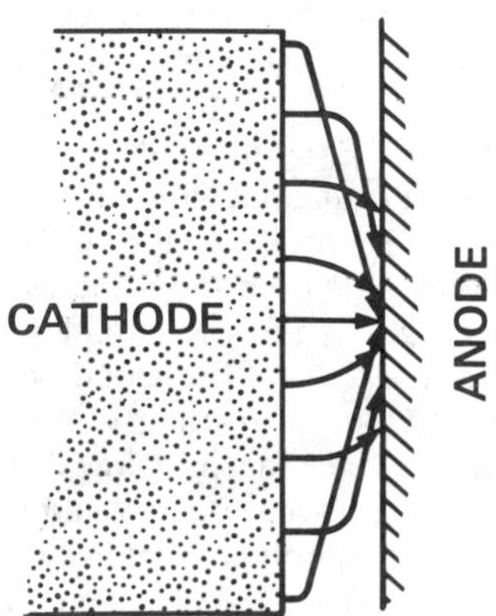

FIG. 2. Parapotential pinched flow.

beam electrons and the attractive magnetic force that arises from the beam current. In vacuum, the electrostatic force dominates, and the beam rapidly expands radially. Electron repulsion can be alleviated by applying a longitudinal magnetic guide field, by supplying neutralizing ions from preexisting plasmas, or by direct ionization of gas by beam electrons. In principle, fractional space-charge neutralization sufficient to provide force balance is possible, but in practice propagating beams very quickly become space-charge neutralized.

With the magnetic force determining particle orbits, propagation is characterized by the strength parameter ν/γ, where ν is the number of electrons in a cross-sectional slab of the beam with a thickness of a classical electron radius. This parameter can be expressed as a ratio of currents, $\nu/\gamma = I/I_A$, where I is the beam current and I_A is the Alfvén–Lawson critical current

$$I_A = Ne\beta c = \frac{4\pi mc\beta\gamma}{\mu_0 e} = 17000\beta\gamma \text{ A}.$$

For $\nu/\gamma < 1$, electron orbits are roughly sinusoidal and propagation is possible (Fig. 3). As beam intensity is increased, a point is reached where the gyrodiameter of beam electrons in the B_θ magnetic field of the current is about equal to the beam radius. Beam electrons turn completely around and flow stops. This limit occurs at $I = I_A$, or $\nu/\gamma = 1$ for a uniform radial current distribution. So when $\nu/\gamma \sim 1$, self-fields strongly influence beam motion.

There are three ways of exceeding this limiting current. The naturally occurring method is "current neutralization," which arises from the tendency of plasma or any conductor to resist sudden changes in magnetic field. A fast-rising beam current induces an opposing axial electric field that drives a partially canceling return current in the plasma. The resulting net current (beam minus plasma current) determines electron orbits, so that ν/γ is replaced by $(\nu/\gamma)_{net}$. For highly conducting plasma, the net current can be reduced to such a low value that beam electrons are not contained and expand to the boundaries of the chamber.

The second method of propagation is to employ a radially nonuniform current distribution. For example, a beam with most of the current flowing in a shell near the edge will propagate with $\nu/\gamma > 1$ by turning beam electrons only in the outer shell. Alternatively, in the Bennett distribution, the current density is peaked on the axis. It can be shown that ν/γ is proportional to the ratio of the magnetic-field energy within the beam to its kinetic energy. Nonuniform beam distributions reduce the magnetic energy, increasing the allowable current that can be propagated.

The third method, application of a longitudinal magnetic guide field, will prevent electrons from turning in their self-field, allowing propagation. The energy density of the field must be at least as great as that of the beam.

FIG. 3. Electron trajectories in propagating beam.

Recent experiments have demonstrated an electrostatic propagation aid. A laser ionizes a channel through a background gas. When the head of an electron beam passes, the electrons are expelled, leaving a background of ions. The remainder of the beam has its charge neutralized by the ions, so the beam propagates along the channel.

The basic features of beam propagation away from the diode have been established in efforts to produce copious x-ray bremsstrahlung from a target and to study the effects of rapid energy deposition in solids. Other applications are the generation of microwaves (1–10 GHz) or infrared to ultraviolet (free-electron laser) by propagation along a periodic magnetic field of suitable choice, and collective acceleration of ions by the space-charge well that exists at the beam front. Recent years have seen rapid development of a wide variety of microwave sources (magnetrons, gyrotrons, backward wave oscillators, virtual cathode oscillators, klystrons) at powers 10^9–10^{10} W for pulse durations $\sim 10^{-7}$ s. Emerging applications for this technology are fusion plasma heating, particle acceleration, and directed energy. Beams are used to excite lasers by exciting electronic and vibrational levels of molecules and excimers, and by initiating chemical reactions.

A major application involves use of the high power and energy of beams for the heating of magnetically confined plasmas in either linear or toroidal geometries. Plasma heating to several-thousand-electron-volt temperatures at particle densities of 5×10^{15} cm^{-3} has been achieved. The heating is due to streaming instabilities. For example, an interaction between beam and plasma electrons causes fluctuating electric fields at the electron plasma frequency. These fields directly heat plasma electrons, and excite oscillations at other frequencies, which then dissipate into particle thermal energy.

Use of beams to create plasma-confining magnetic-field configurations has centered on electron and ion rings, although toroidal systems have shown promise as well. Electron rings are formed by injection of an electron beam transverse to a magnetic field. The electrons gyrate about the field lines and generate a cylindrical current, which in turn produces a magnetic field in its interior in the opposite direction. For a sufficiently large current the direction of field can be reversed, forming a closed set of field lines, a minimum-B magnetic trap. The gyrating electrons radiate their energy rapidly by synchrotron radiation, requiring that ions be used to form the layer in fusion applications.

Pinched electron beams have been applied to inertial-confinement fusion. Self-focused electron beams have been observed with current densities of 10^6–10^7 A/cm^2, resulting in large charge concentration near the diode axis. The electron flow path is long compared with that of ions, which can cross from the anode with little deflection by the magnetic field. Simple arguments show that higher-current generators could have a substantial portion of the current carried by ions that could be focused geometrically for inertial confinement fusion. Ion currents ~ 1 MA have been produced and propagated onto targets from diodes optimized for ion beam production. The basic principle is suppression of electron flow by use of a transverse magnetic field.

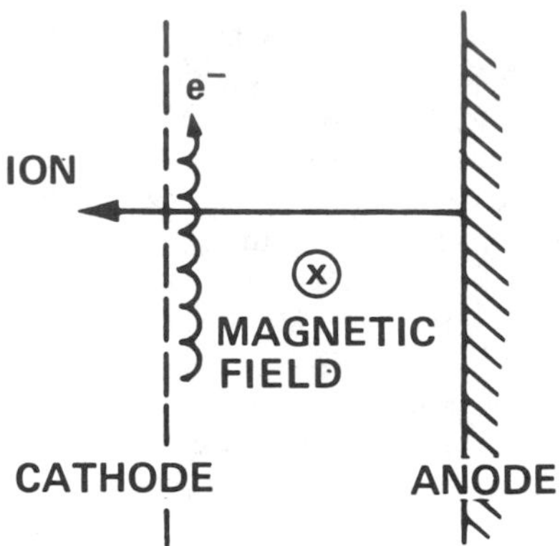

FIG. 4. Ion diode employing external magnetic field for electron current suppression.

$$B_c = \frac{mc}{e} \frac{1}{d} \beta\gamma.$$

Ions cross the gap freely (Fig. 4), undergoing a small deflection, and electrons execute magnetron orbits.

An alternative method of intense ion-beam generation is to prevent pinching with an axial magnetic field and to allow electrons to pass through the anode, losing a small fraction of their energy, and on into another diode. Electrons reflect from the second cathode and return to the first diode. Multiple reflections and the accompanying energy loss result in concentration of the electrons near the anode surface (Fig. 5). The increased space charge there causes vastly enhanced ion emission from both sides of the anode. Currents of ~200 kA have been generated by this method.

Such intense beam diodes are used in an emerging application, production of very high pulse power (~10^{13}–10^{14} W). Such diodes serve as opening switches, rapidly diverting energy from high-energy magnetic stores.

The technological developments that have spawned the rapid growth of electron-beam research and have led to the production of intense ion beams are continuing. Higher-current (several megamperes) and higher-power (10^{13}–10^{14} W) generators are under development.

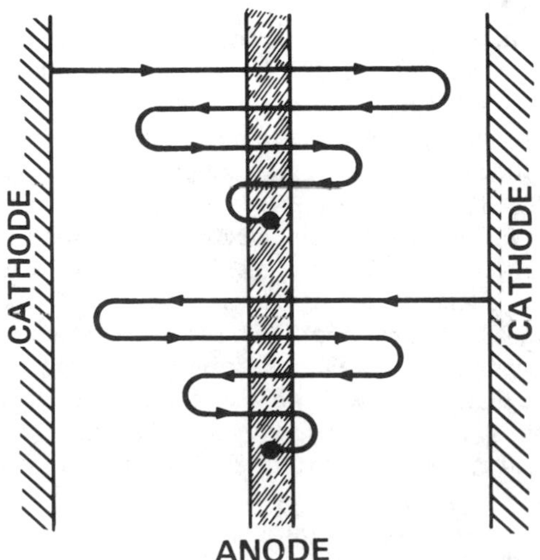

FIG. 5. Double diode arrangement—electron making multiple transits through anode foil.

BIBLIOGRAPHY

G. Benford and D. Book, *Adv. Plasma Phys.* **4,** 125 (1971). Reviews beam-equilibrium models. (A)

H. Fleischmann, *Phys. Today* **28,** 34 (1975). Surveys applications of electron beams. (E)

J. Nation, *Particle Accelerators* **8,** (1979). Discusses ion- and electron-beam generation and technology. (A)

Electron and Ion Impact Phenomena

Edward W. Thomas

An energetic electron or ion impacting on a surface transfers energy to the lattice of the solid and to the free electrons in the solid by a successive series of collisions. As a result, certain particles are ejected from the surface; familiar examples are electron ejection (secondary-electron emission) and ejection of surface atoms (known as sputtering). The projectile itself may suffer sufficient angular deviation in a collisional scattering event so that it becomes directed out of the surface and may be observed as a reflected primary particle. The deposition of energy into the solid will cause various displacements of the lattice structure, known collectively as radiation damage, and it may excite electrons which subsequently decay with the emission of light (luminescence).

The rates at which these phenomena occur are described by a quantity known as a "coefficient," which is the number of relevant events occurring for every primary projectile incident; thus the secondary-electron emission coefficient is the number of secondary electrons ejected for each primary electron incident. Coefficients are strongly dependent on the condition of the surface and the crystal structure of the substrate. For a reliable measurement that can be related to a theoretical concept it is necessary that the target surface be atomically clean, the substrate must be of known crystallographic structure, and the orientation of the surface to the incident projectile direction must be defined.

ELECTRON IMPACT MECHANISMS

Most often studied is the flux of emergent electrons including both scattered primary electrons and ejected secondary electrons. Figure 1 shows the energy distribution of emergent electrons; region I is due mainly to reflected primaries, region III to secondary electrons, and region II includes components from both sources.

Secondary Electron Ejection

The incident primary electron produces excitation and ionization within the solid. Those excited electrons which diffuse to the surface, overcome the potential barrier, and emerge to be detected are called secondary electrons. Figure 1 indicates that true secondary electrons cannot be distinguished from reflected primary electrons; it is conventional

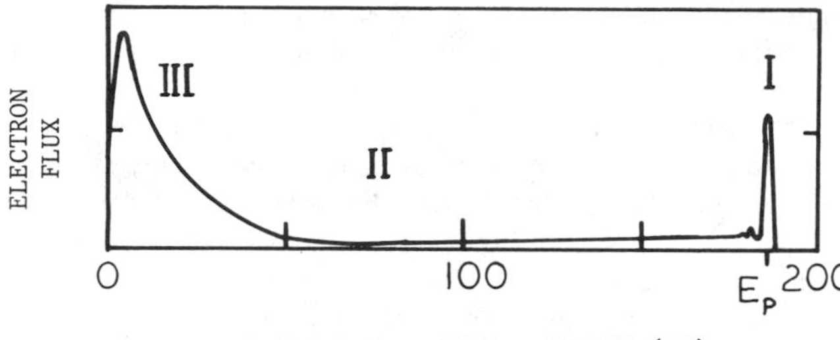

FIG. 1. A typical energy distribution for electrons emerging from a solid as a result of electron bombardment. E_P denotes the energy of the primary electrons which produce the secondary emission.

to regard the secondary electrons to be those with energies below 50 eV (excepting Auger electrons discussed below). Secondary electrons arise primarily from excitation of outer-shell electrons of the solid and they have an average energy of a few electron volts. Typical peak secondary-emission coefficients of metals and semiconductors are 1 to 1.5 (secondary electrons out per primary electron in); for insulators and intermetallic compounds yields may be as high as 10 or 20.

Primary electrons of sufficient energy may eject inner-shell electrons from the atoms of the solid, so creating a vacancy which must be subsequently filled by an electron falling from a higher level. The energy liberated as the vacancy is filled may be transferred to some other outer-shell electron which thereby becomes ionized, and escapes from the solid, a process known as Auger-electron emission. Auger electrons are energetic with a relatively small energy spread; they are observed in the secondary-electron energy spectrum as discrete high-energy peaks, superimposed on the general continuous background of other secondaries and reflected primaries. The Auger-electron energy is characteristic of the atom from whence it was ejected so that analysis of the Auger spectrum, using an electron energy analyzer, can indicate which atomic species are present in the target.

Electron Reflection

The high-energy peak of electrons emerging from the solid (I in Fig. 1) is due to primary electrons scattered or reflected from close to the surface and much of the lower energy tail (II in Fig. 2) is due to primary electrons scattered from deeper in the solid. Small subsidiary peaks slightly below the incident energy are due to primary electrons which have lost discrete amounts of energy by exciting plasma oscillations in the electron gas. Reflection coefficients increase with the nuclear charge of the substrate and do not vary strongly with the energy of the projectile.

Electron-Induced Desorption

Because an incident electron has a small mass it cannot transfer appreciable energy to the heavy substrate nuclei by collisions and therefore cannot directly eject atoms. An incident electron may, however, ionize the electrons which bond a surface atom to the substrate; removal of the electron destroys the bond and the atom is ejected. This phenomenon is observed in the removal of weakly bonded adsorbed molecules and is consequently known as electron-induced desorption.

Luminescence Phenomena

Impact of electrons on a surface gives rise to a variety of light-emission phenomena emanating from the surface, or, if the material is sufficiently transparent, from some distance within.

Broad-band luminescence characteristic of the solid itself is caused by excitation of electrons to a higher level followed by their decay across a forbidden band gap with resulting emission of light. The phenomenon, called cathodoluminescence, is best known for insulators and semiconductors where large gaps exist between bands, but is observed also in metals.

Plasmon radiation is an optical emission from the decay of collective electron oscillations excited in metals; it emanates from the surface. Transition radiation occurs as a result of annihilation of the dipole formed by an electron and its image charge in the metal; this source also is located at the surface. Bremsstrahlung radiation occurs when electrons are decelerated or deflected in the Coulomb field of the atoms and this may emanate from deep in the solid. These three luminescence phenomena are seen only with electrons having energy in excess of tens of keV.

ION IMPACT MECHANISMS

Impact of an ion or atom on a surface exhibits certain complexities associated with the structure of the projectile. Figure 2 represents a potential energy diagram of an ion at a few angstroms from a metal surface. It is possible for an electron (1) to tunnel through the barrier and enter the ground state of the projectile; the excess energy is given to electron (2) causing its ejection by the Auger mechanisms. Thus as an ion approaches a surface it changes its form to a neutral atom and electrons are ejected before the projectile penetrates the surface. Clearly an incident atom cannot undergo this change. After the projectile enters the solid it may lose some of its electrons by collisional ionization but we have no direct monitor of its structure.

Secondary-Electron Emission

Collisional transfer of kinetic energy from the projectile to electrons in the target will result in a secondary-electron emission spectrum similar to that observed with incident primary electrons. There is a large peak of low-energy emergent electrons and subsidiary peaks at high energies due to Auger transitions. One should note that heavy particles transfer energy inefficiently to light electrons and the projectile energy must exceed some thousands of electron volts before the energy transferred to the electrons is sufficient to overcome the surface potential barrier. Thus low-energy incident ions and atoms cannot eject electrons by "kinetic" mechanisms involving transfer of kinetic energy from the projectile.

Incident ions may also eject electrons by "potential"

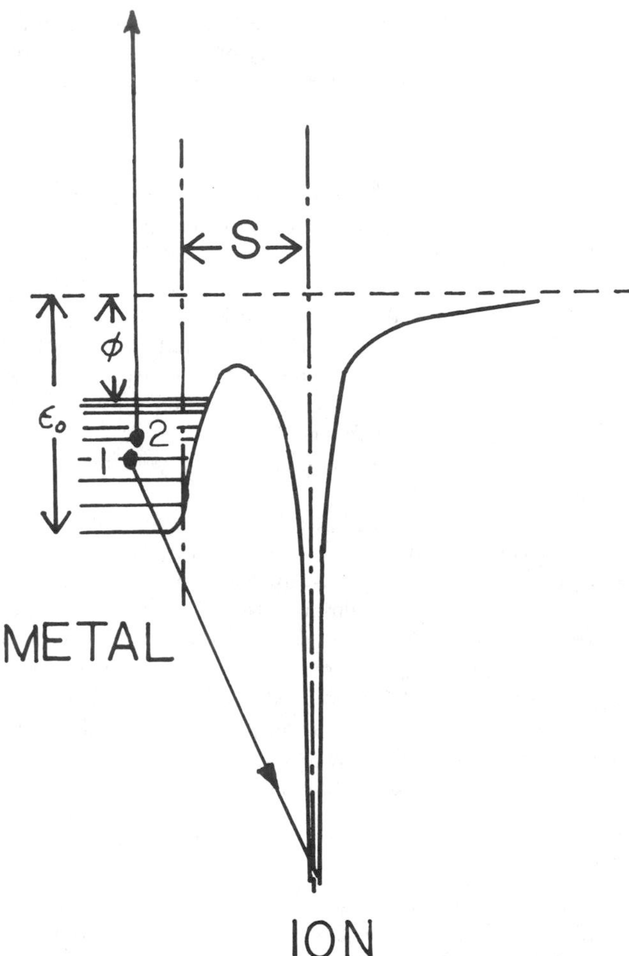

FIG. 2. A potential energy diagram showing an ion at a distance S from a metal surface; the conduction-band electrons lie at energies between the work function ϕ and the bottom of the conduction band ϵ_0. Auger neutralization of the ion occurs by electron 1 falling to the bound state in the ion; electron 2 carries away the excess energy.

mechanisms related to their structure; the Auger effect described above and illustrated in Fig. 2 is such a mechanism and will obviously occur at any impact energy however small. Impact of neutral atoms cannot produce secondary electrons by this mechanism.

Reflection of Ions

Projectile ion collisions with electrons cause energy loss but, because of the substantial mass difference between the particles, no appreciable deviation of the heavy projectile. Only collisions of the projectile ion with lattice atoms can produce the deviation necessary to return the projectile toward the surface. For low-energy impact the projectile recoils from the surface with an energy loss related only to the angle of scattering and the mass of the atom from which it recoiled. Measurement of energy loss may be used to determine the mass of the atoms from which scattering occurs and thereby to analyze the composition of the surface. For high-energy impact there is a continous distribution of recoil energies related to the penetration of projectiles into the solid.

The reflected projectile flux is found to include both ions and atoms. In part the distribution between charge states is related to interaction of the recoiling species with the target surface; Fig. 2 can be used to represent an ion which emerges from a surface and which picks up an electron by the Auger mechanism to become a neutral atom. The distribution of charge states is also partially established while the projectile is in the solid. A small fraction of the reflected projectiles may be excited and will emit normal atomic spectral lines when the projectiles have receded from the surface.

Sputtering

Ion impact on a surface will collisionally transfer energy to the lattice with the result that some lattice atoms may acquire sufficient energy to be ejected or sputtered from the surface. Typically the sputtered particles have an average energy of a few electron volts and are ejected with a coefficient which ranges from 10^{-3} (atoms ejected per ion incident) for a light projectile such as H^+ to 10 or 20 for heavy projectiles such as Pb^+. The ejected particles are found to include both ions and atoms. The distribution between these charge states is related in part to interaction of the emerging particle with the surface. Figure 2 can be used to describe how a sputtered ion can be neutralized as it emerges from the surface. A substantial fraction (up to 10%) of the sputtered particles may emerge as molecules representing the simultaneous ejection of two or more neighboring atoms in the lattice. A fraction of the emerging sputtered particles is found to be excited and will emit atomic line radiation at some distance from the surface.

Luminescence Phenomena

The emission of light from the surface or the bulk occurs through essentially the same mechanisms as for electrons discussed earlier. Intrinsic luminescence is termed ionoluminescence (the counterpart of cathodoluminescence) and is complicated by formation of additional energy levels through radiation damage caused by the ion impact.

Atomic line emission from sputtered or reflected atoms or molecules is not to be termed luminescence as it arises from a region in front of the surface rather than from the bulk material.

CRYSTALLOGRAPHIC EFFECTS

When a simple cubic crystal is "viewed" parallel to the sides of the unit cube it appears highly transparent with the strings of atoms lined up behind one another and large open "channels" exposed between the strings. The majority of projectiles directed along such a channel will lose energy only by collisions with electrons and therefore will penetrate considerable distances. Coefficients for reflection and ejection should be very small. Clearly the magnitude of electron and ion impact coefficients will be closely related to the orientation of the incoming projectile with respect to these crystallographic channels.

APPLICATIONS

Electron and ion impact phenomena find much application in the analysis and modification of surfaces. The utility of the techniques is related to their performance in the vacuum environment where contaminants may be avoided. Under electron bombardment each element in the near surface region produces Auger electrons of characteristic energy, which may be separately detected by use of an electron energy spectrometer; the analysis provides the atomic composition of the surface and is known as Auger electron spectroscopy (AES). The secondary-electron emission mechanism can be used to provide amplification of weak electron currents using materials where the secondary-emission coefficient exceeds unity; this finds application in photomultipliers. Sputtering of a surface by ion impact provides a method for removal of thin layers of material, one atomic layer at a time. Sputtering is used to remove contaminants from a surface (sputter cleaning), to machine a surface (ion milling), or to remove a layer and expose some underlying surface (sputter etching). Mass analysis of the ions sputtered from a surface can provide a qualitative indication of that surface's composition, a technique known as secondary-ion mass spectrometry (SIMS). Energy analysis of slow ions scattered from the surface gives an indication of surface composition (ion scattering spectrometry or ISS).

See also AUGER EFFECT; CHANNELING; LUMINESCENCE; RADIATION DAMAGE IN SOLIDS; SECONDARY ELECTRON EMISSION.

BIBLIOGRAPHY

J. Schou, *Scanning Microscopy* **2,** 607 (1988). A rather complete review of secondary electron emission by electron and proton impact. (A)

R. Behrisch, ed., *Sputtering by Particle Bombardment I,* Topics in Applied Physics, Vol. 47. Springer-Verlag, Berlin, 1981. (A)

R. Behrisch, ed., *Sputtering by Particle Bombardment II,* Topics in Applied Physics, Vol. 52. Springer-Verlag, Berlin, 1983. The two books edited by Behrisch provide a complete compendium of information on all aspects of sputtering phenomena. (A)

S. Datz, ed., *Applied Atomic Collision Physics, Volume 4, Condensed Matter.* Academic Press, New York, 1983. A collection of articles concerning heavy particle impact including reflection (by W. Heiland and E. Taglauer, Chapter 4), inelastic and excitation events (by E. W. Thomas, Chapter 5), and secondary ion ejection (by P. Williams, Chapter 7). (A)

Electron Beam Technology

Dieter P. Kern

Key features that make electron beams attractive for technological applications are that (i) intense electron beams can be easily generated, for example, by using a hot filament, (ii) electron beams can be easily focused by electric and magnetic fields, and (iii) electrons efficiently interact with matter with a range of distances, or effective strengths, which can be controlled by an applied voltage. For example, with modest acceleration potentials of about 20,000 to 200,000 volts and focusing magnetic fields, electron beams can be focused into spots a few angstroms (1 Å = 10^{-8} cm) in diameter. Thus objects can be imaged with essentially atomic resolution either by magnifying projection or by rapidly scanning such beams by means of time-varying electric or magnetic deflection fields at rates up to hundreds of megahertz. Electrons interact strongly with matter and consequently typical electron beam processes are performed in vacuum of about 10^{-3} Pa and better. Electron beam applications can be divided into two categories, those where materials are modified, for fabrication purposes, and the noninvasive techniques used for characterization, such as the various types of electron microscopy, and for display purposes.

ELECTRON BEAM GENERATION AND FOCUSING

An electron optical system typically contains an electron gun and beam forming optics. In the gun, a cathode, mostly a hot tungsten filament, emits electrons which are accelerated toward the anode; the voltage between cathode and anode is in the range 1–200 kV (Fig. 1). The Wehnelt electrode, situated between cathode and anode, is some 100 V more negative than the cathode. It has two purposes: it controls the emission current and it shapes the electric field in such a way that the electrons are collimated to a beam, which passes through an aperture in the anode. Recently also photocathodes and field-emission sources have been used for special applications, particularly in order to increase the brightness (i.e., current density per solid angle) of the electron beam.

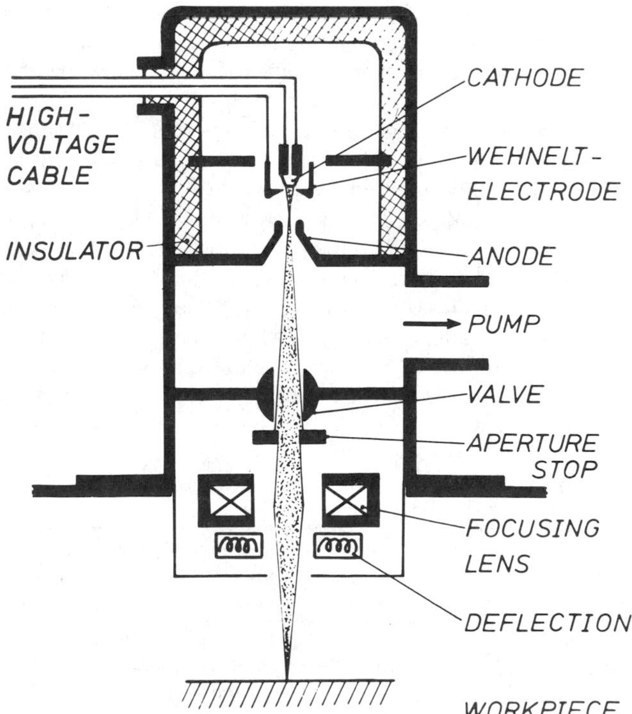

FIG. 1. Schematic diagram of an electron beam gun with a focusing lens and deflection coils.

The beam-forming optics consists of one or more electron lenses. These are typically rotationally symmetric, static electric or magnetic fields, generated by a set of electrodes or by current-carrying coils, mostly surrounded by appropriately shaped ferromagnetic material. With these lenses, the beam can be focused or any plane, e.g., that of a thin film inserted into the beam path, can be imaged. Scanning of the beam can be achieved by time-varying electric or magnetic fields.

ELECTRON BEAM FABRICATION

Electron beam processes utilize the energy of the electron beam to modify the target material by heating or by altering chemical bonds. In most cases a focused beam is used and depending on the application a wide range of power densities, total beam power, acceleration voltage which determines the penetration depth, and beam diameter is available (see Table I).

Electron Beam Melting is used for refining of metals. Refractory metals (e.g., with high melting point) and reactive materials can be kept in molten state in vacuum while impurities evaporate.

Electron Beam Evaporation is commonly used to produce metal and insulator coatings, such as mirrors, antireflection coatings on optical lenses, and in the semiconductor industry. Again high-melting-point and reactive materials, including alloys and compounds, can be handled, since the power is directly supplied to the material to be evaporated while the crucible can be kept cool.

Electron Beam Welding relies on the formation of a thin cavity channel by removing material under the impact of a high–power-density beam. The solidifying melting bath around the channel joins the workpieces together. A remarkable depth-to-width ratio of the welding seams (up to 50:1 and several centimeters deep) can be achieved, while only a small volume around the seam is thermally affected, resulting in low material distortions. This has led to a variety of new construction principles, such as welding of precision machined parts and of disparate materials.

Electron Beam Drilling involves pulsed beams with diameters ranging from 0.01–1 mm and power densities up to 10^{13} W/m^{-2} causing explosive removal of the material via holes and cavities of controlled depth can be produced.

Table I Summary of EB processes and their technical parameters

Process	Acceleration voltage (kV)	Beam power (kW)	Power density (W m^{-2})	Beam diameter (mm)
Melting	15–35	10–2000	10^8–10^9	1–10
Evaporation	10–35	5–500	10^9–10^{10}	0.5–5
Welding	30–170	0.1–100	10^9–10^{11}	0.2–2
Fusion treatment	20–150	1–10	10^8–10^{10}	0.5–5
Drilling	80–170	1–60	10^{10}–10^{13}	0.05–0.5
Lithography	10–25	10^{-8}–10^{-4}	10^5–10^8	10^{-6}–10^{-3}
Polymerization	20–100	1–50	10^4–10^6	10–100

Electron Beam Surface Treatment uses the energy of the electron beam to modify thin surface layers, involving melting, often together with previously deposited additives, and subsequent quenching of the liquid, or crosslinking of the molecules of organic surface coatings (paints). In many cases, surfaces with remarkable properties can be attained.

Electron Beam Lithography is the main application of electron beams for microfabrication in semiconductor technology. In electron beam lithography the substrate is first coated with a thin film of radiation sensitive material, commonly called a resist (in most cases an organic polymer). The electron beam irradiation changes the chemical solubility of the resist in the exposed areas. Subsequently, in a development step, the more soluble part of the resist (either the exposed or the unexposed, depending on the polarity of the resist) gets removed and a pattern stays behind. This remaining resist then serves as a mask to transfer the pattern to the substrate, either by etching, deposition, or ion implantation, for example. Two features make electron beam lithography particularly attractive: the high degree of flexibility, accuracy, and speed with which original patterns can be directly generated with a computer controlled beam and the high-resolution capability of electron optics.

The relative ease with which original patterns can be generated by a scanning focused beam under computer control makes electron beam lithography the main technology for fabrication of masks for other types of lithography such as standard optical projection lithography. This will continue to be true in the future for submicrometer optical lithography and for x-ray lithography as the resolution requirement becomes more demanding. Another application is direct-write lithography for manufacturing of relatively low-volume products, application-specific integrated circuits (ASICs), for which the demand is increasing due to the diversification of applications and sophisticated performance requirements.

Electron beam lithography tools today are of the scanning type, in which a fine beam that can be rapidly turned on and off is steered to delineate a pattern serially. Such scanning tools have adequate speeds to prepare optical masks, which are used many times, but are relatively slow and cannot compete with optical lithography tools in processing throughput. State-of-the-art tools (Fig. 2) can scan a beam at pixel rates of several hundred megahertz, and, using complex variable-shaped spots that can expose about 100 pixels in parallel processing rates of $>10^{10}$/s will be achieved (a pixel is the smallest resolution spot in the pattern). The quick turnaround time of direct writing tools together with the high resolution make such tools extremely suitable for development work even when future high-volume manufacturing is performed with optical lithography.

Areas with very little practical alternative to electron beam lithography are the development of submicrometer integrated circuits, research into limits of technologies, and the search for techniques and devices based on novel effects, which requires structures with 100 nm dimensions and below. At this point of time, electron beam lithography is the only proven technology for sub-0.5-μm and even sub-

FIG. 2. Electron beam lithography tool for manufacturing, including the electron optical column and the control computer (photo IBM).

100-nm work. As an example, Fig. 3 shows a scanning electron micrograph of gates from circuits of field effect transistors (FET). These circuits with gates as short as 70 nm have been fabricated with five levels of electron beam lithography, overlaid with an accuracy of better than 30 nm.

Fundamental resolution limits involve several factors: (i) the electron optical resolution, which can be a fraction of a nanometer, if current density in the beam and therefore exposure speed is not an issue, (ii) the range of secondary electrons with sufficient energy to expose the resist, (iii) the resolution, or effective size of the basic building blocks of the resist material. For poly-methyl-methacrylate (PMMA), the most widely used resist for high-resolution applications, the latter two factors seem to limit resolution to ~ 10 nm, even if the electron-optical resolution is 10 times better. In inorganic resist materials such as AlF_3, patterns with 2 nm minimum dimension have been produced by electron beam lithography. These materials, however, require rather high exposure doses, so that breaking this apparent resolution limit of ~10 nm with highly sensitive material remains one of the challenges for electron beam lithography.

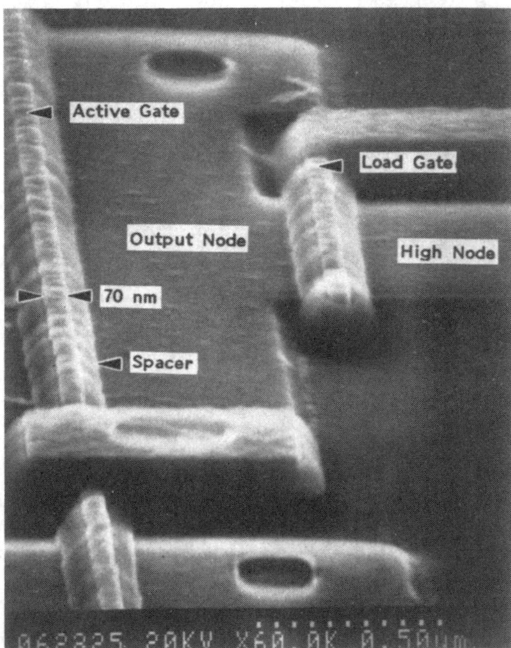

FIG. 3. High-resolution scanning electron micrograph of 70-nm gates in an experimental silicon circuit, fabricated by electron beam lithography.

Normal operation of a scanning tunneling microscope (STM) has little or no effect on the sample. But increasing the voltage and distance between tip and substrate so that a strongly confined beam of electrons with sufficient energy to cause chemical reactions is formed by field emission enables the STM to perform nanolithography. Thin layers of electron beam resists have been exposed, including PMMA, Langmuir–Blodgett films, metal halides, and hydrocarbon contamination. Metallic deposits have been formed by decomposing metal-organic adsorbates and bumps have been created on the surface of metallic glasses by thermally and electrostatically enhancing surface diffusion. Atomic scale resolution has been shown in modifications to a single-crystal germanium surface. These concepts are very new and promising for ultrahigh-resolution processing.

NONINVASIVE TECHNIQUES

Noninvasive techniques exploit the various effects that occur when an electron beam interacts with a sample: the electrons get absorbed in thin layers, depending on the local sample properties (in thin films of low atomic number material, essentially the phase of the electron waves gets shifted), they suffer material-characteristic energy losses, they excite secondary radiation in the form of photons (luminescence, x rays) or secondary electrons, again with sample specific energies and yields.

Cathode Ray Tubes

One or more (e.g., three for color applications) finely focused electron beams are scanned across a phosphor coated screen where light is generated at the points of impact. Major applications are video displays (television, terminals) and oscilloscopes.

Electron Beam Characterization

The scanning electron microscope (SEM) in its various forms has come to play a key role in semiconductor technology, in particular because optical microscopy has poor resolution in the submicrometer regime. Sample preparation is simple—even production wafers or packaging modules can be inspected—since a focused electron beam is scanned across the surface, and secondary radiation is collected to form a point-by-point image. Major new developments are advances in the capability to focus significant currents into small spots, leading to high-resolution images with good signal-to-noise ratio. The advent of field emission sources and sophisticated electron optics has led to commercial instruments with better than 1 nm resolution and to high-resolution images with low-energy electron beams (a few hundred electron volts), so that even insulating samples can be investigated without charging. Together with automation and improved metrology by means of laser interferometry, this opens the way to use SEMs for inspection and critical dimension measurements in semiconductor production lines.

The transmission electron microscope (TEM) has been the driving force for the developments in electron optics earlier this century. In the conventional case, an unfocused electron beam is incident on a very thin sample and, by using the transmitted electrons, magnifying projection electron optics form an image of the sample on photographic film with a resolution of a few angstroms. Thus, essentially atomic resolution has been achieved, in particular with the scanning version of the tool. Lattice imaging, where the contrast is improved as a result of the many atoms lined up in a properly oriented lattice plane, has proven to be very powerful in materials research, in particular interface studies. The analytical capabilities by means of energy loss spectroscopy and filtering, and small-area diffraction, for example, are becoming more and more important. A drawback is the rather complicated sample preparation—thinning to a few tens of nanometers is typically required—and the limited sample size. Recent advances in electron holography promise new possibilities for image evaluation and enhancement.

An exciting new form of electron microscopy has become available with the invention of the scanning tunneling microscope (STM). In an STM, a very fine metal tip is scanned across a sample at a height of only a few atomic spacings. At such spacings, a tunneling current occurs when only a few volts are applied between tip and sample. This tunneling current depends sensitively on the tip-to-sample spacing and is used to measure and control the height of the tip. In this way, scanning micrographs of the surface with a height resolution of one-tenth to one-hundredth of an atom and lateral resolution of approximately one to three atomic distances can be directly produced. Also, the electron energy distribution of the tunneling current depends on the electronic properties of the sample and can be used to select or view different types of surface atoms.

Noncontact electrical testing with focused electron beams is another application that will increase in importance as the complexity of circuits and packaging modules increases and critical dimensions further decrease. The current induced by the beam can directly be detected on the substrate and in this way give information on electrical continuity (specimen current) and electronic properties (EBIC). Another technique measures the energy of secondary electrons and thus the voltage of their point of origin on the specimen. Moreover, the electron beam can be chopped into pulses of picosecond duration, and thus fast changes in voltages on IC lines can be sampled at high temporal resolution.

See also ELECTRON BOMBARDMENT OF ATOMS AND MOLECULES; ELECTRON AND ION BEAMS, INTENSE; ELECTRON AND ION IMPACT PHENOMENA; ELECTRON MICROSCOPY; SCANNING TUNNELING MICROSCOPE.

BIBLIOGRAPHY

A. Septier (ed.), *Focusing of Charged Particles*, 2 vols. Academic Press, New York and London, 1967.

O. Winkler and R. Bakish, *Vacuum Metallurgy*. Elsevier, Amsterdam, London, and New York, 1971.

A. H. Meleka, *Electron Beam Welding*. McGraw-Hill, London, 1971.

G. R. Brewer and J. P. Ballantyne, *Electron Beam Technology in Microelectronic Fabrication*. Academic Press, New York, 1980.

T. H. P. Chang *et al.*, "Nanostructure Technology," *IBM J. Res. Develop*. **32,** 462–493 (1988).

L. Reimer, *Transmission Electron Microscopy*. Springer-Verlag, Berlin, 1984.

O. C. Wells, A. Boyd, E. Lifschin, and A. Rezanowich, *Scanning Electron Microscopy*. McGraw-Hill, New York, 1974.

Electron Bombardment of Atoms and Molecules

Earl C. Beaty

INTRODUCTION

In scientific research and in technology the bombardment of atoms and molecules by electrons occurs under a great variety of physical circumstances. Some are deliberately arranged for a special purpose, such as stripping electrons from gas molecules in order to use a magnetic deflection spectrometer for chemical analysis of the gas, or transferring energy into an atomic or molecular system in order to induce the emission of electromagnetic radiation. For example, the excitation required for the operation of lasers is generally produced directly or indirectly by electron bombardment. For such applications, we seek to understand the effects of electron bombardment in order to obtain the desired result. In other circumstances, the collision of electrons with atoms or molecules is a disruptive influence. For instance, in cathode ray tubes collisions of electrons with residual gas molecules increase the spot size. For these applications, we seek to understand electron bombardment effects in order to reduce their influence. For these and other reasons, there has been a great deal of scientific effort directed at understanding the consequences of electron bombardment of atoms and molecules since the electron was first identified by J. J. Thomson in 1897.

The subject of electron bombardment of atoms and molecules is here restricted to the case of atoms and molecules in the gaseous state. Thus, it is adequate to consider a projectile electron as interacting with only one molecule at a time. Even in a gas as dense as the Earth's atmosphere, it is relatively rare for two molecules to be in close enough proximity that a free electron could interact significantly with them simultaneously. Although interesting and important effects follow electron bombardment of solids and liquids, the considerations are very different.

Many of the effects of electron collisions with atoms and molecules are relatively easy to measure and describe. There is an extensive and long-standing literature describing research in this field. Over the years, the pursuit of more accurate and complete descriptions has caused both the experiments and the descriptions of the data to become rather complex. The following paragraphs attempt to provide an elementary description of the phenomena and an introduction to some of the language normally used in the research literature.

GENERAL CONSIDERATIONS

Among the early observed phenomena that could not be described in the framework of classical mechanics were some effects of electron bombardment of atoms. While many aspects of an electron beam moving through a rarefied gas can be described adequately using classical mechanics, some features demand description in terms of quantum mechanics.

Quantum mechanics now offers a mathematical framework for describing all collision effects. With a small set of experimentally determined atomic constants (the charge and mass of the electron, Planck's constant, and the speed of light) and a great deal of mathematical work, it is possible to compute both qualitatively and quantitatively all of the electron collision effects to high accuracy. While the preceding sentence is universally believed by atomic physicists as a matter of principle, in practice, the mathematics is so complicated that even experts cannot carry it out accurately except in special circumstances. The prescriptions embodied in quantum mechanics are well-defined but so complex that, except in unusual situations, it is necessary to make approximations which lead to errors of unknown magnitude.

Current research in this field involves using a combination of mathematical analysis and experimental results to work out ways of making the simplifying approximations without introducing unduly large errors. Thus in spite of the acceptance of the quantum-mechanical description as accurate, it is still standard practice to describe electron bombardment effects in the context of classical mechanics with suitable *ad hoc* modifications to account for the quantum or wave effects. In judging the adequacy of a particular classical model, it is necessary to appeal to quantum mechanics or to make a direct check with observations.

It is almost always a good approximation to treat a beam of electrons as a stream of particles. Many years ago the wave effects of diffraction and interference were observed in electron beams, establishing an important principle. However, such observations require rather special circumstances such that the classical approximation is ordinarily used in work on electron motion. Similarly, it is a good approximation to treat the motion of the target molecule from a classical point of view. On the other hand, the electronic structure of atoms and molecules cannot be well described in the context of classical mechanics. It is sometimes convenient to speak of the electrons as being in orbit; however, such descriptions quickly prove inadequate.

In the encounter of a free electron with a free molecule, linear momentum is conserved in the classical sense. The encounter may also cause a change in the wave function of the molecule. In almost all circumstances it is appropriate to regard the molecule as being in one of its energy eigenstates both before and after the encounter. If the final state is different from the initial state, the collision is known as inelastic and the energy difference must come from (or go to) the projectile. Thus energy is conserved in the classical sense; any energy absorbed by the molecule is treated as potential energy.

Generally the mechanics of a collision problem are best described in center-of-mass coordinates. For the case of electrons colliding with molecules, the center of mass is sufficiently close to the molecule that little error is made if one considers the molecule to be a fixed target, with the electron being a projectile hurled at the target. The probability of causing some particular effect upon collision is usually specified by giving the cross section for that effect. More specifically, if a beam of electrons of particle current density J (particles per unit area per unit time) is directed at molecular targets having a density N (molecules per unit volume), the rate R of collisions (a collision being defined by some prescribed measurement of changed trajectory, change of energy state, or other specific effect) will be proportional to both J and N. This may be expressed as

$$R = \sigma J N, \tag{1}$$

where σ is the collision cross section. In this expression, R is in units of events per unit volume per unit time, and σ has units of area per molecule per electron. In giving the cross section for an effect, we specify the equivalent area of the target which the projectile must hit to produce the effect. While it is convenient to think in terms of a single electron being projected toward a single target, such detailed control is not actually possible. Furthermore, as a matter of principle, the effects can only be described and measured in a statistical sense.

SCATTERED ELECTRONS

One way to observe the effects of electron bombardment of molecules is to pass a beam of electrons through a sample of gas and note the changes in the electron beam caused by the gas. The electron properties capable of being changed by collision with a molecule are momentum and spin. Spin changes are observable only in highly specialized situations and are not considered further here. Instead of working with the vector quantity, momentum, it is frequently more useful to use the equivalent quantities kinetic energy and the polar coordinates giving the direction of motion.

Description of scattering cross sections is assisted by extending the definition to "differential" cross sections. Suppose the collisional effect to be examined is the scattering of an electron into solid angle $\delta\Omega$ with its final energy in the range δE. The cross section for this effect as defined in Eq. (1) will be proportional to $\delta\Omega$ and δE (for appropriately small $\delta\Omega$ and δE). Let

$$S \equiv \lim_{\delta\Omega, \delta E \to 0} \frac{\sigma}{\delta E \delta\Omega}. \tag{2}$$

S is called a differential cross section, or sometimes a double differential cross section. The integral of S over either energy or solid angle is useful and is also called a differential cross section.

The double differential cross section embodies a complete

description of the electron scattering effects. It is normally defined to include changes in the target as well. Consideration of the energy and momentum of the electron before and after the collision allows deduction of the momentum and energy transferred to the target. However, the target might have more than one way to respond to the changes. A general rule of quantum mechanics is that some energies of excitation cannot be accepted by atoms and molecules with the result that S is zero for some values of the transfer energy. To illustrate this effect let us consider an example. Suppose a narrow beam of 100-eV electrons is passed through helium gas. Examine the electrons which leave the gas sample. (We are imagining that it is possible to restrict the gas to a small volume without having solid objects interfering with the observations.) Most of these electrons will not have been affected, with the result that the original beam will be present with reduced intensity. Electrons will be observed with 100 eV energy but traveling in directions different from the beam. These electrons are said to be elastically scattered. Other electrons will be observed with 80.2 eV energy. These have been inelastically scattered, leaving the helium atom in the first excited state, spectroscopically designated the 2 ^3S state. No electrons will be found with energies in the range 80.3–99.9 eV. (In helium the energy gap between the ground state and the first excited state is particularly large; however, such a gap exists with all atoms and molecules.) The result is that with enough control over the electron energies it is possible to distinguish elastically scattered electrons from others. In considering elastic scattering it is useful to define a cross section that is differential only in solid angle. Such a cross section is equivalent to integrating S as defined in Eq. (2) over a small energy range which includes the energy of incident electrons.

For some purposes it is desirable to know the total scattering cross section σ_T, which is related to S by

$$\sigma_T = 2\pi \int_0^{\pi} \int_0^{\infty} S(E,\theta)\sin\theta \, dE \, d\theta. \tag{3}$$

To be complete, Eq. (3) should include a sum over all the final states of the target; however, the total cross section is most often used when all the scattering is elastic. Measurements of σ_T are most conveniently made by passing a beam of electrons through a gas with magnetic or electric fields applied in such a way that electrons which lose energy or change direction by collisions are removed from the beam. By measuring the transmitted beam current with and without the gas it is straightforward to deduce σ_T. The total scattering cross section σ_T can be considered the effective area of an atom or molecule for all collisional effects. An important observation is that σ_T is finite.

TARGET EFFECTS

As noted above, a collision between an electron and a molecule can do nothing to the electron except change its kinetic energy and direction of motion. The molecule, being a much more complex object, is capable of much more extensive changes. It is useful to speak of the cross section for making a particular change in a target with the understanding that what happens to the projectile is of no concern. Thus we might be interested in the cross section of helium atoms for excitation to the 2 ^1P state by 100-eV electrons. Such a cross section might be inferred from observation of the photons generated as the excited atoms return to the ground state. For atoms, a change of state means a transition to a different energy eigenstate corresponding to a different electronic configuration. For molecules, vibrational and rotational excitations are also possible. For both atoms and molecules the final electronic state may be high enough that a previously bound electron is set free, leaving the target with a net positive charge (i.e., ionized).

Perhaps the most consequential effect of electron bombardment of atoms and molecules is ionization. A major result is that additional free electrons are produced. Gases at ordinary temperatures and pressures are electrically insulating; however, a few free electrons in an electric field may gain enough energy to have ionizing collisions. The result is an electron avalanche which makes the gas electrically conducting.

GASEOUS CONDUCTION

A large array of physical processes is involved in gaseous conduction. A high density of electrons cannot be maintained in the absence of positive ions because the electrons repel each other. But when the source of electrons is the ionization of molecules, the positive ions produced cause an electric field which on the average cancels the electric field due to the electrons. Such a mixture of positive ions and electrons is called a plasma and can be a very good electrical conductor. An applied electric field causes the electrons to move in one direction and the positve ions to move in the other. Since the electrons have much smaller mass they move faster and thus are the primary agent for the movement of charge. The movement of electrons through a gaseous conductor involves a balance of momentum (and energy) gained from the force of the electric field and with that lost from collisions with the gas molecules.

See also ATOMS; ELECTRON BEAM TECHNOLOGY; MOLECULES; QUANTUM MECHANICS; SCATTERING THEORY.

BIBLIOGRAPHY

Earl W. McDaniel, *Collision Phenomena in Ionized Gases*. Wiley, New York, 1964. A relatively elementary survey of ionized gas effects. Included are the effects of electron bombardment of atoms and molecules with a brief survey of relevant classical and quantum theory and a substantial compilation of data.

H. S. W. Massey, *Electronic and Ionic Impact Phenomena, Vols. I and II, Electron Collisions with Molecules and Photo-ionization*, 2nd ed. Oxford University Press, London, 1969. The second edition of this monograph is published in four volumes, the first two of which cover electron collisions with atoms and molecules. The phenomena are reviewed with special attention to experimental and theoretical techniques. Included is a good summary of available data.

N. F. Mott and H. S. W. Massey, *The Theory of Atomic Collisions*. Oxford University Press, London, 1965. A thorough account of the application of quantum mechanics to atomic collisions. Includes a thorough discussion of collisions of electrons with atoms and molecules.

Electron Diffraction

L. Marton† and Michel Van Hove

Diffraction, in light optics, has been defined as "Any departure of the actual light path from that prescribed by geometrical optics" [1]. Contrary to light, where the discovery of diffraction preceded the wave theory, electron diffraction was discovered as a consequence of a deliberate attempt to prove the wave nature of the electron.

In 1924 Louis de Broglie advanced the, for the times, revolutionary view that electrons (or any corpuscles) are accompanied by waves [2]. The origins of that view can be traced back to Einstein's 1905 theory about the wave–particle duality of light. Starting with this theory of Einstein, de Broglie came to the conclusion that electrons should exhibit the same kind of dual nature as light, where the relation between a particle's mass m_0, its velocity v, and its wavelength λ could be expressed as

$$\lambda = \frac{h}{m_0 v}$$

h being the constant of Planck. Today we usually write the de Broglie relation as

$$\lambda = \frac{h}{p}$$

where p is the momentum of the particle; this formulation implies the necessary relativistic correction.

About a year later a young graduate student at the University of Göttingen, Walter Elsasser, surmised that Clinton J. Davisson and Charles H. Kunsman's work on the scattering of slow electrons by metal surfaces [3] may have shown evidence of the diffraction of the electrons by the crystal lattice, thus supporting de Broglie's view of the wave nature of the electron [4].

In 1927 came definite evidence for the particle–wave duality through the discovery of electron diffraction by crystal lattices. This discovery was made almost simultaneously by C. J. Davisson and Lester H. Germer, using slow electrons diffracted by a nickel single crystal [5], and by George P. Thomson and A. Reid, using fast electrons passing through a thin celluloid film [6]. Their work was followed by rapid confirmation and extensive further development. Quite early it was shown that the angular dependence of the diffracted beams satisfied Bragg's law (originally established for x-ray diffraction):

$$n\lambda = 2d \sin\theta$$

where d is the interatomic distance, θ the diffraction angle and n an integer. The de Broglie wavelength λ, as calculated from the relativistically correct equation

$$\lambda = \frac{h}{[2m_0 eV(1 + eV/2m_0c^2)]^{1/2}},$$

was confirmed with good accuracy. In this equation e is the charge of the electron, c the light velocity, and V the potential difference applied to accelerate the electron [7]. It was also shown that, while the Davisson–Germer experiment corresponded to the Laue diagram of x-ray diffraction, the Thomson–Reid experiment was a counterpart of the Debye–Scherrer diagram of x rays.

Because of easier experimental requirements (lesser vacuum, less stringent control of the accelerating potentials) the high-voltage, transmission method (Thomson) developed much faster than the slow-electron reflection technique. The original diffraction cameras, consisting of an electron source and an aperture, collimating the beam before it passed through the specimen, were soon replaced by the arrangement first suggested by A. A. Lebedev [8]. In this arrangement an axially symmetrical magnetic field, the axis being parallel to the electron beam, is used to focus the beam on the recording medium (usually the photographic emulsion). This procedure has the advantage of replacing a well-collimated beam with a divergent beam and still attaining a higher intensity of the recorded pattern. At the same time this device allows the use of a larger area of the specimen. The accelerating potentials used by different authors were generally in the 10- to 100-kV range, resulting in electron wavelengths corresponding to $(12.2\text{–}3.7) \times 10^{-12}$ m (0.0122–0.0037 nm).

In the experiments described so far, diffraction was produced by periodic structures, such as crystal lattices. The reason for this is that the discovery of the equivalent to x-ray diffraction antedated by about 10 years the discovery of diffraction on opaque edges. In light optics we are familiar with the two kinds of edge diffractions: the Fraunhofer diffraction, produced by a plane wave front; and the Fresnel diffraction, produced by all other wave fronts. Because of the extreme shortness of the electron wavelength, the interference fringes, caused by the diffraction effect, are so closely spaced that their observation required a high degree of resolution of the electron optical system used for recording them. By 1940 the electron microscope was sufficiently developed for the observation of Fresnel fringes, and they were described almost simultaneously by H. Boersch [9] and J. Hillier [10]. By that time, all workers in electron microscopy agreed that the resolving power, as in light microscopy, is diffraction limited and obeys the Abbé relation

$$\rho = \frac{0.61\lambda}{n \sin\alpha}$$

where ρ is the least resolved distance, n the index of refraction of the medium surrounding the object, and α the semiangle subtended by the object at the lens.

The year 1940 also marks the beginning of another important link between electron diffraction and electron microscopy. Until that time the dominant interpretation of the origin of contrast in electron microscope images was based on mass-thickness scattering by the specimen [11]. B. von Borries and E. Ruska found, however, variations of intensities in chromium oxide smoke images, which could not be interpreted by scattering alone [12]. This finding was followed by the observation of Bragg reflections in images of crystalline specimens [13,14] and led finally to the theory of diffraction contrast in images of crystalline specimens [15,16].

Introduced in 1939, convergent-beam electron diffraction started being used only in the 1970s [17,18]. Instead of il-

† Deceased.

luminating the sample with a parallel beam, the beam is focused as a convergent cone onto a submicron area in the transmission mode. Advantages are the small-scale analysis area, avoiding twinning and other faults, for example, and microstructure fingerprinting of complex materials.

In the 1980s, electron microscopy started being applied to surfaces, using either samples of submicron thickness with the transmission mode, or submicron-size particles grown on carbon supports in the profile imaging mode [19,20]. In this manner, deviations from the bulk structure were observed in the first one or two atomic layers of the surface. Obtaining ultrahigh-vacuum conditions is a necessity for control of surface composition [20].

For an understanding of intensity relations in electron diffraction and in electron microscope images of crystalline objects, detailed calculations based on the theory of diffraction phenomena are necessary. There are two theories of electron diffraction: the kinematic (or geometric) theory and the dynamic theory. Both theories were developed originally for x-ray diffraction and adapted for use with electrons. The kinematic theory, in the ideal case, describes the interaction of the electron with a single atom and derives the intensities of the resulting beams. The dynamic theory takes into account the multiple scattering resulting from the presence of many atoms. This theory requires a wave-mechanical treatment, since the scattering produces a background that very considerably modifies the intensity distributions. X rays, which produce much less scattering than electrons, can usually be treated by means of the kinematic theory. In the case of electrons, very thin layers of crystals, roughly below 100 atom layers thick, give diffraction phenomena sufficiently well described by the kinematic theory. Above a thickness of 100 atom layers there is so much scattering that we have to have recourse to the dynamic theory [15,16,21]. The thicknesses employed in electron microscopy require the application of the dynamic theory in calculating image intensities [22].

In 1928, however, S. Kikuchi, while observing the diffraction of electrons passing through thin mica sheets, discovered a pattern differing widely from the single-crystal Laue spots or the polycrystalline Debye–Scherrer patterns [23]. He found a system of lines and bands, requiring a modification of the dynamic theory. These Kikuchi patterns consist of a continuous angular distribution of inelastically scattered electrons modulated by lines or band edges at the position of the various Bragg reflections. The energy analysis of the electrons diffracted into a Kikuchi line has shown that, for silicon, the average energy loss was about 450 eV, in contrast to the ten-times-lower loss for the ordinary patterns taken in transmission [24].

Similar effects are found with electrons that have energies on the order of 1000 eV and that have either lost energy to an excitation or have themselves been emitted by, for example, a photon in photoelectron emission. Such electrons radiate outward from the point of interaction and are diffracted by the surrounding lattice. They thereby carry structural information which can modulate the excitation process, as in extended x-ray absorption fine structure (EXAFS) [25]. Near a surface, the electrons themselves can escape to a detector and deliver surface structure information, as is the case with photoelectron diffraction [26].

For three decades high-energy electron diffraction (HEED) remained the only practical approach to the problems accessible by diffraction methods, although L. H. Germer pointed out quite early [27] the advantages of low-energy electron diffraction (LEED). By 1960, however, two of the experimental obstacles to LEED were removed: high-vacuum technology was radically improved to the extent that the surfaces investigated remained stable during the experiment; and the low intensities of the diffraction patterns, which hampered earlier work, could be enhanced by post-acceleration of the diffracted electrons. The usual energy range of LEED is from 20 to 500 eV, corresponding to a wavelength range of (274–55) $\times 10^{-12}$ m (0.274–0.055 nm). In the case of LEED, multiple scattering is largely unavoidable and the dynamic theory therefore a necessity [28]. Due to strong inelastic and elastic scattering at LEED energies, the electron mean free path is small, 5–10 Å, and the electron–atom scattering cross section is large, comparable to the geometrical atomic size. The LEED theory in common use is an adaptation of the theory of electronic band structure, after inclusion of inelastic damping and of x-ray-like Debye–Waller factors to describe vibration effects. The theory is capable of determining structures within atomic and molecular monolayers to an accuracy of 0.01–0.1 Å. Traditionally a trial-and-error search procedure has been applied to structure determination, but advances such as tensor LEED [29] and the direct method [30] permit rapid and computer-efficient determinations.

The discovery of electron diffraction was followed, within a remarkably short time, by application of the new effect to many other problems. Within the framework of this short survey it is necessary to limit ourselves to an enumeration of some of these applications, without extensive discussion. As with x-ray crystallography, structure determinations have been in the forefront. Because of the limited penetration of electrons, surface structures are much more accessible to electron diffraction investigation than to x rays. Thus surface physics benefits both from HEED reflection observations and from observations by means of LEED. For example [28], it was found that many clean metal surfaces exhibit ideal bulk lattice terminations, but show small relaxations in the first few atomic layers. Other metal surfaces and most semiconductor surfaces, by contrast, reconstruct: one or more atomic layers adopt lattices that differ from the bulk lattice. Also, a few hundred structures of atomic adsorption have been solved. They often involve simple bonding in high-coordination "hollow" sites of the substrate. Increasingly many cases of adsorbate incorporation into or adsorbate-induced restructuring of the substrate are being uncovered. Such results are of direct value in such diverse fields as microelectronic device properties, catalysis, corrosion, and tribology.

HEED in transmission is very widely used for the study of thin films of monocrystalline, polycrystalline, or amorphous materials. In the gaseous phase HEED is very useful for the study of molecular structures [29]. Within 10 years after the discovery of electron diffraction, results were published on more than 150 substances and by 1953 Pinsker's book [21] tabulated the structures of about 500 molecular species.

Selected-area diffraction has been extended from the high energies [15] to surface-sensitive energies in the low-energy electron microscope [31]. Thus areas of the order of 0.1–100 μm^2 can be selectively examined. Using computer simulations of emitted photoelectrons, Barton has for the first time shown the possibility of performing holography with electrons [32]. This technique will allow one to generate three-dimensional real-space images of a crystalline lattice with atomic resolution. With less than atomic resolution, only perpendicular to the surface, a screw dislocation has in fact been imaged using reflection–electron holography [33]. Electron diffraction on three parallel single crystals, placed at about 0.035 m distance from each other, was used to create the electron optical analog to the Mach–Zehnder interferometer [34], by means of which for the first time electron interferometer fringes could be observed [35].

See also Crystal Defects; Crystallography, X-Ray; Diffraction; Electron Microscopy; Gratings; Holography; Interferometers and Interferometry; Low-Energy Electron Diffraction (LEED); Quantum Mechanics; Reflection High-Energy Electron Diffraction (RHEED); Surfaces and Interfaces; X-Ray Spectra.

REFERENCES

1. E. U. Condon and H. Odishaw, eds., *Handbook of Physics*, p. 6–78. McGraw-Hill, New York, 1958.
2. L. de Broglie, Thèse, Paris. Masson, Paris, 1924.
3. C. J. Davisson and C. H. Kunsman, *Science* **64**, 522 (1921).
4. W. Elsasser, *Naturwiss.* **13,** 711 (1925).
5. C. J. Davisson and L. H. Germer, *Nature* **119,** 558 (1927); *Phys. Rev.* **30,** 705 (1927).
6. G. P. Thomson and A. Reid, *Nature* **119,** 890 (1927).
7. Extensive tabulation of λ is given by L. Marton, C. Marton, and W. G. Hall in "Electron Physics Tables," Natl. Bur. Stds. Circ. No. 571 (1956).
8. A. A. Lebedev, *Nature* **128,** 491 (1931).
9. H. Boersch, *Naturwiss.* **28,** 709, 711 (1940); *Phys. Z.* **44,** 202 (1943).
10. J. Hillier, *Phys. Rev.* **58,** 842 (1940).
11. L. Marton and L. I. Schiff, *J. Appl. Phys.* **12,** 759 (1941).
12. B. v. Borries and E. Ruska, *Naturwiss.* **28,** 366 (1940).
13. J. Hillier and R. F. Baker, *Phys. Rev.* **61,** 722 (1942).
14. R. D. Heidenreich, *Phys. Rev.* **62,** 291 (1942).
15. R. D. Heidenreich, *Fundamentals of Transmission Electron Microscopy*. Wiley (Interscience), New York, 1964.
16. P. B. Hirsch, A. Howie, R. B. Nicholson, D. W. Pashley, and M. J. Whelan, *Electron Microscopy of Thin Crystals*. Butterworth, Washington, 1965.
17. J. W. Steeds, in *Introduction to Analytical Electron Microscopy* (J. J. Hren, J. I. Goldstein, and D. C. Joy, eds.), p. 387. Plenum Press, New York, 1979.
18. J. M. Cowley and J. C. H. Spence, *Ultramicroscopy* **6,** 359 (1981).
19. L. D. Marks and D. J. Smith, *Nature* **303,** 316 (1983).
20. K. Yagi, K. Takayanagi, and G. Honjo, in *Crystals, Growth, Properties and Applications* (H. C. Freyhardt, ed.), Vol. 7, p. 47. Springer-Verlag, Berlin, Heidelberg, 1982.
21. Z. G. Pinsker, *Electron Diffraction* (translated from the Russian by A. A. Spink and E. Feigl). Butterworths, London, 1953.
22. M. J. Whelan, "Electron Diffraction Theory and Its Application to the Interpretation of Electron Microscope Images of Crystalline Materials," *Adv. Electronics Electron Phys.* **39,** 1 (1975).
23. S. Kikuchi, *Proc. Imp. Acad. Japan,* **4,** 271, 275, 354, 471 (1928); *Jpn. J. Phys.* **5,** 83 (1928).
24. W. Hartl and H. Raether, *Z. Phys.* **161,** 238 (1961).
25. P. A. Lee, P. H. Citrin, and B. Kincaid, *Rev. Mod. Phys.* **53,** 769 (1981).
26. J. J. Barton, S. W. Robey, and D. A. Shirley, *Phys. Rev. B* **34,** 778 (1986).
27. L. H. Germer, *Z. Phys.* **54,** 408 (1929); *Bell Syst. Tech. J.* **8,** 591 (1929).
28. J. B. Pendry, "Low Energy Electron Diffraction," in *Techniques of Physics* (G. K. T. Conn and K. R. Coleman, eds.), Vol. 2. Academic Press, New York, 1974. M. A. Van Hove, W. H. Weinberg, and C.-M. Chan, *Low-Energy Electron Diffraction: Experiment, Theory, and Surface Structure Determination*. Springer-Verlag, Berlin, 1986.
29. L. S. Bartell, "Electron Diffraction by Gases" in *Physical Methods of Chemistry* (A. Weissburger and B. W. Rossiter, eds.), 4th ed., p. 125. Wiley (Interscience), New York, 1972. K. Kuchitsu, "Gas Electron Diffraction" in *Molecular Structure and Properties* (G. Allen, ed.), p. 203. MTP, Oxford, 1972. J. Karle, "Electron Diffraction" in *Determination of Organic Structures by Physical Methods* (F. G. Nachod and J. J. Zuckerman, eds.), p. 1. Academic Press, New York, 1973.
30. J. B. Pendry, K. Heinz, and W. Oed, *Phys. Rev. Lett.* **61,** 2953 (1988).
31. W. Telieps and E. Bauer, *Ultramicroscopy* **17,** 57 (1985).
32. J. J. Barton, *Phys. Rev. Lett.* **61,** 1356 (1988).
33. N. Osakabe, J. Endo, T. Matsuda, A. Tonomura, and A. Fukuhara, *Phys. Rev. Lett.* **62,** 2969 (1989).
34. L. Marton, *Phys. Rev.* **85,** 1057 (1952); L. Marton, J. A. Simpson, and J. A. Suddeth, *Phys. Rev.* **90,** 490 (1953); *Rev. Sci. Instr.* **25,** 1099 (1954).
35. P. J. Rous, M. A. van Hove, and G. A. Somorjai, *Surf. Sci.*, to be published.

Electron Energy States in Solids and Liquids

Joseph Callaway

INTRODUCTION

In an atomic or molecular gas at ordinary temperatures and densities, electrons are bound to specific atoms or molecules and occupy states characteristic of isolated systems. Such states have wave functions which decay exponentially at large distances from the home-base atom or molecule, and the energy levels of the system are reasonably sharp. Interactions between electrons on different atoms are, on the average, quite weak.

In contrast, in condensed matter, interactions cause broadening of the energy levels of electrons in outer atomic shells. It is more correct to describe the electron states as belonging to the system as a whole, rather than to any specific atom or molecule. Such states are called band states, and one says that energy bands are formed. At the normal interatomic distances in solids, only the outer shells have broadened into bands. The energies of tightly bound core states are still quite sharp.

These ideas are, in a rough way, independent of the particular atoms or molecules involved, and are valid whether the condensed phase is solid or liquid. Closer investigation indicates that in molecular crystals, such as solid hydrogen, or in crystals of rare gases, solid argon for example, the band of valence electron states is quite narrow, and a molecular or atomic description is quite appropriate. Also, if the system is strongly disordered, the disorder itself may lead to electron localization. On the other hand, in metals, the broadening is quite advanced and states originating in different atomic levels may be strongly mixed.

It has proved much easier to develop a detailed understanding of electron states in crystals than in glassy solids or liquids, and in pure materials rather than alloys. This is because the existence of long-range structural order makes possible enormous simplifications in the mathematical description of electron states. Ordered systems are also often easier to characterize from an experimental point of view, and many types of experiments are possible only for ordered systems because of increased electron mean free path. We shall therefore devote most of this article to electron states in perfect crystalline solids, and reserve only a few paragraphs toward the end for disordered materials.

ELECTRON STATES IN ORDERED SOLIDS

The essential characteristic which simplifies the description of states in perfect crystals is periodicity. We will ignore here the existence of surfaces. Then there exists an infinite set of direct-lattice vectors $\mathbf{R}_i$,

$$\mathbf{R}_i = n_{i1}\mathbf{a}_1 + n_{i2}\mathbf{a}_2 + n_{i3}\mathbf{a}_3, \tag{1}$$

in which the $\mathbf{a}_i$ are the three primitive lattice vectors and the n_{ij} are integers, such that the crystal is unaltered if all atoms are displaced by any $\mathbf{R}_i$. We shall consider the electron states from a single-particle point of view in which each electron has its own wave function. There are some materials, particularly antiferromagnetic insulators (NiO and CoO are good examples), for which a single-particle description of the electron states is totally inadequate. We will not consider such systems here. The following discussion does apply rather well to metals, common semiconductors (such as Si, Ge, GaAs, etc.), and insulators (such as NaCl) in which the atoms or ions have filled shells of electrons. In these cases, the one-electron wave function is the solution of a Schrödinger equation

$$H\psi = E\psi, \tag{2a}$$

in which the Hamiltonian is periodic,

$$H(\mathbf{r}+\mathbf{R}_i) = H(\mathbf{r}). \tag{3}$$

It is a fundamental property of solutions of the Schrödinger equation for a periodic Hamilton that (Bloch's theorem)

$$\psi(\mathbf{r}+\mathbf{R}_i) = \exp(i\mathbf{k}\cdot\mathbf{R}_i)\psi(\mathbf{r}): \tag{4}$$

If the coordinate of an electron is changed by a direct-lattice vector, the only change in the wave function is multiplication by a phase factor. The quantity $\mathbf{k}$ is called the wave vector and can be regarded as a quantum number specifying the electron state. It is customary to incorporate $\mathbf{k}$ into the notation for the wave function so that it is written $\psi_{\mathbf{k}}(\mathbf{r})$ or $\psi(\mathbf{k},\mathbf{r})$. It follows from (4) that the wave function can be written in the form

$$\psi(\mathbf{k},\mathbf{r}) = \exp\,(i\mathbf{k}\cdot\mathbf{r})\;u(\mathbf{k},\mathbf{r}), \tag{5}$$

in which $u(\mathbf{k},\mathbf{r})$ has the same periodicity as the Hamiltonian,

$$u(\mathbf{k},\mathbf{r}+\mathbf{R}_i) = u(\mathbf{k},\mathbf{r}). \tag{6}$$

Thus, $u(\mathbf{k},\mathbf{r})$ repeats itself in each unit cell of the crystal, and the wave function ψ has the form of a plane wave modulated in each cell by the periodic function u.

It is useful to define a set of reciprocal-lattice vectors $\mathbf{K}_s$ through

$$\mathbf{K}_s = n_{s1}\mathbf{b}_1 + n_{s2}\mathbf{b}_2 + n_{s3}\mathbf{b}_3, \tag{7}$$

where the n_{si} are integers and the $\mathbf{b}$'s are related to the $\mathbf{a}$'s by

$$\mathbf{b}_i\cdot\mathbf{a}_j = 2\pi\,\delta_{ij}. \tag{8}$$

It is easy to find $\mathbf{b}$'s which obey (8); for example,

$$\mathbf{b}_1 = \frac{2\pi(\mathbf{a}_2\times\mathbf{a}_3)}{\mathbf{a}_1\cdot(\mathbf{a}_2\times\mathbf{a}_3)}, \tag{9}$$

and the other b's are obtained by cyclic permutation. All direct- and reciprocal-lattice vectors satisfy

$$\mathbf{K}_s\cdot\mathbf{R}_i = 2\pi\times\text{integer}. \tag{10}$$

The reciprocal lattice may or may not look like the direct lattice; for example, if the direct lattice is simple cubic, so is the reciprocal lattice, but if the direct lattice is body-centered cubic, the reciprocal lattice is face-centered cubic, and vice versa. Equation (10) has the following implication: If $\mathbf{k}$ and $\mathbf{k}'$ are two wave vectors which differ by a reciprocal-lattice vector, $\mathbf{k}' = \mathbf{k} + \mathbf{K}_s$, then

$$\exp(i\mathbf{k}\cdot\mathbf{R}_i) = \exp(i\mathbf{k}'\cdot\mathbf{R}_i) \tag{11}$$

for all $\mathbf{R}_i$. Hence $\psi(\mathbf{k},\mathbf{r})$ and $\psi(\mathbf{k}',\mathbf{r})$ behave identically under all lattice translations according to (4) and may be considered to be the same function. For this reason, wave vectors can be restricted to be in a region of momentum space called the Brillouin zone (abbreviated BZ) such that no two interior points differ by a reciprocal-lattice vector. Two examples of BZ's are shown in Fig. 1.

We adopt the convention that wave vectors $\mathbf{k}$ are restricted to a BZ. When this is done, there will be many solutions of the Schrödinger equation which have the same wave vector but different energies. We designate these different, orthogonal, solutions by an index, usually in the order of increasing energy; thus (2a) is written in a more explicit way as

$$H\psi_n(\mathbf{k},\mathbf{r}) = E_n(\mathbf{k})\;\psi_n(\mathbf{k},\mathbf{r}). \tag{2b}$$

The energies of states with a given value of the index n form a continuous manifold when regarded as functions of $\mathbf{k}$ inside

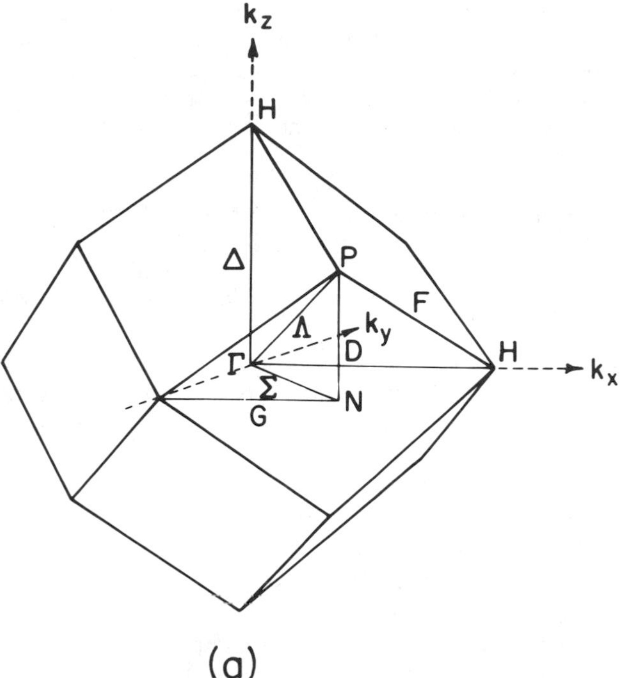

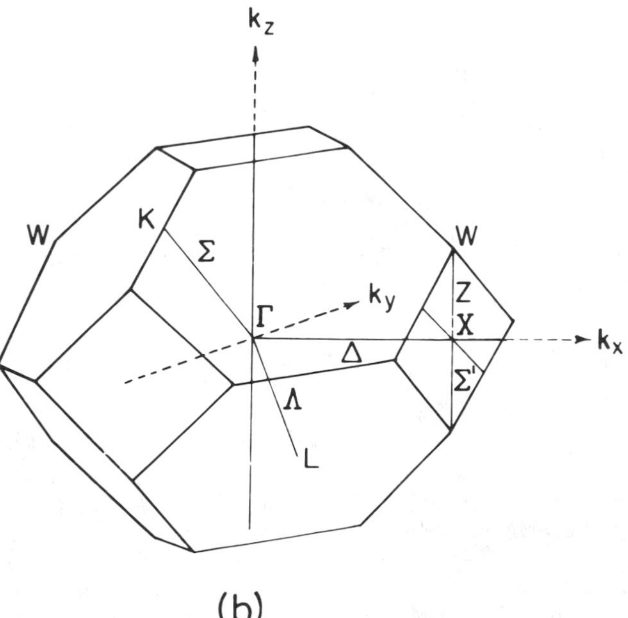

FIG. 1. Brillouin zones for the body-centered-cubic lattice (a), and the face-centered-cubic lattice (b). Points and lines of symmetry are indicated according to the accepted notation.

a BZ. This manifold is called an energy band, and n in (2b) is referred to as the band index.

As an example, calculated energy bands for the semiconductor silicon are shown in Fig. 2.

Band states can be regarded as occupied at $T = 0$ K in order of increasing energy according to the rules of Fermi–Dirac statistics, until the required number of electrons per atom has been accommodated. If the energy of the highest occupied state falls inside an energy band, the substance is a metal. The energy of this state is the Fermi energy E_F. In the (excellent) approximation in which the wave vector $\mathbf{k}$ is treated as continuous, the solutions $\mathbf{k}_F$ of

$$E_n(\mathbf{k}_F) = E_F \tag{12}$$

(for all relevant bands) define the Fermi surface. The Fermi surface can simply be considered to be the surface(s) bounding the occupied region of $\mathbf{k}$ space. There are several experimental methods which can be used to determine the Fermi surface and so check theoretical calculations. Some of the most important ones include the de Haas–van Alphen effect, the radio-frequency size effect, and the magnetic field dependence of the electrical resistivity and of ultrasonic attenuation. Limitations of space preclude much discussion of these techniques here, but additional information can be found in the references.

If there are no allowed states immediately adjacent in energy to the highest occupied state, the material is a semiconductor or an insulator. The minimum energy separation between the highest occupied state and lowest empty state is the "band gap," which determines the temperature dependence of the electrical conductivity of a pure semiconductor. If the minimum separation between full and empty states is determined for states of the same $\mathbf{k}$, one has the "direct" band gap, which is of great importance in the study of optical properties.

It is found in many applications that electrons in solids move in weak applied external electric ($\mathbf{E}$) and magnetic ($\mathbf{B}$)

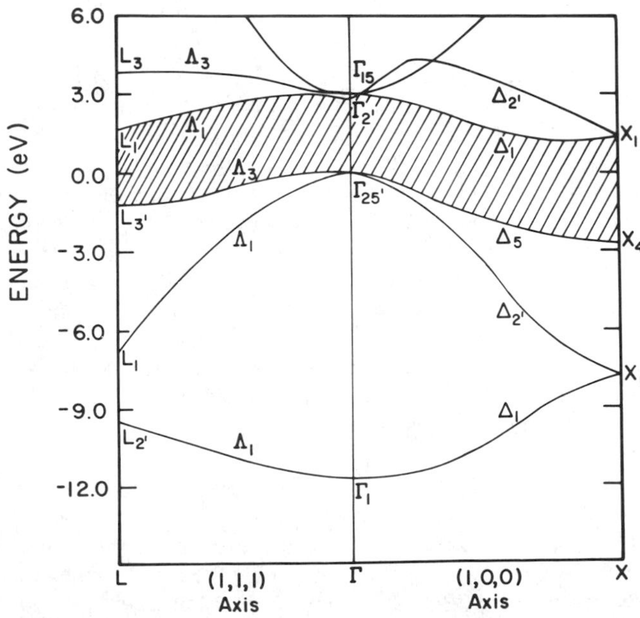

FIG. 2 Energy bands calculated for the semiconductor silicon for two directions of the wave vector $\mathbf{k}$. The points Γ, X, L, etc., on the horizontal axis refer to the Brillouin zone of Fig. 1(b). The cross-hatched region is the band gap. Bands are labeled according to symmetry properties of the wave functions (see references for details).

fields according to the simple equation

$$m^*\left(\frac{d\mathbf{v}}{dt}+\frac{1}{\tau}\right)=-e(\mathbf{E}+\mathbf{v}\times\mathbf{B}), \tag{13}$$

in which e is the magnitude of the electronic charge, τ is a relaxation time which roughly describes the effect of collisions, and m^* is known as the effective mass. The quantity m^* is, in general cases, a tensor; that is, acceleration and force are not necessarily in the same direction. It is possible to determine the effective-mass tensor from the derivatives of the energy band function. Suppose that only a single band need be considered. Then one has

$$\left(\frac{1}{m^*}\right)_{\alpha\beta}=\frac{1}{\hbar^2}\frac{\partial^2 E(\mathbf{k})}{\partial k_\alpha \partial k_\beta}, \tag{14}$$

in which α, β are Cartesian indices. A variety of experimental measurements are available to determine m^*, and values smaller than the free electron mass by factors of the order of 100 are found in some materials (e.g., InSb).

It is possible for regions of $\mathbf{k}$ to be important, where $E(\mathbf{k})$ has negative ("downward") curvature. In this case m^* is negative, but it is customary to consider the mass to be positive and to delete the minus sign in (13) so that one describes the motion of a positive charge. Such effective positive charges are known as holes. The concept of a hole is particularly important in semiconductors, where states near the top of a full valence band are described in this way.

Another important concept is the "density of states." This measures the fractional number of states which have energies in a certain interval. Specifically let $G(E)\,dE$ be the number of states of a given spin direction which have energies between E and $E+dE$. $G(E)$ is given by the expression

$$G(E)=\frac{\Omega}{(2\pi)^3}\sum_n \int d^3k\ \delta(E-E_n(\mathbf{k})), \tag{15}$$

in which the sum includes all bands and the integral runs over the Brillouin zone. The delta function serves as a counter, registering a unit contribution from those bands (n) and wave vectors $(\mathbf{k})$ such that $E_n(\mathbf{k})=E$. The quantity Ω in the multiplicative constant is the volume of the unit cell. Since the volume of the Brillouin zone is $(2\pi)^3/\Omega$, the constant normalizes the integral so that the contribution from each band to the integral of $G(E)$ over all energies is unity. A very simple case in which (15) may be explicitly evaluated is that of a parabolic band characterized by an effective mass m^* as is found for a nearly free electron system. Suppose $E(\mathbf{k})=\hbar^2k^2/2m^*$ (a scalar effective mass) and consider only a single band. The result is

$$G(E)=\frac{\Omega}{4\pi^2}\left(\frac{2m^*}{\hbar^2}\right)^{3/2}E^{1/2}. \tag{16}$$

This expression will not be valid in a region of energies where an energy band intersects a Brillouin-zone face, for then the expression for $E(\mathbf{k})$ must become more complicated, as some derivatives of the energy must vanish. In fact, it can be shown that the density of states must possess sharp structure at such energies, called "van Hove singularities." Further information can be found in the references.

Electrons in a metal produce a contribution to the specific heat that depends linearly on temperature. This contrasts with the contribution from lattice vibrations, which is proportional to T^3 at low temperatures. At sufficiently low temperatures the linear term dominates the cubic one, and one can determine experimentally the coefficient γ in the relation

$$C_v^{(e)}=\gamma T, \tag{17}$$

in which $C_v^{(e)}$ is the electronic contribution to the specific heat per atom. It can be shown that γ is given by

$$\gamma=\frac{1}{3}\pi^2 K^2G(E_\mathrm{F}), \tag{18}$$

in which K is Boltzmann's constant and $G(E_\mathrm{F})$ is the density of states evaluated at the Fermi energy. Hence measurement of γ yields a value for $G(E_\mathrm{F})$ that can be compared with calculations.

CALCULATION AND MEASUREMENT OF ELECTRONIC ENERGIES

There are several rather standard techniques available for the numerical calculation of electronic energies and wave functions from Eq. (2). The most commonly applied methods include (1) the "Green's-function" method, (2) the "augmented-plane-wave" method, (3) the method of "linear combination of atomic orbitals," and linearized versions of the first two (linear muffin tin orbitals method, linearized augmented plane wave method). Space permits only a brief description of some common general features, but details of specific approaches are described in the Bibliography. All of the methods involve the expansion of the unknown $\psi_n(\mathbf{k},\mathbf{r})$ as a combination of some previously specified functions $\Phi_s(\mathbf{k},\mathbf{r})$, which are considered to obey Bloch's theorem explicitly,

$$\Phi_s(\mathbf{k},\mathbf{r}+\mathbf{R}_i)=\exp(i\mathbf{k}\cdot\mathbf{R}_i)\ \Phi_s(\mathbf{k},\mathbf{r}).$$

The simplest set of potentially useful functions are plane waves of the form

$$\exp[i(\mathbf{k}+\mathbf{K}_s)\cdot\mathbf{r}]$$

in which $\mathbf{K}_s$ is a reciprocal-lattice vector. In general, one writes

$$\psi_n(\mathbf{k},\mathbf{r})=\sum_s c_{ns}(\mathbf{k})\Phi_s(\mathbf{k},\mathbf{r}). \tag{19}$$

The sum includes, in principle, an infinite number of terms, but of course only a finite subset can be employed in actual calculations. The coefficients $c_{ns}(\mathbf{k})$ have to be determined so that (2) is satisfied. Equation (19) is substituted into (2); the result is multiplied by $\Phi_t^*(\mathbf{k},\mathbf{r})$, where Φ_t is a general member of the basis set, and integrated over the entire crystal. A set of linear homogeneous equations is obtained:

$$\sum_s H_{ts}(\mathbf{k})c_{ns}(\mathbf{k})=E_n(\mathbf{k})\sum_s S_{ts}(\mathbf{k})c_{ns}(\mathbf{k}), \tag{20}$$

in which the H_{ts} are matrix elements of the Hamiltonian on the basis Φ,

$$H_{ts}(\mathbf{k})=\int \Phi_t^*(\mathbf{k},\mathbf{r})H\Phi_s(\mathbf{k},\mathbf{r})\ d^3r, \tag{21}$$

and the S_{ts} are overlap matrix elements,

$$S_{ts}(\mathbf{k}) = \int \phi_t^*(\mathbf{k},\mathbf{r})\Phi_s(\mathbf{k},\mathbf{r})\, d^3r. \quad (22)$$

The overlap matrix S_{ts} is usually not simply a unit matrix since the functions Φ are not generally orthonormal. However, a variety of standard mathematical techniques exist to enable numerical solution of (20) once the elements of matrices H and S are constructed.

The most important problem encountered in these calculations is the determination of the function $V(\mathbf{r})$, which represents the potential energy of an electron in the crystal. At present most calculations of energy bands are based on density functional theory (see the article by Callaway and March in the Bibliography). This approach gives a procedure by which the contributions from the nuclei of the system, the average electrostatic field of the other electrons, and exchange and correlation interactions can be included in $V(r)$ (the latter only approximately). The fundamental quantity is the electron density which has to be computed self-consistently. That is, one assumes some initial distribution of electrons; computes the corresponding potential function; then solves the Schrödinger equation to find energy bands and wave functions. The new distribution of electrons must now be used and the process repeated until the electron distributions resulting from two successive stages of the procedure agree within some assigned limit of error. This iterative process is quite time consuming. However, reasonably good results are usually obtained in a single-pass, non-self-consistent calculation if the potential energy is computed by solving Poisson's equation for a charge distribution produced by the superposition of neutral-atom charge densities centered on crystal lattice sites.

Many ingenious techniques have been developed to permit the experimental determination of the properties of electron states. The easiest to understand are optical. The absorption of light can produce a transition between an occupied state in band n and an unoccupied state in band l. Because the wave vector of a light wave in the infrared, visible, or near-ultraviolet regions of the spectrum is quite small on a scale appropriate to crystalline lattices, it is an excellent approximation to suppose that the initial and final states have the same wave vector. Such transitions can be represented by a vertical line on an energy band diagram, and are said to be direct. Indirect transitions also occur in which one or more phonons are emitted or absorbed. The interpretation of the spectra of band-to-band transitions is not so simple as in the case of free atoms because both initial and final states belong to continuous bands. Sharp structures that are easily identifiable are associated with the onset of transitions into a band or with the occurrence of van Hove singularities in the density of states. Ultraviolet photoemission spectroscopy, in which the excited electron escapes from the solid, also provides information concerning the distribution of initial states.

Optical investigations can determine the energy differences between states substantially removed from the Fermi energy. States near the conduction-band minimum or valence-band maximum in semiconductors, and states close to the Fermi surface in metals, can be studied experimentally by applying a strong external magnetic field. There are many interesting effects of which cyclotron resonance is the easiest to understand. Consider Eq. (13) in the absence of an electric field and suppose (1) that the relaxation time τ is long enough so that $1/\tau$ can be neglected, and (2) that the effective mass m^* is a scalar rather than a tensor. Then it is easy to verify that an electron will spiral around the direction of the magnetic field with a frequency

$$\omega_c = eB/m^*. \quad (23)$$

The quantity ω_c is called the cyclotron frequency. Electromagnetic waves of frequency ω_c will be absorbed and in this way m^* can be determined. The discussion is easily extended to allow for a tensor effective mass, and for relaxation. Another effect of great importance is the de Haas–van Alphen effect. The magnetic susceptibility of a metal in a strong magnetic field exhibits oscillations as a function of the field whose period is proportional to $1/B$. The observed constant of proportionality can be used to determine the areas of cross sections of the Fermi surface in a direction perpendicular to the magnetic field.

ELECTRON STATES IN DISORDERED MATERIALS

Our ability to describe electron states in extended systems rests on the applicability of Bloch's theorem. If the Hamiltonian is not periodic because of disorder on a microscopic scale, the wave vector **k** is no longer a good quantum number. We can imagine that the disorder (caused either by random atomic positions in an alloy, or by thermal motions as in a liquid) causes electrons to scatter rapidly between states. Even in pure materials, some such scattering is always present, and gives rise to the electrical resistivity of a metal. In a liquid metal, the scattering is so frequent that the mean free path may be only a few atomic separations. If a state of definite **k** were somehow to be prepared initially, it would soon be destroyed.

A straightforward description of the properties of a disordered system in terms of energy bands is therefore not to be expected. The concept of the density of states, however, remains a useful one. The density of states of a disordered system will not exhibit the sharp structure characteristic of ordered systems in which the periodic lattice structure gives rise to van Hove singularities, but need not be a featureless object. An energy gap between filled and empty regions of a density of states may still persist. In some cases, especially highly anisotropic systems such as thin films (effectively two dimensional) or long narrow wires (effectively one dimensional), disorder may lead to complete localization of the electrons, even though bulk, well-ordered samples of the same materials would be normal metals. In certain amorphous solids, particularly semiconductors, disorder may lead to at least partial filling of the energy gap. It has been shown for such materials that states in the vicinity of the gap are localized and therefore will not permit electrical conduction, while states more removed in energy are extended and will conduct. The energy separating conducting and nonconducting states is called the mobility edge.

Much current research is devoted to the development of a quantitative theory of disordered systems. In regard to normal bulk metallic alloys, this work has emphasized the development of approximations in which the actual disordered system is conceptually replaced by an ordered effective medium. The properties of this medium are determined from the scattering properties of the individual atoms by the condition that, on the average, no further scattering should occur (the coherent potential approximation). An effective, non-Hermitian Hamiltonian results, which can be used for the computation of a density of states. Studies of electron states in strongly disordered systems in which electrons are localized, especially those of reduced dimensionality, have to be undertaken by more powerful methods because an effective medium approach is not appropriate. One finds, for example, that the electrical resistance of a disordered one-dimensional system increases exponentially with its length, and at sufficiently low temperatures, the resistance of a given sample may fluctuate as a function of time.

See also CRYSTAL SYMMETRY; CYCLOTRON RESONANCE; DE HAAS–VAN ALPHEN EFFECT; FERMI SURFACE; HAMILTONIAN FUNCTION; LIQUID METALS; LIQUID STRUCTURE; METALS; QUANTUM MECHANICS; RESISTANCE; SCHRÖDINGER EQUATION; SEMICONDUCTORS, CRYSTALLINE; ULTRASONICS.

BIBLIOGRAPHY

A. Elementary

J. S. Blakemore, *Solid State Physics,* 2nd ed. Cambridge University Press, Cambridge, 1985.

G. Burns, *Solid State Physics*. Academic Press, New York, 1985.

R. E. Hummel, *Electronic Properties of Materials*. Springer-Verlag, Berlin, Heidelberg, 1985.

C. Kittel, *Elementary Solid State Physics*. Wiley, New York, 1966.

M. N. Rudden and J. Wilson, *Elements of Solid State Physics*. Wiley, New York, 1980.

B. Intermediate

P. W. Anderson, *Concepts in Solids*. Benjamin, New York, 1963.

N. W. Ashcroft and N. D. Mermin, *Solid State Physics*. Holt, Rinehart and Winston, New York, 1976.

R. J. Elliott and A. F. Gibson. *An Introduction to Solid State Physics and Its Applications*. Barnes and Noble, New York, 1974.

H. Y. Fan, *Elements of Solid State Physics*. John Wiley, New York, 1987.

C. Kittel, *Introduction to Solid State Physics,* 5th ed. Wiley, New York, 1976.

J. M. Ziman, *Principles of the Theory of Solids,* 2nd ed. Cambridge University Press, London, England, 1972.

C. Advanced

J. Callaway, *Quantum Theory of the Solid State*. Academic Press, New York, 1974.

J. Callaway and N. H. March, "Density Functional Methods, Theory and Applications," *Solid State Phys*. **38,** 135 (1984).

W. A. Harrison, *Solid State Theory*. McGraw-Hill, New York, 1970.

W. A. Harrison, *Electronic Structure and the Properties of Solids*. W. H. Freeman and Co., San Francisco, 1980.

W. Jones and N. H. March, *Theoretical Solid State Physics*. Wiley-Interscience, London, 1973.

C. Kittel, *Quantum Theory of Solids*. Wiley, New York, 1963.

P. A. Lee and T. V. Ramakrishnan, "Disordered Electronic Systems," *Rev. Mod. Phys*. **57,** 287 (1985).

O. Madelung, *Introduction to Solid State Theory*. Springer-Verlag, Berlin, Heidelberg, 1978.

G. Mahan, *Many Particle Physics*. Plenum Press, New York, 1981.

J. C. Slater, *Quantum Theory of Molecules and Solids,* Vols. 2 and 3. McGraw-Hill, New York, 1963.

Electron–Hole Droplets in Semiconductors

Jagdeep Shah

A pure semiconductor crystal is devoid of free electrons and holes (vacant electron states) at low temperatures. Excitation of such a semiconductor by photons of appropriate energy creates electrons and holes which interact among themselves via Coulomb interactions. At low densities and temperatures, an electron becomes bound to a hole to form a hydrogen-atom-like complex known as an exciton. In many semiconductors it is possible to create a large density of excitons by intense photoexcitation. In 1968 Keldysh suggested that observations of anomalous photoconductivity in Ge at high excitation intensities might be explained if, at some critical density, the gas of excitons condensed into a metallic liquid phase consisting of electron–hole droplets (EHD). This gas–liquid transition is analogous to condensation of atomic Na vapor into a metallic Na liquid, i.e., an electron is not bound to a particular hole but is free to move independently.

The correctness of this idea was demonstrated by three key experiments. In the first, a rapid increase in the intensity of a luminescence feature of Ge with slight decrease in temperature was interpreted as a first-order phase transition and the line shape of this feature was shown to be that expected for EHD. In the second experiment, small differences in dielectric constants between the metallic droplets and the surrounding exciton gas in an optically excited crystal of Ge scattered a probe laser beam in much the same way as fog scatters light. The angular distribution of the scattered laser light conclusively demonstrated the existence of macroscopic droplets with an average radius of the order of 5 μm. In the third experiment, discrete giant charge pulses were observed from an optically excited reverse-biased *p-n* junction in Ge. These result from dissociation of EHD in the strong electric field at the junction and provided further evidence for the finite size of EHD.

A detailed theory describing this electron–hole liquid has been developed by considering the ground-state energy $E_G(r_s)$ of an electron–hole pair in a degenerate electron–hole plasma as a function of r_s, the interparticle separation in units of the exciton radius. This energy is the sum of the kinetic, exchange, and correlation energies. The kinetic and exchange energies can be calculated analytically, but various approximations have to be used to calculate the correlation energy. Many factors, such as the anisotropy and the multivalley nature of the band structure, affect these energies. However, it has been shown that the sum of the exchange and correlation energies (ϵ_{xc}) is independent of the band structure of semiconductors. $\epsilon_{xc}(r_s)$, in the unit of the excitonic Rydberg (ϵ_x), has been shown to obey the following simple expression for many semiconductors over a large range of r_s: $\epsilon_{xc}(r_s) = (a + br_s)/(c + dr_s + r_s^2)$, where $a = -4.8316$, $b = -5.0879$, $c = 0.0152$, and $d = 3.0426$. Typically, $E_G(r_s)$ has the shape shown in Fig. 1, with a minimum at r_0. If this minimum energy lies below the exciton binding energy, i.e., if $\phi_0 = |E_G(r_0)| - |\epsilon_x|$ is positive, then a condensate can form because it is stable against emission of an exciton at absolute zero degrees ($T = 0$); ϕ_0 is then the binding energy of the condensate. As a result of the Coulomb

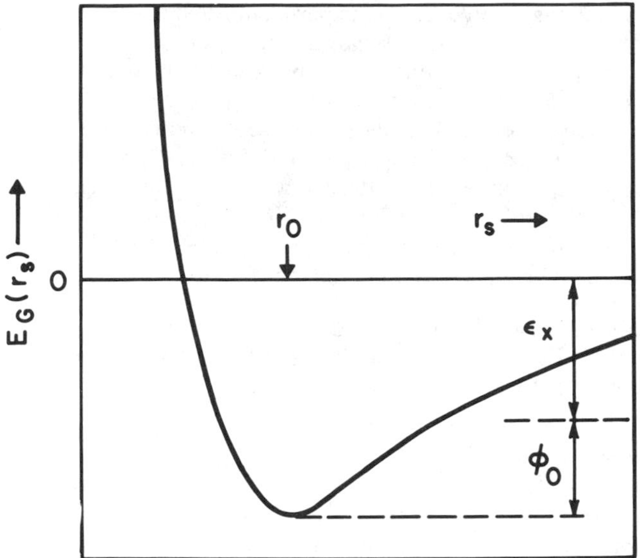

FIG. 1. Schematic representation of ground-state energy per pair versus normalized interparticle separation.

interactions between the carriers the band gap within the EHD is renormalized from E_g to $E_g(r_0)$ as is shown schematically in Fig. 2. The chemical potential $\mu(r_0)$ and $E_g(r_0)$ can be simply calculated from $E_G(r_s)$.

The formation of a metallic state in optically excited semiconductors provides a unique opportunity to explore the properties of quantum metals, and a wide variety of experimental techniques have been used to study this novel state of matter. One technique which stands out above all others is the study of photoluminescence spectra of EHD and excitons under a variety of conditions. It is clear from Fig. 2 that at $T=0$ the energies of the lowest- and the highest-energy emitted photons correspond to $E_g(r_0)$ and $\mu(r_0)$, respectively. Since $E_F{}^e + E_F{}^h = \mu(r_0) - E_g(r_0)$, one can obtain the liquid density n_0 from these measurements. Also, ϕ_0 is obtained by measuring the difference in energy between $\mu(r_0)$ and the low-energy edge of exciton emission. One can also obtain these quantities at $T \neq 0$ by analyzing the line shapes; such measurements show that n_0 and ϕ_0 vary with T in a manner expected for a metal or a degenerate plasma, confirming once again the metallic character of EHD.

When the source of excitation is removed, the EHD luminescence decays with a characteristic decay time. The e-h pairs within a droplet decay by radiative recombination, by nonradiative Auger recombination, and also by thermionic emission into the surrounding gas if the temperature is not too low. The finite lifetime of this metallic liquid phase distinguishes it from other liquids. However, assuming quasi-equilibrium between the gas of excitons and the liquid, one can discuss the gas–liquid coexistence curve in the temperature–density phase diagram of EHD, as for any other gas which undergoes a first-order transition to a liquid. A schematic phase diagram for EHD is shown in Fig. 3. The exciton–EHD system has a critical temperature T_c above which liquid cannot form. The region under the coexistence curve is the unstable region in that a plasma created at any point (n,T) in this region will spontaneously break up into droplets of density n_0 and gas of density n_{ex}. The coexistence curves for Ge and Si, including the values of T_c, have been determined from luminescence measurements. Phase diagrams have also been calculated theoretically.

At low temperatures and densities, the photoexcited e-h pairs form excitons, corresponding to the lower left region in Fig. 3. For $T<T_c$, as the density of the gas phase is increased beyond the density of gas on the coexistence curve, the gas becomes supersaturated and condensation occurs. Experimentally, there is a well-defined excitation intensity (I_{th}) beyond which the EHD becomes observable, i.e., the EHD luminescence increases extremely rapidly for $I>I_{th}$. It has been shown that the threshold for monotonically increasing intensity is considerably larger than that for monotonically decreasing intensity. This hysteresis is a result of

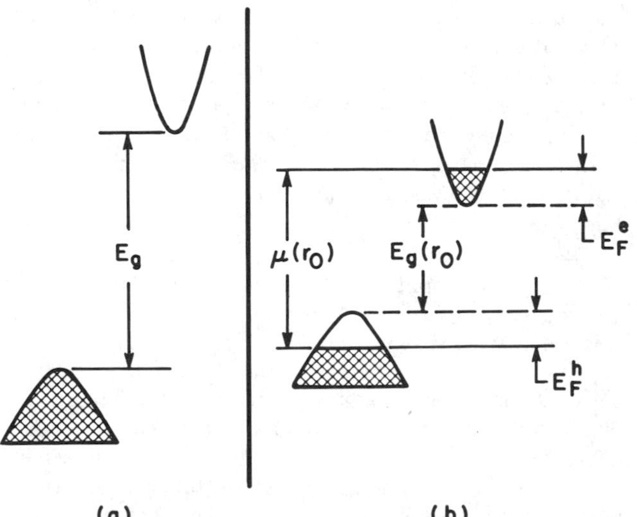

FIG. 2. Schematic energy versus wave-vector representation of the band structure of an indirect-bandgap semiconductor (a) outside and (b) inside EHD. The shaded regions are occupied by electrons. $E_F{}^e$ and $E_F{}^h$ are the electron and hole Fermi energies.

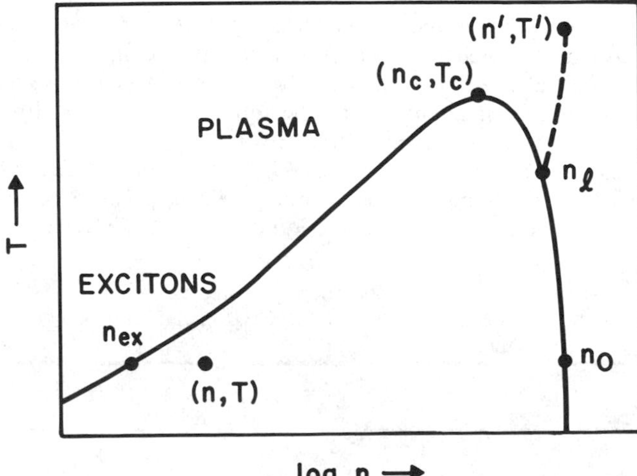

FIG. 3. Schematic phase diagram of EHD.

the surface energy σ of EHD and provides a technique of determining σ.

The theory of nucleation of drops from excitons has been developed by including the effect of finite lifetime of drops in the well-known classical theory of homogeneous nucleation. Density fluctuations of excitons create clusters of excitons or embryos. Embryos with radius R less than a critical radius R_c vanish quickly while embryos with $R>R_c$ continue to grow to a macroscopic size limited by the EHD lifetime. The time to form a macroscopic EHD is determined by the time to form a critical embryo and the time it takes for the critical embryo to collect excitons to grow to its final size. At high supersaturation ($I \gg I_{th}$), the second factor dominates, while at small supersaturation, the first factor dominates. Nucleation theory predicts a very strong dependence of nucleation time on supersaturation close to threshold. Time-resolved luminescence spectroscopy on nanosecond time scales provides a direct confirmation of these predictions. This technique also demonstrates that EHD can form not only by nucleation from a supersaturated gas but also through expansion and cooling of a hot dense plasma, i.e., expansion and cooling of (n',T') first creates a liquid of density n_l (Fig. 3) and then leads to phase separation into EHD and excitons.

The droplets can move through the crystal under exciton or phonon pressure or under external forces such as strain gradients, or electric field, or phonon pressure. Acceleration and deceleration of EHD has been directly observed by Doppler velocimetry. Motion produced by applied electric fields has been used to measure the net charge on EHD in Ge. A remarkable phenomenon has been observed in Ge in the presence of a stress gradient; the EHD accelerate to the point of maximum stress and coalesce into a single large drop whose radius and lifetime are 300 μm and 500 ms, respectively, at high pump intensities. These large drops have also been photographed using infrared image-scanning cameras.

Early work on EHD was done exclusively on Ge and Si. The existence of the metallic liquid phase has now also been established in GaP; the polar character of GaP and CdS has been shown to play an important role in the theoretical calculation of the binding energy in these semiconductors. In addition to the thermodynamic considerations, one must also consider whether the plasma lifetime in direct-bandgap semiconductors such as CdS is sufficiently long to allow the formation of the liquid phase. Other subjects which have attracted recent attention are the influence of the exciton–plasma Mott transition on the EHD phase diagram and interactions of phonons with EHD. The question of EHD in quasi-2D systems such as quantum wells has also been considered. Calculations predict liquid formation under certain conditions. However, no experimental results on EHD in quasi-2D semiconductors have been reported.

Table I Binding energy (ϕ_0), liquid density (n_0), critical temperature (T_c), and lifetime (τ) of EHD in various semiconductors

	ϕ_0 (meV)	n_0 (cm^{-3})	T_c (K)	τ (μs)
Ge	1.8	2.4×10^{17}	6.5	40
Si	8.2	3.3×10^{18}	27	0.15
GaP	14	6×10^{18}	>35	0.035

A study of electron-hole droplets in semiconductors has been a fascinating field of research, and has provided increased understanding not only of semiconductors, but also of phenomena normally associated with other branches of physics of condensed matter. It allows us to investigate, for example, effective ultrahigh densities, quantum and many-body effects in metals, gas–liquid phase transitions, and thermodynamics of multicomponent systems.

See also Electron Energy States in Solids and Liquids; Excitons; Luminescence; Semiconductors, Crystalline.

BIBLIOGRAPHY

V. S. Bagaev, *Springer Tracts Mod. Phys.* **73,** 72 (1985). (I)
C. D. Jeffries, *Science* **189,** 955 (1975). (E)
Ya. E. Pokrovskii, *Phys. Status Solidi (a)* **11,** 385 (1972). (I)
G. A. Thomas, *Sci. Am.* **234,** 28 (June 1976). (E)

A two part review of EHD (*Theory* by T. M. Rice and *Experimental Aspects* by J. C. Hensel, T. G. Phillips and G. A. Thomas) can be found in *Solid State Physics, Vol. 32* (H. Ehrenreich, F. Seitz and D. Turnbull, eds.). Academic Press, New York, 1977 (A); this is an excellent, exhaustive review covering the literature up to mid-1976.

Electron Microscopy

David J. Smith

The electron microscope uses a finely focused beam of energetic electrons to provide a highly magnified image of the specimen region of interest. Since the electron wavelength is typically much less than 1 Å, for example at 100 kilovolts it is 0.04 Å (where 1 Å = 10^{-10} m), it might be anticipated that the electron image would show structural details on the same scale. In practice, however, all existing electron lenses suffer from an unavoidable imaging defect, known as spherical aberration, which causes off-axis electrons to be improperly focused. The microscope resolution can then be expressed in the form

$$D = AC_s^{1/4}\lambda^{3/4}$$

where C_s is the spherical aberration coefficient of the microscope objective lens, λ is the electron wavelength, and A is a constant, varying between about 0.43 and 0.7, which depends on the operating conditions. Typical theoretical values for D range from about 5.0 Å for an accelerating voltage of 20 kV to about 1.2 Å at 1000 kV. The quandary for the microscopist is to choose between operation at higher voltages when the resolution is better but the sample structure is likely to be altered by collisions with the energetic electron beam, or to operate at lower voltage and accept that the resolution will be somewhat degraded. The recent trend has been towards operating voltages of 200–400 kV with corresponding image resolutions in the range of 2.5–1.6 Å. Details of atomic configurations are then easily obtained in low-index projections of many types of inorganic materials. Most biological and organic materials can not, however, withstand the effects of the electron beam and resolution figures are typically worse by an order of magnitude or more.

There are two basic modes of operation of the electron microscope, utilizing either fixed or scanning electron beams.

The conventional transmission electron microscope (CTEM) involves a relatively broad beam of electrons, perhaps 0.5–1.0 micron across, which is transmitted through a suitably thinned specimen. A highly magnified image can then be seen on the final viewing screen or on a television monitor via an image pickup system. By judicious choice of the size and position of the so-called objective aperture, which is normally located within the magnetic field of the objective lens pole-pieces, it is possible to highlight significant aspects of the sample morphology, particularly if it has an irregular or defective crystalline structure. Electron diffraction patterns can also be interpreted to provide useful specimen information.

The scanning electron microscope (SEM) has a finely focused electron beam which is scanned across the sample in a raster-like fashion. Electrons which are scattered by interaction with the sample are synchronously collected by nearby detectors and, after suitable amplification, form highly magnified images of the specimen surface on a viewing monitor. Three-dimensional views of the sample can be obtained by recording successive pairs of (stereo) images which differ only by small changes in the direction of the incident beam. The image resolution of the SEM is closely related to the minimum probe diameter, which mainly depends on the type of electron source. Typical resolutions are perhaps 50 Å but with a field emission electron gun, which has a very sharp metallic tip as its electron source, surface details as fine as 5 Å across are sometimes discernible.

The scanning transmission electron microscope (STEM) also involves rastering of the focused beam across the sample in synchronization with the detection of electrons which have passed through the sample, followed by modulation of the image monitor. The sequential nature of the image formation process again lends itself to online manipulation of the electron signal, in particular to take advantage of the fact that the amount of scattering of the electron beam depends strongly upon the elemental composition of the sample. Local variations in composition or structure, sometimes down to the atomic scale, can be differentiated by large variations in image contrast. Finally, by stopping the probe on a selected region of the sample, elemental information about the irradiated area can be obtained using either the characteristic x rays emitted by the sample or by measuring the energy spectrum of the transmitted electrons.

Overall, the electron microscope represents an instrument suitable for probing the *local* structure and composition of materials at levels which are unapproachable by most other techniques. For example, Fig. 1 shows a grain boundary in a nickel oxide bicrystal. Each black spot represents a row of nickel atoms viewed end-on, and so it is possible to deduce directly the arrangement of atoms in the vicinity of the boundary. Since the microscopic properties of most solids depend to a large extent on such local irregularities the electron microscope is a powerful tool for the characterization of advanced materials.

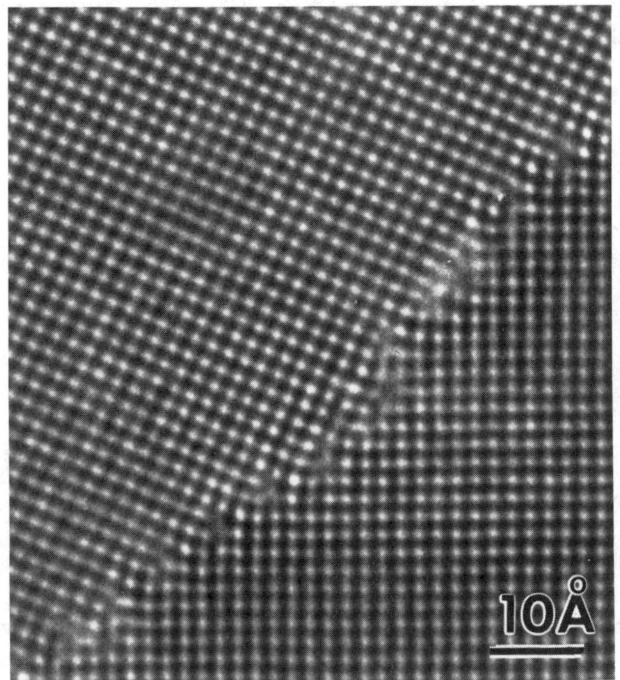

FIG. 1. Atomic-resolution electron micrograph showing a grain boundary between two nickel oxide crystals. Black spots represent rows of nickel atoms viewed end-on (10 Å = 10^{-9} m).

See also ELECTRON DIFFRACTION; FIELD EMISSION.

BIBLIOGRAPHY

J. M. Cowley and D. J. Smith, "The Present and Future of High Resolution Electron Microscopy," *Acta Crystallographica* **A43**, 737 (1987).

A. V. Crewe, "A High Resolution Scanning Electron Microscope," *Scientific American* **224** (4), 1971

Electron Optics *see* Charged-Particle Optics

Electron Spectroscopy *see* Photoelectron Spectroscopy; Charged-Particle Spectroscopy

Electron Spin Resonance

A. Rassat

A free electron with spin $\hbar\mathbf{s}$ and magnetic moment $\boldsymbol{\mu}_s = -g_e\beta_e\mathbf{s} = -\gamma_s\hbar\mathbf{s}$ ($\hbar$ is Planck's constant/2π, β_e is the Bohr magneton$=\hbar|e|/2m$, γ_s is the electron magnetogyric ratio $=g_e|e|/2m$, g_e is Landé factor $=2.0023\ldots$, e is electron charge, m is electron mass) has in a magnetic field $\mathbf{B}_0$ a "Zeeman energy'' $W_Z = -\boldsymbol{\mu}_s \cdot \mathbf{B}_0 = g_e\beta_e \mathbf{B}_0 \cdot \mathbf{s} = \gamma_s \hbar \mathbf{B}_0 \cdot \mathbf{s}$ with two stationary states, α ($m_s = +\frac{1}{2}$) and β ($m_s = -\frac{1}{2}$), separated by $\Delta W = g_e\beta_e B_0$. $\mathbf{B}_0$ also induces precession of the spin at angular velocity $\boldsymbol{\omega}_0 = \gamma_e \mathbf{B}_0$. In a collection of independent free electrons, spins precess in phase, unless there is inhomogeneity δB in the magnetic field (broadening the Zeeman levels by $\delta W_z = g_e\beta_e|\delta B|/2$), which after a time τ induces a dephasing $\delta\phi = \tau\delta\omega = \tau\gamma_e\delta B$. If a small oscillatory magnetic field $\mathbf{B}_1$ [conveniently provided by an electromagnetic radiation of frequency $\nu_0(=\omega_0/2\pi)$ and amplitude $B_1 \ll B_0$] is applied perpendicular to the static magnetic field B_0 so that $h\nu_0 = g_e\beta_e B_0$ (resonance condition), transitions are induced between the α and β levels: (electron spin) resonance occurs. Induced absorption and emission having equal probability, the populations of each level tend to become equal; there is no net absorption or emission of radiation, and resonance cannot be detected. However, in most systems, *relaxation* occurs: some mechanism (''spin–lattice relaxation'') permits thermal equilibrium to be transferred to the spins with a characteristic time constant T_1, the ''spin–lattice relaxation time.'' If this process is fast enough (but not too fast, because of inherent broadening $\Delta\omega = 1/T_1$), resonance should be detected. ''Spin–spin relaxation,'' with characteristic time constant T_2, must also be considered. It does not modify the level population, but reflects the spin-dephasing and level-broadening due to the *local* magnetic field inhomogeneity. (T_2 may be visualized as the time for an average dephasing $\Delta\phi$ between spins to become of the order of 1 rad.) Now, if relaxation is not too slow and B_1 is not too strong ($\gamma_e^2 B_1^2 T_1 T_2 \ll 1$), resonance may be detected.

The basic components of an electron-spin-resonance (ESR) *instrument* are a homogeneous magnetic field, a source of monochromatic electromagnetic radiation, and a device to detect its absorption. For $B_0 \simeq 3300$ G (0.33 T), the frequency is $\nu_0 \simeq 10^{10}$ Hz (wavelength 3 cm, in the microwave range). A klystron is a convenient source. For technical reasons, its frequency is fixed, and resonance condition is obtained by varying the applied static magnetic field B_0. Absorption I of radiation is recorded as a function of B_0 and is maximum at resonance. Figure 1 gives a typical signal. Since linewidth ΔB characterizes the local field inhomogeneity, $1/T_2 = \gamma_e \Delta B$.

Generally, samples are paramagnetic solids or liquids. Let us study first a crystalline sample containing *perfectly oriented* paramagnetic molecules (or ions), having zero orbital angular momentum, and spin $\frac{1}{2}$ (''one unpaired electron''), diluted in a diamagnetic matrix, so that the electronic spins do not interact appreciably. Resonance occurs when $h\nu_0 = g_e\beta_e B$. $\mathbf{B}$, the local field felt by the electron spin, is now the sum of $\mathbf{B}_0$ and of the magnetic fields due to orbital and spin angular momenta (spin–orbit and spin–spin interactions) in the molecule. (An individual electron i has an instantaneous angular momentum $\hbar \mathbf{l}_{iN}$ relative to each nucleus N). Interactions with nuclear spins are separated from other magnetic interactions and give rise to the ''hyperfine structure.'' For instance, if the unpaired electron interacts with one nuclear spin $\hbar I_N$, the local field depends on the $2I_N+1$ possible stationary orientations of the I_N spin relative to $\mathbf{B}_0$: $2I_N+1$ resonance lines are observed. Separating the hyperfine terms, the resonance conditions at one orientation

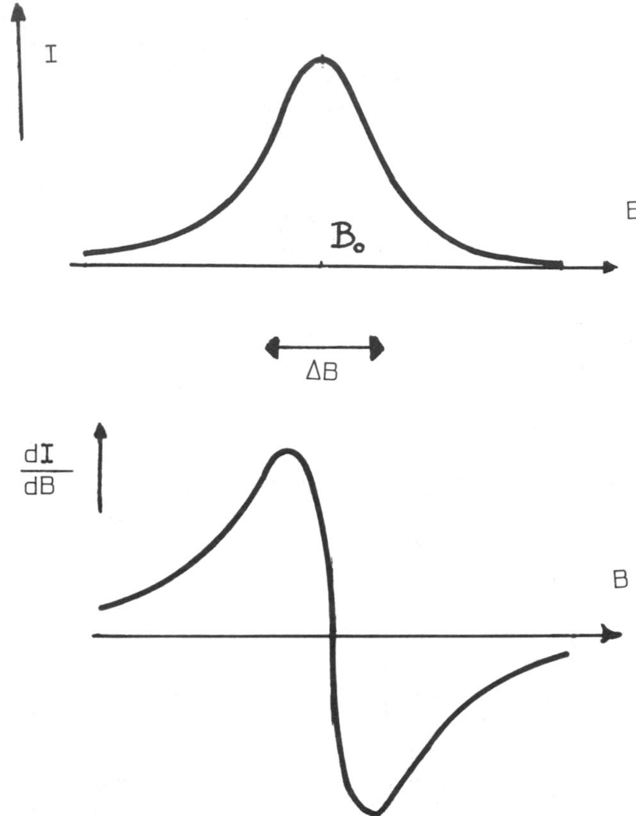

FIG. 1. Resonance curve for a collection of electrons resonating at $g = h\nu_0/\beta_e B_0$. (a) Intensity I of absorption recorded as a function of the applied magnetic field, ν_0 being fixed. (b) Derivative curve dI/dB, the usual presentation of results. Schematically, a single ESR absorption line δ is expected to be Gaussian, $I \simeq \exp[-(B-B_0)^2/(\Delta B)^2]$, in solids, where linewidth ΔB comes from a statistical distribution of local magnetic fields (inhomogeneous broadening), or Lorentzian, $I \simeq 1/[\Delta B^2 + (B-B_0)^2]$, where linewidth ΔB is the damping term of a usual (mechanical) resonance curve (homogeneous broadening); $\Delta B = 1/\gamma_e T_2$.

(θ,Φ) of $\mathbf{B}_0$ relative to the molecule may be written (if $A_N \ll g_e\beta B_0$)

$$h\nu_0 = g(\theta,\Phi)\beta_e B_0 + A_N(\theta,\Phi)m_{IN}$$

$$(m_{IN} = -I_N,\ -I_N+1,\ \ldots,\ +I_N).$$

$g(\theta,\Phi)$ and $A_N(\theta,\Phi)$ are the Landé factor and hyperfine coupling constant of nucleus N with the unpaired electron for direction (θ,Φ) of $\mathbf{B}_0$. If there are different interacting nuclei N, resonance conditions are $h\nu_0 = g\beta_e B_0 + \Sigma_N A_n m_{IN}$. The number of observed lines is often less than $\Sigma_N (2I_N+1)$ because of some coincidences. (Figure 2 shows a spectrum obtained at two orientations for one electron interacting with one spin $I=1$.) If resonance is studied at all possible orientations, results may be given in the form of an ''effective-spin Hamiltonian,'' $\mathcal{H}'_{\text{spin}} = \beta_e \mathbf{B}_0 \cdot \mathbf{g} \cdot \mathbf{s}' + \Sigma_N \mathbf{I}_n \cdot \mathbf{A}_n \cdot \mathbf{s}'$ (nuclear terms may be added). $\mathbf{g}$ and $\mathbf{A}_N$ are second rank tensors ($\mathbf{B}_0 \cdot \mathbf{g} \cdot \mathbf{s}' = \Sigma_u \Sigma_v B_u g_{uv} s_v'$, $u,v = x,y,z$) which by a suitable choice of Cartesian axes (''principal axes'') may be written in diagonal form (as a set of three ''principal values''). Except for symmetry reasons, both tensors will not be diagonal in the same axes. Diagonalization of the matrix representation of this Hamiltonian in the basis of the effective electron spin $|s'=\frac{1}{2}, m_{s'}\rangle$ and nuclear-spin $\Pi_N|I_N, m_{IN}\rangle$ states gives energy levels reproducing exactly the experimental resonance conditions, if **g** and $\mathbf{A}_N$ have correctly been determined. Independently from their experimental determination, the **g** and $\mathbf{A}_N$ tensors could in principle be computed from a magnetic Hamiltonian, in the basis of the (electronic) kinetic and (electronic and nuclear) electrostatic spinless Hamiltonian eigenfunctions. The **g** tensor comes from Zeeman and spin–orbit interactions. The $\mathbf{A}_N$ tensor comes from two interactions: dipolar interaction with nucleus N,

$$H_{\text{dipN}} = \sum_i^{\text{all electrons}} (\gamma_s\gamma_N\hbar^2\mu_0/4\pi) \times [3(\mathbf{I}_N\cdot\mathbf{r}_{iN})(\mathbf{s}_i\cdot\mathbf{r}_{iN})I_{iN}^{-2} - \mathbf{I}_N\cdot\mathbf{s}_i]r_{iN}^{-3}$$

(γ_N is nuclear magnetogyric ratio of nucleus N, μ_0 is free space permeability, $\mathbf{r}_{iN}$ connects electron i and nucleus N); and (Fermi) contact interaction with nucleus N,

$$H_{\text{contact}N} = \sum_i^{\text{all electrons}} (2\mu_0/3)\gamma_s\gamma_N\hbar^2\delta(\mathbf{r}_{iN})\mathbf{s}_i\cdot\mathbf{I}_N$$

[$\delta(\mathbf{r}_{iN})$ is the three-dimensional Dirac "function"]. H_{dip} gives rise to a traceless tensor $\mathbf{A}'$ ("purely anisotropic") and H_{contact} to an isotropic term $a_N\cdot\mathbf{1}$ ($\mathbf{A}_N = a_N\cdot\mathbf{1} + \mathbf{A}_N'$).

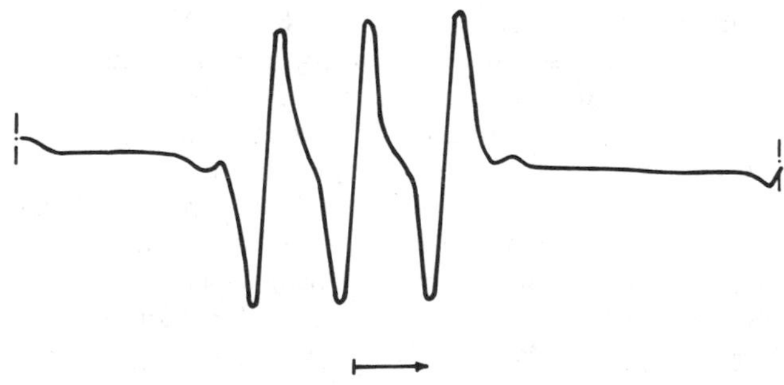

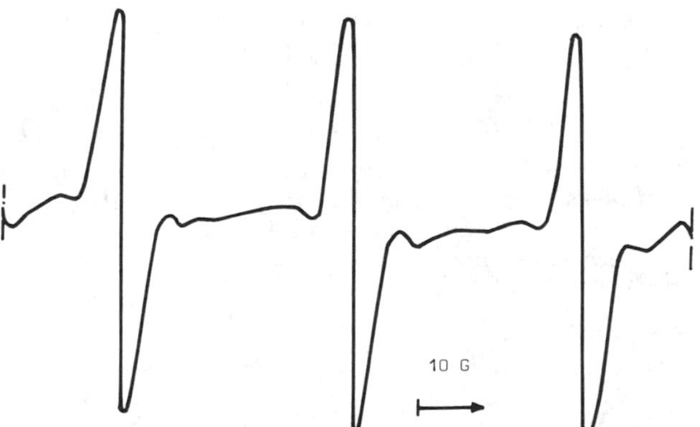

FIG. 2. Observed spectrum at $\nu_0 \sim 10^4$ MHz for a single crystal containing a diluted and oriented nitroxide free radical, at two perpendicular orientations of the magnetic field B_0. In this paramagnetic molecule, the unpaired electron interacts with one nitrogen nucleus (spin $I=1$). The three lines correspond to $m_1 = -1, 0, +1$. $g(\theta,\Phi)$ is measured of the central line $g(\theta,\Phi) = h\nu_0/\beta_e B_0(\theta,\Phi)$. The hyperfine coupling constant $A(\theta,\Phi)$ is measured as the average hyperfine splitting A' and A'' between the lines. If $g\beta_e B_0 \gg A_N$, $A' = A'' = A_N$.

Let us study a different situation for the same paramagnetic species, still in a rigid diamagnetic matrix but so that the orientations of B_0 relative to the molecule have angular distribution. The observed "immobilized spectrum" $I = F(B_0)$ is the superposition of spectra $I(\theta,\Phi) = f(\theta,\Phi,B_0)$ corresponding to the various orientations: $F(B_0) = \int\int f(\theta,\Phi,B_0)$ $\rho(\theta,\Phi)$ $\sin\theta\, d\theta\, d\Phi/4\pi$. For an isotropic distribution ("powder spectrum"), $\rho(\theta,\Phi) = 1$. For an anisotropic distribution, $\rho(\theta,\Phi)$ may be estimated from the experimental spectrum. Anisotropies of **g** and hyperfine **A** terms both affect the "powder spectrum," as shown in Fig. 3.

Let us consider now the same paramagnetic species in *dilute fluid solution*. Magnetic interactions are anisotropic and the energy levels vary as the molecules rotate. For a spherical molecule of radius a, this random tumbling can be characterized by one rotational correlation time τ_c, related to temperature t and viscosity η of the medium by $\tau_c = 4\pi a^3\eta/3kt$ (k is Boltzmann's constant). This random motion has a component at the resonance frequency and induces transitions, thus modifying the spin–lattice relaxation time T_1. Spin–spin relaxation time T_2 is also influenced by random rotation if anisotropic interactions $\delta\omega$ are present: let us simulate this rotation by sudden random jumps between two orientations of resonance field B_0 and $B_0 + \delta B$ (at fixed ν_0), τ_c being the average lifetime at each orientation (thus simulating the correlation time). For spins precessing initially in phase, each jump introduces a dephasing $\delta\phi = \pm\tau_c\,\delta\omega = \pm\tau_c\gamma_e\,\delta B$. If $\tau_c \gg T_2$, the phase difference will be 1 rad after T_2, $1/T_2 \simeq \gamma_e\,\delta B$. (In the limiting case $\tau_c \to \infty$, the "immobilized spectrum" consists of two lines whose separation is of course δB.) If $\tau_c \ll T_2$, dephasing after n random jumps is given by the accumulation of n random dephasing $\delta\phi$: $\overline{\Delta\phi^2} = n\delta\phi^2 = n(\tau_c\delta\omega)^2$. $\Delta\phi \approx 1$ rad is reached after T_2/τ_c jumps: $1/T_2 \approx (\delta\omega)^2\tau_c = \gamma_e^2(\delta B)^2\tau_c$. In this case, the spectrum is averaged out to a single line, of width $1/T_2$, at the average position between B_0 and $B_0 + \delta B$. Spin–spin relaxation may be induced by other mechanisms (such as internal motion in a free radical, or chemical exchange of an electron between molecular systems). The same relationships exist between T_2 and the difference of resonance frequencies and the characteristic lifetime of the processes (see Fig. 4). When different mechanisms operate at the same time, the observed linewidth is the sum of all contributions.

Let us consider now a case of *spin-1* species ("two un-

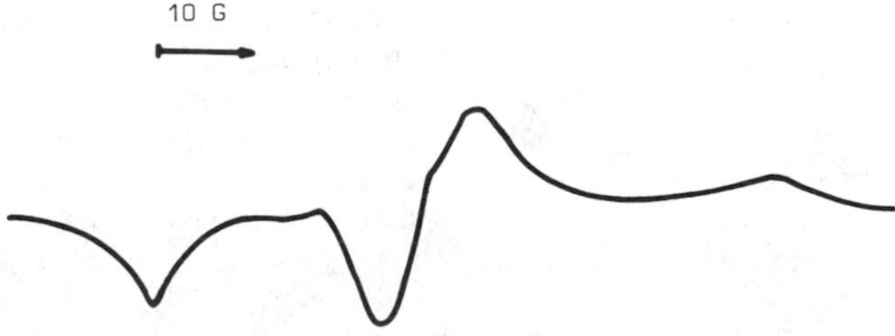

FIG. 3. Spectrum of a $M/100$ frozen solution of a nitroxide radical. Principal axes of **g** and **A** tensors coincide. The spectrum can be computer simulated with the principal values $A_{xx} = 7.1$ G, $A_{yy} = 5.6$ G, $A_{zz} = 32$ G, and $g_{xx} = 2.0089$, $g_{yy} = 2.0061$, $g_{zz} = 2.002$.

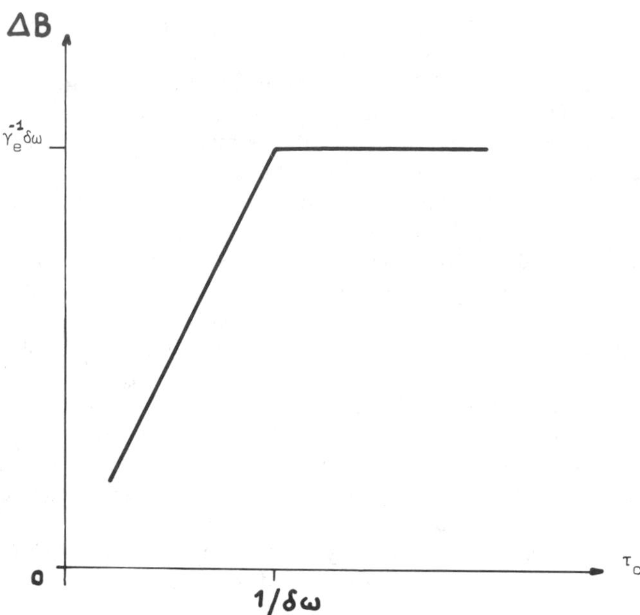

FIG. 4. Schematic ESR linewidth ΔB in solution of a paramagnetic species having an anisotropic interaction $\delta\omega$, as a function of its correlation time τ_c: when $\tau_c \ll 1/\delta\omega$, $\Delta B = \gamma_e^{-1}\delta\omega^2\tau_c$; when $\tau_c \gg 1//\delta\omega$, $\Delta B = \gamma_e^{-1}\delta\omega$.

paired electrons'') diluted in a diamagnetic environment. There are singlet and triplet states, the lowest having ^1E and (average) ^3E energy. The ''exchange integral'' J is by definition $2J = {}^1E - {}^3E$. Electron spin–spin dipolar interaction is nonzero in the triplet state: To first order, it comes from the spin–spin magnetic dipolar interaction

$$H_{\text{dip}} = \sum_i^{\text{all electrons}} \sum_{j<i} [\gamma_s{}^2\hbar^2\mu_0/4\pi][\mathbf{s}_i \cdot \mathbf{s}_j - 3(\mathbf{r}_{ij}\cdot\mathbf{s}_i)(\mathbf{r}_{ij}\cdot\mathbf{s}_j)r_{ij}^{-2}]r_{ij}^{-3}$$

($\mathbf{r}_{ij}$ connects electrons i and j), and to second order, from spin–orbit interactions,

$$H_{\text{LS}} = \sum_N^{\text{nuclei}} \sum_i^{\text{electrons}} \zeta(r_{iN})\boldsymbol{l}_{iN}\cdot\mathbf{s}_i,$$

where $\zeta_N(r_{iN})$ is a characteristic function of nucleus N. The influence of singlet and triplet levels and of the dipolar interaction on ESR spectra can be simulated by an effective-spin Hamiltonian $\mathcal{H}'$ similar to the one used for spin-$\frac{1}{2}$ systems, but dealing with two effective spins $\frac{1}{2}(\mathbf{s}_1{}'$ and $\mathbf{s}_2{}')$, and having ''fine-structure'' terms: an exchange term $\mathcal{H}'_{\text{spin(exch)}} = 2J\mathbf{s}_1{}'\cdot\mathbf{s}_2{}'$ and a dipolar term $\mathcal{H}'_{\text{spin(dip)}} = \mathbf{s}_1{}'\cdot\mathbf{D}\cdot\mathbf{s}_2{}'$. $\mathbf{D}$ is a traceless tensor whose principal values are written $(2D, -D+3E, -D-3E)$ and $D = g_e^2\beta_e^2/r^3$, r being an average distance between the two unpaired electrons (spin–orbit contribution being neglected). As for spin-$\frac{1}{2}$ systems, various situations occur depending on the state (solid or liquid) of the system and of the relative order of magnitude of the Zeeman, exchange, fine, or hyperfine terms: In rigid matrix (or in viscous solutions, if $\tau_c \gg \hbar/D$), the dipolar interaction gives ''fine structure'' and also ''half-field transitions'' at $g \simeq 4$. In solutions, the dipolar interaction is averaged out to zero if $\tau_c \ll \hbar/D$, but contributes (very effectively) to the line broadening. Finally, the case of two interacting spins may also be found for spin $\frac{1}{2}$ not perfectly isolated from their neighbors, i.e., for concentrated solutions of radicals in a diamagnetic matrix or liquid. Pairwise interactions give rise to exchange J and dipolar D energies. In a pair of spin-coupled electrons, the individual electron spins (as any nonconservative observable in a two-level system) oscillate with a period $\tau = h/2J$. By analogy, we may expect this oscillation (''exchange'') to induce a spin–spin relaxation with $1/T_2 = \pi D^2/\hbar J$. This is indeed the case: as the concentration increases, lines broaden (dipolar broadening) and then narrow (exchange narrowing).

Applications of ESR use information related to the spectral moments: Total *intensity* is proportional to the number of paramagnetic species. Concentration is measured by comparison with a reference, because absolute determination is difficult. Under favorable conditions 10^{11} spins may be detected. The *position* of spectral lines permits identification of the paramagnetic species, of its surroundings, and of its orientation. *Linewidth,* related to T_2, depends on a characteristic time τ and some characteristic difference $\delta\omega$. For instance, τ may be the lifetime of some jump process between positions having resonance at ω_0 and $\omega_0 + \delta\omega$, or the rotational correlation time in solution. In this case, the linewidth is a measure of the ''effective volume'' of the species, of the solvent viscosity and temperature, and of anisotropic interactions, among which dipolar interactions ($\simeq r^{-3}$) provide geometrical information. T_1 measurements (saturation, pulses) give information complementary to those obtained from T_2. This information is used, for instance, in chemical (structural and dynamical) analysis of free radicals, of excited triplet molecules, and of transition-metal ions, and in biophysics, especially in the ''spin-label method,'' where a free radical reveals information about the biomolecule to which it is bound.

See also PARAMAGNETISM; RELAXATION PHENOMENA; RESONANCE PHENOMENA.

BIBLIOGRAPHY

Elementary treatment of ESR may be found in quantum-mechanics textbooks, for instance, in D. Park, *Introduction to the Quantum Theory,* p. 151. McGraw-Hill, New York, 1964. (E) or in C. Cohen-Tannoudji, B. Diu, and F. Laloe, *Quantum Mechanics (I),* p. 443. Wiley-Hermann, Paris, 1977. (I)

General Texts

N. M. Atherton, *Electron Spin Resonance: Theory and Applications*. Wiley, New York, 1973. (E,I)

A. Carrington, A..D. McLachlan, *Introduction to Magnetic Resonance*. Harper and Row, New York, 1966.

G. E. Pake and T. Estle, *The Physical Principles of Electron Spin Resonance*. Addison-Wesley, Reading, MA 1973. (I,A)

C. P. Slichter, *Principles of Magnetic Resonance*. Harper & Row, 2nd ed., New York, 1978. (I,A)

J. E. Wertz and J. R. Bolton, *Electron Spin Resonance: Elementary Theory and Practical Applications*. Chapman and Hall/Methuen, New York, 1986. (E,I)

Specialized

A. Abragam and B. Bleaney, *Electron Paramagnetic Resonance of Transition Ions*. Clarendon Press, Oxford, 1970. (A)

L. J. Berliner (ed.) *Spin Labelling: Theory and Applications*, Vol. 1, 1976. Academic Press, New York, vol. 2, 1979. (E,A)

J. R. Bolton, D. C. Borg, and H. M. Schwartz, *Biological Applications of Electron Spin Resonance*. Wiley-Interscience, New York, 1971. (E)

J. A. Weil, M. K. Bowman, J. R. Morton, and K. F. Preston (ed.), *Electronic Magnetic Resonance of the Solid State*. The Canadian Society of Chemistry, Ottawa, 1987. (I,A)

Spin Hamiltonian and Magnetic Interactions

R. McWeeny, *Spins in Chemistry*. Academic Press, New York, 1970. (I)

H. F. Hameka, *Advanced Quantum Chemistry*. Addison-Wesley, Reading, MA, 1965. (A)

J. E. Harriman, *Theoretical Foundations of Electron Spin Resonance*. Academic Press, New York, 1978. (A)

Relaxation

A. Hudson and G. R. Luckhurst, *Chem. Rev.* **69,** 191 (1969). (E)

R. Lenk, *Fluctuations, Diffusion and Spin Relaxation*. Elsevier, Amsterdam, 1986. (A)

A. Manenkov and R. Orbach, *Spin-Lattice Relaxation in Ionic Solids*. Harper & Row, New York, 1966. (A)

L. T. Muus and P. W. Atkins (eds.), *Electron Spin Relaxation in Liquids*. Plenum Press, New York, 1972. (A)

Electron Tubes

John D. Ryder

The Edison effect, observed by Thomas A. Edison in 1883 as an electric current between a metal plate and a heated filament in an evacuated bulb, marked the birth of the electron tube. Current passed only with the plate positive to the filament, and the charge carriers were later identified as negative electrons. Edison patented several ideas for application but did not investigate further.

Since this two-element device or diode was unilateral in conduction properties, it was used as a rectifier or detector of radio signals by J. A. Fleming, ca. 1904. In 1906, Lee deForest, following unsuccessful experiments with gas lamps, placed a wire grid as a control element between filament and plate, making a sensitive relay, the three-element or triode tube.

THERMAL EMISSION OF ELECTRONS

To achieve emission from a surface, the electron must be given sufficient outward directed energy to overcome surface binding forces. The energy needed is called the work function, E_W at 0 K, $=mv^2/2$. Velocity v is the outward-directed velocity component. This required energy depends on cathode material and surface condition; values for a few common emitting materials are given in Table I.

In thermionic emitters the work-function energy is supplied by raising the temperature of the emitter. The emitted current is predicted by an equation proposed by S. Dushman in 1923:

$$I = A_0 S T^2 \epsilon^{-b_0/T} \text{ A}, \qquad (1)$$

where A_0 is the surface constant, S is the emitting area in m^2; T is the surface temperature in degrees Kelvin, $b_0 = 11\,600 E_W = eE_W/k$, the voltage equivalent of temperature.

Table I Values of the emission constants

Material	A_0 ($\times 10^4$)	b_0	E_W (eV)[a]
Cesium	16.2	21 000	1.81
Thorium	60.2	39 400	3.4
Tungsten	60.2	52 400	4.52
Thorium on tungsten	3.0	30 500	2.6
Oxide (BaO + SrO)	0.01	11 600	1.0

[a] 1 eV $= 1.60 \times 10^{-19}$ J.

Thermionic cathodes are (1) filamentary or (2) indirectly heated. In a filament, emission occurs from a wire, heated by the passage of a current. In the indirect form, a heater is used inside a metal cylinder to raise the surface to about 1000°C, sufficient to cause emission from the surface, usually coated with barium and strontium oxides. A short heating time is required for the surface to reach emitting temperature.

Oxide-coated cathodes were normally used, but high-power tubes employed tungsten with a monatomic thorium layer or pure tungsten.

THE DIODE

About 1912, independent studies by Langmuir and Child allowed explanation of the diode volt–ampere curve, Fig. 1. The toe from a to b indicates that some electrons have outward velocities and need no accelerating potential to reach the plate. From b to c the current increases with potential but is less than predicted by the Dushman relation. Temperature saturation, as predicted by the Dushman relation, is reached above c.

The lowered current from b to c is caused by the repulsion of the electronic space charge near the cathode, stopping electrons with low outward velocities. If the anode potential is increased, more electrons are attracted from the space cloud to the anode and the repulsion is weakened. Equilibrium is reached at a space charge density at which the num-

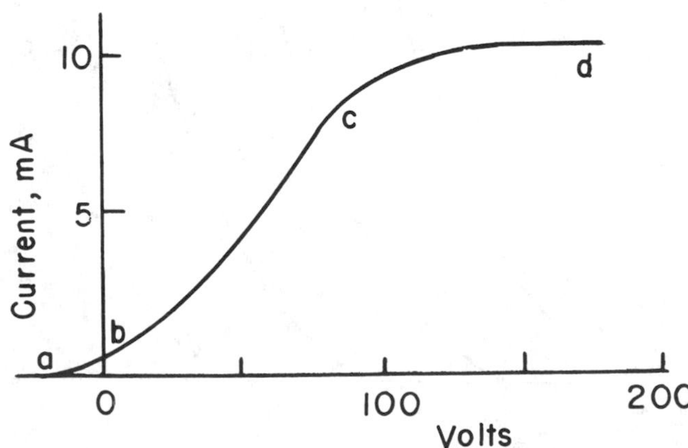

FIG. 1. Volt–ampere diagram for a vacuum diode.

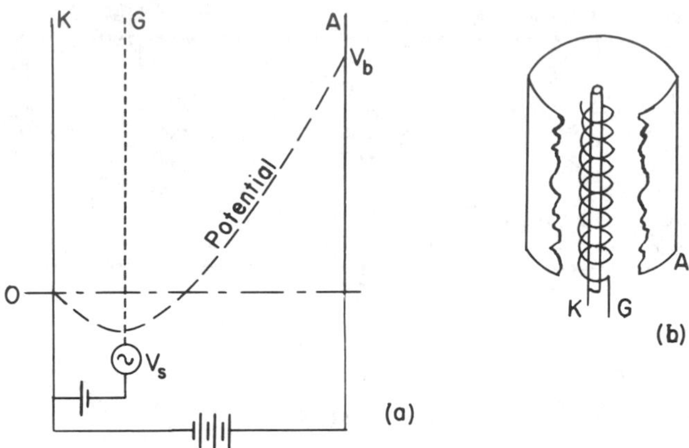

FIG. 2. (a) Space potential with space charge in a triode; (b) typical triode structure.

ber of electrons reaching the anode is balanced by the number leaving the cathode.

The equation relating the current collected to the applied voltage and the geometry of the diode is called the three-halves power law, and for large parallel-plane electrodes is

$$J = \frac{4\epsilon_v}{9}\left(\frac{2e}{m}\right)^{1/2}\frac{V_b^{3/2}}{d^2} = 2.34\times 10^{-6}\frac{V_b^{3/2}}{d^2} \quad \frac{A}{m^2}, \quad (2)$$

where ϵ_v is the space permittivity $= 10^7/4\pi c^2 = 8.85\times 10^{-12}$; m is the electron mass $= 9.106\times 10^{-31}$ kg; d is the cathode–anode spacing, in meters.

THE TRIODE

The potential minimum, established by the space charge near the cathode surface, is illustrated in Fig. 2(a). It acts as a barrier to the lower velocity electrons, and by raising or lowering the potential by the signal v_s, the current can be varied or controlled. The average grid potential is maintained negative to cathode, and the grid current is zero in most operating regions.

The anode current is a function of the grid–cathode voltage v_c; if v_b is changed the potential minimum shifts up or down, so we may say

$$i_b = f(v_c, v_b) \quad (3)$$

and the general triode relations are shown in Fig. 3. To show the effect of small changes, as with a grid signal, Δv_c and Δv_b, we have

$$\Delta i_b = \frac{\partial i_b}{\partial v_c}\Delta v_c + \frac{\partial i_b}{\partial v_b}\Delta v_b. \quad (4)$$

We define

$$g_m = \frac{\partial i_b}{\partial v_c}, \qquad g_p = \frac{\partial i_b}{\partial v_b} \quad (5)$$

and these are the transconductance and the plate conductance, respectively. From Eq. (5) we also have

$$\mu = \frac{g_m}{g_p} = g_m r_p = -\left.\frac{\partial v_b}{\partial v_c}\right|_{i_b=K}, \quad (6)$$

where μ is the amplification factor and r_p is known as the plate resistance. Small ac signals may replace the Δ values and Eq. (4) becomes a circuit equation as

$$I_p = g_m V_g + V_b/r_p. \quad (7)$$

This equation represents a current summation at A in Fig. 4(a); this is known as the current-source equivalent circuit for the triode. The circuit dual is shown in Fig. 4(b) as the voltage-source equivalent circuit. Loads may be connected at the output as shown. The voltage gain is A = (output voltage) (input voltage) and

$$A = -\frac{\mu Z_L}{r_p + Z_L} = -\frac{g_m Z_L}{1 + Z_i/r_p}. \quad (8)$$

The minus sign indicates a 180° shift due to the tube; any additional angle on A is called the circuit phase shift.

The grounded-cathode circuit was most often used because of its high gain, but the cathode follower and the grounded-grid circuits were used to take advantage of particular gain and impedance properties. These circuits appear in Fig. 5.

THE PENTODE

At frequencies of the order of 1 MHz, the triode may become unstable and oscillate in the grounded-cathode circuit. The anode, at a higher signal potential, drives a current to the grid through the grid-plate capacitance, in phase with the

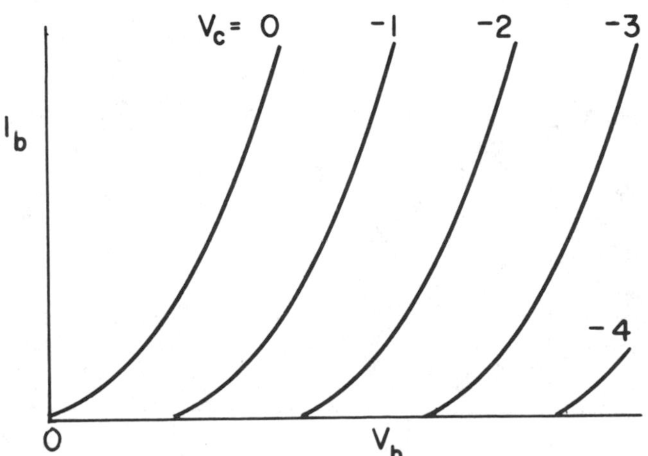

FIG. 3. Voltage–current characteristics for a triode.

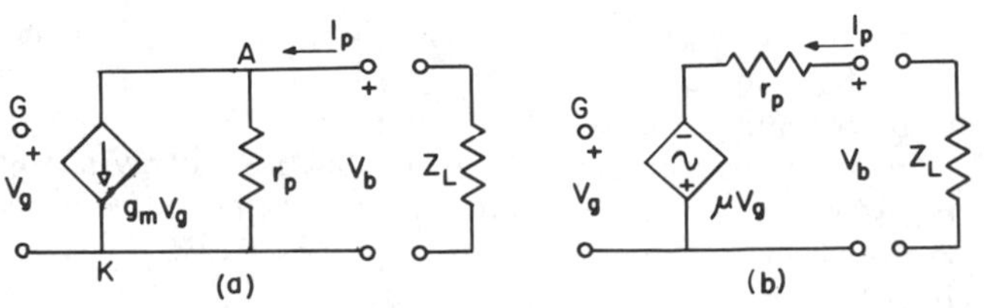

FIG. 4. (a) Current-source equivalent circuit for a triode; (b) voltage-source equivalent circuit. The diamonds are control sources.

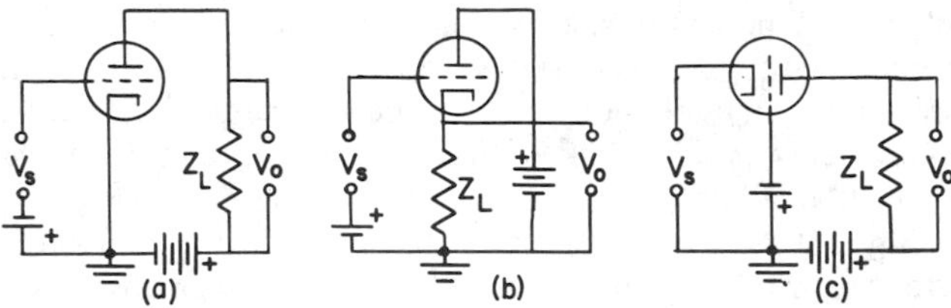

FIG. 5. (a) Grounded-cathode triode circuit; (b) cathode follower; (c) grounded-grid circuit.

signal. This raises the grid voltage, and a cumulative process of regeneration occurs, with instability of gain. Shielding of the grid from the anode by a second and grounded grid reduces this current through C_{gp}. A tube with a screen grid, and often another grid called the suppressor, has a C_{gp} of about 0.004 pF, compared to 2–5 pF for a triode. With five internal elements, the tube is called a pentode.

The first screen is operated at a high dc potential to accelerate the electrons, but is grounded for signal frequencies by a shunt capacitor. The energy of the electrons is sufficient for them to reach the positive anode. The values of μ and r_p are much higher than for a triode, and the low internal capacitance makes the pentode useful to frequencies above 100 MHz.

Output loads of 100–200 kΩ are used, small compared to an r_p of 600–1000 kΩ. Equation (8) with $r_p \gg Z_L$ reduces to

$$A = -g_m Z_L. \tag{9}$$

The equivalent circuit for a pentode becomes Fig. 6(a), with characteristics in Fig. 6(b). Voltage gains of several hundred per stage are possible, with g_m values of 0.01 to 0.04 siemens.

PHOTOCELLS OR PHOTOEMISSIVE TUBES

Electrons are emitted when light of appropriate frequency strikes a specially prepared plate. This photoelectric emission differs from thermionic emission only because the work-function energy is derived by the energy of a photon. The Einstein equation explains the phenomena; photon energy hf must supply the work energy, eE_W J. That is

$$hf = eE_W + mv^2/2. \tag{10}$$

Any excess photon energy appears as kinetic energy of the emitted electron; the photon then disappears. Below some

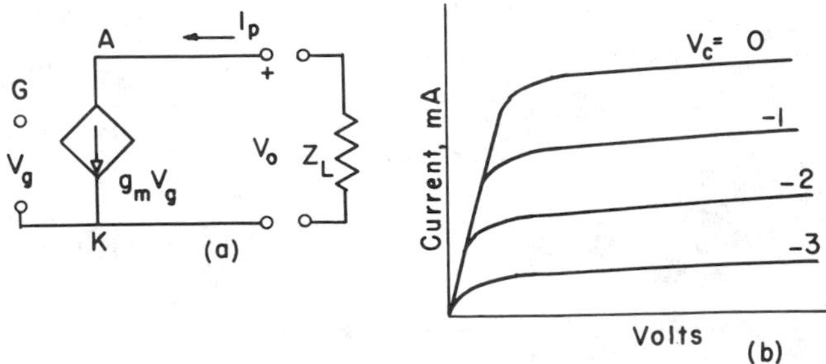

FIG. 6. (a) Equivalent circuit for a pentode; (b) output characteristics of a pentode.

threshold frequency $f_0 = eE_W/h$ the emitted current becomes zero.

Emitting surfaces, usually with cesium or rubidium, are available with peak responses from the infrared to the ultraviolet.

All emitted electrons are collected at voltages over 25 V. The internal resistance is many megohms and the cell approximates a current source. Vacuum cells are linear with light intensity, but gas cells are linear only for small light variations. Ionization by collision with gas atoms gives such cells additional sensitivity, gains of 4 to 7 being usual.

THE GAS DIODE RECTIFIER

By introduction of a gas—argon, xenon, or mercury vapor—into a thermionic diode, the voltage required for conduction is reduced and power efficiency is raised. When the anode–cathode voltage exceeds the gas-ionization potential, electrons acquire sufficient energy to ionize upon collision with gas atoms. The resulting positive ions neutralize the negative space charge due to the electrons, and more electrons are accelerated in a cumulative action leading to an arc discharge. Upon reversal of the anode polarity the ions migrate to the anode and tube walls, recombining with charges there, and the current goes to zero.

The positive ions are heavy and slow moving and their function is to neutralize the negative space charge. It was possible to secure emission from deep cavities, and heat shielding was added to reduce the heating power needed. With mercury vapor the pressure is dependent on temperature, and below 20°C insufficient ions are formed. The tube voltage rises and positive ions bombard the cathode, destroying the emitting surface; xenon or argon were often used, but with reduced efficiency. At temperatures above 80°C the vapor provides insufficient insulation.

Such devices were efficient rectifiers operating with forward voltages of 10–15 V with mercury vapor and handling forward currents of 100 A with reverse voltages as high as 15 000 V. The fragility of the oxide-coated cathodes posed maintenance and application problems.

THE THYRATRON OR GAS TRIODE

The thyratron (door tube) was developed by addition of a grid structure to the gas diode, Fig. 7(a). The grid was a cylinder with baffles pierced by large holes, and the anodes were often graphite blocks.

With the grid negative to cathode, emitted electrons face a repelling field and there is no conduction. With a less negative grid, a few electrons pass through the holes in the baffles, are accelerated toward the anode, and ionize gas atoms. The positive gas ions neutralize the negative space charge, more gas ions are produced, and the discharge proceeds to an arc. The grid potential is the critical factor in starting the discharge, and the relation between critical grid voltage and anode voltage is plotted in Fig. 7(b) for a typical tube.

Positive ions are attracted to the negative grid and build up a positive space charge there. This sheath of ions insulates the grid, and changes in grid potential alter the sheath thick-

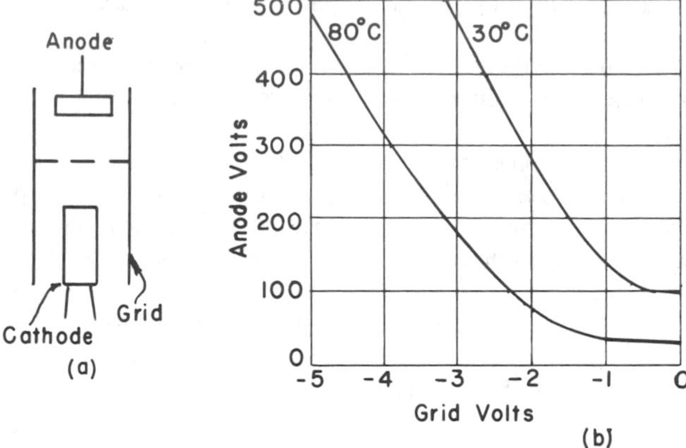

FIG. 7. (a) Cross section of a negative-grid thyratron; (b) critical grid characteristics for a thyratron.

ness but have no effect on the current to the anode. Reduction of the anode potential to zero or a negative value must occur to stop the current. Positive ion current to the grid should be limited by resistance.

The time of deionization was typically 10–1000 μs. This is the time for the grid sheath to dissipate and for the grid to regain control; this fixed an upper frequency limit for thyratron operation. Lowest deionization time was obtained with hydrogen as the gas.

See also THERMIONIC EMISSION; WORK FUNCTION.

BIBLIOGRAPHY

W. L. Chaffee, *Theory of Thermionic Vacuum Tubes*. McGraw-Hill, New York, 1933.

S. Dushman, "Thermionic Emission," *Rev. Mod. Phys.* **2**, 381 (1930).

A. L. Hughes and L. A. Dubridge, *Photoelectric Phenomena*. McGraw-Hill, New York, 1932.

A. W. Hull, "Hot Cathode Thyratrons," *Gen. Elec. Rev.* **32**, 213, 390 (1929).

A. W. Hull and N. H. Williams, "Characteristics of Shielded-Grid Pliotrons," *Phys. Rev.* **27**, 432 (1926).

J. D. Ryder, *Engineering Electronics*, 2nd ed. McGraw-Hill, New York, 1967.

F. E. Terman *et al.*, "Calculation and Design of Resistance-Coupled Amplifiers Using Pentode Tubes." *Trans. AIEE.* **59**, 879 (1940).

Electronics

James A. McCray and Wolfgang Nadler

From the physicist's standpoint electronics may be defined as the use of materials and devices which transport or store electrons for the purpose of recording, transmitting, manipulating, or storing physical information. The variables usually monitored as functions of time are voltage, current, and charge. In actual practice an arrangement of passive components, such as resistors, capacitors, or inductors, and active components, such as field effect transistors (FETs), tubes, or bipolar transistors, make up collectively an electronic circuit which is designed to perform a given function. This circuit may be a discrete component circuit, where the components are individually arranged on a circuit board, or it may be an integrated circuit (IC) where thousands of mainly active but also passive components are intimately arranged within a small silicon "chip." Another possibility is a hybrid combination where most of the circuitry is contained in an IC chip with just a few passive components "outboarded" around the chip. Changing these passive components alters the function of the circuit.

It is the responsibility of the physicist to understand the function of the circuit, that is, how the circuit modifies an input voltage or current, so that the physical information represented by these variables is changed in a known manner.

A very important point that must be remembered is that the function of a circuit depends on the relative values of the parameters of the circuit (for example, *RC* time constants) and the values of the incident waveform (for example, the rise time of a pulse).

Since electronic circuits may be situated at some distance from one another, they must be connected together by transmission lines. Therefore, the properties of the transmission line, such as transit time and characteristic impedance, reflections at its terminations, and voltage and current traveling waves on the line must be understood if the transmitted physical information is not to be distorted.

Electronic circuits may be classified as analog, where for example the magnitude of the information variable is monitored directly, or digital, where the information is contained in a sequence of binary states. A further classification of electronic circuits is to separate them into linear and nonlinear circuits. If the coupled differential equations which describe the variation with time of the relevant charges involved in a circuit are linear in the charges and their time derivatives, then the circuit is described as a linear circuit. If there are any nonlinear terms in the differential equations, then the circuit is called nonlinear.

A given configuration of ideal resistors, capacitors, and inductors in addition to an ideal voltage source constitute an electronic circuit and may be represented by a drawing which is called a circuit diagram. An example of such a diagram that represents an integrating circuit is given in Fig. 1. The physical laws governing the operation of electronic circuits are Maxwell's equations and the equation of con-

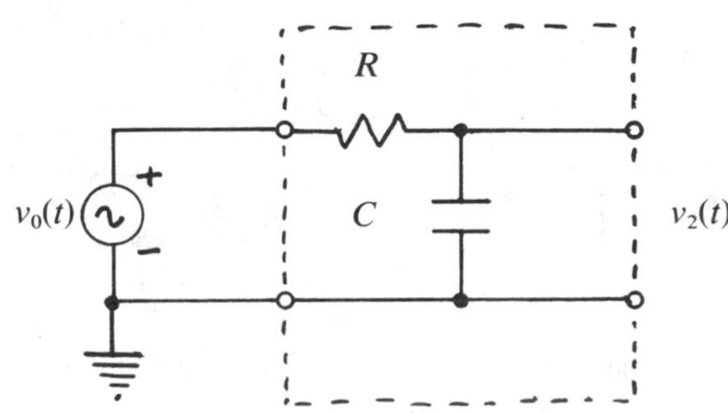

FIG. 1. *RC* integrating circuit.

tinuity. A discussion of these laws and conservation of charge and conservation of energy leads to Kirchhoff's laws.

In order to determine the function of each linear circuit, either experimentally or theoretically, the response, i.e., output voltage or current, is obtained for two standard types of input. The steady-state, frequency-dependent properties of a circuit are obtained by using a sinusoidal input voltage or current. For Fig. 1 the input voltage would be $v_0(t) = A\sin(\omega t + \theta)$, where A, ω, and θ are the amplitude, angular frequency, and phase, respectively, of the sinusoidal input. Instead of using Kirchhoff's laws directly and then solving the differential equations it is easier to use the phasor technique. The properties of the circuit to be determined are the amplitude frequency response (relative *gain* or *attenuation* as a function of frequency) and the frequency-dependent phase response (*phase shift*). Since only the induced phase shift of the circuit is desired, the circuits are tested with the input $v_0(t) = A\sin \omega t$ which is mathematically equal to the imaginary part of the complex function $\hat{v}_0 = Ae^{j\omega t}$, where $j = \sqrt{-1}$. This latter form is called a *phasor* and, when the phasor form of current is used, $\hat{\imath} = Be^{j\omega t}$, the voltage drops given above for resistance, capacitance, and inductance become $\hat{v} = R\hat{\imath}$, $\hat{v} = \hat{\imath}/j\omega C$, and $\hat{v} = j\omega L\hat{\imath}$. These all have the form of Ohm's law if R, $1/j\omega C$, and $j\omega L$ are taken as generalized resistances (impedances). Kirchhoff's laws then become just algebraic equations with complex numbers and the analysis reduces to that for dc circuits. For the integrating circuit shown in Fig. 1 the output voltage phasor $\hat{v}_2$ may be easily found by using the voltage divider ratio. Thus $\hat{v}_2 = \{[1/j\omega C]/[R + 1/j\omega C]\}\hat{v}_0$. The ratio of the two complex numbers $\hat{v}_2/\hat{v}_0$ is called the voltage transfer function $\hat{T}$ and can be written in polar form: $\hat{T}(\omega) = M(\omega)e^{j\phi(\omega)}$. $M(\omega)$ is the amplitude frequency response and $\phi(\omega)$ is the phase response of the circuit. For the circuit of Fig. 1 we have $\hat{T}(\omega) = 1/(1 + j\omega\tau)$, where $\tau = RC$ is called the time constant of the circuit. In polar form this complex number is $\hat{T}(\omega) = [1/(1 + \omega^2\tau^2)^{1/2}] \exp(-j\tan^{-1}\omega\tau)$, so that the amplitude response is $M(\omega) = 1/(1 + \omega^2\tau^2)^{1/2}$ and the phase response is $\phi(\omega) = -\tan^{-1}\omega\tau$. These results are shown in Fig. 2 and illustrate that the circuit can also be called a *low-pass filter* since high frequencies are attenuated, or a lag circuit since the phase shift is negative. When $\omega = 1/\tau$, the amplitude has been decreased by $1/\sqrt{2} = 0.707$ and the power has been decreased by $\frac{1}{2}$. Thus this frequency is a measure of the pass band of the filter and is called the upper half-power point. In terms of decibels the power has fallen off 3 dB.

The transient voltage response of a linear circuit is obtained by using a voltage step input, which is zero up to a specified time origin and then ideally jumps to a value V and remains constant thereafter:

$$v_0(t) = \begin{cases} 0, & t<0 \\ V, & 0 \le t \end{cases}$$

The transient response may be readily found if Laplace transforms are used. For Fig. 1 the result is $v_2(t) = V(1 - e^{-t/\tau})$, which is an exponential rise. Figure 3 illustrates this basic property of electronic circuits. All circuits have some shunt capacitance to ground which must be charged, and, hence, there will be a certain amount of time necessary for the circuit to respond to a step input and this will depend on the capacitances and resistances involved. A measure of this response time is the 10–90% rise time shown in Fig. 3. The rise time depends on the time constant of the circuit, and it is easily shown that $T_R \simeq 2.2\tau$. This result may be combined with the upper half-power-point frequency for this circuit, $f_2 = 1/2\pi\tau$, to yield the relation $f_2T_R \simeq \frac{1}{3}$. This expression is particularly useful since most wide-band amplifier circuits have this circuit as a high-frequency approximation. Thus if the frequency band pass of an amplifier, f_2, is given, then the rise time T_R may be easily determined. The frequency spectrum of noise usually encountered is so-called "white noise" and is uniform with frequency. Thus one would like to minimize the electronic bandwidth used. However, if too small a bandwidth is used, then the rise time of the amplifier $T_R \simeq \frac{1}{3}f_2$ will be too long to measure a desired signal rise time or initial round-off of the signal will result. Thus one must adjust the bandwidth in order to optimize the signal-to-noise ratio.

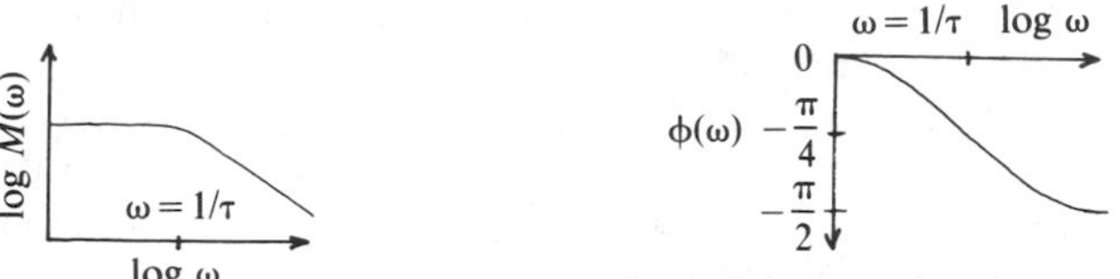

FIG. 2. Amplitude and frequency responses of the RC integrating circuit.

ACTIVE DEVICES

One of the most important features of some electronic circuits is that they are capable of amplification. If an external source of energy is applied to a circuit containing an active device such as a tube, FET, or bipolar transistor, then it is possible to arrange that a low voltage or current control the flow of a large current through the active device. Semiconductor devices are the most important of these. One of the basic features of active semiconductor devices is the p-n junction. This device is produced when a silicon crystal is grown and the doping is changed at a given point from n type to p type, as shown in Fig. 4.

On the n-type side there is effectively an electron "gas" which can diffuse into the p-type side leaving behind at the junction a layer of bound positive charges. Likewise on the p-type side there is effectively a positive "hole gas" which can diffuse into the n-type region leaving behind a layer of negative charges. The dipole layer electric field created causes Ohmic current in the opposite direction to that of the diffusion current. At steady state the currents just balance and a region, called the depletion region, has been estab-

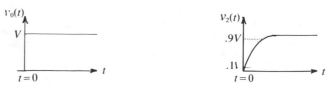

FIG. 3. Response of circuit to step input (rise time).

lished where the potential changes and an electric field exists. This is shown by the middle curve of Fig. 4. If an external voltage is applied across the *p-n* junction so that the positive side is connected to the *n*-type material, then the electric field adds to that already there, resulting in an increased potential barrier and a larger depletion width. It is difficult then for current—conventional positive current—to flow from *p*-type to *n*-type material, and the junction is said to be back-biased. However, if the polarity of the external voltage is reversed, then the additional electric field at the junction subtracts from that already there thus resulting in a lower potential barrier. Current then may more easily flow through the device, and the junction is said to be forward biased. This voltage-dependent directionality of current flow allows rectification of ac currents to dc currents, so the device is called a rectifier or diode and is indicated by the symbol shown in Fig. 4.

p-n junctions may be combined in various ways to form, for example, a voltage-dependent device, the field-effect transistor (FET), or a current-dependent device, the bipolar transistor. These devices are shown in Fig. 5 and both are capable of amplification.

In the FET case a small voltage change at the gate (*G*) produces a large change in channel current from source (*S*) to drain (*D*) by changing the size of the opposite depletion layers. For the bipolar transistor shown a small change in base (*B*) current produces a large change in emitter (*E*) to collector (*C*) current by varying the base–emitter potential barrier. The emitter and collector region are heavily doped while the base region is very thin and lightly doped. Here, again, one can achieve an amplification, in this case, of current.

Feedback is a very important concept in electronics. A voltage, for example, which is proportional to the output voltage of a circuit, is introduced in either a negative or positive sense into the input circuit of an amplifier. Negative voltage feedback is used for stability and positive voltage feedback is used to make oscillator circuits. In modern circuits an integrated-circuit differential amplifer with high gain, high input impedance, and low output impedance is used to form various functional linear circuits. Various passive components, such as resistors, capacitors, and inductors, are "outboarded" around the so-called operational amplifier (op-amp) and the function of the overall circuit depends on the type and arrangement of these components. For example, application of Kirchhoff's laws to the circuit shown in Fig. 6 indicates that this circuit is a constant-amplitude variable 180-deg phase shifter with voltage transfer function $\hat{T}(\omega)=e^{j(-2\tan^{-1}\omega\tau)}$, where $\tau=RC$. The voltage characteristics of the op-amp (triangular symbol) are represented by $\hat{V}_2=\hat{A}(\hat{V}_+-\hat{V}_-)$, where $\hat{V}_-$ and $\hat{V}_+$ are measured with respect to ground. These circuits are called operational amplifier circuits because various analog operations may be performed such as differentiation, integration, summing, etc.

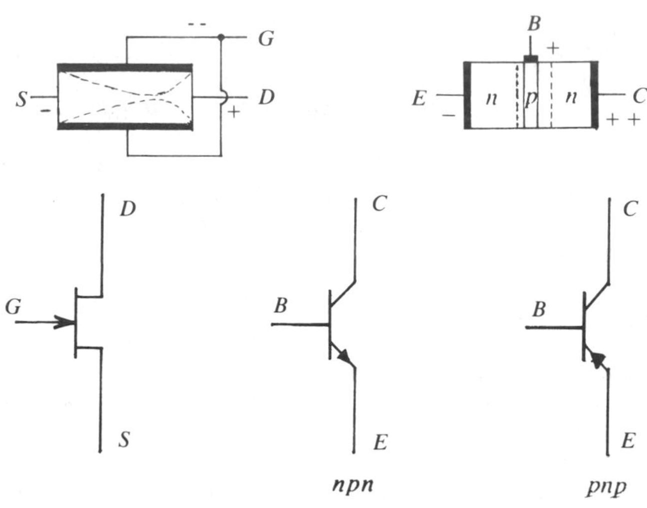

FIG. 5. *n*-channel field effect transistor (FET) and bipolar transistor.

An interesting application of these circuits is to solve nonlinear differential equations, for example, the driven Duffing oscillator equation. Nonlinear effects such as period-doubling leading to chaos can be demonstrated. In fact, it is possible to see and even hear such effects with a simple driven *L–R* diode circuit where the diode is a nonlinear capacitor.

DIGITAL ELECTRONICS

As can be seen from the previous discussion in this article the fidelity of information transfer of analog signals greatly depends on the frequency and phase responses of the circuits and transmission lines through which these signals pass. For many applications this possible distortion of data is intolerable. It is also possible to perform analytical computation with analog circuits and many of the first computers were analog computers, but the accuracy, reliability and possible memory storage were not considered satisfactory.

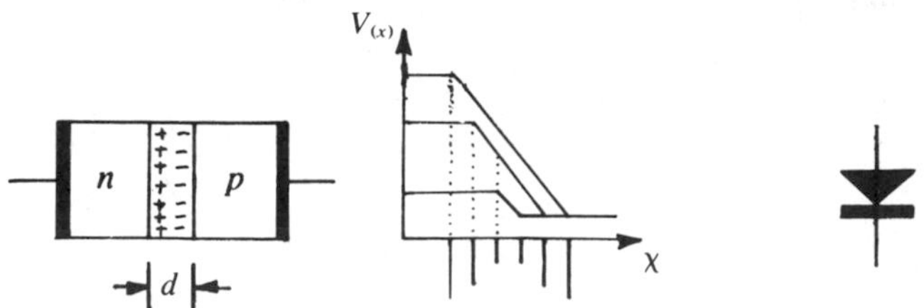

FIG. 4. *p-n* junction diode.

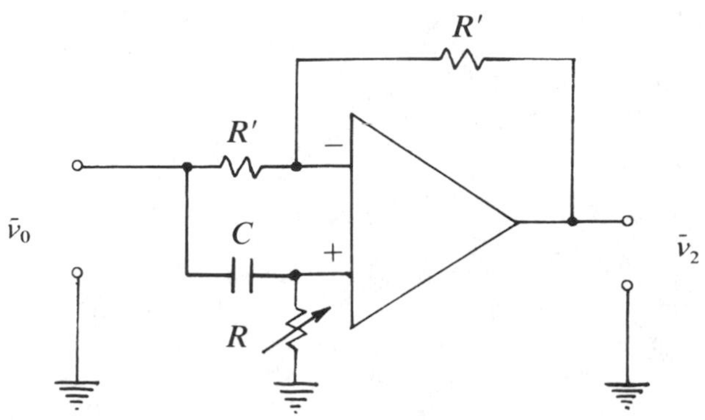

FIG. 6. Variable constant-amplitude op-amp phase shifter.

An alternative approach to data transfer and computation is based upon the binary system of counting. Instead of using the base 10 (decimal system) where a number is expressed in terms of powers of 10 [for example, the number $35486 = N_4N_3N_2N_1N_0$ is $N = N_4 \times 10^4 + N_3 \times 10^3 + N_2 \times 10^2 + N_1 \times 10^1 + N_0 \times 10^0$, where $10^0 = 1$, $10^1 = 10$, $10^2 = 100$, etc.] the number is expressed instead in terms of base 2. As an example, the number $38_{10} = 3 \times 10^1 + 8 \times 10^0$ in base 10 would become $100110_2 = 1 \times 2^5 + 0 \times 2^4 + 0 \times 2^3 + 1 \times 2^2 + 1 \times 2^1 + 0 \times 2^0$ in base 2. The general form is $N = \cdots B_5B_4B_3B_2B_1B_0 = \cdots + B_52^5 + B_42^4 + B_32^3 + B_22^2 + B_12^1 + B_02^0$. The reason for the choice of the binary system is clear when it is realized that to represent one binary digit (bit) a device with only two states (0 or 1) is needed while to represent one decimal digit a device with 10 states (0,1,2,3,4,5,6,7,8,9) would be necessary. There are many electrical two-state devices that can be used for binary representation such as switches, relays, bipolar transistor binaries (flip-flops), MOSFET (metal oxide semiconductor field effect transistor) binaries, magnetic cores, magnetic domains on tapes, disks, or bubbles, etc. Binary numbers can then be stored in a linear sequence of such devices, called a register, and can also be added and subtracted, thus making possible all of the higher mathematical operations such as multiplication, division, differentiation, integration, etc. This is done by using repeated addition and data transfer operations which are expressed in a computer program and stored in the memory of a computer.

Since our analytical thinking processes are based on a two-state logic system (true or false), two-state electronic devices are very readily arranged so that various logical operations can be represented by circuits such as the NOT, AND, NAND, OR, NOR, and exclusive OR circuits. These circuits (gates) and various combinations are available in integrated circuit chips. The packing density of gates on a chip is indicated by the terminology SSI (small-scale integration, < 10 gates), MSI (medium-scale integration, < 100 gates), LSI (large-scale integration, ~ 1000 gates) and VLSI (very large-scale integration, > 1000 gates). The type of two-state devices used to make up the logic circuits gives rise to various logic families such as CMOS (complementary metal oxide semiconductor logic—low power consumption), ECL (emitter-coupled logic—high speed), TTL (transistor–transistor logic), and I^2L (integrated injection logic). Computer chip sets made up of the central microprocessor chip and the coprocessor (math processor) chip such as the Motorola 68020/68881, the Intel 80386/80387, and the Digital Equipment Corporation 78032/78132 chip sets form, when combined with ROMs (read-only memory chips) and RAMs (random-access memory chips) along with several other chips necessary for proper matching of circuits, the basis for the Apple MAC II computer, the IBM AT computer, and the DEC Micro-Vax computer, respectively. The new supercomputers connect miniaturized central microprocessor chips in parallel with very short interconnections in order to achieve a higher through-put and greater overall speed of calculation. High density requires special cooling techniques; for example, the Cray-3 supercomputer is cooled by an inert fluorocarbon liquid.

The general philosophy now would be to obtain analog information from a transducer at an experimental site, then convert this information into digital form as soon as possible with an A/D converter (analog-to-digital), store this information in a digital memory, and subsequently process this information in a digital computer. There is, however, one disadvantage in handling information in digital form. If only one bit of information is lost, then the total information may be incorrect. Modern semiconductor and solid-state technology and circuit-design philosophy have been so successful that device and circuit reliability can be excellent even for very large computers.

Finally it should be noted that all types of functional circuits such as optical couplers, LEDs (light emitting diodes), PLLs (phase-locked-loops), CCDs (charge-coupled devices), sample and hold circuits, etc. have been designed and produced on single chips. Thus before designing or building any circuit it is advisable to check various manufacturer's data and application manuals (see Bibliography) to see if that particular desired function can be obtained with the use of one or two integrated circuit chips.

See also CIRCUITS, INTEGRATED; ELECTRON TUBES; NETWORK THEORY: ANALYSIS AND SYNTHESIS; SEMICONDUCTORS, CRYSTALLINE; TRANSISTORS; TRANSMISSION LINES AND ANTENNAS.

BIBLIOGRAPHY

General Electronics

J. J. Brophy, *Basic Electronics for Scientists*, 4th ed. McGraw-Hill, New York. 1983. (E)

L. O. Chua and R. N. Madan, "Sights and Sounds of Chaos," *IEEE Circuits and Devices Magazine*, January 1988, and references therein. (A)

C. F. G. Delaney, *Electronics for the Physicist with Applications*. Halsted Press, New York, 1980. (I)

A. J. Diefenderfer, *Principles of Electronic Instrumentation*. Saunders College Publishing, Orlando, FL, 1979. (E)

T. M. Frederiksen, *Intuitive Operational Amplifiers*. McGraw-Hill, New York, 1988. (E)

P. Horowitz and W. Hill, *The Art of Electronics*. Cambridge University Press, Cambridge, 1987. (I)

W. G. Jung, *IC Op-Amp Cookbook*, 3rd ed. Howard W. Sams & Co., Indianapolis, 1986. (E)

D. Lancaster, *Active-Filter Cookbook*. Howard W. Sams & Co., Inc., New York, 1988. (E)

J. W. Nilsson, *Electronic Circuits*. Addison-Wesley, Reading, MA, 1985. (I)

R. F. Pierret and G. W. Neudeck, eds., *Modular Series on Solid State Devices*. Addison-Wesley, Reading, MA, 1989. (A)

M. J. Sanfilippo, *Solid-State Electronics Theory—with Experiments*. Tab Books Inc., Blue Ridge Summit, PA, (E)

M. R. Spiegel, *Theory and Problems of Laplace Transforms*. McGraw-Hill, New York, 1965. (I)

W. T. Thomson, *Laplace Transformation*. Prentice-Hall, Englewood Cliffs, N.J., 1962. (A)

T. H. Wilmshurst, *Signal Recovery from Noise in Electronic Instrumentation*. Adam Hilger Ltd., Accord, MA, 1985. (I)

Digital Electronics

P. Antognetti and G. Massobrio, *Semiconductor Device Modeling with Spice*. McGraw-Hill, New York, 1988. (A)

M. Bird and R. Schmidt, *Practical Digital Electronics* (laboratory workbook). Hewlett-Packard Co., Santa Clara, CA, 1974. (I)

J. Blukis and M. Baker, *Practical Digital Electronics* (textbook). Hewlett-Packard Co., Santa Clara, CA, 1974. (I)

P. Burger, *Digital Design—A Practical Course*. Wiley, New York, 1988. (E)

L. A. Glaser and D. W. Dobberpuhl, *The Design and Analysis of VLSI Circuits*. Addison-Wesley, Reading, MA, 1985. (I)

R. L. Goodstein, *Boolean Algebra*. Pergamon Press, London, 1963. (A)

Handbooks, Dictionaries and Tables

S. Gibilisco, ed., *Encyclopedia of Electronics*. Tab Books, Blue Ridge Summit, PA, 1985.

R. F. Graf, *Modern Dictionary of Electronics,* 6th ed. Howard W. Sams & Co., New York, 1984.

M. Kaufman and A. A. Seidman, *Handbook for Electronics Engineering Technicians*. McGraw-Hill, New York, 1976.

J. Marcus and C. Weston, *Essential Circuits Reference Guide*. McGraw-Hill, New York, 1988.

P. A. McCollum and B. F. Brown, *Laplace Transform Tables and Theorems*. Holt, Rinehart and Winston, New York, 1965.

The Radio Amateur's Handbook, published annually by the Headquarters Staff of the American Radio Relay League, Newington, CT.

Magazines and Newspapers

BYTE, A McGraw-Hill publication, BYTE Subscriptions, P.O. Box 551, Hightstown, NJ.

EDN, A Cahners Publication, P.O. Box 5563, Denver, CO.

Electronic Design, VNU Business Publications, Inc., 10 Holland Drive, Hasbrouck Heights, NJ.

Electronic Engineering, Morgan-Grampian House, 30 Calderwood St., Woolwich, London.

EE Product News, P.O. Box 12973, Overland Park, KS.

Electronic Products, Hearst Business Communications, Inc./UTP Division, 645 Stewart Ave. Garden City, NY.

Electronics, VNU Business Publications, Inc., 10 Holland Drive, Hasbrouck Heights, NJ.

HP benchbriefs, Hewlett-Packard Co., 1820 Embarcadero Road, Palo Alto, CA.

Hewlett-Packard Journal, 3200 Hillview Avenue, Palo Alto, CA.

Journal of Physics E: Scientific Instruments, IOP Publishing Ltd., 7 Great Western Way, Bristol.

Nuclear Instruments and Methods in Research A, Elsevier Science Publishers B. V., P.O. Box 211, 1000 AE Amsterdam.

Personal Engineering, Box 182, Brookline, MA.

Review of Scientific Instruments, American Institute of Physics, 335 East 45th St., New York, NY.

Manuals

A great deal of use information may also be obtained from manufacturer's data and application manuals and books from companies such as American Telephone and Telegraph, General Electric, General Instruments, Harris, Intel, Motorola, National Semiconductor, Radio Corporation of America, Signetics, Texas Instruments, TRW, and many others. *IC Master,* Hearst Business Communications, 645 Stewart Ave., Garden City, NY 11530, is quite useful. Information about electronics manufacturers and distributors may be found in sources such as *Electronics Buyer's Guide,* published annually by VNU Business Publications, Inc., 10 Holland Drive, Hasbrouck Heights, NJ 07604; *Electronic Design's Gold Book,* published annually by Hayden Publishing Co., Inc., 10 Holland Dr., Hasbrouck Heights, NJ 07604; and *Electronic Engineers Master,* published semiannually by Hearst Business Communications, Inc., 645 Stewart Avenue, Garden City, NY 11530.

Electro-optical Effects *see* Kerr Effect, Electro-optical

Electrophoresis

Andreas Chrambach

Electrophoresis, the migration of charged particles in an electric field, is a method for separating such particles. In most biological applications, the separation proceeds in aqueous buffers, within 3 to 4 pH units on either side of neutrality, at low (0.01 to 0.03 M) ionic strength and in the presence of an inert polymer network (gel). According to the most common gel types used, the separation method is designated polyacrylamide or agarose gel electrophoresis [1,2].

For separation of particles differing in size but sharing a common surface net charge density, such as nucleic acids or protein subunits derivatized with sodium dodecyl sulfate (SDS-proteins), separation occurs at gel concentrations at which electrophoretic mobility is inversely related to molecular size (i.e., by "molecular sieving"). At any of those gel concentrations, separated species can be characterized by their migration distances which are related to size, commonly expressed as molecular weights (kDa) or basepairs in the case of DNA and numerically evaluated in reference to size standards [3].

For particles differing from one another both in size and in surface net charge density (related to isoelectric point), electrophoretic mobility is a function of both. Separation between such particles is most efficiently carried out by molecular sieving (as above) if size differences predominate. However, in that case molecular weight is related to the rate of change of electrophoretic mobility with gel concentration, rather than the mobility at any one gel concentration. That rate is expressed as the slope of the plot of log(mobility) versus gel concentration ("Ferguson plot"), designated as the retardation coefficient. The "free electrophoretic mobility," related to net charge density, is given by the intercept of the Ferguson plot with the mobility axis. Gel electrophoresis, therefore, allows one to evaluate separately the two elements of electrophoretic mobility, size and net charge, and thus to define the two most important properties of macromolecules [4].

If, in a particular separation problem, the sizes of all components are similar and differences are preponderantly based on surface net charge densities, the most efficient separation is that on the basis of free mobilities. The gel in those separations serves as an anticonvective medium, not a means of sieving. One applicable technique in such cases is gel electrophoresis in a pH gradient ("isoelectric focusing") [5], provided that particles carry both positively and negatively charged groups and are therefore reduced to zero mobility (and arrest of migration) at a pH at which the two balance one another (isoelectric point). Separation under those circumstances leads to an alignment of particles separated in the order of their isoelectric points.

Two-dimensional electrophoresis of proteins (2-D gel electrophoresis) [6] achieves a consecutive separation based on net charge differences and a separation based on size differences by combining an isoelectric focusing gel in the first dimension with a polyacrylamide gel separation in SDS-containing buffer in the second dimension. This technique is important for the analysis of multicomponent protein systems. Each separated component on a 2-D gel after application of a detection procedure gives rise to a spot defined by coordinates of size and isoelectric *p*H. On appropriately large gels, thousands of such separated spots can provide a protein map of the system.

Another important application of nonsieving, anticonvective gels is in the pulsed field gel (PFG) electrophoresis of large DNA [7]. DNA larger than 2×10^4 bp is stretched on agarose gel electrophoresis, migrates by "reptation" at a rate independent of size, or is entangled in the gel network and does not migrate at all. Pulsing of the electric field and variation of the direction of the field during electrophoresis restores both the migration of entangled species and the separation by size. Using PFG electrophoresis, chromosomal-size DNA up to sizes higher than 10 megabasepair has been separated.

Electrophoresis can be applied preparatively. At the microgram and milligram scale, gel electrophoretic separation, sectioning of the gel, and electrophoretic extraction are applicable [8]. At a larger scale, free-flow electrophoretic separation techniques [9] are available, but necessarily suffer in resolving power for being free-mobility separations in the absence of molecular sieving effects in those applications where size and shape differences between species predominate, e.g., in the bulk separation of whole cells [10].

See also RHEOLOGY; ELECTROPHOTOGRAPHY.

REFERENCES

1. A. T. Andrews, *Electrophoresis*. Clarendon Press, Oxford, 1986. (I)
2. R. C. Allen, C. A. Saravis, and H. R. Maurer, *Gel Electrophoresis and Isoelectric Focusing of Proteins*. de Gruyter, Berlin and New York, 1984. (A)
3. D. Rodbard, "Estimation of Molecular Weight by Gel Filtration and Gel Electrophoresis," in *Method of Protein Separation, Vol. 2*, N. Catsimpoolas (ed.). Plenum Press, New York, 1976. (A)
4. A. Chrambach, *The Practice of Quantitative Gel Electrophoresis* (V. Neuhoff and A. Maelicke, eds.). Verlag Chemie, Weinheim, 1985. (I)
5. P. G. Righetti, *Isoelectric Focusing: Theory, Methodology and Applications*. Elsevier, Amsterdam, 1983. (A)
6. M. J. Dunn, "Two-dimensional polyacrylamide gel electrophoresis," in *Advances in Electrophoresis, Vol. 1* (A. Chrambach, M. J. Dunn and B. J. Radola, eds.). Verlag Chemie, Weinheim, 1987. (A)
7. M. V. Olson, "Pulsed-Field Gel Electrophoresis," in *Genetic Engineering* (J. K. Setlow, ed.). Plenum Press, New York, 1989. (A)
8. R. Horuk, "Preparative polyacrylamide gel electrophoresis of proteins," in *Advances in Electrophoresis, Vol. 1* (A. Chrambach, M. M. Dunn, and B. J. Radola, eds.). Verlag Chemie, Weinheim, 1987. (A)
9. M. Bier, "Effective principles for scaling-up of electrophoresis," in *Frontiers in Bioprocessing* (S. K. Sikdar, M. Bier, and P. Todd, eds.). CRC Press, Boca Raton, FL, 1990. (A)
10. A. W. Preece and K. A. Brown, "Recent Trends in Particle Electrophoresis," in *Advances in Electrophoresis, Vol. 3* (A. Chrambach, M. J. Dunn, and B. J. Radola, eds.). Verlag Chemie, Weinheim, 1989. (I)

Electrophotography

L. B. Schein

Electrophotographic (also called xerographic) printers and copiers are based on two well-known but not well-understood physical phenomena: electrostatic charging and photoconductivity in insulators. It is a tribute to the genius of Chester F. Carlson, the inventor of electrophotography, that he was able in 1938 to combine such little-understood physical phenomena into a process that is now at the heart of a rapidly growing, $20 billion industry (the estimated U.S. market in 1988). Commercial products span the range from low-cost personal copiers and laser printers that produce six pages per minute to high-speed printers that produce up to 220 pages per minute. Laser printers are expected to be the dominant printer technology of the 1990s because they are much quieter and faster than impact printers, and can print multiple type fonts and pictures in black and white and in color.

The process of electrophotography is shown schematically in Fig. 1. It is a complex process involving six distinct steps in most cases:

Charge. An electrical corona discharge caused by air breakdown uniformly charges the surface of the pho-

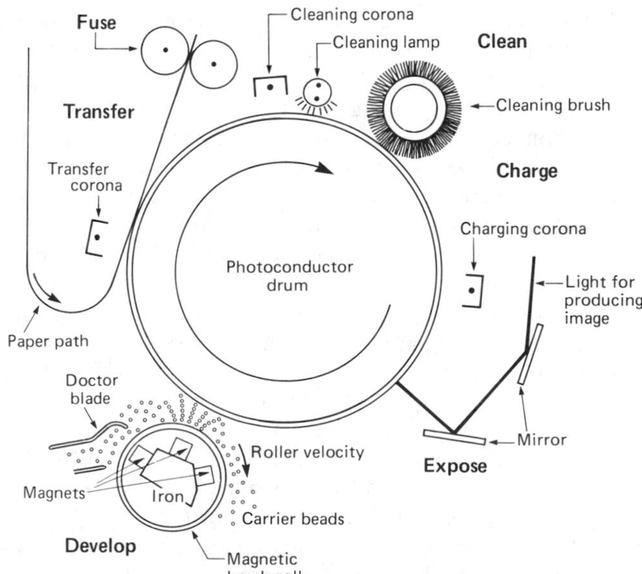

FIG. 1. Electrophotography can be separated into six steps: charge, expose, develop, transfer, fuse, and clean. The diagram locates the steps schematically in a commercial copier.

toconductor, which, in the absence of light, is an insulator.

Expose. Light, reflected from the image (in a copier) or produced by a laser (in a printer), discharges the normally insulating photoconductor producing a latent image—a charge pattern on the photoconductor that mirrors the information to be transformed into the real image.

Develop. Electrostatically charged and pigmented polymer particles called toner, ≈10 microns in diameter, are brought into the vicinity of the latent image. By virtue of the electric field created by the charges on the photoconductor, the toner adheres to the latent image, transforming it into a real image.

Transfer. The developed toner on the photoconductor is transferred to paper by corona charging the back of the paper with a charge opposite to that of the toner particles.

Fuse. The image is permanently fixed to the paper by melting the toner into the paper surface.

Clean. The photoconductor is discharged and cleaned of any excess toner using coronas, lamps, brushes or scraper blades.

Discussed below are the physical phenomena (1) occurring in the photoconductor (during the expose step), (2) associated with the development step, and (3) related to toner charging.

Much of the scientific literature of electrophotography has focused on the microscopic processes occurring inside photoconductors. All photoconductors, either inorganic (amorphous selenium alloys) or organic (molecularly doped polymers), appear to have similar photogeneration and charge-transport characteristics. Photogeneration is usually modeled as the escape, by random diffusion, of a charge carrier from a Coulomb well, whose source is the countercharge. The escape process is aided by temperature and the electric field, which lowers the well height. Charge transport is usually pictured as a hopping process with the holes (or electrons) hopping among dopant molecules (in the organic photoconductor). Recent work has suggested that polaron formation and hopping may be occurring. However, the electric field dependence of the mobility remains puzzling. A remarkable feature of photoconductors involves trapped charges. After repeated charging and discharging, charge may become trapped in the photoconductor which cannot be dissipated by light. The amount of trapped charge must remain small compared to the charges associated with the latent image. It may be shown that very low levels of traps, 5×10^{13} cm^{-3}, or 1 molecule in 10^8, will produce a useless photoreceptor. That any such material exists in nature is remarkable; that it is a glassy or amorphous material is even more surprising given the discussions of bandtail states that are postulated to exist in amorphous materials.

One important example of a development system is the magnetic brush development system, which is used today in almost all copiers and printers operating above 30 cpm, and is shown schematically in Fig. 1. In this system, the toner (approximately 10 μm in diameter) is mixed with carrier beads (200 μm in diameter). The materials are chosen such that the mixing causes the toner and carrier particles to become oppositely charged, as described below, thereby causing the toner to adhere electrostatically to the carrier particles. The carrier beads are made from a soft magnetic material so that they form magnetic "brushes" on a roller that carries them by frictional and magnetic forces past stationary magnets into the development zone. Here, in response to the electric field of the latent image, the toner transfers to the photoconductor from the carrier particles at the end of the brushes.

The conditions for toner development depend upon the type of image and development system. For solid areas or lines, in the development system shown in Fig. 1, toner develops until the local electric field goes to zero. The local electric field is determined primarily by two terms, one due to the latent image and one due to charge buildup on the carrier particles adjacent to the photoconductor as toner develops.

An evolution of development systems was introduced in 1980. This approach, called monocomponent development, eliminates the need for carrier particles. It is therefore smaller and lower cost, important characteristics for personal copiers and printers. The physical principles behind their development characteristics are similar to those discussed above.

In discussing development, we assumed that the toner is charged. The electrostatic charge exchange between the toner and carrier surfaces (triboelectrification) is key to successful image formation. This occurs by the mixing action in the hopper upstream of the developer. The phenomenon is familiar to anyone who has touched grounded metal after walking across a rug in a dry room or whose hair clings to a comb in the winter.

Improving development requires control of the toner charge. Knowledge of the physics of contact electrification is crucial to this effort, yet the physics of electrification of insulators is very poorly understood. At present, the most is known, both experimentally and theoretically, about contact electrification for metal–metal contacts, and successively less for metal–insulator and insulator–insulator contacts. In electrophotography we are interested primarily in insulator–insulator electrification.

A straightforward model for metal–metal contact electrification involves a simple equilibration of Fermi levels. As the metals are separated, the levels remain in equilibrium out to a distance of 10 Å, where the tunneling currents vanish. Quite reasonable semiquantitative agreement between theory and experiment has been achieved.

For metal–insulator or insulator–insulator triboelectrification, experiments between laboratories do not appear to be reproducible and theoretical uncertainties exist concerning the mobility of charges in the insulators and the applicability of the concept of a Fermi level when excess charge can be trapped in nonequilibrium situations for longtimes. Attempts to arrange insulators (along with metals) in an empirical triboelectric series such that materials higher up in the series become positively charged on contact with materials lower down in the series have had limited success. Unfortunately, there is not general agreement on the ordering of insulators and in some cases the series appears to be a ring.

The physics of electrostatic charge exchange remains one

of the least understood branches of solid-state physics. Part of the reason is the difficulty of performing definitive experiments in this area. When the surfaces of two materials are brought into contact and separated, the actual area which made contact is difficult to determine. Whether pure contact or friction is required has not been determined. The precise nature of the surfaces are usually not well defined: dust particles, surface contaminants, and even water layers may be the "surface." Even for "clean" surfaces the nature of intrinsic and extrinsic surface electron states on insulators is not well understood. The magnitude of return currents during separation remains controversial. Finally, the number of surface molecules involved in the charging process is extremely small, on the order of one molecule in 10^5. Nevertheless, progress in understanding insulator–insulator charging would be highly useful, not only for electrophotography but also for a variety of processes that involve electrostatic charging, such as dust precipitation, spray painting, and reduction of sparking.

See also CORONA DISCHARGE; ELECTROPHORESIS; PHOTOCONDUCTIVITY; TRIBOLOGY.

BIBLIOGRAPHY

General Sources

L. B. Schein, *Electrophotography and Development Physics.* Springer-Verlag, New York, 1988.

E. M. Williams, *The Physics and Technology of Xerographic Processes.* Wiley, New York, 1984.

R. M. Schaffert, *Electrophotography.* Focal, New York, 1980.

D. M. Burland and L. B. Schein, *Phys. Today* **39**(5), 46 (1986).

Physics of Electrophotography

J. Mort and D. M. Pai (eds.), *Photoconductivity and Related Phenomena.* Elsevier, Amsterdam, 1976.

J. Mort and G. Pfister (eds.), *Electronic Properties of Polymers.* Wiley, New York, 1982.

L. B. Schein, A. Peled, and D. Glatz, *J. Appl. Phys.* **66,** 686 (1989).

D. M. Pai, *J. Non-Cryst. Solids* **60,** 1255 (1983).

L. B. Schein, K. J. Fowler, G. Marshall, and V. Ting, *J. Imaging Technol.* **13,** 60 (1987).

J. Lowell and A. C. Rose-Inner, *Adv. Phys.* **29,** 1947 (1980).

W. R. Harper, *Contact and Frictional Electrification.* Oxford University Press, Oxford, 1967.

D. A. Sennor, in *Physiochemical Aspects of Polymer Surfaces,* Mittal (ed.), vol. 1, p. 477, Plenum, New York, 1983.

Electrostatics

R. Casanova Alig

Electrostatics is concerned with the effects of positive and negative charges. The fundamental charges are the electron and the proton. They are equal in magnitude. The electron is negative, the proton is positive. Like charges repel; unlike charges attract.

The basic electrostatic law is Coulomb's law: Two electric charges attract or repel each other with a force that is proportional to the product of the charges and that varies inversely with the square of the distance between them.

The unit charge, the coulomb (C), consists of 0.6242×10^{19} electrons or protons. In a vacuum a charge of 1 C repels a like charge, at a distance of 1 m, with a force of about 9×10^9 newtons, (N) or approximately 1 million tons. In physics and industry the charges encountered are typically vastly smaller than the coulomb.

All the Coulomb forces on a charged object define the electric field. The force is proportional to the charge on the object and to the field intensity at its location. Near the center of a parallel-plate capacitor, the field is uniform. At a separation of 1 cm and with 1 V applied, the central field intensity is 100 V/m.

When water is raised to a higher level against gravitational force, energy is stored as potential energy. Likewise, when a charged object is moved against the Coulomb force, it is raised to a higher electric potential, and energy is stored. The volt (V) is the unit of electric potential, measured in joules per coulomb.

The capacitance measures the capacity of a system to store electric energy. The farad (F) is the unit of capacitance, measured in coulombs per volt. The farad is far greater than the amount of capacitance normally encountered; e.g., the capacitance of a parallel-plate capacitor with area 1 m^2 and separation 1 cm is less than 10^{-9} F.

The electric field is related to the volume charge density $\rho(\mathbf{r})$ by the differential equation $\nabla^2\phi = \rho(\mathbf{r})$, where $\phi(\mathbf{r})$ is the electric potential and $\rho(\mathbf{r})$ is the volume charge density. This equation is called Poisson's equation, and when $\rho(\mathbf{r}) = 0$, it is called Laplace's equation. The old definition of electrostatics restricted it to the effects of fixed charges. More recently, dynamic phenomena are included in electrostatics, i.e., $\phi(\mathbf{r})$ and $\rho(\mathbf{r})$ are time dependent, provided the time variation is slow enough to ignore the other phenomena of electrodynamics.

Particles with unequal numbers of electrons and protons are said to be charged. Ions are charged atoms or molecules. Larger charged particles are formed when ions or electrons become attached to bits of solid or liquid matter. Most applications of electrostatics can be grouped into phenomena involving (1) the direction of electron and ion beams or (2) the transfer of larger charged particles.

Bulk matter may contain an internal fixed charge density. To describe the forces on charges inside the material, the charge density is divided by the dielectric constant of the material. The semiconductor *p-n* junction is an example of an internal fixed charge density.

SOURCES

Gaseous atoms and molecules are neutral. They can be ionized, i.e., separated into positive ions and electrons, by an intense nonuniform field, such as that near a wire or point. This is called a corona discharge. The electrons formed in this way join neutral molecules to form negative ions. Ions can also be formed by cosmic rays or radioactive sources. Solid and liquid particles can be charged through the exchange of electrons or ions on contact; this is frictional, or tribo, electricity. Historically this was the earliest experience with electricity; frequently the exchange was made by rubbing cat's fur on amber. Solids can emit electrons. Thermionic emission is the release of electrons by heating, and field emission is their release in an intense electric field.

ATMOSPHERIC ELECTROSTATICS

Cosmic rays and contact electrification give rise to myriad atmospheric electrostatic phenomena. In fair weather the atmosphere is positive, and the earth is negative, resulting in a downward electric field of about 100 V/m. Inside thunderheads the fields can be 10 000 V/m. Local corona discharges produce ions, and when a path of dense ion concentration develops, charge is rapidly transported to lower the field. In this way, lightning occurs in the atmosphere. On a smaller scale, sparks discharge the contact electrification that occurs in daily life, e.g., from walking on rugs. While these discharges from the body are generally only physically annoying, in some environments, such as petroleum tanks or grain elevators, such sparks can be quite dangerous. On a yet smaller scale, discharges of voltages 100 times below the threshold for human sensitivity can damage modern integrated circuits (ICs). People working with unshielded ICs must be grounded and exercise extreme caution to avoid sporadic, latent damage to the ICs.

ELECTRON AND ION BEAMS

A multitude of uses for electron and ion beams exist in research and industry. The most common commercial product is the cathode ray tube, used in television. These beams are focused and directed by electrostatic fields. The design of these fields is called electron optics, because they guide beams much as lenses and prisms guide light. Electron and ion microscopes reveal details extending down to atomic dimensions. Electron-beam lithography is used extensively in manufacturing ICs. In ion implantation, ion beams are used for the detailed control of material composition.

CHARGED PARTICLES

Many electrostatic applications use the charge on particles to move them. For example, in precipitation, ions from a corona charge airborne particles and move them to precipitator walls to be collected. In this way 20 million tons of fly ash are collected each year from coal-burning power stations. Electrostatic coating saves great quantities of paint: corona from the paint gun charges tiny particles; with the target grounded, the electric field guides them to it. Powder coating is a growing art: a refrigerator can be dry-powder coated and the coat later fused or baked on. Xerography, in copying machines, depends on a corona to charge a coated drum's surface and on the triboelectric effect to charge the toner. Charged-particle movement is basic to the manufacture, electrostatically, of a fifth of a billion dollar's worth of sandpaper and grit cloth per year. Separation of many mixtures is achieved electrostatically. In the mining industry, the machines of a single company electrostatically separate minerals from dross at the rate of 10 million tons per year.

See also CHARGED-PARTICLE OPTICS; ELECTRIC CHARGE; ELECTRODYNAMICS; ELECTRON AND ION BEAMS; ELECTRON BEAM TECHNOLOGY; ELECTRON MICROSCOPY; ELECTROPHOTOGRAPHY; FIELD THEORY, CLASSICAL; ELECTRODYNAMICS, CLASSICAL.

BIBLIOGRAPHY

A. D. Moore, ed., *Electrostatics and its Applications*. Wiley, New York, 1973.

J. M. Crowley, *Fundamentals of Applied Electrostatics*. Wiley, New York, 1986.

Elementary Particles in Physics

S. Gasiorowicz and P. Langacker

Elementary-particle physics deals with the fundamental constituents of matter and their interactions. In the past several decades an enormous amount of experimental information has been accumulated, and many patterns and systematic features have been observed. More recently, a highly successful mathematical theory of the electromagnetic and weak interactions has been devised and tested, and a promising candidate theory of the strong interactions has been developed. These theories, which are collectively known as the standard model, are most likely the correct description of Nature, to first approximation, down to a distance scale 1/1000th the size of the atomic nucleus. There are also speculative but encouraging developments in the attempt to unify these interactions into a simple underlying framework, and even to incorporate quantum gravity in a parameter-free "theory of everything." In this article we shall attempt to highlight the ways in which information has been organized, and to sketch the outlines of the standard model and its possible extensions.

CLASSIFICATION OF PARTICLES

The particles that have been identified in high-energy experiments fall into distinct classes. There are the *leptons* (see ELECTRON, LEPTONS, NEUTRINO, MUONIUM), all of which have spin $\frac{1}{2}$. They may be charged or neutral. The charged leptons have electromagnetic as well as weak interactions; the neutral ones only interact weakly. There are two well-defined lepton pairs, the electron (e^-) and the electron neutrino (ν_e), and the muon (μ^-) and the muon neutrino (ν_μ). These particles all have antiparticles, in accordance with the predictions of relativistic quantum mechanics (see CPT THEOREM). There appear to exist "lepton-type" conservation laws: the number of e^- plus the number of ν_e minus the number of the corresponding antiparticles e^+ and $\bar{\nu}_e$ is conserved, and similarly for the muon-type leptons. These conservation laws follow automatically in the standard model if the neutrinos are massless.

In 1976, a third (much heavier) charged lepton, the τ, was discovered. The pattern above suggests that the τ possesses an associated massless neutrino and that the τ-type leptons are also conserved. Fairly compelling indirect evidence from the weak interactions of the τ and the consistency of the standard model indicates that this is true. However, there is no understanding of the mass patterns in Table I or why the observed neutrinos appear to be massless.

In addition to the leptons there exist *hadrons* (see HADRONS, BARYONS, HYPERONS, MESONS, NUCLEON), which have

Table I The leptons. Charges are in units of the positron (e^+) charge $e = 1.602 \times 10^{-19}$ coulomb. The ν_τ has not yet been directly observed.

Particle	Q	Mass
e^-	−1	0.51 MeV/c^2
μ^-	−1	105.7 MeV/c^2
τ^-	−1	1784 MeV/c^2
ν_e	0	<11 eV/c^2
ν_μ	0	<0.25 MeV/c^2
ν_τ	0	<35 MeV/c^2

strong interactions as well as the electromagnetic and weak. These particles have a variety of spins, both integral and half-integral, and their masses range from the value of 135 MeV/c^2 for the π^0 to 11 020 MeV/c^2 for one of the upsilon (heavy quark) states. The particles with half-integral spin are called *baryons,* and there is clear evidence for baryon conservation: The number of baryons minus the number of antibaryons is constant in any interaction. The best evidence for this is the stability of the lightest baryon, the proton (if the proton decays, it does so with a lifetime in excess of 10^{31} yr). In contrast to charge conservation, there is no deep principle that makes baryon conservation compelling, and it may turn out that baryon conservation is only approximate. The particles with integer spin are called *mesons,* and they have baryon number $B = 0$. There are hundreds of different kinds of hadrons, some almost stable and some (known as *resonances*) extremely short-lived. The degree of stability depends mainly on the mass of the hadron. If its mass lies above the threshold for an allowed decay channel, it will decay rapidly; if it does not, the decay will proceed through a channel that may have a strongly suppressed rate, e.g., because it can only be driven by the weak or electromagnetic interactions. The large number of hadrons has led to the universal acceptance of the notion that the hadrons, in contrast to the leptons, are composite. In particular, experiments involving lepton–hadron scattering or e^+e^- annihilation into hadrons have established that hadrons are bound states of point-like spin-$\frac{1}{2}$ particles of fractional charge, known as quarks. Five types of quarks have been identified (Table II) and a sixth quark is predicted. For each type of quark there is a corresponding antiquark. Baryons are bound states of three quarks (e.g., proton = *uud*; neutron = *udd*), while mesons consist of a quark and an antiquark.

Table II The quarks (spin-$\frac{1}{2}$ constituents of hadrons). Each quark carries baryon number $B = \frac{1}{3}$, while the antiquarks have $B = -\frac{1}{3}$. The *t* quark is predicted but not yet observed.

Particle	Q	Mass
u (up)	$\frac{2}{3}$	5 MeV/c^2
d (down)	$-\frac{1}{3}$	10 MeV/c^2
s (strange)	$-\frac{1}{3}$	200 MeV/c^2
c (charm)	$\frac{2}{3}$	1.4 GeV/c^2
b (bottom)	$-\frac{1}{3}$	5 GeV/c^2
t (top)	$\frac{2}{3}$	80–200 GeV/c^2

CLASSIFICATION OF INTERACTIONS

For reasons that are still unclear, the interactions fall into four types, the *electromagnetic, weak,* and *strong,* and the *gravitational* interaction. If we take the proton mass as a standard, the last is 10^{-36} times the strength of the electromagnetic interaction, and will mainly be neglected in what follows. (The unification of gravity with the other interactions is one of the major outstanding goals.) The first two interactions were most cleanly explored with the leptons, which do not have strong interactions that mask them. We shall therefore discuss them first in terms of the leptons.

ELECTROMAGNETIC INTERACTIONS

The electromagnetic interactions of charged leptons (electron and muon) are best described in terms of equations of motion, derived from a Lagrangian function, which are solved in a power series in the *fine-structure constant* $e^2/4\pi\hbar c \equiv \alpha \simeq 1/137$, a small parameter. The Lagrangian density consists of a term that describes the free-photon field,

$$\mathcal{L}_\gamma = -\frac{1}{4} F_{\mu\nu}(x)F^{\mu\nu}(x), \tag{1}$$

where

$$F_{\mu\nu}(x) = \frac{\partial A_\nu(x)}{\partial x^\mu} - \frac{\partial A_\mu(x)}{\partial x^\nu} \tag{2}$$

is the electromagnetic field tensor. $\mathcal{L}_\gamma$ is just $\frac{1}{2}[\mathbf{E}^2(x) - \mathbf{B}^2(x)]$ in more common notation. It is written in terms of the vector potential $A_\mu(x)$ because the terms that involve the lepton and its interaction with the electromagnetic field are simplest when written in terms of $A_\mu(x)$:

$$\mathcal{L}_l = i\bar{\psi}(x)\gamma^\alpha \left(\frac{\partial}{\partial x^\alpha} - ieA_\alpha(x)\right) \psi(x) - m\bar{\psi}(x)\psi(x). \tag{3}$$

Here $\psi(x)$ is a four-component spinor representing the electron, muon, or tau, $\bar{\psi}(x) = \psi^\dagger(x)\gamma^0$, the γ^α ($\alpha = 0, 1, 2, 3$) are the Dirac matrices [4×4 matrices that satisfy the conditions $(\gamma^1)^2 = (\gamma^2)^2 = (\gamma^3)^2 = -(\gamma_0)^2 = -1$ and $\gamma^\alpha\gamma^\beta = -\gamma^\beta\gamma^\alpha$ for $\beta \neq \alpha$]; m has the dimensions of a mass in the natural units in which $\hbar = c = 1$. If e were zero, the Lagrangian would describe a free lepton; with $e \neq 0$ the interaction has the form

$$-eA^\alpha(x)j_\alpha(x), \tag{4}$$

where the current $j_\alpha(x)$ is given by

$$j_\alpha(x) = -\bar{\psi}(x)\gamma_\alpha\psi(x). \tag{5}$$

The equations of motion show that the current is conserved,

$$\frac{\partial}{\partial x_\alpha} j_\alpha(x) = 0, \tag{6}$$

so that the *charge*

$$Q = \int d^3\mathbf{r}\, j_0(\mathbf{r},t) \tag{7}$$

is a constant of the motion.

The form of the interaction is obtained by making the replacement

$$\frac{\partial}{\partial x^\alpha} \rightarrow \frac{\partial}{\partial x^\alpha} - ieA_\alpha(x) \quad (8)$$

in the Lagrangian for a free lepton. This *minimal coupling* follows from a deep principle, *local gauge invariance*. The requirement that $\psi(x)$ can have its phase changed locally without affecting the physics of the lepton, that is, invariance under

$$\psi(x) \rightarrow e^{-i\theta(x)}\psi(x), \quad (9)$$

can only be implemented through the introduction of a vector field $A_\alpha(x)$, coupled as in (8), and transforming according to

$$A_\alpha(x) \rightarrow A_\alpha(x) - \frac{1}{e}\frac{\partial\theta(x)}{\partial x^\alpha}. \quad (10)$$

This dictates that the free-photon Lagrangian density contains only the gauge-invariant combination (2), and that terms of the form $M^2 A_\alpha{}^2(x)$ be absent. Thus local gauge invariance is a very powerful requirement; it implies the existence of a massless vector particle (the photon, γ), which mediates a long-range force [Fig. 1(a)]. It also fixes the form of the coupling and leads to charge conservation, and implies masslessness of the photon. The resulting theory (see QUANTUM ELECTRODYNAMICS, COMPTON EFFECT, FEYNMAN DIAGRAMS, MUONIUM, POSITRONIUM) is in extremely good agreement with experiment, as Table III shows. In working out the consequences of the equations of motion that follow from (3), infinities appear, and the theory seems not to make sense. The work of S. Tomonaga, J. Schwinger, R. P. Feynman, and F. J. Dyson in the late 1940s clarified the nature of the problem and showed a way of eliminating the difficulties. In creating *renormalization theory* these authors pointed out that the parameters e and m that appear in (3) can be identified as the charge and the mass of the lepton only in lowest order. When the charge and mass are calculated in higher order, infinite integrals appear. After a rescaling of the lepton fields, it turns out that these are the only infinite integrals in the theory. Thus by absorbing them into the definitions of new quantities, the renormalized (i.e., physically measured) charge and mass, all infinities are removed, and the rest of the theoretically calculated quantities are finite. Gauge invariance ensures that in the renormalized

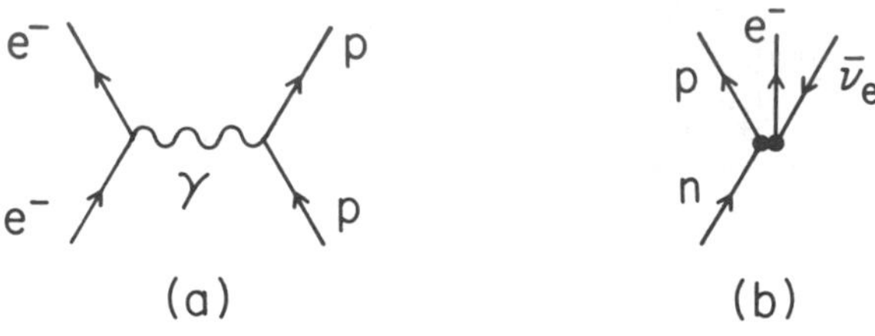

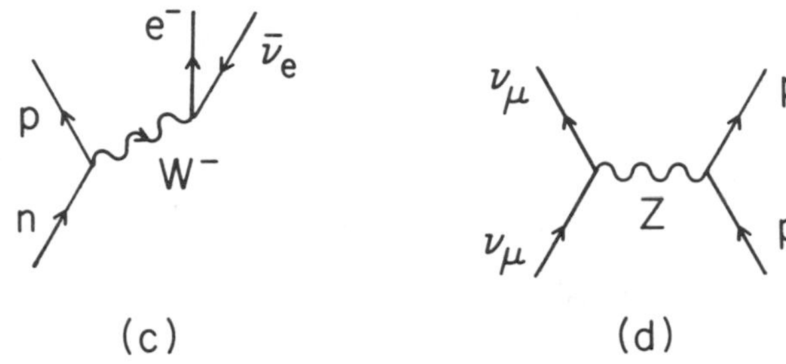

FIG. 1. (a) Long-range force between electron and proton mediated by a photon. (b) Four-fermi (zero-range) description of beta decay ($n \rightarrow pe^-\bar{\nu}_e$). (c) Beta decay mediated by a W^-. (d) A neutral current process mediated by the Z.

theory the current is still conserved, and the photon remains massless.

Subsequent work showed that the possibility of absorbing the divergences of a theory in a finite number of renormalizations of physical quantities is limited to a small class of theories, those involving the coupling of spin-$\frac{1}{2}$ to spin-0 particles with a very restrictive form of the coupling. Theories involving vector (spin-1) fields are only renormalizable when the couplings are minimal and local gauge invariance holds. Thus gauge-invariant couplings like $\bar{\psi}(x)\gamma^\alpha\gamma^\beta\psi(x)F_{\alpha\beta}(x)$, which are known not to be needed in quantum electrodynamics, are eliminated by the requirement of renormalizability.

The electrodynamics of hadrons involves a coupling of the form

$$-eA^\alpha(x)j_\alpha{}^{\text{had}}(x). \quad (11)$$

For one-photon processes, such as photoproduction (e.g., $\gamma p \rightarrow \pi^0 p$), matrix elements of the conserved current

Table III Comparison of theory and experiment in QED. The numbers in parentheses represent the uncertainty in the last digits.

Physical quantity	Experiment	Theory	Difference
Deviation from gyromagnetic ratio, $a_e = (g-2)/2$ for electron	$1\,159\,652\,188.2(4.3) \times 10^{-12}$	$1\,159\,652\,192(131) \times 10^{-12}$	$(-4 \pm 131) \times 10^{-12}$ (0.11 ppm)
The same for the muon	$1\,165\,923(8.4) \times 10^{-9}$	$1\,165\,920.1(2.0) \times 10^{-9}$	$(2.9 \pm 8.6) \times 10^{-9}$ (7.4 ppm)
Lamb shift in hydrogen (MHz)	1057.845(9)	1057.875(10)	-0.030 ± 0.013 (12 ppm)
Hyperfine splitting in hydrogen (MHz)	1420.4057 5	1420.4034(13)	0.0024 ± 0.0013 (0.9 ppm)
Hyperfine splitting in muonium, μ^+e^- (kHz)	4463 302.88 (0.16)	4463 303.3 (1.7)	0.4 ± 1.7 (0.38 ppm)
Photon mass (MeV/c^2)	$<3 \times 10^{-33}$	0	—

$j_\alpha{}^{had}(x)$ are measured to first order in e, while for two-photon processes, such as hadronic Compton scattering ($\gamma p \to \gamma p$), matrix elements of products like $j_\alpha{}^{had}(x) j_\beta{}^{had}(y)$ enter. Within the quark theory one can write an explicit form for the hadronic current:

$$j_\alpha{}^{had}(x) = \frac{2}{3}\bar{u}\gamma^\alpha u - \frac{1}{3}\bar{d}\gamma^\alpha d - \frac{1}{3}\bar{s}\gamma^\alpha s \ldots, \quad (12)$$

where we use particle labels for the spinor operators (which are evaluated at x), and the coefficients are just the charges in units of e. The total electromagnetic interaction is therefore $-eA^\alpha j_\alpha{}^\gamma$, where

$$j_\alpha{}^\gamma = j_\alpha + j_\alpha{}^{had} = \sum_i Q_i \bar{\psi}_i \gamma_\alpha \psi_i, \quad (13)$$

and the sum extends over all the leptons and quarks ($\psi_i = e$, μ, τ, ν_e, ν_μ, ν_τ, u, d, s, . . .), and where Q_i is the charge of ψ_i.

WEAK INTERACTIONS

In contrast to the electromagnetic interaction, whose form was already contained in classical electrodynamics, it took many decades of experimental and theoretical work to arrive at a compact phenomenological Lagrangian density describing the weak interactions. The form

$$\mathcal{L}_W = -\frac{G}{\sqrt{2}} J_\alpha{}^\dagger(x) J^\alpha(x) \quad (14)$$

involves vectorial quantities, as originally proposed by E. Fermi. The current $J^\alpha(x)$ is known as a charged current since it changes (lowers) the electric charge when it acts on a state. That is, it describes a transition such as $\nu_e \to e^-$ of one particle into another, or the corresponding creation of an $e^-\bar{\nu}_e$ pair. Similarly, $J_\alpha{}^\dagger$ describes a charge-raising transition such as $n \to p$. Equation (14) describes a zero-range "four-fermi" interaction [Fig. 1(b)], in contrast to electrodynamics, in which the force is transmitted by the exchange of a photon. An additional class of "neutral-current" terms was discovered in 1973 (see Weak Neutral Currents, Currents in Particle Theory). These will be discussed in the next section. $J^\alpha(x)$ consists of leptonic and hadronic parts:

$$J^\alpha(x) = J^\alpha_{lept}(x) + J^\alpha_{had}(x). \quad (15)$$

Thus, it describes purely leptonic interactions, such as

$$\mu^- \to e^- + \bar{\nu}_e + \nu_\mu,$$

through terms quadratic in J_{lept}; semileptonic interactions, most exhaustively studied in decay processes such as

$$n \to p + e^- + \bar{\nu}_e,$$

$$\pi^+ \to \mu^+ + \nu_\mu,$$

$$\Lambda^0 \to p + e^- + \bar{\nu}_e,$$

and more recently in neutrino-scattering reactions such as

$$\nu_\mu + n \to \mu^- + p,$$

$$\bar{\nu}_\mu + p \to \mu^+ + n,$$

and, through terms quadratic in J^α_{had}, purely nonleptonic interactions, such as

$$\Lambda^0 \to p + \pi^-,$$

$$K^+ \to \pi^+ + \pi^+ + \pi^-,$$

in which only hadrons appear. The coupling is weak in that the natural dimensionless coupling, with the proton mass as standard, is $Gm_p{}^2 = 1.01 \times 10^{-5}$.

The leptonic current consists of the terms

$$J^\alpha_{lept}(x) = \bar{e}\gamma^\alpha(1-\gamma_5)\nu_e + \bar{\mu}\gamma^\alpha(1-\gamma_5)\nu_\mu + \bar{\tau}\gamma^\alpha(1-\gamma_5)\nu_\tau. \quad (16)$$

Both polar and axial vector terms appear ($\gamma_5 = i\gamma^0\gamma^1\gamma^2\gamma^3$ is a pseudoscalar matrix), so that in the quadratic form (14) there will be vector–axial-vector interference terms, indicating *parity nonconservation*. The discovery of this phenomenon, following the suggestion of T. D. Lee and C. N. Yang in 1956 that reflection invariance in the weak interactions could not be taken for granted but had to be tested, played an important role in the determination of the phenomenological Lagrangian (14). The experiments suggested by Lee and Yang all involved looking for a pseudoscalar observable in a weak interaction experiment (see Parity), and the first of many experiments (C. S. Wu, E. Ambler, R. W. Hayward, D. D. Hoppes, and R. F. Hudson) measuring the beta decay of polarized nuclei (^{60}Co) showed an angular distribution of the form

$$W(\theta) = A + B\mathbf{p}_e \cdot \langle \mathbf{J} \rangle, \quad (17)$$

where $\mathbf{p}_e$ is the electron momentum and $\langle \mathbf{J} \rangle$ the polarization of the nucleus. The distribution $W(\theta)$ is not invariant under mirror inversion (P) which changes $\mathbf{J} \to \mathbf{J}$ and $\mathbf{p}_e \to -\mathbf{p}_e$, so the experimental form (17) directly showed parity nonconservation. Experiments showed that both the hadronic and the leptonic currents had vector and axial-vector parts, and that although invariance under particle–antiparticle conjugation C is also violated, the form (14) maintains invariance under the joint symmetry CP (see Conservation Laws). There is evidence that CP itself is violated at a much weaker level, of the order of 10^{-9} of the weak interactions. As will be discussed later, this is consistent with a second-order weak effect, though it is possible that an otherwise undetected superweak interaction also plays a role. The part of $J^{\alpha\dagger}_{had}$ relevant to beta decay is $\sim \bar{u}\gamma_\alpha(1-\gamma_5)d$. The detailed form of the hadronic current will be discussed after the description of the strong interactions.

Even at the leptonic level the theory described by (14) is not renormalizable. This manifests itself in the result that the cross section for neutrino absorption grows with energy:

$$\sigma_\nu = (\text{const}) G^2 m_p E_\nu. \quad (18)$$

While this behavior is in accord with observations up to the highest energies studied so far, it signals a breakdown of the theory at higher energies, so that (14) cannot be fundamental. A number of people suggested over the years that the effective Lagrangian is but a phenomenological description of a theory in which the weak current $J^\alpha(x)$ is coupled to a charged *intermediate vector meson* $W_\alpha{}^-(x)$, in analogy with quantum electrodynamics. The form (14) emerges from the exchange of a vector meson between the currents (see Feynman Diagrams) when the W mass is much larger than the momentum transfer in the process [Fig. 1(c)]. The intermediate vector boson theory leads to a better behaved σ_ν at high energies. However, massive vector theories are still not renormalizable, and the cross section for $e^+e^-\to W^+W^-$ (with longitudinally polarized Ws) grows with energy. Until 1967 there was no theory of the weak interactions in which higher-order corrections, though extraordinarily small because of the weak coupling, could be calculated.

UNIFIED THEORIES OF THE WEAK AND ELECTROMAGNETIC INTERACTIONS

In spite of the large differences between the electromagnetic and weak interactions (massless photon versus massive W, strength of coupling, behavior under P and C), the vectorial form of the interaction hints at a possible common origin. The renormalization barrier seems insurmountable: A theory involving vector bosons is only renormalizable if it is a gauge theory; a theory in which a charged weak current of the form (16) couples to massive charged vector bosons,

$$\mathcal{L}_W = -g_W[J^{\alpha\dagger}(x)W_\alpha{}^+(x) + J^\alpha(x)W_\alpha{}^-(x)], \tag{19}$$

does not have that property. Interestingly, a gauge theory involving charged vector mesons, or more generally, vector mesons carrying some internal quantum numbers, had been invented by C. N. Yang and R. L. Mills in 1954. These authors sought to answer the question: Is it possible to construct a theory that is invariant under the transformation

$$\psi(x)\to\exp[i\mathbf{T}\cdot\boldsymbol{\theta}(x)]\psi(x), \tag{20}$$

where $\psi(x)$ is a column vector of fermion fields related by symmetry, the T_i are matrix representations of a Lie algebra (see Lie Groups, Gauge Theories), and the $\theta_i(x)$ are a set of angles that depend on space and time, generalizing the transformation law (9)? It turns out to be possible to construct such a *non-Abelian gauge theory*. The coupling of the spin-$\frac{1}{2}$ field follows the "minimal" form (8) in that

$$\bar{\psi}\gamma^\alpha\frac{\partial}{\partial x^\alpha}\psi\to\bar{\psi}\gamma^\alpha\left(\frac{\partial}{\partial x^\alpha}+igT_iW_\alpha{}^i(x)\right)\psi, \tag{21}$$

where the W^i are vector (gauge) bosons, and the *gauge coupling constant* g is a measure of the strength of the interaction. The vector meson form is again

$$\mathcal{L}_V = -\tfrac{1}{4}F_{\mu\nu i}(x)F_i{}^{\mu\nu}(x), \tag{22}$$

but now the structure of the fields is more complicated than in (2):

$$F_{\mu\nu i}(x) = \frac{\partial}{\partial x^\mu}W_\nu{}^i(x) - \frac{\partial}{\partial x^\nu}W^i_\mu(x) - gf_{ijk}W_\mu{}^j(x)W_\nu{}^k(x), \tag{23}$$

because the vector fields $W_\mu{}^i$ themselves carry the "charges" (denoted by the label i); thus, they interact with each other (unlike electrodynamics), and their transformation law is more complicated than (10). The numbers f_{ijk} that appear in the additional nonlinear term in (23) are the *structure constants* of the group under consideration, defined by the communication rules

$$[T_i, T_j] = if_{ijk}T_k. \tag{24}$$

There are as many vector bosons as there are generators of the group. The Abelian group U(1) with only one generator (the electric charge) is the local symmetry group of quantum electrodynamics. For the group SU(2) there are three generators and three vector mesons. Gauge invariance is very restrictive. Once the symmetry group and representations are specified, the only arbitrariness is in g. The existence of the gauge bosons and the form of their interaction with other particles and with each other is determined. Yang–Mills (gauge) theories are renormalizable because the form of the interactions in (21) and (23) leads to cancellations between different contributions to high-energy amplitudes. However, gauge invariance does not allow mass terms for the vector bosons, and it is this feature that was responsible for the general neglect of the Yang–Mills theory for many years.

S. Weinberg (1967) and independently A. Salam (1968) proposed an extremely ingenious theory unifying the weak and electromagnetic interactions by taking advantage of a theoretical development (see Symmetry Breaking, Spontaneous) according to which vector mesons in Yang–Mills theories could acquire a mass without its appearing explicitly in the Lagrangian (the theory without the symmetry breaking mechanism had been proposed earlier by S. Glashow). The basic idea is that even though a theory possesses a symmetry, the solutions need not. A familiar example is a ferromagnet: the equations are rotationally invariant, but the spins in a physical ferromagnet point in a definite direction. A loss of symmetry in the solutions manifests itself in the fact that the ground state, the vacuum, is no longer invariant under the transformations of the symmetry group, e.g., because it is a Bose condensate of scalar fields rather than empty space. According to a theorem first proved by J. Goldstone, this implies the existence of massless spin-0 particles; states consisting of these *Goldstone bosons* are related to the original vacuum state by the (spontaneously broken) symmetry generators. If, however, there are gauge bosons in the theory, then as shown by P. Higgs, F. Englert, and R. Brout, and by G. Guralnik, C. Hagen, and T. Kibble, the massless Goldstone bosons can be eliminated by a gauge transformation. They reemerge as the longitudinal (helicity-zero) components of the vector mesons, which have acquired an effective mass by their interaction with the ground-state condensate (the *Higgs mechanism*). Renormalizability depends on the symmetries of the Lagrangian, which is not affected by the symmetry-violating solutions, as was elucidated through the work of B. W. Lee and K. Symanzik and first applied to the gauge theories by G. 't Hooft.

The simplest theory must contain a W^+ and a W^-; since their generators do not commute there must also be at least one neutral vector boson W^0. A scalar (*Higgs*) particle associated with the breaking of the symmetry of the solution is also required. The simplest realistic theory also contains a photon-like object with its own coupling constant [hence the description as SU(2)×U(1)]. The resulting theory incorporates the Fermi theory of charged-current weak interactions and quantum electrodynamics. In particular, the vector boson extension of the Fermi theory in (19) is reproduced with $g_w = g/2\sqrt{2}$, where g is the SU(2) coupling, and $G \simeq \sqrt{2}g^2/8M_W^2$. There are two neutral bosons, the W^0 of SU(2) and B associated with the U(1) group. One combination,

$$A = \cos\theta_W B + \sin\theta_W W^0, \tag{25}$$

is just the photon of electrodynamics, with $e = g\sin\theta_W$. The *weak* (or *Weinberg*) *angle* θ_W which describes the mixing is defined by $\theta_W \equiv \tan^{-1}(g'/g)$, where g' is the U(1) gauge coupling. In addition, the theory makes the dramatic prediction of the existence of a second (massive) neutral boson orthogonal to A:

$$Z = -\sin\theta_W B + \cos\theta_W W^0, \tag{26}$$

which couples to the *neutral current*

$$J_\alpha{}^Z = \sum_i T_3(i)\bar{\psi}_i\gamma_\alpha(1-\gamma_5)\psi_i - 2\sin^2\theta_W j_\alpha{}^\gamma, \tag{27}$$

where $j_\alpha{}^\gamma$ is the electromagnetic current in (13) and $T_3(i)$ [$+\frac{1}{2}$ for u, ν; $-\frac{1}{2}$ for e^-, d] is the eigenvalue of the third generator of SU(2). The Z mediates a new class of weak interactions (see WEAK NEUTRAL CURRENTS),

$$(\nu/\bar{\nu}) + p,\ n \to (\nu/\bar{\nu}) + \text{hadrons},$$

$$(\nu/\bar{\nu}) + \text{nucleon} \to (\nu/\bar{\nu}) + \text{nucleon},$$

$$\nu_\mu + e^- \to \nu_\mu + e^-,$$

characterized by a strength comparable to the charged-current interactions [Fig. 1(d)]. Another prediction is that of the existence, in electromagnetic interactions such as

$$e^- + p \to e^- + \text{hadrons},$$

of tiny parity-nonconservation effects that arise from the exchange of the Z between the electron and the hadronic system. Neutral current–induced neutrino processes were observed in 1973, and since then all of the reactions have been studied in detail. In addition, parity violation (and other axial current effects) due to the weak neutral current has been observed in asymmetries in the scattering of polarized electrons from deuterons, in the induced mixing between S and P states in heavy atoms, and in asymmetries in electron–positron annihilation into $\mu^+\mu^-$, $\tau^+\tau^-$, and heavy quark pairs. All of the observations are in excellent agreement with the predictions of the standard SU(2) $\times$ U(1) model and yield values of $\sin^2\theta_W$ consistent with each other (Fig. 2). Another prediction is the existence of massive $W^\pm$ and Z bosons (the photon remains massless because the condensate is neutral), with masses

$$M_W{}^2 = \frac{A^2}{\sin^2\theta_W}, \qquad M_Z{}^2 = \frac{M_W{}^2}{\cos^2\theta_W}. \tag{28}$$

where $A \sim \pi\alpha/\sqrt{2}G \sim (37\ \text{GeV})^2$. (In practice, a significant, 7%, higher-order correction must be included.) Using $\sin^2\theta_W$ obtained from neutral current processes, one predicts

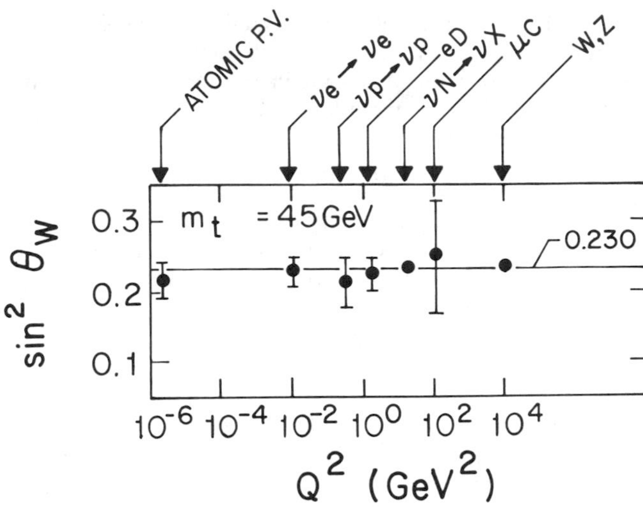

FIG. 2. $\sin^2\theta_W$ from various reactions, as a function of Q^2, the typical four-momentum transfer squared.

$M_W = 80.2 \pm 1.1$ GeV/c^2 and $M_Z = 91.6 \pm 0.9$ GeV/c^2. In 1983 the W and Z were discovered at the new $\bar{p}p$ collider at CERN. The current values of their masses, $M_W = 80.8 \pm 1.3$ GeV/c^2, $M_Z = 91.9 \pm 1.7$ GeV/c^2, dramatically confirm the standard model predictions. The neutral current and boson mass data together establish that the standard (Weinberg–Salam) electroweak model is correct to first approximation down to a distance scale of 10^{-16} cm (1/1000th the size of the nucleus), and yield the world average $\sin^2\theta_W = 0.230 \pm 0.005$. (It is hoped that the value of this one arbitrary parameter may emerge from a future unification of the strong and electromagnetic interactions, but the simplest attempts have not been successful in detail.) It is expected that future accelerators will increase the precision of these electroweak tests by an order of magnitude, and also directly measure the (so far unobserved) self-interactions of the gauge bosons. The major outstanding ingredient is the Higgs boson, which is hard to produce and detect and which could have a mass anywhere from 5 to 1000 GeV/c^2. Many physicists suspect that the elementary Higgs field may be replaced by a dynamical or bound-state symmetry-breaking mechanism.

THE STRONG INTERACTIONS

The strength of the coupling that manifests itself in nuclear forces and in the interaction of pions with nucleons is such that perturbation theory, so useful in the electromagnetic interaction, cannot be applied to any field theory of the strong interactions in which the mesons and baryons are the fundamental fields. The large number of hadronic states strongly suggests a composite structure that cannot be viewed as a perturbation about noninteracting systems. In fact, it is now generally believed that the strong interactions are described by a gauge theory, quantum chromodynamics (QCD), in which the basic entities are quarks rather than hadrons. Nevertheless, prior and parallel to the development of the quark theory a wealth of experimental information concerning the hadrons and their interactions was accumulated. In spite of the absence of guidance from field theory, and in spite of the fact that each jump in available accelerator energy brought a shift in the focus of attention, certain simple patterns were identified.

INTERNAL SYMMETRIES

The first hint of a new symmetry can be seen in the remarkable resemblance between neutron and proton. They differ in electromagnetic properties, and, other than that, by effects that are very small; for example, they differ in mass by 1 part in 700. W. Heisenberg conjectured that the neutron and proton are two states of a single entity, the *nucleon* (see NUCLEON), just as an electron with spin up and an electron with spin down are two states of a single entity, even though in an external magnetic field they have slightly different energies. Pursuing this analogy, Heisenberg and E. U. Condon proposed that the strong interactions are invariant under transformations in an internal space, in which the nucleon is a spinor (see ISOSPIN). Thus, the nucleon is an *isospin* doublet, with $I_z(p)=\frac{1}{2}$ and $I_z(n)=-\frac{1}{2}$, and isospin (in analogy with angular momentum) *is conserved*. In the language of group theory, the assertion is that the strong interactions are invariant under the transformations of the group SU(2), and that particles transform as irreducible representations. The electromagnetic and weak interactions violate this invariance. The expression for the charge of the nucleons and antinucleons,

$$Q=I_z+B/2, \tag{29}$$

shows that the charge picks out a preferred direction in the internal space. (It is now believed that the strong interactions themselves have a small piece which breaks isospin symmetry, in addition to electroweak interactions.)

With the discovery of the three pions (π^+,π^0,π^-) with mass remarkably close to that predicted by H. Yukawa (1935) in his seminal work explaining nuclear forces in terms of an exchange of massive quanta of a mesonic field, the notion of isospin acquired a new significance. It was natural, in view of the small $\pi^\pm-\pi^0$ mass difference, to assign the pion to the $I=1$ representation of SU(2). The invariance of the pion–nucleon interaction under isospin transformations led to a number of predictions, all of which were confirmed. In particular, states initiated in pion–nucleon collisions could only have isospin $\frac{1}{2}$ or $\frac{3}{2}$. Early work on pion–nucleon scattering led to the discovery of a resonance with rest mass 1236 MeV/c^2, width 115 MeV/c^2, and angular momentum and parity $J^P=\frac{3}{2}^+$. This resonance occurred in π^+p scattering, so that it had to have $I=\frac{3}{2}$, and its effects seen in $\pi^-p\rightarrow\pi^-p$ and $\pi^-p\rightarrow\pi^0n$ should be the same as those in $\pi^+p\rightarrow\pi^+p$. This prediction was borne out by experiment.

Formally, SU(2) invariance is described by defining generators I_i $(i=1,2,3)$ obeying the Lie algebra

$$[I_i,I_j]=ie_{ijk}I_k \tag{30}$$

where e_{ijk} is totally antisymmetric in the indices and $e_{123}=1$. The statement that a pion is an $I=1$ state then means that the pion field transforms according to

$$[I_i,\theta_i]=-(\mathcal{I}i)_{ab}\theta_b, \qquad a=1,2,3, \tag{31}$$

where the $\mathcal{I}_i$ are 3×3 matrices satisfying (30). In relativistic quantum mechanics conservation laws must be local, so the conservation law

$$\frac{dI_i}{dt}=0 \tag{32}$$

really follows from the local conservation law

$$\frac{\partial}{\partial x^\mu}\mathcal{I}_i{}^\mu(x)=0 \tag{33}$$

for the isospin-generating *currents*, for which

$$I_i=\int d^3\mathbf{r}\mathcal{I}_i{}^0(\mathbf{r},t). \tag{34}$$

Isospin [and SU(3)] are *global* symmetries: The symmetry transformations are the same at all space-time points, as opposed to the local (gauge) transformations in (20). Hence, they are not associated with gauge bosons or a force.

In the early 1950s a number of new particles were discovered. The great confusion generated by the widely differing rates of production and decay was completely cleared up by M. Gell-Mann and K. Nishijima, who extended the notion of isospin conservation to the strong interactions of the new particles, classified them (and along the way noted "missing" particles that had to exist, and were subsequently found), and discovered that the observed patterns of reactions could be explained by assigning a new quantum number S (strangeness) to each isospin multiplet.

The selection rules were

$$\Delta S=0 \tag{35}$$

for the strong and electromagnetic interactions, and

$$\Delta S=0,\pm1 \tag{36}$$

for the weak interactions. Relation (29) now takes the form

$$Q=I_z+(B+S)/2. \tag{37}$$

[Equation (37) holds for all hadrons except for those involving heavy (c and b) quarks, discovered in the 1970s and 1980s.]

The success of the strangeness scheme immediately started a search for a higher symmetry that would include isospin and strangeness (or hypercharge, $Y=B+S$), and that would, in some limit, include the nucleons and the newly discovered strange baryons in a supermultiplet. The search ended when M. Gell-Mann and Y. Ne'eman discovered that the Lie group SU(3) was the appropriate (global) symmetry. The group is generated by eight operators F_i $(i=1,2,3,\ldots,8)$, of which the first three may be identified with the isospin generators I_i, and (by convention) F_8 is related to hypercharge. The other four change isospin and strangeness. The nucleons and six other baryons discovered in the 1950s fit into an eight-dimensional (octet) representation containing doublets with $I=\frac{1}{2}$ and $Y=\pm1$, and $I=1,0$ states with $Y=0$. Similarly, the $I=1$ pions, the (K^+,K^0) with $I=\frac{1}{2}$, $Y=1$, and $(\overline{K}^0,K^-)$ with $I=\frac{1}{2}$, $Y=-1$, could be fitted into an octet that was soon completed with the discovery of an $I=Y=0$ pseudoscalar meson, the η (see Table IV). SU(3) is only an ap-

Table IV Table of low-lying mesons and baryons, grouped according to SU(3) multiplets. There may be considerable mixing between the SU_3 singlets η', φ, and f' and the corresponding octet states η, ω, f.

Particle	B	Q	Y	I	J^P	Mass (GeV/c^2)	Quark content
π	0	1, 0, −1	0	1	0^-	0.14	$u\bar{d}$, $u\bar{u} - d\bar{d}$, $d\bar{u}$
K	0	1, 0	1	$\frac{1}{2}$	0^-	0.49	$u\bar{s}$, $d\bar{s}$
$\bar{K}$	0	0, −1	−1	$\frac{1}{2}$	0^-	0.49	$s\bar{d}$, $s\bar{u}$
η	0	0	0	0	0^-	0.55	$u\bar{u} + d\bar{d} - 2s\bar{s}$
η'	0	0	0	0	0^-	0.96	$u\bar{u} + d\bar{d} + s\bar{s}$
ρ	0	1, 0, −1	0	1	1^-	0.77	$u\bar{d}$, $u\bar{u} - d\bar{d}$, $d\bar{u}$
K^*	0	1, 0	1	$\frac{1}{2}$	1^-	0.89	$u\bar{s}$, $d\bar{s}$
$\bar{K}^*$	0	0, −1	−1	$\frac{1}{2}$	1^-	0.89	$s\bar{d}$, $s\bar{u}$
ω	0	0	0	0	1^-	0.78	$u\bar{u} + d\bar{d}$
ϕ	0	0	0	0	1^-	1.02	$s\bar{s}$
A_2	0	1, 0, −1	0	1	2^+	1.32	$u\bar{d}$, $u\bar{u} - d\bar{d}$, $d\bar{u}$
$K^*(1420)$	0	1, 0	1	$\frac{1}{2}$	2^+	1.43	$u\bar{s}$, $d\bar{s}$
$\bar{K}^*(1420)$	0	0, −1	−1	$\frac{1}{2}$	2^+	1.43	$s\bar{d}$, $s\bar{u}$
f	0	0	0	0	2^+	1.27	$u\bar{u} + d\bar{d}$
f'	0	0	0	0	2^+	1.53	$s\bar{s}$
N	1	1, 0	1	$\frac{1}{2}$	$\frac{1}{2}^+$	0.94	uud, udd
Λ	1	0	0	0	$\frac{1}{2}^+$	1.12	$uds - dus$
Σ	1	1, 0, −1	0	1	$\frac{1}{2}^+$	1.19	uus, $uds + dus$, dds
Ξ	1	0, −1	−1	$\frac{1}{2}$	$\frac{1}{2}^+$	1.32	uss, dss
Δ	1	2, 1, 0, −1	1	$\frac{3}{2}$	$\frac{3}{2}^+$	1.23	uuu, uud, udd, ddd
$\Sigma(1385)$	1	1, 0, −1	0	1	$\frac{3}{2}^+$	1.39	uus, uds, dds
$\Xi^*(1530)$	1	0, −1	−1	$\frac{1}{2}$	$\frac{3}{2}^+$	1.53	uss, dss
Ω^-	1	−1	−2	0	$\frac{3}{2}^+$	1.67	sss

proximate symmetry of the strong interactions. Mass splittings within SU(3) multiplets and other breaking effects are typically 20–30%.

Most interesting is that the search for partners of the *resonance* $\Delta(1236)$ with $I=\frac{3}{2}$ led to a dramatic confirmation of SU(3). The simplest representation containing an $(I=\frac{3}{2}, Y=1)$ state is the 10-dimensional representation, which also contains $(I=1, Y=0)$ and $(I=\frac{1}{2}, Y=-1)$ states and an isosinglet $Y=-2$ particle. The symmetry-breaking pattern that explained the mass splittings among the isospin multiplets in the octet predicted equal mass splittings. Thus, when the $I=1$ $\Sigma(1385)$ was discovered, predictions could be made about the $I=\frac{1}{2}$ Ξ^*, found at mass 1530 MeV/c^2, and the Ω^-, predicted at 1675 MeV/c^2. The latter mass is too low to permit a strangeness-conserving decay to $\Xi^0 K^-$, so the Ω^- had to be long-lived, only decaying by a chain of $\Delta S=1$ weak interactions with a very clear signature. The dramatic discovery in 1964 of the Ω^- with all the right properties convinced all doubters. [See SU(3) AND HIGHER SYMMETRIES, HYPERONS, HYPERNUCLEAR PHYSICS AND HYPERNUCLEAR INTERACTIONS].

S-MATRIX THEORY

The construction of higher-energy accelerators, the invention of the bubble chamber by D. Glaser, and the combination of large hydrogen bubble chambers, rapid scanning facilities, and high-speed computers into a massive data production and analysis technology, pioneered by L. Alvarez and collaborators, led to the discovery of many new resonances during the 1950s and 1960s. The basic procedure was to measure charged tracks in bubble-chamber pictures, taken in strong magnetic fields, and to calculate the invariant masses $(\Sigma E_i)^2 - (\Sigma \mathbf{p}_i c)^2$ for various particle combinations. Resonances manifest themselves as peaks in mass distributions, and the events in the resonance region may be further analyzed to find out the spin and parity of the resonance. Baryonic resonances were also discovered in phase-shift analyses of angular distributions in pion–nucleon and K–nucleon scattering reactions. The patterns of masses and quantum numbers of the resonances showed that all the mesonic resonances came in SU(3) octets and singlets, and the baryonic ones in SU(3) decimets, octets, and singlets.

There was good evidence that there was no fundamental distinction between the stable particles and the highly unstable resonances: The Δ and the Ω^-, discussed above, are good examples, and theoretically it was found that both stable (bound) states and resonant ones appeared in scattering amplitudes as pole singularities, differing only in their location. Furthermore, the role assigned by Yukawa to the pion as the nuclear "glue"—it was the particle whose exchange was largely responsible for the nuclear forces—had to be shared with other particles: Various vector and scalar mesons were seen to contribute to the nuclear forces, and G. F. Chew and F. E. Low explained much of low-energy pion physics in terms of nucleon exchange. Chew, in collaboration with S. Mandelstam and S. Frautschi, proposed to do away with the notion of any particles being "fundamental." They hypothesized that the collection of all scattering amplitudes, the scattering matrix, be determined by a set of self-consistency conditions, the *bootstrap* conditions (see *S*-MATRIX THEORY), according to which, crudely stated, the exchange of all possible particles should yield a "potential" whose bound states and resonances should be identical with the particles inserted into the exchange term.

Much effort was devoted to bootstrap and *S*-matrix theory during the 1960s and early 1970s. The program had its greatest success in developing phenomenological models for strong interaction scattering amplitudes at high energies and low-momentum transfers, such as elastic scattering and total cross sections. In particular, Mandelstam applied an idea due to T. Regge to relativistic quantum mechanics, which related a number (perhaps infinite) of particles and resonances with the same SU(3) and other internal quantum numbers, but different masses and spins, into a family or *Regge trajectory*. The exchange of this trajectory of particles led to much better behaved high-energy amplitudes than the exchange of one or a small number, in agreement with experiment (see REGGE POLES). Related models had some success in describing inclusive processes (in which one or a few final particles are observed, with the others summed over) and other highly inelastic processes (see INCLUSIVE REACTIONS).

The more ambitious goal of understanding the strong interactions as a bootstrap (self-consistency) principle met with less success, although a number of models and approximation schemes enjoyed some measure in limited domains. The most successful was the dual resonance model pioneered by G. Veneziano. The dual model was an explicit closed-form expression for strong-interaction scattering amplitudes which properly incorporated poles for the Regge trajectories of bound states and resonances that could be formed in the reaction, Regge asymptotic behavior, and duality (the property that an amplitude could be described *either* as a sum of resonances in the direct channel *or* as a sum of Regge exchanges). However, the original simple form did not incorporate unitarity, i.e., the amplitudes did not have branch cuts corresponding to multiparticle intermediate states, and the resonances in the model had zero width (their poles occurred on the real axis in the complex energy plane instead of being displaced by an imaginary term corresponding to the resonance width). Perhaps the most important consequence of dual models was that they were later formulated as *string theories,* in which an infinite trajectory of "elementary particles" could be viewed as different modes of vibration of a one-dimensional string-like object (see STRING THEORY). String theories never quite worked out as a model of the strong interactions, but the same mathematical structure reemerged later in "theories of everything."

Many of the *S*-matrix results are still valid as phenomenological models. However, the bootstrap idea has been superseded by the success of the quark theory and the development of QCD as the probable field theory of the strong interactions.

QUARKS AS FUNDAMENTAL PARTICLES

The discovery of SU(3) as the underlying internal symmetry of the hadrons and the classification of the many resonances led to the recognition of two puzzles: Why did mesons come only in octet and singlet states? Why were there no particles that corresponded to the simplest representations of SU(3), the triplet 3 and its antiparticle 3*? M. Gell-Mann and G. Zweig in 1964 independently proposed that such representations do have particles associated with them (Gell-Mann named them *quarks*), and that all observed hadrons are made of $(q\bar{q})$ (quark + antiquark) if they have baryon number $B=0$ and of (qqq) (three quarks) if they have baryon number $B=1$. They proposed that there exist three different kinds of quarks, labeled *u, d,* and *s*. These were assumed to have spin $\frac{1}{2}$ and the internal quantum numbers listed in Table V. The quark contents of the low-lying hadrons are given in Table IV. The vector meson octet (ρ,K^*,ω) differs from the pseudoscalars (π,K,η) in that the total quark spin is 1 in the former case and zero in the latter. The $(A_2,K^*(1490),f)$ octet are interpreted as an orbital excitation $(^3P_2)$. All of the known particles and resonances can be interpreted in terms of quark states, including radial and orbital excitations and spin.

The first question was answered automatically, since products of the simplest representations decompose according to the rules

$$\mathbf{3}\times\mathbf{3}^* = \mathbf{1}+\mathbf{8}, \qquad \mathbf{3}\times\mathbf{3}\times\mathbf{3} = \mathbf{1}+\mathbf{8}+\mathbf{8}+\mathbf{10}. \tag{38}$$

A problem immediately arose in that the decimet to which Σ(1236) belongs, being the lowest-energy decimet, should have its three quarks in relative *S* states. Thus the Δ^{++}, whose composition is *uuu*, could not exist, since the spin-statistics connection requires that the wave function be totally antisymmetric, which it manifestly is not when the Δ^{++} is in a $J_z=\frac{3}{2}$ state, with all spins up, for example. The solution to this problem, proposed by O. W. Greenberg, M. Han, and Y. Nambu and further developed by W. A. Bardeen, H. Fritzsch, and M. Gell-Mann, was the suggestion that in addition to having an SU(3) label such as (u,d,s)—named *flavor* by Gell-Mann—and a spin label (up, down), quarks should have an additional three-valued label, named *color*. Thus according to this proposal there are really nine quarks:

$$\begin{array}{ccc} u_R & u_B & u_Y \\ d_R & d_B & d_Y \\ s_R & s_B & s_Y. \end{array}$$

Hence, the low-lying (qqq) state could be symmetric in the flavor and spin labels, provided it were totally antisymmetric in the color (red, blue, yellow) labels. More colors could be imagined but at least three are needed. Transformations among the color labels lead to another symmetry, $SU(3)_{color}$. The totally antisymmetric state is a color singlet. The mesons

Table V The *u, d,* and *s* quarks

	B	*Q*	*Y*	*I*	I_z
u	$\frac{1}{3}$	$\frac{2}{3}$	$\frac{1}{3}$	$\frac{1}{2}$	$\frac{1}{2}$
d	$\frac{1}{3}$	$-\frac{1}{3}$	$\frac{1}{3}$	$\frac{1}{2}$	$-\frac{1}{2}$
s	$\frac{1}{3}$	$-\frac{1}{3}$	$-\frac{2}{3}$	0	0

can also be constructed as color singlets, for example,

$$\pi^+ = \frac{1}{\sqrt{3}}(u_R\bar{d}_R + u_B\bar{d}_B + u_Y\bar{d}_Y).$$

The existing hadronic spectrum shows no evidence for states that could be color octets, for example, so the present attitude is that either color nonsinglet states are very massive compared with the low-lying hadrons or that it is an intrinsic part of hadron dynamics that *only color singlet states are observable*.

The first evidence that there are *three* (and not more) colors came from the study of $\pi^0 \rightarrow 2\gamma$ decay. Using general properties of currents, S. Adler and W. A. Bardeen were able to prove that the π^0 decay rate was uniquely determined by the process in which the π^0 first decays into a $u\bar{u}$ or a $d\bar{d}$ pair, which then annihilates with the emission of two photons. The matrix element depends on the charges of the quarks, and a calculation of the width yields 0.81 eV. With n colors, this is multiplied by n^2, and the observed width of 7.8 ± 0.9 eV supports the choice of $n=3$. Subsequent evidence for three colors was provided by the total cross section for e^+e^- annihilation into hadrons (see below), and by the elevation of the $SU(3)_{\text{color}}$ symmetry to a gauge theory of the strong interactions.

The quark model has been extremely successful in the classification of observed resonances, and even predictions of decay widths work very well, with much data being correlated in terms of a few parameters. The ingredients that go into the calculation are (a) that quarks are light, with the (u,d) doublet almost degenerate, with mass in the 300-MeV/c^2 range (one-third of a nucleon), (b) that the s quark is about 150 MeV/c^2 more massive—this explains the pattern of SU(3) symmetry breaking—and (c) that the low-lying hadrons have the simple $q\bar{q}$ or qqq content, without additional $q\bar{q}$ pairs. However, nobody has ever observed an isolated quark (free quarks should be easy to identify because of their fractional charge). It is now generally believed that quarks are *confined*, i.e., that it is impossible, even in principle, for them to exist as isolated states. However, in the 1960s this led most physicists to doubt the existence of quarks as real particles. That view was shattered by the deep inelastic electron scattering experiments in the late 1960s.

DEEP INELASTIC REACTIONS AND ASYMPTOTIC FREEDOM

In 1968 the first results of the inelastic electron-scattering experiments (Fig. 3),

$$e + p \rightarrow e' + \text{hadrons}$$

measured at the Stanford Linear Accelerator Center, were announced. The experiments were done in a kinematic region that was new. Both the momentum transfer squared (that is, the negative mass squared of the virtual photon exchanged) and the "mass" of the hadronic state produced were large. The cross section could be written as

$$\frac{d^2\sigma}{dE'\,d\Omega} = \left(\frac{d\sigma}{d\Omega}\right)_{\text{point}} \left(W_2(x,Q^2) + 2W_1(x,Q^2)\tan^2\frac{\theta}{2}\right), \quad (39)$$

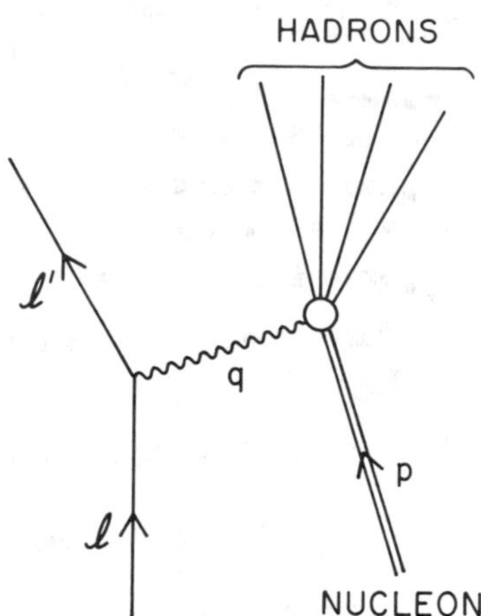

FIG. 3. Kinematics of deep inelastic lepton scattering.

where $(d\sigma/d\Omega)_{\text{point}}$ is essentially the cross section for a collision with a free point particle, and the hadronic part of the process was expressed in terms of certain *structure functions* W_1 and W_2. In (39), E' is the energy of the final electron, θ the scattering angle, $Q^2 = -(p_e - p_e')^2$, and the quantity x is $Q^2/2m_p\nu$, where $\nu = p \cdot q/m_p$ is the electron energy loss. No one knew what to expect for the behavior of W_1 and W_2. On the one hand, cross sections for production of definite resonances (*exclusive* reactions rather than *inclusive* ones) fell as powers of $1/Q^2$; on the other hand, J. D. Bjorken had predicted, on simple grounds of the irrelevance of masses when all the variables were large, that the dimensionless functions $F_1(x,Q^2) \equiv m_p W_1$ and $F_2(x,Q^2) \equiv \nu W_2$ should depend on x alone.

The results spectacularly confirmed Bjorken's conjecture of *scaling*. (Recent results using muons are shown in Fig. 4.) R. P. Feynman interpreted the detailed shapes of the distributions with his *parton model*, in which the proton, in a frame in which it is moving rapidly, looks like a swarm of independently moving point "parts" without any structure. The shape of F_2 can be interpreted as the probability distribution for a parton to carry a fraction x of the proton's momentum (weighted by the square of its charge), while the relation between F_1 and F_2 depends on the parton spin. The observed relation $F_2 \simeq 2xF_1$ establishes that the partons have spin $\frac{1}{2}$. Comparing the structure functions obtained from e and μ scattering (from proton and nuclear targets) with those obtained more recently by weak reactions such as

$$\nu_\mu(\bar{\nu}_\mu)p \rightarrow \mu^-(\mu^+) + \text{hadrons},$$

one can constrain the parton quantum numbers. They are consistent with the assumption that the partons are quarks, and that the proton consists of the three *valence* quarks assigned to it by the naive quark model, supplemented with a *sea* of quark–antiquark pairs. The relative amount of $q\bar{q}$ sea and its composition (e.g., amount of $s\bar{s}$ relative to $u\bar{u}$) are also determined. The mechanism for deep inelastic scattering is the ejection of a single quark by the virtual photon, or by the weak current in the neutrino reactions. The model assumes that the quarks that make up the proton do not interact, and that seems somewhat mysterious. Furthermore,

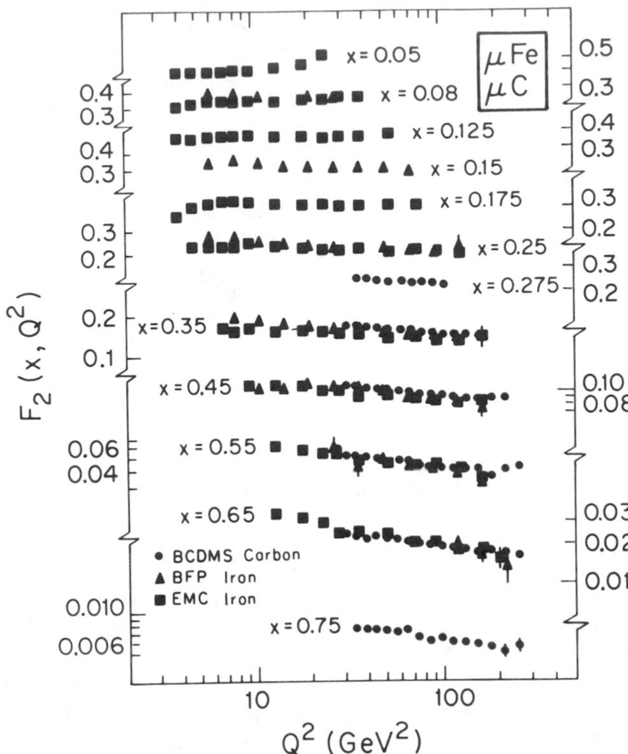

FIG. 4. Nucleon structure functions $F_2(x, Q^2)$ obtained from deep inelastic scattering of muons on iron and carbon as a function of Q^2 for various x, from "Reviews of Particle Properties," *Phys. Lett.* **204B**, 116 (1988). The data agree with the parton model prediction $F_2(x, Q^2) \simeq F_2(x)$, up to small ($\ln Q^2$) corrections. The latter are consistent with the predictions of QCD.

the model of the mechanism suggests that one quark is strongly deflected from the original path. If that is so, where is it?

The problem of how the quarks appear to be noninteracting is answered by quantum field theory. There we find that the coupling strength is really momentum dependent. For example, in quantum electrodynamics, because of the polarizability of the vacuum, the net charge of an electron seen from afar (low momentum transfer) is smaller than the charge as seen close in (large momentum transfer) where it is not shielded by the positrons produced virtually in the vacuum. Quantum electrodynamics is not the right kind of theory for quarks, since the coupling (charge) increases with momentum transfer. It was pointed out by D. Gross and F. Wilczek, by D. Politzer, and by G. 't Hooft that a theory of quarks coupled via Yang–Mills vector mesons will have the property desired for the quarks probed with high-momentum-transfer currents. The requirement of such a high-momentum-transfer decoupling, named *asymptotic freedom,* thus suggests that the "glue" that binds the quarks together is generated by a non-Abelian gauge theory, which has the attraction of being renormalizable, universal (only one coupling constant), and unique, once the number of "colors," that is, the group structure, is determined. The high-energy lepton scattering experiments provide evidence for the existence of some kind of flavor-neutral glue, in that the data are well fitted in terms of quarks, except that only about 50% of the momentum of the initial proton is attributable to quarks.

QUANTUM CHROMODYNAMICS

Quantum chromodynamics (QCD), the modern theory of the strong interactions, is a non-Abelian gauge theory based on the $SU(3)_{color}$ group of transformations which relate quarks of different colors. (The transformations are carried out simultaneously for each flavor, which does not change. The number of flavors is arbitrary.) The gauge bosons associated with the eight group generators, known as *gluons,* can be emitted or absorbed by quarks in transitions in which the color (but not flavor) can change. Since the gluons themselves carry color they can interact with each other as well (Fig. 5). As long as there are no more than 16 flavors, QCD is weakly coupled at large momentum transfers (asymptotic freedom) and strongly coupled at small momentum transfers, in agreement with observations.

For the theory to be renormalizable the gluons must either acquire mass through a Higgs mechanism (spontaneous symmetry breaking) or remain massless. The first type of theory destroys asymptotic freedom (its *raison d'être*), and the second has the difficulty that no massless vector mesons, aside from the photon, have ever been observed. It has been proposed that the theory has a structure such that only color singlets are observable, so that the vector mesons, like the quarks, are somehow *confined.* It has been speculated that the non-Abelian field lines, in contrast to electric and magnetic field lines, do not fan out all over space, but remain confined to a narrow cylindrical region, which leads to an interaction energy that is proportional to the separation of the sources of the field lines, and thus confinement. The linearly extended structure so envisaged is reminiscent of the string models suggested by duality, and thus may yield the spectrum characteristics of rectilinear Regge trajectories and the associated high-energy behavior. At large separations the potential presumably breaks down, with energy converted into ($q\bar{q}$) pairs, that is, hadrons. This would explain why quarks are never seen in deep inelastic scattering or other processes. This picture of quark and gluon confinement has not been rigorously established in QCD, but is strongly supported by calculations in the most promising approximation scheme for strongly coupled theories: viz., *lattice calculations,* in which the space-time continuum is replaced by a discrete four-dimensional lattice (see LATTICE GAUGE THEORY).

QCD is very successful qualitatively, but is hard to test quantitatively. This is partly because the coupling is large

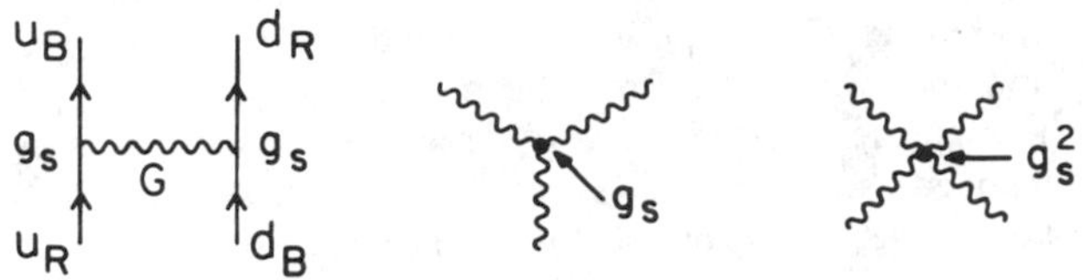

FIG. 5. Fundamental QCD interactions. g_s is the $SU(3)_{color}$ gauge coupling, which becomes small at large momentum transfers.

for most hadronic processes. Also, QCD brings a subtle change in perspective. The "strong interactions" are those mediated by the color gluons between quarks, and they give rise to the color singlet hadronic bound states. The interaction between these states need not be simple, any more than the interactions between molecules (the van der Waals forces) manifest the simplicity of the underlying Coulomb force in electromagnetism. It is hoped, but not conclusively proved, that successful phenomenological models such as Regge theory or the one-boson-exchange potential emerge as complicated higher-order effects (Fig. 6). Similarly, it has not been possible to calculate the hadron spectrum (because of strong coupling, relativistic, and many-body effects), but lattice attempts are promising. *Glueballs* (bound states of gluons) and other nonstandard color singlet states are expected. Candidate mesons exist but have not been unambiguously interpreted. QCD fairly naturally explains the observed hadronic symmetries. Parity and *CP* invariance (except for possible subtle nonperturbative effects) and the conservation of strangeness and baryon number are automatic, while approximate symmetries such as isospin, $SU(3)_{flavor}$, and chiral symmetry (see below) can be broken only by quark mass terms.

More quantitative tests of QCD are possible in high-momentum-transfer processes, in which one glimpses the underlying quarks and gluons. To zeroth order in the strong coupling g_s, QCD reproduces the quark–parton model. Higher-order corrections lead to calculable logarithmic variations of F_1 and F_2 with Q^2, in agreement with the data (e.g., Fig. 4). These experiments will be pushed to much higher Q^2 in the future $e+p$ collider HERA at DESY in Hamburg, hopefully leading to a clean demonstration of the decrease of g_s and therefore (indirectly) the gluon self-interactions.

Another consequence of the quark–parton picture is the prediction that at high energies the cross section for $e^+ + e^- \rightarrow$ hadrons should proceed through the creation (via a virtual photon) of a $q\bar{q}$ pair, which subsequently converts into hadrons through the breakdown mechanism (Fig. 7). Thus the cross section is expected to be point-like, with the modification that the quark charges appear at the production end, so that

$$R \equiv \frac{\sigma(e^+ + e^- \rightarrow \text{hadrons})}{\sigma(e^+ + e^- \rightarrow \mu^+ + \mu^-)} \Rightarrow \sum_{\text{all quarks}} Q_i^2, \quad (40)$$

where the Q_i are the quark charges. At relatively low energies the (u,d,s) contribution is 2/3 per color, that is, 2. Above

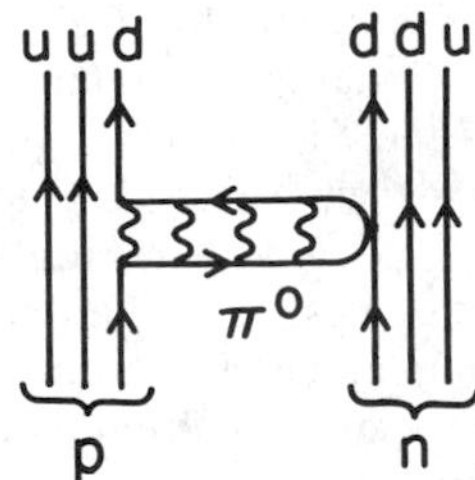

FIG. 6. One-pion exchange in QCD.

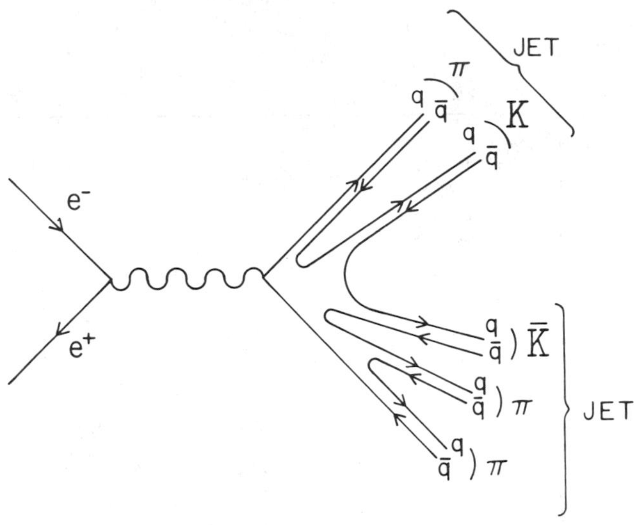

FIG. 7. Schematic picture of hadron production in $e^+ + e^-$ annihilation.

the energy (3–4 GeV) needed to produce a charm quark pair ($c\bar{c}$) one expects $R=10/3$, while above the bottom ($b\bar{b}$) threshold (~10 GeV), $R \sim 11/3$. Calculable higher-order corrections in QCD increase these predictions slightly, as does the new contribution of a virtual Z boson above ~30 GeV. These predictions are in excellent agreement with the data (Fig. 8), strongly supporting QCD and the existence of color.

In large-Q^2 processes at sufficiently high energies it is expected (and observed) that the produced hadrons tend to cluster in reasonably well-collimated *jets* of particles following approximately the direction of the final quarks. For example, the angular distribution of the jets observed in e^+e^- annihilation at SLAC and DESY confirms that the quarks have spin $\frac{1}{2}$, as well as the existence of the intermediate quark–antiquark state. Similarly, experiments at DESY have shown the existence of three-jet events whose characteristics are consistent with the hadronization of a $q\bar{q}$ pair as well as a gluon (gluon bremsstrahlung, analogous with electron bremsstrahlung). These results give fairly convincing evidence for the existence of gluons, and in particular establish their spin as 1. Finally, jets produced in hadronic processes, especially at high-energy proton–antiproton colliders at CERN and Fermilab, probe the strong interactions at extremely high Q^2 (e.g., 10^4 GeV^2). The observations are all consistent with the QCD predictions of underlying hard quark and gluon-scattering processes.

It is believed that at high temperatures and densities confinement would no longer be relevant and that a plasma of quarks and gluons should be possible. It may be possible to create such a state in high-energy heavy-ion collisions, though the signatures are not clear. Presumably, quarks and gluons were unconfined in the very early Universe when the temperature exceeded ~1 GeV.

In view of these various successes, QCD is the prime candidate for the "correct" theory of the strong interactions, even though there has been no single "gold-plated" test. In fact, QCD is the only realistic candidate within the framework of renormalizable field theories.

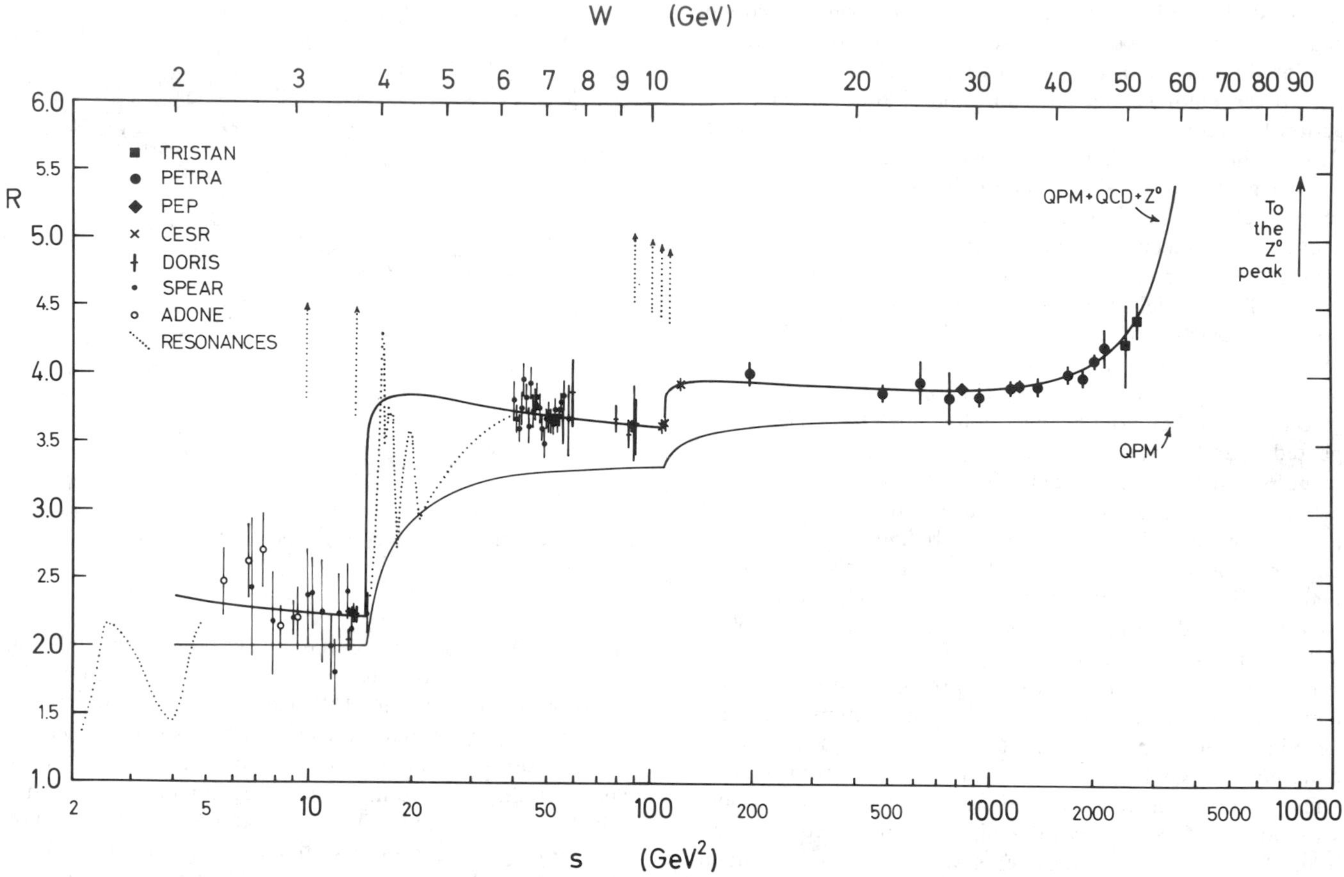

FIG. 8. Data on $R = \sigma(e^+ + e^- \to \text{hadrons})/\sigma(e^+ + e^- \to \mu^+ + \mu^-)$ as a function of the center-of-mass energy $W = \sqrt{s}$ (courtesy Dr. R. Marshall, from Rutherford Appleton Lab. Pub. RAL 88-049). The predictions of the quark parton model and a fit by QCD are also shown. At the highest energies the contribution of an intermediate virtual Z becomes important.

HADRONIC WEAK INTERACTIONS, CURRENT ALGEBRA, AND HEAVY QUARKS

The weak interactions, whether in the phenomenological local current–current form or in the Weinberg–Salam [SU(2) × U(1)] form, need a specification of the hadronic currents, both vector and axial. The experimental evidence showed that the phenomenological form for the part of the current involving the proton and neutron was

$$\bar{n}(x)[(0.9744 \pm 0.0010)\gamma^{\alpha} - (1.227 \pm 0.004)\gamma^{\alpha}\gamma_5]p(x). \quad (41)$$

This form bears a strong resemblance to the leptonic form in (16), and that, paradoxically, is surprising, since, in general, *form factors* due to strong-interaction corrections should appear. For the vector part, for example, the general expectation would be that for low momentum transfers from proton to neutron (as in beta decay)

$$\langle n|V_{\alpha}|p\rangle = G_V(q^2)\bar{n}\gamma_{\alpha}p \quad (42)$$

and the question is, Why should $G_V(0)$ be equal to unity? (The small deviation of the vector coefficient in (41) from unity is due to the Cabibbo angle factor discussed below.) R. P. Feynman and M. Gell-Mann, as well as S. Gershtein and Y. B. Zel'dovich, pointed out that if the vector current were conserved, and if $\int V_0(x)d^3x$ is normalized like the generator of a symmetry group, then the observed result would follow. Feynman and Gell-Mann identified the vector current with the current generating the isospin transformations. This satisfied the conditions that led to $G_V(0) = 1$, and it led to a model-independent characterization of the weak vector current. From this it was possible to predict unambiguously the rate for the decay $\pi^+ \to \pi^0 + e^+ + \nu_e$, and certain "magnetic" corrections to beta-decay spectra, all in excellent agreement with experiment.

In 1963, after the discovery of SU(3), N. Cabibbo generalized this assignment and proposed that the weak current, now also describing the weak interactions of the strange particles, has a form involving the SU(3)-generating currents.

The next important step was to give a general characterization of the axial current. M. Gell-Mann proposed that there exists an additional global symmetry of the strong interactions, also of the SU(3) type, but generated by eight pseudoscalar charges whose associated currents are axial currents, in fact, the weak axial currents. The difficulty that in the symmetric limit a pseudoscalar charge acting on a proton state, for example, yields another "odd" proton, for which there is no experimental evidence, was resolved by Y. Nambu, who made use of the Goldstone mechanism discussed earlier. The proposal was that in the symmetry limit the axial current is conserved, but that the solutions do not obey the symmetry. This leads to the prediction of massless pseudoscalar particles, which would approximately describe the pions and their SU(3) partners. The symmetry generated by the SU(3) generators F_i and the pseudoscalar F_{5i} is called a *chiral* [SU(3) × SU(3)] symmetry, and through the study of weak interactions it has been established that it is a good symmetry, leading to a number of experimentally verified relations involving matrix elements of the weak hadronic currents. In particular, in 1965 S. Adler and W. Weisberger independently derived a relation between the coefficient $G_A(0)$ of the axial current in (41) in terms of other observables, such as the πp cross section. They obtained $G_A(0) \simeq 1.21$, in excellent agreement with experiment. The algebra SU(3) × SU(3) emerges quite naturally in a quark model. With minimal couplings, the currents

$$\bar{q}(x)\gamma^\alpha\lambda_i q(x), \quad \bar{q}(x)\gamma^\alpha\gamma_5\lambda_i q(x), \tag{43}$$

where the λ_i are the SU(3) generalizations of the SU(2) Pauli matrices, are conserved in the limit that the quark masses vanish. The *current* quark masses, which are the masses appearing in the QCD Lagrangian (they may actually be generated in the electroweak sector by the same Higgs mechanism which yields the W and Z masses), break the symmetries associated with the axial currents explicitly and generate small masses for the π, K, and η. Quark mass differences break the vector symmetries and lead to multiplet mass differences. From these effects one can estimate the current masses given in Table II. The u and d masses are extremely small. (The much larger *constituent* masses of order 300 MeV/c^2 in the naive quark model are dynamical masses associated with the spontaneous breaking of chiral symmetry.) Experiments indicate $m_u \neq m_d$, implying a breaking of isospin in the strong interactions, which is however no larger than the (separate) electromagnetic breaking because of the small scale of the masses. The much larger m_s leads to a substantial breaking of SU(3).

The quark model gives a simple expression for the electromagnetic current of the hadrons [Eq. (12)], with the coefficients determined by the quark charges. Similarly, the weak neutral current coupling to the Z is given in (27). The weak current that couples to the charged W is given (for three quarks) by

$$\begin{aligned} J_W{}^\alpha &= (\bar{u}\bar{d}\bar{s})Q_W\gamma^\alpha(1-\gamma_5)\begin{pmatrix} u \\ d \\ s \end{pmatrix} \\ &\equiv (\bar{u}\bar{d}\bar{s})\begin{pmatrix} 0 & 0 & 0 \\ \cos\theta_C & 0 & 0 \\ \sin\theta_C & 0 & 0 \end{pmatrix}\gamma^\alpha(1-\gamma_5)\begin{pmatrix} u \\ d \\ s \end{pmatrix} \\ &= (\bar{d}\cos\theta_C + \bar{s}\sin\theta_C)\gamma^\alpha(1-\gamma_5)u. \end{aligned} \tag{44}$$

This form leads to the Cabibbo theory, and the angle θ_C, the so-called Cabibbo angle, is of magnitude 0.22 rad. Its origin lies in the difference in the ways in which the strong and weak interactions break SU(3) symmetry. This is associated with the quark masses, which are generated by the Higgs mechanism in the standard model. Their values, as well as θ_C and the other *mixing angles* introduced later, are free parameters that are not understood at present.

The weak current in (44) has the property that the change in strangeness and the change in charge are equal ($\Delta S = \Delta Q$), in agreement with experiment. The Cabibbo theory explains a large number of strange particle decays with a universal choice of θ_C. It has one difficulty: If this current is incorporated into a gauge theory of the weak interactions (e.g., the SU(2) × U(1) model) in a manner analogous to the leptonic current, then the neutral intermediate W^0 vector meson couples naturally to a neutral current obtained by commuting Q_W with its adjoint,

$$[Q_W, Q_W{}^\dagger] = \begin{pmatrix} -1 & 0 & 0 \\ 0 & \cos^2\theta_C & \sin\theta_C\cos\theta_C \\ 0 & \sin\theta_C\cos\theta_C & \sin^2\theta_C \end{pmatrix}. \tag{45}$$

Among the neutral currents there will be strangeness-changing currents of the type $(\bar{d}s + \bar{s}d)\sin\theta_C\cos\theta_C$ (we ignore the γ matrices for brevity), which give rise to processes such as

$$K^+ \to \pi^+ + \nu + \bar{\nu},$$
$$K_L{}^0 \to \mu^+ + \mu^-$$

at rates much larger than experimental limits or observations. Thus, a major modification is needed. The solution had actually been proposed before the Weinberg–Salam theory became popular. In 1970, S. Glashow, J. Iliopoulos, and L. Maiani, building on some earlier work of Glashow and J. D. Bjorken, proposed that the number of quark flavors be extended, with a fourth quark c carrying a new conserved quantum number called charm. The c quark is taken to have charge $\frac{2}{3}$, hypercharge $\frac{1}{3}$, and baryon number $B = \frac{1}{3}$. The weak current, constructed to have the form

$$(\bar{u}\bar{d}\bar{s}\bar{c})Q_W\gamma^\alpha(1-\gamma_5)\begin{pmatrix} u \\ d \\ s \\ c \end{pmatrix} \tag{46}$$

with

$$Q_W = \begin{pmatrix} 0 & 0 & 0 & 0 \\ \cos\theta_C & 0 & 0 & -\sin\theta_C \\ \sin\theta_C & 0 & 0 & \cos\theta_C \\ 0 & 0 & 0 & 0 \end{pmatrix}, \tag{47}$$

treats the s quark symmetrically with the d. This implies that the neutral current constructed as in (45) does not have any strangeness-changing (or charm-changing) terms. The smallness of the Cabibbo angle implies that the dominant charged-current transitions are $u \to d$ and $c \to s$, which means that mesons involving c quarks are predicted to usually decay into final states with strangeness (e.g., a $\bar{K}$) rather than into non-strange (e.g., pions) final states.

Notice that if we set $\theta_C = 0$ and replace (u,d,s,c) by $(e^-, \nu_e, \mu^-, \nu_\mu)$, we get the leptonic current. It was this analogy that originally led Bjorken and Glashow to generalize the SU(3) flavor symmetry to the four-flavor SU(4) symmetry. The analogy goes deep. It turns out that renormalizability demands that each lepton pair [e.g., (e^-, ν_e)] must have a compensating quark pair [e.g., (u,d)], so that the existence of the charmed quark is compelling in gauge theories. In the early 1970s it was also realized that the observed mass difference between the two neutral kaons could be accounted for by a calculable higher-order weak effect (Fig. 9) if the c quark existed and had a mass around 1.5 GeV/c^2.

In 1974 in the experimental study of the ratio R in (40), B. Richter and collaborators at the Stanford Linear Accelerator Center, simultaneously with S. Ting and collaborators at Brookhaven National Laboratory, who were studying the reaction

$$p+\text{Be}\rightarrow e^+ + e^- + \text{hadrons},$$

found an extremely sharp resonance at 3097 MeV and, soon after that, another one at 3685 MeV (Fig. 8). It is interesting that these had been anticipated theoretically by T. Appelquist and D. Politzer, who advanced reasons why the $J=1$ $c\bar{c}$ (charmed quark–antiquark) states should be quite long-lived. This interpretation of the $J/\psi(3097)$ and $\psi(3685)$ as $1\,^3S_1$ and $2\,^3S_1$ "charmonium" states was confirmed by the later discovery of $^3P_0$, $^3P_1$, $^3P_2$, and $^1S_0$ states as well as additional $^3S_1$ radial excitations in the vicinity of the ones already discovered. There now exists a complete and well-studied charmonium spectroscopy. The arrangement of levels and the spacing cannot be understood in terms of a simple $1/r$ potential, as for positronium. Rather, a better fit is obtained with a potential of the form $V(r)=A/r+Br$, where the first term represents the short-range (asymptotically weakly coupled) contribution and the second term represents the long-range confining potential. The extreme narrowness of the $^3S_1$ resonances can be understood qualitatively by arguing that a $c\bar{c}\rightarrow u\bar{u}$ (say) transition can only take place through the mediation of three color-carrying gluons, and these couple weakly for large momentum transfers.

Numerous hadrons consisting of a single c quark bound to ordinary quarks or antiquarks have also been identified and their decay modes studied. These include the spin-0 $(c\bar{u})$ D meson, with mass 1865 MeV/c^2, the corresponding spin-1 D^* meson of mass 2010 MeV/c^2, their charged $(D^+=c\bar{d})$ partners, and the $D_s^{\ +}$ $(c\bar{s})$ at 1969 MeV/c^2, as well as charmed baryons with isospin 0, 1, and $\frac{1}{2}$ at 2285, 2455, and 2460 MeV/c^2, respectively. The spectroscopy of the charmed hadrons, as well as additional evidence from deep inelastic neutrino scattering, have thus clearly established the existence of the c quark as needed for a consistent and realistic gauge theory of the weak interactions.

Similarly, if the τ lepton is "normal" in that it has an associated neutrino (as is indicated by the τ properties), then

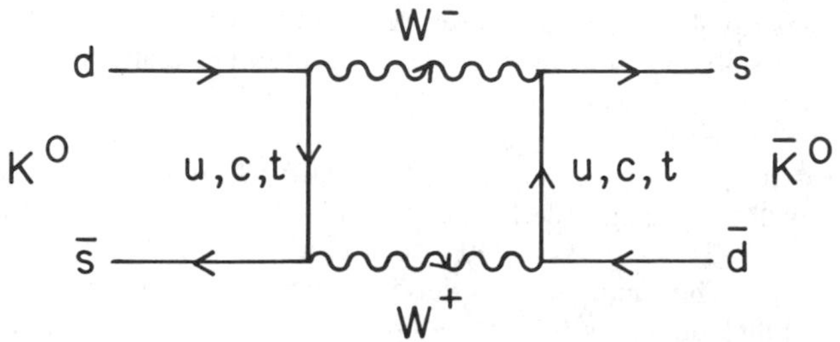

FIG. 9. Higher-order weak diagram leading to the mass difference between $K_s \sim K^0 + \bar{K}^0$ and $K_L \sim K^0 - \bar{K}^0$. The diagram with intermediate t quarks can also account for most of the observed CP violation.

another quark pair, which we label (t,b) for *top* and *bottom*, respectively, with charges $\frac{2}{3}$ and $-\frac{1}{3}$, should exist.

This notion of lepton–quark symmetry was confirmed in 1977 by the discovery by L. Lederman and collaborators at the Fermi National Accelerator Laboratory of two new narrow resonances, the Υ(9460 MeV/c^2) and Υ'(10023 MeV/c^2), in the reaction

$$p+\text{Be}\rightarrow\Upsilon+\text{ hadrons }\rightarrow\mu^+ + \mu^- + \text{ hadrons}.$$

The interpretation of these states as $^3S_1$ states of a new "quarkonium" composed of $b\bar{b}$ quarks with charge $\frac{1}{3}$ was supported by their production in e^+–e^- collisions at DESY and at the Cornell Electron Storage Ring, where additional $^3S_1$ as well as P wave states were also discovered.

The bottom mesons $B^+=u\bar{b}$ and $B^0=d\bar{b}$, with masses 5278 and 5279 MeV/c^2, as well as their antiparticles, have also been identified. The observed weak interactions of the b quark as well as renormalizability and lepton–quark symmetry indicate that a top quark (t) should exist. By the middle of 1989 the t had still not been discovered. Searches in e^+e^- reactions at the TRISTAN collider in Japan imply $m_t>27$ GeV/c^2, while searches at $p\bar{p}$ colliders suggest that $m_t>80$ GeV/c^2. Indirect arguments based on the consistency of weak neutral current data (which would be upset by higher-order corrections associated with a very heavy t) imply the upper limit $m_t<200$ GeV/c^2. Why the t (assuming it exists) is so much heavier than the other fermions is not understood.

In the four-quark model, Q_W in (47) involves a single mixing angle $\theta_C\simeq0.22$, and the higher-order diagram in Fig. 9 can account for the K_L-K_S mass difference. The generalization of Q_W to six quarks involves three mixing angles and one complex phase. In addition to θ_C, the dominant b quark transition $b\rightarrow c$ is described by a mixing angle $\theta_{bc}\sim0.05$ (determined from the b lifetime). The much weaker $b\rightarrow u$ transition, which has not yet been observed, is characterized by the angle $\theta_{bu}<0.01$. These angles suffice to account for the large mixing observed at DESY between the $B^0=d\bar{b}$ and $\bar{B}^0=\bar{d}b$ states, by a second-order diagram analogous to Fig. 9 (with the external s replaced by b) provided the top mass is sufficiently large (>50 GeV/c^2). Very large mixing between the $B_s^{\ 0}=s\bar{b}$ and $\bar{B}_s^{\ 0}=\bar{s}b$ is predicted.

Even after the discovery that the weak interactions violated parity (P) and charge conjugation (C) invariance, it was generally believed that CP invariance was maintained. In particular, the $CP=+1$ state $K_S\sim K^0+\bar{K}^0$ should decay rapidly ($\tau\sim0.89\times10^{-10}$ s) into $\pi^+\pi^-$ or $\pi^0\pi^0$, while the $CP=-1$ state $K_L\sim K^0-\bar{K}^0$ should decay into 3π in $\sim5.2\times10^{-8}$ s. However, in 1964 J. Cronin, V. Fitch, and collaborators working at Brookhaven observed the CP violating decays $K_L\rightarrow2\pi$ with branching ratios of $\sim10^{-3}$. The results could be accounted for by a small CP violating mixing between the K_L and K_S states. One possibility was that this mixing is generated by an entirely new *superweak* $\Delta S=2$ interaction. In 1973, M. Kobayashi and M. Maskawa suggested that CP breaking could be generated by the higher-order diagram in Fig. 9 if there were three (or more) fermion families, implying observable phases in Q_W. The observed K_L-K_S mixing could thus be generated either by the three-family standard model or by the superweak model. However, in 1987 an experiment at CERN observed small differences in the CP violating $K_L \to \pi^0\pi^0$ and $\pi^+\pi^-$ rates (relative to K_S) that required direct CP breaking in the decay amplitude, not just in the state mixing. If confirmed, this would be consistent with the standard model and rule out the superweak theory.

The standard SU(2)×U(1) model has been extensively confirmed. The charged current sector will be tested even more stringently in the future following the (probable) discovery of the t quark, measurement of $B_s{}^0\bar{B}_s{}^0$ mixing, the $b \to u$ transition rate, and searches for CP-violating asymmetries in B meson decays.

PROBLEMS WITH THE STANDARD MODEL

The standard model (QCD plus the Weinberg–Salam electroweak model and general relativity) has been spectacularly successful. Although some aspects (e.g., the gauge self-interactions) have not been tested and it is possible that the elementary Higgs mechanism for symmetry breaking is too naive, the basic structure of the standard model is nevertheless almost certainly correct at some level. However, it contains far too much arbitrariness to be the final story of Nature. One way of seeing this is that the minimal version with massless neutrinos has 21 free parameters, and nobody believes that the ultimate theory of Nature could have so much arbitrariness. In addition, the standard model suffers from:

(a) *The Gauge Problem:* The standard model gauge group is a complicated direct product of three factors with three independent coupling constants. *Charge quantization,* which refers to the fact that the magnitudes of the proton and of the electron electric charges are the same, is not explained.

(b) *Fermion Problem:* The standard model involves a very complicated reducible representation for the fermions. Ordinary matter can be constructed out of the fermions of the first family (ν_e, e^-, u, d). We have no fundamental understanding of why the additional families (ν_μ, μ^-, c, s), (ν_τ, τ^-, t, b), which appear to be identical with the first except that they are heavier, exist. In addition, the standard model does not explain or predict the pattern of fermion masses, which are observed to vary over five orders of magnitude, or mixings.

(c) *Higgs/Hierarchy Problem:* The spontaneous breakdown of the SU(2)×U(1) symmetry in the standard model is accomplished by the introduction of a Higgs field. Consistency requires that the mass of the Higgs boson be not too much different from the weak interaction scale; that is, it should be equal to the W mass within one or two orders of magnitude. However, there are higher-order corrections which change (renormalize) the value of the square of the Higgs mass by $\delta m_H{}^2 \sim m_P{}^2$, where $m_P \equiv (G_N/\hbar c)^{-1/2} \simeq 10^{19}$ GeV/c^2 is the gravity scale. Therefore, $\delta m_H{}^2/M_W{}^2 \gtrsim 10^{34}$, and the bare value of $m_H{}^2$ in the original Lagrangian must be adjusted or fine tuned to 34 decimal places. Such a fine tuning is possible, but extremely unattractive.

(d) *Strong CP Problem:* It is possible to add an additional term, characterized by a dimensionless parameter θ, to the QCD Lagrangian which breaks P, T, and CP invariance. Limits on the electric dipole moment of the neutron require θ to be less than 10^{-9}. However, weak interaction corrections change or renormalize the lowest-order value of θ by about 10^{-3}—that is, 10^6 times more than the total value is allowed to be. Again, one must fine-tune the bare value against the correction to a high degree of precision.

(e) *Graviton Problem:* The graviton problem has several aspects. First, gravity is not unified with the other interactions in a fundamental way. Second, even though general relativity can be incorporated into the model by hand, we have no idea how to achieve a mathematically consistent theory of quantum gravity: attempts to quantize gravity within the standard model framework lead to horrible divergences and a nonrenormalizable theory. Finally, there is yet another fine-tuning problem associated with the cosmological constant. The vacuum energy density $\langle V \rangle$ associated with the spontaneous symmetry breaking of the SU(2)×U(1) model generates an effective renormalization of the cosmological constant $\delta\Lambda = 8\pi G_N \langle V \rangle$, which is about 50 orders of magnitude larger than the observed limit. One must fine-tune the bare cosmological constant against the correction to this incredible degree of precision.

EXTENSIONS OF THE STANDARD MODEL

There must almost certainly be new physics beyond the standard model. Some of the possible types of new physics that have been discussed extensively in recent years are shown in Table VI. Additional gauge bosons, fermions, or Higgs bosons do not by themselves solve any problems, but may exist at accessible energies as remnants of underlying physics such as unified gauge groups. There could well be additional fermion families if they are sufficiently heavy. However, experimental evidence from the cosmological abundance of helium suggests, within the framework of the standard "big-bang" cosmological model, that there cannot be more than three or four species of neutrinos if they are massless (or lighter than ~1 MeV/c^2). The new e^+e^- colliders at CERN and SLAC will yield precise values for the Z width and will therefore be able to count the number of neutrinos and other particles lighter than ≃45 GeV/c^2. Many theories also predict *exotic* heavy fermions which are not simply repetitions of the known families.

Family symmetries are new global or gauge symmetries which relate the fermion families. Another approach to understanding the fermion spectrum is to assume that the quarks and leptons are bound states of still smaller constituents. However, experimental searches for substructure suggest that the underlying particles must be extremely massive (>1000 GeV/c^2). Hence, unlike all previous layers of matter, extremely strong binding would be required. Neither of these ideas has led to a particularly attractive model.

The tuning problem associated with the Higgs mechanism could be solved if the elementary Higgs fields were replaced by some sort of bound-state mechanism. However, models

Table VI Some possible extensions of the standard model

Model	Typical scale (GeV)	Motivation
New *W*s, *Z*s, fermions, Higgs	10^2–10^{19}	Remnant of something else
Family symmetry	10^2–10^{19}	Fermion (No compelling models)
Composite fermions	10^2–10^{19}	Fermion (No compelling models)
Composite Higgs	10^3–10^4	Higgs (No compelling models)
Composite W, Z (G, γ?)	10^3–10^4	Higgs (No compelling models)
New global symmetry	10^8–10^{12}	Strong *CP*
Kaluza–Klein Higgs (0) ⇔ gauge (1) ⇔ Graviton (2) ($d > 4$)	10^{19}	Graviton
Grand unification Strong ⇔ electroweak	10^{14}–10^{19}	Gauge
Supersymmetry/supergravity Fermion ⇔ boson	10^2–10^4	Higgs, graviton
Superstrings Strong ⇔ electroweak ⇔ gravity Fermion ⇔ boson ($d > 4$)	10^{19}	All problems!?

which generate fermion as well as W,Z masses are complicated and tend to predict certain unobserved decays at too rapid a rate. Composite W and Z bosons could be an alternative to spontaneous symmetry breaking, but such theories abandon most of the advantages of gauge theories.

The strong *CP* problem could be resolved by the addition of a new global symmetry which ensures that θ is a dynamical variable which is zero in the lowest energy state. Such models imply a weakly coupled Goldstone boson (the *axion*) associated with the symmetry breaking. Constraints from astrophysics and cosmology limit the symmetry breaking scale to the range 10^8–10^{12} GeV. Much of this window may be closed by the energetics of supernova 1987A, which would be affected by axion emission.

Kaluza–Klein theories are gravity theories in $d>4$ dimensions of space and time. It is assumed that four dimensions remain flat while the other $d-4$ are *compactified* or curled up into a compact manifold with a typical radius of $\hbar/m_P c \sim 10^{-33}$ cm. Gravitational interactions associated with these unobserved compact dimensions would appear as effective gauge interactions and Higgs particles in our four-dimensional world. Although such ideas are extremely attractive for unifying the interactions they have great difficulty in incorporating parity violation and also in achieving a stable configuration for the compact manifold.

Much work has been devoted to grand unified gauge theories (GUTs), in which it is proposed that at very high energies (or very short distances) the underlying symmetry is a single gauge group that contains as its subgroups both SU(3) for color and SU(2) × U(1) for the weak and electromagnetic interactions. The larger symmetry is manifest above a unification scale M_X at which it is spontaneously broken. At lower energies, the symmetry is hidden and the strong and electroweak interactions appear different. The unification scale can be estimated from the observed low-energy coupling constants. The energy-dependent couplings are expected to meet at M_X, as in Fig. 10. Since they vary only logarithmically, M_X is predicted to be extremely large, typically around 10^{14} GeV/c^2. (These heady speculations rest on the assumption that nothing fundamentally different occurs in a leap in energy scale from 1 to 10^{14} GeV. This is extremely unlikely, but then who would have thought that Maxwell's equations are good over distances ranging from astronomical down to 10^{-15} cm?) From the strong and electromagnetic couplings one can predict the weak angle $\sin^2\theta_W$. The simplest GUTs predict a value about 2.5σ too low, but variations can yield the correct value.

In addition to unifying the interactions, the additional symmetries within a grand unified theory relate quarks, antiquarks, leptons, and antileptons. For example, the simplest grand unified theory—the SU(5) model of H. Georgi and S. Glashow—assigns the left-helicity (L) fermions of the first family to a reducible $5^* + 10$-dimensional representation:

$$W^{\pm}\updownarrow \underset{\leftarrow X,Y\rightarrow}{\overset{5^*}{\begin{pmatrix} \nu_e & \\ & \bar{d} \\ e^- & \end{pmatrix}_L}} \quad \underset{\leftrightarrow X,Y\leftrightarrow}{\overset{10}{\begin{pmatrix} & u & \\ e^+ & & \bar{u} \\ & d & \end{pmatrix}_L}} . \tag{48}$$

The neutrino and electron can be rotated into the anti-down quark by the new symmetry generators, as can the positron and the up, down, and anti-up quarks. Because of this relation between the different types of fermions, the grand unified theories naturally explain charge quantization. In addition, there are new gauge bosons associated with these extra symmetries. The SU(5) model contains new bosons which carry both color and electric charge, known as the X and Y bosons, with masses $M_X \sim M_Y \sim 10^{14}$ GeV/c^2. These can

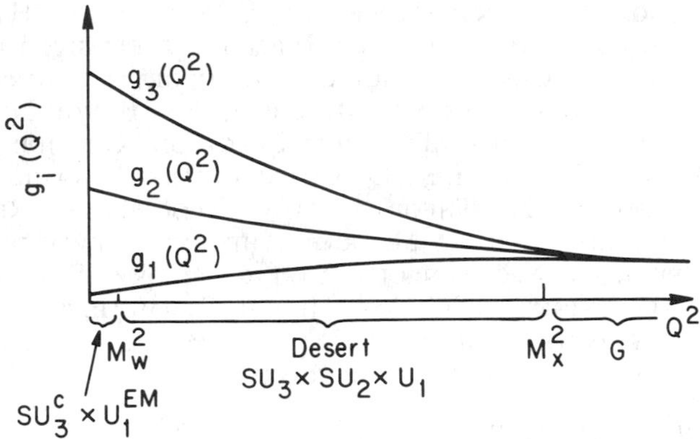

FIG. 10. The momentum dependent normalized coupling constants $g_3 \equiv g_s$, $g_2 \equiv g$, and $g_1 \equiv \sqrt{5/3}\, g'$.

mediate proton decay (and also the decay of otherwise stable bound neutrons). A diagram for $p \to e^+\pi^0$ is shown in Fig. 11. Motivated by such predictions, a number of experiments have searched for proton decay. No events have been observed, and one finds $\tau(p \to e^+\pi^0) > 4 \times 10^{32}$ yr, in conflict with the prediction $4 \times 10^{29 \pm 2}$ yr of the simplest GUTs. However, many extended or nonminimal models predict longer lifetimes, often in the 10^{30}–10^{35}-yr range.

An attractive feature of grand unified theories is that they can naturally explain the cosmological baryon asymmetry. It is an observational fact that our part of the Universe consists of matter and not antimatter: there is approximately one baryon for every 10^{10} microwave photons, but essentially no antibaryons. In GUTs this asymmetry does not need to be invoked as an initial condition on the big bang; rather it can be generated dynamically in the first instant (the first 10^{-35} s or so) after the big bang, by baryon number violating interactions related to those which lead to proton decay.

Grand unified theories predict the existence of superheavy *magnetic monopole* states with masses $M_M \sim M_X/\alpha_G \sim 10^{16}$ GeV/c^2. These may have been produced during phase transitions during the early Universe and could still be left over as relics today. Monopoles could be detected as a component of cosmic rays, but unfortunately we have no firm prediction as to what the flux should be. So far none have been detected.

Grand unified theories offer little or no help with the fermion, Higgs, strong *CP*, or graviton problems, and the simplest versions are ruled out by $\sin^2 \theta_W$ and proton decay.

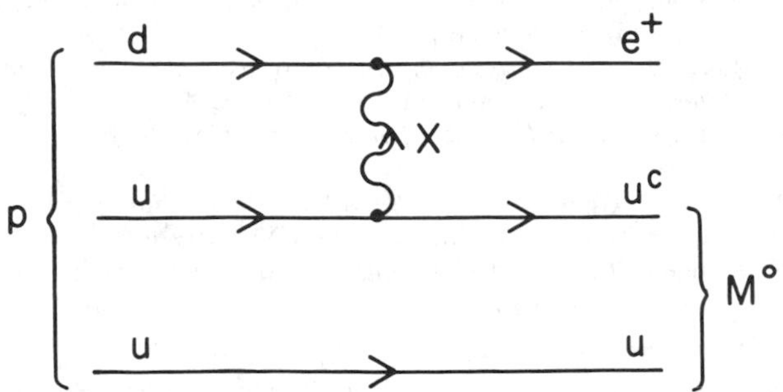

FIG. 11. A typical proton decay diagram.

GUTs are probably not ambitious enough, but it is likely that some ingredients (e.g., related to charge quantization and the baryon asymmetry) will survive.

Supersymmetry (see SUPERSYMMETRY AND SUPERGRAVITY) is a new kind of symmetry in which fermions are related to bosons. Realistic models are complicated in that they require more than a doubling of the number of fundamental particles. For example, one must introduce a spin-0 superpartner of each known fermion, a spin-$\frac{1}{2}$ partner of each gauge boson or Higgs field, etc. The primary motivation is the Higgs problem: higher-order corrections associated with the new particles cancel the unacceptable renormalization of the Higgs mass that occurs in the standard model. If supersymmetry were exact, the new particles would be degenerate with the ordinary particles. However, there exist experimental lower limits of order 80 GeV/c^2 on the masses of most superpartners, so that supersymmetry (if it exists) must be broken. The most successful models assume that supersymmetry is broken at some very large scale (e.g., 10^{11} GeV) in a sector which is only coupled very weakly to the ordinary particles and their partners, implying superpartner masses in the range 10^2–10^4 GeV/c^2. (Larger masses would fail to solve the Higgs problem.)

Supersymmetric grand unified theories lead to larger values for $\sin^2 \theta_W$ (in better agreement with experiment) and a larger unification scale. This leads to a longer proton lifetime into $e^+\pi^0$ from X and Y exchange. However, some versions lead to new mechanisms for proton decay (usually into modes such as $\bar{\nu}K$) mediated by the new superpartners at rates that may be unacceptable.

Just as the promotion of an ordinary internal symmetry to a gauge symmetry implies the existence of spin-1 gauge bosons, the requirement that supersymmetry transformations can be performed independently at different space-time points implies the existence of spin-2 gravitons; i.e., gauged supersymmetry automatically unifies gravity. However, *supergravity* theories do not solve the problems of quantum gravity. Higher-order corrections are still divergent and nonrenormalizable. There is no experimental evidence for supersymmetry. Whether Nature chooses to make use of it remains to be seen.

Superstrings are a very exciting development which may yield finite theories of all interactions with no free parameters. Superstring theories introduce new structure above the Planck scale. The basic idea is that instead of working with (zero-dimensional) point particles as the basic quantities, one considers one-dimensional objects known as strings, which may be open or closed. The quantized vibrations of these strings lead to an infinite set of states, the spectrum of which is controlled by the string tension, given by the square of the Planck scale $\tau \sim m_P{}^2$. When one probes or observes a string at energies much less than the Planck scale, one sees only the "massless" modes, and these represent ordinary particles. The physical size of the string is given by the inverse of the Planck scale $\simeq 10^{-33}$ cm, and at larger scales a string looks like a point particle.

The mathematical consistency of superstring theories requires (in most versions) that there are nine dimensions of space and one of time. Presumably, the extra six space dimensions are curled into a compact manifold of radius

$\sim 10^{-33}$ cm, reminiscent of Kaluza–Klein theories. However, the interactions are due to gauge interactions in the ten-dimensional space. The absence of mathematical pathologies requires an essentially unique gauge group called $E_8 \times E_8$. At energy scales less than the Planck scale an effective particle field theory in four dimensions emerges, with a supersymmetric gauge symmetry based on a subgroup of $E_8 \times E_8$. The effective group, the number of fermions, and their masses, mixings, etc., are all determined by the way in which the extra dimensions are compactified. In principle, superstring theories have no arbitrary parameters or other features, and most likely they yield completely finite (not just renormalizable) quantum theories of gravity and all the other interactions. That is, they are candidates for the ultimate "theory of everything." Unfortunately, there are enormous numbers of possible ways in which the extra dimensions can compactify, and at present we do not possess the principles or mathematical tools to determine which is chosen. It is therefore not clear what the predictions of superstring theories really are (e.g., whether they lead to an effective supergravity GUT) or whether they correspond to the real world.

THE FUTURE

The standard model is extremely successful, but there is almost certainly new physics underlying it. There are many theoretical ideas concerning this new physics. Some of the most promising involve energy scales much larger than will ever be probed by direct experimentation, though they may still lead to testable low-energy predictions. Apart from theoretical ideas and new computational techniques (e.g., lattice calculations), many types of experiments will improve the tests of the standard model and search for manifestations of new physics. These include direct searches for new particles at high-energy accelerators, precision neutral current and charged current electroweak tests and QCD tests, searches for rare decays of muons, kaons, B mesons, etc., that are forbidden or strongly suppressed in the standard model, and searches for neutrino mass (which is predicted to be nonzero in most extensions and which has very important astrophysical and cosmological implications). Other probes include searches for magnetic monopoles and proton decay. Finally, there has been an increasingly close connection between particle physics and cosmology. The dynamics of the early Universe was controlled by the elementary particles and their interactions, and refined studies of the large-scale structure of the Universe and of other relics from the big bang place severe constraints on new physics.

It is impossible to do justice to the subtlety and richness of this field in a brief survey, or to give their due to the thousands of researchers who have painstakingly uncovered the beautiful structure that is emerging. The interplay of imaginative experimentation and daring conjectures has been a source of wonder to all who have witnessed the growth and maturing of elementary-particle physics.

See also Baryons; Compton Effect; Conservation Laws; Currents in Particle Theory; Electron; Feynman Diagrams; Field Theory, Axiomatic; Gauge Theories; Grand Unified Theories; Gravitation; Hadrons; Hypernuclear Physics and Hypernuclear Interactions; Hyperons; Inclusive Reactions; Isospin; Lattice Gauge Theory; Leptons; Lie Groups; Mesons; Muonium; Neutrino; Nucleon; Partons; Positron–Electron Colliding Beams; Positronium; Quantum Electrodynamics; Quantum Field Theory; Quarks; Regge Poles; Renormalization; *S*-Matrix Theory; Symmetry Breaking, Spontaneous; String Theory; Strong Interactions; SU(3) and Higher Symmetries; Supersymmetry and Supergravity; Weak Interactions; Weak Neutral Currents.

BIBLIOGRAPHY

Space limitations preclude an exhaustive bibliography. At best we can provide references to some standard textbooks, monographs, readily available review articles, and popular articles covering some recent developments.

Textbooks

I. Aitchison and A. Hey, *Gauge Theories in Particle Physics: A Practical Introduction*. A. Hilger, Bristol, 1982.

T. P. Cheng and L.-F. Li, *Gauge Theory of Elementary Particle Physics*. Clarendon Press, Oxford, 1984.

S. Gasiorowicz, *Elementary Particle Physics*. Wiley, New York, 1966.

D. H. Perkins, *Introduction to High Energy Physics*. Benjamin, Menlo Park, CA, 1982.

J. Smith and B. de Wit, *Field Theory in Particle Physics*. North-Holland, Amsterdam, 1986.

Monographs

S. L. Adler and R. F. Dashen, *Current Algebra and Applications to Particle Physics*. Benjamin, New York, 1968.

V. D. Barger and D. B. Cline, *Phenomenological Theories of High Energy Scattering*. Benjamin, New York, 1969.

F. Boehm and P. Vogel, *Physics of Massive Neutrinos*. Cambridge University Press, New York, 1987.

S. Coleman, *Aspects of Symmetry*. Cambridge University Press, New York, 1985.

E. D. Commins and P. H. Bucksbaum, *Weak Interactions of Leptons and Quarks*. Cambridge University Press, New York, 1983.

V. de Alfaro, S. Fubini, G. Furlan, and C. Rossetti, *Currents in Hadron Physics*. North-Holland, Amsterdam, 1973.

R. P. Feynman, *Photon–Hadron Interactions*. Benjamin, Reading, MA, 1973.

P. H. Frampton, *Gauge Field Theories*. Benjamin, Menlo Park, CA, 1987.

M. B. Green, J. H. Schwarz, and E. Witten, *Superstring Theory*. Cambridge University Press, New York, 1987.

B. Kayser, *The Physics of Massive Neutrinos*. World, Singapore, 1989.

R. N. Mohapatra, *Unification and Supersymmetry*. Cambridge University Press, New York, 1986.

C. Quigg, *Gauge Theories of the Strong, Weak, and Electromagnetic Interactions*. Benjamin, Reading, MA, 1983.

G. C. Ross, *Grand Unified Theories*. Benjamin, Menlo Park, CA, 1985.

J. C. Taylor, *Gauge Theories of the Weak Interactions*. Cambridge University Press, London and New York, 1976.

S. Weinberg, *The First Three Minutes, A Modern View of the Origin of the Universe*. Basic, New York, 1977.

Advanced Review Articles

E. S. Abers and B. W. Lee, "Gauge Theories," *Phys. Rep.* **9C**, 1 (1973).

U. Amaldi *et al.*. "A Comprehensive Analysis of Data Pertaining to the Weak Neutral Current and the Intermediate Vector Boson Masses," *Phys. Rev.* **D36,** 1385 (1987).

P. Bagnaia and S. D. Ellis, "CERN Collider Results and the Standard Model," *ARNPS* **38,** 659 (1988).

B. C. Barish and R. Stroynowski, "The Physics of the τ Lepton," *Phys. Rep.* **157,** 1 (1988).

J. D. Barrow, "Cosmology and Elementary Particles," *Fund. Cosmic Phys.* **8,** 85 (1983).

M. A. B. Bég and A. Sirlin, "Gauge Theories of Weak Interactions," *Annu. Rev. Nucl. Sci.* **24,** 379 (1974); *Phys. Rep.* **88,** 1 (1982).

J. Bernstein, "Spontaneous Symmetry Breaking, Gauge Theories, the Higgs Mechanism and All That," *Rev. Mod. Phys.* **46,** 7 (1974).

M. S. Chanowitz, "Electroweak Symmetry Breaking," *Annu. Rev. Nucl. Part. Sci.* **38,** 323 (1988).

H. Y. Cheng, "The Strong *CP* Problem Revisited," *Phys. Rep.* **158,** 1 (1988).

P. D. B. Collins, "Regge Theory and Particle Physics," *Phys. Rep.* **1C,** 103 (1971).

B. Diekmann, "Spectroscopy of Mesons Containing Light Quarks (*u,d,s*) or Gluons," *Phys. Rep.* **159,** 99 (1988).

J. F. Donoghue, E. Golowich, and B. Holstein, "Low Energy Weak Interactions of Quarks," *Phys. Rep.* **131,** 319 (1986).

E. Eichten, I. Hinchliffe, K. Lane, and C. Quigg, "Super Collider Physics," *Rev. Mod. Phys.* **56,** 579 (1984).

H. E. Fisk and F. Sciulli, "Charged-Current Neutrino Interactions," *Annu. Rev. Nucl. Part. Sci.* **32,** 499 (1982).

P. J. Franzini, "$B\bar{B}$ Mixing: A Review of Recent Progress," *Phys. Rep.* **173,** 1 (1989).

M. K. Gaillard, B. W. Lee, and J. L. Rosner, "Search for Charm," *Rev. Mod. Phys.* **47,** 277 (1975).

J. Gasser and H. Leutwyler, "Quark Masses," *Phys. Rep.* **87,** 77 (1982).

M. B. Green, "Unification of Forces and Particles in Superstring Theories," *Nature* **314,** 409 (1985).

O. W. Greenberg, "Quarks," *Annu. Rev. Nucl. Part. Sci.* **28,** 327 (1978).

D. E. Groom, "In Search of the Supermassive Magnetic Monopole," *Phys. Rep.* **140,** 323 (1986).

H. E. Haber and G. L. Kane, "The Search for Supersymmetry," *Phys. Rep.* **117,** 75 (1985).

A. Hasenfratz and P. Hasenfratz, "Lattice Gauge Theories," *Annu. Rev. Nucl. Part. Sci.* **35,** 559 (1985).

V. W. Hughes, "Atomic Physics and Fundamental Principles," *Nucl. Phys.* **A463,** 3C (1987).

J. E. Kim *et al.,* "A Theoretical and Experimental Review of the Weak Neutral Current," *Rev. Mod. Phys.* **53,** 211 (1981).

E. W. Kolb and M. S. Turner, "Grand Unified Theories and the Origin of the Baryon Asymmetry," *Annu. Rev. Nucl. Part. Sci.* **33,** 645 (1983).

W. Kwong, J. L. Rosner, and C. Quigg, "Heavy Quark Systems," *Annu. Rev. Nucl. Part. Sci.* **37,** 325 (1987).

P. Langacker, "Grand Unified Theories and Proton Decay," *Phys. Rep.* **72,** 185 (1981); *Comm. Nucl. Part. Phys.* **15,** 41 (1985).

P. Langacker, "Is the Standard Model Unique?," *Comm. Nucl. Part. Sci.* **19,** 1 (1989).

L. McLerran, "The Physics of the Quark–Gluon Plasma," *Rev. Mod. Phys.* **58,** 1021 (1986).

W. Marciano and H. Pagels, "Quantum Chromodynamics," *Phys. Rep.* **36C,** 137 (1978).

W. J. Marciano and Z. Parsa, "Electroweak Tests of the Standard Model," *Annu. Rev. Nucl. Part. Sci.* **36,** 171 (1986).

H. P. Nilles, "Supersymmetry, Supergravity, and Particle Physics," *Phys. Rep.* **110,** 1 (1984).

H. Pagels, "Departures from Chiral Symmetry," *Phys. Rep.* **16C,** 219 (1975).

J. Preskill, "Magnetic Monopoles," *Annu. Rev. Nucl. Part. Sci.* **34,** 461 (1984).

J. R. Primack, D. Seckel, and B. Sadoulet, "Detection of Cosmic Dark Matter," *Annu. Rev. Nucl. Part. Sci.* **38,** 751 (1988).

J. Rich, D. L. Owen, and M. Spiro, "Experimental Particle Physics Without Accelerators," *Phys. Rep.* **151,** 239 (1987).

N. P. Samios, M. Goldberg, and B. T. Meadows, "Hadrons and SU(3): A Critical Review," *Rev. Mod. Phys.* **46,** 49 (1974).

G. Steigman, "Cosmology Confronts Particle Physics," *Annu. Rev. Nucl. Part. Sci.* **29,** 313 (1979).

E. H. Thorndike and R. A. Poling, "Decays of the *b* Quark," *Phys. Rep.* **157,** 183 (1987).

F. Wilczek, "Quantum Chromodynamics: The Modern Theory of the Strong Interaction," *Annu. Rev. Nucl. Part. Sci.* **32,** 177 (1982).

L. Wolfenstein, "Present Status of *CP* Violation," *Annu. Rev. Nucl. Part. Sci.* **36,** 137 (1986).

T. M. Yan, "The Parton Model," *Ann. Rev. Nucl. Sci.* **26,** 199 (1976).

There are also many excellent reviews available in conference and summer school *Proceedings*.

Scientific American Articles

M. A. Bouchiat and L. Pottier, "Atomic Parity Violation," June (1984).

R. A. Carrigan and W. P. Trower, "Superheavy Monopoles," April (1982).

D. Cline, C. Rubbia, and S. Van der Meer, "Search for the Intermediate Vector Boson," March (1982).

D. B. Cline, A. K. Mann, and C. Rubbia, "The Detection of Neutral Weak Currents," December (1974).

D. Freedman and P. van Nieuwenhuizen, "The Hidden Dimensions of Spacetime," March (1985).

H. Georgi, "Unified Theories," April (1981).

M. Green, "Superstrings," September (1986).

A. Guth and P. Steinhardt, "Inflation," May (1984).

H. Haber and G. Kane, "Is Nature Supersymmetric?," June (1986).

G. 't Hooft, "Gauge Theories," June (1980).

J. D. Jackson, M. Tigner, and S. Wojcicki, "The Superconducting Supercollider," March (1986).

L. Krauss, "Dark Matter in the Universe," December (1986).

J. Lo Secco, F. Reines, and D. Sinclair, "Search for Proton Decay," June (1985).

Y. Nambu, "The Confinement of Quarks," November (1976).

C. Quigg, "Elementary Particles and Forces," April (1985).

D. N. Schramm and G. Steigman, "Particle Accelerators Test Cosmological Theory," June (1988).

J. H. Schwarz, "Dual Resonance Models of Elementary Particles," February (1975).

M. Veltman, "The Higgs Boson," November (1986).

A. Vilenkin, "Cosmic Strings," December (1987).

S. Weinberg, "Unified Theories of Elementary Particle Interactions," July (1974).

S. Weinberg, "Proton Decay," June (1981).

E. P. Wigner, "Violations of Symmetry in Physics," December (1965).

F. Wilczek, "The Matter-Antimatter Asymmetry," December (1980).

Other

Resource letter NP-1: New Particles, J. L. Rosner, *Am. J. Phys.* **48** (No. 2), 90 (1980).

Finally, every other April the Particle Data Group's updated "Review of Particle Properties" appears in *Physics Letters B*.

Elements

O. Lewin Keller, Jr.

The concept that one or a few elementary substances could interact to form all matter was originated by Greek philosophers beginning in the sixth century B.C. The atomic hypothesis, which has proven indispensable in the development of our understanding of chemical elements, originated with the philosopher Leucippus and his follower Democritus in the fifth century B.C. These two interacting concepts of elements and atoms have been unsurpassed in their importance to the development of modern science and technology.

Aristotle accepted from the earlier philosophers that air, water, earth, and fire were elements, and he added a fifth—the ether—representing the heavenly bodies. Although all matter was supposed to be formulated from these elements, in Aristotelian thought the elements themselves represented qualities and were nonmaterial. Centuries later, as the alchemists found new transformations, three more elements were added. These were sulfur, mercury, and salt, thought to represent the quantities of combustibility, volatility, and incombustibility, respectively. Aristotle's view of the elements as nonmaterial qualities thus prevailed for about 2000 years. It was only in the seventeenth century that a scientific atomic theory began to emerge from the philosophical theory of Democritus. In 1661 Robert Boyle, who had developed a chemical atomic theory based on the concepts of Democritus, gave the definition of chemical elements as "certain primitive and simple, or perfectly unmingling bodies, which, not being made of any other bodies or one another" are the constituents of chemical compounds. This physical and rather modern definition did not allow Boyle to conceive how such elements might be distinguished from compounds, however. The experimental determination of which substances were actually elements awaited the development of chemistry as a quantitative science. Over one and one quarter centuries after Boyle had given his definition of an element, Antoine-Laurent Lavoisier was able to determine a list of elements based on an experimentally verifiable definition: A chemical element is a substance that cannot be decomposed further into simpler substances by ordinary chemical means. He had determined his elements (about 30) by making careful quantitative studies of decomposition and recombination reactions. However, his list still included some very stable compounds such as silica and alumina, which he had not been able to break down. Also, from a philosophical point of view, it is interesting that he included heat and light! Apparently, the Greek influence was lingering on.

Work of the sort being carried out by Lavoisier soon led to the development of the law of definite proportions, which stated that in any given compound the elements always occur in the same proportions by weight no matter how the compound is synthesized. This law led to the definition of "equivalent weights" of elements as that weight which will combine with or replace a unit weight of some standard element such as hydrogen. John Dalton, in 1808, was the first to postulate an atomic theory that incorporated atomic weight as distinguished from equivalent weight and was capable of explaining the empirically derived laws of chemical combination. Dalton assumed that (1) all atoms of a given element are identical but are different from the atoms of other elements; (2) compounds are formed from these elemental atoms; (3) chemical reactions result from the atoms becoming rearranged in new ways; and (4) if only one compound can be formed from two elements X and Y, then that compound contains one atom of X and one atom of Y. So, for example, Dalton assumed water was HO instead of H_2O. Dalton needed postulate (4) in order to determine atomic weights experimentally, and he decided that nature is simpler than it is. Really, the only new idea introduced by Dalton was that of atomic weight as the ratio of the mass of a given element's atoms to the mass of some standard atom such as hydrogen. Naturally, Dalton's too-simple postulate (4) led to confusion in the atomic weight field, but a considerable resolution of these difficulties by Joseph-Louis Gay-Lussac, Amedeo Avogadro, and Stanislao Cannizzaro resulted in the publication of useful specific results on atomic weights.

The availability of fairly reliable atomic weights for a number of elements allowed chemists to seek relationships among them on a weight basis. In 1869, Dimitri Mendeleev knew of 65 elements with their atomic weights. While looking for relationships among these elements he made one of the most important discoveries in the history of chemistry: The properties of the elements are periodic functions of their atomic weights. This periodic law allowed the arrangement of the elements in a table in order of increasing atomic weight such that the table contains columns and rows of elements. Elements with similar chemical properties, such as silicon, tin, and lead, were found to fall in the same column. Thus there appeared a regular recurrence of chemical and physical properties of the elements from the top to the bottom of the column even though the elements were widely separated in atomic weight. For the first time in history it was shown that the chemical elements form an entity in their interrelationships, and undiscovered elements with predictable properties could be sought to fill up the holes in the table.

The periodic law proposed by Mendeleev was a daring break with the thought of the scientific community in 1869. In fact, Mendeleev's bold predictions of the chemical and physical properties of still undiscovered elements undoubtedly furnished the touch of drama needed to gain acceptance of his system. The predictions are probably the main reason Mendeleev's name is more firmly associated with the periodic law than that of Lothar Meyer, who discovered it independently and simultaneously. The three most famous predictions by Mendeleev concerned eka-aluminum (gallium), eka-boron (scandium), and eka-silicon (germanium). When the elements themselves were discovered, Mendeleev's detailed predictions were found to be amazingly accurate.

Earlier formulations of somewhat limited periodicities had led to obscurity or ridicule. A particularly brilliant insight is represented by the three-dimensional model presented in 1862 to the French Academy by Alexandre E. Beguyer de Chancourtois. In this model the elements were placed on a helix drawn on a cylinder in such a way that the distances of the elements along the helix were proportional to their atomic weights. This procedure resulted in elements of similar properties falling on parallel vertical lines along the surface of the cylinder. De Chancourtois stated, as a result, that "the properties of substances are properties of numbers." His remarkable insight went unnoticed, however, and the *Comptes Rendus* did not publish his table for nearly 30 years. The limited observation by J. A. R. Newlands in 1864 that a periodicity with atomic weight existed after each eighth element was another important foreshadowing of the periodic law. Unfortunately, his "law of octaves" received only ridicule at the time and the Chemical Society refused to publish his paper, although 23 years later they awarded him the Davy Medal for it.

Mendeleev's periodic table naturally emphasized the value of more accurate atomic weights, partly because some elements were not in the proper order. As more accurate values for atomic weights were obtained, it was found that some of the elements persisted in disobeying the periodic law. Nickel is lighter than cobalt, iodine is lighter than tellurium, and potassium is lighter than argon, yet in each case, it was clear from the chemical properties that the lighter element must follow the heavier in the periodic table. Also, further anomalies in atomic weights developed as measurements became more precise. For example, T. W. Richards obtained the perplexing result that the atomic weight of lead depends on its geological source. Difficulties of this sort were finally resolved through the unexpected discovery by Sir J. J. Thompson that all atoms of an element are not identical, but can differ in their atomic weights. In 1912, Thompson was studying a gas discharge of neon in a mass spectrometer he had constructed. Much to his surprise he found that the neon atoms corresponded to two atomic weights, 20 and 22, rather than one atomic weight of 20.2. Thus, one of the most cherished of the concepts of the atomic hypothesis was reluctantly acknowledged to be wrong—all atoms of a given element are *not* identical. After the neutron was discovered, it was possible to understand that these atoms had the same number of protons in the nucleus but a different number of neutrons. Such atoms are called isotopes. Their chemical properties are virtually identical, although not completely so. The atomic weight of an element was redefined as an average for its naturally occurring isotopes. Also, it became the practice to simply assign a number to each element according to its place in the periodic table, since the order was obviously not strictly by atomic weight. This identifying number from the periodic table came to be called the *atomic number* of the element.

After the discovery of the electron in 1897 by Sir J. J. Thompson following experimentation of Sir William Crookes, several models of atomic structure were proposed, including the nuclear model of Hantaro Nagaoka. Nagaoka viewed the positive charge as concentrated at the center of the atom, with the electrons forming rings (like the rings of Saturn) around the central charge. The essential correctness of the nuclear picture of the atom was established by Ernest Rutherford through studies of the scattering of alpha particles (helium ions) by thin metallic foils. Thus atoms were shown to have structure, and another of the basic concepts of the atomic hypothesis was dethroned. Rutherford's experiments also gave the value of the charge on the nucleus for the elements he studied. The value was found to be roughly one half the atomic weight. In 1911 A. van den Broek suggested that the atomic number of an element could be identified with the number of protons in the nucleus. The suggestion of van den Broek was confirmed in a brilliant series of experiments carried out by H. G. J. Moseley in 1913. Moseley found that the atomic number (nuclear charge) of an element can be easily identified with the energy of its characteristic *K* and *L* x rays. Through the characteristic x rays the atomic number of each element could be positively identified with its position in the periodic table. It was thus found that, indeed, the properties of the elements are correlated in a periodic fashion through their atomic numbers rather than through their atomic weights as originally assumed by Mendeleev.

In 1922, Niels Bohr was able to interpret the periodicity of the chemical and physical properties of the elements by atomic number as a reflection of periodicity in electronic structure. In current periodic tables (Fig. 1) there are three short periods (rows) of 2, 8, and 8 elements; three long periods of 18, 18, and 32 elements; and an incomplete long period of 20 elements—a total of 106 elements in all. The progression of properties in a row occurs as each electron is added in going from each atomic number to the next higher atomic number. The electrons are added to the electronic shells of the atoms in accordance with the laws of quantum mechanics. The development of an understanding of the electronic structure of atoms on the basis of quantum theory began with Bohr's formulation in 1913. The culmination of the basic foundation was achieved in 1925 with the discovery of the electron spin by Uhlenbeck and Goudsmit, the discovery of the exclusion principle by Pauli, and the formulation of quantum mechanics by Heisenberg and Schrödinger. The electronic structure of atoms is described in terms of "orbitals," each of which is able to accommodate up to two electrons. The description of these orbitals is given in terms of so-called quantum numbers, which are designated n, l, and m_l. The principal quantum number n can assume the values 1, 2, 3, ...; the azimuthal quantum number l can have the values 0, 1, 2, ..., $n-1$; and the orbital magnetic quantum number m_l can be 0 and any positive or negative integer up to and including l. There are therefore $2l+1$ values of m_l for each value of l. Designations that are now arbitrary have been given to the orbitals. For $l=0$ the orbital is given the designation *s*; for $l=1$, *p*; for $l=2$, *d*; for $l=3$, *f*; and so on. The Pauli exclusion principle states that a maximum of two electrons can occupy one of these orbitals. This is because of the spin magnetic quantum number m_s which can assume the values $\pm\frac{1}{2}$. In order for two electrons to occupy the same orbital, their spins must be paired.

All electrons with the same value of n appear in the same "shell." The shells are called *K*, *L*, *M*, *N*, ... for $n=1$, 2, 3,4, ... It is the orderly filling of these electronic shells that gives the periodicity to the properties of the elements (Fig. 1). The *K* shell is filled at helium with two electrons ($n=1$; $l=0$). The first short period completes the *L* shell at neon by filling the 2*s* and 2*p* subshells ($n=2$; $l=0,1$). The second short period ends at argon where the 3*s* and 3*p* subshells are filled, although the *M* shell is still incomplete. The filling of the five 3*d* orbitals in the *M* shell produces the first transition-element series of 10 elements (the "iron group"),

IA	IIA	IIIA	IVA	VA	VIA	VIIA	VIIIA			IB	IIB	IIIB	IVB	VB	VIB	VIIB	VIII
H 1																	He 2
Li 3	Be 4											B 5	C 6	N 7	O 8	F 9	Ne 10
Na 11	Mg 12											Al 13	Si 14	P 15	S 16	Cl 17	Ar 18
K 19	Ca 20	Sc 21	Ti 22	V 23	Cr 24	Mn 25	Fe 26	Co 27	Ni 28	Cu 29	Zn 30	Ga 31	Ge 32	As 33	Se 34	Br 35	Kr 36
Rb 37	Sr 38	Y 39	Zr 40	Nb 41	Mo 42	Tc 43	Ru 44	Rh 45	Pd 46	Ag 47	Cd 48	In 49	Sn 50	Sb 51	Te 52	I 53	Xe 54
Cs 55	Ba 56	La 57	Hf 72	Ta 73	W 74	Re 75	Os 76	Ir 77	Pt 78	Au 79	Hg 80	Tl 81	Pb 82	Bi 83	Po 84	At 85	Rn 86
Fr 87	Ra 88	Ac 89	Rf 104	Ha 105	106	107	108	109	(110)	(111)	(112)	(113)	(114)	(115)	(116)	(117)	(118)

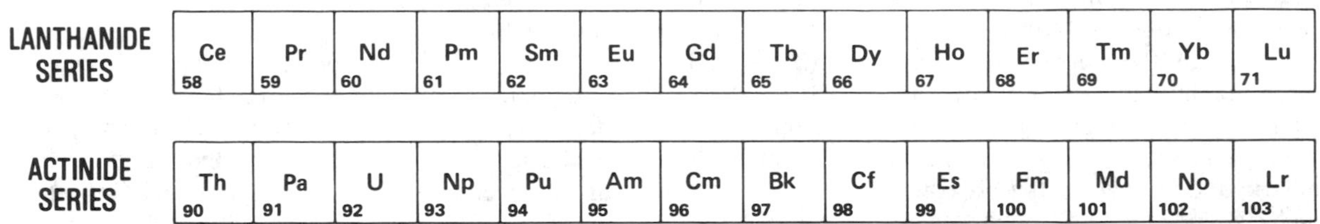

LANTHANIDE SERIES	Ce 58	Pr 59	Nd 60	Pm 61	Sm 62	Eu 63	Gd 64	Tb 65	Dy 66	Ho 67	Er 68	Tm 69	Yb 70	Lu 71
ACTINIDE SERIES	Th 90	Pa 91	U 92	Np 93	Pu 94	Am 95	Cm 96	Bk 97	Cf 98	Es 99	Fm 100	Md 101	No 102	Lr 103

FIG. 1. Periodic table of elements.

which begins at scandium and ends at zinc. The 4*s* subshell is filled at calcium and the 4*p* subshell is filled at the next rare gas, krypton. This leaves the *M* shell full but the *N* shell incomplete at the end of the first long period. The second long period, beginning at rubidium and ending at xenon, leaves the *N* shell complete with an octet of electrons in the *O* shell. The next long period contains 32 elements rather than 18. This results from the appearance of 14 elements characterized by 4*f* electrons in addition to the 18 elements corresponding to the filling of the 5*d* subshell of 10 electrons and the 6*s*, 6*p* octet. The 4*f* or lanthanide elements have extremely similar properties because their valence and outer electronic shells are essentially the same, with the 4*f* orbitals being deeply buried in the atom where they can have little effect. They thus effectively represent one "place" in the periodic table, and this is represented by placing element 57, lanthanum, in the body of the table and placing the other lanthanides at the bottom of the table. The next period, currently pausing at element 106, will presumably end at element 118, the next rare gas. In this period, the 5*f* or actinide elements are built up in a fashion analogous to the lanthanides. They are therefore placed beneath the lanthanides to show this similarity, as first suggested by Glenn T. Seaborg.

Discovery of new elements is still an active and challenging area of chemistry and physics. The heaviest element discovered to date (December 1988) is 109. The difficulties in new element discovery are illustrated by the 11-day bombardment time needed by the group at the Gesellschaft für Schwerionenforschung (GSI) in Darmstadt (FRG) to produce one atom of element 109. Another difficulty is posed by 109's short half-life of several milliseconds. Similar characteristics are found for elements 107 and 108.

Although the production cross sections are so small and the half-lives so short, according to the liquid drop model of the nucleus, these elements should not exist at all. It is, indeed, through their production and study that stabilization of matter by nucleon shell effects has been established in this region. Their enhanced stability holds out the possibility of the existence of a predicted "island of stability" for superheavy elements centering on a nucleus of 110 protons and 184 neutrons. The elements 107–109 could be looked upon as a sort of "causeway" from the main group of elements to the superheavy region. Relativistic effects on the electrons in these superheavy elements will be especially important in causing differences to appear between their chemistry and that of the lighter members of their group. The discovery of superheavy elements with long enough half-lives to study their chemistry would therefore open up a whole new area of the periodic table for the chemist as well as the physicist.

The GSI workers have designed and built a powerful accelerator (the UNILAC) capable of delivering the high intensity beams of chromium and iron needed to bombard bismuth and lead targets to produce a few atoms of elements such as 107, 108, and 109. They have also designed and built ultrasensitive, rapid detection and identification methods. Such equipment and methods are also useful for searching for superheavy elements. The search for superheavy elements is, however, requiring the heaviest transuranium target elements including einsteinium-254.

Other groups that are actively searching for new elements are located at the University of California, Berkeley (USA), and the Joint Institute for Nuclear Research, Dubna (USSR).

The techniques for a valid identification of a new element must, by definition, be capable of determining the atomic number unequivocally. Applicable chemical methods must be capable of yielding definitive results using only one atom at a time. Experience shows that ion exchange, solvent extraction, and gas chromatography can be definitive. Applicable physical methods include (1) the detection of characteristic x rays emitted by the daughter of the new element in coincidence with unique alpha particles from the decay of the new element itself and (2) the observation of "mother–daughter" decay chains. In (2) the new element is detected as it decays (for example, by alpha emission) into a known daughter. Then, through a sequence of alpha decays, a series of known nuclides is produced. Sufficient information for identifying the original nuclide in the chain can thus be obtained by recording the *chain of events* set up by the original event that occurs in the decay of the new, unknown nuclide.

It should also be mentioned that the upper limit of 10^{-14} s for a compound nucleus lifetime is considered to be the shortest lifetime of a composite nuclear system that could be thought of as an element.

See also ATOMS; BOHR THEORY OF ATOMIC STRUCTURE; ISOTOPES; RARE EARTHS; RARE GASES AND RARE-GAS COMPOUNDS; SUPERHEAVY ELEMENTS; TRANSURANIUM ELEMENTS.

REFERENCES

1. M. E. Weeks and H. M. Leicester, *Discovery of the Elements*, 7th ed. Journal of Chemical Education, Easton, Pa, 1968.
2. G. Münzenberg, "Recent Advances in the Discovery of Transuranium Elements," *Rep. Prog. Phys.* **51**, 57–104 (1988).
3. E. K. Hyde, D. C. Hoffman, and O. L. Keller, Jr., "A History and Analysis of the Discovery of Elements 104 and 105," *Radiochim. Acta* **42**, 57–102 (1987).

Ellipsometry

H. Lüth

Ellipsometry is an optical reflectance technique in which the change of the state of polarization of light upon reflection on a surface is measured, rather than the change of intensity. Even though the method dates back to Drude (1889) [1], ellipsometry has recently attracted considerable attention because of a wide variety of applications in modern solid-state optics, especially in the physics of surfaces, interfaces and thin films [2,2a].

The change of the state of polarization upon reflection can be expressed in terms of the ratio of the two complex reflection coefficients $r_{\parallel}$ and $r_{\perp}$ for light polarized parallel and perpendicular to the plane of incidence, r being the ratio of reflected and incident electric field strength. The complex quantity

$$\rho = r_{\parallel}/r_{\perp} = \tan\psi \exp(i\Delta)$$

defines the two ellipsometric angles Δ and ψ that are measured in ellipsometry.

In contrast to a standard reflectivity (intensity) measurement, the two angles Δ and ψ completely determine the two optical constants n (refractive index) and κ (absorption coefficient) of an isotropic reflecting medium. For the mathematical dependence of n and κ on Δ and ψ calculated from Fresnel's formulas, see, e.g., Ref. [3].

The application of ellipsometry in surface physics is favored because in suitable cases an adsorbed species can be detected in a coverage as low as 1/100 of a monolayer. If used at a fixed wavelength λ, measured changes $\delta\Delta$ and $\delta\psi$ due to adsorption of gases can give information about adsorption kinetics (e.g., sticking coefficients) or absolute coverages [4]. The analysis, mostly done by means of computers, is rather involved because the two measured changes $\delta\Delta$ and $\delta\psi$ are not sufficient to determine all parameters (optical constants, thickness) of one or more adsorbed phases. For the study of adsorption processes, therefore, ellipsometry is most effectively used in combination with other methods, such as Auger electron spectroscopy [4].

Surface electronic structure can be successfully investigated by ellipsometric spectroscopy, in which $\delta\Delta$ and $\delta\psi$ caused by surface treatments (e.g., gas adsorption) are taken at various wavelengths over a given spectral range [5]. Transitions between surface states as they can be found by this type of optical spectroscopy are of considerable importance, e.g., for semiconductor device technology and for corrosion and catalysis studies. Another promising development can be expected from an extension of ellipsometric spectroscopy into the infrared in order to study vibration frequencies of adsorbed gases [6]. In the physics of thin films, ellipsometry can be used as a convenient *in situ* technique to follow epitaxial growth.

The conventional experimental setup for ellipsometry consists of two rotatable polarizers and a fixed quarter-wave plate as compensator to compensate the change of polarization of a parallel light beam due to reflection (null ellipsometry) [7]. The parameters Δ and ψ are calculated from component settings at extinction. For spectroscopy or high-speed measurements, automatic ellipsometers are much more convenient: The extinction settings can be done and read out by means of electronically controlled Faraday rotators in a self-compensating ellipsometer [8]. Polarization modulators [9] or rotating analyzers [10] are used in other types of automatic ellipsometers in which the parameter information is contained in the phase and relative amplitude of the ac component of the transmitted light intensity.

See also ABSORPTION COEFFICIENTS; POLARIZED LIGHT; REFLECTION; SURFACES AND INTERFACES.

REFERENCES

1. P. Drude, *Ann. Phys. Chem.* **36,** 532 (1889). (I)
2. "Proc. Symp. Recent Developments in Ellipsometry (Nebraska, 1968)," *Surface Sci.* **16,** (1969); "Proc. 3d Int. Conf. on Ellipsometry (Nebraska, 1975)," *Surface Sci.* **34,** (1976). (I)

2a. R. M. A. Azzam and N. M. Bashara, *Ellipsometry and Polarized Light*. North-Holland, Amsterdam, 1971. (A)

3. G. A. Bootsma and F. Meyer, *Surface Sci.* **14,** 52 (1969). (A)
4. R. Dorn, H. Lüth, and G. J. Russell, *Phys. Rev.* **B10,** 5049 (1974). (I)
5. H. Lüth, *Appl. Phys.* **8,** 1 (1975). (I)
6. M. J. Dignan, B. Rao, M. Moskovits, and R. W. Stobie, *Can. J. Chem.* **49,** 1115 (1971). (A)
7. F. L. McCrackin, E. Passaglia, R. R. Stromberg, and H. L. Steinberg, *J. Res. Natl. Bur. Stand.* **A67,** 363 (1963). (A)
8. H. J. Mathieu, D. E. McClure, and R. H. Muller, *Rev. Sci. Instrum.* **45,** 798 (1974). (A)
9. S. N. Jasperson and S. E. Schnatterly, *Rev. Sci. Instrum.* **40,** 761 (1969). (A)
10. D. E. Aspnes and A. A. Studna, *Appl. Opt.* **14,** 220 (1975). (A)

Energy and Work

Jearl D. Walker

Energy is a certain abstract scalar quantity that an object (either matter or wave) is said to possess. It is not something that is directly observable, although in certain cases, the behavior of the object possessing a particular amount of energy can be observed and the energy inferred. The usefulness of the concept comes from the fact that total energy cannot be eliminated or created in the world, rather energy must be conserved. In predicting the behavior of objects, one uses the conservation of energy to keep track of the total energy and the interchange of energy between its various forms and between objects. The conservation of energy does not account for the method by which energy is transformed between its forms, only that the total amount must remain constant.

Work is the transfer of energy from one object to another by a force from one on the other when that second object is displaced by the force. Often the nature of the first object need not be specified if the force doing the work is known. For example, one can push a box, and the force performing the work is the pushing that the box experiences. The work can be calculated without knowing anything about the person pushing. In general, the work done on an object can be calculated from knowledge of either the applied force or the energy changes involved.

For there to be work in the scientific definition, the object must be displaced, and the displacement must have a component parallel to the force. The work done in such a displacement is the integral

$$W=\int \mathbf{F}\cdot d\mathbf{r}=\int F\cos\theta\, dr,$$

where θ is the angle between the force and displacement vectors. If the applied force is constant and is at a constant angle with respect to a displacement Δx, then

$$W=F\cos\theta\,\Delta x.$$

For rotation of a rigid body about an axis by a torque τ, the work is

$$W=\int \boldsymbol{\tau}\cdot d\boldsymbol{\theta}.$$

If the torque is constant, then

$$W=\tau\Delta\theta.$$

Because work is a scalar quantity, the total work done by several forces on an object can be calculated by finding the work done by each separately, and then algebraically adding the work. Alternately, the total work is that work done by the resultant of the several applied forces.

If the dot product is positive (i.e., if the displacement and the force component along the displacement axis are in the same direction), then the work done on the object is positive and energy is added to the object. If the dot product is negative, then the work is negative and energy is removed from the object. These results do not mean that the conservation of energy is violated. If energy is added to the object, that energy came from the second object responsible for the force used in the work calculation. The total energy of the two objects is constant, and the work done is merely the transfer of the energy (equal to the work done) from one object to the other.

An example of a work calculation is the following. A block initially at rest has two forces acting on it: 3 N to the right and 2 N to the left. It moves to the right and after 2 m, the net work done on it is 2 J. If there are no other forces on it, then this 2 J appears as kinetic energy of the moving block. The right-going force did a positive 6 J of work. The left-going force did a negative 4 J of work.

Kinetic energy is a form of energy associated with the motion of an object, that is, with the magnitude of its linear velocity v or rotational velocity ω. In classical mechanics, kinetic energy is given by

$$K=\tfrac{1}{2}mv^2$$

for linear velocity and by

$$K=\tfrac{1}{2}I\omega^2$$

for rotation of a body with a moment of inertia I about the rotational axis. For an example of the first equation, a mass of 4.0 kg moving with a velocity magnitude of 5.0 m/s has a kinetic energy of 50 J.

To stop an object having a particular kinetic energy, that amount of energy must be removed from the object. In the previous example, the object would come to a stop if an external force such as friction does work on the object such that 50 J of energy is removed.

In the special theory of relativity, the kinetic energy is defined by

$$K=E-E_0$$

where E is the total energy and E_0 is the rest energy of the object. For velocities smaller than about 0.1 the speed of light, this expression can be approximated by the classical one.

Potential energy is an energy associated with the relative position of an object. As with kinetic energy, this energy can be removed from the object by the object doing work on another system or this potential energy can be increased by an external force doing positive work on the object.

Gravitational potential energy is related to the distance r of a point mass m_1 from another point mass m_2:

$$U=G\frac{m_1m_2}{r},$$

where G is the gravitational constant. This classical expression for gravitational potential energy is zero for an infinite separation of the masses and is not defined for zero separation. If an external force is to increase the distance between the objects from r_1 to r_2, that external force must do an amount of work given by

$$W=\Delta U=Gm_1m_2\left(\frac{1}{r_1}-\frac{1}{r_2}\right).$$

If one of the point masses, say m_1, is released and is allowed to accelerate toward the other mass, its gravitational potential energy is converted to kinetic energy as the gravitational force between the masses does work on the moving mass.

The gravitational potential energy and change in that energy near the surface of the earth can be approximated, respectively, by

$$U=mgh,\quad \Delta U=mg\,\Delta h,$$

where a point mass is a vertical distance h above a reference level, and where g is the local gravitational acceleration, approximately 9.8 m/s^2. For example, if an object of mass 4.0 kg is raised 3.0 m near the earth's surface, then the external force causing the rise does work of 117.6 J. Thus the object's gravitational potential energy is increased by that amount. If the object is released and allowed to fall through 3.0 m, the gravitational force does work on it during the fall and converts the gravitational potential energy to kinetic energy. At the end of that fall, the object has kinetic energy of 117.6 J.

With such a definition the amount of potential energy a particular object has depends on an arbitrary choice of reference level from which to measure h. The change of vertical height results in the same change of potential energy regardless of the choice of reference level. Thus, the potential energy change of an object moving in a gravitational field is not arbitrary.

Potential energy is also associated with the compression or elongation of an ideal spring:

$$U=\tfrac{1}{2}kx^2$$

where x is the change in the spring's length and k is the spring constant. This type of potential energy is useful in any situation in which the force on an object obeys Hooke's law, either exactly or approximately.

Electrostatic potential energy between two point charges q_1 and q_2 is given by

$$U=k\frac{q_1q_2}{r},$$

where k is a constant, r is the separation distance, and the charges can be either positive or negative. If both charges have the same sign, then an external force would have to do positive work to bring the charges closer. If the charges are already close and are then released, the electrostatic repulsive force would do work on them to accelerate them away from each other and to convert the electrostatic potential energy into their kinetic energy.

If there is no net external force on a system, its total energy must remain constant. The various forms of the energy of the system can change, but the total amount cannot.

If a net external force does work on a system and then the process is reversed, and if the system returns to its initial value of energy, the net work done is zero and the force is said to be conservative. The energy of the system is thus independent of how the work was done and depends only on the initial and final states of the system. An example of a nonconservative force is friction. The amount of energy lost to friction by a moving body depends on the distance over which the body slides on the surface having friction and therefore the choice of path determines the amount of energy lost. If a force F is conservative, a potential energy U can be defined by

$$F=-\nabla U.$$

In the special theory of relativity the total energy E of an object is related to its relativistic mass m,

$$E=mc^2.$$

Mass and energy are considered to be the different and interchangeable forms of the same thing.

See also CONSERVATION LAWS; DYNAMICS, ANALYTICAL.

Entropy

Laszlo Tisza

The term entropy derives from the Greek expression for "transformation," and was suggested by Rudolf Clausius in 1865. However, the concept was already implicit in his resolution of an impasse in thermodynamics 15 years earlier. The problem was how to account for the fundamental asymmetry in the conversion of work and heat. The dissipation of work into heat is a spontaneous process, whereas the reverse conversion calls for such special arrangments as heat engines, and it is at best of limited efficiency. It was sensed that the underlying factor is the "lower quality" of energy associated with random molecular motion as compared with ordered mechanical energy. However, this contrast was not adequately expressed by the terms of heat and work, since heat was at least partially convertible. The problem was how to express the difference in randomness in terms of measurable quantities.

This complex requirement was filled by the concept of entropy, which was established by Clausius and William Thomson, later Lord Kelvin, by the most careful logic. The resulting practical rules can be stated with relative ease.

Let us inject a small quantity of heat dQ into a system at constant absolute temperature T. The entropy increase of this system is then $dS=dQ/T$.

Suppose that the heat quantity has been extracted from

a reservoir of temperature T' which loses the entropy dQ/T'; then the total entropy change is

$$dS_{tot} = dQ(1/T - 1/T'). \quad (1)$$

This quantity is positive for the natural process in which heat flows "downhill" ($T'>T$) and it would be negative for the impossible process in which heat would flow "uphill" ($T'<T$). Of course, such an uphill flow of heat does occur in an air conditioner, but this happens with the intervention of the compressor, and the dissipated work ensures that the total entropy change is positive.

For the limiting case of $T' \cong T$ the entropy is unchanged; it appears to be conserved. In this ideal case the process can be reversed and we speak of reversible heating.

Let us consider now the situation in which the heat quantity dQ is obtained from the dissipation of mechanical energy. Since the latter is not associated with entropy, we have no compensating entropy decrease, but the entropy increase of the system is the same as before; we say that entropy is a function of the state of the system only, and does not depend on the nature of the process that led up to it. In this respect it differs from the heat quantity dQ, which is not a differential of a state function Q; there is no such thing as a heat content Q of a system. This is sometimes expressed by slashing the symbol of differential:

$$đQ = T\,dS. \quad (2)$$

Note the analogy with the expression of compressional work:

$$đW = -p\,dV. \quad (3)$$

Excess pressure decreases volume and excess temperature increases molecular disorder, as expressed by entropy.

The requirement that the total entropy cannot decrease enables us to distinguish natural irreversible processes, for which the entropy increases, from the impossible reverse processes, for which the entropy would decrease and which do not occur in nature. It is worthwhile to note, however, that the term "irreversible" has a different meaning in thermodynamics from that in the everyday language. The melting of ice by a heating coil is said to be irreversible, but this does not preclude the refreezing of the ice, provided that we supply the necessary entropy to another system.

Although it is remarkable that a simple concept of entropy could be established entirely within classical thermodynamics [1], more elaborate versions emerge in a variety of roles.

By merely postulating the entropy, this concern serves to organize the structure of much of thermodynamics. Having been demonstrated in an austere setting by Gibbs, this idea has proved successful on an elementary level [2,3].

Considered as a measure of molecular disorder, the entropy concept generates a vast range of quantitative calculations within quantum statistics [3–6].

It is possible to elaborate the conceptual connections with quantum mechanics [6], and more specifically the "absolute entropy" of the Third Law yields a variety of nontrivial applications if used with sufficient care [6–8].

The connection with information theory [9] and biology [10] marks the extent to which this concept evolved from its original beginnings.

See also Carnot Cycle; Heat Engines; Heat Transfer; Statistical Mechanics; Thermodynamics, Equilibrium; Thermodynamics, Nonequilibrium.

REFERENCES

1. E. Fermi, *Thermodynamics*. Dover, New York, 1956. (E)
2. H. B. Callen, *Thermodynamics*. Wiley, New York, 1960. (I)
3. David Chandler, *Introduction to Modern Statistical Mechanics*. Oxford Press, New York, 1987. (E)
4. L. D. Landau and E. M. Lifshitz, *Statistical Physics*, 2nd ed. Pergamon, New York, 1969. Third edition with L. P. Pitaevskii. Pergamon, New York, 1980. (I)
5. F. Reif, *Fundamentals of Statistical and Thermal Physics*. McGraw-Hill, New York, 1965. (I)
6. L. Tisza, *Generalized Thermodynamics*. MIT Press, Cambridge, Mass., 1966. (I)
7. J. Wilks, *The Third Law of Thermodynamics*. Oxford, London and New York, 1961. (I)
8. L. Tisza, in *Energy Transfer Dynamics. Studies and Essays in Honor of Herbert Fröhlich*, T. W. Barrett and H. A. Pohl (eds.). Springer-Verlag, Berlin, 1987. (A)
9. L. E. Brillouin, *Science and Information Theory*, 2nd ed. Academic Press, New York, 1962. (I)
10. E. Schrödinger, *What Is Life?* Macmillan, New York, 1945. Doubleday Anchor Books, Garden City, NY, 1956. (E)

Equations of State

Raymond D. Mountain

An equation of state describes the relationship of the directly observable quantities that specify the thermodynamic state of a system. For a fluid the observable quantities are the pressure p, the specific volume v, and the temperature T. The equation of state is then the relationship

$$p = p(v,T). \quad (1)$$

For a solid, it may be necessary to specify the stress and strain components in addition to the fluid variables. For a ferromagnet the applied field **H** and magnetization **M** must be included in the set of variables.

The simplest example of an equation of state is the ideal gas equation

$$pV = nRT. \quad (2)$$

This applies to n moles of a gas of noninteracting point molecules occupying a volume V; thus the ideal-gas form of Eq. (1) is

$$p = RT/V \quad (3)$$

with specific volume

$$v = V/n. \quad (4)$$

$R = 8.314\,471 \pm 0.000\,014$ J mol^{-1} K^{-1} is a universal constant known as the gas constant. Another example of an equation of state is provided by the van der Waals equation

$$p = RT/(v-b) - a/v^2. \quad (5)$$

The constants a and b are intended to take into account the attraction between the molecules and the finite size of the molecules, respectively. With suitable choices for a and b, Eq. (5) provides a qualitative representation of the equation of state for fluids.

Statistical mechanics provides the connection between the thermodynamic properties of a system and the molecular properties of the molecules which make up the system. Given the interaction potential for molecules, the methods of statistical mechanics provide a procedure for calculating the equation of state of a gas. The resulting equation (known as the "virial equation of state") takes the form of an infinite series in powers of v^{-1} with the ideal-gas expression as the first term:

$$p=\frac{RT}{v}\left(1+\frac{B(T)}{v}+\frac{C(T)}{v^2}+\cdots\right). \tag{6}$$

The temperature-dependent quantities $B(T)$ and $C(T)$ are the second and third virial coefficients, and these objects contain information on the interaction of pairs and triples of molecules. $B(T)$ and $C(T)$ are known both theoretically and experimentally for various types of molecular interactions. The higher-order virial coefficients are difficult to obtain accurately. Theoretically this is so because the computations are complex. Experimentally it is due to the need for very precise p-v-T data over a large range of v if one is to extract the virial coefficients by fitting the data with Eq. (6). Even so, the virial equation of state is an important way of describing the equation of state of gases.

Empirical representations of the equation of state of fluids are frequently developed by fitting expressions of the type

$$p=\sum_{i,j} A_{ij}T^i v^{-j} \tag{7}$$

to experimental data. Such expressions have considerable computational utility even though little physical significance can be attached to the coefficients A_{ij}.

A portion of the p-v projection of an equation of state for a fluid is sketched in Fig. 1. Three isotherms (lines of constant temperature) are shown for the critical temperature T_c, for $T_3>T_c$, and for $T_1<T_c$. Several universal features for equations of state for fluids should be noted.

For temperatures greater than T_c, the pressure is a decreasing function of v and the isotherms are smooth functions of v^{-1}. This changes at the critical point (CP). The coexistence of liquid and gas phases requires nonanalytic regions in the equation of state for temperatures less than the critical temperature. This is because p and T are constant for a range of specific volumes. Analytic equations, such as Eqs. (5) or (7), can be used to describe much of the subcritical region *only* if they are augmented with the Maxwell construction to determine the location of the phase boundary. Such equations are not suitable in the near vicinity of the critical point as there the p-v-T relationship is known to involve nonintegral powers of $T-T_c$ and $v-v_c$.

Theoretical studies of the equation of state for gases use the ideal gas as a reference system. A different reference is needed for liquid-state studies. We shall arbitrarily suppose that liquids are fluids with $v<v_c$, the value of the specific volume at the critical point. Since the equation of state for a liquid is dominated by the strong repulsion that molecules experience when close together, the hard-sphere fluid provides a useful reference system for liquid-state studies when the attractive part of the potential is treated as a perturbation.

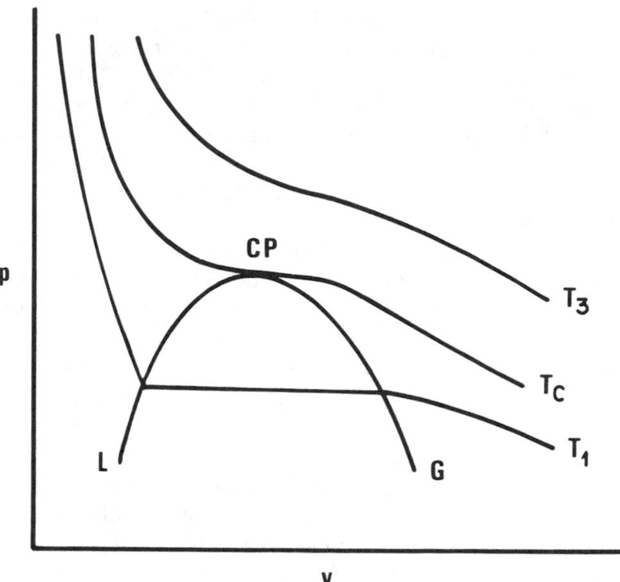

FIG. 1. A sketch of a portion of the p-v projection of the equation of state of a fluid. The curve L-CP-G represents the liquid (L)–gas (G) phase boundary which terminates at the critical point (CP). Three isotherms are indicated for temperatures $T_3>T_c>T_1$. The critical temperature is T_c.

The hard-sphere fluid has been extensively studied using a combination of statistical-mechanical and computer-simulation methods. An accurate representation of the hard-sphere equation of state is

$$p=\frac{RT}{v}\frac{1+y+y^2-y^3}{(1-y)^3}, \tag{8}$$

where $y=\pi N_A d^3/6v$, d is the diameter of the sphere, and N_A is Avogadro's constant. A variety of techniques have been devised to obtain numerical equations of state starting with Eq. (8). For example, and diameter d can be used as a variational parameter to minimize the free energy derived by perturbation theory. Such calculations yield values for the equation of state which are accurate to a percent or so over a wide range of liquid-state temperatures and volumes.

Theoretical studies of the equation of state of solids use a collection of coupled harmonic oscillators as the reference system. Debye's theory provides an example which illustrates the concepts involved. In that theory it is assumed that the crystal can be represented as a collection of harmonic oscillators with a distribution of oscillator frequencies which is quadratic in frequency ν up to some maximum value ν_m. The distribution is identically zero for $\nu>\nu_m$. The maximum frequency is determined by the number of modes possible for the system. The maximum frequency can be expressed as a temperature, the Debye temperature,

$$\Theta=h\nu_m/k, \tag{9}$$

where h is Planck's constant and $k=R/N_A$ is Boltzmann's constant. The explicit form for the equation of state is too complicated to exhibit here. The Debye temperature is commonly used to characterize a solid since it can be readily inferred from heat-capacity measurements. If $T<\Theta$ is satisfied, then not all modes of vibration of the lattice are excited and explicitly quantum-mechanical effects are observed in the thermodynamic properties of the solid. If $T>\Theta$ is satisfied, then the solid behaves much like a classical system of oscillators.

More accurate theories of solids modify Debye's theory in two ways. The first modification is to replace the postulated quadratic distribution of frequencies with one based on the lattice structure and interaction parameters of the crystal. The second modification is important at high temperatures and involves the introduction of anharmonic interactions into the equations used to determine the frequencies of the oscillators. These changes are refinements in the theory rather than departures from Debye's theory.

One other aspect of equations of state should be mentioned. These equations do not provide a complete, local description of the thermodynamics of a system. The thermodynamic potentials, such as $A(v,T)$, the Helmholtz free energy, do. Thus, while the pressure can be obtained by differentiation

$$p(v,T) = -\left(\frac{\partial A(v,T)}{\partial v}\right)_T, \qquad (10)$$

the determination of $A(v,T)$ from $p(v,T)$ requires both a global knowledge of the equation of state and information (the integration constant) not contained in the equation of state. The latter information is obtained from calorimetric studies, a topic outside the domain of this article.

See also CALORIMETRY; PHASE TRANSITIONS; THERMODYNAMICS, EQUILIBRIUM.

BIBLIOGRAPHY

M. R. Moldover, J. P. M. Trusler, T. J. Edwards, J. B. Mehl, and R. S. Davis, "Measurement of the Universal Gas Constant *R* Using a Spherical Acoustic Resonator," *J. Res. Natl. Bur. Stand.* **93,** 85–144, 1988. (A)

A. Münster, *Statistical Thermodynamics,* Vol. 2. Academic Press, New York, 1974. (A)

J. O. Hirschfelder, C. F. Curtis and R. B. Bird, *Molecular Theory of Gases and Liquids.* Wiley, New York, 1954. (I)

J. Kestin and J. R. Dorfman, *A Course in Statistical Thermodynamics.* Academic Press, New York, 1971. (I)

John R. Dixon, *Thermodynamics I: An Introduction to Energy.* Prentice Hall, Englewood Cliffs, N.J., 1975. (E)

Ergodic Theory

Joseph Ford

The ice in a glass of tea placed on a dining room table is always observed to melt and, eventually, the water–tea mixture comes to room temperature. Quite generally on the macroscopic level, a system initiated in a specified disequilibrium state is observed to approach and finally reside in a unique macroscopic equilibrium state having specified values for quantities such as temperature T, volume V, and pressure P. In its most general physical (but not mathematical) sense, classical ergodic theory is that branch of classical particle mechanics that attempts to rigorously explain and predict the observed macroscopic behavior of matter starting from microscopic particle dynamics, namely, Newton's equations. We immediately note that the macroscopic state of the system is specified by giving only a few quantities such as T, P, and V, while the microscopic state is specified by giving the positions and velocities for each of an enormous number of individual particles. Thus many distinct microstates correspond to each macrostate, whether equilibrium or disequilibrium; however, it may convincingly be argued that, for a given isolated mechanical system, the vast majority of microstates correspond to a single equilibrium macrostate. The central problem of ergodic theory then lies in rigorously proving that almost all microstates corresponding to macroscopic disequilibrium evolve with time into the sea of microstates corresponding to macroscopic equilibrium, and in rigorously computing the times required to reach equilibrium.

Although much work remains to be done in this area, equally, much success has been achieved. The flavor of work in this field may be sampled by considering an extremely simple (but unphysical) model that, nonetheless, exhibits many of the characteristics of more physically realistic systems. Consider a system with a large number of particles in which the microstate for each particle is specified by a position q and a speed p, with both q and p confined to lie in the unit interval (0,1). Let the equations of motion for each particle (giving the time evolution of microstates at 1-s intervals) be given by

$$q_{n+1} = 2q_n \quad \text{and} \quad p_{n+1} = p_n/2 \qquad (1a)$$

when $0 \le q_n < \frac{1}{2}$, or by

$$q_{n+1} = 2q_n - 1 \quad \text{and} \quad p_{n+1} = (p_n/2) + (1/2) \qquad (1b)$$

when $\frac{1}{2} \le q_n < 1$. Equations of motion (1) generate an area-preserving transformation that first transforms the unit square into a rectangle having twice the original width and one half the original height, and then it slices the rectangle into a left and a right half and places the right half on top of the left to reform the unit square; in addition, each initial microstate evolves via Eqs. (1) through a unique set of microstates (a trajectory). Unique trajectories and the area-preserving property are generic characteristics of dynamical systems.

Macroscopic equilibrium for this system clearly corresponds to a macroscopically uniform distribution of particles over the (q,p) unit square, and here ergodic theory can rigorously prove that almost all initial particle distributions tend to this unique equilibrium state. In order to understand this intuitively, suppose that all the initial-particle microstates are uniformly spread over an arbitrary, small rectangle (Δq Δp). By Eqs. (1), this small initial rectangle grows exponentially ($\sim 2^n$) in the q direction and shrinks exponentially ($\sim 2^{-n}$) in the p direction, eventually wrapping many times across the unit square and uniformly covering the square as a set of thin, horizontal ribbons. This exponential (called C-system or K-system) approach to equilibrium is characteristic of many physical systems, most notable the hard-sphere gas. Moreover, for Eqs. (1), since every small rectangle spreads uniformly over the whole unit square, this system is mathematically mixing (analogous to stirring a mixture to uniformity) and hence mathematically ergodic (which means, loosely speaking, the time-evolved iterates of almost all initial-particle microstates densely cover the unit square). Finally, the trajectories of this simple system, like those of certain more physically realistic systems, exhibit striking statistical behavior. Indeed, this system can be proved to be a Bernoulli system, which means that if the unit square is divided into N disjoint regions, successive iterates of most initial-particle microstates generated by the deterministic Eq. (1) sequentially appear in the various regions in a "random" sequence that could equally well have been selected by a roulette wheel.

See also STATISTICAL MECHANICS; THERMODYNAMICS, NONEQUILIBRIUM.

BIBLIOGRAPHY

V. I. Arnold and A. Avez, *Ergodic Problems of Classical Mechanics*. Benjamin, New York, 1968. (A)

J. Ford, "How Random Is a Coin Toss?," *Phys. Today* **36** (No. 4), 40 (1983). (E)

J. Ford, "What Is Chaos That We Should Be Mindful of It?," in S. Capelin and P. C. W. Davies, eds. *The New Physics*. Cambridge University Press, Cambridge, 1988. (I)

J. Ford, "Chaos: Solving the Unsolvable, Predicting the Unpredictable!," in M. F. Barnsley and S. G. Demko (eds.), *Chaotic Dynamics and Fractals*. Academic Press, New York, 1986. (I)

Error Analysis

Paul S. Olmstead

Any discussion of error analysis must make its objective clear. In this article, the objective is a description of how to obtain and present data that are sufficiently "trouble free" to satisfy anyone who would wish to analyze them in a different way. It is recognized that all data are subject to error. However, it is the duty of the experimenter to show what he has done to reduce that error to a point that gives essentially "trouble-free" data. The experimenter's first step toward this goal is to record the data in a form that provides all information necessary for repeating the experiment successfully. The following form is suggested:

$$x_i = f(H_i, C_i, t_i)$$

where i is each successive test; f is the function to be either derived or verified; H_i is the identification of the human factor for the ith test; C_i is the identification of the physical factors for the ith test; t_i is the time that the ith test was made; and x_i is the "error" observed on the ith test. When data are presented in this form, any future analyst with a new hypothesis has an opportunity to test it with these data that is equal to that of the original experimenter.

The simplest set of data in this form is where x is assumed to be a constant. (In the very simplest form, the average error could be zero if certain very stringent conditions were met.) Many experimenters assume that all that is required is to find the average of the x_i's. Experience has shown that this may not be so. Most of us have made experiments where

1. a single observation seems inconsistent with the rest, or
2. several observations, low in value, are followed by a group that are high, or
3. the spread of the values is first low but becomes high, or
4. there is a tendency to increase in value during the experiment, or
5. the data show periodic peaks and valleys, etc.

Usually, we associate these observations with something that we have failed to control, such as

1′. misreading a meter—a human error—or
2′. a shift in the conditions of test, or
3′. change in the sensitivity of a meter, or
4′. drift in calibration, or
5′. picking up a stray cyclical field, etc.

Most experimenters can add other symptoms and explanations to these lists. The purpose here is to point out that the things we have observed are related to patterns in the data that some statisticians have studied. They have been concerned with similar findings in other fields as well as physics. Their approach is to calculate how often a similar "presumed error" would be expected in an experiment with only random effects present. If an "error" as large as that observed could happen half the time, it would be considered a normal occurrence. When, then, should we look for "trouble"? How often should we be willing to look for a false trouble? In other fields, scientists are willing to look 5% of the time for a "nonexistent trouble" particularly if they have found that they can locate a similar "real trouble" at least 50% of the time. This ability to locate "real trouble" is much more important than the inability to locate "nonexistent trouble" indicated by a false clue. This means that scientists have more to gain than to lose by applying statistical criteria to their data.

In considering an outlying observation, for example, one statistical test is the ratio of the difference between the largest and the second largest and the difference between the second largest and the smallest. For a ratio of 1.0 to be meaningful at the 5% level, at least 8 measurements are needed; for a ratio of 0.5, at least 16; for a ratio of 0.4, at least 23; etc., making it clear that it is very dangerous to discard data simply because they "look bad." This is a serious matter in the case of three observations where we find that the ratio must exceed 16 at the 5% level and be at least 8 at the 10% level.

A shift in level has several possibilities for statistical test: a long run of observations on

a. one side of the median,
b. either side of the median,
c. each side of the median, or
d. each side of any cut.

For a sample of size 40, the length criteria for these to be significant are a. 8, b. 9, c. 7 and d. 7; for a sample of 100, they become: a. 10, b. 11, c. 8, and d. 9. The criteria do not change markedly with increase in sample size but experimenters will often find that "real trouble" will be associated with lengths of run that are not marginal as judged by the statistical criteria.

The simplest measure of spread in the observations is the range from maximum to minimum in small samples. To make use of this for detecting trouble, control charts for ranges of consecutive groups of four are recommended. For these, the average range is obtained and any range exceeding $2.28\bar{R}$ is considered as indicating variability worth investigating. If control charts for averages of the groups of four are also made, the expected limits about the grand average are $\pm 0.73\bar{R}$. A necessary condition for data consistency is to find 25 consecutive groups of four with their averages and ranges within control limits.

A slow fluctuation or trend in a set of data may be detected by applying a run test or by looking for a run-up or rundown in either individual observations or averages. Such a run of length 5 has a probability of not over 5% for up to 25 observations; for a run of length 6, the same probability exists to 154 observations.

The best method of identifying a fast fluctuation or cycle is to plot the individual observations in the order in which they were obtained and note any periodicity in the occurrence of peaks and valleys. For more information, this should be followed by lag correlation plots (x_i vs x_{i-L}). If a cyclical effect does exist in the data, the various lag plots will approximate a series of elliptical plots with superimposed random variation. Absence of data near the center of the plot is typical.

Up to this point, it has been assumed that the objective has been to determine a universal constant based on data obtained by several observers at various times and with differing experimental setups. From the results, a standard procedure is to be defined with predictable limits of error. A second type of physical experiment is to establish a curve of relationship for which a theoretical expression,

$$y = f(x),$$

has been assumed. If this expression is the median curve expected, a check of less than 15 points over the range of interest should not have a run of length 6 on one side of the curve in more than 1% of the experiments. A run-up or rundown of length 6 is even less likely. Failure to meet these criteria suggests need for modification in the expected curve of assumed relationship.

It should be pointed out that use of statistical concepts in various fields differs and the vocabulary used is seldom identical. This paper considers the problem of a physicist examining data that may contain errors worth investigating. A few simple clues to the possible existence of such errors are shown. How to identify the probable cause of the suspected error will depend on the experimental ingenuity of the physicist. Data that have been examined for the types of error discussed here may not be "trouble free" but they will have passed the first requirement for reaching such a condition.

For those interested in reading more about data analysis as it is practiced in industry, a bibliography is appended. Some of these show the fundamental steps outlined here, some show how particular tests have been derived, some show examples of use, and others extend the inquiry into the determination of correlated relationships. No text covers completely the problems of the physicist. However, one very useful text is *Precision Measurement and Calibration, Selected NBS Papers on Statistical Concepts and Procedures*, Harry H. Ku (ed.), National Bureau of Standards Special Publication 300, Volume 1, issued February 1969.

Standardization in this field is important. In the United States, this is being carried on by Committee E-11 of the American Society for Testing and Materials, 1916 Race Street, Philadelphia, PA 19103.

See also PROBABILITY; STATISTICS.

BIBLIOGRAPHY

General

Precision Measurement and Calibration (Harry H. Ku, ed.). NBS Special Publication 300, Vol. 1, 1969, $9.00 (Sup't Doc., U.S. Printing Office, Washington, D.C. 20402), with particular attention to articles by Churchill Eisenhart, W. J. Youden, John Mandel, R. B. Murphy, and Milton E. Terry.

Elementary

American Society for Testing and Materials, Committee E-11, *Manual on Quality Control of Materials*, STP No. 15, 1960.

American Society for Testing and Materials, Committee E-11, *Use of the Terms Precision and Accuracy as Applied to Measurement of a Property of a Material*, ASTM Standards, Part 41, 1975, pp. 165–182, E 177-71.

Paul S. Olmstead, "How to Detect the Type of an Assignable Cause," *Industrial Quality Control* **9**(3), 32–36 (1952); **9**(4), 22–32 (1953).

Intermediate

Paul S. Olmstead, "Distribution of Sample Arrangements for Runs Up and Down," *Ann. Math. Stat.* **17**, 24–33 (1946).

Paul S. Olmstead, "Runs Determined in a Sample by an Arbitrary Cut," *Bell System Tech. J.* **37**, 55–82 (1958).

Paul S. Olmstead, "Grouping of High (or Low) Values in Observed Data," *Statistica Neerlandica* **26**(3), 29–36 (1972).

Paul S. Olmstead and John W. Tukey, "A Corner Test for Association," *Ann. Math. Stat.* **18**, 495–513 (1947).

Walter A. Shewhart, *Statistical Method from the Viewpoint of Quality Control*. (The Graduate School, Department of Agriculture, Washington, D.C., 1939).

Advanced

ASTM, Committee E-11, "Dealing with Outlying Observations," ASTM Standards, Part 41, 1975, pp. 183–211, E 178-75.

ASTM, Committee E-11, *Manual on Fitting Straight Lines*. STP No. 313, 1962.

ASTM, Committee E-11, *Manual on Conducting an Interlaboratory Study of a Test Method*. STP No. 335, 1963.

Richard L. Anderson, "Serial Correlation," *Ann. Math. Stat.* **13**, 1–33 (1942).

Cuthbert Daniel, "Calibration Designs for Machines with Carry-Over and Drift," *J. Qual. Tech.* **7**, 103–108 (1975).

Wilfrid J. Dixon, "Analysis of Extreme Values," *Ann. Math. Stat.* **21**, 488–506 (1950).

Wilfrid J. Dixon, "Ratios Involving Extreme Values," *Ann. Math. Stat.* **22,** 68–98 (1951).
Jane F. Gentleman and Martin B. Wilk, "Detecting Outliers in a Two-Way Table, 1. Statistical Behavior of Residuals," *Technometrics* **17,**1–14 (1975).
Frank E. Grubbs, "Procedures for Detecting Outlying Observations in Samples," *Technometrics* **11,** 1–21 (1969).
Richard E. Lund, "Tables for Approximate Test for Outliers in Linear Models," *Technometrics* **17,** 473–476 (1975).

Excitons

Carson Jeffries

The term exciton is broadly used to describe elementary localized excited states in solids, with the characteristic feature that the excitation can propagate through the crystal lattice. Two limiting experimental cases are recognized depending on the type of solid. For cubic band structures these can be shown to form a theoretical continuum.

FRENKEL EXCITONS

In molecular crystals such as anthracene, in rare-gas crystals, and in some alkali halides, a particular molecule, atom, or ion may be initially in an electronically excited state. But since the coupling to the adjacent lattice atoms is strong this excitation can be rapidly transferred from site to site without motion of the atoms themselves. Indeed these Frenkel excitons are best described as propagating waves of electric polarization. The excitons are experimentally manifested as optical absorption bands or characteristic luminescence radiation.

MOTT–WANNIER EXCITONS

In semiconducting crystals like Si, Ge, CdS, and Cu_2O, optical excitation promotes an electron from a lower-energy valence band to an upper conduction band, leaving a vacant state, or hole, in the valence band. The electrons and holes are independently mobile and give rise to the electrical conductivity of the semiconductor. However, an electron and a hole have an attractive electrical interaction. They may combine into a hydrogen-like particle—a neutral mobile excited state. The radius of this Mott–Wannier exciton is much larger than the crystal lattice spacing and the binding energy is correspondingly small. In the simplest cases the exciton displays a set of energy levels approximately like $E \propto n^{-2}$, i.e., like atomic hydrogen. As an example the exciton in Ge has a binding energy of 4 meV and a radius of 115 Å, to be compared to 13.6 eV and 0.5 Å for the H atom. The exciton in Ge may be visualized as a rather large, loosely bound particle (an electron and a hole in orbit about their center of mass) drifting through the crystal lattice; the electron and hole ultimately recombine, giving up characteristic luminescence radiation in the infrared. Being neutral, excitons do not contribute directly to electrical conductivity. They can be trapped on impurity atoms and other lattice defects; the radiation from commercial light-emitting diodes is usually due to the decay of trapped excitons.

EXCITON CONDENSATION

At sufficient densities and low temperatures, excitons form macroscopic droplets of a conducting electron–hole liquid in crystals like Ge and Si, much like the condensation of water vapor into fog droplets. In the liquid phase the electrons and holes are no longer bound into excitons but are free to move independently: The medium is a plasma of constant density, displaying both quantum and classical properties. The droplets have surface tension; they move freely through the crystal and can be accelerated to the velocity of sound by crystal strain gradients.

BIBLIOGRAPHY

A. S. Davydov, *Theory of Molecular Excitons.* McGraw-Hill, New York, 1962. (A)
J. J. Hopfield, "Excitons and their Electromagnetic Interactions," in *Quantum Optics*, pp. 340–395, R. J. Glauber (ed.). Academic Press, New York, 1969. (I)
C. D. Jeffries, "Electron–Hole Condensation in Semiconductors," *Science* **189,** 955–964 (1975). (E)
R. S. Knox, *Theory of Excitons.* Academic Press, New York, 1963. (A)
D. C. Mattis and J.-P. Gallinar, *Phys. Rev. Lett.* **53,** 1391 (1984).

Far-Infrared Spectra

Walter G. Rothschild and K. D. Moeller

The far-infrared (FIR) spectrum is the electromagnetic radiation which begins at wavelengths (λ) longer than can be dispersed by prisms [about 20–50 μm; 1 μm = 1 micrometer (or "micron," μ) = 10^{-6} m], and ends where continuous light sources cease to furnish FIR energy ($\lambda \sim 10000$ μm = 1 cm). *Far-infrared instruments* use FIR from the electron plasma of mercury lamps; they eliminate unwanted λ with low-pass transmission or reflection filters. Dispersion is by a mirror with ruled grooves which pass a narrow λ range ("grating spectrometer"). Such an instrument successively scans a total of N spectral events during time T, spending time T/N on each event. A more recent technique is *FIR Fourier transform spectroscopy*. It observes all N events during T—which gains $N^{1/2}$ in signal-to-noise ratio ("Fellgett" or "multiplex advantage")—by taking an "interferogram" and transforming it numerically to the ordinary frequency spectrum. Usually a Michelson interferometer is used: a beam splitter generates two FIR beams, one following an optical path of fixed length, the other's travel being varied by distance x ($-L \leq x \leq L$, $L \sim 10$ cm) through reflection on a mirror which moves, parallel to the FIR beam, back and forth. The two beams are recombined to give the interferogram

$$s(x) = 2\int_0^{\infty} A^2(\nu)\{1+\cos(2\pi\nu x)\}d\nu,$$

where $A(\nu)$ is beam amplitude at frequency ν, and $2\pi\nu x$ is phase angle difference due to the moving mirror. Fourier transformation gives the spectrum

$$S(\nu) = \int_{-\infty}^{+\infty} s(x)\cos(2\pi\nu x)dx.$$

Much thought has gone into writing programs which tell the detector when to observe $s(x)$ at mirror positions 0, x_1, x_2, . . . ("sampling"), to reconcile $|x| \leq L$ with infinity integration range ("apodization"), to do the many integrations rapidly, etc. Until continuously tunable narrow-band FIR sources (or detectors) are realized, Fourier-transform spectroscopy remains the choice for the FIR. (It is a classical technique: Michelson only lacked an on-line computer.) A *typical FIR wavelength* of 100 μm corresponds to the frequency 2.997×10^{12} Hz = 100 cm^{-1} (cm^{-1} = number of waves per cm, imprecisely called "wave number") and to an energy 1.986×10^{-14} erg = 285.8 cal/mol = 0.0124 eV, or an energy-equivalent temperature 143.9 K (room temperature ~300 K). *Far-infrared phenomena* consequently involve transitions between small energy differences or low temperatures. Most such phemonena are found in crystalline and solid states of matter because the multiplicity of three-dimensional directions of forces, the propagation of waves as patterned through the mobility of many closely spaced particles, the long-range order permitting magnetic and ferroelectric phenomena, etc., yield closely spaced energy levels. The breakdown of long-range order in liquids eliminates most of these cooperative effects, and the chaos of positions and orientations in vapors and gases restricts the application of FIR spectroscopy to weak intramolecular and intermolecular effects.

One-phonon effects: A "primitive unit cell" (i.e., smallest building block of a crystal) containing m atoms or ions possesses $3m$ degrees of freedom ("lattice modes"), which can be described by branches of standing waves ("phonon branches") $\psi = A\exp\{i(\mathbf{q}\cdot\mathbf{r} - \omega t)\}$ of angular frequency $\omega = 2\pi\nu$ and phonon wave vector $\mathbf{q}$ in crystal direction $\mathbf{r}$. Their spectra are usually in the FIR. For instance, cubic rock salt Na^+Cl^- ($m = 2$, lattice constant $d = 5.6402$ Å), has six phonon branches with q between 0 ("center of Brillouin zone") and $\pm\pi/2d = \pm 2.79 \times 10^7$ cm^{-1} ("boundary"). In contrast, a FIR photon wave vector K is small: For instance, at 100 μm; $K = 2\pi/\lambda = 628$ cm^{-1}. Its interaction with the lattice modes ("destruction of a FIR photon and creation of a FIR phonon") therefore involves the $q \sim 0$ phonon ("law of energy and wave vector conservation"). Only a doubly degenerate pair and a single phonon branch of the six are FIR-active ("optical branches"). The dipole moment in the pair is "perpendicularly polarized" (oscillates perpendicularly to phonon propagation direction) and thus can couple to the FIR field (always perpendicularly polarized). This pair is the "transverse optic branch" TO, with resonance frequency ω_{TO}. (Two people wiggling between them a long rope, up-down and left-right, illustrate TO.) The three other branches, the doubly degenerate TA ("transverse acoustic") and the single LA ("longitudinal acoustic"), are FIR inactive since no dipole moment is set up during their motion ("compressional waves"). When the photon–phonon interaction in the TO branch exceeds the resonance energy $\hbar\omega_{TO}$ (the latter depends on the forces, masses, and charge distributions in the crystal), the effective dielectric constant becomes negative. According to Maxwell's equations, this means that the FIR field cannot be sustained within the crystal and is thus strongly reflected ("Reststrahlen") in the range $\omega_{TO} < \omega < \omega_{LO}$ or $\lambda_{TO} > \lambda > \lambda_{LO}$ ("forbidden band"). Examples (λ_{TO} given) are 33 μm (NaF), 52 μm (NaCl), 61 μm (KCl), 77 μm (KBr), 150 μm (CsI). For more complicated crystals consisting of multiatomic ions (e.g., NO_3^-) or molecules, there are correspondingly more active phonon branches.

Multiphonon effects: For strongly anharmonic crystal forces, the lattice modes are not independent and several phonon branches interact with a FIR photon. For example, two phonons of frequency ω_i, ω_j lead to summation and difference frequencies $\omega_i \pm \omega_j$ ("two-phonon process") and total wave vector $q = q_i \pm q_j \sim 0$. Since q_i, $q_j \neq 0$, they are not at the center of the Brillouin zone but (usually) at the boundary. Optic modes can, in this way, couple with acoustic modes (FIR inactive in the one-phonon approximation; see above). For instance, $TO - LA = 105$ cm^{-1} (95.2 μm) for cadmium telluride at 110 K. FIR spectroscopy is thus helpful in characterizing the density of states at critical points of phonon branches, leading to a better understanding of compressibility, ionic charges, heat capacity, thermal expansion, dielectric properties, sound velocity, etc.

Impurity-induced absorption: Crystals with impurity atoms or structure faults show additional FIR absorption or reflection because their translational symmetry is broken. *(i) Local modes:* The vibrational motion of the impurity is concentrated within a few neighboring particles and is strongly influenced by them. Such impurity modes afford a probe for studying vibrational energy transfer and relaxation in solids. *(ii) Gap modes:* The impurity mode moves into the energy gap between the acoustic and optic branches. For instance, the gap mode of a Cl^- (chloride) impurity (~0.4 ppm) in K^+I^- (potassium iodide) appears at 77.10 and 76.79 cm^{-1} ($\lambda = 129.7$ and 130.2 μm) showing the ^{35}Cl and ^{37}Cl isotope splitting due to natural chlorine. *(iii) Resonance modes:* The impurity mass, driven by the rest of the lattice motions, performs large-amplitude forced vibrations which extend far into the crystal and can give rise to one-phonon absorption in the acoustic (usually FIR-inactive) branches. Study of FIR impurity absorption affords an important probe of lattice dynamics: The introduction of impurities need not change the crystal symmetry but the local disruption of ideality leads to additional absorption and hence to a greater wealth of available crystal parameters.

Soft-mode behavior: On cooling a crystal toward its Curie temperature (below which it becomes ferroelectric), it is found that certain *TO* branches tend to zero frequency. This means the crystal would disrupt, since $\omega_{LO}^2/\omega_{TO}^2 = \text{const}$ ("Lyddane–Sachs–Teller relation"), unless it undergoes a phase transition to another structure (e.g., cubic→tetragonal). This has been studied extensively for perovskite-type crystals ($CaTiO_3$). It offers electro-optic and FIR laser applications (tuning a soft mode with an electric field).

Spin-wave excitation and crystalline-field-effect spectra: The oscillatory motion of the ordered spins in the antiparallel sublattices of antiferromagnetic insulators in an applied magnetic field can couple to the magnetic component of FIR radiation, leading to "single-" ($q = 0$) or "two-magnon" ($q = \pi/2d$) resonance absorption below the ordering (Néel) temperature. For instance, FeF_2 has a one-magnon resonance at 52.7 cm^{-1}. In certain rare-earth ion garnets, e.g., $5Fe_2O_3 \cdot 3Y_2O_3$ (a mixed iron-ytterbium oxide), temperature-dependent collective $q \sim 0$ spin-wave excitation is observed (e.g., 20 cm^{-1} = 500 μm at 50 K) due to mutual spin precession of the iron and rare-earth sublattices. Temperature-independent FIR single-rare-earth ion transitions also appear, within energy levels determined by the atomic environment ("ligand-field potential") and the iron exchange field.

Semiconductors: Refractive index and absorption coefficients from FIR reflectance data are related to free-carrier concentration and mobility. For instance, in InSb (indium antimonide), free-carrier reflection occurs at $\lambda > 90$ μm, far beyond the *Reststrahlen* band ($\lambda_{TO} = 56$ μm). Of particular interest are ground-state impurity energy levels in regard to photocurrent production for solid-state FIR detectors. Beryllium-doped germanium (Ge) is a good detector for 15–52 μm at 4.2 K, copper-doped Ge is usable to ~130 μm. The photon-excited change in the electronic mobility of InSb is utilized for very-long-wavelength detection. Since this is an "intraband effect," there is no long-λ cutoff due to the gap between impurity level and conduction band. Such devices ("Putley detector") are useful to $\lambda > 3000$ μm.

Cyclotron-resonance measurements of free carriers determine many features of band structure in semiconductors. Typically, they are observed for electron carriers, e.g., in indium arsenide and phosphide, at 23–160 μm using strong magnetic fields, and at much longer λ for the heavier holes, e.g., in cadmium-doped *p*-type Ge. On sweeping of the magnetic field, several distinct absorption peaks appear in the FIR which allow distinction between the mass of the different holes involved in the transitions. *LO* phonons, which do not couple to perpendicularly incident FIR radiation (see above), can do so if a magnetic field is applied to the semiconductor, giving the free electron carriers both longitudinal and transverse components. These couple the longitudinal *LO* to the transverse FIR photon ("collective plasma cyclotron-*LO* modes"). Such magnetoplasma effects are useful, e.g., for the generation of tunable coherent radiation sources in the FIR.

High-temperature superconductivity: Far-infrared properties of ceramic high-T_c superconductors, in particular efforts to determine the superconducting energy gap 2Δ, have been actively pursued since the detection of the effect by Bednorz and Müller. Several reflectivity experiments seemed to have verified the Bardeen–Cooper–Schrieffer weak-coupling value $2\Delta = 3.5k_BT_c$. However, the spectra are complex (sample inhomogeneity, contamination, among others); certainty and consensus are still elusive at this writing.

Astrophysics: Several galactic and interstellar sources emit FIR radiation, for instance the Orion nebula. It is at present believed that the FIR is emitted from dust particles, surrounding the distant source and heated up to ~100 K. Recent far-infrared luminosity observations of nebulae by the IRAS satellite have divulged features that allow tentative attributions as to the chemical nature of the radiating dust particles, such as silicates, silicone carbides, polycyclic hydrocarbons, and ice matrices with chemical inclusions. Airborne far-infrared Fabry–Perot observations have revealed emissions from a large abundance of warm (about 300 K) and dense atomic and molecular (predominantly carbon monoxide) material in galactic star formation regions, in the inner region of our galaxy, in the galactic center, and in external galaxies. It appears that these fine-structure emissions originate from surfaces of molecular clouds excited photoelectrically from OB stars or the galactic interstellar radiation field. These observations open exciting vistas regarding future developments of theories of star formation and other cosmological phenomena.

Chemical physics: FIR is used to study rotational mobility in liquids, the conformational behavior of flexible vapor molecules, aspects of hydrogen bonding, collision-induced absorption in pressurized rare gases and homodiatomic molecules (hydrogen, oxygen, etc.), and rotational spectra of gaseous small molecules (H_2O, HF). The results of such experiments are of greatest interest to chemists.

See also INFRARED SPECTROSCOPY; MOLECULAR SPECTROSCOPY.

BIBLIOGRAPHY

R. J. Bell, *Introductory Fourier Transform Spectroscopy*. Academic Press, New York, 1972. (I/A)

Far Infrared Science and Technology, J. R. Izatt (ed.), SPIE—International Society of Optical Engineering, Bellingham, WA, 1986.

K. D. Moeller and W. G. Rothschild, *Far-Infrared Spectroscopy*. Wiley-Interscience, New York, 1971. (I/A)

W. G. Rothschild and K. D. Moeller, *Phys. Today*, **23**, 44, 1970.

Faraday Effect

Laura M. Roth

In 1845, Michael Faraday discovered that when plane-polarized light passed through lead glass in the direction of a magnetic field, the plane of polarization was rotated. The Faraday effect has since been observed in many media and over a wide range of frequencies. A plane-polarized light wave can be resolved into left and right circularly polarized waves, and the rotation arises from a difference in index of refraction between the two. The effect is usually proportional to the magnetic field, and the rotation per unit field per unit path length is given by the Verdet constant V. For positive V, the rotation is in the sense of the current flowing in the coil producing the field. Thus on reflection, the rotation increases, in contrast to the optical activity of quartz and many organic molecules which is due to an intrinsic left or right handedness of the structure, for which the rotation reverses upon reflection. Related effects are the Voigt effect, the Cotton–Moulton effect, and the Hall effect.

In atoms and in impurity ions in solids such as lead glass, the Faraday effect is related to the strong dispersion which occurs at frequencies near an optical absorption line. In a magnetic field the lines undergo Zeeman splittings, with a differential dispersion for the two circularly polarized components. For paramagnetic ions the rotation is enhanced by population effects. In semiconductors, Faraday rotation occurs which is related to electronic transitions across the energy gap. At frequencies above the absorption edge for interband transitions oscillatory effects are seen in large magnetic fields. The Verdet constant of Ge is -0.16 at 1.5 μm, and of Corning 8363 lead glass is 0.17 at 5000 Å, in min/Oe cm.

In solids, free charge carriers also contribute to the Faraday effect. In a plasma, such as the ionosphere, it is the electrons which contribute at radio frequencies. The rotation due to free carriers depends on their mass and density and is proportional to the square of the wavelength. In solids, the electron and hole effective masses can be measured by infrared Faraday rotation when the damping is too large to observe microwave cyclotron resonance. The density of electrons in intragalactic space is extremely small but the path length is enormous so that Faraday rotation of polarized radio waves, for example from pulsars, has been observed and has been used to estimate galactic magnetic fields. Ionospheric Faraday rotation has been measured by satellite.

In magnetic materials the Faraday rotation is related to the tensor property of the magnetic permeability. For ferromagnets and ferrimagnets the rotation is extremely large and proportional to the magnetization. The corresponding effect in reflection is the magneto-optic Kerr effect. Magnetic semiconductors also have large Faraday rotations. Metallic ferromagnets absorb light, but Faraday rotation has been observed in thin films. Transparent ferromagnets such as yttrium iron garnet (YIG) have large rotations (100°/cm at 7.9 μm for YIG) and are important optical materials. Ferrites and antiferromagnets are good magneto-optic materials at microwave frequencies.

In applications of the Faraday effect the figure of merit is the ratio of rotation to absorption. Faraday rotation is used to observe magnetic domain structure in transparent ferromagnets as well as thin films of metallic ferromagnets. Faraday rotation isolators are used in microwave and optical systems including optical fibers to prevent the highly amplified signal from reflecting back to the source. Other applications are mode converters, magnetic sensors, and optical recording systems.

See also HALL EFFECT; IONOSPHERE; KERR EFFECT, MAGNETO-OPTICAL; MAGNETIC MATERIALS; OPTICAL ACTIVITY; ZEEMAN AND STARK EFFECTS.

BIBLIOGRAPHY

C. L. Andrews, *Optics of the Electromagnetic Spectrum*, Chap. 19. Prentice-Hall, Englewood Cliffs, NJ, 1960. (E)

E. Scott Barr, "Men and Milestones in Optics V: Michael Faraday," *Appl. Opt.* **6**, 631 (1967). (E)

J. K. Furdyna, "Diluted Magnetic Semiconductors," *J. Appl. Phys.* **53**, 7637 (1982). (A)

F. A. Jenkins and H. E. White, *Fundamentals of Optics*, 3rd ed., Chap. 29. McGraw-Hill, New York, 1957. (E)

M. Lambeck, "Image Formation by Magneto-Optic Effects," *Op. Acta* **24**, 643 (1977). (I)

J. G. Mavroides, "Magneto-Optical Properties," in *Optical Properties of Solids* (F. Abeles, ed.). North-Holland, Amsterdam, 1972. (I)

P. A. H. Seymour, "Faraday Rotation and the Galactic Magnetic Field—A Review," *Quart J. R. Astron. Soc.* **25**, 293 (1984). (A)

K. Shiraishi, S. Sugaya, and S. Kawakami, "Fiber Optic Rotator," *Appl. Opt.* **23**, 1103 (1984). (E)

S. Wang, M. Shah, and J. Crow, "Studies of the Use of Gyrotropic and Anisotropic Materials for Mode Conversion in Thin Film Optical Wave Guide Application," *J. Appl. Phys.* **43**, 1861 (1972). (A)

K. A. Wickersheim, "Optical and Infrared Properties of Magnetic Materials," and K. J. Button and T. S. Hartwick, "Microwave Devices," in *Magnetism*, Vol. I (G. T. Rado and H. Suhl, eds.). Academic Press, New York, 1963. (I)

H. J. Zeiger and G. W. Pratt, *Magnetic Interactions in Solids*, Secs. 4.10.3 and 6.10. Clarendon, Oxford, 1973. (A)

Faraday's Law of Electromagnetic Induction

Leslie L. Foldy

The observation that time-varying magnetic fields induce electric fields proportional to the time rate of change of the former was made independently by Michael Faraday and Joseph Henry in the early 1830s. Its quantitative expression is one of the fundamental laws of electromagnetism and is commonly referred to as *Faraday's law of electromagnetic induction.* Its most familiar statement takes the form: *A change, by whatever means, in the magnetic flux linking a closed circuit will result in an electromotive force in the circuit instantaneously proportional to the time rate of change of the linking flux.*

Electromotive force (emf) is here interpreted in the sense that if the circuit is formed of a conducting material obeying Ohm's law, then the emf is the driving force for the electrical current which by Ohm's law is equal to the emf divided by the resistance of the circuit. The change in the linked flux may arise through any one or combination of the following means: (a) motion of a permanent magnet in the vicinity of the circuit, (b) a change of the position of, or the electric current in, a neighboring circuit such as an electromagnet, (c) motion of the original circuit through an externally produced magnetic field, (d) a change in the electric current in the original circuit (self-induction), (e) a change in the shape of the latter. The above statement of Faraday's law encompasses the explanation of the operation of many familiar electromagnetic devices including inductances (chokes), induction or spark coils, transformers, induction motors, generators of various types, etc.

A complete appreciation of Faraday's law is only possible by introducing the field concepts which so strongly influenced Faraday's thinking about electromagnetism. The electric field **E** and the magnetic induction **B** then exist at each point in space independently of the presence of any material substance at these points. The circuit referred to may then be conceived of as a closed mathematical curve, not necessarily material. The electromotive force in this circuit (whether at rest or changing in time with velocity small compared to vacuum light velocity) is expressed as a line integral about this curve in the form

$$\mathscr{E} = \text{emf} = \oint_C \mathbf{E}' \cdot d\mathbf{l},$$

$$\mathbf{E}' = \mathbf{E} + \mathbf{v} \times \mathbf{B},$$

where **E** is the electric field and **B** is the magnetic induction at the element $d\mathbf{l}$ of the circuit and **v** is the velocity of this element. The magnetic flux linking the circuit can be expressed as an integral of the magnetic induction over a surface S bounded by the circuit:

$$\Phi = \text{flux} = \int_S \mathbf{B} \cdot d\mathbf{S}.$$

The quantitative expression of Faraday's law then takes the form

$$\mathscr{E} = -\frac{\partial \Phi}{\partial t}.$$

The minus sign indicates that the direction of the emf is such as to induce a current in the circuit whose magnetic field has the direction required to oppose the change in flux linking the circuit (Lenz's law). The same result written in differential (purely local) form for a stationary infinitesimal circuit becomes

$$\text{curl } \mathbf{E} = -\frac{\partial \mathbf{B}}{\partial t},$$

which is one of Maxwell's equations. This together with another of Maxwell's equations, div $\mathbf{B} = 0$, expressing the nonexistence of magnetic monopoles, can be written as a Lorentz-invariant equation thus demonstrating the consistency of Faraday's law with Einstein's principle of special relativity.

With only minor changes in interpretation, Faraday's law in its differential form has survived the upheavals in physics of the twentieth century: special relativity, general relativity, and quantum mechanics. Recent theoretical attempts to unite weak, electromagnetic, and strong interaction (hadronic) phenomena within a unified theory have seriously raised again the question of the existence of free magnetic monopoles and their associated currents. Experimental discovery of magnetic monopoles would finally herald the necessity of a modification of Faraday's law through the addition of an inhomogeneous term in the mathematical equation expressing it.

The equations above are correct in various unit systems, one of which is: $\mathscr{E}$ in volts, E in volts/meter, Φ in maxwells, B in maxwells/meter^2, distances in meters, and time in seconds.

See also ELECTRODYNAMICS, CLASSICAL; MAGNETIC MONOPOLES; MAXWELL'S EQUATIONS.

BIBLIOGRAPHY

E. W. Cowan, *Basic Electromagnetism*. Academic Press, New York, 1968. (I)

R. P. Feynman, R. B. Leighton, and M. Sands, *The Feynman Lectures on Physics*, Vol. II. Addison-Wesley, Reading, Mass., 1964. (E)

J. D. Jackson, *Classical Electrodynamics*. 2nd ed. Wiley, New York, 1975. (A)

Fatigue

Johannes Weertman

Everyone is familiar with the fact that although a paper clip cannot be broken in two by simply bending it in one direction with the fingers, it is possible to break the clip without much trouble by repeatedly bending it back and forth. This is a simple example of the phenomenon of fatigue—the breaking of material by subjecting it to a cyclic stress whose amplitude is significantly smaller than the magnitude of the stress required to cause failure under static conditions.

In an engineering test of the fatigue properties of a metal the number N of stress cycles that a sample can withstand before failure occurs may be measured as a function of the amplitude S of the cyclic stress. If the experimental data are plotted on a graph of cyclic stress amplitude S versus the log of the number N of cycles to failure to obtain what is called an S–N curve, it is often (especially for steel) found that S decreases monotonically with increasing N from a value equal to the static failure stress (at $N=\frac{1}{4}$) to an apparent limiting value (for N larger than about 10^3–10^5) equal to about one half of the static failure stress. This lower value of S is called the fatigue limit or the endurance limit. A cyclic stress whose amplitude is smaller than the fatigue limit will never produce a fatigue failure, regardless of the number of stress cycles. However, in a number of metals and alloys there is no indication that the S–N curve ever levels off and for these materials a fatigue limit does not exist. Even in a material for which a fatigue limit appears to exist there is no complete assurance that if N is increased to a value larger than any of those used in the tests, fatigue failure may not occur under a cyclic stress whose amplitude is smaller than the apparent fatigue limit. There is virtually no frequency effect in fatigue if the temperature is not high relative to the melting point of the material and an active environment is not present. Whether the cyclic stress is applied at 1 cycle per hour or at 100 cycles per second, failure occurs at the end of the same number of total cycles. Only if the frequency is so very high that heat produced by plastic working cannot be dissipated easily is a frequency effect observed.

In material that contains no preexisting cracks, fatigue failure occurs in two stages. In the first stage small cracks are nucleated. If the material already contains cracks, the first stage of fatigue failure is of course bypassed. Small cracks may be nucleated in the interior of the material by the coalescence of a number of crystal dislocations on intersecting slip planes or, more commonly, at small second-phase particles. It is easier for small cracks to be nucleated at the surface of a metal part or fatigue specimen. The escape at a free surface of the dislocations that create slip bands causes the surface to be roughened. With an electron microscope it can be seen that small notches are produced in the roughened surface that are the equivalent of small surface cracks.

The subsequent growth of a small crack to a large size constitutes the second stage of the fatigue failure process. The rate of growth is given by the Paris equation:

$$da/dN = C(\Delta K)^n$$

where $2a$ is equal to the length of the crack, N now is the number of stress cycles (not the number required to produce failure), C is a constant whose value depends on the material tested, and ΔK is the cyclic stress intensity factor. During each stress cycle a fatigue crack will increase its length by an incremental amount if ΔK is large. If ΔK is small an incremental advance occurs only after many cycles. The term ΔK is a measure of the strength of the stress singularity at the tip of a crack. Depending on the crack geometry and the specimen geometry, the stress intensity factor either is equal to, or is approximately equal to, $\Delta K \cong S(\pi a)^{1/2}$. The exponent n in the Paris equation usually is equal to 4 but it can have values within the range of $n=2$ to $n=6$. In some materials, particularly in aluminum, regularly spaced striations can be seen on the fatigue crack surface. The spacing between the striations is such that it is exactly equal to the experimentally measured crack growth rate, proving that one striation is created in each stress cycle.

Complete failure of a part or test specimen occurs when a crack reaches a length sufficiently large that it will propagate catastrophically under a static stress of the same magnitude as S. The fatigue life of a part is thus determined by how long it takes a fatigue crack to be nucleated if no cracks are present initially, and by how long it takes a small crack to grow to a large size. Generally the growth stage occupies the major fraction of the fatigue life of a specimen or metal part when ΔK is large and the nucleation stage is the more important one when ΔK is small.

When the value of the stress intensity factor ΔK approaches the value required for catastrophic crack propagation under a static load, the fatigue crack propagation rate becomes greater than that predicted by the Paris equation. The Paris equation also breaks down when ΔK is so small that the predicted crack growth rate is smaller than about 3–10 atom distances per stress cycle. In these circumstances fatigue cracks apparently do not propagate at all and thus $da/dN \equiv 0$.

In tests in which a specimen is cycled under conditions of a constant plastic strain amplitude ϵ_p (the plastic strain is equal to the total strain less the elastic strain component ϵ_e) the number N of cycles required to reach failure is given by the Coffin–Manson law:

$$N^q = C_p/\epsilon_p$$

if $\epsilon_p \gg \epsilon_e$. Here q and C_p are constants ($q \simeq \frac{1}{2}$ and $C_p \simeq \epsilon_s/2$ where ϵ_s is the plastic strain at failure under a static load). If the specimen is cycled under a constant elastic strain amplitude and $\epsilon_e \gg \epsilon_p$, the failure equation is the Basquin law $N^{q^*} = C_e/\epsilon_p$, where $q^* \simeq \frac{1}{8}$ and C_e is another constant.

The fatigue life is affected by the environment around a specimen. For example, the fatigue crack growth rate of a crack in aluminum is an order of magnitude slower if the specimen is in a high vacuum than if it is in air of 50% humidity. When the temperature is high, diffusion–controlled cavity or crack-like cavity formation generally is the dominant fatigue mechanism. At high temperatures the fatigue life is a strong function of frequency as well as the wave shape of the cyclic stress. Failure paths tend to become intercrystalline rather than transcrystalline.

A phenomenon called static fatigue is observed in glass and in some inorganic crystalline material. No cyclic loading is involved in this phenomenon. A crack will grow under a static load very slowly until its length reaches the size required for catastrophic failure. A continuously occurring corrosive reaction at the crack tip is required to produce static fatigue.

See also ANELASTICITY; CRYSTAL DEFECTS; ELASTICITY; MECHANICAL PROPERTIES OF MATTER.

BIBLIOGRAPHY

J. T. Barnby, *Fatigue*. Mill and Boon Ltd., London, 1972. (E)

N. E. Frost, K. J. Marsh, and L. P. Pook, *Metal Fatigue*. Oxford (Clarendon Press), London and New York, 1974. (A)

M. F. Kanninen and C. H. Popelar, *Advanced Fracture Mechanics*. Oxford University Press, Oxford, 1985.

Special Technical Publications on fracture and fatigue, published at irregular intervals by the American Society for Testing and Materials, Philadelphia. (A)

Fermi–Dirac Statistics

P. M. Platzman

In classical mechanics physically similar particles are identifiable. We can number the particles at some instant of time and, in principle, follow each one in its motion during some dynamical process. In quantum mechanics there is, even in principle, no possible way of distinguishing or labeling identical particles. This is true because the system is described by a wave function Ψ satisfying some type of Schrödinger equation which, in turn, implies an uncertainty principle and the impossibility of labeling trajectories.

If we consider a system of two identical particles the states of the system obtained by interchanging the two particles must be equivalent. In particular, this means that the wave function $\psi(\xi_1,\xi_2)$ can change only by a phase factor when we interchange the two particles, i.e.,

$$\psi(\xi_1,\xi_2)=e^{i\alpha}\psi(\xi_2,\xi_1)$$

where α is a real number. By repeating the interchange in three-dimensional space we return to the original state so that $e^{2i\alpha}=1$, i.e., $e^{i\alpha}=\pm 1$. This argument can be generalized to any number of particles, i.e., the wave function of identical particles is either symmetrical or antisymmetrical under the interchange of any pair of particles. The symmetrical wave function corresponds to bosons, whereas the antisymmetric wave function corresponds to fermions. Fermions are said to obey Fermi–Dirac statistics.

Using some reasonable arguments regarding the general properties of relativistic quantum mechanics, Pauli (1940) constructed an argument which linked the intrinsic spin of the particle to the symmetry or antisymmetry of its wave function. In particular, he showed that half-integer spin particles were fermions, whereas integer spin particles were bosons. Atoms, the building blocks of ordinary matter, are made up of fermions. However, complex particles containing an even number of fermions, for example, He^4, behave as bosons when one considers processes (scattering, statistical mechanics, etc.) at energies small compared to the binding energy of the fermions.

If we consider a system composed of N noninteracting fermions, then the wave function of the various one-particle stationary states ψ_k with energy ϵ_k which each of the particles may occupy specifies the wave function of the N-particle system. Since the wave function is antisymmetric in the interchange of any pair, it follows that no two fermions may occupy the same one-particle state. This is called the Pauli exclusion principal and was first set forth by Jordan and Wigner (1928).

There are many dramatic experimental consequences of Fermi–Dirac statistics or equivalently, at zero temperature, the Pauli exclusion principle. In particular, all descriptions of atoms, the interpretation of their optical spectra, chemical valence, etc., use the concept of singly occupied one-particle states. Classification of nuclei and the spectra of their excited states are also linked to the exclusion principle.

At a finite temperature T it is easy to show that the thermodynamic potential of a set of noninteracting fermions is given by

$$\Omega=-k_BT\sum_k \ln\{1+\exp[(\mu-\epsilon_k)/k_BT]\}.$$

Here μ is the chemical potential. The average number of particles in each state ϵ_k is

$$\bar{n}_k=\{1+\exp[(\mu-\epsilon_k)/k_BT]\}^{-1}.$$

For free or almost free electrons, $\epsilon_k=\hbar^2k^2/2m$. Near $T=0$, $\mu\cong E_F=\hbar^2{k_F}^2/2m$, where n is the density.

The low-lying excited states of such a degenerate system are quantitatively described by exciting a particle from an occupied state $\bar{n}_k=1$ to an empty state $\bar{n}_k=0$, i.e., creating a particle–hole pair. Landau (1956) pointed out that many interacting-fermion systems (Fermi liquid) should have excitation spectra which could be put in one-to-one correspondence with the noninteracting system. The excitations in this case are called quasiparticle-quasihole pairs. There are many examples in nature of such degenerate Fermi liquids. For example, the Landau theory of Fermi liquids describes most of the low-temperature low-frequency properties of metals: their specific heat, their electrical conductivity, and their magnetic susceptibility. It also describes the properties of such diverse systems as low-temperature liquid He^3, the excited states of heavy nuclei, and the observed properties of white dwarf stars.

Modifications of Fermi liquid theory to include the attractive interactions between quasiparticles leads naturally to the so-called Bardeen–Cooper–Schrieffer theory of superconductivity. This theory describes most known low-temperature superconductors.

See also BOSE–EINSTEIN STATISTICS; HELIUM, LIQUID; QUANTUM FLUIDS; QUANTUM STATISTICAL MECHANICS; QUASIPARTICLES; SUPERCONDUCTIVITY THEORY.

BIBLIOGRAPHY

A more complete discussion of this topic and specific references may be found in:

L. D. Landau and E. M. Lifshitz, *Quantum Mechanics, Non-relativistic Theory*. Pergamon Press, New York, 1981.

L. D. Landau and E. M. Lifshitz, *Statistical Physics*. Addison-Wesley, Reading, MA, 1958.

Fermi Surface

Bernard R. Cooper

The concept of the Fermi surface in metals is intimately related to the existence of lattice periodicity, the most fundamental property of crystalline solids. The existence of the

Fermi surface is a consequence of the presence of lattice periodicity and the fact that electrons obey Fermi-Dirac statistics (i.e., obey the Pauli exclusion principle). Interest in the Fermi surface stems from the fact that it is the most important single entity characterizing the electron states in a metal, and also from the fact that macroscopic thermal, electric, magnetic, and optical properties are determined by, or have important contributions from, the behavior of the electrons at and near the Fermi surface. In this article we define the Fermi surface, particularly relating its existence to the lattice periodicity and to the occupation of electron energy bands in solids. We then briefly describe several of the most important experimental techniques for measuring the properties of the Fermi surface.

For a perfect three-dimensional lattice there are three primitive translations, $\mathbf{a}_1$, $\mathbf{a}_2$, $\mathbf{a}_3$, which define the periodicity of the lattice. That is, the crystalline atomic arrangement looks the same in every respect when viewed from any point $\mathbf{r}$ as when viewed from the point

$$\mathbf{r}' = \mathbf{r} + \mathbf{t}_n, \tag{1}$$

where

$$\mathbf{t}_n = n_1\mathbf{a}_1 + n_2\mathbf{a}_2 + n_3\mathbf{a}_3. \tag{2}$$

Here n_1, n_2, and n_3 are arbitrary integers; and for $\mathbf{a}_1$, $\mathbf{a}_2$, $\mathbf{a}_3$ to be primitive, any two points from which the crystalline atomic arrangement looks the same must be given by (1).

The requirement that the potential acting on an electron moving through a solid is identical at two points $\mathbf{r}$ and $\mathbf{r}'$ related by (1) leads to the requirement that the electronic wave functions obey Bloch's theorem and are labeled by a crystal-momentum quantum number $\mathbf{k}$. So the most general electron wave function in a three-dimensional periodic lattice has the form

$$\psi_k(\mathbf{r}) = u_k(\mathbf{r})e^{i\mathbf{k}\cdot\mathbf{r}}, \tag{3}$$

where u_k is periodic so that

$$u_k(\mathbf{r}) = u_k(\mathbf{r} + \mathbf{t}_n). \tag{4}$$

The crystal momentum $\mathbf{k}$ takes on a discrete set of values determined by periodic boundary conditions. We consider a system in the shape of a parallelepiped of edges N_1a_1, N_2a_2, and N_3a_3; and take the wave function to be identical for locations separated by a translation $\mathbf{t}_N = N_1\mathbf{a}_1 + N_2\mathbf{a}_2 + N_3\mathbf{a}_3$. (That is, we imagine all of space to be filled by identical contiguous parallelepipeds, and require the electronic wave functions to behave identically in each parallelepiped.) Then the discrete values of $\mathbf{k}$ are given by

$$\mathbf{k} = 2\pi\left(\frac{k_1\mathbf{b}_1}{N_1} + \frac{k_2\mathbf{b}_2}{N_2} + \frac{k_3\mathbf{b}_3}{N_3}\right), \tag{5}$$

where k_1, k_2, and k_3 are integers; and $\mathbf{b}_1$, $\mathbf{b}_2$, and $\mathbf{b}_3$ are the primitive reciprocal-lattice vectors given by

$$\mathbf{b}_1 = \frac{\mathbf{a}_2 \times \mathbf{a}_3}{\mathbf{a}_1\cdot(\mathbf{a}_2 \times \mathbf{a}_3)}, \tag{6}$$

with $\mathbf{b}_2$ and $\mathbf{b}_3$ obtained by cyclic permutation of the indices on the $\mathbf{a}$'s.

Reciprocal space is defined as the space in which the vectors $\mathbf{b}_1$, $\mathbf{b}_2$, and $\mathbf{b}_3$ are primitive vectors in the same way that $\mathbf{a}_1$, $\mathbf{a}_2$, and $\mathbf{a}_3$ are in ordinary space. If we consider a vector

$$\mathbf{K}_m = 2\pi(m_1\mathbf{b}_1 + m_2\mathbf{b}_2 + m_3\mathbf{b}_3), \quad \text{with } m_1,\ m_2,\ m_3 \text{ integers}, \tag{7}$$

then because the electron wave functions are of the form given by Eqs. (3) and (4) as required by the Bloch theorem for a periodic lattice, it is not possible to distinguish between the wave function with crystal momentum $\mathbf{k}$ and that with $\mathbf{k} + \mathbf{K}_m$. The set $\mathbf{K}_m$ (or strictly speaking $\mathbf{K}_m/2\pi$) are called reciprocal-lattice vectors and define the reciprocal lattice in the same way that the $\mathbf{t}_n$ of Eq. (2) define the real-space crystal lattice. The wave function shows periodicity with $\mathbf{k}$ in reciprocal space (also referred to as $\mathbf{k}$ space) analogous to the periodicity of $u_k(\mathbf{r})$ in real space. Thus all distinct electronic states can be considered by treating only that part of $\mathbf{k}$ space closer to the origin than to any nonzero $\mathbf{K}_m$. This part of $\mathbf{k}$ space is called the first Brillouin zone, and is constructed by setting up planes that perpendicularly bisect the lines connecting the origin to all $\mathbf{K}_m$, and then taking the volume about the origin enclosed by these intersecting planes. Figure 1 shows the first Brillouin zone for the face-centered-cubic (fcc) structure with conventional labeling of points and lines of high symmetry.

There are $N_1N_2N_3$ discrete values of $\mathbf{k}$ contained within the first Brillouin zone, i.e., one discrete value per real-space lattice point. For $\mathbf{k} = 0$, there will be a number of discrete electronic energy eigenvalues. For instance, for a noninteracting free electron gas these values are given by $E_m = \hbar^2K_m^2/2m$ for K_m taking on all values given by Eq. (7), including 0. As $\mathbf{k}$ varies across the Brillouin zone, the values of E emanating from each E_m form a semicontinuous band with band index m. The fact that $\mathbf{k}$ is a good quantum number

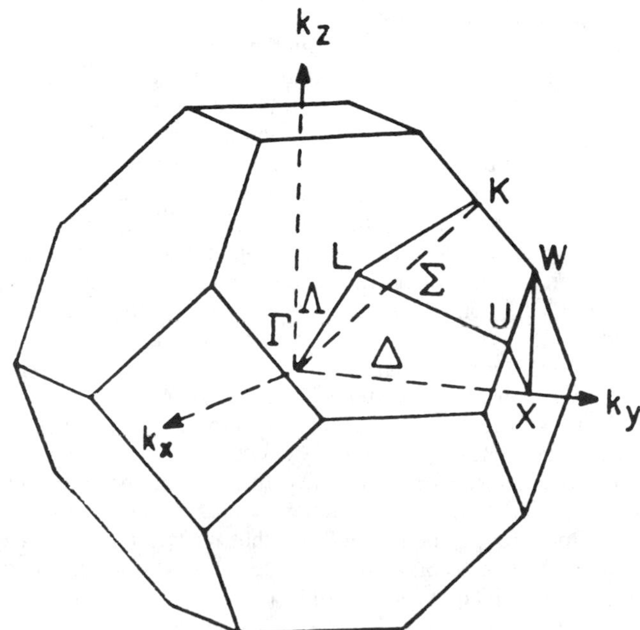

FIG. 1. The first Brillouin zone for the face-centered-cubic structure with points and lines of high symmetry labeled in conventional notation. [after B. Segall, *Phys. Rev.* **125**, 109 (1962)].

and that the band index m serves to lump all other quantum numbers (allowing for degeneracies for **k** at high-symmetry points and lines), together with the fact that electrons obey Fermi statistics, means that each band can hold two electrons (one with spin up and one with spin down) for a general discrete **k** in the Brillouin zone. Thus the first Brillouin zone can hold two electrons per real-space crystal lattice site per band. At absolute zero the band energy states will be filled with electrons up to an energy level, the Fermi energy (E_F), such that all the electrons are accounted for. For a metal there are one or more partially filled bands. Thus there is a surface in reciprocal space separating filled from empty states. This surface is the *Fermi surface*.

The simplest example of a Fermi surface is that for a noninteracting electron gas. The Fermi surface in that case is a sphere, and the radius of that sphere is the Fermi radius. For example, if one had an fcc lattice with one free electron per lattice site, the Fermi sphere would have a volume equal to one-half that of the first Brillouin zone shown in Fig. 1. Copper presents a simple and much studied case for a real metal. The 10 $3d$ electrons per copper atom form filled bands in the solid and do not give a Fermi surface. There is one valence electron per atom, corresponding to the atomic $4s$ state. The band formed from the valence electrons is partially filled and has behavior close to that for free electrons (i.e., the dispersion relationship over much of the Brillouin zone is approximately $E \sim k^2$ with an electron effective mass differing somewhat from the free electron mass); and as shown in Fig. 2, the Fermi surface is basically spherical. However, interaction effects between the s and d electrons give rise to "necks" contacting the hexagonal <111> faces of the Brillouin zone.

The Fermi surface has a symmetry appropriate to the particular crystal. Depending on the location in the Brillouin zone of the lowest-energy states and the number of electrons available for filling states, the Fermi surface can be quite complicated and bear no resemblance to a sphere. There may be disconnected surfaces in different parts of the Brillouin zone. Especially for metals with several valence electrons there may be several Fermi surfaces corresponding to partial filling of different bands. One can view the surface either from the direction of the filled states or from that of the unfilled states, the holes. If the filled states occupied the outer part of the Brillouin zone in Fig. 1, then the Fermi surface for electrons would be concave. For example, we could interchange the filled and unfilled states for the copper Fermi surface of Fig. 2. In such a case it is simpler to deal with the convex surface enclosing the unfilled volume, and one then treats the situation in terms of a *hole surface*.

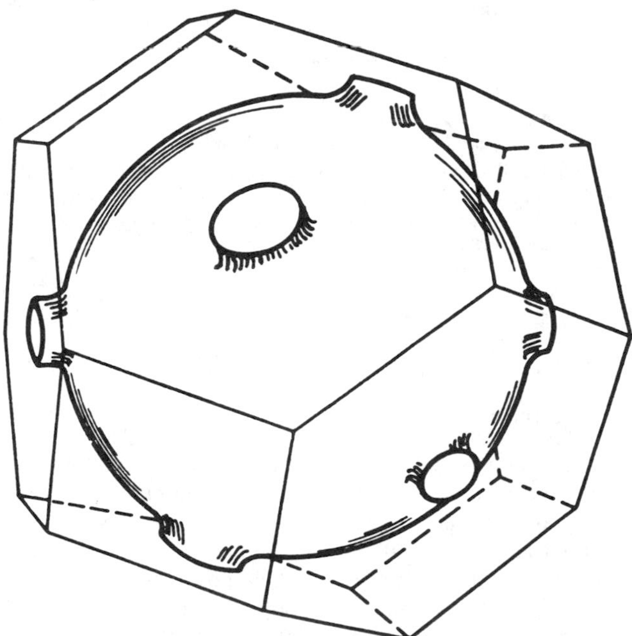

FIG. 2. A sketch of the Fermi surface of copper. The polyhedron represents the Brillouin zone [after B. Segall, *Phys. Rev.* **125,** 109 (1962)].

At finite temperature, following the laws of Fermi–Dirac statistics, the electron states with energy lower than E_F have only a finite probability for being filled, and those above E_F have only a finite probability for being unfilled. However, these probabilities differ significantly from unity only within kT of E_F. Since E_F is typically several electron volts and 1 eV is the energy equivalent of $T = 11,606$ K, this results in a slight blurring of the Fermi surface at finite temperatures.

Because of its relationship to the filling of the energy band states, the Fermi surface characterizes the electronic structure of a metal. Also the behavior of electrons at and near the Fermi surface determines, or gives important contributions to, many macroscopic properties of metals such as the electrical transport properties, magnetic behavior, and specific heat. For example, under the influence of a dc electric field, the only electrons that can respond are those at the Fermi surface, since they can be displaced into adjacent empty states with different lattice momenta. Thus the behavior of the electrons at the Fermi surface determines the electrical transport properties of metals. Because of this fundamental relationship to the electronic structure and to the macroscopic properties of metals, over a period of about 25 years there has been a great deal of experimental and theoretical work aimed at characterizing the Fermi surfaces of metals.

The experiments most directly characterizing the Fermi surface are those measuring the sizes and shapes of the orbits of electrons moving on the Fermi surface in a magnetic field. Changing the direction of the magnetic field causes the observed behavior to be governed by different parts of the Fermi surface thereby allowing one to map out the Fermi surface. Such experiments are possible only at very low temperature in very pure materials since the electron must be able to complete its orbit without scattering for the necessary observations to be possible.

Among the most valuable techniques for measuring important features of the Fermi surface are the following. (1) *The de Haas–van Alphen effect*. This involves measuring the periodic fluctuation of magnetic susceptibility in a varying magnetic field. The period of oscillation gives a direct measure of extremal cross-sectional areas of the Fermi surface normal to the magnetic field. By making measurements at all orientations of the crystal relative to the magnetic field, one can reconstruct the Fermi surface almost exactly. The de Haas–van Alphen effect has probably yielded the most precise Fermi-surface measurements. (2) *Magnetoacoustic attenuation.* This involves ultrasonic attenuation in a magnetic field applied perpendicular to the direction of sound propagation. Variation of the attenuation with magnetic field gives extremal diameters of the Fermi surface for sections normal to the applied field. (3) *High-field magnetoresistance.* This involves observing whether the electrical resistance transverse to a high magnetic field saturates or increases indefinitely on increasing H for varying directions of H. Nonsaturation for a given direction implies a multiply connected Fermi surface, i.e., the geometry in directions such as the necks in Fig. 2 where the Fermi surface intersects the Brillouin-zone boundary. Thus one can map out "neck" directions or directions of other multiply connected Fermi surfaces, as opposed to directions of closed Fermi surface such as the spherical portion in Fig. 2. (4) *Azbel'-Kaner cyclotron resonance.* This involves absorption of microwave energy as a function of varying magnetic field applied parallel to the surface of the metal. The period of oscillation in $1/H$ depends on the integral of the inverse of the component of velocity of the electron normal to $\mathbf{H}$ around the circumference of the orbit. If the shape of the Fermi surface is known by other means, this allows one to measure the electron velocity, i.e., the derivative of electron energy with lattice momentum, at the Fermi energy.

Finally, we should point out that once the Fermi surface is known in detail, as is now the case for most metallic elements, this knowledge can be combined with measurements of macroscopic thermal, electrical, magnetic, and optical properties to obtain information about the interactions of conduction electrons with each other, with collective excitations of various types, and with electromagnetic fields. Thus, once the Fermi surface is known, one can use that knowledge to consider effects beyond those treated by the one-electron theory (i.e., theory based on the model of electrons moving independently in an effective potential) for the electronic structure of metals.

In recent years there has been great interest in qualitative changes of behavior of certain metallic systems from the predictions of one-electron theory as described above. This one-electron theory involves describing the electronic behavior in terms of uncorrelated motion of the itinerant "band" electrons in the solid-state potential. Thus the departures of interest in the electronic behavior are referred to as correlated-electron effects and systems showing such effects are commonly referred to as correlated-electron systems.

The class of correlated-electron system that has gained the most attention are the "heavy-electron metals" often referred to as "heavy-fermion systems." These systems are typically compounds of light rare earths or light actinides (mostly of cerium or uranium) and are primarily characterized by having a low-temperature electronic contribution to the specific heat hundreds of times that corresponding to the band-electron density of states near the Fermi energy as found in one-electron band-theory calculations. There have been a variety of theories trying to explain heavy fermion behavior. These vary from Fermi-liquid-type theories where the Fermi-surface-type description remains essentially intact, to theories which question the existence of a measurable Fermi surface. Furthermore, the nature of the superconductivity in heavy fermion systems is strongly linked to behavior on the Fermi surface, e.g., whether there are anisotropic electron pairing states differing from the usual isotropic BCS pairing state and whether these new states are characterized by the vanishing of the energy gap at points or on lines on the Fermi surface. Thus the experimental observation of a Fermi surface, as for example by the de Haas–van Alphen measurements of Taillefer *et al.* (1987) on UPt_3, was very important in restricting and defining allowable mechanisms and theories. For UPt_3 the Fermi surface is multisheeted. The complicated and difficult measurements involved have called for refinements and improvements in the de Haas–van Alphen measuring techniques, very high-purity single-crystal samples, high magnetic fields, and very low temperatures involving use of magnetic-dilution refrigeration.

See also Cyclotron Resonance; de Haas–van Alphen Effect; Electron Energy States in Solids and Liquids; Heavy-Fermion Materials; Magnetoacoustic Effect; Magnetoresistance.

BIBLIOGRAPHY

R. J. Elliott and A. F. Gibson, *An Introduction to Solid State Physics*. Harper & Row, New York, 1974. (E)

Charles Kittel, *Introduction to Solid State Physics*, 5th ed. Wiley, New York, 1976. (E)

A. P. Cracknell and K. C. Wong, *The Fermi Surface*. Oxford Univ. Press, Oxford, 1973. (I)

J. M. Ziman, *Electrons in Metals*. Taylor & Francis, London, 1970. (I)

Walter A. Harrison, *Solid State Theory*. McGraw-Hill, New York, 1970. (A)

W. A. Harrison and M. B. Webb (eds.), *The Fermi Surface*. Wiley, New York, 1960. (A)

A. B. Pippard, *The Dynamics of Conduction Electrons*. Gordon and Breach, New York, 1965. (A)

Z. Fisk, D. W. Hess, C. J. Pethick, D. Pines, J. L. Smith, J. D. Thompson, and J. O. Willis, "Heavy-Electron Metals: New Highly Correlated States of Matter," *Science* **239**, 33–42 (1988).

L. Taillefer, R. Newbury, G. G. Lonzarich, Z. Fisk, and J. L. Smith, "Direct Observation of Heavy Quasiparticles in UPt_3, via the de H–van A Effect," *J. Magnetism and Magnetic Materials* **63, 64,** 372–376 (1987).

Ferrimagnetism

W. P. Wolf

Ferrimagnets are magnetic materials which exhibit a spontaneous magnetization below a certain temperature, but one which in contrast to *ferro*magnets arises from atomic moments which are not all parallel. In the simplest ferrimagnets there are two sets of antiparallel moments (generally described as *spins*) which are unequal either in number or in the magnitudes of their magnetic moments. Thus, a ferrimagnet represents the general intermediate case between a ferromagnet of parallel spins and an antiferromagnet of equal antiparallel spins.

HISTORY

The concept of ferrimagnetism was formulated by Néel in 1948 to explain the properties of ferrites with the spinel structure and the general composition Fe_2O_3MO, with M = Fe, Mn, Co, Ni, etc. These were known to exhibit a spontaneous magnetization, but the moment extrapolated to T = 0 K did not correspond to that which would be expected if all of the spins were parallel. For example, for iron ferrite (magnetite), with $M = Fe^{2+}$, the two Fe^{3+} moments would give $5\mu_B$ each and the Fe^{2+} moment $4\mu_B$, so that parallel spins would give $14\mu_B$ per molecule of Fe_3O_4. The observed moment, on the other hand, was close to only $4\mu_B$. Néel postulated that the spins aligned antiparallel, with the Fe^{2+} and one of the two Fe^{3+} spins pointing in one direction, and the other Fe^{3+} spin in the opposite direction. Such an arrangement immediately explained the observed moment. The same model also explained the observed moments for the ferrites with $M = Ni^{2+}$, Co^{2+}, and Mn^{2+}.

Néel's insight was based on earlier crystallographic analyses, which had shown that the ferrites which showed the spontaneous magnetization had a so-called inverse spinel structure in which half the trivalent Fe^{3+} ions and all of the divalent M ions occupied the same type of crystallographic site (denoted as the B sites) while the remaining trivalent ions occupied the so-called A sites. It was then only necessary to postulate that there exists a dominant interaction between the spins on the A and B sites which favors antiparallel alignment.

Such a mechanism had already been proposed for antiferromagnets by Kramers, in 1934. It is explained by a form of indirect exchange interaction which involves the nonmagnetic O^{2-} ions.

Néel's simple model was subsequently extended in a number of ways, but the fundamental idea of *sublattices* of spins aligned along different directions is common to all ferrimagnets.

MEAN FIELD THEORY

Temperature Dependence of Magnetization

Néel's theory also explained quantitatively the temperature dependence of the magnetization. For this he generalized the mean field theory originally due to Weiss to explain ferromagnetism.

If we consider a typical spin on, say, the A sublattice, the effect of the exchange interactions with other spins on the A sublattice and with spins on the B sublattice can be approximated by a molecular or mean field

$$\mathbf{H}_A = -N_{AA}\mathbf{M}_A - N_{AB}\mathbf{M}_B, \tag{1a}$$

where $\mathbf{M}_A$ and $\mathbf{M}_B$ are the average sublattice magnetizations and N_{AA} and N_{AB} are proportional to the appropriate exchange interaction constants. Similarly, for a spin on the B sublattice

$$\mathbf{H}_B = -N_{BA}\mathbf{M}_A - N_{BB}\mathbf{M}_B. \tag{1b}$$

The sublattice magnetizations $\mathbf{M}_A$ and $\mathbf{M}_B$ are determined self-consistently using the relation between magnetization and field:

$$M_A = \sum_i n_i g_i \mu_B S_i B_{S_i}(x_A), \tag{2}$$

where

$$x_A = \frac{g_i \mu_B S_i}{k_B T} H_A$$

and

$$B_{S_i}(x_A) = \frac{2S_i+1}{2S_i} \coth\left(\frac{2S_i+1}{2S_i} x_A\right) - \frac{1}{2S_i} \coth\left(\frac{x_A}{2S_i}\right)$$

is the Brillouin function for spins S_i with spectroscopic splitting factor g_i. Here, μ_B is the Bohr magneton, k_B is Boltzmann's constant, T is the temperature, and n_i is the number of spins per unit volume of species i on the A sublattice. The sublattice magnetization $\mathbf{M}_B$ is given by a similar expression involving $\mathbf{H}_B$.

As in the corresponding case for ferromagnetism, it is found that there are nontrivial solutions with $M_A \neq M_B \neq 0$ for T less than some critical temperature, which is now known as the Néel temperature, T_N.

The temperature dependences of $M_A(T)$ and $M_B(T)$ are generally similar, but not identical, depending on the relative values of the coupling constants N_{AA}, N_{AB}, etc., and the other parameters. Correspondingly, the vector sum $\mathbf{M}_S = |\mathbf{M}_A + \mathbf{M}_B|$ can show a variety of unusual variations. Three of these are shown in Figure 1. All three were predicted by Néel in his original paper and all were observed subsequently. Néel's theory also predicted some other variations which were later shown to be unphysical and to correspond to more complex situations with more than two sublattices (see below).

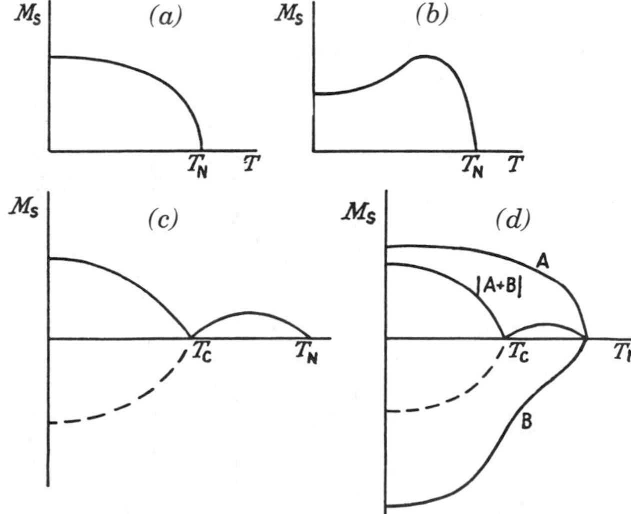

FIG. 1. (a), (b), and (c). Three types of spontaneous magnetization temperature curves predicted by Néel's theory. (d) illustrates the origin of the compensation point T_c arising from the cancellation of two-sublattice magnetizations.

Compensation Temperature

The most unusual and interesting variation is case (c), whose origin is explained in more detail in Figure 1(d). This shows the situation when one sublattice has a larger low-temperature magnetization, but slower initial temperature variation. It is then possible for $|M_A|$ to equal $|M_B|$ at some temperature T_c, the so-called *compensation point*. Since experiments often measure the absolute magnitude of $\mathbf{M}_A+\mathbf{M}_B$, the change of sign at T_c may not always be evident, but it can be observed in weak magnetic fields and has given rise to some important practical applications (see below).

High-Temperature Susceptibility

For $T > T_c$, there is no spontaneous magnetization, but an applied field can induce a paramagnetic moment in the usual way. The temperature dependence of the corresponding low-field susceptibility χ is also predicted by the mean field theory. For a simple two sublattice ferrimagnet, χ is given by

$$\frac{1}{\chi}=\frac{T}{C_A+C_B}+\frac{1}{\chi_0}-\frac{\sigma}{T-\theta}, \quad (3)$$

where

$$C_A=\sum_i n_i g_i^2 \mu_B^2 S_i(S_i+1)/3k_B$$

is the Curie constant for the A sublattice, and C_B is the same for the B sublattice, and χ_0, σ, and θ are constants which depend on the interaction parameters N_{AA}, N_{AB}, etc.

At high temperatures, Eq. (3) tends to the form $1/\chi=T/C+\theta'$, as for an antiferromagnet, consistent with the concept that the dominant interaction favors antiparallel neighboring spins. As the temperature decreases, however, $1/\chi$ will fall below the high-temperature asymptote and at some temperature, T_N, it will reach the value zero. Figure 2 shows experimental results for yttrium iron garnet (YIG) and the fit by an expression of the form of Eq. (3). It can be seen that the agreement is very good, except for temperatures very close to T_N.

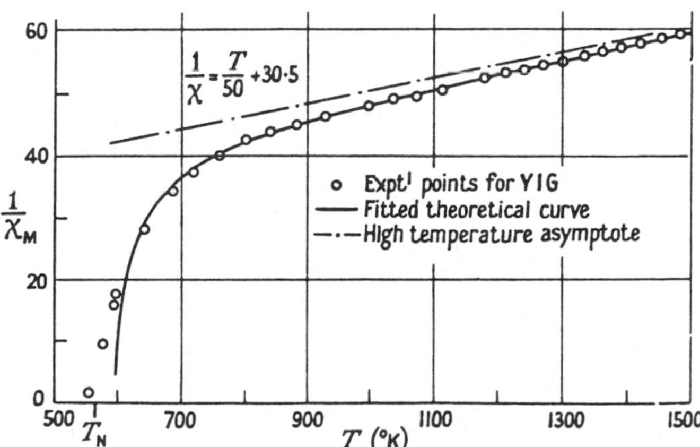

FIG. 2. Temperature variation of the reciprocal of the susceptibility for YIG above its Néel point (after Aléonard and Barbier, 1959).

Extension of the Two-Sublattice Model

The simple Néel model can be extended in a number of ways. The most obvious is the extension to more than two sublattices. Such a situation arises naturally in materials with a crystal structure with more than two inequivalent sites occupied by magnetic ions.

The best known examples of this situation are the rare-earth iron garnets, which have compositions of the form $5Fe_2O_3{\cdot}3M_2O_3$, where M is a rare earth or yttrium. In these materials, the Fe^{3+} ions occupy two crystallographically inequivalent sites, the octahedrally coordinated [*a*] sites and the tetrahedrally coordinated (*d*) sites, which occur in the ratio 4:6. A strong antiferromagnetic a-d coupling aligns the Fe^{3+} spins ferrimagnetically below about 570 K with a spontaneous moment which tends to $(6-4){\cdot}5\mu_B=10\mu_B$ as $T\rightarrow 0$ K. In the yttrium compound (commonly known as YIG for yttrium iron garnet), these are the only magnetic spins and the observed moment indeed tends to $10\mu_B$ at low temperatures, as shown in Figure 3. The small discrepancy can be ascribed to impurities in these (early) results.

The rare-earth ions occupy the eightfold coordinated {*c*} sites and are relatively weakly coupled to the Fe^{3+} on the [*a*] and (*d*) sites. Their presence does little to affect the ordering near T_N but, as the temperature is decreased, the relatively weak exchange field due to the Fe^{3+} sublattices polarizes the rare-earth moments and significantly changes the magnetization, as shown in Figure 3.

The simplest case to understand quantitatively is GdIG (M = Gd). If one assumes that the Gd^{3+} moments, with $7\mu_B$ each, are all antiparallel to the net magnetic moment of the Fe^{3+} spins, we would expect a total moment of $[(6){\cdot}7-(6-4){\cdot}5]\mu_B=32\mu_B$ in excellent agreement with the observed value. It must be noted that the net moment is larger than that of the Fe^{3+} sublattices only at low temperatures, and that there is thus a compensation point, T_c, at

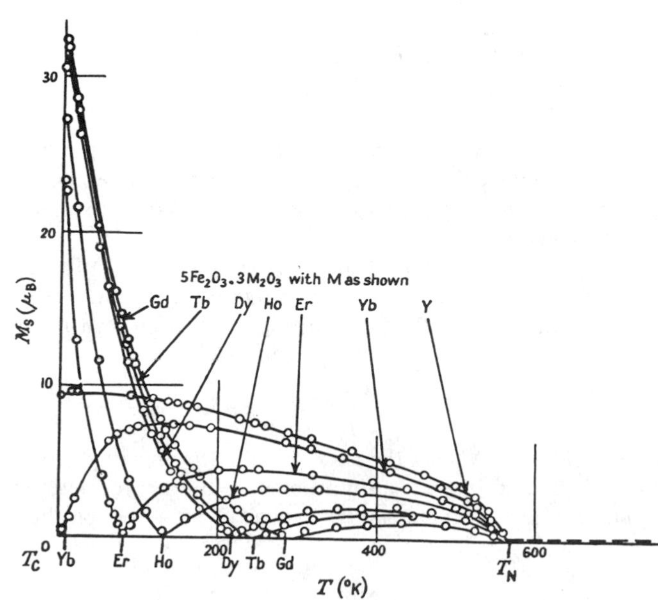

FIG. 3. Temperature variation of the spontaneous magnetization of several rare-earth garnets (after Pauthenet, 1958).

which the two become equal. Above T_c, the Fe^{3+} moment dominates.

For the other rare earths the quantitative moment predictions are complicated by orbital crystal-field effects, but the general behavior is similar. The orbital effects also lead to another complication which was not included in the original Néel theory: the possibility of noncollinear sublattices.

Noncollinear sublattices are generally the result of competing interactions. For rare-earth ions with large angular momenta, the competition is between local anisotropy effects, which favor different directions for different crystallographic sites, and the exchange field which tends to favor a common direction for the spins. For some of the rare-earth ions in the garnets, this leads to six rare-earth sublattices related to one another in threes by cyclic permutation about a [111] axis.

Noncollinear sublattice structures can also arise from competition between *inter*sublattice (N_{AB} and N_{BA}) and *intra*sublattice exchange interactions (N_{AA} and N_{BB}). This possibility was first recognized by Yafet and Kittel, in 1952, who showed that relatively strong antiferromagnetic interactions *within* a given sublattice (say the B sublattice in a two-sublattice ferrimagnet) can actually cause that sublattice to split into two nonparallel sublattices, which then make a common angle with the other (A) sublattice, resulting in a triangular three-sublattice structure.

These ideas can clearly be extended further and in general there is no *a priori* way to determine how many inequivalent sublattices will actually form. However, in practice it is often found that the simple two-sublattice model will adequately describe most ferrimagnets, provided one allows for two conceptual extensions: the possibility of more than one type of spin on a particular sublattice and also the possibility that some of the spins may deviate from the common direction but with an average moment which coincides with that of the rest of the sublattice. The averaging implied by such extensions is readily accommodated in the mean field approximately. Figure 4 summarizes schematically some of the commonly found ferrimagnetic sublattice structures.

(a) (b) (c)

(d) (e) (f)

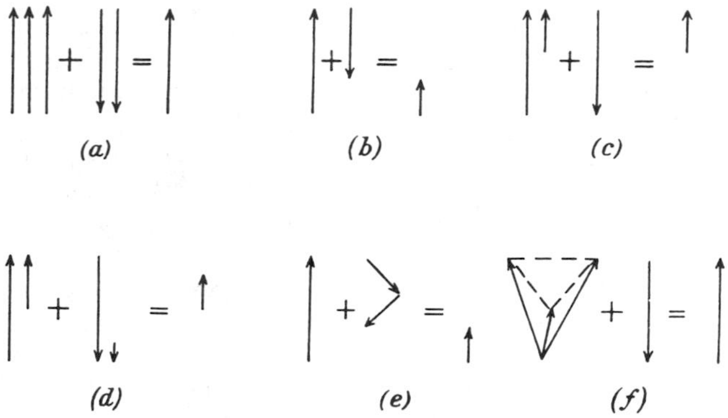

FIG. 4. Illustration of six simple sublattice arrangements which can give rise to a ferrimagnetic moment.

MORE EXACT THEORIES

Of course, the mean field theory is not exact, even for a simple two-sublattice ferrimagnet, and more exact models can be developed. Specifically, for low temperatures and for microwave properties it is appropriate to consider the role of spin waves, quantized excitations with characteristic wavelengths and energies. Mean field theory is also a poor approximation close to T_N, as in other cooperative transitions.

MATERIALS ENGINEERING

Both the spinel and the garnet structures allow wide varieties of substitutions of both magnetic and nonmagnetic ions. Correspondingly, it is possible to synthesize ferrimagnetic materials with specific magnetic properties and to control such parameters as magnetic anisotropy, magnetostriction, magneto-optical parameters, and microwave response, and the temperature dependence of all these properties. This has resulted in a large number of studies of related materials and a large body of experimental data. An extensive compilation may be found in the Bibliography.

Historically, the basic idea of making atomic substitutions to achieve specific properties was closely related to the early development of the theory of ferrimagnetism and it may have served as a useful model for the more general concept of materials engineering which is now widely used.

TECHNICAL APPLICATIONS

Practical applications make use of several distinct properties of ferrimagnetic materials. Some of these have nothing to do with the specifically *ferri*magnetic nature of the spin alignment but rather the almost coincidental fact that most of the earliest ferrimagnets also happened to be good electrical insulators. Consequently, they could be used for magnetic applications involving alternating fields without the complication of eddy current losses associated with most ferromagnetic metals and alloys.

Such applications include radio-frequency transformers and inductances, and microwave devices such as isolators, circulators, and phase shifters. They were also important for early computer memories, using ferrite cores, and for magnetic bubble memories, using thin films of garnets deposited by liquid-phase epitaxy.

Another technical application which remains very important is the use of γ-Fe_2O_3 and its derivatives as the recording medium for audio, video, and computer tapes and for both floppy and hard discs. Although γ-Fe_2O_3 does not have the typical spinel composition, it does, in fact, have a crystal structure closely related to that of the spinels and its formula may be written as $(Fe)[\square_{1/3}Fe_{5/3}]O_4$, where $\square_{1/3}$ denotes a vacancy on one-sixth of the $[B]$ sites in the spinel structure. The corresponding moment should thus be $(5/3-1)\cdot 5\mu_B = 3.3\mu_B$, in excellent agreement with the observed value of $3.2\mu_B$ per formula unit.

Other technically important ferrimagnetic materials include a series of complex hexagonal ferrites with compositions such as $BaFe_{12}O_{19}$ and $Ba_3Co_2Fe_{24}O_{41}$, which have high magnetic anisotropies and, correspondingly, applications at high microwave frequencies. The high anisotropy also leads to a large magnetic remanence and these materials are used extensively as permanent magnets for such applications as small electric motors.

All of these applications depend only on the existence of a net spontaneous moment, but not on the presence of more than one sublattice. Recently, however, there have been some important developments which make use of thermomagnetic hysteresis near compensation points for recording digital information. The materials which have been used are amorphous *metallic* alloys with compositions such as $Tb_xFe_yCo_z$, with various x, y, and z, in which the transition-metal moments align antiparallel to those of the rare earths. The readout is by magneto-optical rotation of reflected light (Kerr effect) which depends on the direction of the remanent magnetization.

CONCLUDING REMARK

The discovery of ferrimagnetic materials which are metals rather than oxides, and moreover amorphous, underscores the generality of the concept of ferrimagnetism, and it suggests that there can be many different systems which will exhibit this type of cooperative magnetism.

See also FERROMAGNETISM; MAGNETIC MATERIALS; MAGNETIC MOMENTS; PARAMAGNETISM.

BIBLIOGRAPHY

References to the specific papers cited in this article may be found in W. P. Wolf, "Ferrimagnetism," *Rep. Prog. Phys.* **24,** 212 (1961), which also contains a general discussion and an extensive bibliography of work prior to that date.

Extensive summaries of experimental data on ferrimagnetic oxides may be found in *Landolt-Börnstein: Numerical Data and Functional Relationships in Science and Technology*, K.-H. Hellwege and A. M. Hellwege, eds., Group 3, Vol. 12, Parts *a–c*. Springer-Verlag, Berlin, 1978, 1980, 1982.

General discussions of ferrimagnetism are given in many textbooks, including:

J. Smit and H. P. J. Wijn, *Ferrites*. Philips Technical Library, Eindhoven, 1959.

S. Chikazumi and S. H. Charap, *Physics of Magnetism*. Wiley, New York, 1964.

A. H. Morrish, *The Physical Principles of Magnetism*. Wiley, New York, 1965.

A. Herpin, *Théorie du Magnétisme*. Presses Universitaires du Magnétisme, Paris, 1968.

D. J. Craik (ed)., *Magnetic Oxides*. Wiley, New York, 1975.

Recent research results may be found in the proceedings of two annual conferences on magnetism, published in the *Journal of Applied Physics* and the *IEEE Transactions on Magnetics*.

Ferroelasticity

S. C. Abrahams

Ferroelasticity is a crystal property that was first fully recognized in 1969, although its effects had long been observed previously in the form of mechanical twinning. A crystal is ferroelastic if it can contain two or more equally stable orientational states, in the absence of mechanical stress, that may be transformed reproducibly from one to the other under the application of stress along an appropriate direction. The minimum stress required in order to transform states is the coercive stress E_{ij}, where i, j denote the effective applied stress and transformed strain directions. The magnitude of E_{ij} generally ranges from 10^4 to 10^8 N m^{-2} and is often strongly dependent upon temperature, pressure, and defect distribution within the sample. The spontaneous strain vector $\mathbf{e}_s$ denotes the distortion direction within the unit cell as developed from a higher symmetry cell and also gives a measure of the distortion. The magnitude of $\mathbf{e}_s$ is typically 10^{-3} or less.

Ferroelastic crystal growth is generally accompanied by the formation of domains in which $\mathbf{e}_s$ is distributed over all directions allowed by the higher symmetry. The resulting twinned crystal can in many cases be detwinned by the application of compressive stress at room temperature; all ferroelastically twinned crystals may be detwinned by raising the crystal temperature above the transition temperature (T_c) at which the higher-symmetry phase forms, then cooling from this paraelastic phase under compressive stress applied along an appropriate direction. This direction is often one assumed by $\mathbf{e}_s$ in the twinned crystal but analysis is required to determine it in each case. Reorientation of the spontaneous strain direction in large transparent single-domain ferroelastic crystals is often readily observable optically as the domain wall induced by the application of compressive stress sweeps through the crystal across the direction of view. Ferroelastic switching of $\mathbf{e}_s$ is always detectable by means of direction sensitive properties within the switching plane. Conoscopic examination of a ferroelastic biaxial crystal, for example, may exhibit readily controlled rotation of the plane containing the optic axes as stress is applied appropriately.

Ferroelastic point groups always form as subgroups of higher-temperature paraelastic point groups. The tetragonal point group *4/mmm*, for example, can give rise to the ferroelastic orthorhombic point groups 222, *mm*2, and *mmm*, with the normal equality of the tetragonal a_1 and a_2 axes replaced by orthorhombic a and b axes such that $e_s=(a-b)/(a+b)$ with $\mathbf{e}_s$ oriented along either [100] or [010].

The reorientation of $\mathbf{e}_s$ under compressive stress is accompanied by a rearrangement of the atomic distribution. The structure within the ferroelastic unit cell is such that all atoms are related pseudosymmetrically, either in pairs or by the identity operation. Thus, for each ith atom at the position $x_iy_iz_i$ there is a jth atom of the same element at the position $x_jy_jz_j$ with $x_iy_iz_i = f(x_jy_jz_j)+\Delta_{ij}$, where Δ_{ij} is an atomic displacement vector between the ith and jth atoms with magnitude on the order of one angstrom and $f(x_jy_jz_j)$ is a transformation that leads to reorientation of the lattice vectors. A typical example is the perovskite-like material $SmAlO_3$, crystallizing in space group *Pbnm*, which has the atomic position relationship

$$x_iy_iz_i=(\tfrac{1}{2}-y_j,\ \tfrac{1}{2}-x_j,\ \tfrac{1}{2}+z_j)+\Delta_{ij}.$$

At room temperature, Δ_{ij} for Sm is 0.18 Å; for Al, Δ_{ij} is zero, and for O, Δ_{ij} has values 0.40 and 0.64 Å. The supergroup of the paraelastic phase is *m*3*m* or *m*3, since the subgroup is *mmm*. With lattice constants at 298 K of $a=5.29\ 108$, $b=5.29\ 048$, and $c=7.47\ 420$ Å, $e_s=5.67\times10^{-5}$. Compressive stress of 50 MN m^{-2} applied along [100] transforms the a axis into $\bar{b}$ and the b axis into $\bar{a}$, as implied by the atomic position relationship which can be expressed in the equivalent form **abc**→$\bar{\mathbf{b}}\bar{\mathbf{a}}\bar{\mathbf{c}}$, i.e., the **a** and **b** basis vectors exchange and reverse sense under the application of stress as illustrated in Fig. 1.

Reported maximum values of Δ_{ij} in a variety of ferroelastic crystals range from about 0.4 to 2.4 Å. The coercive stress required to reorient $\mathbf{e}_s$ in crystals with large Δ_{ij} may exceed the cohesive strength, as is the case for the langbeinite-type structure $K_2Cd_2(SO_4)_3$ in which the maximum value of Δ_{ij} is 0.91 Å even at 418 K, with $T_c=432$ K. A crystal that may be ferroelastically reoriented only by heating to T_c and cooling under stress is a *frozen* ferroelastic.

The thermal dependence of e_s, i.e., (de_s/e_sdT), for a number of ferroelastics has been reported in the range -1 to -20 mK^{-1}. Ferroelasticity is often accompanied by such other properties as ferroelectricity, ferromagnetism, or superconductivity. For materials in which the other property is coupled, ferroelastic reorientation of $\mathbf{e}_s$ may cause reorientation of the second property. A well-known example of coupling between ferroelasticity and ferroelectricity is provided by $Tb_2(MoO_4)_3$ (TMO), where $x_iy_iz_i=(\frac{1}{2}-y_j,\ x_j,\ 1-z_j)+\Delta_{ij}$, so that the a and b axes interchange under the application of compressive stress as the polar c-axis reverses sign simultaneously, with consequent reversal of the spontaneous polarization P_s, i.e., the basis vector transformation is **abc**→**ba**$\bar{\mathbf{c}}$. Similarly, reversal of P_s by the application of an electric field results in the interchange of the a and b crystal axes and hence the reorientation of $\mathbf{e}_s$. The normalized thermal dependence of e_s is identical to that of P_s (see Pyroelectricity, Fig. 2), confirming experimentally that these two properties are fully coupled in TMO.

In the case of a fully coupled ferroelastic–ferromagnetic crystal such as Fe_3O_4 at room temperature, or Mn_3O_4 below T_c, either field reversal of the spontaneous magnetization or the application of compressive stress will simultaneously reorient both spontaneous magnetization and spontaneous strain.

Among the new high-T_c superconductors are some with a slight orthorhombic distortion from tetragonal symmetry; these are generally ferroelastic and may have enhanced properties if detwinned as outlined above, since conductivity losses at domain wall boundaries are eliminated in single-domain crystals.

Ferroelastic devices, based upon the control of domain wall motion by a small electric field, include optical switches, adjustable optical slits, pattern generators including color modulators, and variable acoustic delay controllers. Micropositioners capable of displacements larger than those possible with piezoelectric positioners have also been designed on the basis of the dimensional changes obtained ferroelastically.

See also Elasticity; Ferrimagnetism; Ferroelectricity; Ferromagnetism; Phase Transitions; Pyroelectricity; Superconductivity Theory.

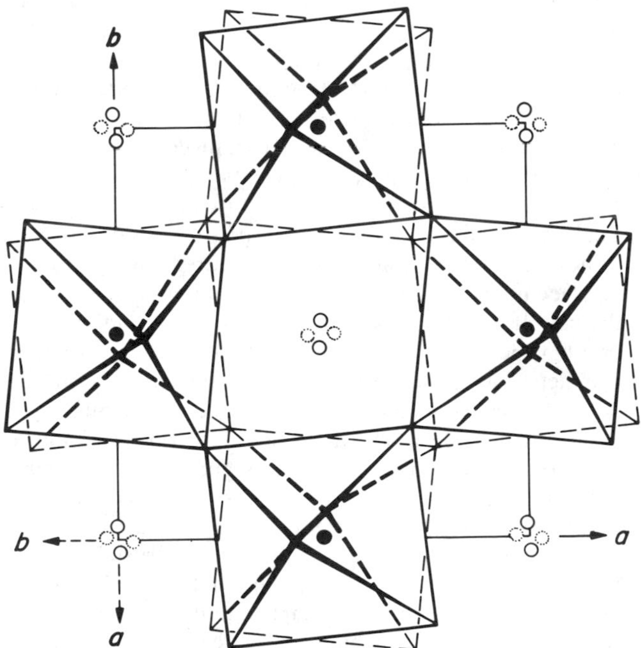

Fig. 1. Unit cell and atomic positions in $SmAlO_3$. Solid lines represent the **abc** orientation and broken lines the $\bar{\mathbf{b}}\bar{\mathbf{a}}\bar{\mathbf{c}}$ orientation following ferroelastic transformation. Octahedra represent O, solid circles Al, and open circles Sm atoms; the lower octahedral edges are omitted for clarity [after S. C. Abrahams, J. L. Bernstein, and J. P. Remeika, *Mat. Res. Bull.* **9**, 1613 (1974)].

REFERENCES

S. C. Abrahams, *Mat. Res. Bull.* **6**, 881 (1971).
K. Aizu, *J. Phys. Soc. Japan* **27**, 387 (1969).
F. Lissalde, S. C. Abrahams, J. L. Bernstein, and K. Nassau, *J. Appl. Phys.* **50**, 485 (1979).

Ferroelectricity

J. D. Axe and M. E. Lines

Ferroelectrics are materials which possess an electric polarization in the absence of an externally applied electric field, together with the property that the direction of the polarization may be reversed by an electric field. The study of the properties of ferroelectric materials and the attempt to understand the nature of the ferroelectric state constitute the field of ferroelectricity.

Any macroscopic collection of matter is, to a very high degree of accuracy, electrically neutral. However, the positive and negative charges which make up the material are not necessarily distributed in a symmetric manner. If the "centers of gravity" ($\pm d$) of the summed positive and negative charges ($\pm q$) do not coincide (i.e., $d\neq0$), then the material is said to possess an *electric dipole moment*, which is a vector quantity of magnitude $2dq$ and direction from $-d$ to $+d$. The dipole moment *per unit volume* is defined as the *dielectric polarization*. Any material develops a dielectric polarization P when placed in an electric field E, but a substance which has such a natural charge separation even in the absence of a field is called a *polar* material and is said to possess a *spontaneous* polarization P_S. Whether or not a material is polar is determined solely by its crystal structure in the sense that only certain classes of crystal structure are compatible with a polar charge distribution. Crystal structures can be divided into 32 classes, or *point groups*, according to the number of rotational axes and reflection planes they exhibit which leave the structure unchanged. Only those which contain a single axis of rotation symmetry without a reflection plane perpendicular to it are polar. Only 10 of the 32 point groups are polar although they do include examples from all crystal symmetry types lower than cubic (e.g., tetragonal, orthorhombic, hexagonal, trigonal, monoclinic, and triclinic).

Under normal circumstances even polar crystals do not display a net dipole moment. As a consequence there are no electric dipole equivalents of bar magnets. This is because the intrinsic electric dipole moment of a polar material is neutralized by "free" electric charge that builds up on the surface, migrating there from within the crystal (by internal conduction) or from the ambient atmosphere. An equivalent buildup of surface "magnetic charge" on bar magnets cannot take place because the free magnetic monopole (which is the magnetic counterpart of electronic charge) is not known to exist in nature. Polar crystals consequently reveal their polarity only when perturbed in some fashion which momentarily disturbs the balance with compensating surface charge. Since spontaneous polarization is in general temperature dependent, the most universally available such perturbational probe is a change of temperature which induces a flow of charge to and from the surfaces. This is the *pyroelectric* effect. All polar crystals are consequently pyroelectric and, as a result, the 10 polar crystal classes are often referred to as the pyroelectric classes.

Another rather obvious perturbational probe of polarization is the electric field itself. Surprisingly, however, the dielectric response of most pyroelectrics to such a field is small, linear, and not significantly different from that of a nonpolar material. The exceptions are those for which the applied field is sufficient in magnitude to reverse the direction of spontaneous polarization. For these the dielectric response is large and uniquely different, a typical variation of polarization P with applied field E being shown in Fig. 1. The highly nonlinear characteristic shape of this curve, which is called a *hysteresis loop*, is therefore the signature of that small subset of polar crystals which can be dielectrically "switched" by externally applied electric fields. It is this particular subset for which the term *ferroelectric* is reserved. More formally, a crystal is said to be ferroelectric if it possesses two or more orientational polar states and can be shifted from one to another of these states by an electric field. Since crystal perfection, electrical conductivity, temperature, and pressure are all factors which affect the ability to reorient polarization, the question of whether or not a particular material is ferroelectric is not one which can be answered by crystallographic determination alone. Also understood in this definition is the fact that the polar character of the orientational states should be absolutely stable in the absence of a field. Materials which mimic ferroelectrics but for which the polar states are only metastable (i.e., decay slowly with time when the field is removed) are defined separately as *electrets*.

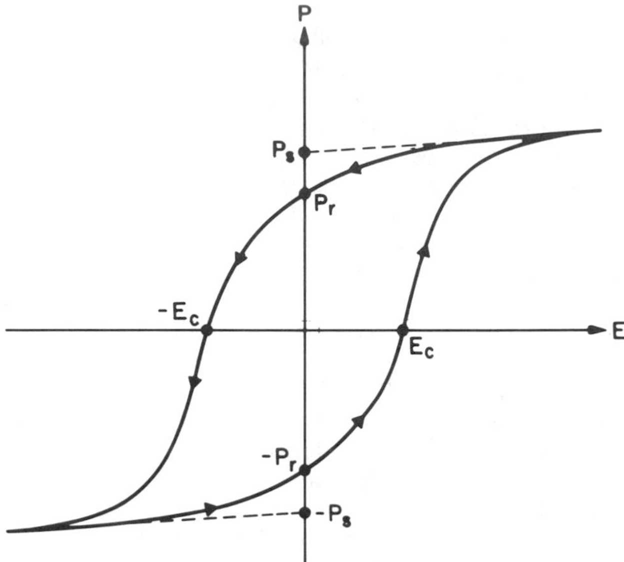

FIG. 1. A typical ferroelectric hysteresis loop showing the variation of polarization P with applied field E as the latter is scanned through a positive and negative cycle in the direction of the arrows. Special values of P and E on the loop, as depicted in the figure, define such ferroelectric parameters as P_S (spontaneous polarization), P_r (remanent polarization) and E_c (coercive field).

Since ferroelectrics can be "switched," or in a sense structurally reoriented, by relatively modest fields (i.e., smaller than those which would produce crystalline dielectric breakdown), the implication is that they have crystal structures which are only slightly removed from some higher-symmetry (usually nonpolar) parent structure. The highest-symmetry phase in terms of which the ferroelectric phase can be described by small internal atomic rearrangements is called the *prototype* phase. In most instances this prototype phase actually exists as the highest-temperature phase of the crystal. For such a case the ferroelectric phase develops from the prototype, as the temperature is lowered, by spontaneous displacements of the equilibrium positions of atoms within the unit cell of the crystal. If this takes place in the absence of a biasing applied electric field, then, because of the high symmetry of the prototype, at least two (and often more) polar atomic displacements are equivalent by symmetry and are distinguished only by their direction of polarization. For example, if the prototype phase is cubic and the incipient ferroelectric phase energetically favors a cubic axis for its polar orientation, then six equivalent such directions exist and, in general, all six of these equivalent crystal structures will coexist in the cooled sample in the form of *ferroelectric domains*. A single-domain sample can be induced by *poling* with an external field which energetically favors one directional-domain over the others.

Quite generally, the stable atomic configuration of any crystal at any particular temperature T is that which minimizes the crystal energy at that temperature. This energy results partly from thermally induced atomic vibrational motion and partly from forces of interaction between the atoms.

The interplay of these two energy components as a function of T can often result in a sudden spontaneous displacement of the equilibrium (i.e., vibrationally averaged) positions of the atoms. These sudden displacements are called *structural transitions* and the temperatures T_C at which they occur *Curie temperatures*. Most such transitions do not result in a separation of the "centers of gravity" of the positive and negative charges and are consequently not ferroelectric. Only those that involve the appearance of a spontaneous polarization are ferroelectric.

It is quite possible for a whole sequence of phases, some ferroelectric, to occur as a function of changing temperature in some systems. It is even possible for one or more ferroelectric phases to occur sandwiched between nonpolar lower- and higher-temperature structures, although such an occurrence is rare. Most often ferroelectrics possess a single low-temperature (polar) phase for which the spontaneous polarization P_S disappears at a single Curie temperature T_C. For most ferroelectrics P_S goes to zero discontinuously at T_C (which is then called a *first-order* transition temperature) although a few examples are known for which P_S goes smoothly to zero at T_C (a *second-order* transition). Second-order ferroelectric transitions are characterized by a linear dielectric response which diverges at T_C while first-order transitions exhibit a marked, but finite, discontinuity at T_C (Fig. 2). Linear dielectric response, which is essentially a measure of the sensitivity of polarization to small applied electric fields, is recorded by a dimensionless quantity called the *dielectric constant*. Most ferroelectrics not only exhibit pronounced peaks in dielectric constant near T_C but tend to maintain large values of this quantity (often in excess of 1000) over wide temperature ranges. Normal nonferroelectric dielectric materials by contrast exhibit dielectric constants which are very much smaller (typically in the range between about 2 and 10). Ferroelectrics are also type characterized as *displacive* or *order–disorder* depending respectively upon whether the prototype phase is nonpolar in a truly microscopic sense (with all atoms vibrating about unique mean positions which together make up a nonpolar configuration) or only in a macroscopic sense (with some atoms hopping between different low-symmetry sites in a manner which averages to a nonpolar configuration only in a statistical sense). Since displacive transitions involve only small oscillations, their physics is simpler, being describable solely in terms of crystal lattice vibrations (or *phonons*).

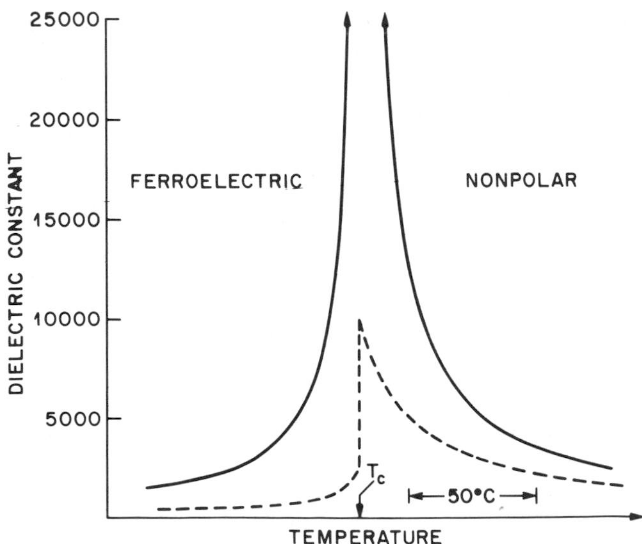

FIG. 2. Typical scans of dielectric constant as a function of temperature near the onset of ferroelectricity at T_c for a second-order (full curve) and first-order (dashed curve) ferroelectric.

When atoms vibrate they do not do so independently. Because of the forces between them they move in a collective fashion which defines a *normal mode* of oscillation. A great many normal modes of different spatial symmetries are present in a typical prototype phase. However, only one such normal mode is involved in an essential manner at a particular displacive structural transition. This mode is the one for which the spontaneous atomic displacements that set in below T_C represent a "time-frozen snapshot" of the vibrational displacement pattern of the mode above T_C. The mode is said to "drive" the transition and, becoming anomalously low in frequency on approach to T_C, is called a *soft mode*. The analogous dynamic anomaly driving an order–disorder transition does not directly involve phonons but concerns the much slower large-displacement intersite hopping motions which characterize this type of transition.

If the soft mode has a polar character, then the phase transition is said to be *intrinsically* ferroelectric and to define a *proper* ferroelectric phase. However, if the driving mode is nonpolar, it is still sometimes possible for the phase below T_C to be ferroelectric. This can occur if there is an interaction of suitable symmetry between the polarization and the nonpolar driving-mode displacement pattern. Such transitions are said to be *extrinsically* ferroelectric and to define an *improper* ferroelectric phase. Only proper ferroelectrics have a dielectric response which becomes anomalously large near T_C; improper ferroelectrics exhibit only a minor (if any) anomaly in dielectric response at T_C.

In a similar fashion, many other physical properties can couple to a greater or lesser extent with a soft polar (or ferroelectric) mode. As a consequence all of these properties undergo extrinsically induced anomalies at a proper ferroelectric transition, anomalies which may become quite large near T_C itself. One universal example involves the appearance of a spontaneous mechanical strain which, if "switchable" by applied stress, defines the phenomenon of *ferroelasticity*. Accompanying the appearance of spontaneous strain is a linear coupling between polarization and stress (and between applied electric field and strain) referred to as *piezoelectric* response. As a result, all ferroelectrics, and indeed all polar materials, are *piezoelectric*. Another common manifestation of extrinsic coupling is the appearance of spontaneous birefringence (the velocity of light becoming different parallel and perpendicular to the polar axis) and more generally a linear dependence of refractive index on applied field known as the *electro-optic* effect. Even magnetic anomalies, in rare instances, can be extrinsically induced by ferroelectric transitions.

Most of the technical applications of ferroelectrics utilize one or more of these anomalous properties in some way. The high dielectric constants of ferroelectric materials make them useful in fabricating high–energy-storage capacitors.

Materials with large piezoelectric response are useful for sensitive sonar detectors or *transducers* (devices that transform pressure variations into voltage variations or vice versa), whereas those with large pyroelectric response have received application as infrared thermal detectors and, more recently, for the recording of infrared images. At present the fastest growing area of technical application centers on the favorable electro-optic properties. These have been put to use in the rapidly developing field of *photonics*, an information technology in which optics and electronics play complementary roles. In this context, important devices include optic switches and light modulators (for signal processing and interconnection), frequency converters, and light valve arrays (in the development of information storage and optical display technologies). In these various capacities some of the more frequently used ferroelectric materials are barium titanate ($BaTiO_3$), lithium niobate ($LiNbO_3$), lithium tantalate ($LiTaO_3$), potassium dihydrogen phosphate (KH_2PO_4), triglycine sulphate (often abbreviated as TGS), gadolinium molybdate [$Gd_2(MoO_4)_3$], bismuth titanate ($Bi_4Ti_3O_{12}$), and various ceramics based on mixtures of lead zirconate and lead titanate (PZT). Although, in principle, it is also possible to use the bistable nature of ferroelectric domains as the basis for high-storage-density computer memories, ferroelectrics in this context have not yet been produced with performance characteristics which can compete with the more reliable magnetic and semiconductor memory elements.

The phenomenon of ferroelectricity was first discovered in 1920 in sodium potassium tartrate tetrahydrate (better known as Rochelle salt) long after the related phenomena of piezoelectricity and pyroelectricity had been recognized and studied in the mid-to-late nineteenth century. For many years Rochelle salt remained the sole known example, and ferroelectricity was considered to be a rarity in nature. By contrast, over 500 other examples have now been discovered so that ferroelectrics are in fact far from being uncommon although, for reasons which are still not fully understood, the overwhelming majority are oxides. Perhaps the simplest (and certainly the most studied) high-symmetry prototype crystal structure which "spawns" ferroelectrics is cubic perovskite, with only five atoms per unit cell. Among its better known ferroelectric family members are barium titanate ($BaTiO_3$), lead titanate ($PbTiO_3$), and potassium niobate ($KNbO_3$). The most thoroughly researched improper ferroelectric is gadolinium molybdate ($Gd_2(MoO_4)_3$). Although most of the presently recognized ferroelectric materials are inorganic crystals, an increasing number of examples are now being discovered among organic polymers and even liquid crystals.

See also CRYSTAL SYMMETRY; DIELECTRIC PROPERTIES; NEUTRON SPECTROSCOPY; PIEZOELECTRICITY; PYROELECTRICITY.

BIBLIOGRAPHY

C. Kittel, *Introduction to Solid State Physics*, 6th ed. Wiley, New York, 1986, Chap. 13. An elementary account of the basic phenomenon.

R. Blinc and B. Zeks, *Soft Modes in Ferroelectrics*. Elsevier, New York, 1974.

M. E. Lines and A. M. Glass, *Principles and Applications of Ferroelectrics and Related Materials*. Clarendon Press, Oxford, 1977.

T. Mitsui, *An Introduction to the Physics of Ferroelectrics*, Gordon and Breach, New York, 1976.

J. C. Burfoot and G. W. Taylor, *Polar Dielectrics and their Applications*. University of California Press, Berkeley, 1979.

Ferromagnetism

A. S. Arrott

Coulomb's law governs the interaction between electrical charges at rest. It is a requirement of the principle of relativity that Coulomb's law be modified when the charges are in motion relative to an observer. Though these modifications are usually very small, depending on the ratio of the products of velocities to the square of the velocity of light, their effects have been observed since antiquity. If the velocity of light were infinity, there would be no magnetism. If it were but ten times larger than it is, electrical technology with its generators, transformers, and motors would be most difficult to achieve. Yet the genius, Faraday, was able to create a language of fields and fluxes that describes these workings of magnetic devices without reference to relativity or the velocity of light. Faraday discovered the law of induction that governs the behavior of the transformer. Not only is the transformer a practical device but also it is useful for studying and teaching the properties of magnetic materials.

Consider a magnetic material in the form of a toroid with two coils of wire wound about its surface as shown in Fig. 1a. The toroid has a mean radius R and rectangular cross-section ab. A current i is passed through the primary coil and a voltage V is measured across the ends of the secondary coil. The voltage is proportional to the rate of change of

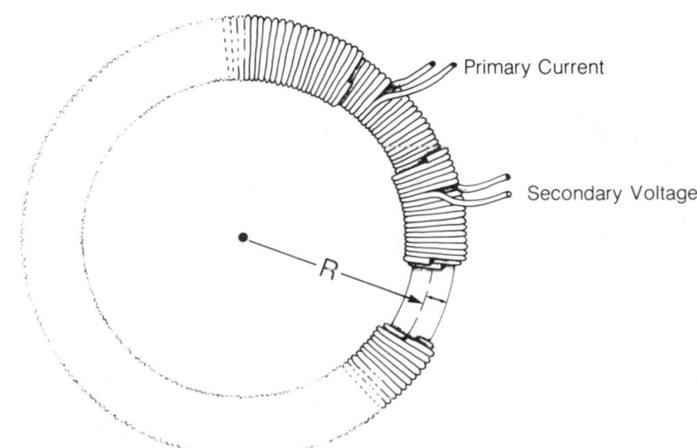

FIG. 1(a). A toroidal transformer. The cross section could be circular, or as assumed in the text a rectangle of area $a \cdot b$.

current according to Faraday's law of induction

$$V = -L_{12}\frac{di}{dt}, \tag{1}$$

where the mutual inductance L_{12} depends on the material of the toroid as well as upon its dimensions. In SI units V is in volts, di/dt is in amperes/second, and L_{12} has the units volt-seconds/ampere which is called the henry. For some materials L_{12} can be much larger than for others. One of these is iron. Materials with properties similar to iron are called ferromagnetic. The least ambiguous definition of ferromagnetism is the existence of a spontaneous magnetic flux density within a material. It is possible to deduce this from the behavior of a transformer, but it is more satisfactory to ascertain this by probing the material with neutrons. From the scattering of neutrons one may map out the magnetic flux density on the scale of atomic dimensions. With a transformer one obtains averages of the magnetic flux density over the volume of the material. If the current were changed at a constant rate from large negative to large positive values, the voltage might vary with current as shown in Fig. 1b. The voltage would be proportional to the mutual inductance. The mutual inductance at large values of the current would approach the value it would have if no iron were present in the core of the toroid. In the absence of iron one could say that the primary current i_p creates a magnetic field strength H given approximately by

$$H \cong N_p i_p / 2\pi R \tag{2}$$

where N_p is the number of turns in the primary winding. The magnetic field strength would be almost parallel to the circulating axis of the toroid. The units of H are amperes/meter. The magnetic field strength is said to polarize the vacuum and create a magnetic flux density of magnitude

$$B = \mu_0 H, \tag{3}$$

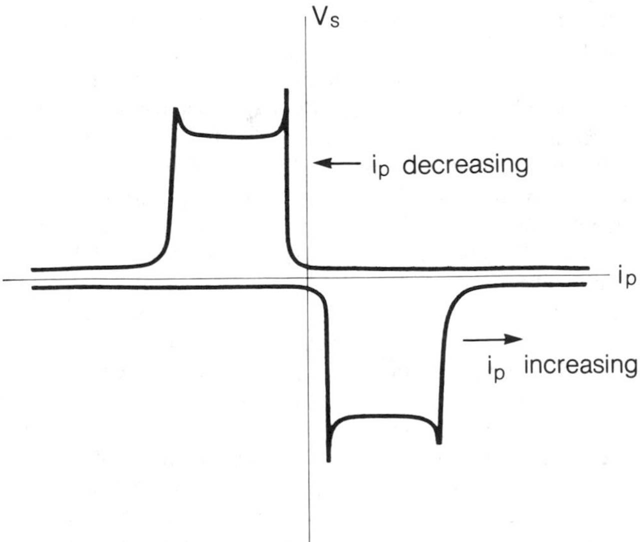

FIG. 1(b). The secondary voltage generated when the primary current increases and decreases linearly in time is plotted against primary current.

where μ_0 is the permeability of the vacuum, a concept of convenience in establishing the SI units of B and H. The flux density is measured in webers/(meter)^2, which are also called teslas. The permeability of vacuum is defined to be $4\pi \times 10^{-7}$ tesla meters/ampere [or newtons/(ampere)^2 or henrys/meter]. The changing current produces a changing magnetic field strength which produces a changing magnetic flux density. The changing magnetic flux $\Phi = abB$ produces the voltage in the secondary winding according to Faraday's law in the form

$$V = -N_s \frac{d\Phi}{dt}, \tag{4}$$

where N_s is the number of turns in the secondary. Actually there are electrons accelerating in the primary giving a push to the electrons in the secondary according to Coulomb's law with relativistic corrections, but Faraday's view is very helpful. The unit of flux is the volt-second or the weber. Magnetic flux is defined as an integral of the magnetic flux density B over an area, here, the cross section of the secondary coil, $a \times b$. The mutual inductance of the toroidal transformer with no iron core is then approximately

$$L_{12} \cong N_p N_s \mu_0 (ab/2\pi R); \tag{5}$$

when the iron is present,

$$L_{12} \cong N_p N_s \mu_R \mu_0 (ab/2\pi R), \tag{6}$$

where μ_R is the relative permeability, which can be as large as 10^6 for iron. As μ_R depends on how the magnetic flux density B varies throughout the toroid and how B changes in time in response to the current in the primary, it will depend on that current and also on the past history of currents applied. At high currents (large H) B uniformly curls about the toroid and has contributions from the current in the primary and from the material of the toroid. This is expressed by the relation

$$B = \mu_0 (H + M_s), \tag{7}$$

where M_s is called the saturation magnetization, H is the field which would be present if the iron were not there, and M_s approaches an almost constant value at large values of H. If a large current i_l is reversed, the direction of B, M_s, and H will be reversed. The change in B, ΔB, will be

$$\Delta B = \mu_0(\Delta H + 2M_s) = \mu_0(N_p/2\pi R)\cdot 2i_l + 2\mu_0 M_s.$$

One can determine ΔB experimentally, calculate ΔH from the current, and then obtain the value of M_s. ΔB can be determined from the time integral of the secondary voltage, that is,

$$-\int V dt = N_s \int \frac{d\phi}{dt} dt = N_s \Delta\phi = N_s ab \Delta B. \tag{8}$$

The voltage in the secondary coil results mainly from the reorientation of the magnetization pattern in the iron as the primary current changes, but also has a direct contribution from the changing current itself. The time integral can be carried out by using an operational amplifier with capacitive feedback as an analog computer. The integration of Fig.

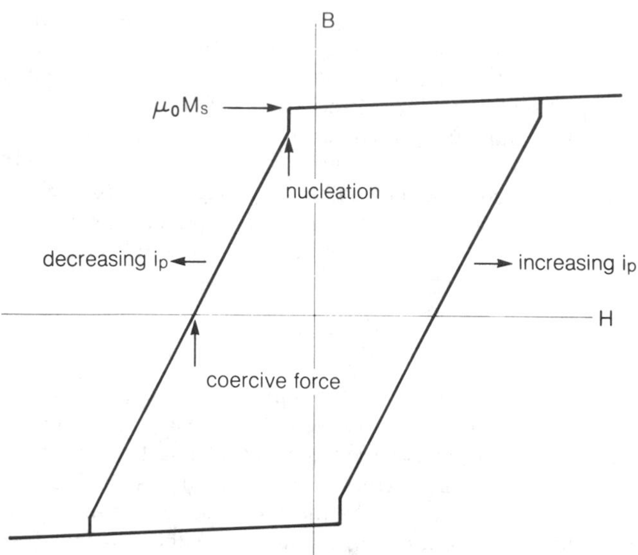

FIG. 1(c). The magnetic flux density, calculated from the integration of Fig. 1b, plotted against the magnetic field, calculated from the primary current.

1b produces Fig. 1c, which is called the major or saturation hysteresis loop. The axes are labeled using average values of $B(=\phi/ab)$ and $H(=N_p i/2\pi R)$. The value of this H for which B is zero is called the coercivity H_c. The value of B for which $H=0$ is called the remanence B_r. For the particular loop shown in Fig. 1(c), $B_r = \mu_0 M_s$.

The study of ferromagnetism divides neatly into attempts to answer two questions: How does the spontaneous magnetization arise in certain materials such as iron, cobalt, and nickel and several of the rare-earth elements as well as in oxides and other compounds of these and a few other metals? And how does the reversal process take place? The latter question is crucial in understanding the technical use of ferromagnetic materials. Though neither question has been fully answered in any single material, the study of magnetism has contributed significantly to understanding the behavior of electrons in solids and to the development of complex technologies based on the behavior of magnetic materials.

The carrier of magnetism is the electron which can be described as possessing an intrinsic angular momentum (spin) as well as a charge. The intrinsic magnetic moment of an electron (the Bohr magneton, $\mu_B = 9.27\times10^{-24}$ joule/tesla) is then a relativistic consequence of a spinning charge. Electrons in atoms combine in such a manner that the spin of the atom is generally less than the sum of the spins of the individual electrons. When atoms combine in solids, the spin of the atom is generally less than when the atoms are separated. The quantum-mechanical symmetry effect known as the Pauli exclusion principle makes it impossible for two electrons to have the same spatial coordinates if they have parallel spins (spins are either parallel or antiparallel for two electrons). Thus electrons with parallel spins avoid one another. This decreases the Coulomb repulsion between them below what it would be on the average for two electrons with antiparallel spins. Thus the net electrostatic binding of electrons to the atom, including the Coulomb attraction to the nucleus, would be greater for the parallel spin electrons. Yet to avoid one another the parallel-spin electrons need more kinetic energy. It is the balance between these effects which determines the net spin of an atom and the spontaneous magnetization of a material.

Calculations of the electronic structure of free atoms are sufficiently accurate to account for Hund's rules, which relate the magnetic moments to the number of electrons on the atom.

The moment of an atom is more likely to be preserved in a solid if the electrons responsible for the moment are not appreciably involved in the binding of the solid. This is the case for elements of the first transition series, the lanthanides, and the actinides. The elements of the first transition series keep their atomic moments better in compounds such as oxides and sulfides than they do in metals.

A free manganese atom has five electrons in orbitals of d symmetry, all with their spins parallel. In some solids manganese exhibits a magnetic moment per manganese atom of $5\mu_B$. If this happened in the metal, it would make a most attractive magnetic material with a saturation magnetic flux density equal to 4.5 tesla. It is technologically unfortunate that this does not happen. The highest spontaneous magnetic flux density at room temperature is found in an alloy of iron and cobalt for which $B_s = 2.4$ tesla. The spontaneous magnetic flux density for the ferromagnetic elements iron, cobalt, nickel, and gadolinium are 2.20, 1.82, 0.64, and 2.45 tesla, respectively, in the limit of low temperatures. At room temperatures iron, cobalt, and nickel have B_s values 2.15, 1.80, and 0.61 tesla, respectively.

In some materials the direction of alignment of the magnetic moment varies from atom to atom, e.g., staggered up–down–up–down. If the net magnetic flux density is zero in any microscopic region, such a material is called antiferromagnetic. If a staggered pattern results in a net magnetic flux density, the material is called ferrimagnetic. The arrangement of magnetic moments in ferrimagnetic and antiferromagnetic materials is studied in detail by neutron diffraction.

At sufficiently high temperatures all of these types of magnetically ordered materials become magnetically disordered and exhibit paramagnetism. The disordering temperature is called the Curie temperature. For gadolinium this Curie temperature is just below room temperature. For iron, cobalt, and nickel the Curie temperatures are ~1040 K (~770°C), ~1390 K (~1120°C), and ~630 K (~358°C), respectively.

Perhaps the behavior of Gd metal is best understood. The seven electrons in the $4f$ shell of Gd are aligned parallel in accordance with Hund's rule. The f electrons on one atom are aligned with respect to the f electrons on neighboring atoms through a polarization of the itinerant conduction electrons which are also responsible for the binding of the metal. The degree of this polarization ($\sim\frac{1}{2}\mu_B$ per atom) and the Curie temperature are reasonably well accounted for by first-principles theory aided by modern computing techniques.

A general framework for discussing ferromagnetism is provided by the Heisenberg model, which assumes a net spin on each atom and an interaction between spins on neighboring atoms. The energy is written as

$$E_{ex} = -\sum_{i=1}^{N}\sum_{j=1}^{n} J_{ij}\mathbf{S}_i\cdot\mathbf{S}_j, \tag{9}$$

where J_{ij} is called the exchange parameter and E_{ex} is called the exchange energy. The sums are over all pairs of n near neighbors on a lattice of N sites. The detailed calculation of the statistical-mechanical properties of the Heisenberg model has been a major activity in theoretical physics for half a century. One attempts to calculate the spectrum of magnetic excitations known as spin waves, the temperature dependence of the spontaneous magnetization, and the nature of magnetic fluctuations near the Curie temperature. Comparisons of the predictions of the model with experimental results are often rather satisfactory for a model with but one parameter $J_{ij}=J$ for the nearest neighbors.

The concept of exchange energy plays an essential role in models of the magnetization reversal process in technical magnetization. In these models the magnetization is taken as a continuum vector field which changes direction by at most a small fraction of a radian from atom to atom and does not change in magnitude. The changing of direction is a magnetic strain which increases the (negative) exchange energy by an amount

$$\Delta E_{ex} = A\int\{[\text{curl }\mathbf{M}_s]^2 + [\text{div }\mathbf{M}_s]^2\}\,dV/M_s^{\,2} \tag{10}$$

where $A = JS^2/a$ (a is an atomic distance) is an exchange constant. The exchange energy, which is responsible for ferromagnetism, resists rapid changes in the magnetization direction. In the toroidal transformer core the magnetization direction tends to curl about the axis of the toroid and to be everywhere parallel to the surfaces. The reason for this is that in addition to the exchange interaction between the spins on neighboring atoms there is a much weaker magnetostatic interaction among all magnetic moments. The magnetostatic interaction is again a relativistic effect between spinning atoms. This energy is a minimum if div $\mathbf{M}_s = 0$ everywhere in the material and $\hat{n}\cdot\mathbf{M}=0$ on all surfaces, where $\hat{n}$ is the normal to the surface. If these conditions are not met, each magnetic moment will experience a demagnetizing field H_D which has as its sources div $\mathbf{M}_s$ and $\hat{n}\cdot\mathbf{M}_s$. The curling pattern satisfies div $\mathbf{M}_s = 0$ and $\hat{n}\cdot\mathbf{M}_s = 0$ while at the same time making only a very small increase in the exchange energy ($|\text{curl }\mathbf{M}_s| = M_s/R$, where R is the radius of the toroid). To reverse the sense of the curling pattern in response to a reversal of current in the primary of the toroid both the exchange energy and the demagnetizing field energy must increase in the process. In addition the process will pass through configurations which are unstable, irreversible changes will occur, and hysteresis will result. A complete theoretical description of this process does not exist. Observations on an iron transformer show that the crystal structure plays an important role in the reversal process. The magnetic moment is not entirely due to the spin of the electrons, but also includes a contribution from the orbital motion of the electrons. This is the origin of the magnetocrystalline anisotropy, an energy which depends on the orientation of the magnetization with respect to the crystal axes. If the toroidal core itself is wound of a thin metallic tape in which all the crystal grains are oriented along the tape, the crystalline anisotropy as well as the exchange energy and the demagnetizing field energy prefers the magnetization to curl about the toroidal axis. All these energies would be the same if the magnetization pattern were completely reversed. But the energy of interaction of the magnetization with the primary current does change if the primary current is reversed with respect to the magnetization pattern. This is the driving force for reversals. The effect of magnetic anisotropy on the reversal process is to inhibit all processes that take place by a gradual rotation of the magnetization everywhere. What happens is that somehow one or more domain walls nucleate and pass through the cross section of the toroid reversing the magnetization as they go. The walls separate regions aligned in opposite directions. The magnetization in a wall rotates from one direction to the other in a distance of a few hundred atoms for iron at room temperature. The wall has exchange energy because of the rapid rotation, anisotropy energy because the magnetization no longer is confined to preferred directions, and generally some demagnetizing field energy because either div $\mathbf{M}\neq 0$ in the wall or $\hat{n}\cdot\mathbf{M}\neq 0$ where the wall intersects the surfaces. A change in the direction of magnetization can also give rise to mechanical strains through the elongation (or shrinkage) of the atomic lattice in the direction of magnetization. This is called magnetostriction, and it also influences the reversal processes. Furthermore, domain walls interact with crystal imperfections, grain boundaries, and surfaces. Presumably imperfections are also necessary for the nucleation of the domain walls. If the field from the current in the primary which is sufficient to nucleate the walls is greater than the field necessary to move the walls, the material has a "square" hysteresis loop. If the reverse is the case, the field necessary to move the walls influences the form of the hysteresis loop.

For the loop shown in Fig. 1c, the nucleation takes place as in a square-loop material, but the walls reach a stable configuration in that nucleation field. Further changes follow the change in the magnetic field, until the walls disappear, leaving the magnetization reversed. In a part of the loop where walls are present, if the direction of field change is reversed, minor hysteresis loops will be followed. Figure 2 shows the behavior of a typical commercial transformer material including a few minor loops. By cycling the magnetic field with decreasing amplitude it is possible to demagnetize the core reaching a state with no net B for zero applied field. The virgin magnetization curve *OABCDEF* is then traced out by increasing this magnetic field monotonically. The initial permeability μ_{RI} and the maximum permeability μ_{RM} are indicated in Fig. 2. The loops *DD'D* or *EE'E* would correspond to efficient operation of a power transformer, while the loop *AA'A* might be employed in a communications transformer. In the first case the area of the hysteresis loop contributes to undesirable energy losses, in the second case the area of the loop corresponds to a decrease in the quality of the signal transmission.

The most used magnetic material is polycrystalline motor-grade iron (99.6% Fe) with μ_{RM} about 5×10^3 and a hysteresis loss, ω_H, about 150 joules per cubic meter of core material for one cycle of a loop with a maximum induction of $B_M = 1.2$ tesla. High-grade power transformers use grain-oriented sil-

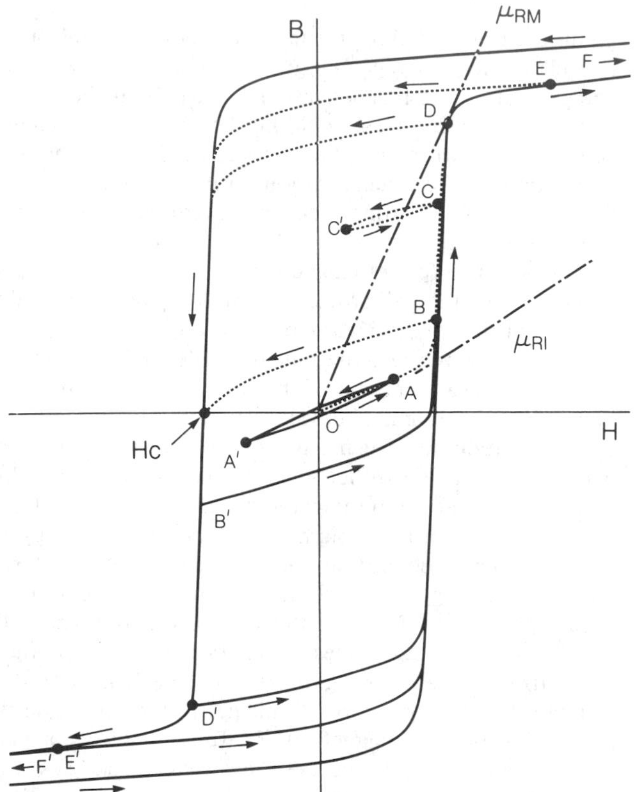

FIG. 2. Hysteresis loops for a commercial transformer material. Portions of the major loop are closed beyond the limits of the diagram (F and F').

icon–iron (3.2 wt% Si) with μ_{RM} values over 7×10^4. It is possible to operate with flux densities near 1.9 tesla for fields of 800 A/m ($\mu_0 H = 10^{-3}$ tesla). Core losses for a loop with $B_M = 1.7$ tesla at 60 Hz can be as little as 200 J/m^3/cycle. Only one-quarter of these losses comes from the area of the hysteresis loop as measured slowly.

An alloy with 79Ni-5Mo-16Fe when suitably heat treated has μ_{RM} over 5×10^5 and a hysteresis loss of less than 0.05 J/m^3 at B_M = 0.5 tesla. As the superior magnetic materials are metallic electrical conductors, it is necessary to use thin laminations with surface insulation to minimize the eddy-current losses. A transformer core as shown in Fig. 1 would be a continuous wound tape. At radio frequencies it is necessary to use low-conductivity materials such as nickel-ferrite which has $\mu_R = 2 \times 10^3$. The best materials in each higher frequency range have lower permeabilities, e.g., μ_R = 10 for $3BaO\cdot2CoO\cdot12Fe_2O_3$ at 1 GHz. Ferrites are used as components for microwave devices.

The preparation of suitable materials for magnetic circuits, motors, generators, transformers, and electromagnets has been an important metallurgical activity for a century. More recently magnetic materials have played a dominant role in the storage of information and are important to the computer revolution. Since mid-century the most familiar permanent magnets are made of barium ferrite and are found pinning notes to metal cabinets or refrigerators in most households. Barium ferrite is very cheap because it is extracted in an almost ready-to-use form from mines. Permanent magnets made of $Fe_{14}Nd_2B$ have made new applications possible in the last decade. These are so strong that they are dangerous to handle.

Permanent magnets are first of all materials for which there is a strong preference for the magnetization to lie in particular directions, but in addition they must be heterogeneous enough to suppress the growth of domains oriented in the directions preferred by the field. A cylinder of such a material with its preferred direction along the axis would retain most of its magnetization parallel to its sides after being magnetized briefly in a strong axial field, despite the strong demagnetizing fields arising from $\hat{n}\cdot\mathbf{M}$ at its ends. In modern permanent magnets the magnetocrystalline anisotropy and the impedance to domain wall motion are sufficient that even quite short cylinders will remain magnetized. Though the stronger permanent magnets are much more expensive, even in terms of stored energy, often design considerations favor the additional cost. The stored energy density in a permanent magnet is a maximum if the remanent magnetization $M_r = B_r/\mu_0$ is maintained against a demagnetizing field $H = -\frac{1}{2}M_r$. The maximum energy product $B_r^2/4\mu_0$ is often quoted in megagauss-oersteds which translates into 0.1 MJ/m^3. This number has increased by a factor of 10 to 4 MJ/m^3 through materials research based on the theory of micromagnetism.

See also FARADAY'S LAW OF ELECTROMAGNETIC INDUCTION; FERRIMAGNETISM; HYSTERESIS; MAGNETIC DOMAINS AND BUBBLES; MAGNETIC MATERIALS; MAGNETS (PERMANENT) AND MAGNETOSTATICS; MAXWELL'S EQUATIONS; PARAMAGNETISM; SPIN.

BIBLIOGRAPHY

Current research in ferromagnetism is reported and reviewed in the *Journal of Magnetism and Magnetic Materials* and in the proceedings of annual magnetism conferences by that journal, the *Journal of Applied Physics*, and the *IEEE Transactions on Magnetism*.

Feynman Diagrams

Thomas Fulton

Feynman diagrams (or graphs) are graphical representations of perturbation-theory calculations involving the scattering and propagation of interacting particles. In the case of relativistic quantum field theory, the particles can interact with external fields or with other particles (through the emission and absorption of virtual quanta). In the many-body problem, the particle–particle interaction can also occur through potentials.

The diagrams serve two principal purposes: to guide actual calculations, and to provide an intuitive physical picture of the process being considered. With their aid, we can ascertain that all contributions to a given order in the perturbation expansion have been counted. Further, a specific integral is associated with each diagram. This integral can be obtained with the use of the "Feynman rules" as indicated in what follows.

As an illustration, consider quantum electrodynamics, involving only electrons (e^-), positrons (e^+), and photons (γ), in the absence of external fields. The diagrams representing the scattering amplitude for e^+–e^- scattering (Bhabha scattering) in lowest order are shown in Fig. 1. When the diagrams are interpreted in space-time, Fig. 1 can be thought of as representing the physical processes that take place in such a scattering, with the various lines corresponding to world lines of particles. Thus, in Fig. 1a, reading from the bottom up, an e^+ and an e^- in some initial state (i) scatter each other into a final state (f) through the exchange of a virtual γ, which then propagates to 2 and creates the final e^+–e^- pair. In Fig. 1b the initial e^+ and e^- annihilate at 1, producing a virtual γ that propagates to 2 and there produces the final e^+ and e^-.

For calculational purposes, it is often more convenient to think of Feynman diagrams in momentum rather than coordinate space. In either case, a general Feynman diagram of arbitrary order (involving $e^\pm$ and γ) consists of external lines (corresponding to incoming and outgoing particles), internal lines representing Feynman propagators or Green's functions, and corners or vertices at which two $e^\pm$ lines and one γ line meet. Figure 1 illustrates a case where all the external lines are $e^\pm$ and the internal line is a photon propagator. Figure 2, corresponding to lowest-order perturbation theory for Compton scattering, has γ's as incoming and outgoing particles as well, and the internal line is an electron propagator. The Feynman rules associate specific functions with each external line, internal line, and vertex; specify sign conventions; and, in momentum space, require conservation of energy and momentum at each vertex. As a consequence, the propagators are in general "off mass-shell" (i.e., the usual relationship, relativistic or nonrelativistic, between energy and momentum of a free particle does not hold). The arrows in Figs. 1 and 2 represent another convention: electrons propagate "forward" and positrons "backward."

A particular advantage of Feynman diagrams in relativistic field theories (which is absent in the many-body applications) is the way in which a given diagram can represent a number of processes. Consider a basic vertex, as in Fig. 3, which is part of a Feynman diagram. (The lines can be external or parts of internal lines.) Line a in Fig. 3 has alternative interpretations as an incoming electron or outgoing positron, line b as an outgoing electron or incoming positron, and line c as an outgoing or incoming photon. Thus, if we are less explicit about the labels $e^\pm$ and i, f, Fig. 1 can also be used to represent e^-–e^- (Møller scattering) as well as e^+–e^- scattering (with Fig. 1b corresponding to the exchange term) and Fig. 2 can describe e^+–e^- annihilation into two photons as well as Compton scattering.

Feynman diagrams of two other types are shown in Figs. 4 and 5. Figure 4 represents one of the radiative corrections to Bhabha scattering—that due to the lowest-order vacuum polarization correction to virtual γ exchange. Figure 5 illustrates a Feynman diagram from weak-interaction theory—the decay $n \rightarrow p + e^- + \bar{\nu}_e$, mediated by an intermediate vector boson W_-. Although charge and fermion-number conservation at vertices are implicit in the Feynman rules for quantum electrodynamics, they are much more apparent in this β-decay process.

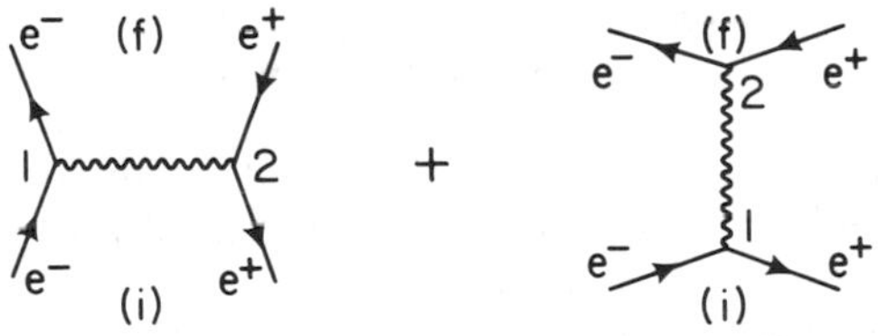

Fig. 1a

+

Fig. 1b

Fig. 2a

+

Fig. 2b

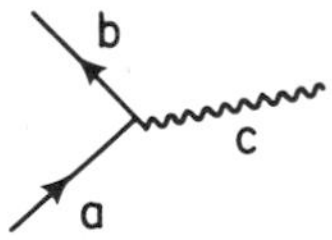

Fig. 3

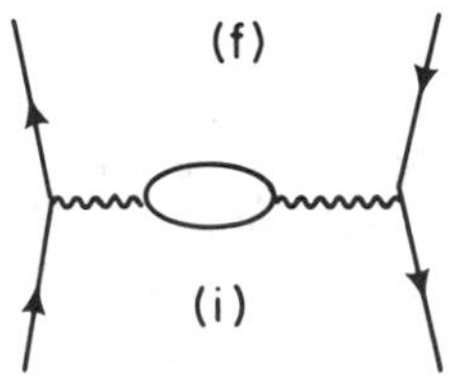

Fig. 4

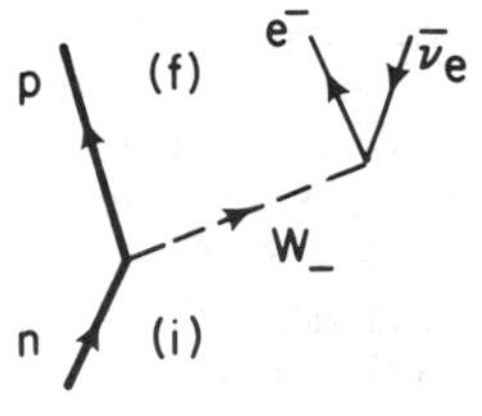

Fig. 5

See also QUANTUM ELECTRODYNAMICS; QUANTUM FIELD THEORY.

BIBLIOGRAPHY

A. A. Abrikosov, L. P. Gorkov, I. E. Dzyaloshinski, *Methods of Quantum Field Theory in Statistical Physics*, Chapters 2 and 3. Prentice-Hall, Englewood Cliffs, NJ, 1963.

J. D. Bjorken and S. D. Drell, *Relativistic Quantum Mechanics*, Appendix B. McGraw-Hill, New York, 1964.

F. J. Dyson, *Phys. Rev.* **75**, 486, 1736 (1949).

R. P. Feynman, *Phys. Rev.* **76**, 769 (1949); **80**, 440 (1950).

J. M. Jauch and F. Rohrlich, *The Theory of Photons and Electrons*, Chapter 8. Addison-Wesley, Reading, Mass., 1955 (2nd ed.: Springer-Verlag, Berlin and New York, 1976).

F. Mandl, *Introducton to Quantum Field Theory*, Chapter 14. Wiley (Interscience), New York, 1959.

R. D. Mattuck, *A Guide to Feynman Diagrams in the Many-Body Problem*. McGraw-Hill, New York, 1967.

J. J. Sakurai, *Advanced Quantum Mechanics*, Appendix D. Addison-Wesley, Reading, Mass., 1967.

S. S. Schweber, *An Introduction to Relativistic Quantum Field Theory*, Chapter 14. Harper, New York, 1961.

Fiber Optics

Walter P. Siegmund

Fiber optics is the branch of physics pertaining to the passage of light through thin filaments of transparent material and includes transmission theory, applications, and manufacturing technology. Although some references of historical interest go back over 100 years, practical development began in the early 1950s [1,2]. In particular, the invention of optically "clad" fibers in 1953 [3] placed fiber optics on a firm foundation. Since then the field has grown steadily both in the range of its applications and in commercial value.

The basic phenomenon involved in fiber optics is the channeling of light by total internal reflections from the fiber walls. In conventional (step-index) fibers the reflections are produced by cladding the walls with a material of lower refractive index to isolate the fibers from one another and ensure a clean reflecting interface. Alternatively the refractive index of the fiber may be graded radially outward from a higher to a lower value to achieve this channeling (graded-index fiber).

The principal characteristics of optical fibers are their diameter (in μm), attenuation (in dB/km), and numerical aperture (*NA*). In step-index fibers the *NA* is simply related to the refractive indices of the fiber core (n_1) and the cladding (n_2) by $NA=(n_1^2-n_2^2)^{1/2}$.

The *NA* of a fiber is a measure of the cone of light which the fiber can accept and, therefore, contributes to its overall transmission efficiency, analogous to the optical "speed" of a camera lens.

In communications applications optical fibers are generally considered as wave guides having discrete transmission modes. To obtain maximum transmission bandwidth through such fibers it is possible to minimize the differences in group velocities of the various modes by means of a precisely controlled graded-refractive-index profile.

Alternatively the fiber may be designed with a sufficiently small core diameter and/or small numerical aperture such that only a single mode is transmitted (single-mode fiber) thus eliminating differences in mode velocities. Such fibers are also of great interest in very long transmission lines.

Materials used in the manufacture of optical fibers include optical glasses, fused silica, and certain plastics. The glasses used in various forms of fiber-optic devices are chosen for their refractive index, transmittance, thermal, and in some cases chemical characteristics. For long flexible fibers transmittance is the principal factor, while for fused rigid-type fiber components such as intensifier tube faceplates, refractive index is the principal factor. In the latter case a core glass having a high refractive index (typically 1.8) is used to obtain an *NA* value near unity which provides the high diffuse light transmittance required for these applications.

For "low-loss" fibers, used in long distance communication, materials of the highest purity are required. One of the principal methods by which such purity is achieved is by chemically depositing silica (or a silica-based glass whose refractive index is modified by additions of boron or germanium oxides) from the vapor phase onto a pure silica substrate (chemical vapor deposition or CVD process). In this process, impurities such as transition metals which are a major cause of optical absorption are effectively eliminated. Fibers drawn from "preforms" made by this process have exhibited an optical attenuation of less than 1.0 dB/km in the near ir spectrum (1.3 μm).

In addition to the low attenuation gradient index and single-mode fibers for long-distance communication cables, fused silica fibers are also being designed for special purposes including polarization maintenance in interferometer applications (such as fiber-optic gyroscopes), as well as low-birefringence single-mode fibers for other sensor applications. In general, optical fibers are finding increased application in a variety of sensors involving either their intrinsic characteristics such as micro-bending losses and birefringence, or as low attenuation relays from remote transducers.

Bundles of coordinated optical fibers (i.e., coherent bundles) capable of image transmission have important applications in medical endoscopes and industrial inspection instruments. In such bundles the individual fibers are positioned identically at both ends of the bundle but are free to flex in between. Thus an image projected on one end appears in recognizable form at the end although it consists of a series of discrete dots when viewed under high magnification. Such bundles are usually made by precisely winding layers or "ribbons" of fibers on a drum or spinning frame and then laminating a large number of such ribbons together for a small portion of their circumference into a tightly packed assembly. After bonding, this portion is cut open and the exposed ends are polished to form the "coherent" ends of the bundle. Continuous winding of the entire cross section of the bundle has also been used but generally does not provide as precise fiber alignment.

Small-diameter, precise, coherent image bundles are also made by the leaching process. In this case, the low refractive index cladding on each fiber is surrounded by a second cladding of an acid-soluble glass. The coherent bundle is made in a series of steps starting with the double-clad preform which is drawn into large single fibers. These are assembled into a precise array containing many thousand fibers and redrawn into a rigid bundle of the desired cross section. With its ends protected against leaching, this bundle is immersed in acid to remove the soluble cladding leaving the bundle highly flexible and precisely "coherent."

One measure of the image quality through such a bundle is the total number of fibers contained i.e., essentially equal to the number of image elements transmitted. For medical or industrial instruments this ranges from about 5,000 to 50,000 fibers depending on the size and type of instrument. The individual fiber size may range from about 8 to 20 μm. For medical instruments (gastroscopes, colonoscopes, bronchoscopes, etc.) the length ranges up to about 200 cm while for industrial instruments up to 450 cm.

A small objective lens at the tip projects an image onto the coherent bundle and the transmitted image is viewed by an eyepiece. Means for remotely focusing the objective lens and articulating the end are often provided in such instruments. In such instruments one or more additional fiber bundles are used as flexible light guides to illuminate the object. These transmit light from a remote light source to the distal tip of the instrument. The fibers in these bundles are not normally coordinated (i.e., they are noncoherent bundles) and thus cannot transmit a recognizable image.

Such noncoherent fiber-optic bundles (light guides) are used in a variety of applications for remote illumination and light sensing. Bundle sizes of 1–6 mm diameter and lengths of $\frac{1}{2}$–2 m are commonly available as stock items; units having special diameters, lengths, and end configurations also are manufactured to meet custom requirements. Fiber size typically ranges from 30–80 μm for glass fibers and up to 250 μm for plastic fibers. Very large fibers (clad rods) of almost any diameter are also available in the trade.

Another important form of image-transmitting fiber optics is the class of "fused" fiber bundles. These consist of a stack of clad glass fibers fused together in which the cladding serves not only as the optical insulation but also as the mortar to cement the fibers together. Usually these are made up of assemblies of "multifibers" or even "multi-multi" fibers (rather than single fibers) made by successive steps of drawing, assembly, and redrawing so as to accumulate the large number of small fibers required in such arrays. The final fusing step is carried out under carefully controlled conditions of temperature and pressure to ensure that all the interstices between fibers are completely sealed off. The fused block or "boule" can then be sliced, ground, and polished into faceplates for use as vacuum-tight end windows in electronic image tubes or CRTs. For these applications the fibers are designed for high numerical aperture (up to $NA=1.0$) and in sizes down to about 4 μm. Such faceplates provide image resolution in excess of 100 line pairs per millimeter. To reduce stray light, which can otherwise degrade image contrast, minute light-absorbing glass fibers are located interstitially among the light-transmitting fibers. These effectively attenuate any stray light moving laterally through the faceplate.

In addition to the faceplate configuration, in which the fibers are precisely parallel, fused bundles may be reshaped with careful heating into "twisters," which rotate the transmitted image about the axis, or into tapers (fiber-optic "cones"), which provide a change in image magnification in the amount of the taper ratio. Such components find important uses in miniature image intensifier tubes for night vision "goggles" and for coupling image intensifier tubes to other electro-optical components such as TV camera tubes or charge-coupled sensor arrays (CCDs).

BIBLIOGRAPHY

A. C. S. Van Heel, "A new method of transporting optical images without aberrations." *Nature* **173** (1954).

H. H. Hopkins and N. S. Kapany, "A flexible fiberscope using static scanning." *Nature* **173** (1954).

A. C. S. Van Heel, *De Ingenieur* **65,** 25 (1953).

W. B. Allan, *Fiber Optics Theory and Practice*. Plenum Press, New York, 1973.

M. K. Barnoski, *Fundamentals of Optical Fiber Communications*. Academic Press, New York, 1976.

The Handbook of Optics, Chap. 13. McGraw-Hill, New York, 1979.

N. S. Kapany, *Fiber Optics*. Academic Press, New York, 1967.

D. A. Krohn, *Fiber Optic Sensors, Fundamentals and Applications*. Instrument Society of America, 1988.

D. Marcuse, *Theory of Dielectric Optical Waveguides*. Academic Press, New York, 1974.

R. Tiedeken, *Fibre-Optics and Its Applications*. Focal Books–Pitman.

Field Emission

L. W. Swanson

Electrons may be extracted from cold conductors or semiconductors by application of a strong electric field. The phenomenon, called *field emission,* which was first reported by R. W. Wood in 1897, occurs at fields of the order of 10^9–10^{10} V/m.

Fowler and Nordheim, in 1928, provided the first generally accepted explanation of field emission in terms of the newly developed quantum mechanics which they applied to the Sommerfeld model for the electronic energy levels in a metal. Application of a high field to the metal (see Fig. 1) produces a triangular potential-energy barrier through which electrons, arriving at the metal surface, may quantum mechanically tunnel. By solving the Schrödinger equation, Fowler and Nordheim obtained the barrier penetration probability $D(E_x)$ as a function of electron kinetic energy in a direction perpendicular to the barrier E_x.

By multiplying $D(E_x)$ by the number of electrons arriving at the surface with energy E_x and integrating over all values of E_x, Fowler and Nordheim derived the "Fowler–Nordheim" equation which relates field emission current I to field F and work function ϕ:

$$I=6.2\times 10^6\{E_F{}^{1/2}/C\ \phi^{1/2}\}F^2 \exp\left\{-\frac{6.8\times 10^9\phi^{3/2}}{F}\right\}, \quad (1)$$

where I is in A/m²; C, ϕ are in volts; F is in V/m; and $C=\phi+E_F$. Equation (1) represents the variation of I with field very well except at high current densities ($>10^{10}$ A/m²), where space-charge effects become noticeable.

In 1937 E. W. Müller invented the field-emission microscope (FEM), which consists of a sharp conducting point directed at a fluorescent screen deposited on conductivized glass that is maintained about 1–10 kV positive with respect

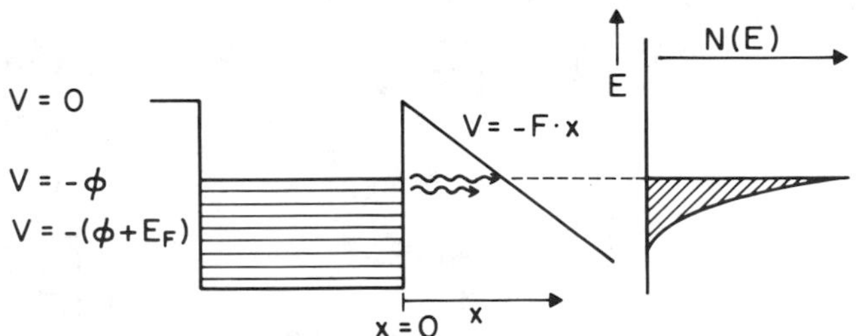

FIG. 1. Sommerfeld model for a metal at $T=0$ K in the presence of a high field. Electrons are able to tunnel through the potential barrier, especially at the level $V=-\phi$ where the barrier is thinner. The energy distribution curve $N(E)$ is shown. Notice the sharp cutoff at $E=-\phi$ due to the absence of electrons above the Fermi level at $T=0$ K.

to the point. The resulting projection microscope (see Fig. 2) has a magnification of upwards of 10^5 and a resolution of 20Å, which is enough to image certain single adsorbed atoms.

From the slope of a plot of $\ln I/F^2$ vs $1/F$, an average work function for the emitter surface may be obtained since, according to Eq. (1), we have $\partial \ln(I/F^2)/\partial(1/F) \propto \phi^{3/2}$. When annealed, field emitters develop hemispherical end forms with large facets of single-crystallographic planes; the work functions of these planes may also be obtained by intercepting the electrons emitted from them in a Faraday cup.

In the hands of Gomer and co-workers, at the University of Chicago, the FEM has been a particularly useful tool for adsorption studies, and it can be used to make measurements of heats of desorption, work function changes, and diffusion energies for adsorbates. For example, in an elegant use of the FEM the relative binding energies of potassium to several planes of tungsten were measured by relating single-plane work functions to surface coverages at different temperatures.

Useful information about emitter band structure, surface states, broadening of adsorbate energy levels, and even vibrational energy levels may be obtained from studies of the total energy distribution (TED) of field-emitted electrons (Fig. 1). An anomalous hump in the energy distribution for the (100) plane of molybdenum and tungsten has been discovered that has been attributed to the existence of a surface state on this plane which disappears when adsorption takes place. A similar structure in the TED from regions containing a single atom of Ca, Sr, or Ba has also been observed. In this case the structure was due to a resonance-enhanced tunneling when the broadened energy levels of the adsorbate atom overlapped the energy levels of the metal just below the Fermi level of the metal. By comparing the TED of hydrogen and deuterium adsorbed on a tungsten field emitter at 78 K, followed by subsequent warming to 300 K, the detection of vibrational energy levels of molecularly adsorbed species has been accomplished.

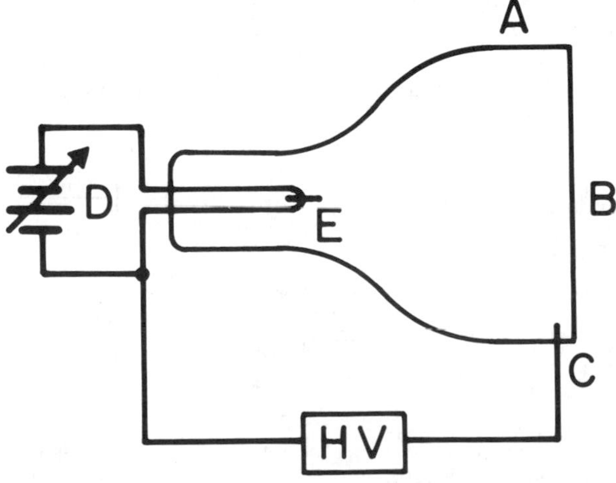

FIG. 2. Simple field-emission microscope. *A* is an evacuated ($p < 10^{-9}$ Torr) glass envelope, *B* is a conductivized fluorescent screen on the inside of the envelope, *C* is an external electrical connection to the screen, and *D* is a low-voltage heating supply for flash cleaning of the field emitter *E*.

The small optical source size and very high current densities of field-emission cathodes make them attractive electron sources for microprobe applications because, for focused beam sizes below about 5,000 Å, field-emission sources provide higher currents than thermionic cathodes. Thus Crewe and co-workers at the University of Chicago were able to develop a field-emission scanning electron microscope (SEM) with a resolution of 5 Å and a beam current one hundred times larger than the corresponding thermionic one; this allows the SEM to scan at a correspondingly higher rate.

Whereas the field-emission process described by Eq. (1) is valid for room temperature and below, the simultaneous application of an electric field and increased temperature to the cathode results in thermal field emission (TFE) at moderate temperatures and Schottky emission (SE) at high temperatures and lower electric field strength. The use of a ⟨100⟩-oriented tungsten field emitter with a zirconium oxide coating results in a low–work-function, thermally stable (up to 1900 K) electron source that can be operated with long life in the TFE or SE modes. The advantage of these operational modes [when compared with the lower-temperature cold field emission (CFE) mode] is the relaxation of the residual gas pressure requirements from 10^{-10} to 10^{-8} Torr while still maintaining a high emission intensity. Typically, the current fluctuations are $<1\%$ and the energy spread of the electron beam ~ 0.4 eV when operated in the SE mode. This results in a high-brightness cathode more practical than the CFE cathode while maintaining emission characteristics favorable to probe forming optics.

The SE cathode is currently finding use in such applications as high-resolution scanning Auger microscopy, high-density information recording, low-voltage scanning electron microscopy, and electron-beam microfabrication of integrated circuits.

In other applications, a CFE cathode array is being used to initiate a field-emission-initiated vacuum arc to produce very large currents in submicrosecond pulses for x-ray generation.

See also ADSORPTION; ELECTRON MICROSCOPY; FIELD-ION MICROSCOPY; TUNNELING; WORK FUNCTION.

BIBLIOGRAPHY

W. P. Dyke, "Advances in Field Emission," *Scientific American* **210** (1), 108 January (1964). (E)

R. Gomer, *Field Emission and Field Ionization*. Harvard University Press, Cambridge, Mass. 1961. (I) An introduction to theory and techniques of field emission.

R. H. Good and E. W. Müller, "Field Emission," *Handbuch der Physik*, Vol. 21, p. 1976, Springer Verlag, Berlin, 1956. (I)

J. W. Gadzuk and E. W. Plummer, "Field Emission Energy Distribution," *Rev. Mod. Phys.* **45,** 487 (1973). (A)

L. W. Swanson and A. E. Bell, "Recent Advances in Field Electron Microscopy of Metals," *Adv. Electr. Electr. Phys.* **32,** 193 (1973). (A)

L. W. Swanson and N. A. Martin, "Field Electron Cathode Stability Studies: Zirconium/Tungsten Thermal Field Cathode," *J. Appl. Phys.* **46,** 2029 (1975).

M. G. R. Thomson, R. Liu, R. J. Collier, H. T. Carroll, E. T. Doherty, and R. G. Murray, "The EBES4 Electron-Beam Column," *J. Vac. Sci. Technol.* **B5**, 53 (1987).

D. W. Tuggle and S. G. Watson, "A Low-Voltage Field-Emission Column with a Schottky Emitter," *Proc. 42nd Electron Microscopy Society of America*, p. 454 (1984). Electron Microscopy Society of America, Woods Hole, MA.

Field-Ion Microscopy

Gary L. Kellogg

Introduced by E. W. Müller in 1951, the field-ion microscope is an instrument which can provide direct images of individual atoms on a solid surface. For nearly 20 years field-ion microscopy had the distinction of being the only technique capable of atomic resolution. Although no longer unique in this respect, the field-ion microscopy remains a powerful research tool for the study of problems in materials and surface science.

The field-ion microscope achieves its high magnification from a nearly radial projection of ions from the apex of a sharply pointed needle to a fluorescent screen. The needle is called a "field emitter" or more commonly a "tip" and has a radius of curvature at the end of several tens of nanometers. Ions are generated above the tip surface in a process known as field ionization. The tip is placed in a vacuum chamber and a low pressure (1–100 mPa) of an "imaging gas" such as helium or neon is introduced. A positive voltage of 5–30 kV is applied to the tip which produces electric fields at the apex of the order of $(1–6)\times 10^{10}$ V/m. At this field strength, the imaging gas atoms in the vicinity of the tip are ionized by an electron tunneling from the gas atom to the surface. The positive "field ions" formed in this manner follow the electric field lines away from the tip surface to the fluorescent screen where they produce image spots. Because the field ions are created preferentially above the protruding surface atoms, the image which appears on the screen consists of a pattern of spots, each spot corresponding to an individual surface atom.

The field-ion microscope is very similar in design to the field-emission microscope invented by Müller in 1936. In the field-emission microscope a negative voltage is applied to the tip and the electrons which tunnel from the surface into the vacuum are imaged on a fluorescent screen. Since the tunneling process is very sensitive to the local work function, different surface regions image with different contrast. The field-emission microscope does not require an imaging gas and can be operated at field strength approximately an order of magnitude lower than the field-ion microscope. However, the large lateral velocity component and the large de Broglie wavelength of the emitted electrons limit the resolution of the field-emission microscope to about 2 nm, not sufficient to achieve atomic resolution. As mentioned above, atomic resolution is achieved in the field-ion microscope with the use of positive ions. The lateral velocity component of the ions can be reduced by lowering the emitter temperature (typically to 77 K or below) and the short de Broglie wavelength of the ions poses a negligible diffraction limit.

An example of a field ion microscope image taken from a Rh tip is shown in Fig. 1. The dark, circular regions seen in the image correspond to flat, low-index planes. These regions are dark because the interior atoms do not protrude sufficiently to produce local field enhancements. The rings of spots which surround these dark regions are images of the edge atoms of individual atomic layers, each larger diameter ring corresponding to a successively deeper atomic layer. From the symmetry of the patterns it is possible to determine the Miller indices of the observed planes. For some of the higher-index planes even the interior atoms are resolved.

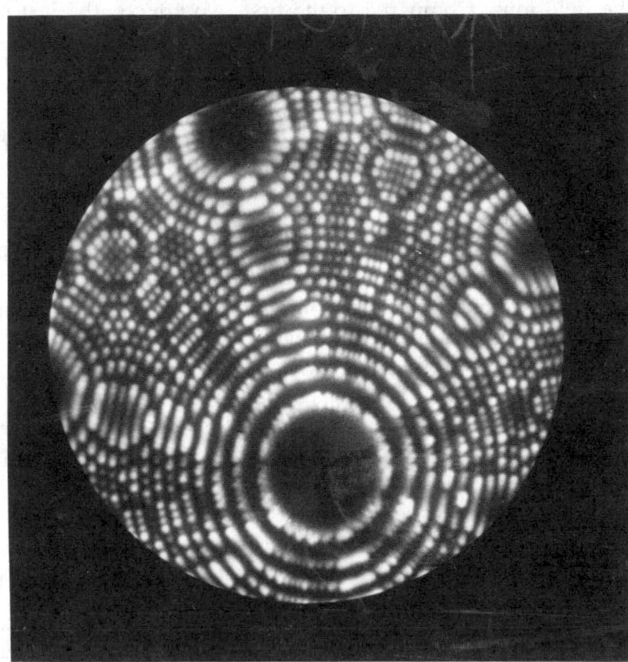

FIG. 1. A field-ion microscope image showing the atomic structure at the apex of a rhodium tip. The image was recorded in 1.3 mPa He at an applied voltage of 6.5 kV. The dark, circular region in the lower central portion of the image corresponds to the flat (111) plane.

Since the field-ion microscope is a point-projection microscope, its magnification is given by

$$M = D/\beta r,$$

where D is the tip-to-screen distance, r is the tip radius, and β is an image-compression factor which accounts for the fact that the tip is not a true sphere in free space. The value of β typically lies in the range from 1.5 to 1.8. By preparing a tip which has a radius of curvature at the apex of a few tens of nanometers, a magnification of several million times can be obtained. Such tips are routinely produced by chemical and electrochemical polishing procedures. It is also important to note that, because the ions are generated at the tip surface, there is no relative motion of the surface with respect to the ion beam. As a result, external vibrations are not magnified as they are in a conventional electron microscope.

The resolution of the field-ion microscope is determined by the spatial extent of the region above the protruding surface atom where ionization occurs (the ionization zone), the lateral velocity component of the emerging ions, and diffraction effects due to the finite de Broglie wavelength of the field ions. Compared to the first two effects, the last is negligible and the resolution is given by

$$\delta = \delta_0 + C(rT/F)^{1/2},$$

where δ_0 is the diameter of the ionization zone (ideally the diameter of the imaging gas atom), C is a constant, r is the tip radius, T is the tip temperature, and F is the applied electric field. The above equation implies that for the best resolution the tip radius should be as small as possible, the tip temperature should be as low as possible, and the applied field should be as high as possible. These conditions are best satisfied with helium as the imaging gas, because of its high ionization potential and low condensation temperature. However, as discussed below, many materials cannot be imaged at the field strengths required to ionize helium. For these, other imaging gases such as neon, argon, or hydrogen are commonly used.

The applied electric field used to image the field ion tip can also be used to remove surface atoms in a controlled, layer-by-layer fashion. At a given field strength, which is typically higher than the ion imaging field, surface atoms are ionized and desorb from the surface. This process is known as field evaporation when the removed atoms are substrate surface atoms or field desorption when the removed atoms are adsorbates. It is quite remarkable that field evaporation in the region of a low-index plane removes the substrate atoms from the edge of the plane inward, one layer at a time. The process is very useful as the final step of sample preparation before field-ion imaging because it leaves behind a surface which is smooth on an atomic scale and free of contaminants. Field evaporation may also be used to probe into the near-surface region of a sample to examine defects which do not extend to the initial surface.

The process of field evaporation is understood qualitatively in terms of classical, one-dimensional energy curves. In his original model Müller viewed field evaporation as the escape of an ion of charge n over a potential energy barrier formed by a superposition of the ionic potential energy curve (approximated by an image potential) and the potential due to the applied electric field. This barrier is known as the Schottky barrier and the activation energy for field evaporation is given by the simple expression

$$Q = Q_0 - (n^3e^3F)^{1/2},$$

where Q_0 is the energy required to remove an ion of charge n from the surface under field-free conditions, e is the elementary charge, and F is the applied field. Corrections which account for the difference in polarizability between the atom and the ion have also been included in subsequent refinements of the model. At a finite temperature the rate of field evaporation is given by an Arrhenius expression of the form

$$k_e = k_0 \exp(-Q/kT),$$

where k_0 is the Arrhenius prefactor, k is Boltzmann's constant, and T is the temperature. Although it is generally accepted that this "image hump" model is not physically realistic, i.e., there is no classical image hump at such close distances, the model is surprisingly accurate at predicting the evaporation fields of most metals. It does not, however, accurately predict the measured field dependence of the activation energy for W and Rh.

In Gomer's "charge exchange" or "intersection" model for field evaporation, ionization occurs at the intersection of the atomic and ionic potential energy curves. If this point lies beyond the maximum of the classical Schottky barrier, the image hump model does not apply. Although the exact form of the ionic potential-energy curve typically is not known, by making reasonable approximations for the potential-energy curves it is possible to reproduce most of the existing experimental data on field evaporation within the framework of this model. More recent quantum-mechanical treatments of field evaporation have also been able to predict accurately the field dependence of the activation energy for field evaporation.

In order to produce a stable image in a field-ion microscope, the evaporation field of the material must be higher than the applied field required for imaging. However, only the most refractory metals have evaporation fields greater than the 4.5×10^{10} V/m required for optimum imaging in He. Many other metals and alloys can be imaged in Ne and Ar, but these heavier ions have a much lower efficiency for exciting the phosphor on the fluorescent screen. A major breakthrough for field-ion microscopy was the development of channel-plate image intensifiers. These intensifiers can convert an ion image to an electron image with gains exceeding 10^3 for a single plate and 10^6 for stacked plates. With a single channel plate and the various imaging gases available it is now possible to obtain stable images of most metals and alloys and even some semiconductors.

The field-ion microscope is primarily a probe of atomic structure. Chemical identification of selected atoms observed in a field-ion microscope image became a possibility in 1967 with the introduction of the atom-probe field-ion microscope. There are several types of "atom probes" currently in use. The most widely used atom probe, shown schematically in Fig. 2, follows the original design of Müller, Panitz, and McLane. The tip is externally adjusted to align the atom(s) of interest with a probe hole in the viewing screen of the field-ion microscope. The superposition of a short-duration (10–100 ns), high-voltage (0.5–3 kV) electrical pulse onto the imaging voltage removes (field evaporates) atoms as n-fold-charged, positive ions. The ions of interest travel through the probe hole and enter a drift tube. The mass-to-charge ratios of the desorbed ions are determined by a measurement of their flight time from the tip through the probe hole and drift tube to a sensitive ion detector. A dramatic improvement in mass resolution can be achieved with the use of an energy-focusing, curved drift tube instead of the straight tube indicated in Fig. 2. The ability to identify selected atoms associated with structural features (e.g., lattice defects) observed in a field-ion microscope image makes the probe–hole atom probe well suited for investigation of metallurgical problems such as impurity segregation to grain boundaries.

Another type of atom probe, known as the magnetic-sector atom probe, uses a magnetic field to separate and identify field-desorbed or field-evaporated ions. Although this type of atom probe has an inherently high mass resolution, it is not commonly used because it suffers from a limited mass

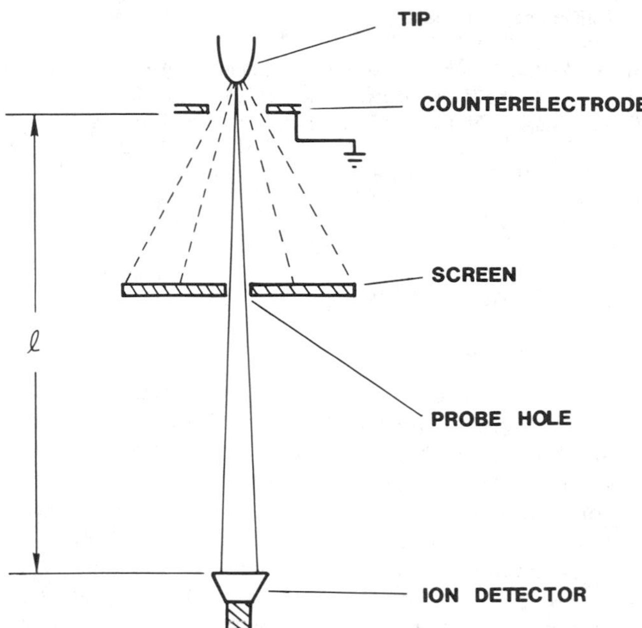

FIG. 2. A schematic representation of the probe–hole, atom-probe field-ion microscope. Ions removed from the tip by a high-voltage pulse are mass analyzed by time-of-flight measurements.

range and requires a large number of ions to obtain a reasonable mass spectrum.

In the imaging atom probe, introduced by Panitz in 1972, the viewing screen of the field-ion microscope is replaced with an imaging detector sensitive to the impact of individual ions. Surface species desorbed from anywhere on the imaged portion of the tip surface are detected and identified by their flight times. The increased number of ions detected per pulse makes the imaging atom probe better suited to the study of surface adsorption and reaction processes than the probe–hole atom probe, but its shorter flight path results in significantly poorer mass resolution. An additional feature of the imaging atom probe is the ability to obtain elemental maps of selected surface species.

Atom-probe investigations of semiconductors and insulators have been greatly facilitated by the introduction of the pulsed-laser atom probe, developed independently by Block and co-workers and by Kellogg and Tsong in 1979. In this version of the atom probe the high-voltage electrical pulse used to stimulate desorption is replaced with a dc voltage and a short-duration laser pulse. The thermal activation provided by the laser pulse initiates the field evaporation or field desorption of surface species. Time-of-flight mass analysis is carried out the same way as in either the probe–hole or imaging atom probe. The use of laser pulses permits analysis of high-resistivity materials which will not transmit short-duration electrical pulses. Elimination of the so-called "energy deficits" associated with high-voltage pulsing also results in a significant improvement in mass resolution in the pulsed-laser atom probe, even better than that obtained with the energy-focusing drift tubes. In addition, the ability to vary the electric field in a continuous manner in field desorption studies makes the pulsed-laser atom probe useful for the investigation of a variety of surface chemical reactions and field desorption phenomena.

The field-ion microscope and atom-probe mass spectrometer have been applied to a wide range of problems in surface and materials science. One of the most unique applications of the field-ion microscope has been the investigation of the diffusion and interaction of individual surface atoms. Pioneered by Ehrlich in 1966, these studies have provided quantitative diffusion parameters and interaction potentials for a variety of atoms on various single-crystal substrates.

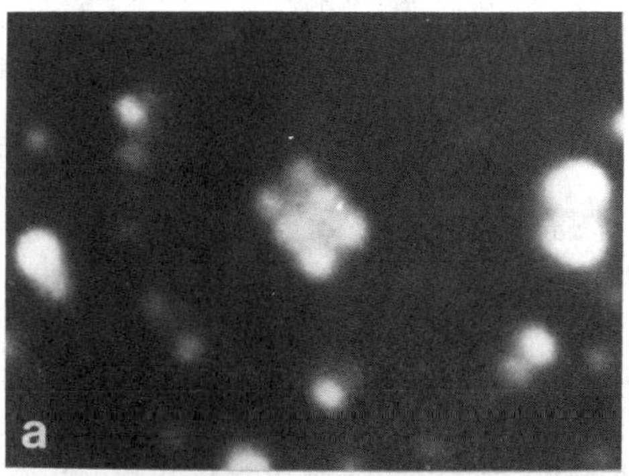

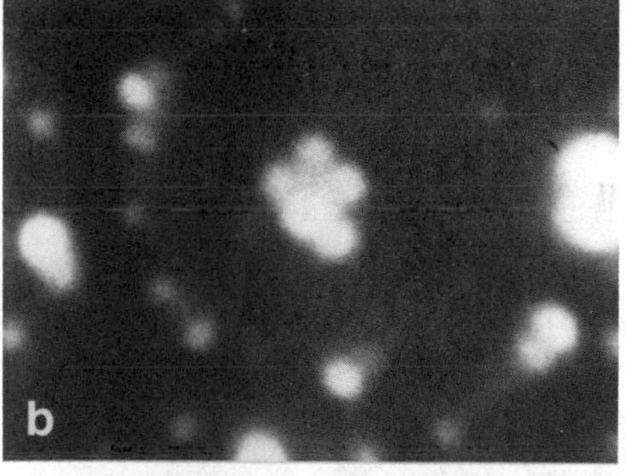

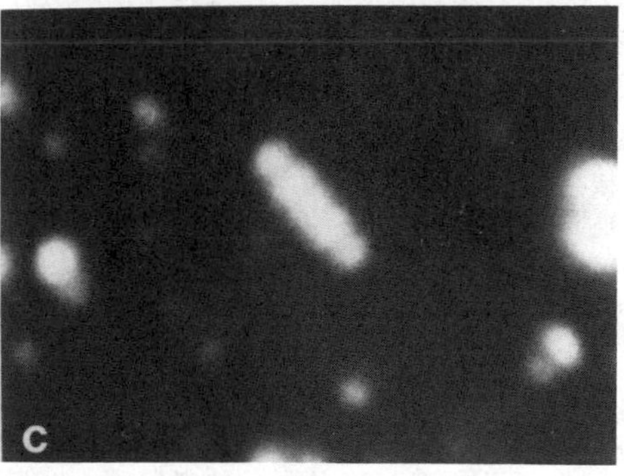

FIG. 3. Field-ion micrographs showing the stability of different cluster configurations on Ir(100). A stable, two-dimensional cluster of six Ir atoms on Ir(100) is shown in (a). In (b) one of the corner atoms has been removed by field evaporation at 77 K producing a metastable, two-dimensional island of five Ir atoms. When thermally equilibrated, the five-atom island transforms to the stable, one-dimensional chain shown in (c). Such studies provide detailed information on the atom–atom interactions involved in cluster nucleation and growth. Figure from P. R. Schwoebel and G. L. Kellogg, *Phys. Rev. Lett.* **61**, 578 (1988).

The clustering of atoms and the formation of epitaxial overlayers have also been investigated in atomic detail with the field-ion microscope. An example of some very recent investigations of cluster nucleation is shown in Fig. 3. The field-ion microscope has also been used to investigate the structure of reconstructed surfaces. These studies have provided unambiguous identification of several reconstructed surface structures as well as information on the atomic forces which cause them.

The most widespread application of the field-ion microscope and atom probe has been in the area of metallurgy. With the field-ion microscope it is possible to examine the microstructure of various lattice defects with atomic resolution. Even point defects, which are beyond the resolution of conventional electron microscopes, can be observed. Prior to the development of the atom probe, the field-ion microscope had been used to examine the imaging characteristics of various alloys and solid solutions, grain boundaries, and precipitates. Computer simulations aided greatly in the interpretation of the field-ion images. The extended capability for chemical analysis offered by the atom probe has led to more quantitative investigations of impurity and solute segregation to interfacial boundaries, surface segregation, ordering and clustering in alloys, and alloy precipitation.

The field-ion microscope and atom probe have had impact on several other scientific disciplines. Defect structures due to radiation damage have been examined at the atomic level in the field-ion microscope. The morphology of biological materials has been examined with field-ion microscopy and related techniques. Compositional variations at semiconductor–metal interfaces have been determined with the atom probe. Various surface chemical reactions and the initial stages of oxide formation on metals have been investigated with atom-probe techniques. Very recently, it has been established that even composite materials such as ceramic-oxide superconductors can be examined at the atomic level with the field-ion microscope and atom-probe mass spectrometer.

See also ADSORPTION; CRYSTAL DEFECTS; ELECTRON MICROSCOPY; SURFACES AND INTERFACES.

BIBLIOGRAPHY

R. Gomer, *Field Emission and Field Ionization*. Harvard University Press, Cambridge, MA, 1961. (I)

J. J. Hren and S. Ranganathan (eds.), *Field Ion Microscopy*. Plenum, New York, 1968. (E)

E. W. Müller and T. T. Tsong, *Field Ion Microscopy, Principles and Applications*. Elsevier, New York, 1969. (I)

E. W. Müller and T. T. Tsong, "Field Ion Microscopy, Field Ionization, and Field Evaporation," in *Progress in Surface Science*, S. Davison (ed.), Vol. 1, part 4. Pergamon Press, New York, 1973. (A)

J. A. Panitz, "Field Ion Microscopy—A Review of Basic Principles and Selected Applications," *J. Phys. E* **15**, 1281–1294 (1982). (I)

G. D. W. Smith, "Field Ion Microscopy and Atom-Probe Microanalysis," in *Metals Handbook*, R. Whan (ed.), pp. 583–602. American Society for Metals, Metals Park, OH (1986). (A)

G. L. Kellogg, "Pulsed-Laser Atom-Probe Mass Spectroscopy," *J. Phys. E* **20**, 125–136 (1987). (E)

M. K. Miller, "Ultrahigh-Resolution Chemical Analysis with the Atom-Probe," *Int. Mater. Rev.* **32**, 221–240 (1987). (I)

T. T. Tsong, "Studies of Surfaces at Atomic Resolution: Atom-Probe and Field Ion Microscopy," *Surf. Sci. Rep.* **8**, 127–209 (1988). (I)

Field Theory, Axiomatic

A. S. Wightman

The phrase "axiomatic field theory" is sometimes used to mean "the general theory of quantized fields" and sometimes to mean any treatment of quantum field theory that has some pretensions to mathematical precision. It is a matter of convention whether constructive quantum field theory, which starts from specific Lagrangian models and constructs solutions of them, is regarded as distinct from axiomatic field theory or as a part of it. Here it will be regarded as distinct but nevertheless will be discussed briefly.

Axiomatic field theory was created in the early 1950s in response to a need for clarification in the foundations of quantum field theory. The great advances in renormalization theory in the late 1940s were based on the use of perturbation theory. This progress made it clear that the ideas of relativistic quantum field theory had more coherence and consistency than had been thought in the 1930s, but it shed little light on what a nonperturbative quantum field theory would be. Axiomatic field theory was regarded as novel in those days because it laid down a set of requirements (axioms) on an acceptable quantum field theory and then investigated what followed from those requirements alone. The consistency of the axioms was clear from the start. They were satisfied by theories of noninteracting fields. The fundamental question that axiomatic field theory left open was whether there are theories of fields in nontrivial interaction that satisfy the axioms.

The early development of the theory is summarized in the books of Jost [1] and Streater and Wightman [2]. The book of Bogolyubov, Logunov, and Todorov [3] contains in addition a summary of an alternative approach created by Lehmann, Symanzik, Zimmermann, and Glaser. In the late 1960s the focus in the study of the general theory of quantized fields shifted from axiomatic field theory to two other areas: the theory of local algebras of bounded operators, sometimes called *local quantum theory,* and constructive quantum field theory. The former, initiated by Haag, Araki, and Kastler, uses the deep theory of operator algebras. It is often called the algebraic approach to quantum field theory [4]. On general grounds it is not obvious that local quantum theory is equivalent to axiomatic field theory. However, that turned out to be true for the models in space-times of two and three dimensions whose existence was established by Glimm and Jaffe and others by the early 1970s [5] using the methods of constructive field theory. By the end of the 1980s constructive quantum field theory had established the existence of solutions for some quantum field theory models in the physically interesting case of four space-time dimensions. However, to prove that the requirements of axiomatic quantum field theory and those of local quantum theory are satisfied for these models remains an open problem. It is plausible that all the axiomatic approaches are, in fact, treating the same objects, and that Lagrangian field theories have solutions satisfying their axioms. That is what the above-mentioned results of constructive field theory state for the models treated.

THE AXIOMS FOR A QUANTUM FIELD THEORY OF A SCALAR FIELD

The basic constituents of a relativistic quantum field theory are, first, a Hilbert space $\mathcal{H}$ whose vectors describe the quantum-mechanical states, and a unitary representation of the inhomogeneous Lorentz (Poincaré) group $\{a,\Lambda\}\to U(a,\Lambda)$ by unitary operators in $\mathcal{H}$ describing the transformation law of states under Poincaré transformation. Second, there are the field operators themselves. For simplicity of exposition, the axioms will be stated only for a theory of a single neutral scalar field ϕ.

Group-theoretical analysis of U shows that the operators $U(a,1)$ representing translations in space-time must be of the form

$$U(a,1)=\exp[i(P\cdot a)]$$

where

$$P\cdot a=P^0a^0-\mathbf{p}\cdot\mathbf{a}$$

and the P^0 and $\mathbf{p}$ are self-adjoint operators whose physical significance is as observables of total energy and total momentum, respectively. The first axiom expresses the physical assumption that the energy is bounded below and that a unique state vector Ψ_0 representing the vacuum exists. This is formally expressed as follows.

1. Spectral Condition. (a) $P^2=(P^0)^2-(\mathbf{P})^2\geq 0$, $\mathbf{P}^0\geq 0$.
(b) There exists a nonvanishing vector Ψ_0 unique up to normalization satisfying

$$U(a,\Lambda)\Psi_0=\Psi_0.$$

The next assumption is about the field, ϕ. The assumption must take into account that the field makes sense as an operator in $\mathcal{H}$ only when it is smeared with an appropriate test function f in a manner formally indicated by

$$\phi(f)=\int d^4x\, f(x)\phi(x).$$

It turns out that although $\phi(f)$ is an operator for each f, $\phi(f)$ cannot in general be applied to every vector of the Hilbert space $\mathcal{H}$. Thus the assumption on $\phi(f)$ must include some kind of specification of the domain on which the $\phi(f)$ are defined. There are several natural choices for the class of test functions. Here only the most commonly used choice will be mentioned: $\mathcal{S}$, the space of all infinitely differentiable complex-valued functions f on space-time that decrease, together with all their derivatives, faster than any negative power of $R^2=(ct)^2+\mathbf{x}^2$ as $R\to\infty$.

2. The Field and Its Domain. For each f in $\mathcal{S}(\mathbb{R}^4)$ there is an operator $\phi(f)$ acting in $\mathcal{H}$ whose domain includes the dense linear set D; $\phi(f)$ and D satisfy

$$U(a,\Lambda)D\subset D,\qquad \Psi_0\in D,$$

$$\phi(f)D\subset D,\quad \phi(f)^*D\subset D,$$

where $\phi(f)^*$ is the Hermitian adjoint of $\phi(f)$. Further,

$$\phi(\alpha f)=\alpha\phi(f),\qquad \phi(f+g)=\phi(f)+\phi(g),$$

$$\phi(f)^*=\phi(\bar{f}),$$

where $\bar{f}$ is the complex conjugate of f. These equations are understood as valid when the operators are applied to vectors in the domain D.

If Φ and Ψ are any two vectors in D, then

$$(\Phi,\phi(f)\Psi)$$

is continuous as a linear functional of the variable f.

The next assumption connects the transformation law of states, U, with the transformation law of a scalar field: $f\to\{a,\Lambda\}f$ where

$$(\{a,\Lambda\}f)(x)=f(\Lambda^{-1}(x-a)).$$

3. Transformation Law of Field

$$U(a,\Lambda)\phi(f)U(a,\Lambda)^{-1}=\phi(\{a,\Lambda\}f),$$

again understood as valid when applied to vectors of D.

The following assumption expresses the idea that measurements of the field ϕ taking place at spacelike separated points cannot influence each other.

4. Local Commutativity. If the support, supp f, of a function f is defined as the closure of the set of all space-time points where f is nonvanishing, then

$$[\phi(f),\phi(g)]=0=[\phi(f),\phi(g)^*]$$

whenever the support of f is spacelike with respect to the support of g, i.e., whenever

$$f(x)g(y)=0$$

for all x and y such that $(x-y)^2\geq 0$.

To state the last axiom, it is necessary to discuss the notion of scattering states within the framework of a theory satisfying assumptions 1–4. This was done early in the development of axiomatic field theory by D. Ruelle [6] following ideas of R. Haag, but only under the assumption of a strengthened form of the spectral condition in which 1 is replaced by 1′.

1′. The Spectral Condition with Mass Gap.
(a) $P^2\geq 0$.
(b) $P=0$ is a simple eigenvalue with unique (up to normalization) eigenvector Ψ_0.
(c) The spectrum of P^2 on $\{\Psi_0\}^\perp$, the orthogonal complement of the vacuum state, lies entirely above m^2 (for some $m>0$) and contains isolated eigenvalues $M_1{}^2$, $M_2{}^2$, ...

The Haag–Ruelle procedure consists in constructing many-particle scattering states out of the single-particle states of mass M_1, M_2, ... provided by assumption (c). There are two sets of such states, corresponding to ingoing and outgoing wave boundary conditions, respectively; they span subspaces of $\mathcal{H}$, designated $\mathcal{H}^{\text{in}}$ and $\mathcal{H}^{\text{out}}$, respectively. Then the last axiom says that the scattering states span the whole Hilbert space.

5. ASYMPTOTIC COMPLETENESS

$$\mathcal{H} = \mathcal{H}^{\text{in}} = \mathcal{H}^{\text{out}}.$$

In a theory with massless particles, the strengthened spectral condition 1′ will be violated and the Haag–Ruelle scattering theory has to be generalized. Nevertheless, scattering states for the massless particles themselves can be defined as was shown by Buchholz [7]. However, it appears that there is still no general rigorous definition of the scattering states for those particles (which for brevity we can refer to as the charged particles) capable of emitting and absorbing the massless particles. The technical difficulty is that there is no eigenvalue of P^2 for such a charged particle but only an end point of a stretch of continuous spectrum. It seems likely that when this technical problem has been overcome, it will still be sensible to require asymptotic completeness in the form shown above.

There is an important phenomenon that may lead to a quantum field theory satisfying 1 or 1′, 2, 3, and 4, but not 5. That is the occurrence of superselection sectors other than the vacuum sector. This will be explained after the axioms of local quantum theory have been introduced.

THE AXIOMS FOR LOCAL QUANTUM THEORY

The basic objects of local quantum theories are algebras $\mathfrak{A}(\mathcal{O})$ of bounded operators attached to regions $\mathcal{O}$ of space-time. The bounded observables that can be measured in $\mathcal{O}$ correspond to self-adjoint elements of $\mathfrak{A}(\mathcal{O})$, and $\mathfrak{A}(\mathcal{O})$ is supposed to be generated by such observables. More precisely, $\mathfrak{A}(\mathcal{O})$ is supposed to be a C^* algebra (see OPERATORS).

The first assumption on the $\mathfrak{A}(\mathcal{O})$ is

1. ISOTONE PROPERTY

$$\text{If} \quad \mathcal{O}_1 \subset \mathcal{O}_2, \quad \text{then} \quad \mathfrak{A}(\mathcal{O}_1) \subset \mathfrak{A}(\mathcal{O}_2).$$

Using Property 1, the quasilocal algebra $\mathfrak{A}$ can be defined:

$$\mathfrak{A} = \overline{\bigcup_{\mathcal{O}} \mathfrak{A}(\mathcal{O})}.$$

Here the regions admitted to the union comprise all bounded open sets and the closure is understood as closure in the sense of C^* algebras.

The transformation law is stated algebraically in local quantum theory.

2. TRANSFORMATION LAW UNDER POINCARÉ GROUP. There is a representation of the Poincaré group, $\{a,\Lambda\} \rightarrow \alpha(a,\Lambda)$, by automorphisms of the quasilocal algebra $\mathfrak{A}$ such that

$$\alpha\{a,\Lambda\}\mathfrak{A}(\mathcal{O}) = \mathfrak{A}(\{a,\Lambda\}\mathcal{O}).$$

Local commutativity is stated in precise analogy with assumption 3 in axiomatic quantum field theory.

3. LOCAL COMMUTATIVITY. If the points of $\mathcal{O}_1$ are all separated from the points of $\mathcal{O}_2$ by spacelike intervals, then the operators of $\mathfrak{A}(\mathcal{O}_1)$ commute with those of $\mathfrak{A}(\mathcal{O}_2)$.

For the finer developments of local quantum theory it is customary to adjoin other axioms, but on the basis of 1–3 we can already discuss a number of significant physical issues. The first concerns the notion of state and, in particular, vacuum state.

A state in the theory of C^* algebras is defined as a complex-valued linear function ω on the algebra $\mathfrak{A}$ such that

$$\omega(A^*A) \geq 0 \qquad \text{for all} \quad A \in \mathfrak{A}$$

and

$$\omega(1) = 1$$

where 1 is the unit element of the algebra. A state ω on $\mathfrak{A}$ determines a Hilbert space $\mathcal{H}_\omega$ and a representation π_ω of $\mathfrak{A}$ by operators in $\mathcal{H}_\omega$: $A \rightarrow \pi_\omega(A)$ such that

$$\omega(A) = (\Omega_\omega, \pi_\omega(A)\Omega_\omega)$$

where Ω_ω is a certain fixed vector of $\mathcal{H}_\omega$ such that $\{\pi_\omega(A)\Omega_\omega\}$ is a dense set in $\mathcal{H}_\omega$. If ω happens to be invariant under a group G of automorphisms α_g of $\mathfrak{A}$,

$$\omega(\alpha_g(A)) = \omega(A) \qquad \text{for all} \quad g \in G,$$

then there exists a unitary representation of the group, $\alpha_g \rightarrow U(\alpha_g)$, such that

$$\pi_\omega(\alpha_g(A)) = U(\alpha_g)\pi_\omega(A)U(\alpha_g)^{-1}$$

and

$$U(\alpha_g)\Omega_\omega = \Omega_\omega.$$

The construction of the Hilbert space $\mathcal{H}_\omega$, the state Ω_ω, and the representations π_ω and U is called the GNS construction after Gelfand, Naimark, and Segal.

The importance of this notion of state and its associated representation is that it makes possible a theory of superselection rules. A superselection rule may be defined in a quantum-mechanical theory by an operator that commutes with all observables. Then the Hilbert space of states breaks up into a sum of subspaces called sectors (or coherent subspaces) within which the superposition principle is valid but between which it is invalid. According to this theory of superselection rules, there are representations of the algebra of observables within the sectors that are unitarily inequivalent. For example, in quantum electrodynamics the algebra of observables has a representation in the charge-1 sector that is inequivalent to the one in the charge-0 (vacuum) sector. In principle, we can find out how many sectors there are by studying the representations of the algebra of observables. In practice, additional assumptions of one kind or another have so far been made [8].

This theory of superselection rules provides an important insight into the axiom of asymptotic completeness. It may happen that the field theory constructed from a given Lagrangian has a Hilbert space $\mathcal{H}_0$ that is the vacuum sector of a larger theory, i.e., a theory with a Hilbert space of states $\mathcal{H}$ that includes $\mathcal{H}_0$ as a proper subspace and that has extra field variables acting in $\mathcal{H}$ but not leaving $\mathcal{H}_0$ invariant. Then we might expect asymptotic completeness to fail in $\mathcal{H}_0$ because there would be multiparticle states with the quantum numbers of the vacuum, some of whose single-particle states are not in the vacuum sector. For example, in a theory of charged particles, the vacuum sector would contain electron–positron pairs even though the states of a single electron or of a single positron are not in the vacuum sector. Thus in order to obtain a Hilbert space in which asymptotic completeness would hold, we would have here to adjoin to $\mathcal{H}_0$ (by direct sum) the Hilbert spaces $\mathcal{H}_{ne}$ describing the other sectors of charge ne, $n=\pm1$, ±2, ... There is convincing evidence that this happens in a variety of model field theories. In particular, it appears that the sine–Gordon theory of a scalar field describes the vacuum sector of the massive Thirring model [9].

PRINCIPAL RESULTS

Since the principal objective of axiomatic quantum field theory has been to put the conceptual and mathematical foundations of relativistic quantum field theory in order, it is not surprising that its impact on particle physics has been mild. What it has done is to make clear how general the arguments are for *CPT* symmetry, the connection of spin with statistics, the LSZ reduction formulas, the connection of broken symmetry with Goldstone bosons, and the occurrence of superselection rules. Its offspring, constructive quantum field theory, has shown the internal consistency of nontrivial theories of interacting fields in space-times of dimensions two and three and established a connection, via Euclidean field theory, between statistical mechanics and quantum field theory.

The ultimate objective of axiomatic field theory is a structure theory for the objects satisfying the axioms. Further progress in constructive field theory seems essential before such a goal can be reached.

See also *CPT* THEOREM; FIELDS; OPERATORS; QUANTUM FIELD THEORY; SPACE-TIME; SYMMETRY BREAKING, SPONTANEOUS.

REFERENCES

1. R. Jost, *The General Theory of Quantized Fields*. American Mathematical Society, Providence, Rhode Island, 1965.
2. R. F. Streater and A. S. Wightman, *PCT, Spin and Statistics and All That*. Benjamin, Advanced Book Program, Reading, Mass., 1964 (2nd printing with additions and revisions, 1978).
3. N. Bogolyubov, A. Logunov, and I. Todorov, *Introduction to Axiomatic Quantum Field Theory*. Benjamin Advanced Book Program, Reading, Mass., 1975.
4. G. Emch, *Algebraic Methods in Statistical Mechanics and Quantum Field Theory*. Wiley (Interscience), New York, 1972.
5. G. Velo and A. Wightman, eds., *Constructive Quantum Field Theory* (*Lecture Notes in Physics* No. 25). Springer-Verlag, Berlin and New York, 1973.
6. D. Ruelle, "On the Asymptotic Condition in Quantum Field Theory," *Helv. Phys. Acta* **35** (1962) 147–174. (See also [1], Chapter VI.)
7. D. Buchholz, "Collision Theory for Massless Fermions and Bosons," *Commun. Math. Phys.* **42,** 269–279 (1975); **52,** 147–173 (1977).
8. S. Doplicher, R. Haag, and J. Roberts, "Local Observables and Particle Statistics I, II," *Commun. Math. Phys.* **23,** 199–230 (1971); **35,** 49 (1974). S. Doplicher and J. Roberts, "Fields, Statistics, and Non-Abelian Gauge Groups," *Commun. Math. Phys.* **28,** 331–48 (1972).
9. S. Coleman, "Quantum Sine–Gordon Equation as the Massive Thirring Model," *Phys. Rev. D* **11,** 2088–2097 (1975).

Field Theory, Classical

Don Weingarten

Classical field theory concerns systems whose (measurable) physical properties at each instant of time are given by a collection of real-valued functions on some region of space. These functions are called fields. The set of fields for each system considered in classical field theory is governed by deterministic rules of time evolution that in most cases permit the values of all fields at any time t_a to be deduced from the values of all fields and a sufficient set of time derivatives at any $t_b<t_a$, or in some more complicated systems, from the values of all fields over the time interval preceding any t_b, $t_b<t_a$. Quantum generalizations of classical field theory lacking such deterministic rules of time development are described in separate articles (QUANTUM FIELD THEORY, QUANTUM ELECTRODYNAMICS).

Classical fields may be either structural or phenomenological. A structural field specifies completely a system's actual microscopic configuration. A phenomenological field, on the other hand, represents only a macroscopic property related in some indirect way to the system's microscopic configuration. Classical field theory may be divided into two branches concerned primarily with structural fields—classical electrodynamics and the general relativistic theory of gravitation—and a third branch concerned mainly with phenomenological fields—the mechanics of continuous media. A typical structural field is the electromagnetic field of electrodynamics; an example of a phenomenological field is the velocity field of continuum mechanics, which may be interpreted as a local average of actual microscopic particle velocities.

In most branches of classical field theory it is possible to identify in a natural way local densities and currents of energy and momentum carried by the fields. These quantities can be combined to form the energy–momentum tensor $T^{\mu\nu}$, where μ and ν run from 0 to 3. The component T^{00} gives the field's energy density, T^{0i} gives the density of the i component of momentum, T^{i0} gives the rate per unit area at which energy is being carried by the field in the i direction, and T^{ij} gives the rate per unit area at which the j component of momentum is being carried by the field in the i direction. For fields interacting in the absence of other sources of energy and momentum, the field energy and momentum currents are conserved:

$$\partial_\mu T^{\mu\nu}=0.$$

Here ∂_μ is an abbreviation of $\partial/\partial x^\mu$, (x^μ) is the four-component vector of special relativity (t,x^1,x^2,x^3), a system of units has been chosen in which the speed of light c is 1, and a summation is performed over repeated indices. These conventions will also be used throughout the remainder of this article.

The dynamical laws which occur in classical field theory are often conveniently stated as principles of stationary action. For both electrodynamics and general relativity a satisfactory action can be found which is the integral of a Lagrangian density $\mathcal{L}(x)$ given by a function of the fields at x and their derivatives of at most second order. For a set of fields φ_i, where i runs from 1 to n, with a Lagrangian density $\mathcal{L}(x)$ of the form $\mathcal{L}[\varphi_i(x),\partial_\mu\varphi_i(x),\partial_\mu\partial_\nu\varphi_i(x)]$, the principle of stationary action

$$\delta \int d^4x\ \mathcal{L}(x)=0 \tag{1}$$

yields the dynamical equations

$$\partial_\mu\partial_\nu \frac{\partial\mathcal{L}(x)}{\partial[\partial_\mu\partial_\nu\varphi_i(x)]}+\partial_\mu\frac{\partial\mathcal{L}(x)}{\partial[\partial_\mu\varphi_i(x)]}-\frac{\partial\mathcal{L}(x)}{\partial\varphi_i(x)}=0. \tag{2}$$

The variation in (1) is required to be zero with respect to each $\varphi_i(x)$ for all i and x. The relations given by (2) are called the Euler–Lagrange equations.

In the remainder of this article we will briefly outline each of the major branches of classical field theory.

ELECTRODYNAMICS

The electromagnetic field of classical electrodynamics is given, in relativistic notation, by a rank-2 antisymmetric tensor field $F^{\mu\nu}$ with indices μ and ν ranging from 0 to 3. The relation between $F^{\mu\nu}$ and the electric and magnetic fields $\mathbf{E}$ and $\mathbf{B}$, respectively, is

$$|F^{\mu\nu}|=\begin{vmatrix} 0 & E^1 & E^2 & E^3\\ -E^1 & 0 & B^3 & -B^2\\ -E^2 & -B^3 & 0 & B^1\\ -E^3 & B^2 & -B^1 & 0\end{vmatrix}.$$

These fields can be measured, in principle, by using the Lorentz force law

$$m\frac{d^2x^\mu}{d\tau^2}=qF^\mu{}_\nu(x)\frac{dx^\nu}{d\tau}$$

giving the acceleration of a particle of mass m and charge q moving in an electromagnetic field with no other forces applied. The second index of $F^{\mu\nu}$ in this equation, and the indices of special relativistic tensors in general, are raised and lowered by contraction with the diagonal metric tensors $\eta^{\mu\nu}=\eta_{\mu\nu}$ with $\eta^{00}=-\eta^{11}=-\eta^{22}=-\eta^{33}=1$. The parameter τ is the particle's proper time, defined by $d\tau^2=dx^\mu\,dx^\nu\,\eta_{\mu\nu}$.

The time development of $F^{\mu\nu}$ is governed by Maxwell's equations, which also restrict the possible field configurations that can occur at a single instant of time. In the presence of an external electric current density J^μ, Maxwell's equations take the form

$$\partial_\nu F^{\mu\nu}=J^\nu, \tag{3}$$

$$\partial_\nu\,{}^*F^{\mu\nu}=0. \tag{4}$$

The field ${}^*F_{\mu\nu}$ is the dual of $F^{\mu\nu}$, defined as $-\frac{1}{2}\epsilon_{\mu\nu\alpha\beta}F^{\alpha\beta}$ where $\epsilon_{\mu\nu\alpha\beta}$ is the completely antisymmetric tensor with $\epsilon_{0123}=1$.

Maxwell's equations and the Lorentz force law are form invariant under Poincaré transformations of the coordinate frame, $x'^\mu=a^\mu{}_\nu x^\nu+b^\mu$, where $a^\mu{}_\nu$ is a Lorentz transformation defined by the condition $a^\mu{}_\lambda a^{\nu\lambda}=\eta^{\mu\nu}$ and b^μ is an arbitrary vector. Under Poincaré transformations, J^μ and $F^{\mu\nu}$ transform as relativistic tensors of ranks 1 and 2, respectively, where a general rank-n tensor transforms by the rule

$$T'^{\mu_1\cdots\mu_n}(x')=a^{\mu_1}_{\nu_1}\cdots a^{\mu_n}_{\nu_n}T^{\nu_1\cdots\nu_n}(x).$$

Equation (4) implies that $F_{\mu\nu}$ can be obtained from a vector potential field A_μ:

$$F_{\mu\nu}=\partial_\nu A_\mu-\partial_\mu A_\nu. \tag{5}$$

If $F_{\mu\nu}$ is considered a function of A_μ and a Lagrangian density is defined by $\mathcal{L}=-\frac{1}{4}F_{\mu\nu}F^{\mu\nu}-A_\mu J^\mu$, then (3) becomes the Euler–Lagrange equations (2) following from the principle of stationary action (1). The remaining Maxwell equations (4) follow automatically once (5) is taken as the definition of $F_{\mu\nu}$.

Under Poincaré transformations A^μ transforms as a tensor field of rank -1 leaving $\mathcal{L}$ form invariant when J_μ is 0. The invariance of $\mathcal{L}$ combined with the Euler–Lagrange equations (3) implies, by Noether's theorem, the conservation relations $\partial_\mu T^{\mu\nu}=0$, $\partial_\mu M^{\mu\nu\lambda}=0$ for the tensor fields

$$T^{\mu\nu}=-F^\mu{}_\delta F^{\nu\delta}+\tfrac{1}{4}\eta^{\mu\nu}F_{\alpha\beta}F^{\alpha\beta},$$

$$M^{\mu\nu\lambda}=x^\nu T^{\mu\lambda}-x^\lambda T^{\mu\nu}.$$

The tensor $T^{\mu\nu}$ may be interpreted as the electromagnetic field's energy–momentum tensor, while $M^{\mu\nu\lambda}$ may be taken to be the relativistic generalization of an angular momentum current. For example, the three-component angular momentum density is $(M^{023}, M^{031}, M^{012})$.

GENERAL RELATIVITY

The physical system considered in general relativity is space-time itself. Points in space-time are specified by a set of four coordinates y^μ, and the measurable property of space-time at each point is its metric, given by the symmetric tensor $g_{\mu\nu}$ or equivalently by its inverse $g^{\mu\nu}$, where the indices μ and ν again run from 0 to 3. A measurement of $g_{\mu\nu}$ can be made, in principle, using the information that an ideal clock following a trajectory successively passing through y^μ and $y^\mu+dy^\mu$ will have $dy^\mu\,dy^\nu\,g_{\mu\nu}$ positive and show an elapsed time of $(dy^\mu\,dy^\nu\,g_{\mu\nu})^{1/2}$ in the course of the displacement dy^μ. At all points in space $g_{\mu\nu}$ has one strictly positive eigenvalue and three strictly negative. Displacements with $dy^\mu\,dy^\nu\,g_{\mu\nu}$ positive are called timelike and those with $dy^\mu\,dy^\nu\,g_{\mu\nu}$ negative are called spacelike. At any point a small displacement purely in the y^0 direction is timelike and small displacements in the y^1,y^2 or y^3 directions are spacelike.

The dynamical laws governing the behavior of $g_{\mu\nu}$ are most easily stated by defining a sequence of intermediate fields. The affine connection is obtained from the metric tensor by

$$\Gamma^\alpha_{\mu\nu}=-\tfrac{1}{2}g^{\alpha\beta}\left(\frac{\partial g_{\beta\mu}}{\partial y^\nu}+\frac{\partial g_{\beta\nu}}{\partial y^\mu}-\frac{\partial g_{\mu\nu}}{\partial y^\beta}\right).$$

Then the curvature tensor is defined by

$$R^{\alpha}_{\beta\mu\nu}=\frac{\partial}{\partial y^{\mu}}\Gamma^{\alpha}_{\beta\nu}-\frac{\partial}{\partial y^{\nu}}\Gamma^{\alpha}_{\beta\mu}+\Gamma^{\alpha}_{\nu\lambda}\Gamma^{\lambda}_{\beta\mu}-\Gamma^{\alpha}_{\mu\lambda}\Gamma^{\lambda}_{\beta\nu}$$

and contracted curvatures are defined by $R_{\beta\nu}=R^{\alpha}_{\beta\alpha\nu}$, $R=R_{\beta\nu}g^{\beta\nu}$. Geometric interpretations of $\Gamma^{\alpha}_{\mu\nu}$ and $R^{\alpha}_{\beta\mu\nu}$ can be found in the 1973 text by Misner, Thorne, and Wheeler listed in the Bibliography. Finally, Einstein's equations for $g_{\mu\nu}$ in the presence of energy and momentum described by the energy–momentum tensor $T_{\mu\nu}$ are

$$R_{\mu\nu}-\tfrac{1}{2}g_{\mu\nu}R=8\pi GT_{\mu\nu} \tag{6}$$

where G is the Newtonian gravitational constant.

As in the case of Maxwell's equations, Einstein's equations both determine the evolution of the field in a timelike direction and constrain the possible values on spacelike hypersurfaces.

Equation (6) is form invariant under twice-differentiable reparametrizations of space-time by a new set of coordinates $y'^{\mu}=y'^{\mu}(y)$ with nonvanishing Jacobian if $g_{\mu\nu}$ and $T_{\mu\nu}$ are both transformed as rank-2 covariant tensor fields. In general, such a tensor $V_{\mu\nu}$ transforms as

$$V'_{\mu\nu}(x')=\frac{\partial x^{\alpha}}{\partial x'^{\mu}}\frac{\partial x^{\beta}}{\partial x'^{\nu}}V_{\alpha\beta}(x).$$

Equation (6) in the absence of an energy–momentum tensor $T_{\mu\nu}$ follows from the variational principle Eq. (1) if the Lagrangian density for the metric tensor is taken to be $\sqrt{-g}R/(16\pi G)$, where g is the determinant of $g_{\mu\nu}$. With this choice, the integrated action in Eq. (1) is form invariant under general coordinate transformations. A Lagrangian density yielding (6) in the presence of an energy–momentum tensor can be found by introducing in the Lagrangian for the fields or matter giving rise to $T_{\mu\nu}$ factors of $g_{\mu\nu}$ and $\Gamma^{\alpha}_{\mu\nu}$ in such a way that the resulting action becomes invariant under general coordinate transformations (see Misner, Thorne, and Wheeler for a discussion of how this can be accomplished), then adding to this Lagrangian density the term $\sqrt{-g}R/(16\pi G)$.

MECHANICS OF CONTINUOUS MEDIA

The motion of a solid, liquid, or gas can be specified in one of two equivalent ways. The material description of motion consists of a field $\mathbf{x}(\mathbf{a},t)$ giving the position $\mathbf{x}$ at time t of the material whose position was $\mathbf{a}$ at a reference time t_0. The spatial description consists of a field $\mathbf{v}(\mathbf{x},t)$ giving the velocity of the material at position $\mathbf{x}$ at time t. These descriptions are related by the differential equation

$$\frac{\partial}{\partial t}\mathbf{x}(\mathbf{a},t)=\mathbf{v}(\mathbf{x},t) \tag{7}$$

with the boundary condition $\mathbf{x}(\mathbf{a},t_0)=\mathbf{a}$.

The thermodynamic properties of a homogeneous medium can be described by a pair of fields giving two conveniently chosen thermodynamic quantities as functions of position and time, for example, a density field $\rho(\mathbf{x},t)$ and a temperature field $T(\mathbf{x},t)$. Any other thermodynamic fields can be obtained from these using the material's equation of state, which by the assumptions of homogeneity is constant throughout the substance. Inhomogeneous media will not be considered here.

The flow of momentum caused by forces which each region of material exerts on neighboring regions is specified by a stress tensor $T^{ij}(\mathbf{x},t)$, where i and j run from 1 to 3, consisting of 9 of the 16 components of the material's energy–momentum tensor $T^{\mu\nu}(\mathbf{x},t)$ described earlier. The flow of heat caused by thermal conduction is given by a field $\mathbf{q}(\mathbf{x},t)$ where $q^i(\mathbf{x},t)$ is the rate of flow of heat per unit area in the i direction. Additional fields which may enter the dynamical equations governing materials include the body force $\mathbf{f}(\mathbf{x},t)$, giving the external force per unit mass acting on the material, and the macroscopic electromagnetic field $F^{\mu\nu}(\mathbf{x},t)$ and current density $J^{\mu}(\mathbf{x},t)$, given by local averages of the corresponding fields of the section on electrodynamics.

The dynamical equations governing a continuous system can be grouped into general laws fulfilled by all materials and constitutive equations which vary from one substance to another. Among the first class of equations are conservation of mass,

$$\frac{D\rho}{Dt}+\rho\nabla\cdot\mathbf{v}=0, \tag{8}$$

and conservation of momentum and energy, which in the absence of electromagnetic effects take the form

$$\rho\frac{Dv^i}{Dt}-\rho f^i-\frac{\partial T^{ji}}{\partial x^j}=0, \tag{9}$$

$$\rho v^i\frac{Dv^i}{Dt}+\rho\frac{DE}{Dt}-\rho v^i f^i-\frac{\partial}{\partial x^j}(v^iT^{ji}-q^j)=0. \tag{10}$$

The operator D/Dt in these equations is called the material derivative, $D/Dt=\partial/\partial t+v^i\,\partial/\partial x^i$, and $E(\mathbf{x},t)$ is the thermodynamic internal energy per unit mass of the material at $\mathbf{x}$ and t with respect to its rest frame. Without an external body-couple field exerting a torque on small regions proportional to their mass, conservation of angular momentum requires T^{ij} to be a symmetric tensor.

The constitutive equations for a substance may include its thermodynamic equation of state, an equation expressing T^{ij} as a function of either $\mathbf{x}(\mathbf{a},t)$ or $\mathbf{v}(\mathbf{x},t)$, and an equation determining $\mathbf{q}$ from the temperature field T. For many materials the constitutive equations for a field $r(\mathbf{x},t)$ actually require the values of another set of fields $s_1(\mathbf{x},t'),\ldots,s_n(\mathbf{x},t')$ not only at $t=t'$ but also over the material's entire preceding history $t'\leq t$. A satisfactory approximation, however, can often be obtained using only $s_1(\mathbf{x},t),\ldots,s_n(\mathbf{x},t)$ and a finite set of their time derivatives.

A typical closed set of conservation laws and constitutive equations are the Navier–Stokes equations for an incompressible Newtonian liquid. These equations apply, for example, to water. Assume ρ and T can be replaced by constants ρ_0 and T_0, respectively, and that T^{ij} obeys the constitutive equation for a Newtonian liquid:

$$T^{ij}=-p\delta^{ij}+\mu\left(\frac{\partial v^i}{\partial x^j}+\frac{\partial v^j}{\partial x^i}\right). \tag{11}$$

The constant μ is the liquid's viscosity and $p(x,t)$ is the pressure field. Equation (11) combined with (8) and (9) forms a closed system of equations for the time development of $p(x,t)$ and $\mathbf{v}(\mathbf{x},t)$. The boundary condition used with these is that the fluid at each boundary point must be at rest with respect to the confining material.

See also ELECTRODYNAMICS, CLASSICAL; FIELDS; GRAVITATION; HYDRODYNAMICS; QUANTUM ELECTRODYNAMICS; QUANTUM FIELD THEORY; RELATIVITY, GENERAL.

BIBLIOGRAPHY

J. D. Jackson, *Classical Electrodynamics*, 2nd ed. Wiley, New York, 1975. (A)

W. M. Lai, D. Rubin, and E. Krempl, *Introduction to Continuum Mechanics*. Pergamon, New York, 1974. (I)

L. D. Landau and E. M. Lifshitz, *The Classical Theory of Fields*, 3d rev. English ed. Pergamon and Addison-Wesley, Reading, Mass., 1971. (A)

L. D. Landau and E. M. Lifshitz, *Fluid Mechanics*. Pergamon and Addison-Wesley, Reading, Mass., 1959. (A)

R. E. Meyer, *Introduction to Mathematical Fluid Dynamics*. Wiley (Interscience), New York, 1971. (A)

C. W. Misner, K. S. Thorne, and J. A. Wheeler, *Gravitation*. W. H. Freeman, San Francisco, 1973. (I–A)

E. M. Purcell, *Electricity and Magnetism* (Berkeley Physics Course), Vol. 2. McGraw-Hill, New York, 1963. (E)

M. Schwartz, *Principles of Electrodynamics*. McGraw-Hill, New York, 1972. (I)

Field Theory, Unified

James L. Anderson

In the early part of this century Einstein achieved in the general theory of relativity a remarkable unification of geometry and gravity by identifying the gravitational field with the Riemannian metric of space-time. This unification in turn served as the impetus for a number of attempts to achieve a further unification by generalizing the geometrical structure of space-time and identifying the electromagnetic field with the additional geometrical elements needed for this generalization. The names of Weyl, Schrödinger, Kaluza, Klein, and especially Einstein and his co-workers figured prominently in the list of authors of such unified field theories (as they came to be called), and indeed Einstein continued to search for such a theory to the end of his life.

Since no unique generalization of the Riemannian geometry of space-time exists, various authors tried different approaches. Einstein for example at one time considered a geometry that utilized an asymmetric metric tensor instead of the symmetric tensor of Riemannian geometry. Other authors introduced an asymmetric affinity into the geometry and identified the electromagnetic field with the antisymmetric components of this affinity. (An affinity or affine connection is used in geometry to define a notion of local parallelism between vectors.) Kaluza, and later Kaluza and Klein, sought the desired generalization by increasing the dimensionality of space-time from four to five but retained its Riemannian character. The additional components of the five-dimensional metric over the ten of four dimensions were then associated with the electromagnetic field.

In judging the degree of unification achieved by these various theories it is helpful to compare them to theories that already possess some degree of unification. Minimal unification is achieved in the Einstein–Maxwell theory of the gravitational and electromagnetic fields in general relativity. In this theory the two fields are represented by two completely independent geometrical objects that transform independently of one another under an arbitrary coordinate transformation. We are not even compelled to introduce the electromagnetic field at all into the Einstein gravitational theory. The only unification that can be said to exist in this theory is the dependence of the field equations for the electromagnetic field on the metric-gravitational field and vice versa. On the other hand the special relativistic theory of the electric and magnetic fields is a highly unified theory. In special relativity a theory of the electric or magnetic field cannot exist by itself; both fields are needed to construct a complete theory. Furthermore, the electric and magnetic fields do not transform independently of one another under a Lorentz transformation. Finally, the group of invariant transformations of special relativity cannot be decomposed into a product of two or more smaller groups.

A comparison of the foregoing theories suggests two possible criteria for a unified theory:

(i) The invariance group of a unified theory cannot be decomposed into a direct product of two or more smaller groups.
(ii) The field variables of a unified theory cannot be decomposed into a direct sum of two or more sets of variables that transform independently of one another under the invariance group of the theory.

Taken together these two conditions are almost certainly too restrictive; even the Einstein gravitational theory fails to satisfy (ii). It does, however, satisfy a somewhat weaker condition that is implied by (ii) but not vice versa.

(ii′) It is not possible to construct a complete theory using a subset of the field variables of a unified theory.

We do not wish to imply here that these are the only criteria for a unified field theory or that there is even any general agreement on what constitutes such a theory. However, they do serve as a convenient yardstick for judging the degree of unification of a theory and we shall so use them here.

While the asymmetric metric or affine theories mentioned earlier fail to satisfy either condition (ii) or (ii′), the five-dimensional theory of Kaluza and Klein does satisfy the latter as well as condition (i). Furthermore, as in the case of general relativity and in contrast to the asymmetric metric and affine theories, the field equations for the metric tensor in this theory are essentially uniquely determined by the condition that they are of second differential order. Unfortunately, in spite of its attractive features, the Kaluza–Klein theory shares with these theories the defect of having no presently testable consequences beyond those of the Einstein–Maxwell theory itself. This fact, coupled with the lack of any physical motivation for unification such as existed in the case of general relativity, was mainly responsible for its not receiving greater acceptance than it has.

In the years that followed the first attempts to construct a unified theory of the electromagnetic and gravitational fields, several developments occurred in physics that cast serious doubt on the possibility of ever constructing a satisfactory unified field theory. One of these developments was quantum mechanics. It soon became clear, especially after the successes of quantum electrodynamics, that any fundamental field theory would have to be a quantum field theory. Because of the difficulties that were encountered in the many attempts to quantize the gravitational field by itself it appeared likely that a unified theory would be even more difficult to quantize.

The second development that had to be taken into account was the discovery of an ever increasing number of fundamental fields in nature. Even if we restrict our attention to the fields that are now believed to be responsible for the fundamental interactions of matter, we have to add to the electromagnetic and gravitational fields those associated with the so-called strong and weak interactions between elementary particles. Since the properties of these four fields are so different from one another—among other things, the former are long-range fields while the latter are short range—it appeared doubtful that they could ever be unified into a single entity. Yet in spite of these difficulties there is good reason to believe that a unified theory of these fields is not only possible but probably even necessary for the construction of a consistent quantum theory of the fundamental interactions.

The search for a unified theory of the elementary interactions is today a very active field of research not only because of the esthetic appeal of such theories, but more important, because of the limited number of such theories we can construct that satisfy conditions such as those given earlier. Since a final theory of these interactions is still to be achieved, such conditions are extremely useful in limiting the possible candidates for such a theory. The first modern unified theory is the non-Abelian gauge theory, developed during the last decade by Steven Weinberg, Abdus Salam, and John Ward, which succeeded in unifying the weak and electromagnetic interactions in spite of their great dissimilarities. It would take us too far afield to describe the details of this theory here. We can say, though, that it has two virtues not shared by the original unified theories: It has testable consequences; and perhaps even more important is that, as Gerard 't Hooft and Martin J. G. Veltman showed in the early 1970s, it can be quantized in a consistent manner even though weak interaction theory by itself cannot.

Recently a new class of unified theories called supergravity theories has been developed. These theories are also non-Abelian gauge theories that include the gravitational interaction and give hope of uniting it and the other basic interactions. They also give hope of constructing a consistent quantum theory of gravity since, for those calculations that we have been able to carry through so far, they give finite results instead of the infinite ones obtained from a quantized version of the Einstein theory. It is, however, still too early to tell if these hopes can be realized and whether we can ultimately construct a single unified quantum field theory of the fundamental interactions found in nature.

See also GAUGE THEORIES; GROUP THEORY IN PHYSICS; MAXWELL'S EQUATIONS; RELATIVITY, GENERAL.

BIBLIOGRAPHY

An eminently readable account of recent developments in unified field theory is given by Daniel Z. Freedman and Peter van Nieuwenhuizen in "Supergravity and the Unification of the Laws of Physics," *Sci. Am.* **238,** 126 (1978).

A technical introduction to non-Abelian gauge theories can be found in Jeremy Bernstein's "Spontaneous Symmetry Breaking, Gauge Theories, the Higgs Mechanism and All That," *Rev. Mod. Phys.* **46,** 7 (1974).

For a technical account of the various classical field theories, see Peter G. Bergmann; *An Introduction to the Theory of Relativity* (Dover Reprints, New York, 1976.

Fields

Cyrus Taylor

The concept of a "field" has been central to the development of modern theoretical physics. The idea was originally introduced as a conceptual alternative to treating forces in terms of action at a distance. Thus, rather than thinking about the earth instantaneously causing the moon to accelerate toward it, one introduces the earth's gravitational field, which is defined throughout space and tells you how a test particle of infinitesimal mass would move. Formulated thus, the field concept is not of much more than philosophical interest; the power becomes apparent only when one assumes that the field obeys mathematical equations governing its evolution. In this sense, the first field theory was Euler's theory of hydrodynamics. He described the motion of a fluid in terms of the velocity at each point (a velocity field). This motion is then governed by a set of partial differential equations. This procedure of abstracting certain important aspects of physical media, representing them as continuous fields, and assuming a suitable set of equations governing them has been a key element in the development of theories of sound and elasticity, and, in the recent past, in the development of the renormalization group governing aspects of phase transitions in condensed matter systems.

The construction of a satisfactory theory of hydrodynamics, in which momentum transport (and hence, forces) are described by partial differential equations, without the notion of an action at a distance, had a profound impact on the development of other branches of physics. Faraday was the first to argue that conventional action-at-a-distance theories would be inadequate to understand electromagnetic phenomena and pointed the way toward the development of a field theory of electromagnetism. This was brought to fruition by Maxwell, who, using symmetry principles, together with mechanical models for wave propagation, proposed a satisfactory and essentially complete theory of isolated electromagnetic phenomena. At the same time, Maxwell pointed the direction towards a more abstract understanding of fields, arguing that the mechanical models he introduced would aid in the understanding of electrical phenomena, but were designed to be illustrative, not explanatory.

The twentieth century has added two principles which underlie all modern field theories of fundamental processes: relativity and quantum mechanics. Indeed, Einstein's formulation of special relativity is, in one sense, an elucidation of symmetries already present in Maxwell's theory of electromagnetism. In contrast, the rise of quantum theory has resulted in new interpretations of the "meaning" of field theories. This began with Schrödinger, who introduced a nonrelativistic equation which describes the evolution of physical systems. This equation is formulated in terms of a "wave function," a field which has the interpretation of a probability amplitude. That is, quantum mechanically, one cannot think of a classical particle as being defined by its position and momentum, but must instead introduce a wave function $\psi(x)$, such that the probability of finding the particle within a distance dx of the position x is $P(x)dx = |\psi(x)|^2 dx$.

Schrödinger's equation is a useful approximation to much of nonrelativistic quantum physics, but requires modification in order to be consistent with relativity. Such modifications were first carried out by Klein and Gordon; the resulting wave equation governs a field having a single component at each point of space-time. It came as a surprise that the interpretation of Schrödinger was no longer tenable. Instead, the field has to be reinterpreted as defining a charge density for an indefinite number of particles rather than a probability density for a single particle. In attempting to circumvent this problem, Dirac introduced a different field theory, involving a four-component "spinor" field, which can be interpreted as defining a probability density for a single electron at the cost of introducing "antiparticles" which formally have negative energy. This, rather than being a drawback, became a virtue when antiparticles were observed. This, in turn, helped motivate the development of second quantization, in which the fields are quantized, and can be decomposed into operators which create and annihilate particles.

In general, fundamental fields are classified according to their properties under the Lorentz group (the relativistic generalization of rotations), and under internal symmetries. As far as the Lorentz group is concerned, particles are classified by their "spin": the Klein–Gordon equation describes a spin-0 object, the Dirac equation, a spin-$\frac{1}{2}$ object, and Maxwell's equations describe a spin-1 particle. Formally, Einstein's theory of general relativity, which also takes the form of a field theory, describes a spin-2 particle, but no existing theory of gravity is satisfactory at the quantum-mechanical level. (One possible resolution of these problems is string theory, which evades technical problems to which other theories of gravity are subject.)

The internal symmetries of particles can be of two types: global or local. A global invariance of a field theory is a simultaneous modification of the field at all points of space-time which leaves the physical properties of the theory invariant. A local symmetry is a transformation of the fields in an infinitesimal neighborhood which leaves the theory invariant. In the latter case, these local symmetries indicate the presence of a gauge invariance, of which the prototypical example is again Maxwell's theory. Generalizations of the gauge invariance of Maxwell's equations, first proposed by Yang and Mills in 1954, form the basis for our modern theory of the electroweak and strong interactions. In this context, it is important to note that 't Hooft and Veltman showed that these theories are renormalizable; that is, that they retain their predictive power after second quantization. It should also be noted that not all of the symmetries of the fundamental laws of physics are actually realized in the physical world: spontaneous breaking of gauge symmetries provides the basis for models of superconductivity as well as for the generation of masses in the standard model of the electroweak interactions.

Field theory is still a rapidly developing subject. The next few years can be expected to see intensive study of the nonperturbative structure of quantum field theories, of quantum formulations of field theories incorporating gravity interacting with matter, the continued development of the areas of overlap between cosmology and particle physics, and the confrontation of theory with experiment as the next generation of particle accelerators is used to test the detailed structure of the standard model.

See also ELECTRODYNAMICS, CLASSICAL; ELEMENTARY PARTICLES IN PHYSICS; FIELD THEORY, CLASSICAL; QUANTUM FIELD THEORY; RELATIVITY, GENERAL THEORY; STRING THEORY.

BIBLIOGRAPHY

A nice review of the early history of field theory is: M. B. Hesse, *Forces and Fields*. Philosophical Library, New York, 1961.

A standard textbook on field theory is: C. Itzykson and J.-B. Zuber, *Quantum Field Theory*. McGraw-Hill, New York, 1980.

A useful textbook discussing the application of field theoretic techniques to critical phenomena is: D. J. Amit, *Field Theory, the Renormalization Group, and Critical Phenomena*. McGraw-Hill, London, 1978.

A nice treatment of the classical theory of the electromagnetic and gravitational fields is: L. D. Landau and E. M. Lifshitz, *The Classical Theory of Fields*. Pergamon, New York, 1979.

Fine and Hyperfine Spectra and Interactions

Daniel Kleppner and G. W. F. Drake

The spectral lines of hydrogen, when observed with high resolution, are found to consist of multiplets which can be ascribed to a doublet splitting of every level except those for which $l = 0$. The multiplet structure, known as the fine structure, is the result of relativistic effects which are neglected in the simple Bohr picture of hydrogen. A yet smaller splitting, the hyperfine structure, arises from the electron interaction with higher multipole moments of the nucleus. In nonhydrogenic atoms, fine structure is often described in terms of various spin–orbit and spin–spin interactions. However, for heavy atoms, these effects become too large to be treated as small perturbations and a complete relativistic calculation of energy levels becomes necessary. The effects also become large in states of high angular momentum, producing a strong mixing of states of the same total angular momentum and parity, but different spin. The hyperfine interaction, in contrast, is generally small enough to be treated as a perturbation irrespective of the atom's complexity.

The fine structure of hydrogen was first described comprehensively by Dirac's relativistic theory of the electron. The major contribution is easily understood by a semiclassical argument. The electron, moving with velocity $\mathbf{v}$, "sees" a motional magnetic field $\mathbf{B}_m = -(\mathbf{v}/c)\times\mathbf{E}$, where $\mathbf{E}$ is the Coulomb field. The motional field gives rise to an interaction $-\boldsymbol{\mu}_e\cdot\mathbf{B}_m$, where $\boldsymbol{\mu}_e$ is the electron's magnetic moment. A straightforward calculation leads to an interaction term of the form

$$\mathcal{H}_{\text{s.o.}} = \zeta(r)\mathbf{L}\cdot\mathbf{S}$$

where $\zeta(r)=(\frac{1}{2}m^2c^2r)dV/dr$. V is the electrostatic energy. The expression includes a factor of $\frac{1}{2}$, the Thomas factor, arising from the relativistic transformation from the accelerating frame of the electron to an inertial frame. The $\mathbf{L}\cdot\mathbf{S}$ interaction, known as the spin–orbit coupling, is the major contribution to the fine structure.

An additional contribution to atomic energies arises from the variation of the electron's mass with velocity. The total energy of the electron is $W=V+(p^2c^2+m^2c^4)^{1/2}$. The kinetic energy is $T=W-V-mc^2\cong p^2/2m-p^4/8mc^2$. The last term,

$$\mathcal{H}_{\text{rel}} = -p^4/8m^3c^2,$$

is the lowest-order relativistic correction to the classical kinetic energy. It shifts fine-structure levels by the same amount, and so does not affect the splitting.

If we treat $\mathcal{H}_{\text{s.o.}}$ and $\mathcal{H}_{\text{rel}}$ as perturbations to the hydrogenic term energy, we eventually obtain the following expression for the fine-structure energy:

$$E_{\text{fs}} = \langle\mathcal{H}_{\text{s.o.}}+\mathcal{H}_{\text{rel}}\rangle = \frac{-hcRZ^2}{n^2}\left[\frac{\alpha^2Z^2}{n}\left(\frac{1}{j+\frac{1}{2}}-\frac{3}{4n}\right)\right],$$

R is the Rydberg constant, n is the principal quantum number, $\hbar j$ is the total electronic angular momentum, Z is the nuclear charge, and $\alpha=e^2/\hbar c\approx 137^{-1}$ is the fine-structure constant. The leading factor is the term energy of a hydrogenic atom. The bracketed expression reveals that the fine structure of hydrogen is smaller than the term energy by approximately $\alpha^2Z^2/n\approx 3\times10^{-5}/n$, which justifies its treatment as a perturbation. The Dirac theory, when taken to lowest order, gives the identical result.

Fine structure splits each angular momentum state into doublets depending on whether $j=l+\frac{1}{2}$ or $j=l-\frac{1}{2}$. For $l=0$, only $j=l+\frac{1}{2}$ is possible. Thus the configurations for hydrogen are of the form ${}^2S_{1/2}$, ${}^2P_{1/2,3/2}$, ${}^2D_{3/2,5/2}$,... . An important feature of hydrogenic fine structure is the degeneracy of all states of a given term having the same j, such as the pairs (${}^2S_{1/2}$, ${}^2P_{1/2}$), (${}^2P_{3/2}$, ${}^2D_{3/2}$).

The fine structure of hydrogen has served as a testing ground for relativistic quantum theory and for quantum electrodynamics. The earliest studies involved optical spectroscopy of the Balmer spectrum ($n>2\rightarrow n=2$) of hydrogen and deuterium. The experiments were inconclusive, however, because of the small fine-structure splitting of the $n=2$ state, and the large Doppler width of hydrogen's spectral line. (The fine structure of the $n=2$ term is only 0.37 cm^{-1} = 11 GHz; the Doppler width at room temperature of the principal Balmer line is 0.2 cm^{-1}.)

Starting in the late 1940s Willis E. Lamb and his colleagues carried out a series of experiments in which the fine structure was studied by radio-frequency spectroscopy of the transitions ${}^2S_{1/2}\rightarrow{}^2P_{1/2}$, ${}^2P_{3/2}$, with a resolution of 10^{-4} GHz. He confirmed that the fine structure of the $2P$ state is accurately given by the Dirac theory, and in fact used his measurements to obtain a more precise value for α. A more dramatic finding, for which Lamb received the Nobel prize, was that ${}^2S_{1/2}$ and ${}^2P_{1/2}$ levels for $n=2$ are not, in fact, degenerate. The ${}^2S_{1/2}$ state was found to be shifted upward by 1060 MHz. The displacement, known as the Lamb shift, is dominantly due to radiative coupling of the electron with the vacuum field. The discovery of the Lamb shift played a central role in the development of quantum electrodynamics. Experiments and calculations on the fine structure and the Lamb shift in hydrogen have undergone continuing refinement. Experiment and theory have been pushed to an accuracy of about 11 kHz, or 1 part in 10^5, and are in good agreement.

Lamb shift measurements are also available for the heavier hydrogenic ions up to one-electron U^{91+}. The most accurate result for He^+ is obtained indirectly from the anisotropy in the angular distribution of Ly-α quenching radiation emitted by the metastable $2\,{}^2S_{1/2}$ state in an electric field. The accuracy of 2.5 parts in 10^5 provides the most sensitive available test of the higher-order quantum electrodynamic contributions to the Lamb shift.

Hyperfine interaction arises principally from the coupling of the nuclear magnetic moment $\boldsymbol{\mu}_I$ with the magnetic field produced by the electron, $\mathbf{B}_e$. From symmetry arguments it can be shown that $\boldsymbol{\mu}_I\propto\mathbf{I}$, and that the hyperfine interaction has the general form

$$\mathcal{H}_{\text{hf}} = a\mathbf{I}\cdot\mathbf{J}.$$

The quantity a, known as the hyperfine constant, has the value for hydrogen

$$a = hcR\frac{Z^3}{n^3}\frac{m}{M}g_I\frac{1}{(l+\frac{1}{2})(j+1)}.$$

This expression neglects small corrections due to nuclear structure and radiative and relativistic effects. m/M is the electron–nuclear mass ratio and g_I is the nuclear g-factor. In the ground state of hydrogen $I=J=\frac{1}{2}$, and $\mathbf{I}\cdot\mathbf{J}$ has the values $\frac{1}{4}$ and $-\frac{3}{4}$. Thus the separation between the hyperfine components is simply

$$\Delta E_{\text{hf}} = a.$$

The hyperfine constant in frequency units, $a/h\approx 1.420$ GHz, has been measured with the hydrogen maser to a precision of 10^{-3} Hz; it is probably the best known physical quantity. The hyperfine separation of hydrogen has played a major role in radioastronomy by giving rise to the 21-cm line which is extensively used for mapping hydrogen radio sources.

In states for which the nuclear and electronic angular momenta are greater than $\frac{1}{2}$, hyperfine structure can arise from an electrostatic quadrupole interaction between the nucleus and the electron. Higher-order magnetic and electric interactions can also occur, though they are generally minute.

The alkalis, which resemble hydrogen in being essentially single-electron atoms, also display doublet structure for all states except those for which $l=0$. The closed-shell core has, however, a major effect on the doublet splitting. Inside the core the potential departs radically from the Coulombic form. The spin–orbit coupling, which varies predominantly as dV/dr, is very sensitive to the core structure, and states which appreciably penetrate the core generally have fine-structure separations which are large compared to that of hydrogen. This is illustrated in Table I, which displays the fine-structure intervals of the lowest p states for the alkalis and hydrogen. Hyperfine separations of the ground states are also listed for comparison. For s states, hyperfine interactions depend on the density of the electron at the nucleus and on the nuclear moment. As Table I shows, although hyperfine separations for the alkalis differ appreciably from that for hydrogen, they are the same order of magnitude.

In certain excited states of the alkalis, both the fine and hyperfine coupling constants are inverted with respect to hydrogen. The mechanism for this involves a spin-dependent exchange interaction between the valence electron and core electrons. For high angular momentum, the interactions assume their normal sign and the systems are well described by hydrogenic theory.

For many-electron atoms, the fine and hyperfine structure can be complex. Systems described by L–S coupling, however, obey a simple rule, the Landé interval rule, which is of great help in identifying spectra. For such systems the spin–orbit interaction can again be written $\zeta\mathbf{L}\cdot\mathbf{S}$, where ζ now represents contributions from each of the electrons. If the total angular momentum can be written in the form $\mathbf{J}=\mathbf{L}+\mathbf{S}$, then the simple vector arguments give $\langle\mathbf{L}\cdot\mathbf{S}\rangle=[j(j+1)-l(l+1)-s(s+1)]/2$. The separation between levels with angular momentum j and $j-1$ is ζj; thus the ratio of separation between adjacent components of a fine-structure multiplet, $(j,j-1)$ and $(j-1,j-2)$, is $j/(j-1)$. (This rule also applies to magnetic hyperfine structure with j replaced by the total electronic and nuclear angular momentum, F.)

Table II shows intervals between the $j=2$, 1, and 0 components of the lowest P terms for atoms with two valence electrons. The intermediate-weight elements obey the rule reasonably, but the very light and very heavy elements depart appreciably. In the case of light elements, fine structure

Table I Fine and hyperfine structure of hydrogen and alkali atoms

	Fine structure		Hyperfine structure[a]	
	State	ΔE (cm^{-1})	State	ΔE (GHz)
H	$2p$	0.37	$1s$	1.42
Li	$3p$	0.44	$2s$	0.83
Na	$4p$	17.2	$3s$	1.77
K	$5p$	57.7	$4s$	0.46
Rb	$5p$	237.6	$5s$	3.03
Cs	$6p$	554.1	$6s$	9.19

[a] Hyperfine structure for most abundant isotope.

Table II Fine-structure splittings of atoms with two valence electrons[a]

	State	ΔE ($^3P_2-{}^3P_1$)	ΔE ($^3P_1-{}^3P_0$)	Ratio
He	$1s2p$	0.076 4261	−0.987 9122	—
Be	$2s2p$	2.35	0.68	3.5
Mg	$3s3p$	40.7	20.1	2.03
Ca	$4s4p$	105.9	52.2	2.03
Sr	$5s5p$	394.2	186.8	2.11
Ba	$6s6p$	878.1	370.6	2.37

[a] Energy in cm^{-1}.

is not well described by a simple $\mathbf{L}\cdot\mathbf{S}$ term; the spin of each electron couples to the spin and orbital moment of the other electron. For helium these effects are so large that the fine structure is actually inverted.

Departures from the interval rule for heavy atoms indicate a breakdown of L–S coupling due to configuration mixing by the Coulomb interaction. In the case of hyperfine structure, departure from the interval rule signals the presence of quadrupole or higher-order interactions.

The theoretical calculation of fine and hyperfine structure from first principles provides a sensitive test of the accuracy of the electronic wave functions used. For helium and the helium-like ions Li^+, Be^{++},..., high-precision variational wave functions constructed from linear combinations of correlated functions of the form $r_1^i r_2^j r_{12}^k e^{-\alpha r_1-\beta r_2}$ are available. Here, $\mathbf{r}_1$ and $\mathbf{r}_2$ are the position vectors of the two electrons and $r_{12}=|\mathbf{r}_1-\mathbf{r}_2|$ is the interelectronic coordinate. Fine structure then arises from matrix elements of the Breit interaction, which includes spin-other-orbit, spin–spin, and Fermi contact interactions in addition to the spin–orbit interaction mentioned above. The corresponding hyperfine structure Hamiltonian is

$$\mathcal{H}_{\mathrm{hf}}=-2\mu_0\sum_{i=1}^{2}\left[-\frac{1}{r_i^3}\mathbf{l}_i\cdot\boldsymbol{\mu}+\frac{1}{r_i^3}\left(\frac{\mathbf{s}_i\cdot\boldsymbol{\mu}-3(\mathbf{s}_i\cdot\mathbf{r}_i)(\boldsymbol{\mu}\cdot\mathbf{r}_i)}{r_i^3}\right)-\frac{8\pi}{3}(\mathbf{s}_i\cdot\boldsymbol{\mu})\delta(\mathbf{r}_i)\right]$$

where $\boldsymbol{\mu}=-g_I(m/M)\mu_o\mathbf{I}$ is the nuclear magnetic moment and $\mu_0=e/2mc$. The above three terms correspond to the electron-orbit–nuclear spin, electron-spin–nuclear spin, and contact terms, respectively. There are numerous small corrections for relativistic, radiative, and finite nuclear mass and size effects. When these are included, the theoretical value for the dominant contact term for the $1s2p$ ^3P state of ^{3}He is -4283.8 ± 0.2 MHz, as compared with the experimental value -4282.72 ± 0.04 MHz. The corresponding values for the orbital term are -28.13 ± 0.02 MHz and -29.85 ± 0.14 MHz. The slight differences are due to higher-order terms not included in the calculation.

The fine-structure intervals of the $1s2p$ ^{3}P state of ^{4}He have been extensively studied, both theoretically and experimentally. When higher-order relativistic effects are included, theory and experiment for the $^3P_1$–$^3P_0$ interval agree to within 0.1 MHz or 3 parts in 10^6. Further progress in the comparison of theory and experiment will require the calculation of two-electron radiative corrections.

Interesting studies of fine-structure splittings have recently been made in the high angular momentum Rydberg states of helium through the measurement of transitions such as 1s10g–1s10h, 1s10h–1s10i, etc., by fast beam microwave/laser resonance techniques. Here, spin–orbit coupling dominates the small singlet–triplet electrostatic splittings to produce four approximately equally spaced levels for each angular momentum. These splittings have fruitfully been analyzed by asymptotic expansion methods which neglect exchange effects and treat the outer electron as moving in the field of a polarizable core consisting of the nucleus and the inner electron. An important feature of these studies is that they are sensitive to long-range Casimir–Polder retardation effects.

A priori calculations of fine and hyperfine structure for many-electron atoms are much more difficult because the splittings are sensitive to the details of the wave function near the nucleus. Simple Hartree–Fock approximations are generally of low accuracy. Only in the case of three-electron atoms such as Li and Be^+ have high-precision correlated variational wave functions of the type described above for helium been obtained. However, considerable progress has been made in the application of many-body perturbation techniques, with accuracies better than $\pm 5\%$ in the hyperfine splittings for atoms as heavy as cesium. This particular case is important as a test of the similar atomic structure calculations required to interpret parity nonconservation measurements in terms of the electroweak interaction and the Weinberg angle.

The term *fine structure* is used in a somewhat different context with respect to molecular spectra. There it refers to the rotational structure of an electronic or vibrational molecular band. The term is descriptive because the structure only becomes apparent when the spectrum is observed with moderate resolution. Molecular spectra can also exhibit hyperfine structure arising from nuclear–electronic interactions. Although magnetic dipole interactions tend to be most important for free atoms, molecular hyperfine structure is often dominated by the electric quadrupole interaction.

Fine structure is generally large enough to be studied by the methods of conventional spectroscopy (an important exception is hydrogen for which most of the precision studies have employed techniques of radio-frequency spectroscopy and atomic beams). Hyperfine structure, on the other hand, is so small that optical observation usually requires high-resolution interferometry. Ground-state hyperfine structure is often studied by molecular-beam electric and magnetic resonance methods which can yield precision in the range of parts per million to parts per billion, or better. Hyperfine-structure studies can provide information on the spin, magnetic dipole moment, and electrical quadrupole moment of the nucleus, as well as the electronic charge distribution in atoms and molecules. One practical application for hyperfine spectroscopy is to atomic clocks; the second is defined in terms of the hyperfine separation of ^{133}Cs, which is taken to be 9 192 631.770 Hz.

Interest in fine and hyperfine spectra has been renewed by the advent of laser spectroscopy, which has allowed the study of excited states that were previously inaccessible, and which overcomes the limitations of Doppler broadening, permitting precision far greater than previously possible. The development of ion traps has opened the way to high-precision measurements in ions as well as atoms. By this method, a hyperfine transition in the 2s ground state of Be^+ has been measured to precision of about one part in 10^{13}—a precision which is exceeded only by similar measurements in neutral cesium.

See also ATOMIC SPECTROSCOPY; LASER SPECTROSCOPY; QUANTUM ELECTRODYNAMICS; QUANTUM MECHANICS.

BIBLIOGRAPHY

H. A. Bethe and E. S. Salpeter, *Quantum Mechanics of One- and Two Electron Atoms*. Academic Press, New York, 1957.

E. U. Condon and G. H. Shortley, *Theory of Atomic Spectra*. Cambridge University Press, London, 1951.

W. R. Johnson, S. A. Blundell, and J. Sapirstein, in *Atomic Physics 11* (S. Haroche, J. C. Gay and G. Grynberg, eds.). World Scientific, Singapore, 1989.

D. Kleppner, in *Atomic Physics and Astrophysics* (N. Chrétien and E. Lipworth, eds.). Gordon and Breach, New York, 1971.

H. G. Kuhn, *Atomic Spectra*, 2nd ed. Longmans, London, 1969.

I. Lindgren, *Rep. Prog. Phys.* **47**, 345 (1984).

G. W. Series, in *The Spectrum of Atomic Hydrogen: Advances* (G. W. Series, ed.). World Scientific, Singapore, 1988; and other articles therein.

J. C. Slater, *Quantum Theory of Atomic Structure*, Vol. II. McGraw-Hill, New York, 1960.

I. I. Sobel'man, *Introduction to the Theory of Atomic Spectra*. Pergamon Press, Oxford, 1972.

C. H. Townes and A. L. Schawlow, *Microwave Spectroscopy*. McGraw-Hill, New York, 1955.

A. van Wijngaarden, J. Patel, and G. W. F. Drake, in *Atomic Physics 11* (S. Haroche, J. C. Gay and G. Grynberg, eds.). World Scientific, Singapore, 1989.

G. Werth, in *Atomic Physics 9* (R. S. van Dyck, Jr. and E. N. Fortson, eds.). World Scientific, Singapore, 1984.

D. J. Wineland, W. M. Itano, J. G. Bergquist, J. J. Bollinger, and J. D. Prestage, in *Atomic Physics 9* (R. S. van Dyck, Jr. and E. N. Fortson, eds.). World Scientific, Singapore, 1984.

Fission *see* Nuclear Fission

Fluctuation Phenomena

Thomas J. Greytak

By fluctuation we usually mean the deviation of some quantity from its mean or most probable value. Almost all the quantities that a physicist might be interested in studying exhibit fluctuations, some on quite a microscopic scale (e.g., the motion of an atom about its equilibrium lattice position in a crystalline solid). This discussion, however, will be limited to macroscopic quantities that can be treated as thermodynamic variables. Fluctuations in these quantities manifest themselves in several ways: They may limit the precision of measurements of the mean value of the quantity, as is the case with thermal noise in electronic circuits or photon shot noise in light beams; they are the cause of some of the familiar features of our surroundings, such as the blue sky; or they may cause spectacular and unexpected effects, such as the critical opalescence of an apparently simple and homogeneous fluid.

FLUCTUATIONS IN A HYDROSTATIC SYSTEM

The simplest example of thermodynamic fluctuations occurs in a hydrostatic system such as a gas, a liquid, or a simple solid under hydrostatic pressure. In such a system of a fixed number of particles, the internal energy U is related to the principal state variables of temperature, entropy, pressure, and volume by the first and second laws of thermodynamics:

$$dU = T\,dS - P\,dV. \qquad (1)$$

Only two of the state variables can be chosen independently; the others are then fixed through the equation of state and an energy equation such as an expression for the heat capacity. Let these variables pertain to a subsystem specified by a certain number N of contiguous particles. N is assumed to be large enough that the definition of local thermodynamic variables makes sense, yet the subsystem must be small enough that its energy is only a very small part of the energy of the total system. In thermal equilibrium the *average* values of the intensive variables T and P are equal to the temperature and pressure of the total system. The average values of S and V will be much smaller than those of the total system and can be related to the average T and P.

At any instant of time there is a finite probability that the subsystem will not be in its most likely state. The probability that a pair of independent variables X and Y will deviate from their mean values by amounts ΔX and ΔY is given by a bivariate Gaussian density that depends only on the three parameters $\langle \Delta X^2 \rangle$, $\langle \Delta Y^2 \rangle$, and $\langle \Delta X\,\Delta Y \rangle$. These quantities are listed in Table I. Notice that the cross correlation between ΔT and ΔV and between ΔS and ΔP vanishes. If either of these pairs of variables were chosen to be the independent ones, their joint probability density would factor into the product of two single Gaussian densities, one for each variable. In this case the variables are statistically independent as well as independent in the thermodynamic sense: a knowledge of the fluctuation in one gives no additional information about the fluctuation in the other. It should be pointed out, though, that ΔT and ΔV or ΔS and ΔP are statistically independent random variables only when taken at the same instant of time. Inspection of the table will show that the ratio of the root mean square fluctuation in any variable to its mean value, $\langle \Delta X^2 \rangle^{1/2}/\langle X \rangle$, is proportional to $N^{-1/2}$. This is why conventional thermodynamics need deal only with the mean values of the variables.

Through expansion of any other thermodynamic variable in terms of two of the principal ones, the fluctuations in that quantity can be found. For example, the fluctuations in the number density, $n \equiv N/V$, are given by

$$\langle \Delta N^2 \rangle = \frac{n^3 kT\kappa_T}{N} \qquad (2)$$

Table I Fluctuations in a hydrostatic system[a]

$\langle \Delta T^2 \rangle = \dfrac{kT^2}{C_V}$	$\langle \Delta V^2 \rangle = kTV\kappa_T$
$\langle \Delta S^2 \rangle = kC_P$	$\langle \Delta P^2 \rangle = \dfrac{kT}{V\kappa_S}$
$\langle \Delta T\,\Delta V \rangle = 0$	$\langle \Delta S\,\Delta P \rangle = 0$
$\langle \Delta T\,\Delta S \rangle = kT$	$\langle \Delta V\,\Delta P \rangle = -kT$
$\langle \Delta T\,\Delta P \rangle = \dfrac{\alpha kT^2}{\kappa_T C_V}$	$\langle \Delta V\,\Delta S \rangle = kTV\alpha$

[a] C_V is the heat capacity at constant volume; C_P the heat capacity at constant pressure. The isothermal compressibility is

$$\kappa_T \equiv \frac{-1}{V}\frac{\partial V}{\partial P}\bigg|_T;$$

the adiabatic compressibility

$$\kappa_S \equiv \frac{-1}{V}\frac{\partial V}{\partial P}\bigg|_S;$$

the expansion coefficient

$$\alpha \equiv \frac{1}{V}\frac{\partial V}{\partial T}\bigg|_P.$$

and the fluctuations in the internal energy can be shown to be

$$\langle \Delta U^2 \rangle = kT^2 C_V + kTV\kappa_T\left(P - \frac{\alpha T}{\kappa_T}\right)^2. \qquad (3)$$

In a similar manner all of the entries in Table I can be found from the ones pertaining to a single pair of variables.

FLUCTUATIONS IN OTHER SYSTEMS

The results for the hydrostatic system can be carried over directly to several other systems that are thermodynamically isomorphic to it. To identify the new variables with those of the hydrostatic case we must examine the increment of work done on the system for a given change in the mechanical variables. For the hydrostatic case [Eq. (1)] the work is $(-P)\,dV$; for a magnetic system with magnetization M and an external or applied magnetic field H it is $H\,dM$; for a dielectric body with polarization $\mathcal{P}$ and electric field E it is $E\,d\mathcal{P}$; for an interface of surface tension $\mathcal{S}$ and area A it is $\mathcal{S}\,dA$; for an elastic rod with tension $\mathcal{T}$ and length L it is $\mathcal{T}\,dL$. For any of these systems we need only make the appropriate substitution of variables in Table I in order to find the fluctuations. For example, for the magnetic system $-P \rightarrow H$ and $V \rightarrow M$, so we find that the fluctuations in T and M are statistically independent and that

$$\langle \Delta T^2 \rangle = \frac{kT^2}{C_M}, \quad \langle \Delta M^2 \rangle = kT\frac{\partial M}{\partial H}\bigg|_T \qquad (4)$$

where C_M is the specific heat at constant magnetization and $(\partial M/\partial H)|_T$ is the isothermal susceptibility.

In a more general system with m independent thermodynamic variables the fluctuations in these variables form a set of m jointly Gaussian random variables. The associated probability density function contains as parameters the m mean square fluctuations $\langle \Delta X^2 \rangle$ and the cross correlations of all possible pairs $\langle \Delta X\, \Delta Y \rangle$. The most important example is probably a binary mixture where $m=3$. If the concentration of one of the components is $c_a \equiv N_a/N$ and the difference in chemical potential of the two components is $\mu' \equiv \mu_a - \mu_b$, then we can show that

$$\langle \Delta c_a\, \Delta T \rangle = \langle \Delta c_a\, \Delta P \rangle = 0, \qquad (5)$$
$$\langle \Delta c_a^{\,2} \rangle = \frac{kT}{N} \frac{\partial c_a}{\partial \mu'}\bigg|_{T,P}.$$

It is interesting to note that for this system we cannot choose a set of three conventional variables that are all mutually statistically independent.

TIME EVOLUTION OF THE FLUCTUATIONS

So far only static fluctuations—those that are evaluated at one instant of time—have been considered. The fluctuations are actually random processes that evolve in time, $\Delta n(t)$, for example. (Density fluctuations will be used for illustration since they are the ones most often studied experimentally, but the discussion that follows could be applied to any thermodynamic variable.) A great deal of information about the temporal behavior of the fluctuations is contained in the time correlation function, $\langle n(t)n(t+\tau) \rangle$. It is usual to consider each spatial Fourier component of the variable separately. If $n_{\mathbf{q}}(t)$ is the complex amplitude of the $\mathbf{q}^{\text{th}}$ Fourier component of the density, the relevant correlation function is $\langle n_{\mathbf{q}}(t) n_{\mathbf{q}}^*(t+\tau) \rangle$. In order to calculate such a quantity we use the Onsager "regression of fluctuations" hypothesis, a version of the fluctuation-dissipation theorem, which states that the time dependence of the equilibrium correlation function is the same as the time evolution of an externally induced density disturbance of wave vector $\mathbf{q}$ which is released at $t=0$. This reduces to a well-posed initial-value problem that can be solved *exactly* once the macroscopic equations of motion are known. The details of this type of calculation for simple fluids have been discussed by Mountain. The general result of such calculations is that the equilibrium correlation function is made up of a sum of contributions, one from each hydrodynamic normal mode of the system, each reflecting the time dependence associated with the macroscopic decay of that mode. In the case of the simple fluid there is a sound-wave contribution to $\langle n_{\mathbf{q}}(t) n_{\mathbf{q}}^*(t+\tau) \rangle$ which oscillates at a frequency $\omega = uq$ (u is the velocity of sound) while it decays toward zero because of processes that damp macroscopic sound waves. There is also a thermal diffusion contribution, which has a simple exponential decay at a rate proportional to the thermal conductivity. For a binary mixture there would be a third contribution to the correlation function, another exponential decay at a rate proportional to the concentration conductivity.

This simple picture for the time dependence of the correlation function is based on the assumption that well-defined hydrodynamic normal modes can exist in the system. In a gas when the mean free path l of the molecules becomes comparable to or greater than q^{-1}, this assumption is no longer valid. The coherent atomic motion characteristic of such modes is dissipated by the distribution of individual molecular velocities. In this case a kinetic theory, such as the Boltzmann equation, must be used to calculate the correlation function. Such calculations are discussed by Yip.

MEASUREMENT OF THE FLUCTUATIONS BY SCATTERING SPECTROSCOPY

A major advance in the study of fluctuations occurred with the realization that all scattering experiments on condensed-matter systems, whether they used light, x rays, neutrons, or electrons, measured essentially the same quantity, the dynamic structure factor $S(\mathbf{q},\omega)$. $S(\mathbf{q},\omega)$ is the space and time Fourier transform of the correlation function for the variable that couples to the scattering probe. For the density fluctuations in a classical system it can be written as

$$S(\mathbf{q},\omega) = \int_{-\infty}^{\infty} d\tau \int_V d^3R\, e^{i(\mathbf{q}\cdot\mathbf{R}-\omega\tau)} \langle n(\mathbf{r},t) n(\mathbf{r}+\mathbf{R},t+\tau) \rangle,$$
$$= \int_{-\infty}^{\infty} d\tau\, e^{-i\omega\tau} \langle n_{\mathbf{q}}(t) n_{\mathbf{q}}^*(t+\tau) \rangle. \qquad (6)$$

Experimentally the scattering geometry singles out one scattering angle, which, through the Bragg condition, determines the wave vector of the fluctuations being observed. A spectrometer then carries out the frequency analysis of the scattered beam. Figure 1 shows the features of $S(q,\omega)$ for a classical gas. The correlation function for such a system was

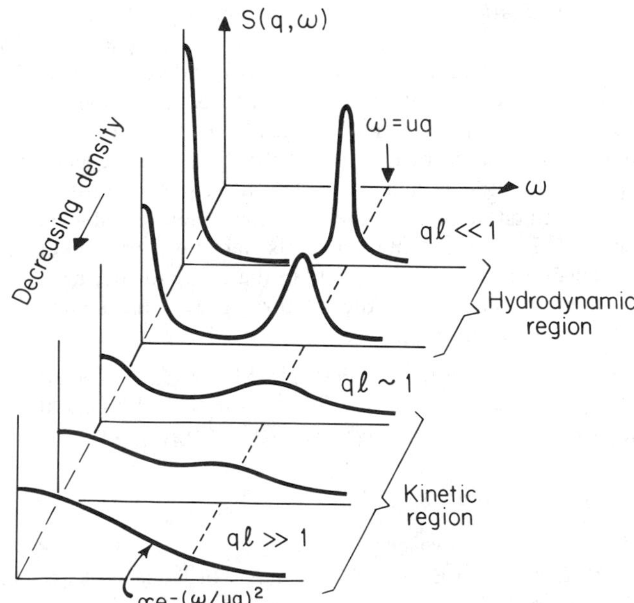

FIG. 1. Dynamic structure factor for density fluctuations in a classical monatomic gas; ℓ is the atomic mean free path and u is the velocity of sound.

discussed in the previous section. A spectrum of this type can be obtained by the Brillouin scattering of laser light from a monatomic gas. Notice that in the hydrodynamic region there are distinct contributions from the two normal modes of the system, sound waves and thermal diffusion, but in the kinetic region such a distinction no longer makes sense.

Light scattering has been used to study the dynamics of density fluctuations in gases, liquids, superfluids, and solids. It has also been applied to concentration fluctuations in solutions and orientation fluctuations in liquid crystals. Neutron scattering has studied fluctuations in the magnetization of ferromagnets and antiferromagnets, and fluctuations in the local atomic order associated with structural phase transitions. Inelastic x ray and electron scattering are also being applied to studies of similar fluctuation phenomena, but often the fluctuations studied are on too microscopic a scale to be classified as thermodynamic fluctuations.

DIVERGENCE OF THE FLUCTUATIONS

It was pointed out earlier that usually we need only deal with the mean of a thermodynamic variable when discussing macroscopic phenomena, since the amplitude of the fluctuations relative to the mean is always inversely proportional to the square root of the number of particles involved. However, in some circumstances, usually at a second-order phase transition, the constant of proportionality (some thermodynamic derivative) diverges. For example, the isothermal compressibility [Eq. (2)] diverges at the gas–liquid critical point, the isothermal magnetic susceptibility [Eq. (4)] diverges at the Curie temperature of a ferromagnet, and the osmotic compressibility $(\partial c_a/\partial \mu')|_{T,P}$ [Eq. (5)] diverges at the consolute point of a binary mixture. In such cases the fluctuations themselves dominate the dynamics of the system and the equations of motion dealing only with the mean values are no longer valid. Exactly how the medium does behave in these circumstances is not well understood. It is the subject of a currently active field of research known as dynamic critical phenomena. The clearest example of this fluctuation-dominated behavior is the critical opalescence of a simple fluid at its critical point. The density fluctuations scatter so much light that the fluid becomes translucent or opaque. It is interesting that this particular phenomenon, which first attracted physicists to the study of thermal fluctuations at the turn of the century, is still the subject of frontier research in physics.

See also CRITICAL POINTS; DYNAMIC CRITICAL PHENOMENA; LIGHT SCATTERING; STATISTICAL MECHANICS; THERMODYNAMICS, EQUILIBRIUM.

BIBLIOGRAPHY

Theory of Thermodynamic Fluctuations

L. D. Landau and E. M. Lifshitz, *Statistical Physics*, Chapter 12. Addison-Wesley, Reading, Mass., 1958.

Mathematics of Random Variables and Random Processes

A. Papoulis, *Probability, Random Variables, and Stochastic Processes*. McGraw-Hill, New York, 1965.

Time Dependence of Fluctuations

R. D. Mountain, *Rev. Mod. Phys.* **28,** 205 (1966).

S. Yip, *J. Acoust. Soc.* **49,** 941 (1971).

Scattering Spectroscopy

W. Marshall and S. W. Lovesey, *Theory of Thermal Neutron Scattering*. Oxford, London and New York, 1971.

P. A. Fleury and J. P. Boon, "Laser Light Scattering in Fluid Systems," *Adv. Chem. Phys.* **24,** 1–93 (1973).

Critical Phenomena

H. E. Stanley, *Introduction to Phase Transitions and Critical Phenomena*. Oxford, London and New York, 1971.

Fluid Physics

Daniel Bershader

WHAT IS A FLUID?

Matter exists in two forms, solid and fluid, the latter, in turn, being divided into the categories of liquids and gases. None of these distinctions is really sharp, but there is a gradation in both spacing and latitude of motion of the atomic components as one proceeds from solids to gases. Intermolecular forces play a dominant role in the closely spaced atomic arrangements in a solid, a smaller role in a liquid, and are much smaller in a gas. As a consequence, matter in the fluid state will generally deform continuously when subjected to a shear stress, no matter how small; such behavior can, in fact, be taken as a working definition of a fluid. For present purposes we will only mention but otherwise ignore more complex materials which show a dual character, such as jellies or paints; these may behave as elastic solids when allowed to stand, but lose their elasticity and behave as a liquid when strongly disturbed by an outside force.

Although fluid properties and related dynamic behavior are ultimately explained in terms of molecular parameters, the largest segment of fluid physics deals with fluid behavior on a macroscopic scale, i.e., large compared with the average molecular spacing. Stated differently, even "local" fluid behavior from a macroscopic view will involve enough molecules so that fluctuations will be smoothed out with no disturbing effect on the measured average values of physical quantities.

If one notes that gases with significant charged-particle content, i.e., plasmas, are included in the fluids category, then it turns out that more than 99.99% of the matter in the universe is in the fluid state. In our own earthly existence, there are many technical subject areas based on fluid physics. They include aerodynamics; thermal hydraulics of nuclear reactors; aerosols and emulsification for fluid transport purposes; meteorological flows; ballistics and atmospheric penetration by hypervelocity vehicles; flow in pipes, turbomachinery, and open channels; energy transfer processes such as magnetohydrodynamic flows, combustion, and propulsion; cavitation associated with behavior of ship propellers or entrance of a solid body into a fluid; flow through porous media in connection with oil extraction and other geophysical applications; and biofluid mechanics, to mention just a few. In what follows we will discuss the more important features of fluid physics which are basic to an understanding and utilization of fluid phenomena.

FLUID PROPERTIES

Here, we treat briefly a few properties that are of special importance for fluid motion. The two which appear most frequently in flow studies are density and viscosity. The fluid density is basic to the analysis of buoyancy and other static fluid phenomena, and contributes to the inertial behavior of fluids in dynamic situations. Thus, drag forces on solid bodies are proportional to fluid density in certain flow regimes and more complex functions of density in other regimes. Heat exchange between immersed bodies and flowing fluids is also a function of fluid density. In flows of gases with heat exchange or at velocities comparable to or larger than the speed of sound in the fluid, the density changes from a constant to a variable. This fact is of special importance because such density variation or "compressibility" couples the flow dynamics with the flow energetics, resulting in a behavior which is quite different from incompressible, isothermal flows.

A fluid bounded by two parallel planes in relative parallel motion will undergo a shearing strain in response to the applied stress. Propagation of the stress is possible because any real fluid, unless very highly rarefied, "sticks" to the wall and exhibits internal friction as well. Such behavior is termed viscosity and stems from the transfer of molecular momentum among flow regimes of different fluid velocity. For a wide range of parameters, a large variety of fluids will strain at a rate which is proportional to the applied stress. Thus, if the planes just mentioned are parallel to the x and z axes of an orthogonal coordinate system and there is steady relative motion in the x direction, then the strain is expressed as a y gradient of the velocity u. The stress in this case may be denoted by τ_{yx} and it is given by

$$\tau_{yx} = \mu \frac{du}{dy}, \tag{1}$$

where the proportionality factor μ is termed the coefficient of viscosity. A so-called Newtonian fluid is one for which μ is constant, and with which the present article is concerned. There are, however, several important non-Newtonian fluids such as blood, paint, and plastics for which μ is not constant, and for which the above relation must be modified in other ways. Note that the fluid pressure does not appear in the above relation, showing how completely different fluid friction is as compared with solid friction. Note further that the relation is also independent of fluid density, which implies an independence from macroscopic inertial fluid features as well.

Viscosity is a function of temperature, increasing for gases and decreasing for liquids. In the temperature range between $-100°C$ and $+300°C$, the viscosity of air is well described by

$$\frac{\mu}{\mu_r} = \left(\frac{T}{T_r}\right)^{0.76}, \tag{2}$$

where μ_r and T_r are reference values.

The ratio of viscosity coefficient to fluid density, μ/ρ, is called the *kinematic viscosity*, usually denoted by ν. Its dimensions of $(\text{length})^2/\text{time}$ consist, in fact, of kinematic quantities and are those of a diffusion coefficient. It is the kinematic viscosity which determines the diffusive spread of fluid straining behavior originating at a local disturbance in an otherwise uniform flow. Such transport is of the same nature as that of heat conduction or of mass diffusion in an inhomogeneous fluid mixture. The latter two properties play an important role, respectively, in fluid heat transfer behavior and in chemically reactive fluid flows involving mixing. The basic metric unit of ν is the *stokes*, which is 1 cm^2/sec, as compared with the metric unit for viscosity μ, the *poise*, denoting 1 dyn s/cm^2. Tabulated values are often given in centipoises or centistokes, which are units 100 times smaller than those quoted above (Table I).

There is a problem in uniquely specifying fluid physical properties associated with viscous, thermal, or mass transport, as just described; it relates to the effect of flow turbulence. The latter, which is discussed more fully later, is characterized by local, rather randomized fluctuations of physical quantities, and includes, as well, a pronounced effect on macroscopic gradients of such quantities as temperature and velocity. The result is that turbulent diffusive transfer may be an order of magnitude greater than so-called transfer of the laminar type, the latter term referring to a flow which involves gradients but which is macroscopically smooth.

An important property of a fluid is its isentropic compressibility, denoted by β_s, a quantity which is essentially a ratio of fractional density change per unit change of applied pressure under conditions of constant entropy:

$$\beta_s = \frac{1}{\rho}\left(\frac{\partial\rho}{\partial p}\right)_s. \tag{3}$$

This quantity determines the speed of small-amplitude disturbances, i.e., sound waves, in the fluid in accordance with

Table I Viscosities of Some Common Fluids

Temperature (°C)		Viscosity (centipoises)
Water	0	1.798
	20	1.0019
Alcohol	20	1.200
Benzene	20	0.652
Glycerin	20	1.490
Liquid solium	250	0.381
	Gases	
Air	0	0.0171
	20	0.0183
	40	0.0190
Argon	20	0.0222
Ammonia	20	0.0098
Carbon dioxide	20	0.0148
Methane	20	0.0109
Oxygen	19	0.0202

the relation

$$a^2 = 1/\rho\beta_s. \tag{4}$$

The value of β_s for water at 20°C and a pressure of 1 atm is 46×10^{-12} cm²/dyn, and the corresponding sound velocity is 1470 m/s, about 4.3 times the corresponding sonic velocity in air.

For most gases of interest, the density and pressure equation relating two states connected by an isentropic change is

$$\frac{p}{p_1} = \left(\frac{\rho}{\rho_1}\right)^\gamma, \tag{5}$$

where γ is the ratio of specific heats at constant pressure and at constant volume, respectively. The equivalent expression for the compressibility is $(\gamma p)^{-1}$, so that the velocity of sound a is now given by

$$a = (\gamma p/\rho)^{1/2}. \tag{6}$$

Recalling that p, ρ, and temperature T are related by the equation of state

$$p = \rho RT, \tag{7}$$

where R is the gas constant per unit mass, we obtain the sound velocity in a gas as a function of temperature only:

$$a = (\gamma RT)^{1/2} = 20.05T^{1/2} \tag{8}$$

for air, with T expressed in degrees kelvin. The sound speed in air at 293 K or 20°C is 343 m/s.

Other principal physical features of fluids which can only be mentioned in the present article, but are adequately treated in the literature, include surface tension and capillarity, thermal conductivity, and additional thermophysical properties such as critical values of pressure, molar volume, and temperature, at which the densities of coexisting liquid and gas phases become identical. At temperatures above the critical temperature a gas cannot be liquified.

FLUID DYNAMICS

First a few remarks on fluid kinematics. In contrast to the so-called Lagrangian description of motion typical of classical physics, the description of fluid flow is termed Eulerian, i.e., instead of following a particle of fixed identity, one relates the changes experienced by a fluid element to the configuration of the flow field. The laws of motion are expressed in terms of the Eulerian or "material" derivative of physical quantities, D/Dt, given by

$$\frac{D}{Dt} = \frac{\partial}{\partial t} + \mathbf{V}\cdot\nabla, \tag{9}$$

i.e., by the sum of the local and convective derivatives.

To express variations in a fluid velocity field, we relate the velocity components (u,v,w) at two nearby points in terms of a Taylor series expansion. Thus, the appropriate first-order terms form a nine-component array $\partial u_i/\partial x_k$.

Application of Newton's second law to a Newtonian fluid, together with certain basic assumptions about the nature of fluid stresses and about viscosity coefficients for shearing and for longitudinal stresses, respectively, yields the following basic relation for the x component of the substantial acceleration:

$$\frac{Du}{Dt} = \frac{\partial u}{\partial t} + \mathbf{V}\cdot \operatorname{grad} u, \tag{10}$$

in a flow field where each component of total velocity $\mathbf{V}$ may be a function of all the space coordinates and the time as well:

$$\rho\frac{Du}{Dt} = f_x - \frac{\partial p}{\partial x} + \frac{\partial}{\partial x}\left[\mu\left(2\frac{\partial u}{\partial x} - \frac{2}{3}\operatorname{div}\mathbf{V}\right)\right] + \frac{\partial}{\partial y}\left[\mu\left(\frac{\partial u}{\partial y} + \frac{\partial v}{\partial x}\right)\right] + \frac{\partial}{\partial z}\left[\mu\left(\frac{\partial w}{\partial x} + \frac{\partial u}{\partial z}\right)\right] \tag{11}$$

with equivalent expressions for the y and z accelerations. Here, (u,v,w) are the velocity components and f_x is the x component of any body force, per unit volume, such as gravity, acting on the fluid. These relations are known as the Navier–Stokes equations. Comparatively few solutions of the general equations for stipulated boundary and thermodynamic conditions are known, but many problems of interest usually permit application of these relations in a simplified form.

Hydrodynamic (liquid) flow and gas flows at speeds considerably below the sonic value and with no heat exchange are essentially incompressible, i.e., the density is constant. In addition, such flows will be isothermal as well in the absence of heat sources or sinks, thus decoupling the dynamic analysis from that of the energetics. Because the temperature does not vary, the viscosity coefficient may then be taken as constant, and the equation for conservation of mass flow,

$$\frac{\partial\rho}{\partial t} + \operatorname{div}(\rho\mathbf{V}) = 0, \tag{12}$$

requires that the quantity div $\mathbf{V}$ vanish. We can now write the x component of the Navier–Stokes equation as

$$\frac{Du}{Dt} = -\frac{1}{\rho}\frac{\partial p}{\partial x} + \nu\left(\frac{\partial^2 u}{\partial x^2} + \frac{\partial^2 u}{\partial y^2} + \frac{\partial^2 u}{\partial z^2}\right) + f_x. \tag{13}$$

For a nonviscous flow, the second-order terms in the above relation disappear and it reduces to the so-called Eulerian equation of motion. Integration of the Euler equation along a streamline (streamlines are the curves which are everywhere tangent to the direction of the velocity vector) for the case of a conservative body force derivable from a potential energy U per unit mass, i.e., $f = -\operatorname{grad} U$, yields

$$\frac{\partial}{\partial t}\int \mathbf{V}\cdot d\mathbf{r} + \frac{V^2}{2} + \int\frac{dp}{\rho} + U = \text{constant along streamline}, \tag{14}$$

where $d\mathbf{r}$ is the differential path along the streamline. Now if the flow is, in addition, irrotational, i.e., $\nabla\times\mathbf{V} = 0$, then the velocity is derivable from a potential ϕ in accordance with $\mathbf{V} = \operatorname{grad}\phi$. Then the equation becomes (in three dimensions)

$$\frac{\partial\phi}{\partial t} + \frac{V^2}{2} + \int\frac{dp}{\rho} + U = \text{const.}, \tag{15}$$

the same constant holding everywhere in the flow, rather than varying from streamline to streamline. For steady, incompressible flow, the above relation takes the form

$$\frac{V^2}{2} + \frac{p}{\rho} + U = \text{const.}, \tag{16}$$

which is the familiar form of Bernoulli's equation.

Incompressible, inviscid flow deals, then, with studies of velocity or pressure distributions in a flow field, the two being related by the Bernoulli equation. The use of velocity potential ϕ, defined by

$$\mathbf{V} = \nabla\phi, \tag{17}$$

or of the stream function ψ is a mathematical aid for the analysis. Thus, an irrotational flow is completely specified by the scalar function ϕ instead of the three velocity components.

On the other hand, the stream function also gives the complete velocity distribution in a two-dimensional flow whether irrotational or not. Its relation to the velocity components, namely

$$u = -\frac{\partial\psi}{\partial y}, \qquad v = \frac{\partial\psi}{\partial x}, \tag{18}$$

is a consequence of the equation of continuity, div $\mathbf{V}=0$. The value of ψ is constant along streamlines and has physical meaning which relates to the mass flow rate in a stream-tube bounded by two streamlines.

In spite of the implied restrictions, two-dimensional, inviscid, incompressible, and irrotational flow fields represent a central area of fluid dynamics with important applications to aerodynamics. The point is that with no shearing stresses acting on a fluid element, the net pressure force passes through its center of mass, and thus there are no turning moments. Therefore, if the fluid is not rotating initially, it will not begin to do so. A body accelerated from rest to a steady velocity in stationary air will experience irrotational motion. Even in the case of a real fluid such as air, flows at typical aeronautical velocities are such that the rotational behavior due to viscosity is limited to narrow regions near the flow boundaries, the so-called boundary layers and near wakes, while the major parts of the flow field may be treated as irrotational in their contributions to the lift and drag on an immersed body.

The use of the velocity potential has been extended to compressible, irrotational flow; it obeys the relation

$$\frac{\partial^2\phi}{\partial t^2} + \frac{\partial}{\partial t}(\nabla\phi)^2 + \tfrac{1}{2}\nabla\phi\cdot\nabla(\nabla\phi)^2 - a^2\nabla^2\phi = 0, \tag{19}$$

where a is the local sound velocity. This is a difficult nonlinear equation but there are two important special cases: For incompressible flows $a \to \infty$, and we have Laplace's equation $\nabla^2\phi = 0$. On the other hand, for small-amplitude motions of a compressible flow, we obtain the wave equation

$$\frac{\partial^2\phi}{\partial t^2} - c^2\nabla^2\phi = 0. \tag{20}$$

Linearization of the ϕ equation has also been utilized to solve for a wide range of small-disturbance, steady supersonic flows, e.g., thin airfoils immersed in a uniform flow field. Assume that the airfoil chord lies along the x axis and that the undisturbed flow far away from the airfoil moves in the x direction at a Mach number M_∞ relative to the airfoil. Then ϕ satisfies the relation

$$(1 - M_\infty^2)\frac{\partial^2\phi}{\partial x^2} + \frac{\partial^2\phi}{\partial y^2} + \frac{\partial^2\phi}{\partial z^2} = 0. \tag{21}$$

For subsonic flow, the compressibility feature is accounted for by an affine transformation, which again reduces the problem to that of solving Laplace's equation. However, when $M_\infty > 1$, the ϕ equation becomes hyperbolic or wavelike. The flow field is now qualitatively different. The disturbance represented by insertion of the airfoil into the flow is propagated along so-called characteristic or Mach lines. The angle μ between the characteristic and flow direction at any point is related to the local flow Mach number M by

$$M = \operatorname{cosec} \mu. \tag{22}$$

A "point" disturbance in such a flow would generate a Mach cone comprised of such characteristics. Existence of the disturbance would be known only within the downstream confines of the conical surface. Physically speaking, we are saying that sound signals cannot travel upstream in a supersonic flow.

FLUID ENERGETICS

Application of energy conservation to the flow of a viscous, heat-conducting fluid yields a relation which is conveniently formulated in terms of change of stagnation or total enthalpy per unit mass h_s, defined by $h_s = h + V^2/2$; in turn, h, the specific static enthalpy, is defined as usual in terms of the specific internal energy e, pressure p, and fluid density ρ by $h = e + p/\rho$. Using the summation convention for repeated indices, we may write the energy equation as

$$\rho\frac{Dh_s}{Dt} = \frac{\partial p}{\partial t} + \frac{\partial}{\partial x_k}(\tau_{ik}u_i - q_k) + f_k u_k + Q, \tag{23}$$

where

$$q_k = -\kappa\frac{\partial T}{\partial x_k}, \text{ the Fourier heat-conduction flux;} \tag{24}$$

f_k = component of body force per unit volume;

Q = power per unit volume exchanged with the outside environment, e.g., radiative energy injected by a laser pulse;

and

τ_{ik} = stress tensor, assumed linearly related to the fluid strain in terms of the viscosity coefficient μ by

$$\tau_{ik} = \mu\left\{\frac{\partial u_i}{\partial x_k} + \frac{\partial u_k}{\partial x_i} - \tfrac{2}{3}\delta_{ik}\nabla\cdot\mathbf{V}\right\}, \tag{25}$$

in which $\delta_{ik} = 1$ for $i = k$ and zero otherwise.

Apart from the term $\partial p/\partial t$ which vanishes in steady flow, the energy relation shows the contributions to the change of total enthalpy of the "work" of the viscous stresses, of heat conduction, of the work done by the body forces, as well as energy transfer with outside systems. In most applications over a wide range of flow conditions, the relation can be suitably simplified to calculate heat transfer rates, temperature distributions, and related quantities.

Some manipulation of the equation yields an energy relation in a form which takes us directly from the first to the second law of thermodynamics. We recall that the differential change of specific entropy s associated with changes in enthalpy and pressure is given by

$$ds = dh - dp/\rho. \tag{26}$$

The energy equation reformulation is

$$\frac{Ds}{Dt} = \frac{Dh}{Dt} - \frac{1}{\rho}\frac{Dp}{Dt} = \frac{1}{\rho}\tau_{ik}\frac{\partial u_i}{\partial x_k} - \frac{1}{\rho}\frac{\partial q_k}{\partial x_k} + \frac{Q}{\rho}. \tag{27}$$

Apart from outside energy addition, we see explicitly that viscous stress "work" (really heat production) and heat conduction increase the entropy and are therefore dissipative in nature. Conversely, in adiabatic flows without viscous or heat conduction effects, we have $Ds/Dt = 0$; i.e., the entropy is constant along streamlines and in that sense the flows are reversible.

SHOCK WAVES

A shock wave is a front across which there is a nearly discontinuous, finite jump in pressure, with corresponding jumps in temperature, density, and other fluid properties. Shadow photographs showing such waves ahead of or in the wake of high-speed projectiles have been widely published (see Fig. 1), and they are commonly seen in the neighborhood of supersonic nozzle exhausts such as rocket nozzles during the launching of space vehicles. Such waves may be changing with time as part of an unsteady flow field, but they are also found in steady flow and may be "made stationary" by a suitable Galilean transformation of the fluid flow equations.

The nonlinear nature of fluid flows is central to the formation of shocks and to the understanding of their behavior. Consider a sinusoidal sound wave of sufficient intensity that (because sound waves propagate adiabatically) the temperature variation across the wave is non-negligible. This means that the higher-temperature portion of the wave will move faster than the lower-temperature portion; thus, the compressive part of the wave will steepen to form a "vertical" pressure front or shock. Shocks, being waves, are formed only in hyperbolic-type, i.e., supersonic, flow fields. In portions of such flows where the pressure is increasing, there is a tendency for the characteristic lines, themselves described by a linear equation, to merge into an envelope, thus creating a shock.

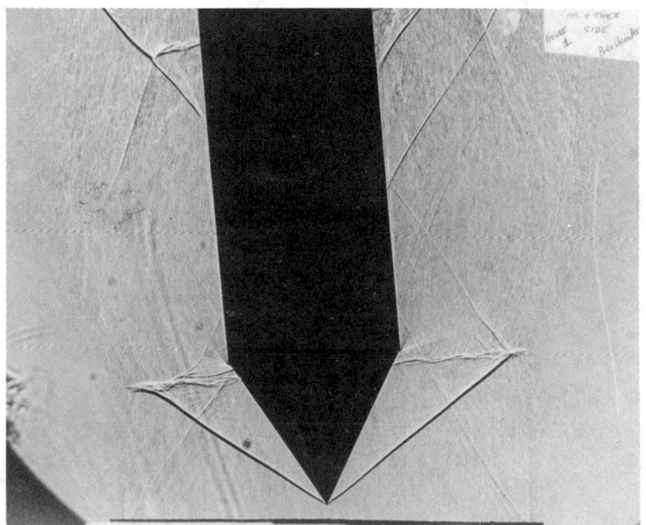

FIG. 1. Shadowgraph showing bow shock wave on a 30° half-angle cone cylinder model immersed in a supersonic flow of Mach number 1.7 at an angle of attack of 5°.

Because shocks are nonlinear, analysis of their behavior is complex, e.g., reflection of a shock wave does not obey any such simple laws as equality of the incident and reflected angles. Indeed, under certain parametric conditions, one finds an unexpected triple shock configuration with a so-called Mach stem. Other types of shock-wave interaction are also more tenuous because linear superposition does not apply. Flows entering a shock at an oblique angle are deflected through a finite angle away from the normal direction. The deflection is due to the decrease of the normal component of the fluid velocity entering the wave. The conservation equations for the normal shock transition of an otherwise steady and uniformly flowing fluid are

$$\begin{aligned} &\text{Mass:} && \rho_1 V_1 = \rho_2 V_2 = \text{const.}; \\ &\text{Momentum:} && p_1 + \rho_1 V_1^2 = p_2 + \rho_2 V_2^2 = \text{const.}; \\ &\text{Energy:} && h_s = h_1 + V_1^2/2 = h_2 + V_2^2/2 = \text{const.}; \end{aligned} \tag{28}$$

where h_s is the total or stagnation enthalpy. For the case of a thermally perfect ($p = \rho RT$) and calorically perfect gas flow ($h = c_p T$, where c_p is constant), the above relations can be manipulated to give the Rankine–Hugoniot relation relating the pressure and density ratios across a shock:

$$\frac{p_2}{p_1} = \left(\frac{\gamma+1}{\gamma-1}\frac{\rho_2}{\rho_1} - 1\right) \Big/ \left(\frac{\gamma+1}{\gamma-1} - \frac{\rho_2}{\rho_1}\right) \tag{29}$$

In the limiting case of very weak shocks with $p_2/p_1 \to 1$, the above relation approaches the isentropic formula [Eq. (5)]. For a shock of nonvanishing strength, the finite jump from state 1 to state 2 is irreversible and the entropy change $s_2 - s_1$ is positive and increases with pressure ratio.

The shock transition for ideal gases increases in strength with the Mach number of the incident flow relative to the shock, M_1; the relation between fractional pressure change and M_1 is

$$\frac{p_2 - p_1}{p_1} = \frac{2\gamma}{\gamma+1}(M_1^2 - 1). \tag{30}$$

Thus, if $M_1 = 3$, the pressure changes by an order of magnitude through the shock for the case of a diatomic gas such as air. The temperature increases by about 70%, and other quantities change correspondingly. The velocity across a normal shock always changes from supersonic to subsonic. The density jump is limited to $(\rho_2/\rho_1)_{\max} = (\gamma+1)/(\gamma-1)$ for the ideal gas case. Thus, for air the density jump can be no larger than 6 according to this formula. In practice, however, gases lose their ideal behavior at the elevated temperature conditions behind very strong shocks, and the density jump for air shocks may approach about 15. Note, by the way, that sonic boom disturbances produced on the ground by

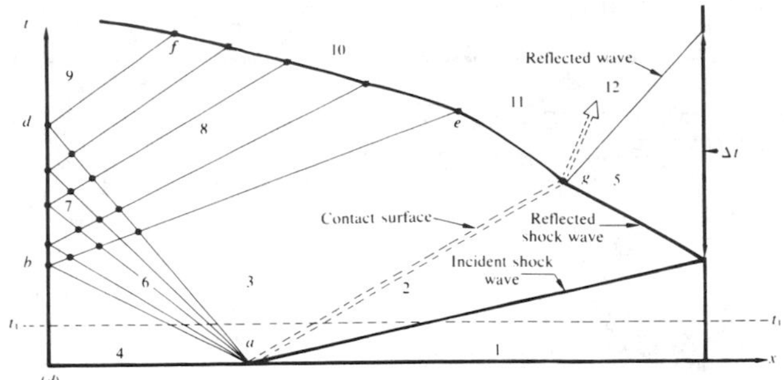

FIG. 2. Wave history in a shock tube given by x–t diagram showing wave paths after breakage of diaphragm at point a at $t = 0$. Note the reflection of the shock and rarefaction waves [latter shown as a fan, regime (6)] from the right- and left-hand ends of the tube, respectively; and the complex interactions which follow. [Reprinted by permission, from Maurice J. Ducrow and Joe D. Hoffman, *Gas Dynamics*, Vol. 2. Wiley, New York, 1977.]

supersonic aircraft passing by are rather weak shocks, with changes of pressure typically under 10%.

Many shocks of common experience are unsteady, decaying with time. This is true of thunder associated with electrical storms, blast waves from explosions, and the bursting of a balloon or pressure vessel. The latter type of process is best illustrated with the help of a wave diagram showing the events in a very useful experimental device called a shock tube, following rupture of the diaphragm separating the low-pressure chamber from the high-pressure section. Figure 2 indicates a shock wave advancing into the low-pressure region while a rarefaction wave travels into the high-pressure section, also accelerating that gas in the positive x direction. The rarefaction wave will reflect from the end wall and begin to overtake the shock. The boundary between the gases in the high- and low-pressure chambers, respectively, called a contact surface, travels with the fluid velocity. Across the contact surface, fluid velocity and pressure are equal, but there is a jump in density, temperature, and entropy. Shock tubes are used for basic studies in chemical kinetics, radiative and convective energy transfer, physics of gases, aerodynamics, magnetohydrodynamics, and plasma behavior.

SUPERSONIC AND HYPERSONIC FLOWS

The existence of shock waves is perhaps the most dramatic feature distinguishing supersonic from subsonic flow. There are other basic physical differences, however, relating especially to the nature of the energy balance at low versus high speeds. In the former case (if we omit the application to heat exchangers), work done on or by the fluid results in a change of velocity, therefore kinetic energy, of the fluid; this is in accordance with the simple form of the Bernoulli relation, Eq. (16). Changes in temperature as well as density are small throughout the flow field. For the high-speed case, however, the thermal energy of the flowing fluid participates in the energy balance as well; that is, the moving fluid can show substantial variations in temperature. Such behavior has important consequences in applications to internal flows (e.g., nozzles and aircraft engine inlets) and also to external flows around aircraft, wind tunnel models, or turbine components. For example, air accelerating from rest at room temperature, say 77°F, to Mach 3 cools off to −267°F. On the other hand, a supersonic plane travelling at Mach 3 in an ambient atmosphere at 77°F will experience a large rise of temperature to 984°F across its bow shock wave.

At appreciably higher Mach numbers, these effects are substantially increased. Thus, above Mach 5, say, a nozzle expansion results in a freezing out of the main components of air, namely, nitrogen and oxygen. At suitably high aircraft speeds, a whole host of high-temperature phenomena occur, relating to the fact that now the internal energy states of the molecules and atoms participate in the energy balance. They include vibrational excitation, dissociation, and ionization (the latter is responsible for the communication blackout which occurs during certain phases of astronaut reentry). Radiation takes place as well; that is, in certain velocity and altitude ranges, the shock layer (flow behind the bow shock) may be observed to glow. At high enough speeds, radiative heat transfer to the vehicle may be comparable or even substantially larger than the convective heat transfer more familiar to the aerothermodynamics engineer. For the Apollo project, the spacecraft returning from the moon entered the earth's atmosphere at about 11 km/s; for this case the peak radiative and convective heat transfers were comparable. As a more extreme example, the ultrahigh-speed entry of the Galileo probe into the Jovian atmosphere will result in peak heating conditions where the radiative heating is about four times as large as the convective heating. The phenomena just described are often categorized by the term *aerophysics*. The corresponding vehicle Mach numbers may be very large, approaching Mach 50 in some applications to transatmospheric flight. In this velocity range, the Mach number designation is somewhat less significant than the basic kinetic energy of the flow relative to the vehicle. However, the term *hypersonics* is widely used for high Mach number flows. *Hypervelocity flows* would be better terminology.

There are several special features of supersonic and hypervelocity flows, relating both to their aerodynamic and to their aerophysical behavior. For example, the existence of shock waves introduces the concept of wave drag which has to be included in the determination of the overall drag coefficient of the aircraft. Another practical problem which affects both drag and heat transfer is that of transition from laminar to turbulent flow in the so-called surface boundary layer; the empirical transition data which hold at supersonic speeds do not easily extrapolate to hypersonic case. Indeed, several aerodynamic phenomena need further study to determine behavior in the hypervelocity regime. For example, three-dimensional flows which are accented at high angles of attack show considerable cross flow over aircraft wings, accompanied by vortex bursting, effects which are important in determining aircraft performance features such as stall. The dependence of such flow configurations on Mach number is a subject of continuing research.

VISCOUS FLOW BEHAVIOR

The participation of viscous effects in the dynamics and energetics of fluid flows has already been indicated in the previous sections dealing with those topics. The relative importance of viscosity in a particular flow situation stems from the magnitude of the viscous terms as compared to other terms in the equations. Such comparison in the case of flow dynamics leads to a nondimensional term called the Reynolds number, which is (apart from a factor of 2) simply a ratio of dynamic pressure $\frac{1}{2}\rho V^2$ to viscous stress $\mu V/L$, L being a typical length, in the flow problem under study. Thus

$$Re = \rho VL/\mu. \tag{31}$$

The Reynolds number is the most important scaling parameter in aerodynamics and a controlling guideline for the overall behavior and analysis of a wide variety of fluid flows. Thus, consider the steady motion of a sphere immersed in a fluid. At low Reynolds numbers, say $Re = 1$ or less, we can ignore the inertial transfer of momentum to the fluid in comparison to the surface shear or fluid "friction." This behavior is known as Stokes flow and is familiar to physicists in connection with the motion of oil drops in Millikan's famous experiment. The drag force F on a sphere of radius a moving with velocity V is, for Stokes flow,

$$F = 6\pi\mu aV. \tag{32}$$

On the other hand, in a Reynolds number range around 10^4, the force is

$$F = c\rho a^2V^2, \tag{33}$$

where c is a constant somewhat less than but of order of magnitude of unity. For $Re > 10^5$, the drag force determination is complicated by the transition from laminar viscous to turbulent flow near the spherical surface and to changes in the location of flow separation from the surface. Note that the Stokes drag depends linearly on viscosity μ and velocity V, but not on density, i.e., not on the inertial feature of the flow. On the other hand, the higher Reynolds number formula does show a dependence on density ρ and a quadratic dependence on V but no dependence on viscosity.

Low-Reynolds-number flows, e.g., glycerine or molasses flowing through a pipe, show viscous effects everywhere in the flow field. However, at high Reynolds numbers, say 10^4 or higher, predominant regions of the flow away from boundaries can be treated on an inviscid basis, while near the boundaries relatively thin layers exist in which both viscous and inertial terms must be taken into account. These so-called boundary layers were first referred to by Prandtl in 1904, and by now have been studied in great detail. To illustrate, if x is the direction of a uniform steady flow parallel to a plane surface with leading edge at $x = 0$ (see Fig. 3), then the angular growth of the laminar boundary layer is given by the formula

$$\frac{\delta}{x} = \frac{A}{\sqrt{Re}}, \tag{34}$$

where δ is the thickness of the layer at a distance x from the leading edge, and A is a constant whose order of magnitude is unity. Since typical Reynolds numbers associated with aircraft and other flight vehicles range around several millions, it is evident that the boundary layers, even when modified by turbulence and other factors, are thin indeed. From a basic conceptual point of view, viscous behavior at the surface generates vorticity which diffuses into the flow at a rate determined by the viscous diffusivity $\nu = \mu/\rho$, while, at the same time, being swept along by the streaming fluid. The equation for δ/x gives the resultant of those two effects.

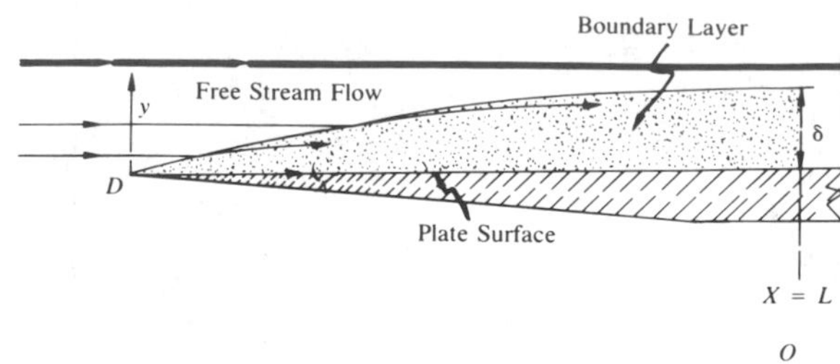

FIG. 3 Schematic diagram of boundary layer development over a flat surface immersed parallel to the flow in a uniform stream. Thickness of layer is much exaggerated for illustrative purposes.

At high speeds, compressibility effects produce temperature gradients and surface heat transfer to accompany the viscous shear in the boundary layer. The comparison of viscous and thermal behavior is represented by the ratio of viscous diffusivity ν to diffusivity associated with heat conduction, $\kappa/\rho c_p$, κ being the Fourier heat-conduction coefficient and c_p the specific heat at constant pressure. The ratio is known as the Prandtl number, denoted by σ:

$$\sigma = c_p\mu/\kappa. \tag{35}$$

For $\sigma = 1$, viscous and thermal boundary layers are equal in thickness, the viscous production of heat being balanced in a sense by the heat conduction effects. For air $\sigma = 0.74$, close enough to $\sigma = 1$ so that the resultant behavior of air boundary layers is not too different from the simplified case of unit Prandtl number.

TURBULENT FLUID FLOW

There are several examples of flow fields which represent solutions of the basic fluid equations but which are not stable with respect to small perturbations. This is especially true of viscous shear flows at high Reynolds numbers. The classical experiments on this problem were performed by Reynolds himself, who observed the rapid breakup of dye filaments in liquids flowing through long straight pipes. The phenomenon represented a transition from laminar to turbulent flow. Stability and transition problems have since received much attention, but should be considered distinct from that of the description and understanding of fully developed turbulent flow.

There is no precise definition of turbulent flow. Many of its properties are unmistakable, however. In contrast to the deterministic nature of a laminar flow, a turbulent flow shows random or stochastic behavior. Physical quantities such as the velocity or vorticity show random fluctuations in time and space, but in such a way that meaningful statistical averaging can be applied to provide at least a partial description of the flow. Associated with these fluctuations is a granularity or "lumpiness" pattern of "eddies" whose size provides a scale for the turbulence. Although such eddies vary in size from the time they are energized and agglomerate to when they disintegrate, the scale of turbulence is in any case much larger than molecular dimensions, i.e., it is macroscopic in that sense.

The largest turbulent eddies are determined by the size of the apparatus, say by the pipe diameter in the case of turbulent pipe flow. These relate to energy injection into the turbulent fluctuations (consider a spoon stirring coffee in a cup). Eddy interactions then take place by a complex nonlinear process which ultimately results in much smaller eddies. The latter lose energy by viscous dissipation. The statistical specification of the sizes of both the large and small eddies utilizes the double correlation coefficient for the velocity components at two points in the flow. In turn, the frequency spectrum of the fluctuations is the Fourier transform of the autocorrelation function of the fluctuating velocities.

Following Reynolds, the analysis of turbulent flows is usually performed by separating the mean fluid properties from the fluctuations with respect to the mean values. In such formulations it is assumed that the scales of motion and energy balance are not changing in such a way as to indicate a really unsteady flow; and therefore the flow is quasisteady in the sense that time averages give definable mean values if the averaging time is long enough. We write

$$\tilde{U}_i = U_i + u_i,$$

where $\tilde{U}_i$ and U_i are the total and mean values of the ith velocity component, and u_i is its fluctuating value. Substituting into the Navier–Stokes equation [see Eq. (11)], and recalling that averages of products such as $\tilde{U}_i\tilde{U}_j$ are given by

$$\overline{\tilde{U}_i\tilde{U}_j} = U_iU_j + \overline{u_iu_j}, \tag{36}$$

we obtain for the momentum equation of ith component of the mean flow (assumed steady and incompressible for simplicity)

$$U_j\frac{\partial U_i}{\partial x_j} = -\frac{\partial P}{\partial x_i} - \frac{\partial}{\partial x_j}(\overline{u_iu_j}) + \nu\nabla^2 U_i, \tag{37}$$

where P is the mean pressure and where the repeated-index summation convention has been used. This differs from the corresponding equation for laminar, steady, incompressible flow by the presence of fluctuation-velocity correlation terms of the type $\overline{u_iu_j}$. It is as if additional stresses $-\overline{u_iu_j}$ were added to those already present in laminar flow. They are indeed termed the Reynolds stresses. The off-diagonal elements of this tensor are shearing stresses which effect transfer of mean momentum by the turbulent motion.

The energy equation for the type of flow just discussed may be written

$$\tfrac{1}{2}U_i\frac{\partial}{\partial x_i}\overline{q^2} = -\overline{u_iu_j}\frac{\partial U_i}{\partial x_j} - \frac{\partial}{\partial x_k}(\overline{u_kq^2}) - \frac{1}{\rho}\overline{u_k\frac{\partial p}{\partial x_k}} + \overline{\nu u_k\nabla^2 u_k}, \tag{38}$$

in which $q^2 = u_iu_i$, the total turbulent kinetic energy per unit mass; p is the fluctuating static pressure; and ν is the kinematic viscosity. The first term represents the rate of convection of turbulent energy by the mean motion, while the second term expresses the rate of production of turbulent energy from mean motion energy by the Reynolds stresses. The third term gives the rate of convection of turbulent energy by the turbulent motion, and the fourth represents turbulent energy transfer resulting from the work of fluctuating pressure gradients. Finally, the last term is equal to the difference between the rate of diffusive transport of turbulent energy by the viscous forces and the rate of dissipation of turbulent energy to heat.

Any turbulence modeling using extensions of the analysis just given must take into account major physical features of turbulent flows. Apart from the irregularities and granularity, the high level of diffusivity in a turbulent flow is of special importance in many practical applications such as fluid flows in processing plants, and dynamical and thermal design of flight vehicles. Transition from laminar to turbulent flow can increase by an order of magnitude or more the rates of momentum, heat, and mass transfer. For predicting this type of behavior, a purely statistical procedure is less useful than a phenomenological approach in which suitable modeling, based partly on dimensional analysis, is employed, typically including constants which have to be determined by experiment. Thus, to explain the highly increased friction in the case of turbulent as compared to laminar flow along a wall, one introduces an eddy viscosity, which can be visualized as corresponding to the product of a length scale times a velocity scale [units (length)^2/time]. The turbulent spreading of jets, for example, has been analyzed by assuming some scaled but constant value of the eddy viscosity across any plane perpendicular to the flow.

A phenomenological approach to turbulent length scales was made by Prandtl who introduced the concept of mixing length of a turbulent lump of fluid, in parallel with the molecular picture of viscosity resulting from "mean-free-path" exchanges of particles among flow layers of different momenta. It is the difference between the original velocity of the fluid lump and that of its "new" environment which is equated with the transverse velocity fluctuation in the fluid flow. Considering a mean flow U_1 in the x direction and principal gradient in the y direction, the mixing-length formulation leads to the following relation for the shearing stress τ_{xy}:

$$\tau_{xy} = \rho l^2\left|\frac{dU_1}{dy}\right|\frac{dU_1}{dy}, \tag{39}$$

where ρ is the density and l is the Prandtl mixing length. From dimensional considerations, we can see that the latter is related to the eddy viscosity ϵ mentioned above by

$$\epsilon = l^2\left|\frac{dU_1}{dy}\right|. \tag{40}$$

Prandtl's formulation has been usefully applied to turbulent boundary layers along solid surfaces and also to turbulent mixing of free flows such as the boundaries of a jet.

Currently, there is considerable interest in the technique of modeling turbulent shear flows with the aid of large-scale computers. The approach is to separate the large- from the small-scale structures. The former show a certain degree of coherent structure, and the attempt is made to model that part of the flow by applying the full Navier–Stokes equations to be solved with the aid of a computer. The smaller-scale or higher wave-number phenomena are treated with a more statistical type of formulation, with the necessity of cutting off the calculation at some upper wave number corresponding to the capabilities of the computer being utilized. Interesting results have been obtained for such flows as turbulent boundary layers along solid surfaces. One advantage of the approach is a better ability to handle the inherently three-dimensional nature of turbulent fluctuations with the help of the computer. It bears repeating, however, that no general solution of the physical equations is known for the case of turbulent flow. As of now, the equations simply do not give the entire story. No adequate model of turbulence currently exists.

Somewhat more recently the theory of chaos has been applied to the dynamics of turbulent fluid flow (as well as to several other scientific problems). Turbulence is an example of dynamical chaos, i.e., apparently random or stochastic fluid behavior. As we know, such motion typically develops from an initially laminar or regular flow representing the solution to a deterministic equation of motion with specified initial and/or boundary conditions. The apparent paradox relates to the extreme sensitivity of the motion to certain types of very small disturbances, amounting essentially to correspondingly small changes in, say, the initial conditions; that is, the regular fluid motion can be considered unstable with respect to such changes which, by the way, should be distinguished from random forcing functions. The growth of such instabilities is also intimately related to the underlying nonlinear character of the equations.

The concepts of dynamical chaos have supplied additional tools for the description of fluid turbulence and, hopefully, are providing a deeper perspective as an aid in the still-elusive attempts to understand the phenomenon. The mechanics of chaos analysis investigates the paths of dissipative systems in phase space, under circumstances where the dissipative energy loss is being balanced, in general, by energy gain from other sources. The phase-space trajectories take the form of orbits which are not closed, i.e., not periodic, but which tend to show a layered, ribbon-like structure. In that sense, the motion can be considered repetitive. The orbits occupy a limited region in phase space and show an inclination to converge to, or are "attracted" to, one or more structures within the region. The sets of such structures are referred to as *strange attractors,* or simply *attractors*.

A parallel concept used in the application of chaos theory to turbulence is that of fractional dimensions or fractals, originally introduced to describe the behavior of a very irregular coastline such as that of Great Britain. More recently, the fractal concept has been applied to the analysis of experimental data obtained from fluid-turbulence studies. In order for an irregular physical process to be treated in terms of fractal behavior, it should exhibit scale invariance. The latter is also a fundamental consideration in turbulence theory. It is probably too much to expect that the concepts of strange attractors, fractals, and others associated with chaos theory will fully clarify the underlying physics of turbulence. However, another "dimension" has now been added to our ways of assessing fluid turbulence, one that is providing further insight into the nature of this ever-challenging phenomenon. For further information on chaos theory, please refer to the article on CHAOS in this volume.

See also BOUNDARY LAYERS; CAPILLARY FLOW; CHAOS; CRITICAL POINTS; EQUATIONS OF STATE; FRACTALS; HYDRODYNAMICS; RHEOLOGY; SHOCK WAVES AND DETONATIONS; SURFACE TENSION; TURBULENCE; VISCOSITY.

Fourier Transforms

Yvon G. Biraud

In the early part of the nineteenth century the French physicist Joseph Fourier [1] studied the propagation of heat in solids [2]. He realized that the solution of this problem was greatly simplified if he decomposed the physical quantities to be calculated into sums of weighted trigonometric functions which bear his name. But the Fourier series are much more than a powerful mathematical tool. They cover the general notion of spectral analysis and that is why they have been widely studied and extended theoretically by mathematicians and used by physicists.

Fourier series had been devised to represent periodic functions and many restrictions appear on the properties these functions must satisfy to insure the convergence of the series [3,4,9,15]. In order to represent nonperiodic functions a new mathematical entity was necessary. It is the so-called Fourier transform (very often denoted as FT). The FT of $f(x)$ is most often defined as

$$F(u)=\alpha \int_{-\infty}^{+\infty} f(x)e^{i\beta ux}\,dx. \tag{1}$$

This formula needs some explanations:

- This integral is taken here in the Riemann sense. Its existence demands several conditions [3,4].
- The argument of the exponential being dimensionless u and x must have inverse dimensions.
- Most often, too, $\alpha = 1$ but $\alpha = 1/2\pi$ or $\alpha = 1/\sqrt{2\pi}$ are sometimes used.
- Usually $\beta = -2\pi$ but sometimes it is set to ± 1. But the choices of the values α and β and of their signs are of no importance for both theory and practice.

In formula (1), $F(u)$ is called the direct Fourier transform of $f(x)$. It is very interesting to note that under precise conditions [3,4], we may define the inverse FT of $F(u)$ as

$$f(x)=\alpha \int_{-\infty}^{+\infty} F(u)e^{\pm i\beta xu}\,du. \tag{2}$$

When this is possible $f(x)$ and $F(u)$ are called a Fourier pair. Many tables of Fourier (and or of Laplace) pairs have been calculated and are very useful [5–8]. Frequently Eq. (1) is rewritten as

$$F_c(u)+iF_s(u)=\alpha\left(\int_{-\infty}^{+\infty} f(x)\cos(2\pi\beta ux)\,dx \pm i\int_{-\infty}^{+\infty} f(x)\sin(2\pi\beta ux)\,dx\right)$$

where the real and imaginary parts are called Fourier cosine and sine transforms, respectively.

Most often in physics the conditions for the existence of the two integrals in Eqs. (1) and (2) are met. But sometimes they are not (mainly when discontinuities appear, when $f(x)$ is null almost everywhere, or when $f(x) \notin L$ or L_2). In these situations the calculation of the transform as a limit may be quite hazardous. Hopefully the introduction of distributions [4,9–11] has allowed a generalization of the notion of functions. This solves the problem of Fourier transforming functions like $e^{i2\pi ux}$, $H(x)$ (Heaviside's step function), $\delta(x-a)$ (Dirac's "impulse function"), the sampling function $Ш(x)=\sum_k \delta(x-k)$... . In this theory all the quantities appearing in physics now possess FTs.

This new mathematical concept allows us to deal with physical situations which were very difficult to handle otherwise such as:

- point masses in mechanics,
- point charge distributions in electrostatics,
- magnetic doublets,
- atoms or ions in crystalline structures [12],
- point sources or narrow slits in optics and astronomy [13],
- particles in quantum mechanics [14].

Moreover the use of the δ distribution, of the sampling distribution $Ш(x)$, and of the Heaviside step function (and their FTs) renders the Z- or Laplace transforms quite outdated or at least obsolete (and so unpractical).

The FTs of functions and distributions have been considerably investigated by both mathematicians and physicists who have derived many properties of this transform. Let us denote

$f(x) \rightleftharpoons F(u)$ a Fourier pair $[F(u)=\mathrm{FT}(f(x))]$

Then if $f(x)$ is real, we may draw up Table I. This table summarizes the most important properties of the FT. Formula (3) is also known as Parseval's or Plancherel's theorem. In Eq. (4) the derivative of $f(x)$ is taken in the distribution sense. Notice that we have described above the FT of functions or distributions of one variable. It may be easily defined and extended to functions operating on 2-D or 3-D spaces (but notice that some theorems established for 1-D do not apply any longer). Two-dimensional FTs are very frequently used in electromagnetism (antenna patterns, optics, astronomy, ...). Notice the existence of transforms related to the FT: Hankel or Mellin transforms [15].

A very important notion which is closely related to FT is the convolution. For two functions $f(x)$ and $g(x)$ it is defined by

$$h(x) = \int f(t)g(x-t)\,dt = \int g(t)f(x-t)\,dt \qquad (9)$$

and also denoted as

$$h(x) = f(x)*g(x) \qquad (10)$$

and may be extended to distributions. Its importance comes from the following statement: the output $h(x)$ of any feasible physical system which is continuous, x-invariant, and linear is related to its input $f(x)$ by a convolution such as in Eq. (9). Because $\delta(x)$ is the convolution unit

$$\phi(x)=\delta(x)*\phi(x) \qquad (11)$$

and if in Eq. (10) $f(x)=\delta(x)$, then

$$h(x)=g(x).$$

Hence $g(x)$ which is characteristic of the linear system is simply its impulse response. Now taking the FT of Eq. (10) provides

$$H(u)=F(u)\cdot G(u) \qquad (12)$$

[where $H(u)$, $F(u)$, and $G(u)$ are the FTs of $h(x)$, $f(x)$, and $g(x)$, respectively] which shows that a linear system simply multiplies the FT of its input by its complex transfer function

Table I

x-space	$\rightleftharpoons$ u-space (FT space)		
$f(x)$	$\rightleftharpoons F(u)$		
Real even	$\Rightarrow$ Real even		
Real odd	$\Rightarrow$ Imaginary odd		
Real (nor even, nor odd)	$\Rightarrow$ Hermitian		
$f(x-a)$	$\rightleftharpoons F(u)e^{-i2\pi au}\,du$	The shift theorem	
$f(ax)$	$\rightleftharpoons (1/\lvert a\rvert)F(u/a)$	The similarity theorem	
$\int \lvert f(x)\rvert^2\,dx$	$= \int \lvert F(u)\rvert^2\,du$	Rayleigh's theorem	(3)
$df(x)/dx$	$\rightleftharpoons i2\pi uF(u)$		(4)
$\delta(x)$	$\rightleftharpoons 1$	$\forall u$	(5)
$e^{i2\pi ax}$	$\rightleftharpoons \delta(u-a)$		(6)
$Ш(x)=\sum_k \delta(x-k)$	$\rightleftharpoons Ш(u)=\sum_k \delta(u-k)$	[see Fig. 1(a) and 1(A)]	(7)
$H(x)$ (Heaviside step function)	$\rightleftharpoons \frac{1}{2}[\delta(u)-i(1/\pi u)]$		
$f(x)\cdot g(x)$	$\rightleftharpoons F(u)*G(u)$	Convolution theorem [see Eq. (10) below]	(8)

$G(u)$. This may be shown in another way. If the input is $e^{i2\pi vx}$, Eqs. (6) and (12) yield

$$H(u) = G(u) \cdot \delta(u - v) = G(v) \cdot \delta(u - v)$$

which proves that complex exponentials are eigenfunctions of linear systems and may be used to determine the variations of attenuation and dephasing of a filter with frequency.

Equation (12) shows all the interest of the FT. The convolution is an integral relation [Eq. (9)] and the value $h(x_0)$, the output for a fixed x_0, is a weighted sum of all the values of the input. Whereas, if we consider the same problem in the Fourier domain, Eq. (12) shows that the value $H(u_0)$ is simply the multiplication of $F(u_0)$ by the known constant $G(u_0)$. This simplification is one of the reasons of the success of the FT and also allows us to face more easily the deconvolution problem which tries to solve Eq. (9): find the unknown $f(x)$ for given $h(x)$ and $g(x)$.

FT and convolution are absolutely essential to understand, modelize and investigate linear filtering in the wide sense. They apply to functions of time and to functions of space variables as well. The conjugate of time is the usual frequency when in the second case it is space frequency. But the same methods apply to both situations (optical filtering or heterodyne, for example).

Nevertheless it must be noticed that FT is not suitable for the study of time-varying spectra. Other techniques like AR or ARMA methods or Wigner–Ville representations are successful in this situation [16,17].

As an illustration, we shall emphasize now the efficiency of the use of distributions, of their FTs, and of the convolution, in solving the problem of the sampling of signals (Shannon's/Nyquist's theorem and interpolation). The introduction of the $\text{Ш}(x)$ distribution (see Table I) simplifies the derivations.

Let s denote the sampling rate on a signal $f(x)$ having a FT $F(u)$. The sampled $f(x)$ is

$$f_s(x) = f(x) \cdot \text{Ш}_s(x) = \sum_k f(ks)\delta(x - ks).$$

Fourier transforming this equation and using Table I yield

$$F_s(u) = F(u) * \frac{1}{s}\, \text{Ш}_{1/s}(u) = \frac{1}{s} \sum_k F(u) * \delta\left(u - \frac{k}{s}\right)$$

$$= \frac{1}{s} \sum_k F\left(u - \frac{k}{s}\right).$$

$F_s(u)$ is the sum of an infinity of shifted versions of $(1/s)F(u)$. With regard to multiplication the Ш distribution acts as a sampler and to the convolution as a replicating symbol. Two situations may arise:

1. $f(x)$ is such that the definition interval of $F(u)$ is $]-\infty, +\infty[$. See Figs. 1(b) and 1(B)
2. $\phi(x)$ is a so-called band limited spectrum function:

$$\Phi(u) \equiv 0 \qquad \forall u \notin [-u_M, +u_M].$$

See Figs. 1(d) and 1(D).

In both situations the FT $F_s(u)$ and $\Phi_s(u)$ are periodical with period $1/s$.

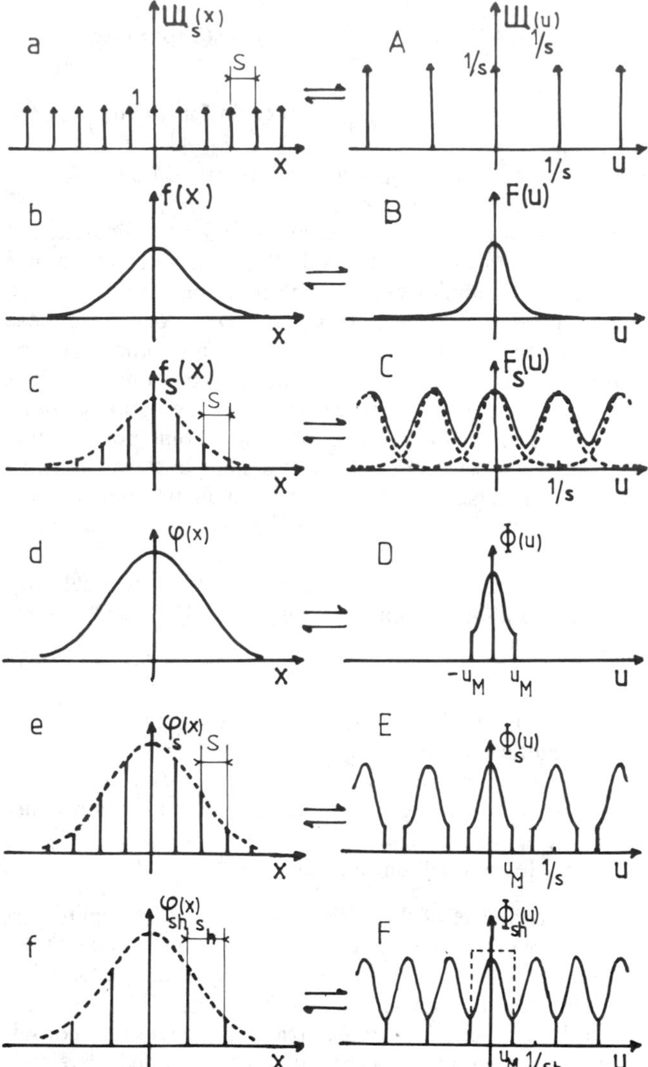

FIG. 1. (a), (A) The sampling function $\text{Ш}_s(x)$ and its FT $\text{Ш}_{1/s}(u)$. (b), (B) A first signal $f(x)$ and its FT $F(u)$ defined on $]-\infty, +\infty[$. (c), (C) $f_s(x)$, the sampled $f(x)$. [—] Its periodical FT: $F_s(u)$; (---) the different $(1/s)F(u - k/s)$. (d), (D) A second signal $\phi(x)$ and its FT $\Phi(u)$ null outside $[-u_M, u_M]$. (e), (E) A first sampling $\phi_s(x)$ and its FT $\Phi_s(u)$ composed of nonoverlapping $(1/s)\Phi(u - k/s)$. (f), (F) (—) The Shannon's sampling $\phi_{sh}(x)$. Its FT $\Phi_{sh}(u)$ composed of just touching patterns $(1/s_h)\Phi(u - k/s_h)$; (---) the window function allowing recovery of $\Phi(u)$.

In case 1, the different patterns $F(u - k/s)$ overlap [Fig. 1(C)], whereas in Case 2, the $\Phi(u - k/s)$ no longer overlap if $u_M \leq 1/2s$ [Fig. 1(E)]. This allows one to sample $\phi(x)$ with a larger period [Fig. 1(f)] which must satisfy

$$s \leq s_h = \frac{1}{2u_M}.$$

s_h is often called Shannon's sampling period. If $s = s_h$, then the different patterns $\Phi(u - k/s_h)$ are just touching [Fig. 1(F)] and $\Phi(u)$ can be recovered by just multiplying by the window function $(1/2u_M)\Pi(u/2u_M)$ $\{\Pi(\tau) = 1,\ \forall \tau \in [-\frac{1}{2}, \frac{1}{2}]$, and is null elsewhere$\}$. Hence

$$\Phi(u)=\Phi_{sh}(u)\,\frac{1}{2u_M}\,\Pi\left(\frac{u}{2u_M}\right).$$

Fourier transforming this relation yields the so-called Shannon's interpolation formula,

$$\phi(x) = \phi_{sh}(x) * \frac{\sin(\pi x/s_h)}{\pi x/s_h}$$

$$= \sum_k \phi(ks_h)\delta(x-ks_h) * \frac{\sin(\pi x/s_h)}{\pi x/s_h}$$

or

$$\phi(x) = \sum_k \phi(ks_h)\,\frac{\sin[\pi(x/s_h-k)]}{\pi(x/s_h-k)}$$

We shall now review different domains which are relevant to the FT. This list does not pretend to be exhaustive but tries to present its most common applications.

First, many situations are concerned with the product of a function $f(x)$ by its shifted version $f(x-a)$. If $f(x)$ is random, the FT of its autocorrelation yields the power spectrum which is of frequent use in noise analyses. In the same field the characteristic function of a random variable may be calculated as the FT of its density probability.

Now if $f(x)$ is no longer random but deterministic, the same process describes the general problem of interferometry. Two different examples:

- In radioastronomy two (or a 2-D array of) antennas measure the visibility function $V(u,v)$ which is Fourier transformed to get the brightness distribution $v(\alpha,\delta)$ of radio sources in the sky (u,v are spatial frequencies, δ and α declination and right ascension).
- In spectroscopy a Michelson interferometer measures the autocorrelation $I(\Delta)$ of the incoming radiation whose spectrum $B(\sigma)$ is the FT of $I(\Delta)$ (Δ: the path difference, σ: the wave number).

Another domain where FT is very useful is the diffraction of electromagnetic waves (radio or optics). Huyghens's principle is very easily applied in the Fourier formalism. FT is eventually the tool for a good understanding of image formation, phase contrast, holography, etc.

Very many phenomena in physics are naturally periodical and their study is evidently relevant to Fourier analysis:

- in acoustics or vibration mechanics;
- in astrometry: rotation of satellites or planets;
- in crystallography: ions or atoms are positioned at the nodes of a more or less regular array;
- in biology or pharmacology FT allows study of the rhythms

At last FT may be also considered as a tool:

- in pure or applied mathematics [18] and statistics [15],
- in electronics,
- in signal processing for which it represents a day-to-day tool [19],
- in image processing (associated with Hadamard, Walsh, or Haar transforms),
- in communications.

This multipurpose characteristic of the FT explains quite well why so much research on FT has been undertaken. The demand for fast processing was the origin of the success of the FFT (fast FT) algorithms [20,21] which now flourish and are theoretically investigated and practically widely used. This is perhaps the best sanction of an almost two-century-old discovery.

REFERENCES

1. J. Herivel, *Joseph Fourier*. Oxford (Clarendon Press), London and New York, 1975.
2. J. Grattan-Guiness, *Joseph Fourier 1768–1830*. MIT Press, Boston and London, 1972.
3. E. C. Titchmarsh, *The Theory of Fourier Integrals*. Oxford (Clarendon Press), London and New York, 1937.
4. J. Arsac, *Fourier Transform and the Theory of Distributions*. Prentice Hall, Englewood Cliffs, NJ, 1966.
5. A. Erdelyi, *Tables of Integral Transforms*. Vol. I. McGraw-Hill, New York, 1954.
6. F. Oberhettinger, *Tabellen zur Fourier Transformation*. Springer-Verlag, Berlin, Gottingen, Heidelberg, 1957.
7. J. Lavoine, *Transformation de Fourier des Pseudo-fonctions*. CNRS, Paris, 1963.
8. G. Harburn, C. A. Taylor, and T. R. Welberry, *Atlas of Optical Transforms*. Bell, London, 1975; Cornell University Press, Ithaca, NY, 1975.
9. L. Schwartz, *Mathematics for the Physical Sciences*. Addison-Wesley, Reading, MA, 1966.
10. I. M. Gelfand *et al.*, *Generalized Functions*. Academic Press, London, 1964.
11. M. J. Lighthill, *An Introduction to Fourier Analysis and Generalised Functions*. Cambridge University Press, Cambridge, 1958.
12. H. Lipson and C. A. Taylor, *Fourier Transforms and X-Ray Diffraction*. Bell, London, 1958.
13. L. Mertz, *Transforms in Optics*. Wiley, New York, London, 1965.
14. P. T. Matthews, *Introduction to Quantum Mechanics*. McGraw-Hill, New York, 1963.
15. R. Bracewell, *The Fourier Transform and its Applications*. McGraw-Hill, New York, 1965 (with an excellent bibliography).
16. P. Flandrin, "Some features of time frequency representations of multicomponent signals," IEEE Int. Conf. on Acoustics, Speech and Signal Processing, ICASSP-84, San Diego, 1984, pp. 41 B4.1–41 B4.4.
17. T. A. C. M. Claasen and W. F. G. Mecklenbräuker, "On the time-frequency discrimination of energy distributions: can they look sharper than Heisenberg?," IEEE Int. Conf. on Acoustics, Speech and Signal Processing, ICASSP-84, San Diego, 1984, pp. 41 B 7.1–41 B 7.4.
18. B. Davies, *Integral Transforms and their Applications*. Springer-Verlag, New York, Heidelberg, Berlin, 1978.
19. E. A. Robinson, *Statistical Communication and Detection*. Griffin, England, 1967.
20. S. W. Cooley and J. W. Tukey, *Math. Comput.* **19**, 297 (1965).
21. H. J. Nussbaumer, *Fast Fourier Transform and Convolution Algorithms*. Springer-Verlag, Berlin, Heidelberg, New York, 1982.

BIBLIOGRAPHY

N. Ahmed and K. R. Rao, *Orthogonal Transforms for Digital Signal Processing*. Springer-Verlag, Berlin, Heidelberg, New York, 1975.

H. S. Carslaw, *Introduction to the Theory of Fourier's Series and Integrals*. Dover, New York, 1930.
D. C. Champney, *Fourier Transforms and Their Physical Applications*. Academic Press, London, 1973.
S. H. Crandall and W. D. Mark, *Random Vibrations in Mechanical Systems*. Academic Press, London, 1963.
J. D. Gaskill, *Linear Systems, Fourier Transforms and Optics*. Wiley, New York, 1978.
J. W. Goodman, *Introduction to Fourier Optics*. McGraw-Hill, New York, 1968.
E. A. Guillemin, *Communication Networks*. Vol. II. Wiley, London, 1947.
H. P. Hsu, *Fourier Analysis*. Simon and Schuster, New York, 1970.
R.E.A.C. Paley and N. Wiener, *Fourier Transforms in the Complex Domain*. American Mathematical Society, New York, 1934.
A. Papoulis, *The Fourier Integral and Its Applications*. McGraw-Hill, New York, 1962.
K. Ramanohan Rao (ed.), *Discrete Transforms and Their Applications*. Van Nostrand Reinhold, New York, 1985.
J. P. G. Richards and R. P. Williams, *Waves*. Penguin, England, 1972.
I. N. Sneddon, *Fourier Transforms*. McGraw-Hill, New York, 1951.

Fractals

Fereydoon Family

Fractal geometry is a mathematical language and a quantitative approach for describing complexity in shapes, forms, and patterns. The word *fractal*, from the Latin *fractus,* was coined by Benoit B. Mandelbrot to denote highly irregular and fragmented objects or mathematical sets. In recent years fractal concepts have been developed into an effective mathematical tool and have been applied to a wide range of phenomena and processes of scientific and practical importance. Although Euclidean geometry and the theory of smooth functions can describe regular shapes and patterns (e.g., lines, planes, and differentiable functions), the concepts of fractal geometry are needed for understanding irregular shapes and forms as well as the behavior of extremely irregular mathematical functions. A major part of the success in the application of fractals is in describing disparate physical phenomena, which evolve through similar fundamental processes and therefore can be characterized by the same set of fractal attributes.

The most profound property of a fractal is its scale invariance or self-similarity. A fractal is a shape or pattern made of parts similar to the whole in some way. The reason is that a fractal object has no characteristic length scale. What this implies is that the essential features of a fractal exist at all length scales. Therefore, magnifying a small piece of a fractal results in a larger object; however, the pattern is similar. One of the fundamental quantitative measures of a fractal is its dimension. Unlike regular geometrical objects, fractals can have a dimension that is generally not an integer. For example, whereas a line has a dimension of 1, a tortuous path, like a river network or a long polymer chain in a solution, has a dimension somewhere between 1 and 2. A square or a triangle has a dimension of 2. But a rough surface, e.g., a mountain or the edge of a broken glass, can have a fractal dimension anywhere from 2 to 3.

The simplest fractals—called self-similar fractals—are invariant under isotropic dilation or contraction. Self-similar fractals can be described by a single fractal dimension. One example is the tree-like pattern shown in Fig. 1. This fractal is made up of squares. One starts with a single square. In the next step, ten squares are put together to construct the square pattern shown in Fig. 1(b). This pattern is the basic building block of the fractal. In each subsequent generation, ten of the objects of the last generation are put together to form a new one having the same overall shape as Fig. 1(b), but with more details. Clearly this process can be continued *ad infinitum* to obtain a very large pattern. Since the same rule is repeated in each iteration, this object is exactly self-similar, having the same shape at any length scale.

In contrast, the shapes and patterns found in nature are usually random fractals. The reason is that they consist of random shapes or patterns that are formed stochastically at any length scale. Because of the randomness, self-similarity of natural fractals is only statistical. Furthermore, natural fractals are self-similar only over a finite range, for example, from atomic sizes to the maximum length in the system. Examples of random fractals includes such objects as the coastlines and mountains, as well as many disordered materials like long polymer chains in solutions, aggregates of colloidal particles, aerosols, and polymeric gels. An example of a random fractal is shown in Fig. 2. This is a viscous fingering pattern formed by injection of a less viscous fluid into a more viscous fluid filling a porous medium. Since the pressure at the interface obeys the Laplace equation, which is a special case of the diffusion equation, this pattern is similar to many other patterns formed in diffusion-limited growth processes.

The fractal dimension of an object or a pattern serves to characterize it and often indicates the manner in which it has been formed. There are several ways to determine the fractal dimension of a fractal, but one of the most widely used methods is to relate the size of the object to its length, which is an extension of our everyday experience of dimensionality for Euclidean objects. Let us consider the fractal tree shown in Fig. 1. In each generation the length of the new pattern is 4 times the length of the previous one, and its mass is 10 times the previous mass. In contrast, if we construct a nonfractal object, like a regular triangle, then every time its length is doubled, its mass is quadrupled. If the length of a regular triangle is increased by a factor of n, then its mass would be $n^2M(r)$, where r is the length of the original triangle and $M(r)$ is its mass. The number 2 is the dimension of the object; a trivial result for a regular two-dimensional triangle. In general, the fractal dimension is the exponent or power to which n is raised. The fractal dimension D is defined by the relation $M(nr) = n^D M(r)$, where $M(nr)$ is the mass after the length has been scaled by a factor of n. For the fractal tree $M(4r) = 10\, M(r)$, because every time r is quadrupled, the mass is increased tenfold. Therefore, the fractal dimension D of the fractal tree can be determined from the relation, $10 = 4^D$, which implies that $D = \log 10/\log 4 = 1.66 \ldots$. This method can be used to determine the fractal dimension of any object or pattern. In the case of a random fractal, one can simply draw circles of increasing radii about a point on the object and determine the mass $M(r)$ within a circle of radius r and compare it with the mass within a circle of radius

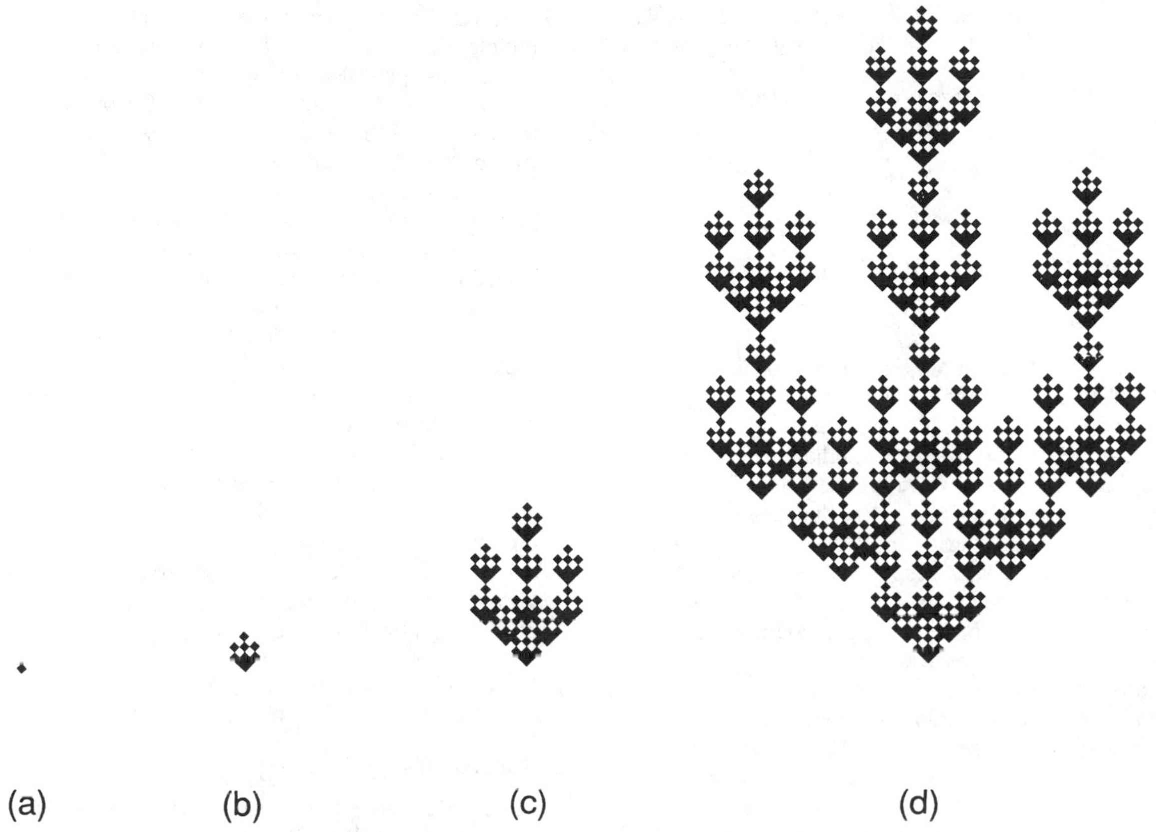

FIG. 1. This tree-like fractal pattern is an example of an exactly self-similar fractal. It is formed by an iterative process which can continue indefinitely.

FIG. 2. An example of a random fractal pattern formed in a viscous fingering experiment where a low-viscosity fluid (air) is pushed into a more viscous fluid (oil) occupying a porous medium.

nr. The fractal dimension D can be calculated from the relation $M(nr) = n^D M(r)$, as before.

Self-similarity is a consequence of invariance under an isotropic dilation. In general, the transformations that rescale an object may not be isotropic. Objects that are invariant under a more general type of rescaling transformations are called self-affine fractals. This class of fractals must be described by more than one fractal dimension. The reason is that a piece of a self-affine fractal must be enlarged anisotropically in order to match larger parts of the object. Self-affine geometry is frequently encountered in the study of rough surfaces, including mountains, thin films, and fracture surfaces.

The advent of fractal geometry has provided scientists and engineers with a fresh mathematical tool for investigating complex phenomena that had been viewed intractable by generations of scientists. Fractal concepts are rapidly becoming a standard tool in many scientific investigations, as witnessed by the rich variety of books published in this field in recent years.

BIBLIOGRAPHY

R. Devaney, *Chaos, Fractals and Dynamics*. Addison-Wesley, Reading, MA, 1989.

F. Family and D. P. Landau (eds.), *Kinetics of Aggregation and Gelation*. North-Holland, Amsterdam, 1984.

J. Feder, *Fractals*. Plenum, New York, 1988.

B. B. Mandelbrot, *The Fractal Geometry of Nature*. Freeman, San Francisco, 1982.

H. E. Stanley and N. Ostrowsky (eds.), *On Growth and Form: Fractal and Non-fractal Patterns in Physics*. Martinus Nijhoff Publishers, Dordrecht, 1986.

T. Vicsek, *Fractal Growth Phenomena*. World Scientific, Singapore, 1989.

The Franck–Condon Principle

B. R. Judd

In the analysis of the photodissociation of diatomic molecules made in 1925, Franck assumed that a transition between two different electronic states takes place so quickly that the positions of the nuclei immediately after the transition are the same as they were just before it. Within a few years, this idea was described in the language of quantum theory by Condon. The variation of the potential energy W of a diatomic molecule with respect to internuclear distance r is illustrated for two electronic states by the curved lines in each of the two parts of Fig. 1. On the left, a molecule in its ground state A absorbs a quantum of radiation and is excited to B, where the two nuclei execute oscillations over those values of r corresponding to the segment BC. If the upper potential curve were to be so flat on its right-hand side that no intersection C occurs, then dissociation would take place. The situation for emission is illustrated in the right part of the figure. A molecule oscillating over a range of r values specified by DE spends most of its time near the extreme positions D and E. The most probable transitions to the lower electronic state correspond to DF and EI, and hence the oscillations corresponding to FG and HI are favored. Quantum mechanics modifies these statements slightly. The most probable values of r lie at points inside the classical ranges of oscillation: the midpoint of the range is the most probable for the lowest vibrational state, as indicated at A. In addition, values of r outside the classical ranges are permitted. Relative transition probabilities between the vibrational states represented by the wave functions ψ_v and $\psi_{v'}$ of two different electronic states depend principally on the square of the overlap integral $\int \psi_v \psi_{v'}\, dr$. The essence of the Franck–Condon principle is that the arrowed lines AB, DF, and EI are vertical: each transition occurs for a well-defined nuclear separation r. The principle can be equally well applied to more complicated molecules, including those exhibiting the Jahn–Teller effect.

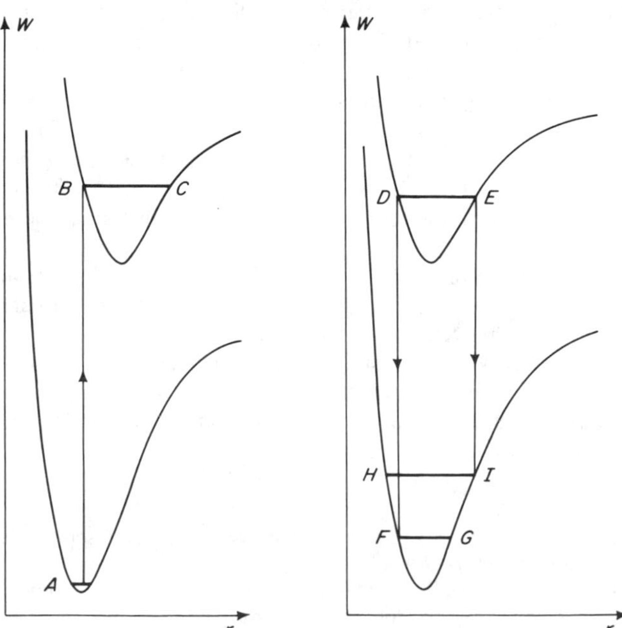

FIG. 1. The Franck–Condon principle illustrated for absorption (on the left) and emission (on the right).

The usefulness of the Franck–Condon principle lies in its capacity to predict the nuclear motion of a molecule after an electronic transition has taken place. If the energy of the transition and also the relevant set of potential-energy curves are known, then the prediction consists of a statement of probabilities for various types of nuclear oscillation (including possibly dissociation). Conversely, of course, an observed nuclear motion can provide information on the shape of the potential-energy curves.

See also MOLECULAR SPECTROSCOPY.

BIBLIOGRAPHY

B. Bak, Elementary Introduction to Molecular Spectra. North-Holland, Amsterdam (1962). (E)

G. Herzberg, Molecular Spectra and Molecular Structure. I. Spectra of Diatomic Molecules. Van Nostrand, New York (1950). (I)

Fraunhofer Lines

Alfred Leitner

Fraunhofer lines are the numerous dark lines in the continuous spectrum of sunlight, first noted by W. H. Wollaston in 1802. He reported seeing a few dark lines—i.e., the most prominent ones—but seems to have paid no further attention to them. They were studied in detail between 1806 and 1826 by J. Fraunhofer, who built the first precision spectrometers, slits, quality prisms, and diffraction gratings ruled with up to 300 lines/mm (7500 lines/in). Some of the dark lines in the solar spectrum became his reference marks for charting the refractive indices of optical glasses. These data enabled Fraunhofer to design and construct outstanding achromatic and otherwise corrected optical instruments, including astronomical refracting telescopes. He determined the wavelengths of a number of Fraunhofer lines to the then unheard of precision of four significant figures. It was Fraunhofer who traced the origin of the spectrum to the sun. He reported that the spectra of other stars, though similar, showed different dark line patterns. Around 1890 H. A. Rowland published wavelength tables and a photographic atlas for about 15,000 Fraunhofer lines in the range between 300 and 650 nm. Charting of this spectrum has continued and is even now being extended into the far ultraviolet as a result of observations from satellites.

The Fraunhofer lines became understood as an *absorption* spectrum by the work of G. Kirchhoff and others in the 1860s. Many of them have since been identified with the spectra of known atoms and ions. As a consequence, the constitution of the solar photosphere is quite well known. From line profiles and from the wavelengths at the line centers it is possible to infer much about what is going on in the solar surface. When observations are confined to small portions of the solar disk, much detail can be culled from spectral line data. The Doppler shift gives evidence about the gross rotational motion of the sun and about local velocity fields. The ambient magnetic field at the solar surface has been calculated from line broadening interpreted as unresolved Zeeman splitting. The far stronger magnetic field in sunspots has also been determined from measurements of Zeeman splitting.

The Fraunhofer spectrum observed from below the Earth's atmosphere contains a few lines whose intensity is strongest at the time of sunset or sunrise, and whose wavelengths display no Doppler shifts due to solar motions (for example: the group of lines in the red, called A, which is due to oxygen). These are due to the earth's atmosphere and are called telluric. The Fraunhofer spectrum is also observed as a discrete emission spectrum of bright lines on a dark field of view during the brief flash from the photosphere in the solar limb just before or just after totality in an eclipse.

In the present-day context, the Fraunhofer spectrum of sunlight forms only a very small part of the great role played by spectroscopy in astrophysics. The interested reader is referred to the large textbook literature in astronomy for further information.

See also Atomic Spectroscopy; Gratings, Diffraction; Sun; Zeeman and Stark Effects.

BIBLIOGRAPHY

Dictionary of Scientific Bibliography, Vol. V, pp. 142–144. Charles Scribner's Sons, New York, 1972. (E)

"Prismatic and Diffraction Spectra," in *The Wave Theory of Light and Spectra*. Arno Press, New York, 1981. (E)

H. A. Rowland, *Astrophysical Journal*, **1–6,** 1895–1898 lists series of Fraunhofer lines between 3 000 and 6 500 Å. (I)

Alfred Leitner, The Life and Work of Joseph Fraunhofer (1797–1826), *American Journal of Physics* **43,** 59–68 (1975). (E)

L. Debouille, L. Neven and G. Roland, *Photometric Atlas of the Solar Spectrum from 3 000 to 10 000*, Institut d'Astrophysique, Université Liége, Liége (1973). (I)

Free Energy

Gabriel Weinreich

In terms of the internal energy U, the temperature T, and the entropy S, the free energy of a thermodynamic system is defined by

$$F \equiv U - TS.$$

In a process whose initial and final states are at the same temperature, the change in free energy is given by

$$\Delta F = \Delta U - T\,\Delta S.$$

If in addition the process is reversible, the last term is equal to the amount of heat transferred, so that the decrease in free energy becomes equal to the work performed by the system. Thus F acts as an "isothermal potential energy." For irreversible processes, $-\Delta F$ represents an upper limit on the work that the system can perform. The free energy does not, however, measure energy stored within the system. For example, in the reversible isothermal expansion of an ideal gas, none of the work comes from inside the system, since the internal energy does not change.

The functional form of the free energy, when given in terms of the temperature and of the variables whose differentials appear in the element of reversible work, comprises a complete thermodynamic description of the system. This puts F into the category of "thermodynamic potentials." For example, the element of reversible work done by a fluid system is $p\,dV$ where p is the pressure and V the volume; accordingly, all thermodynamic properties of such a system are determined as soon as the functional form $F(T,V)$ is known. Specifically, entropy and pressure are the negative partial derivatives of F with respect to T and V, and other quantities are derivable by further differentiations. A general thermodynamic system, for which the element of reversible work is $\Sigma_i\, Y_i\, dX_i$, is similarly completely specified by giving $F(T,X_i)$.

Spontaneous processes at fixed T, X_i always go in a direction that decreases F; hence, any equilibrium state has a minimum value of F when compared to other states of the same T, X_i. This is the sense in which "systems tend toward a state of lowest energy."

In terms of microscopic properties, the free energy is given by

$$F = -kT\log[\sum_n \exp(-\epsilon_n/kT)],$$

where ϵ_n are the energy levels available to the system and k is Boltzmann's constant. The dependence on the macroscopic coordinates X_i enters through the energy levels. The simplicity of the foregoing relation makes the free energy one of the primary links between microscopic and macroscopic descriptions of a system.

Both the nomenclature and the notation for free energy vary considerably. The quantity we have been discussing is associated with the name of Helmholtz, and is sometimes called "work content" or "work function" and denoted by A instead of F. American chemists almost universally use "free energy" to mean the Gibbs potential $G \equiv F + pV$ (which, however, they denote by F).

See also Thermodynamics, Equilibrium.

BIBLIOGRAPHY

G. N. Lewis and Merle Randall, *Thermodynamics* (revised by K. S. Pitzer and Leo Brewer). McGraw-Hill, New York, 1965. (The chemist's approach.) (I)

Max Planck, *Treatise on Thermodynamics*. Dover, New York. 1945. (Translation from the German treatise whose first edition appeared in 1897, giving the classical view.) (I)

Gregory H. Wannier, *Statistical Physics*. Wiley, New York, 1966. (Connections with statistical mechanics.) (I)

Gabriel Weinreich, *Fundamental Thermodynamics*. Addison-Wesley, Reading, Mass., 1968. (A modern approach that emphasizes thermodynamic potentials.) (I)

Friction

Francis E. Kennedy, Jr.

Whenever one body moves tangentially against another, there is resistance to that motion. The resistance is called friction and the resisting force is the friction force. Although there is also friction between a solid surface and a fluid, most of the following discussion will deal with friction between solid bodies. Particular attention will be focused on the contacting surfaces of the solids, since friction is primarily a surface phenomenon.

COEFFICIENT OF FRICTION

If two stationary contacting bodies are held together by a normal force N and if a tangential force is applied to one of them, the tangential force can be increased until sliding occurs. The maximum tangential force at the time of incipient sliding is, in most cases, approximately proportional to the normal force. The ratio of the friction force at incipient sliding to the normal force is known as the static coefficient of friction, f_s. After sliding commences, the friction force always acts in the direction opposing motion and the ratio between that friction force and the normal force is the kinetic coefficient of friction, f_k.

Generally f_k is slightly smaller than f_s and both coefficients are independent of the size or shape of the contact surface. Both coefficients are, however, very much dependent on the materials of the two bodies and on the cleanliness of the contacting surfaces. Typical values of the friction coefficients are given in Table I. For ordinary metallic surfaces, the friction coefficient is not sensitive to the surface roughness. For ultrasmooth or very rough surfaces, however, the friction coefficient can be larger for reasons which will be discussed. If a metal is sliding against a very hard or very soft material, the roughness of the harder material can have an important influence on friction.

The kinetic coefficient of friction, f_k, for metallic or ceramic surfaces is relatively independent of sliding velocity at low and moderate velocities, although there is often a slight decrease in f_k with increasing sliding speed at higher velocities. With polymers and soft metals there may be an increase in friction with increasing velocity or temperature until a peak is reached, after which friction decreases with further increases in velocity or temperature. The decrease in kinetic friction with increasing velocity, which may become especially pronounced at higher sliding velocities, can be responsible for friction-induced vibrations (stick–slip oscillations) of the sliding systems. If the sliding velocity is high enough, the amount of frictional heating may be sufficient to cause surface melting of one of the sliding materials. This can lead to very low friction coefficients. For example, the friction coefficient of ice against itself at a very low velocity (10^{-5} m/s) is about 0.5, whereas at a higher velocity (10^{-2} m/s), f_k is less than 0.05.

Table I Some Typical Friction Coefficients[a]

	f_s		f_k	
Material pair	in air	in vacuo	in air, dry	oiled
Mild steel vs. mild steel	0.75		0.57	0.16
Mild steel vs. copper	0.53	0.5 (oxidized) 2.0 (clean)	0.36	0.18
Copper vs. copper	1.3	21.0	0.8	0.1
Tungsten carbide vs. copper	0.35		0.4	
Tungsten carbide vs. tungsten carbide	0.2	0.4	0.15	
Mild steel vs. PTFE	0.04		0.05	0.04

[a] The friction coefficient values were compiled from several of the references listed in the Bibliography.

BASIC MECHANISMS OF SLIDING FRICTION

Solid surfaces are not perfectly flat and smooth, but have roughness consisting of peaks and valleys produced in the manufacturing process. When two solids are forced against one another, contact occurs not within the entire nominal or apparent contact area, but only at the tips of contacting peaks or asperities on the surfaces. The total real area of contact is generally much smaller than the nominal area, and it is within that small area that friction occurs. There are two primary contributions to sliding friction in the contact region, adhesion and deformation.

Adhesion occurs when the atoms or molecules of the two contacting surfaces approach each other so closely that attractive forces between approaching atoms (or molecules) bond them together. The strength of the bond depends on the size of the atoms (or molecules), the distance between them, and the structure of the contacting materials. In order to separate the materials, the adhesive bonds must be broken. If the applied force is a tangential force, relative motion between the two surfaces requires that all adhesive junctions formed within the real area of contact must be sheared, and the force required to do this constitutes the major part of the friction force.

Another contribution to the friction force occurs whenever one of the two contacting surfaces is softer than the other. In such cases, when the surfaces are forced together by a normal force, the peaks on the harder surface will indent the softer surface to some extent. Relative tangential motion of the surfaces results in the plowing of the softer surface by the harder one, so the indentations become fine grooves in the softer surface. The deformation in the softer surface could be plastic or viscoelastic in nature, with the former occurring in metallic surfaces and the latter in polymers. The force required to displace the softer material ahead of the moving hard asperities is a second contribution to the friction force, the plowing or deformation contribution. For most dry sliding situations, the deformation contribution is considerably smaller than the adhesion term. If one of the surfaces is both harder and rougher than the other, though, or if a lubricant or contaminant surface film is present on the surfaces, the deformation contribution could be significant.

If one considers only the adhesion contribution, one could say that the friction force is approximately the product of the real area of contact times the shear strength of the adhesive junctions within that area. Large differences in friction coefficient, such as those evident in Table I, are generally caused by differences in the shear strength term. Ordinary surfaces are not completely clean, but are covered with oxide films, adsorbed gases, or other contaminant films which reduce both interfacial shear strength and adhesion. Thus, friction of sliding solids is considerably reduced by the presence of lubricants, oxides, or other contaminant films. If ultraclean surfaces are placed together in a vacuum, the interfacial adhesion is very strong and the application of a tangential force causes the junctions to grow, resulting in a very high friction force or even complete seizure.

Within adhesive junctions, the strength of the adhesive bonds is determined by the relative sizes of the surface atoms, the number of atoms that can approach each other closely, and the type of bond formed between the atoms (or molecules). The adhesive bonds and the resulting friction coefficients are greatest when the two surfaces are identical metals, slightly lower with dissimilar but mutually soluble metals, still lower for metal versus nonmetal, and lowest for dissimilar nonmetals.

EFFECT OF LUBRICATION ON FRICTION

When a small amount of lubricant is applied to a surface or when lubricated surfaces slide together at low sliding speeds or with a high applied normal load, the lubricant will not be successful in separating the two solid surfaces. The lubricant can still significantly reduce the friction coefficient by reducing the shear strength of adhesive junctions between the solid surfaces. In this situation, called the boundary lubrication regime, the effectiveness of the lubricant can be improved if the lubricant molecules adhere well to the solid surfaces. This is best accomplished by introducing a lubricant or additive that forms a surface film through physical adsorption, chemisorption, or chemical reaction with the surface. The reduced shear strength of the surface film can lower the friction coefficient by as much as an order of magnitude from the dry friction value.

When a good supply of a viscous lubricant is available, the separation between the surfaces will increase as the sliding speed increases or the normal load decreases. As the separation increases, the amount of solid/solid contact between the surfaces will decrease, as will the friction coefficient. In this "mixed lubrication" regime, friction is determined by the amount of plowing deformation by the harder asperities and by adhesion within the solid/solid contacts. When the surfaces become completely separated by a self-acting or externally pressurized lubricant film, the lubricant regime is hydrodynamic and friction usually reaches a low value governed by fluid shear of the lubricant. If the lubricant is Newtonian, the local shear stress is equal to the product of the lubricant viscosity times the local shear rate of the lubricant film. The shear rate is approximately equal to the difference in velocity between the two surfaces divided by the lubricant film thickness. The total friction force can be obtained by integrating the shear stress over the nominal contact area. Friction coefficients in such cases can be 0.001 or lower, depending on the surface velocities and the lubricant viscosity.

Most thrust bearings take advantage of the low friction resulting from hydrodynamic lubrication, and the lubricant in such bearings could be mineral oil, a synthetic lubricant, gas, water, or other liquids. Unless the lubricant contains long-chain polymer additives, the flow in these bearings is generally assumed to be Newtonian. A difficulty arises in a lubricated contact if the sliding velocity goes to zero, because the fluid film separating the surfaces disappears and friction increases to a value determined by boundary lubrication or solid contact conditions.

In concentrated, nonconformal contacts such as in rolling element bearings, gears or cams, the pressures are often high enough to cause deformation of the surfaces and variations in lubricant viscosity. In such cases, the laws of Newtonian hydrodynamic lubrication are not necessarily obeyed and friction must generally be determined by empirical methods.

See also MECHANICAL PROPERTIES OF MATTER; RHEOLOGY; TRIBOLOGY.

BIBLIOGRAPHY

F. P. Bowden and D. Tabor, *The Friction and Lubrication of Solids*. Part II. Clarendon Press, Oxford, 1964.

F. P. Bowden and D. Tabor, *Friction. An Introduction to Tribology*. Anchor Press/Doubleday, New York, 1973.

D. Dowson, *History of Tribology*. Longman Group Ltd., London, 1979.

K. C. Ludema, "Friction," *CRC Handbook of Lubrication*. Vol. 2 (E. R. Booser, ed.), pp. 31–48. CRC Press, Boca Raton, FL, 1984.

M. J. Neale, *Tribology Handbook*. Butterworth & Co. Ltd., London, 1973.

Galaxies

Alan Dressler

In 1924 Edwin Hubble discovered Cepheid variable stars in M33 and M31, the Andromeda nebula. With one strike he put an end to the debate that had been raging among astronomers for years [1]. The Andromeda nebula was proved to be a giant stellar system—a galaxy—at an immense distance (about 2 million light years) and similar in size to the Milky Way system in which the Sun is situated.

Astronomers now recognize galaxies as the fundamental units in the organization of baryonic matter [2]. A typical galaxy like our own Milky Way is roughly 60 000 light years in diameter. Galaxy sizes range in order of magnitude larger and smaller than this. Galaxy masses, which range from 10 million to 1 trillion times the mass of the sun, are mainly made up of ordinary stars, with smaller contributions from neutral and ionized gas and dust. However, evidence from the motions of stars and extended gas distributions suggests that the luminous structures we recognize as galaxies are condensed regions of more extensive mass concentrations containing, typically, 10 times the visible mass. This "dark halo" may be composed of ordinary baryonic matter in nonstellar form, but it could also be made up of weakly interacting elementary particles like neutrinos (which would have to have a nonzero rest mass). The nature of this dark matter remains a crucial problem in astronomy.

Galaxies are found to cluster on many scales. Rich clusters include environments several orders of magnitude more dense than the average background, which is mainly composed of loose groups. Such clusters are conspicuous, but contain only 5–10% of all galaxies. A surprisingly complex structure has recently been discovered in the large-scale distribution of galaxies. Giant superclusters, containing tens of thousands of galaxies, have diverse topologies, including flat and filamentary forms in addition to basically spherical shapes. These superclusters are interspersed with immense spherical or tube-like voids in which few galaxies are found.

Observations of galaxies now span the electromagnetic spectrum from radio waves to x rays. Starlight is the primary source of continuum light in the ultraviolet, visible, and near infrared; dust heated by these stars produces most of the far-infrared flux. Optical line emission from ionized regions surrounding hot stars provides important information on the densities, temperatures, and atomic abundances in regions of active star formation. Radio emission is commonly observed as the 21-cm line from cool, neutral hydrogen gas. Continuum radio flux arises from synchrotron emission in the remnants of exploded stars called supernovae and in the form of diffuse emission from the interaction of cosmic ray particles with the galactic magnetic field. Thermal radio emission from recombination of ionized hydrogen in star forming regions is also seen. Emission from molecules, most importantly CO, is found in cold, dense gas clouds.

X rays from galaxies arise primarily from binary star systems where mass is being transferred to a white dwarf, neutron star, or black hole companion. Thermal x rays have been observed from hot gas around some galaxies. Both x-ray and radio emission also arise in energetic processes in galactic nuclei, described below. Neutrinos of a bona fide extragalactic origin were first seen with the detection of supernova 1987a in the Large Magellanic Cloud, the closest neighbor galaxy to the Milky Way.

Galaxies come in a variety of morphological forms which have been described simply by Hubble and more elaborately by Sandage, de Vaucouleurs, and van den Bergh. These classification schemes recognize the basic elements of galaxy form as the disk and the spheroid, structural forms that reflect the specific angular momentum of the system. Elliptical galaxies are ellipsoidal or triaxial distributions of stars whose nearly round shapes are supported against gravity by mainly random motions. Similar spheroidal components are also found in the centers of many disk galaxies. In general, spheroidal components, including the so-called "bulges" of these disk galaxies, are made up of old stars and have little cold gas or dust, the raw materials of star formation.

Disk systems often show a pronounced spiral pattern which outlines sites of ongoing star formation, usually associated with cold, dense gas clouds rich in molecules, and ionized gas surrounding young, hot stars. The spiral structure is often due to a resonance phenomena wherein a density fluctuation propagates around the galaxy. Morphological classification systems refer to the tightness of the spiral pattern or its coherence. Spirals with central straight structures called *bars* represent a general class of axisymmetric distortion resulting from dynamical instabilities in stellar systems with high angular momentum. Irregular morphologies, including amorphous and disturbed forms, comprise a few percent of luminous galaxies, but are more common among dwarf galaxies.

The distribution of galaxy luminosities, and, by inference, galaxy mass, has been found to be remarkably similar over a wide range of environment. This *luminosity function* has a power law form which is truncated exponentially at the bright end. Some outstandingly bright "cD" galaxies at the centers of clusters appear to be star piles formed from the aggregation of many galaxies. On the faint end, the luminosity function continues to rise to systems with 10 million solar luminosities. Though these dwarf galaxies are the most common galaxies in the universe, their contribution by mass is far less than that of luminous systems like the Milky Way.

Though the connection of morphology to the history of star formation is self-evident, the mechanism that determined whether a galaxy became spiral rather than an elliptical disk, rather than spheroid dominated, is still unclear. An important clue is the strong correlation of the type of galaxy with the crowdedness of its surroundings. Densely packed clusters of galaxies are mainly composed of spheroidally dominated galaxies like elliptical and SO galaxies; low-density groups and isolated galaxies are usually disk-dominated spirals. Most galaxies are found in lower-density environments, so, overall, spiral galaxies are the most common. The morphology–density relation suggests that galaxy–galaxy interactions, particularly early in a galaxy's life, may determine its ultimate form. The effect on angular-momentum distributions, perhaps connected to the relationship between the luminous material and the dark halo, could be of crucial importance. Also suggestive are observations of merging galaxies which appear to be in the act of forming spheroidal systems from former disk galaxies. This highlights the possibility that morphology may be drastically altered late in the lives of at least some galaxies.

Although a satisfactory, detailed model of galaxy formation has yet to be formulated, and may, in fact, depend strongly on the nature and distribution of dark matter, the role of gaseous dissipation in achieving the high average density of a typical galaxy seems clear. Some 1–2 billion years after the Big Bang, gas clouds the size of galaxies were cooled by radiation and by inverse Compton scattering of cosmic background photons. In this way they dissipated thermal energy, contracting until they reached densities suitable for copious star formation. Larger structures with the mass of tens or hundreds of galaxies were unable to follow this path because their cooling times scale exceeded the dynamical time.

The exact epoch at which galaxy structure became distinguishable and the first stars formed is still not known. A considerable effort has gone into looking directly for *primeval* galaxies at high redshifts, but only rare, nonrepresentative cases have been identified. However, the common ~15 billion year age found for the oldest stars in the Milky Way and its neighboring galaxies, and the approximate coincidence of this time with the present age of the universe, seems to ensure that such a process occurred over a relatively short interval. It is also unknown to what extent the collapse to a galaxy-size structure was a single, coherent process as opposed to the more random coagulation of a number of fragments. In spiral galaxies, at least, the process of turning gas into stars has persisted over the age of the universe, resulting in a steady increase in the abundance of elements heavier than hydrogen. Our own Sun was born in a later generation of star formation some 5 billion years ago, in a spiral arm near the rim of the Milky Way galaxy.

Among the most interesting phenomena in astrophysics are those galaxies which release enormous amounts of energy, often far exceeding the stellar output, from solar-system size regions at their centers. The most luminous examples of these "active galactic nuclei," *radio galaxies* and *quasars,* are important probes of the early universe since their light can be seen over far greater distances, and thus to greater lookback times, than the light of ordinary galaxies. Furthermore, absorption of quasar light by intervening galaxies probes the properties of gas in these objects that are themselves too faint to be observed directly. Successful models of the central energy source of active nuclei invoke accretion onto massive black holes of tens of millions of solar masses. Some evidence based on the motions of stars in the nuclei of nearby galaxies indicates that such massive black holes may be common.

See also ASTROPHYSICS; COSMOLOGY; INTERSTELLAR MEDIUM; MILKY WAY; QUASARS; UNIVERSE.

REFERENCES

1. E. P. Hubble, *The Realm of the Nebula,* p. 93. Oxford University Press, Oxford and London, 1936.
2. D. Mihalas and J. Binney, *Galactic Astronomy.* W. H. Freeman, San Francisco, 1981.

Galvanomagnetic and Related Effects

D. J. Sellmyer and C. M. Hurd

These are transport effects which occur in conductors in the presence of forces produced by electric and magnetic fields, and temperature gradients. The irreversible, cross-coupled processes can be described by the expressions:

$$E_m = \sum_n \rho_{mn}(\mathbf{B}) J_n + \sum_n \alpha_{mn}(\mathbf{B}) \frac{\partial T}{\partial x_n}, \quad (1)$$

$$Q_m = \sum_n \pi_{mn}(\mathbf{B}) J_n - \sum_n \kappa_{mn}(\mathbf{B}) \frac{\partial T}{\partial x_n}, \quad (2)$$

where E_m, J_m, Q_m, and $\partial T/\partial x_m$ represent the mth component of the electric field, electric current density, heat current density, and temperature gradient, respectively. **B** is the magnetic flux density and m and $n = 1, 2, 3$ represent x, y, and z Cartesian coordinates, respectively. The transport-coefficient arrays ρ_{mn}, α_{mn}, π_{mn}, and κ_{mn} are known as the *resistivity, thermoelectric power, Peltier coefficient,* and *thermal conductivity,* respectively.

A large number of experimental situations can be described by Eqs. (1) and (2). By convention, three classes of transport effects are defined: (1) *galvanomagnetic* effects in which there are no temperature gradients, (2) *thermoelectric* effects in which there is no magnetic field present, and (3) *thermomagnetic* effects in which all quantities in Eqs. (1) and (2) are allowed to be nonzero. The thermoelectric effects will not be considered further here and some of the more important galvanomagnetic and thermomagnetic effects, with their names and experimental conditions, are defined for an isotropic conductor in Table I.

The effects are characterized as isothermal or adiabatic depending on whether temperature gradients or heat flows are allowed to exist. In addition, the galvanomagnetic effects are classified according to the three basic relative orientations of applied electric and magnetic fields:

1. *Transverse* effects in a *transverse* magnetic field. **B** is perpendicular to the primary current density **J** flowing in the sample, and the galvanomagnetic effect is measured in the direction mutually perpendicular to **B** and **J**.
2. *Longitudinal* effects in a *transverse* magnetic field. **B** is again perpendicular to **J**, but the galvanomagnetic effect is measured along the direction of **J**.

Table I Galvanomagnetic and related effects in isotropic media

Class	Name	Primary fluxes ($\mathbf{B}=\mathbf{B}_z$)	Quantity measured	Defining conditions
1. Transverse effects in a transverse field	Hall (isothermal)	$\mathbf{J}=\mathbf{J}_x$	E_y/J_x	$J_y=\delta T/\delta x=\delta T/\delta y=0$
	Hall (adiabatic)	$\mathbf{J}=\mathbf{J}_x$	E_y/J_x	$J_y=\delta T/\delta x=Q_y=0$
	Ettingshausen	$\mathbf{J}=\mathbf{J}_x$	$(\delta T/\delta y)/J_x$	$J_y=\delta T/\delta x=Q_y=0$
2. Longitudinal effects in a transverse field	Electrical transverse magnetoresistivity (isothermal)	$\mathbf{J}=\mathbf{J}_x$	E_x/J_x	$J_y=\delta T/\delta x=\delta T/\delta y=0$
	Electrical transverse magnetoresistivity (adiabatic)	$\mathbf{J}=\mathbf{J}_x$	E_x/J_x	$J_y=\delta T/\delta x=Q_y=0$
	Nernst (isothermal)	$\mathbf{J}=\mathbf{J}_x$	$(\delta T/\delta x)/J_x$	$J_y=Q_x=\delta T/\delta y=0$
	Nernst (adiabatic)	$\mathbf{J}=\mathbf{J}_x$	$(\delta T/\delta x)/J_x$	$J_y=Q_x=Q_y=0$
3. Longitudinal effects in a longitudinal field	Electrical longitudinal magnetoresistivity	$\mathbf{J}=\mathbf{J}_z$	E_z/J_z	$\delta T/\delta z=0$
	Unnamed	$\mathbf{J}=\mathbf{J}_z$	$(\delta T/\delta z)/J_z$	$Q_z=0$

3. *Longitudinal* effects in a *longitudinal* electric field. **B** is parallel to **J** and the galvanomagnetic effect is measured along **J**.

Fundamental to an understanding of the galvanomagnetic effects is the Lorentz force equation (in SI units):

$$\mathbf{F} = q(\mathbf{E}+\mathbf{v}\times\mathbf{B}). \tag{3}$$

Here, q is the charge on a charged particle (such as an electron), $\mathbf{v}$ is the particle's velocity, and $\mathbf{v}\times\mathbf{B}$ is the vector product of these two quantities (which is perpendicular to both).

The galvanomagnetic effects arise from a combination of two circumstances: the deflection of the itinerant electrons in a conductor by the Lorentz force, and the constraint on the electron flow produced by the boundaries of the sample. Consider a conducting sample subjected to a steady electric field **E**. Under the influence of the electrostatic force, each itinerant electron has superimposed upon its random thermal motion an acceleration along the direction exactly opposite to **E** (since an electron carries a negative charge). During its mean free path between randomizing collisions, which occur especially with the ions in the substance, each electron therefore contributes to a general current drift along **E**. The Lorentz force produced by a uniform flux density **B**, assumed for generality to be noncollinear with **E**, deflects each moving electron from its unperturbed path. A few will at first be deflected without hindrance, but since electrons are mutually repulsive, and since they are constrained to remain within the bounds of the sample, they eventually pile up against an inside face. This creates an electric field which opposes the Lorentz force. Ultimately, a new dynamical equilibrium is reached among the electrons in which the current drift is maintained along **E** but is now nonuniform because of the established concentration gradients. Since the electrons possess thermal energy as well as charge, a temperature gradient will be set up concomitantly with the electric field. A third effect is the decrease in the effective electrical conductivity that is implied as electrons are deflected from their otherwise forward motion under the influence of **E**. These three features are typical manifestations of a galvanomagnetic effect.

The most informative and most-often measured of the galvanomagnetic and thermomagnetic effects are two in the former class: Hall effect and magnetoresistance. For these effects two field regions are defined. In the presence of a magnetic field the Lorentz force causes the electrons to move in circular or helical paths. The angular frequency of this periodic motion is called the *cyclotron frequency* which is given by

$$\omega_c = qB/m_c, \tag{4}$$

where m_c is the cyclotron mass of the carrier. However, the periodic motion is interrupted by collisions of the carriers with impurity atoms or lattice vibrations, with the average time between collisions, or *relaxation time*, being defined as τ. The two field regions defined above can then be characterized by

$$\omega_c\tau \ll 1 \quad \text{(low-field region)}, \tag{5}$$

$$\omega_c\tau \gg 1 \quad \text{(high-field region)}. \tag{6}$$

Physically, in the low-field region the carriers are prohibited from making a complete cyclotron orbit by collisions. In the high-field region, typically reached in very pure single crystals at liquid-helium temperatures ($T \approx 4$ K), many cyclotron orbits can be transversed before scattering takes place. The behavior of the galvanomagnetic effects in the two field regions is markedly different.

In the low-field region, the transverse Hall effect for a conductor with only one type of carrier present (electrons or holes) gives a transverse voltage *inversely* proportional to the carrier density. This permits the measurement of this density for simple monovalent metals like the alkali and noble metals. But the small size of the Hall effect in metals makes it difficult to measure and interpret. For semiconductors, where the carrier density is several orders of magnitude smaller than for metals, the magnitude of the Hall field, E_y (see Table I), is correspondingly larger. This leads to applications such as magnetometers, susceptibility meters, Hall-effect amplifiers, multiplying elements, displacement transducers, and contactless switches.

In metals and semiconductors where there is more than one type of carrier present—for example, electrons and holes—the theory for the low-field Hall effect becomes more difficult and the information obtained less precise.

In the high-field region the transverse magnetoresistance of certain pure single-current metals can show striking anisotropy as a function of magnetic field direction. These effects have been shown to be related to the topology of the Fermi surface of the metal, the surface bounding the region of occupied electron states in wave vector space. In addition, quantum oscillations in the magnetoresistance are directly related to extremal cross sections of the Fermi surface. Thus these galvanomagnetic effects are important for obtaining fundamental information about the Fermi surfaces of metals.

The Hall effect in ferromagnetic metals exhibits remarkable features. The Hall field per unit current density can be written as

$$E_H = R_0 B + R_1 M, \tag{7}$$

where R_0 is the "ordinary" Hall coefficient, R_1 the "extraordinary" or "anomalous" Hall coefficient, and M the magnetization. In thin-film magnetic structures, measurements of the Hall coefficient have been shown to be a sensitive method for studying the field and temperature dependence of the magnetization. Recently, this technique has been combined with microlithography to detect the magnetization reversal of a single- or nearly single-domain Co-Cr particle with a volume of only about 3×10^{-15} cm^3.

Additional developments of high interest are the discovery of the quantum and anomalous quantum Hall effects. These phenomena occur in quasi-two-dimensional semiconducting structures. The Hall coefficient as a function of field exhibits plateau regions which correspond to a combination of fundamental constants divided by an integer or simple rational fraction. Microscopic theories of these effects involve many-body effects and are an area of active research.

The galvanomagnetic effects have been helpful in furthering our knowledge of conduction processes in both semiconductors and metals. Measurements of the low-field Hall and magnetoresistance effects are made routinely to determine the sign, density, and mobility of the charge carriers in semiconductors. Studies of the dependence of these effects on the magnetic field strength lead to information about the nature of the electron scattering processes in both semiconductors and metals, while studies in the high-field condition, particularly in metals, are an important tool for investigating the substance's electronic structure. Recent research on novel microstructures has led to important new information of great interest in pure and applied physics.

See also Conduction; Hall Effect; Hall Effect, Quantum; Magnetoresistance; Resistance.

BIBLIOGRAPHY

F. J. Blatt, *Physics of Electronic Conduction in Solids*. McGraw-Hill, New York, 1968. (I)

C. L. Chien and C. R. Westgate (eds.), *The Hall Effect and its Applications*. Plenum, New York, 1980. (A)

E. Conwell, "Transport: The Boltzman Equation," in *Handbook on Semiconductors*. Vol. 1 (T. S. Moss, ed.). North-Holland, Amsterdam, 1982. (A)

C. M. Hurd, *The Hall Effect in Metals and Alloys*. Plenum, New York, 1972. (A)

C. M. Hurd, "Galvanomagnetic Effects in Anisotropic Metals," *Adv. Phys.* **23,** 315–433 (1974). (A)

R. E. Prange and S. M. Girvin (eds.), *The Quantum Hall Effect*. Springer-Verlag, New York, 1986. (A)

E. H. Putley, "The Hall Effect and its Applications," *Contemp. Phys.* **16,** 101–126 (1975). (E)

W. A. Reed, "Experimental Methods of Measuring High-Field Magnetoresistance in Metals," in *Methods of Experimental Physics*. Vol. 11 (R. V. Coleman, ed.), pp. 1–31. 1974. (I)

A. C. Smith, J. F. Janak, and R. B. Adler, *Electronic Conduction in Solids*. McGraw-Hill, New York (1967). (I)

B. C. Webb and S. Schultz, "Detection of the Magnetization Reversal of Individual Interacting Single-Domain Particles within Co-Cr Columnar Thin-Films," *IEEE Trans. Magn.* **MAG-24,** 3006 (1988). (A)

Gamma Decay

E. der Mateosioan

Gamma decay is a process in which atomic nuclei change their states to states of less energy with the simultaneous emission of electromagnetic radiations, or photons, called *gamma rays*. The probabilities with which these transitions take place may be vastly different and are strongly dependent, in a complicated fashion, on the angular momenta, energies, and parities of the initial and final energy states of the nuclei involved. Other considerations give rise to "selection rules" and perturbations that may prohibit or affect the probability of certain transitions. Since this process involves the electromagnetic field, familiar to physicists through classical events, the study of gamma decay assumed historical importance because it made possible the understanding and determination of nuclear properties (energy, angular momentum, parity) even though the nuclear forces were not known.

The atomic nucleus is a complex system of particles (nucleons) that may absorb or emit energy according to the rules of quantum mechanics (QM). One of the striking restrictions imposed by QM on the nucleus is that it may not absorb or emit any arbitrary amount of energy; rather, the nucleus may exist only in a set of discrete (not continuous) energy states. It is customary to illustrate this situation by means of an energy diagram called a *nuclear level scheme* (Fig. 1). A permissible nuclear energy state for a particular atom is represented by a horizontal line whose vertical position indicates the energy content of the nucleus when in that state. The energy scale is only a relative one based upon the assumption that the lowest energy state (ground state) has zero energy. Properties of the nucleus when in one of these states, such as its energy, angular momentum, and parity, are written beside the level lines. In a quantum-mechanical system such properties are given by *quantum numbers*. When a nucleus spontaneously changes from one state to another with less energy, it emits the difference in energy between these states in the form of a monoenergetic photon, or electromagnetic radiation. If the energy of the photon is $\hbar\omega$ and the energies of the initial and final states are E_a and E_b, respectively, then

$$E_a - E_b = \hbar\omega \tag{1}$$

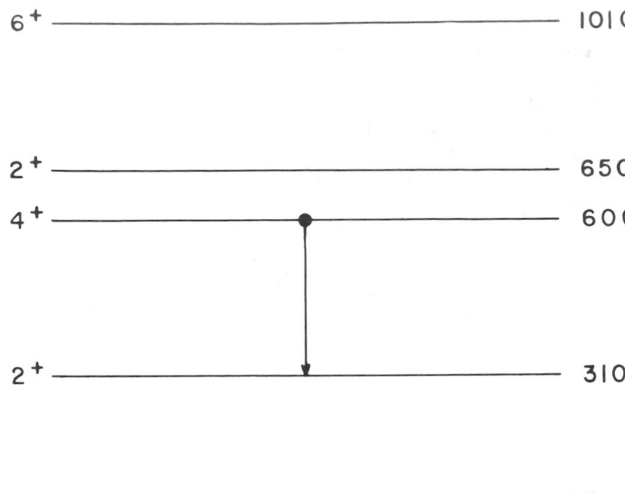

FIG. 1. A nuclear energy level diagram. The horizontal lines represent discrete energy levels in a hypothetical nucleus. To the right of each level appears its energy above the ground state (or lowest level) and to the left, its angular momentum and parity. The vertical line connecting the 4+ and 2+ levels indicates that the nucleus, when in its 4+ state, will make a transition to its 2+ state with the emission of a gamma ray.

and energy is conserved. The transition of the nucleus from one energy state to another by means of the emission of a photon is indicated on the energy-level scheme of Fig. 1 by a vertical line between two levels. A dot on the end of this line indicates the initial state; an arrowhead indicates the final state. Thus, Fig. 1 may be taken to mean that the particular nucleus under consideration was in the 600-keV state (which has 4 units of angular momentum) and it decayed to the 310-keV state (2 units of angular momentum). However, Fig. 1 is interpreted to mean more than this. The transition between the two levels does not take place haphazardly but with a probability that can be predicted by various model-dependent theoretical calculations or the systematics of experimentally determined decay probabilities. Figure 1 predicts that if the nucleus is excited to the 600-keV state, it will decay to the 310-keV state (with a certain probability expressed, when known, reciprocally as a half-life). It is thus seen that Fig. 1 gives information not only on the states of the nucleus but on its behavior in these states as well. A nuclear level scheme with all of the known transitions of the nucleus indicated on it is called a decay scheme. The usual decay scheme in practice can be quite complicated. An example of what one may expect is shown in Fig. 2.

Decay schemes of nuclei are sufficiently unique that if the nuclear gamma rays are detected and their energies determined, it is frequently possible to identify the nuclei emitting the radiations; but more than energy characterizes a photon emitted in a nuclear transition between two specific states of a nucleus. For example, in a mechanical system not only energy but angular momentum is conserved, and a transition between two nuclear levels with different angular momenta may take place only if the emitted photon, or gamma ray, carries that amount of angular momentum ℓ such that

$$\mathbf{J}_a = \mathbf{J}_b + \boldsymbol{\ell}, \tag{2}$$

where $\mathbf{J}_a$ and $\mathbf{J}_b$ are the angular momenta of initial and final states of the nucleus. Equation (2) is a vector equation that describes the addition of angular momenta in quantum mechanics and it results in the selection rule that transitions between states with angular momenta $\mathbf{J}_a$ and $\mathbf{J}_b$ can take place only if

$$|J_a - J_b| \le l \le J_a + J_b. \tag{3}$$

Another selection rule says

$$M_a - M_b = m \tag{4}$$

where M_a and M_b are the Z components of $\mathbf{J}_a$ and $\mathbf{J}_b$ and m is the Z component of the angular momentum associated with the emitted photon. Photon radiation with $l=0$ is forbidden and gamma decay between two states with angular momenta zero cannot take place.

A precise and complete description of the emission (and absorption) of photons by the nucleus is given only in a quantum-mechanical theory of radiation. However, some aspects of the problem may be understood through a comparison of the nucleus to a classical system of charges and currents varying in time and confined to a space small compared to the wavelength of the radiation emitted. The starting point of the classical treatment is Maxwell's equations and it has been found convenient to separate the radiation field of such a classical system into components with definite angular momentum and parity (which see), which process is described as separating the electromagnetic field into its electric and magnetic multipole components. The multipole with $l=1$ is called dipole radiation; that with $l=2$, quadrupole radiation. Depending on whether or not parity changes in the transition, the multipole may be electric or magnetic. Multipole radiations of odd l accompanied by a change of parity and of even l with no change of parity are electric. The reverse combinations of angular momentum and parity change give magnetic multipole radiations.

The probability of multipole radiation depends (in the absence of selection rules and other perturbations) on l and the presence of parity change. For example, magnetic dipole radiation ($l=1$, no parity change) is less probable than electric dipole radiation ($l=1$, parity change). This is true for all l. Multipole radiation probability rapidly decreases as l increases and increases as the energy of the transition increases. It varies also, but to a lesser degree, with the size of the nucleus. The probability of a transition is frequently given in terms of half-life, which equals 0.693 divided by the probability. The fantastic range of values assumed by the probability for gamma decay is dramatically illustrated by ^{113}Cd, in which one level has a transition probability of 7.7×10^{10} sec^{-1} and another level has a probability for gamma emission of 3.5×10^{-13} sec^{-1} (the nucleus in this energy state mostly decays by β-particle emission, the probability for which is 1.6×10^{-9} sec^{-1}). Because of the rapid variation of probability of radiation with l, radiations in gamma decay are most often pure in multipolarity, although radiations that are a mixture of magnetic dipole ($M1$) and electric quadrupole ($E2$) radiations are observed.

In gamma decay the direction in which a photon is emitted is also a function of the multipolarity of the radiation. Normally, an assembly of excited (radioactive) nuclei will be evenly distributed among all directional orientations. If the angular distribution of photons emitted by this array is stud-

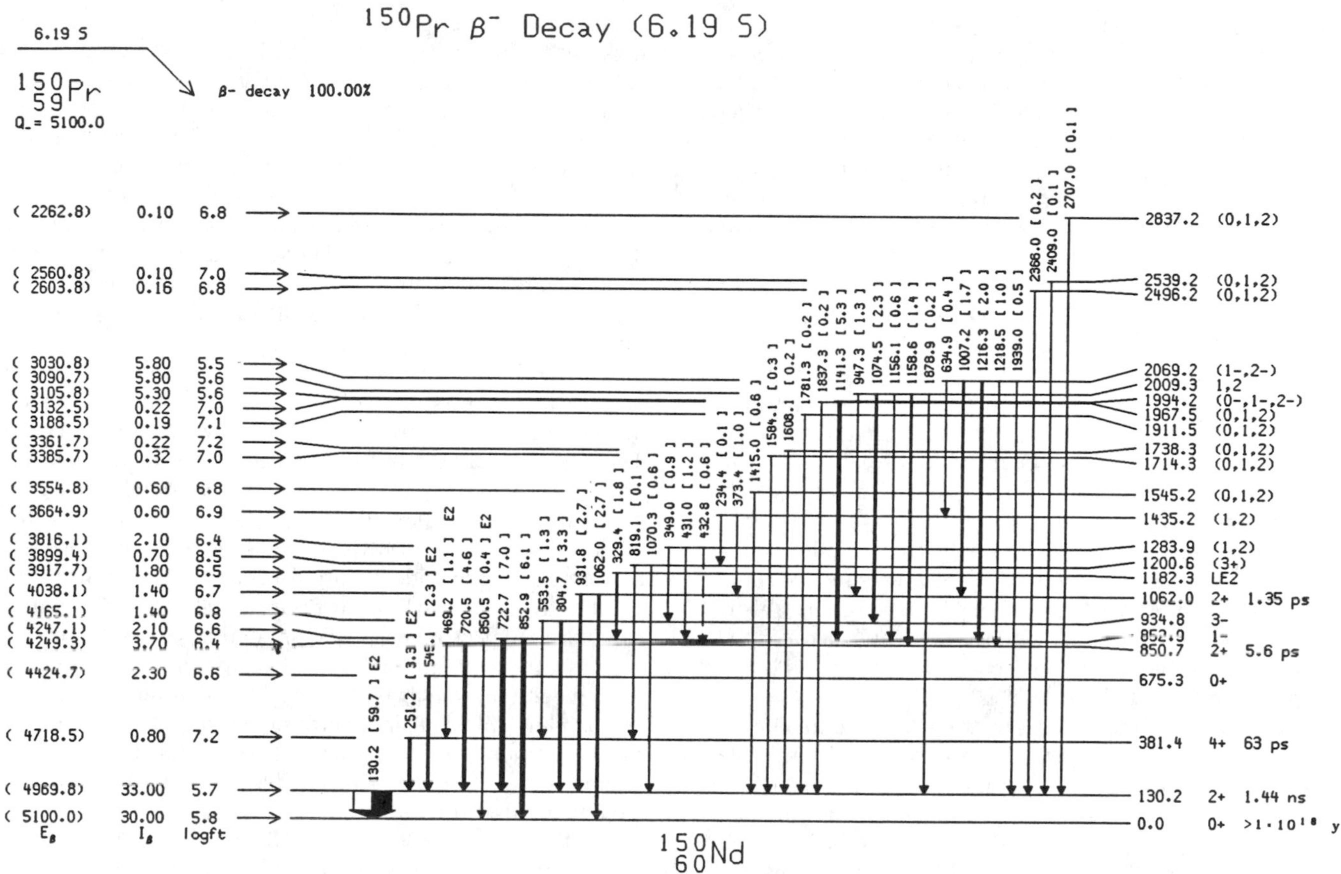

FIG. 2. β decay of ^{150}Pr to ^{150}Nd. Gamma decay of excited states of ^{150}Nd formed in the beta decay of ^{150}Pr is shown. γ-ray information (energy [intensity] multipolarity) appears above each vertical line representing a specific γ-ray transition. The information to the left of the decay scheme is related to the β decay of ^{150}Pr. To the right, the three columns give the energy, spin, and parity, and the half-life of the excited energy levels of ^{150}Nd. In this decay scheme the thickness of a given transition line is related to the relative intensity of that γ ray.

ied, any inherent anisotropy will be washed out. There are conditions in which the angular dependence of multipole radiations may be observed. For example, nuclei excited through charged-particle bombardment or reactions are often aligned in space, and gamma decay on the part of such an assembly of nuclei can show evidence of a nonisotropic angular distribution. If the gamma-decay scheme of the nuclei under investigation is such that more than one photon is emitted in rapid succession by any member of the assembly, it is not necessary to align the nuclei to see the angular distribution of the emitted photons. If two detectors are used to catch two photons emitted by the same nucleus, the detection of the first photon has the effect of selecting out of the assembly of nuclei those that are aligned in such a fashion as to make the detection of the first photon probable in the first detector. If the second photon is detected simultaneously in the second detector, the probability of observing this event will be a function of the angle between the two detectors. This type of experiment is called a coincidence measurement. These types of experiments have verified the theoretical predictions that radiations accompanying gamma decay have an angular distribution. The angular distribution is the same for electric and magnetic multipole radiation of the same l and m. Other differences, however, do exist that allow one to differentiate between these cases (determination of parity).

See also ANGULAR CORRELATION OF NUCLEAR RADIATIONS; GAMMA-RAY SPECTROMETERS; ISOMERIC NUCLEI; MULTIPOLE FIELDS; NUCLEAR PROPERTIES; NUCLEAR STATES.

BIBLIOGRAPHY

R. D. Evans, *The Atomic Nucleus.* McGraw-Hill, New York, 1955.(E)

J. M. Blatt and V. F. Weisskopf, *Theoretical Nuclear Physics.* Wiley, New York, 1958.(I)

Gamma-Ray Spectrometers

Guy T. Emery

Gamma rays from radioactive nuclei have energies that range from a few keV to a few MeV. The energies of gamma rays emitted during the course of nuclear reactions, when these are initiated by low-energy particles, can extend up to about 20 MeV. The major part of gamma-ray spectrometry concerns this energy range; we return later to spectrometers for gamma rays of higher energy. The variety of gamma-ray spectrometers in current use is so great that we can only summarize the properties of the most important types.

Crystal diffraction gamma-ray spectrometers make use of

the wave properties of gamma rays, in particular of constructive interference between amplitudes for elastic scattering from many atoms arranged in a regular crystalline array. Such spectrometers can achieve high resolution and high precision in the determination of gamma-ray wavelengths, and thus energies. The efficiency of such spectrometers is not large, and decreases rapidly as the gamma-ray energy increases. For further details, see X-RAY SPECTRA AND X-RAY SPECTROSCOPY.

All other gamma-ray spectrometers convert the gamma-ray energy into kinetic energy of charged particles, primarily electrons. The principal processes by which this occurs are the photoelectric effect, the Compton effect, and pair production. The most widely used spectrometers are the scintillation spectrometer and the semiconductor spectrometer. In both cases the energy and intensity of the gamma rays are ordinarily determined by the position and number of counts in a peak in the spectrum of voltage (or charge) pulses from the spectrometer, as recorded in a multichannel pulse-height analyzer. The peak usually used corresponds to absorption in the scintillator or semiconductor of the full energy of the gamma ray. A series of photoelectric, Compton, and, for high energies, pair-production and bremsstrahlung events is necessary for such total absorption.

In scintillation gamma-ray spectrometers the kinetic energy of electrons is converted into optical and ultraviolet light. The intensity of the scintillation light is proportional to the electron energy absorbed, and is measured by the output pulse height of a photomultiplier tube optically connected to the scintillating material. Sodium iodide (thallium activated) is the material most widely used for scintillation gamma-ray spectroscopy. The efficiency is large (but the higher the gamma-ray energy, the larger the crystal must be to give good efficiency for the full-energy peak). The resolution is modest; a typical full width at half maximum is 35–40 keV for 662-keV gamma rays. Such spectrometers are well suited for routine use where high resolution is unnecessary. In recent years new scintillating materials incorporating high-Z elements have come into use, especially bismuth germanate and barium fluoride.

Plastic scintillators are also used in gamma-ray spectrometers. Since plastics are composed almost entirely of low-Z elements, the efficiency is much lower (except for very low-energy gamma rays) and a Compton distribution, rather than a full-energy peak, is the most prominent feature associated with each discrete gamma ray. Plastic scintillators are well suited, however, to fast-coincidence measurements.

In semiconductor spectrometers the kinetic energy of the fast electrons from gamma-ray absorption or scattering is dissipated by the promotion of electrons from below the band gap to above. A bias voltage allows the charge thus liberated to be collected. The resolution of semiconductor spectrometers can be very good, since only a few electron volts of energy are needed for each electron liberated. Both germanium and silicon can be used, but germanium spectrometers are more efficient, except for energies below about 100 keV. The initial semiconductor gamma-ray spectrometers used lithium-drifted material for proper charge compensation, but in the last few years ultrapure intrinsic germanium suitable for spectrometers has been available. The resolution of these solid-state spectrometers can be very good; a typical full-width at half maximum for 662-keV gamma rays is 1.5–1.8 keV. Efficiencies of germanium spectrometers can now be as large as about 30% of those of 3-in. by 3-in. NaI scintillation spectrometers for the same gamma-ray energies and source-to-spectrometer distances. For gamma rays of energy above about 2 MeV the primary absorption process is pair production, and it is sometimes more convenient to use the "double-escape peak" (both 511-keV photons from positron annihilation having left the semiconductor without interacting) as the spectral feature that defines the energy and intensity of the incident gamma ray.

With total-absorption spectrometers, whether scintillation or semiconductor, study of low-intensity gamma rays in the presence of gamma rays of higher energy can be difficult. The useful dynamic range of intensity can sometimes be increased substantially by the use of more complicated geometries or multisection detectors. NaI well detectors, for example, provide not only a larger solid angle but also enhancement of the total-absorption peaks relative to background from partial absorption processes. Compton-suppression spectrometers use an auxiliary scintillator surrounding most of the principal detector, and analyze only those events in which no secondary radiation reaches the surrounding shield. Pair spectrometers require detection of 511-keV annihilation photons in two auxiliary scintillators on opposite sides of the principal detector. These techniques have many useful variants.

Mention should perhaps be made of the use of gas-filled proportional counters as total-absorption spectrometers for low-energy gamma rays.

In a rather different style of spectrometer the gamma-ray energy is used to liberate individual electrons, or electron–positron pairs, from a converter, and these electron energies, which differ from that of the initiating gamma ray by an atomic energy, are measured, usually by magnetic analysis. Photoelectric conversion is most often used, although for energies above about 1 MeV pair production or Compton conversion may be useful. Transitions between nuclear states can in general occur by internal conversion as well as by gamma-ray emission, and measurements of the energies of such transitions, and information about their intensities, can be obtained from measurements of the internal conversion electrons. The intensity of internal conversion in general decreases with the transition energy, and increases with the atomic number of the nucleus.

For high-energy gamma rays, above a few tens of MeV, the series of events associated with absorption of the energy of a gamma ray in matter necessarily occupies an extended space, and becomes a "shower." Large volumes of scintillating material are used in some spectrometers for high-energy gamma rays. Others use the Čerenkov light generated by the electrons and positrons of the shower. In the 100-MeV range a thin converter followed by magnetic analysis of the electron and positron of a pair can provide good energy resolution at the cost of high efficiency.

In almost all gamma-ray spectrometers, electronic amplification of small signals, analog-to-digital conversion, and the sorting and storing of information play important roles. Low noise, linearity, and efficiency of these electronic systems are crucial for high-quality results, as is careful calibration.

See also COMPTON EFFECT; GAMMA DECAY; SCINTILLATION AND ČERENKOV RADIATION; SEMICONDUCTOR RADIATION DETECTORS.

BIBLIOGRAPHY

C. E. Crouthamel (ed.), *Applied Gamma-Ray Spectrometry* (Pergamon, New York, 1960), reviews traditional methods and applications. (I)

G. T. Ewan, in *Progress in Nuclear Techniques and Instrumentation*, Vol. 3, pp. 69–157 (Wiley, New York, 1968), covers semiconductor spectrometers. (A)

P. Quitner, *Gamma-Ray Spectroscopy* (Adam Hilger, London, 1972), has "particular reference to detector and computer evaluation techniques." (A)

S. M. Schafroth (ed.), *Scintillation Spectroscopy of Gamma Radiation*, Vol. 1. Gordon & Breach, New York, 1967. (A)

L. C. L. Yuan and C.-S. Wu (eds.), *Nuclear Physics, Part A*, Vol. 5 of *Methods of Experimental Physics* (Academic, New York, 1963; L. Marton, ed.), contains several useful discussions of various types of spectrometers. See especially those by S. J. Lindenbaum and L. C. L. Yuan, p. 162; T. R. Gerholm, p. 582; J. W. M. DuMond, p. 599; G. D. O'Kelley, p. 616; D. E. Alburger, p. 641; and R. Hofstadter, p. 652. (A)

Gauge Theories

William J. Marciano

Gauge theories provide a fundamental description of elementary particles and their interactions. In that framework, forces between particles are mediated by gauge fields associated with an underlying symmetry. A gauge theory based on the symmetry group $SU(3)_C \times SU(2)_L \times U(1)_Y$ accounts for all observed strong, weak, and electromagnetic interaction phenomena. It correctly predicted weak neutral currents as well as the existence and properties of gluons, $W^{\pm}$, and Z gauge bosons. Because of its mathematical elegance, simplicity, and many phenomenological successes, that theory is called the "standard model" of elementary particles, a label that describes its acceptance as a standard against which future discoveries and alternative theories are to be compared.

The concept of local gauge invariance is simply illustrated for a free electrically charged field $\chi(\mathbf{x},t)$ governed by the (nonrelativistic) Schrödinger equation (with units $\hbar = c = 1$)

$$-\frac{\nabla^2}{2m}\chi(\mathbf{x},t) = -i\frac{\partial}{\partial t}\chi(\mathbf{x},t). \qquad (1)$$

The overall phase of a complex field such as χ is not measurable; so Eq. (1) must be and is invariant under global phase transformations of the form $\chi \rightarrow e^{ie\theta}\chi$, $0 \leq \theta < 2\pi$. (The unit of electric charge, e, is introduced in the exponent, so the phase symmetry can be identified with electric charge conservation.) The set of all such phase transformations form a $U(1)$ group. They are global in that the same phase change is made on $\chi(\mathbf{x},t)$ at each space-time point. If one generalizes that symmetry and requires invariance under the set of local gauge transformations,

$$\chi(\mathbf{x},t) \rightarrow e^{ie\theta(\mathbf{x},t)}\chi(\mathbf{x},t), \qquad (2)$$

where $\theta(\mathbf{x},t)$ can vary over space-time points, then Eq. (1) is no longer adequate. It must be modified by the introduction of an electromagnetic potential (gauge field) $(\mathbf{A}(\mathbf{x},t),\phi(\mathbf{x},t))$ such that

$$-\frac{1}{2m}(\nabla - ie\mathbf{A})^2\chi = i\left(\frac{\partial}{\partial t} - ie\phi\right)\chi, \qquad (3)$$

where under the local gauge transformation

$$\begin{aligned} \mathbf{A}(\mathbf{x},t) &\rightarrow \mathbf{A}(\mathbf{x},t) + \nabla\theta(\mathbf{x},t) \\ \phi(\mathbf{x},t) &\rightarrow \phi(\mathbf{x},t) + \frac{\partial}{\partial t}\theta(\mathbf{x},t). \end{aligned} \qquad (4)$$

Equation (3) is invariant under the combined transformations in Eqs. (2) and (4). The specific manner in which the electromagnetic potential is introduced into Schrödinger's equation is called minimal coupling. The terminology gauge field and gauge transformation correspond to the freedom of changing one's standard or measurement gauge for the electromagnetic potential via Eqs. (2) and (4). The gauge field $A_u \equiv (\mathbf{A},\phi)$ itself satisfies Maxwell's equations which are also invariant under Eq. (4).

The principle of local gauge invariance is one of the fundamental precepts of modern physics. It elevates ordinary global symmetries to space-time–dependent gauge symmetries via the introduction of a gauge field. The validity of that approach is borne out by the existence of 12 known gauge fields with the properties specified by the local gauge invariance prescription of the standard model.

Quantum electrodynamics (QED), the theory of interacting electrons and photons, is based on the relativistic version of the principle of local $U(1)$ gauge invariance. The electron field, $\psi(x)$, $x = (\mathbf{x},t)$, is a four-component Dirac spinor which satisfies the Dirac equation ($\partial_\mu \equiv \partial/\partial x^\mu$)

$$i[\partial_\mu - ieA_\mu(x)]\gamma^\mu\psi(x) = m\psi(x), \qquad (5)$$

where m is the electron mass; γ^μ, $\mu = 0,1,2,3$, are 4×4 Dirac matrices; and the repeated index μ is summed over. Each spinor component of ψ transforms according to Eq. (2) while the gauge field $A_\mu(x)$ satisfies Eq. (4) under local $U(1)$ gauge transformations, thus rendering Eq. (5) invariant.

Second quantizing the $\psi(x)$ and $A_\mu(x)$ fields, they become operators on a Hilbert space and represent spin-$\frac{1}{2}$ electrons and spin-1 photons, respectively. In that full framework, QED provides a relativistic quantum-mechanical description of interacting electrons and photons with the interaction specified by the coupling between $\psi(x)$ and $A_\mu(x)$ in Eq. (5). Despite its simplicity and elegance, there is a problem with QED. Products of operators become ill defined when evaluated at the same space-time point. That results in short-distance (or ultraviolet) divergences in perturbative expansions of QED. Fortunately, QED is renormalizable quantum field theory which means that ultraviolet divergences can be consistently absorbed into the difference between bare (i.e., before turning on interactions) and physical parameters (in this case the electron's mass and charge). After renormalization, high-order perturbative calculations are finite and unambiguously given in terms of e and m. Comparison of such predictions with very high-precision experiments confirms the validity of QED at a level better than 1 part in 10^{11}, making it the best tested theory in physics.

The $U(1)$ symmetry underlying QED is an Abelian (commuting) group. Any two transformations commute, $e^{ie\theta_1}e^{ie\theta_2}=e^{ie\theta_2}e^{ie\theta_1}=e^{ie(\theta_1+\theta_2)}$. As shown by C. N. Yang and R. L. Mills in 1954, the concept of local gauge invariance can be extended to non-Abelian (i.e., noncommuting) symmetry groups such as $SU(N)$, the group of unitary $N\times N$ matrices with determinant 1. For example, the $SU(2)$ isodoublet

$$\psi(x) = \begin{pmatrix}\psi_1(x)\\ \psi_2(x)\end{pmatrix}, \tag{6}$$

where each ψ_i, $i=1,2$, is a four-component Dirac spinor and satisfies the free Dirac equation

$$i\partial_\mu\gamma^\mu\psi_i(x)=m\psi_i(x), \tag{7}$$

is invariant under the general set of global $SU(2)$ transformations

$$\begin{aligned}\psi(x)&\rightarrow U\psi(x)\\ U&=\exp(-i\boldsymbol{\tau}\cdot\boldsymbol{\alpha}),\end{aligned} \tag{8}$$

where the α_i, $i=1,2,3$, are real parameters and τ_i are 2×2 Pauli matrices. That symmetry can be (gauged) expanded to local $SU(2)$ gauge invariance with $\boldsymbol{\alpha}\rightarrow\boldsymbol{\alpha}(x)$ by introducing three gauge fields $W_\mu{}^i(x)$, $i=1,2,3$ [i.e., one for each generator of $SU(2)$], which transform as an $SU(2)$ triplet. They are introduced into Eq. (7), analogous to minimal coupling, by replacing the derivative ∂_μ with a covariant derivative operator

$$\partial_\mu\rightarrow D_\mu=\partial_\mu+ig\frac{\boldsymbol{\tau}}{2}\cdot\mathbf{W}_\mu, \tag{9}$$

where, under $U(x)$, $W_\mu{}^i$ transforms as

$$\frac{\boldsymbol{\tau}}{2}\cdot\mathbf{W}_\mu\rightarrow U(x)\frac{\boldsymbol{\tau}}{2}\cdot W_\mu U^{-1}-\frac{i}{g}U^{-1}(x)\partial_\mu U(x). \tag{10}$$

The coupling g characterizes the strength of the interaction of $\psi(x)$ with the gauge field in analogy with the electric charge, e, of QED. In non-Abelian gauge theories, the fields themselves carry $SU(2)$ charge or are self-coupled because they transform as a nontrivial isovector multiplet. That is to be contrasted with QED where the photon carries no charge and is, therefore, not self-interacting. The generalization to $SU(N)$ symmetries, $N=3,4,\ldots$, is straightforward. N^2-1 gauge bosons (corresponding to the number of group generators) are introduced via a covariant derivative analogous to Eq. (9).

Like QED, non-Abelian gauge theories are renormalizable; so unambiguous finite perturbative calculations are possible. However, the fact that non-Abelian gauge fields (unlike the photon) are self-interacting makes their dynamical properties very different. The effective coupling g of non-Abelian gauge theories grows at low energies (large distances) and decreases at high energies (short distances). Those two properties find their natural application in quantum chromodynamics (QCD), an $SU(3)$ gauge theory of strong interactions. In that theory, quarks interact by exchanging gluons [non-Abelian $SU(3)$ gauge quanta]. Because their coupling to gluons grows at long distances, quarks are confined within hadrons (infrared slavery). At very short distances, quark couplings decrease and they have much less interaction with one another. The latter property known as asymptotic freedom is well established by high-energy scattering processes.

The standard model of strong, weak, and electromagnetic interactions is based on gauging the symmetry group $SU(3)_C\times SU(2)_L\times U(1)_Y$ where the subscripts denote special features of a given symmetry. The symmetries act on quark and lepton fields. (Leptons are electrons, muons, neu-

Table I Elementary particles

Particle	Symbol	Spin	Charge	Color	Mass (GeV)	
Electron neutrino	ν_e	$\frac{1}{2}$	0	0	$<1\times10^{-8}$	
Electron	e	$\frac{1}{2}$	-1	0	0.51×10^{-3}	1st
Up quark	u	$\frac{1}{2}$	$\frac{2}{3}$	3	5×10^{-3}	generation
Down quark	d	$\frac{1}{2}$	$-\frac{1}{3}$	3	9×10^{-3}	
Muon neutrino	ν_μ	$\frac{1}{2}$	0	0	$<0.25\times10^{-3}$	
Muon	μ	$\frac{1}{2}$	-1	0	0.106	2nd
Charm quark	c	$\frac{1}{2}$	$\frac{2}{3}$	3	1.25	generation
Strange quark	s	$\frac{1}{2}$	$-\frac{1}{3}$	3	0.175	
Tau neutrino	ν_τ	$\frac{1}{2}$	0	0	<0.035	
Tau	τ	$\frac{1}{2}$	-1	0	1.78	3rd
Top quark	t	$\frac{1}{2}$	$\frac{2}{3}$	3	>77	generation
Bottom quark	b	$\frac{1}{2}$	$-\frac{1}{3}$	3	4.5	
Photon	γ	1	0	0	0	
W boson	$W^\pm$	1	±1	0	80.2 ± 0.4	Gauge
Z boson	Z	1	0	0	91.1 ± 0.1	bosons
Gluon	g	1	0	8	0	
Higgs scalar	H	0	0	0	$\lesssim1000$	

trinos, etc.) The C in $SU(3)_C$ stands for color. Each quark has three color components and $SU(3)_C$ transforms them into one another. $SU(3)_C$, the basis of quantum chromodynamics, is an exact symmetry of nature. There are eight massless gluons which correspond to the eight gauge fields of $SU(3)_C$.

The L on $SU(2)_L$, the weak isospin group, denotes the fact that only left-handed components, ψ_L, of spinor fields,

$$\psi_L \equiv \frac{1-\gamma_5}{2}\psi, \tag{11}$$

transform as doublets under that group. Right-handed spinor components $\psi_R \equiv \frac{1}{2}(1+\gamma_5)\psi$ are isosinglets under $SU(2)_L$, i.e., they are unchanged under $SU(2)_L$ transformations and therefore do not couple to its three gauge fields which we denote by $W_\mu^{\pm} \equiv (W_\mu^1 \mp W_\mu^2)/\sqrt{2}$ and W_μ^3. The fact that only left-handed quarks and leptons couple to those gauge fields makes their (weak) interactions maximally parity violating.

The Y on $U(1)_Y$ stands for weak hypercharge, the charge associated with that Abelian group. That gauge group has one gauge field B_μ which couples to quarks and leptons via their weak hypercharge Y. It couples to left- and right-handed components of those particles differently and therefore also violates parity, but not maximally.

The $SU(2)_L \times U(1)_Y$ part of the standard model is not an exact symmetry. If it were, the $W^{\pm}$, W^3, and B would all be massless gauge bosons. That is not the case. To accommodate electroweak phenomenology, a scalar (spin 0) field is introduced which breaks the symmetry $SU(2)_L \times U(1)_Y$ down to the $U(1)$ symmetry of QED. That breaking gives mass to the $W^{\pm}$ and the combination of fields

$$Z_\mu = W_\mu^3 \cos\theta_W - B_\mu \sin\theta_W \tag{12}$$

which is called the Z boson. The orthogonal combination

$$A_\mu = B_\mu \cos\theta_W + W_\mu^3 \sin\theta_W \tag{13}$$

remains massless and is identified as the photon. The angle θ_W, called the weak mixing angle, is experimentally found to be $\sin^2\theta_W = 0.23$. That leads to the standard model's predictions $m_W \simeq 80$ GeV and $m_Z \simeq 91$ GeV for $W^{\pm}$ and Z boson masses. Those values have been confirmed by measurements at high-energy accelerators.

The full particle spectrum of the standard model is illustrated in Table I. All 12 gauge bosons as well as all quarks and leptons, except the top quark, have been observed. Only the top quark and Higgs scalar, a remnant of the symmetry breakdown, remain to be discovered. Their masses are, therefore, presently unknown; but some theoretical and experimental bounds do exist.

Attempts have been made to embed the standard model into a larger compact simple symmetry group such as $SU(5)$, $SO(10)$, and E_6. Such models are called grand unified theories (GUTs). The idea of a simple unified gauge theory is mathematically appealing; but, so far, there is no experimental support for any GUT. A fairly generic prediction of GUTs is that the proton should be unstable and decay, albeit with a very long lifetime. Experiments have searched for proton decay with negative findings and now give a bound on its lifetime $\tau_p > 10^{33}$ year.

See also ELECTROMAGNETIC INTERACTION; ELEMENTARY PARTICLES IN PHYSICS; FIELD THEORY, UNIFIED; GRAND UNIFIED THEORIES; QUANTUM FIELD THEORY; QUANTUM MECHANICS; WEAK INTERACTIONS.

BIBLIOGRAPHY

E. S. Abers and B. W. Lee, "Gauge Theories," *Phys. Rep.* **9,** 1 (1973).

W. Marciano and H. Pagels, "Quantum Chromodynamics," *Phys. Rep.* **36C,** 137 (1978).

J. C. Taylor, *Gauge Theories of Weak Interactions*. Cambridge, London and New York, 1976.

Gauss's Law

Robert W. Brown

In its narrowest definition, Gauss's law refers to the integral form of a Maxwell equation in electrodynamics: The electric flux through a closed surface S of any volume V is proportional to the electric charge contained in that volume. Generalizations include a result from macroscopic averaging, an "electroweak" extension, and "non-Abelian" gauge theory analysis.

The law in electrodynamics is

$$\int_S \mathbf{E}\cdot\hat{n}\,dS = K\int_V \rho\,dV. \tag{1}$$

In this equation, $\mathbf{E}$ is the electric field, ρ is the charge density, and $\hat{n}$ is a unit vector normal to the surface at the point of integration, pointing away from the enclosed volume. The proportionality constant K depends on the system of electromagnetic units. In Gaussian units, $K=4\pi$. In another popular system, rationalized mks units, $K=\epsilon_0^{-1}$, where $(4\pi\epsilon_0)^{-1} = 10^{-7}c^2$ (c is the speed of light).

The left-hand side of Eq. (1) can be taken as a definition of the flux through S associated with a vector field (a vector whose direction and magnitude may change in space). If we imagine drawing lines in space denoting the direction of $\mathbf{E}$ at any point with the density of such lines proportional to the magnitude $|\mathbf{E}|$, then the flux through S is proportional to the number of lines crossing S. Historically, one thought of the flow of something through S, a picture which helps us to introduce similar statements about other physical phenomena. Equations analogous to Eq. (1) can be developed for heat conduction, gravity, water flow, particle diffusion, quantum-mechanical probability, electric current conservation, and so forth. Aside from sources or sinks (charges, heat generators, masses, etc.), the common denominator is a conservation law prohibiting the disappearance of the "stuff" flowing out or in.

The point is that Eq. (1) implies that static electric field lines cannot end outside of the source region and will spread according to the familar inverse-square law, which brings to mind the usual derivation of Eq. (1) for electrostatics. Consider the Coulomb field of an electric point charge q at rest:

$$\mathbf{E} = \frac{Kq}{4\pi}\frac{\hat{r}}{r^2}. \tag{2}$$

The fact that this field is radial (central) and drops off like r^{-2} leads to

$$\int_S \mathbf{E}\cdot\hat{n}dS = 0 \tag{3}$$

if the charge is outside V. That is, contributions to the integral from opposite areas on S subtending the same solid angle at the point charge cancel. (Observe that $dS = r^2d\Omega/|\cos\theta|$, where θ is the angle between $\hat{n}$ and $\mathbf{E}$.) Now suppose S consists of disconnected inner and outer surfaces with V the volume between them and the charge inside the inner surface [still outside V so Eq. (3) is still applicable]. If the inner surface is a sphere centered around the charge, the corresponding inner contribution to Eq. (3) is $-Kq$ [see Eq. (2)]. By linear superposition of the fields due to an assemblage of charges, Gauss's law (1) follows.

We can convert Gauss's law into differential form. What is needed is the divergence theorem, sometimes called Gauss's theorem, which reads

$$\int_S \mathbf{F}\cdot\hat{n}dS = \int_V \nabla\cdot\mathbf{F}dV \tag{4}$$

for any suitable vector field $\mathbf{F}$. Since this is an integral identity, the class of vectors $\mathbf{F}$ can be generalized to include those whose derivatives are distributions or generalized functions, which actually arise in $\nabla\cdot\mathbf{E}$ for a point charge. In fact, Eqs. (2) and (4) imply the identification

$$\nabla\cdot\mathbf{E} = Kq\delta(\mathbf{r}) \tag{5}$$

in terms of the Dirac delta function. The derivation of Eq. (4), in which partial integration of the right-hand side is the principal step, shows that this theorem applies to an arbitrary number of dimensions with S a hypersurface of V; this is relevant to conservation laws and the investigation of conserved "charges" in four-dimensional space-time. Getting back to Gauss's law, Eqs. (1) and (4) can be combined to give

$$\nabla\cdot\mathbf{E} = K\rho \tag{6}$$

by considering arbitrarily small volumes. This is the promised differential form.

Equation (6) is an equivalent formulation of Gauss's law and one of the Maxwell differential equations. It must be emphasized that no change is required when time-dependent fields and sources are considered. We can use Eq. (1) even if the charges are moving relativistically; the field is to be evaluated at all points on S and the density at all points inside V at the same time in a given Lorentz frame of reference.

Furthermore, the full quantum theory of electrodynamics assumes the correctness of Gauss's law where the electric field and the charge density are now operators in a Hilbert space. Therefore this law is a basic "equation of motion" which governs electromagnetism from microscopic distances (checked down to 10^{-17} cm for electrons and photons) to intergalactic plasmas. It is interesting that Gauss's law is a statement about the masslessness of photons, the electromagnetic quanta. If the photon had a mass λ, the Coulomb potential would be modified by a factor $e^{-\lambda cr/\hbar}$, where $\hbar$ is Planck's constant (essentially a Yukawa potential). Flux lines could disappear as we go away from the sources and Eq. (1) would not be valid.

In fact, generalizations are characterized by nonvanishing divergence. In electroweak theory, there is a partner to the photon, the neutral Z boson, whose huge mass most definitely changes things. In non-Abelian gauge theories, the fields carry "charge." In macroscopic media, the polarization charge is absorbed into a displacement vector $\mathbf{D}$, hiding the fact that $\nabla\cdot\mathbf{E}$ may not be zero.

See also ELECTRODYNAMICS, CLASSICAL; MAXWELL'S EQUATIONS.

BIBLIOGRAPHY

R. P. Feynman, R. B. Leighton, and M. Sands, *The Feynman Lectures on Physics,* Vol. II. Addison-Wesley, Reading, Mass., 1964. (E)

C. Itzykson and J.-B. Zuber, *Quantum Field Theory*. McGraw-Hill, New York, 1980. (A)

J. D. Jackson, *Classical Electrodynamics,* 2nd ed. Wiley, New York, 1975. (I)

Geochronology

Marvin A. Lanphere

INTRODUCTION

Geochronology is the study of time in relation to the history of the earth. The term includes *relative* dating systems based on fossils and *physical* (sometimes called absolute) dating systems based on radioactive decay. These physical dating methods, collectively called radiometric dating, are based on decay of naturally occurring radioactive nuclides in rocks and minerals.

In 1896 the French physicist Henri Becquerel discovered that uranium salts emitted radiation that had properties similar to x rays. This spontaneous emission of radiation was termed "radioactivity," and this property is the basis for all radiometric dating methods. In the next few years many investigators provided evidence on the manner in which one element is produced from another by radioactive decay. In 1905 Ernest Rutherford suggested that the rate of transformation of radioactive nuclides might be used to determine the age of geologic materials. Rutherford in 1906 determined the first mineral ages using the uranium and helium contents of minerals and the disintegration rate of radium, a decay product of uranium. Bertram Boltwood suggested that lead was the final product of the radioactive decay of uranium, and in 1907 he published uranium–lead ages for several minerals. These ages were calculated before the distintegration rate of uranium was accurately known, before isotopes were discovered, and before it was discovered that lead also is a product of the radioactive decay of thorium. But it was clear even then that the phenomenon of natural radioactivity offered great potential in constructing a quantitative scale of geologic time.

PRINCIPLES OF RADIOMETRIC DATING

When a radioactive parent atom decays to produce a stable daughter atom, each disintegration results in one more atom of daughter and one less atom of parent. Several workers in the early 1900s conducted experiments on uranium and thorium salts that led to construction of a general theory to explain radioactivity. It was suggested that atoms of radioactive elements are unstable and a fixed proportion of the atoms spontaneously disintegrate in a given period of time to form atoms of a new element. It was further proposed that the activity or intensity of radiation is proportional to the number of atoms that disintegrate per unit time. The probability of a parent atom decaying in a given period of time is the same regardless of temperature, pressure, or chemical conditions; this probability of decay is called the decay constant. Thus, if one knows the decay constant and can measure the amounts of parent and daughter isotopes in a mineral, then the time since the mineral formed can be calculated. Another measure of the rate of radioactive decay is the half-life which is inversely proportional to the decay constant; the half-life is the time required for one-half of the number of radioactive parent atoms to decay.

A number of long-lived radioactive isotopes have been used in radiometric dating; some of the parent and daughter nuclides and half-lives are shown in Table I. Some radioactive parent nuclides decay directly to stable daughter nuclides in a process known as simple decay. The radioactive uranium and thorium parents decay to radioactive daughter products that also decay, and so forth, until stable lead isotopes are produced in a process known as chain decay. ^{40}K decays to two different daughter elements in a process known as branching decay. All of the decay schemes in Table I, except radiocarbon, are based on accumulation of daughter nuclides. The radiocarbon (^{14}C) method is based on disappearance of the parent nuclide by radioactive decay.

Key elements in radiometric dating are knowing the decay constants accurately and demonstrating that the rate of decay is constant. The decay constants for radioactive elements listed in Table I are known with an accuracy of a few percent from laboratory experiments involving counting the number of particles emitted per unit time by a known quantity of parent nuclide. Because radioactive decay occurs spontaneously in the nucleus of an atom, the decay rate should not be affected by physical or chemical conditions. Many attempts have been made to change radioactive decay rates, but these have not produced any significant changes. A type of decay known as electron capture, in which an orbital electron falls into the nucleus, is affected very slightly by external conditions because the process involves a particle outside the nucleus. Measurements have shown, however, that physical and chemical conditions have no significant effects on the decay constant of isotopes which decay by electron capture.

Table I. Parent and Daughter Isotopes Used in Radiometric Dating

Parent isotope	Daughter isotope	Half-life (years)
Carbon-14 (^{14}C)	Nitrogen-14 (^{14}N)	5.73×10^3
Potassium-40 (^{40}K)	Argon-40 (^{40}Ar)	1.25×10^9
Rubidium-87 (^{87}Rb)	Strontium-87 (^{87}Sr)	4.88×10^{10}
Samarium-147 (^{147}Sm)	Neodymium-143 (^{143}Nd)	1.06×10^{11}
Lutetium-176 (^{176}Lu)	Hafnium-176 (^{176}Hf)	3.59×10^{10}
Rhenium-187 (^{187}Re)	Osmium-187 (^{187}Os)	4.3×10^{10}
Thorium-232 (^{232}Th)	Lead-208 (^{208}Pb)	1.40×10^{10}
Uranium-235 (^{235}U)	Lead-207 (^{207}Pb)	7.04×10^8
Uranium-238 (^{238}U)	Lead-206 (^{206}Pb)	4.47×10^9

There are several fundamental conditions that must be met before a radioactive decay scheme can be used to successfully measure ages of geologic material. These are:

1. The decay of a radioactive parent takes place at a constant rate that is accurately known.
2. The present-day proportion of a radioactive parent nuclide to the total quantity of the same element is the same in all materials. There is no natural fractionation of isotopes of the same element.
3. A rock or mineral may contain nonradiogenic nuclides of the same mass as the radiogenic daughter; in this case a correction must be made for the nonradiogenic nuclides present when the rock or mineral formed.
4. The rock or mineral has been a closed system since the time of formation. That is, there has been no gain or loss of radioactive parent or radiogenic daughter.

RADIOMETRIC DATING METHODS

This review is too short to describe all of the dating methods in use. Only the three most commonly used methods will be described. These are the potassium–argon (K–Ar), uranium–lead (U–Pb), and radiocarbon (^{14}C) methods. These methods have been applied to problems covering the entire span of geological time from the present to the formation of Earth, moon, and meteorites.

The K–Ar Method

The K–Ar method is the radiometric dating technique used most widely by geologists. Potassium is a common element found in many minerals and the half-life of ^{40}K has a value that permits application to very young as well as ancient rocks. ^{40}K has a branching decay to produce the daughter products ^{40}Ca and ^{40}Ar. ^{40}Ca is the most abundant isotope of calcium which also is very abundant in the Earth's crust. Thus, it usually is not possible to correct for the ^{40}Ca initially present and the ^{40}K–^{40}Ca decay is rarely used for dating, except with potassium-rich, calcium-poor salts. ^{40}K also decays to ^{40}Ar, a noble gas, and a K–Ar age commonly is referred to as a gas-retention age. The K–Ar method is used mostly on igneous rocks, such as lava or granite, where generally there is no initial radiogenic ^{40}Ar. While a rock is molten, any radiogenic ^{40}Ar will escape from the liquid, but as the rock solidifies and cools the radiogenic ^{40}Ar is trapped within mineral grains and accumulates as time passes. However, argon makes up approximately 1% of the Earth's atmosphere, and this nonradiogenic argon is present in most minerals and the laboratory apparatus in which samples are processed. The correction for this nonradiogenic argon is easily made and, unless the atmospheric argon is a large proportion of the total argon, the calculated age is not adversely affected.

For the conventional K–Ar method, the quantities of potassium and argon in a sample are measured in separate experiments by different techniques. In a variant of K–Ar dating known as the $^{40}Ar/^{39}Ar$ method the sample is irradiated in a nuclear reactor in order to convert part of the ^{39}K, which is the most abundant isotope of potassium, to ^{39}Ar. Then the ratio of daughter to parent is given by the ratio of ^{40}Ar to ^{39}Ar. Corrections must be made for atmospheric argon and interfering isotopes produced by undesirable neutron reactions with calcium and potassium. If the irradiated sample is totally melted in one experiment, the method gives ages with precisions comparable to conventional K–Ar ages.

An irradiated sample also can be heated in increments to progressively higher temperature and the gas released at each temperature treated as a separate experiment. Data from these experiments are plotted as an age spectrum (apparent age versus percentage of ^{39}Ar released). From the shape of the spectrum one can infer whether the sample is undisturbed, has suffered argon loss since crystallization, or contains excess (initial) argon. Although many age spectra from rocks with complex histories cannot yet be interpreted, this new technique has greatly expanded the usefulness of the K–Ar method.

Conventional K–Ar ages younger than five thousand years have been measured on lavas; these ages overlap the range covered by ^{14}C ages. $^{40}Ar/^{39}Ar$ ages older than four billion years have been measured on lunar rocks and meteorites. The ability to apply the method to the entire span of geologic time underscores the importance of the K–Ar dating method.

The U–Pb Method

The radioactive decays of ^{238}U to ^{206}Pb and ^{235}U to ^{207}Pb yield two independent ages. The decay of ^{232}Th to ^{208}Pb gives a third independent age. An age also can be calculated from the $^{207}Pb/^{206}Pb$ ratio since this ratio changes as a function of time because the half-lives of the two parent uranium isotopes differ by a factor of about 6. If these four ages agree, then this represents the age of the rock. However, the ages often do not agree because lead is often lost from minerals whose crystal structures have been damaged by radioactive decay. Some graphical and numerical procedures have been developed that permit interpretation of discordant U–Pb ages.

On a U–Pb parent–daughter diagram the locus of concordant $^{206}Pb/^{238}U$ and $^{207}Pb/^{235}U$ ages is a curve called "concordia." However, most rock-forming minerals yield U–Pb ages that are discordant to some degree and do not plot on concordia. This is true for zircon, an accessory mineral in igneous rocks which contains uranium when it crystallizes but very little lead. The ages for a suite of zircon samples which have lost different amounts of lead generally form a linear array, called discordia, below the concordia curve. A chord drawn through this array will intersect concordia in two places. The older intercept is considered the original age of the rock. The younger intercept may be interpreted in several ways, but usually it indicates the time at which lead loss occurred. The U–Pb concordia–discordia system is quite resistant to heating and is the dating method of choice in rocks with complex histories.

The Radiocarbon Method

Radiocarbon (^{14}C) dating is quite different from other radiometric dating methods because it depends on disappearance of the parent isotope rather than accumulation of the daughter isotope. The ^{14}C method is used on organic material, and its half-life of 5730 years restricts the method's use to about 50,000 years.

^{14}C is produced in the upper atmosphere by a number of nuclear reactions, the most important of which is the interaction of cosmic ray neutrons and ^{14}N. ^{14}C is radioactive and decays back to ^{14}N. The ^{14}C produced in the atmosphere is incorporated into organic material, and as long as the organism lives, the ^{14}C in the organism and the atmospheric reservoir remain in equilibrium. When the organism dies, the ^{14}C begins to disappear with time because it ceases to be replenished. The ^{14}C age of a sample is given by the amount of ^{14}C left compared to the amount of ^{14}C in a modern sample of known age.

For the ^{14}C dating method to work requires a constant inventory of ^{14}C in the atmosphere. It is known, however, that the production rate of ^{14}C has not been constant with time because of variations in the cosmic ray flux in the upper atmosphere. Deviations from a constant production rate have been documented by dating tree rings from certain long-lived species. This detailed dendrochronology has been determined back to 5400 B.C. This calibration curve is used to adjust measured ages for variations in the ^{14}C production rate. The accuracy of ages older than 5400 B.C. may have uncertainties of $\pm 10\%$ or so. However, the relative differences between older ^{14}C ages are not affected by production rate variation.

CONCLUSIONS

Radiometric dating now is a standard technique in most geological investigations. The use of radioactive decay to measure elapsed time has been applied to problems in the earth sciences since early in the century. However, accurate measurement of decay constants, development of analytical techniques, and availability of adequate instrumentation dates mostly from about 1950. During the past four decades technique development has permitted the measurement of ages with analytical precisions of better than $\pm 0.5\%$ for several of the dating methods. During the past 10 years techniques for measuring K–Ar and U–Pb ages on single mineral grains have been developed. The future promises continued improvements and extension of radiometric dating to a broad range of geological and geophysical problems.

See also COSMIC RAYS, SOLAR SYSTEM EFFECTS; GEOPHYSICS; RADIOACTIVITY.

BIBLIOGRAPHY

G. B. Dalrymple and M. A. Lanphere, *Potassium-Argon Dating*. Freeman, San Francisco, 1969. (I)

G. Faure, *Principles of Isotope Geology*. John Wiley & Sons, New York, 1986. (A)

D. York and R. M. Farquhar, *The Earth's Age and Geochronology*. Pergamon Press, London, New York, 1972. (E)

Geomagnetism

J. A. Jacobs

At its strongest near the poles, the Earth's magnetic field is several hundred times weaker than between the poles of a toy horseshoe magnet—being less than a gauss (G). Thus in geomagnetism we are measuring extremely small magnetic fields and a more convenient unit is the gamma (γ), defined as 10^{-5} G. These are the traditional units used in geomagnetism and are found in the older literature. SI units were officially adopted in 1973: the conversion factor is $1\ \text{G} = 10^{-4}\ \text{Wb/m}^2 = 10^{-4}$ tesla (T). Thus, $1\ \gamma = 10^{-9}\ \text{T} = 1$ nT.

In a magnetic compass the needle is weighted so that it will swing in a horizontal plane, and its deviation from geographic north is called the declination D. Over most of the northern hemisphere the north-seeking end of a magnetized needle balanced horizontally on a pivot will dip downward, the angle it makes with the horizontal being called the magnetic dip or inclination I. Over most of the southern hemisphere, the north-seeking end of the needle points upward.

At points on the Earth's surface where the horizontal component of the magnetic field vanishes, a dip needle will rest with its axis vertical. Such points are called dip poles. Two principal poles of this kind are situated near the north and south geographic poles and are called the magnetic north and south poles. Their positions in 1980 were 77.3°N, 258.2°E and 65.6°S, 139.4°E. They are thus not diametrically opposite, each being about 2500 km from the point antipodal to the other. The total intensity F of the Earth's magnetic field is a maximum near the magnetic poles—its value is just over 60 μT near the northern dip pole and just over 70 μT near the southern dip pole. In some areas such as Kursk, south of Moscow, and Berggiesshubel in Germany, the magnitude of F may exceed 300 μT, but this is entirely due to local concentrations of magnetic ore bodies.

The variation of the magnetic field over the Earth's surface is best illustrated by isomagnetic charts, i.e., maps on which lines are drawn through points at which a given magnetic element has the same value. Figures 1 and 2 are world maps showing lines of equal declination (isogonics) and lines of equal inclination (isoclinics) for the year 1955. It is remarkable that a phenomenon (the Earth's magnetic field) whose origin lies within the Earth should show so little relation to the broad features of geography and geology. The isomagnetics cross from continents to oceans without disturbance and show no obvious relation to the great belts of folding

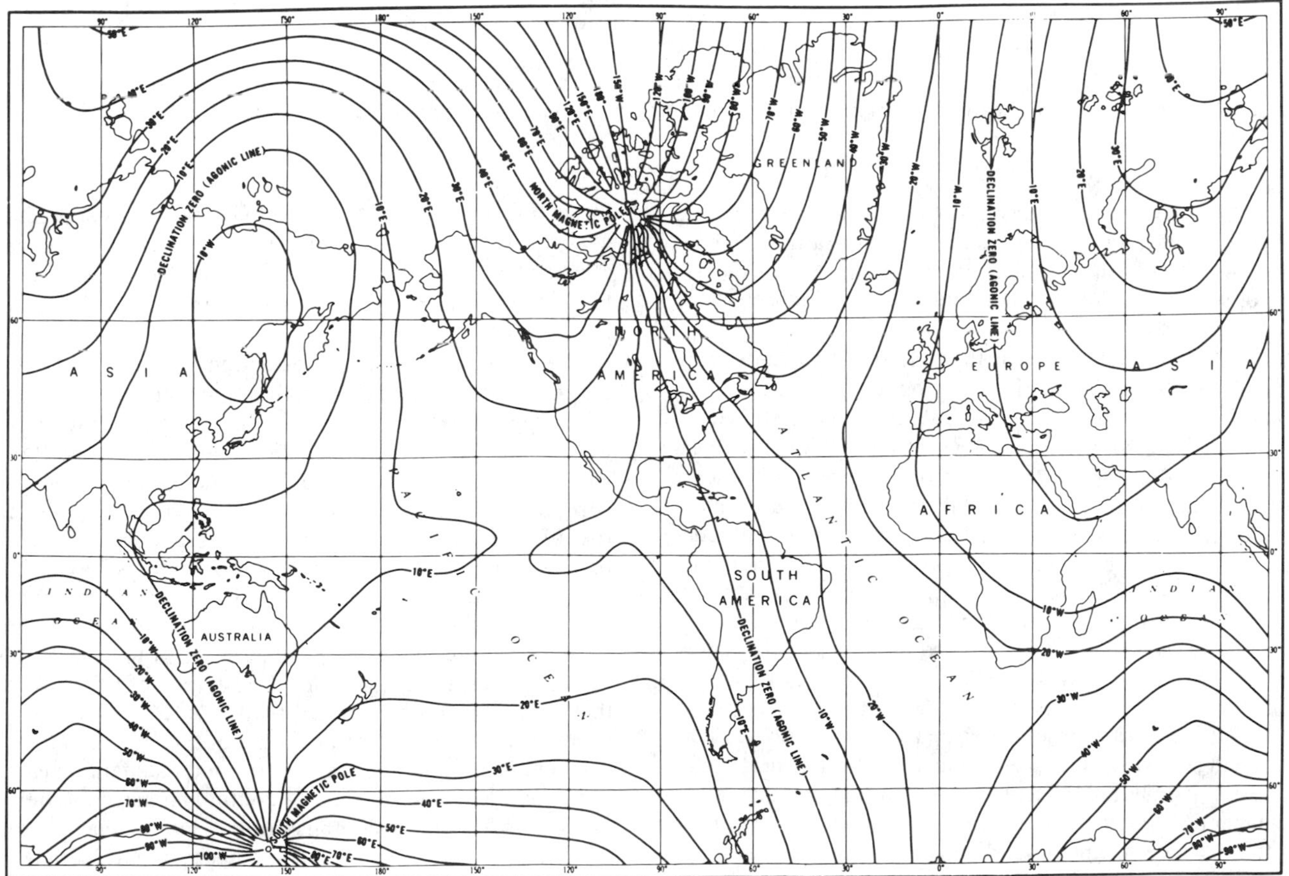

FIG. 1. Simplified isogonic chart of the entire world except polar regions, for 1955. Mercator projection. (Derived from U.S. Navy Hydrographic Office Chart 1706.)

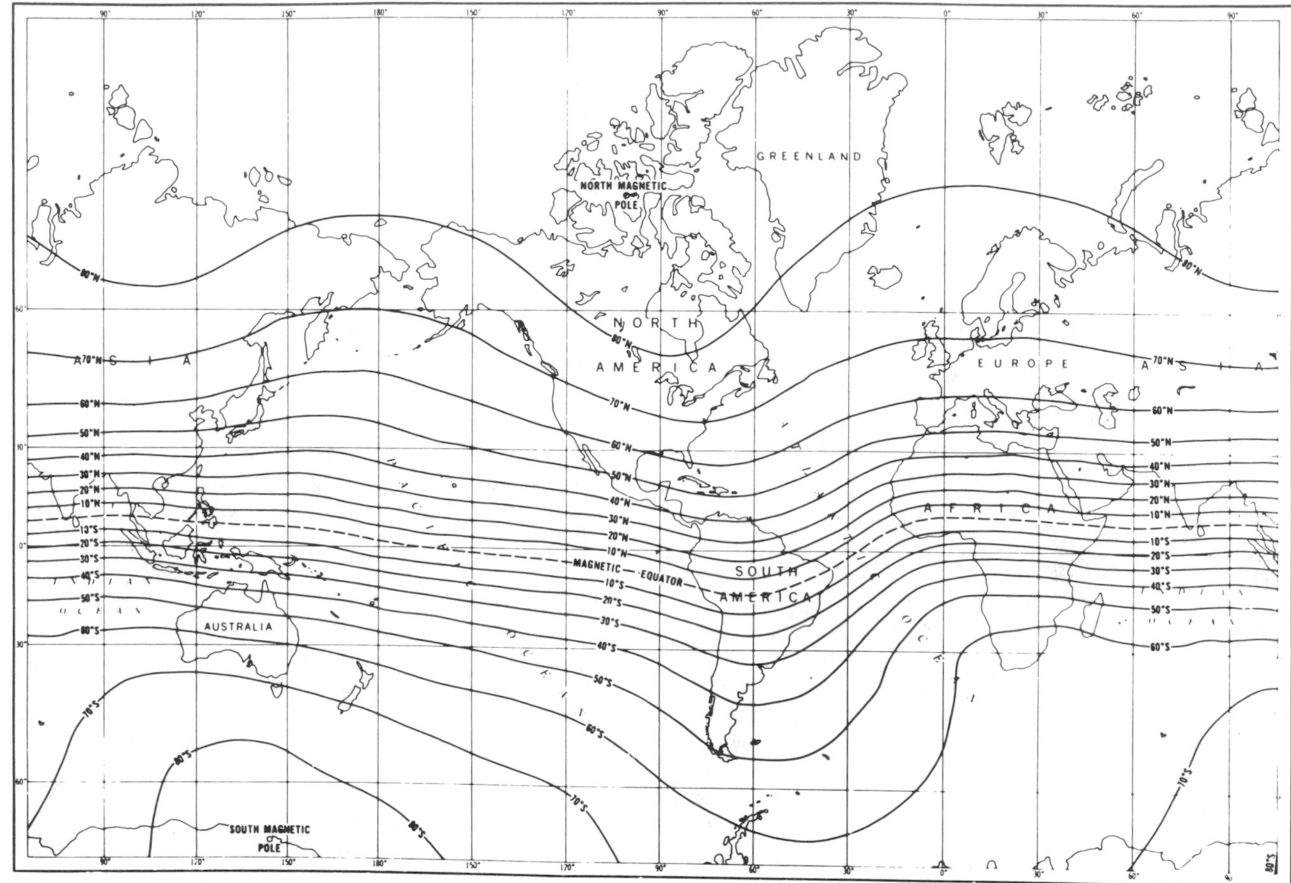

FIG. 2. Simplified isoclinic chart of the entire world except polar regions, for 1955. Mercator projection. (Derived from U.S. Navy Hydrographic Office Chart 1700.)

or to the pattern of submarine ridges. In this respect, the magnetic field is in striking contrast to the Earth's gravitational field and to the distribution of earthquake epicenters, both of which are closely related to the major features of the Earth's surface.

In 1839 Gauss showed that the field of a uniformly magnetized sphere (which is the same as that of a geocentric dipole) is an excellent first approximation to the Earth's magnetic field. This was nearly 250 years after William Gilbert had reached the same conclusion from experimental studies on the variation in direction of the magnetic force over the surface of a piece of the naturally magnetized mineral lodestone which he had cut in the shape of a sphere. The points where the axis of the geocentric dipole which best approximates the Earth's field meet the surface of the Earth are called the geomagnetic poles and are situated approximately at 79°N, 289°E and 79°S, 109°E. The geomagnetic axis is thus inclined at about 11° to the Earth's geographic axis.

In 1635 Gellibrand discovered that the magnetic declination changed with time. This change in the magnetic field with time is called the secular variation and is observed in all magnetic elements. A spherical harmonic analysis of the Earth's surface magnetic field shows that its source is predominantly internal. Superimposed on this field is a rapidly varying external field giving rise to transient fluctuations. Unlike the secular variation, which is also of internal origin, these transient fluctuations produce no large or enduring changes in the Earth's field. They are mostly due to solar effects which disturb the ionosphere and give rise to a number of related upper atmospheric phenomena such as magnetic storms and aurora.

There has been much speculation on the origin of the Earth's main field and no completely satisfactory answer has as yet been given. The only possible source seems to be some form of electromagnetic induction, electric currents flowing in the Earth's fluid, electrically conducting (mainly iron) core. Such currents may have started originally by chemical irregularities which separated charges and thus initiated a battery action. Paleomagnetic measurements have shown that the Earth's main field has existed throughout most of geologic time and that its strength has never differed much from its present value. Electric currents in the Earth would decay by Joule heating in a time of the order of 10^5 yr whereas the age of the Earth is more than 4×10^9 yr, so that the geomagnetic field cannot be a relic of the past, and a mechanism must be found for generating and maintaining electric currents to sustain the Earth's continuing magnetic field. A process that could accomplish this is the familiar action of the dynamo. The dynamo theory suggests that the magnetic field is ultimately produced and maintained by an induction process, the magnetic energy being drawn from the kinetic energy of the fluid motions in the core.

The energy needed to drive the geodynamo could originate in a variety of forms (gravitational, chemical, or thermal), ultimately being converted into heat that flows out into the mantle. However, the probable efficiency of any heat-driven dynamo is very small, being only about 5–10%. The most favored energy source is compositionally driven convection caused by the cooling and freezing of the liquid outer core to form the solid inner core. This separates a heavy fraction (mainly iron) in the inner core, leaving behind in the outer core a lighter, buoyant liquid fraction.

One of the most interesting results of paleomagnetic studies is the discovery that the Earth's field has reversed its polarity many times. Reverse magnetization was first discovered in 1906 by Brunhes in France and examples have since been found in every part of the world. Over the last 45 Myr the field has reversed at least 150 times. The frequency of reversals has however changed throughout geologic time, and there were times when there were almost no reversals; e.g., during the Cretaceous (from about 107 to 85 Myr ago) the field remained normal and during the upper Carboniferous and Permian (from about 290 to 235 Myr ago) the field remained reversed with but few changes in polarity.

Reversals of the Earth's field have played a key role in the development of plate tectonics, which has revolutionarized geologic thinking during the last decade. As new oceanic crust forms at the center of an oceanic ridge and cools through its Curie temperature, the permanent component of its magnetization will assume the ambient direction of the Earth's magnetic field. The striped pattern of magnetic anomalies observed in the ocean basins is thus due to seafloor spreading as the upwelling mantle material is carried away from the ridge in both directions.

See also GEOPHYSICS.

BIBLIOGRAPHY

J. A. Jacobs, *The Earth's Core,* 2nd ed. Academic Press, New York, 1987.

R. T. Merrill and M. W. McElhinny, *The Earth's Magnetic Field.* Academic Press, New York, 1983.

W. D. Parkinson, *Introduction to Geomagnetism.* Scottish Academic Press, 1983.

Geometric Quantum Phase

J. H. Hannay

The geometric quantum phase, or "Berry phase," results from a fundamental feature of quantum states: while the phase of $\langle a|b\rangle\langle b|a\rangle$ is zero, the phase of $\langle a|c\rangle\langle c|b\rangle\langle b|a\rangle$ is *not* zero. Moreover, because each of the states occurs twice, it is independent of their individual, arbitrary, phases. The Berry phase is defined not for this discrete chain of states, but for any *continuous cyclic* chain of states $|n(\mathbf{R})\rangle$, labelled by their position $\mathbf{R}$ in parameter space, as the phase of the product $\langle n(\mathbf{R}_0)|n(\mathbf{R}_\infty)\rangle \ldots \langle n(\mathbf{R}_2)|n(\mathbf{R}_1)\rangle\langle n(\mathbf{R})|n(\mathbf{R}_0)\rangle$. Each term is of the form $\langle n+dn|n\rangle = 1+\langle dn|n\rangle = \exp(id\gamma)$ in the limit $|dn\rangle \rightarrow 0$, where $d\gamma = -i\langle dn|n\rangle = i\langle n|dn\rangle$ (using $d\langle n|n\rangle = 0$). So the Berry phase is $\gamma = \int d\gamma = i\int\langle n|dn\rangle$. Here $\langle n|dn\rangle$ means $\langle n|\nabla n\rangle \cdot d\mathbf{R}$, or it can be read directly as a 1-form: the "phase 1-form" in parameter space.

Schematically, if the parameter space is thought of as the surface of a sphere, the cycle of states could be represented as a loop of points on it, with the (arbitrary) phase of each state indicated by the direction of a little arrow in the surface attached to its representative point. The change of direction between neighboring arrows (the angle between one and its *parallel transported* neighbor) could be likened to the phase of $\langle n(\mathbf{R}+d\mathbf{R})|n(\mathbf{R})\rangle$. Just as the sum of these direction increments around the loop is nonzero for a curved surface, so also the phases add up around the cycle to give a nonzero sum. In both these "holonomy" effects, the sum captures a fundamental quantity enclosed by the loop. On the one hand, the direction change captures the integrated curvature of the surface (the integrated flux of the curvature 2-form). On the other hand, the Berry phase captures, by Stokes's theorem, the integrated flux of a "curl": $\nabla\wedge\langle n|\nabla n\rangle = \langle \nabla n|\wedge|\nabla n\rangle$, or more correctly, of a 2-form $d\langle n|dn\rangle = \langle dn|\wedge|dn\rangle$. Unlike the phase 1-form, the phase 2-form does *not* depend on the arbitrary choice of phases of the states $|n\rangle$. For the simplest circumstance of a quantum system with only two basis states, the sphere construction above is not merely a schematic description but a quantitative one (the sphere being the "Bloch sphere," or "Poincaré sphere" in optics).

The Berry phase is realized physically by adiabatic (i.e., infinitely slow) cyclic change of a quantum Hamiltonian $H(\mathbf{R}(t))$. The quantum adiabatic theorem guarantees that if a system starts in an eigenstate of this Hamiltonian, the nth say, then it will stay in the nth, making no transitions to others on the way. The phases of the initial and final states, however, are not the same. During the excursion the phase changes at a rate $-\hbar^{-1}$ times the energy, $E_n(\mathbf{R}(t))$, of the nth eigenstate of $H(\mathbf{R}(t))$ so there is a "dynamical" phase change given by the integral of this over the duration of the excursion: $-\hbar^{-1}\int E_n(\mathbf{R}(t))dt$. If the Hamiltonian, and therefore the eigenstates, are purely real, this is the only phase change apart from a possible change of sign (see next paragraph). However for a general (Hermitian) Hamiltonian there is an extra phase change, the Berry phase, due to the cycle that $|n(\mathbf{R}(t))\rangle$ has executed. The derivation is straightforward: the instantaneous state, $|\psi(t)\rangle$, is given by $e^{i\chi(t)}|n(\mathbf{R}(t))\rangle$, where $\chi(t)$ is to be determined. Substitution into the Schrödinger equation $H|\psi\rangle = i\hbar|\dot\psi\rangle$ and use of $H|n\rangle = E_n|n\rangle$ yields $E_n e^{i\chi}|n\rangle = i\hbar[i\dot\chi e^{i\chi}|n\rangle + e^{i\chi}|\nabla n\rangle\cdot\dot{\mathbf{R}}]$. Premultiplying by $\langle n|$ gives $\dot\chi = -E_n/\hbar + i\langle n|\nabla n\rangle\cdot\mathbf{R}$. The time integral of the second term is recognized as the Berry phase; "geometrical" because it only depends on the sequence of parameters passed through, not on the duration of each.

In this adiabatic realization of the geometrical phase, the phase 2-form $\langle dn|\ |dn\rangle$ can be expressed as a sum over states $m\neq n$:

$$\langle dn|\wedge|dn\rangle = \sum \langle dn|m\rangle\wedge\langle m|dn\rangle = \sum \langle n|dH|m\rangle\wedge\langle m|dH|n\rangle/(E_n-E_m)^2,$$

as can be verified by differentiating the relations $\langle n|H|m\rangle = 0$, $\langle n|n\rangle = 1$, and using $H|n\rangle = E_n|n\rangle$. This expression shows that the 2-form is singular near "accidental" degeneracy points **R**, generic in parameter space of 3-D (or more), where $E_n(\mathbf{R})$ contacts a neighboring energy $E_m(\mathbf{R})$ $(m = n \pm 1)$. Local analysis shows it to take an inverse square form (like the magnetic field a magnetic monopole would produce). If we imagine a succession of loops in parameter space, enclosing the degeneracy point like the lines of latitude of a globe, the Berry phase factor $e^{i\gamma}$ changes from loop to loop. It starts at unity for the little loop near the north pole and traces out the unit circle in the complex plane, ending at unity again at the south pole. The integrated flux of the phase 2-form through the enclosing surface is the change of γ: $\pm 2\pi$ depending on the direction of the tracing. (The integrated flux for the partner state m is the opposite: $\mp 2\pi$.) Degeneracies thus bring topology into the geometric phase. Indeed if the Hamiltonian is purely real, only topology survives: the geometric phase is zero for any loop except those enclosing degeneracy lines (the reality restriction makes degeneracies lines, not points, in 3-D). For enclosing loops the phase turns out to be π, so the Berry phase factor is a sign change.

One manifestation of the Berry phase arises in the Born–Oppenheimer theory of molecules in which the nuclear coordinates **R** are adiabatically changing parameters as far as the (nondegenerate) electron state $|n(\mathbf{R})\rangle$ is concerned. This provides a standard potential energy function $V_n(\mathbf{R}) = \langle n(\mathbf{R})|H_{\text{elec}}|n(\mathbf{R})\rangle$ felt by the nuclei which affects their motion by adding a term $V_n(\mathbf{R})|N\rangle$ to the Schrödinger equation for the state $|N\rangle$ of the nuclear motion. However the kinetic energy form for the nuclei is also influenced, being not merely $\nabla^2|N\rangle$ but $\langle n|\nabla^2|n\rangle|N\rangle$. This evaluates to $(\nabla - \mathbf{A})(\nabla - \mathbf{A})|N\rangle + \phi|N\rangle$ where the vector $\mathbf{A}(\mathbf{R}) \equiv \langle n|\nabla n\rangle$ acts like a magnetic vector potential corresponding to the phase 1-form, and $\phi(\mathbf{R})$ is a certain additional potential energy. The derived magnetic field $\nabla \wedge \mathbf{A}$ corresponds to the phase 2-form. A similar effect is present in the theory of gauge anomalies in quantum field theory. Other phenomena in which the Berry phase plays a role include magnetic resonance and the quantum Hall effect.

A geometrical holonomy effect analogous to the Berry phase is present in the classical mechanics of integrable (i.e., chaos free) systems. Their "action" variable I is adiabatically conserved, while the change in the "angle" variable under cyclic excursion of the classical Hamiltonian is a dynamical part plus a geometric angle θ. In the semiclassical limit $(\hbar \to 0)$ the geometric phase γ is related to θ by $\theta = -\hbar d\gamma/dI$.

See also Hall Effect, Quantum; Hamiltonian Function; Resonance Phenomena.

BIBLIOGRAPHY

M. V. Berry, *Proc. Roy. Soc.* A *392*, 45 (1984).

M. V. Berry, *Sci. Am.* **259** No 6, 46, December (1988).

A. Shapere, and F. Wilczek, "Geometric Phases in Physics," *Advanced Series in Mathematical Physics*, Vol. 5. World Scientific (1989).

Geophysics

Thomas J. Ahrens

Geophysics in its broadest sense includes the study of the physical processes and properties of the atmosphere and hydrosphere as well as those of the solid earth. Hence atmospheric physics, meteorology, oceanography, and hydrology may be considered as branches of geophysics, although the term more commonly refers to the study of the solid earth (and planets) alone. The major subdivisions of solid-earth geophysics are seismology, geodesy and the earth's gravitational field, geomagnetism and electricity, thermal properties of the earth, rheology, and mineral geophysics. Applied geophysics describes the techniques and theory of geophysical exploration and prospecting; these include seismic methods, gravity and magnetic surveys, electrical and combined electromagnetic measurements, and investigation of surface heat flow. Marine geophysics is a broad subdivision concerned with the properties of the earth beneath the oceans. In the study of geophysics much use is made of knowledge from other branches of science, principally from physics (e.g., elasticity, electromagnetic theory, fluid dynamics, thermodynamics and solid-state physics), but also from geology, chemistry and geochemistry, crystallography, and mineralogy.

The theories of sea-floor spreading and plate tectonics have been the most significant unifying hypotheses in the earth sciences, and have provided explanations for, and relationships between, many geophysical, geochemical, geological, and paleoecological observations. The scientific study of the physics of plate tectonics has been termed "geodynamics." The basic tenet of plate tectonics is that the entire surface of the earth consists of a small number of rigid lithospheric plates, ~150 km thick, that move continuously relative to one another and with respect to the underlying, more plastic, asthenosphere by rotations about axes through the center of the earth. The plate boundaries are defined by narrow belts of seismicity and volcanism. The plates may contain both continental and oceanic regions. The continental component of plates appears not to be destroyed, and has generally grown throughout earth history; however, the configuration of a plate may be changed by collision with another continent (which causes the formation of mountains like the Himalayas) or by rifting to form small continents and new intervening oceans. Oceanic plates are continuously created at mid-ocean ridges and reabsorbed (subducted) at destructive boundaries (trenches). Conservative boundaries, where the plates simply slip past one another horizontally, are termed transform faults. The subduction process only involves the oceanic portion of plates. When a plate sinks in the earth, the intermittent sliding motion of the downgoing plate against the overriding plate gives rise to major earthquakes. Upon bending downwards, the fact that 150 km thick plate is cooler and undergoes phase changes as higher pressures are encountered gives rise to earthquakes within the plate as the plate sinks down to depths in some places of 650 km. Whether subducting plates go deeper than 650 km has been subject to much controversy and is at present unclear. Plate motions are driven by mantle convection, which is part of the process of cooling of the earth. The energy for convection is provided both by radioactivity of the mantle and by the initial gravitation energy of the earth which is released by several processes including contributions from the latent heat of freezing of the inner core of the earth. The rise of molten material from the upper mantle at mid-ocean ridges is also thought to be an intrinsic driving process of plate tectonics.

The original paradigm of plate tectonics became widely accepted in the 1960s after it was recognized that all the continents were part of a super continent which broke some 200 M years ago (Fig. 1), forming the present distribution of plates (Fig. 2). Two additional concepts, which have been accepted, are the presence of mantle plumes and the movement of oceanic and continental terranes by plate tectonics. The study of the geology of China and the easternmost portion of the USSR and Alaska indicates that most of the continental areas of these regions stems from the succession of on the order of 10 oceanic and continental small plates, only 100 km in extent, which have been rafted into their present position by ancient plate tectonic motion.

Mantle plumes appear to be fixed, virtually cylindrical regions of the mantle which are continual sources of molten rock for periods approaching 80 My. The plume sources appear to be deep seated, at least below a depth of 150 km, and a number appear to be associated with mid-ocean ridges. The minimum depth of plume sources is inferred from the observation of the resulting molten material which pierces the oversliding 150 km thick plate. Some geodynamicists have suggested that the ultimate source of the heat which is required from mantle plumes come from a proposed thermal boundary layer at the core mantle boundary (Fig. 3). In the case of, for example, the Hawaiian hotspot, a chain of submerged volcanic islands with successively older rocks, extends east northeast 3,000 km from the present active volcanic Kilauea to Mawaukee sea mount (submerged, submarine volcano). From this point the chain continues

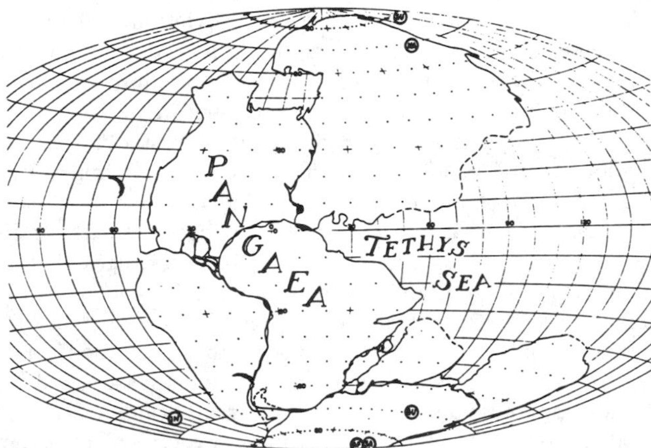

FIG. 1. Reconstruction of the supercontinent Pangea as it may have looked 200 million years ago. The relative positions of the continents are based upon numerical best fits (R. S. Dietz and J. G. Holden, *J. Geophys. Res.* **75**, 4939 (1970)).

north northeast 2,000 km to the intersection of the Kurile and Alaskan trenches (Fig. 4). Mantle plumes are apparently a major source of molten rocks of the earth's interior (largely basalt) and taken with the other two sources of basalt—mid-ocean ridges and subduction volcanoes—provide a framework for understanding the origin of the earth's volcanic rocks. This can be seen in Fig. 5. The geoid, upon removal of the effect of the earth's ellipticity, is dominated by highs (around the Pacific) whose origin can be accounted for with the excess density associated with the subducting oceanic plates.

Seismology is the study of earthquakes, their mechanisms and near source effects, and the propagation of seismic waves through the earth. As predicted by the theory of plate tectonics, the propagating fractures (earthquakes) that produce seismic waves occur principally along zones of preexisting weakness around plate boundaries. The propagation of waves generated by earthquakes can be described using elasticity theory. Three types of waves are generated: *P* (longitudinal), *S* (transverse), and surface waves; in addition, large earthquakes may excite whole-earth vibrations termed free oscillations.

Earthquakes are of major public concern if they occur in populous areas, and seismic risk evaluation and the siting and vibration-resistant design of manmade structures have given rise to the field of earthquake engineering. Although there has been much research conducted on testing a wide range of proposed techniques for predicting the time and place of the occurrence of earthquakes, no reliable technology exists for earthquake prediction. However, it is known that the largest earthquakes are associated with either the subduction process (thrust faulting) or motion on transform faults. It is empirically observed that due to the relative motion on a fault, similar lengths of faults rupture on a quasi-periodic basis. The historical record of periodicity of rupture has recently been augmented backwards in time by as much as 50,000 years using ^{14}C dating methods in many places. What is measured is displacement of ancient soil layers disturbed by faulting. These data have led to the gap theory of earthquake prediction. The theory assigns characteristic earthquakes to sectors of characteristic length on a major fault. Each length can be assigned a repeat rupture period. Typical values range from 30 to 15,000 years, and are the basis of assigning earthquake probabilities to a particular region along a plate boundary. Although the gap theory predicts where earthquakes occur and what magnitude is expected, it cannot predict exactly when an earthquake will occur. Thus the gap theory is most useful in providing information on the degree of shaking which can be expected to be experienced by critical structures, such as bridges, dams, nuclear reactors, and public buildings.

Seismic methods are used to derive both shallow and whole-earth structure. The latter is determined from the travel times and wave forms of seismic waves, and the periods of free oscillations; resultant models give longitudinal velocity (α), transverse velocity (β), and density (ρ) as a function of depth (from which may be derived bulk and shear moduli). These show that the earth consists of a number of layers (Table I).

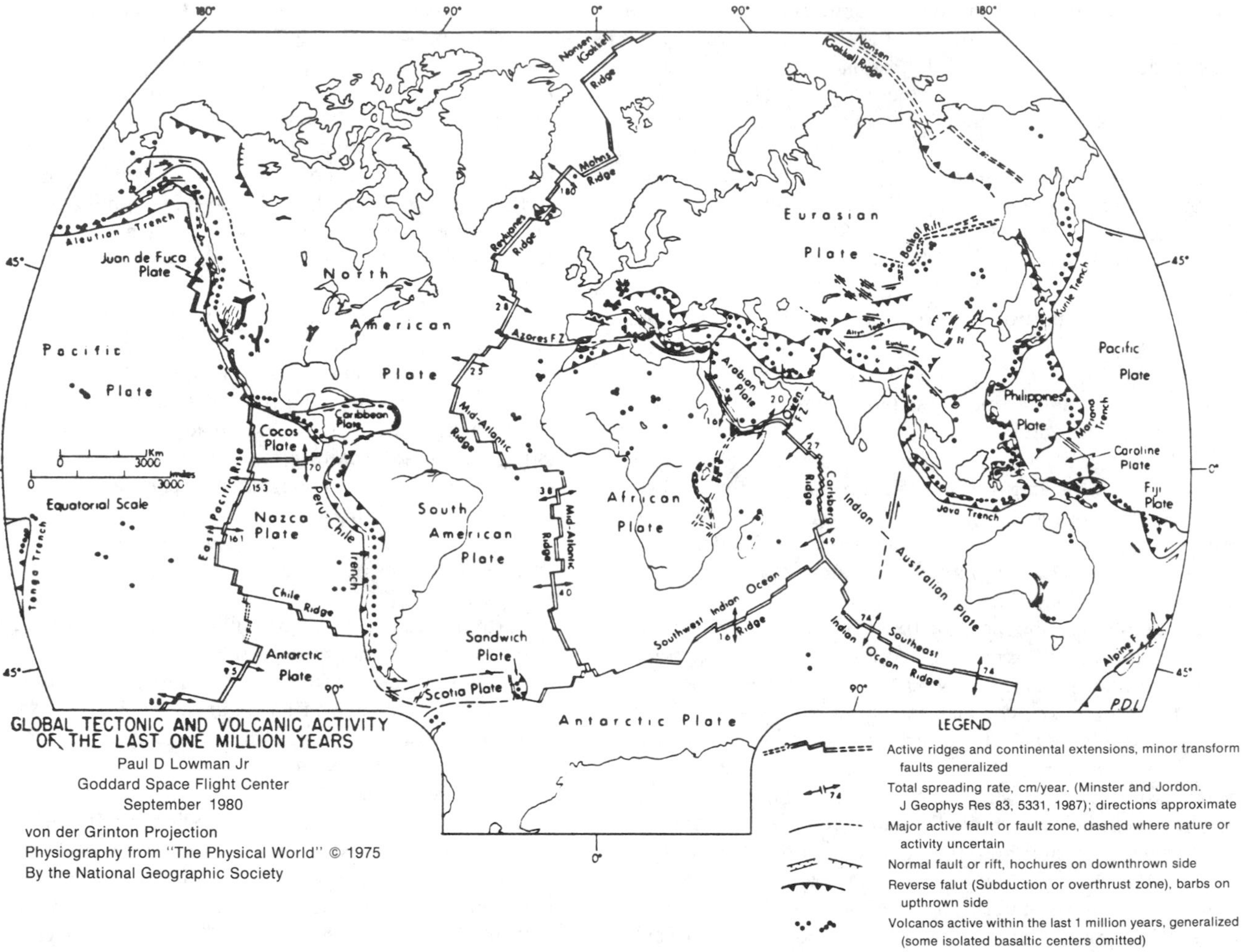

GLOBAL TECTONIC AND VOLCANIC ACTIVITY OF THE LAST ONE MILLION YEARS
Paul D Lowman Jr
Goddard Space Flight Center
September 1980
von der Grinton Projection
Physiography from "The Physical World" © 1975
By the National Geographic Society
LEGEND
Active ridges and continental extensions, minor transform faults generalized
Total spreading rate, cm/year. (Minster and Jordon. J Geophys Res 83, 5331, 1987); directions approximate
Major active fault or fault zone, dashed where nature or activity uncertain
Normal fault or rift, hochures on downthrown side
Reverse falut (Subduction or overthrust zone), barbs on upthrown side
Volcanos active within the last 1 million years, generalized (some isolated basaltic centers omitted)
Eurasian Plate
North American Plate
Pacific Plate
Juan de Fuca Plate
Aleutian Trench
Cocos Plate
Caribbean Plate
Nazca Plate
South American Plate
African Plate
Arabian Plate
Antarctic Plate
Sandwich Plate
Scotia Plate
Philippines Plate
Caroline Plate
Fiji Plate
Indian Australian Plate
Mid-Atlantic Ridge
East Pacific Rise
Chile Ridge
Peru-Chile Trench
Tonga Trench
Kurile Trench
Mariana Trench
Java Trench
Carlsberg Ridge
Southwest Indian Ocean Ridge
Southeast Indian Ocean Ridge
Reykjanes Ridge
Mohns Ridge
Nansen (Gakkel) Ridge
Azores F.Z.
Baikal Rift
Alpine F.
Equatorial Scale
PDL

FIG. 2. Global map of plate boundaries and of tectonic and volcanic activity. (P. D. Lowman, Jr., *Bull. Int. Assoc. Eng. Geol.*, **23**, 37 (1981)).

670
PLUME
CMB
OUTER CORE
CBL
CBL
TBL
INNER CORE

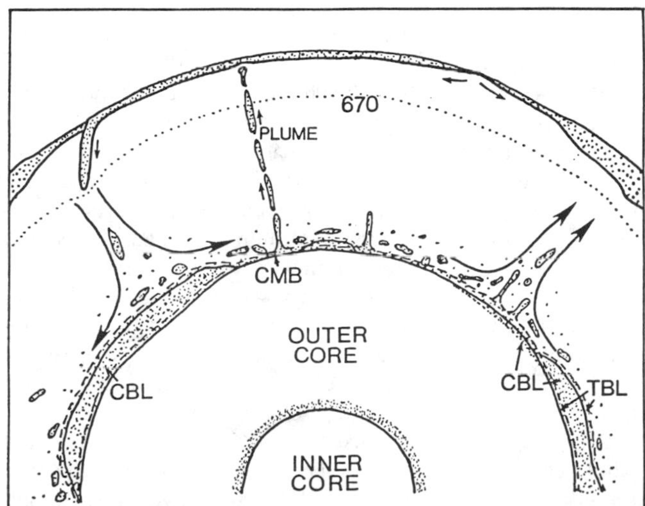

FIG. 3. A schematic model of the core-mantle transition zone. Geophysical measurements suggest a heterogeneous chemical boundary layer (CBL) is embedded in a thermal boundary layer (TBL). Thermal plumes caused by boundary layer instabilities ascend from the core-mantle boundary (CMB) (T. Lay, *EOS, Trans. Amer. Geophys. U.* **70,** 49 (1979)).

Geodesy is concerned with the shape of the earth, and classically has been divided into two parts: determination of the spheroid (the ellipsoid giving the best fit to the sea-level surface) and the measurement of the deviation of the actual sea-level surface (geoid) from the spheroid (Fig. 5). The shape of the earth is basically that of an ellipsoid of revolution slightly flattened at the poles; accurate measurements indicate that it is in fact slightly nonhydrostatic. Determinations of the geoid and spheroid are currently made by a combination of precise surveying and Doppler tracking of artificial satellite orbits. Currently, geologists are attempting to measure directly the absolute distance changes that accompany plate motion by using radio interferometry in combination with the global positioning satellite system. The gravitational potential field for the earth is described via expansion in a series of spherical harmonics (associated Legendre polynomials). The higher-degree harmonics ($n \approx 10$), and to some extent the lower ones, are caused by lateral variations in the upper mantle and transition zone; these may be due to chemical differences, or thermal fluctuations, or may possibly be the direct consequence of convective motion in the mantle. As can be seen in Fig. 5, the geoid, upon removal of the effect of the earth's ellipticity, is dominated

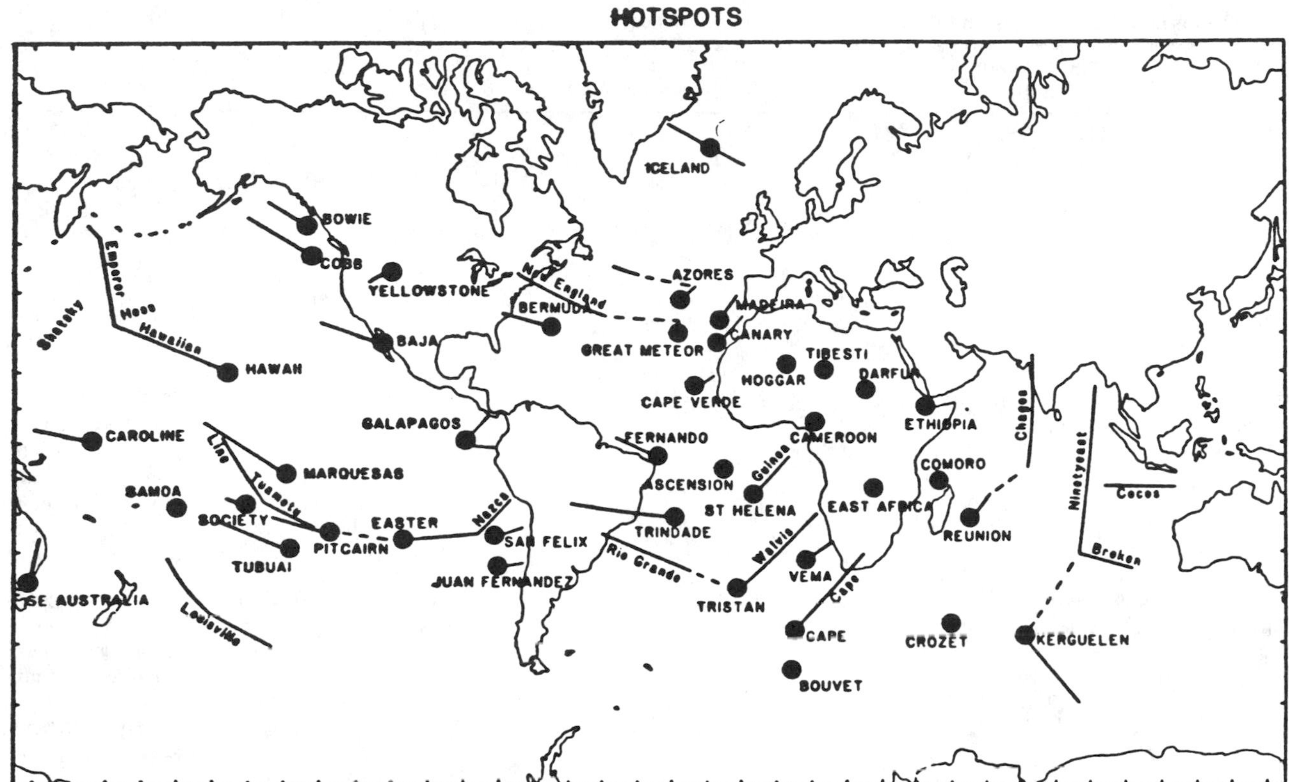

FIG. 4. Global distribution of hotspots (dots) and hotspot traces (lines). (S. T. Crough, Annual Review of Earth and Planetary Science **11**, 165 (1983)).

by highs whose origin can be accounted for by the density which is associated with subduction of plates.

Local fluctuations in the gravitational field are caused by density variations within the crust, and are used in geophysical exploration to locate structures such as dense ore bodies. Spatial gravity anomalies extending over 10^2–10^3 km do not correlate well with topography because surface features and density variations appear to be compensated at depth (isostasy).

The earth's magnetic field, its history and origin, and the behavior of naturally occurring magnetic minerals are studied in geomagnetism. The field is commonly believed to be caused by electrical currents in the fluid, conducting, outer core of the earth, and undergoes secular variations in strength, quasiperiodic fluctuations on a time scale of 10^2–10^4 yr, and reversals of polarity. Measurements of the intensity and direction of magnetism in rocks of known ages enable study of the history of the magnetic field; these have also provided key evidence in favor of the theories of plate tectonics and continental drift.

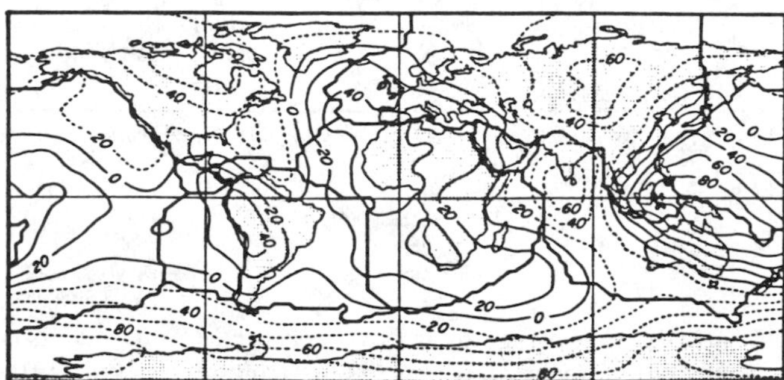

FIG. 5. Geoid height anomaly in meters from the GEM 10B method of Lerch et al. (1979) corrected by removal of the dynamically inferred flattening of Nakiboglu (1980). Solid contours are positive anomalies, dashed contours are negative anomalies. Contour interval is 20 m.

The electrical conductivity of the earth as a function of depth is determined from damping of magnetic-storm–induced fluctuations in the magnetic field, and temporal and spatial variation of magnetic variations induced by air tides. Magnetotelluric methods utilize the interaction of magnetic and electrical field fluctuations, generated by naturally or artificially induced currents, to determine shallow conductivity structure for prospecting purposes or for studies of the lower crust and upper mantle, depending on frequency.

Thermal properties of the earth are generally investigated by measuring heat flow and the thermal diffusivity and radioactivity near the surface, and then extrapolating to greater depths using constraints provided by the measured rock-melting intervals at high pressure and the position of the known phase changes in the earth's upper mantle (Table I).

Solar heating has a negligible effect on the interior of the earth, and the energy released by earthquakes and tidal friction is also small compared with geothermal heat losses. The major sources of heat within the earth, at present, are thought to be the radioactive decay of long-lived isotopes and, to a lesser and more uncertain degree, the release of

Table I Major layers of the earth and their approximate characteristics[a]

Layer	Approx. depth (km)	α (km/s)	β (km/s)	ρ (gm/cm^3)	σ ($\Omega^{-1} m^{-1}$)	η (P)	T (°C)	
Crust	Variable: Continents 0–~35 Oceans 0–~10	~2–7	~1.2–3.8	~2–3.0				Laterally heterogeneous, layered, complex velocity, density, and conductivity variations
Mohorovicic (M–) discontinuity		*Jumps*	*Jumps*	*Jumps*				Usually known as the Moho
Upper Mantle	~35–400 (continents) ~10–400 (oceans)	8–9	4.3–4.7	3.3–3.6	$\sim 10^{-2}$	$\sim 10^{21}$ P	~400–1500	Laterally inhomogeneous, possibly convecting; principal minerals probably magnesium-rich olivine and pyroxene
Mantle transition zone	400–1000	9–11.5	4.7–6.4	3.6–4.7	10^{-2}–10^{2}	$\sim 10^{22}$ P at 1000 km	~500–2000	Mixed-phase region with complex velocity and density gradients. Two major discontinuities: ~400 km, olivine → spinel; ~650 km, spinel → postspinel.
Lower Mantle	1000–2900	11.5–13.8	6.4–7.5	4.7–5.7	$\sim 10^{2}$	$\sim 10^{23}$ or $\sim 10^{26}$ P [b]	2000–~3500 (±500)	Fairly homogeneous; possibly chemically similar to upper mantle
Core–Mantle Boundary (Gutenberg discontinuity)[c]								
Outer core	2900–4980	~8–10	0	9.4–12	$\sim 5 \times 10^{5}$		3800 ±500	Liquid; alloy of iron–nickel with lighter elements, probably sulfur, carbon, or silicon; probably convecting and source of earth's magnetic field
Transition zone[d]	4980–5120							
Inner core	5120–6370	10–11.7	3.4–3.6	15.8–17.2			6300 ±1000	Solid, probably similar to outer core in composition; pressure at center of earth ~3.1 Mbar.

[a] Key: α, P velocity; β, shear velocity; ρ, density; σ, electrical conductivity; η, viscosity; T, temperature. α, β, ρ, and T generally increase with depth within each layer.
[b] Depends on method.
[c] Chemical boundary; pressure ~1.4 Mbar.
[d] Probably phase boundary between liquid outer and solid inner core.

gravitational energy and the decay of short-lived isotopes, such as ^{26}Al.

The present heat flow out of the earth is highly variable (on land an average of 57 mW/m^2), and correlates strongly with the radioactivity and hence heat production of continental rocks. The contribution to land heat flow from the mantle accounts for approximately 50% of the heat flow. In contrast, the average heat flow measured on the oceans is 100 mW/m^2, most of which results from convection in the mantle below.

Below about 100 km the distribution of temperature and heat sources and the mechanism of heat transfer are uncertain. Lattice thermal conductivity mechanisms are thought to dominate throughout the entire mantle as the iron-bearing silicates of mantle become opaque at high pressure. Convection probably occurs in the upper mantle and outer core. To what degree convection now takes place in the lower mantle of the earth is unclear. It appears, however, that thermal convection was more vigorous during the early history of the earth than at present. Higher heat flow is found in regions of volcanism such as mid-ocean ridges and geothermal activity, and geothermal energy may be a source of power in the future.

Rheology is the study of the deformation and flow of matter. When applied to the earth, this generally means the magnitude and time dependence of viscosity within the earth. The study of fracture processes in rock, creep processes, and the motion along faults at depth, as well as the sliding of the lithosphere over the asthenosphere and the subduction of the lithosphere, may also be considered geophysical applications of rheology. Direct measurements of the earth's viscosity are not possible, but estimates can be made by studying the rate of uplift of areas such as the ancient terranes of Canada and Fennoscandia, earlier subjected to down-warping by ice-sheet loading, and provides a determination of the upper-mantle viscosity in the range from 10^{20} to 10^{22} P. Estimates of the lower-mantle viscosity vary from 10^{22} to $\sim 10^{26}$ P.

Mineral geophysics includes the study of mineral and rock properties at high pressures, and the interpretation of velocity and density models of the earth in terms of possible mineral assemblages. (Estimates of the bulk composition of the earth are based on solar and meteoritic abundances and on the composition of crustal and upper-mantle rocks.) The development of theories concerning high-pressure behavior of properties such as density, crystal structure, conductivity, and melting point utilizes results from thermodynamics, elasticity, and solid-state physics. Experimental methods include ultrasonic measurements of velocity as a function of pressure (limited to $\sim$30 kbar), and high-pressure techniques such as the static compression of samples in diamond anvil cells are currently being carried out to pressures of at least 1000 kbar. Using dynamic compression via shock waves and diamond anvil cells, pressures in excess of those at the center of the earth ($\sim$3600 kbar) can be achieved.

Ideally, measurements of compressional and shear velocities, the elastic moduli and their pressure derivatives, density, thermal expansion, specific heat, melting points, viscosity, and electrical and thermal conductivity, all as functions of pressure, are needed for a complete understanding of processes occurring at depth in the earth. However, most of these data are not available, and much present knowledge is in the form of equations of state for density as a function of pressure. Often it is necessary to extrapolate experimentally determined data to higher pressures using theoretical equations of state. Lattice-dynamical, quantum-mechanical, and finite-strain theories are used to construct equations of state.

Experimentally determined empirical relationships such as those between velocity, density, and mean atomic weight have played an important role in estimating properties of materials whose elastic properties are as yet unmeasured, and in interpreting the seismic data for the earth in terms of composition and phase. All earth models show discontinuities in density and velocity at various depths; these are due to changes in chemical composition, but some have been identified with phase changes or rearrangements of crystal structure in the minerals present. Experimental observations of phase changes have been made using x-ray diffraction methods; these studies allow identification of discontinuities in the earth with phase changes.

Phase changes in the earth occur over a band of pressure, and are also "broadened" by the simultaneous transformation of several coexisting minerals. The transition zone between depths of 400 and 1000 km includes two major discontinuities at depths of $\sim$400 km and $\sim$670 km, which have been identified as being olivine [$(Fe,Mg)_2SiO_4$] to spinel and spinel related structures and spinel to a denser assemblage of perovskite plus magnesiowüstite. This latter transition is very sharp and takes place in the earth in a depth interval range of less than about 4 km and thus could be expected to be a good seismic reflector.

Recently, using tomographic methods, lateral variations in seismic velocity of $\pm 2\%$ have been mapped in the earth's mantle. The velocity structure of the upper mantle, especially the outer 100 km, can be related to many surface features and to heat flow. For example, the mid-ocean ridge system and the stable continental interiors indicate hotter and cooler shear wave velocities than does the average earth. The lower mantle also demonstrates similar magnitude compressional and shear wave velocity variability except that it is not correlated with surface features.

The core-mantle boundary is a chemical one; the outer core is thought to consist largely of liquid iron-nickel with a lighter alloying element, probably sulfur, oxygen, or silicon.

The most accurate methods of determining the ages of rocks (cf. GEOCHRONOLOGY) involve radioactive decay processes within the rock, and are studied in isotope geochronology, which forms a link between geophysics, geochemistry, and nuclear physics. Radioactive dating is accomplished by measuring the proportions of parent and daughter isotopes present and, allowing for the initial concentration of daughter isotope, using the measured half-life of the parent to calculate the age. Originally, the uranium and thorium decay series were widely used, but now a wide range of methods have found more general applicability. Recently the samarium method has found wide use to infer the petrogenetic history of igneous and metamorphic rocks. The age of the earth is inferred from the ages of primitive meteorites to be $\sim 4.65 \times 10^9$ yr. However, the oldest dated terrestrial rocks are only $\sim 4.0 \times 10^9$ yr old.

See also ATMOSPHERIC PHYSICS; GEOCHRONOLOGY; GEOMAGNETISM; GRAVITY, EARTH'S; METEOROLOGY; RHEOLOGY; SEISMOLOGY.

BIBLIOGRAPHY

D. L. Anderson, *Theory of the Earth*. Blackwell, Boston, 1989.

J. M. Bird, *Plate Tectonics*, 2d ed., Am. Geophys. U., Washington, D.C., 1980.

W. R. Peltier, *Mantle convection*. Gordon and Breach, New York, 1989.

G. Ranalli, *Rheology of the Earth*. Allen and Unwin, London, 1986.
F. D. Stacey, *Physics of the Earth*, 2d ed. Wiley & Sons, New York, 1979.

Glass

B. Golding

Glasses and amorphous solids comprise a class of materials (i) whose microscopic atomic arrangement exhibits no periodicity or long-range order, in contrast to crystals; and (ii) whose shear viscosity is sufficiently large that macroscopic shapes are maintained for relatively long times, in contrast to fluids. A substance in its amorphous phase is metastable, since its atoms or molecules exist in a higher configurational energy state than in its crystalline phase, but are unable to reach that state because of substantial energy barriers separating the two phases. An amorphous solid state can be achieved (1) by rapid cooling of a viscous fluid, the traditional preparation method for inorganic glasses such as SiO_2 (vitreous silica), As_2S_3, and sulfur, and for organics such as polystyrene and polymethyl methacrylate; or (2) through circumvention of the liquid state by preparation from the vapor phase, as in vacuum deposition, sputtering, gaseous decomposition, etc., as used in the preparation of amorphous silicon and germanium. The term *glass* is often restricted to specify materials formed by process (1).

If, upon cooling, a liquid successfully avoids crystallization, i.e., becomes supercooled, it passes through the glass transition temperature T_g, defined as the temperature at which the substance's shear viscosity exceeds approximately 10^{14} P. On cooling through T_g, the specific volume and entropy of a glass decrease abruptly but continuously from values characteristic of the liquid toward values approaching, but not reaching, those of a crystalline state. The rate of cooling through the glass transition region determines the total change in the substance's extensive thermodynamics properties by influencing the overall degree of structural rearrangement.

The structures of many inorganic glasses can be visualized as made up of small molecular groups (e.g., SiO_4 tetrahedra in silica glass), joined together to satisfy all, or nearly all, bonding requirements (each oxygen joined to two adjacent tetrahedra). This continuous random network differs from a crystalline network in that the bonding angles linking the groups are not fixed but have distributions of values about the crystalline angle. Nevertheless, the short-range order of the nearest atomic coordination shells about a central atom in a crystal is retained in the amorphous phase.

The glassy network of vitreous silica can be modified and the softening temperature lowered for glasses of commercial and technological importance by incorporating other oxides (e.g., Na_2O and B_2O_3) into the structure to form the soda silicates and borosilicates. Such glasses are transparent to visible light because optical excitation of electronic transitions begins in the ultraviolet part of the spectrum, whereas excitation of vibrational modes occurs in the infrared. The minimum optical loss due to absorption and scattering in silica glass fibers occurs typically in the near infrared at a wavelength of approximately 1.5 μ.

A glass is an elastically isotropic solid, characterized by two independent elastic constants, and supports a long-wavelength longitudinal and a doubly degenerate transverse acoustic mode. Oxide glasses are generally brittle, exhibit little plastic deformation, and fracture by crack propagation, whereas many organic polymeric glasses are capable of withstanding strains of several hundred percent before breaking. Heat is carried by sound waves but because of the disordered structure the thermal vibrations undergo strong scattering, resulting in low thermal conductivities.

A glass may be regarded as a collection of intrinsic structural defects. When it is subjected to external stimuli such as time-dependent electric or stress fields, the induced rearrangement of defects creates frequency- and temperature-dependent features in the dielectric and elastic responses, i.e., large dielectric and elastic relaxational loss and dispersion characteristic of a particular glass. In addition to a primary relaxational process occurring near T_g, a secondary structural relaxation appears at temperatures below T_g. The specific heat of glasses reflects the extra atomic degrees of freedom by an enhancement of the T^3 Debye specific heat at low temperatures and the appearance of a quasilinear temperature-dependent defect contribution below 1 K.

See also CRYSTAL DEFECTS; ELASTICITY; GLASSY METALS; LIQUID STRUCTURE; SEMICONDUCTORS, AMORPHOUS

BIBLIOGRAPHY

R. J. Charles, "The Nature of Glass," *Sci. Am.* **217**(3), 127 (1967). (E)
R. H. Doremus, *Glass Science*. Wiley, New York, 1973. (I)
P. A. Fleury and B. Golding, *Coherence and Energy Transfer in Glasses*. Plenum, New York, 1984. (A)
P. A. Lee (ed.), "Disordered Solids," *Phys. Today* **41**(12) (1988). (I)
J. Wong and C. A. Angell, *Glass*. Dekker, New York, 1976. (A)
R. Zallen, *The Physics of Amorphous Solids*. Wiley (Interscience), New York, 1983. (I)

Glassy Metals

Pol Duwez and Takeshi Egami

The crystalline state is characterized by a periodic repetition of a unit cell which contains a certain number of atoms. In the amorphous, or glassy, state the atoms are arranged more or less randomly without any periodicity, as in the liquid state. The transition from the liquid to the crystalline state is very sharp, with various properties changing discontinuously at the transition. By contrast the transition from the liquid to the glassy state is gradual and continuous. When a liquid is cooled, the viscosity of the liquid changes rapidly but continuously, until it becomes so high that the material behaves like a solid. During this transition, or the glass transition, the atoms remain in approximately the same configuration. Therefore the glass transition is not really a phase transition in a rigorous sense, and glassy metals can be considered as supercooled liquid metals.

Fused silica and mixtures of silica with other oxides of Al, Na, Ca, etc., constitute the oldest and most commonly seen class of glassy inorganic solids. Other glassy solids include chalcogenide glasses containing elements of group Va, VIa, and VIIa, and amorphous polymers. Glassy metals belong to the same category of materials as these more traditional glasses and share many properties with them. On the other hand they require much higher cooling rates to form from the liquid state. If a molten liquid alloy is cooled with too slow a rate, it simply crystallizes. Around 1960 new methods were developed to cool liquid alloys at very high rates—up to 1 million degrees per second or more—and under these extreme conditions crystallization of some metallic alloys can be avoided and glassy metals can be formed.

The basic principle involved in achieving very high rates of cooling is the rapid extraction of heat from the liquid by conduction onto a solid substrate of a high-conductivity metal, such as pure copper. Since the rate of heat extraction is governed largely by thermal diffusion, the thickness of the glassy metal foil is limited to rather small values, usually of the order of 50 μm or less. Earlier methods of rapid cooling, or rapid quenching, were the splat-cooling method, in which a droplet of molten metal is propelled by gas pressure against a copper plate, and the piston-and-anvil method, in which a molten droplet is squeezed between a moving piston and a fixed anvil. The methods most widely used today are the melt-spinning and melt-extraction methods. A metal drum rotating at a rate of several thousand rpm takes heat away from the molten metal which comes out of a crucible and is cast into a ribbon or a sheet. In the melt-spinning method the molten liquid alloy is pressured out of a crucible to form a stream and hits the cooling drum. In the melt-extraction method the crucible is close to the surface of the drum, and the melt fills the gap between them. By feeding the melt continuously and cooling the drum by water it is possible to produce sheets of almost unlimited length.

Glassy metals can be produced by other methods as well. The earliest reports of amorphous phase formation are by electrochemical deposition and vapor deposition at low temperatures. The samples obtained by the latter, however, are usually unstable at room temperature. In the earlier studies the researchers really could not distinguish amorphous phases from microcrystalline phases. It took the advent of rapid cooling methods to initiate extensive studies of this new phase. Amorphous thin films can be produced by sputtering deposition and are beginning to be used for magnetic applications as mentioned later.

It is also possible to produce an amorphous phase directly from the crystalline state, without going through a gas or liquid phase. Irradiating a crystal with electrons and ions produces lattice defects, and eventually leads to a collapse of the whole lattice structure into the amorphous state. More interestingly, it was discovered recently that when a mixture of two elements is annealed at a temperature just high enough to promote sufficient atomic diffusion, an amorphous phase can result due to interdiffusion. If powders of two elements are mechanically mixed by a ball-milling machine, the product can again be amorphous. Thus glassy metals can be produced by many thermal or athermal processes.

At least theoretically, everything becomes glassy when it is quenched from the liquid state with a sufficiently high cooling rate. In reality, however, the critical cooling rate necessary for some materials to become glassy is technically too high to attain. For instance, it is generally recognized that pure metals cannot be quenched from the liquid state into a glassy solid, at least not with the presently achieved rates of cooling which are in the range of 10^5–10^7°C/s. So all glassy metals actually are alloys containing two or more elements. Over the last three decades a very large number of alloy systems have been found to form glasses, including the transition-metal–metalloid alloys such as Fe-B and Ni-P, and metallic alloys such as Cu-Zr, Fe-Tb, Au-La, and Al-Ca. It is now clear that glass formation is not limited to particular classes of alloy systems, but is a phenomenon widely found in metallic alloys.

Various criteria have been proposed in attempting to predict glass-forming combinations of elements. However, there is yet no perfectly reliable method of predicting the glass formability. A large variety of chemical interactions found in glass-forming alloy systems imply that chemistry has little to do with glass formation, except that a strongly positive heat of mixing would lead to phase segregation and prevent glass formation. A factor which is definitely important in glass formation is the atomic radii of the elements. A large difference in the atomic size makes the solid solution unstable and makes the glass formation easier. In binary systems (A-B) the minimum composition (in atomic percent) of the second (B) element necessary for glass formation, $c_B^{\min}$, is usually related to the atomic volume of each, v_A and v_B, by $c_B^{\min} \sim 10v_A/|v_A - v_B|$.

In addition, in order for the glass to be stable, the rate of crystallization at room temperature has to be sufficiently low. Glassy metals are always unstable against crystallization when the atomic diffusivity is high enough, and they usually crystallize when they are heated to near the glass-transition temperature, T_g. For many transition-metal–metalloid glasses such as $Fe_{80}B_{20}$, T_g is around 400°C; thus they crystallize at around this temperature. Since T_g is roughly equal to $0.4\bar{T}_m$, where $\bar{T}_m$ is the compositionally averaged melting temperature of the constituent elements, $\bar{T}_m$ has to be above 1000 K for the glass to be stable. The length of time required for crystallization to occur, τ_c, changes exponentially with temperature approximately obeying the Arrhenius law, $\tau_c = \tau_0 \exp(E_c/kT)$, with the apparent activation energy of 2–5 eV. Thus sufficiently below T_g, glassy metals are stable for quite a long time and can be regarded stable in practice.

The atomic structure of glassy metals can be best studied by diffraction experiments, using x rays, neutrons, or electrons. A typical diffraction pattern consists of several broad peaks, rather than a number of sharp Bragg peaks as for crystalline solids. The diffraction peak width, if interpreted in terms of microcrystalline aggregates, corresponds to the grain size of 15–20 Å. It is now well established, however, that the structure of glassy metals cannot be described as a collection of small crystalline grains. The simplest and yet physically most sensible model of liquid and glassy metals, the dense random packing (DRP) model, was proposed by Bernal and Scott. The model just consists of a large number of steel balls randomly packed in a bag. Today the model can be built by a computer, using more realistic interatomic potentials. The calculated distribution of the atomic distances in the model compares very well with those of the glassy metals directly obtained from the diffraction intensities by the Fourier transformation, as shown in Fig. 1.

Because of rapid quenching necessary for production, the structure of glassy metals is not quite settled as quenched, and relaxes when they are annealed at temperatures below T_g. Various properties change during the structural relaxation; some drastically, and others only slightly. For instance diffusivity changes by many orders of magnitude, while the changes in volume and structure are only barely observable. For many applications this relaxation should be avoided by applying preannealing treatment.

Since glassy metals are isotropic, their elastic properties can be described by only two elastic constants, namely, bulk and shear constants. The bulk modulus of glassy alloys is similar to that of crystalline alloys of the same composition, while the shear modulus is significantly (by about 30%) lower than in the corresponding crystalline materials. Thus the average Debye temperature determined by low-temperature specific heat is also lower. Mechanical deformation of crystalline solids occurs via the motion of lattice dislocations. In glassy metals the dislocations cannot be defined in the absence of a lattice; thus they show very high mechanical strength, often of the order of 300 kg/mm^2 or more, which is comparable to that of the strongest steel (piano) wires. At least in the as-quenched state the glassy metals exhibit considerable ductility. However, the ductility is lost when they are annealed, and the glasses become brittle. This embrittlement is largely a consequence of structural relaxation.

Glassy metals show relatively high electrical resistivity, ranging from 50 to 200 $\mu\Omega$ cm. This high resistivity is due to structural disorder and is only weakly dependent upon temperature, with the temperature coefficient being either positive or negative. The resistivity and its temperature dependence of some alloys, particularly the simple metal alloys, can be understood well in terms of the scattering theory by Ziman which was successfully applied to many liquid alloys. At low temperatures the resistivity sometimes shows a minimum similar to that of Kondo alloys, but the details of its mechanism are not well understood. In highly resistive glasses, electrons show a tendency of localization. Some of the glassy metals containing Zr, Mo, and La are superconducting at low temperatures. The critical temperature is somewhat lower than in crystalline alloys of a similar composition, but the critical field, H_{c2}, and the critical current density are higher. The superconducting properties of glassy metals are resistant against radiation damage, and thus they may be useful for applications in nuclear reactors or in accelerators.

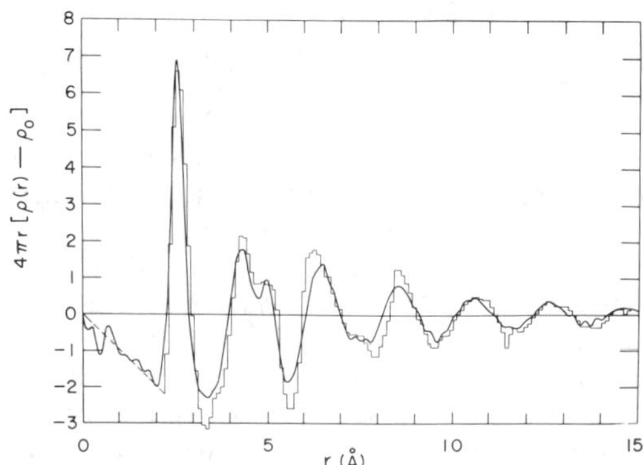

FIG. 1. The distribution of interatomic distances presented in terms of $G(r) = 4\pi r[\rho(r) - \rho_0]$, where $\rho(r)$ is the atomic pair distribution function which describes the probability of finding two atoms at a distance r and ρ_0 is the average number density of atoms in the alloy. The solid line is $G(r)$ determined by x-ray diffraction experiment for amorphous $Fe_{40}Ni_{40}P_{14}B_6$, while the histogram describes that of a DRP model.

Ferromagnetism is not limited to crystalline solids, and many glassy alloys containing Fe, Co, and Ni are ferromagnetic at room temperature and above. The saturation magnetic moment, M_s, and the Curie temperature, T_C, of Co-rich alloys are very similar to the crystalline counterparts, while those of Fe-rich alloys are usually smaller than in the crystalline alloys. This contrast reflects the difference in the electronic structures of these two. In Co-rich alloys the majority-spin d band is full (strong magnetism), so that the magnitude of the magnetic moment depends only upon the electron density, while in Fe-rich alloys both the majority and minority bands are partially full (weak magnetism), and the spin polarization is influenced by structure and other factors.

Their magnetic transition at T_C is well defined and sharp, unless they contain significant compositional fluctuations. The critical behavior in the vicinity of T_C is very similar to the corresponding crystalline alloys, since the atomic-level disorder does not influence the long-range magnetic fluctuations. Away from T_C, disorder has some effects on magnetism, but the compositional disorder which can occur in crystalline solids as well has more significant effects than topological disorder. Changes in the compositional short-range order can affect magnetic properties, such as T_C and field-induced magnetic anisotropy.

Many ferromagnetic glassy alloys show very low coercivity and therefore high permeability, except when they contain rare-earth elements, because they do not have grain boundaries and lattice defects like dislocations. Their permeability after annealing can be equal to or better than that of the crystalline alloy with the highest permeability, Supermalloy. Ferromagnetic glassy alloys are already used in a large number of devices. Their use in distribution power transformers is expected to result in significant energy savings, which could run up to $\$10^8$ or more, and, in switching regulators which supply stable dc power to computers and other devices, since they save much space and weight. Tape-recorder heads with glassy alloys allow good reproduction of high-frequency sounds, and in antitheft devices the mechanical toughness is an important asset, because the magnetic signals do not become deteriorated by handling. On the other hand those containing rare-earth elements show high coercivities, particularly in thin films. Sputter-deposited Fe-Co-Tb thin films are now used for magneto-optical data storage in which information is written by a laser in the form of tiny magnetic bubbles.

Glassy metals are a new class of materials rich in prospect of leading to interesting and important science and applications, some of which have already materialized. They posed a challenge to the conventional theories of solid-state physics which are largely based upon the periodicity and symmetry of crystalline lattices, and are stimulating the growth of new and more general theory of solids. Even though they are not thermodynamically stable, they can be used in many applications just as a number of metastable crystalline solids are currently in use. The glassy metallic state is not an unusual state which is an object of scientific curiosity alone, but represents a general state of matter which can occur in a large number of metallic alloy systems.

See also ALLOYS; CONDUCTION; CRYSTAL GROWTH; GLASS; LIQUID METALS; LIQUID STRUCTURE; VISCOSITY.

BIBLIOGRAPHY

Pol Duwez, *Ann. Rev. Mater. Sci.* **6,** 83–117 (1976).
H.-J. Güntherodt and H. Beck (eds.), *Glassy Metals.* Vols. 1 and 2. Springer-Verlag, Berlin, 1981, 1984.
F. E. Luborsky (ed.), *Amorphous Metallic Alloys.* Butterworth, London, 1983.
T. Egami, *Rep. Prog. Phys.* **47,** 1601–1725 (1984).

Grand Unified Theories

Stephen M. Barr

HISTORICAL BACKGROUND

The first example of a unification of forces was Maxwell's electromagnetic theory. This was not simply a matter of the electric and magnetic fields (**E** and **B**) appearing together in the same equations but, more profoundly, involved a new symmetry principle: Lorentz invariance. A symmetry is a transformation which leaves the form of something—in this case the laws of electromagnetism—unchanged; and the set of such transformations form a mathematical structure called a "group." Certain Lorentz transformations rotate electric and magnetic fields into each other. Thus one can only speak in an absolute sense of a unified entity called the "electromagnetic field."

A point of cardinal importance is that a deep connection exists between forces or "interactions" and symmetries which act locally in space-time. For example, underlying the electromagnetic force is a local symmetry called electromagnetic gauge invariance (Weyl 1929). Einstein's theory of gravity is based on a local symmetry called general coordinate invariance. Other similarities between these two forces are their long-range character (inverse square law) and their coupling to conserved charges (electric charge and energy-momentum, respectively). The other two interactions, the "strong" and "weak" nuclear forces, seemed not to share any of these features. Thus the first attempts were to find a "unified field theory" of electromagnetism and gravity with a larger local symmetry containing both electromagnetic gauge invariance and general coordinate invariance. Such theories can be found (Kaluza 1921, Klein 1927) but suffer from severe difficulties. In the late 1960s and early 1970s it came to be understood that the strong and weak forces like the electromagnetic are based on "gauge" symmetries. This had long been obscured, in the case of the weak interactions by "spontaneous symmetry breaking," and in the case of the strong interactions by "confinement." The local symmetry of gravity is of a somewhat different kind and the early attempts of Einstein and others to unify gravity with other forces are now seen as premature. Modern grand unified theories unify the three nongravitational forces.

The electroweak theory of Glashow, Salam, and Weinberg (1967) is based on the gauge group $SU(2) \times U(1)$. [$SU(N)$ is the group whose elements can be represented by the $N \times N$ unitary matrices with unit determinant. $U(1)$ is the group whose elements can be represented by complex numbers of unit magnitude.] $SU(2)$ is the "weak isospin" symmetry; $U(1)$ is the "weak isospin" symmetry. The vacuum state (or ground state) does not respect this full symmetry, however, but only a $U(1)$ subgroup which is the electromagnetic gauge group. The rest of the symmetries (associated with the weak interactions) are said to be spontaneously broken. In this theory the left-handed components of the electron and electron–neutrino form a multiplet of weak isospin as do the left-handed u and d quarks:

$$\begin{bmatrix} \nu_{eL} \\ e_L^- \end{bmatrix}, \quad \begin{bmatrix} u_L \\ d_L \end{bmatrix}.$$

This means there are $SU(2)$ transformations that rotate $\nu_{eL} \leftrightarrow e_L^-$ and $u_L \leftrightarrow d_L$. A remarkable feature of gauge theory is that there corresponds to every type of gauge symmetry transformation a particle—a "gauge boson"—which by emission or absorption actually performs it on the particles. For the weak interactions these are the $W^\pm$ and Z^0 (Fig. 1). To broken gauge symmetries correspond massive gauge bosons ($M_W \cong 80$ GeV, $M_Z \cong 90$ GeV), to unbroken ones massless gauge bosons ($M_{\text{photon}} = 0$). Symmetry breaking accounts for the short-range and nonconserved charges that distinguish the weak from the electromagnetic forces. Spontaneous symmetry breaking is associated with a mass or energy scale; that of electroweak breaking is about 10^2 GeV ($\sim M_W, M_Z$).

The strongly interacting particles (hadrons, e.g., proton,

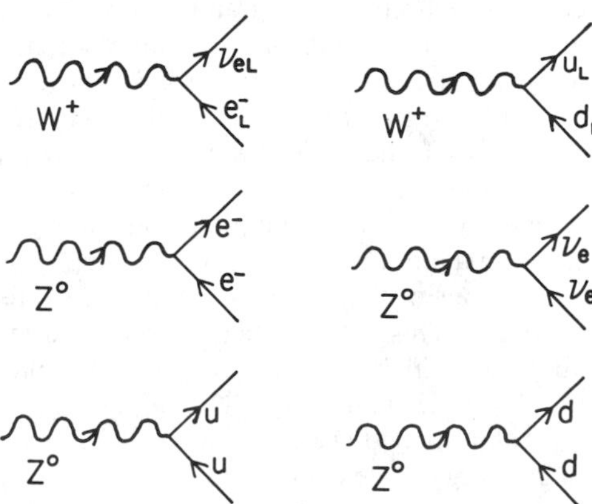

FIG. 1. The absorption or emission of the weak gauge bosons $W^\pm$ causes transitions within particle multiplets.

neutron, pion) are made up of more elementary constituents called quarks and gluons. It is believed that only hadrons and not free quarks or gluons can be directly observed experimentally (confinement). The theory describing the interactions of quarks and gluons is called quantum chromodynamics (QCD) and is based on the gauge group $SU(3)$. The gauge bosons are called gluons and couple to a conserved charge called "color." Each quark is really a multiplet of $SU(3)$ whose components have different color. For example "the u quark" is actually (u^r, u^g, u^b). (r = "red," g = "green," b = "blue"). $SU(3)$ transformations rotate these components among themselves, and, correspondingly, gluon emission or absorption physically accomplishes this.

Together the electroweak theory and QCD comprise the standard model. The elementary spin-$\frac{1}{2}$ particles of this theory are the quarks (which have color) and the leptons (which do not). These come in three sets called families identical except for their masses. The first family consists of the following multiplets:

	multiplet	$SU(3) \times SU(2) \times U(1)$	electric charge
leptons	$\begin{bmatrix} \nu_{eL} \\ e_L^- \end{bmatrix}$,	$(1, 2, -\frac{1}{2})$	$\begin{bmatrix} 0 \\ -1 \end{bmatrix}$
	e_R^-	$(1, 1, -1)$	(-1)
quarks	$\begin{bmatrix} u_L \\ d_L \end{bmatrix}$,	$(3, 2, \frac{1}{6})$	$\begin{bmatrix} \frac{2}{3} \\ -\frac{1}{3} \end{bmatrix}$
	u_R,	$(3, 1, \frac{2}{3})$	$(\frac{2}{3})$
	d_R,	$(3, 1, -\frac{1}{3})$	$(-\frac{1}{3})$

This assortment of particles is strange and complicated, yet also exhibits certain mysterious regularities. We note three: (1) Left-handed $SU(2)$ doublets, right-handed singlets. (2) Quarks have three colors and $\frac{1}{3}$-integral charges. (3) Electric charge is quantized. Also note that since quarks and leptons come in separate multiplets, gauge interactions (except for "anomalies") conserve both lepton number ($L \equiv N_{\text{leptons}} - N_{\text{antileptons}}$) and baryon number ($B \equiv [N_{\text{quark}} - N_{\text{antiquark}}]/3$). Protons as the lightest baryons ($B = 1$) cannot decay.

GRAND UNIFIED THEORIES

The technical definition of a grand unified theory (GUT) is a gauge theory of elementary-particle interactions in which the gauge symmetries of the strong, weak, and electromagnetic forces are contained within a single simple Lie group. ("Simple" is a group-theoretical term.) An apparent problem is that the three standard-model forces have different strengths given by the gauge coupling constants g_1, g_2, and g_3. However, in quantum field theory, coupling "constants" are really functions of the energy scale at which interactions are taking place. This is described by the renormalization group equations, which predict a slow (logarithmic) dependence of g_i ($i = 1, 2, 3$) on energy. Georgi, Quinn, and Weinberg (1974) found that these three couplings if extrapolated to very high-energy scales appear to converge to a common "unified coupling" g_{GUT} at an energy near 10^{15} GeV. This is the unification scale, M_{GUT}.

In the simplest (and first) grand unified theory, that of Georgi and Glashow (1974), the unification group is $SU(5)$ which undergoes two stages of spontaneous symmetry breaking. At the unification scale, M_{GUT}, $SU(5)$ breaks down to the standard model group $SU(3) \times SU(2) \times U(1)$, which itself undergoes the electroweak breaking at the scale M_W. The ratio of these two energy scales, M_W/M_{GUT}, is roughly 10^{-13} and its smallness is very hard to explain satisfactorily in the context of quantum field theory where it receives contributions that are of order $g_i^2/4\pi$ ($\approx 10^{-2}$) from higher-order quantum effects. Different contributions must therefore conspire to cancel to many decimal places which involves a highly unnatural "fine tuning" of parameters. This is called the "gauge hierarchy problem" (Gildener 1976) and is the key theoretical difficulty afflicting unified theories. The fermions of the first family form only two multiplets of $SU(5)$, an enormous simplification (Fig. 2). (Instead of the right-handed particles, e_R^-, u_R, and d_R we display their CP conjugates, the left-handed antiparticles, e_L^+, u_L^c, d_L^c.) Importantly, the quarks and leptons appear together in the same multiplets, so that there are $SU(5)$ gauge transformations, and hence also gauge bosons—called usually X and Y—which convert leptons into quarks and vice versa (Fig. 3). Thus baryon and lepton numbers are violated and the proton decays, for example into $\pi^0 e^+$, $\pi^+ \bar{\nu}$, and $K^0 \mu^+$. As $M_X \cong M_Y \approx M_{\text{GUT}}$ the proton decay rate is suppressed by $(m_p/M_{\text{GUT}})^4$ relative to typical nuclear interaction rates (10^{24} s^{-1}). So $\tau_p = 2 \times 10^{29 \pm 1.7}$ yr in the minimal $SU(5)$ model, with $\tau(p \to e^+ \pi^0)$ about three times that.

Since the renormalization group allows one to compute the standard-model couplings g_1, g_2, and g_3 in terms of only g_{GUT} and M_{GUT} there is a prediction which is that $\sin^2\theta_w$ $[\equiv g_1^2/(g_1^2 + g_2^2)] = 0.214 \pm 0.004$ (at M_W). This agrees fairly well with the experimental value of 0.229 ± 0.0064. Unfortunately, the minimal $SU(5)$ model is ruled out by the Irvine–Michigan–Brookhaven (IMB) experiment which set a limit of $\tau(p \to e^+ \pi^0) > 2.5 \times 10^{32}$ yr.

In models with unification groups other than $SU(5)$ at least two stages of symmetry breaking are required to reach the standard model: $G \xrightarrow[M_{\text{GUT}}]{} G' \xrightarrow[M_{\text{GUT}}']{} \cdots \to SU(3) \times SU(2) \times U(1)$. With more unknown parameters there is no definite prediction for τ_p or $\sin^2\theta_w$. Since the unknown scales of breaking enter as the fourth power in τ_p and only logarithmically in $\sin^2\theta_w$, the IMB bound on τ_p can be satisfied with only a small adjustment of parameters that leaves the $\sin^2\theta_w$ prediction only slightly disturbed. Thus the significance of

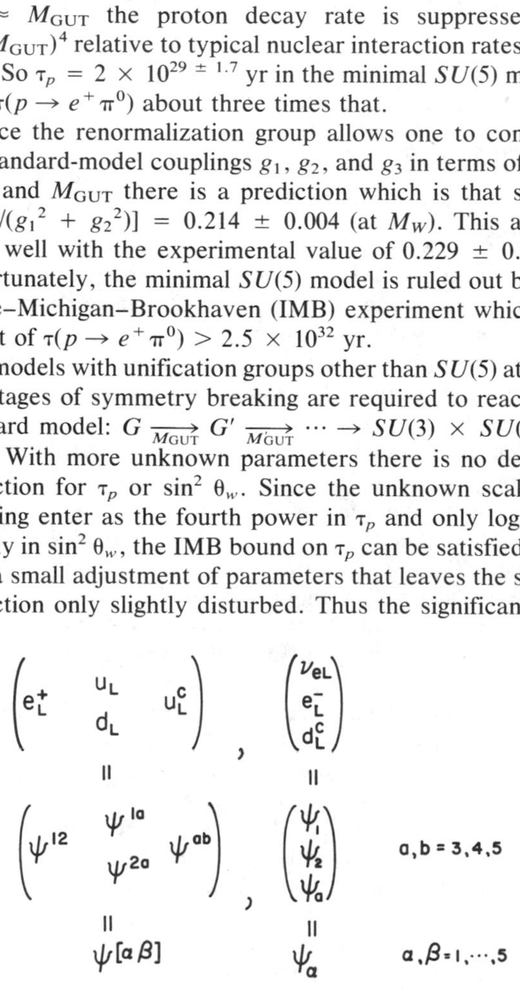

$$\begin{pmatrix} e_L^+ & \begin{matrix} u_L \\ d_L \end{matrix} & u_L^c \end{pmatrix}, \quad \begin{pmatrix} \nu_{eL} \\ e_L^- \\ d_L^c \end{pmatrix}$$

$$\| \qquad\qquad \|$$

$$\begin{pmatrix} \psi^{12} & \begin{matrix} \psi^{1a} \\ \psi^{2a} \end{matrix} & \psi^{ab} \end{pmatrix}, \quad \begin{pmatrix} \psi_1 \\ \psi_2 \\ \psi_a \end{pmatrix} \qquad a, b = 3, 4, 5$$

$$\| \qquad\qquad \|$$

$$\psi^{[\alpha\beta]} \qquad \psi_\alpha \qquad \alpha, \beta = 1, \cdots, 5$$

FIG. 2. The two $SU(5)$ multiplets containing the quarks and leptons of the first family.

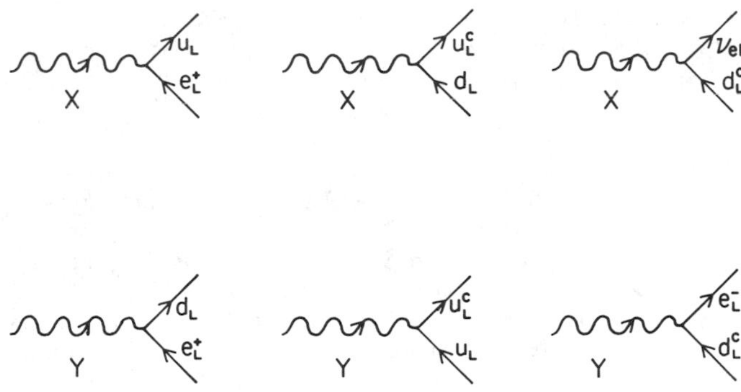

FIG. 3. The superheavy ($\approx 10^{15}$ GeV) gauge bosons of the $SU(5)$ unified theory, X and Y, can cause transitions between leptons and quarks.

the IMB result for grand unification generally [as opposed to the minimal $SU(5)$ model] should not be overstated.

Aside from these quantitative tests, grand unification has some impressive explanatory successes. (1) The peculiar assortment of quarks and leptons in a family follows automatically in grand unified theories from the *smallest* set of fermion multiplets which is complex (a group-theoretic term) and anomaly free. [The so-called Georgi survival hypothesis (Georgi and Glashow 1979) imposes the requirement of complexity. Gauge invariance requires anomaly freedom.] (2) Grand unification implies charge quantization. (3) The "mysterious regularities" of the quarks and leptons follow from simple group-theoretical considerations. For example the tracelessness of the $SU(5)$ charges requires that the total electric charge of a multiplet vanish. So $Q(e^-) + 3Q(d^c) = 0$ or $Q(d) = -\frac{1}{3}$. Many theorists believe that the remarkable group-theoretical "fit" of the standard model within unified models based on $SU(5)$ and related groups is too exact to be easily dismissed as coincidence.

GRAND UNIFIED GROUPS

A unification scheme just slightly older than $SU(5)$ is the $SU(4) \times SU(2) \times SU(2)$ model of Pati and Salam (1974). This is not truly a grand unified model since the group is not simple (it has three factors). Nonetheless a very appealing unification of quarks and leptons is achieved. The first family looks like

$$\begin{bmatrix} \nu_{eL} & u_L^r & u_L^g & u_L^b \\ e_L^- & d_L^r & d_L^g & d_L^b \end{bmatrix} \qquad \begin{bmatrix} N_{eR} & u_R^r & u_R^g & u_R^b \\ e_R^- & d_R^r & d_R^g & d_R^b \end{bmatrix}$$

The left-handed particles are doublets under one $SU(2)$ group and the right-handed particles are doublets under the other—a pleasing left-right symmetry (which can be made mathematically exact). Moreover one can see that leptons become the "fourth color" of quark. Thus as great a degree of quark–lepton unification is achieved in the Pati–Salam model as in $SU(5)$. (Notice that a right-handed neutrino N_R must be present in each family.)

While $SU(5)$ may be the smallest and simplest possibility, $SO(10)$ is in many ways a privileged unification group. It is the smallest which allows the quarks and leptons of one family to be unified into a single multiplet—namely, the 16-component spinor. $SO(10)$ also mathematically contains the Georgi–Glashow and Pati–Salam groups as subgroups—and thus in a sense as special cases. One can make a chart of the most interesting groups for unification (Fig. 4). Of course, ultimately only experiment can tell us which, if any, unified group is correct. But theoretical (or aesthetic) considerations have focused greater attention on certain ones. An interesting approach to understanding why there are families is the idea that all of them are contained in a single irreducible multiplet. For three (or more) families this can be achieved with the groups $SO(N)$, $N > 14$, E_7, and E_8. Note that all of these contain $SO(10)$ as a subgroup. The exceptional series (E_N) is interesting for other reasons both mathematical and physical. Some theorists would like the true unification group to be somehow special or unique. E_8 is the largest exceptional group (with finite-dimensional representations), while there is no largest $SO(N)$ or $SU(N)$ group. If one asks what the smaller members of the E_N series would be, it turns out to be natural to make the (very suggestive) identifications $E_5 \equiv SO(10)$, $E_4 \equiv SU(5)$, and $E_3 \equiv SU(3) \times SU(2)$. E_6 allows the fermions (the quarks and leptons) and the scalar bosons (the Higgs fields required for spontaneous symmetry breaking) of the theory to appear in multiplets of the same size, which is very interesting from the viewpoint of supersymmetry where a fermion–boson unification occurs. On the other hand, since exceptional groups rarely arise in physics, by far the most attention has focused on the unitary and orthogonal groups. However, recently, E_6 and E_8 have aroused great interest since they arise naturally as unification groups in "superstring" theory.

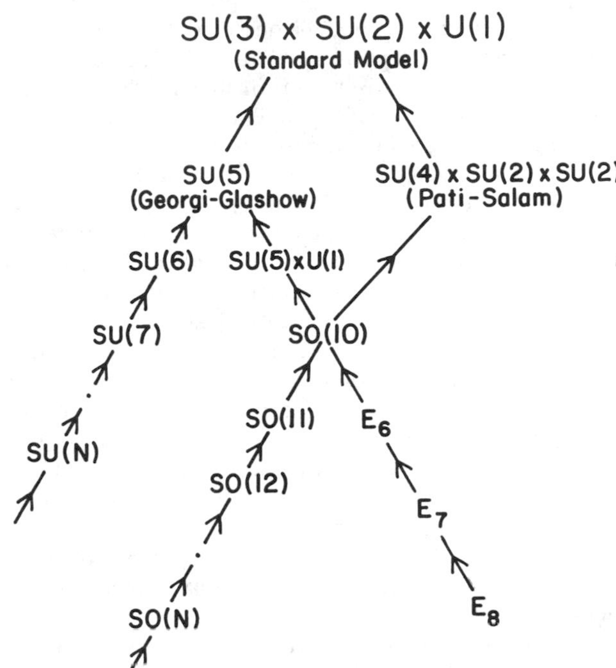

FIG. 4. Groups of most interest for unification of forces. $G \rightarrow H$ means here that H is a subgroup of G.

EXPERIMENTAL TESTS

The main prediction of grand unification is proton decay. However the lifetime is very model dependent and can vary over a range much wider than one can ever hope to test experimentally. (If for example $M_{GUT} \sim M_{Planck}$ one can expect $\tau_p \sim 10^{46}$ yr.) (It is worth mentioning that the large underground detectors built to find proton decay, though so far failing of their original purpose, have made the dramatic and important discovery of neutrinos from the supernova 1987A.) Other baryon number violating processes, such as neutron–antineutron oscillations, are even more model dependent and even less likely to be observed. Most unified models [though not $SU(5)$] predict the existence of right-handed neutrinos and thus (probably) small nonzero neutrino masses, but such masses can arise without unification. Grand unified theories predict that magnetic monopoles should exist whose mass is of order M_{GUT}. These are too heavy to be produced terrestrially, but primordial monopoles produced in the Big Bang have been searched for. Given various astrophysical bounds on such particles, finding them would seem a long shot. One of the few *low-energy* tests of unification would be the discovery of so-called extra Z bosons (which can arise in GUTs with group of rank > 4) and a measurement of their properties.

Unfortunately to date there has been no experimental confirmation of grand unification.

ASTROPHYSICAL IMPLICATIONS

The ideas of unification have had an enormous impact on cosmology. It is now believed that the predominance of matter over antimatter in our universe is due to baryon-number-violating processes which occurred in the early universe. This idea (Sakharov 1967) is a powerful argument in favor of grand unification. Such theories of "baryogenesis" predict that the primordial fluctuations that led to the growth of galaxies were probably nearly "adiabatic" (i.e., uniform entropy/baryon), an important constraint in the theory of galaxy formation. Spontaneous symmetry breaking associated with unified theories would have had dramatic consequences in the early universe when phase transitions connected with such breakings would have occurred. The study of such transitions led (Guth 1981) to the important idea of inflationary cosmology [which was also motivated by attempts to solve the "primordial monopole problem" (Preskill 1979) which arises in unified theories]. Such phase transitions also can lead to "cosmic strings" which are much studied as possible seeds of galaxy formation.

FUTURE

Theory has gone beyond the original grand unified theories in two main respects. First, the attempt to solve the "gauge hierarchy problem" and to achieve a unification of bosons with fermions has led to models with low-energy "supersymmetry." The attempt to include gravity has led to supergravity and superstring unification. Whether any of these beautiful, important and theoretically fruitful ideas describe physical reality awaits experimental confirmation.

See also ELECTROMAGNETIC INTERACTION; ELEMENTARY PARTICLES IN PHYSICS; GAUGE THEORIES; GROUP THEORY IN PHYSICS; INVARIANCE PRINCIPLES; NEUTRINO; PROTON; STRONG INTERACTIONS; SU(3) AND HIGHER SYMMETRIES; SYMMETRY BREAKING, SPONTANEOUS; WEAK INTERACTIONS.

BIBLIOGRAPHY

H. Georgi, "A Unified Theory of Elementary Particles and Forces," *Sci. Am.* 48–61 (1981). (E)

A. Zee, *Fearful Symmetry, The Search for Beauty in Modern Physics*. Macmillan Publishing Co., London, 1986. (E)

T.-P. Cheng and L.-F. Li, *Gauge Theory of Elementary Particle Physics*, pp. 428–453. Oxford University Press, Oxford, 1984. (A)

P. Langacker, "Grand Unified Theories and Proton Decay," *Phys. Rep.* **74,** 185–385 (1981). (A)

A. Zee, *The Unity of Forces in the Universe* (2 vols.). World Scientific, Singapore, 1982.

Gratings, Diffraction

E. G. Loewen

INTRODUCTION

Diffraction gratings are the most widely used elements for dispersing light into its spectral components. They are typically in the form of plane or spherically concave mirror surfaces that contain a large number of straight, parallel grooves. Groove frequencies vary from 20 to 6000 per millimeter. Incident light is diffracted by each of the grooves, but for a given wavelength is visible only in that direction in which light from each groove interferes constructively with that of the others. Depending on the ratio of wavelength to groove spacing, there may be one or more of such sharply defined directions, the *grating orders*. As reflection devices they can be used over the entire electromagnetic domain from 0.1 nm to 1 mm. In the form of transmission gratings, their use is restricted by the transparency of available materials to a band from the near ultraviolet to the near infrared.

Since gratings can be considered a special type of interferometer, it has always been clear that to obtain clean separation of wavelengths requires the grooves to be located with a uniformity of 0.1–0.001 of the wavelength of the light being analyzed. To achieve accuracy of such levels while mechanically impressing the grooves on the grating surface is the ultimate challenge to contemporary machinery.

HISTORY

Joseph Fraunhofer was the first (1823) to rule precision diffraction gratings and use them to isolate solar spectral lines. He derived the basic grating equation:

$$m\lambda = d(\sin\alpha + \sin\beta)$$

where m is the spectral order of wavelength λ, d the groove spacing, and α and β are the angles of incidence and diffraction, respectively, with respect to the normal of the grating surface.

The next major step was taken by Henry Rowland, who built the first ruling engine that produced large gratings of high resolving power (1885), with the grooves diamond burnished onto blanks made of speculum metal. He also invented the concave grating that performed both the imaging and the dispersing functions, eliminating the need for collimating optics required for plane gratings. With that invention he opened up spectroscopy in the entire vacuum ultraviolet.

Albert Michelson devoted decades to the building of large ruling engines, recognizing in 1915 the desirability for long (250-mm) travels to be monitored interferometrically, instead of relying on even the best and most carefully corrected lead screws. George Harrison took advantage of technological advances in isotope lamps, electronics, and servos to bring this concept into being in 1950. At the same time he introduced the idea of the echelle grating, used in high orders at high angles of incidence. With this grating, resolution of over 10^6 is attainable.

R. W. Wood was responsible (1920) for showing how the diffraction efficiency could be greatly improved through the shaping of individual grooves. John Strong introduced the concept of ruling in vacuum-deposited aluminum layers on glass substrates, an idea crucial to the ruling of larger and finer pitch gratings.

Michelson indicated in 1927 the possibility of generating high-precision gratings by recording in a suitable medium the highly uniform interference fringe field generated when two monochromatic collimated beams intersect. He lacked an intense source (i.e., a laser) as well as a grainless photographic recording medium. Labeyree in France, and Rudolph and Schmahl in Germany, began in 1967 to use photoresists and argon ion lasers to make such gratings, usually termed holographic. For some applications these gratings have the advantage of eliminating the effects of residual mechanical errors, but they have restrictions in groove spacing (no coarser than 600 per millimeter) and in diffraction efficiency.

From a user's point of view a very important step forward was the development by 1950 of a plastic casting process with which it was possible to make replicas of master gratings that were optically identical. For the first time precision gratings became available in quantity at low cost, and quickly displaced prisms from most spectroscopic instruments. Compared to prisms, they are cheaper, easier to mount, identical from one to the other, mathematically predictable, and not limited by transparency of substrates. Resolving powers such as 10^5, beyond the reach of prisms, became routine.

MONOCHROMATORS AND SPECTROGRAPHS

Spectrometry requires not only dispersion but also imaging of an entrance slit onto a receiving surface. The latter may be a fixed exit aperture, in which case we speak of monochromators, because only one wavelength is transmitted at a time; or it may be a receiving surface, in which case we speak of spectrographs.

Most monochromator designs are based on collimating the entrance beam so that the collimated light diffracted by a plane grating can be focused stigmatically onto the exit slit. All elements are fixed except the grating, whose angular setting with respect to the incident beam determines the wavelength transmitted. It is possible to use concave gratings instead, which can image without the use of separate collimating optics, but at the expense of nonstigmatic imaging.

Spectrographs are frequently designed around concave gratings because the optics achieves the ultimate in simplicity, i.e., just the grating. The traditional photographic film plate recording of spectra has been largely displaced by electronic array detectors; sensitivity is increased and direct computer data handling becomes straightforward. The grating's usefulness is greatly enhanced by the development of holographic-generating methods that reduce aberrations and flatten the image field. One important consequence is that much more compact systems become possible.

A high-dispersion, high-resolution system with an unusually compact display is possible by combining a plane echelle-type grating with a cross-dispersing grating or prism. The latter conveniently separates the otherwise overlapping high orders and leads to a rectangular display.

Largely due to lasers, a number of nonspectrometric uses have been found for gratings. Examples are beam splitters and recombiners, beam-scanning systems, and pulse compression.

See also FRAUNHOFER LINES; HOLOGRAPHY; INTERFEROMETERS AND INTERFEROMETRY; VISIBLE AND ULTRAVIOLET SPECTROSCOPY.

BIBLIOGRAPHY

R. M. Barnes and R. F. Jarrell, "Gratings and Grating Instruments," in *Analytical Emission Spectroscopy* (E. Grove, ed.), Ch. 4. Dekker, New York, 1968.

G. R. Harrison, R. C. Lord, and J. R. Loofbourow, *Practical Spectroscopy*. Prentice-Hall, Englewood Cliffs, N.J., 1948.

M. C. Hutley, *Diffraction Gratings*. Academic Press, London, New York, 1982.

E. W. Palmer, M. C. Hutley, A. Franks, J. F. Verrill, and B. Gale, "Diffraction Gratings," *Rep. Prog. Phys.* **38,** 975–1048 (1975).

R. W. Sawyer, *Experimental Spectroscopy*. Dover, New York, 1963.

Gravitation

John Friedman

INTRODUCTION

Early History

Our knowledge of gravity is rooted in astronomy and geometry. Like astronomy, geometry has written records spanning four millenia, and also like astronomy, it began as an empirical science. Greek geometers learned their ropes in apprenticeship to Egyptians, whose stretched cords had served as both ruler and compass. From direct measurement, the Egyptians had developed a set of geometric laws that included special cases of the Pythagorean theorem and approximate areas of circles and spheres. Over five centuries the Greeks gradually built their theory of geometry on an experimental, initially Egyptian, foundation. Plato, recognizing the parallel between astronomy and geometry, hoped similarly to unify the rather complicated Babylonian rules governing the recurrence of eclipses and planetary positions: to his students he set the problem

> to find the uniform and ordered movements by the assumption of which the motion of the planets can be accounted for.

Solving Plato's problem required a quantitative understanding of gravity, and it ultimately led, early in the present century, to general relativity, a subtle unification of astronomy and geometry.

The parallel between the two sciences can, in fact, be drawn quite closely if one regards early geometry as a study of space via objects that show extreme spatial regularity, for early astronomy can be called a study of time via objects whose dynamics (behavior in time) show extreme temporal regularity. The importance of astronomy to agricultural societies, in whose religions it played a usually dominant role, was related to the accurate timekeeping the stars provided. In fact, until the development of atomic clocks stellar positions provided the most accurate measure of elapsed time.

Once the Greeks had recognized that the earth was spherical (by 200 B.C. Eratosthenes had measured its circumference to better than 10%), it was possible to explain the apparent daily revolution of the stars from east to west as the result of a real daily rotation of the earth from west to east, and at least a minority of Greek astronomers adopted this view. Although accounting for the simple apparent motion of the stars was thus a straightforward step from a round earth, solving Plato's problem—explaining the apparent motion of the planets—was much more difficult. The key obstacle was that, seen from the earth, the planets move against the background stars in paths that are only roughly repetitive and in which an average west-to-east motion is periodically interrupted by a retrograde motion from east to west. The Greek models of planetary motion reproduce retrograde motion by supposing that the planets move in epicycles (small circles) whose centers themselves move in circles about the earth. This description is in fact correct for a frame of reference centered at the earth: Seen from earth, the sun appears to move yearly in a circle, and the nearly circular motion of each planet about the sun is then approximated by an epicycle whose center (the sun) itself appears to circle the earth. Because planetary orbits are not exactly circles, the most accurate Greek model (Ptolemy's) was burdened by additional small epicycles for each planet.

Copernican Revolution

A simple description of planetary motion requires one to choose a frame of reference centered at the sun—to abandon thereby the apparently obvious fact that the earth is at rest at the center of the celestial sphere. A sun-centered system brings a dramatic conceptual simplification: To an observer at rest relative to the sun, there is no retrograde motion—the need for major epicycles disappears for all five visible planets at once. Although Aristarchos had, by 250 B.C., suggested such a system, subsequent Greek models failed to exploit his idea, and nearly two millenia passed before Copernicus finally contrived a model in which planets moved about the sun (Fig. 1).

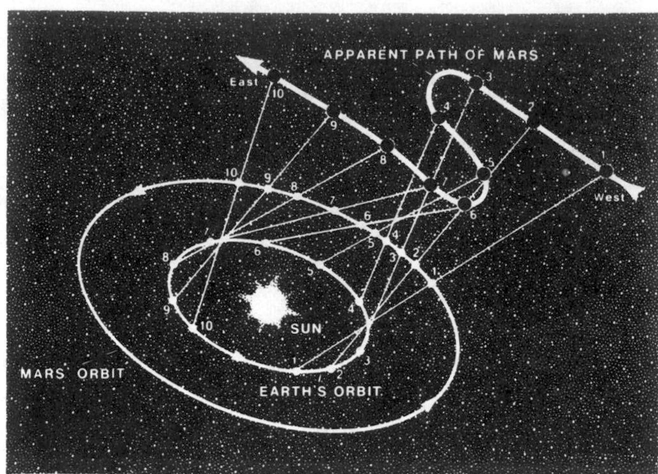

FIG. 1. The motion of planets, complex as seen from earth, is simple in a frame of reference at rest relative to the sun. The motion of Mars seen from earth and sun is depicted above: The earth moves more quickly in its orbit than Mars does, and as we pass Mars, it appears to reverse its direction of motion relative to the background stars.

The subsequent logical development has been rapid. Within a hundred years it was clear that if the earth did not define an absolute rest frame, neither did the sun: After Galileo and Newton it was clear that there was no natural candidate for an object at rest, and so no natural way to decide what it meant to be at the same place at two different times. One lost absolute space, the idea that one can pick out the same point in space at two different times. When the earth is a natural rest frame, one seems to need a force to make an object move. With no natural frame, there is no way to know if an object is moving or at rest. Thus, a force is needed not to make something move—with no force, a puck slides forever on perfect ice—but to change its motion.

With force needed only for acceleration, the force that acts on a planet in circular orbit about the sun must point to the sun, because the acceleration, the change in velocity per unit time, is an arrow pointing to the center of a circular orbit. This was the insight that Kepler, still tied to the idea of absolute space, had missed: from the fact that all objects fell to a round earth, he guessed that gravity was universal, that

> If two stones were placed anywhere in space near to each other and outside the reach of a third like body, then they would come together, after the manner of magnetic bodies, at an intermediate point, each approaching the other in proportion to the other's mass.

Failing to understand Galileo, Kepler still thought a force was needed to keep a planet moving. His spectacular success in deducing the elliptical shape of planetary orbits and the speed of a planet along any orbit was accompanied by a failed attempt to explain these simple motions in terms of the sun's magnetic field. To the next generation, knowing both Kepler and Galileo, it was clear that gravity pointed in the right direction to bend a planet's path, and both Newton and Hooke guessed that the force between two point masses decreased with distance as $1/r^2$ (r the distance between the masses). Armed with his calculus, Newton proved that the single new axiom of a $1/r^2$ force (and Galileo's law that force was proportional to acceleration) implied Kepler's three laws and hence the motion of planets.

Newtonian Gravity

Newton's law governs the acceleration of one particle from the gravity of a second. If the second particle is at a distance r from the first and has mass M, the acceleration of the first particle given by the equation,

$$\frac{d^2\mathbf{x}}{dt^2} = \frac{GM_{\hat{r}}}{r^2}, \tag{1}$$

where the constant G has the value 6.67×10^{-8} kg^{-1} m^3 s^{-2}. This relation describes the geometry of each particle's path in space-time: its path in space as a function of time. For mass distributed through space with density ρ, Eq. (1) takes the form

$$\frac{d^2\mathbf{x}}{dt^2} = -\nabla\Phi, \tag{2}$$

where

$$\nabla^2\Phi = 4\pi G\rho. \tag{3}$$

Using Eq. (2) one can easily prove a result that Newton struggled with for some years, that the acceleration of a ball (a spherically symmetric distribution of mass m) due to the gravity of a second ball of mass M is again given by Eq. (1), with r now the distance from the center of each ball. The corresponding mutual attraction of two spherically symmetric objects is described by the force $F = GMm/r^2$.

At the close of the Copernican revolution, marked by Newton's law of gravity, all but one of the ancient problems of astronomy were explained—the apparent motion of sun, moon, stars, and planets, the seasons, tides, and eclipse cycles, the precession of the equinoxes. Only the "guest stars," apparent new stars whose sightings were recorded once every few centuries, remained mysterious.

THE GEOMETRY OF SPACE-TIME

Space and time are to some extent already blended in the Newtonian framework. By adjoining time to the three spatial coordinates of Descartes, Newtonian dynamics acquired at least the formal trappings of space-time, the set of points or "events," $(t, \mathbf{x})$. More striking, however, is the mixing of space and time that Newton's laws imply: The path of an observer "at rest" and one moving at constant speed are both straight lines in space-time, pointing to the future, but not parallel. Because there is no empirical way to decide who is at rest, one has no unique way of splitting space-time into space + time. If we suppress a dimension and think of space as two dimensional, a plane with two-dimensional objects moving about on it, then space-time can be regarded as a set of photographs of the plane, placed one on top of the other. An observer stacking the photographs aligns them so that successive pictures of an object at rest relative to the observer are directly over one another, forming a vertical line. Thus an observer at rest relative to the earth would place the earth on the $t = 1$ second photograph directly above the point that marks the earth on the $t = 0$ photograph. On the other hand, for an observer at rest relative to the sun, the earth moves at 30 km/s, forming a helix in space-time. Any straight line directed toward the future is a path of an unaccelerated observer, and the fact that there is no absolute space means that there is no preferred direction for such a line.

By the middle of the last century, however, it appeared that absolute space was reestablished, when Maxwell found that light was an electromagnetic wave moving across space with finite speed c. For you can claim to be at rest relative to space—now conceived of as the "aether" through which light propagates—if, when you set off a firecracker, its light at all later times forms a sphere with you at the center. Instead of resurrecting absolute space, however, Maxwell's discovery turned out to destroy absolute time: In 1887, Michelson and Morley found experimentally that the speed of light was, in fact, independent of the motion of its observer (and of its source)! Thus, the new prescription for determining when you are at rest fails: Suppose, for example, you are in the middle of a train car and you try to decide whether you and the train are moving left or right by setting off a firecracker and observing whether you first receive the light reflected from the right wall or the left wall. Then regardless of the (uniform) speed of the train, both reflected signals will hit you at once. What is more, you will thereby deduce (by symmetry) that the light hit left and right walls simultaneously. But if the train is moving to the right, say, relative to an observer on the earth, the observer will report that the light hit the left wall before it hit the right wall. Thus, the two observers disagree on which events took place at the same time, and absolute time has been lost.

What remains? The absolute speed of light and the fact that no information can travel faster than light allows causality to survive. Only if two events occur so close together in time that information does not have time to travel from one to another can different observers disagree on which was first. The expanding sphere of light from a firecracker set off at a point $P = (t, x)$ expands to form a cone in space-time, the "future light cone" of P. Anything that starts at P and travels more slowly than light remains inside this light cone, and the event P can thus affect only future events inside the light cone. (See Fig. 2.)

Observers who move along straight, parallel lines in space-time are again at rest relative to each other, and they agree on what events occurred at the same time. These are their surfaces of simultaneity which slice space-time into space + time. In the absence of gravity, these spatial slices are flat: they continue to obey the laws of Euclidean geometry. For each choice of time-like direction, one can label the surfaces of simultaneity by a coordinate $x_0 = ct$ for which the light cone through a point P, at $t = 0$, is the set of points $(\mathbf{x}, t)$, for which

$$\eta_{\alpha\beta}x^\alpha x^\beta \equiv c^2t^2 - \sum(x^i)^2 = 0, \qquad \alpha = 0,1,2,3,$$

where $\|\eta_{\alpha\beta}\| = \text{diag}(-1,1,1,1)$ and x^i, $i = 1$–3, are Cartesian coordinates of the surface of simultaneity. The matrix η is

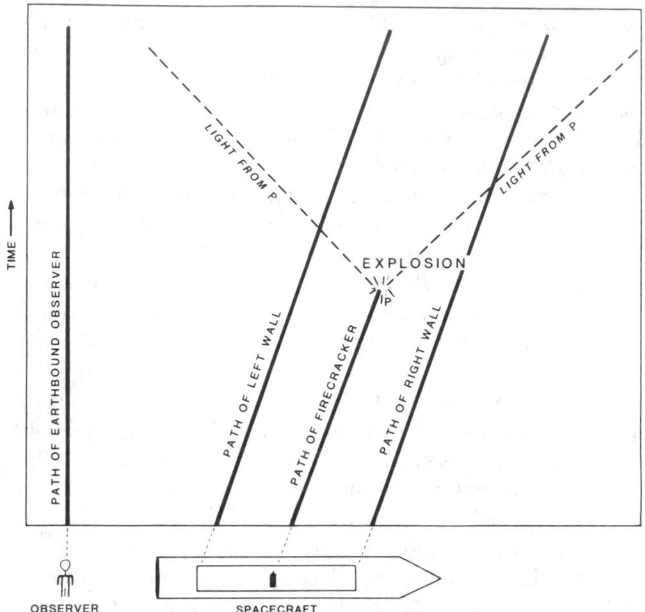

FIG. 2. Event P is the explosion of a firecracker set off midway between the left and right walls of a spacecraft. Seen by an observer on the spacecraft, the light hits the two walls simultaneously: A line joining events L and R is "space," the set of events that is simultaneous for this observer. But to an earthbound observer the light hits the left wall first: L occurs before R. The fact that nothing can move faster than light limits the tilt of line LR relative to the horizontal space of the earthbound observer, and it implies that events can be seen as simultaneous only if light does not have time to travel from one to another.

the flat metric of this "Minkowski" space-time, giving the distance between any two events, x^α and x'^α.

If, for two events, the squared interval is positive, then there is an observer for which the events are simultaneous, and their spatial distance is $|x-x'|\equiv|\eta_{\alpha\beta}(x^\alpha-x'^\alpha)(x^\beta-x'^\beta)|^{1/2}$. If $\eta_{\alpha\beta}x^\alpha x^\beta$ is negative, then a clock moving along the straight line joining the events will measure the time interval $|x-x'|$.

This is the framework of Einstein's special relativity, but Newton's law of gravity is not consistent with it. When Eq. (1) is satisfied on the surfaces of simultaneity of one observer, it fails on those of any other. The problem is that according to Eq. (1), the position of the second particle (M) instantaneously determines the acceleration of the first, contradicting the requirement that no information can travel faster than light. Instead, to find a theory of gravity that agrees with Newton's for small relative velocities and is consistent with special relativity, one must abandon the assumption that space is flat. In its place Einstein realized in part the dream of the mathematician Riemann that particles curve owing not to forces but to the curvature of space.

The idea seems on the face of it absurd. Balls thrown on earth deviate obviously and macroscopically from straight-line motion, but space is experimentally flat near the earth to better than one part in a million. The key, however, is simple. In measuring intervals between events in space-time, one uses the speed of light c to compare times to distances. A time of 1 s has the same magnitude as a distance of one light-second (300,000 km), the distance light travels in 1 s. Thus, in falling for 1 s, a ball drops less than 10^{-7} light-seconds, and its path in space-time barely deviates from the 1-s-long straight line it would have described in the absence of gravity. Similarly, because the earth's speed in its orbit about the sun is $10^{-4}\,c$, in one year it traces out a nearly straight line in space-time, a helix whose radius is 10^{-4} times smaller than its height.

The advent of general relativity at once destroyed Euclidian geometry and Newtonian astronomy, but it showed that a unification of space and time implied a unification of the two sciences as well. Newton's law (1) is replaced by the statement that particles move on geodesics, paths of shortest length on a curved space-time. Equation (2), relating the gravitational potential Φ to the distribution of mass, becomes a limiting case of the equation relating the curvature of space-time to the matter density. The flat metric of Minkowski space is replaced by a metric describing the curved space-time geometry. The way in which mass curves space is given by the equation

$$R_{\alpha\beta}-\tfrac{1}{2}g_{\alpha\beta}R=8\pi GT_{\alpha\beta}, \tag{4}$$

where $R_{\alpha\beta}$ is the Ricci curvature tensor ($R_{\alpha\beta}\equiv g^{\alpha\beta}R_{\alpha\beta}$), constructed from the metric and its first two derivatives. For a collection of particles with mass density ρ moving along paths with unit tangent u^α, the matter tensor $T^{\alpha\beta}$ is $\rho u^\alpha u^\beta$.

Newtonian Limit: Gravity in the Solar System

The sun, planets, and moons that comprise the solar system move with relative velocities small compared to c. In the frame of the solar system, the components of u^α are then approximately given by (1, 0, 0, 0), the metric has the form,

$$\|g_{\alpha\beta}\|=\left\|\begin{matrix} 1+\dfrac{2\phi}{c^2} & & \\ & 1+\dfrac{2\phi}{c^2} & \\ & & 1+\dfrac{2\phi}{c^2}\end{matrix}\right\| \tag{5}$$

and Eq. (4) becomes, in this Newtonian limit, Eq. (2). The solar system's metric deviates from the flat metric η by at most one part in 10^6, and the corrections provided by general relativity to Newtonian gravity are similarly small. At the time it was formulated, the only discrepancy in Newtonian gravity was an inaccurate prediction of the precession of Mercury's elliptical orbit that would arise from the other planets' gravity. The predicted precession disagreed with observations by 43″ of arc/century, and the first triumph of general relativity was to account for that difference.

Not until the past two decades was gravity systematically measured to a precision high enough that one could verify the geometry predicted by relativity. Among the most accurate of these are the Eötvös experiments, tests of the "principle of equivalence," the statement that particles with the same initial velocities move on identical trajectories, independent of the mass or composition of the particles. Galileo's demonstration at the tower of Pisa is the precursor to experimental work of Eötvös, Dicke *et al.*, Braginsky *et al.*, and other recent investigators, searching for a dependence of the gravitational acceleration on the internal structure of two suspended masses. Lack of such a dependence is currently verified at a level of 1 part in 10^{12}.

The implication of these experiments is that in a gravitational field that is nearly uniform over the size of a laboratory, one cannot distinguish the gravitational field from an accelerating laboratory. In a uniformly accelerating laboratory, or in one at rest in the earth's nearly uniform field, free particles will fall with constant acceleration relative to the laboratory. To discover that he is not in uniform acceleration, an observer confined to a closed laboratory must make measurements accurate enough to see a change in the gravitational field from one side of the room to the other, or to detect the energy lost in radiation by a particle at rest relative to the room. The approximate inability to distinguish a slowly varying gravitational field from uniform acceleration is known as the principle of equivalence. Historically one of the guiding principles leading to relativity, its shadow remains in the theory as the statement that any smooth geometry is approximately flat when measured over intervals short compared to its radius of curvature.

A second, related, class of experiments measure the gravitational redshift, the change in frequency of light as it climbs higher in a gravitational field. To conserve energy, the work done by gravity on a photon of energy E that travels upward a height h in the gravitational field of the earth must be the work done on a mass of magnitude E/c^2, namely, $(E/c^2)\,gh$, where g is the acceleration due to the earth's gravity, $g=9.8$ m/s^2. Because the energy of a photon is related to its frequency ω by $E=\omega$, a photon climbing up a height h is redshifted, changing its frequency by $Z=\Delta\omega/\omega=-gh/c^2$. The first successful, high-precision redshift measurements (due to Pound, Rebka, and Snider) were made in 1960–1965. The redshift measures the part of the metric that determines rates of clocks at different heights in a static gravitational field, and hydrogen maser clocks on orbiting satellites have restricted the deviation from general relativity to a part in 10^4.

An additional set of tests involves the bending of light by the sun's gravity and delays in light travel time due to the comparably large curvature of the space-time near the planets. The bending of light by the sun, the most famous of the theory's early predictions, was first observed during the 1919 solar eclipse. Current striking examples of light deflection are gravitational double images of distant quasars, for whom intervening galaxies act as gravitational lenses. Light deflection and light travel times depend on a parameter independent of the redshift, and measurements by Viking spacecraft have limited its departure from the prediction of relativity to 0.001%.

The predicted precession of the orbit of Mercury has been measured to a precision of about 0.5%, by using radar determination of the planet's position: By bouncing radio waves off an object and carefully timing their round-trip time, one obtains solar system positions to within a few meters. The test of precession, however, is limited not only by planetary position, but by our knowledge of the sun's shape: rotation of the solar interior could make the sun slightly oblate, and a departure from spherical symmetry also implies a rate of planetary precession.

Finally, comparison of orbital precession of the earth, and measurements of solid tides of the earth due to the combined gravity of sun, moon, and earth, restrict each of two additional independent parameters to within 0.1% of its predicted value. In summary, the past two decades of solar system tests have left little doubt that general relativity correctly describes gravity for objects that are large compared to atoms and small enough that a particle can escape their gravity with a speed small compared to light.

Gravitational Collapse and Black Holes

The most dramatic predictions of general relativity are in exactly the cases where it is poorly tested, where objects are very small compared to atoms or have a large enough concentration of mass that one must travel at speeds approaching light to escape. In the Newtonian approximation, a particle of mass m can escape the gravitational field of an object if its kinetic energy, $\frac{1}{2}mv^2$, is larger than its gravitational binding energy, $m\phi$. For a spherical star, $\phi=-GM/R$, where R is the distance from the object's center. This means that a particle can escape only when its velocity is given by

$$v>v_{esc}\equiv\sqrt{2GM/R}. \tag{6}$$

At the surface of the earth, $v_{esc}=11$ km/s. From Eq. (6), the escape velocity from the surface of an object exceeds the speed of light when its radius shrinks to within the "Schwarzschild radius,"

$$R_S=2GM/c^2. \tag{7}$$

This turns out to be the correct result in the exact theory as well, if R is defined as the length for which $2\pi R$ is the distance circumference of the object (the fact that space is curved then implies a larger distance from the object's center to its edge). For the sun, the Schwarzschild radius is 1.5 km, vastly smaller than the star's 700,000-km radius. Remarkably, however, stars with masses larger than 10 or 20 times that of the sun seem fated to end their lives by collapsing to within their Schwarzschild radius—to form black holes.

As a star evolves, the atomic nuclei in its core fuse to build up a sequence of heavier elements. The free protons and electrons (hydrogen ions) that largely comprise the initial star gradually fuse together to create the elements from helium through iron that populate the universe. Iron is the most stable nucleus, and in the most massive stars, an iron core gradually builds up as lighter nuclei fuse together. Because iron nuclei cannot fuse, the core supports itself by the pressure of its electrons, each restricted to a distinct state.

There is, however, a maximum mass that such matter can support, related to the finite speed of light. For an ideal gas, the pressure per unit mass (p/ρ) is equal to the kinetic energy per unit mass, equal to v^2, where v is the average speed of a gas molecule. The restriction $v<c$ suggests a limit on the pressure per unit mass of c^2, and this limit is, in fact, implied for any matter by the causality requirement that the speed of sound be less than the speed of light.

But there is no limit on the gravitational attraction per unit mass. As a result, there is a maximum mass that can be supported by electron pressure, the Chandrasekhar limit of 1.4 $M_\odot$, where $M_\odot$ is the mass of the sun. Shortly before the mass of a star's core reaches this limit, as the speed of the electrons approaches the speed of light, the core collapses, with the electrons pushed by gravity onto the iron nuclei. The entire core is thereby converted to neutrons, becoming a neutron star whose mass of nearly $1\frac{1}{2}$ times the mass of the sun is compressed to a radius of about 10 km. The gravitational energy of the collapse blows off the rest of the star in an explosion called a supernova. A closely similar collapse is thought to occur when white dwarfs accrete enough matter from a companion star that they exceed the Chandrasekhar limit. The "guest stars" of ancient astronomy were supernovae, their explanation an achievement of the last half-century.

Like the iron core or white dwarf from which it forms, a neutron star has a maximum mass set by causality—by the finite speed of light. Our lack of knowledge of the behavior of matter at the extreme densities of neutron stars, 10^{15} g/cm^3, means that the largest neutron star mass is known only roughly, but general relativity implies that no spherical configuration of matter at neutron density or higher can have a mass greater than $5M_\odot$. For dense matter with mass greater than this, the escape velocity exceeds the speed of light, and the star collapses to form a black hole.

In the collapse of a star, the escape velocity at the star's surface increases as the star contracts. When the surface passes through the Schwarzschild radius, R_S, the escape velocity exceeds that of light, and no light (or matter) can escape from the region inside a sphere of circumference $2\pi R_S$. In fact, within the classical theory, the matter inside is drawn to a single point. The gravitational field, however, persists and a region of empty space outside the tiny collapsed star is cut off from the surrounding universe: the light cones at the Schwarzschild radius tip inward, preventing anything from emerging (Fig. 3). Events inside the "event horizon" at $r=R_S$ are inaccessible to external observers. When the collapse is not spherical, as in the collapse of a rapidly rotating star, the event horizon is oblate, but a theorem of Penrose again implies, within classical gravity, a singular evolution inside the black hole.

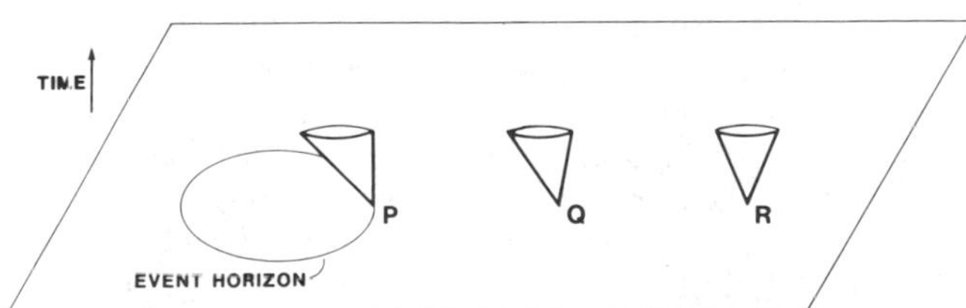

Fig. 3. The diagram above shows the path followed by light emitted at points outside the event horizon of a black hole. One spatial dimension is suppressed, so the horizon of the black hole is a circle. A flash of light emitted at time 0 expands to form a widening sphere, depicted in the diagram as a widening circle. Far from the event horizon (point R), the circle is centered about the point where the light was emitted; the flash from Q is pulled somewhat inward; and the flash at P, on the event horizon, is pulled inward so strongly that even the outward directed light cannot escape—rays directed radially outward remain forever at the same radius.

The experimental evidence for black holes, if not decisive, has become quite strong. Black holes can in principle be observed in binary systems (systems of two stars orbiting their common center of mass) when mass from the ordinary star falls onto the black hole. (The expansion of the ordinary star to form a red giant late in its evolution is commonly the event that triggers the accretion: matter at the outer edge of the giant is closer to the black hole than to the center of its parent star.) The accreting matter moves at speeds comparable to c, and emits a rapidly varying spectrum of x rays. While matter falling onto a neutron star will also emit x rays, observations of the binary system can set a lower limit on the mass of the dense member. In cases where this mass is above $5M_\odot$ there seems to be no way to avoid the conclusion that one is observing a black hole. The best studied black hole candidates of this kind are in the binary systems Cygnus X-1, LMC X-3, and Cir X-1, all x-ray sources for which the candidate is a dense star with mass greater than $8M_\odot$.

Even more remarkable is a second class of black hole candidates in the centers of large galaxies. Observations of the central regions of galaxies show a sharp increase in brightness near the center. To account for the emission with a cluster of stars, the density of stars would have to be exceptionally high, high enough that it seems quite difficult to avoid rapid coalescence to giant stars, leading to gravitational collapse to large black holes. The resulting prediction of sharp central concentrations of mass has been dramatically verified in recent years by radio telescopes electronically joined across the globe (called very long baseline interferometers). The resolution of these telescopes is comparable to a single telescope the size of the earth, and by looking at the velocity of orbiting clouds, they reveal a mass of $5\times10^9\ M_\odot$ in a region the size of our solar system.

Gravitational Radiation

A second implication of the finite speed of information is the prediction of gravitational waves, waves of curvature that travel at the speed c. For nearly flat space-times, with $h_{\alpha\beta}\equiv g_{\alpha\beta}-\eta_{\alpha\beta}$ small, Eq. (4) becomes the wave equation for $h_{\alpha\beta}$:

$$h_{\alpha\beta}\equiv\left[\frac{\partial^2}{\partial x^{0^2}}-\frac{\partial^2}{\partial x^{1^2}}+\frac{\partial^2}{\partial x^{2^2}}+\frac{\partial^2}{\partial x^{3^2}}\right]h_{\alpha\beta}=0.$$

The first evidence for these waves has come in the last two years from observations of a system of two neutron stars that is losing energy at the rate corresponding to its predicted emission of "gravitational radiation." Verification of the predicted power is currently at an accuracy of about 3%, with steady improvement expected as the number of years of observation increases. Direct detection is still in the future, but 20 years of efforts in building gravitational wave detectors are gradually developing a sensitivity for which gravitational wave astronomy can become a reality.

Cosmology

On the largest scale, the visible universe appears to be uniform in its distribution of matter, which on smaller scales is clumped into stars, galaxies, and a hierarchy of clusters of galaxies. Models for the universe, satisfying the field equations (4), are either expanding or contracting, and the discovery by Slipher, Hubble, and Humason of a uniform expansion traced by galactic redshifts should have been an early triumph of the theory. It was not only because Einstein had by then added an additional term $\Lambda g_{\alpha\beta}$ to the equation to allow a static universe. If one follows an expanding universe backward in time, one encounters ever-increasing densities, and a second theorem of Penrose, this time in collaboration with Hawking, again implies a singularity, marking the beginning of the universe, or at least of the part of the universe in which we reside. The explosive expansion immediately after the singularity is termed the "big bang."

There is little doubt that the big bang in fact occurred. The most remote galaxies, over 1×10^{10} light-years distant, are seen as they were over 1×10^{10} years in the past, and as one looks back in time toward the big bang the character of galaxies changes. Finally at redshifts greater than about 4, one sees no galaxies at all: they could not have formed until after the big bang. The key piece of evidence concerns the cosmic microwave background radiation predicted by Alpher, Bethe, and Gamow in 1948. In the hot, dense, early universe, light and matter are in equilibrium, but as the universe expands, the time between successive collisions of a photon with the surrounding matter grows longer, until, after about 1×10^{5} years, the density is low enough that the time for a photon to hit a particle is longer than the time to cross the universe. Light from that time should now be uniformly distributed and should have been redshifted in tandem with the expansion, its wavelength proportional to the radius of the universe. In 1965, the prediction was confirmed, with the spectacular discovery by Penzias and Wilson of isotropically distributed microwave radiation, and its identification with the predicted background by Dicke *et al.* Subsequent satellite observations depict a nearly exact blackbody spectrum at 2.7°K, with deviations from perfect uniformity less than 0.01% (after an overall motion of the earth relative to the rest frame of the radiation is taken into account).

Quantum Gravity

On scales larger than the size of an atom, gravity appears to be accurately described as a deterministic theory of the geometry of spacetime. The theory, however, is not consistent with quantum mechanics, whose probabilities characterize matter on small scales, as well as large. The logical inconsistency of the two theories is exacerbated by the singularity theorems, which show that within black holes, and at early times in the history of our universe, the density of matter is so high that one can no longer do physics without a way to unify the theories.

While a quantum-mechanical geometry must follow the quantum-mechanical probabilities for atoms on an atomic scale, gravitational fields on such scales are far too small to be detected. Only when one looks at distances of the order of the Planck length,

$$l=(c/G^3)^{1/2}=1.6\times10^{-33}\text{ cm},$$

do quantum fluctuations in the geometry become noticeable. At such lengths, even the topology of space-time may fluctuate. Small structures with non-Euclidean topology might conceivably represent elementary particles, and child universes might branch off our own, or the concept of spacetime as smooth geometry might simply fail to describe reality on so small a scale. For somewhat larger distances, however, fluctuations of the geometry can be ignored and a convincing approximate theory has been developed. Among its more striking implications is a predicted evaporation of black holes: The quantum nature of matter alters gravity, allowing the horizon of an isolated black hole gradually to shrink as it radiates its mass in a thermal spectrum of particles.

See also MACH'S PRINCIPLE; NEWTON'S LAWS; RELATIVITY, GENERAL.

BIBLIOGRAPHY

S. W. Hawking, *A Brief History of Time*. Bantam, New York, 1988.
C. W. Misner, K. S. Thorne, and J. A. Wheeler, *Gravitation*. W. H. Freeman, San Francisco, 1973.
R. M. Wald, *Space, Time, and Gravity*. Chicago, 1977.
C. M. Will, *Was Einstein Right?* Basic Books, New York, 1989.

Gravitational Lenses

Jacqueline N. Hewitt

Gravitational lensing occurs when the ray paths of electromagnetic radiation are altered as they pass through a gravitational field. The first measurements of the influence of a gravitational field on the propagation of light took place during the solar eclipse of 1919, when the deflection due to the gravitational field of the sun was measured. After this measurement, there were many articles in the literature of physics and astronomy that discussed the expected properties of images of objects viewed through a gravitational lens. In particular, it was recognized that a gravitational lens is likely to produce distorted, multiple images of the background source. Two short articles that have received considerable attention were those of Einstein [1], in which he showed that perfect alignment of the source and a circularly symmetric lens would produce a circular ring image (now popularly referred to as an "Einstein ring"), and of Zwicky [2], in which he pointed out that the large masses and distances of galaxies and clusters of galaxies implied that gravitational lenses would be observed in extragalactic sources of radiation. The predictions of both articles have been realized: the first gravitational lens, consisting of a distant quasar lensed by a large galaxy and its associated cluster of galaxies, was discovered in 1979 [3], and the first Einstein ring was discovered in 1988 [4]. There are now some 20 gravitational lens systems known, all involving propagation of light across intergalactic distances. They include multiply imaged quasars and galaxies, two Einstein rings, and the luminous arcs which surround many dense clusters of galaxies.

The deflection of a ray of light in the gravitational field of a spherically symmetric mass distribution is given by the following formula, due to general relativity, for the angle θ between the perturbed and unperturbed light ray:

$$\theta = \frac{4GM}{c^2 b},$$

where G is the gravitational constant, M is the mass, c is the speed of light, and b is the impact parameter of the light ray with respect to the mass. In the limit of a weak gravitational field, one can sum the contributions to the deflection angle of a collection of point masses; therefore, the lensing properties of an arbitrary mass distribution can be calculated. Most of the qualitative features of the known gravitational lenses can be reproduced by calculations that invoke a gravitational potential with elliptical symmetry.

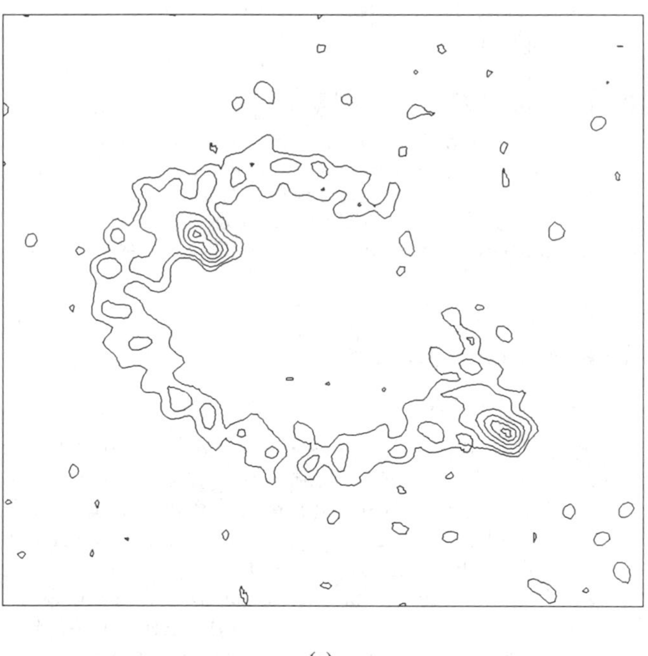

(a)

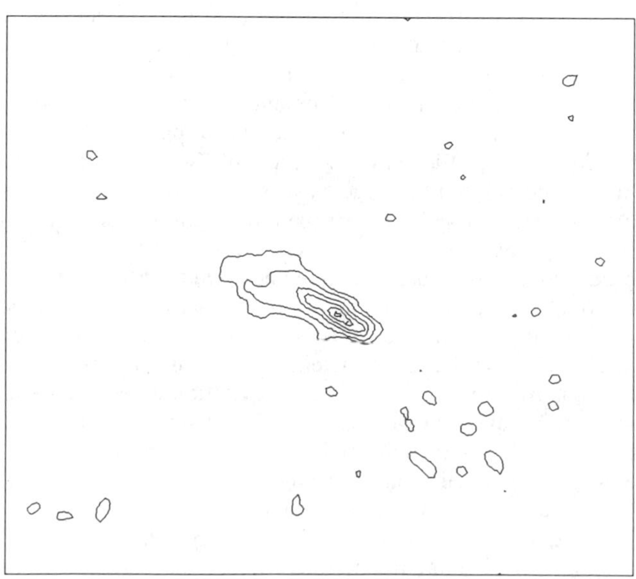

(b)

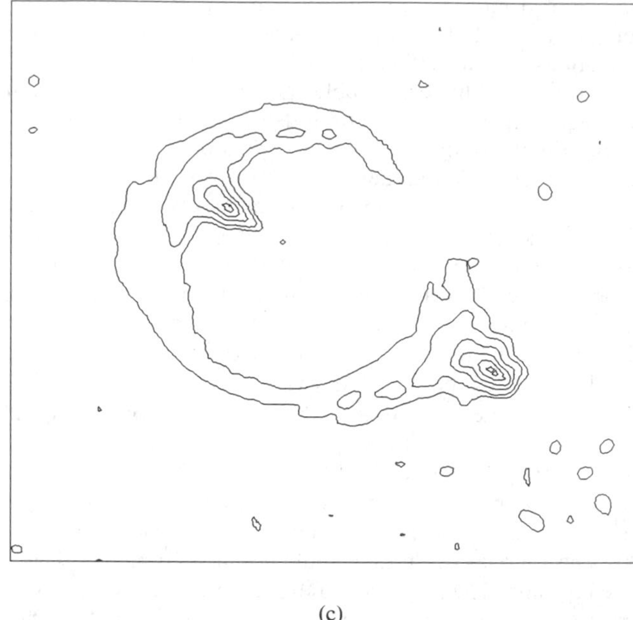

(c)

FIG. 1. (a) A radio image of an "Einstein ring" gravitational lens as observed with the National Radio Astronomy Observatory's Very Large Array radio telescope. The contours represent loci of equal surface brightness. (b) The hypothesized background source responsible for the Einstein ring, viewed without the gravitational lens in the foreground. (c) The calculated appearance of the source in (b) if it were viewed through a gravitational lens characterized by an elliptical potential. The model of (b) and (c) reproduces the major features of the object in (a). Figures (b) and (c) represent calculations presented by C. Kochanek *et al.*, *Monthly Notices of the Royal Astronomical Society*, Volume 238, page 43.

Gravitational lenses provide a new laboratory in which to address problems in astrophysics. They are unique in that they signal the presence of concentrated matter, regardless of whether the matter emits electromagnetic radiation. There is evidence for dark matter in the universe in the dynamics of galaxies and clusters of galaxies. The presence of dark matter on several scales should manifest itself in gravitational lenses. On the smallest scales, from planet-sized bodies to masses of approximately 100 solar masses, the lenses within galaxies act as "microlenses," causing apparent fluctuations in the brightness of images. The time scale of these fluctuations depends on the amount of matter present on these scales, as well as on the geometry of the lens, the relative velocities of the lens and the source, and the size of the source [5]. On intermediate scales, consisting of galaxies and clusters as a whole acting as lenses, the geometric properties of the images give independent measures of the ratio of mass to light in the systems. On the largest scale, that of the universe itself, the frequency of occurrence of gravitational lenses places limits on the quantity of cosmologically distributed dark matter in the form of concentrated matter.

A second application of gravitational lensing is in the determination of Hubble's constant. The different images of a gravitational lens travel along different paths from source to observer, and in general have different propagation times. Therefore, a fluctuation in the brightness, or other property, of the source will be observed at different times in the different images, giving a measure of the time delay between the images. The angle between the images is measured, and this is related to the time delay by the distances between the source, the lens, and the observer. For example, for a point mass lens, the expression relating the angular image separation, $\Delta\theta$, and the time delay Δt, is

$$\Delta t = \frac{(1 + z_L)}{2c} \frac{D_{OL}D_{OS}}{D_{LS}} \left[\beta\Delta\theta + 2 \ln \frac{\Delta\theta + \beta}{\Delta\theta - \beta} \right],$$

where z_L is the redshift of the lens, D_{OL} is the observer–lens distance, D_{OS} is the observer–source distance, D_{LS} is the lens–source distance, β is the angle between the (unlensed) source and the optic axis of the lens measured in units of the Einstein ring radius of the lens, and c is the speed of light. The distances are inferred from measurements of the lens and source redshifts, and scale inversely with Hubble's constant. Therefore, measurements of the image separation, time delay, and redshifts in principle give a measurement of Hubble's constant. In practice, the relationship between the time delay and the image separation may not be clearly determined because of uncertainties in the mass model for the lens. To date, measurements of the time delay have been reported in only one system [6], and the mass model in that system is not well constrained. However, the inferred value for Hubble's constant is consistent with estimates made using other methods.

A third application is simply to use the gravitational lens as a magnifying glass, giving astronomers the means to view distant objects in more detail than otherwise possible. Of particular interest would be the observation of a magnified quasar or active galactic nucleus. The degree of magnification, however, is a very strong function of the alignment of the source and the lens. While several known lens systems appear to offer moderate magnification of the background object, a highly magnified quasar or active galactic nucleus remains to be discovered.

The study of gravitational lenses is relatively young, and it is most likely that their potential has yet to be realized. Gravitational lenses can be applied to many problems in astrophysics, and observational programs aimed at exploiting their properties are under way. New systems are being discovered at a rate of several per year, and many with remarkable symmetry, such as the Einstein rings, are among them.

See also ASTROPHYSICS; COSMOLOGY; GRAVITATION; RELATIVITY, GENERAL THEORY.

REFERENCES

1. A. Einstein, *Science* **84**, 506 (1936).
2. F. Zwicky, *Phys. Rev.* **51**, 290 (1937).
3. D. Walsh, R. F. Carswell, and R. J. Weymann, *Nature* **279**, 381 (1979).
4. J. N. Hewitt, E. L. Turner, D. P. Schneider, B. F. Burke, G. I. Langston, and C. R. Lawrence, *Nature* **333**, 537 (1988).
5. R. Kayser, S. Refsdal, and R. Stabell, *Astron. Astrophys.* **166**, 36
6. C. Vanderriest, J. Schneider, G. Herpe, M. Chevreton, M. Moles, and G. Wlérick, *Astron. Astrophys.* **215**, 1 (1989).

BIBLIOGRAPHY

R. D. Blandford, C. S. Kochanek, I. Kovner, and R. Narayan, "Gravitational Lens Optics," *Science* **245**, 797 (1988).
E. L. Turner, "Gravitational Lenses," *The Fourteenth Texas Symposium on Relativistic Astrophysics* (E. J. Fenyves, ed.). Acdemy of Sciences, New York (1989).
E. L. Turner, "Gravitational Lenses," *Sci. Am.* **259**, 54 (July 1988).

Gravitational Waves

J. Anthony Tyson

Einstein's field equations relating space-time curvature to energy-momentum,

$$R_{\mu\nu} - \tfrac{1}{2} g_{\mu\nu} R = \frac{8\pi G}{c^4} T_{\mu\nu},$$

have radiative solutions: gravitational waves. These waves in the curvature of space-time exist in all gravitational theories, and they propagate at the velocity of light in most post-Newtonian theories of gravity. Attempts to find a general radiative solution have been frustrated by the nonlinearity of the field equations, although some elegant particular solutions valid in the strong-field limit have been found. Gravitational waves are produced by accelerated mass-energy, analogous to the production of electromagnetic waves by accelerated charges. However, there is no gravitational dipole radiation, since linear momentum is conserved: $(d/dt) \int \rho \dot{r}\, d^3r = \dot{p} = 0$. The first nonvanishing radiative multipole is quadrupole, and far from the source the flux falls off as the inverse square of the distance.

Any sources of this radiation must be spherically nonsymmetric and dynamic. In general, the luminosity is proportional to the square of the third time derivative of the mass quadrupole moment:

$$L_{\text{grav}} \approx (G/5c^5)\langle \dddot{Q}_{ij} \dddot{Q}_{ij} \rangle$$

where $Q_{ij} = \int \rho(x_i x_j - \delta_{ij} r^2/3)\, d^3x$. Dynamic quadrupoles radiate and induce dynamic-quadrupole deformations in absorbing media. Although interesting astrophysical sources of intense bursts would correspond to the strong-field limit, we can get a clearer picture of the effects of gravitational waves by going to the weak-field nonrelativistic approximation. For example, imagine two freely falling scientist-astronauts in separate space labs separated initially by a certain distance. Chart their progress by plotting their world lines in the (x, t) plane: If space-time were flat, their world lines would be straight. If a gravitational wave passed by and perturbed this flat space-time, their world lines would become wavy. In fact, they might detect the presence of this wave by exchanging bursts of light and agreeing on their separation distance (now oscillating) as a function of time. Joseph Weber's pioneering experiments to search for astrophysical bursts of gravitational waves basically used this idea, except that an elastic solid rod was sandwiched between the two space-time points. As a gravitational wave passes by, the separation changes—and hence the length of the rod changes—and the wave does work on the rod's elastic forces, imparting energy to the longitudinal vibrational modes of the rod.

In nonrelativistic weak-field systems ($v \ll c$, $R \gg GM/c^2$) the metric tensor is given as a perturbation of the flat Minkowski metric $\eta_{\mu\nu}$: $g_{\mu\nu} = \eta_{\mu\nu} + h_{\mu\nu}$, $|h_{\mu\nu}| \ll 1$. The transverse gravitational wave force f that acts on a mass m is given by $f = -m\,\nabla\Phi = -\Sigma_{ij}\, mR_{i0j0}x_j$, where Φ is the equivalent Newtonian gravitational potential of the wave and R_{i0j0} are the components of the Riemann curvature tensor: $R_{i0j0} = \partial^2\Phi/\partial x_i\,\partial x_j$. The relative change in separation $\Delta l/l$ between two such free masses under action of a gravitational wave is given by $\Delta l/l = h(t)$, where h is the dimensionless field strength related to the flux by $F = (c^3/16\pi G)\langle \dot{h}^2 \rangle$ erg cm^{-2} s^{-1}. For a resonance in an elastic solid, the resulting strain is $\Delta l/l \approx h(\omega_0)\omega_0/2$, where $h(\omega_0)$ is the Fourier component of $h(t)$ at the resonance ω_0. In general, an elastic solid will absorb energy ΔE from the gravitational wave: $\Delta E = \int \sigma(\nu)\epsilon(\nu)\,d\nu$, where $\sigma(\nu)$ is the cross section for absorption and $\epsilon(\nu)$ is the flux spectral density in units of ergs per square centimeter per hertz. Flux spectral density is a natural flux unit for Weber-type detectors, since they respond to the entire gravitational burst energy within a given bandwidth. For a rod of mass m and sound velocity v_s, the integrated absorption cross section for an optimally polarized incident gravitational wave is

$$\int \sigma(\nu)\,d\nu = (8G/\pi c^3)v_s^2 m \text{ cm}^2\text{ Hz}.$$

Present detectors of this type have an integrated absorption cross section in the range 10^{-21} cm^2·Hz. Thus, because of the weakness of the gravitational coupling, these detectors are dominated by internal thermal noise. Clearly, massive detectors with high sound velocity are of maximal efficiency. The flux spectral density of any detected wave burst is then given by $\epsilon(\omega_0) = \Delta E/\int\sigma(\omega)\,d\omega$ if the burst spectral width is large compared to the detector resonance width. The inferred minimum total burst energy (intrinsic luminosity integrated over duration of burst) is $E \gtrsim 2R^2\epsilon(\omega_0)\omega_0$, where R is the distance to the source. A commonly used unit of flux spectral density is the GPU (gravitational pulse unit): 1 GPU $= 10^5$ erg cm^{-2} Hz^{-1}.

The luminosity from hypothetical sources in this nonrelativistic weak-field limit can be calculated exactly. The flux F emitted in a given direction (unit vector n_i) is

$$F = (G/8\pi c^5 R^2) \sum_{j,k} \langle (\dddot{Q}_{jk}^{TT})^2 \rangle_{\text{ret}}$$

where

$$Q_{jk}^{TT} = \sum_{l,m} (\delta_{jl} - n_j n_l) Q_{lm} (\delta_{mk} - n_m n_k)$$

and δ_{ij} is the Kronecker delta function. For example, the luminosity of a rigidly rotating spheroid (peak luminosity at twice the rotation frequency Ω) is given by $L(2\Omega) = 32G\Omega^6 I^2 e^2/5c^5$ erg s^{-1} where I and e are the moment of inertia and equatorial ellipticity. The Taylor–Hulse pulsar may be braking at a rate determined by this luminosity. Intense sources of gravitational waves, however, would be highly relativistic and in strong fields.

Some general remarks are possible regarding these high-luminosity systems. Since $\ddot{Q}_{ij} \sim MR^2/T^3$ is proportional to the time derivative of the nonspherical part of the kinetic energy, the gravitational-wave luminosity L_{grav} is proportional to the square of the internal power flow L_{int} in the quadrupole mode: $L_{\text{grav}} = L_{\text{int}}^2/L_0$, where $L_0 = c^5/G = 3.6 \times 10^{59}$ erg s^{-1} $= 2 \times 10^5 M_\odot c^2$ s^{-1}. Thus, the power output in gravitational waves approaches L_{int} only for $L_{\text{int}} \approx L_0$, hardly a feasible laboratory experiment! If this internal power flow arises from quadrupole deformations in an elastic medium, then $L_{\text{grav,max}} \sim U_{\text{grav}}^2 (v_s/c)^6 L_0^{-1}$ where U_{grav} is the gravitational potential energy and v_s is the sound velocity. Clearly, significant luminosity can occur only for compact systems of high density for which the sound velocity approaches the velocity of light, such as rapidly deforming neutron stars and/or black holes. Maximum luminosity occurs when the system is near its gravitational radius $R_s = 2GM/c^2$, when $L_{\text{grav}} \lesssim L_0$. This curious maximum intrinsic luminosity L_0 is due entirely to the nonlinearity in the field equations: Energy present in gravitational waves contributes to the mass-energy of the system. It has been estimated that in some astrophysically plausible scenarios these systems radiate as much as 30% of their mass-energy in a single burst of gravitational waves. This burst would be relatively broadband, since in this high-luminosity limit the width of the burst spectrum will be the inverse of the radiation reaction time $\tau_r \approx (L_0/L_{\text{int}})\tau \approx (Rc^2/GM)^{5/2}\tau$, where τ is the characteristic time of rotation or free fall. In 1962 Freeman Dyson showed that such bursts were astrophysically plausible, and he suggested that Joseph Weber perform an experiment to search for these bursts.

By 1969 Weber had assembled two aluminum bar detectors at 1660 Hz and was claiming coincident detection of bursts apparently in the direction of the galactic center. Estimates of source luminosity ranged between 10^3 and $10^5 M_\odot c^2$ yr^{-1}. Subsequently, however, several other groups using detectors of greater sensitivity have searched for kilohertz-band bursts and have found none. Such a detector can be treated as a shock-excited oscillator for these bursts. Their output time signature is known and is independent of the detailed dependence of the wave amplitude $h(t)$ in time. An event is detected provided sufficient gravitational wave power is present at the detector resonance to excite the detector above its filtered internal noise. Thus, from information theory, for each detector arrangement there exists an optimal filter for extracting this signal from the detector noise—independent of the source scenario. This convenient situation would break down only for (noisier) detectors whose output time response was fast compared to τ_r. Undaunted, the search for astrophysical sources of gravitational waves goes on. The signal-to-noise ratio for these detectors can be shown to be proportional to $mQ^{1/2}N^{-1/2}$, where m is the detector mass, Q the resonance quality factor, and N the detector noise per unit of electromechanical coupling.

Making a longer detector by replacing the spring in the spring-mass resonant detector with a laser beam can in principle result in higher sensitivity, wider bandwidth, and better time resolution. Recently, groups working at MIT and CalTech have achieved strain sensitivity, at frequencies above a few hundred Hz, of 10^{-18} strain in Michelson interferometer detectors, similar in burst sensitivity to that of the resonant bar antennas.

J. Taylor *et al.* have presented compelling evidence that the binary pulsar 1913+16 is losing orbital energy and angular momentum at approximately the rate expected from gravitational radiation due to the 7.8-hour orbital motion of the two neutron stars. The observed derivative of the orbital period, $(-2.1 \pm 0.4) \times 10^{-12}$ s s^{-1}, is in excellent agreement with theory $(-2.38 \pm 0.02) \times 10^{-12}$ s s^{-1}, using the separately determined masses, orbital period, and eccentricity. Possible contamination of this binary pulsar system by a third body, mass loss, or tidal dissipation has been made unlikely through this confrontation between observation and theory. In the absence of direct detection of gravitational waves, pulsar 1913+16 remains our only experimental check of the general relativistic quadrupole radiation formula. It is an ideal relativistic laboratory for testing theories of gravity.

See also BLACK HOLES; PULSARS; RELATIVITY, GENERAL; SPACE-TIME.

BIBLIOGRAPHY

Jonothan Logan, "Gravitational Waves," *Phys. Today*, March, p. 44 (1973). (E)

P. F. Michelson, J. C. Price, and R. C. Taber, *Science* **237**, 150 (1987). (E)

C. W. Misner, K. S. Thorne, and J. A. Wheeler, *Gravitation*, pp. 943–1044. W. H. Freeman, San Francisco, 1973. (I,A)

P. Saulson, *Rev. Sci. Instrum.* **55**, 1315 (1984). (A)

J. A. Tyson and R. P. Giffard, *Annu. Rev. Astron. Astrophys.* **16**, 521 (1978). (I,A)

J. M. Weisberg and J. H. Taylor, *Phys. Rev. Lett.* **52**, 1348 (1984). (A)

Gravity, Earth's

J. C. Harrison

Newton's law of gravitation states that particles attract each other with a force proportional to the product of their masses and inversely proportional to the square of the distance between them. Gravity at a point on the earth's surface, defined as force per unit mass or equivalently as acceleration of a freely falling body, consists of the net gravitational attraction of the earth's mass together with the centrifugal force of rotation. The latter (about 0.3% at the equator) is not required in space applications where an inertial reference frame is employed. The unit the Gal (after Galileo), equal to 0.01 m/s^2, is frequently used. Other related quantities include astronomical latitude and longitude—the direction of the gravity vector; and the geoid—the equipotential or level surface which, over the oceans, coincides with mean sea level. Gravity varies predominantly with latitude (from 983.2 Gal at the poles to 978.0 Gal at the equator), because of the rotation and nonspherical shape of the earth, and with height above sea level (0.3086 mGal/m). The remaining variations, 400 mGal extreme, 34 mGal root mean square, are due to topography and density variations within the earth.

On land, gravity can be measured by a transportable free-fall apparatus with an absolute accuracy of about 3 μGal [1]. However, more commonly and easily, gravity meters [2] (sensitive spring balances) capable of 30-μGal accuracy worldwide and 3-μGal accuracy locally under carefully controlled conditions are used to measure gravity differences. Heavily damped gravity meters and high-quality accelerometers operating on gyroscopically stabilized platforms are used on ships with an accuracy which can be as high as 1 mGal but is often degraded by navigational uncertainties [3]. These instruments can also be used on helicopters and fixed-wing aircraft with some loss of accuracy and resolution [4, 5]. Absolute levels and scale factors of gravimetric surveys are based on the International Gravity Standardization Net, 1971 [6], which replaced the Potsdam System in use from 1909 to 1971. This network of nearly 500 stations is based on a reduction of many pendulum and gravity-meter measurements, with the pendulum measurements controlling the scale factors of the meters employed and 11 absolute determinations by free-fall methods providing the absolute level. Its worldwide accuracy is 0.1 mGal, although much higher accuracies are attained in some national gravity base networks as a result of great improvements in the accuracy and transportability of free-fall apparatus since 1971. Gravity at a given site varies periodically with time as a result of the tidal attractions of the sun and the moon, of the ocean tides, and of the associated deformation of the Earth. This variation, less than 200 μGal in amplitude, is predictable at most places to within a few tenths of a μGal [7]. Nonperiodic changes due to vertical ground motion, subterranean motion of fluids, and igneous activity have been reported from some sites, usually at the sub-mGal level.

The correlations with latitude and elevation account for much of the variation of gravity, and so it is convenient to work with gravity anomalies, the amount by which gravity differs from a normal (ellipsoidal) model in which these factors are taken into account. This model compares gravity with that outside a body whose surface is (1) an exact ellipsoid of revolution and (2) a gravity equipotential, and (3) whose equatorial radius, geocentric gravitational constant (GM), rate of rotation, and dynamical form factor J_2 (see below) have specified values close to those observed for the actual earth [8]. Gravity γ on this surface is given by

$$\gamma(\phi) = 978032.677(1 + 0.001931851 \sin^2\phi)/(1 - 0.006694380 \sin^2\phi)^{1/2},$$

where ϕ is geodetic latitude. This formula replaces earlier versions adopted in 1930 and 1967. The vertical gradient of normal gravity is close to -0.3086 mGal/m and the difference

$$g_{\text{meas}} - \{\gamma(\phi) - 0.3086h\},$$

where h is the station elevation above sea level (the geoid), the anomaly after correction for height is the *free air anomaly*. (A similar expression but using height above the reference ellipsoid gives the *gravity disturbance*.) The *Bouguer anomaly* is obtained by subtracting the attraction of the topography above sea level from the free air anomaly.

The free air anomaly is strongly correlated with topography on a local scale but not regionally (continents to oceans, mountains to plains, etc.). The local topographic effects are eliminated from the Bouguer anomalies, which are therefore the anomalies normally used in geophysical prospecting. These anomalies are, however, strongly correlated with regional topography because this is largely compensated—that is, mountains are underlain by less dense rocks than oceans, so that the obvious mass excesses and deficits due to the topography are normally canceled out (isostasy). There are many specific hypotheses for the distribution of this compensating mass, and for each of these an isostatic gravity anomaly can be computed. The isostatic effects are often represented spectrally in the form of a response function giving the mGal of free air gravity anomaly associated with 1 m of topographic amplitude as a function of wavelength. This response tends to zero at long wavelengths, tends to $2\pi G\rho$ (where ρ is the rock density) at short wavelengths, and its behavior between these extremes is related to the ability of the lithosphere to support topographic loads.

The gravity field in space is important because of its influence on the orbits of artificial satellites. Accurate prediction of these orbits requires that the gravity field be well known; conversely, gravity field models can be improved by adjusting gravity field parameters and tracking station coordinates to minimize the sum of the squares of the residuals between satellite positions and velocities as predicted from the model and as obtained from tracking data. In geophysics the potential is defined at a point in space as

$$V = G \sum_i \frac{m_i}{r_i} ,$$

where masses m_i are at distances r_i from the point. Force per unit mass is given by grad V. (This sign convention is opposite to that adopted in physics.) In free space, V obeys Laplace's equation and in the vicinity of the earth V can be written in a series expansion in spherical harmonics. A frequently used form is

$$V_{\text{space}} = \frac{GM}{r} \left[1 + \sum_{n=2}^{n\max} \sum_{m=0}^{n} \left(\frac{a_e}{r} \right)^n \overline{P}_{nm}(\sin \phi) \times \{ \overline{C}_{nm} \cos m\lambda + \overline{S}_{nm} \sin m\lambda \} \right] ,$$

where GM is the product of the Newtonian gravitational constant and the mass of the earth, a_e is its equatorial radius, and r, ϕ, and λ are radial distance from the earth's center of gravity and geocentric latitude and longitude, respectively.

The $\overline{C}_{nm}$ and $\overline{S}_{nm}$ are coefficients and the $\overline{P}_{nm}(\sin \phi)$ normalized associated Legendre polynomials with the property that the mean square values of $\overline{P}_{nm}(\sin \phi) \cos m\lambda$ and $\overline{P}_{nm}(\sin \phi) \sin m\lambda$ over a sphere are 1 [9]. Traditionally the zonal terms ($m = 0$) were written in an expansion

$$V_{\text{zonal}} = \frac{GM}{r} \left[1 - \sum_{n=2}^{n\max} J_n \left(\frac{a_e}{r} \right)^n P_n(\sin \phi) \right] ,$$

where the P_n are conventional (not normalized) Legendre polynomials and the J_n are coefficients related to the $\overline{C}_{n0}$ by

$$J_n = -(2n+1)^{1/2} \overline{C}_{n0} .$$

These expansions link the field in space to that on the earth's surface; the earthbound observer adds a centrifugal term $\frac{1}{2}p^2\omega^2$, where p is the perpendicular distance from the axis of rotation and ω the earth's angular velocity.

An anomalous potential T is defined as $V - U$, where U is the potential of the normal (ellipsoidal) model. The geoid height is given by T/g, where g is the mean value of gravity, and the gravity anomaly field by $-(\partial T/\partial r + 2T/r)$ [10]. The Goddard Earth Model GEM-T1 [11], complete to degree and order 36, is the most recent gravity field model based entirely on satellite tracking.

A number of satellites carrying radar altimeters to measure the distance from the satellite to the sea surface have been launched since 1975. If the sallite orbit is known, these measurements, after a number of corrections including one for ocean tides, allow the height of the sea surface relative to the reference ellipsoid to be determined. This surface is almost a gravity equipotential (the geoid) but includes a small (~2 m maximum, ~70 cm rms) signal due to ocean dynamics. These effects cannot be entirely separated but sea surface altimetry has vastly extended our knowledge of the marine gravity field and will be important in oceanography. The radial orbital errors, initially of the order of 5 m and reduced to 70 cm by the early 1980s, are larger than those in the altimeter (~10 cm for SEASAT, ~3 cm for GEOSAT). However, they are dominantly of long wavelength and a strategy of removing biases on the basis of discrepancies at the crossovers of ascending and descending tracks has been successful in reducing their influence to about 25 cm. SEASAT was launched in June 1978 into a repeat orbit which gave a spacing of about 160 km between ascending tracks at the equator but it unfortunately failed after 3 months of operation. GEOSAT has been repeating the SEASAT orbit since late 1987, yielding a more precise and complete data set and a better understanding of the temporal variations in sea surface height. Mean gravity anomalies over $0.5° \times 0.5°$ blocks have been recovered with a precision of about 3.5 mGal for most of the world's ocean from the SEASAT and earlier GEOS 3 altimetry. In addition some important relative information can be extracted at finer detail. The GEOSAT results are expected to be even better. A useful summary is given in [12] and an attractive map of the gravity field of the world's oceans has been published [13].

Spherical harmonic expansions of the gravity have been made to degree 180 ($1° \times 1°$ resolution) and higher by combining data from satellite tracking, altimetry, and terrestrial measurements [14]. The long-wavelength information (degree less than 25) is derived from satellite tracking whereas that at shorter wavelengths comes largely from altimetry over the oceans and from terrestrial measurements on land. Of the continents, only Europe, North America, and Australia are adequately surveyed.

The interpretation of gravity anomalies in terms of the causative density variations is not unique and there generally exist an infinite number of possibilities that must be narrowed using other geological or geophysical information. Features less than a few hundred kilometers in size usually correlate well with geological structure and are very helpful in modeling this structure. The longer-wavelength features originate in the mantle and at the core–mantle boundary, and show little correlation with surface features; their origin is not understood, although they are of great interest in connection with convective and other geodynamic models.

See also GEOPHYSICS.

REFERENCES

1. G. Peter, R. E. Moose, C. W. Wessells, J. E. Faller, and T. M. Niebauer, "High-Precision Absolute Gravity Observations in the United States," *J. Geophys. Res.* **94,** 5659–5674 (1989).
2. J. C. Harrison, "Gravity Sensors," in *Geoscience Instrumentation* (E. A. Wolff and E. P. Mercanti, eds.). Wiley, New York, 1974.
3. L. J. B. LaCoste, "Measurement of Gravity at Sea and in the Air," *Rev. Geophys.* **5,** 447–526 (1967).
4. S. Hammer, "Airborne Gravity Is Here," *Geophysics* **48,** 213–223 (1983).
5. J. M. Brozena and M. F. Peters, "An Airborne Gravity Study of Eastern North Carolina," *Geophysics* **53,** 245–253 (1988).
6. *The International Gravity Standardization Net, 1971.* International Association of Geodesy Special Publ. No. 4, 1974.
7. T. F. Baker, R. J. Edge, and G. Jeffries, "European Tidal Gravity: An Improved Agreement Between Observations and Models," *Geophys. Res. Lett.* **16,** 1109–1112 (1989).
8. H. Moritz, "Geodetic Reference System 1980," *Bull. Geodesique* **62,** 348–358 (1988).
9. W. M. Kaula, "Determination of the Earth's Gravitational Field," *Rev. Geophys.* **1,** 507–551 (1963).
10. H. Moritz, *Physical Geodesy,* Chap. 2.14. W. H. Freeman and Company, San Francisco and London, 1967.
11. J. G. Marsh *et al.,* "A New Gravitational Model for the Earth from Satellite Tracking Data: GEM-T1," *J. Geophys. Res.* **93,** 6169–6215 (1988).
12. B. C. Douglas, D. C. McAdoo, and R. E. Cheney, "Oceanographic and Geophysical Applications of Satellite Altimetry," *Rev. Geophys.* **25,** 875–880 (1987).
13. W. F. Haxby, *Gravity Field of the World's Oceans,* National Geophysical Data Center, NOAA, Boulder, Colorado, 1987.
14. R. H. Rapp and J. Y. Cruz, Spherical harmonic expansions of the Earth's gravitational potential to degree 360 using 30′ mean anomalies. Report no. 376, Dept. of Geodetic Science and Surveying, The Ohio State University, 1986.

Group Theory in Physics

J.-P. Antoine

ORIGINS

As far back as we know, people have been fascinated by symmetry. From megalithic monuments to Mycenian jewelry, from Egyptian pyramids to Greek temples, symmetrical forms were clearly perceived and intuitively understood as manifestations of harmony and perfection. When modern science developed, this innate inclination became incorporated into the work of the great pioneers of physics, whose beautiful theories and equations were thought of as reflecting the harmony of the world (Kepler). However, the technical study of symmetry properties requires a level of mathematical sophistication not available before the late nineteenth century when, following E. Galois, the abstract theory of groups was developed by mathematicians like G. Frobenius, I. Schur, W. Burnside, E. Cartan, and H. Weyl. Classical physics had not created much motivation for it; from Newton to Lagrange or Maxwell, the aim was to derive and solve the partial differential equations describing the structure of matter. Only in some works of Euler do we see a first approach to the idea of a group and its role in physics. Thus physics followed (and in some cases preceded) analysis rather than algebra.

Group theory first entered physics with crystallography. Given an arbitrary crystal, it is an easy exercise to figure out all symmetry operations that leave it unaffected: reflection through certain planes or inversion with respect to the center, rotations around given axes through the center (only the angles $2\pi/n$, with $n=2,3,4$, or 6, will be compatible with the periodicity of the crystalline lattice), or any combination of these. A systematic investigation shows there exist exactly 32 different combinations of symmetry properties, and accordingly crystals are subdivided into 32 *crystal classes*. For a given class, any two symmetry operations can be combined to give a third one (and this product is associative), there is one that does nothing at all (the identity), and every operation has an inverse (the combination of the two being the identity); in other words, the symmetry operations form a *group* with a *finite* number of elements.

Each crystal class corresponds to such a group, which is called a *point group*, since it consists of symmetry operations around a fixed point; thus there are 32 different point groups. Combining these with the lattice translations for each of the 14 different types of lattices (the Bravais lattices), we obtain, after some hard work, 230 possible invariance groups for crystal lattices, the so-called *space groups*. This remarkable achievement, due to E. V. Fedorov (1885) and A. Schönflies (1891), illustrates the primordial role of group theory in physics, namely, to organize data in a rational fashion.

GROUPS AND REPRESENTATIONS

Let there be given an arbitrary physical system, for which we know a symmetry group G. In order to exploit this information we need another concept, that of a (linear) *group representation*. An element $g\varepsilon G$ transforms a given configuration of the system into another one: the symmetry group is realized in the space V of configurations. Quite often, the latter has or can be given the structure of a vector space. Every $g\varepsilon G$ will then be represented by a nonsingular transformation or operator $L(g)$, acting on V in such a way that the operators $\{L(g)\colon g\varepsilon G\}$ obey essentially the same algebraic relations as the group elements themselves: $L(g_1)L(g_2)=L(g_1g_2)$, $L(g^{-1})=[L(g)]^{-1}$, $L(e)=1$ (e is the neutral element of G, 1 is the identity operator). We say in this case that the map $L\colon g\rightarrow L(g)$ is an n-dimensional (linear) representation of G, where n is the dimension of V $(1\leq n\leq\infty)$.

If the space V possesses an inner product and the latter is invariant under G, the representation L is called *unitary* (if V is a complex vector space) or *orthogonal* (if V is real). So the representation L describes completely the action of the symmetry group on the system. For instance, the configurations may be decomposed into subsets V_j (vector subspaces of V), each of which is transformed into itself by all the operations of G; the representation is then said to be *reducible*. If L is unitary or orthogonal, it can be decomposed into a direct sum of subrepresentations L_j: $L=\oplus_j L_j$, corresponding to $V=\oplus_j V_j$; if a given V_k no longer contains an (nontrivial) invariant subspace, then L_k is called *irreducible*. With this machinery, we can describe as follows the common pattern of most applications of group theory in physics. First, identify the symmetry group G of the system under study; next determine the representation L of G in the space of configurations V, and decompose L into a direct sum of irreducible representations L_j. At this stage, in general, the analysis of the system will be enormously simplified, since the number of configurations available in V_j is much smaller than the one in V. Accordingly, we can evaluate the quantities of interest, such as normal frequencies or transition probabilities, rather easily. A typical example of this procedure is the study of the normal modes of vibration of a mechanical system, that is, a system composed of coupled harmonic oscillators; this is the lowest approximation of the classical description of a molecule or a crystal. In all such cases the symmetry group of an object is an essential element for simplifying its investigation.

THE PRINCIPLE OF RELATIVITY AND THE ASSOCIATED GROUPS

Geometrical symmetries, as discussed earlier, are very useful, but group theory enters physics in a much more fundamental way through the *principle of relativity*. The latter asserts that two observers will describe a physical process by the same equations whenever they are at rest or in uniform linear motion with respect to each other (such observers are called equivalent). "Observer" here means simply a reference frame in space-time. (Notice that the existence of *inertial* frames is an independent assumption.) In other words, a transformation of space-time that maps any observer into an equivalent one cannot affect the description of physical processes. These transformations obviously form a group, called the *relativity group* G^{rel} of the theory. Thus, technically, space-time is a homogeneous manifold on which the group G^{rel} acts. The importance of the principle of relativity is that it guarantees (because of the homogeneity of space-time) that experiments can be repeated whenever and wherever we wish, i.e., that science is possible.

However, as stated earlier, the principle of relativity does not determine G^{rel} uniquely; an additional postulate is needed. Three possibilities have been used. We can say that two obervers are equivalent only if they are at rest with respect to each other; G^{rel} is then the Euclidean group $\mathscr{E}(3)$, which consists of all rotations and translations in three-dimensional space as met in ordinary vector calculus. If, in addition, equivalent observers are allowed to be in relative uniform motion, but if time remains absolute, we have the Galilei group $\mathscr{G}$, which is the relativity group of Newtonian mechanics. If, instead, we require that light propagate with the same speed for every (inertial) observer (which is incompatible with an absolute time), we obtain the inhomogeneous Lorentz group, also called the Poincaré group $\mathscr{P}$. This is the relativity group of Einstein's theory of special relativity, which adapts classical mechanics to the symmetry properties inherent in electromagnetism. The principle of relativity can be extended to observers in gravitational fields, but the interpretation of general relativity in group-theoretical terms is no longer straightforward.

Once the relativity group of the theory G^{rel} has been determined, the principle of relativity must be put into action: It asserts that all laws of physics are relations between *tensors* associated to the chosen G^{rel}. Tensors are usually introduced through multilinear algebra, but they can also be defined in a purely group-theoretical language. Each of the relativity groups $\mathscr{E}(3)$, $\mathscr{G}$, $\mathscr{P}$, has a defining matrix representation M; a tensor of order (p,q) may then be defined as an element of the vector space of the representation $M(p,q)=M\otimes\cdots\otimes M\otimes M^*\otimes\cdots\otimes M^*\equiv M^{\otimes p}\otimes M^{*\otimes q}$, the tensor product of p factors M and q factors M^* (M^* is the contragredient representation defined by the relation $\langle M^*(g)\phi|\psi\rangle=\langle\phi|M(g)\psi\rangle$ for any vectors ϕ, ψ in the representation space). Thus the classification of irreducible tensors of order (p,q) amounts to the decomposition of $M^{(p,q)}$ into its irreducible constituents. At this point, however, a new element enters, for this reduction involves the study not only of representations of G^{rel}, but also of those of the group S_n of *permutations*, which act on the indices of the tensors. As a result, we can write explicitly, in principle, all possible forms for the laws of physics in a theory governed by a given relativity group. This is of course an enormous simplification, which is achieved by postulating the principle of relativity from the outset. In practice, further restrictions (based on simplicity, for instance) are used.

One example of such an enterprise is the derivation of all covariant wave equations for a given relativity group. For instance, the Dirac equation is the unique first-order covariant equation describing a relativistic ($G^{\text{rel}}=\mathscr{P}$) spin-$\frac{1}{2}$ particle of mass $m>0$. It should be mentioned that all these equations are classical, although historically they have been discovered in a quantum context. (Second) quantization is the reinterpretation of their solutions as operators acting on the Hilbert space of the theory.

INVARIANCE AND CONSERVATION LAWS

The crystallographic groups are very special; they have either a finite or a denumerably infinite number of elements; such groups are called *discrete*. In contrast, the three relativity groups $\mathscr{E}(3)$, $\mathscr{G}$, and $\mathscr{P}$ depend *continuously* on some parameters (e.g., rotation angles or translations). Such groups are called *Lie groups*, after the Norwegian mathematician Sophus Lie, who first investigated them systematically. In a Lie group, there exist infinitesimal transformations, arbitrarily close to the identity. The law of group multiplication can be linearized in terms of their coefficients, called *infinitesimal generators*. The latter form a vector space known as the *Lie algebra*. Conversely (and this was

Lie's main achievement), the group can be reconstructed from its algebra. The latter plays an important role in physics, for it contains the essential information that can be extracted from the symmetry of a system, namely, the conservation laws that it satisfies. This crucial connection was formulated in the following fundamental theorem by the German mathematicians David Hilbert (1916) and Emmy Noether (1918): If a Lagrangian theory is invariant under an N-parameter Lie group of transformations (in the sense that the Lagrange function $\mathscr{L}$, or, more generally, the action integral $I=\int\mathscr{L}\,dt$, is invariant), then the theory possesses N conserved quantities. This theorem, which is a consequence of a variational principle, holds in any Lagrangian theory, whether the Lagrange function depends only on finitely many coordinates (point mechanics), or is a functional of fields (classical field theory). It is one of the cornerstones of classical physics and, by the correspondence principle, of quantum physics as well.

As an example, consider the relativity groups introduced earlier. The Euclidean group $\mathscr{E}$ is a six-parameter Lie group. Accordingly, any Lagrangian system invariant under $\mathscr{E}$ possesses six constants of the motion, which span the Lie algebra, namely, the components of the total momentum and those of the total angular momentum. Both $\mathscr{G}$ and $\mathscr{P}$ have 10 parameters and contain $\mathscr{E}$ as a subgroup. Hence systems invariant under either $\mathscr{G}$ or $\mathscr{P}$ have four additional constants of the motion, namely, the total energy (the Hamiltonian) and three others that guarantee that the center of mass moves as a *free* particle.

It should be noted that those conservation laws, although fundamental, are not sufficient in general for a complete integration of a classical mechanical problem. There are, however, some problems that can be integrated in an elementary way, without solving any differential equations of motion, thanks to additional symmetry properties (called *hidden dynamical* symmetry). The two main cases are the n-dimensional oscillator, which has a hidden $SU(n)$ symmetry [$SU(n)$ is the group of all $n \times n$ unitary matrices of determinant 1]; and Kepler's problem, which is invariant under a hidden four-dimensional rotation group $SO(4)$. In the latter case, the three additional constants of motion have an immediate physical interpretation, namely, the components of the major semiaxis of the trajectory; hence, not only does the orbit lie in a fixed plane (angular momentum conservation), but its position in the plane does not change in time. This fact has an obvious importance for astronomy!

SYMMETRIES IN QUANTUM MECHANICS

So far we have discussed symmetry principles of classical physics only, but it is in quantum theory that group-theoretic ideas have found their most fertile ground, with a spectacular development as the result.

The basic reason is the *linearity* of the theory; indeed, the first axiom of quantum mechanics asserts that the states of a system are represented by unit rays in a Hilbert space $\mathscr{H}$. Thus, contrary to classical physics, quantum theory has a built-in linear structure through which symmetries are automatically realized by group representations, thanks to fundamental theorems of E. P. Wigner and V. Bargmann. More precisely, if a symmetry is defined as a map of the states into themselves that preserves all transition probabilities, i.e., the modulus of the inner product between any two states, then every group G of symmetries of the system is realized, up to phase factors, by a unitary representation U of G into $\mathscr{H}$.

For exploiting this theorem, we proceed along the general lines described earlier. The unitary representation U can be decomposed into irreducible constituents U_j, corresponding to subspaces $\mathscr{H}_j$. Most quantities of interest are given by matrix elements of certain observables that have a simple behavior under G (e.g., vector and tensor operators, when G is a rotation group). If A is one of these observables, a matrix element $\langle\phi|A|\psi\rangle$, with $\phi\varepsilon\mathscr{H}_j$, $\psi\varepsilon\mathscr{H}_k$ can be evaluated using only properties of the symmetry group; either it will vanish (selection rule), or it will depend to a large extent on the subrepresentations U_j and U_k only, but *not* on the individual states ϕ, ψ (this is the idea of the famous Wigner–Eckart theorem).

Furthermore, many observables themselves can be derived from the symmetry group via Noether's theorem and the correspondence principle. For instance, invariance under $\mathscr{E}$ (translations and rotations) yields total momentum and total angular momentum; time translation gives the Hamiltonian; Galilei invariance yields position observables. In all cases, these observables belong to the Lie algebra of the symmetry group and are realized in the state space by self-adjoint operators.

APPROXIMATE SYMMETRIES

So far we have discussed only *exact* symmetries, but we can go beyond that and, in the spirit of perturbation theory, consider *approximate* symmetries. This concept is useful whenever the Hamiltonian can be split into two terms where the first is invariant under a given group and the second is a small correction, invariant only under a subgroup; in addition, the symmetry-breaking term is assumed to have well-defined transformation properties under the full approximate symmetry group. Then we can resort to the general procedure described earlier for computing matrix elements. The procedure can be repeated, leading to a hierarchy of approximate symmetries, more and more badly broken, corresponding to a descending chain of subgroups. This idea, which might be traced back, for instance, to the analysis of the Zeeman effect, has been remarkably successful.

In summary, besides implying fundamental conservation laws, the symmetries of a system, exact or approximate, provide an extremely powerful tool for computing physical quantities. This explains the considerable development of group-theoretical methods in the various domains of physics starting in the late twenties under the impulsion of such great physicists or mathematicians as Heisenberg, Pauli, Weyl, van der Waerden, Wigner, and Bargmann.

APPLICATIONS TO ATOMS, MOLECULES, AND SOLIDS

The first applications of group theory were to atomic physics, where in most cases the bewildering complexity of spectroscopic data resisted analysis. The basic fact is the rotational symmetry of a free atom: in the center-of-mass frame, the Hamiltonian is invariant under all rotations around the origin. This $SO(3)$ symmetry guarantees that the total angular momentum of the atom is conserved.

Consider first a simplified hydrogen atom: a spinless electron bound in a Coulomb potential (i.e., to a spinless proton). Schrödinger's equation can then be solved exactly in polar coordinates; the energy levels obtained, indexed by an integer $n=1,2,\ldots$, are highly degenerate: the nth level accommodates angular momenta $l=0,1,2,\ldots,n-1$, and for each l there are still $(2l+1)$ different states indexed by $m=-l, -l+1,\ldots,l$. This so-called *accidental degeneracy* of the Coulomb potential has a group-theoretical explanation, exactly as in the classical case: for a given level n, the irreducible representations $D^{(l)}$ $(l=0,1,\ldots,n-1)$ of $SO(3)$ can be grouped to form a single irreducible representation of $SO(4)$. This fact was recognized by Pauli in 1926 and exploited by him to solve the Coulomb problem in a purely algebraic way. If the central potential is not exactly Coulombic, the energy levels depend on both n and l. Following Pauli, we now take into account the electron spin: According to the exclusion principle, each state $|n,l,m\rangle$ can accommodate two, and only two, electrons. The total angular momentum of the atom is now the (vector) sum of the orbital angular momentum and the spin, and can take values $j=l\pm\frac{1}{2}$; this law of addition is nothing but the decomposition into irreducible constituents of the *tensor product* of two $SU(2)$ representations: $D^{(l)}\otimes D^{(1/2)} = D^{(l+1/2)}\oplus D^{(l-1/2)}$.

Notice that the usual rotation group $SO(3)$ describes properly *integer* angular momenta only; for describing half-integer ones we need $SU(2)$, which is closely related to it [technically, $SO(3)$ is the quotient of $SU(2)$ by a two-element subgroup].

Consider now a many-electron atom. The Schrödinger equation can no longer be solved exactly, and we must have recourse to approximation. The crudest one is the so-called *central field approximation*, in which each electron moves in an average, central potential created by all the other electrons. This allows us to find an energy spectrum $E(n,l)$ for each electron, and the global spectrum by addition. The result is the description of an atom called the *shell model*: All electrons with the same values of n and l are said to form a "shell," and we can list all possible configurations within each shell. This model explains the periodic classification of the elements (Mendeleev table). Group theory enters this process in two ways. First, the addition of all individual angular momenta and spins is just, as before, the decomposition of a tensor product of many representations $D^{(l)}$ or $D^{(1/2)}$ of $SU(2)$. Then, also, the permutation groups enter, through the Pauli exclusion principle. A further step was taken by Racah, who introduced in this context an *approximate symmetry*. The idea is to consider the single-electron eigenfunctions in a given shell (n,l) as the basis of a representation of the much larger $SO(8l+5)$. A classification of all configurations is then obtained by studying this representation and its restriction to a chain of smaller and smaller subgroups. At the same time, transformation properties under these larger groups may be exploited for computing various matrix elements, and this is useful for the next approximation, namely when electron spins and spin–orbit coupling (which is a truly relativistic effect) are introduced as a perturbation. By this method we can obtain a good picture of the complete energy spectrum of the heaviest atoms. It can be fairly said that without the enormous simplifying power of group theory, the unraveling of atomic spectra would be all but impossible.

A different extension of the notion of symmetry has been considered, first in the case of the hydrogen atom. Some operators that describe transitions between different levels, such as the dipole operator, generate, if combined with the generators of $SO(4)$, two still larger groups, $SO(4,1)$ and $SO(4,2)$. A single irreducible representation of each of these contains *all* the representations of $SO(4)$ corresponding to all energy levels. These groups are variously called *dynamical symmetry groups*, *noninvariance groups*, or *spectrum generating groups*; they are not invariance groups (although they contain the true symmetry group as a subgroup), but can describe transitions as well. The use of such groups has proved very fruitful.

The usefulness of group theory does not stop at isolated atoms, but can be extended to their interaction with radiation (emission, absorption, scattering). The key remark here is that creation and annihilation operators of the radiation field obey commutation relations identical to those of the Lie algebra of a unitary group. This fact can be used for solving all problems with a Hamiltonian at most quadratic in these operators. These Hamiltonians cover a large part of quantum optics, notably lasers and other coherent phenomena, including the intriguing squeezed states of light. Here again group theory has provided new insights as well as powerful techniques.

Going one step higher, to molecules, a similar story can be repeated. The rotation group, however, is restricted to the geometrical symmetry group of the molecule (known from chemistry). The simplest case is a diatomic molecule; it retains an axis of symmetry, and is accordingly invariant under $SO(2)$. More complicated molecules are invariant only under a (finite) crystallographic group: e.g., the benzene ring (hexagonal symmetry) or the ammonia molecule. Here too group theory provides an efficient tool, first for classifying the stable configurations and energy levels of the molecule, then also for analyzing interactions with radiation (such as Raman scattering).

Then, of course, we may return to crystals. The quantum theory of solids encompasses all the knowledge accumulated from classical crystallography and applies it to dynamical problems: energy bands in solids, and the behavior in a perfect solid of foreign particles, such as electrons (semiconductors and metals), phonons or photons (optical or electromagnetic properties of solids), or ions (impurities), including their movement under the influence of external electric or magnetic fields. In each case, the use of crystal symmetry is crucial for solving the problem.

NUCLEAR PHYSICS

A free nucleus can be viewed as a bound state of N particles (protons and neutrons) interacting through two-body forces; its geometrical invariance group is again $SO(3)$. To understand its structure better we use exactly the same method as for atoms, which leads to the shell model of nuclei. To a first approximation, each of the N particles may be visualized as moving in the average field of the others. But here this average field is approximately a harmonic potential [which has an $SU(3)$ invariance]. We then build an approximate symmetry group again, this time a unitary group of large dimensions, and use it to estimate the energy levels of the nucleus.

A NEW KIND OF SYMMETRY

All the symmetries discussed so far are geometrical in origin; but in nuclear physics, and even more so in elementary-particle physics, internal symmetries play an essential role. The simplest case is *charge conjugation,* which exchanges particle and antiparticle—this is still a discrete symmetry. Then, as Heisenberg was the first to realize (1932), protons and neutrons may be considered as two different states of the same entity, the *nucleon*. They differ only through electromagnetic, but *not* through strong (i.e., nuclear), interactions. The typical quantum system with only two possible states is a spin-$\frac{1}{2}$ particle; hence Heisenberg attributed to the nucleon a new *internal* degree of freedom, later called *isospin*. Mathematically it is isomorphic to ordinary spin: the nucleon is a particle with spin $\frac{1}{2}$ and isospin $\frac{1}{2}$. Hence the Hamiltonian of a nucleus must be invariant under the isospin $SU(2)$ group. If only Wigner and Majorana forces are present, it will also be invariant, to a good approximation, under the ordinary spin $SU(2)$, hence under the enveloping group, $SU(4)$. This approximate symmetry, proposed by Wigner (1937), has given good clues for understanding nuclear ground states.

THE CLASSIFICATION OF ELEMENTARY PARTICLES

Nucleons are not the only particles subject to strong interactions. Since the discovery of the π meson in 1947 the list of strongly interacting particles, or hadrons, has increased steadily, to more than 400 entries at present! Here again group theory has provided a unique tool for organizing such overwhelming experimental data.

First, all the hadrons can be grouped in isospin multiplets. Then another internal degree of freedom, called the *hypercharge* Y (or the closely related *strangeness*) was introduced by Gell-Mann; it satisfies the famous relation $Q=T_3+\frac{1}{2}Y$ where Q is the electric charge and T_3 the third component of the isospin. After many unsuccessful attempts, Gell-Mann (and others independently) discovered in 1962 how to merge the isospin and the hypercharge into a new internal symmetry group $SU(3)$. $SU(3)$ is an approximate symmetry that permits the classification of all known hadrons in multiplets corresponding to irreducible representations and thereby predicts many new particles. Here again a use of the Wigner–Eckart theorem has led to correct predictions of many physical properties, such as mass differences, magnetic moments, or branching ratios for various decay modes. The most spectacular success was the discovery (1964) of the quasi-stable particle Ω, whose mass was correctly predicted by Gell-Mann. Various extensions of this model have been proposed, but none of them proved really satisfactory.

Instead the classification problem took a new direction. Shortly after $SU(3)$ was introduced, Gell-Mann (and, independently, G. Zweig) suggested that all known hadrons could be thought of as bound states of three elementary building blocks, the so-called *quarks,* and their antiparticles. This model, naive at first but steadily refined, has been remarkably successful for describing the dynamical properties of hadrons (although the quarks themselves have never been found—this is the famous *confinement* phenomenon). More recently, the discovery of totally new particles has forced the theorists to enlarge their model. First, in 1974–1975, came the charmonium family ($\psi,\psi',\psi'',\ldots$), whose properties can be best understood by the existence of a new quantum number called *charm*: this demands a fourth quark. Similarly, the upsilon family, discovered in 1977, requires a fifth quark, the so-called *bottom* quark. More important, the phenomenological quark model is now incorporated, and understood, within a genuine theory, the so-called *standard model* (see below).

UNDERSTANDING DYNAMICAL PROPERTIES

However, group-theoretical ideas were not confined to classification. As far back as 1958, Feynman and Gell-Mann had proposed that part of the (conserved) electromagnetic current and the (almost conserved) weak current are an isospin triplet (the CVC, or conserved vector current, hypothesis). A corresponding but weaker hypothesis (PCAC) was soon made for the axial currents. These assumptions were extended to $SU(3)$ by Cabibbo, who thus obtained very good predictions of weak decay rates. These ideas finally led to Gell-Mann's *algebra of charges*. To each current there corresponds a charge (space integral of the zeroth component of the current); the vector charges transform like the adjoint representation of $SU(2)$, or $SU(3)$ if hypercharge-changing currents are included; this implies that the commutator of a vector charge with an axial charge is again an axial charge. Then Gell-Mann postulated that the commutator of two axial charges be a vector charge, i.e., the algebra of all charges closes under commutation to the Lie algebra of $SU(2)\otimes SU(2)$, or $SU(3)\otimes SU(3)$. From this so-called *chiral symmetry* a large number of predictions were obtained, simply by taking matrix elements of the commutation relations between adequate states and using the Wigner–Eckart theorem again. Finally, going one step further, Gell-Mann postulated that the currents themselves satisfy a local $SU(3)\otimes SU(3)$ algebra: this is the famous *current algebra*. These developments represent a remarkable evolution since the early applications of group theory. The precise structure of the various hadronic currents is unknown, but only their symmetry properties are important: the line of thought is exactly opposite to the one originally used, e.g., in atomic physics!

GAUGE THEORIES AND ALL THAT

By far the most promising development, however, is the emergence of the so-called *gauge theories,* which are based on an extension of the concept of symmetry. Within a field theory, an internal symmetry is said to be *global* if the action of the symmetry group on the field $\phi(x)$ is independent of the space-time point x; this is the concept used so far. The symmetry is called *local* if the action is, in addition, allowed to vary from point to point; such a theory is called a gauge field theory, and the local symmetry group is called a gauge group. The idea goes back to H. Weyl in 1918, who treated electromagnetism as a gauge theory based on the commutative gauge group $U(1)$. A noncommutative theory, based on $SU(2)$, was proposed by C. N. Yang and R. L. Mills in 1954, but the gauge concept did not become popular until the Dutch physicist G. t' Hooft proved in 1971 that a noncommutative gauge field theory is renormalizable (i.e., susceptible of giving consistently finite predictions). The key point in his proof was that a clever use of group identities produced sufficiently many cancellations between potentially divergent terms. Since then, gauge theories (and with them, differential geometry) have invaded the whole field of particle physics, with rather remarkable results.

The most important aspect is that a gauge symmetry must necessarily be exact; this eliminates lots of arbitrary parameters and gives the theory a much greater coherence (and elegance too). In particular the form of the interaction Lagrangian is uniquely determined. On the other hand, the interaction in a gauge theory is mediated by *massless* particles; the canonical example is the photon, corresponding to the fact that electromagnetism is a $U(1)$ gauge theory.

This mechanism extends to the other interactions. On the one hand, the model of Weinberg, Salam, and Glashow, based on the gauge group $SU(2)\otimes U(1)$ (which, incidentally, requires a sixth quark), gives a unified description of weak and electromagnetic interactions; it has accumulated excellent experimental support and is by now almost universally accepted (although a key ingredient—particles known as Higgs bosons—has not been found so far). On the other hand, a new theory of strong interactions, called *quantum chromodynamics* (QCD), has emerged: it is also a gauge theory, with gauge group $SU(3)$, that generalizes the old quark model by assuming that every quark comes in three types (or *colors*). Notice that, in both cases, all the gauge particles, except the photon, acquire nonzero masses by a subtle mechanism based on the notion of *spontaneously broken symmetry* (this term describes the situation where the ground state of a system has *less* symmetry that the Hamiltonian—the canonical example is the Heisenberg ferromagnet).

Taken together, these two models constitute what is now known as the *standard model*, covering simultaneously all three types of interactions. In fact it may fairly be called a *theory* instead of a mere model, although it contains a large number of free parameters, notably the masses of the fundamental fermions (quarks and leptons). This implies that, in the standard model, the isospin symmetry and even more the flavor $SU(3)$ symmetry are in fact *accidental*. On the other hand, the color $SU(3)$ is truly fundamental, because it reflects dynamical properties, as did already current algebra and chiral symmetry.

RECENT DEVELOPMENTS

Where do we go now? Although the answer to that question is certainly confused (things become clear only with hindsight, of course), one aspect is certain: the role of group theory is more central than ever!

A first direction to be mentioned is that of the so-called *grand unified theories* (GUTs), which try to unify all four interactions, including gravity. Various schemes have been proposed, based on groups like $SU(5)$, $SO(10)$, etc., but all predict new particles and the decay of the proton—both unseen so far.

Another interesting development is *supersymmetry* (which leads to *supergravity* when gravitation is included), a theory that seeks to unify bosons and fermions in a common framework. This idea has opened a new branch of (super) mathematics, namely, analysis (including group theory) with *anticommuting* variables. As a physical model, supersymmetry is very elegant (although it is a badly broken symmetry), but it has received so far no experimental confirmation whatsoever; in particular, it predicts the existence of many new particles with fancy names (photinos, gluinos, etc.), none of which has been seen. So the question remains totally open.

A totally different idea yet has emerged from gauge theory, and that one is much more promising. The original idea of gauge invariance introduced by Weyl was that the theory should be insensitive to a redefinition of length standards—in other words, that it should be invariant under reparametrization. This idea has proven extremely successful in Einstein's general relativity (which is also a gauge theory, albeit of a very special type). On the other hand, two-dimensional models have been popular among field theorists for a long time: they are easier to solve than their 4-dimensional, real world, counterparts, and they present striking (and probably deep) similarities with various models in statistical physics. Applying the idea of reparametrization in a 2-dimensional world leads to *conformal* invariance, which has become an extremely successful concept in high-energy physics and statistical mechanics as well. First of all, it lies at the basis of the theory of strings or superstrings, according to which the basic constituents of matter are no longer point-like, but 1-dimensional, string-like, objects. Various models have been proposed, and they are all based on heavy use of (unexpected) groups like $SO(32)$ or the exceptional $E(6)$, $E(7)$, $E(8)$. Second, the conformal group in two dimensions is no longer a Lie group, since it has infinitely many parameters. Thus *infinite-dimensional Lie algebras* have entered physics. First there is the Virasoro algebra, closely related to the conformal algebra. Next, combining the idea of reparametrization invariance with the formulation of classical string theory, one is led to a whole class of simple, infinite-dimensional Lie algebras, the so-called Kac–Moody algebras (discovered, independently, by the mathematicians V. Kac and R. V. Moody). The representation theory of those algebras is by now well under control, and they play a central role both in quantum string theory and in various models of classical statistical mechanics (conformal invariance is the link between the two). It is interesting to notice that Kac–Moody algebras were encountered previously as invariance algebras of some nonlinear differential equations giving rise to soliton solutions, such as the famous Korteweg–de Vries equation describing waves in shallow water.

CONCLUSION

Clearly our present understanding of elementary particles is sketchy and the whole picture is confusing. Yet, whatever theory finally emerges, it seems fair to say that group theory has grown into one of the essential tools of contemporary physics. Besides its fundamental role in relativity, it has provided physicists with a remarkable analyzing power for exploiting known symmetries, and thereby with a considerable predictive capability, precisely in cases where the basic physical laws are unknown. One of the striking aspects is its versatility: going from the rather restrictive study of exact symmetries to that of approximate ones, more and more badly broken, group theory has pervaded all fields of physics, often in a fundamental way. Except for calculus and linear algebra, no mathematical technique has been so successful.

See also ELEMENTARY PARTICLES IN PHYSICS; GAUGE THEORIES; GRAND UNIFIED THEORIES; INVARIANCE PRINCIPLES; ISOSPIN; LIE GROUPS; RELATIVITY, SPECIAL; STRING THEORY; *SU*(3) AND HIGHER SYMMETRIES; SUPERSYMMETRY AND SUPERGRAVITY.

BIBLIOGRAPHY

H. Weyl, *The Theory of Groups and Quantum Mechanics*. Dover, New York. 1950 (original, 1st German ed., 1928).

E. P. Wigner, *Group Theory and Its Application to the Quantum Mechanics of Atomic Spectra*. Academic Press, New York, 1959 (original, 1st German ed., 1931).

F. J. Dyson, *Symmetry Groups in Nuclear and Particle Physics* (a lecture note and reprint volume). Benjamin, New York, 1966.

E. M. Loebl (ed.), *Group Theory and Its Applications*. Vols. 1–3. Academic Press, New York, 1968–1975.

M. B. Green, J. H. Schwarz, and E. Witten, *Superstring Theory*. Vols. I–II. Cambridge University Press, Cambridge, 1987.

A. Böhm, Y. Ne'eman, and A. O. Barut (eds.), *Dynamical Groups and Spectrum Generating Algebras*. Vols. I–II. World Scientific, Singapore, 1988.

Gyromagnetic Ratio

Jack H. Freed

The gyromagnetic ratio (or magnetogyric) ratio γ is defined as the ratio of the magnetic moment $\boldsymbol{\mu}$ to the angular momentum $\mathbf{J}$ for any system. Specifically, one introduces for electrons with electron spin angular momentum $\mathbf{J}=\hbar\mathbf{S}$ (where $\hbar=h/2\pi$ and h is Planck's constant) the electron gyromagnetic ratio γ_e by $\boldsymbol{\mu}=-\gamma_e\hbar\mathbf{S}$, where the negative sign represents the fact that the spin and moment are oppositely directed. For a nucleus, which is a composite particle with total spin $\mathbf{I}$, one introduces the nuclear gyromagnetic ratio γ_I by $\boldsymbol{\mu}_I=\gamma_I\hbar\mathbf{I}$. Both $\boldsymbol{\mu}_I$ and $\mathbf{I}$ (or $\boldsymbol{\mu}_e$ and $\mathbf{S}$) must be regarded as quantum-mechanical operators. In the case of nuclear moments, the defining equation for γ_I must be considered as implying that the expectation values of $\boldsymbol{\mu}_I$ and $\mathbf{I}$ are taken for the given state (usually the ground state) of the nucleus. It then follows from symmetry considerations expressed in the Wigner–Eckart theorem that $\boldsymbol{\mu}_I$ and $\mathbf{I}$ may be taken as collinear, with $\gamma_I\hbar$ the proportionality constant.

A classical spinning spherical particle with mass m and charge e can be shown to give rise to a magnetic moment $e\hbar/2mc$, where c is the velocity of light. For an electron, this moment is known as the Bohr magneton, $\beta=0.9274096\times10^{-20}$ erg/G. But one has $\gamma_e\hbar=g_s\beta$, where g_s is the anomalous g value of the electron spin, which was first derived from the relativistic Dirac equation to be exactly 2. Schwinger showed how to correct this for quantum-electrodynamic effects to first order in $\alpha-c^2/\hbar c$ to give $g_s=2(1+\alpha/2\pi)=2.0023$. The most accurate theoretical calculation, due to Sommerfield, gives $g_s=2.0023192768$, which is in excellent agreement with the value of 2.002319244, obtained by Wilkinson and Crane by electron-beam experiments. Earlier atomic-beam measurements by Kusch on the hydrogen atom, for which corrections must be made for the relativistic mass change due to binding, yielded $g_s=2.002292$.

The nuclear gyromagnetic ratio, representing a composite nuclear property, may be measured very accurately by molecular-beam techniques. Also useful are nuclear magnetic resonance and optical, microwave, electron paramagnetic, Mössbauer, and electric quadrupole resonances. One typically introduces the nuclear magneton $\beta_N=0.5050951\times10^{-23}$ erg/G and the nuclear g value; thus $\gamma_I\hbar=g_I\beta_N$. One finds for the proton $g_P=5.585564$ with $\gamma_P=2.6751965\times10^4$ rad/s G (corrected for diamagnetism of H_2O).

See also MAGNETIC MOMENTS; NUCLEAR MOMENTS.

BIBLIOGRAPHY

H. Kopfermann, *Nuclear Moments*. Academic Press, New York 1958.

C. M. Lederer, J. M. Hollander, and I. Perlman, *Tables of Isotopes*. Wiley, New York, 1967.

N. F. Ramsey, *Molecular Beams*. Oxford, London, 1956.

B. N. Taylor, W. H. Parker, and D. N. Langenberg, *Rev. Mod. Phys*. **41**, 375 (1969).

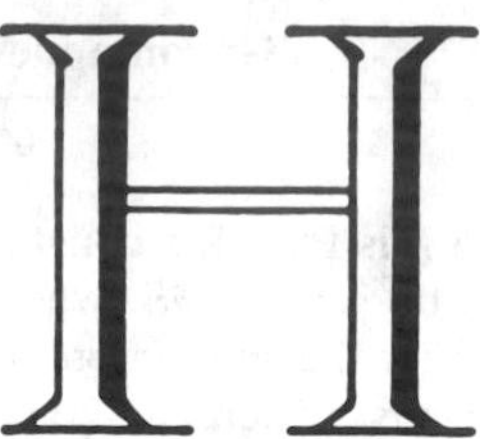

H Theorem

Mark Kac† and G. W. Ford

Originally the *H* theorem, or more precisely the Boltzmann *H* theorem, referred to the following statement first enunciated by L. Boltzmann in a famed memoir in 1872:

In a spatially homogeneous dilute (in the sense that only binary collisions need be considered) gas of molecules interacting through central forces, let $f(\mathbf{v},t)d^3v$ denote the number of molecules having velocity $\mathbf{v}$ within the (three-dimensional) volume element d^3v; then setting

$$H = \int d^3v f(\mathbf{v},t) \log f(\mathbf{v},t),$$

one has that for any initial distribution *H* decreases monotonically with increasing time, with *H* constant only for the Maxwellian equilibrium distribution

$$f(\mathbf{v}) = \left(\frac{m}{2\pi kT}\right)^{3/2} \exp\left(-\frac{mv^2}{2kT}\right).$$

This is a rigorous consequence of the Boltzmann equation (first derived in the aforementioned memoir of 1872):

$$\frac{\partial f}{\partial t} = \int d^3v_1 \int d\Omega\, gI(g,\Omega)\{f(\mathbf{v}',t)f(\mathbf{v}_1',t) - f(\mathbf{v},t)f(\mathbf{v}_1',t)\},$$

where the velocity variables refer to the four velocities of a binary collision $(\mathbf{v},\mathbf{v}_1)\rightleftarrows(\mathbf{v}',\mathbf{v}_1')$, $g = |\mathbf{v}_1 - \mathbf{v}| = |\mathbf{v}_1' - \mathbf{v}'|$ is the magnitude of the relative velocity which in an elastic collision is invariant, and $I(g,\Omega)$ is the differential cross section for a collision in which the relative velocity turns through angle θ into the element of solid angle $d\Omega$.

Since in equilibrium $-H$ is proportional to the thermodynamic entropy and $-H$ increases in time, Boltzmann saw his theorem as a mechanistic derivation (at least for dilute gases) of the second law of thermodynamics for irreversible processes. Subsequent critical analyses showed that Boltzmann's derivation of his equation did not follow from the mechanics of collision processes alone but required in addition a nonmechanical statistical assumption equating the actual number of collisions of a certain type with its average (*Stosszahlansatz*). This objection was answered by the statistical method of Boltzmann and the Boltzmann–Gibbs statistical mechanics that is at the heart of much of present-day physics. The modern view is that Boltzmann's *H* theorem embodies the essential features of the irreversible approach to equilibrium.

Since Boltzmann's time the *H* theorem has been generalized to a wide variety of mechanical systems, both classical and quantum mechanical. Gradually the term *H* theorem has acquired a much broader meaning, referring essentially to any statement about the approach to equilibrium which asserts that an appropriate quantity decreases (or increases) with time.

†Deceased

See also Entropy; Statistical Mechanics; Thermodynamics, Equilibrium; Thermodynamics, Nonequilibrium.

BIBLIOGRAPHY

S. G. Brush, "Kinetic Theory. Vol. 2. Irreversible Processes," in *Selected Readings in Physics*. Pergamon Press, New York. 1966. Here one finds an English translation of the 1872 Boltzmann memoir.

J. R. Dorfman and H. van Beijeren, "The Kinetic Theory of Gases," in *Statistical Mechanics. Part B: Time-Dependent Processes* (B. J. Berne, ed.). Plenum Press, New York, 1977. An account of more recent developments.

Martin J. Klein, *Paul Ehrenfest*. North-Holland, Amsterdam and American Elsevier, New York, 1970. Contains the best critical and historical treatment of the *H* theorem and its role in elucidating the statistical nature of the second law of thermodynamics.

George E. Uhlenbeck and George W. Ford, *Lectures in Statistical Mechanics*. American Mathematical Society, Providence, RI, 1963. Chapter IV contains a succinct derivation of the Boltzmann equation and the *H* theorem.

Hadrons

Martin L. Perl

Hadrons such as protons, neutrons, and mesons are the largest family of subnuclear particles. Hadrons have three related properties. (1) They interact with each other through the strong force or strong interaction. Since the strong force also holds protons and neutrons together in a nucleus, its older name is the nuclear force. The simplest measure of the strong force is that when two hadrons collide at high energy (above several GeV) the total cross section is 10–60 mb (1 mb is 10^{-27} cm^{-2}). (2) In the collision of two hadrons at high energy, additional hadrons are usually produced, up to 50 additional hadrons when the energy is in the thousand GeV range. (3) Hadrons are roughly spherical, with radii of the order of 10^{-13} cm.

Each hadron is composed of elementary particles called quarks and gluons. The quarks contribute to the mass of the hadron and determine other hadron properties such as electric charge and spin. Depending on the types of quarks, the hadron may also possess the properties called strangeness, charm, or beauty. The gluons carry the strong force which holds the quarks together inside the hadron, in the same quantum-mechanical manner that photons carry the electro-

Table I Properties of some hadrons. The quark content is in the order of the electric charge. The quark symbols are u=up, d=down, s=strange, c=charm, and b=beauty also called bottom. The bar means antiquark.

Name	Symbol	Mass (GeV/c^2)	Electric charge	Quarks in hadron	Spin	Lifetime (s)
Charged pion	π^+	0.140	+1, −1	$u\bar{d}$, $\bar{u}d$	0	2.6×10^{-8}
Neutral pion	π^0	0.135	0	$u\bar{u}$ and $d\bar{d}$	0	8.7×10^{-17}
Charged kaon	$K^\pm$	0.494	+1, −1	$\bar{s}u$, $s\bar{u}$	0	1.2×10^{-8}
Neutral kaon	K^0	0.498	0	$\bar{s}d$	0	5.2×10^{-8} and 8.9×10^{-11}
Rho meson	ρ	0.770	+1, 0, −1	$u\bar{d}$, $u\bar{u}$ and $d\bar{d}$, $\bar{u}d$	1	4.3×10^{-24}
Proton	p	0.938	+1	uud	$\frac{1}{2}$	Stable
Neutron	n	0.940	0	udd	$\frac{1}{2}$	898
Charged D meson	$D^\pm$	1.869	+1, −1	$c\bar{d}$, $\bar{c}d$	0	9×10^{-13}
Neutral D meson	D^0	1.865	0	$c\bar{u}$	0	4×10^{-13}
Psi	ψ	3.097	0	$c\bar{c}$	1	1.0×10^{-20}
Charged B meson	$B^\pm$	5.271	+1, −1	$\bar{b}u$, $b\bar{u}$	0	About 10^{-13}
Neutral B meson	B^0	5.276	0	$\bar{b}d$	0	About 10^{-13}
Upsilon	Υ	9.460	0	$b\bar{b}$	1	1.5×10^{-20}

magnetic force. Both quarks and gluons are less than 10^{-16} cm in size. Thus the hadron volume of 10^{-13} cm radius is mostly empty space containing a few quarks and gluons. Yet the strong force makes it difficult for hadrons to penetrate each other.

There are two types of hadrons. The baryons such as protons and neutrons contain three quarks or three antiquarks. The mesons contain one quark and one antiquark.

More than 100 different kinds of hadrons have been found. A few prominent examples and some of their properties are listed in Table I. All hadrons, except for the proton, are unstable. Some hadrons decay into other hadrons through the strong interaction, for example, $\rho^0\rightarrow\pi^+ + \pi^-$. Other hadrons decay through the electromagnetic interaction, $\pi^0\rightarrow\gamma+\gamma$, or through the weak interaction, $\pi^+\rightarrow e^+ + \nu_e$. The number of different kinds of hadrons is expected to be very large because there seems to be no upper limit on their mass. Hadrons with masses greater than 10 GeV/c^2 have been identified. There is a rough rule that the larger the mass, the shorter the lifetime and hence the more unstable the particle. However, when hadrons decay through the weak interaction their lifetimes are relatively longer. Examples in Table I are the charged pions, K mesons, neutron, D mesons and B mesons.

See also ELEMENTARY PARTICLES IN PHYSICS; HYPERONS; MESONS; NUCLEON; PARTONS; QUARKONIUM; QUARKS; STRONG INTERACTIONS.

BIBLIOGRAPHY

F. Close, M. Marten, and C. Sutton, *The Particle Explosion*. Oxford University Press, New York, 1987.

Y. Ne'eman and Y. Kirsch, *The Particle Hunters*. Cambridge University Press, Cambridge, 1986.

G. Kane, *Modern Elementary Particle Physics*. Addison-Wesley, Redwood City, CA, 1987.

Hadron Colliders at High Energy

R. Ronald Rau

INTRODUCTION

Interactions from high-energy colliding beams were first observed in 1963. They were produced by counter-rotating beams of electrons and positrons, each of 250 MeV, in a small ring in which the particles circulated for long periods. It was built by Italian scientists at the Frascati Laboratory. In the succeeding years, perhaps 20 colliders have been built and used for experiments, some of which have revolutionized the understanding of high-energy particle physics. Just three of these have produced hadron–hadron collisions, either proton–proton (pp) or proton–antiproton ($p\bar{p}$). The first hadron collider was the ISR (Intersecting Storage Rings) pp machine at CERN, which reached an energy of 31 GeV per beam or 62 GeV center of mass (c.m.) energy. This state of the art collider "proved" the importance of hadron colliders to elementary-particle physics research. The ISR was decommissioned in December 1983 after having produced collisions between beams of alpha particles as well as protons.

Two hadron colliders are in operation and these are the forefront machines for experiments with hadrons at very high energies. Both are $p\bar{p}$ colliders. The S$p\bar{p}$S began operations at CERN in 1981 and now operates at 630-GeV c.m. energy. It was at this collider that the charged, $W^\pm$, and neutral, Z^0, weak intermediate vector bosons were discovered in 1983. The second, which began operating for experiments in 1987, is located at Fermi National Laboratory (FNAL). Its c.m. energy is 1800 GeV.

These two colliders exist due to the ingenuity of high-energy and accelerator physicists. Both colliders are modifications to previously existing single-beam accelerators, originally built for fixed-target operations, i.e., their high-energy beams of protons were extracted from the machines striking targets, such as liquid hydrogen, in which the hydrogen (proton) was at rest in the laboratory. In the late 1970s, the major problem for $p\bar{p}$ colliders was to produce enough antiprotons and then to increase their space density so as to attain a large interaction rate between the beams of protons and antiprotons. The problem was solved at CERN by the invention of "stochastic cooling" for bunches of antiprotons (applicable to other hadrons as well). This process reduces the relative momentum between antiprotons and increases their space density. With high-density bunches of $\bar{p}$ available, counter-rotating beams of protons and antiprotons could be stored in one accelerator ring and produce large numbers of interactions. This approach, $p\bar{p}$ collisions using one ring and modifying an existing accelerator complex, provided a minimal cost route to reaching very high-c.m.-energy hadron–hadron collisions.

At CERN the SpS accelerator (~250 GeV) was converted to the S$p\bar{p}$S which could accelerate and collide protons and antiprotons in the original accelerator ring. At FNAL the Tevatron collider uses a new ring of superconducting magnets which was installed in the original machine tunnel. The new ring stores counter-rotating bunches of protons and antiprotons. Since the superconducting magnets produced a magnetic field of 4.5 T (tesla), the energy of each beam was raised to ~900 GeV. The original synchrotron (400 GeV) is used as part of the injection system for the Tevatron collider.

The operation of the $p\bar{p}$ colliders is complex, as the example of the Tevatron will show. In order to produce $\bar{p}$'s, protons are accelerated in the original accelerator ring to 120 GeV. The beam is then extracted, striking an external target. $\bar{p}$'s are collected by a magnet focusing system and injected into a "debunching" ring, where the bunch structure is removed and the $\bar{p}$'s are "cooled," as mentioned earlier. They are then transferred to a concentric "accumulating" ring, in which $\bar{p}$'s are collected from many repetitions of this operation, lasting several hours. After sufficient $\bar{p}$'s are collected, they are injected into the main ring, orbiting in a sense opposite to protons, accelerated to 150 GeV, then transferred into the Tevatron. Six bunches of $\bar{p}$'s are stored, followed by six bunches of protons circulating opposite to the antiprotons. Both beams are then accelerated to 900 GeV and brought into collision in the experimental regions, providing collisions at a center-of-mass energy of 1800 GeV. In the Tevatron, there are provisions for four collision areas, but only two are currently available for large detector facilities. Figure 1 is a schematic showing the machine elements of the complete Tevatron collider and Table I lists some of the relevant parameters. The S$p\bar{p}$S, at CERN, is similar in operating principle.

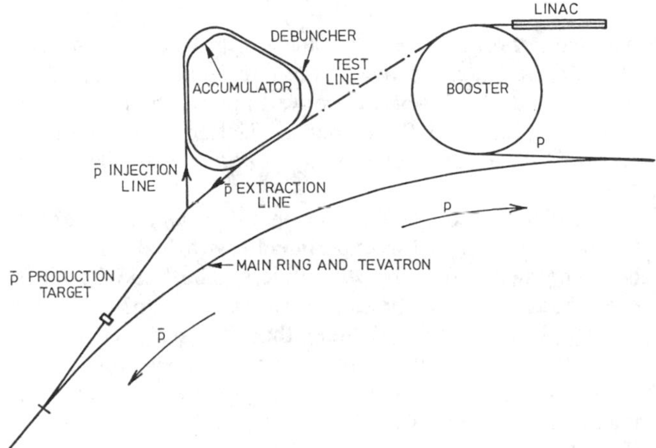

FIG. 1. This schematic drawing shows the elements of the injection system for the FNAL Tevatron collider.

Table I Parameters: Tevatron collider

Injection	
Linear accelerator	200 MeV
Booster	8 GeV
Main ring	150 GeV
$\bar{p}$ debuncher and accumulator	8.9 GeV
Maximum energy per beam	900 GeV
Circumference	6.3 km
Dipole field	4.5 T
No. bunches/beam	6
No. $\bar{p}$/bunch	3×10^{10}
No. p/bunch	7×10^{10}
Peak luminosity	2.06×10^{30} cm^{-2} s^{-1}
Beam lifetime	10–15 h

WHY COLLIDING BEAMS?

The key parameter in particle physics research has always been the available c.m. energy. As higher-energy accelerators were constructed, startling discoveries were made. For example, in 1952, at the Chicago synchrotron with its proton beam of 1 GeV energy, the π^+p cross section was observed to increase unexpectedly and dramatically up to the maximum π^+ energy available (~140 MeV). In a short time the π^+p, I (isotropic spin) = $\frac{3}{2}$ resonance at a mass of 1232 MeV was discovered. At the Brookhaven National Laboratory 3-GeV Cosmotron, the experimental proof was obtained, in 1953, for the associated production of the newly discovered "strange" particles. In 1955, the antiproton was produced at the 6-GeV Bevatron at Lawrence Berkeley Laboratory. The 30-GeV AGS, at Brookhaven, provided several surprises: 1962, the discovery of a second neutrino, always associated with the muon; 1964, CP (C = charge conjugation, P = parity) is violated in neutral kaon decay; 1974, the fourth quark, "charm" (simultaneously found at SLAC). Note that at each energy increase, new phenomena have been observed.

Recall, however, that these discoveries were made in fixed-target collisions. At relativistic energies the c.m. energy, $E_{c.m.}$, increases only with the square root of the projectile energy,

$$E_{c.m.} \cong \sqrt{2m_pE_1}, \tag{1}$$

where m_p is the mass of the target proton, and E_1 ($\gg m_p$) is the projectile energy in the laboratory frame of reference.

On the other hand, when particles of equal and opposite

momenta collide in the laboratory, the c.m. energy increases linearly with the energy of the particles. It therefore follows that, with the current accelerator technology, colliding beams offer the only practical method of achieving c.m. energies of hundreds or thousands of GeV in hadron collisions. The example of the discovery of the weak intermediate vector bosons ($W^{\pm}$, mass 81 GeV, and Z^0, mass 92.4 GeV), at CERN in 1983, will illustrate this argument. In the late 1970s, the SpS, at CERN, accelerated protons to 270 GeV and upon their striking a fixed target provided 23 GeV c.m. energy. In order for a fixed-target proton accelerator to achieve 92 GeV c.m. energy, a beam of about 4300 GeV would be required. However, by modifying the SpS to the S$p\bar{p}$S collider at 540 GeV c.m. energy, more than enough energy was available to produce the Z^0 and even the $W^{\pm}$, which must be produced in pairs.

PARAMETERS AND TECHNICAL CONSIDERATIONS

It is natural to ask whether it is possible to achieve useful interaction rates with colliding beams since, in comparison to liquid hydrogen or solid targets used in fixed-target accelerators, these beams have very low particle density. The luminosity, L (cm^{-2} s^{-1}), is the parameter connecting the interaction rate, R, and the relevant cross section, σ, through the relation

$$R = \sigma L. \tag{2}$$

At the Tevatron collider energy, $\sigma \approx 50$ mb (50×10^{-27} cm^2), and with $L = 2 \times 10^{30}$ cm^{-2} s^{-1}, $R = 10^5$ interactions per second, which is indeed a useful event rate. For bunched beams, assuming N_b particles in each bunch and head-on collisions, L is given by

$$L = \frac{N_b^2 k f}{4\pi\sigma^2}, \tag{3}$$

where k is the number of bunches in the beam, f is the revolution frequency, and σ is the root mean square beam size. The beams, at the collision points, are assumed circular transverse to their motion. Equation (3) suggests how to maximize the luminosity: decrease the beam size σ; increase the number of particles per bunch and the number of bunches in each beam (f is fixed by the circumference of the machine). Naturally there are limitations on the range of values for these parameters.

There are two components contributing to beam size. One arises from the intrinsic nature of the beam itself, the other from the linear focusing properties of the collider ring magnet system. The beam size in one dimension can be written

$$\sigma = \sqrt{\epsilon\beta^*} \tag{4}$$

corresponding to one root mean square beam height or width. ϵ is the emittance of the beam (units are mm-mrad). β^* is the value of the betatron amplitude function (or beta function) $\beta(s)$ at the collision point and s is the longitudinal coordinate around the ring.

$\beta(s)$ is characteristic of the linear focusing properties of the magnet structure around the ring (referred to as the "lattice") and is independent of the beam itself. β^* is made small so as to maximize the luminosity. However, decreasing β^* requires that $\beta(s)$, at some other point such as at a quadrupole focusing magnet, increases. The aperture of the magnet thus becomes a limitation on β^*. In hadron machines, it is possible to achieve β^* values as small as 0.5–1 m.

For a linearly focusing lattice at a fixed central momentum, the emittance, ϵ, is an invariant characteristic of the beam. Even for changing momentum there is an invariant of the motion, termed the normalized emittance,

$$E_N = \beta\gamma\epsilon, \tag{5}$$

where β and γ are the usual relativistic variables. Combining Equations (4) and (5) we see that the beam size shrinks as $\sqrt{\beta\gamma}$ or $\sqrt{(\text{momentum})}$. Although the emittance, E_N, is an invariant for a linear machine, in a real accelerator or collider, there are effects which tend to make the emittance grow. Sources of some of these effects are (1) nonlinearities present in the magnet system, (2) electromagnetic fields generated by the circulating beam (self-beam effects) or those generated by the "other" beam (beam–beam effects), and (3) scattering processes: intrabeam scattering, i.e., multiple Coulomb scattering of protons (hadrons) within bunches in the circulating beam, or beam–gas multiple Coulomb scattering. The cumulative effect of these processes is a kind of "second law," embodied in Liouville's theorem, which states that for a given current the real beam emittance, at any energy, cannot be less than the initial emittance. (E_N can be decreased by "cooling" processes, which seem to "violate" Liouville's theorem, but in actuality do not. $\bar{p}$ "cooling" mentioned earlier is such a case.) In fact, any collider design must ensure that the growth time of the emittance is large compared to the time needed for the performance of useful physics. That this was indeed possible was first demonstrated at the ISR where useful circulating beams, lasting 24 hours or more, were routinely available for physics experiments. E_N is initially determined by the injection system and the source of particles; hence these sources must be designed to minimize E_N in order to achieve high luminosity.

There are limitations on N_b, the number of particles in each bunch. For example, when two beams collide head on, the particles in a bunch moving clockwise experience a defocusing by the electromagnetic field of the colliding bunch moving counterclockwise, and vice versa. This is expressed by a "tune shift," ΔQ, which is written

$$\Delta Q = \frac{N_b r_p}{4\pi E_N}. \tag{6}$$

r_p is the classical radius of the proton. If ΔQ becomes too large, individual beam particles experience forces which eventually remove them from the beam. From experience, a maximum value of $\Delta Q \cong 0.004$ for each collision point, in hadron colliders, appears to be a safe limit.

Similarly, k, the number of bunches in the ring, is limited. Two different effects illustrate limitations. For head-on collisions, if the distance between bunches, d_b, is less than the field-free space on either side of the collision point, then bunches in one beam have close encounters with several bunches in the oppositely moving beam and this acts to increase the effective beam–beam tune shift. Second, if the luminosity is high so that the effective number of interactions per beam crossing is ≥ 1, then when d_b is short, a particle detector could record several interactions from succeeding beam crossings. These multiple interactions would be difficult to separate in the subsequent analysis of data. This is not a problem with current colliders ($d_b \cong 1000$ m), but it is an important consideration for the next generation of super-high-energy colliders ($d_b \cong 5$–10 m).

FUTURE FACILITIES

In Construction

Two hadron colliders are currently being built. UNK, a *pp* collider, is under construction at the Serpukov Laboratory, in the USSR. This machine will use the existing 70-GeV accelerator as the source of protons for a new 400-GeV booster ring, which will in turn inject protons into two rings of superconducting magnets. The energy is 3000 GeV (3 TeV) per beam: 6 TeV c.m. energy. The booster, when finished, will be used initially as a fixed-target facility until the superconducting magnet rings are available in the mid-1990s. The design luminosity is 4×10^{32} cm^{-2} s^{-1}.

The second collider, HERA, is a hybrid. It will produce collisions of 30-GeV electrons with 820-GeV protons: c.m. energy of 320 GeV. (In $p\bar{p}$ or *pp* colliders the c.m. is at rest in the laboratory, but in HERA it will move, in the proton beam direction, with high velocity, $\beta_{c.m.} = 0.93$.) HERA is located at DESY Laboratory in Hamburg, West Germany, and is scheduled to be ready for experimental research in early 1991. The design is for a luminosity greater than 10^{31} cm^{-2} s^{-1}. Superconducting dipole magnets at 4.7 T will be used in the proton ring.

Proposed Facilities

Table II lists three hadron–hadron colliders which are in the proposal stage. The SSC (Superconducting Super Collider) is the highest-energy hadron (*pp*) collider to be considered seriously, 20-TeV protons in each beam or 40 TeV c.m. energy. The design is 83 km in circumference, employing 10 000 superconducting magnets, with the dipole magnets designed to operate at 6.6 T. This project, estimated to cost \$5–\$6 billion, has been approved by the Congress and has received initial construction funds in the 1990 budget. The SSC should be operational before the year 2000.

A less ambitious project is the LHC (Large Hadron Collider) which CERN has proposed at 16 TeV c.m. energy, using 10-T superconducting magnets. The two proton rings would be installed in the LEP tunnel which is 27 km in circumference. LEP is a large electron–positron collider which began operation in late summer 1989 at CERN: 100 GeV c.m. energy. The estimated cost for LHC is significantly less than for the SSC since the injector system, all tunnels, experimental areas, etc., are available from LEP and the energy is roughly one-half that proposed for the SSC. This proposal is not yet formalized by submission to the member states of CERN.

RHIC (Relativistic Heavy Ion Collider) is designed for the collisions of ions, ranging from protons to gold, including collisions between different species of ions, e.g., protons and gold. This unique collider would be installed in an existing 3.8-km tunnel, at Brookhaven National Laboratory, and use existing accelerators for the injection process. RHIC would use simple superconducting magnets operating at 3.5 T. The expected cost is under \$300 million and it could be built in five to six years. The proposal is included for funding in the president's budget for 1991.

Table II Parameters: proposed hadron colliders

	SSC	LHC	RHIC
Beam particles	*pp*	*pp*	AuAu, *pp*, *p*Au
Injection energy	1 TeV	0.45 TeV	15 GeV/nuc. (Au) 30 GeV (*p*)
Max. beam energy	20 TeV	8 TeV	100 GeV/nuc. (Au) 250 GeV (*p*)
Circumference (km)	83.6	26.7	3.8
Dipole field (T)	6.6	10	3.5
No. bunches/ring	16 470	3 564	57
Particles/bunch	10^{10}	2.6×10^{10}	10^9 (Au) 10^{11} (*p*)
Luminosity (cm^{-2} s^{-1})	10^{33}	1.4×10^{33}	2×10^{27} (Au) 3×10^{32} (*p*)
Beam lifetime (h)	>24	>20	≈10

INTERACTION OF COLLIDER AND EXPERIMENTS

Colliders differ from large accelerators (synchrotrons) in significant respects. The need for high-field dc magnet operation and longtime stability of the beams are examples which are briefly touched upon elsewhere in this article. However, the crucial difference is the close interrelationship between the design of each experiment and the collider.

Since collisions between the circulating beams occur within the vacuum chamber of the collider, experiments become an integral part of the machine and must be designed about the specific properties of the beam and with consideration for the geometry of the collider and the stability of the beam.

The vacuum system provides a clear example of how the experimental needs affect the machine design. The beams must be able to circulate for many hours with minimal loss of beam quality or luminosity. Protons scattering from the residual gas is one such loss mechanism and implies the need for a far better vacuum than in conventional accelerators. For example, a vacuum of 10^{-10} torr or better is required to minimize background particles, against which the experimental equipment must discriminate. The collider design must also include long field-free regions on either side of the beam-collision points into which the very large detectors must fit.

CONCLUSION

For producing interactions above several hundred GeV c.m. energy, only *pp* colliders currently have the required high luminosity (>10^{32} c.m.$^{-2}$ s^{-1}). $p\bar{p}$ colliders have significantly lower maximum luminosity. Beyond the next round of high-energy colliders (SSC and LHC), new technology will be needed to keep costs within reason. In the past, new ideas allowed higher energies to be obtained. For example, the synchrotron principle, 1944, the alternating gradient or strong focusing principle, 1952, and the use of superconducting magnets, 1985. Will the cycle repeat?

See also ANTIMATTER; BARYONS; CENTER-OF-MASS SYSTEM; POSITRON-ELECTRON COLLIDING BEAMS; SYNCHROTRON.

BIBLIOGRAPHY

General

E. D. Courant and H. S. Snyder, "Theory of the Alternating Gradient Synchrotron," *Ann. Phys. (NY)* **3,** 1 (1958).

M. S. Livingston and J. P. Blewett, *Particle Accelerators.* McGraw-Hill, New York, 1962.

C. Pelligrini, "Colliding-Beam Accelerators," *Annu. Rev. Nucl. Sci.* **20,** 1 (1972).

J. D. Lawson and M. Tigner, "The Physics of Particle Accelerators," *Annu. Rev. Nucl. Part. Sci.* **34,** 29 (1984).

W. Scharf, *Particle Accelerators and Their Uses.* Harwood Academic Publishers, 1986 (2 volumes).

D. E. Johnson, "Accelerator Issues at the SSC," *J. Mod. Phys.* **A3** (No. 11), 2503 (1988).

Specific Accelerator Components

F. T. Cole and F. E. Mills, "Increasing the Phase-Space Density of High Energy Particle Beams," *Annu. Rev. Nucl. Part. Sci.* **31,** 295 (1981).

R. Palmer and A. V. Tollestrup, "Superconducting Magnet Technology," *Annu. Rev. Nucl. Part. Sci.* **34,** 247 (1984).

The Fermilab Antiproton Source Design Report, Fermi National Accelerator Laboratory, Batavia, IL, February, 1982.

Proceedings

The Proceedings of various conferences and schools contain the current information on accelerator theory and practice. A few are given here.

Proceedings of the Particle Accelerator School.

Proceedings of the CERN Accelerator School.

Proceedings of the International Conference on Magnet Technology.

Proceedings of the 12th International Conference on High Energy Accelerators, Fermilab, August 11–16, 1983, and the subsequent "13th," held at Novosibirsk, USSR, August 7–11, 1986.

R. A. Carrigan, F. R. Huson, and M. Month (eds.), *The State of Particle Accelerators and High Energy Physics,* AIP Conference Proceedings No. 92. American Institute of Physics, New York, 1982.

M. Month, P. Dahl, and M. Dienes (eds.), *The Physics of High Energy Particle Accelerators,* AIP Conference Proceedings No. 105 and No. 127. American Institute of Physics, New York, 1983 and 1985.

Design Proposals

SSC Conceptual Design Report, SSC Central Design Group Report No. SSC-SR-2020 (1986).

The Large Hadron Collider in the LEP Tunnel, CERN 87-05 (1987).

Conceptual Design of the Relativistic Heavy Ion Collider RHIC, BNL Report 52195 (1989).

Hall Effect

P. R. Emtage

In 1879, E. H. Hall found that when a metal strip which bore a current was placed in a magnetic field, as in Fig. 1a, a voltage was produced across the strip. This transverse voltage is the simplest and most widely useful of the galvanomagnetic effects, since the number and nature of the current carriers can often be found from it.

The Hall voltage comes from a transverse electric field that cancels the sideways deflection of the current carriers by the magnetic field. Suppose the current is carried by electrons of charge $-e$ and drift velocity v_x in the direction of the current. In a magnetic field B_z they are acted on by a transverse force f_y—the Lorentz force—which causes the cyclotron motion and is at right angles to both field and velocity,

$$f_y = ev_xB_z.$$

The electron trajectories are therefore bent toward one side of the conductor; a surface charge forms on the sides and causes an electric field E_y that opposes further deflection—see Fig. 1b. Equilibrium is reached when the total force is zero,

$$ev_xB_z - eE_y = 0. \tag{1}$$

The transverse field E_y should be expressed in terms of measurable quantities, such as electric current, rather than electron velocity. If there are n electrons per unit volume, the current density is

$$j_x = -nev_x,$$

the negative sign being inserted because the charge on the electron is negative. Accordingly from Eq. (1) we find

$$E_y = Rj_xB_z, \tag{2}$$

wherein R is defined as the Hall coefficient, and from the above argument is given by

$$R = -1/ne. \tag{3a}$$

Note that, had we supposed the current to be carried by p positive carriers of charge $+e$, we should have found

$$R = 1/pe. \tag{3b}$$

The sign of the Hall coefficient therefore says whether the current is borne by positive or by negative charges, and its magnitude yields the density of current carriers.

The above results are so simple that they should be regarded with the gravest suspicion. Equations (3) hold only in very high magnetic fields, when each electron makes several cyclotron revolutions between collisions. In low fields no unique electron drift velocity v_x exists, and therefore Eq. (1) cannot be satisfied for all groups of electrons simultaneously. The total transverse current must, however, be zero, and therefore a well-defined low-field Hall coefficient does exist, and in simple cases has the form

$$R = -r/ne,$$

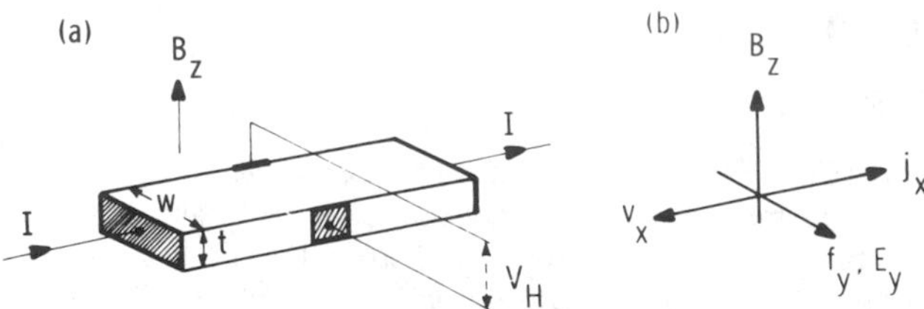

FIG. 1. (a) Configuration in which Hall effect is found. (b) Fields, etc., in the system.

r being a factor of order unity. It is common to define a "Hall density of carriers" through

$$n_H \equiv -1/Re. \tag{4}$$

Where other evidence suggests that the electron structure is in fact simple, this definition is a useful guide to the electron density.

MEASUREMENTS

The Hall coefficient R is defined from Eq. (2), and measurements of it are carried out on bar-shaped samples such as that in Fig. 1a. Let the bar have width w and thickness t; if the total current is I, the current density is $j_x = I/wt$. The Hall voltage V_H is measured between electrodes on the sides, and is related to the Hall field through $E_y = V_H/w$. Therefore,

$$R = E_y/j_xB_z = tV_H/IB_z.$$

In practice it is always necessary to find the mean of V_H from four measurements with field and current both forward and reversed, so as to eliminate errors resulting from misalignment of the electrodes, thermoelectric effects, and other galvanomagnetic effects, as well as inhomogeneities or anisotropy in the sample.

Values of the Hall coefficient at room temperature for a variety of elemental conductors are shown in Table I; the last column shows the apparent number of free electrons per atom, n_H being defined by Eq. (4). The results for the polyvalent metals are distressingly irregular, but there is some order amidst the chaos: all the metals have high carrier densities, and the monovalent metals have nearly one free electron per atom; the semimetals show much lower carrier densities, as expected; while Ge, a semiconductor with no free electrons at all at low temperatures, is nearly destitute of current carriers.

The outstanding oddity of Table I is that the Hall coefficients are in many cases positive. It was found in Eqs. (3) that a positive Hall coefficient should correspond to positively charged current carriers; the results therefore conflict with the electron theory of conduction, and for a resolution of this enigma we must turn to the band theory of electrons.

BAND THEORY

There is not space here for more than the briefest description of the band theory. In a semiconductor the valence electrons occupy a complete set or band of states that spans a finite range of energies; no electron in this set can be moved to any other valence state without violating the Pauli exclusion principle, so the valence electrons are "rigid" and can carry no current. There must, however, be states of higher energy, the conduction bands, within which an electron can move—otherwise, the material would be impervious to high-energy electrons.

Table I Hall Coefficients R (at room temperature) and Apparent Number of Free Electrons per Atom, n_H/n_a, for Some Elemental Conductors

Element	Group[a]	$R \times 10^{10}$ (m^3/C)	n_H/n_a
Na	Ia (M)	−2.1	1.17
Cu	Ib (M)	−0.536	1.38
Be	IIa (M)	+2.44	−0.21
Al	IIIa (M)	−3.0	0.31
Cr	IVb (TM)	+3.55	−0.21
Pt	VIII (TM)	−0.244	3.87
As	Va (SM)	−70	1.9×10^{-2}
Sb	Va (SM)	+250	-7.6×10^{-3}
Ge	IVa (SC)	-8×10^{8}	1.7×10^{-9}

[a] In this column M = metal, TM = transition metal, SM = semimetal, SC = semiconductor.

In a semiconductor the conduction band is separated from the valence band and is normally empty, as in Fig. 2a. In many semimetals the conduction and valence bands overlap somewhat, as in Fig. 2b; to minimize energy, electrons in the upper part of the valence band "fall" into the lower part of the conduction band. In most polyvalent metals the overlap of bands is large, and no clear distinction between conduction and valence bands can be drawn.

The peculiarities of the band picture can be seen if we imagine an electron in state i removed from a filled valence band, as in Fig. 2a; conduction can now occur since other valence electrons can move into the vacant state. Denote by Σ a sum over all valence states, and by Σ' a sum over all states except i. The charge on each valence electron is $-e$; let the energy of each electron relative to the band maximum be $-\epsilon_v$. The energy ϵ_h and charge q_h of the "hole" in the band are

$$\epsilon_h = \Sigma'(-\epsilon_v) - \Sigma(-\epsilon_v) = \epsilon_i,$$

$$q_h = \Sigma'(-e) - \Sigma(-e) = +e.$$

The hole in the valence band therefore acts as a "particle" of positive energy and positive charge, and the Hall coefficient due to its motion is positive. It is from the motion of

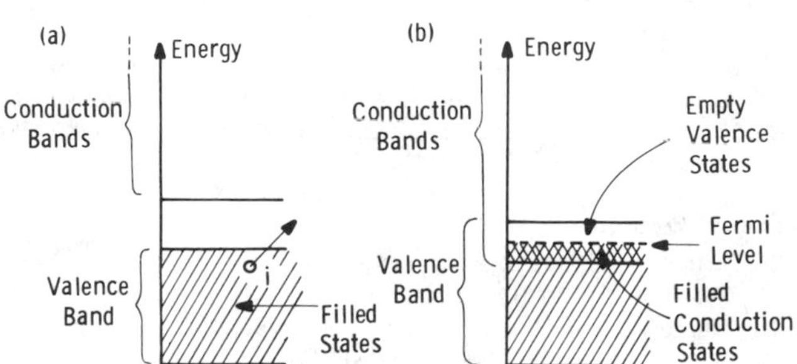

FIG. 2. Electron bands in (a) a semiconductor or insulator; (b) a semimetal.

electrons in incompletely filled bands that so many materials derive positive Hall coefficients.

See also CONDUCTION; ELECTRON ENERGY STATES IN SOLIDS AND LIQUIDS; GALVANOMAGNETIC AND RELATED EFFECTS.

BIBLIOGRAPHY

A. A. Abrikosov, *Introduction to the Theory of Normal Metals*. Academic Press, New York, 1972. (A)

C. Kittel, *Introduction to Solid State Physics*. Wiley, New York, 1966. (E)

E. H. Putley, *The Hall Effect and Related Phenomena*. Butterworth, London, 1960. (I)

Hall Effect, Quantum

Klaus von Klitzing

The most exciting result of the quantum Hall effect (QHE) is the fact that from measurements on microelectronic devices a new type of electrical resistance can be deduced [quantized Hall resistance (QHR)] with a value which is independent of the material and microscopic details of the conductor. High-precision measurements in different countries demonstrated that within the experimental uncertainty of about 2×10^{-7} the same value R_K is found for the quantized Hall resistance, and on the basis of a recommendation of the Comité Consultatif d'Electricite (September 9, 1988) the best value for R_K is given as $R_K = 25\,812.807\ \Omega$ with an uncertainty of $\Delta R_K = \pm 0.005\ \Omega$. Up to now, all theories of the QHE show that the quantized Hall resistance is identical with the fundamental constant h/e^2 (h = Planck constant, e = elementary charge) and the table of recommended values of fundamental constants (1986) uses the expression quantized Hall resistance for the unit h/e^2 and gives the value $25\,812.805\,6 \pm 0.0012\ \Omega$. The quantity h/e^2 is directly proportional to the Sommerfeld fine-structure constant (the proportionality constant is a fixed number which is known without any uncertainty) and therefore the title of the first publication in 1980 about the quantized Hall effect was "New method for high precision determination of the fine structure constant . . .".

Since January 1, 1990 all calibrations of resistance are based on a fixed value of the quantized Hall resistance with $R_{K-90} \equiv 25\,812.807\ \Omega$ (von Klitzing constant). Experimentally, the quantized Hall resistance is observed in Hall-effect measurements at low temperature T and high magnetic fields B (typically T = 2 K and B = 10 T) on two-dimensional electronic systems. Two dimensional means that the electrons are free to move within a plane but have a fixed energy for the motion in the direction perpendicular to the plane.

Normally silicon field-effect transistors or GaAs–AlGaAs heterostructures are used as two-dimensional electronic systems for the investigation of the QHE. The electrons in these devices are confined in such a thin layer close to the interface Si–SiO_2 or GaAs–AlGaAs that the energy of the electrons for the motion perpendicular to the interface becomes quantized into well-separated electric subbands E_i. At low temperatures only the lowest electric subband E_0 is occupied with electrons and the energy of the electrons can be written

$$E = E_0 + \frac{\hbar^2 k_\parallel^2}{2m}. \tag{1}$$

The second term in this equation characterizes the free motion of the electrons (effective mass m and momentum $\hbar k_\parallel$) within the plane parallel to the Si–SiO_2 or GaAs–AlGaAs interface (x-y plane). The free motion of the electrons within the x-y plane is drastically changed if a strong magnetic field B_z is applied perpendicular to the two-dimensional electronic systems. The motion of the electrons on closed cyclotron orbits leads to a quantization in the energy (Landau levels) comparable with the discrete energies of electrons in the hydrogen atom. Under this condition the energy spectrum of the electrons becomes discrete,

$$E_n = E_0 + (n + \tfrac{1}{2})\hbar\omega_c, \qquad n = 0, 1, 2 \ldots. \tag{2}$$

The cyclotron energy $\hbar\omega_c = eB_z/m$ is typically of the order of 10 meV at B_z = 10 T. Each cyclotron orbit occupies an area of h/eB_z within the x-y plane which leads to a degeneracy factor per unit area for each Landau level of $N = eB_z/h$. Since the area of a cyclotron orbit is relatively small (the cyclotron radius at B_z = 10 T is only 8 nm), a very large number of electrons (2.4×10^{11} cm^{-2} at 10 T) can occupy each Landau level.

The classical Hall effect which relates the Hall voltage U_H (measured perpendicular to the magnetic field B_z and the current direction I) to the carrier density n becomes independent of the magnetic field and the dimension of the sample if for a two-dimensional electron gas, the carrier density agrees exactly with an integer number i of fully occupied Landau levels. Under this condition the Hall voltage has the value

$$U_H = \frac{h}{ie^2} I, \qquad i = 1, 2, 3, \ldots. \tag{3}$$

where the proportionality constant between the Hall voltage and the current through the sample is the quantized Hall resistance R_H.

Figure 1 shows a typical result of Hall effect and resistivity measurements on a two-dimensional electron gas (2DEG). A GaAs–$Al_{0.3}Ga_{0.7}As$ heterostructure was used in this experiment, because a fully quantized energy spectrum, necessary for the observation of the quantized Hall resistance, is already obtained at a relatively low magnetic field: the cyclotron energy for electrons in GaAs is much larger than in silicon for the same magnetic field.

The Hall voltage U_H and the voltage drop U_p between the potential probes are measured under constant current conditions (I = const) as a function of the magnetic field applied perpendicular to the plane of the 2DEG. The oscillation in $R_x = U_p/I$ originates from variations in the filling factor of the energy levels E_n in a magnetic field. Whenever R_x becomes zero, an integer number i of energy levels E_n is occupied and the Hall resistance adopts the value $R_H = h/e^2 i$. Maxima in R_x are observed at approximately half-filled energy levels.

The simple theory discussed before cannot explain the flat regions in the $U_H(B)$ measurements (Hall plateaus). For a

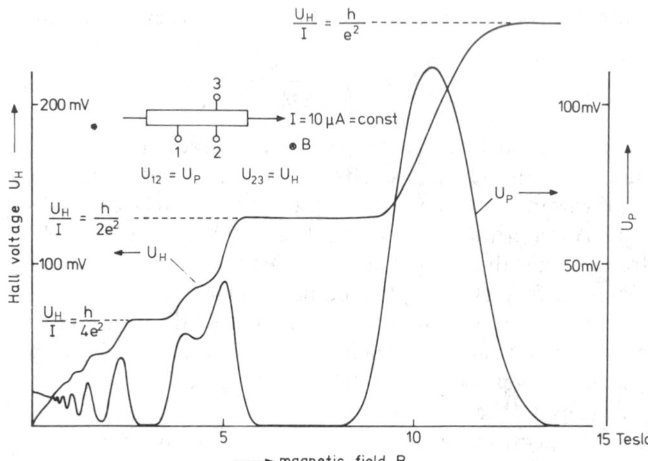

FIG. 1. Hall voltage (U_H) and resistivity ($\rho_{xx} \sim U_p$) data for a GaAs–$Al_{0.3}Ga_{0.7}As$ heterostructure as a function of the magnetic field B at a temperature of $T = 1.6$ K. Steps in the Hall voltage are visible with resistance values $R_H = h/e^2i$ at magnetic fields where U_p becomes zero.

fixed surface carrier density n only at well-defined magnetic field values $B_i = hn/ei$ should the Hall resistance be expressed by $R_H = h/ie^2$. A number of theoretical papers have discussed the phenomena of Hall *plateaus* and the authors conclude that potential fluctuations within the area of the sample (for example due to ionized impurities) lead to a localization of electrons which stabilize the Hall resistance at the quantized values. The plateaus should disappear for an ideal system. Experimentally the width of the plateaus becomes smaller if the number of impurities is reduced (higher mobility of the electrons). Simultaneously, new phenomena related to the fractional quantum Hall effect become more pronounced which lead to Hall plateaus not only at integer values of i in the equation $R_H = h/ie^2$ but also at "fractional values" of i like $i = \frac{1}{3}, \frac{1}{5}, \frac{1}{7}$.

The origin of these additional Hall plateaus are new gaps in the energy spectrum due to the electron–electron interactions. The appearance of (integer or fractional) Hall plateaus is always connected with a vanishing energy dissipation (vanishing voltage drop U_p in the current direction I as shown in Fig. 1). The existence of such supercurrents forms the basis of different theories which are able to explain the quantum Hall effect in a more general way. The experimental finding that a quantized longitudinal resistance $R = h/ie^2$ ($i = 2, 4, 6, \ldots$) is also observed *without* magnetic field if the width of the two-dimensional system is so small (≈100 nm) that an additional quantization in the energy spectrum of the electrons becomes effective (one-dimensional channel for the electron motion) favors an interpretation of the quantum Hall effect on the basis of one-dimensional channels along the edges of the sample. In this picture the interior of the two-dimensional sample is only a reservoir of localized electronic states. It seems that the edges of the sample are extremely important for an explanation of the quantum Hall effect but at present there exists no experimental evidence that the current in QHE experiments flows exclusively along the edges of the devices.

BIBLIOGRAPHY

R. E. Prange and S. M. Girvin (eds.), *The Quantum Hall Effect*. Springer-Verlag, New York, 1987.

K. von Klitzing, "The quantized Hall effect" (Nobel Lectures in Physics 1985), *Rev. Mod. Phys.* **58,** 519 (1986).

Hamiltonian Function

R. H. Good, Jr.

The Hamiltonian is the function that, through the way that it depends on its arguments, specifies the time development of a system.

Consider first a classical (nonquantum) system with coordinates q_i and momenta p_i. The Hamiltonian equations of motion are

$$\frac{dq_i}{dt} = \frac{\partial H}{\partial p_i}, \qquad \frac{dp_i}{dt} = -\frac{\partial H}{\partial q_i}. \tag{1}$$

Knowing the functional dependence of the Hamiltonian on coordinates and momenta, $H(q_i,p_j)$, we have here a set of differential equations that determine the time dependence of the coordinates and momenta. An example is a system with one coordinate, one momentum, and

$$H(q,p) = (p^2/2m) + V(q). \tag{2}$$

The Hamiltonian equations of motion lead to

$$\frac{dq}{dt} = \frac{p}{m}, \qquad \frac{dp}{dt} = -\frac{\partial V}{\partial q}. \tag{3}$$

The first equation serves to define the momentum in terms of the velocity; if the momentum is eliminated, then

$$m\frac{d^2q}{dt^2} = -\frac{\partial V}{\partial q}, \tag{4}$$

which is Newton's second law for the motion of a particle with coordinate q in a potential field $V(q)$. In general the Hamiltonian equations of motion are equivalent to Newton's equations. The Hamiltonian equations apply to a variety of problems and they permit general discussions of equations of motion and transformation between systems, the detailed descriptions of the systems being relegated to the functional dependence of the Hamiltonian. Because of the formal elegance of the equations and the discussions they lead to, they are often called the canonical equations of motion. Another aspect of the role of the Hamiltonian as determining the evolution in time is the theorem

$$\frac{du}{dt} = [u,H]. \tag{5}$$

Here $u(q_i,p_j)$ is any function of the coordinates and momenta and the right-hand side is the Poisson bracket of the two functions, defined by

$$[u,H] = \sum_i \left(\frac{\partial u}{\partial q_i}\frac{\partial H}{\partial p_i} - \frac{\partial u}{\partial p_i}\frac{\partial H}{\partial q_i}\right). \tag{6}$$

Next consider a quantum-mechanical system. Here also there are classical-type coordinates q_i, momenta p_i, and a Hamiltonian $H(q_i,p_j)$ that describes the system to be studied. Now, however, these are all operators, the coordinates and momenta satisfying

$$[p_i,q_j] = -i\hbar\delta_{ij} \tag{7}$$

where the brackets indicate the commutator $[p_i,q_j]=p_iq_j-q_jp_i$. In a quantum-mechanical system there may also be nonclassical coordinates, as for the spin of a particle. In the Schrödinger picture the state of the system is described by a time-dependent wave function $\psi(q_i,t)$. The operators act on the wave function, and the commutation rules, Eqs. (7), are realized by identifying p_i as $-i\hbar\, \partial/\partial q_i$. The Hamiltonian again determines the time evolution of the system, here through the Schrödinger equation

$$H\psi = i\hbar \frac{\partial\psi}{\partial t}. \tag{8}$$

For example, if the system has the Hamiltonian of Eq. (2), the equation for the time development of the wave function is

$$-\frac{\hbar^2}{2m}\frac{\partial^2\psi}{\partial q^2} + V\psi = i\hbar\frac{\partial\psi}{\partial t}. \tag{9}$$

The Hamiltonian will be a Hermitian operator in some sense, so that the wave function can be normalized. In the example above H is Hermitian in the sense that

$$\int \psi_1{}^*H\psi_2\, dq = \int (H\psi_1)^*\psi_2\, dq \tag{10}$$

where ψ_1 and ψ_2 are any two wave functions that go to zero at the limits of the integration, and then it follows from Eq. (8) that

$$\frac{d}{dt}\int \psi^*\psi\, dq = 0, \tag{11}$$

so that the normalization condition

$$\int \psi^*\psi\, dq = 1 \tag{12}$$

can be assigned independent of time. In the Heisenberg picture the time dependence is moved from the wave function to the operators. For any operator $u_S(q_i,p_j)$ in the Schrödinger picture the Heisenberg operator is defined by

$$u_H(t) = e^{-Ht/i\hbar}u_S e^{Ht/i\hbar} \tag{13}$$

and the Heisenberg operator has time dependence such that

$$\frac{du_H}{dt} = (i\hbar)^{-1}[u_H,H]. \tag{14}$$

Here again the Hamiltonian determines the time development. There is a close parallel between classical mechanics, as expressed by Eq. (5), and quantum mechanics in the Heisenberg picture, as expressed by Eq. (14). The Poisson bracket and $(i\hbar)^{-1}$ times the commutator play corresponding roles.

In classical mechanics and nonrelativistic quantum mechanics the Hamiltonian is the total energy of the system. For example, when Eq. (2) applies, H is the kinetic energy plus the potential energy. In relativistic quantum mechanics this interpretation cannot be applied straightforwardly. In Dirac's theory of the electron–positron, for example, the Hamiltonian for a free particle is

$$H = c\boldsymbol{\alpha}\cdot\mathbf{p} + mc^2\beta, \tag{15}$$

where $\boldsymbol{\alpha}$ and β are 4×4 matrices satisfying

$$\alpha_i\alpha_j + \alpha_j\alpha_i = 2\delta_{ij},$$
$$\alpha_i\beta + \beta\alpha_i = 0, \qquad \beta^2 = 1.$$

Consider eigenstates of the operator $\mathbf{p}$ and use the same symbol for the eigenvalue. The eigenvalues of H are found to be $\pm(p^2c^2+m^2c^4)^{1/2}$. We cannot identify H as the energy operator without further discussion, in view of the negative eigenvalues. One way out of the difficulty is Dirac's hole theory. He suggested that H be identified as the energy operator and that the vacuum consists of all the positive-energy states empty, all the negative-energy states filled. A physical electron is the vacuum plus a particle in a positive-energy state; a physical positron is the vacuum minus a particle in a negative-energy state. Both electron and positron then have positive energy. Another way out of the difficulty is to abandon the interpretation of H as the energy operator. We can define $|H|$ as the operator $(p^2c^2+m^2c^4)^{1/2}$, having always positive eigenvalues, and identify it as the energy operator. The operator $H/|H|$ has eigenvalues $+1$ and -1, corresponding to electron states and positron states.

See also DYNAMICS, ANALYTICAL; KINEMATICS AND KINETICS; QUANTUM MECHANICS; SCHRÖDINGER EQUATION.

BIBLIOGRAPHY

H. C. Corben and P. Stehle, *Classical Mechanics*, 2nd ed. Wiley, New York, 1960.

P. A. M. Dirac, *The Principles of Quantum Mechanics*, 4th ed. Oxford, London and New York, 1958.

H. Goldstein, *Classical Mechanics*, 2nd ed. Addison-Wesley, Reading, Mass., 1980.

Heat

J. W. Morris, Jr.

In thermodynamics the term heat is used to denote the quantity of energy exchanged through thermal interaction between the system of interest and its environment. If, for example, a flame is applied to a cool body of water, the energy content of the water increases, as evidenced by its increased temperature, and we say that heat has passed from the flame to the water. If energy losses from the water to the atmosphere may be ignored, the heat transferred is numerically equal to the energy gained by the water.

In more complex processes, which may involve mechanical as well as thermal interaction, the heat transferred is more difficult to identify. In fact, much of the history of thermodynamics concerns the slow evolution of clear concepts of heat, energy, and entropy, and of the distinctions between them. In the resulting science of thermodynamics heat is defined and measured indirectly in terms of the objectively measurable quantities energy and mechanical work.

Specifically, thermodynamics postulates the conservation of energy (the first law of thermodynamics) and distinguishes two general ways in which the conserved quantity, energy, may be transferred from one system to another. First, the interacting systems may do work on one another through the action of the forces imposed by their mechanical, electromagnetic, or chemical interaction. Second, the energy may be exchanged without sensible work and is in this case called heat. The change in the energy (ΔE) of a system is objectively measurable; since energy is a state function (i.e., is uniquely determined by the thermodynamic state of the system) the net change in energy may be found by identifying the initial and final states of the system and computing the difference between their associated energies. The net work (W) done on a system is also objectively measurable through the rules supplied by the science of mechanics. The heat, in contrast, is not directly measurable. Thermodynamics rather defines the heat supplied to a system as the difference between its energy change and the work done on it:

$$Q \equiv \Delta E - W. \quad (1)$$

The concept of a thermal interaction is then defined, in an equally indirect way, as a nonmechanical interaction that results in the transfer of heat.

Since heat and work are forms of the same physical quantity, energy, and since heat is not directly measurable, we might wonder why it is important to retain the concept of heat in modern physics. The answer is contained in the second law of thermodynamics, which defines a new state function, the entropy (S) of the system. The change in the entropy of a system is sensitive to the manner in which energy is supplied; it does not necessarily change when work is done on the system, but necessarily increases when heat is added. The entropy change governs the reversibility of thermodynamic processes. When two systems exchange energy solely in the form of work the thermodynamic process may, in theory, be reversed to reestablish their initial states. However, when the thermodynamic process involves an exchange of heat or the conversion of work into heat via friction, the process generally cannot be reversed without causing a corresponding irreversible change in some third system.

See also ENERGY AND WORK; ENTROPY; THERMODYNAMICS, EQUILIBRIUM; THERMODYNAMICS, NONEQUILIBRIUM.

Heat Capacity

Norman Pearlman

An intrinsically positive, extensive property of matter, defined as the limiting ratio of heat input ΔQ to temperature increment ΔT, is the heat capacity

$$c = \lim_{\Delta T \to 0} \Delta Q / \Delta T.$$

Constraints, such as constant volume or pressure, are usually indicated by subscripts. Tabulated values are ordinarily heat capacity per mole. The quantity normally measured is c_p, which is related to c_v by

$$c_p - c_v = VTB\beta^2$$

(V is volume, β the thermal expansion coefficient, B the bulk modulus). This is related to energy differences,

$$\Delta E = \int_{T_1}^{T_2} c_v(T)\, dT,$$

and, according to the constraint, to entropy differences,

$$\Delta s_{p,v} = \int_{T_1}^{T_2} \frac{c_{p,v}(T)}{T}\, dT,$$

thereby yielding information on the excitations concerned. For instance, the temperature-independent molar heat capacity

$$C_v = (\nu/2)R$$

(ν is an integer; R the gas constant per mole, $R = N_A k_B$, with N_A Avogadro's number and k_B Boltzmann's constant) observed for many gases and solids at room temperature corresponds to the classical equipartition energy $k_B T/2$ per degree of freedom, of which there are ν per molecule. In solids, $\nu = 6$ accounts for lattice energy in terms of normal-mode harmonic oscillators. Their energy, and hence also their heat capacity, vanishes exponentially as T approaches zero, as required by the third law of thermodynamics. For the quadratic oscillator frequency distribution assumed in Debye's theory,

$$(C_v/R)_{\text{lattice}} = (12\pi^4/5)(T/\theta_D)^3, \quad T \ll \theta_D$$

(θ_D, the Debye temperature, is $h\omega_m/k_B$, with ω_m the maximum oscillator frequency). It is possible to relate θ_D to lattice elastic properties. The electrons in normal metals form a degenerate Fermi gas, and so their heat capacity contribution can be observed only at low temperatures, where $(C_v)_{\text{lattice}}$ has become negligible. The electronic contribution, $(C_v)_{\text{electronic}}$, is found to be linear in T, as calculated in free-electron theory,

$$(C_v/R)_{\text{electronic}} = \gamma T/R = (\pi^2/2)(T/T_F).$$

The Fermi degeneracy temperature T_F is related to the Fermi level E_F by $T_F = E_F/k_B$, and thereby to the density of electron states per electron at the Fermi level,

$$D(E_F) = \frac{3}{2} E_F$$

and hence also to the electron effective mass, which is proportional to $D(E_F)$. In superconductors, below the transition temperature T_c the ground state is separated from excited states by an energy gap E_g which, according to BCS theory, is zero at T_c and increases to about $3.5 k_B T_c$ at 0 K. In consequence, there is a jump in $(C_v)_{\text{electronic}}$ at T_c:

$$(C_v)^{\text{supercond}}_{\text{electronic}}(T_c)/\gamma T_c = 2.43$$

according to BCS theory. Also, the heat capacity vanishes exponentially rather than linearly as T approaches zero:

$$(C_v)^{\text{supercond}}_{\text{electronic}}(T)/\gamma T_c = a e^{-bT_c/T}$$

where both a and b vary with T. While in a superconductor the ground state and the energy gap are collective properties of all the electrons due to their interaction with the lattice, a number of interactions (magnetic, crystal field, etc.) involving isolated degrees of freedom also produce a gap, ΔE, between the ground state and the first excited state. In terms of the temperature parameter $T_S = \Delta E/k_B$, the heat capacity of such a mode (called a "Schottky anomaly") is

$$c/k_B = (T_S/T)^2 e^{T_S/T}/(1+e^{T_S/T})^2.$$

Unlike the contributions described earlier, this is not monotonic. It has a peak at $T=0.416T_S$, decreasing, as $(T_S/T) \times e^{-T_S/T}$ for $T<T_S$, and as $(T_S/2T)^2$ for $T>T_S$. Since the excited state is empty at $T=0$ K while its occupancy equals that of the ground state for $T \gg T_S$, the total area under c/T is the entropy difference corresponding to this rearrangement, or $k_B \ln 2$.

See also FERMI–DIRAC STATISTICS; LATTICE DYNAMICS; THERMODYNAMICS, EQUILIBRIUM.

BIBLIOGRAPHY

C. Kittel, *Introduction to Solid State Physics*, 5th ed. Wiley, New York, 1976.

M. W. Zemansky, *Heat and Thermodynamics*. McGraw-Hill, New York, 1968.

Heat Engines

M. Garbuny

Systems that receive heat and convert it into mechanical work in a periodic process are called heat engines. The heat is supplied to a working substance externally, as in the steam turbine, or internally, as in the gasoline engine. At constant volume the working substance, which may be either a gas or a liquid, experiences an increase of its internal energy U, where U represents the sum of the molecular kinetic energies due to thermal motion and of the potential energies caused by molecular fields. However, if the working substance is allowed to expand its volume V against a moving piston or turbine blade at a pressure P, it will convert part of the received heat into work $W = \int P(V)\, dV$. The extent of this conversion is governed by the first and second laws of thermodynamics.

The first law, a manifestation of the energy conservation theorem, asserts that heat δQ_{in} supplied to a system is, in the limit of lossless operation, just equal to the sum of the internal energy increase in the system and the work performed by the system or, in differential form, $\delta Q_{in} = dU + P\, dV$. In executing a cycle, a system returns to its starting point, i.e., to the original values of pressure, volume, and temperature T. Therefore the contribution of heat to the internal energy after a complete cycle is zero, and the first law then yields the relation $W = Q_{in} - Q_{out}$, where Q_{out} is the heat not converted into work. A *thermal efficiency* η can now be defined as the ratio W/Q_{in}, i.e., $\eta = 1 - Q_{out}/Q_{in}$.

The existence of residual heat Q_{out} after the completion of a thermal engine cycle is a requirement imposed by the second law. The mathematical formulation of this law introduces the *entropy* S defined by the differential $dS = \delta Q_r/T$, where δQ_r is an amount of heat converted *reversibly*, so that reversing the heat transfer reestablishes conditions inside and outside the system to those before the conversion. The second law asserts that dS is an exact differential, so that after a cycle completed by any thermal system with any working substance, $\oint \delta Q_r/T = 0$. Thus an engine cannot completely convert a heat input Q_{in} into work W. If Q_{in} is obtained from a heat source at a temperature T_H, then for reversible conversion into work, the system requires a heat sink at a lower temperature T_L, so that $\oint \delta Q_r/T = (Q_{in}/T_H) - (Q_{out}/T_L) = 0$. Thus $Q_{in}/T_H = Q_{out}/T_L$. With the first law, this yields the *Carnot* thermal efficiency

$$\eta_r = 1 - T_L/T_H. \tag{1}$$

According to Eq. (1), the thermal efficiency of a reversible cycle depends only on the temperatures T_H of the source and T_L of the sink. An ideal cycle operating reversibly between two heat reservoirs at T_H and T_L was first proposed in 1824 by Sadi Carnot. The operation of the cycle is shown by the pressure–volume (P–V) diagram in Fig. 1a and is most easily visualized as applied to a gas passing through appropriate engine components such as cylinders with pistons and heat exchangers. Starting at point 1, the gas is compressed *adiabatically*, i.e., isolated from heat exchange with the surroundings (constant entropy), until its temperature has increased from T_L at point 1 to T_H at point 2 as the result of converting compressional work into heat. In the process 2–3, the gas is brought into thermal contact with a heat reservoir

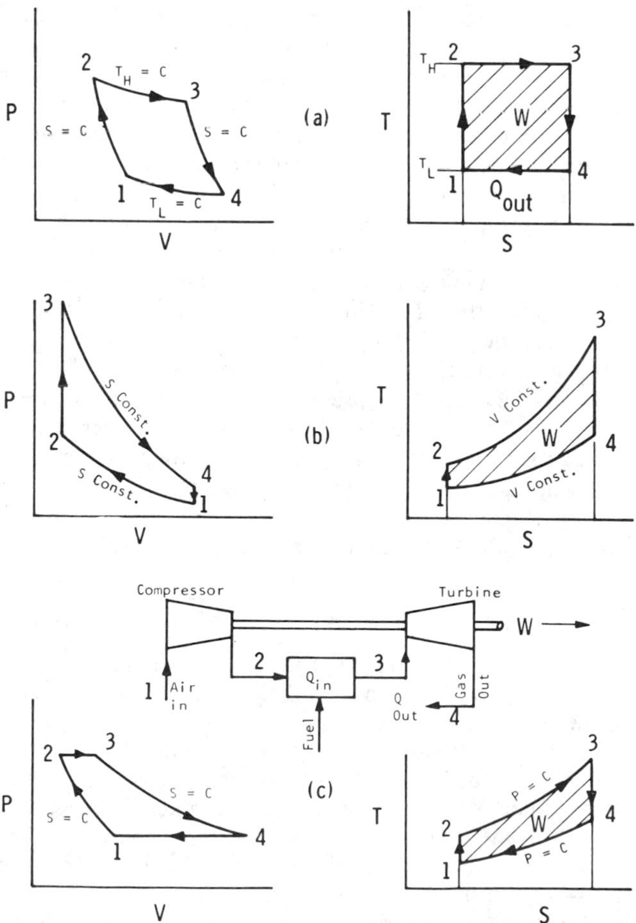

FIG. 1. P–V and T–S diagrams of thermal engines: (a) Carnot cycle; (b) Otto cycle (gasoline engine); (c) Brayton cycle (gas turbine).

at temperature T_H and expands isothermally, i.e., without change of temperature. At point 3, the gas continues its expansion adiabatically until, because of the partial conversion of heat into work, it has cooled to the temperature T_L at point 4. In the final process 4–1, the gas is brought into thermal contact with a heat reservoir at temperature T_L and compressed isothermally, so that at point 1 it assumes the values of pressure, volume, and temperature it had at the start of the cycle. If the four steps are performed reversibly, the gas receives heat Q_{in} isothermally at T_H and rejects heat Q_{out} isothermally at T_L, while the two adiabatic processes contribute nothing to a heat exchange with the environment of the gas. The conditions required for achieving the Carnot efficiency η_r according to Eq. (1) are therefore fulfilled. The area enclosed by the *P–V* diagram of steps 1–2–3–4–1 is representative of the work $\oint P\,dV$ performed by the system, but since the diagram yields no information on the isothermal heat exchange, it provides no indication of the thermal efficiency. That information is obtainable, however, from the temperature–entropy (*T–S*) diagram shown at the right in Fig. 1a. Since, by definition, $\delta Q_r = T\,dS$, the area under a curve $T(S)$ for which dS increases indicates heat Q_{in} received, while that under a curve for which dS decreases indicates heat Q_{out} rejected. Therefore the area $\oint T\,dS$ enclosed by 1–2–3–4–1 represents the work W produced reversibly, and the ratio of this area to that under $T(S)$ for $dS>0$ (i.e., 2–3) is equal to the thermal efficiency. For the Carnot cycle the efficiency is read directly as $(T_H - T_L)/T_H$.

Since the efficiency of a reversibly operating thermal engine depends only on T_H and T_L, that of the Carnot cycle cannot be exceeded. In fact, however, reversibility of the various processes can only be approached by infinitely slow and frictionless operation. In actual operation, the upper temperature of the gas must be somewhat lower than T_H to produce the required heat transfer. To invert the process by isothermal compression, the gas temperature must be somewhat higher than T_H. Thus somewhat less work is gained than must be applied to reverse the process, but complete reversibility is then not achieved. Similar considerations apply to the other three processes of the Carnot cycle, with the effect that the areas enclosed by indicator diagrams are larger than those utilized by the gas, so that the actual efficiencies are smaller than the ideal η_r. The susceptibility of an engine to losses incurred by irreversible processes is large for small values of a merit factor called the *work ratio* r_W,

$$r_W = \frac{W_{out} - \Sigma W_{in}}{W_{out}} = \frac{W}{W_{out}}. \qquad (2)$$

The work $W = \oint \delta Q_r$ ideally produced is the difference of the work output W_{out} and the sum ΣW_{in} of energies required for isothermal and adiabatic compression. W_{out} and W_{in} are the work contributions expected from the ideal Carnot cycle, but because of losses, the actual work required is always larger than W_{in} and the actual output is smaller than W_{out}. Therefore, if r_W is small, the practically achieved $(r_W)_{pract}$ may be negative. The work ratio of a Carnot engine using a gas is quite small, since the work contributions of adiabatic and isothermal compression and expansion are large compared to the net work generated. Despite the thermally ideal cycle, Carnot gas engines are not practical. An improvement results in a Carnot engine using steam. In this case the steps 2–3 and 4–1 in the *T–S* diagram of Fig. 1a represent, respectively, transitions from the liquid to the vapor phase in a boiler and from the vapor to the liquid phase in a condenser. Both processes occur at constant temperature and pressure. Step 1–2 is adiabatic compression of the liquid requiring relatively little work and step 3–4 represents adiabatic expansion of the vapor, which performs work on a piston or turbine. Practical implementation, however, is difficult for a variety of reasons. Therefore, practical steam turbines (and the older steam engines) use the *Rankine cycle*, which differs from the Carnot steam cycle mainly in step 1–2 of the *T–S* diagram. This step consists here of compressing the liquid to the pressure at point 2 and heating it at that pressure to the saturation (boiling) temperature T_H at point 2. Important alternatives of the Rankine cycle include the use of "superheated" steam and operation above the triple point. Although the thermal efficiency of the Rankine cycle is less than that of the Carnot cycle, practical efficiencies of 40–45% result from the much larger work ratio.

In addition to steam (or vapor) power cycles, in which the working substance alternates between the liquid and vapor states, there exists the general class of gas power cycles, in which the working substance, although usually injected as a fluid, remains in the gaseous state throughout the cycle. These cycles are particularly suitable for engines of relatively small weight, and for powers below about 10^4 kW, such as those used for ground and air transportation. Most of these engines operate by internal combustion, and their design must be primarily aimed at optimizing the conversion of the available chemical energy rather than at approaching a Carnot cycle. Nevertheless, combustion should begin at the highest, and heat rejection at the lowest, possible temperature. The conventional thermodynamic engineering approach to an analysis of these cycles is the air-standard approximation, which treats the working substance as an ideal gas and combustion as simple heat addition.

The Otto cycle, represented by the *P–V* and *T–S* diagrams of Fig. 1b, is the model for the operation of the gasoline engine. After an intake stroke (not shown), the gas is adiabatically compressed in step 1–2, receives heat (by spark-initiated explosion) at constant volume in step 2–3, expands adiabatically against a piston (power stroke) in step 3–4, and rejects heat (after passing through the exhaust valve) at constant volume in step 4–1. The cycle is simple because two steps involve only heat, but not work, and the two other steps involve work, but not heat. The air-standard analysis (assuming temperature-independent specific heats, c_P and c_V) yields $T_2/T_1 = T_3/T_4$ and $\eta_r = 1 - r^{1-\kappa}$ for a compression ratio r and $\kappa = c_P/c_V$.

The Brayton cycle represents the model for the operation of the basic gas turbine process as shown in Fig. 1c. In step 1–2, air drawn into the compressor is compressed adiabatically. In step 2–3, the air is mixed with fuel in the combustion chamber and receives heat at constant pressure. Step 3–4 represents adiabatic expansion, which drives the turbine. In step 4–1, heat is rejected at constant pressure. The air-standard analysis yield for the thermal efficiency, at a pressure ratio r_p, $\eta_r = 1 - r_p^{(1-\kappa)/\kappa}$.

The diesel engine is in many respects similar to the gasoline engine, but operates without spark ignition at very high compression ratios. After the intake, air is compressed adiabatically and reaches a temperature exceeding the ignition point of the fuel that is injected when the piston begins to reverse its motion. Fuel is injected at such a rate that its consumption and the expansion of the volume occur at approximately constant pressure. The rest of the cycle proceeds as in steps 3–4 and 4–1 of the Otto cycle. The *T–S* diagram is a hybrid of the Otto and Brayton cycles in that heat is received at constant pressure and rejected at constant volume.

See also CARNOT CYCLE; THERMODYNAMICS, EQUILIBRIUM.

BIBLIOGRAPHY

J. B. Jones and G. A. Hawkins, *Engineering Thermodynamics*. 2nd ed., Wiley, New York, 1986. (I)

E. F. Obert and R. L. Young, *Elements of Thermodynamics and Heat Transfer*. 2nd ed., reprint of 1962 edition, Krieger, Melbourne, FL, 1980. (I)

M. W. Zemansky and R. Dittman, *Heat and Thermodynamics*. 6th ed. McGraw-Hill, New York, 1981. (I)

R. Becker, *Theory of Heat*, 2nd ed., revised by G. Leibfried, Springer-Verlag, Berlin, 1967. (A)

Heat Transfer

Alan J. Chapman

INTRODUCTION

Heat is defined as that energy exchange which occurs as the result of a temperature difference or a temperature gradient. The term heat transfer is used to denote that body of knowledge which attempts to describe, physically and mathematically, the rate of exchange of heat between bodies, various parts of bodies, fluids, etc., when the appropriate physical conditions are specified. The two fundamental mechanisms of heat transfer are known as *conduction* and *radiation*. However, a third mechanism is customarily identified when the conduction process is coupled with the macroscopic motion of a fluid medium; it is termed *convection*. Although problems of physical interest exist in which only one of these mechanisms is present, there are many for which a complex simultaneous accounting for two or more must be made. Perhaps the most complex heat transfer problems are those in which the situation is further complicated by the simultaneous occurrence of a phase change process such as condensing, boiling, solidification, or sublimation.

CONDUCTION

Heat conduction is the exchange of heat from one body at a given temperature to another body at a lower temperature with which it is in contact, or between portions of a body at different temperatures, by the transfer of kinetic energy of motion through molecular impact. The transfer of energy takes place from the more energetic (higher-temperature) molecules to the less energetic (lower-temperature) molecules. In a gas the molecules are spaced relatively far apart and their motion is random, leading to an energy transfer that is less than that observed in a liquid, in which the molecules are more closely packed. Amorphous solids exhibit conduction much like liquids; however, in dielectric solids, which form a crystal lattice, the transfer of energy by conduction is enhanced by the vibratory motion of the lattice structure itself. In solids that are also electrical conductors, heat conduction is further enhanced by the drift of free electrons within the lattice structure.

Macroscopically, the above-described transfer of heat by conduction is hypothesized to be described by *Fourier's conduction law*:

$$\frac{q}{A} = -k\frac{\partial T}{\partial n}, \tag{1}$$

in which the heat transfer rate q per unit of area A is presumed to be proportional to the gradient of the temperature T in the direction normal to the area n. The proportionality constant k is identified as the *thermal conductivity*. The thermal conductivity is a physical property of the matter composing the body under consideration and is a function of the composition of the matter, the phase in which it exists, and such state variables as pressure and temperature.

In general, the solution of a conduction problem consists of finding, in a body of given geometry and initial temperature distribution, the temperature at any point at any subsequent time. This solution is obtained from the *heat conduction equation*, which results from the combination of Fourier's law and the energy conservation principle:

$$\nabla\cdot(k\nabla T) + q^* = \rho C_p\frac{\partial T}{\partial t} \tag{2}$$

where q^* represents the volumetric internal production of heat, if any; ρ the material density; C_p the specific heat of the material; and t the time. When internal heat generation is absent and the thermal conductivity is taken as constant we have

$$\nabla^2 T = \frac{\rho C_p}{k}\frac{\partial T}{\partial t}. \tag{3}$$

A special case of common interest is that of the *steady state*, in which the temperatures are independent of time. In this instance, if internal heat generation is absent and the material properties are constant, the heat conduction equation reduces to the familiar Laplace equation:

$$\nabla^2 T = 0. \tag{4}$$

A complete solution to a heat conduction problem involves the solution of Eq. (2), or (4), subject to an initial known temperature distribution (if nonsteady) and whatever boundary conditions apply. Typically, the boundary condition commonly encountered is that of either a specified temperature on the body surface or convection from the surface into a bounding moving fluid of known temperature. The latter condition is usually represented by *Newton's law of cooling*, which states

$$\frac{q}{A} = -k\left(\frac{\partial T}{\partial n}\right)_s = -h(T_s - T_\infty) \tag{5}$$

in which the subscript "s" denotes the body surface, n the coordinate direction normal to the surface, and T_∞ the temperature of the fluid far away from the surface. The quantity h is known as the surface *heat transfer coefficient* and is the primary quantity sought in the analysis of convection.

CONVECTION

Heat convection is the term applied to the heat transfer mechanism that occurs when a fluid of some known temperature flows past a solid surface at some different temperature. The convection is termed *natural*, or *free*, when the fluid motion results from buoyant forces created by temperature (and, hence, density) differences in the fluid. It is called *forced* convection when the fluid motion is produced by imposed pressure differences or by motion of the surface itself. The flow of heat between the surface and the fluid is usually expressed in terms of the heat transfer coefficient h defined in Eq. (5).

The determination of the heat transfer coefficient necessitates the solution of the governing equations of a viscous, heat-conducting fluid. These equations in their general form are not solvable analytically, and the problem is usually simplified by introduction of the simplifications provided by boundary layer theory. In this simplified form, the fundamental equations that need to be solved near the body surface are noted in the following equations (for an incompressible fluid), where the coordinate x is measured parallel to the body surface, y is measured normal to x, and the symbols v_x and v_y denote the fluid velocities in those directions.

Conservation of mass:

$$\frac{\partial v_x}{\partial x} + \frac{\partial v_y}{\partial y} = 0. \tag{6}$$

Equation of motion:

$$v_x \frac{\partial v_x}{\partial x} + v_y \frac{\partial v_x}{\partial y} = -\frac{1}{\rho}\frac{dp}{dx} + \frac{1}{\rho}\frac{\partial \tau}{\partial y} + g\beta(T_f - T_\infty). \tag{7}$$

Conservation of energy:

$$v_x \frac{\partial T_f}{\partial x} + v_y \frac{\partial T_f}{\partial y} = \frac{1}{\rho C_p}\left(\tau \frac{\partial v_x}{\partial y}\right) - \frac{1}{\rho C_p}\frac{\partial}{\partial y}\left(\frac{q}{A}\right). \tag{8}$$

In these equations, ρ represents the fluid density, p the pressure, C_p the specific heat, and β the coefficient of volume expansion. The symbol T_f is used to denote the fluid temperature, to distinguish it from temperatures within the solid surface, and it is generally a function of position within the fluid (i.e., a function of x and y). In the momentum equation and energy equation, the quantities τ and q/A are used to denote the shear stress and heat flux, respectively, within the fluid domain. For *laminar* flow these are usually expressed in terms of the gradients of velocity and temperature within the fluid by introduction of the following hypothesized rate laws:

$$\tau = \rho\nu \frac{\partial v_x}{\partial y}, \tag{9}$$

$$\frac{q}{A} = -\rho C_p \alpha \frac{\partial T_f}{\partial y}, \tag{10}$$

where ν and α represent the kinematic viscosity and thermal diffusivity, respectively. In the instance when *turbulent* flow exists in the fluid, it is customary to augment these rate coefficients with their turbulent counterparts, the so-called eddy viscosity and eddy diffusivity.

In the equation of motion, Eq. (7), the last term represents buoyant forces arising from temperature gradients in the fluid and may be absent in cases of pure forced convection. Likewise, in the energy equation, Eq. (8), the first term on the right-hand side represents the rate of dissipation of kinetic energy into thermal energy through viscous action, and it may in certain cases be negligible.

In general, the solution to a particular convection problem consists of the simultaneous solution of Eq. (6), (7), and (8), under the appropriate boundary conditions, for the fluid temperature $T_f(x,y)$ from which we eventually ascertain the sought-for heat transfer coefficient by application of Fourier's law at the surface:

$$h = -\frac{k(\partial T_f/\partial y)_s}{(T_f - T_\infty)}, \tag{11}$$

the subscript "s" denoting the surface.

Usually, the solutions just referred to are expressed in dimensionless form. If Eqs. (6)–(11) are nondimensionalized by introducing a characteristic velocity U and a characteristic dimension L, we can deduce that the above-implied dependence of h on the other parameters may be expressed as

$$N_{NU} = f_n(N_{RE}, N_{GR}, N_{EC}, N_{PR}). \tag{12}$$

In Eq. (12) the quantities shown are dimensionless groupings of variables that bear the following designations:

Nusselt number: $N_{NU} = hL/k$;

Reynolds number: $N_{RE} = UL/\nu$;

Grashoff number: $N_{GR} = \dfrac{L^3 g\beta(T_s - T_\infty)}{\nu^2}$;

Eckert number: $N_{EC} = \dfrac{U^2}{C_p(T_s - T_\infty)}$;

Prandtl number: $N_{PR} = \nu/\alpha$.

A vast number of solutions for convection problems comprising a wide variety of surface geometries and hydrodynamic boundary conditions have been carried out and are reported in the literature. An equally vast number of empirical convection solutions have been deduced in the form of Eq. (12) by conducting appropriate experiments.

The foregoing discussion applies to an incompressible fluid for which the density is taken to be constant. When fluid compressibility is important, as might be the case in aerodynamic applications, the governing equations become more complex. However, the same dimensionless correlation implied in Eq. (12) still holds.

RADIATION

Heat transfer by thermal radiation constitutes a completely different mechanism from those discussed earlier for convection and conduction. The basic mechanism is that of electromagnetic radiation in a wavelength band usually taken to extend from 0.1 to 100 μm. Thermal radiation emitted by a heated surface is found to be a complex function of the surface temperature, the nature of the surface, the wavelength of the emission, and the particular direction of concern. If the "hemispherical" emission is considered, wherein the radiation emitted in all directions is summed together, the *monochromatic emissive power* W_λ is defined as the total energy emitted, per unit time and area, at the wavelength in question. The symbol λ is used to denote the wavelength. If the surface in question is a perfect *blackbody* (i.e., absorbs all radiant energy incident upon it), the monochromatic emissive power $W_{b\lambda}$ is known to be a function of wavelength and temperature as described by Planck's radiation law. When integrated over all wavelengths to obtain the *total* emissive power W_b, Planck's equation yields the Stefan–Boltzmann law:

$$W_b = \sigma T^4. \tag{13}$$

For nonblack surfaces, the monochromatic emissive power is found to be less than that for a blackbody at the same temperature. The ratio of these emissive powers is defined as the *monochromatic emissivity* ϵ_λ:

$$\epsilon_\lambda = W_\lambda / W_{b\lambda}. \tag{14}$$

We may also define a *monochromatic absorptivity* α_λ as the fraction of an incident radiation at a given wavelength that is absorbed by a nonblack surface. Kirchhoff's law states that, at equal surface temperatures, the monochromatic emissivity and monochromatic absorptivity are equal:

$$\epsilon_\lambda = \alpha_\lambda. \tag{15}$$

Generally, these quantities are functions of the wavelength λ. However, if a surface exhibits the characteristic by which $\epsilon_\lambda = \alpha_\lambda$ is constant, the surface is termed *gray*.

In a similar fashion, the total emissivity ϵ of a nonblack surface is defined as the ratio of the energy emitted, per unit time and area, at *all* wavelengths to the same quantity for a black surface at the same temperature. Thus, for the nonblack surface

$$W = \epsilon \alpha T^4. \tag{16}$$

Similarly, there is defined a total absorptivity α, which represents the fraction of incident radiation absorbed by a nonblack surface at all wavelengths. Kirchhoff's law in the total sense,

$$\epsilon = \alpha, \tag{17}$$

will follow from Eq. (15) only if the surface is gray, or if the incident radiation is from a black surface at the same temperature as the receiving surface.

A frequently encountered problem of practical interest is that of determining the radiant exhange between a collection of finite surfaces, black or gray, maintained at different temperatures. In such instances it is necessary to know the fraction of the radiation leaving one surface that is intercepted by another. This fraction is expressed by the *shape factor* F_{1-2}, in which 1 denotes the originating surface and 2 the receiving surface. If we presume that Lambert's law of diffuse radiation governs the spatial distribution of the emitted energy—that is, the radiant energy density varies as the inverse square of the distance from the source and as the cosine of the angle between the surface normal and the direction in question—we may deduce that the shape factor between two surfaces A_1 and A_2 of given shape and orientation is

$$F_{1-2} = \frac{1}{A_1} \int_{A_1}\int_{A_2} \frac{\cos\theta_1 \cos\theta_2 \, dA_1 \, dA_2}{\pi r_{12}^2}, \tag{18}$$

in which θ_1 and θ_2 represent the angles between the surface normals to dA_1 and dA_2 and the line connecting them, r_{12}. The shape factor is a geometric factor that depends only on the shape and orientation of A_1 and A_2. It is generally very difficult to determine analytically because of the twofold double integrals in Eq. (18). Values for many surface combinations are reported in the literature, and computer techniques are available for complex geometries.

For an enclosure consisting of a number of finite black surfaces, maintained at specified temperatures, the heat exchanged between any pair of the surfaces (say A_i and A_j maintained at T_i and T_j) is

$$q_{i-j} = A_i F_{i-j} \sigma (T_i^4 - T_j^4) \tag{19}$$

and the total energy given up by one surface is then

$$q_i = \sum A_i F_{i-j} \sigma (T_i^4 - T_j^4), \tag{20}$$

in which the summation is taken over all the surfaces in the enclosure. (In the event that a true enclosure does not exist, the vacant space may be represented by a surface at zero temperature to simulate an enclosure.) The use of Eqs. (19) and (20) requires the prior determination of the shape factors of all the surface-pair combinations in the enclosure. This may be very time consuming.

If the surfaces of an enclosure are gray instead of black, the effect of multiple reflections must be included. Likewise, if the heat flux of a surface is specified (such as zero) instead of the temperatures, the problem becomes much more complex. In such instances highly developed techniques exist that yield formulations similar to those in Eqs. (19) and (20) where F_{i-j} is replaced with complex exchange coefficients ϕ_{i-j} involving the F's, the surface emissivities, etc.

See also ABSORPTION COEFFICIENTS; BLACKBODY RADIATION; BOUNDARY LAYERS; LATTICE DYNAMICS; TURBULENCE.

BIBLIOGRAPHY

All three modes of heat transfer; applications

Alan J. Chapman, *Heat Transfer,* 4th ed. Macmillan, New York, 1984. (I)

E. R. G. Eckert and R. M. Drake, *Analysis of Heat and Mass Transfer*. McGraw-Hill, New York, 1972. (A)

J. P. Holman, *Heat Transfer,* 5th ed. McGraw-Hill, New York, 1981. (E)

F. P. Incropera and D. P. Dewitt, *Fundamentals of Heat and Mass Transfer*. Wiley, New York, 1985. (I)

M. Jakob, *Heat Transfer*. Wiley, New York, 1949 (Vol 1), 1957 (Vol. 2). (A)

Conduction

V. S. Arpaci, *Conduction Heat Transfer*. Addison-Wesley, Reading, Mass., 1966. (A)

H. S. Carslaw and J. C. Jaeger, *Conduction of Heat in Solids*, 2nd ed. Oxford, London and New York, 1959. (A)

P. J. Schneider, *Conduction Heat Transfer*. Addison-Wesley, Reading, Mass., 1955. (I)

Convection

W. M. Kays and M. E. Crawford, *Convective Heat and Mass Transfer*. 3rd ed. McGraw-Hill, New York, 1980. (A)

H. Schlichting, *Boundary Layer Theory*, 7th ed. McGraw-Hill, New York, 1979. (A)

Radiation

H. C. Hottel and A. F. Sarofim, *Radiative Transfer*. McGraw-Hill, New York, 1967. (A)

R. Siegel and J. R. Howell, *Thermal Radiation Heat Transfer*. 2nd ed. McGraw-Hill, New York, 1981. (A)

E. M. Sparrow and R. D. Cess, *Radiation Heat Transfer*. Brooks-Cole, 1966. (I)

Heavy-Fermion Materials

Douglas E. MacLaughlin

Heavy-fermion or heavy-electron compounds are metals which exhibit enormously enhanced effective conduction electron masses [1]. The classical heavy-fermion metals contain 4*f* or 5*f* elements (usually Ce or U, e.g., $CeAl_3$, UBe_{13}). The heavy-electron mass is inferred from the large enhancement of the electronic specific heat C_e; more specifically, the coefficient γ of the linear term in C_e is a factor of 100–1000 larger than in ordinary metals. (Observed values of γ for various rare-earth compounds are distributed between ~10 and ~1000 mJ K^{-2} mol^{-1}; a value greater than ~200–400 mJ K^{-2} mol^{-1} is usually arbitrarily taken to define a heavy-fermion system.) This mass enhancement (which reflects the interaction between electrons and their environment and is not a "real" change of the electron mass) is associated with strong hybridization between conduction and *f* electrons. At high temperatures the effect of hybridization is weak, and the properties of heavy-fermion systems are those of a collection of weakly-interacting localized *f* ions embedded in a normal metal, but at low temperatures heavy-fermion materials exhibit a rich variety of ground states and unusual phenomena [1,2]. Heavy-fermion systems have been discovered which are superconducting, which order magnetically, and which do not undergo a phase transition but appear to approach a metallic Fermi-liquid-like "normal" ground state at zero temperature. Experimental and theoretical interest has been particularly attracted to heavy-fermion superconductivity, which offers the possibility of exotic Cooper pairing mechanisms and previously unknown pairing symmetries [3], and to the effect of extremely strong correlations between conduction electrons characteristic of these materials.

The characteristic energy scales of all of the heavy-fermion ground states are very low (0.1–10 MeV, or 1–100 K in temperature units) compared to typical Fermi energies of ordinary metals (1–10 eV, 10^4–10^5 K). In Fermi liquids one expects on very general grounds an effective characteristic or degeneracy energy ϵ_0 (Fermi energy for a weakly interacting Fermi gas) to be given by $\epsilon_0 = p_0^2/2m^*$, where m^* is the effective mass and p_0, an effective quasiparticle momentum at the Fermi surface, is determined by the electron concentration (Luttinger theorem). $p_0/\hbar$ is typically 1–10 $Å^{-1}$ independent of m^*. Therefore, an enhanced electron mass and a reduced energy scale are equivalent descriptions of heavy-fermion behavior. This characteristic energy (or equivalent characteristic temperature $T_0 = \epsilon_0/k_B$) has been associated with the "Kondo" temperature T_K of a single magnetic impurity in a normal metal, in part because heavy-fermion metals behave approximately as "concentrated Kondo systems" for $T \gtrsim T_0$.

It is hard to understand the stability of heavy Fermi liquids even in the normal state, because the narrow bandwidths characteristic of nearly localized electronic states should be costly in energy of localization. The various ground states of heavy-fermion systems are in fact not very stable and are easily modified, e.g., by application of pressure or by doping with impurities.

It was realized some time ago [4] that the observed large discontinuity of the specific heat in heavy-fermion superconductors at the superconducting critical temperature T_c, shown in Fig. 1 for $CeCu_2Si_2$, implies that the heavy electrons themselves are involved in the superconducting Cooper pairing. (Figure 1 also shows the extremely enhanced linear specific heat in the normal state of $CeCu_2Si_2$, compared to the isostructural compound $LaCu_2Si_2$ which has no 4*f* electrons.) The fact that nearly localized electrons experience strong Coulomb repulsion when on the same site makes a conventional (BCS) phonon-mediated attraction suspect, since the latter is a local interaction and would be seriously weakened by the tendency of the electrons to avoid each other. Instead, Cooper pairing might take advantage of an interaction which is attractive at intermediate electron separations. Such an interaction, together with the on-site repulsion, would favor a non-*s*-like Cooper pair state with a node at the origin of the relative coordinate.

Experimental evidence on the nature of the pairing interaction is difficult to obtain. Consequences of unconventional Cooper pair states are somewhat less obscure, and the experimental evidence by and large supports claims for unconventional superconductivity in particular systems. This evidence includes [2,5] (a) signs of an extremely anisotropic superconducting energy gap, (b) rapid depression of T_c by nonmagnetic impurities, and (c) in two systems, $U_{1-x}Th_xBe_{13}$ and UPt_3, evidence for multiple superconducting phases. All of these phenomena are predicted for unconventional Cooper pairing [5]. Group-theoretical considerations put the various kinds of energy-gap anisotropy into two general classes: point nodes and line nodes (zeros) of the gap function $\Delta(\mathbf{k})$ on the Fermi surface. At low temperatures each of these node geometries gives rise to characteristic power-law temperature dependences of quantities, such as the ultrasonic attenuation coefficient and the nuclear spin–lattice relaxation rate, which depend on the number of thermal excitations across the gap. These power laws are more or less easily distinguished from the activated tem-

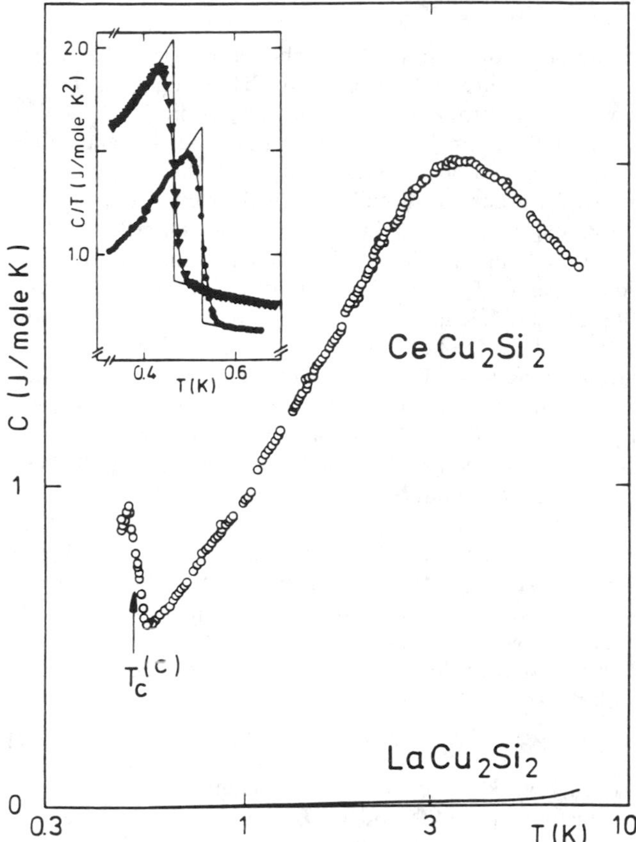

FIG. 1. Temperature dependence of the molar specific heat C of $CeCu_2Si_2$ in zero magnetic field. The arrow gives the superconducting transition temperature $T_c = 0.51 \pm 0.04$ K. Inset: C/T vs. T near T_c for two other $CeCu_2Si_2$ samples, showing the large jump. From Ref. 4, with permission.

perature dependence $\exp(-\Delta/k_B T)$ expected for a nonzero gap, and have been verified in a number of experiments.

The recent discovery of strong antiferromagnetic (AF) spin fluctuations in several heavy-fermion superconductors is consistent with unconventional pairing mediated by such spin fluctuations, as is the observation in UPt_3 of an unusual phase boundary in the H–T plane. A case has been made, however, for conventional but highly anisotropic pairing [1]. Some authors [6] have speculated that a relation exists between superconductivity in heavy-fermion metals and in high-temperature copper oxides, involving spin fluctuations of the rare-earth ions in the former and Cu^{2+} ions in the latter.

The other heavy-fermion ground states are scarcely less puzzling. Thermodynamic measurements in "normal" heavy-fermion materials suggest extremely low characteristic energies, as noted above. In particular the magnetic susceptibility tends toward a constant Pauli-type value $\chi(0)$ at low temperatures which is also enhanced compared to normal metals. Surprisingly, the so-called "Wilson ratio" $\chi(0)/\gamma(0)$ is seldom far from the value for free-electron metals. Various electron spectroscopies [photoemission spectroscopy (PES), bremsstrahlung isochromat spectroscopy (BIS), and their variants] yield information on the single-particle Green's function, which is in turn affected by interactions between the added hole or electron and the rest of the system. There is evidence for the onset of "coherence" effects (consequences of well-defined quasiparticles with long mean free paths) at a temperature T_{coh} well below the characteristic or "Kondo" temperature T_0. Coherence effects are most notable in transport properties, but also appear to influence the neutron-scattering form factor and NMR properties at low temperatures. The nature of the coherence is not well understood. The best evidence for the Fermi-liquid nature of the normal ground state is the observation of de Haas–van Alphen (dHvA) oscillations in the magnetic susceptibility of heavy-fermion metals at extremely low temperatures [2]. The dHvA effect is due to quantization of quasiparticle orbits in a magnetic field and its effect on the population of orbits at the Fermi surface, and its very existence indicates the presence of a Fermi surface and long-lived charged fermion quasiparticles in these sytems. For UPt_3 the dHvA experiments yield a Fermi surface geometry in good agreement with band structure calculations, but the measured cyclotron masses are about an order of magnitude larger than the calculated values.

Some heavy-fermion compounds undergo a transition to a magnetically ordered state. This magnetism has some experimental features in common with the itinerant AF (spin density wave) state of chromium and related systems, but differs in that the antiferromagnetism is always commensurate and of a simple structural symmetry. In addition, the ordered magnetic moment per f ion is typically much smaller than a Bohr magneton. Magnetic order and superconductivity appear to coexist in at least three compounds: UPt_3, URu_2Si_2 (with moderately heavy band electrons), and $U_{1-x}Th_xBe_{13}$, $0.018 < x < 0.045$. The latter alloys are unique in that the magnetic ordering (Néel) temperature T_N is less than the superconducting transition temperature: $T_N < T_c$. This is hard to understand on the basis of an itinerant AF state because superconductivity should produce a gap over essentially the entire Fermi surface. No low-lying quasiparticles would then remain to form a spin density wave. New evidence from magnetic resonance and neutron diffraction experiments indicates very weak magnetism in several normal and superconducting heavy-fermion materials. Essentially no first-principles theory of magnetic order in heavy-fermion systems has appeared to date.

See also DE HAAS–VON ALPHEN EFFECT; KONDO EFFECT; METALS; QUASIPARTICLES; SUPERCONDUCTING MATERIALS; TRANSITION ELEMENTS.

REFERENCES

1. For a review see P. Fulde, J. Keller, and G. Zwicknagl, in *Solid State Physics* (H. Ehrenreich and D. Turnbull, eds.), vol. 41. (Academic Press, New York, 1988). (I-A)
2. *Proceedings of the 6th Int. Conf. on Crystal Field Effects and Heavy Fermion Physics, Frankfurt, July 1988* (W. Assmus, P. Fulde, B. Lüthi, and F. Steglich, eds.). North-Holland, Amsterdam, 1988; reprinted from *J. Magn. Magn. Mat.* **76–77** (1988). (I-A)
3. See M. Sigrist and T. M. Rice, in Ref. 2, p. 487. (A)
4. F. Steglich, J. Aarts, C. D. Bredl, W. Lieke, D. Meschede, W. Franz, and J. Schäfer, *Phys. Rev. Lett.* **43**, 1892 (1979). (E-I)

5. See P. A. Lee, T. M. Rice, J. W. Serene, L. J. Sham, and J. W. Wilkins, *Comments Cond. Mat. Phys.* **12,** 99(1986). (I-A)
6. Z. Fisk, D. W. Hess, C. J. Pethick, D. Pines, J. L. Smith, J. D. Thompson, and J. O. Willis, *Science* **239,** 33 (1988). (E-I)

Helium, Liquid

R. A. Guyer

The liquid-helium systems, liquid ^{4}He, liquid ^{3}He, and liquid ^{3}He-^{4}He mixtures, are among the most extensively studied systems in physics. The primary reason for this is that the helium systems do not solidify down to arbitrarily low temperature unless pressures in excess of about 25 atm are applied. Liquid ^{3}He and ^{4}He can be carried to low temperatures and for any kind of interaction can be studied where the thermal energy, $k_B T$, is comparable to or less than the interaction energy; e.g., for liquid ^{4}He at $T \approx T_\lambda = 2.17$ K, where the normal-liquid to Bose superfluid transition occurs, or for liquid ^{3}He at $T \approx T_A = 2.7$ mK, where the Fermi-liquid to *A*-phase superfluid transition occurs, etc. The only other laboratory system which remains liquid (or gaseous) to arbitrarily low temperature is the electron gas that inhabits the ion lattice of a metal, but the properties of this gas are enormously complicated by its environment.

The helium systems are so exceptional because the helium–helium interaction (which is the same between 4-4 pairs, 3-3 pairs, and 3-4 pairs) is very weak, of order 10 K, and the mass of the atoms is very small. Thus the force tending to bond the atoms is relatively weak and the zero-point energy, proportional to $\hbar^2/m$, which is inimical to bonding and localization, is relatively large. So in the absence of a substantial external pressure the systems are liquid.

The most important temperatures encountered in describing these liquids are the temperatures at which quantum-statistical effects set in for it is below these temperatures that these liquids exhibit properties not seen in classical liquids. Liquid ^{4}He is composed of composite particles which are bosons; the Bose character of this system becomes important when the thermal de Broglie wavelength $\lambda(T) = h/p$ ($p^2/2m = k_B T$) becomes comparable to the interparticle spacing $\rho^{-1/3}$, where ρ is the number density. Liquid ^{3}He is composed of particles that are fermions; it is for $k_B T \leq k_B T_F$ [$\lambda_T(T_F) \approx \rho^{-1/3}$] that the Fermi character of the constituents leads to liquid behavior unlike that of classical liquids. For both liquid ^{4}He and liquid ^{3}He, $T_B \approx T_F \approx 1$ K.

Helium-four was first liquefied by Kamerlingh Onnes in 1908. This liquid and liquid ^{3}He, which is relatively difficult to obtain in large quantities, have been studied intensively for the past 60 years. With the discovery of superfluidity in ^{3}He at $T \leq 3$ mK, by Osheroff, Richardson, and Lee in 1972, ^{4}He was left behind, while during the next decade attention turned to the ^{3}He superfluid. Liquid ^{3}He and ^{4}He, of interest because of their unusual low-temperature properties, are currently regarded as well understood. These fluids continue to be the focus of serious research activity that more and more makes use of their properties to probe other systems.

LIQUID ^{4}He

The phase diagram of ^{4}He is shown in Fig. 1. At suitably high temperature and low pressure ^{4}He is a gas; as the temperature is lowered (at, e.g., 1 atm) a gas–liquid phase transition (of the normal variety) occurs at about 4 K and a more or less conventional liquid results. As the temperature is lowered further the liquid passes through $T_\lambda(P)$ at about 2.2 K and becomes liquid helium II, the superfluid. Above $T_\lambda(P)$ the liquid has properties similar to those of a normal liquid; below T_λ it has many unusual properties. For example, liquid ^{4}He in a bucket at the same temperature as the bath and with the bottom of the bucket having an unglazed ceramic plug will remain in the bucket at $T > T_\lambda(P)$ but pour through the bottom of the bucket as T is lowered below $T_\lambda(P)$; and at $T < T_\lambda(P)$ a temperature pulse will propagate through the liquid as a temperature wave (see Fig. 2). These phenomena are understood in terms of a phenomenologic model, the two-fluid model first introduced by Tisza, which has been very successful in describing macroscopic ^{4}He phenomena. The essential element of this model is that below T_λ the fluid, which has been unexceptional at $T > T_\lambda$, can be described as being made up of two components, a normal component with density $\rho_n(T)$ and a superfluid component with density $\rho_s(T)$ (Fig. 3):

$$\rho = \rho_n(T) + \rho_s(T). \qquad (1)$$

ρ is only weakly temperature dependent because of thermal expansion, but $\rho_s(T)$ and $\rho_n(T)$ have striking temperature dependences. The two components of the density have different properties: the normal component is like a normal liquid in that it carries entropy and has viscosity; the superfluid component carries no entropy and flows without viscosity. Superfluid ^{4}He phenomena are described by the linearized two-fluid equations (ignoring viscosity and nonlinear terms)

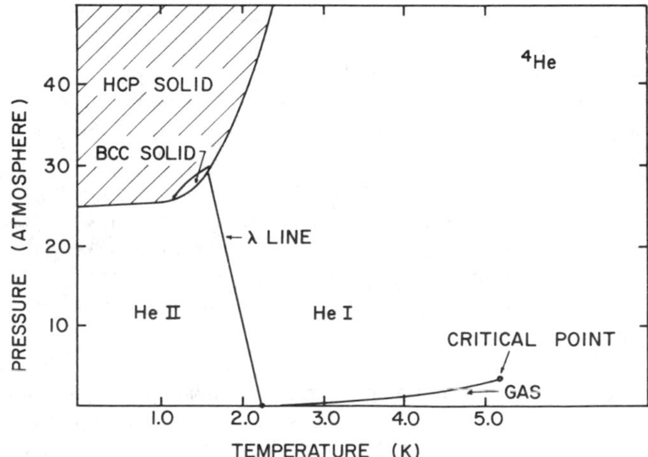

FIG. 1. Phase diagram of ^{4}He. ^{4}He remains a liquid as $T \to 0$ K unless pressures in excess of 25 atm are applied. At temperatures less than about 2 K, $P < 25$ atm, the liquid changes from a normal liquid, He I, to a superfluid, He II. Below $T_\lambda(P)$ the fluid is described as a combination of "normal" and superfluid components; it exhibits many unusual properties.

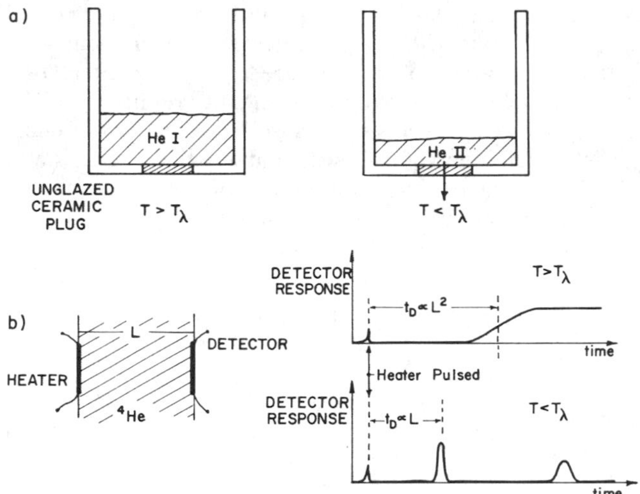

FIG. 2. Superfluid phenomena. Among the many superfluid properties of the liquid are those illustrated here. Experiment 1: liquid helium is placed in a bucket that contains an unglazed ceramic plug in the bottom. With the system (bucket, liquid, etc.) at $T > T_\lambda$ the liquid helium remains in the bucket. When the system is carried to $T < T_\lambda$ the fluid pours through the bottom of the bucket. Experiment 2: a heat pulse is propagated from a heater (a piece of carbon resistor board) through liquid helium to a detector of the same material. The temperature at the detector is recorded as a function of time from initiation of the heater pulse.

$$\rho = \rho_n + \rho_s, \tag{2}$$

$$\mathbf{j} = \rho_n \mathbf{v}_n + \rho_s \mathbf{v}_s, \tag{3}$$

$$\frac{\partial \rho}{\partial t} + \nabla \cdot \mathbf{J} = 0, \tag{4}$$

$$\frac{\partial \mathbf{j}}{\partial t} = -\nabla P, \tag{5}$$

$$\frac{\partial \mathbf{V}_s}{\partial t} = -\nabla G, \tag{6}$$

where $\mathbf{V}_n$ and $\mathbf{V}_s$ are the velocities of the two components, $\mathbf{j}$ is the momentum density, and P and G are the pressure and Gibbs free energy. The physical content of these equations is that the total fluid density is driven by the forces that normally drive a fluid but that the superfluid component obeys a special equation of motion, Eq. (6). Application of the two-fluid model with many elaborations (beyond the linear regime, with viscosity) leads to a remarkably good description of the properties of liquid ^{4}He.

In 1938 London suggested that the basic physical event that presages the superfluidity is the occurrence of Bose condensation in the liquid. Below T_λ the system goes into a single quantum-mechanical state, the Bose condensed ground state, with excitations above the Bose condensed ground state corresponding to the normal fluid. Although this picture is strictly correct only for the dilute Bose gas, it is believed that the essential elements remain unchanged at the densities of the real liquid. London's suggestion gives the two-fluid model some microscopic underpinnings. A complete understanding of the model comes from the work

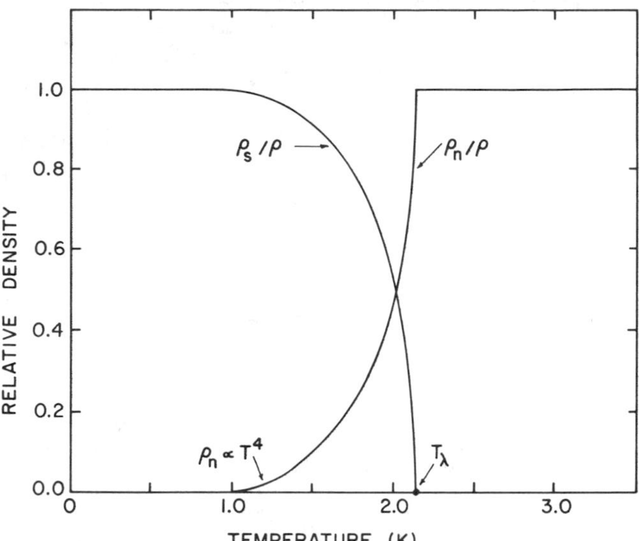

FIG. 3. Superfluid density. The superfluid component and normal fluid component of the total density are strong functions of T for $T < T_\lambda$. As $T \to 0$ K, $\rho_s \to \rho$; as $T \to T_\lambda$, $\rho_n \to \rho$. The normal fluid density is a measure of the degree of thermal excitation of the system above the ground state.

of Landau, who emphasized the special character of the ground state and excitations above it. Excitations above the ground state are identified with the normal component of the fluid and are called the phonons and rotons; they have been carefully studied in neutron scattering experiments. (See Fig. 4.) As $k \to 0$ the excitations are long-wavelength longitudinal phonons,

$$\epsilon(k) \to \hbar C_l |k|; \tag{7}$$

at $k \approx k_0$ the excitations are single-particle-like and are called rotons,

$$\epsilon(k) \to (\hbar^2/2\mu)(k - k_0)^2. \tag{8}$$

The specific character of the excitation is observable in thermodynamic measurements at $T < T_\lambda$, e.g., in the specific heat, and in two-fluid phenomena since $\rho_n(T)$, $S_n(T)$, etc., as well as the transport coefficients that enter the two-fluid equations, are given in terms of them. The normal density $\rho_n(T)$ is a measure of the density of excitations; at $T \ll 0.5$ K,

$$\rho_n(T) = \frac{2\pi^2}{45} \frac{k_B{}^4 T^4}{\hbar^3 C^5} \tag{9}$$

due to the phonons.

^{3}He

The ^{3}He phase diagram is shown in Fig. 5. Like ^{4}He, the liquid ^{3}He system can be carried to arbitrarily low temperature at pressures less than 30 atm. In the temperature range $T \approx 1$ K the liquid phase diagram has none of the exotic features of the ^{4}He phase diagram. When the system is carried to $T \lesssim 1$ K, just as in ^{4}He, it is approaching the temperatures at which quantum statistics become important. Unlike ^{4}He, this does not bring on superfluidity but rather the continuous

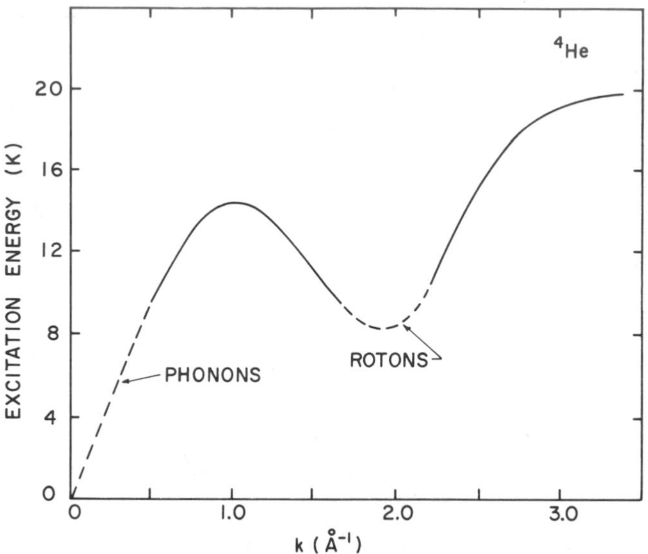

FIG. 4. Excitation spectrum of ^{4}He. The excitation spectrum of ^{4}He is directly observable in neutron scattering experiments. The excitation spectrum is found to contain two branches: a phonon branch as $k \to 0$, $\epsilon_k \propto k$, and a roton branch at $k \approx k_0$, $\epsilon_k \propto (k-k_0)^2$. There are also single-particle-like excitations at $k > k_0$, $\epsilon_k \propto k^2$. At $T \neq 0$ these excitations are present in the system in numbers that depend on the temperature; they are the normal component.

evolution of a nondegenerate Fermi system to a degenerate Fermi system.

Although in principle the ^{3}He liquid is a "can of worms," it was suggested by Landau that quite simple behavior might occur. The ^{3}He system at $T \lesssim 1$ K, like the ^{4}He system, is strongly interacting in the sense that the strength of the pair interaction, $\epsilon = 10$ K, is large compared to the temperature (this interaction has a very repulsive short-range core) and

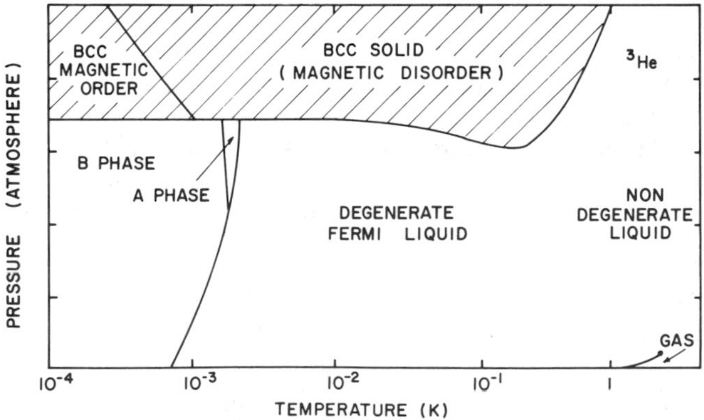

FIG. 5. Phase diagram of ^{3}He. ^{3}He remains a liquid as $T \to 0$ K unless pressures in excess of 30 atm are applied. At temperatures less than about 1 K the liquid goes from a nondegenerate Fermi system to a degenerate Fermi system (this "transition" is not sharp). The liquid is a Fermi liquid for 3 mK$\lesssim T<1$ K. At temperatures below about 3 mK the Fermi liquid undergoes a transition to one of the two superfluid ^{3}He states. The details of phase diagram in this temperature range are a sensitive function of magnetic field. An interesting and not well-understood magnetic transition also occurs in solid ^{3}He in this temperature range.

the system is dense so that the typical ^{3}He atom is in repeated strong interaction with many other particles. Landau suggested that almost all of the repeated strong interaction among particles goes into determining the ground state. The little bit that is left over is seen in weak interactions of quasiparticle excitations of the system above the ground state. Specifically, the excitation properties of the system can be described by an aggregate of fermion quasiparticles whose energy relative to the ground state is

$$\epsilon(\mathbf{p}) = \epsilon_0(\mathbf{p}) + \int f(\mathbf{pp}')\delta n(\mathbf{p}')\, d\mathbf{p}', \tag{10}$$

where

$$\epsilon_0(p^0) = \mu + \left(\frac{\partial \epsilon}{\partial p}\right)_0 (p - p_0). \tag{11}$$

The quasiparticles are single particles like, in leading approximation, the term $\epsilon_0(\mathbf{p})$ of Eq. (10). They have a finite lifetime by virtue of their interactions through $f(\mathbf{pp}')$ with other quasiparticles that occupy the system with density $\delta n(\mathbf{p})$. As the quasiparticles are fermions their mutual interactions go to zero as the temperature is lowered as $(T/T_F)^2$ so that they are long lived as $T \to 0$ K. Lifetimes and mean-field effects for the quasiparticles depend upon the Fermi-liquid parameters which are employed to parametrize the phenomenologic interaction $f(\mathbf{pp}')$; these parameters and the other characteristics of the quasiparticle excitation spectrum are determined by the many experiments on ^{3}He. For example, the specific heat as a function of temperature varies as T, $C = \gamma T$, like a degenerate ideal Fermi gas (the magnitude of the constant γ determines one of the "Fermi-liquid" parameters); the magnetic susceptibility behaves like that of a Pauli paramagnet but with a magnitude 10 to 20 times as great as an ideal Fermi gas (this enhancement of the susceptibility is understood in terms of the behavior of the Fermi-liquid parameter which measures the degree to which the polarization of one atom affects another). The Landau theory of a Fermi liquid provides a scheme for correlating the many data that are available on liquid ^{3}He. Furthermore, the structure of the physical ideas in the theory leads to the prediction of new phenomena, e.g., zero sound, the observation of which phenomena was a crowning achievement of the extensive theoretical and experimental attack on liquid ^{3}He.

^{3}He-^{4}He MIXTURES

The phase diagram of liquid ^{3}He-^{4}He mixtures at 1 atm is shown in Fig. 6. When ^{3}He is mixed into ^{4}He at concentrations less than that at the tricritical point, x_T, the resulting solution is a normal liquid at $T > T_\lambda(x_3, P)$ and undergoes a normal liquid to superfluid transition at $T = T_\lambda(x_3, P)$. At temperatures below $T_\lambda(x_3, P)$ the liquid behaves like a superfluid and is describable by the two-fluid model. The ^{3}He atoms in the fluid join the usual normal component and the Bose condensed ground state continues to be identified with the superfluid. At suitably low ^{3}He concentrations ^{3}He atoms see the ^{4}He as an inert medium and behave like single particles with dispersion relation

$$\epsilon_k = \epsilon_0(x_3, P) + \hbar^2 k^2/2m^*, \qquad (12)$$

where m^* is an effective mass that measures the departure in the behavior of the ^{3}He atom in ^{4}He from the behavior of a free ^{3}He atom, and $\epsilon_0(x_3, P)$ is an energy which measures the bonding of the ^{3}He atom to the ^{4}He medium. The normal component of the superfluid is then made up of three parts, phonons, rotons, and ^{3}He atoms:

$$\rho_n = \rho_P(T) + \rho_R(T) + \rho_3(T), \qquad (13)$$

with $\rho_3(T) \approx x_3 m^* N/V$. As $T \to 0$ K, $\rho_P(T)$ and $\rho_R(T) \to 0$ and the normal component is dominated by the ^{3}He excitations; this occurs at $T \lesssim 0.5$ K. Then, in conventional two-fluid-model experiments on the mixtures, e.g., second sound, osmotic pressure, etc., one learns about the properties of the ^{3}He normal component, e.g., m^*, the ^{3}He-^{4}He interactions, etc.

In 1962 Edwards showed, from an analysis of scant thermodynamic data, that one should expect that as $x_3 \to 0$, $\epsilon_0(x_s, P) \to 0$; i.e. a single ^{3}He atom would prefer to be in a ^{4}He medium than in a ^{3}He medium. There is a natural tendency for mixing of ^{3}He in ^{4}He at low ^{3}He concentrations. This is seen in Fig. 6 as the finite solubility of ^{3}He in ^{4}He at $T = 0$ K for concentrations less than about 6%. Because of this finite solubility it is possible to carry low-concentration ^{3}He in ^{4}He mixtures to very low temperature; the ^{3}He phase does not separate out (compare to solid ^{3}He-^{4}He mixture diagram for a contrast) but the ^{4}He phonon-roton excitations disappear completely. The dilute gas of fermions that now inhabit the ^{4}He medium have a degeneracy temperature of

$$k_B(T_F(x_3)) = k_B T_F x_3^{2/3} \ll 1 \text{ K}:$$

$$k_B T_F = (\hbar^2/2m^*)(3\pi^2\rho)^{2/3};$$

$$k_B T_F(5\%) \approx 70 \text{ mK}.$$

At $T < T_F(x_3)$ the dilute mixtures are systems of degenerate fermions. As such these systems constitute a second class of Fermi systems which have been explored extensively by the full range of experimental techniques and against which the Landau theory of Fermi liquids can be tested. Furthermore, the dilute mixtures can be employed to operate a "dilution refrigerator" that makes the temperature range $10 \leq T \leq 100$ mK readily accessible.

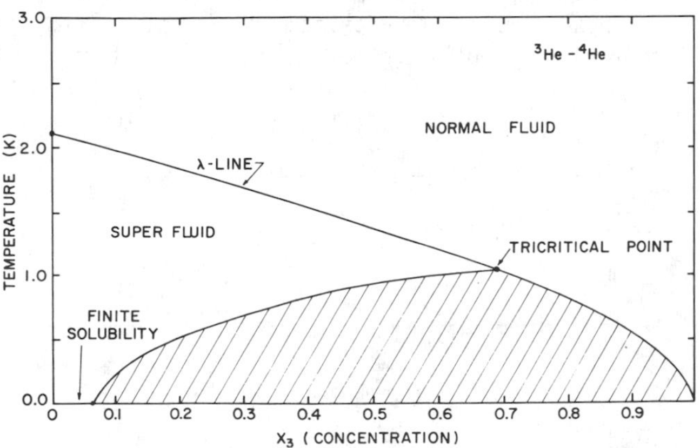

FIG. 6. The phase diagram of ^{3}He-^{4}He mixtures. The ^{3}He-^{4}He mixtures are phase separated at temperatures less than about 1 K. But the unique feature of the mixture phase diagram is the finite solubility of ^{3}He in ^{4}He as $T \to 0$ K up to concentrations of order 6%. This feature makes it possible to study dilute ^{3}He in ^{4}He, i.e. dilute Fermi liquids, to arbitrarily low temperature.

^{3}He BELOW 3 mK

In 1972 Osheroff, Richardson, and Lee encountered a "glitch" in P vs T on the melting curve of ^{3}He near 2.7 mK during an experiment in which they were trying to elucidate the properties of solid ^{3}He. Within a few months Osheroff, Richardson, and Lee had identified the glitch with a modification of the behavior of the liquid ^{3}He in the sample chamber and had identified several important characteristics of liquid ^{3}He below 2.7 mK, i.e., superfluid ^{3}He. A ^{3}He superfluid had been expected on theoretical grounds (by analogy with the electron superfluid in metals and the hadron superfluids in nuclei and dense matter) but the experimental search for this phenomenon had driven the theory before it to lower and lower temperatures.

The phase diagram of liquid ^{3}He in the millidegree range is shown in Fig. 5. There are two ^{3}He superfluids, the A superfluid ^{3}He-A at $2.2 \leq T \leq 2.7$ mK on the melting curve and the B superfluid ^{3}He-B at $T < 2.2$ mK on the melting curve. Also shown is the solid ^{3}He magnetic transition at 1 mK. Unlike ^{4}He which is composed of bosons, the ^{3}He liquid is composed of fermions and the occurrence of superfluidity requires a coupling that will tend to bond ^{3}He pairs. The ^{3}He-^{3}He interaction is strongly repulsive at short distances so that one expects an effective attractive interaction to occur in an $l \neq 0$ relative angular momentum state for a pair. Experimental evidence points to an $l = 1$ pairing state. With $l \neq 0$ the ^{3}He pairs may have a variety of spin states, unlike the $l = 0$ pairing in superconductors that have antiparallel spins only. The A phase is composed of $l = 1$, with $\uparrow\uparrow$ pairs and $\downarrow\downarrow$ pairs behaving as two weakly coupled spin fluids; the B phase has all three $l = 1$ spin states in equal number. Because of the nature of the microscopic pairing a correct characterization of the ^{3}He superfluids is substantially more complex than characterization of ^{4}He superfluids; e.g., a piece of the fluid is described by **I**, a measure of the angular momentum of the pairs in the fluid. Similarly a piece of the fluid can be described by **S**, a measure of its spin, and **d**, the order parameter. Because of the weak dipolar interaction between ^{3}He spins the energy of the system depends on the relative orientation of **S**, **I**, and **d**. Furthermore, **S** couples directly to a magnetic field; **I** is influenced by walls, etc. As a consequence a great deal about the ^{3}He superfluids has been learned from experiments that probe the relationship of **S**, **I**, and **d** and the influence of **H**, walls, etc. Experiments on superfluid ^{3}He have explored these relationships and have provided strong evidence in support of an $l = 1$ ground state and in support of the Leggett theory of the dynamics of **S**, **I**, and **d**. The understanding of superfluid ^{3}He advanced in a few years (1972–1982) to a level comparable to that of our understanding of superfluid ^{4}He. Superfluid ^{3}He and ^{4}He are currently regarded as well understood.

USES OF LIQUID HELIUM

Having a well-understood physical system in hand permits the possibility of employing this system as a probe.

(1) As a consequence of the superfluidity of ^{4}He, thin films (thickness of order 10 Å) of ^{4}He support vortices having quantized circulation, + or −, that interact with one another through a log r potential. Thus, thin films of ^{4}He have behavior that is due to the $d=2$ neutral Coulomb gas of vortices that resides in them. It was in experimental studies of thin ^{4}He films (Bishop and Reppy, 1978) that compelling evidence for the Kosterlitz–Thouless scenario for $d=2$ phase transitions (a scenario built on the neutral Coulomb gas model) was found.

(2) The fourth sound mode of ^{4}He in packed powders (a mode made possible because the normal fluid is held immobile in the confined space available to it between the powder particles) can be used to study the properties of an elastic fluid, the superfluid component of ^{4}He, in a porous medium. Experimental and theoretical studies of the ^{4}He system, ^{4}He in Vycor glass, have been important in elucidating the properties of a correct description of complex fluid–matrix systems; e.g., ^{4}He in Vycor glass is analogous to oil in a rock.

(3) The single-particle states appropriate to describing particles in a thin film, e.g., a thin film of ^{3}He, are strongly influenced by the film's size (thickness). The magnetic properties of the film depend on the occupation of these states by the ^{3}He particles, à la Pauli. Thus the magnetization of a thin ^{3}He film is a sensitive probe of the geometry of the space in which it resides. Should this space be fractal the magnetization of the ^{3}He will show it.

The quantum fluids, well known, are valuable probes of the unknown.

See also BOSE–EINSTEIN STATISTICS; FERMI–DIRAC STATISTICS; QUANTUM FLUIDS; QUASIPARTICLES; SECOND SOUND.

BIBLIOGRAPHY

E. E. Keller, *Helium 3 and Helium 4*. Plenum Press, London, 1969.
J. C. Wheatley, *Am. J. Phys.* **36,** 181 (1968).
J. C. Wheatley, *Physics Today* **29,** No. 2, 32 (1976).
J. Wilks, *The Properties of Liquid and Solid Helium*. Clarendon Press, Oxford, 1967.

Helium, Solid

Henry R. Glyde

INTRODUCTION

Helium was first solidified at the famous Kamerlingh Onnes low-temperature physics laboratories in Leiden by W. H. Keesom [1] on June 25, 1926. The initial experiments by Sir Francis Simon at Oxford University and by Keesom and their collaborators focused on the melting curve, the specific heat, and the thermal conductivity of solid helium as a test of our early understanding of solids. These measurements showed, for example, that the Lindemann criterion of melting does not hold in solid helium. This pioneering work up to 1957 is elegantly and beautifully reviewed by Domb and Dugdale [2], a review that stands today along with those of Guyer [3], Koehler [4], Glyde [5], and Roger et al. [6] as an excellent introduction to solid helium.

The pair potential $v(r)$ between helium atoms is now precisely known [7,8]. It is weakly attractive at large separation, $r \gtrsim 3$ Å^{-1}, with a maximum well depth ϵ = 10.95 K. At close approach $r \lesssim \sigma = 2.63$ Å, $v(r)$ becomes steeply repulsive with hard-core radius σ defined by $v(\sigma)=0$. The potential parameters σ and ϵ of the rare gases are compared in Table I. The potential seen by a helium atom lying between two atoms in a linear lattice is depicted in Fig. 1. The well shape is clearly dominated by the repulsive core of $v(r)$, which is wide and anharmonic.

Since helium is light, its thermal wavelength is long, e.g., at $T=4.2$ K, $\lambda_T \sim 10$ Å for ^{4}He. Helium is therefore difficult to localize. Localization leads to a high kinetic or zero point energy. Since $v(r)$ is weak, helium does not solidify under attraction via $v(r)$. Rather, it solidifies only under pressure and then solidifies due to the hard core of $v(r)$, much as billiard balls form a lattice under pressure. The total energy E is small and positive, except at the lowest pressures.

The phases of helium are sketched in Fig. 2. At $p \approx 35$ bar, the lighter isotope ^{3}He solidifies into an expanded body-centered-cubic (bcc) structure, having $V=24.4$ cm^3/mole with interatom spacing $R=3.75$ Å. At higher pressure, both ^{3}He and ^{4}He are compressed into the close-packed (fcc and hcp) phases (e.g., at $p=4.9$ kbar, fcc ^{4}He has V = 9.03 cm^3/mole, $R=2.77$ Å). At very high pressure ($p=156$–223 kbar), ^{4}He solidifies at room temperature in the hcp phase [10] (at $p=223$ kbar, $V=3.42$ cm^3/mole). This is therefore evidence for a transition from the fcc to the hcp phase somewhere along the melting line ($10 \leq p \leq 150$ kbar) and possibly other transitions [11] A discussion of phases can be found in the books by Keller [12], Wilks [13], and Wilks and Betts [14].

HELIUM, THE QUANTUM SOLID

The basic character of solid helium is revealed in the specific heat, C_v^P, due to exciting collective vibration of the atoms about their lattice points, the phonons (p). At low temperature (T), C_v^P is well described in all phases by the

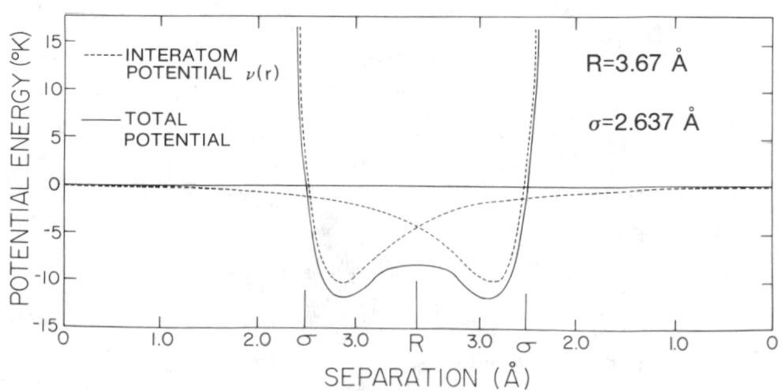

FIG. 1. Total potential seen by a helium atom in a linear solid due to its nearest neighbors (solid line). R is the interatom space of the linear lattice. Dashed line is the interatomic potential, $v(r)$. Note that the second derivative of the total potential at R is negative.

Table I Comparison of solid ^{3}He (at $V=24$ cm^3/mole) and solid ^{4}He (at $V=21.1$ cm^3/mole) with the heavier rare-gas crystals. The interatomic potential parameters are the core radius σ [$v(\sigma)=0$] and the well depth ϵ for the following potentials: He, HFD-B (Ref. 7); Ne, HFD-C2 (Ref. 8); Ar, HFD-C (Ref. 8); Kr, HFD-C (HFGKK) (Ref. 8). For Xe we quote σ and ϵ from Barker et al. (Ref. 9).

Rare-gas crystal	Debye temperature θ_D (K)	Melting temperature T_M (K)	Debye zero point energy $E_{ZD}=\frac{9}{8}\theta_D$	Lindemann parameter $\delta=\langle u^2\rangle^{1/2}/R$	Potential parameters σ (Å)	ϵ (K)	de Boer parameter Λ
^{3}He(bcc)	19	0.65	21	0.368	2.637	10.95	0.325
^{4}He(bcc)	25	1.6	28	0.292	2.637	10.95	0.282
Ne	66	24.6	74	0.091	2.758	42.25	0.061
Ar	84	83.8	95	0.048	3.357	143.22	0.019
Kr	64	161.4	72	0.036	3.579	199.9	0.011
Xe	55	202	62	0.028	3.892	282.35	0.0065

traditional Debye expression

$$C_v^P=3R\left(\frac{4\pi^4}{5}\right)\left(\frac{T}{\theta_D}\right)^3. \quad (1)$$

Here R is the gas constant and θ_D is a free parameter, the Debye temperature, obtained by fitting to the observed C_v^P. θ_D is the characteristic energy of the phonons. From Table I, θ_D is clearly large compared to the melting temperature (T_M) in solid helium.

In Debye's model the zero point vibrational energy is

$$(E_Z)_D=2\langle T\rangle_D=\tfrac{9}{8}\theta_D. \quad (2)$$

This energy, a purely quantum effect, clearly dominates thermal energies (T_M) so that solid helium may be called a quantum solid. From θ_D we may also evaluate the mean square vibrational amplitude $\langle u^2\rangle$ of the atoms about their lattice points. In the Debye model at $T=0$ K this is

$$\langle u^2\rangle_D=109.2\left(\frac{1}{M\theta_D}\right)\text{Å}^2, \quad (3)$$

where M is the atomic mass. From Table I we see that $\langle u^2\rangle$ and the Lindemann [15] ratio $\delta\equiv\langle u^2\rangle^{1/2}/R$ are large in solid helium.

Simply from θ_D we see that the atoms are not well localized. The atoms therefore explore a wide range of the potential well depicted in Fig. 1. For the dynamics, the hard core is the important part of the potential. Large-amplitude vibration also means a highly anharmonic crystal.

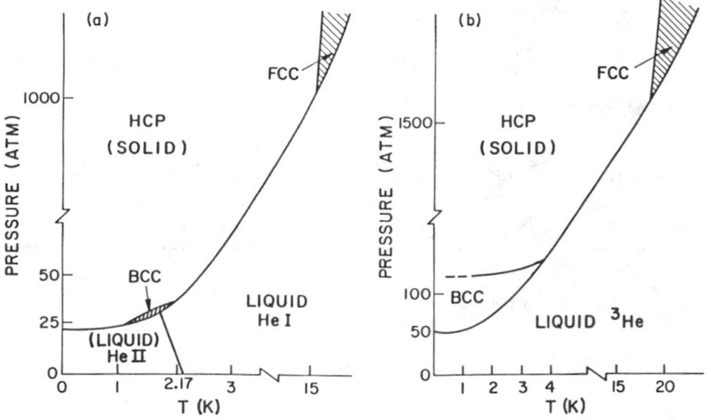

FIG. 2. Phase diagram of helium.

The large zero point vibration in solid helium results from the uncertainty principle of quantum mechanics. This states that we cannot simultaneously localize an atom at a point (a lattice point here) and make its momentum vanish. There is a trade-off between reducing the momentum (energy) and localizing the particle. To keep the zero point energy manageable each nucleus has a large vibrational distribution about its lattice point. This distribution may be described at $T=0$ K by a wave function with wavelength given by the famous de Broglie wavelength, λ. The quantumness of a solid was articulated by de Boer [16] in his famous parameter

$$\Lambda=\frac{1}{2\pi}\frac{\lambda}{\sigma}=\frac{h}{\sigma(2m\epsilon)^{1/2}} \quad (4)$$

which is the ratio [18] of the de Broglie wavelength, $\lambda=h/p$ [with momentum $p=(2m\epsilon)^{1/2}$], to the minimum separation of atoms in the crystal, σ. If Λ is comparable to σ, the solid is highly quantum (see Table I). Also, Λ is clearly mimicked by the Lindemann ratio δ (see Table I). A large Λ therefore means very anharmonic vibration. Hence very anharmonic vibrations remaining at $T=0$ K can also be used to identify a quantum solid.

The large wave function of atoms means that wave functions of atoms on adjacent lattice sites overlap. This leads to tunnelling and exchange of atoms between sites. In solid ^{3}He, which has a nuclear spin $\frac{1}{2}$, this exchange can be traced in nuclear magnetic resonance experiments and in thermodynamic properties. The characteristic energies in helium are total energy $E\sim1$ K, phonons $\theta_D\sim25$–100 K, and exchange $J\sim10^{-3}$ K.

GROUND STATE AND DYNAMICS

A key to understanding helium is finding a suitable wave function to describe the vibrational distribution of the atomic nuclei. Nobel laureate Max Born and his student D. J. Hooten first recognized that this vibrational distribution and the phonon dynamics must be determined consistently [2]. Nosanow [17] in 1966 proposed a Gaussian function for the distribution with corrections to account for short-range correlations between pairs of atoms induced by the hard core of $v(r)$. Koehler [18] and Horner [19] extended this to correlated Gaussian functions, the wave function for a harmonic crystal, plus corrections for short-range correlations. The short-range correlations can be described by a Jastrow pair function or by use of a Brueckner T-matrix method, both developed initially to describe short-range correlations between nucleons in nuclei. The self-consistent determination of the phonon frequencies and lifetimes and this vibrational distribution constitutes the self-consistent phonon (SCP) theory [20]. This theory is exhaustively reviewed [4,5,21,22].

Phonons in solid helium are most directly observed by inelastic neutron scattering. Measurements on all phases show that, although solid helium is highly anharmonic, phonons are well defined and have long lifetimes [23]. The expanded bcc ^{4}He phase is so anharmonic that phonon energies can be determined at low wave vector (Q) only (see Fig. 3). At higher Q, the scattering intensity contains fascinating interference effects between the one-phonon and multiphonon scattering contributions that can be well described using the SCP theory [23]. The more compressed fcc phase of helium is significantly less anharmonic and can be quite well described without short-range corrections.

Ground-state properties of ^{4}He are now most accurately evaluated using variational Monte Carlo (MC) and Green's function Monte Carlo (GFMC) methods combined with accurate potentials. Properties such as the total energy E, the static structure factor $S(Q)$, the pair correlation function $g(r)$, and momentum distribution $n(p)$ have been evaluated. In Table II we quote [26] GFMC values of the kinetic energy $\langle T\rangle$, the potential energy $\langle V\rangle$, and $E = \langle T\rangle + \langle V\rangle$ for fcc ^{4}He. The evaluated properties of fcc and hcp ^{4}He are the same within MC error. Clearly, both $\langle V\rangle$ and $\langle T\rangle$ are large with a large cancellation to yield a small E.

The $\langle T\rangle_{\rm MC}$ is much larger than $\langle T\rangle_{\rm D}$ predicted by a Debye model, $\langle T\rangle_{\rm D}=(\frac{9}{16})\theta_{\rm D}$ (see Fig. 4). This difference is a direct manifestation of the anharmonic hard core of $v(r)$. Interactions via the hard core introduce high-energy tails in response functions. These high-energy components raise the average $\langle T\rangle$ above that predictable by an adjusted Debye model. High-energy tails are common to all phases of helium.

The atomic kinetic energy $\langle T\rangle$ has been measured by exciting new applications of neutron scattering at high wave vector (Q) transfer [27]. At $Q \geq 10$ Å^{-1}, the observed coherent dynamic structure factor $S_c(Q,\omega)$ is well approximated by the incoherent $S_i(Q,\omega)$. $\langle T\rangle$ is related to the second moment of $S_i(Q,\omega)$ by

$$\langle T\rangle_{\rm obs} = \frac{3\hbar}{4\omega_R}\int_{-\infty}^{\infty} d\omega(\omega-\omega_R)^2 S_i(Q,\omega).$$

Assuming a shape for $S_i(Q,\omega)$, the second moment may be obtained from the observed width of $S_c(Q,\omega)$. $S_i(Q,\omega)$ will be a Gaussian in ω if $n(p)$ is a Gaussian or if $S_i(Q,\omega)$ approximately satisfies the impulse approximation. GFMC evaluations find $n(p)$ very close to a Gaussian [26]. The $\langle T\rangle_{\rm obs}$ obtained in this way shown in Fig. 4 lie somewhat above but in good agreement with MC evaluations in fcc ^{4}He and with evaluations using the SCP theory in bcc ^{4}He. Neutron scattering at high Q offers exciting possibilities to determine $n(p)$ and other single-atom properties in condensed helium [27].

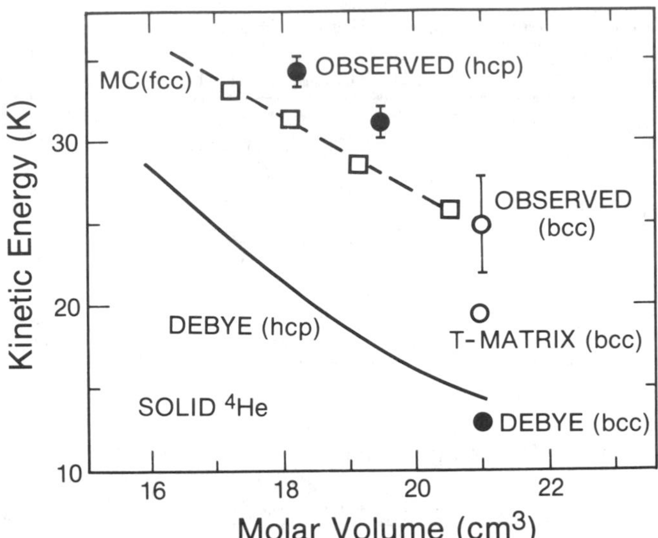

FIG. 4. Atomic kinetic energies in solid ^{4}He. The solid and open circles with error bars are observed values in hcp and bcc ^{4}He, respectively. The squares with dashed line are the Monte Carlo (MC) values $\langle T\rangle_{\rm MC}$ in fcc ^{4}He from Table II. The open circle is a calculated value using self-consistent phonon/T-matrix methods. The solid line and solid dot are Debye model values, $\langle T\rangle_{\rm D}=\frac{9}{16}\theta_{\rm D}$, calculated using observed $\theta_{\rm D}$ values.

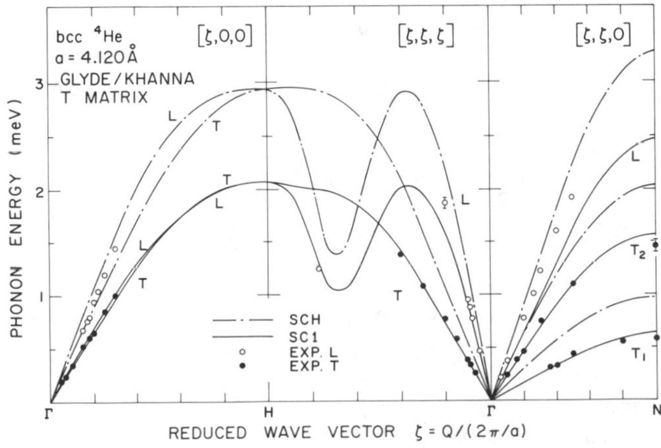

FIG. 3. The phonon–frequency dispersion curves observed in bcc ^{4}He by Minkiewicz *et al.* (Ref. 24) and calculated by Glyde and Goldman (Ref. 25) in the self-consistent harmonic (SCH) and complete first-order self-consistent (SC1) approximations.

MAGNETIC PROPERTIES

Solid bcc ^{3}He displays rich magnetic properties at low temperature ($T \leq 10$ mK) where phonons and thermal vacancies

Table II Total, kinetic, and potential energies from GFMC calculations (Ref. 26) in the fcc solid phase computed using the HFDHE2 potential. All energies are in kelvins per particle

Density (Å^{-3})	Volume (cm^3/mole)	E	$\langle T\rangle$	$\langle V\rangle$
0.0294	20.5	−5.61 ± 0.03	25.70 ± 0.07	−31.31 ± 0.07
0.0315	19.1	−4.96 ± 0.03	28.3 ± 0.1	−33.2 ± 0.1
0.0335	18.0	−3.87 ± 0.03	31.8 ± 0.1	−35.7 ± 0.1
0.0353	17.1	−2.70 ± 0.06	33.3 ± 0.2	−36.0 ± 0.1

[28] are frozen out. Solid ^{3}He is then an ideal collection of spin-$\frac{1}{2}$ nuclei on a lattice. The nuclear wave functions on adjacent lattice sites overlap leading to exchange and tunneling of nuclei between sites. Because the lattice is expanded and $v(r)$ has a steep repulsive core (which restricts simple pair exchange between nearest neighbors), there are several exchange processes involving pairs, triplets, four atoms and higher which have similar exchange rates [29] (of $\sim 10^7$ s ~ 1 mK). This leads to competing antiferromagnetic and ferromagnetic properties. It is a tribute to human ingenuity that this complex magnetic solid has recently been quite well understood on a microscopic basis.

The magnetic phase diagram of bcc ^{3}He is depicted in Fig. 5. Early studies of the paramagnetic phase are reviewed by Guyer *et al.* [30]. These showed that even the paramagnetic phase could not be explained using a simple Heisenberg exchange Hamiltonian. At low magnetic field ($B \leq 0.46$ T), there is a transition from the paramagnetic to an antiferromagnetically ordered phase [31]. Nuclear magnetic resonance [32] and neutron-diffraction data [33] are consistent with this being a tetragonal [32] U2D2 ordering displayed in Fig. 6. The antiferromagnetic resonant frequency, Ω_0, of the U2D2 ordering agrees better with experiment [34] than do other orderings not excluded by neutron and NMR data. At higher field, $B \gtrsim 0.46$ T, there is a transition to a second ordered phase believed to be a canted normal antiferromagnetic (CNAF) phase (see Fig. 6). As B is increased, the canting angle decreases until it goes to zero at an upper critical field [34], $B_{C2} = 21.7 \pm 1$ T. For $B \gtrsim 10$ T, the phase diagram is not well known.

To explain the magnetic properties of the ordered and paramagnetic phases, Roger, Hetherington, and Delrieu [6], and Roger [35] proposed an exchange Hamiltonian

$$H_{ex} = -J_{1N} \sum_{ij} P_{ij} + T_1 \sum_{ijk} P_{ijk} - K_p \sum_{ijkl} P_{ijkl}. \quad (5)$$

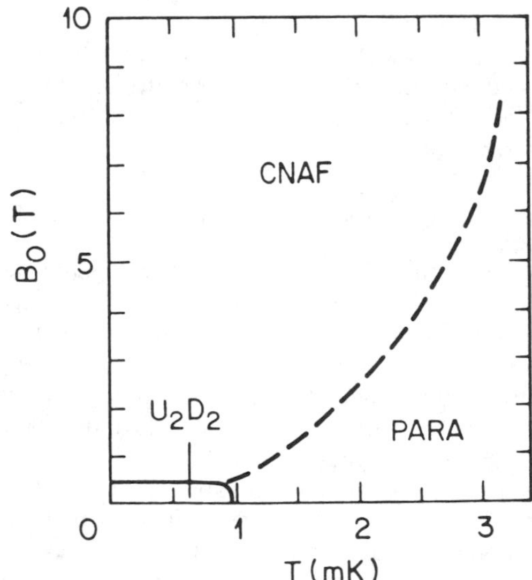

FIG. 5. Magnetic phase diagram of bcc ^{3}He. U2D2 is the low magnetic field ordered antiferromagnetic phase, CNAF is the high field ordered canted normal antiferromagnetic phase, and PARA is the paramagnetic phase.

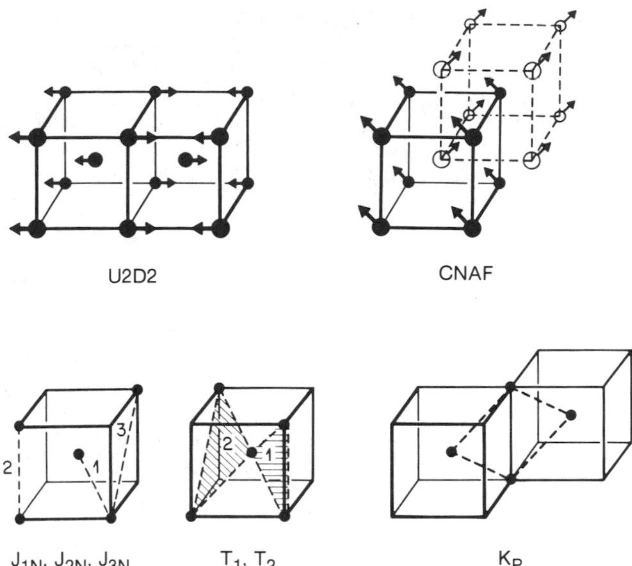

FIG. 6. The tetragonal up-up down-down (U2D2) and cubic canted normal antiferromagnetic (CNAF) structures. Exchange processes involving two, three, and four atoms.

H_{ex} includes nearest-neighbor pair exchange P_{ij}, triplet exchange P_{ijk}, and a four-atom rotation process P_{ijkl} where $P_{ij} = \frac{1}{2}(1 + \boldsymbol{\sigma}_1 \cdot \boldsymbol{\sigma}_j)$ is the pair interchange operator. From the symmetry of the fermion wave function, exchange involving even (odd) numbers of nuclei leads to antiferromagnetic (ferromagnetic) ordering. This symmetry is reflected in the signs in H_{ex}. Triplet exchange (ferromagnetic) therefore competes with two- and four-atom exchange but antiferromagnetic exchange dominates. With H_{ex}, Roger *et al.* [6] and Roger [35] were able to obtain good agreement with a wide range of properties treating J_{1N}, T_1, and K_p as adjustable parameters.

In a remarkable application of path integrals, Ceperley and Jacucci [36] (CJ) evaluated the exchange energy of many exchange processes within 10%. They confirmed that the exchange processes in H_{ex} (5) are the dominant terms and obtained values for the coefficients of $J_{1N} = -0.46$ mK, $T_1 = -0.19$ mK, and $K_p = -0.27$ mK at $V = 24.12$ cm^3/mole. Other processes are significant and including these Godfrin and Osheroff [37] have been able to obtain impressive agreement with experiment, especially the Curie–Weiss constant. This agreement confirms the basic H_{ex} in (5) and the path integral evaluation of the coefficients. The reviews by Roger *et al.* [6] and by Cross and Fisher [38], and the papers by Godfrin and Osheroff [37] and in the *Canadian Journal of Physics,* Vol. 65, 1987, provide an excellent modern introduction to this field.

See also CRYSTAL BINDING; CRYSTAL DEFECTS; INTERATOMIC AND INTERMOLECULAR FORCES; RARE GASES AND RARE-GAS COMPOUNDS; TUNNELING.

REFERENCES

1. W. H. Keesom, *Comm. Kamerlingh Onnes Lab., Leiden* No. **184b** (1926); *Helium,* p. 180. Elsevier, Amsterdam, 1942.
2. C. Domb and J. S. Dugdale, *Prog. Low Temp. Phys.* **2,** 338 (1957).

3. R. A. Guyer, *Solid State Phys.* **23,** 413 (1969).
4. T. R. Koehler, in *Dynamical Properties of Solids,* Vol. II (G. K. Horton and A. A. Maradudin, eds.). North-Holland, Amsterdam, 1975.
5. H. R. Glyde, in *Rare Gas Solids,* Vol. I (M. L. Klein and J. A. Venables, eds.). Academic Press, New York, 1976.
6. M. Roger, J. H. Hetherington, and J. M. Delrieu, *Rev. Mod. Phys.* **55,** 1 (1983).
7. R. A. Aziz, F. R. W. McCourt, and C. C. K. Wong, *Mol. Phys.* **61,** 1487 (1987).
8. R. A. Aziz, in *Inert Gases* (M. L. Klein, ed.). Springer-Verlag, Berlin, Heidelberg, 1984.
9. J. A. Barker, M. L. Klein, and M. V. Bobetic, *IBM J. Res. Dev.* **20,** 222 (1976).
10. H. K. Mao, R. J. Hemley, Y. Wu, A. P. Jephcoat, L. W. Finger, C. S. Zha, and W. A. Bassett, *Phys. Rev. Lett.* **60,** 2649 (1988).
11. J. M. Besson and J. P. Pinceaux, *Science* **206,** 1073 (1979).
12. W. E. Keller, *Helium-3 and Helium-4.* Plenum, New York. 1969.
13. J. Wilks, *The Properties of Liquid and Solid Helium.* Oxford University Press, Oxford, 1967.
14. J. Wilks and D. S. Betts, *Liquid and Solid Helium.* Oxford University Press, Oxford, 1987.
15. F. A. Lindemann, *Phys. Z.* **11,** 609 (1911).
16. J. de Boer, in *Progress in Low Temperature Physics,* Vol. 2 (J. C. Gorter, ed.). North-Holland, Amsterdam, 1957. We use the definition introduced by Newton Bernardes, *Phys. Rev.* **120,** 807 (1960), which is $(2\pi\sqrt{2})^{-1}$ times de Boer's definition.
17. L. H. Nosanow, *Phys. Rev.* **146,** 120 (1966).
18. T. R. Koehler, *Phys. Rev. Lett.* **17,** 89 (1966); **18,** 654 (1967).
19. H. Horner, *Z. Phys.* **205,** 72 (1967).
20. P. F. Choquard, *The Anharmonic Crystal.* Benjamin, New York, 1967.
21. H. Horner, *J. Low Temp. Phys.* **8,** 511 (1972); *Dynamical Properties of Solids,* Vol. I (G. K. Horton and A. A. Maradudin, eds.). North-Holland, Amsterdam, 1974.
22. C. M. Varma and N. R. Werthamer, in *The Physics of Liquid and Solid Helium,* Vol. I (K. H. Bennemann and J. B. Ketterson, eds.). Wiley, New York, 1976.
23. H. R. Glyde and E. C. Svensson, in *Methods of Experimental Physics,* Vol. III (K. Sköld and D. L. Price, eds.). Academic Press, New York, 1987.
24. V. J. Minkiewicz, T. A. Kitchens, G. Shirane, and E. B. Osgood, *Phys. Rev.* **A8,** 1513 (1973).
25. H. R. Glyde and V. V. Goldman, *J. Low Temp. Phys.* **25,** 601 (1976).
26. P. A. Whitlock and R. M. Panoff, *Can. J. Phys.* **65,** 1409 (1987).
27. R. O. Simmons, *Can. J. Phys.* **65,** 1401 (1987).
28. P. R. Granfors, B. A. Frass, and R. O. Simmons, *J. Low Temp. Phys.* **67,** 353 (1987), and references therein.
29. A. K. McMahan and J. W. Wilkins, *Phys. Rev. Lett.* **35,** 376 (1975).
30. R. A. Guyer, R. C. Richardson, and L. I. Zane, *Rev. Mod. Phys.* **43,** 532 (1971).
31. W. P. Halperin, C. N. Archie, F. B. Rasmussen, R. A. Buhrman, and R. C. Richardson, *Phys. Rev. Lett.* **32,** 927 (1974).
32. D. D. Osheroff, M. C. Cross, and D. S. Fisher, *Phys. Rev. Lett.* **44,** 792 (1980).
33. A. Benoit, J. Bossy, J. Flouquet, and J. Schweizer, *J. Phys. Lett.* **46,** L923 (1985).
34. D. D. Osheroff, H. Godfrin, and R. R. Ruel, *Phys. Rev. Lett.* **58,** 2458 (1987).
35. M. Roger, *Phys. Rev. B* **30,** 6432 (1984).
36. D. M. Ceperley and G. Jacucci, *Phys. Rev. Lett.* **58,** 1648 (1987).
37. H. Godfrin and D. D. Osheroff, *Phys. Rev. B* **38,** 4492 (1988).
38. M. C. Cross and D. S. Fisher, *Rev. Mod. Phys.* **57,** 881 (1985).

Hidden Variables

Michael Horne

The quantum-mechanical state of a system provides, in general, statistical predictions on the results of measurements. A controversy long existed as to whether the dispersions predicted by quantum mechanics indicate that the theory is incomplete and whether it is possible to reproduce the quantum-mechanical statistics by suitable ensembles of dispersion-free states for which the results would be determined individually. The dispersion-free states are commonly referred to as hidden variables, and a theory which uses dispersion-free states to explain quantum-mechanical statistics is called a hidden variable theory. If dispersion-free states could actually be prepared in the laboratory, quantum mechanics would be observably inadequate. However, the controversy focused not on whether the hidden variables should be observable but on whether the mere hypothesis of their existence is compatible with the structure of quantum mechanics. Another facet of the controversy focused on whether hidden variables are needed, i.e., whether their existence solves any pressing dilemmas in physics.

To describe the results of research on the existence question, it is helpful to recognize, within the general family of hidden variable theories, two distinct classes which may be called *contextual* theories and *noncontextual* theories. A noncontextual hidden variable theory would be related to quantum mechanics in much the same way that classical mechanics is related to classical statistical mechanics. Each individual system is specified by a state λ from a phase space Γ of hidden states. Each quantum-mechanical state ψ is associated with a probability density $\rho_\psi(\lambda)$ on the space Γ; that is, the state ψ is associated with an ensemble of the λ states. The value of each observable A is specified by a function $f_A(\lambda)$ mapping the hidden states λ into the real line; that is, the result of measuring the observable A is predetermined by the hidden state of the individual system. Clearly, agreement with quantum mechanics requires that the range of the function $f_A(\lambda)$ be the set of eigenvalues of the operator $\mathbf{A}$ associated with the observable A. And finally, the statistical predictions of quantum mechanics are reproduced: for each observable A and state ψ

$$|\langle\psi|a_i\rangle|^2 = \int_{\Gamma(a_i)} \rho_\psi(\lambda)d\lambda,$$

where $|a_i\rangle$ is the eigenvector of $\mathbf{A}$ with eigenvalue a_i and $\Gamma(a_i)$ denotes the subspace in Γ where $f_A(\lambda) = a_i$. The central feature of this scheme is quite simple: each individual system in effect possesses a definite value for every observable.

It is now known that noncontextual hidden variable states are incompatible with the formal structure of quantum mechanics. For many years it was widely believed that von Neumann had already demonstrated the incompatibility in 1932, but in 1966 his proof was examined by Bell and shown to be inadequate. Bell and others noted, however, that incompatibility does follow as a corollary to a 1957 theorem of Gleason. The proof presupposes a Hilbert space of dimension greater than two, which permits the existence of noncommuting operators $\mathbf{B}$ and $\mathbf{C}$, each commuting with a third operator **A**, such that one can in principle simultaneously observe either A and B or A and C. The proof then shows from quantum mechanics that there exists no function $f_A(\lambda)$ which specifies the value of the observable A and which is independent of whether B or C is measured simultaneously with A. Thus the above noncontextual hidden variable scheme is impossible. However, the proof does not exclude *contextual* hidden variable theories: a necessary condition for a hidden variable interpretation of quantum mechanics is that the function specifying the value of an observable A depends not only on the hidden state λ but also on the context of the whole apparatus, e.g., on whether the apparatus measures A and B or A and C.

Contextual theories do not suffice to salvage the hidden variable program because it turned out that any such theory which agrees with all the statistical predictions of quantum mechanics must have an undesirable property of nonlocality. This fact was shown by Bell, who worked out the implications of an argument initiated in 1935 by Einstein, Podolsky, and Rosen (EPR). In their argument, EPR consider a composite system consisting of two spatially separated, yet correlated, subsystems and argue persuasively that the quantum-mechanical description is incomplete unless hidden variables exist. Consider, for example, a pair of spin-$\frac{1}{2}$ particles produced somehow with total spin angular momentum zero (singlet spin state) and moving freely in opposite directions. Measurements can be made, say with Stern–Gerlach magnets, on selected components of the spins of the two particles. If measurement of some component of the spin of particle one yields the result "up" or $+1$, then, according to quantum mechanics, measurement of the same component of the spin of particle two must yield the result "down" or -1, and vice versa. Now assume the two measurements are made at widely separated places and consider the following premises of EPR:

1. Since at the time of measurement the two systems no longer interact, no real change can take place in the second system in consequence of anything that may be done to the first system.
2. If, without in any way disturbing a system, we can predict with certainty (i.e., with probability equal to unity) the value of a physical quantity, then there exists an element of physical reality corresponding to this physical quantity.
3. Every element of the physical reality must have a counterpart in the [complete] physical theory.

Since, in our example, we can predict in advance the result of measuring any component of the spin of one particle by previously measuring the same component of spin of the other remotely located particle, it follows from premises 1 and 2 that the value of any spin component of either particle is an element of physical reality. That is, the value of every spin component of either particle must actually be predetermined. And reflection will convince the reader that the only way premise 3 may be fulfilled is through the existence of a hidden variable theory. Because of premise 1, the hidden variable theories required by EPR have the property called *locality*.

In 1964 Bell investigated the local hidden variable theories and found that the whole class is incompatible with certain statistical predictions of quantum mechanics. It is significant that he was able to expose an *observable* discrepancy between quantum mechanics and the local hidden variable theories. Consider, as does Bell, the pair of spin-$\frac{1}{2}$ particles in the singlet state discussed above. Measurements can be made on the **a** component of the spin of particle one and on the **b** component of the spin of particle two. If the unit vectors **a** and **b** are not parallel, all combinations of results are possible: $++$, $+-$, $-+$, and $--$. For given directions **a** and **b**, quantum mechanics predicts a probability for the occurrence of each of these outcomes: $P_{++}(\mathbf{a},\mathbf{b})$, $P_{+-}(\mathbf{a},\mathbf{b})$, $P_{-+}(\mathbf{a},\mathbf{b})$, and $P_{--}(\mathbf{a},\mathbf{b})$, where the first and second subscripts refer to the results for the first and second particle. The specific statistical quantity considered by Bell is the expectation value of the product of the two results:

$$E(\mathbf{a},\mathbf{b}) = P_{++}(\mathbf{a},\mathbf{b}) - P_{+-}(\mathbf{a},\mathbf{b}) - P_{-+}(\mathbf{a},\mathbf{b}) + P_{--}(\mathbf{a},\mathbf{b}).$$

The quantum mechanical prediction for this is

$$E_{\rm qm}(\mathbf{a},\mathbf{b}) = -\mathbf{a}\cdot\mathbf{b}.$$

Bell proves that for any local hidden variable theory the expectation value satisfies the inequality

$$|E_{\rm hv}(\mathbf{a},\mathbf{b}) - E_{\rm hv}(\mathbf{a},\mathbf{c})| - E_{\rm hv}(\mathbf{b},\mathbf{c}) - 1 \leq 0,$$

where **a**, **b**, and **c** are any three directions. The quantum-mechanical function $E_{\rm qm}$ does not, in general, satisfy the inequality. For example, choosing **a**, **b**, and **c** coplanar, letting the angle between **a** and **b** be 60°, the angle between **b** and **c** be 60°, and the angle between **a** and **c** be 120°, insertion of $E_{\rm qm}$ into the hidden variable inequality yields the contradiction $\frac{1}{2} \leq 0$.

Bell's theorem concerning the local hidden variable theories is a profound development in the history of hidden variable discussions. First, the theorem proves that the EPR premises are untenable if quantum mechanics is correct. Second, the theorem makes it possible to design experiments for testing the entire family of local hidden variable theories, on the chance that quantum mechanics may break down for widely separated, yet correlated, pairs of systems. Of course, the suggestion of such a breakdown is a radical proposal in view of the success of quantum mechanics. But on the other hand, the premises of EPR are plausible and some physicists may prefer experimental evidence before giving them up. Many experiments have been designed and carried out, the majority yielding results in good agreement with quantum mechanics and in violation of local hidden variable predictions. An important feature of one recent experiment is the use of time-varying polarization analyzers. In this experiment, the settings of each analyzer are effectively changed during the flight of the particles, thereby substantially reducing the possibility that one particle or its analyzing apparatus "knows" the setting of the distant analyzing apparatus. Although the experiments so far are not sufficiently close to ideal to be conclusive, experimental evidence is accumulating that the EPR premises and the hidden variables they imply are untenable.

See also QUANTUM MECHANICS; QUANTUM STATISTICAL MECHANICS.

BIBLIOGRAPHY

For the EPR argument that local hidden variables are needed, see *Phys. Rev.* **47,** 777 (1935); for Bohr's reply, see *Phys. Rev.* **48,** 696 (1935).

For an explicit hidden variable model of elementary wave mechanics, see D. Bohm, *Phys. Rev.* **85,** 166, 180 (1952).

For criticism of early noncontextual incompatibility proofs and for proof based on Gleason's work, see J. Bell, *Rev. Mod. Phys.* **38,** 447 (1966); for simpler proofs, see F. J. Belinfante, *A Survey of Hidden-Variables Theories.* Pergamon Press, New York 1973.

For Bell's theorem concerning local theories, see *Physics (Long Island City)* **1,** 195 (1964); for proposal of an experimental test, see J. Clauser *et al., Phys. Rev. Lett.* **23,** 880 (1969); for the experiment, see S. Freedman and J. Clauser, *Phys. Rev. Lett.* **28,** 938 (1972).

For general discussion of experimental tests, see J. Clauser and M. Horne, *Phys. Rev. D* **10,** 526 (1974).

For extensive discussion of the implications of Bell's work, see B. d'Espagnat, *Conceptual Foundations of Quantum Mechanics,* 2nd ed. Benjamin, New York 1976; M. Redhead, *Incompleteness, Nonlocality and Realism.* Clarendon, Oxford, 1987.

For a comprehensive review of experimental tests through 1978 and their implications, see J. Clauser and A. Shimony, *Rep. Prog. Phys.* **41,** 1881–1927 (1978).

For the experiment with time-varing analyzers, see A. Aspect *et al., Phys. Rev. Lett.* **49,** 1804 (1982).

For collections of reprints, including many of the papers listed in this bibliography, see J. A. Wheeler and W. H. Zurek (eds.), *Quantum Theory and Measurement.* Princeton University Press, Princeton, NJ, 1983; J. S. Bell, *Speakable and Unspeakable in Quantum Mechanics.* Cambridge University Press, Cambridge, 1987; and L. E. Ballentine (ed.), *Foundations of Quantum Mechanics Since the Bell Inequalities.* American Association of Physics Teachers, College Park, MD, 1988.

High-Field Atomic States

James E. Bayfield

An atom placed in a sufficiently strong electric, magnetic, or electromagnetic field is in some high-field atomic quantum state. The strength of the field must be high enough for the alteration of the energy of the atom not to be a small perturbation. Thus by definition, high-field states are quite different from those of isolated atoms. The electron density distribution within the atom is controlled by both the nuclear Coulomb field and the applied field. While some high-field states can be considered to be stationary, others involve an explicit time evolution of the electron's energy.

ATOMS IN STRONG STATIC FIELDS

One early interest in high-field states was in astrophysics, as atoms can be in the intense static magnetic fields of neutron stars. An analysis of the spectra of such atoms yields the value of the magnetic field. The interaction Hamiltonian or energy of interaction of a hydrogen atom with a static magnetic field is

$$\frac{e}{mc}\mathbf{A}\cdot\mathbf{p} + \frac{e^2}{2mc^2}\mathbf{A}^2,$$

where $\mathbf{p}$ is the electron's momentum and the quantity $\mathbf{A}$ is related to the position $\mathbf{r}$ of the electron relative to the nucleus and to the magnetic field strength $\mathbf{B}$ by

$$\mathbf{A}(\mathbf{r}) = \tfrac{1}{2}\mathbf{B}\times\mathbf{r}.$$

The Zeeman effect occurs at low fields $\mathbf{B}$ and is a perturbative effect arising from the first term in the interaction energy. High-field states are formed when the second term exceeds the first. Whereas the classical physics of the Zeeman effect involves a field-induced precession of the electron's orbit plane about the magnetic field direction, the high-field states classically can involve a keen competition between the electron orbiting around the nucleus and orbiting around the magnetic field direction. This competition produces a mixture of possible orbits in the classical limit, with some being periodic, some quasiperiodic, and some chaotic. The periodic classical orbits play a major role in determining the stationary high-field quantum states that produce the spectrum of the atom in the field. On the other hand, one quantum manifestation of the presence of chaotic orbits is an alteration of the distribution of quantum energy level spacings in a certain characteristic way.

Whereas a free electron classically orbits in a circle when in a static magnetic field, a static electric field accelerates such an electron along the field direction. Thus magnetic high-field states tend to have the electron's localization near the nucleus maintained, whereas electric high-field states tend to involve ultimately a removal of the electron from the region of the nucleus, a process called field ionization. When both electric and magnetic high fields are present, some high-field states are slowly ionizing while others ionize rapidly.

ATOMS IN HIGH-STRENGTH ELECTROMAGNETIC WAVES

When an atom is exposed to a monochromatic electromagnetic wave of high-field strength and of frequency comparable to the classical electron orbit frequency of the unperturbed atom, a large number of photons can be absorbed from the wave before the electron leaves the vicinity of the nucleus. More photons may be absorbed than the minimum energetically required for photoionization, a phenomenon called above-threshold ionization. The many-photon absorption has been studied for the case of highly excited atoms, where the classical electron orbit is very large and the electron orbit frequency is in the microwave region, much lower than the optical frequencies corresponding to lowest quantum atomic energy levels. In the classical limit, the possible electron dynamics again spans the range from periodic to chaotic. Although the high-field states cannot be stationary since the applied electromagnetic field is time-varying, in the absence of field ionization there exist "quasi-energy" states that display the time periodicity of the field. When the electric field strength of the wave reaches a frequency-dependent threshold value of the order of 10% of the mean atomic Coulomb field strength, a many-photon field ionization process does rapidly occur that in the classical limit involves the chaotic electron orbits only. During the ionization process the time evolution of electron energy exhibits a near-classical growth characteristic of a diffusion process. As a result the ionization threshold fields are near-classical. At ionizing field strengths the quasienergy level separations again exhibit the distribution function characteristic of chaos in the classical limit. This situation is altered when the frequency of the wave is higher than twice the initial electron orbit frequency, where destructive quantum wave interference suppresses the near-classical time evolution and forces the ionization threshold field to higher values.

See also CHAOS; DYNAMICS, ANALYTICAL; ORDER–DISORDER PHENOMENA; PHOTOIONIZATION; QUANTUM MECHANICS; ZEEMAN AND STARK EFFECTS.

BIBLIOGRAPHY

H. Hasegawa, M. Robnik, and G. Wunner, in "New Trends in Chaotic Dynamics of Hamiltonian Systems," *Prog. Theoret. Phys.* Supplement, 1989. (A)

J. E. Bayfield, *Comments At. Mol. Phys.* **20,** 245 (1987). (A)

G. Casati, B. V. Chirikov, D. L. Shepelyansky, and I. Guarneri, *Phys. Rpt.* **154,** 77–123 (1987). (A)

High-Pressure Techniques *see* Ultrahigh-Pressure Techniques

High Temperature

John W. Hastie, David W. Bonnell, and Joan B. Berkowitz

High temperature is a pervasive condition, occurring as a natural intra- and extraterrestrial phenomenon and in many aspects of ancient and modern science and technology. The human sense of temperature derives from the fact that biological processes (as we know them) are confined to a very narrow temperature range. A person coming into contact with an object at a much higher (or lower) temperature than that of his own body will describe the object as "hot" (or "cold"). In a thermodynamic sense, an object will feel hot when the direction of heat flow is from the object to the individual. The observation that heat will only flow spontaneously (i.e., without expenditure of work) from a hot object to a cold object, and never in the reverse, is in essence one statement of the second law of thermodynamics. In general, if two bodies are brought into contact and there is no heat flow from one to the other, they are said to be at the same temperature. This forms the basis for more formal definitions of temperature, and for the establishment of a scale by which to measure temperature quantitatively. The realization of the measurement of temperature is the purpose of the International Temperature Scale of 1990 (ITS-90), which replaces IPTS-68(75), the previous standard termperature scale. The four temperature scales in use are Kelvin (K), Celsius (°C), Fahrenheit (°F), and Rankine (°R). They are interrelated as: $t(°C) = T(K) - 273.15$; $t(°F) \cong 9/5t(°C) + 32$; $t(°R) \cong 9/5T(K)$. The Kelvin and Rankine temperature scales are referred to as *absolute* thermodynamic scales, because their lower reference point is absolute zero. The Kelvin scale is currently realized in practice through the ITS-90, and the other scales are derivative. The Rankine scale is of historic interest because many tabulated engineering data are based on that scale. The Fahrenheit scale is used mainly in the United States, and for nonscientific purposes. Implicit in the concept of temperature is the concept of equilibrium. A system not at equilibrium cannot be said to have a temperature. However, in many globally nonequilibrium situations, if one carries out measurements on an appropriate scale, either in space or time, then a local equilibrium can be identified with which a temperature can be associated. This concept provides the basis for dealing with the real world of matter and energy flows. For example, in a system with a temperature gradient and with mass transport occurring, there is no three-dimensional volume at constant temperature. However, if the energy and mass transport are occurring under equilibrium conditions, a temperature still exists at every point in the system.

From a human physiological perspective, high temperatures are those in excess of the range of stability of living systems. From a materials or chemical perspective, the influence of high temperatures can often be defined from the thermodynamic relationship for the Gibbs energy change for a physicochemical process, $\Delta G = \Delta H - T\Delta S$, where ΔH is the heat energy, or *enthalpy*, change, ΔS is the entropy change, and T is the absolute temperature. In many practical situations, ΔG provides a measure of the thermal stability of a system. The more negative ΔG is, the more stable the products of the process are. Often, and particularly in the range from 1000 to 5000 K, both the ΔH and $T\Delta S$ terms are influential in determining the sign and magnitude of ΔG. This interplay between the opposing stability factors of bond formation (negative ΔH) and bond breaking (positive ΔS) leads to complex, often unpredictable, behavior as a function of temperature. Thus, the usual insight at ambient temperature regarding heat energy as the driving force can be counter-intuitive in understanding high-temperature processes. From a scientific viewpoint, high-temperature phenomena are often associated with entropy-dominated processes.

From an engineering and technology point of view, high temperatures are usually those temperatures sufficiently in excess of normal ambient temperatures to change significantly the properties of physical and chemical systems. While ambient temperatures are usually considered to fall around room temperature, 298 K, the recent development of "high-temperature" superconductors, which exhibit zero electrical resistance at temperatures reported as high as about 100 K, is an example of the nebulous and changing nature of our concept of high temperature.

The properties of greatest interest at high temperatures are either those obtained under equilibrium conditions, such as thermodynamic stability and physical, molecular, and mechanical properties, or under kinetic (dynamic or nonglobal equilibrium) conditions. The significant changes in conditions of equilibrium include changes in phase (solid, liquid, gas), and changes in molecular composition. Physical properties include density, surface tension, viscosity, thermal and electrical conductivity, and the like. Molecular properties include lattice structure and bond energies, and spectroscopic parameters such as vibrational frequencies, electronic states, and so on. Basic thermodynamic properties include heat of formation, entropy, heat capacity and equation of state. Mechanical properties of solids and glasses include a variety of correlative properties (creep, deformation, strain, fracture resistance) which are all measures of cohesive strength, i.e., surface and grain-boundary energy. Microstructure is also important to understanding chemical, mechanical, and physical properties of solids. The significant changes in kinetic interactions include changes in the rates at which thermodynamically favorable processes occur and changes in the mechanisms (i.e., the ways) by which chemical reactions take place. In addition to chemical kinetic interactions, rates of energy transfer, particularly among rotational, vibrational, and electronic quantum states, and their conversion of kinetic or translational energy, are of key significance to the concepts of equilibrium and temperature. In a gas, translational energy and thermodynamic temperature, T, are related by $\frac{1}{2}m\bar{v}^2 = 3/2\ kT$, where m is molecular mass, $\bar{v}$ is average velocity, and k is the Boltzmann constant, 1.38×10^{-16} erg/K.

Significant changes in properties are generally observed at temperatures above 1000 K. At temperatures above 3000 K, the only metals that are still stable in the solid phase at ordinary pressures are rhenium, tantalum, and tungsten. The normal boiling points of some common metals are: aluminum, 2720 K; copper, 2855 K; gold, 2980 K; silver, 2450 K; tin, 2960 K; and zinc, 1181 K. Above these temperatures the only stable phase of the metal at atmospheric pressure will be gaseous. Temperatures above 3000 K are reached in both industrial and natural processes. In the manufacture of silicon carbide (one of the common abrasives), temperatures in the center of the fabricating furnace are above 3600 K. Temperatures within the earth range from about 1000 K in the mantle regions, 100 km below the surface, to possibly 6900 K in the metal core. The core temperature of Jupiter is estimated to be 10 000–20 000 K. Some stars have surface temperatures as high as 100 000 K. Our own sun has a surface temperature of about 6000 K, with temperatures of its coronal gases reported as high as 2 000 000 K. In thermonuclear reactions (e.g., nuclear fusion and stellar cores), temperatures can be more than 100 000 000 K. In astrophysically catastrophic events, such as supernovas, violent galaxies, quasars, and the standard model of the big bang, or the bizarre objects of gravitational collapse, neutron stars and black holes, temperatures are often measured in units of 10^9 K. The maximum realizable temperature in nature, limited by the onset of hadron creation from the available energy, have been estimated to be of the order of 2×10^{12} K. Experimental nuclear temperatures from nucleus-nucleus collisions in accelerators of over 1×10^{12} K have been reported.

The remainder of the article is devoted to a number of illustrative examples of the properties of physical and chemical systems at temperatures higher than 1000 K. Since the dominant physicochemical properties, and the methods of temperature generation and measurement, vary considerably between 1000 and 10^{12} K, the temperature range is subdivided for purposes of discussion as follows: 1000–5000 K; 5000–10 000 K; 10 000–50 000 K; and more than 50 000 K.

THE TEMPERATURE RANGE FROM 1000 TO 5000 K

In the 1000 to 5000 K temperature range, at a pressure of one bar (=0.1 MPa, ≈one atmosphere), all condensed phases (solids and liquids) become unstable against a gaseous phase. There are no known solids stable at one bar above 4300 K, the approximate melting point of a mixture of tantalum carbide and hafnium carbide. There are stable liquids over the entire range, although they have not been studied extensively. The normal boiling point (1 bar) of tungsten for example, is about 6150 K. This temperature range is one of the most significant for high-temperature science and technology. It is particularly important to the processing and performance of present day high-performance materials, such as high strength alloys, advanced ceramics, high-temperature structural composites, semi- and superconductor materials—as well as traditional materials, such as steels, aluminum, and other metals and refractories. Modern technologies, such as chemical vapor deposition used in semiconductor production and the relatively new area of diamond films, rely heavily on processes occurring in this temperature range.

Although the intermolecular forces responsible for the stability of solids and liquids begin to weaken as the temperature is increased from 1000 to 5000 K, chemical valence forces that bind atoms into molecules are still of considerable importance in the gas phase. In fact, complex molecular species that are unstable at room temperature have been found under conditions of equilibrium in high-temperature vapors.

In the late 1940s and early 1950s, there was considerable controversy over the correct value for the heat of sublimation of graphite to gaseous atoms; i.e., the heat of the reaction

$$\mathrm{C(graphite)} \rightarrow \mathrm{C}(g).$$

The sublimation of graphite had been studied experimentally by a number of investigators, and the heats of sublimation obtained from seemingly good data tended to cluster around one of two values—140 kcal/mol or 170 kcal/mol. The controversy was settled in favor of the higher value when the composition of the equilibrium vapor over graphite at temperatures between 2400 and 2700 K was analyzed directly with a mass spectrometer. The vapor was found to contain not only gaseous carbon atoms, $C(g)$, but also $C_2(g)$ and $C_3(g)$ gaseous molecules. In fact, for this temperature range, the concentration of $C_3(g)$ molecules in the vapor is roughly six times that of carbon atoms (and continues to increase up to the graphite triple point, where solid, liquid, and gas are all in equilibrium, at a pressure of about 100 bar and a temperature of approximately 4500 K). The low value for the heat of sublimation had been calculated from experimental data, with the assumption that the vapor consisted almost entirely of atomic carbon.

Through the use of the high-temperature mass spectrometer, many complex gas-phase molecules have been discovered. For example, about one-third of the vapor molecules in equilibrium with molten table salt at 1200 K are in the form of the dimer, Na_2Cl_2, with the remainder in the form of the monomer, NaCl. Electron diffraction studies in the vapor phase and other spectroscopic evidence show that the dimer molecules are planar tetragons of the form

$$\begin{array}{ccc} & Na^{+} & \\ Cl^{-} & & Cl^{-} \\ & Na^{+} & \end{array}$$

The first electron diffraction studies that were carried out were interpreted on the assumption that the vapor was entirely monomeric. The calculated Na–Cl bond distance was significantly larger than that obtained from later microwave spectra. The discrepancy was resolved based on the fraction of dimer in the vapor, as determined by mass spectrometry. These types of complex molecules have the ability to combine with other metal halide, oxide, sulfide, etc. species to form molecules of even greater complexity, e.g., $MnFe_2Cl_8$. The exceptional thermodynamic stability of these complexes allows their use as vapor transport agents in modern metallurgical processes. Similar schemes are typically used in various forms of chemical vapor deposition (CVD) in the burgeoning semiconductor industry. CVD generally exploits the differing stabilities/reactivities of molecules over a thermal gradient to provide highly controlled film deposition. A relatively recent discovery of unusually stable clusters of atoms or molecules (e.g., C_{60}) has been made, using techniques similar to those applied to complex species of the type just mentioned. These clusters represent an intermediate state of matter between the gas and condensed phase. The formation of smoke from a fuel-rich hydrocarbon flame is a common example, where high-temperature cluster formation occurs as an intermediate step. The research area of cluster physics and chemistry is currently developing at a rapid pace.

There are many ways to achieve temperatures in the 1000 to 5000 K range. The one most commonly known is electrical resistance heating (used in both the electric toaster and the tungsten-filament light bulb). Other traditional techniques include combustion processes, induction heating, electron bombardment, and radiant heating, including solar furnaces and arc imaging furnaces. In recent years, high-power laser beams have been used as heat sources, particularly where localized heating and short heating times are desired. A serious difficulty in high-temperature research is the chemical reactivity of container materials with the system of interest. These reactions are generally prevalent at high temperatures, and have increasing rates (i.e., speed) as the temperature is raised. Various methods of "containerless" heating have been used in recent years. Electromagnetic levitation/heating and electrical pulse heating of conductors have been used to generate temperatures to above the melting temperature of tungsten (3695 K). Extending these methods using microgravity levitation and auxiliary heating is a promising future technique. The use of pulsed laser heating has also been used to alleviate the containment difficulty by providing heat only to a localized hot spot on the surface of the sample. The use of thermocouples for high-temperature measurement is also complicated by the materials interaction problem, as well as the tendency of insulating sleeves to be conductive, particularly above 2500 K. The most common temperature measurement method for the 1000 to 5000 K range is optical pyrometry, calibrated against a tungsten strip lamp. This method is the basis for the high temperature region above 1234.93 K of the ITS-90.

THE TEMPERATURE RANGE FROM 5000 TO 10 000 K

In the temperature range from 1000 to 5000 K, stable molecules exist in the gas phase, and for some systems, larger and more complex vapor molecules appear as the temperature is increased. However, as temperatures are raised above 5000 K, a point is reached where no molecules at all can exist. This temperature has been estimated to lie somewhere between 8000 and 12 000 K.

Consider nitrogen, for example. At 5000 K and a pressure of one bar, N_2 molecules predominate in the gas phase, much as they do at room temperature, but with a higher average translational energy, and more importantly, higher vibrational energies; about 1% of the molecules will have dissociated into nitrogen atoms because their internal vibrational energy exceeded the N—N bond strength. At 6000 K, more than 10% of the original nitrogen (N_2) molecules will be split into atoms. At 10 000 K, nitrogen atoms will predominate in the vapor phase, with molecules accounting for less than 1% of the particles present.

Temperatures above 5000 K are both difficult to achieve and difficult to measure. Temperature is generally defined by the equilibrium established among the accessible states of the system under consideration and the resulting average of the distribution of energy (*equipartition* of energy). However, the methods available for bringing systems to temperatures in excess of 5000 K often do not result in thermal equilibrium; i.e., evidence exists that all parts of the system, on the atomic and molecular level, may not be at the same temperature.

Flames have been used to achieve temperatures in the lower part of the range of interest. A one-bar carbon subnitride (C_4N_2)-oxygen flame, burning to carbon monoxide and nitrogen, yields temperatures around 5260 K. Theoretical flame temperatures are calculated from tabulated thermodynamic data by assuming that the heat released in the combustion reaction goes into raising the temperature of the combustion products. The calculations assume that equilibrium is achieved, and depend on a knowledge of the nature of the product species. The reaction zone of a flame is usually only a steady state region, and can often be far from equilibrium. The simplest frequently used method for measuring flame temperatures is the "spectral-line reversal" method, which makes use of the fact that certain chemical species at high temperatures emit visible light at well-defined wavelengths. If light from a calibrated continuum source (preferably a thermal equilibrium light emitter) is passed through a flame containing sodium vapor, the two sodium spectral lines (589.0 and 589.6 nm) will appear dark in absorption if the background source is at a higher temperature than the flame, and bright in emission if the background source is at a lower temperature than the flame. When the (known) source temperature matches the flame temperature, the lines disappear. For the highest-temperature flames, however, well-calibrated sources are not generally available.

Spectral methods to measure gas temperatures that do not require a calibrated source have been developed, but the measurements often reflect the temperature of abnormally excited (i.e., having nonequilibrium energy distribution) species rather than mean average temperatures. Observed spectral lines reflect the transition of a chemical species from one allowed energy level to another. If two spectral lines of the same species (molecule, radical, atom, ion, etc.) are excited at different energy levels, E_1 and E_2, and if the respective transition probabilities are P_1 and P_2, then the *equilibrium* ratio of the intensities of the two lines will be given by the ratio of Boltzmann factors, $P_1 e^{-E_1/kT}/P_2 e^{-E_2/kT}$, defining the distribution between energy levels 1 and 2, where k is the Boltzmann constant, T is the thermodynamic temperature (K), and e is the natural logarithm base. On a quantum scale, the overall Boltzmann distribution, which defines the equilibrium populations among system energy levels, is a definition of temperature. Those systems which do not obey the distribution are considered to be nonequilibrium by definition. The use of the Boltzmann distribution to derive temperatures for distributions, such as the state population inversions characteristic of lasing, have led to the concept of negative temperatures. To avoid infinite energies being associated with creating such a condition, the state spectrum must have an upper limit, the states at negative temperature must be inaccessible to the rest of the system states which are at a positive temperature, *and* the states at negative temperature must be in thermal equilibrium with the appropriate Boltzmann distribution for the assigned negative temperature. Often, when different degrees of freedom of a system have different equilibrium distributions, distinct temperatures can be assigned to the various distributions, particularly if they form isolated subsystems, as in the above example. If the transition probabilities are determined by measuring spectral-line intensities in sources of known temperature, then the ratio of the spectral-line intensities in a source of unknown temperature can be used to calculate the temperature. Use has been made of spectra arising from electronic transitions in atoms (such as iron) and from rotational and vibrational transitions in diatomic molecules (such as OH or CN). Measured rotational temperatures of OH in the reaction zone of flames have ranged from 5400 K in a one-bar oxyacetylene flame to 10 000 K in a low-pressure oxyhydrogen flame containing a trace of acetylene. These temperatures are much higher than the theoretical flame temperatures, and probably reflect the fact that the OH radicals can be formed in a highly excited nonequilibrium rotational distribution.

Temperatures up to 10 000 K (and even higher in some cases) can be achieved, for times on the order of microseconds, in shock waves. A shock wave can be produced in a so-called shock tube by the sudden bursting of a diaphragm separating a gas at low pressure from one at high pressure. A compression wave, which rapidly steepens to form a shock front, is generated in the low-pressure gas, and temperatures at the front rise abruptly. Shock waves are always produced in explosions or detonations. The energy released at the shock front goes immediately into increasing the kinetic energy (or mean square speed) of the gas molecules, which is equivalent to increasing translational temperatures.

For a gas at equilibrium, almost by definition, translational, rotational, vibrational, and electronic temperatures will all be equal. At the shock front, however, there may be some delay in achieving equilibrium between the translational energies of the molecules and their internal energies. Equilibrium is brought about by the interconversion of energy from translational to internal modes during molecular collisions. This process of energy equilibration is known as "relaxation," and has been studied for a number of gases in shock tubes. Generally, transfer of vibrational and electronic energy is relatively slow, which can lead to these internal state systems becoming isolated. Dissociation of molecules to equilibrium concentrations of atoms and ions has also been studied in shock tubes and is mechanistically complex, depending not only on collision frequency, but also on the energy of the colliding particles and the distribution of that energy between translational and internal modes. Equilibrium temperatures are, of course, lower than initial translational temperatures.

Direct laser heating of surfaces is used for industrial processes such as welding and cutting, yielding industrial process temperatures estimated to be above 5000 K. Surface interaction with very high power infrared or visible lasers ($>10^7$ W/cm^2), or with near-UV lasers, produces a luminous plume which contains, in addition to atoms and molecules, an abundance of ions having translational energies which correspond to temperatures of 8000 K and above. Use of lower power levels, with visible or infrared lasers, can produce approximately equilibrium high temperatures. This appears to be true even for short pulse-length ($\approx$10 ns) lasers. The bulk temperature of the surface can vary from near ambient for very low power levels to values above 5000 K. The majority of reported measurements have used spectroscopic probes, which are generally only sensitive to excited species; mass spectrometric determinations have found that vaporization of neutral species with apparent thermal equilibrium among emitted species can be the major process.

Gas discharges and electric arcs have been used for studies between 5000 and 10 000 K, but since they are also capable of achieving higher temperatures, they are discussed in the next section.

THE TEMPERATURE RANGE FROM 10 000 TO 50 000 K

At 10 000 K, atoms will predominate over molecules in most gases at equilibrium, and concentrations of ions can be appreciable. As temperatures are increased further, ions and free electrons become the dominant species, and gases that were electrically insulating at lower temperatures become electrically conducting. In nitrogen at one bar pressure, for example, the concentrations of nitrogen atoms, singly ionized nitrogen ions (N^+), and free electrons become equal at approximately 14 000 K. Doubly ionized nitrogen (N^{2+}) begins to form at about 20 000 K, and N^{3+} becomes significant at about 34 000 K. By 30 000 K, neutral nitrogen atoms have virtually disappeared.

The devices most commonly used to produce temperatures in the 10 000 to 50 000 K range are known generically as electric discharge devices. Examples include dc glow discharges, induction and microwave plasmas, electric arcs, and plasma jets or torches. The devices differ in many ways, but they all have a source of electrical power and a coupling mechanism for delivering the power to a plasma environment (i.e., a glowing environment characterized by the presence of positive ions and free electrons). The nature of the plasma produced (e.g., electron and gas temperatures, degree of gas ionization, deviations from equilibrium among the species present, electric field strengths, gas density, ratio of plasma volume to device volume) is highly sensitive to the particular device and its mode of operation. The basic principle of the electric discharge, however, is that electrons accelerated by large electric fields transfer energy from the power supply to the gas under study. Collisions between accelerated electrons and gas molecules result in the production of ions and dissociated free radicals or atoms. Laser-based welding systems produce a laser-opaque plasma by direct vaporization/ablation of the surface, followed by further heating of the plasma by direct transfer of laser energy to the plasma.

The mean kinetic energies, and hence the temperatures, of species present in glow discharge and laser-generated plasmas can be very different. Electrons can reach 30 000 K while neutral gas molecules in the same discharge can be present effectively at 300 K. In a hydrogen discharge, both atoms and molecules can have velocity distributions characteristic of room temperature, while the fraction of atoms present is that which would be expected for hydrogen at equilibrium at 4000 K. It is possible, however, to generate plasmas in which there is at least local thermal equilibrium—i.e., regions where the concentrations of atoms, ions, and electrons, and the distribution of energy among particle vibrations, rotations, and electronic excitations, can be described by a single temperature. Such equilibrium plasmas are generally formed at higher pressures (one bar and above) than for plasmas where there is a wide divergence in particle temperatures.

TEMPERATURES OVER 50 000 K

As temperatures are increased beyond 50 000 K, atoms lose more and more electrons in a stepwise fashion until a stable gas at 10 million to about 10 billion kelvins consists of bare nuclei and free electrons. It is this type of plasma environment that exists in the cores of "normal" stars and within the fireball of a nuclear explosion. It is also the environment of the nuclear fusion reactors that are looked upon as possible power sources of the future.

The stability of atomic nuclei increases rapidly with atom mass from hydrogen (whose nucleus is a proton) to neon (atomic mass 20), peaking at iron (atomic mass 56). The fusion of lighter nuclei to form heavier nuclei therefore results in a high release of energy. For technical and economic reasons, the combination of deuterium and tritium nuclei (heavy hydrogen nuclei with atomic masses of 2 and 3, respectively) to form a 3.5-MeV α particle (a helium nucleus, atomic mass 4) and a 14-MeV neutron (atomic mass 1) is still considered to be the best reaction for generation of fusion power. The deuterium and tritium (D-T) have to be heated to a temperature between about 50 000 000 K and 1 000 000 000 K (depending on the plasma density), which requires a large energy input; clearly if a fusion reactor is to be practical, there must be an energy payback (i.e., the energy output must exceed the energy required to build and operate the system). The deuterium and tritium nuclei formed in the plasma will of course be positively charged, and therefore mutually repulsive. Both high temperatures and plasma confinement schemes are needed to promote sufficient closeness of approach between nuclei so that fusion can occur.

Current research centers around two quite different techniques. The more traditional is electromagnetic/electrostatic confinement, in which strong magnetic and/or electrical fields are used to confine and heat a fusion plasma. Recent successes with tokamak-type devices have brought this method close to "break-even," as evidenced by the detection of very high neutron fluxes characteristic of fusion reactions. The other method is inertial confinement, in which a very high-energy laser beam is divided, and the resulting beams are all directed at a tiny sphere containing a D-T mixture. Laser impact simultaneously implodes and heats the pellet to (near) fusion temperatures. The NOVA laser system at Lawrence Livermore Laboratory is a major U.S. facility of this type. Currently, both methods are within about a factor of two of the 5×10^{14} keV cm^{-3} sec temperature $\times$ density $\times$ confinement-time product, which represents the break-even fusion energy objective. The D-T ignition point is about an order of magnitude higher. Both techniques are reaching the point of "scientific feasibility"—where fusion plasmas can be expected to be achieved in the laboratory with realistic apparatus. The question of economic viability of fusion power still hinges on a variety of materials science issues, such as development of an acceptable "first wall" material and reactor construction materials to withstand the expected mechanical, thermal, and radiation fatigue problems. A variety of approaches are being considered, representing a rapidly growing area of high-temperature research and development.

Temperatures even hotter than the several billion kelvin fusion reaction temperatures of highly evolved class O stars are the province of the interiors of supernova explosions, violent galaxies, quasars, and gravitationally collapsed objects such as "White Dwarf" stars and neutron stars. Above 10 billion kelvins, the plasma requires a relativistic treatment to describe it. At the pressures and temperatures present inside neutron stars, the "electron pressure" limit is exceeded, and electrons are forced into nuclei to create an object of neutrons. These conditions are of keen interest to theoreticians, and are experimentally accessible in high energy nucleon-colliding particle accelerators. Temperatures in collider events are now nearing nature's limits, where additional energy goes into creating matter (hadrons), rather than increasing the temperature of the nuclear fluid.

Although many scientific and technological areas have benefitted greatly from high temperature research, even more interesting and challenging problems still remain in each of the temperature regions.

See also Alloys; Combustion and Flames; Geophysics; Levitation, Electromagnetic; Maxwell–Boltzmann Statistics; Shock Waves and Detonations; Stellar Energy Sources and Evolution; Superconductors; Thermodynamics;

THERMOMETRY; ULTRAHIGH-PRESSURE TECHNIQUES; VAPOR PRESSURE

BIBLIOGRAPHY

Bibliography on the High Temperature Chemistry and Physics of Materials (M. G. Hocking and V. Vasantasree, current eds.) Vol. 33. Imperial College of Science and Technology, London, 1989.

M. Bass, "Laser Heating of Solids," in *Physical Processes in Laser-Materials Interactions* (M. Bertolotti, ed.), pp. 77–115. Plenum, New York, 1983.

B. D. Blaustein (ed.), *Chemical Reactions in Electrical Discharges*. American Chemical Society, Washington, DC, 1969.

D. W. Bonnell, R. L. Montgomery, B. Stephenson, P. C. Sundareswaran, and J. L. Margrave, "Levitation Calorimetry," in *Specific Heat of Solids* (C. Y. Ho and A. Cezairliyan, eds.), pp. 265–298. Hemisphere Publishing Corp., Washington, DC, 1988.

L. E. Brus, R. W. Siegal, et al. "Research Opportunities on Clusters and Cluster-Assembled Materials: A Department of Energy, Council on Materials Science Panel Report," *J. Mater. Res.* **4**(3), 704–736 (1989). Brus and Siegal were panel cochairmen.

G. Chaudron and F. Trombe (eds.), *Les Hautes Températures et leurs utilisations en physique et en chimie*, Vols. 1–2. Masson et Cie Editeurs, Paris, 1973.

W. A. Chupka and M. G. Inghram, "The Thermodynamics of Carbon Molecules as Determined in the Mass Spectrometer." *Mem. Soc. Roy. Sci. Liege, Quatrième Sér.*, **15**, 373–377 (1954).

Committee on High Temperature Science and Technology, *High Temperature Science: Future Needs and Anticipated Developments*. National Academy of Sciences, Washington, DC, 1979.

H. P. Furth, "High-Temperature Plasma Physics," in *Physics in a Technological World. XIX General Assembly International Union of Pure and Applied Physics*, pp. 315–345. American Institute of Physics, New York, 1988.

I. Glassman, *Combustion*. Academic Press, New York, 1977.

R. Goulard (ed.), *Combustion Measurements: Modern Techniques and Instrumentation*. Academic Press, New York, 1976.

W. Greiner and H. Stöcker, "Hot Nuclear Matter," *Sci. Am.* **252**(1) 76–87 (1985).

J. W. Hastie (ed.), *Characterization of High Temperature Vapors and Gases*. Government Printing Office, Washington, DC, 1979.

J. W. Hastie, *High Temperature Vapors: Science and Technology*. Academic Press, New York, 1975.

J. W. Hastie (ed.), *Sixth International Conference on High Temperatures: Chemistry of Inorganic Materials*. Humana Press, Clifton, NH, 1990. The conference was held in 1989.

High Temperature Technology (IUPAC International Symposium on High-Temperature Technology). Butterworth, Washington, DC, 1964.

C. Kittel, *Thermal Physics*. Wiley, New York, 1969.

G. N. Lewis, M. Randall, K. S. Pitzer, and L. Brewer, *Thermodynamics*. McGraw-Hill, New York, 1961.

B. W. Mangum, "Special Report on the International Temperature Scale of 1990: Report of the 17th Session of the Consultative Committee on Thermometry," *J. Res. Natl. Inst. Std. Technol.* **95**, 69–77 (1990).

J. L. Margrave, ed., *The Characterization of High Temperature Vapors*. Wiley, New York, 1967.

G. de Maria and G. Balducci (eds.), *Fifth International Conference on High Temperature and Energy-Related Materals*. Pion Ltd., London, 1989. The conference was held in 1987.

Physics through the 1990's, Vols. 1–9. National Academy Press, Washington, DC, 1986

H. Preston-Thomas, "The International Temperature Scale of 1990 (ITS-90)," Metrologia **27**, 3–10 (1990).

Symposium on High Temperature and Materials Chemistry, LBL-27905. Lawrence Berkeley Laboratory Materials and Chemical Science Division, Berkeley, CA, 1989.

Temperature: Its Measurement and Control in Science and Industry. American Institute of Physics, New York, 1941–1989. The volumes for 1941–1971 are proceedings of the Symposium on Temperature; the volumes for 1982–1989 are proceedings of the International Temperature Symposium.

E. T. Turkdogen, *Physical Chemistry of High Temperature Technology*. Academic Press, New York, 1980.

S. Woosley and T. Weaver, "The Great Supernova of 1987," *Sci. Am.* **261**(2), 32–40 (1989).

K. M. Young, "Summary Abstract: New Diagnostic Approaches for High-Temperature Plasmas," *J. Vac. Sci. Technol. A* 6(3), 2061–2062.

History of Physics

Lewis Pyenson

Physics in the sense of Aristotle, the study of the material representation of natural phenomena, may be identified in all civilizations and cultures. If the historian were to write a history of physics based on Aristotle's definition, he would reveal many unfamiliar cosmologies that provide unusual taxonomies for natural knowledge. This essay has a more modest focus. It concerns physics understood as a scientific discipline with a recognizably modern syllabus and institutional locus. Beyond preliminary remarks the following text does not address the history of physical thought before the crystallization of the modern discipline of physics, around 1830. The restriction may be deceiving, for even over the past six generations physics has been expressed through ideologies and institutions that no longer exist. Physics in the world we have just lost is a mixture of recognizable equations, slightly unusual life-styles, and unfamiliar methodological preoccupations. In focusing on modern physics it must be remembered that many of the social forms and philosophical prejudices associated with physics during the past 160 years have roots in previous centuries; indeed, one of the most exciting problems in the history of science has long been the transformation in physical world views between 1600 and 1800. Notwithstanding the traditional focus of interest of historians of physics, reconstructing the recent past provides a much-needed perspective on the decisions and pressures that confront physics today, and in part for this reason the discussion here will concentrate on developments in the late nineteenth and the twentieth centuries.

Given the persistence of classical learning in medieval Europe and the continual stimulation provided by Islamic institutional developments, it is difficult to speak of a renaissance in natural knowledge in the same way that the term has come to be applied in literature and art. To a considerable extent natural philosophy in the sixteenth and seventeenth centuries was a reaction against classical wisdom and the established institutions—the universities—that consecrated it. The vehicle for circulating new ideas beyond the university was the printed book. By 1500 the book had ceased to be an innovation concerned only with disseminating the classics, and it interacted with the nonverbal tradition of the mechanical arts. Engineers, navigators, cartographers, engravers, watchmakers, surveyors, and architects availed themselves of the new medium and contributed to it in increasing numbers. The printing press was also open to the hermetic tradition of magic and occult learning which influenced the labors of Johannes Kepler, Robert Hooke, and Isaac Newton. Finally, the press promoted new learned disciplines, such as chemistry, that were in part based on a pedagogical tradition beyond the universities.

Geometrical astronomy was a high art when the posthumously published work of Nicholas Copernicus suggested that a heliocentric astronomical model might provide a simpler picture of the heavens, a point of view that by 1600 was widely discussed throughout northern Europe. When in that year Giordano Bruno died at the stake for his Copernican heresies, however, only about a dozen writers had come out in print against geocentrism. It was left to Galileo, at the beginning of the seventeenth century, to invent modern mechanics and physical astronomy. Although he taught the traditional wisdom at several universities. Galileo limited his own classical inspiration to Archimedes. Instead of following any master, he urged reasoning through experimentation. Galileo was best at synthesizing numerous unrelated phenomena, and for this task his superb physical intuition allowed him to set aside irrelevant, but persistent, experimental irregularities. Galileo's work is usually designated as the beginning of the "scientific revolution." During the seventeenth century intense concern with creating a unified picture of the physical world was reflected in the algebraic interpretation of mechanics and the final acceptance of the differential calculus. At the same time, new experimental and inductivist methods were explored by the earliest scientific societies. Conventional disciplinary lines became blurred as new mathematical and experimental approaches slowly penetrated universities at Leiden, Jena, and Edinburgh. The seventeenth-century natural philosopher was different from a scientist. He sought to investigate the entire natural world. At the same time, he navigated currents that we would call nonscientific. The natural philosopher might equally be a magus casting horoscopes, a hermeticist believing in occult traditions, or a divine obsessed with biblical prophecy.

Isaac Newton was an exemplary seventeenth-century natural philosopher. Frank Manuel has argued that Newton acquired Puritan religious sympathies during Oliver Cromwell's Protectorate and that his later religious convictions, personal temperament, and scientific prejudices were conditioned by an early fixation on his mother. Newton entered Trinity College, Cambridge, as a poor subsizar who was required to perform menial tasks. It appears as if throughout his life, even after having been knighted and having served as Master of the Mint and President of the Royal Society, he felt insecure about his low birth. At the urging of his friends, Newton set down his three laws of motion and derived Johannes Kepler's three laws of planetary motion from the gravitational attraction of the sun. The publication of these results in 1685 provided conclusive evidence against Cartesian celestial vortex mechanics, which had been unable to account for Kepler's laws. The mature Newton saw his youthful work as the prelude to more important labors in theology, alchemy, and speculation about the ultimate composition of matter. He fretted over biblical numerology and believed that he had discovered a way to produce "philosophical mercury."

Newton's endeavors did not directly result in the creation of the modern *physics* discipline. At the end of the eighteenth century, mathematicians transformed Newton's geometrical representation of mechanics into more familiar expressions in differential and integral calculus. The experimental study of matter was refined by mineralogists, apothecaries, and especially professional chemists. Observation of nature was the business of anatomists, botanists, geologists, and astronomers. Engineering was the product of a largely independent tradition of mathematical practitioners—those who had for many generations designed and built optical systems, navigational equipment, and chronometers. Distinct from these activities, Newtonian natural philosophy around 1800 constituted speculative preoccupation with the essence of physical substances such as heat and light. For example, the followers of Pierre Simon de Laplace attempted to interpret physical processes through Newtonian mechanics by the hypothesis of short-range forces. Late-eighteenth-century natural philosophers who called themselves physicists did study mechanics, light, heat, acoustics, and occasionally electricity, but few systematic treatments of physics as a unified field of knowledge existed before the second third of the nineteenth century.

Modern physics may be traced directly to curriculum innovations at several German universities around 1830. The most influential program was founded at the University of Königsberg by Carl Gustav Jacobi in collaboration with Carl Neumann and the astronomer Wilhelm Friedrich Bessel. The Königsberg physics seminar included training in both theory and experiment. Partial differential equations from Joseph Fourier's new science of heat provided the theoretical language of physics; precision measurement, itself not a critical innovation, spanned a wide range of physical phenomena and for the first time took account of significant figures. Within a generation the Königsberg innovation had spread to universities throughout Prussia and the other German states. The high value accorded all branches of scholarly research in Germany provided incentives for talented youth to pursue physical science despite the disappointingly low salaries often paid university professors. At mid-century, education ministries of the German states were recognized leaders in processing large numbers of students. However autocratic these authorities may seem to us now, at the time they made competent and even brilliant choices for professorships in physics and mathematics, and they managed to provide at least some funds for the earliest laboratories devoted entirely to physics.

After mid-century the concept of a physics discipline was conveyed from Germany to France and Great Britain, where it interacted with national traditions. French and British investment in physics manpower and fixed assets became comparable with that of Germany by 1900, but the theoretical part of the discipline assumed a distinct character in each of the three countries. Since the eighteenth century France had maintained a strong tradition of mechanics, a subject that united astronomy, analysis, and engineering. A good deal of theoretical physics in France was treated by mathematicians in the discipline of mechanics. In Britain university physics was not clearly distinguished from either mathematics or chemistry until the end of the century.

At the beginning of the twentieth century neither France nor Great Britain had a system of higher education to match that of Germany, and it was above all in universities that physics flourished. Because physics was seen as abstract learning, it occupied a natural place in the curricula of elite institutions of higher education. In nineteenth-century Britain and Germany (although not in France), the ordinary practitioners of physics came from the upper classes in proportionately greater numbers than did other physical scientists. Physicists were educated in the best secondary schools and were under less pressure than chemists or geologists to practice their profession in industry or commerce. It might even be maintained that the life-style of nineteenth-century physicists resembled that of classical philologists more closely than that of chemists. The great democratization of higher education in the United States and the Soviet Union during the first part of the twentieth century significantly transformed the class structure of the physics profession. Vestiges of the former elite status of physicists may be found in the disproportionate influence that physicists still exert on university affairs and in the carefully cultivated image of the physicist as an omnivorous intellectual.

It would be a mistake to imagine that elite nineteenth-century physicists devoted themselves exclusively to a search for synthetic representations of fundamental physical laws. The research of some was directed to a deeper understanding of properties of certain chemical elements or minerals, to routine collection of meteorological data, or to technical improvements in electrical or mechanical devices. The published work of other physicists was indistinguishable from that of physical chemists, applied mathematicians, or engineers. At least in terms of the problems that came to dominate physics in the early years of the twentieth century, however, three threads may be identified in nineteenth-century physical discourse: electrodynamics, thermodynamics, and kinetic theory. During the last half of the nineteenth century, each comprised a circumscribed, though not entirely self-contained, set of theories and experiments.

Although various theories of electric current had been explored by physical scientists of the Laplacian school in the period around 1830, electrodynamics emerged as a science when William Thomson (Lord Kelvin) and James Clerk Maxwell succeeded in expressing Michael Faraday's intuitive picture of an electromagnetic field in mathematical language. Maxwell formulated a set of partial differential equations to describe the electromagnetic field. For over a generation his achievement was not clearly recognized by physicists on the continent, who continued to use action-at-a-distance formulations of electromagnetism in the tradition of André-Marie Ampère and Wilhelm Weber. Heinrich Hertz, who first demonstrated the existence of electromagnetic waves in 1887, was instrumental in gaining acceptance for Maxwellian electrodynamics in Germany. Hertz argued that the mathematical form of physical laws, in particular those of mechanics and electrodynamics, did not have to be based on physical models. In the middle 1890s Hendrik Antoon Lorentz reduced Hertz's formulation of Maxwell's equations to the now-familiar expressions.

Thermodynamics, a second important area of nineteenth-century physics, emerged from the disintegration of the caloric theory (where heat was viewed as a fluid) in part through the work of Joseph Fourier and Sadi Carnot. The conservation of energy, a principle independently announced at nearly the same time by four scientists, formed the basis of the first law of thermodynamics. Beginning around 1850, Rudolf Clausius and others elaborated a general criterion for physical processes in the second law of thermodynamics. To express the second law succinctly Clausius formulated a new quantity, *entropy*. The second law of thermodynamics was of great practical advantage for chemists. The change in entropy for any conceivable chemical reaction could be calculated easily once the heats of formation of the initial and final products were known. During the last third of the nineteenth century, some chemists such as Marcellin Berthelot refused to acknowledge the validity of the second law, and they proposed a variety of alternative rules for explaining reactions. Nevertheless, by 1900 the two laws of thermodynamics seemed as secure to most physicists as Newton's laws and Maxwell's equations.

Kinetic theory was the third important development in nineteenth-century physics. Around the middle of the century many physicists believed that if heat were equivalent to mechanical motion, as the first law of thermodynamics requires, then heat might reasonably be produced by the vibrations and collisions of molecules, the microscopic building blocks of matter that some physicists and chemists assumed to exist but about which few detailed investigations had been carried out. From data on the specific heats of gases and the gas laws, Clausius, Maxwell, and others were able to calculate molecular dimensions. For Maxwell and those who followed, the temperature and heat of a macroscopic quantity of gas was related to the *average* velocity of gas molecules. At the end of the century, Ludwig Boltzmann was able to interpret the laws of thermodynamics through an elastic-sphere model of gas molecules, and Josiah Willard Gibbs formulated a statistical mechanics.

Around 1900 serious doubts arose about the completeness of classical mechanics, and partisans of each of the three threads of nineteenth-century physics offered their specialty as the basis for a new picture of the world. Lorentz, Emil Wiechert, and Wilhelm Wien thought that some modification of Maxwell's equations might provide the basis for all physical laws. Their optimism stimulated much work on the theory of the recently discovered electron, then believed to be the smallest physical particle. Although the so-called electromagnetic view of nature was short-lived, intense and mathematically exacting work on electron dynamics during the period around 1900 sharpened the analytical tools of atomic theorists and conditioned contemporary physicists within the intellectual orbit of German-speaking Europe for the reception of Einstein's special theory of relativity. Indeed, until around 1911 many physicists and mathematicians thought that special relativity was equivalent to the electron theory of Lorentz. The electron theory provided inspiration for Hermann Minkowski's ideas of four-dimensional space-time and David Hilbert's version of the covariant field equations of general relativity. The laws of thermodynamics, too, furnished several physical scientists with a world view called energeticism. According to Georg Helm, Pierre Duhem, and Wilhelm Ostwald, all the laws of mechanics and electrodynamics would emerge from thermodynamics. At the end of the century atomism was held in disrepute by some physicists, for it postulated physical objects that could not be observed; most continued to make use of the atomic hypothesis, nevertheless, in their daily research. Marian Smoluchowski and Jean Perrin then gave reason to current belief.

Albert Einstein accomplished more than any other physicist attempting to unify these three threads of nineteenth-century physics. It is ironic that Einstein, who in some ways had little in common with the professional scientists of his day, has come to be seen as the greatest representative of theoretical physics. He held teaching professorships only at Zurich and Prague between 1909 and 1913. Many physicists felt that Einstein wasted the last 40 years of his life in a search for a field theory to unify electromagnetism and gravitation. Unlike most physicists, he never accepted Niels Bohr's Copenhagen interpretation of indeterminism in quantum mechanics. Einstein reported to his collaborator Leopold Infeld that he was considered an old fool at the Institute for Advanced Study in Princeton, where he worked from 1933 to his death in 1955. Einstein's unpretentious and direct lifestyle contrasted with the formality, extravagance, and chauvinism of powerful professors and laboratory directors. He never sanctioned the use of physics for national ends. In December 1914 he wrote to his good friend Paul Ehrenfest about the war: "The international catastrophe has imposed a heavy burden upon me as an internationalist. In living through this 'great epoch,' it is difficult to reconcile oneself to the fact that one belongs to that idiotic, rotten species which boasts of its freedom of will. How I wish that somewhere there existed an island for those who are wise and of good will! In such a place even I should be an ardent patriot." Notwithstanding his "convinced" pacifism, Einstein vigorously opposed the rise of fascism in Europe.

Einstein was born and raised in southern Germany. He dropped out of school at age 15 to join his parents in Milan when his father moved there from Munich. Within a year he sat for entrance examinations in the engineering faculty at a school near Milan that offered advanced instruction in German, the Swiss Federal Institute of Technology in Zurich. He failed the nonscientific part of the examination and was advised to spend a year at a secondary school in nearby Aarau. One year later he entered the Zurich Polytechnic, this time in the section for preparing secondary-school teachers. He studied in Zurich for four years, although upon receiving his teaching diploma in 1900 he could not find a university assistantship in physics. After a lean two years he was hired as a scientific examiner in the Patent Office at Berne. During the years at Berne, Einstein developed the special theory of relativity, as well as his ideas on molecular motion and the quantum theory of light. Soon after he became associate professor of theoretical physics at the University of Zurich in 1909, Einstein was recognized as one of the leading theoretical physicists in German-speaking Europe.

Fig. 1. Theoreticians. Albert Einstein and Niels Bohr, photographed by Paul Ehrenfest.

Along with other theoretical physicists Einstein believed that mathematics and experiment provided tools for constructing theories in physics, but that the theoretician required above all physical insight. Einstein's early master in thermodynamical reasoning was Boltzmann, and the laws of thermodynamics provided him with a logical model in much of his early work. The young Einstein felt himself attracted to Ernst Mach's critical approach to physical theories, but he was not a consistent follower of Mach's philosophy. Later, in the period around 1920, he came to appreciate the ideas of Immanuel Kant. From his youth (his family ran various electrotechnical enterprises), Einstein had special affinity for scientific instruments and apparatus, but contemporary experiments seem not to have influenced the course of his theoretical work. In formulating the special theory of relativity, Einstein sought to eliminate the apparent contradiction between classical electromagnetic theory and classical mechanics. He argued that the notions of absolute motion and ether were not essential elements of physical reality. Before he settled on the covariant field equations late in 1915, Einstein occasionally wrote that observation might disprove the general theory of relativity. Nevertheless, by the period around 1914 his faith in general relativity became fixed. He wrote to his good friend Michele Angelo Besso in 1914 about an attempt to verify the gravitational deflection of light during a solar eclipse: "Now I am fully satisfied, and I no longer doubt the correctness of the whole system, whether the observation of the eclipse will succeed or not. The reasonableness of the matter is too evident." As he grew older aesthetic criteria from pure mathematics played an ever more important role in his formulation of physical laws.

Einstein's contemporary Ernest Rutherford has epitomized the modern experimental physicist. To our sophisticated eyes his ingenious equipment may seem small and improvised, but using it, Rutherford had a hand in revealing more about the nature of submicroscopic physical reality than perhaps any other experimentalist of the twentieth century. He was an expansive and commanding man, gifted with penetrating physical insight and the talent for demonstrating his intuitions. Around his laboratories in Montreal, Manchester, and Cambridge gathered a brilliant constellation of twentieth-century physical scientists.

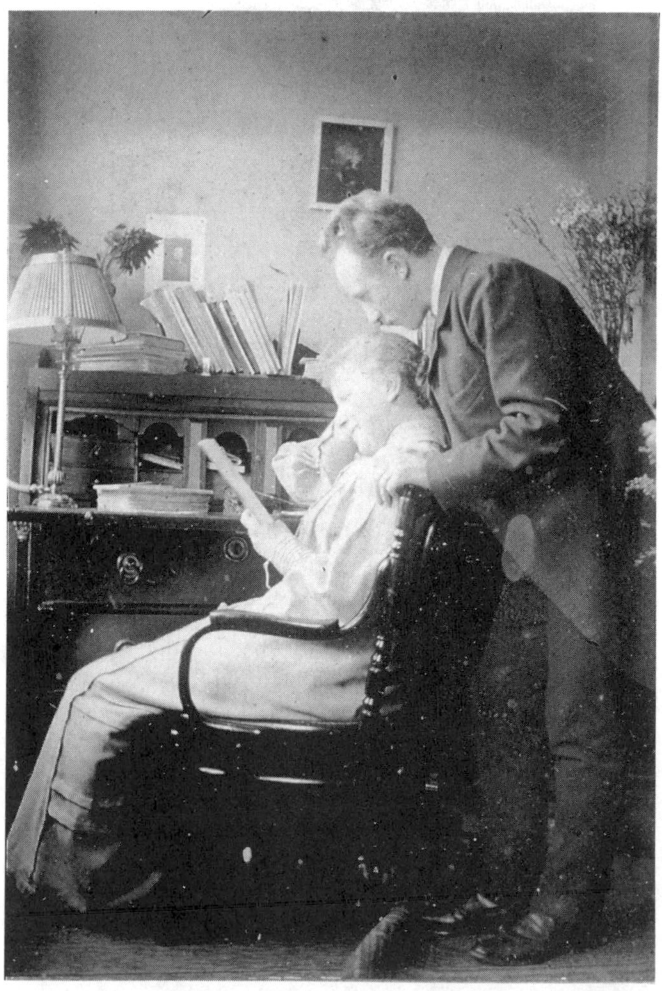

Fig. 2. A Couple of Experimentalists. Jacob Clay and Tettje Jolles, 1906, in Leiden. Both studied with Heike Kamerlingh Onnes. After more than a decade of work on low-temperature physics and the philosophy of science, he became the first physics professor at the Bandung Institute of Technology, on Java. There, in the 1920s, the Clays carried out research into cosmic rays and solar radiation. (Courtesy of Mevr. Nelke van Osselen-Clay.)

Fig. 3. Master and Disciple. Ernest Rutherford and J. J. Thomson, June 1934. (Bainbridge Collection, Niels Bohr Library, American Institute of Physics).

Rutherford received early university physics training in his native New Zealand. He went to Cambridge in 1895 when the first choice for a scholarship declined to accept the award. Working under J. J. Thomson at the Cavendish Laboratory, he examined the ionizing radiation produced by x rays, which had just been discovered by Wilhelm Conrad Röntgen at Würzburg. Rutherford quickly showed that the spontaneous radiation discovered in uranium by Henri Becquerel was the same as the ionizing radiation that he had studied, and that the uranium radiation had two components, which he called alpha and beta rays. In 1898 the 27-year-old Rutherford was appointed to a research chair endowed by the Macdonald tobacco fortune at McGill University in Montreal. There he showed that radioactivity was solely a property of atoms, discovered atomic transmutation through radioactive decay, and identified alpha rays as helium atoms with a positive charge twice that carried by an electron. In 1907 he went to fill Arthur Schuster's chair at Manchester University, where in 1911 he demonstrated the existence of atomic nuclei by observing the recoil of alpha particles that had been directed at a thin gold foil. In 1919 Rutherford succeeded his old teacher J. J. Thomson at the Cavendish. During the interwar period Cambridge was radioactive with Rutherford-dominated illuminati.

By the eve of the first World War quantum physics and relativity together with new ray and radioactivity phenomena were widely discussed in German periodicals. The theoretical side of these developments only slowly penetrated to Great Britain, France, and the United States. The acceptance of the relativity theories and quantum physics cannot be summarized easily, but the most important factor in their reception may be that theoretical physics was institutionalized principally in greater German-speaking Europe. Theoretical physicists often sought to synthesize all fundamental laws in what they called a physical world picture. Used by Max Planck, Einstein, and others, the term world picture—*Weltbild*—indicated a comprehensive structure of the physical universe that was based on a number of fundamental physical principles or prejudices. A physical world picture was considerably more than a model useful for explaining a limited range of phenomena. Theoretical physics was framed in the language of mathematics and it explained or addressed experimental propositions, but many theoretical physicists saw a world picture as the necessary basis of physical theories.

Theoretical physics beyond German-speaking Europe was carried out by and addressed to a broad community of physical scientists. Theoretically minded physicists in Britain studied to succeed in the Cambridge mathematics tripos, examinations that emphasized mathematics as well as mathematical problems in mechanics. The British physics community showed little interest in quantum theory or the theoretical side of atomic physics. Furthermore, distinguished physicists like Oliver Lodge and Joseph Larmor were hostile to relativity, and they refused to abandon the ether. Many others thought that the principle of relativity itself was nonsense. French theoreticians were trained as mathematicians, lesser copies of Henri Poincaré. They focused for the most part on problems in classical mechanics. Young theoreticians who studied quantum theory and relativity, such as Perrin, Paul Langevin, and Edmond Bauer, were exceptional.

The first World War disrupted physics throughout Europe. No different from men in other professions, H. G. J. Moseley, Karl Schwarzschild, and Friedrich Hasenöhrl marched to their death. Funds and personnel bled away from major centers of research at Manchester, Göttingen, and Leipzig. Dozens of young French physicists, graduates of the Polytechnical School of Paris and junior officers in the army, led hopeless charges against German machine-gun emplacements. German physicists travelling in Allied lands—Erich Hupka passing through Ceylon, Peter Pringsheim visiting Australia—were interned. When Allied forces overran German institutions, Bruno Meyermann and Georg Angenheister spent the duration under various forms of confinement. Laboratories and observatories—in Lebanon's Beqa'a Valley, on Tahiti, at Tsingtao, in Belgium—suffered violation and ruin. Loyal British subjects—Australia's T. H. Laby and Canada's J. C. MacLennan—volunteered their talents to professional killers. And through it all came brilliant elaborations of general relativity, atomic physics, and meteorology by physicists whom fortune or conscience had granted respite from the slaughter. Only a small flow of information passed between belligerent powers through Niels Bohr's Copenhagen, Paul Ehrenfest's Leiden, and Svante Arrhenius's Stockholm.

FIG. 4. Astrophysicist in Paradise. Milan R. Stefánik on Tahiti, 1910, for observing Halley's comet. A Slovak educated in Prague, Štefánik moved to France and across the decade before the First World War travelled widely on French astrophysical missions. During the war he commanded Czechoslovak troops under French colors; then he became the first minister of war in Czechoslovakia. (Emil Purghart, ed., *Štefánik vo fotografii. 2: Vydanie* [Orbis, Prague, 1938].)

The greatest part of quantum mechanics was forged by German-speaking physicists during the years after the war. In 1913 Bohr's atomic model had combined quantum constraints with classical mechanics and succeeded in interpreting the gross features of atomic spectra. Nevertheless, ten subsequent years of atomic model building failed to explain spectral fine structure. Arnold Sommerfeld and his students at Munich formulated complicated, although never completely satisfying, atomic models by quantizing the action principle to describe electrons orbiting a central nucleus in precise paths. What we know today as matrix mechanics emerged when Werner Heisenberg abandoned classical models in favor of formalism from a quantum theory of radiation that had been used by Rudolf Ladenburg to explain dispersion of light in gases. By the spring of 1926, P. A. M. Dirac, Max Born and Norbert Wiener, and Erwin Schrödinger had proposed alternative formulations of quantum mechanics, although all were soon demonstrated to be equivalent to matrix mechanics.

In 1927 Heisenberg announced the principle of uncertainty, according to which the product of certain pairs of operators that correspond to physical quantities has as a lower limit a fixed value determined by Planck's constant. This principle later gave rise to a large philosophical literature about modern physics, causality, free will, and human destiny, but at the time many physicists quickly assimilated it as a fundamental limitation on physical knowledge. Paul Forman suggested in 1971 that physicists embraced indeterminism in response to a hostile environment in Weimar Germany. As the nation careened through political and economic crises, many nonscientists developed a strong antipathy to logic, reason, and deterministic science. These highly valued attributes of culture in Imperial Germany were placed at the root of Weimar Germany's troubles. German physicists had long sought to precipitate crises in physics, and during the early 1920s increasing numbers of them hoped for a clean break with the past. A flurry of papers asserted the necessity of a quantum mechanics predicated on abandoning classical principles. It was even argued that energy might not rigorously be conserved in all physical processes. German physicists looked to the uncertainty principle in a perhaps unconscious attempt to appease critics of physical science and to rehabilitate their own image as important contributors to Weimar culture.

If the theoretical underpinning of physics today is a product of nineteenth- and early twentieth-century German scientific culture, contemporary experimental physics has been most strongly conditioned by events in the United States and to a lesser extent in the Soviet Union. Teams were able to mount expensive experimental research with Rutherford at Cambridge, with Enrico Fermi at Rome, and with Frédéric Joliot-Curie at Paris, but it was in the United States that experimental physics emerged on a massive scale. Requiring many experts working for years to construct enormous experimental installations, big physics in America was guided by an elite circle of science brokers who channeled money from many sources into selected research programs. European physicists watched while their American colleagues helped themselves from a groaning table.

Big physics depended on funding from three sources. The first source, industry, had furnished university laboratories in physical chemistry and mechanics in Germany and France since the end of the nineteenth century, and industrial laboratories such as those at Eindhoven and Schenectady cautiously began to fund basic research. Private foundations were a second source of support. Some of these, such as the Teyler's Stichting in Haarlem, the Royal Institution in London, and the dozens of Jesuit congregations, had long played a key role in financing physical research, but the generation of enormous private fortunes late in the nineteenth century gave new meaning to scientific philanthropy. Empires in physics arose from the generosity of plutocrats and robber barons, although it is an open question if results at the most lavishly appointed institutions—the Carnegie Institution of Washington, the Nice Observatory in France, the Nobel Institute in Stockholm—came close to satisfying expectations. A third source of funding was government largesse distributed outside the usual channels that supported higher education. Governments had long favored research at their own institutions, such as the Imperial Institute of Physics and Technology at Berlin and the Naval Observatory in Washington. By the 1920s national supervisory agencies were established in Canada and much of Europe to direct government funds to authors of the cleverest grant proposals. The rise of accelerator physics depended on consummate skill in obtaining money from all three sources, and during the lean 1930s enterprising physicists exploited lucrative technologies generated by their research. Ernest O. Lawrence financed part of his cyclotron research at Berkeley by dispensing cancer radiation therapy.

The second World War and its aftermath produced an awesome respect for the knowledge wielded by physicists. The period 1939–1945 could be called, with some accuracy, the "physicists' war." Although the world pictures of physicists could not directly power the engines of industry or the wheels of commerce, physicists were held to guide basic research that yielded useful gadgets, and they possessed the key—the "secret," many believed—to the enormous energies of the atom. Respect was tempered with official sanctions, and none of the great powers hesitated to discipline physicist heroes for entertaining politically subversive notions. Nevertheless, by the 1950s most reasonable funding requests from physicists were honored. Driven forward by military and space research, physics flourished in the lavish environment of the late 1940s and 1950s.

About mid-twentieth century, national military establishments funded the lion's share of physics-related work. Armies and navies, long interested in purchasing or pioneering new technologies, supported a great deal of pure research. The integrity of pure learning—the proud cornerstone of science in the nineteenth-century German universities—fell to crass expediency. In the United States, the Soviet Union, and elsewhere, expediency has been a way of life for more than a generation, for what escaped the military's grasp has been caught up by governmental regulation and control. The pattern is familiar to students of physics in France, where for nearly 200 years generals and admirals, sporting academic laurels, have captained major research boards and institutions, and where physics has long tacked to political winds.

The postwar development of big physics resulted in a new kind of physics laboratory. Only during the last third of the nineteenth century had laboratories especially designed for physics research and teaching first been erected. The principal desideratum in their construction was an attempt to provide space for undisturbed measurement: Whole wings were built without iron, and massive piles were fixed to resist vibrations. By around 1910 these laboratories had outlived their design. Vibrations from new trolley lines made precision mechanical adjustments almost impossible. Furthermore, many measurements of spectra, radioactivity, and phenomena using cathode-ray tubes did not require vibration-free, magnetic isolation. Accelerators brought an end to the self-contained physics laboratory. Just as geophysics and astrophysics had left the physics laboratory by 1914, so the apparatus of atomic and nuclear physics moved to new quarters during the 1930s.

In the period after the second World War the accelerator laboratory was much larger than even the biggest prewar government or industrial research installations in physics. Big physics centered around big machines, much as astronomy had long depended on powerful telescopes. Large teams of specialists conducted experiments conceived by several principal investigators. It became common to see papers authored by tens of physicists who were affiliated with a handful of institutions. In addition to resident and visiting

FIG. 5. A Modern Coeducational Teaching Laboratory, 1893–1894. West Physics Building, University of Michigan. Built with high ceilings and adequate plumbing, outfitted with a spectroscope and variable-resistance box, such a laboratory would not have seemed out of place for elementary undergraduate instruction as late as the 1960s. (Physics Department, University of Michigan.)

physicists, accelerator laboratories employed many service and auxiliary personnel. Rutherford's talented mechanics—Kay, Baumbach, and Niedergesass—were reflected in scores of engineers who built and serviced high-energy accelerators. Routine analysis of experimental data, often recorded in cloud-chamber photographs, was performed by nonphysicists. Housewives and bohemians, modern counterparts of Justus von Liebig's laboratory janitor Aubel, scanned hundreds of thousands of photographs for evidence of elementary-particle interactions. New elementary particles were discovered by nonphysicists who stood near the bottom of the social hierarchy in large national laboratories.

Big physics contributed to increasing estrangement between theoreticians and experimentalists. In the 1950s Enrico Fermi was sometimes held to be the last physicist at ease equally with experiment and theory. During the course of the twentieth century increasingly fewer physicists mastered both the yin and the yang of their discipline. Twentieth-century theoretical physicists found that they could speak more easily with mathematicians and theoretically minded chemists and astronomers than with other physicists interested in experimental problems. Experimentalists drew closer to nuclear chemists and those scientists interested in designing equipment using optical, electrical, and vacuum technologies. The unity of experiment and theory continued to supply the foundation for physics education after the war. The two were indeed combined by many physicists who undertook research in physical optics, solid-state electronics, and low-temperature phenomena, but the physics profession nevertheless continued to drift even farther into theoretical and experimental camps.

Modern physics has so far been described within the developed countries of Western Europe and North America. This emphasis should not be surprising, for English, German, and French cultures ruled over intellectual activity through the first third of the twentieth century, and their

FIG. 6. Physicists Preparing the Big Eye. Enrique Gaviola (with beret) and John Strong inspecting the aluminum coating on the 100-inch mirror for Mt. Wilson in California, June 1935. Gaviola, a native Argentinian who had obtained a doctorate at Berlin in 1926, was on leave from the University of Buenos Aires; in 1940 he became director of the La Plata Observatory in Argentina. Strong, at the California Institute of Technology, was the supervisor of Gaviola's Guggenheim fellowship. (Courtesy of Dr. Gaviola.)

FIG. 7. Van de Graaff Generator Constructed by the Department of Terrestrial Magnetism of the Carnegie Institution of Washington. By the end of the 1930s projects were underway to produce particle accelerators many orders of magnitude greater in cost and energy output. This photograph shows the Van de Graaff on 18 May 1935. (Courtesy of the Department of Terrestrial Magnetism of the Carnegie Institution of Washington.)

impact in science has survived even when the physical presence of military garrisons and cultural missions has disappeared. At the same time, by the beginning of the twentieth century indigenous physics had developed in perhaps a dozen regions previously dominated by a foreign scientific culture. By around 1920 the discipline of physics was taught and successfully practiced in popular languages in Poland, Finland, Norway, Hungary, and Flanders. Among those who issued from emerging twentieth-century physics communities in Europe were Leopold Infeld, Gunnar Nordström, Vilhelm Bjerknes, Leo Szilard, and Marcel Minnaert.

By the eve of the first World War physics had been established in several non-European cultures. The most unusual implantation occurred in Japan. The Meiji regime that came to power in 1868 formulated a national policy of adopting Western learning. Japanese science students went abroad for extended periods of study and, upon returning to Japan, taught their specialty in a Western language. The first generation of Japanese physicists was predisposed to acquire learning from a foreign culture, for all had been trained in Chinese as Confucian scholars. Some Japanese physicists sought to preserve their culture from destruction at the hands of the Western barbarians. In 1888 the young physicist Nagaoka Hantarō wrote in English to Tanakadate Aikitsu, then studying in Glasgow: "We must work actively with an open eye, keen sense, and ready understanding, indefaticably [*sic*] and not a moment stopping. . . . There is no reason why the whites shall be so supreme in everything, and as you say, I hope we shall be able to beat those *yattya hottya* [pompous] people in the course of 10 or 20 years: I think there is no use of observing the victory of our descendents over the whites with the telescope from *jigoku* [hell]." Nagaoka need not have been alarmed. Japanese physics had grown to maturity by 1922 when Einstein spent ten weeks in Japan visiting with his colleague and translator Ishihara Jun. Within twenty years the Japanese had research and teaching institutions in physics throughout their Asian and Pacific empire.

When the Japanese conquered European colonies in Asia, they found themselves in possession of laboratories and observatories that in some cases surpassed the standards of establishments in Japan. Particularly impressive were the French Jesuit observatory at Zikawei (Xujiahui) near Shanghai, directed by physicist Pierre Lejay, and the Dutch observatories and faculties on Java, which had been home to physicists Willem van Bemmelen and Jacob Clay, among others. After China and Indonesia freed themselves from colonial rule, each country slowly built upon a heritage of excellence in physical research; significant programs, in place during the 1960s, have since translated into vital enterprises in nuclear, astrophysical, and geophysical specialties.

Physics also came to countries with established, national scientific communities, such as Argentina. Eager to create a technical elite, in 1904 the Argentine minister of education, Joaquín V. González, renovated the moribund provincial University of La Plata. González advertised throughout Germany the availability of well-paying faculty positions. He succeeded in attracting Emil Bose, editor of the physics journal *Physikalische Zeitschrift*; Jakob Johann Laub, Einstein's first scientific collaborator; and Richard Gans, a recognized authority on magnetism. The La Plata program declined after Bose died and Laub and other German scientists left for positions elsewhere. Before the first World War there were enough physicists for critical discussions in the Buenos Aires region, and distinguished scientists such as Walther Nernst came for short lecture tours. The foreigners began to publish in Spanish, but they continued to measure professional aspirations by circumstances in Germany. When Einstein lectured at La Plata in 1925 he was well received by a German community whose numbers had been augmented after the war, but Argentine physics was a pale reflection of German forms. The wholesale transplantation of foreign physicists has been repeated many times throughout Africa and Asia among former colonies governed by unenlightened despots and dominated by ravaged economies and oppressive social structures.

The latest discoveries in physics have been made at a time when many physicists in the industrial Western countries are adjusting their professional and personal aspirations to new economic and social conditions. Research in all fields of science and scholarship suffered a major decline in financial support when the postwar economic expansion foundered in

Fig. 8. Danish Feminist in Argentina. Margrete Bose (*née* Heiberg), recipient of an advanced degree in chemistry from the University of Copenhagen (*Magistra Scientiarum i Kemi,* 1901) and an alumna of Walther Nernst's Göttingen, here instructing students as a professor of physics at the National University of La Plata, *circa* 1914. (Courtesy of Walter B. L. Bose.)

the middle 1960s. Fueled by American investment in exotic military and aerospace hardware, inflation and economic stagnation wrought havoc with research budgets. Physics continued to receive relatively greater support than other fields because it could claim to be directly relevant to military and technological goals. Even though only a fraction of young physicists found permanent academic or research positions in their own specialties, unemployment among American physicists remained considerably lower than the national average. Though many universities instituted programs cultivating narrow specialties in the physical sciences, well-trained physicists still commanded the respect of industry, government, and neighboring disciplines, a favored status that they had enjoyed for nearly a century.

Changes in the physics discipline also stemmed from the growth of disillusionment among young people with physics and with scientific world views. In the late 1960s physics acquired the image of an activity estranged from human values. Many critics observed that physicists were working with extravagant or destructive technologies. Others argued that abstract physics contributed to the anomie of industrial societies and to the grinding poverty of former colonies. As evidence that physics was suffering from a profound malaise, critics cited the absence of major synthetic formulations over the preceding generation. They argued that some problems in applied physics—for example, research in nuclear fusion power—were being suppressed by powerful interest groups in favor of other, less promising investigations. An undercurrent of discontent with both the Copenhagen interpretation of quantum mechanics and mechanical explanation in general, blossomed into alternative visions of physical reality. By the 1970s prophets of a new holism—David Bohm, René Thom, Ilya Prigogine, James Lovelock—contributed to a physics counterculture. The holists married physics to biology in an ecological vision of natural processes: Molecular biology (which by this time had come to rival physics for popular attention), subatomic physics, and the vast cosmos came under the rule of indeterminism, nonlinearity, and intangible emotion.

The holists emboldened nonphysicists in the 1970s to question the cost of big physics, believing that it was the same sort of cultural pursuit as music or art. High-energy physics seemed to confirm an increasingly popular notion that the discipline related less to seeking physical truths than to formulating an aesthetic view of the world. This understanding received encouragement from Thomas Pynchon and Stanislaw Lem, whose metaphorical use of physics continued a tradition upheld earlier by Paul Valéry, Henry Adams, and Robert Musil. Other authors were stimulated by the discovery of new astrophysical objects, and artists and musicians found new means of expression in laser optics and electronic data processing. Physicists tacked to the winds of public opinion. Departing from the moral justifications of Wilhelmian physicists and the arguments of technical utility used by early twentieth-century American physicists, in the late 1960s and early 1970s many physicists pointed to the aesthetic value of their research. Aesthetes like Hubert Reeves, Carl Sagan, and Stephen Hawking retained a large following during the 1970s and 1980s, and their paens have contributed to public support for large projects—colossal particle accelerators, enormous telescopes, and ambitious space missions.

The holists and the aesthetes contended for hegemony beyond the confines of nation and language. In the last decade of the twentieth century, progress in physics has ceased to be practised only within national institutions financed by the caprice of national resolve. International laboratories and research efforts—CERN, Ariane, and extensive mountaintop observatories—configure the cutting edge of research. Research teams increasingly have partners or entire laboratories in foreign lands, just as professors sport permanent overseas appointments. During the past hundred years, disciplinary coherence came to physics from the designs of the nation-state. What shall happen if this bond dissolves in a new and more humane world?

Note: Quotations in the text from: Otto Nathan and Heinz Norden, eds., *Einstein on Peace*. Schocken, New York, 1960; Lewis Pyenson, "Einstein's Early Scientific Collaboration," *Historical Studies in the Physical Sciences, 7,* 83–124 (1976); Kenkichiro Koizumi, "The Emergence of Japan's First Physicists, 1868–1900," *ibid., 6,* 3–108 (1975).

See also ACCELERATORS, LINEAR; ACCELERATORS, POTENTIAL-DROP LINEAR; ATOMIC SPECTROSCOPY; BOHR THEORY OF ATOMIC STRUCTURE; CLOUD AND BUBBLE CHAMBERS; DYNAMICS, ANALYTICAL; ELECTRODYNAMICS, CLASSICAL; ELECTROMAGNETIC INTERACTION; ELECTROMAGNETIC RADIATION; ELEMENTARY PARTICLES IN PHYSICS; ENTROPY; HEAT; KEPLER'S LAWS; KINETIC THEORY; LIGHT; LORENTZ TRANSFORMATIONS; MAXWELL'S EQUATIONS; MICHELSON–MORLEY EXPERIMENT; NEWTON'S LAWS; PHILOSOPHY OF PHYSICS; QUANTUM MECHANICS; RADIOACTIVITY; RELATIVITY, GENERAL; RELATIVITY, SPECIAL THEORY; STATISTICAL MECHANICS; THERMODYNAMICS, EQUILIBRIUM; UNCERTAINITY PRINCIPLE.

BIBLIOGRAPHY

Students of the history of physics may begin with the index to the *Dictionary of Scientific Biography* (16 vols.; Scribner's, New York, 1970–1980) and several sets of cumulative indexes compiled from the annual bibliographies published by the history of science journal, *Isis* (founded in 1913 by George Sarton). Nineteenth-century and early twentieth-century physicists are often well represented in the various editions of J. C. Poggendorff's bio-bibliography and in the catalogue to scientific papers published by the Royal Society of London. The Center for the History of Physics of the American Institute of Physics publishes an occasional newsletter with a selected bibliography. The journal *Historical Studies in the Physical and Biological Sciences* also notes recent books.

For students interested in the history of physics beyond the North-Atlantic world, the standard sources may be supplemented by bibliographical notices in periodicals with national or regional focus, such as: *Historia Scientiarum* and related publications of the History of Science Society of Japan; the *History of Australian Science Newsletter* and the *Historical Records of Australian Science; Scientia Canadensis*; the *Boletín Informativo* of the Sociedad Latinoamericana de Historia de las Ciencias et la Tecnología, as well as the Society's journal *Quipu*; the *Revista* of the Sociedade Brasileira de História da Ciência; *Indian Journal of History of Science*.

Location of published material has been made easier by on-line library catalogues. Notable authors of recent books and monographs treating themes germane to the preceding essay include: Klaas van Berkel, Alan Beyerchen, Jed Z. Buchwald, David Cahan, Elisabeth Crawford, Paul Forman, Robert Friedman, John L. Heilbron, Roderick W. Home, Christa Jungnickel, Daniel Kevles, Mario Mariscotti, Russell McCormmach, Walter A. McDougall, Kathryn Olesko, Dominique Pestre, Chikara Sasaki, Terry Shinn, Vladimir P. Vizgin, Spencer Weart, and José F. Westerkamp. It should be emphasized that the field of the history of science has produced a great number of important additions to the bibliography provided in the first edition of this encyclopedia.

Holography

P. Hariharan

Holography is a technique that makes it possible to store and reproduce three-dimensional images. Holographic images can be produced, in principle, with any type of wave. However, this article will only deal with optical holography, excluding related techniques such as acoustical and microwave holography, whose applications have not developed to the same extent.

Holography was invented by Dennis Gabor in 1947 in an attempt to improve the resolution of images obtained with an electron microscope. Within a few years of its invention, it appeared destined for obscurity because of two serious problems: the desired image was overlaid by a spurious image in line with it and there was no suitable source of coherent light for making holograms. However in the early 1960s, Leith and Upatnieks demonstrated a new approach to holography based on communications theory that made it possible to obtain images of good quality. At about the same time the invention of lasers provided an intense source of coherent light. These developments triggered a major research effort in holography.

While Gabor's original goal has not yet been achieved, the last two decades have seen several breakthroughs which have established holography as a practical technique with a remarkably wide range of applications.

BASIC CONCEPTS

The unique characteristic of holographic imaging is the idea of recording both the amplitude and the phase of the light waves coming from the object. Since all available recording media respond only to light intensity, it is necessary to convert the phase information into variations of intensity. This is done by using a coherent, monochromatic source and adding to the wave front from the object a reference wave front of known amplitude and phase derived from the same source.

To record a hologram, the object is illuminated with light from a laser as shown in Fig. 1, and a photographic film is placed so that it receives light reflected from the object. The first difference from ordinary photography is that no lens is needed. Each point on the object reflects light to the entire photographic film. The second difference is that a portion of

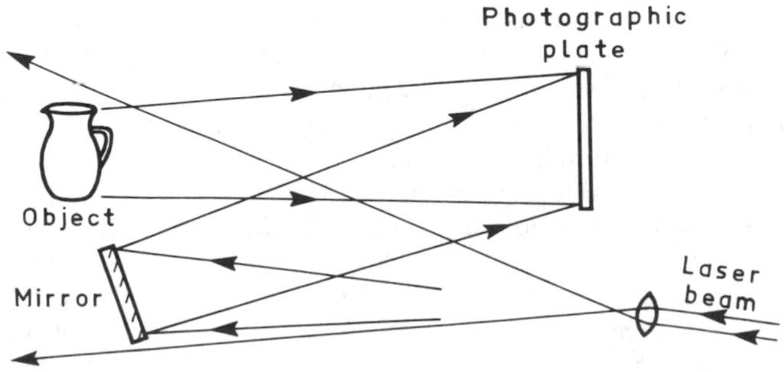

FIG. 1. Schematic of the optical setup used to record a hologram.

the light from the laser is incident directly on the film. This is called the reference beam. The film records the interference pattern between the light waves reflected from the object and the reference beam. This record is called the hologram of the object.

Since the light waves reflected from the object have a very complex form, the interference pattern making up the hologram is quite irregular. It is also so fine that its details can only be seen under a microscope. To the naked eye the hologram looks like a grey blur which bears little or no resemblance to the object. However, all the details of the light waves reflected by the object are recorded in it.

To view the image, the hologram is illuminated with a beam of monochromatic light as shown in Fig. 2. The light waves diffracted by the hologram are a replica of the light waves that came originally from the object. An observer looking through the hologram sees a lifelike image of the object in the same position in which it was when the hologram was made.

This reconstruction presents a natural 3-D appearance and has all the visual properties of the original object. The hologram looks like a window through which the image can be seen. When the observer moves his head, the perspective of the image changes, as shown in Fig. 3. If an object in the foreground masks something, the observer can look around it.

If we try to photograph the image, we find that it also exhibits depth-of-focus effects. With the camera lens wide open, it is possible to focus on only one plane in the image; other planes in front of this plane and behind it are out of focus. To bring the whole image into focus, it is necessary to stop down the camera lens just as with the original object.

An interesting feature of the hologram is that if it is broken, each part still reproduces the entire image. This is because each point on the hologram has received light from all parts of the object. The only change is that the viewing window through which the image is seen is smaller. For the same reason, holograms can be scratched quite badly with little effect on the image, which appears some distance behind the hologram.

Several practical problems have to be faced when making holograms. In the first place, the spacing of the interference fringes recorded in making a hologram is extremely small—usually less than a micrometer. Special ultrafine-grain films are needed, which are correspondingly slow. In addition, it is necessary to see that the object and, in fact, the whole

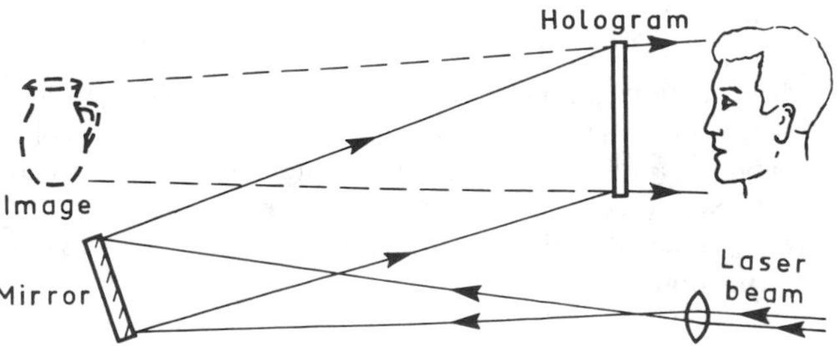

FIG. 2. Viewing the holographic image.

FIG. 3. Views from two positions of the image reconstructed by a hologram, showing different perspectives.

setup, does not move by more than 0.1 μm during the exposure. For these reasons, holograms are normally recorded on a massive steel or granite table floated on an antivibration system. Alternatively, it is possible to use a pulsed ruby laser which produces a flash of light lasting only a few nanoseconds.

APPLICATIONS

Holography has many fascinating applications. Current trends suggest that the areas of greatest interest are art and advertising, high-resolution imaging, information storage and security coding, holographic optical elements, and holographic interferometry.

Art and Advertising

Some very spectacular applications of holography as a medium for art and advertising were demonstrated at quite an early stage. These included the production of a hologram of the Venus de Milo and portraits of living human beings with pulsed lasers. However, these early holograms had several drawbacks which stood in the way of their wider acceptance. In the first place, they needed a laser or a bright point source of monochromatic light to illuminate them. Even then, the reconstructed image was dim and could be viewed only in subdued light. Finally, the image was reconstructed in a single color, that of the source used to illuminate the hologram.

Some progress was made toward the solution of these problems in the early 1960s. This period saw the development of white-light imaging using reflection holograms, the development of new recording materials such as photopolymers and dichromated gelatin, and improved processing techniques for commercial photographic materials.

A major advance was the invention in 1969 of the rainbow hologram by Benton. This is what may be called a second-generation hologram. A hologram of a solid object is recorded initially in the conventional way. When this primary hologram is properly illuminated, it projects a real image of the object. The aperture of this primary hologram is then deliberately limited by a horizontal slit and a second hologram is made of the projected real image.

When this second hologram is illuminated with monochromatic light, it forms, in addition to an image of the object, an image of the slit which is projected out into the viewing space. With white light, the slit image is spread out into a spectrum in the vertical plane. The observer can position his eyes at any point on this spectrum (hence the name "rainbow" hologram); he then sees a bright, sharp image in the color corresponding to that part of the spectrum.

Two further advances followed. One was the white-light holographic stereogram of Cross. This was a composite rainbow hologram built up from a number of views of the subject from different angles in the horizontal plane. These views can be recorded with a motion picture camera in white light and subject movement presents no problems. Another was the production of multicolor rainbow holograms. In this case, the final hologram is built up by successive exposures to the real images projected by three primary holograms, made either with different laser wavelengths or with the reference beam incident at different angles. When this multiply-exposed hologram is illuminated with a white-light source, the observer sees a full-color three-dimensional image reconstructed from the three color separations. These techniques have made it possible to produce multicolor holographic stereograms up to 2 m wide and 1 m high.

Another valuable technique which was developed during the 1970s was the production of copies of holograms by embossing on a thin sheet of plastic backed with an evaporated metal coating. Embossed holograms can be produced cheaply in large quantities and give a bright image when viewed by reflection. They have opened up a rapidly expanding field of applications ranging from novelties and greeting cards to book and record covers.

Holographic movies have also attracted considerable attention. The major problem here is that the holograms have to be small for economy, while the image has to be large enough for viewing by a number of people. The most advanced system so far is that developed by Komar in the USSR, in which a series of holograms are recorded with a pulsed laser on 70-mm film using a lens with an aperture of about 200 mm. The same lens is then used to project the reconstructed images on a holographic screen which forms a number of images of the pupil of the lens in the viewing space. Each of these imaged pupils constitutes a viewing zone within which an observer can see a three-dimensional image.

High-Resolution Imaging

One of the interesting features of a hologram is that it can reconstruct an image with very high resolution over a great depth. Because of this, one of the first successful applications of holography was in the study of aerosols. Measurements on moving microscopic particles distributed over an appreciable volume are not possible with a conventional microscope because of its very limited depth of field. Pulsed laser holography makes it possible to store an image of the whole particle field at a given instant. The stationary image reconstructed by the hologram can then be studied in detail with a microscope.

Holography can also be used to obtain an image which is unaffected by lens aberrations and even by the presence of aberrating media in the optical path. The object is illuminated with coherent light and a hologram is recorded of the aberrated image wave using a collimated reference beam. When the hologram is illuminated by the same collimated reference beam reflected back along its original path, it reconstructs the conjugate to the object wave, which has the same phase errors as the original object wave but with the opposite sign. Accordingly, when this conjugate wave propagates back through the optical system, these phase errors cancel out exactly, resulting in an aberration-free image. Such an image can be produced in real time if a photorefractive crystal (e.g., bismuth silicon oxide) is used as the recording material.

One possible application of holographic high-resolution imaging is in the production of photolithographic masks for semiconductor devices. Holographic systems could overcome the problem of designing and producing lenses which can give very high resolution over a large field.

Information Storage

The fact that information stored in a hologram is not easily corrupted by local surface damage led at an early stage to research on holographic memories. Due to the lack of a suitable recording material, this application now appears to have been overtaken by other techniques. However, holographic information storage still appears to have advantages for some specialized applications. One is the reduction of the space required for archival storage of documents, far below that possible with microfilm. Others are for storage of multicolor material, such as motion pictures, and for storing three-dimensional information.

An application with considerable commercial importance is security coding for credit and identity cards. A simple method adopted by some of the major credit cards involves an embossed hologram incorporated in the card to provide an additional safeguard against forgery.

In the cardphone, now widely used in Britain, the credit card has a number of holograms imprinted on its surface. These are illuminated by a light source in the cardphone and counted to work out how many units the user has left. As the call progresses, the holographic patterns are erased one by one by a focused heat source in the cardphone.

Holographic Optical Elements

Diffraction gratings formed by recording an interference pattern in a photoresist layer are commonly known as holographic gratings. Holographic gratings are free from ghosts and exhibit very low levels of scattered light. They have replaced conventional ruled gratings for many spectroscopic applications.

Holographic scanners are now widely used in point-of-sale terminals. Typically, the scanner consists of a disc with a number of holograms recorded on it. Rotating the disc about an axis perpendicular to its surface causes the reconstructed image spot to scan the image plane in the desired pattern.

Holographic optical elements (HOEs) are much lighter than lenses and mirrors, since they can be produced on thin substrates. Another advantage is that several holograms can be recorded in the same layer to produce multifunction elements. In addition, holograms can correct system aberrations. One of the most successful applications of HOEs has been in head-up displays for high-performance aircraft. The optical system in such a head-up display projects an image of the instruments along the pilot's line of vision so that he can monitor critical functions while looking straight ahead through the windscreen.

An attractive possibility is the use of computer-generated holograms as HOEs. With these it should be possible to realize quite unusual optical systems. However, the main application of computer-generated holograms so far has been in interferometric tests of aspheric optical surfaces. The hologram corresponds to the interferogram that would be obtained if the wave front from the desired aspheric surface were to interfere with a tilted plane wave front. This hologram is placed in the plane in which the surface under test is imaged. The superposition of the actual interference pattern and the computer-generated hologram then produces a moiré pattern which gives the deviation of the actual wave front from the ideal wave front.

A potentially very interesting application of HOEs is to provide optical interconnections between integrated circuits; these could overcome many of the limitations of conventional wire interconnections.

Holographic Interferometry

Holographic interferometry makes use of the fact that a hologram can store an accurate, three-dimensional image of an object. This makes it possible to compare the shape of the object when it is stressed with a hologram image of its shape in its normal condition.

To do this, the hologram is replaced, after processing, in exactly the same position in which it was recorded and is illuminated with the original laser beam, so that it reconstructs an image which coincides with the object. If, then, a stress is applied to the object, so that its shape changes very slightly, the observer receives two sets of light waves which have traversed slightly different paths. These waves reinforce each other at points in the field of view where the separation of the surfaces of the object and the image corresponds to an optical path difference of zero or a whole number of wavelengths and cancel each other at some points in between. As a result, the image is covered with a pattern of bright and dark fringes which, as shown in Fig. 4, map in real time the displacements of the surface of the object. Very accurate measurements can be made of these changes since the fringes are not affected by surface roughness, and the contour interval corresponds to a movement of the surface of approximately half a wavelength (a fraction of a micrometer).

Alternatively, two holograms can be recorded on the same photographic plate, one of the object in its normal state and

FIG. 4. A hologram reconstructs an image of the unstressed object. The interference fringes map the changes in the shape of the object.

the other while it is under stress. In this case, interference fringes are generated by the two reconstructed images, so that the hologram is a permanent record of the changes in the shape of the object. With a pulsed laser it is possible to study transient deformations and rotating objects.

Holographic interferometry has found wide application in nondestructive testing to detect cracks and areas of poor bonding and is now an important part of the quality-control process in the aerospace industry. Turbine blades, load-bearing structures in aircraft, and even tires are routinely subjected to holographic inspection. The automobile industry is also using this technique to improve engine performance and reduce noise from transmissions and door panels. Other applications are in studies of plasmas and in medical research to measure the deformations of implants under stress.

Holographic interferometry can also be used to produce an image modulated by a fringe pattern corresponding to contours of constant elevation with respect to a reference plane. Three basic methods of holographic contouring are available; these involve changing either the wavelength used or the angle of illumination or, alternatively, the refractive index of a medium in which the object is immersed.

Yet another application of holographic interferometry has been in studies of vibrating surfaces. One way is to use stroboscopic illumination, but a widely used alternative is time-average holographic interferometry. In this technique a hologram of the vibrating object is recorded with an exposure which is long compared to the period of vibration.

The scattered light from the object now exhibits a time-varying phase modulation, and hence contains frequency-shifted components. Only the component at the original frequency can interfere with the reference wave. As a result, the brightness at the image at any point is multiplied by a factor $J_0^2(\xi)$, where ξ is proportional to the amplitude of vibration. The image appears overlaid with dark fringes corresponding to the zeros of this function which can be used to map the vibration amplitude.

Figure 5 shows a set of time-average holograms of some

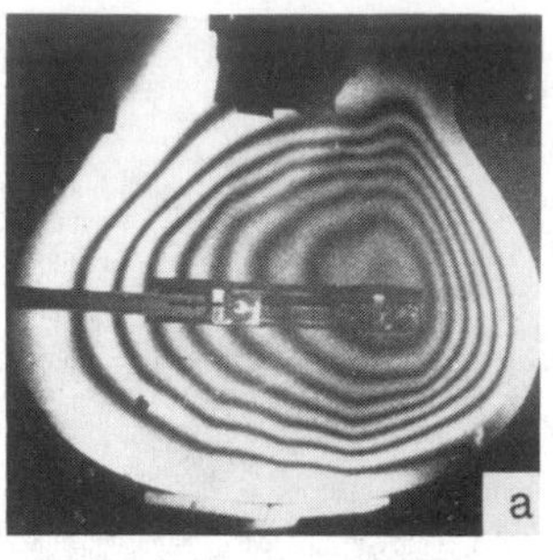

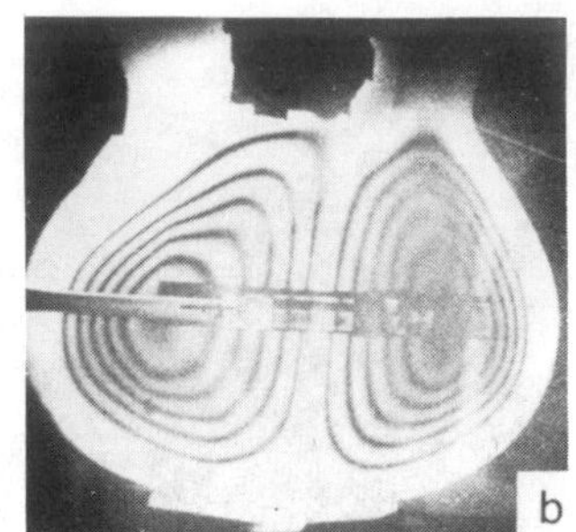

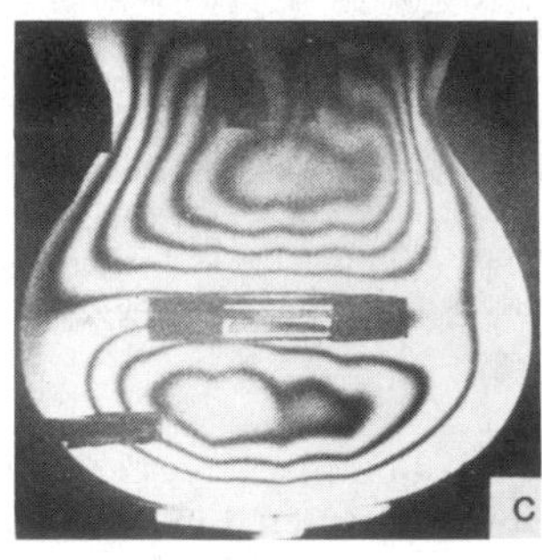

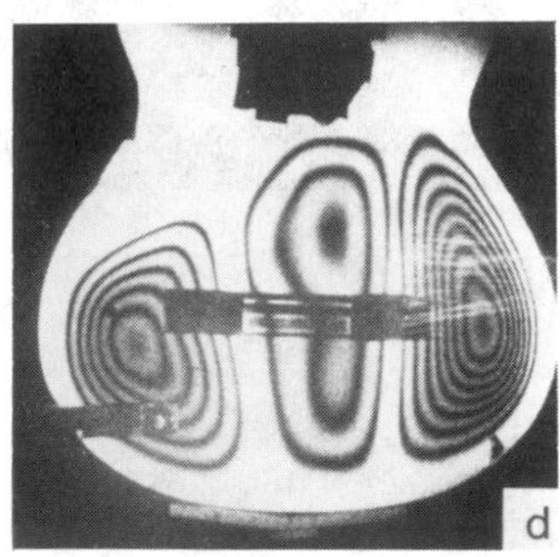

FIG. 5. Time-average holograms showing the vibration modes of the soundboard of an acoustic guitar at frequencies of (a) 195, (b) 292, (c) 385, and (d) 537 Hz.

of the modes of vibration of the soundboard of an acoustic guitar. Such studies have led to instruments with improved acoustic performance.

Digital phase-stepping techniques now make possible very accurate measurements of the surface displacements at a uniformly spaced network of points covering the object. Because of the speed with which data can be acquired and manipulated, digital phase-stepping techniques open up many new applications for holographic interferometry in quantitative stress analysis.

CONCLUSION

It is clear from this survey that holography has already established itself in several rapidly expanding areas. We can look forward confidently to other completely new applications in the next few years.

See also GRATINGS, DIFFRACTION; INTERFEROMETERS AND INTERFEROMETRY; LASERS; LIGHT; OPTICS, PHYSICAL; WAVES.

BIBLIOGRAPHY

E. N. Leith and J. Upatnieks, *Sci. Am.* **212,** 24 (June 1965). (E)

R. J. Collier, C. B. Burckhardt, and L. H. Lin, *Optical Holography*. Academic Press, New York, 1971. (A)

H. M. Smith, *Principles of Holography*. Wiley, New York, 1976. (A)

E. N. Leith, *Sci. Am.* **235,** 80 (October 1976). (E)

H. J. Caulfield (ed.), *Handbook of Optical Holography*. Academic Press, New York, 1979. (A)

C. M. Vest, *Holographic Interferometry*. Wiley, New York, 1979. (A)

G. Saxby, *Holograms*. Focal Press, London, 1980. (I)

P. Hariharan, *Optical Holography: Principles, Techniques and Applications*. Cambridge University Press, Cambridge, 1984. (A)

Hot Atom Chemistry

Peter P. Gaspar

Hot atom chemistry is the study of the chemical events, such as bond making, energy transfer, and electron exchange, that occur after excitation of an atom by a nuclear transformation.

The initial excitation leading to the formation of a hot atom can be induced by bombardment with high-energy particles such as protons or neutrons, and such transformations often result in high recoil energies that lead to bond rupture. Other processes, such as thermal neutron activation, and radioactive decay by beta emission, electron capture, or internal conversion, often endow daughter atoms with low recoil energies. Nevertheless, bond rupture can occur in these processes, despite their low recoil energy, as a result of the formation of vacancies in the inner shells of electrons. The vacancies are filled by electrons from outer shells, but these nonradiative atomic events, known as Auger processes, result in the loss of as many as 10 additional electrons, leading to highly positively charged atoms in excited electronic states. If these atoms are bound in molecules, the electrons will be redistributed, leading to several positively charged atoms in the same molecule, which immediately flies apart due to electrostatic repulsion. Photoionization with monochromatic x rays can also bring about inner electron-shell vacancy cascades by a non-nuclear process.

Thus, hot atoms are generally formed as ions carrying multiple positive charges and a recoil energy of from tens to millions of electron volts. The recoiling atom will lose energy to its surroundings through collisions with atoms and molecules in which electron exchange can also occur. When the hot atom is sufficiently deexcited so that its collisions can lead to the formation of new chemical bonds, a variety of chemical reactions can take place. At this stage the hot atom is typically electrically neutral, but monopositive ions can also reach the epithermal energy range, from 1 to 100 eV.

The production of hot atoms by nuclear recoil allows the study of chemical reactions of atoms and molecules possessing far more energy than the threshold for thermally induced processes. It is virtually the only technique known for producing superexcited molecules containing tens of eV of internal excitation in a bath of room-temperature molecules. Through the study of the decomposition of molecules formed with high internal energies through reactions of hot atoms, the distribution of recoil energies of the hot atoms undergoing reaction can be deduced.

One limitation on the nuclear recoil technique is the small number of hot atoms produced. In a typical experiment employing an accelerator or a nuclear reactor, 10^6 to 10^{10} hot atoms are produced over a 1-h period, each with a chemical lifetime of only on the order of 1 μs. The instantaneous concentrations of hot atoms are therefore so low, perhaps one hot atom among 10^{20} normal molecules, that it is difficult to detect a hot atom until it has incorporated itself via its reactions in a chemically stable molecule. Detection of hot atoms via their radioactive decay is facilitated by the production of short-lived isotopes.

Nevertheless, much has been learned about the chemistry of free atoms from the end products of hot atom reactions. Since new bonds tend to be formed one or two at a time in the collisions of hot atoms with molecules, novel molecules in which the valence of the hot atom is not saturated are produced as reactive intermediates in the reactions of hot atoms that must make more than two bonds to satisfy their bonding capability. Many new chemical reactions of free atoms and reactive intermediates have been discovered in hot atom experiments. Nuclear techniques for producing hot atoms complement such instrumental methods as molecular and ion beam experiments that can provide more direct information about reaction kinetics and dynamics.

When hot atoms with high recoil energies are produced by nuclear transformations in the gas phase, they can escape the molecule debris of atoms, radicals, and ions formed in the collisional energy loss cascade that brings the hot atom down to the energy range of chemical reactions. The much shorter range of recoiling atoms in the liquid and solid state sometimes results in chemical reactions of hot atoms with their own debris.

From the earliest days of hot atom chemistry in the 1930s, practical use has been made of the chemical consequences of nuclear transformations. Recoil processes facilitate the isolation of radioisotopes in high specific activities by delivering the product of a nuclear transformation in a different chemical form from that of the target nucleus. Thus, an organic halide could be irradiated with neutrons, and the radioactive halide atoms freed by nuclear recoil dissolved in aqueous solution without dissolving the stable halide bound in the parent molecules, an effect known as the Szilard–Chalmers reaction.

Molecules can be labeled directly by the reactions of hot atoms, but since these reactions can be quite unselective, the use of hot atom reactions to provide labeled molecules has been limited to such simple systems that the number of reaction channels is small. This has led to the practice of employing recoil reactions to form small molecules such as $^{11}CO_2$ and $H^{11}CN$ which are converted by chemical synthesis into larger molecules, no mean feat with half-lives as short as 20 min for ^{11}C. Hot atom chemistry has made important contributions to nuclear medicine by providing biomolecules containing short-lived isotopes for physiological and diagnostic studies.

The processes following nuclear recoil in solids are of interest because of the short range of the hot atom in the solid state, and have important implications for materials science and reactor technology. Solid-state hot atom chemistry can lead to the synthesis of simple inorganic and organic compounds that can be converted into a variety of useful molecules.

In situ methods have been directed at answering the most basic questions in solid-state hot atom chemistry: What fraction of the recoiling atoms remains in the chemical form of the target atom? What is the nature of the recoil species produced? What reactions do the recoiling atoms undergo? Mössbauer emission spectroscopy has revealed changes in charge state, ligands, and ligand geometry of atoms followng nuclear recoil on the time scale of 10^{-9}–10^{-7} s. Measurements of the perturbed angular correlation of emitted γ-rays caused by the coupling of nuclear moments with extranuclear electromagnetic fields can provide information about the local environment of a recoiling atom on a nanosecond time scale following a nuclear transformation.

See also ATOMS; RADIOACTIVITY.

BIBLIOGRAPHY

T. Tominaga and E. Tachikawa, *Modern Hot-Atom Chemistry and Its Applications*. Springer-Verlag, Berlin, 1981.

T. Matsuura, ed., *Hot Atom Chemistry: Recent Trends and Applications in the Physical and Life Sciences and Technology*. Kodansha, Tokyo and Elsevier, Amsterdam, 1984.

"Hot Atom Chemistry" (special issue), *Radiochim. Acta* **43** (1988).

Hot Cells and Remote Handling Equipment

J. M. Davis

In nuclear terminology, "hot" means radioactive, and a "cell" is an enclosure that provides containment for the radioactive material and shielding from penetrating radiation. Remote handling equipment includes mechanical, electrical, and hydraulic devices with which workers outside the hot cell can handle, process, and examine radioactive material inside the cell.

HOT CELLS

The design of hot cells and remote handling equipment varies greatly, depending on the kinds and amounts of radioactive materials to be handled, the processes or examinations to be conducted, and the production efficiency desired. A small steel box with a window, one ball-and-socket tong, and a syringe-actuated pipette is a hot cell for dispensing medical isotopes. A 450-m-long "canyon" with 1.2-m-thick concrete walls and inside bridge crane is a hot cell (see Fig. 1) for a reactor fuel reprocessing plant.

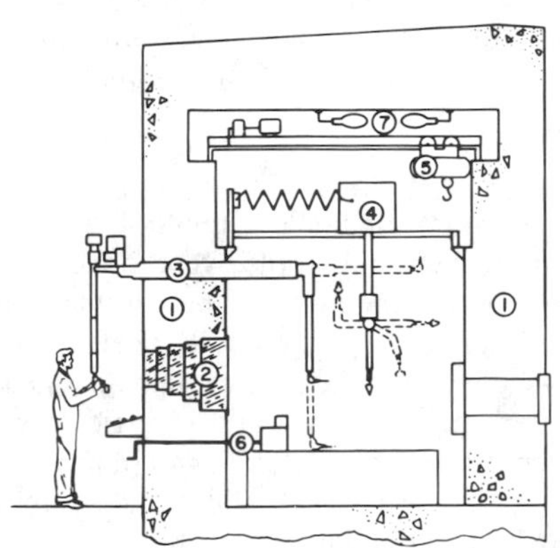

FIG. 1. A typical hot cell.

Design bases common to all hot cells, regardless of size or purpose, follow:

1. *Containment Integrity*. The radioactive particulate or gaseous effluent(s) from the hot cell must be ALARA (*a*s *l*ow *a*s *r*easonably *a*chievable), and in no case greater than allowed by Federal regulation. This is usually achieved by tight construction, operation at negative pressure, and highly efficient air filtration.
2. *Radiation Protection*. The hot cell itself must provide sufficient shielding to reduce radiation exposure of workers to ALARA, and never more than permitted by Federal regulation. This requires correct construction materials. These can vary from thin sheet-metal walls and plate glass or plastic windows for alpha and weak beta emitters to thick concrete, (①, Fig. 1) steel, or lead walls and shielding glass windows (②, Fig. 1) for gamma and neutron emitters.
3. *Consequences of Malfunction and Accidents*. The designer of hot cells and associated remote handling equipment must be able to predict normally expected malfunctions and accidents, and provide for recovery without breaching containment or significantly adding to occupational radiation exposure. He must also predict unexpected and more serious accidents, as well as the maximum credible accident, and provide for protecting workers and the general public against the consequences. Since some malfunction or accident scenarios may involve the possibility of breaching hot cell containment, the room containing the hot cell must also be at negative pressure and have exhaust air filters. The room thus becomes an emergency secondary containment system.

REMOTE HANDLING EQUIPMENT

The type, design, and use of remote handling equipment varies greatly with the nature and quantities of radioactive materials being handled, and with the processes or examinations being conducted. The major basis for decision is the ALARA principle of controlling occupational radiation exposure. If the materials to be handled emit little or no penetrating radiation, and if there is insignificant hazard from heat, corrosion, or machinery, then manipulations can be safely accomplished by hand through long rubber gloves affixed to openings in a thin-walled containment cell (glove box). Otherwise, remote handling becomes necessary.

General Purpose

During the growth of nuclear technology, a few general-purpose remote handling tools were developed to a high degree of excellence and now can be purchased from several vendors. Perhaps the most generally useful of these tools is the master–slave manipulator. Through an ingenious system of cables, steel tapes, gears, and electric drives operating through a penetration in a hot cell wall of any thickness, this manipulator (③ in Fig. 1) duplicates the movements and forces of the human hand inside the hot cell. Given sufficient light (⑦ in Fig. 1) and vision, a skilled operator can thread a needle or operate an engine lathe with a pair of these manipulators.

Somewhat less dexterous, but far more powerful, is the rectilinear power manipulator (④, Fig. 1). When rail-mounted on a trolley and bridge inside the hot cell, it can reach and manipulate within nearly the entire volume. With interchangeable jaws, special attachments, and power tools, it is capable of performing a great variety of work that is beyond the reach or capacity of master–slaves. Control is by console.

Perhaps the oldest and simplest of the remote handling tools is the in-cell bridge crane (⑤, Fig. 1). Equipped with inching controls and variable speeds, and operated by a person skilled in its use, it becomes a marvelously versatile manipulator.

Special Purpose

Remote handling equipment for special purposes in the hot cell performs all of the required functions that cannot or should not be done with general-purpose equipment, for reasons of inaccessibility, overstress, economy of operator time, etc. (⑥, Fig. 1). Material transfer can often be better done by conveyor, pneumatic tube, or even toy electric train, than by crane or master–slave. Most repetitive motions can easily be automated or computer controlled.

Requirements for all special-purpose equipment are reliability, simplicity, accessibility, visibility, and ease of repair or replacement by remote handling.

OTHER CONSIDERATIONS

1. *Vision*. Vision is usually afforded by shielding glass windows (②, Fig. 1) designed and furnished by vendors who are specialists in this field. Closed-circuit television is also proving increasingly useful.
2. *Maintenance, Repair, Replacement*. Careful analysis and preplanned solutions to anticipated problems will pay off handsomely in reduced downtime, process interruption, and occupational radiation exposure.
3. *Quality*. The best quality of equipment, materials, and workmanship that money can buy will prove to be the least expensive approach to building, equipping, and operating a hot cell.
4. *Decommissioning*. Cleanup and disposition of the hot cell, when obsolete or not needed, must be considered in the initial design.

See also RADIATION INTERACTION WITH MATTER; RADIOACTIVITY; RADIOLOGICAL PHYSICS.

BIBLIOGRAPHY

Title 10, Code of Federal Regulations, Parts 20, 30, 40, 50, 70, 71.

Proc. 9th–26th Conf. Remote Systems Technology. American Nuclear Society, 1961–1978.

Proc. Int. Symposium High Activity Hot Laboratories Working Methods, Grenoble, France, June 1965, published by ENEA/OECD.

Remote Handling of Mobile Nuclear Systems. U. S. Atomic Energy Commission, 1966.

Hubble Effect

Kenneth R. Lang

As early as 1917 V. M. Slipher of the Lowell Observatory had shown that the majority of spiral nebulae are receding from our Galaxy with tremendous velocities as large as 1,800 kilometers per second. Slipher had inferred these large velocities by observing the difference between the wavelengths, λ_0, of the spectral lines of the nebulae and the wavelengths, λ_L, of the same lines observed in terrestrial laboratories. According to the Doppler effect, the radial velocity of the nebula, V, is given by

$$V = \frac{c(\lambda_0 - \lambda_L)}{\lambda_L} = cz,$$

where $c = 2.997\,924\,56 \times 10^5$ kilometers per second is the velocity of light, and the parameter z is called the redshift of the nebula. Six years after Slipher's pioneering work, C. Wirtz looked for correlations between these velocities and other observable properties of the nebulae. Wirtz found that when suitable averages of the available data were taken an approximate linear dependence of velocity and apparent magnitude was visible. Because the observed light intensity of an object falls off as the square of its distance, this meant that the more distant objects had larger outward velocities. Two years after Wirtz argued for a velocity–distance relation, Arthur Eddington showed that when matter is placed in the de Sitter solution for Einstein's field equations it will expand; and Hermann Weyl showed that the redshift of this matter will, to first order, increase linearly with distance. By 1923 it was clear that Einstein's field equations suggested an expansion of the Universe, and that the Wirtz interpretation of Slipher's observations suggested that the spiral nebulae participate in this expansion.

It was not until an additional six years later that Edwin Hubble put the expected linear dependence of velocity on distance in a quantitative observational framework. When the observed radial velocities, V, were compared with the estimated distances, D, of the spiral nebulae, the now famous Hubble law

$$V = H_0 \times D$$

was found to hold for each spiral nebula. Here the parameter H_0 is called Hubble's constant. Hubble's original velocity–distance plot, from which the Hubble law was inferred, is given in Fig. 1.

Hubble's constant essentially measures the rate of recession of the nebulae, and can therefore be used to infer an age parameter of the Universe through the relation age $= 1/H_0$. Hubble obtained a value of $H_0 = 550$ kilometers per second per megaparsec for his recession constant, and this value of the constant gives an age parameter of 2×10^9 years.

Today we know that Hubble's basic ideas were correct, but that his estimates of distance were too small. Many of the distance calibrations were based on Cepheid variable stars, and involved the assumption that all Cepheids obeyed the same relationship between period of light variation and luminosity. But, as Walter Baade showed in 1952, the period–luminosity relation for Cepheids in spiral arms is quite different from that for Cepheids in galactic nuclei or in globular clusters. When this difference was taken into account, the extragalactic distance scale was revised, and Hubble's constant was revised downward to about 180 kilometers per second per megaparsec. During the past 15 years Allan Sandage and his colleagues at the Mount Wilson and Palomar Observatories have published at least five successively smaller values for Hubble's constant, the most recent value being about 50 kilometers per second per megaparsec. This value of Hubble's constant results in a Universe age parameter of about 20×10^9 years.

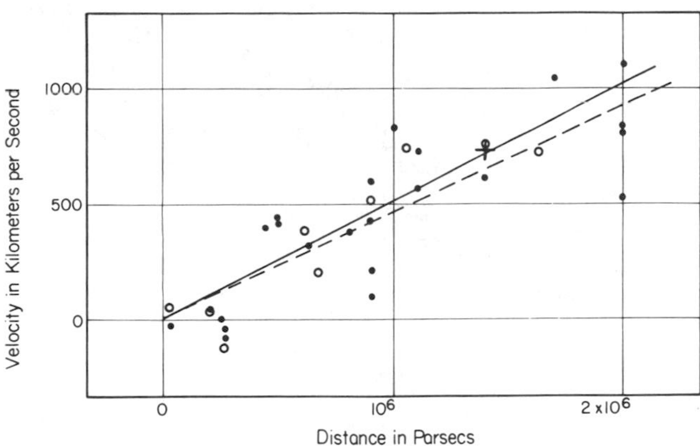

FIG. 1. The velocity–distance relationship for galaxies, as plotted by E. P. Hubble in 1929. The solid line represents the relation from individual galaxies (dots), the dashed line the relation when the galaxies are combined into groups (circles). The cross indicates the mean for 22 galaxies of uncertain distance.

In order to test the Hubble effect for the more distant objects, Milton Humason, Nicholas Mayall, and Allan Sandage reformulated Hubble's law in the form

$$\log(V) = 0.2m + B,$$

where m is the apparent magnitude of the object and the constant B depends on the absolute magnitude of the object and Hubble's constant. They then examined the log velocity-apparent magnitude relation for 474 nebulae using data obtained during a 25-year period with the Mt. Wilson, Palomar, and Lick observatories. In order to test the velocity–distance relation, redshifts were corrected for the solar motion and the observed magnitudes were corrected for absorption within our Galaxy and for the effects of Doppler shifts on the radiation spectrum. The linearity of the corrected log velocity-apparent magnitude relation was then tested by a linear least-squares fit. Within the accuracies of the data, the relation is linear and the slope is 0.2. It has since been shown that this slope is exactly the expected result for a homogeneous, isotropic, expanding Universe that obeys the laws of Einstein's general theory of relativity.

As illustrated in Figure 2, the author and his colleagues have recently shown that the Hubble-diagram plots for normal galaxies, radio galaxies, and quasistellar objects are all consistent, within the errors, with that expected for a homogeneous, isotropic, expanding Universe. The presently available data suggest that all extragalactic objects with redshifts ranging from 0.001 to about 3 participate in the expansion of the Universe. Unhappily, the great dispersion in the observed data makes it impossible to determine the de-

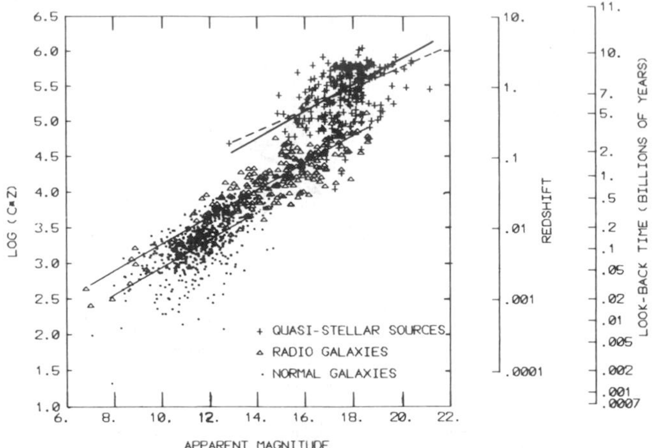

FIG. 2. The author's composite Hubble diagram, in which are plotted 663 normal galaxies, 230 radio galaxies, and 265 quasistellar objects. The vertical axis at right gives the look-back time in billions of years, calculated on the assumption of a Hubble constant of 50 kilometers per second per megaparsec. For normal galaxies and radio galaxies, the solid line denotes the least-squares fit to the data. For the quasistellar objects, the dashed line denotes the least-squares fit, while the solid line represents the theoretical Hubble relation that would hold for a homogeneous, isotropic, expanding universe.

tails of the expansion. The data suggest instead that the distant quasistellar objects are about a factor of 10 brighter than the nearer radio galaxies, which are in turn about a factor of 10 brighter than the very nearby normal galaxies. Because the light with which we view distant objects left them some time ago, this change in brightness suggests that younger extragalactic objects are brighter than older ones.

See also ASTROPHYSICS; COSMOLOGY; GALAXIES; UNIVERSE.

BIBLIOGRAPHY

W. Baade, *Pub. Astron. Soc. Pacific* **68,** 5 (1956). (I)

A. S. Eddington, *The Mathematical Theory of Relativity*. Cambridge University Press, London, 1923. (I)

E. P. Hubble, *Proc. Natl. Acad. Sci.* **15,** 168 (1929). (I)

E. P. Hubble, *The Realm of the Nebulae*. Yale University Press, New Haven, 1936. Reprinted by Dover Publications, New York, 1958. (E)

M. L. Humason, N. U. Mayall, and A. R. Sandage, *Astron. J.* **61,** 57 (1956). (I)

K. R. Lang, *Astrophysical Formulae*. Springer-Verlag, New York-Berlin-Heidelberg, 1974. (A)

K. R. Lang and O. Gingerich, *Source Book in Astronomy and Astrophysics 1900–1975*. Harvard University Press, Cambridge, Mass., 1979. (I)

K. R. Lang, S. D. Lord, J. M. Johanson, and P. D. Savage, *Astrophys. J.* **202,** 583 (1975). (I)

K. R. Lang and G. S. Mumford, *Sky and Telescope* **51,** 83 (1976). (E)

A. R. Sandage, *Astrophys. J.* **178,** 25 (1972). (I)

A. R. Sandage and G. A. Tamman, *Astrophys. J.* **196,** 313 (1975). (I)

V. M. Slipher, *Proc. Amer. Philos. Soc.* **56,** 403 (1917). (I)

S. Weinberg, *Gravitation and Cosmology: Principles and Applications of the General Theory of Relativity*. Wiley, New York, 1972. (A)

H. Weyl, *Phys. Zeit.* **24,** 230 (1923). (A)

C. Wirtz, *Astron. Nach.* **215,** 349 (1921). (I)

Hydrodynamics

Mark Nelkin

Hydrodynamics is the study of the mechanics of fluid flow. More recently called fluid mechanics or fluid dynamics, it has become an extensively developed discipline of applied science. This is hardly surprising since fluid flows are such an important part of our natural and man-made environment. Of greater interest here, however, is the currently growing interest in hydrodynamics as an active research subject in physics. We will see that there are many problems for which a basic physical understanding is still lacking.

The equations of motion of a fluid, the Navier–Stokes equations, are relatively simple partial differential equations. The molecular properties of the fluid enter only through the values of certain transport coefficients, such as the viscosity. When properly scaled, these equations describe the same motions in air that they do in water. This is well confirmed experimentally, so that we can be confident that the phenomena we wish to study are consequences of the underlying hydrodynamic equations. These phenomena, however, occur in bewildering variety, and there are very few of them that we can calculate directly. This is a consequence of the nonlinearity of the underlying equations and of our inability to predict, from nonlinear equations of motion, even the qualitative features of the resulting motion. This same situation exists in many other parts of physics, but seldom do we have so much confidence that we have the correct equations of motion. Thus hydrodynamics becomes a very important testing ground for our developing understanding of nonlinear phenomena. (For an elementary but elegant expansion of this viewpoint see Ref. 1.)

We turn now to a more quantitative description. The velocity of a fluid can be described by a vector function of space ($\mathbf{x}$) and time t, written as $\mathbf{u}(\mathbf{x},t)$. The density of the fluid is given by a scalar function $\rho(\mathbf{x},t)$ of the same variables. The conservation of mass is expressed by the equation of continuity,

$$\frac{\partial\rho}{\partial t}+\nabla\cdot(\rho\mathbf{u})=0. \tag{1}$$

For many fluid flows of interest, typical flow speeds are very small compared to the speed of sound in the fluid. Under these conditions the density of the fluid is nearly constant, and Eq. (1) can be replaced by

$$\nabla\cdot\mathbf{u}=0. \tag{2}$$

Such fluid flows are called incompressible, and Eq. (2) is sometimes called the incompressibility condition. It is important to recognize that Eq. (2) expresses a property of a fluid flow, and not of a fluid.

The momentum balance for a moving fluid element is given by

$$\rho\left[\frac{\partial \mathbf{u}}{\partial t}+(\mathbf{u}\cdot\nabla)\mathbf{u}\right]=-\nabla p+\mathbf{f}_{\text{ext}}+\mathbf{f}_{\text{vis}} \tag{3}$$

where $p(\mathbf{x},t)$ is the pressure, $\mathbf{f}_{\text{ext}}$ is the force per unit volume due to external forces such as gravity, and $\mathbf{f}_{\text{vis}}$ is the force per unit volume due to internal viscous friction. Of all the terms in Eq. (3) only $\mathbf{f}_{\text{vis}}$ depends on the molecular nature of the fluid. For an ordinary isotropic classical fluid, $\mathbf{f}_{\text{vis}}$ can be simply expressed in the form

$$\mathbf{f}_{\text{vis}}=\eta\,\nabla^2\mathbf{u}+\left(\frac{1}{3}\eta+\eta_v\right)\nabla(\nabla\cdot\mathbf{u}) \tag{4}$$

where η is the ordinary shear viscosity and η_v is a volume coefficient of viscosity. These are to be taken as empirical properties of the fluid that depend on temperature and density. For gases they can be calculated from kinetic theory. For liquids theory is more complicated and less reliable.

Fluids that obey the viscous friction law of Eq. (4) are called Newtonian. Air and water and most other gases and liquids are Newtonian under usual conditions. At high enough frequencies, however, all fluids exhibit viscoelastic effects that make them non-Newtonian. Other fluids, such as polymer solutions, may exhibit non-Newtonian effects even at ordinary frequencies. Anisotropic fluids such as liquid crystals have a more complicated behavior. Finally, superfluids such as helium II below 2.17 K have a totally different hydrodynamics.

To close our system of equations we need an equation for the energy balance of a moving fluid element. This will contain the temperature $T(\mathbf{x},t)$. We also need the equation of state $\rho(p,T)$. The problem simplifies greatly, however, if we make the approximation of incompressible flow, $\rho=$constant. By use of Eq. (2), Eq. (4) becomes

$$\frac{\partial \mathbf{u}}{\partial t}+(\mathbf{u}\cdot\nabla)\mathbf{u}=-\frac{1}{\rho}\nabla p+\frac{1}{\rho}\mathbf{f}_{\text{ext}}+\nu\,\nabla^2\mathbf{u}, \tag{5}$$

where we have introduced the kinematic viscosity $\nu=\eta/\rho$. For air under standard conditions $\nu=0.15$ cm^2/sec, and for water $\nu=0.01$ cm^2/sec. For incompressible flow this is the only property of the fluid that enters.

Equation (5) is the Navier–Stokes equation of hydrodynamics. It is to be supplemented by the incompressibility condition of Eq. (2). Taking the divergence of Eq. (5) and using Eq. (2), we have

$$\nabla\cdot(\mathbf{u}\cdot\nabla)\mathbf{u}=-\frac{1}{\rho}\nabla^2 p+\frac{1}{\rho}\nabla\cdot\mathbf{f}_{\text{ext}}. \tag{6}$$

The pressure $p(\mathbf{x},t)$ can be obtained by integrating this equation over space subject to the appropriate boundary conditions. Thus the pressure term in the Navier–Stokes equation is a quadratically nonlinear contribution that is spatially nonlocal. The subtle interplay of the two nonlinear terms with the linear viscous term gives the Navier–Stokes equation very special and interesting properties.

Under almost all conditions the appropriate boundary condition is that the velocity vanish at a solid surface. The fluid sticks to the solid. As stated nicely by Feynman, "You will have noticed, no doubt, that the blade of a fan will collect a thin layer of dust—and that it is still there after the fan has been churning up the air. In spite of the fact that the fan blade is moving at high speed through the air, the speed of the air relative to the fan blade goes to zero right at the surface. So the very smallest dust particles are not disturbed." Exceptions to this boundary condition can occur in highly rarified gases, and in attempts to extend hydrodynamic ideas to motion at the molecular level, but these are unusual cases. In the two-fluid model of liquid helium, the normal-fluid component obeys the usual boundary condition, but the superfluid slips freely at a surface.

In most flows there is a well-defined largest speed U and a well-defined largest length L. It is then convenient to rewrite the Navier–Stokes equations in dimensionless form. We measure velocities in units of U, distances in units of L, times in units of L/U, and pressures in units of ρU^2. Leaving out any external forces for simplicity, the dimensionless equation of motion becomes

$$\frac{\partial \mathbf{u}}{\partial t}+(\mathbf{u}\cdot\nabla)\mathbf{u}=-\nabla p+\frac{1}{R}\nabla^2\mathbf{u}. \tag{7}$$

The parameter

$$R-UL/\nu \tag{8}$$

is called the Reynolds number, and completely defines the characteristics of the flow. The length scale L, velocity scale U, and fluid viscosity ν can be varied, but if the Reynolds number is the same, the flow will be the same. This is well confirmed experimentally even for very complicated turbulent flows.

When the Reynolds number is small, corresponding to large viscosity or low speed or small length scales, the linear viscous term dominates the nonlinear terms, and the problem is straightforward and well understood. Consider for example the flow past a sphere of radius a. At small Reynolds numbers the drag force on the sphere is given by the famous Stokes formula

$$F=6\pi\eta aU \tag{9}$$

where U is the free-stream speed far from the sphere. The physics of this situation is best seen by looking at the rate of energy dissipation

$$P=FU=6\pi\eta aU^2. \tag{10}$$

The motion of the sphere through the fluid creates a velocity field extending a considerable distance from the sphere. The energy dissipation is accounted for by the decay of the fluid kinetic energy into heat through action of the viscosity.

For another simple example consider a cylindrical pipe of diameter D whose axis is in the z direction. In cylindrical coordinates (r,θ,z) the flow field is given by

$$u_z(r)=\frac{1}{16\eta}\left(\frac{dp}{dz}\right)(D^2-4r^2), \tag{11}$$

where dp/dz is the pressure gradient along the pipe.

As a third simple example, consider two concentric cylinders with the outer cylinder stationary, and the inner cylinder rotating at angular frequency ω (couette flow). The flow field, again in cylindrical coordinates, is given by

$$u_r = u_z = 0, \quad u_\theta(r) = \frac{\omega R_1^2}{r} \frac{(R_2^2 - r^2)}{(R_2^2 - R_1^2)} \tag{12}$$

where R_1 and R_2 are the radii of the inner and outer cylinders, respectively.

These three examples are illustrative in their differences as well as in their similarities. The theoretical results of Eqs. (9), (11), and (12) all agree with experiment at sufficiently low Reynolds numbers. They all disagree with experiment at high Reynolds numbers, but in different ways and for different reasons. For the flow past a sphere, Eq. (9) is an approximate result in which the nonlinear terms in the equation of motion have been explicitly neglected. Corrections can be calculated as a power series in the Reynolds number, but this perturbation expansion is itself valid only for small Reynolds numbers. At high Reynolds number, the flow becomes more complicated, and eventually becomes turbulent. For the flow in a pipe or between rotating cylinders, the solutions we have given are exact steady solutions of the Navier–Stokes equations at all Reynolds numbers. Whether these solutions are observed is a question of hydrodynamic stability.

For pipe flow the Poiseuille profile of Eq. (11) is the only known steady solution, but for large enough Reynolds numbers this solution is unstable. The observed flow when the stability of the Poiseuille flow is lost is fully developed turbulence. Under some conditions the laminar Poiseuille flow may be metastable, much as a supercooled liquid can be metastable against solidification. The transition to turbulence and the detailed nature of turbulent pipe flow have been extensively studied experimentally. The transition is reasonably well understood, but the fully turbulent state is very complicated. A fundamental theoretical understanding of fully developed turbulence is still lacking. There is, however, no reason to believe that turbulent pipe flow cannot be understood in principle starting from the Navier–Stokes equations. It is just an extremely difficult problem, both mathematically and computationally. Qualitatively it is not surprising that high Reynolds number flow should be turbulent. The Navier–Stokes equations can be thought of as a nonlinear dissipative dynamical system with a very large number of degrees of freedom. When the dissipation is weak, such systems generically exhibit chaotic time dependence. The essential feature is sensitivity to initial conditions. Two solutions that are close together in the appropriate phase space at some initial time will be far apart at much later times. In such a situation, a statistical description is the natural one. Turbulent flows exhibit a subtle mixture of randomness and regularity that is beyond our present ability to calculate, but turbulence is an active field of research, and considerable progress can be expected. For more about turbulence, see the article in this volume.

The Couette flow problem exhibits a more interesting structure before the onset of turbulence. At a critical Reynolds number the steady flow of Eq. (12) becomes unstable. The stable flow is still steady, but has a complex spatial structure consisting of toroidal rolls, the so-called Taylor vortices discovered by G. I. Taylor in 1923. At a second critical Reynolds number this structured steady flow becomes unstable to a transverse wave motion of the Taylor vortices. The stable flow is then still laminar, but is now structured in space and periodic in time. The behavior up to this point can be understood by hydrodynamic stability theory. With a further increase in Reynolds number, the flow becomes more complicated, and appears to show a sudden transition to turbulence. This transition is quite different from that in pipe flow, and is currently the subject of active experimental and theoretical research. At much higher Reynolds numbers the turbulence becomes stronger. The fully developed turbulence is quite similar in Couette and in pipe flow, but the pattern of transitions to reach it is quite different.

The flow past an obstacle such as a sphere or a cylinder is again of a different character. In contrast to Couette and Poiseuille flow, in this case the flow is perturbed in a negligible fraction of the total space. At sufficiently large distances cross-stream or upstream, the velocity field will approach its uniform unperturbed value. Downstream perturbations may extend farther, but these also will eventually die out. At large distances upstream or cross-stream the flow will be slowly varying and streamlined. A variety of elegant mathematical methods of classical hydrodynamics are applicable in this region, but the problem of how to join on to the rapidly varying flow near the obstacle makes the application of these methods to real problems quite subtle. Near the obstacle the flow varies rapidly, and the singular perturbation of the viscous term is important. In a thin layer near the obstacle useful approximate solutions including this singular perturbation can be obtained by boundary layer theory. The asymptotic solution obtained by boundary layer theory near the obstacle, and from potential flow theory far away, must be joined together using the method of matched asymptotic expansions. All of this works in precise form when the flow is laminar, but useful empirical methods exist also for the case when the boundary layer and wake are turbulent.

There is an immense amount known about fluid mechanics, and there are a tremendous variety of qualitatively interesting phenomena. The spirit of the subject as a part of physics is perhaps best given by Landau and Lifshitz in their book on fluid mechanics [2], in almost every chapter of which are several examples of phenomena whose understanding and application are of interest in modern physics.

See also BOUNDARY LAYERS; CHAOS; FLUID PHYSICS; NONLINEAR WAVE PROPAGATION; TURBULENCE; VISCOSITY.

REFERENCES

1. *The Feynman Lectures on Physics,* Vol. II, Chapters 40 and 41. Addison-Wesley, Reading, Mass., 1964. (E)
2. L. D. Landau and E. M. Lifshitz, *Fluid Mechanics (Course of Theoretical Physics,* Vol. 6, translated from the Russian by J. B. Sykes and W. H. Reid). Pergamon, New York, 1987. (I)

BIBLIOGRAPHY

G. K. Batchelor, *An Introduction to Fluid Dynamics.* Cambridge, London and New York, 1970. (E)

C. C. Lin, *The Theory of Hydrodynamic Stability.* Cambridge, London and New York, 1955. (I)

D. J. Tritton, *Physical Fluid Dynamics*. Van Nostrand Reinhold, New York, 1977. (E)

Advanced work appears principally in *J. Fluid Mech.*, *Phys. Fluids A*, and *Phys. Rev.* **A.** (A)

Hydrogen Bond

George C. Pimentel

INTRODUCTION

In the preponderant majority of its compounds, the element hydrogen is found attached to one atom, whether in a diatomic or polyatomic molecule. The bond energies (from 40 kcal/mol in RbH to 134 kcal/mol in HF) identify the attachments as chemical bonds and form the basis for the classical bonding rule that hydrogen has a ''valence'' of one, i.e., it can bond to one other atom.

In many molecules, however, there is evidence of additional attractive interaction that specifically links the hydrogen atom in a group (or molecule) A–H to another atom B in the same or a different molecule. In most cases, this additional interaction involves bond energies of a few kilocalories per mole, well above van der Waals interactions but well below chemical-bond energies. Since the hydrogen atom has already expended its classical bonding capacity, this interaction is given a special name; it is called a *hydrogen bond* (hereafter, called H bond). It is usually designated $A\text{–H}\cdots B$. H bonds are formed by the hydrohalides and by molecules that contain hydrogen that is bonded to oxygen or to nitrogen but not to carbon (examples are alcohols, phenols, carboxylic acids, amides, and amines). Such an A–H group can form H bonds to any other molecule that contains an effective electron donor, B. (Examples are oxygen- and nitrogen-containing compounds including ethers, ketones, amines, etc.).

MACROSCOPIC MANIFESTATIONS OF HYDROGEN BONDING

Two readily observed properties that can reveal H bonding are the melting and boiling points. Figure 1 shows the trends in these properties for the sixth and seventh column hydrides. The anomalous values for ice and solid HF are symptomatic of specially strong intermolecular forces in these condensed phases. It is now recognized that H_2O and HF form relatively strong H bonds.

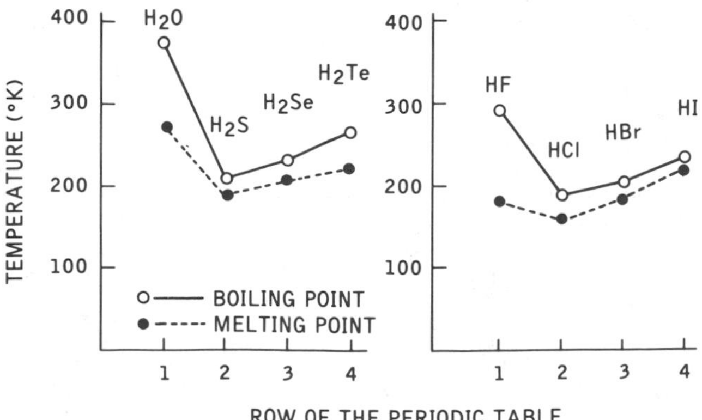

FIG. 1. H-bond formation by H_2O and HF revealed by melting and boiling point trends.

The entropy of vaporization demonstrates another facet of this interaction. Most substances have about the same value of $\Delta S_{\text{vap}} = \Delta H_{\text{vap}}/T$ (Trouton's rule). In contrast, hydrogen-bonded substances all display ΔS_{vap} values higher than 21 cal/deg mol (H_2O, 26.1; CH_3OH, 25.0; NH_3, 23.3 cal/deg mol). The excess over the Trouton-rule norm indicates that there is special order in the liquid state when hydrogen bonding occurs. This special order is associated with the fixed, linear orientation in the $A\text{-H}\cdots B$ link.

This fixed orientation gives H-bonding liquids specially high dielectric constants. These high dielectric constants reflect the ability of the H bonds to reorient in response to an electric field.

The specially strong intermolecular attractions caused by H bonding are evidenced in the gas phase through the *PVT* behavior. Specially high deviations from the ideal gas law are always observed when H bonding can be expected. Thus, the experimental second virial coefficients of methanol and of water (−1220 and −976, respectively) are each over twice as negative as those calculated from the Berthelot equation (−550 and −421, respectively). The Berthelot-equation estimate does not include the H-bond part because it is based on the critical constants, and H bonds do not persist at very high temperatures.

The viscosities of H-bonding substances tend to be higher than those of other substances of comparable molecular weight. This is particularly true if the molecule has two H-bonding sites, as for water and for polyhydroxy alcohols. For example, the viscosities of diethyl ether and *n*-octane are 0.23 and 0.54 centipoise, respectively. In contrast, the hydrogen-bonding liquids *n*-octanol, ethylene glycol, and glycerol have viscosities of 10.6, 19.9, and 1490 centipoise, respectively.

This catalogue of observable properties that are affected by H bonding could be extended to include density, surface tension, heat of solution, solubility, conductance, ferroelectric behavior, refractive index, thermal conductivity, acoustic conductivity, and adiabatic compressibility.

SPECTROSCOPIC MANIFESTATIONS

Spectroscopic techniques furnish effective means for detecting and studying H bonds. Infrared absorption spectra are most used because there are quite characteristic spectral changes caused by H bonding. There are four characteristic vibrational modes of the $A\text{–H}\cdots B$ group that can be examined, as pictured in Fig. 2.

The A–H stretching mode ν_{AH} has been most studied. For example, the vibration ν_{OH} of monomeric CH_3OH (in dilute CCl_4 solution) absorbs at 3642 cm^{-1} with a bandwidth near 20 cm^{-1}. At higher concentrations, H bonding causes the formation of dimers, $(CH_3OH)_2$, trimers, $(CH_3OH)_3$, and higher polymers, $(CH_3OH)_n$. The involvement of the O–H group in the H bond causes its absorption band to shift to lower frequencies, to broaden, and to become much more intense (see Fig. 3). The extent of each of these changes is found to increase as the H-bond strength increases, so these spectral properties, $\Delta\nu$, $\nu_{1/2}$, and B/B_1 (B_1 being the integrated absorption of the monomer) furnish useful criteria for ordering substances according to hydrogen bonding strength. Furthermore, systematic concentration studies in an inert solvent (such as CCl_4 or *n*-hexane) furnish equilibrium data. Such measurements carried out at various temperatures lead to ΔH^0 and ΔS^0 for H bond formation. All of these valuable results make the infrared study of H bonds one of the most fruitful techniques available.

The other vibrations, ν_b, ν_t, and ν_{AB}, are less studied and they are somewhat less useful diagnostically. The bending mode ν_b shifts upward slightly (<0.2%) and is not intensified. The torsional mode ν_t can exhibit as spectacular shifts (to higher frequencies), broadening, and intensification as those of ν_{AH}. Rather less is known about ν_{AB} because the low frequencies (<250 cm^{-1}) have made them less accessible, but there is significant potential for their future study and application.

The nuclear-magnetic-resonance (NMR) spectrum also furnishes a rather definitive criterion for detecting H bonding and for classifying it as to strength. The characteristic NMR proton frequency shifts significantly toward lower fields upon H-bond formation. Figure 4 shows how distinctive the bonding perturbation of the NMR resonance can be. In pure ethanol, the OH proton NMR resonance is at much lower magnetic field than that of the CH_2 protons. Dilution to 1.0 *M* in CCl_4 causes the OH proton resonance to move over close to the CH_2 signal. Further dilution to less than 0.1 *M* reveals the monomeric OH proton resonance to be at higher field than that of the CH_3 protons!

THERMODYNAMIC PROPERTIES

The energy and entropy changes (ΔH^0 and ΔS^0) that accompany H-bond formation can be deduced from almost any of the manifestations mentioned above. Some of the most reliable values come from the spectroscopic measurements. In general, H-bond energies are in the range 2–7 kcal/mol,

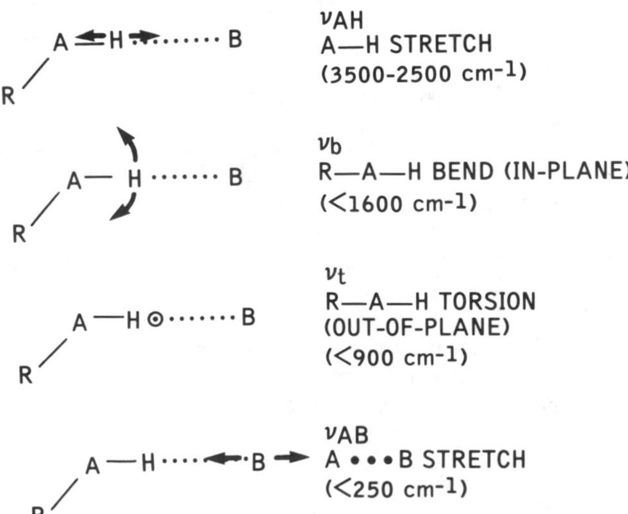

FIG. 2. Vibrational modes of a hydrogen bonded system A–H···B.

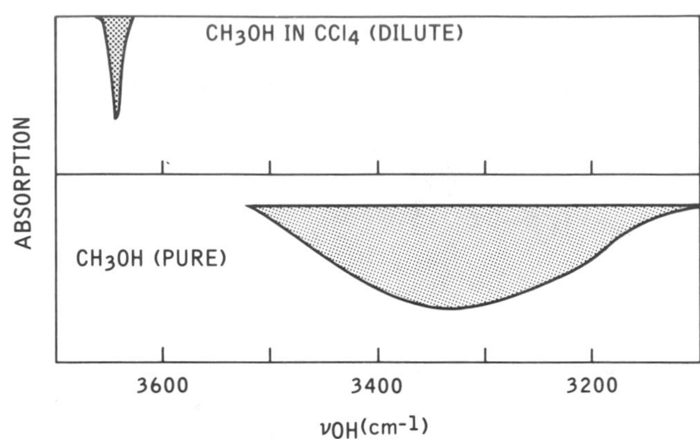

FIG. 3. The O–H stretching vibration in the infrared spectrum of methanol: (a) monomeric CH_3OH in carbon tetrachloride; (b) H-bonded CH_3OH.

with the notable exception of H bonds to negative ions, as in the bihalide ions, HF_2^- (−37 kcal/mol) and HCl_2^- (−14 kcal/mol).

Entropy changes $\Delta S°$ on hydrogen-bond formation correlate with the $\Delta H°$ values. The reason is obvious: As $\Delta H°$ becomes more negative, the H bond becomes more rigid, and hence its positional randomness decreases.

STRUCTURAL PROPERTIES

X-ray and neutron-diffraction studies of crystals and electron-diffraction studies of gases provide us with information about the geometry of H bonds. Data of this type, which are voluminous, lead to some important generalizations:

(i) In crystals, the A–H bond is almost always within 15° of collinearity with A···B.

(ii) The A···B distance R_{AB} is less than the sum of the van der Waals radii of A and B, usually by about 0.1 to 0.2 Å. As the H-bond strength increases, R_{AB} decreases.

(iii) The A–H distance R_{AH} is always longer in A–H···B than in the non-H-bonded A–H parent molecule. For the very strongest H bonds, as in HF_2^-, the proton is symmetrically placed between the A and B atoms.

(iv) H bonds figure prominently in the crystal lattices of

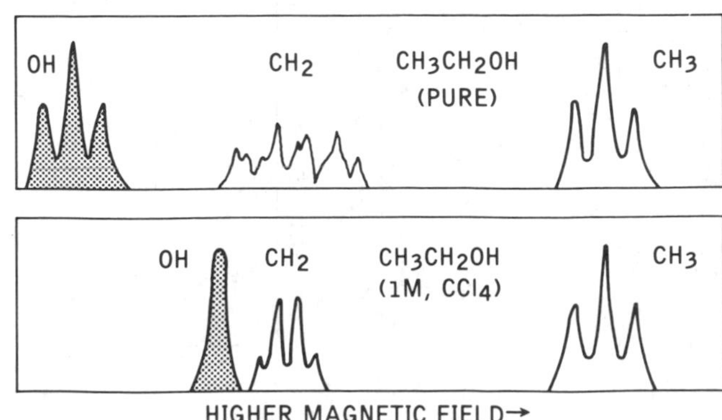

FIG. 4. The effect of H bonding on the NMR resonance of the OH proton in ethanol.

all molecules that have the capability of forming intermolecular H bonds to other, like molecules.

These generalizations are of great value in predicting and deducing the molecular structures of biologically important molecules (proteins, DNA, RNA, etc.) in which H bonding is a structural element.

THEORY

Early explanations of the H bond were essentially electrostatic in character, and these persist today. However, no correlation is found between the H-bond energy and the dipole moments of either acid or base. There are other deficiencies, such as inability to explain the intensity effects in the *A*–H stretching and bending modes or the zigzag arrangement of hydrogen halide molecules in their crystal lattices in preference to the parallel orientation preferred by a dipolar array. Any calculational successes obtained using the electrostatic theory must be attributed, at least in part, to the parametric freedom associated with charge placement.

The molecular-orbital description gives a satisfactory qualitative explanation of the existence of the H bond. Axial molecular orbitals are constructed from the hydrogen 1*s* orbital and the terminal, axially directed *p* orbitals. If the hydrogen atom is symmetrically placed (as in HF_2^-), there results one bonding, one nonbonding, and one antibonding molecular orbital. With four electrons to be accommodated, only the first two orbitals need to be occupied, so two half-order bonds result. The nonbonding orbital places the charge on the terminal atoms, those with high electronegativity.

The most extensive *ab initio* calculations on H bonds utilize the Hartree–Fock approximation. These calculations depend upon cancellation of electron-correlation effects since configuration interaction has not been included. Yet the completeness of the basis sets has approached the point that useful calculations of the rather small H-bond energies can be made. These studies seem to corroborate the charge distribution implied by the molecular orbital approach just described.

Incorporation of configuration interaction in calculations of H-bond energies and geometries must be high on the agenda of today's theorists because of the crucial role played by hydrogen bonding in biological molecules such as the proteins and DNA. Both intramolecular H bonds within the DNA molecule and intermolecular H bonds involving the aqueous environment play a role in determining primary, secondary, and tertiary structures of these macromolecules. Accurate theory is needed to predict such structures because these structures determine biological function.

SIGNIFICANCE

The mere existence of the oceans on earth is, of course, attributable to the H bonding that elevates the boiling point of water (see Fig. 1). Thus we are afforded the vast thermostat that moderates the climate and that provided the primordial brew in which life first appeared. Water's ionizing solvent properties, too, can now be seen to involve H bonding. The H-bond energies of bifluoride and bichloride ions imply an HOH$\cdots$Cl$^-$ bond energy near 20 kcal. Four such

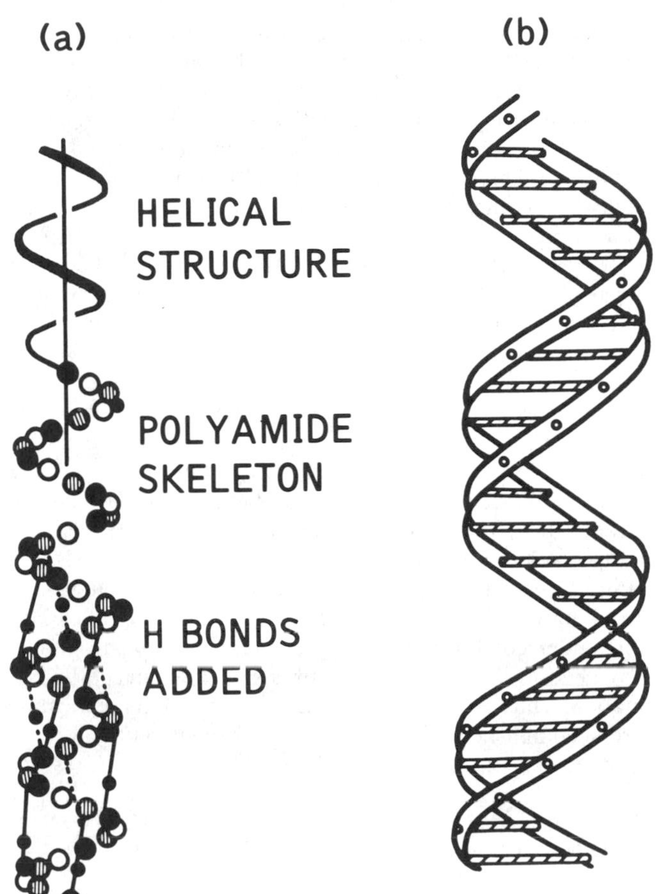

FIG. 5. (a) A schematic representation of the α-helix in protein: the helical configuration is determined and maintained by H bonds. (b) A schematic representation of DNA: the spiraling ribbons denote sugar–phosphate chains and the bars represent matched pairs of bases linked by H bonds.

H bonds to Cl$^-$ account nicely for the heat of aquation of a chloride ion, a heat that is essential to the stability of electrolyte solutions. Finally, all of the sugars, carbohydrates, cellulose, proteins, and other stuff of which plants and animals are made are generously sprinkled with ether linkages, hydroxyl groups, amides, and esters—groups that raise aqueous solubility and that provide compatibility with the aqueous milieu in which they must function.

The structure of a protein even more intimately depends on H bonding. Its skeleton is a long polymer of amide linkages formed from α-amino acids. This skeleton coils into a specially configured helical structure, called the α-helix, which is maintained by H bonds between successive coils (see Fig. 5a). Even more ornate in their dependence on H bonding are the structures of nucleic acids, including DNA. Figure 5b pictures diagrammatically the double-stranded helix that represents the basic makeup of DNA. The spiraling ribbons denote sugar-phosphate chains and the bars represent matched pairs of bases that bond the chains together by H bonds. Each base pair includes a purine molecule linked to a pyrimidine partner that matches in a lock and key H-bonded configuration. Despite the size of a DNA molecule (molecular weight can be as high as 10^9), it contains only four different amine bases, adenine, thymine, cytosine, and guanine (abbreviated A, T, C, and G). Adenine and thymine have geometrically fixed capacities to form hydrogen bonds that match each other so well that an adenine base can recognize a thymine molecule and bond to it in strong preference to any other bases present. Cytosine and guanine match in a similar way.

This recognition capability is incorporated when two sugar-phosphate strings twist into the double-helix structure. Because of the specificity of the A–T and C–G hydrogen bonds, the helix can form only if the sequence of bases on the first string is matched in perfect complementarity to the sequence on the second string. The order in which these base pairs occur creates an information code in the molecule. This information can be copied (replicated) to produce duplicate DNA molecules through enzymatic synthesis. The double strands unzip their hydrogen bonds to expose a single strand that serves as a sequence guide in enzymatic synthesis of a complementary copy. This process involves making and breaking of relatively weak H bonds which can be carried out under conditions that do not break the much stronger covalent bonds that carry and preserve the molecular information. Thus, the genetic coding in DNA and its replication are accomplished through an elegant orchestration of weak H bonds and strong covalent bonds.

See also Chemical Bonding; Interatomic and Intermolecular Forces; Molecular Structure Calculations.

BIBLIOGRAPHY

S. Bratoz, *Adv. Quantum Chem.* **3,** 209 (1966).

D. Hadzi (ed.), *Hydrogen Bonding, Symposium on Hydrogen Bonding.* Pergamon Press, New York, 1957.

G. C. Pimentel and A. L. McClellan, *The Hydrogen Bond.* Freeman and Co., San Francisco, Calif., 1960.

G. C. Pimentel and A. L. McClellan, *Hydrogen Bonding, Ann. Rev. Phys. Chem.* **22,** 347 (1971).

J. A. Pople, W. G. Schneider, and H. J. Bernstein, *High Resolution Nuclear Magnetic Resonance,* Chap. 15. McGraw-Hill, New York, 1959.

N. D. Sokolov, *Ann. Chim. (Paris)* **10,** 497 (1965).

Hypernuclear Physics and Hypernuclear Interactions

A. R. Bodmer

INTRODUCTION

Hypernuclear physics deals with the interactions of hyperons. These are "strange baryons" which are heavier than neutrons (n) or protons (p) referred to collectively as nucleons (N). Strangeness, represented by the quantum number S, is conserved in strong interactions but is violated in weak interactions which are responsible for the slow decay of hyperons with a lifetime of about 10^{-10} s. The lightest hyperon is the neutral Λ with $S=-1$, spin $\frac{1}{2}$, and a mass of 1116 MeV, i.e., about 1.2 times that of a nucleon. The free Λ decays are mostly $\Lambda \rightarrow p+\pi^-$ or $n+\pi^0$ where π denotes a pion (the lightest meson). The next heavier hyperon is the Σ also with $S=-1$ and spin $\frac{1}{2}$; it exists in three charge states Σ^+, Σ^-, Σ^0 (positive, negative, and neutral) and is about 80 MeV heavier than the Λ.

In the quark model the baryons are composites of three (valence) quarks (plus "sea" quarks). Nucleons are composites of the lightest up (u) and down (d) quarks, whereas hyperons contain one or more strange (s) quarks which carry strangeness $S=-1$ and which are about 150 MeV heavier than the u, d quarks. Thus a proton $p \equiv (uud)$ is a composite of two u quarks and a d quark whereas a $\Lambda \equiv (uds)$ is a composite of a u, d, and s quark.

Hypernuclei are "strange" nuclei with one or more hyperons bound to an (ordinary) nucleus. Most observed hypernuclei are Λ hypernuclei with one Λ, but Σ hypernuclei have also been observed and there is evidence for $\Lambda\Lambda$ hypernuclei with two Λs. The ground states of Λ hypernuclei decay via the weak interaction with lifetimes about the same as that of a free Λ, i.e., about 2×10^{-10} s. Thus for most purposes the low-lying hypernuclear states may be considered as completely stable, just as are the β-decaying states of ordinary nuclei. Striking differences between ordinary and hypernuclei arise because the Λ is distinct from the nucleons, with no Pauli principle between it and these. The Λ can then go into deeply bound orbits already occupied by nucleons. In particular for the ground state the Λ occupies the 1s orbit which has zero orbital angular momentum.

THE STRONG HYPERON–NUCLEON FORCE

Because Λs and Σs are short-lived one cannot produce beams of these at low energies. The information about the Λ-N and Σ-N forces is thus much less than about the N-N force and has been obtained from the scattering of Λs and Σs by protons in a hydrogen bubble chamber. Nevertheless, these data show quite clearly that the Λ-N nuclear force is strong and about half as attractive as the N-N force. A distinctive feature of the ΛN, ΣN systems is their strong coupling to each other because the Λ and Σ may be rapidly converted into each other since they have the same strangeness. Thus above the threshold energy of about 80 MeV one has the nuclear reaction $\Lambda+N \rightarrow \Sigma+N$. However, even for the ΛN system below the ΣN threshold, where no Σ can be produced, this "$\Lambda\Sigma$" conversion can strongly influence the net ΛN force because of the rather small $\Lambda\Sigma$ mass difference. Thus the time–energy uncertainty principle allows the Λ to turn into a Σ for a few percent of the time, proportional to $(M_\Sigma - M_\Lambda)^{-1}$.

Theoretical models of these forces invoke the exchange of mesons for larger separations, greater than about 0.5×10^{-13} cm, and the interaction of quarks for shorter distances. For a meson of mass m, the force range is about $\hbar/mc$. Because there is only a single neutral Λ (its isospin is zero), the exchange of a pion—the lightest meson—between Λ and N, or another Λ, is strongly inhibited. Thus this long-range pion-exchange force is strongly suppressed in the Λ-N (and Λ-Λ) system and mostly only heavier mesons contribute. The inner part of the N-N force is known to be strongly repulsive, and can be explained as mostly due to the interaction of the quarks in the two nucleons when these overlap at small separations. The short-range character of the hyperon–nucleon forces is not established, although quark models suggest a repulsion for the Λ-N and Σ-N forces. Since strange quarks are now involved, knowledge about this part of the force could give important insight into the strong interactions of strange quarks. It is clearly important to obtain more data about the hyperon–nucleon forces. This could be done by studying the interactions of a nucleon and a hyperon produced, for example, by the inelastic scattering of energetic electrons by nuclei using the CEBAF electron accelerator projected to come on-line in a few years time.

HYPERNUCLEI AND STRONG INTERACTIONS

The earlier data (1950s and 1960s) are from emulsion and bubble chamber experiments in which hypernuclei are an end product of negative kaon–nucleus interactions, the negative kaon (K^- meson) with $S=-1$ transmitting its strangeness to a nucleon to produce a Λ or Σ (e.g., $K^- + n \rightarrow \pi^- + \Lambda$). Especially through observation of the decay products of a hypernucleus one has determined accurate ground-state binding energies, the spin of very light hypernuclei, and certain decay characteristics for individually identified hypernuclei up to a mass number (total number of nucleons plus Λ) of $A=14$ corresponding to hypernuclear nitrogen, ${}^{14}_{\Lambda}\text{N}$.

More recently, extensive experiments with K^- beams of well-defined low energy have produced many excited states of hypernuclei by means of the strangeness-exchange (K^-, π^-) reaction $K^- + {}^AZ \rightarrow \pi^- + {}^A_{\Lambda}Z$, where ${}^AZ$ is the target nucleus and ${}^A_{\Lambda}Z$ is the final hypernucleus. The fundamental process is $K^- + n \rightarrow \pi^- + \Lambda$. The Λ may be produced with a controllable and small momentum and thus various states of the final hypernucleus may be produced. In particular, the Λ can be substituted with large probability in place of a loosely bound neutron in the original nucleus, producing a "strangeness-analog" or "substitutional" state where the Λ is in the same higher angular momentum orbit as the replaced neutron, whereas for the ground state the Λ has to occupy the $1s$ orbit. In this way many excited states of relatively light (p-shell) hypernuclei have been identified. In some cases γ rays (photons) emitted by the decay of an excited state to the ground state have also been observed. Σ hypernuclei have also been observed in (K^-, π^-) reactions.

Very recently beams of positive pions have been used to produce hypernuclei through the (π^+, K^+) reaction: $\pi^+ + {}^AZ \rightarrow {}^A_{\Lambda}Z + K^+$ with detection of the K^+ meson (strangeness $= +1$), the basic process being $\pi^+ + n \rightarrow K^+ + \Lambda$. In this reaction the Λ is in general produced with a larger momentum than in the (K^-, π^-) reaction and is thus less favorable for the production of substitutional-type states. However, this larger momentum makes the (π^+, K^+) reaction much more favorable for the production of hypernuclei with the Λ in deeply bound states, in particular the $1s$ state. Also since the K^+ is absorbed much less inside a nucleus than the K^-, the (π^+, K^+) reaction allows the production and observation of much heavier hypernuclei than the (K^-, π^-) reaction, with the heaviest hypernucleus studied so far being ${}^{89}_{\Lambda}\text{Yb}$ with $A=89$.

The following are some of the principal results and problems of (strong) hypernuclear physics.

The lightest Λ hypernuclei are more weakly bound than ordinary nuclei because the Λ-N force is less attractive than the N-N force. There is in fact no two-body bound ΛN (or ΣN) system, whereas a neutron and proton are bound together as a deuteron. The lightest hypernucleus is the hypertriton ${}^3_{\Lambda}\text{H}$ (bound state of $np\Lambda$) with the Λ only just bound. For heavier hypernuclei a dramatic effect is that the energy needed to separate a Λ from a hypernucleus in its ground state is much greater than that ($\simeq$8 MeV) needed to separate a nucleon from a nucleus. This nucleon, because of the Pauli principle, must go into the last unfilled shell and thus has a large (Fermi) kinetic energy which largely compensates the attractive potential energy due to its interactions with the other nucleons. The Λ, however, can go into deeply bound orbits already occupied by nucleons; in particular it can go into the $1s$ orbit to produce the hypernuclear ground state. For this orbit the Λ kinetic energy can be quite small and the resulting binding energy quite large, even though the potential energy of the Λ is less than that for a nucleon. For very heavy hypernuclei the $1s$ state of the Λ is bound by about 30 MeV which is then also the ground-state Λ separation energy. The recent (π^+, K^+) experiments, made for a large range of nuclei, show in fact a very well-defined sequence of deeply bound single-particle-like states which are very well explained as a result of the Λ moving in a potential well generated by the nucleons of the hypernucleus. In a given hypernucleus these states correspond to increasing orbital angular momentum of the Λ, with the $1s$ state being the most deeply bound. These results also show very clearly that the Λ maintains its identity inside a nucleus and does not dissociate into its constituent quarks.

More detailed calculations of the ground-state energies of Λ hypernuclei which use Λ-N forces consistent with Λp scattering show the need for strongly repulsive many-body, in particular three-body, forces involving the Λ and at least two nucleons. Thus the free Λ-N force is effectively weakened in a nucleus. One possible mechanism for such a weakening is the inhibition of the $\Lambda\Sigma$ conversion process by the presence of other nucleons. Recent calculations have shown that three-body forces between nucleons are important also for ordinary nuclei.

A noteworthy result obtained from the excited states of hypernuclei (in particular of ${}^9_{\Lambda}\text{Be}$ and ${}^{16}_{\Lambda}\text{O}$) observed in the ($K^-$, π^-) experiments is that the Λ-nucleus spin-orbit force is much weaker than the nucleon–nucleus one which plays such a vital role for the nuclear shell model. The (K^-, π^-) experiments have also led to the observation of excited Σ hypernuclear states with remarkably narrow widths in view of the strength of the (strong) $\Sigma + n \rightarrow \Lambda + n$ reaction by which these states can decay. These narrow widths have so far received no satisfactory explanation.

Recently an observation of the $A=4$ helium hypernucleus ${}^4_{\Sigma}\text{He}$ ($\Sigma^+ p2n$) has been reported. There is now a whole family of $A=4$ nuclei: the ordinary ${}^4\text{He}$ nucleus (the alpha particle), the Λ hypernuclei ${}^4_{\Lambda}\text{H}$ ($\Lambda 2np$) and ${}^4_{\Lambda}\text{He}$ ($\Lambda n2p$) in both the ground and an excited state, and now, if confirmed, also ${}^{4}_{\Sigma}$He, with all the hypernuclei being much less bound than ^{4}He. The excited states of ${}^{4}_{\Lambda}$H and ${}^{4}_{\Lambda}$He are observed by their γ decay to their respective ground states, with the Λ flipping its spin relative to the spin of the nucleons. Analysis of the ground-state and excited-state energies of both ${}^{4}_{\Lambda}$H and ${}^{4}_{\Lambda}$He give quite detailed information about the charge-symmetry breaking of the Λ-N force, i.e., about the difference between the Λ-p and Λ-n forces. The Λ-p force is found to be slightly more attractive than the Λ-n force. The lack of any spin dependence in the difference is in strong disagreement with meson-exchange models, suggesting the possible importance of quark effects.

Two emulsion events of $\Lambda\Lambda$ hypernuclei have been reported. The binding energy of such a $\Lambda\Lambda$ hypernucleus, where the two Λs are bound to an ordinary nucleus, can then give information about the Λ-Λ force. Analysis of the ${}^{10}_{\Lambda\Lambda}$Be event (two alpha particles + two Λs) indicate that the Λ-Λ force is about as attractive as the corresponding Λ-N force. Proposed experiments to produce $\Lambda\Lambda$ hypernuclei are clearly very much needed.

An exciting and fundamental question is the existence and properties of bound or resonant states of six quarks with a baryon number of 2. In particular, theoretical models suggest that the most strongly bound such state is the so-called H dibaryon$\equiv$($uuddss$) with the same quantum numbers as two Λs, namely, $S=-2$ (and zero spin). The H could possibly be quite strongly bound with respect to two Λs, and could then decay only by weak interactions and would be long-lived. Experiments to search for the H have been proposed and are under way. An exciting possibility is the existence of superdense hadronic "nuclei" of very large strangeness which are more strongly bound than ordinary nuclei. Such "strangelets" would be multiquark objects with roughly equal numbers of u, d, s quarks, and would thus have a large negative strangeness equal to the number of s quarks and a small (positive) charge (Bodmer, Witten). Their properties would be quite unusual, e.g., the ability to grow by "eating up" nucleons. It has been suggested that very rapidly rotating very compact pulsars may then be strange quark stars rather than ordinary neutron stars.

An intriguing situation where hypernuclear interactions may play an important role is in the deeper interior of neutron stars (pulsars). For fairly massive neutron stars (about the mass of the sun or greater) the matter not too deep in the interior is expected to be mostly in the form of neutrons. However, deeper in the interior the pressure due to gravitation may be sufficient to compress the matter so much beyond ordinary nuclear densities that it may become energetically favorable for some or even most of the neutrons to transform into hyperons. The interior of massive neutron stars may thus consist of a baryonic "soup" containing many species of hyperons whose effect would be to soften the equation of state with possible observational consequences.

WEAK INTERACTIONS

The weak decays of the ground states of Λ hypernuclei are a unique source of information about the weak Λ-N force. The free pionic decay $\Lambda \rightarrow N+\pi$ leaves the final nucleon with only a small kinetic energy ($\simeq$5 MeV). This decay will however be strongly suppressed inside a hypernucleus since most of the states for the final nucleon are now unavailable being occupied by the nucleons of the hypernucleus as a result of the Pauli principle. The decay of the Λ then proceeds mainly by the nonmesonic strangeness-changing weak interaction: $\Lambda+N \rightarrow N+N$ whereby the Λ loses its strangeness turning into a nucleon, the net result being two energetic nucleons for which there is little reduction in the available final states. The large energy release is a signature of the nonmesonic decay. Thus recent measurements on ${}^{12}_{\Lambda}$C show the nonmesonic decay rate to be about 20 times the pionic rate. In spite of the very different decay mechanism from that of a free Λ the lifetime of ${}^{12}_{\Lambda}$C is observed to be about the same as the free Λ decay time, i.e., about 2.5×10^{-10} s. For even heavier hypernuclei the nonmesonic rate is about 100 times the pionic rate, dramatic evidence for Pauli suppression of the latter. Experimentally, the neutron nonmesonic rate ($\Lambda+n \rightarrow n+n$) is about the same or even greater than the proton rate ($\Lambda+p \rightarrow n+p$). A puzzle is that theoretical models predict the neutron rate to be much smaller than the proton rate.

See also HYPERONS; NEUTRON STARS; NUCLEAR FORCES; NUCLEAR STRUCTURE; NUCLEON; PULSARS; WEAK INTERACTION.

BIBLIOGRAPHY

R. H. Dalitz, *Nuclear Interactions of the Hyperons*. Oxford University Press, London and New York, 1965.

A. Gal, "Strong Interactions in Λ-Hypernuclei," in *Advances in Nuclear Physics, Vol. 8* (Michel Baranger and Erich Vogt, eds.). Plenum Press, New York and London, 1975.

B. Povh, *Annual Review of Nuclear Science,* Vol. 28, p. 1. Annual Reviews, Inc., Palo Alto, CA, 1978.

Proceedings of the 1985 International Symposium on Hypernuclear and Kaon Physics, *Nucl. Phys.* **A450** (1986).

Proceedings of the 1988 International Symposium on Hypernuclear and Low-Energy Kaon Physics, *Nuovo Cimento* (1989).

A. R. Bodmer, *Phys. Rev.* **D4** (1971) 1601; E. Witten, *ibid.* **D4** (1984) 272.

Hyperons

David O. Caldwell

Hyperons are particles that are baryons, like the proton, but they are heavier than the proton and possess a different property, designated by the quantum number of "strangeness." Baryons are fermions (i.e., they obey Fermi–Dirac statistics and have half-integer intrinsic angular momentum, or spin) and hadrons, meaning that they interact via the strong interactions.

Of the first two "strange" particles observed in cosmic rays in 1947, one was a hyperon. The particles were considered strange because they were produced copiously (i.e., via a strong interaction) and decayed slowly (i.e., via a weak interaction), and yet the two processes appeared to be the inverse of each other and both should have been strong. That is, the decay should have taken about 10^{-23} s, instead of the observed 10^{-10} s, a decided anomaly. As was proved in 1954 when strange particles could be produced in large numbers at the Cosmotron accelerator, their apparently strange behavior was due to their possessing a new quantum number, which has been called strangeness or hypercharge. Since this quantum number is conserved by the strong interactions, nonstrange particles (like nucleons—such as the proton, p—and π mesons) can produce strange particles only in pairs having opposite values of the quantum number. Thus the Λ^0 hyperon (the first observed) has strangeness $S=-1$, whereas the K^0 meson (the other particle seen first in 1947) has $S=+1$, making the sum $S=0$ for both sides of the production equation, $\pi^- + p \to \Lambda^0 + K^0$. On the other hand, the decay $\Lambda^0 \to p + \pi^-$ can occur only by the weak interaction because the new quantum number is not conserved by the process (i.e., $S=-1 \to S=0$). In the cosmic-ray observations the Λ^0 and K^0 were not seen in the same interaction because they were observable only by their short-lived charged-particle decays (e.g., $K^0 \to \pi^+ + \pi^-$) and not when $\Lambda^0 \to n + \pi^0$ or $K^0 \to \pi^0 + \pi^0$. In addition, half the time the K^0 has a much longer-lived decay, which was not then observable. Thus the apparently strange behavior is not really strange, but was the first observation of the effects of a new quantum number.

The hyperons differ from the strange mesons not only in having the opposite sign (for particles, as opposed to antiparticles) of their new quantum number S, but also in possessing another property called baryon number, B. Baryons are part of the wider class of fermions (unlike the mesons, which are bosons), but both protons and electrons are fermions. Since the proton is not observed to decay into the much less massive positive electron, or positron, the difference between the particles is characterized by assigning $B=1$ to the proton, whereas an electron has an analogous lepton number, describing its membership in its own family. The baryons, as opposed to the leptons (electron, muons, and tauons, and their neutrinos), are strongly interacting particles, or hadrons. It is sometimes convenient to utilize B and replace S with the hypercharge quantum number, referred to above, which is $Y=S+B$.

Because their decays could be observed directly, the first hyperons discovered were those that decayed by the weak or electromagnetic (e.g., $\Sigma^0 \to \Lambda^0 + \gamma$) interactions. Some of the properties of these so-called "stable" hyperons are given in Table I. Note that all the hyperons have a larger mass than that of the proton (938.3 MeV), a fact that was the original source of their name. Note also that hyperons exist with larger values of the strangeness quantum number, the Ξ having $S=-2$ and the Ω^- having $S=-3$.

Table I Properties of the "Stable" Hyperons

Particle	Mass (MeV)	Mean life (s)	Strangeness	Spin
Λ^0	1115.6	2.63×10^{-10}	-1	$\frac{1}{2}$
Σ^+	1189.4	0.80×10^{-10}		
Σ^0	1192.6	7×10^{-20}	-1	$\frac{1}{2}$
Σ^-	1197.4	1.48×10^{-10}		
Ξ^0	1314.9	2.9×10^{-10}	-2	$\frac{1}{2}$
Ξ^-	1321.3	1.64×10^{-10}		
Ω^-	1672.4	0.8×10^{-10}	-3	$\frac{3}{2}$

All hadrons are made up of quarks, mesons being a quark–antiquark combination and baryons being constituted of three quarks. One type of quark has $S=-1$, so the Λ and Σ possess one of these, the Ξ has two, and the Ω^- has all three quarks of the strange variety. The strange quark has charge $-\frac{1}{3}$, and the remainder of the quarks in the hyperons are the charge $+\frac{2}{3}$ up quark and the charge $-\frac{1}{3}$ down quark. Thus the Λ^0 hyperon consists of one strange, one up, and one down quark. While baryons with charm or bottom quarks can have strange quarks as well, these particles are not usually called hyperons.

There are a large number of hyperons which decay via the strong interaction. These particles are massive enough so that they have available other particles into which they can decay while still conserving the strangeness quantum number. At present, about 13 Λ-like, 9 Σ-like, 5 Ξ-like, and 1 Ω-like heavier particles are known. Their quantum numbers, masses, partial decay rates, and in the case of the particles listed in Table I, even their magnetic moments give detailed and strong support to the existence of a quark structure.

See also Baryons; Elementary Particles in Physics; Strong Interactions; *SU*(3) and Higher Symmetries; Weak Interactions.

BIBLIOGRAPHY

R. H. Dalitz, *Rep. Prog. Phys.* **20**, 163 (1957). (On early history.)

Particle Data Group, *Phys. Lett.* **B204**, 1 (1988), or later versions, which appear every two years, alternately in *Reviews of Modern Physics*. (On hyperon properties.)

Hysteresis

R. A. Dunlap

Hysteresis is a phenomenon in which the state of a system does not reversibly follow changes in an external parameter. Examples of hysteresis include the dependence of strain or stress for many materials and the resonant response of nonlinear systems. Most commonly, however, the term hysteresis refers to the behavior of the magnetic induction B or, analogously, the magnetization M as a function of the applied magnetic field intensity H in ferromagnetic and ferrimagnetic materials, or the behavior of the polarization as a function of applied electric field in ferroelectric materials.

For a typical ferromagnetic material, the hysteresis in B as a function of H is illustrated in Fig. 1. This hysteresis loop arises principally from irreversible domain wall motion in the material which results from structural imperfections such as grain boundaries, dislocations, etc. The hysteresis loop is measured by applying a cyclic magnetic field. Beginning with an initially unmagnetized sample in zero field, the B-H behavior will follow the curve oa as H is increased. This portion of the curve is the initial magnetization curve. In sufficiently large H, B reaches its saturation value, B_s. As H is reduced

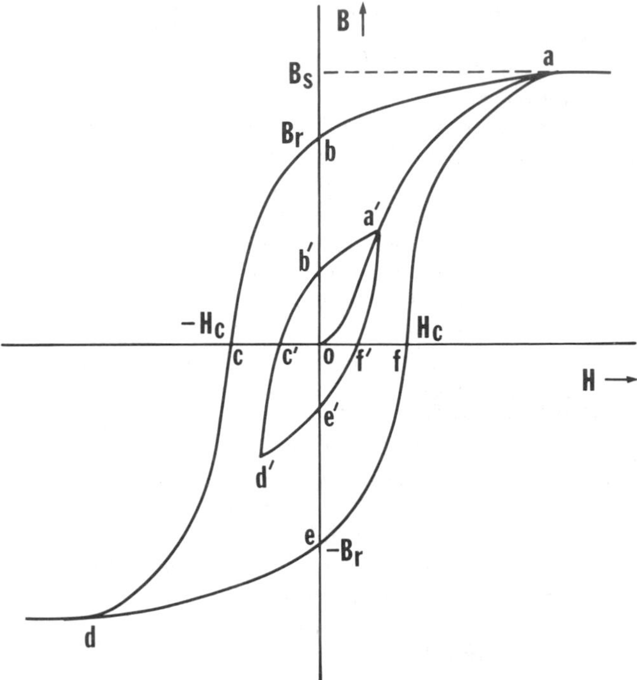

FIG. 1. Hysteresis loop of a ferromagnetic material.

along the portion of the curve ab, the magnetic induction decreases to a remanent value, B_r, at $H=0$. The direction of the applied field is then reversed and the magnetic induction follows the portion of the curve bc, reaching zero at $H=H_c$, the coercive field or coercivity. Reverse saturation is obtained at point d on the curve. The portion of the curve *defa* results from applying progressively more positive H and is typically symmetric with the *abcd* portion of the curve. Repeated cycling of the applied magnetic field will repeatedly trace out the hysteresis loop *abcdefabc*... . The so-called minor hysteresis loop $a'b'c'd'e'f'a'b'c'$... is obtained by cycling the applied field with an amplitude which is insufficient to attain magnetic saturation. A family of such curves exists for different driving field amplitudes.

The area enclosed by the hysteresis loop is the energy loss W per cycle of the applied field and is an integral over the loop of the form

$$W = \oint H\,dB, \tag{1}$$

where H is in A/m, B is in Wb/m^2 (or tesla), and W is in J/m^3. Materials with small H_c and B_r have correspondingly small W and are referred to as soft magnetic materials. Soft magnetic materials with sufficiently large B_s have applications as transformer cores, electric motor stators and rotors, tape recorder heads, etc. One of the best soft magnetic materials known is "Supermalloy," a Ni-Fe-Mo alloy which has $H_c \approx 0.16$ A/m and $W \approx 0.8$ J/m^3. Commercial applications, such as power distribution transformer cores, typically use Fe-Si alloys which have $H_c \approx 8$ A/m and $W=70$ J/m^3, although the development of low-loss amorphous alloys for such uses is in progress.

The so-called "hard" magnetic materials have large H_c and B_r, and a correspondingly large value of W, and form the class of permanent magnets. The maximum of the product BH in the second quadrant of the hysteresis loop (i.e., the portion of the curve bc) is a convenient measure of the energy density of a permanent magnet. Commercially available Co-Sm magnets have $BH_{\max}$ as large as ~150 kJ/m^3. Nd-Fe-B magnets have been reported with $BH_{\max}$ up to about 400 kJ/m^3, and are currently being developed for commercial applications.

See also FERROELECTRICITY; FERROMAGNETISM; MAGNETIC DOMAINS AND BUBBLES; MAGNETIC MATERIALS; MAGNETS (PERMANENT) AND MAGNETOSTATICS.

BIBLIOGRAPHY

R. M. Bozorth, *Ferromagnetism*. Van Nostrand, New York, 1951.

B. D. Cullity, *Introduction to Magnetic Materials*. Addison-Wesley, Reading, MA, 1972.

A. H. Morrish, *The Physical Principles of Magnetism*. Wiley, New York, 1965.

I

Ice

Hermann Engelhardt

Ice, the frozen form of liquid water, is abundant on the earth's surface, in the planetary system, and in interstellar space. If all the ice presently existing on earth melted, the sea level would rise about 66 m. In some planets and in most moons, ice is the major constituent. Pluto is 80% ice; Jupiter's moons Ganymede and Callisto and Saturn's Titan contain 40% ice. Ice is also present in many other moons, in the planetary ring systems, and in comets. The mass of ice stored in the terrestrial polar ice sheets and in alpine glaciers is closely related to the climate.

All of the natural ice on earth is hexagonal ice, ice Ih, as manifested in six-cornered snow flakes (see Fig. 1). At lower temperatures and at pressures above 2 kbar many other ice phases with different crystalline structures exist. No other known substance exhibits such a variety of forms. The phase diagram of ice shows the conditions of stability for the ice phases (Fig. 2; Table I). The equilibrium line between water and ice Ih has negative slope, which is a consequence of the solid having a lower density than the liquid, unlike most other substances and the high-pressure ice phases. The equilibrium lines extend as metastable phase boundaries into the area of stability of other ice phases. Ice IV exists only as a metastable phase within the stability fields of ice III, V, and VI. After forming by nucleation from supercooled water, it quickly transforms into one of the other phases unless it is quenched to low temperatures.

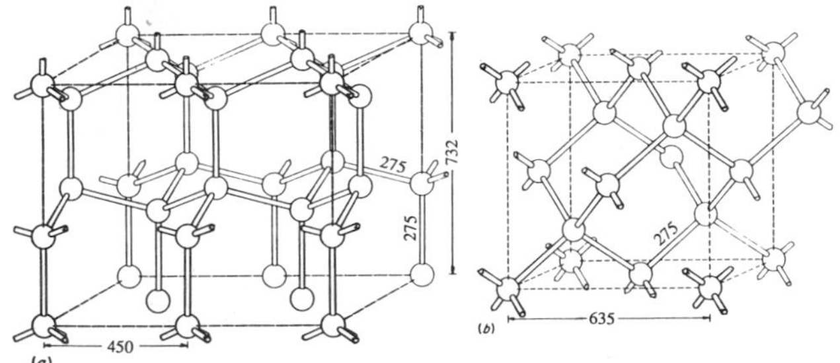

FIG. 1. The structure of (a) ice Ih and (b) ice Ic. Note the tetrahedral coordination and the openness of the structure. On each hydrogen bond, shown by a rod joining the oxygen atoms, lies one proton in an asymmetric position (not shown). Bond lengths, 275 pm, are indicated. The hexagonal c axis is labeled 732 pm, and one of the hexagonal a axes is labeled 450 pm. In (b) the cubic unit cell is outlined with dashed lines; dimensions are in pm at 110 K.

The individual water molecule remains intact as the fundamental building unit in all the ice phases except ice X as

Table I Structural Data on the Phases of Ice

Ice phase	Crystal system	Space group	Cell dimensions[a] (pm, deg)	No. of molecules in a unit cell	No. of nearest neighbors	Distance of nearest neighbors (pm)	O···O···O angles[a] (deg)	Hydrogen positions	Density[a] ($Mg\ m^{-3}$)
Ih	Hexagonal	$P6_3/mmc$	$a=450$; $c=732$	4	4	275	109.3–	Disordered	0.93
		$Fd3m$					109.6		0.93
Ic	Cubic	—	$a=635$	8	4	275	109.6	Disordered	0.94
a_I	Amorphous	—	—	—	4	277		Disordered	1.17
a_{II}	Amorphous		—	—	4	280		Disordered	
II	Rhombohedral	$R\bar{3}$	$a=778$; $\alpha=113.1$	12	4	275–284	80–128	Ordered	1.18
III	Tetragonal	$P4_12_12$	$a=673$; $c=683$	12	4	276–280	87–141	Disordered	1.15
IV	Rhombohedral	$R\bar{3}c$	$a=760$; $\alpha=70.1$	16	4	279–292	87.7–127.8	Disordered	1.27
V	Monoclinic	$A2/a$	$a=922$; $b=754$; $c=1035$; $\beta=109.2$	28	4	276–287	84–128	Disordered[b]	1.24
VI	Tetragonal	$P4_2/nmc$	$a=627$; $c=579$	10	4	280–282	76–128	Disordered[c]	1.33
VII	Cubic	$Pn3m$	$a=341$	2	8	295	109.5	Disordered	1.56
VIII	Tetragonal	$I4_1/amd$	$a=480$; $c=699$	8	8	296 and 280[d]	109.5	Ordered	1.56
IX	Tetragonal	$P4_12_12_1$	$a=673$; $c=683$	12	4	276–280	87–140	Ordered	1.16
X	Cubic	$Pn3m$	$a=283$	2	8	245	109.5	Symmetric	2.51

[a] Cell dimensions and density are at 110 K and atmospheric pressure. Ice X at 44 GPa and 25°C.
[b] Becomes partially ordered when quenched to low temperatures. A more fully ordered phase V′ has space group $P2_1/a$.
[c] When cooled slowly to −150°C, ice VI′ is partly ordered with orthorhombic space group *Pmmn*. A fully ordered ice VI″ has space group *Pn*.
[d] Four of the nearest neighbors are hydrogen-bonded to central molecule at a distance of 296 pm; four are nonbonded at 280 pm.

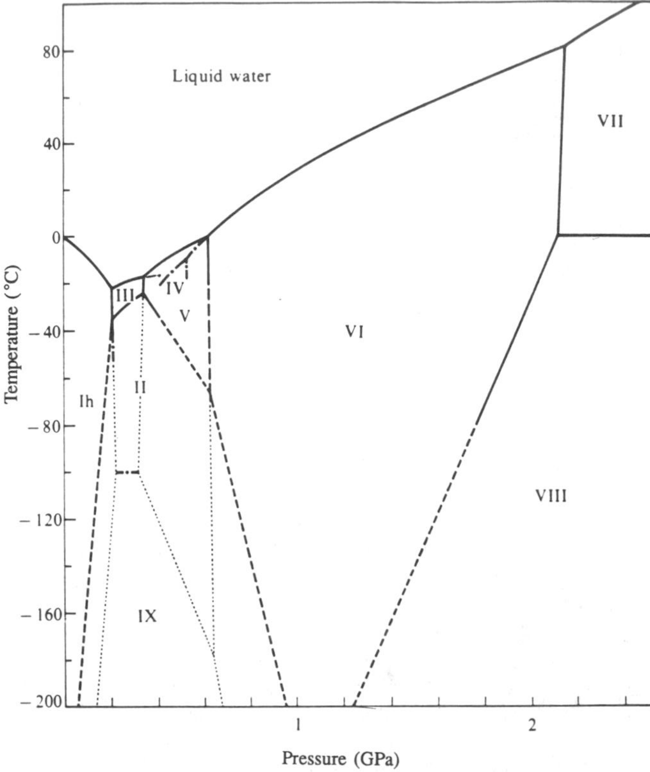

FIG. 2. The phase diagram of ice: ——, measured stable equilibrium lines; ·–·–·–·, measured metastable lines; – – –, extrapolated or estimated stable lines; ····, extrapolated or estimated metastable lines. Ice IV is metastable in the region of stability of ice V. The line above ice IV is the equilibrium line between liquid water and ice IV. Not shown are ice Ic, the amorphous phases, and the ordered versions of ice V and ice VI. Ice X exists above 44 Gpa.

noted later. Since there are equal numbers of protons and of lone electron pairs in the water molecule, each proton can match up with an electron pair of a neighboring molecule forming a hydrogen bond (H-bond). Each water molecule forms H-bonds to four nearest neighbors in a tetrahedral arrangement with one proton occupying each H-bond. The tetrahedral bond geometry explains the openness and relatively low density of the ice Ih structure. In ice Ih the O···O···O angles are nearly the same as the perfect tetrahedral angle (109.5°), which matches fairly well the H–O–H angle of the water molecule, little changed from the angle in the free molecule (104.4°). In the higher-pressure phases the tetrahedral bonding geometry is distorted; the bond angles are bent from perfect tetrahedral and the H-bonds are stretched. The distances to nonbonded near neighbors become shorter the higher the pressure. In ice VII the nonbonded neighbors come as close as the nearest neighbors. Ice VII and ice VIII can be visualized as two ice Ic structures completely intertwined with one another. A general feature repeated in all of the ice structures is H-bonded rings. In ice I and II the smallest rings are of six molecules, but in higher-pressure phases rings of five and four molecules occur.

Since the H-bond lengths are in the range of 274 pm (ice Ih) to 296 pm (ice VIII) and the O–H distances are only 98.5 pm (ice Ih), the position of the protons in the H-bonds is asymmetric. There are six different orientations possible for a water molecule in its tetrahedral bonding environment, each corresponding to a different arrangement of protons in its four H-bonds. If all the possible orientations of the water molecule at each lattice site are equally realized, the phase is proton-disordered; if one orientation is preferred the phase is proton-ordered. Most ice phases have a high- and low-temperature modification which are distinguished by their degree of proton order. Ice VIII is the proton-ordered version of ice VII and ice IX is the proton-ordered version of ice III. Partially proton-ordered modifications exist for ice V and ice VI. Ice II is a completely proton-ordered phase for which the proton-disordered version is unstable and is not observed. In most cases the existence of proton order or disorder affects the crystallographic symmetry of the ice phase. The cubic symmetry of fully disordered ice VII changes to tetragonal symmetry on transformation to ordered ice VIII.

If water vapor is condensed on a cold substrate between −80° and −130°C, a cubic modification, ice Ic, is formed. with two associated protons (D defect). The ions and ori- is related to the structure of ice Ih in the same way that cubic diamond is related to hexagonal diamond, the cubic and hexagonal forms having almost the same density. Below −130°C a noncrystalline amorphous solid known as low-density amorphous ice, ice a_I, appears. A high-density amorphous phase, ice a_{II}, with a density of 1.31 Mg/m^3 can be made at 77 K by compressing ice Ih to 10 kbar.

At a pressure of 44 GPa, the H-bonds in ice VII are so shortened by compression to 245 pm that the protons leave their asymmetric positions in the bonds and move to the centers, forming the structure of ice X, in which, because of the symmetric H-bonding, discrete H_2O molecules are no longer identifiable.

Many of the physical properties of ice are unique and relate to the special features of structure, especially the H-bonding (Table II). Certain physical properties arise from defects in the ideal ice structure. Important among the point defects are the H_3O^+ and OH^- ions and the D and L bond defects—sites where H-bonds are broken by violation of the rule that there be one and only one proton along each O···O bond line. The violation is caused when a water molecule undergoes rotation from one of its six possible orientations to another without compensating adjustments by the adjacent molecules to which it is H-bonded. Such a rotation interchanges some of the molecule's protons and electron pairs, with the result that some of the H-bonds are converted to nonbonded O···O contacts with no protons (L defect) or with two associated protons (D defect). The ions and orientational defects, together with substitutional or interstitial impurity atoms or ions, have important influence on the electrical and mechanical properties of ice and in some cases completely dominate them. Also important are crystal dislocations (line defects) and subgrain or grain boundaries (surface defects), the latter in polycrystalline ice.

The electrical conduction in ice is carried by the protons, whose motion is closely tied to the motion of ionic defects. An ionic defect can move from one lattice site to the next by a small jump of a proton from one off-center (asymmetric) position to the other in an H-bond that connects the two

Table II Physical Properties of Ice Ih at 0°C

Density	0.917 Mg m^{-3}
H-bond length	276.5 pm
Adiabatic compressibility	0.119 GPa^{-1}
Isothermal compressibility	0.33 GPa^{-1}
Melting point	273.15 K
Melting point depression	−74 K GPa^{-1}
Specific heat	2.01 kJ kg^{-1} K^{-1}
Heat of melting	334 kJ kg^{-1}
Thermal conductivity	2.2 W m^{-1} K^{-1}
Linear expansion coefficient	55×10^{-6} K^{-1}
Cubical expansion coefficient	166×10^{-6} K^{-1}
Vapor pressure	610.7 Pa
Static dielectric constant	96.5
High-frequency dielectric constant	3.2
Dielectric relaxation time	20 μs
Activation energy	
for dielectric relaxation	55 kJ mol^{-1}
Refractive index	1.31
Electric dc conductivity	
of ice single crystals at −10°C	1.1×10^{-8} Ω^{-1} m^{-1}
of polycrystalline glacier ice	10^{-5}–10^{-6} Ω^{-1} m^{-1}
Proton mobility	0.8×10^{-4} m^2 V^{-1} s^{-1}
Acoustic velocity	
longitudinal wave	3828 m s^{-1}
transverse wave	1951 m s^{-1}
Velocity of radio waves	170 m μs^{-1}

lattice sites. Since the small proton jumps occur collectively along a favorably oriented chain of H-bonds, the ionic defects and the effective charge carried by them have a very high mobility in the ice structure, 10 times greater than the mobility of ions in normal ionic conductors. Protonic conduction plays a major role in biological charge transfer processes across membranes. Proton conduction in ice and H-bonded materials is analogous to electron conduction in semiconductors. Ice Ih has an unusual high static dielectric constant (96.6 at 0°C compared to 88 for liquid water). The reorientation of the molecular dipoles in ice is facilitated by the mobility of orientational L and D defects, which are generated by thermal activation. The dielectric relaxation time is extremely slow compared to water (20 μs in ice and 10 ps in water at 0°C).

Ice is transparent to visible light. It has the lowest index of refraction for the sodium D line of any known crystalline material. It is doubly refracting, uniaxial, optically positive with very small birefringence. The proton-disordered phases have a broad infrared absorption band for the fundamental intramolecular bending and stretching vibrations (near 3220 cm^{-1} and 1650 cm^{-1} for ice Ih). The infrared band for hindered rotations of the water molecules in ice Ih is centered around 840 cm^{-1}. The translational lattice vibrations absorb in the range 350–50 cm^{-1} (with peak for ice Ih at 229 cm^{-1}). The proton-ordered phases show distinct narrow peaks in their infrared and Raman spectra. From infrared and Raman spectroscopy on the ice phases, much has been learned about the intermolecular coupling mechanism, lattice dynamics, and the properties of H-bonds.

For electromagnetic waves with frequencies from 5 to 300 MHz the loss of energy by absorption in ice is sufficiently small that they can penetrate large ice masses great distances. Radio waves are reflected by inhomogeneities in the ice and at material boundaries, especially at the ice–water and ice–rock interfaces, and these waves can therefore be utilized to examine the internal structure of glaciers and to determine the depth and bottom topography of large polar ice sheets and ice shelves.

Ice is a viscoelastic material with a nonlinear flow law. When shear stress is applied to a single crystal of ice, it undergoes plastic shear strain easily parallel to the basal plane, which is perpendicular to the hexagonal c axis. In other directions the stress needed to produce plastic shear deformation is much higher. When polycrystalline ice is subjected to stress, it immediately deforms elastically, followed by transient creep, and finally steady viscous flow called secondary creep is reached. For high stresses in excess of 400 kPa the creep curve accelerates which is called tertiary creep. Several physical processes are responsible for these deformations: movement of dislocations, sliding along grain boundaries, and recrystallization. The steady-state secondary creep rate $\dot{\epsilon}$ in secondary creep is related to the stress σ by the flow law

$$\dot{\epsilon}=A\sigma^n \exp(-Q/kT),$$

where A is a temperature-independent constant, Q is the activation energy for creep, n is the nonlinear exponent, k is the Boltzmann constant, and T is the absolute temperature. Values of A, Q, and n depend to some extent on the grain size, grain orientation distribution, and impurity content of the polycrystalline material. At temperatures between 0 and −10°C and for stresses between 100 and 250 kPa, $A=5\times10^{-15}$ s^{-1} kPa^{-3}, $Q=139$ kJ mol^{-1}, $n=3$. At stresses lower than 100 kPa, $n=2$.

The surface of ice Ih shows unique properties. Near the melting point, the surface contains many dangling broken bonds that promote the existence of a liquid-like layer. Consequences of the surface properties are sintering of snow to cohesive snowballs, recrystallization of snow to firn and its transformation to glacier ice, and the low friction of many materials on ice, which is useful for sled riding, skiing, and skating. Regelation is a unique ice property: melting of ice under pressure, coupled with adjacent refreezing of meltwater at lower pressure, is the mechanism by which a loop of wire can be pulled slowly through an ice block without cutting the block in two.

There is a whole class of solids, clathrates, where the ice forms a H-bonded host lattice that encages a great variety of small guest atoms or molecules like argon or methane. Ice VI, VII, and VIII can be viewed as self-clathrates, where two equal ice lattices interpenetrate each other but are not H-bonded to each other. Ice-like structures or structurally arranged water molecules are encountered in hydrates and hydrated compounds including all macromolecules that would not be biologically active without their structured water or ice-like shells. Even water can be viewed as a partially broken down structure of ice.

H-bonds, so pervasive in the crystalline ice phases, play a major role in many substances from glues and mortar to the life-supporting structure of proteins. The properties of ice, and H-bonds, are recognized in many aspects of physics, chemistry and biology, in several branches of geophysics, including glaciology (dynamics of large ice masses), meteorology (cloud physics), and in oceanography (sea ice), in planetary sciences, and in astronomy.

See also HYDROGEN BOND; LATTICE DYNAMICS; PHASE TRANSITIONS; WATER.

BIBLIOGRAPHY

N. H. Fletcher, *The Chemical Physics of Ice*. Cambridge University Press, Cambridge, 1970. (An intermediate and advanced text that illustrates many aspects of solid-state physics using ice as an example.)

P. V. Hobbs, *Ice Physics*. Clarendon Press, Oxford, 1974. (A comprehensive treatment and documentation of ice at an intermediate and advanced level.)

B. Kamb, "Crystallography of Ice," in *Physics and Chemistry of Ice* (E. Whalley, S. J. Jones, and L. W. Gold, eds.). Royal Society of Canada, Ottawa, 1973. (An advanced text.)

E. Whalley, "The Hydrogen Bond in Ice," in *The Hydrogen Bond* (P. Schuster, G. Zundel, and C. Sandorfy, eds.). North-Holland, Amsterdam, 1976. (An advanced text.)

W. S. B. Paterson, *The Physics of Glaciers*. Pergamon Press, Oxford, 1981. (An intermediate text on applied ice physics.)

Journal of Glaciology, published by the International Glaciological Society, Cambridge. (A professional journal.)

Impedance *see* Resistance

Inclusive Reactions

T. Ferbel

Collisions at high energies generally involve the production of many elementary particles. To establish which features of multiparticle production reflect the basic dynamics requires a detailed study of such production processes. One approach is to examine as many complete reactions as are accessible to experimental investigation. For example, in a collision between particle A and particle B we might study the specific channels A + B→A + B + C, A + B→C + D, A + B→A + C + D + G, or the like (where C, D, and G represent produced particles). Reactions such as these, in which all final-state particles are measured, are termed *exclusive* reactions. A complementary approach to the problem is based on a form of statistical sampling of all exclusive channels using *inclusive* reactions. An inclusive experiment measures the probability for the production of a specified configuration of particles in a collision, independent of whatever else might be produced. Thus the simplest (albeit trivial) inclusive measurement, yielding what might be termed a zero-particle inclusive cross section, is one in which no particles at all are specified, but rather a sum is taken over all possible final states. The rate for the zero-particle inclusive reaction A + B→ANYTHING is consequently determined by the total cross section (a Lorentz-invariant quantity, σ_T) for the interaction of particle A with particle B.

A more informative inclusive reaction is a single-particle inclusive process of the kind A + B→C + ANYTHING, in which the cross section for the production of particle C is measured regardless of whatever is produced along with C. This type of reaction is characterized by a function, $d\sigma/(d^3p/E)$, proportional to the probability for emitting particle C, of momentum **p** and energy E, into a Lorentz-invariant differential element of momentum space d^3p/E. Further information pertaining to multiparticle production can be obtained by examining a two-particle inclusive reaction of the kind A + B→C + D + ANYTHING; in this case the production of C and D is characterized by a Lorentz-invariant two-particle differential cross section, $d\sigma/[(d^3p_C/E_C)(d^3p_D/E_D)]$. Although we might expect that additional dynamic information could be obtained from studies of higher-order inclusive reactions, this does not seem to be the case; it appears that inclusive cross sections of order greater than 2 may be calculable from lower-order data for total collision energies in the center-of-mass (c.m.) system (E_T*) of at least up to ~50 GeV.

Inclusive production cross sections depend in general on the nature of the colliding particles, on their states of polarization, on the energy available in the center-of-mass frame, and on the momentum vectors of the specified final-state particles. In the absence of polarization, the cross section for any specific inclusive reaction can depend only on E_T* and on the longitudinal and transverse momenta of the specified particles (longitudinal being along the collision axis). Transverse momenta (p_T) of most particles produced in hadronic reactions tend to be small and their distribution only weakly dependent on E_T*. (The average p_T value for pions is ~0.4 GeV/c.) Longitudinal momenta (p_L) can be large, and are often comparable to incident momenta. Two p_L-like variables have been found to be particularly useful in describing inclusive reactions. These are (1) the Feynman x variable, $x = 2p_L^*/E_T^*$, where the asterisks signify c.m. quantities; and (2) the rapidity variable $y = \frac{1}{2}\ln[(E+p_L)/(E-p_L)]$. The range of x is limited between -1 and $+1$, while the allowed range of y values grows logarithmically with E_T*.

It is often convenient to distinguish two regimes of particle production in "soft" high-energy collisions (i.e., collisions that do not ostensibly involve constituents). One of these is the "fragmentation region": here, the produced particles have momenta comparable to those of the incident particles (i.e., large $|x|$ values) and can be thought of as fragments from the breakup of the target or projectile. The other regime is the "central region"; here particles are produced with comparatively small momenta in the c.m. frame and therefore do not appear to be associated with either the target or the projectile particle.

Total cross sections for hadrons interacting with other hadrons vary substantially at low energies ($E_T^* < 3$ GeV), but at intermediate incident momenta ($E_T^* < 100$ GeV) the dependence is much weaker. For $E_T^* > 100$ GeV, σ_T appears to grow logarithmically (almost doubling every decade). It has been suggested that a similar simplification may obtain in the case of inclusive reactions for large E_T*. In particular, the hypothesis of limiting fragmentation (HLF), motivated in part by a diffraction picture of particle production, asserts that at fixed p_T the invariant cross section (or this cross section divided by σ_T) for producing a particle of any finite p_L, as measured in the rest frame of the target or the projectile particle, approaches a finite, energy-independent limit as E_T^* grows.

An alternative statement of limiting behavior in inclusive reactions is the scaling hypothesis, which is based on an analogy between hadron emission in high-energy collisions and the radiation of photons by an accelerating charge. This hypothesis states that for fixed p_T and large E_T^*, the invariant cross section for any x value approaches a limit that is E_T^* independent. Yet another formulation of asymptotic behavior is available in a macroscopic picture of high-energy collisions that involves a short-range order (SRO) hypothesis. This view of particle production suggests that among the particles involved in a scattering event only those with similar rapidities are correlated. Thus at asymptotic energies, particle production and correlations among produced particles should depend not on rapidity but rather on differences in the y values of produced particles. At asymptotic E_T^*, HLF and scaling yield the same predictions in the fragmentation region. The scaling and the SRO hypotheses also provide predictions for limiting behavior in the central production domain.

A comparison of the x dependence of invariant inclusive cross sections for pion production at fixed p_T and varying E_T^* indicates that up to $E_T^* \approx 900$ GeV, scaling appears to hold surprisingly well at small p_T, particularly at large $|x|$ values (fragmentation region). However, small but significant deviations are observed near $x \approx 0$ (central region). If inclusive cross sections were to start scaling at some E_T^*, then the invariant cross sections would develop plateaus in the rapidity variable and, as E_T^* is increased, those plateaus would grow in extent as $\ln E_T^*$. It is easy to show that as a result of this, the average number of particles of any particular species produced in an inelastic collision would rise linearly with $\ln E_T^*$. This is only approximately what happens for $E_T^* \gtrsim 10$ GeV. In fact, the average number of charged particles ($\langle n \rangle$) produced in inelastic p–p collisions for $E_T^* \gtrsim 10$ GeV is well represented by $\langle n \rangle = 0.5 \ln^2 E_T^* + 0.9 \ln E_T^* + 1$. (Pi mesons account for about 90% of all produced particles, and K mesons for about 10%.) The values of $\langle n \rangle$ for other types of incident particles differ only somewhat from that in p–p reactions. In fact, except for an overall normalization (given approximately by σ_T), inclusive particle production in the central region appears to be independent of the incident channel.

At small p_T, where there is no fundamental theory of particle production (such as, e.g., QCD at large p_T), the rate at which inclusive cross sections (or their normalized values) might approach their asymptotic forms and relationships between different reactions can be examined by means of a generalization of the optical theorem. The usual optical theorem relates the amplitude for the two-body elastic-scattering process $A+B \rightarrow A+B$ to the total cross section (zero-particle inclusive reaction $A+B \rightarrow$ ANYTHING); there is an analogous relationship connecting cross sections for higher-order inclusive reactions to multiparticle elastic-scattering amplitudes ($A+B \rightarrow C+$ ANYTHING is related to $A+B+\bar{C} \rightarrow A+B+\bar{C}$ scattering, where $\bar{C}$ is the antiparticle of C). Although experiments on multiparticle scattering are not feasible, models can be used to calculate these "elastic"-scattering amplitudes and such calculations provide the phenomenology for relating different inclusive reactions. There has been substantial success in this regard in the application of Regge models to inclusive reactions.

High-energy collisions typically produce a great number of particles, but the number of exclusive channels that are simple enough to be amenable to experimental probing at high energies is rather small. It therefore appears that, of the two approaches available for investigation of high-energy phenomena, the inclusive technique may be the more valuable one for gaining an understanding of the dynamics of multiparticle production.

See also ELEMENTARY PARTICLES IN PHYSICS; HADRONS; STRONG INTERACTIONS.

BIBLIOGRAPHY

H. Bøggild and T. Ferbel, *Ann. Rev. Nucl. Sci.* **24,** 451 (1974). (I–A)

D. Horn and F. Zachariasen, *Hadron Physics at High Energies*. Benjamin, Advanced Book Program, Reading, Mass., 1973. (I–A)

J. G. Rushbrooke, in *Proceedings of the International Symposium on Multiparticle Dynamics* (J. Granhans, ed.). World Scientific Publisher, Singapore, 1985.

Induction *see* Faraday's Law of Electromagnetic Induction

Inertial Fusion

Steven W. Haan and Heiner W. Meldner

Inertial fusion (IF) is one of the two principal schemes proposed for producing technologically useful energy from nuclear fusion. In IF the burning plasma is confined by its own inertia, the burn time being limited by hydrodynamic disassembly. For electromagnetically confined fusion see PLASMA CONFINEMENT DEVICES.

In IF, beams of radiation or ions (designated driver beams) are used to implode targets containing fusionable material, in order to compress and heat them sufficiently so that thermonuclear reactions ensue. Generation of electric power from such a process is potentially attractive, as large amounts of energy could be produced from readily available fuel with relatively little radioactive waste. Also, an IF facility producing explosions of a few hundred MJ could be of value to defense research, both for studies of effects of such explosions on military hardware and for basic studies of plasma processes under conditions relevant to thermonuclear weapons.

IF was proposed in the early 1960s, very quickly following the invention of the laser in 1960. Work by Stirling A. Colgate, Ray E. Kidder, John H. Nuckolls, Ronald F. Zabawski, and Edward Teller in that period already defined many of the crucial concepts as currently understood. Most of their work was initially classified and was not published until the early 1970s.

In the following we consider the requirements for an implosion intended to produce significant yield efficiently. Basic features of this implosion determine many characteristics of the drivers, described in subsequent sections here. They also determine the overall scale for high-yield IF, which requires drivers somewhat larger than any currently existing.

Details of the present discussion are limited to "direct drive" targets, in which the driver energy is deposited directly in an ablator. In another class of targets, "indirect drive," symmetry is enhanced by converting the driver energy to thermal x rays which compress the fuel regions. Details regarding indirect drive are currently classified.

IMPLOSION CHARACTERISTICS

A target generally includes two nested spherical shells (see Fig. 1). The outer shell, the ablator, is heated by energy from the driver. Ablation creates pressure at the outer surface, which causes inward acceleration, and ultimately compression, of the remainder of the pellet. The interior shell is fusionable material, usually equimolar deuterium and tritium (D-T). Schemes have been proposed using other fusion fuels such as D-^{3}He, but such fuels require higher temperatures for the reactions.

For fusion to occur, the thermal kinetic energy of the D and T ions must be high enough to overcome the Coulomb barrier. The Maxwellian-averaged D-T reaction rate, $\overline{\sigma v}$, is therefore strongly temperature dependent. Significant energy is produced only when temperatures of a few keV are reached. Each reaction produces a 14-MeV neutron and a 3.5-MeV α particle. The neutrons generally escape. (In a reactor, these would be absorbed in a heat-transfer medium, and in a lithium supply for tritium reproduction.) If the volume of hot D-T in the fuel (the hot spot) has column density (ρr) of order 0.3 g/cm^2, the αs are stopped in the hot spot providing further heating.

Self-sustaining burn can ensue, provided there is a positive

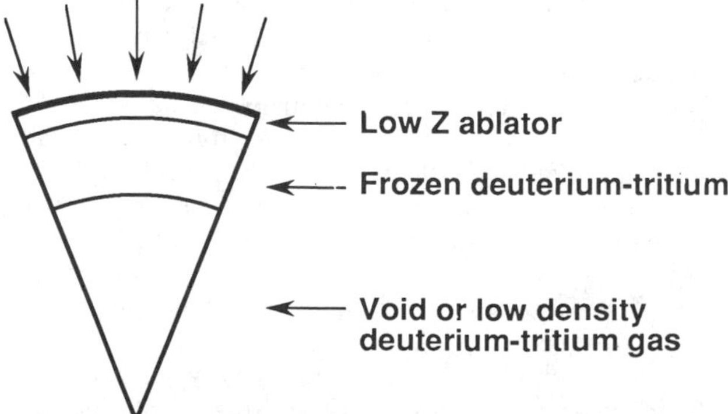

FIG. 1. Sectional view of a typical direct drive IF target. Light from a powerful laser impinges on a spherical target consisting of a low atomic number ablator enclosing a shell of fusible deuterium–tritium fuel. About 60 individual beams, each focused to a spot size about equal to the capsule diameter, are required to produce sufficiently symmetric illumination.

energy balance of four processes: α deposition and PdV work from the implosion, which deposit energy into the fuel, and bremsstrahlung radiation and thermal conduction, which remove energy. The balance of these four processes depends on the implosion velocity of the igniting material, the temperature, and the ρr of the hot spot. Ignition requires temperatures of about 5 keV, implosion velocities of about 3×10^7 cm/s, and ρr about 0.3 g/cm^2. Once such conditions are achieved, the burn rate increases further, and the ignited fuel "bootstraps" via accelerating α deposition up to temperatures of tens of keV.

Subsequent dynamics are controlled by the disassembly. The net energy produced per mass of fuel depends on the density, since the reaction rate is proportional to ρ, on the temperature via $\overline{\sigma v}$, and on the confinement time, which is essentially the time it takes a rarefaction wave to travel in from the perimeter. These variables enter in such a way that the burn-up fraction depends only on the density–confinement time product, or equivalently on the total ρr.

Significant burn-up (30%) requires $n\tau$ of about 2×10^{15} cm^{-3} s, or ρr of about 3 g/cm^2. Clearly there is advantage to having ρr larger than the 0.3 g/cm^2 required for ignition, but heating additional fuel to ignition temperature wastes driver energy. So the optimal imploded configuration has two distinct fuel regions: a hot spot with ρ-$r\cong0.3$ g/cm^2 and $T\cong5$ keV to produce ignition, and a surrounding "main fuel" layer, at lower temperature so that it can be efficiently compressed to increase the total ρr to about 3 g/cm^2.

For a given mass of D-T in radius R, $\rho r\propto1/R^2$. So the required ρr values could in principle be achieved by compression of an arbitrarily small mass of D-T and correspondingly small energy investment. However, while the energy produced by burning a mass of D-T at a given ρr is proportional to the mass, the energy required to compress a dense Fermi gas to density ρ is proportional to $m\rho^{2/3}$. So the gain (energy out/energy in) drops for smaller masses, which require proportionately more energy for compression. Target gains of order 100 are required to produce a significant net system gain given a driver efficiency of order 10–30%. The mass required to produce such a target gain depends on how efficiently the driver energy is coupled to compression, how much energy is used to heat the hot spot, and other factors, but for most configurations the requirement that the gain be about 100 sets the main fuel mass at about 10^{-3} g. Another related factor is the compression that can be realistically achieved. This is limited by the symmetry of the implosion. Plausible convergence ratios (initial outer radius/final hot-spot radius) are in the range 30–50. The main fuel itself then converges in linear dimensions by a factor of about 10, corresponding to compressions of order 1000. Given an initial density of 0.25 g/cm^3 for cryogenic D-T, a mass of 2 mg would produce a ρr of 3 g/cm^2. This corresponds to about the same mass as is set by the requirement of gain 100. These two issues, fortunately compatible, set the overall scale for high-gain IF: the target contains a few mg of D-T, at initial radius R of a few mm. The energy produced is between 100 MJ and 1 GJ. In a reactor designed to produce about 1 GW of electricity, targets of this scale would have to be shot at a rate of 2 to 30 per second, depending on target yield and reactor efficiency. The energy in the assembled DT is about

100 kJ; depending on the coupling efficiency, the driver energy E is required to be in the range 1–10 MJ. The pulse length, $t \cong R/(3\times10^7$ cm/s), is about 10 ns, so that the power E/t is a few times 10^{14} W, and the intensity $E/4\pi r^2 t$ is a few times 10^{14} W/cm^2.

Additional requirements are determined by considering how the ignition configuration can be assembled. The main fuel must be compressed and accelerated by the ablation pressure; the kinetic energy thereby achieved is then converted into further compression, as well as compression and PdV heating of the hot spot. If the hot-spot mass is a few percent of the main fuel mass, the kinetic energy in the latter at the velocity required for ignition (3×10^7 cm/s) is also adequate to produce heating of the hot spot to several keV. Thus the object of the first part of the implosion is to accelerate the main fuel to about 3×10^7 cm/s, while keeping its entropy very low so that its pressure is close to the minimum imposed by Fermi degeneracy. Minimal entropy requires that the fuel be initially liquid or solid. A series of shocks may pass through the fuel, but they must be relatively weak. Other sources of energy deposition in the main fuel must be minimized.

Achieving the needed velocity requires high ablation pressure, low mass per unit shell area, and/or large distance over which the acceleration can act. Thus there is a tradeoff between aspect ratio (radius/thickness) and required ablation pressure, for the given final velocity. The aspect ratio of the compressed shell in flight is limited by hydrodynamic stability considerations to be in the range of 30 to perhaps 80. (The achievable upper limit is a topic of current research.) This requires that the ablation pressure be in the range 50–200 Mbar. Such a large pressure cannot be applied immediately to cold D-T, but four or five shocks in succession can produce adequate pressure while maintaining sufficiently low entropy. Each shock is launched by an increase in driver intensity, and the increases are timed so that the shocks coalesce just inside of the high-density fuel layer.

Another important issue is the efficiency with which the incoming driver energy is converted to ablation pressure and then to kinetic energy. The ablation efficiency depends on the driver type used. The driver energy is usually absorbed somewhere outside the actual ablation surface, and then conducted or otherwise transported to the ablation front. Differences in the radii of absorption and ablation help to improve symmetry, but reduce the efficiency.

Other ignition configurations have been considered that are more complicated than the hot-spot/main-fuel configuration described here. If the igniting hot spot is surrounded by a shell of material with high atomic number, and hence high density and high opacity, the energy balance for ignition improves. Such a shell can also be driven to very high velocity by collision with another, heavier, shell converging from a larger radius. However, these more complicated configurations generally involve interfaces that are hydrodynamically unstable. They also risk contamination of the fuel with high atomic number material, which would increase the energy losses dramatically. Another important alternative configuration is the "exploding pusher," in which the entire shell is heated rapidly and about half of it implodes while the other half explodes. Such targets have low gains, but have been very useful for physics experiments since they produce high temperatures, albeit at very low ρr.

POTENTIAL DRIVERS AND RELATED TRADE-OFFS

Four driver systems are being actively considered at the time of this writing: two short-wavelength laser systems, light ions, and heavy ions.

High-energy lasers are most readily available in the infrared, but research in the 1970s found that infrared radiation–plasma interactions generate energetic electrons which produce copious preheat and seriously degrade the implosion. The most important such process is stimulated Raman scattering, in which the incident infrared resonantly decays into a scattered electromagnetic wave and an electron plasma wave, which produces the energetic electrons. A similar process, stimulated Brillouin scattering, is the decay into an electromagnetic wave and an ion acoustic wave; this process can degrade the absorption or change its location, since nearly all the incident energy can be transferred to the scattered light wave. In more recent experiments, these deleterious plasma processes have been sharply reduced by using laser light with wavelengths 0.35 μm or shorter. The short-wavelength laser light is absorbed primarily by inverse bremsstrahlung (collisional heating). In all cases, the interactions are concentrated near the surface of so-called critical density, at which the incident frequency equals the plasma frequency and inside of which the light cannot propagate. To a large degree, the advantage of shorter-wavelength illumination is the higher critical density, so that the plasma is more collisional where the light is being absorbed. In indirect drive the laser is incident on high atomic number material, and the efficiency of its conversion to thermal x rays is a key issue.

Of the two candidate short-wavelength laser systems, the most thoroughly investigated is doped Nd glass, pumped via optical flashlamps. This system emits 1.06-μm infrared, which is frequency-tripled in optically nonlinear crystals before hitting the target. In most current glass lasers the infrared pulse is generated by passing through a series of slabs, individually pumped, which serve as amplifiers. Several systems of this type have been built, the largest of which can now produce 120 kJ. This type of laser, at larger scale, would adequately function as a driver for a very low repetition-rate high-gain experimental facility. With respect to electricity production, it has yet to be demonstrated that glass lasers can be sufficiently efficient and that they can be built to sustain the thermal loads resulting from a repetition rate of order 10 s^{-1}.

The other candidate laser system is the KrF gas laser. A gaseous mixture of Ar–Kr–F_2 is pumped either by electronic discharge or by an electron beam, and the system lases at a wavelength of 0.28 μm wavelength via decay of Kr^+F^- ion pairs to a ground state of dissociated covalent Kr and F. Efficient amplification requires a pulse length of >100 ns, and so the short pulse required for IF must be created by a pulse-compression scheme. For example, a synthetic pulse can be created from a sequence of short pulses which propagate through the gain meduim in slightly different directions. After amplification, the pulses are individually temporally delayed by passing them through different path lengths before they are overlapped on the target.

Light-ion beams are also being investigated as driver candidates. They evolved from efforts to drive IF with electron beams, which are unacceptable because the energy is deposited over too large a penetration depth. Light ion drivers (Li^+ at 30 MeV) appear to be relatively inexpensive, but are difficult to focus and pulse shape.

Finally, heavy-ion beams could become the most attractive driver option. Singly to triply charged ions with atomic mass 130 or higher would be accelerated to about 10 GeV. The target coupling physics appears to be much simpler than with lasers and has desirable scaling with intensity. The system appears to be efficient, and a 10 s^{-1} repetition rate is feasible. Although cost scaling to a large facility is very attractive, it is also the principal difficulty at this time: a small, "experimental" heavy ion driver is much more expensive than a comparable laser or light-ion machine.

CURRENT RESEARCH

Most experimental facilities are generally doped Nd glass lasers, frequency multiplied. Experiments, details of which are classified, have also been done using nuclear explosions in underground tests by the United States and possibly other countries.

The basic physics of the implosions is being tested at various laboratories, with both direct and indirect drive. High-density compressions and high temperatures have been generated, and it is generally agreed that the spherically symmetric implosion hydrodynamics is well understood. Detailed simulations with large computer codes are used to model the experiments.

The effects of deviations from spherical symmetry are the topic of much current research, both theoretical and experimental. It is difficult to estimate the effect of irradiation asymmetry and hydrodynamic instabilities. They are important because the shell as described above is hydrodynamically unstable on the outside while being accelerated, and on the inside while being decelerated. Instability growth probably determines the acceptable values for the aspect and convergence ratios, and thus is related to the required size and other properties of the driver. Also, since inhomogeneities in driver intensity can couple to the instabilities, smooth beams are very important and work is being done to improve beam quality and test the consequences.

Laser/matter interactions are another important area of current research. Hot electron production has been reduced with short-wavelength lasers, but questions remain regarding how the phenomena may change in plasmas with larger length scales and other features of high-gain designs. The efficiency of absorption, and for indirect drive, conversion to x rays, are very important in establishing the size and properties of a high-gain driver.

Experiments to date indicate that high-gain targets would work within the parameter range described above; the major research efforts now are related to minimizing the driver size and expense within this range, so that a driver producing ignition and burn can be built and operate perhaps in the late 1990s.

ACKNOWLEDGMENT

This research was supported by the U.S. Department of Energy, Lawrence Livermore National Laboratory under Contract No. W-7405-ENG-48.

See also BREMSSTRAHLUNG; FERMI–DIRAC STATISTICS; FLUID PHYSICS; LASERS; NUCLEAR FUSION; PLASMAS; PLASMA WAVES; SHOCK WAVES AND DETONATIONS.

BIBLIOGRAPHY

R. S. Craxton, R. L. McCrory, and J. M. Soures, *Sci. Am.* **255,** No. 2, 68 (1986).

J. I. Duderstadt and G. A. Moses, *Inertial Confinement Fusion.* Wiley, New York, 1982.

J. F. Holzrichter, E. M. Campbell, J. D. Lindl, and E. Storm, *Science* **229,** 1045 (1985).

J. H. Nuckolls, L. L. Wood, A,. R. Thiessen, and G. B. Zimmerman, *Nature* **239,** 139 (1972).

E. Storm et al., "High Gain Inertial Confinement Fusion: Recent Progress and Future Prospects," Lawrence Livermore National Laboratory Report UCRL-99383 (1989).

Infrared Spectroscopy

David A. Dows

Infrared spectroscopy is primarily a method for the study of the vibrations of atomic nuclei in molecules and crystals. When electromagnetic radiation in the infrared region of the spectrum falls on a sample, energy may be transferred to or from the moving nuclei. The resulting removal or addition of electromagnetic energy is measured to record the spectrum. The infrared region of the electromagnetic spectrum lies at longer wavelengths, or lower energy, than the visible light region. It is common to refer to the near infrared (about 0.7–2μm), the middle infrared (about 2–50 μm), and the far infrared (greater than 50 μm) wavelength regions.

Energy level spacings for molecular and crystal vibrations lie in the 0–0.5-eV energy range, and transitions between them which can absorb or emit light lie in the infrared. These transitions can also be observed in light scattering, where a monochromatic incident light beam, usually from a visible laser, gives rise to frequency-shifted lines in the scattered light (the Raman spectrum); the shift is just the vibrational energy difference. They can also be observed as vibronic structure in the electronic spectrum (in the visible and ultraviolet regions of the spectrum).

It is usual in infrared spectroscopy to describe transition energies by giving the wave number of the transition, obtained from the Bohr frequency condition $\bar{\nu}=(E_1-E_2)/hc$, where E_1 and E_2 are the upper and lower state energies, h is Planck's constant, and c is the velocity of light. The units of $\bar{\nu}$ are cm^{-1}, but $\bar{\nu}$ is often called the "frequency" of the transition (the frequency, measured in hertz, is really equal to $c\bar{\nu}$). The vibrational "frequency" ranges from about 10 to 4000 cm^{-1}, and the corresponding wavelength of infrared light from about 1000 to 2.5 μm.

Since the frequencies are determined by the molecular or crystal potential function, or by its parameters, the "force constants,'' study of the infrared or Raman spectrum most immediately yields information about the strengths and stiffness of chemical bonds and valence angles or about the binding of atoms in crystals. In addition, due to the characteristics of the physical processes leading to interaction with light and thus to spectroscopic phenomena, there exist ''selection rules'' limiting the number of transitions which can be observed. These selection rules (which may be different for the infrared and the Raman spectra) depend on the symmetry of the molecule or crystal, and their existence often adds to the determination of molecular structure or crystal symmetry.

INFRARED SPECTROSCOPIC METHOD

Infrared spectroscopy, first developed in the early years of this century by Coblentz, experienced a long period of dormancy followed by a dramatic resurgence at the end of the Second World War due to development of new technologies. In an ideal spectrometer, strong radiation from a perfectly monochromatic, tunable light source would pass through the sample to a sensitive, noise-free detector. In reality, a typical spectrometer uses continuous radiation, from a heated source, which passes through the sample, is dispersed by a monochromator into narrow frequency bands, and is measured by a thermal or photoconductive detector. The wartime development of electronic systems, particularly the phase-sensitive detector and low-noise amplifier, together with more recent photoconductive detectors and bolometers brought infrared spectroscopy to the status of a routine tool in physics and chemistry. The recently developed injection diode laser seems almost an ideal light source, which would obviate the need for a monochromator, but it is at present restricted to special applications. An alternative to the monochromator is also in wide use; the light from the sample is passed through a Michelson interferometer, which records the Fourier transform of the spectrum as its movable mirror is scanned. The Fourier-transform method of spectroscopy has a large advantage in light-gathering power over the monochromator method, but requires a small computer to make the Fourier analysis and cannot as directly study a single infrared wavelength as a function of time.

When a narrow band of radiation selected by the spectrometer is of a frequency given by the Bohr condition, absorption of the light by the sample may take place. For interaction with the radiation field the sample must have an electrical polarity change resonating with the field frequency. In practice it is necessary that the electric dipole of the molecule (or the electric polarization of the crystal) change during the course of the vibration in order that the transition between energy levels of the vibration appear in the infrared spectrum.

The requirement of an oscillating dipole leads to the conclusion that homopolar molecules (e.g., H_2, Cl_2) and crystals (e.g., diamond) will not have infrared spectra. Heteropolar diatomics (e.g., HCl, CO) have infrared spectra, though the strengths of the characteristic absorptions vary greatly. Polyatomic molecules and almost all crystals absorb infrared radiation, though their various vibrations vary in absorption strength. If symmetry is present in the absorber, some vibrations may not carry an oscillating dipole. In crystals, second-order effects are much more important than in molecules, and cause extensive complex absorptions not simply given by the basic selection rules.

Gases, liquids, and solutions to be studied in the infrared are ordinarily contained in cells ranging from 10 or more to 0.1 or less millimeters in thickness. The cell windows are usually cut from crystals of salts (e.g., KBr, NaCl) which do not absorb near- or middle-infrared radiation. In the case of crystalline materials, where it is difficult to prepare samples of the requisite thinness (*ca.* 0.001 cm), most routine analytical work is done in mixtures where a small amount of finely powdered sample is intimately mixed with an oil (a ''mull'' spectrum) or mixed with a powdered alkali halide and pressed into a transparent pellet. It is also possible to study the spectrum in the reflected light from the sample surface. And the emission of infrared radiation from heated samples (flames, shock waves) is an important technique.

The advent of the computer and of affordable integrated circuit and chip technology has profoundly modified modern infrared instrumentation. In turn, this has opened new areas of exploration. Primarily, the computer has pushed Fourier-transform spectroscopy to the fore to such extent that the purchase of new grating spectrometers for infrared studies is no longer an option for many industrial and university laboratories—save for specific purposes such as infrared observations from satellites, research on phenomena that stretch over narrow-band regions, and infrared band profile analysis (see below), to name some. The advantage of Fourier-transform infrared spectroscopy over a grating instrument (''Fellgett'' or ''multiplex'' advantage) not only yields a gain in signal-to-noise level normalized to the observation time, but also an *n*-fold doubling of resolution by a mere doubling of the time of observation if the original signal strength is demanded. Further, the ease of data handling (smoothing, transforming, storing and retrieving, analyzing, and plotting) is appreciable since sampling is, *a priori*, done at precise discrete intervals.

New technology of lens construction in combination with the intensity-gathering advantages of infrared Fourier-transform spectroscopy have spawned a method of observing the diffusely scattered infrared radiation (''DRIFT''). The technique has been very useful in exploring vibrational modes of samples, such as powders and entire flat panels, which do not lend themselves to being pressed into pellets (possible modification by the applied high pressures) or otherwise being decimated (art objects). It is also very promising for *in situ* adsorption/desorption studies on amorphous materials.

VIBRATIONAL MOTIONS STUDIED BY INFRARED SPECTROSCOPY

The molecular motions studied by the infrared technique are understood by considering well-established theories of classical and quantum mechanics [1,2] as are those of atoms in crystals [3].

In a molecule containing N atoms, $3N-6$ linearly independent coordinates ($3N-5$ in linear molecules) describe the relative positions of the atoms with respect to one another.

The displacements of these coordinates from their equilibrium values form the basis for description of the molecular vibrations. When an atom is displaced from equilibrium, a force, approximately described by Hooke's law, resists the motion, and a vibration results. A typical molecular vibration frequency is 1000 cm^{-1} (3×10^{13} Hz), and with a typical atomic mass of 10^{-26} kg the Hooke's-law force constants are found to be of the order of 10^5 dyn/cm (100 newtons/m). This number is characteristic of the magnitude of force constants opposing the stretching of a chemical bond.

Polyatomic molecules require a large number of Hooke's-law force constants to represent their force fields; the theoretical treatment of the vibrational problem, called a "normal coordinate analysis" is straightforward though tedious. Seldom is there enough experimental data to determine all of the force constants in a general force field for a polyatomic molecule, and a number of approximate fields (the "valence force field," the "Urey–Bradley field", etc.) have been adopted by different investigators. Though a degree of arbitrariness is involved, the force fields and vibrational spectra of many polyatomic molecules have come to be quite well understood, and this understanding has contributed greatly to modern pictures of chemical bonding and molecular structure.

In a crystalline solid, the translational symmetry, based on the crystalline unit cell, greatly simplifies what would appear to be an intractable theoretical problem. Description of the crystalline vibrational motions is reduced to a separate description of the motions of atoms in one unit cell and a statement of the phase relationship between unit cells. A given possible phase relationship corresponds to a point in the "Brillouin zone." For a crystal containing N atoms in its unit cell, there are $3N$ vibrational coordinates describing each unit cell, and therefore $3N$ vibrational motions at any single point in the Brillouin zone.

The requirement of an oscillating polarization for interaction of a sample with infrared radiation takes on a special importance for crystals. If a vibrational motion of a single unit cell generates an oscillating polarization, then the motion can in principle be observed as a transition in the infrared spectrum. But the wavelength of the infrared radiation is very long (of the order of micrometers) compared to the dimensions of the unit cell (nanometers); thus for a particular vibrational motion to absorb radiation, a very large number of unit cells must oscillate in phase with one another to avoid cancellation of the interaction by interference. The phase relationship in which all unit cells oscillate exactly in phase with one another corresponds to the "center of the Brillouin zone." Thus the crystalline modes which can cause infrared transitions are limited to the $3N$ at the zone center (since the frequencies of the modes change slowly with change of location in the zone, the limitation to the zone center, though not rigorous, is in practice correct).

Of the zone-center modes, three correspond to motions where all atoms in the cell have the same displacement, i.e., to a shift of the whole cell; since all cells move in phase, this motion corresponds to a translation of the whole crystal, and has zero frequency. Thus the infrared spectrum can, in the simplest approximation, contain $3N-3$ frequencies. This number itself can be further curtailed if the crystal has symmetry beyond the translational.

Quantum mechanics tells us that the various normal vibrational modes can be treated independently to a good approximation, and that the energy of the molecule or crystal is a sum of terms, one for each mode. If the normal-mode frequencies are ν_i, then the energy of vibrational motion is $\Sigma_i(n_i+\frac{1}{2})hc\bar{\nu}_i$, where n_i is the quantum number for the ith mode. Light-induced transitions in the infrared (or Raman) spectrum are most commonly those where only one of the n_i changes by one unit. Thus the various normal frequencies appear directly in the infrared spectrum. Breakdown of the independence of the normal vibrations, and of the simple rule governing changes in n_i, can come as a result of anharmonic terms in the force law governing the displacements of atoms from equilibrium. The effects of anharmonicity in gaseous molecules are generally small, but in crystals quite strong effects occur. Of particular interest is the observation of "second-order effects" including spectroscopic transitions in which two or more n_i change simultaneously. In these transitions the two vibrational modes involved need not be zone-center modes, and as a result it is possible to observe indications of all the vibrations in the Brillouin zone in some crystals.

APPLICATIONS OF INFRARED SPECTROSCOPY

Infrared spectra of gases are due to simultaneous transitions involving both vibrational and rotational energy levels of the molecules. By study of these lines it is possible to locate the band origin, and thus determine the pure vibrational frequency, and also to determine the moments of inertia of the molecule, which control the rotational line spacings. From the moments of inertia, and with spectra of different isotopic species of the molecule, it is possible to determine the structure of the molecule, as in microwave spectroscopy.

In order to make a full analysis of the rotational spectrum high resolution is needed; for small, light molecules a great deal of detail may be obtained with good grating spectrometers, but for heavier molecules injection-diode laser spectrometers reveal a wealth of fine rotational structure which must be studied to define the rotational problem completely. At lower resolution this fine rotational structure is blurred, leaving only a "rotational envelope" which may have a number of maxima. Some information on moments of inertia may still be derived from the rotational contour, but the multiple maxima may cause confusion if several vibrational transitions are closely spaced.

An upsurge of interest in irreversible mechanisms of the condensed phase, almost entirely due to the theoretical work of R. G. Gordon, has created an entire field of experimental and theoretical research on the dynamics of molecular motion by applying the principles of linear response theory and of the fluctuation–dissipation theorem to vibrational–rotational spectra of condensed-phase molecules. The principle, which relates the frequency distribution of a spectral profile to the time-dependent correlation function of the observed transition moment tensor (one is the

Fourier transform of the other), requires precise data in the wings of a band. Hence, Fourier-transform spectroscopy with its (arbitrary) cutoff methods (''apodization'') is not useful here. The vast amount of effort spent in this field has generated a large base of quantitative knowledge on the nature of rotational and vibrational motion (''relaxation,'' ''fluctuations'') of molecules that are constantly perturbed by their neighbors. Rather precise notions on the complicated aspects of ''hindered'' and ''free'' rotation, the statistics and dynamics of environments local to the observed oscillator, the clustering of molecules, the persistence of local order within a (macroscopically) isotropic medium were among the fruits of these studies. In addition, new theories on vibrational resonant energy transfer, equilibrium renewal processes by Markov chaining, collision-induced effects, conformation changes, short-lived phonon modes, etc. were engendered by these principles. Finally, the difficulty of preparing a suitable and well-defined ''sample'' of a liquid of random-motion molecules spurred, in combination with experimental data from this kind of analysis, the use of computer simulations or ''experiments'' for the solutions of Newton's equations of motion (molecular dynamics calculations). These, in turn, have led to workable and transferable intermolecular potential functions, thereby allowing the separation of a particular dynamic process from the panoply of concurring mechanisms.

Infrared spectroscopy has contributed heavily to solid-state physics as well as to the chemical analysis of solid materials. In general the forces between atoms in crystals are somewhat weaker than chemical bonds, and as a result the vibrational transitions of solid materials occur at lower frequencies than those involving distortion of the chemical bonds within molecules. For this reason much solid-state information is obtained in the far-infrared spectrum, and many details of the spectroscopic study of solid-state phenomena are given in the article on far-infrared spectral in this encyclopedia. Molecular crystals are characterized by containing strongly bonded molecular groups of atoms which interact with weaker forces to form the crystal. In this case the internal or molecular vibrations cause transitions in the middle infrared while the external motions of the molecules, derived from rotation and translation, give rise to lattice vibrations in the far infrared.

To obtain a crystalline sample thin enough (from 1 to 10 μm in thickness) for infrared transmission studies is quite difficult. In some cases thin crystals can be grown, but a common and powerful method of preparing solid samples involves mixing the sample with a larger amount of powdered potassium bromide, KBr, which is a transparent crystal often used for windows in the mid-infrared. The mixture is then pressed into a transparent disk which is studied in the normal way. In addition to allowing a convenient control of the amount of the sample, this method reduces the anomalous reflection and scattering effects which occur in solid-state spectra by enclosing the sample particles in a medium (the KBr) which has about the same index of refraction. Generally there is little difference in vibrational frequency between gas, liquid, and solid (hydrogen-bonded systems are an important exception), and the KBr pellet technique is of great value in the qualitative analysis of synthetic mixtures or of unknown solids.

Methods of analysis based on vibrational spectra involve both the frequency specificity and the intensity of vibrational transitions. Their use for qualitative characterization and for quantitative measurement of the amounts of materials is widespread.

Qualitative study of a sample for chemical content, as in air pollution, characterization of mineral samples, detection of impurities in solid samples, or study of planetary atmospheres, often relies on the presence of characteristic group frequencies. Because of the similarities in strengths of bonds which are similar chemically (e.g., carbon–hydrogen bonds, carbonate groupings), there are found to exist quite definite ''group frequencies'' which are present in the spectrum whenever the chemical group is present in the sample. Extensive tables of such group frequencies are available.

Molecular vibration frequencies depend primarily on *intra*molecular force constants, and vibrational spectra are commonly rather independent of the state of the sample, with gases, liquids, solids, or solutions yielding quite similar frequencies. The analyst takes advantage of this fact to adapt his techniques to the other problems which face him, and he can often combine data taken in two or more phases to identify a molecule. However, the effects of *inter*molecular forces are appreciable, and these forces may be studied by infrared spectroscopy. Of particular chemical interest are studies of association and hydrogen bonding, while in solid materials the small shifts in frequency of an impurity or dopant may give a clue to understanding multiple sites available to the guest particle.

Quantitative analysis by infrared absorption spectroscopy involves the use of the absorption law common to all absorption spectroscopic techniques, $\log (I_0/I) = \alpha cl$, where the logarithm of the ratio of incident to the transmitted light is dependent on an absorption coefficient (α, a property of the sample and of the frequency of the light), the sample concentration (c), and cell length (l). The basic law is not always followed exactly, but corrections derived from measurements on standard samples make the analytical procedure an accurate one. Use of highly monochromatic light, for example of laser origin, to single out a particular absorption feature of a sample (e.g., a single rotational line of the molecular of interest in a gas mixture) can give a highly sensitive, specific analytical method for a wide variety of problems. Pollution control, for example, relies heavily on infrared measurements. The concentrations of reaction intermediates and products in such inhospitable environments as flames and plasmas can be followed by observations of the infrared spectrum, often in emission; the dependence of line intensities in the gas phase on temperature makes the careful quantitative measurement of line intensities a useful thermometer for high-temperature systems.

Many lasers have been constructed in which the energy of a chemical reaction, deposited in excited vibrational levels of the reaction products, emerges as infrared radiation. The best-known example is the powerful hydrogen fluoride laser, deriving its energy from the gas phase reaction $H_2 + F_2 = 2HF$. The reaction product, HF, is produced in excited vibrational levels, and laser action can be stimulated. Conversely, the study of infrared radiation, often in laser-amplifier configurations, gives detailed evidence of the exact states which result in chemical reactions. The information on microscopic chemical reaction kinetics which emerges is revolutionizing the study of kinetic processes.

See also FAR-INFRARED SPECTRA; INTERATOMIC AND INTERMOLECULAR FORCES; MICROWAVE SPECTROSCOPY; MOLECULAR SPECTROSCOPY; MOLECULES; RAMAN SPECTROSCOPY.

BIBLIOGRAPHY

G. Herzberg, *Molecular Spectra and Molecular Structure II. Infrared and Raman Spectra of Polyatomic Molecules*. D. Van Nostrand, New York, 1945.

E. B. Wilson, J. C. Decius, and P. C. Cross, *Molecular Vibrations*. McGraw-Hill, New York, 1955.

M. Born and K. Huang, *Dynamical Theory of Crystal Lattices*. Oxford, London, 1954.

R. W. Ramirez, *FFT, Fundamentals and Concepts*. Prentice-Hall, Englewood Cliffs, NJ, 1985. (I)

R. G. Gordon, *Adv. Magn. Resonance* **3,** 1 (1968). (A)

B. J. Berne and G. D. Harp, *Adv. Chem. Phys.* **17,** 63 (1970). (A)

W. G. Rothschild, *Dynamics of Molecular Liquids*. Wiley, New York, 1984. (A)

Insulators

R. A. Anderson

INTRODUCTION

An insulator, or dielectric, could be conveniently defined as a material that does not conduct electricity, if not for the fact that all known materials conduct a measurable amount of electricity. Nevertheless, insulators are readily distinguished from conductors. At room temperature, the conductivity of a good insulator typically is more than 20 orders of magnitude smaller than that of a good conductor, and the difference becomes larger at lower temperatures.

The physics that underlies the vast difference between the electrical conductivity of insulators and conductors is the first topic considered. Following that, discussions are devoted to properties of insulators, including the importance of the insulator–electrode contacts in determining conduction processes. This article concludes with a brief description of some of the ways an insulator loses its ability to insulate when electrically stressed to the point of breakdown.

BAND STRUCTURE AND CONDUCTIVITY

A medium consisting of isolated atoms or molecules cannot conduct electricity because no charged particles are free to flow in response to an electric field. However, if atoms are brought close together, as in a crystalline solid, electrons can move from one atom to another. It would appear, then, that all crystalline solids should conduct electricity.

The distinction between insulators and conductors originates in the quantum-mechanical behavior of electrons. As a consequence of the periodic structure of a crystalline solid, electron energies are restricted to well-defined energy bands, normally separated by bands of forbidden energies. Each band of allowed energies consists of a very large number of closely spaced allowed momentum values, each of which can be occupied by, at most, two electrons having opposite spins. For this reason, there is a finite number of electrons that can be included within an energy band, and the lower-lying energy bands are completely or partially filled with electrons.

Electrical conductivity is crucially dependent on the degree to which the energy bands are filled with electrons. A completely empty band obviously cannot contribute to electrical conductivity. On the other hand, no charge can be transported by a completely filled band: Because all available momentum values are occupied, the net momentum must remain zero in the presence of an electric field. In a partially filled band, however, the boundary between occupied and unoccupied allowed momenta can become displaced in response to an electric field. Charge is transported because the net momentum is no longer zero. The charge carrier in a nearly filled band is the absence of an electron, called a hole. A hole behaves very much as though it were an electron of reversed charge in a nearly empty band.

In most cases for a material to be an insulator, all energy bands must be either full or empty. If one or more energy bands are partially filled, the material usually is a metallic conductor. A further characteristic of insulators is that the gap between filled (valence) and empty (conduction) bands is large enough (usually >2 eV) that few electrons are thermally excited across this band gap. If, instead of a band gap, there is a small overlap between these bands, electrons at the top of the valence band spill over into the bottom of the conduction band. Such materials, of which antimony is an example, are known as semimetals. In the case of a pure semiconductor, such as germanium, the band gap is so small that a portion of the valence-band electrons are thermally excited into the conduction band, thereby creating electrons and holes which can conduct electricity. Such a material is an insulator at very low temperatures. Other semiconductors owe their conductivity to impurities that either remove electrons from the valence band or add them to the conduction band. While there is no natural demarcation between insulators and semiconductors, it is convenient to place the boundary at a conductivity of 10^{-6} S/m. At the other extreme, fluorinated ethylene–propylene copolymer, a very good insulator, has a room-temperature conductivity less than 10^{-17} S/m.

At this point the reader may be wondering how the unlikely situation of exactly enough electrons to fill the valence bands can occur at all. The reason it occurs frequently is as follows. Each repeating unit of the crystal structure contributes one allowed momentum value, which can accommodate two electrons, to each energy band. Therefore, bands can be filled without leftover electrons if each repeating unit contains an even number of valence electrons, a situation that is not at all unlikely.

In the preceding discussion, we have considered only crystalline solids. However, the short-range atomic order in amorphous materials is sufficient to establish energy bands similar to those found in solids having highly periodic structure. Examples of insulators, semiconductors, and conductors are found in both crystalline and amorphous solids, as well as in liquids.

INSULATOR–ELECTRODE CONTACTS

The conductivity of an insulator is strongly influenced by interactions between the insulator and the electrodes attached to it. When an insulator and a conductor are brought into contact, electrons diffuse either into or out of the insulator, depending on the particular materials involved, to establish thermodynamic equilibrium.

If electrons diffuse into the insulator, the insulator–electrode contact is called Ohmic, or injecting. Because an accumulation of electrons injected into the conduction band at the insulator surface is established, which can supply current to the bulk, conduction is limited by the bulk conductivity of the insulator. The bulk conductivity itself depends on whether or not the injected charge extends throughout the thickness of the insulator. If this charge is everywhere larger than the intrinsic charge in the conduction band, then the conductivity reflects the properties of the insulator–electrode contacts rather than the intrinsic conductivity, and the conduction process is referred to as space-charge limited.

A blocking contact occurs when electrons diffuse out of the insulator to establish thermodynamic equilibrium. In this case, conduction is limited by the rate at which electrons can cross the interface, and the conduction process is referred to as electrode limited. Current–voltage characteristics depend on whether electrons are thermally excited over, or tunnel through, the potential barrier at the insulator surface.

Contacts also may be either blocking or injecting for holes, which introduces the complexity of recombination of electrons and holes if both charge carriers are flowing simultaneously.

TRAPS

All known insulators contain localized energy levels in the band gap called traps, which tend to capture and immobilize the few charge carriers present in the valence and conduction bands. Traps originate from impurity atoms as well as naturally occurring defects. Because trapping affects recombination as well as the flow of carriers, essentially no conduction process escapes being influenced.

PHOTOCONDUCTIVITY

The absorption of light by an insulator may be accompanied by an increase in the electrical conductivity during illumination. Some insulators also display an increased conductivity in the dark as the result of previous exposure to light. These phenomena are called primary and secondary photoconductivity, respectively. Primary photoconductivity occurs when absorbed photons have sufficient energy to excite electrons into the conduction band from traps or from the valence band. Secondary photoconductivity is ascribed to light-induced changes in the occupation of traps near the electrodes, giving rise to local electric fields which assist the injection of electrons into the insulator. A well-known application of photoconductivity is the xerographic duplicating process.

Photoconductivity provides a means of investigating the mobility and lifetime of charge carriers in insulators. For example, a thin layer of carriers can be generated near an electrode by strongly absorbed light, so that the conductivity due to either electrons or holes can be measured.

ALTERNATING-CURRENT CONDUCTIVITY

The amount of alternating current flowing through a capacitor depends on the electrical polarizability of the insulator between the capacitor electrodes. Polarization, however, does not respond instantaneously to changes in applied voltage. Rather, several separate processes are involved, each of which proceeds at a different rate and contributes a frequency dependence to the polarizability of the insulator. Alternating-current measurements are a useful probe of slow processes such as chain motion in polymers and drift of mobile ions, as well as faster electronic processes, and can provide a wealth of information when performed at a series of different temperatures.

ELECTRICAL BREAKDOWN

The most important application of insulators is electrical insulation, without which electricity would be of little use. Failure of insulators under electrical stress, therefore, remains a topic of research interest.

Ideally, electrical breakdown can be classified as either thermal or electronic. Thermal breakdown occurs when the conductivity increases rapidly enough with increasing temperature to permit thermal runaway. Electronic breakdown involves an instability of the conduction electrons, such as a runaway drift velocity which leads to an ionization avalanche, or enhanced electron injection due to trapped charge near an electrode, as in secondary photoconductivity. The ultimate stage of a destructive electronic breakdown event is probably thermal.

Some insulators are able to withstand very high electric fields. Thin sheets of polyethylene terephthalate, a widely used capacitor dielectric, have a dielectric strength near 4×10^8 V/m. The highest known values exceed 10^9 V/m, such as reported for layers of silicon dioxide grown on silicon during microelectronic device fabrication.

In practice, breakdown often occurs at stresses well below the intrinsic dielectric strength of an insulator, and the mechanisms leading to breakdown are complex and poorly understood. For example, insulation on ac power transmission cables is susceptible to breakdown phenomena collectively called treeing, in which branching networks of tubular voids resembling trees slowly develop. Eventually the insulation is penetrated and a damaging discharge occurs. Trees tend to start growing wherever the electric field is enhanced: at cracks or voids in which gas discharges occur, or at metallic inclusions or electrode irregularities. Cable insulation in a moist environment often develops trees that are filled with water.

We consider, finally, another class of breakdown phenomena called surface flashover, in which discharges follow the insulator surface rather than penetrating the bulk. Insulator surfaces exposed to vacuum are particularly susceptible. Although vacuum surface flashover often appears to be the consequence of an avalanche of electrons which multiply on the insulator surface by secondary emission, mechanisms responsible for flashover in other cases are unresolved.

See also CONDUCTION; DIELECTRIC PROPERTIES; ELECTRON ENERGY STATES IN SOLIDS AND LIQUIDS; PHOTOCONDUCTIVITY.

BIBLIOGRAPHY

A. R. Blythe, *Electrical Properties of Polymers*. Cambridge University Press, Cambridge, England, 1979. (I)

H. Fröhlich, *Theory of Dielectrics*. Oxford, London and New York, 1958. (A)

T. J. Gallagher, *Simple Dielectric Liquids*. Oxford, London and New York, 1975. (A)

N. E. Hill, W. E. Vaughan, A. H. Price, and M. Davies, *Dielectric Properties and Molecular Behavior*. Van Nostrand, Princeton, NJ, 1969. (A)

C. Kittel, *Introduction to Solid State Physics*. 5th ed. Wiley, New York, 1976. (E)

M. A. Lampert and P. Mark, *Current Injection in Solids*. Academic, New York, 1970. (A)

N. F. Mott and R. W. Gurney, *Electronic Processes in Ionic Crystals*, 2nd ed. Oxford, London and New York, 1948. (I)

E. H. Nicollian and J. R. Brews, *MOS (Metal Oxide Semiconductor) Physics and Technology*, Chap. 11. Wiley, New York, 1982. (A)

J. J. O'Dwyer, *The Theory of Electrical Conduction and Breakdown in Solid Dielectrics*. Oxford, London and New York, 1973. (A)

J. G. Simmons, "Conduction in Thin Dielectric Films," *J. Phys. D: Appl. Phys.* **4**, 613 (1971). (A)

J. M. Ziman, *Principles of the Theory of Solids*, 2nd ed. Cambridge University Press, Cambridge, 1972. (I)

Integrated Circuits *see* Circuits, Integrated

Interatomic and Intermolecular Forces

J. A. Barker and John P. McTague

Molecules, and atoms which are not chemically bonded to one another, exert forces on one another which are attractive at large distances and repulsive at short distances. This is what is generally meant by the phrases "intermolecular forces" and "interatomic forces," which are therefore not normally used to describe the "chemical" or "valence" forces acting between atoms bonded together in the same molecule. It is the existence of these intermolecular forces which explains the existence of matter in solid, liquid, and gaseous states, and the stability of different states of matter in particular regimes of temperature and pressure (phase diagrams and equations of state). Modern statistical mechanics has made it possible to predict accurately the properties of solids, liquids, and gases starting from information on intermolecular forces.

The repulsive forces are dominant at distances comparable to and smaller than the size of the molecules, a few Å (1 Å $=10^{-10}$ m) for nonpolymeric molecules. In fact, it is the repulsive forces which determine the size and shape of the molecule. The attractive forces, which hold together the molecules in solids and liquids, are of longer range and fall off proportionally to R^{-7} for nonpolar molecules, where R is the distance between the molecules. The corresponding potential energy of interaction is proportional to $-R^{-6}$ (the force is the derivative of the potential energy with respect to distance). These forces are consequences of the electrostatic interactions between the electrons and nuclei and are described satisfactorily by quantum mechanics (before the advent of quantum mechanics, classical mechanics and electrodynamics were unable to describe these phenomena since they predicted that both static and dynamic charge distributions were unstable). The nonelectrostatic interactions between nuclei arising from the "strong" and "weak" nuclear forces are of much shorter range than the electrostatic and electrodynamic interactions, and completely negligible at distances comparable with molecular dimensions. The gravitational interaction is also negligible compared with the electrostatic interactions at such distances, though it falls off much more slowly with distance (potential energy proportional to $-\mathrm{R}^{-1}$), and provides the dominant interaction between electrically neutral and nonmagnetic bodies at macroscopic distances. Magnetic contributions to intermolecular forces are usually unimportant.

It is almost universal to consider the interaction between pairs of molecules as a sum of separate elementary interactions. This is strictly valid only when the molecules are sufficiently far apart that their electronic charge do not overlap. From a practical point of view, this requires intermolecular distances R about 1 Å greater than the nearest-neighbor spacings in solids. In this range the permanent electrostatic, induction, and dispersion interactions can be related to the properties of the individual molecules.

The permanent electrostatic interactions reflect the more or less asymmetrical average distribution of the electronic charge within the molecule, which can be described in terms of the charge (zero for neutral molecules) and the dipole, quadrupole, and higher permanent multipole moments of the molecule. The dipole moment describes the separation of positive and negative charge in the molecule and the quadrupole and higher moments describe in successively higher detail the average distribution of the charge. The dipole–dipole interaction, and all higher multipole interactions, give an average of zero when averaged over orientations of the molecules. Both dipole and quadrupole forces can play a significant role in determining the relative orientation of molecules in molecular crystals, but do not play a predominant role in the total binding energy of the solid.

Induction forces arise because a permanent molecular dipole or quadrupole moment produces electric fields which induce charge distortions, and therefore dipole and higher moments, in neighboring molecules. Those forces play only a small role in molecular orientation and binding energies.

None of the above-named forces can exist for spherical systems, such as rare gases, for which all multipole moments are zero. Nevertheless their binding energies can be as large as those for polar molecules. F. London (1930) showed that, owing to the constant motion of the electrons in an atom or molecule, there are instantaneous electrical moments, which can set up transient electrical fields at neighboring molecules. These electric fields cause the electrons in the neighbor to correlate their motion with those of the first molecule, leading to dynamic dipole–dipole, dipole–quadrupole, quadrupole–quadrupole, etc., attractive interactions. From the point of view of molecular quantum mechanics these are "electron correlation" effects, which are not included in, for example, a Hartree–Fock self-consistent field calculation. The leading (dipole–dipole) contribution to the energy varies as $-R^{-6}$, with higher-order terms in $-R^{-8}$, $-R^{-10}$, etc. (At very long range they are modified by relativistic effects, becoming proportional to $-R^{-7}$, etc.) These forces, variously called London, van der Waals, or dispersion forces, are universal and account for the major part of the binding energy of molecular solids, liquids, and even such weakly bound dimers as Ar_2, which cannot have ordinary chemical bonds.

The above-mentioned long-range forces, which are all directly related to properties of the isolated molecules, do not explain the repulsion at very short distances. This additional force comes into play when the electronic charge distributions of different molecules overlap. They have their origin in the quantum-mechanical exchange interaction, which is also involved in chemical bonding. In fact, the repulsive interaction may be looked upon as "antibonding." Since these energies are of the order of chemical interactions it is not surprising that only a very small amount of electron overlap is required to cancel out completely the weak long-range attractive forces, as illustrated in Fig. 1.

The short-range repulsive energy varies approximately as $\exp(-R/R_0)$, where R_0 depends on the pair of atoms involved, but is typically of the order of 0.1–0.2 Å. The extremely steep repulsion suggests that a useful approximation is to consider the atomic cores to be impenetrable spheres,

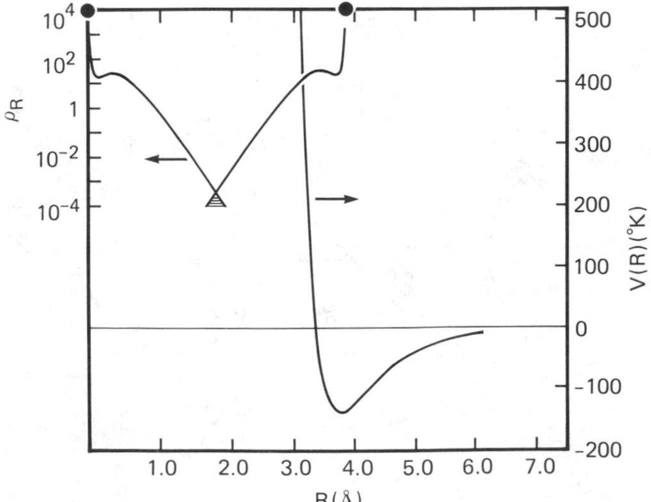

FIG. 1. Potential energy of a pair of argon atoms versus internuclear separation R. The electron density ρ (electrons/Å^3) is also shown for the equilibrium internuclear distance.

whereas the effect of the attractive forces can be approximated as a uniform background potential. These approximations are remarkably successful in correlating the thermodynamic properties and geometrical arrangement of molecules in both liquid and solid phases. The energies of collections of molecules are approximately, but not exactly, pairwise additive. For liquid and solid densities the additional term is of the order of 5–10% of the pair energy, where it is known.

There is detailed information on the nature of the intermolecular potential energy function for only a few relatively simple atomic and molecular pairs, in particular the rare gases. For the simplest case, that of two helium atoms, the interatomic potential has been derived with impressive accuracy from *ab initio* quantum-mechanical calculations, and the results agree very closely with empirical estimates. This provides a convincing validation both of our understanding of the origin of the intermolecular forces and of the empirical methods used to study them.

For systems involving more electrons accurate *ab initio* calculations are so far prohibitively time-consuming. However for heavier rare-gas pairs the interatomic potentials are known very accurately from empirical studies which use theoretically motivated mathematical forms together with a wide range of experimental data to determine the values of a number of adjustable parameters. The experimental data which are useful include the deviation of real gases from ideal gas behavior ("virial coefficients"); the viscosity, thermal conductivity, and diffusion coefficients of dilute gases; spectroscopic data on energy levels of bound pairs; and measurements of scattering cross sections for pairs of atoms and molecules. Earlier potentials which were derived from measurements of only one property were much less satisfactory.

For more complicated cases, such as small organic molecules, mathematically simple models involving a minimum of adjustable parameters, often two, are commonly used. The model most commonly used in calculations of bulk properties is the Lennard–Jones 6-12 potential:

$$V(R) = 4\epsilon[(\sigma/R)^{12} - (\sigma/R)^6],$$

where ϵ is the attractive energy at the potential minimum and σ the intermolecular distance where $V=0$. The parameters are fixed to reproduce selected bulk properties. These phenomenological potentials have often been used as effective potentials for approximate calculations even for systems where the potential is more accurately known. This is because of their simplicity and the scaling properties of V/ϵ and R/σ which enable the properties of one substance to be related to those of another at a different temperature and density (principle of corresponding states). In quantitative studies where more accurate potentials are available this is no longer necessary.

The interactions of large molecules are often modeled as a sum of atom–atom interactions between the atoms of one molecule and those of the other, with the atom–atom interactions described by the Lennard–Jones potential. Several sets of "transferable" Lennard–Jones paremeters for atom pairs relevant to organic materials are available in the literature. These atom–atom potentials are useful in predicting both thermodynamic and structural properties of organic solids and liquids and of polymeric materials including those of biological importance such as proteins. For flexible polymeric molecules it is necessary to include atom–atom potentials for nonbonded atoms in the same molecule, since these intramolecular interactions are important in determining the structure adopted by the molecule.

See also Electric Moments; Equations of State; Magnetic Moments; Molecular Structure Calculations; Molecules; Polarizability; Statistical Mechanics.

BIBLIOGRAPHY

G. C. Maitland, M. Rigby, E. B. Smith, and W. A. Wakeham, *Intermolecular Forces, Their Origin and Determination*. Clarendon Press, Oxford, 1981. (I)

M. L. Klein and J. A. Venables (eds.), *Rare Gas Solids*, Vol. 1. Academic Press, New York, 1976. (I)

J. O. Hirschfelder, C. F. Curtiss, and R. B. Bird, *The Molecular Theory of Gases and Liquids*. Wiley, New York (1964 corrected printing). (I)

J. A. Barker and D. Henderson, "What Is 'Liquid'? Understanding the States of Matter," *Rev. Mod. Phys.* **48,** 587 (1976). (I)

H. Margenau and N. R. Kestner, *Theory of Intermolecular Forces*, Vol. 18. Pergamon Press, New York, 1969. (A)

J. N. Israelachvili, *Intermolecular and Surface Forces with Application to Colloidal and Biological Systems*, Academic Press, New York, 1985. (I)

Califano (ed.), *International School of Physics "Enrico Fermi," Course LV*. Academic Press, New York, 1975. (A)

Interferometers and Interferometry

K. M. Baird

Light is propagated according to laws of wave motion, and when two or more beams of light are combined under appropriate conditions, they will sometimes reinforce, sometimes tend to nullify, each other, like waves in and out of phase. This can cause cyclical intensity changes, corresponding to changes in the difference in the phases of the light waves in the respective beams, giving rise to an "interference" pattern of light and dark contours, usually called fringes, whose interval corresponds to a change of one cycle in the phase difference—i.e., a change of one wavelength (λ) in the relative distance traveled by the beams. This effect is the basis of interferometry, widely used to obtain information about the path traversed by a light beam and to study the characteristics of the light itself.

With natural light the phase due to different atoms varies independently and very rapidly, so that interference cannot be observed on the combination of beams that have come from different parts of a light source. Observable interference fringes are produced by the superposition of a very large number of nearly identical patterns, each due to a pair or set of beams that have come from the same atom by different paths; an interferometer provides the means for doing this by division of light into separate beams, which are then recombined under suitable conditions regarding polarization imaging, and relative path lengths. For laser-produced light, atoms are stimulated to emit in phase with one another and the restrictions are relaxed by many orders of magnitude; for example, the paths traveled by the interfering beams can differ in length by hundreds of meters, as compared with the few decimeters possible with natural sources.

Since the wavelength of visible light is about 0.5 μm and it is quite possible to observe changes as small as 10^{-3} fringe, interferometry provides an extremely sensitive means for observing small displacements, microtopography of surfaces, etc. Furthermore, even natural light will produce fringes over a length of up to 10^6 wavelengths, so that physical lengths can be determined in terms of wavelengths with a precision approaching $1:10^9$. This fact and the very precise reproducibility of some wavelengths accounts for the great importance of interferometry in metrology, as exemplified by the definition of the international meter in terms of a wavelength of light. Similarly, interferometry makes possible great precision and resolution in spectroscopic measurements.

There exist a very large number of types of interferometer, differing in the arrangement of reflectors, lenses, etc. necessary for the separation and recombination of the light beams in a manner convenient for specific applications. They can be classified in a number of ways—according to whether two or a multiplicity of beams interfere, whether the beams are separated by division of the area of a wave front or by division of its amplitude at a partially transmitting reflector, etc. These will be illustrated in the following description of the most widely used types of interferometer that can be considered as basic examples.

TWO-BEAM INTERFEROMETERS

One of the most important interferometers is the Michelson, which is described in some detail with reference to

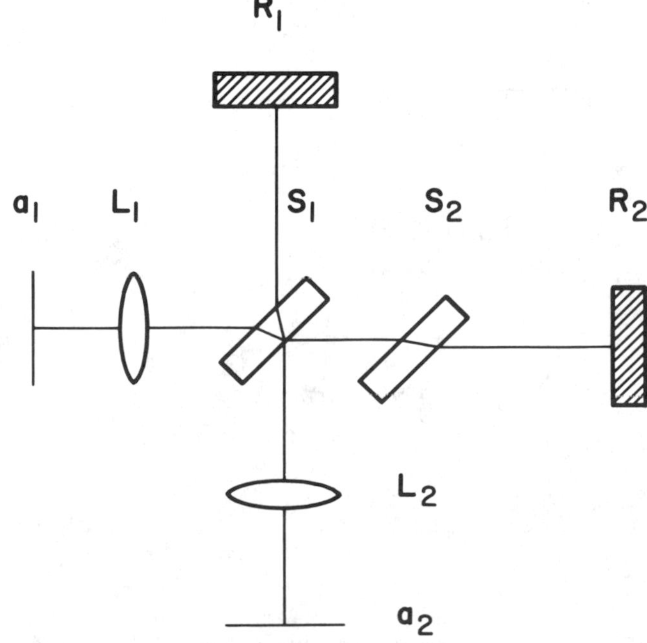

Fig. 1. Twyman–Green form of the Michelson interferometer.

later modifications that included lenses and apertures; this type employs two beams and division of amplitude. Referring to Fig. 1, we see that light from given points at a_1 is rendered parallel by the lens, L_1, and split into two beams at the partially transmitting reflector, S_1. The beams travel to mirrors R_1 and R_2, respectively, and thence back to S_1 where they recombine to form a beam that is focused at a_2 where an aperture can be used to block unwanted light such as that reflected by the second surface of S_1. The optional plate S_2 is to provide symmetry in the light paths.

The recombining beams produce interference fringes that can be observed in three ways:

1. If the reflectors are plane and adjusted so that the recombined beams are collinear, the phase difference between wave fronts from a given point in the plane of a_1 will be constant across the whole wave front and the intensity of the recombined beam is given by

$$I = I_0 \cos^2 \left[\frac{2\pi}{\lambda} (S_1R_1 - S_1R_2) \cos\theta \right]$$

where θ is the angle a beam makes with the mirror normal. Beams perpendicular to the mirrors, which become focused at the center of a_2, have the greatest phase difference; obliquely traveling beams, which are focused at points removed from the center of a_2, have lesser phase differences and there results a pattern of circular fringes having a sinusoidal intensity distribution.

2. If under the foregoing conditions apertures are placed at the centers of a_1 and a_2 and are made small compared with the central circular fringe, the total light passed will vary sinusoidally with displacement of one of the reflectors, provided the light is of a sufficiently narrow spectral band. This type of observation is particularly suited to photoelectric detection and is useful for length metrology and for Fourier analysis of spectra.

3. Viewed through a small aperture at a_2, the reflectors will appear uniformly illuminated because of the constant phase difference. If now a variation of phase difference across the wave front is introduced, say, by an inclination of one of the reflectors, or by a distortion resulting from reflection or transmission by an object placed in one of the beams, a pattern of interference fringes showing contours of equal optical path length will be seen "localized" near the plane of R_1. This mode of observation, devised by Twyman and Green, is applied to the testing of optical components, microtopography of surfaces, refractive index studies, length metrology, etc.

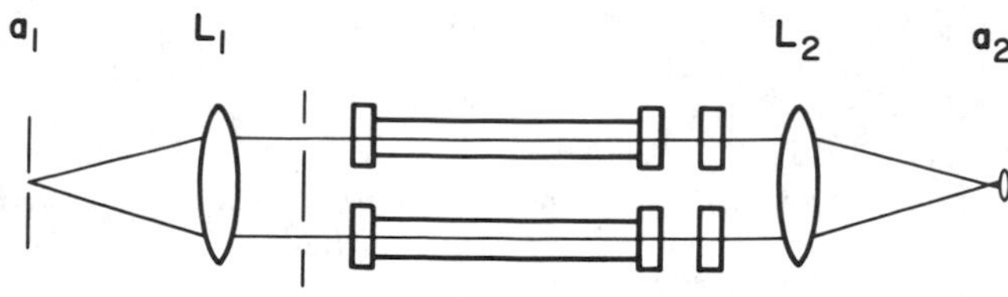

FIG. 2. Rayleigh interferometer.

The basic features described above are common to a large number of similar interferometers modified for special applications by the use of arrangements such as a prism beam splitter, separated beam splitter and combiner, retro reflectors (R_1 and R_2), curved wave fronts, unsymmetrical beam sizes, and a shear displacement of the wave fronts.

Another class of two-beam interferometer separates the beams by division of the wave-front area. Important examples are Michelson's stellar interferometer and the Rayleigh, shown in Fig. 2, which is used for refractive index measurement. Light scattering, diffraction, and polarization are other methods used for separation of the beams.

MULTIPLE-BEAM INTERFEROMETERS

This important class is typified by the Fabry–Perot interferometer or "etalon," of great importance in spectroscopy and, in modified forms, in lasers. In Fig. 3, light from given points in the plane a_1, rendered parallel by passage through L_1, is partially transmitted by the plane-parallel reflecting surfaces, R_1 and R_2, which may have reflectances as high as 99%. Light in the space between R_1 and R_2 is multiply reflected back and forth, resulting in a series of wave fronts leaving R_2 with successive retardations of twice the separation of the mirrors. The light is brought to a focus at the plane a_2, where interference can be observed in much the same manner as described in connection with Fig. 1. An essential difference in the form of the fringes arises, however, because of the multiple reflections. These create a standing-wave system between R_1 and R_2 that is relatively powerful when the retardation of successive wave fronts is an exact multiple of λ, under which conditions the combination of reflectors has a transmittance of nearly 100%. The standing-wave power, and consequently the transmittance, falls off rapidly with departure from this condition as given by the Airy function $A(n) = T^2/[(1-R)^2 + 4R \sin^2(\pi n)]$ where R is the reflectance, T the mean transmittance of the reflectors, and $n = 2(\mu t/\lambda) \cos\theta$, the order of interference, μt being the optical separation of the reflectors. The very sharp "resonance" effect in this class of interferometer produces fringes that are very narrow compared to the spacing between them.

In the Fabry–Perot, the fringes are observed either as circles formed at a_2 or as a change in flux passing through a small aperture at a_2 while the optical separation μt or the wavelength λ is varied. This form has had its greatest use in spectroscopy, where it has been used for wavelength measurement and very high-resolution spectrometry; it is also used in a form, originated by Connes, having curved reflectors.

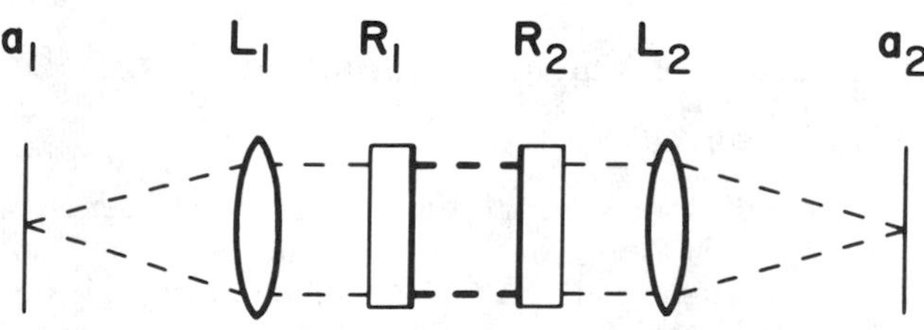

FIG. 3. Fabry–Perot interferometer.

The Fizeau form of interferometer makes use of the fringes localized at the reflectors observed through the small aperture at a_2. It has had its greatest application in microtopography, sometimes in conjunction with a microscope to provide great resolution in three dimensions.

Multiple-beam interferometry using division of amplitude is also the basis of the Lummer-Gehrcke plate. The echelon of Michelson formerly used in spectroscopy and the diffraction grating can be considered multiple-beam interferometers displaying division of aperture.

MISCELLANEOUS

The fringes observed in the interferometers described are formed by the use of single very narrow bands of the spectrum and represent contours of either equal path length or equal angle. Another type of fringe can be observed when light in a broad band of the spectrum passes successively through an interferometer and a spectroscope, in which case they are said to represent contours of equal chromatic order.

Interferometry has a history of well over a hundred years but continues to be actively developed by the application of newly available sources and detection and recording techniques.

It is applied throughout the spectrum from radio waves (for astronomy) to x rays (for measuring fundamental constants) and it has been enormously influenced by the development of the laser, which has made possible a wide variety of new techniques, such as optical heterodyning, holography, and laser speckle, that themselves justify book-length treatment.

See also GRATINGS, DIFFRACTION; HOLOGRAPHY; WAVES.

BIBLIOGRAPHY

K. M. Baird and G. R. Hanes, *Applied Optics and Optical Engineering,* Vol. 4, pp. 309–361. Academic Press, New York, 1967. (I)

J. C. Dainty, *Laser Speckle and Related Phenomena*. Springer-Verlag, Berlin and New York, 1975. (A)

Encyclopaedic Dictionary of Physics, Vol. 3, pp. 873–893. Pergamon Press, New York, 1961. (E)

P. Hariharan, *Optical Interferometry*. Academic Press, Sydney and Orlando, 1985. (A)

G. Hernandez, *Fabry–Perot Interferometers*. Cambridge University Press, London and New York, 1986. (A)

Elliott R. Robertson and James M. Harvey, *The Engineering Uses of Holography*. Cambridge University Press, London and New York, 1970. (I)

Current issues of journals of the Optical Society of America. (A)

Intermediate Valence Compounds

Jon M. Lawrence

Intermediate valence, or homogeneous mixed valence, occurs in certain rare-earth compounds such as $CePd_3$, SmS, $EuPd_2Si_2$, TmSe, and $YbAl_2$ where the rare-earth valence is intermediate between two integral values, i.e., fractional. There are three key characteristics of these solids: the crystal density is unusually sensitive to changes in temperature and/or pressure; the low-temperature susceptibility is finite; and the coefficient of the linear term in the low-temperature specific heat is enhanced (the heavy-fermion compounds are a related class of materials).

Rare-earth elements have an electronic configuration consisting of a filled xenon core, 2–4 bonding (valence) electrons in $5d$ and $6s$ states, and 0–14 $4f$ electrons. The configuration can be expressed as $(5d6s)^z4f^n$, where z is the valence and n the $4f$ count. In the solid state the bonding electrons form energy bands. The $4f$ orbital has a small radius and does not contribute directly to the bonding. In ordinary rare-earth compounds the $4f$ behaves as a localized orbital, with a well-defined magnetic moment. Ordinarily the energy separation $E=E_n-E_{n-1}$ between configurations $(5d6s)^z4f^n$ and $(5d6s)^{z+1}4f^{n-1}$ is large (5–10 eV); for compounds of the elements Ce, Sm, Eu, Tm, and Yb the separation can be much smaller (0–2 eV). Under these circumstances an interaction which hybridizes the two configurations via a process where the conduction electrons hop on and off the rare-earth sites can cause them to fluctuate between the $4f^n$ and $4f^{n-1}$ configurations. This leads to a quantum-mechanical ground state which contains an admixture of both configurations. The strength of this mixing interaction is given by $\Gamma = V^2N(E_n)$, where V is the matrix element of the interaction and $N(E_n)$ is the number of conduction electron states per unit energy interval. The degree of admixture depends on the ratio E/Γ; when this is small, both states contribute equally and the valence is nonintegral (near $z + \frac{1}{2}$). In these compounds the valence has the same fractional value at each rare-earth site. This is referred to as homogeneous mixed valence to distinguish it from inhomogeneous mixed valence where the rare-earth element has different values of integral valence at different lattice sites. This latter case occurs in compounds such as Sm_3S_4 and Fe_3O_4.

The radius of the $4f$ ion in the $4f^n$ state is larger than that of the $4f^{n+1}$ state. The lattice constant of a given compound will thus be larger for the $4f^n$ case than for the $4f^{n+1}$ case. At ambient pressure, SmS contains divalent Sm; but at elevated pressure, SmS converts to an intermediate valence phase. The lattice constant is intermediate between that expected for divalent and trivalent SmS (Fig. 1). A lattice constant deviation is a signature of mixed valence. The fractional valence can be estimated by assuming the lattice constant varies linearly with valence.

This ability of the $4f$ ion to change size lies behind the above-mentioned extreme sensitivity of the crystal density (or lattice constant) of many mixed valence compounds to changes in temperature and pressure. In two cases (SmS and elemental Ce metal) this leads to an unusual kind of phase transition, namely, an *isomorphic* transition where at the transition pressure the lattice constant undergoes a large (5%) discontinuous decrease without any change in the crystal symmetry; further at a critical temperature and pressure the compressibility and thermal expansion coefficients diverge. These are valence transitions where the $4f$ count decreases causing the ions to shrink. In other materials no such phase transition is observed, but the thermal expansion is large and highly temperature dependent.

The isomer shift measured in a Mössbauer experiment

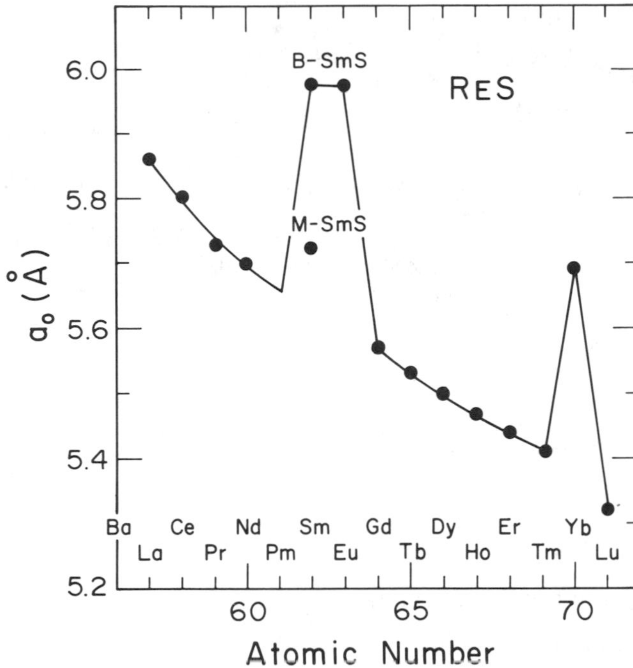

FIG. 1. The lattice constants versus atomic number for the rare-earth sulfides. In B-SmS, EuS, and YbS the rare earth is divalent; in M-SmS the Sm is in an intermediate valence state; for all other cases the rare earth is trivalent.

measures the electron density at the nucleus. This differs for the two configurations $4f^n$ and $4f^{n-1}$ because the extra $4f$ electron increases the screening of the nucleus seen by the outer valence electrons and hence decreases the probability for finding them at the nucleus. For homogeneous mixed valence Sm and Eu compounds, a single isomer shift is observed intermediate between that observed for divalent and trivalent compounds. For inhomogeneous mixed valence compounds two isomer shifts are observed, at the values expected for the integral valence states.

The lattice constant and isomer shift measure properties of the ground state and therefore yield one (intermediate) value for homogeneous mixed valence compounds. Other experiments measure properties of highly excited states, and the signatures of both integral valence states are observed. An important example is x-ray absorption. The L_{III} absorption edge occurs when the photon energy equals the value necessary to excite a $2p$ deep core electron to the valence band. For a rare-earth ion in the $4f^{n-1}$ configuration about 8–10 eV more energy is required to unbind the $2p$ electron than for the $4f^n$ case. Again, this is due to screening effects by the extra $4f$ electron. In mixed valence compounds *two* absorption edges are observed, at the values expected for the two integral valence cases. This is because each component of the hybridized wave function couples to a different integral-valent excited state. The intensity of the two absorption peaks can be used to estimate the valence; indeed, this is presently considered to be the best way to determine valence.

A very important property of these compounds is that the thermodynamic behavior is a universal function of a scaled temperature. For several compounds the magnetic susceptibility $\chi(T)$ (Fig. 2) has the same shape: a maximum at a temperature T_*; a low-temperature value proportional to C/T_*, where C is the Curie constant (the upturn at low temperature is not an intrinsic property but is an impurity effect); and a high temperature law $C/(T + 2T_*)$. For all cases the product $T\chi(T)/C$ has exactly the same dependence on T/T_*; this property is called *scaling*. Simultaneously the specific heat at low temperatures has a linear temperature dependence with coefficient γ proportional to $1/T_*$. The thermal expansion coefficient typically shows a maximum near the characteristic temperature. The inelastic neutron-scattering spectrum (which measures the energy required to flip a $4f$ spin) has a maximum at an energy $k_B T_*$. The characteristic temperature T_* has values between 10 and 2000 K for different compounds and can be varied in a given compound by pressure or alloying. There is a correlation between T_* and valence; e.g., in cerium compounds, large T_* corresponds to smaller f count.

For a conventional rare earth with a well-defined local moment the susceptibility should diverge as $T \rightarrow 0$. For intermediate valence materials the susceptibility is finite as $T \rightarrow 0$; for Ce, Sm, and Yb, $\chi(T)$ is intermediate between the values expected for the integral valence cases (Fig. 2b). The local moment is *quenched*; finite susceptibility (Pauli paramagnetism) is characteristic of nonmagnetic metals. This and the fact that the low-temperature specific heat has a linear temperature dependence implies that the low-temperature behavior is that of a *Fermi liquid*. That is, despite the extremely complicated interactions between the conduction electrons and the $4f$ ions, in the ground state the conduction electrons behave as though they are noninteracting.

The theory which describes mixed valence compounds is based on the Anderson model. This describes a single $4f$ impurity embedded in a nonmagnetic host. The energy separation $E = E_n - E_{n-1}$, the hybridization V, and the density

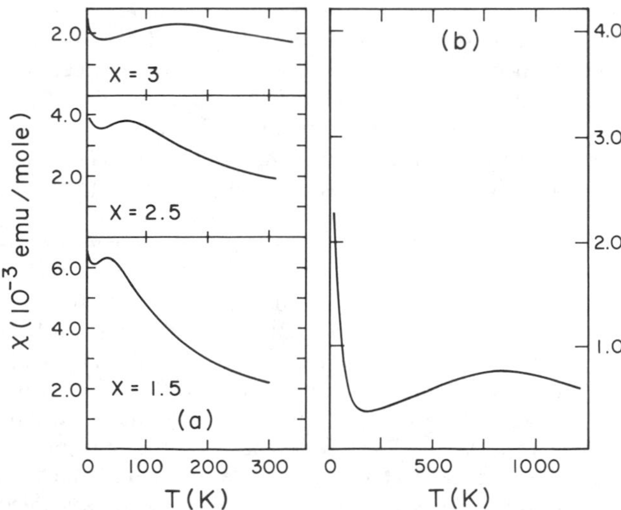

FIG. 2. (a) Susceptibility for three concentrations x of $CeIn_{3-x}Sn_x$. (b) Susceptibility of CeN. The line indicates the susceptibility for a trivalent Ce ion; the susceptibility of a tetravalent cerium ion is essentially zero.

of states $N(E_n)$, as well as the energy separation $U = E_{n+1} - E_n$, are taken as model parameters. For the rare earths U is typically 6 eV; V is of order 0.1 eV and E varies from close to zero for Sm, Eu, Tm, and Yb mixed valence compounds to 2 eV for cerium. The model predicts the above-mentioned universality; the characteristic temperature T_* is proportional to the so-called Kondo temperature T_K which is computed in the model. This is the energy scale for fluctuations of the 4f spin. The local moment is quenched by these spin fluctuations whose ultimate origin lies in the fluctuations between the two integral-valent configurations. The ground state is thus a singlet and the enhanced specific heat coefficient arises from thermal excitation of spin fluctuations. The predictions of the model depend on the orbital degeneracy of the rare-earth integral valence states (e.g., 6 for the Ce $J = 5/2$ state and 8 for the Yb $J = 7/2$ state). The susceptibilities shown in Fig. 2 are computed precisely in the model, as are the inelastic neutron-scattering spectra, the linear coefficients of specific heat, and the dependence of the valence (and hence lattice constant, or thermal expansion) on temperature. Confidence in the application of the Anderson model to these materials has been significantly increased by recent photoemission and inverse photoemission experiments. These essentially measure the model parameters E, U, and V, eliminating unknown parameters and allowing direct calculation of the thermodynamic quantities.

The Anderson model describes an impurity, whereas mixed valence compounds consist of a periodic lattice of 4f sites. For temperatures of order T_K or larger the model adequately describes the data, implying that the 4f ions behave as a set of noninteracting impurities. At lower temperatures this is not the case, and interactions among the 4f ions affect the behavior. Properties such as susceptibility, specific heat, and neutron spectra show deviations from the single-impurity predictions at low temperatures. For Ce, Yb, and Eu compounds the resistivity (which is large and finite for an Anderson impurity) vanishes as $T \rightarrow 0$. This buildup of interactions, the vanishing of the resistivity, and the onset of Fermi liquid behavior is referred to as *coherence*. Such coherence is observed most convincingly in de Haas–van Alphen experiments, which show that the wave functions in the ground state are not random (as they would be for impurity scattering) but fully periodic. The description of coherence on the theoretical level requires an Anderson *lattice* model. This is an extremely difficult problem at the present time. Future work on the mixed valence problem will stress the development of coherence, which is as yet poorly understood.

The ground states of Sm and Tm compounds are not always metallic. In the intermediate valence state, SmB_6 appears to be an insulator with an extremely small energy gap, of order 0.01 eV as compared to 1–10 eV in ordinary semiconductors. One expectation is that the Anderson lattice model can lead to the development of a hybridization-induced energy gap at low temperatures; hence the model should prove capable of describing these unusual semiconductors as well as the Fermi liquid compounds.

See also FERRIMAGNETISM; HEAVY-FERMION MATERIALS; PHASE TRANSITIONS; RARE EARTHS.

BIBLIOGRAPHY

C. M. Varma, *Rev. Mod. Phys.* **48,** 219 (1976).

J. M. Lawrence, P. S. Riseborough, and R. D. Parks, *Rep. Prog. Phys.* **44** 1, (1981).

Valence Fluctuations in Solids (L. M. Falicov, W. Hanke, and M. B. Maple, eds.). North-Holland, Amsterdam, 1981.

Valence Instabilities (P. Wachter and H. Boppart, eds.). North-Holland, Amsterdam, 1982.

Proceedings of the 4th International Conference on Valence Fluctuations, Aug. 1984, Cologne, FRG, edited by E. Muller-Hartmann, B. Roden, and D. Wohlleben, *J. Magn. Magn. Mater.* **47&48** (1985).

Theory of Heavy Fermions and Valence Fluctuations (T. Kasuya and T. Saso, eds.). Springer-Verlag, Berlin, 1985.

Theoretical and Experimental Aspects of Valence Fluctuations and Heavy Fermions (L. C. Gupta and S. K. Malik, eds.). Plenum Press, New York, 1987.

Internal Friction in Crystals

Andrew V. Granato

A description of the mechanical properties of crystals is formally similar to that for electrical or magnetic properties, with the elastic constant playing a role analogous to that of the dielectric constant or the permeability.

The mechanical behavior (response to applied forces, or strain produced by a given stress) of a crystal is determined by the interatomic forces between the atoms as well as by the motion of defects in the crystal. The interatomic forces act quickly and may be taken to be instantaneous for mechanical tests, whereas the displacements or strain from defect motion requires a finite time. A perfect or ideal crystal can be defined as one for which there are no defects providing energy-loss mechanisms, or internal friction, for mechanical excitations. The elastic strain response ϵ_{el} of an ideal elastic crystal to an applied stress σ is given by elasticity theory as $\epsilon_{el} = \sigma/G$, where G is the elastic modulus of the crystal. When this stress–strain relation is combined with Newton's laws for a volume element of the crystal, plane-wave solutions without attenuation, or energy loss, of the equations of motion are found.

Real crystals contain defects. When the defect contribution to the strain is linear with a unique equilibrium value for a given stress, but is achieved only after the passage of a characteristic time, the crystal is called anelastic. For a sinusoidal applied external stress, the in-phase (real) part of the delayed response (ϵ_R) leads to an effective modulus change ΔG, while the out-of-phase (imaginary) part (ϵ_I) leads to an energy loss, or internal friction Q^{-1}. These are given by

$$\frac{\Delta G}{G} = \frac{G\epsilon_R}{\sigma_0} \quad \text{and} \quad Q^{-1} = \frac{G\epsilon_I}{\sigma_0}, \tag{1}$$

respectively. The response is often of the Debye type for which

$$\frac{\Delta G}{G} = \frac{\Delta_R}{1+(\omega\tau)^2} \quad \text{and} \quad Q^{-1} = \frac{\Delta_R\omega\tau}{1+(\omega\tau)^2} \tag{2}$$

where Δ_R is the relaxation strength, ω is the angular frequency of the applied stress, and τ is the relaxation time for the delayed anelastic component of the strain.

The classic example of an internal friction peak, which occurs when $\omega\tau = 1$, is the Snoek relaxation of carbon in bcc iron. In bcc crystals interstitial impurities normally occupy octahedral sites and have tetragonal symmetry. Depending on the particular site chosen, the tetragonal axis may lie along any one of the cube axes x, y, or z. When a stress is applied along one of the Cartesian x, y, or z directions, the equivalence of the sites is lost, and the carbon atoms will jump by diffusion, preferentially occupying the lowest energy sites. This changes the length of the crystal in the direction of the stress, leading to an anelastic strain with a characteristic delay determined by the time of diffusion of the carbon atom between the sites. When a stress is applied along the ⟨111⟩ direction, no effect is observed, since this stress deforms the x, y, and z directions by the same amount. The relaxation strength is determined by the number of interstitials and the direction of the applied stress.

The overall behavior can be classified conveniently with the help of simple models, as illustrated in Fig. 1. For a crystal containing point defects and dislocations, there is a point-defect strain ϵ_p and a dislocation strain ϵ_d in addition to the elastic strain ϵ_{el}. The point-defect strain may be paraelastic (stress-induced rotation between equivalent positions of permanent dipoles with the help of thermal fluctuations, as in the Snoek effect), or diaelastic (induced dipole, which produces an internal strain, as with interstitials in metals). Also, the effects of phonons and electrons can be included within the formalism as point defects.

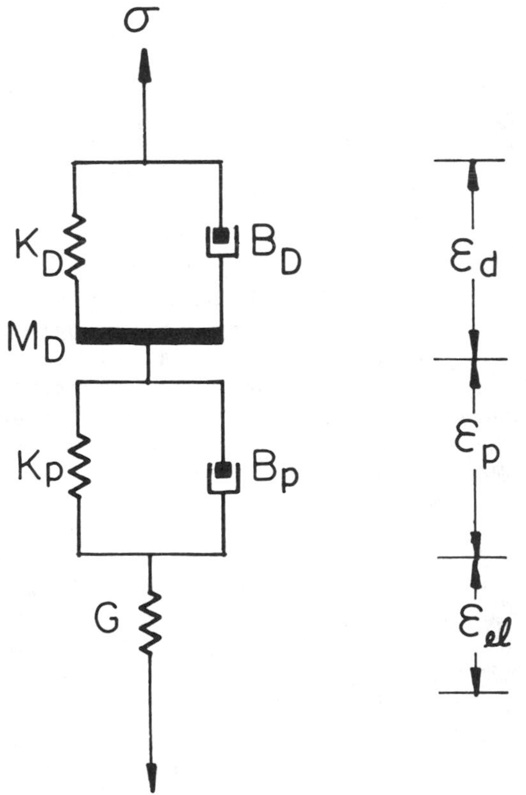

FIG. 1. Schematic representation of strain in a specimen containing point defects and dislocations.

The dislocation strain is a diaelastic effect requiring a mass term in the general case and is given by

$$\epsilon_d = \Lambda b y \tag{3}$$

where Λ is the dislocation density, b the Burgers vector, and y the average dislocation displacement. The displacement y is given by the dislocation equation of motion

$$M\ddot{y} + B\dot{y} + Ky = b\sigma \tag{4}$$

where $M \simeq \rho b^2$ is the dislocation mass per unit length, B is the dislocation viscosity, and K is a restoring-force constant for the dislocation. Comparing Eqs. (3) and (4), we obtain the model values M_d, B_d, and K_d from M, B, and K by dividing the latter values by Λb^2. For small enough B, the response ϵ_d has a resonance character, but for large B, inertial effects can be ignored and we have a relaxation with

$$\Delta_R = \frac{\Lambda G b^2}{K_d} \quad \text{and} \quad \tau = \frac{B_d}{K_d}. \tag{5}$$

The coefficients B and K are phenomenological, can arise from many sources representing the dislocation interactions with other defects, and are related to the corresponding quantities appearing for the point-defect strain ϵ_p. In general, any mechanism that damps ultrasonic shear waves in a crystal containing no dislocations will also provide a viscous drag on dislocations, since a moving dislocation generates a shear strain rate in the crystal. Sources of drag B are phonons, electrons, and point defects, while sources of K are the dislocation tension and immobile aligned dipolar point defects. Dislocation internal-friction effects have been detected over a frequency range from 10^{-2} to 10^{12} Hz, and over a point-defect concentration range from 10^{-12} to 10^{-2}.

See also ANELASTICITY; CRYSTAL DEFECTS; ELASTICITY.

BIBLIOGRAPHY

R. DeBatiste, *Defects in Crystalline Solids*. North-Holland, Amsterdam, 1972.

A. S. Nowick and B. S. Berry, *Anelastic Relaxation in Crystalline Solids*. Academic, New York, 1972.

R. Truell, C. Elbaum, and B. B. Chick, *Ultrasonic Methods in Solid State Physics*. Academic, New York, 1969.

Internal Rotation *see* Inversion and Internal Rotation

Interstellar Medium

Dennis J. Hegyi

The interstellar medium refers to the matter and radiation lying between the stars within the region of space associated with a galaxy. Most interstellar matter is found in the planes of spiral galaxies; observations of elliptical galaxies have revealed relatively little interstellar material. Until quite recently, almost all data on the interstellar medium have been obtained from optical and radio studies of our own galaxy, a spiral, and, in fact, even more locally, from observations in the solar neighborhood. However, new observations are just beginning to reveal information about the interstellar medium in nearby galaxies. Using mostly data from our galaxy, an overview of the interstellar medium will be presented by discussing the following: the hierarchy of structure, chemical composition, cosmic rays, the galactic magnetic field, and the interaction between the components.

Hydrogen is the main constituent of both the interstellar medium and the stars. Hydrogen's widespread abundance, coupled with the ability of radio telescopes to detect interstellar hydrogen in absorption and emission using the 21-cm hyperfine transition [1], has made it possible to trace out the distribution of hydrogen over most of the galaxy. Beginning with the grossest features and proceeding to those on smaller scales, the hydrogen is confined to a thin flat disk, the galactic plane, whose thickness [2] inward of the sun (solar galactic radius $\simeq 10$ kpc; 1 pc $=3.26$ light-years $=3.09\times10^{18}$ cm) is about 230 pc and decreases in thickness to less than 100 pc at a galactic radius of less than 300 pc. Beyond the galactic radius of the sun, the galactic disk is known to be warped, possibly because of tidal distortion by a nearby galaxy, the Large Magellanic Cloud [3].

The observations show that neutral hydrogen (H I) in the disk has spiral-like structure. This pattern is superimposed on the overall differential rotation of the galaxy. To account for the persistence of this structure, which is not smeared by differential rotation, it has been suggested [4,5] that the spiral features are density waves propagating in a uniform-density disk. This is consistent with observations showing the interarm regions to be only a factor of 2 less dense than the spiral arms.

The predominant motion of the hydrogen in the galactic plane is differential rotation, for which the angular velocity is a decreasing function of radius. The linear velocity reached at the solar position is approximately 225 km s^{-1} [6,7] and increases slightly at larger galactic radii as a consequence of the massive dark halo inside of which the disk is located. Deviations from circular motion in the plane are typically less than 10 km s^{-1}. Away from the galactic plane, H I clouds have been observed with velocity components ranging from 100 to 200 km s^{-1}, moving toward the plane [8]. It appears that these clouds are from the galaxy–Magellanic Cloud system and have reached these high velocities by having been gravitationally accelerated by the galaxy from distances larger than 20 kpc from the galactic disk [9].

The structure of the interstellar gas in the galactic plane is usually described as cloud-like. The reference to clouds is meant to invoke a picture not of matter in a void, but rather of a density enhancement in a more tenuous background. Since the clouds are not gravitationally stable, they are expected to dissipate due to internal thermal pressure on a time scale of about 10^7 yr. To stabilize the clouds, Spitzer [10] proposed a hot ($\sim10^6$ K) tenuous intercloud medium in pressure equilibrium with the clouds. The existence of the hot component of the interstellar medium is now well established from observations of the diffuse x-ray background and observations of O VI absorption lines [11]. The present picture of the interstellar medium [12] consists of cool clouds in a mixed intercloud medium consisting of both warm and hot gas in pressure equilibrium. The energy needed to heat and compress the interstellar medium comes from a combination of stellar winds and supernovae while the advance of shock fronts forms the clouds by compressing the initially cool material.

Parameters describing individual H I clouds lie in the following ranges: size, 0.01–50 pc; density, 10–10^7 particles cm^{-3}; mass, $(0.1\text{–}10^6)M_\odot$; and kinetic temperature, 10–100 K. Taken collectively, approximately 5–10% of the mass of the galactic plane resides in neutral hydrogen. The distribution function for interstellar cloud masses is consistent with a single power-law slope [14] extending from cloud masses of less than $0.01M_\odot$ to $10^6M_\odot$. For cloud masses higher than $(10^2\text{–}10^3)M_\odot$, the distribution shifts from atomic to molecular clouds because the more massive clouds can more effectively shield the molecules from photodissociation.

Hydrogen is also found in an ionized state (H II) in the radiation fields surrounding hot stars ($T_{star} > 20\,000$ K) (see, e.g., Spitzer [15]). These emission nebulae (which are among the most photogenic astronomical objects, containing very delicate shadings of light and dark) sometimes exhibit a sharp outer boundary beyond which is H I. The abrupt transition between H II and H I occurs when the ionizing flux due to the hot central star has dropped to a low enough level so that hydrogen is mostly in the ground state. Because H II requires an ionizing source, it is not distributed as widely nor as uniformly throughout the galaxy as H I [16].

Not only do H I and H II exist in interstellar clouds, but a significant amount of H_2 has been found, and it may be extremely abundant in the interior of dark clouds where it is difficult to observe. Molecular hydrogen is believed to form on the surfaces of grains. At low temperatures an appreciable fraction of the hydrogen atoms that collide with a grain adhere, then move about the grain surface occasionally finding another hydrogen, and coalesce to form H_2. Because H_2 is only weakly bound to a grain, it can escape, leaving the grain as gaseous H_2. Molecular hydrogen does not exhibit the 21-cm hyperfine transition and must be detected in absorption in the far ultraviolet from above the atmosphere.

It is generally believed that the present chemical composition of the interstellar medium is relatively similar to the solar composition, since the sun formed from interstellar material only 4.6 billion years ago, a short time (~25%) relative to the age of the galaxy. Nevertheless, this point of view does require some explanation when confronted with data that ostensibly show 25 elements to be depleted [17] by factors as high as 10^{-4}. It is argued that a significant amount, if not all, of the depleted material must be condensed out onto dust grains [18]. These are particles ~0.1 μm in size that account for the obscuration in certain directions in space. They are most evident as the cause of the dark band in the center of the Milky Way visible to the unaided eye on a dark summer night. Supporting condensation onto dust grains is an observation [19] that shows that at the higher temperatures encountered in the intercloud medium, about

5000 K in this case, the calcium–hydrogen ratio is an order of magnitude larger than the value obtained from the cooler interstellar clouds. Presumably, the elevated temperature reduces the probability of a calcium atom sticking to a grain. Also supporting grain condensation is the observed correlation between the amount of gas and dust in a particular direction in space [13]. Since grains contribute about 1% of the mass of the interstellar medium, they must be quite rich in heavy elements.

In addition to H_2, the interstellar medium contains other molecules, which were detected first in the interstellar medium. The first molecules to be discovered, CN, CH, and CH^+, were found as optical interstellar absorption lines by Adams and Dunham in 1937 [20]. The next interstellar molecule to be discovered, the OH radical, detected in the radio-frequency spectrum in 1963, precipitated an avalanche of other detections. Now, well over 50 molecules have been discovered, including molecules as complex as $(CH_3)_2O$, dimethyl ether, and C_2H_5OH, ethanol. Almost all of these molecules have been found at millimeter wavelengths. The principal reason for the productivity of the millimeter techniques for detecting molecules is that most molecules have millimeter rotational transitions, and these are about the only lines that can be excited in the cold interior of interstellar clouds. For molecules other than hydrogen, there is no agreement about the formation mechanism. They might form in two-body reactions in the gas phase [21] or on the surfaces of grains [22].

The isotopic abundance ratios of the elements in the interstellar medium can be interpreted to provide a variety of different types of information. For example, the interstellar deuterium to hydrogen ratio is sensitive to the density in baryons at the epoch of primordial nucleosynthesis, when the universe was less than 3 minutes old. Observations of the deuterium to hydrogen abundance ratio have been obtained from Ly-α absorption features in the spectrum of nearby stars using the International Ultraviolet Explorer satellite [23] and yield a range of values consistent with cosmological expectations. Also, isotopic abundance ratios can be used to reveal information about the cycling of interstellar material through stars and galactic chemical evolution. Using both radio and optical techniques many workers have collected data on the $^{12}C/^{13}C$ ratio, for example [24], finding galactic gradients such that the ratio varies from about 25 at the galactic center to about 50 [25] in the local interstellar medium with variations to above 100 that complicate the interpretation. The corresponding solar value for the $^{12}C/^{13}C$ ratio is 89 and probably reflects the interstellar value when the sun was formed 4.6 billion years ago.

The interstellar medium is also filled with cosmic rays. These are energetic particles consisting mostly of protons and heavier nuclei with an electron component amounting to about 1% of the proton flux. Cosmic rays ranging in energy from less than 10^6 eV to more than 10^{20} eV are observed. However, it is only at energies from 10^9 eV to about 10^{17} eV that the cosmic ray data is believed to be characteristic of the galactic cosmic radiation. At lower energies the solar wind influences fluxes, whereas at higher energies extragalactic cosmic rays can penetrate the galactic magnetic field and, conversely, galactic cosmic rays can escape. The solar wind, the plasma flowing from the solar corona, carries with it frozen-in magnetic field lines that affect the diffusion of cosmic rays in the solar cavity [26,27]. The cosmic ray element abundances are somewhat similar to the solar abundances except for Li, Be, and B. The abundances of these elements can be accounted for by the fragmentation of heavier nuclei as they spiral around galactic magnetic field lines, a path traversing about 5 g cm^{-2} of matter.

The origin of cosmic rays is a topic that has been discussed for years. At present, the most common viewpoint is that low-energy cosmic rays are accelerated in supernova shocks. For energies above 10^{14} eV, the acceleration is believed to occur in pulsar magnetospheres. There, the energy density in cosmic rays, about 10^{-12} erg cm^{-3}, is significant and comparable to the energy density of the 2.8 K cosmic blackbody radiation, of starlight in the galactic plane, of turbulence in the interstellar medium, and of the galactic magnetic field. Whether this is a coincidence or has fundamental significance is an unanswered question.

The existence of the galactic magnetic field has been deduced from observation of the polarization of starlight [28,29]. To explain the consistency of magnitude and direction of the polarization of light from stars viewed through the same interstellar cloud, a galactic magnetic field has been postulated that aligns elongated grains in the cloud, establishing a preferred scattering direction. The value of the galactic magnetic field, approximately 1–3 μG [30], may be obtained most directly by use of the Faraday effect, the rotation of the plane of polarization of radiation, and the Zeeman splitting of the 21-cm line. Somewhat less directly, magnetic field values may be obtained from synchrotron radiation due to electrons spiraling in the galactic magnetic field. The topology of the galactic magnetic field is of interest to understand the origin of the field. If the field is generated by a self-starting dynamo [31], the magnetic field lines in the disk are expected to be closed and circular, but if they are primordial they are expected to have a bisymmetric configuration open to intergalactic space [32]. The observations appear to favor a bisymmetric spiral pattern [33].

After enumerating the various constituents of the interstellar medium, it is not yet possible to proceed significantly further by discussing the interactions among the components. Any complete theory must begin with the formation of our galaxy, approximately 10^5 yr after the "big bang" when the universe consisted of a homogeneous gas containing only hydrogen and helium. It must follow the matter through cycles of star formation and nuclear burning to account for the formation of heavy elements, and stellar mass ejections, including supernovae, which disperse these elements throughout the interstellar medium. Since, for example, the star formation rate is dependent upon the composition of the matter, magnetic field strength, and cosmic ray flux, no individual aspect of the system can be treated without taking all the other components into account. Consequently, any advance in understanding of the interstellar medium is related to general progress in astrophysics.

See also COSMIC RAYS—ASTROPHYSICAL EFFECTS;

Cosmic Rays—Solar System Effects; Cosmology; Galaxies; Milky Way; Solar Wind.

REFERENCES

1. H. I. Ewen and E. M. Purcell, *Nature* **168,** 356 (1951).
2. W. B. Burton, *Galactic and Extra-Galactic Radio Astronomy* (G. L. Verschuur and K. I. Kellerman, eds.). p. 82. Springer-Verlag, Berlin and New York, 1974.
3. C. Hunter and A. Toomre, *Astrophys. J.* **155,** 747 (1969).
4. B. Linblad, *Stockholm Obs. Ann.* **22,** 5 (1963).
5. C. C. Lin, C. Yuan, and F. H. Shu, *Astrophys. J.* **155,** 721 (1969).
6. L. Blitz, *Ap. J.* **231,** 115 (1979).
7. J. E. Gunn, G. R. Knapp, and S. D. Tremaine, *Astrophys. J.* **84** (1979).
8. G. L. Verschuur, *Annu. Rev. Astron. Astrophys.* **13,** 257 (1975).
9. I. F. Mirabel, *Astrophys. J.* **256** (1982).
10. L. Spitzer, *Astrophys. J.* **120,** 1 (1954).
11. E. B. Jenkins, *IAU Colloquium 81, The Local Interstellar Medium* (Y. Kondo, F. C. Bruhweiler, and B. D. Savage eds.), p. 155. NASA Conf. Pub. 2345, 1984.
12. R. M. Cretcher, *Astrophys. J.* **254,** 82 (1982).
13. L. Spitzer and E. B. Jenkins, *Annu. Rev. Astron. Astrophys.* **13,** 133 (1975).
14. J. M. Dickey and R. W. Garwood, *Astrophys. J.* **341,** 201 (1989).
15. L. Spitzer, *Diffuse Matter in Space,* p. 112. Wiley-Interscience, New York, 1968.
16. F. J. Lockman, *Astrophys. J.* **209,** 429 (1976).
17. D. C. Morton, *Astrophys. J. Lett.* **193,** L35 (1974).
18. G. B. Field, *Astrophys. J.* **187,** 453 (1974).
19. L. M. Hobbs, *Astrophys. J. Lett.* **206,** L117 (1976).
20. W. S. Adams and T. Dunham, *Pub. Astrom. Astrophys. Soc.* **9,** 5 (1937).
21. E. Herbst and W. Klemperer, *Astrophys. J.* **185,** 505 (1973).
22. W. D. Watson and E. Salpeter, *Astrophys. J.* **175,** 659 (1972).
23. W. B. Landsman, R. C. Henry, H. Moos, and J. L. Lansky, *Astrophys. J.* **285,** 801 (1984).
24. P. G. Wannier, *Annu. Rev. Astron. Astrophys.* **18,** 399 (1980).
25. P. Crane and D. J. Hegyi, *Astrophys. J. Lett.* **326,** L35 (1988).
26. J. R. Jokipii, *Rev. Geophys. Space Phys.* **9,** 27 (1971).
27. H. K. Volk, *Rev. Geophys. Space Phys.* **13,** 547 (1975).
28. W. A. Hiltner, *Science* **109,** 165 (1949).
29. J. S. Hall, *Science* **109,** 166 (1949).
30. G. L. Verschuur, in *Galactic and Extragalactic Radio Astronomy* (G. L. Verschuur and K. I. Kellerman, eds.), p. 179. Springer-Verlag, Berlin and New York, 1974.
31. M. Stix, *Astron. Astrophys.* **47,** 243 (1976).
32. J. H. Piddington, *Astrophys. Space Sci.* **59,** 237 (1978).
33. Y. Sofue, M. Fujimoto, and R. Wielebinski, *Annu. Rev. Astron. Astrophys.* **24,** 459 (1986).

Invariance Principles

Feza Gürsey

Invariance principles came to occupy their preeminent position in contemporary physics chiefly through the impact of three successive waves. The first was relativity, which emphasized the invariance of the fundamental laws of physics under the Poincaré transformations relating equivalent inertial systems and the invariance of the laws of gravitation under Einstein's general coordinate transformations (See Relativity, Special Theory; Relativity, General) in space-time. The second was the study of the invariance properties of the Hamiltonian in quantum mechanics, which led to an explanation of degeneracies in energy of the different states of a physical system. The pioneering work of Weyl, Wigner, and Fock and the extensive use of Noether's theorem showed the importance and usefulness of symmetry considerations. The third wave came with particle physics, where invariance principles turned out to be indispensable tools for the understanding of selection rules in particle reactions, for the classification of particle states, and for the restriction of the possible forms of the interaction Hamiltonian or the scattering matrix.

VARIOUS KINDS OF INVARIANCE

Invariance principles may be classified in different ways: (a) Those related to *space-time* (like parity, time reversal, Poincaré invariance, conformal invariance, or general relativistic invariance) and those related to *internal* coordinates or internal quantum numbers (like charge conjugation, gauge invariance, charge independence, etc.). (b) Those associated with *discrete* or *continuous* groups of transformations. For instance T (time reversal), P (parity), C (charge conjugation), and their combinations, as well as invariance under discrete translations in crystals, fall in the first category, while translations, rotations, scaling transformations, isospin, or unitary transformations are examples in the second category. (c) Those which are *exact* and those which are *approximate*. Poincaré invariance, TCP invariance, invariance under fermion-number gauge transformations (and perhaps separate baryon- and lepton-number transformations), gauge invariance related to electric charge are all thought to be exact. On the other hand, invariance principles related to C, P, and T separately, charge independence, unitary symmetry are approximate in varying degrees. (d) Those corresponding to a *local* (x-dependent) or *global* (x-independent) continuous transformations. For example, the fermion conservation law is a global invariance principle, whereas invariance of electromagnetism is local. (e) Those which are *manifest,* being associated with an explicit symmetry of the Lagrangian of the system, and those which are *hidden* as in the case of spontaneous symmetry breaking where the Lagrangian is expressed in terms of physical fields which are not covariant with respect to the invariance group. Topological quantum numbers that arise in special solutions like monopole or soliton solutions to field equations provide another example of hidden symmetry not contained explicitly in the Lagrangian.

INVARIANCE GROUPS AND CONSERVED QUANTITIES

In order to illustrate the strong links between invariance principles and symmetry groups, let us characterize a system by a Lagrangian $L(q_i,\dot{q}_i)$ where q_i are the generalized coordinates and $\dot{q}_i$ (their time derivatives) the generalized velocities. The equations of motion are given by the variational principle

$$\delta \int_{t_1}^{t_2} L\, dt = 0 \qquad (1)$$

which tells us that the action is minimized (or extremized) by the actual motion. The Hamiltonian is $H(p,q)=p_i\dot{q}_i-L$ where the momenta p_i are given by $\partial L/\partial \dot{q}_i$. If we consider a physical observable $\Omega=\Omega(p,q)$ as a function of the canonical variables p_i and q_i, the time variation of Ω is given by

$$\frac{d\Omega}{dt}=\frac{\partial\Omega}{\partial q_i}\frac{dq_i}{dt}+\frac{\partial\Omega}{\partial p_i}\frac{dp_i}{dt}. \tag{2}$$

Now the equations of motion derived from Eq. (1) can be written in the Hamiltonian form

$$\frac{dq_i}{dt}=\frac{\partial H}{\partial p_i}, \quad \frac{dp_i}{dt}=-\frac{\partial H}{\partial q_i} \tag{3}$$

so that we can write

$$\frac{d\Omega}{dt}=\{\Omega,H\}=\frac{\partial\Omega}{\partial q_i}\frac{\partial H}{\partial p_i}-\frac{\partial H}{\partial q_i}\frac{\partial\Omega}{\partial p_i} \tag{4}$$

by means of a Poisson bracket.

The quantum-mechanical case is similar. Ω and H are then Hermitian operators that are functions of the Hermitian operators p_i and q_i that obey Heisenberg's commutation relations. In Dirac's formulation of quantum mechanics, the Poisson bracket is replaced by i times the commutator, i.e.,

$$\frac{d\Omega}{dt}=i[\Omega,H]=i(\Omega H-H\Omega). \tag{5}$$

Conservation in time of the observable Ω implies the vanishing of its bracket (Poisson or Dirac) with the Hamiltonian. In particular, the Hamiltonian H is a constant of motion if L does not depend on t explicitly. Since H is the sum of kinetic and potential energies both in classical and quantum mechanics, this means conservation of total energy when L is invariant under time translations. This is an example of *Noether's theorem* according to which invariance of the Lagrangian under a one-parameter transformation implies the existence of a conserved quantity associated with the generator of the transformation. Another example is provided by invariance under space translations. Let q_i represent the six coordinates $\mathbf{r}_1$ and $\mathbf{r}_2$ of a two-point system. If the potential energy is invariant under space translations $\mathbf{r}\rightarrow\mathbf{r}+\mathbf{a}$, it must be a function of $\mathbf{r}_1-\mathbf{r}_2$ only. Using for Ω in Eq. (4) or (5) the three components of the total momentum $\mathbf{p}=\mathbf{p}_1+\mathbf{p}_2$, we find $dp/dt=0$, so that the total momentum is conserved. Thus momentum conservation is associated with the principle of invariance under the translation group. Using the commutators of $\mathbf{r}_1$, $\mathbf{r}_2$ with $\mathbf{p}_1$, $\mathbf{p}_2$ in the quantum-mechanical case we can show readily that

$$\exp(i\mathbf{p}\cdot\mathbf{a})f(\mathbf{x}_1,\mathbf{x}_2)\exp(-i\mathbf{p}\cdot\mathbf{a})=f(\mathbf{x}_1+\mathbf{a},\mathbf{x}_2+\mathbf{a}). \tag{6}$$

This shows that the conserved quantity $\mathbf{p}$ is also the generator of space translations.

In a similar way, conservation of angular momentum follows from the isotropy (invariance under rotations) of the Lagrangian.

According to Poisson's theorem, if **H** has vanishing brackets with quantities C_r, it also has vanishing brackets with $\{C_r,C_s\}$ (or $[C_r,C_s]$ in the quantum-mechanical case). Thus the conserved quantities C_r have a Lie algebraic structure, so that they can be associated with the generators of a Lie group. There are N conserved quantities corresponding to the dimension N of the group. For example, there are eight conserved quantities (three of them being the angular momenta) in the case of the three-dimensional harmonic oscillator, the Lagrangian of which is invariant under the group $SU(3)$. In the case of the hydrogen atom (or the Kepler) problem, the invariance group is $SO(4)$ leading to six conserved quantities, namely, the three angular momenta and the three components of the Runge–Lenz vector.

If we now turn to a discrete invariance principle, we can consider a system invariant under the parity operator with the two eigenvalues ± 1. A parity noninvariant term in the Lagrangian will induce transitions among states $P=1$ and $P=-1$ causing a parity-violating decay of the system in the course of time. This is what happens in a weak interaction such as beta decay. Strong and electromagnetic interactions, insofar as they can be separated from weak interactions, obey P (and C) invariance. Again, while the usual weak interactions are invariant under CP (and T) but not under C and P separately, a special kind of weak decay associated with neutral kaons exhibits a small violation of CP (or T) invariance. Such a violation of time reversal is ascribed to superweak forces.

GAUGE INVARIANCE AND INTERNAL SYMMETRIES

In the domain of internal symmetries, the best known (and oldest) invariance principle is associated with the conservation of electric charge. This exact conservation law and the vanishing of the photon mass are also related to invariance under a local gauge transformation which affects simultaneously the electromagnetic potential A_ν and the wave function (or field operator) ψ of the charged particle (e.g., the electron) so that the Lagrangian remains invariant under the transformation

$$A_\nu\rightarrow A_\nu-\frac{1}{e}\frac{\partial\Lambda}{\partial x^\nu}, \qquad \psi\rightarrow\exp(i\Lambda)\psi, \tag{7}$$

where Λ is an arbitrary function of the space-time variables x^ν and e is the charge of the particle. The electromagnetic field tensor $F_{\mu\nu}$ and the charge current j_ν are gauge invariant. Then the charge $Q=\int j^0\, d^3x$ has a vanishing bracket with the Hamiltonian and represents an exactly conserved quantity. The photon field associated with the charge remains massless in virtue of Eq. (7). The fact that the field ψ is charged is expressed by the quantum-mechanical relation

$$[Q,\psi]=e\psi. \tag{8}$$

The principle of local gauge invariance associated with electric charge has been generalized by C. N. Yang and R. L. Mills in 1954 to charges Q_i that are the generators of a Lie group G [say $SU(2)$]. The G invariance is supposed to hold independently at every space-time point x. In the color gauge theory of strong interactions, G is taken to be the color group $SU(3)$. The electron field is then replaced by colored quark fields and the photon is replaced by eight massless gluons. In the $SU(2)$ example, ψ is a doublet and we can use the 2×2 Pauli matrices $\boldsymbol{\tau}$ to write a generalization of Eq. (7), i.e.,

$$\boldsymbol{\tau}\cdot\mathbf{A}_\nu \to S\boldsymbol{\tau}\cdot\mathbf{A}_\nu S^{-1} - \frac{1}{g} S\left(\frac{\partial S^{-1}}{\partial x^\nu}\right), \qquad \psi \to S\psi, \tag{9}$$

where $S=\exp[\frac{1}{2}\boldsymbol{\tau}\cdot\boldsymbol{\omega}(x)]$. In the case of global gauge transformations the parameters Λ or $\boldsymbol{\omega}$ become constant and the inhomogeneous terms in the transformation laws of A_ν or $\mathbf{A}_\nu$ drop out. Established field theories such as gravitation, electromagnetism, and quantum chromodynamics for the quark–gluon system are all based on exact local gauge invariance principles and involve massless boson fields whose interaction with other fields is universal and determined by the gauge principle.

INVARIANCE PRINCIPLES ASSOCIATED WITH SPONTANEOUSLY BROKEN SYMMETRIES

Systems can be constructed in such a way that the Lagrangian is invariant under a group G, but the ground state (vacuum) which minimizes the energy is not invariant under G. Then the excited states will not be G covariant. The action of G on such states will transform them into other degenerate states. Then we speak of a spontaneous symmetry breakdown. If the ground state is still invariant under a subgroup H of G, the generators of H are associated with massless boson fields, whereas the other fields corresponding to the remaining generators of G acquire a mass. When the Hamiltonian is reexpressed in terms of physical noncovariant fields the H symmetry is manifest, whereas the remainder of the G symmetry is hidden. A prototype of spontaneous symmetry breakdown is the ferromagnet that has a ground state with all its atomic spins aligned in one direction although interatomic forces are rotationally invariant. In this case, rotational invariance is restored by noting that the magnet pointing in an arbitrary direction through rotation of its initial position provides an equally good ground state. The totality of all rotated magnets forms an isotropic set. Another example is the generalization of electrodynamics based on a local $U(1)$ group to a unified gauge field theory of weak and electromagnetic interactions based on the Weinberg–Salam group $SU(2)\times U(1)$. A realistic theory is obtained by allowing the group to be spontaneously broken so that out of the four generators of the group one remains associated with a massless boson (the photon), while the other three correspond to massive weak bosons that mediate charged and neutral weak interactions.

USEFULNESS AND POWER OF INVARIANCE PRINCIPLES

Invariance principles are of great practical value in determining the general form of physical observables. Wave functions and transition amplitudes related to a Hamiltonian invariant under G can be decomposed into irreducible representations of G with coefficients that are functions of invariants only. For example, a rotationally and translationally invariant quantity depending on $\mathbf{r}_1$ and $\mathbf{r}_2$ can only be a function of $|\mathbf{r}_1-\mathbf{r}_2|^2$. Projection operators for spin 0 and 1 in a system composed of two spinors can only be a function of the invariant combination $\boldsymbol{\sigma}^{(1)}\cdot\boldsymbol{\sigma}^{(2)}$ and because of the algebraic properties of Pauli matrices they must be linear in this expression. Invariants of discrete translations in a lattice are periodic functions. When a scaling invariance principle holds at high energy and momentum transfer, the structure functions occurring in the cross sections can only be functions of dimensionless combinations of kinematic variables called scaling variables. Homogeneity and isotropy of the universe at large determine the form of the cosmological metric (the Robertson–Walker metric).

SOME RECENT DEVELOPMENTS

In two-dimensional dynamical systems (like spin models) scaling invariance near the critical point leads to conformal invariance, hence to a new invariance principle related to the behavior of the system under holomorphic mappings. This also holds for the system represented by a two-dimensional field theory. Such conformal field theories have a very high symmetry described by infinite algebras like the Virasoro algebra and Kac–Moody algebras. The invariants of these algebras are in turn related to critical exponents and other characteristics of the dynamical systems they represent. Since string theory also makes use of conformal field theories, the principle of two-dimensional conformal invariances has become a keystone of modern mathematical physics.

Another development concerns a possible symmetry (broken in nature) between fermions obeying the Fermi–Dirac statistics and bosons obeying the Bose–Einstein statistics. It is called supersymmetry. If a supersymmetric invariance principle were fundamental, then the Poincaré group would be extended to the super Poincaré group that would imply the existence of new particles in the role of superpartners of known fundamental particles. Although there is no experimental evidence for such an extension of the particle spectrum, supersymmetric theories are attractive because of their improved convergence properties and the possibilities they offer for the unification of all fundamental forces.

Following the lead of Einstein, physicists would like to derive the fundamental dynamical laws for all fundamental interactions from general invariance principles comprising and generalizing relativity and gauge principles and perhaps including symmetries between bosons and fermions (supersymmetry) and the conformal invariance of a world sheet substructure. If the dream comes true, all of fundamental physics could be reduced to invariance principles.

See also CONSERVATION LAWS; CPT THEOREM; QUANTUM MECHANICS; RELATIVITY, GENERAL; RELATIVITY, SPECIAL THEORY; STRING THEORY; SUPERSYMMETRY AND SUPERGRAVITY.

BIBLIOGRAPHY

A. Belavin, A. Polyakov, and A. Zamolodchikov, "Infinite Conformal Symmetry in Two-dimensional Quantum Field Theory." *Nucl. Phys. B* **241,** 333 (1984).

J. P. Elliott and P. G. Dawber, *Symmetry in Physics,* Vols. I and II, Oxford University Press, New York, 1979.

D. Friedan, Z. Qiu, and S. Shenker, "Conformal Invariance, Unitarity and Critical Exponents in Two Dimensions," *Phys. Rev.. Lett.* **52,** 1575 (1984).

T. D. Lee, *Symmetries, Asymmetries and the World of Particles*. University of Washington Press, Seattle, 1988.
C. Quigg, *Gauge Theories of the Strong, Weak and Electromagnetic Interactions*. Benjamin Cummings, New York, 1983.
J. J. Sakurai, *Invariance Principles and Elementary Particles*. Princeton University Press, Princeton, NJ, 1964.
J. Wess and J. Bagger, *Introduction to Supersymmetry*. Princeton University Press, Princeton, NJ, 1983.
H. Weyl, *Space-Time-Matter*. Dover, New York, 1951.
E. P. Wigner, *Symmetries and Reflections*. MIT Press, Cambridge, MA, 1970.

Inversion and Internal Rotation

Lawrence C. Krisher

Both inversion and internal rotation, despite their names, belong to the category of molecular dynamics called molecular vibrations. In molecular spectroscopy we find an energy hierarchy by which general molecular dynamics can be usefully described in decreasing magnitudes, as electronic transitions, molecular vibrations, and molecular rotations. Vibrational transitions ordinarily fall in the infrared region of the spectrum around 1000 cm^{-1} and rotations in the microwave region near 1 cm^{-1}. Incongruously enough, both inversion phenomena and internal rotations have been extensively studied from observations made in the microwave region, since they often represent relatively low-energy (low-frequency) vibratory motions which can be strongly coupled to the overall molecular rotations.

Inversion is a vibration characterized by a potential function which has a double minimum, idealized as a parabola with a small bump at the bottom. Because of the rather atypical nature of the allowed energy levels for these potential functions, a number of common molecules, the most well-known perhaps being ammonia, exhibit pure vibrational transitions at microwave frequencies. For ammonia, NH_3, the classical motion involves the passing back and forth of the nitrogen atom through the plane formed by the three equidistant hydrogen atoms. The vibrational levels are shown in Fig. 1, and for high values of the quantum number v can be seen to exist more or less like harmonic oscillator levels, with a dissociation limit at the top, not indicated in the diagram. The bump at the bottom is sufficiently high, for molecules such as ammonia, to stabilize quantum states, in pairs, below the top of the classical barrier. The "inversion doublet" spectrum involves jumps between members of these pairs, and corresponds quantum mechanically to the nitrogen atom tunneling through the barrier. The lowest inversion doublet, labeled 0.8 cm^{-1} in Fig. 1, actually represents a large set of microwave transitions near that energy, the detailed frequencies of which depend on the rotational J and K quantum numbers for the molecule, although the latter do not change during the transition. In effect, the rotational state, specified by J and K, indicates the height on the barrier at which the tunneling is to take place. At ordinary temperatures the most intense member of this set is the 3,3 transition ($J=3$, $K=3$) at 23,870 MHz, which was used for the first maser oscillator and time standard.

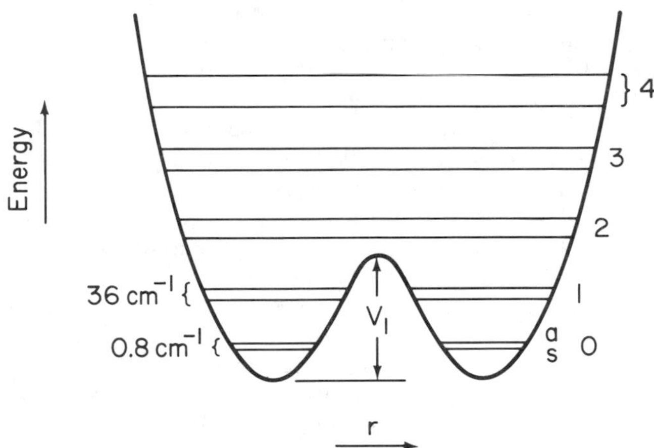

FIG. 1. Potential-energy curve for the ammonia molecule (NH_3) showing the allowed inversion energy levels, labeled in pairs with the appropriate vibrational quantum number. The energies are not to scale; the doublet separations are greatly exaggerated. The molecular coordinate r indicates the distance of the nitrogen atom from the center of the triangle formed by the hydrogen atoms.

The concept of internal rotation involves motion around the type of chemical bond known as a single bond, such as that between the carbon atoms in the ethane molecule, CH_3–CH_3. The idea of internal rotation, the twisting of one of the methyl groups with respect to the other, when hindered by a potential-energy function such as the threefold barrier shown in Fig. 2, actually represents a vibrational degree of freedom, the torsional vibration. This type of motion for ethane and similar organic chemicals, long thought to be virtually free, is hindered by potential barrier heights of about 1000 cm^{-1}, sufficient to stabilize several vibrational

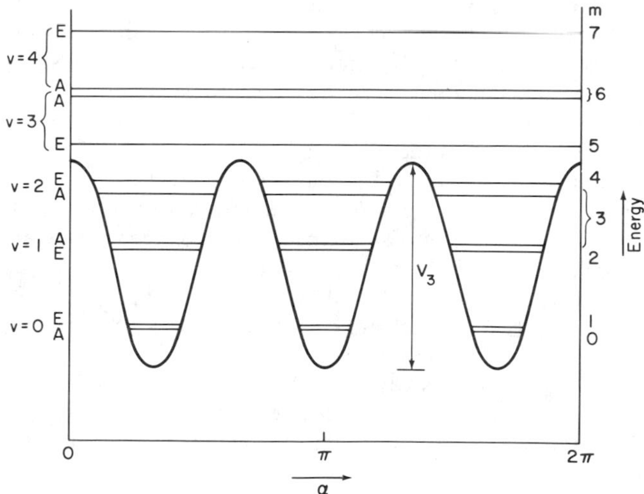

FIG. 2. Potential-energy diagram for a threefold symmetric internal rotation, such as that presented by a methyl group (CH_3–), showing the allowed vibrational (torsional) levels. The vibrational quantum numbers are indicated on the left with the corresponding "free rotor" or m quantum numbers at the right. The molecular coordinate α is an angle describing the degree of internal rotation; for a methyl group it indicates the position of the three hydrogen atoms with respect to the remaining part (framework) of the molecule.

states. For more complex hydrocarbons, this type of barrier also stabilizes the definite molecular geometries which are known as the conformations of the carbon chain, and which give rise to rotational isomers, or *rotamers*.

The finite threefold barrier for ethane-type molecules also imposes a pairing of the torsional vibration levels, with non-degenerate A and doubly degenerate E symmetry, as shown in Fig. 2. Transitions can occur between states with like symmetry, and for the lower levels are typically in the far infrared, at 100–300 cm^{-1}. The height of the torsional barrier can also be derived very accurately from a complex quantum-mechanical treatment of the coupling between the torsional vibration and the pure rotational motion of the molecule. For reasonably high internal barriers such as those found for the ethanes, the expected "pure" rotational transitions appear in the microwave region as close-spaced doublets. In effect, a molecule in a specific low torsional state, say $v=0$, and of symmetry species A will have its own slightly different rotational constants and rotational spectrum from one in the $v=0$ state but of symmetry E. The doublet splittings (frequency differences) afford a very accurate, if indirect, measurement of the barrier height. It should be carefully noted that the microwave energies do not effect transitions from one torsional state to another, nor are they energetic enough to allow jumps over the barrier to yield new conformations. The microwave measurements indicate the two different rotational behaviors of the A and E vibrational species, which exist because of the presence of the barrier.

INVERSION SPECTRA

Ammonia is usually singled out because of the intensity of the microwave transitions and its historical importance. The first microwave transition, observed by Cleeton and Williams in 1934, involved the lowest inversion levels of this molecule. Following the development of radar in the 1940s, as microwave technology became available to spectroscopists, ammonia was one of the earliest molecules studied in detail, and in the 1950s was used by C. H. Townes and co-workers as the working substance for the first molecular oscillator, known as the *maser*.

The presence of the bump at the bottom of the potential, shown in Fig. 1, gives rise to a pairing of the inversion energy levels, with wave functions of alternating symmetry. The molecular symmetry allows direct transitions between these pairs, only two of which are bound by V_1, the lowest at about 0.8 cm^{-1} accounting for the microwave spectrum. The inversion frequency is very sensitive to the height of the barrier, as is found for most tunneling phenomena, and several semiempirical aproaches lead to a value for V_1 of about 2080 cm^{-1}.

The NH_3 molecule can also rotate, of course, but because of its very small moment of inertia, allowed transitions between rotational levels occur at much higher energies, in the far infrared. The inversion energies add to and subtract from these to form a doublet structure, and thus a rather special case of a vibration–rotation band. In addition, the "pure" inversion transition frequency shows a strong dependence on the particular rotational state in which the molecule is found. This gives rise to the large number of intense microwave inversion transitions, with frequencies ranging from 17,000 to 40,000 MHz, dependent on the values of the rotational quantum numbers J and K, which remain unchanged for the transition. From the form of the dependence of the inversion frequency on the potential, V_1, Costain has developed an exponential power series expansion which fits these transitions very accurately through $J=16$, as

$$\begin{aligned}\nu(MHz) = 23{,}785.88 \exp\{&-6.36996\times 10^{-3}J(J+1)\\ &+8.88986\times 10^{-3}K^2+8.6922\times 10^{-7}J^2(J+1)^2\\ &-1.7845\times 10^{-6}J(J+1)K^2+5.3075\times 10^{-7}K^4\}.\end{aligned}$$

Quantum-mechanical considerations prevent the number $K=0$ from being used in this formula.

Many other trigonal–pyramidal molecules should in principle exhibit this "umbrella inversion," but only the deuterated ammonias such as ND_3 and the phosphine molecule PH_3 appear to have realistically observable frequencies. Such molecules as $AsCl_3$, because of mass effects and much higher values for V_1, have the inversion doublets collapsed to essentially zero frequency, as do more complex trigonal molecules such as CH_3Cl.

Beyond these symmetric umbrella inversions, there are many other vibrational phenomena which have double-minimum potentials. In larger molecules, a trivalent nitrogen atom may often be regarded as tunneling back and forth through a molecular plane, which motion is compensated in a complex way by the remainder of the molecule. Another example is provided by "ring-puckering" in molecules such as cyclopentene and trimethylene oxide. In cyclopentene, for example, the carbon atom furthest from the double bond is not coplanar with the other four, and the pertinent vibration allows it to move to an equivalent position on the opposite side of the plane. In more complex molecules, however, other low-frequency vibrational motions are often possible which, along with the molecular rotations, may be coupled to the inversion in such a way as to prevent a separable quantum mechanical description.

INTERNAL ROTATION SPECTRA

The ethane molecule, CH_3–CH_3, is often regarded as the prototype exhibiting the property of hindered internal rotation, largely because of interest in the thermodynamic properties of hydrocarbons. The bond between the two carbon atoms represents the chemist's classic "single bond," of Pauling hybridization type sp^3, as contrasted to "double-bond" molecules of the ethene series, with Pauling hybridization sp^2, which afford separable isomers (the *cis-trans* species) for many substituents.

Careful thermodynamic work in the late 1930s, along with developing spectroscopic techniques, indicated that internal rotatory motion in the ethane case was also hindered by a periodic energy barrier of about 1000 cm^{-1}, with bound torsional vibration states as shown in Fig. 2. Quantum-mechanical considerations for this case forbid transitions between the "doublets" (the selection rules are $A\leftrightarrow A$, $E\leftrightarrow E$), and torsional transitions, such as between the $v=0$ and the $v=1$ levels, are found in the far infrared at about 290 cm^{-1}.

Interestingly enough, the barrier heights hindering such motion can often be determined very accurately by a somewhat indirect method, which essentially involves only the rotational spectrum in the microwave region near 1 cm^{-1}. This requires analysis of the coupling between the internal and overall rotations, actually a vibration–rotation effect, in which the quantity V_3 appears as a parameter.

Analogues of ethane, and, more generally, series with threefold symmetric internal rotors, have been extensively studied. A methyl group or any other group with C_3 rotational symmetry provides a simplification of the dynamics, in that the overall moments of inertia of a molecule remain unchanged regardless of the detailed orientation of the internal rotor coordinate, labeled α in Fig. 2.

The model for a molecule such as acetaldehyde, CH_3CHO, assumes that the molecule is composed of a rigid threefold symmetric internal rotor (the CH_3– group) connected to a rigid framework (the CHO group) and is able to rotate about the intervening bond. All vibrations other than the torsion are thus assumed to be uncoupled and to provide a rotational average set of moments of inertia. The shape of the internal potential is assumed to be sinusoidal, and a small number of other parameters appear which can in principle be determined from the molecular structure.

The Hamiltonian for this model is expressed by Herschbach as

$$\mathcal{H} = \mathcal{H}_r + F(p - \mathcal{P})^2 + (V_3/2)(1 - \cos 3\alpha),$$

where $\mathcal{H}_r$ is the usual rigid-rotor expression involving the three rotational constants, F is a derived "internal rotational constant," and the operator $p - \mathcal{P}$ represents the relative angular momentum of the top (CH_3– group) and the framework. Expansion of the second term gives Fp^2 which can be combined with the V_3 term to give Mathieu's equation, and $F\mathcal{P}^2$ which, since $\mathcal{P}$ is a linear function of the overall molecular angular momenta, can be combined with $\mathcal{H}_r$ to give the new "effective rigid-rotor" constants. The cross term $-2F\mathcal{P}p$ is treated by perturbation theory in a basis set composed of the modified rigid-rotor functions times Mathieu functions. Contributions from the cross term lift the A,E degeneracy of a given torsional level and, for a certain range of barrier height, give rise to two sets of slightly different effective rotational constants. Because of the selection rule $A \leftrightarrow A$, $E \leftrightarrow E$, the rotational spectrum thus appears as close-spaced doublets, with frequency splittings $(\nu_A - \nu_E)$ which are a sensitive function of the barrier height V_3.

This PAM (for "principal-axis method") treatment is especially useful for high barriers and for those cases where A-E splittings in the ground torsional state, $v = 0$, are too small to observe. One must then measure the pure rotational spectrum of molecules in a higher torsional state, and if there are A-E splittings, the barrier can be calculated. An alternative treatment called the IAM (for "internal-axis method") is often used for molecules which have lower barriers or which have other low-frequency vibrational modes besides the torsion. For very low barriers it is more convenient to treat the barrier energy itself as a perturbation to an otherwise "free internal rotor," using a model which employs the m quantization scheme indicated in Fig. 2 at the right.

INTERNAL BARRIERS AND CONFORMATIONS

Molecules for which barriers have been determined are chemically quite diverse but tend to fall into series, the members of which differ by the progressive change of one substituent. Substituents with widely different electronegativities appear to have remarkably little effect on the barrier values through a given series. Thus the ethanes, when progressively fluorinated to give CH_3CH_2F, CH_3CHF_2, continue to have barriers near 1000 cm^{-1}. Analogous methyl silane molecules, CH_3SiH_3, CH_3SiH_2F, etc., all have barriers near 500 cm^{-1}, and the methyl germanes at about 350 cm^{-1}. For the "acetyl series" in which the methyl group rotates against a double–single bond structure, substitution of any halogen for the acetaldehyde hydrogen leads to very little change in the barrier from 380 cm^{-1}. Acetic acid appears to be an anomaly with an intermediate barrier value of about 170 cm^{-1}. Much lower barriers are found for molecules such as $CH_3C{\equiv}CSiH_3$ in which the "central bond" is a linear group, and for pure sixfold barrier cases such as CH_3NO_2, which has a barrier of about 2 cm^{-1}.

For internal rotors which lack threefold symmetry, the presence of the barrier allows one to describe in detail the preferred molecular conformations, indicated by the coordinate α. The overall moments of inertia and thus the pure rotational spectra of such molecules are now strongly dependent on α, and species at potential minima (which no longer need be threefold symmetric) are called *rotamers*. As shown in Fig. 3, the *cis-trans* terminology is no longer sufficient, and the term *gauche* is applied to the asymmetric forms which occur as a result of the threefold bonding.

It is interesting to compare these barrier energies with those of conventional chemical bonds. The energy of for-

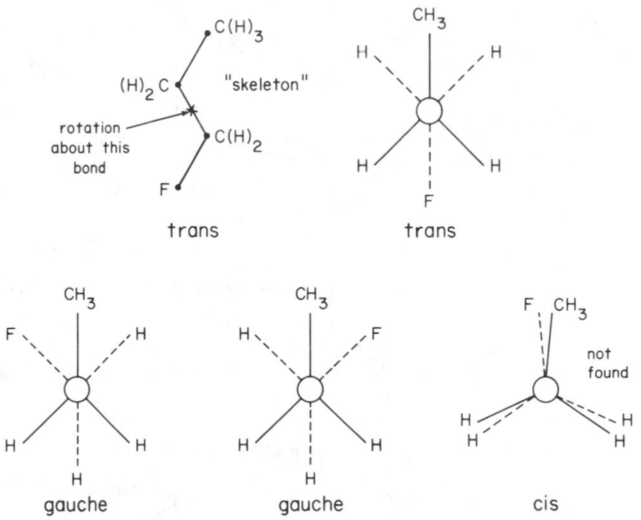

FIG. 3. Conformations of the molecule 1-fluoropropane $CH_3CH_2CH_2F$. The first sketch shows the carbon–fluorine chain or "skeleton" in the *trans* conformation, accompanied by the appropriate Newman diagram. The second row shows the two inertially equivalent *gauche* forms, and the unstable *cis* form which represents the "top of the hill" for the internal rotation potential. Also note that both the *trans* and *gauche* forms are "staggered" whereas the *cis* form is "eclipsed."

mation of a chemical single bond, such as that between the two carbon atoms in the ethane molecule, is about 30,000 cm^{-1}. Also, the energy of an ethene type *cis-trans* conversion, which can be thought of as an internal rotation about a chemical double bond, is about 10,000 cm^{-1}. The rotamer barriers, typically less than 1000 cm^{-1}, are thus a fraction of a conventional bond energy. The stable species shown in Fig. 3 can exist, however, through many hundreds of overall rotations, and be observable with distinct values for I_a, I_b, and I_c. These single-bond conformations when stabilized in condensed phases are also of great interest to molecular biologists. An entire issue of the *Journal of Physical Chemistry* (Vol. 91, No. 21, October 8, 1987), is concerned with "dynamical stereochemistry," including these conformational effects in the synthesis of complex organic compounds.

From the study of the conformations of small molecules, two powerful generalizations have emerged. When one threefold group, such as a methyl group or any sp^3 hybridized carbon, is bonded to another, the stable internal conformation is "staggered," as shown in Fig. 3 for the *trans* and the two *gauche* forms of 1-fluoropropane. The *cis* form represents a conformation corresponding to a maximum in V, and is said to be "eclipsed." Secondly, when a threefold sp^3 group is bonded to an sp^2 carbon, such as in the molecule propylene or in the acetyl series, the energy minima occur for those conformations in which a methyl hydrogen eclipses the double bond of the sp^2 group.

The internal-rotation coupling phenomenon is an important example from a larger area of vibration–rotation interactions, which serves to indicate the difficulties sometimes encountered when one tries to employ the often convenient and useful idea of the separation of molecular dynamics.

See also MICROWAVE SPECTROSCOPY; MOLECULES; TUNNELING.

BIBLIOGRAPHY

Inversion

T. M. Sugden and C. N. Kenney, *Microwave Spectroscopy of Gases*. Van Nostrand, London, 1965 (I)

C. H. Townes and A. L. Schawlow, *Microwave Spectroscopy*. McGraw-Hill, New York, 1955. (A)

Internal Rotation

C. C. Lin and J. D. Swalen, *Rev. Mod. Phys.* **31,** 841 (1959). (A)

E. B. Wilson, Jr., *Advances in Chemical Physics,* Vol. II, p. 367. Wiley-Interscience, New York, 1959. (I)

J. E. Wollrab, *Rotational Spectra and Molecular Structure*. Academic Press, New York, 1967. (I)

R. B. Bernstein, D. R. Herschbach, and R. D. Levine, *J. Phys. Chem.* **91,** No. 21 (1987), see page 5366ff.

Ion Beams *see* Electron and Ion Beams, Intense

Ion Impact Phenomena *see* Electron and Ion Impact Phenomena

Ionization

M. E. Rudd

Ionization is the process in which a neutral atom or molecule is given a net electrical charge. Some atoms can accept an electron to form a negative ion in the process called *attachment*. One or more electrons can be removed from an atom or molecule forming a positive ion.

Of all the basic atomic processes, ionization requires the greatest energy. The amount of energy required to remove the least tightly bound electron from a neutral atom or molecule is called the *first ionization potential I* and is usually measured in electron volts (eV). The additional energy needed to remove the next electron is the *second ionization potential*, etc. The energy by which an excess electron is bound in a negative ion is known as the *electron affinity*. Tables of electron affinities, ionization potentials, and binding energies of inner-shell electrons are available for most atoms and many molecules.

Because ions and charged radicals are chemically highly reactive, much of the radiation damage to matter caused by high-energy photons or particles is a result of ionization, caused both by the primary radiation and by secondary products.

It is an empirical fact that approximately one ion–electron pair is formed for every 30 eV of energy lost by a fast charged particle traversing nearly any material. This makes it possible to determine the approximate energy of a particle by counting the ions formed in a detector which is sufficiently thick to stop the incoming particle.

PHOTOIONIZATION

Electromagnetic radiation of sufficiently short wavelength will cause ionization. The energy of an individual photon, $h\nu$, must equal or exceed the first ionization potential of the atom or molecule for this to take place. For most atoms the

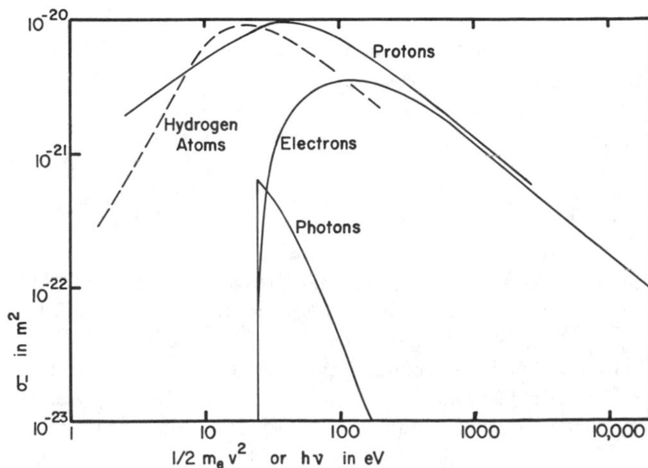

FIG. 1. Cross section for ejection of electrons from a helium atom by various incident particles. The abscissa is the incident energy for photons or electrons; for heavy particles it is the energy of an electron of equal velocity.

ionization threshold is in the ultraviolet region of the spectrum. The cross section (which is a measure of the probability that a given process takes place) for photoionization rises abruptly at the threshold and for most atoms falls off at higher energies as shown in Fig. 1. Photoionization in the upper atmosphere caused by ultraviolet radiation from the sun is largely responsible for the presence of the ionosphere.

Two-photon ionization can also occur in which a single photon has insufficient energy to ionize an atom but enough to raise it to a real or virtual excited state. If a second photon arrives before the atom decays back to the ground state, the additional energy may then be enough to cause ionization. Three-photon and higher multiphoton ionizations have also been observed. For these processes to take place the radiation field must be sufficiently intense so that successive photons interact with a single atom within a time span shorter than the decay time. Such radiation fields are usually obtained with lasers.

ELECTRON COLLISIONS

If an electron makes a collision with an atom or molecule, enough energy may be transferred to one of the electrons to eject it. To do this, the incident electron must have an energy at least equal to the first ionization potential I. As shown in Fig. 1, the cross section for this process increases rapidly above threshold to a maximum and then falls off monotonically at high energies. In atoms with more than one subshell, additional thresholds occur at higher energies yielding structure in the ionization function.

The cross section varies with impact energy E as $E^{-1}\log E$ at high energies. This prediction of the Bethe–Born approximation has been well verified experimentally. The dependence of the cross section on energy near threshold is of the form $(E-I)^n$. The semiclassical treatment by Wannier yields $n=1.127$, a value verified by experiment.

HEAVY-PARTICLE COLLISIONS

Because of the added possibilities of electron removal from the projectile (called *stripping* or *electron loss*) and the transfer of electrons (variously called *electron transfer, charge transfer*, and *charge exchange*), ionization by heavy atomic particles is a more complex process than photon or electron ionization. For example, the following are possible final charge states when a proton collides with a helium atom:

$$H^+ + He \rightarrow H^+ + He^0, \quad (a)$$

$$H^+ + He^+ + e^-, \quad (b)$$

$$H^0 + He^+, \quad (c)$$

$$H^+ + He^{2+} + 2e^-, \quad (d)$$

$$H^0 + He^{2+} + e^-, \quad (e)$$

$$H^- + He^{2+}. \quad (f)$$

Of these six processes, b, d, and e contribute to the production of free electrons; b, c, d, e, and f contribute to the production of positive ions; and c, e, and f are electron-transfer reactions. Note that while reactions b and d are the only ones that might fairly be called simple ionization, other reactions contribute to the charging of the target or to the release of free electrons.

In most experiments involving the collision of a beam of particles with a stationary gas target, what is measured are the positive and negative charges left behind or the change of charge state of the projectile. If we call the negative charge production cross section σ_-, the cross section for production of slow positive ions σ_+, and the electron-transfer cross section σ_c, then one can readily show from the basic reactions listed above that $\sigma_+ = \sigma_- + \sigma_c$.

For collisions of fast neutral atoms with other neutral atoms a similar analysis can be made. If the cross section for producing negative ions is small compared to the other cross sections (as it usually is), then the approximation $\sigma_- = \sigma_+ + \sigma_s$ results where σ_s is the electron-loss cross section.

Figure 1 shows cross sections for ejection of electrons from collisions of protons or neutral hydrogen atoms with helium atoms. Note that at high energies, equal-velocity protons and electrons yield the same ionization cross sections.

Cross sections for ionization from various electronic shells of large numbers of targets have been and are being measured as a function of many parameters. These include the collision energy, the excitation and charge state of the target (and the projectile in the case of heavy-particle impact), the impact parameter (or distance of closest approach), and the energy and angular distributions of the ejected electrons and of the scattered and recoil particles.

As a result of such measurements and the concomitant theoretical investigations, a number of different mechanisms of ionization have been identified. Some of these are as follows.

1. Direct Coulomb Ionization

If a charged particle such as an electron or proton makes a sufficiently close collision with an atom, enough energy may be transferred to one (or more) of the atomic electrons to remove it from the atom. This transfer of energy is through the Coulomb interaction between the electron and the incident charged particle and can occur for inner-shell as well as outer-shell electrons.

This mechanism of ionization may be described classically either in a binary-encounter model (taking account only of the interaction between the charged projectile and one electron of the target) or by the use of Monte Carlo methods to solve the three-body problem. Quantum mechanically, calculations usually use the Born approximation.

In the case of a collision between two neutral atoms, the electric fields of the electrons and the nucleus tend to screen each other for large distances of separation, but at close approaches, the Coulomb field of the nucleus and the electrons separately may cause ionization.

2. Electron Promotion

A pair of colliding atoms can be thought of as a temporary molecule with an internuclear distance which decreases, reaches a minimum, and then increases to an infinite separation. The molecular energy levels must change accordingly, and if the collision is slow enough, the electronic wave functions adjust adiabatically except near crossings of the energy levels. At such crossings electrons readily transfer from one level to another and upon separation electrons may be "promoted" to higher excited levels or into the continuum. Provided there are crossings at reasonably large internuclear separations, this mechanism is very efficient in producing inner-shell vacancies. This picture is called the *molecular-orbital promotion model* or the *Fano–Lichten model*.

3. *Autoionization and the Auger Effect*

When an atom is put into an excited state with an energy above the ionization potential, instead of decaying by the emission of a photon it is usually more likely to make a transition to a continuum state, i.e., a state of the ion plus a free electron with kinetic energy. This spontaneous ejection of an electron is known as *autoionization*. Electrons ejected by this mechanism have discrete energies characteristic of the initial and final states of the atom.

Autoionizing states can be produced by photons, electrons, or heavy particles. Excited states with energies above the ionization potential may result, e.g., from the simultaneous excitation of two electrons. The excitation may also be caused by the removal of an inner-shell electron. In this case the resulting transition involving the filling of the vacancy and the ejection of an electron is known as the *Auger effect*. This competes with x-ray emission, the branching ratio in favor of the latter being called the *fluorescence yield*. Successive Auger transitions in a multishell atom can leave the outer shell completely or nearly completely stripped of electrons.

4. *Electron Transfer*

A collision between a fast positive ion and a neutral atom or molecule may result in a transfer of an electron to the ion. Even though no electron is released, this reaction leaves a positive ion behind and therefore can be thought of as a mechanism for producing ionization.

It has been found that the transferred electron may end up not only in the ground state of its new host atom or in an excited state, but can also be transferred to a continuum state of the projectile. Electrons ejected by this mechanism are distinguishable from those in continuum states of the target atom only insofar as their angular and energy distribution is characteristic of the moving frame of reference rather than that of the stationary target.

5. *Electron Shakeoff*

When an electron is ejected from a multielectron atom, the remaining electrons suddenly find themselves in a different central potential. After the ensuing readjustment there is a probability that a given electron will find itself in a new state. If this is a higher excited state, the process is called *shakeup*, but if it is a continuum state it is known as *shakeoff*. Theoretically, the sudden approximation says that the probability of a transition is the square of the overlap integral of the wave functions for the initial and final states. The outermost electrons are the ones most likely to be ejected by shakeoff.

6. *Penning Ionization*

If one atom has a greater excitation energy than the ionization potential of a second atom with which it makes a collision, the energy released by the deexcitation of the first atom can go to ionize the second. This process, known as *Penning ionization*, normally requires that the excited state be a metastable (i.e., long lived) one so that the atom can retain its excitation energy long enough to make a collision.

7. *Dissociative Ionization*

Photon, electron, or heavy-particle impact on a molecule can cause it to break up or *dissociate* into fragments which may be charged. Dissociative ionization also occurs when ionic solids such as salts are dissolved in certain solvents. This occurs when the ions of the solute are more strongly attracted to the dipoles of the solvent molecules than to the oppositely charged ions and therefore are separated from them.

See also AUGER EFFECT; COLLISIONS, ATOMIC AND MOLECULAR; PHOTOIONIZATION.

BIBLIOGRAPHY

J. W. Gallagher, C. E. Brion, J. A. R. Samson, and P. W. Langhoff, "Absolute Cross Sections for Molecular Photoabsorption, Partial Photoionization, and Ionic Photofragmentation Processes," *J. Phys. Chem. Ref. Data* **17,** 9–153 (1988). (I)

J. B. Hasted, *Physics of Atomic Collisions,* 2nd ed. American Elsevier, New York, 1972. (I)

Wolfgang Lotz, "Electron Binding Energies in Free Atoms," *J. Opt. Soc. Am.* **60,** 206–210 (1970). (E)

T. D. Mark and G. H. Dunn, eds, *Electron Impact Ionization.* Springer-Verlag, Wien and New York, 1985. (I)

H. S. W. Massey, E. H. S. Burhop, and H. B. Gilbody, *Electronic and Atomic Impact Phenomena,* 2nd ed. Oxford, London, 1969–1975 (5 volumes). (I)

E. W. McDaniel, *Collision Phenomena in Ionized Gases.* Wiley, New York, 1964. (I)

N. F. Mott and H. S. W. Massey, *Theory of Atomic Collisions,* 3rd ed. Oxford, London, 1965. (A)

M. E. Rudd and J. H. Macek, "Mechanisms of Electron Production in Ion-Atom Collisions," *Case Studies At. Phys.* **3,** 47–136 (1972). (I)

M. E. Rudd, Y.-K. Kim, D. H. Madison, and J. W. Gallagher, "Electron Production in Proton Collisions: Total Cross Sections," *Rev. Mod. Phys.* **57,** 965–994 (1985). (I)

H. Tawara and T. Kato, "Total and Partial Ionization Cross Sections of Atoms and Ions by Electron Impact," *At. Data Nucl. Data Tables* **36,** 167–353 (1987). (E)

U. Wille and R. Hippler, "Mechanisms of Inner-Shell Vacancy Production in Slow Ion-Atom Collisions," *Phys. Rep.* **132,** 129–260 (1986). (I)

Ionosphere

J. Weinstock

The ionosphere is an upper region of the atmosphere where gas particles are ionized, thereby forming a plasma that contains free electrons and positive ions. The presence of free electrons enables the ionosphere to reflect high-frequency waves and control radio communication around the earth. The study of this region, whose altitude extends from approximately 50 km to several hundred km, is called aeronomy.

The electrons and ions of the ionosphere are produced by shortwave solar radiation and cosmic particles that penetrate the upper atmosphere and ionize molecules and atoms. The dominant ionization process can be described by the photoionization formula

$$A + h\nu \rightarrow A^{+} + e^{-},$$

where A is an atmospheric molecule or atom, ν is the frequency of a photon of solar ultraviolet radiation, h is Planck's constant, A^{+} is the positive ion produced, and e^{-} is the free electron. With a knowledge of the atmospheric composition of the various particles A, the solar radiation flux, absorption cross sections, and ionization efficiencies, it is possible to compute the ion production rate q at each altitude. When q is balanced against electron loss processes, such as recombination, there results a steady-state value of electron number density n_e. Since the atmospheric composition varies with altitude, it follows that n_e varies with altitude.

Large values of n_e occur from roughly 50 km to several hundred km. This (ionospheric) range is divided into three principal regions, or layers: D, below about 90 km; E, between 90 and 150 km; and F, above 150 km. Each region has either a local maximum of n_e with altitude or a slow variation of n_e with altitude. The slow variation is referred to as a ledge. The F region has both an absolute maximum of n_e, the F_2 layer, and a bottom-side ledge, the F_1 layer. Typical maximum (noontime) values of n_e are: D region, 10^3; E region, 10^5; and F region, 10^6, in units of electrons/cm^3.

These values of n_e vary with the intensity of incident solar radiation and, consequently, vary with time of day and of year. There are also less predictable variations due to solar disturbances. Superimposed on the smooth variations of n_e are small-scale irregularities that are caused by atmospheric waves and plasma waves and are somewhat random. These have been studied intensively because they are difficult to predict.

The F region makes possible communication by high-frequency radio waves since it reflects (back down to earth) radio waves below a critical frequency. This critical frequency ω^*, the plasma frequency, is given by

$$\omega^* = (4\pi n_e^* e^2/m)^{1/2},$$

where n_e^* is the maximum value of n_e in the ionosphere, e is the electronic charge, and m is the electron mass. Radio waves above this frequency propagate through the ionosphere into outer space. Additionally, there is a lower bound to radio wave frequencies because of absorption in the D region. High-frequency radio communications are also influenced by the mentioned small-scale irregularities in n_e which can cause static. Very-short-scale irregularities can sometimes be utilized for communication in a mode called *scatter propagation,* by scattering VHF or UHF radio waves back down to earth. An advantage of VHF is that it is less susceptible to interference than ordinary radio waves. In the past two decades, it has become possible to modify (100-km) sections of the ionosphere with powerful (e.g., 2 MW) ground-based radio wave transmitters. The intense radio waves heat electrons and excite short-scale plasma turbulence. This turbulence has been used for scatter propagation to transmit photographs from Texas to California.

Most recently, the general circulation of the lower ionosphere (also referred to as the middle atmosphere) has been studied intensively. The circulation pertains to motions of neutral particles. The coupling of these neutral motions to charged particle motions and emissions is presently being investigated with renewed interest.

Aeronomy has advanced greatly in the past three decades by the use of rockets, satellites, and radar, in addition to classic radio reflection and chemiluminescent emission techniques. The ionosphere remains an active field of study.

See also Atmospheric Physics; Electromagnetic Radiation; Ionization; Photoionization; Plasmas.

BIBLIOGRAPHY

J. A. Ratcliffe, *Sun, Earth, and Radio.* Weidenfeld and Nicolson, London, 1970. (E)

H. Risbeth and O. K. Garriot, *Introduction to Ionospheric Physics.* Academic Press, New York, 1969. (I)

Ising Model

Fa Yueh Wu

It is an empirical fact that the magnetization of a permanent magnet diminishes in strength as the magnet is heated, disappearing completely above a certain temperature, called the Curie point. The Ising model is a simple physical model for explaining this phenomenon from a microscopic point of view. Historically, the model was first proposed by Lenz in 1920 and later studied by his student Ernst Ising. For this reason, it is also referred to in the literature as the Lenz–Ising model.

In the microscopic picture, the atoms of a magnetic substance are themselves tiny magnets. The existence of a spontaneous magnetization in the bulk is then explained as a result of an alignment of these tiny magnets, or "spins," due to their mutual interactions. To construct a simple model which simulates this situation, Lenz assumed that (1) the spins are arranged on a regular lattice, (2) each spin can point in only one of two directions, "up" or "down," and (3) there exists an interaction energy $-J$ between two neighboring spins which point in the same direction, and an energy J if they point in opposite directions. This describes the Ising model. Since a physical system attains its lowest energy at absolute zero temperature, for positive J the Ising model will be in a configuration with all spins aligned at absolute zero. It is then hoped that the spin alignment would persist, at least partially, up to some nonzero but finite temperature identified as the Curie point.

The mathematical formulation of the model is as follows: associate a two-valued variable σ_i to the ith spin such that $\sigma_i = +1$ (or -1) denotes an up (or down) spin. The interaction energy between two spins i and j can then be written as $-J\sigma_i\sigma_j$ and the total energy of the system is

$$E = -J \sum_{\langle i,j \rangle} \sigma_i\sigma_j,$$

where the summation is taken over all neighboring pairs $\langle i,j \rangle$. Then, according to the principles of statistical mechanics, the thermodynamics of the system can be derived from the partition function

$$Z = \sum_{\text{configurations}} e^{-E/kT}$$

where k is the Boltzmann constant and T is the absolute temperature. The summation is taken over all possible spin configurations. The mathematical problem is to find a closed-form expression for Z. Once this is done, the Curie point, if any, will manifest itself in this expression as a point of mathematical nonanalyticity in the variable T.

In his dissertation summarized in a 1925 paper, Ising solved this problem for a one-dimensional system (spins arranged on a chain) and found no Curie point. The existence of a spontaneous magnetization, which is related to the occurrence of a Curie point, in two- and three-dimensional systems was established by Peierls in 1936, and in 1944 Onsager published the exact solution of a two-dimensional model. For a square lattice of N spins, Onsager obtained the following expression for the partition function Z in the limit of infinite N:

$$\lim_{N\to\infty} \frac{1}{N} \ln Z = \ln 2 + \frac{1}{2\pi^2} \int_0^\pi d\theta \int_0^\pi d\phi \ln \left[\cosh^2 \left(\frac{2J}{kT}\right) - \sinh \left(\frac{2J}{kT}\right) (\cos\theta + \sin\phi) \right].$$

Analysis of this expression indicates that the two-dimensional Ising system possesses a unique Curie point T_C located at

$$\sinh(2J/kT_C) = 1$$

near which the specific heat diverges as $-\ln|1 - T/T_C|$. The spontaneous magnetization of this Ising model, whose derivation was first given by Yang in 1952, is

$$M = [1 - \sinh^{-4}(2J/kT)]^{1/8}, \qquad T \leq T_C$$
$$= 0, \qquad T \geq T_C.$$

While the problems of the three-dimensional Ising model and the two-dimensional model in a magnetic field remain unsolved to this date, their thermodynamic properties have been studied extensively by numerical analyses and computer simulations. The subject of the Ising model has found applications in many diverse areas of research including chemistry, biology, and materials science, and has over years developed into a major field of research.

See also FERROMAGNETISM.

BIBLIOGRAPHY

S. G. Brush, "History of the Lenz-Ising Model," *Rev. Mod. Phys.* **39,** 883 (1967). (E)

C. Domb, "On the Theory of Cooperative Phenomena in Crystals," *Adv. Phys.* **9,** 149 (1960). (I)

C. Domb, "Ising Model," in *Phase Transitions and Critical Phenomena, Vol III* (C. Domb and M. S. Green, eds.). Academic Press, New York, 1974. (I)

E. Ising, "Beitrag zur Theorie des Ferromagnetismus," *Z. Physik* **31,** 253 (1925). (I)

B. M. McCoy and T. T. Wu, *The Two-Dimensional Ising Model.* Harvard University Press, Cambridge, MA, 1973. (A)

L. Onsager, "Crystal Statistics, I. A Two-Dimensional Model with an Order–Disorder Transition," *Phys. Rev.* **65,** 117 (1944). (I)

C. N. Yang, "The Spontaneous Magnetization of a Two-Dimensional Ising Model," *Phys. Rev.* **85,** 809 (1952). (I)

Isobaric Analog States

Hanns L. Harney

The term "isobaric analog states" implies that we are discussing states of equal mass numbers having the same intrinsic configuration but different charges. The basic example is that the proton is the isobaric analog of the neutron.

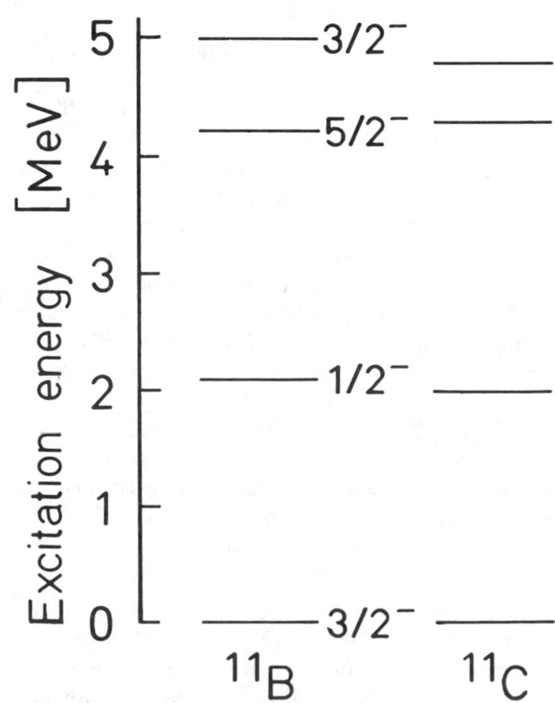

FIG. 1. Level diagrams for ^{11}B and ^{11}C.

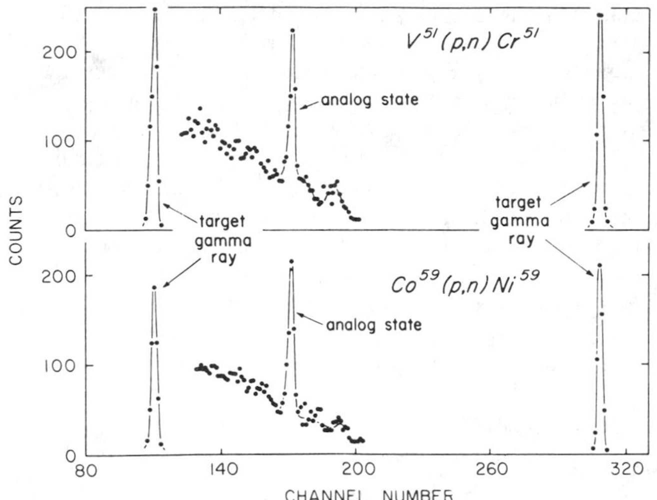

FIG. 2. Neutron spectra showing isobaric analog states. From Anderson *et al.*, 1962.

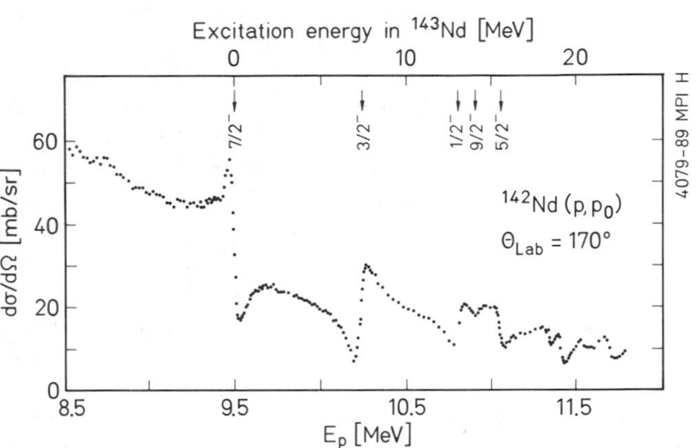

FIG. 3. Differential cross section for elastic proton scattering on ^{142}Nd showing resonances in the compound system ^{143}Pm. They are the isobaric analogs of low-lying states in ^{143}Nd as indicated by the arrows. From Grosse *et al.*, 1970.

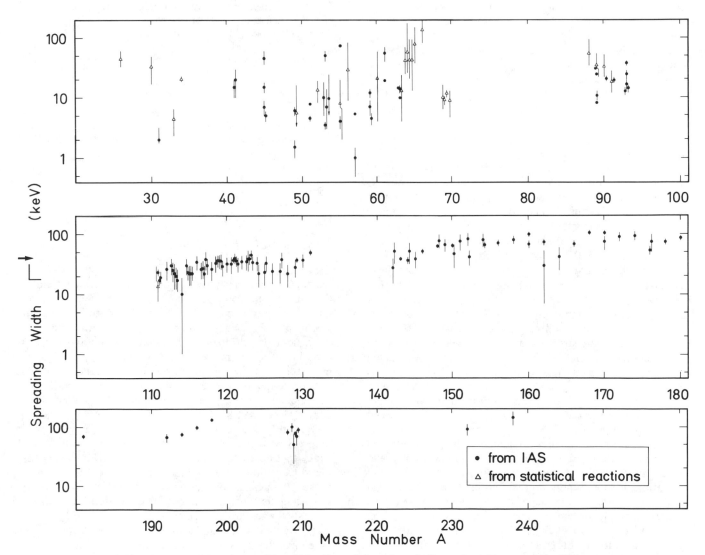

FIG. 4. Experimental spreading widths. Extended version of a figure from Harney *et al.*, 1985.

Protons and neutrons can be treated as the two orientations of the isobaric spin $\frac{1}{2}$ (in short isospin) of the nucleon. This concept, which is analogous to ordinary spin, has been widely used to classify the states of light nuclei since the pioneering work of Wigner in 1937 (see Wilkinson in *Isospin in Nuclear Physics,* 1969), but its full impact on nuclear science was only recognized with the discovery of isobaric analog states in heavy nuclei.

As an example from light nuclei, we show in Fig. 1 the energy levels of the mirror nuclei ${}^{11}_{5}B_6$ and ${}^{11}_{6}C_5$. We place their ground states on the same level, i.e., we disregard their Coulomb energy difference which is non-nuclear in origin. One then sees a one-to-one correspondence in quantum numbers and a close correspondence in the spacings between the states of these two nuclei. The states in ^{11}C and ^{11}B are isobaric analogs of each other. The small differences in the energy spacings are attributed to the nonconservation of isospin produced by the Coulomb force which breaks isospin since it affects only protons. On the other hand, the close correspondence between levels in light isobaric nuclei is not surprising because the Coulomb energy of the last proton is, in general, less than the nuclear binding energy and, hence, the nuclear properties which are charge independent dominate and preserve isospin symmetry. Conversely, it seemed likely that isospin symmetry would be destroyed in heavy nuclei, when the Coulomb energy significantly exceeded the nucleon binding energy.

In 1961, Anderson and Wong published experiments that led to the unexpected conclusion that isospin is a valid quantum number even in heavy nuclei. A study of the (p,n) reaction on target nuclei with masses between 48 and 93 revealed an intense neutron group in each spectrum as shown in Fig. 2. Conceptually this (p,n) reaction is very simple: the incident proton exchanges its charge with one of the (excess) neutrons of the target, leaving the nuclear configuration of the target unchanged. Thus the isobaric analog state of the target is formed and the resultant neutron is emitted with an energy loss (the Q value of the reaction) equal to the Coulomb energy, i.e., the energy required to deposit one unit of charge on the target nucleus. The surprise was that these isobaric analog states are so narrow. Let their isospin be $T_>$. They are located in a region of excitation energy (indeed above nuclear binding energy) where the density of states with isospin $T_< = T_> - 1$ is very large. Any isospin violating force spreads their strength out over the $T_<$ states within some energy range $\Gamma^{\downarrow}$. Hence, although an isobaric analog state may be strongly mixed into the local $T_<$ states, this mixing occurs only within an energy range $\Gamma^{\downarrow}$ remaining small and therefore the state stands out from the background of $T_<$ states. In this sense isospin is approximately conserved even in heavy nuclei.

The isobaric analog states populated in the last-mentioned experiments occur above the nucleon binding energy and especially their proton decay does not violate isospin conservation. It should therefore be possible to excite them via proton scattering, where they should show up as resonances in the compound system. This was indeed discovered by Fox and co-workers in 1964. Figure 3 shows an especially beautiful example from a different source. The resonances occur in the compound system ($^{142}Nd + p$)$\rightarrow$$^{143}Pm$ and are isobaric analogs of the ground and first excited states of ^{143}Nd. If one removes the Coulomb energy difference between ^{143}Pm and ^{143}Nd, similar to Fig. 1, the spectrum of resonances agrees (according to spins, parities, and energy spacings) with the spectrum of ^{143}Nd sketched at the top of Fig. 3.

The spreading of the isobaric analog states over the fine structure of nearby $T_<$ states has been impressively demonstrated by the high-resolution proton scattering experiments of Bilpuch *et al.* (1976). A compilation of spreading widths $\Gamma^{\downarrow}$ is given in Fig. 4. Up to the heaviest nuclei, $\Gamma^{\downarrow}$ does not exceed $\approx$100 keV. Here, the Coulomb force provides an energy difference of $\approx$19 MeV between isobaric analog states, but still only a very limited isospin breaking.

BIBLIOGRAPHY

F. D. Anderson and C. Wong, *Phys. Rev. Lett.* **7,** 250 (1961) (A); *Phys. Rev. Lett.* **8,** 442 (1962). (A)

E. G. Bilpuch, A. M. Lane, G. E. Mitchell, and J. D. Moses, *Phys. Rep.* **28C,** 145 (1976). (A)

Joseph Cerny (ed.), *Nuclear Spectroscopy and Reactions.* Academic Press, New York, 1974. See the articles by G. M. Temmer, Part B, p. 61 (E,I), P. von Brentano, John G. Cramer, Part B, p. 89 (E,I), and D. Robson, Part D, p. 179 (E,I).

W. R. Coker and C. F. Moore, *Phys. Today* **22** (No. 1), 53 (1969). (E)

J. D. Fox, C. F. Moore, and D. Robson, *Phys. Rev. Lett.* **12,** 198 (1964). (A)

E. Grosse, K. Melchior, H. Seitz, P. von Brentano, J. P. Wurm, and S. A. A. Zaidi, *Nucl. Phys. A* **142,** 345 (1970). (A)

H. L. Harney, A. Richter, and H. A. Weidenmüller, *Rev. Mod. Phys.* **58,** 607 (1986). (A)

D. H. Wilkinson (ed.), *Isospin in Nuclear Physics.* North-Holland, Amsterdam, 1969. (E,I,A)

Isomeric Nuclei

Gertrude Scharff-Goldhaber

The term isomers [from the Greek ισο (same) and μερος (part, share)] refers to two or more aggregates consisting of the same components, but in different arrangements. It was introduced into nuclear physics in analogy to its use in molecular physics. Whereas in molecules the difference in arrangement has to do with the structure and/or spatial distribution of the atoms within a molecule (e.g., two isomeric molecules may be mirror images of each other), the most noticeable distinction between two or more nuclear isomers [i.e., nuclei with the same atomic number (Z) and mass number (A)] is the half-life ($\tau_{1/2}$). One of the isomers, representing the ground state of a given nuclide (N,Z), may be stable, i.e., $\tau_{1/2} = \infty$. The other isomer or isomers are metastable states that decay to one or more lower-lying states by "isomeric transitions," emitting energy quanta in the form of γ rays, or internal-conversion electrons from the various atomic shells. The percentages of conversion electrons from each shell and subshell are accurately described by electromagnetic theory. The ratio of conversion electrons to γ rays is referred to as the internal-conversion coefficient α. Competition with the isomeric transition by β radiation, electron

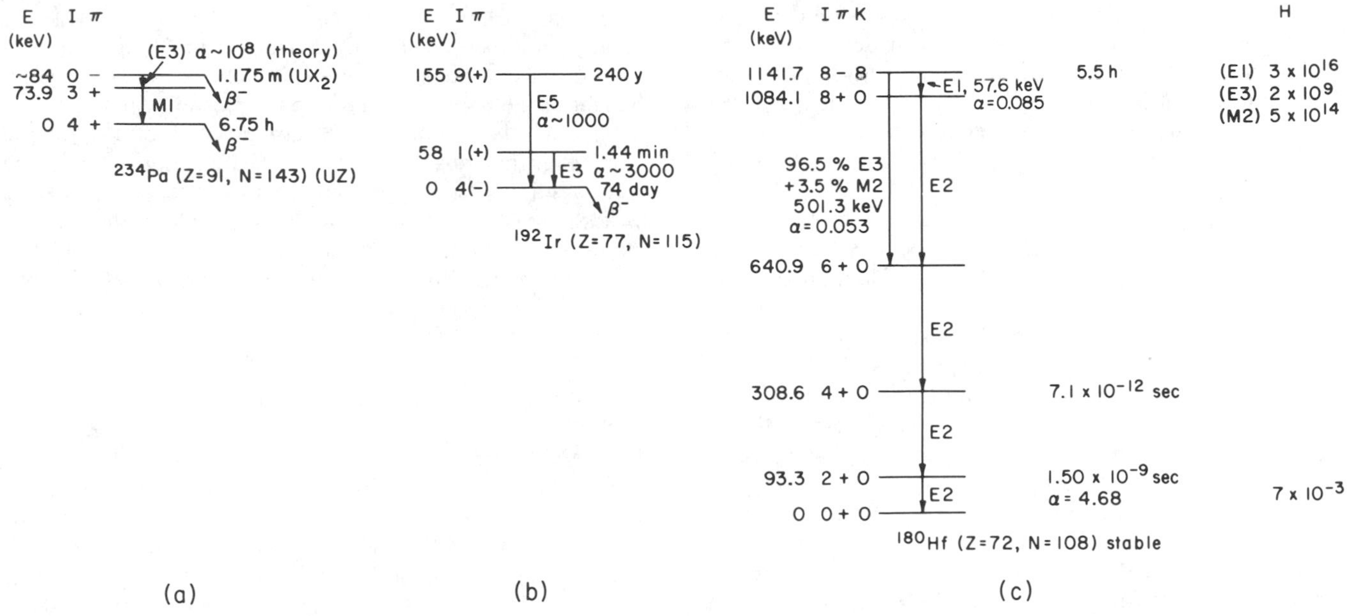

FIG. 1 Three examples of isomerism. At the left of each diagram displaying the level scheme of the three nuclides are shown the energy E in keV, the spin I, and the parity π. (Since for ^{192}Ir only the parity *changes* for the two transitions have been determined, the probable parities are given in parentheses.) The arrows representing isomeric transitions originating from metastable states are labeled with the corresponding multipole order ($E3$, etc.). At the right of each level scheme the half-lives of the states and the conversion coefficients α are stated.

(a) The first pair of isomers, discovered in nature in the decay of uranium (^{238}U), occurs in protactinium (^{234}Pa). The ~10-keV energy of the isomeric transition is based on indirect evidence. Theory predicts that $\alpha \sim 10^8$, i.e., γ rays are too weak to be observed.

(b) Triple isomerism in ^{192}Ir: The 74-day ground-state activity, which is efficiently produced by neutron capture in natural iridium, plays an important role in industry and medicine. The 240-year isomer is ~10^{11} times weaker than the ground state at the time of production.

(c) K isomerism in ^{180}Hf: The 5.5-h $E1$ transition of 57 keV is exceedingly retarded, as indicated by the hindrance factor H in the last column, whereas the $E2$ transitions within the rotational band, especially the 93.3-keV transition, are strongly enhanced. Since the 501-keV transition is less K-forbidden than the 57-keV transition, it can compete, in spite of the higher multipole orders.

capture, α radiation, etc., may occur. The half-life of an isomeric transition increases by a large factor for each unit of the difference in angular momentum of the isomeric state and the final state. Although in principle all excited states of a given nuclide might be considered isomers, in practice only states with "measurable" half-lives are given this name. An isomer whose half-life $\tau_{1/2} \gtrsim 1$ s is frequently referred to as a "long-lived" isomer. The first, and for a long time only, pair of isomers was found by Otto Hahn in 1921 among the decay products of ^{238}U in the element protactinium ($Z=91$). They are ^{234m}Pa (then named UX_2) with $\tau_{1/2}=1.17$ min, and ^{234}Pa (the ground state of the nuclide, then named UZ) with $\tau_{1/2}=6.75$ h. Although both isomers decay mainly by β emission into ^{234}U, ^{234m}Pa decays in 0.1% of the cases to the ground state via a "two-step isomeric transition." The energy quantum released in the first of the two transitions, which determines the half-life of the state, can so far only be estimated (~10 keV). It is followed by a 73.9-keV transition (Fig. 1, part a).

In the early 1950s the study of isomeric transitions contributed importantly to the establishment and detailed testing of the shell model of the atomic nucleus. While this model, conceived in close analogy to the atomic shell model, had been shown to describe the angular momenta (or spins) I of the nuclear *ground states* of odd-A nuclei, the study of the "multipole order" l of isomeric transitions as a function of N and Z proved that the shell model was valid also for a large number of *excited states*. Here $l \geq I_{\text{init}} - I_{\text{final}}$, where the I's are the spins of the initial and final states of the transition. In addition, it is of importance whether the wave functions of the initial and final states have the same parity (π), which may be even (+) or odd (−), or whether the parity changes. If $l=1$, and no parity change takes place, the transition is said to be a magnetic dipole ($M1$); if $l=1$ and the parity changes, it is an electric dipole ($E1$). The lowest magnetic and electric multipoles—the most common ones found—are shown in Table I. The multipole order of an isomeric transition can be determined in various ways, e.g., by deter-

Table 1

	Dipole		Quadrupole		Octupole		Hexadecapole		32-pole	
l	1		2		3		4		5	
$\Delta\pi$	no	yes	no	yes	no	yes	no	yes	no	yes
	$M1$	$E1$	$E2$	$M2$	$M3$	$E3$	$E4$	$M4$	$M5$	$E5$

mining α, or ratios of the numbers of conversion electrons from the subshells of an atomic shell, or by determining the angular correlation of the gamma rays with those of a transition following it in the decay. For the isomeric transition in the "odd–odd" nucleus ^{234}Pa referred to earlier, it is believed that $l=3$ and $\Delta\pi$ yes, i.e., that the transition is $E3$ (Fig. 1, part a).

An interesting long-lived triplet of isomers is known in ^{192}Ir ($Z=77$). Here the ground state, which decays mainly by β decay to ^{192}Pt ($Z=78$) with a 74.2-day half-life, has spin 4. The 58-keV first excited state, with $I=1$, lives 1.44 min. It decays to the ground state by an $E3$ transition, i.e., the parity changes. The second excited state, at 155 keV, with $I=9$, decays to the ground state by an $E5$ transition. This state has a half-life of approximately 240 yr (Fig. 1, part b). In nature, there exists an isomer, ^{180m}Ta, with an abundance of 0.012% and a half-life $>10^{17}$ yr.

The shell model, which is based on the assumption that the nucleus has a spherical shape, makes predictions for the half-life of an isomeric transition of a single proton, depending on the transition energy E (in MeV), the multipole order l, the absence or presence of parity change, and the mass number A: for electric transitions,

$$[\tau_e(1+\alpha)]^{-1} = B(El)E^{2l+1}A^{2l/3}$$

and for magnetic transitions,

$$[\tau_m(1+\alpha)]^{-1} = B(Ml)E^{2l+1}A^{(2l-2)/3}$$

where $B(El)$ and $B(Ml)$, the electric and magnetic reduced transition probabilities, are estimated to be as in Table II. Whereas these predictions were found to be in amazingly good agreement with observed $M4$ transition half-lives (within a factor of 2), considerable discrepancies were found for most other multipoles. Some of these have been attributed to the presence of large numbers of nucleons outside closed shells, which bring about a change of the nuclear shape from spherical to spheroidal [usually prolate (like a football), sometimes also oblate (like a pancake)]. The difference in the length of the axis of symmetry from that of the two other axes of the spheroid is expressed by the deformation parameter (β). The following deviations from the foregoing shell-model predictions for lifetimes of isomeric transitions are noteworthy.

(a) The $E2$ transitions taking place within a rotational band, especially that built on the ground state, are strongly enhanced. For the largest values of β, their half-lives are about 200 times shorter than the individual-particle shell model predicts. An example of an enhanced $E2$ transition to the ground state of the even–even nucleus ^{180}Hf is given in Fig. 1, part c.

Table II

Multipolarity	$B(El)$	Multipolarity	$B(Ml)$
$E1$	1.0×10^{14}	$M1$	2.9×10^{13}
$E2$	7.4×10^{7}	$M2$	8.4×10^{7}
$E3$	3.4×10^{1}	$M3$	8.7×10^{1}
$E4$	1.1×10^{-5}	$M4$	4.8×10^{-5}
$E5$	2.5×10^{-12}	$M5$	1.7×10^{-11}

(b) Many transitions, in particular $E1$, $M1$, and $M2$, are greatly retarded because of "K isomerism." This name refers to the quantum number K, which measures the component of the nuclear spin along the axis of symmetry. For a rotational band built on the ground state of a deformed even–even nucleus, $K^{\pi}=0^{+}$. If the transition takes place from a state with a high spin I whose direction coincides with the axis of symmetry, $K=I$, to a state in the ground-state band, the retardation (compared with shell-model predictions), measured by the *hindrance factor* H, follows (within very wide limits) the rule

$$\log H \approx 2(\Delta K - l).$$

Let us take as an example the 5.5-h isomer of ^{180}Hf (Fig. 1, part c) with $K=I=8^{-}$, which decays to the state $K=0$, 8^{+} by a 57.5-keV $E1$ transition. Hence, according to the rule, $\log H \approx 2(8-1)=14$; i.e., $\tau_{1/2}$ is expected to be $\sim10^{14}$ times longer than the individual-particle model predicts. As seen from Fig. 2, which displays the observed distribution of $\log H$ versus $\Delta K - l$, the experimental value of H for this isomer (labeled 180) is even somewhat larger, namely, 3.1×10^{16}.

(c) A second type of retarding effect may be due to a shape change of the nucleus, e.g., from oblate to prolate, or to a change representing a jump in deformation to almost twice

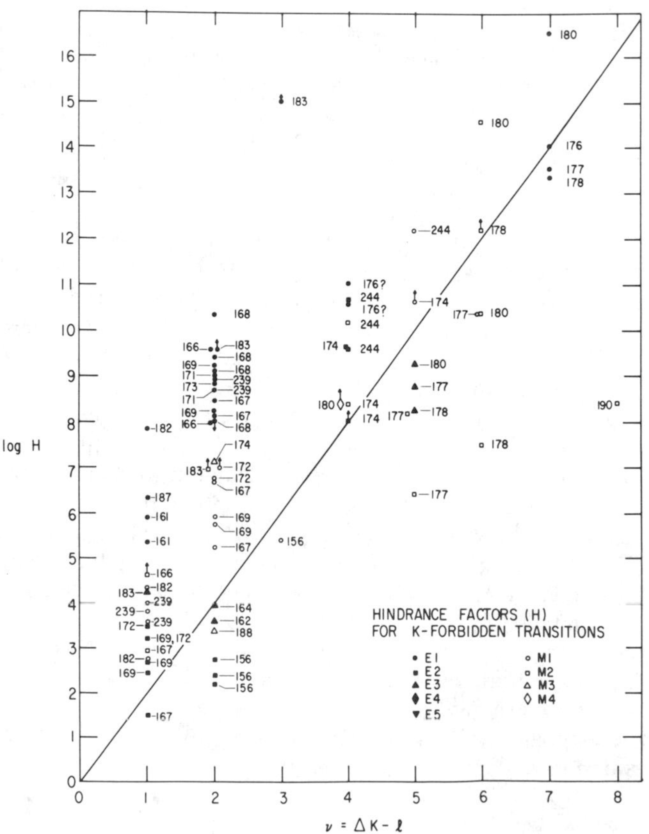

FIG. 2 Log$H$ versus $\nu=\Delta K - l$, where H is the hindrance factor and ν denotes the degree of K forbiddenness. The symbols indicate the multipole order of the transition in a K isomer characterized by its mass number. The points with arrows show lower limits for log$H$. (Prepared by G. T. Emery and G. Scharff-Goldhaber.)

that of the ground state, as has been observed in the so-called fission isomers occurring in the heaviest elements known. A fission isomer is a metastable state of a heavy nucleus that decays largely by fission. Little is known about the isomeric transition lifetimes from these states.

See also GAMMA DECAY; NUCLEAR PROPERTIES; NUCLEAR STRUCTURE.

BIBLIOGRAPHY

D. E. Alburger, *Encyclopaedic Dictionary of Physics*, Vol. 4, pp. 101–102. Pergamon, New York, 1961.

Y. Y. Chu and G. Scharff-Goldhaber, "Decay of ^{234}Th to the ^{234}Pa Isomers," *Phys. Rev.* **C17**, 1507 (1978).

J. Godart and A. Gizon, "Niveaux de ^{234}Pa atteints par la désintegration de ^{234}Th," *Nucl. Phys.* **A217**, 159 (1973).

M. Goldhaber and R. D. Hill, "Nuclear Isomerism and Shell Structure," *Rev. Mod. Phys.* **24**, 179–239 (1952).

M. Goldhaber and A. W. Sunyar, "Classification of Nuclear Transition Rates," in *Alpha-, Beta- and Gamma-Ray Spectroscopy* (K. Siegbahn, ed.), pp. 931–949. North-Holland, Amsterdam, 1965.

S. A. Moszkowski, "Theory of Multipole Radiation," in *Alpha-, Beta- and Gamma-Ray Spectroscopy*, (K. Siegbahn, ed.), pp. 863–886. North-Holland, Amsterdam, 1965.

H. C. Pauli, K. Alder, and R. M. Steffen, "The Theory of Internal Conversion," *The Electromagnetic Interaction in Nuclear Spectroscopy*, pp. 341–440. North-Holland, Amsterdam, 1975.

G. Scharff-Goldhaber, "Multipole Order and Enhancement Factor of Long-Lived Isomeric Transition in ^{192}Ir.," *Bull. Am. Phys. Soc.* **22**, 545 (1977).

G. Scharff-Goldhaber, C. B. Dover and A. L. Goodman, "The Variable Moment of Inertia Model and Nuclear Collective Motion," *Ann. Rev. Nucl. Sci.* **26**, Section 3 (1976).

G. Scharff-Goldhaber and M. McKeown, "Triple Isomerism in ^{192}Ir," *Phys. Rev. Lett.* **3**, 47 (1959).

R. Vandenbosch and J. R. Huizenga, *Nuclear Fission*, pp. 59–76. Academic, New York, 1973.

Isospin

Sir Denys Wilkinson

Isospin, also called isotopic or isobaric spin, is a concept that arises only for systems of particles, hadrons, that interact with each other through the nuclear force. The most familiar of such hadronic systems are the atomic nuclei, comprised of neutrons (n) and protons (p)—together called nucleons (N)—but hadrons include the mesons whose exchange between the nucleons generates the N-N force and also include the excited isobaric states or resonances into which nucleons can be raised under appropriate bombardment.

The essence of the matter is that hadrons exist in families, of varying numbers of members, called multiplets, all members having identical quantum numbers (spin, parity, charge conjugation, G-parity, etc., as appropriate) and closely similar masses; the various members are distinguished essentially only by their electric charges. Thus the neutron and proton form a family of two and have charges of 0, +1 but differ in mass by only 0.14%; the lightest of the mesons, the pion (π), forms a family of three members of charge −1, 0, +1 whose masses span a range of only 3.3%; the lightest of the nucleon resonances, the Δ, forms a family of four of charge −1, 0, +1, +2 whose masses differ over all by only about 1%, and so on. These mass splittings, of order 1%, within the multiplets, separating the members of different electric charge, are of the same order as the factor by which the distinguishing electrical force is weaker than the nuclear force that gives the family its definition in terms of its common quantum numbers. This suggests that from the point of view of the dominating nuclear force that defines the particles the multiplet members are identical each to the others and that the charge acts merely as a kind of label to distinguish the members; since electrical charge carries electrostatic energy and energy by $E=mc^2$ implies mass, the different charge states therefore also have slightly different masses.

This idea of identity from the viewpoint of the nuclear force with distinction only by the electric force is strongly reinforced by the observation that the interaction between the members of a multiplet (called an isobaric or charge multiplet because of the near identity of the masses of the various charge states that we have noted) does not depend on the charges of the multiplet members involved after allowance for the direct effect of the electric charges, if any, upon each other and for other electromagnetic interactions such as those associated with the magnetic moments, if any. In other words, the nuclear force that defines the particles does not depend on the charge state of the members. Thus the np force between a neutron and a proton in the $^1S_0$ state is the same to within about 2% as that, pp, between two protons in the same spectroscopic state after subtraction of the electromagnetic effects, while the similar force between two neutrons and between two protons is the same to within 1%: $nn \simeq np \simeq pp$. We can even understand the fact that the np strong force is slightly different (2%) from the pp strong force: The strong force is generated by the exchange of mesons between the nucleons; the allowed charges for the exchanged mesons depend on the charges of the nucleons that are interacting—thus two protons can exchange only neutral mesons if they are to remain nucleons because the exchange of a meson of charge ±1 would involve one of the nucleons becoming of charge +2 which does not exist for the nucleon, but a neutron and a proton can exchange either neutral mesons or mesons of charge ±1, the latter simply effecting an interchange of the charge label between the nucleons; but since the charged and neutral mesons themselves have slightly different masses the forces that they engender by their exchange will also be slightly different. It is quite possible that if we could correct for this slight charge difference *within* the meson multiplets, itself presumably due to the electric force, then the np and pp strong forces would be found to be identical.

This indifference of the nuclear force to the charge states is called the principle of charge independence. It extends to forces *between* isobaric multiplets as well as *within* multiplets; thus in the interaction between pions and nucleons we have only one strength, the πN force, which is experimentally practically the same for all charge combinations: $\pi^+p \simeq \pi^+n \simeq \pi^0p \simeq \pi^0n \simeq \pi^-p \simeq \pi^-n$.

This idea of charge independence of the nuclear force, the charge merely acting as a label, leads us to ask how we might most conveniently define that label in mathematical terms. Since the multiplet families are of different sizes, two members for N, three for π, four for Δ, etc., and some with only a single member such as the η and ω mesons, we naturally look for a labeling that will not only tell us the charge state but also how many members there are in the family. A ready-made analogy is to hand in angular momentum. A particle of intrinsic spin s (e.g., $s=0$ for an α particle, $s=\frac{1}{2}$ for an electron, $s=1$ for a deuteron, $s=\frac{3}{2}$ for the Δ, etc.) is allowed by the ordinary rules of quantum mechanics to assume $2s+1$ alternative orientations in space distinguished by the projections s_z of the angular momentum along whatever axis, z, we choose for our consideration. The particle is identical in its properties in each of its $2s+1$ possible states but the states are distinguishable physically. Thus a proton with its spin "up," $s_z=+\frac{1}{2}$, is no different intrinsically when it is turned upside down, $s_z=-\frac{1}{2}$, but we can tell the difference by an appropriate test such as passing it through an inhomogeneous magnetic field which will bend the trajectory up or down depending on s_z. So we invent an analogous quantity, t, for the hadrons and call it isospin by analogy with the ordinary spin s. We imagine it to exist in its own, totally fictitious, isospace within which it can take up its $2t+1$ alternative orientations each of which, t_z, labels the charge state of the particle. t tells us the number, $2t+1$, of members in the family and t_z tells us the charge of the individual member. Thus for the nucleon, with two members to the family, $t=\frac{1}{2}$; conventionally in nuclear physics, $t_z=+\frac{1}{2}$ represents the neutron and $t_z=-\frac{1}{2}$ represents the proton (to emphasize that it is only a convention, high-energy physicists have it the other way around). Because of charge independence a nucleon with its isospin "up," a neutron, is not essentially changed when it is turned upside down in isospace, then merely changing its charge label and becoming a proton distinguished not by any "nuclear" property but only by its charge which totally changes its chemistry but does not affect its physics. So for the other families so far mentioned: η and ω have $t=0$; π has $t=1$; Δ has $t=\frac{3}{2}$.

But now comes the critical point. Ordinary angular momentum is a conserved quantity: In the absence of external torques the overall angular momentum of a mechanical system remains constant no matter how violent the forces at play *within* that system. This principle is congruent with the remark that the mechanical properties of the system are independent of its orientation in space: a proton with its spin up, $s_z=+\frac{1}{2}$, is identical to a proton with its spin down, $s_z=-\frac{1}{2}$, so that an *ensemble* of protons of total spin $\mathbf{S}=\Sigma\mathbf{s}$ may exist in $2S+1$ different orientations $S_z=\Sigma s_z$ in space, with identical properties, no matter how violent the interactions between the protons that may result in a constant turning of each other upside down, i.e., a constant interchange of their individual s_z values. S is then a constant of the motion, a conserved quantity, a "good quantum number." (This illustration assumes no orbital angular momentum to be present and so ignores the Pauli principle, but this is irrelevant to our illustration.) Consider now an *ensemble* of, say, nucleons. Because of charge independence a nucleon in its neutron state, $t_z=+\frac{1}{2}$, is not changed in its interactions when in its proton state, $t_z=-\frac{1}{2}$, so the *ensemble* with $\mathbf{T}=\Sigma\mathbf{t}$ may exist in $2T+1$ different orientations $T_z=\Sigma t_z$ in isospace with identical properties no matter how violent the interactions that turn the nucleons upside down in isospace, viz., that interconvert neutrons and protons, and T, as S, will be a good quantum number. So for N neutrons and Z protons we have $T_z=\frac{1}{2}(N-Z)$, but this must belong to a multiplet of $2T+1$ nuclear states, identical to each other in physical properties but being found in $2T+1$ different chemical elements, the only restriction being $\frac{1}{2}(N+Z)\geq T\geq|T_z|$. The actual value of T for a given multiplet is determined by the Pauli principle, viz., the antisymmetrization of the overall wave function in the three variables of space, spin, and isospin, and, crudely speaking, is a measure of the number of successive conversions of, say, neutrons into protons that can be made before two neutrons or two protons are placed into identical quantum states. We might offhand expect the most symmetrical systems, of lowest T, to lie lowest in energy, and although this is usually the case it does not necessarily have to be so and sometimes is not. So any nuclear state must be accompanied by $2T$ other nuclear states of identical physical properties in $2T$ other chemical elements. Illustrations of this are given in the energy level diagrams of Figs. 1 and 2 for both odd and even mass systems. Note particularly that all members of an isospin multiplet must have the same ordinary spin J and parity π labeled J^π.

With perfect overall charge independence the masses of the members of an isomultiplet would be identical. But we do not have perfect charge independence because of the Coulomb force, the effect of which differs for differing members of the multiplet. This has been subtracted out in the figures by drawing the members of the lowest multiplet all at the same level so that the figures effectively show relative excitation energies within the different chemical elements.

The figures demonstrate the utility of isospin as a label for nuclear states. The other aspect of isospin, namely, its conservation, is also of great importance. Nothing in our argument says that the N neutrons and Z protons have to con-

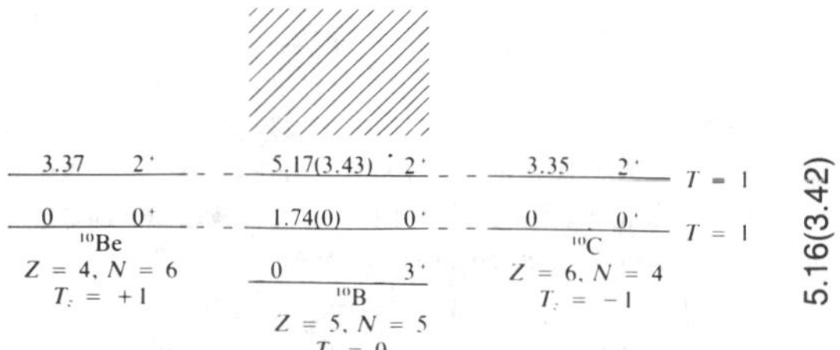

FIG. 1. Some low-lying levels of $A=10$. The numbers on the left of the lines are the excitation energies in MeV; on the right of the lines are the J^π values. In brackets for ^{10}B are the excitation energies relative to the lowest $T=1$ state. The unlabeled states of ^{10}B (and the ground state) are of $T=0$. $T=0$ states are dense in the hatched region (the next $T=1$ states lie about 2 MeV higher up). The lowest $T=1$ states have been drawn at the same level to display the symmetry; in practice, because of the extra Coulomb energy in the more highly charged nuclei, the ground state of ^{10}C is some 1.4 MeV heavier than its companion state at 1.74 MeV in ^{10}B, which in turn is some 0.7 MeV heavier than the ground state of ^{10}Be.

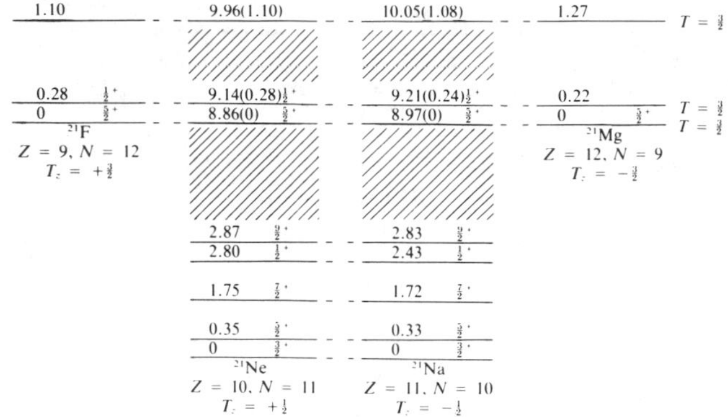

FIG. 2. Energy levels of $A=21$ with labeling conventions as for $A=10$. The states without a T label, dense within the hatched regions, are of $T=\frac{1}{2}$. Where J^{π} is not shown it is unsure. (Note the energy gap, not to scale, between the $T=\frac{1}{2}$ states at about 2.8 MeV and the onset of the $T=\frac{3}{2}$ states; it contains at least 40 $T=\frac{1}{2}$ states and probably many more.)

stitute or continue to constitute a single nucleus. Thus if two nuclei A and B engage in a nuclear reaction: $A+B \rightarrow C+D$, then, because of charge independence and exactly as with ordinary angular momentum, the overall isospin T must be conserved: $\mathbf{T}=\mathbf{T}_A+\mathbf{T}_B=\mathbf{T}_C+\mathbf{T}_D$. As an illustration consider the reaction $^{12}C+d \rightarrow ^{10}B+\alpha$. The target ^{12}C which is in its ground state has $T=0$ as also do the incident deuteron and the product α particle. So the left-hand side of the equation has $T=0$ and so must the right-hand side; since $T_{\alpha}=0$ we can only form those states of ^{10}B that have $T=0$. By reference to Fig. 1 we should therefore expect to be able to form ^{10}B in, for example, its ground state but not in those at excitation energies of 1.74 and 5.16 MeV: It is indeed found that production of these latter states is very considerably suppressed.

Similarly in the decay of excited nuclear states. Consider the lowest $T=\frac{3}{2}$ state of ^{13}N at 15.06 MeV. Energetically this can emit a proton (of isospin $\frac{1}{2}$) leaving the ground state of ^{12}C which, however, has $T=0$ so that this is forbidden and indeed this emission takes place about 1000 times more slowly than we might otherwise have expected.

The fact that the isospin selection rules, which also exist in a somewhat subtler form for gamma-ray emission and β decay, are well, but not perfectly, obeyed reflects the role of the Coulomb force, which prevents *de facto* overall charge independence from being perfect; the energy of the T multiplet *does* depend slightly on its T_z value, its orientation in isospace, because, for example, the Coulomb energy of the ground state of ^{10}C is bigger than that of the 1.74-MeV state of ^{10}B which is bigger than that of the ground state of ^{10}Be (see Fig. 1 where these differences have been subtracted out). T, unlike ordinary angular momentum, cannot therefore be a perfect quantum number but it is good enough to be of very great utility not only in nuclear physics, from which most of the present illustrations have been drawn, but also for high-energy or particle physics.

This discussion has been conducted in terms of the analogy between isospin and the rotational invariance of ordinary space. In an alternative symmetry language the multiplets that have been discussed here are $SU(2)$ multiplets: The algebras of three-dimensional rotation and $SU(2)$ symmetry are identical so the discussions are completely equivalent.

See also ELECTROMAGNETIC INTERACTION; ELEMENTARY PARTICLES IN PHYSICS; HADRONS; NUCLEAR PROPERTIES; NUCLEAR REACTIONS; NUCLEAR STATES; ROTATION AND ANGULAR MOMENTUM; STRONG INTERACTIONS.

Isotope Effects

W. Alexander Van Hook

The term isotope effect (IE) denotes a difference in some molecular or atomic property consequent to a change of mass or mass distribution caused by isotopic substitution. Radioactive and other specific nuclear effects are excluded. IEs are of theoretical interest because within the framework of the Born–Oppenheimer approximation the electronic properties of a system of atoms or molecules can be separated from those involving nuclear motion. Solution of the electronic part of the problem determines an electronic potential surface on which nuclear motion occurs. The electronic structure determines most of the chemistry and much of the physics of the molecule or process under description. Investigation of isotopic substitution amounts to a study of how mass and/or mass distribution affects motion on the (isotope-independent) electronic potential surface. Alternatively, experimental data on IEs can be regarded as a tool to gain information about the potential-energy surface. We note that isotope effects are expected to be small; some representative values are given in Table I. IEs on equilibrium constants or vapor pressure amount to several percent or more for H/D substitution (around room temperature) but are smaller for more massive isotopic isomers ($^{12}C/^{13}C$, for example). IEs on the rate constants of chemical reactions are typically larger. Isotope effects on a given property are normally much much smaller than the corresponding differences between chemically distinct systems. Part of the practical interest in isotope effects stems from the World War II and postwar efforts at preparing sizable quantities of separated isotopes.

A convenient starting point for a theoretical understanding of isotope effects is to consider isolated molecules in the gas phase. Nuclear motion can be described in terms of small displacements from the equilibrium nuclear positions. The potential energy is approximated with an expansion in terms of displacement coordinates where the leading (quadratic) term is the largest, but cubic and higher-order terms can also be important. Even so, the cubic and higher terms are often neglected and this leads to a description of the molecule in terms of a set of $3n-j$ harmonic oscillator frequencies ($j=5$ for a linear, 6 for a nonlinear, or 3 for a monatomic molecule). These frequencies are often described as the set of solutions which follow from diagonalization of a matrix, $\bar{F}\bar{G}$, where $\bar{F}$ is a $(3n-j)$-dimensional potential-energy (force constant) matrix and $\bar{G}$ a kinetic-energy matrix which takes due ac-

Table I Some representative isotope effects

System	Process	Temperature (K)	Light/Heavy
$H_2+2DI \leftrightarrows D_2+2HI$	Equilibrium (gas)	600	$1.21 = \bar{K}_H/\bar{K}_D$
$^{15}NH_3+{}^{14}NH_4{}^+ \leftrightarrows {}^{14}NH_3+{}^{15}NH_4{}^+$	Equilibrium (solution)	298	$1.034=\bar{K}_{14}/\bar{K}_{15}$
$^{10}BF_3(g)+\text{donor}\cdot{}^{11}BF_3(l) \leftrightarrows {}^{11}BF^3(g) + \text{donor}/{}^{10}BF_3(l)$ [donor $=(CH_3)_2O$]	Equilibrium	303	$1.026=K_{10}/K_{11}$
$^{36}Ar-{}^{40}Ar$	Vapor pressure	87	$1.006=P_{36}/P_{40}$
H_2O-D_2O	Vapor pressure	373	$1.054=P_{H_2O}/P_{D_2O}$
$C_6D_6-C_6H_6$	Vapor pressure	353	$0.976=P_H/P_D$
$H_2+Cl\cdot \rightarrow HCl+H\cdot$ $D_2+Cl\cdot \rightarrow DCl+H\cdot$	Reaction rate	298	$9.6 = K_H/K_D$
$(CH_3)_3CCl \rightarrow (CH_3)_3C^+ + Cl^-$ $(CD_3)_3CCl \rightarrow (CD_3)_3C^+ + Cl^-$	Reaction rate	298	$2.4 = K_H/K_D$

count of mass and mass distribution. From the paragraph above we recognize $\bar{F}$ as isotope independent whereas $\bar{G}$ reflects the known mass differences. Given $\bar{F}$ and $\bar{G}$ for a set of isotopic molecules one can readily calculate a complete set of frequencies and isotopic frequency shifts which can then be employed to evaluate partition functions, partition function ratios, and, therefore, thermodynamic properties via the methods of a statistical thermodynamics. In the calculation of the thermodynamic properties it is necessary to properly account for the j zero-frequency rotations and translations, which are usefully described in the classical limit in the gas, but which take on nonzero values in the condensed phase and are sometimes treated there in an harmonic approximation. It is to be noted that the isotopic frequency shifts are small, at least for heavier isotopes, and it therefore becomes convenient to formulate the isotopic partition function ratios (which are near unity) in powers of $1/T$. In this connection it has been long established that IEs vanish in the classical limit, and Jacob Bigeleisen and Maria G. Mayer in 1947 pointed out the convenience of defining a reduced partition function ratio: q/q(classical). Their approach and notation has since been widely employed.

The methodology outlined above has been applied to a wide variety of special cases including gas-phase chemical equilibrium constants (Table I), isotope effects on rate constants in chemical reactions, vapor pressure, molar volume, and other condensed phase isotope effects, etc. In suitable cases, corrections for such effects as anharmonicity, rotation–vibration interaction, centrifugal distortion, the effect of anharmonicity on the zero-point energy, and corrections for deviations from the Born–Oppenheimer approximation have been made. Most analyses to date have considered systems in thermal equilibrium so that Maxwell–Boltzmann distribution functions are appropriate for use in the calculations. Isotope effects for systems not in thermal equilibrium are also of considerable interest although not discussed in this article. Non-Maxwellian distributions are important in the treatment of laser methods of isotope separation, IEs on fast unimolecular reactions, etc.

The theory outlined above for equilibrium readily lends itself to the interpretation of data on isotope effects on rate constants of chemical reactions. The analysis is based on the absolute rate (activated complex) approach of Eyring and others. The theoretical construct readily allows assumptions about the nature of the transition state and reaction path to be tested in a reasonably straightforward way. This area has been treated in detail with extensive computer model calculations. It has been demonstrated that isotope effects essentially probe force-constant changes at the position of isotopic substitution, transition state minus reactant. The effects are normal (light faster than heavy) if force constants in the "transition state" (read "product" if isotope effects on equilibria are being considered) are smaller than in the reactant, and inverse in the opposite case. Unusual temperature dependences can occur and corrections for quantum-mechanical tunneling can be important.

The IE on equilibrium between condensed and vapor-phase species as expressed by the vapor–pressure isotope effect (VPIE) has been measured and interpreted for a wide variety of compounds. The VPIE is directly related to the mean square force or Laplacian of the intermolecular potential. The available data on argon, for example, are in excellent agreement with recent sophisticated calculations. For polyatomic molecules both normal and inverse IEs are observed together with complicated temperature dependences. The inverse effects (commonly seen for protio-deuterocarbons) are a consequence of the net red shift in internal frequencies, gas to liquid, due to the influence of the intermolecular van der Waals forces. In many cases detailed correlation with spectroscopic information has been possible. Differences in vapor pressure between equivalent isomers such as *ortho, meta,* and *para* dideuterobenzene or *cis, trans,* and *gem* dideuteroethylene have allowed the importance of rotation–vibration interaction in the liquid phase to be assessed. Studies on solutions of isotopic isomers (C_6H_6 in C_6D_6, H_2 in D_2, deutero/protio polystyrene, etc.) have been made demonstrating significant nonideality. Such effects owe their origin to isotopic differences in vibrational amplitudes (in turn described in terms of an isotope-independent intramolecular potential function). In cases where the temperature can be forced low enough, i.e., solutions of He^3/He^4 or H_2/D_2, or the cumulative effect per substituted atom is large enough, i.e., mixtures of H/D isomers of certain polymers, the excess free energy from this effect can be large enough to cause phase separation.

In addition to the effects described above, IEs on density,

molar volume, transition temperatures, superconductivity, crystal structure, and gas-phase virial coefficients have been investigated. IEs on molecular properties like polarizability, dipole moments, chemical shifts in magnetic resonance, etc., are understandable in terms of amplitude differences. These arise from the fact that the mean square displacement of any internal coordinate (even in the harmonic approximation) is nonzero and is temperature and mass dependent. The effects are important for those molecular properties which must be averaged over vibrational motion.

See also Kinetics, Chemical; Molecular Structure Calculations.

BIBLIOGRAPHY

J. Bigeleisen, *Science* **147,** 463 (1965); J. Bigeleisen, M. W. Lee, and F. Mandel, *Annu. Rev. Phys. Chem.* **24,** 407 (1973); *Acc. Chem. Res.* **8,** 179 (1975).

E. Buncel and C. C. Lee (eds.) *Isotopes in Organic Chemistry.* Elsevier, Amsterdam, Vols 1 (1975)–7 (1987).

C. J. Collins and N. S. Bowman (eds.), *Isotope Effects in Chemical Reactions.* Van Nostrand–Reinhold, New York, 1970.

T. Ishida, W. A. Van Hook, and M. Wolfsberg (eds.), "Festschrift in honor of J. Bigeleisen." *Z. Naturforschung A,* **44** (1989).

G. Jancso and W. A. Van Hook, *Chem. Rev.* **74,** 689 (1974): W. A. Van Hook, *J. Chem. Phys.* **83,** 4107 (1985): R. R. Singh and W. A. Van Hook, *J. Chem. Phys.* **88,** 2969 (1987); *J. Chem. Phys.* **87,** 6088, 6097 (1987); *Macromolecules* **20,** 1855 (1987).

L. Melander and W. H. Saunders, *Reaction Rates of Isotopic Molecules.* John Wiley and Sons, New York, 1980.

I. B. Rabinovitch, *Effect of Isotopy on Physico Chemical Properties of Liquids.* Consultants Bureau, New York, 1970.

M. Wolfsberg, *Annu. Rev. Phys. Chem.* **20,** 449 (1969); *Acc. Chem. Res.* **5,** 225 (1972).

Isotope Separation

William Spindel and Takanobu Ishida

Most elements exist in nature as a mixture of isotopes differing in mass as a result of differences in the number of neutrons in their atomic nuclei. Isotopes of an element have very similar physical and chemical properties that make their separation difficult, but some possess distinctly different nuclear properties that make those isotopes particularly useful.

Enrichment plants for isotopes such as uranium-235 (^{235}U) and deuterium are of industrial scale yielding, e.g., hundreds of tons per year; others yield product at rates ranging only from grams per day to grams per year [1]. However, the scale of production does not necessarily reflect the relative importance of the separated isotopes. Uranium-235 and deuterium are essential materials for the nuclear power industry. Deuterium oxide (heavy water) is a coolant and moderator in heavy-water reactors, which use natural abundance uranium rather than enriched ^{235}U as fuel. Deuterium is also an essential ingredient of fuels for all types of nuclear fusion schemes currently under development. Lithium-6 may be a useful fuel in fusion reactors and is also a starting material for producing tritium. Boron-10 is used in control rods for fission reactors.

Carbon-13, nitrogen-15, and oxygen-17 are produced in plants whose typical capacities are grams per day. They are powerful tracers for various fields of research in the physical and, particularly, the life sciences because they are nonradioactive and thus physiologically harmless. They occur in nature at low abundances. For this reason relatively low enrichments yield significant signals, and their nuclear magnetic moments make it possible to use them to determine molecular environments surrounding the atoms by means of nuclear magnetic resonance spectroscopy.

Critical needs exist for enriched isotopes of virtually every element in the periodic table, although the required quantities of some may only be on the order of grams per year [2]. Some highly enriched stable isotopes are vital to fundamental research in nuclear physics and chemistry, solid state physics, geoscience, and biology and medicine. Many separated stable isotopes are used as nuclear targets in research reactors or particle accelerators to produce particular radioisotopes, which are then used as radiotracers or radiopharmaceuticals. High isotopic enrichment of the target is the key to minimizing interference in measurements and possible physiological side effects because of the presence of other radioisotopes, which may result from isotopic impurities in the target.

A variety of methods has been developed especially adapted to separating isotopes. Some methods, such as gaseous diffusion, gas centrifuge, separation nozzle, and electromagnetic separation, are obviously based directly on mass differences; others, such as thermal diffusion, chemical exchange, chromatographic and ion exchange, distillation, and photochemical methods, result from less obvious mass-dependent changes in atomic and molecular properties.

GENERAL COMMENTS

Although the common principles underlying isotope separation are relatively few, a large variety of processes has been considered. Examination of processes for separating deuterium turned up 98 potential candidates, and one study group evaluated at least 25 known processes other than gaseous diffusion and gas centrifuges [3] for separating uranium isotopes. All individual processes have some merit for the separation of isotopes, but evaluating them depends on several factors: whether isotopes of light-weight (e.g., deuterium), intermediate-weight (e.g., nitrogen), or heavy-weight elements (e.g., uranium) are to be separated; and whether the quantities needed are grams (as in research) or tons (as for use in power reactors). The choice depends on the properties of the element, the degree of separation needed, and the scale and continuity of the demand; even for a given isotope, there is no one best method. For large-scale applications availability of feed materials, capital costs, and power requirements may be the overriding considerations, while for laboratory needs simplicity of operation or versatility may be primary.

More than 30 years ago Manson Benedict and Thomas Pigford (in the first edition of [1]) drew general conclusions that are still valid regarding isotope separation methods:

1. the most versatile means for the production of small quantities of isotopes is the electromagnetic method;

2. the simplest and most inexpensive means for small-scale separation of many isotopes is thermal diffusion;
3. distillation and chemical exchange are the most economical methods for large-scale separation of the lighter elements; and
4. gaseous diffusion and the gas centrifuge are most economical for large-scale separation of the heaviest elements.

In addition, laser-induced isotope separation (LIS) has become an extremely attractive method for separating isotopes of many elements, as a result of recent advances in laser technology [4], and probably the most economical option for future large-scale production of isotopes of heavy elements such as Pu and U.

Since isotopic molecules have very similar properties, the degree of separation achieved in a single separating unit device (e.g., for distillation, a vessel containing a multicomponent liquid and its vapor phase in equilibrium with each other) is extremely small. The degree of separation achieved in such a unit is measured by a property of the separating process called the elementary separation factor. If one envisions a separating unit for any process as a "black box" (Fig. 1) that separates a single feed stream into a product stream—somewhat enriched in the desired isotope—and a waste stream—somewhat depleted in that component—the elementary separation factor, α, is defined as the abundance ratio of the isotopes in the product stream divided by the same ratio for the waste stream:

$$\text{elementary separation factor:}\quad \alpha = \frac{[x/(1-x)]_{\text{product}}}{[x/(1-x)]_{\text{waste}}}\,.$$

Here, x is the mole fraction of the desired isotope. If α is unity, no separation occurs. For typical (other than electromagnetic and laser isotope separation) processes and most elements other than hydrogen, α is at most only several percent different from unity, and often only tenths of a percent away from unity. A single separating stage can thus increase the enrichment by only a few percent or less, relative to the enrichment level of its feed.

To achieve a useful enrichment it is thus necessary to construct a stack of separating stages in which the product stream from a lower stage (processing less enriched material) is fed to the next higher stage, whose product is in turn fed to the stage above, and so on. In addition, the waste streams from upper stages usually contain desired isotope concentrations higher than the natural abundance, so they are recycled as part of the feed to an appropriate intermediate stage. Such a stack of interconnected stages is called a separation cascade. Furthermore, when, as is often the case, the desired isotope has a low relative abundance in nature, it is a minor component of the natural-abundance feed material going into the separation plant. Consequently, the section of such an enriching cascade near the feed-point must handle considerably higher material flow rates than those close to the product-end. The higher-flow section should therefore contain larger flow-capacity separative elements or a number of elements connected in parallel.

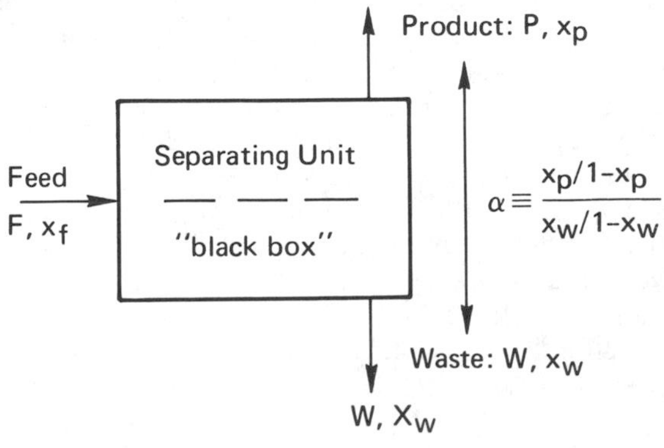

FIG. 1. An elementary separating stage.

Material conservation (see Fig. 1) requires that the total amount of material fed into the stage be balanced by the amounts leaving in the product and waste streams, and further that the quantities of the desired isotope in the two exit streams equal the amount of the isotope that entered the stage:

$$\text{total material balance:}\quad F = P + W;$$

$$\text{isotopic material balance:}\quad Fx_f = Px_p + Wx_w.$$

Here F, P, and W symbolize the molar flows in the feed, product, and waste streams, and x again represents the mole fraction of the desired isotope. These two material-balance conditions, together with the elementary separation factor, govern the steady-state operation of a separating unit. In fact, a similar set of three equations can be applied to either an element or a stage, and the performance of an entire cascade is governed by the overall material balance equations and the cumulative effect of the separative capability of the individual elements.

More detailed theoretical treatments of cascade parameters will be found in the references (see particularly [1,3,5–7]). Now to illustrate the principles just outlined, the article will provide brief descriptions of several isotope-separating processes used (or proposed) on the industrial or laboratory scales.

SEPARATION PROCESSES

Gaseous Diffusion [*1,6–9*]

Most of the current ^{235}U enrichment capacity of the world is still provided by gaseous diffusion plants. In each stage of the process, gaseous uranium hexafluoride (UF_6, the only volatile compound of uranium at ordinary temperature) flows through a diffusion barrier with very fine holes from a high-pressure chamber into a lower-pressure region. This phenomenon is gaseous effusion, although the process is usually termed gaseous diffusion. The lighter $^{235}UF_6$ molecules have a slightly higher mean speed than their ^{238}U counterparts; therefore, the gas passing through the barrier is slightly richer in ^{235}U than the portion remaining behind. The mean speeds of the molecules are in inverse ratio to the square roots of the molecular weights of the isotopic molecules: $(^{238}UF_6/^{235}UF_6)^{1/2} = (352/349)^{1/2} = 1.0043$, the elementary separation factor for $^{235}U/^{238}U$ enrichment by gaseous diffusion.

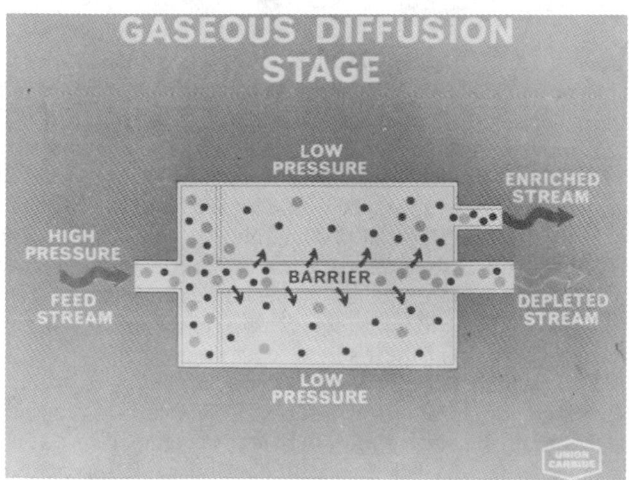

FIG. 2. A gaseous diffusion stage.

The low separation factor per stage requires the use of many enriching stages in a countercurrent cascade to produce a useful degree of enrichment. Figure 2 depicts schematically the operation of a stage, and Fig. 3 shows the arrangement of three stages in a cascade. The gas in the enriched stream is at a lower pressure, so it must be recompressed before it is fed to the next higher stage, accounting for the high electrical power usage of such enrichment plants; about 70% of the operating expenses of a gaseous diffusion plant (exclusive of feed material) is for electricity. An "ideal cascade" to separate natural uranium feed containing 0.71% ^{235}U into a product containing 3% and waste containing 0.2% requires 1272 separating stages of varying size and capacity arranged in series as in Fig. 3. (To produce product containing 90% ^{235}U, at the same feed and waste compositions, would require 3919 stages.) Figure 4 shows the arrangement of stages in an "ideal" cascade, one in which the compositions of streams from the upper and lower stages at each enrichment level are equal (Fig. 3).

Many of the plant parameters that make large contributions to the cost of enriched product (other than the cost of

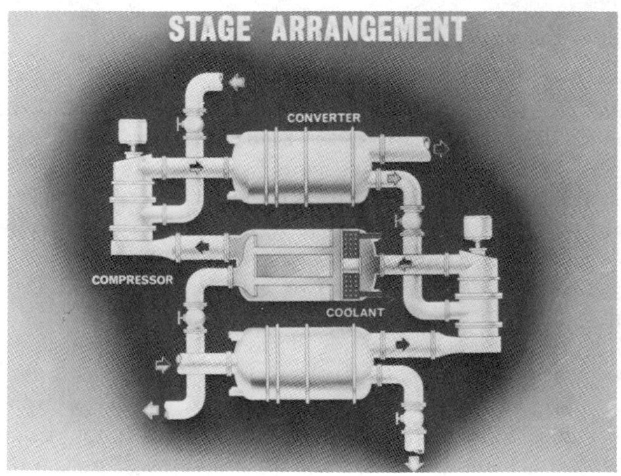

FIG. 3. Arrangement of gaseous diffusion stages.

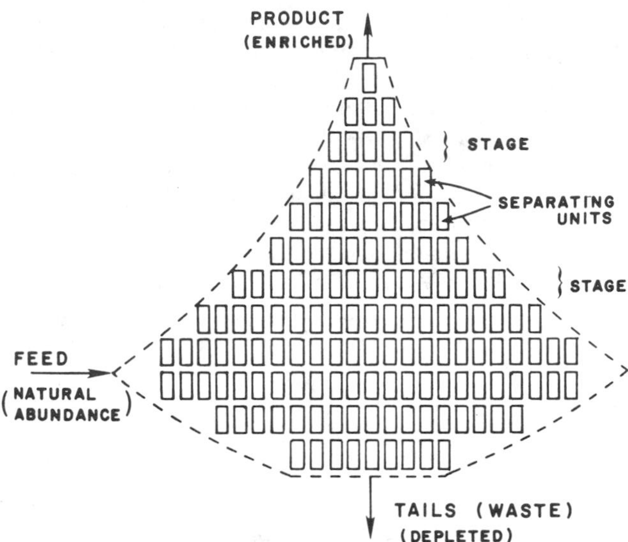

FIG. 4. A schematic representation of relative flow rates of ideal cascade.

feed material) are proportional to a simple mathematical function of the isotopic abundances and the quantities (moles or kilograms) of the feed, product, and waste streams entering and leaving the plant. This function—applicable to an ideal separating cascade and called the separative work [1,6,7]—is a measure of the value added by the plant to the material processed by the cascade. It is proportional to the quantity of the enriched product, and it increases with the change in the enrichment levels of product and waste from that of the feed. It does not, however, depend on the stage separation factor. The cost of an enriched product is proportional to this separative work, which is a measure of the quantity of separation achieved, and an additional factor—a function of the separation factor and thus dependent on the efficiency of the separation process. Thus the relative merits of various separation processes can be discussed in terms of their costs per separative work unit (SWU). The kilogram of uranium is the conventional unit for the SWU for ^{235}U.

For a plant with a single waste, product, and feed stream, there are three mole fraction variables and three quantity variables, of which only four can be specified independently because of the two material balance requirements. Thus, for instance, two plants each taking the same fixed feed concentration and yielding the same amount of product enriched to the same isotopic concentration and the same waste assay have the same SWUs, no matter what separation principle and engineering design they operate on. A plant of a given SWU capacity may be run (a) to produce either a small quantity of material at high enrichment or a larger quantity at lower enrichment; and (b) to maintain a given quantity and quality of product by extracting more ^{235}U from the feed (lower waste assay) when natural uranium is in short supply and discarding wastes of higher assay when feed material is plentiful.

The gaseous diffusion plants in the world have a combined annual capacity of about 50 million SWUs, or 50 thousand metric tons, of uranium enrichment, of which the United States and the Soviet Union have, respectively, capacities of 27 million SWUs/yr and about 10 million SWUs/yr [9].

Overall, the advantages of the gaseous diffusion process are its proven reliability and demonstrated ability (over more than 40 years) to operate at a capacity factor of over 99%. Among its disadvantages are its high energy consumption—about 2200 kwh/SWU—an inevitable consequence of the need to recompress the gas many times—and its need for large installations to make economical operation possible.

Gas Centrifugation [1,6–11]

When gaseous uranium hexafluoride spins in a centrifuge at a high rotational speed, the heavier ^{238}U molecules move preferentially toward the periphery, leaving the inner zone enriched in ^{235}U. Modern developments in this method (see e.g., Fig. 5) are all variants of the countercurrent gas centrifuge design of Gernot Zippe. UF_6 gas enclosed within a rapidly rotating cylinder is subject to centrifugal acceleration thousands of times greater than gravity. At peripheral speeds above 500 m/s, made possible in current centrifuges using rotors composed of high-performance carbon fiber- or aramide-reinforced polymers, the ^{235}U content at the center of the cylinder can be as much as 18% greater than at the periphery. A system of rotating baffles and stationary scoops induces longitudinal countercurrent gas flow with light gas

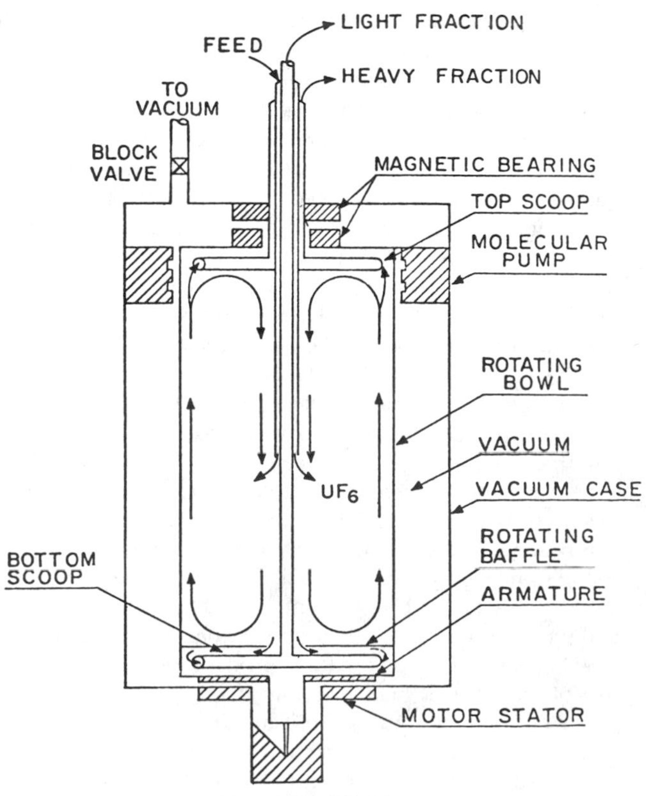

FIG. 5. Gas centrifuge.

(rich in $^{235}UF_6$) flowing upward near the axis and heavy gas (depleted in ^{235}U) flowing downward near the periphery. This sets up a cascade of multiple separative elements in a single centrifuge and, in a sufficiently long centrifuge, a ^{235}U concentration ratio as high as 2 between the top and bottom; the enriched product is then withdrawn from the top of the cylinder. Only seven stages would be needed in an ideal cascade to produce 3% ^{235}U product and waste at 0.25%.

The intrinsic separation factor in centrifuges is proportional to the difference in the masses of isotopic molecules rather than the ratio (or its square root) of the masses. Therefore, this process is best suited to a system such as $^{235}UF_6$–$^{238}UF_6$, which has a mass difference of three mass units. Unfortunately, uranium hexafluoride condenses at pressures higher than one-sixth atmosphere at room temperature. This, coupled with the fact that at the periphery of a centrifuge the pressure is millions of times the pressure near the axis, necessarily makes the throughput of gas (the amount of material processable in a given time) by an individual centrifuge quite low. Thus, a centrifuge plant with a capacity of 3×10^6 kg SWU/yr producing 3% ^{235}U would need 6×10^4 centrifuges, each 10 m long, to do the separative work produced by 1200 gaseous diffusion stages. On the other hand, the electrical energy consumed would be only 90 kwh/SWU as opposed to 2200 kwh/SWU for gaseous diffusion.

Among centrifuge plants existing or designed around the world are a plant with a capacity of about 1×10^6 SWUs per year built and operated by a joint British, Dutch, West German venture, a plant with an annual capacity of 2.2×10^6 SWUs per year designed by the U.S. Department of Energy at Portsmouth, Ohio, and in Japan a prototype plant with 2×10^5 SWUs per year capacity built and operating, and another with a capacity of 1.5×10^6 SWUs per year scheduled for completion in 1991.

Laser Isotope Separation [1,4,6–9,12,13]

This process takes advantage of the differences in absorption spectra of different isotopic species. By using sufficiently monochromatic light of an appropriate wavelength, it preferentially excites a particular isotopic species to an upper energy level. The excited species must then be separated from its isotopic partners before it exchanges isotopes or before it loses excitation by some physical or chemical process that need not be isotopically selective. The technique was successfully used to separate small quantities of mercury isotopes almost three decades ago. Attention has been focused on these methods by the recent development of narrow-band, tunable lasers of high intensity and high repetition rate. Laser methods have been notably successful in achieving separation of isotopes of elements such as hydrogen, boron, chlorine, sulfur, and bromine, but the greatest importance of the method is likely to lie in its potential as a next generation process for uranium isotope separation.

One such process, called atomic vapor laser isotope separation (AVLIS), is illustrated in Fig. 6. Atoms of uranium heated in a high-vacuum chamber by a high-energy beam of electrons vaporize and flow upward through a region between a pair of negatively charged plates. The space between the plates is illuminated by light from a system of lasers pass-

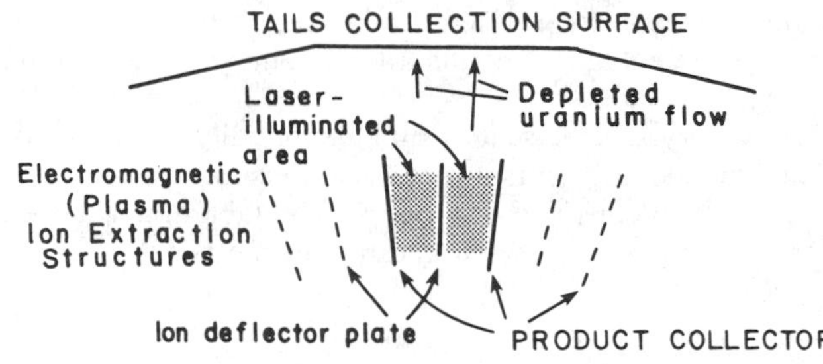

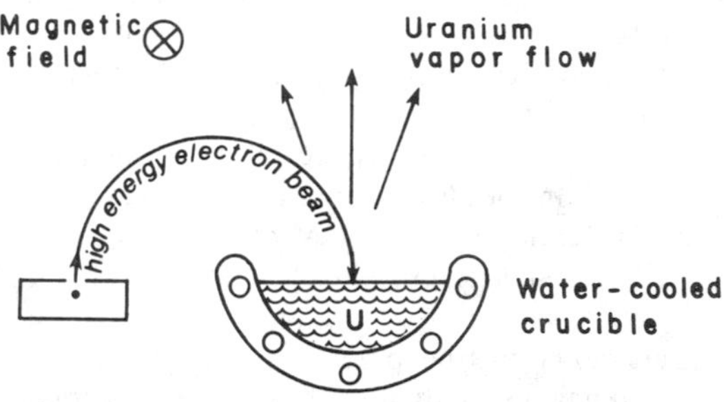

FIG. 6. Cross-sectional view of an atomic vapor laser isotope separation module.

ing through in the direction perpendicular to the page. Laser photons with an appropriate wavelength selectively excite ^{235}U atoms but leave ^{238}U atoms unexcited. At least one additional laser beam is used to ionize the excited ^{235}U atoms. The ionizing photons have energies sufficiently high to ionize the already excited ^{235}U atoms but insufficient to ionize unexcited ^{238}U atoms. The ^{235}U ions are then collected by negatively charged plates, while ^{238}U atoms continue their upward flow until condensed on the ceiling surface. It is a major attraction of the method that under ideal conditions, a complete separation of ^{235}U and ^{238}U in a single LIS stage is theoretically possible.

The future of the LIS process depends on successful development of laser systems. Since even the largest energy difference between the atoms of ^{235}U and ^{238}U corresponds only to a change on the order of 1 in 5×10^4, the excitation laser photon energy must be tuned more closely than this to avoid accidental excitation of ^{238}U atoms. Lasers generate pulses (or packets) of photons at a fixed repetition rate. The lasers for the LIS must have a sufficiently rapid repetition rate to catch nearly all the ^{235}U atoms passing between the collection plates; ^{235}U atoms left un-ionized by the lasers will not be collected by the plates, thus reducing the effective separation factor. Only a fraction of laser photons passing through an LIS chamber is actually absorbed by uranium, and it takes one mole of absorbed photons to excite one mole of uranium atoms. Thus, LIS requires a high-power (high photon production rate), rapid-repetition, finely tunable laser. In addition, laser photons of high energy are expensive to produce. Since a higher density of uranium atoms in the illuminated region increases the likelihood of collision between excited ^{235}U atoms and ^{238}U atoms, which can produce excited ^{238}U atoms at the expense of de-excited ^{235}U, the throughput of LIS processes is necessarily low.

In spite of these technical difficulties, LIS for ^{235}U [13] holds high promise because of the theoretical possibility of its extremely high separation factor. The laser isotope separation processes other than the AVLIS for ^{235}U that have been developed include the molecular LIS for ^{235}U, in which UF_6 is working material, and "Special Isotope Separation" for converting fuel-grade [7–19% ^{240}Pu] into weapon-grade [less than 7% ^{240}Pu] by removing unwanted isotopes such as ^{240}Pu from ^{239}Pu.

Separation Nozzle [1,3,6–9]

This aerodynamic process achieves partial separation of isotopes in a flowing gas stream subjected to high linear or centrifugal acceleration, or both. A cross section of an improved separation nozzle developed by E. W. Becker and his co-workers in Germany is shown in Fig. 7. A feed gas, typically consisting of a mixture of about 5 vol% UF_6 and 95% H_2 at a pressure of about 1 atm, passes through a nozzle with first a convergent, then a divergent cross section into a lower-pressure region. The change in cross section accelerates the mixture to supersonic speed, and the curved groove produces centrifugal acceleration. The gas adjacent to the curved wall is preferentially enriched in ^{238}U, and the knife edge downstream divides the stream into a (waste) portion depleted in ^{235}U (region C) and a portion enriched in the light isotope (region B).

The separation factor increases with the mean speed of the gas molecules and with their centrifugal acceleration. Therefore, a larger pressure ratio between the feed and product gases, a narrower throat width at the entrance, and a smaller radius of curvature of the curved groove all contribute to increasing the separation factor. But the resulting miniaturization of nozzle elements reduces the throughput and makes fabrication of the elements a formidable task. A separation factor of 1.015 has been achieved with a mixture of 5% UF_6–95% H_2 flowing through a pressure ratio of 3.5. The dilution of UF_6 with a gas of low molecular weight yields (1) a higher sonic velocity and thus a higher separation factor, and (2) higher diffusion rates of UF_6 molecules, which allow operation at higher product pressures and correspondingly increased uranium throughput. The Becker nozzle process offers a higher separation factor than gaseous diffusion, but the 20-fold dilution of UF_6 also makes power consumption

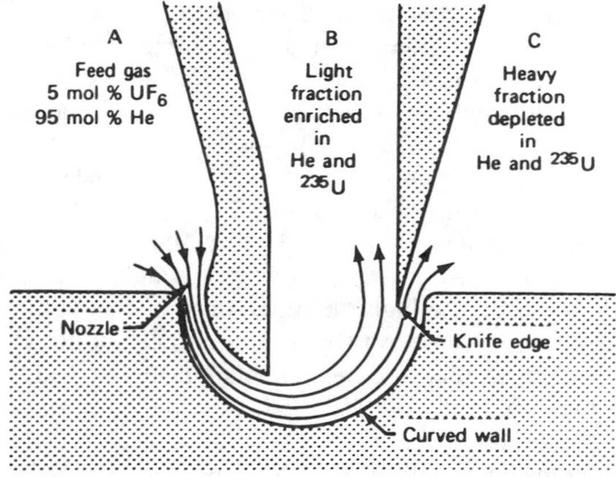

FIG. 7. Operating principle of the separating nozzle.

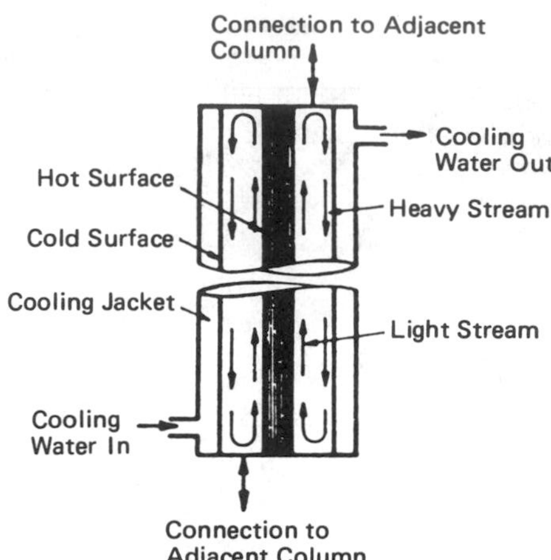

FIG. 8. Thermal diffusion column.

(about 3×10^3 kwh/SWU) higher than for gaseous diffusion. A demonstration cascade of about 0.3×10^6 SWU/yr has been operated in Brazil with technical assistance from West Germany.

Electromagnetic Separation [*1–3, 14*]

In this process, the principle is essentially that of a large-scale mass spectrometer, and its use dates back to the World War II Manhattan District Project when such machines were used to separate ^{235}U in kilogram quantities. Since the end of World War II, the electromagnetic separators (called Calutrons for *Cal*ifornia *U*niversity Cyclo*tron* because they were originally developed at the University of California by E. O. Lawrence) have been used for separating an amazing variety of isotopes, both stable and radioactive. A retrospective paper by L. O. Love [14] describes activities at Oak Ridge National Laboratory (ORNL) in this area over three decades. See Ref. 2 for a discussion of needs for various isotopes of practically every element that can be separated by this versatile method, and a description of the current ORNL electromagnetic separation program.

Thermal Diffusion [*1,3,6*]

The thermal diffusion effect—namely, that a concentration gradient is produced in a gaseous mixture subjected to a temperature gradient (e.g., in a vessel with walls at different temperatures)—was transformed from a laboratory curiosity into a useful and simple method for separating fluid mixtures (including mixtures of isotopes) by the invention of the thermal diffusion column by Clusius and Dickel. The operating principle of such a column is illustrated in Fig. 8. A fluid is confined in a long vertical cylinder between an inner wall, kept at a higher temperature, and an outer wall, kept at a lower temperature. A countercurrent convective flow is set up in the fluid, as indicated in Fig. 8, descending along the cool wall and rising adjacent to the heated wall. Since horizontal slices of the column at different levels along the column length act as unit separation stages, the countercurrent vertical flows multiply the small (less than 1%) isotopic concentration difference between the two walls, so that a single column contains a large number of separating stages.

Laboratory-scale cascades suitable for daily separation of milligram quantities of isotopes in gaseous form are remarkably simple to construct and operate. Notably, the isotopes of the noble gases are separated by this method.

Chemical Exchange [1,3,5–6,15–16]

Isotopic exchange equilibria, which lead to nonequipartition of isotopes between different chemical species, result from differences in internal vibrational frequencies of the isotopic forms. The magnitude of the separative effect depends on the isotopic mass difference between the isotopes ($\Delta m/m$), and on the difference in free energy (produced by the mass difference) in a pair of different chemical species [3,5,15]. The first criterion leads to the conclusion that chemical exchange is most effective for separating isotopes of the lighter elements and becomes increasingly difficult as one proceeds to heavier elements (>sulfur). The second criterion leads one to expect a large separation effect between two molecules, one in which the isotopic atom is unbonded or weakly bonded, and another in which it is connected by a maximum number of strong bonds. Unfortunately, such compounds do not necessarily undergo isotope exchange rapidly, and exchange processes often require a catalyst to promote the reaction at an effective rate.

Reactions that have been used to produce quantities of separated isotopes include the exchanges: H_2O(liq)/HDS(gas); H_2O(liq)/HD(gas); NH_3(liq)/HD(gas); $H^{14}NO_3$(aq)/^{15}NO(gas); $^{12}CN^-$(aq)/$H^{13}CN$(gas); $^{10}BF_3$-ether(complex)/$^{11}BF_3$(gas), and $H^{32}SO_3^-$(aq)/$^{34}SO_2$(gas).

The elementary exchange effects are multiplied in long exchange columns similar to those used for distillation. In contrast to distillation, where reflux at the ends is effected by a boiler (at the bottom) to evaporate the liquid, and a condenser (at the top) to liquefy the vapor, exchange columns use chemical reactors at each end to interconvert the exchanging species. Figure 9 illustrates the analogy between the gas–liquid exchange process and distillation.

Among the above listed exchanges, those involving exchange of hydrogen and deuterium have been and are being used in plants producing thousands of tons of D_2O annually [16]; the exchange between nitric acid and nitric oxide (termed the Nitrox process) has been widely used throughout the world in simple laboratory scale cascades to produce grams per day of nitrogen-15 enriched to purities up to 99.8% from the natural abundance of 0.37%. Operating characteristics for the systems are described in the references, particularly [1,3,5–6,16].

Distillation [*1,3,5–6*]

H. C. Urey discovered deuterium by concentrating it sufficiently through distillation of liquid hydrogen and, for the first time, detecting it spectroscopically. The method, which is one of the common chemical engineering processes for separating chemical substances, has been used for producing deuterium by the distillation of liquid hydrogen or water.

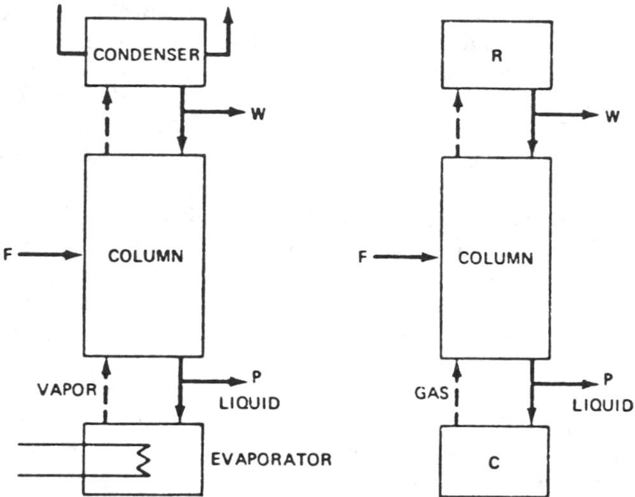

FIG. 9. Analogy between distillation and gas–liquid chemical exchange. R = reflux converter; C = chemical reactor or electrolytic cell; F = feed; P = product; W = waste.

Isotopes of carbon, nitrogen, and oxygen have been produced by distillation of their compounds, carbon monoxide, nitric oxide, and water.

The energy consumed by distilling water to produce heavy water is relatively high, due partly to the fact that heavy water is less volatile than light water. This makes distillation less competitive than chemical exchange processes for primary enrichment of heavy water. However, its high throughput (because the working material is liquid), reasonably high separation factor (1.027 at 100°C), design simplicity, and operational reliability make it the preferred process for final enrichment—from several percent deuterium to 99.9% D. This illustrates the comment made earlier in this article that, even for a given isotope, there is no one best separation method.

The authors acknowledge with thanks many helpful suggestions from Robert C. Rooney. One of us, TI, acknowledges support by the U.S. Department of Energy, Office of Basic Energy Sciences (DE-FG02-88ER13855).

See also ISOTOPE EFFECTS.

REFERENCES

1. M. Benedict, T. H. Pigford, and H. W. Levi, *Nuclear Chemical Engineering*, 2nd Edition. McGraw-Hill, New York, 1981. (A)
2. *Separated Isotopes: Vital Tools for Science and Medicine*. Office of Chemistry and Chemical Technology, National Research Council, Washington, DC, 1982. (E)
3. W. Spindel, in *Isotopes and Chemical Principles*, ACS Symposium Series, Vol. 11, Chap. 5. American Chemical Society, Washington, DC, 1975. (E)
4. V. S. Letokhov and C. Bradley Moore, Laser Isotope Separation, in *Chemical and Biochemical Applications of Lasers* (C. Bradley Moore, ed.). Academic Press, New York, San Francisco, London, 1977. (I)
5. J. Bigeleisen, in *Isotope Effects in Chemical Processes*, Advances in Chemistry Series, Vol. 89, Chap. 1. American Chemical Society, Washington, DC, 1969. (I).
6. S. Villani, *Isotope Separation*. American Nuclear Society, Hinsdale, IL, 1976. (A)
7. S. Villani (ed.), *Uranium Enrichment*. Vol. 35 of Topics in Applied Phys., Springer-Verlag, Berlin, Heidelberg, New York, 1979. (A)
8. M. Benedict (ed.), *Developments in Uranium Enrichment*, AIChE Symp. Ser. 169, Vol. 73. American Institute of Chemical Engineers, New York, 1977. (I)
9. M. Benedict, in AIChE Symp. Ser. 235, Vol. 80, pp 149–156. American Institute of Chemical Engineers, New York, 1983. (E)
10. D. R. Olander, *Sci. Am.* **239,** 37 (August 1978). (E)
11. S. Whitley, *Rev. Mod. Phys.* **56,** 41 (1984); **56,** 67 (1984). (A)
12. R. N. Zare, *Sci. Am.* **236,** 86 (February, 1977). (E)
13. J. A. Paisner and R. W. Solarz, Resonance Photoionization Spectroscopy (pp. 175–260), and J. L. Lyman, Laser-Induced Molecular Dissociation: Applications in Isotope Separation and Related Processes (pp. 417–506), in *Laser Spectroscopy and Its Applications* (L. J. Radziemski, R. W. Solarz, and J. A. Paisner, eds.). Marcel Dekker, New York, 1987.
14. L. O. Love, *Science* **182,** 343 (1973). (E)
15. J. Bigeleisen, *Science* **147,** 463 (1965). (E)
16. H. K. Rae (ed.), *Separation of Hydrogen Isotopes*, ACS Symposium Series, Vol. 68. American Chemical Society, Washington, DC, 1978. (I)

Isotopes

R. C. Barber

INTRODUCTION

An isotope is one of two or more different species of the same chemical element, having the same atomic number Z but different mass number A. Thus, for a given element the nuclei of isotopes possess the same number of protons but differ in the number of neutrons. The term, proposed by Frederick Soddy (1913), derives from the Greek *isos* (equal) and *topos* (place) and reflects the fact that such species occupy the same position in the periodic table of the elements. For this reason the term isotope implies the existence of at least two different species of a given element. The general term for a particular atomic species characterized by an atomic number Z and mass number A is nuclide, although the term isotope has frequently been used loosely with this meaning.

Within each of the known elements at least three nuclides have been observed, some of which may be stable with the remainder being radioactive. However, studies of the isotopic nature of elements are normally confined to those nuclides that are either stable or sufficiently long-lived to exist in significant amounts in nature. The occurrence and relative abundances of such species have been studied by mass spectrometry.

The largest number of stable isotopes, ten, occurs in tin ($Z = 50$, $A = 112$, 114, 115, 116, 117, 118, 119, 120, 122, 124). Xenon possesses nine isotopes; tellurium and cadmium each have eight. A total of 62 elements possess naturally occurring isotopes; a further 20 elements are mononuclidic, that is, in each case only one nuclide of the element is stable. In more precise terms, the Commission on Atomic Weights and Isotopic Abundances of IUPAC has defined an element to be mononuclidic if it has one and only one nuclide that is stable or has a half-life greater than 3×10^{10} years. Therefore the Commission considers Be, F, Na, Al, P, Sc, Mn, Co, As, Y, Nb, Rh, I, Cs, Pr, Tb, Ho, Tm, Au, and Bi to be mononuclidic, while Pa is not. Among the elements below bismuth ($Z=83$) only two, technetium and promethium, do not have any stable nuclide.

For those elements that possess naturally occurring isotopes, the relative abundances of these isotopes, in most cases, are remarkably independent of the source of the material. Accordingly, the chemical atomic weight, which is the average value of the masses of all of the stable isotopes, weighted according to the relative abundances of each, is found to be a constant for most materials. Situations in which relative abundances do vary are then of special interest; some of these are discussed later.

DISCOVERY

Against the background of the idea, generally accepted in the latter part of the nineteenth century, that a given number of atoms of a particular element (Avogadro's number) had a mass characteristic of the element (the atomic weight), the first speculation recognizably consistent with the modern idea of isotopes was made by Sir William Crookes. Addressing the Chemical Section of the British Association at Birmingham in 1886 he suggested that "when we say the atomic weight of, for instance, calcium is 40, we really express the fact that, while the majority of calcium atoms have an actual atomic weight of 40, there are not a few which are represented by 39 or 41, a less number by 38 or 42 and so on." He further considered the possibility that "these heavier and lighter atoms may have been in some cases subsequently sorted out by a process resembling chemical fractionation. This sorting out may have taken place in part while atomic matter was condensing from the primal state of intense ignition, but also it may have been partly effected in geological ages by successive solutions and precipitations of the various earths."

Subsequently he developed this idea in connection with his pioneer work with the rare earths, subdividing a sample of "yttria" into a number of components by fractional precipitation. These components, which he called "meta-elements," possessed different phosphorescent spectra and differed a little in atomic weight but closely resembled each other in chemical properties. With advances in chemical techniques, however, one after another of these components was identified as a distinct element, each having its characteristic atomic spectrum and atomic weight. Thus the yttria studied by Crookes proved to be a mixture of elements and the idea of meta-elements was abandoned.

By the end of the nineteenth century, the increased precision in the determinations of atomic weights, coupled with the lack of variations in these values, had further reinforced the general assumption that the atomic weight of an element was a fundamental property. Two problems remained unsolved: Why were so many atomic weights nearly integers on the scale O = 16, and why were there a few notable cases of fractional atomic weights (e.g., Cl = 35.5)?

The first experimental evidence that chemically identical substances might possess different physical properties emerged from the studies of radioactivity in the elements heavier than bismuth (i.e., $Z>83$). In particular these studies exploited the fact that minute quantities of a particular substance could be detected and identified by observing the type of radiation and the half-life. While studying the decay chains of uranium, B. B. Boltwood (1906) identified a new element, which he called ionium and which, when mixed with thorium, could not be separated by any known chemical process. Shortly thereafter, similar behavior was observed in mixtures of radiothorium (a disintegration product of thorium) and thorium by H. N. McCoy and W. H. Ross, and in mixtures of mesothorium 1 (also in the thorium series) and radium by W. Marckwald and by F. Soddy. Further evidence of the chemical identity of ionium and thorium came from the examination of the visible emission spectrum of a mixture of the two by A. S. Russell and R. Rossi (1912). No new lines attributable to ionium were observed when compared to the spectrum of pure thorium.

Soddy, convinced that in spite of the unique radioactive decay properties and distinct atomic weights, the chemically inseparable species were in fact the same elements, undertook a survey of the radiochemistry of the heavy elements and proposed the existence of isotopes, along with a general scheme that reconciled the chemical properties with the decay data and notably with the recently established identity of the α particle. Moreover he saw no reason to restrict the incidence of isotopes to the radioactive elements and recognized that it might account for fractional atomic weights.

Clear evidence of the existence of stable isotopes emerged from the investigations by J. J. Thomson of the nature of positive rays in gas discharge tubes. The positive rays from a gas discharge were collimated, passed through an arrangement of parallel and coterminous electric and magnetic fields, and allowed to fall on a photographic plate. In such apparatus all particles having the same value of e/M, regardless of energy, strike the plate at some point on a particular parabola. Thomson observed that, in addition to the intense line for neon at atomic weight 20, there was a much fainter line that accompanied it at mass 22. This result led Thomson to encourage F. W. Aston, then a research student at Cambridge, to pursue the study of stable isotopes.

Aston initially attempted to enrich neon in its suspected isotopes but then turned his attention to the construction of a "mass spectrograph" (1919), which both focused and analyzed positive-ion beams. With this instrument he demonstrated conclusively that neon consists of a mixture of isotopes and shortly thereafter confirmed the isotopic nature of chlorine, mercury, nitrogen, and the noble gases.

Independently and simultaneously, A. J. Dempster, working at the University of Chicago, devised a "mass spectrometer" (1918), which he used to discover, and make accurate abundance determinations for, the isotopes of magnesium and, shortly thereafter, lithium, potassium, calcium, and zinc.

Thus with the work of Aston and Dempster the isotopic nature of the stable elements was unequivocally demonstrated and an important instrument in the study of isotopes, the mass spectrometer, was introduced.

ISOTOPES AND NUCLEAR STRUCTURE

The work of Aston indicated early that the atomic masses of the various isotopes are close to integers. This is confirmed in the modern values for atomic masses, including many nuclides not known to Aston, all of which lie within 0.1 u of the value given by the mass number.

Initial attempts to account for this behavior in terms of nuclear electrons were recognized to be unsatisfactory and were discarded when the neutron, a neutral particle with a mass of ~1 u, was discovered by Chadwick in 1932 and identified as a nuclear particle. Differing numbers of neutrons were seen to have no effect on the electronic structure, which determined chemical properties, but altered the mass in steps of ~1 u and led to very different nuclear structure and properties.

Aston also realized that the study of divergences from "whole numbers" could be highly rewarding inasmuch as it was a measure of nuclear stability. In this connection, he realized that the definition of atomic weights used in chemistry at the time, wherein oxygen in its "natural" abundance of isotopes was taken to be 16 amu, was imprecise and adopted instead $^{16}O = 16$ amu exactly as his standard.

This double standard persisted until 1960 when the standard for the "chemical" scale was found to be insufficiently precise for modern chemical work, while the standard for the "physical" scale was generally inconvenient for use. Accordingly, in a coordinated action, both the International Union of Pure and Applied Chemistry and the International Union of Pure and Applied Physics adopted in that year the "unified" scale of atomic mass, with the symbol u, where $^{12}C = 12$ u exactly. Thus for this scale, 12 g of ^{12}C contains Avogadro's number of atoms, 6.022 136 7 (36) $\times 10^{23}$ atoms.

All atoms are stable with respect to spontaneous decomposition into the constituent particles. Thus, as required by special relativity, the mass of the aggregation of protons and neutrons in the form of a tightly bound nucleus, with the relatively loosely bound electrons surrounding it, must be less than the sum of the masses of the dissociated, free particles. That is, we may express the mass of an atom, $M(A,Z)$, as

$$M(A,Z) = A[1 + f(A,Z)] \qquad (1)$$

where $f(A,Z)$ is the packing fraction, a quantity that for stable nuclides lies in the range 8×10^{-3} to -8×10^{-4}. This approach to the description of the average binding energy of the nucleus was first used by Aston, who defined the packing fraction and presented sufficient experimental data to determine the general outline of the way that $f(A,Z)$ varies with mass number. In hindsight, this remarkable work indicated the large changes in energy to be expected from the fusion of light nuclei or from the fission of very heavy nuclei following neutron capture.

In more recent work, the mass of an atom is usually expressed in some variation of the form

$$M(N,Z) = Z[m_p + m_e] + Nm_n - (\mathrm{BE})_{\mathrm{nucleus}} - (\mathrm{BE})_{\mathrm{electron}} \qquad (2)$$

where N is the number of neutrons; m_p, m_e, and m_n are the masses of the proton, electron, and neutron, respectively; and $(\mathrm{BE})_{\mathrm{nucleus}}$ and $(\mathrm{BE})_{\mathrm{electron}}$ represent the total binding energies of the nuclear particles and of the atomic electrons, respectively, with the former being much larger than the latter. The variation of average nuclear stability with nuclear size is usually expressed in terms of the binding energy per nucleon, $(\mathrm{BE})_{\mathrm{nucleus}}/A$, and given as a function of the mass number. This quantity increases quickly from 1.1 MeV for ^2H to ~6.5 MeV at $A = 10$ and increases thereafter to a maximum value of 8.793 MeV for ^{58}Fe. From this value it declines to 7.6 MeV in uranium.

Evidence of the enhanced stability associated with closed nuclear shells may be seen in the fine detail of the variation of the binding energy per nucleon. It may be seen more clearly in a comparison, among isotopes of a given element, of the systematic changes in the energy required to remove the last pair of neutrons. Abrupt and dramatic decreases in this energy are seen as the "magic" numbers ($N = 28$, 50, 82, 126) are exceeded. In a similar way it can be demonstrated that the same magic numbers of protons are also configurations of enhanced stability.

Current values for atomic masses are based on data that derive from two major sources: precise values for mass differences between the members of doublets in mass spectra obtained with large, high-resolution mass spectrometers; and precise determinations of the energies of particles taking part in nuclear reactions, i.e., in a reaction

$$a + X \rightarrow Y + b + Q$$

where Q is the net change in rest mass, converted to energy. Data of both types may be regarded as determinations of the mass-difference connections between the various nuclides. Such values outnumber the actual mass differences, and so constitute an overdetermined set for which the method of least squares is appropriate for deducing "best" values for

Table I Naturally Occurring Nuclides

Atomic no. Z	Element	Atomic weight[a] (u)	Mass no. A	Atomic mass[b] (u)	Isotopic composition[c] (%)	Notes*
1	H	1.00794 (7)	1	1.007 825 035 (12)	99.985	*d,f*
			2	2.014 101 779 (24)	0.015	
2	He	4.002602 (2)	3	3.016 029 31 (4)	0.000138	*e*
			4	4.002 603 24 (5)	99.999862	
3	Li	6.941 (2)	6	6.015 121 4 (7)	7.5	*d–g*
			7	7.016 003 0 (8)	92.5	
4	Be	9.012182 (3)	9	9.012 182 2 (4)	100	
5	B	10.811 (5)	10	10.012 936 9 (3)	19.9	*d,f,g*

Table I (*continued*)

Atomic no. Z	Element	Atomic weight[a] (u)		Mass no. A	Atomic mass[b] (u)		Isotopic composition[c] (%)	Notes*
				11	11.009 305 4	(4)	80.1	
6	C	12.011	(1)	12	12.000 000 0	(0)	98.90	*d,f*
				13	13.003 354 826	(17)	1.10	
7	N	14.00674	(7)	14	14.003 074 002	(26)	99.634	*f*
				15	15.000 108 97	(4)	0.366	
8	O	15.9994	(3)	16	15.994 914 63	(5)	99.762	*d,f*
				17	16.999 131 2	(4)	0.038	
				18	17.999 160 3	(9)	0.200	
9	F	18.9984032	(9)	19	18.998 403 22	(15)	100	
10	Ne	20.1797	(6)	20	19.992 435 6	(22)	90.51	*f,g*
				21	20.993 842 8	(21)	0.27	
				22	21.991 383 1	(18)	9.22	
11	Na	22.989768	(6)	23	22.989 767 7	(1)	100	
12	Mg	24.3050	(6)	24	23.985 042 3	(8)	78.99	*e*
				25	24.985 837 4	(8)	10.00	
				26	25.982 593 7	(8)	11.01	
13	Al	26.981539	(5)	27	26.981 538 6	(8)	100	
14	Si	28.0855	(3)	28	27.976 927 1	(7)	92.23	*f*
				29	28.976 494 9	(7)	4.67	
				30	29.973 770 1	(7)	3.10	
15	P	30.973762	(4)	31	30.973 762 0	(6)	100	
16	S	32.066	(6)	32	31.972 070 70	(25)	95.02	*d,f*
				33	32.971 458 43	(23)	0.75	
				34	33.967 866 65	(22)	4.21	
				36	35.967 080 62	(27)	0.02	
17	Cl	35.4527	(9)	35	34.968 852 73	(7)	75.77	
				37	36.965 902 62	(11)	24.23	
18	Ar	39.948	(1)	36	35.967 545 52	(29)	0.337	*d–f*
				38	37.962 732 5	(9)	0.063	
				40	39.962 383 7	(14)	99.600	
19	K	39.0983	(1)	39	38.963 707 4	(12)	93.258	
				40	39.963 999 2	(12)	0.0117	*h* 1.28×10^9 a
				41	40.961 825 4	(12)	6.730	
20	Ca	40.078	(4)	40	39.962 590 6	(13)	96.941	*e,f*
				42	41.958 617 6	(13)	0.647	
				43	42.958 766 2	(13)	0.135	
				44	43.955 480 6	(14)	2.086	
				46	45.953 689	(4)	0.004	
				48	47.952 533	(4)	0.187	
21	Sc	44.955910	(9)	45	44.955 910	(14)	100	
22	Ti	47.88	(3)	46	45.952 629 4	(14)	8.0	
				47	46.951 764 0	(11)	7.3	
				48	47.947 947 3	(11)	73.8	
				49	48.947 871 1	(11)	5.5	
				50	49.944 792 1	(12)	5.4	
23	V	50.9415	(1)	50	49.947 160 9	(17)	0.250	*h* $>4 \times 10^{16}$ a
				51	50.943 961 7	(17)	99.750	
24	Cr	51.9961	(6)	50	49.946 046 4	(17)	4.345	
				52	51.940 509 8	(17)	83.789	
				53	52.940 651 3	(17)	9.501	
				54	53.938 882 5	(17)	2.365	
25	Mn	54.93805	(1)	55	54.938 047 1	(16)	100	
26	Fe	55.847	(3)	54	53.939 612 7	(15)	5.8	
				56	55.934 939 3	(16)	91.72	
				57	56.935 395 8	(16)	2.2	
				58	57.933 277 3	(16)	0.28	
27	Co	58.93320	(1)	59	58.933 197 6	(16)	100	
28	Ni	58.69	(1)	58	57.935 346 2	(16)	68.27	
				60	59.930 788 4	(16)	26.10	
				61	60.931 057 9	(16)	1.13	

Table I (*continued*)

Atomic no. Z	Element	Atomic weight[a] (u)		Mass no. A	Atomic mass[b] (u)		Isotopic composition[c] (%)	Notes*
				62	61.928 346 1	(16)	3.59	
				64	63.927 967 9	(17)	0.91	
29	Cu	63.546	(3)	63	62.929 598 9	(17)	69.17	*d*
				65	64.927 792 9	(20)	30.83	
30	Zn	65.39	(2)	64	63.929 144 8	(19)	48.6	*f*
				66	65.926 034 7	(17)	27.9	
				67	66.927 129 1	(17)	4.1	
				68	67.924 845 9	(18)	18.8	
				70	69.925 325	(4)	0.6	
31	Ga	69.723	(1)	69	68.925 580	(3)	60.1	
				71	70.924 700 5	(25)	39.9	
32	Ge	72.61	(2)	70	69.924 249 7	(16)	20.5	
				72	71.922 078 9	(16)	27.4	
				73	72.923 462 6	(16)	7.8	
				74	73.921 177 4	(15)	36.5	
				76	75.921 401 6	(17)	7.8	
33	As	74.92159	(2)	75	74.921 594 2	(17)	100	
34	Se	78.96	(3)	74	73.922 474 6	(16)	0.9	
				76	75.919 212 0	(16)	9.0	
				77	76.919 912 5	(16)	7.6	
				78	77.917 307 6	(16)	23.6	
				80	79.916 519 6	(19)	49.7	
				82	81.916 697 8	(23)	9.2	
35	Br	79.904	(1)	79	78.918 336 1	(26)	50.69	
				81	80.916 289	(6)	49.31	
36	Kr	83.80	(1)	78	77.920 396	(9)	0.35	*e–g*
				80	79.916 380	(9)	2.25	
				82	81.913 482	(6)	11.6	
				83	82.914 135	(4)	11.5	
				84	83.911 507	(4)	57.0	
				86	85.910 616	(5)	17.3	
37	Rb	85.4678	(3)	85	84.911 794	(3)	72.165	*e,f*
				87	86.909 187	(3)	27.835	*h* 4.7×10^{10} a
38	Sr	87.62	(1)	84	83.913 430	(4)	0.56	*e,f*
				86	85.909 267 2	(28)	9.86	
				87	86.908 884 1	(28)	7.00	
				88	87.905 618 8	(28)	82.58	
39	Y	88.90585	(2)	89	88.905 849	(3)	100	
40	Zr	91.224	(2)	90	89.904 702 6	(26)	51.45	*e,f*
				91	90.905 643 9	(26)	11.22	
				92	91.905 038 6	(26)	17.15	
				94	93.906 314 8	(28)	17.38	
				96	95.908 275	(4)	2.80	
41	Nb	92.90638	(2)	93	92.906 377 2	(27)	100	
42	Mo	95.94	(1)	92	91.906 808	(4)	14.84	*f*
				94	93.905 085 3	(26)	9.25	
				95	94.905 841 1	(22)	15.92	
				96	95.904 678 5	(22)	16.68	
				97	96.906 020 5	(22)	9.55	
				98	97.905 407 3	(22)	24.13	
				100	99.907 477	(6)	9.63	
43	Tc							
44	Ru	101.07	(2)	96	95.907 599	(8)	5.52	*e,f*
				98	97.905 287	(7)	1.88	
				99	98.905 938 9	(23)	12.7	
				100	99.904 219 2	(24)	12.6	
				101	100.905 581 9	(24)	17.0	
				102	101.904 348 5	(25)	31.6	
				104	103.905 424	(6)	18.7	
45	Rh	102.90550	(3)	103	102.905 500	(4)	100	

Table I (*continued*)

Atomic no. Z	Element	Atomic weight[a] (u)		Mass no. A	Atomic mass[b] (u)		Isotopic composition[c] (%)	Notes*
46	Pd	106.42	(1)	102	101.905 634	(5)	1.020	*e,f*
				104	103.904 029	(6)	11.14	
				105	104.905 079	(6)	22.33	
				106	105.903 478	(6)	27.33	
				108	107.903 895	(4)	26.46	
				110	109.905 167	(20)	11.72	
47	Ag	107.8682	(2)	107	106.905 092	(6)	51.839	*e*
				109	108.904 756	(4)	48.161	
48	Cd	112.411	(8)	106	105.906 461	(7)	1.25	*e,f*
				108	107.904 176	(6)	0.89	
				110	109.903 005	(4)	12.49	
				111	110.904 182	(3)	12.80	
				112	111.902 757	(3)	24.13	
				113	112.904 400	(3)	12.22	*h* 9×10^{15} a
				114	113.903 357	(3)	28.73	
				116	115.904 755	(4)	7.49	
49	In	114.82	(1)	113	112.904 061	(4)	4.3	*e*
				115	114.903 882	(4)	95.7	*h* 5×10^{14} a
50	Sn	118.710	(7)	112	111.904 826	(5)	0.97	
				114	113.902 784	(4)	0.65	
				115	114.903 348	(3)	0.36	
				116	115.901 747	(3)	14.53	
				117	116.902 956	(3)	7.68	
				118	117.901 609	(3)	24.22	
				119	118.903 311	(3)	8.58	
				120	119.902 199 1	(29)	32.59	
				122	121.903 440 4	(3)	4.63	
				124	123.905 274 3	(17)	5.79	
51	Sb	121.753		121	120.903 821 2	(29)	57.3	
				123	122.904 216 0	(24)	42.7	
52	Te	127.60	(3)	120	119.904 048	(21)	0.096	*e,f*
				122	121.903 050	(3)	2.60	
				123	122.904 271 0	(22)	0.908	*h* 1.2×10^{13} a
				124	123.902 818 0	(18)	4.816	
				125	124.904 428 5	(25)	7.14	
				126	125.903 309 5	(25)	18.95	
				128	127.904 463	(4)	31.69	
				130	129.906 229	(5)	33.80	
53	I	126.90447	(3)	127	126.904 473	(5)	100	
54	Xe	131.29	(2)	124	123.905 894 2	(22)	0.10	*e–g*
				126	125.904 281	(8)	0.09	
				128	127.903 531 2	(17)	1.91	
				129	128.904 780 1	(21)	26.4	
				130	129.903 509 4	(17)	4.1	
				131	130.905 072	(5)	21.2	
				132	131.904 144	(5)	26.9	
				134	133.905 395	(8)	10.4	
				136	135.907 214	(8)	8.9	
55	Cs	132.90543	(5)	133	132.905 429	(7)	100	
56	Ba	137.327	(7)	130	129.906 282	(8)	0.106	*e,f*
				132	131.905 042	(9)	0.101	
				134	133.904 486	(7)	2.417	
				135	134.905 665	(7)	6.592	
				136	135.904 553	(7)	7.852	
				137	136.905 812	(6)	11.23	
				138	137.905 232	(6)	71.70	
57	La	138.9055	(2)	138	137.907 105	(6)	0.09	*e,f,h* 1.05×10^{11} a
				139	138.906 347	(5)	99.91	
58	Ce	140.115	(4)	136	135.907 140	(50)	0.19	*e,f*
				138	137.905 985	(12)	0.25	

Table I (*continued*)

Atomic no. Z	Element	Atomic weight[a] (u)		Mass no. A	Atomic mass[b] (u)		Isotopic composition[c] (%)	Notes*
				140	139.905 433	(4)	88.48	
				142	141.909 241	(4)	11.08	h >5 × 10^{16} a
59	Pr	140.90765	(3)	141	140.907 647	(4)	100	
60	Nd	144.24	(3)	142	141.907 719	(4)	27.13	e,f
				143	142.909 810	(4)	12.18	
				144	143.910 083	(4)	23.80	h 2.1 × 10^{15} a
				145	144.912 570	(4)	8.30	
				146	145.913 113	(4)	17.19	
				148	147.916 889	(4)	5.76	
				150	149.920 887	(4)	5.64	
61	Pm							
62	Sm	150.36	(3)	144	143.911 998	(4)	3.1	e,f
				147	146.914 894	(4)	15.0	h 1.0 × 10^{11} a
				148	147.914 819	(4)	11.3	h 8 × 10^{15} a
				149	148.917 180	(4)	13.8	h >1 × 10^{16} a
				150	149.917 273	(4)	7.4	
				152	151.919 728	(4)	26.7	
				154	153.922 205	(4)	22.7	
63	Eu	151.965	(9)	151	150.919 847	(4)	47.8	e,f
				153	152.921 225	(4)	52.2	
64	Gd	157.25	(3)	152	151.919 786	(4)	0.20	e,f,h 1.1 × 10^{14} a
				154	153.920 861	(4)	2.18	
				155	154.922 618	(4)	14.80	
				156	155.922 118	(4)	20.47	
				157	156.923 956	(4)	15.65	
				158	157.924 099	(4)	24.84	
				160	159.927 049	(4)	21.86	
65	Tb	158.92534	(3)	159	158.925 342	(4)	100	
66	Dy	162.50	(3)	156	155.924 277	(8)	0.06	f,h 2 × 10^{14} a
				158	157.924 403	(5)	0.10	
				160	159.925 193	(4)	2.34	
				161	160.926 930	(4)	18.9	
				162	161.926 795	(4)	25.5	
				163	162.928 728	(4)	24.9	
				164	163.929 171	(4)	28.2	
67	Ho	164.93032	(3)	165	164.930 319	(4)	100	
68	Er	167.26	(3)	162	161.928 775	(4)	0.14	
				164	163.929 198	(4)	1.61	
				166	165.930 290	(4)	33.6	
				167	166.932 046	(4)	22.95	
				168	167.932 368	(4)	26.8	
				170	169.935 461	(4)	14.9	
69	Tm	168.93421	(3)	169	168.934 212	(4)	100	
70	Yb	173.04	(3)	168	167.933 894	(5)	0.13	
				170	169.934 759	(4)	3.05	
				171	170.936 323	(3)	14.3	
				172	171.936 378	(3)	21.9	
				173	172.938 208	(3)	16.12	
				174	173.938 859	(3)	31.8	
				176	175.942 564	(4)	12.7	
71	Lu	174.967	(1)	175	174.940 770	(3)	97.41	f
				176	175.942 679	(3)	2.59	h 2.7 × 10^{10} a
72	Hf	178.49	(2)	174	173.940 044	(4)	0.162	h 2 × 10^{15} a
				176	175.941 406	(4)	5.206	
				177	176.943 217	(3)	18.606	
				178	177.943 696	(3)	27.297	
				179	178.945 812 2	(29)	13.629	
				180	179.946 545 7	(3)	35.100	
73	Ta	180.9479	(1)	180	179.947 462	(4)	0.012	h >1.6 × 10^{13} a
				181	180.947 992	(3)	99.988	

Table I (*continued*)

Atomic no. Z	Element	Atomic weight[a] (u)		Mass no. A	Atomic mass[b] (u)		Isotopic composition[c] (%)	Notes*
74	W	183.85	(3)	180	179.946 701	(5)	0.13	
				182	181.948 202	(3)	26.3	
				183	182.950 220	(3)	14.3	
				184	183.950 928	(3)	30.67	
				186	185.954 357	(4)	28.6	
75	Re	186.207	(1)	185	184.952 951	(3)	37.40	
				187	186.955 744	(3)	62.60	*h* 5×10^{10} a
76	Os	190.2	(1)	184	183.952 488	(4)	0.02	*e,f*
				186	185.953 830	(4)	1.58	
				187	186.955 741	(3)	1.6	
				188	187.955 830	(3)	13.3	
				189	188.958 137	(4)	16.1	
				190	189.958 436	(4)	26.4	
				192	191.961 467	(4)	41.0	
77	Ir	192.22	(3)	191	190.960 584	(4)	37.3	
				193	192.962 917	(4)	62.7	
78	Pt	195.08	(3)	190	189.959 917	(7)	0.01	*h* 7×10^{11} a
				192	191.961 019	(5)	0.79	
				194	193.962 655	(4)	32.9	
				195	194.964 766	(4)	33.8	
				196	195.964 926	(4)	25.3	
				198	197.967 869	(6)	7.2	
79	Au	196.96654	(3)	197	196.966 543	(4)	100	
80	Hg	200.59	(3)	196	195.965 807	(5)	0.14	
				198	197.966 743	(4)	10.02	
				199	198.968 254	(4)	16.84	
				200	199.968 300	(4)	23.13	
				201	200.970 277	(4)	13.22	
				202	201.970 617	(4)	29.80	
				204	203.973 467	(5)	6.85	
81	Tl	204.3833	(2)	203	202.972 320	(5)	29.524	
				205	204.974 401	(5)	70.476	
82	Pb	207.2	(1)	204	203.973 020	(5)	1.4	*d–f,h* 1.4×10^{17} a
				206	205.974 440	(4)	24.1	
				207	206.975 872	(4)	22.1	
				208	207.976 627	(4)	52.4	
83	Bi	208.98037	(3)	209	208.980 374	(5)	100	
84	Po							
85	At							
86	Rn							
87	Fr							
88	Ra							
89	Ac							
90	Th	232.0381	(1)	232	232.038 050 8	(23)	100	*e,h* 1.4×10^{10} a
91	Pa							
92	U	238.0289	(1)	234	234.040 946 8	(24)	0.0055	*e–h* 2.44×10^{5} a
				235	235.043 924 2	(24)	0.7200	*h* 7.04×10^{8} a
				238	238.050 784 7	(23)	99.245	*h* 4.47×10^{9} a

* The general symbol for the year is a.

[a] Values are those recommended by the Commission on Atomic Weights and Isotopic Abundances, Inorganic Chemistry Division, IUPAC, in *Pure Appl. Chem.* **60,** 841 (1988). Uncertainty is in parentheses.

[b] Values are given by A. H. Wapstra and G. Audi, *Nucl. Phys.* **A432,** 1 (1985). Uncertainty is in parentheses.

[c] Values are those recommended by the Commission on Atomic Weights and Isotopic Abundances, IUPAC, in *Pure Appl. Chem.* **56,** 675 (1984).

[d] Variations in isotopic composition prevent a more precise atomic weight being given.

[e] Specimens having anomalous isotopic composition are known in which the atomic weight may exceed the uncertainties given.

[f] Element has a highly anomalous composition in certain specific geological specimens.

[g] Commercially available element may vary in isotopic composition for inadvertent or undisclosed reasons.

[h] Radioactive isotope; half-life is given.

the atomic masses and the mass differences. The values of the atomic masses given in Table I have been derived in this manner.

The distribution and relative abundances of the naturally occurring nuclides also reflect certain features of the nuclear force.

The 287 nuclear species that occur in nature in appreciable quantities (Table I) may be classified as follows: 168 with even Z and even N; 57 with even Z and odd N; 53 with odd Z and even N; and 9 with odd Z and odd N. From this it is seen that there is a strong tendency for protons to pair with protons and neutrons to pair with neutrons. The distribution also suggests that odd-neutron configurations are similar to odd-proton ones and that the nuclear force is charge independent. This is further supported by the tendency toward equal numbers of neutrons and protons in light nuclei.

Additionally, the shell nature of the nuclear force is reflected in the incidence of stable nuclides. The existence of a large number of stable isotopes is consistent with a particularly stable arrangement of protons. For example, tin, with $Z=50$, has the largest number of stable isotopes. Moreover, for a particular isotope having a magic number of neutrons, the isotopic abundance will be larger than would otherwise be expected. In barium, which has seven stable isotopes, the heaviest one, ^{138}Ba, for which $N=82$, has a relative abundance of 71.9%.

Examination of Table I shows that there is a stable nuclide for every value of $Z \leqq 83$, except for $Z=43$ (Tc) and for $Z=61$ (Pm). At least one stable nuclide occurs for each value of $A \leqq 209$ except for $A=5$ and $A=8$.

In general, even-Z elements have many more isotopes than their nearby odd-Z neighbors. This is especially evident above $A=16$. Only in one case ($Z=19$, K) throughout the entire table does an odd-Z element have more than two isotopes. By contrast, among the even-Z elements, only Be ($Z=4$) exists in a single stable form, and all elements from oxygen and above ($Z \geqq 8$) have at least three stable isotopes except the heavy radioactive element Th ($Z=90$).

Mattauch's rule, reflecting the properties of β decay, excludes the existence of stable isobars (nuclides with the same value of A) whose atomic numbers differ by unity. In each of the ten groups of naturally occurring adjacent isobars, one has been found to be radioactive, as indicated in Table I.

Of the 287 naturally occurring species appearing in the table, 23 are now known to be radioactive, decaying with long half-lives, comparable at least to the age of the earth (4.5×10^9 years). Of these, seven have half-lives between 10^{11} years and 10^{15} years, whereas seven have half-lives from 10^{15} years to 10^{17} years.

NATURAL ISOTOPIC ABUNDANCES

The relative amounts of various elements have been seen to vary widely, depending on the location in the solar system. However, for many elements the relative isotopic abundances seem to be remarkably constant. Moreover, for these elements the same abundances are found in terrestrial, lunar, and meteoritic samples. Accordingly, it is presumed that the observed relative abundances reflect the process by which the elements were originally formed in the solar system. For this reason the reproduction of the observed isotopic abundances by any proposed mechanism of nucleosynthesis constitutes an important requirement of the theory.

Elements in which variations in isotopic abundances do occur are therefore of special interest. Such changes may take place as a result of the very small differences in the physical properties of isotopes that are reflected in physical and chemical processes. Fractionation, or the alteration of relative isotopic abundances in this way, is invariably small for each such process. By contrast, nuclear transformations involving particular nuclides may result in dramatically altered abundances. The latter form the basis of what is sometimes called nuclear geology.

As indicated in Table I, the half-lives for the decays ^{40}K to ^{40}Ar (and ^{40}Ca), ^{87}Rb to ^{87}Sr, ^{232}Th to ^{208}Pb, ^{235}U to ^{207}Pb, and ^{238}U to ^{206}Pb are well known and are appropriate to the geological time scale. In each case the parent nuclide is present in rocks of geological interest and decays to form a particular stable daughter isotope whose abundance may be determined relative to other isotopes of that element. We can thus identify the amount of the daughter isotope that is a product of the decay and deduce from it the time interval since the mineralization of the sample.

The formation of ^{3}He, the lighter and very rare isotope of helium, is believed to take place in rocks by the nuclear reactions

$$^{6}\mathrm{Li}(n,\alpha)^{3}\mathrm{H}, \qquad ^{3}\mathrm{H}(\beta^{-})^{3}\mathrm{He},$$

and in the atmosphere by

$$^{14}\mathrm{N}(n,^{12}\mathrm{C})^{3}\mathrm{H} \quad \text{or} \quad ^{14}\mathrm{N}(n,3\alpha)^{3}\mathrm{H} \quad \text{and} \quad ^{3}\mathrm{H}(\beta^{-})^{3}\mathrm{He}.$$

In rocks the energetic neutrons come from the spontaneous fission of ^{238}U and from a variety of (α,n) reactions on various elements; in the atmosphere the neutrons are present in the cosmic rays. Variations in the relative abundance of ^{3}He reflects both the formation process and the retention of the helium. Values range widely from the one shown in Table I with observations as large as 7.7%.

A third example, in which natural abundances of isotopes have been drastically altered by nuclear reactions, is the natural nuclear reactor at the Oklo quarry in Gabon. The structure of the geological formation was such that natural uranium was sustained in a critical reaction for a significant length of time. The ore is correspondingly found to be anomalously depleted in ^{235}U, the isotope that is effectively responsible for slow-neutron-induced fission. Also, as a result of fission, anomalous isotopic abundances have been observed for Kr, Zr, Mo, Ru, Pd, Ag, Sn, Sb, Te, Xe, Ce, Nd, Sm, Eu, and Gd.

Similarly, variations in the abundances of Kr and Xe can be related to the accumulation of fission-product gases in uranium or thorium minerals arising from both neutron-induced and spontaneous fission.

Among the processes based on the differences in physical properties between isotopes (apart from nuclear properties) that may lead to isotopic fractionation, the most significant are diffusion, evaporation, and chemical exchange reactions. These three processes, which depend primarily on the relative differences in mass, are most pronounced among the light elements, especially hydrogen, carbon, nitrogen, and oxygen.

The rate of diffusion of a gas through a porous membrane is inversely proportional to the square root of the molecular weight (Graham's law). In a more elaborate geological setting, considerable variation in the $^{14}N/^{15}N$ ratio has been observed in natural gas fields and related to crude oils as a result of molecular flow through porous rock and surface diffusion.

In evaporation, the vapor is enriched in lighter isotopes whereas a corresponding enrichment in heavier isotopes occurs in the residual liquid. This has been observed in the $^{2}H/^{1}H$ and $^{18}O/^{16}O$ ratios for precipitation compared to ocean water and to isolated bodies of water from which extensive evaporation has taken place.

In chemical exchange reactions, the differences in molecular vibrational energies, moments of inertia, etc., of structurally similar molecules under equilibrium conditions lead to fractionation. For example, carbon is involved in the carbonate–atmospheric CO_2 exchange reaction

$$^{13}CO_2 + {}^{12}CO_3^{2-} \rightleftharpoons {}^{12}CO_2 + {}^{13}CO_3^{2-},$$

for which the equilibrium constant differs somewhat from unity and is temperature dependent. By making use of the empirical values of the equilibrium constant as a function of temperature and of the measured $^{13}C/^{12}C$ ratios from $CaCO_3$ in the shells of marine organisms, it is possible to determine paleo-temperatures.

Both carbon and sulfur participate in reaction cycles in which biological processes are involved, and the isotopic fractionation in these processes has been studied very extensively.

For carbon the $^{13}C/^{12}C$ ratio can be used to distinguish materials originating as limestone carbon, organic carbon of marine origin, and carbon from land plants. These variations are presumed to be determined by the influence of the carbon isotopic exchange reaction for the limestone, by the preferential assimilation of ^{12}C in photosynthesis by both marine and land plants, and by kinetic effects for the various reactions involving carbon in the atmosphere.

The $^{32}S/^{34}S$ ratio determined in meteoritic samples has been found to be remarkably invariant and is presumed to represent an average value for all terrestrial material. However, individual terrestrial deposits are found to vary widely, primarily as a result of the action of bacteria that reduce sulfate to produce H_2S or free sulfur, either of which is depleted in ^{34}S with a corresponding enrichment of ^{34}S in the sulfate.

SEPARATED ISOTOPES

Inasmuch as nuclear structure and properties are unique to particular nuclides, there are many experiments in nuclear physics where it is desirable or essential to use material enriched in, or consisting solely of, a particular isotope. Such a case would be the studies of a reaction on a particular nucleus. Similarly, nuclear properties determine the use of certain isotopes in reactor applications where large quantities of the material are required, e.g., deuterium in heavy-water-moderated natural uranium reactors, or fuel rods enriched in ^{235}U for light-water-moderated reactors.

Stable isotopes may be used to examine biochemical reactions by the introduction of material with an abnormal isotopic composition that can be monitored at subsequent stages. The method is suitable for reactions involving oxygen and nitrogen where radioactive isotopes suitable for tracer work do not exist.

A variety of techniques by which differing degrees of isotope separation are achieved are summarized here. The processes of gaseous diffusion, evaporation, and chemical exchange, mentioned in the previous section, have been exploited on a large scale. Other processes in which separation takes place include the following:

1. **Centrifuging.** The sample in a centrifuge experiences an effective force that depends on the mass; hence the separation factor depends on the ratio of the masses.
2. **Electrolysis.** When a water solution of NaOH has been electrolyzed, the residual electrolyte is enriched in ^{2}H.
3. **Electromagnetic separation.** Mass spectrometers with ion sources capable of producing large elemental ion beams may be used to achieve high degrees of separation for small quantities of virtually any isotope.
4. **Laser separation.** The optical spectra of atoms and molecules depend on the mass, volume, and shape of the nuclei involved. Laser light may be used to excite and then ionize a given isotope. The ions are then separated from residual material by an electric field.

ATOMIC WEIGHT

Current values of the atomic weights of the elements as recommended by the Commission on Atomic Weights of the Inorganic Chemistry Division of IUPAC are given in Table I. The determination of isotopic abundances by mass spectroscopic methods combined with the very precisely known atomic masses usually yields the most precise values of atomic weights. In certain unfavorable cases, however, chemical methods are of comparable precision, as described in the detailed IUPAC report on atomic weights of the elements (1987).

In Table I, attention is drawn to the fact that there are commercially available materials whose isotopic composition has been altered from that of the natural material. Materials that vary in isotopic abundances according to the source are so indicated in the table.

See also Atomic Spectroscopy; Atoms; Elements; Geochronology; Isotope Effects; Isotope Separation; Nuclear Properties; Radioactivity.

BIBLIOGRAPHY

F. W. Aston, *Mass Spectra and Isotopes*. Edward Arnold, London, 1942. (I)

H. E. Duckworth, R. C. Barber, and V. S. Venkatasubramanian, *Mass Spectroscopy*, 2nd Edition. Cambridge University Press, Cambridge, 1986. (A)

IUPAC Inorganic Division, Commission on Atomic Weights and Isotopic Abundances, "Element by Element Review of Their Atomic Weights," *Pure Appl. Chem.* **56,** 695 (1984). (A)

IUPAC Inorganic Division, Commission on Atomic Weights and Isotopic Abundances, "Atomic Weights of the Elements 1987," *Pure Appl. Chem.* **60,** 841 (1988). (I)

K. Rankama, *Isotope Geology*. McGraw-Hill, New York, 1954. (I)

A. Romer, *Radiochemistry and the Discovery of Isotopes*. Dover, New York, 1970. (E)

M. A. Preston, *Physics of the Nucleus*. Addison-Wesley, Reading, MA, 1962. (A)

F. W. Walker, D. G. Miller, and F. Feiner, *Chart of the Nuclides,* 13th Edition. Available from General Electric Company, Nuclear Energy Operations, 175 Curtner Ave., M/C 684, San Jose, CA 95125. (E,I)

F. A. White and G. M. Wood, *Mass Spectrometry, Applications in Science and Engineering*. John Wiley and Sons, New York, 1986. (A)

Jahn–Teller Effect

B. R. Judd

The Jahn–Teller effect refers to the tendency of molecular systems to distort when electronic degeneracy is present. The electronic energy of such a system almost always depends linearly on a displacement coordinate d of some kind, while the elastic energy resisting the distortion is invariably a quadratic function of d. Consequently, the total energy of the system can be lowered if a distortion takes place. There are two exceptions to this general rule. Linear molecules (like CO_2) do not exhibit the required linearity of the electronic energy with respect to d, and may or may not distort. In addition, all molecular systems possessing an odd number of electrons necessarily exhibit Kramers degeneracy, a phenomenon that pairs every quantum-mechanical state with its time-reversed companion. Such twofold degeneracy cannot be split by distortions of the nuclear frame, and thus the inducement to undergo a distortion is absent. The possible existence of the Jahn–Teller effect was pointed out to Teller by Landau in 1934, and a systematic study of the symmetry types for which the effect can be expected was carried out by Jahn (in collaboration with Teller) and presented in 1936. In that form it is referred to today as the "static" Jahn–Teller effect. It is considered responsible for the comparatively low symmetry of the sites of many transition-metal ions in crystals. The paradoxical result that a Hamiltonian with a definite symmetry may lead to systems with lower symmetry is resolved by noting that, in principle, the many equivalent systems of lower symmetry are connected in virtue of quantum-mechanical tunneling. Given enough time (perhaps an eternity), all systems will be represented in a way consistent with the original symmetry. The apparent preference of Nature for broken symmetries has been commented on in recent years by Teller in the wider context of elementary particles.

Because the electronic energy depends on d, there is a coupling between the electronic states and vibrations. Thus the absorption line of an electron trapped in an oxygen vacancy in CaO, which corresponds to an excitation from an s state to a p state, exhibits a broad absorption band on its low-wavelength side. Some of the structure can be associated with the excitation of vibrational modes of the octahedron of calcium ions surrounding the defect site. This is referred to as the "dynamic" Jahn–Teller effect. It is often accompanied by a partial quenching of electronic properties such as the spin-orbit coupling, a phenomenon known as the Ham effect after its discoverer.

The static Jahn–Teller effect provides good examples of phase differences between physically identical states of the same system (referred to as Berry's phase). In the theoretical limit of strong coupling between the p orbital and the displacements of the calcium ions in the example mentioned above, it can be shown that a rotation by 180° of the p orbital (which interchanges the positive and negative lobes) brings the calcium ions back to their original positions. The system is physically unchanged, but the rotation produces a phase reversal in the wave function of the entire system.

See also Geometric Quantum Phase; Symmetry Breaking, Spontaneous; Tunneling.

BIBLIOGRAPHY

R. Englman, *The Jahn–Teller Effect in Molecules and Crystals*. Wiley-Interscience, New York, 1972. (I)

Yu. E. Perlin and M. Wagner (eds.), *The Dynamical Jahn–Teller Effect in Localized Systems*. North-Holland, Amsterdam, 1984. (A)

E. Teller, "The Jahn–Teller Effect—Its History and Applicability," in *Group-Theoretical Methods in Physics* (L. L. Boyle and A. P. Cracknell, eds.). North-Holland, Amsterdam, 1982. (E)

Josephson Effects

D. J. Scalapino

Brian Josephson won the Nobel Prize in 1973 for theoretical work he carried out on the properties of two superconducting metals separated by a thin insulating oxide layer. This structure, a sandwich in which the two superconducting films are the bread with the oxide analogous to a very thin slice of cheese separating the metal films, is now often called a Josephson tunnel junction, and the phenomena associated with the superconducting current flow between the metals are called the Josephson effects. Figure 1 shows a typical junction configuration. Here two cross strips of Pb are separated by a thin oxide layer 10–20 Å in thickness.

In order to understand the Josephson phenomena it is essential to realize that the superconducting state exhibits quantum-mechanical properties on a macroscopic scale. In a superconducting metal, pairs of electrons are bound together by an attractive interaction mediated by the motion of the ionic lattice. According to the basic laws of quantum mechanics, the motion of the center of mass of the pairs is described by a wave. Now, ordinarily the quantum-mechanical waves of the many electrons which make up a metal differ from one another, so that on a macroscopic scale we are not directly aware of the phase properties of the electron wave functions. However, in the superconducting state, the state of lowest free energy corresponds to a situation in which all of the pairs have the same center-of-mass wave function. With this vast number of electron pairs having the same wave structure, it becomes possible to observe the relative change in phase $\Delta\varphi$ of the center-of-mass wave function.

The first Josephson equation states that if there is a phase difference $\Delta\varphi$ between the pair waves in the superconductors on the two sides of a Josephson junction, pairs will tunnel through the oxide giving rise to a supercurrent

$$I = I_1 \sin \Delta\varphi. \tag{1}$$

This is the dc Josephson effect. It predicts, for example, that if a phase difference $\Delta\varphi = \pi/2$ exists between the superconductors on either side of a Josephson junction, then a current I_1 will flow across the junction in the absence of a voltage difference. Naturally, one needs leads on the junction to supply and remove this current, or excess charge will build up giving rise to a voltage which produces further effects which we will discuss below.

Suppose leads from a current source are attached across the junction shown in Fig. 1 and another set of high-impedance leads are used to monitor the voltage V across the junction. The low-temperature $I(V)$ characteristic of the junction will look like that shown in Fig. 2. At low temperatures, the current I passing through the junction can be increased at zero voltage to a critical value beyond which the junction will switch (dashed line of Fig. 2) to a state with a finite voltage. Then as the current is reduced, the voltage will decrease toward zero as shown by the arrow on the solid line in Fig. 2. The zero-voltage current corresponds to the dc Josephson effect and was first reported by P. W. Anderson and J. M. Rowell. The magnitude of the critical current $I_1(T)$ is temperature dependent, vanishing above the superconducting transition temperature T_c and approaching a constant $I_1(0)$ at low temperatures. For a junction composed of the same type of superconducting metal on both sides, $I_1(0)$ is equal to the current which is carried by the junction in the normal (nonsuperconducting) state at a voltage $\pi\Delta/2e$, where Δ is the superconducting energy gap and $I_1(T)/I_1(0) = \tanh(\Delta/2kT)$. A typical Josephson junction will have a current density of order amperes per square centimeter while a very strongly coupled junction can have a current density several hundred times larger.

If a voltage difference V exists across the junction, quantum mechanics implies that the relative difference $\Delta\varphi$ in the pair phase between the two superconductors changes at a rate set by$2eV/\hbar$:

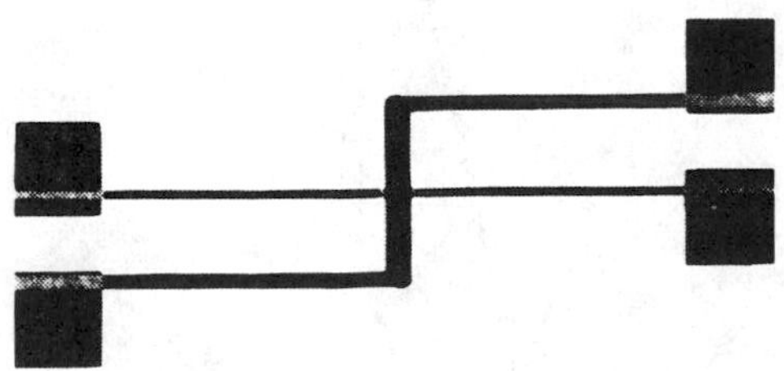

FIG. 1. A typical Josephson tunnel junction formed by two overlapping Pb strips separated by an oxide layer. The glass slide on which this junction is mounted is about an inch square.

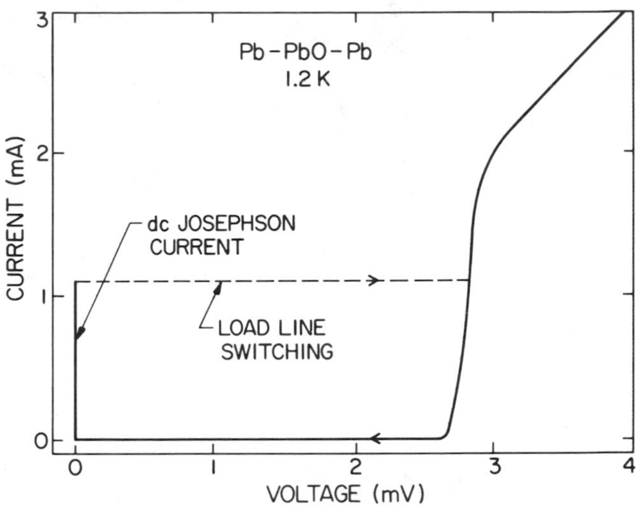

FIG. 2. The $I(V)$ characteristic of a Pb-PbO-Pb Josephson junction showing the dc Josephson current and the quasiparticle current onsetting at a voltage $2\Delta/e$.

$$\frac{\partial}{\partial t}\Delta\varphi = \frac{2eV}{\hbar}. \tag{2}$$

Here e is the electron charge, $\hbar$ is Planck's constant divided by 2π, and the 2 arises from the fact that we are dealing with electron pairs. If the voltage is constant at a value V_0, then the phase difference will increase with time, $\Delta\varphi(t) = 2eV_0t/\hbar + \Delta\varphi(0)$, and a current $I = I_1 \sin(2eV_0t/\hbar + \Delta\varphi)$ will oscillate back and forth across the oxide at a frequency $f = 2eV_0/h$. Using the values of e and h, one finds that this frequency is approximately 500 MHz/μV. This is the ac Josephson effect.

In order to observe the ac Josephson effect, Josephson suggested applying an rf field to the junction and looking for current steps in the dc $I(V)$ characteristic. With both a dc bias voltage V_0 and an rf bias voltage $V_1 \cos \omega t$ across the junction,

$$\frac{\partial\Delta\varphi}{\partial t} = \frac{2e}{\hbar}(V_0 + V_1 \cos \omega t)$$

and

$$I = I_1 \sin\left(\frac{2e}{\hbar}V_0t + \frac{2eV_1}{\hbar\omega}\sin \omega t + \Delta\varphi(0)\right).$$

This current has a dc step whenever $2eV_0 = n\hbar\omega$. Figure 3 shows this step structure for a Josephson junction irradiated by a microwave field with frequency $f \sim 12$ GHz. With the approximate value of 500 MHz/μV, this implies a step spacing of order 25 μV. The ac Josephson effect was first observed in this way by S. Shapiro.

The frequency of the ac Josephson current can be measured with high precision. If, in addition, a careful measurement of the applied voltage V_0 is made, the ratio of the fundamental constants $2e/h$ can be obtained. Measurements of this type were carried out by Taylor, Parker, and Langenberg, who found that $2e/h = 4.835976(12) \times 10^{14}$ Hz/V_{NBS}. Here V_{NBS} is the NBS volt. Given a value of $2e/h$ and the

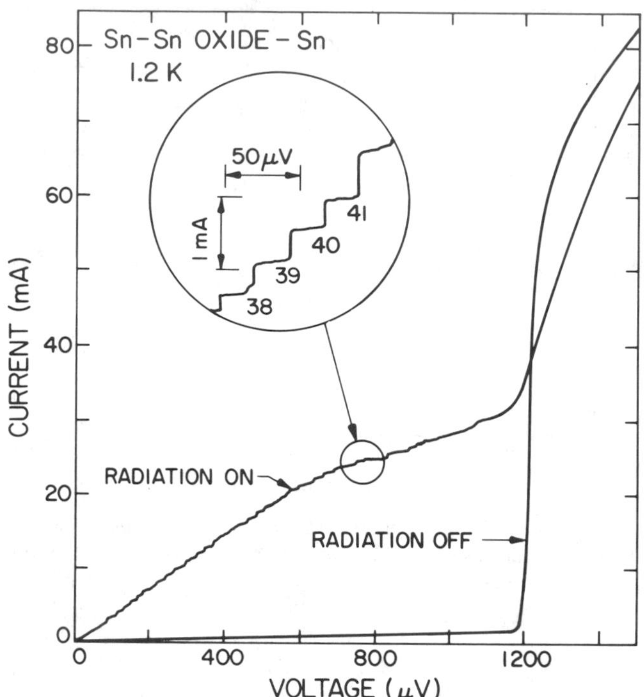

FIG. 3. The $I(V)$ characteristic of a Josephson junction in the presence of a microwave field of frequency f shows vertical steps separated by $hf/2e$.

ability to measure frequency, the Josephson relation $f=2eV_0/h$ can be used to determine V_0. This type of measurement is, in fact, used to define the present NBS voltage standard.

The final Josephson relation concerns the behavior of a junction in a magnetic field. If a magnetic field H is applied parallel to the oxide layer, it again follows from quantum mechanics that the relative pair phase difference between the superconducting film on opposite sides of the junction varies with the position in the plane of the junction. Specifically, the change of $\Delta\varphi$ with distance is greatest as one moves in the plane of the junction perpendicular to the magnetic field. For identical superconducting films which are thick compared to the penetration depth λ of the magnetic field into the superconducting film, the change of $\Delta\varphi$ over a distance Δx perpendicular to the field H is $(2e/\hbar c)H(2\lambda+l)\Delta x$. Here l is the thickness of the oxide and $\hbar$ is Planck's constant divided by 2π. This means that the current density varies with position. For a junction lying in the x-y plane with a uniform field H in the oxide parallel to the y axis, the current density is

$$j(x)=j_1 \sin\left(\frac{2eH(2\lambda+l)x}{\hbar c}+\Delta\varphi(0)\right),$$

with $j_1=I_1/A$, where A is the area of the junction.

This effect of a magnetic field H on the current density is illustrated in Fig. 4. In part a the magnetic field is zero, and the current density is uniform. As the field is increased, parts b–d, the current density oscillates in a direction perpendicular to H. The value H_0 is just such that $2eH_0(2\lambda+l)\cdot L/\hbar c=2\pi$, where L is the length of the junction perpendicular to the field direction. This corresponds to one flux unit $hc/2e$ passing through the junction. When H is equal to H_0, Fig. 4c, the current density makes one complete oscillation and no net current flows across the junction. In Fig. 4d the field has been increased to $3H_0/2$ so that 3/2 wavelengths fit into the junction. In this case a net current smaller than that shown for Figures a or b can flow across the junction.

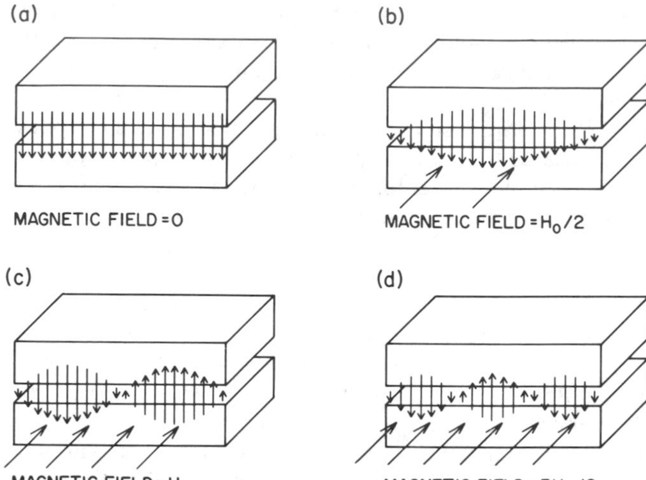

FIG. 4. Effect of a magnetic field on the current density in a Josephson junction.

A plot showing how the critical junction dc current depends on H is shown in Fig. 5. This single-slit-like interference pattern, first observed by J. Rowell, shows in a vivid manner the de Broglie wave properties of electron pairs in superconductors. Thus, by monitoring the Josephson current, it is possible to tell when $(2\lambda+l)LH/(hc/2e)=2\pi n$. The flux quantum $hc/2e$ corresponds to approximately 2×10^{-7} G/cm^2. Sensitive instruments can measure changes of parts in a thousand in the maximum Josephson current. Thus, Josephson junctions provide the means for making flux detectors with sensitivities of 10^{-10} G/cm^2.

Besides their use as magnetometers and voltage standards, Josephson junctions are now used in a variety of devices

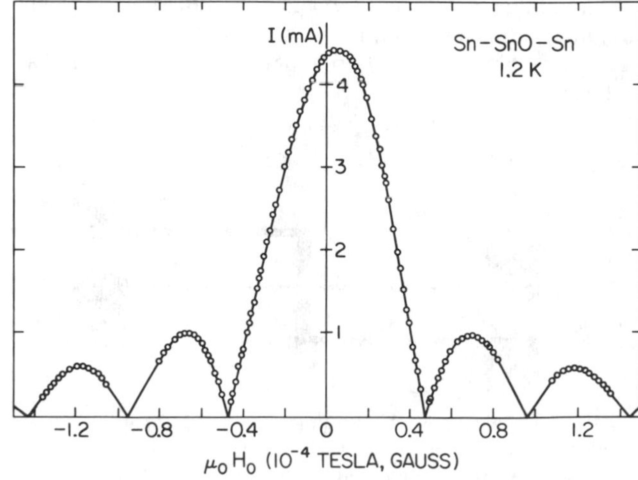

FIG. 5. The critical Josephson dc current versus magnetic field.

such as picovoltmeters and high-frequency electromagnetic detectors and mixers. Perhaps their most important applications will be as switching elements in which the junction switches from a zero-voltage current-carrying state to a finite-voltage current-carrying state (see the dashed line of Fig. 2) as the critical current I_1 is exceeded. This switching can occur in times of order tens of picoseconds and with power dissipation less than microwatts. While equivalently fast semiconductor switching elements can achieve similar switching times, their power dissipation is many orders of magnitude larger. Thus the density with which one can pack Josephson junctions can be much higher, reducing the time for electromagnetic signals to pass between them. At present, work on high-speed microcircuit technology using Josephson junctions is an active area of research.

ACKNOWLEDGMENT

The author thanks the University of Pennsylvania tunneling group for the experimental figures used. Support from the Office of Naval Research and the J. S. Guggenheim Foundation is gratefully acknowledged.

See also SUPERCONDUCTIVE DEVICES; SUPERCONDUCTIVITY THEORY.

BIBLIOGRAPHY

P. W. Anderson, in *Lectures on the Many-Body Problems* (E. R. Caianello, ed.). Academic Press, New York, 1964.

P. W. Anderson, *Physics Today*, (November, 1970).

B. D. Josephson, *Phys. Lett.* **1,** 251 (1962).

D. N. Langenberg, D. J. Scalapino, and B. N. Taylor, *Scientific American*, **214**(5), 30 (May, 1966).

Kepler's Laws

Edward H. Kerner

One of the giant decisive strides on the way toward that watershed of physics that is the Newtonian world view was taken by Johannes Kepler (1571–1630) in his *Astronomia Nova* of 1609, containing (besides much else) his first two laws:

I. the planets circulate around the sun in elliptical orbits with the sun at one focus;
II. the radius vector from the sun to a planet sweeps out equal areas in equal times;

and in his *Harmonices Mundi* of 1619, stating the third law,

III. "The ratio which exists between the periodic times of any two planets is precisely the ratio of the $\frac{3}{2}$th power of the mean distances"

(i.e., the ratio of the $\frac{3}{2}$ power of the two semimajor elliptical axes).

The laws were, with great and skillful computational labor, drawn out of the large body of pretelescopic planetary observations of the master observer Tycho Brahe (1546–1601) with whom the theoretically bent Kepler worked in Prague for a time prior to Tycho's death, continuing thereafter for years to mine Tycho's data. It is remarkable that Tycho right off gave Kepler the problem of Mars' motion, notoriously intractable since Hipparchus (and thought at first by Kepler to be manageable in two weeks work), but, as it happens, with an orbit far enough from circular to be clearly discernable as such from the accuracy (by eye observation) then available. Though the Copernican controversy was in full furore, Kepler adopted the heliocentric position definitively. However, the Platonic conception of the "perfectness" of the sphere and circle completely dominated all thinking, including Kepler's; to break this domination was to be a principal achievement. After some 70 attempts to fit Mars' orbit to a circle, employing Tychonian observations-in-opposition (for which Sun–Earth–Mars are well aligned and for which uncertainties of Earth's orbit are then minimized), Kepler's best fit to selected observations was within 2 min of arc (or about a sixteenth of the moon's angular breadth) of them, viz., within the limit of direct eye judgment. But then still other observations-in-opposition were off by as much as 8 min. These were fateful minutes: for Kepler was sure that Tycho, "that most diligent observer," could not have been that much in error: and he had perforce to give up circles, later saying ". . . These eight minutes alone have pointed the way to the reformation of the whole of astronomy."

It took another year for Kepler to realize that the Martian egg in the sky was indeed no mere oval but exactly an *ellipse*; he had gotten the equation of his oval but did not recognize it for what it was (the times were before analytic geometry); then he cast it aside, thought independently about ellipses after trials with other figures, and thereto reinvented his selfsame oval equation! With charming candor he wrote, "Why should I mince my words? The truth of Nature, which I had rejected and chased away, returned by stealth through the back door . . .."

As regards the third law, which within 70 years came to be the critical linkup to Newton's law of gravitation and equations of motion, it was a dream of Kepler's for 25 years to pierce through the (to him) tantalizing data on orbit sizes and periods to the simple rule he felt with intuitive certainty must be controlling them.

The state of Tycho's art may have been providential in another way. Had Tycho had a telescope, his observations might have been *too* good, in that deviations from Keplerian ellipses due to perturbations from other planets might well have come into range, blurring and bothering possibly fatally the simplicity of the coarser view. Similar good fortune perhaps attended the study centuries later of another Kepler-system-in-little, that of a planetary electron wheeling about a heavy attracting proton, when Balmer in 1885 discerned from the hydrogen-spectrum data of Ångström and of Huggins—nicely set between coarseness and fineness—the simple rule for the spectral wavelengths that, stamped decisively with integers, foreshadowed quantum physics.

While uncovering the geometrodynamical peaks of the world-machine, Kepler was also acutely sensitive all along to the question of how the machine could operate as pure clockwork from fundamental universal law applying both terrestrially and cosmically, recognizing the need for a "true doctrine concerning gravity" and for a comprehensive theory of motion. But, wholly in thrall to the Aristotelian (mis-) conception about inertia, by which a body had to be shoved along to sustain even uniform motion (otherwise coasting to rest), his quest after the clockwork had to go unfulfilled. Even so, Kepler had the brilliant physical intuition that the Sun as center of both light and motion produced its "moving effect" on planets so as to be weakened "through spreading from the Sun in the same manner as light"; but then, in thought of spreading in a *plane*, Kepler adduced an inverse *first-power* law that was convenient to his Aristotelian theorizing.

On the Newtonian base, the laws of Kepler fall into place swiftly and simply. If we assume for simplicity that the Sun's mass M is so great that it may be taken to be at the center of an inertial frame, a planet of mass m at radius vector $\mathbf{r}$ moves according to the equation (G is the universal gravitational constant)

$$m\ddot{\mathbf{r}} = -\frac{GMm}{r^2}\hat{r}$$

(note too the powerful assertion that inertial mass on the left is equal to gravitational mass on the right, exemplifying the principle of equivalence that undergirds the later Einsteinian gravitation). Two conservation laws follow readily, reflecting the rotational invariance and time-translational invariance of the equation of motion:

$$\mathbf{r}\times m\dot{\mathbf{r}}=\mathbf{L}=\text{conserved vector angular momentum},$$

$$\tfrac{1}{2}m(\dot{r}^2+r^2\dot{\theta}^2)-\frac{GMm}{r}=E=\text{conserved energy}$$

(plane polar coordinates r,θ for the position of m). The scalar content of the first is $mr^2\dot{\theta}=L$ or $\frac{1}{2}r(rd\theta)=dS$ (the area swept out in time dt) $=(L/2m)dt$. Hence Kepler's second law (which was actually discovered first) with its physically vivid portrayal that m at close approach to M must be speeding along, while at the far reach of orbit it must be relatively lolling. In the statement of energy conservation, one may introduce $\dot{\theta}=L/mr^2$ and then go to the geometry of the orbit by $dr/dt=(dr/d\theta)\dot{\theta}=(dr/d\theta)(L/mr^2)$ whose differential equation then integrates to the polar equation of an ellipse with focus at M,

$$\frac{1}{r}=\frac{GMm^2}{L^2}+\left(\frac{2mE}{L^2}+\frac{G^2m^4M^2}{L^4}\right)^{1/2}\cos\theta$$

provided $E<0$. The third law follows from integration of the first, $S=\pi ab=(L/2m)T$ (T is the period and a and b are semimajor and semiminor axes), upon using the geometry of the preceding ellipse for $b=b(a)$, giving $T=2\pi(GM)^{-1/2}a^{3/2}$.

Keplerian "harmony" in more recent times has been embraced in the perhaps wider term "symmetry," and in the Kepler problem a further symmetry beyond those of space rotations and time translations shows itself in the conserved Runge–Lenz vector

$$\mathbf{M}\equiv\dot{\mathbf{r}}\times\mathbf{L}-GMm\hat{r}.$$

It is orthogonal to $\mathbf{L}$, lying therefore in the plane of motion, and it is an expression of the fact that the Kepler orbits are closed. The higher symmetry of the Kepler problem, with all of E, $\mathbf{L}$, $\mathbf{M}$ conserved, is that of the four-dimensional orthogonal group usually designated as $O(4)$.

The Kepler problem has continued to course like some deep inevitable stream in Nature's ocean or man's mind down through the centuries from Kepler and Newton on celestial planes, to Balmer, Bohr, and Schrödinger at the atomic level, to Einstein and Schwarzschild again cosmically but on a grander geometrorelativistic base for gravitation, to Dirac and Lamb in our own day once again atomistically. Through all the refinements and revolutions of later times, Kepler's laws and Newton's synthesis stand steadfastly as foundation stones.

See also CONSERVATION LAWS; GROUP THEORY IN PHYSICS; NEWTON'S LAWS.

BIBLIOGRAPHY

Alexander L. Fetter and John D. Walecka, *Theoretical Mechanics of Particles and Continua*. McGraw-Hill, New York, 1980.

Arthur Koestler, *The Watershed: A Biography of Johannes Kepler*. Anchor Books, Garden City, New York, 1960.

L. Pars, *Treatise on Analytical Dynamics*. Wiley, New York, 1965.

Kerr Effect, Electro-optical

Robert E. Hebner, Jr.

The electro-optic Kerr effect is the name given to a modification of the index of refraction of a material by an electric field E. In general, the field induces a difference between two components of the index of refraction, Δn, which can be expressed as

$$\Delta n=\alpha+\beta E+\gamma E^2+\cdots. \tag{1}$$

For the purpose of this description, the fact that the elements of this equation are tensors has been neglected and only scalar quantities are considered. If only the quadratic term is nonzero, Eq. (1) describes the Kerr effect. Similarly, if the first-order term is the only one which is nonzero, Eq. (1) describes the Pockels effect.

For the Kerr effect, Eq. (1) can be written as

$$n_{\parallel}-n_{\perp}=\lambda BE^2, \tag{2}$$

where $n_{\parallel}$ and $n_{\perp}$ are the indices of refraction of the electro-optic material for light polarized parallel and perpendicular to the electric field, respectively; λ is the wavelength of the transmitted light; B is the Kerr coefficient of the material; and E is the applied electric field.

Measurements of the difference between the indices of refraction are usually made using a laser—a convenient monochromatic light source.

The Kerr effect can be modeled as dealing with two different electric fields: one called the orienting field and the other the sensing field. To appreciate the distinction between the two, it is useful to consider, qualitatively, the behavior of a system on a molecular scale. Assume that the molecules in question have some electrical anisotropy, e.g., a dipole moment. If a low-intensity, linearly polarized light beam is passed through the material, no net effect of the anisotropy is observed because the dipoles are randomly oriented. The electric field associated with this light beam is the sensing field. If a second field, with sufficiently high intensity, is applied, a detectable alignment of the dipoles results. This second field is the orienting field. It should be emphasized that the distinction between the orienting and sensing field is based solely on intensity.

It should be recognized that the Kerr effect is primarily due to molecular reorientation for orienting fields with temporal variations slow compared to molecular relaxation time and due to molecular deformation for high-frequency orienting fields, e.g., high-power laser beams.

Values for Kerr coefficients range from 10^{-21} m/V^2 for helium gas, to 10^{-12} m/V^2 for nitrobenzene, to 10^{-7} m/V^2 for bentonite in water. These depend on temperature, wavelength of the light, and frequency of the orienting field.

The Kerr effect has been used in a variety of applications. It is the basis for a very accurate measurement technique for high-voltage pulses. The Kerr effect has been used to measure the electric field and space-charge distributions in insulating materials. Measurements of the Kerr effect have provided insight into molecular structure and into molecule–molecule interactions in the liquid and gaseous states. Optical shutters and digital light beam deflectors based on the Kerr effect have been constructed using a variety of electro-optic materials.

See also DIELECTRIC PROPERTIES; KERR EFFECT, MAGNETO-OPTICAL; OPTICAL ACTIVITY; POLARIZED LIGHT; REFRACTION.

BIBLIOGRAPHY

Elementary

F. A. Jenkins and H. E. White, *Fundamentals of Optics*, 3rd ed. McGraw-Hill, New York, 1976.

R. W. Wood, *Physical Optics*. Dover Publications, New York, 1967.

Intermediate

E. Fredericq and C. Houssier, *Electric Dichroism and Electric Birefringence*. Clarendon Press, Oxford, 1973.

R. E. Hebner *et al.*, Proc. IEEE **65,** 1524 (1977).

C. G. LeFevre and R. J. W. LeFevre, *Rev. Pure Appl. Chem.* **5,** 261 (1955).

C. T. O'Konski, *Molecular Electro-Optics: Part 1—Theory and Methods*. Marcel Dekker, New York, 1976.

Kerr Effect, Magneto-optical

J. F. Dillon, Jr.

When polarized light is reflected from the surface of a magnetic material, careful measurements show that there are small changes in polarization state attributable to the magnetization. This phenomenon is called the magneto-optical Kerr effect after its discoverer, the Scottish physicist John Kerr (1824–1907). Though usually thought of relative to visible light, the effects may be seen with radiation anywhere in the electromagnetic spectrum. They are closely related to the magneto-optical rotation of the axis of linear polarization seen on transmission through a magnetized material (Faraday rotation) and to first-order Raman scattering from magnetic excitations. All three can be attributed to off-diagonal terms in the dielectric or magnetic susceptibility tensors. For instance in the dielectric case which is dominant for visible light, the tensor $\dot{\epsilon}$ in the relation $\mathbf{D}=\epsilon_0\dot{\epsilon}\mathbf{E}$ has the following form for cubic and isotropic materials (to first order in **M**):

$$\epsilon=\epsilon_0\begin{bmatrix} 1 & -iQM_z & iQM_y \\ iQM_z & 1 & -iQM_x \\ -iQM_y & iQM_x & 1 \end{bmatrix}$$

In this, ϵ_0 is the dielectric constant in the absence of magnetization and M_i are the components of the magnetization. The complex quantity Q is a magneto-optical parameter which varies with material, frequency, and temperature. A quantum-mechanical theory for Q is based on the differing probabilities for transitions between ground and excited states for the two signs of circular polarization. Note that the off-diagonal terms are first order in **M**, and thus the effects which arise from them change sign when **M** is reversed. This tensor expresses the fact that in magnetized materials a radiation field **E** not parallel to the magnetization can give rise to dielectric polarization perpendicular to both **E** and the static **M**. This new time-varying polarization then radiates,

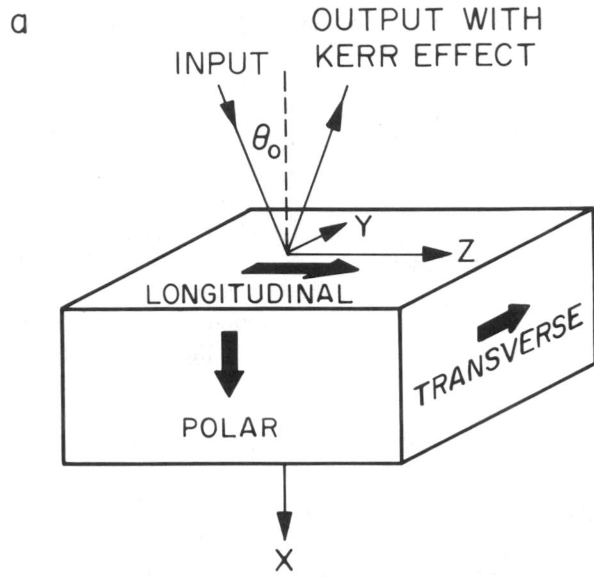

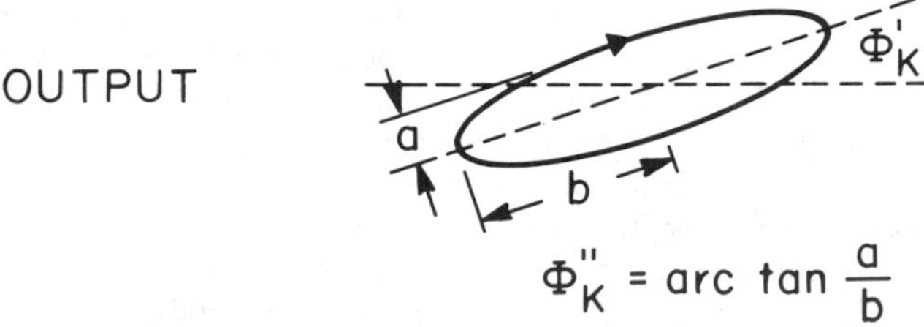

FIG. 1. (a) Magnetization orientations for the three magneto-optical Kerr effects. (b) For longitudinal and polar cases, incident light is linearly polarized with **E** parallel (p) or perpendicular (s) to the plane of incidence. The polarization of the reflected light may be an ellipse with its major axis slightly rotated from the original s- or p-axis. The sketch shows the definitions of the complex Kerr rotation.

Table I Magneto-optical Kerr Effect—Selected Values

Material	T (K)	λ_0 (nm)	Longitudinal[a] p-polarization Φ'_p (deg)	Φ''_p (deg)	Polar[b] Φ' (deg)	Φ'' (deg)	Transverse[a] ΔR_{pp} (10^{-2})
Fe[c]	300	633	0.14	0.07	0.63	−0.47	0.54
EuO[c]	4.2	550	2.35	0.00	1.60	2.01	0.39
Mn Bi[c]	300	633	0.95	0.23	2.47	0.00	2.
$Y_3Fe_5O_{12}$[c]	300	430	0.001	0.02	0.002	.05	0.001
U_3As_4[d]	15	2250	—	—	5.9	4.0	—
Amorphous $Tb_{0.21}Fe_{0.79}$[e]	300	600	—	—	0.37	0.30	—

[a] Incident from air at $\theta=45°$.
[b] Incident from air at $\theta=0°$.
[c] J. M. Judy, *Ann. NY Acad. Sci.* **189,** 239 (1972).
[d] J. Schoenes and W. Reim, *J. Magn. Mat.* **54–57,** 1371 (1986).
[e] R. Allen and G. A. N. Connell, *J. Appl. Phys.* **54,** 2353 (1982).

and the net polarization of the radiation field is altered. These first-order magneto-optical effects may be derived by solving the appropriate boundary-value problem using these dielectric and magnetic susceptibility tensors in Maxwell's equations.

Since reflection from a surface in general will produce changes in polarization state even without the presence of a magnetization, the magneto-optical Kerr effect is traditionally defined for three geometries in which nonmagnetic effects play no part. All three are illustrated in Fig. 1a. Light with an angle of incidence θ_0 from the normal is reflected at an equal angle. These two beams define the plane of incidence. In the absence of a magnetization, light which is linearly polarized with **E** in this plane (p-polarization or direction) or normal to it (s-polarization) is reflected without change of polarization. On the other hand, if the material has a magnetization, the output state may be elliptical as shown exaggerated in Fig. 1b. The angle between the major axis of this ellipse and the original s- or p-direction is called the Kerr rotation Φ'_K. The shape of the ellipse is given by the Kerr ellipticity Φ''_K as defined in Fig. 1b. Φ'_K is taken as positive if the major axis of the ellipse appears to have rotated in a counterclockwise sense to an observer looking into the beam. Φ''_K is positive if the tip of optical frequency **E** is seen to go around the ellipse in a counterclockwise sense by such an observer. However, different sign conventions have been used. Together, this rotation and ellipticity are taken to be the real and imaginary components of a complex Kerr rotation, i.e., $\hat{\Phi}_K=\Phi'_K+i\Phi''_K$.

The longitudinal (meridional) Kerr effect pertains to the geometry in which the static **M** lies both in the surface and in the plane of incidence. In the dielectric tensor $M_x=M_y=0$, $M_z\neq 0$. Both rotations and ellipticities are different in magnitude and often in sign for incident p- and s-polarizations.

The polar Kerr effect is seen when the magnetization lies normal to the surface. Only M_x is nonzero. Measurements are usually made with light incident normally, and so there is no distinction between s- and p-polarizations.

In the transverse (equatorial) geometry (only M_y nonzero) it is found that the reflection of s-polarized light is not changed at all on reversing M_y, but that there are changes in the reflectivity of p-polarized light. The effect can be seen even with unpolarized incident light.

In Table I are given representative values of the longitudinal $\Phi'_K(p)$ and $\Phi''_K(p)$ for p-polarized light, and of the polar Φ'_K and Φ''_K. The last column gives ΔR_{pp}, the fractional change in the intensity of reflected p-polarized light on the reversal of magnetization in the transverse case. Conditions must be specified for every value in Table I since all of these quantities vary with the refractive index of the incident medium, the angle of incidence, the temperature, and the wavelength. They can be drastically modified or enhanced by thin covering layers of other materials. Though polar Kerr rotation and ellipticity for Fe are about 0.5°, there are a few materials with much larger values. Usually the polar rotation is much larger than the longitudinal effect.

In magnetics research, the Kerr effects are used to visualize the distribution of magnetization, i.e., to see magnetic domains, at the surface of a sample. It is often valuable to construct magneto-optical hysteresis loops by measuring rotation or ellipticity as field is cycled. The effects are used to measure the off-diagonal elements of the dielectric tensor, and thus to obtain valuable spectroscopic information on energy levels associated with magnetization. An exceedingly important technological application of the polar Kerr effect has been developed in recent years, the erasable magneto-optical memory. In these computer memories, information is stored in the orientation of the magnetization in small spots on a magnetic film. These bits of information may be read optically using the polar Kerr effect. The materials now used in this technology are amorphous alloys of transition metals and the rare earths such as the amorphous TbFe shown in Table I.

See also FARADAY EFFECT; FERROMAGNETISM; MAGNETIC DOMAINS AND BUBBLES; REFLECTION.

BIBLIOGRAPHY

R. Carey and E. E. Isaac, *Magnetic Domains and Techniques for Their Observation*. Academic Press, New York, 1966.

R. P. Hunt, *IEEE Trans. Magn.* **MAG-5,** 700 (1969). Theory.

J. H. Judy, *Ann. NY Acad. Sci.* **189,** 239 (1972). Phenomenological theory and Kerr effects of many materials.
M. H. Kryder, *J. Appl. Phys.* **57,** 3913 (1985). Magneto-optical recording technology.
M. R. Parker, *Physica* **86-88B,** 1171 (1977). A review article on magneto-optical Kerr effects.

Kinematics and Kinetics

E. C. G. Sudarshan

Physics is the study of changes and of invariance: In the stability of matter it sees the dynamics of atomic systems, while in the movement of a stream it sees the unfolding of a system by a family of canonical transformations. The simplest changes pertain to *motion,* to changes of configuration. The study of motion, without reference to the causes or effects of motion or the nature of the object moving, is *kinematics*.

When the moving entity has no structure, we refer to a particle, a position as a function of time, a *trajectory*. In general, the position is a vector **r** in three dimensions. The time rate of change of position is the *velocity* **v** and its rate of change is the *acceleration* **a**.

$$\mathbf{a}=\frac{d}{dt}\mathbf{v};\quad \mathbf{v}=\frac{d}{dt}\mathbf{r}.$$

For constant acceleration

$$\mathbf{v}(t)=\mathbf{v}(0)+t\mathbf{a},$$
$$\mathbf{r}(t)=\mathbf{r}(0)+t\mathbf{v}(0)+\tfrac{1}{2}t^2\mathbf{a}.$$

The trajectory is a parabola with its axis parallel to the direction of acceleration, suitable for describing terrestrial projectiles and electrically charged particles in constant electric fields.

Motion along a circle of constant radius and with constant speed is accelerated since the direction of velocity changes. If the radius of the circle is R and the speed v, the acceleration is v^2/R directed toward the center. Such orbits describe possible motions of planets and satellites, and of electrically charged particles in constant magnetic fields and in the field of a central electric charge.

A *rigid body* may be pictured as a collection of particles with invariable mutual distances. The motion of a rigid body can then be obtained as the motion of the particles under these constraints. Alternatively, we may use the position of the center of mass and the orientation angles of the rigid body as coordinates. The center of mass moves like a mass point; the changes of the orientation involve *angular velocities*. When the center of mass remains fixed, we have the rigid rotator. More generally, any system of particles subject to any number of constraints between their coordinates can be dealt with in terms of *generalized coordinates*, not all necessarily Cartesian position variables.

So far motion has been treated as secondary to configuration, as a derived quantity. It is possible to view motion as a quality in the same sense as position; and profitable to do so when we recognize that motion is seen against (or from) a *reference frame* and that when the frame is changed, motion is also changed along with position. Dynamical quantities undergo well-defined transformations under changes of reference frames. Since the composition of frame changes is a frame change, they can be seen to form a group; the transformation of the dynamical quantities furnish realizations of this group. But to see it thus requires that positions and motions are to be treated on par. Another reason so to treat them is the change in motion produced by collisions or more generally under interaction. We have to take account of the quantity of motion or momentum. It is traditional to denote generalized coordinates by q_j and generalized momenta by p_j. The number of distinct values that j takes is called the number of *degrees of freedom*.

The trajectory of a particle may be seen either as a particular solution to a differential equation of motion or as an abstract general solution, which is mapped into a particular trajectory by specifying the initial state (coordinates and momenta). The latter view leads to the generalization, to a statistical description, as well as the passage to quantum physics. In quantum physics coordinates and momenta cease to be numerical variables, but become noncommuting *operators* with characteristic commutation relations. For Cartesian coordinates and momenta these are

$$q_jp_k-q_kp_j=(ih/2\pi)1$$

where h is Planck's constant (6.5×10^{-26} erg s). For a rigid rotator the angular momenta J_1,J_2,J_3 satisfy the *commutation relations*

$$J_1J_2-J_2J_1=(ih/2\pi)J_3\text{(cyclically)}.$$

Such commutation relations together with equations of motion consitute *quantum kinematics*.

The causes, effects, and interplay of motions constitute *kinetics*. The traditional starting point of mechanics is the set of Newton's three *laws of motion*: (i) A body continues in its state of rest or of uniform motion unless acted upon by an external force. (ii) Force equals product of acceleration induced times the mass of the body. (iii) Action and reaction are equal and opposite. The first law is a characterization of free motion; the second a quantitative definition of force. The third law states that the total momentum of a system under interaction is additive and invariable.

The energy of an isolated system is a scalar invariant consisting of an additive *kinetic energy* (quadratic in the momentum) and a nonadditive *potential energy*. The specification of these energy functions is the seed which generates the complete kinetics given the kinematics.

In *Lagrangian mechanics* one starts with a function $L(q,\dot{q})$ of the *generalized coordinates* and velocities and defines the generalized momenta p_j by

$$p_j=\frac{\partial L}{\partial q_j}.$$

These equations may be derived as the Euler–Lagrange variational equations for the *action principle*

$$\delta \int L(q,\dot{q})dt = 0.$$

For a system where the potential energy is independent of the generalized velocities the Lagrangian is the difference between the kinetic and potential energies.

In the Lagrangian formulation in terms of the action principle the consequences of *invariance properties* of the system are seen immediately. If we have invariance of the action with respect to, say, the position of the system, when we choose these quantities as generalized coordinates, the Lagrangian is independent of them; and by virtue of the equations of motion the corresponding generalized momentum is invariant. An invariance becomes translated into an invariant dynamical quantity: Symmetry and conservation become related. Invariance with respect to time origin corresponds to conservation of energy; indifference to the choice of space origin implies constancy of the space momentum (Newton's third law!); irrelevance of spatial orientation is equivalent to a conserved angular momentum (Kepler's second law). Conversely, conserved dynamical variables entail symmetries of the dynamical system: For motion under the inverse square law the axes of the ellipse remaining fixed implies the existence of a symmetry group changing the eccentricity of the ellipse without changing the energy. These ideas have important consequences for modern particle physics.

The clear separation between kinematics and dynamics is already blurred by including the "boost" transformation to moving frames. In this process the kinematic quantity of momentum and the dynamic quantity of energy undergo transformations into linear combinations of themselves. However, if a system obeyed Newton's First Law of Motion, it will continue to do so. There are other transformations to noninertial frames which affect the dynamical picture more drastically. We now turn to two different cases of this.

First is the transformation to rotating coordinates. A particle at rest in the original inertial frame has an apparent motion along a circle as seen from the rotating coordinate system. This is no longer an inertial motion. If the angular velocity is given by the (axial) vector $\boldsymbol{\omega}$ any vector dynamical quantity $\mathbf{A}$ has time derivatives $d\mathbf{A}/dt$ and $d'\mathbf{A}/dt$ with respect to the inertial and rotating coordinate system related by

$$\frac{d\mathbf{A}}{dt} = \frac{d'\mathbf{A}}{dt} + \boldsymbol{\omega}\times\mathbf{A}; \qquad \frac{d\boldsymbol{\omega}}{dt} = \frac{d'\boldsymbol{\omega}}{dt}.$$

Specializing to the radius vector $\mathbf{r}$ we relate the velocities in the two frames

$$\mathbf{v} = \frac{d\mathbf{r}}{dt}, \qquad \mathbf{v}' = \frac{d'\mathbf{r}}{dt}$$

by

$$\mathbf{v} = \mathbf{v}' + \boldsymbol{\omega}\times\mathbf{r}.$$

But for accelerations we have a more complicated relation:

$$\mathbf{a} = \mathbf{a}' + \boldsymbol{\omega}\times(\boldsymbol{\omega}\times\mathbf{r}) + 2\boldsymbol{\omega}\times\mathbf{v}' + \frac{d\boldsymbol{\omega}}{dt}\times\mathbf{r}.$$

The last term vanishes for steady rotation. The second term,

$$\boldsymbol{\omega}\times(\boldsymbol{\omega}\times\mathbf{r}) = \omega^2\mathbf{r} - (\mathbf{r}\cdot\boldsymbol{\omega})\boldsymbol{\omega},$$

is the "centrifugal" acceleration which could be pictured as if there is a centrifugal force on *all* particles *proportional to their mass*. The term $2\boldsymbol{\omega}\times\mathbf{v}'$ is the Coriolis acceleration, which is perpendicular to both the axis of rotation and the velocity as measured in the rotating system. It vanishes for particles at rest or seen to be moving parallel to the axis of rotation. The corresponding apparent Coriolis force on a particle is *velocity dependent* and *proportional to its mass*.

These noninertial frame-dependent forces are thus proportional to the mass of a particle betraying their kinematic origin. Viewed in this light, the fact of gravitational force on a particle being proportional to its mass suggests the natural way of understanding *gravitation is as a kinematically induced force*. This idea is implemented in the General Theory of Relativity in which a gravitating mass causes a curvature in the space around it; more precisely, the *stress tensor* composed of the momentum and energy flows and their densities is the *source of space-time curvature*.

Hamiltonian mechanics starts with a collection of phase-space variables q,p. The equations of motion are now

$$\dot{p}_j = -\frac{\partial H}{\partial q_j},$$

$$\dot{q}_j = +\frac{\partial H}{\partial p_j},$$

where $H(q,p)$ is the *Hamiltonian* (energy) function. Hamiltonian mechanics can be extended to systems with more general phase-space variables ω, like the rigid rotator. For any two phase-space functions $F(\omega)$ and $G(\omega)$ a new function $E(\omega)$ is defined called the *Poisson bracket* of F and G:

$$E = \{F,G\}$$

with the properties

$$\{F,G\} = -\{G,F\},$$

$$\{c_1F_1 + c_2F_2, G\} = c_1\{F_1,G\} + c_2\{F_2,G\},$$

$$\{\{F_1,F_2\},F_3\} + \{\{F_2,F_3\},F_1\} + \{\{F_3,F_1\},F_2\} = 0,$$

where c_1 and c_2 are any constants. The last property is called the Jacobi identity. With this structure the phase-space functions constitute a *Lie algebra*, which is associated with the Lie group of canonical transformations of phase-space functions and dynamical variables:

$$\phi(\omega) \rightarrow \phi(\omega;\tau),$$

with

$$\frac{\partial\phi(\omega;\tau)}{\partial\tau} = \{\phi(\omega;\tau), F(\omega)\}.$$

By virtue of the Jacobi identity, we have

$$\{F(\omega;\tau),G(\omega;\tau)\}=E(\omega;\tau).$$

The dynamical structure is therefore preserved under canonical transformations; the *transformation is a change of perception*.

We may adjoin the derivation property of the Poisson brackets,

$$\{F_1\cdot F_2,G\}=F_1\cdot\{F_2,G\}+\{F_1,G\}\cdot F_2,$$

to construct all Poisson brackets from the fundamental brackets:

$$\{\omega^\mu,\omega^\nu\}=\epsilon^{\mu\nu}(\omega).$$

For canonical variables

$$\{q_j,q_k\}=\{p_j,p_k\}=0,$$

$$\{q_j,p_k\}=\delta_{jk},$$

so that

$$\{F(q,p),G(q,p)\}=\sum_j\left(\frac{\partial F}{\partial q_j}\frac{\partial G}{\partial p_j}-\frac{\partial F}{\partial p_j}\frac{\partial G}{\partial q_j}\right).$$

For the rotator

$$\{J_1,J_2\}=J_3\ \text{(cyclically)}$$

so that

$$\{F(J),G(J)\}=\sum_{jkl}\frac{\partial F}{\partial J_j}\frac{\partial G}{\partial J_k}\epsilon_{jkl}J_l.$$

Quantum dynamics is obtained by identifying the Poisson bracket with a multiple of the commutator,

$$\{F,G\}=(ih)^{-1}(FG-GF),$$

and identifying the Hamiltonian operator of the system.

Recognition that canonical transformations are changes of perception entails the dynamical transformations due to change of reference frames forming a canonical transformation group. This is the *relativity group* built up from the following:

(1) change of space origin: displacement generator **P**;
(2) change of time origin: time evolution generator H;
(3) change of space orientation: rotation generator **J**;
(4) change to moving frames: boost generator **K**.

These generators are dynamical variables, which have, respectively, the physical significance of linear momentum **P**, energy (Hamiltonian) H, angular momentum **J**, and moment of energy **K**. The Poisson bracket relations deduced from the structure of the frame transformations are

$$\{J_j,P_k\}=\sum_l\epsilon_{jkl}P_l,$$

$$\{J_j,J_k\}=\sum_l\epsilon_{jkl}J_l,$$

$$\{J_j,K_k\}=\sum_l\epsilon_{jkl}K_l,$$

$$\{K_j,P_k\}=\delta_{jk}H,$$

$$\{K_j,H\}=P_j,$$

$$\{K_j,K_k\}=\sum_l\epsilon_{jkl}J_l,$$

with all other Poisson brackets vanishing. All relativistic systems are realizations of this *Poincaré group*; the simplest realization is by a particle with $H=(m^2+p^2)^{1/2}$, $\mathbf{K}=\mathbf{q}H$, $\mathbf{J}=\mathbf{q}\times\mathbf{p}$, where m is the mass of the particle. In this case we have a simple geometrical interpretation of the free-particle trajectory as a *world-line*: the quantities $\mathbf{q}(t),t$ transform as the components of a four-vector. A collection of interacting particles which have both the canonical structure and world-line trajectories cannot have any interaction; this is the *no-interaction theorem* of classical dynamics.

For many purposes the velocity of light is very large compared with the velocities of interest. The *Galilean group* of transformations is adequate to describe them. The Lie algebra consists of $\mathbf{P},H,\mathbf{J},\mathbf{G}$, with the new generator **G** (the moment of mass) with the Poisson bracket relations

$$\{J_j,G_k\}=\sum_l\epsilon_{jkl}G_l,$$

$$\{G_j,G_k\}=0,$$

$$\{G_j,H\}=P_j,$$

$$\{G_j,P_k\}=M\delta_{jk},$$

where M has vanishing brackets with all generators (and may be identified with the constant total mass of the system). The generator **G** is the moment of mass $m\mathbf{q}$ for a particle.

The dynamical description by Newtonian/Lagrangian or Hamiltonian equations has three limitations: first, it is a short-time description and does not deal with *asymptotic behavior*; second, the equations of motion may become inapplicable or indeterminate for certain configurations causing *catastrophes*; third, it deals only with a finite number of degrees of freedom.

The initial configuration can be determined only with a finite accuracy; the predictions of the dynamical equations ought to be such that neighboring initial configurations should lead to neighboring final configurations. Dynamics should not only map configuration space points into points but also map *"neighborhoods" into neighborhoods*. For most dynamical systems this property is true only for short times; for sufficiently long times two arbitrarily near configurations can become separated widely. A simple example is provided by a cluster of particles of varying speeds moving along (1) a straight line and (2) a circle. In the first case we have a Hubble law: proportionality between velocity and distance from origin. In the second case although such correlations obtain for short times, if we wait long enough there is no correlation between the (angular) positions of the particles and their speeds. This simple example also illustrates the importance of global (topological) properties on apparently local correlations for the asymptotic development of the system.

The inertia of a particle participating in a nonlinear interaction is a dynamical variable and it could vanish for certain dynamical configurations. This in turn leads to unpredictable behavior leading to catastrophes.

The limitation to systems with a finite number of degrees of freedom must be removed if the dynamics of fluids, elastic bodies, and fields is to be considered. Natural generalizations of the Lagrangian, the action principle, the Hamiltonian, and the transformation group obtain; but since there are an infinity of fundamental dynamical variables, their algebra is not locally compact and considerable care is needed in handling realizations. One important consequence is the existence of states of the system which do not admit the original relativity transformations meaningfully. The restriction to finite time behavior has to be removed if we have to study the longtime asymptotic behavior of dynamical systems, especially in the context of the foundations of statistical mechanics. A discussion of these fascinating topics is beyond the scope of this article.

See also CATASTROPHE THEORY; CENTER-OF-MASS SYSTEM; CHAOS; CONSERVATION LAWS; DYNAMICS, ANALYTICAL; ENERGY AND WORK; GRAVITATION; GROUP THEORY IN PHYSICS; HAMILTONIAN FUNCTION; INVARIANCE PRINCIPLES; LIE GROUPS; NEWTON'S LAWS; RELATIVITY, GENERAL; RELATIVITY, SPECIAL THEORY.

Kinetic Theory

R. N. Varney

Kinetic theory is a term applied to the study of gases by assuming that gases are composed of molecules subject to rapid random motions colliding with one another and with the container walls, analyzing these motions, and deducing various properties of the gases such as their molal heat capacities, the pressure they exert, their viscosity, coefficient of diffusion, and rate of effusion through apertures. The adjective *kinetic* refers to the fact that the properties of the gases are deduced from the *motions* of the molecules. The term *theory* refers to the *deduction* of the properties of the gases *without* reference to *any* known experimental properties, not even the ideal gas law (q.v.).

The atomic nature of matter, as a hypothesis, dates back to antiquity. It is generally accepted that the concept of *molecules* as distinct from atoms and their role in the behavior of gases dates from 1811 with the formulation of Avogadro's hypothesis which states that equal *volumes* of gases at the same temperature and pressure contain equal numbers of *molecules*. The key feature of the hypothesis lies in Avogadro's recognition of molecules as distinct from atoms.

The great and key developments of the kinetic theory of gases were made largely during the next 100 years despite the fact that some highly respected scientists (Ostwald, Mach) denied the existence of molecules. The reality of their existence, as opposed to the hypothesis, was finally proved to the satisfaction of even the disbelievers by the experimental work on Brownian motion by Jean Perrin in 1908 coupled with the analysis of it by A. Einstein.

In 1843, Joule applied the hypothesis of the intense motions of molecules in gases to explain the pressure exerted by a gas. His simplified assumptions were that (i) the molecules all had the same mass, m; (ii) they all had the same speed, designated by c; (iii) an experimental volume of gas V contained n molecules; (iv) the molecules bombarded the walls of the container with speed c and momentum mc, always striking the walls normally and rebounding elastically, hence with the same speed but reversed direction, and traveled unimpeded to the opposite wall of the rectangular container where the bombardment and rebounding occurred again. Momentum $2mc$ was transferred to the wall at each impact, the factor of 2 arising because the momentum changed from $+mc$ to $-mc$ upon impact. The time to cross the box of length x from one wall across and back was $t=2x/c$, so that the transfer of momentum per *unit time* was $2mc/t=2mc/(2x/c)=mc^2/x$. By Newton's second law, the rate of change of momentum equals the force f exerted by one molecule on one wall. (v) Joule assumed that of the n molecules in the box, $\frac{1}{3}n$ at each instant traveled in the x direction so that the force of all of these molecules on one wall was $(n/3)(mc^2/x)$. If the wall had dimensions y by z and area yz, then the pressure on the wall, $p=f/A$, became

$$p=\frac{1}{3}\frac{nmc^2}{xyz}.$$

Since the product xyz is the volume of the box V, the result is

$$pV=\tfrac{1}{3}nmc^2.$$

Before commenting about this result of Joule's it may be noted that when the derivation was later redone with far more rigorous and realistic assumptions to include the collisions of molecules with one another, the random direction of hitting the walls, a wide distribution of values of the velocity c, and even mixtures of molecules with various mass values, the result was unchanged except for replacing c^2 by $\overline{c^2}$, the average squared speed, and the replacing of p by the partial pressure of each homogeneous constituent gas. This feature of kinetic theory of giving a correct or nearly correct result with highly simplified assumptions recurs in many other derivations.

While neither n nor m could be known to Joule, the product nm was the mass of gas in the volume V and was subject to experimental measurement. With a known pressure p, known volume V, and known mass of gas, the value of c or at least a mean value of c could be found. For N_2 gas at room temperature it proved to be about 5×10^4 cm s^{-1}.

A series of deductions was at once made from Joule's equation.

(i) Since Boyle's law states that at constant temperature T, pV is constant, it follows that the mean speed c is *not* a function of p.

(ii) Since the ideal gas law gives $pV = RT$, it follows that $\frac{1}{3}Mc^2 = RT$, where M is the molecular weight of the gas, in kg if joules and n s^{-1} are the units used.

(iii) The kinetic energy of all the molecules of total mass M is $\frac{1}{2}Mc^2$ and hence this kinetic energy is equal to $\frac{3}{2}RT$. The quantity $\frac{3}{2}RT$ has been known from thermodynamics to be the internal energy of one mole of any monatomic gas. Its derivative $(\partial(\frac{3}{2}RT)/\partial T)_V$ is the molal heat capacity at constant volume, C_V, and is equal to $\frac{3}{2}R$, a result widely verified for monatomic gases such as He, Ar, and Ne.

A new concept now enters the subject of kinetic theory from the domain of mechanics, called *the equipartition of energy*, one that has been more disputed than any other in kinetic theory. In a simple form it may be stated that each degree of freedom of the gas possesses the same amount of energy. This idea was tacitly assumed by Joule when he pictured one-third of all the molecules to be moving in the x direction, one-third in the y direction, and one-third in the z direction, all with the same speed c. He thus assumed that the same amount of kinetic energy was to be found in each of these components of motion or *degrees of freedom*. There would seem to be little reason to doubt this assumption; if it were false, it would seem that a wind must be blowing in a favored direction to cause the unbalance. Thus $\frac{1}{3}(\frac{1}{2}Mc^2) = \frac{1}{2}RT$ must be the kinetic energy to be ascribed to each degree of translatory motion.

The concept, as a hypothesis, was immediately extended in two respects: The first was that if the gas consisted of a mixture of molecules having different masses, then each molecular type would have the same kinetic energy per degree of freedom as each other type. It followed at once that the more massive molecules, having larger M, must have slower speeds of agitation c in order that the kinetic energies would all be the same. The second extension was that if the molecules of a gas were not simply single pointlike atoms but had molecular structure so as to have moments of inertia about principal axes and kinetic energies of rotation $\frac{1}{2}I\omega^2$ about each axis, where I is the moment of inertia about that axis, then each degree of rotational freedom would also possess the same kinetic energy $\frac{1}{2}RT$. A simple diatomic molecule composed of two identical atoms, as shown in Fig. 1, has the same moment of inertia about an axis in the x direction and the y direction, and essentially no moment of inertia at all about an axis in the z direction. The total kinetic energy of a mole of such molecules may then be written

$$E_K = \tfrac{1}{2}nm(c_x^2 + c_y^2 + c_z^2) + \tfrac{1}{2}nI(\omega_x^2 + \omega_y^2);$$

but by the equipartition of energy theorem, each of the five terms is the same, each is equal to $\frac{1}{2}RT$, and the total kinetic energy becomes $\frac{5}{2}RT$. The molal heat capacity at constant volume C_V should then be $\frac{5}{2}R$, and this has also been abundantly demonstrated to be true. However, it definitely fails to be true when quantum limitations become applicable. Thus when hydrogen gas is cooled to nearly its boiling temperature, its heat capacity at constant volume drops from $\frac{5}{2}R$ to $\frac{3}{2}R$ because the equipartition energy is less than one quantum of rotational energy of the hydrogen molecule.

The result most commonly associated with the term kinetic theory is the law of distribution of molecular velocities,

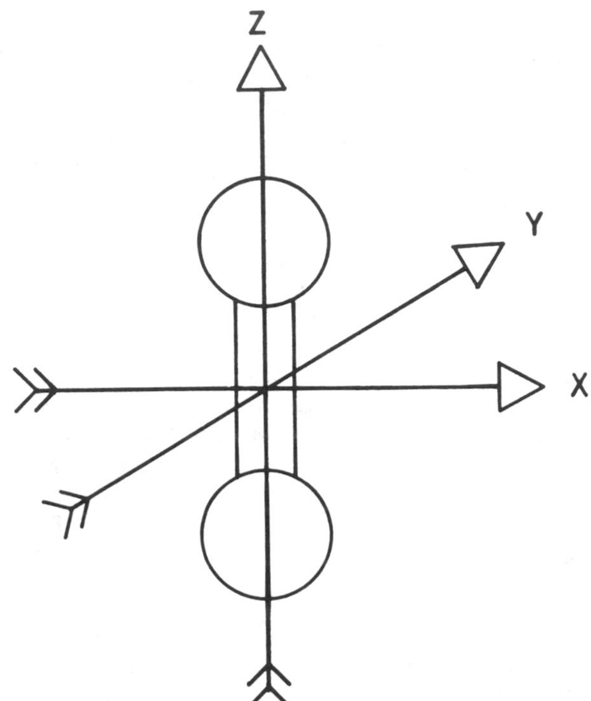

FIG. 1. A diatomic molecule oriented along the z direction has equal moments of inertia about axes through the center of mass parallel to the x and the y axes. The z-axis moment of inertia is negligible.

a law that describes how many molecules out of each mole of gas have molecular velocities in any selected range of values. It may equally well be described as telling the probability that one molecule will have a velocity in any defined range of values. Perhaps the most remarkable feature of the distribution is that it exists, i.e., that the random motions settle down to a distinct pattern whereby only a few molecules have very high or very low speeds and the number in any range in between remains essentially constant.

The expression *Maxwell distribution* is often used for the distribution law, as J. C. Maxwell gave one of the earliest derivations of it. A derivation leading to the same result by L. Boltzmann is generally regarded as more rigorous but both derivations are criticized on the grounds of containing unproved, if experimentally valid, assumptions. (The basic issue has been whether the law can be derived solely from considerations of Newtonian mechanics or whether a theorem of probability is required in addition. The latter has proved to be the case.) The criticisms do not alter the validity of the law nor its enormous importance in physics but only the philosophical perfection of kinetic theory in reaching an important result entirely free of experimental aspects.

The Maxwell distribution law is illustrated graphically in Fig. 2. In this form, the x axis shows the speed (regardless of direction) of molecules, and the y axis shows the probability that a molecule will have this speed within a narrow but specified range Δc. (If there were an infinite number of molecules, it would be possible to speak of a probability of an exact speed, but since there is only a finite number, it is not possible to speak of an exact speed or the probability

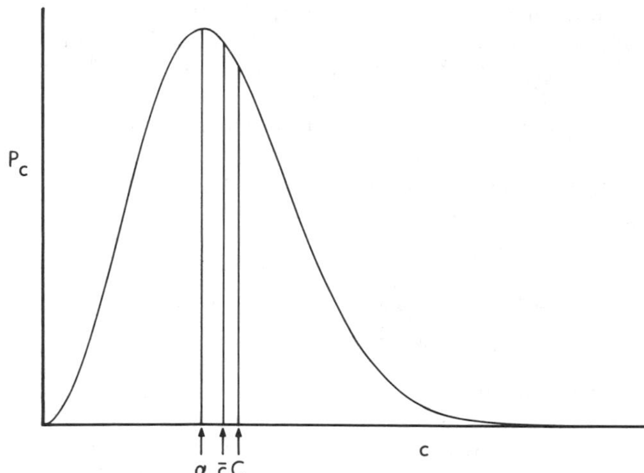

FIG. 2. Probability P_c that a molecule has a speed between c and $c+\Delta c$. Shown in the figure are the most probable speed α, the average speed $\bar{c}$, and the root-mean-square speed $C=(\overline{c^2})^{1/2}$.

of it.) The curve is represented by the equation,

$$P_c=\frac{4}{\alpha^3\pi^{1/2}}c^2e^{-c^2/\alpha^2}\Delta c.$$

Here P_c is plotted on the y axis and is the probability of a speed c, shown on the x axis, in a range of speeds Δc. The symbol e is the base of natural logarithms, and α, a constant, proves to be the value of c at the highest point of the curve, i.e., at the maximum value of P_c. It is thus correct to say that α is the most probable speed of a molecule.

Some important features of the equation and its graph may be noted. The curve is *not* symmetric around the most probable speed α, since it terminates at zero on the low side but extends to infinity on the high side. By use of the mean-value theorem of integral calculus, it is possible to determine the average speed of molecules, $\bar{c}$. It is shown in Fig. 2 and has the value $\bar{c}=2\pi^{-1/2}\alpha$, some 12.8% greater than α. In addition, because of the appearance of c^2 in the Joule derivation, there is interest in the average value of c^2, written $\overline{c^2}$. Its square root is called the *root-mean-square speed,* often abbreviated rms speed, and is commonly designated by C. The value of C is $C=\frac{3}{2}^{1/2}\alpha$ or some 22.5% greater than α. It is also marked on the curve of Fig. 2.

Some caution is necessary in using the distribution law. Thus if it is desired to find the probability that a molecule has a certain component of velocity *in a certain direction,* say, for example, parallel to the x axis, the equation is

$$P_x=\frac{1}{\alpha\pi^{1/2}}e^{-c^2/\alpha^2}\Delta c,$$

and is pictured in Fig. 3 for positive directions only. Since it is equally likely that a molecule should have a negative as well as a positive component of velocity in any given direction, the symmetric negative half of the curve is not drawn. The most probable component of velocity *in a given direction* is zero.

Finally, it is possible to calculate the probability that a molecule has a given energy, within a narrow band of energies, and this variation of the distribution law is often called

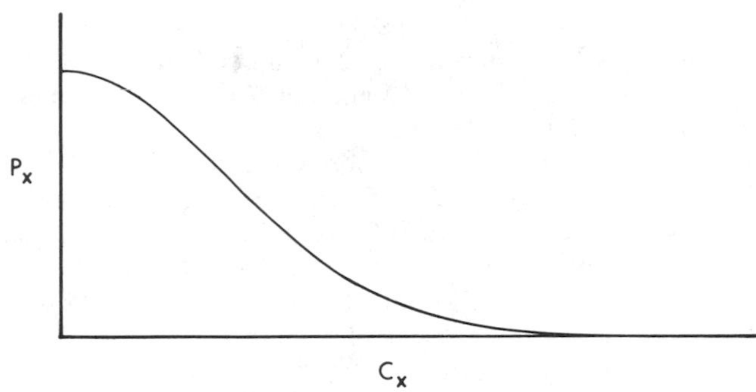

FIG. 3. Probability P_x that a molecule has an x component of velocity between c_x and $c_x+\Delta c_x$. Only positive values of c_x are shown, but the curve extends symmetrically to negative values of c_x. The most probable x component of velocity is seen to be zero.

the *Maxwell–Boltzmann law*. It is usually written in a somewhat different form from the law of distribution of speeds, and the symbols used are presented first.

P_E is the probability of a molecular kinetic energy between E and $E+\Delta E$. If the molecules are in a potential-energy field such as the earth's gravitational field, this potential energy is introduced as E_p. It is usually set equal to zero for a standard condition such as sea level in the gravitational case. Finally, recourse to Joule's derivation is taken wherein the mean molecular kinetic energy arises from

$$\tfrac{1}{2}nm\overline{c^2}=\tfrac{3}{2}RT$$

or

$$\tfrac{1}{2}m\overline{c^2}=\tfrac{3}{2}(R/n)T=\tfrac{3}{2}kT.$$

The molal gas constant R, divided by Avogadro's number n, is called Boltzmann's constant and is represented by k. The best value of k is 1.3806×10^{-23} J K^{-1}. Since $\frac{3}{2}kT$ is the average kinetic energy of a molecule, this quantity is used instead of α or a most probable kinetic energy. The Maxwell–Boltzmann law then is

$$P_E=\frac{2}{\pi^{1/2}(kT)^{3/2}}E^{1/2}e^{-(E+E_p)/kT}\Delta E$$

and it is plotted in Fig. 4.

The curve has several distinctive characteristics. Thus its slope at the origin is infinite whereas the speed-law curve has zero slope at the origin. The curve has a maximum, and

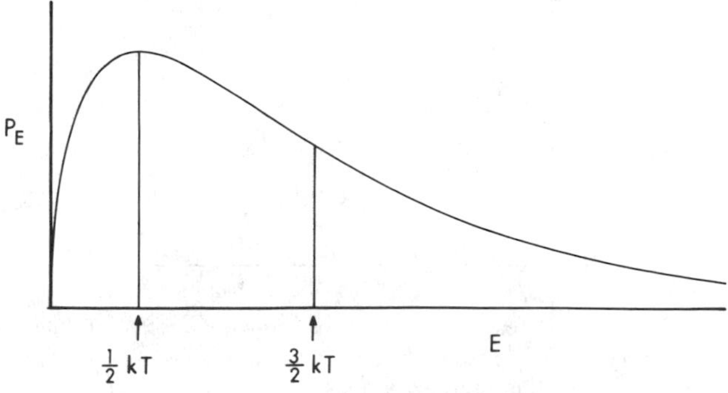

FIG. 4. Probability P_E that a molecule has kinetic energy between E and $E+\Delta E$. The most probable energy is $\frac{1}{2}kT$ and the average kinetic energy is $\frac{3}{2}kT$.

the most probable kinetic energy proves to be $\frac{1}{2}kT$. At first glance this is surprising as a calculation of $\frac{1}{2}m\alpha^2$ with α the most probable speed gives kT and not $\frac{1}{2}kT$. The average energy $\frac{3}{2}kT$ is also indicated on the graph.

Direct experimental verification of the distribution law of molecular velocities was achieved in the late 1920s and early 1930s when the techniques of molecular beams (q.v.) had been developed. Indirect verification of a sort was achieved earlier by studies of the Doppler broadening of spectroscopic lines occurring from the speed of the radiating atoms toward or away from the spectroscope.

One of the great achievements of kinetic theory has been the analysis of interdiffusion of gases. The experimental problem in its simplest form is to expose some ammonia in one corner of a draft-free room and determine the time interval until the odor is detectable several meters away. The molecular velocity of 5×10^2 m s^{-1} obtained from application of the Joule derivation might suggest that the odor should be noted 2 m away in 4 ms, a time much too short to correspond with observations that are in fact measurable in minutes. The first analysis at the hands of R. Clausius generated the concept of the *mean free path* whereby the ammonia molecules travel only some 10^{-5} cm between collisions with gas molecules at atmospheric pressure, in time intervals of about 10^{-10} s between collisions, and describe a long and intricate path in the process of acquiring a displacement of several meters.

Assuming that a molecule has just had a collision at the origin of coordinates, the probability that at a distance r from the origin it still has not had a new collision is given by $P_0=e^{-r/l}$. The symbol l, a constant in the equation, proves to be the mean value of r and hence is called the *mean free path,* commonly abbreviated mfp. The curve of P_0 vs r is shown in Fig. 5. Somewhat more enlightening is the curve marked P_1 in the same figure, which is the probability that the molecule *will* have collided *once* after traveling the distance r. Note that P_0 is the probability that the molecule *will not* have collided after traveling a distance r, and P_1 is the probability that it *will* have collided *just once*. The curve for

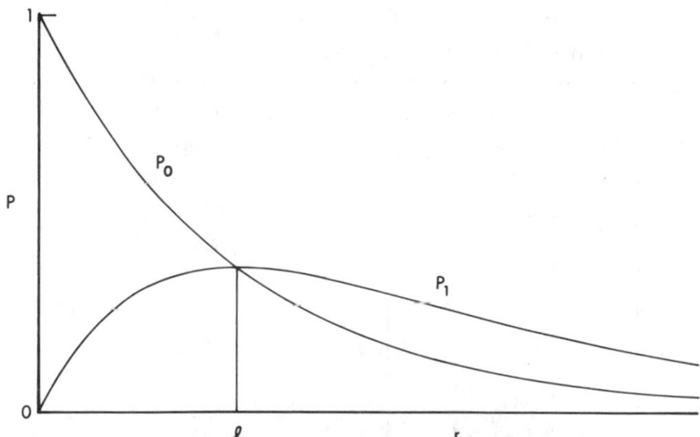

FIG. 5. Two curves, P_0 and P_1, are shown. P_0 is the probability that a molecule travels a distance r without a collision, and P_1 that the molecule having traveled a distance r has had exactly one collision—no more, no less. The value of $r=l$ is the mean free path. P_0 has the value unity at $r=0$ and the value e^{-1} or 0.368 at $r=l$. P_1 has its peak value at $r=l$.

P_1 drops at distances $r>l$, since the likelihood of two or more collisions rises. The value of P_1 is given by $P_1=(r/l)e^{-(r/l)}$.

From the mean free path l and the average molecular speed $\bar{c}$, J. C. Maxwell deduced expressions for the coefficients of viscosity, of thermal conduction, and of diffusion in gases. The derivations required calculation, respectively, of the transfer of momentum, of energy, and of mass through the gases, and the results proved to be

$$\text{Coefficient of viscosity} \quad =\eta=\tfrac{1}{3}\rho\bar{c}l,$$

$$\text{Thermal conductivity} \quad =\kappa=\tfrac{1}{3}C_v\rho\bar{c}l=\eta C_v,$$

$$\text{Coefficient of diffusion} \quad =D=\tfrac{1}{3}\bar{c}l=\eta/\rho.$$

In these equations, ρ is the gas density, C_v is the molal heat capacity at constant volume, and the other symbols have been defined previously. The equations agree with experimental measurements to 5% or better over a large range of the variables, limited only by the requirement that the mean free path l be short compared with the main apparatus dimensions. Greater refinement is obtained by replacing the product $\bar{c}l$ by the average $\langle cl(c)\rangle$, that is, by multiplying the velocity by the path length for that velocity and then averaging.

Derivation of the value of the coefficient of viscosity is presented here in simplified form by way of illustrating kinetic-theory analysis. It is assume that by some mechanism of moving belts, a velocity gradient $\partial u/\partial z$ is created across a gas. Here u refers to a flow velocity in the x direction. Then across a mean free path l in the z direction, there is a difference of x velocity $l\partial u/\partial z$, and $ml\partial u/\partial z$ is the difference of momentum across such a free path. If it is now assumed that there are n molecules per cm^3 and that $\frac{1}{3}n$ are moving at any instant in the z direction, with velocity $\bar{c}$, it follows that $\frac{1}{3}n\bar{c}$ molecules per cm^2 per second travel across a plane at constant z. Hence the product $\frac{1}{3}n\bar{c}$ times $ml\partial u/\partial z$ is the momentum transferred per cm^2 per second across such a plane, and by Newton's second law, this must be the force per cm^2 acting across a plane at constant z. The coefficient of viscosity is defined by the relation

$$\eta=-\frac{f/A}{\partial u/\partial z},$$

so that it emerges that $\eta=\frac{1}{3}nm\bar{c}l$. Since nm is the mass of gas per cm^3, or ρ, Maxwell's result above is obtained.

The value derived for η, the coefficient of viscosity, depends on the product ρl, and these two quantities change oppositely with change of gas pressure, annulling any change in η with p. The original prediction of this fact, made by J. C. Maxwell, seemed unbelievable at first but has been verified over a large range of pressures. The reason for the surprise is traceable to an attempt to compare the viscosity of gases with that of liquids, a comparison that fails because in liquids, the viscosity arises from intermolecular forces, while in gases it arises from the collisions and the transfer of momentum. This difference causes η to rise with temperature in gases but fall in liquids, an experimentally substantiated distinction.

The kinetic theory of gases would probably not have a place in this encyclopedia if its major underlying hypothesis, the existence of molecules, had not been proved beyond any lingering doubts in 1908 by the work of Perrin and of Einstein on Brownian motion. A by-product of the 1908 work was the first reliable evaluation of Avogadro's number, the number of molecules in a mole. There are today no less than 20 methods for evaluating this number (see L. B. Loeb in the bibliography). The best current value is believed to be $6.022045(31)\times 10^{26}$ molecules per kg mole.

A precise distinction between kinetic theory and statistical mechanics cannot be made. The latter is usually more highly formalized but utilizes the same basic laws of mechanics and collisions. Statistical mechanics is generally regarded as the aspect of kinetic theory that deals with interpreting thermodynamic quantities like entropy and free energy by molecular collision analysis.

See also EQUATIONS OF STATE; STATISTICAL MECHANICS; VISCOSITY.

BIBLIOGRAPHY

L. B. Loeb, *Kinetic Theory of Gases*. Dover, New York, 1961. This reference is highly readable, contains many applications, and is notable for balancing out excessive mathematical rigor in favor of physical comprehensibility.

A. I. Khinchin, *Mathematical Foundations of Statistical Mechanics*. Dover, New York, 1949. This book, translated from Russian, is essentially mathematical and contains the proof that the Maxwell–Boltzmann distribution law cannot be derived solely from Newton's second law without the aid of probability theory. It is not easily readable below advanced levels.

C. Schaefer, *Einführung in die theoretische Physik*, Vol. 2, part 1. Akad. Verlagsges, Leipzig, 1937. This reference is cited for the benefit of readers who prefer German to English. It is comparable with Loeb's book, going somewhat further in statistical mechanics.

Kinetics, Chemical

Richard M. Noyes

Chemical kinetics is the study of the rates of those processes by which systems move toward equilibrium states determined by chemical thermodynamics. The importance of the subject is its use to elucidate the mechanism by which chemical change takes place in any system.

A system is *homogeneous* if it consists of a single uniform phase; the rate of chemical change per unit volume is then a function only of temperature, pressure, and the concentrations of various component species. The system is *heterogeneous* if at least one additional phase is present; the rate may then depend on factors such as distribution and extent of subdivision of that phase.

Let us postulate a homogeneous uniform system in which extent of chemical change can be described in terms of a single balanced equation such as

$$aA+bB+\cdots \rightleftarrows cC+dD+\cdots, \tag{1}$$

where capital letters are chemical species and lowercase letters are rational numbers. The *rate* of this reaction v is defined by

$$v=-\frac{1}{a}\frac{d[A]}{dt}=-\frac{1}{b}\frac{d[B]}{dt}=\frac{1}{c}\frac{d[C]}{dt}, \tag{2}$$

where $[A]$ denotes concentration of A in units such as mol dm^{-3}.

If such a system is far from thermodynamic equilibrium, an equation of the form

$$v=k[A]^{\alpha}[B]^{\beta}[C]^{\gamma}[D]^{\delta}[P]^{\pi} \tag{3}$$

may be valid over a wide range of conditions if temperature and either pressure or volume are maintained constant. The coefficients α, β, ..., may be positive, negative, or zero, and the concentration of species P may affect the rate even though it does not appear in Eq. (1) for net chemical change. Such a reaction is said to be α *order* in A, β order in B, and $(\alpha+\beta+\gamma+\delta+\pi)$ order overall. Often orders are small integers and the equation can be integrated in closed form. Thus, if the rate depends only on $[A]$ and if $a=\alpha=1$, the solution becomes $\ln([A]_0/[A])=kt$.

The *reaction rate constant* k is a function of temperature but not of species concentrations. It can often be fitted empirically to the Arrhenius equation

$$k=Ae^{-E/RT}, \tag{4}$$

where the *pre-exponential factor* A and *activation energy* E are disposable parameters, while R and T are the gas constant and absolute temperature, respectively.

It is generally believed that net chemical change in any system is the consequence of a number of *elementary processes* each of which involves simultaneous interaction of one, two, or at most three molecules. If an elementary process involves single isolated molecules of a species, the rate is first order in that species. If the process involves collisions between pairs of molecules, it is second order. The order of an elementary process is thus determined by its molecularity. However, the empirical observation of a simple reaction order does not prove the reaction takes place in a single step with the indicated molecularity.

At chemical equilibrium, $v=0$ and all reactants and products are at finite concentrations. Therefore, Eq. (3) cannot remain valid as the system approaches equilibrium. In general, the exact expression for v contains another term of opposite sign that is of negligible magnitude when the system is far enough from equilibrium. Because the equation must reduce to $v=0$ for all conditions of thermodynamic equilibrium, the allowable coefficients in the second term are strongly restricted. However, they are not uniquely determined except that $[P]^{\pi}$ must appear in both terms or neither because this species does not affect the position of equilibrium.

For many chemical reactions, an equation of the form of (3) can not describe the data over a significant range of concentrations no matter how far the system is from equilibrium. Such reactions must involve more than a single elementary process to accomplish net chemical change. Detailed measurements of the dependence of v on various concentrations provide insight to the molecular *mechanism* of the overall reaction. Such measurements can never unequivocally establish the way in which elementary processes accomplish net chemical change, but they can permit many otherwise plausible mechanisms to be rejected. If a comparatively simple explanation remains consistent with all measurements designed to test it, we develop an ever increasing confidence that it does describe true molecular behavior. When hypothetical mechanisms are being tested in this way, it is helpful to recognize that elementary processes may be *consecutive* involving formation of intermediates that subsequently react by unique paths, or they may be *competitive* so that species have different probable fates depending on experimental conditions.

The subject of chemical kinetics cannot be covered in detail here. Reference 1 illustrates how theories of elementary processes can be related to rates of collisions of gas molecules, and Ref. 2 illustrates an alternative viewpoint based on statistical thermodynamic partition functions. Reference 3 is an excellent presentation of the overall subject. Reference 4 illustrates how kinetic measurements are used to elucidate chemical mechanism, and Ref. 5 develops the importance of energetic considerations. Reference 6 develops the concepts of consecutive and competitive processes and is found in one of two volumes that offer detailed information about kinetic techniques.

See also THERMODYNAMICS, EQUILIBRIUM.

REFERENCES

1. L. S. Kassel, *The Kinetics of Homogeneous Gas Reactions*. Chemical Catalog Company, New York, 1932.
2. S. Glasstone, K. J. Laidler, and H. Eyring, *The Theory of Rate Processes*. McGraw-Hill, New York, 1941.
3. S. W. Benson, *The Foundations of Chemical Kinetics*. McGraw-Hill, New York, 1960.
4. A. A. Frost and R. G. Pearson, in *Kinetics and Mechanism,* 2nd ed., Chap. 12, pp. 285–387. Wiley, New York, 1976. Unfortunately, this material is omitted in a later edition.
5. S. W. Benson, *Thermochemical Kinetics*. 2nd ed. Wiley, New York, 1976.
6. R. M. Noyes, in *Investigations of Rates and Mechanisms of Reactions,* 4th ed., Part I, Chap. V. pp. 373–423. C. F. Bernasconi (ed.) of *Techniques of Chemistry*. Vol. VI, A. Weissberger (ed). Wiley, New York. 1986.

Klystrons and Traveling-Wave Tubes

J. A. Arnaud

Klystrons and traveling-wave tubes (TWT) are electron tubes that can amplify (or generate with the help of a suitable feedback mechanism) electromagnetic waves in the microwave range of frequency, from about 1 to 200 GHz. The operation of these devices can be understood on the basis of a classical theory, the electrons being viewed as point charges subjected to the electric force, and the electromagnetic waves as solutions of the Maxwell equations satisfying the boundary conditions imposed by the conductors.

The klystron tube was first demonstrated by the Varian brothers in 1937. As shown schematically in Fig. 1a, a klystron consists of an electron gun; a modulating cavity, called

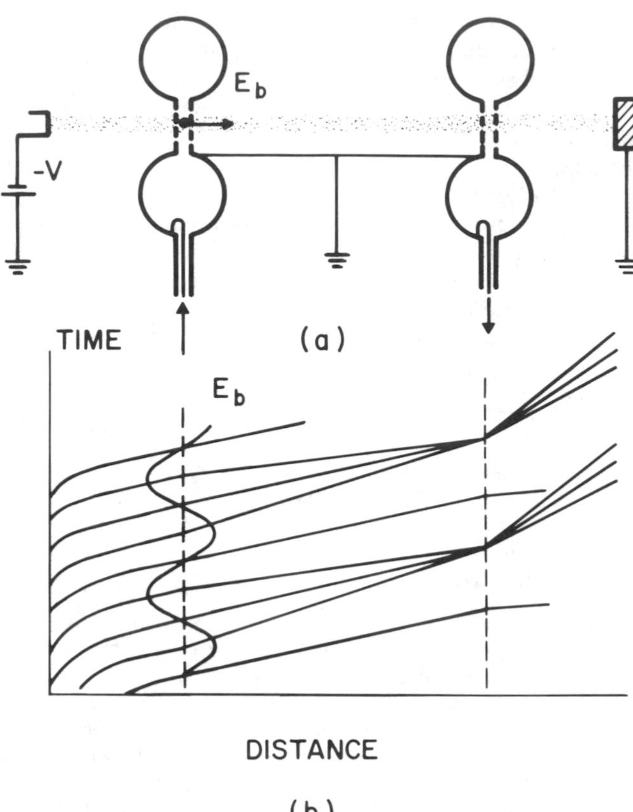

FIG. 1. Schematic view of a two-cavity klystron amplifier.

the buncher; and a collecting cavity, called the catcher. The electrons originating from the gun are accelerated by a dc field. When they reach the buncher they have almost all the same velocity, except for a small spread due to the cathode temperature. When a particular electron of the beam traverses the buncher, it is accelerated or decelerated, depending on whether the axial electric field in the buncher is negative or positive, respectively. Some electrons traverse the buncher at a time when the field is equal to zero. Their trajectory, then, is unaffected. Because the buncher modulates the velocity of the passing electrons, the klystron is called a velocity-modulation device. In Fig. 1b the positions of a few electrons are shown as a function of time. It is easily seen that the electrons emitted within a given period of field oscillation tend to meet at a distance from the modulating cavity that is inversely proportional to the strength of the modulating field. The velocity-modulated electrons are said to form "bunches." At the particular location where the electrons are most concentrated the electron beam carries a large alternating current whose fundamental frequency is equal to that of the modulating field. This alternating current, flowing through the catcher, generates a power that may be considerably larger than the power used to modulate the beam. In fact, according to the mechanism of operation just described, electron bunches could be obtained with arbitrarily small modulating fields. The gain of klystrons could be increased to arbitrarily large values merely by increasing the distance between the catcher and the buncher. This is not the case, however, because electrons repel each other. In order to obtain a large gain, it is necessary to introduce additional cavities between the buncher and the catcher. These cavities are not coupled directly to each other, but only to the electron beam.

Klystrons generate electromagnetic waves when a suitable feedback mechanism is provided. In the "reflex" klystron, the modulated electron beam is reflected back into the buncher with the help of a negatively biased electrode called the reflector. The reflex klystron therefore incorporates a single cavity, which plays the role of both the buncher and the catcher. The frequency of oscillation is primarily determined by the resonant frequency of that cavity, but it can be changed slightly by changing the reflector voltage. In spite of increasing competition from solid-state sources, particularly Gunn and Impatt devices, reflex klystrons remain invaluable as low-noise local oscillators, particularly in the 30- to 170-GHz frequency range. High-power multicavity klystrons, which can generate peak powers of tens of megawatts, are commonly used as sources of electromagnetic energy in linear accelerators.

The traveling-wave tube, invented and demonstrated during the second World War by Kompfner, has some similarity to the multicavity klystron described earlier. However, the traveling-wave tube makes use of a metallic structure that can guide electromagnetic waves independently of the electron beam. The traveling-wave tube is inherently a broadband device because of the absence of any resonant structure required for its basic operation. The guiding structure that was first considered and that remains the most useful is the helix, shown schematically in Fig. 2a. The phase velocity of waves guided by the helix is approximately equal to the velocity of light in free space, c, times the ratio of the helix period to its perimeter. A typical value for the phase velocity is $c/10$. The wave guided by the helix is amplified when it interacts with an electron beam whose velocity exceeds

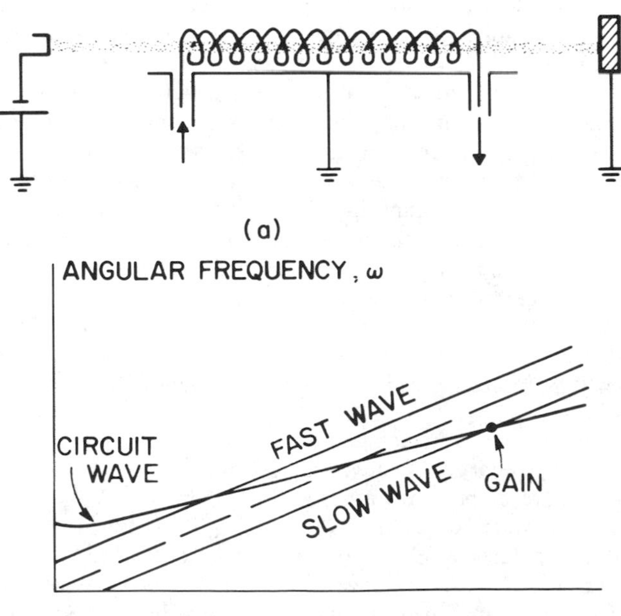

FIG. 2. Schematic view of a helix-type travelling-wave amplifier.

slightly the phase velocity of the guided wave. When this near-synchronism condition is met, the electrons are submitted to an almost constant axial field. On the average, they supply energy to the electromagnetic wave. In a precise theory, the fact that electrons repel each other must be taken into account. When waves are excited on electron beams (in the absence of any circuit), the electronic fluid tends to oscillate at the so-called plasma angular frequency ω_p as a result of the electrostatic forces. If the angular frequency of the electron wave in the laboratory frame of reference is ω, the apparent wave number can be either $\beta_+ = (\omega + \omega_p)/v_e$ (slow wave) or $\beta_- = (\omega - \omega_p)/v_e$ (fast wave), where v_e denotes the average electron velocity. From the fact that the electron-wave energy in the electron-beam frame of reference is positive, and that the group velocity $d\omega/d\beta_+ = v_e$ in the laboratory frame exceeds the phase velocity ω/β_+ of the slow wave (see Fig. 2b), it can be shown quite generally that the energy of the slow wave in the laboratory frame is negative. Approximately, this means that as the amplitude of the slow wave grows, the total energy of the electron beam decreases. Thus, when the slow electron wave is synchronous with the circuit wave, the electron beam supplies energy to the circuit wave, both the circuit and the slow electron waves growing exponentially as a function of distance. This is the condition for maximum gain. If, however, the circuit wave is synchronous with the *fast wave*, the circuit wave supplies energy to the electron beam and is attenuated. The stability of traveling-wave tubes with respect to reflections from the load is considerably improved when a properly located attenuation is introduced on the circuit: Reflected waves are attenuated but the gain is not very much reduced.

Traveling-wave tubes are commonly used to amplify weak signals up to a level of a few tens of watts in the 3- to 30-GHz range of frequency, for example in communication satellites. Below 10 GHz and for powers less than a few watts, field-effect transistors can successfully compete with TWT. Gigawatts of peak power have been reported in TWT having beam voltages of a few megavolts.

Two important modifications of the conventional TWT should be mentioned. In this analysis it has been implicitly assumed that the circuit group velocity, as well as the circuit phase velocity, is in the forward direction. However, some circuits, particularly periodic circuits such as the inderdigital line, may in fact support waves whose phase and group velocities are in opposite directions. When such circuits are made to interact with electron beams, a continuous feedback takes place that results in an oscillation of the device. The electromagnetic energy flows in the direction opposite to the electron flow. This is the principle of the backward-wave oscillators. These devices are commonly used as electronically tunable sources of radiation. They provide milliwatts of power up to 1000 GHz.

The second modification concerns the electron beam. Electron beams can be kept confined in static electric and magnetic fields perpendicular to each other and to the direction of propagation. Crossed-field tubes, such as the magnetron, usually exhibit higher efficiencies than conventional traveling-wave tubes, but they are more noisy. Some magnetrons operate at voltages up to 1 MV.

A recently developed tube is the gyrotron which operates

in the relativistic regime and can generate up to 100 kW, cw, at 80 GHz.

The free-electron laser is a related device whose frequency range extends to optics.

See also MICROWAVES AND MICROWAVE CIRCUITRY.

BIBLIOGRAPHY

General

C. Susskind, *The Encyclopedia of Electronics*, Reinhold, New York, 1962. (E)

History

E. L. Ginzton, "The $100 Idea; How Russel and Sigurd Varian, with the Help of William Hansen and a $100 Appropriation, Invented the Klystron," *IEEE Spectrum* **12** (2) (1975). (E)

R. Kompfner, *The Invention of the Traveling-Wave Tube*, San Francisco Press, San Francisco, 1964. (E)

Technical

J. F. Gittins, *Power Traveling-Wave Tubes,* Elsevier, New York, 1965. (I)

V. Granatstein and I. Alexeff, *High Power Microwave Sources,* Artech House, 1987.

E. Okress, ed., *Crossed-Field Microwave Devices,* Academic Press, New York, 1961. (I)

Theory

J. R. Pierce, *Traveling-Wave Tubes*, Van Nostrand, Princeton, NJ, 1950. (A)

J. E. Rowe, *Nonlinear Electron–Wave Interaction Phenomena*, Academic Press, New York, 1965. (A)

Kondo Effect

Alan S. Edelstein

The Kondo effect is the name given to describe the temperature dependence of restivity in certain alloys containing a small concentration of a magnetic impurity. The resistivity of these alloys as a function of decreasing temperature exhibits a minimum, followed by a logarithmic increase at lower temperatures and finally becomes temperature independent at still lower temperatures. The effect was named after J. Kondo who in 1964 performed a perturbation-theory calculation which correctly predicted the logarithmic behavior. Though first observed in 1939 in "pure" Cu, the effect was later understood to be associated with residual impurities in the Cu. These investigations started a research area that is still active and that has evolved into two closely related research areas, valence fluctuations and heavy fermions. These two research areas are discussed separately in this volume. It will be pointed out below that heavy-fermion systems are concentrated Kondo systems, i.e., concentrated systems which exhibit the Kondo effect.

The reasons for the interest in and longevity of the field are threefold. First, since the resistivity in Kondo's calculation increased as ln T, it would become infinite at low temperatures. Instead of this behavior, experiments on dilute alloys containing magnetic impurities showed that the low-temperature resistivity is temperature independent and relatively small. The qualitative difference between the prediction of perturbation theory and experiment is an indication that the low-temperature properties and the ground state of the system cannot be obtained from perturbation theory. New theoretical techniques had to be developed to determine these quantities. Second, as discussed below, the magnetic moments of the impurities in Kondo systems become unstable at low temperatures. Thus it is necessary to understand the Kondo effect in order to understand magnetic moment formation in metals. Third, the discovery of a continuing variety of new phenomena, such as f-band superconductivity, has served to sustain the field.

It is useful to contrast the Kondo effect with the usual temperature dependence of a metal. Usually the resistivity of a metal, $\rho(T)$, monotonically decreases with decreasing temperature to a temperature-independent value, ρ_0, at low temperatures. The resistivity ρ_0 is due to nonmagnetic impurities and other things which break the perfect periodicity of the lattice. The resistivity increases with increasing temperature since there are increasingly larger thermal vibrations which also disrupt the lattice periodicity. Based upon calculations using first-order perturbation theory, it was thought that magnetic impurities would only increase ρ_0 and thus shift $\rho(T)$ upward. Instead of this behavior, Kondo systems show a logarithmic increase in their resistance with decreasing temperature in a temperature region which is approximately a decade wide. The temperature where this increase occurs, which is system dependent, often is below 10 K. This behavior is illustrated in Fig. 1 for LaB_6 containing various percentages of Ce as the impurity.

Let us consider the scattering of conduction electrons off a magnetic impurity to see how the temperature dependence of $\rho(T)$ arises. Scattering can alter an electron's momentum (both its direction and magnitude) and can cause the conduction electron's spin to flip. Since scattering from nonmagnetic impurities can alter a conduction electron's momentum, clearly spin-flip scattering must be essential in giving rise to the Kondo effect. In a perturbation theory treatment of scattering, one attempts to express the state of the system after the scattering in terms of the initial state. It turns out that this is accomplished by considering repeated scattering off the impurity. Because of the Heisenberg uncertainty principle, the uncertainty of the lifetime of the intermediate states between successive scatterings gives rise to an uncertainty in the electron's energy when it is in these intermediate states. Pauli's exclusion principle, which states that there can only be one electron in a given state, places an important constraint on the occupancy of intermediate states.

In metals at zero temperature, all electronic states below the Fermi energy, E_F, are occupied, and all the higher-energy states are empty. At an absolute temperature T, there is a band of states of width k_BT (where k_B is Boltzmann's constant) centered at E_F where there are empty states below E_F and occupied states above E_F. For nonspin-flip scattering, when one sums over the allowed intermediate states, there are two competing effects which cancel the temperature dependence of the scattering. Kondo's calculation showed that

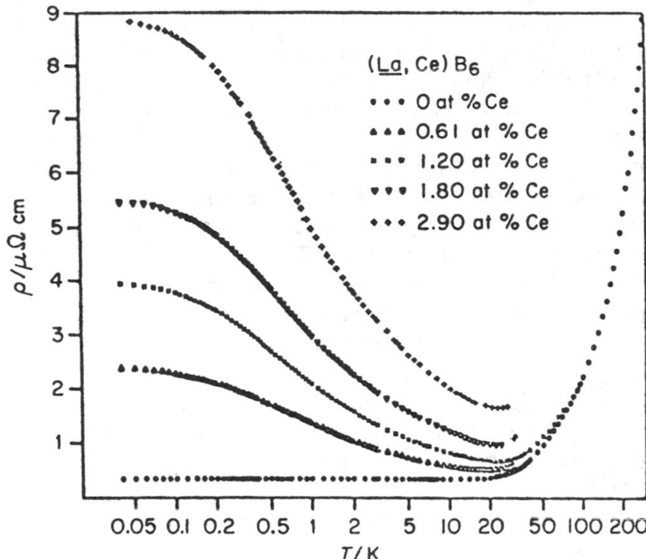

FIG. 1. Electrical resistivity of LaB_6 and four $(La,Ce)B_6$ samples as a function of temperature (after Samwer and Winzer 1976) (from "Exact Results in the Theory of Magnetic Alloys," by A. M. Tsvelick and P. B. Wiegmann, from *Advances in Physics,* Vol. 32, published by Taylor & Francis Ltd, copyright 1983. Used by permission).

these two competing effects no longer cancel in the case of spin-flip scattering.

In his calculation Kondo assumed that the interaction between a conduction electron and the impurity had the form

$$E_{\text{int}} = -2J\mathbf{S}\cdot\mathbf{s}, \quad (1)$$

where J is an energy and is called the exchange integral and $\mathbf{S}$ and $\mathbf{s}$ are the spins of the magnetic impurity and the conduction electron, respectively. Using the second Born approximation in a perturbation calculation, Kondo found that scattering off magnetic impurities increases the resistivity by an amount $\Delta\rho$ given by

$$\Delta\rho = c\rho_m[1 + N(E_F)J\ln(k_BT/D)], \quad (2)$$

where c is the impurity concentration, ρ_m is the contribution to the resistivity of a single impurity calculated using the first Born approximation, and $N(E_F)$ is the density of states of the conduction electrons evaluated at the Fermi energy. For simplicity, Kondo assumed that the conduction-electron density of states is independent of E and of width $2D$. One sees from Eq. (2) that $J<0$ is necessary for the Kondo effect to occur, i.e., for the resistivity to increase at low temperatures. One also sees that $\rho\rightarrow\infty$ for $T\ll D/k_B$ and $J<0$. This is an unphysical result. There is a limit, the unitary limit, which sets an upper bound on the amount of scattering from a single impurity. The fact that Eq. (2) is incorrect at low temperatures indicates that the method used in its derivation, i.e., perturbation theory, may be inapplicable. This turns out to be the case. Sophisticated methods, including renormalization-group and Bethe-*Ansatz* techniques, have been used to derive the low-temperature properties correctly. K. Wilson received the Nobel prize for his renormalization group work which included the solution of the spin-$\frac{1}{2}$ Kondo problem.

Thus there is special interest in the low-temperature properties. There is a temperature, called the Kondo temperature, below which the system gradually condenses into its ground state. No sharp phase change occurs at T_K, since an isolated impurity does not have enough degrees of freedom to permit a sharp phase transition. The value of T_K is given by

$$k_BT_K = \tilde{D}[\,|J|\,N(E_F)]^{1/2}\exp[-1/|J|\,N(E_F)], \quad (3)$$

where $\tilde{D}$ is an effective bandwidth which is of order D. The value of T_K is important in that it determines the energy and temperature scale; i.e., $T/T_K>1$ and $E/k_BT_K>1$ correspond to high temperatures and energies, respectively. Much work has been done to develop a theory that is valid at all temperatures. The most promising work is based on renormalization-group techniques. Even with this approach, exact calculations have been limited to a few properties such as the susceptibility and have not been generalized to spin values greater than $\frac{1}{2}$.

Since the Kondo temperature T_K sets the energy scale, it also determines the temperature scale for experimental properties. Thus, one expects these properties to be approximately functions of just T/T_K. Experimentally the value of T_K for a given system is determined by fitting experimental data by theoretical predictions. The experimental values for T_K obtained in this way can vary drastically depending on the impurity and host. For dilute Mn in Cu, $T_K<0.01$ K, while for dilute Fe in Cu, $T_K\approx30$ K.

The exchange form of the interaction energy, Eq. (1), can, in a certain limit, be derived from the more fundamental formulation of P. W. Anderson. Anderson was interested in magnetic moment formation of dilute 3d transition-element impurities in a nonmagnetic host. In the dilute limit one need only calculate the contribution of a single, isolated impurity. He considered the energy of the total system to consist of a conduction–electron contribution, an impurity contribution, and an interaction term between the impurities and the conduction electrons. The impurity contribution was taken to be a sum over σ of $\epsilon_d n_{d\sigma} + Un_{d\sigma}n_{d-\sigma}$, where ϵ is the energy of the d state, $n_{d\sigma}$ is the number of d electrons of spin σ occupying the impurity site, and U, which is positive, is the Coulomb repulsion between d electrons of opposite spin. For a nondegenerate d orbital, double occupancy by electrons with the same spin is not allowed because of the Pauli exclusion principle. The interaction term allows an electron to hop on or off the impurity site without changing its spin. A spin flip occurs if an electron with spin σ hops off the site and is replaced by an electron with spin $-\sigma$. This interaction causes the original sharp impurity level to have a width Δ proportional to $|V_{dk}|^2$, where $|V_{dk}|$ is the magnitude of the matrix element involved in the hopping. For $\Delta\ll U$ one can transform the interaction energy in Anderson's model to the form of Eq. (1). The exchange integral that then appears in Eq. (1) is $J = J_d + J_{\text{eff}}$ where J_d is a positive quantity called the direct exchange integral and J_{eff} is given by

$$J_{\text{eff}} = \frac{|V_{dk}|^2U}{E_{df}(E_{df}+U)}, \qquad E_{df} = E_d - E_f < 0. \quad (4)$$

In Anderson's model, the conditions that most favor magnetic moment formation are $\Delta \ll U$ and $E_{df}/U = -\frac{1}{2}$. As mentioned earlier, the total exchange integral must be negative in order for the Kondo effect to occur. Since J_d is positive, this requires that $E_{df}(E_{df} + U) < 0$ and sufficiently small in magnitude that $-J_{\text{eff}} > J_d$. Notice that the conditions required for the Kondo effect to occur are different from those favoring moment formation. Thus it is not surprising that the ground state of Kondo systems is a nonmagnetic singlet.

For dilute, stable magnetic moments the susceptibility χ per impurity of a substance having a magnetic moment μ per impurity is given by Curie's law, $\chi = \mu^2/3k_BT$. In Kondo systems for $T<T_K$, the interactions between the local moment and the conduction electrons reduce the value of χ below the Curie's law prediction. One can define an effective moment $\mu_{\text{eff}}^2 = 3kT\chi$ as a measure of how the magnetic response of the impurity is reduced. This reduction might be anticipated from Eq. (1). Since $J<0$, the total energy of the system is minimized if the conduction electrons' spins are aligned opposite to the impurity spins. This antiparallel arrangement reduces the effective moment. In fact, for Kondo systems for $T \ll T_K$, χ approaches a constant χ_0 and hence $\mu_{\text{eff}}^2 \to 0$. Numerical calculations using the renormalization group show that the distribution of allowed energy states in the limit $T \ll T_K$ is similar to that which would exist if $|J|$ in Eq. (1) were to become infinite. Since $T \to 0$ is like $|J| \to \infty$ in Eq. (1), the coupling between a spin-$\frac{1}{2}$ impurity and the conduction electrons becomes very strong. This implies for $T \ll T_K$ that there is always a conduction electron paired with its spin antiparallel to that of the impurity. Thus, the ground state is a nonmagnetic singlet. The spin of the impurity is compensated by the spins of the conduction electrons in a dynamic way. There is not a static compensating conduction electron cloud.

Evidence against the existence of a static compensating conduction electron cloud was provided by neutron-scattering measurements on Y:Ce alloys and nuclear-magnetic-resonance measurements on Cu:Fe alloys. The measurements on Y:Ce alloys showed that the spatial distribution of the Ce moment was the same as that of Ce^{3+} and did not appear to be modified by the conduction electrons for $T<T_K$. The nuclear-magnetic-resonance measurements were performed on the Cu nuclei, and it was found possible to detect the separate resonance signals from Cu nuclei which were different distances from the Fe impurities. Signals from the first through fourth nearest neighbors of the Fe impurities have been separated from the main resonance of most of the Cu nuclei. From the temperature dependence of these signals it is possible to conclude that the magnetization or spin density has the form $\sigma(r, T) = f(r)\chi(T)$, i.e., it does not change its shape for $T<T_K$ at the position of these neighbors.

For $0<T \ll T_K$ the renormalized J is large but finite. In this case real transitions out of the ground state are impossible since they require a finite amount of energy. Virtual transitions are possible and the impurity is polarizable. This leads to a repulsive interaction between the conduction electrons and to a finite value for the susceptibility for $T \ll T_K$. Hence $\mu_{\text{eff}} \to 0$ as $T \to 0$.

One can also understand why $\mu_{\text{eff}} \to 0$ as $T \to 0$ by considering the time that an impurity spin will remain correlated with itself, which is called the intrinsic spin correlation time, τ. For Kondo impurities it turns out that $\tau \approx h/k_BT_K$, where h is Planck's constant. The susceptibility of a dilute impurity is the time average of $\{S(0)S(t)\}$ from 0 to a time $t \approx h/k_BT$. If a spin only stays correlated with itself a time τ, then for $T \gg T_K$ the susceptibility is approximately given by Curie's law, while for $T \ll T_K$ the susceptibility is given by $\chi \approx \mu^2\tau/3h$. Hence, $\mu_{\text{eff}} \propto (T\tau)^{1/2}$ for $T \ll T_K$.

A great variety of both microscopic and macroscopic experiments have been performed on Kondo systems in which the impurity, the host, the concentration of impurities, and such parameters as the temperature and the magnetic field were varied. For most properties, the single-impurity contribution must be proportional to the impurity concentration. For many systems one must employ concentrations less than 100 ppm to observe the single-impurity contribution. Early experimental results were often complicated by interaction effects.

Table I shows the qualitative behavior for the resistivity, susceptibility, and specific heat for Kondo systems for $T \gg T_K$ and $T \ll T_K$. The temperature dependence of these properties gradually changes from its high- to its low-temperature behavior. There is a broad specific-heat anomaly, which occurs over several decades in T/T_K, that is centered at approximately $T_K/3$. The entropy change derived from the area under a plot of C/T vs. T, where C is the impurity specific heat, is approximately equal to $R \ln(2S + 1)$. This is consistent with the ground state being a singlet. Note that the specific heat is proportional to cT/T_K, where c is the impurity concentration for $T \ll T_K$. Thus, if c/T_K is large, then the low-temperature specific heat will have a large term which is linear in T. Heavy-fermion systems are defined as systems which have such a large linear term. The mechanism which gives rise to this term in heavy-fermion systems is believed to be large values of c/T_K. Theoretical calculations predict that χ is a univeral function of T/T_K. It is interesting that the low-temperature susceptibility of several Kondo systems can be fitted by this function of T/T_K. This agreement is somewhat surprising since these systems do not satisfy the conditions of the calculations, e.g., they are spin-$\frac{1}{2}$ systems.

Magnetic impurities drastically affect the properties of superconductors. The electrons in superconductors condense into states in which pairs of electrons are bound and have their spins oppositely directed. Magnetic impurities break this symmetry property and cause a large depression in the superconducting transition temperature T_c. Kondo impurities cause an especially large depression in T_c and gapless superconductivity at very low impurity concentrations. The

Table I Temperature Dependence of Kondo Impurity Contributions to Some Macroscopic Properties[a]

	Resistivity	Susceptibility	Specific heat
$T \gg T_K$	See Eq. (2)	$\mu^2/3k_B(T + T_K)$	$\propto 1/T^2$
$T \ll T_K$	$\propto(1 - bT^2/T_K^2)$	$\chi_0(1 - dT^2/T_K^2)$	$\propto cT/T_K$

[a] b and d are constants.

impurities' tendency to inhibit superconductivity is greatest for $T \approx T_K$. Thus, if T_K is smaller than the superconducting transition temperature of the pure system, it is possible for the system to undergo a transition into the superconducting state at a temperature T_{c1} above which the impurity has its full effect and then undergo a second transition out of the superconducting state at a temperature $T_{c2} \approx T_K$. These effects have been observed in (La,Ce)Al_2 and (La,Th):Ce alloys.

See also CONDUCTION; DIAMAGNETISM AND SUPERCONDUCTIVITY; HEAVY-FERMION MATERIALS; INTERMEDIATE VALENCE COMPOUNDS.

BIBLIOGRAPHY

G. Grunier and A. Zawadowski, *Rep. Prog. Phys.* 37, 1497 (1974); theoretical review. (A)

A. J. Heeger, *Solid State Physics,* (F. Seitz, D. Turnbull, and H. Ehrenreich, eds.), Academic Press, New York, 1969; experimental review. (I)

C. Kittel, *Introduction to Solid State Physics,* 4th ed., p. 660. Wiley, New York, 1971. (E)

J. Kondo, *Solid State Physics* (F. Seitz, D. Turnbull, and H. Ehrenreich, eds.), Vol. 23, p. 184. Academic Press, New York, 1969; theoretical review. (I)

P. Nozières, *J. Low Temp. Phys.* **17,** 31 (1974); Fermi-liquid description of the problem. (I)

C. Rizzuto, *Rep. Prog. Phys.* **37,** 147 (1974); experimental review. (I)

P. L. Rossiter, *Aust. J. Phys.* **39,** 529 (1986). (I)

K. Samwer and K. Winzer, *Z. Phys. B* **25,** 269 (1976).

H. Suhl (ed.). *Magnetism,* Vol. V. Academic Press, New York, 1973; both theoretical and experimental reviews which include the effect on superconductivity. (I and A)

A. M. Tsvelick and P. B. Wiegmann, *Adv. Phys.* **32,** 453 (1983). (A)

K. G. Wilson, *Rev. Mod. Phys.* **47,** 773 (1975); review of renormalization theory. (A)

Laser Spectroscopy

Steven Chu

The interaction of laser light with matter has dramatically extended the capabilities of optical spectroscopy. These developments are due to the synergism between technological advances in lasers, a deeper understanding of how light interacts with matter, and new spectroscopic techniques that have emerged from this understanding.

Some of the technological capabilities of lasers are summarized below. The frequency range of lasers has been extended from the far-infrared to the x-ray region of the electromagnetic spectrum range. There is almost complete coverage of the ultraviolet to mid-infrared portion of the spectrum with broadly tunable light sources. The frequency control of lasers now rivals the precision of the best microwave sources: stability of the most precisely controlled lasers is better than one part in 10^{-13}. If linked to suitable atomic references, they have the potential to be stable to better than one part in 10^{-18}. Lasers can deliver high energy pulses in excess of 10^{12} watts, and because photons interact extremely weakly with each other without the aid of intervening matter, this power can be concentrated into volumes of less than 100 μm in diameter. Even with small laser systems, electric fields can be made far stronger than atomic electric fields. Pulses as short as a few femtoseconds (10^{-15} seconds) can be generated in precisely regular pulse trains or as a single high intensity burst. Miniature solid state diode lasers are now commonly used in laser spectroscopy as well as in optical communications, laser printing, and information retrieval.

DOUBLE RESONANCE

The interaction of two beams of monochromatic light with atoms or molecules, satisfying a double resonance condition, can be used to eliminate Doppler broadening and allow the much narrower homogeneous linewidth to be resolved. When only a single transition is involved, the fields can be of the same frequency, but must propagate in opposite directions (saturation spectroscopy, Lamb-dip effect). When two transitions sharing a common level are excited, the two fields may be of different frequencies and can propagate either in the same direction (resonant Raman effect) or in opposite directions (two-photon absorption). Energy and momentum conservation or perturbation theory show that the Doppler shift $k_{1,2}v_z$ is eliminated for the Lamb-dip case when $\omega_1=\omega_2=\Omega_1=\Omega_2$, for the resonant Raman case when $\Omega_1-\Omega_2=\omega_1-\omega_2$, and for the two-photon absorption case when $\Omega_1+\Omega_2=\omega_1+\omega_2$. Here, $\omega_{1,2}$ is the transition frequency of laser beams 1 and 2, $\Omega_{1,2}$ is the laser frequency, $\mathbf{k}_1-\mathbf{k}_2$ is the propagation vector of light, and v_z is the molecular velocity component along the laser beam direction. To lowest order, the intensity of the double-resonance signal depends on the product of intensities of the two applied fields (nonlinearly) and exhibits a sharp Doppler-free tuning behavior.

APPLICATIONS OF DOUBLE RESONANCE TECHNIQUES

One significant application of the Lamb dip has been the precise measurement of the speed of light. As a result, the speed of light is now a defined quantity, $c=299\ 792\ 458$ m/s, that ties the cesium time standard to the length standard. The frequency of a continuous wave (cw) 3.39-μm He-Ne laser has been frequency locked ($\Delta v/v=10^{-13}$) to the Lamb dip of a CH_4 sample located inside the laser cavity. Figure 1 shows the tuning behavior of this laser, a nonlinear resonance 400 Khz wide riding on top of a 260-Mhz-wide Doppler profile. The 88 THz optical frequency corresponding to the 3.39 μm wavelength was directly tied to the cesium microwave standard by generating a chain of microwave and laser oscillators linked by ultrahigh-frequency mixing diodes. Lamb-dip measurements of the hydrogen $2S$-nD transitions in atomic hydrogen using a cw tunable dye laser have produced a new Rydberg value, $R_\infty=109\ 737.315\ 714\ (19)$ cm^{-1}.

Two-photon spectroscopy has been used to measure the $1S\rightarrow 2S$ transitions in fundamental atoms such as hydrogen,

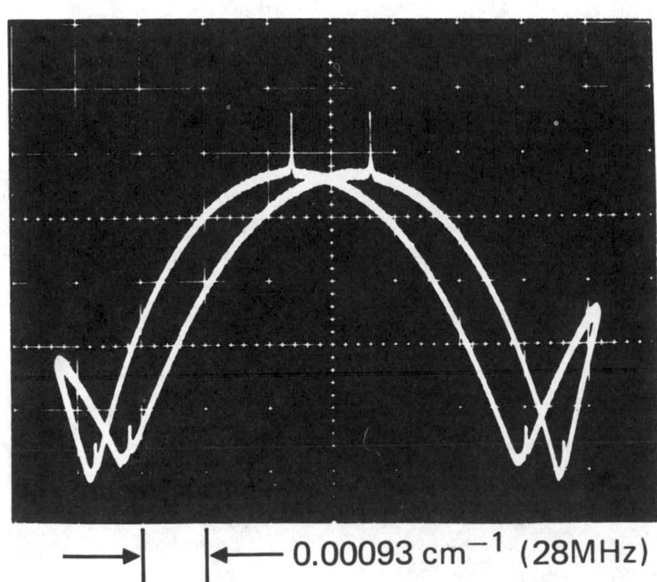

FIG. 1. The tuning characteristic of a 3.39-μm He–Ne laser that contains an intracavity CH_4 sample (by permission of The American Physical Society).

positronium ($e^+ - e^-$) and muonium ($\mu^+ - e^-$). The precise energy measurements of these intervals provide some of the best tests of quantum electrodynamics for a bound system.

LASER COOLING OF TRAPPED IONS AND ATOMS

Neutral atoms and trapped ions can be cooled to temperatures on the order of ten microkelvins with laser light. At those temperatures, the atoms have velocity spreads on the order of 10 cm/sec. Both the first and second order Doppler shifts of the particles are greatly reduced and the available measurement time is greatly increased. These techniques have yielded linewidths in the microwave regime of less than 10 milliHz, and optical regime of less than 30 Hz, and will probably form the basis of our next generation time standards. Tunable lasers can be phase-locked to a reference Fabry-Perot cavity, and the length of the cavity can then be stabilized by signals derived from laser cooled samples of ions or atoms. Laser cooling techniques are also broadly applicable to spectroscopic measurements where the signal appears as a small frequency shift. Tests of mass anisotropy, nonlinearity in quantum mechanics, time reversal invariance, and charge neutrality of atoms are examples of work where laser cooling is playing a role. Laser spectroscopy of cooled ions and atoms has also opened up the study of single and few ion systems, ultracold plasmas, ultracold collisions, and the formation of dilute quantum gases.

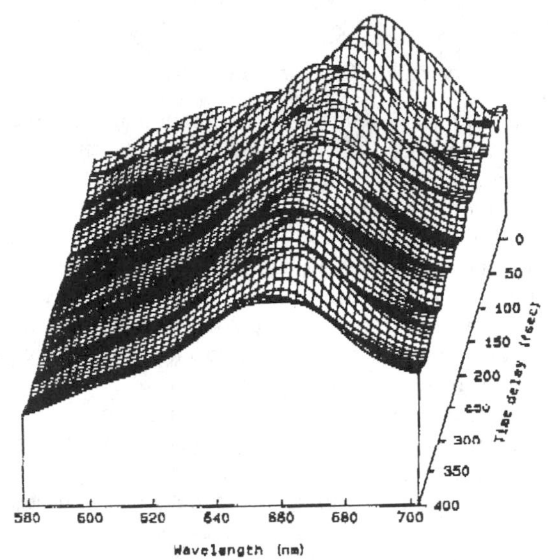

FIG. 2. Three dimensional plots of the differential absorption spectrum as a function of time delay and wavelength for the dye Nile Blue. The amplitude and peak of the induced absorption oscillates with a period of 60 femtoseconds. (From C. V. Shank, et al., p. 302 of *Laser Spectroscopy IX,* eds. M. S. Feld, J. E. Thomas, and A. Mooradian, Academic Press, Boston, 1989.)

COHERENT TRANSIENTS

Nonlinear optical processes of a coherent nature can be observed in the time domain and are capable of yielding detailed dynamic information about relaxation processes as well as Doppler-free spectra. Coherent transient phenomena, such as optical nutation, free induction decay, and photon echoes arise in a sample excited by a sequence of resonant laser pulses. These phenomena are completely analogous to effects in pulsed NMR spectroscopy. Each transient effect may be used to select and examine a specific dephasing process. The laser field prepares the sample by placing the transition levels in quantum-mechanical superposition, and the resulting dipoles radiate in accordance with the coupled Schrödinger–Maxwell wave equations. The detection of the emitted radiation then yields information about the environment of the oscillators.

PICOSECOND AND FEMTOSECOND SPECTROSCOPY

Extremely short pulses of light are used to examine events such as vibrational and rotational relaxation of molecules, the thermalization of photo-excited electron plasmas in a semiconductor, and nonthermal processes such as desorption of adsorbates from surfaces. Subpicosecond snapshots of chemical reactions, studies of the primary photochemical behavior of retinal in rhodopsin, and ultrafast electrical pulses are now possible with the short pulse technology. Most experiments use some variation of the "pump-probe" technique, where the first optical pulse is used to trigger an event with the absorption of one or more photons. A second pulse optically delayed with respect to the first pulse then probes the system spectroscopically. Figure 2 shows the induced transmittance change for the molecule Nile Blue using a 6 femtosecond pump and probe. Coherent transient techniques such as photon echoes or four-wave mixing are also applicable in time domain spectroscopy. The short laser pulses can also be used to generate very short bursts of x rays, phonons, electrons, and electrical pulses.

See also LASERS; NONLINEAR OPTICS.

BIBLIOGRAPHY

L. Allen and J. H. Eberly, *Optical Resonance and Two Level Atoms*. Wiley, New York, 1975.

S. Chu and C. Wieman, ed., Special Issue of *J. Opt. Soc. Am. B,* **6,** "Laser Cooling and Trapping of Atoms," 1989.

W. Demtröder, *Laser Spectroscopy*. Springer-Verlag, Berlin, 1988.

M. Feld, J. E. Thomas, and A. Mooradian, eds., *Laser Spectroscopy IX*. Academic Press, Boston, 1989.

G. R. Fleming and A. E. Siegman, eds., *Ultrafast Phenomena V*. Springer-Verlag, Berlin, 1986.

S. Haroche, J. C. Gay, and G. Grynberg, *Atomic Physics 11*. World Scientific, Singapore, 1988.

Y. Prior, A. Ben-Reuven, and M. Rosenbluth, eds., *Methods of Laser Spectroscopy*. Plenum Press, New York, 1986.

Y. R. Shen, *Principles of Nonlinear Optics*. Wiley, New York, 1984.

A. Siegman, *Lasers,* University Science Books. Mill Valley, Calif., 1986.

A. Yariv, *Quantum Electronics,* 3rd ed. Wiley, 1988.

Lasers

H. J. Zeiger and P. L. Kelley

The laser is a device for the generation of coherent, nearly single-wavelength (and -frequency), highly directional electromagnetic radiation emitted somewhere in the range from submillimeter through ultraviolet and x-ray wavelengths. The word laser is an acronym for "light amplification by stimulated emission of radiation." The principle of the laser is similar to that of the maser (see MASERS), which is somewhat arbitrarily defined as a device operating somewhere in the range from the radio or microwave region down to millimeter wavelengths.

Quantum theory shows that matter can exist in only certain allowed energy levels or states (see QUANTUM MECHANICS). In thermal equilibrium, lower energy states of matter are preferentially populated, since occupation probability is proportional to $e^{-\mathscr{E}/kT}$, where $\mathscr{E}$ is the state energy, T the temperature, and k the Boltzmann constant. An excited state can decay spontaneously (i.e., with only zero-point electromagnetic radiation present) to a lower energy state, emitting a quantum or wave packet of electromagnetic radiation (photon) with transition frequency $\nu = \Delta\mathscr{E}/h$, where $\Delta\mathscr{E}$ is the energy difference between the two states and h is Planck's constant. The presence of radiation at frequency ν can cause a transition from the lower energy state to the upper energy state with the absorption of a photon and coherent (phase preserving) decrease in the electromagnetic field energy. A transition from the upper state to the lower state can also be induced by the radiation at the same time a photon wave packet is emitted in step (i.e., is emitted coherently) with the stimulating radiation wave; this stimulated emission process is the reverse of the absorption process. If the matter can be forced out of thermal equilibrium to a sufficient degree, so that the upper state has a higher population than the lower state (population inversion), then more stimulated emission than absorption occurs, leading to coherent growth (amplification or gain) of the electromagnetic wave at the transition frequency.

A laser generally requires three components for its operation: (1) an active medium with energy levels that can be selectively populated; (2) a pumping process to produce population inversion between some of these energy levels; and usually (3) a resonant electromagnetic cavity structure containing the active medium, which serves to store the emitted radiation and provide feedback to maintain the coherence of the electromagnetic field (see Fig. 1). In a continuously operating laser, coherent radiation will build up in the cavity to a level set by the decrease in inversion required to balance the stimulated emission process with the cavity and medium losses. The system is then said to be lasing, and radiation is emitted in a direction defined by the cavity.

An important aspect of the laser involves the design of resonators to accommodate the characteristics of the active medium and the diffractive properties of the radiation, and at the same time meet requirements such as low angular beam divergence and high efficiency. The electromagnetic field in a resonator has well-defined modes that have patterns both transverse to and along the cavity axis. Waveguiding with index-of-refraction profiles or reflecting walls can be usefully employed in some cases (most notably in semiconductor lasers) not only to confine the radiation to the amplifying medium but to force the laser to operate in a single transverse mode. More frequently, however, laser resonators are open in the sense that the transverse structure is defined only by axial mirrors or lenses (see Fig. 1). Open resonators formed with convergent optics ("stable" resonators) generally have the lowest diffraction losses, while planar resonators have higher losses, and resonators formed with divergent optics ("unstable" resonators) have the highest losses. Figure 2 shows a few of the lowest-order transverse mode distributions for a stable resonator. There is generally one transverse mode of a cavity that has the largest net gain (product of amplifier gain and the transmittance of the remainder of the cavity). This is the transverse mode that oscillates first and is typically the lowest-order (TEM_{00}) mode (see Fig. 2). Single transverse mode lasers, in particular lasers operating in the TEM_{00} mode, have nearly optimal "brightness" in the sense that the beam divergence is near the minimum value for the size of the laser output mirror; such a laser is said to have near-diffraction-limited output. The optimal resonator for a particular laser (e.g., "stable" or "unstable") is determined by the geometry of the gain medium, the desired cavity length, and the single-pass gain. Cavity modes have an axial periodicity that is determined by the cavity length; the frequency spacing between the axial modes is the inverse of the round-trip time for radiation in the cavity. The gain in a laser is peaked at a transition frequency determined by the energy levels of the active medium, and laser operation tends to occur at the axial mode frequency closest to the gain peak.

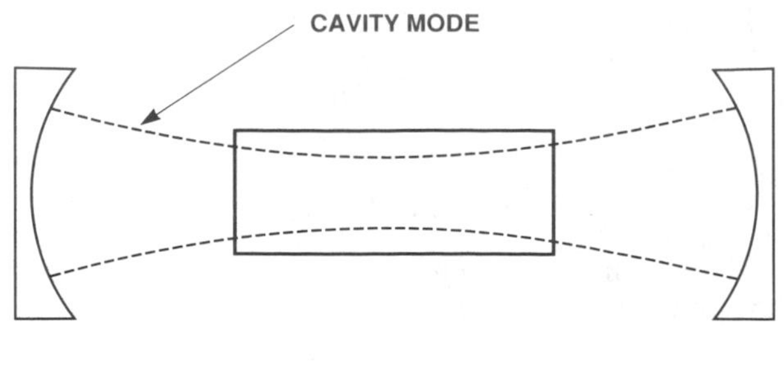

FIG. 1. A simplified schematic of a laser oscillator. The mirrors at the ends of the laser form an open resonator. Stable modes that consist of electromagnetic waves which travel back and forth in the resonator are amplified by the active laser material. At the radiative steady state, the gain due to amplification balances the loss due to intracavity absorption, mirror reflection losses, and diffraction beyond the edge of the mirrors. The pumping system is not shown, nor are ancillary intracavity elements that are often used for temporal (including frequency selection) and spatial control of the laser output. Usually, one of the mirrors is partially transmitting so that some of the highly directional radiation leaves the cavity through the mirror. The dashed lines are approximately characteristic of the transverse extent of the lowest-order transverse mode.

A rich variety of physical systems has been exploited to

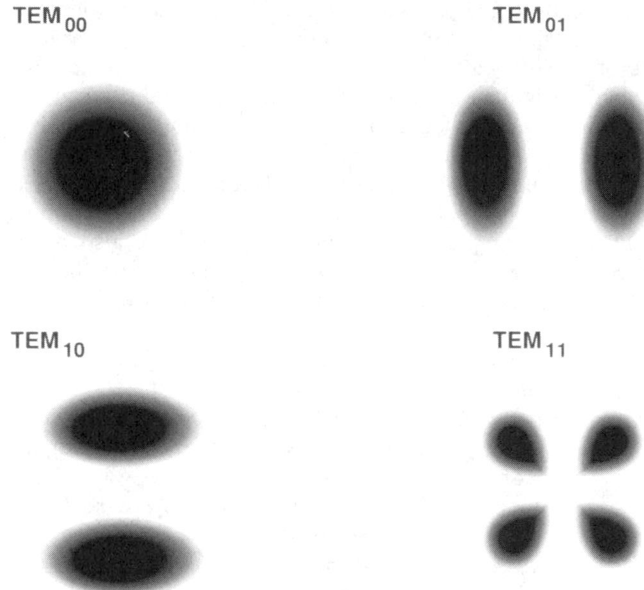

FIG. 2. The lowest-order transverse modes for a "stable" resonator with square symmetry. The TEM_{mn} notation indicates that the modes have nearly transverse electric and magnetic fields with m nodes vertically and n nodes horizontally.

produce laser radiation over a five-decade range of wavelengths. We can only briefly describe here some of the most significant of these.

SOLID-STATE LASERS

In 1960 the first laser was successfully operated. It used as the active medium a rod of ruby (single-crystal Al_2O_3 with the ion Cr^{3+} substituted for a small fraction of the Al^{2+} ions). In this laser, a xenon flashlamp pumps the chromium ions from their ground state to a broad band of states, from which they rapidly decay nonradiatively to a long-lived state at an energy about 14,422 cm^{-1} (693.4 nm) above the ground state.* It is the narrow (~11 cm^{-1}) emission line from this level back to the ground state that gives rise to the ruby laser emission. When lasing occurs, the emission narrows in frequency width to less than 1 cm^{-1} and leaves the ruby rod with an angular spread of a few milliradians.

The ruby laser is a three-level system (see Fig. 3), and therefore requires depopulation of the ground state by more than 50% to obtain population inversion. For this reason, pumping of the laser requires a high-energy source, and continuous-wave operation of the system is difficult to achieve. In the free-running mode, as opposed to the Q-switched mode (see below), the output radiation of the ruby laser fluctuates rapidly over a time of about 1 ms and, for a 1-cm-diameter × 10-cm-long laser rod, is emitted in pulses of about 1 J.

The high-threshold pump power requirement for a three-level laser is greatly relaxed if a four-level system is used.

* The state is actually a doublet split by about 38 cm^{-1}. It is usually the lower-energy transition of this R-line pair that lases.

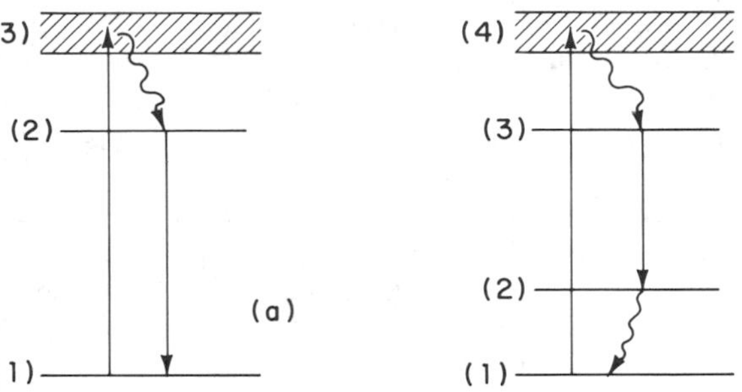

FIG. 3. Schematic representation of an optically pumped laser: (a) three-level system; (b) four-level system.

In the four-level system, population inversion is produced, but radiative transitions do not terminate on the ground state (see Fig. 3). If the final state is not significantly populated at the operating temperature, then population inversion can be maintained with only moderate pump power, and such lasers can usually be easily operated on a continuous basis.

One type of four-level solid-state laser is based solely on the electronic levels of ions in a crystal or other solid state host, with the laser transition terminating on an excited electronic level of the ion. The rare earth ions Nd^{3+}, Ho^{3+}, Er^{3+}, and Tm^{3+} are the most frequently used active ions in such lasers. Extensively developed lasers of this type are the Nd-glass laser and the Nd:YAG (yttrium aluminum garnet, $Y_3Al_5O_{12}$) laser, both of which have their most important transitions in the vicinity of 1.06 μm. Nd:YAG lasers have produced in excess of 1 kW in continuous operation, several joules with low-repetition-rate (10 Hz) pulsed operation, and about 10 mJ with high-repetition-rate (1 kHz) pulsed operation. Large Nd-glass lasers have produced tens of joules at repetition rates of several Hz, and several kilojoules at very low repetition rates (~1 per hour). Rare-earth–doped glass fiber lasers are of potential utility in optical communications. Er^{3+} fiber lasers have given outputs of the order of 100 mW, and their characteristics as amplifiers of fiber optical signals have been studied.

A second type of four-level solid-state laser involves fewer electronic levels of the active ion; instead, laser transitions are employed that terminate on an excited vibrational level of the ion in the host lattice. The basic optical pumping–emission cycle is similar to that of a molecular dye laser (see section below on the dye laser). One important class of these solid-state lasers uses transition-metal ions such as Cr^{3+}, Ti^{3+}, Ni^{2+}, and Co^{2+}. As with the ruby (Al_2O_3) laser, the alexandrite ($BeAl_2O_4$) laser employs the Cr^{3+} ion; in the latter case, however, the laser transition can terminate on a variety of final vibrational states. This laser is tunable from 700 to 818 nm and has pulsed output energies similar to the ruby laser. The Ti:Al_2O_3 laser has even broader tunability, covering a range from 660 nm to beyond 1.1 μm; it has operated continuously at power levels up to 17 W with Ar-laser pumping, and pulsed at energies of hundreds of millijoules when pumped by doubled Nd:YAG laser radiation. Lasers based on color centers in alkali halide crystals operate on a principle similar to dye lasers (see below) and transition-metal lasers. Using different types of F-centers in various alkali halides, wavelength coverage over a range from 0.82 to 3.3 μm can be obtained, with continuous output powers ranging from tens of milliwatts to over a watt. Stability of F-centers can be a problem, and storage at low temperature is required for several of these lasers.

The energy stored in a pumped solid-state laser medium can be delivered as a giant pulse in a time much shorter than the lifetime of the upper laser energy level by the use of Q-switching (quality factor switching). This technique, which was first used in the ruby laser, has been applied to many other laser systems. It makes use of the idea that if the resonant cavity structure is maintained at a very lossy (low-Q) level while the active medium is pumped, a high level of population inversion can be reached. If the cavity is then suddenly switched to a high-Q state, stimulated emission occurs rapidly, and radiation is emitted in a short pulse. In the case of the Nd:YAG laser, this pulse is of the order of 10 ns and can be obtained by the use of a saturable dye absorber, an electro-optic element, or an acousto-optic element as the switch in the laser cavity.

Because solid-state lasers have spectrally broad gain regions, many equally spaced resonant cavity axial modes can lie inside the gain bandwidth when cavity lengths are of the order of tens of centimeters. If the electric fields from a number of resonant cavity axial modes are locked in phase (mode-locked) by modulating a parameter such as dielectric constant or loss, the axial mode fields interfere coherently and constructively for short times of the order of the reciprocal of the lasing bandwidth. The result is a train of high-intensity, narrow pulses of light, with spacing in time determined by the length of the cavity. Mode locking has been applied to many solid-state laser systems such as Nd:YAG (20-ps pulses), and $TiAl_2O_3$ (200-fs pulses). Mode locking has also been used in many other laser systems (see, e.g., the section below on dye lasers).

GAS LASERS

A number of methods can be used to produce population inversion in gaseous media. Inversion can exist between some of the energy levels of the constituents in a gas discharge. The first such system, demonstrated not long after the announcement of the ruby laser, was the He–Ne laser, now a standard item in optics laboratories. This system makes use of a discharge in He at a pressure of about 1 Torr, with an admixture of Ne at about 0.1 Torr. The discharge excites He atoms to their first excited level, about 160,000 cm^{-1} above their ground state. This excitation is readily transferred by collisions to a Ne atomic level with nearly the same excitation energy above the ground state (resonant transfer). These excited states decay radiatively to lower energy Ne states, giving rise to continuous laser emission in the red at 15,820 cm^{-1} (632.8 nm) with output power in the range of 10^{-2} W. Other transitions produce strong emission at 8,680 cm^{-1} (1.15 μm) and 2,957 cm^{-1} (3.39 μm). In a pure Ne discharge, excitation would occur to many Ne levels, and population inversion would not occur as effectively.

An important gas-discharge laser system is based on the energy levels of the argon ion (Ar^+). By a complex series of steps, argon-ion–electron collisions in the discharge lead to population inversion and lasing at a number of frequencies near 20,500 cm^{-1} (488 nm). Continuous output at power levels of tens of watts in the blue-green make this device especially useful as a spectroscopic source in Raman scattering, and for pumping continuously operating tunable dye lasers.

Other intense laser sources arise from atomic transitions of metal ions in a pulsed He discharge (the Cu and Cd vapor lasers, for example). Gas-discharge lasers have also been made to operate in the UV, but special problems arise in this frequency range, and these are discussed below in a separate section.

Gas lasers with output at longer wavelength make use of the vibration–rotation energy levels of molecules (see MOLECULAR SPECTROSCOPY). The energy-level structure of a free molecule in a gas is shown schematically in Fig. 4. In addition to the electronic-state energy-level structure characteristic of atoms, there is vibration–rotation structure associated with the relative motion of the nuclei. The spacing of vibrational energy levels corresponds to frequencies in the infrared, and it is in this region of the electromagnetic spectrum that molecular gas-discharge lasers are especially important.

The most efficient and powerful of the molecular gas-discharge lasers is the CO_2 laser. One version of this laser makes use of a dilute mixture of CO_2 in an N_2 discharge. The N_2 molecules are excited by collisions with electrons to their first excited vibrational state, from which the excitation is resonantly transferred by molecular collisions to preferentially excite CO_2 molecules to a particular vibrational state. These molecules, in turn, undergo radiative transitions to lower vibrational levels, and the presence of the rotational structure gives rise to a cluster of many lines that can lase,

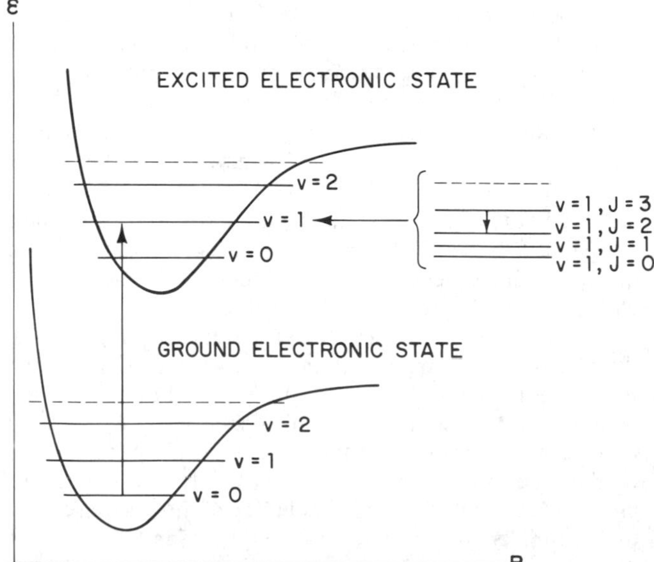

FIG. 4. Diagram of the potential energy curves of a molecule as a function of nuclear coordinate R. The vibrational and rotational energy levels are labeled respectively by v and J (the rotational structure is shown only for $v = 1$). The classical turning points of the motion are the intersections of the vibrational levels with the potential energy curves. The downward-directed arrow indicates a pure rotational transition.

grouped near a frequency of about 944 cm^{-1} (10.6 μm), and 1042 cm^{-1} (9.6 μm). The electric-discharge CO_2 laser is quite efficient (better than 10% electrical power converted to laser power), and is capable of producing power outputs greater than 1 kW in a continuous mode of operation. Other important molecular gas-discharge lasers make use of vibration–rotational or pure rotational transitions of H_2O, CO, and HCN, and produce emission at (78 μm, 119 μm), 5.3 μm, and (337 μ, 311 μm), respectively.

A number of other methods of generating laser radiation using molecular or atomic energy levels have been devised. Powerful pulses of laser radiation at 1.3 μm from excited iodine atoms have been produced by flash photolysis (UV photodecomposition) of CH_3I (methyl iodide). In the gas-dynamic laser, a nonthermal distribution of molecular vibrational energy levels is produced by the rapid expansion of a hot gas through a nozzle. This method has produced continuous emission of tens of kilowatts at 10.6 μm from CO_2 gas. In the chemical laser, two reacting molecular species in a gas produce a product that is left in an excited vibration–rotation state, and returns to the ground state radiatively. An example is the HF (DF) laser, which lases in the 2.5-μm to 3.5-μm region when H_2 (D_2) and F_2 gases combine chemically. Other types of lasers produced by excitation of gases are TEA (transverse electric discharge-excited) lasers, *e*-beam (electron-beam-excited) lasers, and UV-preionized electric discharge lasers.

CO_2 laser radiation has been used to pump other gases, yielding far-infrared emission. If there is a coincidence between a CO_2 laser line and a vibration–rotation transition in another gas, an excited level can be populated directly or by collisional transfer of excitation. Pure rotational transitions can take place radiatively to an unpopulated level, producing far-infrared laser radiation (see Fig. 4). Such gases as NH_3 (output at 291 μm) and CH_3OH (output at 164 μm and 205.3 μm) have been made to lase by this method.

DYE LASERS

It has long been known that certain organic dyes, when illuminated with visible or UV radiation, fluoresce strongly, but at lower frequencies. This so-called Stokes shift of the fluorescence can be understood in terms of the change of equilibrium internuclear position with electronic excitation, the rapid vibrational relaxation within an electronic state, and the Franck–Condon principle (see Franck–Condon Principle; Molecular Spectroscopy). This principle states that an electronic transition in a molecule takes place so rapidly that the nuclear coordinates can be regarded as nearly fixed. A quantum-mechanical calculation shows that the probability of finding the system at a given value of nuclear coordinate is largest near the classical turning point of the vibrational level (which occurs at the points where the vibrational level intersects the electronic potential curve). It then follows that the probability of occurrence of an electronic-vibrational transition is greatest when the initial- and final-state turning points are nearly the same.

An optical pumping–fluorescence cycle for a molecular system is indicated schematically in Fig. 5(a). It can be seen that after absorption of a photon and nonradiative vibrational

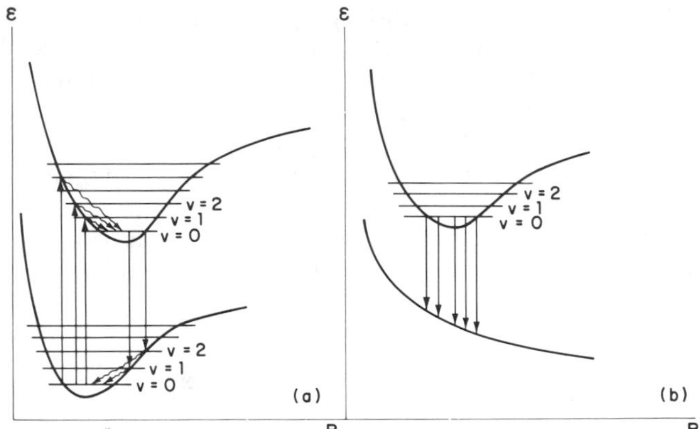

FIG. 5. Light emission from molecular systems: (a) bound–bound system; (b) bound–free system.

cascading to the lowest vibrational level, the excited electronic state can decay by fluorescence at a lower frequency, and the molecule returns to the ground state by nonradiative vibrational cascading. Since the final state in the radiative transition process is unoccupied, the pumping requirements for such a laser system need not be too severe. Discrete vibrational structure is washed out in a liquid, but the general outlines of the cycle indicated in Fig. 5(a) are still preserved in dye fluorescence in a liquid. This process is virtually identical to that occurring in tunable transition-metal solid-state lasers such as $Ti^{3+}:Al_2O_3$ or $Cr^{3+}:BeAl_2O_3$; in the solid, the substitutional ion's coordinates with respect to neighboring host ions play the same role as the internuclear coordinates in the dye molecules.

A problem that arises in the use of a dye system in a laser is the existence of long-lived electron spin triplet states into which the usual excited electron spin singlet state can relax nonradiatively; for rhodamine 6G this relaxation time is $\sim 10^{-7}$ s. This process interferes with laser action in the desired frequency range, but it can be largely circumvented by circulating the dye solution through the resonant cavity structure. Dye lasers can be flashlamp pumped, or pumped with Ar-ion, frequency-doubled Nd:YAG, or N_2 gas laser radiation. Dye systems fluoresce over a wide band of frequencies from the near infrared through the visible, and are well adapted to use in a continuous laser operation mode with wavelength tuning over as much as 40 nm for a single dye; a battery of dyes placed in optical cavity structures gives tunable laser radiation over a range from roughly 10,000 cm^{-1} (1 μm) to 25,000 cm^{-1} (0.4 μm).

Because of the broad bandwidth of the dye laser fluorescence line and, in the case of pulsed dye lasers, the low Q of the resonant structure provided by a simple mirror cavity, the dye laser emission line is fairly broad, but can be greatly narrowed without great loss in power output by the use of a diffraction grating in place of one of the mirrors in the cavity structure. Even narrower linewidths may be obtained by using intracavity frequency-selective elements (etalons); single longitudinal mode operation can be obtained in both pulsed and continuous lasers. Dye jet fluctuations limit the stability of continuous dye lasers; however, stabilities of hundreds of kilohertz are readily obtained, and stabilities of a fraction of a hertz have been achieved. Thus, the dye laser is an important source of radiation for spectroscopy. Since the dye laser has a spectrally broad gain region with many equally spaced resonant cavity modes falling inside the gain bandwidth, it is well suited for mode-locked operation (see section above on solid-state lasers). Using continuous mode-locked dye lasers, pulse trains having pulses of less than 10 fs duration have been obtained and used for studies of fast electronic processes in solids and organic molecules.

SEMICONDUCTOR LASERS

In 1962, it was reported that a forward biased semiconductor diode of GaAs radiated efficiently at about 11,800 cm^{-1} (850 nm). In the following year a number of groups reported the observation of laser emission in this frequency region from suitably prepared diode structures. While these semiconductor devices are solids, they are differentiated from those lasers involving optically active ions in ionic hosts on account of their markedly different physical and technological characteristics. To understand how these devices function, it is necessary to consider the nature of the electronic energy states in a semiconductor (see SEMICONDUCTORS, CRYSTALLINE).

A periodic crystal has bands of allowed energy levels separated by forbidden energy gap regions. In an intrinsic semiconductor at low temperatures, there are just enough electrons present to fill the uppermost occupied energy band (valence band), leaving the next higher band (conduction band) empty. In an *n*-type semiconductor, impurity atoms (donors) are present that contribute electrons to the conduction band; in a *p*-type semiconductor, there are impurity atoms (acceptors) present that can bind electrons, leaving behind missing electrons (holes) in the valence band.

Figure 6 shows a schematic of a *p–n* junction, fabricated by forming *p*- and *n*-type semiconductor layers in intimate contact. When a voltage is applied in the forward direction, electrons flow across the junction from the *n* region by dropping into empty hole states in a region of the junction about 1 μm thick. They may do this with the emission of radiation (hole–electron recombination radiation) in a frequency region in the vicinity of the energy gap frequency ($\nu \cong \mathscr{E}_g/h$); when the injection current density is sufficiently high, population inversion and gain will be induced in this frequency region. In addition to injection pumping of semiconductor

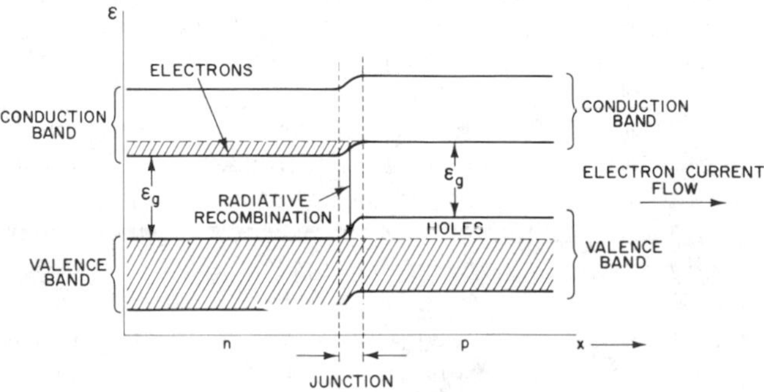

FIG. 6. Schematic showing the emission of radiation from a forward-biased, light-emitting diode.

lasers, a number of other methods not requiring the fabrication of a junction have been used successfully, including electron-beam pumping and optical pumping. Pumping with high-energy electrons (~200 keV) results in thick (~100 μm) gain regions and has been used to produce output throughout the visible from CdS_xSe_{1-x}.

In the diode (or injection) laser, a cavity structure (typically a fraction of a millimeter in dimensions) is provided by plane-parallel, cleaved faces at right angles to the junction plane, and laser emission occurs in that plane in one or more modes of the cavity when the diode injection current (and hence the spontaneous emission) reaches a threshold value. Since stimulated emission occurs in a narrow area near the junction (a few micrometers in the plane of the junction and a fraction of a micrometer perpendicular to the junction), the angular spread of emitted radiation, as expected from consideration of diffraction, is fairly large. The technology of diode lasers has undergone considerable development with the primary goals of achieving room-temperature operation, low thresholds, high output powers, improved mode quality, wavelength diversity, and long lifetimes. Progress has included improvements in electrical and optical confinement, and closer coupling of the gain region to the heat sink. In 1969, room-temperature continuous operation was achieved in a GaAlAs double heterostructure laser. This laser structure consists of a small region of *p*-type GaAs sandwiched between *n*-type layers of the alloy $Al_xGa_{1-x}As$ ($x < 1$). With further development, device lifetimes of tens of years were obtained, with output powers in the tens of milliwatts range. These improvements opened up applications of considerable significance, in particular fiber-optical communication and optical data storage. GaAlAs diode lasers have also been operated continuously at room temperature with output in the range of several watts, and linear arrays of diodes in the form of multiple stripes in a 1-cm bar cleaved from a wafer have given a total output of over 50 W. In addition, electrical to optical power conversion efficiencies of greater than 50% have been obtained.

In late 1976, continuous laser operation at room temperature was obtained in the quaternary alloy system, InGaAsP. Because laser operation is much less sensitive to dislocation effects in this system, long-lifetime devices were readily achieved. When operating at 1.3 and 1.55 μm, these lasers are matched to low-loss (<1 db km^{-1}), low-dispersion fiber optics, and are currently used in high-data-rate, long-distance communication.

The compositionally tuned lead salt lasers ($Pb_xSn_{1-x}Te$, PbS_xSe_{1-x}) operate at cryogenic temperatures. The alloy systems are particularly interesting, since changes in composition change $\mathscr{E}_g$, and therefore change the frequency of laser emission. Furthermore, since $\mathscr{E}_g$, and hence the dielectric constant at near-bandgap wavelengths, are sensitive functions of temperature in these small-gap semiconductor systems, cavity mode frequencies and the gain peaks are tunable, giving rise to tunable laser output. These tunable sources have been used extensively for high-resolution infrared spectroscopy in the 5–20-μm region.

In the visible region, progress has recently been made using the quaternary alloy AlGaInP. Room-temperature operation has been obtained continuously at 640 nm and pulsed at 603 nm; nitrogen-temperature continuous operation has been achieved at 583.6 nm, which is in the yellow region of the spectrum.

UV AND X-RAY LASERS

To produce stimulated emission in the UV and x-ray region, special problems must be considered. Short lifetimes of inverted populations become important, since the dependence of spontaneous radiative lifetime varies with frequency ν as ν^{-3}. For example, the nitrogen-gas-discharge laser, which radiates in the near UV (337 nm), can only operate in a pulse mode and must be pumped by a powerful intermittent source. Another problem is associated with the difficulty of devising resonant structures, since the reflectivity of materials for electromagnetic radiation becomes very small in the vacuum UV and the x-ray region. A different conceptual scheme, involving directionally amplified spontaneous emission from an inverted population (superradiance), must frequently be used in these spectral regions, where the directional amplification is achieved by the geometry of the pumped region.

The nitrogen- and hydrogen-discharge lasers make use of radiative transitions between two bound electronic levels (bound–bound transitions). On the other hand, there are molecular systems (excimers) in which radiative transitions occur between a bound excited state and a free or very weakly bound ground state [bound–free transitions; see Fig. 5(b)]. Xe and Kr form the excited molecular states Xe_2^* and Kr_2^*, although the diatomic molecules are unstable in their ground states. These gases at high pressure, when pumped by powerful electron-beam sources, emit superradiantly at 172 nm from Xe_2^* and 145.7 nm from Kr_2^*. At somewhat longer wavelengths, rare-gas–halide excimers such as ArF* at 248 nm, KrF* at 193 nm, and XeCl* at 308 nm have been operated, both by electron-beam pumping and by transversely excited discharge. These lasers can produce multijoule pulses at 100-Hz repetition rates and are commercially available.

X-ray lasers at wavelengths shorter than a few tens of nanometers present a more difficult challenge. Nevertheless, several groups have achieved laser operation in this region. A multijoule visible laser can directly pump the gain medium through inversion produced in the laser-generated plasma, or indirectly by using the x rays from a laser plasma to pump a separate x-ray laser medium. The first approach has been used to achieve operation near 21 nm in Se XXV, which has a $1s^2 2s^2 2p^6$ Ne-like electron configuration. Current research involves efforts to improve the very low efficiency of x-ray lasers as well as studies of new potential x-ray laser systems.

FREE-ELECTRON LASERS

Relativistic electrons traveling in a periodically alternating transverse magnetic field (wiggler) can be stimulated to give up radiation to a copropagating electromagnetic field of wavelength $\lambda = \lambda_w/2\gamma^2$, where λ_w is the wiggler period and γ is the ratio of the electron energy to its rest mass energy. This phenomenon can be pictured as stimulated inverse Compton scattering from the electromagnetic field of the wiggler seen in the electron rest frame. To obtain efficient conversion over reasonable distances with high output power, the electron beam must be very monoenergetic, high current, and have very low angular divergence. Efficient operation also requires either the use of an electron storage ring or a tapered period wiggler. Free-electron lasers have been operated at several wavelengths ranging from the visible to millimeter waves. While free-electron lasers do not have the wavelength restrictions imposed on other lasers, and do not have the inhomogeneities characteristic of many laser media, they require electron beams of a current level and quality which are difficult to achieve simultaneously. Further, the electron beam and wiggler requirements tend to become more demanding as wavelength decreases. These systems have the large sizes which are associated with relativistic electron-beam sources.

FREQUENCY CONVERSION

The frequency coverage of lasers can be extended using nonlinear optical techniques. Harmonic generation, frequency mixing, optical parametric oscillation, and stimulated Raman scattering are used for this purpose. The article on nonlinear optics describes these processes.

See also ELECTRON ENERGY STATES IN SOLIDS AND LIQUIDS; FRANCK–CONDON PRINCIPLE; MASERS; MOLECULAR SPECTROSCOPY; NONLINEAR OPTICS; QUANTUM THEORY OF MEASUREMENT; SEMICONDUCTORS, CRYSTALLINE; TRANSISTORS; ULTRASHORT OPTICAL PULSES.

BIBLIOGRAPHY

J. Hecht and R. Teresi, *Laser, Supertool of the 1980's*. Ticknor & Fields, New Haven and New York, 1982. (E)

A. J. Siegman. *Lasers*. University Science Books, Mill Valley, CA, 1986. (A)

O. Svelto, *Principles of Lasers*, 3rd ed. Plenum Press, New York and London, 1989. (I)

M. J. Weber, (ed.), *Handbook of Laser Science and Technology*. Vols. I & II. CRC Press, Boca Raton, 1982. (A)

A. Yariv, *Quantum Electronics*, 3rd ed. John Wiley and Sons, New York, 1988. (A)

Lattice Dynamics

Richard F. Wallis

1. INTRODUCTION

Lattice dynamics concerns the vibrations of atoms of a solid about their equilibrium positions and the effect of these vibrations on the properties of the solid. In the adiabatic approximation [1] the total wave function of the solid, $\Psi(\mathbf{r},\mathbf{R})$, is written as the product of the electronic wave function, $\psi(\mathbf{r},\mathbf{R})$, and the nuclear wave function, $\chi(\mathbf{R})$, where $\mathbf{r}$ represents all the electronic coordinates and $\mathbf{R}$ all the nuclear coordinates. The term adiabatic signifies that the nuclear motion does not induce electronic transitions. The vibrational motion of the atoms is determined by the nuclear potential energy, which is the electronic energy eigenvalue. The latter is a function of the nuclear coordinates **R**.

2. LATTICE-DYNAMICAL COUPLING CONSTANTS

We assume that the crystal is in a nondegenerate electronic ground state and that the atoms execute small vibrations about their equilibrium positions. The nuclear potential energy, $\Phi(\mathbf{R})$, may then be expanded in powers of the displacement components of the nuclei from their equilibrium positions:

$$\begin{aligned}\Phi(\mathbf{R}) = \Phi(\mathbf{R}^{(0)}) &+ \sum_{l\kappa\alpha} \Phi_\alpha(l\kappa) u_\alpha(l\kappa) \\ &+ \tfrac{1}{2} \sum_{l\kappa\alpha} \sum_{l'\kappa'\beta} \Phi_{\alpha\beta}(ll',\kappa\kappa') u_\alpha(l\kappa) u_\beta(l'\kappa') \\ &+ \tfrac{1}{6} \sum_{l\kappa\alpha} \sum_{l'\kappa'\beta} \sum_{l''\kappa''\gamma} \Phi_{\alpha\beta\gamma}(ll'l'',\kappa\kappa'\kappa'') \\ &\times u_\alpha(l\kappa) u_\beta(l'\kappa') u_\gamma(l''\kappa'') + \cdots .\end{aligned} \tag{2.1}$$

In Eq. (2.1), $u_\alpha(l\kappa) = R_\alpha(l\kappa) - R_\alpha^{(0)}(l\kappa)$ is the αth Cartesian component of the displacement from equilibrium of the κth nucleus in the lth primitive unit cell and the superscript zero denotes the equilibrium configuration. The equilibrium condition requires that the quantities $\Phi_\alpha(l\kappa) = 0$. The $\Phi_{\alpha\beta}(ll',\kappa\kappa')$ are the *harmonic* coupling constants and the $\Phi_{\alpha\beta\gamma}(ll'l'',\kappa\kappa'\kappa'')$, . . . , are the *anharmonic* coupling constants. The coupling constants are partial derivatives of the potential energy and therefore satisfy symmetry conditions such as $\Phi_{\alpha\beta}(ll',\kappa\kappa') = \Phi_{\beta\alpha}(l'l,\kappa'\kappa)$. Other conditions [1,2] satisfied by the coupling constants are those imposed by infinitesimal translational invariance and infinitesimal rotational invariance. For the harmonic coupling constants, these conditions take the respective forms

$$\sum_{l'\kappa'} \Phi_{\alpha\beta}(ll',\kappa\kappa') = 0 \tag{2.2}$$

and

$$\sum_{l'\kappa'} \{\Phi_{\alpha\beta}(ll',\kappa\kappa') X_\mu(l'l,\kappa'\kappa) - \Phi_{\alpha\mu}(ll',\kappa\kappa') X_\beta(l'l,\kappa'\kappa)\} = 0, \tag{2.3}$$

where $X_\mu(l'l,\kappa'\kappa) = R_\mu^{(0)}(l'\kappa') - R_\mu^{(0)}(l\kappa)$. Analogous equations are satisfied by the anharmonic coupling constants. Additional conditions on the coupling constants are provided by utilizing the point symmetry and translational symmetry of the crystal [2].

Typically, the nuclear potential energy of a crystal has several physical origins. These include the Coulomb interaction between electrically charged ions, van der Waals interactions, interactions associated with chemical bonds, and the short-range quantum-mechanical repulsion between atoms due to the Pauli principle.

3. EQUATIONS OF MOTION AND NORMAL COORDINATE TRANSFORMATION

The classical equations of motion for the nuclei can be written as

$$\partial_\kappa \ddot{u}_\alpha(l\kappa) = -\frac{\partial\Phi(\mathbf{R})}{\partial u_\alpha(l\kappa)}, \tag{3.1}$$

where the double dot above the u represents a second time derivative. Making use of Eq. (2.1) and retaining only the harmonic terms, we obtain

$$M_\kappa \ddot{u}_\alpha(l\kappa) = \sum_{l'\kappa'\beta} \Phi_{\alpha\beta}(ll',\kappa\kappa') u_\beta(l'\kappa'). \tag{3.2}$$

The set of equations (3.2) corresponding to the various values of l, κ, and α describe a system of coupled harmonic oscillators. Their solution is facilitated by the substitution $u_\alpha(l\kappa) = M_\kappa^{-1/2} w_\alpha(l\kappa)\exp(i\omega t)$, which leads to

$$\omega^2 w_\alpha(l\kappa) = \sum_{l'\kappa'\beta} D_{\alpha\beta}(ll',\kappa\kappa') w_\beta(l'\kappa'), \tag{3.3}$$

where the quantities $D_{\alpha\beta}(ll',\kappa\kappa') = (M_\kappa M_{\kappa'})^{-1/2}\Phi_{\alpha\beta}(ll',\kappa\kappa')$ are the elements of the *dynamical matrix*. For a nontrivial solution, the determinant of coefficients of the $w_\alpha(l\kappa)$ in Eq. (3.3) must be zero:

$$|D_{\alpha\beta}(ll',\kappa\kappa') - \omega^2\delta_{ll'}\delta_{\kappa\kappa'}\delta_{\alpha\beta}| = 0, \tag{3.4}$$

where the δ_{ij}'s are Kronecker deltas. Equation (3.4) is applicable to any crystal, perfect or imperfect, and specifies the normal-mode frequencies of the crystal. If the crystal contains N atoms, there are $3N$ normal modes and hence $3N$ solutions for ω to Eq. (3.4). We shall distinguish the various normal modes by the index s, $1 \le s \le 3N$. Typically, the normal-mode frequencies range from zero up to the infrared region.

For a macroscopic crystal, Eq. (3.4) is a polynomial equation in ω^2 of very high order. Calculations are simplified for a perfect crystal if we introduce periodic boundary conditions, justified by Ledermann's theorem [2], and write

$$w_\alpha(l\kappa) = W_{\alpha\kappa}(\mathbf{k})\exp[i\mathbf{k}\cdot\mathbf{R}^{(0)}(l)], \tag{3.5}$$

where **k** is the wave vector. Periodic boundary conditions require that the atoms at corresponding points on opposite faces of the crystal have the same displacements. There are $3N$ values of **k** that occupy the first Brillouin zone and that specify the independent normal modes [2]. The analog of Eq. (3.4) becomes

$$|D_{\alpha\beta}(\mathbf{k},\kappa\kappa') - \omega^2\delta_{\alpha\beta}\delta_{\kappa\kappa'}| = 0, \tag{3.6}$$

where

$$D_{\alpha\beta}(\mathbf{k},\kappa\kappa') = \textstyle\sum_{l'} D_{\alpha\beta}(ll',\kappa\kappa')\exp ei\mathbf{k}\cdot[\{\mathbf{R}^{(0)}(l') - \mathbf{R}^{(0)}(l)]\}$$

is independent of l. If there are r atoms per unit cell, the determinant in Eq. (3.6) is $3r \times 3r$ in size, far smaller than that in Eq. (3.4) and therefore more suitable for computation.

Let us denote the eigenvectors and eigenvalues of $D(\mathbf{k},\kappa\kappa')$ by $e_\kappa(\mathbf{k}j)$ and $\omega^2(\mathbf{k}j)$, respectively, where j is the branch index. The normal coordinates, $Q(\mathbf{k}j)$, can be introduced through the transformation

$$w_\alpha(l\kappa) = N^{-1/2} \sum_{\mathbf{k}j} e_{\alpha\kappa}(\mathbf{k}j) Q(\mathbf{k}j)\exp[i\mathbf{k}\cdot\mathbf{R}^{(0)}(l)]. \tag{3.7}$$

The Hamiltonian for the harmonic crystal can be written as

$$H = \tfrac{1}{2}\sum_{\mathbf{k}j} \{|\dot{Q}(\mathbf{k}j)|^2 + \omega^2(\mathbf{k}j)|Q(\mathbf{k}j)|^2\}, \tag{3.8}$$

which consists of the sum of contributions from independent harmonic oscillators. In the quantum-mechanical theory in the absence of anharmonicity, the nuclear wave function is a product of harmonic oscillator functions. The energy eigenvalue can be written as

$$E = \sum_{kj} \hbar\omega(\mathbf{k}j)[n(\mathbf{k}j) + \tfrac{1}{2}], \quad (3.9)$$

where $n(\mathbf{k}j)$ is the harmonic oscillator quantum number for the mode $\mathbf{k}j$. The quantum of excitation energy for a normal vibrational mode is called a phonon.

4. *PHONON DISPERSION CURVES*

The solutions of Eq. (3.6) determine the phonon dispersion curves, i.e., the normal-mode frequencies $\omega(\mathbf{k}j)$ as functions of the wave vector $\mathbf{k}$ for the various branches j. As a simple example, we give the normal-mode frequencies for a monatomic linear chain with atomic mass M and nearest-neighbor interactions characterized by the coupling constants

$$\Phi_{xx}(l,l+1) = \Phi_{xx}(l,l-1) = -\tfrac{1}{2}\Phi_{xx}(l,l) = -\sigma: \quad \omega(k) = (4\sigma/M)^{1/2}\sin(ak/2), \quad (4.1)$$

where a is the lattice constant. In this case we have only one atom per primitive unit cell and only one branch. For a crystal with r atoms per unit cell, there are $3r$ branches of which three are acoustical branches with $\omega(\mathbf{k}j)\to 0$ as $|\mathbf{k}|\to 0$ and $3r-3$ are optical branches. In crystals of high symmetry with $\mathbf{k}$ in a high-symmetry direction, the branches can be classified as longitudinal (displacements $\mathbf{u}$ parallel to $\mathbf{k}$) or transverse ($\mathbf{u}$ perpendicular to $\mathbf{k}$). The phonon frequencies associated with a given branch are restricted to a finite range or "band." There may be gaps between the various phonon bands.

Phonon dispersion curves can be determined experimentally by means of the coherent inelastic scattering of cold neutrons or by the thermal diffuse scattering of x rays. Inelastic neutron scattering [3] is the more widely used method.

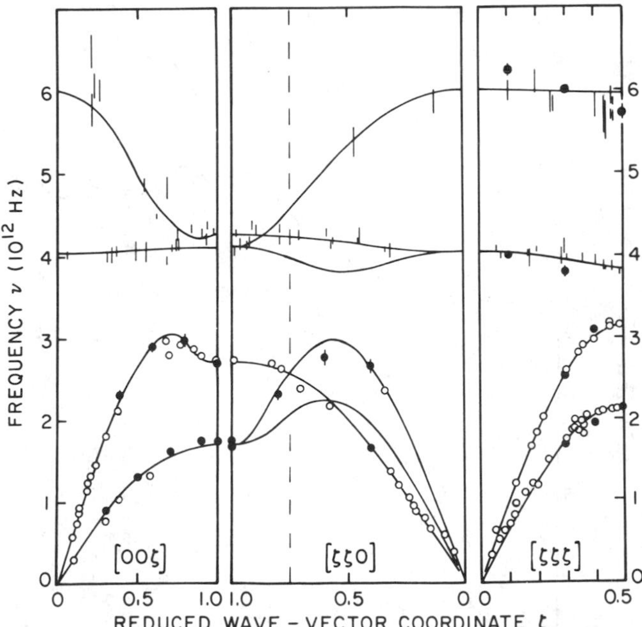

FIG. 1. Phonon dispersion curves for NaBr in the [001], [110], and [111] directions (from Ref. 4). The experimental data are indicated by bars and circles. The solid curves are calculated from a lattice-dynamical model and $\nu = \omega/2\pi$.

If the incident and scattered neutron energies are E_0 and E_s, respectively, the energy, $\hbar\omega(\mathbf{q}j)$, of the phonon created or destroyed in the scattering process is given by conservation of energy as

$$E_0 - E_s = \pm\hbar\omega(\mathbf{q}j), \quad (4.2)$$

where the plus sign refers to the creation and the minus sign to the destruction of the phonon. In addition, conservation of momentum requires that

$$\mathbf{k}_0 - \mathbf{k}_s = \mathbf{q} + \mathbf{G}, \quad (4.3)$$

where the wave vectors of the incident and scattered neutrons are $\mathbf{k}_0$ and $\mathbf{k}_s$, respectively, and $\mathbf{G}$ is a reciprocal lattice vector. From experimental values of E_0, E_s, $\mathbf{k}_0$, and $\mathbf{k}_s$, we can determine $\omega(\mathbf{q}j)$ and $\mathbf{q}$ and, hence, the phonon dispersion curves. Results for NaBr [4] are shown in Fig. 1. The gap between the acoustical and optical branches is evident.

5. *SPECIFIC HEAT*

The specific heat associated with the lattice vibrations of a crystal can be calculated in the harmonic approximation by taking the thermal average, $\langle E\rangle$, of the energy given by Eq. (3.9). Replacing the sums over $\mathbf{k}$ and j by an integral over frequency ω, we obtain for the specific heat

$$C_v(T) = \left(\frac{\partial\langle E\rangle}{\partial T}\right)_V = k_B\beta^2 \int_0^{\omega_m} d\omega\, g(\omega)\hbar^2\omega^2 \frac{e^{\beta\hbar\omega}}{(e^{\beta\hbar\omega}-1)^2} \quad (5.1)$$

where $\beta = 1/k_BT$, k_B is Boltzmann's constant, ω_m is the maximum phonon frequency, and $g(\omega)$ is the phonon frequency distribution. In general, $g(\omega)$ must be evaluated numerically using a computer. However, for the case of an isotropic elastic continuum, $g(\omega)$ takes the simple form

$$g(\omega) = \frac{\Omega}{2\pi^2}\left(\frac{2}{c_t^3} + \frac{1}{c_l^3}\right)\omega^2, \quad 0 \le \omega \le \omega_m; \qquad = 0, \quad \omega > \omega_m, \quad (5.2)$$

where c_t and c_l are the speeds of transverse and longitudinal sound waves and Ω is the volume of the crystal. Substitution of Eq. (5.2) into Eq. (5.1) gives

$$C_v(T) = 9Nk_B\left(\frac{T}{\Theta}\right)^3 \int_0^{\Theta/T} dx\, \frac{x^4e^x}{(e^x-1)^2}, \quad (5.3)$$

where $\Theta = \hbar\omega_m/k_B$ is the Debye temperature [5]. At low temperatures, $T \ll \Theta$, we obtain

$$C_v(T) \cong (12Nk_B\pi^4/5)(T/\Theta)^3, \quad (5.4)$$

which is the "Debye T^3 law." A tabulation of some Debye temperatures is given in Table I. The T^3 law is observed experimentally at sufficiently low temperatures; at higher temperatures, however, deviations from Eq. (5.3) appear because a real crystal is not an elastic continuum. The correct frequency distribution is in general a complicated function of ω and exhibits Van Hove singularities [6], which arise from maxima, minima, or saddle points on the constant-frequency surfaces in $\mathbf{k}$ space. The frequency distribution for copper [7] is shown in Fig. 2.

Table I. Debye Temperatures of Some Cubic Crystals[a]

Solid	Debye temperature (K)
Sodium	157
Copper	342
Lead	102
Nickel	427
Silicon	647
NaCl	321
LiF	732

[a] From J. S. Blakemore, *Solid State Physics,* 2nd ed., p. 130. Saunders, Philadelphia, PA, 1974.

Table II. Lattice-Dynamical Parameters of Some Cubic Crystals

Solid	ϵ_∞	$\epsilon(0)$	ω_T (cm^{-1})	ω_L (cm^{-1})
LiF[a]	1.9	8.9	307	662
NaCl[a]	2.25	5.9	164	264
KI[a]	2.7	5.1	101	139
ZnS[b]	5.14	8.67	271	352
GaAs[b]	10.9	12.9	269	292
InSb[b]	15.6	17.7	185	197
PbS[b]	17.2	202	65	223

[a] Ref. 8.
[b] Ref. 9.

6. *OPTICAL PROPERTIES*

An ionic crystal can interact with the electromagnetic field through the vibrations of its ions, since the latter are electrically charged and undergo acceleration during a lattice vibration. Optical phenomena are simplest in cubic crystals, which are optically isotropic and can be described by the complex dielectric constant $\epsilon(\omega)=\epsilon_1(\omega)+i\epsilon_2(\omega)$. For a crystal such as NaCl, with two atoms per primitive cell, we can show that [1,8]

$$\epsilon(\omega)=\epsilon_\infty+\frac{4\pi e^{*2}}{\mu\Omega_0}\left[\frac{\omega_T^2-\omega^2+i\omega\gamma}{(\omega_T^2-\omega^2)^2+\omega^2\gamma^2}\right], \qquad (6.1)$$

where ω_T is the transverse optical phonon frequency of long wavelength, γ is the damping constant, e^* is the transverse effective charge, μ is the reduced mass of the two ions in the unit cell, Ω_0 is the volume of the unit cell, and ϵ_∞ is the limiting value of $\epsilon(\omega)$ for $\omega \gg \omega_T$.

The real part of the dielectric constant, $\epsilon_1(\omega)$, can be written in the absence of damping in the simple form

$$\epsilon_1(\omega)=\frac{\epsilon_\infty(\omega_L^2-\omega^2)}{(\omega_T^2-\omega^2)}, \qquad (6.2)$$

where $\omega_L=[\omega_T^2+(4\pi e^{*2}/\mu\Omega_0\epsilon_\infty)]^{1/2}$ is the longitudinal optical phonon frequency of long wavelength. Without damping, the dielectric constant has a pole at ω_T and a zero at ω_L. If we set $\omega=0$ in Eq. (6.1), we obtain

$$\frac{\epsilon(0)}{\epsilon_\infty}=\frac{\omega_L^2}{\omega_T^2}, \qquad (6.3)$$

which is the Lyddane–Sachs–Teller relation. Values of the lattice-dynamical parameters for several crystals are given in Table II.

The complex index of refraction, $n-i\kappa$, is defined by $(n-i\kappa)^2=\epsilon(\omega)$. The reflectivity at normal incidence, R, is given by

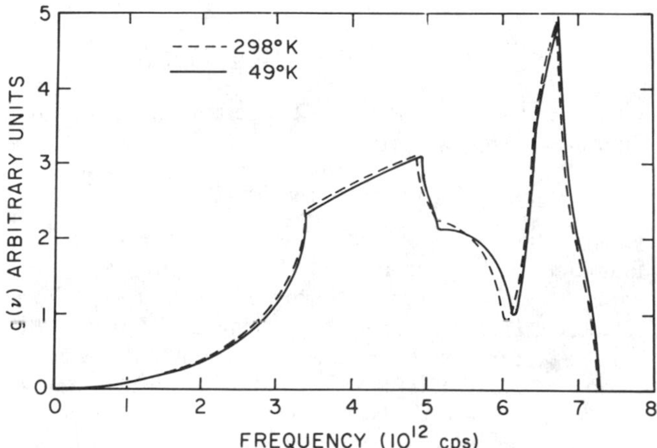

FIG. 2. Frequency distribution function for copper (Ref. 7). Note that $\nu=\omega/2\pi$.

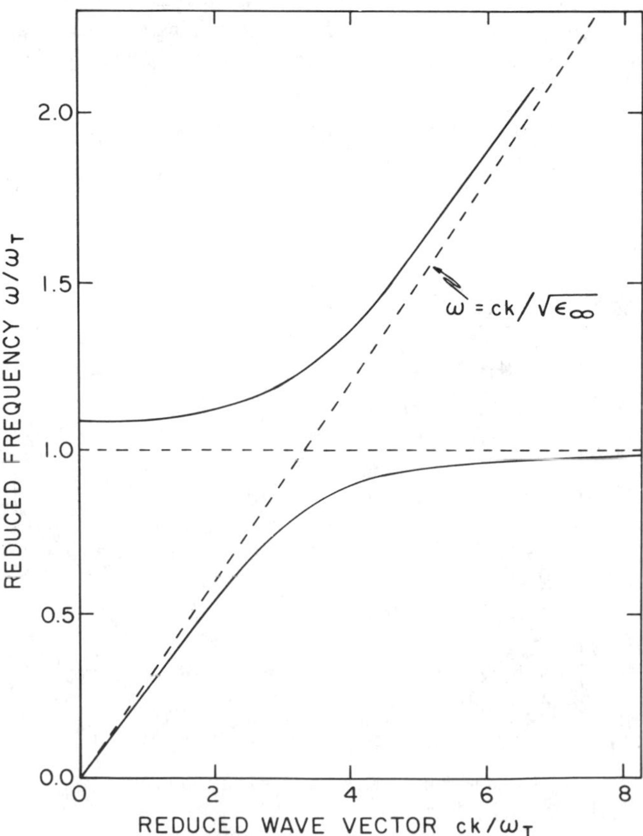

FIG. 3. Polariton dispersion curves for GaAs.

$$R = \frac{(n-1)^2+\kappa^2}{(n+1)^2+\kappa^2} \tag{6.4}$$

and is large between ω_T and ω_L, the region of *Reststrahlen*. The absorption coefficient, $\eta = 2\omega\kappa/c$, is strongly peaked close to ω_T if γ/ω_T is small. The coupling between the electromagnetic field and the transverse optical phonon of long wavelength leads to coupled modes called *polaritons* whose dispersion relation is given by $c^2k^2 = \omega^2\epsilon(\omega)$. Experimental information about the polariton dispersion relation can be obtained by Raman scattering [10]. The dispersion curves for polaritons in GaAs are shown in Fig. 3. Note that there are two branches, one above and one below the *Reststrahlen* region.

7. *IMPURITY MODES*

Associated with an impurity atom in a crystal are changes in atomic mass and coupling constants compared to the perfect crystal. These changes lead to modifications of the normal-mode eigenvectors and frequencies. In accordance with Rayleigh's theorem [2], a decrease (increase) in an atomic mass or increase (decrease) in a coupling constant produces either no change or an increase (decrease) in the normal-mode frequencies. Of particular importance are modes whose frequencies lie in regions that are forbidden for the frequencies of the perfect crystal. Such modes are termed localized modes because the atomic displacement amplitudes decrease exponentially from the impurity site. For an isotopic impurity of mass M' replacing a host crystal atom of mass M, the localized-mode frequency, ω_i, is specified by the equation [11]

$$\epsilon\omega_i^2 \int_0^{\omega_m} \frac{g(\omega)\,d\omega}{\omega_i^2-\omega^2} = 1, \qquad \omega_i > \omega_m, \tag{7.1}$$

where $\epsilon = (M-M')/M$ and $g(\omega)$ is the frequency distribution of the unperturbed crystal normalized to unity. Typically, a localized mode appears if M' is sufficiently light compared to M.

For a crystal with two atoms per unit cell there are optical modes of vibration whose frequencies can be separated from those of the acoustical modes by a gap (Fig. 1). If an atom of the host crystal is replaced by an impurity atom of mass M', a localized impurity mode can appear with frequency above the maximum frequency of the optical modes or in the gap between the optical and acoustical modes. Whether an impurity mode exists and where its frequency lies depends on which host atom is replaced and the relation of the impurity mass M' to the host atom masses M_1 and M_2 [12].

8. *SURFACE MODES*

A crystal surface may be regarded as a defect in which certain coupling constants are set equal to zero. Modes localized at a surface may be derived from both acoustical and optical branches. Long-wavelength surface acoustic modes are known as Rayleigh waves [12], whose speed, c_R, in an isotropic medium is given by

$$16\left(1-\frac{c_R^2}{c_l^2}\right)\left(1-\frac{c_R^2}{c_t^2}\right) = \left(2-\frac{c_R^2}{c_t^2}\right)^4. \tag{8.1}$$

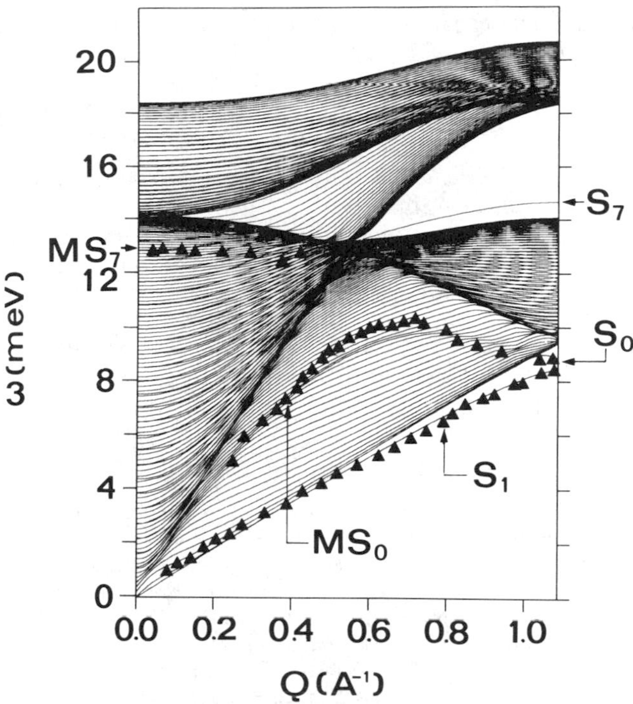

FIG. 4. Dispersion curves for surface and bulk modes of silver with a (110) surface. The continuous curves are theoretical results (Ref. 15) and the triangles are experimental results (Ref. 16).

The speed of Rayleigh waves is always less than c_t and c_l.

In recent years complete surface-phonon dispersion curves have been measured for many crystal surfaces using the techniques of electron energy loss spectroscopy [13] and inelastic helium atom scattering [14]. The surface-mode frequencies appear below the bulk continuum or in gaps within the bulk continuum. An example is shown in Fig. 4 for the silver (110) surface [15] where the Rayleigh mode is indicated by S_1, the gap surface mode by S_7, and so-called resonance modes by MS_0 and MS_7.

9. *ANHARMONICITY*

The cubic, quartic, and higher-order terms in the displacements in Eq. (2.1) give rise to anharmonic effects such as thermal conductivity, thermal expansion, broadening of optical absorption lines, and ultrasonic attenuation. The coefficient of thermal conductivity, K, for an isotropic crystal can be shown [17] to have the form

$$K = \tfrac{1}{3}Cs\lambda, \tag{9.1}$$

where C is the specific heat per unit volume, s is the average speed of sound, and λ is the phonon mean free path. The coefficient K increases as T^3 at very low temperatures, reaches a maximum, and then decreases as $1/T$ at higher temperatures.

Anharmonicity also leads to a volume dependence of the Debye temperature described by the Grüneisen constant, γ, defined by

$$\gamma = -\frac{V}{\Theta}\frac{d\Theta}{dV}. \tag{9.2}$$

The thermal expansion coefficient, α, can be expressed in terms of γ,

$$\alpha = \frac{1}{V}\left(\frac{\partial V}{\partial T}\right)_P = \frac{\kappa\gamma C_v}{V}, \qquad (9.3)$$

where κ is the compressibility and C_v is the specific heat at constant volume V.

In certain circumstances it is possible to have vibrational waves propagating in an anharmonic lattice without damping. Such waves are known as solitons [18].

See also Crystal Defects; Crystal Symmetry; Nonlinear Wave Propagation; Phonons; Quasiparticles; Solitons.

REFERENCES

1. M. Born and K. Huang, *Dynamical Theory of Crystal Lattices*. Oxford, London and New York, 1954.
2. A. A. Maradudin, E. W. Montroll, G. H. Weiss, and I. P. Ipatova, *Theory of Lattice Dynamics in the Harmonic Approximation*. Academic Press, New York, 1971.
3. B. N. Brockhouse, in *Phonons and Phonon Interactions* (T. A. Bak, ed.), p. 221, Benjamin, New York, 1964.
4. J. S. Reid, T. Smith, and W. J. L. Buyers, *Phys. Rev.* **B1,** 1833 (1970).
5. P. Debye, *Ann. Phys. (Leipzig)* **39,** 789 (1912).
6. L. Van Hove, *Phys. Rev.* **89,** 1189 (1953).
7. R. M. Nicklow, G. Gilat, H. G. Smith, L. J. Raubenheimer, and M. K. Wilkinson, *Phys. Rev.* **164,** 922 (1967).
8. E. Burstein, in *Phonons and Phonon Interactions* (T. A. Bak, ed.), p. 276. Benjamin, New York, 1964.
9. E. Burstein, A. Pinczuk, and R. F. Wallis, *The Physics of Semimetals and Narrow Gap Semiconductors* (D. Carter and R. Bate, eds.), p. 251. Pergamon, Oxford, 1971.
10. C. H. Henry and J. J. Hopfield, *Phys. Rev. Lett.* **15,** 964 (1965).
11. P. G. Dawber and R. J. Elliott, *Proc. Roy. Soc. London* **A273,** 222 (1963).
12. Lord Rayleigh, *Proc. London Math. Soc.* **17,** 4 (1887).
13. S. Lehwald, J. M. Szeftel, H. Ibach, T. S. Rahman, and D. L. Mills, *Phys. Rev. Lett.* **50,** 518 (1983).
14. R. B. Doak, U. Harten, and J. P. Toennies, *Phys. Rev. Lett.* **51,** 578 (1983).
15. A. Franchini, G. Santoro, V. Bortolani, and R. F. Wallis, *Phys. Rev.* **B38,** 12139 (1988).
16. G. Bracco, R. Tatarek, F. Tommasini, V. Linke, and M. Persson, *Phys. Rev.* **B36,** 2928 (1987).
17. C. Kittel, *Introduction to Solid State Physics,* 4th ed., p. 224. Wiley, New York, 1971.
18. M. Toda and M. Wadati, *J. Phys. Soc. Japan* **34,** 18 (1973).

Lattice Gauge Theory

Michael Creutz

Supercomputers have recently become a crucial tool for the quantum field theorist. Applied to the formalism of lattice gauge theory, numerical simulations are providing fundamental quantitative information about the interactions of quarks, the fundamental constituents of those particles which experience nuclear interactions. Perhaps most strikingly, these simulations have provided convincing evidence that the interquark forces can prevent the isolation of these constituents.

Quarks are the primary constituents of particles subject to the strong nuclear force. Their basic interactions are believed to follow from a generalization of the gauge theory of electromagnetism. Instead of a single photon, this theory involves eight spin-1 quanta, referred to as gluons. Furthermore, these eight gluons are themselves charged with respect to one another. This introduces subtle nonlinear effects which appear even in the pure glue theory.

One particularly important consequence of these nonlinearities is that the quark interactions weaken at small separations. This phenomenon, known as "asymptotic freedom," is essential to many of the successes of the simple quark model. As long as the quarks remain near each other, their interactions are small.

In contrast, the behavior of the gauge fields changes dramatically as the quarks are pulled apart. The experimental nonobservance of free quarks has led to the conjecture of the phenomenon of "confinement," wherein interquark forces increase and remain strong as quarks are pulled apart to arbitrary separations. In this picture, it requires an infinite amount of energy to separate a single quark from the other constituents of a physical particle. This explains why free quarks are not produced in nature.

Standard field-theoretical tools are severely hampered in the regime of large distances where these effects come into play. Perturbation theory, the historic mainstay of quantum field theory, begins with free particles and then treats their interaction as a small correction. With confinement, however, the fundamental constituents become increasingly strongly interacting as their separation is increased. In this domain the conventional perturbative approach fails totally.

Lattice gauge theory, originally formulated by K. Wilson, provides a novel framework for calculations in this regime. This approach replaces the relativistic continuum of space and time with a discrete space-time lattice. The quarks move through this scaffolding by a sequence of discrete hops between nearest-neighbor sites. The gluon fields lie on the bonds connecting these sites.

This lattice is a mathematical trick, introduced for calculational purposes only. It should not be taken as requiring a crystalline basis for physical space. At the end of any calculation, one should consider a continuum limit, wherein the lattice spacing is extrapolated to zero. In this limit observable quantities, such as the masses of particles and the forces between them, should approach their physical values.

The lattice artifice, however, has several advantages. First, by replacing an infinite number of space-time points in any given volume by a finite number, the field-theoretical system becomes mathematically considerably simpler and better defined. Continuum quantum field theories are notorious for the appearance of formally infinite quantities. These divergences involve short-distance singularities and must be exorcised by a renormalization procedure. A space-time lattice provides a particularly convenient regulator of such divergences. Indeed, the lattice spacing represents a minimum length and singularities arising from wavelengths shorter than this distance are automatically excluded.

Second, this formulation makes no assumptions on calculational schemes to be applied. Other techniques for controlling the singular behavior of a field theory are usually formulated directly in terms of some calculational method. For example, conventional discussions of renormalization regulate the divergences only after they are encountered in the perturbative expansion. On the lattice the theory is mathematically well defined at the outset.

Finally, the lattice formulation gives a system particularly well suited to numerical simulation. While there are several analytic techniques which have been applied to the strongly interacting lattice gauge problem, numerical simulations by Monte Carlo techniques currently dominate the field. These simulations have given compelling evidence that the confinement phenomenon does indeed occur in the standard gauge theory of the nuclear force. In addition, the approach is now giving quantitative predictions for long-range hadronic properties not accessible to more traditional theoretical methods.

In the Wilson approach, the gauge degrees of freedom are represented by matrices, one of which is associated with each lattice bond. To describe the physical theory of quarks, these are 3×3 unitary matrices with determinant 1; thus, they are elements of the group $SU(3)$. The interactions of these degrees of freedom are most concisely summarized in the "action"

$$S = -\tfrac{1}{3} \sum_{p} \mathrm{ReTr}\, U_p.$$

Here the sum is over all elementary squares, or "plaquettes," p, and U_p is the product of the link variables around the plaquette in question. The latter represents the flux of the gauge fields through the corresponding tile. For slowly varying fields, the above sum reduces to the conventional gauge-theory action as the lattice spacing goes to zero.

The numerical techniques used for lattice gauge simulations are borrowed directly from statistical mechanics. Indeed, there is a deep mathematical relationship between quantum field theory and classical statistical mechanics in four dimensions. In this relationship, the strength of the quark coupling to the gauge fields corresponds directly to temperature and the action S corresponds to the classical energy. Thus, a study of confinement and long-distance quark interactions is equivalent to a study of a high-temperature statistical model.

In the Monte Carlo approach to studying such a lattice system, one attempts to create an ensemble of field configurations with a Boltzmann-like distribution where the probability for any given configuration C takes the form

$$P(C) \propto e^{-\beta S(C)}.$$

Here β is proportional to the inverse of the gauge coupling squared. Thus one wants configurations typical of "thermal equilibrium."

The procedure begins with the storage of some initial values for all the lattice fields in the computer memory. These are then updated with pseudorandom changes on the field variables, thus mimicking thermal fluctuations. The structure of such a program is quite simple. On the outside is a set of nested loops over all the system variables. These loops surround calls to the random number generator, so as to simulate a thermal coupling to these degrees of freedom. The field changes are constructed with a bias toward lower values of S so as to obtain the appropriate thermal weighting of configurations.

Having the values of all fields at his disposal, the physicist is free to calculate any quantity of interest. There will, of course, be statistical errors coming from the thermal fluctuations. In addition, there will be errors coming from the requisite extrapolation to the continuum limit and from the practical requirement of working with a finite volume. It is attempts to reduce these uncertainties that have driven the theorists to the most powerful computers available.

Despite the inevitable uncertainties, several important results have been extracted. Perhaps the most dramatic of these is the measurement of the confinement force and how it relates to the weaker interactions of the quarks at short distances. Then there are successful studies of the mass spectra of the bound states of quarks. These calculations are being refined to give information on the distributions of the quarks and on the strong-interaction effects on other processes, such as weak decays.

In addition, there have been quantitative studies of physics at temperatures sufficiently high that strongly interacting particles are created by thermal fluctuations. Here lattice gauge calculations have provided strong evidence that the vacuum undergoes a phase transition at a temperature of $kT \sim 200$ MeV. This transition is from a phase of ordinary matter, made up of quarks bound into the familiar nuclear particles, to a new plasma phase where the quarks and their attendant gluon fields form a thermal gas. Indeed, the lattice approach gives the best estimates for the temperature of the transition to this phase, which will be looked for in future accelerator experiments.

Until recently, the bulk of the numerical simulations in lattice gauge theory considered the full dynamics of the gauge fields but only included the quarks as fixed sources. In this way the confinement potential has been mapped out. A more refined approach allows the primary "valence" quarks to carry kinetic energy, but still ignores the creation of matter–antimatter quark pairs by quantum fluctuations. Most of the hadronic spectrum calculations have been done in this approximation.

Thus far, only limited results have been obtained beyond this valence approximation. To proceed in this direction, one must allow for the possibility of unlimited numbers of virtual quarks being created by quantum fluctuations. The most difficult part of the problem is the inclusion of the effects of the Pauli exclusion principle in the simulations. The development of algorithms to treat the dynamical quarks appropriately forms an area of intense current research. Promising new schemes combine Monte Carlo methods with ideas from molecular-dynamics simulations and stochastic processes.

In conclusion, computer simulations of lattice gauge theory provide a powerful tool for the study of nonperturbative phenomena. The technique provides a first-principles approach to calculating particle properties as well as details of the phase transition to a quark gluon plasma.

See also Gauge Theories; Hadrons; Quarks; Strong Interactions.

FURTHER READING

Elementary

M. Creutz, *Phys. Today* **36** (No. 5), 35 (1983).

Intermediate

C. Rebbi, *Sci. Am.* **248,** 54 (1983).

M. Creutz, *Comments Nucl. Part. Phys.* **10,** 163 (1981).

Advanced

K. Wilson, *Phys. Rev.* **D10,** 2445 (1974).

M. Creutz, *Quarks, Gluons, and Lattices.* Cambridge University Press, Cambridge, 1983.

X. Li, Z. Qiu, and H. Ren (eds.), *Lattice Gauge Theory Using Parallel Processors.* Gordon and Breach, New York, 1987.

Leptons

Martin L. Perl

The known leptons are a class of elementary particles having two defining characteristics:

1. They are fermions. That is, they have spin $\frac{1}{2}$ and obey Fermi–Dirac statistics.
2. They have no strong interactions among themselves or with any other known particles. Table I lists the known leptons and their properties. The name *lepton*, meaning small or light in Greek, was coined when the heaviest known lepton was the muon which is lighter in mass than all hadrons. However, the tau discovered in 1975 is heavier than many hadrons.

The electron is, of course, the best known of the leptons, being a constituent of all ordinary matter. An atom consists of electrons moving in orbits around a nucleus. The symbol for the electron which has one unit of negative electric charge is e^-. Associated with the electron is its antiparticle, the *positron* with one unit of positive charge, denoted e^+. The e^+ has the same mass and spin as the e^-.

Also associated with the electron is the *electron-neutrino*, ν_e, and the *electron-antineutrino*, $\bar{\nu}_e$. These neutrinos are neutral, that is have zero electric charge, and they may have zero mass. The e^-, e^+, ν_e, and $\bar{\nu}_e$ constitute the electron family of leptons. For convenience only the e^- and ν_e are listed in Table I. The tabulated properties of the antiparticles e^+ and $\bar{\nu}_e$ are the same except that the e^+ has opposite charge to the e^-.

Two more lepton families have been discovered. The muon family consists of the negative muon, μ^-, and its associated neutrino, ν_μ, plus the associated antiparticles, μ^+ and $\bar{\nu}_\mu$. Similarly, the tau family consists of τ^-, ν_τ and the antiparticles τ^+ and $\bar{\nu}_\tau$. In Table I the three families are listed in order of increasing mass of the charged lepton, which is also the historical order in which the families were discovered.

The families are also called generations, 1st, 2nd, and 3rd for convenience. There are believed to be connections between these lepton generations and the three quark generations: e^-, ν_e corresponding to the u, d quark pair, μ^-, ν_μ corresponding to the c, s quark pair and τ^-, ν_τ corresponding to the assumed t, b quark pair.

The three lepton generations are separated from each other and from all quarks in collision and decay processes by a well-established but unexplained lepton conservation law. For example, a positron, e^+, can annihilate an electron, e^-, producing two photons

$$e^+ + e^- \rightarrow \text{photon} + \text{photon}.$$

But a positive muon, μ^+, or positive tau, τ^+, does not annihilate an electron. In another example, an electron-neutrino, ν_e, can interact with a neutron to produce an electron and a proton

$$\nu_e + \text{neutron} \rightarrow e^- + \text{proton},$$

but the ν_e does not produce a μ^- or a τ^-. Analogously for a muon neutrino, ν_μ,

$$\nu_\mu + \text{neutrino} \rightarrow \mu^- + \text{proton},$$

but the ν_μ does not produce an e^- or a τ^-.

The electron, positron, all neutrinos, and all antineutrinos are stable, that is, they do not decay. The muon and tau are unstable, decaying through the weak interaction. The negative muon has only one major decay mode,

$$\mu^- \rightarrow \nu_\mu + e^- + \bar{\nu}_e.$$

The corresponding decay mode for the positive muon is

$$\mu^+ \rightarrow \bar{\nu}_\mu + e^+ + \nu_e.$$

The tau because of its greater mass has many decay modes such as

Table I Properties of the Known Leptons. Only the particles are listed. The antiparticles e^+, $\bar{\nu}_e$, μ^+, $\bar{\nu}_\mu$, τ^+ and $\bar{\nu}_\tau$ have the same mass and lifetime as their corresponding particles e^-, ν_e, μ^-, ν_μ, τ^-, ν_τ.

Generation	1	2	3
Charged lepton name	Electron	Muon	Tau
Charged lepton symbol	e^-	μ^-	τ^-
Charged lepton mass (MeV/c^2)	0.511	105.7	1784. ± 3.
Charged lepton lifetime (s)	Stable	2.197×10^{-6}	3.0×10^{-13}
Neutrino name	Electron neutrino	Muon neutrino	Tau neutrino
Neutrino symbol	ν_e	ν_μ	ν_τ
Neutrino mass upper limit	<18 eV/c^2	<0.25 MeV/c^2	<35 MeV/c^2
Neutrino lifetime	Stable	Stable	Stable

$$\tau^- \rightarrow \nu_\tau + e^- + \bar{\nu}_e,$$

$$\tau^- \rightarrow \nu_\tau + \mu^- + \bar{\nu}_\mu,$$

$$\tau^- \rightarrow \nu_\tau + \pi^-,$$

$$\tau^- \rightarrow \nu_\tau + \pi^- + \pi^0,$$

$$\tau^- \rightarrow \nu_\tau + \pi^- + \pi^+ + \pi^-,$$

with corresponding modes for the τ^+. The symbol π stands for a pion.

All of the leptons have sizes less than 10^{-16} cm; they may be much smaller. There is no known natural law which limits the number of different leptons. Table I lists all leptons discovered as of 1988. It is also possible that other kinds of leptons may exist, leptons with zero spin for example, but no other kinds have been found.

See also CONSERVATION LAWS; ELECTRON; ELEMENTARY PARTICLES IN PHYSICS; NEUTRINO; POSITRON; POSITRON–ELECTRON COLLIDING BEAMS; WEAK INTERACTIONS.

REFERENCES

1. F. Close, M. Marten, and C. Sutton, *The Particle Explosion*. Oxford University Press, New York, 1987.
2. Y. Ne'eman and Y. Kirsh, *The Particle Hunters*. Cambridge University Press, Cambridge, 1986.
3. G. Kane, *Modern Elementary Particle Physics*. Addison-Wesley, Redwood City, CA, 1987.

Levitation, Electromagnetic

G. Wouch and A. E. Lord, Jr.

Levitation is the act of holding up an object with no visible support. The generation of eddy currents in an electrical conductor by a time-varying magnetic field is the basis of electromagnetic levitation. (It is also the basis of induction heating, most metal detectors at the airport and beach, some geological surveys, and eddy-current nondestructive testing of metal parts. Transformer cores are laminated in order to break up the eddy currents and hence reduce loss.)

Electromagnetic levitation can be described in very simple terms. A time-varying magnetic flux generates eddy currents (in a conductor) and Lenz's law says that the magnetic field of these circulating currents must be in such a direction as to oppose the flux. If the flux varies spatially, then there is a force on the object

$$\mathbf{F} = \mathbf{m} \cdot \nabla \mathbf{B} \tag{1}$$

where **m** is the total magnetic moment included in the metal object and $\nabla\mathbf{B}$ is effectively the gradient of the applied magnetic field (actually $\nabla\mathbf{B}$ is a dyadic). Thus, for example, if at a particular instant **B** is in, say, the Z direction, **m** will be in the $-Z$ direction and if $dB/dZ<0$, Eq. (1) gives a positive or lifting force.

Now let us be a bit more analytical [1] with a very simple

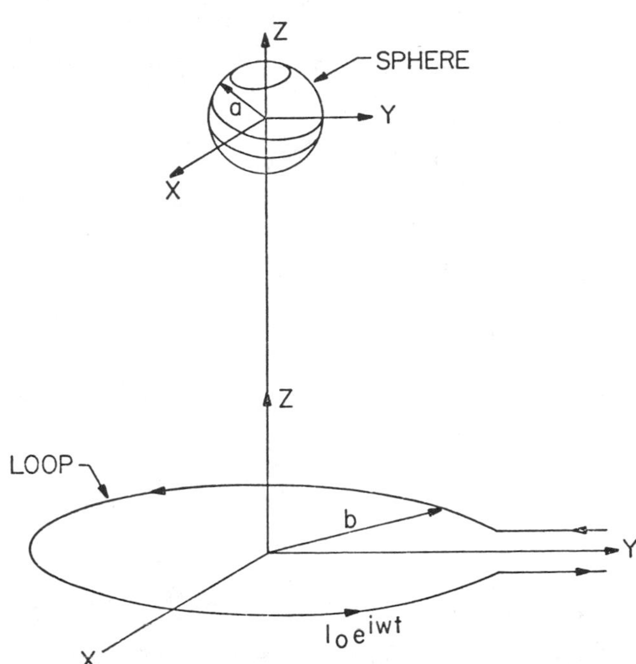

FIG. 1. Sphere positioned on the axis of a single current loop.

model, but one that is quite realistic. Consider the situation shown in Fig. 1. A small metal sphere of radius a is located on the axis of a circular loop of wire of radius b. The wire is carrying a current $I_0e^{i\omega t}$. A most important quantity in the problem is the ratio of the skin depth δ to the radius a. The skin depth (here taken as the effective depth of penetration of the magnetic field) is given by [2]

$$\delta = (2/\omega\sigma\mu_0)^{1/2}, \tag{2}$$

where ω is the circular frequency, σ is the electrical conductivity, and μ_0 is the permeability of free space ($=4\pi\times10^{-7}$ in SI units). A typical induction heating unit also used for levitation studies operates at 400 kHz and a typical metallic conductivity is $10^7\ \Omega^{-1}\ \mathrm{m}^{-1}$. This gives $\delta = 4\times10^{-4}$ m $=0.4$ mm. Usual metal spheres being levitated are at least 1 cm in diameter. Thus $\delta/a \ll 1$; this also applies to most metal-detector situations where the frequency may be lower but the objects to be detected are usually somewhat larger. Thus the calculation here will be limited to the regime of $\delta/a \ll 1$. No essential physics is lost in this way.

In the extreme case of $\delta/a \ll 1$ the eddy currents induced on the surface of the sphere completely shield the interior from magnetic flux. Hence

$$\mathbf{B} = \mu_0(\mathbf{H} + \mathbf{M}), \tag{3}$$

where **B** is the magnetic induction, **H** is the total magnetic field, and **M** is the magnetization in the sphere. Now

$$\mathbf{H} = (H_0 e^{i\omega t} - H_{\text{demag}})\hat{k}, \tag{4}$$

where for a sphere the demagnetizing field is given by [3]

$$\mathbf{H}_{\text{demag}} = \tfrac{1}{3}\mathbf{M}. \tag{5}$$

Thus setting $\mathbf{B}=0$ we obtain

$$0 = \mu_0(H_0 e^{i\omega t} + \tfrac{2}{3}M), \tag{6}$$

and so

$$\mathbf{M} = -\tfrac{3}{2}H_0 e^{i\omega t}\hat{k}. \tag{7}$$

The magnetic moment **m** of the sphere is given by

$$\mathbf{m} = (\tfrac{4}{3}\pi a^3)(-\tfrac{3}{2}H_0 e^{i\omega t}\hat{k}) = -2\pi a^3 H_0 e^{i\omega t}\hat{k}. \tag{8}$$

LEVITATION FORCE

The force on a dipole in a nonuniform field is given by [4]

$$\mathbf{F} = \mathbf{m}\cdot\nabla\mathbf{B}. \tag{9}$$

The axial magnetic field of a single loop is the well-known result [5]

$$\mathbf{B} = (I_0/2)[\mu_0 b^2/(b^2+z^2)^{3/2}]e^{i\omega t}\hat{k} \tag{10}$$

and

$$\nabla\mathbf{B} = -[3b^2 I_0\mu_0 z/(b^2+z^2)^{5/2}]e^{i\omega t}\hat{k},\hat{k}. \tag{11}$$

From Eqs. (8), (9), and (11), the levitation force is

$$\mathbf{F} = [3\pi a^3 b^4 z/2(b^2+z^2)^4]\mu_0 I_0^2 e^{2i\omega t}\hat{k} \tag{12}$$

Taking the real part, averaging over a period, and using $I_{rms} = I_0/\sqrt{2}$ gives the average force as

$$\langle\mathbf{F}\rangle = [3\pi a^3 b^4 z/2(b^2+z^2)^4]\mu_0 I^2_{rms}\hat{k}. \tag{13}$$

where I_{rms} is the root-mean-square current. This result agrees exactly with the general treatment of levitation forces given by Okress *et al.* [6] and Fromm and Jehn [7] in the case where $\delta/a \ll 1$. A typical levitation situation might be

$$a = 0.5\ \text{cm} = 5\times10^{-3}\ \text{m},$$
$$b = 2.5\ \text{cm} = 2.5\times10^{-2}\ \text{m},$$
$$z = 0.5\ \text{cm} = 5\times10^{-3}\ \text{m};$$

this yields the numerical result

$$\langle F\rangle = 8.12\times10^{-8} I^2_{rms}. \tag{14}$$

A current of $I_{rms} = 1000$ A will give a levitation force of 0.08 N or about 7.2 g, which is about twice the weight of the above sphere assuming a density of 8 g/cm^3. A current of 707 A is sufficient to levitate the sphere of radius a and density 8 g/cm^3 at a distance of 0.5 cm above the plane of the loop. Normally a coil of n turns is used so that I_{rms} to first order can be replaced by NI_{rms} in the equation. For a three-turn coil, then the I_{rms} required would be 236 A to levitate the above sphere at that distance. Those numbers are quite realistic. The change in inductance of the coil due to the sphere and the power absorption in the sphere is also worked out in Ref. 1.

The eddy currents induced in the conductor by the time-varying electromagnetic field heat it and may melt it. It can be shown [7] that for $a/\delta \gg 1$, the average power absorbed by the sphere producing Ohmic heating is given by

$$P = (\tfrac{4}{3})\pi\sigma\mu_0^2\omega^2\delta^3 a^2 H_0^2.$$

Research work conducted jointly by the authors has resulted in some new accomplishments of electromagnetic levitation. This work has primarily been directed at obtaining independent control over levitation force and heating. The achievement of this has made possible some interesting and perhaps useful laboratory investigations of solidification phenomena [8–12]. Containerless solidification of melts, which avoids contamination by crucibles, has been accomplished by utilizing an auxiliary heating source, i.e., electron beam, laser, or thermal imaging source, to obtain independent control of levitation force and heating power [8–10]. Containerless solidification of melts has also been achieved by utilizing the microgravity environment of space, where objects are naturally levitated [8,11–13]. In that environment only very weak electromagnetic fields are required to confine objects, and specimens can be melted and solidified as slowly or as rapidly as desired while containerless. The utilization of the space environment for containerless experiments represents a definite advance in this field of research. By the use of weak fields to confine melts (either electromagnetic or acoustic), physical and chemical phenomena can be studied with negligible disturbance of the melt. This is in contrast to terrestrial levitation, where the large forces required to levitate appreciably disturb the melt, i.e., electromagnetic stirring and agitation. The microgravity environment of space thus allows almost complete independence of control over all of the parameters, i.e., heating, confinement, stirring, for containerless experiments. This, coupled with the microgravity environment itself, has permitted new studies of phenomena in melts and during solidification. As an example of this, in the microgravity environment, the rates of rising and settling of constituents of different densities in a melt and gravity-driven convection are greatly reduced. Hence the time for agglomeration and separation from the melt of particles or a dispersed phase in a melt is greatly extended, if other agitations are minimized (i.e., electromagnetic stirring). This has been observed both in containerless and noncontainerless experiments in space [8,11–13].

Figure 2 is a schematic of an advanced levitation apparatus combining electromagnetic levitation with electron beam heating [14]. Figure 3 is a sketch of an electromagnetic containerless processing payload, flown twice on sounding rockets [10], using electromagnetic fields to confine the specimen and inductively melt it while floating in the microgravity environment. Independent control over confinement and heating is achieved here by the use of the space environment and adjustment of the strength of the electromagnetic field. The apparatus has a high-power mode for heating and a low-power mode for containerless solidification. Figure 4 shows the levitation coil assembly used in the terrestrial levitation apparatus shown in Fig. 2. A shaped molybdenum ring shown in Fig. 5 above the coil is used to achieve stable levitation of molten metals. Eddy currents induced in this ring by the alternating magnetic field of the levitation coil produce a magnetic field that presses down on the levitated specimen, preventing it from jumping out of the coil. Active rings carrying current or passive rings carrying only the induced currents are used to obtain stable levitation. The thick molybdenum ring above the stabilizing ring is used to collimate the electron beam and prevent it from striking the coil.

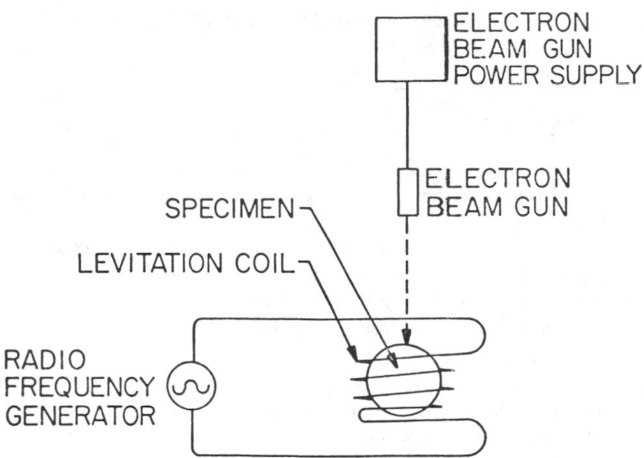

FIG. 2. Schematic of terrestrial containerless melting and solidification apparatus.

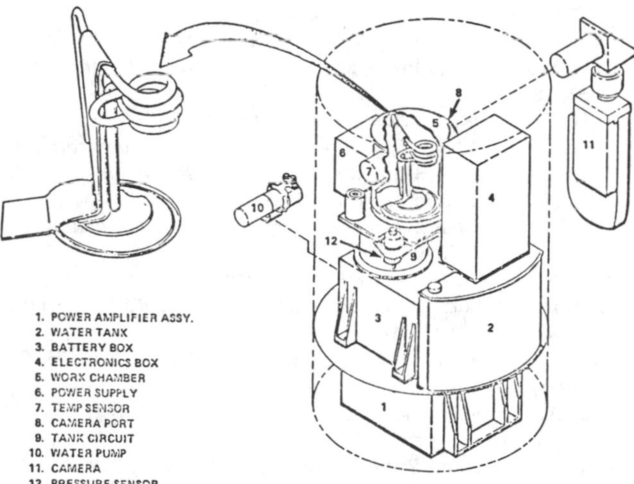

FIG. 3. Electromagnetic containerless processing payload schematic.

FIG. 4. Levitation coil assembly with insertion pedestal extended and specimen positioned in the coil for levitation.

FIG. 5. Levitation coil with passive stabilizing ring and electron beam collimator above it.

Figure 6 shows a levitated solid tungsten sphere. Figure 7 shows a sequence where tungsten is melted while levitated, using an electron beam [8].

Electromagnetic levitation is used in laboratory investigations of the physical and chemical properties of materials. Examples of these investigations are the measurement of the thermodynamic properties of metals [15–18], the study of gas–metal interactions [19,20], observations of supercooling and nucleation phenomena in liquid metals [21,22], determination of the densities and surface tensions of liquid metals [23], and the study of evaporative purification of liquid metals [24].

It must be mentioned that objects can also be levitated by intense acoustic fields [25], electrostatic fields [26], magnetostatic fields [27], and gas streams [28]. Other forms of electromagnetic levitation include utilizing the radiation pressure of laser light waves [29] and microwave confinement [29]. Many ingenious schemes have also been utilized to neutralize the effects of gravity such as obtaining up to 3 s of free fall in the NASA drop towers [30], up to 20 s of free fall during KC135 airplane ballistic trajectory flights [31], and sounding rocket flights [8,13].

FIG. 6. Levitated solid tungsten ball at a temperature of 2873 K.

FIG. 7. Levitation melting sequence. Topmost view is solid tungsten sphere levitated at a temperature of 2873 K. The dark band is a turn of the levitation coil; center view is when melting is begun by allowing electron beam to strike the specimen from above; bottom view is completely molten, levitated tungsten specimen at a temperature of about 3773 K.

See also MAGNETS (PERMANENT) AND MAGNETOSTATICS.

REFERENCES

1. G. Wouch and A. E. Lord, Jr., "Eddy Currents: Levitation, Metal Detectors and Induction Heating," *Am. J. Phys.* **46,** 464 (1978).
2. J. R. Reitz and F. J. Milford, *Foundations of Electromagnetic Theory,* p. 305. Addison-Wesley, Reading, Mass. 1967.
3. J. R. Reitz and F. J. Milford, in Ref. 2, p. 213.
4. J. R. Reitz and F. J. Milford, in Ref. 2, p. 243.
5. J. R. Reitz and F. J. Milford, in Ref. 2, p. 156.
6. E. C. Okress, D. M. Wroughton, G. Comenetz, P. H. Brace, and J. C. R. Kelly, "Electromagnetic Levitation of Solid and Molten Metals," *J. Appl. Phys.* **23,** 545 (1952).
7. E. Fromm and H. Jehn, "Electromagnetic Forces and Power Absorption in Levitation Melting," *Br. J. Appl. Phys.* **16,** 653 (1965).
8. G. Wouch, "Containerless Melting and Solidification of Metals and Alloys in the Terrestrial and Space Environments," Ph.D. Thesis, (Drexel University, Copyright, 1978).
9. G. Wouch *et al., Rev. Sci. Instruments* **46**(8), 1122 (1975).
10. G. Wouch, R. T. Frost, and A. E. Lord, Jr., *J. Crystal Growth* **37,** 181 (1977).
11. G. Wouch *et al., Nature* **274,** 235 (1978).
12. G. Wouch, *NBS Sponsored Conference on Applications of Space Flight in Materials Science and Technology,* at NBS, Gaithersburg, Maryland, April 20–21, 1977.
13. G. Wouch *et al., General Electric Company, Final Report, NAS8-31963,* June, 1977.
14. L. S. Nelson, *Nature,* **210,** 410 (1966).
15. A. K. Chadhuri, D. W. Bonnell, L. A. Ford, and J. L. Margrave, *High Temperature Sci.,* **2,** 203 (1970).
16. J. A. Treverton and J. L. Margrave, *J. Phys. Chem.* **75,** 3737 (1971).
17. J. A. Treverton and J. L. Margrave, *J. Chem. Thermodyn.* **3,** 473 (1971).
18. L. A. Stretz and R. G. Bautista, *Metall. Trans.* **5,** 921 (1974).
19. O. C. Roberts, D. G. C. Robertson, and A. E. Jenkins, *Trans. Metall. Soc., AIME,* **245,** 2413 (1969).
20. LL. A. Baker, N. A. Warner, and A. E. Jenkins, *Trans. Metall. Soc., AIME* **230,** 1228 (1964).
21. E. T. Turkdogan, *Trans. Metall. Soc., AIME* **230,** 740 (1964).
22. E. T. Turkdogan and K. C. Mills, *Trans. Metall. Soc., AIME* **230,** 750 (1964).
23. S. Y. Shiraishi and R. G. Ward, *Canadian Metall. Quart.* **3,** 118 (1964).
24. B. F. Oliver, *Trans. Metall. Soc., AIME* **227,** 996 (1963).
25. R. R. Whymark, "Acoustic Positioning for Containerless Processing," *Proc. Third Space Proc. Symp., Skylab Results, V. II,* April 30–May 1, 1974, NASA M-74-5.
26. A. T. Nordsieck, "Free-gyro systems for navigation or the like," *U.S. Patent 3,003,356* (A. Nov. 1, 1954: P. Oct. 10, 1961).
27. F. T. Backers, "A Magnetic Journal Bearing," *Philips Tech. Rev.* **22**(7), 232 (April 5, 1961).
28. R. A. Happe, "Oxide Glass Processing," in Ref. 25.
29. A. H. Boerdijk, "Technical Aspects of Levitation," *Philips Res. Rep.* **11**(1), 45 (Feb., 1956).
30. H. F. Wuenscher, "Space Processing Experiments on Sounding Rockets," in Ref. 25.
31. R. T. Frost *et al., Free Suspension Processing Systems for Space Manufacturing, General Electric Company, Final Report Contract NAS8-26157,* June 15, 1971.

Lie Groups

M. Hamermesh

A *group* is an algebraic structure with the following properties: The group consists of a set of elements $g,h,k...$, together with an operation of binary combination such that to each ordered pair of elements g and h of the set there is associated a unique element of the set, p, which is often called the *product* of g and h, so that we write $gh=p$. This multiplication is associative. There is an element e in the set such that its product with any element of the set in either order simply reproduces that element: $eg=ge=g$ for any g in G. Also, for each g in G there exists an element g' that is the *inverse* of g: $gg'=g'g=e$. The inverse of g is usually denoted by g^{-1}. Using these algebraic properties we can deduce a rich structure of theorems concerning groups with a finite or countable number of elements. Among these groups are the crystal symmetry groups that have important applications to the physics of the solid state.

If we add to the algebraic structure some notion of *nearness* or *continuity* of the elements of the group, we obtain a *continuous* group. The most general form of this extension is the *topological* group, where the elements of the group G form a topological space. A special case of this extension, which includes practically all the groups that have physical applications, is the case where the elements of the group G can be labeled by a finite number r of continuously varying parameters $a_1,...,a_r$, so that two elements of the group $R(a)=R(a_1...a_r)$ and $R(b)$ are near if the Euclidean distance $[\Sigma_{i=1}^r(a_i-b_i)^2]^{1/2}$ between the points a and b is small. If we take the product of the elements $R(a)R(b)$ we get an element $R(c)$ of the group, where the parameters c: $(c_1...c_r)$ are functions of the parameters a and b:

$$c_i=\varphi_i(a_1,...,a_r;b_1,...,b_r)=\varphi_i(a;b) \qquad (i=1,...,r). \quad (1)$$

Similarly, the parameters of the inverse of the group element $R(a)$ will be functions of the parameters a. If all these functions are *analytic*, the group is said to be an *r-parameter Lie group*.

Of particular interest in physics are Lie groups of transformations. For example, the assumption that space is homogeneous means that the transformations $x'=x+a$, $y'=y+b$, $z'=z+c$ must leave all physical statements unchanged for all real a, b, c. Similarly, isotropy of space implies the equivalence of descriptions in rotated systems. In classical mechanics such symmetry statements lead to Noether's theorem: Invariance of the Hamiltonian under a group of transformations implies the existence of corresponding constants of the motion. For example, translation invariance implies conservation of linear momentum of an isolated system.

In quantum mechanics similar results hold, but because we have a superposition of states in quantum mechanics the results of symmetries are much more profound. If the Hamiltonian H of a system is invariant under a group G of transformations g, there exists a set of unitary operators $U(g)$ that commute with H:$UHU^{-1}=H$. As a result, if ψ is an eigenstate of H belonging to eigenvalue E, then $U(g)\psi$ is also an eigenstate with the same eigenvalue E. Unless there is some accidental degeneracy, the states belonging to a given eigenvalue form a subspace in Hilbert space and are transformed among themselves by the operators $U(g)$. We are then led to the notions of irreducibility, the splitting of multiplets when a perturbation is applied (the lowering of the symmetry means that the group G' is a subgroup of G and the states that previously were degenerate may no longer transform into one another), and selection rules on transitions.

For most of the problems of physics we need not deal with the Lie group as a whole. If we consider the transformations of the group G that are near the identity element of the group, we can, because of analyticity, regard them as being generated by integration of infinitesimal transformations, $U(g)=1+\epsilon\hat{g}$, where 1 is the identity element and ϵ is a small parameter. For an r-parameter Lie group we can construct r independent basis operators $\hat{g}_i$ $(i=1,...,r)$. By the fundamental theorem of Lie these operators form an algebra: They can be added, and multiplied by real constants, and the product $\hat{g}_i\hat{g}_j-\hat{g}_j\hat{g}_i$, the commutator $[\hat{g}_i,\hat{g}_j]$ of any two elements, is a linear combination of the basis operators:

$$[\hat{g}_i,\hat{g}_j]=\sum_{k=1}^{r} c_{ij}^k\hat{g}_k, \quad (2)$$

where the c_{ij}^k are the structure constants of the Lie group. We thus arrive at a new algebraic structure, the Lie algebra corresponding to the original Lie group. For the translation group the Lie algebra consists of three basis elements, p_x,p_y,p_z, that commute with one another. The p_i are simply the momentum operators along the three axes. For the three-dimensional rotation group, the Lie algebra has three basis elements, the angular momentum operators for x, y, z. Thus we can, instead of working with the continuous manifold of the Lie group, work with the finite-dimensional Lie algebra. We can associate with each element of the Lie algebra $\hat{g}$ a unitary operator $U(\hat{g})$ that acts on the states of the system. In this way we recover the simple theory of representations.

See also GROUP THEORY IN PHYSICS; INVARIANCE PRINCIPLES; OPERATORS; QUANTUM MECHANICS; ROTATION AND ANGULAR MOMENTUM.

BIBLIOGRAPHY

M. Hamermesh, *Group Theory and Its Applications to Physical Problems*. Addison-Wesley, Reading, Mass., 1962.

M. Tinkham, *Group Theory and Quantum Mechanics*. McGraw-Hill, New York, 1964.

H. Lipkin, *Lie Groups for Pedestrians*. North-Holland, Amsterdam, 1965.

R. Gilmore, *Lie Groups, Lie Algebras, and Some of Their Applications*. Wiley (Interscience), New York, 1974.

B. Wybourne, *Classical Groups for Physicists*. Wiley, New York, 1974.

M. Gourdin, *Basics of Lie Groups*. Editions Frontières, Paris, 1982.

J. F. Cornwell, *Group Theory in Physics*, Vol. 2. Academic Press, New York, 1984.

D. H. Sattinger and O. L. Weaver, *Lie Groups and Algebras with*

Applications to Physics, Geometry, and Mechanics. Springer-Verlag, New York, 1986.

P. Olver, *Applications of Groups to Differential Equations*. Springer-Verlag, New York, 1986.

Ligand Fields *see* Crystal and Ligand Fields

Light

Victor L. Granatstein

NATURE OF LIGHT

Light is energy in the form of radiation that is preferentially received by the human eye and causes the sensation of vision. It can be regarded as composed either of photons in the energy range $(2.5 \text{ to } 5.2)\times 10^{-19}$ J or equivalently of electromagnetic waves in the wavelength range 800 nanometers (nm) down to 380 nm. Undoubtedly, it is no coincidence that the spectrum of solar radiation reaching the earth's surface is peaked in this same spectral range where humans and other earth animals possess sensitive receptors.

The foregoing definition speaks of both photons and electromagnetic waves and thus brings to mind the question whether light is composed of particles or of waves. The answer to this question is that, strictly speaking, light is like neither the waves nor the particles that we are familiar with from everyday experience. Rather, light is something more subtle that displays both wavelike and particlelike properties. The behavior of light is adequately described by the theory of quantum electrodynamics. This theory uses Maxwell's equations to determine the pattern of fields in an electromagnetic wave and interprets the square of these fields as a probability density function for the photons that compose the light radiation. The photons themselves have particlelike characteristics (e.g., each photon has associated with it a discrete value of energy); however, they also have properties that are very different from the particles of normal experience (e.g., it is postulated that photons have zero rest mass, which implies that they always travel at a constant speed independent of their energy).

ELECTROMAGNETIC WAVES

Light forms a small part of the continuous spectrum of electromagnetic radiation, which encompasses radio waves and infrared radiation at wavelengths longer than light, and ultraviolet, x-ray, gamma-ray, and cosmic-ray radiation at progressively shorter wavelengths. At all wavelengths, electromagnetic radiation is made up of a magnetic field **H**, which exerts a force on moving electric charges, and an electric field **E**, which exerts a force on charges even when they are stationary.

Maxwell's equations may be combined to form a wave equation for the fields. In a medium that is homogeneous, stationary, and free of electric charges, this equation takes the form

$$\nabla^2\mathbf{E} - \mu\epsilon\frac{\partial^2\mathbf{E}}{\partial t^2} - \mu\sigma\frac{\partial\mathbf{E}}{\partial t} = 0 \tag{1}$$

where ∇^2 is the Laplacian operator ($\nabla^2 = \partial^2/\partial x^2 + \partial^2/\partial y^2 + \partial^2/\partial z^2$ in Cartesian coordinates) and t denotes time. The properties of the medium through which the wave propagates are given by three parameters: (1) the magnetic permeability μ; (2) the dielectric constant ϵ; and (3) the conductivity σ.

In free space, $\sigma=0$, while $\epsilon_0 = 10^7/4\pi c^2$ F/m and $\mu_0 = 4\pi\times 10^{-7}$ H/m (mks units). Then, Eq. (1) takes the form

$$\nabla^2\mathbf{E} - \frac{1}{c^2}\frac{\partial^2\mathbf{E}}{\partial t^2} = 0. \tag{2}$$

A simple solution is obtained by restricting spatial variation to the z direction and specifying **E** to point in the y direction. Then Eq. (2) yields

$$E_y = e_0 \sin 2\pi[\nu t - (z/\lambda) + \phi]$$

where $\lambda\nu = c$, and e_0 (the wave amplitude) and ϕ (the phase angle) are constants. This expression describes a propagating plane wave that is sinusoidal in both space and time. The spatial periodicity is characterized by the wavelength λ while the time variation has a frequency ν. Any one crest of the wave moves forward a distance λ in a time $1/\nu$, so that the speed of wave propagation is $\lambda\nu = c$.

The electric field in our example is transverse to the direction of propagation. This is a general characteristic of electromagnetic waves. The time-varying electric field is accompanied by a time-varying magnetic field **H**, which is oriented so that the direction of wave propagation is given by $\mathbf{E}\times\mathbf{H}$. In free space, the amplitude of the magnetic field is related to the amplitude of the electric field by $H = (\mu_0/\epsilon_0)^{1/2}E$.

THE SPEED OF LIGHT

In vacuum, light waves travel at a constant speed c. The value of the constant c has been determined by measuring the time taken by a modulated light beam to cover a known distance. The presently accepted value of c is 2.99792458×10^8 m/s $\pm$ 400 m/s.

A photon always travels at the speed c. In a material medium, however, the photons are absorbed and reemitted by molecules, a process that slows down the rate of travel of a light wave. Macroscopically, this retardation is represented in Maxwell's equations by having the parameters μ, ϵ, and σ depart from their free-space values. We can define a refractive index $n = (\epsilon\mu/\epsilon_0\mu_0)^{1/2}$, and in a weakly conducting medium the light wave propagates with phase velocity c/n, and has its fields exponentially attenuated through absorption as $\exp(-k_i z)$ where $k_i = (\mu/\epsilon)^{1/2}\sigma/2$.

In general, the refractive index n is a function of wavelength, so that two light beams of different colors (different frequencies) propagate through materials at two different speeds. This phenomenon is known as dispersion and it is important to distinguish between phase velocity and group velocity in a dispersive medium. Any realizable measurement of the speed of light involves measurement on a wave packet, i.e., a wave that persists only for a finite time T. Such a wave packet is not monochromatic but covers a range of frequencies $\Delta\nu \approx 2/T$. As a wave packet travels through a dispersive medium the phase relation between the different frequency components changes. However, this change is relatively slow for an almost monochromatic wave in which $\Delta\nu \ll \nu$, and the wave packet progresses as a recognizable entity. The measured speed of the packet will be the group velocity $-\lambda^2\, \partial\nu/\partial\lambda$ rather than the phase velocity $\lambda\nu$.

REFRACTION AND REFLECTION

When an electromagnetic wave in one medium impinges on a second medium it will be partly reflected and only a portion of the wave will be transmitted into the second medium. Moreover, the direction of wave propagation will be changed in the second medium; i.e., the light "rays" will be bent or refracted. This bending is the basis for simple optical devices such as lenses.

The rules for reflection and transmission may be derived by a straightforward application of electromagnetic theory. The results are conveniently stated in terms of the configuration shown in Fig. 1. A plane electromagnetic wave is incident at an angle θ_i on a plane interface between two transparent media. The propagation vector $\mathbf{k}_i$ of this incident wave and the normal $\hat{m}$ to the interface lie in and define the plane of incidence. The propagation vectors of the reflected and transmitted waves (denoted by $\mathbf{k}_r$ and $\mathbf{k}_t$, respectively) lie in this plane of incidence.

The direction of $\mathbf{k}_r$ is given by Snell's law of reflection, viz., $\theta_r = \theta_i$. The direction of $\mathbf{k}_t$ is given by Snell's law of refraction,

$$\sin\theta_t = \sin\theta_i[(\mu_1\epsilon_1)/(\mu_2\epsilon_2)]^{1/2} = n_{12}\sin\theta_i$$

where n_{12} is the relative refractive index of the two media.

When the refractive index in medium 1 is greater than that in medium 2 (i.e., $n_{12}>1$), as happens when light is refracted as it passes from glass into air, the refracted "ray" grazes the surface if $\sin\theta_i = 1/n_{12}$. For $\sin\theta_i > 1/n_{12}$ the incident wave is totally reflected.

The magnitudes of the transmitted and reflected electric fields are given by the Fresnel coefficients. For the incident wave polarized with its *electric field in the plane of incidence* the field magnitudes are

$$e_{r\parallel} = -e_{i\parallel}\tan(\theta_i - \theta_t)/\tan(\theta_i + \theta_t)$$

and

$$e_{t\parallel} = e_{i\parallel} 2\sin\theta_t\cos\theta_i/\sin(\theta_i + \theta_t)\cos(\theta_i - \theta_t),$$

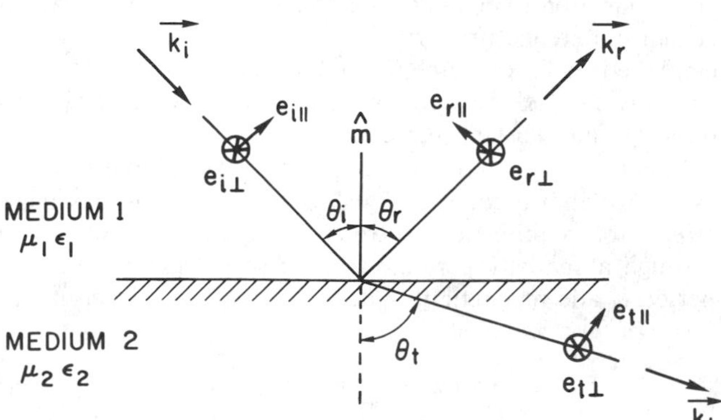

FIG. 1. Reflection and refraction at a planar interface.

while for the incident *electric field perpendicular to the plane of incidence*

$$e_{r\perp} = -e_{i\perp}\sin(\theta_i - \theta_t)/\sin(\theta_i + \theta_t)$$

and

$$e_{t\perp} = e_{i\perp} 2\sin\theta_t\cos\theta_i/\sin(\theta_i + \theta_t).$$

For the case of normal incidence ($\theta_i = 0$), $e_r = e_i(n_{12}-1)/(n_{12}+1)$ and $e_t = e_i 2n_{12}/(n_{12}+1)$ independent of polarization.

From the Fresnel coefficients given above, it is clear that $e_{r\parallel}$ vanishes when $\theta_i + \theta_t = \pi/2$. The angle of incidence at which this occurs is known as Brewster's angle and is given by $\tan\theta_i = (n_{12})^{-1}$. When randomly polarized light is incident on an interface at the Brewster angle, the reflected light will be linearly polarized with the electric field perpendicular to the plane of incidence; also, light polarized with the electric field in the plane of incidence will be totally transmitted.

DISPERSION, ANISOTROPY, AND NONLINEARITY

In the preceding discussion of refraction, the refractive index was assumed to be constant for the sake of simplicity. In general, the refractive index will be a function of wavelength (dispersion), of polarization and direction of wave propagation (anisotropy), and of light intensity (nonlinearity).

Dispersion is easily observed. In most "transparent" media, the refractive index increases regularly toward the blue end of the spectrum. Thus white light passing through an air–glass interface at some oblique angle will be dispersed into a spectrum of colors, with the rays of blue light ($\lambda = 400$ nm) being bent more than the rays of red light ($\lambda = 640$ nm). A glass prism is useful in displaying the spectral content of white light or in measuring the wavelength of monochromatic light as in a prism spectrometer. The Kramers–Kronig relation allows us to calculate the light absorption properties of a medium when its dispersion is known.

Anisotropy is displayed strongly by certain crystalline materials such as calcite. A beam of linearly polarized light normally incident on a calcite plate may be resolved into two components with mutually perpendicular electric fields that will propagate through the calcite with different phase velocities. If the plate is of such a thickness that the phase difference between the two components is 90° after traversing the plate, the linearly polarized wave will have been converted into an elliptically polarized wave. Such a quarter-wave plate is a useful device for studying the polarization of light.

The application of a strong steady electric or magnetic field to a medium can significantly affect the interaction of a light wave with the molecules composing the medium. For example, Kerr demonstrated that strong anisotropy could be induced in glass by applying an external electric field. Of course, if external electric and magnetic fields can affect the propagation of light through matter, then the self fields of the light beam can have a similar effect if they are sufficiently strong. In recent years, with the development of lasers, light beams have been created with sufficiently great intensity to produce nonlinear effects such as harmonic generation.

DIFFRACTION AND INTERFERENCE

Usually, nonlinear effects can be neglected. In that case, when two or more light waves are present at the same place and at the same time, their fields add *algebraically* and in a linear way, as given by the principle of superposition,

$$\mathbf{E}=\mathbf{E}_1+\mathbf{E}_2+\mathbf{E}_3+\cdots+\mathbf{E}_N. \quad (3)$$

This implies the possibility of forming interference fringes, which are an observable confirmation of the wavelike properties of light.

A simple experimental demonstration of interference fringes can be made using the apparatus in Fig. 2, which constitutes one form of Young's double-slit experiment. A single-filament lamp is used as the light source. A filter (e.g., a piece of colored glass) is placed in front of the lamp to make the light more nearly monochromatic. This quasimonochromatic light is now passed through two closely spaced slits in an opaque plate.

Semicylindrical waves spread out from each slit. The spreading of a light beam that is restricted by a narrow slit is a manifestation of the phenomenon of diffraction; the trajectories of photons passing through a slit spread over an angle ϕ given by $\sin\phi \approx \lambda/w$ where w is the slit width. Thus for narrow slits the spread angle is considerable and the light beams from the two slits overlap.

The circular lines in Fig. 2 represent the crests of the waves. The intersection of two lines indicates the arrival at that point of two waves with the same phase or with phase difference equal to an integral multiple of 2π. Here the fields add according to Eq. (3) to yield maximum brightness. On the other hand, when a trough of one wave coincides with a crest of the second wave the fields cancel, resulting in darkness.

Through this interference between the two waves, a pattern of bright and dark bands or fringes is formed that may be recorded on a photographic plate placed far from the slits ($D_2 \gg d$). Bright fringes occur for displacement from the centerline of $x=p\lambda D_2/d$ and dark fringes for $(p+\frac{1}{2})\lambda D_2/d$ where p is an integer.

In a typical experiment, the double slits might be separated by $d=0.2$ mm with the distance from the slits to the photographic plate $D_2=1$ m. Then, the spacing between adjacent bright fringes is $\Delta=5000\lambda\approx 3$ mm, and the fringes are readily observable. Thus, although the spatial period of light waves is too small to be seen directly, interference techniques can effectively magnify the periodicity so that it becomes observable.

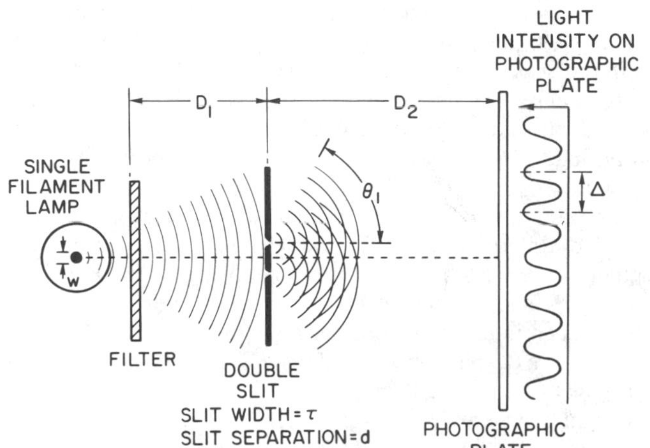

FIG. 2. Experimental arrangement for observing interference fringes (not to scale).

COHERENCE AND LASERS

The production of fringes indicates that light waves from the two slits have a high degree of spatial coherence (i.e., their phase difference remains constant during the period of observation.) However, the light waves will be highly coherent and fringes will be produced only when d, the slit separation, is small. For strong fringe formation $d<\lambda D_1/\tau$ where τ is the thickness of the filamentary source. When d becomes large, the light waves will not interfere and their intensities will add (law of photometric summation) instead of their fields, producing a low-level illumination without fringes. The angular width of an inaccessible source can be determined by studying fringe visibility as d is varied; Michelson used this method to obtain the angular diameter of a star.

In addition to measuring coherence between light at two points separated transversely to the direction of wave propagation, measurements can also be made of the mutual coherence between light at points separated along the direction of propagation. This allows us to define a coherence length l_c such that strong coherence is measured when the separation between the two sampling points is less than l_c. A related coherence time can be defined as $t_c=l_c/c$. Measurement of l_c or t_c yields information on the frequency spread of the wave $\Delta\nu$, viz., $\Delta\nu\approx t_c^{-1}$.

Light beams produced by lasers are highly coherent and monochromatic. Strong spatial coherence extends over the beam cross section, and the coherence length along the direction of propagation may be many meters long. In conventional light sources even when filters are employed as in Fig. 2, coherence lengths are at best on the order of centimeters.

A description of laser physics is beyond the scope of this article. To be very brief, lasers are sources that are based on in-phase light emission from an ensemble of excited atoms or molecules at a resonant frequency characteristic of the particular atomic or molecular species. Because of the excellent coherence properties of laser beams they are being employed in myriad unique and important applications in such diverse fields as communications, weaponry, physics research, and corneal surgery.

Finally, we take note of the recent advent of free-electron lasers which are coherent light sources based on passing a stream of relativistic electrons with injection velocity v through a spatially periodic transverse magnetic field with period l_w and magnetic flux density $\mathbf{B}_\perp$. Such a laser amplifies light at a wavelength

$$\lambda=l_w(1+a_w^2)(1-v/c)^{-1}, \quad (4)$$

where $a_w=e\mathbf{B}_\perp l_w/2\pi mc$ in MKS units. Thus, the wavelength is continuously tunable by varying the electron velocity.

Free-electron lasers are also characterized by unusually high power and efficiency.

This article has attempted to treat a number of important topics in sufficient detail so as to be meaningful to a wide readership. It has not been possible to give a comprehensive treatment of the subject of light, for a more detailed treatment of both topics surveyed in this article and of the many topics that were omitted, see the Bibliography.

See also DIFFRACTION; ELECTROMAGNETIC RADIATION; FARADAY EFFECT; KERR EFFECT, ELECTRO-OPTICAL; LASERS; MAXWELL'S EQUATIONS; PHOTONS; POLARIZED LIGHT; REFLECTION; REFRACTION.

BIBLIOGRAPHY

G. Birnbaum, *Optical Masers*. Academic Press, New York, 1964. (I)

M. Born and E. Wolf, *Principles of Optics*, 4th ed. Pergamon, New York, 1970. (A)

R. Ditchburn, *Light*, 2nd ed. Wiley (Interscience), New York, 1963. (I)

F. A. Jenkins and H. E. White, *Fundamentals of Optics*. McGraw-Hill, New York, 1957. (I)

M. Kerker, *The Scattering of Light and Other Electromagnetic Radiation*. Academic Press, New York, 1969. (A)

T. C. Marshall, *Free-Electron Lasers*. Macmillan Publishing Company, New York, 1985.

M. G. J. Minnaert, *The Nature of Light and Color in the Open Air*. Dover, New York, 1954. (E)

J. W. Simmons and M. J. Guttman, *States, Waves and Photons; a Modern Introduction to Light*. Addison-Wesley, Reading, Mass., 1970. (A)

S. Tolansky, *Curiosities of Light Rays and Light Waves*. American Elsevier, New York, 1965. *Revolution in Optics*. Pelican Books, Harmondsworth, England, 1968. (E)

Light Scattering

R. W. Detenbeck

The scientific study of light scattering began in 1869 with the experiments of Tyndall, who sent a beam of white light through a suspension of fine particles. The scattered light was bluish and, observed at right angles to the beam, was highly polarized. These effects suggested to Tyndall that the blue color and polarization of light from the sky are produced in the scattering of sunlight by atmospheric dust. In a subsequent, theoretical treatment Rayleigh calculated the scattering from a random collection of small (compared with a wavelength), widely separated spheres. He predicted the observed polarization phenomena, as well as the inverse-fourth-power dependence of scattered intensity on wavelength which gives the sky its blue color. That the blue color and polarization were most striking in the clearest skies led Rayleigh to conclude that the molecules of air itself, not added dust particles, are responsible for the blue sky.

Wavefronts propagate straightforward within a homogeneous medium, and the effect of the medium is described in terms of its average electromagnetic properties. Light scatters when it passes through a medium that contains inhomogeneities on the scale of a wavelength. These may be embedded particles, as in Tyndall's scattering experiments, or the molecules of the medium itself, as in Rayleigh's theory of the blue sky. However, Rayleigh's molecular theory applies only to gases, where it is proper to add the *intensities* of light scattered by individual molecules. In condensed matter the medium is quite homogeneous on the scale of optical wavelengths, and the coherent addition of scattered *amplitudes* cancels radiation in all but the forward direction. Using a different model suggested by Smoluchowski, Einstein in 1910 described the weak light scattering from liquids in terms of thermodynamic density fluctuations within small volume elements of a continuous medium.

Descriptions of light-scattering phenomena can be classified as static or dynamic. Static descriptions include time-averaged angular distributions of scattered intensity and polarization. One interesting case is the scattering of linearly polarized light from a single particle, illustrated in Fig. 1. If the scattering takes place within a surrounding medium of index n, and the vacuum wavelength of the incident light beam is λ_0, the incident wave vector has magnitude $k_0 = 2\pi n/\lambda_0$. The scattered light has wavelength λ_S, and the magnitude of its wave vector is $k_S = 2\pi n/\lambda_S$. The incident beam propagates in the direction of the vector $\mathbf{k}_0$, and the scattered light travels from the particle to the detector in the direction of $\mathbf{k}_S$. The two wave vectors determine the scattering plane, in which the scattering angle between them is denoted by θ. In a common arrangement the incident light is polarized with its electric field aligned along the normal to the scattering plane, often called the vertical direction and denoted here by the unit vector $\mathbf{e}_V$. The detector may select either the vertically polarized component or the horizontal component, with electric field aligned along $\mathbf{e}_H$.

Electromagnetic scattering from a particle remains an interesting theoretical problem in itself. The exact solution for the scattering from an isolated sphere was obtained by Mie in 1908, but the only other exact solution for a finite particle, an ellipsoid of revolution, awaited the work of Asano and

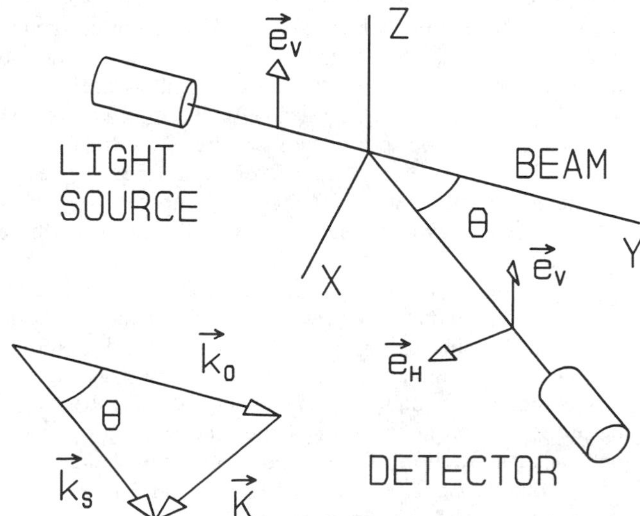

FIG. 1. Light scattering from a single scattering center located at the origin of the coordinate system.

Yamamoto in 1975. Numerical approximation techniques, which can calculate the scattering in almost any particular case, have been developed to run on modern computers. Moreover, they can be tested experimentally by scaling the electromagnetic wavelength and a model of the scatterer to centimeter sizes. However, many of the most important applications of light scattering to other fields (e.g., astronomy, atmospheric and ocean physics) deal with naturally occurring collections of particles differing in size, shape, orientation, and composition. In such cases one cannot afford a simple sum over all parameters, and theoretical ingenuity and intuition are required. In more complicated situations, such as dense atmospheric clouds, where the light is multiply scattered, even more powerful approximation techniques are needed.

Light carries momentum and energy. Therefore, when it is scattered or absorbed, there is a momentum transfer to the scatterer. A small particle can be levitated in a focused laser beam. Even single atoms can be manipulated by radiation forces. Laser cooling of atoms exploits the enhancement and strong frequency dependence of the scattering of light when its frequency is near that of an atomic resonance. The Doppler shift of near-resonance radiation can be exploited to favor scatterings which slow moving atoms in a viscous manner. Temperatures less than 1 mK for small numbers of atoms have been obtained.

Frequency shifts in the scattered light carry information about dynamic processes involving the scatterers. Brillouin in 1922 predicted a doublet in the frequency distribution of the light scattered from thermally excited sound waves in solids. Although experiments were very difficult with the light sources available at the time, Gross observed the Brillouin doublet in liquids in the early 1930s. Meanwhile, Raman and others had observed large frequency shifts associated with light scattered by molecular excitations in condensed matter.

The intense, coherent light from laser sources developed in recent decades is ideally suited to light-scattering studies of time-dependent phenomena. The spatial scale of accessible phenomena is determined by the wavelength of light. The scattering geometry is that of Fig. 1. The scattering vector $\mathbf{K} = \mathbf{k}_0 - \mathbf{k}_S$ is the difference between the incident and scattered wave vectors. In Rayleigh and Brillouin scattering the fields may be treated classically. The scattering process selects spatial Fourier components of the scattering fluctuations with a wave vector $\mathbf{K}$. Each temporal frequency component ω_S of the scattered light is produced by a corresponding frequency component Ω of the fluctuations, where $\omega_S = \omega_0 \pm \Omega$. One associates Brillouin scattering in condensed matter with sound waves of gigahertz frequencies, inaccessible to acoustic measurements. Frequency shifts in Rayleigh scattering are caused by translational and rotational motions of particulate scatterers or by nonpropagating collective modes in condensed matter. In Rayleigh/Brillouin scattering the frequency shift of the scattered light is often such a small fraction of the incident frequency that k_S is nearly equal to k_0. Then the vector-addition triangle of Fig. 1 is isosceles, and the direction of $\mathbf{K}$ is along the bisectrix of $-\mathbf{k}_0$ and $\mathbf{k}_S$. The magnitude is $K = (4\pi n/\lambda_0)\sin(\frac{1}{2}\theta)$; the scattering angle θ selects a particular value of K. The maximum value of K occurs for backward scattering, where $K_{max} = 2k_0$. For light from a He–Ne laser, k_0 is about 10 μm^{-1}, and K_{max} corresponds to a spatial resolution of the order of 0.1 μm.

When the frequency shifts in dynamic light scattering are very small, the scattering process is called quasielastic. Light scattering from swimming microorganisms, diffusing macromolecules, and critical opalescence falls into this category. The narrow bandwidth of the incident laser beam permits measurement of these very small frequency shifts, but in some cases they are beyond the resolution capabilities of tuned optical elements. Then the frequency spectrum can be measured by beating the scattered light against a sample of the incident laser light, by beating the various Fourier components against each other, or by equivalent intensity-autocorrelation techniques.

Raman scattering is discussed in terms of quantum excitations. Momentum conservation requires that the scattering excitation accept a momentum equal to $\hbar K$. The corresponding frequency ω_S of the outgoing photon, for a change of excitation energy $\pm\Delta E$, is determined by energy conservation: $\hbar\omega_S = \hbar\omega_0 \pm \Delta E$. Typically, Raman scattering involves optical phonons of wavelength $2\pi/K$ in solids, or molecular excitations in fluids, but it is also valuable for the study of other, more exotic excitations in condensed matter.

The intense optical fields of high-power lasers produce nonlinear scattering effects, such as stimulated Raman and Brillouin scattering, in addition to the linear phenomena discussed above.

See also AEROSOLS; ATMOSPHERIC PHYSICS; BRILLOUIN SCATTERING; DYNAMIC CRITICAL PHENOMENA; FLUCTUATION PHENOMENA; LIGHT; NONLINEAR OPTICS; POLARIZED LIGHT; RAMAN SPECTROSCOPY; RAYLEIGH SCATTERING.

BIBLIOGRAPHY

B. Berne and R. Pecora, *Dynamic Light Scattering*. Wiley, New York, 1976. (A)

C. F. Bohren and F. R. Huffman, *Absorption and Scattering of Light by Small Particles*. John Wiley, New York, 1983. (A)

W. Bragg, *The Universe of Light*. Dover, New York, 1959. (E)

B. Chu, *Laser Light Scattering*. Academic Press, New York, 1974. (A)

I. L. Fabelinskii, *Molecular Light Scattering*. Plenum Press, New York, 1968. (A)

M. Kerker, *The Scattering of Light*. Academic Press, New York, 1969. (A)

M. Minnaert, *Light and Colour in the Open Air*. G. Bell and Sons, London, 1940 (reprinted by Dover, New York, 1954). (E)

D. W. Schuerman, (ed.), *Light Scattering by Irregularly Shaped Particles*. Plenum Press, New York, 1979. (A)

D. J. Wineland and W. M. Itano, "Laser Cooling," *Phys. Today* **40**(6), 34 (1987). (I)

H. C. van de Hulst, *Light Scattering by Small Particles*. John Wiley, New York, 1957 (reprinted by Dover, New York, 1981). (I)

H. C. van de Hulst, *Multiple Light Scattering*, Vols. 1 and 2. Academic Press, New York, 1980. (A)

Light-Sensitive Materials

H. E. Spencer

Many materials are sensitive to visible and ultraviolet radiation. They include gases, liquids, and solids, and exhibit a wide variety of properties. They possess one common feature, however, in that absorption of visible and ultraviolet photons produces excited electronic states. The fate of the excited states determines the fate of the exposed materials, which may undergo chemical or physical change.

These states can best be discussed in terms of two somewhat different types of materials. In molecular materials, the ground-state properties are characteristic of individual molecules, and the excited states also are molecular in nature. In many solids, however, individual molecules or ions interact strongly and both the ground and excited electronic states are characteristic of the collection of ions or molecules in the solid.

In molecular organic materials, both the ground and excited states, in spectroscopic notation, are usually singlets. Deactivation of the excited singlet back to the ground state, mostly by a radiationless path, but sometimes accompanied by fluorescence, usually dominates unless internal conversion to a triplet state is energetically and kinetically possible. The long lifetime of the triplet state coupled with its high energy relative to the ground state results in a high probability of its reaction with another species.

A specific example of a molecular organic material that has been studied extensively is benzophenone, $C_6H_5COC_6H_5$, in solution at room temperature. It absorbs weakly in the violet and near-ultraviolet region. The excited singlet converts with near unit efficiency to a triplet state that can react with numerous compounds. For instance, if isopropyl alcohol is present in solution, the benzophenone triplet abstracts hydrogen atoms from the alcohol. This is but an initial step; the final products are acetone and benzopinacol. If isobutylene is present instead of isopropyl alcohol, a triplet benzophenone molecule and an isobutylene molecule combine to form a new molecule, an oxetane. Such photochemical reactions often provide the easiest ways of synthesizing certain complex molecules.

Many molecular inorganic materials are photosensitive, also. The gas NO_2 is one, and its photochemistry is important in the chain of events whereby the atmospheric pollutant O_3 is formed. NO_2 is generated as a by-product of high-temperature combustion such as is found in automobile engines. Absorption in NO_2 of ultraviolet radiation of 300- to 400-nm wavelengths from the sun produces an excited state that rapidly decomposes into NO and O. The latter reacts with O_2 to form O_3.

Other examples of photosensitive inorganic materials are found among coordination compounds of transition metals. For example, $K_3Fe(C_2O_4)_3 \cdot 3H_2O$, both as a solid or in solution, is decomposed by visible and ultraviolet radiation. In solution, products consisting of Fe(II) complexes and CO_2 form with high quantum efficiency. One of the most commonly employed actinometers is an acidic solution of $K_3Fe(C_2O_4)_3 \cdot 3H_2O$.

In many solids, absorbed ultraviolet and visible radiation energetically raises electrons from the valence band to the conduction band, leaving behind mobile holes. Often these electronic carriers are detected in photoconductive or photovoltaic experiments. The electrons and holes commonly recombine, often in a radiationless manner, but sometimes accompanied by luminescence, particularly at low temperatures. Proper doping with impurities enhances this luminescence.

Silver halides exemplify photosensitive solids of scientific and commercial importance. The chloride, bromide, iodide, and mixtures of these salts form the basis of many photographic systems. Values of extinction coefficients are high in the ultraviolet region and become lower as they extend into the visible portion. After absorption, the hole–electron pairs that do not recombine reduce silver ions to metallic silver and oxidize halide ions to halogen. For photographic applications of a silver salt, microcrystals of micron or submicron linear dimensions are usually dispersed in a gelatin coating on an inert polymeric substrate. To decrease recombination and thereby increase sensitivity, small amounts of impurities, such as silver, gold, silver sulfide, and gold sulfide, are added to the microcrystals. Sensitizing dyes often are also adsorbed to the microcrystals. Absorption of light by the dyes produces conduction-band electrons in the silver halides, in the usual case.

These photoelectrons form silver centers. By a repeated cycle of trapping of the photoelectrons, and neutralization by interstitial silver ions, silver centers nucleate and grow. Those of sufficient size to act as catalyst sites when the exposed film is immersed in a developer solution are called latent-image centers. In the developer, microcrystals that have been exposed and contain latent-image centers are reduced to metallic centers at a much greater rate, in general, than those that have not been exposed. The final image is either the reduced silver or a dye formed by the reaction of the oxidized developer and a suitable reagent. Many variations of this basic process are utilized.

BIBLIOGRAPHY

D. R. Arnold, N. C. Baird, J. R. Bolton, J. C. D. Brand, P. W. M. Jacobs, P. deMayo, and W. R. Ware, *Photochemistry, an Introduction*. Academic, New York, 1974. (I)

J. G. Calvert and J. N. Pitts, Jr., *Photochemistry*. Wiley, New York, 1966. (A)

T. H. James, ed., *Theory of the Photographic Process*, 4th ed. Macmillan, New York, 1976. (A)

Albert Rose and Paul K. Weimer, "Physical Limits to the Performance of Imaging Systems," *Phys. Today* **42**(9), 24–32 (1989). (E)

Tadaaki Tani, "Physics of the Photographic Latent Image," *Phys. Today* **42**(9), 36–41 (1989). (E)

Lightning

Martin A. Uman

Lightning is a transient, high-current electric discharge whose path length is generally measured in kilometers. The source of most lightning is the electric charge separated in the common thunderstorm, although forms of lightning occur due to charge separation in snowstorms, in sandstorms, in the clouds generated by some volcanoes, and near thermonuclear explosions. Most thunderstorm-produced lightning occurs within the cloud (intracloud discharges). Cloud-to-ground lightning (sometimes called streaked or forked lightning) has been studied more extensively than other forms of lightning because of its practical interest and because it is easily photographed. Cloud-to-cloud and cloud-to-air lightning are less common than intracloud or cloud-to-ground discharges.

The typical lightning between cloud and ground is initiated in the cloud and results in the lowering of tens of coulombs of negative charge to ground in about 0.2 s. The total discharge is called a *flash* and is composed typically of three or four component *strokes*, each lasting about 1 ms with a separation time of roughly 50 ms. Lightning often appears to flicker because the eye resolves the individual luminous stroke pulses. The process that initiates the first stroke in a flash, the *stepped leader*, is sketched in Figs. 1 and 2. In the idealized model of cloud charge shown in Fig. 1a, P and N are of the order of many tens of coulombs of positive and negative charge, respectively, and p is a few coulombs of positive charge. The local discharge in the cloud base (Fig. 1b) allows negative charge (electrons) to be funneled toward the ground in a series of luminous steps of typically 1-μs duration and 50-m length with a pause time between steps of about 50 μs (Fig. 1c–f, Fig. 2a). A fully developed stepped leader has about 5 C of negative charge on it, has traveled to ground in about 20 ms with an average velocity of 1.5×10^6 m/s and an average current of about 100 A, and has an electric potential with respect to ground of about -10^8 V. The intermittent leader steps have a pulse current of about 1 kA and are brighter than the channel above. The stepped leader branches in a downward direction during its trip to ground.

When the stepped leader is near ground, its electric field causes upward-moving discharges to be launched from the ground (Fig. 2b). When one of these discharges contacts the leader some tens of meters above the ground, the leader bottom is connected to ground potential. The leader is discharged by virtue of a ground potential wave, the *return stroke*, which propagates continuously up the leader channel at a velocity of typically one third the speed of light (Fig. 2c–e), the trip taking less than 100 μs. The return-stroke channel carries a peak current of typically 20 kA with a time to peak of a few microseconds. Currents measured at the channel base fall to half of peak value in about 50 μs, and

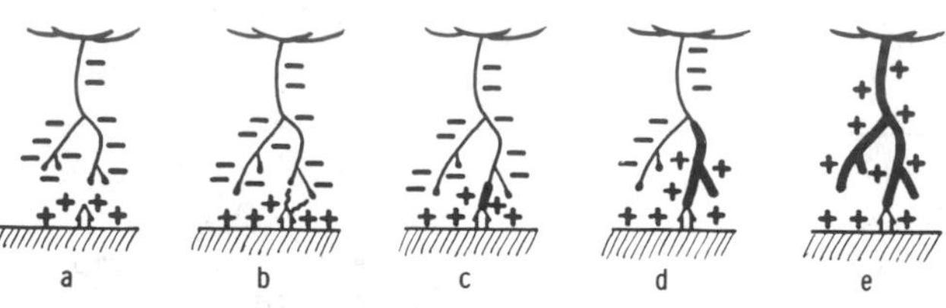

Fig. 2. Return-stroke initiation and propagation. (a) Final stage of stepped-leader descent; (b) initiation of upward-moving discharges; (c)–(e) return-stroke propagation from ground to cloud. Scale of drawing is distorted for illustrative purposes. (Adapted from M. A. Uman, *All About Lightning,* Dover, New York, 1986.)

currents of the order of hundreds of amperes may flow for milliseconds or longer. The return stroke energy input heats the leader channel to near 30 000 K with the result that the tortuous high-pressure channel expands, creating shock waves that become the thunder we hear.

After the stroke current has ceased to flow, the flash may end. On the other hand, if additional charge is made available to the decaying channel top in a time less than about 100 ms, a continuous or *dart leader* (Fig. 3) will traverse the defunct return-stroke channel at about 2×10^6 ms, depositing charge along the channel by virtue of its 1-kA current and carrying cloud potential earthward once more. The dart leader thus sets the stage for the second (or any subsequent) return stroke. Dart leaders and strokes subsequent to the first are generally not branched. Some leaders begin as dart leaders and end their trip to ground as stepped leaders. A drawing of a streak photograph and a still photograph of a lightning flash are given in Fig. 4.

While most lightning to ground lowers negative charge, about 10% of the worldwide ground flashes are initiated by downward-moving positive leaders, and the resultant flash lowers positive charge. Positive ground discharges are more common at the higher latitudes and in winter. They generally have only one return stroke followed by a relatively long period of continuous current flow between cloud and ground.

In addition to being initiated by downward leaders, lightning can occasionally be initiated from high structures by upward-going stepped leaders which can be either positively or negatively charged. The upward-going leaders branch in an upward direction. Upward-going leaders can be artificially initiated by firing from ground under a charged cloud small rockets trailing a few hundred meters of wire, either connected or unconnected to the earth. The wire serves to concentrate and enhance the electric field of the cloud charge

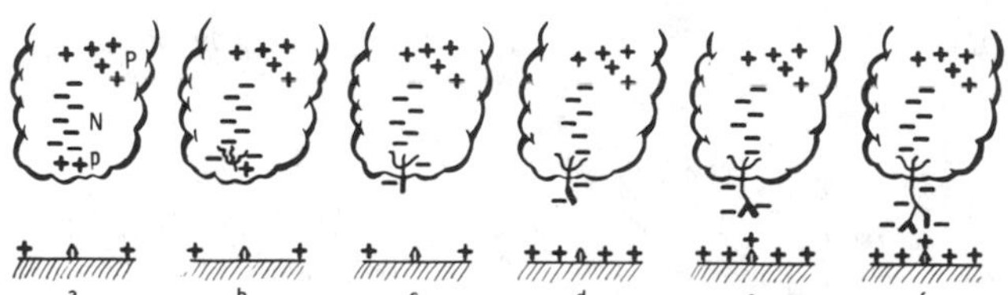

Fig. 1. Stepped-leader initiation and propagation. (a) Cloud charge distribution prior to lightning; (b) discharge in lower cloud; (c)–(f) stepped-leader progression toward ground. Scale of drawing is distorted for illustrative purposes. (Adapted from M. A. Uman, *All About Lightning,* Dover, New York, 1986.)

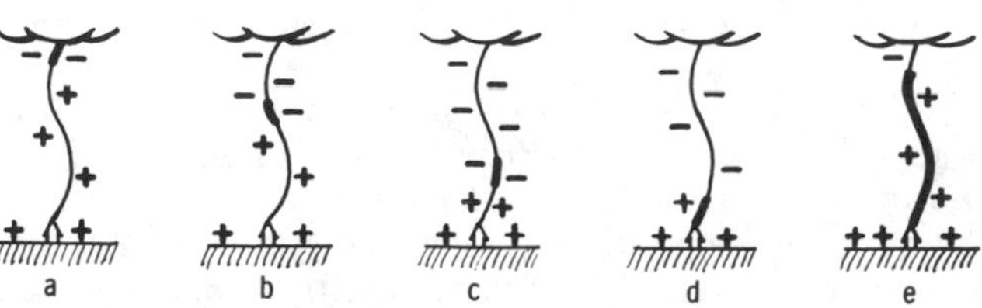

Fig. 3. Dart leader and subsequent return stroke. (a)–(c) Dart leader deposits negative charge on defunct first-stroke channel; (d)–(e) return stroke propagates from ground to cloud. Scale of drawing is distorted for illustrative purposes. (Adapted from M. A. Uman, *All About Lightning,* Dover, New York, 1986.)

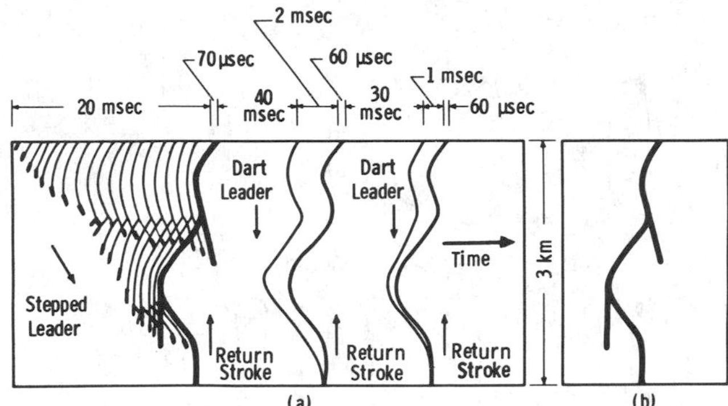

FIG. 4. (a) The luminous features of a typical lightning flash as would be recorded by a camera with relative motion (horizontal and continuous) between lens and film, a so-called streak camera. Scale of drawing is distorted for illustrative purposes. (b) The same lightning flash as recorded by an ordinary camera. (Adapted from M. A. Uman, *Lightning*, McGraw-Hill, New York, 1969.)

to the point of lightning initiation. For both natural and artificially initiated upward lightning, the initial upward discharge may be followed by dart leader–return stroke sequences similar to those in normal downward-initiated lightning.

Intracloud discharges take place between two charge centers in the cloud, have a total duration about equal to that of ground discharges, and appear to the camera as a continuous luminosity with brighter pulses superimposed. It is thought that a leader moves through the cloud generating weak return strokes (called K-changes) when contacting pockets of opposite charge. The charge neutralized in a cloud discharge is about the same as in a ground discharge.

Heat and *sheet* lightning are due to cloud illumination by normal lightning in situations where the storm is too distant for thunder to be heard (generally over 25 km). Rocket lightning is the name given to the very long air discharges that often travel along the bases of the clouds. *Ribbon* lightning occurs when the cloud-to-ground discharge channel is shifted horizontally by the wind in the time between strokes. *Bead* lightning is the name given to the form of lightning in which the channel to ground breaks up into a "string of pearls" which persists in luminosity for a longer-than-normal flash duration. *Ball* lightning is the name given to the mobile luminous spheres that have been observed during thunderstorms, often in the vicinity of strokes to ground. A typical ball lightning is the size of a grapefruit and has a lifetime of a few seconds. At present there is no completely satisfactory explanation for ball lightning.

Lightning has been tentatively identified on the planet Jupiter from spacecraft television pictures and has been postulated to exist on Venus, primarily from radio noise measurements.

See also ARCS AND SPARKS; ATMOSPHERIC PHYSICS; CORONA DISCHARGE.

BIBLIOGRAPHY

R. H. Golde, *Lightning Protection*. Edward Arnold, London, 1973. (A)

R. H. Golde, ed., *Lightning*, Vol. 1: *Physics of Lightning*; Vol. 2: *Lightning Protection*. Academic Press, New York, 1977. (A)

D. J. Malan, *Physics of Lightning*. English Universities Press, London, 1963. (A)

B. Schonland, *The Flight of Thunderbolts*, 2nd ed. Oxford (Clarendon Press), London and New York, 1964.

M. A. Uman, *Lightning*. McGraw-Hill, New York, 1969. (A)

M. A. Uman, *All About Lightning*. Dover, New York, 1986. (E)

M. A. Uman, *The Lightning Discharge*. Academic Press, San Diego, CA, 1987. (A)

P. E. Viemeister, *The Lightning Book*. Doubleday, Garden City, NY, 1961; MIT Press (paperback), Cambridge, Mass., 1972. (E)

Liquid Crystals

P. G. de Gennes and J. Prost

CLASSIFICATION AND EXAMPLES

In a crystal the building blocks (atoms or molecules) are regularly arranged in a periodic lattice. In a liquid they are completely disordered. With suitably chosen molecules, we can find other states of organization, with less symmetry than in a crystal but more symmetry than in a liquid: they are called *mesomorphic phases* or, more loosely, liquid crystals. The main types known at present are the following [1].

1. *Nematics* [2] are fluids of rodlike molecules with their centers disordered but with a common axis of alignment for the rods (Fig. 1a). A typical molecule giving rise to a nematic phase is methoxybenzilidenebutylaniline (MBBA):

CH_3O— ⬡ —CH=N— ⬡ —C_3H_7

The sequence of phases for MBBA is

$$\text{solid} \xrightarrow{10^\circ\text{C}} \text{nematic} \xrightarrow{43^\circ\text{C}} \text{isotropic liquid.}$$

2. *Smectics A* and *smectics C* [3] are layered systems, each layer behaving like a two-dimensional *liquid*. The A type is optically uniaxial (Fig. 1b), whereas the C type is biaxial (Fig. 1c). Both types can be obtained with certain long organic molecules, such as terephthal-*bis*-*p*-butylaniline (TBBA):

C_4H_9— ⬡ —N=CH— ⬡ —CH=N— ⬡ —C_4H_9

The sequence of phases of TBBA is complex: the main phases are

$$\text{crystal} \xrightarrow{113^\circ} \text{B} \xrightarrow{144^\circ} \text{C} \xrightarrow{172^\circ} \text{A} \xrightarrow{200^\circ} \text{N} \xrightarrow{236^\circ} \text{I}$$

(where I stands for isotropic and N for nematic, and B is another smectic type, to be discussed later). This shows that one compound may have many states in between the crystal

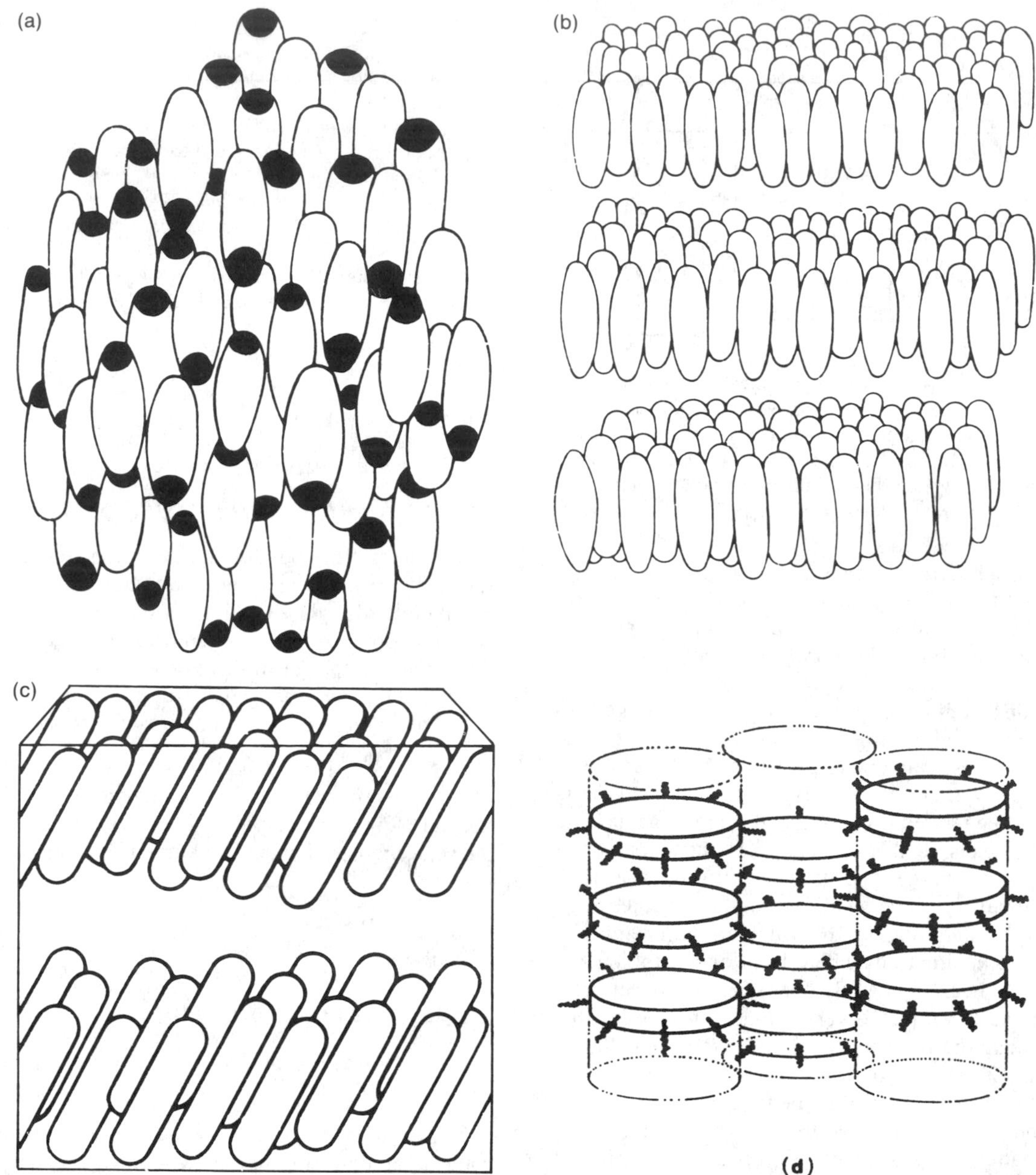

FIG. 1. (a) Nematic phase: the molecules are purposely represented with one black end; note that as many molecules are "up" as are "down"; even if the molecules carry a permanent dipole, the phase is not ferroelectric. (b) Smectic-A phase. (c) Smectic-C phase. Note that in parts (b) and (c) each layer is disordered (liquid-like). (d) Thermotropic "columnar" phase: the disks represent the rigid part of a "plate-like" molecule and the wavy lines alkyl chains. Note the absence of periodicity along a column.

and the liquid! This situation is not exceptional; the number of known molecules giving rise to mesomorphic phases is of the order of 2000.

In pure materials like TBBA, the main physical variable that allows us to explore the sequence of phases is the temperature; for this reason the resulting liquid crystals are called *thermotropic*. Another group is obtained with mixtures, where the phases can be obtained by changes in concentration (lyotropic liquid crystals). A major example of lyotropic smectics is found with mixtures of lipid and water [4]. A lipid molecule contains a polar head (hydrophilic) and a hydrocarbon tail (hydrophobic). Many lipid-water systems show a lamellar phase (Fig. 2a) that has the symmetry of a smectic A.

What is the origin of the smectic layers? For lyotropic systems they appear as a natural way of having the heads close to the water and the tails far from it. For thermotropic systems the situation is less clear, but something similar may happen: the molecules usually have an aromatic—rigid—part (Ar) plus terminal chains that are aliphatic (Al). It may be that (Ar)–(Ar) attractions dominate the system's behavior, and they favor a layered system.

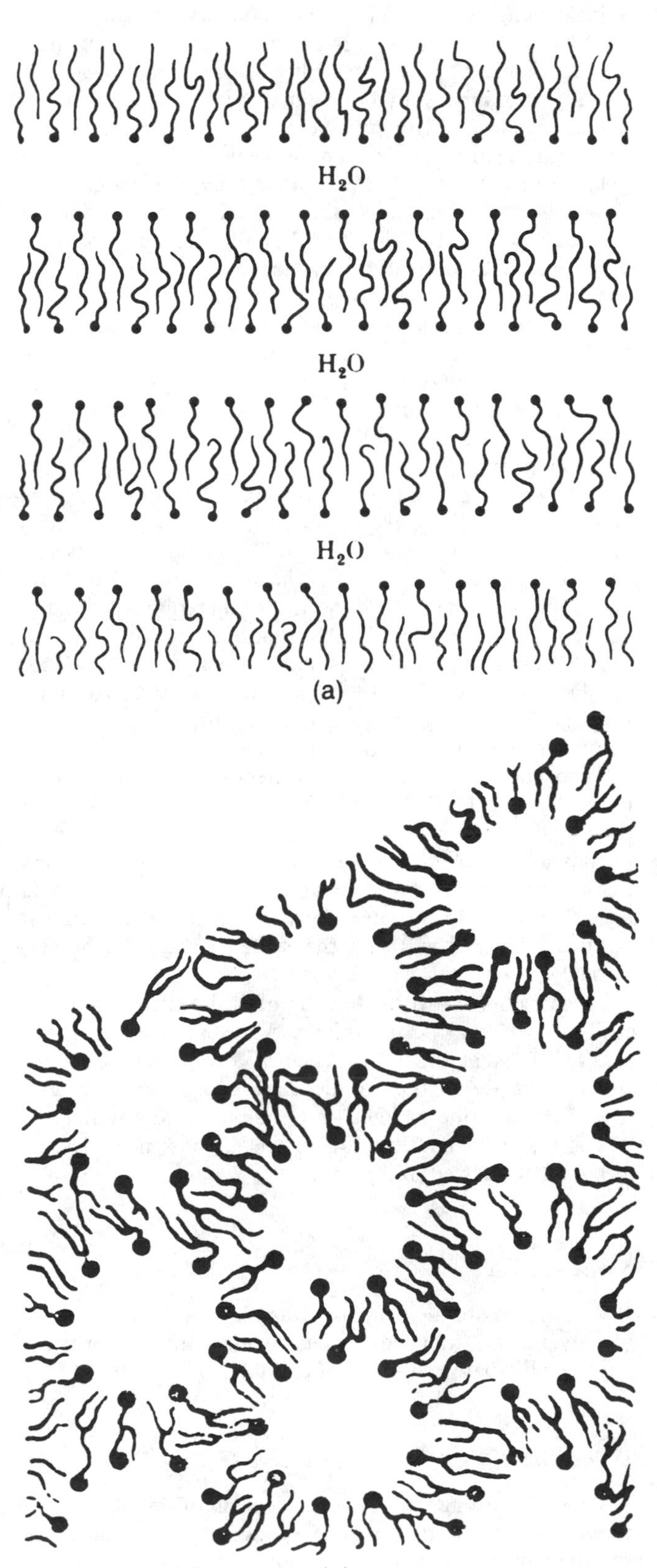

FIG. 2. (a) Lamellar phase of lipid + water systems (smectic A). (b),(c) Hexagonal phases of lipid–water systems (canonic).

3. *Smectics A* and *C* are not the only smectic phases. *Hexatic* smectics are also stacks of liquid layers. Like the A type, they are optically uniaxial, but the pair correlation function exhibits a sixfold macroscopic modulation. Locally, there is a triangular order like in a crystal, but a high enough defect density destroys the periodicity without decorrelating the triangles directions.

Smectics I, F, and *K* are hexatics in which molecules are tilted with respect to the layers. In terms of symmetry, they are not different from smectics C.

More recently, an "incommensurate" smectic has been reported. A real-space picture is not easy to give: imagine two conventional smectic A systems with incommensurate layer spacing; mix them together so that the two percolate through each other, this gives an approximate image of an incommensurate smectic.

4. Many other phases have been called smectics historically, smectics B in particular. The transition from normal solid to smectic B is often associated with the "melting" of aliphatic chains at the end of the molecule while other parts (the aromatic region in thermotropic systems, or the polar heads in lipid water) remain well ordered: this gives a structure in which successive layers are nearly uncoupled. X rays show that they should be classified as crystals. The same remark holds for smectics D, E, and G. The most interesting is the D phase which is a cubic gel with several hundred molecules per unit cell.

5. "Canonic" phases [5], also called columnar phases, have periodicity in two directions and not in the third; they are found with lipid water and also with thermotropic systems. In the latter case, disc-like molecules pack on top of each other, forming columns; however, the lack of a fixed repeat distance between the centers of gravity of the molecules gives a liquid-like order along the column axis (Fig. 1d). In the former case, one component is associated in rods, each rod being a one-dimensional fluid tube (Fig. 2b,c). The macroscopic symmetry is the same. There are many possible variations of this basic arrangement among which about a dozen are currently known.

The physics of all these liquid crystals has expanded rapidly during the last twenty years, and cannot be summarized easily. We have chosen here to omit all further discussion of properties at the molecular level (for which there exist many reviews [9,10] and to focus on certain macroscopic features, which are most unusual.

OPTICAL PROPERTIES OF NEMATICS

Nematics are optically uniaxial, but they flow like liquids. Their optical properties are interesting mainly because the optical axis may be *rotated by weak external agents:* the effects of the walls of the container, of magnetic and electric fields, and of flows are most important in practice.

1. *Alignment by walls:* It is possible to fix the orientation of the optical axis in a nematic slab by suitable treatment of the container surfaces: rubbing in one direction, or covering the surface with suitable detergents, or evaporating metallic films at an oblique angle. It is thus possible to prepare single-domain samples with well-defined boundary conditions.

2. *Alignment by magnetic fields H:* Nematic fluids have a weak diamagnetic susceptibility χ, and χ is anisotropic. Usually, this anisotropy favors an optical axis parallel to the magnetic field H. With suitable orientation of H, there may be an interesting competition between wall alignment and field alignment, first studied long ago by Freedericksz [6]. Typically, for a slab 50 μm thick, the fields involved are of the order of 10^3–10^4 Oe. The field effects can be monitored very directly by optical observations.

3. *Flows* are strongly coupled to the molecular alignment [7]. In a simple shear flow the long axis of the molecules tends usually to become nearly parallel to the lines of flow. In many cases, a competition can be set up between wall alignment and flow alignment, leading to unusual mechano-optic effects and to instabilities even at extremely low Reynolds numbers.

4. *Electric fields E* have remarkable effects on nematics. Alignments similar to the H effect discussed above provide display devices which are sensitive to voltages as low as a few volts. Optical anisotropy or linear dichroism provide the necessary coupling to light. Multiplexing requires the use of active addressing schemes. More interesting from a fundamental standpoint is the instability obtained with slightly conducting fluids ($\sigma \approx 10^{-9}$ mho): (a) The molecular alignment is slightly deformed by a thermal fluctuation; (b) this reacts on the conduction–current pattern J, and makes J inhomogeneous; (c) the inhomogeneous J builds up a space charge Q; (d) the field E acting on Q gives a bulk force in the liquid; (e) the result is a flow pattern, which distorts the alignment as explained in item (3) and *enhances* the original deformation—hence the instability [8]. When V is high enough, a state of turbulent flow is reached, the alignment is strongly perturbed, and the system scatters light very strongly.

These examples show that the optical axis can be tilted easily by external perturbations. As usual in physics, this also implies that, in the absence of any perturbation, the *spontaneous fluctuations* of the axis are important: they give rise to a scattering of light (in the absence of any field E) that is much higher than for normal liquids: nematics are *turbid* when viewed in thicknesses $\gtrsim$ 1 mm.

SMECTICS A

Smectics A are the simplest example of a system with liquid layers, and they have unusual mechanical properties, intermediate between those of crystals and liquids.

Deformability of the Layer Structure

The ideal arrangement is a succession of flat sheets. But the sheets can be deformed very easily. The distance between them is essentially fixed, but they can be moved over large regions since successive layers can slip freely over one another. This shows up in various problems.

Textures: If the boundary conditions do not allow for the ideal flat arrangement, the layers build an interesting deformed texture. The simplest example is a jelly-roll shape (with a singular line on the axis). But there are more complex patterns, where the singular lines become ellipses and hyperbolas ("focal conics") [11].

Undulation modes: An oscillation with a wave vector parallel to the layers does not alter the interlayer distance; thus, the restoring force for these particular modes is extremely weak, and they have a large thermal amplitude, giving rise to a special type of light scattering.

Instability under traction: A smectic slab (typical thickness 100 μm) with the layers parallel to the slab plane is rapidly pulled along the normal to the layers; as soon as the relative displacement of the limiting plates exceeds a few hundred angstroms, the smectic layers *fold* to fill the available space without changing their thickness (Fig. 3).

Flows in a Smectic Structure

At first sight, a smectic appears very viscous—very much like a soap. But the real situation is more subtle. Flow *parallel* to the layers involves only weak friction, very much like in a conventional fluid. Flow *through* the layers requires a pressure gradient, just like the flow of water through a filter. But here the filter and the flowing fluid are one and the same species! The hydrodynamic consequences of these strange features have now been observed in certain weak flows. Strong flows usually imply the creation and motion of many defects (dislocations, focal conics) and are just beginning to be understood.

Fluctuations play an unusual role in smectics: elastic moduli and viscosities become wave-vector and frequency dependent in the long-wavelength long-time limit.

CANONIC PHASES

They again have unusual mechanical properties bearing some similarities with those of smectics.

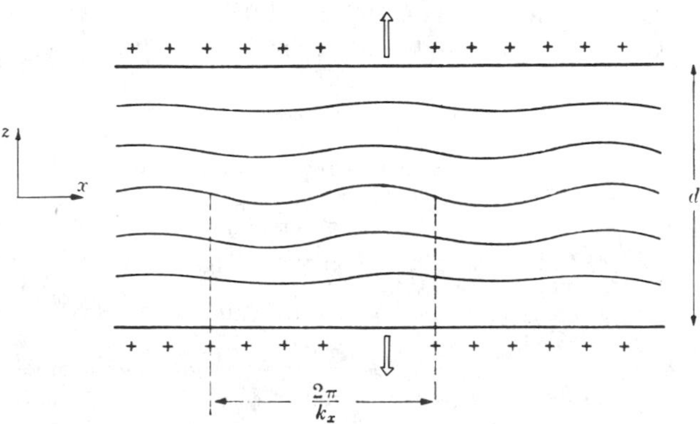

FIG. 3. Instability of smectic-A layers under traction.

Deformability of the Structure

The ideal arrangement is that of straight columns. However, bending the columns, keeping the distance between them constant, is possible: this costs only nematic energy and thus occurs easily.

Textures

The equivalent of the smectic "focal conics" are developable domains. The simplest example is obtained by wrapping the columns around a cylinder of revolution perpendicularly to its axis.

Instabilities

A slab of canonic sample with the columns parallel to the slab behaves pretty much like the smectic of Fig. 3. The novelty is now that one can get a buckling instability when the columns are perpendicular to the slab and under compressive stress.

Flows have not been studied, but pulling on a columnar material produces thin strands (a few microns diameter, a few millimeters long), almost perfectly oriented.

CHIRALIZATION

Another remarkable example of the deformability of liquid-crystal structures is connected with the effect of solute molecules that are not identical to their mirror image (chiral molecules). A nematic N doped with such molecules becomes twisted into a helical form N* (Fig. 4a), usually known as a *cholesteric* phase [12]. The pitch of the helix is typically in the micron range and thus gives rise to Bragg scattering of optical light. For this reason many cholesterics show beautiful iridescent colors. Also, the pitch (and thus the color) can be very sensitive to various external agents (temperature, fields, chemical contaminants, etc.) and this has been the basis of certain applications [13]. When added to a smectic A, chiral molecules cannot induce a macroscopic twist (because this twist is incompatible with the condition of constant layer thickness). But in a smectic C, chiralization does give a twisted phase C* (Fig. 4b). The C* systems have Bragg reflections and optical rotations very much like N*. They have one further property that was recognized recently [14]: each C* layer is *ferroelectric*, with a permanent moment parallel to the layers and normal to the tilt direction. The ferroelectric moment is usually rather small (of the order of 10^{-2} debye per mole), but it can reach important values (almost one debye per mole) for well-chosen molecules. The existence of a moment in a system that can flow leads to remarkable cross effects. In particular, a thin slab (≈ 1 μm) of C*, in which the layers are perpendicular to the slab, provides a novel, highly multiplexable and fast display. In this geometry, the helical precession is unwound by the boundaries, and only two states of polarization (up and down) are possible. A modest electric field commutes the states.

On the whole, we see that liquid cystals seem to have a relatively broad spectrum of future applications.

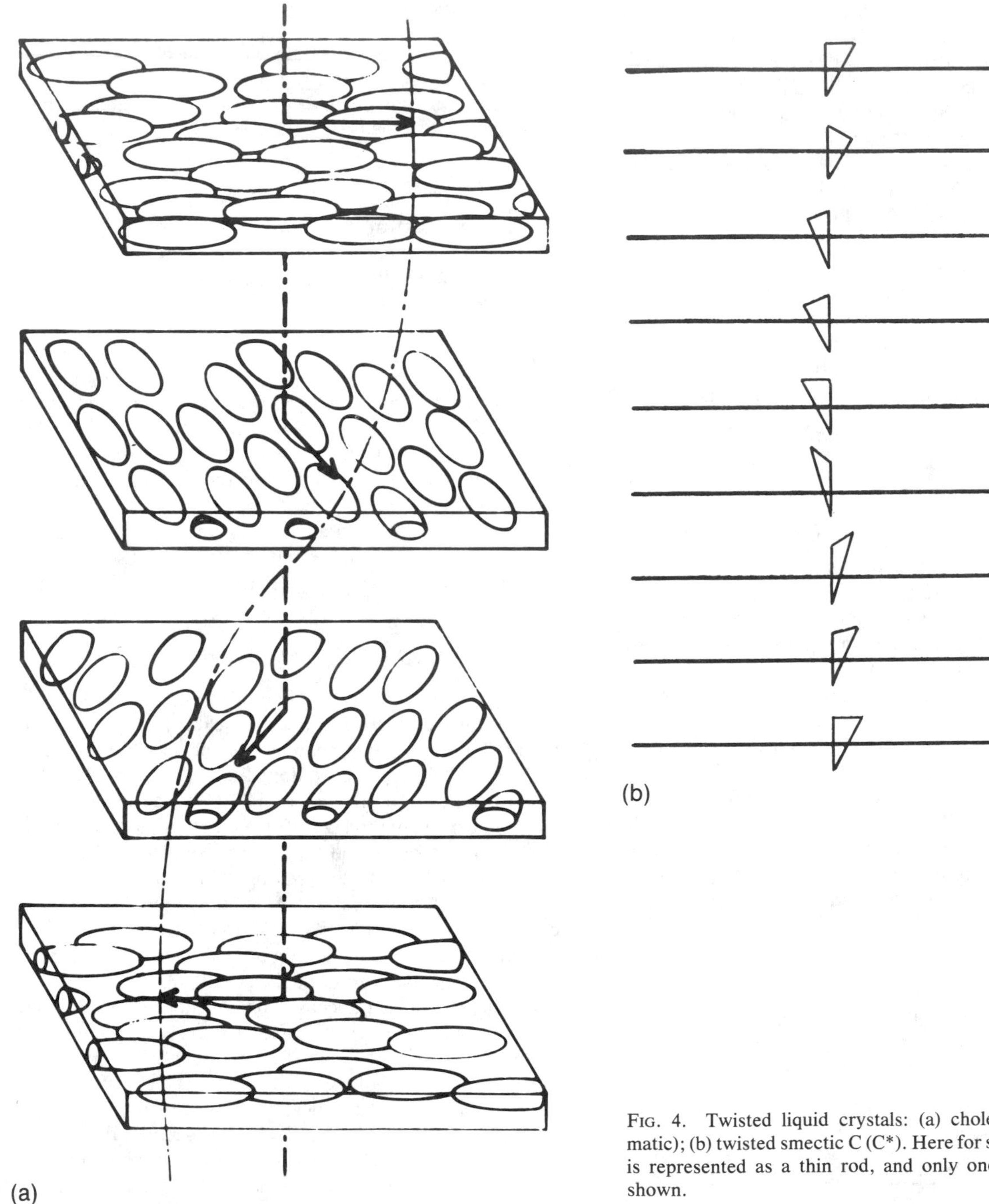

FIG. 4. Twisted liquid crystals: (a) cholesteric N* (twisted nematic); (b) twisted smectic C (C*). Here for simplicity each molecule is represented as a thin rod, and only one molecule per layer is shown.

See also LIQUID STRUCTURE; VISCOSITY.

NOTES AND REFERENCES

1. The classification of liquid crystals has been constructed mainly by G. Friedel [*Ann. Phys.* **18,** 273 (1922)] using optical observations, and by H. Sackmann and D. Demus [*Mol. Cryst.* **2,** 81 (1966)] using miscibility criteria and textural features.
2. From the greek *nema*, thread; this refers to certain threadlike defects (disclinations) that are found in poorly oriented nematics (see Ref. 6).
3. From the greek *smegma*, soap; certain lamellar phases of the soaps (salts of fatty acids) are smectics.
4. For a review of the polymorphism of lipid–water systems, see V. Luzzati in *Biological Membranes* (H. Chapman, ed.), Academic Press, New York, 1968.
5. From the greek *canon*, rod; this terminology was introduced by F. C. Frank.
6. The continuum elastic theory on which the interpretation of phase effects is based is due to C. Oseen, H. Zocher, and F. C. Frank; see F. C. Frank, *Disc. Faraday Soc.* **25,** 19 (1958).
7. The equations of "nematodynamics" were constructed by J. L.

Ericksen and F. M. Leslie. A more general formulation covering the smectics as well can be found in P. Martin, O. Parodi, and P. Pershan, *Phys. Rev.* **A6,** 2401 (1972).
8. The complete explanation is due to W. Helfrich, *J. Chem. Phys.* **51,** 4092 (1969).
9. G. Gray, *Molecular Structure and the Properties of Liquid Crystals.* Academic Press, New York, 1962. A. Saupe, *Angew. Chem. (Int. Ed.)* **7,** 97 (1968). G. H. Brown, W. Doane, and V. Neff, *C.R.C. Crit. Rev. Solid State Sci.* **1,** 303 (1970). P. G. de Gennes, *The Physics of Liquid Crystals,* Oxford, London and New York, 1974.
10. For a recent perspective on the field, see the *Proceedings* of the 5th International Conference on Liquid Crystals (Stockholm, 1974) published by *J. Phys. (Paris), Colloq.* **C1,** (1975).
11. The focal conics are always with sets of equidistant surfaces, such as wave fronts in optics. G. Friedel realized this very early, and from observations of focal conics under the microscope he was able to infer—long before any x-ray work (see Ref. 1)—that smectics A were made up of equidistant fluid layers.
12. Because it was found first with cholesterol esters.
13. J. Fergason, *Sci. Am.* **211,** 77 (1964).
14. R. B. Meyer, L. Liébert, L. Strzelecki, and P. Keller, *J. Phys. (Paris), Lett.* **36,** L69(1975).

Liquid Metals

N. E. Cusack

Among the new insights into liquid-metal physics which began to emerge in the 1960s was the realization that several strands of theory were converging. Simple nonmetallic liquids like argon had become well understood through steady progress in liquid-state theory. This theory was ready for use for liquid metals if their interatomic forces resembled, at least in some ways, those of nonmetals. Striking developments in the physics of crystalline metals were showing that this was indeed conceivable, so a fusion of metal and liquid-state theories began, and continues still. In another perspective, liquid metals can be seen as one class of structurally disordered, or noncrystalline, materials which also includes all other liquids and glasses and amorphous semiconductors. Thus, liquid metals pose problems which arise, *mutatis mutandis*, in crystalline metals, metallic glasses, and nonmetallic liquids. To see what these problems are let us describe a metal in general terms.

The general character of metals results from the dense fluid of free electrons that permeates them. Indeed a pure liquid metal can be seen as a mixture of (i) a fluid of positively charged ions and (ii) a gas of negative electrons containing one electron per ion in sodium, two in calcium, and so on according to valency. There will, therefore, be about 10^{28} electrons/m^3, and these are loosely called "free electrons" because they can move through the entire system. The free-electron gas conducts electricity well and is also responsible for thermal conduction, metallic reflectivity, and all the other typically metallic properties.

To understand a liquid metal we must tackle three problems which are not really separate. They concern the forces between the ions and the electrons, the forces between one ion and another, and the way electrons behave in a disordered (noncrystalline) environment. The following three sections touch on these.

FORCES BETWEEN IONS AND ELECTRONS

The ions contain electrons tightly bound to nuclei. The free electrons flow through the ions under the control of two major influences: (a) the electrostatic attraction of the positive ions and (b) a complicated response to those electrons bound to the nuclei. A powerful development of the modern wave-mechanical theory of solids made it possible to treat influence (b) mathematically as if it were an electrostatic influence like (a) except that it is a repulsion, which partially cancels the attraction (a). The net effect of the two is quite small and is called a *weak pseudopotential.*

An electron passing through an ion may therefore be thought of as deviated by the appropriate pseudopotential. However, the ion does not suddenly exert its influence as the electron crosses a well-defined ionic boundary. The ions are positively charged and the electron gas is mobile, and therefore the ions exert a long-range influence on the gas by attracting electrons to themselves. Meanwhile the free electrons are also repelling each other, and a complex balance of forces sets in whose ultimate effect is to envelop each ion in an electron cloud. This cloud, being negative, neutralizes the effect of the positive ion rather completely at large distances (say, several atomic diameters) and less completely at smaller distances.

The cloud of electrons piled up near an ion is called the ion *screening cloud* and the total effect on an approaching electron of the pseudopotential and the screening cloud is called the *screened pseudopotential.*

In summary we may therefore say that from the point of view of one electron passing through a liquid metal it encounters the screened potentials of the ions (the nearest ones being felt more strongly) and is scattered by them. The total effect of all the ions acting through their weak screened pseudopotentials on all the electrons is to produce the very electrons clouds that screen the ions. This idea underlies most current thinking about liquid-metal physics.

Pseudopotentials can be calculated from first principles, i.e., from the known nature of the ions, and they enable a good account to be given of the conductivity and other electronic properties in the following way.

Perfectly regular arrays of atoms in space are known theoretically not to have any electrical resistance. Cold crystalline metals are good approximations to this and their resistance is indeed very low. As they are heated, their electrical resistance rises because the ions vibrate about their ordered positions, thus acquiring *thermal disorder.* On melting, the array of ordered positions dissolves and *structural disorder* sets in as illustrated in Fig. 1. This causes a further rise of resistance by a factor of about 2 at the melting point. Further heating increases the resistance of liquid metals still more. Thus, although the resistance of liquid metals is higher than that of the corresponding crystals, it is not catastrophically so and liquid metals are good conductors. They also have thermoelectric powers and thermal conductivities comparable with those of solid metals.

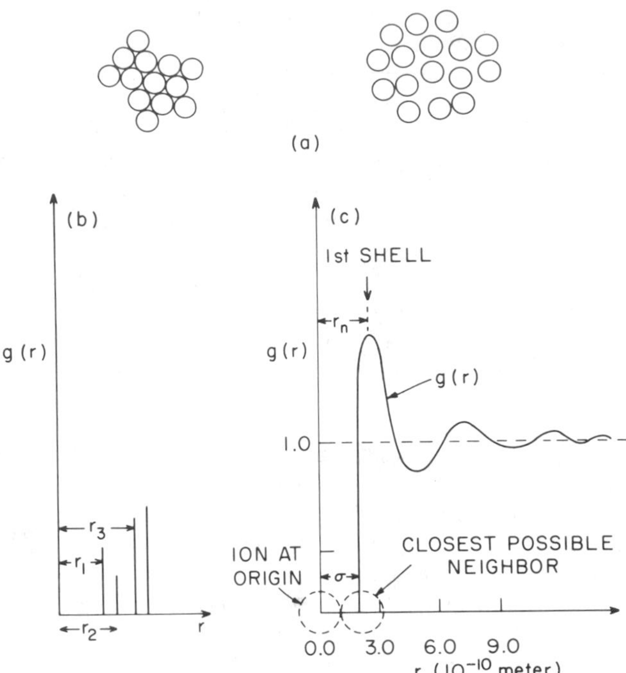

FIG. 1 (a) Schematic regular and irregular arrays of atoms shown in two dimensions. (b) Radial distribution function for a perfect regular or crystalline array. Atoms occur only at definite distances r_1, r_2, etc. (c) Radial distribution function $g(r)$ for a liquid metal. Atoms occur at any distance greater than σ (the effective hard-sphere diameter) with varying probability; r_n is the mean distance of nearest-neighbor shell, which typically accommodates 11–12 neighbors on average.

These properties are now well understood on the basis of wave mechanics. As explained earlier, the electrons are deflected when they collide with the ions because the latter have weak screened pseudopotentials. This is in terms of particle language. In the wave language of wave mechanics, the electron waves are scattered by the ions to a degree depending on the pseudopotential, and the scattered wavelets from the different ions may be in or out of phase, thus reinforcing or cancelling one another. To find the net effect of all the wavelets is a typical problem in diffraction theory, well understood in, for example, x-ray crystallography. The combination of pseudopotential and diffraction theory, employing the structural knowledge of Fig. 1c, leads to expressions for electrical resistance, thermoelectric power, and thermal conductivity in good agreement with observations on many pure liquid metals and alloys.

FORCES BETWEEN IONS

Any two positive ions repel one another electrostatically. But in a metal they have another, indirect, mutual influence because they both exert force on the same electron gas. The latter contribution must depend on the density of the system because this determines the density of the electron gas. This contrasts with the intermolecular forces in nonmetals which are independent of density to a good approximation. However, for a *fixed density*, it turns out that the total ion–ion interaction in a metal can be treated just like the forces in nonmetallic liquids. In particular, the ions have pairwise interactions. The exact form of the ion–ion potential function will vary from metal to metal and differ from those of nonmetals; it depends on the pseudopotential. But once it is known, it can be inserted into liquid-state theory to calculate thermodynamic properties such as free energy and entropy. This is technically complicated but has been done with some success for pure liquid metals and some binary alloys, thus bringing these materials into the theory of liquids.

It is even possible to use liquid-state theory and the ion–ion potentials to say something about structure. But generally the details of the interaction are not well enough known to calculate structure accurately from first principles. The structure of crystals is of course revealed in full by x-ray crystallography. This is not so for liquids. What x-ray and neutron diffraction studies can do for liquids is to provide a quantity known as the radial distribution function, denoted by $g(r)$ and defined as follows: choose any one ion as origin; then in a spherical shell of radius r, the average number density of the ions is $g(r)$ times the average number density in the whole liquid. Figure 1 contrasts the structures of liquid and crystalline metals. The function $g(r)$ exceeds unity at the first closely packed shell of neighbors and is zero at smaller radii because the neighbors cannot penetrate the origin atom. Note that $g(r)$ is an average. The ions are constantly moving about and interchanging neighbors like a milling crowd; $g(r)$ represents the average appearance of numerous instantaneous snapshots of the crowd. To think of the arrangement of ions in a liquid metal as a jumble of billiard balls in a bag is not far from the truth, and a hypothetical system of hard spherical atoms is called the "hard-sphere model" of a liquid. Because hard spheres are comparatively simple objects, it is possible to treat the hard-sphere model mathematically and to derive its thermodynamic properties by statistical mechanics and computing.

It is a somewhat remarkable fact that the structure and some of the thermodynamic properties of liquid metals can be represented quite well by those of the hard-sphere model provided that the diameter of the hypothetical spheres is selected by a theoretical process that takes the electron–ion and ion–ion interactions into account.

ELECTRONS IN DISORDERED MATTER

The free or conduction electrons in metals are of course not perfectly free—they interact with the ions. But when, as assumed in the preceding two sections, the electron–ion pseudopotential is weak, the electrons are nearly free, and to treat them as wholly so between collisions with the ions is an acceptable approximation. This is valid for many metals for which the division of their electrons into those bound in the ions and those that are free in the electron gas is clear; examples are the alkali metals, Ca, Al, Pb, and Bi. In transition metals of the iron group the situation is more obscure for the 3d shell of electrons; these interact strongly with the ions but are partially free as well. There are other circumstances, such as particular compositions of binary liquid alloys, where anomalous electrical and thermodynamical properties show clearly that strong electron–ion interactions render the free-electron concept dubious.

The obvious question then is: how do electrons behave under strong forces from the ions? For crystals the answer is available in the Bloch or band theory of solids (q.v.) and the regular order of the crystal structure is essential for the solution. In disordered liquids or glasses, the problem is much more difficult and apt to lead quite rapidly to mathematical technicalities and unanswered questions. The following difficulties are characteristic: (i) the actual ionic arrangement in liquids is unknown—the radial distribution function gives only limited information; (ii) even if a typical configuration were wholly known, it would remain difficult to solve Schrödinger's equation for the electron motion and to average the solution over all possible configurations of the moving ions; (iii) the multiple collisions of an electron with the ions make the simple electron diffraction theory of electrical resistance untenable and more powerful, but very complex theories have to be tried instead. By now there are several theoretical and computational attempts to deal with these problems and it is fairly clear that the electron energy spectrum in a molten metal will normally be similar to that of its crystal but with the sharpness removed from peaks and discontinuities, a general smoothing out of features, and a removal of any anisotropy.

In alloys, strong electron–ion interactions are an expression of binding forces and a conceptual approach that emphasizes chemical bonds and valency can be helpful in the absence of complete descriptions of the electron motion from physical first principles. For example, the extraordinarily low conductivity of the liquid alloys of special compositions Li_4Pb and CeAu—the latter actually being nonmetallic!—testify to the enduring challenge of dealing fully with the physics of liquid metals outside the weak pseudopotential regime. The same applies to liquid metals of which the density is greatly reduced by heating under pressure to high temperatures near the liquid–vapor critical point. This process induces a gradual disappearance of metallic properties.

In the late 1980s the phenomena mentioned in the preceding paragraph are at the frontiers of the subject. The emphasis now is less on describing liquid metals as such and more on understanding what conditions of density, temperature, and composition make a disordered material metallic, semiconducting, or insulating and how its electrons are behaving to cause these properties.

See also CONDUCTION; ELECTRON ENERGY STATES IN SOLIDS AND LIQUIDS; GLASSY METALS; LIQUID STRUCTURE; METALS; MOLTEN SALTS; SEMICONDUCTORS, AMORPHOUS.

BIBLIOGRAPHY

N. E. Cusack, *The Physics of Structurally Disordered Matter*. Adam Hilger, Bristol, 1987. (I, A)

T. E. Faber, *Theory of Liquid Metals*. Cambridge University Press, Cambridge, 1972. (I, A)

M. Shimoji, *Liquid Metals,* Academic Press, London, New York, San Francisco, 1977. (I, A)

Liquid Structure

B. Alder

VAN DER WAALS EQUATION

Over 100 years ago van der Waals proposed that a liquid could be distinguished from a gas by a simple modification of its equation of state. The equation of state constitutes the functional relation between the pressure p, volume v, and temperature T of a material, which for a perfect gas was then known to be $pv=RT$, where R is a universal constant. The modification that was required came about because the volume occupied by a liquid is considerably less than that of a gas. Hence the volume b taken up by the molecules themselves could no longer be ignored, and furthermore, when the molecules on the average are close together, they attract each other, so that the pressure required to contain the liquid is reduced. Thus, he derived

$$(p+a/v^2)(v-b)=RT,$$

where the constant a is a measure of the cohesive force between the molecules of the particular gas.

One of the great achievements of this formulation is that it led to what was then by no means generally accepted, namely, the continuity of the states of matter, which conceives of the liquid and gas phases as being part of a single model. The van der Waals equation is formally a cubic equation in the volume with three real roots below a certain temperature and only one above that temperature, called the critical temperature. The largest and smallest of the three volume roots at a given pressure and temperature describe the liquid in equilibrium with its vapor (gas), the coexistence curve, while the middle root has no physical meaning. Above the critical temperature the liquid cannot exist.

What van der Waals did not realize at the time was that his concept of the continuity of states of matter could be extended using his model to include the solid state. This can be demonstrated by rearranging his equation to

$$\frac{pv}{RT}=\frac{v}{v-b}-\frac{a}{vRT}$$

so that the $v/(v-b)$ term can be recognized as a purely geometrical factor, describing the packing of the particles without consideration of their attractive force. Such particles can be idealized as hard spheres and the $v/(v-b)$ term then represents a crude approximation to their equation of state. To this day there exists no adequate theory for the equation of state of hard spheres that accounts for their complex arrangements at all possible volumes; however, numerical simulation computer results have been obtained for this system. These results convincingly demonstrate that hard spheres for a range of volumes beyond close-packing arrange themselves in an ordered solid structure, beyond which they undergo melting. Thus the van der Waals equation, modified by the correct hard-sphere properties, also describes the melting curve, including the temperature where all three phases of matter coexist, called the triple point.

VIRIAL EXPANSION

Subsequent to the van der Waals model there have been many attempts to describe the liquid structure in terms of either a modified gas or a modified solid. None of these have been entirely successful, because both of these states of matter are too different from a liquid. The motivation for such models is primarily the availability of simple theories for the perfect gas and the harmonic solid, and the unavailability of any equivalent idealization of a liquid that leads to a simple, manageable mathematical theory. Thus the perfect-gas model was modified to account for the clustering of molecules leading to the so-called virial expansion. Each successive term in this expansion accounts for the properties of a cluster larger by one molecule than the previous one, the liquid being thought of as a gigantic cluster. The hope, not yet realized, was that one could derive some general properties of the infinite cluster relevant to liquids, or could at least prove the divergence of the virial series at the condensation point of the gas.

LATTICE MODELS

On the other hand, it was thought that an adequate description of a liquid could be obtained by introducing holes (missing molecules at lattice sites) into a perfect crystal to account for the larger volume and increased mobility of molecules in a liquid. There has never been the slightest bit of encouragement from either experimental or quantitative considerations that such holes exist. The larger volume available to the molecules in a liquid, over that in a solid, seems to be democratically shared by all the molecules. For example, the entropy of a liquid, which is a measure of its order, is such that it requires an unreasonable number of holes in view of the energetic requirements to create such holes. The mechanism of mobility in a liquid is not by a large jump of a molecule adjacent to a vacancy into the hole, as in a solid, but overwhelmingly by a large number of much smaller steps. Such lattice models impose both a local geometrical arrangement and a long-range order, known, from x-ray diffraction studies, to be absent from a liquid.

MELTING

Similarly, there is at present no clear evidence in the solid phase just prior to melting of any liquid-like behavior. The search for such premelting phenomena in the solid phase is motivated by an effort to try to understand the universal mechanism by which all solids melt or reach their stability limit. The only generally valid observation on when melting occurs was made long ago by Lindemann, who observed that solids melt at a temperature at which they are about 30% expanded from their volume at absolute zero or, equivalently, when the root mean square excursion of a particle from its lattice site exceeds about 10% of the interatomic distance. This geometrical criterion is quantitatively confirmed by the hard-sphere simulations. The universality of melting is then ascribed to the general similar nature of the steep repulsive forces between all atoms separated by close distances, which can be well approximated by hard spheres.

This leaves open, however, the question of the mechanism of instability leading to melting. A number of instability criteria for the solid phase, such as the disappearance of shear stress, have been investigated, but none has been shown to coincide with melting. In two-dimensional systems, constructed by polystyrene spheres confined to a plane, melting has been associated with the dissociation of dislocation pairs. By the same mechanism, helium films have been proposed to lose their superfluidity when vortex pairs start to dissociate. However, computer simulation of hard disks does not confirm such an instability criteria; it is rather a metastability criterion very close to melting. It is, of course, evident that at best only the least stable mode will coincide with melting, while all others will lead to metastable extensions of the solid phase into the liquid. The most conservative view is simply that melting occurs whenever the chemical potentials or Gibbs free energy of the solid and liquid are the same, and this need not be associated with an instability mechanism.

The melting curve exhibits a maximum whenever the liquid has a smaller volume than the solid, as the thermodynamic relation known as the Clausius–Clapeyron equation quantitatively demonstrates. The cause of the denser liquid is associated with the fewer constraints on the configurational arrangement imposed by the liquid structure than by the solid. The ice–water transition is a case in point; the ice crystal is constrained to a low-coordination-number (4), tetrahedral, regular structure by the highly directional interaction between the molecules, whereas in water the coordination number does not have to be exactly 4, and is in fact larger, and hence the fluid phase is denser.

Because the entropy of melting, $\Delta S/R$, has been observed to be of the order of 1 for many simple fluids, the notion that this entropy could be identified with the communal entropy, which is exactly 1, seemed attractive. The communal entropy derives from the difference between N particles, each confined to a volume per particle around a lattice site, and hence distinguishable, and these same N particles permitted to roam over the entire volume of the system, and hence indistinguishable, which leads to $N!$ additional identical configurations. A more detailed examination of the problem, however, leads to the conclusion that the entropy of melting is primarily due to the expansion of the system and only a small fraction can be identified as communal entropy.

CRITICAL POINT

The melting curve at high temperature and pressure can be predicted on quite general grounds to continue indefinitely; that is, there is no solid–fluid critical point, contrary to the gas–liquid coexistence curve. The argument, attributed to Landau, depends on the observation that if there were such a critical point, one could find paths in the phase diagram by which a substance could go continuously from a state of long-range order (solid) to one of short-range order (liquid), which is not allowable. On the other hand, this argument does not rule out the existence of a gas–liquid critical point, since only the degree of short-range order varies across it. This same argument eliminates critical points for phase transitions within the solid phase unless the two solids in equilibrium happen to have the same crystal structure. There is at least one such known case, cerium, where the atom in the two different phases has a different internal electronic structure.

The liquid–gas critical point has been found experimentally for all liquids, including liquid metals, not to be of the van der Waals type, that is the coexistence curve is not quadratic but closer to cubic. The universality of this behavior has been ingeniously and quantitatively described by the renormalization theory. For cesium the critical point coincides with the metal–insulator transition.

RADIAL DISTRIBUTION FUNCTION

Because the liquid does not have a predominant structure but a wide range of possible arrangements of the atoms, the only practical description is in terms of statistically averaged distributions. There have been many attempts, particularly for water, to pick out a few of the more stable configurations and represent the structure as an average over them weighted by their relative energy. Even for water, however, there are so many configurations that have nearly the same energy or probability of occurrence that the task becomes hopeless. The x-ray or neutron diffraction pictures obtained from liquids give no hint of any definite arrangements of atoms but yield a set of smeared-out bands of rings. From the Fourier transform of this intensity pattern the radial distribution function is experimentally determined. It quantitatively describes the structure of a liquid in terms of the probability of finding another particle at a given distance from a central particle. The rapidly damped oscillations of this function with the period of about a lattice spacing show that the short-range order in a liquid extends to about three neighboring shells, beyond which deviations from the average density are no longer discernible. The accurate calculation of this function is the central problem in the theory of liquids, since using this weighting function for the appropriate pairwise property allows the determination of any thermodynamic property. For example, the potential part of the internal energy is calculated as an average of the pair intermolecular potential. For the calculation of the pressure of a liquid under normal conditions this radial distribution function must be exceedingly accurately known, since the pressure is the result of the near cancellation of the kinetic pressure with the potential part of the pressure determined from the average of the derivative of the potential (virial).

Theoretical expressions for the radial distribution function describing the structure of a liquid are derived from the partition function. Mathematical transformation of this partition function leads to a hierarchy of equations that express the pair radial distribution function in terms of triplet distribution functions, which in turn are expressed in quadruplet distribution functions, etc., *ad infinitum*. To close this hierarchy, the higher-order distribution functions must be expressed in terms of lower-order ones. For example, the triplet distribution function, describing the probability of three atoms being separated by three given distances, is expressed as a triple product of pair distribution functions, each for a separation by one of the three distances. This neglect of triplet correlations is called the pairwise superposition approximation and leads to a semiquantitative theory of the pair radial distribution function. Different closure approximations, but physically less transparent, have led to more quantitative predictions.

Most of the foregoing theories for the radial distribution are mathematically quite complex and of uncertain accuracy, so that the need for an alternative method led to the development of a computer simulation method—the so-called Monte Carlo method. As applied to hard spheres, for example, the particles are placed initially in some nonoverlapping configuration in a box and then randomly displaced, one at a time. Upon displacement, a check is made whether in the new position the sphere is closer than a diameter to any other sphere (overlaps). If yes, the displacement is rejected and the particle is placed in its former position. If no overlap occurs, a new configuration has been generated and the process is repeated. Averaging over a sufficient number of such configurations leads, via the evaluation of the radial distribution function, to an exact calculation of any thermodynamically required average.

PERTURBATION ON STRUCTURE

These computer calculations, and particularly those for hard spheres, have led to a great deal of insight into the structure of simple liquids and a greater understanding of the validity of the van der Waals equation. It can be shown that the addition of a weak and long-range attractive potential to the hard-sphere potential leads rigorously to the van der Waals equation. Since this attractive potential is unrealistic, another way to obtain the van der Waals equation is to add a short-range attractive potential to the hard-sphere potential as a perturbation. But then it is necessary to show that the higher-order perturbation terms are negligible. The first-order perturbation term represents the mean number of hard-sphere particles within the range of the attractive forces and leads rigorously to the van der Waals equation, while the higher-order perturbation terms represent higher-order fluctuations in the number of particles about the mean.

If the added attractive potential is sufficiently long range or so weak that first-order perturbation suffices, the attractive potential does not change the structure of the fluid from that of pure hard spheres. This is a good approximation for a normal liquid, since the particles are so closely packed that they rarely escape outside the range of their attractive forces. The validity of the van der Waals equation at much lower density can be traced to the structural observation that the fluctuations of the number of particles within the range of the attractive forces does not change significantly with density over the entire fluid density range. For this reason the higher-order perturbation terms can be neglected.

COMPLEX LIQUID MIXTURES

Under this topic quantum fluids such as helium, metals, liquid crystals, fluids made up of molecules as big as polymers and microemulsions are considered. All these topics are under active investigation but much progress has been made already, primarily via computer simulations. For helium a Monte Carlo path integral simulation has succeeded in describing the thermodynamic properties of the so-called λ transition between the normal and superfluid phase. For metals, the electrons must be dealt with quantum mechanically as well; however, since they are fermions instead of bosons, as in helium, a stable algorithm has not as yet been developed. Good results are, however, obtained by treating the electrons approximately by the density functional theory, used so successfully in solid-state band theory, and by treating the ion cores classically.

The notion that the formation of liquid crystals required anisotropic attractive potentials between the molecules was dispelled by a computer simulation of hard ellipsoids. In that simulation a phase transition from an isotropic liquid phase to a nematic phase in which the ellipsoids are only rotationally but not translationally ordered was found. It is believed that most of the many possible other rotationally ordered phases, such as a smectic phase, can be found for particles with other purely repulsive anisotropic potentials. On the other hand, the details of the attractive forces appear to be essential to the formation of lamellar phases from microemulsions in such mixtures as oil, water, and soaps. Such systems by suitably adjusting concentrations, temperature, and attractive potentials assume layer-like structures with alternate layers of water and oil with soap layers sandwiched between each interface. They are modeled by a soft ellipsoid (the soap molecules) which attracts water molecules on one end and oil molecules on the other, each represented by soft spheres. Lattice model versions of this system have been able to analytically predict this complex phase diagram qualitatively. Simpler mixtures of molecules have been successfully described by the van der Waals theory with appropriate values of a and b for each species. Statistical mechanical theories of simpler mixtures more realistically represent each molecule by a spherical soft potential originating at the site of each atom in the molecule, leading to what is called the interaction site potential. When dealing with polymers, that is, a highly viscous system of entangled very long molecules, the simulation procedure uses the reptation process to sample more efficiently the possible configurations of the polymer in the melt. Reptation, as the name implies, is a process by which a snake moves, namely, the head moves forward as the tail contracts.

TRANSPORT PROPERTIES

Computer simulation of hard spheres has shown that the fundamental assumption of molecular chaos by which collision becomes eventually uncorrelated is not valid at any finite density. Thus, strictly speaking, the Boltzmann transport equation is not valid although the quantitative corrections are exceedingly small. The persistence of velocity between successive collisions is caused by the generation of hydrodynamic vortex modes in which the momentum of a particle is fed back into its motion by the vortex it created in the medium. The phenomenon is well understood by graph theoretical means as well as mode–mode coupling (hydrodynamic) arguments. The latter lead to the unexpected conclusion that hydrodynamics is quantitatively valid on a microscopic scale, that is, on a few-molecular-diameters and a few-collision-times scale. The slow decay of the stress is not of the same origin as that of the velocity. That slow decay is not, at present, well understood, but leads to the rapid increase of the viscosity near melting and is the key to the understanding of glass formation.

The structure and dynamics of liquids is hence primarily determined by the steep repulsive forces. The attractive forces, whose range is typically about half that of the repulsive forces, do not alter the structure significantly. The hard-core computer simulation results, unfortunately in numerical form, are hence accurate idealizations of the liquid structure, even for metals and liquid crystals.

See also Equations of State; Fluid Physics; Monte Carlo Techniques; Order-Disorder Phenomena.

BIBLIOGRAPHY

J. P. Hansen and I. R. McDonald, *Theory of Simple Liquids*. Academic Press, London, 1986.

J. S. Rowlinsen and F. L. Swinton, *Liquids and Liquid Mixtures*. Butterworth, London, 1982.

Lorentz Transformations

Allen E. Everett

The Lorentz transformation equations give the relations between the space and time coordinates of a single event as measured by observers in two different inertial reference frames in motion relative to one another. Let S and S' be the two reference frames, and v be their relative speed. We can think of a "reference frame" as being a rigid framework of measuring rods (e.g., meter sticks), which allow us to determine the spatial coordinates of any point in space relative to some origin, together with a set of identical clocks distributed throughout space and attached to the measuring rods. The clocks must be properly sychronized so that observers moving with the reference frame agree that all of the clocks show the same time, even though they are at different locations. An inertial frame may be defined, somewhat loosely, as one in which the stars have, on the average, a constant velocity. Hence, given one such frame, the set of all inertial frames consists of all possible reference frames in uniform linear motion relative to the original one. For simplicity we will suppose that the coordinate axes in S and S' are chosen to be parallel to one another, and that the velocity of the origin of S' as seen from S is along the positive x axis. Furthermore we suppose that the observers in the two reference frames agree to set their clocks in such a way that, at the instant the origins of the frames coincide, the clocks located at their origins both read zero. Let us suppose that some event, for example the detection of some particle or of a light pulse, occurs at the point with spatial coordinates x, y, z in S at a time t as measured on the clock fixed in S at the point where the event occurs. The Lorentz transformation equations giving the coordinates x', y', z', and t' of the same event in S' are

$$x' = (x - vt)\,(1 - v^2/c^2)^{-1/2}, \tag{1a}$$

$$y' = y, \tag{1b}$$

$$z' = z, \tag{1c}$$

$$t' = (t - vx/c^2)(1 - v^2/c^2)^{-1/2}, \tag{1d}$$

where c is the speed of light. If we solve this set of equations for the unprimed quantities in terms of the primed, we obtain the identical set of equations except that v is replaced by $-v$. Hence, S and S' enter the transformation equations on an equal footing, the only difference arising from the fact that the velocities of S' relative to S and S relative to S' are the negatives of one another.

These transformation equations have a crucial property. It follows from them by simple algebra that

$$x'^2 + y'^2 + z'^2 - c^2t'^2 = x^2 + y^2 + z^2 - c^2t^2. \tag{2}$$

Suppose that the event in question is the detection of a light pulse that was emitted at the origin at $t = t' = 0$. Then, if the speed of light in S is c, we will have $x^2 + y^2 + z^2 - c^2t^2 = 0$. But it then follows from Eq. (2) that the light pulse will also be observed to have speed c in S'. Thus the Lorentz transformations imply that the speed of light is the same in all inertial frames, in accordance with the experimental results of the Michelson–Morley experiment and the basic assumptions of the special theory of relativity. The Lorentz transformation equations (1) in fact follow from the requirement that Eq. (2) be satisfied and the assumptions that the form of the transformations is independent of the choice of origin of the coordinate system, and that all directions in space are equivalent.

The quantities x, y, z, and t are said to be the components of a 4-vector. There are other examples of 4-vectors, whose components by definition transform according to Eqs. (1). An important case is the 4-momentum, with x, y, z, and t replaced by p_x, p_y, p_z, and E/c^2, where **p** and E are momentum and energy. The form of the relativistic expressions for E and **p** is in fact determined by the requirement that the components of the 4-momentum transform between inertial frames by Eqs. (1).

The unusual and unintuitive properties of the Lorentz transformations, such as that the time of an event is different in different reference frames, result from the experimentally based requirement that the transformations leave the speed of light unchanged. These properties become important only for speeds comparable to c. For $v \ll c$, Eqs. (1a) and (1d) reduce to the simpler equations $x' = x - vt$, $t' = t$, which are called the Galilean transformations and which yield predictions in agreement with everyday experience, based on the behavior of objects with speeds small compared to c.

See also MICHELSON–MORELY EXPERIMENT; RELATIVITY, SPECIAL THEORY; TWIN PARADOX.

BIBLIOGRAPHY

P. G. Bergmann, *Introduction to the Theory of Relativity*. Prentice-Hall, Englewood Cliffs, NJ, 1942. (I)

M. Born, *Einstein's Theory of Relativity*. Dover, New York, 1962. (E)

A. Einstein, *Relativity, the Special and the General Theory* (translated by R. W. Lawson). Crown, New York, 1961. (E)

A. P. French, *Special Relativity*. Norton, New York, 1968. (E)

H. Goldstein, *Classical Mechanics*, 2nd ed. Addison-Wesley, Reading, Mass., 1980. (I)

W. Pauli, *Theory of Relativity* (translated by G. Field). Pergamon, New York, 1958. (A)

J. L. Synge, *Relativity: The Special Theory*. North-Holland, Amsterdam, 1965. (A)

Low-Energy Electron Diffraction (LEED)

C. B. Duke

Low-energy electron diffraction (LEED) is the coherent reflection of electrons in the energy range $5 \lesssim E \lesssim 500$ eV from the uppermost few atomic layers of crystalline solids. A schematic diagram of an apparatus for measuring the configuration and intensities of the diffracted beams is presented in Fig. 1. Diffraction occurs because the wavelength λ (in angstroms) is related to the accelerating voltage V (in volts) via

$$\lambda = 12.2638/V^{1/2}, \tag{1}$$

leading to λ of the order of atomic dimensions for V in the "low-energy" range. The surface sensitivity of the diffraction process arises from the short inelastic-collision mean free paths, $5 \lesssim \lambda_{ee} \lesssim 10$ Å, for substantial energy losses ($w \gtrsim 5$ eV) by these low-energy electrons. Since most electron detectors exhibit an energy resolution of $\Delta E \lesssim 0.5$ eV, the detection only of elastically scattered electrons effectively discriminates against reflected electrons emanating from depths greater than λ_{ee}. The detection of electrons that have experienced only a single energy-loss process also measures the consequences of scattering events occurring within distances of about λ_{ee} of the surface.

The nature of electron diffraction from crystals can be understood most simply by regarding a crystal as composed

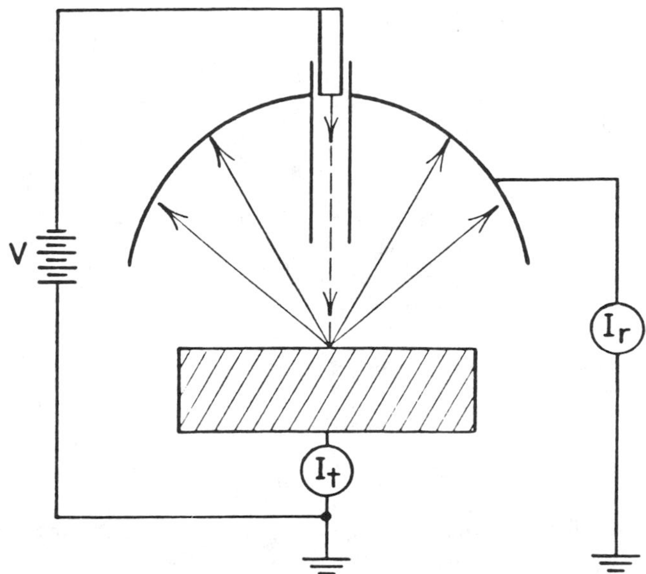

FIG. 1. Schematic diagram of a low-energy electron diffraction apparatus (i.e., $V \sim 100$ V) illustrating that the current drawn from the gun, $I = I_t + I_r$, either passes through the target to ground (I_t) or is reflected back from the target (I_r). [After C. B. Duke, *Adv. Chem. Phys.* **27**, 1 (1974).]

of geometrically equivalent layers of atoms parallel to a given surface. An incident electron of wave vector **k** (i.e., momentum $\mathbf{p} = \hbar\mathbf{k}$) is diffracted from the array of these two-dimensional gratings stacked together to form the crystal. The periodic translational symmetry of these diffraction gratings is manifested in the electron–solid scattering cross sections by virtue of the reflected electrons emerging from the solid in a series of diffracted beams as indicated in Fig. 2. The configuration of the diffracted beams relative to the beam of incident electrons may be predicted from the translational symmetry of the system, which leads to the momentum conservation law

$$\mathbf{K}'_{\parallel} = \mathbf{k}_{\parallel} + \mathbf{g}, \tag{2a}$$

$$k_{\parallel} = (2mE/\hbar^2)^{1/2} \sin\theta , \tag{2b}$$

for electrons of energy E incident on a planar surface at an angle θ relative to its exterior normal. The vectors **g** are the reciprocal lattice vectors of the two-dimensional atomic diffraction gratings, as indicated in Fig. 2. The only information conveyed by the configuration of the diffracted beams is the translational symmetry parallel to the surface of the two-dimensional atomic diffraction gratings that make up the solid. Determination of the unit-cell structure within each of these gratings and of the packing sequence of the gratings relative to each other requires analysis of the diffracted intensities.

The observation in 1927 of diffracted beams corresponding to the parallel-momentum conservation law (Fig. 2) was one of the original demonstrations of the wave nature of the electron, a fact that led to the recognition in 1937 of Clinton Davisson as a Nobel laureate. The potential for the use of LEED as a technique for surface crystallography was recognized in Davisson and Germer's original 1927 paper. This application of LEED was not achieved, however, until the early 1970s. The reason for this 40-year delay between recognition and achievement lies in the nature of the electron–solid interaction. The fact that it is strong (i.e., $\lambda_{ee} \sim 5$ Å) is necessary for LEED to be a surface-sensitive spectroscopy. Because it is strong, however, a single-scattering (Born-approximation) calculation, patterned on the analysis of x-ray diffraction from bulk solids, does not suffice to describe the scattered intensities. Since analysis of these intensities is required for surface crystallography, the accomplishment of the latter awaited the construction of an adequate renormalized quantum field theory of the electron-diffraction process that properly describes the occurrence of multiple elastic and inelastic electron-scattering events. Such theories were forthcoming during 1969–1971. Surface crystallography via LEED intensity analysis subsequently became routine for low-index faces of clean metals and semiconductors, as well as for ordered adsorbed overlayers, when computer programs based on these theories became fast and inexpensive on modern supercomputers. The structures thereby attained have led to many new results in surface physics including the discoveries that the top layer spacing on most metal surfaces is contracted, that semiconductor surfaces exhibit new types of chemical bonding that exhibit neither molecular nor bulk analogs, and that surface relaxations of atomic layer spacings exhibit an oscillatory behavior extending several layers inward for both metals and semiconductors.

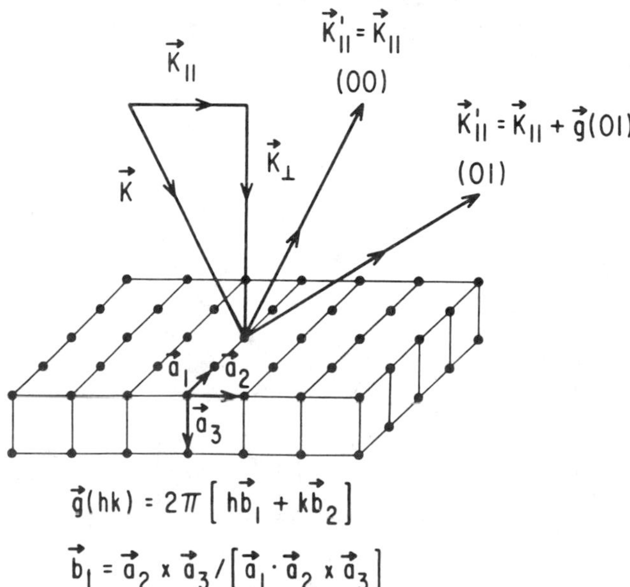

FIG. 2. Schematic illustration of an incident electron beam of wave vector $\mathbf{k} = \mathbf{k}_{\perp} + \mathbf{k}_{\parallel}$, scattered elastically from a single crystal into a state characterized by the wave vector $\mathbf{k}' = \mathbf{k}_{\parallel} + \mathbf{k}_{\perp}'$. The construction of the reciprocal lattice associated with the single-crystal surface is also shown. The vectors $\mathbf{g}(hk)$ designate the reciprocal lattice vectors associated with the lowest-symmetry Bravais net parallel to the surface. [After C. B. Duke, *Adv. Chem. Phys.* **27,** 1 (1974).]

That inelastically scattered electrons as well as elastically scattered ones emerge from the surfaces of crystalline solids in a series of beams was also noted in the original Davisson and Germer paper. The application of inelastic low-energy electron diffraction ("ILEED") for quantitative surface characterization awaited the construction in 1971 by Duke and Laramore of a complete theory of the inelastic diffraction process. Analyses of ILEED intensities from the low-index faces of clean metals have been utilized to determine the energy–momentum relation of electronic surface excitations (e.g., surface plasma oscillations).

Measurements of the configuration and intensity of LEED beams are carried out under ultrahigh-vacuum conditions ($p \lesssim 10^{-8}$ N/m^2) in order to prevent contamination of the surface during the experiment. A typical apparatus consists of the vacuum chamber and pumps, electron gun, electron detector, and goniometer (i.e., a sample holder permitting precise control of the orientation and temperature of the specimen). Most LEED measurements are performed, however, in conjunction with other experiments which determine the elemental composition and electronic structure of the surface under examination. Thus, LEED is best viewed as one of a number of spectroscopic techniques utilized in concert to characterize solid surfaces. Specifically, LEED, ion scattering, and scanning tunneling microscopy are the primary techniques for the determination of the atomic geometries of crystalline surfaces and ordered overlayers thereon. ILEED is only one of many techniques for the assessment of the electronic structure of surfaces, the most prominent of which is angle-resolved photoemission spectroscopy.

See also CRYSTALLOGRAPHY, X-RAY; DIFFRACTION; ELECTRON DIFFRACTION; SCANNING TUNNELING MICROSCOPY.

BIBLIOGRAPHY

History and General Usage

C. B. Duke, *Crit. Rev. Solid State Sci.* **4,** 541 (1974). (E)
C. B. Duke, *Appl. Surf. Sci.* **11/12,** 1 (1982). (E)
P. M. Marcus and F. Jona, *Appl. Surf. Sci.* **11/12,** 20 (1982). (E)
G. Thompson, *Contemp. Phys.* **9,** 1 (1968). (E)

Electron–Solid Interactions

C. B. Duke, *Crit. Rev. Solid State Sci.* **4,** 371 (1974). (E)

LEED Theory and Applications to Surface Crystallography

C. B. Duke, *Adv. Chem. Phys.* **27,** 1 (1974). (I)
C. B. Duke, in *The Chemical Physics of Solid Surfaces and Heterogeneous Catalysis* (D. A. King and D. P. Woodruff, eds.), Vol. 5, pp. 69–118. Elsevier, Amsterdam, 1988. (I)
A. Kahn, *Surf. Sci. Rept.* **3,** 193 (1983). (I)
J. B. Pendry, *Low Energy Electron Diffraction Theory and Its Application to Determination of Surface Structure.* Academic Press, London, 1974. (A)
M. A. Van Hove and S. Y. Tong, *Surface Crystallography by LEED.* Springer-Verlag, Berlin, 1979. (A)

Surface Resonances in LEED

R. O. Jones and P. J. Jennings, *Surf. Sci. Rept.* **9,** 165 (1988). (I)

Role of LEED in Surface Science

F. W. de Wette (ed.), *Solvay Conference on Surface Science.* Springer-Verlag, Berlin, 1988. (A)
M. A. Van Hove and S. Y. Tong (eds.), *The Structure of Surfaces.* Springer-Verlag, Berlin, 1985. (A)

Luminescence (Fluorescence and Phosphorescence)

Arthur M. Halpern

Luminescence is the spontaneous emission of optical radiation (infrared, visible, or ultraviolet light) by matter. This phenomenon is to be distinguished from incandescence, which is the emission of radiation by a substance by virtue of its being at a high temperature (> 500°C) (cf. BLACKBODY RADIATION). Luminescence can occur in a wide variety of matter and under many different circumstances. Thus, atoms; organic, inorganic, or organometallic molecules; polymers; organic or inorganic crystals; and amorphous substances all luminesce under appropriate conditions. The most straightforward way in which luminescence can be produced is by the application of nonionizing radiation (visible or ultraviolet light) to some form of matter. For example, a solution of the organic dye rhodamine-B emits orange light if it is irradiated with green light (or shorter-wavelength light). Many papers, fabrics, and laundry detergents contain "optical brighteners" which emit blue or purple light when exposed to ultraviolet radiation. The production of light in this manner is called *photoluminescence*.

Other common types of luminescence and their modes of production are as follows:

Radioluminescence (or scintillation) is produced by ionizing radiation. Some polymers contain organic molecules which emit visible light when exposed to such radiation as x rays, γ rays, or cosmic rays, and thus act as detectors for high-energy radiation. In many biomedical applications scintillation counters are used. The scintillators are organic molecules that emit visible light when they are excited by the decay products of certain radioactive isotopes such as ^{3}H (tritium), ^{14}C, and ^{18}O.

Electroluminescence is produced by electric fields or plasmas. An example is the gas-discharge tube. These are lamps that are filled with a gas such as neon, argon, xenon, nitrogen, etc., or the vapor of metals such as mercury or sodium. The lamps contain electrodes, which are connected to an electrical power supply. The applied voltage is high enough to ionize the atoms and produce a low-pressure plasma, or ionic (current-carrying) gas. The recombination of metal ions and electrons in the plasma leads to the production of electronically excited states of the atoms and hence the emission of light, usually in the form of very narrow lines throughout the electromagnetic spectrum. The common fluorescence lamp is a discharge tube in which mercury atoms are electrically excited and then transfer this energy to a material (phosphor), which coats the inside surface of the tube. The phosphor, after being excited via collisions with fast-moving electrons, emits most of the observed light. Another example is lightning in which the static electricity generated by a moving air mass ionizes or removes electrons from the nitrogen and oxygen molecules. Charge recombination produces energized molecules, and nitrogen, formed in such a state, emits radiation that can be seen as visible light. Another example of electroluminescence is the familiar light-emitting diode.

Chemiluminescence is produced as a result of a chemical reaction usually involving an oxidation–reduction process, in which, simply viewed, electronic charge is transferred from one species to another. One common example is the so-called luminol reaction. When luminol, an amino-substituted phthalazine derivative, is treated with base and an oxidizing agent, nitrogen molecules are produced along with intense blue light. This light comes from the excited singlet state of a product, the aminophthalate molecule. In this example, the light emitted represents the release of energy from the reaction.

Bioluminescence is the result of certain oxidation processes (usually enzyme-catalyzed) in biological systems. The firefly and its larva, the glowworm, are well-studied examples; light emission for the firefly serves as a mating signal. These organisms produce a molecule called luciferin and an enzyme known as luciferase. When these molecules are in the presence of oxygen and adenosine phosphate, yellow-green light is produced with very high efficiency. Many other organisms such as bacteria, fungi, and certain marine species are bioluminescent. It is thought that chemiluminescence is a now-defunct evolutionary mechanism that was developed by some anaerobic organisms for the disposal of oxygen.

Thermoluminescence is visible light produced (usually after heating) by a substance initially irradiated by some other means (e.g., optical, x ray, or γ ray).

Although there is a very wide range of circumstances in

which luminescence is observed, there is one common aspect to this phenomenon. When a luminescent atom or molecule emits a photon (a quantum of light), there is a change in its electronic structure. When electrons are confined to a region of space which is defined by the positions of the atomic centers of a molecule (or just the nucleus, in the case of an atom), a set of discrete energy levels is imposed on the system. The way in which the electrons fill up, or occupy, these levels (or orbitals) determines the physical and chemical properties of the system. Thus, the electronic structure of a molecule or atom is a representation of how the electrons are arranged in the orbitals.

For example, the ground state of a system corresponds to the situation in which the electrons occupy the lowest possible energy levels. The Pauli exclusion principle requires that not more than two electrons occupy the same orbital. Electronic excited states refer to other arrangements in which the electrons occupy the orbitals in such a way that the total energy is higher than the ground state. In this context, then, luminescence is the release of energy via photon emission and is a consequence of a transition from an electronically excited state to some other, lower-energy state, usually the ground state.

Because the electron possesses an intrinsic spin angular momentum, the picture describing electronic states is more complicated. It is found experimentally that when an electron is confined to an atom or molecule (or when it is under the influence of a magnetic field), its spin angular momentum is quantized and can thus have only one of *two* possible values; e.g., + or − $h/2$ (where h is Planck's constant, a universal quantity). The + or − value indicates the sense of electron spin, or rotation, i.e., clockwise or counterclockwise. If two electrons populate the same orbital, they must have different spin values; in such a case, the electrons are said to be paired. In a representation of the electronic structure of an excited state, one of the electrons is "promoted" to occupy a higher (usually vacant) energy level. Thus the excited state is depicted as having two half-filled orbitals: the vacated, half-populated lower one, and the half-filled upper one. The electrons in these half-filled orbitals can have opposite values (or spin quantum numbers), or the same ones; the spin quantum numbers of these unpaired electrons are no longer restricted by the Pauli exclusion principle. The former situation is referred to as a *singlet* state, while the latter denotes a *triplet* state. These terms stem from the observed behavior of atomic or molecular spectra under the influence of a magnetic field. This is an important distinction because the magnetic quantum numbers associated with the unpaired electrons greatly affect the characteristics of the luminescence that is produced by the excited states.

It is observed experimentally that a triplet state has a lower total energy relative to the corresponding singlet state. The reason for this state ordering is that in a triplet state, the two unpaired electrons are prevented from being too close to each other, and this space restriction reduces the electrostatic repulsion between them. Moreover, it is known that in the presence of a magnetic field, the triplet state splits into three sublevels whose separation is proportional to the strength of the magnetic field; hence the term "triplet" state.

Thus in cases where spin quantum numbers play an important role in determining state energies, one may distinguish between two different types of luminescence: *fluorescence*, in which emission occurs with no net change in spin quantum numbers, and *phosphorescence*, in which emission is accompanied by a change in spin quantum numbers. These distinctions hold true for species possessing "light" atoms, i.e., atoms derived from the first or second row of the periodic table. For "heavy" atoms (e.g., ≥ third row), or molecules containing such atoms, one may often speak of emission in a general sense as "luminescence." The relative energies of the ground state and various excited singlet and triplet states are schematically portrayed in Fig. 1. In this diagram, each state is also represented by an orbital-energy-level scheme which is filled with (in this example) four electrons. The upward and downward arrows symbolize the different spin quantum numbers of the electron.

Because most organic molecules are ground-state singlet systems (all electrons paired), the direct production of an excited singlet state (via photon absorption) is considered to be a spin-allowed transition. In contrast, the direct production of a triplet state is a spin-forbidden transition because during such a process there would have to be a net change in the spin angular momentum of the system. This has the consequence that singlet–singlet transitions occur with much higher probability relative to singlet–triplet transitions. For organic molecules containing "light atoms" such as H, C,

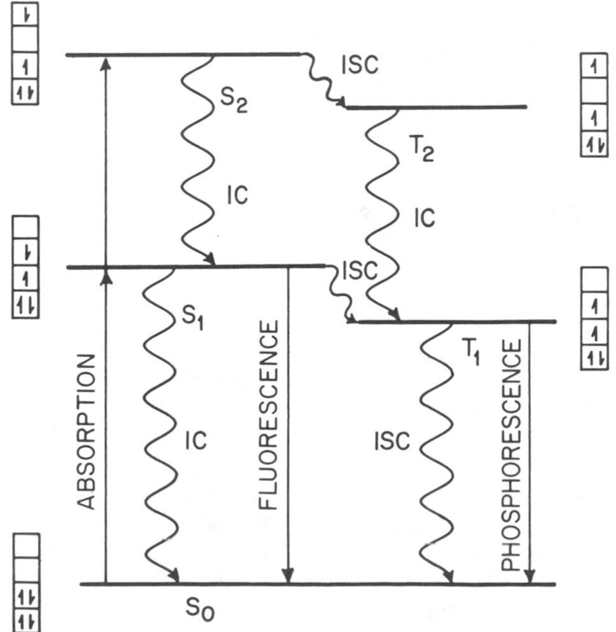

FIG. 1. Schematic electronic-state diagram for a four-electron system. S_0, S_1, and S_2 refer to the ground, first, and second excited singlet states, respectively, while T_1 and T_2 denote the first and second excited triplet states. Alongside each of the electronic state positions is an orbital diagram in which the direction of the arrow symbolizes the electron's spin value. The straight arrows connecting states describe radiative transitions. For example, S_1 or S_2 can be reached from S_0 via photon absorption, and the ground state is produced from S_1 or T_1 by photon emission. The wavy arrows portray radiationless transitions: internal conversion (IC), or intersystem crossing (ISC).

N, O, S, P, and others, the ratio of a singlet–singlet *vis-à-vis* singlet–triplet transition probability is typically 10^8.

Thus, in photoluminescence, an excited singlet state (e.g., S_1 or S_2 in Fig. 1) can be directly produced by photon absorption. This excited state can then undergo what are called radiationless transitions. Through these processes, energy initially deposited in the molecule is dissipated by various means, and thus electronic energy is converted into vibrational (i.e., thermal) energy, or chemical energy (if bonds are broken or rearranged). The spin-forbidden radiationless transition (i.e., singlet → triplet or triplet → singlet) is often called *intersystem crossing* (ISC), and the spin-allowed transition (singlet → singlet or triplet → triplet) is referred to as *internal conversion* (IC) (see Fig. 1). It is generally observed that the rate of a radiationless process decreases sharply as the energy gap between the two electronic states increases. Because for most molecules, the energy differences between the *excited* states are much larger than the gap between the lowest excited state and the ground state, the lowest excited state is usually formed with very high efficiency after optical excitation into any of the higher states. Thus, with very few exceptions, fluorescence and phosphorescence take place from the lowest excited singlet and triplet states (S_1 and T_1), respectively.

For most inorganic molecules or organic molecules that contain "heavy atoms" (e.g., bromine and iodine atoms, metal atoms or ions, etc.) the spin-forbidden processes are much less restrictive, and luminescence is frequently called "fluorescence" even if it is associated with states with different spin quantum numbers. In such cases, the emission is much shorter lived than the typical phosphorescence lifetime of a molecule containing "light atoms." In this sense, then, the distinction between fluorescence and phosphorescence is made on the basis of the luminescence lifetime, which is determined from the rate with which the intensity of emitted light is diminished after a sample is photoexcited with a very short-lived pulse of light. Technically, the luminescence lifetime is the time required for the emission intensity to fall to 36.788% ($1/e$, where e is the base of the natural logarithm) of its initial value after pulse excitation.

For most organic molecules, however, the phosphorescence lifetimes are generally much longer than fluorescence lifetimes (see above). Because the triplet state is so long-lived, it has a high probability of undergoing a reaction by which excitation energy is dissipated by some means. If this process occurs as a result of an interaction (e.g., via a collision) with another molecule, it is called *quenching*. One interesting and important consequence of the quenching of a triplet state by another molecule is energy transfer in which the excitation energy is "picked up" by the quencher (the energy acceptor), The quencher, now enriched, can undergo some chemical or physical reaction just as it would if it were directly excited. If energy transfer takes place in a system, the donor luminescence is diminished (because it is quenched) and in some cases the acceptor emits light. This process is called *sensitized* luminescence and can be observed in singlet–singlet or in triplet–triplet energy transfer (i.e., sensitized fluorescence or phosphorescence, respectively). In certain systems where there is a high concentration of molecules, such as in crystals, excitation energy can "migrate" over fairly large distances before it is dissipated. If the crystal contains an impurity (or a site defect) having a lower excited-state energy, the impurity acts as an energy trap, and it may luminesce with its own characteristics.

A common mode of quenching of triplet states is the interaction with molecular oxygen. This process may take place with high efficiency not only because the triplet state is intrinsically long-lived, but also because in the ground state, the oxygen molecule has two unpaired electrons (triplet). The consequence of this is that the overall quenching reaction is a spin-allowed process. In some cases, this has been shown in the production of electronically excited (singlet state) oxygen. This energy-enriched, long-lived form of molecular oxygen can act as a powerful oxidizing agent toward some suitable substrate. The application of light in the presence of oxygen is known to result in the oxidative denaturation of biological systems. This is called photodynamic action and is known to involve the excited triplet state as a sensitizer.

Triplet states also have a high reactivity toward other molecules that might be present in very small amounts (such as impurities). For this reason, phosphorescence in fluid media is usually very weak because of the high mobility of the molecules and the correspondingly high probability of an encounter between the triplet state and a quencher. The phosphorescence intensity of a system can be significantly enhanced by cooling the system so that the bulk solvent viscosity increases. This results in a decrease in the encounter probability, the (excited) triplet state and a potential quencher. Alternatively, phosphorescence may be observed at room temperature if the phosphorescent molecules are dissolved in an immobilizing medium, such as a rigid polymeric "glass," a micelle, or some other type of isolating matrix.

See also Atomic Spectroscopy; Chemiluminescence; Electroluminescence; Thermoluminescence.

BIBLIOGRAPHY

R. S. Becker, *Theory and Interpretation of Fluorescence and Phosphorescence*. Wiley–Interscience, New York, 1969. (I)

J. B. Birks, *Photophysics of Aromatic Molecules*. Wiley–Interscience, London, 1970. (A)

M. D. Lumb (ed.), *Luminescence Spectroscopy*. Academic Press, London, 1978. (A)

C. A. Parker, *Photoluminescence in Solutions*. Elsevier, Amsterdam, 1968. (I)

P. G. Seybold, "Luminescence," *Chemistry* **42,** 6 (1973). (E)

G. H. Schenk, *Absorption of Light and Ultraviolet Radiation*. Allyn & Bacon, Boston, 1973. (E)

N. J. Turro, "The Triplet State," *J. Chem. Educ.* **46,** 2 (1969). (E)

F. Williams (ed.), *Luminescence of Crystals, Molecules, and Solutions*. Plenum Press, New York, 1973. (A)

Mach's Principle

Hubert F. Goenner

The Newtonian laws of motion are valid only in inertial frames of reference (IF). Newton tried to explain the privileged status of IF's by the concepts of absolute space and time. Bodies moving unaccelerated with respect to absolute space constitute (part of) an IF. Acceleration with regard to absolute space is recognized by the effects of inertial (e.g., centrifugal or Coriolis) forces. An IF is realized by the axes of a gyroscope or the plane of a Foucault pendulum. It so happens that such IF's are at rest, approximately, with respect to the fixed stars.

Mach criticized Newton's concept of motion with respect to absolute space for being unobservable. He replaced it by relative motion with regard to the fixed stars, which at his time represented the bulk of matter in the universe. Thus he pointed out a material system in place of a mathematical construct and suggested a change in the role of the fixed stars from that of marker-points for a frame of reference to that of participants in an interacting physical system. Mach claimed that the motion of each individual body is determined by all other bodies in the universe.

Einstein seems to have coined the name Mach's principle (MP) in context with his general theory of relativity (GR). For him it expressed the idea that the metric is determined entirely by gravitating matter. Nowadays, each connection between the local dynamics of matter and the global structure of the universe is considered an expression of MP, especially the determination of local IF's through the gravitational action of distant matter (stars, systems of nebulas, cosmos). Effects linking the inertial mass of a body with proximate or remote matter or predicting rotating matter to generate (gravitationally) Coriolis and centrifugal fields are called Machian.

Because of the empirically found proportionality of inertial mass and gravitational mass, a homogeneous gravitational field may be simulated by a field of inertial forces induced by use of an accelerated frame. Through this fact and his principle of equivalence Einstein abolished the preferred role of IF's. In GR, the same quantity, i.e., the metric of space-time, is used to describe both gravitation and inertia. Nevertheless, gravitation stands out by being linked to the curvature of space-time. To the extent that the motion of bodies is governed by the metric (geodesic equation), which itself depends, via the field equations of GR, on energy and momentum of matter, the local dynamics of a body is influenced by the matter elsewhere. Therefore, GR is considered by some to express MP. GR also predicts the dragging along of the plane of the Foucault pendulum in the sense of the earth's rotation. It has not yet been possible to test this tiny effect. Moreover, GR neither incorporates the relativity of all motion nor provides a unique formulation of what the inertia of a body is.

MP has encouraged some to derive cosmological facts from local phenomena and to invent new theories of gravitation. To others, the transformation of the inertial mass, by MP, from an intrinsic quantity into one reflecting the gravitational interaction of matter in universe means an unwanted complication of physics.

See also MASS; RELATIVITY, GENERAL THEORY.

BIBLIOGRAPHY

R. S. Cohen and R. J. Seeger, eds., *Ernst Mach, Physicist and Philosopher* (Boston Studies in the Philosophy of Science, Vol. 6). Reidel, Dordrecht, 1970. (A)

R. H. Dicke, "The Many Faces of Mach," in *Gravitation and Relativity* (H. Y. Chiu and W. F. Hoffmann, eds.). Benjamin, New York, 1964. (A)

A. Einstein, *The Meaning of Relativity*. Princeton Univ. Press, Princeton, N.J., 1950. (E)

H. F. Goenner, "Machsches Prinzip und Theorien der Gravitation," in *Grundlagenprobleme der modernen Physik* (J. Nitsch et al., eds.). B.I.-Wissenschaftsverlag, Mannheim, 1981. (I)

D. W. Sciama, *The Unity of the Universe*. Faber and Faber, London, 1959. (E)

J. A. Wheeler, "Mach's Principle as Boundary Condition," in *Proceedings of the 1962 Conference on the Theory of Gravitation* (Waszawa; L. Infeld, ed.). Gauthier-Villars, Paris, 1964. (A)

Magnetic Circular Dichroism

Philip J. Stephens

Magnetic circular dichroism (MCD) is one manifestation of the optical activity induced in matter by an applied magnetic field. Magnetic optical activity occurs when the optical properties of matter differ for left and right circularly polarized radiation because of an external magnetic field. The effect of a magnetic field on the optical properties of matter was first observed by Faraday in 1846, when he discovered the rotation of the plane of polarization of linearly polarized light when passed through a piece of glass to which was applied a magnetic field parallel to the light propagation direction. This Faraday rotation is due to the induction of a difference in refractive index for right and left circularly polarized light (n_R and n_L, respectively) according to the equation

$$\phi = (\pi/\lambda)(n_L - n_R)l, \qquad (1)$$

where ϕ is the rotation in radians, λ is the wavelength of the light, and l is the path length traversed. MCD is a companion phenomenon to Faraday rotation in which one observes the difference in absorption coefficient for right and left circularly polarized light due to a longitudinal magnetic field. In terms of the absorbance A, defined by

$$I/I_0 = 10^{-A}, \quad A = \log_{10}(I_0/I), \tag{2}$$

where I_0 and I are the light intensity before and after passing through a sample, respectively, the MCD of a sample is

$$\Delta A = A_L - A_R. \tag{3}$$

When the sample obeys the Beer–Lambert law, we have

$$\Delta A = (\Delta\epsilon)cl, \tag{4}$$

where $\Delta\epsilon = \epsilon_L - \epsilon_R$ is the differential molar extinction coefficient of the absorbing species, c is its concentration in moles per liter, and l is the path length in cm. At low fields, ΔA and $\Delta\epsilon$ are usually proportional to the magnetic field H and are frequently normalized to a standard field of 10 kG (1 tesla).

MCD is observable in principle in any spectral region where a sample absorbs radiation. In practice, MCD arising from electronic excitations, in which an electron is excited from a ground state to a state of higher energy, is generally studied. MCD measurements are, consequently, carried out almost exclusively in the near infrared, visible, and ultraviolet spectral domains. MCD is measured as follows (Fig. 1): Light is first monochromated and linearly polarized. It then passes through a device which rapidly modulates the polarization between right and left circular polarizations. This phase-modulated light beam then passes through the sample (placed in a magnetic field) onto a suitable detector. If the sample exhibits MCD, the intensity at the detector contains a component oscillating at the light modulation frequency, which is detected with optimal sensitivity using phase-sensitive detection techniques. In this way, ΔA values of as little as 10^{-5} can be measured.

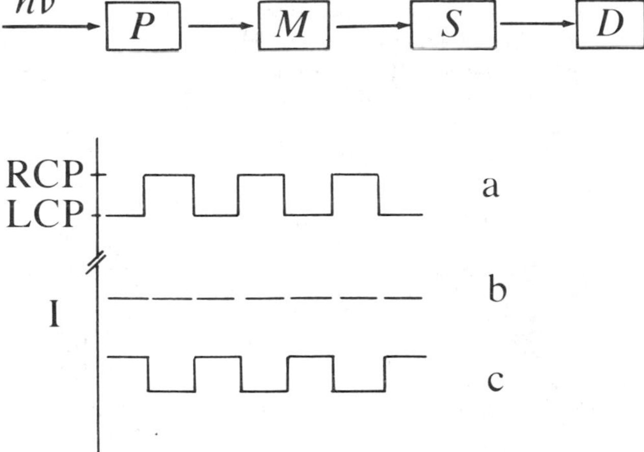

FIG. 1. The measurement of MCD: P is a linear polarizer, M is a modulator, S is the sample, and D is a detector. In a, the polarization of the light beam after M and prior to S is shown as a function of time; RCP and LCP denote right and left circular polarizations, respectively. In b, the corresponding intensity of the light is shown. In c, the light intensity after passing through S is illustrated, in the case that $A_R > A_L$. The oscillating component of the intensity is related to the MCD, ΔA.

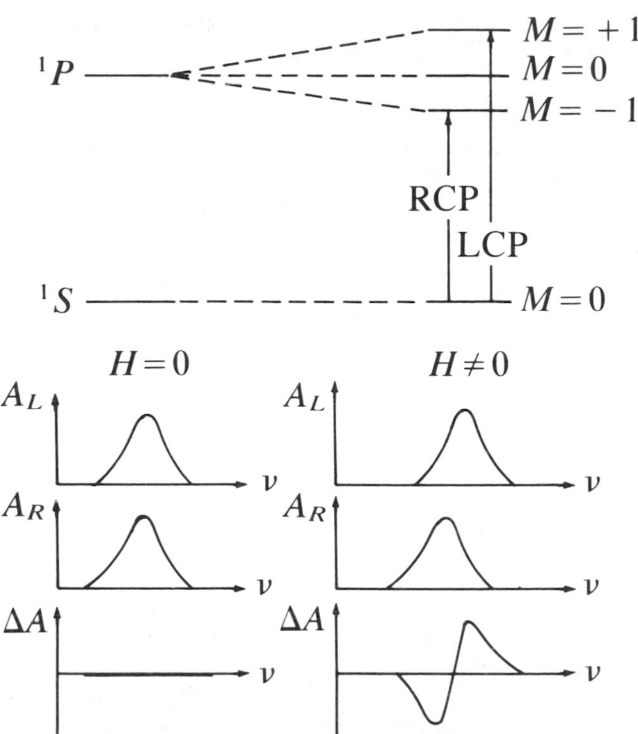

FIG. 2. The origin of MCD: In the presence of the magnetic field the ^1P excited state is split into three components ($M = +1, 0, -1$). Transitions from the ground ^1S ($M=0$) state are allowed to the $M=-1$ and $M=+1$ components for right and left circularly polarized light, respectively. In the absence of a magnetic field these occur at identical frequencies with identical intensity and there is then no net circular dichroism ($\Delta A = 0$). In the presence of the field, however, the transitions are displaced by the Zeeman splitting and at any frequency $A_L \neq A_R$; there is, therefore, MCD ($\Delta A \neq 0$).

MCD arises from the effect of a magnetic field on matter and has a close relationship to other magnetic phenomena, particularly the Zeeman effect (the splitting of spectral lines by a magnetic field) and magnetic susceptibility. A simple example, involving a hypothetical atom, can be used to illustrate how MCD arises from the Zeeman effect. Consider an atom whose ground state is ^1S and an electronic transition to a ^1P excited state (Fig. 2). In the magnetic field the excited state is split into three components. For circularly polarized light, transitions to only two of these are allowed, one for right and one for left circular polarization. Since these are at different frequencies (because of the Zeeman splitting) the absorption for left and right circularly polarized light is different—there is MCD.

MCD has been used to study the electronic states and transitions of molecules, of impurities in crystals (e.g., metal ions, color centers), of magnetic solids, and of biological systems (such as proteins and nucleic acids). For a detailed review of the applications of this phenomenon the reader is referred to the articles in the bibliography.

See also Absorption Coefficients; Faraday Effect; Zeeman and Stark Effects.

BIBLIOGRAPHY

A. D. Buckingham and P. J. Stephens, *Ann. Rev. Phys. Chem.* **17,** 399 (1966).

C. Djerassi, E. Bunnenberg, and D. L. Elder, *Pure Applied Chem.* **25,** 57 (1971).

P. N. Schatz and A. J. McCaffery, *Quart. Rev. Chem. Soc.* **23,** 552 (1969); **24,** 329 (1970).

P. J. Stephens, *Ann. Rev. Phys. Chem.* **25,** 201 (1974).

P. J. Stephens, in *Electronic States of Inorganic Compounds*, p. 141 (P. Day, ed.). Reidel, Dordrecht, 1975.

P. J. Stephens, *Adv. Chem. Phys.* **35,** 197 (1976).

D. M. Dooley and J. H. Dawson, *Coord. Chem. Rev.* **60,** 1 (1984).

Magnetic Cooling

O. V. Lounasmaa

Adiabatic demagnetization of a paramagnetic salt is the oldest method for reaching temperatures significantly below 1 K. This technique was introduced by W. F. Giauque in 1933. The use of nuclear instead of electronic magnetic moments for cooling was pioneered by N. Kurti in 1955. Since 1970, dilution refrigerators have largely replaced paramagnetic salts, while nuclear cooling has proven its worth in numerous experiments, notably in studies involving superfluid ^3He and nuclear ordering in metals. Cerium magnesium nitrate (CMN), however, is still occasionally used for electronic magnetic cooling. For information beyond this brief article, the reader may consult a textbook [1] and several papers with comprehensive lists of references [2–4].

PARAMAGNETIC SALTS

Consider a solid with some paramagnetic ions. In salts suitable for magnetic cooling, the interaction energy ϵ of these ions with their crystalline environment and with each other is small in comparison with the thermal energy kT at temperatures near 1 K. Every ion is thus relatively free in a system of randomly oriented dipoles. This results in a spin entropy $S = R \ln(2J+1)$ per mole, where R is the gas constant and J is the total angular momentum quantum number. In practice, only salts of the iron or the rare-earth groups of ions are important for refrigeration. Their paramagnetism is caused by the magnetic moment associated with electrons in the incompletely filled $3d$ or $4f$ shells.

In zero magnetic field the molar entropy of the salt is a slowly varying function of the temperature until ϵ becomes an appreciable fraction of kT. Spontaneous order of the dipoles occurs at a temperature $\theta \approx \epsilon/k$, characteristic of the salt; this reduces the entropy. At sufficiently low temperatures the system approaches a ground state of zero entropy as required by the third law of thermodynamics. At high temperatures, usually when >1 K, the entropy rises above $R \ln(2J+1)$ as the lattice contribution becomes appreciable.

At $>\theta$ the entropy of the spin system can be lowered by an external magnetic field B which orients the dipoles. For significant polarization, i.e., for large reductions in entropy, the magnetic interaction energy between a dipole and the external field should be comparable to or larger than kT.

The qualitative features of the entropy versus temperature diagram of a paramagnetic salt are illustrated in Fig. 1. We assume that, by pumping on a bath of liquid ^4He or ^3He, point X at the initial temperature T_i and zero magnetic field has been reached. The salt is then magnetized isothermally to an external field B_i (1 T in Fig. 1) along the path $X \rightarrow Y$. The latent heat of magnetization, represented by the rectangular area $XYZ''X''$, is absorbed by the helium bath. Next, the salt is thermally isolated by pumping away the exchange gas and then demagnetized adiabatically to $B \rightarrow 0$ along the path $Y \rightarrow Z$. Considerable cooling results; the temperature T_0 after the demagnetization depends on T_i, B_i, and θ. For CMN, the weakest paramagnetic salt, $T_0 \approx 2$ mK.

The external heat leak then begins to warm up the system along the $B=0$ curve. The amount of heat that the salt absorbs during this part of the experiment is represented in Fig. 1 by the darkly shaded area, which is considerably smaller than the heat of magnetization that was removed at T_i. If demagnetization is stopped at a nonzero external field B_f (0.25 T in Fig. 1), the final temperature T_f is higher than T_0 but the amount of heat absorbed (darkly and lightly shaded areas in Fig. 1) during warming along the $B = B_f$ curve from Z' to X' is larger. The warming-up rate is thus slower.

"BRUTE FORCE" NUCLEAR MAGNETIC COOLING

The basic principle of nuclear cooling is the same as for paramagnetic salts (see Fig. 1), but there are very important differences in practice. Because nuclear magnetic moments are about 2000 times smaller than their electronic counterparts, it is more difficult to produce significant changes in the nuclear polarization by external means. For a 5% reduction of the spin-entropy before demagnetization, using copper as the working substance, one must start with temperatures below 20 mK and magnetic fields in excess of 5 T. Fortunately, these conditions can be produced with dilution refrigerators and superconducting magnets.

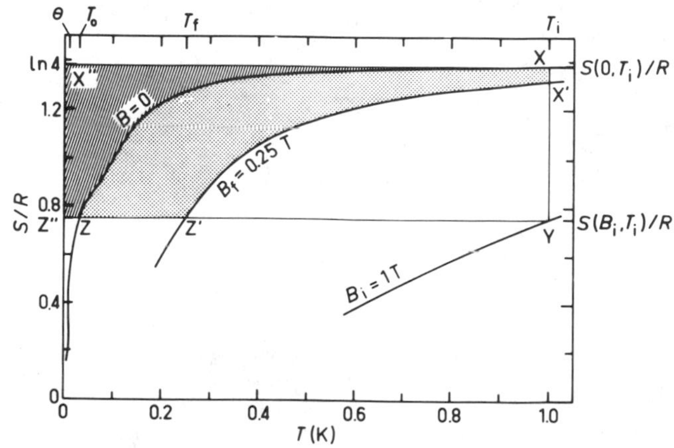

Fig. 1. Entropy diagram of chromic potassium alum illustrating magnetic cooling: $T_i = 1$ K, $T_f = 0.25$ K, $T_0 = 35$ mK, $\theta = 10$ mK. For more details, see text.

The main advantage of nuclear cooling is the very low temperature that can be reached. Nuclei order themselves spontaneously, owing to their mutual interactions, at $\theta \approx \epsilon/\kappa \ll 1$ μK. Because spontaneous ordering is the limit of any cooling process, temperatures in the nanokelvin region can be produced by nuclear demagnetization.

The nuclear spin-entropy is given by the approximate expression

$$S = R\ \ln(2I+1) - \tfrac{1}{2}\Lambda(B/T)^2,$$

where I is the nuclear spin and Λ is the Curie constant. The only variable on the right side of this equation is B/T. Consequently, during an isentropic process, such as adiabatic demagnetization from an initial field B_i to a final field B_f, this ratio remains constant, i.e.,

$$B_i/T_i = B_f/T_f.$$

In nuclear cooling the sample is first magnetized isothermally by raising the externally applied field from 0 to $B_i > 5$ T at the low initial temperature $T_i < 20$ mK. The precooling dilution refrigerator and the nuclear stage are then thermally separated by means of a superconducting heat switch, whereafter the magnetic field is adiabatically reduced to B_f. The nuclear spins thereby cool to the temperature

$$T_f = (T_i/B_i)(B_f{}^2 + b^2)^{1/2};$$

here b represents the effective field between the nuclei which was ignored in the previous equation. Thus, for example, if one starts from $B_i = 8$ T and $T_i = 16$ mK, and demagnetizes to $B_f = 0$, the final temperature in an ideal cooling process with copper nuclei ($b = 0.3$ mT) would be $T_0 = 0.6$ μK. In actual experiments, losses always occur and T_f is higher than calculated from the above equation. This conceptually simple method is often called "brute force" nuclear cooling.

COOLING OF CONDUCTION ELECTRONS

In a nuclear coolant it is meaningful to speak about two distinct equilibrium temperatures at the same time; these are the nuclear spin temperature T_n and the common lattice and conduction electron temperature T_e. During nuclear cooling experiments these two quantities may differ by many orders of magnitude. The nuclei reach local thermal equilibrium among themselves in a time characterized by τ_2, the spin–spin relaxation time, whereas the approach to equilibrium between nuclear spins and conduction electrons is governed by the spin–lattice relaxation time τ_1. At low temperatures, $\tau_2 \ll \tau_1$, which makes a separate nuclear spin temperature meaningful and real.

For most metals τ_1 is on the order of seconds at 10 mK, whereas for insulators τ_1 is days or even weeks. The short spin–lattice relaxation time in metals is due to conduction electrons which act as intermediaries between the nuclear spins and the lattice. Only electrons near the Fermi surface contribute; their number is proportional to T_e, making τ_1 proportional to $1/T_e$, i.e.,

$$\tau_1 = \kappa/T_e,$$

where κ is the so-called Korringa constant. In practice, only copper, indium, silver, thallium, and scanduim have been used for brute force nuclear cooling. For these metals, in low magnetic fields, κ has the values of 0.5, 0.086, 10, 0.0044, and 0.7 s K, respectively.

The unavoidable external heat leak $\dot{Q}_e$ into the conduction electron system has an important effect on the equilibrium between T_e and T_n. If $\dot{Q}_e$ is high, the spin–lattice relaxation process is not sufficiently rapid for cooling the conduction electrons adequately, and $T_e - T_n$ will be large. The relationship

$$(T_e/T_n) - 1 = \mu_0 \kappa \dot{Q}_e / \Lambda(B^2 + b^2)$$

describes the situation quantitatively.

Numerical calculations show that often, in order to stay cold for sufficiently long times and to obtain significant refrigeration of conduction electrons or an external specimen, nuclear demagnetization should not be carried out all the way to $B_f = 0$ but should be stopped instead at some intermediate field (see Fig. 1). In particular, the lowest T_e is reached by demagnetizing to

$$B_f(\mathrm{opt}) = (\kappa \dot{Q}_e/\Lambda)^{1/2};$$

in this case $T_e/T_n = 2$. The lowest conduction electron temperature, 12 μK, has been reached in copper by F. Pobell's group at the University of Bayreuth [3] and by G. R. Pickett and A. M. Guénault at the University of Lancaster [4].

NUCLEAR COOLING BY INDIRECT METHODS OF POLARIZATION

A different procedure is the hyperfine-enhanced nuclear cooling method, originally developed by K. Andres at Bell Laboratories [2]. In van Vleck paramagnetic lanthanide alloys, in which the rare-earth ion occupies a crystal-field–split singlet ground state, an applied magnetic field mixes excited states with the ground state. A large electronic polarizability is thereby induced, which enhances the external magnetic field at the sites of the rare-earth nuclei. In the best specimens, usually intermetallic compounds of praseodymium, the field is boosted by a factor varying from 10 to 20. In $PrNi_5$, with modest precooling conditions ($B_i = 2.0$ T and $T_i = 20$ mK), a reduction of 25% in the nuclear spin entropy can be achieved. Using this compound, temperatures as low as 0.5 mK have been reached. For work between 2 and 4 mK, $PrCu_6$ is an excellent coolant.

In nuclear demagnetization experiments it is sometimes sufficient just to lower the temperature of the nuclei; this is possible when τ_1 is long. Dynamic polarization techniques, developed by A. Abragam and M. Goldman at Saclay [5], have been employed for cooling the nuclear spin system of some insulators, like CaF_2 and LiH, below 1 μK. Ferromagnetic and antiferromagnetic ordering has been observed at positive and negative spin temperatures. Using this method, a high initial nuclear magnetization, up to 90%, must first be achieved with the help of suitable electronic magnetic impurities in the sample. Demagnetization is then performed in the rotating frame. The nuclei retain the low temperature for a considerable length of time because of their high degree of thermal isolation from the lattice.

CASCADE NUCLEAR REFRIGERATION

The very lowest temperatures, in the low picokelvin range, have been obtained in the nuclear spin system by means of cascade magnetic cooling techniques. Figure 2 illustrates a two-stage nuclear refrigerator [2]. A typical experiment is performed as follows: The large first nuclear stage, made of 1 kg of copper, is magnetized to 8 T and cooled, by the dilution refrigerator, to 15–20 mK. The superconducting heat switch between these two precooling stages is then opened, and the field is reduced to 0.1 T, whereby the upper nuclear stage cools to 200 μK. Toward the end of this first demagnetization, the small second nuclear stage, made of the 2-g copper or silver specimen itself, is magnetized to 7 T.

After a sufficient waiting time for equilibrium, typically 1.5 h, the demagnetization of the first stage is continued to 20 mT, which produces a conduction electron temperature $T_e = 50$ μK in copper. The specimen is then demagnetized in about 20 min from 7 T to zero field. The nuclear spins in the second stage are thereby cooled to about 30 nK in copper or 40 1 mK in silver, whereas the conduction electrons remain at 50 μK (200 μK for silver). By monitoring the nuclear spin system, using ac susceptibility measurements or Bragg-reflected neutrons, several nuclear antiferromagnetic phases have been discovered in copper; in zero external field the Néel temperature is 58 nK [6]. The lowest temperature ever produced and measured is 600 pK; this was achieved at the Helsinki University of Technology in the nuclear spin system of a silver specimen. Warm-up to 3 nK took 3 hours. By quickly reversing the external field at the end of demagnetization, negative spin temperatures, corresponding to population inversion of the nuclear evergy levels, have been achieved in silver. So far, the lowest negative temperature produced and measured is −4 mK.

Cascade nuclear refrigerators in which the first stage is made of $PrNi_5$ and the second of copper have also been constructed [3,4,7].

COOLING OF ^3He

When superfluid ^3He is being cooled, the whole low-temperature assembly may be separated into three thermal reservoirs; the nuclear spin system, the conduction electrons, and liquid ^3He. Because of heat leaks and thermal barriers, each subsystem will reach, in dynamic equilbrium, a different thermodynamic temperature: T_n, T_e, and T_3, respectively.

Heat transfer between the experimental cell and the ^3He sample is determined by the Kapitza thermal boundary resistance, defined by $R_K = \Delta T/\dot{Q}_3$, where ΔT is the temperature drop. T_3 adjusts itself to such a value that the heat flow through the boundary just balances the heat leak $\dot{Q}_3$, i.e.,

$$\Delta T = T_3 - T_e = R_K \dot{Q}_3.$$

This equation shows that for reaching low ^3He temperatures R_K and $\dot{Q}_3$ should be reduced as much as possible.

Decreasing the heat leak is difficult because frequently one does not know the origin of $\dot{Q}_3$. The cryostat, however, should be rigid and mechanical vibrations must be prevented.

Assuming that heat transfer at the liquid–solid interface occurs via phonons only, one would expect a $1/T^3$ temperature dependence for R_K. The boundary resistance below 1 mK would then be prohibitively high. Fortunately, an additional thermal coupling mechanism, important below 5

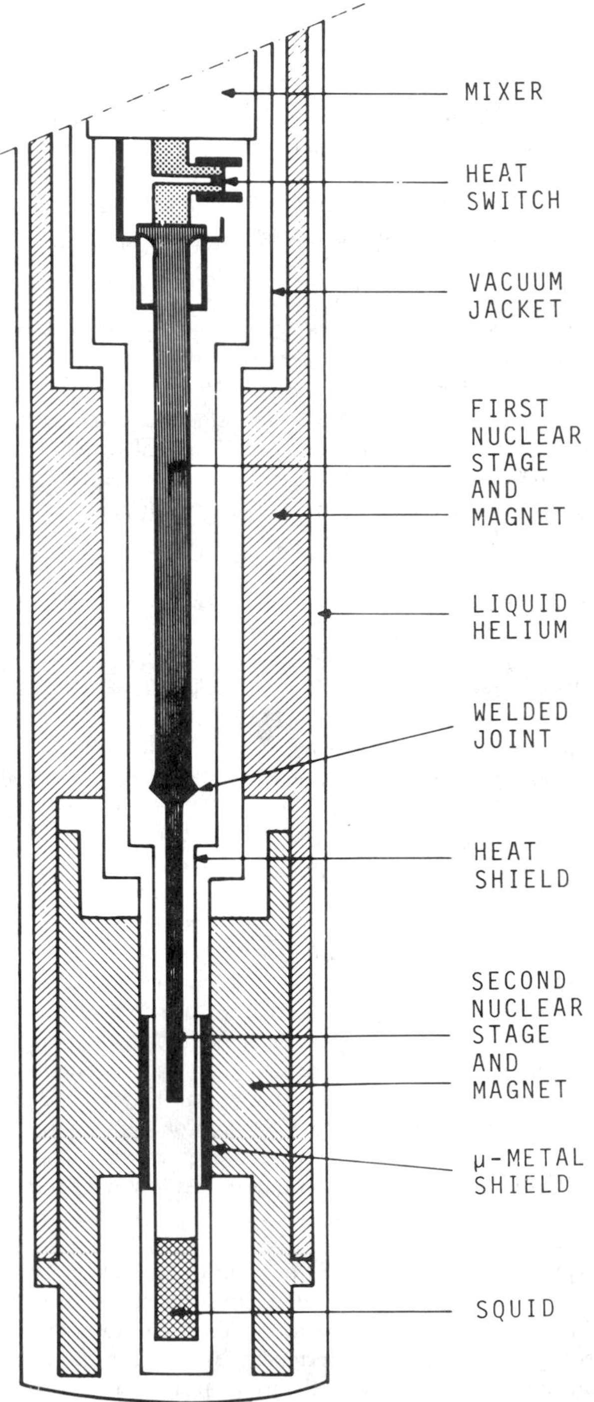

FIG. 2. The cascade nuclear refrigerator in Helsinki [2]. With this apparatus, using silver as the working substance, conduction electrons have been cooled to 200 μK and nuclear spins to 600 pK; both can be considered, within the respective subsystems, as equilibrium temperatures because the spin–lattice relaxation time is many hours.

mK, exists between liquid ^3He and several metals. The extra conductance is believed to arise from a dipolar interaction between the nuclear spins of ^3He and the localized electronic impurity moments on the surface of the metal. The boundary resistance then becomes proportional to some lower power of $1/T$, depending on the material and the temperature.

To reduce the boundary resistance further, it is necessary to increase the surface area between liquid ^3He and the nuclear coolant as much as possible. For this reason copper or silver sinter, made of 0.1 μm diameter or even smaller powder, is put into the experimental cell. The surface area can be increased to 10 m^2 per 1 cm^3 of the sinter.

Several laboratories, in the United States and elsewhere, have cooled superfluid ^3He well below 1 mK; the current record, 100 μK, is held by A. M. Guénault and G. R. Pickett of Lancaster University [4]. Solid ^3He has been cooled to 43 μK by the Tokyo group of H. Ishimoto [7].

CONCLUSION

Figure 3 is an illustration of successive low temperature "records" during the years 1970–1990, all obtained by nuclear magnetic cooling. Over the 20-year period, progress has been very rapid. There are good reasons for trying to reach even lower temperatures; for example, the expected superfluid transition in dilute ^3He awaits to be discovered.

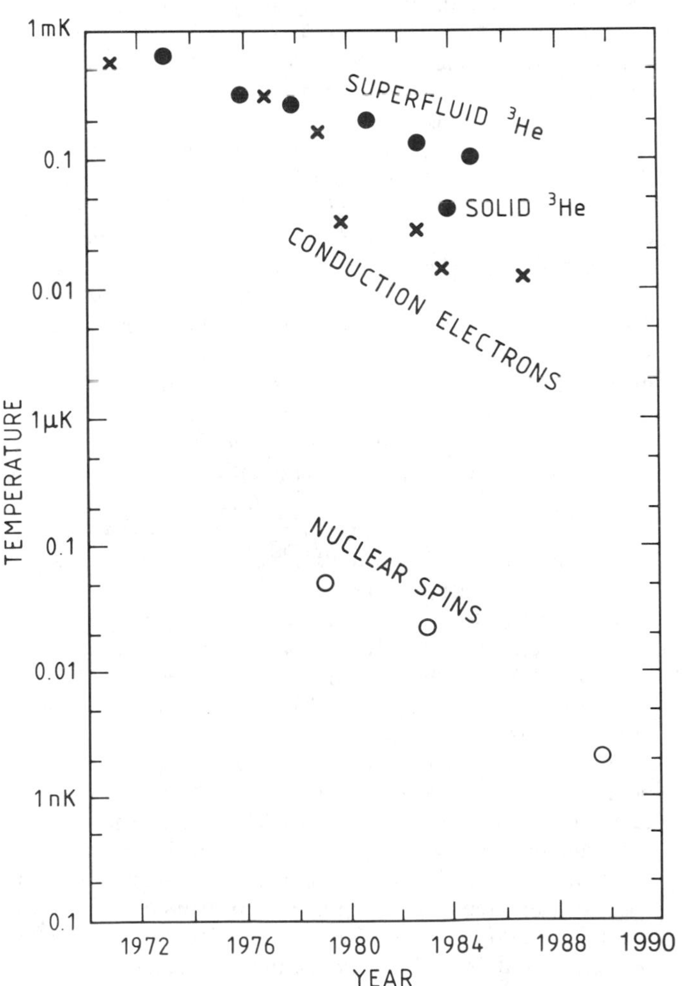

FIG. 3. Progress in nuclear cooling for the 20-year period 1970–1990, as illustrated by the three series of low temperature "records."

See also CRYOGENICS; HELIUM, LIQUID; PARAMAGNETISM.

REFERENCES

1. O. V. Lounasmaa, *Experimental Principles and Methods below 1 K*, Chapters 5 and 6. Academic Press, London, 1974, reprinted 1988.
2. K. Andres and O. V. Lounasmaa, *Prog. Low Temp. Phys.* **8,** 222 (1982).
3. K. Gloos *et al.*, *J. Low Temp. Phys.* **73,** 101 (1988); F. Pobell, *La Recherche* **19,** 784 (1988).
4. G. R. Pickett, *Rep. Prog. Phys.* **51,** 1295 (1988).
5. A. Abragam, *Proc. R. Soc. London, Sect. A* **412,** 225 (1987).
6. T. A. Jyrkkiö et al., *J. of Low Temp. Phys.* **74,** 435 (1989); O. V. Lounasmaa and A. S. Oja, *Phys. Rep.* (to be published).
7. H. Ishimoto et al., *J. Low Temp. Phys.* **55,** 17 (1984); Y. Takano et al., *Phys. Rev. Lett.* **55,** 1490 (1985).

Magnetic Domains and Bubbles

G. P. Vella-Coleiro

A magnetic material may be thought of as an aggregation of atomic magnetic moments, or tiny magnets, which are acted upon by a large internal force known as the exchange interaction. This interaction is a quantum-mechanical concept and although it is electrostatic in origin it has no classical analog. Its effect is to couple adjacent atomic magnetic moments parallel to one another in a ferromagnetic material, and antiparallel in an antiferromagnetic one, thus forming regions where the magnetic moments are aligned in the same direction. These regions are known as magnetic domains. The magnitude of the exchange interaction can be appreciated by noting that in iron, a typical ferromagnetic material, it is equivalent to an internal magnetic field with a strength of 4×10^8 A-turns/m (5×10^6 Oe) at room temperature, which is several orders of magnitude larger than commonly encountered magnetic field strengths.

Ordinarily, a sample of ferromagnetic material does not form a single magnetic domain, but rather a collection of domains with the magnetization of adjacent domains pointing in different directions. If the sample were to form a single domain, as shown schematically in Fig. 1a for a rectangular bar, there would be a substantial amount of magnetostatic energy associated with the magnetic field emanating from the surfaces, this energy being proportional to the square of the magnetic field strength integrated over the volume in which it resides. By a break up into two oppositely oriented domains, as in Fig. 1b, the magnetostatic energy is reduced, and further reductions occur with subsequent subdivisions. Each domain is separated from adjacent domains by a transition region, known as a domain wall, which is discussed below. Magnetic energy is associated with a domain wall, and the process of subdivision halts when the reduction in

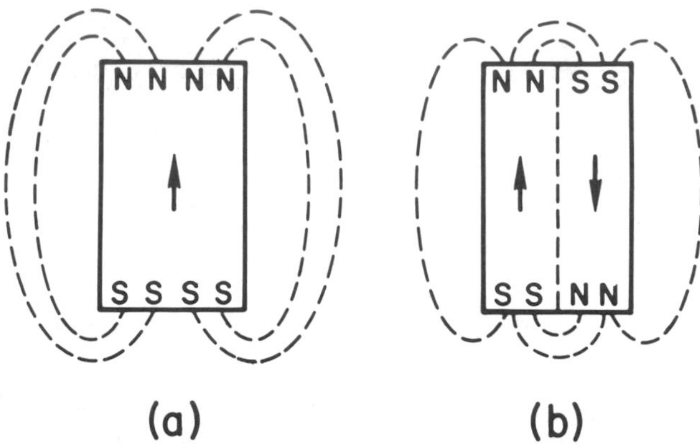

FIG. 1. Magnetostatic energy is associated with the magnetic field (broken lines) which surrounds a magnetic domain due to the magnetic poles (indicated by N and S) on the surfaces. This energy is highest for a single domain, as in (a), and it is approximately halved by subdividing the sample into two domains, as in (b). Further reductions in magnetostatic energy occur as the number of domains increases.

magnetostatic energy is counterbalanced by the energy of the additional domain walls which would be introduced.

In a single-crystal magnetic material the domains are not oriented in arbitrary directions, but they line up parallel to well-defined crystalline axes. This effect is due to the phenomenon of magnetocrystalline anisotropy which causes certain directions in the crystal to be energetically favored over others. For example, in single-crystal iron, whose atoms have a cubic arrangement, the magnetic energy is lowest when the magnetic moments in a domain are oriented parallel or antiparallel to a cube edge direction of the atomic structure, which is therefore known as the easy axis of magnetization. Since in a cubic lattice there are three equivalent cube edges, the domain structure in iron can be rather complicated, the angle between adjacent domains depending on the orientation of the sample. By contrast, a material such as cobalt can exhibit a very simple domain pattern. Single-crystal cobalt has a hexagonal structure and at room temperature the easy direction of magnetization is parallel to the hexagonal axis. Thus a plate of cobalt cut with its faces perpendicular to the hexagonal axis forms domains oriented parallel or antiparallel to the plate normal. A similar situation exists in specially prepared ferrite crystals known as garnets, which have achieved great technological importance recently due to their suitability for use in magnetic bubble devices (see below). The domain pattern in such a crystal, made visible by means of the Faraday effect, is shown in Fig. 2a. It consists of serpentine domains with the direction of magnetization alternating between up and down. In the absence of an external magnetic field the areas of the up and down domains are equal, resulting in no net magnetization.

As noted above, the domain walls which separate adjacent domains contain energy, predominantly from two sources, the exchange interaction and the magnetocrystalline anisotropy. The exchange energy density associated with a domain wall is small if the wall is broad, occupying many lattice sites, since this results in adjacent magnetic moments within the wall being nearly parallel. In this case, however, the magnetocrystalline energy density would be large since the wall would contain many magnetic moments pointing in a direction other than an easy axis. Reducing the wall width reduces the number of these moments, thus reducing the magnetocrystalline energy. The equilibrium wall width is therefore determined by a balance between the exchange and the magnetocrystalline energies. The former would like all the moments to be parallel, which would result in a very broad wall, while the latter prefers the moments to lie parallel to an easy axis, which would result in a wall only one lattice constant wide. Somewhere in between these two extremes lies a compromise wall width which minimizes the sum of the two energies. Analysis of the wall structure in an infinite magnetic medium shows that the minimum energy occurs at the wall width which makes the exchange energy density equal to the magnetocrystalline energy density.

Another source of energy which is very important in determining the behavior of domains is an external magnetic field. Just like a magnetic compass needle which orients itself parallel to the earth's magnetic field, magnetic domains try to orient so that the magnetic moments are parallel to an external field. Because of the competing forces arising from the magnetocrystalline anisotropy and the magnetostatic energy, complete orientation only occurs when the magnetic field exceeds a critical value which depends on the magnetic properties of the material and on its shape. For field strengths lower than the critical value, the domains whose magnetic moments are most nearly parallel to the external field are favored over others and thus occupy a larger volume. In the example of Fig. 2a, application of a low-strength magnetic field perpendicular to the plane of the magnetic film, say in the up direction, causes the domains pointing up to expand in area at the expense of those pointing down. The process continues until the critical magnetic field strength is reached, when the sample becomes a single domain pointing up. The situation is modified somewhat by the presence of coercivity which inhibits changes in the net magnetization for external fields lower than some value which depends on the material and on its preparation, such as heat treatment. Values of the coercive field range from less than 10 A-turns/m (0.1 Oe) for materials such as that of Fig. 2, to a few hundred thousand ampere-turns per meter for permanent magnet materials.

Another type of domain can occur if the stripe domains do not extend from one edge of the sample to another, but are isolated, for example by cutting with a magnetized needle at both ends (several of these "island" domains can be seen in Fig. 2a). In the presence of a sufficiently strong magnetic field oriented antiparallel to the magnetic moments in these domains, they shrink into magnetic bubbles, which are actually cylindrical domains with their axes perpendicular to the plane of the magnetic film, see Fig. 2b. Bubble domains are stable over a diameter range of approximately 3 to 1, in a range of applied field strengths of approximately one-tenth the saturation moment of the material. Their stability is determined by an interplay between the magnetostatic energy, the total wall energy, and the external field energy. The magnetostatic energy decreases with increasing radius, while the other two increase, and a stable bubble domain is formed

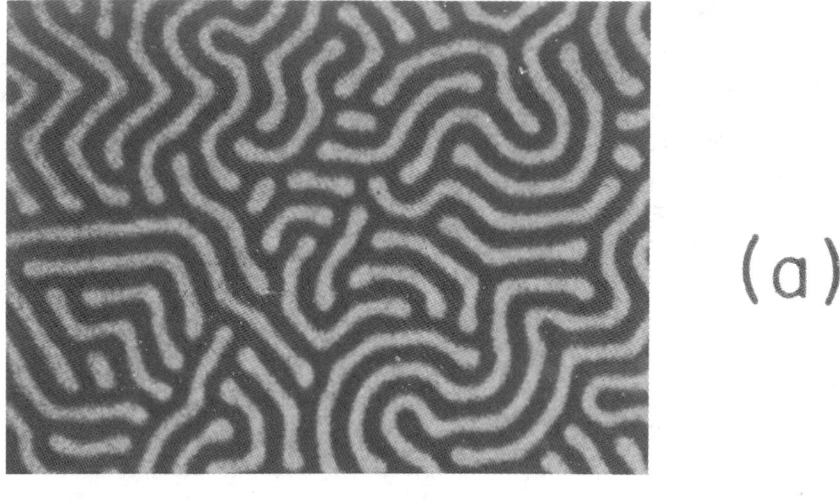

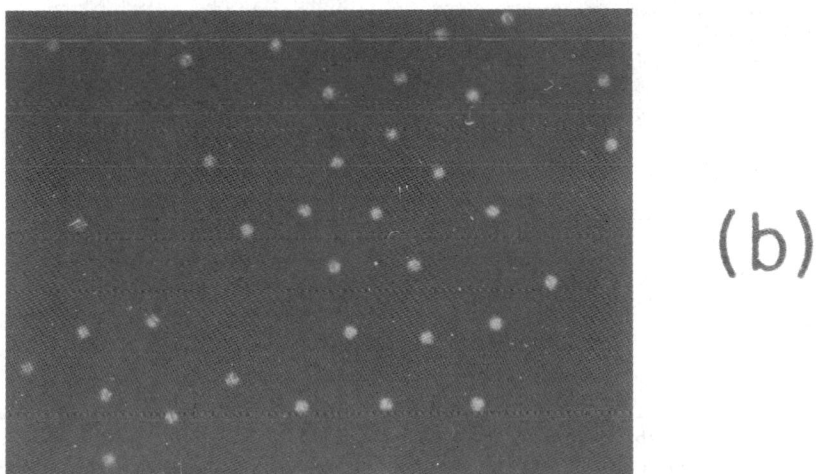

FIG. 2. Photomicrographs of the domain patterns in a garnet film with easy axis of magnetization perpendicular to the film plane. The yttrium–samarium–iron garnet film had a thickness of approximately 6.5 μm, and it was grown epitaxially on a nonmagnetic gadolinium gallium garnet substrate. The domains were observed in a microscope by means of polarized light transmitted perpendicular to the plane of the sample. The plane of polarization of the light was rotated by the magnetization in the domains (Faraday effect), the up domains rotating the plane of polarization in the opposite direction from the down domains. After traversing the sample the light was passed through an analyzer which was adjusted to extinguish the light transmitted through one set of domains. Thus one set of domains appears dark, the other bright. Photograph (a) was obtained in the absence of a magnetic field and shows serpentine domains with magnetization perpendicular to the plane of the film. In (b) a magnetic field was applied in the direction opposite to the magnetization in the bright domains, with a magnitude sufficient to cause the "island" domains to become bubbles. The bubble diameter was approximately 6 μm.

at the radius where the sum of the three energies is a minimum. As the magnetic field is increased, the bubble radius decreases until a point is reached where a further increase in magnetic field energy can no longer be compensated by a decrease in magnetostatic and wall energies; then the bubble becomes unstable and collapses. Conversely, as the magnetic field strength is reduced, the bubble radius increases up to the point where a distortion of the circular shape of the bubble to an elliptical one would result in a decrease of the magnetostatic energy by a larger amount than the increase in wall and magnetic field energies. The bubble thus becomes unstable toward such distortions and the domain runs out into a stripe.

The current interest in magnetic bubbles stems from their potential applicability to memory systems capable of handling large amounts of data at high speed. In a manner analogous to magnetic tape or disc reorders, binary information can be coded in terms of the presence or absence of a domain. Unlike the tape or disc units, however, where the domains are stationary and data are accessed by moving the magnetic material past read/write heads, in a bubble device the bubble material is stationary. Access to the data is

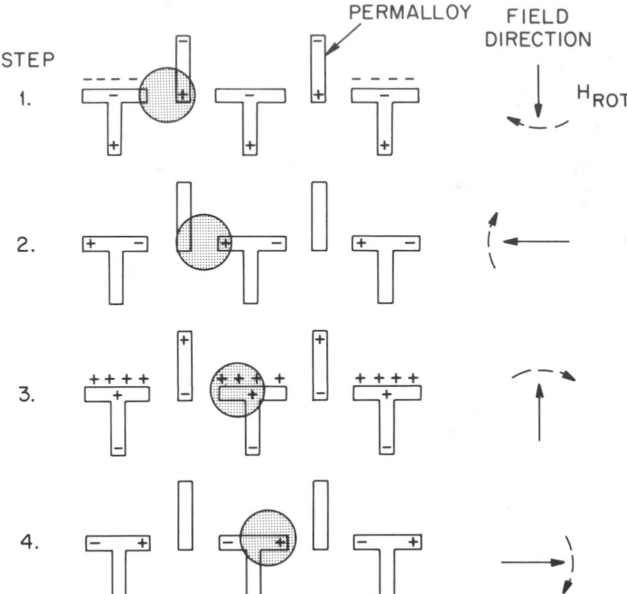

FIG. 3. Schematic representation of bubble propagation in a Permalloy structure consisting of T and I elements. The thin Permalloy elements are in close proximity to the bubble material, to which a perpendicular magnetic field is applied to stabilize the bubble domains, represented by the discs. A rotating in-plane magnetic field H_{rot} is also applied. As the in-plane field rotates from the direction shown in step 1 to step 2, the magnetization of the permalloy elements, indicated by the + and − signs, changes and causes the bubble to move to the right. Futher motion occurs as the field rotates to positions 3 and 4, and when one cycle of rotation is complete the bubble will be in a location equivalent to that in step 1, but one circuit period to the right. A stream of bubbles can thus be propagated in the plane of the film, data being encoded according to the presence or absence of a bubble in the stream.

achieved by translating the bubbles within the material by means of externally applied magnetic fields. In the presence of a nonuniform magnetic field, a bubble domain moves in the direction in which the magnetic field strength is least since this results in a reduction of the total energy. In a practical bubble device, a nonuniform magnetic field is obtained by means of an array of high-permeability Permalloy elements in close proximity to the magnetic material, in conjunction with a magnetic field in the plane of the material (see Fig. 3). The in-plane field is separate from, and in addition to, the perpendicular field which is required to stabilize the bubble domains. The Permalloy elements are magnetized by the in-plane field, thereby creating localized nonuniform fields which exert forces on the bubble domains, causing them to seek regions where the local magnetic field has the least value. As the direction of the in-plane field rotates, these regions move along the Permalloy elements, carrying the bubbles with them. A coded stream of bubbles can thus be made to propagate past a detector, which makes use of the magnetoresistive properties of a Permalloy element to produce an electrical signal in response to the passage of the bubble domains. In-plane fields rotating at rates well in excess of 10^5 Hz can be produced readily by means of a pair of orthogonal solenoids, and materials which support bubbles with submicrometer and larger diameters are available. Photolithographic and electron-beam techniques have been developed to produce the Permalloy patterns required for bubble propagation and detection, thus forming the basis for high-speed, high-storage-density magnetic bubble memories.

An alternative to the use of Permalloy elements for bubble propagation is to implant the bubble material with ions such as H^+, He^+, or Ne^+ at energies in the neighborhood of 100 keV, which produces a change in the magnetic properties of a thin layer near the surface. Ion implantation produces stress in the crystal lattice, causing it to expand slightly in the direction perpendicular to the film plane. In the plane of the film, however, the lattice is constrained by the material underneath, which is much thicker than the implanted layer. The resulting in-plane compressive stress, in conjunction with the inverse magnetostrictive effect, produces a change in the easy direction of magnetization of the implanted layer. By a judicious choice of the magnetic properties of the bubble material, it is possible to produce a surface layer in which the magnetization lies in the plane of the film. Photolithographic techniques are used to implant only selected areas, so that regions analogous to the Permalloy elements shown in Fig. 3 are produced. In the presence of an in-plane rotating field, these regions cause the bubbles to propagate. An advantage of ion implantation over the use of Permalloy elements is the relative ease with which micrometer-sized features can be produced, making it possible to manufacture bubble memories with very high bit density.

Another type of bubble memory, which has received considerable attention, utilizes an array of bubbles, called a bubble lattice, which is stabilized by the magnetic interaction between bubble domains. For domains of similar polarity this interaction results in a repulsive force between domains which, when closely spaced, form a two-dimensional hexagonal array with a spacing approximately equal to twice the bubble diameter. Information coding is achieved by using two types of bubble which differ only in the detailed structure of their domain walls. One bubble type has a wall which contains Bloch lines, i.e., 180° reversals of the magnetization direction, while the other does not. These two types of bubble can be differentiated by the fact that they propagate at different angles with respect to a field gradient, which forms the basis for data retrieval in this type of bubble memory.

Additional information on magnetic domains and bubbles may be found in the references listed below. Reference 1 is a very readable, nonmathematical account of ferromagnetic domains and their observation. Reference 2 contains a more advanced discussion of domain theory as well as detailed descriptions of several techniques used for studying domains. References 3 and 4 are textbooks on magnetism which contain several chapters on domains. A review of the theory and device aspects of magnetic bubbles may be found in Ref. 5, while more detailed information is contained in Refs. 6, 7, and 8. The applications of ion implantation in bubble memories are described in detail in Refs. 9–11.

See also FERRIMAGNETISM; FERROMAGNETISM; HYSTERESIS; MAGNETIC MOMENTS.

REFERENCES

1. E. A. Nesbitt, *Ferromagnetic Domains*. Williams and Wilkins, Baltimore, 1962. (E)
2. R. Carey and E. D. Isaac, *Magnetic Domains and Techniques for their observation*. Academic Press, New York, 1966. (A)
3. A. H. Morrish, *The Physical Principles of Magnetism*. Wiley, New York, 1966. (I)
4. S. Chikazumi, *Physics of Magnetism*. Wiley, New York, 1966. (I)
5. A. H. Bobeck and H. E. D. Scovil, "Magnetic Bubbles," *Scientific American*, 78–90 (June 1971). (I)
6. A. H. Bobeck and E. Della Torre, *Magnetic Bubbles*. North-Holland, Amsterdam, 1975. (A)
7. T. H. O'Dell, *Magnetic Bubbles*. Wiley, New York, 1974. (A)
8. A. H. Eschenfelder, *Magnetic Bubble Technology*. Springer-Verlag, New York, 1980.
9. J. C. North, R. Wolfe, and T. J. Nelson, "Applications of Ion Implantation to Magnetic Bubble Devices," *J. Vac. Sci. Technol.* **15**, 1675 (1978). (I)
10. K. Yoshimi, "Ion-implanted Bubble Devices," *Japan Ann. Rev. Electron., Computers & Telecommun.* **10**, 43 (1983). (I)
11. P. Gerard, "Implantation of Bubble Garnets," *Thin Solid Films*, **114**, p. 3 (1984). (A)

Magnetic Fields, High

Simon Foner

The term "high" magnetic fields here is restricted to fields above those available in "conventional" electromagnets where the first 20 kG (2 T) can be produced by iron-core magnets with a small magnetizing field, and therefore small power consumption. Although an iron-core-type construction can be employed for high fields, most high fields do not involve iron for various technical reasons. The volume occupied by the iron core can be used more effectively by replacing it with current-carrying conductors. The development of "high" fields has progressed from the late 1920s when pulsed field technology in the range of 30 T was developed and used in experiments by Kapitza, and the mid-1930s when Bitter developed water-cooled continuous high-power magnets for experiments to 10 T. Since then there has been a continuing growth of high-field research and high-field applications.

The total energy stored in a magnetic field in free space is given by

$$E=\tfrac{1}{2}\int \mu_0 H^2\, dv=\tfrac{1}{2}LI^2, \tag{1}$$

where H is the magnetizing field, μ_0 is the vacuum permeability, v is the volume over which the field is generated, L is the inductance, and I is the current of the coil generating the field. High fields involve high energy densities. For a fixed energy, in order to generate higher fields the field volume must be reduced. Alternatively, for given field volume, increasing the field requires increasing the energy as H^2. Inherent in the conventional generation of magnetic fields by means of normal conductors is the dissipation produced by Joule heating. For fixed volume the power required to generate the field increases as H^2. Below a certain magnetic field level, superconducting conductors (which have no Joule heating, but require cooling to low temperatures) may be used to reduce the power requirement. Generation of high fields by means of superconductors is considered later.

Associated with the increase in energy density at high fields are increased forces. The pressure generated in the high-field volume also varies as H^2, so that at very high fields one may exceed the strength of conventional (and unconventional) structural materials. For example, the pressure which varies as H^2 reaches 9.8×10^8 N/m^2 (100 kbar) at $H=50$ T (500 kG). The generation of very high fields is limited by the combination of the conductivity of the conductor, the mechanical strength of the conductor and supporting structure, the power available and the power dissipation, the conductor strain which varies as the product IHR where R is the radius of the winding, and the cost of the magnet which increases more rapidly than VH^2. The combination of these factors places practical limitations on production of very high fields.

One way of avoiding the large powers required for production of continuous high magnetic fields with normal conductors is to utilize a transient field so that large amounts of energies are used, but intermittently for short pulses. This can be done by storing energy electrically in a capacitance bank or an inductance, or mechanically in rotating machinery, and then for short periods transferring the stored energy to the magnet to producing a transient field. One problem with transient (pulsed) fields is that in some experiments thermal equilibrium is difficult to achieve. Furthermore, the short time that is available in a given experimental situation reduces resolution and sensitivity. In addition, a rapidly changing magnetic field produces eddy currents in a conducting material which produce large forces as well as heating effects.

Use of superconducting materials for magnets in the field range where these superconductors have zero resistance can reduce the power requirements substantially, particularly for very large-volume field requirements. Research in high magnetic fields is limited by magnet technology. This technology is strongly dependent on new materials development; conductors with higher conductivity than copper, or higher strength, or both can lead to substantially higher fields. Because very high-field systems are expensive, these are generally developed as central facilities for a broad user community.

CONTINUOUS HIGH MAGNETIC FIELDS

Continuous high magnetic fields are produced by (1) water-cooled or cryogenically cooled normal conductors, (2) superconducting conductors, which have no resistive losses below their critical field, and (3) a combination hybrid magnet composed of a water-cooled central section which generates the highest fields, surrounded by a lower-field superconducting magnet. Since 1936, when Bitter pioneered water-cooled high-power magnets, many facilities have been built for research in high magnetic fields worldwide.

In 1961, Kunzler of Bell Laboratories discovered that Nb_3Sn, a type II superconductor, exhibited high current-carrying capacities in high fields and still remained superconducting. During the last 20 years development of such type II superconductors has resulted in two commercially available materials, NbTi and Nb_3Sn, which are useful as conductors for generating high fields with no appreciable resistive losses. NbTi is a ductile alloy that is now produced as a reliable multifilamentary superconducting material which can be used to fabricate magnetic fields up to 9 T when cooled to liquid helium (4.2 K). Nb_3Sn is a relatively brittle compound and is less well developed, but has been used in various conductor configurations to generate fields up to 18 T. Current technology is developing superconducting materials for applications of even higher fields for research-type magnets.

In 1986, Bednorz and Müller discovered an oxide of LaBaCuO which had a superconducting transition temperature T_c above 30 K. This finding was startling because the highest T_c values in NbTi, Nb_3Sn, and NbGeAl were 9, 18, and 23 K, respectively, and theories suggested a maximum of 30 K for superconductors. This discovery led to a worldwide rush to search for new materials for high-temperature superconductors and a reexamination of possible applications. The highest T_c achieved to date (1989) is 125 K, and many different oxide systems have been discovered between 20 and 125 K. These new discoveries suggest that high-field magnets operating at liquid nitrogen (77 K) may be feasible in the future. The challenge is to develop practical conductors from high-temperature superconductors which can carry large currents in these high magnetic fields. Recent studies indicate that extensive research and development will be required for several years before it will be known whether these new superconductors will be useful for such magnets.

The "hybrid" magnet concept was suggested for generation of high magnetic fields in the mid-1960s. A hybrid magnet is composed of an inner water-cooled magnet, which operates at high field and high current density, surrounded by an outer large-bore superconducting magnet operating at a lower field, and producing a large-volume field. A minimal power requirement is needed for cooling this superconductor at 4.2 K. The two fields add resulting in a higher field with a much lower total power requirement than needed for a totally water-cooled magnet. With a given power source, it is possible always to make a higher field with the hybrid approach than would be possible by either the water-cooled or superconducting approach alone. However, there is a substantial investment required for the large outer superconducting magnet structure and its refrigeration needs. The superconductor is always limited to a field below its upper critical field H_{c2}, as well as by a rapid reduction in the critical current density which can be carried by the superconductor as the field is increased toward H_{c2}. Nonsuperconducting magnets generally require a disproportionate amount of power for useful field generation in the bore. The hybrid magnet concept maximizes the field for a given available power. The present record high fields are produced at the Francis Bitter National Magnet Laboratory, Massachusetts Institute of Technology. A field of 31.8 T (318 kG) is generated in a 33-mm-bore room-temperature access and 35.3 T in a 2-mm gap of holmium pole pieces centered in the bore at 4.2 K. The central water-cooled section dissipates 9 MW of power; a comparable, completely water-cooled 30-T magnet would require about 30 MW. Several laboratories are building higher-field hybrid magnets and projections for major new facilities in Japan and the United States, and expect to attain fields of 40 and 45 T, respectively.

PULSED MAGNETIC FIELDS

Transient fields can generate fields above those feasible by dc means. The time that the field can be maintained is limited by several factors: the melting point of the conductor, the mechanical strength of the conductor and supporting structure, the heat capacity of the magnet, and the energy available. Sources of energy for small-volume fields can be capacitor banks (for energies up to a few MJ). For larger energies, rotating machinery becomes more economical. Even larger transient energy sources are available from electrochemical and chemical stored energy, e.g., batteries or explosives.

Fields of about 75 T are achievable with beryllium–copper or maraging steel structures. These are generated in reproducible coil structures for short times (generally in the range of a millisecond or less) from high-voltage capacitive discharges. With a sufficiently short discharge time, efficient generation of high fields is feasible even for relatively high-resistance conductors. Longer pulse times (10 ms or more) and fields up to about 50 T (500 kG) can be attained with multilayer multiturn copper coils when the magnet is cooled initially to 77 K. Recently, fields above 68 T were generated in such wire-wound magnets using a new high-conductivity, high-strength, metal-matrix copper–niobium microcomposite. It is expected that new materials will permit efficient generation of long-pulse fields of 70 to 80 T in the near future.

Above 100 T (1 MG), the magnet conductors move during the pulse, and the moving conductors are utilized to generate transient fields. The Cnare technique is such a method for generating fields of the order of 200 T. The procedure is to trap magnetic flux in a thin conducting metal liner, then generate (outside the liner) a rapid pulsed field. If the pulse is sufficiently rapid so that the penetration depth of the outside field into the liner is small compared to the liner thickness, the liner is compressed by the outside field whereas the trapped flux cannot diffuse out of the liner. As the volume of trapped flux reduces, the field increases as indicated in Eq. (1). The field is changing most rapidly as the liner approaches its minimum diameter. Eventually the pressure generated inside the liner by the trapped flux becomes sufficiently large that it then can overwhelm the external pressure. This limits further reduction in volume (and increase in field). Fields well above 100 T have been produced for short times on the order of microseconds by the Cnare technique.

If larger fields are required, then explosive energy sources may be used. A very large-mass container may be used for protection around the explosive or a special explosive test site may be used. Starting with ~100 kG flux trapped in a conducting liner, compression of the liner by explosives has produced fields up to 1500 T (15 MG). Explosive flux compression devices may also be used for efficient conversion of very large amounts of chemical energies to electrical energy on a one-shot basis.

A single-turn pulsed magnet can produce fields over 200

T for a few microseconds using a very high-voltage, low-inductance capacitor bank. Although the magnet is destroyed during the pulse, the specimen can often be preserved by careful design.

HIGH MAGNETIC FIELD RESEARCH

Figure 1 shows, on a logarithmic scale, the ranges of fields produced by various techniques. Laboratory fields and the technique for producing them are indicated qualitatively. Hybrid techniques have extended the continuous range to ~30 T to date (1989), and explosive techniques should extend the megagauss range. Astronomical environments (e.g., neutron stars) which are only accessible to us indirectly have much larger energy densities than those produced on earth. This is shown by the asterisk in that figure. Estimates, based on astrophysical observations of x-ray cyclotron emission of electrons from Hercules X-1, indicate a magnetic field of approximately $\sim 5\times10^8$ T (5×10^{12} G) at the surface of this neutron star.

The highest field reported to be generated on earth (by expensive compression) is about 2.5×10^3 T (25 MG). It should be noted that, in addition to generation of extremely high fields, ultrahigh fields also generate extremely high pressures comparable to those of the explosives used to drive the liners. The yield strength of most metals is exceeded at fields of about 50 T. At ~100 T the temperature of a conductor approaches the melting point. At a few MG the surface of the conductor evaporates and intense shock waves are produced, and at ~500 T (5 MG) the internal volume behaves as if it is filled with detonating high explosive. In the range of 10^3 T (10 MG), the energy density of the field becomes larger than the binding energy of most solids, and the pressures approach those near the center of the earth. Powerful lasers have produced (for very short times) fields of about 10^3 T (10 MG). The internal fields in solids, e.g., exchange fields, can range to 10^3 T (10 MG), and the hyperfine fields in solids can approach 2×10^3 T (20 MG).

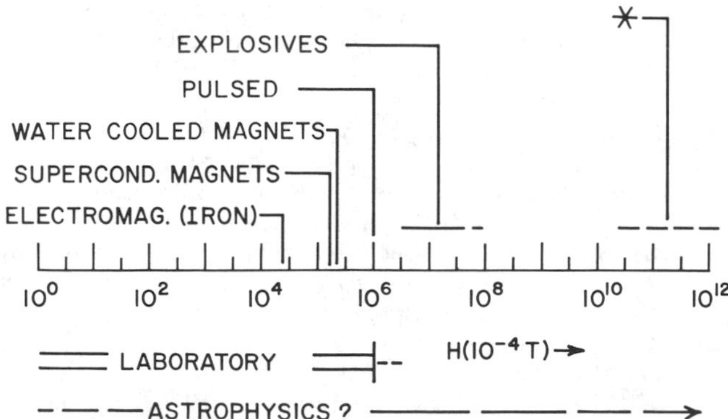

FIG. 1. High magnetic field generation. The units of the horizontal logarithmic scale are in 10^{-4} T (1 G $=10^{-4}$ T). The vertical bars correspond to estimates of the present upper limit achieved by each technique for generation of high fields. Horizontal dashed lines suggest the range of fields for a given technique. Laboratory fields extend above 100 T (1 MG) for single-shot devices and about 1000 T for explosive techniques of field generation. The combination of water-cooled (inner) and superconducting (outer) hybrid magnet achieving over 30 T (300 kG) is shown. The limits of high fields observed in astrophysics may be well above 10^8 T.

High fields permit studies of the properties of matter under extreme conditions. External fields can be made comparable to those found in the nucleus, and comparable to the magnetic and electric interactions in matter. Laboratory fields can be used to generate the lowest temperatures by means of adiabatic demagnetization, and to compete effectively with internal fields for ordered and disordered magnetic systems. In addition, the spin and orbital motion of electrons are strongly affected by the applied fields and many basic properties of matter can be explored when the field can be employed as an independent variable. The corresponding energy densities can be extremely high [5×10^5 J/cm^3 at $\sim10^3$ T (10 MG)].

The generation of very high fields over large volumes is costly. The costs are orders of magnitude smaller than those which one encounters in high-energy physics experiments, but they are comparable to many other modern tools currently used in other research areas. Because high-field technology is specialized, very large continuous fields are most cost-effective when centralized in efficient multiuser facilities. At present, high-field laboratories exist in England, France, Germany, Holland, Japan, Poland, Russia, and the United States, and several new facilities are in the planning stage.

LARGE-SCALE APPLICATIONS OF HIGH MAGNETIC FIELDS

There are many high-field applications. These all involve large-volume systems at fields well above those achievable by efficient iron-core magnets. The use of high-field superconductors to achieve the required high flux densities is often a practical means to achieve these high fields. These applications include:

1. *Power generation and conversion.* Superconducting ac and dc generators and motors have been demonstrated. These are compatible with conventional systems, are more economical, smaller in volume, and would permit higher power capacity than that achieved with conventional technology. Magnetohydrodynamic (MHD) generators using large superconducting magnets in order to achieve increased efficiency of power generation for peak loading have been studied. However, no industrial scale units have been built.

2. *Fusion power generation.* High-temperature plasma containment by high magnetic fields is another application of magnet technology to power generation. Experiments with tokamak-type toroidal devices have employed liquid-nitrogen-cooled copper magnets to generate fields for times of a few seconds. Several large toroidal machines have been built in Europe, Japan, and the United States with continuous increases in plasma density times containment time, and at increasing temperatures of the contained plasma. The alternative, magnetically contained plasmas in a linear mirror machine, also has been developed. A large mirror machine employing some of the largest superconducting magnets was built at Livermore; however, the mirror concept has been abandoned in favor of the toroidal machine in the United States.

3. *Energy storage devices.* Sufficiently large-volume, high-field magnets can store enormous energies as indicated in Eq. (1). If the magnets employ high-field superconductors, the losses can be minimized. The economics of scale suggest structures enclosing volumes as large as $\sim 10^6$ m^3 with field levels of 5 T and energy storage capacities of 10 GWh. A large superconducting energy storage device (SCES) using NbTi at 4.2 K is being constructed in the United States.

4. *High-energy accelerators.* Some of the largest superconducting-magnet systems are high-energy accelerators for research. The Energy Doubler of the Fermi National Accelerator Laboratory operates at 4 T and doubles the energy of the Fermi Laboratory iron-core accelerator using the same machine tunnel. Nearly 800 superconducting magnets are distributed over a circumference of about 4 miles. A much larger accelerator, the superconducting supercollider (SSC), is planned for construction in the United States with a circumference of 53 miles, a field of 6.8 T, and almost 10,000 magnets. Another application of large-volume high fields for high-energy physics involves large superconducting magnets for bubble chambers.

5. *Nuclear magnetic resonance.* Two important applications of high-field superconducting magnets involve nuclear magnet resonance (NMR). Very high homogeneity magnets are used for NMR of large molecules for research in chemistry and biology. Commercial magnets and systems are available up to 15 T or more. Very large-volume lower-field systems (1.5 to 4 T) have been developed on a worldwide commercial scale for whole-body imaging of the soft tissues in living systems (humans) for medical diagnostics. This is the major current market for superconductors and more than 1000 units have been delivered.

6. *Industrial processing.* Large-volume high-field magnets also appear to be promising for rapid magnetic filtration. High fields with high-field gradients produce a force on ferromagnetic or weakly paramagnetic particles in order to separate them from the remaining fraction.

7. *Research magnets.* With improved materials, higher-field superconducting laboratory magnets have become available commercially. Superconducting magnets up to 15 T or more can be used in a laboratory environment on a routine basis. The cost is a very nonlinear function of field reflecting the technical limits of the materials. At this time, relatively inexpensive magnets are produced for fields of 9 to 10 T using NbTi.

It is not surprising that the applications of high magnetic fields are so broad. The worldwide development of modern technology has made use of smaller-scale electromagnetic devices in machines and control mechanisms during the last century. This iron-based magnet technology has been a dominant contributor to our electronic age. For flux densities less than 2 T (20 kG) very efficient field generation is feasible with iron-core magnet devices. No new developments in magnetic materials promise flux densities much larger than those already achieved during the iron technological age. However, much higher flux densities can be achieved with modern high-field superconducting magnets. This will lead to smaller-volume, high-efficiency devices with inherently large savings in energy and cost, assuming compatible cryogenic features. If normal conductors are used, large-volume, high-field applications would involve large losses. Superconducting materials reduce resistive losses to negligible values. If magnetic containment of high-temperature plasmas results in the practical approach to fusion power, superconducting magnet systems will be essential for reasonable efficiency.

Within the last few years, high-field research has led to the development of superconducting materials of great value for many practical applications. High magnetic fields are expected to continue to play an important role in research and technology in the future.

See also Electrodynamics, Classical; Magnetic Materials; Magnets (permanent) and Magnetostatics; Superconducting Materials; Superconductivity Theory.

BIBLIOGRAPHY

D. Bruce Montgomery, *Solenoid Magnet Design*. John Wiley and Sons, New York, 1969. (A) A definitive text of magnet design which treats the magnetic and mechanical aspects of resistive and superconducting systems. No iron circuits are considered—the text is directed toward high-field structures.

B. B. Schwartz and S. Foner, "Large Scale Applications of Superconductivity," *Phys. Today* **30,** 34 (1977). (E)

Megagauss Technology and Pulsed Power Applications (C. M. Fowler, R. S. Caird, and D. J. Erickson, eds.). Plenum, New York, 1987. (A) The most recent (4th) megagauss conference proceedings.

Report of NSF Panel on Large Magnetic Fields. Publications Office, U.S. National Science Foundation, Washington, DC, 1988. (E) This study reviews research opportunities of high magnetic fields and gives an assessment of the present and projected magnet technology, and recommendations for future high-field developments in the United States. See also *Phys. Today* (1988).

Strong and Ultrastrong Magnetic Fields and Their Applications (F. Herlach, ed.). Springer-Verlag, Berlin, 1985. (A) A comprehensive text of transient field generation techniques including the theoretical background, techniques for generation of pulsed fields (including implosion techniques for megagauss fields), and the measurements of these fields and research in high fields. Extensive reference listings of research in high fields up to 1985 are also given.

Superconducting Applications: SQUIDS and Machines (B. B. Schwartz and S. Foner, eds.). Plenum, New York, 1977. (I) Reviews worldwide efforts in superconducting quantum interference devices (SQUIDS). An extensive updated review of large-scale applications of superconductors up to 1976 is also included and supplements *Superconducting Machines and Devices—Large Systems Applications*.

Superconducting Machines and Devices—Large Systems Applications (S. Foner and B. B. Schwartz, eds.). Plenum, New York, 1974. (I) A text which gives reviews of all major large-scale applications of superconductivity using high-field superconducting magnet technology.

Superconductor Materials Science (S. Foner and B. B. Schwartz, eds.). Plenum, New York, 1981. (I) Reviews metallurgy and fabrication of conventional practical superconductors (NbTi and Nb_3Sn) and other processing high-field superconductors. The large-scale high-field applications in the above two books include large magnet systems for research and technology, energy generation (fusion and MHD) energy storage, ac and dc generators and motors, magnetic levitation for high-speed ground transportation, and industrial processing.

Research in high magnetic fields appears in various scientific journals. Proceedings of International Conferences on High Magnetic Fields and Their Applications are useful as a control source of activities. These have been held at Cambridge, Massachusetts (1961), Oxford, England (1963), Grenoble, France (1966), Nottingham, England (1969), and Grenoble, France (1974); see *Colloques Internationaux C.N.R.S., No. 242,* Physique sous Champs Magnetiques Intenses, Anatole France, 75700, Paris, France. There is also a series of conferences on semiconductors in high magnetic fields; see, for example, *Physics in High Magnetic Fields* (S. Chikazumi and N. Miura, eds.). Springer-Verlag, Berlin, 1981. (A)

Technical aspects of magnet design appear in *Proceedings of International Conferences on Magnet Technology* every two years. The most recent (11th) was held in Tsukuba, Japan, 1989. (A)

The Applied Superconductivity Conference takes place every two years. Recent technical advances in superconductors and applications are presented in the extensive Proceedings. The latest (1988) Proceedings are *IEEE Trans. Magn.* **MAG-25** (1989). (A)

Magnetic Materials

C. D. Graham, Jr.

INTRODUCTION

All materials are magnetic, in the sense that they develop some magnetization when subjected to an applied magnetic field. However, the term *magnetic material* usually refers to a material that can have a relatively large magnetization, and especially to a material that is used in a practical engineering device because of its magnetization.

Engineering magnetic materials are customarily divided into two categories: *soft* and *hard*. The names refer to the ease with which the level of magnetization can be changed by an applied field, not to the mechanical hardness. Soft magnetic materials can be easily magnetized and demagnetized; they have low values of coercive field (see Fig. 1) and are used when the magnetization must change with time, as in alternating-current devices generally. Hard magnetic materials resist changes in their magnetization (have high values of coercive field), and so are used for permanent magnets.

Materials with magnetic properties intermediate between hard and soft are sometimes called semihard materials, but since their principal application is in magnetic recording of analog and digital information (audio, video, and numeric), it seems better to call them *recording materials*.

SOFT MAGNETIC MATERIALS

The element with the highest saturation magnetization at room temperature is iron, which is also low in cost and widely available. It is therefore natural that iron and iron-based alloys are the most commonly used soft magnetic materials. For direct-current applications such as electromagnets for lifting scrap iron or for providing laboratory fields, solid iron parts may be used. Unfortunately, carbon, which is the most common alloying or impurity element in commercial iron, has bad effects on the soft magnetic properties, especially in increasing the coercive field. Iron with an unusually low carbon content is therefore required. For many

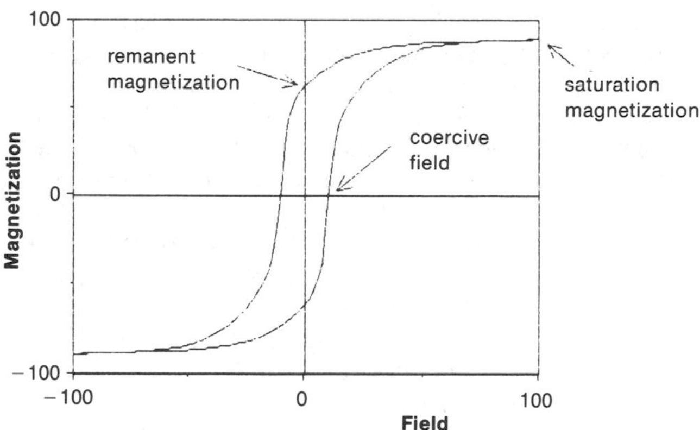

FIG. 1. Schematic hysteresis loop of a ferromagnetic or ferrimagnetic material.

years, the standard material was produced by the Armco Steel Company under the name of ARMCO iron. ARMCO iron is no longer made, but similar materials are available from other suppliers.

Higher saturation magnetization than pure iron can be achieved in iron–cobalt alloys, although these materials are expensive and rather difficult to produce. The usual composition is about 50% cobalt, with 1 or 2% molybdenum added to prevent or reduce atomic ordering and so reduce brittleness.

Devices to be operated at power frequencies (50 or 60 Hz) are made of layers or *laminations* of relatively thin sheet, in order to reduce the magnitude of the circulating eddy currents induced by the changing magnetic flux. These eddy currents generate heat and reduce the efficiency of the device. The special low-carbon steel sheet for magnetic applications is called *electrical steel*. It is often sold in a cold-rolled condition to increase the strength and hardness during manufacturing and punching operations, and is then annealed to restore the magnetic properties.

Improved magnetic properties can be obtained by adding alloying elements that increase the electrical resistivity of the steel. The usual choice is silicon, in amounts less than about 3% (larger amounts make the material brittle). In addition to increasing the resistivity, silicon reduces the magnetic anisotropy of iron. The disadvantage is that the saturation magnetization is decreased.

Silicon steel can be treated to produce a strong crystallographic texture or preferred orientation, which gives the sheet strongly directional (anisotropic) properties. This *grain-oriented* steel is widely used for medium and small power transformers; nonoriented grades of similar composition are used for large motors and generators. The sheet thickness is generally near 0.25 mm (0.01 in.) for use at 50 or 60 Hz. Ordinary electrical steel (without silicon) is used for small motors and transformers, where efficiency is of less concern than low cost.

A relatively new class of materials for power frequency use are the amorphous alloys or metallic glasses (see GLASSY METALS). These are alloys of 70 to 85 at. % iron + nickel + cobalt, with the balance comprising mainly boron and silicon, sometimes with a small amount of carbon. They are made by rapid solidification from the melt. These alloys are noncrystalline and highly homogeneous, with high electrical resistivity, and excellent soft magnetic properties. Their magnetization depends on the composition, with a maximum value about two-thirds that of pure iron. The maximum thickness is about 0.025 mm (0.001 in.). Amorphous alloys are somewhat more expensive than conventional sheet steels, but are useful where very low power losses are necessary or at frequencies up to several kilohertz.

In cases where very high magnetic permeability is required, especially at frequencies higher than 60 Hz, the usual choice is a nickel–iron alloy of about 80% iron, often containing a few percent of molybdenum to improve the mechanical and magnetic properties. These alloys have various commercial names, but are generically called *Permalloy*. At the 80Fe–20Ni composition, the magnetic anisotropy is very small, and the saturation magnetization still reasonably large. Nickel–iron alloys can be rolled to thicknesses below 0.025 mm (0.001 in.) if necessary; they require careful heat treatment and are easily damaged by bending.

Above about 10 kHz, the eddy-current losses in any metallic material become prohibitive, so that nonmetallic magnetic materials must be used. The customary choices are *ferrites,* which are ferrimagnetic oxides having the spinel structure (see FERRIMAGNETISM). Various standard grades are available, usually with some nickel, zinc, or manganese replacing iron in the basic Fe_3O_4 composition. Parts are made by pressing the oxide powder to the desired shape and sintering at elevated temperature to bond the individual particle by solid-state diffusion. The resulting parts are brittle ceramics, generally containing some remaining pores. The saturation magnetization is about one-quarter that of iron, but the power losses are reasonably low up to frequencies of 1 to 10 MHz.

PERMANENT MAGNET MATERIALS

Uses

Permanent magnets are used to create a steady magnetic field in some region of space, without the continuing expenditure of power. This field (or more precisely the field gradient) may be used to exert a force on another magnetic object, as in various latches and holding devices, scrap metal sorters, and torque-coupling devices. The field may be used to exert a force on a current-carrying wire, as in loudspeakers and motors; or the force may act on electric charges moving in a vacuum, as in a traveling-wave tube or a synchrotron (where the magnet assemblies that provide special field geometries are known as *wigglers* and *undulators*). The magnetic compass is based on a permanent magnet, and permanent magnets can be used to provide the field required for nuclear magnetic resonance in medical scanning technology.

Measure of Quality

Unlike the case of soft magnetic materials, there is a single index of quality for permanent magnet materials. This is the *maximum energy product,* defined as the area of the largest rectangle bounded by the positive *B* and negative *H* axes of the hysteresis loop and the *B–H* curve (see Fig. 2). The value of the energy product [units are megagauss-oersted (MG Oe) in cgs or kJ/m^3 in SI] is inversely proportional to the volume of magnetic material required to produce a given field in a given volume of space.

Materials

The original permanent magnets were natural Fe_3O_4 magnetites, known as *lodestones*. After the industrial revolution, these were replaced by hardened steel, and slow improvements were made by adding various alloying elements to the steel. Energy products were less than 8 kJ/m^3 (1 MG Oe). The first major advance occurred about 1930 and led to the development of a family of iron–aluminum–nickel–cobalt alloys, generally known by the generic name *Alnico*. Alnicos are generally made by casting, followed by a heat treatment that produces a very fine-scale two-phase structure. The two phases have very different magnetic properties, so that the structure can be thought of as an array of elongated magnetic particles in a nonmagnetic matrix. Alnico, which is still manufactured in limited quantities, usually has an energy product about 48 kJ/m^3 (6 MG Oe), a dramatic improvement over magnet steels.

A second new class of permanent magnets was developed about 1950, based on a family of iron oxide materials with hexagonal crystal symmetry. These hexagonal ferrites have nominal composition $MO_6(Fe_2O_3)$, where *M* is barium or strontium. They have relatively low values of magnetization, but are low in cost, low in weight, and have quite high values of coercive field. This permits the production of short cylindrical magnets magnetized parallel to the short direction, a configuration not practical with steel or Alnico magnets. As a result, ferrite magnets are widely used in many consumer products where low cost is important and extreme miniaturization is not required.

Theoretical work just before and after World War II led

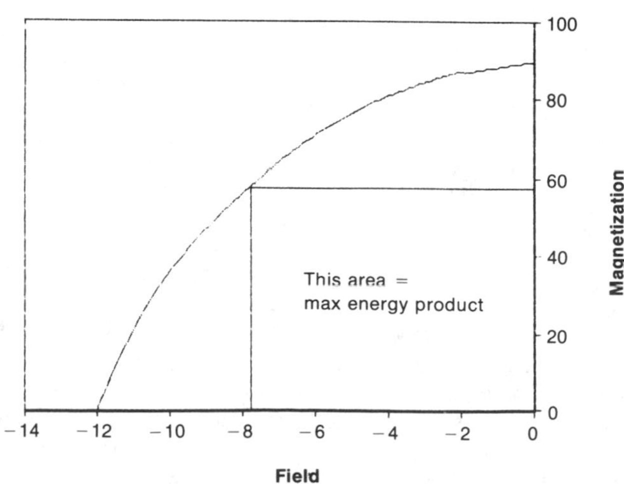

FIG. 2. Second-quadrant (field negative, magnetization positive) portion of hysteresis loop of a permanent magnet material, showing definition of maximum energy product.

to the idea of making new permanent magnet materials from very small single-domain magnetic particles (of order 10 to 100 nm) with a strong uniaxial anisotropy (see MAGNETIC DOMAINS AND BUBBLES). Iron and iron–cobalt particles in approximately cylindrical shape were produced by electrodeposition and made permanent magnets of respectable quality. An alternate approach was to use spherical particles, which are easier to make, but to choose a material with a strong uniaxial crystal anisotropy. The first successful magnet of this type was $SmCo_5$ (Sm is samarium, one of the rare-earth elements). $SmCo_5$ can be made with energy product values of 160 kJ/m^3 (20 MG Oe), easily twice the best previous values. To make full use of the intrinsic properties of the material, each crystal or grain must be oriented with its hexagonal axis parallel to the direction of magnetization. This is done by grinding the bulk material to a particle size such that each particle is a single crystal, placing the loose powder in a strong magnetic field to rotate each particle to the desired orientation, and compacting the powder while the field is still applied. The particles are then bonded by high-temperature sintering, just like ferrites. $SmCo_5$ is an excellent permanent magnet material. Aside from its cost, its major disadvantage is its mechanical brittleness and the strong tendency of rare-earth elements to oxidize.

Better magnetic properties are obtained using magnets based on a related compound, Sm_2Co_{17}, with substantial amounts of added iron and copper plus small amounts of other elements. These materials require a rather elaborate heat treatment in order to obtain optimum properties and are more difficult to magnetize initially than $SmCo_5$, but they can reach energy products of 480 kJ/m^3 (60 MG Oe) or more.

The most recent permanent magnet material is a previously unknown compound of composition $Fe_{14}Nd_2B$ (Nd is neodymium, another rare earth), which has tetragonal rather than hexagonal crystal symmetry. It can be prepared in the same way as the other rare-earth magnets, or alternatively by very rapid solidification from the liquid, grinding the resultant ribbons, and hot pressing at about 700°C to final shape. Energy products of 320 kJ/m^3 (40 MG Oe) or more can be attained, with better mechanical properties than the other rare-earth alloys. The price is also somewhat lower. The major disadvantage of FeNdB is the low Curie temperature of about 310°C, which makes the magnetic properties at room temperature fairly temperature dependent. This is a major disadvantage in some applications, but of little importance in others.

It has gradually become clear that the theory of single-domain particles, on which much development work has been based, is not directly relevant to real permanent magnets. The actual origin of the high coercive fields remains a subject for debate and experiment.

Bonded Magnets

Most of the permanent magnets described above can be ground into coarse powder and embedded in a polymer or rubber (or even metal) matrix. Such a composite material is called a *bonded magnet,* or, in Japan, a *plastic magnet*. The magnetic properties are degraded by the inclusion of the nonmagnetic matrix material, but complicated shapes can be easily made and the magnets are not brittle. They may even be flexible.

RECORDING MATERIALS

The third major class of magnetic materials is used for magnetic recording of analog and digital information. Audio- and videotape cassettes, and computer storage disks (floppy and hard), are common examples. The requirements for magnetic materials for use in recording are difficult to specify in any simple way, since various recording techniques have been proposed and used. Most magnetic recordings are made on multilayer structures consisting of a substrate of a flexible or rigid polymer, a metal, or a glass; a surface coating of some magnetic material; and a protective surface layer that may include a lubricant. Information is stored on a track along the length of a tape or around the circumference of a disk by magnetizing the surface layer parallel to the surface and in a positive or negative direction along the length of the track, using a small recording head that applies a strong localized field to the moving tape or disk. This technique is known as longitudinal recording.

The magnetic material used for longitudinal recording must have a coercive field high enough to resist demagnetization by the field from adjacent regions of the tape (or from any other source) but low enough to permit recording of a new signal over the previous one. The material must also have high enough magnetization to provide an adequate playback signal. The most widely used material is gamma ferric oxide, γ-Fe_2O_3, in the form of very small elongated crystals, about 100 nm in diameter and perhaps 500 nm long. Often the particles are made with a surface layer enriched in cobalt, which increases the coercive field substantially and so permits higher recording densities (more information per unit area of recording surface). Also used, especially for video recording, is CrO_2, a ferromagnetic oxide. The so-called "metal tape" uses particles of metallic iron or iron–cobalt, protected from oxidation by an inert surface layer.

Rigid disks (hard disks) used for digital information storage often employ a continuous thin-film magnetic layer rather than an assembly of particles. Usually a nickel–cobalt–phosphorus composition is used, which may be produced by chemical deposition, electroplating, or vacuum sputtering.

Magneto-optical materials

In magnetic recording, there is constant pressure to increase the density of recorded information. In conventional magnetic recording, the limit on density is the physical separation between the magnetic surface and the recording head, and present technology is approaching the practical limit of this dimension. Accordingly, there have been extensive efforts to find and develop alternate recording technologies. Most of these make use of magneto-optical behavior, usually the Kerr effect, which is the rotation of the plane of polarization of polarized light when it is reflected from a magnetized material. Using solid-state lasers as a light source, very high recording densities can be achieved, with no physical contact between the reading head and the magnetized surface. A rather special combination of properties is required in the magnetic material: it must reflect light, provide measurable rotation of the plane of polarization, be magnetized perpendicular to the surface, be recordable and erasable multiple times without degradation, and (of course) be stable against oxidation and corrosion. The technology is still under development; most current products use a continuous thin film of rare earth plus transition metal, such as terbium–iron.

PHYSICS OF MAGNETIC MATERIALS

Physicists are working on a wide variety of other magnetic materials quite independent of their possible usefulness. Three topics of special interest at the time of writing (1989) are spin glasses, multilayers, and superconductors.

Spin glasses are usually dilute alloys of magnetic atoms in a nonmagnetic matrix, such as cobalt in copper. The magnetic atoms are located at random positions and so have varying and sometimes competing magnetic interactions. The resulting magnetic behavior is complex and raises difficult theoretical questions.

Current technology allows the preparation of multiple-layered structures with individual layers whose thickness can be varied down to a few atomic diameters. The layers can be variously magnetic, antiferromagnetic, nonmagnetic, superconducting, etc. The resulting properties provide a rich field for experimental and theoretical physics and may lead to new applications.

High-temperature superconductors (with transition temperatures above 77 K) are a major current research topic. The magnetic behavior of these materials, both above and below the superconducting transition temperature, may provide important information about the possible mechanism of superconductivity.

See also FERRIMAGNETISM; FERROMAGNETISM; GLASSY METALS; MAGNETIC DOMAINS AND BUBBLES; MAGNETOSTRICTION; RARE EARTHS; SUPERCONDUCTIVITY THEORY.

BIBLIOGRAPHY

C. W. Chen, *Magnetism and Metallurgy of Soft Magnetic Materials*. North-Holland, Amsterdam, 1972.

S. Chikazumi, *Physics of Magnetic Materials*. McGraw-Hill, New York.

B. D. Cullity, *Introduction to Magnetic Materials*. Addison-Wesley, Reading, MA, 1972.

Malcolm McCaig and Alan G. Clegg, *Permanent Magnets in Theory and Practice*. Pentech Press, London, 1987.

John C. Mallinson, *The Foundations of Magnetic Recording*. Academic Press, San Diego, CA, 1987.

R. M. White, *Physics of Magnetic Recording*. IEEE Press, New York.

For access to current work in magnetic materials, consult the Proceedings of two major annual conferences:

Annual Conference on Magnetism and Magnetic Materials, jointly sponsored by the American Institute of Physics and the Institute of Electrical and Electronic Engineers (IEEE). Published in *J. Appl. Physics*, usually a special issue in April. Emphasis on basic science, although considerable coverage of applications as well.

Intermag, sponsored by the Magnetics Society of the Institute of Electrical and Electronic Engineers (IEEE). Published in the *IEEE Magnetics Transactions*, usually in September. Emphasis on practical applications, especially (in recent years) on magnetic recording.

There is also an International Conference on Magnetism largely devoted to the physics of magnetism, held every three years. The Proceedings are published, but each Conference publication is separate.

Magnetic Moments

R. K. Nesbet

The magnetic dipole moment of an infinitesimal current loop carrying electric current J (in units of electrostatic charge times velocity) is defined as [4]

$$d\boldsymbol{\mu} = \frac{1}{c} J\, d\mathbf{S}, \tag{1}$$

where $d\mathbf{S}$ is the oriented surface element enclosed by the current loop and c is the velocity of light. The magnetization $\mathbf{m}$, or magnetic moment per unit volume, in volume element $d\tau$ is defined by

$$d\boldsymbol{\mu} = \mathbf{m}\, d\tau. \tag{2}$$

For distributed currents, the electric current density j is defined as net current crossing a unit element of surface locally perpendicular to j within the distribution. When the electric charge and current densities are time independent, the vector fields $\mathbf{m}$ and $\mathbf{j}$ can be related by considering surface currents in a plane locally perpendicular to $\mathbf{m}$, which can be taken to have a single component $m_z(x,y)$. Each element $dxdy$ in the surface of constant z can be assumed to define a current J circulating about its periphery. All current between z and $z+dz$ is included in J. The net current through surface element $dzdx$ perpendicular to the current plane is obtained as the difference of adjacent loop currents, using Eqs. (1) and (2):

$$\begin{aligned} j_y dzdx &= J(x,y) - J(x+dx,y) \\ &= -c\frac{\partial m_z}{\partial x}\, dzdx. \end{aligned} \tag{3}$$

Similarly, the net current through $dzdy$ is

$$\begin{aligned} j_x dzdy &= J(x.y+dy) - J(x,y) \\ &= c\frac{\partial m_2}{\partial y}\, dzdy. \end{aligned} \tag{4}$$

In arbitrary geometry, these equations are

$$\mathbf{j} = c\nabla\times\mathbf{m}. \tag{5}$$

Since Eq. (5) implies

$$\nabla\cdot\mathbf{j} = 0, \tag{6}$$

the equation of continuity for the electric charge density ρ associated with $\mathbf{j}$,

$$\nabla\cdot\mathbf{j}+\frac{\partial\rho}{\partial t}=0, \tag{7}$$

requires the charge density to be time independent for Eq. (5) to be valid. Stationary charge and current densities will be assumed here.

Each elementary particle or aggregate with intrinsic angular momentum (spin) $\hbar\mathbf{J}$ carries an associated magnetic moment

$$\boldsymbol{\mu}=\gamma_J\hbar\mathbf{J}, \tag{8}$$

where γ_J is the gyromagnetic ratio. The magnetic moment of an electron is defined in terms of the Bohr magneton

$$\mu_e=e\hbar/2m_ec, \tag{9}$$

while nuclear magnetic moments are defined in terms of the nuclear magneton

$$\mu_N=e\hbar/2m_pc. \tag{10}$$

Here $-e$ is the charge of the electron, m_e is the electron mass, and m_p is the proton mass.

The magnetic moment of an electron is $\boldsymbol{\mu}_s+\boldsymbol{\mu}_l$, where

$$\boldsymbol{\mu}_s=-g_e\mu_e\mathbf{s}, \qquad \boldsymbol{\mu}_l=-\mu_e\mathbf{l}. \tag{11}$$

Here g_e is the electronic g factor, approximately

$$g_e\cong2.00232, \tag{12}$$

and $\hbar\mathbf{s}$ and $\hbar\mathbf{l}$ are the spin and orbital angular momentum operators, respectively. The total magnetic moment operator for the electrons in an atom is

$$\boldsymbol{\mu}=-\mu_e(\mathbf{L}+g_e\mathbf{S}), \tag{13}$$

where $\mathbf{S},\mathbf{L}$ are total $\mathbf{s},\mathbf{l}$, respectively. In a state of total angular momentum

$$\hbar\mathbf{J}=\hbar\mathbf{L}+\hbar\mathbf{S}, \tag{14}$$

the atomic magnetic dipole moment, in the limit of weak spin-orbit coupling, is

$$\boldsymbol{\mu}_J=-g\mu_e\mathbf{J}. \tag{15}$$

Here g is the Landé factor

$$g=1+\frac{J(J+1)+S(S+1)-L(L+1)}{2J(J+1)} \tag{16}$$

if g_e is approximated by 2.0.

The magnetic dipole moment of a nucleus is defined by the operator for each nucleon,

$$\boldsymbol{\mu}=\mu_N(g_l\mathbf{l}+g_s\mathbf{s}) \tag{17}$$

where for a proton,

$$g_l=1, \qquad g_s/2\cong2.793, \tag{18}$$

and for a neutron,

$$g_l=0, \qquad g_s/2\cong-1.913. \tag{19}$$

Tables of nuclear magnetic dipole moments have been published by Ramsey [5,6] and by Lederer and Shirley [3].

Rotational magnetic moments of molecules have also been discussed by Ramsey [6]. These moments involve coupling of electronic excitation to rotation of the nuclear framework of a molecule. They are of the order of magnitude of the nuclear magneton because the molecular rotation constant contains the reciprocal of the nuclear mass of the molecule.

In atoms and molecules, magnetic hyperfine-structure energy level splittings arise from the interaction between nuclear magnetic moments and the electronic magnetic moment distribution, which is stationary but nonuniform in a quantum-mechanical stationary state. The energy of interaction of stationary current density distributions due to physically distinct particle distributions is given by Maxwell's equations in the form

$$W_{12}=-\frac{1}{c^2}\iint\frac{1}{r_{12}}\mathbf{j}_1(1)\cdot\mathbf{j}_2(2)\,d\tau_1d\tau_2, \tag{20}$$

$$W_{12}=-\iint\frac{1}{r_{12}}(\nabla_1\times\mathbf{m}_1)\cdot(\nabla_2\times\mathbf{m}_2)\,d\tau_1d\tau_2, \tag{21}$$

in terms of the equivalent magnetizations. If $\mathbf{m}_1$ and $\mathbf{m}_2$ vanish outside the region of integration, this result can be transformed by partial integration to

$$W_{12}=\int\int_{r_{12}>\epsilon\to0}\frac{1}{r_{12}}(-\nabla_1\cdot\mathbf{m}_1)(-\nabla_2\cdot\mathbf{m}_2)\,d\tau_1d\tau_2 - \frac{8\pi}{3}\int\mathbf{m}_1\cdot\mathbf{m}_2\,d\tau. \tag{22}$$

The infinitesimal volume integral excluded from the first term here is $+(4\pi/3)\int\mathbf{m}_1\cdot\mathbf{m}_2\,d\tau$, in the limit $\epsilon\to0$. This is included in the second term, which vanishes if the particle distributions are spatially separated. The second term contributes to hyperfine structure as the Fermi contact interaction between a nuclear magnetic moment and the electronic spin density at that nucleus.

The first term in Eq. (22) is of the same form as the Coulomb interaction, except that electric charge densities ρ are replaced by magnetic monopole densities $(-\nabla\cdot\mathbf{m})$. The integrated monopole moment of any isolated system vanishes, since

$$\int(-\nabla\cdot\mathbf{m})\,d\tau=0 \tag{23}$$

by the divergence theorem, if $\mathbf{m}$ vanishes outside a finite volume. The magnetic dipole moment is

$$\mathbf{u}=\int(-\nabla\cdot\mathbf{m})\mathbf{r}\,d\tau=\int\mathbf{m}\,d\tau, \tag{24}$$

in agreement with the definition of $\mathbf{m}$.

A magnetic scalar potential can be defined by

$$\psi_2(1)=\int\frac{1}{r_{12}}(-\nabla_2\cdot\mathbf{m}_2)\,d\tau_2. \tag{25}$$

Then the first term of Eq. (22) becomes

$$W^m = \int (-\nabla\cdot\mathbf{m}_1)\psi_2\, d\tau, \tag{26}$$

analogous to the electrostatic interaction energy

$$W^e = \int \rho_1\phi_2\, d\tau. \tag{27}$$

This analogy provides definitions of magnetic multipole moments and polarizabilities formally equivalent to their electrostatic counterparts. Details of magnetic multipole interactions relevant to hyperfine-structure interactions have been worked out [1,2,7].

Important differences between electric and magnetic interactions arise from the different nature of the source terms ρ and $-\nabla\cdot\mathbf{m}$. For a wave function of definite parity (inversion symmetry), ρ is a scalar quantity (even parity) while $-\nabla\cdot\mathbf{m}$ is pseudoscalar (odd parity). Spherical tensor multipole moments are defined by integrating the source density multiplied by r^l times a spherical harmonic of order l. Since the spherical harmonic has the parity of l, electric multipole moments of a system with definite parity are nonzero only for even values of l, whereas magnetic multipole moments are nonzero for odd values. Hence, elementary particles or systems with definite inversion symmetry can only have intrinsic electric monopole, quadrupole, etc., moments, and can only have intrinsic magnetic dipole, octupole, etc., moments.

See also ELECTRODYNAMICS, CLASSICAL; ELEMENTARY PARTICLES IN PHYSICS.

REFERENCES

1. H. B. G. Casimir and G. Karreman, *Physica* **9**, 494 (1942).
2. V. Jaccarino, J. G. King, R. A. Satten, and H. H. Stroke, *Phys. Rev.* **94**, 1798 (1954).
3. C. M. Lederer and V. S. Shirley (eds), *Table of Isotopes*, 7th ed., Appendix VII. Wiley, New York, 1978.
4. W. K. H. Panofsky and M. Phillips, *Classical Electricity and Magnetism*. Addison-Wesley, Reading, MA, 1955.
5. N. F. Ramsey, *Nuclear Moments*. Wiley, New York, 1953.
6. N. F. Ramsey, *Molecular Beams*. Oxford University Press, New York, 1956.
7. C. Schwartz, *Phys. Rev.* **97**, 380 (1955).

Magnetic Monopoles

Alfred S. Goldhaber

A magnetic monopole is a hypothetical entity, the source of a magnetic field like the familiar Coulomb field of an electric charge. All magnets known up to now are made of magnetic dipoles. Cutting through such a dipole would yield two dipoles rather than isolated monopoles of opposite polarity. In classical and then quantum electrodynamics, magnetism has been described as a by-product of the motion of electric charge. This explanation accounts for the absence of free magnetic poles, since they represent the only kind of magnetic phenomenon that cannot be produced by circulating electric current. Although magnetic poles are not required by experiment or by electrodynamic theory, they are allowed by the theory, under certain conditions.

In classical nonrelativistic mechanics the equation of motion for interaction of an electric charge with a magnetic pole may be solved by exploiting the existence of a conserved total angular momentum

$$\mathbf{J} = \mathbf{L} + \mathbf{s}.$$

Here $\mathbf{L}$ is the orbital angular momentum of the charge–pole system, while $\mathbf{s}$ is a vector of constant magnitude directed from charge to pole, interpreted as the "angular momentum stored in the electromagnetic field":

$$\mathbf{s} = \int d^3r\, \frac{\mathbf{r}\times(\mathbf{E}\times\mathbf{B})}{4\pi c}.$$

If electric and magnetic charges are measured in the same units, then the magnitude of $\mathbf{s}$ is simply the product of the electric charge q and magnetic pole strength g divided by the speed of light c. Consequently, the projection of the conserved total angular momentum $\mathbf{J}$ in the direction from charge to pole is gq/c. In quantum theory an angular momentum operator must have quantized values for its projection in any direction. This gives the basis for a simple and rigorous derivation of Dirac's conclusion that the product of any pair of magnetic and electric charges must obey the quantization condition

$$gq/\hbar c = N/2.$$

For odd N a spinless charge and a spinless pole would combine to make a system with half-integer spin. Further, if both charge and pole were bosons, it may be shown that the composite would be a fermion, and thus would obey the familiar connection of spin and statistics.

Charge–pole interactions have another unusual aspect. Reflection in space or time shows motions that may be impossible, but would be possible if all poles changed sign. Thus pole strength must be treated as a pseudoscalar quantity in order for parity and time-reversal symmetries to hold.

If even one monopole exists in the universe, then the quantization condition implies that all electric charges must be multiples of a smallest unit of charge, in agreement with observation. In addition to this conceptual point, there is an important quantitative conclusion, namely, that the smallest nonzero monopole strength would have the enormous value $e/2\alpha$, where e is the charge of the electron and α is the fine-structure constant 1/137.036 ....

The enormous pole strength required by the quantization condition in turn implies that a monopole must have a geometric size two orders of magnitude greater than its Compton wavelength, and therefore may be described in some approximation by classical dynamics. However, ordinary electrodynamics cannot accommodate such internal structure, since the expression of the magnetic field as the curl of a vector potential is inconsistent with spread-out magnetic charge. Thus there must be an appropriate modification of electrodynamics at distances smaller than some characteristic length L if monopoles are to occur, and the pole mass must be greater than g^2/L. Successes of the standard model of electroweak interactions (which like electrodynamics is incompatible with spread-out monopoles) imply a lower bound on monopole mass of at least 10 TeV, well beyond the reach of any existing accelerator.

Combining classical electrodynamics and general relativity leads to a stable magnetic monopole which is also a black hole, having a mass of 0.1 mg (20 orders of magnitude larger than that of a proton) and a radius of 10^{-34} m (19 orders of magnitude smaller than that of a proton). The mass and size should be little affected by quantum corrections, since they are both an order of magnitude larger than the corresponding Planck units which set the scale for quantum gravity. A similar statement applies to lighter, larger monopoles arising in more complex, "unified" models (ranging from Kaluza–Klein gravitoelectrodynamics to minimal $SU(5)$).

Because electrons and other charged Dirac particles in the presence of a monopole have radial wave functions which diverge as 1/r, remarkable effects on fermion dynamics are possible, including deeply bound fermion states, as well as enhancement of the proton decay rate beyond that which might be observed in vacuum. The enhancement could be anywhere from imperceptible to strong (rate of 10^{23} s^{-1} for a monopole in nuclear matter).

A minimum-strength Dirac pole would feel a force of 2 MeV/m in a 1-G magnetic field. This means that a pole traveling in interstellar space could acquire enough energy from the galactic field to pass through the earth without being stopped by ionization energy loss. The drain on magnetic field energy coming from monopoles produced in the galaxy has been used to set an upper limit on the flux of high-energy poles at the earth of less than 10^{-11} m^{-2} s^{-1}. This limit could be less stringent by several orders of magnitude if the monopoles were very massive. It has even been suggested that, rather than draining magnetism, monopoles in great quantity may actually generate magnetic fields through monopole plasma oscillations, but it is not clear that such oscillations would have sufficient stability to account for the observed galactic fields. The characteristic properties of strong binding to ferromagnetic materials, easy acceleration in magnetic fields, large ionization loss in matter, and ability to induce an emf always of one sign as it passes through a conducting circuit have all been used in experiments designed to search for monopoles. So far extremely low limits (10^{-20} per square meter per second) have been established on monopole cosmic rays below 10^7 GeV, as well as comparable limits (cross section $\sigma \lesssim 10^{-40}$ m^2) on production rates at accelerators up to 10^3 GeV proton beam energy. A report in 1975 that a particular cosmic-ray track appeared compatible with a nonrelativistic monopole having twice the minimum Dirac strength has since been withdrawn. A report in 1982 of an induction signal corresponding to a minimum-strength monopole has not been reproduced more than once, despite an enormous increase in collection time and area for such detectors.

See also Dynamics, Analytical; Electrodynamics, Classical; Elementary Particles in Physics; Gauge Theories; Maxwell's Equations; Quantum Electrodynamics; Symmetry Breaking, Spontaneous.

BIBLIOGRAPHY

B. Cabrera, *Phys. Rev. Lett.* **48,** 1378 (1982). (I)

R. A. Carrigan, Jr., *Phys. Teacher* **13,** 391 (1975). (E)

R. A. Carrigan, Jr., and W. P. Trower, *Sci. Am.* **246** (No. 4), 106 (1982). (E)

P. A. M. Dirac, *Proc. Roy. Soc. London* **A133,** 60 (1931): *Phys. Rev.* **74,** 817 (1948). (A)

P. Goddard and D. I. Olive, *Rep. Prog. Phys.* **41,** 1357 (1978). (A)

A. S. Goldhaber, *Phys. Rev. Lett.* **36,** 1122 (1976); *Phys. Rev.* **D16,** 1815 (1977). (A)

A. S. Goldhaber and W. P. Trower, *Am. J. Phys.* (1989). (I)

C. Kittel and A. Manoliu, *Phys. Rev.* **B15,** 333 (1977). (I)

I. R. Lapidus and J. L. Pietenpol, *Am. J. Phys.* **28,** 17 (1960). (E)

P. B. Price, E. K. Shirk, W. Z. Osborne, and L. S. Pinsky, *Phys. Rev.* **D18,** 1382 (1978). (I)

M. Turner, E. N. Parker, and T. Bogdan, *Phys. Rev. D* **26,** 1296 (1982). (I)

T. T. Wu and C. N. Yang, *Phys. Rev.* **D12,** 3845 (1975); **D14,** 437 (1976). (A)

Magnetic Ordering in Solids

D. Sherrington

Just as the atoms in a crystalline solid are arranged periodically in space, so too if the atoms (or ions) in such a solid carry microscopic magnetic moments (which requires that they have nonzero orbital and/or intrinsic spin angular momentum) those magnetic moments may be aligned (ordered) in a spatially periodic fashion. Again, just as a solid melts on heating to form a liquid phase in which the atoms are disordered and free to move without long-range correlations, so too, when the temperature is raised sufficiently, long-range magnetic order is lost and the individual magnetic moments are no longer frozen in orientation, either individually or relative to one another. The characteristic temperatures at which the transition from magnetic order to disorder occurs vary widely from a few millidegrees to several hundred degrees Kelvin.

The simplest such magnetic order is ferromagnetism in which the average direction of all the magnetic moments is the same and macroscopic magnetization results, but many more complex magnetic structures are possible. Restricting discussion first to cases where all the moment-bearing atoms are identical, the next simplest ordering is that of (collinear) antiferromagnetism in which the atomic magnetic moments order periodically, alternating between opposite directions, for example, as shown in Fig. 1. In contrast to ferromagnets, antiferromagnets exhibit no macroscopic magnetization, the average magnetization over length scales much greater than the lattice spacing being effectively zero. However, the new periodicity associated with the magnetic structure is clearly exhibited in neutron diffraction. Since neutrons interact with the microscopic magnetic moments, antiferromagnetic order shows up as extra diffraction directions caused by the increase in the unit-cell size. Neutron diffraction also serves

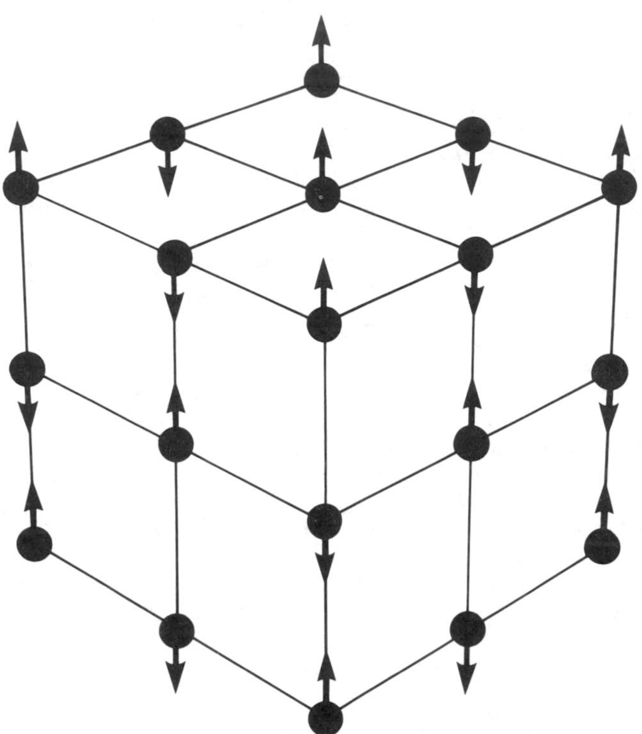

FIG. 1. Example of antiferromagnetic order.

to demonstrate the existence of noncollinear periodic order in certain materials, such as the heavy rare-earth metals which exhibit a wealth of helical and oscillatory orders, usually incommensurate with the atomic lattice. Examples are illustrated in Fig. 2.

In fact, it is not essential that all the atoms are magnetic. Often only a subset carry magnetic moments, but these can order as described; for example in MnO, only the Mn atoms carry moments but they order as a periodic antiferromagnet. When there is more than one type of magnetic atom, they can all order but contribute differently to the macroscopic magnetization; ferrimagnets are such a case where two moment-bearing atomic types order ferromagnetically within each type but antiferromagnetically relative to one another, leading to a reduced overall magnetization.

In all cases the ordering is a consequence of energy (or free energy) minimization. There are several driving mechanisms for these orderings, but all are quantum mechanical in origin, arising from so-called exchange contributions to electronic energies; direct magnetic interactions are much weaker. However, most can be modeled as though the interactions were between the magnetic moments, with the energy of a pair of moments dependent on their relative orientation but with the effective strength determined electronically. Among the specific mechanisms are: (i) *direct exchange:* due to the Coulomb energy difference between adjacent atoms with oppositely directed moments, caused by the changes in electronic charge distribution necessitated by the Pauli exclusion principle (which prevents two electrons from being in the same quantum state); (ii) *superexchange* in insulators: here the interaction is mediated by an intermediate nonmagnetic atom which shares electrons with the magnetic atoms on either side; (iii) *indirect exchange* in metals and semiconductors: in this case mobile conduction electrons carry polarization from one magnetic moment to another and the resultant ordering period reflects features of the electronic structure which may be incommensurate with the atomic lattice, such as Fermi wavelengths. In insulating solids the resultant effective pairwise interactions most often favor antiparallel (antiferromagnetic) order, but for metals both ferromagnetic and antiferromagnetic exchange occur. The local crystalline electric field due to the surrounding ions can further influence the particular order, by leading to preferences or dislikes for certain directions or by quenching orbital contributions to the magnetic moments. At finite temperatures entropic effects compete with energetic ones, always destroying the order when the temperature is raised sufficiently.

In fact, magnetic order sometimes occurs even when the fixed ions of a solid do not carry magnetic moments. In metals, magnetic order can be due to spontaneous spin polarization of conduction electrons, caused by the combination of Coulomb repulsion and the Pauli principle. This so-called itinerant-electron magnetism is the origin of the ferromagnetism of transition metals such as Ni, Co, and Fe, but can also yield periodic orders without macroscopic magnetization, as for example in Cr where there is a spin-density wave structure, the average local magnetization of the mobile electrons oscillating periodically in magnitude and direction. Another interesting situation is one where in isolation the ions of a solid have singlet ground states, so that in the absence of interaction they would be nonmagnetic. Magnetic order can nevertheless be induced by exciting the atoms to higher moment-bearing states if by so doing they gain a binding energy due to the resulting interactions which is greater than the cost of the excitation. Both itinerant-electron and induced-moment magnetism involve a competition between energetic mechanisms favoring and disfavoring magnetic correlation, with a minimum relative strength needed for purely electronic magnetic ordering to occur. However, even when such purely electronic ordering does not occur, ordering due to weaker nuclear magnetic moments can arise; an example is found at measurable temperatures in Pr (which is just not quite able to induce spontaneous atomic magnetic order).

Thus far, we have implicitly assumed perfect compounds. However, in some substitutional alloys one finds situations in which the periodicity of orientation is maintained over long distances even if a significant fraction of the magnetic atoms are randomly replaced by nonmagnetic atoms. It is remarkable that such ordering memory can sometimes be maintained at very high levels of dilution; for example, helical order persists to less than 1% of magnetic atoms (Gd) in $Y_{1-x}Gd_x$.

In alloys and amorphous solids another interesting type of magnetic order is often found. This is spin-glass order, which is the magnetic analog of the atomic amorphous (or glassy) solid state in which atoms are frozen over long times but without long-range periodicity. In a spin glass, magnetic moments are frozen in their average local orientation but are not arranged periodically relative to one another. There are

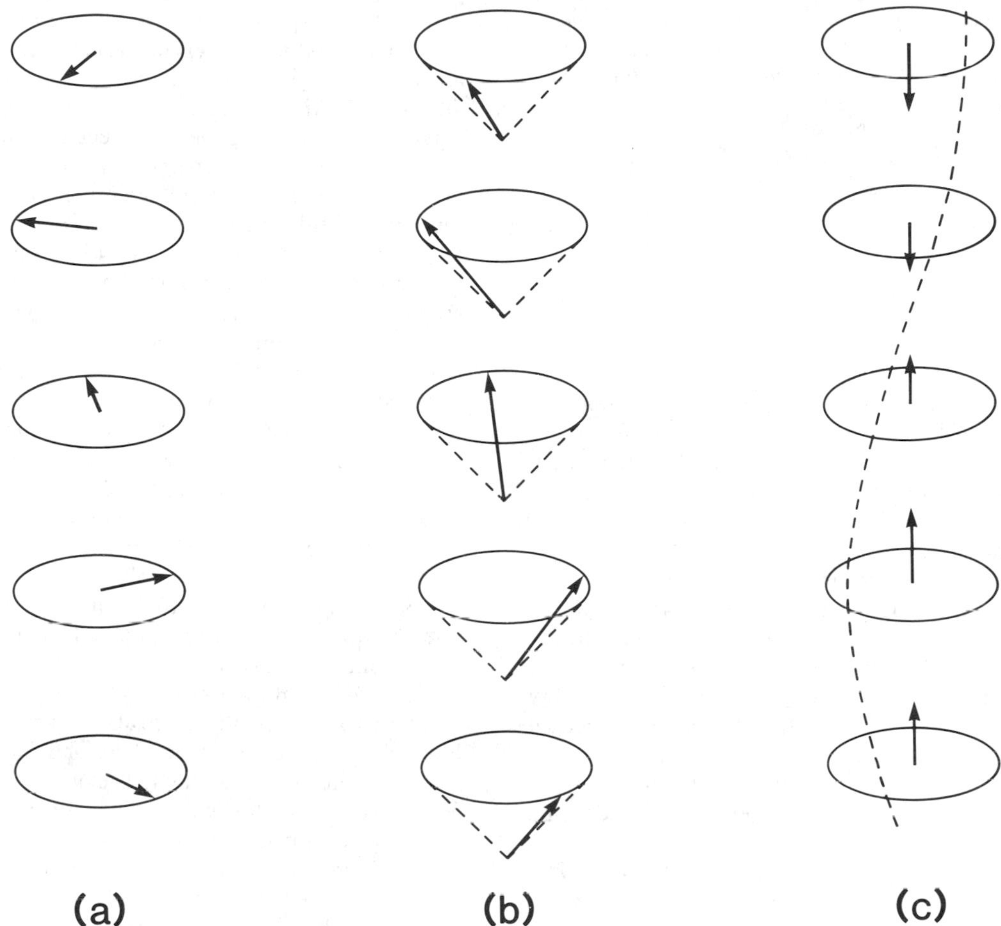

FIG. 2. Examples of magnetic order found in heavy rare-earth metals; (a) spiral, (b) conical spiral, (c) sinusoidally modulated. These metals have hexagonal lattice structures. The figures depict the variation in the magnetic order in the c direction, showing just one atom at each level. In each case the complete order is of ferromagnetically ordered layers in the planes perpendicular to the c axis.

several pieces of evidence for this behavior, but among the simplest are the following: nuclear magnetic spectroscopy can be used to demonstrate, via the hyperfine interaction between nuclear and electronic spins, the existence of frozen electronic moments; and the absence of periodicity in the magnetic freezing can be shown by neutron diffraction. The spin-glass ordering also shows up in a singularity in the magnetic susceptibility at the ordering temperature.

Spin glasses exhibit a number of other interesting features, particularly concerning their history dependence and their response to changes in their environment. One classic example concerns the magnetization response to an applied field at a low temperature. In a normal magnet one expects this to be independent of whether the field is applied before or after the temperature is lowered. However, in spin glasses these two procedures lead to very different magnetizations. This observation and others has led to the conclusion that spin glasses have, under any set of applied external conditions, many macroscopically equivalent but microscopically distinct sets of states, which, furthermore, evolve chaotically as the external conditions are changed.

The origins of the magnetic-ordering forces in spin glasses are believed to be similar to those in periodic magnets. The difference in their behavior compared with periodic magnets is a consequence of a competition between different ordering instructions, ferromagnetic and antiferromagnetic, frustrating the possibility of satisfaction of all, which combines with the inequivalency of different magnetic sites, due to the spatial disorder, to yield many alternative, imperfect compromises.

Ferromagnets are of great practical importance. All the types of magnetic order discussed here are of fundamental interest in solid-state physics as externally observable manifestations of microscopic interactions. Simple models for magnetic order have also been valuable as canonical simple models incorporating the fundamentals germane to the general study of phase transitions and critical phenomena. The occurrence of disorderly frustration in many problems in operational research, computer science, and neural modeling has led to an importance of studies of spin glasses well beyond the realm of solid-state physics.

See also FERRIMAGNETISM; FERROMAGNETISM.

BIBLIOGRAPHY

K. Binder and A. P. Young, "Spin Glasses: Experimental Facts, Theoretical Concepts and Open Questions," *Rev. Mod. Phys.* **58,** 801–976 (1986). (A)

J. Crangle, *The Magnetic Properties of Solids*. Arnold, London, 1977. (I)
C. Kittel, *Introduction to Solid State Physics*. John Wiley, New York, 1976. (E)
E. W. Lee, *Magnetism*. Penguin, New York, 1963. (E)
G. T. Rado and H. Suhl (eds.), *Magnetism*, Vols. 1–5. Academic, New York. (A)

Magnetoacoustic Effect

J. D. Gavenda

In pure metals at liquid-helium temperatures the conduction electrons provide the dominant mechanism for attenuation of ultrasonic waves. The variation in this attenuation caused by an applied magnetic field **B** is referred to as the magnetoacoustic effect. In recent years much has been learned about the properties of conduction electrons in metals through use of this experimental technique.

An ultrasonic wave propagating through a metal sets up an oscillating electric field in the neighborhood of the vibrating ions. The conduction electrons move about so as to rapidly screen this field and, in the process, absorb energy from the sound wave which is then transferred to the thermal motion of the lattice via collisions. The attenuation of the sound wave can be quite large (as much as 60 dB/cm at 100 MHz) for both longitudinal and transverse bulk modes and for surface modes.

Electronic attenuation is important when the mean free path l of the electrons is at least as large as the sound wavelength λ. This condition is easily satisfied in very pure (at least 99.999%) metals at liquid-helium temperatures, for ultrasonic frequencies f greater than about 10 MHz.

When $l \gg \lambda$, only those electrons moving perpendicular to the direction of propagation $\hat{\mathbf{q}}$ of the wave are effective absorbers. The explanation lies in the much higher velocity of the conduction electrons than that of the sound wave. In metals the electrons which play a role in transport processes are those moving with the Fermi velocity v_F. Since v_F is typically several hundred times the sound velocity v_s, the sound wave can be regarded as stationary in the metal as far as the electrons are concerned. If an electron is moving in the direction $\hat{\mathbf{q}}$, it experiences an alternating electric field which does no net work on it. However, if it moves perpendicular to $\hat{\mathbf{q}}$ along a constant phase plane, it will stay in a constant electric field which does a net amount of work on it.

When a magnetic field is applied to the metal, the situation is drastically changed. The electrons are forced to move in orbits with their direction of motion continually changing. On the portions of their orbits where they move essentially parallel to a sound wave front they absorb energy from the sound field; elsewhere the net interaction is very small. The result is a magnetic-field-dependent attenuation which may be monotonic, oscillatory, or resonant in character, depending on the electronic properties of the metal and on the geometry of the experiment.

The most common experimental geometry has the magnetic field **B** perpendicular to $\hat{\mathbf{q}}$. In a metal such as potassium which has a spherical Fermi surface, i.e., free-electron behavior, the projections of electron orbits on planes perpendicular to **B** will be circles with radii given by $R = p_\perp/eB$, where $p_\perp$ is the component of electron momentum perpendicular to **B** and e is the charge on the electron. As **B** is varied in intensity, an electron will absorb energy coherently from the sound field when the diameter of its orbit equals $n\lambda$, where n is an integer. Since the diameter is inversely proportional to B, the attenuation will have oscillations which are periodic when plotted against $1/B$. These periodic variations in attenuation are called *geometric oscillations* to distinguish them from other magnetoacoustic effects. Given values of λ, e, and B, one can solve for the value of $p_\perp$ for the electrons causing the oscillations. If the sound frequency is 50 MHz, geometric oscillations will be observed for values of B ranging from about 0.01 to 0.1 T.

For a spherical Fermi surface of radius p_0, we have $p_\perp = p_0 \times \sin\theta$, where θ is the angle between **B** and the electron's velocity. Thus, for a given value of B the diameters of the orbits will vary from zero to a value determined by p_0. The effects of all the orbits are summed to give the total attenuation, but those having θ near 90° will dominate, since there are more of them than in any other angular interval.

Geometrical oscillations are also observed in metals which have nonspherical Fermi surfaces, hence noncircular electron orbits in a magnetic field. The periods of the oscillations are determined by the extremal diameters of the orbits in the direction of propagation, so a Fermi surface can be calipered by measuring periods for a variety of directions of **B** and $\hat{\mathbf{q}}$.

In a metal such as copper with a Fermi surface which contacts the Brillouin zone boundaries, open orbits are possible when **B** is perpendicular to the directions ($\hat{\mathbf{G}}$, for example) in $\hat{\mathbf{k}}$ space for which contact occurs. The open orbit in real space will have a net displacement in a direction perpendicular to both **B** and $\hat{\mathbf{G}}$. If $\hat{\mathbf{q}}$ is parallel to this displacement, sharp resonance-type peaks in the attenuation will occur when the period of the open orbit equals $n\lambda$. *Open-orbit resonances* are narrow peaks, in contrast to the sinusoidal geometric oscillations, because all electrons on open orbits will have precisely the same period, determined by the crystal lattice spacing.

In an extremely pure metal it is possible to have an electron relaxation time τ comparable to, or greater than, the period of the sound wave, $1/f$. Then it is no longer valid to neglect the motion of the sound field with respect to the electron velocity. The sound field at a particular point in the metal will have changed appreciably by the time an electron on a large-diameter closed orbit returns to its starting point. When the sound frequency f equals the cyclotron frequency $f_c = eB/2\pi m_c$, *acoustic cyclotron resonance* will be observed if τ is greater than $1/f$. The latter condition is so stringent that acoustic cyclotron resonance has not been widely employed in the investigation of metals.

When the magnetic field is applied parallel, rather than perpendicular, to the direction of propagation of a transverse mode, *Doppler-shifted cyclotron resonance* may be observed. The sound field experienced by the electrons has its frequency Doppler-shifted by their drift velocity v_B along the magnetic field. Energy is absorbed from the sound field when the electron cyclotron frequency f_c is an integral multiple of the Doppler-shifted sound frequency $f - v_B/\lambda$.

For sufficiently large B the cyclotron frequencies of all the electrons will exceed their Doppler-shifted frequencies and none of them will be able to absorb energy from the sound field. The value of B beyond which there is no attenuation is called the Kjeldaas edge field. It can be related to the Gaussian curvature of the Fermi surface in the direction of $\hat{\mathbf{q}}$. For nonspherical Fermi surfaces, attenuation peaks will result if there are bands of electrons with the same value of $m_c v_B$ which will thus all experience Doppler-shifted cyclotron resonance at the same value of B.

Virtually all of the experiments involving the magnetoacoustic effect have utilized pulsed rather than continuous ultrasonic waves. This allows the use of propagation delay to discriminate between stray electromagnetically coupled signals and the actual acoustic signal arriving at the receiving transducer. In pure metals with an electron mean free path approaching the length of the ultrasonic pulse in the specimen, an interesting phenomenon called *anomalous sound propagation* may occur. The electrons absorb energy while passing through the wave packet and then, when they enter a region where there was no sound field, they are no longer in local equilibrium with the crystal lattice. This leads to an electrical current which oscillates with the ultrasonic frequency and which then excites lattice vibrations. The net result is that ultrasonic energy is carried into the region surrounding the wave packet at the electron Fermi velocity. The range of the effect is, of course, limited by the electron mean free path.

The phenomena discussed so far can all be explained by semiclassical transport theory. For magnetic fields above about 5 T quantum phenomena related to the de Haas–van Alphen effect show up in the magnetoacoustic effect. Since the magnitude of the attenuation is proportional to the number of electrons at the Fermi energy, quantum oscillations in the density of states will manifest themselves as oscillations in the attenuation. For $\mathbf{B}$ parallel to $\hat{\mathbf{q}}$ these effects can be particularly pronounced and are termed *giant quantum oscillations*. Although related to the conventional de Haas–van Alphen effect, giant quantum oscillations are caused by just those electrons which experience Doppler-shifted cyclotron resonance. This selectivity makes it possible to single out specific cross-sectional areas of the Fermi surface for measurement.

All of these phenomena cause variations in the velocity of the sound wave as well as in the attenuation. In other words, both the real and imaginary parts of the complex wave propagation vector exhibit magnetoacoustic effects. However, since the small variations (of the order of 10^{-6} to 10^{-4}) are difficult to measure, and because there has not been much theoretical work to exploit these effects, little use has been made of velocity measurements.

The variety of experimental arrangements of propagation, polarization, and magnetic field direction is limited only by the ingenuity of the experimentalist, but the interpretation of the data is relatively straightforward only for the cases described above.

See also CYCLOTRON RESONANCE; DE HAAS–VAN ALPHEN EFFECT; FERMI SURFACE; ULTRASONICS.

BIBLIOGRAPHY

J. D. Gavenda, "Study of the Electronic Properties of Metals by Ultrasonics," *Prog. Appl. Mat. Res.* **6,** 41–68 (1954). (E)

A. R. Mackintosh, "The Interaction of Long-Wavelength Phonons with Electrons," in *Phonons and Phonon Interactions* (Thor A. Bak, ed.), pp. 181–220. Benjamin, New York, 1964. (I)

J. D. Gavenda, "Anomalous Sound Propagation by Conduction Electrons," *Comments Solid State Phys.* **9,** 49–54 (1979). (I)

A very complete review of all aspects of ultrasonic wave propagation in metals has been published by J. Mertsching, "Theory of Electromagnetic Waves in Metals (III)," *Phys. Status Solidi* **37,** 465–522 (1970). (A) Part I of this series is also relevant: *Phys. Status Solidi* **14,** 3–61 (1966). (A)

Magnetoelastic Phenomena

R. L. Melcher

The magnetic and elastic properties of both crystalline and amorphous solids are, in general, coupled through the magnetoelastic interaction. This coupling arises from the elastic strain dependence of the interactions experienced by the individual magnetic moments associated with the ions or nuclei of the solid. Through these strain-dependent interactions a stress applied to the solid can influence its magnetic behavior and, conversely, a change in the magnetic state of the solid can alter its elastic properties. Numerous physical phenomena, both static and dynamic, result from the magnetoelastic coupling and a variety of experimental techniques have been developed to investigate them.

One of the more familiar phenomena associated with magnetoelastic coupling is static magnetostriction which refers to the change in size or shape of a magnetically ordered solid when its state of magnetization is changed by an applied magnetic field. In addition, magnetoelastic coupling can strongly influence the elastic moduli and concomitantly the dispersion and absorption of elastic waves in a solid. Resonant magnetoelastic coupling occurs in those cases when the magnetic and elastic modes of a solid are tuned to the same frequency. The establishment of thermal equilibrium in a magnetic solid is governed by the coupling of the magnetic and elastic degrees of freedom. Magnetoelastic coupling is important not only in magnetically ordered media (e.g. ferro-, ferri-, and antiferromagnetic) but also in paramagnetic materials including nuclear paramagnetism.

Microscopically, magnetoelastic coupling arises from the strain dependence of any of several interactions which influence the magnetic behavior of the material. Often the dominant contribution to the magnetoelastic interaction is the strain dependence of the crystalline electric field. For the transition-metal ions, this crystal field acts on the electronic orbital motion which in turn is coupled to the spin magnetic moments via the spin–orbit interaction. For the rare-earth ions, the spin–orbit interaction is so strong that the spin and orbital moments cannot be considered separately and the crystalline electric field acts on the moments associated with the total angular momentum of the ion. These types of single-ion magnetoelastic coupling are equally valid for paramagnetic impurities in diamagnetic host lattices and for magnetically ordered media. As the concentration of magnetic ions increases, two-ion contributions to the magnetoelastic coupling can become important. These include the strain dependence of the dipolar and exchange interactions which couple two magnetic ions. The single-ion and two-ion magnetoelastic interactions in magnetically ordered media are not always easily distinguished experimentally and theoretical estimates of the relative magnitudes are used to determine which dominates. In magnetically dilute materials only the single-ion magnetoelastic coupling is significant.

The magnetic moments associated with nuclear spin are coupled to the lattice strain via one of several types of interaction. In magnetically ordered materials, the strain can couple first to the electronic moments which are coupled to the nuclear dipole moments through the hyperfine interaction, thus resulting in an indirect nuclear spin magnetoelastic coupling. In metallic materials the dynamic interaction of the strain with the conduction electrons can cause the generation of an internal magnetic field which couples to the nuclear magnetic dipoles. Finally, nuclear spins which possess an electric quadrupole moment are coupled to the elastic strain through strain-induced changes in the electric field gradient at the nucleus due to the surrounding lattice.

Although the elastic properties of a nonmagnetic solid can be fully described by the symmetric strain tensor, the magnetoelastic interaction has been shown to depend not only on the symmetric strain but also on the antisymmetric elastic rotation tensor. These rotational contributions arise from the torques exerted on the magnetic moments in an anisotropic material by local lattice rotations and/or the applied field. The total interaction Hamiltonian of a magnetoelastic solid takes the general form

$$\mathcal{H} = \mathcal{H}_M(\mathbf{M},\mathbf{H}) + \mathcal{H}_{ME}(\mathbf{M},e_{ij},\omega_{ij}) + \mathcal{H}_E(e_{ij}),$$

where $\mathcal{H}_M(\mathbf{M},\mathbf{H})$ describes the magnetic properties of a rigid solid, $\mathcal{H}_E(e_{ij})$ describes the elastic properties of a nonmagnetic solid, and $\mathcal{H}_{ME}(\mathbf{M},e_{ij},\omega_{ij})$ is the magnetoelastic interaction which couples the two. Here, **M** represents the magnetic variables required to describe magnetic properties of the system, **H** is an applied magnetic field, and e_{ij} and ω_{ij} represent the symmetric elastic strain tensor and the antisymmetric elastic rotation tensor, respectively. Phenomenological descriptions of magnetoelastic phenomena rely heavily on the symmetry properties of the elastic strain, of the surrounding of the magnetic ion or nucleus, and of the interaction itself. Through application of these symmetries, the number of coupling constants required to characterize the magnetoelastic system is reduced to a minimum. Once the explicit form of the Hamiltonian $\mathcal{H}$ is known, both the static and dynamic properties of the coupled system can be derived.

The static properties of the system are calculated from the thermodynamic free energy which is deduced from $\mathcal{H}$. These properties include the magnetostrictive distortion of the lattice, the magnetoelastic contributions to the magnetic anisotropy and the magnetoelastic contributions to the elastic stiffness moduli. In most cases the contributions to the elastic moduli are small; however, there are cases in which the coupling completely dominates the elastic behavior and can even result in lattice instabilities and structural phase transitions which reduce the symmetry of the lattice.

The dynamic properties of the system are calculated with the aid of the coupled equations of motion which can be derived from the Hamiltonian $\mathcal{H}$. The dynamic coupling of the magnetic and elastic modes of the system enables one to use elastic waves as a probe of the magnetic system and the magnetoelastic interaction. Sensitive techniques have been developed to measure the resulting absorption and dispersion of elastic waves. These measurements provide valuable information regarding the coupling between the lattice and the magnetic degrees of freedom and how this coupling leads to the establishment of thermal equilibrium in the solid.

Magnetostrictive transducers, which convert high-frequency electromagnetic radiation into elastic waves, and magnetoelastic delay lines, whose delay time can be varied by changing the biasing magnetic field, are two technological applications which are made possible by magnetoelastic interactions.

See also Crystal and Ligand Fields; Diamagnetism and Superconductivity; Elasticity; Magnetic Moments; Magnetoacoustic Effect; Magnetostriction.

BIBLIOGRAPHY

C. Kittel, *Phys. Rev.* **110,** 836 (1958). (I)

R. C. Le Craw and R. L. Comstock, "Magnetoelastic Interactions in Ferromagnetic Insulators," in *Physical Acoustics,* Vol. 3B, Warren P. Mason (ed.). Academic Press, New York, 1965. (E)

Edmond B. Tucker, *Physical Acoustics,* Vol. 4A, Warren P. Mason (ed.). Academic Press, New York, 1966. (I)

D. I. Bolef, *Physical Acoustics,* Vol. 4A, Warren P. Mason (ed.). Academic Press, New York, 1966. (I)

R. L. Melcher, "Rotational Invariance and Magnetoelastic Phenomena in Anisotropic Media," in *Proceedings of the International School of Physics, Enrico Fermi,* Course LII, Varenna, Italy, E. Burstein (ed.). Academic Press, New York, 1973. (A)

R. L. Melcher, "The Anomalous Elastic Properties of Materials Undergoing Cooperative Jahn-Teller Phase Transitions," in *Physical Acoustics,* Vol. 12, Warren P. Mason and R. N. Thurston, (eds.). Academic Press, New York, 1976. (I)

Magnetohydrodynamics

P. H. Roberts

INTRODUCTION

Magnetohydrodynamics (MHD, for short) is that branch of continuum mechanics which studies the flow of an electrically conducting fluid in the presence of an electromagnetic field. The subject is also occasionally called hydromagnetics and is very often called *magnetofluid dynamics.* In addition to some significant technological uses, it finds important applications in cosmical physics, a few of which will be described below. MHD processes usually take place on such a long time scale, compared with the time taken by light to cross the system under study, that relativistic effects may be safely ignored. The electromagnetic field is then governed by the pre-Maxwell theory, in which displacement currents are absent, and in which the energy density of the electric field is negligible compared with that of the magnetic field. The fluid flow obeys the Navier–Stokes equations.

At the root of magnetohydrodynamics lies the mutual interaction of fluid flow (**u**) and magnetic field (**B**). On the one hand, the field and its associated electric current density [$\mathbf{j}=$ curl $\mathbf{H}=$ curl $(\mathbf{B}/\mu)$, where μ is the permeability] create a *Lorentz force* ($\mathbf{j}\times\mathbf{B}$, per unit volume) which seeks to accelerate the fluid in a direction perpendicular to both the electric current and the magnetic field. On the other hand, as in the motion of any conductor across a magnetic field, the fluid flow creates an *electromotive force* ($\mathbf{u}\times\mathbf{B}$) perpendicular to both fluid velocity and magnetic field, and proportional to each; this emf can create currents that modify the electromagnetic field. Magnetohydrodynamics is, however, not merely the hydrodynamics of fluids driven by Lorentz forces, nor yet the electrodynamics of moving fluids. To the marriage of fluid mechanics and electromagnetism are born new physical effects, particularly the Alfvén wave, that are not present in these subjects taken singly.

ELECTROMAGNETIC INDUCTION IN FLUIDS

The two ingredients of MHD will at first be considered separately, it being initially supposed that the fluid motion, **u**, is given.

It is well known in conventional electromagnetic theory that magnetic fields and currents on a length scale L are significantly changed by Ohmic resistance only over time scales comparable with or larger than the *electromagnetic diffusion time* $\tau_\eta=L^2/\eta=\mu\sigma L^2$, where σ is the electrical conductivity and $\eta=1/\mu\sigma$ is the *magnetic diffusivity*. For $t\ll\tau_\eta$, electrical resistance is ineffective, and the material behaves as a perfect conductor. Then, in the frame moving with a particular fluid element, the electric field $\mathbf{E}'$ must vanish, since otherwise it would drive an infinite current. In the laboratory frame the electric field **E** is thus equal and opposite to the induced emf $\mathbf{u}\times\mathbf{B}$.

A central theorem, due to Alfvén, states that perfectly conducting fluid contained within any tube of magnetic flux remains in that tube for all time; i.e., the magnetic field is "frozen" to the moving fluid.

Alfvén's theorem provides a vantage point from which many MHD processes are easily viewed. For example, an interstellar gas cloud from which a star forms is generally threaded by a weak magnetic field ($\sim 10^{-10}$ T) whose lines of force are (Alfvén's theorem) brought together as the protostar condenses. The associated Maxwell stresses (see below) oppose the collapse but, under the assumption that they are overcome, the star may be born with a strong magnetic field. If, as seems possible, τ_η exceeds the entire lifetime of the star, its magnetism is a fossilized relic of its birth. If possible eruption of field from its interior is ignored, the net (unsigned) flux leaving the star will not change, although

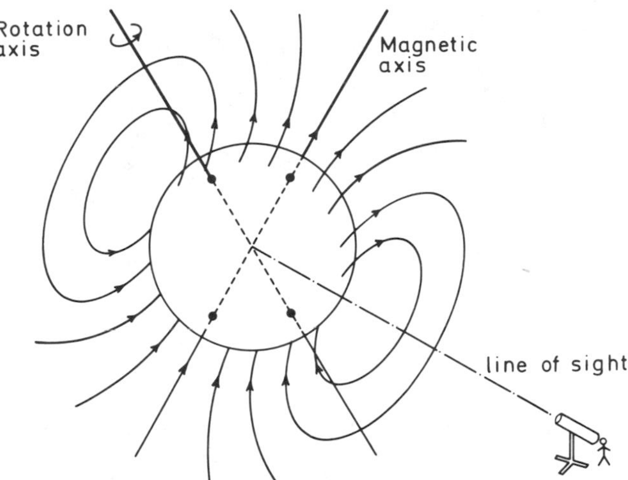

FIG. 1. Oblique rotator model of magnetic variability. This idealized model is symmetric about a magnetic axis, and the lines of force emerging and returning to one particular magnetic meridian are shown. The rotation axis and line of sight are also indicated.

it may be redistributed over its surface by internal motions (Alfvén's theorem). In particular, its lines of force will be carried round by a rotation, and the star will present a continuously changing but periodic magnetic face to the observer (the "oblique rotator" model of magnetic variability; see Fig. 1).

The magnetic energy of a system is increased not only by motions that compress flux tubes as explained above, but also by flows that stretch them; indeed, this is the only amplification mechanism available in an incompressible fluid. A shearing motion can extend a magnetic field and align it with the flow. Zonal shears arise readily in highly rotating systems such as the solar convection zone, and can create zonal fields that greatly exceed the meridional "poloidal" field from which they were spun (see Fig. 2). Sunspot pairs mark areas where an erupted segment of such a "toroidal" flux rope crosses the photosphere into the solar atmosphere.

Violations of Alfvén's theorem arise through the finite conductivity of real fluids. Substantial diffusion of field lines

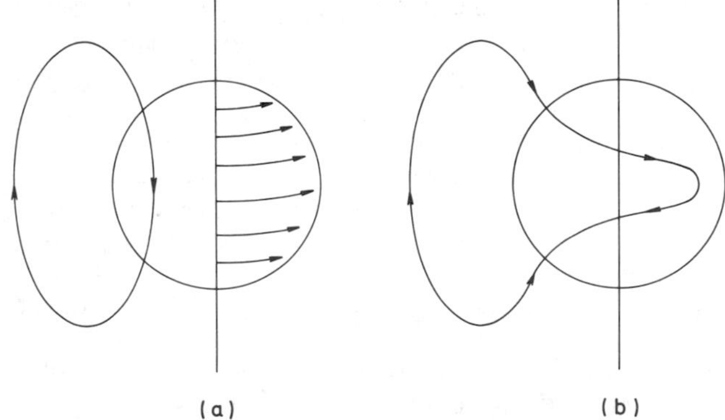

FIG. 2. In (a), a field lying in meridian planes is shown on the left of an axis of symmetry, and a zonal motion, sheared in latitude and confined to the conducting sphere, is shown on the right. This motion creates a zonal component of field inside the fluid, as indicated in (b).

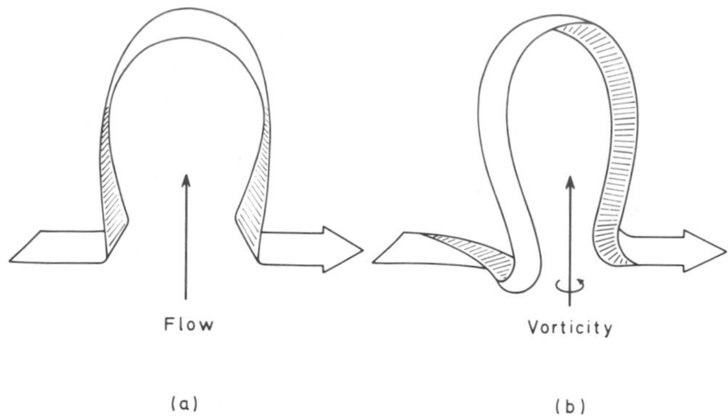

FIG. 3. The effect of helical flow on a ribbon of magnetic flux. A localized motion shapes the ribbon into an Ω; see (a). The twist of the (correlated) vorticity turns all but the feet of the Ω into a plane perpendicular to the original field (b). The combined effect of many such loops creates, by Ampère's law, an average current antiparallel to that field (or parallel if the helicity is negative).

through fluid occurs over times $t \gtrsim \tau_\eta$. Thus, for example, the fossil-field theory is powerless to explain how the main geomagnetic field, which apparently emanates from the earth's core ($\tau_\eta = 10^4$ yr), can have persisted over all geologic time. In this and many similar cases it is thought that, despite the short-circuiting unavoidable in a homogeneous body like the earth's core, the magnetic field is sustained by the emf it itself induces, as in a commercial self-excited dynamo. The lack of the deliberate asymmetry of the man-made machine may in the earth's core be compensated by an absence of mirror symmetry in the fluid motions, arising especially through the action of Coriolis forces. A non-mirror-symmetric flow generally possesses *helicity*, i.e., a correlation between its velocity and vorticity fields. The way an eddy of helical turbulence can contribute to the mean electric current ($\bar{\mathbf{j}}$) induced from the mean magnetic field ($\bar{\mathbf{B}}$) is indicated in Fig. 3. The process is called *the* α *effect*, from the coefficient α usually used to describe it ($\bar{\mathbf{j}} = \alpha\bar{\mathbf{B}}$). The study of magnetism self-excited by given fluid flows is usually called kinematic *dynamo theory* [1].

Other situations in which Alfvén's theorem is poorly obeyed are those where the distance $u\tau_\eta$ through which the field would be convected over the time τ_η is small compared with L; i.e., when the *magnetic Reynolds number*

$$R_m = uL/\eta = \mu\sigma uL$$

is small. A particular system may possess several scales of differing R_m, and differing electromagnetic response. For example, even when $R_m \gg 1$, motions may convect oppositely directed flux into close proximity. The concomitant currents and large field gradients define a new length scale $l \ll L$ and a small R_m. During times of order l^2/η, the field lines outside the high-current region move with the fluid, but where they cross the region, diffusion can alter their overall topology by separating them from the fluid and reconnecting them in a new way, processes sometimes called "severing" and "coalescence" of field lines (see Fig. 4). The new topology may allow substantial shortening of field lines outside the region, to the accompaniment of a transfer of magnetic to kinetic energy. Solar flares are often interpreted as regions of large current density in the solar atmosphere, where reconnection of field lines is under way.

Turbulence provides another example of a multiple-scale system. The lines of force of a field $\mathbf{B}_0$, which would otherwise be of large length scale L, are stretched and crinkled by the (given) small-scale motions. This process is similar to, though more complicated than, the convection of a scalar field, such as heat or concentration of a contaminant, by a turbulent flow. The stretching tends to enhance the field energy and the crinkling to increase the Ohmic dissipation. It is often argued that in the absence of helicity in the motions, crinkling will be more potent than stretching and that $\mathbf{B}_0$ will decay more rapidly than it would have done had the conductor been stationary. In the simplest picture, the effect of the turbulence on $\mathbf{B}_0$ is recognized by adding to η a *turbulent diffusivity*, η_T. In many astrophysical plasmas, $\eta_T \gg \eta$ and $\tau_\eta = L^2/\eta_T \ll L^2/\eta$. It is hard to account for the reversal of the global field of the sun during its 22-yr cycle except by saying that (helical) turbulence destroys and recreates the field by ac-dynamo action over a period of $L^2/\eta_T \sim 10$ yr rather than $L^2/\eta \sim 10^7$ yr.

If, as a result of dynamo action for example, the magnetic field grows to where the Lorentz force $\mathbf{j}\times\mathbf{B}$ is comparable with the controlling forces in the equation of motion, the basic assumption of this section, that $\mathbf{u}$ is determined without reference to $\mathbf{B}$, must be reconsidered. Indeed the Lorentz

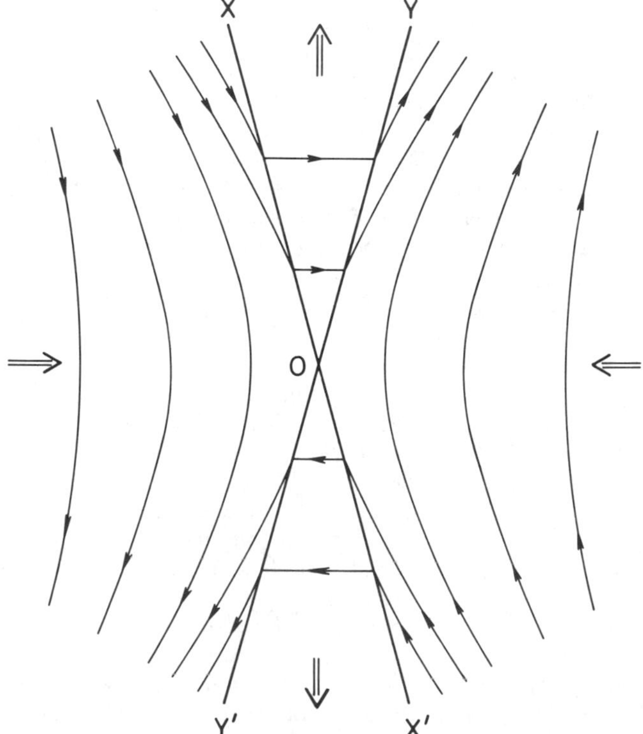

FIG. 4. Magnetic field line reconnection associated with a neutral "point," O, as described by the Petshek mechanism. In this cross section of a two-dimensional model, motions from right and left (double arrows) carry oppositely directed field lines toward O, in the vicinity of which they are reconnected with the new topology indicated. They are then carried away to the upper and lower halves of the figure. XOX' and YOY' are hydromagnetic shocks initiated at O across which the direction of flow and field change abruptly.

forces will prevent indefinite field growth. When Lorentz forces are comparable with inertial forces ($\rho\mathbf{u}\cdot\nabla\mathbf{u}$, where ρ is the fluid density), the magnetic and kinetic energy densities are of the same order, and $A \sim 1$ where A is the *Alfvén number,* or "magnetic Mach number": $A^2 = (\frac{1}{2}\rho u^2)/(B^2/2\mu)$. The Lorentz force is comparable with the pressure gradient when magnetic and internal energy densities are of the same order, and $\beta = \mu p/B^2 \sim 1$. It has been supposed in this section that $A \gg 1$ and $\beta \gg 1$.

MAGNETOSTATICS

It is now supposed that the magnetic field $\mathbf{B}$ and its associated current density $\mathbf{j}$ are given. It will be helpful to interpret the associated Lorentz force $\mathbf{j} \times \mathbf{B}$ by the equivalent Maxwell stresses: an isotropic *magnetic pressure* $B^2/2\mu$ together with a tension B^2/μ along field lines.

The sum of magnetic pressure and kinetic pressure (p) is called the *total pressure* (P). It often fills the role played in the nonmagnetic theory by p, as for example when providing the restoring force for magnetoacoustic waves traveling perpendicular to a uniform field. Another example is provided by a flux tube inside the sun. Hydrostatic balance requires P to be about the same inside and outside the tube but, since the interior magnetic field and magnetic pressure are larger, the kinetic pressure is smaller. In a stellar interior radiative transfer diffuses heat rapidly, and the gas inside the tube will be at the same temperature T as outside; therefore, by the gas law ($p \propto \rho T$), its density will be less. The flux tube acquires in this way a *magnetic buoyancy*, which may raise it until it partially erupts into the solar atmosphere. Near the photosphere, turbulence is an efficient convector of heat, but its effectiveness may be diminished by the magnetic field of the spot. The lower kinetic pressure of the spot compared with the surrounding photosphere would then imply a smaller temperature and explain the characteristic relative darkness by which the spot is seen and named.

A quiescent solar prominence is often modeled as a magnetostatic structure. Other magnetostatic equilibria arise in the terrestrial search for controlled thermonuclear reactors operating by the fusion of a hot deuterium plasma. Contact of such a plasma with walls would release impurities that would enhance radiation losses by bremsstrahlung, thus cooling the plasma and quenching the reactions. This would be prevented if the plasma could be confined by "walls" of magnetic field lines, impenetrable over containment times less than τ_η because of Alfvén's theorem. The magnetic walls could be moved inward to "pinch" (i.e., compress) the plasma and raise it to reaction temperatures. Containment fields can be provided in principle by a current distribution flowing within the plasma itself. For example, the toroidal pinch consists of a "doughnut" of plasma around whose surface an axial current flows. The associated lines of force are hoops that pass through the hole in the doughnut and surround the pinch, so exerting a compressing magnetic stress upon it. For a number of reasons, some of which are discussed below, these ideas meet with practical difficulties. The controlled thermonuclear reaction field has, however, added much new understanding of MHD processes. One much studied, and theoretically simpler, model is the linear pinch, a straightened-out toroidal pinch with ends; because of these, the plasma in this pinch is not properly confined magnetically.

An important special class of magnetohydrostatic equilibria are the *force-free fields*, in which field and current are parallel ($\mathbf{j} \times \mathbf{B} = 0$). These model tenuous, but highly conducting, plasmas (such as the lower solar atmosphere) in which $\beta \ll 1$. The field of minimum magnetic energy continuous with a given radial field emerging from the photosphere is, by a well-known result of potential theory, curl free ($\mathbf{j} = 0$). In contrast, the high conductivity of the solar atmosphere implies that the two feet where a field line enters and leaves the photosphere are always attached to that field line. Under this stronger constraint, the field of minimum energy is force free, its energy being, of course, larger than the unconstrained minimum. It can, however, be reduced by the flares which (see above) reconnect field lines with different pairs of feet, thus relaxing the force-free field toward the potential field.

ALFVÉN WAVES AND MAGNETOHYDROSTATIC STABILITY

Each side of the MHD interaction is needed to understand *Alfvén* or *magnetohydrodynamic waves*. These are most readily pictured in an incompressible, inviscid, perfectly conducting fluid pervaded by a uniform field $\mathbf{B}_0$. The total pressure is not affected by the wave, and only the stress B_0^2/μ stretching field lines need be considered. The absence of electrical resistance implies that every flux tube retains its contents as it moves (Alfvén's theorem). A tube of unit cross-sectional area therefore behaves like a string of mass ρ per unit length stretched under tension B_0^2/μ. Like the stretched string, it can, when plucked, transmit a transverse, nondispersive wave, moving with the *Alfvén velocity*

$$V_A = B_0/(\mu\rho)^{1/2}.$$

Its polarization is the common direction of the flow and field ($\mathbf{B} - \mathbf{B}_0$) associated with the wave; this is perpendicular to $\mathbf{B}_0$ and to the electric field of the wave. The mean energy density of the wave is equipartitioned between the motion and the magnetic field. Its transmission with velocity V_A gives the correct mean (Poynting) energy flux. Wave energy is completely reflected from perfectly conducting or insulating walls.

When dissipative effects are considered in the commonly occurring case of small *magnetic Prandtl number*, $P_m = \nu/\eta$ (where ν is the kinematic shear viscosity), the wave is attenuated over the characteristic distance $V_A\tau_\eta$, where $\tau_\eta = l^2/\eta$ and l is its wavelength. Waves on the same scale as the system ($l = L$) can cross it without serious attenuation only if the *Lundquist number*

$$Lu = V_A L/\eta = \mu\sigma V_A L$$

is large, i.e., if the Alfvénic time scale $\tau_A = L/V_A$ is small compared with τ_η.

The magnetohydrostatic equilibria considered earlier were implicitly of large Lu, for they expressed a balance between the Lorentz and the remaining forces that lasted over times large compared with τ_A, and diffusion of $\mathbf{j}$ and $\mathbf{B}$ drastically alters the Lorentz force in time τ_η.

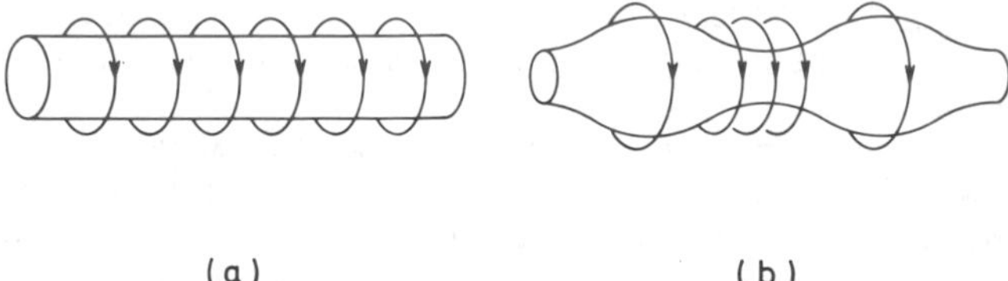

FIG. 5. The linear pinch (a) and the sausage mode of instability (b) to which it is prone.

When the balance of forces in a magnetohydrostatic equilibrium is disturbed, it responds on the dynamic time scale τ_A. If the disturbance increases indefinitely at this rate, the equilibrium is often said to be *dynamically unstable*.

Most instances of magnetic confinement are dynamically unstable. Since few reactions can occur in the time τ_A, the design of this type of thermonuclear reactor is difficult. The simple linear pinch defined earlier is subject to "necking instability," during which its surface develops an axisymmetric, volume-preserving deformation of increasing amplitude (see Fig. 5). The cross section is alternately widened and narrowed, the constricted portions, or "necks," resembling the links joining sausages, so suggesting the alternative name, the sausage mode, for the instability.

Energy methods provide a powerful tool with which to attack questions of dynamic stability. They explain in a general way why the unstable motions are, even in a compressible medium, usually divergenceless; were they not so, energy that could have been used to drive the motions would be uselessly squandered on increasing the internal energy. In a similar way, since energy must usually be expended to bend field lines, the displacements most dangerous to the stability of a system are often those that permute field lines without substantially bending them. These are called *interchange instabilities*. The most stable are often "minimum-B" configurations, in which the smallest B occurs within the plasma, and $B=|\mathbf{B}|$ increases outward in all directions from that point up to the surface of the plasma.

Because of micro-instabilities, even dynamical stability would not guarantee a viable configuration. Also, perhaps surprisingly, diffusive processes can *de*stabilize a system that would be stable in their absence. Particularly significant are the *resistive instabilities* arising from magnetic diffusion. These grow on time scales intermediate between τ_A and τ_η. For example, the *tearing mode* has a characteristic growth rate of $\tau_A^{2/5}\tau_\eta^{3/5}$. The instabilities succeed in severing and reconnecting the field lines in thin magnetic diffusion layers; the reconnected lines can shorten, thereby freeing magnetic energy to drive the instability.

Astrophysics also provides a number of interesting examples of dynamic and resistive instabilities. For instance, the toroidal field wound up in the solar convection zone by zonal shear motions may, as it strengthens, be subject to a "watchspring instability," in which a coil of the field releases energy by jumping, like an overwound watchspring, out of the plane of shear. A similar eruption into a galactic halo of the field wound by shear in the plane of a galactic spiral is sometimes called a "galactic hernia."

The discussion of this section has ignored all questions of *magnetogasdynamics*, a term used in preference to magnetofluiddynamics when the compressibility of the fluid is important. In reality, the disturbed magnetostatic equilibria would respond not only on the Alfvénic time scale but also on two time scales defined by the "fast" and "slow" magnetoacoustic waves; shocks might develop. Space does not permit further discussion of these fascinating topics here; the interested reader is referred to a text by Jeffrey [2]. As noted above, the stability of a system is often decided by its response to divergenceless motions, and it seems generally true that the incompressible fluid provides a good basis for understanding MHD processes and their applications.

LABORATORY AND ENGINEERING MAGNETOHYDRODYNAMICS

Although modest values of R_m can be attained in the liquid sodium magnetically pumped in large volumes around a fast reactor to extract heat from its core, most earthbound experience of magnetohydrodynamics is obtained from the flow of liquid metals, such as Hg, Na, or NaK, under conditions of comparatively small R_m, in which τ_η is small compared with all other relevant time scales (such as L/u) and in which Alfvén's theorem can be misleading. Nevertheless, the Lorentz force can have a profound effect on the fluid flow.

If $R_m \ll 1$, the motionally induced field $\mathbf{B}-\mathbf{B}_0$ is small compared with the externally applied field $\mathbf{B}_0$ required to prevent the diffusive demise of $\mathbf{B}$ in the time τ_η. The associated current $\mathbf{j}$ ($\sim\sigma u B_0$) creates from $\mathbf{B}_0$ the Lorentz force $\mathbf{j}\times\mathbf{B}_0$ ($\sim\sigma u B_0^2$), which by Lenz's law has the overall effect of decelerating the flow in a characteristic *magnetic response time* of $\tau_m=\rho/\sigma B_0^2$. The electric field in a system evolving slowly compared with this time is quasi-electrostatic.

The relative strength of the Lorentz and viscous forces is approximately M^2, where M is the *Hartmann number*

$$M=LB_0(\sigma/\rho v)^{1/2}.$$

Even when $M\gg 1$, viscous forces are always important at bounding walls where they bring the flow to relative rest in a *Hartmann layer* of thickness $\delta_H=(\rho v/\sigma B_0^2)^{1/2}$. If in the mainstream outside such boundary layers no other forces act, a state of steady flow is, paradoxically, one of magnetohydrostatic balance between the driving and the Lorentz forces, the flow merely serving to induce the currents necessary to bring this about [3]. The physical idea does not determine conditions in the mainstream uniquely. In this sense the Hartmann layer *controls* the mainstream. The structure of the layer depends on the magnetic Prandtl number $P_m=v/\eta=\mu\sigma v$. Thus even when $R_m\gg 1$ and $M\gg 1$, the mainstream, although approximately an inviscid ($v=0$) perfectly conducting ($\eta=0$) flow, depends on the ratio $v/\eta=P_m$.

The Hartmann layer is not the only kind of MHD boundary layer, but it is the most important. Other types exist, typically when $\mathbf{B}$ is parallel to the boundary, that passively adapt the mainstream to the wall instead of exerting a controlling influence on it.

A strong uniform field creates a form of two-dimensionality. Flow across field lines induces Lorentz forces that oppose the motion (see above), but motion along the field lines generates no currents or forces. Thus, for example, a strong magnetic field imparts a two-dimensional structure to MHD

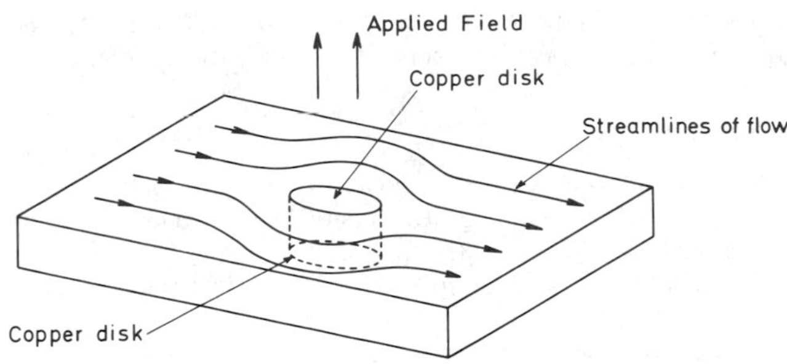

FIG. 6. The controlling influence of boundary layers is illustrated in the flow down a duct in opposite faces of which two disks of high conductivity have been placed. The right circular cylinder they define has generators parallel to the applied magnetic field. Slow flow down the duct avoids the cylinder almost as though it were solid.

turbulence, and turbulent transport coefficients such as thermal conductivity are then anisotropic, being smaller perpendicular to the field than along it. Striking experimental confirmation of such "two-dimensional theorems" has been demonstrated in a rectangular duct across which a strong uniform field is applied perpendicular to two of its sides. By an increase in the electrical conductivity of (say) two facing disks embedded in these sides, the Hartmann layers on them are structurally changed and a pressure-driven flow down the duct is constrained to avoid the entire cylinder whose generators are the field lines through the rims of the disks, almost as though that cylinder were a solid obstacle actually situated within the duct (Fig. 6).

A strong vertical field also highly constrains convective overturning, as for example in the Bénard layer in which a horizontal slab of fluid (depth L) is heated from below. The applied temperature gradient (γ) is usually measured by the *Rayleigh number* $Ra = g\bar{\alpha}\gamma L^4/\nu\kappa$, where g is the acceleration due to gravity, $\bar{\alpha}$ is the coefficient of volume expansion, and κ is the thermal diffusivity. If Ra exceeds a critical value Ra_c, a tessellated pattern of convection rolls forms of horizontal wave number a/L. When $M=0$ it is found that $Ra_c \sim 1708$ and $a_c \sim 3.16$, corresponding to cells "as broad as they are deep." If M is increased, so are Ra_c and a_c, the latter because of the increasing tendency of the motions toward two-dimensionality, the former because of the greater

Table I Typical values of parameters in magnetohydrodynamic systems of various scales[a]

System	Earth's core	Solar convection zone	Solar corona	Magnetic stars	Galactic disk	Ionized H (10^5 K, 42 Pa)	Hg	Na	NaK
Length scale, L (m)	10^6	10^8	10^8	10^9	10^{20}	10^{-1}	10^{-1}	10^{-1}	10^{-1}
Velocity scale, U (m/s)	10^{-4}	10^2	10^5	10?	10^4	1	10^{-1}	10^{-1}	10^{-1}
Field scale, B_0 (T)	10^{-2}	10^{-1}	10^{-4}	1	10^{-9}	10^{-1}	1	1	1
Density scale, ρ (kg/m^3)	10^4	10^{-2}	10^{-15}	10^3?	10^{-21}	10^{-7}	10^4	10^3	9×10^2
Magnetic diffusivity η (m^2/s)	1	(10^7)	1	1	(10^{24})	15	1	8×10^{-2}	0.3
Electromagnetic diffusion time $\tau_\eta = L^2/\eta$ (s)	10^{12}	(10^9)	10^{16}	10^{18}	(10^{16})	7×10^{-4}	10^{-2}	10^{-1}	3×10^{-2}
Magnetic Reynolds number $R_m = UL/\eta$	10^2	(10^3)	10^{13}	10^{10}	(1)	7×10^{-3}	10^{-2}	1	3×10^{-2}
Alfvén velocity $V_A = B_0/(\mu\rho)^{1/2}$ (m/s)	10^{-1}	10^3	10^6	10	10^4	3×10^5	8	30	30
Alfvén number $A = U/V_A$	10^{-3}	10^{-1}	10^{-1}	1	1	3×10^{-6}	10^{-2}	3×10^{-3}	3×10^{-3}
Lundquist number $Lu = V_A L/\eta$	10^5	(10^4)	10^{14}	10^{10}	(1)	10^3	1	40	10
Magnetic response time $\tau_m = \eta/V_A^2$ (s)	10^2	(10)	10^{-12}	10^{-2}	(10^{16})	10^{-10}	10^{-2}	9×10^{-5}	3×10^{-4}
Kinematic viscosity v (m^2/s)	10^{-2}?	(10^7)	10^{16}	10^{-6}?	(10^{24})	10	10^{-7}	8×10^{-7}	6×10^{-7}
Magnetic Prandtl number $P_m = v/\eta$	10^{-2}?	(1)	10^{16}	10^{-6}?	(1)	0.7	10^{-7}	10^{-5}	2×10^{-6}
Hartmann number $V_A L/(v\eta)^{1/2}$	10^{-6}?	(10^4)	10^6	10^{13}	(1)	2×10^3	3×10^3	10^4	7×10^3

[a] Astrophysical quantities are given representative values which in some cases (?) are extremely uncertain. Quantities based on turbulent transport coefficients are in parentheses.

Ohmic and viscous opposition to such motions, which must be overcome by the buoyancy force before convection can take place. If M is sufficiently large, and the diffusivity ratio $q=\kappa/\eta$ is sufficiently small, the layer may exhibit overstability, i.e., a periodic oscillation of the motion in each cell. In the case of steady convection, it is found that $Ra_c \sim (\pi M)^2$ and $a_c \sim (\pi^4/2)^{1/6} M^{1/3}$ as $M \rightarrow \infty$.

When motion follows field lines, as in an aligned flow, the Alfvén velocities are Galilean transformed into $u \pm V_A$. Thus if $A>1$ both waves move downstream, whereas if $A<1$ one moves upstream. A highly conducting solid body placed in an aligned flow of small A therefore signals its presence ahead by an upstream wake. Because $Lu \ll 1$ in most experiments with liquid metals, such precursors are rapidly damped and hard to detect; indeed clear-cut laboratory demonstrations of Alfvén radiation are hard to give except in laboratory plasmas.

MAGNETOHYDRODYNAMICS OF ROTATING SYSTEMS

The importance of rotation in geophysical and astrophysical contexts has already been stressed in connection with the α effect. The influence of Lorentz forces on flows of rotating fluids is no less striking, and is thought to be of considerable significance in these large-scale applications of MHD.

When angular velocity ($\boldsymbol{\Omega}$), as measured by the *Taylor number* $Ta=(2\Omega L^2/\upsilon)^2$, is large and M is zero, flows that are slow and steady in the rotating frame are constrained to be highly two-dimensional with respect to the direction of $\boldsymbol{\Omega}$ (the Taylor–Proudman theorem; see [4,5]). For instance, when the *Prandtl number* $Pr=\upsilon/\kappa$ is sufficiently large, it is found that convection in the Bénard layer with $\boldsymbol{\Omega}$ vertical is steady, and that $Ra_c \sim 3(\pi^2 Ta/2)^{2/3}$ and $a_c \sim (\pi^2 Ta/2)^{1/6}$ for $Ta \rightarrow \infty$.

When both Ta and M are large, the rotational and magnetic tendencies toward two-dimensionality oppose each other. Lorentz forces "release the constraint" of the Coriolis forces on the motion. For example, they facilitate Bénard convection. For $Ta \rightarrow \infty$ (and $q<1$) the overall minimum of Ra_c occurs when the *Elsasser number* $El=M^2/Ta^{1/2}=\sigma B_0^2/2\Omega=\tau_m/2\Omega$ is of order 1. It is then found that $a_c=O(1)$; i.e., the cells are again as broad as they are deep. The corresponding reduction in Ohmic dissipation implies that Ra_c is *less* than when $M=0$; in fact it is $O(Ta^{1/2})$ rather than $O(Ta^{2/3})$. [Note that a_c and γ_c are now independent of ν; i.e., viscosity is not influential when $Ta \gg 1$ and $El = O(1)$.]

It appears that a rotating conductor can transmit heat *more* readily when it can self-excite an internal magnetic field by dynamo action than when it cannot. This provides a heuristic thermodynamic reason for the all-pervasive presence of magnetism on astrophysical scales.

The MHD of rotating fluids is a rapidly growing field. For further details, see [4,5].

Typical magnitudes of times, velocities, and dimensionless parameters that have been defined in this article are shown in Table I for a number of MHD systems.

See also Boundary Layers; Electrodynamics, Classical; Hydrodynamics; Plasma Confinement Devices; Plasmas; Plasma Waves; Turbulence.

REFERENCES

1. H. K. Moffatt, *Magnetic Field Generation in Electrically Conducting Fluids*. Cambridge University Press, Cambridge, London, and New York, 1978. (I)
2. A. Jeffrey, *Magnetohydrodynamics*. Wiley (Interscience), New York, 1966. (I)
3. A. G. Kulikovskii, "Slow Steady Flows of a Conducting Fluid at Large Hartmann Numbers," *Fluid Dynamics* **3**(2), 1–5 (1968). (I)
4. P. H. Roberts and A. M. Soward (eds.), *Rotating Fluids in Geophysics*. Academic Press, New York, 1978. (I)
5. J. A. Jacobs (ed.), *Geomagnetism*, Vol. **2**. Academic Press, London and Boston, 1987. (I)

BIBLIOGRAPHY

Shercliff (1965) has written a comprehensive and comprehensible text full of physical insight; it is highly recommended. Hughes and Young (1966) and Sutton and Simon (1965) provide thorough treatments of laboratory and engineering MHD. Cowling (1976), Parker (1979), Priest (1984) and Zeldovich *et al.* (1983) discuss various astrophysical applications. Bateman (1978) and Friedberg (1987) analyze MHD processes in the context of plasma physics and especially of containment devices.

G. Bateman, *MHD Instabilities*. MIT Press, Cambridge, MA, 1978. (I)

T. G. Cowling, *Magnetohydrodynamics*. Adam Hilger Ltd., Bristol, UK, 1976. (I)

J. P. Friedberg, *Ideal Magnetohydrodynamics*. Plenum, New York and London, 1987. (I)

W. F. Hughes and F. J. Young, *Electromagnetodynamics of Fluids*. Wiley, New York, 1966. (I)

E. N. Parker, *Cosmic Magnetic Fields, their Origin and their Activity*. Clarendon Press, Oxford, 1979. (I)

E. R. Priest, *Solar Magnetohydrodynamics*. Reidel, Dordrecht, 1984. (I)

P. H. Roberts, *An Introduction to Magnetohydrodynamics*. Elsevier North-Holland, New York, 1967. (I)

J. A. Shercliff, *A Textbook of Magnetohydrodynamics*. Pergamon, New York, 1967. (I)

G. W. Sutton and A. Sherman, *Engineering Magnetohydrodynamics*. McGraw-Hill, New York, 1965. (I)

Ya. B. Zeldovich, A. A. Ruzmaikin, and D. D. Sokolov, *Magnetic Fields in Astrophysics*. Gordon and Breach, New York, 1983. (I)

Magneto-optical Effects *see* Kerr Effect, Magneto-optical; Faraday Effect

Magnetoresistance

W. A. Reed

When a conductor carrying a current is placed in a magnetic field its resistance changes (usually increases). This increase is called its magnetoresistance and is due to the Lorentz force acting on the electrons combined with a distribution in the electrons' velocities. The magnetoresistance is only one of several galvanomagnetic effects and a general discussion of all these effects is given by Jan [1].

The relation between the electric field, current density, and magnetic field is given by

$$E_i = \rho_{ij}(B)J_j, \tag{1}$$

where ρ_{ij} is the resistivity tensor. The element $\rho_{ij}(B)J_j$ can be expanded in powers of **B** such that

$$\rho_{ij}(B) = \rho_0 + \rho_{ijk}B_k + \rho_{ijkl}B_kB_l. \tag{2}$$

The magnetoresistance terms of the tensor are contained in the diagonal elements of $\rho_{ij}(B)$ which, from symmetry considerations, only contain terms with even powers of **B**. When the magnetic field is parallel to the current, the resistance change (proportional to ρ_{iiii}) is called the *longitudinal magnetoresistance*, whereas when the magnetic field is perpendicular to the current, the change in resistance (proportional to ρ_{iijj} or ρ_{iikk}) is called the *transverse magnetoresistance*. The magnetoresistance is usually expressed as $\Delta\rho/\rho_0$, where $\Delta\rho = \rho(B) - \rho_0(0)$ is the change in resistivity in the magnetic field, and ρ_0 is the resistivity in zero magnetic field at the temperature of measurement. The magnitude of the transverse magnetoresistance is a function of both the magnetic field and the resistivity and generally follows a reduced Kohler's rule which states that:

$$\frac{\Delta\rho}{\rho_0} = FB\frac{\rho_\Theta}{\rho_0}, \tag{3}$$

where ρ_Θ is the resistivity at the Debye temperature and F is a function depending only on the geometrical configuration and the metal [2].

Theories of magnetoresistance generally fit into one of five regimes [3,4] depending on the value of $\omega_c\bar{\tau}$, where ω_c, the cyclotron frequency (eB/m^*c), is 2π times the reciprocal of the time it takes an electron to travel once around its orbit on the Fermi surface, and $\bar{\tau}$ is the average relaxation time of the electrons. These regimes are as follows:

(1) The low-field regime ($\omega_c\bar{\tau} \ll 1$) is where the time between electron collisions is short and the electron traverses only a small fraction of the Fermi surface before scattering. In this regime the anisotropy of the scattering mechanisms dominates the magnetoresistance. Experimentally the condition ($\omega_c\bar{\tau} \ll 1$) is satisfied at high temperatures where phonon scattering is significant and/or in materials with nontrivial amounts of defects or impurities. In this regime the size of $\Delta\rho/\rho_0$ in metals is generally about 0.1 to 1%. "Many-valley" models have been used to successfully fit low-field galvanomagnetic data of semiconductors and semimetals [5] with Fermi surfaces constructed from a few ellipsoids, but these models have not been successfully applied to the more complicated Fermi surfaces usually found in metals.

(2) The intermediate-field regime ($\omega_c\bar{\tau} \approx 1$) is a regime for which few if any theories of magnetoresistivity exist and is therefore of little interest experimentally.

(3) The high-field regime ($\omega_c\bar{\tau} \gg 1$) is where the electrons are able to complete many cyclotron orbits between collisions. Thus it is the topology of the Fermi surface and not the details of the scattering mechanisms that dominates the character of the magnetoresistance [6,7]. To obtain the high-field regime experimentally the measurements are made at low temperatures (T<4 K), in high magnetic fields ($B > 10^4$ G), and on high-purity single crystals [8]. In metals the high-field magnetoresistance can saturate at $\Delta\rho/\rho_0 \approx 1$ when the metal is not compensated (the number of electrons does not equal the number of holes) or when the current and magnetic field are perpendicular to an open orbit on the Fermi surface. However, if the metal is compensated or an open orbit is parallel to the current, the transverse high-field magnetoresistance can become greater that 10^6. Because $\Delta\rho/\rho_0$ is so sensitive to the topology of the cyclotron orbits, the high-field magnetoresistance has been extensively used to study the Fermi surfaces of metals [6,7].

(4) The quantum oscillation regime ($\hbar\omega_c \gg kT$ and $\omega_c\bar{\tau} \gg 1$) is where the separation of the quantized energy levels is greater than the thermal smearing. Experimentally this regime is the same as the high-field regime except that the requirements on temperature, magnetic field, and crystal perfection are more stringent. The oscillations in $\Delta\rho/\rho_0$, which are periodic in $\mathbf{B}^{-1}$, provide information similar to that gained from de Haas–van Alphen measurements, i.e., the extremal cross-sectional areas of the Fermi surface. In many metals the spacing between Landau levels may become greater than the energy gap between bands and the phenomenon of magnetic breakdown can occur. Since magnetic breakdown can change both the topology of the cyclotron orbits and the state of compensation, it profoundly affects both the oscillatory and monotonic magnetoresistance [9].

(5) The quantum limit regime ($\hbar\omega_c \gg \epsilon_F$ where ϵ_F is the Fermi energy) is almost impossible to obtain experimentally in metals except for possibly bismuth, so that experimental studies have been restricted to semiconductors [10].

The phenomenon of negative magnetoresistance is usually associated with measurements on samples which either are magnetic or contain magnetic impurities, or on samples where the mean free path of the electrons is comparable to the sample dimensions. In these cases the decrease in the magnetoresistance is a result of the magnetic field reducing the amount of scattering by some mechanism such as aligning the spin of magnetic impurities or rotating domains in magnetic samples, or reducing the number of collisions with the sample's walls in size-effect measurements. In addition to the usual negative magnetoresistance effects numerous anomalous effects have been reported, usually in the longitudinal magnetoresistance. These have most often proved to be artifacts of the experiment [11].

The magnetoresistive effect has been used in a number of sensing and measuring devices. Typically these devices use a magnetic material, often a film, as the magnetoresistive element since magnetic materials can have large internal (**B**) fields for small applied fields (**H**). Examples of magnetoresistive devices are magnetic bubble domain sensors [12], and position and angle sensors in automobiles [13].

See also CONDUCTION; CYCLOTRON RESONANCE; DE

HAAS–VAN ALPHEN EFFECT; FERMI SURFACE; GALVANOMAGNETIC AND RELATED EFFECTS; MAGNETIC DOMAINS AND BUBBLES.

REFERENCES

1. J.-P. Jan, in *Solid State Physics,* Vol. 5 (F. Seitz and D. Turnbull, eds.), pp. 1–96. Academic Press, New York, 1957.
2. J. M. Ziman, *Electrons and Phonons,* p. 491. Oxford University Press, London, 1962.
3. R. G. Chambers, in *The Fermi Surface* (W. A. Harrison and M. B. Webb, eds.), pp. 100–124. Wiley, New York, 1960.
4. Colin M. Hurd, *The Hall Effect in Metals and Alloys,* pp. 1–50. Plenum Press, New York, 1972.
5. Arthur C. Smith, James F. Janak, and Richard B. Adler, *Electronic Conduction in Solids.* McGraw-Hill, New York, 1967.
6. E. Fawcett, *Adv. Phys.* **13,** 139 (1964).
7. W. A. Reed and E. Fawcett, *Science* **146,** 603 (1964).
8. W. A. Reed, in *Methods of Experimental Physics,* Vol. 11 (R. V. Coleman, ed.), pp. 1–32. Academic Press, New York, 1974.
9. R. W. Stark and L. M. Falicov, in *Progress in Low Temperature Physics,* Vol. 5 (C. J. Gorter, ed.), pp. 235–286. North-Holland, Amsterdam, 1967.
10. E. N. Adams and T. D. Holstein, *J. Phys. Chem. Solids* **10,** 154 (1959).
11. W. A. Reed, E. Blount, J. A. Marcus, and A. J. Arko, *J. Appl. Phys.* **42,** 5433 (1971).
12. G. S. Almasi, G. E. Keefe, Y. S. Lin, and D. A. Thompson, *J. Appl. Phys.* **42,** 1268 (1971).
13. A. Peterson, *Electronik* **3,** 99–102 (1985).

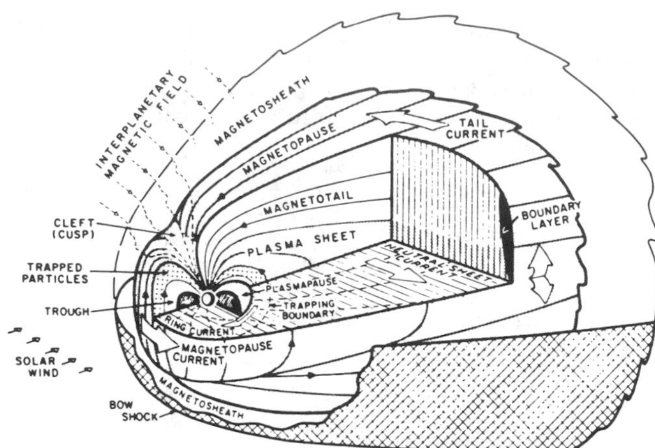

FIG. 1. A three-dimensional view of the magnetosphere drawn by W. J. Heikkila, University of Texas. The magnetopause current provides the $\mathbf{J} \times \mathbf{B}_M$ force which holds off the solar-wind dynamic pressure. The polar cleft is an opening in the geomagnetic field which permits the direct penetration of magnetosheath plasma onto the ionosphere; this plasma precipitation produces the dayside auroras. The tail boundary layer, which is now known to encircle the tail, is a region of anti-sunward plasma flow on recently merged field lines. The tail magnetopause current closes through the plasma sheet and produces the extended tail magnetic field. The trapping boundary denotes the outermost extension of the Van Allen radiation belt of trapped particles, i.e., particles which execute complete drift orbits around the earth. The trough is a region of very low plasma density.

Magnetosphere

F. V. Coroniti

The term magnetosphere refers to the spatial region surrounding a compact object in which the plasma dynamics are controlled by the object's magnetic field. In the solar system, spacecraft have discovered magnetospheres around Mercury, Earth, Jupiter, Saturn, Uranus, and Neptune. Astrophysical magnetospheres include magnetic white dwarfs, neutron stars that are either radio or x-ray pulsars, and possibly the supermassive black holes that are in the nuclei of active galaxies and quasars. The energy source for the magnetosphere can be either external (e.g., the solar wind as for Earth) or internal (e.g., the rotation of the central object as for Jupiter and radio pulsars). The terrestrial magnetosphere is the best studied, and so we will discuss it in some detail before commenting briefly on the magnetospheres of the other planets.

THE TERRESTRIAL MAGNETOSPHERE

The magnetosphere is formed by the interaction of earth's magnetic field and the solar wind, which is the supersonic, super-Alfvénic, and magnetized plasma expansion of the solar corona. The geomagnetic field presents a blunt obstacle to the solar wind, so that a hydromagnetic bow shock forms to divert the flow around the magnetosphere. The location of the magnetopause, which separates the solar wind and magnetosphere, is determined by balancing the solar-wind dynamic pressure with the magnetic pressure $B_M^2/2\mu_0$ of the magnetospheric field B_M. The various magnetospheric boundaries and regions are depicted in Fig. 1; spatial scales are listed in Table I, and typical thermal plasma parameters are given in Table II. We will now discuss several dynamical processes that determine the basic magnetospheric structure.

MAGNETIC FIELD MERGING AND MAGNETOSPHERIC CONVECTION

The magnetosphere is a driven hydromagnetic system whose internal dynamics are controlled by the input of solar-wind energy through magnetic field merging. As the post-shock magnetosheath solar wind impacts the dayside magnetopause, the solar-wind magnetic field $\mathbf{B}_{sw}$ comes into

Table I Location of boundaries and scale size of regions[a]

	Location	Spatial extent
Bow shock	13–17	
Magnetosheath		3–5 thick
Magnetopause	10–12 (dayside)	
Magnetotail boundary layer		1–3 thick
Magnetotail		~30–40 wide, ~1000 long
Plasma sheet		2–4 thick
Plasmasphere	4–6 (dusk), 2–3.5 (dawn)	

[a] In earth radii: $1.0R_E = 6.378 \times 10^6$ m.

Table II Thermal plasma properties

Region	Density (cm^{-3})	Electron temp. (eV)	Ion temp. (eV)	Magnetic field (10^{-5} G)	Flow speed (km/sec)
Solar wind	5–20	20–40	10–20	5–15	350–1000
Magnetosheath	20–40	100–200	~1000	20–40	100–200
Magnetotail boundary layer	0.01–0.1	10–50	500–5000	10–30	200
Plasma sheet	0.1–1.0	200–2000	$(10–50) \times 10^3$	10–20	0–1000
Ring current	5–20	~1000	~1.0	100–500	10
Plasmasphere	100–1000	~1.0	0.1–0.2	>100	<1.0
Ionosphere	10^4–10^6	0.1–0.2		0.3–0.6	<1.0

contact with $\mathbf{B}_M$. Unless $\mathbf{B}_{sw}$ is parallel to $\mathbf{B}_M$, the two fields can merge at a magnetic neutral line, so that $\mathbf{B}_M$ becomes connected to $\mathbf{B}_{sw}$ (Fig. 2); clearly, a large southward component of $\mathbf{B}_{sw}$ maximizes the merging interaction.

In hydromagnetics the magnetic field moves with the plasma, so that the Lorentz force $\mathbf{E} + \mathbf{v} \times \mathbf{B} = 0$ in the laboratory frame. The solar-wind flow around the magnetosphere stretches the merged field lines into a long magnetic tail of open flux behind the Earth. About 3×10^{15} J of solar-wind flow energy is stored as magnetic energy in the tail. In steady state the motional electric field $\mathbf{E} = -\mathbf{v} \times \mathbf{B}$ (out of page in Fig. 2) appears throughout the magnetosphere and corresponds to a potential drop of 50–200 kV across the system. In the tail the open field lines flow toward the plane of symmetry, where they reconnect at the tail neutral line; during quiet times the neutral line is located about $100R_E$ (R_E = Earth radius) behind the Earth. The reconnected, but stretched, field lines accelerate the plasma back toward the Earth's nightside, thus injecting plasma into the inner magnetosphere. As the flow penetrates the dipole region it decelerates and the plasma is heated by adiabatic compression.

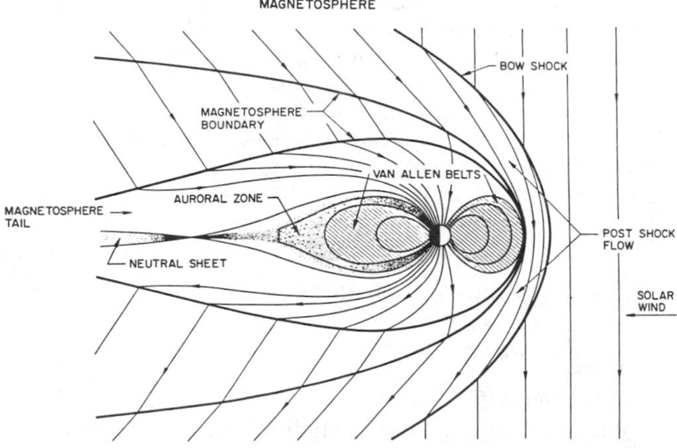

FIG. 2. The topology of the interconnected solar-wind and magnetospheric magnetic fields drawn for the optimum merging configuration of a southward solar-wind field. After dayside merging the solar-wind flow drags the field lines into the tail while the foot of the field in the ionosphere flows anti-sunward across the polar cap. In the tail field lines flow toward the neutral sheet and reconnect. Reconnected or plasma sheet field lines flow toward the earth and connect to the nightside auroral zone ionosphere. The field lines flow around the earth and back to the dayside magnetopause.

The field lines then flow around the Earth and return to the dayside magnetopause, thus completing the internal hydromagnetic convection or circulation system. Typical circulation times are 1–3 h.

Not all of the Earth's field lines participate in the convection system. Near the Earth the field lines are forced to corotate by the ionosphere and do not generally become open to the solar wind. These plasmaspheric field lines fill with cool, dense plasma that diffuses upward from the ionosphere. The plasmapause separates the corotating and convecting field lines.

TIME-VARYING CONVECTION AND MAGNETOSPHERIC SUBSTORMS

In steady state the dayside merging rate equals the tail reconnection rate. However, $\mathbf{B}_{sw}$ is rarely steady in direction. When $\mathbf{B}_{sw}$ has zero southward or a northward component for several hours, the magnetosphere is relatively quiet and convection is slow. Occasionally $\mathbf{B}_{sw}$ shifts and stays strongly southward for about 1 h. The enhanced dayside merging rate produces a transient change in the convection state and structure of the magnetosphere that culminates in the magnetosphere's most striking phenomenon—the magnetospheric substorm.

For about 10–60 min following the southward shift, the dayside merging rate exceeds the tail reconnection rate, and the tail magnetic energy increases. Eventually tail reconnection must increase to balance the dayside merging rate. Recent observations indicate that tail reconnection onsets explosively with the formation of a new neutral line 10–20 R_E behind the Earth within the closed field line region of the plasma sheet. On the ground, tail reconnection onset corresponds to the breakup or expansion phase of the auroral substorm. The auroras, which are recombination radiation caused by the precipitation of 1–10-keV electrons, brighten and then expand poleward and westward. In the nightside ionosphere an electrojet current of about 10^6 A flows between 100 and 120 km altitude, causing a negative depression or bay in the geomagnetic field. The expansion phase lasts 0.5–1.0 h and may have a total energy dissipation rate exceeding 10^{11}–10^{12} W.

During the middle to late expansion phase, previously open tail field lines reconnect at the near earth neutral line, thus disconnecting the distant plasma sheet from the Earth. The solar wind now exerts a tailward Maxwell stress on the distant plasma sheet, which accelerates a large volume of plasma (a plasmoid) down the tail and back into the solar wind. During recovery phase, the near-earth neutral line retreats toward its quiet time position near $100R_E$, and the aurora moves equatorward and becomes fainter as the electron precipitation fluxes decrease. Although substorms occur at irregular intervals, a typical moderately disturbed period has 4–8 substorms per day.

MAGNETIC STORMS

A large solar flare produces a major perturbation in the solar wind that may double the wind velocity and raise $\mathbf{B}_{sw}$ and the density by a factor of 5. During the disturbance $\mathbf{B}_{sw}$ often remains southward for 8–12 h; the enhanced internal convection and energy input results in a magnetic storm. The increased solar-wind dynamic pressure moves the magnetopause inward to $(5–6)R_E$. During the storm's main phase, large fluxes of 20–50-keV protons and singly ionized oxygen and helium from the ionosphere are injected deep within the radiation belts, where they form a westward ring current that produces a worldwide depression in the geomagnetic field. Convection is so strong that the plasmapause may move to within $2R_E$. The storm recovery phase begins when $\mathbf{B}_{sw}$ shifts northward and convection slows. The ring-current ions decay by charge exchange with the neutral hydrogen geocorona over a period of 1–2 days. The storm dissipates a total energy of 10^{16} J.

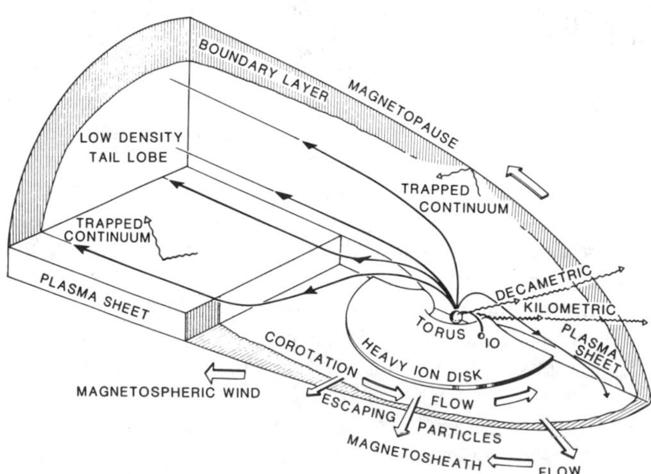

FIG. 3. The Jupiter–Io magnetosphere. Volcanos and vents on Io inject SO_2 neutrals which are ionized and form a heavy-ion torus. Field-aligned currents which couple Io and the torus to Jupiter's ionosphere generate decametric and kilometric radio waves which escape from the magnetosphere. Outward radial diffusion transports the torus ions into the middle magnetosphere and populates a cold heavy-ion disk. A hot ion plasma sheet surrounds the heavy-ion disk. Plasma sheet ions escape into the magnetosheath as a magnetospheric wind. Dayside magnetic field reconnection creates a long magnetic tail with extremely low-density lobes. Electromagnetic continuum radiation is trapped in the low-density magnetospheric cavity.

OUTER-PLANET MAGNETOSPHERES

Unlike the Earth (and Mercury) the magnetospheres of the giant planets are dominated by the rapid planetary rotation. Merging between the planetary and solar-wind magnetic fields does create a long magnetotail (Jupiter's tail may extend 3–4 AU behind the planet), but solar-wind-driven internal convection influences only the outermost magnetospheric regions.

At Jupiter, the powerful injection of heavy molecules (e.g., SO_2) by Io's volcanos provides the dominant source of magnetospheric plasma. The molecules are dissociated and ionized by hot electrons. The ions are then accelerated to the local corotation velocity by the planetary rotationally induced electric field, forming a dense torus in the region near Io (Fig. 3). Approximately 10^{13} W are dissipated in this ion-acceleration process, which is a sufficient energy input to power the Jovian aurora, decametric radiation, and synchrotron radiation. Under the strong centrifugal force, the heavy ions diffuse radially outward, and when no longer confined by the magnetic field, form a planetary wind that stretches the field lines into a disk configuration; the heavy ions eventually escape into the solar wind. The magnetospheres of the other outer planets have much weaker internal sources of plasma, and are therefore relatively quiescent.

See also AURORA; IONOSPHERE; MAGNETOHYRODYNAMICS; PLASMAS; PLASMA WAVES; RADIATION BELTS; SOLAR WIND.

BIBLIOGRAPHY

S. I. Akosofu, *Solar and Magnetospheric Substorms*. Springer-Verlag, Berlin and New York, 1969.

F. V. Coroniti, "Explosive Tail Reconnection: The Growth and Expansion Phases of Magnetospheric Substorms," *J. Geophys. Res.* **90,** 7427 (1985).

E. W. Hones, Jr., "Transient Phenomena in the Magnetotail and their Relation to Substorms," *Space Sci. Rev.* **23,** 393 (1979).

A. T. Y. Lui, *Magnetotail Physics*. The Johns Hopkins University Press, Baltimore and London, 1987.

F. L. Scarf, F. V. Coroniti, C. F. Kennel, and D. A. Gurnett, "Jupiter and Io: A Binary Magnetosphere," *Vistas Astron.* **25,** 263 (1982).

Magnetostriction

B. S. Chandrasekhar

If the magnetization of a solid object is changed, for example, by changing the magnitude of a magnetic field applied to it, then the size of the object changes slightly. This is the phenomenon of magnetostriction, discovered in 1842 by J. P. Joule, eponym for our unit of energy. The effect is small, with the change in length between the unmagnetized and magnetized states being typically a few parts in 10^5; but the effect is both an important strand in the fabric of our basic understanding of solids, and of use in several practical applications.

The origins and detailed characteristics of magnetostriction depend on whether the solid under consideration is ferromagnetic or not. Ferromagnetic magnetostriction has been investigated for a longer time and far more extensively than the other kind, and it also has provided all the practical applications.

The physical origin of magnetostriction in a ferromagnet like iron or cobalt ferrite is as follows. The energy of a magnetized single crystal depends on the direction of the magnetization, essentially because of the interaction between the spin moments of the electrons which cause the magnetization and the orbital motion of the electrons around the nuclei. This anisotropy energy, for any given direction of the magnetization, is itself dependent on the deformation of the crystal, a deformation which produces its own contribution of an elastic energy to the total energy of the crystal. The minimizing of the sum of these magnetoelastic and elastic contributions to the energy leaves the crystal with a resultant strain, the magnetostriction.

An unmagnetized crystal consists of many microscopic magnetized domains pointing in different directions so that the net magnetization is zero. The application of a strong enough external field, along a direction whose direction cosines referred to the cubic crystal axes (assuming such symmetry) are α_1, α_2, and α_3, will produce a single domain extending over the crystal and magnetized along that direction, and will also change the length l_0 of the crystal in that direction magnetostrictively by δl. Then one finds that

$$\frac{\delta l}{l_0} = \lambda_0 + 3(\lambda_1 - \lambda_0)(\alpha_1^2\alpha_2^2 + \alpha_2^2\alpha_3^2 + \alpha_3^2\alpha_1^2),$$

where λ_0 and λ_1 are the magnetostriction constants for a given material. The equation takes a different form for a noncubic crystal. For an isotropic polycrystalline sample, one would measure a single magnetostriction constant λ_p, which is approximately equal to $(2\lambda_0 + 3\lambda_1)/5$.

Magnetostriction constants decrease with increasing temperature, and vanish at the Curie point. Some typical experimental values for cubic materials at room temperature are given below:

Substance	$10^6\lambda_0$	$10^6\lambda_1$	$10^6\lambda_p$
Fe	+21	−21	−7
Ni	−46	−24	−34
$FeO \cdot Fe_2O_3$	−20	+78	+40
$Co_{0.8}Fe_{0.2}O \cdot Fe_2O_3$	−590	+120	—

In nonferromagnetic substances, where the magnetization is induced by and is proportional to the applied field, a similar magnetoelastic coupling, i.e., a dependence of magnetization on strain, leads to a magnetostriction which is quadratic in the field. Typically, a strain $\delta l/l_0$ of 10^{-6} is produced by a field of 10 T ($=10^5$ G). In pure conductors at low temperatures the magnetization shows an oscillatory dependence on field because of the Landau quantization of the orbital motion of the conduction electrons in the field (the de Haas–van Alphen effect). In this regime the magnetostriction also shows an oscillatory behavior with magnetic field, with amplitudes as large as 10^{-5}. These effects have been used to analyze the electronic structure and its response to a lattice strain, the latter being related to the electron–phonon interaction, in a number of metals and semiconductors.

Magnetostriction and its inverse effect, viz., the dependence of magnetization on strain, have found several applications using ferromagnets. Magnetostrictive acoustic transducers have been used in echo sounders, fish finders, sonar, and certain particle detectors used in high-energy physics. They have been employed, because of their ability to transmit considerable ultrasonic energy into the surrounding medium, in ultrasonic cleaning and in processes such as homogenization, emulsification, and particle dispersion and agglomeration. In the field of medicine, ultrasonic transducers are used in physical therapy and in imaging. Magnetostrictive delay lines, filters, and memories have been used in information processing.

See also DE HAAS-VAN ALPHEN EFFECT; MAGNETIC DOMAINS AND BUBBLES; MAGNETOELASTIC PHENOMENA.

*BIBLIOGRAPHY**

W. J. Carr. Jr., "Secondary Effects in Ferromagnetism," in *Handbuch der Physik, XIII/2,* pp. 274–340. Springer-Verlag, Berlin, 1966. Phenomenological theory.

B. S. Chandrasekhar and E. Fawcett, "Magnetostriction in Metals," *Adv. Phys.* **20,** 775–794 (1971). An advanced review of the effects in nonferromagnetic substances.

R. M. Hornreich, H. Rubinstein, and R. J. Spain, "Magnetostrictive Phenomena in Metallic Materials and Some of their Device Applications," *IEEE Trans. Magn.* **MAG-7,** 29–48 (1971).

W. Joss, R. Griessen, and E. Fawcett, "Electron States and Fermi Surfaces of Homogeneously Strained Metallic Elements," in K. H. Hellwege and J. L. Olsen (eds.), *Landolt-Börnstein: Numerical Data and Functional Relationships in Science and Technology,* Group 3, Vol. 13, P8. b, pp. 1–258. Springer-Verlag, Berlin, 1983. A comprehensive review, *inter alia,* of experimental data on nonferromagnetic magnetostriction.

J. Kanamori, "Anisotropy and Magnetostriction of Ferromagnetic and Antiferromagnetic Substances," in *Magnetism,* Vol. I, Chap. 4 (G. T. Rado and H. Suhl, eds.). Academic Press, New York, 1963. Advanced treatment, mainly theoretical.

Y. Kikuchi, "Magnetostrictive Materials and their Applications," *IEEE Trans. Magn.* **MAG-4,** 107–117 (1968). A review at intermediate level.

Landolt–Börnstein: Zahlenwerte und Funktionen aus Physik, Chemie, Astronomie, Geophysik und Technik, Band 2, Teil 9 (K. H. Hellwege and A. M. Hellwege, eds. Springer-Verlag, Berlin, 1962), and Band 4, Teil 3 (E. Schmidt, ed. Springer-Verlag, Berlin, 1957), provide formulas for different crystal classes as well as graphical and tabular data. *Landolt-Börnstein: Numerical Data and Functional Relationships in Science and Technology,* Group 3, Vol. 4 (K. H. Hellwege and A. M. Hellwege, eds., 2 parts. Springer-Verlag, Berlin, 1970), and Vol. 12 (K. H. Hellwege and A. M. Hellwege, eds., 3 parts. Springer-Verlag, Berlin, 1978, 1980, 1982) contain an extensive collection of data.

* Each of these articles gives extensive references to the original literature.

Magnets (Permanent) and Magnetostatics

A. E. Berkowitz

Permanent magnets are useful because they produce a constant magnetic flux density in their vicinity. This occurs because permanent magnet materials (PMM) can sustain a significant intensity of magnetization in the absence of any external applied magnetic field. In this article, some of the more important properties of PMM will be introduced and their magnetostatic behavior will be reviewed.

Figure 1 shows schematically the intensity of magnetization **M** (dashed arrow) and lines representing the flux density in and around a rod of PMM in zero applied field after a saturating field has been applied in the direction of **M**. The flux lines are shown as emanating from the north and south pole distributions at the ends of the magnet. The concept of magnet poles is an entirely fictitious but extremely useful analytical tool for calculating field strengths around magnetized media. Actually, the fields result from divergences in **M**.

The flux density B outside a magnetized body is the gradient of a scalar potential Φ, i.e.,

$$\mathbf{B} = -\nabla\Phi,$$

where

$$\Phi = \frac{\mu_0}{4\pi}\int \frac{\mathbf{M}\cdot\mathbf{r}^0\, d\tau}{r^2}$$

The integration is over the magnetized body, r is the distance to the point at which **B** is being calculated, $\mathbf{r}^0$ the corresponding unit vector, and $\mu_0 = 4\pi \times 10^{-7}$ H/m is the vacuum permeability. Volume and surface pole densities (ρ_M and σ_M, respectively) are introduced by a transformation of Φ to the form

$$\Phi = \frac{\mu_0}{4\pi}\int \frac{\rho_M}{r}\, d\tau + \int \frac{\sigma_M}{r}\, dS$$

where

$$\rho_M = -\nabla\cdot\mathbf{M}, \qquad \sigma_M = \mathbf{n}\cdot\mathbf{M};$$

dS denotes integration over the surface of the magnetized body with **n** the unit vector normal to the surface. **B** may now be written as

$$\mathbf{B} = \frac{\mu_0}{4\pi}\int \frac{(d\rho_M)}{r^2}\,\mathbf{r}^0,$$

where $d\rho_M = \rho_M\, d\tau$ or $\sigma_M\, dS$.

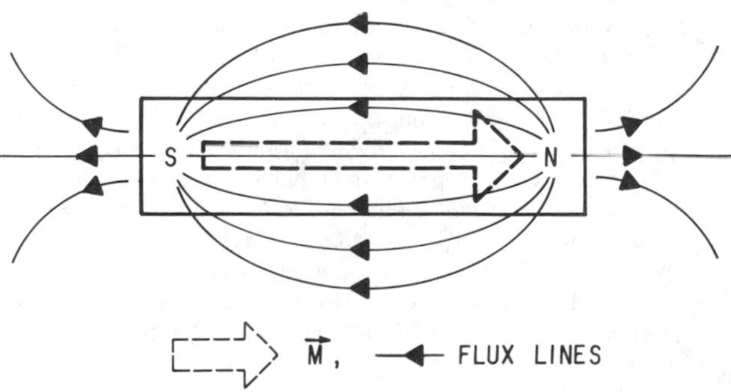

Fig. 1. Magnetization **M** (shown as dashed arrow) and resulting magnetic flux lines in a magnetized permanent magnet in zero applied field.

Thus the pole strength and the resulting external magnetic flux depend on variations in **M**. We would expect that the maximum pole strength would occur at the surface of a uniformly magnetized PMM. Uniform magnetization of PMM generally does not exist because, as depicted in Fig. 1, the fields produced by surface poles *within* the PMM are antiparallel to **M**. These internal fields can reduce **M** by demagnetizing the PMM. In the demagnetized state, which is the lowest energy state, the PMM is subdivided into regions called domains whose dimensions are of the order of 1 μm. In each domain, **M** has the saturation value corresponding to the composition of the PMM, but the orientation of **M** among the various domains is such that the net **M** of the PMM vanishes. Thus the measures of a useful PMM are twofold. First, **M** must be high enough so that variations in **M** produce large pole strengths. Second, the PMM must be able to withstand the effects of the internal demagnetizing field as well as the effects of external fields directed opposite to **M**.

To examine the latter property, we consider the second quadrant of the hysteresis loop of a PMM as shown in Fig. 2. Figure 2 indicates the response of **M** and the internal flux density $\mathbf{B} = \mu_0\,\mathbf{H} + \mathbf{M}$ of the PMM to an external applied field **H**. Also indicated are B_R, the remanent magnetization in zero applied field, and H_c, the coercive force. $H_c(M)$ and $H_c(B)$ are the fields required to reduce **M** and **B**, respectively, to zero. It may readily be shown that the maximum external field produced by a PMM is proportional to $(BH)_{\max}$, which is called the energy product. Thus, from Fig. 2 it is evident that a useful PMM has high B_R and H_c. In terms of the discussion relating to Fig. 1, we note that a high B_R implies a large value of **M**, whereas H_c is a measure of the ability of a PMM to withstand demagnetizing fields.

There are several principal mechanisms responsible for high H_c in PMM. The first arises from the manner in which the net magnetization increases in response to an external field. In the demagnetized state, the direction of **M** varies from one domain to another. As an increasing field is applied, those domains whose **M** is parallel, or nearly parallel, to **H** grow at the expense of the other domains. The regions between adjacent domains are called domain walls. The magnetization of a material may, therefore, be viewed as the growth of domains by the motion of domain walls. If it requires a large amount of energy (supplied by a magnetic field) to move a domain wall from one region of a PMM to another, then a high H_c results. Impediments to wall motion arise from microstructural heterogeneity in some PMM. Defects, impurities, precipitates, grain boundaries, etc., can be energy barriers to wall motion. This occurs because there is an energy density associated with a domain wall. This energy density depends on intrinsic magnetic properties of the PMM. If a domain wall is intercepted by a nonmagnetic microstructural feature whose dimension is of the same order as the domain wall width, the energy of the wall is lowered, and the wall is "pinned." Alternatively, a PMM in the state

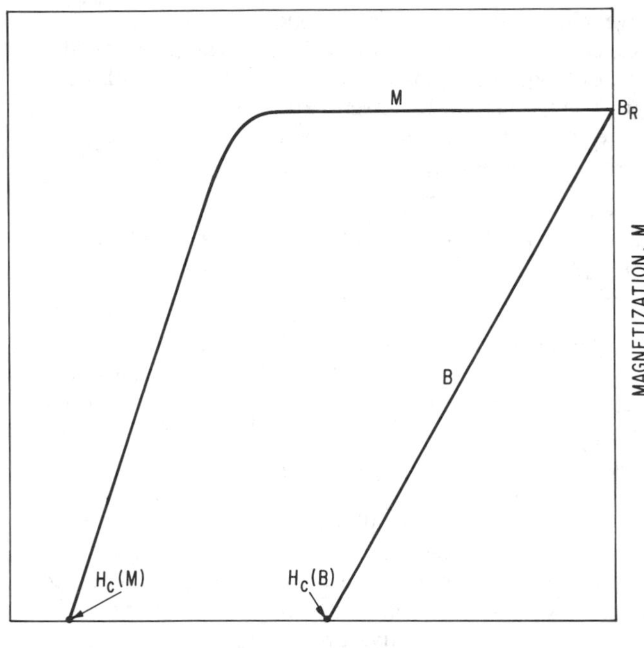

FIG. 2. Second quadrant of the hysteresis loop of a permanent magnet. Field **H** is applied opposite to direction of **M**.

of remanent magnetization may have very few domains. H_c then depends on the energy required to nucleate a domain of reversed magnetization. Again, this energy is a function of intrinsic magnetic properties. One or both of these mechanisms are considered to be responsible for the high H_c and energy products of the strongest current PMM, namely, $Fe_{14}Nd_2B$, Co_5R, and $Co_{17}R_2$, where R stands for a rare earth, usually Sm.

Another prominent mechanism for high H_c in PMM is the existence of single magnetic domains. When a particle of magnetic material is small enough, i.e., <0.1 μm, it is always uniformly magnetized in zero applied field. This is the definition of a single-domain particle. If it requires a high field to reverse the magnetization of an assembly of these particles, then the assembly will have a high coercive force. In all magnetic materials, the magnetization prefers to lie in certain crystallographic directions. It requires a finite energy to rotate the magnetization from an "easy" crystallographic direction to a "hard" one. This energy is called the magnetocrystalline anisotropy energy. If this anisotropy energy is high in a given material, then an aggregate of single-domain particles will have a high H_c. This is the operative mechanism in barium oxide ferrite PMM. Another type of anisotropy energy can be provided by the shape of single-domain particles. If the particles are long and thin, the energy will be minimum when the magnetization is along the longitudinal axis. This can be seen from Fig. 1 to arise from the fact that the internal demagnetizing field is minimum in this case because of the r^{-n} $(n > 1)$ dependence of the demagnetizing field. The particles in magnetic recording tape are dominated by this mechanism, as are some of the Alnico magnets. In the latter case, the elongated particles are precipitates in a heterogeneous alloy.

BIBLIOGRAPHY

Magnetostatics

W. F. Brown, Jr., *Magnetostatic Principles in Ferromagnetism.* North-Holland, Amsterdam, 1962.

Permanent Magnets

H. Zijlstra, "Permanent Magnets; Theory," in *Ferromagnetic Materials,* Vol. 3 (E. P. Wohlfarth, ed.), p. 37. North-Holland, Amsterdam, 1982.

M. McCaig and A. G. Clegg, *Permanent Magnets in Theory and Practise,* 2nd ed. Pentech Press, London, 1987.

J. D. Livingston, "Nucleation Fields of Permanent Magnets," *IEEE Trans. Magn.* **MAGN-23,** 2109 (1987).

Many-Body Theory

D. C. Mattis

To most physicists today the many-body problem has connotations of a highly theoretical, quantum-mechanical study of systems of large numbers of interacting particles, such as electrons in a metal, with densities greater than 10^{20} particles/cm^3, or nucleons within the nucleus of an atom or in the nuclear matter of stars, with densities approaching 10^{40} particles/cm^3. But the question may be asked, *how many* bodies are required before we have a problem [1]? In Newtonian mechanics, the three-body problem was insoluble. With the birth of general relativity around 1910 and of quantum electrodynamics in 1930, the two- and one-body problems became insoluble; and within modern quantum field theory, the problem of zero bodies (the vacuum) has become insoluble!

Of course, one goal of many-body physics is to discover to what extent a complex system may be approximated by a collection of individual particles moving more or less independently of one another in the collective field of all of them. A separate, distinct goal is to discover to what extent such "quantum fluids" satisfy appropriate hydrodynamical equations. On closer inquiry, we observe that the interaction of each individual particle with the remaining fluid affects important properties such as its mass, because of the inertia of the "cloud" of interactions that accompany it. Further, properties of pairs of particles, such as their mutual interaction potential, are profoundly modified by the motion of the remaining particles. Thus, when the self-consistent equations of the fluid are derived, these changes must be taken into account *a priori*. It was not until the late 1950s that a proper theoretical framework—the use of second quantization and introduction of the Greens functions of quantum field theory and statistical mechanics—permitted a systematic development of this program.

The mounting popularity of this formerly recondite subject stems from successful applications to the theory of metals, both normal and superconducting; nuclear matter; and quantum fluids such as helium (^{3}He and ^{4}He). But the new techniques have not led to many *exact* results, nor have they been without notable antecedents in classical and in quantum mechanics. It is therefore most instructive to start by an examination of some early versions of the many-body problem.

THE CLASSICAL MANY-BODY PROBLEM

It is well known that Newton's equations of motion for three or more bodies, interacting by inverse-square-law forces, admit no closed-form solution. The difficulties are partly related to a physical question: Is the system in a bound state? For it is possible that one or more of the objects acquires energy from the others, achieves escape velocity, and is ejected from the system. Note also that the potential energy function satisfies Laplace's equation, so that there are no configurations of static equilibrium. Partly, also, the difficulties arise because of the detailed nature of the information that is required. Given a set of initial conditions, perhaps the shape of a galaxy, we wish to integrate the equations of motion of the celestial bodies either backward to the origins of time or forward to the ultimate end—but round-off error and long computation times endemic to multidimensional and numerical integration do not permit many results of this kind to be obtained. Thus the tendency in astrophysics is increasingly to forget about the many-body problem and concentrate on the study of hydrodynamics of "cosmic fluids."

When, however, the forces are not pure inverse-square law [2] it becomes possible for the *total* potential energy function $V(\mathbf{r}_1,\ldots,\mathbf{r}_N)$ of N interacting objects to have configurations of stable equilibrium, at $\mathbf{r}_i=\mathbf{r}_i^0$. Taylor's series expansion of this function up to terms quadratic in the displacements $x_i=x_i-x_i^0$ yields a system of coupled oscillators that can always be solved. The normal modes of regular one-, two-, and three-dimensional lattice structures, first studied by Hamilton [3] in 1839, form the basis for the theory of sound propagation (or "phonons" in the quantized version [4]) and that of the elastic properties of solids. The equations of coupled oscillators can be solved subject to arbitrary initial conditions, even when the masses of the particles are unequal—e.g., when one of the masses is much greater than the rest, as in Brownian motion [5]. Whereas for finite N the correlation functions are all periodic, with the period being Poincaré's recursion time t_P, in the limit $N\rightarrow\infty$ when the sum over normal modes must be replaced by an integral, we obtain that $t_P\rightarrow\infty$ and that all initial correlations decay inexorably. This explanation of *irreversibility*, then, is one important result of proceeding to the large-N limit.

If coupled oscillators represent the one extreme of a permanently bound system, the hard-sphere gas represents the other of a permanently *un*bound assembly that must be held together by external pressure or by mathematically imposed boundary conditions. In one dimension, where it is known as Tonk's gas of impenetrable rods, the constraints of conservation of energy and momentum together with the limitations of one-dimensional motion permit two colliding rods merely an *exchange* of energies and momenta upon impact, and thus the initial distribution of velocities is permanently maintained. However, the time evolution of N hard disks in two dimensions, or of hard spheres in three, is quite different, since (depending on impact parameter) each collision alters the relative momenta $\mathbf{p}_i-\mathbf{p}_j$. Indeed, after a (finite) characteristic time the entire system "forgets" its initial conditions to approach a well-defined state of thermodynamic equilibrium [6]. Computer simulations of the final state [7] have identified *two* different equilibrium phases: a close-packed solid phase and a low-density gas phase, with the lack of a liquid presumably due to the absence of attractive forces in the model. In any case, the study of such examples has established that the many-body problem is at the root of the science of thermodynamics.

A VARIETY OF COMPLEMENTARY QUANTUM METHODS

Bohr's 1913 theory of the hydrogen atom failed for helium and the heavier elements because it lacked the framework for the many-electron problem. The successor Hartree method improperly distinguished among in-principle indistinguishable electrons. Finally, use of determinantal wave functions by Hartree, Fock, and Slater allowed the construction of a one-particle self-consistent dynamics compatible with the exclusion principle. By the 1930s a substantially correct picture of the electron waves in atoms, molecules, and solids had emerged; the complementary, contemporaneous Thomas–Fermi method sacrificed the individual particles in favor of collective properties of the total *electron fluid* [8]. However, it also became apparent that a significant part of the energy of open-shell atoms, molecules, and metals, of the order of $\frac{1}{2}$ eV per electron, could not be explained by an independent-particle model; this energy came to be known as the "correlation energy," which many-body theory assumed the task of explaining.

The process of selecting relevant operators and establishing an "effective Hamiltonian" governing their interactions is the starting point. Because correlation energy is related to the electrostatic cost of density fluctuations, the construction of an effective Hamiltonian with the use of particle-density operators is indicated. Consider a metal containing noninteracting electrons having all possible momenta $\mathbf{p}$ within a Fermi sphere of radius p_F. It is convenient to take this state as the "vacuum," and for the elementary particles to take electrons only for $p>p_F$, introducing "holes" as the quasiparticles inside the Fermi sphere. The appropriate second-quantized operators a_p, $a_p\dagger$ destroy/create an electron (and are thus defined to vanish inside the Fermi sphere), b_p, $b_p^{\dagger}$ destroy/create a hole (and vanish outside p_F). All fermion operators are subject to the anticommutation relations

$$\{a_p,a_{p'}\}\equiv a_pa_{p'}+a_{p'}a_p=\{b_p,b_{p'}\}=\{a_p,b_{p'}\}=0,$$
$$\{a_p,a_{p'}^{\dagger}\}=\{b_p,b_{p'}^{\dagger}\}=\delta_{p,p'} \quad (1)$$

to ensure that the Pauli principle is exactly obeyed by any state constructed with their help. The particle-density operators

$$\rho_q=\sum_{j=1}^{N} e^{iq\cdot R_j}$$

are, in this formalism,

$$\rho_q=\sum_k (a_{k+q}^{\dagger}a_k+a_{k+q}^{\dagger}b_{-k}^{\dagger}+b_{-k-q}a_k-b_{-k}^{\dagger}a_{-k-q}), \quad (2a)$$

$$\approx\sum_k (a_{k+q}^{\dagger}b_{-k}^{\dagger}+b_{-k-q}a_k). \quad (2b)$$

Terms neglected in approximating (2a) by (2b) are only involved in the scattering of free electrons or holes, of which there are expected to be very few when the potential that creates these free particles is itself weak. We now study the "two-body" interaction potential:

$$\mathcal{H}' = \frac{1}{2V}\sum_q U_q \rho_q \rho_{-q} \quad (V = \text{volume}) \tag{3}$$

with $U_q = 4\pi e^2/q^2$ the Fourier transform of e^2/r or, for neutral weakly interacting fermions, $U_q \approx$ constant for small q, and 0 at large q. The unperturbed kinetic energy operator is

$$\mathcal{H}_0 = \sum \frac{p^2 - p_F^2}{2m}(a_p^\dagger a_p - b_p^\dagger b_p). \tag{4}$$

Examination of the equations of motion of operators such as $a^\dagger b^\dagger$ or ab indicates that no new operators appear if, to approximations (2b), we join

$$\begin{aligned}[b_{-k'+q'}a_{k'}, a_k^\dagger b^\dagger_{-k+q}] &\equiv b_{-k'+q'}a_{k'}a_k^\dagger b^\dagger_{-k+q} \\ &\quad - a_k^\dagger b^\dagger_{-k+q} b_{-k'+q'} a_{k'} \\ &\approx \delta_{k,k'}\delta_{q,q'} \quad (|k| > p_F, |k-q| < p_F)\end{aligned} \tag{5}$$

in addition to the exact commutations

$$\begin{aligned}&[a_k^\dagger a_k, a_{k'}^\dagger b_{k''}^\dagger] = \delta_{k,k'} a_{k'}^\dagger b_{k''}^\dagger \quad \text{and} \\ &[b_k^\dagger b_k, a_{k'}^\dagger b_{k''}^\dagger] = \delta_{k,k''} a_{k'}^\dagger b_{k''}^\dagger.\end{aligned} \tag{6}$$

With this algebra, operators carrying momentum q do not mix with operators carrying $q' \neq q$. This constitutes the so-called random-phase approximation (RPA) [9]. The problem is now effectively linearized, i.e., reduced to coupled oscillators, and all that remains is to find the normal modes. One method consists of constructing operators $\Omega_q \equiv \Sigma_k (F_k a_k^\dagger b^\dagger_{-k+q} + G_k b_{-k} a_{k-q})$ that satisfy the commutator equations of motion of exact normal modes,

$$[\mathcal{H}_0 + \mathcal{H}', \Omega_q] = \omega \Omega_q. \tag{7}$$

Equating coefficients of the ab operators on both sides of this equation leads to a set of linear equations in the F's and G's, the solution of which reduces to a proper-value equation first enunciated obtained by Bohm and Pines [9];

$$1 = \frac{2U_q}{V} \sum_{\substack{p<p_F \\ |p+q|>p_F}} \frac{(p\cdot q/m) + (q^2/2m)}{\omega^2 - (p\cdot q/m + q^2/2m)^2}. \tag{8}$$

This equation possesses two types of solutions: continuum solutions $\omega_{pq} = p\cdot q/m + q^2/2m + O(1/V)$, which interlace [10] the excitation spectrum of $\mathcal{H}_0$, and bound-state "collective" solutions above the highest value in the continuum, $p_F q/m + q^2/2m$. For the latter, now that the summand in (8) is finite, the sum may be replaced by an integration that is evaluated by elementary methods to yield the solutions displayed in Fig. 1. For the Coulomb potential, the quantized unit of this bound-state mode is denoted "plasmon," a stable high-energy charge fluctuation. For the weak finite-ranged potential with finite $U_{q=0}$, the same manipulations yield the "phonon" spectrum $\omega \propto q$, also indicated in the figure.

Landau's 1956 theory of Fermi liquids [11] concerns the energies of quasiparticles on either side of the Fermi level, particularly their dependence on the states of occupation of

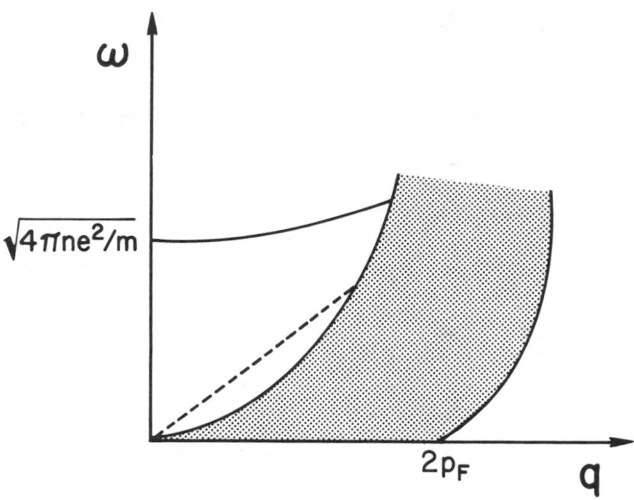

FIG. 1. The excitation spectrum of a Fermi gas (schematic). Shading indicates the continuum and coincides with the excitation spectrum of the unperturbed Hamiltonian, Eq. (4). Solid line shows *plasmon* branch (for $U_q = 4\pi e^2/q^2$), dashed line the *phonon* branch of the weakly imperfect neutral Fermi gas (U_q is constant at small q, vanishes at large q).

the other particles. Originally developed for the study of liquid ^{3}He, this theory was ultimately shown to provide the justification, to all orders in the perturbation expansion, of many important features of the independent-particle model. Thus, violent departures from free-particle behavior, e.g., the solidification of ^{3}He (under pressure) or the superconductivity of an electron gas (at low temperature), now have to be ascribed to the failure of perturbation theory to converge, which is a good sign of a genuine phase transition. Subtler examples concern the metal–insulator phase transition that is supposed to occur as atoms in a metal are slowly separated. Beyond a critical interatomic separation of the order of several angstroms, the Coulomb fluctuation energy greatly exceeds the covalent binding responsible for the band structure of the itinerant electrons, and it "pays" for atoms to become neutral and for the dc conductivity to vanish abruptly [12]. A much simplified model due to Hubbard [13] has been studied in connection with such problems; expressed in terms of operators that create or destroy particles at discrete lattice sites R_i, it is

$$\mathcal{H} = -\sum_{\substack{i,j \\ s=\pm 1}} T_{ij} c_{is}^\dagger c_{js} + U \sum_i c_{i+}^\dagger c_{i+} c_{i-}^\dagger c_{i-}. \tag{9}$$

The Fourier transform of the transfer matrix T_{ij} yields the Bloch energies of electrons in the conduction band. The interaction U has been simplified to where it only connects pairs of (necessarily opposite-spin) electrons on the same site. For T_{ij} transferring only between nearest neighbors, in one dimension, the model is *exactly* soluble [14]. In any number of dimensions, rigorous upper and lower bounds to the ground-state energy may be computed [14], as illustrated in Fig. 2. Unfortunately, none of the exact results have yet confirmed the variety of phase transitions (magnetic, insulator, etc.), at various values of the electron density and

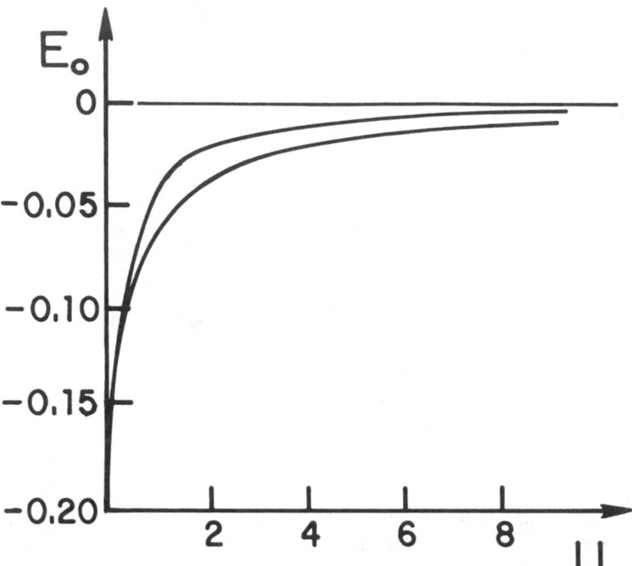

FIG. 2. Rigorous upper and lower bounds to the ground-state energy [14] of the Hubbard model in a half-occupied three-dimensional simple cubic lattice. Aside from indicating much steeper slope for $U<1$ than for $U>1$ (in units of the bandwidth), this sort of calculation appears incapable of determining changes of phase, for which the much more difficult study of Green functions or correlation functions is required.

parameter U, that have been predicted by a number of intuitive or approximate methods.

Bardeen, Cooper, and Schrieffer's 1957 theory of superconductivity [15] shows that the various phenomena of superconductivity come from a most spectacular breakdown of perturbation theory. Formulated in terms of Cooper-pair destruction operators $a_{ps}a_{-p-s}$ or $b_{ps}b_{-p-s}$ and their Hermitian conjugate creation operators, this theory showed that however weak a phonon-induced attractive matrix element $-V_0$ connecting such pairs in the vicinity of the Fermi surface might be, there would always be energies $|p^2-p_F{}^2|/m<V_0$ for which this interaction was, in fact, *strong* compared to the unperturbed energies. The BCS result consequently finds the ground state and the elementary excitations to be nonanalytic in the coupling constant, with an essential singularity at $V_0=0$.

Somewhat related results could already be seen in a 1947 study [16] of the superfluidity of a weakly imperfect Bose gas modeling ^{4}He. Here the field operators a_p, $a_p{}^\dagger$ satisfy commutation relations $[a_p, a_{p'}^\dagger]\equiv a_pa_{p'}^\dagger - a_{p'}^\dagger a_p=\delta_{p,p'}$ and the Hamiltonian is

$$\mathcal{H}=\sum\frac{p^2}{2m}a_p{}^\dagger a_p+\frac{1}{V}\sum U_p a_{k_1-p}^\dagger a_{k_2+p}^\dagger a_{k_2}a_{k_1}. \qquad (10)$$

In the unperturbed ideal Bose–Einstein condensed state, all particles occupy the $p=0$ momentum state, so that for weak interactions we may continue to treat $a_0\sim a_0{}^\dagger\sim n_0{}^{1/2}$ as macroscopic parameters and neglect their fluctuations. The largest parts of the remaining interaction Hamiltonian are therefore

$$\frac{1}{V}U_0n_0{}^2+\frac{2n_0}{V}\sum_{p\neq0}U_pa_p{}^\dagger a_p + \frac{n_0}{V}\sum_{p\neq0}U_p(a_p{}^\dagger a^\dagger_{-p}+a_{-p}a_p)+O(n_0{}^{1/2}). \qquad (11)$$

Both this operator and the kinetic energy are quadratic in the field operators, and thus a difficult problem has once again been reduced to a simpler one in coupled harmonic oscillators. Again, we seek normal modes and obtain, as for the weakly interacting Fermi gas, a linear excitation spectrum $\omega\propto q$ in the long-wavelength limit $q\to0$. But the two results differ significantly in that there is now no continuum of quasiparticle excitations: The entire excitation spectrum of the weakly interacting Bose gas consists of collective modes.

Many flaws have been noted in the foregoing theory. (a) Neither the van der Waals attractive potential nor the repulsion of the hard cores at short distances admits of a finite Fourier transform, so that the formalism is defective. (b) A study of the dilute limit [17] indicates that it is the scattering length of the interaction, and not the potential, that enters into a correct theory. (c) Penrose and Onsager have noted that, for hard spheres, not more than 8% could be condensed into the $p=0$ state, making n_0 a very poor approximation to $n=N/V$ in any event. (d) The theory seems unable to predict the existence or nature of *rotons* [18], which have been so important in the phenomenology of superfluidity.

Although normal-mode approximation methods seem to do even worse in two dimensions or one dimension, for reasons associated with the amount of phase space available to long-wavelength fluctuations, by a fortunate compensation many of the interesting many-body problems have been able to be solved in one dimension, through a variety of special methods [19].

GREEN'S FUNCTION METHODS

A vital simplification was brought into quantum electrodynamics by the introduction in 1949 of the graphical methods of Feynman diagrams for the systematic evaluation of various physical quantities in given orders of perturbation theory. It is not certain who first appreciated the relevance of such methods to the many-body problem [20]; however, by the 1960s they had become quite prevalent. One of the essential ingredients in this calculus is the "propagator," also known as the causal Green function [21],

$$G(p,E)=-i\int_{-\infty}^{\infty}dt\,e^{iEt}\langle\Psi_0|T\{c_p(t)c_p{}^\dagger(0)\}|\Psi_0\rangle \qquad (12)$$

in which Ψ_0 is the exact ground state of the interacting system, T is the symbol for time ordering, and the time dependence of the operators is given by $c_p(t)=[\exp(i\mathcal{H}t)]c_p(0)[\exp(-i\mathcal{H}t)]$. All observables—occupation numbers, kinetic energy, interaction energy, etc.—can be obtained by suitable manipulations of (12). It is therefore an important quantity to calculate, and a relatively simply one, since only the operators of the pth particle are explicitly involved. We can express G in terms of G_0, the trivially calculable *unperturbed* Green function,

$$G_0=(E-\epsilon_p+i\delta_p)^{-1}$$

where $\epsilon_p=$ one-particle energy and

$$\delta_p=\pm 0^+ \quad \text{for} \quad |p|\gtrless p_F,$$

by means of Dyson's equation:

$$G(p)=G_0(p)+G_0(p)\,\Sigma\,(p)G(p), \tag{13}$$

which has the formal solution

$$G=\{E-[\epsilon_p+\Sigma\,(p)]+i\delta_p\}^{-1}. \tag{14}$$

The proper self-energy part $\Sigma(p)$ must be calculated separately; its real part yields the effect of interactions on the mass of the quasiparticle, while its imaginary part tells about the lifetime. To first order, $\Sigma(p)$ is given by (a) in Fig. 3, for which the rules for the evaluation of such diagrams yield [21]

$$\Sigma_1(p)=\tfrac{1}{2}\left(\frac{1}{2\pi}\right)^3\int d_3q\,U_q-\frac{1}{(2\pi)^3}\int_0^{|q|<q_F} d_3q\,U_{|q+p|}, \tag{15}$$

which is entirely equivalent to Hartree–Fock theory. Higher-order diagrams yield further corrections, including the finite *lifetime*. By examining these corrections to all orders, Luttinger was able to prove [22] that subject to the convergence of perturbation theory, in three dimensions the imaginary part of $\Sigma(p)$ vanished as $(p^2-p_F{}^2)^2$ as $p\to p_F$, establishing that the quasiparticles near the Fermi surface were sharp and well defined. Use is often made of the so-called Lehmann representation

$$G(p,E)=\int_0^\infty d\epsilon\left[\frac{A^+(p,\epsilon)}{E-(\epsilon_F+\epsilon)+i0^+}+\frac{A^-(p,\epsilon)}{E-(\epsilon_F-\epsilon)-i0^+}\right] \tag{16}$$

in which it can easily be shown that $A^\pm>0$. Eq. (16) shows that the causal G has a complicated analytic structure as a function of the complex variable E; this makes the theory awkward to generalize to finite temperature T.

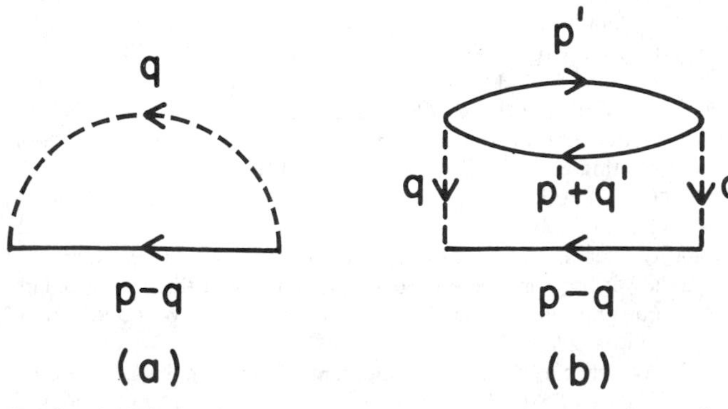

FIG 3. (a) First-order contribution to proper self-energy of a quasiparticle shown in diagrammatic form. The rules for evaluating it [1] yield Eq. (15). (b) A simple second-order diagram that contributes to the finite lifetime of a quasiparticle [22].

The causal Green's functions relevant to the various components of a many-body system (electrons, plasmons, phonons, etc.), the equations which they satisfy, and their interrelationships constitute the principal subject of such textbooks as are cited in [9]. Additionally, once perturbation theory ceases to be the overriding consideration, analytically more satisfactory Green's functions can be used, which easily carry over to finite T. There exist at least four species [23]: advanced and retarded, which are analytic in the lower and upper half-planes of complex E, respectively; each of these can be a commutator or anticommutator, the former being most useful for the study of bosons (plasmons, phonons, magnons, etc.) and the latter for Fermi–Dirac-type quasiparticles (electrons, holes, etc.). They are

$$G_r(t'-t'')=-i\theta(t'-t'')\langle[A(t'),B(t'')]_\eta\rangle,$$
$$G_a(t'-t'')=i\theta(t''-t')\langle\text{same}\rangle \tag{17}$$

where $\theta(t)$ is the unit step function, and averages are with respect to the grand canonical ensemble:

$$\langle op\rangle=\frac{1}{Z}\,\mathrm{Tr}\{ope^{-\beta(\mathcal{H}-\mu N)}\},\quad Z\equiv\mathrm{Tr}\{e^{-\beta(\mathcal{H}-\mu N)}\} \tag{18}$$

and $\eta=\pm1$ is used to distinguish commutator and anticommutator Green's functions,

$$[A,B]_\eta\equiv AB-\eta BA. \tag{19}$$

Basically, it is the Fourier transform $G(E)$ that we need:

$$G(E)=\frac{1}{2\pi}\int_{-\infty}^{\infty} dt\,G(t)e^{iEt}, \tag{20}$$

for we can show [23] that the thermal-averaged correlation functions are given by the discontinuity of the corresponding Green functions on the real axis:

$$\langle B(t')A(t)\rangle=i\int_{-\infty}^{\infty} dE\,\frac{e^{-iE(t-t')}}{e^{\beta E}-\eta}[G(E+i0^+)-G(E-i0^+)] \tag{21}$$

where $G(E\pm i0^+)=G_{r,a}(E)$. By appropriate choices of the operators A and B *all* physical observables can be examined in this manner, once the corresponding Green function is known. The method of choice for obtaining it involves the *equations of motion*:

$$EG(E)=\frac{1}{2\pi}\langle[A(0),B(0)]_\eta\rangle+G^{(1)}(E) \tag{22}$$

where the equal-time commutator is usually a known constant, and

$$G^{(1)}(E)\equiv-\frac{i}{2\pi}\int_0^\infty dt\,e^{iEt}\langle[C(t),B(0)]_\eta\rangle. \tag{23}$$

This is itself a Green function, with the operator $C(t)$, given by a commutator,

$$C(t) \equiv [A(t), \mathcal{H} - \mu N], \quad (24)$$

taking the place of A in $G^{(1)}$.

We may then study the equation of motion of $G^{(1)}$, which in turn yields a new function $G^{(2)}$, and continue this process Q times until either of two conclusions: After the Qth iteration no new functions appear, in which case the set of $Q+1$ linear equations in $Q+1$ Green's functions is closed and may be straightforwardly solved; or the hierarchy is terminated by our approximation of the $G^{(Q)}$. If the approximation consists of replacing $G^{(Q)}$ by its (unperturbed) value in the absence of interactions, then the system of Q equations in the remaining functions is essentially equivalent to Qth-order perturbation theory. But vastly more interesting results have been obtained when $G^{(Q)}$ is treated as a function, generally nonlinear, of the remaining Green functions on the basis of some physically motivated arguments. The system of equations, now nonlinear, has been found to exhibit such behavior as phase transitions at various temperatures and coupling constants, etc.; i.e., to mimic the rich variety of behavior that has come to be associated with the many-body problem.

See also PHONONS; PLASMONS; QUANTUM FIELD THEORY; QUANTUM FLUIDS; QUANTUM STATISTICAL MECHANICS; QUASIPARTICLES; SUPERCONDUCTIVITY THEORY; THREE-BODY PROBLEM, CLASSICAL.

REFERENCES

1. Here we quote loosely from Chapter 1 in R. D. Mattuck's delightful *Guide to Feynman Diagrams in the Many-Body Problem* (McGraw-Hill, New York, 1967). The difficulties (and successes) of the *few*-body ($n=1,2,3,4,\ldots$) problem are the subject of D. C. Mattis, *Rev. Mod. Phys.* **58,** 361 (1986).
2. The potential function for two rare-gas atoms is not r_{ij}^{-1} but, approximately, $V(r_{ij})=Ar_{ij}^{-12}-Br_{ij}^{-6}$; the potential function of open-shell atoms may be more complicated because of covalent bonding and other quantum effects. That an equilibrium configuration always exists (i.e., that collapse of matter is impossible) has been proved by F. J. Dyson and A. Lenard, and most recently and elegantly by E. Lieb and W. Thirring in *Phys. Rev. Lett.* **35,** 687 (1975).
3. Volume 2 of *The Collected Mathematical Papers of Sir William Rowan Hamilton* (Cambridge Univ. Press, London and New York, 1940). Recent advances are discussed in A. Maradudin, E. Montroll, and P. Weiss's "Theory of Lattice Dynamics in the Harmonic Approximation," *Solid State Phys.* Suppl. No. 3 (F. Seitz, *et al.*, eds.; Academic Press, New York, 1963), as well as in the conference proceedings "Molecular Dynamics and the Structure of Solids" (L. Carter and J. Rush, eds.; National Bureau of Standards Special Publication 301, Washington, D.C., June 1969).
4. A familiar topic in all introductory solid-state textbooks. But recent interesting developments in "quantum solids," i.e., solid ^{3}He or ^{4}He isotopes or mixtures, are not so well known; see a recent report by B. Bertman and R. Guyer in *Sci. Am.* August 1967, p. 85.
5. R. J. Rubin, "Statistical Dynamics of Simple Cubic Lattices, Model for the study of Brownian Motion," *J. Math. Phys.* **1,** 309 (1960).
6. A rigorous demonstration that a many-body system subject to arbitrary initial conditions approaches thermal equilibrium (i.e., satisfies the famous "ergodic hypothesis" to a greater or lesser extent) exists indeed *only* for the hard-sphere gas. The immensely difficult proof is the work of Y. Sinai, *Dok. Akad. Nauk SSSR* **153,** 1261 (1963).
7. B. J. Alder and T. E. Wainwright. *J. Chem. Phys.* **31,** 459 (1959) and **33,** 1439 (1960). Amazingly, many of their results have been duplicated by an approximate theory in which the coordinates r_i, p_i of the individual particles are sacrificed in favor of an equal number of "collective coordinates," chosen to be the Fourier components of the particle density. These were treated as approximately independent dynamical variables and used in the construction of a self-consistent equation of state by J. Percus and G. Yevick, *Phys. Rev.* **110,** 1 (1958), following a procedure known as RPA that had been found useful in the quantum many-electron problem. This topic is treated elsewhere in the present text.
8. This method, largely discarded since the 1930s, has lately been reexamined and extended by E. H. Lieb and B. Simon, "The Thomas–Fermi Theory of Atoms, Molecules and Solids," *Adv. Math.* **23,** 22–116 (1976).
9. For a discussion of this topic and a collection of many relevant reprints see David Pines's *The Many-Body Problem* (Benjamin, Advanced Book Program, Reading, Mass., 1962). A pedagogic and reasonably up-to-date approach to the many-body problem, containing numerous applications to condensed matter physics (electron gas, superconductivity, superfluidity, etc.) is found in G. D. Mahan's massive (1000-page) book, *Many-Particle Physics*. Plenum, New York, 1981. The similar (but terser) *Quantum Theory of Many-Particle Systems* by A. L. Fetter and J. D. Walecka (McGraw-Hill, New York, 1971) covers much the same topics, together with the nuclear many-body problem ("nuclear matter").
10. The slight shift in the excitation spectrum of the continuum may be interpreted by scattering theory to be associated with the decay in time of a prepared initial state. This decay is better studied by examining the individual quasiparticles (i.e., the individual electron or hole) than by examining the couple that make up the elementary excitation of the system. For this, the propagator formalism is required (see Ref. 1). Note that the finite lifetime of elementary excitations does not of itself limit the electrical conductivity, since the scattering is among excitations *all* of which carry momentum q.
11. The original papers in English translation are reprinted in Ref. 9. A simple textbook approach is provided by D. Pines and P. Nozières, *The Theory of Quantum Liquids*, Vol. 1 (Benjamin, New York, 1966).
12. Experimentally, this effect has been seen in metal vapors. Theoretically, a projection-operator method to reduce charge fluctuations to an energetically optimal amount was introduced by M. Gutzwiller, *Phys. Rev.* **134,** A923 (1964), following ideas that date back to E. Wigner, *Phys. Rev.* **46,** 1002 (1934) and *Trans. Faraday Soc.* **34,** 678 (1938). N. F. Mott has contributed many papers to the theory of collective insulator, and there is a nice characterization of this topic in W. Kohn's "Theory of the Insulating State," *Phys. Rev.* **133,** A171 (1964).
13. J. Hubbard, *Proc. Roy. Soc. (London)* **A276,** 238 (1963), **277,** 237, and **281,** 401 (1964).
14. E. Lieb and F. Wu, *Phys. Rev. Lett.* **20,** 1445 (1968). Upper and lower bounds to the ground-state energy of this Hamiltonian in one and three dimensions were computed by W. Langer and D. Mattis, *Phys. Lett.* **36A,** 139 (1971).
15. An especially clear development is found in J. R. Schrieffer's *Theory of Superconductivity* (Benjamin, New York, 1964).
16. N. N. Bogoliubov, apparently following a suggestion of D. Landau to have the two-fluid hydrodynamic theory of superfluidity placed on a more microscopic basis; the relevant papers are reprinted in Ref. 9.

17. T. D. Lee, K. Huang, and C. N. Yang, *Phys. Rev.* **106,** 1135 (1957); also E. Lieb, *Phys. Rev.* **130,** 2518 (1963).
18. Discussed by R. P. Feynman, *Phys. Rev.* **94,** 262 (1954); R. P. Feynman and M. Cohen, *Phys. Rev.* **102,** 1189 (1956).
19. See the text and reprint volume covering many areas of the many-body problem and statistical mechanics by E. Lieb and D. Mattis, *Mathematical Physics in One Dimension* (Academic Press, New York, 1966).
20. The earliest referenced papers in this general area include K. Husimi, *Proc. Phys. Math. Soc. Japan* **22,** 264 (1940); H. Koppe, *Ann. Phys.* **9,** 423 (1951); H. B. Callen and T. A. Welton, *Phys. Rev.* **83,** 34 (1951), L. van Hove, *Physica (Utrecht)* **21,** 901 (1955); T. Matsubara, *Prog. Theor. Phys.* **14,** 351 (1955); followed by others too numerous to list.
21. Indicated by a directed solid line in diagrams such as Fig. 3. The rules for evaluating these diagrams and the resulting physical observables may be found in Ref. 1.
22. J. M. Luttinger, "Analytic Properties of Single-Particle Propagators for Many-Fermion Systems," *Phys. Rev.* **121,** 942 (1961).
23. For an exceptionally clear introduction to the use of these retarded and advanced Green functions consult D. N. Zubarev in *Sov. Phys. Usp.* **3,** 320 (1960).

Masers

James P. Gordon

The maser is a device for coherent amplification or generation of electromagnetic waves by stimulated emission from an ensemble of material particles such as atoms or molecules. The word maser is an acronym for Microwave Amplification by Stimulated Emission of Radiation, coined in 1954 by Charles H. Townes and his students at Columbia University to celebrate the operation of the first such device, the molecular beam ammonia maser. The word has now become a generic label for coherent wave amplifiers and oscillators operating on the above principle, particularly in the frequency range up to about 3×10^{11} Hertz (1 mm wavelength). Higher frequency devices of a similar nature are called LASERS, wherein Light has been substituted for Microwave.

Masers have proved particularly useful as low-noise amplifiers for radio-astronomy and space communications, and as "atomic clock" oscillators of exceptional frequency stability. Their optical counterparts, which were first called optical masers, have achieved great importance in a wide variety of roles in science and technology.

Stimulated emission is a concept arising from the quantum theory of the interaction of electromagnetic radiation with matter. It was first introduced by Einstein in 1916-1917 to provide a consistent basis for thermodynamic equilibrium of a system containing both material particles—atoms, molecules, etc.—and radiation. The molecules transfer energy to and from the radiation field at certain discrete frequencies by making transitions among their various allowed energy states. The state energies W and transition frequencies f obey the relations

$$W_b - W_a = hf_{ab}, \tag{1}$$

where the subscripts a and b indicate two different states, and h is Planck's constant. Energy is conserved in these processes; hence the absorption of radiation results in upward (in energy) particle transitions, while the emission of radiation results in downward transitions. The downward transitions can occur spontaneously whether or not there is radiation present; the resulting incoherent spontaneous emission provides the irreducible noise in maser amplifiers. The presence of ambient radiation stimulates the upward and downward transitions with equal probability. The coherent stimulated emission process is the essential ingredient of a maser.

When thermodynamic equilibrium exists, there must be more particles in states of lower energy than in those of higher energy. The net absorption that results is necessary to balance spontaneous emission and so to maintain equilibrium. In a maser, thermodynamic equilibrium is upset so that net amplification rather than absorption occurs. The most common way of doing this is by inverting the populations of two energy levels so that the states of the upper level are more populated than those of the lower level. This process is known as "pumping"—stimulated emission exceeds absorption, and an incident electromagnetic wave near the transition frequency between the two levels is amplified.

THEORETICAL CONSIDERATIONS

Any material exhibiting a relatively strong electric or magnetic resonance absorption line is a candidate for conversion to maser action if a suitable pumping mechanism can be found. Examples come from many areas of spectroscopy. The ammonia maser uses the strong electric dipole inversion resonances of NH_3 molecules occurring near 24 GHz. The hydrogen maser uses the magnetic dipole hyperfine transition of atomic hydrogen near 1.420 GHz. The ruby maser uses the paramagnetic resonance transitions of the CR^{+3} ions in a ruby crystal. This and other masers using electron paramagnetic resonances can be tuned by application of a d-c magnetic field at a rate of about 2 MHz per gauss; maser amplification has been obtained with these materials at frequencies ranging from 300 MHz to 75 GHz. The nuclear magnetic resonance of protons in water has been used to make a maser operating in the earth's magnetic field at a frequency near 2 kHz.

The response of any material to the presence of an electromagnetic field can be described by its electric and magnetic polarizations, **P** and **M**, respectively. For a field of angular frequency ω and for a dipolar material, the polarizations are related to the fields by

$$\mathbf{P} = \epsilon_0\chi_e\mathbf{E} \text{ and } \mathbf{M} = \chi_m\mathbf{H}, \tag{2}$$

where the χ's are dimensionless quantities, respectively the electric and magnetic susceptibilities. [Note: We use S.I. units, and the usual engineering convention that the time varying fields are expressed as the real parts of (**E** or **H**) exp($j\omega t$). The physics literature usually uses exp($-j\omega t$); this difference means a sign change of imaginary quantities, which can be a source of confusion to the unwary.]

These susceptibilities are the most important material parameters in determining the properties of a maser. In quantum theory they are the consequence of induced transitions. They are generally complex quantities, reflecting the components of polarization both in phase with and in quadrature with the field. Conventionally, we write

$$\chi = \chi' + j\chi'', \tag{3}$$

where χ' and χ'' are real. The real part, χ', affects the electromagnetic wave velocity in the material, while the imaginary part, $j\chi''$ is proportional to the rate at which the material absorbs energy from the field. In the range of field strengths where χ is not field dependent, χ' and χ'' are derivable from one another via the Kramers-Kronig relations.

A typical form of χ for a narrow resonance line is the Lorentzian

$$\chi_L(\omega) = \frac{j\chi''(\omega_m)}{1 + j(\omega - \omega_m)T_2} \tag{4}$$

where ω_m is the material resonance frequency, and T_2 is the lifetime of the polarization in the absence of any field. Often, real resonance lines are sums of Lorentzian components, each with a slightly different resonance frequency due to variations in the local surroundings of the molecules, or to motional Doppler shifts. The resulting absorption lines then often display nearly Gaussian shapes if this "inhomogeneous" broadening is greater than the lifetime broadening.

In general, quantum theory gives the following expression for the imaginary parts of the electric and magnetic susceptibilities resulting from a transition from a lower energy level A to a higher energy level B:

$$\chi''_{AB}(\omega) = P\left(\frac{n_B}{g_B} - \frac{n_A}{g_A}\right)\frac{2\pi^2 S_{AB}}{3h}\, g(\omega) \tag{5}$$

where P is ϵ_0^{-1} for the electric case or μ_0 for the magnetic case, n_A and n_B are the population densities of atoms in the two levels, g_A and g_B are their degeneracies†, $S_{AB} \equiv \Sigma\Sigma_{a,b} \mu^*_{ab} \cdot \mu_{ab}$ is the sum, over all the degenerate states of both levels, of the squared magnitudes of the electric or magnetic dipole moment transition matrix elements, and finally $g(\omega)$ is the normalized line shape function, whose integral over ω is equal to unity. A Lorentzian line, for example, has the line shape function

$$g_L(\omega) = \frac{T_2/\pi}{1 + (\omega - \omega_m)^2 T_2^2}. \tag{6}$$

The frequency width at half height of this shape is $\Delta\omega_m = 2/T_2$. Note that the maximum of $g(\omega)$ at $\omega = \omega_m$ is approximately the reciprocal of the line width. Note also that the susceptibility is proportional to the difference in state populations of the two levels; absorption (negative χ'') results when $n_A/g_A > n_B/g_B$, or gain when $n_B/g_B > n_A/g_A$. The frequency integrated susceptibility is a material property independent of the line width. Relatively narrow resonance lines are needed to provide useful gain.

The frequency integrated electric susceptibility, per molecule, tends to be considerably larger (about 10^4 times) than the magnetic susceptibility. On the other hand, magnetic resonances are less susceptible to excessive broadening due to random fields from neighboring material, and so much higher particle densities (liquids and solids as opposed to gases) can be achieved. Thus the solid state paramagnetic masers can have larger susceptibilities; in addition they are tunable, as discussed above, and can handle higher field intensities before their gain is saturated (see below).

† The degeneracy, or statistical weight, of an energy level is the number of independent quantum states of the same energy. Spatial degeneracy is one ubiquitous type of degeneracy.

PUMPING MECHANISMS

The population inversion necessary to a maser has been achieved in several ways. In the ammonia and hydrogen masers, a molecular (atomic) beam, in a vacuum chamber, traverses a region of spatially varying d-c field (electric for ammonia, magnetic for hydrogen) which focuses the upper level particles into the interior of a microwave resonator and simultaneously defocuses the lower level particles. Maser action then occurs in the microwave resonator. Masers using stationary media require three or more energy levels to operate continuously. The Cr^{+3} ions of ruby have four pertinent energy levels, for example, resulting from the different quantized orientations of the magnetic dipole moment associated with the total electron spin (3/2) of the ion. Consider the case of three non-degenerate ($g = 1$) levels. Various relaxation processes (emission and absorption of phonons, for example) normally transfer the particles among the available states so that, at temperature T, the usual Boltzmann relation $n_i/n_j = \exp(hf_{ij}/kT)$ obtains, the lower levels always having the greater populations. Here f_{ij} is the transition frequency between levels i and j. Pumping this system consists of applying an intense microwave field at the transition frequency between the outer two levels. If the transition rate between these outer levels induced by the pump exceeds their relaxation rates, then they will achieve substantially equal populations. This is called saturating the transitions. The population of the intermediate level then comes to an equilibrium value determined by its relaxation rates to the two pumped levels. If its relaxation rate to the upper level is the faster, it will equilibrate to a population greater than that of the upper level, and hence also greater than that of the lower level. On the other hand, if the intermediate level relaxes faster to the lowest level, it will achieve a population smaller than that of the other two, and hence a population inversion with respect to the higher level. A variation on this scheme is called push-pull pumping, and requires four levels. Here the first and third levels are pumped to saturation, and also the second and fourth. This scheme results in inversion of the second and third levels. If the system can be tuned so that the two pumped transitions occur at the same frequency, then only a single pumping frequency is required. Both of these methods have been used successfully in ruby masers.

With regard to these multilevel pumping mechanisms, it is important to note that at a microwave frequency of 10 GHz, the temperature hf/k is only about 0.5 °K. Thus at room temperature, the level population differences are quite small. In addition, the interlevel relaxation rates are so fast at room temperature that the resonance lines broaden and disappear. Hence the ruby maser, for example, is always operated at liquid helium temperature. While this necessity restricts such masers to uses where no simpler amplifier will do, it has the advantage of reducing the maser's noisiness nearly to the minimum allowed by nature. The maser's sensitivity to incoming signals is greater than that of any other amplifier.

NEGATIVE RESISTANCE, NEGATIVE TEMPERATURE, AND NOISE

The electromagnetic circuits of masers are of two types, resonators and traveling wave structures, with the active medium filling all or part of the circuit. Resonators maximize the interaction of the electromagnetic field with the maser medium, but their amplification bandwidth narrows rapidly at high gain. They have mainly been used for maser oscillators. The traveling wave structures used are of the slow wave type, meaning that the group, or energy propagation, velocity of electromagnetic waves in the structure is considerably less (by a factor of 100, say) than the velocity of light. This again serves to keep the field in contact with the active medium longer, increasing the gain per unit length of amplifier.

At low frequencies a simple resonator is made by connecting together an inductor and a capacitor. At microwave frequencies, a typical resonator might be a section of transmission line or waveguide one half-wavelength long, shorted at both ends, with some provision for coupling energy to an external transmission line. A slow wave structure is made essentially by coupling a series of resonators together so that energy passes relatively slowly from one resonator to the next.

The effects of inserting a maser-active medium into a resonator or traveling wave structure can be described most easily by considering the equivalent lumped circuit elements. For example, the impedance Z (voltage/current ratio) of an inductor with an inserted piece of material of magnetic susceptibility χ is $Z_L = R_L + j\omega L(1+\eta\chi)$, where R_L is the (small) series resistance of the inductor, and η, called the filling factor, is the effective fraction of the inductor's field volume occupied by the material. From Equation (4) one may recognize that if χ has the Lorentzian form, the impedance added by the material has the form of that of a parallel *RLC* circuit connected in series to the inductor. This analogy has been fruitfully pursued. The added impedance has both resistive and reactive components; explicitly

$$Z_L = R_L - \omega L\eta\chi'' + j\omega L(1+\eta\chi'). \tag{7}$$

One sees that if the medium is active (positive χ'') it acts as a negative resistance. The quantity $\omega L/R_L$ is called the quality factor, or Q, of the inductor. Thus if $\eta\chi'' > 1/Q$, the inductor stops being lossy and instead can provide gain to a circuit. Exactly analogous arguments apply to the admittance of a capacitor containing an active material.

Generally, the Q of a resonant circuit is the ratio of 2π times its stored energy to its energy loss per cycle. The quantity $\eta\chi''(\omega_m)$, the maximum value of $\eta\chi''$, is called Q_m^{-1}, the reciprocal quality factor for the maser material. Typical numbers for a ruby maser, for example, might be a circuit Q of the order of 10^3 to 10^4, and a material Q of the order of 10^2.

One can pursue these ideas to construct resonators, slow wave circuits, etc., composed of such active elements. For a resonator coupled to a transmission line, the reciprocal resonator Q has three parts:

$$Q_{\text{total}}^{-1} = Q_{\text{int}}^{-1} + Q_{\text{ext}}^{-1} - Q_m^{-1} \tag{8}$$

where Q_{int}^{-1} accounts for internal losses, Q_{ext}^{-1} for the loss to the output, and Q_m^{-1} for the material Q. When $Q_m^{-1} > Q_{\text{int}}^{-1}$, the resonator begins to provide gain to a signal field incident from the line. When $Q_m^{-1} \gg Q_{\text{int}}^{-1} + Q_{\text{ext}}^{-1}$, then the gain from the active medium exceeds all losses, and spontaneous oscillation ensues. Just below the oscillation threshold, high gain can be achieved, but one can show that the half-power bandwidth of the gain is inversely proportional to the gain, according to the relation

$$G\Delta\omega = 2(\Delta\omega_0)Q_{\text{ext}}^{-1}/(Q_{\text{int}}^{-1} + Q_{\text{ext}}^{-1}) \tag{9}$$

where $\Delta\omega_0^{-1} = \Delta\omega_m^{-1} + \Delta\omega_{\text{res}}^{-1}$, and $\Delta\omega_{\text{res}}$ is the bandwidth of the resonator $\Delta\omega_{\text{res}} = \omega_{\text{rest}}(Q_{\text{int}}^{-1} + Q_{\text{ext}}^{-1})$. Here G is the amplitude gain. Thus the initial band $\Delta\omega_0$ in the gain-bandwidth product is in effect the smaller of the resonator or material bandwidth.

A feature of primary importance to the use of masers as frequency standards is the precise frequency at which oscillation takes place. When threshold is exceeded, the oscillation frequency is that for which the total circuit reactance (the imaginary part of Z) is zero, and so it is affected by the real part of the susceptibility, χ'. Straightforward circuit analysis yields the relation

$$\frac{\omega - \omega_{\text{res}}}{\Delta\omega_{\text{res}}} + \frac{\omega - \omega_m}{\Delta\omega_m} = 0 \tag{10}$$

for the oscillation frequency ω. That frequency is therefore intermediate between the resonator and material frequencies, closer to the center of the sharper resonance. Thus, for the best frequency standard, one wants the narrowest material resonance possible, combined with the broadest resonator resonance consistent with meeting the threshold condition. Then one attempts to keep the resonator tuned as closely as possible to the material resonance. Typically, the maser oscillation frequency can be set with an accuracy of the order of 10^{-3} of the material resonance bandwidth.

Traveling wave structures maintain relatively broader bandwidths up to high gain, because the individual resonators of which they are composed need not individually operate as close to threshold. The expression

$$\frac{\Delta\omega}{\Delta\omega_m} = \left(\frac{3}{G_{db} - 3}\right)^{1/2} \tag{11}$$

can be shown to apply in this case, where G_{db} is the power gain in decibels. For a gain of 30 *db*, say, the bandwidth of the traveling wave amplifier is reduced only by a factor of 3, whereas the single resonator amplifier would suffer a bandwidth reduction greater than 15.

Low noise is the other primary feature of masers. Because masers use dipolar rather than charged particles to provide gain they are much less sensitive to fluctuations in the number of particles. The primary unavoidable noise source in a maser is the spontaneous emission from the excited particles. To analyze noise in any circuit, one must ally, with every resistive element at temperature T, a series noise voltage source whose mean square voltage in a narrow band δf is $4RW(f)\delta f$, where $W(f)$ is the thermodynamic equipartition energy $hf/[\exp(hf/kT)-1]$. If the resistance is due to a material susceptibility such as equation (5), the source of this noise is the spontaneous emission from the upper state particles. The quantity $[\exp(hf/kT)-1]^{-1}$ is equal to the level population ratio $(n_B/g_B)/(n_A/g_A-n_B/g_B)$, according to Boltzmann's law. An active material, with $n_B/g_B>n_A/g_A$, has, in effect, a negative temperature $T_m=-(hf/k)\ln(n_Bg_A/n_Ag_B)$. The concept of negative temperature is very useful, and has been shown to have thermodynamic validity when restricted to interactions covering a finite range of frequencies. The expression above for the noise voltage source remains unchanged on passage to negative temperatures. The values of R and $W(f)$ both become negative, but their product stays positive. Note, however, that as T gets small and positive, $W(f)$ goes to zero, but as T gets small and negative (particles mainly in the upper state) then $|W(f)|$ goes to hf.

The consequence to maser amplifiers is as follows. For a maser amplifier with power gain G, one can write the expression $P_{out}=GP_{in}+(G-1)P_n$ for the power output P_{out}. Here P_n is the effective input noise power at high gain. One can verify that cascaded amplifiers, each with the same P_n, have the same P_n in combination, the overall gain being the product of the individual amplifier gains. For a maser with negligible circuit losses, one finds $P_n=|W(f)|\delta f$, with the minimum value $hf\Delta f$. Actual masers have approached this ideal value to within a factor of 100. The effective input noise is often characterized by an effective input noise temperature, defined by equating P_n to that produced by an input termination at the positive temperature T_{eff}. This yields

$$T_{eff}=(hf/k)/\ln[1+hf/|W(f)|], \qquad (12)$$

whose minimum value, for the ideal maser, is $T_{eff.min}=hf/k\ln 2$. At microwave frequencies $T_{eff.min}$ is a fraction of a degree Kelvin; actual masers, operating at 4.2 °K (liquid helium) have achieved noise temperatures less than 10 °K.

USEFULNESS AND IMPORTANCE OF MASERS

The original ammonia maser used a state-selected beam of about 10^{14} NH_3 molecules per second to provide amplification at the inversion resonance at 23.87 GHz as they traversed a microwave cavity resonator. The cavity was structured so that the usual Doppler broadening of a gaseous resonance line was largely suppressed, yielding a material bandwidth of about 7 kHz. The maximum power output was about a nanowatt. This type of maser did not make a practical amplifier, but was used to verify predictions of oscillator stability and low noise amplification. As oscillators, ammonia masers achieved frequency stability over a few minutes of about 0.02 Hz (1 part in 10^{12}), a record at that time, and were used in 1958–59 in an improved test of special relativity, looking for an ether drift and confirming none down to about 10^{-3} of the earth's orbital velocity, or 30 m/sec. The most stable atomic clock is now the hydrogen maser, which is discussed below.

Solid state paramagnetic traveling wave maser amplifiers have been made with a variety of substances, but perhaps most used has been pink ruby (Cr^{3+} ions at about 0.05% concentration in Al_2O_3 crystals) which combines nearly ideal mechanical and electromagnetic properties. The characteristics of one such maser, built at Bell Telephone Laboratories in 1960–61, were 36 *db* gain, 13 MHz bandwidth, and a noise temperature of 9 °K. This maser, operating at 2.39 GHz, used a comb type slow wave circuit, consisting of a series of metal fingers, each of which formed a resonant structure, projecting into a waveguide. The magnetic field in this structure is circularly polarized at the location of the amplifying crystals, and the magnetic susceptibility was arranged to be large for the sense of circular polarization of only the forward wave. In this way, and by putting in a lossy material (actually also ruby with a much higher Cr^{+3} concentration which does not work as a maser) to interact with the backward wave, a highly unidirectional and stable amplifier was made. It was used with a horn-reflector antenna at Crawford Hill to receive reflected signals from the early Project Echo experimental communication satellite. Careful sky noise measurements using this system resulted in 1965 in the discovery by A. Penzias and R. W. Wilson of the 3-K background which has been interpreted as residual red-shifted radiation left over from the beginning of the universe, in support of the "big bang" theory. Many such masers have been used in radio-astronomy, measuring the very weak (effective antenna temperatures of the order of hundredths of a degree Kelvin) signals from stellar objects. Many kinds of molecules have been found in interstellar space by observing their microwave emission or absorption spectra. Because the tenuous matter of space, particularly near strong radiation sources, is not in thermal equilibrium, the conditions for maser action can exist in space also. Since 1965, a number of masers have been discovered there, beginning with one driven by the hydroxyl (OH) molecule in the Great Nebula of Orion. Others include water (H_2O) and nitrous oxide (NO). These masers are characterized by emissions with unusual ratios of line strengths resulting from gain, and strongly fluctuating intensities.

The maser most in use today is the hydrogen atom maser, conceived and built in 1961 by N. F. Ramsey and his associates at Harvard. A beam of hydrogen atoms created in an r-f discharge passes through a magnetic state selector, which directs those atoms in the upper state of the 1.420 GHz hyperfine transition into a coated quartz storage bulb interior to a microwave cavity resonator. The atoms can bounce many times off the walls of the bulb without causing relaxation of the hyperfine transition, resulting in a material bandwidth of a small fraction of one Hz. Because of this exceedingly narrow bandwidth, the gain is sufficient even for this weak magnetic dipole transition to give self-sustained oscillations. Such oscillators are the most stable in existence. They have proved important as clocks in astronomical studies. One has been sent up to height of about 10 km in a successful test of the gravitational equivalence principle. A number are in use at the U.S. Naval Observatory in Washington D.C. They have demonstrated frequency stability of 5 parts in 10^{16} for times of about four hours. Their absolute frequency accuracy is about 2 parts in 10^{12}, the main source of error being uncertain phase shifts which occur when the atoms bounce off the walls of their storage bulbs. Future improvements in these numbers are likely.

BRIEF HISTORY

The maser idea was conceived in 1951 by Charles H. Townes at Columbia University, and slightly later by the Russian physicists Nicolai Basov and Alexander M. Prokhorov. The first maser was built by Townes and his students James P. Gordon and Herbert J. Zeiger in 1954. The idea for multilevel pumping which allowed masers using stationary substances was advanced in 1956 by Basov and Prokhorov, and independently, in a more practical embodiment, by Nicolaas Bloembergen at Harvard University. H. E. Derrick Scovil, George Feher, and Harold Seidel built the first of these at Bell Telephone Laboratories in 1957.

See also ASTRONOMY, RADIO; LASERS; MICROWAVES AND MICROWAVE CIRCUITRY.

BIBLIOGRAPHY

J. P. Gordon, "The Maser," *Sci. Am.* **199,** 42, December 1958. (E)

T. K. Ishii, *Maser and Laser Engineering*. Krieger, Melbourne, Florida, 1980. (A)

A. E. Siegman, *Microwave Solid State Masers*. McGraw-Hill, New York, 1964. (A)

Joseph Weber, ed., *Masers, A Collection of Reprints with Commentary*, Gordon and Breach, New York, 1967. (A)

Mass

Johannes A. Van den Akker

Mass is the measure of the gravitational and inertial properties of matter. Once thought to be conceivably different, *gravitational mass* and *inertial mass* have been shown [1] to be the same to one part in 10^{11}.

Inertial mass is defined through Newton's second law, $\mathbf{F}=m\mathbf{a}$, in which m is the mass of a body, $\mathbf{F}$ is the force acting upon it, and $\mathbf{a}$ is the acceleration of the body induced by the force. If two bodies are acted upon by the same force (as in the idealized case of connection with a massless spring), their instantaneous accelerations will be in inverse ratio to their masses. Although this idealized experiment suggests a means for determining mass, exact measurement is done through weighing, for which a number of techniques exist [2]. (Local weight, which is local gravitational force modified by spin of the earth, is in exact proportion to mass.)

The international unit of mass is the kilogram—the mass of the International Prototype Kilogram of platinum–iridium, which is preserved by the International Bureau of Weights and Measures at Sèvres, France. A number of copies of the prototype are distributed throughout the world.

The equivalence of mass and energy, deduced by Einstein [3] from his special theory of relativity, is given by $E=mc^2$. When m is in kilograms and c, the velocity of light in a vacuum, is in meters per second, E is in joules. A single mass–energy conservation principle replaces the old conservation principles. However, for the purposes of ordinary ("nonrelativistic") calculations of engineering and science it is necessary, convenient, and accurate to conserve mass and energy separately. The equivalence of mass and energy must be called into play wherever one is concerned with, e.g., fast particles and nuclear processes.

If the reader is concerned about the relativistic increase in mass of a high-speed spacecraft, it is of interest to calculate that, for each metric ton, this is, at escape velocity 11,180 m/s, 7×10^{-7} kg, or 0.7 mg. One might well be more concerned about the collection of and/or erosion by cosmic dust and meteorites.

See also CONSERVATION LAWS; MACH'S PRINCIPLE; NEWTON'S LAWS.

REFERENCES

1. R. H. Dicke, "Experimental Relativity," in *Relativity, Groups, and Topology* (Les Houches Lectures, 1963). Gordon and Breach, New York, 1964, cited by E. L. Hill, in *Handbook of Physics*, 2nd ed., p. 2–47 (E. U. Condon and H. Odishaw, eds.). McGraw-Hill, New York, 1967. (I)
2. National Bureau of Standards. *NBS Special Publication 700-1. "A Primer for Mass Metrology,"* by K. B. Jaeger and R. S. Davis. November 1984. (E)
3. A. Einstein, *Ann. Phys.* **17,** 891 (1905). (A) See also A. Einstein, *The Meaning of Relativity*, 5th ed., pp. 43–47. Princeton University Press, Princeton, N.J. 1956. (I,A)

Mass Spectroscopy

Alfred O. Nier

Mass spectroscopy made its start in the early 1900s when J. J. Thomson employed his first "parabola" apparatus. He made use of the discovery by Goldstein in 1886 that positive rays are produced in an electrical discharge and the discovery by Wien in 1898 that these rays could be deflected by a magnetic field. In Thomson's apparatus a positive-ray beam, produced in a discharge tube, passed through a combination of electric and magnetic fields which resulted in a separation of the beam into components which, when they struck a detector such as a photographic plate, gave parabolic images on the plate, the shape of the parabolas depending on the mass-to-charge ratio (m/e) of the particles (ions) in the beam.

The method was improved upon by Aston, who in 1919 built the first *mass spectrograph*. In it a focused spectrum of lines, corresponding to the m/e ratios of the ions in the beam, was produced on a photographic plate. In the 20 years that followed, Aston determined the isotopic composition of many of the elements and established that the exact masses of nuclei were not integral multiples of a common number. The latter discovery, when coupled with the Einstein energy relation, $E=mc^2$, gave the first determinations of the binding energies of nuclei throughout the atomic table.

Dempster, 1922, built the first *mass spectrometer*. Here, ions were accelerated after their formation and deflected through 180° in a magnetic field. An electrical detector, such as an electroscope, mounted behind a slit in the focal plane, measured only one mass at a time. Mass spectra were obtained by varying an electrical potential or the magnetic field in the instrument [see Eq. (5)].

In the early 1930s the stage was set for modern mass spectroscopy. Vacuum techniques were steadily improving, the electrometer vacuum tube replaced the quadrant electrometer and electroscope for measuring small currents, and the magnetically collimated electron beam for producing ions was introduced. These developments, together with the wider application of electronic circuitry, gradually transformed the field from an art to a science, although a generous component of the former still remains when unusual applications are undertaken.

A large proportion of mass spectrometers in use today employ magnetic deflection to perform mass separation. If, in a region where the magnetic induction is B, ions of mass m and charge e move with speed v in a direction normal to the lines of magnetic induction, they experience a force

$$f = Bev. \tag{1}$$

This force results in a circular trajectory for the particle, the radius r being found by equating the force to the product of the mass and central acceleration:

$$Bev = mv^2/r. \tag{2}$$

Then, if the particles acquired their velocity by falling through a potential difference V, the kinetic energy gained equals the potential energy lost,

$$\tfrac{1}{2}mv^2 = eV. \tag{3}$$

If Eq. (3) is combined with Eq. (2) to eliminate v, one obtains

$$m/e = B^2r^2/2V, \tag{4}$$

which can be written as

$$m^*/e^* = 4.82 \times 10^7 r^2B^2/V, \tag{5}$$

where m^* is the mass expressed in atomic mass units (amu), e^* is the charge expressed in number of electronic charges on the ions (i.e., $e^* = 1$ for singly charged ions, 2 for doubly charged, etc.), B is the magnetic induction in teslas, r is the radius in meters, and V is the potential difference in volts.

A typical mass spectrometer used for gas analyses (Fig. 1) has a continuously pumped vacuum chamber, historically called the mass spectrometer tube. The gas or vapor to be analyzed flows from a volume through a pinhole "leak" into the mass spectrometer tube. The equilibrium pressure in the ion source region of the tube is of the order of 10^{-5} Torr. The pumping system frequently consists of a mechanical vacuum forepump followed by either an oil or a mercury diffusive pump and a cold trap maintained at dry-ice or liquid-air temperature. In some installations the pumping is accomplished with a titanium sputter pump.

An electron beam, collimated by a weak magnetic field perpendicular to the figure, is produced by a heated tungsten or rhenium filament, not shown. The energy of the electrons may be varied, but typically it has a value around 75 eV. Electrons, striking gas moleules, knock off one or more electrons, producing positive ions. In some cases negative ions may be formed in an attachment process. When a polyatomic molecule is struck, fragment ions as well as singly or multiply charged molecular ions are formed.

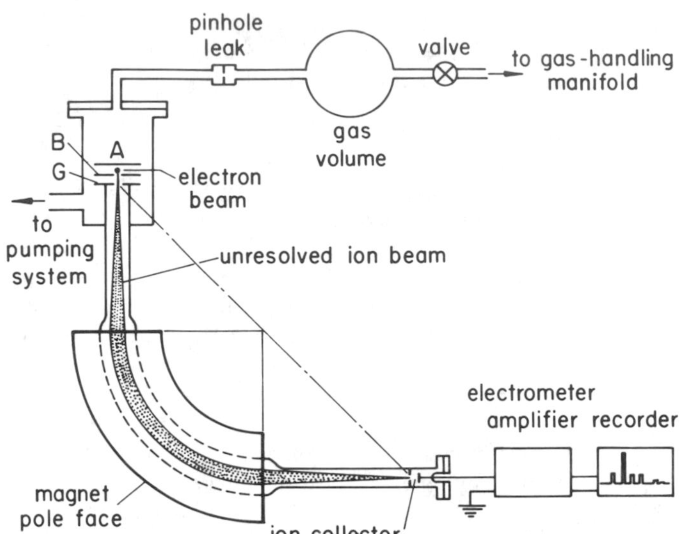

FIG. 1. Schematic drawing of a magnetic deflection mass spectrometer. In typical instruments radii of curvature 5–30 cm are used. With pressures of 10^{-5} Torr in the ionizing region and an electron beam of 100 μA, analyzed ion currents of 10^{-11}–10^{-10} A are observed at the ion collector.

The ions formed are accelerated toward a slit in plate B (Fig. 1) by a potential difference of a few volts between A and B. Further energy is given the ions by allowing them to fall through an electrical potential of several hundreds or thousands of volts applied between B and the grounded plate G. Because of the finite width of the slits in B and G, the ion beam diverges as shown as it enters the field-free region of the vacuum housing. The beam passes between the wedge-shaped poles of a magnet and on to the ion collector behind the ion-collector slit. If the slit in G, the apex of the magnet wedge, and the ion collector slit are on a straight line, the diverging beam is focused as shown. Most instruments in use employ either 60° or 90° deflection. In a modification of this geometry the edges of the magnet poles are tilted to form a 36° wedge magnet so that ions enter and leave the magnetic field at 27° to the normal instead of 90° as shown in Fig. 1. It has been shown that twice the dispersion is obtained, and there is also focusing of ions in a direction perpendicular to the plane of the figure. This geometry has been employed to obtain higher resolution and sensitivity without increasing the overall size of the instrument.

As may be seen from Eq. (4) or (5), if either B or V is varied continuously, a mass spectrum is obtained. Figure 2 is an example of such a spectrum for krypton.

Compounds of many elements, especially the alkali metals or alkali earths, give off ions if heated. A heated filament coated with such materials and placed at the position of the electron beam in Fig. 1 serves as a convenient ion source. In some applications ions are produced by field ionization or by sputtering techniques, using fast ions or fast atoms for removing and ionizing molecules from the surfaces of materials to be studied.

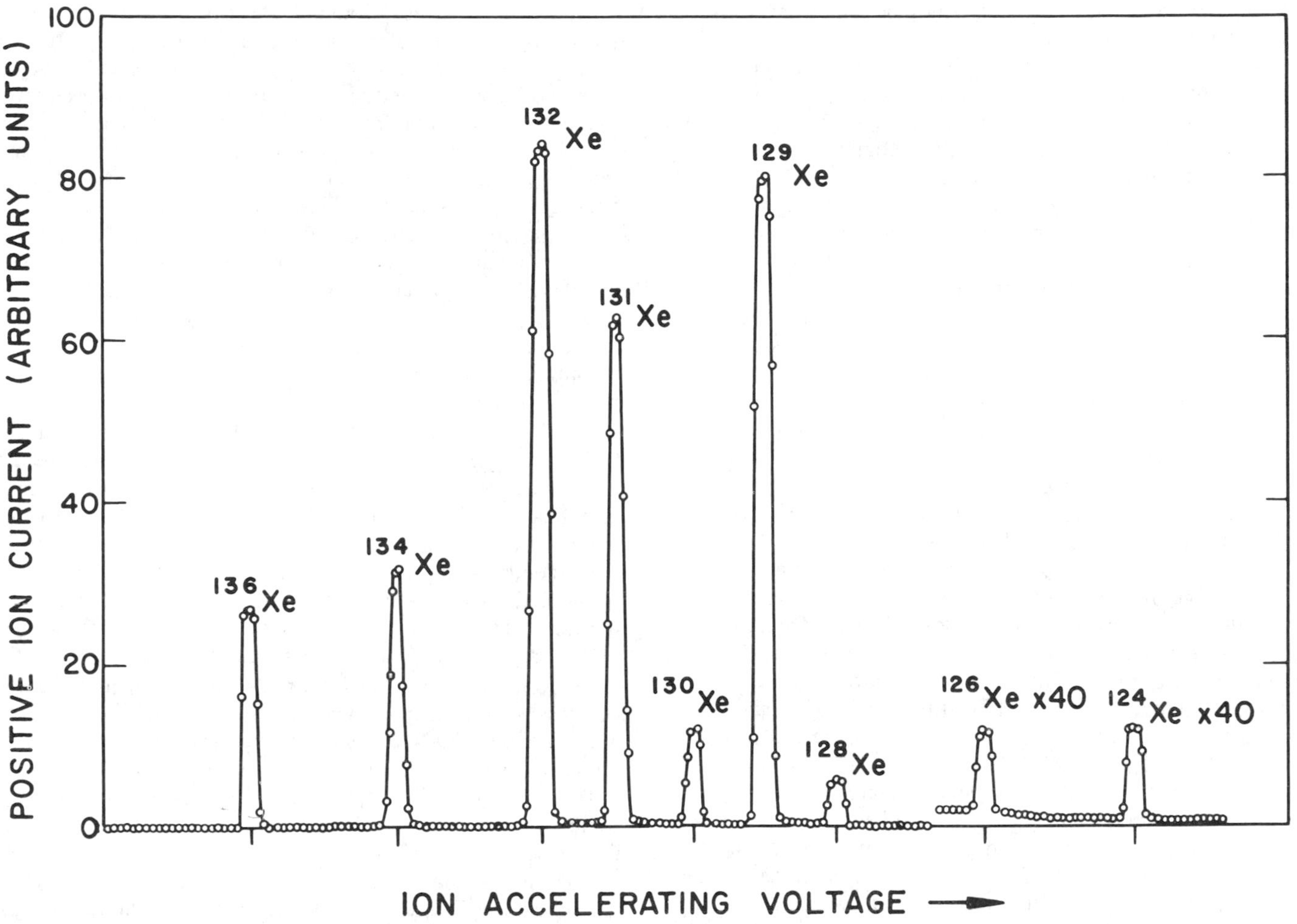

FIG. 2. Mass spectrum of krypton obtained with an instrument such as shown in Fig. 1.

When an electron multiplier rather than an electrometer tube amplifier is employed as a detector, single ions may be detected and the sensitivity of the mass spectrometer is very much increased.

If very high mass resolution or precision in determining atomic masses is desired, the geometry described above, having only an angle-focusing property, does not suffice, since the ions will generally have an energy spread as they leave the ion source. For high resolutions one employs what is called double focusing, i.e., angle plus energy focusing. This may be accomplished by sending the ions leaving the ion source through a cylindrical electric analyzer in tandem with a magnetic field such as employed in the single-focusing instruments just described. During the 1930s and 1940s when there was a strong interest in determining atomic masses precisely, a number of different geometries evolved. In much of the work at the time, ions were produced in an electrical discharge or spark rather than by electron impact as described above, and because of the large energy spread of the ions energy focusing was an absolute necessity. A number of different geometrical arrangements evolved. In one of these, the Mattauch–Herzog geometry, ions pass through a 31.8° cylindrical capacitor followed by a 90° magnet, as shown in Fig. 3. This instrument has the property that there is nearly perfect focusing for a wide range of masses along a straight line, so that a photographic plate may conveniently be used as a detector. Originally developed for measuring atomic masses precisely, instruments employing this geometry have effectively been used for analytical purposes

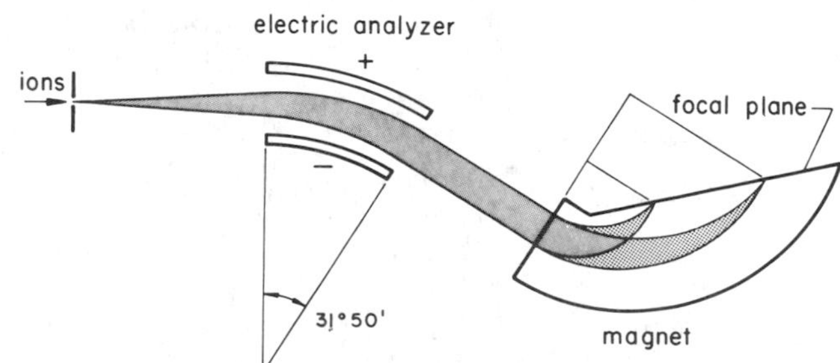

FIG. 3. Double-focusing mass spectroscope employing Mattauch–Herzog geometry. Ions having both an energy and angular spread leave the ion source and are deflected in a cylindrical electric analyzer before entering the magnetic analyzer. If the deflection in the electric analyzer is 31°50′ and in the magnetic analyzer is 90°, ions having different m/e values will focus along a line as shown. In some applications a photographic plate is employed as a detector; in others, several slits with collectors are used.

in the analyses of solids where it may be important to cover a wide range of masses.

Another commonly used double-focusing geometry, the Nier–Johnson, employs a 90° electric analyzer followed by a 60° or 90° magnetic analyzer and a single-slit collector as in Fig. 1. Instruments based on this geometry are used extensively for determining molecular structure and analyzing mixtures of heavy organic molecules, making use of the deviation of exact atomic masses from whole numbers. Figure 4 shows the separation obtained for two complex compounds differing in mass by only one part in 21,000.

Mass separation of ions may also be accomplished by employing time-varying electric or magnetic fields. Instruments based on such designs are sometimes called dynamic mass spectrometers. Many different arrangements have been devised—some with considerable success. Best known of these is the quadrupole (also called *Massenfilter*) perfected by Paul and his co-workers. In this, four parallel cylindrical rods have their axes on the corners of a square and ions travel along the line of symmetry between the four rods. A potential, varying sinusoidally with time, is superimposed upon a steady potential between opposite pairs of rods, and filters out all ions except those of the desired mass. Quadrupoles are used extensively in gas-analysis instruments. Mass spectrometers are frequently used in conjunction with gas or liquid chromatographs for analytical purposes. The technique is far more powerful than a mass spectrometer or chromatograph used by itself.

In Fourier transform mass spectrometry, ions are trapped in a cell mounted in a magnetic field and subjected to a high-frequency electric field causing them to undergo ion cyclotron resonance, with the frequency of resonance depending sharply on the mass of the ions present. Sweeping the radiofrequency field produces a complex mass spectrum which can be analyzed by Fourier transform methods. The method has considerable potential as a high-resolution means for studying complex chemical reactions and other analytical problems.

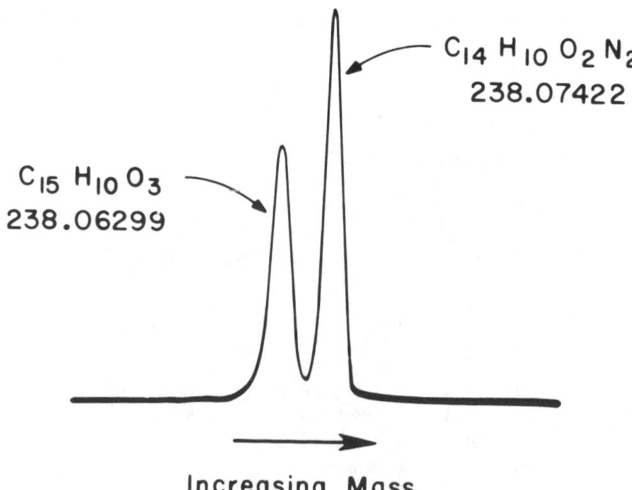

FIG. 4. Mass doublet at mass number 238 showing separation of two compounds which differ in mass by only one part in 21,000. (Courtesy J. H. Beynon.)

Mass spectroscopes, originally developed for determining the isotopic nature of elements and for measuring atomic masses exactly, have found application in virtually every field of pure and applied science. From the study of products of ionization resulting from the electron impact of gases, not only has a great deal been learned about the structure of molecules but techniques have been perfected for the analyses of complex mixtures of gases and vapors. Dilution studies employing separated stable isotopes such as ^{15}N or ^{18}O, to mention but a few cases, have made possible tracer studies in chemistry and biochemistry not practical with radioactive tracers. Instruments tuned to a particular gas such as helium are extensively used as leak detectors in the field of vacuum technology; sensitivities to leaks as low as 10^{-13} cm^3 (STP) per second have been achieved.

Configurations in which two double-focusing or quadrupole mass spectrometers are placed in tandem have been used effectively in determining molecular structure of complex heavy molecules. Ions emerging from the first instrument pass through a high-pressure gas cell before entering the second instrument. Energy transfers in collisions made in the cell lead to fragmentation patterns in the spectra of the ions leaving the second instrument which clarify structure details not possible with a single instrument.

Geology and cosmology have been particular beneficiaries of mass spectroscopy. Systematic studies of the relative abundances of the isotopes of common lead in terrestrial and meteoritic samples have provided essential data for determining the age of the solar system. Investigation of the isotopic composition of the lead, argon, or strontium resulting from radioactive decay of parent elements has made possible the age determination of the minerals from which the elements were extracted. Anomalous isotopic abundances observed in elements such as xenon apparently are related to the very early history of the solar system.

Miniaturized mass spectrometers carried on sounding rockets and earth satellites have been used extensively to study the neutral and ionic composition of the earth's upper atmosphere. *In situ* measurements of the Martian atmospheric composition were made with mass spectrometers carried to Mars in 1976. It is expected that mass spectrometers will play an important part in missions to other planets and comets.

See also GEOCHRONOLOGY; ISOTOPES.

BIBLIOGRAPHY

F. W. Aston, *Mass Spectra and Isotopes*. Longmans, Green and Co., Boston, Mass., 1942. (I)

J. H. Beynon, *Mass Spectrometry and its Applications to Organic Chemistry*. Elsevier, New York, 1960. (I)

H. E. Duckworth, *Mass Spectrometry*. Cambridge University Press, Cambridge, 1958. (I)

C. A. McDowell (ed.), *Mass Spectrometry*. McGraw-Hill, New York, 1963. (I)

K. Ogata and T. Hayakawa (ed.), *Recent Developments in Mass Spectroscopy*. University of Tokyo Press, Tokyo, 1970. (A)

John Roboz, *Mass Spectrometry*. Interscience, New York, 1968. (I)

F. A. White and G. M. Wood, *Mass Spectrometry—Applications in Science and Engineering*. Wiley, New York, 1986.

Matrices

Edgar A. Kraut

This article summarizes some of the most important features of matrices. References containing proofs, additional details, and applications are listed in the bibliography.

MATRIX ALGEBRA

Definitions

A *matrix* is a rectangular array of real or complex numbers arranged in an orderly table of rows and columns; an example is the $(n\times m)$ matrix containing nm elements:

$$A=a_{ij}=\begin{pmatrix} a_{11} & a_{12}\cdots a_{1m}\\ a_{21} & a_{22}\cdots a_{2m}\\ \vdots & \vdots \quad \vdots\\ a_{n1} & a_{n2}\cdots a_{nm}\end{pmatrix}$$

A *row vector* is one row of a matrix:

$$(a_{n1},a_{n2},\cdots,a_{nm}).$$

This row vector is also a $(1\times m)$ matrix.

A *column vector* is one column of a matrix:

$$\begin{pmatrix} a_{11}\\ a_{21}\\ \vdots\\ a_{n1}\end{pmatrix}$$

This column vector is also an $(n\times 1)$ matrix.

Matrix Sum and Difference: If $A=a_{ij}$ and $B=b_{ij}$, then $C=A\pm B$ is defined as

$$c_{ij}=a_{ij}\pm b_{ij}.$$

An example is

$$\left[\begin{pmatrix} c_{11} & c_{12}\cdots c_{1m}\\ c_{21} & c_{22}\cdots c_{2m}\\ \vdots & \vdots \quad \vdots\\ c_{n1} & c_{n2}\cdots c_{nm}\end{pmatrix}=\begin{pmatrix} a_{11} & a_{12}\cdots a_{1m}\\ a_{21} & a_{22}\cdots a_{2m}\\ \vdots & \vdots\cdots\vdots\\ a_{n1} & a_{n2}\cdots a_{nm}\end{pmatrix}\right.$$

$$\left.\pm\begin{pmatrix} b_{11} & b_{12}\cdots b_{1m}\\ b_{21} & b_{22}\cdots b_{2m}\\ \vdots & \vdots\cdots\vdots\\ b_{n1} & b_{n2}\cdots b_{nm}\end{pmatrix}=\begin{pmatrix} a_{11}\pm b_{11}\cdots a_{1m}+b_{1m}\\ a_{21}\pm b_{21}\cdots a_{2m}\pm b_{2m}\\ \vdots \quad \cdots \quad \vdots\\ a_{n1}\pm b_{n1}\cdots a_{nm}\pm b_{nm}\end{pmatrix}\right]$$

Scalar Multiplication: $B=\lambda A$ where λ is a scalar means $b_{ij}=\lambda a_{ij}$ for each i and j; for example,

$$4\begin{pmatrix} 2 & 1 & 0\\ 3 & 1 & 9\end{pmatrix}=\begin{pmatrix} 8 & 4 & 0\\ 12 & 4 & 36\end{pmatrix}.$$

Matrix Product: For $C=c_{ij}$, $A=a_{ij}$, and $B=b_{ij}$ the matrix product $C=AB$ is defined when the number of columns in A equals the number of rows in B. Then A and B are said to be conformable and their product C is defined by

$$c_{ij}=\sum_{k=1}^{r} a_{ik}b_{kj}, \qquad i=1,2,\ldots,n;\, j=1,2,\ldots,m.$$

The $(n\times m)$ matrix C is the product of the $(n\times r)$ matrix A and the $(r\times m)$ matrix B. In general the matrix product is not commutative, i.e., $AB\neq BA$. The rule for the matrix product states that element c_{ij} is the scalar product of the ith row vector of A with the jth column vector of B. The following are some examples.

$$\begin{pmatrix} 1 & 2\\ 3 & 4\end{pmatrix}\begin{pmatrix} 1 & 3\\ 6 & 2\end{pmatrix}$$

$$=\begin{pmatrix} 1\times1+2\times6 & 1\times3+2\times2\\ 3\times1+4\times6 & 3\times3+4\times2\end{pmatrix}=\begin{pmatrix} 13 & 7\\ 27 & 17\end{pmatrix}$$

$$\begin{pmatrix} a_{11} & a_{12}\\ a_{21} & a_{22}\end{pmatrix}\begin{pmatrix} b_{11} & b_{12}\\ b_{21} & b_{22}\end{pmatrix}=\begin{pmatrix} a_{11}b_{11}+a_{12}b_{21} & a_{11}b_{12}+a_{12}b_{22}\\ a_{21}b_{11}+a_{22}b_{21} & a_{21}b_{12}+a_{22}b_{22}\end{pmatrix}$$

$$\begin{pmatrix} a_{11} & a_{12}\\ a_{21} & a_{22}\end{pmatrix}\begin{pmatrix} x_1\\ x_2\end{pmatrix}=\begin{pmatrix} a_{11}x_1+a_{12}x_2\\ a_{21}x_1+a_{22}x_2\end{pmatrix};$$

$$[a_1,\ldots,a_n]\begin{bmatrix} b_1\\ b_2\\ \vdots\\ b_n\end{bmatrix}=a_1b_1+a_2b_2+\cdots+a_nb_n;$$

$$\begin{bmatrix} a_1\\ a_2\\ \vdots\\ a_n\end{bmatrix}[b_1b_2,\ldots,b_n]=\begin{bmatrix} a_1b_1 & a_1b_2\cdots a_1b_n\\ a_2b_1 & a_2b_2\cdots a_2b_n\\ \vdots & \vdots \quad \vdots\\ a_nb_1 & a_nb_2\cdots a_nb_n\end{bmatrix}$$

The last two examples represent, respectively, the product of a $(1\times n)$ matrix with an $(n\times 1)$ matrix, and the product of an $(n\times 1)$ matrix with a $(1\times n)$ matrix. The former is a (1×1) matrix that is simply a scalar number. The latter is a special kind of $(n\times n)$ matrix called a *dyadic*.

Direct Product: If A is an $(n\times n)$ matrix and B is an $(m\times m)$ matrix, then the direct product of A and B is written as

$$C=A\times B$$

where C is a matrix having nm rows and nm columns defined by

$$C=A\times B=\begin{bmatrix} a_{11}B & a_{12}B\cdots a_{1n}B\\ \vdots & \cdots\ \vdots\ \vdots\\ a_{n1}B & a_{n2}B\cdots a_{nn}B\end{bmatrix}$$

For example,

$$\begin{pmatrix} a_{11} & a_{12}\\ a_{21} & a_{22}\end{pmatrix}\begin{pmatrix} b_{11} & b_{12}\\ b_{21} & b_{22}\end{pmatrix}$$

$$=\begin{pmatrix} a_{11}\begin{pmatrix} b_{11}b_{12}\\ b_{21}b_{22}\end{pmatrix} & a_{12}\begin{pmatrix} b_{11}b_{12}\\ b_{21}b_{22}\end{pmatrix}\\ a_{21}\begin{pmatrix} b_{11}b_{12}\\ b_{21}b_{22}\end{pmatrix} & a_{22}\begin{pmatrix} b_{11}b_{12}\\ b_{21}b_{22}\end{pmatrix}\end{pmatrix}$$

$$=\begin{pmatrix} a_{11}b_{11} & a_{11}b_{12} & a_{12}b_{11} & a_{12}b_{12}\\ a_{11}b_{21} & a_{11}b_{22} & a_{12}b_{21} & a_{12}b_{22}\\ a_{21}b_{11} & a_{21}b_{12} & a_{22}b_{11} & a_{22}b_{12}\\ a_{21}b_{21} & a_{21}b_{22} & a_{22}b_{21} & a_{22}b_{22}\end{pmatrix}$$

Rank r(A) of Matrix A: $r(A)$ is the number of rows or columns in the largest nonvanishing determinant contained in A. If A is an $(n \times n)$ matrix and $r(A)=n$, then A is nonsingular. Otherwise A is singular.

The *trace* of a matrix $A=a_{ij}$ the sum of the elements on its main diagonal. For example,

$$\mathrm{Tr}(A)=\sum_{k=1}^{n} a_{kk}$$

Useful properties of the trace of a matrix are

$$\mathrm{Tr}(AB)=\mathrm{Tr}(BA);$$

$$\mathrm{Tr}(A\times B)=\mathrm{Tr}(A)\cdot\mathrm{Tr}(B).$$

The *determinant det|A| of matrix* $a_{ij}=A$ satisfies

$$\det|A|\delta_{ik}=\sum_{j=1}^{n} a_{ij}A^{kj}$$

(expansion by cofactors of the *i*th row), and

$$\det|A|\delta_{ik}=\sum_{j=1}^{n} a_{ji}A^{jk}$$

(expansion by cofactors of the *i*th column) where

$$\delta_{ik}=\begin{cases}1, & i=k,\\ 0, & i\neq k.\end{cases}$$

Here A^{ij} is the cofactor $[(-1)^{i+j}\times(\text{minor of } a_{ij} \text{ in } \det|a_{ij}|)]$ of a_{ij} in $\det|a_{ij}|$ (Laplace development).

The determinant remaining after striking out the *i*th row and *j*th column in $\det|a_{ij}|$ is the *minor of element* a_{ij}. Useful properties of determinants include

1. If all elements of a row or column vanish, $\det|A|=0$.
2. If a pair of rows (or columns) are proportional, $\det|A|=0$.
3. Interchanging a pair of rows (or columns) changes sign of the determinant.
4. Multiplying each element of a row (column) by the same constant multiplies the value of the determinant by this factor.
5. Addition of a constant multiple of any one row (column) of a determinant to any other row (column) leaves its value unchanged.
6. Transposing rows with columns leaves the value of a determinant unchanged: $\det|a_{ij}|=\det|a_{ji}|$.
7. $\det|ABC\cdots Z|=\det|A|\cdot\det|B|\cdots\det|Z|$.

The *identity matrix I* is the matrix satisfying $IA=AI=A$ for arbitrary A. Examples are

$$I=\begin{pmatrix}1&0\\0&1\end{pmatrix},\quad I=\begin{pmatrix}1&0&0\\0&1&0\\0&0&1\end{pmatrix},\quad I=\delta_{ij}$$

where

$$\delta_{ij}=\begin{cases}1, & i=j,\\ 0, & i\neq j.\end{cases}$$

Diagonal Matrix $D=D_i\delta_{ij}$: Diagonal matrices have nonvanishing components only along the main diagonal. They always commute with one another.

Transpose $A^T=a_{ji}$ *of Matrix* $A=a_{ij}$: The matrix obtained from A by interchanging rows and columns is the transpose. Transposing a product of several factors reverses the order of the transposed factors; for example,

$$C=AB,\qquad C^T=B^TA^T,$$

$$Z=ABC,\qquad Z^T=C^TB^TA^T.$$

The matrix obtained by replacing each element of matrix A by its cofactor in $\det|A|$ is the *cofactor matrix* A^c:

$$A=\begin{pmatrix}a_{11}&a_{12}\\a_{21}&a_{22}\end{pmatrix},\qquad A^C=\begin{pmatrix}a_{22}&-a_{21}\\-a_{12}&a_{11}\end{pmatrix}.$$

The matrix obtained by transposing the cofactor matrix A^C of matrix A is the *adjoint matrix* A^{CT}. It has the properties

$$A^{CT}=A^{TC},\qquad AA^{CT}=A^{CT}A,$$

$$I\det|A|=AA^{CT},\qquad I\det|A|=A^TA^C.$$

Inverse A^{-1} *of* A: For nonsingular ($\det|A|\neq 0$) square $(n\times n)$ matrices,

$$AA^{-1}=A^{-1}A,$$

$$(AB)^{-1}=B^{-1}A^{-1},$$

$$(ABC\cdots Z)^{-1}=(Z^{-1}\cdots C^{-1}B^{-1}A^{-1}).$$

A *complex matrix* is one with complex numbers as matrix elements:

$$A=\begin{pmatrix}1+2i&3-2i\\4&2i\end{pmatrix}.$$

The *complex conjugate* of A is $\bar{A}$:

$$\bar{A}=\begin{pmatrix}1-2i&3+2i\\4&-2i\end{pmatrix},$$

$$Z=ABC\cdots XY,\qquad \bar{Z}=\bar{A}\bar{B}\bar{C}\cdots\bar{X}\bar{Y}.$$

The transpose of the complex conjugate of A is the *Hermitian conjugate* A^*:

$$A^*=(\bar{A})^T=\overline{(A^T)}=\bar{A}^T,$$

$$a_{ij}=A,\qquad A^*=\bar{a}_{ji};$$

$$Z=ABC\cdots XY,$$

$$Z^*=Y^*X^*\cdots C^*B^*A^*.$$

MATRIX GEOMETRY

Definitions

The *linear operator A* is an abstract quantity that has the properties

$$A(kX)=k(AX),\qquad A(X+Y)=AX+AY$$

where k is any scalar. For example, a matrix A is a linear operator that transforms a column vector x into a new column vector $y=Ax$ having different direction and magnitude. Matrices are linear operators because of the way in which the rules for matrix addition and multiplication have been defined.

A *unitary matrix* rotates a vector while leaving its magnitude unchanged. The matrix A is unitary if $A^{-1}=A^*$. For purely real matrices the corresponding condition is $A^{-1}=A^T$ and the matrix A is called orthogonal.

The matrix H is called *Hermitian* if $H=H^*$. It is also called self-adjoint or self-conjugate. If H is real and Hermitian, then $H=H^T$ and H is symmetric, i.e., $H_{ij}=H_{ji}$. Hermitian matrices have real eigenvalues and their eigenvectors, which correspond to different eigenvalues, are orthogonal.

The *eigenvectors of a matrix* are those vectors whose direction is unchanged under matrix multiplication. For example, the vector x is an eigenvector of A if Ax is parallel to x, i.e., $Ax=\lambda x$ where λ is a scalar.

The *eigenvalue problem* is to find all the eigenvectors x and all the eigenvalues λ for a given matrix A by solving the linear homogeneous set of equations $(A-\lambda I)x=0$. The λ's are determined from the condition $\det|A-\lambda I|=0$ for the existence of a nontrivial solution to the homogeneous system $(A-\lambda I)x=0$. Eigenvectors of A are also called characteristic or proper vectors of A and $\det|A-\lambda I|=0$ is known as the secular, characteristic, or eigenvalue equation for matrix A.

Similar Matrices: In a given coordinate system, a matrix A is a linear operator rotating a vector x into a vector $y=Ax$. If a rotation of coordinates S is introduced, then the components of x and y are changed to x' and y'. The relation between (x,y) and (x',y') is expressed through the transformation of coordinates $x=Sx'$ and $y=Sy'$. The matrix A' that rotates the vector x' into the vector $y'=A'x'$ is said to be *similar* to A. For example, the matrices A and A' are similar if $A'=S^{-1}AS$ (similarity transformation). *Properties* of similar matrices include

$$\mathrm{Tr}(A')=\mathrm{Tr}(A),$$

$$\det(A')=\det(A),$$

$$\det|A'-\lambda I|=\det|A-\lambda I|.$$

Frequently, we seek a coordinate transformation S that makes A' a diagonal matrix. It can be shown that the matrix S whose column vectors are the eigenvectors of A has exactly this property. Similar matrices have the same eigenvalues. The matrix S can always be found for any matrix A satisfying $A^*A=AA^*$. Therefore, Hermitian and unitary matrices can always be diagonalized by a similarity transformation.

MATRIX COMPUTATIONS

The total operation count (toc) for a numerical computation is defined as the total number of elementary arithmetic operations required to perform the computation.

Thus toc = moc + aoc + doc + soc where these terms represent respectively, the total number of multiplications, additions, divisions, and subtractions required to perform the computation. The total operation count for the matrix product Ax of an $(n\times m)$ matrix A and an $(m\times 1)$ vector x is given by toc $=n[m+(m-1)]=n(2m-1)$ as the following discussion shows.

The numerical evaluation of the matrix product Ax requires the formation of the dot product of the n rows of A with the m-dimensional column vector x. Each dot product requires m arithmetic multiplications followed by $(m-1)$ additions. Exactly n dot products must be formed. The total number of multiplications moc $=nm$ and the total number of additions aoc $=n(m-1)$. This gives toc = moc + aoc $=nm+n(m-1)$ or toc $=n(2m-1)$.

For large n and m, toc and moc are large numbers of the same order, $O(\text{toc})=O(\text{moc})=O(mn)$, because

$$\text{toc/moc}=n(2m-1)/nm=2-(1/m)$$

$$\lim \text{toc/moc}\rightarrow 2,$$

therefore toc/moc $=O(1)$ and $O(\text{moc})=O(mn)$. Simply stated: The total operation count for Ax is of the same order as the total multiplication count $O(nm)$.

Many problems in physics require the computation of matrix products of matrices of very large order. Anything that can be done to reduce the number of arithmetic operations required to form such products makes numerical computations more tractable and can, in some cases such as many-body simulations in molecular dynamics and plasma physics, lead to significant progress in the numerical simulation of complex interacting particle systems (Greengard, 1988).

A sufficient condition for reducing the operation count for the matrix product Ax is that the matrix A be factorable into the product of an $(n\times p)$ matrix B and a $(p\times m)$ matrix C such that $A=BC$. For example: Let $Ax=BCx$. The matrix product Cx generates a column vector having p rows and has an operation count that is $O(pm)$. The matrix product $B(Cx)$ generates a column vector having n rows with an operation count that is $O(np)$. The total number of operations is then $O(np+pm)$. Now if p is held fixed while n and m are allowed to increase without bound, $O(np+pm)\mapsto O(n+m)$ and the total operation count for computing Ax has been reduced from $O(nm)$ to $O(n+m)$ by computing Ax by the process BCx. The crucial part of the argument, of course, is that the dimension p can be bounded while n and m increase without bound. This depends on the physics of the systems to which this factorization is applied. Many interesting examples can be found in the book by Greengard and the references contained therein.

The proliferation of microcomputers, workstations, desktop supercomputers, etc., has stimulated research in numerical matrix methods (Coleman and Van Loan, 1988; Golub and Van Loan, 1988; Dongarra, Bunch, Moler, and Stewart 1979). Iterative methods of matrix inversion, in particular, have received much attention because of many important practical applications. An excellent historical summary of the development of the conjugate gradient and Lanczos algorithms for solving linear systems of equations and eigenproblems is contained in the review article by Golub and O'Leary (1989). The generalized conjugate residual algorithm (GCR) and variants thereof have been shown to be important for the iterative solution of linear problems involving complex nonsymmetric matrices whose Hermitian parts $(A+A^*)/2$ are positive definite (Eisenstat, 1983). These types of linear problems arise frequently in the solution of Fredholm integral equations of scattering theory and in many other branches of physics and engineering.

BIBLIOGRAPHY

E. Bodeweg, *Matrix Calculus*, 2nd ed. North-Holland, Amsterdam, 1959. (E)

Thomas F. Coleman and Charles F. Van Loan, *Handbook for Matrix Computations (Frontiers in Applied Mathematics 4)*. Society for Industrial and Applied Mathematics, Philadelphia, 1988.

Jack J. Dongarra, James R. Bunch, Cleve B. Moler, and G. W. Stewart, *Linpack User's Guide*. Society for Industrial and Applied Mathematics, Philadelphia, PA, 1979.

Stanley C. Eisenstat, "A Note on the Generalized Conjugate Gradient Method," *SIAM J. Numer. Anal.* **20,** 358–361 (1983).

G. Forsythe and C. B. Moler, *Computer Solution of Linear Algebraic Systems*. Prentice-Hall, Englewood Cliffs, NJ, 1967. (I)

R. A. Frazer, W. J. Duncan, and A. R. Collar, *Elementary Matrices*. Cambridge, London and New York, 1960. (I)

F. R. Gantmacher, *The Theory of Matrices*, Vols. I and II. Chelsea, New York, 1959. (A)

F. R. Gantmacher, *Applications of the Theory of Matrices*. Wiley (Interscience), New York, 1959. (A)

Gene H. Golub and Dianne P. O'Leary, "Some History of the Conjugate Gradient and Lanczos Algorithms" *SIAM Rev.* **31,** 50–102 (1989).

Gene H. Golub and Charles F. Van Loan, *Matrix Computations*. The Johns Hopkins University Press, Baltimore, 1988.

Leslie Greengard, *The Rapid Evaluation of Potential Fields in Particle Systems*. MIT Press, Cambridge, MA, 1988.

R. T. Gregory and D. L. Karney, *A Collection of Matrices for Testing Computational Algorithms*. Wiley (Interscience), New York, 1969. (I)

P. Hlwaiczka, *Matrix Algebra for Electronic Engineers*. Hayden, New York, 1965. (I)

A. S. Householder, *The Theory of Matrices in Numerical Analysis*. Blaisdell, Boston, 1965. (A)

E. A. Kraut, *Fundamentals of Mathematical Physics*. McGraw-Hill, New York, 1967. (E)

P. Lancaster, *Theory of Matrices*. Academic Press, New York, 1969. (I)

H. Margenau and G. Murphy, *The Mathematics of Physics and Chemistry*, 2nd ed., Vol. I, Chapters 4, 5, and 10. Van Nostrand-Reinhold, Princeton, NJ, 1956. (E)

S. Perlis, *Theory of Matrices*. Addison-Wesley, Reading, Mass., 1952. (E)

R. S. Varga, *Matrix Iterative Analysis*. Prentice-Hall, Englewood Cliffs, NJ, 1962. (A)

J. H. Wilkinson, *The Algebraic Eigenvalue Problem*. Oxford (Clarendon) London and New York, 1965. (A)

Maxwell–Boltzmann Statistics

N. D. Mermin

A collection of N noninteracting (or weakly interacting) identical particles in thermal equilibrium at temperature T obeys Maxwell–Boltzmann statistics ("classical statistics") if the mean number of particles occupying any one particle level of energy ϵ, is proportional to $\exp(-\epsilon/k_B T)$, where k_B is Boltzmann's constant.

The Maxwell–Boltzmann statistics are valid only at temperatures and densities for which the probability of any given level being occupied is very small. In the case of a gas the validity of Maxwell–Boltzmann statistics requires low enough densities and high enough temperatures for the mean interparticle distance to be large compared with the de Broglie wavelength of a particle with the classical thermal velocity.

When the probability of each one-particle level being occupied is small, the mean number of particles in each level is numerically equal to that probability. Thus the Maxwell–Boltzmann statistics can also be characterized as assigning to each level a thermal-equilibrium probability of occupation proportional to $\exp(-\epsilon/k_B T)$ (the Maxwell–Boltzmann distribution). One of the most familiar applications is that which ascribes a Gaussian ("Maxwellian") velocity distribution to the molecules of a classical ideal gas: $P(\mathbf{v}) \propto \exp(-\frac{1}{2}mv^2/k_B T)$. Note, though, that this velocity distribution is a general feature of a classical system in thermal equilibrium regardless of the strength of interactions, whereas the Maxwell–Boltzmann statistics are limited to the treatment of noninteracting (or weakly interacting) particles.

One can "derive" the Maxwell–Boltzmann mean occupation number by (incorrectly) treating as distinct two N-particle states that only differ by an interchange of the levels occupied by a pair of particles. It is known, however, that such an interchange of identical particles does not yield a new physical state. The errors induced by so miscounting are unimportant if every level has a mean occupation number small compared with unity. In this case the mean occupation numbers given by the correct (Fermi–Dirac or Bose–Einstein) quantum statistics reduce to the Maxwell–Boltzmann form.

The Rayleigh–Jeans catastrophe can be viewed as a result of the misapplication of the Maxwell–Boltzmann distribution to the photon gas comprising the radiation in a cavity; the Planck distribution law follows from the application of the correct Bose–Einstein statistics. Similarly, the anomalies associated at the turn of the century with the electronic and lattice contributions to the specific heats of solids can be viewed from a modern perspective as arising from the erroneous use of the Maxwell–Boltzmann distribution to describe the electron and phonon gases instead of the correct Fermi–Dirac and Bose–Einstein distributions.

Examples in which the Maxwell–Boltzmann distribution gives an accurate description (in addition to the classical gas) are the distribution of electrons and holes in nondegenerate semiconductors, the low-temperature distribution of optical phonons in a solid, and the distribution of rotons in liquid ^{4}He below the lambda point. In the last two examples, entities which the statistics describe are not the particles themselves, but rather the weakly interacting "elementary excitations" or "quasiparticles" that characterize the low-lying excited states of what is in general a strongly interacting quantum system.

The Maxwell–Boltzmann distribution should not be confused with the Gibbs distribution, which asserts that the thermal-equilibrium probability of finding an N-particle system in an *N-particle stationary state* of energy E is proportional to $\exp(-E/k_B T)$. The Gibbs distribution is generally valid for classical or quantum systems, whether or not they are weakly interacting. The Maxwell–Boltzmann, Fermi–Dirac, and Bose–Einstein distributions (which characterize the occupation of *single-particle levels* for weakly interacting particles or excitations) all follow from the Gibbs distribution in the appropriate special cases.

See also BOSE–EINSTEIN STATISTICS; FERMI–DIRAC STATISTICS; QUASIPARTICLES; STATISTICAL MECHANICS.

BIBLIOGRAPHY

P. M. Morse, *Thermal Physics*. Benjamin, New York, 1965.
F. Reif, *Statistical and Thermal Physics*. McGraw Hill, New York, 1965.
Gregory H. Wannier, *Statistical Physics*. Wiley, New York, 1966.

Maxwell's Equations

Philip Stehle

Maxwell's equations describe the behavior of the electromagnetic field. In free space, vacuum, there are two basic field quantities, the electric field $\mathbf{E}(\mathbf{r},t)$ and the magnetic induction $\mathbf{B}(\mathbf{r},t)$, whose values at a given place and time can be determined by observing the force on a test charge q moving with velocity $\mathbf{v}$;

$$\mathbf{F}=q\left(\mathbf{E}+\frac{1}{c}\mathbf{v}\times\mathbf{B}\right). \tag{1}$$

Here c is a constant with the dimensions of a velocity that turns out to be the speed with which electromagnetic waves (light waves) propagate in free space. The sources of the fields are electric charges and electric currents.

The equations, written out in customary vector notation, are

$$\nabla\times\mathbf{B}-\frac{1}{c}\frac{\partial\mathbf{E}}{\partial t}=\frac{4\pi}{c}\mathbf{j}, \tag{2}$$

$$\nabla\cdot\mathbf{E}=4\pi\rho, \tag{3}$$

$$\nabla\times\mathbf{E}+\frac{1}{c}\frac{\partial\mathbf{B}}{\partial t}=0, \tag{4}$$

$$\nabla\cdot\mathbf{B}=0. \tag{5}$$

$\mathbf{j}$ is the current density and ρ is the charge density. Gaussian units have been used.

The first of Maxwell's equations is a statement of Ampère's law giving the magnetic field produced by a current, including the displacement current introduced by Maxwell. The second, in conjunction with Eq. (1), includes Coulomb's law of force between charges. The third gives Faraday's law of induced emf. The fourth describes the nonexistence of magnetic charge.

The two homogeneous equations [(4) and (5)] show that the fields can be expressed in terms of a scalar potential $\phi(\mathbf{r},t)$ and a vector potential $\mathbf{A}(\mathbf{r},t)$;

$$\mathbf{B}=\nabla\times\mathbf{A} \qquad \mathbf{E}=-\frac{1}{c}\frac{\partial\mathbf{A}}{\partial t}-\nabla\phi. \tag{6}$$

The fields $\mathbf{E}$ and $\mathbf{B}$ are not affected by making a gauge transformation on the potentials consisting of

$$\mathbf{A}\to\mathbf{A}'=\mathbf{A}+\nabla\chi, \qquad \phi\to\phi'=\phi-\frac{1}{c}\frac{\partial\chi}{\partial t}. \tag{7}$$

If the potentials are chosen so that the Lorentz condition

$$\nabla\cdot\mathbf{A}+\frac{1}{c}\frac{\partial\phi}{\partial t}=0 \tag{8}$$

is satisfied, then the potentials satisfy the wave equations

$$\frac{1}{c^2}\frac{\partial^2\phi}{\partial t^2}-\nabla^2\phi=4\pi\rho, \tag{9}$$

$$\frac{1}{c^2}\frac{\partial^2\mathbf{A}}{\partial t^2}-\nabla^2\mathbf{A}=\frac{4\pi\mathbf{j}}{c}. \tag{10}$$

In this form the meaning of c as a propagation speed is clear.

As proposed by Maxwell these equations described the behavior of stresses in a material medium, the luminiferous ether. The present interpretation is that the electric and magnetic fields are physical quantities without reference to a material medium, an interpretation demanded by the special theory of relativity. In classical electrodynamics these fields are taken as numerical functions of space-time. In quantum electrodynamics they are taken as linear operator-valued functions on a suitable state space and are subject to certain canonical commutation relations. At the present time quantum electrodynamics is consistent with all experiments despite the complications of the renormalization procedures that must be used in making calculations.

In the presence of matter such as dielectric and magnetic materials the physical situation is more complicated. From a microscopic viewpoint the matter merely contributes new sources of the fields, and these must be included in the sources on the right-hand sides of Eqs. (2) and (3). To the extent that these sources are induced by the field, and that we are interested only in the average fields over regions containing many atoms of the material medium, the effects of these sources can be incorporated into the fields on the left to form new fields, $\mathbf{H}$ and $\mathbf{D}$, with only the externally impressed sources on the right. The equations then can be written in the macroscopic form

$$\nabla\times\mathbf{H}-\frac{1}{c}\frac{\partial\mathbf{D}}{\partial t}=\frac{4\pi}{c}\mathbf{j}_{\text{ext}}, \tag{11}$$

$$\nabla\cdot\mathbf{D}=4\pi\rho_{\text{ext}}, \tag{12}$$

$$\nabla\times\mathbf{E}+\frac{1}{c}\frac{\partial\mathbf{B}}{\partial t}=0, \tag{13}$$

$$\nabla\cdot\mathbf{B}=0 \tag{14}$$

together with the "constitutive equations"

$$\mathbf{D}=\epsilon\mathbf{E}=\mathbf{E}+4\pi\mathbf{P}, \qquad \mathbf{B}=\mu\mathbf{H}=\mathbf{H}+4\pi\mathbf{M}, \tag{15}$$

where ϵ and μ are the dielectric constant and the magnetic permeability, and $\mathbf{P}$ and $\mathbf{M}$ the electric and the magnetic polarizations, respectively. The last equations can be generalized to include anisotropic media in which $\mathbf{D}$ is not necessarily parallel to $\mathbf{E}$ nor $\mathbf{B}$ to $\mathbf{H}$, and to include nonlinear media in which ϵ and μ depend on the fields. The presence of these nonlinear effects has consequences such as harmonic production in intense light beams and makes solid-state electronic devices possible.

Gaussian units have been used above. In the SI system of

units the vacuum is treated on the same footing as a material medium so that **E** is distinguished from **D**, and **B** is distinguished from **H**, even in vacuum. Also the unit of charge is rationalized so that the inevitable factor of 4π is eliminated from the field equations to reappear in Coulomb's law. The homogeneous equations are unchanged. The equations with sources become

$$\nabla \times \mathbf{H} - \frac{\partial \mathbf{D}}{\partial t} = \mathbf{j} \text{ [SI]}, \tag{16}$$

$$\nabla \cdot \mathbf{D} = \rho \text{ [SI]} \tag{17}$$

with the constitutive equations for the vacuum

$$\mathbf{D} = \epsilon_0 \mathbf{E}, \qquad \mathbf{B} = \mu_0 \mathbf{H}. \tag{18}$$

The wave equations for the potentials are

$$\epsilon_0 \mu_0 \frac{\partial^2 \phi}{\partial t^2} - \nabla^2 \phi = \rho \text{ [SI]}, \tag{19}$$

$$\epsilon_0 \mu_0 \frac{\partial^2 \mathbf{A}}{\partial t^2} - \nabla^2 \mathbf{A} = \mathbf{j} \text{ [SI]} \tag{20}$$

with the speed of light being given by

$$c = \frac{1}{\sqrt{\epsilon_0 \mu_0}}. \tag{21}$$

By the 1983 definition of the meter, these constants are defined to be

$$c \equiv 299\ 792\ 458 \text{ m/s}$$

$$\epsilon_0 \equiv 8.854\ 187\ 817 \times 10^{-12} \text{ F/m},$$

$$\mu_0 \equiv 4\pi \times 10^{-7} \text{ N/A}^2.$$

See also ELECTRODYNAMICS, CLASSICAL; QUANTUM ELECTRODYNAMICS.

BIBLIOGRAPHY

M. Born and E. Wolf, *Principles of Optics*. Pergamon, New York, 1965.

J. D. Jackson, *Classical Electrodynamics*, 2nd ed., Wiley, New York, 1975.

L. Landau and E. Lifschitz, *Classical Theory of Fields*. Addison-Wesley, Reading, MA, 1951.

L. Landau and E. Lifschitz, *Electrodynamics of Continuous Media*. Addison-Wesley, Reading, MA, 1960.

Mechanical Properties of Matter

George A. Alers

INTRODUCTION

Mechanical properties describe the change in shape of matter when external forces are applied. Examples include the simple bending of a beam, the propagation of sound waves, the permanent deformation of metals into useful shapes, and the flow of liquids or gases around obstacles. For matter in the liquid and gaseous states, the usual force is the hydrostatic pressure and the deformation is a change in volume and a flow of the material from one point to another. For matter in the solid state, both tensile and shearing type forces are available to produce the corresponding elongations and shear distortions.

DEFORMATION OF SOLIDS

A solid is said to be *elastic* if the amount of deformation is directly proportional to the applied force. This implies that the process is completely reversible and independent of the way in which the final force or its direction of application has been achieved. A solid is said to be *anelastic* if there are small, additional displacements that depend on the rate with which the forces are applied. For most metals and ceramics, these time-dependent effects are small, but they play the important function of giving rise to the dissipation of vibrational energy that causes the damping of vibrations in machines and oscillating mechanical systems. In materials such as plastics and rubbers, the time-dependent contributions to the deformation are large and these types of solid matter are often called *viscoelastic* materials. The term *plastic solid* or *plastic deformation* describes those materials in which the displacement is a nonlinear, irreversible function of the applied force. Examples are the continuous and permanent deformation of metals by large forces and the response of organic polymers and glasses at elevated temperatures.

Mathematical Description of Elasticity in Solids

The forces that deform solids are described in terms of a *stress tensor* σ_{ij}, which represents a force F_i in direction i applied to an area ΔA_j whose normal is in the j direction; that is,

$$\sigma_{ij} = \lim_{\Delta A_j \to 0} F_i / \Delta A_j$$

and has the units of force per unit area. Obviously, when $i = j$, the force and the area normal are parallel, hence the stresses σ_{11}, σ_{22}, and σ_{33} are called *normal stresses* and work to move the area ΔA parallel to its normal. When $i \neq j$, a *shear stress* results in which the area ΔA is slipped or sheared in either of the two orthogonal directions that are perpendicular to the normal.

The deformations are described in terms of a *strain tensor* ϵ_{ij}, which measures the change in separation between two points in a solid divided by their original separation. Thus, a bar whose original length L becomes stretched to a length $L + \Delta L$ is said to have a strain of magnitude $\Delta L/L$. Thus, the strain has no units. For the general case, the deformation is described by a *displacement vector* which has Cartesian components u, v, and w and whose magnitude and direction describe the displacement of each point in the solid. This means that when a point at position X is displaced in the X direction by an amount $u(X)$ and its neighboring point at $X + \Delta X$ is displaced by $u + (\partial u / \partial X)\Delta X$, the strain in the X direction is $\partial u / \partial X$. The *normal strains* along all three coordinate axes are, therefore,

$$\epsilon_{xx}=\frac{\partial u}{\partial x}=\epsilon_{11},\quad \epsilon_{yy}=\frac{\partial v}{\partial y}=\epsilon_{22},\quad \epsilon_{zz}=\frac{\partial w}{\partial z}=\epsilon_{33},$$

and describe the deformation of a cube into a parallelepiped with unequal but orthogonal edges. To describe the shearing of a cube into a parallelepiped without orthogonal edges, the shear strains are

$$\epsilon_{xy}=\epsilon_{yx}=\frac{1}{2}\left(\frac{\partial u}{\partial y}+\frac{\partial v}{\partial x}\right)=\epsilon_{12}=\epsilon_{21},$$

$$\epsilon_{yz}=\epsilon_{zy}=\frac{1}{2}\left(\frac{\partial v}{\partial z}+\frac{\partial w}{\partial y}\right)=\epsilon_{23}=\epsilon_{32},$$

$$\epsilon_{zx}=\epsilon_{xz}=\frac{1}{2}\left(\frac{\partial w}{\partial x}+\frac{\partial v}{\partial z}\right)=\epsilon_{31}=\epsilon_{13}.$$

Geometrically, the shear strain ϵ_{ij} measures the angle by which the edges of a square in the ij plane is tilted away from 90° to form a parallelogram in that plane.

For elastic solids, the stress and strain are linearly related by a constant whose value is called the *elastic constant* of the material. Since the stress and strain are tensors of second rank, the constant of proportionality is a fourth rank tensor. Thus,

$$\sigma_{ij}=\sum C_{ijkl}\epsilon_{kl}$$

and

$$\epsilon_{ij}=\sum S_{ijkl}\sigma_{kl}$$

where the C_{ijkl} are called the *elastic moduli* and the S_{ijkl} are called the *elastic compliances*. It has become customary to contract this notation using the symmetry properties of the elastic tensors by applying the rule 11→1, 22→2, 33→3, 23→4, 13→5, and 12→6 so that stress and strain become six-component vectors and the elastic constants become second rank tensors with 36 elements. Thus,

$$\sigma_i=\sum C_{ij}\epsilon_j \quad \text{or} \quad \epsilon_i=\sum S_{ij}\sigma_j,\ i,j=1 \text{ to } 6.$$

The inherent symmetry of the stress and strain relations (invariant to interchange of i and j) reduces the number of independent elastic tensor components to 21 for the most general case. Further reductions occur when the symmetry of the atomic structure of the particular material is introduced. Orthorhombic crystals have nine constants, tetragonal crystals have six constants, hexagonal crystals have five constants, and cubic crystals have three constants. *Isotropic solids,* which exhibit no variation of elastic properties with direction, have only two independent elastic moduli. These two constants appear in the literature with different names and describe different common distortions. The Lamé constants, λ and μ, are related to the C_{ij} tensor elements by $C_{11}=C_{22}=C_{33}=\lambda+2\mu$, $C_{12}=C_{23}=C_{13}=\lambda$, $C_{44}=C_{55}=C_{66}=\mu$, all other $C_{ij}=0$. The Young's modulus E and the Poisson's ratio ν describe the distortions of a long thin rod subjected to a tensile stress along its length:

$$E=\frac{\text{tensile stress}}{\text{lengthwise strain}}=\frac{\mu(2\mu+3\lambda)}{\lambda+\mu},$$

$$\nu=\frac{\text{lateral contraction (strain)}}{\text{lengthwise strain}}=\frac{\lambda}{2(\lambda+\mu)}.$$

The bulk modulus B and the shear modulus G describe the change in volume resulting from the application of a hydrostatic pressure and the shear strain resulting from a shear stress, respectively.

Most common materials are aggregates of tiny, single-crystal grains. If the orientations of these grains are randomly distributed, the solid will exhibit isotropic elastic properties and only two elastic constants are needed to describe its response to external forces. In this case, these two elastic constants are simply determined by measuring the distortions of rod-shaped specimens subjected to known longitudinal stretching or twisting forces. Specialty materials such as single crystals and composites are generally anisotropic and more than two elastic constants must be considered in order to solve their elasticity problems. In addition, the rolling and drawing processes used to fabricate common plate and wire materials introduce nonrandom distributions of the grains so textures are present and the complexities of anisotropic elasticity should be used if accurate descriptions of the elastic deformations of these solids are required. In all these anisotropic cases, the many elastic constants needed to characterize the materials are most efficiently measured by ultrasonic techniques in which the velocity of compressional and shear waves propagating along the principal symmetry directions are measured. A simple relation that equates the square of the sound velocity to the ratio of the elastic modulus to the density provides very accurate values for the many elastic constants of anisotropic materials.

Internal Friction and Damping in Elastic Solids

For a perfectly elastic solid in which stress and strain are linearly related and completely reversible, the energy stored in the solid when under stress is exactly returned when the stress is removed. Thus, a vibrating solid will continue to oscillate between deformed and undeformed states indefinitely. To describe the dissipation of elastic energy that causes the vibrations of a solid to die away, anelastic or time-dependent relaxation phenomena must be introduced. The common way of doing this is to generalize the relation between stress and strain with the addition of the first time derivatives of the stress and the strain:

$$\sigma+\tau_1\frac{d\sigma}{dt}=M_R\left(\epsilon+\tau_2\frac{d\epsilon}{dt}\right)$$

The coefficients τ_1 and τ_2 have the physical significance of relaxation times which describe the time constant with which the stress or strain relaxes after the sudden application of a fixed strain or stress, respectively. The coefficient M_R is called the relaxed modulus because it corresponds to the ratio of stress to strain after relaxation has occurred. Another modulus, M_U, which defines the ratio of stress to strain at times short compared to τ_1 or τ_2, is called the unrelaxed modulus. If the stress and strain are periodic functions of time with angular frequency ω, then the tangent of the phase angle θ by which the strain lags behind the stress is given by

$$\tan\theta=\frac{M_U-M_R}{M}\,\frac{\omega\tau}{1+(\omega\tau)^2},$$

where $M=(M_U M_R)^{1/2}$ and $\tau=(\tau_1\tau_2)^{1/2}$. The quantity $\tan\theta$ is a measure of the degree of damping in a vibrating system and is related to the logarithmic decrement δ, the elastic energy dissipated per cycle ΔW from a maximum stored elastic energy of W, the quality factor Q of the vibrating system at resonance, and the $1/e$ decay distance for the attenuation of a propagating wave of wave length λ, by the equations

$$\tan\theta=\frac{\delta}{\pi}=\frac{1}{2\pi}\frac{\Delta W}{W}=Q^{-1}=\frac{1}{\pi}\alpha\lambda.$$

These relationships show that the damping is large when the difference between the unrelaxed and relaxed modulus is large and that the maximum damping occurs when the vibrational frequency ω is equal to the reciprocal of the relaxation time τ.

The physical origin of the relaxation process may be a redistribution of molecular segments on a polymer; the diffusion of atoms between adjacent sites in a crystal; the rearrangement of heat, electric polarization, or magnetization within the sample; and many other stress-induced, microscopic motions. Some phenomena have broad ranges of relaxation times so there is always a relaxation process that matches the vibrational frequency and thus a high level of damping is observed over a broad range of frequencies. Some relaxation processes are thermally activated and the relaxation time depends on the temperature with a form

$$\tau\propto\exp(H/RT),$$

where H is an activation energy and R is the gas constant.

Strength of Solid Materials

For small stresses, the elastic properties of solids discussed above provide a sufficient description. At higher stresses, irreversible phenomena take place which cause either the fracture of the part or its permanent distortion. The maximum stress for which there is no permanent deformation or fracture is called the *elastic limit* or the *proportional limit*. *Brittle* materials fracture into separate pieces abruptly at this stress while *ductile* materials exhibit a time-dependent extension which is not recovered upon unloading. A graph of stress versus strain for a ductile material shows a deviation from linear behavior at the elastic limit and a rapidly increasing degree of strain for each additional increment of applied stress. The stress level at which the strain deviates from the linear relationship between stress and strain by 0.2% is often called the *yield stress* (or more exactly the 0.2% yield stress) and is used in engineering design to designate the maximum stress to which the material can be safely subjected before any permanent set is produced. The slope of the graph of stress versus strain above the yield stress is called the *work-hardening coefficient* and the maximum value of the stress achieved during the deformation is called the *ultimate stress*. For static loads that exceed the elastic limit, the extension or strain will continue to increase as a function of time giving rise to the *creep* of the solid.

When viewed on a microscopic scale, these irreversible changes in shape of a solid can be described in terms of the motion of irregularities in the atomic structure of the material called *dislocations*. A dislocation is a line in the crystal lattice about which the atom rows are distorted from their normal crystallographic positions. By moving from one point in a specimen to another, a dislocation produces a unit of permanent shear distortion in the specimen. The yield stress can be viewed as the stress level at which a detectable number of dislocations become mobile. To strengthen a solid (i.e., to make the yield point occur at a higher stress level) alloying with foreign atoms or precipitating second phases of other chemical species or different phases of the parent material within the solid provides obstacles which retard dislocation motion. When the dislocations are too firmly fixed, the solid will be elastic to very high stresses but then it is likely to break abruptly by the sudden growth of a crack.

DEFORMATION OF GASES AND LIQUIDS

The formal definitions of stress and strain needed to describe the deformation of elastic solids are greatly simplified when gases and liquids are described. Since these forms of matter do not support shear stresses (except for viscous effects) the stress tensor elements σ_{ik} for which $i\neq k$ are zero and those principal stresses for which $i=k$ are equal to one another since the medium is isotropic. Thus for liquids and gases

$$\sigma_{11}=\sigma_{22}=\sigma_{33}=-P,$$

where P is the pressure and has a negative sign because it conventionally acts inward. Likewise the strain components with $i\neq k$ vanish and those with $i=k$ add to describe the relative volume change produced by the pressure P. Thus, for liquids and gases, the relation connecting stress and strain becomes

$$P=-B(\Delta V/V),$$

where B is the bulk modulus and V is the volume of matter involved. For liquids, the bulk modulus B is only slightly smaller than it is for solids and for many problems of flow, an assumption of incompressibility ($B=\infty$) can adequately describe the flow behavior.

For gases, the compressibility is high (B very small) and the volume changes that accompany pressure changes are large enough that they must be considered. In fact, the bulk modulus of a gas cannot usually be considered as a constant independent of volume, and the question of whether or not the deformation occurs at constant temperature (isothermal) or at constant heat content (adiabatic) has a marked effect.

The usual problem found in the mechanics of those liquids which can be assumed incompressible is to calculate the velocity of each volume element in the material and to construct the path followed by a representative point in a flow pattern. Thus, the velocity vector replaces the displacement vector as the principal variable to be related to the external forces. Since the velocity of the volume element is important, its inertia as measured by its density, ρ, and its acceleration, $d^2\mathbf{R}/dt^2$ (where $\mathbf{R}$ is its position vector), occupy a prominent position in the equations of motion. In general, a volume element of an incompressible fluid acted upon by a body force $\mathbf{G}$ per unit mass and a local pressure P obeys the equation of motion

$$\rho \frac{d^2\mathbf{R}}{dt^2} = \rho\mathbf{G} - \text{grad } P.$$

If the velocity vector $\mathbf{V} = d\mathbf{R}/dt$ that satisfies this equation is such that curl $\mathbf{V} = 0$, the flow is said to be *irrotational*. Furthermore, if the body force $\mathbf{G}$ can be derived from a potential function ϕ, then Bernoulli's equation follows which states that

$$\frac{V^2}{2} + \phi + \frac{P}{\rho} = \text{constant},$$

where V is the magnitude of the velocity at the point where ϕ and P satisfy the equation. This equation is useful because it can relate the velocity and pressure at a location in the stream where they are not known to a more distant location where they can be determined easily.

In order to describe a *viscous* liquid, the equations of motion for a volume element must be modified by introducing, in addition to the pressure, a shear stress whose magnitude is proportional to the velocity gradient across the element

$$P_{ij} = -\eta \frac{\partial V_i}{\partial X_j}$$

which introduces the coefficient of viscosity η relating the shear stress P_{ij} to the rate of change of the i component of the velocity with respect to a change in the j position coordinate.

See also ANELASTICITY; CRYSTAL DEFECTS; DEFORMATION OF CRYSTALLINE MATERIALS; ELASTICITY; FATIGUE; HYDRODYNAMICS; INTERNAL FRICTION IN CRYSTALS; RHEOLOGY.

BIBLIOGRAPHY

H. B. Huntington, *The Elastic Constants of Crystals*. Academic Press, New York and London, 1958.

C. Kittel, *Introduction to Solid State Physics*, 3rd ed., Chaps 3–5 and 19. Wiley, New York, 1966.

W. T. Read, Jr., *Dislocations in Crystals*. McGraw-Hill, New York, 1953.

Arnold Sommerfeld, *Mechanics of Deformable Bodies*. Academic Press, New York, 1950.

Mechanics, Classical *see* Dynamics, Analytical

Mesonic Atoms *see* Muonic, Mesonic, and Other Exotic Atoms

Mesons

Malcolm Derrick

"Meson" is a generic name for a group of strongly interacting particles having baryon number 0. Approximately 100 different mesons are known [1]. They have masses ranging upward from the π meson with a mass of 140 MeV through the ψ family with masses in the range 3–4 GeV, to the Υ family, near 10 GeV in mass. All are unstable and decay, sometimes by a cascade process, to the lowest mass states that are accessible. Since such decays may be via the strong, the weak, or the electromagnetic interaction, a wide range of lifetimes is encompassed.

All mesons have definite properties specified by quantum numbers such as spin, parity, and isospin. These symmetry properties of the mesons can be understood if they are considered to consist of bound pairs of elementary building blocks called quarks (q). Mesons consist of a quark–antiquark pair ($q\bar{q}$), bound together by the strong force, which is mediated by the exchange of gluons. At least six quarks are thought to exist, arranged in three doublets: up (u), down (d); charm (c), strange (s); top (t), bottom (b). All but the top quark have been experimentally identified. The c, s, and b quarks carry a charm, strange, or bottomness property that is conserved in the strong interaction. The quantum numbers of the mesons are specified by the intrinsic properties of the quarks and the spectroscopic state of the ($q\bar{q}$) spin and orbital angular momentum. The mesons can also carry a so-called flavor property, which is the flavor of the heavy quark constituent: strangeness (s), charm (c), or bottomness (b).

Since most mesons decay strongly with lifetimes of $\sim 10^{-24}$ s, their existence must be inferred by looking for peaks in the invariant mass distributions of the decay products. High-mass mesons involving the u and d quarks cascade down and eventually give π mesons. Similarly, mesons with strangeness, charm, or bottomness decay to a K ($\bar{s}d$), D ($c\bar{d}$), or B ($\bar{b}d$) meson plus π mesons. These lowest-lying states then decay weakly with lifetimes in the range 10^{-8}–10^{-10} s. A few mesons (π^0, η, η') decay via the electromagnetic interaction with lifetimes intermediate between those decaying weakly and those decaying strongly.

The higher-mass mesons form a series of excited states of the $q\bar{q}$ system in relative orbital momenta 0, 1, 2, etc. In fact, a striking regularity is observed between masses and spin with groups of mesons exhibiting a linear relationship on a plot of square of mass versus spin. The straight lines connecting such groups are termed Regge trajectories. The higher-mass mesons are called Regge recurrences of the lower-mass mesons.

In strong-interaction theory, the gluons that are the carriers of the interquark force also interact strongly and so should be able to stick together to form particles. Such mesons are called glueballs. Although several candidates are known, no particle has so far been unambiguously identified as a glueball.

Mesons can also be classified in families having the same spin and parity but different strangeness and isospin. For example, the eight lowest-lying pseudoscalar states can be arranged in a hexagon on a plot of strangeness (S) against the third component of the isospin (T_3), as shown in Fig. 1. In these mesons, the u, d, and s quarks are in a relative S wave and have their spins antiparallel. The central point has two particles, the π^0 ($T = 1$, $T_3 = 0$), and the η ($T = 0$, $T_3 = 0$). Similar classifications can be made of the vector, $J^P = 1^-$, and tensor, $J^P = 2^+$, particles, and there is also experimental evidence that families of scalar 0^+, axial vector 1^+, and 3^- mesons exist. Such arrangements of particles are those of the group $SU(3)$.

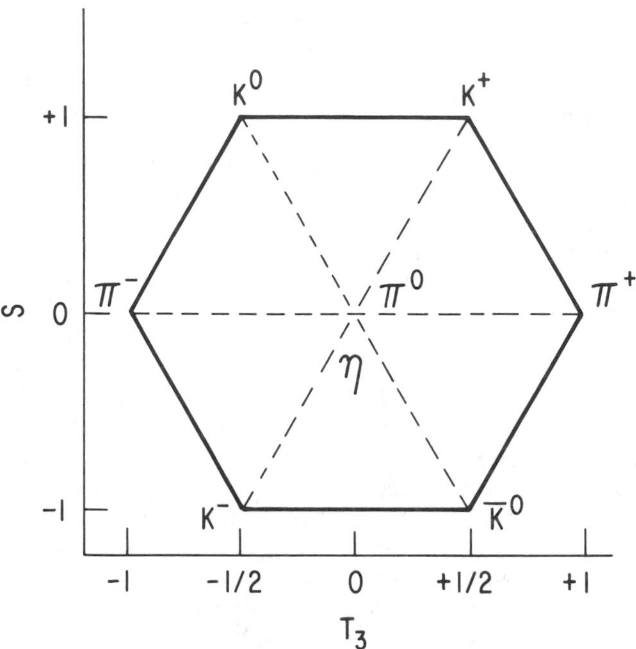

FIG. 1. Arrangement of the pseudoscalar mesons in an octet of $SU(3)$.

The lowest-mass meson, now called the π meson, was predicted 40 years ago by Yukawa as the carrier of the nuclear force. Such an idea was clearly much too simplified, and the exact role that the mesons play in holding nucleons together inside nuclei is still not completely understood. It is possible, however, to understand the size of the proton as measured, say, by electron scattering, if we consider that the photon that probes the proton structure couples to those mesons that have the same quantum numbers as the photon and that are in the meson cloud of the nucleon, i.e., the $T = 1$, $J^P = 1^-$ ρ meson, and the $T = 0$, $J^P = 1^-$ ω and ϕ mesons. The ρ and ω mesons have masses of ~700 MeV and the ϕ mass is 1016 MeV. These particles determine the momentum-space form factor of the nucleon, which on Fourier transformation gives a size of 10^{-13} cm, as observed.

Those mesons having the quantum numbers of the photon can also be seen as peaks in the total cross section for e^+e^- annihilation at center-of-mass energies equal to the meson masses. Such experiments are carried out using electron–positron storage rings. As the energy of these rings is increased to the 3–4-GeV range, two very narrow states are seen, called the $\psi(3100)$ and $\psi'(3700)$. This pattern is repeated in the 9–10-GeV mass range where three narrow states, called the Υ family, are seen. These states are much narrower than would be expected if a strong decay were fully allowed, yet are wider than appropriate for a weak or electromagnetic decay. We can understand the observed narrow widths if they are formed of quarks carrying a new quantum number.

The ϕ meson is thought to be predominantly made of an $s\bar{s}$ quark pair; similarly, the ψ mesons consist of a $c\bar{c}$ pair and the Υ of a $b\bar{b}$ pair. These particles act as if the quarks were moving in a potential that is weak for small distances but increases in strength as the separation between the quarks increases.

See also ELEMENTARY PARTICLES IN PHYSICS; QUARKONIUM; QUARKS; REGGE POLES; STRONG INTERACTIONS; $SU(3)$ AND HIGHER SYMMETRIES.

REFERENCES

1. Particle Data Group, *Phys. Lett.* **B204** (1988).

BIBLIOGRAPHY

B. Diekmann, "Spectroscopy of Mesons Containing Light Quarks (u, d, s) or Gluons," *Phys. Rep.* **159,** 99 (1988).
W. Kwong, J. L. Rosner, and C. Quigg, "Heavy Quark Systems," *Annu. Rev. Nucl. Particle Sci.* **37,** 325 (1987).
E. H. Thorndike and R. A. Poling, "Decays of the b Quark," *Phys. Rep.* **157,** 183 (1988).

Metal–Insulator Transitions

N. F. Mott

Solid materials are divided rather sharply into metals, which contain free electrons able to carry a current, and nonmetals such as sodium chloride or glass, which are insulators. It might be said that there are intermediate materials, semiconductors, containing only a few electrons, but in fact, for low temperatures at which bound electrons cannot be freed by heat motion, the division is still sharp. For metals the electrical conductivity tends to a finite value as the temperature is lowered, depending on the purity and degree of perfection if the metal is crystalline; for nonmetals, including semiconductors, the reverse is the case, the conductivity tending to zero. For the purpose of this article a metal is defined as a material that has this former property; this includes materials such as the "metallic" oxides (for instance, CoO_2 and the new high-temperature superconductors such as $La_{2-x}Sr_xCuO_4$) and also very heavily doped semiconductors. All these must contain electrons which are free to move, even at the lowest temperature.

Before the advent of quantum mechanics, it was quite unclear why the electrons in metals were free and in nonmetals not free, though it was known from the Hall effect, soon after the discovery of the electron, that the number of free electrons in a metal such as silver was near to one per atom. However, the formulation of wave mechanics by Schrödinger and its application to electrons in solids by Bloch, Peierls, and Wilson gave a convincing explanation in the early 1930s. Essentially it was assumed that in both metals *and* insulators the electrons are in a sense free, that is to say that they could be described in crystals by the Bloch wave functions

$$\psi = e^{ikx}u_k(x,y,z), \qquad (1)$$

which are now a very familiar part of solid-state physics.

Here e^{ikx} represents an electron moving with the wave vector k in the x direction, and u_k is a function having the periodicity of the crystal lattice. These wave functions describe an electron with an infinite mean free path; scattering, and with it a finite electrical resistivity, arises only because of deviations from a perfect lattice due to thermal vibrations or impurities. Moreover, the effect of one electron on another was entirely neglected. The insulating property arose because, first of all, the electrons described by the wave function (1) suffer Bragg reflection within the crystal, and this leads to the gaps of forbidden energy, in which no solution of the Schrödinger equation of type (1) exists; and also because the electrons obey Fermi–Dirac statistics, according to which not more than two electrons can occupy one quantum state. The "gaps" lead to the conclusion that the number $N(E)dE$ of states allowed for an electron in the range of energy dE can have the forms shown in Fig. 1. The quantity $N(E)$ is known as the density of states. Bands of energy states, containing (for a cubic lattice) two states per atom, are formed, which may or may not overlap. For a monovalent metal, for instance silver, the lowest band is half filled (Fig. 1a). For a nonmetal a filled band is separated from an empty band by the gap ΔE. For a divalent metal (for instance calcium) the first and second bands must overlap as in Fig. 1c.

From this model it was clear that, if two overlapping bands could be separated, for instance by pressure or by change of composition in an alloy, a "metal–insulator transition" would occur; and this has been achieved, for instance, for the divalent metal ytterbium under pressure. However, there was little if any discussion in the literature on this subject before World War II, apart from a seminal paper by Wigner [1] in 1938 which will be described later. The author's [2] paper of 1949, however, was the first to point out that if two bands overlap, a *small* number of electrons and of "holes" in the valence band is not possible, because of the Coulomb attraction between them, which would lead to their forming bound pairs (excitons). It is a result of quantum mechanics that, if two particles attract with a force that behaves as const/r^2 at large distances r, their ground state is always bound. A discontinuous change in the number of electrons and holes could thus be predicted, from zero to a value high enough to screen out the Coulomb attraction sufficiently to prevent pair formation. The prediction of a discontinuous change turned out to be correct, but the nature of the state in which pairs formed was obscure. These were called "excitonic insulators," and after many rather speculative papers about them, which have not been very fruitful in predicting properties of actual materials, a seminal paper by Brinkman and Rice [3] made it possible to look at the problem in the following way. The paper starts with the properties of "electron–hole droplets"; it is known that if electrons and holes (that is vacancies in a valence band) are formed in crystalline germanium, they condense into metallic droplets, and that these droplets have lower energy than bound pairs (excitons). In a droplet in equilibrium the number n of electrons and holes per unit volume is given by

$$n^{1/3}a_{\mathrm{H}} \sim 0.1,$$

where a_{H} is the hydrogen radius $\hbar^2\kappa/me^2$, with κ the dielectric constant. The corresponding energy gained is

$$E_0 \sim 0.1me^4/\hbar^2\kappa^2. \tag{2}$$

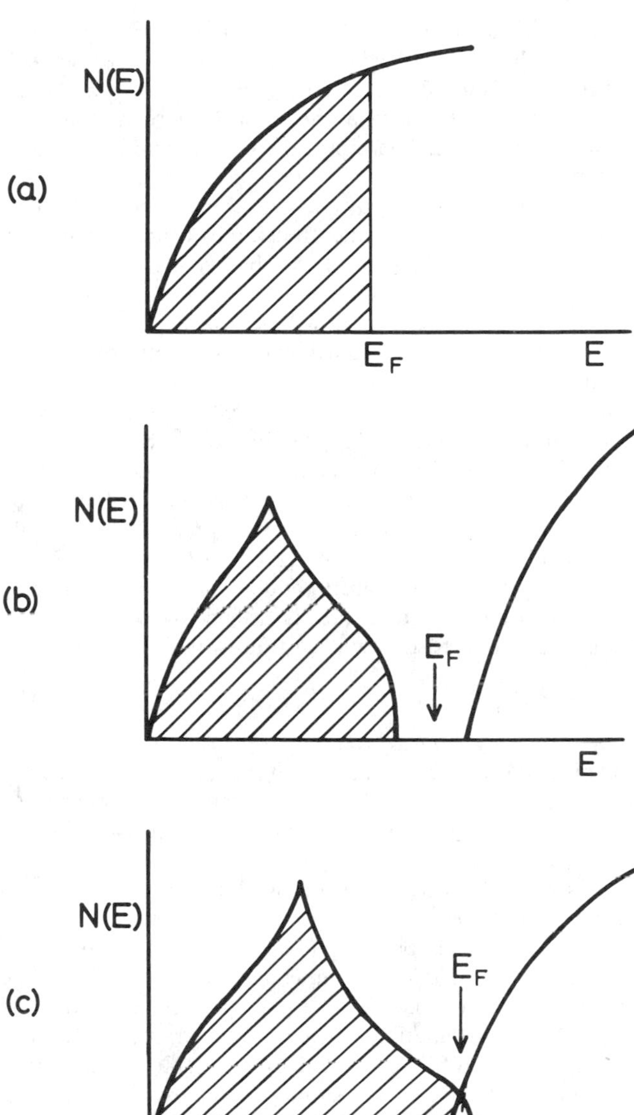

FIG. 1. Density of states in crystalline solids: (a) monovalent metal, (b) nonmetal with gap ΔE, (c) semimetal.

The numerical constants depend both on the band form and also on the approximation made in any calculation of this kind. The application to metal–insulator transitions is as follows: If the band gap ΔE decreases to the point where $\Delta E = E_0$, then an electron–hole gas will form discontinuously, filling the whole volume. When $\Delta E > E_0$, the material is a nonmetal; as soon as $\Delta E = E_0$, it becomes metallic, even though the bands do not overlap.

It would be very interesting if this discontinuous change from metal to nonmetal could be observed experimentally. This is difficult to do. One problem is that, if an attempt is made to change the gap by, for instance, hydrostatic pressure, there will necessarily be a discontinuous change in volume, and the transition will occur in an unstable region. This is because, if one plots free energy F against volume, there will be a kink in the curve at the point P where the transition occurs as in Fig. 2. Therefore, the discontinuous change in volume will be from A to B in the diagram. In the same way, if one changes the gap by varying the composition of an alloy, the region where the transition occurs is unstable. However, it might be possible to observe it in quenched alloys. This has been done when the bands are what we call Hubbard bands; their nature will now be explained.

Many materials, particularly transition-metal oxides, are antiferromagnetic insulators (e.g., NiO, V_2O_3). Each transition metal contains one or more electrons, oriented in the same direction by Hund's rule. For these Slater [4] first pointed out that the antiferromagnetic lattice could split the d band, making possible, in NiO for instance, a full and an empty band. These are now called the upper and lower Hubbard bands, after the work of Hubbard [5] in 1964. If the widths of the two bands are B_1 and B_2, and the tight binding approximation is used so that both bands are symmetrical about their centers, they will overlap when

$$\tfrac{1}{2}(B_1+B_2)=U. \tag{3}$$

U is the energy required to take an electron from, say, Ni^{2+} and place it on another Ni^{2+} ion. If there is only one electron per atom (as in Ti in $TiBr_3$), this is called the Hubbard U, given by the average of e^2/r_{12} over one atom, r_{12} being the distance between the two electrons.

This kind of transition is called a "Mott transition." It can occur with change of composition, temperature, or, when the centers are donors in semiconductors, with magnetic field. Since the metal and nonmetallic phases must have free energies close together for the phenomenon to occur, materials showing this kind of transition are rare. It occurs in the series [6] $Ni(S_{1-x}Se_x)_2$ and $(V_{1-x}Ti_x)_2O_3$. In the latter case, at $x=0.5$ the low-temperature conductivity jumps from zero to $10^4\ \Omega^{-1}\ \mathrm{cm}^{-1}$.

Equation (3) is only approximate because it does not take into account the discontinuous change in the number of carriers, or the volume or structure change that can take place at the transition.

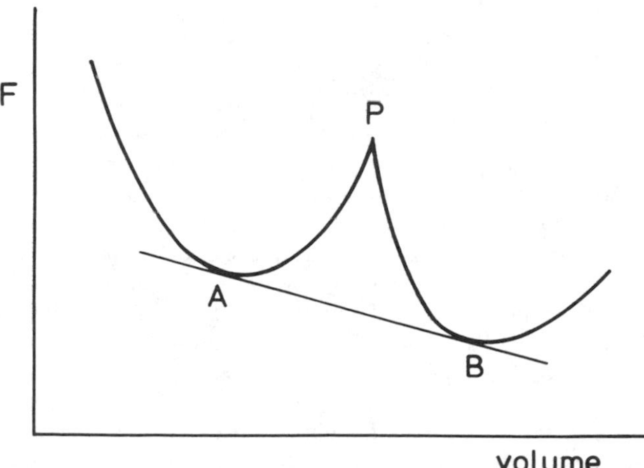

FIG. 2. Free energy F as a function of volume near a metal–insulator transition.

Other transitions not related to disorder are those such as that in VO_2; at low temperatures the vanadiums form pairs, giving rise to a band structure, without antiferromagnetism, which allows a full and an empty band. At 340 K there is a transition to a metallic state; according to Zylbersztejn and Mott [7] the Hubbard U plays a major role in determining the condition for the transition.

Antiferromagnetic insulators with one electron (or hole) per center are rare. La_2CuO_4 and other oxide superconductors when doped are some of the few known. Insulators such as $TiBr_3$ and TiBrO in which each Ti atom should be in the state $3d^1$ do not show antiferromagnetism but a small paramagnetism independent of temperature. It is possible that these are the "resonance valence bond" insulators proposed by Anderson.

A totally different form of transition can take place when a condensed electron gas exists in a noncrystalline environment, when parameters such as composition, stress, or magnetic field are changed. Examples are heavily doped semiconductors, where for a condensed electron gas the electron energies are in an impurity band, and alloys such as amorphous Si_xNb_{1-x}. Our understanding of these transitions depends on Anderson's [8] paper of 1958 on the "Absence of Diffusion in Certain Random Lattices," and the concept of a mobility edge (Mott [9]). According to Anderson, disorder can turn the states in a conduction band into traps (localized states). A mobility edge is a value E_c of the electron energy separating localized from nonlocalized states. If the Fermi energy E_F of the condensed gas lies below the mobility edge, conduction at low temperature is by variable-range hopping, with the conductivity given as a function of temperature T by

$$\sigma=\sigma_0\exp\{-(T_0/T)^{\nu}\} \tag{4}$$

with ν lying between $\frac{1}{4}$ and $\frac{1}{2}$; if it lies above, the conductivity remains finite as $T\rightarrow0$, varying only weakly with temperature. If through change of the parameters E_F passes through E_c, a metal–insulator transition takes place (See Fig. 3).

The transition of this kind most fully investigated is that in doped semiconductors and takes place in an impurity band. Experimentally it is found over a wide variety of substances that it takes place for a concentration n of donors given by [8]

$$n^{1/3}a_H\simeq0.27, \tag{5}$$

where a_H is the hydrogen radius $\hbar^2\kappa/m_{\mathrm{eff}}e^2$, κ being the background dielectric constant. Curiously, treatments of the problem as an Anderson transition and as a Mott transition lead approximately to the same equation, with the constant little different from that in Eq. (5). While it is generally agreed that for compensated samples the transition is of Anderson type, it is not so clear that this is so for uncompensated samples [9]. However, for the many-valley conduction bands of silicon and germanium the evidence favors the Anderson type.

For such transitions the zero-temperature conductivity should rise, according to most theorists, with composition

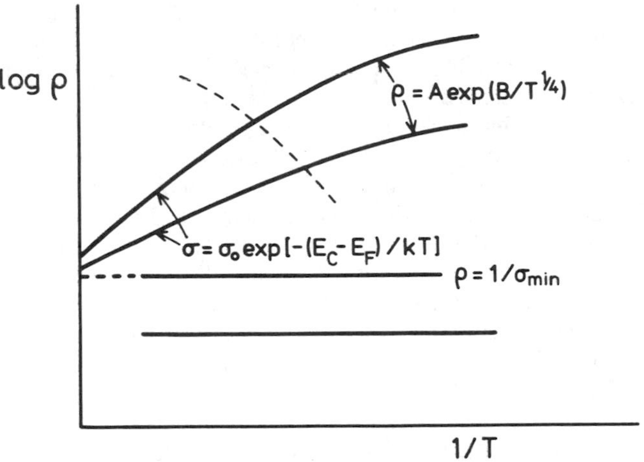

FIG. 3. Log(resistivity) as a function of $1/T$ for a system in which $E_c - E_F$ can be varied, by changing the composition or otherwise. σ_{min} is the "minimum metallic conductivity."

as $(n-n_c)^\nu$ with $\nu = 1$, and this is widely observed. However, for uncompensated many-valley materials $\nu = \frac{1}{2}$, the explanation of this is still in some doubt.

It should be noted that the first of the new high-temperature superconductors, $La_{2-x}Sr_xCuO_4$, is an antiferromagnetic insulator when $x = 0$, doped with Sr (or Ca or Ba) until it becomes metallic, the transition probably of Anderson type.

Several other kinds of transition have been described in the literature. In 1938, Wigner [1] suggested that at low densities electrons in a metal could crystallize under the influence of their own repulsion, producing a nonconducting state. While this has not been observed with certainty, the "Verwey transition" occurring in magnetite (Fe_3O_4) has been extensively investigated. At low temperature, the d electrons, forming Fe^{2+} and Fe^{3+}, are ordered and conductivity is activated; at 119 K there is a first-order transition to a state in which they are not, with a sharp increase in the conductivity.

See also ELECTRON ENERGY STATES IN SOLIDS AND LIQUIDS; ELECTRON–HOLE DROPLETS IN SEMICONDUCTORS; EXCITONS; INSULATORS; METALS.

REFERENCES

1. E. Wigner, *Trans. Faraday Soc.* **34,** 678 (1938).
2. N. F. Mott, *Proc. Phys. Soc. A* **62,** 416 (1949).
3. W. F. Brinkman and T. M. Rice, *Phys. Rev. B* **7,** 1508 (1973).
4. J. C. Slater, *Phys. Rev.* **82,** 538 (1957).
5. J. Hubbard, *Proc. R. Soc. London, Ser. A* **277,** 237 (1964).
6. See articles in *Metallic and Non-metallic States of Matter,* referenced in the Bibliography.
7. A. Zylbersztejn and N. F. Mott, *Phys. Rev. B* **11,** 4983 (1975).
8. P. W. Anderson, *Phys. Rev.* **109,** 1492 (1958).
9. N. F. Mott, *Adv. Phys.* **16,** 49 (1967).
10. P. P. Edwards and M. J. Sienko, *Phys. Rev. B* **17,** 2575 (1978).
11. N. F. Mott, *Philos. Mag. B* **58,** 369 (1988).

BIBLIOGRAPHY

Articles on the subject, including one by the author, are to be found in: *The Metallic and Non-metallic States of Matter* (P. P. Edwards and C. N. R. Rao, eds.). Taylor and Francis, London, 1985 and in "Sir Nevill Mott Festschrift"; Localization and Metal–Insulator Transitions"; Vol. 3 (H. Fritzsche and D. Adler, eds.). Plenum Press, New York and London, 1985.

Metallurgy

Bruce Chalmers

In earlier days the term metallurgy was used to refer to the technology of the sequence of processes by which a useful object was produced from metal which usually was, at the outset, a component of a chemical compound found in the earth's crust as an ore. More recently, the term has come to describe the newly emerging branch of science that is concerned with exploring and understanding the structure of metals and alloys at the microscopic and submicroscopic levels, the relationship between the structure of a metal (or alloy) and its properties, and the way in which the structure of a particular sample is determined by the chemical composition and its thermal and mechanical history, and the exploitation of this understanding to produce the properties that are called for by the increasingly stringent demands of sophisticated technology.

It is, therefore, convenient to divide the whole field of metallurgy into process metallurgy and physical metallurgy. The former deals with the technology of extracting metals from their ores, by chemical, thermal or electrochemical processes, refining and alloying, and producing a metallic object by casting the molten metal in a mold of the appropriate shape and then, if necessary, deforming it mechanically by forging, rolling, extrusion, or otherwise to bring it to the required final shape.

Physical metallurgy is chiefly concerned with the properties of the final product as distinct from its chemical composition and its shape. The science of physical metallurgy is based on the concept that the properties of an alloy depend only on its structure, and that the structure is determined only by the chemical composition and the thermal and mechanical history of the sample. The structure is studied at two levels: the microscopic and the atomic. At the microscopic level, the structure is defined by the phases that are present and by the shapes and sizes of the crystals of each phase; at the atomic level, the structure relates to the basic crystal structure of each phase, the disposition of the atoms of the various elements within that crystal, and the ways in which the crystals depart from this ideal structure. Thus an alloy may consist of two or more distinct types of crystal, each with its own characteristic structure and composition; one or two of these crystal types may form a continuous phase (the matrix) while the others may be in the form of discrete particles, which could be needlelike, platelike, or equiaxed in shape. These characteristics can be determined by optical microscopy, as the typical size of the crystal is larger than the limit of resolution of the optical microscope, 1 μm. The study of structure at this level is called "metallography.'' Each type of crystal has its characteristic crystal structure (face centered cubic, body centered cubic, etc), and all the elements present in the alloy are to be found in each crystal, although usually not in the same proportion as in the alloy as a whole; for example, an alloy of iron and carbon (steel) containing 0.2% (by weight) of carbon consists of crystals of iron containing about 0.002% carbon (ferrite), and crystals of iron carbide, Fe_3C (cementite) with about 6.7% of carbon. Thus a phase may have the basic crystal structure of one of the elements that is present, with atoms of other elements present in it in solid solution. It is then described as a ''terminal phase.'' The solute atoms may substitute for atoms of the parent phase at sites which, in the perfect crystal, would be occupied by atoms of the parent element. The substitutional atoms may be dispersed randomly within the crystal, or they may exhibit an ordered arrangement, which can either be of short range, consisting of very small ordered domains, or long range, in which the regularity of site selection extends over many atoms. Substitutional solid solutions do not occur over extended ranges of composition when the diameters of the atoms differ by more than about 15%. The solid solution may also be interstitial, when the solute atoms are accommodated between the atoms that define the structure of the crystal; this usually occurs only when the solute atoms have a diameter not more than about one-half of that of the host. A phase may, on the other hand, have a structure that differs from those of all of its component elements; it is then described as an ''intermediate phase.'' Some intermediate phases have compositions in which atoms of the constituent elements are present in simple proportions (these are referred to as intermetallic compounds); other intermediate phases owe the stability of their structures to geometrical factors that exist when the sizes of the atoms are related in specific ways; others, to a preferred ratio of the number of valence electrons to the number of atoms; and for some, electronic and geometrical factors combine to stabilize the structure. The study of the equilibrium distribution of the elements that are present between the various phases, as a function of temperature, leads to the construction of ''phase diagrams'' or ''equilibrium diagrams.'' The equilibrium diagram is a thermodynamic statement of the most stable distribution of the elements that are present; the phases that are actually present, and the distribution of the atoms in them as well as the sizes and shapes of their crystals, however, depend on the thermal history of the sample, because the processes that allow the structure to approach equilibrium are either the diffusion of atoms within the solid, which is inherently a slow process, or a cooperative change of structure, which does not take place without a large driving force in the form of a substantial departure from equilibrium.

''Heat treatment'' is the term that describes the sequence of processes designed to produce the nonequilibrium structure that provides the required properties. For example, steel is heat treated by first heating it to a temperature at which the equilibrium structure is face-centered-cubic crystals of iron with carbon in interstitial solid solution. If the steel were cooled slowly, so as to maintain equilibrium, it would transform into body-centered-cubic crystals of iron containing very little carbon, and crystals of iron carbide (Fe_3C); if the steel is ''quenched,'' that is, rapidly cooled by plunging into water, the separation into two phases is suppressed because there is insufficient time for the carbon to diffuse to form crystals of iron carbide, and instead the carbon remains where it is and a phase, martensite, which does not appear in the phase diagram, is formed. Martensite is an extremely hard and brittle material, and for most purposes for which hardened steel is needed it is necessary to temper the martensite by heating it to an intermediate temperature. During tempering the structure of the steel is allowed to change toward equilibrium to a carefully controlled extent to achieve the required compromise between hardness and toughness. Another example in which heat treatment provides a controlled approach to equilibrium is the age hardening of some aluminum alloys; the alloy is first ''solution treated'' at a relatively high temperature at which the alloying elements, such as copper, go into solid solution. Slow cooling would allow the copper largely to separate in the form of crystals of $CuAl_2$; rapid cooling prevents this, but subsequent aging at an intermediate temperature allows the copper to achieve the first stage of its separation, in the form of ''zones'' that endow the alloy with very high strength.

It has been long known that, while an ideally perfect crystal of a metal should be stronger than any known materials, real crystals are almost always very easily deformed; the reason is the presence in the crystal of imperfections called dislocations, which permit the sliding of one part of the crystal over another to take place progressively, by the motion of the dislocations, which is easy, rather than simultaneously, which would require a very high stress. In order to strengthen a metal, therefore, it is necessary to restrict the movement of the dislocations; there are several ways in which this can be done, of which the two most important are by the introduction of very large numbers of dislocations and by the introduction of obstructions which inhibit their motion. When a metal is deformed, the dislocations that are present not only move, but they also increase in number; typically $1 cm^3$ of a metal crystal would contain 100 km of dislocations, but if it were deformed, for example by rolling, this could increase to 10^7 km/cm^3. Dislocations interact with each other in such a way as to increase the force required to move them, and an effect of deforming a metal or alloy, therefore, is to increase its resistance to further deformation, that is, its hardness. This is the process of work hardening, which can be negated by annealing, which is heating to a temperature at which the dislocations are reduced in extent as a result of the mobility of the atoms. The second method of hardening a metal is by making it heterogeneous on a sufficiently fine scale. Concentrations of solute atoms that strain the crystal locally, or precipitated particles of a second phase, provide obstacles to the motion of dislocations so that the strength is increased. A third method of hardening an alloy is to convert it into a crystal structure which is inherently resistant to the passage of dislocations, as is the case with martensite. The strength depends not only on the internal structure of the individual crystals, but also on the shape, size, and distribution of the crystals of the various phases that are present; it follows that the composition of an alloy and the processes, such as casting, that determine the shape, size, and distribution of the crystals, interact with subsequent processes such as work hardening and heat treatment to determine the properties of the final product.

Some of the most challenging demands on the skill of the physical metallurgist are in the development of alloys that will withstand high stresses for long times at high temperatures in a highly oxidizing environment. This requires an understanding of the processes which allow slow deformation to take place over long periods of time at high temperatures and the ways in which these types of deformation can be stopped or retarded, and it requires the selection of materials that are highly resistant to oxidation.

See also ALLOTROPY AND POLYMORPHISM; ALLOYS; CRYSTAL DEFECTS; CRYSTALLOGRAPHY, X-RAY; METALS.

BIBLIOGRAPHY

R. W. Cahn (ed.) *Physical Metallurgy*. North Holland, Amsterdam, 1965.

J. Wulff, H. F. Taylor, and A. J. Shaler, *Metallurgy for Engineers*. Wiley, New York, 1952.

Metals

Stanley Engelsberg

A large majority of elements in the periodic table crystallize in the metallic state. The properties that essentially characterize a metal are its high electrical and thermal conductivity, its ductility and malleability, and its luster. These properties may in part be deduced from the type of binding that characterizes a metal.

In molecular and ionic crystals the binding can be discussed using classical concepts. That is, once we have the forces between the pairs of particles, which in most cases are derived on the basis of quantum mechanics, we can just add together the potentials of all the pairs of particles to obtain the total energy. In the case of metallic binding the picture is completely different. Here the electrons in the conduction band are no longer tied to a single atom, but effectively overlap all the ions in the crystal. The calculation is quantum mechanical by nature in all of its stages. If we consider the wave functions for the atoms in the metal, we find that at the known interatomic spacings, the last maximum in the radial part of each atomic wave function almost touches that for its nearest neighbors. Because of this overlap in the metal, the crystal's properties are quite distinct from those of, say, a gas of the constituent atoms. On the other hand, in ionic and molecular crystals, many of the physical properties (e.g., optical, dielectric, and magnetic) may be predicted from those of the constituent atoms, ions, or molecules.

One of the important ideas in the Wigner–Seitz theory of the binding of metals was to join the atomic wave functions smoothly with zero slope at the position between nearest neighbors. This new boundary condition on the wave function for the lowest electron state of the conduction band lowers the kinetic energy contribution and is an important part of the cohesive energy of metals.

In nonmagnetic metals and in the absence of an external field, this lowest state can accommodate two electrons, one with spin along the direction of quantization, the other opposed. To obtain the other allowed states for the electrons in the conduction band we must solve the problem of the motion of electrons in a periodic potential, which is the subject matter of the band theory of solids. The electrons fill the lowest energy states up to the Fermi energy.

That there are available energy states just above the last filled state means that the application of the smallest electric field will accelerate the electrons. For a perfect pure crystal, in the absence of atomic vibrations, which can be made exceedingly small at very low temperatures, the electrical conductivity of a metal would be infinite. At low temperatures the resistivity of a metal is governed by its purity and perfection. That is the case until the metal reaches its superconducting transition temperature, below which temperature there is no measurable resistance. There are still a number of metals that have not been found to be superconducting at the lowest attainable temperatures. Most notably these include the best high-temperature conductors, i.e., the alkali metals and the noble metals. Those metals that undergo transitions to a magnetically ordered state are also nonsuperconducting.

That there are states available to the electrons just above the Fermi surface means that no matter how small the energy of a photon of light that you shine on a metal, the light can be completely absorbed. This is the basic reason for the opacity of metals in the visible and lower-energy part of the spectrum. It is directly connected with the good electrical conductivity of metals.

At all finite temperatures the electrons in fact do not just fill up the allowed energy states, stopping at the Fermi energy. In addition they have a thermal smearing in a small energy interval about the Fermi energy of width $k_B T$ (Boltzmann's constant times the absolute temperature) that is about 1/40 of an electron volt at room temperature. That there is a finite density of states (number of allowed states per unit energy interval) for the electrons at the Fermi surface leads to an electronic contribution to all of the equilibrium thermodynamic properties of metals. For example, there is an electronic contribution to the specific heat of metals, which is linearly proportional to the temperature. There is also a small contribution to the magnetic susceptibility of metals, which is paramagnetic for elements including and to the left of the transition elements in the periodic table. Starting with the noble metals and for those following, a diamagnetic susceptibility is observed. The oscillations of the diamagnetic or orbital part of the susceptibility as a function of the magnetic field give a good deal of information about the metals' Fermi surface. In a large number of metals that have incomplete 3*d* or 4*f* shells, the electronic interactions lead to ordered magnetic phases.

See also CONDUCTION; CRYSTAL BINDING; DE HAAS–VAN ALPHEN EFFECT; DIAMAGNETISM; ELECTRON EN-

ERGY STATES IN SOLIDS AND LIQUIDS; FERMI SURFACE; FERRIMAGNETISM; FERROMAGNETISM; MAGNETIC MATERIALS; PARAMAGNETISM; SUPERCONDUCTING MATERIALS; SUPERCONDUCTIVITY THEORY.

BIBLIOGRAPHY

N. Ashcroft and D. Mermin, *Solid State Physics*. Holt, Rinehart & Winston, New York, 1976.

F. Seitz, *Modern Theory of Solids*. McGraw-Hill, New York, 1940.

Metals, Liquid *see* Liquid Metals

Meteorology

John L. Stanford

ATMOSPHERIC SCIENCES (METEOROLOGY)

In the past, meteorology has often been considered related to weather forecasting, whereas atmospheric physics has been more concerned with individual processes occurring in the atmosphere. Today, the distinction is blurred and the field is more properly known as atmospheric science. A wide variety of scientists trained in physics, mathematics, computer science, chemistry, oceanography, and weather prediction pool their talents to focus on understanding the complex behavior of the atmosphere, compressed by gravity on the surface of a spherical, rotating planet, with friction, land–sea contrasts, mountains, complicated radiative absorption and emission, intricate chemistry effects, and fluid dynamical instability processes.

WEATHER OBSERVATIONS

A worldwide network of reporting stations provide simultaneous ("synoptic") surface measurements several times a day of temperature, moisture content, atmospheric pressure, wind, and other weather conditions (such as precipitation and visibility). Conditions above the surface are sampled twice a day at 0000 and 1200 UTC using small instrument packages called radiosondes attached to helium-filled balloons. The balloons typically reach altitudes of 30 km, radioing back data at selected pressure (altitude) levels as they ascend. Wind measurements are derived from observation of the balloon motion.

Satellite observations of the state of atmosphere are made continuously by polar-orbiting (about 14 orbits per day) and geosynchronous (stationary over one location) satellites. The data are telemetered to earth where temperature profiles and other characteristics are derived through statistical inversion and other analysis techniques.

The satellite and balloon data are merged and used as the starting point for sophisticated computer models of the atmosphere. Weather forecasts are based on the output of these models. The satellites provide especially important data over largely uninhabited areas, particularly the vast tracts of the Southern Hemisphere oceans.

WEATHER FORECASTING

Though the World Weather Watch of participating nations has developed an excellent network for exchanging conventional weather observations, the fact remains that these measurements are crude and widely separated in time and space. The observational networks are not sufficiently dense to resolve weather phenomena of dimensions smaller than approximately 500 km, and existing computer methods are not able to specify accurately smaller-scale phenomena such as fronts, tornadoes, squall lines, and small hurricanes. Consequently, the initial analysis of weather patterns is inaccurate. Forecasting future weather trends beyond a few hours or days is increasingly in error because of the complex interactions of the environment (see below under chaotic dynamics). Weather forecasts in the Northern Hemisphere and tropics have considerable utility for periods up to 48 h for weather systems having dimensions of 1000 km or more (large cyclones and anticyclones). However, small-scale features embedded in these systems may cause hour-to-hour weather variations that are difficult to predict. The exact location of severe thunderstorms and tornadoes cannot be forecast accurately, although the general area of severe storm activity may be predicted up to 24 h in advance. Accurate forecasts of infrequent events such as heavy snow, sleet, and damaging winds are usually limited to periods not exceeding 24 h.

CURRENT RESEARCH

Remote Sensing

New observational instruments and techniques are evolving rapidly as a consequence of advances in space technology, computers, and electro-optical sensors. Conventional measurements of atmospheric properties are being supplemented by both active and passive remote sensing devices from earth-bound and space platforms.

Active remote-sensing systems include radar, lidar, and sodar. *Radar* is the system most widely used in synoptic meteorology. It emits high-frequency (wavelengths 0.1–10 cm) radio waves (microwaves) which penetrate clouds, but are reflected by precipitation. The time it takes the wave to reach the precipitation and return gives the distance to the precipitation. The evolution, motion, and structure of the precipitation patterns can be monitored at distances out to several hundred kilometers. *Lidar* is similar to radar; the radiation is in the form of visible or infrared laser beams. These beams of light are extremely narrow and can detect microscopic particles, aerosols, in the atmosphere. Lidars can also detect turbulent regions containing temperature fluctuations. The useful range of lidars is only a few kilometers. *Sodars* are acoustic devices that emit sound impulses which are then reflected from regions of turbulent temperature fluctuations. Sodars are proving useful in measuring the thickness of the friction layer near the surface, thus giving a measurement of the diffusion of pollutants in urban areas. Active radar transmitters have been placed on earth-orbiting weather satellites. Information about wave heights, ocean currents, thickness, and age of sea ice, snow cover, and winds are obtained.

Passive remote-sensing systems are most commonly placed on earth-orbiting or geostationary satellites. The most familiar satellite observations are photographs taken with visible light, showing cloud patterns, water areas, mountains, vegetation changes, and snow cover. Infrared radiometer instruments provide images both day and night. The amount of radiation recorded is a measure of the temperature of the radiating object—the ground, clouds, or the atmosphere; thus, the horizontal and vertical temperature distribution of the atmosphere and ocean surface can be inferred.

Other passive radiometric measurements from satellites include water vapor, carbon dioxide, ozone, aerosol content, and other "greenhouse gases" such as CH_4 and N_2O.

The Upper Atmosphere Research Satellite (UARS) is a major research instrument with 10 separate sensor packages designed to provide detailed information about the structure and composition of the stratosphere (10–50 km altitude) and mesosphere (50–80 km). The data are especially important for assessing and monitoring of ozone depletion and the greenhouse effect (see below).

The complexities of the atmosphere offer exciting challenges to new generations of atmospheric scientists in the coming decades. Among the major issues are the following.

Chaotic Dynamics and Implications for Weather Forecasting

Stimulated by MIT Professor and meteorologist E. N. Lorenz, the field of research known as nonlinear dynamics and chaos is causing a revolution in thinking in all of science. Chaotic dynamics has particular relevance to weather forecasting since the atmospheric state depends on highly nonlinear fluid dynamical equations of motion which can exhibit chaotic behavior. As mentioned earlier, computer forecast models begin with the atmospheric state at a particular time and project ahead in time. If the system is chaotic, the end result (the forecast) is extremely (exponentially) sensitive to the starting conditions, so much so that it may not even prove possible to predict weather accurately beyond a week or two, due to the fundamental complexities of chaotic dynamics.

Ozone Depletion

Ozone absorbs solar ultraviolet radiation, protecting life on earth and also influencing large-scale wind patterns. It is thus vital to understand the ozone depletion caused by increasing emissions of anthropogenic gases (chlorofluorocarbons, CFCs). CFCs, widely used as refrigerants and in production of plastic foam products, have long lifetimes (around 100 years) and migrate to the stratosphere where chlorine is photochemically released. Complicated chemical interactions, documented in the discovery of the "ozone hole" in extreme cold conditions over Antarctica, then destroy ozone catalytically.

Greenhouse Effect

Potentially of even greater importance are slow climate changes due to "dirtying the infrared window" by increased concentrations of atmospheric trace gases such as CO_2, CH_4, N_2O, and CFCs. Understanding the detailed radiative balance between solar input and the earth's infrared thermal radiation back to space is crucial to protecting our environment. Serious climate changes may occur if the emissions of these gases continue to increase unabated, but present understanding is not adequate to give accurate guidance to policymakers. Major uncertainties relate to the role of clouds and oceans in climate-system feedbacks.

Acid Rain

The sources and atmospheric transport related to "acid rain," suspected to be due to industrial activity but incompletely understood, represent another important challenge.

See also ATMOSPHERIC PHYSICS.

BIBLIOGRAPHY

R. A. Anthes, H. A. Panofsky, J. J. Cahir, and A. Rango, *The Atmosphere*. Merrill Publishing Co., New York, 1978.

J. T. Houghton, *The Physics of Atmospheres*, 2nd ed. Cambridge University Press, Cambridge, 1986.

Metrology

John A. Simpson

Metrology is the science of measurement, and if broadly construed would encompass the bulk of experimental physics. The term is usually used in a more restricted sense to refer to that portion of measurement science used to provide, maintain, and disseminate a consistent set of units; to provide support for the enforcement of equity in trade by weights and measures laws; or as an adjunct to quality control in manufacturing.

A measurement is a series of manipulations of physical objects or systems according to defined protocols that result in a number. The objects or systems involved are test objects, measuring devices, and computational operations. The objects and devices exist in and are influenced by some environment. The value obtained is purported to represent uniquely the magnitude, or intensity, of some quantity embodied in the test object. This number is acquired to form the basis of a decision affecting some human goal or satisfying some human need that depends on the properties of the test object.

In order to attain this goal of useful decision making, metrology has focused on the task of assuring that the value obtained for a given quantity of a given object is functionally identical wherever and whenever the measurement process is repeated. Only then can all parties to the decision work from a concordant data base. Such a universally reproducible measurement is called a proper measurement.

An analysis of the logical conditions that must be satisfied to achieve a proper measurement shows that three independent arbitrary axioms must be universally agreed upon:

1. All parties must agree upon and have access to a common unit in which the results will be expressed.
2. There must be an agreed-upon physically realizable method of obtaining a continuous scale of magnitude based on the unit.
3. There must be an agreed-upon physically realizable method of determining when the quantity of interest, as embodied in a physical object or system, is equal to, less than, or greater than, some fixed point on this realized scale.

The principal activity of metrologists consists of generating, propagating, testing, and applying to an object or system of interest sets of these measurement axioms for all quantities and all useful magnitudes of those quantities.

Historically the units (or standards) were man-made artifacts, for example, kilogram platinum weights and painstakingly fabricated meter bars. These artifacts were carefully preserved in national laboratories and infrequently compared with secondary standards. These secondary standards served to disseminate the unit to the required working standards. This formal hierarchical system is characteristic of classical metrology. Fundamental to the success of such a system is the development, at each transfer, of realistic estimates of uncertainty. Control of the uncertainty, as indicated by the statistical indices such as standard deviation of repeated measurements or transfers, provided assurance that the measurement was proper enough for the intended use. This practice was extended in modern times to the concept of measurement as a production process whose product is a number to which industrial quality-control techniques can be applied. These techniques, already well developed at the time of application to metrology, made use of the power of statistical analysis, experimental design, sampling theory, and tests of closure to develop reliable indices of measurement quality. Measurement programs using these tools are often called *measurement assurance programs*, especially when closure of results around the hierarchy is employed as a final test of a measurement system.

There have been continuing efforts in metrology directed toward making the ultimate standards (SI base units), which are defined by the International Committee of Weights and Measures (CIPM), constants of nature, which, in theory at least, would make them independently realizable. Initial efforts, for example, resulted in the second being defined in terms of a multiple of a natural frequency of the cesium atom and the meter in terms of the vacuum wavelength of the radiation corresponding to the transition between two atomic levels of a krypton isotope. Most recently, the meter was again redefined, indirectly this time by defining c, the velocity of light, as $2.997\ 924\ 58 \times 10^8$ m/s. Hence, the meter is now defined as the distance traveled by light in $1/c$ s. Of the SI base units, only the unit standard of mass, the kilogram, despite serious efforts in a number of National Standards Laboratories, still remains a pure artifact standard. This desire to use natural phenomena as units has brought about a fundamental change in the logical structure of metrology. The units are no longer arbitrary since they are now linked by the equations of physics. The defined velocity of light directly links time and length and also ties these quantities into the electrical units since c is the ratio of electrostatic to electromagnetic units.

The desire for a consistent set of fundamental physical constants in turn puts constraints on both the quantity scales and the methods of determining equality since the total measurement procedure must yield numbers that preserve the self-consistency of the redundant set of fundamental constants. National metrology laboratories and academic institutions have programs devoted to the measurement of these fundamental constants that are the interface between metrology and the rest of physics.

Desirable as this change is for scientific purposes, these new units and the methods that transfer them are often not as appropriate for more mundane purposes as were their predecessors. For purposes of maintaining equity in trade and ensuring interchangeability of manufactured parts, the added complexity and expense entailed are unjustified. The problems arise because these new unit standards differ markedly from the usual objects or systems embodying the quantity of interest. In length, for example, a line scale is a convenient secondary standard. In the days of the artifact, meter direct comparison was a simple physical operation. Use of the meter, as now defined, requires an optical frequency heterodyne link from the frequency of the second-defining cesium atomic transition in the infrared to the frequency of a line in the visible spectrum. This operation, since the wavelength in vacuum equals the defined c divided by the frequency, yields a standard wavelength suitable for interferometry. The visible line almost universally used is the He–Ne laser line at $632\ 991.399 \times 10^{-12}$ m. This line is an internationally defined secondary standard. Comparing a line scale to a wavelength demands an elaborate interferometer requiring not only temperature correction for the coefficient of expansion of the scale (typically 0.25 m/°C), but also correction for the effect of temperature on the local air density, since the air density affects the correction of wavelength in air to the vacuum unit. Modern interferometers can make such corrections under most conditions, making use of environmental sensors and internal computing capability, but at the expense of complexity and cost.

In practice, therefore, the first transfers from the unit are usually performed only in the National Metrology Laboratories of the industrial nations. Transfers from the secondary wavelength standards are done in primary standards laboratories. Below this level in the hierarchy, artifact standards are still used.

Metrology in support of weights and measures activity has been uninfluenced by these changes and continues to depend on the maintenance and use of simple artifact standards. In the United States this activity is carried out in state laboratories, which maintain standards supplied to them by the National Institute of Standards and Technology (formerly National Bureau of Standards). These standards are restricted to those quantities used in trade and commerce. Elsewhere in the world this application of metrology is a federal function and extends to the broader field of consumer protection for health and safety in addition to trade equity. This activity has given rise to a subdiscipline called legal metrology, which deals with the development and enforcement of laws that involve measurement and measurement devices. This activity is coordinated by the International Organization for Legal Metrology, a treaty organization to which the United States is a party.

By far the greatest activity in metrology is that performed in the service of quality control. Manufacturing establishments of any size maintain standards laboratories and/or metrology laboratories. These laboratories maintain the company master standards, gauges, and measuring instruments, which are periodically calibrated against the national standards. The working measuring equipment on the shop floor is calibrated by the metrology laboratory on a scheduled basis. In the mechanical industries this measurement equipment tends to be simple tools, such as micrometers, line scales, gauge blocks, and static gauges of simple geometrical form such as "plugs" (cylinders) and rings. These are used to "set" the more elaborate on-line measuring machinery, which is either pneumatic or, more recently, electronic. In this manner the measurements made for quality control are considered "traceable" to national standards. Such "traceability" is required of all measurements on items produced for the Department of Defense, and more recently has been invoked by organizations such as the Environmental Protection Agency in connection with their regulatory activities.

Metrology as a discipline is not currently taught at the baccalaureate level or beyond in any academic institution in America. Its practitioners are usually physicists, mechanical engineers, or persons who entered the field after experience in the field of quality inspection.

See also CONSTANTS, FUNDAMENTAL; INTERFEROMETERS AND INTERFEROMETRY; MASS; SYMBOLS, UNITS, AND NOMENCLATURE; TIME.

BIBLIOGRAPHY

D. M. Anthony, *Engineering Metrology*. Pergamon Press, Oxford, 1986. Classic metrology, mechanical, elementary level.

Rudolph Carnap, *Philosophical Foundations of Physics* (Basic Books, New York, 1960), develops the logical foundations of metrology in some detail.

Rexmond C. Cochrane, *Measures for Progress* (U.S. Department of Commerce, National Bureau of Standards, Washington, D.C., 1966). A history of the National Bureau of Standards, including development of metrology in the United States.

Francis T. Farago, *Handbook of Dimensional Metrology*, 2nd ed. Industrial Press, New York, 1982. Very complete.

Society of Manufacturing Engineers, *Handbook of Industrial Metrology*. Prentice-Hall, Englewood Cliffs, NJ, 1967. General text, very practical in tone.

The Michelson–Morley Experiment

L. S. Swenson, Jr.

One of the most famous optical experiments ever performed, this attempt to measure "the relative motion of the Earth and the Luminiferous Ether" gradually became known as a notorious failure that was widely supposed to have prompted, if not caused, the advent of relativity theory. First conceived and performed in the 1880s, repeated and polished in the early 1900s, and then elevated, repeated, and finally completed in the 1920s, the so-called Michelson–Morley–Miller "aether-drift" experiment gained a reputation beyond its character as a test in physical optics. It became a celebrated part of the rationale for and the debate over Einstein's ideas of 1905 and 1915.

Albert A. Michelson, a U.S. Naval officer on leave in 1880 to study physics in Europe, initially designed his new precision instrument (later called an interferometer) as a means for observing a second-order effect of the earth's motion, presumably absolute, against the background of an interstellar luminiferous medium. Stimulated by Maxwell's speculations and encouraged by Helmholtz's advice, Michelson conceived a way of comparing the velocity of light at right angles on a laboratory bench so that, in accord with the undulatory theory of light and A. J. Fresnel's explications, it ought to be possible to measure the absolute velocity of the earth's motion through the luminiferous ether.

The original apparatus (Fig. 1) was a rigid cruciform of brass optical arms about a meter in length and crossed eccentrically so that a single beam of sodium light could be split into two pencils, raced over equal paths at right angles to each other, and then recombined into a pattern of interference fringes. Estimating fractional shifts in these fringes at eight cardinal azimuths about noontime in or near Berlin in early April 1881, Michelson was disappointed to find no displacement of fringes beyond what he attributed to experimental design error. These results he interpreted as negating the hypothesis of a *stationary* ether filling the void of interstellar space and as contradicting the received theory of astronomical aberration.

Over the next few years, Simon Newcomb, H. A. Lorentz, Oliver Lodge, William Thomson (later Lord Kelvin), J. Willard Gibbs, Lord Rayleigh, and other leading physicists took an interest in this experiment and its proper interpretation. Meanwhile, Michelson settled at a new academic post in Cleveland, Ohio, and was encouraged to pursue these problems with his new neighbor, a senior chemist, Edward W. Morley. Together they first designed an elaborate repetition of H. Fizeau's 1859 "water-drag" experiment in order to test Fresnel's convection coefficient. They constructed a hydrodynamic apparatus to measure the velocity of light moving with and against a monitored flow of distilled water. This crucial test of the "influence of the motion of the medium on the velocity of light" turned out in 1886 to corroborate Fizeau's confirmation of Fresnel's theory, so Michelson and Morley moved on with confidence to try for a definitive "aether-drift" test, this time seeking merely the earth's orbital velocity.

Despite many personal tribulations, Michelson and Morley made ready their apparatus (Fig. 2) in June 1887 for what would become the "classic" experiments. The new optical parts were set atop a massive stone slab mounted on an annular bearing floating on a circular bed of mercury. This apparatus was built upon a brick pier in a basement room close to bedrock foundation, and carefully protected against mechanical and thermal disturbances. The best times for ob-

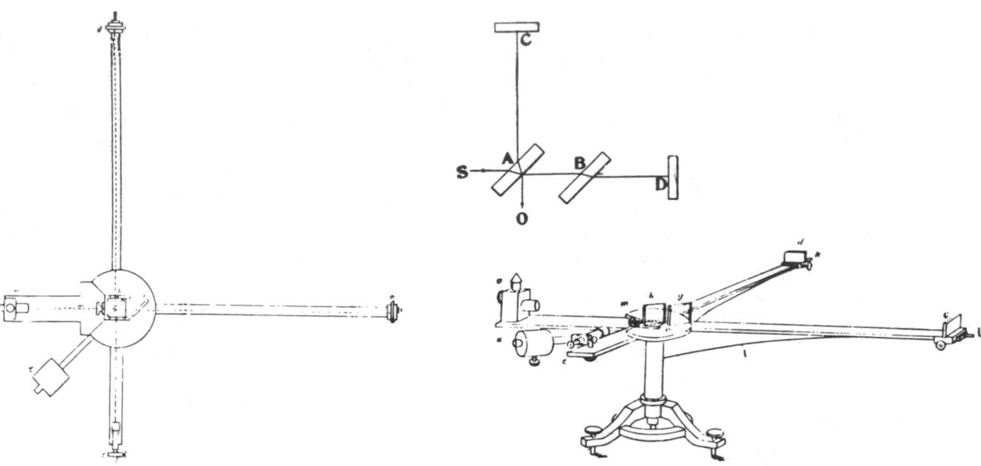

FIG. 1. These drawings describe Michelson's original interferometer as designed in 1881 and made in Germany. The brass apparatus with two arms, each a meter long, was supposed to be able to detect a relative "aether wind" that could be analyzed to reveal the resultant velocity of the earth's many motions through space. (A. A. Michelson, *Am. J. Sci.* 3d ser., **22,** 252–253 (1881), and Univ. of Chicago Press, 1902.)

serving were recalculated from astronomical theory, but the sun's proper motion was ignored. The optical path was greatly increased and 16 cardinal azimuths were to be marked.

Finally from 8 through 12 July 1887 Michelson and Morley took the observations that were in time to become known as the null results of the ether-drift experiment. Only 36 turns of the interferometer, covering a total of 6 hours duration over a 5-day period, were used for the data of record. Having expected a significant set of visible fringe shifts, Michelson and Morley were severely disappointed to find after data reduction at most no more than one twentieth of the predicted shift. They interpreted these results to mean that "the relative velocity of the earth and the ether is probably less than one-sixth of the earth's orbital velocity, and certainly less than one-fourth." Thus the received theories of astronomical aberration were called even further into serious question.

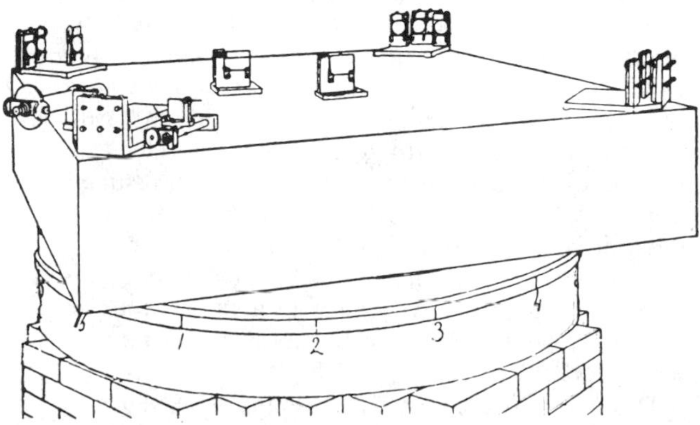

FIG. 2. This is the prime illustration of the classic "aether-drift" apparatus developed by Michelson and Morley in 1887 and tested in Cleveland, Ohio. The optical bench atop the massive stone slab floating on mercury could achieve a path length about ten times its former value, but as before, results were null. (A. A. Michelson and E. W. Morley, *Am. J. Sci.* 3d ser., **34,** 277 (1887).)

Although Michelson and Morley were confused and disappointed over these results, they were elated over the sensitivity of the instrument. They abandoned all further ether-drift tests in favor of a series of experiments to establish the feasibility of using wavelengths of light as actual and practical standards of length. They failed to perform the promised seasonal tests, and soon they parted, never to resume their collaboration.

In response to the Michelson–Morley experiment G. F. FitzGerald and H. A. Lorentz independently suggested the possibility of material contractions of "rigid rods" in motion through the ether. As the electromagnetic world view captured the imagination of many physicists toward the end of the 1890s, the theoretical speculations of Lorentz, Henri Poincaré, Ernst Mach, Max Planck, J. J. Thomson, and Joseph Larmor gained wide currency. Many references to the Michelson–Morley experiment already were raising its reputation higher than its actual character. Michelson devised a large vertical interferometer outside his laboratory building at the University of Chicago in 1895, again with no significant results.

Meanwhile back in Cleveland, Morley and Dayton C. Miller, Michelson's successor, began a series of revised interferometric tests in 1902 that they hoped would clarify the theory and their own assumptions as well as solve the relative motion problem. But these tests of different materials for contractions as measured by interferometry were also inconclusive. In addition to other second-order tests, which also failed to measure the earth's motion relative to the supposed ether, the Morley–Miller experiments merely reinforced the reputation of the Michelson–Morley experiments.

In Berne in 1905 a young patent examiner and physicist was composing a set of papers that would eventually have profound influences on the world of physical scientists. Albert Einstein's third paper of that miraculous year, "On the Electrodynamics of Moving Bodies," carried a critique of the idea of simultaneity and an enhanced appreciation for the postulates of the velocity of light and relativity. This paper declared the luminiferous ether to be superfluous. By inference it banished the Newtonian ideas of absolute time and space because these concepts prove to be operationally meaningless.

After the development of Einsten's general theory of relativity in 1915, the earlier work of 1905 became known as the special or restricted theory of relativity. Although Einstein did not refer specifically to Michelson's work in 1905, he did defer to those leaders of the profession who, like Lorentz and Poincaré, were almost reverential toward the Michelson–Morley test. Thus by implication in the 1920s the Michelson–Morley experiment became an embarrassment to its authors as it became a primary pedagogical tool for explaining Einstein's theories of relativity.

D. C. Miller recognized the need to return to the original Michelson experiment and to finish it for all seasons of the year and at a high altitude with many new precautions against systematic errors. With Michelson's encouragement at Mount Wilson in southern California, Miller worked sporadically from 1921 to 1926 to perfect the Michelson–Morley experiment. Even before finally finishing in 1926, Miller announced that he had found the "absolute motion" of the earth: 200 km/s toward the head of Draco!

This challenge aroused Michelson himself to respond with new and different tests and with vastly improved optical equipment. Many other experimentalists also joined the fray from 1926 to 1930, but no one else was able, it seems, to corroborate Miller's findings. Therefore, Michelson's original interpretation of his null results for ether-drift experiments were reconfirmed several times over, and the data obtained by Miller were generally considered anomalous. Reanalyses as late as 1955 by R. S. Shankland and others found temperature fluctuations after all to have been the cause of these spurious data. Yet contemporary cosmology and cosmogony continue to generate interest in much more sophisticated versions of the Michelson–Morley experiment.

See also RELATIVITY, SPECIAL THEORY.

BIBLIOGRAPHY

Stanley Goldberg and Roger H. Stuewer (eds.), *The Michelson Era in American Science, 1870–1930*. American Institute of Physics, New York, 1988. See especially Parts II and III.

Gerald Holton, "Einstein, Michelson, and the 'Crucial Experiment,'" *ISIS* **60**, 133–197 (1969). (I)

Dorothy Michelson Livingston, *The Master of Light: A Biography of Albert A. Michelson*. Scribner's, New York, 1973. (E)

Dayton C. Miller, "The Ether-Drift Experiment and the Determination of the Absolute Motion of the Earth," *Rev. Mod. Phys.* **5**, 203–234 (1933). (A)

Special Issue: Michelson–Morley Centennial, *Phys. Today* **40** (No. 5), 9–69 (1987); with articles by A. I. Miller, L. S. Swenson, J. D. Jackson, J. Stachel, A. E. Moyer, M. P. Haugan, and C. M. Will.

Loyd S. Swenson, Jr., *The Ethereal Aether: A History of the Michelson–Morley–Miller Aether-Drift Experiments, 1880–1930*. Univ. of Texas Press, Austin, 1972. (I)

Microscopy, Optical

Gordon S. Kino

The compound optical microscope is said to have been invented by a Dutch spectacle maker, Zacharias Janssen, around 1590. Galileo's microscope came a little later in 1610. Despite its long history, the optical microscope is still being developed in response to the need to observe submicron features in semiconductors and other materials. Further developments of the optical microscope also continue to be made for observing internal features of transparent biological samples and for obtaining evidence of their biological activity.

MAGNIFICATION

The eye can distinguish between two points in space only if they subtend an angle greater than 1.5′ at the eye; this corresponds to two points approximately 0.1 mm apart at a distance of 25 cm. To observe microscopic objects perhaps only 0.5 μm apart (a micron is one-thousandth of a millimeter), with the eye focused at the normal reading distance of 25 cm, it is necessary to display an image magnified by a factor of 200 or more. The compound microscope fulfills this purpose by using two lenses, the objective and the eyepiece, to give two stages of magnification. The lens nearest to the object, the objective, gives a magnified image of the object at a plane a distance L from the objective. In turn, this image is further magnified by the eyepiece. The objective typically has a magnification in the range of 5–150 and the eyepiece a magnification in the range of 5–20, so the total magnification may be as large as 3000 but as low as 25. Often, several objective lenses are mounted on a rotating turret so that an overall view of a sample can be taken at low magnification and then a high-magnification lens can be rotated into position to observe a feature of interest.

The length L from the first image plane to the second focus of the objective (approximately the position of the diaphragm at the back of the lens) is known as the tube length. The tube length is standardized by several manufacturers at 160 mm and is marked on the barrel of the lens, along with the numerical aperture (discussed later) and the magnification. Many modern objectives have an infinite tube length, i.e., the image of the object is focused at infinity; this makes standardization and interchangeability of lenses easier. An additional lens, the tube lens, is employed with these lenses to focus an image on the first image plane.

IMAGE QUALITY

The quality of the microscopic image is determined by the magnification, limitations caused by diffraction, distortion due to aberrations of the lenses, and contrast.

Diffraction

Geometrical optics predicts that a point in the object plane of a perfectly corrected optical system will be imaged as a point. It is apparent that an out of focus image where the rays have not converged to a point must be blurred. However, diffraction effects spread out even the ideal point image and create a pattern, called an Airy disk, consisting of a central core surrounded by concentric dark and bright rings. As the separation between two points on an object is reduced, their Airy disks begin to merge. Rayleigh's criterion is based on the idea that two neighboring points are just distinguishable if one point is located on the first dark fringe of the second point's Airy disk. On this basis, the minimum detectable separation d between two neighboring points is $d = 0.61\ \lambda/\mathrm{NA}$, where λ is the optical wavelength. The numerical aperture NA is a figure of merit for the objective lens, defined in terms of the sine of the half-angle θ subtended at the object by the entrance pupil of the lens, where $\mathrm{NA} = n \sin\theta$ and n is the refractive index of the medium between the objective and the object.

When the medium between the objective and the sample is air, the practical limit for NA is about 0.95; thus, at the center of the visible spectrum, where the wavelength $\lambda = 0.5\ \mu\mathrm{m}$, the transverse resolution is $d \cong 0.3\ \mu\mathrm{m}$. With oil of refractive index 1.52 as the operating medium, the maximum numerical aperture is raised to 1.4 and reduces the transverse resolution to $d \cong 0.2\ \mu\mathrm{m}$.

OPTICAL ABERRATIONS

Prior to the latter half of the nineteenth century, microscope designers lacked the knowledge to rid microscope optics of the degrading effects of lens aberrations. The two primary types of aberrations are *spherical aberration,* which causes the outer rays from the lens to converge to a point on a different plane from the inner rays (thus limiting the resolution), and *chromatic aberration* which causes rays of different colors to come to a focus at different points (thus showing color fringing). By the end of the century, design theory had reached a level where microscope optics with numerical apertures up to 1.4 could be designed to perform at nearly the theoretical diffraction limit—at least in the center of the field of view. By employing high-speed computers, along with modern glass technology and manufacturing methods, this near-perfect state of correction has been extended to the entire field of view and the entire visible optical range in the modern professional microscope.

IMAGE CONTRAST

Most biological specimens are prepared for examination as thin sections or smears and observed with a *transmission microscope* in which light is passed through the object. Opaque specimens, such as metallurgical samples, are typically observed with a *reflection microscope* in which the light is reflected from the object. Often specimens show very poor contrast between adjacent areas, even though the areas differ chemically or morphologically. Consequently, specimen recognition is often limited by lack of image contrast rather than by diffraction or insufficient magnification. Routinely, contrast-enhancement techniques, such as chemical staining, are used with dyes which fix themselves to different specimen constituents. Acid etching is used to make specimen details visible in opaque metallurgical specimens.

In biological applications, fluorescence techniques are now commonly employed in which light at one wavelength excites fluorescence at a longer wavelength. Since specific fluorescent dyes attach themselves to certain types of cells or parts of cells, fluorescence is a powerful technique for observing biological activity on a microscope scale.

Phase-contrast techniques, which were first introduced by Zernike, rely upon refractive index and thickness differences between specimen constituents, producing phase differences in the image. *Optical phase contrast microscopes* employ optical components designed to convert these phase differences into differences in image intensity. The interference, phase contrast, and differential interference contrast microscopes are the three most important microscopes designed for this purpose. Yet another type of microscope. the *polarizing microscope,* has important applications for the examination of crystalline and other materials with anisotropic material properties.

THE CONFOCAL SCANNING OPTICAL MICROSCOPE

The *confocal scanning optical microscope* (CSOM) was invented by Minsky in 1957 and a working version was demonstrated by Davidovidts and Egger in 1969. The CSOM offers the distinct advantage over the standard reflection microscope that out-of-focus features of an image disappear, i.e., the intensity drops off as the image is defocused. In the standard microscope, out-of-focus features blur but the intensity does not change as the image is defocused. This property allows structures with height differences comparable to a wavelength to be imaged independently; thus, quantitative measurements of height, profiles, or three-dimensional reconstructions can be made. An additional benefit of the CSOM is that the resulting images also tend to have sharper edges and better contrast than those obtained using a standard microscope. When examining biological materials, a confocal microscope can be used to optically cross section transparent specimens. One layer of a transparent biological material can be observed without having its features obscured by glare from layers in front of or behind the area of interest, or by reflections from the glass slide on which the sample is mounted.

The principle of operation of the CSOM is to illuminate the objective through a pinhole. An image of the pinhole in the form of a small spot is formed on the object. This spot, in turn, forms a reflected image on the original pinhole. If the object is in focus, the light passes through the pinhole to a detector. If the object is not in focus, the light reflected from it is defocused at the pinhole and very little light passes through the pinhole; thus the image disappears.

If a highly reflective sample, such as a mirror, is scanned axially through the focal plane, a profile of the range resolution can be observed directly by the detector. The range resolution of the microscope is given by the approximate formula for the distance d_z between points where the signal drops to half its maximum level at focus by the formula $d_z = 0.45\lambda/(1 - \cos\theta)$. Typically this distance is of the order of 0.5 μm.

Images may be formed by scanning the object back and forth in a raster scan and the variation of output intensity is displayed on a video screen. Alternatively, the optical beam may be scanned by moving mirrors. Typically, these scanning techniques form one frame of the image in a few seconds.

At about the same time that Davidovidts and Egger were building their microscope, Petran and Hadravsky, in Czechoslovakia, invented a different type of confocal microscope, the *tandem scanning microscope,* which operated in real time and produced an image which can be directly observed with the naked eye. Their idea was to use a Nipkow disk to fulfill the requirements of point illumination and point detection. The Nipkow disk, invented in Germany in 1884, was used in the first form of mechanically scanned television and was easily adapted to provide the raster scan in the confocal microscope. In the form used in the microscope, a Nipkow disk is an opaque disk into which many thousands of pinholes have been drilled or etched in a set of interleaved spiral patterns. Each illuminated pinhole on the disk is imaged by the objective to a diffraction-limited spot on the sample. The light reflected from the sample can be seen in the eyepiece after it has passed back through the Nipkow disk pinholes. Several thousand points are simultaneously illuminated on the disk achieving, in effect, several thousand confocal microscopes, all running in parallel. Spinning the disk fills in the spaces between the holes and creates a real-time confocal image which can be directly observed with the naked eye, as with the standard microscope.

Confocal microscopes of both types are rapidly being developed for use in biological applications and for the inspection and metrology of semiconductor materials. Most of the standard modes of microscopy, such as fluorescence and phase contrast imaging, can be used with the confocal microscope.

See also DIFFRACTION; OPTICS, GEOMETRICAL.

BIBLIOGRAPHY

Peter Gray (ed.), *Encyclopedia of Microscopy and Microtechnique*. Van Nostrand Reinhold, New York, 1973.

Rudolf Kingslake (ed.), *Applied Optics and Optical Engineering*, Vols. I–V. Academic Press, New York, 1965–1969.

Michael Spencer, *Fundamentals of Light Microscopy*, Cambridge University Press, Cambridge, 1982.

Tony Wilson and Colin Sheppard, *Theory and Practice of Scanning Optical Microscopy*. Academic Press, New York, 1984.

Microwave Spectroscopy

James R. Durig

Microwave spectroscopy is the detection and interpretation of the transitions between rotational energy levels of molecules which occur in the microwave and millimeter-wave regions of the electromagnetic spectrum. Every molecule is characterized by its own unique energy-level diagram and on the atomic scale this energy-level diagram is dependent on the electron distribution, atomic masses, and interparticle distances. When a source of electromagnetic radiation impinges upon a molecular sample, certain transitions between energy levels may occur, provided the source emits radiation of an energy which corresponds to the difference between two energy levels. The allowed transitions are determined by a set of selection rules which follow from the symmetry of the molecule and the corresponding properties of its wave functions. A molecule may undergo changes in the vibrational (oscillation of atoms about equilibrium positions without displacement of the center of mass), electronic, nuclear, and rotational (molecular movement about the center of mass as a fixed point while the atoms maintain their positions with respect to one another) energies, and each of these energies is characterized by appropriate quantum numbers. Pure rotational transitions occur between two levels in the same vibrational and electronic state, i.e., two levels having identical vibrational and electronic quantum numbers but different rotational quantum numbers. Since the rotational energies of a molecule are generally much smaller than vibrational energies, it is usually a good approximation to treat these two energies separately. Transitions between two vibrational levels occur in the infrared spectral region. Even larger energies are normally required to produce the electronic transitions characteristic of the visible and ultraviolet spectral regions. By averaging the relatively rapid vibrational and electronic motions, the molecule may be treated as a rigid rotor, i.e., a model which exhibits only rotational degrees of freedom with an average vibrational-electronic structure. Energy contributions which are usually much smaller than the rotational energies, such as magnetic dipole interactions, nuclear electric quadrupole interactions, and perturbations resulting from relatively small external electric and magnetic fields (which influence the rotational energy levels only slightly), can be neglected in the first approximation and these contributions to the total rotational energy can then later be evaluated by perturbation methods.

The assumption of the complete separation of the different energies of the molecule is equivalent to neglecting all interaction terms in the total energy Hamiltonian and results in the total wave function being the product of the contributions of each of the different energies:

$$\Psi^{(0)} \sim \Psi^{(0)}_{\text{rot}} \Psi^{(0)}_{\text{vib}} \Psi^{(0)}_{\text{etc.}},$$

where the corresponding zeroth-order Hamiltonian is

$$\sum_i H^{(0)}_i = H^{(0)}_{\text{rot}} + H^{(0)}_{\text{vib}} + H^{(0)}_{\text{etc.}} + \cdots.$$

In some cases, however, this approximate separation breaks down. It frequently occurs when the molecule has a vibrational energy spacing for a low-frequency vibration (i.e., an internal torsional mode or a ring-puckering vibration of a four- or five-membered ring) which approaches the rotational energy spacing. These near degeneracies can produce strong perturbations of the energy levels and mixing of the separated wave functions so that the energy can no longer be approximated by perturbation theory. In this case an explicit secular equation which involves all of the interacting energy levels must be solved.

In order for a molecule to exhibit pure rotational absorption spectra it must have a permanent dipole moment and have sufficient vapor pressure to give an observable microwave spectrum. There are a large number of molecules which meet these two criteria and have rotational energy transitions in the frequency range between 7.5 and 40.0 GHz, the region most frequently studied. In the vapor state at low pressures the effect of molecular collisions is quite small because the period of rotation is quite short in comparison with the interval between collisions. Therefore a molecule may execute several rotations before it is disturbed and possibly transferred to a different rotational energy level by the collision with another molecule. The pressure at which interaction becomes appreciable depends on the intermolecular forces which may vary from one molecule to another, but 0.01 Torr is a rough upper limit of the pressure where the effect of collisions on the rotational energy levels can be ignored. Rotations in solids or liquids are sometimes examined to obtain information about intermolecular forces but if one wishes to study rotational energy levels, one must limit the studies to molecules in the gas phase.

Except under the influence of collisions, a molecule can transfer from one rotational energy level to another only with emission or absorption of radiation. These transitions between rotational states comprise the pure rotational spectrum of the molecule. Transitions between the rotational states are restricted. It is found that the transition probability is zero for many combinations and in reaching a preliminary understanding of the spectrum it is clearly of greatest importance to know which transitions are forbidden ones. This information is obtained from the considerations of the wave functions for the rotations but it is beyond the scope of this article.

Assuming that the rotational energy is separable from the total energy, a molecule can be treated as a rigid rotator whose energy is a function of the moments of inertia of the molecule. The moment of inertia of a rigid molecule about any axis passing through the center of mass is defined by $I=\Sigma_i m_i r_i^2$, where r is the perpendicular distance from the axis and m_i is the mass of the ith nucleus. It is a theorem of mechanics that the locus of points formed by plotting $I^{-1/2}$ radially from the center of mass in the direction of the axis of rotation is the surface of a triaxial ellipsoid known as the momental ellipsoid. The three mutually perpendicular axes of this ellipsoid coincide with the three principal axes of the molecule around which the rotation is dynamically balanced. These axes of the ellipsoid are usually labeled as a, b, and c in order of decreasing length; therefore a is the major axis, b is the intermediate axis, and c is the minor axis of the ellipsoid. Since the axial lengths are proportional to $I^{-1/2}$, it follows that $I_a<I_b<I_c$. The generalized rotation of a molecule has components along the three perpendicular directions (i.e., three degrees of freedom) and therefore the total energy of rotation is $E_R=(P_a^2/2I_a)+(P_b^2/2I_b)+(P_c^2/2I_c)$, where P_a, P_b, and P_c are the components of the total angular momentum P along the principal axes a, b, and c. The total angular momentum is given by $P^2=P_a^2+P_b^2+P_c^2$. The quantization of the rotation restricts the rotational energy of the molecules to well-defined values allowed by wave mechanics. In the rigorous approach the eigenvalues of the operator which correspond to the classical Hamiltonian for the energy of rotation must be obtained. However, this treatment is beyond the scope of this article, but there are a number of postulates which can be used to change the classical equations to the quantum conditions. Before considering these quantum restrictions, it is useful to describe the simplification of the classical energy expression which results from the symmetry of the momental ellipsoid.

If the three principal moments of inertia are all equal, the rotational energy expression becomes $E_R=(P_a^2+P_b^2+P_c^2)/2I=P^2/2I$. Molecules in this class are known as spherical tops (e.g., methane, CH_4) and since they have no permanent dipole moments, they have no ordinary microwave spectrum. For linear molecules the moment of inertia about the internuclear axis is zero and $I_b=I_c$. Since I_a is zero, the component of the momentum P_a along the a axis is also zero so that the total angular moment reduces to $P^2=P_b^2+P_c^2$, and the classical rotational energy for linear molecules becomes $E_R=P^2/2I_b$. A necessary and sufficient condition for a molecule to be a symmetric top is that it has one axis of threefold or higher symmetry. Thus, symmetric tops have two moments of inertia which are equal but different from the third one. The top is described as prolate (e.g., chloromethane, CH_3Cl) if $I_a<I_b=I_c$ or oblate (e.g., trifluoroborane, BF_3) if $I_a=I_b<I_c$. For a prolate symmetric top the rotational energy expression becomes $E_R=(P^2/2I_b)+(P_a^2/2)([1/I_a]-[1/I_b])$, and if the top is oblate the subscript c replaces a in this expression. If all three moments of inertia of a molecule are unequal, then the molecule is classified as an asymmetric top and the original classical expression for the energy cannot be further simplified. The majority of molecules fall into this class. Thus molecules can be separated into four classes on the basis of their rotational motions, and the quantum restrictions are then applied to these four cases.

For the quantization of the total angular momentum, only values of $[J(J+1)]^{1/2}h/2\pi$, where $J=0,1,2,\ldots$, are permitted. Additionally, for symmetric-top molecules, the axial component of the total angular moment is restricted to values of $\pm Kh/2\pi$, where $K=0,1,2,\ldots,J$ with the restriction that K cannot exceed J since it is the axial component of J. With these restrictions the quantized energy levels for a prolate symmetric top become $E_R=[J(J+1)h^2/8\pi^2 I_b]+K^2\{1/I_a-1/I_b\}h^2/8\pi^2$. If both sides of the equation are divided by h, the energy expression becomes $(E_R/h)=J(J+1)B+K^2(A-B)$, where $A=h/8\pi^2 I_a$ and $B=h/8\pi^2 I_b$. A and B are known as the rotational constants and they are proportional to reciprocal moments of inertia. When the top is oblate the subscript c replaces a and the formula for the energy levels of an oblate top is then $E_R/h=J(J+1)B+K^2(C-B)$. From the convention that $I_a<I_b<I_c$, it follows that $A>B>C$ and the quantity $C-B$ must be negative. Although the axial component of the angular momentum of a symmetric top may be positive or negative, the energy depends on the magnitude but not the sign so that every energy level for which $K\neq 0$ is doubly degenerate.

For linear molecules, the classical expression with the introduction of the quantum restriction becomes $E_R/h=J(J+1)B$. This formula also applies to spherical-top molecules. In asymmetric-top molecules the K degeneracy of the symmetric top is lifted and there are $2J+1$ levels of different energy for each value of J. The quantization of the classical energy is too complicated to be treated here but the energy levels are designated as $J_{K_{-1}K_{+1}}$ where K_{-1} is the limiting K quantum number for the prolate top and K_{+1} is the limiting K quantum number for an oblate top. A thorough examination of the rotational transitions usually allows the assignment of the microwave spectrum, and the determination of the rotational constants (A, B, and C), from which the moments of inertia are calculated and then related to the molecular structure. In order to calculate meaningful structural parameters, several schemes have been devised to obtain the bond angles and distances from the microwave spectrum. These structural parameters and their relationships to the equilibrium molecular structure are dependent on the methods used in the structure determination and they differ from the parameters obtained by the electron diffraction technique. Nevertheless, the largest proportion of the microwave studies are for the purposes of obtaining structural information for molecules in the vapor state.

For the most part molecules are not rigid rotors. Centrifugal and Coriolis forces may act on the rotating molecule to alter its energy levels and its microwave spectrum. Low-frequency vibrations in the molecule such as the internal torsional vibrations associated with a methyl group or the ring-puckering vibrations of four- and five-membered rings can interact with the overall rotation of the molecule. The angular momentum of the overall rotation may couple with the internal angular momentum associated with these internal vibrations to perturb the rotational energy levels. Analysis of the energy levels then allows one to obtain information on these perturbations. The vibrational perturbations on the microwave spectrum can be utilized to determine the magnitudes and shapes of barriers to internal rotation and ring-puckering motions. The intensities of the microwave lines are indicative of the relative energy-level populations and, by assuming a Boltzmann population distribution, the intensities of the rotational lines resulting from excited vibrational states can provide the approximate frequency of the vibrational motions.

Microwave spectroscopy also provides a very accurate method for the determination of the molecular dipole moments. When an external electric field is applied to the molecule, the rotational energy levels are perturbed through an interaction called the Stark effect. An accurate measurement of the Stark lobes (splittings) usually provides a better value for the dipole moment than that obtained by any other physical method. If a molecule contains a nucleus with a spin greater than $h/4\pi$, e.g., nuclei with spin quantum numbers $=1,\frac{3}{2},2,\frac{5}{2},\ldots$, information on the electric field gradient along the axes through this nucleus may be obtained. Such nuclei possess a nuclear quadrupole moment which measures the local charge asymmetry at these nuclei. This charge asymmetry can couple the nuclear spin axis, which is also the axis of the quadrupole moment, to the molecular axis through the electric field gradient at the nucleus. This coupling splits the rotational energy levels and produces a spectrum with what is called "hyperfine" structure. Analysis of this hyperfine structure can then lead to estimates of double-bond character, the extent of hybridization, and the percentages of ionic and covalent bonding in molecules.

See also INVERSION AND INTERNAL ROTATION; MOLECULAR SPECTROSCOPY; ZEEMAN AND STARK EFFECTS.

BIBLIOGRAPHY

J. C. D. Brand and J. C. Speakman, *Molecular Structure: The Physical Approach*. Edward Arnold Ltd., London, 1960.

W. Gordy and R. L. Cook, *Microwave Molecular Spectra*. Wiley, New York, 1970.

W. A. Guillory, *Introduction to Molecular Structure and Spectroscopy*. Allyn and Bacon, Boston, 1977.

C. H. Townes and A. L. Schawlow, *Microwave Spectroscopy*. McGraw-Hill, New York, 1955.

J. E. Wollrab, *Rotational Spectra and Molecular Structure*. Academic Press, New York, 1967.

Microwaves and Microwave Circuitry

Fred J. Rosenbaum

INTRODUCTION

Microwaves are coherent electromagnetic waves whose frequencies lie in the range from about 500 MHz to, perhaps, 300 GHz. The range may be divided in terms of wavelengths as centimetric waves (3–30 GHz), millimetric waves (30–300 GHz) and submillimetric waves (beyond 300 GHz). At lower frequencies conventional circuit theory is adequate to describe the properties of electrical networks, while at infrared wavelengths, and above, geometrical or wave optics is appropriate. Microwaves span this gap and so have aspects of both conventional circuit theory and of wave phenomena. In fact, microwave techniques are needed when the physical dimensions of circuits or systems approach the wavelength of the highest-frequency components of the signals being processed.

Microwaves are propagated in free space in applications such as radar, and point-to-point and satellite communications. In order to generate and to detect and process these signals the microwave energy must be guided in microwave circuits to obtain the desired functions. The principal forms of waveguides in common use are (a) coaxial cables, (b) stripline or microstrip (microwave forms of printed circuits with metal conductor patterns supported on dielectric substrates), (3) closed (rectangular or cylindrical) metallic waveguides and (d) dielectric waveguides which localize or trap electromagnetic radiation in the vicinity of dielectric discontinuities, without the use of conductors.

PROPERTIES OF TRANSMISSION LINES

Wave propagation in any electromagnetic waveguide that has a uniform cross section perpendicular to the direction in which the energy travels can be described in terms of the transmission-line equations:

$$\frac{dV}{dz} = -ZI,$$

$$\frac{dI}{dz} = -YV,$$

where V and I are voltage and current variables associated with the transverse electric and magnetic fields of the particular waveguide. The field quantities are assumed to vary sinusoidally in time; z is the coordinate variable in the direction of propagation. The quantity Z represents the series impedance *per unit length* along the line, while Y is the shunt admittance *per unit length*. Their form and magnitude depend on the type of transmission line, its dimensions and cross-sectional shape, the nature of the material medium filling it, e.g., free space, air, etc., and the frequency and electromagnetic mode of operation.

Solutions to the transmission-line equations are

$$V(z,\omega) = V^+ e^{j\omega t - \gamma z} + V^- e^{j\omega t + \gamma z},$$

and

$$I(z,\omega) = Z_c^{-1}[V^+ e^{j\omega t - \gamma z} - V^- e^{j\omega t + \gamma z}]$$

where $j = \sqrt{-1}$; ω is the radian frequency; γ is the complex propagation constant, $\gamma = \sqrt{ZY} = \alpha + j\beta$; and Z_c is the characteristic impedance of the transmission line, $Z_c = \sqrt{Z/Y}$.

The phase of the wave traveling along the line is given by $\omega t \pm \beta z$. Since time is an increasing variable, to keep pace with a point of constant phase one must move along the line with a velocity $v_p = \omega/\beta$, called the phase velocity. Thus the $-$ signifies waves traveling in the increasing z direction, while the $+$ corresponds to a backward-going wave. The wave amplitudes are likewise identified: V^+ for the forward wave V^- for the backward wave.

If a generator is connected to the input to the transmission line, a forward wave V^+ is established. If the line is terminated in a complex impedance Z_L, then some of the signal incident on this load may be reflected toward the generator. The ratio of reflected to incident voltage amplitudes is known as the voltage reflection coefficient ρ,

$$\rho = \frac{V^-}{V^+} = \frac{Z_L - Z_c}{Z_L + Z_c}.$$

When the line is terminated in a load impedance identical with its characteristic impedance, there are no reflections and the line is said to be matched.

The reflected wave from a mismatched load interferes with the incident wave, periodically, giving rise to a stationary standing wave along the line, whose amplitude varies sinusoidally with time. The ratio of the voltage, at a measurement location ($z = -l$), and the current at the same point gives the input impedance to the line at that point, when the line is terminated in Z_L; viz.,

$$\frac{V(-l,\omega)}{I(-l,\omega)} = Z_{\text{in}} = Z_c \times \frac{Z_L + jZ_c \tan\beta l}{Z_c + jZ_L \tan\beta l}.$$

Here we have assumed that the transmission line has no losses and so $\gamma = j\beta$.

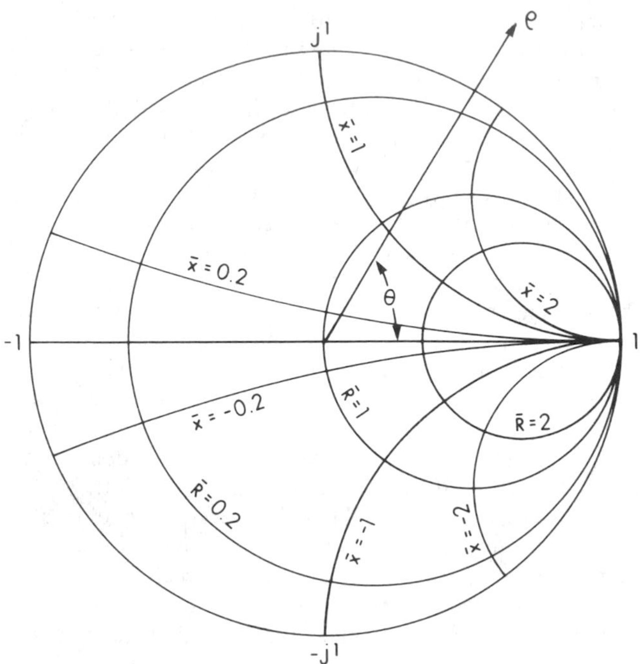

FIG. 1. Smith chart: Constant normalized resistance and reactance circles in the reflection coefficient plane.

The previous result illustrates the most distinctive feature of transmission lines, from a microwave-circuit viewpoint. The input impedance of a transmission line varies with length and with the frequency of operation,* i.e., $\beta = \omega/v_p$. This gives rise to the important property of impedance transformation. By choosing the appropriate load impedance Z_L, characteristic impedance Z_c, and line length l, arbitrary input impedances can be realized over narrow frequency bands.

The voltage reflection coefficient can be measured at any point along the line, not merely at the location of the termination. Thus

$$\rho(-l) = \frac{Z_{\text{in}}(-l) - Z_c}{Z_{\text{in}}(-l) + Z_c}.$$

Noting that all the quantities may be complex numbers, this is recognized as a bilinear transformation between the complex variables ρ and Z_{in}. Thus, the input impedance and the corresponding reflection coefficient can be mapped in the same space, giving rise to an invaluable design and display tool, the Smith chart (see Fig. 1). The orthogonal circular trajectories are lines of constant resistance (conductance) and reactance (susceptance), respectively, normalized to Z_c. Radial distance measured from the center gives $|\rho(-l)|$. Its angle is found by measurement from the positive abscissa. Rotation of a point of constant $|\rho|$ about the center gives the variation of input impedance that is obtained when either the length of line or the frequency is changed. The circumference of the chart is calibrated in fractional wavelengths.

* The phase velocity itself may be a function of frequency. In this case the line is said to be dispersive. Lines whose phase velocity is constant with frequency propagate electromagnetic waves whose field components lie entirely in planes transverse to the direction of propagation. These are the so-called transverse-electromagnetic waves or TEM waves.

MICROWAVE-CIRCUIT THEORY

Microwave-circuit design is the art of realizing networks of transmission-line elements for the precise control of microwave signal amplitude or phase over prescribed bandwidths. The problem has three parts: (a) the system problem—defining the performance desired of the network; (b) the network synthesis problem—determining electrical elements of prototype networks that approximate (to some order) the desired performance; and (c) the realization problem—finding equivalent microwave networks and determining their required physical dimensions. Many types of system components exist including impedance transformers, filters, power splitters and combiners, and phase or amplitude comparators.

The field of distributed networks is relatively young compared to lumped-network theory. It was stimulated by the need for microwave techniques prior to and during World War II, when radar was being developed. Subsequently, the field has matured and grown to the point where extremely sophisticated signal-processing networks can be implemented.

An important development was made by Richards in 1948, who showed that a simple frequency transformation could be employed to reduce the network functions of a distributed network to a ratio of polynomials, thus making the whole of lumped-element network theory applicable to the microwave problem.

The fundamental distributed network is a section of transmission line of length l (or electrical length $\beta l = 2\pi l/\lambda_g$, where λ_g is the guided wavelength). If there are no losses, then by replacing $j\tan \beta l$ with a complex frequency variable S, the input impedance to the line may be written

$$\frac{Z_{in}}{Z_c} = \frac{Z_L + SZ_c}{Z_c + SZ_L}.$$

If the line is terminated by a short circuit, then $Z_{in} = SZ_c$. Since this is analogous to the impedance of an inductor (SL), a short-circuited stub will be taken as a "distributed inductor," with $L = Z_c$. Similarly, for an open-circuited stub ($Z_L \to \infty$), $Z_{in} = Z_c/S$. This element is referred to as a "distributed capacitor" with $C = 1/Z_c$.

The Richards transformation, $S = j\tan \beta l$, is useful for the synthesis of distributed networks containing transmission lines of equal time delay (or equal length if they are constructed identically). Quarter-wavelength lines ($l = \lambda_g/4$) are commonly used. The response of a distributed network is readily found from the response of the "corresponding" lumped-element network. For example, consider the lumped-element low-pass filter and the corresponding distributed network shown in Fig. 2. When the frequency response of the lumped filter is compared with the behavior of the distributed one as a function of S/j, both exhibit a low-pass characteristic. When examined in the frequency domain, however, the response of the distributed network is periodic because of the periodicity of the tangent function. It will have interlaced pass bands and stop bands rather than a simple low-pass response.

A further difficulty in realizing distributed networks results from limitations on the size and hence characteristic impedance levels of transmission lines that can be obtained

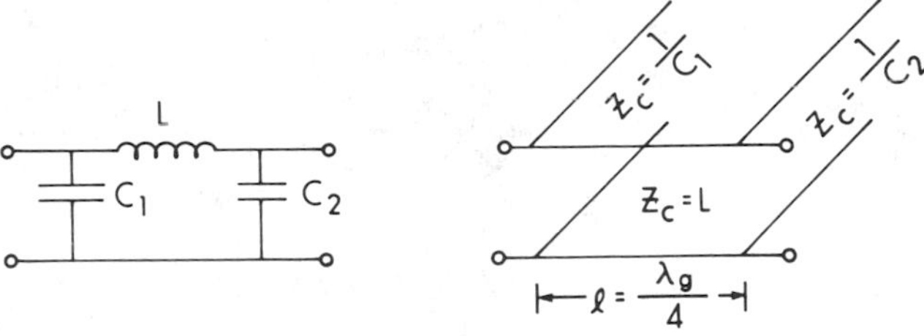

FIG. 2. Lumped-element low-pass filter and equivalent distributed filter. The parallel lines represent transmission lines.

in practice, with reasonable losses; effectively $200 > Z_c > 10$ Ω. Thus designs requiring elements with impedance ratios in excess of 20:1 are impractical. This limitation often prevents the direct realization of designs incorporating cascades of transmission-line elements. Another approach is to develop several realizations for the same network function. This is possible since the synthesis procedure is not unique. In 1952 K. Kuroda presented four basic equivalences of distributed networks which facilitate the design of many filter structures and other components.

COUPLED LINE ELEMENTS

Coupled transmission lines can be used to realize certain network functions. Electromagnetic energy can be transferred between two (or more) transmission lines provided that their phase velocities are nearly identical and that they are in sufficiently close proximity so that their electric and magnetic fields can be shared. Jones and Bolljahn present a circuit analysis of two coupled lines (a four-terminal network).

To use coupled lines as an element in a cascade network, where the output of one is the input to the next, it is necessary to convert the coupled pair into an equivalent two-port network. This can be achieved by terminating two of the ports by reactive elements, typically open or short circuits. By this method 10 different two-port networks can be realized. This approach allows the design of coupled-line (comb-line or interdigital) filters, which are widely used in practice.

If three of the ports of two coupled lines are terminated in matched load ($Z_L = Z_c$), then a signal entering the remaining port can be selectively coupled to two of the ports, and isolated from the other one. This device is known as a directional coupler. Its use permits the separation of the forward and backward-going signals described earlier. In systems, it allows a single channel to be used for two-way communication.

Such microwave four-port junctions can be realized with the coupled port providing a signal that is 90° out of phase with the input signal (quadrature coupler) or 180° out of phase with it (180° coupler). Combinations of these components with other two- and three-port junctions devices such as filters and power splitters permit the realization of complicated microwave-signal-processing networks.

MICROWAVE CONTROL DEVICES

In addition to the components previously discussed, microwave networks often contain other passive devices as

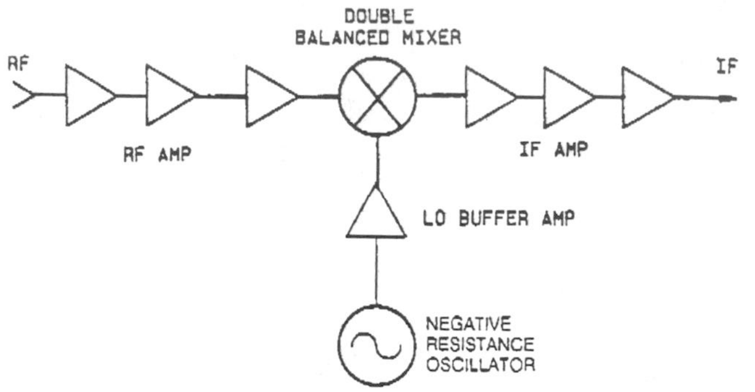

DOUBLE BALANCED MIXER
RF
RF AMP
IF AMP
IF
LO BUFFER AMP
NEGATIVE RESISTANCE OSCILLATOR

FIG. 3. Block diagram of 3–6-GHz monolithic convertor subsystem which includes three microwave amplifier stages, a transistor local oscillator and buffer amplifier, a double balanced mixer, and three stages of IF gain. (Courtesy, Pacific Monolithics Inc., Sunnyvale, CA.)

well as active ones. Among the most significant are nonreciprocal elements realized by incorporating magnetic materials such as ferrites with microwave-transmission-line elements. Devices such as load isolators (one-way transmission lines) and circulators can be realized. A circulator is a three-port device which couples a microwave signal incident at one port to only one of the two output ports. Energy incident at the next port is coupled only to the third one, and not to the first, and so on. These devices find application in systems where a transmitter and a receiver must share a common antenna, for example.

Variable-phase-shift networks and switches can be implemented using ferrites or *p-i-n* diodes properly located in microwave circuits.

MICROWAVE-TRANSMISSION MEDIA

Microwave networks can be implemented with a variety of transmission-line media. Metallic waveguides of rectangular or cylindrical cross sections are used where high-power levels must be carried, where circuit quality factor, Q, must be high, or where losses must be kept low. Metal obstacles, such as irises and tuning screws, are used to provide the reactance elements needed in filters. Coaxial transmission lines are used extensively in filter and coupler applications.

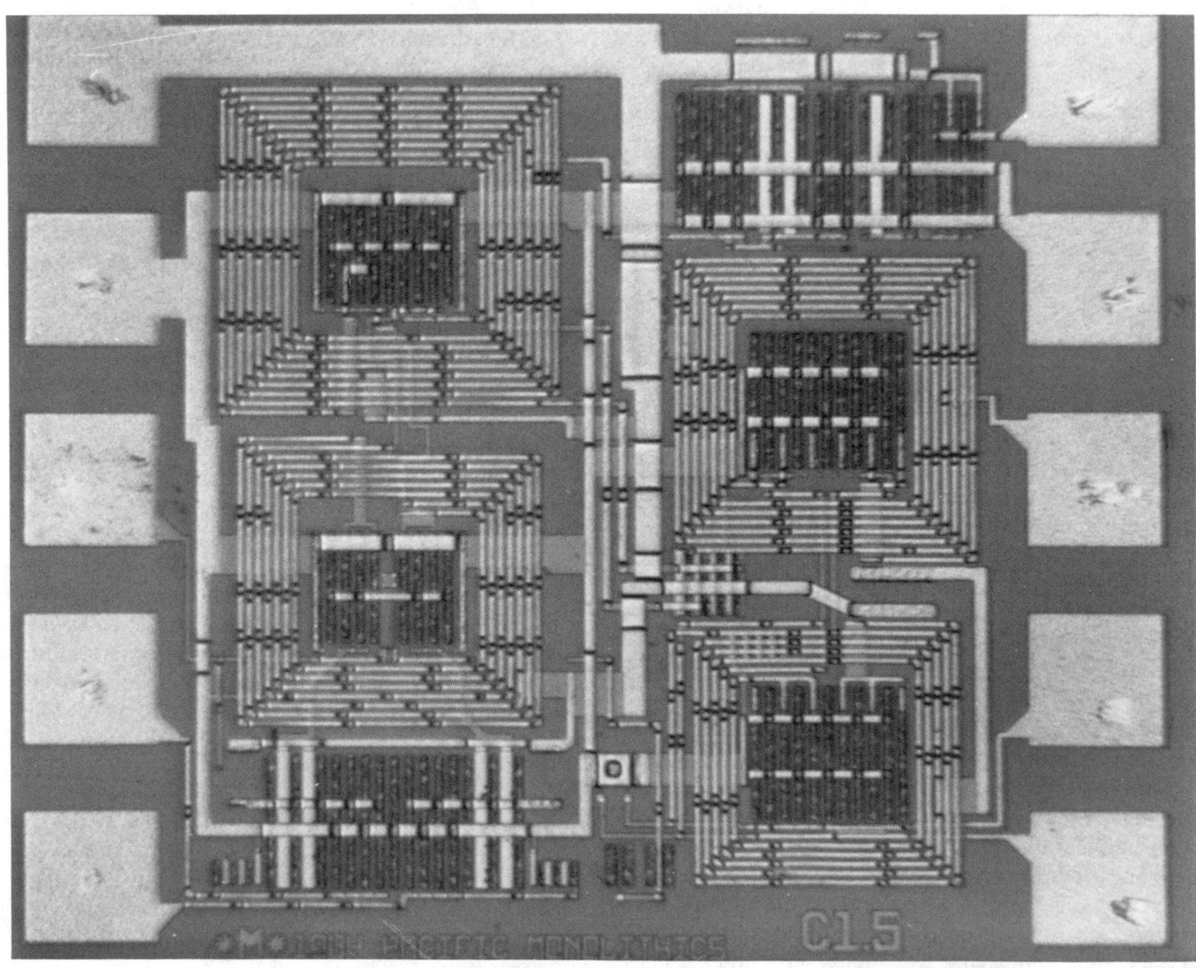

FIG. 4. GaAs MMIC microwave receiver whose block diagram is shown in Fig. 3. The chip's actual dimensions are 0.635 mm × 0.914 mm. The IF amplifier is shown in the upper right-hand corner. The rectangular spiral structures are inductors and transformers. They are constructed on tiny pillars which raise them above the surface of the GaAs in order to provide insulation and to reduce parasitic capacitance. (Courtesy, Pacific Monolithics Inc., Sunnyvale, CA.)

Their Q and power-handling capacity are somewhat lower than corresponding waveguide components.

Microwave printed circuits are realized using stripline or microstrip transmission lines. Stripline employs a conductor pattern etched or printed on one side of a thin dielectric slab. The other side is a metallic ground plane. A similar dielectric slab and ground plane are located over the conductor pattern to form a shielded sandwich-type construction. Microstrip is a medium which uses only a single dielectric slab with the conductor pattern on one side and the ground plane on the other. Although the circuit losses are higher in these transmission lines, complicated networks can be fabricated with them inexpensively and reproducibly.

ACTIVE MICROWAVE DEVICES AND MONOLITHIC MICROWAVE INTEGRATED CIRCUITS

In order to provide certain system functions it is often necessary to generate, amplify, detect, and translate signals to other frequencies. Microwave semiconductor devices have been developed for these purposes. Schottky-barrier diodes, employing metal–semiconductor junctions, are used as mixers and detectors, and as variable-capacitance diodes (varactors) for tuning oscillators. The transferred electron effect in GaAs and InP is employed in Gunn diodes, which convert dc power directly to microwave- or millimeter-wave signals. These devices are used in oscillators that can be operated from a few GHz to about 150 GHz with appropriate device design and circuit techniques. The phenomenon of avalanche breakdown is employed in IMPATT (Impact Ionization Avalanche Transit Time) diodes.

Microwave amplifiers are constructed from silicon bipolar junction transistors (BJJ) for use at the low frequencies (below about 6 GHz), whereas the GaAs field effect transistor (FET) provides excellent performance up to 40 GHz (and beyond). In this device, a narrow "gate" (0.25 μm) in the form of a Schottky-barrier diode is used to control the charge flowing between two Ohmic contact strips in a thin (0.25-μm) heavily doped channel of GaAs. The region under the gate can be fully or partially depleted of carriers by adjusting the bias voltage on the gate. A variation of this device, called HEMT (high electron mobility transistor) uses heterojunctions of GaAs and GaAlAs to introduce a two-dimensional electron gas into a nearly insulating channel of GaAs. The conduction electrons travel through the channel with a high saturated drift velocity without experiencing significant impurity scattering. This permits high-frequency operation with very low added noise.

High-frequency active components and multifunction subsystems are now being fabricated in GaAs using semiconductor batch processing techniques. This technology, called microwave monolithic integrated circuits (MMIC), incorporates FETs, diodes, and lumped and distributed circuit elements to provide complex circuit functions and complete subsystems like the microwave receiver whose block diagram is shown in Fig. 3. A photograph of the actual chip is presented in Fig. 4. The receiver converts microwave signals in the frequency range 3–6 GHz to an intermediate frequency range of 50 MHz–1.5 GHz. The most prominent features on the chip are rectangular spiral inductors surrounding the FET amplifiers. This chip has been widely used in satellite TV receivers. In MMIC technology device physics, materials science, microwave engineering, and systems design all come together on the surface of the GaAs chip.

See also KLYSTRONS AND TRAVELING-WAVE TUBES; NETWORK THEORY: ANALYSIS AND SYNTHESIS; RADAR; TRANSMISSION LINES AND ANTENNAS.

BIBLIOGRAPHY

S. F. Adam, *Microwave Theory and Applications*. Prentice-Hall, Englewood Cliffs, N.J., 1969. (E,I)

Advances in Microwaves. Academic Press, New York, 1966 (also supplements). (A)

R. E. Collin, *Foundations for Microwave Engineering*. McGraw-Hill, New York, 1966. (I)

H. Howe, Jr., *Stripline Circuit Design*. Artech House, Dedham, Mass., 1974. (I)

E. M. T. Jones and J. T. Bolljahn, "Coupled Strip-Transmission Line Filters and Directional Couplers," *IRE Trans. Microwave Theory Tech*. **MTT-4,** 75–81 (1956). (I)

K. Kuroda, "Derivation Methods of Distributed Constant Filters from Lumped Constant Filters," text for lectures at Joint Meeting of Konsoi Branch of Institute of Elec. Commun. of Elec. and of Illumin. Engrs. of Japan, p. 32, October 1952 (in Japanese). (I)

K. Kurokawa, *An Introduction to the Theory of Microwave Circuits*. Academic Press, New York, 1969. (A)

B. Lax and K. J. Button, *Microwave Ferrites and Ferrimagnetics*. McGraw-Hill, New York, 1962. (A)

G. L. Matthaei, L. Young, and E. M. T. Jones, *Microwave Filters, Impedance-Matching Networks and Coupling Structures*. McGraw-Hill, New York, 1964. (I,A)

R. S. Pengelly, *Microwave Field-Effect Transistors—Theory, Design, and Application*. Wiley, New York, 1972.

R. A. Pucel (ed.), *Monolithic Microwave Integrated Circuits*. IEEE Press, Piscataway, NJ, 1985.

S. Ramo, J. R. Whinnery, and T. Van Duzer, *Fields and Waves in Communication Electronics*. Wiley, New York, 1965. (I)

P. I. Richards, "Resistor Transmission-Line Circuits," *Proc. IRE.*, **36,** 217–220 (1948). (I)

L. N. Ridenour (ed.), *M.I.T. Radiation Laboratory Series*, 28 vols. (see Vols. 8–11). McGraw-Hill, New York, 1948. (I,A)

P. H. Smith, *Electronic Applications of the Smith Chart*. McGraw-Hill, New York, 1969. (E,I)

Transactions on Microwave Theory and Techniques. Institute of Electrical and Electronics Engineers, New York, 1952. (A)

H. A. Watson (ed.), *Microwave Semiconductor Devices and Their Circuit Applications*. McGraw-Hill, New York, 1969. (I,A)

R. J. Wenzel, "Exact Design of TEM Microwave Networks Using Quarter-Wave Lines," *IEEE Trans. Microwave Theory Tech*. **MTT-12,** 94–111 (1964).

L. Young (ed.), *Microwave Filters Using Parallel Coupled Lines*, Artech House, Dedham, Mass., 1972. (I,A)

L. Young (ed.), *Parallel Coupled Lines and Directional Couplers*, Artech House, Dedham, Mass., 1972. (I,A)

Milky Way

Bart J. Bok

The luminous band of the Milky Way encircles the heavens in very nearly a great circle. It marks the midplane of our home galaxy, the Milky Way. Most of the stars and associated gas and cosmic dust are within a highly flattened disk centered on the midplane, with a diameter of the order of 30 000 parsecs (1 parsec $= 3.09 \times 10^{18}$ cm). The majority of the stars are within 500 parsecs of the midplane. The average distance from the midplane is larger for the older groups of stars. Our sun is an inconspicuous star located only about 10 parsecs from the midplane, but at a distance of 9000 parsecs from the center of our galaxy. Almost all of the young stars and star clusters (of Population I) are found within the disk. The total mass of the stars and gas in our galaxy is about 2×10^{11} solar masses. The flattened main body of the system is embedded in a thin spheroidal halo, which contains the older varieties of star clusters—the globular clusters—and stars of so-called Population II. The most conspicuous of the Population-II stars are variable stars of the RR Lyrae variety with periods of light variation of the order of 1 day and less. There is some evidence that our galactic halo contains a very massive and extended component made of nonluminous "dark matter" whose nature remains unknown. The radius of this dark component may be larger than 100 000 parsecs; its total mass may be 10^{12} solar masses, or greater.

It is generally thought that our galaxy was formed about 1.5×10^{10} years ago from a huge blob of slowly rotating gas. Initially, conditions must have been such as to favor formation of Population-II objects, notably globular star clusters. Because of a gradual dissipation of individual cloud velocities, especially through collisions between interstellar clouds, the gas and associated cosmic dust began to settle to the central plane of our galaxy, which is the sole place where formation of young stars (Population I) is now occurring. The youngest known stars of Population I, the massive, hot, blue-white O and B stars, have ages since formation of only a few million years. Our sun—an intermediate-age Disk Star—was formed about 5×10^9 years ago.

We recognize three major divisions in our galaxy, the flattened *disk region*, the *halo*, and the central *nuclear bulge*, which reaches as far as 4000 to 5000 parsecs from the galactic center. The nuclear bulge is perhaps bar shaped as opposed to spherical.

We cannot actually observe the central parts of our galaxy in the visible, ultraviolet, and near-infrared parts of the spectrum. The center itself is hidden from view by dense clouds of cosmic dust. However, the inner central parts can be studied to some extent in the far infrared and, without difficulty, at radio wavelengths. The direction toward the galactic center as viewed from earth lies in the constellation of Sagittarius and the most clearly marked central radio-star cloud is generally referred to as Sagittarius A. This central cloud is observable as well at infrared wavelengths between 2 and 100 μm. Observations of the galactic center itself indicate the presence there of a condensed mass, possibly a completely collapsed object known as a black hole, of 3×10^6 solar masses.

Our sun and the great majority of the stars in our vicinity move around the galactic center in nearly circular orbits at a rate of about 225 km s^{-1}. Our sun and its neighbors should complete one circuit around the galactic center in about 2.5×10^8 years. In other words, our sun should have made about 20 such galactic revolutions since the time of its birth as a star accompanied by a planetary system. A star with an age of $2–3 \times 10^6$ years since formation will have completed since its birth only one one-hundredth of a galactic revolution.

In the past 75 years or so Milky Way research has become one of the most active areas for investigation. We can now assign distances with errors not exceeding 20% to all observable objects in our galaxy. Crosswise motions, called *proper motions*, can be measured (in seconds of arc per year) for hundreds of thousands of stars, especially for the stars within 1000 parsecs of our sun. Line-of-sight velocities—*radial velocities*, measured in kilometers per second—can be obtained for all stars within reach of our spectrographs. For the gas clouds of our Milky Way system, radial velocities are obtainable by optical as well as by radio means. The kinematical properties of our galaxy are hence quite well known. The general rotation of our highly flattened galaxy has been studied in a variety of ways. The law of variation of circular velocity with distance from the galactic center is well established. The apparent failure of the circular velocity to decrease at large distances provides evidence for the existence of the extended dark component of the galactic halo.

Two current major areas of research on our galaxy are the study of its spiral structure and work on star formation.

It is now known that our galaxy resembles some of the spiral galaxies in the universe of galaxies in that it has a tightly wound system of spiral arms. Two major arms and a system of spiral spurs are known to exist in our galaxy. They are recognizable as spiral features in cosmic dust, in interstellar gas, and in very young, extreme Population-I stars. Close to the central galactic plane, the density of the gas associated with the spiral features is about five to eight times the density of the gas in the interarm medium. The dust appears to be mostly concentrated at the insides of the spiral arms. The density-wave theory—proposed in the early 1960s by C. C. Lin and F. H. Shu—explains the piling up of the gas and dust in and near the spiral arms. An accompanying shock wave assists in furthering additional concentration of gas and dust—presumably sufficiently strong to trigger star formation. Radio evidence for the presence of the shock wave comes from observations of synchrotron radiation observable at the insides of spiral arms.

In the past decade studies of star formation in our galaxy have come more and more to the fore. It is now recognized that star formation is actively taking place deep inside the large complexes of cosmic dust and associated gas. Most of the gas inside such complexes is molecular hydrogen, for which carbon monoxide (CO), observable at radio wavelengths near 2.6 mm, is an excellent tracer. Star formation is also recognized as taking place in the regions surrounding emission nebulae with their associated very hot blue-white O and B giant and supergiant stars. The process appears to be especially prominent at the interface between the emission region, rich in ultraviolet light, and a huge molecular cloud (H_2) often found adjacent to the gas cloud of the emission region. Cosmic dust seems to play a major role. Finally, we have evidence for protostar formation in some very peaceful, isolated, roundish, dark nebulae, which are most likely in gravitational collapse. Almost without exception, these *globules* are found to be centers for concentration of carbon monoxide, and hence of molecular hydrogen. The kinectic temperatures inside the globules are very low mostly between 5 and 12 k.

The triggering of the processes of star formation is probably provided in part by the spiral shock waves associated with the density waves that produce spiral structure. Another possible source of shock waves leading to protostar (or cluster!) formation is the shock waves generated as a by-product of supernova explosions.

It should always be borne in mind that our galaxy possesses two companion galaxies, the Large and the Small Magellanic Clouds, observable only from southern latitudes. Star formation is relatively more active in these galaxies than in our own. The relative absence in the Magellanic Clouds of cosmic dust and the presence of huge amounts of atomic neutral hydrogen (9% and 25% of total mass, respectively!) show that star formation is a far-flung process not necessarily limited to clouds of cosmic dust and molecular hydrogen.

See also ASTROPHYSICS; BLACK HOLES; GALAXIES; INTERSTELLAR MEDIUM; STELLAR ENERGY SOURCES AND EVOLUTION.

BIBLIOGRAPHY

Any one of the many general text books on astronomy gives an account of the structure and arrangement of our Milky Way system. Two of the Harvard Books on Astronomy, published by the Harvard University Press (*The Milky Way*, by B. J. and P. F. Bok, 5th ed. 1981, and *Atoms, Stars and Nebulae*, by L. H. Aller, Revised Edition, 1972), deal with the subject at hand. For further advanced reading, we urge the reader to turn, first, to the magazines *Sky and Telescope* or *Scientific American*, next to the *Annual Reviews of Astronomy and Astrophysics*, published since 1962 by Annual Reviews, Inc. of Palo Alto, California. Finally, the Symposium and Colloquium Volumes, published under the auspices of the International Astronomical Union, include several volumes that deal with Milky Way problems.

Molecular Spectroscopy

David R. Lide, Jr. and Alfons Weber

The field of molecular spectroscopy covers a broad range of phenomena associated with the interaction of electromagnetic radiation with molecules. Emission and absorption of radiation are the processes of primary concern although certain types of scattering (e.g., the Raman effect) are also included. The feature which distinguishes this field from other studies of optical effects in materials is the emphasis on wavelength or frequency dependence. By careful study of the variation of absorption and emission with wavelength, a considerable amount of information can be obtained on the microscopic structure of the substance involved. Thus, molecular spectroscopy has become one of the primary tools for the determination of molecular structure and, to a growing extent, the study of molecular dynamics.

The existence of discrete energy levels is basic to the science of spectroscopy. The Bohr condition,

$$\nu = c/\lambda = (E_2 - E_1)/h,$$

relates the frequency ν or wavelength λ of the emitted or absorbed quantum of radiation (photon) to the initial and final energy levels involved in the transition (here c is the speed of light and h is Planck's constant). Measurement of ν or λ for a number of spectroscopic transitions thus yields quantitative data on the manifold of energy levels which characterize the molecule.

The term "molecular" in this context is interpreted rather broadly. The objects of study may vary from virtually isolated molecules in low-pressure gases to crystals in which the surroundings of the absorbing or emitting center have a major influence on the spectrum. However, there is a general implication that the features of the spectrum can be correlated with the characteristics of individual molecules or segments of molecules (e.g., repeating units of polymer chains).

The primary basis for subdividing this very broad field is in terms of the wavelength of the radiation. The experimental techniques which are utilized lead to a natural division into the subareas of microwave, infrared, and visible-ultraviolet spectroscopy. The approximate ranges covered by each, and the molecular processes which contribute most strongly to the spectrum in each region, are given in Table I. The boundaries between these regions are not sharply defined, and there is a certain amount of overlap between the experimental techniques, as well as the molecular processes which contribute to the spectrum.

Molecular spectra are also classified according to whether they are observed in emission, in absorption, or in scattering processes (Raman effect). Emission spectra are studied by exciting molecules to higher energy levels through the use of electric discharges, high temperatures, or some other means of imparting energy, and then observing the radiation emitted as the molecules return to a lower energy state. In absorption spectroscopy the sample is exposed to radiation, and the absorption of this radiation by the molecules (raising them to higher energy states) is detected by measuring the fraction of radiation transmitted. In the Raman scattering process molecules are transferred in a two photon event from a low energy to a higher energy state (Stokes Raman scattering) or from a high energy to a low energy state (anti-Stokes Raman scattering). In principle, the same basic information can be obtained by either of these approaches, but experimental considerations determine the relative convenience of these types of experiments. In particular, emission spectra are difficult to observe at longer wavelengths (i.e.,

Table I Ranges and molecular processes for the various spectral regions

Spectral region	Wavelength range	Frequency range	Predominant molecular process
Microwave	10 cm–0.5 mm	3–600 GHz	Rotation
Infrared	500 μm–1 μm	0.6–300 THz	Vibration
Visible-ultraviolet	1000 nm–50 nm	0.3–6 PHz	Excitation of electrons

in the microwave and infrared region) while Raman spectra, originating as a result of a second order quantum process, are generally weaker than absorption or emission spectra.

HISTORY

Studies of the emission spectra of molecules in the visible and near-ultraviolet regions began in the latter half of the 19th century. These were often observed in conjunction with atomic spectra, but were differentiated by their closely spaced, regular series of features, in contrast to the sharp, isolated lines characteristic of atomic spectra. The term "band spectra" was attached to these patterns and is still in common use. The regularity of the patterns attracted much interest, and spectroscopists such as Deslandres established various empirical relations before the end of the 19th century. The development of the quantum theory made it possible to relate these band spectra to the structure of the molecules which were responsible.

Infrared absorption by gases and liquids became an active subject of research in the early part of the 20th century. The first studies were empirical in nature, but early workers (particularly W. W. Coblentz) recognized that the infrared absorption spectrum provided a unique fingerprint of each molecule and that certain gross structural features of the molecule could be correlated with particular features in the spectrum. The discovery of the Raman effect in 1928 provided a valuable supplementary technique to infrared spectroscopy. Finally, the microwave region became accessible to experimental research after 1945, so that the full spectral range of interest in molecular studies can now be covered by readily available instruments.

The invention of the laser in 1960 has had a profound influence on the practice of molecular spectroscopy, in many cases replacing many of the classical techniques. Moreover, as a result of the high degree of coherence of laser radiation, many new "nonlinear" and time-resolved spectroscopies have been developed. The uses of the laser in molecular spectroscopy are still being expanded.

EXPERIMENTAL TECHNIQUES

An investigation of a molecular spectrum requires measurement of both the intensity and the wavelength (or frequency) of the radiation involved. The intensity measurement is often done in relative terms although there are situations where absolute measurements are required. In essence, one needs a detector or sensor which responds to electromagnetic radiation in the wavelength range of interest and whose response function is at least roughly known. The type of detector depends, of course, on the spectral region. The photographic plate has been the traditional detector in the visible and ultraviolet regions, although photomultipliers, solid state detectors, and charge coupled devices (CCDs) are now in common use. Simple thermocouple sensors may be used in the infrared, but photoconductive devices (semiconductors whose electrical characteristics change sharply on exposure to infrared radiation) and liquid helium cooled bolometers offer many advantages. The photoacoustic detector, which uses a sensitive microphone to transduce the photon energy absorbed by a molecular gas into an acoustical pressure wave, is a very sensitive device free of some of the shortcomings of other detectors. Microwave radiation is most easily measured with rectifying diodes—devices that are conceptually similar to rectifiers employed for alternating currents at lower frequency. Modern research in solid-state physics has led to a large number of detectors in every spectral region, which have special characteristics desirable for various applications.

Measurement of wavelength requires either a monochromatic source of radiation (in absorption measurements) or a dispersive device which can separate the wavelengths in an emission spectrum (or the incident radiation used to observe an absorption spectrum). Prisms and diffraction gratings are the traditional dispersive elements used in the ultraviolet, visible, and infrared regions. Modern instrumentation usually employs gratings rather than prisms, in order to provide higher resolving power and to avoid the limitations of absorption by the prism material. Fourier-transform spectroscopy utilizes a different approach in which the wavelengths are effectively separated by measuring interference patterns. Here the primary measurement is the Fourier transform of the spectrum (i.e., light intensity as a function of displacement of the elements of the interferometer), and the spectrum itself is obtained by computer processing of these data.

In the microwave region monochromatic radiation sources are available in the form of klystrons, backward wave oscillators, and other types of vacuum tubes, as well as solid-state devices. These provide an output in a narrow band of frequencies, which can be made as sharp as desired, and the frequency can be varied smoothly. Thus, no dispersive element is needed to carry out absorption spectroscopy.

Continuously tunable laser sources, of both the pulsed and cw (continuous wave) variety, have greatly modified the experimental practice of molecular spectroscopy, in many cases replacing the traditional dispersive spectrograph or spectrometer. Commercially available, tunable cw diode lasers are now used throughout the infrared to wavelengths as long as 30 μm. More specialized continuously tunable lasers for the infrared, operating over narrower wavelength ranges, are the color center laser and the difference frequency laser. The CO_2 and CO gas lasers emit radiations corresponding to individual infrared rotation-vibration lines. These lasers are therefore only discretely, or line, tunable. With microwave side band modulation impressed on these radiations a modest tunability is available spanning approximately $\frac{2}{3}$ of the gap between neighboring rotation-vibration lines. The CW dye laser is, with appropriate dyes, capable of spanning the visible and near infrared region. Tunable lasers for the ultraviolet result from the mixing of tunable visible or near infrared laser radiations in a nonlinear optical crystal to produce sum frequencies or harmonics. Since cw laser radiations have nearly always a very narrow bandwidth (less than 0.001 cm^{-1}), molecular absorption spectra are obtained by simply placing the substance directly into the path of the laser beam and tuning the laser over the absorption band. The radiation transmitted by the substance then falls onto an appropriate detector. Such a system, composed of a tunable laser and a detector, constitutes a basic laser spectrometer. The conventional dispersive spectrometer is then no longer needed except occasionally as a secondary optical device, e.g., as a band pass filter to isolate a single laser mode from the several that may be present in the laser output. Though the laser has often resulted in the abandonment of the traditional dispersive spectrograph, or spectrometer, these instruments are still the only ones suitable for spectroscopy in spectral regions for which tunable laser radiations do not yet exist or are under development, or when only continuum light sources are available, e.g., synchrotron radiation, or radiation emitted by hydrogen or deuterium lamps, etc. Figure 1 gives schematic descriptions of various experimental arrangements used in molecular spectroscopy.

Pulsed lasers generally yield higher powers than cw lasers. Their greatest advantage is their ability to produce very sharp pulses ranging in duration from several nanoseconds to picoseconds (1 ps $= 10^{-12}$ s) and even femtoseconds (1 fs $= 10^{-15}$ s). This characteristic permits investigations of intramolecular processes such as energy flow, isomerization, relaxation of excited states, molecular dissociation dynamics, and other phenomena by following the evolution of these processes in real time.

The impact made by the laser on the practice of molecular spectroscopy is due to its two principal characteristics, namely high power over a narrow spectral line width and the coherence of the radiation. The first of these by itself has played a large role in problems of conventional spectroscopy, such as the improvement of signal to noise ratio and the vastly enhanced spectral resolution in the recorded spectra. The resolution actually achieved is, however, most often not determined by the instrument, i.e., the laser spectrometer, but by the broadening of spectral lines beyond their natural width. Various causes are responsible for this broadening. The most important is the thermal motions of the molecules in a gas, giving rise to the Doppler broadening of spectral lines. The second property, the coherence of the laser radiation, has not only permitted the achievement of resolution beyond the Doppler limit, but is also responsible for the various nonlinear spectroscopies not feasible with incoherent sources.

BROADENING OF SPECTRAL LINES

Spectral lines possess an intrinsic width, the "natural" linewidth, which is determined by the spontaneous decay

Classical Spectroscopy

Laser Spectroscopy

(a) S AC D Sp

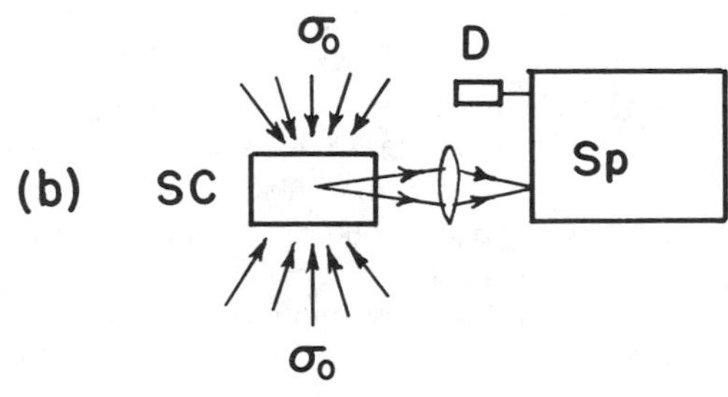

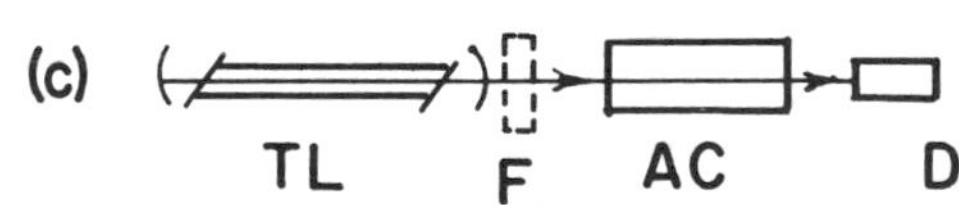

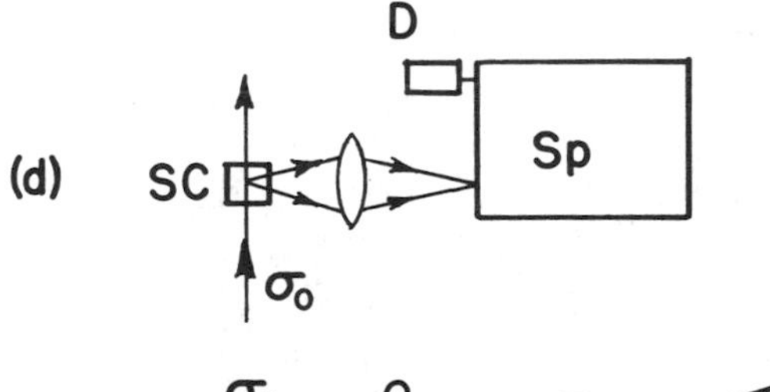

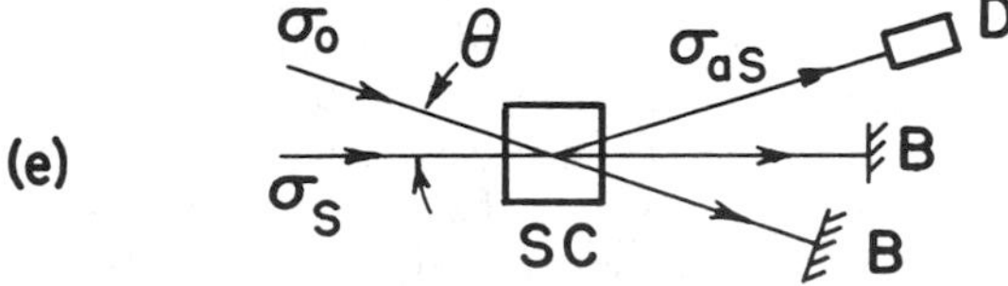

FIG. 1. Schematic arrangements for experiments in classical and laser spectroscopy. Sp = classical spectrometer (prism, grating, Fourier transform), D = detector, S = broadband continuum light source, AC = absorption cell, SC = scattering cell, TL = tunable laser, F = filter, B = beam stop. Classical spectroscopy: in part (a), removal of the absorption cell permits the study of the emission spectrum (discrete line spectrum or continuum) of source S; part (b) shows the arrangement used to record Raman or fluorescence spectra excited by incoherent monochromatic radiation of wavenumber σ_o. Laser spectroscopy: Arrangement (c) constitutes a laser absorption spectrometer; (d) laser radiation of fixed wavenumber σ_o is used to generate spontaneous, incoherent Raman scattering, of for laser induced fluorescence (LIF) spectroscopy; (e) concept of the CARS experiment, σ_o = fixed wavenumber of pump beam, σ_s = wavenumber probe beam tuned to the Stokes transition such that $|\sigma_o - \sigma_s| = \nu_i$ = molecular vibration frequency, σ_{as} = emitted coherent aanti-Stokes beam. After passing through the scattering cell the pump and probe beams are absorbed by the beam stops.

rate of the excited state. This line width is given by $\delta\nu_{nat} = (2\pi\tau)^{-1}$ where τ is the lifetime of the excited state. For rotation and rotation-vibration spectra of stable molecules $\delta\nu_{nat}$ is ~10 to 1000 Hz. Resolution at this level is rarely achieved in molecular spectroscopy nor is it needed except in very special cases. Various mechanisms are responsible for broadening the spectral lines beyond their natural width. These are molecular collisions (Lorentz broadening), collisions with the walls of the containing vessel, transit time broadening determined by the time taken by a molecule to travel through a light beam, power broadening resulting from the interaction with a high intensity laser beam which stimulates a high transition rate, Stark and Zeeman effects, and the Doppler broadening. Except for the broadening due to the Doppler effect, these broadenings, including the natural lifetime broadening, are classified as "homogeneous," while the Doppler broadening is said to be "inhomogeneous."

The translational motions of molecules give rise to absorption or emission frequencies which are displaced from their "rest frequencies" ν_0 by amounts $\Delta\nu$ which are determined by the individual molecular velocities. To lowest order (first-order Doppler effect) these shifts are given by $\Delta\nu = \nu_0(v_z/c)$ where v_z is the component of the molecular velocity along the direction of observation and c is the speed of light. For a molecular gas in thermodynamic equilibrium the RMS average of these displacements is determined by the Maxwell–Boltzmann distribution function. The width of an observed spectrum line, for emission or absorption of radiation, is therefore due to a set of narrow spectral lines, each of at least natural width, spread over a Gaussian intensity profile centered about the rest frequency ν_0. The Full width between half maximum (FWHM) intensity points of this line profile is given by the relation $\delta\nu_{Doppler}$ (FWHM) = $(2\nu_0/c)\,(2RT/M)^{1/2} = 7.16\times10^{-7}\,\nu_0(T/M)^{1/2}$, where R is the universal gas constant, T is the absolute temperature, and M is the molecular weight. (A more general relationship, which also includes an angular dependence, applies to scattering, i.e., Raman spectra.) In addition to line broadenings, there are also often line asymmetries and line shifts in observed spectra. Extensive efforts are therefore expended to minimize the effects of line broadenings, shifts, and asymmetries in order to improve on the accuracy of measurement and to discern detail that may be obscured due to the overlap of closely spaced spectrum lines. Thus, experiments are performed with low-pressure gases to minimize the collision broadening effects; absorption cells are usually of sufficient dimensions to render the molecule-wall collision effect negligibly small, etc. The Doppler broadening can be reduced by simply cooling the gas. This, however, can be done only within limits, these being determined by the condensation of the gas on the walls of the container. Special techniques are therefore employed to reduce the (first order) Doppler broadening further and in fact eliminate it in principle.

DOPPLER FREE SPECTROSCOPY

Spectroscopy free of Doppler broadening, also known as sub-Doppler spectroscopy, is achieved by several different techniques. For several of these it is essential that the molecular velocity v_z along the direction of observation be zero, or very nearly so. This can be accomplished geometrically by the use of collimated molecular beams, a method already in use for atomic spectroscopy in the 1930s. Free supersonic jet expansions of a gas into a region of high vacuum are also used. For absorption spectroscopy with a molecular beam, the laser beam crosses the molecular beam at right angles. Though this condition of orthogonality is not strictly met in the use of free supersonic jet expansions, a very substantial decrease in the Doppler linewidth is nevertheless achieved. The reduction in the Doppler linewidth is equivalent to a thermodynamic temperature of 1–10K, far below the condensation temperature of molecular gases.

The coherence of laser radiation is the basis for sub-Doppler spectroscopy, without relying on the geometrical effect of collimated molecular beams. Various techniques are available, among which saturation spectroscopy, two-photon absorption spectroscopy, and polarization spectroscopy are the most prominent ones. Moreover, the coherence property has given birth to various types of new "coherent" Raman spectroscopies, such as stimulated Raman spectroscopy, Raman gain spectroscopy, inverse Raman spectroscopy, CARS (Coherent Anti-Stokes Raman Scattering) spectroscopy, and others. Of these, the CARS technique has proven to be the most widely used, both for high-resolution spectroscopy of gases, as well as for the spectroscopy of liquids and solids (see Fig. 1d).

Because the experimental configurations for stimulated Raman gain and for CARS spectroscopy are "forward" scattering arrangements, or nearly so, the widths of the Raman or CARS lines so obtained are substantially less than those obtained with the conventional 90° scattering configuration (Fig. 1b), and are close to the Doppler widths of the corresponding infrared transitions.

Figure 2 shows an example of the improvement in resolution that has been achieved in the infrared with laser sources. The figure shows the infrared absorption spectrum of the ν_3 rotation-vibration band near 10.6 μm (948 cm^{-1}) of the SF_6 molecule. SF_6 is a spherical top-rotator molecule and, on account of its high symmetry, one would expect a very simple band structure consisting of *P*-, *Q*-, and *R*-branches. Instead, due to various internal interactions, the structure of the band is quite complex. The top spectrum, shown in part (A), was obtained with a special type of high resolution grating spectrometer ("Grille" spectrometer), with a resolution of 0.07 cm^{-1}, a gas pressure of 0.1 Torr, and room temperature. The band is unresolved, showing only a few strong *Q*-branches above the broad continuum. This unresolved continuum is due to a very narrowly spaced rotational fine structure of the ν_3 fundamental band, as well as the superposition of several unresolved "hot" bands on the fundamental. The width between the (ill-defined) half-intensity points is roughly 11 cm^{-1}, or 330 GHz. With the instrumental resolution of 0.07 cm^{-1}, this band remains unresolved even at a lower temperature of 150K. Part (B) shows a narrow 0.128 cm^{-1} = 3850. GHz wide portion in the *R*-branch of (A), recorded with a semiconductor diode laser at the same pressure (0.1 Torr) and room temperature as in (A). The laser linewidth, which is taken to be the measure of the

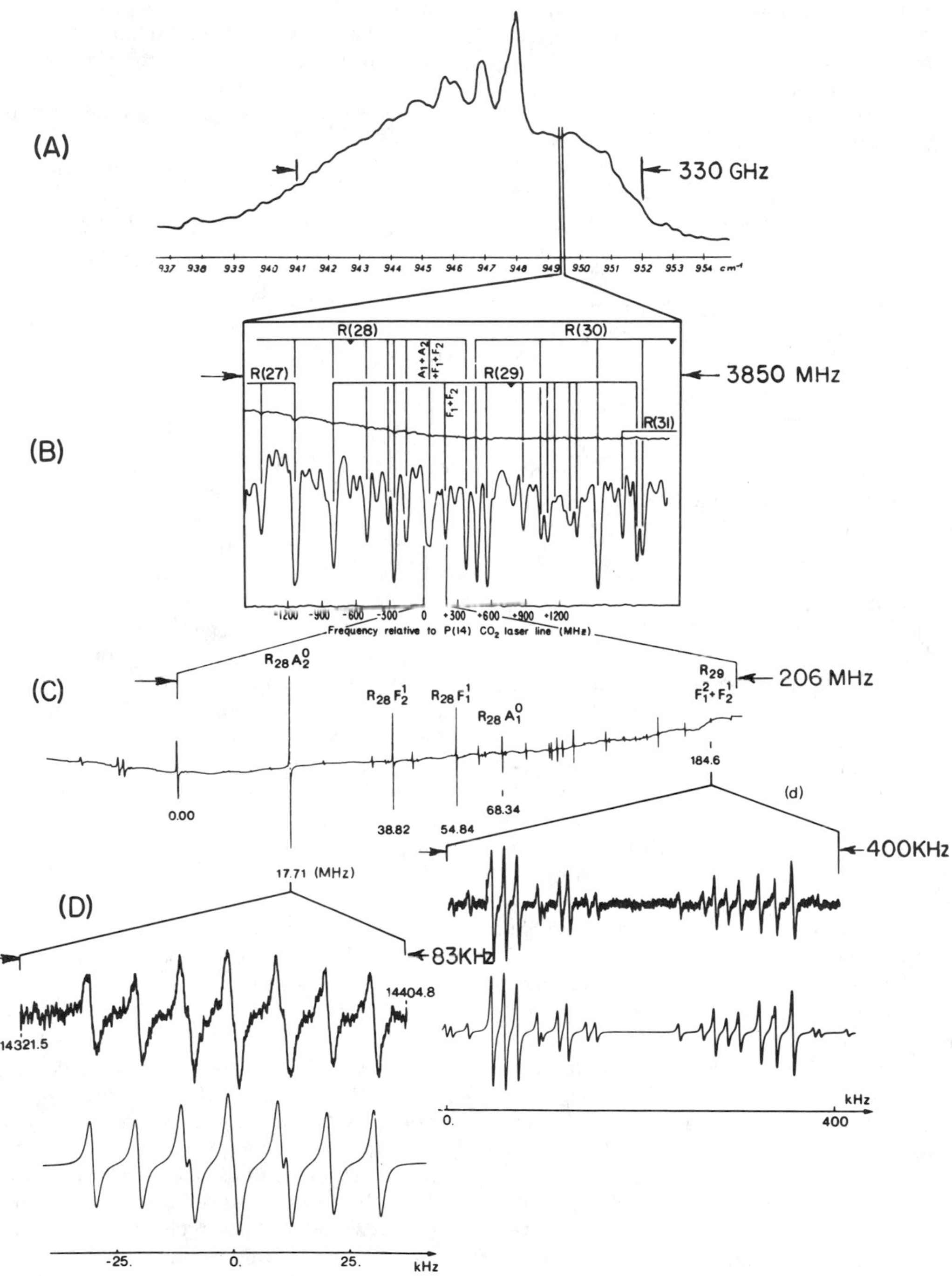

Fig. 2. The ν_3 rotation-vibration band of SF_6 at 10 μm (948 cm^{-1}) showing successive stages of improved resolution. (A) Spectrum recorded with a classical grating spectrometer (see Fig. 1a) capable of resolving 0.07 cm^{-1}. (B) Spectrum recorded with a tunable diode laser (see Fig. 1c) revealing the rotational fine structure; gas pressure (0.1 Torr) and temperature (room temp.) are the same as in (A). The experimental resolution of ~0.01 cm^{-1} is determined by the Doppler line broadening. (C) Saturated absorption spectrum at sub-Doppler resolution obtained with a free-running (unstabilized) CO_2 laser; the resolution is ~20–40 KHz. (D) Doppler free absorption spectrum obtained with a stabilized CO_2 laser; the resolution is ~1 KHz. The gas pressure in (C) and (D) was 10^{-5} Torr; the spectra are derivatives of the absorption line profiles. (Figure provided by Dr. C. Bordé).

instrumental resolution, is less than 10^{-4} cm^{-1} or 3 MHz. The actual resolution in (B) is, however, limited by Doppler broadening of the individual transitions to ~0.01 cm^{-1} (300 MHz). This level of resolution is also achieved by commercial high resolution Fourier transform spectrometers. Part (C) shows further resolution to ~20–40kHz achieved by sub-Doppler saturation spectroscopy, with a free-running CO_2 waveguide laser spectrometer tuned over a range of 520 MHz about the $P(14)$ CO_2 laser line, which is centered at 949.4793 cm^{-1}; the resolution is between 20 and 40 kHz. The R_{26} A_2^0 line, located at 17.71 MHz from the center of CO_2 laser line, and the unresolved R_{29} $(F_1^2+F_2^1)$ doublet at 184.6 MHz from the CO_2 laser line are further resolved into hyperfine structure components by means of a frequency-stabilized CO_2 waveguide laser. These structures are shown in part (D) (top = observed, bottom = calculated). The resolution is ~1 kHz (~3.3 10^{-8} cm^{-1}). (The symbols A_2^0, F_1^2, and F_2^1 are symmetry labels).

Instead of the usual absorption line profiles such as shown in (B), the spectra in (C) and (D) are the first order derivatives of the line profiles. (For both (C) and (D) the gas pressure was 10^{-5} Torr).

REACTIVE MOLECULES, IONS

Highly reactive species, e.g., free radicals and also ions, can be studied in the gas phase as well. An important aid for the identification and vibrational assignment as well as the determination of the geometry of these species is the method of matrix isolation spectroscopy. Reactive species are codeposited with an inert gas on a cryogenic substrate to form a dilute solution in an optically transparent solid medium, i.e., the matrix formed by the condensed inert gas such as argon, neon, nitrogen, and so on. The species may alternatively be produced *in situ* in the matrix by photochemical decomposition or ionization of codeposited stable molecules. Absorption, fluorescence, and Raman spectra of radicals and ions can be obtained by this method. Since molecular rotation is quenched in the solid, only electronic and vibrational transitions are observed.

There are several methods available to record the gas phase spectra of molecular ions. Since molecular ions constitute only a small part of a gas plasma discharge, the bulk being neutral molecular species, the spectrum of the ions is blended with the spectra of the neutrals. The identification and analysis of the spectra of ions is thereby made very difficult. An effective method of discriminating against the spectra of the neutral species is the technique of velocity modulation spectroscopy. In a plasma discharge, molecular ions are accelerated by the electric field and experience a drift that is superposed on the random thermal motions of the ions and neutrals. By alternating the electric field, the ion drift velocity also alternates its direction. Molecular absorptions from a laser beam, which is collinear with the direction of the electric field and is tuned to the ion rest frequency ν_0, will be shifted in and out of coincidence with ν_0 due to the Doppler effect. The absorption of the laser radiation by the ions is thereby modulated. Neutral species do not move in response to the electric field, and thus do not drift in and out of resonance with ν_0. The modulation of the transmitted laser power is thus produced by ion absorption only, the presence of the more numerous neutrals having no effect, even if the absorption of a neutral species happens to coincide exactly with ν_0. This technique has proven to be very effective and supplements other methods of spectroscopy of molecular ions.

DETERMINATION OF WAVELENGTHS, WAVENUMBERS, AND FREQUENCIES

The important quantity in molecular spectroscopy is the transition frequency ν between two quantum states whose energy difference $\Delta E = h\nu$. Direct measurements of this frequency are only feasible in the radio frequency, microwave, and submillimeter wave regions of the spectrum. The determination of the energy difference ΔE in the infrared, visible, and higher frequency regions is therefore based on the measurement of wavelength and use of the relation $c = \nu\lambda$ where c is the speed of light in vacuum. The reciprocal of the wavelength is the wavenumber, $\sigma = 1/\lambda$, so that $\Delta E = hc/\lambda = hc\sigma$. Since the product hc is a constant the wavenumber is equivalent to the energy difference. In a similar manner since the frequency $\nu = c\sigma$, the wavenumber is often loosely called the frequency. In practice the wavelengths or wavenumbers of spectrum lines in an unknown spectrum are determined by means of standard reference spectra for which the wavelengths or wavenumbers have been precisely determined by independent measurements based on the use of Fabry–Perot interferometers or Fourier transform spectrometers. These reference spectra are recorded either simultaneously or sequentially with the unknown spectrum. Unless the spectra are recorded with a vacuum spectrograph or spectrometer, the dispersion of air must be accounted for in the determination of the wavenumbers. Reference spectra are furnished by the emissions of hollow cathode lamps with Fe, Th, or U cathodes or rare gas discharge lamps filled with Ne or Ar gas, or by the pure rotational or rotational-vibrational infrared absorption spectra of simple molecular gases such as CO, HCN, CH_4, H_2O, and others. The visible molecular absorption spectrum of molecular iodine contains many closely spaced lines whose wavenumbers have been accurately determined by Fourier transform spectrometers. Direct measurement of unknown wavelengths is also accomplished by means of wavemeters, which in turn are calibrated by a single reference wavelength such as that of the He-Ne laser line at 632.8 nm or other possible visible or ultraviolet laser wavelengths. In all these methods the determination of wavelengths or wavenumbers is based on a length measurement, e.g., the separation of the plates in a Fabry–Perot interferometer or the displacement of a moving mirror in a Fourier transform interferometer or wavemeter. The spectroscopy resting on such determination may be termed "wavelength based spectroscopy." Frequency based spectroscopy has, until recently, been limited to the submillimeter wave, microwave, and lower-frequency regions of the spectrum. Direct frequency measurements of molecular absorption lines in the infrared are now feasible so that a direct

Table II Conversion Factors

Units	cm^{-1}	Joules/molecule	electron volts	Hz
1 cm^{-1}	1	$\{hc\}$ $1.9864475(12) \times 10^{-23}$	$\{hc/e\}$ $1.23984244(37) \times 10^{-4}$	$\{c\}$ $2.99792458 \times 10^{10}$
1 Joule/molecule	$\{1/hc\}$ $5.0341125(30) \times 10^{22}$	1	$\{1/e\}$ $6.2415064(19) \times 10^{11}$	$\{1/h\}$ $1.50918897(90) \times 10^{33}$
1 electron volt	$\{e/hc\}$ 8065.5410(24)	$\{e\}$ $1.60217733(49) \times 10^{-12}$	1	$\{e/h\}$ $2.41798836(72) \times 10^{14}$
1 Hz	$\{1/c\}$ $3.335640952 \times 10^{-11}$	$\{h\}$ $6.620755(40) \times 10^{-34}$	$\{h/e\}$ $4.1356692(12) \times 10^{-15}$	1

link of infrared reference spectra to the primary Cs frequency standard is now available.

DETAILS OF MOLECULAR SPECTRA

The energy of a molecule is conveniently divided into three additive parts representing (a) the motion of the electrons relative to the nuclear framework; (b) the internal vibrations of the nuclei; and (c) the rotation in space of the molecule as a whole. The magnitudes of these three energy components follow the order: electronic>vibrational>rotational. Such a division is only a first approximation, since each type of motion influences the others to some extent. However, it is very useful in describing the relation between molecular structure and the type of spectrum observed.

The quantitative description of molecular energy levels and spectra is most commonly given in terms of cm^{-1} units because of the historical practice of using the cgs unit of the reciprocal of the wavelength, $1/\lambda$, as the most meaningful measure of a spectral line. Thus, the energy of a molecule is written, to a first approximation, as

$$E = hc(T_e + G_v + F_J),$$

where T_e represents the electronic contribution to the energy, G_v is the vibrational contribution, and F_J is the rotational contribution. In this expression T_e, G_v, and F_J have units of cm^{-1}, and the quantity hc has the value 1.9864×10^{-23} J cm. Since the joule is an inconveniently large unit for describing energy on a molecular scale, it is customary to quote simply the value of T_e, G_v, or F_J and refer to this as an "energy" although the dimensions are reciprocal length. This practice has the additional important advantage that changes in the accepted values of c and h do not affect spectroscopic data reported in the literature or listed in tables. The quantities T_e, G_v, and F_J are called the electronic, vibrational, and rotational "terms" respectively.

Spectroscopists working in different regions, i.e., the microwave, infrared, visible, and ultraviolet, and so on, often use different units to represent the energy. Among these the electron volt and the hertz are the most commonly used other representations. The conversions between the various units employed is given in Table II. These are based on the 1986 adjusted values of the fundamental physical constants. The equivalence between the electron volt and the wavenumber (cm^{-1}), for example, is that 1 eV = 8065.5410 cm^{-1}, while that between the joule/molecule and the hertz is 1J/molecule = $1.50918897 \times 10^{33}$ Hz. There is no single system of units that is employed—mixtures of cgs, MKS, and SI units are often used, lengths of chemical bonds are given in Ångstrom units (1Å = 10^{-8} cm), Hooke's law stretching force constants are reported in millidynes/Å, and the calorie/mol often appears when the spectroscopy is related to chemical processes, etc.

ELECTRONIC SPECTRA

When two or more atoms combine to form a stable molecule, the discrete energy states of the separated atoms correlate, in an often complex manner, with a set of states of the molecule. Since the motion of the electrons is much more rapid than that of the nuclei, one can define an electronic energy for each configuration of the nuclei; this includes the kinetic energy of the moving electrons and the potential energy resulting from the various Coulomb attractions and repulsions. Thus, the electronic energy of each state can be visualized in terms of a potential surface which is a function of all the nuclear coordinates. In the simplest case of a diatomic molecule, the surface reduces to a curve representing electronic energy as a function of internuclear distance (see Fig. 3). When a given potential curve has a minimum, the state is referred to as stable, in the sense that the molecule can exist as an entity for a finite time. If there is no minimum, the state is unstable or repulsive.

Electronic states are characterized by the symmetry properties of the electronic wave functions. One important feature in a diatomic molecule is the component of the net angular momentum of all the electrons along the internuclear axis. States are labeled Σ, Π, Δ, . . . corresponding to values of 0, 1, 2, . . . , respectively (in units of $\hbar$) of this component. In polyatomic molecules, a similar classification can be made on the basis of the symmetry elements (axes, planes) which the nuclear framework possesses. The net spin S of the electrons determines the multiplicity of the states, which is indicated by a superscript $2S+1$ before the symmetry label. Thus, the ground state of the CO molecule is designated as $X\,{}^1\Sigma$, indicating a complete pairing of electron spins ($S=0$) and no net angular momentum along the internuclear axis. (X is the conventional label for the lowest-energy stable electronic state). The first excited state of CO is called a ${}^3\Pi$, and there are numerous states with higher energy, including the ${}^1\Pi$, ${}^3\Sigma$, ${}^3\Delta$, etc., states.

The pattern of electronic states is usually complicated, and the energy does not follow any simple relation (an exception

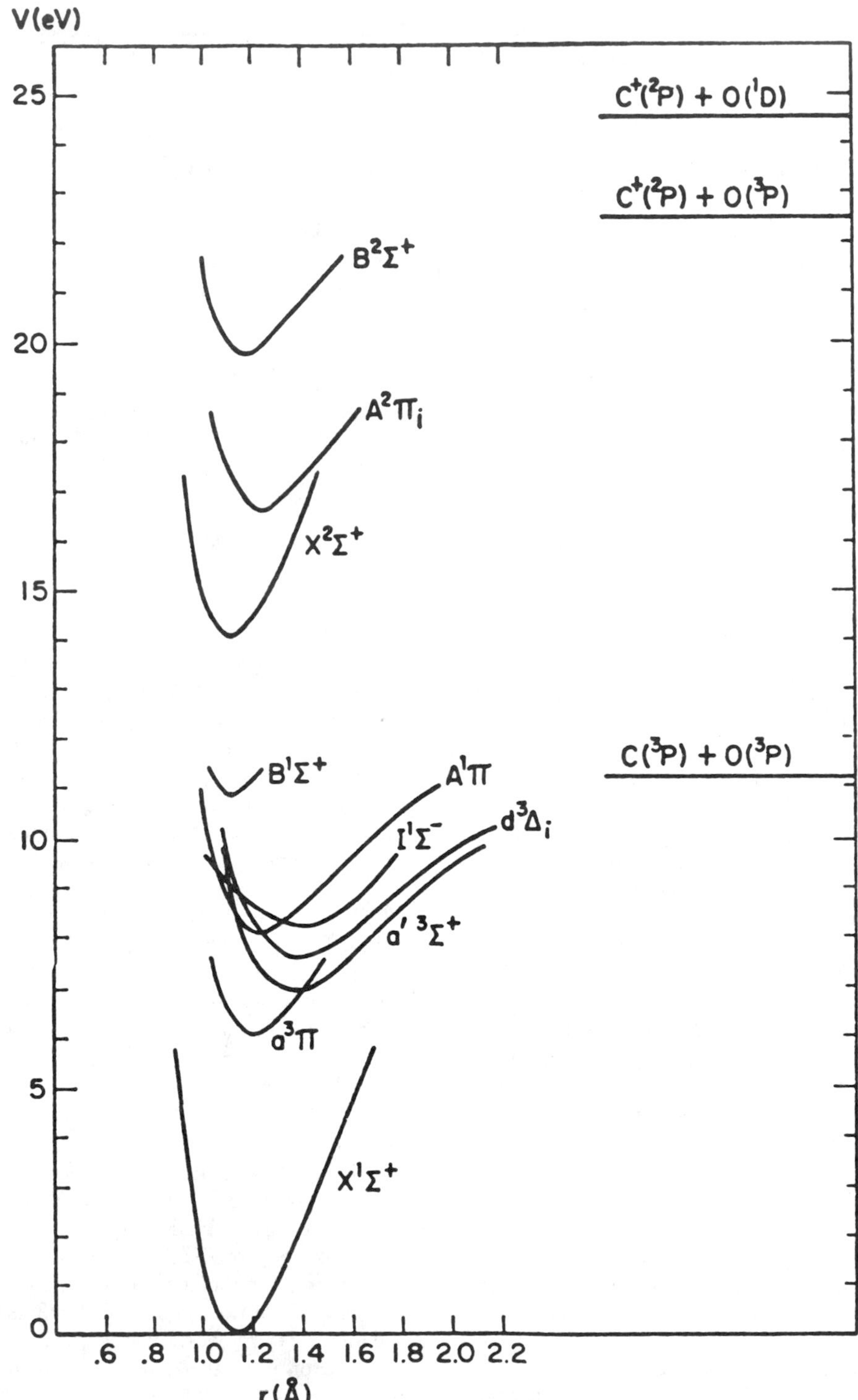

FIG. 3. Potential-energy curves for the carbon monoxide molecule. The asymptotes at the right are labeled by the states of the C and O atoms into which the CO molecule in a given electronic state will dissociate.

is Rydberg states, which result when one electron is so highly excited that it behaves rather like the electron in a hydrogenic atom). In many common molecules the first excited state is 5 eV or more above the ground state (but there are exceptions such as O_2 where states as low as 1 eV exist). Since an energy separation of 5 eV corresponds to a transition in the ultraviolet region, many molecules show no electronic absorption spectrum until one goes to fairly short wavelengths. However, the emission spectrum is normally much richer because excited states may lie closer together in energy. Some molecules show emission spectra throughout the ultraviolet, visible, and near-infrared regions.

Spectroscopic transitions between electronic states do not appear as single lines, but have a complex, banded structure. This is due to vibrations of the nuclei. The potential curves in Fig. 3 can be each considered as a potential well in which the nuclei oscillate. This oscillation is quantized, producing a discrete set of energy levels. A molecule in a given vibrational level of one electronic state can in general undergo a transition to any one of the vibrational levels of another electronic state. Thus, each electronic transition appears as a series of "bands" which may extend over a fairly broad region of the spectrum. An example is shown in Fig. 4. These bands are often roughly equally spaced since the vibrations are harmonic to a first approximation. The vibrational contribution to the energy is discussed in more detail below.

When a given band is observed at higher resolution, it is found to consist of many single lines, often forming a regular pattern. This results from changes in the rotational state, which will be discussed later.

Radiative transitions between electronic states are governed by certain selection rules. Symmetry considerations play an important part in determining which transitions will appear in a spectrum. The quantitative intensities are a function of the change of electric dipole moment in going from one state to the other. Within a given electronic transition, the relative intensities of the different bands are determined by so-called "Franck–Condon factors." These factors measure the overlap between vibrational wave functions in the initial and final electronic states, and thus specify the relative probabilities of different vibrational transitions.

The band system of the diatomic phosphorus molecule P_2 results from transitions between the vibrational and rotational energy levels of the ground electronic state $X\ {}^1\Sigma_g^+$ and the upper (excited) electronic state $C\ {}^1\Sigma_u^+$. The analysis of the complete band system (see below) provides the data from which the potential energy curves of these two electronic states are constructed. The minima of these curves define the equilibrium separation, r_e, between the phosphorus nuclei. The energy difference between the horizontal asymptote of $X\ {}^1\Sigma_g^+$ and the $v''=0$ vibrational level is the dissociation energy D_0 of P_2 in its ground electronic state. The energy difference between the vibrational level $v''=0$ and the minimum of the potential curve is the zero point energy of the oscillation. (The same definitions apply to the corresponding quantities of the upper state $C\ {}^1\Sigma_g^+$).

The potential curves constructed from the experimental data, the so-called RKR (Rydberg–Klein–Rees) curves, are said to be the "true" potentials of a diatomic molecule. Various analytic functions have been proposed to approximately represent the true potentials. Among these is the Morse potential $U(r-r_e)=D_e[1-e^{-\beta(r-r_e)}]^2$ where D_e is the dissociation energy referred to the potential minimum and r_e is the equilibrium separation of the two nuclei; the quantity β is a function of D_e, the reduced mass of the molecule, and the harmonic frequency of vibrations ω_e (see below). Though other, more elaborate functions give a closer fit to the true potentials, the Morse potential is widely used on account of its simple form.

FIG. 2. Electronic emission spectrum of the aluminum oxide molecule in the visible region. The coarse structure is the vibrational band structure, while the fine lines represent the rotational structure.

Electronic spectra observed in the liquid and solid state show less detail than gas-phase spectra. The rotational fine structure is lacking because collisions strongly perturb the free rotation that can occur in the gas phase. Vibrational band structure may still be observed, but the bands are broader, and overlapping of bands reduces the amount of information which can be obtained.

Electronic spectra of organic molecules provide valuable information on chemical bonding. Certain types of bonds tend to produce spectra in specific wavelength regions, corresponding to excitation of one of the electrons forming the bond to a higher energy state. Thus, double and triple bonds (C═C, C═O, C≡N, etc.) show characteristic electronic transitions in the near-ultraviolet and visible parts of the spectrum. Such groups are known as chromophores because they can lead to a characteristic color of the compound (where the transition is in the visible region). Since other nearby structural groups can perturb these chromophores, the electronic spectrum is a useful tool for studying these features of molecular structure.

VIBRATIONAL SPECTRA

As has already been mentioned, the surface describing the electronic energy of a stable state can be considered as a potential to which the nuclei are subjected. This leads to a set of vibrational energy levels for each electronic state. While the surface has no simple form in general, if the wells in this surface are deep enough, the region near the bottom can be described by a power series expansion. In the simplest case of a diatomic molecule, this is equivalent to approximating the potential curve as a parabola, with higher terms in the expansion added to improve the approximation when necessary. In this first approximation, then, the nuclei move as though they were subject to a Hooke's law force (i.e., restoring force proportional to displacement from equilibrium position) and undergo simple harmonic motion. The vibrational contribution to the energy of a diatomic molecule is, in this approximation.

$$G_v=\left(\frac{1}{2\pi c}\right)(k/\mu)^{1/2}(v+\tfrac{1}{2})=\omega_e(v+\tfrac{1}{2})$$

where k is the "force constant" of the bond, μ is the reduced mass, and the vibrational quantum number v takes values 0, 1, 2, . . . The constant ω_e is often referred to as the vibrational "frequency" although its units are cm^{-1}

The resulting vibrational levels are equally spaced with separation ω_e. Values of ω_e can be as high as 4400 cm^{-1} (for the H_2 molecule), but are much lower for bonds between two heavy atoms (e.g., 380 cm^{-1} in NaCl).

The pattern of vibrational levels is more complex in a polyatomic molecule. A molecule of N atoms has $3N-6$ vibrational degrees of freedom ($3N-5$ if its configuration is linear). Fortunately, the vibrational motion can be reduced to a set of $3N-6$ "normal modes" constructed by taking linear combinations of the individual atomic displacements. In a normal mode of oscillation all atoms vibrate in phase, with exactly the same frequency, and reach their maximum displacements from their equilibrium positions simultaneously. The normal coordinate Q_i associated with each of these modes describes a complex set of motions of the N atoms; in many cases, however, a given Q_i can, to a rough approximation, be identified with an easily visualized motion such as stretching of a single chemical bond or deformation (bending) of an angle formed by two adjacent bonds. In this way one can develop a physical interpretation of the vibrational energy levels which is of great value in analyzing vibrational spectra.

The energy levels of a polyatomic molecule are described in the harmonic approximation by

$$G^v = \sum_{i=1}^{3N-6} \omega_i(v_i + \tfrac{1}{2}).$$

The ω_i may range from 3000–4000 cm^{-1} (for bond stretching modes involving a hydrogen atom) to a few hundred cm^{-1} for deformation modes.

Symmetry considerations are of major importance in constructing the normal modes and carrying out quantitative calculations. Each molecule may be assigned to a point group on the basis of the symmetry operations (reflections, rotations, etc.) which leave the equilibrium configuration invariant. The normal modes are then classified according to the irreducible representations (species) of this point group.

Figure 5 shows the normal modes of the carbon dioxide and water molecules. In the equilibrium configuration the CO_2 molecule has axial symmetry about the line that joins the nuclei, 180° rotational symmetry about an axis passing through the C atom and perpendicular to the internuclear axis, reflection symmetry in a plane perpendicular to the line joining the nuclei and with the C atom in this plane, and reflection symmetry with respect to an infinite number of planes that contain the line joining the nuclei. The symmetry of this, and similar linear molecules, is specified by the symbol $D_{\infty h}$; such molecules are said to "belong" to the point group $D_{\infty h}$. There are four normal modes: Q_1 with an oscillation frequency $\nu_1 = 1388$ cm^{-1}, Q_{2a} and Q_{2b} with frequencies $\nu_{2a} = \nu_{2b} = \nu_2 = 677$ cm^{-1}, and Q_3 with $\nu_3 = 2349$ cm^{-1}. The symbols Σ_g^+, Π_u, and Σ_u^+ respectively are symmetry labels for the so-called vibrational species, i.e., the irreducible representations of the point group $D_{\infty h}$, used in classifying the normal modes according to their behavior under the various symmetry operations that are feasible in the point group $D_{\infty h}$. The two modes Q_{2a} and Q_{2b} have exactly the same vibrational frequency. Their distinction is that the oscillations are 90° out of phase; they occur in two orthogonal planes. These two modes therefore constitute a vibrationally "degenerate" pair; there is only one observable frequency for this pair, namely $\nu_2 = 677$ cm^{-1}. Vibrational degeneracies (double and triple) also occur in nonlinear molecules having high symmetry. (The ν_3-band of SF_6 shown in Fig. 2 is that of a triply degenerate vibration of species F_{1u}.) The three nuclei of the water molecule define a plane and all oscillatory

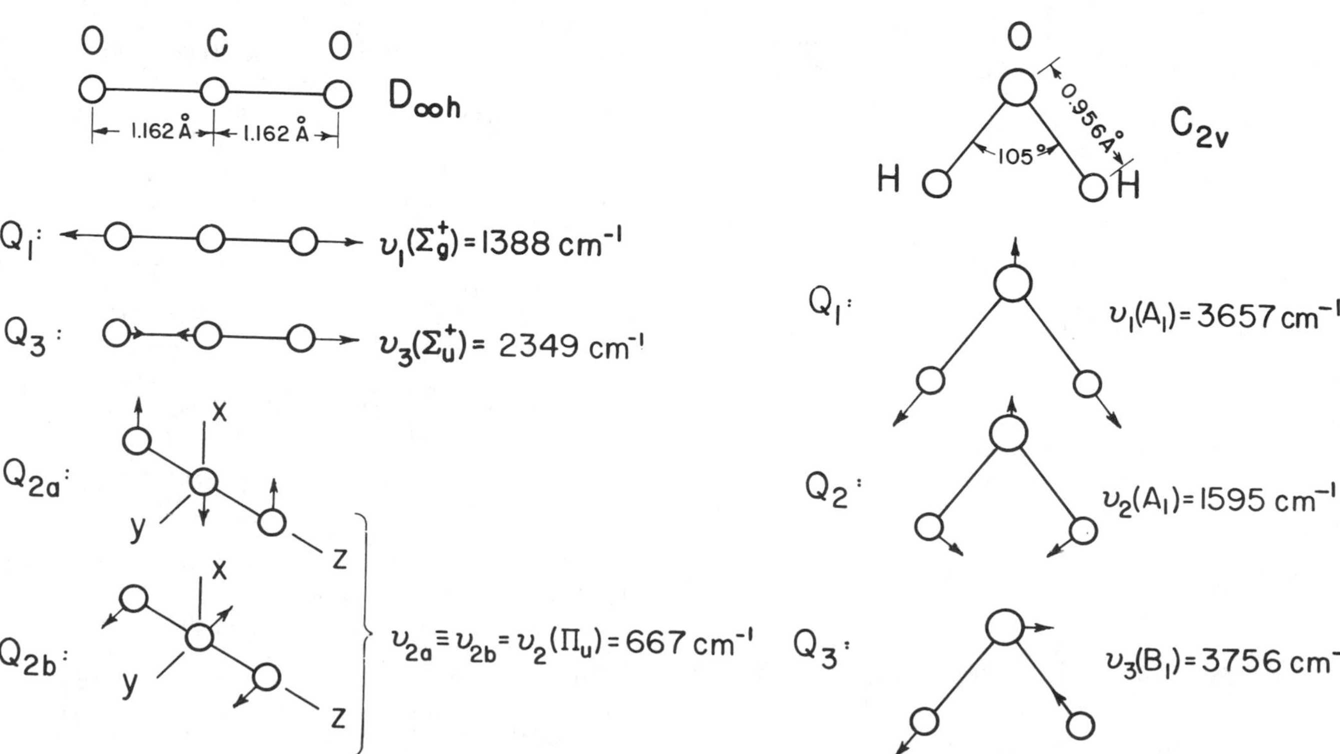

FIG. 5. Equilibrium configurations aand normal modes of vibration of the CO_2 and H_2O molecules together with their observed frequencies. The quantities in the parentheses are the vibrational symmetry species.

displacements of the nuclei about their equilibrium positions are confined to this plane. There are three normal modes, two being of the totally symmetric species A_1 and one of the antisymmetric species B_1; all three modes are nondegenerate.

This harmonic approximation of molecular vibrations can be improved by adding higher terms to the power-series expansion of the potential energy. In diatomics the energy expression then becomes

$$G_v = \omega_e(v+\tfrac{1}{2}) - \omega_e x_e(v+\tfrac{1}{2})^2 + \omega_e y_e(v+\tfrac{1}{2})^3 + \ldots .$$

The first anharmonic term $\omega_e x_e$ is typically of the order of 1% of ω_e, and the higher terms are correspondingly smaller. The $\omega_e x_e$ term is almost always positive, so that the levels become closer together as the vibrational quantum number v increases. A similar expansion of G^v may be carried out in polyatomic molecules.

In the harmonic approximation, radiative transitions can occur only between levels differing by unity in one of the quantum numbers v_i. As a further constraint, for absorption and emission spectra the electric dipole moment must change during the motion represented by the particular normal mode. Thus, a homonuclear diatomic molecule does not show a vibrational spectrum, and certain modes of many polyatomic molecules (especially those belonging to highly symmetric point groups) are "inactive," i.e., no transitions are possible. Thus, the Q_1 mode of CO_2 is "forbidden," or inactive in the infrared absorption or emission spectrum since there is no change in the electric dipole moment during this oscillation. The Q_1 mode is, however, "allowed" in the Raman spectrum. Therefore, in this simple approximation the vibrational spectrum of a polyatomic molecule consists of a small number of bands ($3N-6$ or $3N-5$ at most) corresponding to unit changes in each of the quantum numbers v_i. These are called the "fundamental frequencies," or simply "fundamentals" of the molecule.

Relaxation of the harmonic approximation allows further transitions to occur, and these make a significant contribution to most observed vibrational spectra. First, the anharmonic terms in the energy expression (e.g., $\omega_e x_e$) destroy the exact coincidence between level spacings, so that a transition between levels $v=1$ and $v=2$ will be slightly displaced from the transition between $v=0$ and $v=1$. In an absorption spectrum this leads to so-called hot bands, whose intensities are less than those of the corresponding fundamentals because of the lower population, at thermal equilibrium, of the excited levels. Anharmonicity of the vibrations also permits overtone and combination bands to appear. The former correspond to transitions in which a given v_i changes by more than one unit, while a combination band occurs when two or more normal modes are simultaneously excited (e.g., v_i and v_j both change by one or more units). Overtone and combination bands are generally weaker than fundamentals, but intensities can vary widely.

Typical vibrational spectra extend throughout the infrared region. The fundamentals usually fall in the 500–3500 cm^{-1} region (although larger molecules will have fundamentals below 500 cm^{-1}). Overtones have been observed as high as 20,000 cm^{-1} (500 nm wavelength).

The first objective in analyzing a vibrational spectrum, whether observed as a pure vibrational spectrum or a vibration-rotation spectrum in the infrared region, or as a band structure in an electronic spectrum, is to determine the fundamental frequencies and explain the remaining spectral features as overtones or combination bands. By use of a suitable model for the vibrational potential function, these fundamentals can then be related to basic structural parameters of the molecule. In the harmonic approximation, these parameters are the force constants describing bond stretching, angle bending, and various interactions among such motions. With further effort, some of the anharmonic terms in the potential energy can also be extracted from the special data.

The analysis of the spectrum is greatly facilitated by the fact that certain types of fundamentals tend to have about the same value, whatever the molecule involved. Thus, almost any molecule containing a C═O bond will have a normal mode whose major component involves the stretching of that bond. Such a mode usually gives rise to a fundamental in the neighborhood of 1700 cm^{-1}. (The CO_2 molecule has two C═O vibrations, the totally symmetric and the antisymmetric stretching modes with frequencies $\nu_1 = 1388\ cm^{-1}$ and $\nu_3 = 2349\ cm^{-1}$ respectively, whose mean value is 1868 cm^{-1}; see Fig. 5.) The recognition of these characteristic "group frequencies" is very important for interpreting infrared spectra and in deducing the structure of a molecule from its spectrum.

The concept of normal modes is extremely useful in describing the spectrum of fundamental vibrations of a polyatomic molecule. In this model the atoms execute infinitesimal displacements about their equilibrium positions in consequence of which the vibrations are harmonic. For overtone and combination vibrations, however, this model becomes less useful, especially for high levels of vibrational excitation. This fact is already discernible from the potential curves for the CO molecule shown in Fig. 3, where the RKR and also the Morse potentials deviate increasingly from the harmonic potential the higher the vibrational level. Indeed this anharmonicity is necessary for the dissociation of the molecule to occur. The dissociation of a polyatomic molecule is most often the result of the breaking of a single bond. Thus, the dissociation processes of, for example, $CO_2 \rightarrow CO + O$ and $H_2 \rightarrow OH + H$ imply that the vibrational energy not be shared by all atoms in the molecule, but that the energy of excitation is localized in that bond that is ultimately broken. The spectra involving highly excited vibrational states of a polyatomic molecule are consequently better described by a "local mode" model.

Infrared bands of gas-phase molecules show a further structure because of molecular rotation. When individual rotation lines can be resolved, they give additional information on the nature of the vibrational transition and the geometric structure of the molecule. Fig. 2B shows the rotational fine structure of the ν_3 infrared band of SF_6 (the further superfine and hyperfine structures shown in C and D result from various intramolecular interactions). Unless the molecular rotations are absent, as in a solid, or highly quenched, as in a liquid, pure vibrational transitions are observed only in special cases (e.g., the totally symmetric ν_1-band of SF_6 represents a pure vibrational transition). The fundamental in-

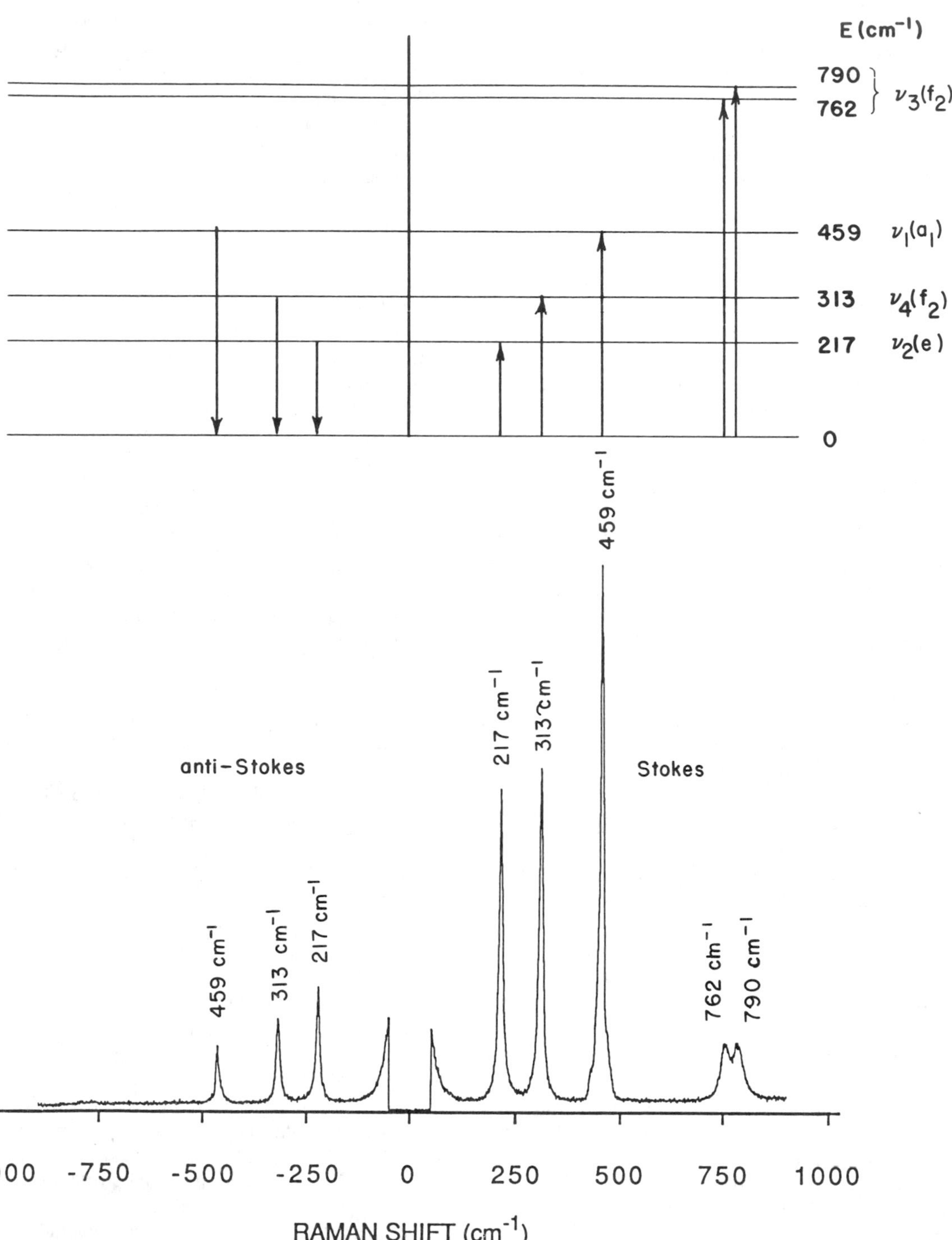

FIG. 6. The vibrational Raman spectrum of liquid CCL_4 obtained with the arrangement indicated in Fig. 1d. The arrows indicate the overall changes from initial to final vibrational states; these changes result from scattering via the virtual states as is indicated in Fig. 9a. The Fermi doublet at 762 cm^{-1} and 790 cm^{-1} is the result of a Fermi resonance interaction between the fundamental ν_3 and the combination tone $\nu_1 + \nu_4$. Letters in parentheses are the symmetry labels of the vibrational fundamentals. The very strong Rayleigh line at zero Raman shift has been blocked out. Stokes Raman shifts are taken as positive. Wavelength increases toward the right. (Figure of spectrum supplied by Dr. M. Bell).

frared absorption bands of the CO and CO_2 molecules are generally composed of P, Q, and R-branches, corresponding to transitions with change in the rotational quantum numbers $\Delta J = -1, 0,$ and $+1$ respectively. For both CO and the ν_3-band of CO_2, there is no Q-branch. Such bands in linear molecules are called parallel bands since the atoms oscillate along the principal molecular axis. The ν_2-mode of CO_2 has a Q-branch; such bands in linear molecules are called perpendicular bands. (See the mode pictures for CO_2 in Fig. 5.) Since there is no change in dipole moment for the ν_1-mode of CO_2, this mode is "forbidden" in infrared absorption; it is however, "allowed" in the Raman spectrum. (For centrosymmetric molecules, including $D_{\infty h}$ molecules such as CO_2, infrared and Raman activities are mutually exclusive.) Even when such resolution is not possible, as for example in part (A) of Fig. 2 or in the Raman spectrum of liquid CCl_4 shown in Fig. 6, useful information is conveyed by the band.

Vibrational and rotation-vibrational spectra are also obtained by the Raman scattering process. The Raman spectrum is obtained by measuring the scattered light from a sample exposed to an intense monochromatic light source. The monochromatic exciting radiation can be of any wavelength; radiations from the deep ultraviolet to those of the near infrared Nd-YAG laser are used, the most commonly employed, however, being the 488 and 514 nm radiations from an argon laser. Figure 6 shows a Raman spectrum of CCl_4 obtained with the laser excitation technique shown in Fig. 1(d). The scattered light contains, in addition to the unmodified radiation of wavenumber σ_0, longer and shorter wavelengths appearing at wavenumbers σ_S and σ_{a-S} respectively, these being the Stokes and anti-stokes Raman bands. The wavenumber differences $|\sigma_0 - \sigma_{S,i}| \equiv |\sigma_0 - \sigma_{a-S,i}| = \nu_i$ are the vibrational frequencies of the scattering molecule. Since the selection rules and intensity relations in the Raman effect are quite different from those applying to infrared spectra, the two types of spectra are complementary to one another and are both desirable for a full understanding of the vibrations of a molecule. The Raman scattering process is illustrated by the energy level scheme shown in Fig. 7a.

The scattering process may be viewed as the monochromatic excitation of the molecule from the initial level i to a virtual state v accompanied by the emission of radiation associated with the transition from the virtual state v to the final level f. If the energy difference $E_f - E_i > 0$ then the scattering is called Stokes Raman scattering, whereas when $E_f - E_i < 0$, the process is called anti-Stokes Raman scatter-

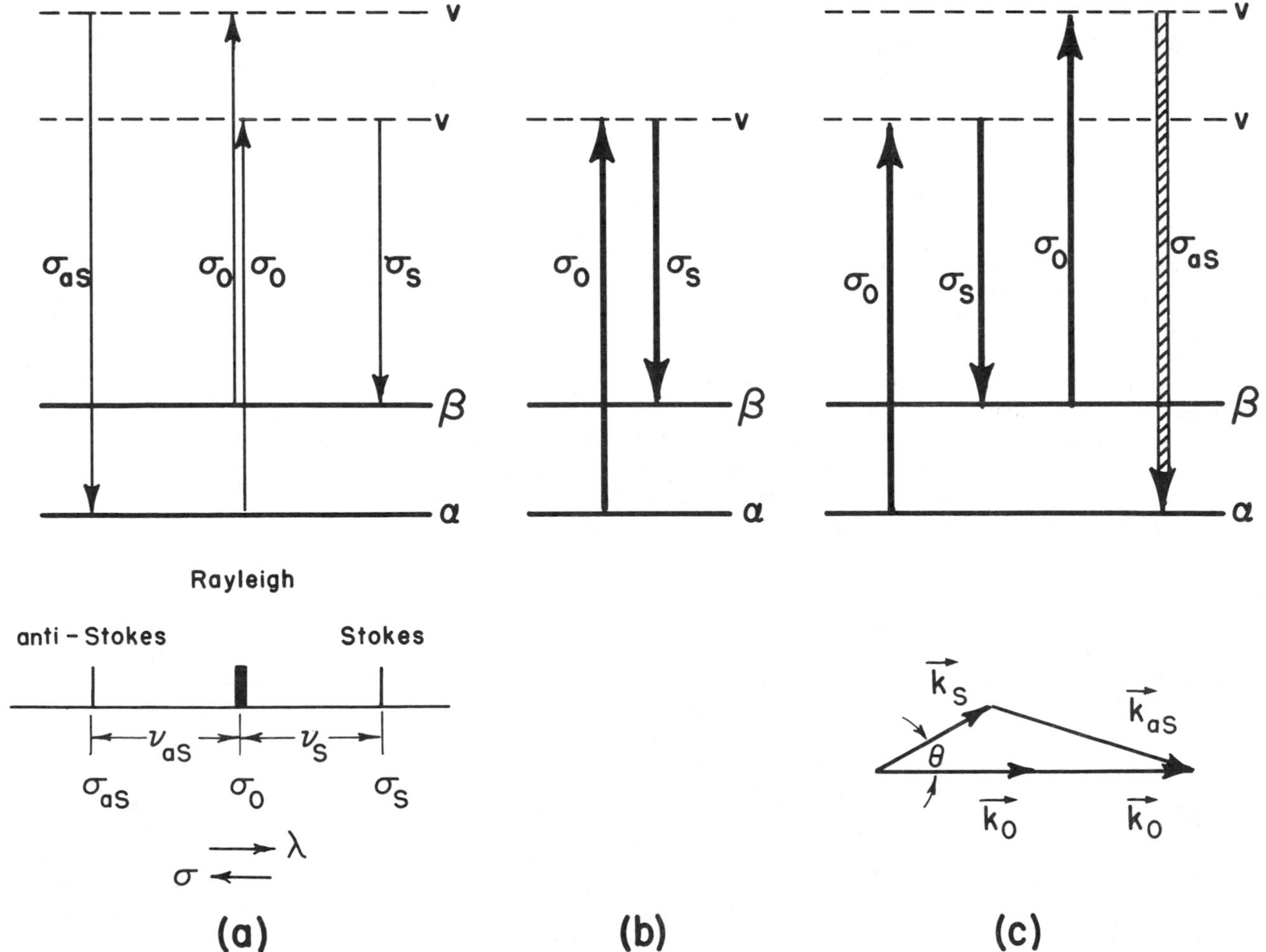

FIG. 7. Energy level diagrams showing transitions for (a) the spontaneous Raman; (b) the SRS, and (c) the CARS processes.

ing. If the initial and final states are the same, there is no energy change in the process, which is then called Rayleigh scattering. In this process, there are no fixed phase relationships between the scatterings produced by the individual molecules of the medium (solid, liquid, gas). The Raman scattering, which is a spontaneous event, is therefore said to be "incoherent." The Stokes and anti-Stokes spectra are symmetrically disposed with respect to the Rayleigh line, with the intensity of the anti-Stokes side determined by the Boltzmann population of the excited state. This is clearly seen in the vibrational spectrum of CCl_4 in Fig. 6. Infrared transitions are governed by the matrix elements of the molecular dipole moment; rotational spectra require the existence of a permanent dipole moment while the derivatives of the dipole moment function with respect to the normal coordinates govern the appearance of the vibration spectrum. For Raman spectra, on the other hand, it is the dipole moment that is *induced* by the electric field of the incident light wave that is important. This *induced* electric moment is a function of the external electric field according to the relation

$$\boldsymbol{\mu}_{\text{ind}} = \alpha \mathbf{E},$$

where α is the polarizability (a second rank tensor) of the molecule. Its matrix elements govern the appearance of the rotational Raman spectrum while the vibrational Raman spectrum is determined by the matrix elements of the derivatives of α with respect to the normal coordinates.

ROTATION AND ROTATION-VIBRATION SPECTRA

In addition to producing a fine structure in electronic and vibrational spectra, rotation of molecules in the gas phase leads to a directly observable spectrum in the microwave and far-infrared regions, as well as by means of Raman scattering. In the first approximation a molecule may be treated as a rigid body with moments of inertia determined by the equilibrium molecular geometry and masses of the atoms. The rotational energy of such a body is quantized, and the pattern of levels is a function of the moments of inertia. In the simplest case of a diatomic (or linear) molecule, this energy is given by the term value expression

$$F_J = BJ(J+1)$$

where $B = \hbar/4\pi cI$ is called the "rotational constant" (in cm^{-1} units), I is the moment of inertia about an axis perpendicular to the line joining the nuclei, and J is the quantum number of total angular momentum. Selection rules for molecules with a permanent electric dipole moment permit transitions between adjacent levels, and so the frequency of a radiative transition (in absorption or emission) is given by

$$\nu = F_{J+I} - F_J = 2B(J+1).$$

The rotational spectrum of a linear molecule thus consists of a series of equidistant lines spaced from one another by an amount equal to $2B$. Since many rotational levels are populated at normal temperatures, a large number of these lines will be observed.

More generally, a rigid molecule is characterized by three moments of inertia, referred to as the "principal moments." If the three moments are equal, the molecule is called a spherical rotor (e.g., CH_4, SF_6). If two moments are equal but the third different, one has a symmetric rotor (e.g., CH_3Cl); and an asymmetric rotor results when all three are different (e.g., H_2O, CH_2Cl_2). The point group, which was important in discussing the vibrational spectrum, also determines the type of rotor.

The energy levels of a spherical rotor follow the same expression as those of a linear molecule, and the levels of a symmetric rotor can be written

$$F_{JK} = BJ(J+1) + (A-B)K^2.$$

Here A is the rotational constant corresponding to the moment of inertia about the unique axis, and K is the projection of the total angular momentum on that axis. The levels of an asymmetric rotor cannot be written in closed form, but must be calculated by solving a secular equation.

The rigid rotor model for a diatomic molecule which predicts equally spaced rotational lines is, however, only a first approximation and has to be replaced by the more general vibrating-rotator model. In this model the interaction between vibration and rotation makes the rotational constants functions of the vibrational state v so that, to lowest order of approximation, $B_v = B_e - \alpha_e(v+\frac{1}{2})$ where α_e is the rotation-vibration interaction constant. Even though α_e is quite small compared to B, this effect has as a consequence that the rotational lines in a vibration-rotation band or in an electronic band spectrum are not equally spaced. A more general extension applies to the rotation constants of polyatomic molecules.

A further effect is due to centrifugal distortion. As a molecule rotates in a given vibrational state it stretches depending on the amount of rotation. Accordingly the rotational term value for a diatomic (or linear) molecule is represented by

$$F_\alpha(J) = B_vJ(J+1) - D_vJ^2(J+1)^2 + \cdots$$

The centrifugal distortion constant D_v is given to a good approximation by

$$D_v = \frac{4B_v^3}{\omega_v^2}$$

where B_v is the rotation constant for the vibrational state v and ω_v is the (harmonic) frequency of vibration in that state. The size of the centrifugal distortion effect is quite small since $D_v < 10^{-4}B_v$ for most molecules. Similar generalizations obtain for symmetric and asymmetric top rotors.

The general selection rule for radiative transitions (emission or absorption) restricts the change in J to one unit, or $\Delta J = J' - J'' = \pm 1$. Additional selection rules apply to each type of rotor. Moreover, a pure rotational (infrared or microwave) spectrum exists only for molecules possessing a permanent (electric) dipole moment (paramagnetic molecules may show a magnetic dipole spectrum). However, except for homonuclear diatomic molecules, this limitation does not apply to rotational fine structure of vibrational bands.

The selection rule for rotational Raman spectra of linear (and diatomic) molecules is $\Delta J = J' - J'' = 0, \pm 2$ and, as before, additional selection rules apply to different types of rotors. Except for spherical top molecules pure rotational Raman spectra always exist. Fig. 8 shows, for example, the pure rotational Raman spectrum of acetylene, C_2H_2, in its ground vibrational state, $v=0$. The observed 3:1 intensity alternation of the J=odd to J=even rotational lines is due to the statistical weights of the rotational levels. This is determined by the spin of the identical nuclei located in symmetrically equivalent positions.

Rotational spectra may extend from essentially zero frequency through the microwave region to 1000 GHz (33 cm^{-1}) or higher; many light molecules such as H_2O have rotational spectra extending well into the infrared region (300–400 cm^{-1}). For molecules other than hydrogen and its isotopes, rotational Raman spectra have been observed as far as 150 cm^{-1} away from the Rayleigh line (see Fig. 7) the extent being determined by the Boltzmann population of the higher lying rotational levels.

Analysis of a rotation as well as rotation-vibration spectra of a molecule requires finding a set of rotation constants

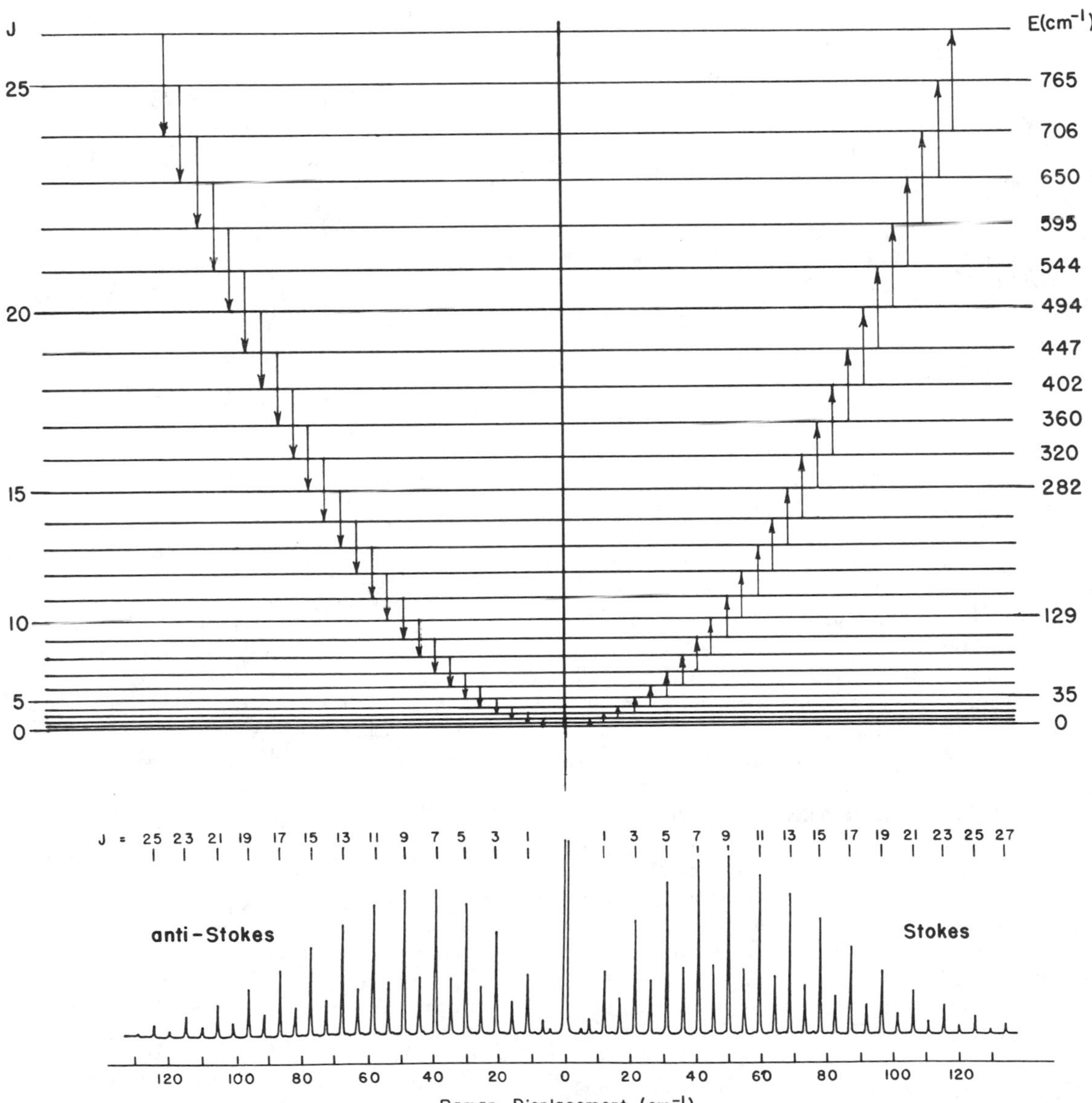

FIG. 8. Pure rotational Raman spectrum of gaseous acetylene, C_2H_2. The top shows the rotational energy levels with the arrows indicating the overall changes from initial to final ratational states; these changes result from scattering via the virtual states as is indicated in Fig. 9a. The rotational quantum number J is that of the lower rotational state. Raman displacements from the Rayleigh line are given as absolute values. Wavelength increases toward the right. (Adapted from H. Finsterhölzl, *et al.*, J. Raman Spectrosc. *6*, 13 (1977).)

(and, in higher approximations, other molecular constants) which reproduce the frequencies and intensities of the observed spectral lines. From the rotation constants one obtains a set of principal moments of inertia which, together with the point group symmetry deduced from the vibrational spectrum, provide information on the geometric structure of the molecule. In favorable cases the equilibrium rotation constants and thus the equilibrium principal moments of inertia can be obtained from which very accurate interatomic distances and interbond angles can be determined. Microwave rotation spectra also furnish other molecular properties, such as electric dipole moments (from Stark effects of rotational lines); nuclear quadrupole coupling constants (from hyperfine structure); barriers to internal rotation; and various parameters appearing in the vibrational potential function.

PERTURBATIONS AND INTERACTIONS

In addition to the vibration-rotation interaction and centrifugal distortion effects, there are others that manifest themselves in the observed spectra. The existence of these perturbations and interactions is often governed by the symmetry properties of the interacting states.

An effect which is often observed in vibrational spectra is Fermi resonance. Two vibrational levels, belonging to different vibrations, which have nearly the same energy and which are also of the same vibrational species may perturb each other. The classical examples are the cases of CO_2 and CCl_4. In CO_2 the level $v_2=2$ ($2\nu_2=2\times667=1334$ cm^{-1}) lies near the fundamental $\nu_1=1337$ cm^{-1}. Since the overtone $2\nu_2$ has symmetry species $\Delta_g+\Sigma_g^+$ it can interact with ν_1 which is of species Σ_g^+. The ν_1 band, which appears only in the Raman spectrum, is therefore shifted away from its unperturbed position at 1337 cm^{-1} upward to 1388 cm^{-1} while $2\nu_2$, which is also active in the Raman effect, is shifted from its unperturbed location at 1334 cm^{-1} downward to 1285 cm^{-1}. In carbon tetrachloride the observed doublet at 762 and 790 cm^{-1} in the Raman spectrum (see Fig. 6) results from a Fermi resonance interaction of the fundamental $\nu_3=775$ cm^{-1}, with the combination band $\nu_1+\nu_4=772$ cm^{-1} since both ν_3 and $\nu_1+\nu_4$ have the same species F_2.

Another type of perturbation results from the interaction of a vibrational angular momentum with that due to the rotation of the molecule in space. The two components of the degenerate pair Q_{2a} and Q_{2b} of CO_2, for example (see Fig. 5), superpose to produce an angular momentum along the internuclear axis. The coupling of this vibrational (also called internal) angular momentum with that of the overall rotation causes a splitting of each rotational line of the ν_2 band into an "*l*-type" doublet.

The vibrating nuclei in a rotating molecule are always subject to a Coriolis force by virtue of which a given normal mode may interact with another. Whether or not vibrational modes can interact by Coriolis coupling is determined by their symmetry species. In the example of the CO_2 molecule (see Fig. 5), Coriolis coupling exists between ν_2 and ν_3, but not between ν_1 and ν_2 or ν_1 and ν_3. The strength of the Coriolis coupling depends on the difference between the vibrational energies of the two modes; the closer the levels the stronger the interaction in consequence of which the levels are mutually repelled.

Another prominent coupling is one that leads to the predissociation of a molecule. For example, in the potential energy diagram of the P_2 molecule, the electronic states $^1\Sigma_u^+$ and $^3\Pi_u$ intersect, with the horizontal asymptote of $^3\Pi_u$ below that of the $^1\Sigma_u^+$ state. Vibrational levels of $^1\Sigma_u^+$, which lie above the asymptote of $^3\Pi$, are metastable. Any molecule excited to one of these levels of $^1\Sigma_u^+$ will transfer into the continuum above the $^3\Pi$ asymptote and thereby undergo "predissociation." Such predissociation may occur over a narrow range of rotational levels and often manifests itself in a sudden termination of the rotational structure of a given vibrational band.

See also Far-Infrared Spectroscopy; Infrared Spectroscopy; Laser Spectrosopy; Microwave Spectroscopy; Molecules; Visible and Ultraviolet Spectroscopy.

BIBLIOGRAPHY

F. T. Arecchi, F. Strumia, and H. Walther, *Advances in Laser Spectroscopy*, NATO ASI Series B, No. 95. Plenum, New York, 1981. (I)

R. P. Bauman, *Absorption Spectroscopy*. Wiley, New York, 1962. (E)

R. J. Bell, *Introductory Fourier Transform Spectroscopy*. Academic Press, New York, 1972. (E).

P. R. Bunker, *Molecular Symmetry and Spectroscopy*. Academic Press, New York, 1979. (A)

A. D. Buckingham (ed.), *MTP International Review of Science, Physical Chemistry Series Two*. Butterworths, London, 1975 (E), *Molecular Structure and Properties*, Vol. 2, contains the following pertinent articles: "Molecular Structure Determination by High Resolution Spectroscopy" (D. R. Lide, Jr.); "The Properties of Molecules from Molecular Beam Spectroscopy" (J. S. Muenter and T. R. Dyke).

W. Demtröder, *Laser Spectroscopy*. Springer Verlag, Berlin, 1981 (E)

G. L. Besley, *Coherent Raman Spectroscopy*. Pergamon, Oxford, 1981. (A)

W. Gordy and R. L. Cook, *Microwave Molecular Spectra*, 3rd ed. John Wiley, New York, 1984. (A)

P. R. Griffiths and J. A. deHaseth, *Fourier Transform Infrared Spectrometry*. Wiley, New York, 1986. (E)

M. D. Haromny, *Introduction to Molecular Energies and Spectra*. Holt, Rinehart, and Winston, New York, 1972. (E)

A. B. Harvey (ed.), *Chemical Applications of Non-Linear Raman Spectroscopy*. Academic Press, New York, 1981. (I)

G. Herzberg, *Electronic Spectra of Polyatomic Molecules*. Van Nostrand, New York, 1966. (A)

G. Herzberg, *Infrared and Raman Spectra of Polyatomic Molecules*. Van Nostrand, New York, 1945. (A)

G. Herzberg, *Spectra of Diatomic Molecules*. Van Nostrand, New York, 1950. (A)

G. Herzberg, *An Introduction to Molecular Spectroscopy*. Cornell University Press, Ithaca, New York, 1971. (E)
J. M. Hollas, *Modern Spectroscopy*. Wiley, 1987. (E)
G. W. King, *Spectroscopy and Molecular Structure*. Holt, Rinehart, and Winston, New York, 1964. (I)
H. W. Kroto, *Molecular Rotation Spectra*. Wiley, 1975. (I)
V. S. Letokhov and V. P. Chebotayev. *Non-Linear Laser Spectroscopy*. Springer Verlag, Berlin, 1977. (I)
M. D. Levenson, *Introduction to Non-Linear Laser Spectroscopy*. Academic Press, New York, 1982. (I)
D. A. Long, *Raman Spectroscopy*. McGraw-Hill, New York, 1977. (E)
D. Papousek and M. R. Aliev, *Molecular Vibration-Rotational Spectra*. Elsevier, Amsterdam, 1982. (A)
L. J. Radziemski, R. W. Solarz, and J. A. Paisner, (eds.), *Laser Spectroscopy and its Application*. Dekker, New York, 1987. (I)
J. I. Steinfeld, *Molecules and Radiation: An Introduction to Modern Molecular Spectroscopy*, 2nd Ed. MIT Press, 1985. (E)
W. S. Struve, *Fundamentals of Molecular Spectroscopy*. Wiley, New York, 1989. (E)
C. H. Townes and A. L. Schawlow, *Microwave Spectroscopy*. McGraw-Hill, 1955. (A)
A. Weber (ed.), *Raman Spectroscopy of Gases and Liquids*. Springer Verlag, 1979. (I)
A. Weber (ed.), *Structure and Dynamics of Weakly Bound Molecular Complexes*, NATO ASI Series C, No. 212. Reidel, Dordrecht, 1987. (I)
D. Williams (ed.), *Methods of Experimental Physics*. Academic Press, New York, 1974. (I) *Molecular Physics*, Vol. 3, contains the following pertinent chapters: 2.1 Microwave Spectroscopy (D. R. Lide, Jr.); 2.2 Infrared Spectroscopy (W. E. Blass & A. H. Nielsen); 2.3 Electronic Spectroscopy (C. W. Mathews); and 3.3 Spontaneous Raman Scattering (D. H. Rank and T. A. Wiggins).
E. B. Wilson, J. C. Decius, and P. C. Cross, *Molecular Vibrations*. McGraw-Hill, New York, 1955. (A)
J. E. Wollrab, *Rotational Spectra and Molecular Structure*. Academic Press, New York, 1967. (I)
L. A. Woodward, *Introduction to the Theory of Molecular Vibrations and Vibrational Spectra*. Oxford, 1972. (E)

Molecular Structure Calculations

Per-Olov Löwdin

INTRODUCTION

Some of the recent developments in the large-scale molecular calculations of molecular ground states, low-lying excited states, molecular geometry and conformations, as well as intermolecular energy surfaces will be reviewed briefly. All studies of molecular stationary states are, in principle, based on the time-independent Schrödinger equation $\mathcal{H}\Psi=E\Psi$, where $\mathcal{H}$ is the Hamiltonian operator

$$\mathcal{H}=\sum_k \frac{p_k^2}{2m_k}+\sum_{k<l}\frac{e_ke_l}{r_{kl}}, \tag{1}$$

which is the sum of the kinetic energy of all particles involved and their Coulomb interaction, with each classical momentum p_k replaced by the differential operator

$$p_k=\frac{h}{2\pi i}\left(\frac{\partial}{\partial x_k},\frac{\partial}{\partial y_k},\frac{\partial}{\partial z_k}\right). \tag{2}$$

This Hamiltonian works on a Hilbert space of type L^2 having a binary product of the form

$$\langle\Phi_1|\Phi_2\rangle=\int\Phi_1{}^*\Phi_2 dX. \tag{3}$$

The energy $\langle\mathcal{H}\rangle_{av}$ of a system described by the wave function Φ is given by the expectation value $\langle\mathcal{H}\rangle_{av}=\langle\Phi|\mathcal{H}\Phi\rangle/\langle\Phi|\Phi\rangle$. Most calculations of stationary states are based on the fact that the eigenvalue problem $\mathcal{H}\Psi=E\Psi$ is equivalent to the variational principle $\delta\langle\mathcal{H}\rangle_{av}=0$.

Current molecular calculations are further based on the Born–Oppenheimer approximation [1], which starts from the assumption that the atomic nuclei have infinite masses and fixed positions in space. The electronic structure is then determined by the electronic Schrödinger equation $\mathcal{H}_{el}\Psi_{el}=E_{el}\Psi_{el}$, where the electronic wave function Ψ_{el} has to be antisymmetric in the electronic coordinates x_1, x_2, ..., x_N to satisfy Pauli's exclusion principle. Once the electronic structure has been found, one may correct for the nuclear motion.

In studying the electronic structure of a molecular system, there are essentially three different types of approaches, referred to as *ab initio*, semi-*ab initio*, and semiempirical calculations. The nomenclature is not entirely proper, since quantum mechanics represents the quintessence of about 150 years of experimental research, so all calculations are essentially "empirical" in their origin.

The experimental discovery of the atomic numbers Z in the Rutherford atomic model of 1911 implies, however, that one may describe a molecule by means of the atomic numbers of its constituents. One can then study the many-electron Hamiltonian $\mathcal{H}_{el}$ derived from Eq. (1) in terms of the electronic charge e, the electronic mass m, and Planck's constant $\hbar=h/2\pi$, and express all physical results in these quantities as units—a system often referred to as "atomic units." Since the Schrödinger equation $\mathcal{H}\Psi=E\Psi$ in the Born–Oppenheimer approximation with fixed nuclei becomes a purely mathematical problem, one may speak of "*ab initio*" calculations, since no more experimental quantities have to be specified—until one introduces the nuclear kinetic energy which requires a knowledge of the nuclear masses involved. Even some of the approximate methods based on the variation principle $\delta\langle\mathcal{H}\rangle_{av}=0$ belong to this class.

In the semi-*ab initio* calculations, it is usually convenient to introduce into the theory some mathematical approximations that depend on one or more parameters which, in turn, are determined by other calculations.

In the semiempirical methods, the variation principle is often used once or several times, but, instead of evaluating all the binary products occurring in the theory, one may try to determine some of them from known experimental data in order to be able to predict other experimental results.

Today, all three approaches require large-scale computations by means of modern electronic computers, and many man-years have gone into the construction of the computer programs. In order to coordinate the computer efforts on an international basis, a special Quantum Chemistry Program Exchange (QCPE) has been set up at the Department of Chemistry, Indiana University, Bloomington, IN.

The molecular calculations of all three types are under continuous development depending on the fact that one gets better programs, new algorithms, and larger computers. There is hence a tendency for so-called standard programs to become obsolete, and it is hence always a good idea to check with QCPE and the leading research centers about the latest developments. It is also very likely that many of the programs mentioned below have become more or less obsolete when this text is being consulted by a reader.

AB INITIO CALCULATIONS

Some molecular structure calculations of this type are essentially based on three related approaches: the Hartree–Fock (HF) method, the multiconfigurational self-consistent-field (MC-SCF) scheme, and the configurational-interaction (CI) method. They are all approximate methods based on the variation principle $\delta\langle\mathcal{H}\rangle_{\text{av}}=0$ for finding the optimum solutions of each type. Other current methods involve elements from perturbation theory, partitioning technique, or resolvent methods.

In all these schemes, the antisymmetric electronic wave function $\Psi=\Psi(x_1,x_2, \ldots, x_N)$ is built up from products or Slater determinants of one-electron functions or spin-orbitals $\psi_1(x)$, $\psi_2(x)$, ..., $\psi_M(x)$, which are usually chosen orthonormal. If $M=N$, the total wave function is approximated by a single Slater determinant $D=(N!)^{-1/2}|\psi_k(x_i)|$, and the variation principle then gives the *Hartree–Fock scheme* [2], in which the best spin-orbitals are determined by the one-electron Schrödinger equation:

$$H_{\text{eff}}(1)\psi_k(1)=\epsilon_k\psi_k(1), \tag{4}$$

where the effective Hamiltonian is given by the expression

$$H_{\text{eff}}(1)=\frac{p_1^{\,2}}{2m}-e^2\sum_g\frac{Z_g}{r_{1g}} + e^2\int dx_2\,\frac{\rho(x_2,x_2)-\rho(x_2,x_1)P_{12}}{r_{12}}. \tag{5}$$

Here $\rho(x_1,x_2)=\Sigma_k\psi_k(x_1)\psi_k{}^*(x_2)$ is the Fock–Dirac density matrix [3] satisfying the relations $\rho^2=\rho$, $\rho^\dagger=\rho$, and tr $\rho=N$, whereas P_{12} is the exchange operator having the property $P_{12}f(x_1)=f(x_2)$. Equation (4) represents a system of nonlinear integrodifferential equations, which is usually solved iteratively by starting from an estimate of the solutions ψ_k and using the cycle

$$\psi_k\rightarrow\rho\rightarrow H_{\text{eff}}\rightarrow\psi_k \tag{6}$$

until the process becomes self-consistent, i.e., there is no change in the significant figures of the wave functions ψ. Since, in the beginning, one was particularly interested in the fields associated with H_{eff}, the approach has become known as the *self-consistent-field* (SCF) method.

In a molecular system, the solutions ψ_k are molecular orbitals (MOs) in the sense of Mulliken, which are extended over the entire molecule. They may often be conveniently constructed by taking linear combinations (LCs) of the atomic orbitals (AOs) involved: $\phi_1(x)$, $\phi_2(x)$, ..., $\phi_{M'}(x)$, i.e.,

$$\psi_k(x)=\sum_{\mu=1}^{M'}\phi_\mu(x)c_{\mu k}, \tag{7}$$

which is the basis for the MO-LCAO approach. The atomic orbitals $\{\phi_\mu\}$ are usually not orthogonal but have a metric matrix of the type $\Delta_{\mu\nu}=\langle\phi_\mu|\phi_\nu\rangle=\int\phi_\mu{}^*(x_1)\phi_\nu(x_1)dx_1$, which leads to the so-called nonorthogonality problem.

If one introduces the row vectors $\boldsymbol{\psi}=\{\psi_1,\psi_2, \ldots, \psi_N\}$ and $\boldsymbol{\phi}=\{\phi_1,\phi_2, \ldots, \phi_M\}$, and the rectangular matrix $\mathbf{c}=\{c_{\mu k}\}$ of order $M\times N$, one may write relation (7) in the form $\boldsymbol{\psi}=\boldsymbol{\phi}\mathbf{c}$. This gives also $\rho=|\boldsymbol{\psi}\rangle\langle\boldsymbol{\psi}|=|\boldsymbol{\phi}\mathbf{c}\rangle\langle\boldsymbol{\phi}\mathbf{c}|=|\boldsymbol{\phi}\rangle\mathbf{c}\mathbf{c}^\dagger\langle\boldsymbol{\phi}|=|\boldsymbol{\phi}\rangle\mathbf{R}\langle\boldsymbol{\phi}|$, where $\mathbf{R}=\mathbf{c}\mathbf{c}^\dagger$ is the charge- and bond-order matrix [4] of the molecule; hence one has

$$\rho(x_1,x_2)=\sum_{\mu,\nu=1}^{M}\phi_\mu(x_1)R_{\mu\nu}\phi_\nu{}^*(x_2), \qquad R_{\mu\nu}=\sum_k c_{\mu k}c^*_{\nu k}. \tag{8}$$

In many applications [5], the charge orders $R_{\mu\mu}$ and the bond orders $R_{\mu\nu}$ are used as fundamental "molecular indices" to determine certain molecular properties. The quadratic matrix $\mathbf{R}$ of order M satisfies the relations $\mathbf{R\Delta R}=\mathbf{R}$, $\mathbf{R}^\dagger=\mathbf{R}$, tr $(\boldsymbol{\Delta}\mathbf{R})=N$, and the last relation has been used in various forms for "population analysis" of the charge clouds of the N-electron system.

In the Roothaan scheme [6], the optimum coefficients $\mathbf{c}=\{c_{\mu k}\}$ are determined by means of the variation principle, which leads to a system of equations of the type

$$\sum_\nu(F_{\mu\nu}-\epsilon\Delta_{\mu\nu})c_\nu=0, \tag{9}$$

where $F_{\mu\nu}=\langle\phi_\mu(1)|H_{\text{eff}}(1)|\phi_\nu(1)\rangle$ are the elements of the Fock matrix $\mathbf{F}=\{F_{\mu\nu}\}$ obtained by substituting the relations (8) into the effective Hamiltonian (5). The associated secular equation has M roots of which the N lowest correspond to occupied spin-orbitals, whereas the $M-N$ highest correspond to so-called "virtual solutions." The system (9) is usually solved by an iterative procedure based on an estimate of the rectangular matrix $\mathbf{c}$ and a cycle of the form $\mathbf{c}\rightarrow\mathbf{F}\rightarrow\mathbf{c}$, which is repeated until one obtains self-consistency. In Mulliken's nomenclature, this approach is referred to as the ASP-MO-LCAO-SCF method, where the symbol ASP (antisymmetrized product) indicates that one is using a single Slater determinant. It has been emphasized [7] that the method may be simplified by using the matrix $\mathbf{R}=\mathbf{c}\mathbf{c}^\dagger$ and an iterative cycle of the form $\mathbf{c}\rightarrow\mathbf{R}\rightarrow\mathbf{F}\rightarrow\mathbf{c}$.

As atomic orbitals $\boldsymbol{\phi}=\{\phi_\mu\}$, one may use atomic Hartree–Fock orbitals, atomical natural orbitals, Slater-type orbitals (STOs), or Gaussians of the type $\exp(-nr^2)Y_{lm}(\theta,\varphi)$, as first suggested by Boys [8]. The bottleneck in the early calculations of $\langle\mathcal{H}\rangle_{\text{av}}$ was connected with the Coulomb terms in the Hamiltonian (1), which lead to "molecular integrals" of the form

$$\left| \nu\kappa \left| \frac{e^2}{r_{12}} \right| \mu\lambda \right|$$
$$= e^2 \int\int \frac{\phi_\nu^*(1)\phi_\kappa^*(2)\phi_\mu(1)\phi_\lambda(2)}{r_{12}}\, dx_1 dx_2$$
$$= e^2 \int\int \frac{\phi_\nu^*(1)\phi_\mu(1)\phi_\kappa^*(2)\phi_\lambda(2)}{r_{12}}\, dx_1 dx_2$$
$$= (\nu\mu|\kappa\lambda), \tag{10}$$

where the last notation, introduced by Mulliken, refers to the two charge densities involved. Since there are four atomic indices occurring in these expressions, one deals with one-, two-, three-, and four-center integrals. Using (10), one obtains, e.g., for the elements of the Fock matrix

$$F_{\mu\nu} = \int \phi_\mu^*(1) \left\{ \frac{p_1^2}{2m} - e^2 \sum_g \frac{z_g}{r_{1g}} \right\} \phi_\nu(1) dx_1 + e^2 \sum_{\kappa,\lambda} R_{\kappa\lambda}\{(\mu\nu|\lambda\kappa) - (\mu\kappa|\lambda\nu)\}. \tag{11}$$

From the point of view of the asymptotic behavior of the wave functions, the exponential-type orbitals (ETOs) closely related to the STOs seem to form the most feasible basis, and the multicenter integrals are then evaluated by using the Legendre or Neumann expansions of $1/r_{12}$ and by expanding the ETOs on each center in terms of spherical harmonics around one and the same point by the α-function technique, so that all integrals are reduced to one-center integrals [9]. Even if many calculations have been successfully carried out by using this approach, full computational programs for arbitrary small-size molecules are not yet available.

Instead, many of the leading research centers have preferred to evaluate the many-center integrals analytically by means of Gaussians, and the pioneering large-scale programs of this type were developed by Pople and his collaborators [10] in the early 1970s, and since then numerous molecular programs based on Gaussians have been developed all over the world. Most of these programs are available through the QCPE. For references the reader is referred to the survey articles and books quoted in the section entitled "Literature." In the late 1980s, molecular SCF-programs for supercomputers were constructed which could calculate the electronic structure of molecules for both closed and open shells having about 80 atoms and thousands of basis functions, and in 1988 some of these programs had been condensed to run on fairly small "personal computers."

If the total electronic Hamiltonian $\mathcal{H}$ has certain symmetry properties, they are usually automatically reflected by the exact eigenfunctions Ψ of the Schrödinger equation. This is unfortunately not always the case for the approximate eigenfunctions Φ derived from the variation principle $\delta\langle\mathcal{H}\rangle_{av}=0$, and this means that both the original HF scheme and some related methods may be afflicted by a certain "symmetry dilemma" [11]. With respect to symmetry, one hence distinguishes between the restricted Hartree–Fock (RHF) scheme, the unrestricted Hartree–Fock (UHF) scheme, and the general Hartree–Fock scheme (GHF), where the one-electron functions may be arbitrary mixtures of α and β spin functions. In these SCF schemes, a symmetry-adapted solution Φ may correspond to a "local" minimum of $\langle\mathcal{H}\rangle_{av}$, whereas the absolute minimum may correspond to a mixture of various symmetry types.

In order to solve this symmetry problem, the projected Hartree–Fock (PHF) method has been constructed [12]. A symmetry property of the Hamiltonian is represented by a projector Q satisfying the relations $\mathcal{H}Q=Q\mathcal{H}$, $Q^2=Q$, $Q^\dagger=Q$, and the corresponding eigenfunctions Ψ fulfill the relation $Q\Psi=\Psi$. In the approximate theory based on the variation principle $\delta\langle\mathcal{H}\rangle_{av}=0$, the relation $Q\Phi=\Phi$ is usually a constraint which will raise the energy above the absolute minimum. In the PHF method, the exact solution is approximated by a projection of a single determinant D, i.e., $\Psi\approx QD$, and this approach preserves very much the simplicity of the original Hartree–Fock method. It contains as special cases the idea of treating electronic correlation by permitting different orbitals for different spins (DODS) as well as the alternant-molecular-orbital (AMO) method [13].

In the study of the total energy in the Hartree–Fock scheme, it has turned out that $\langle H\rangle_{av}$ has not only maxima and minima but also saddle points of a complicated character, and the study of these *instabilities* has for three decades been a very hot field of research [14]. It has further been mathematically proven that the solutions to the GHF equations do *exist* [15], which is by no means obvious. In connection with the method of complex scaling, it has finally been shown that the Hartree–Fock method may be extended also to non-Hermitan Hamiltonians [16].

If one starts from M one-electron functions $\psi_1(x)$, $\psi_2(x)$, ..., $\psi_M(x)$, with $M>N$, one may form $\binom{M}{N}$ different Slater determinants D_K or "configurations" and approximate the total wave function Ψ by means of an expansion of the type

$$\Psi \approx \sum_K D_K C_K, \tag{12}$$

which may be referred to as a "superposition of configurations." The coefficients C_K are again determined by the variation principle $\delta\langle\mathcal{H}\rangle_{av}=0$, which leads to the equation system

$$\sum_L (\mathcal{H}_{KL} - E\delta_{KL})C_L = 0, \tag{13}$$

where $\mathcal{H}_{KL}=\langle D_K|\mathcal{H}|D_L\rangle$. For historical reasons, this approach is often called the configurational-interaction (CI) method. If $M\rightarrow\infty$ and the set $\{\psi_k\}$ becomes complete, the relation (12) should in principle give the exact solution Ψ.

If the number M is fairly low $(M>N)$, it may be convenient to optimize the basic one-electron functions ψ_1, ψ_2, ..., ψ_M, which leads to the multiconfigurational (MC) SCF method [17]. Recently there has been considerable progress in the practical development of these MC-SCF methods, and several fine programs are now available through QCPE.

In the original CI method, one keeps the basis ψ_1, ψ_2, ψ_3, ..., ψ_M fixed but increases the number of functions, so that the set becomes more and more complete. In recent applications, the orthonormal basis $\{\psi_k\}$ is usually expanded in terms of fixed atomic orbitals $\{\phi_\mu\}$ having the metric matrix $\Delta_{\mu\nu}=\langle\phi_\mu|\phi_\nu\rangle$.

As in the SCF methods, the problem of the evaluation of the molecular integrals $(\nu\mu|\kappa\lambda)$ is now usually handled by using Gaussians, and it is no longer considered a "bottleneck." One of the first simplifications of this approach was achieved by the "direct CI method" developed by Roos [18], and—by concentrating the computational efforts on the vectors (and not the matrices) involved—one could handle calculations involving more than 100 000 configurations.

It is evident that the CI expansion (12) is invariant under unitary transformations of the orbital basis $\{\psi_k\}$ of order M, or under linear transformations of the atomic orbital basis $\{\phi_\mu\}$. The formal connection between the CI expansion and the unitary group was discovered by Jordan [19]. After the successful classification of the irreducible representations of the unitary group by Gelfand [20], several quantum chemists [21] used the "Gelfand symbols" to study and simplify the CI expansion in the unitary group approach (UGA). It was shown by Shavitt [22] that a graphical representation of this approach now known as the graphical unitary group approach (GUGA) was a powerful computational tool, and—after a certain class of the "outer graphs" were successfully treated by Siegbahn [23]—one could carry out the first molecular calculations involving more than one million (10^6) configurations [24]. Many fine molecular CI programs are now available through QCPE.

The ultimate goal of the CI approach is, of course, to reach the *natural expansion* [25] of the total wave function Ψ, which is characterized by the property of fastest convergence. The importance of the *natural orbitals* in this connection has been stressed by Davidson [26], and various types of natural orbitals have now also become useful as basis functions in large-scale CI calculations.

As the power of the computers steadily increases, and one is using larger and larger basis sets, one becomes increasingly aware of the existence of *approximate linear dependencies* [27]. If the set of atomic orbitals $\boldsymbol{\phi}=\{\phi_\mu\}$ has p linear dependencies, the metric matrix $\boldsymbol{\Delta}=\langle\boldsymbol{\phi}|\boldsymbol{\phi}\rangle$ has necessarily p vanishing eigenvalues, and vice versa. If the set $\boldsymbol{\phi}$ is linearly independent, the matrix $\boldsymbol{\Delta}$ is positive definite, and the smallest eigenvalue $\mu_{\min}$ is called the "measure of linear independence." If $\mu_{\min}\approx 10^{-m}$ is smaller than the accuracy used in the computer, the set $\boldsymbol{\phi}$ is linearly dependent from all practical points of view, and one speaks of an approximate linear dependency.

Since orthonormal sets $\Psi=\{\psi_k\}$ do not suffer from this difficulty, one could think that it would be possible to remove the approximate linear dependencies by some orthonormalization procedure, for instance, the *symmetric* or *canonical* orthonormalizations [28] defined by the formulas:

$$\Psi=\Phi\langle\Phi|\Phi\rangle^{-1/2}, \qquad \chi=\mu^{-1/2}\Phi U, \tag{14}$$

respectively, where U is the unitary transformation which brings the metric matric $\Delta=\langle\Phi|\Phi\rangle$ to diagonal form μ, so that $U^\dagger\,\Delta U=\mu$ and $\Delta^{-1/2}=U\mu^{-1/2}U^\dagger$. Again one finds that it is the smallness of some of the eigenvalues $[\mu_k]$ which remains the essential problem, and that one can remove the approximate linear dependencies only by removing the canonical functions χ_k associated with the eigenvalues μ_k which are too small to lead to significant results in the calculations.

All the methods described above are essentially based on the idea of the existence of *molecular orbitals* as developed by Mulliken and Hund [29]. A somewhat different approach is based on the *valence bond* (VB) method, which is a generalization of the original Heitler–London function [30] for the H_2 molecule. The VB method is closely associated with the concept of the chemical bond, and, for a long time, the calculation of the matrix elements of the Hamiltonian with respect to two VB structures represented a very difficult problem [31]. During the last two decades, the VB method has had a remarkable renaissance [32] and some molecular structure programs of this type are now available from QCPE. The theoretical basis for this approach seems still to be under rapid development.

Another *ab initio* method which requires large number crunchers is the *Monte Carlo* method, which transforms the Schrödinger equation into a diffusion equation which may be solved by using some principles from *probability* theory—hence the name. This approach invented by Metropolis has been adapted to molecular structure calculations by Clementi [33] and has led to some remarkable results.

The *ab initio* methods mentioned above have a comparatively straightforward theoretical background but require very large number crunchers to be effective in molecular structure calculations. There is also a different group of methods, which usually requires a much deeper theoretical analysis but which instead may give results even with more modest computational resources. If the former starts from the *expectation value* of the Hamiltonian $\langle H\rangle_{\rm av}=\langle\Phi|H|\Phi\rangle/\langle\Phi|\Phi\rangle$ and the variation principle $\delta\langle H\rangle_{\rm av}=0$, the latter are associated with the *transition value* defined by the relation $\langle H\rangle_{\rm tr}=\langle\psi|H|\Phi\rangle/\langle\psi|\Phi\rangle$ and the bivariation principle [34]. Here the function ψ is a *reference function* in the Hilbert space, which is usually chosen normalized. Even if the transition value has no direct physical interpretation, it is a very strong mathematical tool for the study of the eigenvalue problem, and—even if it is in principle unbounded—it has usually some interesting "bracketing properties."

Instead of the Schrödinger equation, one studies the solution to the *inhomogeneous* equation $(H-z\cdot 1)\Psi_z=a\psi$, where z is a complex variable and the coefficient $a=a(z)$ is chosen so that the solution satisfies the *intermediate normalization* $\langle\psi|\Psi_z\rangle=1$. In order to solve the eigenvalue problem, one is essentially looking for the values of z corresponding to $a(z)=0$. If one introduces the *resolvent* operator $R(z)=(z\cdot 1-H)^{-1}$, which is bounded as long as one is outside the eigenvalues E so that $|z-E|\geq\rho$, one gets directly the solution

$$\Psi_z=aR(z)\psi=R(z)\psi/\langle\psi|R(z)|\psi\rangle, \tag{15}$$

and one can then find the eigenvalues by looking for the poles of the function $W(z)=\langle\psi|R(z)|\psi\rangle$ in the complex plane by various methods including contour integration. Since $R(z)$ satisfies the formula $R(z)=z^{-1}+z^{-1}HR(z)$, it is sometimes referred to as a *propagator* [35] and its kernel as a *Green's function*. For $z=E$, expression (15) takes the form ∞/∞, but its limiting value may be found by using the resolvent identity [36]:

$$R(z)\psi/\langle\psi|R(z)|\psi\rangle\equiv(1-PH/z)^{-1}\psi, \qquad P=1-|\psi\rangle\langle\psi|, \tag{16}$$

which is valid for any value of z. Formula (16) gives the relation between the resolvent method and the *partitioning* technique, in which the Hilbert space is partitioned into two parts by means of the projectors $O = |\psi\rangle\langle\psi|$ and $P = 1 - O$. The operator $W = (1 - PH/z)^{-1}$ is often referred to as the wave operator, and one has $\Psi_z = W\psi$. If one substitutes this expression into the transition formula and the expectation value, one obtains respectively

$$f(z) = \langle H\rangle_{tr} = \langle\psi|H|\Phi\rangle/\langle\psi|\Phi\rangle = \langle\psi|HW|\psi\rangle/\langle\psi|W|\psi\rangle, \quad (17)$$

$$\langle H\rangle_{av} = \langle\Phi|H|\Phi\rangle/\langle\Phi|\Phi\rangle = \langle\psi|W^{\dagger}HW|\psi\rangle/\langle\psi|W^{\dagger}W|\psi\rangle, \quad (18)$$

where the function $f(z)$ has the interesting property that, for any real z, there is at least one true eigenvalue E in the interval between z and $f(z)$ [37]. If the Hamiltonian has the form $H = H_0 + \lambda V$, and one expands the wave operator $W(z)$ in power series in λ, one can easily derive the various forms of *perturbation theory*. If one starts from the Hartree–Fock function D as the reference function, this leads to the *Møller–Plesset perturbation theory* (MPPT) [38], and various forms of this approach are now available for molecular calculations. By introducing the so-called *reaction operator* t and the unperturbed wave operator W_0 through the formulas

$$\begin{aligned} t &= V + V(z\cdot 1 - PH_0)^{-1}PV \\ t^{-1} &= V^{-1} - (z\cdot 1 - PH_0)^{-1}P \\ W_0 &= (1 - PH_0/z)^{-1} \end{aligned} \quad (19)$$

where, in the second relation, it is assumed that V^{-1} at least formally exists, one obtains the following closed forms for the wave operator W and the "bracketing" operator $\Omega = HW$ associated with the energy $\langle H\rangle_{tr}$:

$$W = V^{-1}tW_0, \qquad \Omega = HW = H_0W_0 + W_0{}^{\dagger}tW_0. \quad (20)$$

We note that one may evaluate the reaction operator t to any accuracy desired by means of "inner projections" using basis sets of finite order M, and that, instead of a power series in λ, one may obtain *rational approximations* for the wave function and the energy [39].

If one considers wave operators of the exponential form $W = e^T$, one obtains immediately the connection with the *coupled-cluster many-body perturbation theory* (CC-MBPT). In the applications of conventional perturbation theory, one had observed that—even if the correct energy expression should be related to the size of the system and the number of electrons N—there were often spurious terms, which were proportional to N^2, N^3, N^4, etc. In the exact theory these terms ought to disappear, and that this is the case is shown by the famous linked-cluster theorem [40]. Ordinarily this theorem is studied by means of Feynman diagrams, but it can also be expressed in the following elementary way [41]: if the Hamiltonian has the form $H = H_a + H_b + \lambda V_{ab}$, where H_a and H_b are many-particle Hamiltonians for the subsystems (a) and (b), and the interaction parameter λ goes to zero, the wave function Ψ for a nondegenerate eigenvalue E should take the product form $\Psi = \Psi_a\Psi_b$, whereas the energy should become a sum $E = E_a + E_b$; even for $\lambda \neq 0$, the results should reflect this property. In connection with molecular calculations, the importance of the "size extensiveness" of the CC-MBPT has been stressed particularly by Bartlett [42].

In the original Hartree–Fock theory, the *Brillouin theorem* [43], which says that $\langle D_{s.e.}|H|D\rangle = 0$ (s.e. = singly excited), played a very fundamental role as an equivalent to the Hartree–Fock equations. The theorem is easily extended [44] to any class of specific trial wave functions $\{\Phi\}$, and it states that, if $\epsilon\psi$ is a permitted variation within the class, one has

$$\{\langle\psi|H|\Phi\rangle - E\langle\psi|\Phi\rangle\} = 0, \quad (21)$$

where Φ is the optimum wave function and $E = \langle\Phi|H|\Phi\rangle/\langle\Phi|\Phi\rangle$ is the optimum energy. This means that, if $\langle\psi|\Phi\rangle \neq 0$, one has also $\langle\psi|H|\Phi\rangle/\langle\psi|\Phi\rangle = \langle\Phi|H|\Phi\rangle/\langle\Phi|\Phi\rangle$, i.e., the transition value equals the expectation value. The extended Briuollin theorem (21) is valid not only in the HF theory, but also in the RHF method as well as in the MC-SCF, CI, and similar approaches.

In connection with the Hellmann–Feynman theorem and its generalizations [45], one has also tried to incorporate the *analytic derivates* of the energy into the various computational schemes [46], and this is currently one of the hottest objects of research.

The technique for evaluating intermolecular potential energy surfaces to be used in the theory of chemical reactions has essentially followed the development of the general quantum-mechanical methods for treating molecular systems, and for a survey of the field up to 1989, the reader is referred to the proceedings from the Tromsø conference [47].

It is evident that the development of *ab initio* methods for calculating molecular structure has been largely dependent on the development of the giant electronic computers. It has been said that the efficiency of the number crunching methods, as the HF, the MCSCF, and the CI(GUGA) methods, has increased by at least a factor of 100 in going from scalar computers to supercomputers. At the same time, it seems very likely that, if one had stayed with the scalar computers but went over to the theoretically more sophisticated resolvent techniques and related methods, one would also have gained a factor of 100 in efficiency. Due to the inertia of the tremendous efforts and man-years invested in the current molecular programs, it will probably take a very long time before one has succeeded in programming the more sophisticated methods on the supercomputers in order to reach an efficiency increase of a factor 100 × 100 = 10 000, but the possibility certainly exists and the prospects for the future of the field of molecular structure calculations are certainly most promising.

SEMI-AB INITIO CALCULATIONS

In the Hartree–Fock scheme, the handling of the "exchange operator" in the effective Hamiltonian (5), i.e., the part of the Coulomb term which contains the permutation P_{12}, represents a comparatively difficult problem. In 1951, Slater [48] suggested that this operator should be approximated by a local potential, and after studying the free-electron model he suggested the form

$$V_{ex}^{\uparrow}(x_1) = -6\alpha[(3/4\pi)\rho^{\uparrow}(x_1)]^{1/3}, \quad (22)$$

where the arrow ↑ indicates that the exchange potential is going to work only on electrons having this type of spin, and α is a coefficient which, according to Gasper [49], should have the value $\alpha = \frac{2}{3}$. This new SCF approach has been called the $X\alpha$ method ($X=$ exchange) and we note that, whereas the use of the form (22) leads to an approximation of the original Hartree–Fock scheme, the $X\alpha$ method contains an adjustable parameter α which makes it possible to incorporate not only exchange but also correlation effects into the theory; for more details, refer to the survey paper by Slater [50]. In the principal applications the $X\alpha$ method turns out to be an excellent "orbital generator" for all the SCF schemes, and, in addition, it gives theoretical results in very good agreement with experiments in the study of the electronic structure of molecules and crystals when starting from the atomic α values involved.

In the numerical work, the $X\alpha$ method is often combined with the multiscattering (MS) wave method to obtain further simplification of the scheme. In the MS$X\alpha$ scheme, each atom is surrounded by a polyhedron as in the cellular method for treating a solid. In the so-called muffin-tin model, each polyhedron is approximated by a sphere, i.e., each atom is surrounded by a sphere and the entire molecule is enclosed in an outside sphere. In some applications, even "overlapping spheres" have been used with excellent results. For more details, the reader is referred to a survey by Johnson [51].

The $X\alpha$ method has been used extensively to predict excitation energies and transition probabilities in various molecules to be compared with the results of the ESCA experiments, using "electron spectroscopy for chemical analysis," carried out by Siegbahn and his co-workers [52].

SEMIEMPIRICAL METHODS

In all the molecular *ab initio* calculations based on the HF, the MC-SCF, the CI, or the VB methods, one is always dealing with a large number of *energy matrix elements* associated with the effective or the many-electron Hamiltonian which are evaluated numerically starting from some given basis set. In the semiempirical methods, these matrix elements are not evaluated from "first principles" but are instead considered as parameters which are determined from experimental data available. The theory is then used to predict the results from other experiments, and it hence becomes essentially a tool for correlating one set of empirical data with another.

As a rule, one tries to keep the number of parameters as small as possible, since one usually does not have too much experimental data to start from. This is often accomplished by considering some of the energy matrix elements as large and essential, whereas others are considered small and are entirely neglected. This gives a certain amount of ambiguity to the semiempirical schemes, and this means that even their most successful applications are often considered with some suspicion by the *ab initio* people who are unable to reproduce their fine results. There is no question, however, that the semiempirical schemes are highly useful for predicting the outcome of experiments within the limited field for which they are constructed and that they provide valuable guidance for the experimentalists as to where to look for their next results. At the same time, the limitations are severe, and the current trend is to use the semiempirical methods only on molecular systems which are so large that they cannot conveniently be handled by the *ab initio* programs. However, many molecular scientists are certainly looking for a revival of the semiempirical approach.

In the study of the Kekulé structures of benzene by quantum-mechanical methods, the principle of "chemical resonance" was discovered early [53]. In the hands of Pauling [54], the theory of resonance between various chemical structures was later extensively developed essentially as a semiempirical form of the valence-bond (VB) theory. It is still considered as one of the most powerful tools for predicting molecular structures and properties without going into large-scale computations, but the reasons for its success are not understood fully.

Most of the conventional semiempirical methods for studies of molecular structure are modifications of the Hartree–Fock scheme in which the matrix elements $F_{\mu\nu}$ of the effective Hamiltonian are considered as parameters to be determined from empirical data available. All of these schemes go back to Hückel's fundamental investigations [55] of the properties of the π electrons in aromatic molecules and conjugated systems, which led to a new understanding of their basic features. In the Hückel scheme, certain simple rules for the determination of the parameters $\alpha_\mu = F_{\mu\mu}$ and $\beta_{\mu\nu} = F_{\mu\nu}$ were developed. The Hückel scheme has been used extensively by the Pullmans in their encyclopedic studies of the conjugated systems in biochemistry [5].

In connection with the conjugated systems, Fukui and his collaborators pointed out the importance of the *frontier electrons* [56] in the highest occupied molecular orbital (HOMO) and their relation to the lowest unoccupied molecular orbital (LUMO) for various types of molecular reactivity, and particularly for charge-transfer reactions.

In order to generalize this approach to include *all* the electrons of a molecular system, Hoffman [57] later developed the so-called extended Hückel method (EHM), which has turned out to be of essential practical importance. In retrospect, it is interesting to observe that the studies of the symmetry properties involved in this method as well as Fukui's idea of the importance of the frontier electrons form the background for the development of the famous Woodward–Hoffman rules [68] in the theory of organic reactions.

In an attempt to include certain correlation effects in the Hückel scheme, the Pariser–Parr–Pople (PPP) method [69] was developed. In connection with the treatment of those matrix elements in the scheme which are so small that they should be neglected, they introduced the idea of "zero differential overlap" between the atomic orbitals involved. It was later shown [70] that this feature could be rather well explained if one introduced symmetrically orthonormalized atomic orbitals defined by the formula $\varphi = \phi\langle\phi|\phi\rangle^{-1/2}$. The same principle is fundamental also in the modifications of this approach known as the CNDO (complete neglect of differential overlap) method, the INDO (intermediate neglect of differential overlap) method, and the MINDO (modified intermediate neglect of differential overlap) method. All of these methods have been used to study the electronic structure of very large molecules, particularly those of importance in molecular biology.

Under the leadership of Dewar and others, the semiempirical methods for studying the electronic structure of very large molecules are still very important, and for a survey of the more recent developments the reader is referred to a review article by Zerner [61].

It should also be observed that the CNDO, INDO, MINDO, etc., methods require large-scale calculations, but that the computer times involved are one or two orders of magnitude less than in the corresponding *ab initio* calculations. However, with the rapid development of the modern electronic computers—particularly through the incorporation of microcomputers having larger and faster memory capacities than ever before—it is evident that, in the future, larger and more complicated molecules will be studied by the *ab initio* methods. It is hoped that this development will lead not only to more numbers of higher and higher accuracy but also to a better conceptual understanding of molecular structure.

LITERATURE

It has been mentioned already in the beginning that many fine programs for molecular structure calculations are available from the Quantum Chemistry Program Exchange (QCPE) at Indiana University, which also publishes a special QCPE bulletin. In addition, there are many commercial molecular programs available. Today there exist large-scale molecular programs which—starting from the ordinary chemical formula of a molecule—will calculate its various configurations including bond distances and bond angles and present its electronic structure in terms of beautiful, colored molecular graphics. It should only be remembered, however, that, if such a program is built on the Hartree–Fock method, it does not include the correlation error, which means that it may give a reasonable approximation without being 100% reliable. It is hence always worthwhile to check the theoretical background of a program.

With the increase in the power of the electronic computers, the development in this field goes very fast. For a survey of the development, the reader is referred to the proceedings of a series of international conferences: the international congresses in quantum chemistry in Kyoto 1979 [62], in Uppsala 1982 [63], in Montreal 1985 [64], and in Jerusalem [65], as well as the conferences in Girona 1988 [66], in Dubrovnik 1988 [67], and in Tromsø 1989 [68]. Surveys of various parts of the development may be found in papers and books by Schaefer [69], Löwdin [70], Clementi [71], and others.

See also Atomic Spectroscopy; Chemical Bonding; Molecular Spectroscopy; Molecules; Quantum Mechanics.

REFERENCES

1. M. Born and J. R. Oppenheimer, *Ann. Phys.* **84,** 457 (1927).
2. D. R. Hartree, *Proc. Cambridge Phil. Soc.* **24,** 89 (1928); J. C. Slater, *Phys. Rev.* **34,** 1293 (1929); *Phys. Rev.* **35,** 210 (1930); V. Fock, *Z. Phys.* **61,** 126 (1930).
3. P. A. M. Dirac, *Proc. Cambridge Phil. Soc.* **26,** 376 (1930); **27,** 240 (1931).
4. C. A. Coulson and H. C. Longuet-Higgins, *Proc. Roy Soc. (London) A* **191,** 39; **192,** 16 (1947); **193,** 447, 456; **195,** 188 (1948).
5. For a survey, see, e.g., B. Pullman and A. Pullman, *Quantum Biochemistry*. Interscience, New York, 1963.
6. C. C. J. Roothaan, *Rev. Mod. Phys.* **23,** 69 (1951); G. G. Hall, *Proc. Roy. Soc. (London) A* **202,** 336 (1950); **205,** 541 (1951); **213,** 113 (1952).
7. P. O. Löwdin, *Phys. Rev.* **97,** 1490 (1955).
8. S. F. Boys, *Proc. Roy. Soc. (London) A* **200,** 542 (1950); **258,** 402 (1960).
9. C. A. Weatherford and H. Jones (eds.), *ETO Multicenter Molecular Integrals*. D. Reidel, Dordrecht, Holland, 1982; see also H. Jones and C. A. Weatherford, *Theochem,* **58,** 233 (1989).
10. W. J. Hehre *et al.,* QCPE **11,** 236 (1973).
11. P. O. Löwdin, *Rev. Mod. Phys.* **35,** 496 (1963).
12. P. O. Löwdin, *Phys. Rev.* **97,** 1474, 1490, 1509 (1955); *Quantum Theory of Atoms, Molecules, Solid-State,* Slater Dedicatory Volume, p. 601 (P. O. Löwdin, ed.). Academic Press, New York, 1966.
13. For a survey of some applications, see R. Pauncz, *Alternant Molecular Orbital Method*. Saunders, Philadelphia, 1967. See also I. Mayer, *Adv. Quantum Chem.* **12,** 189 (1980), J. L. Calais, *ibid.* **17,** 225 (1985).
14. D. J. Thouless, *The Quantum Mechanics of Many-Body Systems*. Academic Press, New York, 1961; W. Adams, *Phys. Rev.* **127,** 1650 (1962); J. Cizek and J. Paldus, *J. Chem. Phys.* **47,** 3976 (1967); **53,** 821 (1970); H. Fukutome: *Prog. Theor. Phys.* **40,** 1156 (1972). For a survey, see *Int. J. Quantum Chem.* **20,** 955 (1981); J. L. Calais, *Adv. Quantum Chem.* **12,** 225 (1985); P. O. Löwdin, *Proc. Ind. Acad. Sci. (Chem. Sci.)* **96,** 121 (1986).
15. E. Lieb and B. Simon, *Comm. Math. Phys.* **53,** 185 (1977).
16. P. Froelich and P. O. Löwdin, *J. Math. Phys.* **24,** 88 (1983).
17. J. Frenkel, *Wave Mechanics, Advanced General Theory,* particularly pp. 460–462. Clarendon Press, Oxford, 1934; J. C. Slater, *Phys. Rev.* **91,** 528 (1953); P. O. Löwdin, *Phys. Rev.* **97,** 1474 (1955).
18. B. Roos, *Chem. Phys. Lett.* **15,** 153 (1972); see also B. Roos, in *Computational Techniques in Quantum Chemistry* (G. H. F. Diercksen *et al.,* eds.). Reidel, Dordrecht, Holland, 1978.
19. P. Jordan, *Z. Phys.* **94,** 531 (1955).
20. I. M. Gelfand and M. L. Zetlin, *Dokl. Akad. Nauk SSSR,* **71,** 825, 1017 (1950); I. M. Gelfand and M. I. Graev, *Izv. Akad. Nauk SSSR, Ser. Mat.* **29,** 1329 (1965) {Amer. Math. Soc. Transl. **64,** 116 (1967).}
21. J. Paldus and J. Cizek, *Adv. Quantum Chem.* **9,** 105 (1975); F. A. Matsen, *Adv. Quantum Chem.* **11,** 223 (1978); J. Paldus, *J. Chem. Phys.* **61,** 3321 (1974); *Int. J. Quantum Chem.* **S9,** 165 (1975).
22. I. Shavitt, *Int. J. Quantum Chem.* **S11,** 131 (1977); **S12,** 5 (1978).
23. P. Siegbahn, *J. Chem. Phys.* **72,** 1647 (1980).
24. P. Saxe, D. J. Fox, H. F. Schaeffer, and N. C. Handy, *J. Chem. Phys.* **77,** 5584 (1982).
25. P. O. Löwdin, *Phys. Rev.* **97,** 1474 (1955); *Advances in Chemical Physics,* Vol. II, p. 207 and 279 (I. Prigogine, ed.). Interscience, New York, 1959; B. C. Carlson and J. M. Keller, *Phys. Rev.* **121,** 659 (1961); A. J. Coleman, *Rev. Mod. Phys.* **35,** 668 (1963); For some applications, see P. O. Löwdin and H. Shull, *Phys. Rev.* **101,** 1730 (1956).
26. E. R. Davidson, *Reduced Density Matrices in Quantum Chemistry*. Academic Press, New York, 1976.
27. P. O. Löwdin, *Adv. Phys.* **5,** (1956), particularly pp. 46–49; *Ann.*

Rev. Phys. Chem. **11,** 107 (1960); *J. Appl. Phys.* **S33,** 251 (1962); D. McLean, in Ref. 40, p. 87, particularly p. 103; G. Hall, *The New World of Quantum Chemistry,* p. 137 (B. Pullman and R. G. Parr, eds.). Reidel, Dordrecht, 1976. Introductory lecture at Oji International Seminar of Theories and *Ab Initio* Computations of Molecular Electronic Structure, Hokkaido Univ., Tomakomai 1976.

28. P. O. Löwdin, *J. Chem. Phys.* **18,** 365 (1950); *Adv. Phys.* **5,** 1 (1956), particularly p. 49; *Adv. Quantum Chem.* **5,** 185 (1970); for a generalization, see H. Kashiwagi and F. Sasaki, *Int. J. Quant. Chem.* **S7,** 515 (1973).
29. R. S. Mulliken, *Phys. Rev.* **32,** 186 (1928); F. Hund, *Z. Physik* **51,** 759 (1928); see also J. E. Lennard-Jones, *Trans. Faraday Soc.* **25,** 668 (1929), and F. Bloch, *Z. Physik* **52,** 335 (1929); **57,** 545 (1929).
30. W. Heitler and F. London, *Z. Phys.* **44,** 455 (1927).
31. W. Heitler and G. Rumer, *Nachr. Ges. Wiss. Gottingen* **277** (1930); G. Rumer, *Nach Ges. Wiss. Gottingen* **337** (1932); L. Pauling, *J. Chem. Phys.* **1,** 280 (1933); F. A. Matsen, A. A. Cantu, and R. D. Poshusta, *J. Phys. Chem.* **70,** 1558 (1966); M. Kotani, K. Ohno, and K. Kayama, "Quantum Mechanics of Electronic Structure of Simple Molecules," in *Encyclopedia of Physics,* vol. 37. Springer-Verlag, Berlin, 1961, especially pp. 118–142; F. A. Matsen, Spin-Free Quantum Chemistry, in *Advances in Quantum Chemistry,* vol. 1, pp. 59–113 (P. O. Löwdin, ed.). Academic, New York, 1964. Matsen reproduces Kotani's tables in an appendix and shows how to derive nondiagonal matrix elements from the diagonal elements given by Kotani, and P. O. Löwdin, *Colloq. Inter. Centre Natl. Rech. Sc. (Paris)* **82,** 23 (1958); for an elegant solution, see H. Shull, *Int. J. Quantum Chem.* **3,** 523 (1969).
32. R. McWeeny, *Proc. R. Soc. (London) A* **253,** 242 (1959); *Rev. Mod. Phys.* **32,** 335 (1960); *Phys. Rev.* **126,** 1028 (1962); F. A. Matsen, *J. Phys. Chem.* **68,** 3282 (1964); F. A. Matsen, A. A. Cantu, and R. D. Poshusta, *J. Phys. Chem.* **70,** 1558 (1966); F. A. Matsen, *J. Phys. Chem.* **70,** 1568 (1966); F. A. Matsen and A. A. Cantu, *J. Phys. Chem.* **72,** 21 (1968); W. A. Goddard III, *Phys. Rev.* **157,** 81 (1967); **169,** 120 (1968); *J. Chem. Phys.* **48,** 1008, 5337 (1968); W. E. Palke and W. A. Goddard III, *J. Chem. Phys.* **50** (1969); P. O. Löwdin and O. Goscinski, *Int. J. Quantum Chem.* **3S,** 533 (1970); W. A. Goddard III and R. C. Ladner, *J. Am. Chem. Soc.* **93,** 6750 (1971); W. A. Goddard III *et al., Acct. Chem. Res.* **6,** 368 (1973); M. Simonetta and A. Gavezzotti, *Adv. Quantum Chem.* **12,** 103 (1980); see also a forthcoming review in *Theochem.* (1991).
33. For a survey, see E. Clementi *et al.* in *Modern Techniques in Computational Chemistry,* p. 363 (E. Clementi, ed.). ESCOM Science Publishers, Leiden, 1989.
34. P. O. Löwdin, *J. Math. Phys.* **24,** 70 (1983).
35. The resolvents of the Hamiltonian and the Liouvillian were first studied as propagators in nuclear physics; the propagator methods were introduced into quantum chemistry by Linderberg and Öhrn: J. Linderberg and Y. Öhrn, *Proc. R. Soc. London S A* **285,** 445 (1965); Y. Öhrn and J. Linderberg, *Phys. Rev.* **139,** A1063 (1965); J. Linderberg and Y. Öhrn, *Chem. Phys. Lett.* **1,** 295 (1967); J. Linderberg and Y. Öhrn, *Propagators in Quantum Chemistry.* Academic, New York, 1973. For some more recent surveys, see, e.g., P. O. Löwdin, *Int. J. Quantum Chem. Symp. QC* **16,** 485 (1983); *Adv. Quantum Chem.* **17,** 285 (*Academic Press, New York,* 1985).
36. P. O. Löwdin, *Phys. Scripta* **3,** 1 (1985).
37. P. O. Löwdin, *J. Mol. Spectrosc.* **10,** 12 (1963); **13,** 326 (1964); **14,** 112 (1964); **14,** 119 (1964); **14,** 131 (1964); *J. Math. Phys.* **3,** 969 (1962); **3,** 1171 (1962); **6,** 1341 (1965); *Phys. Rev.* **139,** A357 (1965); *J. Chem. Phys.* **43,** S175 (1965); *Int. J. Quantum Chem.* **2,** 867 (1968); **S4,** 231 (1971); **5,** 685 (1971) (together with O. Goscinski); *Phys. Scripta* **21,** 229 (1980); *Adv. Quantum Chem.* **12** (1980); *Int. J. Quantum Chem.* **21,** 69 (1982).
38. C. Møller and M. S. Plesset, *Phys. Rev.* **46,** 618 (1934). For a survey over the development of this approach, see, e.g., J. A. Pople *et al., J. Chem. Phys.* **157,** 479 (1989); *J. Chem. Phys.* (to be published, 1990), and references therein.
39. See, e.g., P. O. Löwdin, in *Quantum Chemistry: Basic Aspects, Actual Trends,* p. 1 (R. Carbó, ed.). Elsevier, Amsterdam, 1989.
40. K. A. Brueckner, *Phys. Rev.* **97,** 1353 (1955); **100,** 36 (1955); H. A. Bethe, *Phys. Rev.* **103,** 1353 (1956); J. Goldstone, *Proc. R. Soc. (London), Ser. A* **238,** 511 (1957).
41. P. O. Löwdin, *J. Math. Phys.* **3,** 1171 (1962), particularly p. 1180.
42. For surveys, see, e.g., R. J. Bartlett, *Ann. Rev. Phys. Chem.* **32,** 359 (1981); S. A. Kucharski and R. J. Bartlett, *Adv. Quantum Chem.* **18,** 281 (1986); M. R. Hoffmann and H. F. Schaeffer III, *Adv. Quantum Chem.* **18,** 207 (1986); R. J. Bartlett, *J. Phys. Chem.* **93,** 1697 (1989).
43. L. Brillouin, *Actualités Sci. Ind.* **71** (1933).
44. P. O. Löwdin, *Proc. Ind. Acad. Sci. (Chem. Sci.)* **96,** 121 (1986).
45. H. Hellmann, *Quantenchemie,* p. 285. Deuticke, Leipzig 1937. R. P. Feynman, *Phys. Rev.* **105,** 1490 (1957). See also P. O. Löwdin, *J. Mol. Spec.* **13,** 331 (1964).
46. See, e.g., P. Pulay, *J. Mol. Phys.* **17,** 197 (1969); *Applications of Electronic Structure Theory,* p. 153 (H. F. Schaefer III, ed.). Plenum, New York, 1977; *Int. J. Quantum Chem. Symp. QC* **17,** 257 (1983); R. J. Bartlett, in *Geometrical Derivatives of Energy Surfaces and Molecular Properties* (P. Jorgensen and J. Simons, eds.). Reidel, Dordrecht, 1986; *NATO ASI Ser. C* **166,** 35 (1986); J. A. Pople, *ibid.,* p. 109. H. F. Schaefer, III, in *New Horizons in Quantum Chemistry* (J. Jortner and B. Pullman, eds.). Reidel, Dordrecht, 1989.
47. Proceedings of the Tromsø Conference on "The Calculation of Intermolecular Energy Surfaces" at the University of Tromsø, Norway, June 20–22, 1989 (P. N. Skancke, ed.), to be published in *Int. J. Quantum Chem.* **38** (1990).
48. J. C. Slater, *Phys. Rev.* **81,** 385 (1951); **82,** 538 (1951).
49. R. Gaspar, Acta Phys. **3,** 263 (1954).
50. J. C. Slater, *Advances in Quantum Chemistry,* Vol. 6, p. 1 (P. O. Löwdin, ed.). Academic, New York, 1972.
51. K. Johnson, *Advances in Quantum Chemistry,* Vol. 7, p. 143 (P. O. Löwdin, ed.). Academic, New York, 1973.
52. K. Siegbahn *et al., ESCA, Atomic, Molecular, and Solid-State Structure Studies by Means of Electron Spectroscopy.* Almqvist & Wiksell, Stockholm, 1967.
53. J. C. Slater, *Phys. Rev.* **37,** 481 (1931); L. Pauling, J. Chem. Phys. **1,** 280 (1931).
54. L. Pauling, *The Nature of the Chemical Bond,* 3rd ed. Cornell University Press, Ithaca, New York, 1960.
55. E. Hückel, *Z. Phys.* **60,** 423 (1930); **70,** 204 (1931); **72,** 310 (1931).
56. K. Fukui, T. Yonezawa, and H. Shingu, *J. Chem. Phys.* **20,** 722 (1952); K. Fukui, T. Vonezawa, C. Nagata, and H. Shingu, *J. Chem. Phys.* **22,** 1433 (1954); K. Fukui, *Theory of Orientation and Stereoselection.* Springer-Verlag, Heidelberg, 1970, 1975.
57. R. Hoffman, *J. Chem. Phys.* **39,** 1397 (1963).
58. R. B. Woodward and R. Hoffman, *J. Amer. Chem. Soc.* **87,** 395, 2046, 2511 (1965); *The Conservation of Orbital Symmetry.* Verlag Chemie GmbH and Academic Press, New York, 1970.
59. R. Pariser and R. G. Parr, *J. Chem. Phys.* **21,** 466, 767 (1953); **23,** 711 (1955); J. A. Pople, *Trans. Faraday Soc.* **49,** 1375 (1953); *Proc. Phys. Soc. (London) A* **68,** 81 (1955); *J. Phys. Chem.* **61,** 6 (1957). For a survey of the development of this method up to now, see "Paper Symposium on the PPP-Method," *Int. J. Quantum Chem.* **37,** 311 (1990).
60. For a survey, see I. Fischer-Hjalmars. *Advances in Quantum Chemistry,* **2,** 25 (P. O. Löwdin, ed.). Academic Press, New York.

61. M. Zerner, in *Reviews in Computational Chemistry,* Vol. 2 (D. Boyd and K. B. Lipkowitz, eds.). McGraw-Hill, New York, to be published.
62. *Horizons in Quantum Chemistry* (K. Fukui and B. Pullman, eds.). Reidel, Dordrecht, 1980; *Int. J. Quantum Chem.* **18,** 1 (1980).
63. *Horizons in Quantum Chemistry* (P. O. Löwdin and B. Pullman, eds.). Reidel, Dordrecht, 1983; *Int. J. Quantum Chem.* **23,** 1 (1983).
64. *Int. J. Quantum Chem.* **29,** 177 (1986).
65. *Horizons in Quantum Chemistry* (J. Jortner and B. Pullman, eds.). Reidel, Dordrecht, 1989; *Int. J. Quantum Chem.* **36,** 187 (1989).
66. *Quantum Chemistry: Basic Aspects, Actual Trends* (R. Carbó, ed.). Elsevier, Amsterdam, 1989.
67. *Electronic Structure of Molecules* (Z. B. Maksic, ed.). IUPAC Publications, Blackwell, Oxford, 1989.
68. See Ref. 47.
69. H. F. Schaefer, "Molecular Electronic Structure Theory 1972–75," *Ann. Rev. Phys. Chem.* (1976); see also *Methods of Electronic Structure Theory* (H. F. Schaefer, ed.). Plenum, New York (1977).
70. P. O. Löwdin, *Adv. Quantum Chem.* **12,** 263 (1980); "On the State of the Art of Quantum Chemistry," *Int. J. Quantum Chem.* **29** (1986); "Some Aspects on the History of Computational Quantum Chemistry in View of the Development as the Supercomputers and Large-Scale Parallel Computers," in *Supercomputer Simulations in Chemistry* (M. Dupuis, ed.). Lecture Notes in Chemistry No. 44. Springe-Verlag, Berlin, 1986.
71. *Modern Techniques in Computational Chemistry,* MOTECC-89 (E. Clementi, ed.). ESCOM, Leiden, 1989.

Molecules

Frank E. Harris

A molecule is a collection of (one or more) atoms which are bound together by their mutual interactions for long enough to be observed as an entity. There is an enormous range in molecular stabilities and lifetimes, with some molecules existing only for the duration of an experiment lasting 10^{-12} s or less, while others may remain intact for billions of years. Molecules are sometimes classified by the number of atoms they contain (e.g., monatomic, diatomic, triatomic); those with three or more atoms are generally termed *polyatomic*. Molecules range in size from monatomic species (as are found in gaseous helium or argon) to macroscopic aggregates (such as single crystals of diamond or quartz, and *polymers*), and in complexity from simple atoms to proteins, enzymes, and nucleic acids. Every known kind of atom is found in at least one diatomic (or larger) molecule; some atoms (e.g., carbon and hydrogen) are found as constituents of millions of different molecules. Molecules containing carbon (with a few exceptions) are called *organic*; all others are *inorganic,* but these terms no longer imply a connection with living organisms. Chemists usually regard the definition of *molecule* as including all observable ionic species. *Ions* differ from (neutral) molecules in that they contain more or fewer electrons than protons and thus carry an electrostatic charge.

In bulk matter, molecules are usually identified as those collections of atoms which by their grouping and its persistence can be distinguished. For example, solid hydrogen may be described as a regular arrangement of hydrogen molecules. However, in some solids (e.g., sodium chloride) each single crystal is a giant molecule (for sodium chloride containing regularly spaced sodium and chloride ions), and some molecules are altered when placed in solution, either by accretion of solvent molecules (as when carbon dioxide is dissolved in water) or by dissociation into ions (as when hydrogen chloride is dissolved in water, yielding hydrogen and chloride ions).

DESCRIPTION

Molecules may be grossly characterized by their elemental composition and electrostatic charge (if any). This information is compactly represented in the *chemical formula*, consisting of the symbols of the different kinds of atoms in a molecule, with right subscripts indicating the number of each kind of atom. By convention, a subscript "1" is dropped. If the molecule has a nonzero charge, it is indicated (in units of the electronic charge) as a right superscript (examples: CO_2, SO_4^{-2}, NH_4^+, $CHCl_3$, C_2H_6O). When the state of aggregation is important, it must be shown in the formula (e.g., CH_3 vs C_2H_6, O_3 vs O_2 vs O).

A further characterization of a molecule is provided by its bond structure. For most (but not all) molecules it is appropriate to regard pairs of individual atoms as connected together by *bonds* representing the existence of significant forces holding these atoms together. It then makes sense to specify the connection pattern exhibited by a molecule. It is sometimes useful to regard a pair of atoms as connected by two or three bonds (a *double* or a *triple* bond). The bond structure can be shown by representing each atom by its symbol and drawing a connecting line to indicate each bond. Examples are shown in Fig. 1. Bond structure diagrams convey information only as to bondedness, but not as to spatial arrangement of the atoms, so the diagrams of Fig. 1 must not be interpreted as indicating bond lengths, orientations, or angles. Chemical formulas are frequently arranged to show bond structure. For example, the formulas of the molecules in Fig. 1 are usually written HCl, Br_2, $Cu(NH_3)_2Cl_2$, C_4H_8, CH_3CCH, CH_3OCH_3, CH_3CH_2OH, CH_3CHO, CH_2CHOH. It is possible for two or more molecules to have the same elemental composition but different bond structures; such molecules are called *structural isomers* (often, simply *isomers*). The sixth and seventh molecules of Fig. 1 are structural isomers, as are the eighth and ninth. The notions of this paragraph assume the bond structure to be reasonably stable, as it is for most molecules. Occasionally, however, one or a few atoms can move between two bonding situations so easily that a pair of isomers rapidly interconverts. Such isomer pairs (example: the eighth and ninth molecules of Fig. 1) are known as *tautomers*.

More detail may now be added by specifying the molecular geometry, i.e., the bond lengths and angles or orientations, as determined from the position of the atomic nuclei. In most molecules the bond lengths and angles between adjacent bonds have definite values; when possible, the molecular geometry often exhibits symmetry. Typical geometries are shown in Fig. 2; note that bond structures, even if conven-

```
                       H                  H     H
                       |                  |     |
                   H - N - H          H - C --- C - H
                       |                  |     |
H—Cl  Br—Br       Cl - Cu - Cl            |     |
                       |              H - C --- C - H
                   H - N - H              |     |
                       |                  H     H
                       H
                                          H  H
   H                 H       H            |  |
   |                 |       |        H - C - C - O - H
H - C - C ≡ C - H  H - C - O - C - H      |  |
   |                 |       |            H  H
   H                 H       H

   H  O              H - O
   |  ||                 |
H - C - C          H - C = C
   |  |                |   |
   H  H                H   H
```

FIG. 1. Bond structures of some molecules.

iently written in a plane, may in fact be three dimensional. Bond lengths and angles show considerable regularity. For example, most C–C and C=C bonds have respective lengths close to 1.54 and 1.34 Å. And when carbon forms bonds to four other atoms, the bonds are usually arranged more or less tetrahedrally. Facts such as these are of great importance in organic chemistry.

The molecular geometry, though definite, is not entirely rigid. Bond lengths oscillate to a small extent about their mean values, as do also bond angles. These types of oscillations are referred to as *vibrations*. There is one further type of internal motion which does not involve changes in bond lengths or angles, namely the *internal rotation* of one portion of a molecule relative to the remainder, about the bond connecting the two parts. A situation of this type is illustrated in Fig. 3a. Internal rotation about ordinary (i.e., single) bonds takes place so freely that under ordinary conditions we cannot identify distinct molecules differing only by an internal rotation angle. However, internal rotation cannot take place about double bonds, and the other atoms directly connected to a pair of double-bonded atoms ordinarily all lie in the same plane. This fact enables the existence of distinct molecules with the same bond structure, as shown in Figs. 3b and 3c. Such molecules are called *geometric isomers*. Another situation exhibiting geometric isomerism is illustrated by the square planar compounds of formula $Pt(NH_3)_2Cl_2$; the two different possible structures are shown in Figs. 3d and 3e.

The geometry of some molecules is sufficiently asymmetric that they show *chirality*, i.e., they can exist in "left-" and "right-handed" forms which are mirror images of each other but are not identical. These forms are called *enantiomers*, or *optical isomers*. A structure exhibiting optical isomerism is shown in Fig. 4. Optical isomers have identical properties in all respects except when in a chiral environment, a circumstance which most frequently occurs during chemical reaction of two enantiomeric molecules. Most molecules in biological systems are enantiomeric, so that optical isomerism is of vital importance in biological and pharmaceutical chemistry.

A few molecules have properties and geometrical arrangements which are not consistent with the notion of bonds connecting pairs of atoms. Such molecules are said to have *nonclassical* structures. Two examples of such structures, illustrated in Fig. 5, are B_2H_6 (best explained in terms of multicenter bonds involving both boron atoms and a central hydrogen) and ferrocene (in which the iron may be considered bound to each ring as a whole).

ELECTRONIC STRUCTURE

At a finer level of detail, a molecule consists of its atomic nuclei and the accompanying electrons. Because of their smaller mass, the electrons move very rapidly in comparison with the nuclei (this is the *Born–Oppenheimer approximation*), so that the nuclei effectively respond only to the average electrical force produced by the electron distribution (this is the *Hellmann–Feynman theorem*). A molecule may thus be described by its nuclear positions and its electron distribution; the actual and most stable molecular geometry corresponds to the nuclear positions which minimize the molecular energy (the potential energies of interaction of nuclei and electrons, plus the kinetic energy of the electrons). The minimum-energy geometry also has the property that the total electrical force on each nucleus vanishes, confirming the stability of the nuclear arrangement. A more detailed understanding of molecules therefore requires consideration of the factors determining their electron distributions.

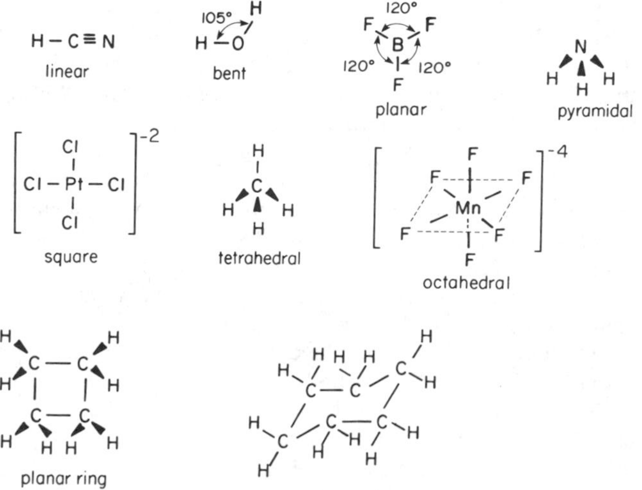

FIG. 2. Typical molecular geometries.

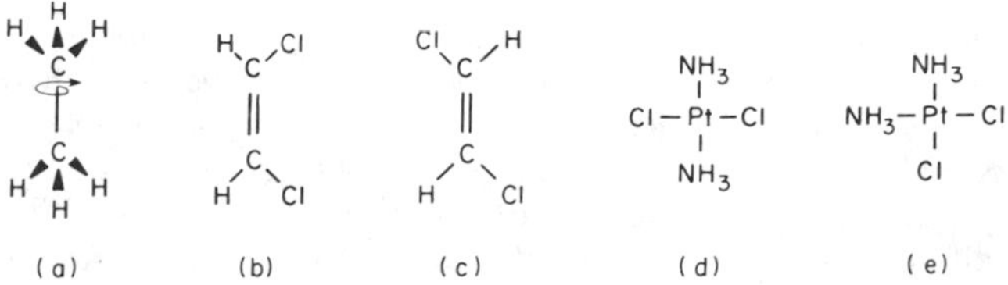

FIG. 3. (a) A molecule in which internal rotation can occur; (b), (c) molecules in which it cannot occur (these molecules are geometric isomers); (d), (e) geometric isomers in which Pt, both N, and both Cl atoms are in square planar configurations.

FIG. 4. A structure exhibiting optical isomerism. *W*, *X*, *Y*, and *Z* represent different atoms. The two molecules are not identical, but are mirror images of each other.

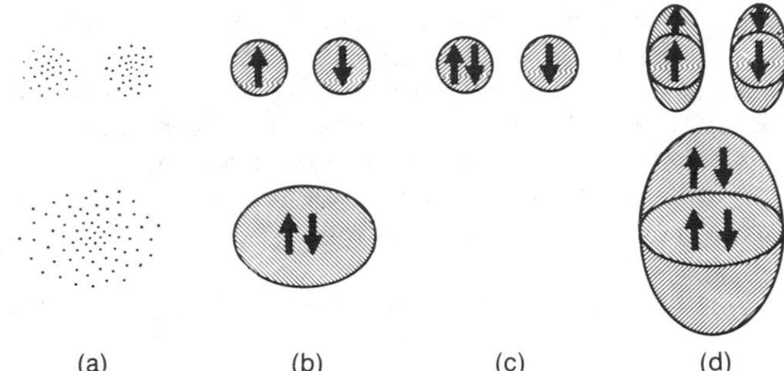

FIG. 6. Electron distributions: (a) of the H–H covalent bond; (b) before and after covalent bond formation; (c) in a case where covalent bond formation cannot occur; (d) before and after formation of a double bond (different shadings identify different distributions). Drawings (b), (c), and (d) are only schematic; arrows denote spin orientations.

The behavior of electrons in molecules can be described by *quantum mechanics*; classical (Newtonian) mechanics cannot be used because the *de Broglie wavelengths* of the electrons are comparable with the molecular dimensions. The main relevant quantum-mechanical ideas are as follows: (i) Electrons are characterized by their entire distributions (called *wave functions* or *orbitals*) rather than by instantaneous positions and velocities; an electron may be considered always to be (with appropriate probability) at all points of its distribution (which does not change with time), (ii) Other things being equal, the kinetic energy of an electron decreases as the volume occupied by the bulk of its distribution increases, so *delocalization* lowers its kinetic energy, (iii) The potential energy of interaction between an electron and other charges is as predicted by classical physics, using the appropriate distribution for the electron; an electron distribution is therefore attracted by nuclei (and its potential energy decreases as the average electron–nuclear distance decreases). An actual (minimum-energy) electron distribution represents the best compromise between concentration near the nuclei (to reduce the potential energy) and delocalization (to reduce the kinetic energy).

It is also necessary to note the existence of *electron spin*, with the following main consequences: (i) In addition to its spatial distribution, each electron can be classified as having one of two spin orientations; and (ii) no two electrons of the same spin orientation may have the same distribution (this is the *Pauli exclusion principle*). In fact, no part of an electron's distribution can be identical with the entire distribution of another electron of the same spin orientation. These facts are responsible for the atomic shell structure; in a many-electron atom, each minimum-energy orbital can be occupied by at most two electrons (one of each spin orientation).

A bond will form between two atoms when the electron distribution of the combined atoms yields a significantly lower energy than the separate-atom distributions. An example is the *covalent bond*, in which two electrons, one originally on each atom, change their distributions so that each extends over both atoms (see Fig. 6a). Since both electrons now have the same distributions, they must have different spin orientations. The extension can be a favored process if the atoms are the right distance apart, because the new distribution then places the electrons on the average as close to the nuclei as before, but enables each electron to be far more delocalized (see Fig. 6b). There will be somewhat more repulsion between the electron distributions than before, but the dominant effect of the delocalization will be a kinetic energy lowering. The potential and kinetic energies will then no longer be at an optimum compromise and the entire electron distribution will need to become more contracted in response to the weakened (lowered) kinetic energy. It is now useful to consider the *virial theorem*, which requires the equilibrium average electronic potential energy to be minus two times the average electronic kinetic energy. This theorem leads to the conclusion that the equilibrium electronic distribution will be more contracted than those of the original bonding atoms, with an increase in the magnitude of the (negative) potential energy and half as large an increase in the (necessarily positive) kinetic energy. Thus, the bonding delocalization is accompanied by a shrinkage in the overall scale of the electron distribution, resulting in a kinetic energy increase, offset by twice as large a potential energy decrease.

A covalent bond will always be possible unless the Pauli exclusion principle prohibits the extension of the atomic orbitals to cover a pair of atoms. This prohibition will occur if there is a second electron with the same spatial distribution as that to be extended, because then the unextended distribution would describe an electron of the same spin orientation as one of the electrons of the extended distribution

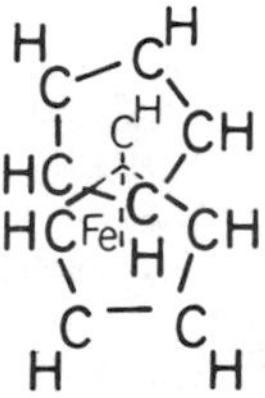

FIG. 5. Nonclassical structures: (a) diborane, B_2H_6; (b) ferrocene, $Fe(C_5H_5)_2$.

(see Fig. 6c). Thus, covalent bonds can be formed using electrons originally in singly occupied atomic orbitals. Some pairs of atoms each have two (or three) electrons in singly occupied orbitals, eligible to form covalent bonds. In such cases two (or three) covalent bonds (each with its own electron distribution) may form between the same atoms; these are the double (and triple) bonds referred to earlier. A double bond is shown in Fig. 6d.

Another electronic rearrangement which may lower the energy is the transfer of an electron from one atom to another. The resulting ions are then attracted to each other, forming an *ionic bond*. This process can occur when the atom receiving the electron has an unoccupied orbital of lower energy than that of an occupied orbital of the donor atom. Covalent bonding is energetically favored over ionic bonding unless there is a large energy difference between the atomic orbitals involved.

A third bonding process, having characteristics of both the first two, is the formation of a *coordinate covalent bond*. This is a covalent bond in which both electrons come from orbitals on one atom; its formation requires the existence of a low-energy unoccupied orbital on the other atom, into which the bonding electron distribution can extend. Coordinate covalent bonding is often an energetically favored process.

Totally reliable prediction of bond formation is difficult, as it depends on detailed quantum-mechanical calculations. However, most bonding can be understood with reference simply to the electronic shell structure of each atom involved, particularly the occupancy (and vacancies) in its outermost, or *valence*, shell. Covalent bonding can take place until this shell is fully utilized. Ionic bonding is most probable between an atom having a nearly empty valence shell [the valence-shell electron(s) of such atoms are easy to remove] and an atom with a nearly full valence shell (in such atoms the unoccupied valence-shell orbitals have unusually low energies). Except for hydrogen and helium, the valence shells of all atoms consist of four orbitals capable of containing up to eight electrons, and the foregoing discussion can be summarized in the *octet rule*, namely, that stable molecular structures are most often characterized by having eight electrons in valence-shell orbitals on, or partly on, each atom. The equivalent of the octet rule for hydrogen and helium (each of which has only one valence-shell orbital) is that two electrons are often on, or partly on, these atoms. Exceptions to the octet rule are most frequently exhibited by atoms beyond the third period, as many such atoms have inner-shell orbitals at energies comparable to those of the four valence-shell orbitals.

A convenient way of showing valence-shell electron distributions is by *electron dot formulas*, using dots on (or between) atomic symbols to represent electrons on an atom (or in covalent bonds). Some typical dot formulas are shown in Fig. 7. The structures not satisfying the octet rule (Cl, BH_3, CH_3, CH_2) are in fact quite unstable and reactive. The octet rule suggests correctly that the stable hydrogen–oxygen and hydrogen–nitrogen neutral molecules will have formulas H_2O and NH_3. The ions H_3O^+ and NH_4^+ are examples of the coordinate covalent binding of a hydrogen nucleus respectively to H_2O and NH_3. The octet-rule structure of SO_4^{-2} explains why that very stable ion has two more electrons than the highly unstable neutral molecule SO_4 (not shown). The dot formula for Cl^- shows the stable ion created when Cl takes part in ionic bonding. The dot formulas for CH_3CHO and CH_3CCH show why these molecules were drawn with multiple bonds in Fig. 1.

Dot formulas can be constructed to determine the probable stabilities and electron distributions of molecules for which this information is not already known. The number of dots to use is given by the total number of valence electrons in the molecule (the sum of the numbers of valence-shell electrons of the atoms, plus or minus, if an ion, the number of electrons added or removed). The dots are then arranged, if possible, in accordance with the octet rule. If several structures are possible, those are preferred which keep the individual atoms more nearly electrically neutral (for this purpose bonded electrons may be assumed half on each bonded atom). Stability is less likely when no octet-rule structure exists.

For some molecules there exist several different electronic structures which differ but little, or not at all, in energy. Two examples are given in Fig. 8. In such molecules the actual electron distribution is an average of the possible structures (called *mesomers* or *resonance structures*), and it can be shown that the averaging is equivalent to a further delocalization of the electron distribution, and hence to an energy lowering. This energy lowering, which makes the molecule more stable, is termed *resonance energy*. The resonance

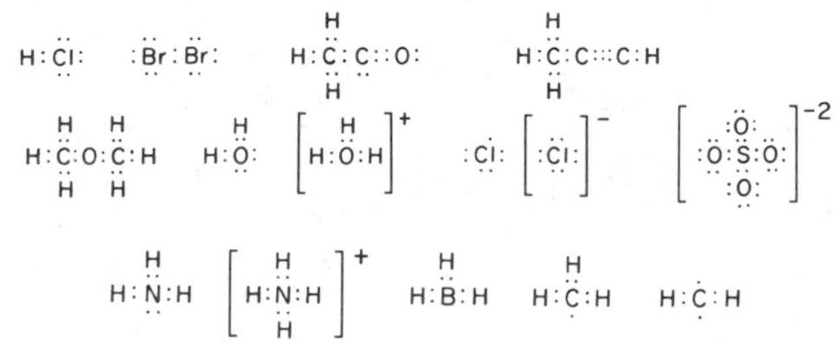

H:Cl: :Br:Br: H:C:C::O: H:C:C:::C:H

H:C:O:C:H H:O: [H:O:H]$^+$:Cl: [:Cl:]$^-$ [:O:S:O:]$^{-2}$

H:N:H [H:N:H]$^+$ H:B:H H:C:H H:C:H

FIG. 7. Electron dot formulas.

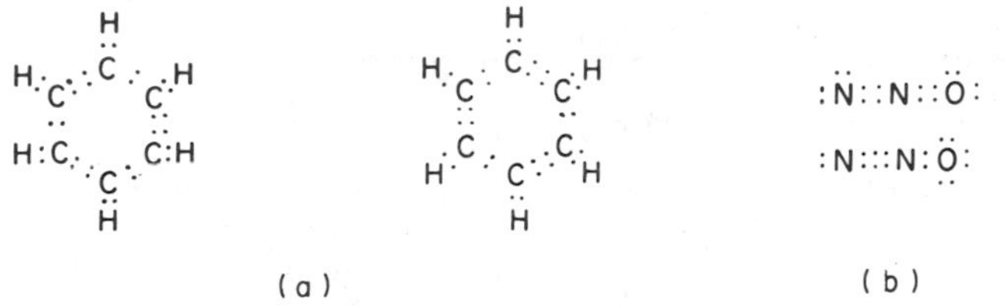

(a) (b)

FIG. 8. Mesomeric electron distributions: (a) benzene, C_6H_6; (b) nitrous oxide, N_2O.

energy tends to be larger for molecules with many mesomers, and for molecules in which the mesomers have completely equivalent structures (as in the first example of Fig. 8).

The methods of quantum mechanics also permit the investigation of other aspects of molecular structure, and in particular can explain bond lengths and bond angles. By studying energy as a function of nuclear position, and by calculating the moments of inertia of a molecule, it is possible to characterize its vibrational and rotational properties. It has also become possible to understand many finer details of the electron distributions and to elucidate further relationships between the electron distribution of a molecule and its physical and chemical properties.

See also CHEMICAL BONDING; CRYSTAL BINDING; INTERATOMIC AND INTERMOLECULAR FORCES; INVERSION AND INTERNAL ROTATION; MOLECULAR STRUCTURE CALCULATIONS; TUNNELING.

BIBLIOGRAPHY

J. A. Campbell, *Chemical Systems*. W. H. Freeman, San Francisco, 1970. (E)

C. A. Coulson, *Valence*, 2nd ed. Oxford University Press, London, 1961. (A)

E. S. Gould, *Inorganic Reactions and Structure* (revised edition). Holt, Rinehart and Winston, New York, 1962. (I)

B. H. Mahan, *College Chemistry*. Addison-Wesley, Reading Mass., 1966. (E)

J. N. Murrell, S. F. A. Kettle, and J. M. Tedder, *Valence Theory*, 2nd ed. Wiley, New York, 1965. (A)

Molten Salts

George J. Janz

The term "molten salts" is generally associated with liquids formed by the fusion of crystalline salts of relatively high melting points. In the solid state such salts are virtually nonconductors (i.e., insulators), whereas in the molten state, most salts conduct electricity so well that the terms liquid electrolytes and molten salts have come to be used almost interchangeably. A comparison with some metals, aqueous electrolytes, and molecular solvents is shown in Table I; the electrical conductance in metals is electronic, whereas in aqueous electrolytes, molecular "solvents," and molten salts, it is ionic. Eutectic mixtures are used to gain the molten state at lower temperatures so as to minimize the problems of corrosion and containment. For example, the LiCl–KCl eutectic mixture (mp. 325°C) is liquid at temperatures greater than 200°C below the melting temperatures of the pure components (KCl, mp. 772°C; LiCl, mp. 610°C), and this system has seen very wide application in molten salts studies.

The variety of inorganic compounds is large and ranges from salt systems that are predominantly ionic to those that are essentially covalent. Mercuric chloride is an example of the latter; on melting, it forms essentially what may be termed a "high-temperature water analogous melt" owing to the molecularities of these two fluids and, as a consequence, the similarities in behaviors of these two systems as solvents for ionic salts. Arranging this wide variety of melts in an ordered manner is important and various schemes have been advanced. One of the more useful is illustrated in Table II. In this approach one draws on structural input and the principles of the well-known phase rule of equilibrium thermodynamics to arrange the systems broadly into groups of increasing complexity. The range of liquids thus formed by fusion of such one-component and multicomponent salt systems is large indeed, but, through such schemes, may be approached in an orderly manner and this can be used in guiding the selections of salt systems for basic studies and as candidate materials in both established technology and in areas still emerging. A partial list of such applications would include:

Chemical: as reaction media for halogenation, oxidation, cracking, condensation, isomerization reactions, and catalysis.

Chemical and environmental engineering: as heat transfer fluids; as reaction media for clean atmosphere processes, such as the removal of SO_2 and SO_3 in emissions of coal burning plants, or sulfide ore smelters; as reaction media for the dissolution of plastic materials for metal recovery, such as gold (from computer chips), or clean copper wire (from coated wires), or silver (from waste photographic film).

Energy: as electrolytes in the high-temperature fuel cells (such as molten carbonate cells), and in the concepts of the superbatteries (e.g., sodium/sulfur; lithium/sulfur), or in high-energy thermal batteries; in advanced nuclear energy concepts, as components in reactors; and in energy storage, as phase change materials for the retention/release of thermal energy in advanced concepts for solar energy utilization, or in power utility stations.

Electrometallurgy and materials science: in metal extractive electrolysis, such as the cryolite process for aluminum extraction, or the sodium chloride electrolysis for winning metallic sodium (and gaseous chlorine); in coating of metal surfaces by electroplating, or by metalliding processes; in the development of glass materials of new compositions, such as the fluoride glasses.

Solid-state technology: as media for single crystal growth; as heat-sensitive detectors, i.e., as in thermal switches; molten semiconductors; as component materials in the search for superconductors.

The list could be extended, but the preceding is sufficient to illustrate this facet of the subject.

Table I Comparison of molten salts with some metals, aqueous electrolytes, and molecular "solvents"

	Metals		Aqueous electrolytes	
	Copper Cu	Gold Au	$NaCl$ (0.1 N)	KCl (0.1 N)
t (°C)	20	20	25	25
Specific electrical conductivity (Ω^{-1} cm^{-1})	5.8×10^5	4.1×10^5	1.1×10^{-3}	1.3×10^{-3}
	Molecular "Solvents"		**Molten Salts**	
	Water H_2O	Carbon tetrachloride CCl_4	$NaCl$ (m. 800°C)	KNO_3 (m. 335°C)
t(°C)	25	25	807	347
Specific electrical conductivity (Ω^{-1} cm^{-1})	0.5×10^{-6}	4×10^{-18}	3.60	0.66
Density (g cm^{-3})	1.0	1.59	1.55	1.86
Viscosity (cP)	1.0	0.97	1.38	2.70
Surface tension (dyn cm^{-1})	72.0	28.0	117	110
Appearance	Clear colorless liquids		Clear colorless liquids	

For conversion between SI and other units:
viscosity: 1 mN s m^{-2} = 1 cP = 1 mPa s
surface tension: 1 mN m^{-1} = 1 dyn cm^{-1}

Table II Molten salts classification system

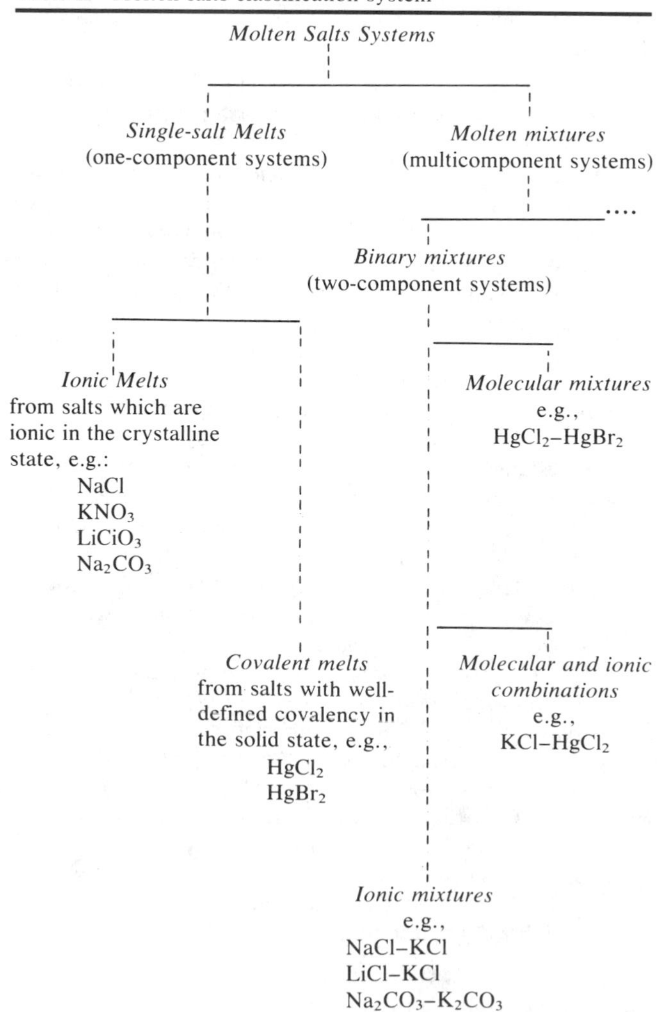

The physics and chemistry of the inorganic molten state have been advanced through studies with a wide range of physicochemical and electrochemical techniques. See Table III. An understanding of the thermodynamic principles governing multicomponent molten salts systems has followed directly with the theoretical advances, such as conformal ionic solution theory (CIST), and its extensions, and through the techniques of experimental thermodynamics, such as solution calorimetry, and equilibrium emf, and/or vapor pressures, in these areas of high-temperature studies.

Table III Experimental techniques used in molten salts studies

Technique	Technique
Spectroscopic	*Computer Simulation*
X-ray diffraction	Simulation of molten salt
Neutron diffraction	Dynamics and equilibrium properties
Microwave	
Nuclear magnetic resonance	*Thermodynamic*
Electron paramagnetic resonance	Density
Raman	Compressibility
Infrared	Cryoscopy
Visible and UV	Electromotive force
Transport	Enthalpy changes
Diffusion	fusion
Viscosity	mixing
Thermal conductivity	reaction
Electrochemical	vaporization
Transport numbers	Heat capacity
Electrical conductance at	Phase equilibria
atmospheric pressures and	Refractive index
superatmospheric pressures	Surface tension
Electroanalytical	Vapor pressure
polarographic	Volume change during fusion
amperometric	*Relaxation Spectroscopy*
chronopotentiometric, etc.	Brillouin light scattering
Dielectric constant	

In the computer-simulation studies of molten state dynamics, the molten halides are being used as model systems since the anions and cations have spherical symmetries, and the external electron shell structures ape those of the rare gases. Neutron-scattering measurements, combined with the techniques of isotopic substitution and radial distribution function (RDF) analyses, are yielding structural information of considerable detail. In related studies, the nature of nearest-neighbor interactions in such molten systems is being probed by the techniques of vibrational spectroscopy using salts with polyatomic anions as *in situ* "sensors." Nitrates are examples.

The principles of acid–base chemistry have been extended to the molten inorganic state to advance the basic understanding of corrosion chemistry, reactive gas "solubilities," oxidation-reduction reactions, the stabilities of various species, such as are possible in the chloroaluminates as a function of overall melt composition. In the past decade "room-temperature" melts have attracted attention for exploratory studies, as electrolytes, and as solvent media for nonaqueous solution chemistry.

BIBLIOGRAPHY

G. J. Janz, "Physical Properties and Structure of Molten Salts," *J. Chem. Educ.* **39,** 59 (1962). (E)

M. Blander, *Molten Salts Chemistry*. Wiley (Interscience), New York, 1964. (A)

B. R. Sundheim, *Fused Salts*. McGraw-Hill, New York, 1964. (A)

G. J. Janz, *Molten Salts Handbook*. Academic, New York, 1967. (I)

B. W. Hatt and D. H. Kerridge, "Industrial Applications of Molten Salts," *Chem. in Britain* **15,** 78 (1979). (I)

H. H. Emons and W. Voight, "Geschmolzene Salze: Vielseitig Nutzbar," *Wissensch. Fortschr.* **30,** 289 (1980). (I)

D. G. Lovering and R. J. Gale (eds.), *Molten Salts Techniques*. Plenum Press, New York, 1983. (A)

G. Mamantov and R. Marassi (eds.), *Molten Salt Chemistry, An Introduction and Selected Applications*. D. Reidel, Boston, 1987. (A)

Moment of Inertia

Frank R. Tangherlini

The moment of inertia of a body depends on the axis of rotation and the distribution of matter about that axis and it is therefore not an unambiguous property of a body such as the translational inertia or "mass." Thus the moment of inertia is part of a more general description of the rotational–inertial properties of a body known as the inertia tensor. For simplicity, the moment of inertia about an axis will be defined and studied first, and then the necessary generalization to the inertia tensor will be given.

DEFINITION

The moment of inertia I about an instantaneous axis of rotation may be defined as that quantity which when multiplied by the instantaneous angular velocity ω yields the angular momentum L about the axis of rotation. Thus the moment of inertia is defined physically by the equation

$$L = I\omega. \tag{1}$$

This equation exhibits an important analogy with the Newtonian definition of linear momentum as the product of inertia and velocity. Historically, the moment of inertia was first introduced in dynamics by Huygens in his analysis of the pendulum, while the name itself is due to Euler. If the axis of rotation is fixed, and the torques about that axis are negligible, L will be a constant of the motion. Hence if I changes, ω will also have to change in order to keep the product constant. Familiar examples of time-varying I resulting in a variable angular velocity are provided by the diver, the ice-skater, and the centrifugal governor. Although other definitions of I are to be found, that provided by Eq. (1) has the advantages that it does not require the body to be rigid, or that it be rotating slowly, and therefore it is consistent with special relativity. Another useful definition of I, albeit restricted to rigid bodies, is obtained from the equation $I\alpha = \Gamma$, where α is the angular acceleration and Γ is the torque.

To calculate the moment of inertia I about a given axis, the body is regarded as a system of N particles of masses $m_i = 1, 2, \ldots, N$, at perpendicular distances r_i from the axis of rotation. It is further assumed that each particle is undergoing an instantaneous "rigid" rotation about the axis so that its instantaneous speed is $v_i = \omega r_i$. Then the magnitude of the linear momentum of an individual particle is $p_i = m_i v_i = m_i r_i \omega$, and hence the angular momentum of an individual particle about the axis of the rotation is $r_i m_i v_i = m_i r_i^2 \omega$. The total angular momentum is therefore,

$$L = \sum_{i=1}^{N} m_i r_i^2 \omega, \tag{2}$$

and hence by Eq. (1), the moment of inertia about the axis is

$$I = \sum_{i=1}^{N} m_i r_i^2. \tag{3}$$

Before proceeding further, the following comments on Eq. (3) will be found useful: (a) If the body is not rigid and/or the axis of rotation is not fixed, the quantities r_i may be time dependent and velocity dependent, and hence, in general, $I = I(t)$ or $I = I(\omega)$ or $I = I(\omega, t)$. (b) In the standard, Newtonian treatment of I, the special relativistic dependence of inertia on speed is neglected. However, if one does not neglect this, one finds a dependence of I on ω^2 in the direction of increasing I with increasing ω^2. Although this ω^2 dependence is under ordinary circumstances a small correction, in contemporary relativistic astrophysical applications such as in rotating neutron stars, it could be important. Likewise, one can anticipate the importance of such corrections in the possible relativistic rotating machinery and scientific instrumentation of the future. For example, for a hoop about an axis through the center and perpendicular to its plane, $I = MR^2(1 - \omega^2 R^2/c^2)^{-1/2}$. Note also that an upper limit to ω is given by c/R. Furthermore, one should not overlook the fact that special relativity predicts that rigid bodies do not exist because the speed of propagation of elastic disturbances is bounded by the speed of light. (c) Although Eq. (3) is based on purely deterministic, "classical" mechanics, the moment of inertia concept continues to hold in the microscopic, quantum world of molecules, atoms, and nuclei. Indeed, the study of the remarkable behavior of the moments of inertia of "deformed nuclei" represents a very active area of contemporary research—in particular, the "backbending" of nuclear moments of inertia. Also it should be remarked that a moment of inertia appears in a time-space component of the metric tensor in the Lense–Thirring perturbative solution to Einstein's field equations. This term is associated with a predicted "dragging" of the inertial frame around a rotating body. This would make it possible for an orbiting gyroscope (Schiff–Fairbank experiment) to determine, say, the moment of inertia of the spinning earth around its axis of rotation, which could then be compared with estimates of it made from geophysical data.

Returning now to Eq. (3), in the case of a continuous distribution of mass, the moment of inertia is calculated from an integral over the mass distribution:

$$I = \int r^2 dm = \int r^2 \rho dV, \tag{4}$$

where ρ is the mass density and dV is the three-dimensional volume element. Homogeneity (ρ=constant), symmetry, choice of coordinates, and choice of axis in conjunction with the parallel-axis theorem given below, frequently lead to simplifications in the evaluation of the sum and/or integral in Eqs. (3) and (4). It has also been found convenient to define the radius of gyration about an axis. All the mass of the object is regarded as concentrated in an infinitesmally thick ring or "hoop" at a radius k from the axis in question, so that $I = Mk^2$, and hence

$$k^2 = \frac{I}{M} = \frac{1}{M}\int r^2 dm, \tag{5}$$

where M is the total mass of the body, $M = \int dm$. It should be pointed out that although the term "moment of inertia" is used in various problems in the theory of elasticity, such as the bending of rods, in this domain, the term means the second moment of the cross-sectional area and, therefore, it has the dimensions of (length)^4. For example, the "moment of inertia" about any axis passing through the center of a circular cross section of radius R is given by $\pi R^4/4$. Also it should be noted that nuclear physicists use the symbol I to denote angular momentum or spin, and the symbol J with a suitable subscript to denote the moment of inertia about some axis.

A simplification of moment of inertia calculations as well as new insight is provided by the parallel-axis theorem of Jacob Steiner: The moment of inertia about an axis parallel to an axis through the center of mass is equal to the moment of inertia about the axis through the center of mass plus the moment of inertia of the center of mass about the parallel axis, hence

$$I_{\parallel} = I_{\text{c.m.}} + MR^2, \tag{6}$$

where $I_{\parallel}$ is the moment of inertia about the parallel axis, and $I_{\text{c.m.}}$ is the moment of inertia about the axis through the center of mass. As an example, consider a uniform sphere rolling without slipping along a horizontal plane. For the sphere, $I_{\text{c.m.}} = \frac{2}{5}MR^2$, and hence about the parallel axis tangent to the sphere where it is in rolling contact with the plane, $I_{\parallel} = \frac{2}{5}MR^2 + MR^2 = \frac{7}{5}MR^2$.

To prove Eq. (6), choose rectangular coordinates (x,y) in the planes perpendicular to the rotation axis, so that $r_i^2 = x_i^2 + y_i^2$. Now let $x_i = X + x_i'$, $y_i = Y + y_i'$, where X, Y are the coordinates of the center of mass, and x_i', y_i' are the coordinates relative to the center of mass. Then $\Sigma\, m_i r_i^2 = \Sigma\, m_i\, (X + x_i')^2 + (Y + y_i')^2$. Upon expanding this sum and using $\Sigma\, m_i x_i' = \Sigma\, m_i y_i' = 0$, the result follows.

The moment of inertia appears in the kinetic energy of a rotating body. In the nonrelativistic case, the kinetic energy of each particle undergoing a rigid rotation is given by $T_i = \frac{1}{2} m_i r_i^2 \omega^2$, and hence the total kinetic energy is

$$T = \sum_{i=1}^{N} T_i = \sum_{i=1}^{N} \frac{1}{2} m_i r_i^2 \omega^2 = \frac{1}{2} I\omega^2. \tag{7}$$

The appropriate generalization to three dimensions will be given below. Some authors prefer to define the moment of inertia starting from the expression for kinetic energy (7). Another useful form of Eq. (7) involves eliminating ω from Eqs. (1) or (2) and writing the kinetic energy in the form

$$T = \frac{1}{2}\frac{L^2}{I}. \tag{8}$$

Once again, this expression is not only applicable in the macroscopic world, but with suitable operator substitutions in the quantum domains as well. However, Eqs. (7) and (8) are only valid for nonrelativistic rotating systems. To get a feeling for the "numbers" note that for a spherical neutron star assumed of uniform density of one solar mass ($M_{\odot} \cong 2 \times 10^{30}$ kg) and radius, say, 10 km, $I_{\text{c.m.}} \cong 0.8 \times 10^{38}$ kg m^2. On the other hand, in studying nuclear moments of inertia, one works with $J/\hbar^2$ in MeV^{-1}.

INERTIA TENSOR

The above considerations will now be generalized to the case of N particles of linear momenta $\mathbf{p}_i$, and position vectors $\mathbf{r}_i$, referred to an arbitrary axis. The total angular momentum vector $\mathbf{L}$ is given by

$$\mathbf{L} = \sum_{i=1}^{N} \mathbf{r}_i \times \mathbf{p}_i. \tag{9}$$

Assuming an instantaneous rigid rotation $\mathbf{v}_i = \boldsymbol{\omega} \times \mathbf{r}_i$, where $\boldsymbol{\omega}$ is the common angular velocity of all the particles about the instantaneous axis of rotation, we can write

$$\mathbf{L} = \sum_{i=1}^{N} m_i \mathbf{r}_i \times (\boldsymbol{\omega} \times \mathbf{r}_i). \tag{10}$$

Upon reducing the vector product, there results

$$\mathbf{L} = \sum_{i=1}^{N} m_i [r_i^2 \boldsymbol{\omega} - \mathbf{r}_i(\boldsymbol{\omega}\cdot\mathbf{r}_i)], \tag{11}$$

where $r_i^2 = \mathbf{r}_i \cdot \mathbf{r}_i$.

From this relation, it follows that the relation between the angular velocity $\boldsymbol{\omega}$ and the angular momentum is not one of direct proportionality, but rather one must introduce the inertia tensor I, which in Cartesian tensor notation is given by

$$I_{jk}=\sum_{i=1}^{N} m_i(r_i^2\delta_{jk}-x_{ij}x_{ik}), \tag{12}$$

where δ_{jk} is the Kronecker-delta symbol, $\delta_{jk}=0$ for $j\neq k$ and $\delta_{jk}=1$ for $j=k$, and $r_i^2=x_{i1}^2+x_{i2}^2+x_{i3}^2$. In Cartesian coordinates, $x_1=x$, $x_2=y$, $x_3=z$, the angular momentum and moment of inertia tensor take the form,

$$\begin{aligned} L_x&=I_{xx}\omega_x+I_{xy}\omega_y+I_{xz}\omega_z,\\ L_y&=I_{yx}\omega_x+I_{yy}\omega_y+I_{yz}\omega_z,\\ L_z&=I_{zx}\omega_x+I_{zy}\omega_y+I_{zz}\omega_z,\end{aligned} \tag{13}$$

where

$$\begin{aligned} &I_{xx}=\sum m_i(y_i^2+z_i^2),\ I_{xy}=-\sum m_ix_iy_i,\ I_{xz}=-\sum m_ix_iz_i,\\ &I_{yx}=-\sum m_iy_ix_i,\ I_{yy}=\sum m_i(x_i^2+z_i^2),\ I_{yz}=-\sum m_iy_iz_i,\\ &I_{zx}=-\sum m_iz_ix_i,\ I_{zy}=-\sum m_iz_iy_i,\ I_{zz}=\sum m_i(x_i^2+y_i^2).\end{aligned} \tag{14}$$

The quantities I_{xx}, I_{yy}, and I_{zz} are the moments of inertia about the x, y, and z axes, respectively, and the quantities $I_{xy}=I_{yx}$, $I_{xz}=I_{zx}$, and $I_{yz}=I_{zy}$ are nowadays called the "products of inertia," although in all the earlier literature the products of inertia were defined with the opposite sign, i.e., $-I_{xy}$, $-I_{xz}$, $-I_{yz}$ were called the products of inertia. Note that some contemporary texts still retain the older definition.

The perpendicular-axis theorem establishes a useful relation between I_{xx}, I_{yy}, and I_{zz} for plane, infinitesmally thin distributions or laminas. Choose the coordinates so that the lamina is in the x-y plane. By assumption, the body has no mass distribution in the z plane, and then $\sum m_iz_i^2=0$. Hence, referring to the expression for the inertia tensor, it follows that

$$I_{xx}+I_{yy}-I_{zz}. \tag{15}$$

This is the perpendicular-axis theorem which may be stated as follows: The moment of inertia of a plane lamina about an axis perpendicular to the plane is equal to the sum of the moments of inertia about two mutually perpendicular axes in the lamina plane. If the plane lamina possesses sufficient symmetry, $I_{xx}=I_{yy}$ and $I_{xx}=I_{zz}/2$. This result simplifies many calculations.

A basic theorem of mechanics (Chasles' theorem) states that the most general motion of a rigid body consists of a translation of some point in the body, plus a rotation about some axis through that point. If one point is therefore kept fixed (imagine a spinning top supported on its tip), the kinetic energy relative to an inertial frame can be written

$$T=\frac{1}{2}\sum_{i=1}^{N} m_i(\boldsymbol{\omega}\times\mathbf{r}_i)^2. \tag{16}$$

This expression is also valid for a nonrigid body all of whose particles are undergoing an instantaneous rotation about some axis. Upon rewriting the above using the vector identity $(\boldsymbol{\omega}\times\mathbf{r}_i)\cdot(\boldsymbol{\omega}\times\mathbf{r}_i)=\boldsymbol{\omega}\cdot\mathbf{r}_i\times(\boldsymbol{\omega}\times\mathbf{r}_i)$ it becomes

$$T=\frac{1}{2}\sum m_i\boldsymbol{\omega}\cdot\left[\mathbf{r}_i\times(\boldsymbol{\omega}\times\mathbf{r}_i)\right]=\frac{1}{2}\boldsymbol{\omega}\cdot\mathbf{L}=\frac{1}{2}\boldsymbol{\omega}\cdot\overleftrightarrow{\mathbf{I}}\cdot\boldsymbol{\omega} \tag{17}$$

where $\overleftrightarrow{\mathbf{I}}$ is another notation for the moment of inertia called the moment of inertia dyadic, and $\overleftrightarrow{\mathbf{L}}=\overleftrightarrow{\mathbf{I}}\cdot\boldsymbol{\omega}$.

Using the moment of inertia tensor $\sum m_i(r_i^2\delta_{jk}-x_{ij}x_{ik})$, the above expression may also be written

$$T=\frac{1}{2}\sum_{j,k} \mathrm{I}_{jk}\omega_j\omega_k, \tag{18}$$

or, expanding in rectangular coordinates,

$$T=\frac{1}{2}(I_{xx}\omega_x^2+I_{yy}\omega_y^2+I_{zz}\omega_z^2+2I_{xy}\omega_x\omega_y+2I_{xz}\omega_x\omega_z+2I_{yz}\omega_y\omega_z). \tag{19}$$

If $\mathbf{n}$ is a unit vector in the direction of the instantaneous axis of rotation, so that $\boldsymbol{\omega}=\omega\mathbf{n}$, then we may write

$$T=\frac{1}{2}\omega\mathbf{n}\cdot\overleftrightarrow{\mathbf{I}}\cdot\mathbf{n}\omega=\frac{1}{2}I\omega^2, \tag{20}$$

where $I=\mathbf{n}\cdot\overleftrightarrow{\mathbf{I}}\cdot\mathbf{n}$ or in tensor notation,

$$I-\sum_{i=1}^{N} m_i\,[r_i^2-(\mathbf{r}_i\cdot\mathbf{n})^2], \tag{21}$$

and this may be used as an alternative definition of the moment of inertia of the body about an instantaneous axis of rotation.

The moment of inertia tensor can be simplified by noting that because the tensor is real and symmetric, it is possible by means of an orthogonal transformation, i.e., a rotation of the coordinate system, to bring the tensor to "principal axes." Thus, under a transformation of the form $\omega_j=\sum a_{jl}\omega_l'$, where the ω_l' are the angular velocities referred to the new coordinate system, and the a_{jl} are the coefficients of the orthogonal transformation, one finds from Eq. (18) that the tensor components I_{jk} transform as

$$I'_{lm}=\sum_{j,k} a_{jl}a_{km}I_{jk}. \tag{22}$$

This is the transformation law for a second-rank Cartesian tensor. The transformation leaves certain properties unchanged; for example, I'_{lm} is also symmetric, $I'_{lm}=I'_{ml}$.

Of particular interest is the fact that for suitable choice of a_{jl}, one can make I'_{lm} diagonal, that is, $I'_{lm}=0$ for $l\neq m$. One refers to such a transformation as a principal-axis transformation; it is not necessarily unique if the body is sufficiently symmetric. Referred to principal axes the tensor takes the form, in matrix notation,

$$\mathrm{I}'_{lm}=\begin{pmatrix} \mathrm{I}_1 & 0 & 0\\ 0 & I_2 & 0\\ 0 & 0 & I_3\end{pmatrix}. \tag{23}$$

The diagonal quantities, I_1, I_2, and I_3, are called the principal moments of inertia, and for a rigid body, they are constants associated with the body which depend on the origin of the coordinates and their labels. To obtain these quantities most directly, given the values of the I_{jk} relative to some arbitrary orientation of the coordinates, one solves the cubic "eigenvalue" equation obtained from expanding the determinant:

$$\begin{vmatrix} I_{11}-I_e & I_{12} & I_{13}\\ I_{21} & I_{22}-I_e & I_{23}\\ I_{31} & I_{32} & I_{33}-I_e\end{vmatrix}=0. \tag{24}$$

On physical grounds, the roots which are the principal moments satisfy $I_1>0$, $I_2>0$, $I_3>0$. Because of the positive-definiteness of the roots, the subdeterminants formed from the components of the I_{ij}, satisfy the following inequalities:

$$|I_{11}|>0,$$
$$\begin{vmatrix} I_{11} & I_{12} \\ I_{21} & I_{22} \end{vmatrix}>0,$$
$$\begin{vmatrix} I_{11} & I_{12} & I_{13} \\ I_{21} & I_{12} & I_{23} \\ I_{31} & I_{32} & I_{33} \end{vmatrix}>0. \tag{25}$$

This is a useful way of checking that the components of the inertia tensor are physically reasonable. In idealized problems, such as that of the infinitely thin needle, one of the principal moments of inertia is allowed to vanish.

After the transformation to principal axes, the kinetic energy takes the form

$$T=\frac{1}{2}(I_1\omega_1{}^2+I_2\omega_2{}^2+I_3\omega_3{}^2), \tag{26}$$

while the angular momentum can be written

$$L_1=I_1\omega_1,\quad L_2=I_2\omega_2,\quad L_3=I_3\omega_3. \tag{27}$$

Hence, upon elimination of the angular velocities, in favor of the angular momenta, the kinetic energy becomes

$$T=\frac{1}{2}\frac{L_1{}^2}{I_1}+\frac{1}{2}\frac{L_2{}^2}{I_2}+\frac{1}{2}\frac{L_3{}^2}{I_3}. \tag{28}$$

The analogy with translational kinetic energy is clear. However, it should be emphasized that the three momenta L_1, L_2, L_3 do not form a canonical set and, consequently, there is greater complexity in treating the dynamics of rotation.

See also DYNAMICS, ANALYTICAL; KINEMATICS AND KINETICS; ROTATION AND ANGULAR MOMENTUM.

BIBLIOGRAPHY

C. W. Allen, *Astrophysical Quantities*. Athlone Press, London, 1973.

H. C. Corben and P. Stehle, *Classical Mechanics*. Wiley, New York, 1950.

R. P. Feynman, *Lectures on Physics*. Addison-Wesley, Reading, Mass., 1962.

J. M. Eisenberg and W. Greiner, *Nuclear Models*, 3rd ed. North-Holland, Amsterdam, 1987.

A. P. French, *Newtonian Mechanics*. W. W. Norton & Co., New York, 1971.

H. Goldstein, *Classical Mechanics*, 2nd ed. Addison-Wesley, Reading, Mass., 1980.

D. Halliday and R. Resnick, *Physics*, 3rd ed. Wiley, New York, 1978.

W. Hauser, *Introduction to the Principles of Mechanics*. Addison-Wesley, Reading, Mass., 1965.

C. Kittel, W. D. Knight, and M. A. Ruderman, *Mechanics*, 2nd ed., Berkeley Physics Course 1. McGraw-Hill, New York, 1973.

L. D. Landau and E. M. Lifschitz, *Mechanics*, 2nd ed. Addison-Wesley, Reading, Mass., 1969.

L. D. Landau and E. M. Lifschitz, *Theory of Elasticity*. Addison-Wesley, Reading, Mass., 1959.

J. B. Marion, *Classical Dynamics of Particles and Systems*, 2nd ed. Academic, New York, 1970.

E. J. Routh, *Elementary Rigid Dynamics*. Macmillan, New York, 1891.

R. B. Sears, M. Zemansky, and H. D. Young, *University Physics*, 7th ed. Addison-Wesley, Reading, Mass., 1987.

A. G. Webster, *The Dynamics of Particles and Rigid, Elastic and Fluid Bodies*. Dover, New York, 1959.

E. T. Whittaker, *A Treatise on the Analytical Dynamics of Particles and Rigid Bodies*. Dover, New York, 1944.

Momentum

Arno Bohm

Newton (1685/1686) introduced a "quantity of motion" reflecting both mass and velocity. This quantity is now called momentum or linear momentum $\mathbf{p}$ and is defined by

$$\mathbf{p}=m\mathbf{v},$$

where $\mathbf{v}$ is the velocity, the change of position per unit time $d\mathbf{x}/dt$, and m is the mass. $\mathbf{p}$ has a direction and a magnitude; it is a vector. If the motion is not a translation but a rotation, then the "quantity of motion" is not linear momentum but angular momentum. For a mass point, orbital angular momentum is given by $\mathbf{j}=\mathbf{x}\times\mathbf{p}$; in the general case of an extended object, angular momentum is defined by

$$\mathbf{j}=I\cdot\boldsymbol{\omega},$$

where $\boldsymbol{\omega}$ is the angular frequency and I is the tensor of inertia of the extended object [1]. Momentum and angular momentum are conserved,

$$\frac{d\mathbf{p}}{dt}=0,\quad \frac{d\mathbf{j}}{dt}=0,$$

if no external forces act on the physical system. If a force $\mathbf{F}$ is applied to the system, then $d\mathbf{p}/dt=\mathbf{F}$.

For velocities comparable with the velocity of light c, Newtonian mechanics is modified according to Einstein's theory of special relativity and momentum is given by

$$\mathbf{p}=\frac{m_0\mathbf{v}}{\sqrt{1-v^2/c^2}}$$

where m_0 is the rest mass [2].

For microphysical systems Newtonian mechanics is modified into the quantum mechanics of Bohr, Heisenberg, Born, Schrödinger, and Dirac and momentum is then no more a number but like all physical observables an operator in a linear space [3]. The three components P_j ($j=1,2,3$) of the momentum $\mathbf{P}$ fulfill together with the three components of the position Q_j the Heisenberg or canonical commutation relations

$$[P_j,P_k]\equiv P_jP_k-P_kP_j=0,\qquad [Q_j,Q_k]=0,$$
$$[P_j,Q_k]=\frac{\hbar}{i}\,\delta_{jk}\mathbf{1},$$

where $\hbar$ is the Planck constant, i is the imaginary unit, $\delta_{jk}=\mathbf{1}$ for $j=k$ and $\delta_{jk}=0$ for $j\neq k$ and $\mathbf{1}$ is the unit operator in the linear space of physical states. Angular momentum in quantum mechanics is an operator whose components fulfill the commutation relation $[J_k,J_l]=i\epsilon_{klm}J_m$, $\epsilon_{klm}=1$ for $k,l,m=1,2,3$ or every even permutation thereof, $\epsilon_{klm}=-1$ for any odd permutation of 1,2,3, and $\epsilon_{klm}=0$ if any two of the k,l,m are equal. The value of momentum, like the value of all quantum physical observables, depends in quantum mechanics upon the state represented by a state vector ψ of the linear space. The average value of many measurements is given by the expectation value $p_i=(\psi,P_i\psi)$ which is the scalar product of the state vector $P_i\psi$ and the state vector ψ.

Momentum (and angular momentum) are connected with motions and their most general definition is: The momentum operators are the generators of the translation group and the angular momentum operators are the generators of the rotation group [4].

For systems with interaction, the notion of momentum undergoes a modification. If we have a particle with charge e in the magnetic field $\mathbf{B}=\nabla\times\mathbf{A}$, where $\mathbf{A}$ is the vector potential, then the operator P_i which fulfills the canonical commutation relations, the canonical momentum (which is the generator of translation), is no more $m\,dQ_i/dt$ as for the interaction-free case. Instead

$$m\frac{dQ_i}{dt}\equiv\Pi_i=P_i-\frac{e}{c}A_i;$$

Π_i is called the kinematical or covariant momentum. The operators Π_j do not commute any more but fulfill the anomalous commutation relation

$$[\Pi_j,\Pi_k]=\frac{i\hbar e}{c}\epsilon_{jkl}B_l,$$

where B_l is the magnetic field. Covariant momenta do not only appear in electromagnetic theory but in any gauge theory [5].

See also Dynamics, Analytical; Group Theory in Physics; Kinematics and Kinetics; Newton's Laws; Relativity, Special Theory.

REFERENCES

1. V. D. Barger, and M. Olsson, *Classical Mechanics: A Modern Perspective*. McGraw-Hill, New York, 1973.
2. W. Rindler, *Essential Relativity*. Springer-Verlag, New York, 1977.
3. A. Bohm, *Quantum Mechanics*. Springer-Verlag, New York, 1986.
4. L. C. Biedenharn and J. D. Louck, *Angular Momentum in Quantum Physics*. Addison-Wesley, Reading, MA, 1981.
5. C. Quigg, *Gauge Theories of Strong, Weak and Electromagnetic Interaction*. Benjamin/Cummings Publishers, Redwood City, CA, 1983.

Monte Carlo Techniques

M. H. Kalos and K. E. Schmidt

Digital computers have become an indispensable tool in many branches of physics; among the numerical techniques available, Monte Carlo methods play a special role because of their combination of immediacy, power, and breadth of application.

A Monte Carlo calculation is one that explicitly uses random variates. There are two complementary reasons for the utility of Monte Carlo: the most obvious is in the study of systems in which random events occur naturally; the other, to be discussed later, is that random sampling can be an efficient numerical method for many-dimensional integrations.

SIMULATION OF NATURAL STOCHASTIC PROCESSES

The transport of particles and photons can be regarded as a sequence of random processes. For neutral particles (e.g., neutrons, photons, neutral pions) the sequence is very clear: the processes that generate the particles are random, the transport is a succession of flights along straight-line paths (whose length is a random variable depending on the geometry and cross sections of the medium) followed by interaction with the isolated scattering centers in the medium. The outcome of these interactions—the kind of radiation produced, the energy, direction, and any other relevant variables—is, to some extent, random. A Monte Carlo simulation of the process may be constructed if we know (or posit) the probability distribution function that governs each step. Using such functions, we can imitate each step with numerical random variates, construct the correct sequence of such events, and measure the appropriate statistics for results. The source is sampled from knowledge of the particle production rate. The straight-line paths are sampled using information about cross sections. The interactions can be simulated from assumptions about differential cross sections.

Numerical procedures may be constructed to produce random variables having any given distribution. It is then a straightforward matter to produce a program for a digital computer that simulates the whole process. The method is easy to apply and powerful in the sense that any complexity of geometry of the medium or of the interaction laws may be treated without approximation.

The method can, of course, be applied to charged particles as well. Although the treatment of very frequent Coulomb collisions presents technical complications, useful calculations have been carried out.

Simulations of this kind have been applied to reactor and nuclear weapons design, to the design of radiation detectors, to the transport of electrons in semiconductors, and to evaluation of the effects of cascades of high-energy particles inside nuclei and in various macroscopic media.

There are many other problems in physics in which the same kind of simulation can be applied directly from random evolution models. Diffusion problems are immediate examples. In solid-state physics, assumptions about the interactions among and motions of atoms and defects can be usefully translated into Monte Carlo simulations of equilibrium behavior and transport processes. Radiation damage and models of the evolution of alloys have been studied this way. The evolution of magnetic systems is analogous. Research on polymer chains and on percolating networks has been carried out by Monte Carlo methods as well.

MONTE CARLO METHODS IN NUMERICAL ANALYSIS

Monte Carlo methods are also useful in situations where they can provide numerical solutions to mathematically posed problems. One example is that of doing an integral of the general form

$$F = \int f(x)p(x)\, dx,$$

where

$$p \geq 0 \quad \text{and} \quad \int p(x)\, dx = 1.$$

(Here x can be a point in a space of any dimensionality.) Under very general conditions, it can be shown that if we draw from $p(x)$ a sequence of samples x_i, $i=1,\ldots,N$ (i.e., the probability that $x_i<X$ is $\int_{-\infty}^{X} p(x)\, dx$; a brief indication of how this may be done will be given later; it is useful but not necessary that successive x_i be independent), then the mean value

$$\bar{F} = \frac{1}{N}\sum_{i=1}^{N} f(x_1) \mapsto F$$

as N increases. Thus Monte Carlo provides a general method of doing integrals, one that is straightforward in many dimensions. Furthermore, it can be shown that if the integral $\int f^2 p\, dx$ exists, then the statistical error of the Monte Carlo method decreases as $N^{-1/2}$, faster in the total number of points at which $f(x)$ must be evaluated than ordinary finite-difference methods for multidimensional spaces. Monte Carlo as a method for numerical quadrature has been applied to a number of problems, including the calculation of cross sections in high-energy physics, the calculation of integrals that appear in variational estimates of the quantum many-body problem, the integration over initial conditions in the classical or semiclassical calculation of atomic and molecular interactions, and the evaluation of path integrals.

The most widely used example of such quadrature is the computation of averages for canonical ensembles of interacting particles. Such an ensemble can be found in any of many possible states. The Boltzmann distribution gives the probability that a system is in that spatial state (the velocity variables are distributed independently in a Maxwellian and do not affect the spatial variables) with potential energy U:

$$\phi_j = \exp\left(-\frac{U_j}{k_B T}\right) \Big/ \sum_k \exp\left(-\frac{U_k}{k_B T}\right),$$

where k_B is Boltzmann's constant and T is the absolute temperature. Then the average potential energy is

$$\langle U \rangle = \sum_j \phi_j U_j.$$

This and related many-dimensional sums (or, more usually, integrals) can be readily estimated by Monte Carlo if p_j can be sampled. A simple, elegant device (described later) has been found to carry out such sampling, and this kind of Monte Carlo has found very broad applications in statistical physics, chemistry, biology, and other sciences.

The method discussed above for numerical quadrature is, in a sense, the inversion of the more usual mathematical problem of formulating the expectation of a stochastic process as an integral. Other such inversions can be found that lead to interesting applications. Finite-difference approximations to differential equations often lead to equations that describe random walks on lattices. The diffusion or heat equations can be solved in this way. It has also been suggested for the Schrödinger equation.

The expected behavior of objects undergoing continuous random walks with birth and death satisfy integral equations. The inversion (in the sense of the preceding paragraph) leads to a simple way of solving integral equations with positive integrable kernels, a way that is practical in many dimensions.

The Schrödinger equation can be solved in this way if a known (or formal) Green's function is used to rewrite the differential equation as an integral equation. This method has proved capable of giving results that are analytically exact and numerically precise enough to improve the accuracy of variational studies of quantum many-body problems.

SAMPLING RANDOM VARIATES

No Monte Carlo calculation can be carried out without some basic random variate. It has been conventional to start with "random numbers" that are independent random variates uniformly distributed on (0,1). That is, if ξ is such a variate, then

$$\Pr\{\xi<X\} = \begin{cases} 0 & \text{if} \quad X<0, \\ X & \text{if} \quad 0\leq X\leq 1, \\ 1 & \text{if} \quad X>1. \end{cases}$$

Such random variates can be generated from natural stochastic processes (e.g., radioactive decays), but it has proved possible and more practical to use artificial (i.e., "pseudo-random") numbers generated by various computing methods.

However, Monte Carlo calculations require more than uniform random variates, for example, in simulation of radiation transport. It is possible to use uniform variates to generate others with any given distribution. Space precludes more than a few examples. If an event occurs with probability p_0, then a program can emulate that choice by considering the event to occur if ξ [uniform on (0,1)] is less than p_0.

A general technique is to map one random variate into another. Thus, let

$$x = -\ln\xi, \qquad 0<x<\infty.$$

Since x is a decreasing function of ξ,

$$\Pr\{x \geq X\} = \Pr\{\xi \leq e^{-X}\} = e^{-X}.$$

This is a useful way of sampling an exponential distribution. Mapping can be used quite generally, sometimes with numerical tables, to sample distributions including those derived from empirical data.

It is interesting to discuss briefly the "Metropolis" method of sampling the Boltzmann distribution for the coordinates of an interacting many-body system. If a system has ergodic dynamics, consistent with detailed balance—i.e., if the system can go from state i to state j with probability p_{ij} such that

$$\phi_i P_{ij} = \phi_j P_{ji}$$

for some function ϕ_j—then the system will come to equilibrium in which the probability of being observed in state j is ϕ_j. The Metropolis method consists of adopting artificial dynamics in which one or more particles are moved at a time according to rules such that

$$P_{ij}/P_{ji} = \exp[-(U_j - U_i)/k_B T].$$

Although many such rules are possible, the most widely used is to move a particle chosen at random uniformly and at random in a cube centered at that particle, and with probability

$$q = \min\{1, \exp[-(U_j - U_i)/k_B T]\}$$

accept the new state (j). With probability $1-q$, the system is returned to the previous state (i).

While the trial state change from i to j is, in a sense, immaterial, dramatic changes in the rate of equilibration can occur. The best known example is in magnetic systems where the Ising model is often used. There the magnetic domains are modeled by simple spins with two states, and single spin flips are a possible move. In the neighborhood of a critical point, the approach to equilibrium becomes very slow (in an infinite system, infinitely slow), and the computing time to get small statistical errors grows very large. This effect is called critical slowing down. The generic solution is to use moves $i \rightarrow j$ that change appropriate collective variables. The detailed solution for the Ising systems has recently been found.

VARIANCE REDUCTION

When the Monte Carlo calculation is posed, at least in part, as a form of quadrature, techniques can be found to reduce the statistical error for one trial (or, more important, per unit of computing time).

An elementary example is that of the integration $\int p(x)f(x) \times dx$. Clearly,

$$F = \int p(x)f(x)\,dx$$
$$= \int [p(x)f(x)/g(x)]g(x)\,dx.$$

If $g(x)/\int g(y)\,dy$ is sampled at random given x_i, $1 \leq i \leq N$, then F is approximately

$$F \approx \frac{1}{N}\sum_i \frac{f(x_i)p(x_i)}{g(x_i)} \int g(y)\,dy.$$

Suppose $g = fp$; then the bracket is identically N and the Monte Carlo estimate is exact, whatever values of x and N were used. In this case there is no statistical error. By choosing g close to fp, the error of the integration can be reduced by large ratios.

A similar, though somewhat more complicated, transformation can be applied to Monte Carlo solutions of linear integral equations in which a linear functional is required or in which an eigenvalue (as in the Schrödinger equation or the fission reactor criticality problem) must be estimated. This transformation in principle gives the result with zero statistical error and in practice gives the possibility—which has been realized—of very large reduction in computing time.

A wide variety of other error-reduction methods have been formulated. One such device is to compute numerically the expected result of some random event that leads directly to a quantity to be tabulated instead of awaiting the successful outcome of the random selection. In other stratagems, correlations are introduced between different estimators in such a way as to reduce the final error.

These error-reducing methods can be applied to Monte Carlo calculations that arise originally as simulations of natural stochastic systems. It is necessary, however, to cast the problem, at least in part, in a mathematical framework. It is straightforward to formulate simulations of particle transport as the solutions of integral equations whose kernel describes the processes of scattering and flight.

HYBRID ALGORITHMS

Hybrid algorithms combine Monte Carlo sampling with deterministic methods. An example is diffusion analyzed using the Langevin method where the medium surrounding the diffusing particle is modeled by a random force. Another application is the use of molecular dynamics to calculate thermodynamic properties in the canonical ensemble. Normally, molecular dynamics methods calculate thermodynamic quantities with constant total energy, i.e., the canonical ensemble. However, by randomly resampling particle velocities, the system is modeled to be in contact with a reservoir at constant temperature, and canonical ensemble quantities can be calculated.

CONTEMPORARY APPLICATIONS

Monte Carlo methods are widely used in theoretical research and as an adjunct to experimental physics. Indeed, a large fraction of the capacity of supercomputers involves Monte Carlo in some form. In experimental physics, especially high-energy physics, the behavior of detectors or proposed experiments is extensively preanalyzed using Monte Carlo methods. The numerical evaluation of predictions of quantum chromodynamics is the subject of very large computations worldwide, and involves very large Monte Carlo or hybrid Monte Carlo molecular dynamics methods. Similar calculational methods, effectively the stochastic evaluation of path integrals in discretized imaginary time, are widely used in calculations on quantum problems in condensed matter physics, including those that relate to theories of high-temperature superconductivity. "Quantum Monte Carlo" is now a recognized technique in electronic structure including applications to atomic and molecular structure. The methods of Monte Carlo transport theory are routinely applied to astrophysics and to the behavior of electrons in solids.

Finally, a class of stochastic optimization methods, including "simulated annealing," has been brought into use in many fields. References are given below to most of these application areas.

SUMMARY

Monte Carlo methods can be readily applied to the simulation of physical systems in which natural stochastic processes occur. In addition, they lend themselves to the numerical solution of many-dimensional integrals, sums, and integral and matrix equations. The distinction between simulation and quadrature is often hard to draw clearly. When problems to be solved are stated in a more formal way, powerful methods for reducing the statistical error, and hence the computer time needed, can be applied.

See also FLUCTUATION PHENOMENA; PROBABILITY; STATISTICS; STOCHASTIC PROCESSES.

BIBLIOGRAPHY

General

J. M. Hammersley and D. C. Handscomb, *Monte Carlo Methods*. Methuen, London, 1964. (Although dated, a good general introduction to the Monte Carlo method.)

M. H. Kalos and P. W. Whitlock, *Monte Carlo Methods, Vol. I: Basics*. Wiley, New York, 1986. (An up-to-date general reference work on Monte Carlo methods.)

Generation of Random Variates

C. J. Everett and E. D. Cashwell, *A Third Monte Carlo Sampler*. Los Alamos National Laboratory report: LA-9721-MS, 1983. (Available from National Technical Information Service, Springfield, VA. An update of the previous two Monte Carlo Samplers by these authors on sampling many probability distributions.)

D. E. Knuth, *The Art of Computer Programming*, 2nd ed., Vol. 2. Chapter 3. Addison Wesley, Reading, MA, 1973. (A good summary on random number generation and testing.)

Applications to Radiation Transport

M. J. Berger, "Diffusion of Fast Charged Particles," in *Methods in Computational Physics* (B. Alder, S. Fernbach, and M. Rotenberg, eds.), Vol. 1. Academic, New York, 1963. (Review of methods for simulating transport of charged particles.)

L. L. Carter and E. D. Cashwell, *Particle-Transport Simulation with the Monte Carlo Method* (TID-26607). (Available from National Technical Information Service, Springfield, VA. A good review of both theory and practice.)

G. Goertzel and M. H. Kalos, "Monte Carlo Methods in Transport Problems," in *Progress in Nuclear Energy* (D. J. Hughes, ed.), Ser. I. Vol. 2. Pergamon, New York, 1958. (The discussion of applications is dated, but the treatment of the basic theory is still useful.)

C. Jacobini and L. Reggiani, "The Monte Carlo Method for the Solution of Charge Transport in Semiconductors with Application to Covalent Materials," *Rev. Mod. Phys.* **55**, 645 (1983). (Good review of transport Monte Carlo in semiconductors.)

J. Spanier and E. M. Gelbard, *Monte Carlo Principles and Neutron Transport*. Addison-Wesley, Reading, MA, 1969. (Fundamentals of Monte Carlo as well as a discussion of several problems in radiation transport.)

High-Energy Physics

T. W. Armstrong *et al.*, *Nucl. Sci. Eng.* **49**, 82 (1972). (Description of methodology and results for Monte Carlo calculation of high-energy nucleon-meson cascades; comparison with experiments.)

I. Montvay, "Numerical Calculation of Hadron Masses in Quantum Chromodynamics," *Rev. Mod. Phys.* **59**, 263 (1987). (A good review of Monte Carlo methods in quantum chromodynamics.)

Applications to Statistical Mechanics

K. Binder (ed.), *Monte Carlo Methods in Statistical Physics* (Vol. 7 of *Topics in Current Physics*). Springer-Verlag, Berlin, 1979. (A collection of articles on many aspects of statistical physics.)

K. Binder (ed.), *Applications of the Monte Carlo Method in Statistical Physics*, 2nd ed. (Vol. 36 of *Topics in Current Physics*). Springer-Verlag, Berlin, 1987. (A continuation of the previous book, this contains articles on additional topics in statistical physics. A further volume in this series is in preparation.)

Critical Slowing Down

R. H. Swendsen and J.-S. Wang, "Nonuniversal Critical Dynamics in Monte Carlo Simulations," *Phys. Rev. Lett.* **58**, 86 (1987). (This gives a method for relieving the critical slowing down problem.)

Simulated Annealing

P. J. M. Laarhoven and E. H. L. Aarts, *Simulated Annealing Theory and Applications*. Reidel, Boston, 1987; E. H. L. Aarts, *Simulated Annealing and Boltzmann Machines: A Stochastic Approach to Combinatorial Optimization and Neural Computing*, Wiley, New York, 1987. (Reviews of the simulated annealing optimization method.)

Quantum Systems

J. E. Gubernatis (ed.), Proceedings of the Conference on Frontiers of Quantum Monte Carlo, held at Los Alamos National Laboratory, Los Alamos, NM published in *J. Stat. Phys.* **43**, 729–1242 (1986). (A collection of articles on many aspects of quantum Monte Carlo. This conference was given in honor of N. Metropolis' seventieth birthday; the proceedings also contains some historical information.)

M. H. Kalos (ed.), *Monte Carlo Methods in Quantum Problems*, NATO Advanced Research Workshop Series C, Vol. 125. Reidel, Boston, 1984. (A collection of articles on quantum Monte Carlo.)

Mössbauer Effect

G. J. Perlow

This effect which bears the name of its discoverer, Rudolph Mössbauer, was first described in a publication in 1958. It is the recoil-free emission and absorption of nuclear gamma rays and is a very useful special case of nuclear resonance fluorescence. The recoil-free process results in emission of γ-rays with natural or nearly natural linewidth which allows the observation in the γ-ray spectrum of the interaction between the nucleus and its atom. The phenomenon occurs only in solids or extremely viscous liquids.

The phenomenon is illustrated in Fig. 1. A γ ray, shown as resulting from a nuclear excitation following the β decay of a radioactive parent, can be removed from a collimated beam by resonantly re-exciting another of the same species. In an *idealization* of the process, the emitter and the target nuclei are constrained from moving. The energy distribution of both the emitted radiation and the absorber cross section has the natural width $\Gamma_0 = \hbar/\tau$, where τ is the mean lifetime of the excited state. Resonant cross sections are high, of order $\lambdabar^2$, where $\lambdabar =$ wavelength/2π. For energies less than about 100 keV, the nuclear events dominate over excitation processes in the electronic shells of the atom and can be observed readily. Such nuclear states may typically have mean lifetimes of a few nanoseconds or longer, and therefore widths less than $\approx 10^{-6}$ eV. Indeed some are very much narrower.

The situation has been over-idealized. There is no method for rigidly constraining the nuclei. Before Mössbauer's discovery, the following description would be appropriate: A γ-ray with energy E carries momentum E/c, and the emitting nucleus (in fact the whole atom) recoils with energy $R = E^2/2Mc^2$. This energy is lost to the transition. A similar shortfall of energy occurs at the absorber, which cannot absorb the quantum unless it also recoils. In addition there is thermal motion which, because of the Doppler effect, spreads both the emission intensity and the absorption cross section out to a width $\Gamma \approx (RK_BT)^{1/2}$, where T is the Kelvin temperature and K_B is Boltzmann's constant. The resonant absorption is then due only to the small overlap between the two distributions.

Mössbauer's remarkable discovery was that the spectrum of radiation emitted from nuclei in *solids* is actually a composite of the idealized and the unconstrained case. The same is true for the distribution of the absorption cross section. For relatively low-energy transitions, some fraction f of the events occur without recoil and as if the nucleus were exactly at rest. The emission or absorption widths have the natural value. Except for a small but important chemical effect there is no energy shift. The sharp line is superimposed on a spectrum of recoil events but on the scale in which one uses Mössbauer spectroscopy, the latter forms a featureless and flat background. The recoil-free fraction f depends sensitively on the energy, and on the temperature and chemical form of the sample.

A technique, also due to Mössbauer, that displays the narrow recoil-free line is the method of velocity or Doppler spectroscopy. If the source and absorber are moved relative to one another with relative velocity v (taken as positive when the two are approaching), then the energy as seen by the absorber is shifted by Ev/c. By varying v, typically a few millimeters per second, the resonance can be scanned. The transmitted intensity is a minimum at the velocity of exact resonance, and on either side of resonance the background is due to recoil events and to irrelevant radiation that had been incompletely rejected by the detector.

Figure 2 is a velocity spectrum for ^{182}W taken at $T = 4.2$ K. The transition is between the first excited and the ground state. Here $E = 100$ keV, $\tau = 2.0$ nsec, and $\Gamma_0 = 3.3 \times 10^{-7}$ eV or 1.0 mm/sec. Because the width enters into the spectrum twice, the minimum observable value (for lines of the Lorentz shape as here) is $2\Gamma_0 = 2.0$ mm/sec. The radiative parent, 115-day ^{182}Ta, is made by neutron irradiation of a tantalum metal foil. Its β decay produces the excited emitter nucleus ^{182}W. The latter is then present as a dilute impurity in a tantalum host lattice. The absorber is a foil of tungsten metal

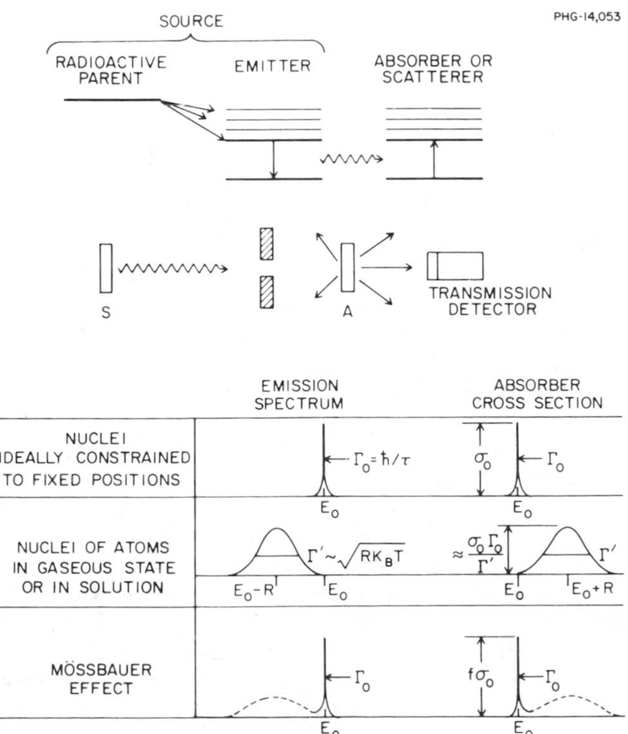

FIG. 1. Description of the Mössbauer effect.

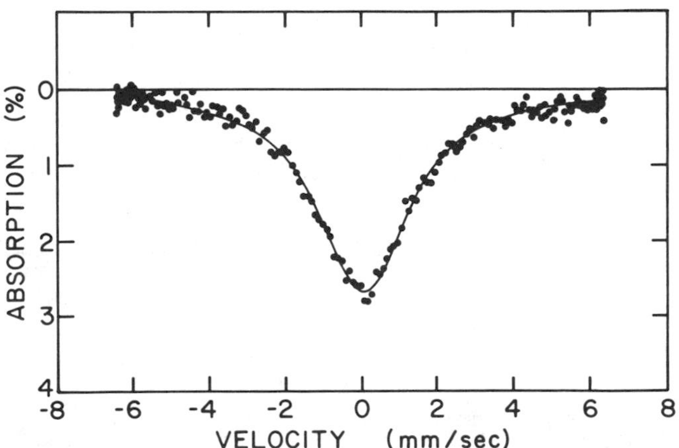

FIG. 2. The Mössbauer effect in ^{182}W. The scatter of the points is due to the statistics of counting. The smooth curve is a Lorentzian, fitted to the data by least squares. The observed width is 3 mm/sec to be compared to the natural width $2\Gamma_0 = 2.0$ mm/sec. The excess is caused by absorber thickness.

in which only the mass-182 isotope whose relative abundance is 26.4% is effective. It is important to note that the substance studied is ^{182}W, not the parent ^{182}Ta. Following a radioactive decay, the daughter atom adjusts its electronic shells to its new nuclear charge in a time short compared to τ in almost all of the cases studied so far.

There are at present 110 transitions in 89 isotopes of 45 elements for which the Mössbauer effect has been observed. A condensed list is given in Table I. The most important is the 14.4-keV transition in ^{57}Fe, because of the key role played by iron in materials science, magnetism, chemistry, and biology. Some of the characteristics of this resonance are $\tau = 140$ nsec, $2\Gamma_0 = 9.3 \times 10^{-9}$ eV or 0.19 mm/sec. The cross section at resonance is $\sigma_0 = 2.4 \times 10^{-18}$ cm^2. This is about 500 times larger than the cross section for all electronic processes in the atom for radiation of this energy. The internal conversion coefficient of the state is about 8, i.e., the state decays eight times more often by internal atomic processes than by γ-ray emission. Most transitions with energies appropriate for Mössbauer spectroscopy decay more often by internal conversion than by radiation. The target nucleus is more of a true absorber than a scatterer.

Although there is an interesting classical analog, the recoil-free phenomenon is basically quantum mechanical.

Before the emission (or absorption) of the quantum, the lattice is in some complicated internal state. An impulse is given to a single atom, and if that causes a change of energy state, the result is a shift of the γ-ray energy. In the contrary case the internal lattice state remains unchanged, and except for an exceedingly small energy shift due to the macroscopic recoil of the sample as a whole, the line is neither shifted nor broadened. It is the fraction f of these in which we are interested. The details of the calculation are complex and are omitted here.

Various treatments lead to the following instructive intermediate formula for the recoil-free fraction:

$$f = e^{-k^2\langle x^2\rangle_{av}}. \tag{1}$$

Here $k = 1/\lambda\!\!\!^-$ and x is the momentary displacement of the nucleus along the direction of observation of the gamma ray. The displacement is due to lattice vibrations—the recoil does not specifically enter. The $\langle\ \rangle_{av}$ brackets denote a time average over a period long compared to τ. The energy enters via k, since $E = c\hbar k$, and because $\langle x^2\rangle_{av}$ increases with temperature, the Mössbauer fraction is increased with lower E and lower T. Even at $T = 0$, however, $\langle x^2\rangle_{av}$ is not zero because there is an unremovable "zero-point" motion. The fraction f is very small if, in the course of a few nuclear lifetimes, the atom moves more than a wavelength or so from its initial position. Thus the Mössbauer effect is characteristic of solids, where the atomic motion is bounded, or of liquids of such extremely high viscosity that the atom, although free, does not diffuse rapidly enough. For an emitter dissolved in water at room temperature, the typical diffusion distance is perhaps 1,000 times too great. Liquids in which the diffusion is small enough to display the Mössbauer effect appear quite solid by ordinary inspection.

Table I Elements having at least one Mössbauer isotope

Potassium	Iodine	Terbium	Osmium
Iron	Xenon	Holmium	Iridium
Manganese	Cesium	Erbium	Platinum
Nickel	Barium	Thulium	Gold
Zinc	Lanthanum	Ytterbium	Mercury
Germanium	Praseodymium	Lutetium	Thorium
Krypton	Neodymium	Hafnium	Protactinium
Technetium	Promethium	Tantalum	Uranium
Ruthenium	Samarium	Tungsten	Neptunium
Tin	Europium	Rhenium	Plutonium
Antimony	Gadolinium	Dysprosium	Americium
Tellurium			

Equation (1) also shows that the effect is favored for heavier stiffer lattices as these have smaller values of $\langle x^2\rangle_{av}$ at a given temperature.

A slight variant of Eq. (1) due to Waller and Debye describes the coherent part of x-ray and neutron scattering from crystals and predates the Mössbauer effect by many years. The quantity f is sometimes called the Debye–Waller factor. Formulas for it in which $\langle x^2\rangle_{av}$ is evaluated with the Debye model of lattice vibrations, can be found in the literature of x-ray diffraction. In an approximation suitable for $T < 0.3\Theta$, where Θ is the Debye temperature of the material, the recoil-free fraction is

$$f = \exp\{-(3R/2K_B\Theta)[1 + (2\pi^2/3)(T/\Theta)^2]\}, \tag{2}$$

which reduces to

$$f = e^{-3R/2K_B\Theta} \quad \text{for } T \ll \Theta. \tag{3}$$

In Eq. (2), $(T/\Theta)^2$ comes from the temperature-dependent atomic motion, while in Eq. (3) only the zero-point motion of the atom remains. It is the only term of importance at liquid helium temperature for almost all materials. Equation (3) predicts the value $f = 0.93$ for ^{57}Fe in its own lattice at low temperature.

APPARATUS

The typical experiment is done in transmission geometry as in Fig. 1. The radioactive source is oscillated so that its velocity is a triangular wave or a sinusoid. The gamma radiation of interest is detected in a scintillation counter, a proportional counter, or a semiconductor detector and after amplification is selected by a single-channel analyzer. The counts are stored in successive memory locations in a computer or in a multichannel analyzer, whose address is advanced and reset in synchronism with the velocity drive, so that each channel corresponds to a different velocity. The process is repeated cyclically until a spectrum has accumulated. The velocity drive is usually an electromechanical device. (However, the spectra shown here were taken in the author's laboratory with a mechanical system.) A cryostat for lowering the temperature of source and absorber separately or together may be required. About 10 Mössbauer nuclides can be used at room temperature. Most of the remainder require liquid helium temperature (4.2 K).

There are many variants on this description. In some experiments, chiefly on magnetic materials, the absorber may be placed in a magnetic field or possibly brought to temperatures below 1 K. It may be an oriented single crystal.

It, or the source, may be subjected to high pressure. In the typical Mössbauer transition, the excited state decays more often by internal conversion than by radiation, so that electrons are emitted at resonance. Typically they have short range. By counting electrons, particularly those that have suffered minimum energy loss, rather than gamma rays, the velocity spectrum represents only the properties of the surface region. Experiments are sometimes done by observation of the resonantly scattered radiation, generally for reasons of intensity, in the backward direction.

Most of the interesting spectra show multiplet structure, a composite of splittings of the energy states of source and absorber. Usually, it is some property of the absorber that is being investigated, and for this it is advantageous to have a single-line source. If the radioactive parent occupies a lattice site with cubic symmetry in a non-magnetic lattice, its daughter generally does so also and meets this requirement.

There are alternatives to a radioactive source. The emitting nucleus may be produced by a nuclear reaction in an accelerator target which is then the source, or it may be produced in one target and implanted downstream by recoil into a different host lattice. Implantation of radioactive material by an isotope separator into specialized hosts is a way of obtaining alloys that cannot be produced otherwise. Here it is often the source that is being studied. Another very interesting technique involves synchrotron radiation, partially monochromatized by Bragg diffraction, and then filtered further by resonant scattering. The method is in its infancy at this time, and although the expected counting rates are only comparable to those of radioactive sources, the radiation is highly collimated and has precise timing. There will be special classes of experiments for which the synchrotron source is eminently suited.

Let us consider with the aid of illustrations some of the kinds of measurements that can be made with Mössbauer spectroscopy.

THE LINE POSITION

On the scale of Mössbauer measurements, the energy of a nuclear transition depends on the structure of the whole atom and thereby on the electronic structure of the lattice. Some of the electrons, those in s states, have non-vanishing density, $\rho(0)$, at the nucleus. The electronic charge within the nuclear volume causes the nuclear ground state whose mean-square proton charge radius is $\langle r^2\rangle_{av}$ to be increased in energy above that of a point nucleus by an amount proportional to $\rho(0)\langle r^2\rangle_{av}$. An excited nuclear state with the same charge radius would suffer the same increase and the transition energy would be unaltered, but if a difference $\Delta\langle r^2\rangle_{av}$ exists, there will be an energy change proportional to $\rho(0)\Delta\langle r^2\rangle_{av}$. This is a chemical shift since for a given nuclear transition the quantity $\rho(0)$ is sensitive to chemical state. It is generally called the isomer shift in analogy to the isotope shift of atomic spectroscopy whose origin is similar. It is a useful indicator of ionic charge state because even if s electrons are not themselves transferred in forming a particular chemical bond, the gain or loss of, for example, p or d electrons changes the shielding of s electrons and thereby $\rho(0)$. In Fig. 1 we note that although the emitter and the target atoms are in different hosts, the shift is barely observable. The reason is that both of the nuclear states are members of the same rotation band and have very nearly the same intrinsic structure. In contrast, the shift between high-spin ferrous and ferric compounds in ^{57}Fe spectroscopy is typically about 0.8 mm/sec.

There are two other interesting causes of line shift. The first is the Pound-Josephson effect. It is a second-order Doppler shift proportional to $\langle u^2\rangle_{av}$, the mean square thermal velocity of the atom about its equilibrium position. It can be thought of as a relativistic time dilatation which lowers the γ-ray frequency by a factor $(1-\langle u^2\rangle_{av}/c^2)^{1/2}$. It shows up as a temperature effect, but is also present whenever one compares the transition in two hosts of different Debye temperatures. It is generally quite small compared to the isomer shift, except in isotope line ^{67}Zn where the isomer shifts are small and the linewidth very narrow. The second is the gravitational red shift. An excited nucleus at a height z above an absorber has an extra potential energy gzE/c^2 because of the mass associated with the excitation by the Einstein relation $E=mc^2$. This appears as an addition to the transition energy. It is a red shift if the absorber is above the source, a blue shift otherwise. For the 93-keV radiation from ^{67}Zn, the shift is Γ_0 for each 4.5 m of height. The effect can also be observed in the centrifugal field produced by a rapidly spinning rotor which holds source and absorber at different radii.

MULTIPLET STRUCTURE OF THE SPECTRUM—MAGNETIC EFFECTS

A nuclear state with angular momentum quantum number $I\geq 1/2$ possesses a magnetic dipole moment $\boldsymbol{\mu}$. In a magnetic field $\mathbf{B}$, the Zeeman interaction $-\boldsymbol{\mu}\cdot\mathbf{B}$ splits the state into its $2I+1$ magnetic substates. If there is no electric quadrupole interaction (see below), these are equally spaced in energy between $-\mu B$ and $+\mu B$. The γ-ray transitions between the components of the excited state and those of the ground state are observed as a multiplet spectrum.

An important source of magnetic fields is internal to the atom. An unpaired electron density at the nucleus is a magnetization because the electron has an intrinsic magnetic moment. The density and its radial variation produce a magnetic field at the nucleus, much as the magnetization and its divergence produce a field in macroscopic materials. It is called the Fermi contact field. In the ions of the transition metals, there can be a net spin in the partially filled d shell. The d electrons have no density at the nucleus but produce a contact field by polarizing s-electron shells. The orbital motion also produces a magnetic field. It is of somewhat minor importance in d-shell ions, but large in the rare earths.

Internal magnetic fields of whatever origin are generally called "hyperfine fields." In order for them to be observed by Mössbauer spectroscopy, they must persist in direction for a time long compared to the precession time of the nuclear spin. In a ferromagnet or in an antiferromagnet the direction of the internal field is preserved in space and its Zeeman effect shows up in the spectrum. We illustrate with two cases, ferromagnetic iron (Fig. 3) and antiferromagnetic IrF_6 (Fig. 4). In the latter case the ordering temperature lies between the values of 4.2 and 27 K where the two spectra

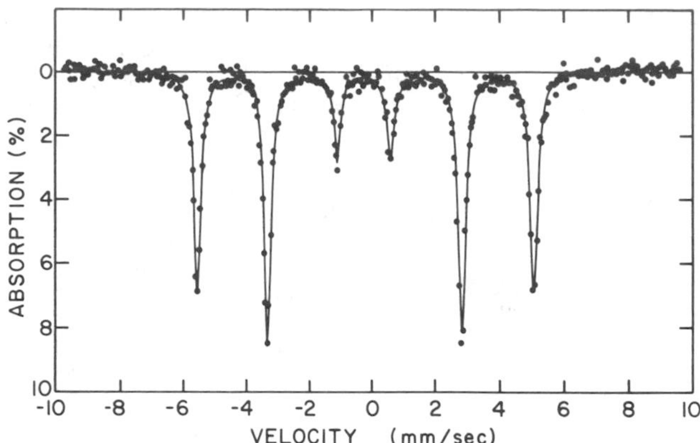

FIG. 3. Velocity spectrum of ^{57}Fe at room temperature. The source is ^{57}Co in a copper host. The absorber is 7.5 μm of pure, annealed, iron foil of normal isotopic abundance (2.2% ^{57}Fe). An average of several runs yields $B = 332.3 \pm 0.2$ kG for the hyperfine field, and $\mu_{ex}/\mu_{gnd} = 1.7132$ for the ratio of the excited to ground-state magnetic moments. Other experiments show that B is negative.

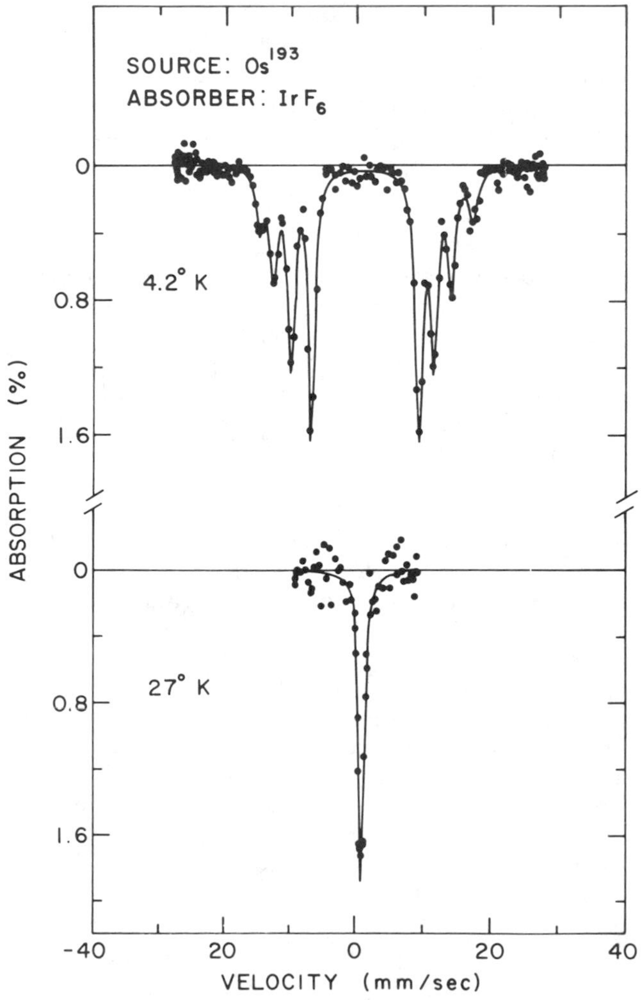

FIG. 4. Spectrum of ^{193}Ir as IrF_6, below and above the antiferromagnetic ordering temperature. Comparison of the split spectrum with one for Ir in iron shows that the observed ratio of magnetic moments depends on the host material because of the existence of a large hyperfine anomaly.

shown are taken. There are eight possible transitions between the magnetically split $I = 3/2$ and $I = 1/2$ states that characterize the Mössbauer transitions in both isotopes. In ^{57}Fe the internal nuclear structure is such that the radiation is almost purely magnetic dipole ($M1$), and two of the possible eight lines (those with $\Delta m = \pm 2$) are forbidden. The ^{193}Ir decay is a mixed multipole ($E2$ and $M1$) and all eight lines are seen. The line intensities depend on the mixing ratio and on the angle between the internal field and the direction of observation. The isomer shifts between source and absorber are visible.

Hyperfine fields may lie either parallel or antiparallel to the bulk magnetization. The behavior of the magnetic splitting with external magnetic field sorts out the two cases.

As the discipline has progressed, interest has developed in compounds and alloys of greater and greater complexity. The simple six-line spectrum of Fig. 3 is seldom seen. Instead there is a complex structure that is sorted out by curve-fitting techniques. A rather less extreme case is shown in Fig. 5, which is a velocity spectrum of a thin single crystal of yttrium iron garnet (YIG) in a magnetic field. It displays 18 lines. (YIG is an example of a ferrimagnet; the magnetization of the sublattices does not cancel out as in the normal antiferromagnet, and the sample has a net magnetization.)

An ion in the paramagnetic condition has an internal field that fluctuates in direction due to interaction of its electronic structure with neighboring spins and with thermal vibrations of the lattice. A typical ferrous ion, for example, shows no magnetic splitting above its ordering temperature as the field averages to nearly zero in the time of a Larmor precession. A ferric ion, however, because it has no orbital angular momentum, interacts less strongly with the lattice. If it is somewhat dilute and the temperature is low, a split spectrum occurs that reflects the electronic state of the ion in its surroundings. Intermediate temperatures and dilutions result in spectra of considerable complexity. They are examples of relaxation processes about which there is considerable theoretical literature. Relaxation spectra are seen also in rare-earth and actinide isotopes. Many applications of Mössbauer spectroscopy to biology arise from analysis of such spectra in iron-bearing proteins.

Related to the relaxation process is the existence of superparamagnetism. In bulk magnetic material, there is a temperature below which magnetic ordering exists. However, if the material is in the form of very small particles, the magnetic hyperfine splitting is not seen until a lower temperature. The lattice magnetization (or that of the sublattice for antiferromagnets) fluctuates between directions of minimum energy in the particle and averages to nearly zero when the fluctuation rate is high enough. As the temperature is lowered, the rate decreases and a distribution of hyperfine fields is observed. At low enough temperature, the spectrum of the bulk material is found. The particle size and shape and the magnetocrystalline anisotropy determine the variation of the spectrum with temperature.

ELECTRIC EFFECTS

Nuclear states with $I \geq 1$ have electric quadrupole moments, that is to say the charge does not have spherical sym-

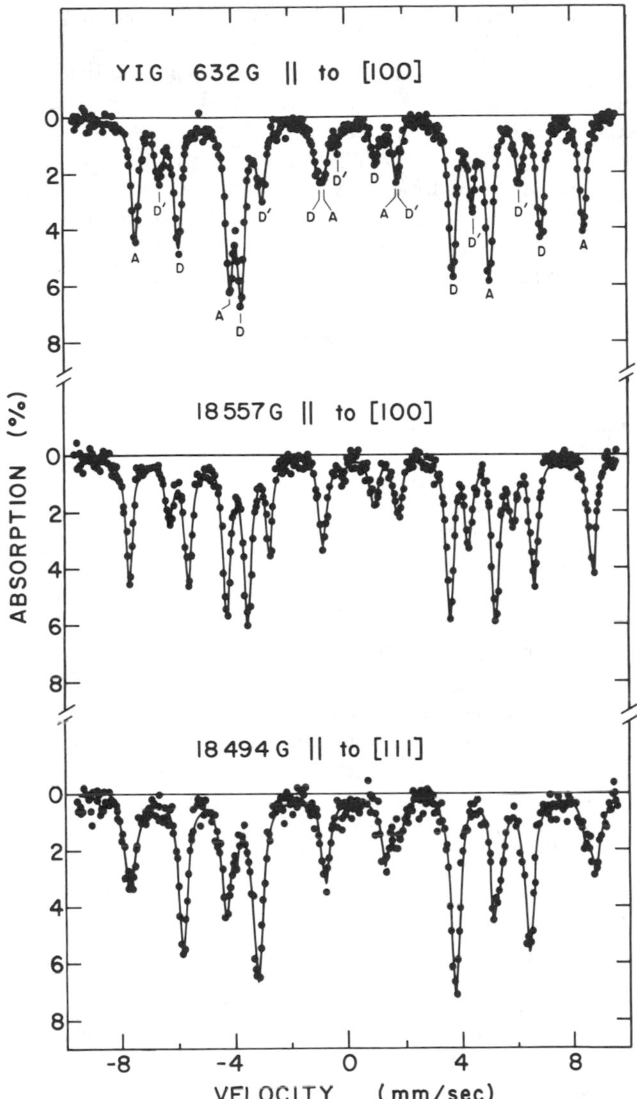

FIG. 5. Velocity spectra of single-crystalline YIG in weak and strong magnetic fields with the 14.4-keV line of ^{57}Fe. The crystal face is perpendicular to [111] and contains [100] and [111] directions. There are electric field gradients whose symmetry directions reduce the possible number of different spectra from seven to three if the magnetic field is along either [111] or [100]. The hyperfine fields of the d and d′ spectra (Fe in tetrahedral sites) decrease in the strong field, and that of the A (octahedral) site increases. Within error the change is equal to the change in applied field. The internal field is opposite to the magnetization, and the magnetization of this ferrimagnetic material has the direction of the tetrahedral site moments. From unpublished work of G. J. Perlow, R. G. Peterson, and E. G. Spencer.

metry. The quadrupole moment interacts with the gradient of the electric field to split an otherwise degenerate manifold of substates, or, if there is already magnetic splitting, to alter the intervals. While the electric field is quite small at a nucleus, the same cannot be said of its derivatives. The largest are produced by the electronic structure of the parent atom. Electric charges on other atoms also produce field gradients at the nucleus of the atom in question but their effect is smaller because of the greater effective distance. However, an antishielding effect (Sternheimer shielding) magnifies their influence. If the charge distribution has cubic or spherical symmetry, the field gradients with respect to any three mutually perpendicular axes must be equal, and by Laplace's equation which states that their sum vanishes, each must do so separately. Thus electric quadrupole splitting is a property of less-than-cubic symmetry.

In the case of nuclei with odd mass number, I is an odd half-integer, and the states are always left with a twofold degeneracy by a quadrupole interaction. Thus instead of $2I+1$ different energy levels one sees half this number. However, if the nucleus has even mass number there need be no degeneracy if the local symmetry is low enough. We illustrate quadrupole splitting with Fig. 5(a) for ^{129}Xe in the compound XeF_4. The excited $I=3/2$ state splits into a doublet, while the ground $I=1/2$ state is unsplit. The resulting spectrum has two lines. The structure of XeF_4 is a square plane of fluorines about a central xenon. The analysis of the spectrum combined with other information shows that the bonds are made by transferring about $0.75e$ from the $5p_{x,y}$ orbitals of Xe to each F, thereby reducing the initial spherical symmetry of the shell. Figure 5(b) is the spectrum of $XeCl_4$ produced in the β decay of ^{129}I in $KICl_4$ and examined at the instant of formation with the aid of an unsplit absorber. The intensity asymmetry is caused by premeditated alignment of

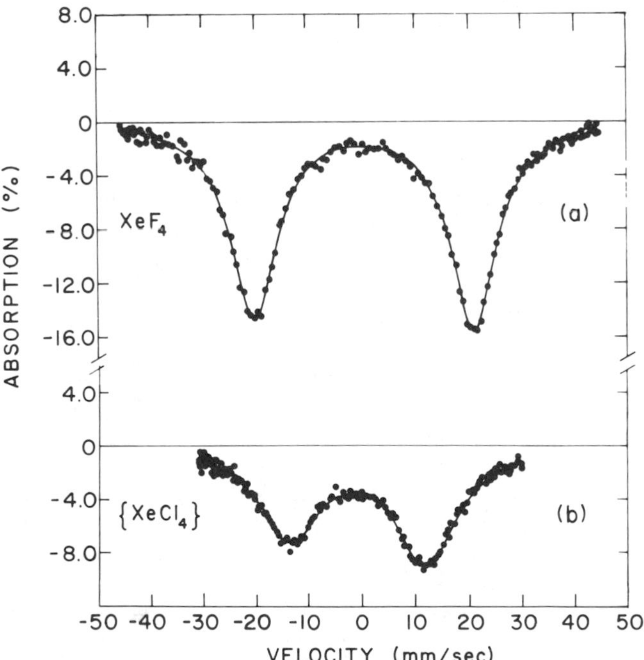

FIG. 6. Quadrupole splitting in XeF_4 and $XeCl_4$. In the former an unsplit source of ^{129}I as sodium paraperiodate ($Na_3H_2IO_6$) is used. In the latter, the unsplit absorber is sodium perxenate (Na_4XeO_6). These both have an octahedron of oxygen ions about the central atom and hence locally cubic symmetry. After the decay of the ^{129}I, both ions are in fact identical, as may be affirmed by the absence of isomer shift when comparing them in the same experiment. The $XeCl_4$ is made in the beta decay process and can be deduced to have the square planar ligands of the ICl_4^- ion of the source material.

the crystals in the source. The mere existence of the lines shows that the species is stable for at least 10^{-9} sec at 4.2 K.

POLARIZATION

If the splitting is magnetic, and the magnetization direction is kept fixed (in a ferromagnet it suffices to apply a modest external field to orient the domains), then the multiplet lines are polarized. The polarization of a line in a source may be measured by using a polarized line in an absorber. Linear polarization is found in directions perpendicular to the magnetization, circular polarization in the parallel directions, and elliptical in between.

Polarization is also a characteristic of emission or absorption of quadrupole components if the sample is a single crystal.

APPLICATIONS

We conclude with a brief discussion of some of the uses to which Mössbauer spectroscopy has been put.

Nuclear Physics

The main thrust here has been in evaluating electromagnetic moments of excited states. If both excited and ground state are split, the ratio of magnetic or quadrupole moments of the two is obtained directly from the spectrum. If the ground-state moment is available from other data, the hyperfine field or the electric field gradient is obtained. If the line is narrow enough for an external magnetic field to be useful, the magnetic moment of the excited state is determined redundantly. Isomer shifts have been used to determine the change in nuclear radius between ground and excited state. The measurements are mainly somewhat ahead of the theory. An exception is in the case of some of the rotational states, where the shifts are very small, and the theory is concerned with small perturbations of the same intrinsic state that are caused by the rotation.

Magnetism and Phase Transformations

This is a vast area. Extensive studies have been made of a great variety of alloys and compounds mainly with ^{57}Fe, but also with a number of other Mössbauer isotopes. In fact, a nominally diamagnetic test probe like ^{119}Sn samples the polarized band structure of a magnetic alloy and shows a transferred hyperfine field. Similar transference is also displayed in nonmetals, for example, CrI_3 studied with the Mössbauer effect in ^{129}I and ^{127}I. Phase transitions between two magnetic phases, or between a magnetic and a paramagnetic phase, are readily determined by observing the hyperfine field with varying temperature. In most materials there are various sites for the magnetic ion. These lead to a number of hyperfine fields. The structure in favorable cases can be sorted out by using the full bag of tricks of the discipline; single crystals, applied fields, variable temperatures, angular effects, and comparisons of the deduced spectral parameters with known cases. Of considerable interest are magnetic phenomena associated with dilute impurities in a magnetic host; for example, the very large ionic magnetic moment associated with an iron impurity in certain hosts which was discovered by measuring the hyperfine field as a function of applied external magnetic field. Study of the Kondo condensation phenomenon has been done similarly. The magnetic structure of the electronic ground state of dilute rare earth impurities has been an active field.

Amorphous materials are actively studied. If they are magnetic, there are distributions of hyperfine fields as well as isomer shift and quadrupole interactions which gives information on the range of the local ordering. In "spin glasses" the crystal lattice possesses order whereas the moment distribution is amorphous. Nonmagnetic amorphous or glassy materials are of current interest and also amenable to insights from Mössbauer spectroscopy, frequently by use of dopants like ^{119}Sn.

Chemistry

In structural chemistry, a Mössbauer spectrum can often decide between two possible structures, or can assign the correct site symmetry for the ion being studied. As an analytic technique, it can detect a suitable element nondestructively, determine a valence state, frequently determine the actual compound by comparison with known parameters, and can with some error tell the quantity. In complex molecules, it can identify particular structural groupings by isomer shift, quadrupole splitting, or magnetic hyperfine field.

The electronic state of an ion and the populations of the orbitals involved in the bonds can be obtained in favorable cases. It is also useful for studying the chemical species formed in radioactive decay. There is a growing collection of systematic data on variation of Mössbauer parameters over a series of related compounds. The greatest concentration of such studies has been on iron and tin compounds, but a considerable fraction of the elements in Table I have also been used. It is possible to investigate a limited area of solution chemistry by Mössbauer spectroscopy, by working with frozen solutions.

There is strong modern interest in surface chemistry in part motivated by the importance of catalysis and of surface oxidation. The ideal detection method, conversion-electron counting, is mainly suitable for Fe, but, for example, other nuclides adsorbed or deposited on substrates having large surface areas can be studied in the usual transition geometry.

Some rare earths form compounds of intermediate valence which is seen by the change of isomer shift with temperature or pressure.

Biology

Iron plays a dominant role as the active element in a variety of biologically important proteins where it exists as Fe^{II} or Fe^{III} in either high- or low-spin forms, distinguishable by their spectra. In the heme proteins, for example, the difference between the oxygenated and deoxygenated state is a change between low- and high-spin Fe^{II}, while ligands other than oxygen can result in the other two possibilities. The analysis of the state of the irons and their electronic structure, including magnetic properties, represents one of the main lines of Mössbauer spectroscopy of biological materials. The phenomenon of superparamagnetism is often seen in biological materials and can be used, for example, to determine the sizes of iron clusters.

Magnetite, a magnetic iron oxide, has been found by Mössbauer spectroscopy in certain bacteria and animals where its interaction with the earth's field provides a directional sense for the organism.

A recent trend has been to use the Mössbauer effect as a tool in diagnostic medicine. Studies to date have concentrated on the spectral variations found in hemoglobin from patients with certain diseases.

With the accumulation of a body of information on the parameters associated with certain structures, the Mössbauer effect can be used as a technique for deciphering unknown ones.

Various

The Mössbauer effect has found use in archaeology, mainly because iron is a component of clays. Pottery can be traced to its source and the firing techniques can be determined. Dating can be done in favorable cases. Studies of meteorites, minerals, volcanic debris, and lake and river sediments have been made.

Studies of lattice dynamics can be made by examining the variation of the recoil-free fraction and other parameters with temperature. Rayleigh scattering by the electrons in a solid can be either recoil-free or not. The recoil-free fraction is described by Eq. (1) with k replaced by $2k \sin \theta/2$, where θ is the scattering angle. The scatterer can be any substance as the scattering is not resonant. The Mössbauer equipment selects the recoil-free part. Rayleigh scattering can be used for example to study diffusion in a solid near its melting point.

It was early recognized that if diffusion of a resonant atom occurred on a slow enough scale, it would show up as a line broadening. This has been observed in some biological materials. A similar effect is observed when an atom jumps about in a cage of its neighbors. This has been observed in radiation-damaged material and in metals with dissolved hydrogen. One observes a line of normal width along with the broadened one.

Various experiments of a fundamental nature have been done with Mössbauer spectroscopy. We mentioned the gravitational and centrifugal redshifts. An ether drift experiment with a source and absorber mounted on a rotor has shown the ether drift to be less than a few centimeters per second. Experiments to detect an absence of invariance under time reversal have been done by polarization measurements in mixed multipole transitions. Experiments have been done to measure parity mixing in a nuclear transition.

Experiments that involve simultaneous time and energy measurements illustrate the ideas of complementarity and lead to interesting phenomena. The radiation that is transmitted through a resonant absorber has its amplitude distribution altered in the process. If one makes a lifetime measurement of the state using the filtered radiation and a previous γ ray to determine zero time, the decay is no longer exponential. This has been called "time filtering." A related set of experiments starts out by vibrating a source with a period shorter than the lifetime of the nuclear state. Then one observes a frequency-modulation spectrum, containing a central line (the "carrier") and symmetric sidebands on either side of it separated (in velocity units) by the modulation frequency. If the source is counted, the relative phases of the components is such that they produce no effect due to interference, but if there is any resonant absorption, a periodic variation, "quantum beats," is seen in the counting rate. The harmonic content of the beats is very sensitive to the energy shift between source and absorber and has been used to measure the second-order Doppler shift corresponding to temperature changes of a few degrees centigrade. Conversely, if a small interval of the modulation cycle is used to gate the counter, dispersion curves are seen in which there is an apparent generation of intensity at certain parts of the spectrum. Although these experiments were done with ^{57}Fe, the long-lived transition in ^{67}Zn and its concomitant narrow line ($2\Gamma = 0.31$ μm/sec) have permitted very interesting subsequent demonstrations, for example, transient phenomena. In all of these experiments the "absorber" is to be thought of as a resonant optical medium with absorption.

Another example of quantum beats occurs in the Bragg and near-Bragg scattering of synchrotron radiation from a single crystal of an iron compound. Because the radiation comes in narrow time bursts, the Zeeman components of the excited Fe state are coherently excited with fixed relative phases. A complicated interference pattern is seen in the resulting time spectrum. It can be analyzed to give the crystal parameters. One also sees a large enhancement of the elastic resonant scattering over the inelastic at the Bragg angle. The effect had been predicted theoretically.

Resonant dispersion is seen in the velocity spectrum of ^{169}Tm. The transition is $E(1)$—relatively rare in low-energy decays. There is interference between the internal conversion amplitude following absorption and the photoelectric amplitude. The former changes phase over the resonance whereas the latter does not. Constructive interference on one side of the line and destructive on the other results in an asymmetric line shape, again with an appearance of generated intensity on one side. The effect has been seen for other multipoles, but much diminished because the photoeffect is very largely $E(1)$.

A variety of other Mössbauer measurements based on the coherence properties of the narrow-band radiation include total reflection, Faraday rotation, and amplitude modulation of the radiation.

The classical description of the Mössbauer effect mentioned earlier regards the recoil-free line as a frequency-modulation carrier, while the sidebands form a continuum due to the thermal vibration of the atom. The treatment leads directly to Eq. (1) as an approximation for small amplitude vibrations.

See also FERROMAGNETISM; GAMMA-RAY SPECTROMETERS; KONDO EFFECT; MAGNETIC MATERIALS; RARE

EARTHS; RELAXATION PHENOMENA; TRANSITION ELEMENTS.

REFERENCES

1. H. Frauenfelder, *The Mössbauer Effect*. W. A. Benjamin, New York, 1963. A review and a collection of reprints of many of the early important papers.
2. A. Abragam, *L'Effet Mössbauer et Ses Applications a L'Etude des Champs Internes*. Gordon and Breach, New York, 1964. A theoretical treatment of the electron–nuclear interactions and a discussion of some experiments.
3. *Chemical Applications of Mössbauer Spectroscopy*, V. I. Goldanskii and R. H. Herber (eds.). Academic, New York, 1968. A collection of articles by various authors.
4. T. C. Gibb, *Principles of Mössbauer Spectroscopy*. Wiley, New York, 1976; *Mössbauer Spectroscopy*. U. Gonser (ed.), Topics in Applied Physics, Vol. 5. Springer-Verlag, New York, 1975.
5. *Mössbauer Isomer Shifts*, G. K. Shenoy and F. E. Wagner (eds.). North-Holland, New York, 1976.
6. *Applications of Mössbauer Spectroscopy*, R. L. Cohen (ed.). Academic, New York, 1976 (Vol. 1) and 1980 (Vol. 2).
7. (a) *Mössbauer Effect Data Index 1958–1965*, A. H. Muir, K. J. Ando, and H. M. Coogan (ed.). Interscience, New York, 1966. (b) *Mössbauer Effect Data Index*, J. G. Stevens and V. E. Stevens (eds.). IFI/Plenum, New York. Yearly volumes for 1971–1976. (c) *Mössbauer Effect Reference and Data Journal*, edited by J. G. Stevens and V. E. Stevens, *et al.* Published by the Mössbauer Effect Data Center, University of North Carolina, Asheville, N.C. Ten issues a year and an annual index. A continuation of items (a) and (b) above since 1977.

These volumes are an invaluable guide to the literature. Articles are referenced by isotope and by author. Titles and some descriptive material are also given.

A series of International Conferences on Applications of the Mössbauer Effect (ICAME) takes place regularly. A few are listed here: ICAME, *J. Phys. (Paris), Colloq.* **35**, C-6 (1974): **37**, C-6 (1976); **40**, C-2 (1978); **41**, C1 (1980).

ICAME Hyperfine Interactions **27/28/29** (1986); **40**, **41**, and **42**.

Multipole Fields

Yukap Hahn

Electromagnetic fields **E** and **B** associated with a localized static or time-dependent source are often described in terms of multipole fields. The method of multipole expansion is especially useful when the typical size d of a source is small compared with the characteristic wavelength, and the charge or current distribution within the source is dominated by a few spherical harmonics.

The fields **E** and **B** generated by a source ρ or **J** satisfy, in general, Maxwell's equations, which are a set of first-order equations that couple **E** and **B** and depend on the source distribution ρ and **J**. For convenience of solving the set, the fields are often expressed in terms of scalar and vector potentials Φ and **A**. As **E** and **B** are invariant under a gauge transformation, this freedom can be used to simplify the equations that Φ and **A** satisfy. An explicit solution for Φ and **A** can then be obtained readily in a form of multipole expansion, so long as the physical geometry of the source distribution is not too complicated.

(A) Thus, let us first consider the potentials generated by a localized static source $\rho(\mathbf{r}')$ or $\mathbf{J}(\mathbf{r}')$. Using the completeness of the spherical harmonics $Y_{lm}(\theta,\phi)$, where (θ,ϕ) are the spherical angles for the observation point **r**, we may write (in the Coulomb gauge) the scalar potential outside the source region:

$$\Phi(\mathbf{r}) = \int \frac{\rho(\mathbf{r}')}{|\mathbf{r}-\mathbf{r}'|}\, d^3r' = \sum_{l=0}^{\infty}\sum_{m=-l}^{l} \frac{4\pi}{2l+1} Q_{lm} \frac{Y_{lm}(\theta,\varphi)}{r^{l+1}}, \tag{1}$$

where Q_{lm} are the electric multipole moments defined by

$$Q_{lm} \equiv \int Y^*_{lm}(\theta',\varphi')r'^{l}\rho(\mathbf{r}')\, d^3r'. \tag{2}$$

For example, we have the monopole field for $l=0$ with $Q_{00}=\int \rho(\mathbf{r}')\, d^3r'/\sqrt{4\pi}$ corresponding to the static Coulomb potential, and the dipole field for $l=1$, $m=0$ with $Q_{10}=(3/4)^{1/2}\int z'\rho(\mathbf{r}')\, d^3r'$ along the z axis, etc.

The vector potential can be expressed in a similar way in terms of the Green's function $|\mathbf{r}-\mathbf{r}'|^{-1}$ and in the Coulomb gauge $\nabla\cdot\mathbf{A}=0$, as

$$\mathbf{A}(\mathbf{r}) = \frac{1}{c}\int \frac{\mathbf{J}(\mathbf{r}')}{|\mathbf{r}-\mathbf{r}'|}\, d^3r' = \frac{1}{c}\sum_{l,m} \frac{4\pi}{2l+1} \frac{Y_{lm}(\theta,\varphi)}{r^{l+1}} \mathbf{M}_{lm}, \tag{3}$$

where the magnetic multipole moments $\mathbf{M}_{lm}$ are defined by

$$\mathbf{M}_{lm} \equiv \int Y^*_{lm}(\theta',\varphi')r'^{l}\mathbf{J}(\mathbf{r}')\, d^3r'. \tag{4}$$

For example, a steady-state current distribution satisfies $\nabla\cdot\mathbf{J}=0$ from the continuity equation, so that the $l=0$ contribution to **A** vanishes. For $l=1$, we have the magnetic dipole potential $\mathbf{A}_{1m}$ with $\mathbf{M}_{1m}$, or equivalently

$$\mathbf{A}_1 = (\mathbf{m}\times\mathbf{r})/r^3$$

with

$$\mathbf{m} = \frac{1}{2c}\int \mathbf{r}'\times\mathbf{J}(\mathbf{r}')\, d^3r'.$$

The **E** and **B** fields are then evaluated with the above Φ and **A** using the definitions $\mathbf{E}=-\nabla\Phi$ and $\mathbf{B}=\nabla\times\mathbf{A}$.

(B) Next consider the case of a time-dependent source which is capable of emitting radiation. Here the Lorentz gauge is often more convenient and the retardation effect is taken into account correctly in relativistic cases. For each Fourier frequency component for the sources,

$$\rho_\omega(\mathbf{r}') = \frac{1}{\sqrt{2\pi}}\int \rho(\mathbf{r}',t)e^{i\omega t}\, dt$$

and (5)

$$\mathbf{J}_\omega(\mathbf{r}') = \frac{1}{\sqrt{2\pi}}\int \mathbf{J}(\mathbf{r}',t)e^{i\omega t}\, dt,$$

we have directly the multipole field expansion for the **E** and **B** fields as

$$\mathbf{E}_\omega = \sum_{l=1}^{\infty} \sum_{m=-l}^{l} [a_M{}^\omega(l,m) g_l(kr) \mathbf{X}_{lm} + (i/k) a_E{}^\omega(l,m) \nabla \times f_l(kr) \mathbf{X}_{lm}] \quad (6a)$$

and

$$\mathbf{B}_\omega = \sum_{l=1}^{\infty} \sum_{m=-l}^{l} [a_E{}^\omega(l,m) f_l(kr) \mathbf{X}_{lm} - (i/k) a_M{}^\omega(l,m) \nabla \times g_l(kr) \mathbf{X}_{lm}]. \quad (6b)$$

In (6), $\mathbf{X}_{lm}$ are the vector spherical harmonics defined by

$$\mathbf{X}_{lm} = \frac{1}{[l(l+1)]^{1/2}} \mathbf{L} Y_{lm}(\theta,\phi), \quad (7)$$

which is completely orthogonal to **r**; it is proportional to the irreducible tensor $\mathbf{T}^m_{ll}$, while $\nabla \times \mathbf{X}_{lm}$ corresponds to a linear combination of $\mathbf{T}^m_{ll\pm 1}$. The radial functions f_l and g_l are to be determined by a set of equations involving the sources, and assume, e.g., an asymptotic form $h_l^{(1)}(kr) \sim (-i)^{l+1} e^{ikr}/kr$ for the outgoing wave boundary condition. The coefficients of the expansion (6) are then given by

$$a_E{}^\omega(l,m) = \frac{4\pi k^2}{i[l(l+1)]^{1/2}} \int Y^*_{lm} \cdot \left\{ \rho \frac{\partial}{\partial r} [r j_l(kr)] + \frac{ik}{c} (\mathbf{r}\cdot\mathbf{J}) j_l(kr) \right\} d^3r \quad (8a)$$

and

$$a_M{}^\omega(l,m) = \frac{4\pi k^2}{i[l(l+1)]^{1/2}} \int Y^*_{lm} \left[\nabla \cdot \left(\frac{\mathbf{r}\times\mathbf{J}}{c} \right) j_l(kr) \right] d^3r, \quad (8b)$$

where ρ and **J** are related by $i\omega\rho = \nabla\cdot\mathbf{J}$. (We have neglected here possible presence of intrinsic magnetic sources.) For an electric dipole radiation with $l=1$, we have the power radiated to a solid angle $d\Omega$, averaged over a frequency interval, given by

$$\frac{dP_E{}^\omega(1,m)}{d\Omega} = \frac{c}{8\pi k^2} |a_E{}^\omega(1,m)|^2 \cdot |\mathbf{X}_{1m}|^2, \quad (9)$$

with $|\mathbf{X}_{10}|^2 = (3/8\pi) \sin^2\theta$ and $|\mathbf{X}_{1\pm 1}|^2 = (3/16\pi)(1+\cos^2\theta)$. Of course $a_E{}^\omega$ and $a_M{}^\omega$ are closely related to Q and **M**.

(C) As stated above, the multipole expansion is a useful procedure in determining the potentials and the fields when $kd \lesssim 1$, or the deviation of the source distribution from a given sphericity is small, so that a few low l values are relevant; the power radiated or the field energy for a given lth multipole behaves roughly as $(kd)^{2l+2}$. On the other hand, when the radiation involved has relatively a short wavelength or a typical source size is large so that $kd \gg 1$, the multipole expansion approach may not be very effective. Alternative approaches are then preferred; for example, (a) the Kirchhoff's integral representation of the potentials and the fields and its approximate solution using the Kirchhoff approximation, and (b) semiclassical approach of the JWKB type in which the eikonal functions (classical action) are evaluated along a straight path, as in the geometrical optics.

See also Electric Moments; Electromagnetic Radiation; Magnetic Moments; Magnetic Monopoles; Maxwell's Equations; Rotation and Angular Momentum.

BIBLIOGRAPHY

H. A. Bethe and E. E. Salpeter, *Quantum Mechanics of One and Two Electron Atoms*. Academic, New York, 1957.

R. D. Cowan, *The Theory of Atomic Structure and Spectra*. University of California Press, Berkeley, 1981.

S. DeBenedetti, *Nuclear Interactions*. Wiley, New York, 1964.

J. D. Jackson, *Classical Electrodynamics*. Wiley, New York, 1975.

E. J. Konopinski, *Electromagnetic Fields and Relativistic Particles*. McGraw-Hill, New York, 1981.

J. B. Marion, *Classical Electromagnetic Radiation*. Academic, New York, 1965.

R. G. Newton, *Scattering Theory of Waves and Particles*, 2nd ed. Springer-Verlag, New York, 1982.

W. K. H. Panofsky and M. Phillips, *Classical Electricity and Magnetism*. Addison-Wesley, Reading, MA, 1962.

M. E. Rose, *Multipole Fields*. Wiley, New York, 1955.

B. W. Shore and D. H. Menzel, *Principles of Atomic Spectra*. Wiley, New York, 1968.

Muonic, Mesonic, and Other Exotic Atoms

Clyde E. Wiegand

Negative particles other than electrons can be implanted in ordinary atoms. Muons, pions, kaons, antiprotons, and sigma hyperons have been used to make exotic atoms of most of the stable chemical elements. The atoms are formed in excited states when the particles are stopped in matter. For example, a beam of K^- mesons (kaons) stopped in carbon will make kaonic carbon atoms. Although the time that an exotic atom lives is only a few microseconds for muons and about 10^{-10} s for the other particles, these are sufficiently long times to study their properties. The beams of particles must come from accelerators and their intensities are limited to around a million per second for muons and pions and to a few hundred per second of the others.

The first of the new atoms to be studied were pionic (1952). Then came muonic (1953), kaonic (1966), sigmonic (1968), and antiprotonic (1970) atoms. These atoms are loosely referred to as mesonic atoms, but in current nomenclature only those of pions and kaons are strictly mesonic.

It is not known exactly how a meson cascades through the cloud of electrons to its host nucleus. Measurements are begun only after the meson has passed through the innermost shell of electrons. In this region the negatively charged meson feels the full force of the positively charged nucleus. The system is similar to a miniature hydrogen atom and the equations of the hydrogen system apply with only a few modifications. Mesons find themselves in Bohr orbits of principal quantum number n and angular momentum l. In early stages of the cascade toward the nucleus, energy is given off in the form of electrons ejected from the electron cloud (Auger effect). As the quantum jumps get larger in energy, x rays become the dominant means of shedding the electric energy due to the coming together of the negative meson and the positive nucleus. It is primarily through the measurements of the x rays that properties of the atoms are discovered.

For spin-0 particles the energy states of hydrogen-like atoms are given approximately by

$$E_n = -(\tfrac{1}{2})mc^2\alpha^2Z^2/n^2, \tag{1}$$

where m is the reduced mass of the orbiting particle $[m_0 \times m_{\text{nucleus}}/(m_0 + m_{\text{nucleus}})]$, c is the velocity of light, $\alpha = e^2/\hbar c = 1/137$, and Z is the nuclear charge. For spin-$\frac{1}{2}$ particles there are additional terms that give the fine-structure splitting, which depends on the particle's magnetic moment and its (n,l) states. As an example, we calculate the x-ray energy for a kaon transition from $n=3$ to $n=2$ in $^6_3$Li atoms:

$$E_{3\to 2} = \left(\frac{1}{2}\right)456\left(\frac{1}{137}\right)^2 3^2\left[\frac{1}{2^2} - \frac{1}{3^2}\right] = 15.2\ \text{keV}$$

in agreement with the spectrum of Fig. 1. Conversely, measuring the energies of x-ray lines gives the mass of the orbiting particle. This has been used to obtain the most accurate mass values for kaons and pions.

All the exotic atoms behave similarly in the early stages of their existence, but only muons continue their cascade down to the ground state $n=1$. From the ground state muons either decay with a half-life of 2.2×10^{-6} s or are absorbed by the nucleus $(\mu^- + p \to n + \nu)$ in a time of $4 \times 10^{-7}(82/Z)$ s. This reaction is so weak that muons can spend appreciable time within the nuclei. All the other particles react strongly with nuclear matter and thus behave differently.

Muonic atoms have yielded extensive results on the distribution of nuclear charge and nuclear moments. Their study revealed that nuclei are smaller than had previously been assumed and that the effect of vacuum polarization is important.

The strongly reacting (hadronic) particles—pions, kaons, antiprotons, and sigmons—disappear near the nuclear surface. They encounter neutrons or protons in reactions exemplified by

$$\begin{aligned} &\pi^- + N \to N + N, \\ &K^- + N \to \Sigma^- + \pi^{+,0}, \\ &\Sigma^- + p \to \Lambda + n, \\ &\bar{p} + N \to \pi\text{'s} + \gamma, \end{aligned} \tag{2}$$

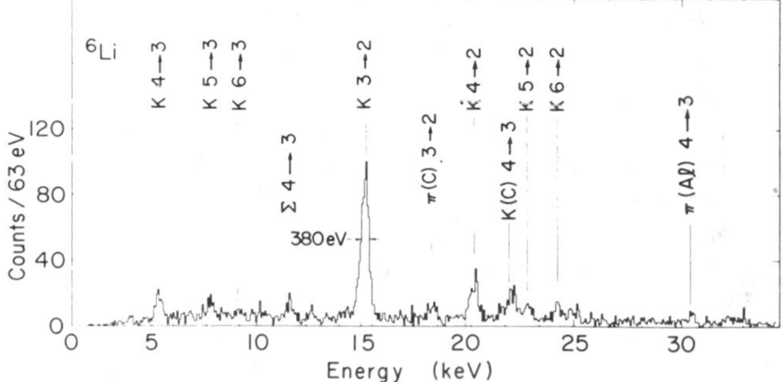

FIG. 1. Kaonic x-ray spectrum of lithium-6. A line from sigmonic atoms appears at 11.9 keV.

where N stands for neutron n or proton p, and Λ is the neutral hyperon.

Experiments show that only 10–50% of the hadrons make measurable x rays. Presumably the remainder encounter nuclear matter before reaching the region where x-ray emission dominates. Some hadrons reach low-n orbits for certain Z, but as Z increases their orbits shrink closer to the nucleus and their x rays disappear. Figure 2 shows how kaonic transitions for $n=6$ to 5 disappear as Z increases. Before experimental studies of kaonic atoms began it was expected that data on their x-ray intensities would give detailed information on the distribution of neutrons near the nuclear surface. Recall that muonic atoms yield the distribution of nuclear charge (protons) but give no direct information on the whereabouts of neutrons. Although kaons react with almost the same strength in encounters with neutrons as with protons, and although nuclear absorption rates have been measured through broadening of x-ray lines, there are complications that prevent the determination of neutron distributions.

Figure 2 shows a remarkable dip in the x-ray intensities from around $Z=20$ to $Z=34$ where kaons were not expected to encounter nuclear matter. The same behavior occurs for higher-n transitions, $n=7\to6$ and $8\to7$. Apparently most of the kaons were captured or forced into orbits of low l where they were absorbed by the nuclei before completing their cascade to low-n orbits. This is an atomic phenomenon induced by the electron structure of the elements. Its mechanism is not understood.

Equation (2) indicates that reactions of kaons with nucleons can make Σ^- particles. This is the source of sigmonic atoms. About 10% of K^- stopped in medium-Z elements result in the emission of Σ^- of which many stay in the target, slow down, and make sigmonic atoms and x rays. Sigmonic-atom x-ray lines were found in the kaonic x-ray spectra as shown in Fig. 1. Sigmonic x-ray lines are doublets due to the spin and magnetic moment of Σ^-. Even though the line intensities are low (a few percent of kaonic lines) and the splitting is too small to be completely resolved, limits have been set on the previously unknown value of the Σ^- magnetic moment $(-1.157 \pm 0.025$ nuclear magnetons).

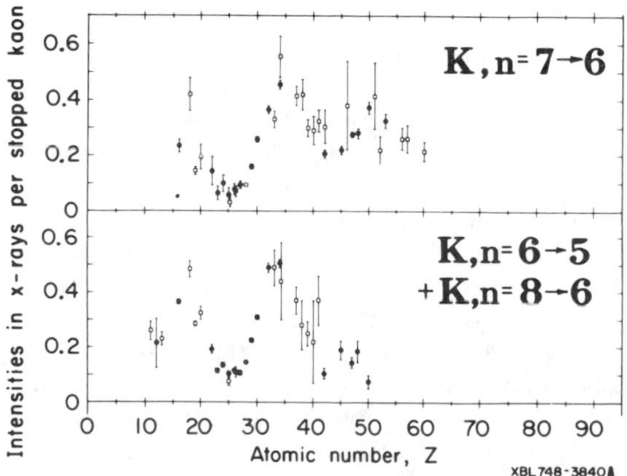

FIG. 2. Intensities of kaonic x rays vs Z. For $n=6$ to 5 transitions and Z larger than 50, kaons were absorbed at the nuclear surface. The dip in intensities around $Z=25$ is the result of atomic effects not yet understood.

Symmetry theories require that antiprotons have the same mass as protons and that their magnetic moments have the same magnitude but opposite sign. The measured mass from x-ray energies was $m_{\bar{p}} = 938.22 \pm 0.04$ MeV/c^2 ($m_p = 938.2723 \pm 1.0003$).

X-ray spectra of antiprotonic atoms of heavy elements show well-resolved doublets due to the spin and magnetic moment of $\bar{p}$. Experimental results were $\mu_{\bar{p}} = -2.795 \pm 0.019$ nuclear magnetons ($\mu_p = 2.7928474$).

There are two other hadronic particles that could form exotic atoms: omega and xi. They have not been observed because of the scarcity of Ω^- and Ξ^-.

See also ATOMS; BOHR THEORY OF ATOMIC STRUCTURE; HADRONS.

BIBLIOGRAPHY

C. S. Wu and L. Wilets, *Ann. Rev. Nucl. Sci.* **19,** 527 (1969).
G. Backenstoss, *Ann. Rev. Nucl. Sci.* **20,** 467 (1970).
R. Seki and C. Wiegand *Ann. Rev. Nucl. Sci.* **25,** 241 (1975).
C. Wiegand, *Sci. Amer.* **227** (5), 102 (1972).

Muonium

Vernon W. Hughes

Muonium is the atom consisting of a positive muon, μ^+, and an electron, e^-. It is the simplest system involving the muon and the electron, and its study has yielded precise information about the interaction of the muon and the electron. The electromagnetic interaction of the muon and the electron has been determined through the precise measurement of the hyperfine structure interval $\Delta\nu$ in the ground ($n=1$) state, and the muon magnetic moment has been determined from the Zeeman effect in this state. In addition, the fine structure and Lamb shift in the $n=2$ state and the $1S$–$2S$ energy interval have been measured. Limits to anomalous weak interactions involving the muon and the electron, in particular to a coupling of muonium (μ^+e^-) to antimuonium (μ^-e^+), have been established. Muonium can be regarded as a light isotope of hydrogen, and its atomic and molecular interactions have been studied in gases and in condensed matter [1–4].

Parity nonconservation in the production and decay of the muon provides an important tool for the study of muonium. The decay of a positive pion at rest ($\pi^+ \to \mu^+ + \nu_\mu$) produces a positive muon with its spin in the direction opposite to its linear momentum. Furthermore, the decay of a positive muon ($\mu^+ \to e^+ + \nu_e + \bar{\nu}_\mu$) occurs with an angular asymmetry favoring positron emission in the direction of the muon spin. Hence polarized muonium can be formed, and changes in muon polarization, which accompany changes in muonium state associated with magnetic resonance transitions or collisions, can be observed through the change in the angular distribution of the decay positrons.

Muonium was discovered through the observation of its characteristic Larmor precession frequency in an external magnetic field [5,6]. The energy-level diagram of the ground ($n=1$) state in an external magnetic field H (Fig. 1) is obtained from the relevant part of the Hamiltonian,

$$\mathcal{H} = a\mathbf{I}_\mu \cdot \mathbf{J} + \mu_B{}^e g_J \mathbf{J} \cdot \mathbf{H} - \mu_B{}^\mu g_\mu{}' \mathbf{I}_\mu \cdot \mathbf{H} \qquad (1)$$

in which $a = h\Delta\nu$ is the hyperfine structure interval, $\mathbf{I}_\mu$ is the muon spin operator, $\mathbf{J}$ is the electron spin operator, $\mu_B{}^e$ ($\mu_B{}^\mu$) is the electron (muon) Bohr magneton, g_J ($g_\mu{}'$) is the electron (muon) gyromagnetic ratio in muonium ($\simeq 2$) and $\mathbf{H}$ is the external static magnetic field. The solution for the energy levels is given by the Breit–Rabi formula

$$W_{F=1/2\pm1/2, M_F} = -\tfrac{1}{4}a - \mu_B{}^\mu g_\mu{}' H M_F \pm \tfrac{1}{2}a[1 + 2M_F x + x^2]^{+1/2}, \qquad (2)$$

in which F is the total angular-momentum quantum number, M_F is the associated magnetic quantum number, and $x = (g_J \mu_B{}^e - g_\mu{}' \mu_B{}^\mu)H/a$. Muonium is formed directly in its ground state through an electron-capture reaction when a muon slows down in a gas. Because the incident muons which originate from pion decays are polarized, there is an unequal population of the four hyperfine-structure (hfs) magnetic substates, and the characteristic Larmor precession frequency of the state $(F, M_F) = (1,1)$,

$$f_L \simeq \mu_B{}^e H/h = (1.40 \text{ MHz G}^{-1})H, \qquad (3)$$

can be observed in the time dependence of the positron angular distribution. The fraction of the muons that form muonium depends on the stopping gas and is large for the heavy rare gases and small for the light rare gases [7].

In a long series of microwave magnetic resonance experiments of increasingly high precision, the Yale and Chicago groups have measured the transition frequencies between the energy levels shown in Fig. 1, both at weak external magnetic field ($\simeq 1$ mG) and at strong magnetic field (6–14 kG). These experiments determine both the hfs interval, $\Delta\nu$, and the ratio of muon and proton magnetic moments, μ_μ/μ_p. From the latest weak-field measurements of the two groups [8–10], the weighted average value of $\Delta\nu$ is

$$\Delta\nu = 4\ 463\ 302.9(11) \text{ kHz } (0.25 \text{ ppm}). \qquad (4)$$

From the latest strong-field measurement by the Yale–Heidelberg group at Los Alamos [11], it is

$$\Delta\nu = 4\ 463\ 302.88(16) \text{ kHz } (0.036 \text{ ppm}), \qquad (5)$$

$$\mu_\mu/\mu_p = 3.183\ 346\ 1(11) \ (0.36 \text{ ppm}). \qquad (6)$$

The theoretical value for $\Delta\nu$ is calculated on the assumption that both the electron and muon are structureless Dirac particles with the conventional coupling to the electromagnetic field. A perturbation-theory solution of the quantum electrodynamic relativistic bound-state two-body problem can be written [12]

$$\Delta\nu_{\text{theor}} = \Delta\nu_F(1+R), \qquad (7)$$

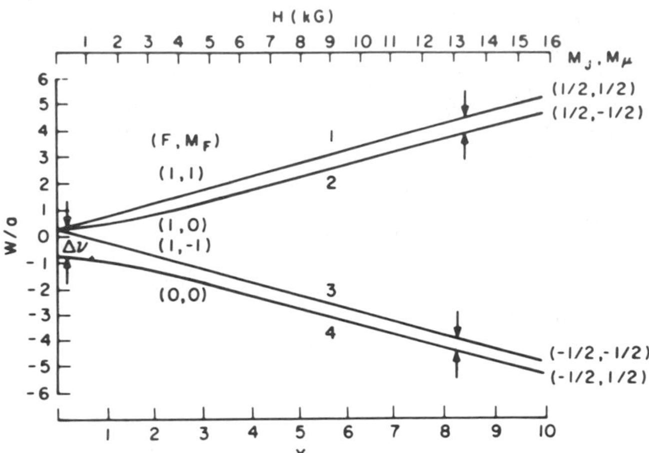

FIG. 1. Energy-level diagram for the ground ($n=1$) state of muonium in a magnetic field. The levels are designated by their weak-field and also by their strong-field quantum numbers (M_J and M_μ are the magnetic quantum numbers associated with the electron and muon-spin angular momenta, respectively).

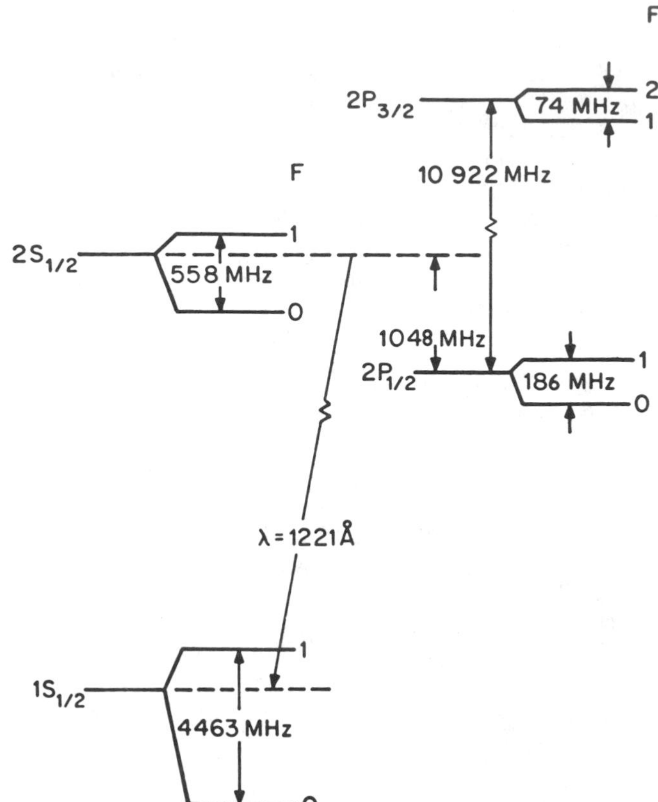

FIG. 2. Energy-level diagram of the $n=1$ and $n=2$ states of muonium.

where $\Delta\nu_F$ is the leading term or Fermi value of $\Delta\nu$ and R is a correction term expressed in powers of the small quantities α and m_e/m_μ, and includes terms up to the orders α^3 and $(m_e/m_\mu)\alpha^2 \ln\alpha$, in which α is the fine-structure constant and m_e/m_μ is the ratio of electron mass to muon mass. Using known values of the fundamental atomic constants [13], we obtain [12]

$$\Delta\nu_{\text{theor}} = 4\ 463\ 303.6(1.8)\ \text{kHz (0.4 ppm)}, \qquad (8)$$

where the uncertainty arises principally from the 0.3-ppm error in μ_μ/μ_p [11,13]. The agreement of $\Delta\nu_{\text{exp}}$ of Eq. (5) and $\Delta\nu_{\text{theor}}$ of Eq. (8) is excellent and provides an important confirmation of the modern theory of quantum electrodynamics and of the behavior of the muon as a heavy electron. The weak interaction axial vector–axial coupling between μ^+ and e^- via Z exchange is calculated to be 1.0 kHz [14] and hence is smaller than both the experimental and theoretical errors in $\Delta\nu$.

Recently, measurements of the fine structure and Lamb shift in the excited $n=2$ state of muonium (Fig. 2) have been made [15–17] by a method similar in principle to the measurements on hydrogen [18]. The experimental values for the $2^2S_{1/2}$ to $2^2P_{1/2}$ Lamb shift and for the $2^2P_{3/2}$ to $2^2P_{1/2}$ fine-structure interval are 1065 ± 10 MHz and 9895 ± 30 MHz, respectively, in good agreement with the theoretical values. The $1S$–$2S$ transition in muonium has been detected by laser spectroscopy as a two-photon transition [19].

The muonic helium atom, an atom analogous to muonium consisting of a ^{4}He nucleus, a μ^-, and a e^-, has been discovered [20] and its hyperfine structure in the ground state has been measured [21,22]. Agreement with theory is satisfactory within the theoretical uncertainty [23].

Recent searches have been made in several experiments [24,25] for the conversion of muonium, μ^+e^-, to antimuonium, μ^-e^+, a process beyond the standard model which may arise from a weak-interaction violating additive muon-number conservation [26]. The most recent direct search for the spontaneous conversion of μ^+e^- to μ^-e^+ involves direct detection of both μ^- and e^+ and establishes the limit to a direct coupling of μ^+e^- to μ^-e^+ through a V–A interaction corresponding to the coupling constant $G_{M\bar{M}} \lesssim 0.1 G_F$ in which G_F is the universal Fermi coupling constant [25]. Some speculative modern theories beyond the standard model [27] predict that the spontaneous conversion of M to $\bar{M}$ should occur.

The atomic and molecular interactions of muonium in gases can be expected to be very similar to those of hydrogen [30]. These interaction cross sections can be studied through their depolarizing effects on muonium [28,29]. Electron spin–exchange collisions with paramagnetic molecules such as NO and O_2 can be particularly well identified because of the dependence of the cross section on external magnetic field. The study of muonium in condensed matter has also become an extensive field of research [31,32].

See also ATOMS; ELECTRON; QUANTUM ELECTRODYNAMICS.

REFERENCES

1. V. W. Hughes, *Annu. Rev. Nucl. Sci.* **16,** 445 (1966).
2. V. W. Hughes and T. Kinoshita, *Muon Physics I* (V. W. Hughes and C. S. Wu, eds.), p. 11. Academic Press, New York, 1977.
3. V. W. Hughes, *Fundamental Symmetries* (P. Bloch, P. Pavlopoulos, and R. Klapisch, eds.), p. 287. Plenum, New York, 1987.

4. V. W. Hughes, in *The Hydrogen Atom* (G. F. Bassani, M. Inguscio, and T. W. Hänsch, eds.), p. 171. Springer-Verlag, Heidelberg, 1989.
5. V. W. Hughes *et al., Phys. Rev. Lett.* **5,** 63 (1960).
6. V. W. Hughes *et al., Phys. Rev. A* **1,** 595 (1970); **2,** 551 (1970).
7. R. D. Stambaugh *et al., Phys. Rev. Lett.* **33,** 568 (1974).
8. D. E. Casperson *et al., Phys. Lett.* **59B,** 397 (1975).
9. D. Favart *et al., Phys. Rev. A* **8,** 1195 (1973).
10. H. G. E. Kobrak *et al., Phys. Lett.* **43B,** 526 (1973).
11. F. G. Mariam *et al., Phys. Rev. Lett.* **49,** 993 (1982).
12. T. Kinoshita and J. Sapirstein, in *Atomic Physics 9* (R. S. Van Dyck, Jr. and E. N. Fortson, eds.), p. 38. World Scientific, Singapore, 1984.
13. E. R. Cohen and B. N. Taylor, *Rev. Mod. Phys.* **59,** 1121 (1987).
14. M. A. B. Bég and G. Feinberg, *Phys. Rev. Lett.* **33,** 606 (1974); **35,** 130 (1975).
15. C. J. Oram *et al., Phys. Rev. Lett.* **52,** 910 (1984); C. J. Oram, in *Atomic Physics 9* (R. S. Van Dyck, Jr. and E. N. Fortson, eds.), p. 75. World Scientific, Singapore, 1984.
16. A. Badertscher *et al., Phys. Rev. Lett.* **52,** 914 (1984); A. Badertscher *et al.,* in *Atomic Physics 9* (R. S. Van Dyck, Jr. and E. N. Fortson, eds.), p. 83. World Scientific, Singapore, 1984.
17. S. H. Kettell, Ph.D. Thesis, Yale University, 1990.
18. F. M. Pipkin in *The Hydrogen Atom* (G. F. Bassani, M. Inguscio, and T. W. Hänsch, eds.), p. 30. Springer-Verlag, Heidelberg, 1989.
19. S. Chu *et al., Phys. Rev. Lett.* **60,** 101 (1988).
20. P. A. Souder *et al., Phys. Rev. Lett.* **34,** 1417 (1975); P. A. Souder *et al., Phys. Rev. A* **22,** 33 (1980).
21. H. Orth *et al., Phys. Rev. Lett.* **45,** 1483 (1980).
22. C. J. Gardner *et al., Phys. Rev. Lett.* **48,** 1168 (1982).
23. K. N. Huang and V. W. Hughes, *Phys. Rev. A* **26,** 2330 (1982).
24. A. Olin *et al.,* in *Nuclear Weak Process and Nuclear Structure* (M. Morita, H. Ejiri, H. Ohtsubo, and T. Sato, eds.), p. 164. World Scientific, Singapore, 1989.
25. B. E. Matthias *et al.,* in *Nuclear Weak Process and Nuclear Structure* (M. Morita, H. Ejiri, H. Ohtsubo, and T. Sato, eds.), p. 157. World Scientific, Singapore, 1989.
26. G. Feinberg and S. Weinberg, *Phys. Rev. Lett.* **6,** 381 (1961).
27. R. N. Mohapatra, in *Quarks, Leptons and Beyond* (H. Fritzsch, R. D. Peccei, H. Saller, and F. Wagner, eds.), p. 219. Plenum, New York, 1985.
28. R. M. Mobley *et al., J. Chem. Phys.* **44,** 4354 (1966).
29. R. M. Mobley *et al., J. Chem. Phys.* **47,** 3074 (1967).
30. D. Walker, *Muon and Muonium Chemistry*. Cambridge University Press, Cambridge, 1983.
31. J. H. Brewer *et al., Muon Physics III* (V. W. Hughes and C. S. Wu, eds.), p. 4. Academic Press, New York, 1975.
32. A. Schenk, *Muon Spin Rotation Spectroscopy*. Adam Hilger, Ltd., Bristol, 1985.

Musical Instruments

A. H. Benade and Gabriel Weinreich

Only a very small fraction of the immense variety of systems capable of producing sound are used as musical instruments; those that are have rather special properties that may appear unimportant to the physicist but are readily apparent to the player and hence (sometimes directly but more often indirectly) to the listener. For this reason, the physics of musical instruments is curiously complex and far-ranging, and the following discussion can do no more than illustrate some of the physical questions that come up in such a context.

THE VIBRATING STRING

The vibrating behavior of a real string is influenced in a musically appreciable way not only by the tension T and linear density μ, but also by stiffness, by the nature of the anchorages to which its ends are attached, by various forms of damping, and by variation of parameters along its length. The wave equation for free transverse oscillations may be schematized thus:

$$\mu\left(\frac{\partial^2\psi}{\partial t^2}\right)=T\left(\frac{\partial^2\psi}{\partial x^2}\right)+[\cdots].$$

Here the bracket on the right contains *musically relevant* terms involving all space derivatives of the displacement ψ up through at least the fourth, along with mixed space and time derivatives, up through at least the third, arising from internal damping and air viscosity effects. Large-amplitude nonlinear effects that can be musically significant are also present. It is apparent that the number of boundary conditions is considerably larger than in the case of the "ideal" string, and that questions of orthogonality and of the interpretation of eigenvalues can become quite subtle.

PLUCKED STRING

If a string is pulled to one side at a point x_e by a narrow object exerting unit force and is then released, the time-varying deflection observed at a point x_0 is

$$D(x_0,x_e,t)=\sum D_n=\sum A_n\psi(\kappa_n x_0)\psi(\kappa_n x_e)\exp(j\omega_n t).$$

Here the ψs are sine-like eigenfunctions with κ_ns chosen so as to fit the boundary conditions at the string ends. The corresponding (complex) frequencies ω_n are found from a dispersion formula determined by the differential equation. For an "ideal" undamped string of length L fixed at both ends, $\kappa_n=n\pi/L$ and $\omega_n=\kappa_n(T/\mu)^{1/2}$. The stiffness of a real string systematically raises the mode frequencies by a factor of $(1+BN^2)$, where B depends on string length, tension, radius, and Young's modulus. The frequencies are further perturbed by string anchorages which have a finite admittance, that is, are not absolutely fixed. The shift of the ω_ns due to this cause can have a real part of either sign, depending on whether the support is mass-like or spring-like at the frequency in question. If the support is itself part of a multiresonant system, such as a soundboard, the resulting string frequency shifts can be quite irregular. By contrast, the sign of the imaginary part of the frequency shift must always be such as to cause decay rather than growth of the free oscillation. (In most, or perhaps all, musically important cases the damping of a freely vibrating string is due almost entirely to the finite admittance of the supports, rather than to internal mechanisms.)

The initial shape of the plucked string is essentially triangular with a slight rounding in a region of width δ_e at the plucking point due to the width of the plectrum joined with the effects of string stiffness. The theory of Fourierlike analysis shows that the A_ns decrease as $1/\kappa_n^2$ for $\kappa_n\delta_e<\pi$ and more rapidly than $1/\kappa_n^3$ for progressively larger values of $\kappa_n\delta_e$.

The velocity spectrum components observed by the magnetic pickup of an electric guitar are proportional to $\partial D_n/\partial t$, with x_0 the location of the pickup. Similarly, the spectrum components of the force which drives the soundboard of an acoustic guitar or a harpsichord are proportional to $\partial D_n/\partial x$ evaluated at the bridge end of the string. In both cases the Fourier-like series for the "signal" converges with a power of κ_n one less than that for the deflection amplitudes themselves. For a pickup of width δ_0 there is an additional band limiting depending on the magnitude of $\kappa_n\delta_0$ exactly as before.

The spectral content of the sound produced by such an instrument can be viewed as the output of a complex linear filter, representing the coupling of the acoustic radiation field to the soundboard (whose physics we shall touch on below), when the above "signal" is applied at its input.

STRUCK STRING

If the initial excitation of the string comes from being struck by a hammer, as in the piano, the situation is rather more complicated; and the "classical" description, in which the initial state of the string is described as having zero displacement and a delta-function of transverse velocity at the hammer position, is never correct, even in the limit in which the mass of the hammer is made to vanish. The reason is that for times less than is required for a wave to return from the (nearer) string end, the hammer sees a string impedance which is resistive (as is generally true for unterminated transmission lines). As a result, a very light hammer (light, that is, compared to the string mass) quickly comes to rest, and remains in contact with the string until the first reflection casts it off again a finite time later. In a piano, this limit (for which the hammer rebound velocity can be shown to be exactly $1/e$ times the impact velocity) is most closely approached in the deep bass, but for most of that instrument's compass the hammer mass is at least comparable to the string mass, leading to contact times at least as long as an unperturbed string period. The resulting behavior depends in an important way on the nonlinear elastic properties of the hammer, because of which the piano sound at high and low dynamic levels differs much more than merely in overall amplitude.

BOWED STRING

A string excited by bowing presents by far the most complicated physical string problem, since here the string is never free, and the dimensionless coupling parameter (which compares the maximum amount of energy exchanged between bow and string in various parts of the cycle to the total stored string energy) is by no means small. In such a situation there is no *a priori* reason to assume that the vibration frequency will have any simple relation to a free-string frequency, and indeed, if the bowing is done without any skill the sound obtained may not even be periodic. The "zero-order" motion of a correctly bowed string was experimentally discovered by Helmholtz to consist of a localized "kink" which travels back and forth on the string with the speed of transverse waves. Recent experiments and computer simulations have thrown a great deal of light both on the corrections to this model and on the conditions for its stability, but the topic continues to be an active area of research.

EXCITATION OF SOUNDBOARDS AND MEMBRANES

We turn next to the motion of a two-dimensional wave medium excited by a localized driving force. The basic wave equation is the two-dimensional generalization of the string equation discussed earlier. For membranes such as drum and banjo heads the tension times the two-dimensional part of ∇^2 dominates the space behavior of the ψs, whereas the wooden soundboards of pianos and the plates of guitars and violins are governed by the two-dimensional part of ∇^4 times a board stiffness factor. In the former case, the average spacing of mode frequencies is inversely proportional to frequency; in the latter, it is independent of frequency. In either case, there is a tendency for higher modes to have higher damping and hence higher resonance widths, so that they become more difficult to resolve. For musical purposes it is seldom necessary to deal individually with more than the lowest half-dozen vibrational modes; above this, one needs only their statistical behavior in frequency and coordinate space. Boundary conditions are of all sorts, and the shapes of the vibrating regions are diverse.

In instruments such as guitars and violins the air enclosed by the box plays an important dynamical role. Since one dimension of the cavity is much smaller than the two others, the air acts as a two-dimensional system whose dispersion formula is membrane-like. Its modes are, however, strongly coupled to those of the wood; in addition, motion of air in and out of the *f*-holes (sound hole) of the violin (guitar) is extremely important in determining the amount and directional pattern of radiated sound. Because of the radiative cancellation between wood motion and air motion, the extreme base range of a violin radiates as a dipole rather than a monopole; but at somewhat higher frequencies the monopole moment is strongly enhanced by the air motion.

STRING-TO-SOUNDBOARD COUPLING

In instruments where the strings vibrate freely after an initial excitation, such as the piano or guitar, an increase in string-to-soundboard coupling will (a) increase the sound volume for a given string amplitude, (b) decrease the "sustaining power" of the instrument by enhancing the rate at which string energy is lost to the soundboard, (c) probably increase the inharmonicity of the normal modes as discussed above. In many instruments, the apparent conflict between the musical desirability of (a) and undesirability of (b) is lessened by the presence of almost-degenerate modes that differ in decay time. In a piano, for example, the modes polarized parallel to the soundboard typically have a much slower decay; in addition, the presence of more than one string per note allows "antiparallel" vibrations for which the force exerted on the soundboard is considerably lessened. The combination of a strong initial "prompt sound" (even if it decays quickly) with a sustained "aftersound" (even if it is weak) is apparently perceived as, paradoxically, *both* loud *and* sustained. Additional subtle effects arise from the fact that the strings belonging to a given note of the piano are individually tunable.

For the bowed instruments, the string–soundboard coupling affects the range of, and responsiveness to, bowing parameters such as normal force, bow speed, and bow placement. The consequences for the musical usefulness of an instrument can be very great.

WIND INSTRUMENTS

A wind instrument consists of an air column coupled at the "blowing end" to a device capable of feeding energy into the oscillations of the column. In the "reed" instruments, this device is a valve actuated by the pressure difference between the inside of the air column and the player's oral cavity (in which a steady overpressure is maintained), and designed so as to present a "negative resistance" to the air column (that is, such that an increase of internal pressure gives rise to an inward, rather than outward, differential flow). Since a reed is actuated by pressure, it is most effective at frequencies at which the pressure amplitude is large and the flow amplitude is small, that is, the input impedance of the air column is at maximum. Equivalently, one can say that the playing frequencies will be near the eigenfrequencies which the corresponding air column would have if it were closed at that end.

In woodwinds such as clarinets and oboes, the resonant frequency of the reed system is typically far above the playing frequency, and the negative-resistance characteristic arises from its static, geometrical structure. In brasses such as trumpets and French horns, on the other hand, the player's lips act as the reed; their own resonant frequency is adjusted by the player to be just below the operating frequency, and the phase shift required to produce a negative resistance arises dynamically from the inertia-limited vibration of the lips. Instruments of the flute family (which includes flue organ pipes and recorders) have their input flow steered by the air-column flow itself into and out of the embouchure hole. The dynamical situation is complex, involving the propagation of waves on the moving air jet, but the general behavior of such a system can be summarized by describing it as a valve for which the flow controls the pressure rather than vice versa. As a result, the playing frequencies are near the eigenfrequencies which the same column would have if the blowing end were open, rather than closed.

At the "far" end, the sounding part of a wind instrument is typically terminated either by a rapidly flaring bell (as in the brasses, or the woodwinds playing their lowest note), or by the section of pipe in which the open tone holes form a quasiperiodic lattice. At low frequencies, either one presents a low-impedance termination, ensuring strongly resonant behavior. At the same time, the bell or the tone-hole proportions determine a "cutoff frequency" (roughly 1500 Hz for trumpets, oboes, and clarinets) above which heavy radiation damping occurs, destroying the resonant behavior in this higher-frequency region.

WIND-INSTRUMENT ENERGETICS

The playing amplitude level of a wind-instrument tone and its related spectral content are stable functions of the structure and blowing conditions of the instrument. This suggests (and even cursory experimental investigation confirms) that the feedback mechanism supporting the multicomponent oscillation has an inherent nonlinearity associated with the flow-control process. For a reed whose own resonant frequency is high, we can schematize this nonlinear relationship between the instantaneous input flow u into the mouthpiece and the controlling pressure p within it thus:

$$u = Ap + Bp^2 + Cp^3 + \cdots + \beta u^2.$$

Here $A, B, C, \ldots$, are reed valving parameters, while the term with coefficient β represents the effect of the Bernoulli force on the reed tip, which is of particular importance in the double-reed woodwinds.

The existence of steady musical tones implies that both $u(t)$ and the corresponding $p(t)$ in the mouthpiece are representable by Fourier series built of harmonics of a playing frequency ω_p. Since the behavior of the air column is (almost) linear, we also have, at each component frequency $n\omega_p$, a relation between the corresponding pressure and flow amplitudes in terms of the column's input impedance Z_n at that frequency. By simultaneously solving these column response equations with the equation above that characterizes the reed, one finds (in agreement with experiment) that (a) energy is fed into the net oscillatory system by any component for which $|Z_nA|>1$; (b) heterodyne action couples the various components and passes energy from one to another; (c) the playing frequency is one which maximizes the aggregate energy production by making a "best fit" of the harmonic components to the (two to four) air-column resonances that lie below cutoff; (d) near threshold the oscillation is nearly sinusoidal, with p_n growing as $p_1{}^n$ during a crescendo until at a (physical) mezzo-forte level the pressure spectrum envelope in the mouthpiece caricatures the envelope of the input impedance peaks; and (e) overall amplitude stability is assured during buildup of an oscillation in part by the increasing dissipation due to heterodyne production of above-cutoff frequency components (for which $|Z_nA|<1$), and in part by increasing quasiturbulent damping in the air column itself, which can reduce the Z_ns by a factor of 5 or more at the loudest playing levels.

PHYSICAL DETERMINANTS OF INSTRUMENT QUALITY

In the case of wind instruments, it is thought that the frequency alignment of the impedance peaks—that is, the degree to which those peaks lie at frequencies which are exact integral multiples of a playing frequency—is a crucial determinant of the ease with which the instruments play, and their responsiveness to variation of blowing parameters as they are controlled by the player. Because the frequency deviations in question are very small, the evidence for this view is indirect although quite strong. For example, if one notes how the playing frequency drifts on a crescendo or diminuendo, one can draw conclusions on how the impedance maxima are (mis)placed; on the other hand, theory predicts how they will shift if small amounts of material are removed from the pipe, or a tone hole is slightly modified. If a "correction" based on such considerations results in a clear improvement of performance, it can be taken as strong evidence for the truth of the original hypothesis.

In the case of string instruments such as violins, the level of understanding is still very low, and the various attempts to attack the problem scientifically have been more in the nature of duplicating vibrational properties of instruments which are known to be good than in understanding fundamentally how those properties function so as to produce the instrument's desirable properties. In the case of violins especially, the problem is exacerbated by a tremendously emotional mystique regarding the great Italian instruments produced some centuries ago (even though they have all been drastically rebuilt since), and by the staggering prices which those instruments command. As a result, alleged discoveries of the "secret of Stradivarius" continue to be publicized with depressing regularity. The truly scientific understanding of these most fascinating physical systems remains a challenge to interested researchers.

See also ACOUSTICS; VIBRATIONS, MECHANICAL.

BIBLIOGRAPHY

Thomas D. Rossing, "Resource Letter MA-2: Musical Acoustics," *Am. J. Phys.* **55**. 589–601 (1987). An outstanding annotated bibliography to the whole field, at all technical levels. This paper is, incidentally, included in Thomas D. Rossing, Ed., *Musical Acoustics: Selected Reprints*. American Association of Physics Teachers, College Park, MD, 1988.

Network Theory: Analysis and Synthesis

Louis Weinberg

INTRODUCTION

Network theory is the study of the behavior and design of electrical *networks* formed by the interconnection of electrical *elements*. *Network analysis* (also designated by its old name, *circuit analysis*) and *network synthesis* form the two parts of network theory. In analysis the problem is to determine the voltages and currents in specified elements of an existing network. This is done by the application of a set of simple laws, which are in essence straightforward rules for finding a solution. The network may contain *voltage* and *current sources,* both *independent* (or *ideal*) and *dependent* (or *controlled*), and a finite number of elements like *resistors, inductors, capacitors, gyrators,* and *transformers,* each of which is a model of an *ideal* component; a combination of elements connected together can represent any *lumped* physical component. There also exist *distributed* systems, which are not treated in this article; network models for such systems require an additional element, namely, a transmission-line element called the *unit element*.

In the inverse problem of network synthesis, a performance or behavior characteristic is specified and the problem is to design a network that achieves the desired performance. The performance is often given in terms of a mathematical function like an *impedance*, or a *matrix* of impedances, or by an equivalent characterization widely used in physics, namely, the *scattering matrix*. The concept of impedance is one of the most important and fundamental concepts of network theory.

Thus, in brief, it may be said that in analysis we are given a network and an excitation and we must find the response; in synthesis, given a response and a specified excitation, we must design a network.

Since network theory is a blend of physics, mathematics, and engineering, it has benefited from contributions in each of these fields. The work of G. R. Kirchhoff on networks and combinatorial topology, more commonly called *graph theory*, was a contribution of the first magnitude. Graph theory is essential for the analysis of networks; recently a generalization of graph theory (and also of matrix theory and vector spaces) called *matroid theory*, discovered by H. Whitney, has been applied to analysis (and synthesis). The characterization of network elements by a *volt–ampere relation* is based on the work of G. S. Ohm, M. Faraday, and J. C. Maxwell. Many network concepts such as the *normal modes* (or *natural frequencies*) of a system and the characterization of electric and magnetic energy by what are (in mathematical terms) *positive definite quadratic forms* are taken over from classical dynamics; in fact, the analysis of networks is analogous to the theory of small vibrations in dissipative mechanical systems.

Network synthesis, in contrast to analysis, is a relatively modern development. Its birth date may be given as 1924 with the publication of the *Reactance Theorem* by R. M. Foster. This paper signaled a revolutionary change in network design; an essentially cut-and-try procedure for the design of networks was replaced by an exact synthesis procedure that satisfied a set of necessary and sufficient conditions. Then followed synthesis contributions by many other research workers, culminating in the similar procedures of H. Piloty and S. Darlington. Their results, arrived at independently and published almost simultaneously, made possible the modern design of precision *filters*, which are of crucial importance in carrier telephony, radar, radio, television, and, in fact, in almost all of electrical engineering. The preceding contributions were in the area of synthesis of *passive* networks, which, roughly speaking, are networks that cannot deliver more power than is furnished to them by external sources. With regard to networks that can amplify power, designated as *active* networks, such as those containing transistors in *feedback* configurations (where by feedback is meant that a portion of the output is fed back to the input to achieve better sensitivity, noise, and other performance characteristics), fundamental work was done by H. W. Bode and H. Nyquist.

In solving the synthesis problem, the properties of networks derived by analysis are essential; a theory of network analysis must precede the development of network synthesis. In addition, use is made of the theory of differential equations, matrices, functions of a complex variable, Fourier and Laplace transforms, linear algebra, vector spaces, graph theory, and matroids.

In addition to its applications throughout electrical engineering, network theory is also applied in analogous fields like acoustics, optics, and mechanics. Furthermore, network concepts have enriched other fields of knowledge. The impedance function, introduced by network theorists, is a major contribution to the field of linear systems; so are the positive real function, the positive real matrix, and the scattering matrix. Network theorists also posed the problem of the realizability of networks and other physical systems in terms of their satisfaction in the time domain of specified physical properties like causality, linearity, passivity, and time invariance; they proved, for example, that causality is implied by linearity and passivity, a result now used in many other fields. The input–output concept and the ubiquitous "black box," a concept now found in all of science, that is, a closed system whose precise contents we are unaware of and even unconcerned about, but whose performance can be specified and measured at external pairs of terminals, or *ports,* have, in addition to their use in the exact sciences, yielded rich rewards in such fields as biology and economics; the Harvard economist, W. W. Leontief, in fact, calls his theory an input–output analysis of economic systems. The feedback concept is essential for the understanding of life processes and the theory is used with precision even by some psychiatrists.

There are also other areas, some of them rather unusual like geometry and probability, that have made use of network theory. The geometric problem of "Squaring the Rectangle," that is, dissecting a rectangle into two or more unequal squares, was solved ingeniously by the use of a resistance-network model, planar graph theory, and the application of Kirchhoff's laws. Another elegant result uses resistance networks and energy concepts (formulated by Lord Rayleigh in another context) to solve the infinite absorbing Markov chain, that is, the infinite random walk in *d*-dimensional space. A third fundamental result that was derived by use of network theory is the *principal partition of a graph,* which has many consequences and applications, one of which is that the minimum number of equations necessary for solving a system is not given by the rank of its graph or its nullity, but by a new concept called the *topological degrees of freedom,* which gives what has been called the *hybrid rank* of the system.

A significant part of network synthesis may be considered applied function theory. The assimilation by network theory of parts of function theory has continued to the present with the use of functions of two or more complex variables and matrices of such functions to represent the system functions of *n* ports composed of mixed lumped-distributed networks or of multidimensional digital filters. Alternatively, for the synthesis of lumped-distributed *n* ports, a single complex variable has been used but the class of functions is then extended from rational functions with real coefficients to functions given by quotients of exponential polynomials with real polynomials as coefficients.

It should be mentioned, finally, that with the advent of the high-speed digital computer, digital networks are being designed to do filtering and control tasks previously performed by electrical networks. Instead of using elements like resistors, inductors, and capacitors, these digital networks are made up of *multipliers, summers,* and *delay* elements; the input and output are sequences of numbers instead of continuous functions of time. This subject of digital networks, which has become exceedingly important, is not pursued further in this article except to mention that there is a one-to-one correspondence between electrical networks and digital networks in the sense that the theory applicable to the analysis and design of one type of network can be translated to apply to the other.

NETWORK ELEMENTS

Two accessible terminals at which voltage or current sources may be applied or responses measured constitute a *terminal pair* or *port*. The terminals are considered as a pair in that the current flowing *into* one terminal is equal to the current flowing *out of* the other terminal. A network with one terminal pair—or a *one-port network*, or more briefly, a *one-port*—may be completely described by one network

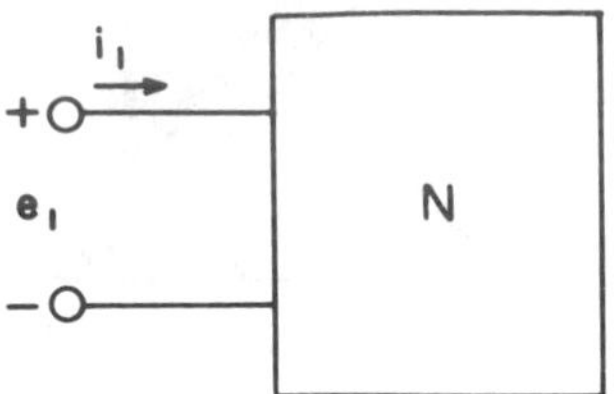

FIG. 1. A one-port network.

function, namely, a *driving-point* function; a *two-port* is described by three independent functions, two driving-point functions and one transfer function, if the two-port obeys the reciprocity theorem (see p. 793, end of column 2); if the two-port is nonreciprocal, then four independent functions are required for a complete characterization. A one-port and a two-port are shown in Figs. 1 and 2, respectively.

The three basic one-port elements are the *resistor, inductor,* and *capacitor*. These may be linear or nonlinear, time varying or time invariant. We consider the linear, time-invariant elements, in which case each of the elements may be characterized by a *constant* parameter. The elements are shown in Fig. 3 and their *volt–ampere* relations are given by

$$\begin{aligned} e_R &= Ri_R, & i_R &= Ge_R, \\ e_L &= L\frac{di_L}{dt}, & i_L &= \frac{1}{L}\int e_L dt, \\ e_C &= \frac{1}{C}\int i_C dt, & i_C &= C\frac{de_C}{dt}, \end{aligned} \tag{1}$$

where R is the *resistance* in ohms (and $G = 1/R$ is the *conductance* in mhos), L is the *inductance* in henrys, and C is the *capacitance* in farads.

The basic two-port elements are the *ideal transformer* and the *gyrator*. The former is shown in Fig. 4 and is defined by the two equations

$$\begin{aligned} \frac{e_2}{e_1} &= \frac{n_2}{n_1}, \\ \frac{i_1}{i_2} &= -\frac{n_2}{n_1}, \end{aligned} \tag{2}$$

whereas the gyrator is defined by

$$\begin{aligned} v_1 &= \alpha i_2, \\ v_2 &= -\alpha i_1, \end{aligned} \tag{3}$$

where α is a constant called the *gyration ratio*. The ideal

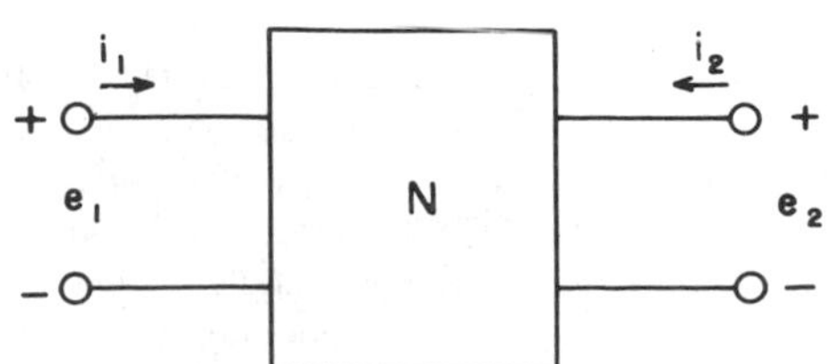

FIG. 2. A two-port network.

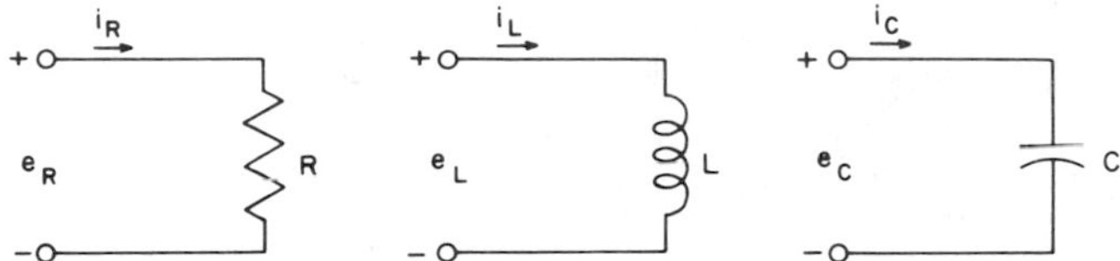

FIG. 3. Symbolic representation of the resistor, inductor, and capacitor.

transformer obeys the reciprocity theorem, that is, it is a reciprocal element, whereas the gyrator is nonreciprocal (Fig. 5).

The number of ports of a network constitute a measure of its complexity; for example, a two-port reciprocal network requires, as was stated above, three independent functions for its complete characterization; a three-port reciprocal network, however, requires six functions. The terminals of a network may also be considered separately, and the network description given in terms of the total number of terminals rather than ports. Thus a two-port is a specialized form of four-terminal network. As a four-terminal network, that is, when the terminals are not considered as pairs, it is a more-complicated structure than a two-port and requires six independent functions to characterize it when it is reciprocal, and nine when it is nonreciprocal.

In addition to the passive elements given above, which may store or dissipate energy, we need sources or generators to furnish energy to a network. An *independent* (or *ideal*) *voltage* source supplies a specified voltage, which may be a constant or a function of time, where the voltage is independent of the current drawn from the source. An *independent current* source is defined by a specified current which is independent of the voltage across the source. A voltage source is shown in Fig. 6 and a current source in Fig. 7.

When we wish to represent devices like a transistor, which can amplify power, we need an additional type of source called a *dependent* or *controlled* source. Such a source may be a current source (or a voltage source) whose current (voltage) is dependent on either a current or a voltage in another element or part of the network. Thus we may have four different types of dependent sources. Unlike the independent sources, which are one-port elements, the dependent sources are two-port elements. A current-controlled voltage source is shown in Fig. 8.

These elements can be used to model any active or passive lumped structure which is assumed to be linear and time invariant. By making the passive elements nonlinear and time varying we may model nonlinear, time-varying networks.

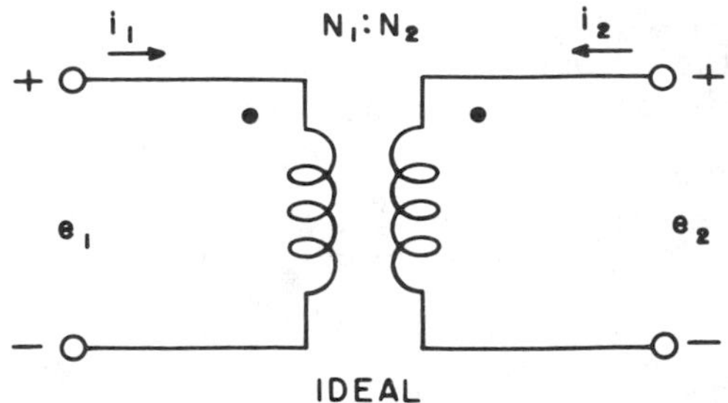

FIG. 4. Representation of an ideal transformer.

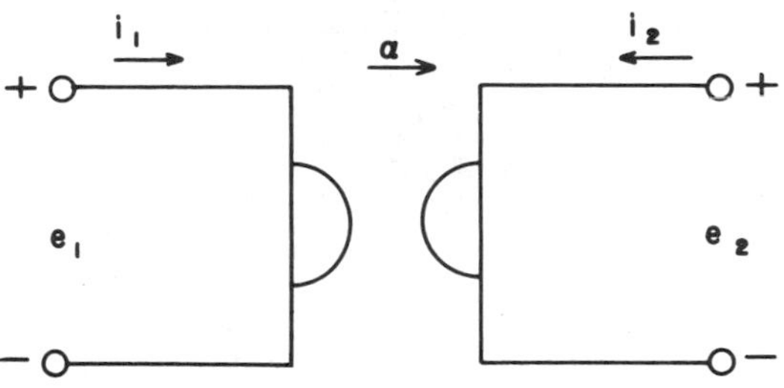

FIG. 5. Representation of a gyrator.

KIRCHHOFF'S LAWS

The basic laws underlying network analysis are Kirchhoff's two laws and the volt–ampere relations given in the preceding section, where the relation for a resistance may be recognized as Ohm's law. Kirchhoff's laws specify the topological or interconnection constraints imposed by the network structure and the conservation of energy. These laws, of course, correspond to the results of experiments in the physical world. It is important to recognize, moreover, that an alternative procedure is to assume them as basic axioms from which all of network theory can be derived logically.

Kirchhoff's laws may be stated as follows:

(1) *Kirchhoff's current law*: the algebraic sum of the currents at a network node equals zero.
(2) *Kirchhoff's voltage law*: the algebraic sum of the voltage drops around a closed path equals zero.

As an example of the application of Kirchhoff's voltage law we write the equations for the two meshes of the network in Fig. 9:

$$\begin{aligned} R_1 i_1 + L\frac{di_1}{dt} - L\frac{di_2}{dt} &= e, \\ -L\frac{di_1}{dt} + R_2 i_2 + L\frac{di_2}{dt} + \frac{1}{C}\int i_2 dt &= 0. \end{aligned} \tag{4}$$

We may then solve these equations for the unknowns by using the theory of differential equations or by the use of transforms.

Kirchhoff's laws can be used to derive two theorems that are useful in network analysis, the *reciprocity theorem* and *Thevenin's theorem*. The reciprocity theorem states that the current produced in branch i by a unit emf in branch j if all other emf's are equal to zero is equal to the current produced in branch j by a unit emf in branch i if all other emf's are equal to zero. Thevenin's theorem states that any combination of emf's and linear circuit elements connected between two terminals is equivalent to a single emf in series

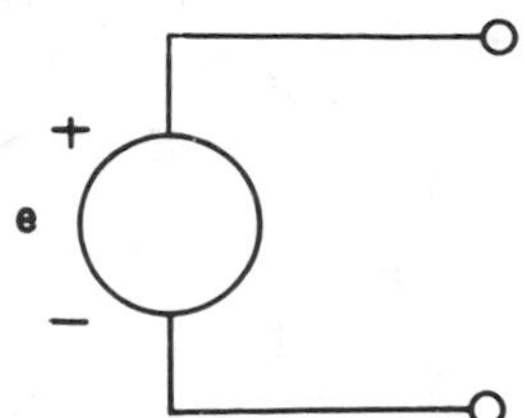

FIG. 6. Ideal voltage source.

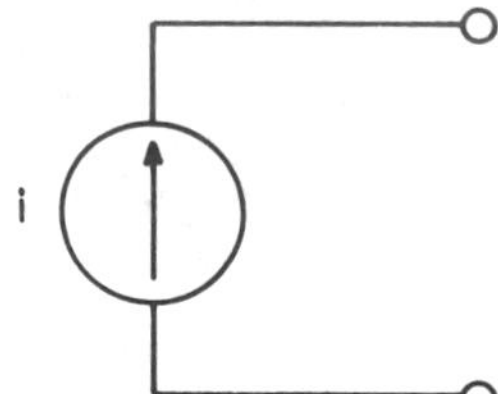

FIG. 7. Ideal current source.

with a single impedance connected between the same two terminals; the value of the emf is the value of the potential difference that would be measured between the terminals when no current is flowing between them, and the value of the impedance is the value that would be measured between them if all independent emf's were equal to zero.

SYSTEM (OR NETWORK) FUNCTIONS

In carrying out the analysis or synthesis of networks we deal with system functions. A *system function* (also called *network function*) is defined as a function of the complex variable $s=\sigma+j\omega$ representing the ratio of the Laplace transform of a *response*, or *output*, variable to the Laplace transform of an *excitation*, or *input*, variable. The principal reason for use of the Laplace transform is that it leads to the replacement of the differential circuit equations by algebraic equations, which are usually easier to handle; the actual currents may then be obtained by application of the inverse Laplace transform. The procedure also has the advantage that the frequency characteristics of the network are obtained by setting $s=j\omega$ in the solution. In the formation of this ratio it is assumed that the initial energy storage of the network is zero.

There are two system functions of interest, the *driving-point function* and the *transfer function*. Each of these may have the dimension of *impedance*, that is, a ratio of a voltage to a current, or *admittance*, while the transfer function may, as another alternative, be dimensionless. For a driving-point function the input and output are measured at the *same* port, whereas for a transfer function the input and output are measured at two different ports.

We illustrate this using Eq. (4). Letting

$$\mathcal{L}[i_1(t)]=I_1(s),$$

$$\mathcal{L}[i_2(t)]=I_2(s),$$

$$\mathcal{L}[e(t)]=E(s),$$

where $\mathcal{L}[\ ]$ indicates the Laplace transform of the function in the brackets, and assuming zero initial conditions, we

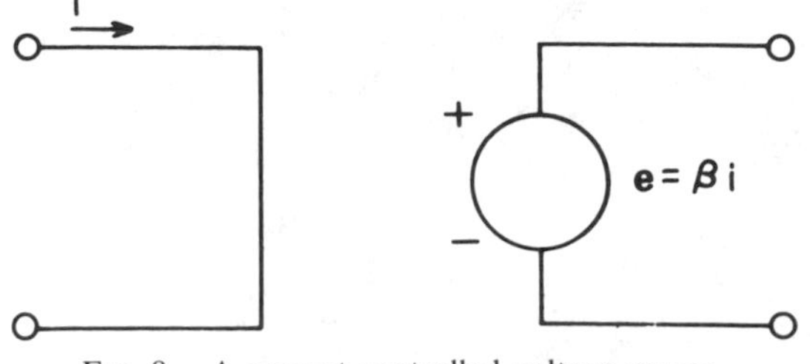

FIG. 8. A current-controlled voltage source.

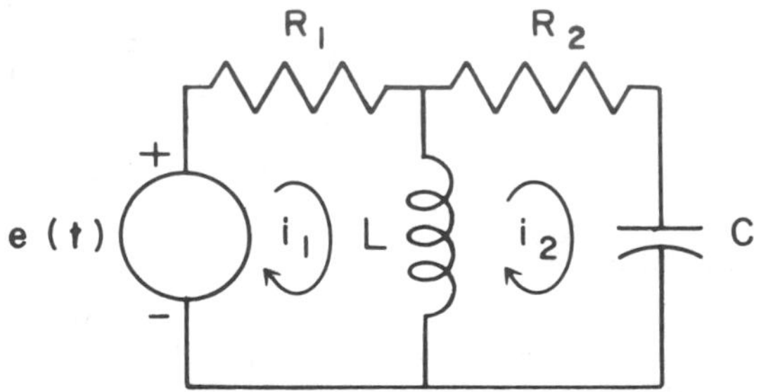

FIG. 9. Two-mesh network.

have

$$(R_1+Ls)I_1-LsI_2=E,$$

$$-LsI_1+\left(R_2+Ls+\frac{1}{Cs}\right)I_2=0.$$

Solving for I_2, we obtain

$$I_2=\frac{s^2}{(R_1+R_2)s^2+(R_1R_2/L+1/C)s+R_1/LC}E,$$

which can be written as the transfer admittance

$$\frac{I_2}{E}=\frac{s^2}{(R_1+R_2)s^2+(R_1R_2/L+1/C)s+R_1/LC}.$$

It is noted that the transfer admittance is a quotient of two polynomials, that is, a rational function.

It can be easily shown that every system function is a rational function of s.

TWO-PORTS AND n-PORTS

Two linear equations completely characterize the behavior of a two-port at its accessible ports; an n-port network requires n equations for its complete description. There are many different sets of equations that may be used, the type used depending on the choice of independent variables. In general, the equations of one set may be derived from those of any other set by appropriate linear transformations; the set used in a particular problem depends mainly on convenience. Two convenient sets are given by

$$E_1=z_{11}I_1+z_{12}I_2,\qquad E_2=z_{21}I_1+z_{22}I_2, \tag{5}$$

and the inverse set of equations

$$I_1=y_{11}E_1+y_{12}E_2,\qquad I_2=y_{21}E_1+y_{22}E_2. \tag{6}$$

The z's are called *open-circuit impedances*, whereas the y's are called *short-circuit admittances*; either set completely characterizes the network. The definitions of these system functions are contained in the equations. From them we obtain

$$z_{11}=\left.\frac{E_1}{I_1}\right|_{I_2=0},\quad z_{21}=\left.\frac{E_2}{I_1}\right|_{I_2=0},$$

$$z_{12}=\left.\frac{E_1}{I_2}\right|_{I_1=0},\quad z_{22}=\left.\frac{E_2}{I_2}\right|_{I_1=0}, \tag{7}$$

and

$$y_{11}=\left.\frac{I_1}{E_1}\right|_{E_2=0}, \quad y_{21}=\left.\frac{I_2}{E_1}\right|_{E_2=0},$$

$$y_{12}=\left.\frac{I_1}{E_2}\right|_{E_1=0}, \quad y_{22}=\left.\frac{I_2}{E_2}\right|_{E_1=0}. \tag{8}$$

The above equations make clear why the impedances are designated as open-circuit functions whereas the admittances are short-circuit functions: each z_{ik} represents a voltage response when all ports are open-circuited and a unit current is applied to the input port; each y_{ik} represents a current response to a unit voltage input when all ports are short-circuited. For example, to evaluate the open-circuit transfer impedance z_{21}, a current source I_1 is used as the input to port 1 and I_2 is made to equal 0, that is, port 2 is open-circuited. Since the internal impedance of a current source is infinite (or an open circuit), it is clear that open-circuit conditions exist at both ports. An analogous discussion applies to the short-circuit admittances.

Extension to the n-port network is straightforward. The set of n equations is given in matrix form by

$$E]=[z]\,I], \tag{9}$$

where

$$[z]=\begin{bmatrix} z_{11} & z_{12} & \cdots & z_{1n} \\ z_{21} & z_{22} & \cdots & z_{2n} \\ \cdot & \cdot & \cdots & \cdot \\ z_{n1} & z_{n2} & \cdots & z_{nn} \end{bmatrix}, \tag{10}$$

the elements of the matrix are defined by

$$z_{ik}=\left.\frac{E_i}{I_k}\right|_{I_1=I_2=\cdots I_{k-1}=I_{k+1}=\cdots=I_{n-1}=I_n=0} \tag{11}$$

and $E]$ and $I]$ denote column matrices. Again it is clear that *all* ports are open-circuited.

Since the equations

$$[z]\,I]=E] \tag{12}$$

may be obtained by solving for $E]$ in

$$[y]\,E]=I] \tag{13}$$

whenever $[y]$ is nonsingular—that is, the two sets of equations are, as was stated previously, inverse sets—the important relation

$$[z]=[y]^{-1} \tag{14}$$

is obtained.

Another method of characterizing a two-port is by use of the *general circuit parameters* or *chain matrix*. This method is important because one often realizes a given system function with a network formed by the cascade connection of two or more simpler networks. Such a network is shown in Fig. 10, where it is specified that I–I′ is the imput port.

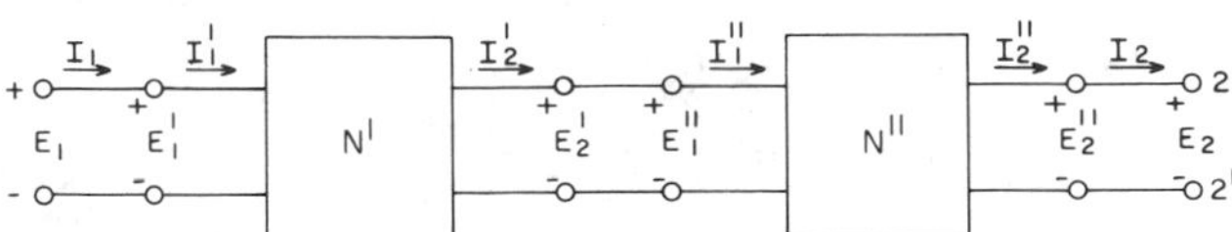

FIG. 10. A cascade connection of two-port networks.

From the figure it is evident that

$$\begin{aligned} E_1&=E_1', & I_1&=I_1', \\ E_2'&=E_1'', & I_2'&=I_1'', \\ E_2''&=E_2, & I_2''&=I_2. \end{aligned} \tag{15}$$

For this form of interconnection the output voltage and output current of the first network are common, respectively, with the input voltage and input current of the second network. Because of this, the matrix that is useful is the *chain matrix*, which is the matrix relating the input and output variables in

$$\begin{bmatrix} E_1 \\ I_1 \end{bmatrix}=\begin{bmatrix} A & B \\ C & D \end{bmatrix}\begin{bmatrix} E_2 \\ I_2 \end{bmatrix}. \tag{16}$$

The parameters of the chain matrix are therefore given by

$$A=\left.\frac{E_1}{E_2}\right|_{I_2=0}, \quad C=\left.\frac{I_1}{E_2}\right|_{I_2=0},$$

$$B=\left.\frac{E_1}{I_2}\right|_{E_2=0}, \quad D=\left.\frac{I_1}{I_2}\right|_{E_2=0}. \tag{17}$$

It is important to note that A, B, C, and D are not system functions but the reciprocals of system functions. For example,

$$C=1/z_{21}. \tag{18}$$

The essential property of the chain matrix, which the reader can easily prove for himself, is that for a network composed of two two-ports connected in cascade, the matrix is equal to the product of the chain matrices of the component two-ports. Thus for the network in Fig. 10, we have

$$\begin{aligned} \begin{bmatrix} E_1 \\ I_1 \end{bmatrix} &= \begin{bmatrix} A' & B' \\ C' & D' \end{bmatrix}\begin{bmatrix} A'' & B'' \\ C'' & D'' \end{bmatrix}\begin{bmatrix} E_2 \\ I_2 \end{bmatrix} \\ &= \begin{bmatrix} A & B \\ C & D \end{bmatrix}\begin{bmatrix} E_2 \\ I_2 \end{bmatrix} \end{aligned} \tag{19}$$

The above process may be repeated for this two-port network cascaded with another two-port. Therefore the chain matrix of a composite two-port formed by the cascade connection of two or more component two-ports is given by the matrix product of the chain matrices of the component networks.

It is observed that in general, that is, for a nonreciprocal two-port, four independent functions are needed for a complete characterization. However, if the two-port obeys the reciprocity theorem, then

$$z_{12}=z_{21}, \qquad (20)$$
$$y_{12}=y_{21},$$

that is, the matrix is symmetric. Thus for the reciprocal n-port there are only $n(n+1)/2$ independent functions. For the chain matrix, reciprocity gives the important relation

$$AD-BC=1, \qquad (21)$$

so that again only three independent functions specify the matrix.

Another set of parameters is the *hybrid* parameters, which are useful for the analysis of transistor circuits. They are called hybrid because, like the chain matrix, they use sets of *mixed* variables; that is, voltages at some ports and currents at other ports are expressed in terms of the corresponding currents at the first ports and voltages at the other ports. These parameters are often referred to as the h matrix and the g matrix.

For the two-port the h parameters are defined by the set of equations

$$\begin{bmatrix} E_1 \\ I_2 \end{bmatrix} = \begin{bmatrix} h_{11} & h_{12} \\ h_{21} & h_{22} \end{bmatrix} \begin{bmatrix} I_1 \\ E_2 \end{bmatrix} \qquad (22)$$

and the g parameters by

$$\begin{bmatrix} I_1 \\ E_2 \end{bmatrix} = \begin{bmatrix} g_{11} & g_{12} \\ g_{21} & g_{22} \end{bmatrix} \begin{bmatrix} E_1 \\ I_2 \end{bmatrix}. \qquad (23)$$

From these equations it is clear that the elements of the h matrix are defined by

$$h_{11}=\left.\frac{E_1}{I_1}\right|_{E_2=0}, \quad h_{12}=\left.\frac{E_1}{E_2}\right|_{I_1=0},$$
$$h_{21}=\left.\frac{I_2}{I_1}\right|_{E_2=0}, \quad h_{22}=\left.\frac{I_2}{E_2}\right|_{I_1=0}, \qquad (24)$$

and those of the g matrix are defined similarly.

It should be noted that like the z's and the y's, the h's and the g's are all system functions, that is, the ratio of a response to an input. For example, h_{11} is the *short-circuit* driving-point impedance at port 1, that is, the input impedance at port 1 with the terminals of port 2 shorted; thus $h_{11}=1/y_{11}$. The h and g parameters are valuable mainly because their use eliminates accuracy problems that arise in making measurements specified by the other sets of parameters.

NATURAL FREQUENCIES, CHARACTERISTIC POLYNOMIALS, AND STABILITY

An important property of a network (or other physical system) is its *natural behavior*. By this is meant that part of the system response whose form is dependent only on the values of the elements and their interconnection and independent of the applied force. This behavior is characterized by means of the *natural frequencies* of oscillation (or *normal modes* or *free vibrations*) of the network. In addition, the behavior may be described as the response to a *unit-impulse* excitation.

Description of the natural behavior in terms of the impulse response as a function of time serves to relate the natural frequencies to the *critical frequencies* of a system function, where by critical frequencies we mean the finite zeros and poles of the rational function that is the system function. The inverse Laplace transform of the system function shows that the impulse response includes terms of the form $A_k e^{s_k t}$, where A_k is a constant and the s_k's are the zeros of the denominator polynomial. Since the s_k's are the natural frequencies, we conclude that the *finite poles of the system function are the natural frequencies*. The denominator is thus called the *characteristic polynomial* of the network described by this system function.

Passive networks must of course be stable; that is, they have no internal energy sources, and therefore an impulse response that increases without bound is impossible. Thus, since a pole in the right half of the s plane corresponds to an exponentially increasing time function, we see immediately that a system function of a passive network can have no poles in the right half-plane. This is not true for an active network, as we show later.

For a two-port network such as is shown in Fig. 11, we cannot properly speak of its stability until we have specified the input and output ports and the types of excitations. Then the natural behavior may be conveniently specified in terms of the *appropriate characteristic polynomial*, that is, there is more than one characteristic polynomial for a two-port.

Since the open-circuit driving-point impedance z_{11} describes the two-port in Fig. 11 with an open-circuit at each port, the denominator of z_{11} is the characteristic polynomial for both ports open-circuited. Now consider the reciprocal of z_{11} as a system function, that is, as an output divided by an input. Thus the denominator of the admittance $y \equiv 1/z_{11}$ is the characteristic polynomial for port 2 open-circuited and port 1 short-circuited. A short circuit is required for port 1 since defining the admittance y as a system function requires a voltage source as the input.

In describing an active network, one frequently uses the expressions *short-circuit stable* and *open-circuit stable*. These may be explained in terms of characteristic polynomials. Though a passive structure is stable for any passive terminations connected to its ports, this is not true for active networks. The stability of an active network may be specified in terms of characteristic polynomials.

Consider the active network shown in Fig. 12; suppose a passive termination is to be connected to the accessible port and we wish to determine whether the complete system will be stable. If the input impedance $Z_{\text{in}}=p(s)/q(s)$ has all its *zeros* in the left half-plane, N is said to be *short-circuit* stable, whereas if some zeros are in the right half-plane, it is short-circuit unstable; if a short-circuit is connected to the

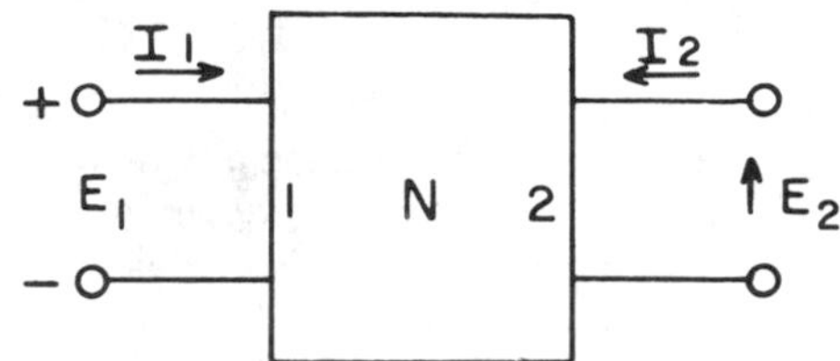

FIG. 11. Two-port network.

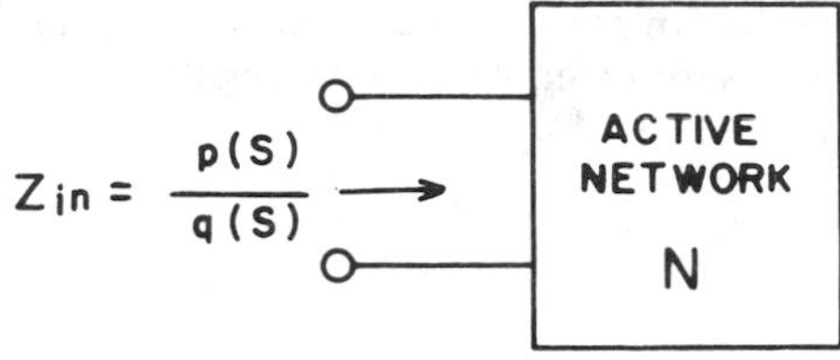

FIG. 12. Active one-port network.

port of a short-circuit stable network, it will not oscillate. Hence, a voltage source may be connected to the port without causing instability. If the input impedance has all its poles in the left half-plane, it is *open-circuit* stable; a current source may thus be connected to the port to yield a stable system. It is therefore clear that depending on the locations of the critical frequencies of Z_{in}, the network may be open-circuit stable, or short-circuit stable, or both, or neither. Since there is no necessary relation between the zeros and poles of an active driving-point impedance, the stability upon being driven by a voltage source, for which the characteristic polynominal is $p(s)$, is independent of the stability upon being driven by a current source, for which the characteristic polynomial is $q(s)$.

It is also possible that N is neither short-circuit stable nor open-circuit stable, but is short-circuit stable when $Z_1=p_1(s)/q_1(s)$ is used as a termination. In this case, though $p(s)$ and $q(s)$ both have zeros in the right half-plane, none of the zeros of Z_1+Z_{in}, that is, the numerator of this sum, namely, pq_1+p_1q, lies in the right half-plane. We thus see that the stability properties can be determined in terms of the appropriate characteristic polynomial.

SYNTHESIS OF DRIVING-POINT AND TRANSFER FUNCTIONS

In this section we introduce briefly some aspects of the synthesis problem for passive networks, mainly, the synthesis of driving-point functions.

Since the numerator and denominator of a driving-point function both represent characteristic polynomials, such a function can have no zeros or poles in the right half-plane. However, these conditions are not enough to guarantee realizability. For example, though the driving-point function

$$Z=\frac{s+3}{s^2+2s+4} \tag{25}$$

clearly satisfies them, it can be shown to be unrealizable. The necessary and sufficient condition for realizability is that the function be a *positive real* rational function. A function of s is defined to be positive real if (1) it is real for s real, and (2) its real part is nonnegative for $\mathrm{Re}(s)\geqq 0$. This concept of positive realness is one of the most important in all of synthesis.

The reader can now test the function in Eq. (25) and demonstrate for himself that it is not positive real.

Often we are interested in realizing a driving-point function that contains only two kinds of elements. For example, in the design of equalizer networks for control systems, we desire an *RC* network, that is, a network containing only resistors and capacitors; for the design of lossless filters we are interested in an *LC* coupling network. *RC* and *LC* driving-point functions must not only be positive real but must satisfy additional conditions. These are given below.

The necessary and sufficient conditions for a function

$$\begin{aligned} Z&=H\frac{p(s)}{q(s)} \\ &=H\frac{s^m+a_{m-1}s^{m-1}+\cdots+a_1s+a_0}{s^n+b_{n-1}s^{n-1}+\cdots+b_1s+b_0} \end{aligned} \tag{26}$$

to be realizable as the driving-point impedance of a network containing only resistors and capacitors are:

(a) All the zeros and poles are simple, that is, of first order, and lie on the negative real axis or at the origin of the complex plane.
(b) The zeros and poles alternate on the nonpositive real axis.
(c) The critical frequency of smallest magnitude, i.e., the one at or nearest the origin, is a pole.
(d) The constant multiplier H is positive.

It is a remarkable fact that the conditions for an *LC* driving-point function may be derived from those for the *RC* function. Alternatively, they may be presented as given below.

The necessary and sufficient conditions for the realization of a function as the driving-point function of an LC network are the following:

(a) The poles and zeros are simple and occur only on the j axis.
(b) The poles and zeros alternate on the j axis.
(c) The constant multiplier is positive.

Using the above theorems we can see that

$$f_1=\frac{3(s+1)(s+5)}{(s+2)(s+6)} \tag{27}$$

is not realizable as an *RC* driving-point impedance. However, its reciprocal satisfies the stated conditions and thus f_1 is realizable as an *RC* driving-point *admittance*.

We observe that the function

$$f_2=\frac{4(s^2+3)(s^2+8)}{s(s^2+5)(s^2+10)} \tag{28}$$

satisfies the conditions for realizability as an *LC* driving-point function; it may be realized as an impedance or an admittance.

There are easy methods for realizing driving-point functions of two-element types of networks, namely, partial-fraction expansions and continued-fraction expansions, and thus a physicist familiar with simple function theory may learn these procedures quickly.

The realization of transfer functions is often accomplished by realizing associated driving-point functions. Thus the realization of driving-point functions is extremely important for practical applications. We do not pursue the synthesis of transfer functions further; the reader may consult the Bibliography.

We mention one further relation that adds to the importance of *LC* driving-point functions, often called *reactance functions*. If we wish to test a given polynomial $p(s)$ to see whether it is the characteristic polynomial of a stable system, that is, has all its zeros in the left half-plane, we merely check to see whether the ratio of its even and odd parts satisfies the conditions for an *LC* driving-point function. Thus if

$$p(s)=m(s)+n(s), \tag{29}$$

where $m(s)$ and $n(s)$ are the even and odd parts, respectively, then $p(s)$ has all of its zeros in the left half-plane if and only if $m(s)/n(s)$ is a reactance function without any common factors; that is, reduced to the lowest terms it has the same degree as $p(s)$.

BIBLIOGRAPHY

J. Bruno and L. Weinberg, "Generalized Networks: Networks embedded on a Matroid, Parts I and II," *Networks* **6,** 53–94 (1976); **6,** 231–271 (1976).

D. C. Youla, L. J. Castriota, and H. J. Carlin, "Bounded Real Scattering Matrices and the Foundations of Linear Passive Network Theory," *IRE Trans. Circuit Theory,* **CT-6,** 102–124 (1959).

W. T. Tutte, "Squared Rectangles," in *IBM Scientific Computing Symposium on Combinatorial Problems*. pp. 3–9. T. J. Watson Research Center, Fishkill, NY, 1964.

P. G. Doyle and J. L. Snell, *Random Walks and Electric Networks*. Mathematics Association of America, Washington, DC, 1984.

A. J. Riederer and L. Weinberg, "Synthesis of Lumped-Distributed Cascades with Lossy Transmission Lines," *IEEE Trans. Circuits Syst.* **CAS-31,** 485–500 (1984).

C. A. Desoer and E. S. Kuh, *Basic Circuit Theory*. McGraw-Hill, New York, 1969.

E. A. Guillemin, *Introductory Circuit Theory*. Wiley, New York, 1953.

L. Weinberg, *Network Analysis and Synthesis*. R. E. Krieger, Huntington, NY, 1975.

Neutrino

Frederick Reines

The neutrino, according to our present view, is a stable elementary particle that has a spin $\frac{1}{2}$ parallel or antiparallel to its direction of motion, can carry momentum and energy, exists in four and possibly six states ($\bar{\nu}_e$, $\bar{\nu}_e$, $\bar{\nu}_\mu$, $\bar{\nu}_\mu$, $\bar{\nu}_\tau$, $\bar{\nu}_\tau$), has no charge or rest mass, and has an exceedingly weak interaction with matter.

THE NEUTRINO CONCEPT

The concept of the neutrino (ν) was put forward by W. Pauli (1930) to account for the apparent loss of energy in the process of nuclear beta decay. He suggested that a hitherto unknown particle, later named the neutrino, was produced in beta decay and that it carried away, unobserved, the missing energy-momentum. E. Fermi in his classic paper on the subject (1934) developed the theory of beta decay in analogy with the process of photon emission from atoms, incorporating the neutrino idea of Pauli. The interaction postulated by Fermi to account for beta decay predicted cross sections for neutrinos on matter to be extremely small, as hypothesized by Pauli. For example, in inverting the process of neutron decay

$$n \rightarrow p+e^-+\bar{\nu}, \tag{1}$$

that is,

$$\bar{\nu}+p \rightarrow n+e^+, \tag{2}$$

the cross section is $\sim 10^{-43}$ cm^2 for a 3-MeV $\bar{\nu}$.

Neutrino interactions and beta decay are now recognized as manifestations of the class of interactions called "weak" (as opposed to the strong, electromagnetic, or gravitational interactions). Detailed studies of the weak interaction via the neutrino are under way at high-energy accelerators and, to a modest degree, at nuclear reactors in the hope that its structure can be elicited and possible connections with other fundamental interactions revealed. The idea of using the neutrino as a probe of the strong interaction in much the same way as the electron has been used is also being exploited.

Because of the tiny cross sections for neutrino interactions the neutrino can penetrate astronomical thicknesses of matter, with profound consequences for the energy transport in stars and hence for their evolution. A further consequence of this great penetrability is the possibility that neutrinos emerging directly from stellar interiors might be used to reveal the conditions in those otherwise experimentally inaccessible regions.

DISCOVERY (ν_e, ν_μ)

Although the concept of the neutrino was generally accepted because of its success in accounting for nuclear beta decay, it was recognized that a logical noncircular proof of its existence required a direct observation of a reaction induced by a neutrino at a location other than that of its parent nucleus. This observation was made by C. L. Cowan, F. Reines, and collaborators from the Los Alamos Scientific Laboratory (1956) using a large liquid scintillation detector in the vicinity of a powerful fission reactor. Antineutrinos ($\bar{\nu}$) associated with the negative beta decay of neutron-rich fission fragments interacted with protons (p) in the liquid scintillator to produce a neutron (n) and a positron (e^+), as indicated in Eq. (2). The reaction was signaled by the characteristic delayed coincidence associated with the positron and neutron-capture pulses.

It was known that other weak decay processes analogous to nuclear beta decay existed. In particular, the shape of the muon decay spectrum as well as its rate was understood in terms of the emission of two neutrinos:

$$\mu^+ \rightarrow e^+ + \nu + \bar{\nu} \tag{3}$$

where $\tau_{1/2}=2.2$ μs and the e^+ end point is 52 MeV. The principle of lepton conservation (the leptons include $e^\pm$, $\mu^\pm$, ν, and $\bar{\nu}$), which required the presence of ν and $\bar{\nu}$ in Eq. (3), suggested a possible decay mode $\mu^+ \rightarrow e^+ + \gamma$, which was not found. This puzzle was resolved in 1962 through the discovery by M. Schwartz and collaborators at the Brookhaven National Laboratory of a second type of neutrino, that associated with the muon. These investigators used a large composite spark-chamber–scintillator detector in the neutrino beam that resulted from the decay of pions produced by protons from the Brookhaven accelerator. They found that, unlike the situation with neutrinos from nuclear beta decay, which produced an electron as in Eq. (2), the particle produced by accelerator neutrinos was the muon. It was therefore necessary to enlarge the list of leptons to include four kinds of neutrinos: ν_e, $\bar{\nu}_e$, ν_μ, and $\bar{\nu}_\mu$, where the subscript indicates the particle produced by neutrino interactions and the overbar denotes the antiparticle. Rewriting (1), (2), and (3) appropriately, we have

$$n \rightarrow p + e^- + \bar{\nu}_e, \tag{1'}$$

$$\bar{\nu}_e + p \rightarrow n + e^+, \tag{2'}$$

$$\mu^+ \rightarrow e^+ + \nu_e + \bar{\nu}_\mu. \tag{3'}$$

The discovery (in 1975) of the production and decay of the τ lepton by M. Perl and collaborators at the Standard Linear Accelerator is interpreted to imply the existence of yet another pair of neutrinos, ν_τ and its antiparticle $\bar{\nu}_\tau$.

INTERACTIONS AT HIGH ENERGIES

Following these initial discoveries there has been a massive and concerted effort, primarily at high-energy accelerators, to make use of the neutrino as a probe of the weak interaction. Total cross sections, multiplicity, and spectra of products have been used to search for a propagator of the weak force that would rescue a purely local interaction from the logical dilemma of exceeding the maximum cross section allowed by the conservation of probability (unitarity limit). No such propagator, the so-called W particle, has been detected within the mass range available to date ($m_W > 10$ GeV/c^2). The relationship with the various quark–parton models of elementary particles has been studied with some success and evidence is accumulating for the existence of additional heavy leptons. It has been conjectured that there may even be found other neutrinos associated with heavy leptons.

NEUTRAL CURRENTS

One of the most exciting discoveries in neutrino physics, the existence of a neutral current, was made at CERN (1973) with a bubble chamber. Unlike inverse beta decay, a neutrino interacting via the neutral current is not absorbed with the production of a charged lepton thus changing the charge of the target nucleus. Instead the neutrino is scattered, exciting the nucleus and emerging with a diminished energy. This process has been interpreted by S. Weinberg and A. Salam as implying a connection between the weak and electromagnetic interactions which they cast in a renormalizable form in the manner of quantum electrodynamics.

Neutral-current interactions that have been observed at accelerators (since 1974) are of two types, purely leptonic and semileptonic, the latter involving a nucleon (N):

$$\nu_\mu + e^- \rightarrow \nu_\mu + e^-, \tag{4}$$

$$\nu_\mu + N \rightarrow N' + \nu_\mu. \tag{5}$$

Electron-type neutrinos (from a nuclear reactor) have also been seen to interact with nucleons (1979) via the neutral-current reaction

$$\bar{\nu}_e + d \rightarrow n + p + \bar{\nu}_e \tag{6}$$

and with electrons in the purely leptonic elastic scattering process (1976)

$$\bar{\nu}_e + e^- \rightarrow \bar{\nu}_e + e^-. \tag{7}$$

The latter process is especially interesting because it proceeds via a mixture of charged plus neutral currents (mediated by intermediate bosons, $W^\pm$ or Z^0), as shown by the Feynman diagrams given in Fig. 1.

Both semileptonic and purely leptonic neutrino reactions are remarkably well described by the Weinberg–Salam theory with a mixing parameter $\sin^2\theta_w = 0.2$.

SOLAR μ_e

Recognizing the unique probe provided by the neutrino for processes occurring deep within stars, physicists have considered ways in which to observe these messengers. In particular, R. Davis, Jr., of the Brookhaven National Laboratory, and his collaborators have over the past 20 years devised and operated a large detector with which they have searched for neutrinos (μ_e) predicted to emanate from the center of the most accessible star, our sun. The detector, located 1500 m underground to reduce backgrounds due to cosmic rays, is a 400 000-liter tank of perchloroethylene (C_2Cl_4) in which solar neutrinos are predicted to cause the inverse beta reaction

$$\nu_e + {}^{37}\text{Cl} \rightarrow {}^{37}\text{Ar} + e^- \tag{8}$$

at a rate of approximately one per day. The technique consists of quantitatively removing the product ^{37}Ar. The results in six years of running (through January 1979) did not agree with standard solar models, which predict a value approximately three times that observed, assuming all nonsolar backgrounds have been taken into account.

Despite the complexity of the sun it is possible to predict the low-energy part of the solar ν_e flux at Earth with good precision (6×10^{10} cm^2 s^{-1}) assuming only that the main energy source derives from the fusion reaction

$$p + p \rightarrow d + e^+ + \nu_e, \tag{9}$$

where the $\bar{\nu}_e$ end point is 0.42 MeV, and the measured value

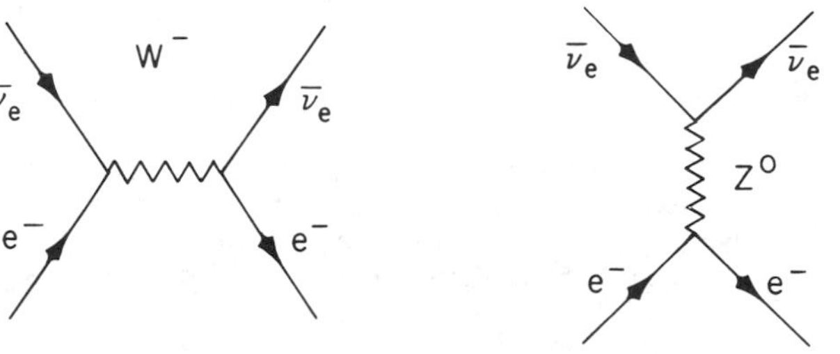

FIG. 1. Feynman diagrams representing mediation of weak-current interactions by intermediate bosons.

of the solar constant. Since this assumption of the nuclear origin of stellar energy generation is central to all astrophysical theory, we can expect the search for solar neutrinos using a suitable target such as ^{71}Ga (threshold 0.24 MeV) to command much attention until the puzzle is solved.

NEUTRINOS IN COSMIC RAYS AND COSMOLOGY

According to our ideas of nuclear and elementary-particle reactions, the earth receives neutrinos of all energies from <1 to 10^{17} eV. These arise from a variety of processes ranging from neutrinos left over from the "big bang" to those produced as decay secondaries from proton–proton collisions in interstellar and intergalactic space. In 1965, the lower part of the spectrum of those high-energy (10^8- to 10^{10}-eV) neutrinos produced as decay products of the collisions between cosmic-ray primaries and the Earth's atmosphere were detected using large-area (20- to 200-m^2) detectors deep (~2 miles) underground. A characteristic of such investigations is the large effective mass (10^3–10^9 tons) of the detector arrays required. An exciting result was obtained in 1987 when two very large imaging water Čerenkov detectors [Irvine/Michigan/Brookhaven Collaboration (5000 tons) and the Kamiokande collaboration (~2000 tons)] designed to search for proton decay observed neutrinos produced by the supernova SN1978A. This historic observation marked the birth of observational neutrino astronomy demonstrating the essential suitability of supernova models. This intriguing field of neutrino astronomy is very much at the beginning.

See also Astronomy, Neutrino; Beta Decay; Cosmic Rays—Astrophysical Aspects; Cosmology; Leptons; Stellar Energy Sources and Evolution; Weak Interactions; Weak Neutral Currents.

BIBLIOGRAPHY

J. N. Bahcall and R. Davis, Jr., "Solar Neutrinos: A Scientific Puzzle," *Science* **191,** 264 (1976). (E)

E. D. Commins, *Weak Interactions*. McGraw-Hill, New York, 1973. (I)

A. De Rújula, H. Georgi, S. L. Glashow, and H. R. Quinn, "Fact and Fancy in Neutrino Physics," *Rev. Mod. Phys.* **46,** 391 (1974). (A)

E. Fermi, "Versuch einer Theorie der β-Strahlen," *Z. Phys.* **88,** 161 (1934). An English translation may be found in *Am. J. Phys.* **36,** 1150 (1968). (A)

IBM Collaboration (R. M. Bionta *et al.*) and Kamiokande Collaboration (K. Hirota *et al.*), *Phys. Rev. Lett.* (1987).

L. M. Lederman, "Resource Letter Neu-1: History of the Neutrino," *Am. J. Phys.* **38,** 129 (1970). (E–A)

F. Reines and C. L. Cowan, Jr., "Neutrino Physics," *Phys. Today* **10** (8), 12 (1957). (E,I)

F. Reines and J. P. F. Sellschop, "Neutrinos from the Atmosphere and Beyond," *Sci. Am.* **214** (2), 40 (1965). (E)

S. Weinberg, "Recent Progress in Gauge Theories of Weak, Electromagnetic and Strong Interactions," *Rev. Mod. Phys.* **46,** 255 (1974). (A)

The literature on neutrinos is large and growing rapidly. The interested reader may wish to consult reports of frequent conferences devoted to the subject, and articles in *Physics Today, Scientific American,* and *Reviews of Modern Physics*.

Neutron Diffraction and Scattering

S. M. Shapiro and G. Shirane

INTRODUCTION

Thermal neutrons are one of the most useful probes of condensed matter such as solids, liquids, polymers, and biological systems. There are five characteristics of the neutron that make it unique in the studies of condensed matter: (i) The energy–wavelength relation of thermal neutrons causes a unique match between the wavelength of the neutron and the interatomic spacings of the solid, and between the energy of the neutron and the characteristic phonon energies in a solid. For example, 10-meV neutrons have a velocity of 1.4 km/s and a wavelength of 2.9 Å. (See Table I.) (ii) The neutron interacts with a solid primarily by a neutron–nuclear interaction. The scattering efficiency of neutrons by different elements varies irregularly through the periodic table. This is seen in Table II where the scattering lengths, b, of the elements are given. Most of the values fall between $\pm 1\times 10^{-12}$ cm. This contrasts with x rays and electrons where the scattering is by the lighter electrons and the scattering efficiency increases as Z^2, where Z is the number of electrons. Thus, it is very difficult to study light elements in the presence of heavy elements. (iii) Different isotopes have different scattering lengths and frequently, different signs of the scattering lengths. This is exploited in contrast measurements where the same material is studied, but with one element replaced by its isotope. Particularly this is used in studies of polymers and biological materials where hydrogen, which has a negative scattering length, can be selectively replaced by deuterium, which has a positive scattering length. (iv) The neutron interacts weakly with matter and so bulk samples, transparent or opaque, can be studied. This has a technical advantage that extreme sample environments can be placed around the sample and the neutron will penetrate into the sample without substantial loss of intensity. (v) The neutron has a spin and a magnetic moment which

Table I Properties and wavelength, frequency, velocity, and energy relationships for neutrons

Rest mass	$m_n = 1.675\times10^{-24}$ g
Magnetic moment ($S=\frac{1}{2}$)	$\mu_n = 1.913$ nuclear magnetons
Charge	0
	Value at 10 meV
Wavelength: $\lambda(\text{Å}) = \dfrac{9.044}{\sqrt{E(\text{meV})}}$	2.86 Å
Wave vector: $k(\text{Å}^{-1}) = \dfrac{2\pi}{\lambda(\text{Å})}$	2.197 Å^{-1}
Frequency: $\nu(\text{Hz}) = 2.418\times10^{11}E(\text{meV})$	2.418 THz
Wave number: $\dfrac{\nu(\text{s}^{-1})}{3.0\times10^{10}(\text{cm/s})}$	80.65 cm^{-1}
Velocity: $v(\text{km/s}) = 0.6302k\ (\text{Å}^{-1})$	1.38 km/s
Temperature: $T(°\text{K}) = 11.605E(\text{meV})$	116.05 K

Table II Coherent Neutron-Scattering Amplitudes in Units of 10^{-3} cm [Prepared by L. Koester and W. B. Yelon (1981)]

Element	Isotope	b
H	^{1}H	−0.3741
	^{2}H	0.6674
	^{3}H	0.494
He	^{4}H	0.326
	^{3}He	0.574
Li		−0.203
	^{6}Li	−0.187+0.026i
	^{7}Li	−0.220
Be	^{9}Be	0.779
B		0.535
	^{10}B	0.0−.11i
	^{11}B	0.66
C		0.6648
	^{12}C	0.6653
	^{13}C	0.62
N	^{14}N	0.937
	^{15}N	0.644
O	^{16}O	0.5805
	^{17}O	0.578
	^{18}O	0.584
F	^{19}F	0.565
Ne	^{20}Ne	0.455
Na	^{23}Na	0.363
Mg		0.5375
	^{24}Mg	0.549
	^{25}Mg	0.362
	^{26}Mg	0.492
Al		0.3449
Si		0.4149
P		0.513
S		0.2847
Cl		0.9579
	^{35}Cl	1.17
	^{37}Cl	0.308
Ar		0.1884
	^{36}Ar	2.42
	^{40}Ar	0.183

Element	Isotope	b
K		0.367
	^{39}K	0.379
	^{41}K	0.258
Ca		0.49
	^{40}Ca	0.49
	^{44}Ca	0.18
Sc		1.23
Ti		−0.3438
	^{46}Ti	0.473
	^{47}Ti	0.349
	^{48}Ti	−0.584
	^{49}Ti	0.100
	^{50}Ti	0.593
V		−0.0382
Cr		0.3635
	^{50}Cr	−0.450
	^{52}Cr	0.491
Mn		−0.373
Fe		0.954
	^{54}Fe	0.42
	^{56}Fe	1.01
	^{57}Fe	0.23
Co		0.253
Ni		1.03
	^{58}Ni	1.44
	^{60}Ni	0.28
	^{61}Ni	0.76
	^{62}Ni	−0.87
	^{64}Ni	−0.38
Cu		0.7718
	^{63}Cu	0.67
	^{65}Cu	1.00
Zn		0.568
	^{64}Zn	0.52
	^{66}Zn	0.60
	^{68}Zn	0.60
Ga		0.729
Ge		0.8193
As		0.658

Element	Isotope	b
Se		0.797
	^{76}Se	1.22
	^{78}Se	0.824
	^{80}Se	0.748
Br		0.679
Kr		0.785
Rb		0.708
Sr		0.702
	^{86}Sr	0.568
	^{88}Sr	0.716
Y		0.775
Zr		0.716
Nb		0.7054
Mo		0.695
Tc		0.68
Ru		0.721
Rh		0.593
Pd		0.591
Ag		0.597
	^{107}Ag	0.764
	^{109}Ag	0.419
Cd		0.5−.16i
	^{113}Cd	−0.8−.75i
	^{114}Cd	0.64
In		0.406−.05i
	^{113}In	0.539
Sn		0.6228
	^{116}Sn	0.58
	^{118}Sn	0.58
	^{120}Sn	0.60
Sb		0.564
Te		0.58
I		0.528
Xe		0.489

Element	Isotope	b
Cs		0.542
Ba		0.525
La		0.827
Ce		0.484
	^{140}Ce	0.470
	^{142}Ce	0.45
Pr		0.445
Nd		0.769
	^{142}Nd	0.77
	^{144}Nd	0.24
	^{146}Nd	0.87
Pm	^{147}Pm	1.26
Sm	^{149}Sm	−1.9+4.5i
	^{152}Sm	−0.50
	^{154}Sm	0.80
Eu		0.60
	^{153}Eu	0.82
Gd		0.95
	^{160}Gd	0.915
Tb		0.738
Dy		1.69
	^{160}Dy	0.67
	^{161}Dy	1.03
	^{162}Dy	−0.14
	^{163}Dy	0.50
	^{164}Dy	4.94
Ho		0.808
Er		0.803
	^{166}Er	1.23
Tm		0.705
Yb		1.24
Lu		0.73
Hf		0.77
Ta		0.691

Element	Isotope	b
W		0.477
	^{182}W	0.83
	^{183}W	0.43
	^{184}W	0.76
	^{186}W	−0.119
Re		0.92
Os		1.07
	^{188}Os	0.78
	^{190}Os	1.14
	^{192}Os	1.19
Ir		1.06
Pt		0.95
Au		0.763
Hg		1.266
Tl		0.879
Pb		0.9401
Bi		0.853
Po		
Rn		
Ra		1.0
Ac		
Th		0.984
Pa		0.91
U		0.842
	^{235}U	0.98
	^{238}U	0.855
Np		1.06
Pu	^{239}Pu	0.77
	^{240}Pu	0.35
	^{242}Pu	0.81
Am		0.83
Cm		0.95

Coherent scattering length for bound atoms. Complex values correspond to neutron wavelength of 1 Å. For references and complete data see: "Summary of Neutron Scattering Lengths," L. Koester, H. Rauch, M. Herkens, K. Schröder, K.F.A. — Report Jül-1755, 1981.

interacts very efficiently with the magnetic moment of a solid.

The traditional sources for neutron scattering are research reactors where a continuous stream of neutrons is produced in the fission reaction. Recently, accelerator-based spallation sources, in which high-energy protons bombard heavy nuclei and shake loose neutrons, have become available. In both cases, the fast neutrons are slowed down by collisions with molecules in the moderator. This moderation is less complete in the spallation sources and, consequently, they have more high-energy neutrons (in the eV range) than the reactor-based sources. The high-flux reactors have continuous neutron fluxes on the order of 10^{15} neutrons/cm^2 s. The pulsed sources can achieve this flux but the neutron pulses have a time width of typically 25 μs and a repetition rate of about 100 Hz. Thus, the time-averaged flux in a pulsed source is less than on a steady-state reactor-based source. Because of the pulsed nature of these sources, time-of-flight techniques to utilize the different wavelengths are commonly used on spallation sources.

In the scattering process of neutrons impinging on a target crystal, the laws of conservation of momentum and energy must be obeyed:

$$\mathbf{k}_i - \mathbf{k}_f = \tau + \mathbf{q} = \mathbf{Q},$$

$$\frac{h^2}{2m}(k_i{}^2 - k_f{}^2) = \hbar\omega;$$

here $\mathbf{k}_i$ and $\mathbf{k}_f$ are the initial and final wave vectors, respectively, of neutrons and $\mathbf{Q}$ is the total momentum transferred to the sample. When $\mathbf{q}=0$ and $\hbar\omega=0$, the scattering process is totally elastic and we have Bragg scattering at τ, a reciprocal lattice vector. Measurements of intensities of many Bragg reflections constitute a structural determination of the solid. The change of neutron energy during the scattering process corresponds to the excitation energy $\hbar\omega$ with a particular momentum $\mathbf{q}$. The object of an inelastic neutron scattering experiment is to measure the excitation energy $\hbar\omega$ and its intensity as a function of $\mathbf{q}$.

ELASTIC SCATTERING

The uniqueness of neutrons as a probe of structures has been adequately demonstrated for over 50 years. The importance of neutron diffraction has been evidenced again, recently, in the field of high-temperature superconductivity. The understanding of high T_c is certainly the most challenging problem in the field of condensed matter science today and its deciphering begins with a knowledge of the structure. One of the first structural determinations performed on the $YBa_2Cu_3O_{7-x}$ compounds was by use of a pulsed neutron source. Figure 1 shows the diffraction pattern for the low-temperature orthorhombic (Fig. 1a) and high-temperature tetragonal phases (Fig. 1b). The solid line is a fit to an assumed structure and the structural parameters were obtained by using the Rietveld profile refinement technique. In this experiment a white beam of neutrons impinges upon the sample and the diffracted neutrons are detected at a fixed angle. The different d spacings are separated by measuring the time of flight for neutrons of different energies. The neutron diffraction studies are easily able to locate the oxygen sites and measure their occupancies. This is important since the oxygens are thought to play a key role in the occurrence of superconductivity.

The magnetic scattering is truly a unique feature of neutron studies of crystals. The magnetic scattering amplitude, p, of an atomic spin S is given by

$$p = \left(\frac{e^2\gamma}{mc^2}\right) Sf = 0.54Sf \times 10^{-12} \text{ cm},$$

where f is a form factor for unpaired electrons due to their finite spatial distribution (the nuclear scattering lengths are independent of the scattering angle). We can see from Table II that the bs are of the same order of magnitude as the ps. An additional factor which enters into the cross-section formula is the angle α between $\mathbf{S}$, the spin direction, and the scattering vector $\mathbf{Q}$. Intensities are proportional to $\sin^2\alpha$. Thus, by a comparison of the intensities and positions of several magnetic peaks one can determine the magnetic spin arrangement. This is demonstrated beautifully in the studies of high-T_c superconductors where neutron diffraction definitively established the antiparallel arrangement of the spins on the Cu^{2+} atoms. Figure 2 shows a partial diffraction pattern of two classes of high-T_c materials. In the La_2CuO_4 compound, Fig. 2a, the weak peaks (labeled with a subscript M) are due to the ordering of the Cu^{2+} spins with a moment of 0.4 Bohr magnetons on each Cu site. Figure 2b shows the data for $YBa_2Cu_3O_6{}^{+x}$ and demonstrates the sensitivity of the magnetic ordering to the oxygen content.

One important application of the magnetic scattering is the production and utilization of fully polarized neutron beams. If the magnetic moment of a monochromator crystal is aligned perpendicular to the scattering vector, the interference effects between the nuclear and magnetic scattering produce the cross sections

$$\sigma = (b \pm p)^2,$$

where the $\pm$ sign corresponds to the two spin states of the incident unpolarized beam. Certain crystal monochromators such as Heusler alloys have $b=p$ and one spin state is not diffracted. The resultant diffracted beam is, therefore, completely polarized. Among the many advantages of polarized-beam studies is the extremely high sensitivity obtainable which allows a measurement of a very small magnetic amplitude. Small induced magnetic moments can be measured which allow detailed spin distributions to be mapped out. In addition, by analyzing the polarization of the diffracted beam it is straightforward to separate the magnetic scattering from the nuclear scattering.

INELASTIC SCATTERING

The most versatile instrument for measuring dispersion curves of excitations in solids is the triple-axis spectrometer. This instrument allows for scans along any direction in $\hbar\omega$–Q space. The three axes consist of a monochromator, sam-

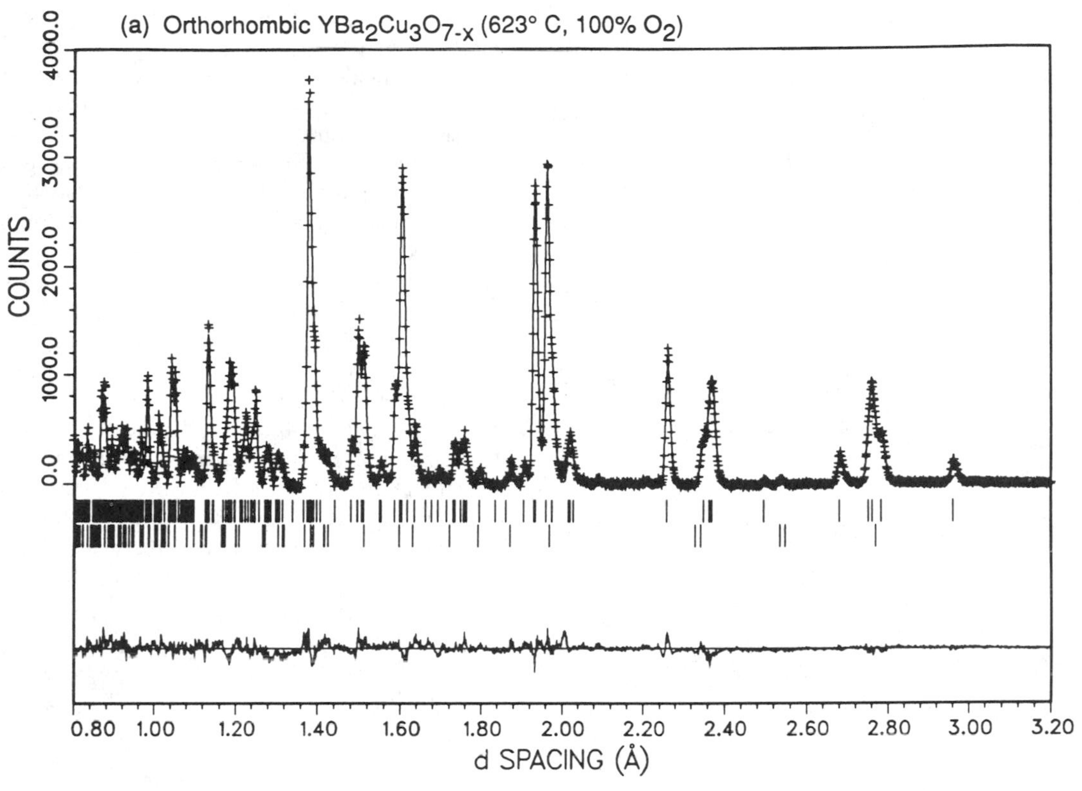

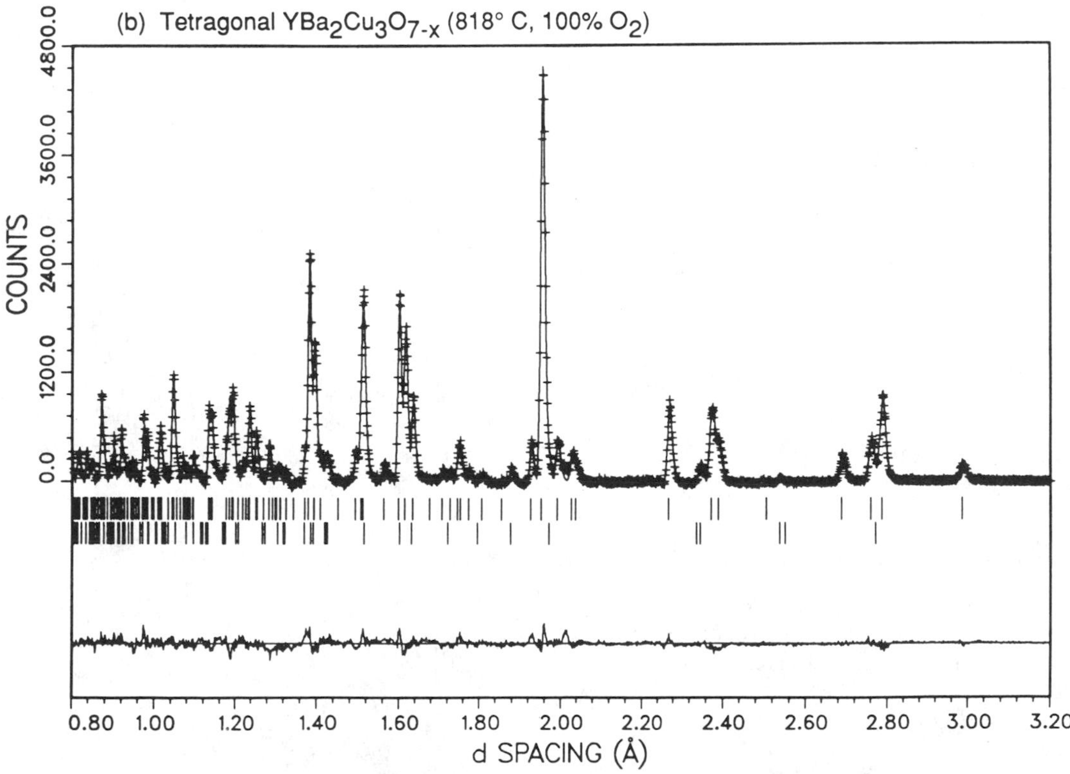

FIG. 1. Rietveld refinement profile of the neutron powder diffraction data for $YBa_2Cu_3O_{7-x}$ in the (a) orthorhombic phase at 623 °C and (b) tetragonal phase at 818 °C. Plus marks (+) are the raw neutron powder diffraction data and the continuous line is the calculated profile. Tick marks below the curves indicate the positions of the allowed $YBa_2Cu_3O_{7-x}$ reflections (upper tick marks) and a minor CuO (less than 3%) impurity phase (lower tick marks). [From J. D. Jorgensen *et al.*, *Phys. Rev.* **B36,** 3608 (1987).]

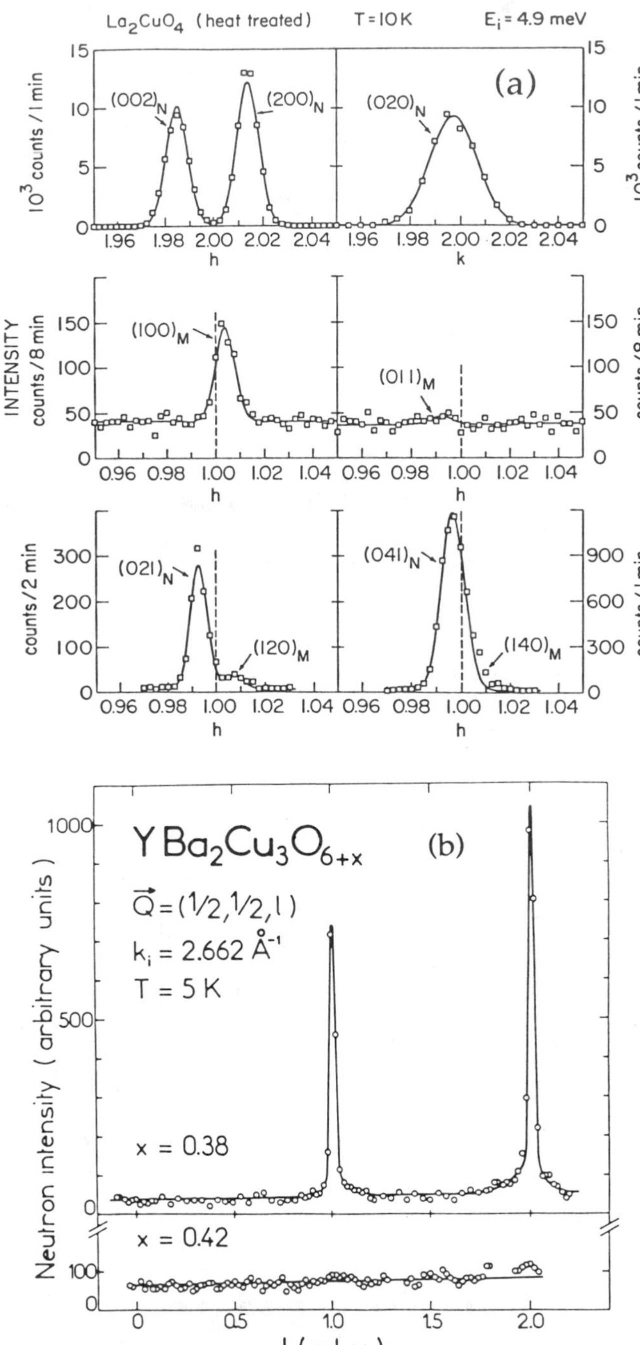

FIG. 2. (a) Neutron diffraction pattern measured at T = 10 K for various Bragg reflections in the La_2CuO_4 crystal heat treated at 400 °C. [From T. Freltoft *et al.*, *Phys. Rev.* **B36**, 826 (1987).] (b) Diffraction pattern measured on a single crystal of $YBa_2Cu_3O_{6+x}$ at T = 5 K showing the magnetic Bragg peaks present for $x = 0.38$ and absent for $x = 0.42$. [From J. Rossat-Mignod *et al.*, *J. Phys.* (*Paris*), Colloque **49**, C8-2119 (1988).].

ple, and analyzer axis. Most frequently, the momentum transfer Q is kept constant and the energy is scanned. The time-of-flight instruments at reactor or pulsed sources are frequently used to measure density of states or localized excitations where there is little dispersion. These measurements are most useful in studies of liquids, amorphous systems, or polycrystalline samples where the direction of momentum transfer is irrelevant.

Phonon dispersion relations are now widely explored for most elements (including solid ^{4}He) and many compounds. These data in turn have stimulated theoretical interest and provide the major experimental input in the field of lattice dynamics. Thus, the dynamical characteristics of solids are now reasonably well understood and they constitute an important part of our current knowledge of condensed matter. As an illustration we use the results of studies on the high-T_c superconductors. Figure 3 shows the phonon dispersion curves measured on a 3-axis instrument for $La_{2-x}SrCuO_4$, the nonsuperconducting parent system. One interesting feature is the decrease in the energy of a transverse optic phonon, measured at the [ζζ0] zone boundary, with decreasing temperature. The frequency of this mode appears to go to zero at the structural transition temperature where the symmetry changes from tetragonal to orthorhombic. In fact, the eigenvectors associated with this mode are just those required to transform the system from the tetragonal to the

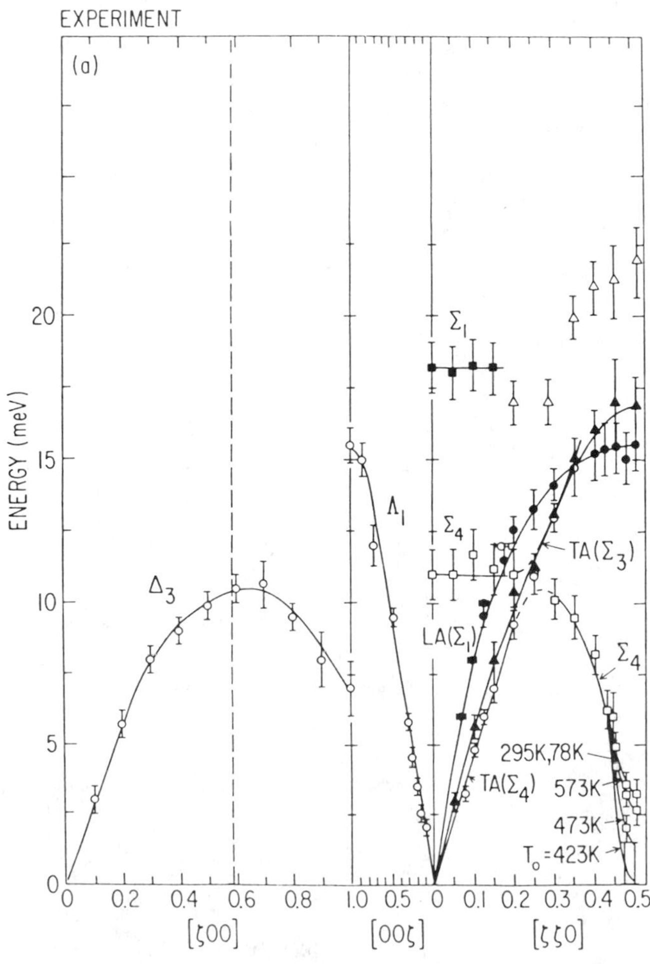

FIG. 3. Low-energy (0–25-meV) phonons measured in $La_{2-x}Sr_xCuO_4$. The dispersion curves are only weakly temperature dependent with the exception of Σ_4 branch near the ($\frac{1}{2}\frac{1}{2}$0) zone boundary. [From P. Böni *et al.*, *Phys. Rev.* **B38**, 185 (1988).]

orthorhombic phase. Both superconducting and nonsuperconducting samples exhibit a similar structure transformation and the role that this lattice instability plays in the appearance of superconductivity is still being explored.

The inelastic scattering technique has also been applied to the study of magnetic fluctuations. As in phonons, the spin fluctuations have been explored for a wide variety of ferromagnetic, antiferromagnetic, and paramagnetic systems. A particularly simple case studied years ago is the two-dimensional (2D) antiferromagnet K_2NiF_4 where the magnetic exchange interaction is strong within the plane and very weak between the planes. This is directly demonstrated by a large spin-wave dispersion within the plane and very little dispersion for directions out of the plane. There is a renewed interest in this class of magnetic compounds because they are isostructural with the high-T_c material La_2CuO_4. The inelastic neutron scattering on this compound readily showed the two-dimensional nature of the magnetic properties, but the energy scale is much larger and the nature of the magnetic interactions is quite different from the prototype K_2NiF_4 material. The novel 2D magnetic state observed in the high-T_c materials has been called a quantum spin fluid. Much of the effort on the present studies of high-T_c compounds is to establish the relationship between the magnetic properties and the superconductivity. It is clear that neutron scattering will play a pivotal role in this endeavor.

The major developments in the technique of neutron scattering at research reactors is the use of cold neutrons with specialized instruments, e.g., neutron spin echo and backscattering, to extend the measurements down to the nanoelectron-volt region. This has proven useful in the study of relaxation times in macromolecular systems such as biological or polymer materials. The use of pulsed neutron sources has extended the energy scale up to the electron-volt regime important in the study of single-particle excitations, molecular spectroscopy, and magnetism in the transition elements and actinides. Now, neutron scattering can measure energies over 9 orders of magnitude. Another new area of development is the application of neutron reflectivity measurements to the study of surfaces. Several phenomena analogous to those observed in classical optics such as reflection, refraction, and interference can also be observed with slow neutrons. Information on surface organization, described by a refractive index profile, can be extracted from reflectivity measurements.

See also FERRIMAGNETISM; FERROMAGNETISM; LATTICE DYNAMICS; MAGNETIC ORDERING IN SOLIDS; NEUTRON SPECTROSCOPY; PARAMAGNETISM; PHASE TRANSITIONS; PHONONS.

BIBLIOGRAPHY

C. Kittel, *Introduction to Solid State Physics*. Wiley, New York, 1986.

G. E. Bacon, *Neutron Diffraction*. Oxford University Press, London, 1975.

G. L. Squires, *Introduction to the Theory of Thermal Neutron Scattering*. Cambridge University Press, Cambridge, 1978.

S. W. Lovesey, *Theory of Neutron Scattering from Condensed Matter*. Oxford University Press, London, 1984.

W. G. Williams, *Polarized Neutrons*. Oxford University Press, London, 1988.

C. G. Windsor, *Pulsed Neutron Scattering*. Taylor and Francis, London, 1981.

Methods of Experimental Physics, Vol. 23: Neutron Scattering (D. L. Price and K. Sköld, eds.). Academic Press, New York, 1987.

Neutron Spectroscopy

B. N. Brockhouse and B. D. Gaulin

Slow-neutron spectroscopy, the study of slow neutrons scattered inelastically by matter, is now recognized as the most generally useful experimental tool that is available for the study of atomic motions in condensed matter and of the low-lying electronic states in magnetic materials. Largely, the experiments employ thermalized neutrons extracted from nuclear reactors. However, spallation neutron sources are in use to an increasing extent. These spallation sources produce neutrons by allowing a high-energy proton beam, from a linear accelerator, to impinge on a heavy-metal target.

THEORY AND PRACTICE

The special usefulness of thermal neutrons comes about because they have, at once, wavelengths (λ) of the order of the interatomic spacings in condensed matter, say 1–4 Å, and (corresponding) energies of ~80 to ~5 meV, that is, of the normal order of the characteristic quantum energies which are excited in condensed matter at ordinary temperatures.

Neutrons scatter from condensed matter by virtue of two interactions:

1. The neutron–nuclear interaction via nuclear forces leads to (S-wave elastic) scattering cross sections (σ) and scattering amplitudes (b), with $\sigma = 4\pi b^2$, that are characteristic of the particular nuclide (in the particular spin state of the combined neutron–nuclide system), and that must be experimentally determined for each nuclide. In most situations the appropriate scattering amplitude is $b_{\text{coh}} \equiv \langle b \rangle$, the average being taken over the distribution of isotopes and spin states of the scattering nuclides of a particular type. Correspondingly, for each element there exists the *coherent scattering cross section* $\sigma_{\text{coh}} = 4\pi\langle b \rangle^2$, which is involved in interference properties of the scattering, and the *incoherent scattering cross section* $\sigma_{\text{inc}} = 4\pi[\langle b^2 \rangle - \langle b \rangle^2]$. These quantities, of course, depend on the isotopic constitution of the specimen, and (in principle, at exceedingly low temperature) on the temperature. Tables of b_{coh} and other quantities for the elements and for many separate isotopes are available. The interaction between a neutron and most nuclei is described by a $\langle b \rangle$ value assigned to the nucleus which is real, positive, and $\sim 10^{-13}$–10^{-12} cm. Strongly neutron-capturing isotopes (e.g., ^3He, ^6Li, ^{10}B, ^{149}Sm, and ^{157}Gd) possess $\langle b \rangle$ values with an important imaginary component. ^1H, ^{51}V, and Mn are examples of nuclei possessing negative coherent scattering amplitudes. There is no very clear dependence between the $\langle b \rangle$ value of an element or isotope and other

properties of the atom in question. This means that elements close to each other on the periodic table, and indeed different isotopes of the same chemical species, may have drastically different coherent "visibilities" to neutrons.

2. The neutron's magnetic moment (-1.91 nuclear magnetons) interacts with any local magnetic field in the specimen, in particular with the fields of any atomic magnetic moments such as those connected with iron group transition elements and with rare-earth and actinide ions. The resulting magnetic scattering amplitudes (b_{mag}) are of the same general order as the nuclear scattering amplitudes (b), but are calculable from theory. In addition to the direct proportionality to the magnetic moment which it exhibits, b_{mag} is also sensitive to the size and shape of the distribution of the magnetic moment upon the scattering atom, and to the relative orientations of the neutron and atomic magnetic moments, and to vector momentum transfer to the neutron. Since the latter quantities are to some degree controllable, the scattering patterns may contain a wealth of information about the magnetic constitution of the specimen.

Thus, slow neutrons are sensitive to almost all nuclear (atomic) excitations, whether in solids, liquids, or gases, as well as to many magnetic (electronic) excitations. With neutrons from a reactor, experiments are sensitive to excitations below, say, 100 meV; spallation neutron sources allow study of excitations with considerably greater energies. An example of such a high-energy excitation spectrum is shown in Fig. 1.

Because of the necessity of maintaining adequately narrow resolution in the incoming (E_0) and outgoing (E') energies and in the angular acceptance of the collimators and detector, the experiments are often intensity limited. Consequently, the bulk of the work has been done at the major neutron sources of the large national laboratories although important contributions have come from other national institutions. The major reactor sources are at Brookhaven and Oak Ridge in the United States, Chalk River in Canada, Harwell in Great Britain, and the multination Institut Laue-Langevin in France; the major spallation sources are at Argonne in the United States and the Rutherford–Appleton Laboratory in Great Britain. Many experiments are seriously hampered by low intensities and long counting periods (days or weeks). (Considered as a source of quanta, even a powerful reactor is a rather dim light bulb.) Because of the massive shielding required against fast neutrons leaking from the reactor or accompanying the beam of slow neutrons, the apparatus tends to be massive. Modern neutron facilities, such as the Institut Laue-Langevin in France, often make use of neutron-reflecting beam guides which allow the slow neutron beams to be transported relatively far from the source of the neutrons, to a so-called guide hall. The guide hall houses many neutron spectrometers in an area where the background and hence shielding problems are very much reduced. Because of the low flux densities available from reactors and because of the comparative transparency of most materials to slow neutrons, use of large specimens (of the order of cubic centimeters) is both required and possible. For studies of crystalline substances single crystals offer enormous advantages over polycrystalline specimens; indeed, use of single crystals is often essential to the experiment. Two major exceptions are for density-of-states measurements and for excitations whose energies are independent of wave vector, such as transitions between crystalline electric field levels.

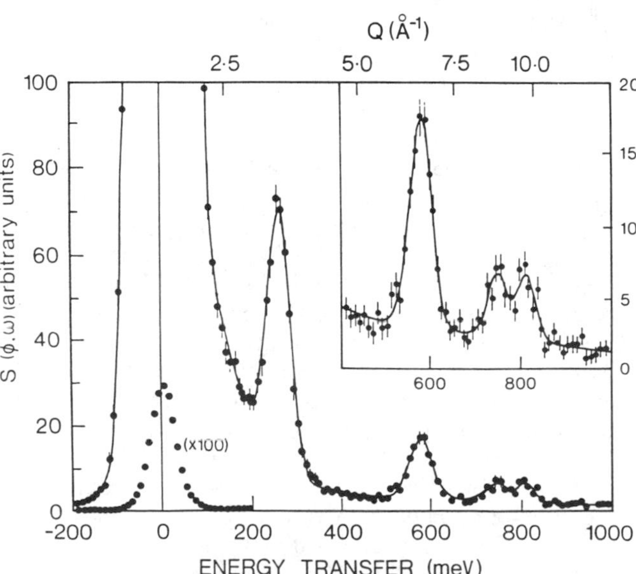

FIG. 1. The neutron scattering cross section of praseodymium measured by Taylor, Osborne, and coworkers at the ISIS spallation neutron source of the Rutherford–Appleton Laboratory. The sample temperature is 5 K, and transitions between crystalline electric field levels are observed at energies in excess of 800 meV.

Neutron energies are determined from the velocity (v) or from the wavelength (λ) as

$$E=\tfrac{1}{2}m_n v^2=\frac{2\pi^2\hbar^2}{m_n\lambda^2}=\frac{\hbar^2k^2}{2m_n},$$

but not directly. Velocities are determined from the times of flight of the neutrons over known path lengths as measured by mechanical/electronic apparatus. Wavelengths are determined by Bragg scattering from large single-crystal monochromators; beryllium, copper, germanium, Heusler alloy, or silicon as well as pyrolitic graphite are most often employed. The crystals may be optimized for this purpose by special metallurgical treatments. Special polycrystalline filters can be used (e.g., beryllium, a low pass filter for neutrons with $\lambda > 4.05$ Å, or pyrolitic graphite, a filter with a more complicated spectrum). These elements and others are used in various combinations in the apparatus to select and define E_0 and E' for neutrons (and therefore, also, $\mathbf{k}_0$ and $\mathbf{k}'$).

The neutrons (of final energy E') are detected in gas or scintillation counters following capture of the neutrons by means of various nuclear reactions, particularly ^{3}He(n,p)^{3}H, ^{6}Li($n,^3$H)^{4}He, ^{10}B(n,α)^{7}Li, and ^{235}U(n,fission). The incoming monoenergetic beam of energy E_0 is sampled (monitored) by a "thin" detector of some sort to establish the incoming flux and thus to enable normalization of the scattering patterns and interpretation of them as partial differential cross sections.

The natural variables of a neutron-scattering experiment

are two: the vector momentum transfer from the neutron to the specimen,

$$\hbar\mathbf{Q} \equiv \hbar(\mathbf{k}_0 - \mathbf{k}'),$$

and the energy transfer,

$$\hbar\omega \equiv E_0 - E' = \frac{\hbar^2}{2m_n}(k_0{}^2 - k'^2).$$

In the first Born approximation the partial differential scattering cross section of the specimen per unit solid angle (Ω) is given by an expression of the form

$$\frac{d^2\sigma}{d\Omega\, d\omega} = \frac{k'}{k_0} S(\mathbf{Q},\omega).$$

For most situations the first Born approximation is very good because of the relative smallness (in certain senses) of the interactions between the neutron and the (nuclear and magnetic) scattering centers of the specimen. The only important exceptions are the study of the Laue–Bragg elastic coherent scattering by large and relatively perfect crystals (the phenomenon of primary extinction and related effects) and study of the *neutron optics* of materials generally.

Because of the Van Hove relationship, data from neutron-scattering experiments are now usually presented in terms of the reduced variables ($\mathbf{Q}$,ω) rather than in terms of the primary laboratory variables, namely, E_0, E', angle of scattering (ϕ), crystal orientation angles with respect to $\mathbf{Q}$, etc. Modern apparatus is now computer controlled and is usually programmed so that the original measurements are made in terms of $\mathbf{Q}$ and ω (e.g., the "constant-**Q**" method used with triple-axis crystal spectrometers). Figure 2 shows a wave vector diagram (and optical analog) for the constant-**Q** method in which the energy distribution of initially "orange" scattered neutrons is studied at constant wave vector transfer **Q**. The initial and final energy ("color") is determined by Bragg reflection from a single-crystal monochromator.

The functions $S(\mathbf{Q},\omega)$ are Fourier transforms of a class of time-dependent (Van Hove) correlation functions $G(\mathbf{r},t)$, which themselves can be the object of experiments. Materials possessing well-defined normal modes which are either atomic or magnetic in nature display scattering functions, $S(\mathbf{Q},\omega)$, which have "spectral lines" substantially of the form $\delta(\mathbf{Q}-q)\delta(\omega-2\pi\nu)$. A major type of experiment involves direct extraction of the dispersion relations for these excitations [e.g., frequency/wave vector $\nu(q)$ dispersion relations for phonons or magnons in crystals] by consideration only of energy (ω) or momentum ($\mathbf{Q}$) conservation. These "spectral lines" have some intrinsic widths and are broadened by instrumental resolution. Often the intrinsic widths of the excitations can be obtained, and thus the "lifetimes" of the excitations induced by anharmonicity, impurities, etc., can be measured. Also, $S(\mathbf{Q},\omega)$ can be written as

$$S(\mathbf{Q},\omega) = \frac{1}{\pi}\langle n(\omega)+1\rangle\, \mathrm{Im}\chi(\mathbf{Q},\omega),$$

where $\langle n(\omega)+1\rangle = [1-\exp(-\hbar\omega/k_B T)]^{-1}$ is the thermal population factor, and $\mathrm{Im}\chi(\mathbf{Q},\omega)$ is the imaginary part of the dynamic susceptibility of the system. This expression for

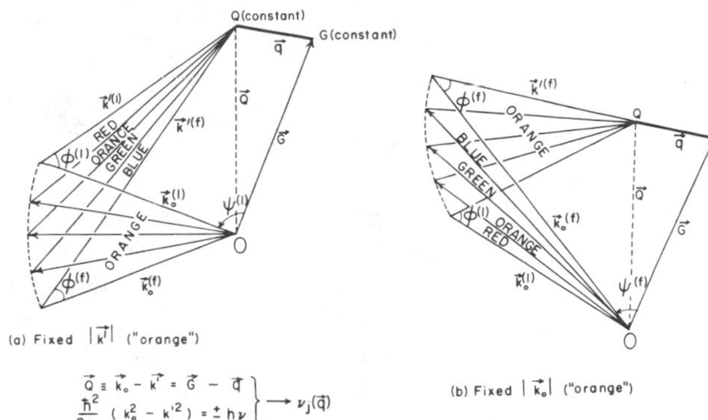

FIG. 2. Reciprocal-space diagrams for a triple-axis crystal spectrometer in two modes of operation: In (a), the incident wave vector k_0 is fixed (in the laboratory system) and the analyzing spectrometer is then set to detect a series of different scattered wave vectors $|\mathbf{k}'|$, while the angle of scattering ϕ and the orientation ψ of the single-crystal specimen in the plane are varied in such a way as to keep the wave vector transfer $\mathbf{Q} \equiv \mathbf{k}_0 - \mathbf{k}'$ fixed (in the crystal system of coordinates). Peaks in the detected neutron intensity indicate the phonon (or magnon) frequencies ν_i for the preselected wave vector $\mathbf{q} = \mathbf{G} - \mathbf{Q}$, where $\mathbf{G}$ is a reciprocal lattice vector of the crystal. In (b) the outgoing wave vector $\mathbf{k}'$ is fixed and $\mathbf{k}_0$ varied with similar results. A simple optical analog is also indicated in both diagrams. Both experiments would detect transverse phonons in preference to longitudinal phonons, because $\mathbf{q}$ is nearly perpendicular to $\mathbf{Q}$.

$S(\mathbf{Q},\omega)$ is most useful in cases where the excitation spectrum of a material cannot conveniently be described by well-defined normal modes, as in many disordered materials.

The neutron's magnetic moment may also be polarized. Polarization analysis allows for model-independent discrimination between scattering which is nuclear and that which is magnetic in origin. Discrimination in the (usual) unpolarized case depends on the contrast of the scattering amplitudes concerned.

SUBJECT MATTER AND HISTORY

The field is too extensive to be reviewed adequately here. A bibliography published in 1974 listed about 8000 references to experimental work and to directly related theory in the twin fields of neutron diffraction and spectroscopy. Since that time the field has flourished. The application of neutron techniques to problems of interest to solid-state physics in particular and materials science in general is usually considered essential. The scientists who have at one time or another worked on these subjects now number in the thousands, with important contributions being made by hundreds of individuals.

The early work on neutron scattering (1935–1950), which established the fields of *neutron diffraction* and *neutron optics* and led to that of *neutron spectroscopy,* is associated preeminently with the name of Enrico Fermi. Other principal figures on the experimental side were E. O. Wollan and C. G. Shull (Oak Ridge National Laboratory), J. R. Dunning, W. W. Havens, Jr., and J. J. Rainwater (Columbia University), D. J. Hughes (Argonne National Laboratory), D. G. Hurst (Chalk River Nuclear Laboratories), and G. E. Bacon (A. E. R. E. Harwell). On the theoretical side, principal figures were F. Bloch, O. Halpern, M. H. Johnson, R. Weinstock, G. Placzek, and (later) L. Van Hove and R. P. Feynman. Obviously, these lists are very incomplete.

The first experiments on energy transfers in inelastic scattering date from 1951 (P. A. Egelstaff and R. D. Lowde at Harwell, and Hurst and B. N. Brockhouse at Chalk River). The field of neutron spectroscopy in a practical sense dates from about 1955 (Brockhouse and A. T. Stewart at Chalk River; B. Jacrot at Saclay; Hughes, H. Palevsky, and R. S. Carter at Brookhaven National Laboratory). Other names that should be mentioned include D. G. Henshaw at Chalk River, and K.-E. Larsson and R. Stedman at A-B Atomenergi, Stockholm. An international conference on the subject was held at Stockholm in September 1957, and a series of six larger conferences were sponsored by the International Atomic Energy Agency between 1960 and 1977. Commencing in 1979 at Julich and continuing to the present day, a triannual International Conference on Neutron Scattering has been held as a satellite conference of the International Conference on Magnetism. The proceedings of these conferences form an important part of the literature on the subject.

Experiments include a considerable variety of types, mainly in the physics and chemistry of solids, liquids, and gases. Some studies of biological interest have also been made. Among the subjects dealt with are the following:

Experimental dispersion curves for the lattice vibrations in most solid phases of most elements (including crystalline ^{4}He) and in a wide variety of compounds and alloys having several atoms per unit cell. The anharmonic properties of lattice vibrations, and the effect of impurities in various concentrations on lattice vibrations. Lattice vibrational properties of several highly anisotropic, low-effective-dimensionality materials such as intercalated graphite. Lattice vibrational properties of hydrogen in some metals. Collective excitations (phase and amplitude modes) in several modulated structures.

Experimental dispersion curves for the excitations in liquid ^{4}He ("phonons" and "rotons") at different temperatures and pressures. The spectrum of collective excitations in ^{3}He and mixtures of ^{3}He and ^{4}He. (The experiments involving ^{3}He are particularly difficult because of the large capture cross section of ^{3}He.)

The correlations between the atomic motions in a variety of "classical" liquid elements and compounds, amorphous substances, and some dense gases.

"Librons" or rotational excitations in molecular crystals.

"Soft modes" and critical effects, especially in ferroelectric and related crystals, and "mode-softening" related to structural phase transitions.

The electron–phonon interaction via the lattice vibrations in superconducting and near-superconducting metals.

All the foregoing measurements rely on the scattering of the neutrons by the nuclei; in magnetic materials analogous situations obtain. The scattering of neutrons by electromagnetic interaction with the electrons allows a nearly equal variety of experiments:

The dispersion curves for the spin waves, "magnons," and related crystalline electric field spectra in a wide variety of ordered and partially ordered magnetic systems containing transition element, rare-earth, or actinide atoms with a wide variety of magnetic structures—ferromagnetic, antiferromagnetic, ferrimagnetic, helical magnetic structures, etc.—sometimes as functions of temperature up to the Curie or Néel point.

Studies of magnetic excitations in some magnetic materials with magnetic or nonmagnetic impurities (such as "spin glasses" and "random-field" materials); also studies of magnetic excitations in some amorphous magnetic materials.

Studies of nonlinear magnetic excitations such as "solitons" and "vortices" in several highly anisotropic, low-effective-dimensional magnetic materials. Also, studies of the excitation spectrum of several realizations of "quantum spin chains."

The behavior of the magnetic excitations in the critical regime of temperature (near the Curie or Néel point) in several materials; also the excitation spectrum in paramagnetic materials.

The magnetic excitations of several materials in which magnetism and superconductivity coexist. The domain structure in certain type-II superconductors.

The first experiments making use of polarization analysis of inelastic scattering were performed at Oak Ridge and demonstrated that the magnon carried one unit of angular momentum. Polarized neutrons have not been used for inelastic scattering experiments to the same extent as for neutron diffraction, largely because of intensity considerations. However, recently (~1978 and later) progress in producing higher-quality polarizing monochromator crystals and associated instrumentation has resulted in new spectrometers being built which are capable of overcoming, at least in part, the poor intensity problem. Using this instrumentation, elegant experiments have been performed which directly determined the eigenvectors of magnetic excitations.

A large number of studies of elastic scattering have given information on the time-independent atomic and magnetic structures of crystals, and some studies of quasielastic and of total differential scattering on the asymptotic and instantaneous structures of liquids have been conducted. These works are reviewed elsewhere in this volume (*see* NEUTRON DIFFRACTION AND SCATTERING).

See also FERRIMAGNETISM; FERROMAGNETISM; LATTICE DYNAMICS; MAGNETIC ORDERING IN SOLIDS; PHONONS; SCATTERING THEORY.

BIBLIOGRAPHY

Included are an elementary review and a collection of historical essays (Bacon); a monograph on magnetic critical scattering (Collins); an extensive bibliography covering the years 1932–1974 (Larose and Vanderwal); an advanced theoretical treatise (Lovesey); a collection of technical review articles related to spallation neutron sources (Newport, Rainford, and Cywinski); a collection of general technical review articles (Sköld and Price); and a readable text at a graduate student level (Squires).

G. E. Bacon, *Neutron Diffraction*, 3rd ed. Oxford, London and New York, 1975.
G. E. Bacon (ed.), *Fifty Years of Neutron Diffracton: The Advent of Neutron Scattering*. Adam Hilger, Bristol, 1986.
M. F. Collins, *Magnetic Critical Scattering*. Oxford, London and New York, 1989.
A. Larose and J. Vanderwal, *Scattering of Thermal Neutrons: A Bibliography (1932–1974)*. IFI/Plenum, New York, 1974.
S. W. Lovesey, *Theory of Neutron Scattering from Condensed Matter*, Volumes 1 and 2. Clarendon Press, Oxford, 1984.
R. J. Newport, B. D. Rainford, and R. Cywinski (eds.), *Neutron Scattering at a Pulsed Source*. Adam Hilger, Bristol, 1988.
K. Sköld and D. L. Price (eds.), *Methods of Experimental Physics, Volume 23, Neutron Scattering*, Parts A, B, and C. Academic Press, Orlando, FL, 1986.
G. L. Squires, *Introduction to the Theory of Thermal Neutron Scattering*. Cambridge University Press, Cambridge, 1978.

Neutron Stars

Gordon Baym

Neutron stars are highly condensed stellar objects produced in supernova explosions, the end point in the evolution of more massive stars. Neutron star masses M are on the order of that of the sun ($M_\odot$), whereas typical radii are only 10 km; the matter they contain, primarily neutrons, is thus the densest found in the universe. The average interior density is greater than that in a large atomic nucleus, $\rho_0 = 3\times10^{14}$ g/cm^3. (By comparison, white dwarf stars, which are of similar masses, have radii of at least several thousand kilometers, and central densities in the range of 10^5–10^9 g/cm^3.) Support against gravitational collapse in a neutron star is provided by the quantum-mechanical Fermi (or zero-point) pressure of the neutrons and other particles in the interior, in the same way that white dwarfs are supported by electron zero-point pressure. (Ordinary stars, on the other hand, are supported by thermal gas pressure.)

Neutron stars were first proposed by Baade and Zwicky in 1934 in their pioneering paper on supernovae, and considerable theoretical work on their properties, beginning with calculations by Oppenheimer and Volkoff in 1939, was carried out prior to their actual observation. It was not until the discovery in 1967 of pulsars—stars that appear to blink on and off—and their identification by Gold as rotating neutron stars, that the existence of neutron stars was established. Since that time the important astrophysical role of neutron stars as highly energetic compact sources has been recognized. In addition to pulsars, neutron stars are found in luminous compact binary x-ray sources (in which a neutron star orbits a more normal companion star), and gamma- and x-ray burst sources; one such binary, Cygnus X-3, is likely to be a principal source of ultrahigh energy ($>10^{15}$ eV) cosmic radiation in the galaxy. To date well over 400 neutron stars have been observed in the galaxy as radio pulsars, and some 20 in compact x-ray sources, whereas about 100 gamma-ray burst sources are detectable each year.

The astrophysical energy source of neutron stars, some 10 times as powerful as thermonuclear burning, is their immense gravitation. The gravitational acceleration ($g = GM/R^2$, where G is Newton's gravitational constant) at the surface is about 10^{11} times that on Earth; gravitational tidal forces would make it impossible for any normal object larger than about 10 cm to reach the surface of a neutron star without being torn apart. The gravitational binding energy GmM/R of a particle of mass m at the surface is about one-tenth of its rest energy mc^2. (Nuclear binding energies are, in comparison, at most 0.9% of the rest energy of matter.) The energy emitted in compact x-ray sources arises directly from the release of gravitational energy of matter that is accreted onto the neutron star from the companion star. The energy source of pulsars—the kinetic energy of rotation of the neutron star—also comes from the release of gravitational binding energy, for as the stellar core in a supernova collapses under gravity to form a neutron star, conservation of angular momentum requires that its rotational rate and rotational energy increase.

Neutron stars characteristically are strongly magnetized. Magnetic fields have been inferred from line spectra in gamma-ray burst sources to be of order 10^{12} G near the surface. This value is consistent as well with theoretical understanding of the slowing down of pulsar rotation. Pulsars, as a consequence of their strong magnetic fields, are surrounded by a highly ionized plasma in which the pulsed radiation originates. The large fields also play a crucial role in the accretion of plasma in compact binary x-ray sources; fields in these systems range from 10^9 to 10^{12} G. The extent to which magnetic fields decay as a neutron star ages remains an open question.

At birth, the interior temperature of a neutron star is about 10^{11} K, and within the first few days it cools by neutrino emission to less than 10^{10} K. Throughout most of its early life, the interior temperature is in the range 10^8–10^9 K, while the surface temperature is one-tenth or less of the interior temperature. Although the surface temperature is high, the extremely small surface area and generally large distance from Earth of neutron stars makes the thermal radiation from the surface difficult to observe with x-ray satellites.

The matter inside neutron stars has a relatively low temperature compared with its characteristic microscopic energies of excitation, typically of the order of MeV (million electron volts). Furthermore, nuclear processes in the early moments of a neutron star take place sufficiently rapidly compared with the cooling of the star that the matter essentially comes—via strong and electromagnetic interactions, as well as weak interactions (which transform protons into neutrons and vice versa)—into its lowest possible energy state.

The cross section of a neutron star interior is shown in Fig. 1. The density of matter increases with increasing depth in the star. Beneath an atmosphere, compressed by gravity to less than 1 cm height, is a crust, typically ~1 km thick, consisting, except in the molten outer tens of meters, of a lattice of bare nuclei immersed, as in a normal metal, in a sea of degenerate electrons. The matter in the outer part of the crust is expected to be primarily ^{56}Fe, the end point of thermonuclear burning processes in stars.

With increasing depth, the electron Fermi (or zero-point) energy rises. Beyond the density 8×10^6 g/cm^3, it is so high (>1 MeV) that ^{56}Fe can capture energetic electrons. In the capture process, as occurs during the formation of the neutron star in a supernova, protons in nuclei are converted into neutrons via the weak interaction $e^- + p \rightarrow n + \nu$. The produced neutrino ν escapes the nascent neutron star, lowering the energy of the system. (Neutrinos generated in the formation and neutronization of the neutron star are eventually responsible for ejection of the mantle in the supernova.) The matter becomes more neutron rich and rearranges into a sequence, with increasing density, of increasingly neutron-rich nuclei, reaching ^{118}Kr at a mass density $\rho_d = 4.3 \times 10^{11}$ g/cm^3. The nuclei present deep in the crust, although unstable in the laboratory, cannot undergo beta decay via the inverse reaction $n \rightarrow p + e^- + \bar{\nu}$ (where $\bar{\nu}$ is an antineutrino) because the electron would, by energy conservation, have to go into a state that is already occupied, a process forbidden because electrons obey the Pauli exclusion principle. Beyond the density ρ_d, called the "neutron drip" point, the matter becomes so neutron rich that not all the neutrons can be accommodated in the nuclei, and the matter, still solid, becomes permeated by a sea of free neutrons in addition to the sea of electrons. Finally, at a density of about ρ_0 at which the nuclei are essentially touching, the matter dissolves into a uniform liquid composed primarily (95%) of neutrons, some protons and electrons, and a few muons. The neutrons are most likely superfluid and the protons superconducting; the electrons, however, are normal.

The states of matter at high pressures deep in the interior are less well understood. With increasing density heavier baryons can live stably in the star. Several interesting phenomena are possible (see Fig. 1). One is that pi mesons are spontaneously produced and form a superfluid "pion-condensed" state; such condensation would greatly enhance the cooling of neutron stars by neutrino emission. The matter may similarly undergo an analogous "kaon condensation." At ultrahigh densities, when the nucleons are strongly overlapping, matter is expected to dissolve into "quark matter," in which the quarks that make up the baryons become free to run throughout the system; such a quark matter core in a neutron star would also enhance its cooling rate, but whether the transition to quark matter is actually reached in neutron stars is uncertain. The possibility of a separate branch of "quark stars" composed essentially of quark matter is a subject of inquiry. More speculatively, if the absolute ground state of matter is "strange quark matter," in which a large fraction of the quarks have nonzero strangeness, then normal matter could be transformed into such matter in neutron stars, giving rise to "strange quark stars."

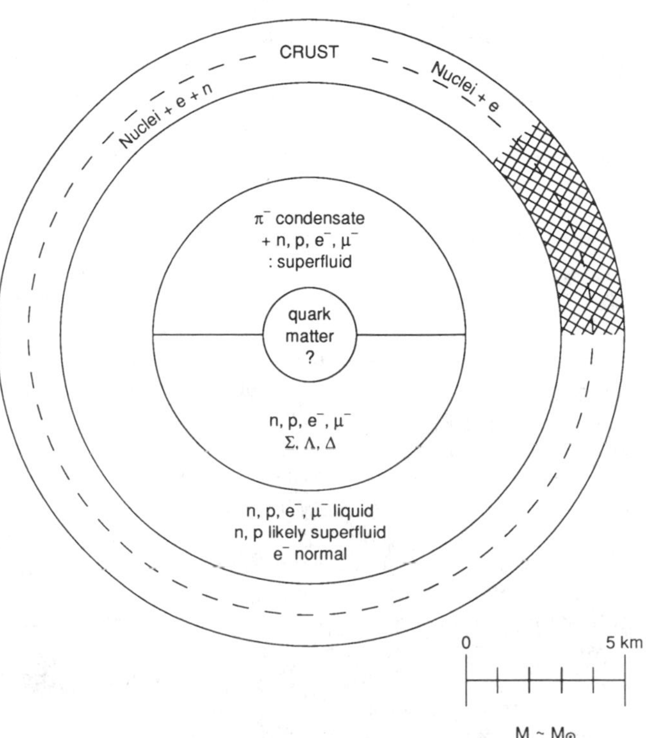

FIG. 1. Schematic cross section of a neutron star, showing the outer crust, the point of neutron drip (broken line), the liquid interior, the possibilities of higher-mass baryons and pion-condensed matter in the core, and a possible quark matter inner core.

The structure of neutron stars, including radii as a function of mass, and the range of masses for which they are stable, is determined by the equation of the state of the matter they contain. A knowledge of the maximum mass M_{max} that a neutron star can have is important in distinguishing possible black holes from neutron stars by observations of their mass. The uncertainty in the present limit, $M_{\text{max}} \sim 3M$, calculated on the basis of physically plausible equations of state, reflects our uncertainty of the properties of matter at densities much greater than ρ_0.

The ever-growing body of observational information on neutron stars provides increasingly stringent constraints on theories of their structure. The masses of the two orbiting neutron stars in the binary pulsar PSR1316+16 have been accurately measured as $1.44M_\odot$ and $1.38M_\odot$, while masses of neutron stars in compact x-ray sources can be inferred through a combination of optical and x-ray observations; those determined so far, typically $\sim 1.4M_\odot$, are consistent with present theories of neutron-star structure and formation in supernovae. Further information on their structure, e.g., moments of inertia as well as radii, can be extracted from measurements of periods and luminosities over time. In addition, sudden speedups and smaller fluctuations in pulsar repetition frequencies give clues to the internal structure of neutron stars, and support the idea that the interior is superfluid. The recent supernova 1987A in the Large Magellenic Cloud, which is expected to have left a neutron star in the center of the remnant, provides an unprecedented opportunity in the coming years to study the birth and development of a neutron star.

See also FERMI–DIRAC STATISTICS; PULSARS; STELLAR ENERGY SOURCES AND EVOLUTION.

BIBLIOGRAPHY

G. Baym and C. J. Pethick, *Annu. Rev. Nucl. Sci.* **25,** 27 (1975); *Annu. Rev. Astron. Astrophys.* **17,** 415 (1979). (On properties of matter in neutron stars; extensive references; A.)

D. J. Helfand and J. H. Huang (eds.), *The Origin and Evolution of Neutron Stars.* D. Reidel, Dordrecht, 1986. (Modern status of neutron stars; A.)

F. K. Lamb, in *Pulsars,* Proceedings of the IAU Symposium 95 (W. Sieber and R. Wielebinski, eds.). D. Reidel, Dordrecht, 1981.

(Overview, in useful compendium volume, of deductions of neutron star properties from observations; A.)

R. M. Manchester and J. H. Taylor, *Pulsars*. W. H. Freeman, San Francisco, 1977. (Describes neutron stars and pulsars; I.)

P. Murdin and L. Murdin, *Supernovae*. Cambridge University Press, Cambridge, 1985. (On supernovae and pulsars; E.)

J. H. Taylor and D. R. Stinebring, *Annu. Rev. Astron. Astrophys.* **24,** 285 (1985). (On pulsar observations and statistics; I.)

S. L. Shapiro and S. A. Teukolsky, *Black Holes, White Dwarfs and Neutron Stars*. Wiley, New York, 1983. (Comprehensive text on physics of compact stars and black holes; I.)

Newton's Laws

Thornton Page

Sir Isaac Newton spoke with assurance (as he usually did) when he said "If I have seen farther than others, it is because I stood on the shoulders of giants." We now think that the giants he referred to were Nicholas Copernicus, René Descartes, Tycho Brahe, Johannes Kepler, and Galileo Galilei. Newton was also a giant in rational, mathematical thought. Like Galileo, he performed many experiments; like Kepler, he employed mathematics; like both, he recognized the importance of "natural law"—exact mathematical relationships between observable quantities such as mass, velocity, acceleration, time, temperature, color, direction, and position. He allowed for the possibility of errors in measurement and deviations from perfect experimental conditions, but felt at heart that Nature systematically follows exact laws, in which he was very nearly right.

By Newton's laws we mean his three laws of motion, partly borrowed from Galileo, and his law of gravitation, similar to the ideas of Kepler and Hooke. (There is also his law of cooling: The rate of temperature decrease of a hot body is proportional to the difference between its temperature and that of its surroundings.) These laws of motion and gravitation refined two basic concepts—mass and force—and were combined with Galileo's equally precise concepts of motion (velocity and acceleration) to describe the "system of the world." All this was summarized in his 1687 book *Philosophiae Naturalis Principia Mathematica* (*The Mathematical Principles of Natural Philosophy*), which reads like a textbook in geometry, with "definitions," "axioms," "corollaries," and proofs that include a whole new branch of mathematics—the differential and integral calculus.

However, Newton was more than a pure mathematician; he included "Natural Philosophy" in the title of his *Principia,* and related its mathematics closely to real things. "It is the quantity of matter that I mean hereafter under the name 'body' or 'mass,' " he says in Definition 1. In the discussion of his definitions he says, " . . .in philosophical disquisitions we ought to abstract from our senses and consider things themselves, distinct from our [nonperfect] measures of them. For it may be that there is no body really 'at rest.' But we can distinguish relative rest and motion. . . ." These abstractions from what we see and measure allowed Newton to combine observations of falling weights with motions of the Moon and planets in a mathematically perfect theory of mechanics, and to set the stage for Einstein's relativity theory over 200 years later.

His three laws of motion were stated as "axioms" in the *Principia*:

"Law 1. Every body perseveres in its state of rest or of uniform motion in a right line unless it is compelled to change that state by forces imposed thereon." This recognizes that we cannot distinguish rest from constant velocity; if the force (F) is zero, the acceleration (a) of a mass (m) is zero, velocity (v) remains constant. Law 2 extends this to an impressed force greater than zero:

$$F = ma,$$

a formula that defines units of force in terms of mass and acceleration. Our present systems of units (cgs, mks, SI, etc.) are based on Newton's reasoning.

"Law 3. To every action there is always opposed an equal reaction. . . ." "If you press a stone with your finger, the finger is also pressed by the stone. . . ." This is the basis of the derived law of conservation of momentum (mv). If the Earth pulls on the Moon, then the Moon must pull on the Earth—one basis for measuring the mass of the Moon.

Long before publishing the *Principia,* Newton got these ideas clearly in mind, and was led to the law of gravitation after seeing an apple fall while he was in a "comtemplative mood," according to his niece, who was with him at Woolsthorpe, his birthplace in Lincolnshire, England. He had been thinking of the Moon's motion around the Earth. Apparently, the apple's fall toward the Earth became related in his mind to the Moon's acceleration toward the Earth, implying a force on the Moon similar to the force on the apple, both toward the Earth. Newton generalized this by assuming a gravitational force between every pair of masses, proportional to the product of the masses (mM). The acceleration of the apple being larger than the Moon's showed that the force must be smaller at larger distances (r), in fact, proportional to $1/r^2$. Geometrically, this inverse square law is consistent with the three-dimensional nature of space, so the law of gravitation emerged as

$$F_g = GMm/r^2$$

where G was a constant yet to be determined. Note that the law depended on the laws of motion as axioms. Edmund Halley urged Newton to summarize all this in one book, but Newton delayed publication for several years so that it could be mathematically perfect. During this period he worked out some details, including the proof that a spherical body such as the Earth or Moon acts gravitationally on outside bodies as if all its mass were concentrated at a point at its center.

The success of Newton's theory of gravitation was immediate and complete: It explained the motions of planets around the sun as expressed in Kepler's laws, which Newton derived from his laws, and it further predicted the perturbations in one planet's orbit by the gravitational attraction of nearby planets. In fact, two more planets (Neptune and Pluto) were discovered from perturbations of their neighbors. It explained the Earth's tides and predicted their long-term effects on the Moon's orbit and the Earth's rotation period. After Henry Cavendish measured the minute gravitational force between spheres of known mass, the gravitational constant G was known, and the masses of all astronomical bodies with satellites could be determined in grams; the sun's from planetary orbits, the planets' from satellite orbits, the masses of double stars from their orbits, and the masses of star clusters, galaxies, and clusters of galaxies. (There are still some discrepancies in the masses of clusters of galaxies, and as yet no good gravitational evidence of the masses of comets.)

With the help of computers and Doppler radar, Newton's laws have provided highly accurate trajectories of space probes; conversely, the space-probe trajectories have been used to improve determination of the masses of planets and satellites as far away as Saturn, to discover the lunar mascons (high-density blocks in the lunar crust), and to establish the hydrostatic mass distribution in Jupiter.

Masses of molecules, atoms, and subatomic particles were determined in the early 1900s, using Newton's laws of motion and impressed electric or magnetic forces. Gravitational forces were found to be overshadowed by strong and weak forces in the atomic nucleus. However, gravitational forces become dominant under the extreme conditions in compact stars. Here Newton's laws are no longer correct, and must be replaced by the field equations of Einstein's general relativity, which lead to the recent concept of the black hole, where the gravitational field is so strong that space-time is "curved back into itself," allowing no light or particles to leave the hole, although its gravitational field still remains outside. (Astronomers have recently obtained observational evidence supporting the existence of black holes.)

Newton's law of gravitation is slightly incorrect near the sun, where the effects of the solar mass on space-time modify the orbit of Mercury. (In 1918 this was the first evidence for general relativity—the advance of Mercury's perihelion.) At the extreme distances of observable galaxies, clusters of galaxies, and quasars, Newton's theory also fails, and must be replaced by general relativity if we are to understand the observed motions detected by Doppler shifts in optical and radio spectra. With these exceptions, however, Newton's four laws describing motion and gravitation apply remarkably well to bodies of sizes ranging from 1 μm to 100,000 light years or 10^{20} km (galaxies), separated by comparable distances.

In the 1980s, theoreticians suggested that Newton's law of gravitation should be modified by changing the inverse square of distance to a power different from 2, and also introduced the possibility that the constant G has changed with time. By the end of 1988, no observational evidence has supported these theoretical possibilities, although some experiments suggest that there may be an additional short-range force added to Newton's inverse square law at distances of a few meters. Several NASA projects have confirmed Einstein's general relativity to high precision as the successor to Newton's law of gravitation.

See also BLACK HOLES: DYNAMICS, ANALYTICAL; GRAVITATION; GRAVITY, EARTH'S; MASS; RELATIVITY, GENERAL THEORY.

BIBLIOGRAPHY

G. Abell, *Exploration of the Universe,* 2nd ed. Holt, Rinehart & Winston, New York, 1969.

W. Bixby and G. de Santillana, *The Universe of Galileo and Newton*. Harper & Row (American Heritage). New York, 1964.

Isaac Newton, *The Principia Mathematica of Isaac Newton* (F. Cajori, transl.). Univ. of California Press, Berkeley, 1934.

L. Page, *Introduction to Theoretical Physics*. Van Nostrand, Princeton, NJ, 1928.

T. Page and L. W. Page, *Wanderers in the Sky*. Macmillan, New York, 1965.

T. Page and L. W. Page, *Beyond the Milky Way*. Macmillan, New York, 1969.

T. Page and L. W. Page, *Space Science and Astronomy*. Macmillan, New York, 1976.

Allen Sandage, Mary Sandage, and Jerome Kristian (eds.), *Galaxies and the Universe*. University of Chicago Press, Chicago, 1975.

Jay M. Pasachoff and Marc L. Kutner, *University Astronomy*. W. B. Saunders, Philadelphia, 1978.

Noise, Acoustical

T. Embleton

The term acoustical noise has two common interpretations. Physically, it may refer to a sound having pseudorandom amplitude and phase as a function of time, the spectrum of which encompasses the audio-frequency range. Pink noise describes such a sound with equal energy per constant percentage bandwidth (e.g., octave) and white noise indicates such a sound with equal energy per unit bandwidth.

Subjectively, noise may refer to "any loud, discordant or disagreeable sound or sounds" (*Webster's New World Dictionary, College Edition*). This definition implies the attributes of unwantedness or of annoyance on the part of some listener.

An approximate relation exists between the mechanical power of any device and the noise power that it radiates. A small fraction of the mechanical power, usually between 1 and 100 parts per million, is inadvertently converted into sound. In industrialized countries where the use of mechanical devices has increased rapidly in recent years (as jet aircraft have largely replaced less powerful and quieter propeller-driven aircraft, motor vehicles have become more numerous, and many powered devices are widespread that were rare or even unknown a few years ago) noise has become a major environmental concern.

Noise control is an important field of activity with many facets: the accurate measurement of sound levels from individual sources and the statistical description of the noise climate due to many sources; the understanding of how noise is generated and propagated from source to receiver, often in an acoustically complicated environment; the practice of reducing the noise output by modification of a source; the legislative control of noise; and psychoacoustical and sociological studies of how individuals and communities react to given situations involving noise.

Instruments for measuring and analyzing most physical aspects of noise are well advanced. Small, lightweight, battery-operated sound-level meters are widely used, and in their simplest form consist of a microphone, amplifiers, frequency-weighting networks, and an output meter. The more precise of these are capable of measuring sound levels to within 1 decibel (dB). A-weighting is the most widely used of the several frequency-weighting networks commonly available, because it correlates well with many human physiological and psychological reactions to noise. (A-weighting has a maximum sensitivity at frequencies of 1–5 kHz and has a progressively decreasing sensitivity at frequencies below 500 Hz and above 8 kHz.) More elaborate instruments, often portable, are widely used for analysis of noise into octave, fractional octave, or narrow frequency bands. This more detailed information is usually required for the scientific or engineering control of noise at the source, and occasionally for regulatory purposes. Integrating-averaging sound-level meters monitor sound levels continuously and can indicate several parameters of time-varying signals, for example, the maximum and minimum sound levels or the equivalent-energy steady sound level during time periods that can be chosen from 1 s to 24 h or even longer.

In order to describe the varying noise levels that constitute the noise climate at a given location, instruments have been developed to sample the noise level (usually A-weighted) at fixed intervals of time. In this way the statistical descriptors, L_x, are obtained, indicating the noise level L in decibels that is exceeded x percent of the time; L_1 or L_3 often taken to indicate the maximum noise levels occasionally attained, and L_{90} or L_{99} the ambient noise level. While useful to describe an existing situation, for which the set of L_x can be measured, several sets of L_x cannot be easily and accurately combined to predict the statistics of noise levels in more complicated situations; for example, knowing the statistics L_x for noise levels near highways and rail tracks does not allow one directly to compute the statistics of noise levels near the intersection of a highway and a rail track.

Another portable device is the noise dose meter, worn by persons working in noisy locations. This instrument sums the noise exposure of the wearer, taking into account both the length of time of exposure and the noise level. Such instruments usually show the exposure as a percentage of the daily dose allowed according to some preset criterion.

A general noise control procedure is to enclose the noise source either completely or partially, taking certain precautions concerning the absorption of sound, though this is often not a practical solution for a variety of reasons, as for example when frequent accessibility is necessary. Noise reduction by redesign of the source usually requires expert knowledge of what mechanisms are producing the noise, how it is radiated, how this can be reduced (many noise-control technques are specific to certain processes), and what are the nonacoustical constraints such as weight, cost, performance requirements, etc. Once noise escapes the vicinity of the source, the intensity and spectrum shape of that which reaches the receiver depends on the possible propagation paths. Inside a building, noise may be either airborne or structure borne, and the noise level depends on numerous factors such as the shape and size of rooms, whether they are reverberant or absorbing, how the absorption is distributed, the efficiency of vibration isolation in solid structures, etc. Outdoors the noise level at the receiver depends on the acoustical impedance of the ground surface, whether or not it is flat, the height of the source and receiver above the ground, and their distance apart. Wind velocity and temperature, and their variation with height, and also molecular absorption, are all important at distances of the order of 100 m or greater.

Certain noise sources can be controlled using active methods for the reduction of sound levels. In essence the sound is detected, processed in near real time, and a related signal is radiated with appropriate changes of amplitude and phase so as to interfere with the original signal at the location of the receiver. Such techniques have limitations in practice: either to low frequencies (below about 500 Hz) because of the need to process the signal in a time comparable to or less than the period of the sound field; to one-dimensional fields such as sound traveling along a duct; or to repetitive signals for which the processor can cancel one event having "learned" the signal waveform from several previous events.

The effects of noise on human activity are extensive. Depending on its frequency content, intensity, time duration, and circumstances under which it is heard, noise can be beneficial—as in masking sounds of high message content that would otherwise be distracting, annoying, or hazardous. Among its annoying effects are interference with speech and the performance of mental tasks, and degradation of the environment of the listener. In most circumstances psychological reaction to noise increases with increasing amounts of noise but there is considerable variability among individuals, depending on past experience, the nature of the noise, and expectation of quietening it. Long or repeated exposure to high levels of noise presents a risk of hearing damage; the extent of damage and its effect on hearing depend in a complex way on duration, intermittency, frequency content, and intensity, and also on the physiological susceptibility of the individual. Noise disturbs sleep, either by awakening or by shifting the depth of sleep, to an extent that depends on the intensity and duration of the noise—whether this is merely annoying or has more serious consequences is not yet established except in extreme cases.

See also ACOUSTICS; ACOUSTICS, ARCHITECTURAL.

BIBLIOGRAPHY

L. L. Beranek (ed.), *Noise and Vibration Control*. McGraw-Hill, New York, 1971.

C. M. Harris (ed.), *Handbook of Noise Control*, 2nd ed. McGraw-Hill, New York, 1979. (I)

L. E. Kinsler and A. R. Frey, *Fundamentals of Acoustics*, 2nd ed. Wiley, New York, 1962. (E)

P. M. Morse and K. U. Ingard, *Theoretical Acoustics*. McGraw-Hill, New York, 1968. (A)

A. D. Pierce, *Acoustics: An Introduction to its Physical Principles and Applications*. McGraw-Hill, New York, 1981.

Nonlinear Wave Propagation

Mark J. Ablowitz

Wave propagation is essentially the study of how information, or some type of signal, is transmitted. The generality in the notion of what actually constitutes a wave, or wave phenomena, necessarily makes the study rather broad and of interest to scientists and mathematicians alike. Applications abound: for example, water waves, acoustic waves, laser beams, traffic flow, and electromagnetic waves. Here we discuss some of the mathematical theories and techniques that have been developed for nonlinear waves. By nonlinear wave propagation we refer to those phenomena whose underlying equations (and hence the basic physical mechanisms) are nonlinear.

The discipline of nonlinear wave propagation essentially began in the nineteenth century with the examination of surface waves by Stokes (1847). For deep-water gravity waves, he found that the phase speed of the wave was amplitude dependent. The larger the amplitude, the faster the wave train moves. Thus, if we call c the phase velocity, k the wave number, and a the amplitude, Stokes found that $c=c(k,a)$. In linear problems, $c=c(k)$ only. This is called a nonlinear dispersive ($c\neq 1$) wave.

At approximately the same time, Riemann was also studying nonlinear wave phenomena. In fact, in his classic paper (1858) on gas dynamics, he discovered a method to solve a certain system of nonlinear partial differential equations (PDEs). The ideas have been generalized, and new methods developed for a class of equations called hyperbolic systems. These are PDEs that have distinct real characteristics. A discussion of these ideas can be found in Courant and Friedrichs (1948) and Courant and Hilbert (1961). Perhaps the most outstanding new phenomenon of this nonlinear theory is the appearance of a shock wave. A shock is the abrupt change in the dependent variables(s) of the underlying equation(s). In gas dynamics these variables are the pressure and the velocity.

In recent years there have been significant advances in the understanding of nonlinear dispersive wave problems. One important step was the understanding of the effects of nonlinearity on wave interactions. These ideas are often called weak interaction theory, because here the nonlinearity is assumed to be small. Very generally, we consider an equation of the form

$$L(u)=\epsilon N(u) \tag{1}$$

where ϵ is very small, and L and N are linear and nonlinear operators, respectively, that depend on the particular physical problem. Equation (1) is assumed to have elementary dispersive wave solutions $u_i-A_i \cos\theta_i$, $\theta_i=k_i x-\omega(k_i)t$ (assuming one spatial dimension and time) in the linear limit $\epsilon=0$; $\omega(k_i)$ is called the dispersion relation and is related to phase speed by $c_i=\omega_i/k_i$. These solutions do not obey the full equation (1). However, approximate solutions can be constructed by using multiple-scale perturbation methods (see, e.g., Kevorkian and Cole, 1985). It turns out that, in general, the nonlinearity creates resonant interactions between the solutions u_i. This in turn causes the amplitudes A_i to depend weakly on time [typically $A_i=A_i(\epsilon t)$]. The effect can be substantial. A mode that initially is absent can "feed" off the others, and eventually become equally important. References regarding physical applications are found in ocean waves [Phillips (1974)] and nonlinear optics [Bloembergen (1965)].

In 1965 Whitham discovered how to develop certain types of approximate solutions to fully nonlinear partial differential equations. The starting point is that for certain equations a single periodic uniform nonlinear "mode" is an exact solution, e.g., the Stokes water-wave solution. Whitham showed how to generate simplified approximate equations that govern gradually modulated nonuniform waves (e.g., a wave in a slowly varying medium). The methods can be viewed as a generalization of the WKB method to nonlinear partial differential equations and can be put into an elegant variational formulation. The theory provides information as to how energy propagates and gives a concept of group velocity in a fully nonlinear problem. In the linear theory, the group velocity is the velocity at which energy is transferred. Although the linear theory was applied only to a single-phase ($\theta=kx-wt$) mode, the ideas can be extended to multiphase modes, as noted by Ablowitz and Benney (1970). In 1980 Flaschka, Forest, and McLaughlin showed how modulated multiphase waves could be completely characterized for a particular equation [see Eq. (2)].

The foregoing theories have led to important new discoveries. For example, Benjamin (1967) showed, using weak interaction theory (the result can also be deduced using Whitham's technique), that the classical Stokes water wave was unstable. This surprising fact (and the very nature of the instability) has stimulated a great deal of research (see also Benney and Newell, 1967).

At approximately the same time, Zabusky and Kruskal were studying special aperiodic solitary-wave solutions of the physically interesting Korteweg–deVries (KdV) equation (1895),

$$u_t+6uu_x+u_{xxx}=0. \tag{2}$$

KdV describes weakly nonlinear and weakly dispersive one-dimensional waves such as water waves in shallow channels. A solitary-wave solution to Eq. (2) is given by

$$u_s(x,t)=2k^2 \operatorname{sech}^2 k(x-4k^2t-x_0), \tag{3}$$

where k and x_0 are constant. They discovered the remarkable fact that two such waves with different amplitudes interact elastically. Specifically, the initial velocity, before interaction, and the asymptotic final velocity, after interaction, are equal! They named this solitary wave a soliton, in conceptual analogy to elementary-particle interactions. Indeed, in 1975, some physicists suggested that an elementary particle is actually a soliton.

This result shortly preceded the pioneering work of Gardner, Greene, Kruskal, and Miura (GGKM) (1967, 1974) in which they developed a method to solve the initial-value problem for Eq. (2), for given $u(x,0)$ decaying sufficiently rapidly as $|x|\to\infty$. The key step is to associate with the KdV equation the linear Schrödinger eigenvalue problem

$$v_{xx}+(\lambda+u(x,t))v=0. \tag{4}$$

Remarkably, Eq. (4) had been studied intensively by theoretical physicists in the 1950s. By using concepts and methods of direct and inverse scattering (see, e.g., Faddeev, 1963) GGKM found that the solution of Eq. (2) obeyed a linear integral equation [Gelfand–Levitan (1951)].

Previous to KdV, Hopf (1950) and Cole (1951) showed that the Burgers (1948) equation describing weak shock waves,

$$u_t + uu_x = u_{xx}, \quad (5)$$

could be reduced to the solution of the linear heat equation by using the transformation $v_x = uv$. Although Eq. (4) is, in some sense, a generalization of this, the implicit nature and the "inverse" concepts are quite novel.

The work of Zakharov and Shabat (1971) showed that the methods of KdV were applicable to still another physically relevant evolution equation (the nonlinear Schrödinger equation). Subsequently, Ablowitz, Kaup, Newell, and Segur (1974) developed a technique termed the inverse scattering transform (IST) by which a class of nonlinear evolution equations can be isolated and solved, many of which are physically relevant and apply to a wide variety of physical problems such as the sine–Gordon and modified KdV equations.

Moreover, researchers in the 1970s have demonstrated that the IST method applies to an even broader class of nonlinear partial differential equations, as well as nonlinear differential-difference, partial difference, and singular integro-differential equations. A review of some of this work can be found in the monograph of Ablowitz and Segur (1981).

Significantly these ideas can be extended to certain nonlinear multidimensional evolution equations. The two-dimensional generalization of KdV is the so-called Kadomtsev–Petviashvili (1970) equation

$$(u_t + 6uu_x + u_{xxx})_x = -3\sigma^2 u_{yy} \quad (6)$$

which arises in water waves, plasma physics, acoustic waves, etc. It has been shown that Eq. (5) is linearized by

$$\sigma v_y + v_{xx} + u(x,y,t)v = 0. \quad (7)$$

New concepts and methods have enabled researchers to apply the IST technique to these and other multidimensional nonlinear equations. Not only have new solutions to Eq. (6) been obtained but novel methods for multidimensional inverse scattering have also been developed [see for example Ablowitz and Fokas (1983), Ablowitz and Nachman (1986)].

See also CHAOS; OPTICS, NONLINEAR; SOLITONS.

BIBLIOGRAPHY

M. J. Ablowitz and D. J. Benney, *Stud. Appl. Math.* **49,** 225 (1970). (A)

M. J. Ablowitz, D. J. Kaup, A. C. Newell, and H. Segur, *Stud. Appl. Math.* **53,** 255 (1974). (A)

M. J. Ablowitz and H. Segur, Solitons and the Inverse Scattering Transform, SIAM Studies in Applied Mathematics, 1981.

M. J. Ablowitz and A. S. Fokas, in *Lecture Notes in Physics, 189* (K. B. Wolf, ed.). Springer-Verlag, Heidelberg, 1983.

M. J. Ablowitz and A. I. Nachman, *Physica* **18D,** 223–241, 1986.

T. B. Benjamin, *Proc. R. Soc. London, Sect A* **299,** 59 (1967). (A)

D. J. Benney and A. C. Newell, *J. Math. Phys. (N.Y.)* **46,** 133 (1967). (A)

N. Bloembergen, *Nonlinear Optics*. Benjamin, New York, 1965. (I)

J. M. Burgers, *Adv. Appl. Mech.* **1,** 171 (1948). (A)

J. D. Cole, *Q. Appl. Math.* **9,** 225 (1951). (A)

R. Courant and K. O. Friedrichs, *Supersonic Flow and Shock Waves*. Wiley (Interscience), New York, 1948. (I)

R. Courant and D. Hilbert, *Methods of Mathematical Physics, Vol. II*. Wiley (Interscience), New York, 1961. (A)

L. Faddeev, *J. Math. Phys.* **4,** 72 (1963). (A)

H. Flaschka, M. G. Forest, and D. W. McLaughlin, *Commun. Pure Appl. Math.* **33,** 739 (1980).

C. Gardner, J. Greene, M. D. Kruskal, and R. Miura, *Phys. Rev. Lett.* **19,** 1095 (1967). (A)

C. Gardner, J. Greene, M. D. Kruskal, and R. Miura, *Commun. Pure Appl. Math.* **27,** 97 (1974). (A)

I. M. Gelfand and B. M. Levitan, *Am. Math. Trans. (2),* **1,** 253 (1951). (A)

E. Hopf, *Commun. Pure Appl. Math.* **3,** 201 (1950). (A)

B. B. Kadomtsev and V. I. Petviashvili, *Sov. Phys. Dokl.* **15,** 539–541 (1970).

J. Kevorkian and J. D. Cole, *Perturbation Methods in Applied Mathematics*. Springer-Verlag, New York, 1985.

D. J. Korteweg and G. deVries, *Philos. Mag.* **39,** 422 (1895). (A)

O. M. Philips, *Ann. Rev. Fluid Mech.* **6,** 93 (1974). (I)

B. Riemann, *Göttingen Abhandl.* **7,** 43 (1858). (A)

G. G. Stokes, *Camb. Trans.* **8,** 441 (1847). (A)

G. B. Whitham, *Proc. R. Soc. London, Sect. A* **283,** 238 (1965a). (A)

G. B. Whitham, *J. Fluid Mech.* **22,** 213 (1965b). (A)

G. B. Whitham, *Nonlinear Waves*. Wiley (Interscience), New York, 1974. (I)

N. J. Zabusky and M. D. Kruskal, *Phys. Rev. Lett.* **15,** 240 (1965). (I)

V. E. Zakharov and A. B. Shabat, *Zh. Eksp. Teor. Fiz.* **61,** 118 (1971) [*Soviet Phys. JETP* **34,** 62 (1972)]. (A)

Novel Particle Acceleration Methods

R. H. Siemann

Accelerators are key instruments for many branches of science, and to some extent advances in these sciences are determined by accelerator performance. This connection is clearest at the energy frontier of particle physics. The accelerators working at this frontier are based on two underlying ideas: colliding beams to reach a high center-of-mass energy ($E_{\text{c.m.}}$) and the long-term confinement of particle beams in storage rings. Because of different advantages, particle physics experiments are performed on hadron colliders and electron–positron colliders. As the SSC design shows, a storage ring can be used to increase the energy of hadron colliders by an order of magnitude. However, this is not possible for electron–positron colliders where synchrotron radiation is the limiting phenomenon. This has led to the development of linear colliders as an alternative to storage rings. Much of the work on novel acceleration methods has concentrated on linear colliders where the requirements are: high acceleration gradient to keep the collider length reasonable, high efficiency for conversion of AC mains power to beam power, and beam emittance (phase space volume) preservation. Other applications could have different requirements.

Acceleration techniques can be separated into three broad categories of near field, far field, and media based. The most familiar is *near-field* accelerators; an example is the disk-loaded waveguide used in electron linacs (linear accelerators). Nearby boundaries give an electromagnetic wave the properties needed to accelerate particles, which are a phase velocity equal to the beam velocity and a component of electric field in the direction of propagation. Because nearby boundaries are important, the transverse dimensions of the accelerator scale with the wavelength of the accelerating wave (λ). Appropriate fabrication methods lead to different looking structures with the same underlying physics; a disk-loaded waveguide is natural for *S*-band and a grating for a CO_2 laser.

Choice of wavelength involves trade-offs among collider requirements and power source properties. Factors that favor short wavelengths are:

1. Limiting gradient. Figure 1 shows the limits on gradient. Typical gradients are below these limits because of stored energy, power, and efficiency considerations.
2. Stored energy and average power. Stored energy and average power (both per unit length) are proportional to $G^2\lambda^2$ because transverse dimensions are proportional to λ and fields to the gradient, G.
3. Energy extraction efficiency. The energy extraction efficiency is the ratio of the energy extracted by the beam to the stored energy. For small efficiency, it is proportional to $N/\lambda^2 G$, where N is the number of particles in a beam bunch.
4. Peak power. The peak power is proportional to $G^2\lambda^{1/2}$ because waveguide filling times vary as $\lambda^{3/2}$.

Emittance preservation favors long wavelengths. Beam-generated fields, called wakefields, interact with the beam and increase its emittance. Emittance growth can be limited by reducing errors, using strong focusing, damping collective motion (BNS damping), and choosing an accelerator structure with small wakefields. The latter means a structure with apertures far from the beam, i.e., a long wavelength. The most troublesome wakefields scale approximately as λ^{-3}.

Parametric analyses have led groups with "short-term" goals, colliders with $E_{\text{c.m.}} = 0.5$–2.0 TeV and construction in the late 1990s, to concentrate on $\lambda \sim 1$–3 cm. Depending on other parameters the peak power requirement is 25–600 MW/m of accelerator. Over 150 MW has been obtained from a pulsed klystron at $\lambda = 10$ cm, but the capability of any design

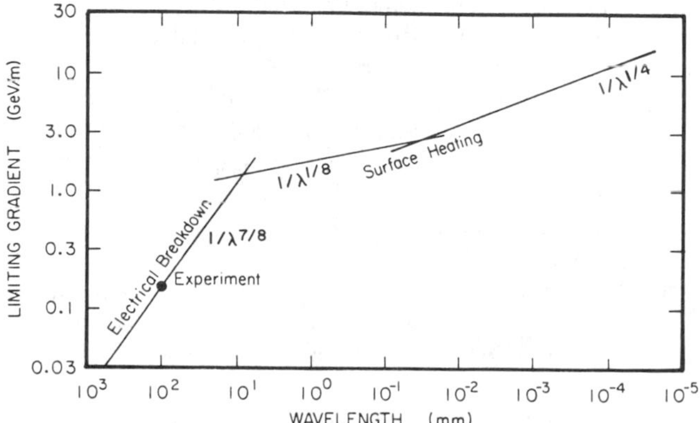

FIG. 1. Limitations on gradient as a function of wavelength due to electric field breakdown and surface heating in a SLAC-type disk-loaded waveguide (reprinted with permission of P. Wilson).

falls as λ^2, and rf amplifiers for 1–3 cm wavelength are a major research area. Among the devices developed at a prototype stage are:

1. a free-electron laser (FEL) that has reached a peak power greater than 1 GW. The original "two-beam" concept was an FEL powered accelerator where the FEL beam was recovered and reaccelerated to increase efficiency.
2. a relativistic klystron that has produced 170 MW. A related device is being considered for a two-beam accelerator.
3. a gyroklystron that could reach 300 MW.
4. a low-power klystron combined with pulse compression to increase peak power.

Research on near-field acceleration for the longer term is aimed at understanding fundamental limits of particle acceleration, and applicability of results will be affected strongly by advances in particle physics, experience with colliders, and progress in related fields such as laser development. The work includes calculations, small-scale experiments, and "proof-of-principle" demonstrations of acceleration with lasers, wakefields, and switched power.

The laser-driven accelerators are conceptually similar to the longer-wavelength rf linacs. The small wavelength makes it natural to choose open structures such as gratings, rows of droplets, and arrays of columns etched in silicon. Some of these will be studied at a laser acceleration facility being developed by Brookhaven and Los Alamos National Laboratories.

A short pulse, rather than a monochromatic energy source, is used in switched power and wakefield accelerators. A power supply, an energy storage device, and an array of fast switches provide the pulse in the former. The switches are arranged around the outer circumference of a radial transmission line, and as the pulses propagate toward the center there is a substantial increase, perhaps a factor of 20, in voltage. In a wakefield accelerator the pulse comes from a ring-shaped, low-energy beam traveling at the outer radius of the radial transmission line; gradients up to 200 MeV/m are possible. Wakefield acceleration has been demonstrated in a prototype at the Deutsches Elektronen-Synchrotron.

Particles in *far-field* accelerators are given periodic deflections that result in motion transverse to the average direction of propagation. This transverse motion can couple to an electromagnetic wave and give acceleration without using nearby boundaries to modify the wave. The inverse free-electron laser (IFEL) is a far-field accelerator with equations of motion identical to the FEL. The difference between the two is the particle injection phase that determines whether particles are accelerated and field energy decreased (an IFEL), or particles are decelerated and field energy increased (a FEL). A disadvantage of the IFEL is that it is limited in energy due to synchrotron radiation from the transverse motion, and, therefore, it is a specialized accelerator that could be used as an injector to a high-energy collider, a low-energy collider, or for applications outside particle physics.

Most research on *media-based* acceleration has concentrated on plasmas which promise gradients in excess of 1

GeV/m. In the plasma beatwave accelerator, a laser with two output frequencies differing by the plasma frequency (ω_p) drive a beatwave, a plasma oscillation with radial and longitudinal electric fields. Other properties of the beatwave are that one-quarter of a wavelength is both focusing and accelerating, the group velocity equals zero, the phase velocity equals $c[1-(\omega_p/\omega)^2]^{1/2}$, where ω is the average laser frequency, and, as a result, there is slippage between the beatwave and the beam that limits the length of an acceleration stage. A very rough rule of thumb in the regime of linear plasma behavior is $G_{max}(\text{GeV/m}) \sim 10^{-12}\omega_p(\text{s}^{-1})$; large gradients and long acceleration stages are incompatible.

Beatwaves with a gradient between 0.3 and 1 GeV/m have been observed in plasmas about 1 mm long at the University of California (Los Angeles). These experiments also observed saturation of the beatwave due to other plasma modes. This is typical of the effects that are not included in simple, linear models and are under study. Other fundamental issues are saturation due to relativistic effects, beam loading and efficiency, laser self-focusing, and acceleration of injected particles.

The plasma wakefield accelerator is closely related; the difference is that the plasma oscillation is driven by a low-energy electron beam. Properties such as gradient, efficiency, and acceleration stage length depend on the intensity, energy, and density profile of the drive beam. Plasma wakefield acceleration has been measured at Argonne National Laboratory; in the first experiment the gradient was about 1 MeV/m which is consistent with experimental parameters.

Among other media-based acceleration concepts discussed in the literature are ones based on (i) the inverse Čerenkov effect, (ii) laser guiding techniques used for intense, low-energy electron beams, and (iii) anomalous x-ray transmission in crystals. The latter is a far-future concept combining ideas from plasma and near-field acceleration. A crystal lattice forms an iris-loaded "waveguide" for x rays, and positively charged muons and protons would be accelerated with a gradient up to 10^9 V/cm. The "power source" is a 40-keV x-ray generator with a peak power of 3×10^9 W!

The ultimate value of this and other novel concepts will depend on many factors, and important, unanticipated uses may develop as ideas evolve. For example, lenses based on the physics of the beatwave and plasma wakefield accelerators are being considered as focusing elements at the collision point of linear colliders, and the work stimulated by the search for high gradients may find application there.

BIBLIOGRAPHY

P. B. Wilson, *AIP Conf. Proc.* **130,** 560 (1985). P. B. Wilson, SLAC-PUB-3674(1985) is a later version with some errors corrected. (I)

R. H. Siemann, *Ann. Rev. Nucl. Part. Sci.* **37,** 243 (1987). (E)

The most detailed articles in this field are proceedings of advanced accelerator conferences:

P. Channell (ed.), *AIP Conf. Proc.* **91** (1982).

C. Joshi and T. Katsouleas (eds.), *AIP Conf. Proc.* **130** (1985).

F. E. Mills (ed.), *AIP Conf. Proc.* **156** (1987).

C. Joshi (ed.), *AIP Conf. Proc.* (to be published).

S. Turner (ed.), CERN 87-11 (1987).

Nuclear Emissions *see* Angular Correlation of Nuclear Radiation

Nuclear Fission

F. Plasil and H. W. Schmitt

I. INTRODUCTION

The process of nuclear fission was discovered unexpectedly in 1938 by Hahn and Strassmann. Since its most characteristic feature is the release of a large amount of energy, coupled with the release of neutrons that make it possible to sustain a chain reaction, its practical implications quickly became apparent as evidenced by the construction of the atomic bomb and the subsequent exploitation of fission in power reactors. In view of this, nuclear fission can be regarded as one of the scientific discoveries that have had profound impacts on the history of people. From the point of view of basic research, fission is a fascinating process involving a large-amplitude rearrangement of nucleons, much like the division of a charged drop of liquid. It is on the basis of the liquid-drop model that Bohr and Wheeler provided a satisfactory account of many of the early, observed properties of fission. As experimental investigations continued, however, it became clear that many observed phenomena, notably the predominant division into fragments of unequal size, lay outside of the predictive capabilities of the liquid-drop model, and it became increasingly obvious that single-particle effects often play decisive roles. Thus, the process of fission provides us with a unique opportunity to study both macroscopic and microscopic nuclear processes, as well as their interaction with each other.

In this short treatment, we can only describe the qualitative aspects of fission and the nature of present-day understanding of the fission process. The interested reader is referred to the Bibliography for further information and study. Below, we first summarize the global fission properties of actinide nuclei, which provide us with the bulk of fission data. Following the description of experimental observables, the liquid-drop model is introduced, together with its modification by single-particle shell effects, which have led to two-peaked fission barriers and to the understanding of most of the observed experimental features. After a discussion of the observed phenomena that are directly related to the existence of the double-peaked fission barrier, we consider the important role of nuclear dynamics in the fission process. We close with a discussion of the fission of relatively light nuclei, including the fission of rotating systems produced in heavy-ion reactions, and with a brief statement on future challenges.

II. FISSION OF ACTINIDE NUCLEI

As an example, consider the nuclear fission reaction induced when a thermal neutron (0.025 eV) is incident on a ^{235}U nucleus, i.e., $^{235}\text{U}(n_{th},f)$. (This reaction has perhaps been the subject of more study than any other fission reaction.) The compound nucleus ^{236}U is formed in an excited state with excitation energy $E^*\cong6.5$ MeV. As the compound

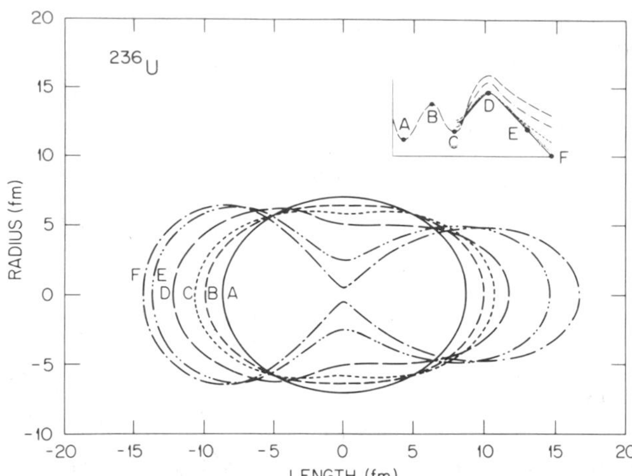

FIG. 1. A possible sequence of shapes of a fissioning nucleus. In the inset, points corresponding to these shapes are shown schematically on the minimum potential-energy curve (Section III). [Reproduced, with permission, from the sixth item in Part D of the Bibliography.]

nucleus begins to fission, this excitation energy transforms into deformation, and a sequence of shapes such as those shown in Fig. 1 may ensue. At the "scission point," i.e., the point of separation, the product nuclei, called "fission fragments" or "fission products," are free to accelerate by mutual Coulomb repulsion. It is this Coulomb potential energy and its subsequent transformation into fragment kinetic energy that accounts for most (80 to 90%) of the energy released in fission. The slowing down and ionization of these high-energy fragment nuclei by atomic processes in the surrounding medium produces the heat which can then be used as a useful energy source in nuclear reactors.

The two fragment nuclei are produced in deformed excited states and decay by neutron and gamma-ray emission. An average of 2.42 neutrons is produced in the ^{235}U(n_{th},f) reaction; in a reactor using ^{235}U as fuel, it is these neutrons that sustain the chain reaction, causing other ^{235}U nuclei to fission, and so on. Energies of 15–30 MeV are accounted for by neutron emission from the fragments. An average energy of approximately 7 MeV is accounted for by gamma emission, and the interaction of gamma rays with atoms in the surrounding medium also contributes to heat generated in a nuclear reactor. Finally, following decay by neutron and gamma-ray emission, the nuclei produced in fission reach their ground states. Almost all of the produced nuclei, however, are very neutron-rich relative to the beta stability line and are, therefore, radioactive. They decay by beta and gamma-ray emission with half-lives ranging from milliseconds to $>10^5$ years. It is the residual long-lived radioactivity of these nuclei that causes the unique waste disposal problems associated with nuclear power reactors.

In summary, for a total energy release of, say, 200 MeV, such as occurs in the fission of the isotopes of U, Pu, etc., fragment kinetic energy accounts for 160–180 MeV; neutron emission accounts for 15–30 MeV, where about 2 MeV per neutron is in kinetic energy and the remainder represents binding energy; and gamma-ray emission accounts for about 7 MeV. The specific energies associated with a particular fission event are determined by the specific division of the protons and neutrons of the compound nucleus, i.e., by the atomic charges and masses of the fragment nuclei, and by the excitation energies with which the fragments are formed. Broad distributions of these quantities occur.

Yield distributions of the masses of fission fragments formed in the thermal-neutron-induced fission of various iso-

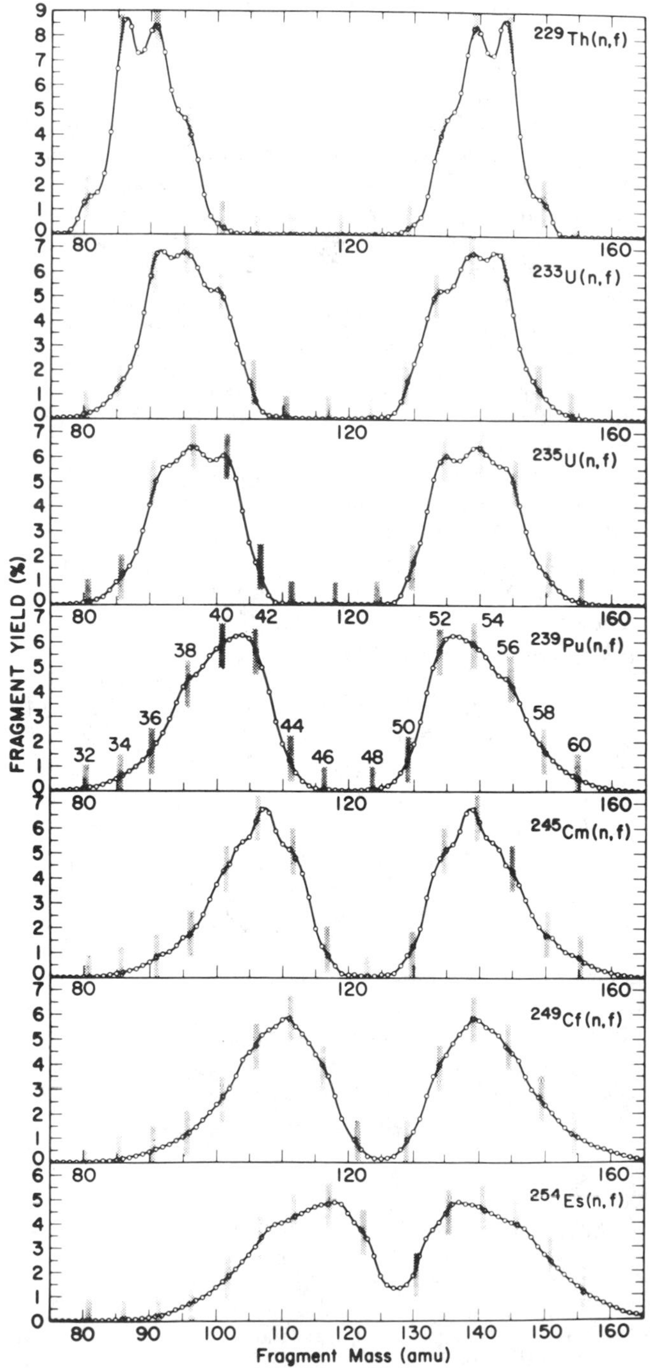

FIG. 2. Fission fragment mass distributions for thermal-neutron-induced fission. Masses at which the even nuclear charges occur are shown as shaded bars with numerical values indicated in the figure for ^{239}Pu(n_{th},f). [Reproduced, with permission, from the second volume of the first item of Part A of the Bibliography.]

topes of thorium, uranium, plutonium, californium, and einsteinium are shown in Fig. 2. The dominant feature of these distributions is that the most probable mass divisions are asymmetric. Thus, for example, the most probable mass division for $^{235}U(n_{th},f)$ occurs at approximately 140 amu/96 amu. Note that the masses shown in Fig. 2 are "primary fragment" masses, i.e., masses before neutron emission, and that the distributions are symmetric about a fragment mass equal to one-half the compound-nucleus mass.

The liquid-drop model proposed by Bohr and Wheeler (Part C of Bibliography) has been the traditional cornerstone in discussions and analyses of the fission process. This model, while successful in many respects, predicts symmetric mass divisions to be most probable. The observation that asymmetric mass divisions are, in fact, most probable could not be explained for a long time, and remained one of the mysteries of nuclear physics for over three decades. The observed asymmetric mass distributions are now understood in terms of nuclear-shell effects.

The overall distributions of fragment charges follow closely those of fragment masses. The even nuclear charges (Z values) are indicated by shaded vertical lines in the figure for $^{239}Pu(n_{th},f)$. The charge distribution corresponding to a particular fragment mass is rather narrow, with a standard deviation of generally less than 0.6.

The heavy-fragment peak remains centered at mass = 140 amu (see Fig. 2), for all compound-nucleus masses from 230 to 255 amu. Average masses of the lighter fission fragments are directly correlated with compound-nucleus masses. It is also interesting to note that the low-mass edge of the heavy-fragment peak occurs uniformly in the 130–135 amu range.

The average fragment total kinetic energies (sum of the energies of both fragments) are plotted as a function of heavy-fragment mass in Fig. 3, for some of the same fissioning systems whose mass distributions are shown in Fig. 2. Recalling that fragment kinetic energy arises from mutual Coulomb repulsive forces between the fragments, the observed total kinetic energy increase with increasing compound-nucleus mass for a given fragment mass is expected.

The most interesting feature of Fig. 3 is that the maximum

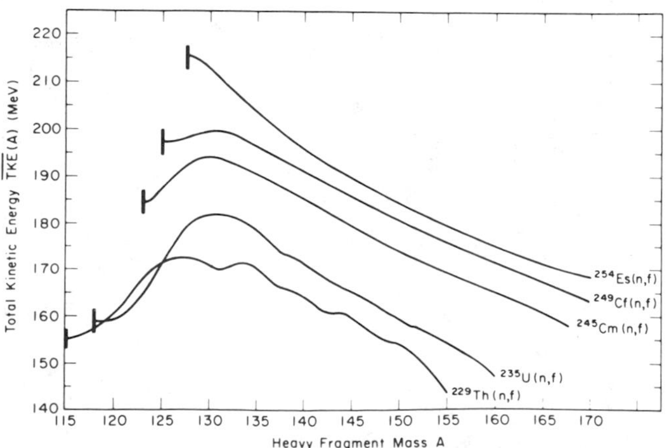

FIG. 3. Average fragment total kinetic energy (both fragments) as a function of the mass of the heavy fragment. Vertical lines indicate the symmetric fission masses. [Reproduced from the second volume of the first item in Part A of the Bibliography.]

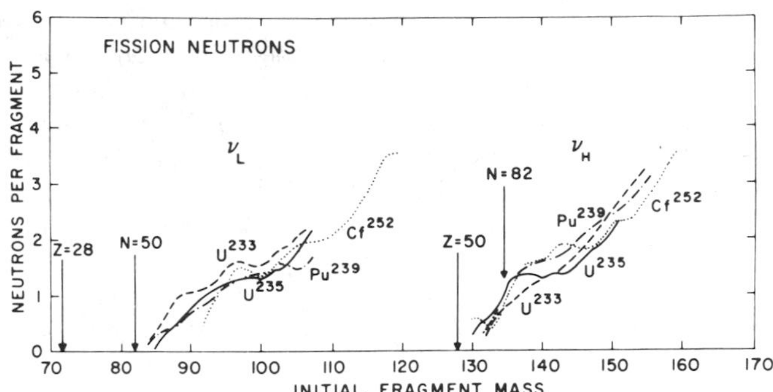

FIG. 4. Average number of neutrons emitted as a function of fragment mass for several low-excitation fission examples. Arrows indicate the fragment masses corresponding to proton and neutron magic numbers. [Reproduced from the first volume of the first item in Part A of the Bibliography.]

total kinetic energy in each case (except $^{229}Th(n_{th},f)$) occurs in the heavy-fragment range of 130–132 amu. It should be noted (see Fig. 2) that the atomic number $Z = 50$ occurs at (or near) fragment mass 130 amu, corresponding to neutron number $N = 80$. The so-called "magic numbers" for nuclei, i.e., those proton and neutron numbers causing nuclei to exhibit extraordinary stability, include $Z = 50$, $Z = 82$, $N = 50$, and $N = 82$. Thus, fragments of mass 130–132 amu are nearly doubly magic with $Z \cong 50$ and $N \cong 82$, and can be expected to exhibit properties of special stability. In particular, they are likely to be formed with more nearly spherical shapes than nearby nonmagic nuclei. The separation of fragment centers at scission will then be reduced, resulting in maximum Coulomb repulsion potential energy and, thus, in maximum total fragment kinetic energy. Furthermore, due to their greater stability and compact shape, the nearly double-magic fragment nuclei have excitation energies that are less than those of nearby fragment nuclei. Since neutron emission from fission fragments reflects their excitation energy, a minimum is expected to occur in the region of doubly magic nuclei (~130 amu). This effect is seen in Fig. 4, which shows the average number of neutrons emitted per fragment, as a function of fragment mass for the same fissioning systems considered above.

From the characteristics of fragment mass and total kinetic energy distributions, as well as from neutron emission data discussed above, it follows that the shell structure of fragment nuclei plays an important role in determining the properties of the fission process. Thus, theoretical descriptions of the fission of the heavy nuclei discussed above clearly need to take nuclear structure effects into account.

III. BASIC FISSION MODELS

The liquid-drop model has formed the basis of theoretical considerations in fission since it was first proposed by Bohr and Wheeler in 1939, within months after the discovery of nuclear fission. In this model the nucleus is assumed to behave as a charged liquid drop endowed with a surface tension, and calculations of shapes, energies, and of other properties are straightforward in principle, complicated only by the choice of shape parametrization and by the addition of other effects such as those of surface diffuseness.

The potential energy of any nucleus is a function of its shape parameters. Following the "fission direction" in the multidimensional potential-energy surface of the liquid-drop model, the potential energy increases smoothly from a minimum at zero deformation (sphere) to a maximum (the fission barrier) at an elongated shape, then decreases again as the nucleus elongates further, develops a neck, and finally divides into two fragments. The point of maximum potential energy in the fission direction is called the saddle point, and the potential energy from this point decreases in the fission direction (and in the direction of return to the original ground state), but increases in all other directions provided that the nucleus is not too light. Figure 5 shows liquid-drop-model (LDM) potentials for several fissile nuclei. As the charge Z and mass A of a fissioning system increase, the fissility parameter, which is equal to the Coulomb energy of a uniformly charged sphere divided by twice its surface energy and which is proportional to Z^2/A, also increases. As this occurs, the barrier height decreases and finally disappears at a fissility value of unity (see the three solid curves), indicating that nuclei with Z^2/A greater than a critical value cannot exist within the framework of this model. Beyond this point, the disruptive Coulomb force exceeds the cohesive surface tension force even for a spherical nucleus.

While the basic liquid-drop model provides a broad description of the fission process, because shell effects are neglected, it is unable to reproduce the features discussed in the previous section, as well as well-known properties of ground-state nuclei, such as the ground-state deformation of actinide nuclei. As a consequence of this situation, the ability of the LDM to describe many of the gross features of fission combined with the important influence of nuclear shell effects, the idea of a macroscopic–microscopic approach, in which smooth trends are taken from the LDM and local fluctuations from a microscopic model, was introduced in 1963 by Swiatecki. It was Strutinsky, however, who in 1966 developed, independently, a quantitative method for treating single-particle shell and pairing corrections to the underlying liquid-drop potential-energy surface. In Strutinsky's approach (Part D of the Bibliography) the nuclear potential energy is the sum of the LDM potential energy and a normalized "shell-correction" energy, where the latter is the difference between the calculated shell-model energy and the energy obtained when the nucleon energy levels are assumed to be broadened and the level density smoothed in the neighborhood of the Fermi surface.

Figure 6 shows the potential energy of a fissile actinide nucleus, with the deformed ground state and a two-peaked barrier which results when shell effects are taken into account. Specifically, such two-peaked–barrier features in the potential-energy surface are caused by nonuniformities in the single-particle levels at deformations where the LDM potential is rather flat as a function of deformation. Also shown in Fig. 6 is the LDM potential (for reference) together with schematic representations of a number of phenomena which result from the existence of a two-peaked barrier. Some of these effects are discussed in the next section.

The Strutinsky method has been extremely successful and has been included, with various refinements, in numerous calculations, with the result that most of the observations are now understood either qualitatively or semiquantitatively. One important conclusion that has emerged is that the route of minimum potential energy for a fissioning nucleus does not necessarily preserve maximum symmetry of shape. Axial asymmetry is indicated for the first barrier and mass asymmetry for the second barrier. This is shown for the case of ^{240}Pu in Fig. 7. The lowest potential energy, shown by the solid line, results when axial symmetry is in-

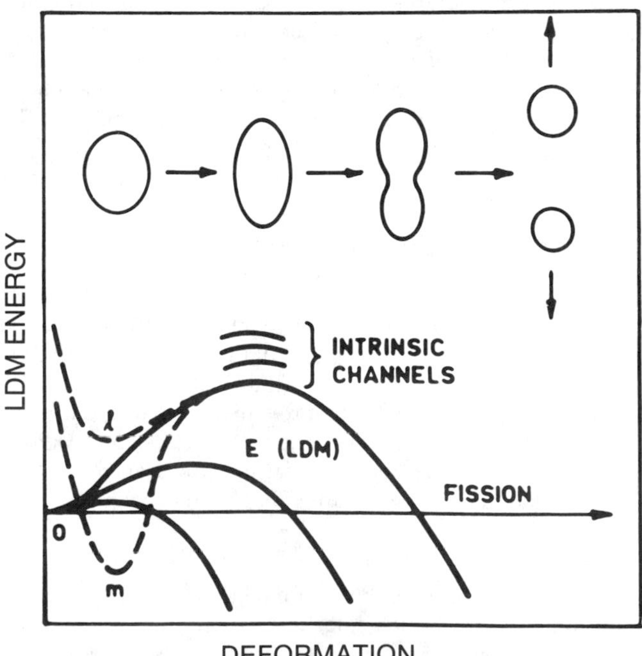

FIG. 5. Liquid-drop-model representation of the nuclear deformation energy in fission. The three solid curves correspond to three values of the fissility parameter Z^2/A; the dashed curves are theoretically corrected curves accounting for the ground-state deformations of actinide nuclei. [Reproduced from the second volume of the first item in Part A of the Bibliography.]

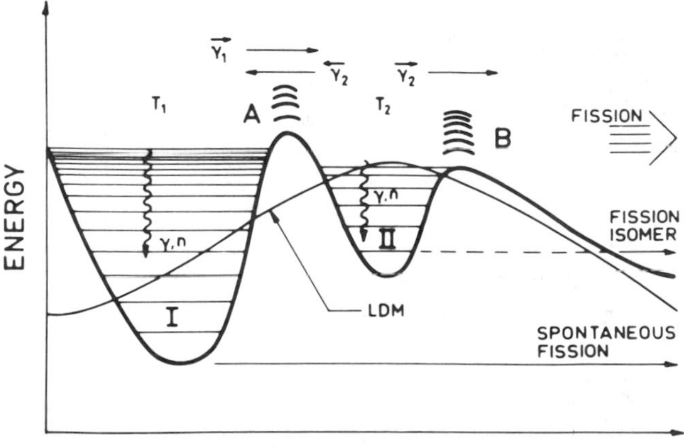

FIG. 6. Schematic representation of the double barrier in the potential energy for fission. Several types of transitions and phenomena associated with the double barrier are shown. [Reproduced from the second volume of the first item in Part A of the Bibliography.]

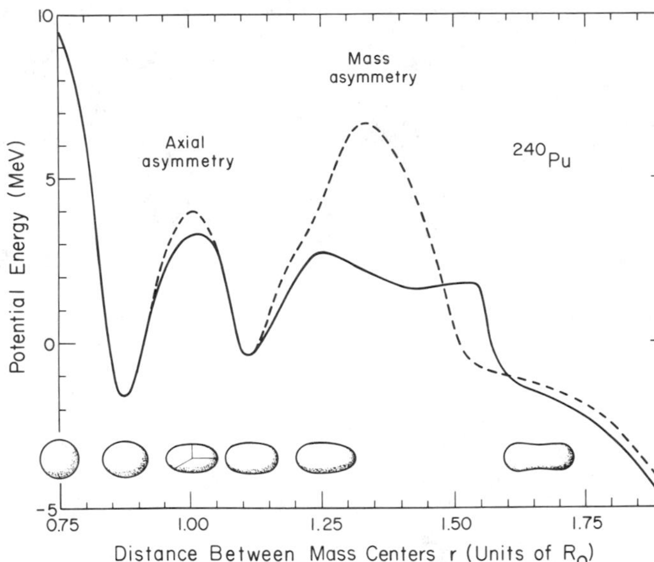

FIG. 7. Fission barrier computed with the macroscopic–microscopic method, illustrating the effects of axial asymmetry and mass asymmetry. [Reproduced from the fourth item in Part D of the Bibliography.]

cluded at the first peak and mass asymmetry is included at the second peak. The dashed curve gives the energies constrained to axial and reflection (mass) symmetry. Mass asymmetric second saddle points, such as the one shown in Fig. 7, are intimately connected with the observed asymmetric fission fragment mass distributions in heavy actinide nuclei discussed in the previous section.

IV. SPONTANEOUSLY FISSIONING ISOMERS AND OTHER OBSERVATIONS ATTRIBUTABLE TO THE TWO-PEAKED BARRIER

An important discovery in fission in 1962 revealed the existence of nuclear isomers that were found to decay by spontaneous fission. Only by 1969, however, had enough evidence been accumulated to establish that these isomers are shape isomers, in support of Strutinsky's theory introduced three years earlier. Since that time, fission isomers have been studied extensively, and their properties have been well determined. They are found to exist in the isotopes of elements from uranium ($Z=92$) through berkeleium ($Z=97$) with neutron numbers ranging from ~141 to ~151. Half-lives range from fractions of a nanosecond to milliseconds, and the isomer excitation energies, which are generally 2–4 MeV above ground state, have been measured. Some moments of inertia and spins also have been determined. The fragment mass and energy distributions have been shown to be similar to those for low-excitation prompt fission in at least two cases.

Figure 6 shows several levels within the second minimum (labelled II). Spontaneously fissioning isomers are produced most commonly in medium-energy reactions, e.g., (d,p), (d,n), $(d,2n)$, etc., whereby the residual nucleus is left in one of the excited states in the second minimum. The nucleus then proceeds via gamma decay to the ground state of the second minimum and subsequently fissions via the penetration of the second potential-energy barrier. Fission–gamma-decay competition takes place at all stages of the process, with the outcome dependent on the relative magnitude of the two barriers.

In addition to fission isomerism, two other phenomena are also related to the nature of the double barrier, and appear as structures in the fission cross section versus energy below and at the fission threshold. First, the fission cross section as a function of compound-nucleus excitation energy is a smooth curve in the absence of a double barrier. The double barrier, however, permits the existence of collective vibrational levels within the second minimum, and the presence of these states (especially near the top of the barrier) gives rise to one or more broad resonances in the fission cross section, which are superimposed on the smooth threshold curve. A number of such cases have been observed. Second, in those cases in which the second minimum is sufficiently deep, the statistical model of nuclear energy levels is used to explain the occurrence of distinct clustered groups of resonances observed in high-resolution neutron-induced fission studied well below the fission barrier. The ^{237}Np(n,f) reaction, for example, exhibits such structure at kilovolt neutron energies, while the fission threshold occurs at a neutron energy of 0.5 MeV. Each group of resonances is interpreted as being due to resonant coupling between states in the first minimum, whose density is high, and a single state in the shallower second minimum, where the level density is smaller.

Recently, certain experimental observations have been attributed to the existence of not two, but three fission barriers. While a three-peaked barrier may be required to explain measured results in some cases, the vast majority of available data is adequately described quantitatively by the microscopic–macroscopic theory that results predominantly in two-peaked barriers, with a strongly developed well nestling between the barrier peaks. It is this structure that has led to the large range of remarkable phenomena observed in nuclear fission, nuclear reactions, and nuclear spectroscopy. For an excellent and complete discussion of the two-peaked barrier and its experimental consequences, see the third item in Part E of the Bibliography.

V. NUCLEAR DISSIPATION AND THE DYNAMICS OF FISSION

Following the introduction of the macroscopic–microscopic method in the 1960s, the next major theoretical advance, which occurred during the 1970s and early 1980s, involved the dynamics of the fission process. Fission provides us with an excellent opportunity to investigate the fundamental nuclear property of kinetic energy dissipation (friction), since it involves the entire nucleus in a dynamic evolution during its descent from the saddle point to scission. At first it was believed that the dissipation mechanism in fission results from two-body collisions, such as those responsible for viscosity in fluids. But it was then realized that the long mean free path of nucleons inside a nucleus arising from the Pauli exclusion principle leads to a very different picture of nuclear dissipation.

In one of the early attempts to elucidate the effect of a

long nucleon mean free path on fission dynamics, it was assumed that the velocity distribution of nucleons striking a moving container wall is completely random. This has led to a result in which nuclei are predicted to follow a highly overdamped motion, and in which the descent from saddle to scission is very slow, reminiscent of the division of a drop of honey. This conclusion, however, turned out to be at variance with a number of observations, notably the systematics of fission fragment kinetic energies. The puzzle remained unsolved until the mid-1980s, when Nix and Sierk proposed another macroscopic approach, in which long mean-free-path dynamics and two-particle dynamics are merged. The new theory is valid for intermediate excitation energies, above which effects of nucleon pairing have disappeared and below which the nucleon mean free path exceeds the nuclear diameter. Dissipation then takes place primarily in the surface region of the fissioning nucleus via two distinct mechanisms. The first is the one-body process with a properly adjusted magnitude, and the second involves two-body collisions which are no longer suppressed by the Pauli exclusion principle as one passes through the nuclear surface to the exterior. In addition, special consideration has been given to dumbbell-like shapes, which are characteristic of a fissioning nucleus, as it descends from the saddle point to scission. (For relatively light nuclei the saddle-point shapes themselves have dumbbell-like configurations.) The thin neck of dumbbell shapes forms an effective "window" between the two halves of the fissioning nucleus, and the transfer of nucleons through this window is considered explicitly in the calculations. The combination of these mechanisms is known as the "surface-plus-window" dissipation, and the resulting dynamical motion is found to be only somewhat overdamped. The calculations provide a satisfactory description of several observed phenomena, including the mean values of fission fragment energies of all fissioning nuclei.

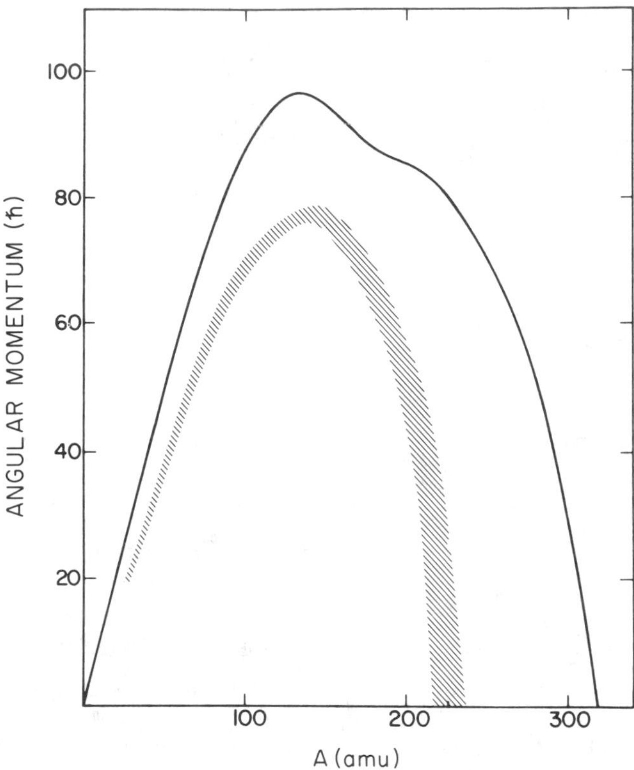

FIG. 8. Limits of stability of rotating nuclei as determined by the rotating liquid-drop model. The solid line gives the value of angular momentum at which the fission barrier of stable nuclei of mass number A is predicted to vanish. The hatched area indicates the region of competition between fission and particle emission. Fission dominates above the hatched area. [Reproduced from the second item in Part G of the Bibliography.]

VI. FISSION OF RELATIVELY LIGHT NUCLEI

The fission of actinide nuclei has been described in Section II, and it was shown in Sections III and IV that many of the observed properties result from the microscopic modification of the underlying macroscopic potential-energy surface. As one considers lighter nuclei, fissility decreases because of the rapid increase in fission barriers, which, in turn, is caused by the increase in the macroscopic (liquid-drop) potential energy. Thus, the shell-model microscopic effects, which are superimposed on the liquid-drop potential-energy surface, play a role of decreasing importance. Furthermore, because of the low fissility of nuclei lighter than actinide nuclei, fission can only be induced by supplying the nucleus with a significant amount of excitation energy to enable it to traverse the higher fission barrier. This additional excitation energy, which is usually supplied by means of bombardments with accelerated charged particles or nuclei, also helps to decrease the role played by shell effects. Consequently, fission of nuclei lighter than actinide nuclei tends to reflect the smooth features associated with the liquid-drop model. Thus, both fragment mass and kinetic energy distributions have symmetric Gaussian shapes and are peaked at symmetric mass divisions. The widths of the distributions increase with increasing excitation energy, which increases with the bombarding energy of the fission-inducing projectiles.

For nuclei in the transition region, such as radium, fission distributions exhibit a curious combination of both liquid-drop and shell-effect characteristics. For example, the fragment mass distribution resulting from the fission of ^{226}Ra has three peaks: a prominent peak centered at symmetric mass divisions reflecting features of the liquid-drop potential-energy surface, flanked by two smaller satellite peaks in the wings, reflecting microscopic shell-effect perturbations. Similar features are found in even lighter nuclei (e.g., ^{201}Tl) provided that the excitation energy is sufficiently low.

In recent years, the study of reactions induced by means of accelerated nuclei (heavy ions) has been an intense area of investigation. Off-center collisions of nuclei produce rotating systems endowed with a large amount of angular momentum. The resulting centrifugal force acts in concert with the disruptive Coulomb force to counteract the cohesive nuclear binding force, which is represented by the surface tension in the liquid-drop model. On the basis of the rotating liquid-drop model (RLDM), which includes angular momentum effects, it was predicted in the early 1960s that all nuclei, regardless of their mass, can become unstable toward disintegration by fission, provided that they are endowed with a sufficient amount of rotational energy. In RLDM calculations, it is possible to obtain, for all nuclei, values of angular momentum (related to the rotational energy) at which the fission barrier is reduced to zero. This vanishing fission barrier determines the ultimate limit of stability of all rotating nuclei and the RLDM calculations, shown in Fig. 8, have been widely used in predicting consequences of various heavy-ion-induced reactions. While the RLDM provides us with reasonable estimates of the fission barriers of rotating nuclei, calculations in good agreement with experimental results were not available until the mid-1980s, when effects of the finite range of the nuclear force and the diffuseness of the nuclear surface were added to the RLDM. The new theory was then found to describe successfully a wide range of heavy-ion-induced fission measurements.

VII. FUTURE CHALLENGES

Fifty years following the discovery of fission, the primary challenges for future investigation lie in the theoretical area. The macroscopic–microscopic method provides us with a general understanding of observed phenomena by merging two diverse theoretical approaches. A full self-consistent microscopic calculation, even within the constraint of an effective two-nucleon interaction treated in the nonrelativistic approximation, remains to be done. The ultimate challenge, a fully relativistic description of fission based on quantum chromodynamics with explicit quark and gluon degrees of freedom taken into account, is well beyond the reach of current capabilities in theoretical physics.

See also GAMMA DECAY; NUCLEAR REACTIONS; NUCLEAR REACTORS.

BIBLIOGRAPHY

A. Collections of Papers

Proceedings of the Symposium on the Physics and Chemistry of Fission, I. Salzburg, II. Vienna, III. Rochester, IV. Jülich. International Atomic Energy Agency, Vienna, 1965, 1969, 1973, 1979.

Proceedings of the International Conference on Fifty Years Research in Nuclear Fission, Berlin, 1989. Nucl. Phys. **A502** (Oct. 1989).

Proceedings of the Conference on Fifty Years with Nuclear Fission, Gaithersburg, 1989. American Nuclear Society, 1989.

B. Discovery of Fission

O. Hahn and F. Strassmann, *Naturwissenschaften* **27,** 11 (1939).

L. Meitner and O. R. Frisch, *Nature* **143,** 239 (1939).

C. General Theory and Liquid-Drop Model

N. Bohr and J. A. Wheeler, *Phys. Rev.* **56,** 426 (1939).

D. L. Hill and J. A. Wheeler, *Phys. Rev.* **89,** 1102 (1953).

S. Cohen and W. J. Swiatecki, *Ann. Phys.* (N.Y.) **22,** 406 (1963).

J. R. Nix and W. J. Swiatecki, *Nucl. Phys.* **71,** 1 (1965).

J. R. Nix, *Nucl. Phys.* **A130,** 241 (1969).

D. Macroscopic–Microscopic Method

W. J. Swiatecki, in *Proceedings of the International Conference on Nuclidic Masses,* Vienna, 1963. Springer-Verlag, Berlin, 1964.

V. M. Strutinsky, *Yad. Fiz.* **3,** 614 (1966); *Sov. J. Nucl. Phys.* **3,** 449 (1966).

V. M. Strutinsky, *Ark. Fys.* **36,** 629 (1971); *Nucl. Phys.* **A95,** 420 (1967); **122,** 1 (1968).

P. Möller and J. R. Nix, in *Proceedings of the Symposium on the Physics and Chemistry of Fission,* III, Rochester. International Atomic Energy Agency, Vienna, 1973.

J. R. Nix, *Annu. Rev. Nucl. Sci.* **22,** 65 (1972).

M. G. Mustafa, U. Mosel, and H. W. Schmitt, *Phys. Rev.* C **7,** 1519 (1973).

E. Fission Isomers and Other Effects of the Two-Peaked Barrier

S. M. Polikanov, V. A. Druin, V. A. Karnaukhov, V. L. Mikheev, A. A. Pleve, N. K. Skobelev, V. G. Subbottin, G. M. Ter-Akopyan, and V. A. Formichev, *Sov. Phys. JETP* **15,** 1016 (1962).

S. Bjornholm and V. M. Strutinsky, *Nucl. Phys.* **A136,** 1 (1969).

S. Bjornholm and J. E. Lynn, *Rev. Mod. Phys.* **52,** 725 (1980).

F. Nuclear Dissipation and Dynamics of Fission

J. Blocki, Y. Boneh, J. R. Nix, J. Randrup, M. Robel, A. J. Sierk, and W. J. Swiatecki, *Ann. Phys.* (N.Y.) **113,** 330 (1978).

J. R. Nix and A. J. Sierk, in *Proceedings of the International School-Seminar on Heavy-Ion Physics,* Dubna, 1986. Joint Institute for Nuclear Research Report, JINR-D7-87-68 (1987).

J. R. Nix, in *Proceedings of the Conference on Fifty Years with Nuclear Fission,* Gaithersburg, 1989, Vol. 1. American Nuclear Society, 1989.

G. Fission of Relatively Light Nuclei and Heavy-Ion-Induced Fission

J. P. Unik, J. G. Cunninghame, and I. F. Croall, in *Proceedings of the Symposium on the Physics and Chemistry of Fission,* II, Vienna. International Atomic Energy Agency, Vienna, 1969.

F. Plasil, in *Proceedings of the International Conference on Reactions Between Complex Nuclei,* Nashville, 1974. North-Holland, Amsterdam, 1974.

S. Cohen, F. Plasil, and W. J. Swiatecki, *Ann. Phys.* (N.Y.) **82,** 557 (1974).

A. J. Sierk, *Phys. Rev.* C **33,** 2039 (1986).

F. Plasil, *Pramana–J. Phys.* **33,** 145 (1989).

J. O. Newton, Australian National University preprint ANU-P/1024, September 1988; Soviet Journal of Particles and Nuclei, to be published.

Nuclear Forces

Ronald Bryan

Forces between aggregates of nucleons (a nucleon is a proton or neutron) are largely due to simple pairwise forces between individual nucleons. The force between each pair of nucleons has the following main features: (1) It vanishes for nucleon–nucleon (*NN*) separations greater than about 4 fm (1 fm $= 10^{-13}$ cm); (2) it is very strong for *NN* separations smaller than 1 fm (much stronger than the Coulomb potential between proton pairs at this separation); and (3) it is very complicated (e.g., much more complicated than the Coulomb force).

To a reasonably good approximation, the *NN* interaction at low energies can be described in terms of the Schrödinger equation,

$$-(\hbar^2/M)\nabla^2\psi + V\psi = i\hbar\ \partial\psi/\partial t, \tag{1}$$

where $\hbar$ is Planck's constant/2π, $M=939$ MeV/c^2 is the nucleon mass (we ignore the slight difference between the neutron and proton masses), V is the potential, and ψ is the two-nucleon wave function.

The potential depends on the two nucleons' relative separation $\mathbf{r}$, their relative momentum $\mathbf{p}$, and the relative orientation of their intrinsic spins $\mathbf{S}^{(a)}$ and $\mathbf{S}^{(b)}$. Since each nucleon has spin $S^{(i)}=\frac{1}{2}\hbar$ ($i=a$ or b) with z component $S^{(i)}=\pm\frac{1}{2}\hbar$, there are four possible spin combinations for the two nucleons. Transitions from one such spin combination to any other are in principle possible, so V must have $4\times4=16$ elements to allow for these. The four possible spin states can be accommodated by writing ψ in Eq. (1) as a four-component wave function (more precisely, a spinor); then V appears as a 4×4 matrix. Since there are three possible combinations of nucleons (pp, nn, and np), there are in fact three such matrices.

The features enumerated as (1) and (2) in the opening paragraph are manifested in each term of the potential (matrix), and in particular are evident in the term corresponding to two protons interacting with total spin $\mathbf{S}=\mathbf{S}^{(a)}+\mathbf{S}^{(b)}=0$ (the singlet state) and relative orbital angular momentum $L=0$. This potential [1] is illustrated in Fig. 1, and is labeled V. The Coulomb potential, $V^{\text{Coul}}=e^2/r$, is also plotted (e is the charge of each proton) and serves as a useful basis of comparison. On the average, V is seen to be much stronger than V^{Coul} at distances of 2 fm or less, but to vanish for distances of about 4 fm or more. (At this distance V^{Coul} also appears to vanish, but is in fact nonzero; indeed it is nonzero out to classical distances. At 1000 fm, V^{Coul} is just as strong as V at 10 fm, and at 10^{13} fm$=1$ cm, as strong as V at 40 fm.)

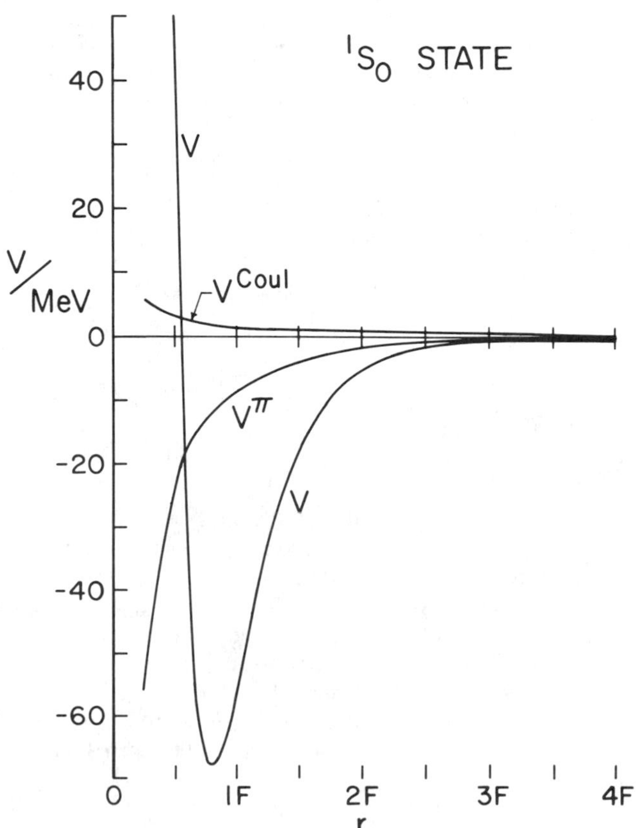

FIG. 1. Plot of the NN one-pion-exchange potential V^π and the full NN potential V versus r, for the $^1S_0$ state. For comparison the Coulomb potential is also plotted.

The reason why V falls off so fast beyond 3 or 4 fm is at the root of our understanding of nuclear forces and provided the basis for Yukawa's predication of the π meson as the source of the nuclear force [2]. Simply put, the nuclear force at long range is understood to be due to the exchange of a π meson (or pion) between two nucleons, as illustrated in Fig. 2. Nucleon a emits the pion and nucleon b absorbs it, and vice versa (not shown). In a reference system where the center of mass of the nucleons is at rest, nucleon a emerges with the same energy E that it has before the exchange; nucleon b likewise emerges with its energy unchanged. However, the *momentum* of nucleon a does change, by an amount $\mathbf{q}=\mathbf{p}'-\mathbf{p}$ (the momentum transfer), which momentum is absorbed by nucleon b. Evidently the π meson carries energy $E_a=0$ and momentum $\mathbf{q}\neq0$ from a to b. But this meson cannot be physical, for the Einstein mass–energy–momentum relationship dictates that $E_q=(q^2c^2+m^2c^4)^{1/2}$ for a real pion with $m_\pi=140$ MeV/c^2 (the rest mass of the π meson) and $c=3\times10^{10}$ cm/s. Thus the energy of the exchanged (or "virtual") meson, being 0, is at least $m_\pi c^2=140$ MeV too low. The Heisenberg uncertainty principle, $\Delta E\Delta t\sim\hbar$, does permit such energy discrepancies ΔE, but only for a sufficiently short time Δt. Thus, if $\Delta E\sim m_\pi c^2$, then $\Delta t\sim\hbar/\Delta E=\hbar/m_\pi c^2=5\times10^{-24}$ s. The distance such a virtual pion can travel in time Δt is then of the order of $c\Delta t=\hbar c/m_\pi c^2$. This determines the range of the nuclear force [3]. Numerically, $\hbar c/m_\pi c^2=(197.3\text{ fm}\cdot\text{MeV})/(140\text{ MeV})=1.4$ fm. Thus this

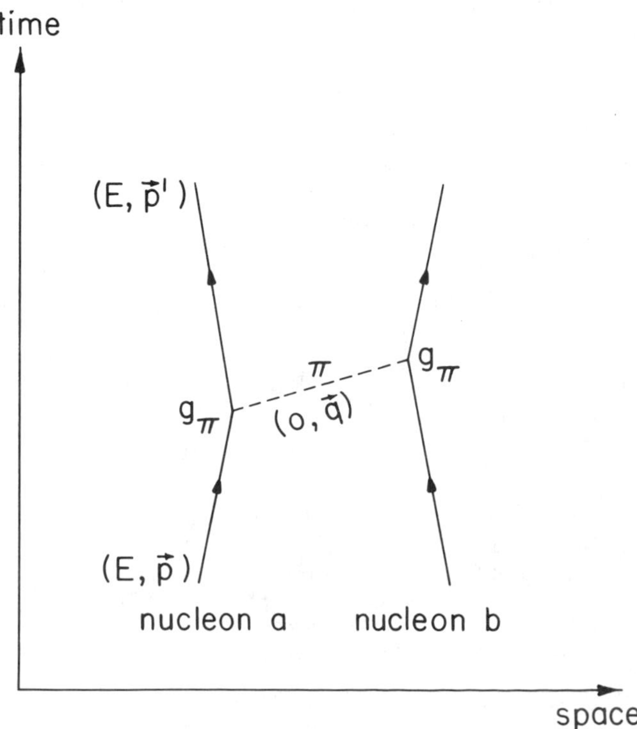

FIG. 2. Feynman diagram for one-pion exchange.

order-of-magnitude estimate of the range of the nuclear force agrees with the observed range [feature (1) in the opening paragraph]. A plot of the exact proton–proton potential due to the exchange of one pion appears in Fig. 1 (for the case of two protons in a spin $S=0$ configuration).

The complete expression for the one-pion-exchange potential V^{π} is rather more complicated than Fig. 1 suggests; through order q^2/M^2 it is

$$V = \boldsymbol{\tau}_a\cdot\boldsymbol{\tau}_b g_{\pi}^2 \left(\frac{m_{\pi}}{2M}\right)^2 \left[\frac{1}{3}\frac{e^{-\mu r}}{r}(\boldsymbol{\sigma}_a\cdot\boldsymbol{\sigma}_b) + \left(\frac{1}{3}+\frac{1}{\mu r}+\frac{1}{\mu^2 r^2}\right)\frac{e^{-\mu r}}{r} S_{ab}\right], \tag{2}$$

where $1/\mu=\hbar/m_{\pi}c=1.4$ fm is the pion Compton wavelength, $g_{\pi}^2/\hbar c\sim 14$ is the square of the pion–nucleon coupling constant, $\boldsymbol{\sigma}_a$ and $\boldsymbol{\sigma}_b$ are the Pauli spin matrices for nucleons a and b, respectively, $\boldsymbol{\tau}_a$ and $\boldsymbol{\tau}_b$ are the 2×2 isospin matrices for nucleons a and b, respectively, and

$$S_{ab}=3(\boldsymbol{\sigma}_a\cdot\hat{\mathbf{r}})\,(\boldsymbol{\sigma}_b\cdot\hat{\mathbf{r}}) - \boldsymbol{\sigma}_a\cdot\boldsymbol{\sigma}_b$$

is the usual tensor operator, where $\hat{\mathbf{r}}$ is a vector of unit length in the direction of $\mathbf{r}$. V^{π} exhibits the properties of the nuclear force enumerated as items (1)–(3) in the first paragraph (although it is by no means the only contributor to the NN potential). The short range [item (1)] is brought about by the factor $\exp(-\mu r)$ which causes the nuclear force to essentially vanish for r greater than $3/\mu$ or so; the great strength [item (2)] is apparent from the size of the pion–nucleon coupling constant squared, $g_{\pi}^2/\hbar\approx 14$, to be compared with its Coulomb counterpart, $e^2/\hbar c=1/137.0$, the photon–proton (or photon–electron) coupling constant squared [although in the nuclear case the factor $(m_{\pi}/2M)^2=0.005$ seriously reduces the strength of $g_{\pi}^2/\hbar c$]; and the complicated nature of the nuclear force [item (3)] is suggested by V^{π}, although in fact it is quite a bit less complicated than the full nuclear force. The factor $\boldsymbol{\sigma}_a\cdot\boldsymbol{\sigma}_b$ is 1 (-3) when the nucleons are in a spin $S=1$ (0) configuration; as for S_{ab}, if we substitute the magnetic moment of each nucleon, $\boldsymbol{\mu}_a$ or $\boldsymbol{\mu}_b$, in place of $\boldsymbol{\sigma}_a$ or $\boldsymbol{\sigma}_b$, then S_{ab}/r is exactly the interaction energy arising from classical magnetism (the interaction is that of two bar magnets separated a distance $\mathbf{r}$, as in Fig. 3). In fact, $\boldsymbol{\mu}_a$ and $\boldsymbol{\mu}_b$ are proportional to the intrinsic spins $\boldsymbol{\sigma}_a\hbar/2$ and $\boldsymbol{\sigma}_b\hbar/2$. However, in this case the analogy goes no further, because in V^{π} the interaction is due to π-meson exchange, not the photon exchange of electromagnetism. Finally, the factor $\boldsymbol{\tau}_a\cdot\boldsymbol{\tau}_b$ in V^{π} relates to isobaric spin or isospin (often misnamed isotopic spin), a concept analogous to ordinary intrinsic spin that has proved to be amazingly useful in predicting the nuclear force. We assign isospin $\frac{1}{2}$ to nucleons, with the third component $\frac{1}{2}$ for protons and $-\frac{1}{2}$ for neutrons (or vice versa in some conventions). Two nucleons are in a total isospin-1 state if they are both protons or neutrons; an np combination has isospin 1 (0) if the isospin wave function is symmetric (antisymmetric) under interchange of the two nucleons. For nucleons in an isospin state 1 (0), $\boldsymbol{\tau}_a\cdot\boldsymbol{\tau}_b=1\,(-3)$.

We now turn to the complete nucleon–nucleon potential. The most important contributions for NN separations greater than 1 fm or so are sketched in Fig. 4. Besides one-pion exchange, there are two-pion-exchange and ω-exchange graphs. Three-pion exchange may also play an important role. By extending Wick's argument we can show that 2π exchange has about one-half the range of one-pion exchange, 3π exchange about one third the range, etc. The ω, which is a neutral vector meson ($J^P=1^-$) of mass 782 MeV/$c^2=5.6m_{\pi}$, has even less exchange range. In the 2π-exchange graphs the blobs surrounding the nucleon lines represent all processes that can take place that result in two external nucleon lines and two internal pion lines, e.g., $N+\pi\rightarrow\Delta\rightarrow N+\pi$, where Δ is the $I=3/2$, $J^P=3/2^+$ isobar of mass 1232 MeV/c^2. Two exchanged pions can also interact in flight, particularly in states of relative orbital angular momentum 0 and 1, in which case their contribution to the NN potential is strongly modified. They appear as scalar and vector mesons, respectively, and provide intermediate-range attraction and repulsion, respectively.

Empirical studies show that the NN force obeys certain invariance principles but is nearly as complicated as possible within these restrictions. The invariance principles obeyed are parity conservation (to about 1 part in 10^6), time-reversal symmetry (to one part in at least 10^5), and isospin conservation (only to 1 or 2 parts in 100, because it is broken by the electromagnetic interaction). The most general form for the potential allowed by these symmetries is

$$V=V^{(0)}+V^{(1)}\boldsymbol{\tau}_a\cdot\boldsymbol{\tau}_b, \tag{3}$$

with

$$V^{(i)} = V_C^{(i)}+V_{LS}^{(i)}\mathbf{L}\cdot\mathbf{S}+V_T^{(i)}S_{ab}+V_{\sigma\sigma}^{(i)}\boldsymbol{\sigma}_a\cdot\boldsymbol{\sigma}_b + V_Q^{(i)}[(\boldsymbol{\sigma}_a\cdot\mathbf{L})\,(\boldsymbol{\sigma}_b\cdot\mathbf{L})+(\boldsymbol{\sigma}_b\cdot\mathbf{L})\,(\boldsymbol{\sigma}_a\cdot\mathbf{L})],\quad i=0,\ 1, \tag{3a}$$

where $\mathbf{L}=\mathbf{r}\times(\hbar/i)\nabla$ is the orbital angular-momentum operator. We see that there are five terms for each of the isospin combinations, or 10 terms in all. This is all that remains of the 3×16 terms in V after the invariance principles are satisfied. The terms as they appear from left to right in Eq. (3a) are denoted the central, spin–orbit, tensor, spin–spin, and

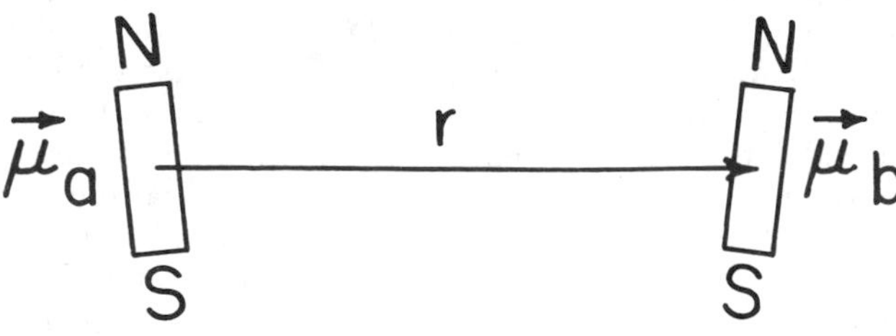

FIG. 3. Interaction between two magnetic dipoles.

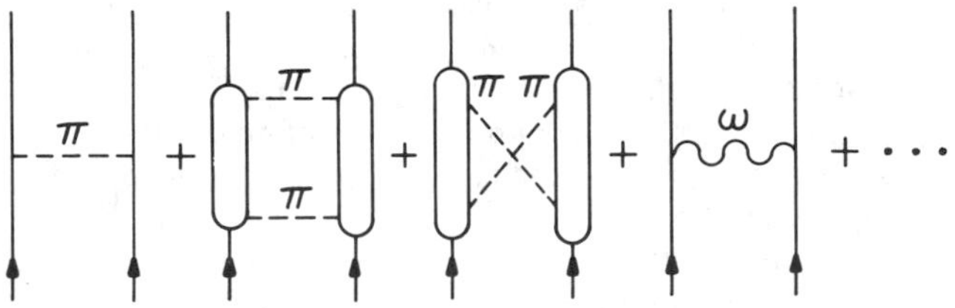

FIG. 4. Principal Feynman graphs for the intermediate- and long-range NN force.

quadratic spin–orbit potentials. In phenomenological work the $V_C^{(i)}$, $V_{LS}^{(i)}$, etc., are usually taken to be functions of r^2 alone, but in general they are also functions of ∇^2 (indeed, are generally nonlocal).

For *NN* separations of less than 1 fm, the potentials are less well understood, but the principal features appear to include a strong attraction in $V_{LS}^{(0)}$, and a repulsion in $V_C^{(0)}$. Such features are characteristic of ω exchange [the ω acts as a heavy photon, giving a repulsion between like charges (nucleons) and the usual *LS* attraction] and in fact led Breit to predict this meson prior to its discovery [4].

Since the nucleon is composed of three quarks, a more fundamental understanding of the nuclear force at distances <1 fm might in principle be provided by quantum chromodynamics, a theory widely believed to give the correct description of interacting quarks and gluons. However, computational difficulties have thus far prevented exact predictions.

The principal source of information about the nucleon–nucleon force has been *pp* and *np* elastic-scattering experiments where the projectile particle's laboratory energy has ranged from 0 to 450 MeV. More detailed information on the *NN* potential may be found in Ref. [5] and references therein.

REFERENCES

1. R. A. Bryan and A. Gersten, *Phys. Rev.* **D6,** 341 (1972); Fit "D".
2. H. Yukawa, *Proc. Phys. Math. Soc. Japan* **17,** 48 (1935); reprinted in *Progr. Theoret. Phys. (Kyoto) Suppl.* **1,** 1 (1955).
3. This argument is due to G. C. Wick, *Nature* **142,** 994 (1938).
4. G. Breit, *Proc. Natl. Acad. Sci. U.S.A.* **46,** 746 (1960).
5. R. Machleidt, K. Holinde, and Ch. Elster, *Phys. Rep.* **149,** 1 (1987).

Nuclear Fusion

J. P. Freidberg

INTRODUCTION

Commercial electric power plants based on the nuclear fusion reaction would satisfy the world's energy needs for as far into the future as one can imagine. Such plants would be environmentally safe, particularly with regard to the issue of waste products, and utilize a fuel that is inexpensive, virtually inexhaustible, and easily extractable from the ocean. In spite of the immense complexity of the task, it is this vision that has motivated the international fusion effort, an effort involving thousands of scientists and engineers, for over three decades.

As its name implies, nuclear fusion is a nuclear reaction in which light elements combine (i.e., fuse together) to form heavier elements, releasing large amounts of energy in the process. The other well-known form of nuclear energy is fission. Here, heavy elements are split into smaller fragments, also releasing large amounts of energy. Nuclear reactions involving intermediate-mass elements are much more difficult to initiate because the binding energy of such elements is substantially higher, peaking at $A=56$ (iron).

Nuclear fusion is familiar in at least three situations: (1) as the energy source for normal stars, our sun in particular, (2) in thermonuclear weapons, and (3) in controlled thermonuclear fusion for ultimate use in electric power production.

In the sun, the most important reaction involves hydrogen as the fuel:

$$4\mathrm{H}+2e\rightarrow{}^4\mathrm{He}+2\nu+26.7\ \mathrm{MeV}. \tag{1}$$

This reaction is extremely slow ($\sim10^{10}$ yr/reaction) as one of the two subchains that constitute Eq. (1) requires the weak interaction force to convert protons to neutrons. Consequently, although there are enormous reserves of hydrogen in the ocean, there is no practical way to produce nuclear energy on earth by this reaction.

Fusion power and thermonuclear weapons depend upon nuclear reactions utilizing the isotopes of hydrogen, namely, deuterium and tritium. These processes involve only the strong nuclear force and hence have overall reaction rates many orders of magnitude larger than Eq. (1). The four major reactions of current interest are

$$\mathrm{D}+\mathrm{D}\rightarrow{}^3\mathrm{He}+n+3.3\ \mathrm{MeV}, \tag{2}$$

$$\mathrm{D}+\mathrm{D}\rightarrow{}^3\mathrm{T}+p+4.0\ \mathrm{MeV}, \tag{3}$$

$$\mathrm{D}+{}^3\mathrm{He}\rightarrow{}^4\mathrm{He}+p+18.3\ \mathrm{MeV}, \tag{4}$$

$$\mathrm{D}+\mathrm{T}\rightarrow{}^4\mathrm{He}+n+17.6\ \mathrm{MeV}. \tag{5}$$

The two D–D reactions have about equal cross sections and are desirable because of the huge resources of deuterium in the ocean (see Fig. 1 for various cross sections). Specifically, D occurs naturally as 0.015 atm% of the hydrogen in sea water which translates into billions of years of energy at the current rate of usage. The difficulty with the D–D reaction is that it has the smallest cross section (by approximately two orders of magnitude) of those listed.

The D–^{3}He reaction has a substantially higher cross section and the desirable feature that, except for a small amount of secondary reactions, the energy produced occurs in charged particles. This feature offers the possibility of directly converting nuclear energy into electrical energy without the need of a thermal cycle, thereby eliminating a factor $\eta_T\approx\frac{1}{3}$ from the overall efficiency. The absence of neutrons in the output products also minimizes radiation and heat load problems on the first wall surrounding the reacting region. The main difficulty with the D–^{3}He reaction is the absence of naturally occurring ^{3}He on earth.

The D–T reaction has the largest cross section of any of the fusion reactions listed above [$\sigma(100\ \mathrm{keV})\approx5$ barns]. However, tritium is a radioactive isotope with a half-life of 12.3 yr and no known reserves on earth. Also, most of the energy appears in the neutron (14.1 MeV). Nevertheless, the overriding issue at the current state of fusion research is the difficulty of achieving sufficient reactions to produce net power. Consequently, the D–T reaction, with the largest cross section, is the primary focus of the fusion program. Probably, the first fusion reactors will be based on this reaction. The neutron wall loading problems while difficult are not expected to be impossible. The problem of supplying tritium as a fuel is solved by surrounding the reacting region

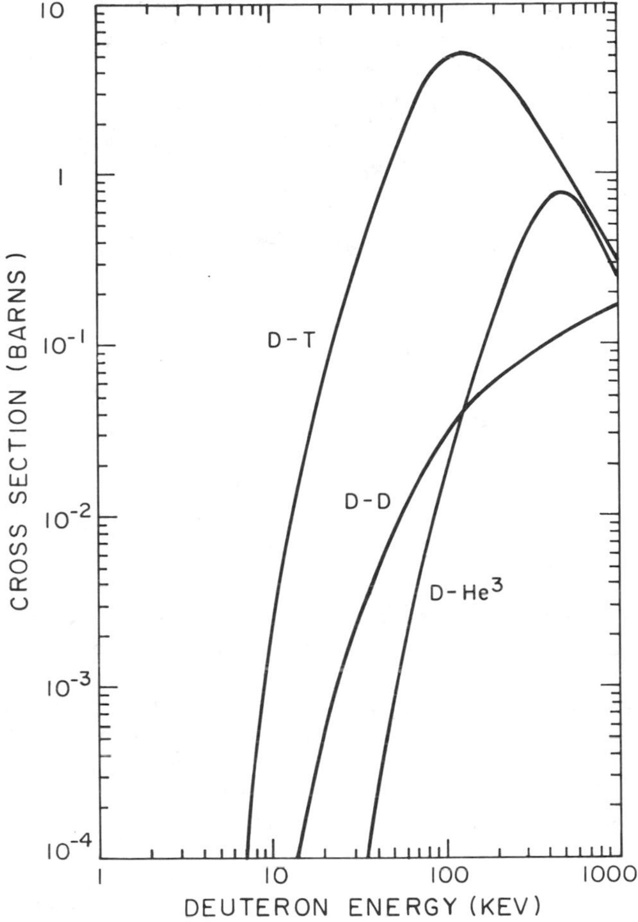

FIG. 1. Cross sections for various reactions.

with a blanket containing ^{6}Li and then breeding T by the reaction

$$n+{}^6\text{Li}\rightarrow{}^4\text{He}+\text{T}+4.8\ \text{MeV}. \tag{6}$$

The neutron appearing in Eq. (6) originates as one of the output products from the D–T reaction. The known reserves of lithium on earth are adequate to supply T for several hundred years of energy, a relatively long time, but still far short of the original promise of fusion. The conclusion is that D–T reactors may be the first but not the ultimately most desirable power plants. However, during the evolution of D–T reactors enough knowledge and expertise should be acquired to enable the development of D–D or other advanced fuel reactors.

CONDITIONS FOR FUSION REACTIONS TO OCCUR

The cross sections for nuclear fusion reactions are typically on the order of the cross-sectional area of the nucleus. In addition, for the reactions of interest, the fusing particles are positively charged nuclei (i.e., D, T, or ^{3}He nuclei). Hence, for a reaction to occur, the particles must have sufficient kinetic energy to overcome the repulsive force of the Coulomb potential. A simple estimate of this energy is obtained by equating the kinetic energy of an incoming particle with the Coulomb potential energy evaluated at $d=d_n$, the nuclear diameter:

$$\frac{1}{2}mv^2=\frac{Z_1Z_2e^2}{4\pi\epsilon_0\, d_n}. \tag{7}$$

For charge numbers $Z_1=Z_2=1$, the kinetic energy is typically 200 keV. Higher values of Z_1 and Z_2 require higher particle energies and are therefore less efficient for producing power. This illustrates why so much attention has been focused on the isotopes of hydrogen which have $Z=1$. In practice, nuclear reactions can occur at energies below 200 keV because of quantum-mechanical tunneling effects.

The kinetic energy of fusing particles far exceeds the ionization potential for neutral atoms (e.g., 13.7 eV for hydrogen). Consequently, fusion fuel is not a gas of neutral atoms, but a fully ionized hot plasma.

Contrary to first intuition, it is not possible to maximize the efficiency of a fusion reactor by injecting a monoenergetic beam of D at the peak of cross-section curve into a stationary plasma of T. Elastic Coulomb collisions between D and both the background electrons and T quickly slow down and Maxwellianize the D before many D–T fusion collisions take place. Thus, a fusion plasma normally contains D, T, and a charge-neutralizing background of electrons, all with Maxwellian distribution functions equilibrated at the same temperature T.

The temperature T is determined as part of the physics and engineering design optimization and normally has the value $T\approx 15$ keV, well below the peak of the cross-section curve. The implication is that most of the fusion reactions taking place in a reactor occur on the tail of the distribution function. Existing magnetic fusion experiments normally operate at several keV, although in special circumstances temperatures as high as 30 keV have been achieved.

SELF-SUSTAINED OPERATION

The basic operating conditions for a fusion reactor are determined by the overall power balance in the system. There must be a sufficiently large number of particles confined for a sufficiently long time at a sufficiently high temperature to produce more fusion energy than is consumed in maintaining the plasma against energy losses. In a magnetic fusion plasma, the dominant energy losses are heat conduction, particle diffusion, and radiation. Usually heat conduction is most severe. If τ is defined as the characteristic time for plasma cooling, then $p_l \equiv 3nT/\tau$ represents the power density loss.

For a magnetically confined plasma, the ^{4}He particle produced by the D–T reaction has a positive electrical charge and thus remains confined in the magnetic field. In a relatively short time, it gives its energy to the plasma through Coulomb collisions. If the ^{4}He energy given to the plasma balances the heat loss due to thermal conduction, the plasma is self-sustaining and is said to be ignited. The power density contained in fusion-produced ^{4}He particles is $p_\alpha=Q_\alpha n^2\overline{\sigma v}/4$, where $n^2\overline{\sigma v}/4$ is the velocity-averaged D–T reaction rate and $Q_\alpha=3.5$ MeV is the ^{4}He energy per reaction. Equating $p_\alpha=p_l$ yields the ignition requirement

$$n\tau = \frac{12T}{Q_\alpha \overline{\sigma v}}. \tag{8}$$

At $T = 15$ keV, $n\tau \approx 3 \times 10^{20}$ m^{-3} s. Existing magnetic fusion experiments in the United States, Europe, and Japan have achieved values of $n\tau \approx 10^{20}$ m^{-3} s, at somewhat lower values of T on the order of several keV.

Assuming ignition is achieved, the reactor produces electricity from the energy in the nonmagnetically confined fusion neutrons (14.1 MeV), the heat generated by breeding T with ^{6}Li (4.8 MeV), and the heat recovered from plasma losses (3.5 MeV if $p_l \approx p_\alpha$). Assuming a thermal conversion efficiency $\eta_T = \frac{1}{3}$, then the net electrical power out of the plant is $P_E = 2.1 p_\alpha V_p$, where V_p is the volume of the plasma. Fusion power plants are expected to be relatively large with $P_E \approx 1000$ MW.

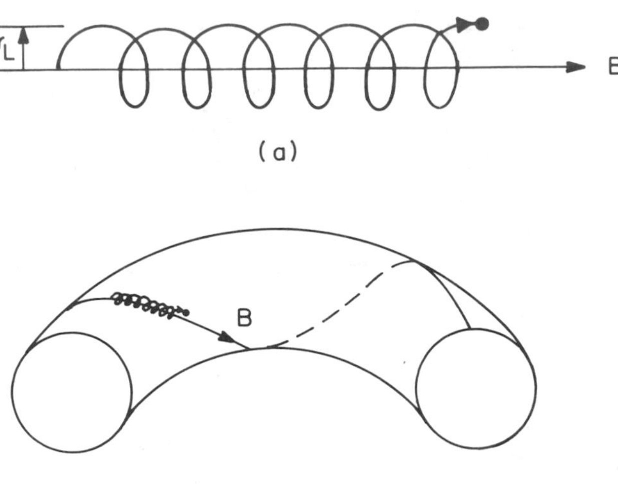

FIG. 2. Magnetic confinement of particles in (a) a linear geometry and (b) a toroidal geometry.

METHODS OF CONFINEMENT

Equation (8) implies that a fusion plasma must have a sufficiently long energy confinement time if significant electrical power is to be produced. There are three basic confinement methods: gravitational, inertial, and magnetic. Gravitational confinement is the mechanism applicable in stars. It is obviously not useful on earth where the gravitational forces are so much weaker.

Inertial confinement has successfully produced uncontrolled thermonuclear fusion in weapons. Here, a sphere of hot thermonuclear fuel resists expansion for an inertial time scale $\tau_i \approx R/v_T$ where R is the radius and v_T is the thermal velocity. If sufficient density is packed into a large enough sphere at a high enough temperature, then the sphere can ignite before disassembling. The condition for ignition is very similar to Eq. (8) and requires $nR \gtrsim 3 \times 10^{26}$ m^{-2} for D–T at $T = 15$ keV.

This same process can be used to produce controlled thermonuclear reactions, thus offering the possibility of the commercial production of electricity. The idea is to dramatically reduce the size of the fueling sphere to that of a small pellet by compressing the density by a factor of $\sim 10^4$. The compression is produced by means of powerful lasers or high-energy particle beams. Since the total mass of fuel scales as $M \sim nR^3$, the condition $nR \approx 3 \times 10^{26}$ m^{-2} at ignition implies that $M \sim n^{-2}$. A compression of 10^4 reduces the mass, and hence the explosive energy release, by a factor $\sim 10^8$ to a level at which it can be contained in a reaction chamber.

The third and most sophisticated method of confinement utilizes magnetic fields to contain the hot plasma and is the primary focus of this discussion. Magnetic confinement is based on the realization that a fusion plasma is a hot, fully ionized gas consisting of freely moving positive and negative electrically charged particles. When such a particle is placed in a strong magnetic field, it executes a helical orbit about a field line (see Fig. 2a). It is confined perpendicular to the field in a circular orbit of Larmor radius $r_L \approx v_\perp/\omega_c$, and cyclotron frequency $\omega_c = eB/m$. Under thermonuclear conditions, $r_L \lesssim 0.2$ cm. Along the field line, the particle moves freely and thus has no parallel confinement.

This suggests that a toroidal magnetic configuration would be an effective container for a fusion plasma. As shown in Fig. 2b the magnetic lines lie on a set of nested toroidal surfaces. The particles are confined within one r_L perpendicular to the surface and are free to move indefinitely along the field lines which never leave the surface. The confinement must be quite good for fusion to succeed, as an ion typically moves 10^6 m along a field during an energy confinement time.

In practice, when a straight magnetic field is bent into a torus, the curvature and gradients in the field modify the particle orbits causing perpendicular drifts and additional complexity. Many years of fusion research have led to an enormous increase in the knowledge of magnetic confinement, and a narrowing down of those configurations most likely to succeed as a fusion reactor. Currently, the three most promising concepts are the tokamak, the stellarator, and the reversed field pinch. The tokamak is the clear leader in terms of performance and investment in the international fusion program.

A TOKAMAK FUSION REACTOR

A schematic diagram of a fusion reactor based on the tokamak confinement configuration is illustrated in Fig. 3. The major components are as follows. The reacting region consists of a hot (15 keV) D–T plasma. It is axisymmetric and typically has a dimension $a \approx 2$ m. The reacting region also contains the magnetically confined ^{4}He particles born from the D–T reactions and whose energy balances the plasma heat conduction losses. The plasma is initially heated to its ignition temperature by a combination of Ohmic heating due to the plasma currents and externally supplied auxiliary power in the form of rf waves or high-energy neutral beams.

Surrounding the plasma is a blanket and shield whose overall dimension is usually $b \approx 1.5$ m. The blanket has several functions. It must contain sufficient ^{6}Li (in either liquid or solid salt form) to breed tritium. It must convert the kinetic energy of the 14.1-MeV fusion neutrons and the 4.8 MeV from ^{6}Li breeding into heat, ultimately used to drive steam turbines. The blanket must also provide some neutron multiplication since the ^{6}Li reaction requires one fusion neutron

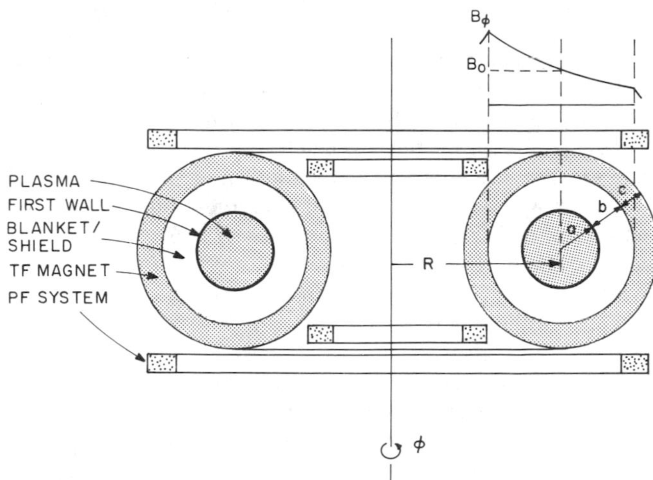

FIG. 3. Schematic diagram of a tokamak reactor.

to produce one T. In practice the breeding will not be perfect and additional neutrons are required to overcome these losses. The primary function of the shield is to protect the magnets and the plant personnel from neutron radiation.

Encompassing the blanket and shield is the toroidal field (TF) magnet that almost certainly will be a superconducting magnet. The joule losses from copper magnets are sufficiently high that they lead to an unfavorable power balance in the plant. Economic considerations require that the magnet operate at as high a field as possible. Current superconducting technology indicates that $B \approx 12$ T is the maximum achievable field in large magnets. This field occurs on the inside leg of the magnet and corresponds to $B_0 \approx 6$ T at the center of the plasma. The magnet design at these high fields is dominated by stress considerations. The TF coils thus require a substantial amount of structural support. Typical designs have an overall magnet thickness of $c \approx 1$ m and a major radius of $R \approx 7$ m. Superconducting magnet technology is an active area of engineering research, and in the future one can anticipate magnets with $B = 15$ T and $B_0 = 7.5$ T. This would lead to considerably improved fusion reactor designs.

The final major component of the reactor is the poloidal field (PF) magnet. The poloidal magnetic field is necessary to hold the plasma in equilibrium. It causes the magnetic field lines to wrap around the torus poloidally as well as toroidally as shown in Fig. 2b. If the poloidal field was zero, then **B** would be purely toroidal. The drifts induced in the particle orbits by the gradients in the toroidal field would cause the plasma to accelerate radially outward along the R axis, leading to plasma–wall contact of several microseconds. The poloidal field prevents this catastrophic motion from occurring.

In equilibrium, the plasma carries a net toroidal current as part of the overall PF field. This dc current can be driven inductively by transformer action, although obviously only for a finite time. For very long pulse or steady-state operation the toroidal current must be driven noninductively. One method is to launch traveling rf waves into the plasma which propagate along the toroidal direction. These waves drag electrons with them by wave–particle interactions thereby producing a net current. Typically $I = 20$ MA is required in a reactor.

A reactor of the type just described produces approximately 1000 MW of electricity. Cost estimates, which are clearly very tentative at this stage, suggest that the cost/watt of a fusion reactor would be comparable to or slightly higher than that of a conventional power plant.

PHYSICS ISSUES

The successful development of a D–T tokamak reactor requires the solution of several quite difficult physics problems. First, there is a maximum pressure that can be stably confined in equilibrium. If this pressure limit is violated, violent magnetohydrodynamic (MHD) instabilities are excited leading to rapid quenching of the plasma, and potentially serious mechanical damage to the first wall. The maximum stable pressure is given by the Troyon limit

$$n(10^{20}\ \mathrm{m}^{-3})T(\mathrm{keV}) \lesssim 0.4\,\frac{I(\mathrm{MA})B_0(\mathrm{I})}{a(\mathrm{m})}\,, \tag{9}$$

where I is the toroidal current and B_0 is the toroidal field on axis.

For a given B_0 and a, the pressure limit cannot be raised by increasing I to large values. Above a critical current, MHD kink instabilities are excited, also leading to a catastrophic termination of the plasma. In an elongated plasma, as would be found in a reactor, the maximum current is given by $I(\mathrm{MA}) \lesssim 5B_0a^2/R$ corresponding to a pressure limit of

$$n(10^{20}\ \mathrm{m}^{-3})T(\mathrm{keV}) \lesssim 2\,\frac{B_0^2(\mathrm{T})a(\mathrm{m})}{R(\mathrm{m})}\,. \tag{10}$$

The macroscopic MHD instabilities just discussed represent important limitations on reaching the high densities required for ignition, $n\tau \gtrsim 3\times10^{20}\ \mathrm{m}^{-3}$. For a reactor, Eq. (10) implies that $n \lesssim 1.4\times10^{20}\ \mathrm{m}^{-3}$.

A second important physics problem is the achievement of large values of the energy confinement time. Heat conduction losses arise because the plasma particles are not perfectly confined by the magnetic field. Coulomb collisions, which occur with a frequency $\nu \ll \omega_c$, cause the particles and energy to slowly diffuse across the magnetic field. These classical losses are in fact quite tolerable. The difficulty is that fusion plasmas almost always exhibit a significant level of density, electric field, and magnetic field fluctuations driven by high-frequency, short-wavelength plasma microinstabilities. This turbulence leads to anomalous transport, often as much as two orders of magnitude greater than that predicted by classical Coulomb collisions. At present there is no universally agreed upon scaling relation for τ. There are over a dozen different empirical and theoretical scaling relations suggested to explain the data in existing tokamaks. Many predict values of τ in the range of 1–3 s for a typical fusion reactor, which in conjunction with the MHD pressure limit, is just sufficient to achieve ignition. The understanding of energy confinement has been, and will continue to be, one of the most challenging physics issues in the magnetic fusion program.

The third important physics problem is the initial heating

of the plasma to $T = 15$ keV (170 million degrees K) in order to ignite. The initial energy is provided by a combination of Ohmic heating and auxiliary power. Ohmic heating is technologically simpler but is usually only effective in achieving values of T of several keV, a consequence of the fact that the plasma resistivity $\eta \sim 1/T^{3/2}$ decreases with temperature. Auxiliary power in the form of rf waves at the ion cyclotron or electron cyclotron frequency may be the most desirable heating method in a reactor. High-energy neutron beams have been successful in achieving temperatures of $T \sim 30$ keV in existing experiments, but may be too technologically complicated in a reactor environment. Even so, existing experience suggests that reactor-compatible heating techniques are available to heat a plasma to ignition. The largest uncertainty is the experimental observation that the plasma energy confinement time degrades as additional auxiliary power is supplied. Thus, the amount of auxiliary power required is strongly coupled to the anomalous transport issue. A reactor may require 50 MW of rf power.

ENGINEERING ISSUES

There are a large number of state-of-the-art engineering problems facing the development of a fusion reactor. Among them are the need for (1) high-field superconducting magnets, (2) a method of burn control to stabilize operation once ignition has been achieved, (3) a proper blanket design (e.g., liquid lithium or a lithium salt) to ensure sufficient and safe tritium breeding, (4) a high-precision equilibrium feedback circuit in the PF system to maintain the plasma in position, and (5) a proper interface between the plasma edge and the external environment to allow fueling and ^{4}He ash removal without impurity contamination.

In addition to these, there are two other issues which are perhaps more fundamental. The first is the question of steady-state operation. From an engineering view a steady-state reactor is highly desirable. Pulsed reactors, even with day-long pulses, would be exposed to serious structural and material damage resulting from thermal cycling. However, achieving steady state in a tokamak is difficult because of the need for a dc toroidal plasma current. As previously discussed, rf current drive can generate a noninductive dc current, but the rf equipment is technologically complicated and expensive, a situation compounded by the fact that the efficiency of generating dc current is much lower than that of just heating the plasma.

The other issue is that of the first wall which is bombarded by 14.1-MeV neutrons, lower-energy neutrons, γ rays, x rays, synchrotron radiation, bremsstrahlung radiation, and plasma ions and electrons. The 14.1-MeV neutron wall loading is probably most severe. The first wall will typically absorb about 10% of the total energy produced in the reactor. Clearly a significant amount of materials research is required to develop materials capable of withstanding this abuse. The situation is further complicated by the fact that the first wall is located deep within the reactor. Thus, replacement will involve a major shutdown of the plant for an extended period of time.

The last two issues discussed represent two of the most difficult engineering problems facing the tokamak concept.

OVERALL ASSESSMENT OF TOKAMAK FUSION

The tokamak concept is the leading contender to produce a fusion reactor based on the D–T reaction, a position achieved by virtue of its superior achievements in terms of $n\tau$ and T. Even so, there remain very challenging physics and engineering problems before success can be claimed.

Assuming these problems are overcome, one can ask for a description of the general properties of such a reactor. Environmentally it would be very attractive. A fusion reactor would make no contribution to acid rain, the greenhouse effect, or other air pollutants generated by fossil fuel power plants. Although the reactor uses tritium and produces activated material because of neutron bombardment, the problems associated with radioactivity are smaller by orders of magnitude than those in a fission plant. Furthermore, it is physically impossible for the reactor to have a thermal runaway induced meltdown.

In terms of fuel, the known reserves of lithium imply that D–T reactors can provide energy for hundreds of years into the future. If D–D reactors can be developed, there is sufficient deuterium in the ocean for billions of years at the current rate of usage.

Economically one can see intuitively that the cost of fusion power will not be low. The need for a large superconducting magnet system and a thick blanket constructed in a relatively complicated toroidal geometry suggests that the capital cost of the nuclear island in a fusion reactor will be at least as great as, if not greater than, that of a fission reactor.

Perhaps the most difficult and least tangible problem to assess is the complexity of a fusion power plant. A fusion reactor will indeed be complex. In addition to the already sophisticated power controls present in any plant, a fusion reactor will have superconducting TF magnets, a complicated PF system with feedback control, and a large supply of high-technology rf equipment and a tritium handling system. Furthermore, a number of physics and engineering problems require solutions at the state-of-the-art of our capabilities. Whereas there is a high probability of success on any given problem, the satisfactory simultaneous synthesis of many new state-of-the-art solutions is by no means guaranteed. Too much complexity may lead to low plant availability.

Although the list of problems facing fusion is large, one should keep in perspective that the progress in fusion research since the inception of the program has been truly remarkable: gains in $n\tau$ and T by factors of 10^4! Current experiments are factors of only 2–3 away in terms of the $n\tau$ and T required in a reactor. The intellectual challenges and potential environmental advantages of fusion will certainly keep scientists and engineers motivated for the next decade and beyond. Whether an attractive fusion reactor can be built, however, remains a question for the future.

BIBLIOGRAPHY

F. F. Chen, *Plasma Physics and Controlled Fusion*. Plenum Press, New York, 1984.

D. R. Cohn and L. Bromberg, "Advantages of High Field Tokamaks for Fusion Reactor Development," *J. Fusion Energy* **5**, 161 (1986).

R. W. Conn, "The Engineering of Magnetic Fusion Reactors," *Sci. Am.* **249,** 60 (1983).
F. J. Dolan, *Fusion Research.* Pergamon Press, New York, 1980.
S. Glasstone and R. H. Lovberg, *Controlled Thermonuclear Reactions.* Robert E. Krieger, Huntington, NY, 1975.
L. Lidsky, "The Trouble with Fusion," *Technol. Rev.* **86,** 32 (1983).
D. J. Rose and M. Clarke, Jr., *Plasma Physics and Controlled Fusion.* MIT Press, Cambridge, MA, 1965.

Plasma Physics and Nuclear Fusion Journals

Fusion Technology, American Nuclear Society, LaGrange, IL.
Journal of Fusion Energy, Plenum Press, New York, NY.
Journal of Plasma Physics, Cambridge University Press, New York, NY.
Nuclear Fusion, IAEA, Vienna, Austria.
Physics of Fluids, American Institute of Physics, New York, NY.
Plasma Physics and Controlled Fusion, Pergamon Press, New York, NY.

Nuclear Magnetic Resonance

R. E. Walstedt

Nuclear magnetic resonance (NMR) is the effect of a resonant rotating (or alternating) magnetic field, imposed at right angles to a typically much larger static field, to perturb the orientation of nuclear magnetic moments. First applied to molecular beam studies, the term has come to refer specifically to the study of nuclear magnetism in bulk matter, with the distinct feature of detection by purely electromagnetic methods. The first such observations were made in 1946, for which F. Bloch and E. M. Purcell received the Nobel Prize in 1952. Since that time, the method has developed extensively and become an important investigative technique in many areas of science, including condensed matter physics, materials science, chemistry, geology, biology, and medicine.

The majority of stable nuclei possess a quantized angular-momentum $I\hbar$, where the *nuclear spin quantum number* I is an integer multiple of $\frac{1}{2}$ and $h=\hbar/2\pi$, h being Planck's constant. The associated coaxial magnetic moment μ is given by $\mu=\gamma\hbar I$, which defines the *nuclear gyromagnetic ratio* γ. Nuclei for which $I=0$ have no magnetic moment and are thus magnetically inert. If it were only for orbital motion of the protons within the nucleus, γ would always have the value $\gamma^*=e/2Mc$, with e and M denoting the charge and mass, respectively, and c being the velocity of light. Although it leads for most nuclei to the observed order of magnitude of μ, this value is based on far too simple a model, since both protons and neutrons have spin$=\frac{1}{2}$ and an associated magnetic moment which must also be taken into account. Actually, γ is found to have a different characteristic value for each nuclear species, varying from the largest value, $\gamma/2\pi=42.5774$ kHz/mT (kHz = kilohertz, mT = millitesla) for the proton to the smallest, $\gamma/2\pi=0.72919$ kHz/mT for ^{197}Au.

Nuclear moments are fundamentally quantum objects. While classical treatment of many aspects of nuclear magnetism is possible, quantitative calculations usually require full quantum-mechanical theory. A simple example is nuclear paramagnetism. In bulk matter this is entirely analogous to the more familiar electronic paramagnetism, but is far weaker owing to the smallness of nuclear moments. The nuclear susceptibility is given under all but the most unusual circumstances by a simple Curie law, where the degree of nuclear polarization p can be written $p\approx\gamma\hbar(I+1)B_0/3k_BT$ ($p\ll1$). B_0 is the field at the nucleus and k_B is Boltzmann's constant. For example, in a typically large laboratory magnetic field $B_0=10$ T obtained with a superconducting magnet, the preceding formula gives for protons at room temperature $p=0.000\,034$. Except at ultralow temperatures, nuclear paramagnetism makes only a negligible contribution to bulk magnetic susceptibilities.

The observation of such minuscule magnetic polarizations is rendered possible by the use of radiofrequency excitation and resonance detection techniques. The essence of this methodology is that the nuclear moments $\boldsymbol{\mu}_i$ in the specimen under observation are rotated so as to lie at some angle with respect to the applied field $\mathbf{B}_0$. Under these conditions the moments experience a torque $\boldsymbol{\mu}_i\times\mathbf{B}_0$, which causes the angular momentum (i.e., $\boldsymbol{\mu}_i$ itself) to precess around $\mathbf{B}_0$ in precise analogy with the motion of a gyroscope in a gravitational field. The nuclei precess at the *Larmor precession frequency* $\omega_L=\gamma B_0$, where γ is the previously defined gyromagnetic ratio. For strong fields, ω_L is in the radio-frequency range. For example, in a field $B_0=1$ T ($=10^4$ Gauss), protons precess at a frequency $f_L=\omega_L/2\pi=42.5774$ MHz. However established, coherently precessing nuclear moments can be caused to induce a voltage at frequency f_L in a surrounding coil of wire, which can then be amplified by standard radiofrequency electronic techniques and recorded.

Many techniques have been developed for the observation of NMR signals. These are divided into two principal classes, *pulsed* and *continuous wave* (cw) methods. In recent years the pulsed methods have gained wider acceptance because of their greater flexibility and efficiency. In pulsed NMR the signal following a single excitation pulse is known as a *free-induction decay*. The excitation of free induction signals is accomplished by means of an applied rotating (or alternating) magnetic field of a frequency at or near the Larmor frequency of the nuclei to be studied, directed along an axis perpendicular to $\mathbf{B}_0$. An alternating field can be decomposed into two counterrotating components, one of which is very far from resonance and, hence, negligible. We may therefore envision the effect of these fields by picturing the motion from the viewpoint of a reference frame rotating in step with the resonant field component B_1, which is then static. If B_1 is applied at exact resonance, then it can be shown that the nuclei precess about $\mathbf{B}_1$ *as though the static field were absent.* Again, borrowing from the terminology of gyroscopic motion, the rotation of nuclei about the field $\mathbf{B}_1$ is generally referred to as *nutation*. If B_1 is applied long enough to rotate the nuclear moments into a right angle with $\mathbf{B}_0$ and then turned off, this is referred to as a $\pi/2$ pulse. A pulse twice as long, i.e., a π pulse, will invert the nuclei to the direction $-\mathbf{B}_0$. The maximum free induction signal is achieved with a $\pi/2$ pulse.

The excitation pulses described above are only useful if they can be made short compared with certain nuclear precession damping effects, known as *relaxation processes*. This is possible in a large number of cases of interest. Thus, nuclei in solids and liquids are typically very strongly decoupled from their surroundings, making them useful probes of local magnetic behavior. There are two principal types of relaxation process affecting NMR measurements. The first is longitudinal or spin-lattice relaxation, which is typified by the recovery of the nuclear polarization following the application of an excitation pulse. The time constant for the polarization to approach its equilibrium value is known as T_1. In such a T_1 process, the nuclear spins exchange energy quanta of magnitude $\hbar\omega_L$ with the "lattice" or environment in which they are immersed. Depending on these circumstances, values of T_1 range from the submicrosecond scale up to times of hours and beyond. Transverse relaxation is the term used to describe the decay in time of a free induction signal following pulsed excitation; it is characterized by a time constant T_2^*. There are several contributions to transverse decay: dipolar interactions among the nuclear moments; spin-lattice relaxation processes as described above; static distributions of Larmor frequencies of whatever origin; and in cases where the nuclear spin quantum number $I>\frac{1}{2}$, nuclear electric quadrupole interactions with electric field gradients generated by ionic and electronic charges in their local environment. The latter are almost always static in nature. The various contributions to transverse relaxation are also known as sources of *line broadening*, since they give rise to a finite width for the NMR line given roughly by $\Delta\omega \sim 1/T_2^*$. (The width in field units is $\Delta H = \Delta\omega/\gamma$.)

Whenever the static line-broadening effects mentioned above are significant, a case encountered frequently in practice, it is then possible to excite a *spin echo*. This is accomplished by applying a second pulse following the (typically) $\pi/2$ initial excitation pulse, for example in a $\pi/2-\pi$ sequence, with a time interval τ separating the pulses. The second pulse has the effect of reversing the static dephasing that occurs after the first pulse, so that at time τ following the second pulse the macroscopic transverse nuclear moment which existed just after the first pulse is partially re-formed. There occurs then a *spin echo signal*, which has a time profile resembling two free-induction decay shapes placed back to back. Spin echoes also decay in amplitude as the time τ between pulses is increased, but only via the dynamic relaxation mechanisms mentioned earlier. The echo decay time constant is usually termed T_2. One frequently encounters cases where $T_2 \gg T_2^*$ in NMR studies of both solids and liquids, as well as in biological systems. In these cases the spin-echo technique is useful in revealing the dynamic decay processes. In many instances the static broadening effects are large, so that T_2^* is very short. The free induction decay may then be unobservable, so that the spin-echo method becomes critical to the success of the experiment.

There are also cases where the dynamic relaxation processes are strong, such as solids with large, fluctuating atomic magnetism. Then one may also find that T_2 is unobservably short (e.g., $T_2 < 1\ \mu s$) and pulsed NMR methods are simply inapplicable. For these cases it is still possible to use continuous irradiation techniques, where the nuclear moments are driven with a small rotating field near their resonance frequency. The driving field produces, near resonance, a small transverse component of nuclear magnetization, which is then observed by the usual inductive pickup method. The "absorption" signal recorded in such an experiment is maximum when the nuclei are driven at their resonance frequency (e.g., the Larmor frequency ω_L), and this maximum is inversely proportional to the linewidth. When the NMR line is narrow and the signal strong, the cw technique has the advantage of being simple to implement. It is popular, therefore, in applications such as field monitors using protons in liquids, etc.

The applications of NMR technique to the various fields of modern science are far too numerous to review here in any detail. We content ourselves with a few illustrative examples. It must also be borne in mind that the elements of technique described in the foregoing paragraphs are rudimentary, and that numerous elaborations upon these methods and the ones mentioned below have been and continue to be developed. As with nearly all fields of experimental study, NMR technique has undergone continuous improvement through advances in digital electronics technology as well as in the computer field. We mention here one development which has been pivotal in the improvement of NMR data. This is the principle of signal-to-noise ratio improvement through repetitive averaging or accumulation of signal waveforms. Implementation of this simple scheme has depended entirely on the availability of digital waveform storage technology and has undergone continuing improvement since the technique was first introduced. This method for improving NMR signal quality underlies many of the applications described below.

One of the earliest applications of NMR was to the monitoring and control of the strength of magnetic fields. This idea has been further extended to studies of geomagnetism through the development of the earth's field magnetometer. The earth's magnetic field ($\sim$0.5 G) gives protons a Larmor frequency of $\sim$2 kHz. Even at such a low frequency the earth's field can be measured to an accuracy of many decimal places with suitably designed NMR apparatus, giving scientists a powerful tool to study detailed behavior of the earth's magnetism.

NMR has had very wide application to the study of the physics and chemistry of solids, including metals, semiconductors, magnetic solids, and organic materials. In this context the nuclei act as microscopic probes which sense local magnetic polarization through NMR frequency shifts (such as the Knight shift in metals, which arises through the Pauli paramagnetism and the contact hyperfine interaction of *s*-band conduction electrons), as well as low-frequency magnetic fluctuations through the spin-lattice relaxation times T_1. A partial list of physical phenomena addressed includes static and dynamic conduction–electron paramagnetism in simple as well as *d*-band metals; spin waves and magnetic fluctuations in ordered magnetic materials; the metal–insulator transition; diffusion in intercalates and superionic conductors; static and dynamic charge density wave phenomena; spin freezing in spin glasses; and lastly a highly diverse array of frequency shift and spin-lattice relaxation effects in heavy-fermion systems and conventional and high-temperature superconductors. For studies at ultralow temperatures, NMR has provided a means of thermometry (using the Curie law for polarization discussed earlier) and an incisive probe of the superfluid phases of ^{3}He.

A very different set of problems is encountered in NMR

studies of solids of interest to the fields of organic chemistry and materials science, such as polymers, amorphous systems, complex molecular solids, etc. Such systems are usually nonmetallic and nonmagnetic. For nuclei having $I=\frac{1}{2}$, the width of the associated NMR lines is dominated by dipolar fields coming from neighboring nuclear moments. For nuclei with $I>\frac{1}{2}$, there will in general be quadrupolar broadening as well, as was mentioned earlier. Solids such as this exhibit complex NMR spectra, however, owing to the presence of small (~1–100 ppm) shifts in nuclear resonance frequencies known as *chemical shifts*. These shifts arise from second-order orbital magnetic effects, are almost always positive, and are unique to each chemical site in the material in which the nuclear species investigated may be found. In many instances the chemical shift spectra are masked by the above-mentioned broadening effects. In an effort to improve the spectral resolution in such cases, a class of techniques has emerged based on the principle known as *magic angle spinning*. The basic idea here is that the dipolar interaction between neighboring nuclei, which is responsible for broadening in high-field NMR, is proportional to $1-3\cos^3\theta$, where θ is the angle between the static field $\mathbf{B}_0$ and the bond axis $\mathbf{r}_{ij}$ joining interacting nuclear moments. Thus, any pair of spins for which $\mathbf{r}_{ij}$ lies at $\theta \sim 54.7°$ will have a vanishing dipolar interaction. By mechanical spinning of the sample about an axis oriented at this "magic angle" relative to $\mathbf{B}_0$, the nonvanishing components of dipolar field will have zero effect on a time average. The spinning frequency has to be greater than the dipolar interaction frequency (typically ~ several kHz) in order to be effective. Using spinning frequencies as high as 20 kHz, significant improvement in resolution has been achieved in many cases.

A similar class of problem is that of complex molecules to be studied in a liquid environment. Owing to the tumbling motion which molecules undergo in liquids, dipolar and quadrupolar couplings average to zero and cease to contribute to the broadening of NMR lines. The resulting NMR spectra will be very sharp, but can also be exceedingly complex, depending on the nature of the system under investigation. The technique almost universally employed in such cases is *Fourier transform NMR spectroscopy*. Rather than scan over the line as is done in continuous irradiation methods, a $\pi/2$ pulse of sufficient intensity is applied to the system so that all components of the NMR line for a single nuclear species are uniformly excited. The free induction signal which follows will have a highly complex time dependence as all the frequency components undergo constructive and destructive interference with each other. It is straightforward to show, however, that if this complex decay waveform can be digitally recorded and subsequently Fourier transformed, a faithful representation of the full spectrum will result. The best resolution is obtained by working at very high fields and by trimming the field to be homogeneous over the sample region to a very high order.

A further improvement in technique devoted to the resolution of complex NMR spectra is the method of *two-dimensional NMR*. This method is a simple extension of the Fourier transform scheme to include a sequence of two $\pi/2$ pulses separated by a variable time t_1. Time t_1, the "evolution" time, is varied from zero up to values slightly greater than the free induction lifetime. After the second pulse, the signal is recorded over a time interval t_2 with a similar range as t_1. The data are then Fourier transformed over both time axes t_1 and t_2 to generate frequency scales f_1 and f_2, and a two-dimensional contour diagram of signal intensity against frequencies f_1 and f_2 is plotted. Depending on additional treatment with irradiation, magnetic field gradients, etc., during t_1 and/or t_2, a variety of effects can be achieved. A simple example will illustrate the possibilities. Consider a system containing two nuclear species which are coupled by an interaction which splits the NMR spectrum of either one into two or more components. Suppose further that there are several inequivalent sites for the species under observation. One can then apply, during time t_2, a "decoupling" radio-frequency field to the species *not* observed, which renders its nuclear–nuclear coupling ineffective. The two-dimensional spectrum will then be broken up into multiplets of intensity, such that the projection on the f_2 axis will show the chemical shift spectrum *without the internuclear splittings*. The projection on the f_1 axis will be the normal spectrum. Many variations on this basic scheme have been developed.

An application of NMR technique which has enormous potential for the fields of biology and medicine is that of *NMR imaging of specimens having spatial structure*. There are many imaging schemes which have been worked out using NMR. The method has been developed to the point where cross-sectional images of the human body can be generated in just a few minutes with enough resolution to be of diagnostic value in medicine. One method of imaging can be described in terms of the pulse sequence employed for two-dimensional NMR which was discussed in the previous paragraph. A three-dimensional object to be imaged is placed in a magnetic field with a field gradient along one axis, e.g, the z axis. A pulsed radio-frequency magnetic field is then applied, exciting the nuclear spins in a planar cross section of small but finite thickness oriented perpendicular to the z axis. This is the first pulse of the two-phase sequence described above. During time t_1 a field gradient is switched on along the x axis, and after the second pulse, this gradient is switched to lie along the y axis. Barring excessive intrinsic structure in the spectrum of the species excited, the effect of these gradients will be to "encode" the spatial variation of NMR intensity into a spectrum of resonance frequencies, along one planar axis and then the other. On performing the Fourier transforms, this encoding will then result in a two-dimensional image of the spatially distributed NMR intensities. The method is not limited to simple intensity, but can be employed to image relaxation times, diffusion rates, etc., depending on the flexibility of the apparatus. The great advantage of NMR imaging in the biological field is that it is noninvasive, in contrast with, e.g., computerized tomography, which uses potentially hazardous x rays and gives comparatively poor resolution of soft tissue. In smaller-scale applications, spatial resolution of features on the scale of a few microns have been achieved with NMR imaging. The ultimate resolution appears to be limited only by the digital memory size and the time spent to acquire the data.

See also FINE AND HYPERFINE SPECTRA AND

INTERACTIONS; GEOMAGNETISM; NUCLEAR MOMENTS; NUCLEAR QUADRUPOLE RESONANCE; RESONANCE PHENOMENA.

BIBLIOGRAPHY

A. Abragam, *The Principles of Nuclear Magnetism*. Oxford (Clarendon Press), London, 1961.

F. A. Bovey, *High Resolution NMR of Macromolecules*. Academic Press, New York and London, 1972.

A. Carrington and A. D. McLachlan, *Introduction to Magnetic Resonance*. Harper & Row, New York and London, 1967.

T. C. Farrar and E. D. Becker, *Pulse and Fourier Transform NMR*, Academic Press, New York and London, 1971.

L. M. Jackman and F. A. Cotton, *Dynamic Nuclear Magnetic Resonance Spectroscopy*. Academic Press, New York, 1975.

P. Mansfield, *NMR Imaging in Biomedicine*. Academic Press, New York, 1982.

J. Schraml and J. M. Bellama, *Two-Dimensional NMR Spectroscopy*. John Wiley and Sons, New York, 1988.

C. P. Slichter, *Principles of Magnetic Resonance*. Harper & Row, New York, 1963.

K. Wüthrich, *NMR in Biological Research*. North Holland/American Elsevier, Amsterdam and New York, 1976.

Nuclear Moments

O. Häusser

The interaction of the nuclear system with electromagnetic fields can be conveniently discussed in terms of various nuclear moments. These moments specify the time-averaged distributions of electric charges and currents in the nucleus and provide sensitive tests of nuclear structure theories.

The basic mechanical moment of the nucleus is the *total angular momentum* (or spin) **I**, expressed in units of Planck's constant $\hbar$. It is the result of coupling together the orbital angular momenta **l** and the intrinsic spins **s** of the individual constituents. The total spin may assume $2I+1$ projections, with values $m=I, I-1, \ldots, -I$, along any chosen direction in space.

The lowest-order magnetic moment is the *magnetic monopole*, whose possible properties have been discussed by Dirac. No conclusive experimental evidence for the existence of magnetic monopoles has yet been found.

All nuclei with total spin $I \geq \frac{1}{2}$ possess a *nuclear magnetic dipole moment* $\boldsymbol{\mu}$, usually expressed in units of nuclear magnetons, $\mu_n = e\hbar/2M_pc$, where M_p is the mass of the proton. The interaction energy of $\boldsymbol{\mu}$ with a static magnetic field **B** is

$$W_{\text{mag}} = -\boldsymbol{\mu}\cdot\mathbf{B} = -\frac{\mu}{I\hbar}\mathbf{I}\cdot\mathbf{B} = -\frac{g\mu_n}{\hbar}\mathbf{I}\cdot\mathbf{B}$$

where $g=(\mu/\mu_n)/I$ is the gyromagnetic ratio or g factor. The energy eigenvalues for the $2I+1$ orientations of the total spin are $W_{\text{mag}}(m) = -mg\mu_n|B| = -m\omega_L\hbar$, as illustrated in Fig. 1a. Nuclei in each magnetic substate precess about **B** with a unique Larmor precession frequency ω_L.

The measured g factors associated with the intrinsic spins of proton and neutron, g_s(proton) = 5.5856 and g_s(neutron) = −3.8263, deviate strongly from those for Dirac particles ($g=2$ and 0, respectively) and indicate the composite nature of nucleons. The g factors for orbital motion, g_l(proton)~1 and g_l(neutron)~0, differ strongly from the g_s values. Nuclear g factors are thus helpful in determining how **l** and **s** of the individual nucleons couple to the total spin **I**. Nuclear g factors do not appreciably exceed those for the free nucleons, because the individual contributions in nuclei tend to cancel each other as a result of the Pauli exclusion principle. Another property of magnetic moments is their sensitivity to localized currents that originate from an exchange of mesons between nucleons, especially of π mesons.

The lowest-order electric moment is the *nuclear charge*, which is quantized in units of e. A finite *electric dipole moment* $\boldsymbol{\mu}_e$ can only exist if the nuclear system is not invariant with respect to spatial inversion (parity transformation) or to a combination of spatial inversion and charge conjugation (replacing particles by their antiparticles). Present experimental limits on the electric dipole moment of the neutron, $|\mu_e/e| \lesssim 10^{-25}$ cm, are beginning to test extensions of the standard model of quarks and leptons and their modified interactions. Significant deviations of nuclear charge distributions from sphericity are described to lowest order by the *nuclear electric quadrupole moment*, $Q=e^{-1}\int r^2(3\times\cos^2\theta-1)\rho d\tau$ where θ is the azimuthal angle of the charge element $\rho d\tau$ with respect to a direction for which the projection of the total spin has its maximum value, $m=I$. Q is nonzero for all nuclei with $I \geq 1$. The interaction energy of Q with an axially symmetric electric field gradient eq ($eq=\partial E_z/\partial z$, and $\partial E_x/\partial x = \partial E_y/\partial y = -eq/2$) is $W_Q = \hbar\omega_Q[3m^2 - I(I+1)]$, where $\omega_Q = e^2qQ/4I(2I-1)\hbar$. The energy splitting is quadratic in m (see Fig. 1b), with the lowest precession frequency being $\omega_0 = 3\omega_Q$ for integer, and $\omega_0 = 6\omega_Q$ for half-integer, values of I.

Electric quadrupole moments can be useful in identifying important components of internucleon forces. For example, the nonzero quadrupole moment of the deuteron can be ascribed to a spin-dependent force (tensor force) acting between proton and neutron in addition to a strong central force. The sign of Q is negative for nuclei consisting of a spherical closed shell plus one particle, and positive for nuclei with nearly closed shells. An outside particle "polarizes" the core, thus enhancing the particle contribution to Q. This effect is caused by the long-range component of the nuclear force, which is also responsible for the sizable collective deformations ($Q\sim$ several times 10^{-24} cm^2) in nuclei between closed shells, especially in rare earth and actinide nuclei.

Powerful methods have been developed for the *measurement of nuclear moments*. The experiments determine either the energy splittings ΔW or spin precession frequencies ω in known electromagnetic fields. The fields have to be sufficiently large that the ΔW exceed considerably the natural linewidth of the nuclear state; thus with currently available fields only states with mean lives longer than $\sim 10^{-12}$ s can be studied. Sources of magnetic fields are atomic electrons ($\lesssim$50 kT = 500 MG), internal fields in ferromagnetics ($\lesssim$800 T), and externally applied fields ($\lesssim$15 T). Electric field gradients of sufficient magnitude are provided by atomic electrons, by solids with a noncubic crystal structure, and by

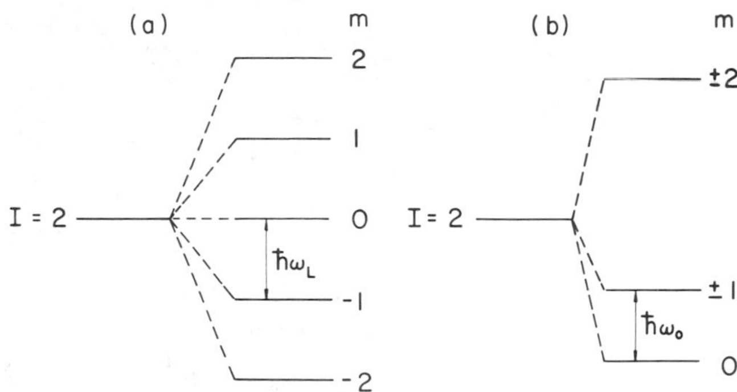

FIG. 1. Energy splitting of the substates of a nuclear state with total spin 2 (a) in a magnetic field, and (b) in an axially symmetric electric field gradient.

high-velocity heavy-ion beams (Coulomb reorientation). The ΔW can be deduced from the hyperfine structure of atomic transitions (optical spectroscopy) or of low-energy (<150 keV) nuclear gamma transitions (Mössbauer spectroscopy). They can be measured accurately by resonant absorption of radio-frequency radiation, as in NMR, NQR, atomic beam resonance, optical pumping, and low-temperature nuclear orientation. The accuracy of experimental methods is sufficiently high that in some cases the atomic hyperfine splittings resulting from higher-order nuclear moments such as *magnetic octupole moments* (nonzero for spins $I \geq \frac{3}{2}$) or *electric hexadecapole moments* (nonzero for $I \geq 2$) have been determined. For short-lived nuclear states the precession of the total spin **I** can often be observed directly by measuring the angular distribution of emitted radiation (particles or β or α rays) as a function of time.

Applications of the experimental methods to other branches of science arise from the high accuracy (10^{-4}–10^{-8}) that can be attained, especially by nuclear resonance methods. Nuclear moments may serve as sensitive probes to detect small changes in the nuclear environment in molecules, liquids, and solids. The information on electromagnetic fields at specific nuclei has applications in solid-state physics, chemistry, biology, and medicine.

See also BEAMS, ATOMIC AND MOLECULAR; FINE AND HYPERFINE SPECTRA AND INTERACTIONS; GYROMAGNETIC RATIO; MAGNETIC MOMENTS; MAGNETIC MONOPOLES; MÖSSBAUER EFFECT; NUCLEAR MAGNETIC RESONANCE; NUCLEAR QUADRUPOLE RESONANCE.

BIBLIOGRAPHY

A. Abragam, *The Principles of Nuclear Magnetism,* 1961, Oxford, London and New York.

P. A. M. Dirac, "The Theory of Magnetic Poles," *Phys. Rev.* **74,** 817 (1948).

G. H. Fuller, "Nuclear Spins and Moments," *J. Phys. Chem. Reference Data* **5,** 835 (1976).

W. Happer, "Optical Pumping," (1972) *Rev. Mod. Phys.* **44,** 169.

O. Häusser, "Coulomb Reorientation," in *Nuclear Spectroscopy and Reactions,* Part C (J. Cerny, ed.). Academic, New York, 1974.

O. Häusser and I. S. Towner, "Hyperfine Interaction Studies in Nuclear Physics," in *Hyperfine Interactions of Radioactive Nuclei* (J. Christiansen, ed.). Springer-Verlag, Berlin, 1983.

H. Kopfermann, *Nuclear Moments.* Academic, New York, 1958.

N. F. Ramsey, *Molecular Beams.* Oxford, London and New York, 1955.

N. F. Ramsey, "Electric Dipole Moments of Particles," *Ann. Rev. Nucl. Part. Sci.* **32,** 211 (1982).

E. Recknagel, "Magnetic Moments of Excited States," in *Nuclear Spectroscopy and Reactions,* Part C (J. Cerny, ed.). Academic, New York, 1974.

R. M. Steffen and K. Alder, "Extranuclear Perturbations of Angular Distributions and Correlations," in *The Electromagnetic Interaction in Nuclear Spectroscopy* (W. D. Hamilton, ed.). North-Holland, Amsterdam, 1975.

C. P. Slichter, *Principles of Magnetic Resonance,* Harper & Row, New York, 1963.

Nuclear Polarization

H. O. Meyer and W. Haeberli

Nuclear polarization or nuclear orientation refers to the spatial distribution of the intrinsic spins of nuclear particles in an assembly containing a large number of particles of the same species. Typical assemblies under consideration might be the protons emerging from an accelerator in a given time or the ^{3}He nuclei contained in a certain volume of ^{3}He gas. If the spin axes ("spins") of the particles in the assembly are oriented randomly, the assembly is *unpolarized,* whereas an assembly of particles whose spins have a preferred orientation is called *polarized* or *oriented.*

DESCRIPTION OF POLARIZATION

The intrinsic spin of nuclear particles is described by a spin quantum number, s, which is either half-integer ($\frac{1}{2}$, $\frac{3}{2}$, etc.) or integer (0, 1, 2, etc.). Examples of $s = \frac{1}{2}$ particles are the electron, muon, proton, neutron, triton, and the ^{3}He nucleus. Nuclei with an even number of protons and an even number of neutrons, such as the alpha particle, have spin $s = 0$. The deuteron and, e.g., the ^{6}Li nucleus, have spin $s = 1$. According to the rules of quantum mechanics, if a measurement is made of the component m_s of the spin along an arbitrarily chosen direction ("quantization axis"), only two answers will be found in the case of $s = \frac{1}{2}$ particles, namely, $m_s = +\frac{1}{2}$ (spin along the quantization axis, or spin "up") and $m_s = -\frac{1}{2}$ (spin "down"). There are three possible answers for spin-1 particles namely, $m_s = +1$, 0, or -1. In general, m_s has $2s + 1$ possible values, ranging from $m_s = -s$ to $m_s = +s$. This quantization into so-called magnetic substates is vividly demonstrated in the Stern–Gerlach experiment.

The quantitative description of polarization is simplest for particles of spin $\frac{1}{2}$, in which case the component of polarization in a given direction **x** is defined as $P_x = f_+ - f_-$, where f_+ is the fraction of all particles whose spin is along **x** and f_- is the fraction of particles whose spin is opposite to **x**. For an assembly completely polarized along the **x** direction, $P_x = 1$ or $P_x = -1$. In general, the polarization of an $s = \frac{1}{2}$ assembly is specified by the three Cartesian components P_x, P_y, P_z of the *polarization vector* **P**. The magnitude P of this vector measures the degree of polarization of the assembly. Sometimes the polarization is expressed as a percentage, e.g., $P = 30\%$ means $P = 0.3$.

For an unpolarized assembly of spin-1 particles, the magnetic substates $m_s = 1, 0$, and -1 are equally populated, i.e., $f_1 = f_0 = f_{-1} = \frac{1}{3}$, regardless of which direction the quantization axis is chosen. If the population of the $m_s = 1$ state is enriched at the expense of the $m_s = -1$ state, the system is said to have *vector* polarization; its value is defined analogous to $s = \frac{1}{2}$ polarization. If the number of particles in the $m_s = 0$ state is enriched or depleted, the assembly acquires *tensor* polarization. The system can at the same time be vector and tensor polarized. In general, five independent components of a second-rank tensor are required to describe the polarization of an assembly of spin-1 particles, in addition to the three vector components. When the spins in the $m_s = 0$ state are uniformly distributed in a plane perpendicular to the quantization axis, i.e., when the system has *axial symmetry*, the assembly is called *aligned* and the degree of alignment is measured by $1 - 3f_0$. In general, the description of assemblies with arbitrary spin s requires tensors of up to rank $2s$.

POLARIZATION OBSERVABLES

Nuclear reactions are studied by bombarding the nuclei in a target with a beam of particles and observing the rate at which reaction products impinge on a detector placed at a specific angle θ with respect to the beam direction, as shown in Fig. 1. The probability for a reaction to occur (per incident particle, per number of atoms in a unit area of the target, and per unit solid angle of the detector) is measured by a quantity called the reaction cross section. If an unpolarized target is bombarded by a polarized beam, the cross section, in general, depends on the polarization of the beam. The change in cross section that arises when the initially unpolarized incident beam is changed to a state of specified polarization, divided by the cross section observed with the unpolarized beam, is a dimensionless observable called the *analyzing power*. If in the arrangement shown in Fig. 1 the incident beam is vector polarized, the number of particles observed at the same angle θ to the left (n_L) or to the right (n_R) will be different. The left–right asymmetry, defined as $(n_L - n_R)/(n_L + n_R)$, equals the product of two factors: the component of the vector polarization **P** perpendicular to the reaction plane (see Fig. 1) and the *vector analyzing power*, A, of the reaction process. More complicated changes in the reaction cross section result from tensor polarization and lead to the measurement of *tensor analyzing powers*. If, instead of the beam, the target is polarized, the statements just made apply equally well.

Even when an unpolarized target is bombarded with an

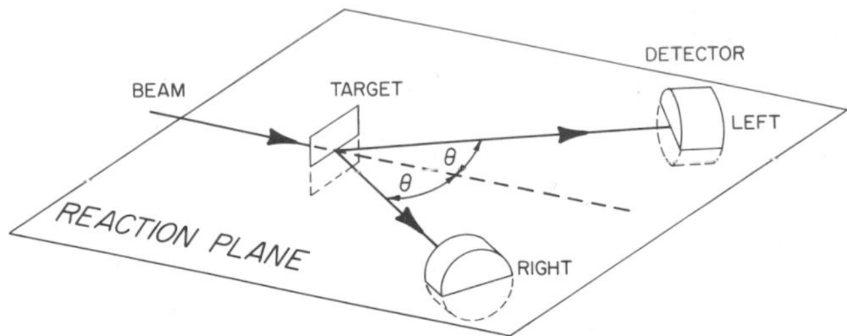

FIG. 1. Symmetric detector arrangement used to observe spin effects in a nuclear reaction or scattering. The reaction plane is a plane that contains both the incoming and outgoing beam directions.

unpolarized beam, the outgoing reaction products are usually polarized (unless, of course, they have spin 0). As an example, consider the elastic scattering of $s = \frac{1}{2}$ particles observed by symmetrically placed detectors as shown in Fig. 1. The incident unpolarized beam contains equal fractions of spin-up and spin-down particles. If we assume that the analyzing power of the target is positive, spin-up particles are scattered preferentially to the left whereas spin-down particles are scattered preferentially to the right. This leads to unequal populations of the substates of the particles arriving at the detector. The polarization P of the scattered beam can be observed by replacing the detector by a *polarimeter,* consisting of a target of known analyzing power followed by an appropriate arrangement of detectors. It follows from general symmetry principles (time-reversal invariance) that the polarization in elastic scattering equals the analyzing power ($P = A$ theorem). In reactions that conserve parity the vector polarization of the outgoing particles is perpendicular to the reaction plane. If the reaction products have spin larger than $\frac{1}{2}$, they may possess *tensor polarization* as well as *vector polarization.*

The observables that can be measured when both beam *and* target are polarized are called *spin-correlation parameters. Polarization-transfer coefficients,* on the other hand, describe how the polarization of the outgoing particles changes in response to a change in the beam polarization (with an unpolarized target) or to a change in the target polarization (with an unpolarized beam). *Spin-rotation* and *depolarization parameters,* which are in use in the description of elastic scattering of $s = \frac{1}{2}$ particles, are special cases of polarization-transfer coefficients. Assuming that both beam and target can be polarized and that the polarization of both outgoing products can be observed, a very large number of different polarization experiments can be conceived; many of these, however, are related when time reversal invariance and parity conservation are assumed.

Polarization observables depend on the type of process, the bombarding energy, and the angle of emission. There are theoretical upper limits on their value (e.g., the vector analyzing power is between -1 and 1) and it is not uncommon to observe values close to these limits. Polarization observables represent *relative* measurements (ratio between two count rates) and are thus less susceptible to systematic errors than cross-section measurements: in some experiments analyzing powers have been determined with an uncertainty of less than one part in 10^7.

POLARIZED BEAMS

The first polarization experiments were carried out with polarized beams produced in nuclear reactions or by nuclear scattering. Much larger intensities and better control over the beam polarization have become available by the development of *polarized ion sources*. One scheme that is in use employs a beam of neutral, polarized *atoms*, produced by substate separation in an inhomogeneous magnetic field or by optical pumping methods. The polarization of the atomic electrons is then transferred to the nucleus by use of the hyperfine interaction. Subsequent ionization (both negative and positive ions are available) prepares the beam for injection into an accelerator. In this manner, polarized beams of protons, deuterons, ^{3}He, ^{6}Li, ^{7}Li, and ^{23}Na have been prepared. Another scheme that has been applied to produce polarized protons and deuterons uses hydrogen atoms in the metastable $2s$ state ("Lamb shift" polarized ion source). Beams of *heavy ions* of moderate polarization have been produced by simply sending the unpolarized beam through a sequence of tilted, thin foils.

Care is needed when accelerating polarized beams in high-energy circular accelerators (synchrotrons) because the bending and focusing fields may cause resonant excitation of spin precession, thus destroying the beam polarization. Methods have been developed to avoid such resonances. One of these is the use of special spin precessors ("Siberian Snakes").

Beams of fast, *polarized neutrons* are produced by suitable nuclear reactions. With both beam and target unpolarized this yields neutron polarizations of typically 30%. A significant enhancement of the polarization can be obtained when the reaction is initiated by a polarized charged-particle beam. Very low-energy ("cold") neutrons can be polarized through the interaction with the polarized atomic electrons in magnetized materials.

Intense beams of *polarized electrons* are produced using photoemission from a GaAs surface using circularly polarized light, or by photoionization of polarized ^{6}Li atoms. In circular electron accelerators the emission of synchrotron radiation causes a slow buildup of polarization in the orbiting beam. Polarized electrons have also been produced by ionization of metastable helium atoms polarized by optical pumping, and by field emission of polarized electrons from magnetized materials.

POLARIZED TARGETS

Nuclei in solids cooled to very low temperature (less than 0.1 K) may be oriented by the application of strong external magnetic fields (several teslas). The nuclear polarization may be enhanced by the presence of paramagnetic centers in the sample and/or by the simultaneous application of electromagnetic radiation. Targets containing highly polarized protons and deuterons have been produced in this way. A variety of polarized nuclei have been prepared in a solid matrix for the purpose of nuclear physics studies, including ^{6}Li, ^{7}Li, ^{13}C, ^{15}N, ^{19}F, and many of the rare-earth nuclei. A famous application for the use of polarized ^{60}Co was the observation of the beta decay asymmetry which was experimental proof for parity violation by the weak force.

Optical pumping can be used to polarize atoms in alkali metal vapors (^{6}Li, ^{7}Li, ^{23}Na, etc.). Hydrogen and deuterium atoms, as well as certain noble gases (^{3}He, ^{21}Ne, ^{129}Xe), have been polarized by spin-exchange collisions with alkali atoms polarized by optical pumping. Optical pumping has also been used to polarize ^{151}Eu and ^{153}Eu nuclei contained in a jet of inert carrier gas. Polarized atomic beams, prepared in polarized ion sources, have served as a target of polarized hydrogen and deuterium atoms. Such targets have the important advantage of being free of admixtures of unwanted nuclei, but they have extremely low density and their main application is in conjunction with intense beams as internal targets in storage rings.

PHYSICS INVOLVING POLARIZATION

The measurement of spin observables yields information on the spin dependence of nuclear forces, and on the nature of the interaction between nuclear particles. In many cases polarization measurements allow one to determine unambiguously spin and parity of excited states of nuclei. Spin-correlation parameters explore the dependence of the interaction on the relative orientation of the spins of the colliding particles and allow the decomposition of the reaction amplitude into pieces that can be more directly related to the ingredients of a model. Often polarization observables can be found that are particularly sensitive to a specific aspect of the interaction.

Symmetry laws such as time-reversal invariance, parity conservation, and charge symmetry predict the cancellation of certain polarization observables. Careful measurements of small, nonzero values of these parameters has yielded information on violations of these symmetries.

Tests of quantum chromodynamics and insight into the makeup of nucleons from quarks and gluons is provided by polarized high-energy electron scattering and by the study of the interaction between polarized high-energy nucleons.

BIBLIOGRAPHY

W. Haeberli, "Sources of Polarized Ions," *Annu. Rev. Nucl. Sci.* **17,** 373 (1967).

W. Haeberli, "Polarized Beams," in *Nuclear Spectroscopy and Reactions* (J. Cerny, ed.), Part A, p. 151. Academic Press, New York, 1974; P. Catillon, "Polarized Targets," *ibid.*, p. 193.

W. J. Thompson and T. B. Clegg, "Physics with Polarized Nuclei," *Phys. Today* **32**(2), 32 (1979).

Polarization Phenomena in Nuclear Physics, Proceedings of the Fifth International Symposium 1980 in Santa Fe (G. G. Ohlsen *et al.,* eds.). AIP Conf. Proc. No. 69. American Institute of Physics, New York, 1981.

J. Kessler, *Polarized Electrons*. Springer-Verlag, Berlin and New York, 1985.

Polarization Phenomena in Nuclear Physics, Proceedings of the Sixth International Symposium 1985 in Osaka (M. Kondo *et al.*, eds.), *J. Phys. Soc. Japan* (Suppl.) **55** (1986).

N. J. Stone and H. Postma, *Low Temperature Nuclear Orientation*. Springer-Verlag, Berlin and New York, 1986.

High-Energy Spin Physics, Proceedings of the International Conference (K. J. Heller, ed.). AIP Conf. Proc. No. 187, Vol 2. American Institute of Physics, New York, 1989.

Nuclear Properties

Rafael Kalish

The scattering of α particles has revealed (Rutherford, 1911) that practically all the mass of an atom is concentrated in the atomic nucleus, which is located in the center of the atom, is positively charged, and has a rather well-defined radius of the order of 10^{-14} m (about 4 orders of magnitude smaller than the atomic radius). The discovery of the existence of several isotopes for the same chemical element (by Thompson in 1912) and the direct observation of the neutron (by Chadwick, 1932) have led to the conclusion that the major constituents of the nucleus are protons and neutrons, together referred to as nucleons (*see* NUCLEON). The great similarity between protons and neutrons regarding their mass (1.67×10^{-27} kg$\simeq$939 MeV/c^2), spin ($\frac{1}{2}$), statistics (Fermi–Dirac), and interaction (strong interaction) suggests that both may be considered as two different states of the same particle—the nucleon—differing only by the projection of their isospin (*see* ISOSPIN).

The (positive) charge of any nucleus equals the atomic number Z of its element (the number of electrons in the neutral atom) multiplied by the elementary electronic charge. The mass of any nucleus is very close to an integral multiple of the mass of a nucleon and is specified by its mass number A (closely related to the atomic weight of the element, for monoisotopic elements). Since the total mass of the nucleus is made up of the masses of its Z protons and N neutrons, $A=N+Z$. A nuclide of chemical element X with Z protons and $N=A-Z$ neutrons is described by the symbol ${}^{A}_{Z}\mathrm{X}_N$.

The forces responsible for binding the nucleons in the nucleus belong to the category of "strong interactions," the details of which are not yet fully understood, and are the subject of studies in low-energy as well as in high-energy nuclear physics (*see* NUCLEAR FORCES; STRONG INTERACTIONS; ELEMEMTARY PARTICLES IN PHYSICS). The other forces active inside nuclei are the electromagnetic force, which is much weaker than the strong nuclear force and is responsible for the electromagnetic properties of nuclei, and the even weaker "weak-interaction" force, which is responsible for β decay, i.e., the transformation of the neutrons into protons and vice versa. The forces acting inside the nucleus couple together under their influence determine the properties of nuclei, such as their mass, stability with respect to radioactive decay, size, shape, angular momentum, electromagnetic moments, energy levels, and interaction with other nuclei as revealed in nuclear reactions. Since nucleons tend to group in shell structures similar to the atomic shells that electrons around the nucleus assume, most properties of nuclei depend on whether these nuclear shells are filled (at "magic numbers"), are close to being filled, or are about half filled. Nuclei belonging to the first two groups reveal the nuclear single-particle properties, whereas those belonging to the latter reveal the collective properties caused by the motion of many nucleons in phase with each other (*see* NUCLEAR STRUCTURE).

NUCLEAR MASSES; NUCLEAR BINDING ENERGIES

Nuclear masses are measured in unified mass units, abbreviated by u (or amu); 1 u is defined as 1/12 the mass of a ^{12}C *atom*. 1 u$=1.6605655(86)\times10^{-27}$ kg$=931.4812(52)$ MeV/c^2.

The total mass of the nucleus is related to its total energy by the relativistic mass–energy equivalence $E=Mc^2$ (c is the speed of light). The nuclear *binding energy* B is defined as the difference between the sum of the masses of the individual nucleons of which a nuclide is composed (Z protons with mass M_p and N neutrons with mass M_N) and its observed mass, multiplied by c^2:

$$B=(ZM_p+NM_n-M)c^2. \tag{1}$$

It represents the energy required to break the nucleus up into its constituents, or conversely, the energy released when the nucleus is put together from individual nucleons. Since one normally deals with atoms rather than with bare nuclei, the binding energy is usually defined as including the Z electrons in their lowest atomic state. This new definition introduces only minor changes to the binding energy as defined in Eq. (1).

Nuclear masses are usually quoted in terms of the *mass excess* (sometimes also called mass defect), defined as $\Delta M=M-A$ (*see* BINDING ENERGY).

Nuclear masses can be determined experimentally by the use of mass spectrometers or by measuring energies involved in nuclear reactions or decay (*see* CHARGED-PARTICLE SPECTROSCOPY; NUCLEAR REACTIONS).

A semiempirical expression for the dependence of the nuclear mass on A and Z, $M(A,Z)$, has been proposed by Weizsäcker (1935) and reproduces the general trend of nuclear masses rather well. In the derivation of this "mass formula" or its equivalent "binding-energy" formula [Eq. (2)], it is assumed that the binding energy per nucleon is, to first order, independent of the number of nucleons A [b_1A in Eq. (2)]; account is taken of the fact that the binding of nucleons near the nuclear surface is reduced ($-b_2A^{2/3}$); a symmetry term is included in the formula—it makes nuclei asymmetric in the number of protons and neutrons less bound than symmetric nuclei $\{-b_3[(A-2Z)^2/A]\}$; a Coulomb term takes into account the repulsion between protons that makes proton-rich nuclei less bound $[-b_4(Z^2/A^{1/3})]$; finally, a pairing term ($+b_5$) is added to account for the fact that nuclei with even numbers of protons and/or neutrons are more stable than those with odd unpaired nucleons of the same kind. The semiempirical binding-energy (mass) formula thus takes the form

$$B=b_1A-b_2A^{2/3}-b_3\frac{(A-2Z)^2}{A}-b_4\frac{Z^2}{A^{1/3}}+b_5. \tag{2}$$

The parameters b_1 to b_5 are fitted to reproduce best the experimental values of the nuclear binding energies. Figure 1 shows the gross as well as the fine structure of the binding energy per nucleon. Also shown is the fit of B/A to the data,

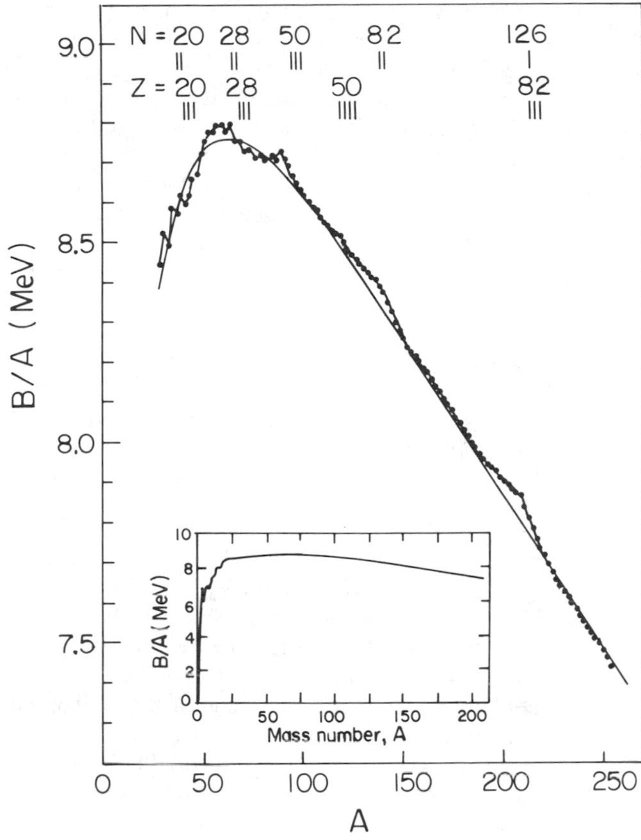

FIG. 1. The binding energy per nucleon as a function of A. The connected points are experimental binding energies, and show some deviation from a smooth behavior at nuclei having a magic number of protons or neutrons. The smooth curve represents the semiempirical mass formula with the best-fitting parameters (From A. Bohr and B. R. Mottelson, *Nuclear Structure*, Vol. I, p. 168)

as calculated from Eq. (2). The best values for the parameters are

$$b_1 \approx 16 \text{ MeV}, \quad b_2 \approx 17 \text{ MeV}, \quad b_3 \approx 25 \text{ MeV},$$

$$b_4 \approx 0.7 \text{ MeV}, \quad b_5 \approx \pm \frac{12}{A^{1/2}} \text{ MeV}.$$

($b_5 > 0$ for nuclei with even Z and N, $b_5 = 0$ for odd A nuclei, and $b_5 < 0$ for nuclei with odd Z and odd N.) The strong deviations of some empirical binding energies from the prediction of the binding-energy formula seen in Fig. 1 are due to shell effects that are not included in the derivation.

NUCLEAR SIZES

The experimental determination of nuclear sizes depends somewhat on the method of measurement employed. Two different classes of measurements exist. The first probes the nuclear spatial extension via electromagnetic interactions and is therefore sensitive to the proton distribution in nuclei. The second utilizes nuclear interactions and is thus sensitive to the total nuclear matter distribution. Among the methods used in the first class are experiments in which the diffraction pattern of charged particles (mostly electrons) scattered off nuclei is measured and experiments in which the effect of the final nuclear charge distribution on atomic transitions (optical transitions, x rays due to transitions of electrons or muons between atomic levels, etc.) are determined (*see* MUONIC, MESONIC, AND OTHER EXOTIC ATOMS). In the second class mostly scattering of strongly interacting particles (neutrons, protons, heavy nuclei, π mesons, etc.) off nuclei is measured, and the distribution of nuclear matter is inferred from these results. Results of all of the above-mentioned experiments indicate that nuclei are of nearly spherical shape, that nuclear matter has approximately constant density, and hence that the volume of nuclei is roughly proportional to the total number of nucleons (A). The nuclear radius can thus be written as

$$R = r_0 A^{1/3}. \tag{3}$$

A more refined look at scattering data (mostly obtained with electrons) shows that the nuclear surface is not sharp but is somewhat diffuse. A charge distribution (density of protons) of the form

$$\rho(r) = \rho_0 \left\{ 1 + \exp\left[\frac{(r-R)}{a}\right]\right\}^{-1}, \tag{4}$$

which allows for a diffuse nuclear surface, reproduces the scattering data rather well, as can be seen in Fig. 2. The proton density well inside the nucleus is ρ_0, while a is the nuclear surface "thickness" parameter. The nuclear radius R of Eq. (3) is defined as the radius at which the nuclear density drops to half of the center value. Systematic studies of nuclear sizes show that ρ_0, r_0, and a are roughly independent of A and assume the values $\rho_0 \approx 0.17$ nucleon/fm^3 (this corresponds to 3×10^{17} kg/m^3, which is of the order of the density of neutron stars), r_0 (electric)≈ 1.1 fm, $a = 0.54$ fm (1 fm $= 10^{-15}$ m).

Results of measurements of the second class are less definitive but seem to indicate that the nuclear matter extends somewhat farther into space than the electric charge distribution does, r_0 (nuclear)≈ 1.3 fm.

NUCLEAR SHAPES

Although most nuclei are roughly spherical, there are regions in the chart of nuclei where the nuclei assume permanent deformations. The nuclear shapes, like their radii, can be divided into shapes of the nuclear mass distribution and those of the charge distribution; the latter are discussed later in the section dealing with nuclear electromagnetic moments. The presence of nonspherical mass distributions of nuclear matter is most pronounced in nuclei in the regions $150 < A < 190$ and $A > 230$, which exhibit clear rotational spectra. They can only be explained if spheroidal shapes are assumed for these nuclei. Deformations as high as 0.7 (minor- to major-axis ratio) have been inferred. The sudden change in rotational spectra observed to occur in rapidly rotating deformed nuclei indicates that the nuclear shape may change abruptly when some critical angular momentum is reached, probably because of the breakdown of the pairing force, which tends to hold nucleons paired with their spins antiparallel. Most recently, a new kind of collective nuclear

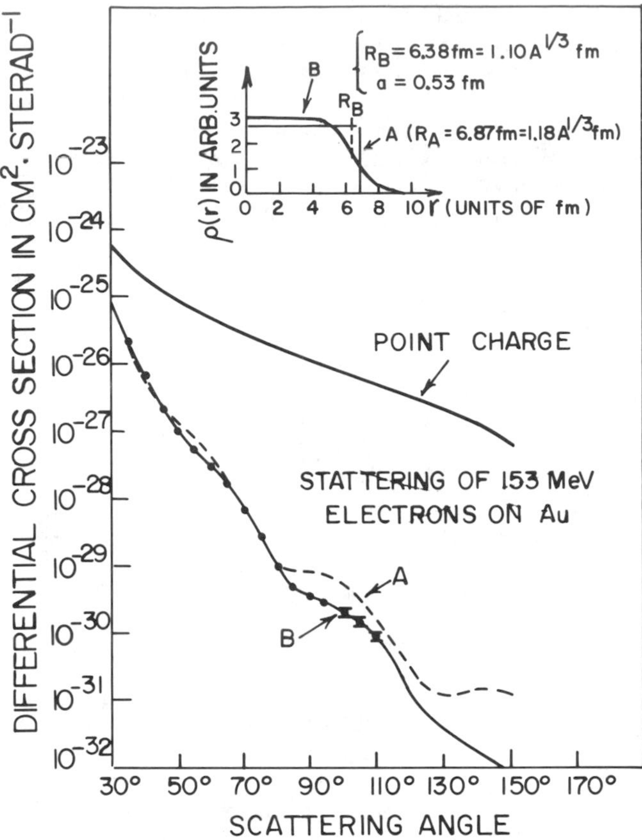

FIG. 2. Angular distribution of 153-MeV electrons scattered off Au. The connected points are experimental data, the smooth lines represent the results of calculations in which different expressions for the nuclear radius have been used as indicated in the insert. (From A. Bohr and B. R. Mottelson, *Nuclear Structure*, Vol. I, p. 159.)

excitation has been observed in some deformed nuclei. It can be explained if the nuclear shape is assumed to consist of the proton and neutron ellipsoids slightly tilted with respect to each other, and both wobble about a common axis (*see* NUCLEAR STATES; ROTATION AND ANGULAR MOMENTUM).

NUCLEAR QUANTUM NUMBERS: SPIN (ANGULAR MOMENTUM), PARITY, AND ISOSPIN

The nucleus consists of nucleons with intrinsic spin quantum number $\frac{1}{2}$ that move in orbitals under the influence of the nuclear potential. The total angular momentum of a nucleus is given by the vector sum of all angular momenta of its constituents:

$$|\mathbf{I}_I| = \left| \sum_{i=1}^{A} (\mathbf{l}_i + \mathbf{s}_i) \right| = \hbar[I(I+1)]^{1/2}. \tag{5}$$

Here ℓ_i and $\mathbf{s}_i$ are the orbital and intrinsic angular momenta of the ith nucleon, and I is the total angular momentum quantum number of the nucleus, usually called the *nuclear spin*.

A nuclear state, like any other quantum-mechanical system, is described by a wave function characterized by a set of quantum numbers. Two of these quantum numbers are I, the total nuclear angular momentum, and its projection on the Z axis, M_I. They carry information about the internal structure of the state since their value depends on the particular way the nucleons inside the nucleus couple to yield that state. The nuclear wave function has a well-defined parity, which is a measure of symmetry of the wave function under reflection of all coordinates through the origin. The parity $\pi = +$ or $-$ is thus another quantum number describing the nuclear state. Finally, a state is sometimes also characterized by its isotopic spin (or isospin) quantum number (T) and its projection (T_z), which are measures of the symmetry of the particular wave function under the operation of interchanging protons and neutrons. Only the spin and the parity, denoted by I^π, are usually quoted in describing a nuclear state (*see* ROTATION AND ANGULAR MOMENTUM; PARITY; ISOSPIN; QUANTUM MECHANICS).

NUCLEAR ELECTROMAGNETIC MOMENTS

Since nuclei are composed of nucleons having electromagnetic (EM) moments (electric charge and magnetic dipole moment), they themselves have EM moments, the most important of which are the magnetic-dipole moment and the electric-quadrupole moment. An EM moment of a nuclear state is defined as the expectation value of the appropriate moment operator. Electromagnetic moments are therefore very sensitive to the internal structure of the state under study and their experimental determination is of great importance in understanding the structure of nuclei.

The *magnetic (dipole) moment* of a nuclear state is given by

$$\boldsymbol{\mu} = \frac{e\hbar}{2M_p c} \int \psi_N^* \left[\sum_{k=1}^{z} (\mathbf{l}_k{}^p + g_p \mathbf{s}_k) + \sum_{k=z+1}^{A} (\mathbf{l}_k{}^n + g_n \mathbf{s}_k) \right] \psi_N \, d\tau \tag{6}$$

where ψ_N is the wave function of the state under study, $\mathbf{l}_k{}^p$ and $\mathbf{l}_k{}^n$ are the orbital angular moments of the kth proton or neutron, $g_p = 5.58$ and $g_n = -3.82$ are the proton and neutron g factors. Nuclear magnetic moments are measured in units of the *nuclear magneton*

$$\mu_N = \frac{e\hbar}{2M_p c} = 5.050824(20) \times 10^{-27} \frac{\text{J m}^2}{\text{Wb}}.$$

The *nuclear g factor* (or gyromagnetic ratio) is defined by

$$\boldsymbol{\mu} = g\mathbf{I} \tag{7}$$

(*see* GYROMAGNETIC RATIO).

The actual measurement of a magnetic moment requires the determination of the effect caused by the interaction of the nuclear-dipole moment with a magnetic field. This can be observed in many ways. Among them are the splitting of optical lines in atomic spectra (hyperfine structure); resonance experiments in which the nuclear moment is flipped in external magnetic fields; or experiments involving nuclear radiation from the state under study, which reveal the precession of the nucleus in the magnetic field or the magnetic substate splitting caused by it (*see* NUCLEAR MAGNETIC RESONANCE; ANGULAR CORRELATION OF NU-

CLEAR RADIATION; MÖSSBAUER EFFECT; FINE AND HYPERFINE SPECTRA AND INTERACTIONS).

Systematic studies of magnetic moments of odd nuclei for which the unpaired nucleon is the main contributor to the sum of Eq. (6) have shown that all measured moments fall between two limits corresponding to parallel ($I=l+\frac{1}{2}$) or antiparallel ($I=l-\frac{1}{2}$) coupling of the orbital and spin angular momenta of the odd nucleon (see Fig. 3). Even–even nuclei in their ground states have spin 0 and thus have no magnetic moment. Many nuclei in their excited states have magnetic moments that indicate that they are undergoing some collective rotational or vibrational motion.

The *electric-quadrupole moment* (EQM) of a state with wave function ψ_N is a tensor quantity with Cartesian components given by

$$Q_{ij}=\sum_{k=1}^{Z}\int e(3x_ix_j-\delta_{ij}r^2)_k|\psi_N|^2d\tau \tag{8}$$

It represents the deviations from spherical shape of the spatial charge distribution of the nucleus. Positive values for the quadrupole moment indicate prolate nuclear shapes, whereas negative values for Q correspond to oblate deformations. A measurement of Q of a certain state thus yields direct information about the shape of the nucleus while in that state.

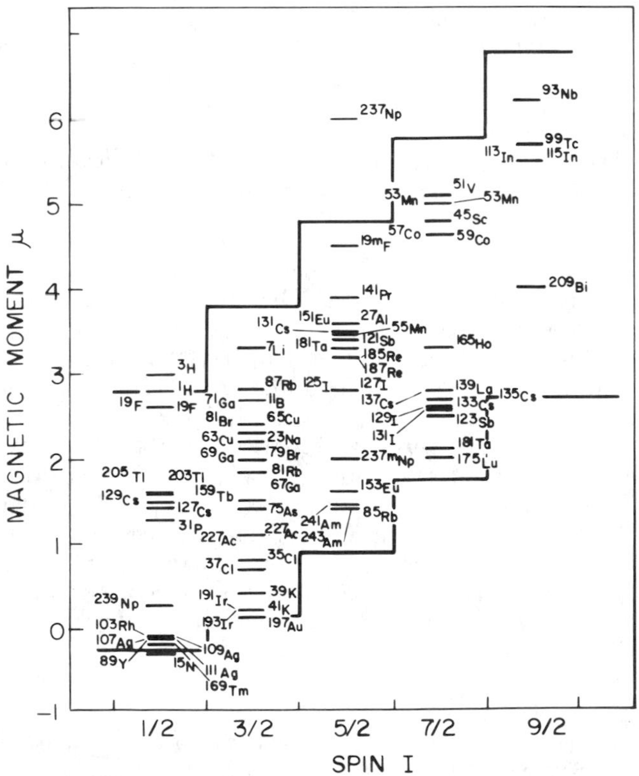

FIG. 3. Magnetic moments for odd-proton nuclei as a function of the nuclear spin. Practically all experimentally determined moments (shown by horizontal short lines) lie between the two "Schmidt lines," which represent the results of extreme single-particle calculations where only the odd proton is assumed to contribute to the total magnetic moment of the nucleus. A similar graph can also be drawn for odd-neutron nuclei. (Based on a figure from M. A. Preston and R. K. Bhaduri, *Structure of the Nucleus*, p. 76.)

In order to measure an electric-quadrupole moment, it is necessary to observe the result of the interaction of the EQM with an electric field gradient (EFG) present at the nucleus. Such EFGs can be achieved by a close collision with a charged particle, as in Coulomb excitation, or by the crystalline or electronic environment surrounding the nucleus under study.

Systematic measurements of nuclear-quadrupole moments have revealed the existence of regions in the chart of nuclei where very large deformations in the nuclear charge distribution are observed (see Fig. 4). The study of these deformations has led to the understanding of the collective structure of nuclei on one hand, and of nuclear shell structure on the other (nuclei having nucleon numbers close to magic numbers exhibit no deformation).

Recent accurate measurements of magnetic-dipole moments have shown that effective g factors for protons and neutrons, different from those measured for the free nucleon, should be used in Eq. (6). Furthermore, effective charges, differing from 0 and 1, should be associated with the neutron and protons, respectively, while inside nuclear matter.

Recent refined measurements of nuclear charge distributions have shown that electric moments of orders higher than quadrupole may be required to explain the Coulomb excitation probabilities measured for some deformed nuclei (*see* NUCLEAR MOMENTS).

NUCLEAR ENERGY LEVELS, TRANSITION PROBABILITIES

Under stable conditions nuclei are found in their energetic *ground state,* which is the state with minimum energy. The nucleus, being a quantum-mechanical system, has discrete energy levels into which it can be excited by various kinds of nuclear reactions or as a result of radioactive decay. An excited state corresponds to an internal rearrangement of the nucleons inside the nucleus. The quantum numbers, energies, EM moments, and interconnecting matrix elements (transition probabilities) of excited states are therefore of prime importance in the understanding of nuclear structure.

Nuclear energy levels are experimentally studied using various techniques involving nuclear reactions or nuclear decay. The excitation (or decay) probability of an excited level is related to its lifetime, as also is the width of the resonance through which the level can be excited. The experimental techniques most commonly used in studying excited states involve the spectroscopy of charged particles following nuclear reactions or the spectroscopy of the radioactive decay products (γ rays, conversion electrons, α and β particles). (*See* NUCLEAR REACTIONS; ALPHA DECAY; BETA DECAY; GAMMA DECAY.)

NUCLEAR STABILITY AND NUCLEAR DECAY

As indicated earlier, in the section on nuclear masses, nuclei are unstable unless their mass is smaller than that of the sum of their constituents. An unstable nucleus may undergo any one of the following radioactive decays, which will convert it into another nuclide:

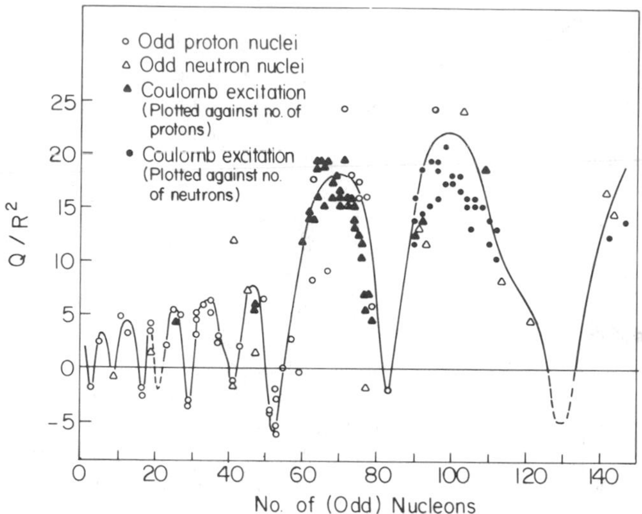

FIG. 4. Experimental values of electric-quadrupole moments divided by R^2 plotted against neutron or proton numbers. Regions where the quadrupole moments are close to 0 (no deformation) characterize nuclei having a closed neutron or proton shell. [From J. Rainwater, *Rev. Mod. Phys.* **48,** 385 (1976).]

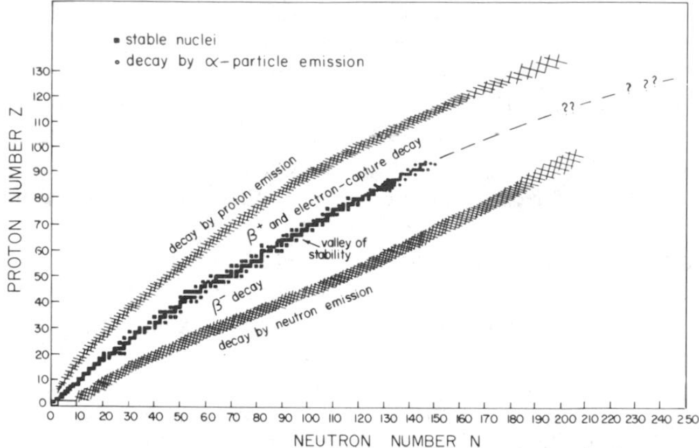

FIG. 5. Stable and alpha-decaying nuclei plotted on a Z-versus-N graph. The region where stable nuclei cluster is called the valley of stability; on its continuation, around $Z\sim114$ and 126, superheavy nuclei may exist, as shown in the figure by question marks.

Alpha decay	${}^A_Z X_N \to {}^{A-4}_{Z-2}Y_{N-2} + {}^4_2He_2$
Beta-minus decay	${}^A_Z X_N \to {}_{Z+1}{}^A Y_{N-1} + \beta^- + \bar{\nu}$
Beta-plus decay	${}^A_Z X_N \to {}_{Z-1}{}^A Y_{N-1} + \beta^+ + \nu$
Electron capture	${}^A_Z X_N + e^- \to {}_{Z-1}{}^A X_N + \nu$
Spontaneous fission	${}_Z X^A_N \to$ 2 heavy fragments ($\sim A/2$) + neutrons

All decay modes conserve the total number of nucleons and the total charge (as well as some other quantities), and follow Poisson statistics with a typical decay probability λ. A characteristic mean life (τ) is associated with each radioactive decay (or with the parent level), which is a measure of the decay probability, $\tau=1/\lambda$ (*see* RADIOACTIVITY; ALPHA DECAY; BETA DECAY; NUCLEAR FISSION).

A graphical presentation of all stable nuclei is given in Fig. 5, in which the region of stable nuclei in the N-versus-Z plot (called the *valley of stability*) is shown. The following points related to the figure are worth noting: (i) The valley of stability bends over from $N\simeq Z$ for light nuclei to $N>Z$ for the heavier ones. This deviation from symmetry for the heavier nuclei reflects the mutual Coulomb repulsion of the protons, as accounted for in the fourth term in Eq. (2). (ii) There are regions where a large number of either stable isotopes (constant Z) or stable isotones (constant N) occur. These are associated with magic numbers characteristic of closed-shell configurations, which yield particularly stable nuclei. (iii) Very few odd–odd stable nuclei exist in nature, in agreement with the last ("pairing") term in Eq. (2). Also shown in the figure are the regions where β^+ and β^- decay occurs, as well as where α-unstable nuclei lie. It should be noted that all nuclei heavier than ${}^{208}Pb$ ($Z=82$, $N=126$) are alpha unstable.

Very recently some evidence for the possible existence of very heavy, presumably stable, nuclei with $Z\sim114$ and 126, has been presented. The discovery of these "superheavy" nuclei may prove the existence of the long-predicted "island of stability" of superheavy nuclei.

Further properties of nuclei are revealed through their interaction with other nuclei, with elementary particles, and with electromagnetic radiation (*see* NUCLEAR REACTIONS; PHOTONUCLEAR REACTIONS). The wealth of information regarding the properties of nuclei has led to deeper understanding of the internal structure of nuclei (*see* NUCLEAR STRUCTURE).

See also ALPHA DECAY; ANGULAR CORRELATION OF NUCLEAR RADIATION; BETA DECAY; BINDING ENERGY; CHARGED PARTICLE SPECTROSCOPY; ELEMENTARY PARTICLES IN PHYSICS; FINE AND HYPERFINE SPECTRA AND INTERACTIONS; GAMMA DECAY; GYROMAGNETIC RATIO; ISOSPIN; MÖSSBAUER EFFECT; MUONIC, MESONIC, AND OTHER EXOTIC ATOMS; NUCLEAR FISSION; NUCLEAR FORCES; NUCLEAR MAGNETIC RESONANCE; NUCLEAR MOMENTS; NUCLEAR REACTIONS; NUCLEAR STATES; NUCLEAR STRUCTURE; NUCLEON; PARITY; PHOTONUCLEAR REACTIONS; QUANTUM MECHANICS; RADIOACTIVITY; ROTATION AND ANGULAR MOMENTUM; SPIN; STRONG INTERACTIONS; SUPERHEAVY ELEMENTS.

BIBLIOGRAPHY

A. Bohr and B. R. Mottelson, *Nuclear Structure*, Vols. I and II. Benjamin, Advanced Book Program, Reading, Mass., 1969, 1975. (A)

M. A. Preston and R. K. Bhaduri, *Structure of the Nucleus*. Addison-Wesley, Advanced Book Program, Reading, Mass., 1975. (I)

B. L. Cohen, *Concepts of Nuclear Physics*. McGraw-Hill, New York, 1971. (E)

H. A. Enge, *Introduction to Nuclear Physics*. Addison-Wesley, Reading, Mass., 1966. (E)

Nuclear Quadrupole Resonance

Richard G. Barnes

The term nuclear quadrupole resonance (NQR) refers to that particular form of nuclear magnetic resonance (NMR) in which transitions are induced between the quantized energy levels of the nuclear spin system which result from the interaction of the nuclear electric quadrupole moment with the gradient of the electric field at the nucleus, rather than from the interaction of the nuclear magnetic dipole moment with the magnetic field at the nucleus. In classical terms, the quadrupole moment executes a kind of complicated precessional motion about the major field-gradient axis analogous to the precession of the magnetic dipole about the field direction. Just as in NMR, the transitions are induced by an oscillating radio-frequency magnetic field acting on the nuclear magnetic dipole moment and can be detected by the same or closely similar experimental instrumentation employed to detect NMR signals, but without the requirement of an applied static magnetic field.

The nuclear quadrupole moment Q measures the extent to which the nuclear charge distribution $\rho(\mathbf{r})$ departs from spherical symmetry. It is defined as the average value of $\frac{1}{2}(3z^2 - \mathbf{r}^2)\rho(\mathbf{r})$ over the nuclear volume and has the dimensions of area. More precisely, Q is the expectation value of $\frac{1}{2}(3z^2 - \mathbf{r}^2)\rho(\mathbf{r})$ in the nuclear angular momentum substate for which $m_I = I$, where I is the nuclear spin angular-momentum quantum number and m_I is the quantum number for the z component of the spin ($m_I = -I, -I+1, \ldots, I-1, I$). The nucleus must have spin $I \geq 1$ in order to have $Q = 0$. Nuclei with $I = \frac{1}{2}$ have magnetic dipole moments, $\boldsymbol{\mu}$, but not quadrupole moments, and those for which $I = 0$ have neither. The quadrupole moment is positive for elongated (cigar-shaped) nuclei, negative for compressed (doorknob-shaped) nuclei, and zero for spherical nuclei ($I = 0$). Values of Q are usually expressed in units of 10^{-24} cm^2 (barns) and range from 0.003 barn for the deuteron (^{2}H) to ~5 barns for heavy nuclei such as tantalum (^{181}Ta).

Just as a magnetic dipole moment experiences an orienting torque, $\boldsymbol{\tau} = \boldsymbol{\mu} \times \mathbf{H}$, in a magnetic field $\mathbf{H}$, and a potential energy $E = -\boldsymbol{\mu} \cdot \mathbf{H}$, the electric quadrupole moment experiences an orienting torque in the gradient of an electric field and an energy proportional to the product of Q and the field gradient. The latter quantity is actually a second-rank tensor, composed of the nine second derivatives, $V_{xx} = \partial^2 V/\partial x^2$, $V_{xy} = \partial^2 V/\partial x\, \partial y$, etc., of the electrostatic potential V at the nucleus due to all charges external to the nucleus. Analogously to the moment of inertia tensor, the electric-field-gradient (EFG) tensor is diagonal in its principal-axis coordinate system (i.e., $V_{xy} = V_{yz} = \text{etc.} = 0$). Furthermore, since V arises from charges external to the nucleus, it satisfies Laplace's equation at the nucleus: $V_{xx} + V_{yy} + V_{zz} = 0$. This constraint means that only two independent parameters are required to characterize the EFG tensor, and these are conventionally taken to be the "field gradient" V_{zz} and the asymmetry parameter, $\eta = (V_{xx} - V_{yy})/V_{zz}$. Choosing the EFG coordinate axes so that $|V_{zz}| \geq |V_{yy}| \geq |V_{xx}|$ means that $0 \leq \eta \leq 1$ always. At a point of cubic symmetry, however, $V_{xx} = V_{yy} = V_{zz} = 0$, so that then the EFG vanishes identically. Hence, the nucleus must be located at a lattice (or molecular) site of less-than-cubic symmetry in order that it experience a quadrupole interaction.

Since the nuclear charge distribution is axially symmetric about the nuclear symmetry axis and also symmetric with respect to a plane perpendicular to that axis at its midpoint (as is a cigar or doorknob), the quadrupole orientation energy is insensitive to 180° reversal of the nuclear orientation. This is quite different from the behavior of the magnetic dipole moment $\boldsymbol{\mu}$ whose energy changes sign when $\boldsymbol{\mu}$ is reversed and indicates that the quadrupole energy levels must be independent of the sign of the z-component quantum number m_I. This is indeed the case, and the quantum-mechanical expression for the energy levels of the nuclear quadrupole moment depends on m_I^2 rather than on m_I as in the case of the nuclear magnetic dipole. The resonance frequencies of the transitions between the levels satisfy the magnetic-dipole radiation selection rule, $|\Delta m_I| = 1$. When $\eta = 0$, the spacings of the levels are such that the transition (resonance) frequencies are in the ratios 1:2:3:etc. For spin $I = \frac{3}{2}$, there is just one resonance frequency, $\nu = eQV_{zz}/2h$; for spin $I = \frac{5}{2}$ there are two frequencies, $\nu_1 = \nu_Q$ and $\nu_2 = 2\nu_Q$; for spin $I = \frac{7}{2}$ there are three frequencies, and so on, with $\nu_Q = 3eQV_{zz}/2I(2I-1)h$ being always the lowest frequency (h being Planck's constant). When $\eta \neq 0$ the number of energy levels and resonance frequencies is unchanged, but these become complicated functions of eQV_{zz} and η. Except when $I = \frac{3}{2}$, at least two resonance frequencies always occur, and hence V_{zz} and η can be determined from frequency measurements alone. When $I = \frac{3}{2}$, $\nu = eQV_{zz}(1 + \eta^2/3)^{1/2}/2h$ and application of an external magnetic field (Zeeman effect) is needed to separate V_{zz} and η. However, in many cases of interest, η is very small and has little influence on the resonance frequency.

Nuclear quadrupole resonance frequencies that have been detected by conventional radio-frequency methods range from about 0.5 MHz up to roughly 1000 MHz. The higher figure represents a strong interaction, indeed. By comparison, the same magnetic resonance (NMR) frequency for protons would require an applied magnetic field strength of 25 T (present-day commercially available NMR spectrometer systems operate at proton resonance frequencies as high as 600 MHz). Smaller quadrupole interactions are usually observed via their perturbation of the NMR spectrum (i.e., in an applied magnetic field), or by means of various double-resonance methods. Since the quadrupole interaction strength, and hence the resonance frequencies, are frequently initially unknown, resonances must often be searched for overwide frequency ranges. One radio-frequency spectrometer that has been much employed in this manner is the superregenerative oscillator detector, quite similar in fact to those used in "walkie-talkie" and other short-range communication systems. These have high sensitivity, but do not necessarily yield a faithful representation of the resonance line shape. So-called "pulse" NMR techniques are now commonly employed in NQR spectroscopy.

For a quadrupole resonance to be detectable, it must not be spread out over too great a frequency range, i.e., it must be relatively narrow. This means that all resonant nuclei must experience very nearly the same EFG. This occurs readily in molecular crystals in which the EFG at a particular nucleus is determined essentially entirely by *intramolecular* charge distributions (covalent bonds within individual molecules). Accordingly, the greatest number of NQR measurements have dealt with the nuclei of chlorine (^{35}Cl), bromine (^{79}Br and ^{81}Br), iodine (^{127}I), and nitrogen (^{14}N), reflecting the fact that these are all abundant isotopes, with typical resonance frequencies in convenient ranges of the radio-frequency spectrum for most compounds, especially solid organic compounds. NQR measurements have also been made on the nuclei of many other elements, both metallic and nonmetallic, in a variety of solid compounds. For example, the noncubic metals gallium (e.g., ^{69}Ga), indium (^{115}In), lanthanum (^{139}La), arsenic (^{75}As), antimony (e.g., ^{121}Sb), and bismuth (^{209}Bi) have been studied in this way.

From the standpoint of solid-state or molecular science, the quantity of interest in NQR measurements is the electric-field-gradient tensor, characterized by V_{zz} and η, since this arises from the charge distribution external to the nucleus. However, since the resonance frequencies always involve the product of these with the nuclear quadrupole moment Q, an essential first step is the determination of Q itself. This is not so straightforward as in the parallel case of the magnetic dipole moment. In that case energies leading to resonance frequencies in the radio-frequency region are readily attained in magnetic field strengths of the order of 1 T easily achieved even with iron-core laboratory magnets. The magnetic field strength can be measured by classical means (the current balance), and since frequencies can be measured very accurately with modern frequency counters, the magnetic moment can be determined. In contrast, the electric field gradient giving rise to the quadrupole energy is "built into" the molecule or crystal, and is not in the range of influence of externally applied fields. This is readily seen from the fact that typical EFG strengths fall in the range 10^{20}–10^{23} V m^{-2}, reflecting the essentially atomic origin of the EFG (the EFG due to a single electronic charge at a distance of 1 Å is $V_{zz} = -2e/4\pi\epsilon_0 r^3 = 2.9\times10^{21}$ V m^{-2}). Hence, determination of Q always entails a calculation of V_{zz}. The simplest and most reliable situations for such calculations are atoms themselves, and correspondingly, the best measurements of nuclear quadrupole moments have been made on free atoms in atomic-beam experiments.

In molecular crystals the principal source of the EFG is the electron bond between atoms in a molecule. Taking chlorine as an example, we note that the chlorine atom lacks one electron of having a filled $3p$ electron shell. The missing electron appears as a positive hole in an otherwise spherical electronic environment about the ^{35}Cl nucleus and causes an EFG proportional to the average value of r^{-3} for a $3p$ electron. Chemically, chlorine is highly electronegative: that is, it has a high affinity for an electron to complete the $3p$ shell. When it is bonded to another atom, the EFG due to the missing electron is reduced in proportion to the degree to which a shared electron from the partner atom is drawn into the chlorine $3p$ shell. In molecular chlorine, Cl_2, the two electrons of the covalent bond between the atoms are shared equally, and to a good approximation the EFG is unaltered from its atomic value. When Cl is bonded to other atoms, however, the EFG and hence the NQR frequency are reduced, the reduction being greatest for bonds between Cl and metals which tend to lose their valence electrons readily. Conversely, when Cl is bonded to the even more electronegative element fluorine (F), as in FCl, the EFG and resonance frequency are increased as a second $3p$ electron is partially withdrawn from the chlorine. In highly ionic solids the principal contribution to the EFG arises from the charges on the surrounding ions, and in metals it is necessary to consider both the EFG arising from the lattice of ion cores and the contribution from the nonspherical distribution of conduction electrons.

In many cases useful information regarding chemical bonding may be obtained from the relative behavior of the NQR frequencies and the "coupling constants," eQV_{zz}/h, of a particular nucleus in a series of related compounds, even though the quadrupole moment Q may not be known with high accuracy. Moreover, since V_{zz} and η are sensitive functions of the molecular or solid structure, the NQR frequencies are sensitive to structural phase transformations, molecular reorientations, and diffusive motions in solids, and these aspects may also be studied without necessarily exact knowledge of Q. For example, since the discovery of high-temperature superconducting materials based primarily on copper oxide, the NQR of ^{63}Cu and ^{65}Cu in these has been actively studied as a source of information on site symmetries and bonding. Finally, from the standpoint of nuclear physics, determination of the quadrupole moments of nuclear and isomeric states is most readily achieved by measurement of the quadrupole interaction energy in solids in which the EFG at the nuclear site has been calibrated by NQR measurements with a stable nuclide.

See also Crystal and Ligand Fields; Nuclear Magnetic Resonance; Nuclear Moments; Resonance Phenomena.

BIBLIOGRAPHY

I. P. Biryukov, M. G. Voronkov, and I. A. Safin, *Tables of Nuclear Quadrupole Resonance Frequencies*. IPST Press, Jerusalem, 1969. (I)

T. P. Das and E. L. Hahn, *Nuclear Quadrupole Resonance Spectroscopy*. Academic Press, New York, 1958. (I)

H. G. Dehmelt, "Nuclear Quadrupole Resonance," *Am. J. Phys.* **22,** 110 (1954). (I)

E. A. C. Lucken, *Nuclear Quadrupole Coupling Constants*. Academic Press, New York, 1969. (I)

J. C. Raich and R. H. Good, Jr., "Discussion of Quadrupole Precession," *Am. J. Phys.* **31,** 356 (1963). (I)

G. K. Semin, T. A. Babushkina, and G. G. Yakobson, *Nuclear Quadrupole Resonance in Chemistry*. Wiley, New York, 1975. (I)

J. A. S. Smith (ed.), *Advances in Nuclear Quadrupole Resonance*, Vols. 1 through 8. Heyden, London, 1974–86. (A)

Nuclear Reactions

John P. Schiffer

The process in which an atomic nucleus interacts with various projectiles, whether single nucleons, other atomic nuclei, photons, electrons, or mesons, is called a nuclear reaction. A major fraction of our knowledge of the nucleus and of nuclear structure has been obtained from nuclear reactions.

The kinematics of a nuclear reaction is determined by the incident energy of the projectile and the energy release. In a reaction $A+B \rightarrow C+D+$energy, this is called the "Q value" (i.e., $A+B \rightarrow C+D+Q$). For reactions with a negative Q value there is a threshold energy ($=-Q$) at which the reaction first becomes possible. All discussion of the kinematics of a nuclear reaction is usually in the center-of-mass coordinate system in which the total momentum is zero.

The techniques of nuclear-reaction studies set the limits on our experimental knowledge; accelerators are needed to produce energetic beams of particles and detectors are needed to observe reaction products.

Nuclear reactions have been studied over the past four decades with ever improving accelerators: electrostatic accelerators, accelerators in which magnetic fields are used to allow rf electric fields to interact with the beam repeatedly (cyclotrons and synchrotrons), and linear accelerators. Superconducting magnets, and more recently superconducting rf accelerating structures, are playing an increasingly important role. Storage rings in which beams of nuclei are cooled and stored for long periods (seconds to hours) are also coming into use. Detection technology has also undergone several revolutions in magnetic spectrographs and considerable evolution in gas, scintillation, and solid-state detectors.

Compound and direct reactions form an important classification scheme for low-energy nuclear reactions in terms of the time scale on which it occurs. If in a process

$$A+B \rightarrow C \rightarrow D+E$$

the intermediate state C is long lived compared to the time it would take A and B to pass through the range of their mutual interaction, then one speaks of a compound nuclear reaction where C is the compound system. If the process is of the type

$$A+B \rightarrow C+D$$

with no well-defined intermediate state, one speaks of a direct reaction.

Compound reactions are characterized either by well-defined resonances corresponding to intermediate states or by fluctuations as a function of incident energy; in either case $\Delta E \approx \hbar/\tau$ where ΔE is the width of the fluctuation and τ is the characteristic lifetime. Angular distributions of the products from compound reactions show no memory of the direction of formation; they are symmetric about 90° when averaged over a sufficient energy interval and may be written as sums of Legendre polynomials.

Direct-reaction cross sections vary smoothly with incident energy, no faster than is allowed by the transit time of the projectile through the target. Angular distributions are asymmetric about 90° and generally sharply forward peaked (except when energies are comparable to the Coulomb repulsion) and exhibit a Bessel-functionlike diffraction pattern, where the order of the Bessel function is determined by the angular momentum transfer. Direct processes become important at higher energies, for nucleons at $E \gtrsim 10$ MeV.

Elastic scattering is the simplest form of nuclear reaction, where the internal coordinates of target and projectile remain unchanged. At low energies, between charged particles, the process is Rutherford scattering; at energies sufficient to overcome the Coulomb repulsion, the nuclear interaction starts playing a role. At energies sufficiently high that compound resonances are not resolved, the interaction is represented by a potential with a real and imaginary part, the so-called optical potential. All reactions other than elastic scattering are treated as absorption and are represented by the imaginary potential. This, strictly empirical, form of potentials as a function of the interaction distance r is the so-called Woods–Saxon shape $V=V_0\{1+\exp[(r-R)a]\}^{-1}$, with the real and imaginary parts having different parameters. The optical potential for nucleons scattered by nuclei is closely related to the shell-model potential, and a spin-orbit potential is needed to fit polarization data. For more complex projectiles absorption dominates.

Inelastic scattering corresponds to a scattering process in which the internal excitation energy of the nucleus is changed by the projectile but no nucleons are transferred. Direct inelastic scattering is an important technique for exciting the *collective* multipole modes of the nucleus; inelastic proton and alpha-particle scattering have been particularly useful in exciting low-lying vibrations as well as higher multipole giant resonances. A special category of inelastic scattering is *Coulomb excitation;* it is a process which occurs at low energies where the projectile cannot reach close enough to the nucleus to get within the reach of nuclear forces, but can nevertheless excite the nucleus through the gradient of the Coulomb field. This process is particularly effective in exciting quadrupole vibrations or rotations. The probability of Coulomb excitation is directly related to the probability of the inverse transition by γ radiation; the reduced transition probability is the $B(E\lambda)$ $B(M\lambda)$ for electric or magnetic transitions of multipolarity λ and is measured in single-particle or Weisskopf units. At higher energies more detailed information on the radial shape of the transition probability becomes available, particularly from electron scattering.

The **single-nucleon transfer reactions** are an important class of direct reactions which have had much to do with establishing the validity of the shell model. A *stripping reaction* adds a nucleon to the target; the reaction $A+d \rightarrow B+p$ or $A(d,p)B$ is an example for neutrons—the (d,n) or $(^3\text{He},d)$ reaction for protons. The angular distributions of these reactions will give the orbital angular momentum (l) of the transferred particles. Some details of the angular distribution (j dependence), but more importantly polarization measurements, will give the total angular momentum (j). The absolute cross section is proportional to the overlap between the "target-plus-nucleon" and the final-state wave functions. This overlap, or reduced transition probability, is called the spectroscopic factor, and contains the significant structure information, other than the quantum numbers.

"Single-particle states" are excited preferentially in stripping. The *pickup reaction* is the inverse; it removes a nucleon from the target nucleus. The (p,d) reaction is a prototype. The spectroscopic factor is again the reduced quantity which here characterizes the extent to which a given final state is a "hole state" in the target.

Of the direct **two-nucleon transfer reactions** the ones involving the transfer of two identical nucleons have been especially useful. The (p,t) and (t,p) reactions, transferring a pair of neutrons with their spins coupled to zero, preferentially populate "pairing states." These reactions form the basis of our knowledge of these states; the description is in close analogy to the theory of superconductivity based on the pairing between electrons.

The **charge-exchange** direct reactions have also been of special interest because of their selectivity. The (p,n) reaction strongly populates an especially simple state, the so-called "isobaric analog state." This is seen when the target has a neutron excess $N-Z>0$. When $N-Z=1$, the reaction populates the ground state of the mirror nucleus: $A[Z=(A-1)/2, N=(A+1)/2]\,(p,n)B[Z=(A+1)/2, N=(A-1)/2]$ in analogy with superallowed Fermi β decay. But when $N-Z>>1$, it is found that the reaction populates a highly excited state analogous to the ground state of the target. This "isobaric analog state," though it may occur at very high excitation energy, is still quite sharp; isospin is a very good symmetry in nuclei.

Other transfer reactions may clearly occur. The α-transfer reactions (^{6}Li,d) or (d,^{6}Li) are of particular interest since the α particle is such a symmetrical and tightly bound piece of nuclear matter. More complex processes also occur, particularly in heavy-ion reactions.

The **distorted-wave Born approximation** (DWBA) has been remarkably successful in describing direct nuclear reactions. Original attempts at using the plane-wave Born approximation were modified to include the distortions of both incident and outgoing waves by the optical potentials (including the Coulomb field) that are needed to describe elastic scattering. The DWBA fits the shapes of angular distributions and the absolute cross sections; it is the reaction mechanism and geometry that needs to be removed from experimental angular distributions to obtain the reduced transition probability [$B(E\lambda)$ or spectroscopic factor, etc.] that is of nuclear structure interest.

Multistep reactions have been explored in more detail recently. Conceptually they are used to describe direct reactions in which two distinct processes take place. Thus in a stripping reaction the projectile may inelastically excite a collective mode as well as transfer a nucleon. A two-nucleon transfer reaction may be described as proceeding by the transfer of the nucleons one at a time. Questions of time scale and orthogonality have caused much confusion in this area. It is clear that multistep reactions provide a transition between direct and compound processes.

In **neutron-induced reactions** the Coulomb repulsion is absent, and thus nuclear reactions are possible even at very low energies. The reaction cross sections with reactor neutrons at thermal energies (~0.02 eV) fluctuate wildly from target to target. This has led to the resonance theory of nuclear reactions which gives $\sigma \sim [(E-E_R)^2+\Gamma^2/4]^{-1}$. The distribution of resonances produced with slow neutrons is essentially statistical, though repulsion between levels of the same quantum numbers has been predicted and found. The width Γ of a resonance may be divided into an internal reduced width (γ^2) and an external penetrability P_l ($\Gamma=2P_l\gamma^2$). For slow neutrons the interaction is s wave ($l=0$) and $P_0 \sim E^{1/2}$. The reduced width is analogous to the spectroscopic factor in stripping. In heavy nuclei the spacing of resonances (D) is very close, a few eV. The average reduced width per unit energy is called the strength function ($\equiv\langle\gamma^2\rangle/D$) and this exhibits smooth maxima and minima as a function of the size of the target. These giant or size resonances in the strength function arise from approximate eigenstates in the complex optical potential that describes the neutron nucleus interaction.

The *capture* process is one in which the incident projectile is captured by the target and no particles, other than photons, are emitted. Capture is the dominant reaction mode with slow neutrons other than elastic scattering (and fission in very heavy nuclei).

Isobaric analog resonances are the isobaric analog states that were discovered in the direct (p,n) reactions, and also showed up as compound resonances in *proton* scattering. They completely dominate the elastic scattering in regions where the density of normal resonances is far too high to be resolved; a complete replica of all the low-lying bound states of the parent nucleus is seen in these resonances.

Statistical reactions are the decay mode of the compound nucleus at high excitation energies. With nucleons, deuterons, alphas, and heavy ions at energies well above the Coulomb barrier more than 90% of the cross section goes into forming a compound nucleus. In a highly excited compound system with a very large number of available decay channels, particles will "evaporate," the mode of decay depending only on phase space, kinematic (penetrability) factors, and the density of final states. Neutron evaporation dominates [e.g., (α,3n), (α,4n), (α,5n) depending on the energy]; proton and α emission are less probable because of the Coulomb barrier. When fission is possible, it competes effectively with neutron emission as a deexcitation process.

Other projectiles may induce nuclear reactions. Examples are **photonuclear reactions** where an energetic photon is absorbed by a nucleus, which in turn decays by emitting a particle. This process is characterized by a huge maximum in the cross section, the dipole giant resonance, which occurs at ~20 MeV in lighter nuclei and ~10 MeV in heavy ones.

Electron scattering: With improving techniques for both the acceleration of electrons and the detection of particles, electromagnetic probes have become the precision tool of nuclear physics. Because the electromagnetic interaction is known precisely, and is relatively weak—so that the nucleus is almost transparent to an electron or photon—one can interpret the data with considerably greater precision than from hadronic nuclear reactions. However, the cross sections are small and experiments have been difficult; this field has begun to come into its own in the decade of the 1980s with accelerators at MIT, Saclay, France, and Amsterdam. A major new facility in Virginia, CEBAF, is scheduled for completion in the 1990s.

Elastic electron scattering reveals the size and shape of the charge and current distribution of the nucleus, while inelastic scattering probes the overlap between the nuclear ground state and specific excited states—providing a radial form factor for each transition that is reduced to the single number represented by the transition probability [$B(E\lambda)$, $B(M\lambda)$] mentioned above in connection with Coulomb excitation. *Quasifree* $(e,e'p)$ scattering, where an inelastically scattered electron is measured in coincidence with a knocked-out proton and the momenta of both are determined accurately, is a precision tool for probing the occupation of proton orbits in the nucleus, in analogy with transfer reactions, and also provides precise quantitative information on the radial shape of these orbits. More generally, these reactions probe the momentum distribution of nucleons in the nucleus.

Pion-induced reactions such as elastic and inelastic scattering, charge exchange, and double-charge exchange (pions come in three charge states π^-, π^0, π^+) have all provided valuable information. In the energy region of 100–200-MeV kinetic energy the interaction of pions with nucleons is dominated by a single resonance, the Δ, which is the first excitation of the nucleon (corresponding to three quarks with their spins aligned). The properties of the Δ are such that positive pions have a much larger cross section for interacting with protons than with neutrons; the converse is true for negative pions. Pions, therefore, provide important tools for investigating differences between the proton and neutron structures in nuclei. Since a pion is a "boson" (its spin is zero, unlike the spin-$\frac{1}{2}$ fermions: neutrons, protons, or electrons) the number of pions is not conserved. This means that a pion may be absorbed within a nucleus, giving up its rest mass to the excitation of the nucleons. After considerable effort this process is still not well understood. The investigation of pion-induced reactions has been pursued very actively since the late 1970s at the "meson factories" in Los Alamos, Zürich, and Vancouver. Even though virtual pion exchange is a major ingredient of the nucleon–nucleon force that holds nuclei together, our understanding of the propagation of real pions in nuclei is still very limited.

Kaons, the mesons that carry a strange quark, provide the opportunity of embedding a Λ (a baryon with one of the quarks of a nucleon replaced by a strange quark) into the nucleus. Because of the relatively long lifetime of a Λ, nuclei called hypernuclei are formed in which three types of particles—protons, neutrons, and Λs—interact. The structure of these hypernuclei is of considerable interest but the information available [from reactions such as $A(K^-,\pi^-)_\Lambda A$, or $A(\pi^+,K^+)_\Lambda A$] is still rather sparse.

Heavy-ion reactions are induced by projectiles heavier than the alpha particle—for example, ^{12}C ions or ^{238}U ions. At low energies, Coulomb barrier effects, of course, become much more important than for lighter projectiles. The separation between compound and direct reactions is much as for lighter projectiles; many of the same phenomena are seen—but because of the larger mass at a given velocity the momenta and angular momenta are large. Heavy-ion reactions are particularly useful in exploring the high-spin states of nuclei. The structure of rapidly rotating but otherwise cold nuclei has received concentrated attention in the 1980s and has been helped by the development of Compton-suppressed Ge detectors that are used in arrays in multiple coincidence. New regimes of superdeformation (where the nuclei are deformed spheroids with the ratio of axes in the vicinity of 2:1) have been revealed through such heavy-ion reactions. A new reaction mode has been observed with heavy ions, the so-called "*deep inelastic*" or "*strongly damped*" process. Here the outgoing reaction products are similar to the incident ones in charge and mass—but the kinetic energy (above the Coulomb barrier) is completely dissipated in the collision with some exchange of nucleons; the two fragments separate again and are observed with the minimal kinetic energies that can be generated from their mutual (Coulomb) repulsion. Heavy-ion reactions are being pursued very actively since the late 1970s at a number of laboratories.

Relativistic heavy-ion reactions have been studied since the late 1970s. Kinetic energies of particles are usually quoted in energy per nucleon or MeV/u, so that one may readily compare the velocity regime for various projectiles. In reactions above about a hundred MeV/u the nuclei fragment into a shower of nucleons and, increasingly, pions and the equation of state of hot nuclear matter may be studied. At energies above 1 GeV/u the number of pions per collision is large, and interferences between the pions may be used to determine the size of the hot emitting region using the "Hanbury Brown and Twiss" effect of astronomy. The interest to study the properties of hot hadronic matter has motivated this field with experiments first at Berkeley, then at Brookhaven, CERN, and soon in Darmstadt, and the construction of a major new facility RHIC is planned for the 1990s to help explore this physics. In particular, the theoretical expectation that at some (not very specifically defined) density hadronic matter should undergo a phase transition in which the quarks would be deconfined and a quark–gluon plasma may be reached has driven these studies to higher and higher energies.

See also Hypernuclear Physics and Hypernuclear Interactions; Isobaric Analog States; Isospin; Nuclear Properties; Nuclear Scattering; Nuclear States; Photonuclear Reactions.

BIBLIOGRAPHY

J. Cerny (ed.), *Nuclear Spectroscopy and Reactions*. Academic Press, New York, 1974.

Nuclear Reactors

Ferdinand J. Shore

INTRODUCTION

In late 1942, the first nuclear reactor made by man began to operate in Chicago at a power level of $\frac{1}{4}$ W through the efforts of E. Fermi and his collaborators on the Manhattan project. Some two billion years earlier, a natural reactor involving a nuclear chain reaction appears to have run spontaneously in Gabon, Africa. Reactors now in use range in thermal power from 1 kW to more than 3 GW, and in neutron flux up to more than 10^{15} neutrons/cm^2/s.

Nuclear reactors are comprised of a core containing the fuel which supports the chain reaction, perhaps a moderator to thermalize neutrons, a coolant to extract fission heat, thermal and biological shields, and a control system.

The large and stable flux of neutrons available in the core and the high-intensity beams which may be brought through the shield have provided science with a major tool for the study of neutron-induced nuclear reactions, radiation damage, and solid-state phenomena. Nuclear medicine and industry have profited from the radioisotopes produced. Other users of fission heat are naval vessels, such as battleships, submarines, and icebreakers with infrequent refueling opportunities, some earth satellites, and power reactors which provide commercial electricity. The present discussion focuses mainly on power reactors.

THE CHAIN REACTION

The chain reaction results from neutron-induced fission of appropriately arranged fissile nuclei such as ^{235}U and ^{239}Pu. In fission, the compound nucleus which is formed breaks into two fragments and simultaneously a few neutrons may be emitted. The total mass energy of the new system is less than the original, and about 200 MeV of energy is liberated. The nascent neutrons induce fission in nearby fissile nuclei, thereby repeating the sequence which can continue for generation after generation if other processes do not remove too many neutrons.

NUCLEAR PROPERTIES

Several nuclear reactions (n,x) can result when a neutron interacts with a nucleus to form a compound state which decays by process x. If σ_x is the cross section for process x, for fission $x=f$; for elastic scattering $x=n_e$; for inelastic scattering $x=n_1$; and $x=c$ denotes resonant capture with emission of gamma rays.

For ^{235}U σ_f is 582 barns (1 barn $=10^{-24}$ cm^2) at 0.025 eV (thermal neutrons) and it falls with energy E as $E^{-1/2}$; but it is only about 1.9 barns in the MeV range (fast neutrons). For ^{238}U, on the other hand, σ_f is *zero* below 0.9 MeV and is only about 1 barn in the MeV region; moreover, many large capture resonances exist below fast-neutron energies and σ_c is relatively large. The ^{239}U which is formed by such capture undergoes two successive beta decays to become ^{239}Pu, which fissions like ^{235}U at all energies. Thus the ^{238}U, which is said to be fertile, is converted into fissile ^{239}Pu. Similarly, ^{232}Th may be converted to fissile ^{233}U.

Neutrons emitted within 10^{-14} s after fission are said to be prompt, and in one fission may number from 0 to 8. The mean number ν born per fission event increases slightly with the energy of the incident particle. At thermal energy, ν is 2.47 for ^{235}U and 2.03 for ^{239}Pu.

Delayed neutrons, on the other hand, originate from the highly excited fragments that arise when fission occurs. The fragments of differing mass form a distribution with two peaks, centered at mass numbers 95 and 139 for ^{235}U. Because they are neutron rich, the fragments are radioactive. More than 50 sequences involving beta and gamma decay have been identified. This radioactivity from several hundred nuclides is the major hazard from reactors, and must be dealt with to ensure biological safety. A few of the nuclides decay by neutron emission: these give rise to the so-called delayed neutrons, which have discrete energies and half-lives determined by the precursor nuclide. For ^{235}U there is 0.0064 delayed neutron per fission neutron and these have half-lives ranging from 0.23 to 56 s. The mean time between generations of *prompt* neutrons is about 10^{-3} s in a thermal reactor; but when delayed neutrons are included it is two or three orders of magnitude larger. When the reactor becomes "critical," the latter becomes important to the control process.

NEUTRON MODERATION

Since σ_f for neutrons which are slow is much larger than for fast, the probability of fission is greatly enhanced by providing a moderator which contains low-mass nuclei. The latter allow neutrons to lose large amounts of energy in elastic scattering events; however, to be useful σ_c must be small. The ratio of slowing-down power to macroscopic absorption cross section, called the moderating ratio, is a figure of merit. Elastic collision by a 2-MeV fission neutron with a low-mass nucleus reduces the neutron energy to thermal energy in 18 collisions with hydrogen, 22 with deuterium, 114 with carbon; but it takes over 2000 with uranium. The latter would be the case in a fast reactor with little moderator, whereas in thermal reactors hydrogen, deuterium, or carbon is used. For power reactor moderators, most frequently used are water, graphite, and heavy water with respective moderating ratios of 67, 170, and 5820.

REACTOR TYPES

Reactors may be classified in many ways, often according to the relative proportions of fissile and fertile nuclides used, the predominant energy at which fission occurs, and the physical nature and distribution of materials employed to extract heat or to moderate neutrons. Heterogeneous reactors have the fuel distributed in lumps throughout the core, whereas homogeneous reactors have it uniformly dispersed. The medium used to extract thermal energy from the core may be a liquid, gas, or a molten metal such as sodium. A reactor is called a breeder if the fissile nuclei created are the same kind and greater in number than those consumed, such as with ^{239}Pu fuel surrounded by a ^{238}U blanket. If ^{235}U fuel is used to convert ^{238}U, it is called a converter. A power breeder converts ^{238}U into ^{239}Pu, simultaneously generating power and more *fissile* nuclei than were initially present. Thermal reactors capitalize on the large value of σ_f at slow-neutron energies, whereas fast reactors operate mainly with fast neutrons and little moderation.

The purpose of a power reactor is to provide heat to make steam for a turbine-driven electric generator. The heat is carried from the reactor fuel by a coolant medium, a liquid or a gas. Power reactors in worldwide use by the beginning of 1988, if classified by the type of coolant employed, fall into four categories: 81% use water (LWR); 11% use gas (GR); 7% use heavy water (HWR); and the remainder use liquid metals. Each group can be further subdivided according to moderator, neutron spectrum, and fuel type. There follows a brief description of only the most important kinds of thermal neutron reactors.

The LWR uses water as the moderator to promote thermalization of neutrons for better interaction with the fuel which is uranium oxide enriched to about 3% in ^{235}U. The fuel rods, composed of ceramic UO_2 pellets sealed inside metallic tubes, are made into bundles and are securely fastened apart to allow flow of coolant water to extract the fission heat. The bundles are arranged vertically in a rigid configuration called the core structure, and this is contained inside a massive steel pressure vessel, typically 8 to 9 in. thick. The pressure vessel resides within a massive concrete and steel housing, the containment structure, which is designed to contain radioactivity which may escape from the pressure vessel if it should rupture. Two types of LWR are used, pressurized water (PWR) and boiling water (BWR). The PWR operates at a pressure of about 150 atm, a water temperature of about 318°C, and thermal efficiency of 34%. Superheated water passes through coils in a steam generator and returns to the core without boiling. Steam made in the secondary loop of the steam generator then passes through a steam turbine, is condensed, and is recycled without coming into contact with the core water. By contrast, in the BWR the core water directly becomes steam which rises to the top of the pressure vessel and passes to the turbine to be condensed and returned to the core. It operates at about 70 atm pressure, 278°C, and has a thermal efficiency of 33%.

In the Soviet Union, water is used as a coolant for power reactors; however the moderator is graphite and the construction is very different. In a common design (RBMK-1000), about 1500 tons of graphite are penetrated by 1700 tubes which contain the fuel, and the water which is pumped through these tubes forms steam to drive the turbogenerator. In addition to making electricity, fuel may be cycled continuously; when it is processed, plutonium may be extracted. With this dual function there is no containment structure above the fuel tubes because easy access is required.

Gas-cooled power reactors have been developed in Great Britain which use a graphite moderator, natural uranium metal fuel, and CO_2 gas coolant. The heated gas is cycled through a steam generator back to the reactor. Although the temperature of the gas is higher than the core water of a PWR, the gas heat-exchange efficiency is smaller and the thermal efficiency is 25%. A later version, the advanced gas reactor (AGR), uses uranium oxide fuel enriched to 2.2% ^{235}U. It employs stainless-steel tubing for the fuel rods which allows substantially higher gas temperatures, and the thermal efficiency is 41%. An American version, the high-temperature gas reactor (HTGR), has helium gas for heat exchange, a graphite moderator, and highly enriched uranium and thorium in the form of tiny carbide-encased spheres for fuel. The HTGR uses a massive reinforced-concrete prestressed pressure vessel instead of steel as in the LWRs.

The heavy-water reactor developed by Canada and called the CANDU uses D_2O for both coolant and moderator. Fuel pellets formed from natural uranium oxide are sealed inside zirconium and small bundles of these are arranged inside horizontal steel pressure tubes through which the D_2O circulates. The pressure tubes are spaced inside a vessel, the calandria, which is filled with D_2O moderator at low pressure. The coolant at high pressure (100 atm) and temperature (293°C) is thus contained inside numerous small-diameter pressure tubes rather than inside a massive pressure vessel. The thermal efficiency is about 29%.

MULTIPLICATION FACTOR

Let us suppose that N_0 neutrons are simultaneously injected into a small lump of uranium. Because of the (n,x) reactions mentioned above, not all neutrons will produce fission. Suppose that there are N_0k neutrons present after the first fission generation which takes a mean time τ. Then in the second generation there are N_0k^2 present at 2τ, etc. The leakage of neutrons depends on the scattering cross section, lump dimensions, and geometry. After a very large number of generations r at a time $t=r\tau$, there are $N=N_0k^r=N_0k^{t/\tau}=N_0\exp[(t/\tau)\ln(k)]$ neutrons present. For a very small leaky lump, $k\ll1$; but there will be a size lump where k may be close to unity. The rate of change of N with time for large r and small τ is $(N\ln k)/\tau$. As the lump size increases, the outer parts are responsible for scattering back, into the middle, neutrons that otherwise would escape; thus, to a degree, the outer parts act as a reflector and enhance the interior neutron flux.

When $0<k<1$, $\ln k$ is negative and the neutron population will die away with exponential time constant $\tau/\ln k$. When $k=1$, the number present remains constant; and when $k>1$, the population increases exponentially with time.

The same type behavior is observed for a reactor when k_{eff} stands for the effective multiplication factor for the whole system. At $k_{eff}=1$ the reactor is said to be "critical."

REACTIVITY

The departure of k_{eff} from unity is called the reactivity ρ. It is defined as $\rho=(k_{eff}-1)/k_{eff}$, which is approximately $k_{eff}-1$, if $k_{eff}\approx1$. When the reactivity is equal to the fraction β of neutrons which are delayed, the reactor is said to be "prompt critical." The small time constant of the prompt neutrons now dominates the multiplication process, and the chain reaction is on the verge of a very rapid growth. When $\rho<\beta$, the time constant is dominated by the slow neutrons, for it is these which are making the reactor critical. The period for growth of power is now many seconds and the reactor may be easily controlled. Thus, a critical reactor is always operated so that $\rho<\beta$. The time for neutron flux or power to change by a factor of e is called the reactor period, and it changes for the different regimes of reactivity.

NONLEAKAGE PROBABILITY

If the medium discussed above were infinitely large, no neutrons could leak away and the multiplication constant $k=k_\infty$ would depend on the fuel and the moderator, but not on the size of the system. In a reactor, if P is the probability that a neutron will not be lost by leakage, then the multiplication factor is $k_{eff}=Pk_\infty$.

FOUR-FACTOR FORMULA

The size of k_∞ may be estimated by considering the fate of N fission neutrons liberated in the medium, say from thermal fission of ^{235}U in the presence of ^{238}U. Because fast neutrons create some fission in the ^{238}U, there will be somewhat more neutrons than would have arisen from the ^{235}U alone. With this enhancement there are $N\epsilon$ fast neutrons created, where ϵ is called the fast fission factor. Now suppose that for each fast neutron produced p is the number which reach thermal energy and are then absorbed. This is called the resonance escape probability and gives the chance that the neutrons will not be absorbed in neutron resonances which have zero fission cross section. Thus $N\epsilon p$ thermal neutrons are available to induce fission.

If f is the fraction of thermal neutrons absorbed by the fuel, called the thermal utilization factor, the number absorbed is $N\epsilon pf$. If η is the number of neutrons produced by each thermal neutron absorbed, for N initial neutrons the number generated is $N\epsilon pf\eta$. Accordingly, $k_\infty = N\epsilon pf\eta/N$; thus, $k_\infty = \epsilon pf\eta$. The factors in the formula, some measured, some calculated, depend on whether the reactor is heterogeneous or homogeneous. Only if k_∞ is greater than unity can a chain reaction be possible and this gives guidance in the design of a thermal reactor.

REACTOR CONTROL

Factors which influence the reactivity of the reactor, hence allow it to be controlled, include the addition or removal of fuel, moderator, reflector, or a neutron absorber. To vary the multiplication factor and ultimately to hold it constant, control rods are inserted in the core. These may introduce positive or negative reactivity according to the purpose at hand; thus, there may be shim rods with positive reactivity to bring the reactor up to power, regulating rods to provide small but fast changes, and safety rods which allow very rapid insertion of strong neutron absorbers in order to shut off the chain reaction. The safety rods are usually made of steel impregnated with strong neutron absorbers such as boron and cadmium. They are made fail safe so that should control power fail, they will return to the core, if not within it, usually by gravity. Slowly withdrawing a rod with negative reactivity will allow k_{eff} to increase and this is detected with appropriate neutron detectors and period meters. Alternatively, slowly inserting a shim rod with positive reactivity will also allow k_{eff} to increase. Various combinations are used in practice.

Factors which affect the continuous operation of the reactor include the depletion of the fissile fuel and the growth of fission products which include some very strong neutron absorbers, and hence are poisons to the neutron balance. A particularly troublesome case is the formation from ^{135}I of ^{135}Xe, with 6.2 h half-life, and the subsequent decay of ^{135}Xe with half-life of 9.2 h. The ^{135}Xe absorption cross section is huge, and once a reactor is shut down it may take some days to start up again after the ^{135}Xe poison has decayed sufficiently.

In western countries such as the United States the reactivity of the fuel and moderator configuration is designed to have a negative temperature coefficient, so that power increases initiate decreased reactivity. This ensures that an escalating chain reaction cannot become a nuclear bomb. In the LWGR reactors used in Russia of the type designated RBMK-1000 in which the coolant water forms steam to drive the turbogenerator, loss of coolant water causes escalation of the chain reaction, the temperature coefficient is positive, and a large power surge is possible. Operation at low power level is difficult to control with this design.

SHIELDING

Tremendous levels of radioactivity exist in the reactor from fission-fragment decay and neutron activation of the materials present. In a high-flux reactor, massive shields 3.5 m thick surround the pressure vessel to attenuate the neutrons and ionizing radiations. Secondary radiations generated in the shield by the attenuation processes increase the required thickness of the shielding materials, such as iron and concrete, which are normally used. In addition in western-type power reactors, a massive containment structure of steel and concrete surrounds the pressure vessel to keep radioactive material from reaching the biosphere if the core is damaged and the pressure vessel is breached.

HUMAN SAFETY

Throughout the world in 1988 more than one-sixth of the electrical energy came from nuclear power plants, although in some countries skepticism about plant safety slowed the growth of nuclear power. In the United States no new plants have been ordered since 1978. Nuclear accidents in 1979 at Three Mile Island (TMI), Pennsylvania, and in 1986 at Chernobyl, U.S.S.R., provide information about health effects from major nuclear accidents.

At TMI more than half the PWR-reactor core melted because of insufficient cooling caused by a stuck valve and operator error; yet, aside from noble gases, only very small amounts of radioactivity entered the biosphere outside of the containment structure. According to the Pennsylvania Department of Health no significant health effects due to exposure to radiation have been found.

On the other hand, in the Chernobyl disaster at the RBMK-1000 reactor which lacked a containment structure, essentially all of the noble gases, more than 20% of ^{131}I, and 13% of ^{137}Cs, as well as other fission products in the reactor, were injected into the air and ultimately detected around the world. This happened during an experiment, in preparation for which the operators had disabled the automatic shutdown system, and had violated several other basic operational rules. Locally 31 deaths resulted, 2 from steam and hydrogen explosions, and the rest from burns and radiation sickness. It has been estimated that ultimately within 30 km of the plant there may be perhaps 100 additional deaths. This will be difficult to detect since it is less than the statistical fluctuation normally expected for cancer deaths $(24{,}000 \pm 155)$ for the population from within 30 km. Similarly within Russia and the rest of the world, the incremental deaths may not be detectable epidemiologically. However, extraordinary costs resulted from resettlement of thousands of people, entombment of the reactor, and decontamination of the region around the plant.

NUCLEAR FUEL

A power reactor of 1000 MW electrical capacity uses about 3 kg of fissile fuel per day. Present American power reactors are light-water moderated and use fuel enriched to 3% ^{235}U. But natural uranium contains only 0.71% ^{235}U and 99.29% ^{238}U. After some decades of growth, the power industry worldwide may find it difficult to obtain fuel which is cost effective if supply becomes difficult. However, fissile fuel from uranium can be multiplied 60 to 100 times by breeding ^{239}Pu.

BREEDING

The neutron excess, $\eta - 1$, must be significantly larger than unity for practical breeding. From the excess neutrons the next generation needs one, some are lost, and the rest may breed. A fast reactor using ^{239}Pu with $\eta - 1 = 1.94$, or possibly a slow reactor using ^{233}U converted from ^{232}Th, has enough neutrons to breed. The doubling time is the time it takes for the breeder to produce a surplus of fissile material equal to that in the initial fuel. Breeding may be economic if the doubling time is not too great.

See also NUCLEAR FISSION; NUCLEAR PROPERTIES; NUCLEAR REACTIONS.

BIBLIOGRAPHY

A. M. Weinberg and E. P. Wigner, *The Physical Theory of Neutron Chain Reactors*. University of Chicago Press, Chicago, IL, 1958.

R. V. Meghreblian and D. K. Holmes, *Reactor Analysis*. McGraw-Hill, New York, 1960.

S. Glasstone and A. Sesonke, *Nuclear Reactor Engineering*. Van Nostrand, New York, 1967.

B. L. Cohen, "The nuclear reactor accident at Chernobyl, USSR," *Am. J. Phys.* **55**, 1076 (1987).

Nuclear Scattering

Herman Feshbach

The scattering of various projectiles by nuclei is one of the principal methods by means of which the structure of nuclei and nuclear dynamics can be studied. It was through the scattering of alpha particles that the nucleus was discovered by Rutherford. In a typical experiment, a beam of particles, the "projectiles," strikes a target which may be in the form of a solid foil, though often liquid and gaseous targets are used. When the projectile strikes an atomic nucleus in the target, the "target nucleus," a scattering is said to have occurred. When the scattering does not result in any change in the internal energy of either the incident projectile or the target nucleus, the scattering is said to be *elastic*. If, on the other hand, internal energy is gained by the target nucleus or by the projectile so that at the end of the collision either or both are excited, the scattering is said to be *inelastic*. If the projectile and concomitantly the target nucleus do not maintain their identity through the exchanges of mass and/or charge, a *reaction* is said to have occurred. The nature of the collision can be determined by observing the emergent particle or the residual nucleus. The measurements include measuring the energy of the emerging particle, its direction of travel, as well as its mass and charge. The decay of the residual nucleus by the emission of gamma rays, by beta decay, and in some cases by the emission of heavier particles such as nucleons or alpha particles helps to determine the nature of the collision.

KINEMATICS

The variables which are important for the collision process are shown in Fig. 1a where the collision is pictured in the zero-momentum frame in which the incident projectile and target nucleus have equal and opposite momenta. This is to be contrasted with the laboratory frame in which the target nucleus is at rest and the incident projectile has all the initial momentum as in Fig. 1b. The relevant variables are the angle of the scattering ϑ, and the initial momentum $\mathbf{p}_i$ and the final momentum $\mathbf{p}_f$ of the projectile. The energy available for the collision, E, in the zero-momentum frame is related in the nonrelativistic limit to the projectile energy in the laboratory frame E_l by $E=[m_n/(m_p+m_n)]E_l$, where m_n is the nuclear mass and m_p is the projectile mass. Relativistically, this relation is replaced by $s=2W_l m_n+m_n^2+m_p^2$, where W_l is the laboratory energy including the rest mass. The available energy in the zero-momentum frame is $\sqrt{s}$.

CROSS SECTION

The differential cross section $d\sigma$ is defined as the number of particles scattered per unit time into a solid angle $d\Omega$ at an angle ϑ divided by the flux, the incident number of particles per unit area per unit time. The total cross section σ is the total number of particles scattered divided by the flux. Hence

$$\sigma = \int \frac{d\sigma}{d\Omega}\, d\Omega.$$

The angular distribution of scattering particles is $d\sigma/d\Omega$. If the inelastic collision leads to a part of the energy spectrum of the residual nucleus which is continuous, then it is possible to define a double differential cross section $d^2\sigma/d\Omega\, d\epsilon$ giving the number of particles scattered per unit time into a solid angle $d\Omega$ with an energy between ϵ and $\epsilon + d\epsilon$.

Since the maximum angular momentum brought into the reaction is $p_i R$, where R is the radius of the interaction re-

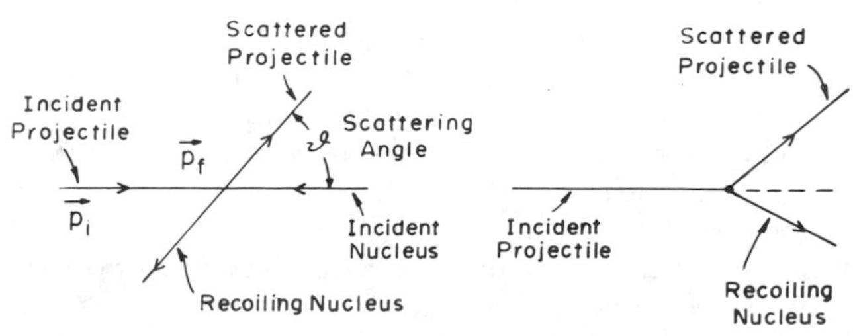

FIG. 1.

gion, and the maximum which leaves is $p_f R$, the highest power of cos ϑ which appears in $d\sigma/d\Omega$ is the lesser of the integers closest to $2p_i R/\hbar$ and $2p_f R/\hbar$ for the collision of spinless particles. For very low energies such that $p_i R < \hbar$, $d\sigma/d\Omega$ is independent of cos ϑ so that the angular distribution is spherical. As p_i increases, the angular distribution will become more complex since higher powers of cos ϑ will be present in the angular distribution.

The nature of the interaction between the projectile and target is reflected in the effects of the collision. If the interaction is strong, one can expect, subject to the kinematical restrictions of the preceding paragraph, an enhanced probability for larger angular deflections. If the interacting particles have intrinsic properties such as spin, isospin, or strangeness, the collision can result in changes in the projectile's intrinsic properties which are complemented by corresponding changes in the target nucleus. A large variety of projectiles have been used including those whose main interaction is electromagnetic such as electrons and muons and those referred to as hadrons whose interaction with nucleons is strong. These include neutrons, protons, and nuclei ranging from the lightest such as ^{3}H, ^{3}He, and alpha particles to heavier nuclei, referred to as heavy ions. The hadron list is augmented by the "exotic" particles such as pions, kaons, and antiprotons. The hadrons possess the intrinsic degrees of freedom of spin and isospin whereas the kaons possess strangeness as well, with the consequence that collisions involving the exchange of angular momentum, charge, and strangeness will occur.

ELASTIC SCATTERING

The elastic scattering of *electrons* by nuclei is a consequence of their mutual electromagnetic interaction. One can consider the electron accelerator providing the electron beam to be a large electron microscope. It is thus possible from the angular distribution to determine the electric charge and current density of nuclei to the extent determined by the wavelength of the incident electrons. At lower energies only the charge radius can be ascertained, whereas at the higher energies, where for example electrons of 500 MeV have a wavelength/2π of 0.4×10^{-13} cm, it becomes possible to obtain considerable information on the graininess and shape of nuclei.

The elastic scattering of nucleons at lower energies presents an important dilemma. With good energy resolution the cross section fluctuates rapidly. When the energy resolution is sufficiently broad, the total cross section and angular distribution vary smoothly with energy and mass number. The latter data can be explained by assuming that the incident nucleon moves in a *mean field* whose parameters are determined from the data. This model is referred to as the *optical model*. This mean field exists in spite of the strong interactions between the incident nucleon and the nucleons of the target nucleus. The dilemma is resolved by introducing the concept of *interaction time*. The optical model describes only the prompt portion of the scattering, that is, the component of the scattering for which the interaction time is short. A long interaction time is required before the fluctuations evidence themselves. The range of interaction times is large so that one can establish a hierarchy of interaction times with characteristic associated phenomena. The isobar analog resonance and the giant resonances are examples for which the interaction time is intermediate.

The optical model describes the interaction in terms of a complex potential which is inserted into the Schrödinger equation: $\nabla^2\psi + (2m/\hbar^2)(E - V_{\text{opt}})\psi = 0$. In its initial application only a central potential $V_{\text{opt}} = -[V(r) + iW(r)]$, where r is the distance between the incident particle and the center of mass of the target, was used. The forms of V and W and their strengths are varied so as to give agreement with the experimentally observed elastic scattering for a variety of nuclei and over a range in projectile energy. This form is adequate for spin-0 projectiles such as the alpha particle incident upon spin-0 nuclei. When either the projectile or target nucleus has a spin, for example if the projectile is a proton, it is necessary to add a spin–orbit potential in order to explain the scattering of polarized projectiles. This potential permits the exchange of angular momentum between the intrinsic spin of the projectile and its orbital angular momentum. When both projectile and target nuclei have spin, there is evidence for the presence of small terms in the optical potential involving the spin of the projectile and the spin of the target nucleus. The optical potential can also involve the isospins of the target and projectile permitting an exchange of charge between them during the collision. The importance of this term in the optical potential was emphasized by the discovery of the *isobaric analog resonance* in the elastic scattering of protons by nuclei. When the projectile has a large enough isospin, e.g., the pion, double-charge exchange in pion collisions becomes possible. When the projectile is a kaon, the interaction will permit strangeness exchange leading to the formation of hypernuclei.

The qualitative relationship between the optical model and the underlying nucleon–nucleon interaction has been established for the lower-energy domain. High-energy hadron projectiles probe the neutron density of nuclei. So far it has been possible to determine that density in the surface and near-surface regions. The elastic scattering of 1-GeV protons by ^{4}He indicates the presence of three-body forces. In the intermediate energy region the central part of the potential $V(r)$ decreases, finally changing sign, that is, becoming repulsive, at high energies. In this intermediate range of a few hundred MeV nucleon energy, the spin-dependent forces are most important. Recently the optical model using the Dirac rather than the Schrödinger equation has been employed resulting in a substantial economy in empirical parameters.

The elastic scattering of pions is unique in that the pion combines with a nucleon in the nucleus to form an excited and unstable state of the nucleon, the Δ. The Δ propagates in the nucleus decaying back into a nucleon plus a pion. The properties of the Δ inside the nucleus turn out to be substantially modified by its nucleon environment and therefore differ from those of a free Δ.

INELASTIC SCATTERING

The energy, angular momentum, and parity of a state of a target nucleus excited by inelastic scattering can be determined from the energy loss of the incident projectile, its angular distribution, and the angular distribution of the gamma ray emitted when the excited state decays back to the ground state. This is most easily accomplished in the case when the projectile–nucleus interaction is electromagnetic, as in inelastic electron scattering and Coulomb excitation by heavy ions. In this case it is possible to determine the nuclear shape as well as the transition electric and magnetic multipole moment densities involved in the excitation. It is often possible to associate a particular energy region, in which inelastic scattering is enhanced, with a giant multipole resonance. Inelastic scattering of protons, deuterons, α particles, and heavy ions has also been used to excite these resonances. Surface vibrations are most readily excited using particles which do not easily penetrate into the nucleus such as α particles and deuterons. On the other hand, proton projectiles which penetrate rather readily excite a much greater variety of nuclear states. The optical model can be extended so as to describe phenomenologically both elastic and inelastic scattering.

See also ANGULAR CORRELATION OF NUCLEAR RADIATION; HADRONS; ISOBARIC ANALOG STATES; ISOSPIN; NUCLEAR PROPERTIES; NUCLEAR REACTIONS; NUCLEAR STATES; PHOTONUCLEAR REACTIONS; SCATTERING THEORY.

BIBLIOGRAPHY

H. Enge, *Introduction to Nuclear Physics*. Addison-Wesley, Reading, MA, 1966. (I)

H. Feshbach, "The Complex Potential Model," in *Nuclear Spectroscopy, Part B* (Fay Ajzenberg-Selove, ed.). Academic Press, New York, 1960. (A)

P. E. Hodgson, *Nuclear Reactions and Nuclear Structure*. Clarendon Press, Oxford, 1971. (A)

Nuclear States

D. Kurath

A nucleus, considered as a many-body system of nucleons bound together by specifically nuclear forces, can exist in a series of stationary states of different energy. In addition to its excitation energy E, each of these states is characterized by a definite total angular momentum J and parity π. In addition to the strong nuclear interactions, nuclei also interact with the electromagnetic field and the still weaker field associated with beta-decay processes. The transitions between nuclear states are determined by this hierarchy of interactions. A state will decay to a state of lower energy by emission of a nucleon or group of nucleons if this is energetically possible; if this is not possible, it will decay electromagnetically by emission of a gamma ray.

Once it reaches the ground state for a given number of protons Z and neutrons N, it will reach the beta-stable ground state for that mass number by the beta-decay process, either changing a neutron into a proton or vice versa. Thus all states other than the beta-stable ground states have a finite mean life τ. Each excited state thus has an energy width Γ, which is related to the mean life by the relationship $\Gamma\tau = \hbar$.

As one expects from a many-body system, the spacing of nuclear energy levels varies both with excitation energy and number of nucleons. In general the low-lying levels are more widely spaced, representing the excitation of few degrees of freedom; and while the levels are still discrete at energies around the threshold for nucleon emission, they become overlapping resonances at higher energies. The levels are most widely spaced for nuclei of even Z and even N, and the level spacing becomes smaller for heavier nuclei although there are a few exceptions due to nuclear structure.

Many reactions are used to probe the nature of these states. Mean lives can often be determined by measuring the gamma-decay time. Transition probabilities from the ground states can be measured by inelastic scattering of electrons, nucleons, alphas, and pions. Transfer reactions like the well-known (d,p) stripping reactions reveal the orbital angular-momentum difference between initial and final states by the angular distribution and give the total angular-momentum transfer if polarization is measured. There is a large body of experimental information about the properties of discrete nuclear states as well as the more highly excited overlapping states. Nuclear models have been proposed to correlate and interpret this information.

Ideally one should solve the Schrödinger equation for the many-nucleon system using the measured nucleon–nucleon interaction as input. However, the very strong short-range nature of this interaction makes this a difficult procedure, and it has been carefully done only in the case of infinite nuclear matter. Instead, for finite nuclei the independent-particle model has provided the best description of nuclear states. The model was proposed on an empirical basis, and calculations similar to the nuclear matter case indicate that the model is not theoretically unreasonable although it has not been derived. The basic assumption of the model is that each nucleon in the nucleus moves in a common potential field which represents the average effect of the interactions with all the other nucleons. This average potential produces an energy spectrum of groups of fairly closely spaced energy levels, separated from other groups by relatively large energy gaps. This results in a shell model and the introduction of a strong spin–orbit coupling term in the potential led to the modern *jj*-coupling shell model. Each energy level is labeled by n (related to the number of radial nodes), l (the orbital angular momentum), and j (the total single-nucleon angular momentum). Then, consistent with the Pauli exclusion principle, up to $2j+1$ identical nucleons can occupy this level. An added assumption of the model is that an even number of nucleons in such a level has lowest energy if coupled to angular momentum zero. With this model one obtains the ground state of A nucleons by filling the levels in order and coupling angular momenta such that the resultant angular momentum is $J=0$ for an even-Z even-N nucleus and $J=j$ for an odd-A nucleus where j comes from the last unfilled level. These features and the major-shell energy gaps when compared with experimental observations established the validity of the *jj*-coupling shell model.

Within one particle of the closed shells resulting from the large energy gaps, the excitation spectra of such nuclei exhibit the level order of the levels within the shells. However, when a number of nucleons are in unfilled levels above a $J=0$ core, such a configuration mixes with other nearby configurations through the residual interaction between nucleons. (This is the remainder of the original interaction which cannot be included in the average single-particle potential, but is small compared to the shell gaps.) The residual interaction can be crudely estimated, but it must be fine-tuned empirically. The resulting calculations provide a good description of the low-lying states of nuclei up to mass 40 and in other regions of the periodic table near closed shells.

When both neutrons and protons have a good number of particles away from closed shells, a better description is obtained by considering the average potential to have an ellipsoidal rather than a spherical shape. The low-lying states are then interpreted as arising from rotations or vibrations of this shape, a collective action of the nucleons beyond the closed spherical core. Again one has single-particle levels, this time in a deformed potential well. These single particles are strongly coupled to the rotations and vibrations of the deformed field. This model provides an excellent picture of the low-lying states for heavy nuclei in several regions of the periodic table.

Other forms of collective behavior are known to be quite general. The best known of these is the giant electric dipole ($E1$) resonance. This resonance arises from a vibration of neutrons with respect to protons, with the strength concentrated in a small region of excitation energy. Other collective multipole resonances exist such as $E2$ and $E3$ resonances as well as magnetic resonances $M1$ and $M2$ resulting from collective spin interactions. These resonances are generally at such excitation energy that they are superimposed on a background of incoherent overlapping levels, known as compound-nucleus states.

Compound-nucleus states, first observed as resonances in slow-neutron capture, occur at relatively large excitation energies E. The density of these states (number per unit energy interval) and their mean lives are several orders of magnitude larger than is predicted by their description in terms of a few neighboring configurations. To account for the observed properties of these states it is necessary to assume that E is shared by many nucleons. Thus, as E becomes larger, configurations with energy near E will involve holes in shells that are fully occupied at lower excitation energies. In this way the residual interaction becomes increasingly effective since interactions between nucleons in partially filled shells are not restricted to the same extent by the Pauli exclusion principle. The density of relevant configurations thus increases with E and they are more thoroughly mixed by the stronger effective residual interactions. This explains the large density of compound-nucleus states. The small observed values of $\Psi_{\lambda c}$, the decay width of the compound-nucleus state λ into a given two-body channel c, are also easily understood. The square root of the decay width is a measure of the amplitude of the configuration corresponding to channel c in the wave function of the state λ. Since there are very many configurations, each with roughly equal amplitude, in this wave function $\Gamma_{\lambda c}$ is expected to be small.

The large density of compound-nucleus states makes it impractical to study the positions and widths of the individual states. Physically significant results are obtained by averaging in some way over many states and this, in turn, implies a statistical theory. The global statistics of these states (the state density and the strength function) are adequately described in terms of single-particle models. For the state density this result is based on the fact that this density is roughly equal to the configuration density when proper account is taken of degeneracies. The strength function is defined as $\langle\Gamma_{\lambda c}\rangle/D$ where D is the average distance between resonances λ (of the same J and π) and $\langle\ \rangle$ denotes an average over resonances λ within a given energy interval. Since a given configuration c, which represents single-particle motion with respect to a relevant target, is distributed over the wave functions of many neighboring resonances, some vestige of this single-particle motion should be evident in the spectrum of resonances averaged over an energy interval large enough to dampen fluctuations due to individual resonances. The observed behavior of strength functions is well represented by a single-particle (optical) potential.

In summary there is a great variety of nuclear states. At low energy they represent excitation of few degrees of freedom and they offer a useful laboratory for testing fundamental theories such as the presence of parity violation in the weak interaction. At higher energy the compound-nucleus states are understood in a statistical manner. Superimposed on this region are collective states understood as cooperative behavior of the nucleons caused by strong features of the residual interaction. Experiments with heavy-ion accelerators, wherein a sizeable projectile strikes a sizable target, have produced highly deformed states with as much as 60 units of angular momentum which then lose energy and angular momentum by gamma decays as a rotational band. This behavior is again superimposed on a background of compound-nucleus states at high excitation energy. The many intriguing features of nuclear structure are revealed by the study of nuclear states.

See also NUCLEAR PROPERTIES; NUCLEAR STRUCTURE; WEAK INTERACTIONS.

BIBLIOGRAPHY

Aage Bohr and Ben R. Mottelson, *Nuclear Structure; Single-Particle Motion, Vol. I.* Benjamin, New York, 1969; *Nuclear Deformations, Vol. II.* Benjamin, Reading, MA, 1975.

R. D. Lawson, *Theory of the Nuclear Shell Model.* Clarendon Press Oxford, 1980.

Nuclear Structure

J. P. Davidson

Initial conjectures about the structure of the nucleus of the atom were put forward soon after Ernest Rutherford proposed the nuclear atom in 1911 and Henry Mosley showed, in 1914, that the nuclear charge (Z) was equal to the atomic number. W. D. Harkins, in 1917, was probably the first person to suggest a reason why the nuclei of even-Z atoms were the most stable. In 1922, he proposed a model based upon a detailed investigation of the then-known nuclear properties. Unfortunately, his model (like others of the time) required negative electrons to be permanent nuclear constituents. However, the laws of quantum mechanics (discovered shortly afterward) prohibit the confining of an electron within such a small volume. Thus, the dilemma of nuclear structure had to await the discovery of the heavy neutral particle. Ten more years were to pass before James Chadwick discovered such a particle, the neutron. From this time, progress was rapid, and, within 4 or 5 years, the general outline of our present understanding had been developed, including models emphasizing single-particle properties, collective effects, and statistical effects. Furthermore, some progress on the nature of the specific nuclear force had been made.

It is known that each nucleus consists of Z protons and N neutrons tightly bound together by a force much stronger than either gravitational or electromagnetic forces and of very short range (about 10^{-13} cm—the magnitude of nuclear dimensions). While it is strongly attractive ($\sim$100 MeV) for these ranges, it must also be very strongly repulsive for shorter ranges in order to prevent the collapse of the system. Thus our current idea of the specific nuclear force is that it is charge independent (i.e., the force acting between any two pairs of nucleons is the same regardless of their charge state) and contains a strong repulsive core in order to prevent collapse. The force contains spin–orbit and tensor as well as central terms and is represented by a class of potentials known by such names as the Hamada–Johnston and Reid soft-core potentials. It is hoped that the use of such potentials will allow the calculation of many fundamental nuclear properties, such as the binding energy per particle (about 8 MeV), as well as of the characteristics of the several successful phenomenological models. A good deal is known about the properties the potential should possess, but even if its precise mathematical expression were known, the solution of the general nuclear structure problem would not be possible, since even in the classical case the solution of the general three-body problem is not available in closed form. Thus theoretical investigations follow one of two paths. Either we attempt to solve, by various approximate means, the many-body nuclear problem using the best available nuclear potential, or we use a phenomenological model that emphasizes those particular nuclear properties believed to play a central role in systems under investigation.

For instance, there has long been known a set of nucleon numbers called the "magic numbers" (N or Z = 2, 8, 20, 28, 50, 82, and N = 126) where many nuclear properties show a discontinuity when plotted as a function of N or Z. Some properties showing such discontinuities are the neutron separation energy, neutron-capture cross sections, the mass parameter associated with the isobaric energy in the semiempirical mass formula, the sign of the static electric-quadrupole moment, etc. The reasons for these discontinuities were understood only with the advent of the shell model about 1949, while the reason nuclei in certain regions have electric-quadrupole moments far too large to be explained by such single-particle models is best understood by using models that emphasize the cooperative effects of a many-nucleon system.

Once the question of the constituents of the nucleus has been resolved, the next most fundamental property of nuclear structure is the size and shape of the nucleus itself. In general, because of the strong attractive forces between nucleons, we might expect the neutron and proton distributions to be essentially identical. Thus the mass distribution is most easily determined experimentally by measuring the nuclear charge distribution. This, in turn, is best done in two essentially different experiments: high-energy ($E \gtrsim 200$ MeV) electron scattering (Coulomb excitation) and analysis of x rays from muonic atoms (in particular the K, L, M, and N x rays). The data in both types of experiment are usually fitted by a deformed Fermi charge distribution of the form

$$\rho(r,\theta)=\rho_0\{1+\exp[(r-R(\theta))/a]\}^{-1},$$

where the deformation term is given by

$$R(\theta)=c[1+\beta_2 Y_{20}(\theta)+\beta_4 Y_{40}(\theta)+\cdots].$$

Here β_2 is a measure of the quadrupole deformation which can be quite large in the deformed regions whereas β_4 is a measure of the hexadecapole deformation. The spherical harmonics are denoted by $Y_{l0}(\theta)$. Making use of large amounts of data one obtains reasonable values for the half-density radius, $c=(1.18A^{1/3}-0.48)$ fm. The parameter a is related to the skin thickness t by the relation $t=4a \ln 3$. The data support a slowly varying value from $a=0.455$ fm near $A=10$ to $a=0.510$ at $A=240$. The deformation term $R(\theta)$ is, of course, zero for spherical nuclei and varies from nucleus to nucleus in the several deformed regions.

The neutron and proton distributions, while essentially the same, are best related by measuring $\Delta=\langle r_n^2\rangle-\langle r_p^2\rangle$ and this is most easily done from the analysis of π- and K-mesonic x rays. Because the π and K mesons are strongly interacting particles, they probe the outermost layers of nuclei. The interpretation of the data, while difficult, is consistent with very small Δ ($\cong$0–0.2 fm).

It is interesting to note that, for the heavy nucleus ^{208}Pb, the current evidence suggests that the central density is somewhat less than the maximum density (at $r \cong 5$ fm) consistent with the notion that the repulsion of the protons "drags" along the neutrons and thus reduces the central density.

The shape of the average nuclear potential well is closely related to, and influences the shape of, the nucleus. In a general way, this well deepens as A increases and in such a way that the nucleons are bound with an energy of about 8 MeV per particle. For nuclei with A near the magic numbers the well is essentially spherical whereas in regions removed from these shell closures the well becomes deformed and is best represented by a prolate spheriod. As one moves up in energy from the bottom of the well (be it spherically symmetric or deformed), the walls broaden out more or less parabolically until, at a radius depending on Z, they hook over and join smoothly onto the $1/r$ Coulomb potential. Thus a barrier is presented to prevent the escape or entry of charged particles (i.e., alpha decay or charged-particle scattering). However, for the very heavy nuclei, the actinides and beyond, the results of neutron experiments as well as neutron and charged particle induced fission indicate that the well becomes "double-" and even "triple-humped." These new wells represent in turn greater and greater deformations. The depths of the second and third wells increase as A does and a nucleus which finds itself in such a high-energy state can decay by several unusual modes. It can decay rapidly within a given well by gamma emission and then more slowly decay into an inner well (shape-changing gamma decay proceeds more slowly than non-shape-changing decay). Alternatively, after reaching the lowest state in an outer well it can decay by spontaneous (isomeric) fission with a lifetime considerably longer than might be expected from such a state (for ^{242}Am, $\tau_{1/2f} = 15$ ms).

The shape of the nuclear charge distribution and hence of the nucleus itself has important ramifications for nuclear structure models. If the nucleus is spherical, it possesses rotational symmetry about any axis through its center, so that rotations will not be measurable for such a system. Because of this spherical symmetry each particle's total angular momentum, **j**, is a good quantum number. If it is deformed from spherical, then particles moving within its associated potential well will not have total angular momentum as a good quantum number but the system as a whole should show a low-energy spectrum proportional to $I(I+1)$, where I is the total nuclear angular momentum eigenvalue. Expanding the nuclear charge distribution in spherical harmonics as given by the equation before, the even terms are related to the static electric moments (quantum-mechanical symmetry conditions make the expectation values of the odd terms identically zero). The first (monopole) term is proportional to the total charge Z, while the second term is related to the quadrupole moment Q_0, the third (even) term to the hexadecapole moment H_0, etc. Since a measurable ground-state quadrupole moment requires a nuclear spin $I \gtrsim 1$, all even–even-A nuclei have zero ground-state quadrupole moments (since $I_{\text{gnd}} = 0$ for all such nuclei). Thus a direct measurement of the ground-state quadrupole shape is only possible for odd-A nuclei. The reduced transition probabilities, $B(E2)$, also depend on the nuclear charge shape and so can be used to determine the quadrupole shape in all nuclei. These moments show strong shell efffects, changing sign passing through a magic number and being very large in the so-called deformed regions, roughly where $150 \lesssim A \lesssim 190$ and $A \gtrsim 225$. The nuclear deformation for even–even-A nuclei is evidenced either by the rotational signature of the low-lying energy levels $[E \propto I(I+1)]$ or through Coulomb excitation. The next electric moment is H_0, the hexadecapole, which determines the deformation parameter β_4. A number of these have been measured in the deformed regions both by Coulomb excitation and by the analysis of muonic x-ray hyperfine spectra. The values for H_0 and β_4 determined by these two methods are in reasonable agreement although the trends with A are not. Currently, no higher electric moments or deformations have been determined.

Static magnetic-multipole moments occur only for the odd terms in the magnetic multipole of expansion, so that the largest and most important is the first or magnetic dipole. Again angular momentum selection rules require that $I \gtrsim \frac{1}{2}$, so the even–even-A nuclei in their ground states display no magnetic-dipole moments. The dipole magnetic moment operator, being a vector, is proportional to the total angular momentum operator, which in turn is the vector sum of orbital and spin angular momenta of each particle. Thus,

$$\boldsymbol{\mu} = \mu_0 g_J \mathbf{I} = \mu_0 \sum_{i=1} (g_i^l l_i + g_i^s \mathbf{s}_i).$$

Since the simplest model (the extreme single-particle model) of odd-A nuclei assumes that all of the nucleons but the last one couple to zero angular momentum, then the magnetic properties of the nucleus are produced by this last particle. If we further assume that its gyromagnetic ratios have the free-nucleon values ($g^l = 0$ or 1, $g^s = -3.826$ or 5.587 for neutron or proton, respectively), then the nuclear moment is quickly calculated. It is remarkable that except for a few light nuclei (^{1}H,^{3}He,^{13}C,^{15}N, and ^{237}Np) the measured values lie between the calculated or Schmidt limits (first calculated by T. Schmidt in 1937). The nuclei with magic number ± 1 come closest to these limits. For the others, numerous explanations have been advanced for the discrepancies. For instance, in self-conjugate nuclei ($N=Z$), only an isoscalar contribution is present, which can be obtained by averaging the moments of corresponding states in mirror nuclei. On the other hand, the deviations for deformed nuclei can be reasonably accounted for by using one of the collective models, which also explain the large electric-quadrupole moments.

The highest moment that has been measured is the magnetic octupole, but only about a dozen have been measured. The angular momentum selection rules require $I \gtrsim \frac{3}{2}$ and the moments calculated using the extreme single-particle model (sometimes called the Schwartz limits) do not describe the measured values as well as the Schmidt limits describe the dipole moments. A collective-model calculation fares somewhat better, although the nuclei in question are generally not considered deformed.

The very large body of nuclear data has given rise to a number of models that are used either to explain and give insight into the workings of nuclear matter or to catalog and relate data from apparently dissimilar origins. We shall outline three or four of the simplest and most used models, leaving to the literature the many others, not a few of which are simply variations on a theme. Because of the diversity of the properties and associated data, each model must emphasize certain properties to the exclusion of others and thus the assumptions of the model delineate the area of its success.

Perhaps the oldest, and conceptionally the simplest, is the statistical model in which numerous nuclear properties are distilled into an equation of state for the total energy of the nucleus. Often called the Bethe–Weizsäcker semiempirical mass formula, it takes the general form

$$\begin{aligned} M(N,Z,\text{shape}) &= NM_n + ZM_{\text{H}} + \text{volume energy} \\ &\quad + \text{surface energy} + \text{Coulomb energy} \\ &\quad + (\text{odd–even}) \text{ correction.} \end{aligned}$$

In this formula M_n and M_{H} are the masses of the neutron and hydrogen atom, respectively. In the earlier formulations, the volume energy was taken to be simply proportional to the volume and hence A, the surface energy proportional to

$A^{2/3}$, and the Coulomb energy proportional to the number of the proton pairs divided by a nuclear length, $Z(Z-1)/A^{1/3}$. Finally, the odd–even correction was taken to be $\pm\delta/2A$ for odd (+) or even (−) nuclei and zero for odd-mass nuclei. In the earlier work, the constants of proportionality were fixed by the experimental data and were the same for all nuclei. For instance, the volume and surface terms were written

$$\text{volume energy} = -a_v A,$$

$$\text{surface energy} = a_s A^{2/3}$$

with a_v and a_s of the order of 15 MeV and $\delta \simeq 2770$ MeV.

When the important role played by the nuclear shells was recognized, a shell function was added to the mass formula. This evidences the shell structure by oscillations with nodes at the magic numbers and by an amplitude which increases with N or Z.

The fact that many of the heavier nuclei are deformed was also incorporated into the mass formula by including the dependence on nuclear shape as well as nuclear composition, so that the volume and surface terms are now expressed as

$$\text{volume energy} = -\bar{a}_v A,$$

$$\text{surface energy} = \bar{a}_s A^{2/3} F(\text{shape}),$$

while the Coulomb energy is written

$$\text{Coulomb energy} = [C_3 G(\text{shape})/A^{1/3} - C_4/A]Z^2$$

with F(shape) and G(shape) different functions of the nuclear shape; and the parameters $\bar{a}_v$ and $\bar{a}_s$ are now identical functions of nuclear composition,

$$\bar{a}_j = c_j\left[1-\kappa\left(\frac{N-Z}{A}\right)^2\right], \qquad j=v \text{ or } s.$$

Here c_j and κ are constants determined by the data.

In their most recent work, Myers and Swiatecki have added an "adjustable" exponential term of the form

$$\Delta E = -CA\bar{\epsilon}e^{-\gamma A^{1/3}},$$

for which they provide no physical reason but which does reduce the errors with experiment. With this addition, they call the complete mass formula the "finite-range droplet model." Here C and γ are more adjustable parameters whereas $\bar{\epsilon}$ is an expansion parameter.

Altogether Myers and Swiatecki used almost 1500 pieces of experimental data to fix the nine parameters of their formula (the six given here—C_3 and C_4 are not independent—plus three more from the shell function). These six, often called the liquid parameters, were determined to be $c_v = 15.677$ MeV, $c_s = 18.56$ MeV, $C_3 = 0.717$ MeV (so $C_4 = 1.21129$ MeV), $\kappa = 1.79$, $C = 230$ MeV, and $\gamma = 1.27$.

Beyond about $A = 50$ (roughly vanadium–iron), the binding energies $[M-(NM_n+ZM_H)]$ are well-reproduced, errors decreasing from about 1% for A near 50 to one-tenth of that for A near 250. For light nuclei the fit is much poorer (for helium the error is 30%), but then the systems are really not large enough to be represented by a statistical model.

This semiempirical mass formula has many uses, only some of which will be touched on. For instance, beta decay with the emission of an electron will occur when

$$M(N,Z,\text{shape}) > M(N-1,Z+1,\text{shape}).$$

Often the shape of the daughter is sufficiently like that of the parent that the shape functions may be ignored. Positron decay will occur when

$$M(N,Z,\text{shape}) > M(N-1,Z+1,\text{shape}) + 2m_e,$$

while only K capture (the capture of an orbital electron) will occur when $M(N,Z,\text{shape})$ is in the range

$$M(N+1,Z-1,\text{shape}) < M(N,Z,\text{shape}) < M(N+1,Z-1,\text{ shape}) + 2m_e.$$

In the first two cases it is clear that the greater the inequality, the greater the maximum energy of the emitted particle. Similar analysis gives information on alpha stability, and the decay rate for even–even-A nuclei follows immediately. However, for odd-A nuclei the ground-state–ground-state transition rates are not necessarily given by the mass formula directly.

Questions of stability against fission can also be dealt with by the mass formula. From a most elementary physical point of view it is clear that as a charged liquid drop is deformed away from spherical, the surface area increases (a sphere has the property of encompassing a given volume by the least area); hence the surface energy increases but the Coulomb energy decreases. If $E_s{}^0$ and $E_c{}^0$ are the surface and Coulomb energies for the spherical shape and if ϵ is the eccentricity of the spheroid into which the sphere is deformed, then the change in energy of the system is, in first order,

$$\Delta E \propto (2E_s{}^0 - E_c{}^0)\epsilon^2,$$

which if negative means the nucleus is unstable against spontaneous fission. Using the fitted values for c_s and C_3, we find the instability criterion to be

$$Z^2/A \gtrsim 52.$$

This is satisfied for all known nuclei (for ^{254}Fm it is about 40). However, a more accurate relation can be derived from the complete mass formula and is fairly successful in predicting trends of fission half-lives.

Another interesting use of the formula is the extrapolation to values of N and Z far beyond those of known nuclei. For instance, if $Z = 126$ is a magic number ($N = 126$ is; see above), then we can extrapolate the mass formula into a region where $Z = 126$ and N is the next magic number. There is evidence that $N = 184$ might be just such a number. Thus the region near $A = 310$ could yield nuclei that are stable against spontaneous fission. Such an extrapolation shows that at this doubly magic nucleus the ground-state mass is depressed to the point that a fission barrier of 9–10 MeV occurs. Thus the half-life for spontaneous fission may be sufficiently long (of the order of milliseconds), depending on the width of the barrier, that such a "superheavy" nucleus might be formed in a heavy-ion collision. Similar extrapolation to a nearby, non-doubly-magic nucleus yields a spontaneous fission half-life of 10^{-22} s. The existence of such "islands" of stability and their exact position can only be determined by experiment. However, the mass formula has opened up exciting possibilities that are still not resolved. In passing it should be noted that although such a superheavy might be rather stable against spontaneous fission, it would still be unstable against alpha and beta decay.

A very successful, yet conceptually simple, model that can be used to explain certain features of nuclear structure is the shell model. It emphasizes the properties of a single nucleon (or a few nucleons) that is assumed to move in some average potential well formed by the remaining nucleons. Thus, the quantum-mechanical many-body problem is replaced by a one-particle problem that is readily solved once the potential function is chosen. While superficially it would seem that a single-particle model would be completely unable to represent any properties of such a strongly interacting many-body system as the nucleus, we must remember that the Pauli principle inhibits the interaction of a nucleon moving within the nuclear volume, since most of the states into which it can scatter are filled. Thus, although the system is bound together by a very strong interaction, the mean free path is very long and therefore the single-particle shell model is a useful approximation to certain nuclear properties.

When we pick the single-particle potential, we fix the Schrödinger equation and can then study the details of the model. The simplest (and often used) potential is the three-dimensional harmonic oscillator. If the wave equation is separated in spherical polar coordinates, the angular momentum l becomes a good quantum number. Thus, the order in which the states fill is given (other potentials frequently used, such as the square well and the Woods–Saxon, yield similar results). Using the notation principal quantum number, angular momentum letter, we find the order of the states to be $0s$,$1p$,($2d$,$2s$), ($3f$,$3p$), ($4g$,$4d$,$4s$), ..., where the states grouped together by ( ) are degenerate. Since the Pauli principle permits $2(2l+1)$ particles per state, we find the predicted magic numbers to be 2,8,20,40,70, ..., instead of 2,8,20,50,82,126, ...

It was the contribution of Mayer and Jensen that solved this order problem and contributed to the success of the model. They pointed out that a spin–orbit interaction proportional to $-2\mathbf{l}\cdot\mathbf{s}$, in addition to the central potential, would remove the degeneracy and split the levels with different j. The negative sign lowers the state with $j=l+\frac{1}{2}$ with respect to the state with $j=l-\frac{1}{2}$ (the opposite is true for the atomic problem). With this addition the order of the levels is now $0s_{1/2}$; $1p_{3/2}$, $1p_{1/2}$; $2d_{5/2}$, $2s_{1/2}$, $2d_{3/2}$; $3f_{7/2}$, $3p_{3/2}$, $3f_{5/2}$, $3p_{1/2}$, $4g_{9/2}$; ..., yielding the magic numbers 2,8,20,50,82,126, ... as observed. Now the ground-state properties are simply described. Since all even–even-A nuclei have observed zero ground-state spin, it is assumed that the nucleons couple up pairwise to zero spin; the entire set then couples to zero total angular momentum. For odd-A nuclei, the even–even core couples to zero spin and the odd particle then has the properties of the shell it is in [angular momentum $j=l\pm\frac{1}{2}$, parity $=(-1)^l$], which then become the properties of the nucleus itself. The few odd–odd-A nuclei will have an angular momentum that results from the coupling of $\mathbf{j}_p$ and $\mathbf{j}_n$. These follow a set of empirical rules called Nordheim's rules and the result is that the model is very successful indeed in predicting the nuclear ground-state properties throughout the periodic table. Low-lying excited states are also successfully predicted by simply promoting the odd particle (for odd-A nuclei, say) to the next higher unfilled state. The model also predicts successfully ground-state magnetic-dipole moments (see above), the "islands of isomerism," and a number of electric-quadrupole moments near magic number or closed-shell nuclei. It is also capable of explaining pickup and stripping reactions. In certain regions, however, notably the rare-earth and actinide regions, it is singularly unsuccessful in predicting anything but the sign of this electric moment. The magnitude is far too large to be explained by any sort of single-particle model. Indeed, any phenomenon related to the collective effects of the bulk of the nucleons (such as the "giant" photonuclear resonance) is not explained by the model and requires a model embodying large-scale cooperative effects for its explanation.

In order to deal with the large electric-quadrupole moments we need only assume that the even–even core of nuclei possessing these large moments be deformed from spherical. Then a classical calculation will yield the quadrupole moment as a linear function, in first order, of the deformation parameter, usually β. The deformed core then becomes a simple model for even–even-A nuclei. Although they possess no measurable ground-state quadrupole moment (since $I_{\text{gnd}}=0$), the consequence of deformation is reflected in the structure of their low-lying excited states. Since in a classical rotating system the Hamiltonian is proportional to the square of the angular momentum, the energy levels will be proportional to $I(I+1)$. This structure is amply confirmed, to within a few percent, in all deformed even–even-A nuclei up to the highest spins measured, $I\approx 22$. (It can be shown by elementary group-theoretic analysis that the low-lying levels that form the ground-state band in even–even-A nuclei can have only even values of I.) If the nucleus is not perfectly rigid, it can also execute vibrations about the equilibrium shape. Such vibrations will increase the rms radius and thus the moment of inertia $\mathcal{I}$. Since the Hamiltonian is inversely proportional to $\mathcal{I}$, these vibrations will tend to lower the rigid-rotor values of the energy levels. To a rather good approximation, then, the ground-state band levels are given by

$$E(I)=\frac{\hbar^2}{2\mathcal{I}_0}I(I+1)-bI^2 2(I+1)^2,$$

where $\mathcal{I}_0$ is the rigid moment of inertia and b is the vibrational parameter. The inertial parameter varies from about 20.3 keV in ^{152}Sm down to 8.3 keV in ^{232}Th and even lower for heavier actinides. The b parameter varies from 0.083 to 0.01 keV, respectively.

That the system can vibrate as well as rotate implies more than a slight lowering of each energy level. To see this we recall the surface expansion due to Bohr, which, for quadrupole shapes in lowest order, is given by

$$R(\theta,\phi)=R_0\left[\alpha_0+\beta\sum_{\mu=-2}^{2}\epsilon_\mu Y_{2\mu}(\theta,\phi)\right],$$

β being the quadrupole deformation parameter and the ϵ_μ satisfying $\sum\epsilon_\mu{}^2=1$. From the symmetry of a spheroid we have that $\epsilon_{\pm 1}=0$, $\epsilon_{-2}=\epsilon_{+2}$; and the ϵs are functions of an asymmetry parameter γ. Thus the fluid has two degrees of vibrational freedom, β and γ. (Note here the connection through the liquid drop with the statistical model.) If we assume that the low-energy structure of the drop is associated with irrotational flow, then the moments of inertia as well as the vibrational properties as functions of β and γ are determined. The beta vibration is associated with zero angular momentum motion and so represents a "breathing" mode familiar to molecular physics. The gamma vibrations are associated with an $I=2$ motion and can be thought of as waves running around the surface of the drop. Upon quantization of such a system we find that the ground-state band contains no phonons of vibration (i.e., $n_\beta=n_\gamma=0$) and so the states have the $I(I+1)$ spacing. If a beta vibration, but not a gamma vibration, is excited ($n_\beta=1, n_\gamma=0$), then we should observe an excited band (a beta band) similar to the ground-state band showing the $I(I+1)$ spacing with $I=0,2,4,\ldots$, but with levels closer together since $\langle r^2\rangle$ (and hence $\mathcal{J}$) is greater in the first excited oscillator state than in the ground state. On the other hand, excitation of only a gamma vibration ($n_\beta=0, n_\gamma=1$) yields a more complicated structure, since now the vibration has spin 2. The level sequence is again proportional to $I(I+1)$ but now $I=2,3,4,\ldots$. Such sequences are observed in all deformed even–even-A nuclei and a great deal can be learned by observing the gamma rays decaying within and between various bands. This model is extremely successful in predicting relative decay rates of all such $E0$ and $E2$ transitions.

What now of the deformed odd-A nuclei? Since the shell model couples the odd particle to a spherical core, we can now couple the particle to a deformed core. In the harmonic-oscillator approximation this amounts to taking the potential to be an anisotropic oscillator. Since the system no longer displays spherical symmetry, but does display axial symmetry, the particle angular momentum **j** is no longer a good quantum number, although its projection on the symmetry axis, Ω, is. This step now unifies the shell and collective models forming the unified or Nilsson model. The energy levels calculated with such a deformed potential and including a spin–orbit term are called Nilsson levels. The level structure is much richer than for even–even-A nuclei, for a rotational band is built on each particle state with a given Ω. The rotational energy levels again follow the $I(I+1)$ rule but the values of I are such that $I=\Omega, \Omega+1, \Omega+2, \ldots$. Again beta and gamma bands are found, as in even–even-A nuclei; however, in special cases the bands can become more complex. If $\Omega=\frac{1}{2}$, then certain terms associated with the "decoupling" parameter come into play that can replace the simple $I(I+1)$ scheme by one of closely spaced doublets. Furthermore, if two nearby bands are such that $|\Omega_1-\Omega_2|=1$, then they can strongly interact through the Coriolis coupling and again cause what appears to be a distorted band structure. An example of a decoupled band is to be seen in ^{169}Tm, where the lowest doublet is $I=\frac{1}{2}, \frac{3}{2}$ separated by 8.4 keV; then the $I=\frac{5}{2}, \frac{7}{2}$ doublet is at 118 keV separated by 21.0 keV; then the $I=\frac{9}{2}, \frac{11}{2}$ doublet is at 332 keV with 36 keV separation, etc.

Another interesting phenomenon is that of "backbending." If we plot the moment of inertia $\mathcal{J}$ against the rotational frequency squared [or in fact $(\hbar\omega)^2$], it starts off with a rather gentle and uniform slope until it gets to energy values associated with angular momenta of about $16\hbar$. At this point the curve bends suddenly upward and back toward smaller values of $(\hbar\omega)^2$. When plotted this way, the "backbending" effect is very obvious, since $\mathcal{J}$ changes by nearly a factor of 2 in some cases, while the angular momentum only changes by $4\hbar$ or $6\hbar$. This phenomenon has been intensively studied and a reasonable understanding of it shows that it is related to band crossings of the associated Nilsson levels.

Finally, there has been discovered recently a band of rotational levels far above the bands related to the deformed ground state and near the yrast region. (An yrast state is that state of lowest energy for a given angular momentum—the yrast line is that line connecting the yrast states and the yrast region that region which lies above the yrast line.) These systems are rotational bands with the highest states having unusually large angular momenta ($I\approx 40\hbar-60\hbar$) and which have very much larger deformation parameters than the ground-state bands. These states are associated with what are called "superdeformed shapes" with deformation parameters $\beta_{SD}\approx 0.50$–0.60 and average quadrupole moments of about (10–20)eb and moments of inertia of the order of $100\hbar^2$/MeV. So far they have been reported in the $A=135$, 150, and 190 mass regions. In these regions the ground-state deformations, β_{GS}, are in the range 0.12–0.40 and the average quadrupole moments range from about 3.0eb to 11eb. It is interesting to note that these superdeformed shapes, which coexist with smaller ground-state deformations, were predicted some 20 years before they were discovered. These "stable" superdeformed shapes occur for much the same reason that the stable, spherical nuclear shapes do at closed shells. This is related to the existence of large gaps in the single-particle levels for zero deformation at the magic numbers. Superdeformation occurs where there are similar large gaps in the single-particle states for which the nuclear shape's semiaxes are in the ratio of either 2:1:1 or 3:2:2.

We thus see that each new artifact leads to more refined models and calculations, which lead in turn to new predictions. In due course these are verified, or not, by experiment and the process continues.

See also Binding Energy; Multipole Fields; Nuclear Forces; Nuclear Moments; Nuclear Properties; Nucleon; Resonances, Giant; Strong Interactions; Superheavy Elements.

BIBLIOGRAPHY

A. Bohr, *Rev. Mod. Phys.* **48,** 365 (1976). (E)

A. Bohr and B. R. Mottelson, *Nuclear Structure,* Vol. I (*Single Particle Motion*), Vol. II (*Nuclear Deformations*). Benjamin, Reading, MA., 1969 and 1975 (A); *Annu. Rev. Nucl. Sci.* **23,** 363 (1973). (I)

J. P. Davidson, *Collective Models of the Nucleus.* Academic Press, New York, 1968. (A)

A. E. S. Green, T. Sawada, and D. S. Saxon, *The Nuclear Independent Particle Model.* Academic Press, New York, 1968. (I)

R. D. Lawson, *Theory of the Nuclear Shell Model.* Clarendon Press, Oxford, 1980. (I)

B. R. Mottelson, *Rev. Mod. Phys.* **48,** 375 (1976). (E).

W. D. Myers, *Droplet Model of Atomic Nuclei*. IFI/Plenum, New York, 1977. (A)

M. K. Pall, *Theory of Nuclear Structure*. Van Nostrand Reinhold, New York, 1983. (I)

Z. Szymański, *Fast Nuclear Rotation*. Clarendon Press, Oxford, 1983. (A)

Nucleon

Owen Chamberlain

The name nucleon is used in physics to designate either a proton or a neutron. The term is useful because there are numerous similarities between protons and neutrons. Thus, it is advantageous to describe the proton and the neutron as two different states of the same basic particle, the nucleon.

The proton has been known since Rutherford presented his model of the atom in 1911. The proton is the nucleus of the ordinary hydrogen atom with mass number 1. The proton is electrically charged—positively. It bears one basic unit of electric charge, just as the electron bears one unit of negative electric charge. (Proton and electron together constitute a hydrogen atom. The total electric charge of the hydrogen atom being zero, we say that the hydrogen atom is electrically neutral.) The proton mass is 1.673×10^{-24} g (938.3 MeV/c^2).

Neutrons were discovered by James Chadwick in 1932 as fragments knocked out of nuclei. Chadwick showed that experiments carried out by Curie and Joliot could be interpreted as showing the existence of neutrons. The neutron was observed to be a neutral particle—with no electric charge—whose mass is about the same as that of a proton. (The neutron mass is 1.675×10^{-24} g or 939.6 MeV/c^2.)

Since the discovery of the neutron, the atomic nucleus has been described as made up of neutrons and protons. As an example we may take carbon-12, which is the most common form of carbon atom. It is often designated as ^{12}C, which is read "carbon twelve" even though the superscript 12 is written to the left of the symbol C for carbon. It has a mass about 12 times the mass of the hydrogen atom, suggesting that its nucleus contains neutrons and protons adding up to 12. The carbon nucleus has 6 units of electric charge, which tells us that the nucleus contains 6 protons. Thus, we say that the nucleus of ^{12}C is composed of 6 neutrons and 6 protons.

It may be helpful at this point to review the meanings of several terms that are rather similar. The word *nucleus* is used for the heavy, positively charged object at the center of any atom. The plural is *nuclei*. The term *nucleon* means neutron or proton. We say that nuclei, then, are composed of *nucleons*.

Already in 1932, in the first year of our knowledge of the neutron's existence, Heisenberg was using the concept of the nucleon to describe either a neutron or a proton. The usefulness of the concept is based on the great similarity between neutrons and protons in their strong (or nuclear) interactions. To enunciate this resemblance clearly we must differentiate between two kinds of interactions among nucleons. First, there are the strong interactions—the strong forces that hold nuclei together and that act only over short distances, about 10^{-13} cm. Then there are the electromagnetic interactions—the longer-range forces that include the electric (Coulomb) repulsion of two protons. The latter interactions are connected with the electric charge. In light and medium nuclei the strong interactions are more important than the electromagnetic, and so we may use light nuclei as evidence of the relative strength of the forces between various nucleons. A small correction can be made for the known Coulomb repulsion of the protons for each other.

As far as the strong interactions are concerned, proton–proton (p–p) forces are identical to neutron–neutron (n–n) forces. As an indication of how this is determined, consider the nuclei ^{3}H and ^{3}He. One is composed of 2 neutrons and 1 proton, the other of 2 protons and 1 neutron. One of these nuclei would become the other if every neutron could be changed into a proton and every proton changed to a neutron. Therefore, that these two nuclei have almost equal energies—the small difference being understood as just due to the electric Coulomb repulsion among protons—leads us to conclude that n–n forces are the same as p–p forces as far as only the strong interactions are concerned.

We can go a step further and show that all three types of strong interactions (n–n, n–p, and p–p) are the same. Since nuclei (below their breakup energy) can exist only in states at certain discrete energies, we speak of various "energy states." If n–n, n–p, and p–p forces are truly the same there should be energy states in different nuclei that have the same energy (apart from a Coulomb correction), but it takes care and patience to sort out the precise states in different nuclei that have just the same internal motions of the nucleons. This has been done, for example, in the nuclei boron-12, carbon-12, and nitrogen-12. Three states—one in each nucleus—do have the same energy, except for a small Coulomb correction. This and other examples convince us that our three interactions are really the same, so that neutrons and protons can well be regarded as forms of the same underlying particles—nucleons.

While this helps us to understand the convenience of using the single term "nucleon" to describe neutrons and protons, there are other similarities that are important to this usefulness. We list some of them here, without a full explanation of each.

1. The proton and neutron masses are very close to each other. The small difference of about 1 MeV is thought to be due to electromagnetic interactions but is difficult to calculate.
2. Each has spin of $\frac{1}{2}$ unit.
3. Neutrons and protons each obey the Pauli principle, meaning they each obey Fermi–Dirac statistics.
4. A consistent description of particles can be made in which neutrons and protons have the same intrinsic parity (even parity).

All these similarities are consonant with the idea that neutrons and protons are really two states of the same basic particle, the nucleon.

Other particles, called antinucleons (antiprotons and antineutrons), can exist. The antiparticles are in a sense counterparts to the particles. The antiparticles have in each case the same mass and same amount of spin angular momentum as the corresponding particles, but have opposite electric charge. Thus a proton has charge e, but an antiproton charge $-e$. For every electrically charged particle, an antiparticle exists. Some electrically neutral particles (e.g., the neutron) have antiparticles (the antineutron). A particle and its antiparticle can, when coming close together, annihilate each other—a process in which they literally eat each other up. The energy represented by the masses of these particles appears after the annihilation as the energy of lighter particles. For example, antineutron and neutron can annihilate to give several pions.

Since 1974 the physicists who study the particles of which all matter is made have come to believe that the nucleon is itself composed of three constituent particles called quarks. (In this view, all strongly interacting particles are composed of quarks and antiquarks.) The arguments for a nucleon made up of three quarks are rather indirect, since quarks have not been identified experimentally. In the simpler forms of the theory the quarks have electric charges such as $\frac{1}{3}e$ and $-\frac{2}{3}e$ that are not integer multiples of the charge of proton. If real quarks have electric charge, but have it in smaller amounts that the proton, they should be observable if they are being created in collision processes between particles. Up to the present time only charges that are integral multiples of the proton charge have been observed. According to present-day theory, quarks are difficult or impossible to produce singly. The known strongly interacting particles are believed to be composed of integral multiples of three quarks, meaning that the quark numbers of each are 0, 3, 6, 9, etc.; when antiquarks are involved, the quark numbers can also be -3, -6, -9, etc.

See also ANTIMATTER; BARYONS; ELEMENTARY PARTICLES IN PHYSICS; FERMI–DIRAC STATISTICS; HADRONS; HYPERONS; ISOSPIN; NUCLEAR FORCES; QUARKS.

BIBLIOGRAPHY

G. E. Brown and A. D. Jackson, *The Nucleon–Nucleon Interaction*. American Elsevier, New York, 1976. (A)

Geoffrey F. Chew, Murray Gell-Mann, and Arthur H. Rosenfeld, "Strongly Interacting Particles," *Sci. Am.* **210** (2), 74 (1964). (E)

Sheldon Lee Glashow, "Quarks with Color and Flavor," *Sci. Am.* **233** (4), 38 (1975). (E)

David Halliday, *Introductory Nuclear Physics*. Wiley, New York, 1955. (I)

Henry W. Kendall and Wolfgang Panofsky, "The Structure of the Proton and the Neutron," *Sci. Am.* **224** (6), 60 (1971). (E)

Yoichiro Nambu, "The Confinement of Quarks," *Sci. Am.* **235** (5), 48 (1976). (E)

R. E. Peierls, "The Atomic Nucleus," *Sci. Am.* **200** (1), 75 (1959). (E)

Steven Weinberg, "Unified Theories of Elementary-Particle Interaction," *Sci. Am.* **231** (1), 50 (1974). (E)

Richard Wilson, *The Nucleon–Nucleon Interaction*. Wiley (Interscience), New York, 1963. (I)

Nucleosynthesis

James W. Truran

Nucleosynthesis studies are aimed at determining the physical mechanisms responsible for the production of the elemental abundance patterns observed in nature and identifying the astrophysical sites in which these mechanisms can operate. Theoretical efforts are guided by our knowledge both of elemental abundances and of nuclear reaction properties. It is evident, therefore, that nucleosynthesis is a relatively young science. The occurrence in nature of natural radioactive elements, recognized early in this century, made clear the fact that the age of the elements is finite. Subsequent advances in nuclear physics and in spectroscopy have provided the tools and the boundary conditions, respectively, essential to nucleosynthesis studies. A substantial body of observational data is provided by our detailed knowledge of the elemental and isotopic abundances of solar system matter and our increasing knowledge of abundances in interstellar gas, cosmic rays, stars, and other galaxies.

Diverse astrophysical environments provide temperature and density conditions under which nuclear transformations and concomitant nucleosynthesis can proceed. High-speed computers now allow detailed studies of stellar evolution, supernova explosions, and the early evolution of the universe itself. Built upon these numerical models, substantial progress has been made over the past 30 years in our understanding of the mechanisms of formation of the elements.

The earliest spectral studies of the sun and stars were able to establish only that the elements of which they and the earth are composed are qualitatively the same. It was therefore quite reasonable to assume the universe to be of uniform chemical composition. Such uniformity strongly suggests a single-event theory of the origin of the elements; this led physicists to search within the framework of cosmology for a set of physical conditions which could produce the present distribution of abundances. In 1948, R. A. Alpher, H. A. Bethe, and G. Gamow proposed such a theory of nucleosynthesis tied to a cosmological "big bang." This model envisioned the early state of matter as a compressed neutron gas. The expansion of this neutron fluid, following an initial explosion, would result in the decay of some of the neutrons into protons and electrons. Subsequent neutron captures, interspersed with beta decays, could then build up all of the heavier elements. In a continuation of this work, Alpher noted that the abundance peaks in the heavy-element region (mass number $A>60$) corresponded to regions in which the neutron-capture probabilities were small. This correlation made apparent the need for a neutron-capture process in the synthesis of the heavy elements, and such a process has indeed been incorporated into modern theories. The current status of the big-bang theory is reviewed below.

Even as these studies proceeded, advances in our understanding of nuclear physics were preparing the way for revised and perhaps definitive theories of element origin. The essential role played by thermonuclear processes in providing an energy source sufficient to account for stellar lifetimes of billions of years was established in the late 1930s by the calculations of H. A. Bethe and C. F. von Weizsäcker. The fusion of four protons to one helium nucleus ("hydrogen burning"), taking place in approximately 10% of the sun's mass, provides several orders of magnitude more energy than has yet been released in gravitational contraction. Not immediately recognized that subsequent stages of nuclear burning in stellar interiors might contribute to the formation of many of the heavier elements observed in nature.

The recognition that nucleosynthesis is a continuing process in stellar interiors followed the discovery by P. W. Merrill, in 1952, of the presence of the element technetium in the atmospheres of red-giant stars. As technetium has no stable isotopes (the longest-lived isotope has a half-life of $<2\times10^6$ years), its presence in abundances sufficient to be observed suggests that element synthesis has taken place quite recently. The view that nucleosynthesis proceeds in stars over the lifetime of our galaxy received further support with the discovery, in the mid-1950s, that there exist abundance differences between certain broad classes of stars, in the sense that the ratio of the abundances of the heavy elements to hydrogen is variable and that this ratio is correlated with the age of the star. The presence in stars of anomalous abundances of elements which are produced by specific nuclear-burning mechanisms provides somewhat more direct evidence for nucleosynthesis in stars. The presence of technetium in red-giant stars, for example, implies that element synthesis by neutron capture has occurred. The high abundances of carbon in "carbon stars" are attributed to helium burning in the interior, assuming that processed matter has been carried to the surface by convection.

These observations are consistent with a rather straightforward model of galactic evolution and nucleosynthesis. Assuming the primordial gas to be composed predominantly of hydrogen and helium, products of the big bang, the first generation of stars will be formed with this composition. The evolution of these stars is characterized both by element synthesis during various stages of nuclear burning and, ultimately, by the return of processed matter to the interstellar gas by wind-driven mass loss or explosive (supernova) ejection. In this manner, the heavy-element content of the interstellar medium will be increased. Subsequent generations of stars will be formed from gas enriched in these heavy elements. Since the late 1950s, significant progress in our understanding of nucleosynthesis mechanisms has proceeded within this theoretical framework. A brief summary of the current status of nucleosynthesis theories follows.

COSMOLOGICAL NUCLEOSYNTHESIS

Interest in the subject of the universal synthesis of the elements was revived after the 1965 discovery, by A. A. Penzias and R. W. Wilson, of the 3-K background microwave radiation. Its identification as the relic of a cosmological big bang remains the most satisfactory theoretical interpretation. This implies that the earliest moments of the universe were characterized by extreme temperatures and densities.

Given an initial temperature above 10^{11} K, it follows that even weak interactions will achieve thermodynamic equilibrium. Equilibrium concentrations of neutrons and protons are maintained via the reactions

$$p+e^-\rightarrow n+\nu_e,$$

$$n+e^+\rightarrow p+\bar{\nu}_e.$$

Following expansion and cooling, the critical temperature for nucleosynthesis of $\sim10^9$ K is realized; production of heavy nuclei is restricted at higher temperatures by nuclear photodisintegrations and at lower temperatures by the inability of charged particles to penetrate their mutual Coulomb barriers on the required time scale. The neutron-to-proton ratio emerging at this era is approximately 1/7. Deuterium formation first proceeds by

$$n+p\rightarrow d+\gamma.$$

Deuterium destruction leading to the production of ^{4}He then takes place rapidly following reactions with neutrons, protons, and with itself, and, subsequently, with the intermediate reaction products tritium and ^{3}He. Most of the initial neutrons are ultimately utilized in the production of ^{4}He, yielding a helium mass fraction $\sim$0.25.

One rather firm conclusion follows from theoretical studies: given the observational limit on the present universal baryon density of the universe and assuming that the microwave radiation is correctly interpreted as a residual 2.7-K radiation temperature, no substantial production of nuclei heavier than ^{4}He is possible in the cosmological big bang. The production of heavier elements would require either a significantly higher present-day density or a lower present-day temperature. The triple-alpha reaction ($3\,^4\text{He}\rightarrow{}^{12}\text{C}$), which serves successfully to bridge the mass gaps at $A=5$ and 8 (no stable nuclei exist with these nucleon numbers) in stellar interiors, is ineffective at the prevailing densities in the expanding fireball at the appropriate burning temperature. It is currently believed that primordial nucleosynthesis contributes only to the present-day abundances of hydrogen, deuterium, ^{3}He, ^{4}He, and ^{7}Li. Our knowledge of the relative abundances of these nuclear species in primordial galactic matter allows constraints to be imposed both on cosmology and on particle physics.

STELLAR ENERGY GENERATION AND NUCLEOSYNTHESIS

Stellar evolution is characterized by a sequence of alternate stages of gravitational contraction to higher temperatures (and densities) and thermonuclear burning of the available fuels at these temperatures. The various burning stages in the lives of stars were first defined in 1957 in the now classic papers by E. M. Burbidge, G. R. Burbidge, W. A. Fowler, and F. Hoyle and by A. G. W. Cameron. By far the greatest fraction of the active burning lifetime of a star is spent converting hydrogen into helium (hydrogen burning) in its core, releasing approximately 7 MeV per nucleon. Nuclear transformations can provide at most only another $\sim$1.1 MeV per nucleon, assuming burning proceeds all the way to ^{56}Fe. This is found to occur only in stars more massive than about 10 solar masses ($M_\odot$). Less massive stars develop dense cores in which electron degeneracy pressure becomes dominant; following the loss of often a substantial fraction of their envelope mass, they evolve to white dwarf stars of helium, carbon–oxygen, or oxygen–neon–magnesium composition. Nuclear transformations in massive stars build toward ^{56}Fe in a succession of stages: helium burning to ^{12}C and ^{16}O proceeding via the reactions $3\,^4\text{He}\rightarrow{}^{12}\text{C}$ and $^{12}\text{C}(\alpha,\gamma)^{16}\text{O}$; carbon and oxygen burning by the heavy-ion reactions

$$^{12}C + ^{12}C \rightarrow ^{20}Ne + ^{4}He$$
$$\rightarrow ^{23}Na + p$$
$$^{16}O + ^{16}O \rightarrow ^{28}Si + ^{4}He$$
$$\rightarrow ^{31}P + p,$$

to intermediate mass nuclei; and silicon burning proceeding by a complex sequence of photonuclear and charged-particle-induced reactions to nuclei in the immediate vicinity of iron. Since ^{56}Fe is the most tightly bound nucleus (per nucleon), further processing of this matter cannot provide a nuclear energy source.

The formation of nuclei heavier than iron is due primarily to neutron-capture reactions. Two distinct environments are required, defined by the conditions that the rates of neutron capture are either slow (*s*-process) or fast (*r*-process), compared to electron decay rates in the vicinity of the valley of beta stability. Production of the most neutron-rich isotopes of heavy nuclei and of nuclei heavier than lead both require more substantial neutron fluxes (*r*-process), as may be realized in supernova environments. Less extreme neutron fluxes, realized during helium shell burning in red-giant stars, can provide an appropriate *s*-process environment. The presence of technetium in the atmospheres of red giants testifies to the operation of this nucleosynthesis mechanism.

From the point of view of nucleosynthesis, stable phases of stellar evolution serve as the likely source of the ^{12}C and ^{16}O existing in our galaxy and of the designated *s*-process heavy elements. Concentrations of nuclei from neon to iron, formed during stable burning phases, can be substantially altered as a result of their shock ejection in supernova events.

SUPERNOVA NUCLEOSYNTHESIS

While the supernova phase constitutes only an extremely small fraction of the lifetimes of stars in restrictive mass ranges, theoretical studies nevertheless suggest that supernova environments represent the likely site of the synthesis of the bulk of the heavy elements (mass number $A \gtrsim 20$) observed in nature. From the point of view of nucleosynthesis, one has the advantage that constructive nuclear transformations proceed in the shock-induced ejection of both core and envelope matter, ensuring that the resulting abundance distributions will not be distorted by subsequent evolution. Hydrodynamic studies of supernova ejection mechanisms predict very promising conditions for the operation of two distinct thermonuclear processes: (1) the synthesis of elements through the vicinity of iron by charged-particle reactions and (2) the neutron-capture synthesis of heavier nuclei.

The characteristics of these environments are apparent from a consideration of the final phases of evolution of massive stars ($\gtrsim 10M_{\odot}$). Massive stars have been shown to proceed in stages of nuclear burning to a configuration consisting of an iron core, with overlying shells dominated by silicon, magnesium, oxygen, carbon, and helium, respectively, and an extended hydrogen envelope. Lacking further nuclear fuel, the iron core is compelled to shrink under gravity. The fate of massive stars is then dependent on existence of some mechanism capable of utilizing the gravitational energy released in core contraction in effecting the ejection of the overlying matter. If this cannot occur, gravitational collapse of the entire star to a black hole is inevitable.

Until recently, definitive theoretical statements concerning nucleosynthesis in supernovae were simply not possible. Observationally, the identification of supernovae with massive star progenitors was suggested by their occurrence in regions of active star formation, and their association with neutron-star remnants was, in some instances, confirmed (e.g., Crab nebula pulsar). However, theoretical studies were unable to establish the precise nature of the mechanism of mass ejection and, therefore, specific predictions concerning the nucleosynthesis conditions accompanying supernovae events were highly uncertain.

The recent outburst of Supernova 1987A in the Large Magellanic Cloud, the closest visual supernova event since that recorded by Kepler in 1604, has changed the situation greatly, by providing an extraordinary opportunity to test theoretical models. One significant outgrowth of observational studies of this supernova has been the unambiguous identification of the stellar progenitor, the star Sanduleak -69 202. It is estimated that this supergiant star had an initial mass of approximately $20M_{\odot}$, of which $\sim 1.4M_{\odot}$ was incorporated into a neutron star remnant and $\sim 5M_{\odot}$ was ejected in the form of neavy nuclei, ranging in mass from oxygen to iron and nickel. Neutrino detections, indicating the release of $\sim 3 \times 10^{53}$ ergs in neutrinos, confirmed the collapse to a neutron star. The exponentially falling light curve of SN 1987A, with a lifetime consistent with ^{56}Co ($\tau_{1/2} = 77.8$ days), together with a knowledge of the total observed luminosity of the supernova and the distance to the Large Magellanic Cloud, allow the determination of the total mass ejected in the form of nuclei of mass $A = 56$: approximately $0.075M_{\odot}$ of matter in the form of ^{56}Ni ($\tau_{1/2} - 6.1$ days), which decays through ^{56}Co to ^{56}Fe. In general, supernovae involving massive stars are expected to eject substantial amounts of matter in the form of such nuclei as oxygen, neon, magnesium, silicon, calcium, chromium, iron, and nickel, in essentially solar proportions, as a consequence of the nuclear burning of matter at temperatures in the range $\sim 2.6 \times 10^{9}$ K. For some range of stellar masses, expansion and cooling of highly neutronized matter from the immediate vicinity of the neutronized core may provide appropriate conditions for the (*r*-process) synthesis of the heavy elements. A single supernova event can therefore be responsible for the formation of a broad range of nuclear species from oxygen to uranium—in roughly solar proportions.

See also Cosmology; Stellar Energy Sources and Evolution.

BIBLIOGRAPHY

W. D. Arnett, J. Bahcall, R. Kirshner, and S. E. Woosley, "Supernova 1987," *Annu. Rev. Astr.* **27,** 269 (1989).

H. A. Bethe and G. Brown, "How a Supernova Explodes," *Sci. Am.*, **252** (5), 60 (1985).

D. N. Schramm, "The Age of the Elements," *Sci. Am.*, **230** (1), 69 (1974).

R. J. Taylor, *The Origin of the Chemical Elements*. Wykeham Publications, London, 1972.

V. Trimble, "The Origin and Abundances of the Chemical Elements," *Rev. Mod. Phys.* **47,** 877 (1975).

J. W. Truran, "Nucleosynthesis," *Annu. Rev. Nucl. Part. Sci.* **34,** 53 (1984).

S. E. Woosley and T. Weaver, "The Great Supernova of 1987," *Sci. Am.*, **261** (2), 32 (1989).

Operators

Konrad Osterwalder

An operator A from a set E to a set F is a law that uniquely assigns to any element ϕ in a subset D_A of E an element $\psi = A\phi$ in F. The subset D_A is called the *domain* of A. The subset R_A of F which consists of all elements ψ of the form $\psi = A\phi$, ϕ in D_A, is the *range* of A. An operator $\tilde{A}$ is an *extension* of A if the domain $D_{\tilde{A}}$ contains D_A and if $\tilde{A}\phi = A\phi$ for all ϕ in D_A.

An operator A from E to F is called one to one if for all ϕ_1, ϕ_2 in D_A, $\phi_1 \neq \phi_2$ implies $A\phi_1 \neq A\phi_2$. If A is one to one, then there is a uniquely defined operator A^{-1} from F to E given by $A^{-1}\psi = \phi$ whenever $A\phi = \psi$. A^{-1} is called the *inverse* of A; its domain $D_{A^{-1}}$ is R_A, its range $R_{A^{-1}}$ is D_A.

Suppose from now on that E and F are *Banach spaces* {a Banach space is a complete, normed linear space; e.g., the set of all functions ϕ on the interval $[a,b]$ with the finite $\|\phi\|_p = (\int_a^b |\phi(x)|^p\ dx)^{1/p}$ is a Banach space denoted by $L^p([a,b])$, $p \geq 1$}. An operator A from E to F is said to be *densely defined* if D_A is dense in E. A is called *linear* if D_A is a linear subspace of E and if $A(a_1\phi_1 + a_2\phi_2) = a_1A\phi_1 + a_2A\phi_2$ for all ϕ_1, ϕ_2 in D_A and arbitrary (real or complex) numbers a_1, a_2. In the following we restrict our attention to linear operators. A linear operator A is *bounded* if its norm

$$\|A\| = \sup_{\substack{\phi \in E \\ \|\phi\| = 1}} \|A\phi\|$$

is finite. Otherwise A is said to be *unbounded*. For unbounded operators the notion of the graph is useful. The *graph* Γ_A of A is defined as the set of all pairs $\langle \phi, A\phi \rangle$ with ϕ in D_A. Γ_A is thus a subspace of $E \times F$ and A is called a *closed operator* if its graph is closed. A is called *closable* if it has a closed extension. The *closure* $\bar{A}$ of A is the smallest closed extension of A (i.e., $D_{\bar{A}} \subset D_{\tilde{A}}$ for all extensions $\tilde{A}$ of A.

Let A be a linear operator from E to F with $D_A = E$. Then A is bounded if and only if its graph Γ_A is closed (*closed graph theorem*). Thus a closed operator which is unbounded cannot be defined on all of E. Many of the difficulties with unbounded operators are due to this fact. For example, let $E = L^p([0,1])$ and $F = L^q([0,1])$ and define the operator A from E to F to be *multiplication* by a function $\alpha(x)$, i.e., $(A\phi)(x) = \alpha(x)\phi(x)$. If α is itself an element of $L^r([0,1])$ such that $q^{-1} = p^{-1} + r^{-1}$, then A is a bounded operator and its domain can be taken to be all of E. This follows from *Hölder's inequality* which says that $\|\alpha\phi\|_c \leq \|\alpha\|_r \cdot \|\phi\|_p$. On the other hand, if we choose $\alpha(x) = x^{-1}$ and D_A equal to all continuously differentiable functions on $[0,1]$ whose first derivative vanishes at $x = 0$, then A is unbounded, but densely defined and closable.

An operator A from a Banach space E to a Banach space F with $D_A = E$ is called *compact* if the image $A\phi_1, A\phi_2, \ldots$ of any bounded sequence $\phi_1, \phi_2, \ldots$ of vectors in E contains a Cauchy subsequence. (A sequence of vectors $\psi_1, \psi_2, \ldots$ is Cauchy if $\|\psi_n - \psi_m\|$ becomes arbitrarily small for n and m sufficiently large.) A is *degenerate* if it has finite rank, i.e., if its range R_A is finite dimensional. A degenerate operator is necessarily compact. The norm limit of a sequence of degenerate operators is compact, but not necessarily degenerate.

Let A be a closed, linear operator on the Banach space E (i.e., an operator from E to E). Suppose there is a complex number λ and a nonzero vector ϕ in D_A such that $A\phi = \lambda\phi$. Then ϕ is called an *eigenvector* of A and λ is the corresponding *eigenvalue*. The set of all eigenvectors ϕ to a given eigenvalue λ is a subspace of E, called the *eigenspace* of λ; its dimension is called the *multiplicity* of λ. More generally, if $A - z$ has a bounded inverse $R(z) = (A - z)^{-1}$, then z is said to belong to the *resolvent set* ρ_A of A and $R(z)$ is called the *resolvent* of A. It follows that for any z in ρ_A, $D_{R(z)}$ is all of E and $R_{R(z)}$ is D_A. The complementary set σ_A of ρ_A (in the complex plane) is called the *spectrum* of A. The spectrum of A contains all the eigenvalues of A, but it may contain much more; it is, however, always a closed subset of the complex plane. If A is bounded, then neither ρ_A nor σ_A is empty. ρ_A contains the exterior of the circle $|z| = \mathrm{spr}A$, where $\mathrm{spr}A$ is the *spectral radius* of A and is given by $\mathrm{spr}A = \lim_{n\to\infty} \|A^n\|^{1/n} = \inf_{n \geq 1} \|A^n\|^{1/n}$. The resolvent $R(z)$ is (piecewise) *holomorphic* for z in ρ_A and it satisfies the *resolvent equation* $R(z_1) - R(z_2) = (z_1 - z_2)R(z_1)R(z_2)$ for z_1, z_2 in ρ_A.

If A is *compact*, then σ_A is a countable set with no limit points except possibly $z = 0$. All nonzero points in σ_A are eigenvalues of A with finite multiplicity (*Riesz–Schauder theorem*).

From now on we assume that A is a densely defined linear operator on a *Hilbert space* $\mathcal{H}$. [A Hilbert space is a Banach space equipped with a scalar product $(\cdot,\cdot)$ such that $\|\phi\| = (\phi,\phi)^{1/2}$. For example, $L^2([a,b])$ is a Hilbert space if we define $(\cdot,\cdot)$ by $(\phi,\psi) = \int_a^b \bar{\phi}(x)\psi(x)dx$.] We define the adjoint A^* of A as follows: D_{A^*} is the set of all ϕ in $\mathcal{H}$ for which there is a ψ in $\mathcal{H}$ such that $(A\eta,\phi) = (\eta,\psi)$ for all η in D_A. For such a ϕ in D_{A^*} we define $A^*\phi = \psi$. A^* thus defined will always be a closed operator. Furthermore, A is closable if and only if D_{A^*} is dense, in which case the closure $\bar{A}$ of A is equal to $(A^*)^*$. If A^* is an extension of A, then A is called *symmetric* or *Hermitian;* if $\bar{A} = A^*$, then A is called *essentially self-adjoint*; if $A = A^*$, then A is called *self-adjoint*. The distinction between symmetric and self-adjoint is crucial for unbounded operators. If A is bounded, then $D_{\bar{A}} = D_{A^*} = \mathcal{H}$ and A symmetric implies $\bar{A} = A^*$ self-adjoint. In quantum mechanics and in quantum-field theory one often is given a formal Hamiltonian H describing some physical system (e.g., a partial differential operator whose domain is not specified). It is usually not hard to find a dense domain D_H on which H is symmetric. But the description of the dynamics of the system requires a self-adjoint Hamiltonian (see below, Stone's theorem). Hence we have to study the question whether H has self-adjoint extensions, and if there are several of them, which one is correct from the point of view of physics. The first question is partially answered by the following analysis. We suppose that A is a closed symmetric operator on $\mathcal{H}$. The dimensions n_+ and n_- of the null spaces $N_{(A^*+i)}$ and $N_{(A^*-i)}$, respectively, are called the *deficiency indices* of A. (The *null space* or *kernel* N_B of a linear operator B is the set of all ϕ in D_B with $B\phi = 0$.) Then A is self-adjoint if and only if $n_+ = n_- = 0$. A has self-adjoint extensions exactly if $n_+ = n_-$. If one of n_+ and n_- equals zero and the other one is different from zero, then A has no symmetric extension (except for A itself) and the operator is called *maximal symmetric*. The situation is particularly simple if A is closed, symmetric and *semibounded*, i.e., there is a constant c such that $(A\phi,\phi) \geq c\|\phi\|^2$ for all ϕ in D_A. Then $n_+ = n_-$ and A does have self-adjoint extensions.

An operator U on $\mathcal{H}$ is called an *isometry* if it preserves the norm, i.e., if $\|U\phi\| = \|\phi\|$ for all ϕ in $\mathcal{H}$. If U is an isometry and the range R_U is all of $\mathcal{H}$, then U is said to be *unitary*. In this case $U^* = U^{-1}$. Let $U(t)$ be a strongly continuous one-parameter group of unitaries; i.e., for each real t, $U(t)$ is a unitary operator, $U(s)U(t) = U(s+t)$ for all s and t, and $\lim_{t\to t_0} U(t)\phi = U(t_0)\phi$ for all ϕ in $\mathcal{H}$. Then by *Stone's theorem* there is a unique self-adjoint operator H such that $U(t) = e^{itH}$; H is the *infinitesimal generator* of $U(t)$. On the other hand, if H is a given self-adjoint operator, then $U(t) = e^{itH}$ is a strongly continuous one-parameter group of unitaries with H as its infinitesimal generator.

This explains why self-adjointness is crucial in quantum mechanics. Let H be the Hamiltonian operator of a physical system whose state at time $t = 0$ is given by the wave function ψ_0. The wave function at time t, denoted by ψ_t, is a solution of the Schrödinger equation

$$i\hbar \frac{\partial \psi_t}{\partial t} = H\psi_t$$

and is formally given by

$$\psi_t = e^{(1/i\hbar)tH}\psi_0.$$

According to what has been explained above, for this to be a well-defined solution it is necessary and sufficient that H is self-adjoint.

An operator P satisfying $P^2 = P$ is called a *projection (operator)*. Self-adjoint projections are necessarily bounded and of norm 1. They are called *orthogonal projections*. (In an infinite-dimensional space there are unbounded projections!) Let A be a compact self-adjoint operator on $\mathcal{H}$. Then there exists a sequence of orthogonal projections $P_1, P_2, \ldots$ with $P_iP_j = 0$ for all $i \neq j$ and real numbers $\lambda_1, \lambda_2, \ldots$ with $\lim_{n\to\infty} \lambda_n = 0$, such that $A = \sum_{n=1}^{\infty} \lambda_n P_n$ (*spectral theorem, compact case*). This representation, called the *spectral representation of A*, allows us to define functions of A by $\hat{f}(A) = \sum_{n=1}^{\infty} f(\lambda_n)P_n$. Here we have assumed that f is a sufficiently nice function on the reals. (For example, we can now write e^{itA} as $\sum_{n=1}^{\infty} e^{it\lambda_n}P_n$.) There is a *spectral theorem* for arbitrary self-adjoint operators; however, it involves more mathematical sophistication.

A *decomposition of the identity* is a one-parameter family of orthogonal projections $P(\lambda)$, defined on a finite or infinite interval $[\alpha,\beta]$, such that $P(\alpha) = 0$, $P(\beta) = \mathbb{1}$ (the identity operator on $\mathcal{H}$), $\lim_{\lambda \nearrow \lambda_0} P(\lambda) = P(\lambda_0)$, and $P(\lambda_1)P(\lambda_2) = P(\min(\lambda_1,\lambda_2))$. Then for any ϕ in $\mathcal{H}$, $(P(\lambda)\phi,\phi)$ is a left-continuous, nondecreasing function (of finite variation). The spectral theorem says that for any self-adjoint operator A on $\mathcal{H}$ there is a decomposition of the identity $P(\lambda)$ such that for all ϕ in D_a, ψ in $\mathcal{H}$, $(A\phi,\psi) = \int \lambda d(P(\lambda)\phi,\psi)$ or in short $A = \int \lambda dP(\lambda)$. On the other hand, every decomposition of the identity $P(\lambda)$ defines a unique self-adjoint operator A through $A = \int \lambda dP(\lambda)$. The domain D_A of A consists of all those vectors ϕ in $\mathcal{H}$ for which $\int \lambda^2 d(P(\lambda)\phi,\phi)$ is finite. As an example we take $\mathcal{H}$ to be $L^2(\mathbb{R})$ and choose $P(\lambda)$ to be multiplication by the Heaviside step function $\theta(\lambda - x)$ ($=1$ for $x<\lambda$, $=0$ for $x>\lambda$), i.e.,

$$(P(\lambda)\phi)(x) = \begin{cases} \phi(x) & \text{for} \quad x<\lambda \\ 0 & \text{for} \quad x>\lambda \end{cases}.$$

Then $A = \int \lambda dP(\lambda)$ is multiplication by x, i.e., the "position operator" of quantum mechanics.

See also EIGENFUNCTIONS; QUANTUM MECHANICS.

BIBLIOGRAPHY

N. I. Akhiezer and I. M. Glazman, *Theory of Linear Operators in Hilbert Space*. Frederick Ungar, New York, 1961 and 1963. (E)

N. Dunford and J. T. Schwartz, *Linear Operators*, I,II,III. Wiley-Interscience, New York, 1958–1971. (A)

P. R. Halmos, *Introduction to Hilbert Space and the Theory of Spectral Multiplicity*. Chelsea, New York 1951. (E)

L. Hörmander, *Linear Partial Differential Operators*. Springer-Verlag, New York, 1969. (A)

T. Kato, *Perturbation Theory for Linear Operators*. Springer-Verlag, New York, 1966. (A)

M. Reed and B. Simon, *Methods of Modern Mathematical Physics*, Vols. I–IV. Academic Press, New York, 1972–1978. (I)

R. D. Richtmyer, *Principles of Advanced Mathematical Physics, I,II*. Springer-Verlag, New York, Heidelberg, Berlin, 1978.

M. H. Stone, *Linear Transformations in Hilbert Space*, Vol. 15. Ann. Math. Soc. Colloq. Publ., Providence, R.I., 1932. (I)

J. von Neumann, *Mathematical Foundations of Quantum Mechanics*. Princeton University Press, Princeton, N.J., 1955. (I)

K. Yosida, *Functional Analysis*. Springer-Verlag, New York, 1974. (A)

Optical Activity

Robert J. Meltzer

A beam of plane polarized light passing through certain substances may remain plane polarized but with its plane of polarization rotated. Substances which cause such a phenomenon are said to be *optically active*.

Some substances are optically active only as crystals. These substances are optically active because of the asymmetric arrangement of the atoms or molecules in the crystal. A common crystal of this kind is quartz.

Other substances are active in the crystalline, liquid, vapor, or dissolved state because the atoms in the molecule are arranged asymmetrically. The most important molecules of this kind are carbon compounds. The spatial configuration of a molecule with an asymmetric carbon is different from its mirror image. The two different configurations are called *enantiomorphs*, or *stereoisomers*. The rotation of the plane of polarization by the two kinds of molecule will be in different directions, but of the same amount. A mixing having equal quantities of both configurations is called *racemic*.

Separation of dissolved or liquid enantiomorphs can be achieved using a chromatographic column. The columns which can perform this function are called *chiral* columns. The separation is made possible by the presence of optically active materials on the surface of the column support material.

If an observer facing the source sees the plane of polarization rotated clockwise, the optical rotation is said to be right-handed and the material is *dextrorotatory*; if counterclockwise the rotation is left-handed and the material is *levorotatory*. Dextrorotatory materials are designated as D-; levorotatory as L-.

The plane of polarization is unchanged if light which has passed through an active substance is then reflected back along the same path.

For solutions of optically active materials the amount of rotation is very nearly linearly proportional to the path length and the concentration of the solute. The proportionality constant is the *specific rotation*, α, and is defined for a 10-cm path. If p is the path length in centimeters, the rotation, θ, will be given by $\theta = \alpha Cp/10$, where C is the concentration in grams of active solute per cubic centimeter of solution. The *molecular rotation* is the specific rotation multiplied by the molecular weight of the active molecule. Strict linearity holds only if the concentration is not too high or the range of concentration is not too wide. The specific rotation is also affected by temperature, solvent, and wavelength of the incident polarized light.

The specific rotation is approximately, but not exactly, inversely proportional to the square of the wavelength. A substance obeying this law is said to have a plain dispersion curve, since it has neither a maximum nor a minimum. Plain dispersion curves are called *positive* dispersion curves or *negative* dispersion curves according to whether they produce more right-hand rotation or more left-hand rotation with decreasing wavelength.

Plain dispersion curves are found in substances without an optically active absorption band in the observed wavelength region. In wavelength regions where there are optically active absorption bands the dispersion curve will be *anomalous* and may show several peaks or troughs in the observed region. In general, anomalous curves are far more useful for elucidating the structure of complex organic molecules than plain curves. Optically active absorption bands are frequently the result of an inherently symmetric absorbing functional group which has been asymmetrically perturbed by its molecular environment. The optical activity induced into the symmetric group thus may act as a probe to explore molecular geometry.

The amount of rotation of the plane of polarization is measured using an instrument called a *polarimeter*. If the amount of rotation is measured as a function of wavelength, the instrument is a *spectropolarimeter*. All such instruments in common use today use photoelectric systems of one sort or another and have sufficient sensitivity to determine rotations as small as a tenth of a millidegree.

The proportionality of optical rotation to concentration is used commercially to measure the concentration of sucrose solutions in an instrument called a *saccharimeter*.

The theoretical explanation for optical rotation was first made by Fresnel who showed the effect can be considered to be the result of a difference in refractive index for left- and right-hand circular polarized light, i.e., circular birefringence. It is also possible for a substance to have different transmittance for left- and right-hand circular polarized light. The effect is called *circular dichroism* or the *Cotton effect*. Circular dichroism spectra are related to anomalous optical rotation spectra and are also used for exploring complex organic structures.

See also MAGNETIC CIRCULAR DICHROISM; POLARIZATION.

BIBLIOGRAPHY

F. J. Bates *et al.*, *Polarimetry, Saccharimetry, and the Sugars*. National Bureau of Standards Circular C440. U.S. Government Printing Office, Washington, DC, 1942.

N. Purdie and K. A. Swallows, "Analytical applications of polarimetry, optical rotatory dispersion, and circular dichroism," *Anal. Chem.* **61**(2), 77A–78A, 80A, 82A, 84A–89A (1989).

W. J. Lough (ed.), *Chiral Liquid Chromatography*. Blackie, Glasgow, U.K., 1989.

R. S. Longhurst, *Geometrical and Physical Optics*, pp. 476–490. Wiley, New York, 1957.

A. Sommerfeld, *Optics*, pp. 158–164, Academic Press, New York, 1964.

Optical Pumping

Robert A. Bernheim

The use of electromagnetic radiation at optical frequencies to produce a state selection of an assembly of atoms or molecules is called *optical pumping*. In such an experiment, the sample under irradiation is converted from an initial equilibrium distribution, usually thermal, over the available energy levels of the system to a final nonequilibrium distribution. The nonequilibrium or "pumped" state of the sample and its optical behavior can be used to study various physical properties of the system, such as its relaxation dynamics, or to perform spectroscopy on the species. In addition, optical pumping is one technique by which a medium can be prepared for subsequent laser action.

In 1949 and 1950 various optical excitation schemes for producing an orientation of atomic angular momenta were proposed by J. Brossel and A. Kastler, and by 1953 successful optical-pumping experiments had been performed on the $^3P_1$ state of the mercury atom and on the ground state of the sodium atom. Optical pumping developed quickly as a spectroscopic technique, especially when combined with radio-frequency and microwave radiation in double-resonance experiments, and today includes optical–optical double-resonance spectroscopy. The specific energy levels that can be populated or "pumped" include magnetic sublevels, hyperfine states, specific molecular vibration–rotation levels, individual components of λ-doubled levels, as well as separate ro-vibrational levels in the same or different electronic states. The information that can be obtained from these experiments includes determinations of nuclear spins, nuclear magnetic and nuclear electric-quadrupole moments, a precision value for the electron magnetic moment, atomic and molecular hyperfine interactions, g factors, excited-state lifetimes, and the excited electronic state structure of the species. For molecules this would also include the molecular constants. The first comprehensive study of quantum beat phenomena was performed with optical-pumping techniques. The inventions of the maser and laser have their roots in some of the early optical-pumping experiments, and other devices such as optically pumped magnetometers, "atomic clocks," and optically pumped lasers are direct applications.

The early optical-pumping experiments were mainly concerned with the magnetic sublevels of electronic states of atoms. These sublevels could be pumped in a variety of ways. Common to many of these experiments is the conservation of angular momentum in the atomic excitation process. By controlling the photon polarization it is possible to select the appropriate excitation for the desired optical-pumping process and can result in a spatial orientation of the species. Optical detection in spectroscopic experiments and the optical monitoring of the distribution of pumped species over the available energy levels can be carried out in a variety of ways that include observation of the absorption of pumping radiation, the change in wavelength and/or polarization of fluorescent radiation, and the appearance of quantum beats or light modulation in the experiment. In atomic beams, changes in beam intensity and beam temperature may be used to monitor the optical-pumping process. Today, there exists a vast array of laser spectroscopic techniques that make use of optical-pumping concepts. Several examples will be illustrated.

ATOMS IN EXCITED STATES

The optical pumping of an excited electronic state of an atom is illustrated by the experiments that can be performed on the Zeeman-split sublevels of the $^3P_1$ state of mercury. Consider the experimental configuration illustrated in Fig. 1a. A cell containing mercury vapor is irradiated along x by the Hg 253.7-nm resonance radiation with the electric vector polarized along the z-axis, which is also parallel to an applied magnetic field H. The fluorescence can be monitored along y. A π-type excitation of the mercury vapor is produced that

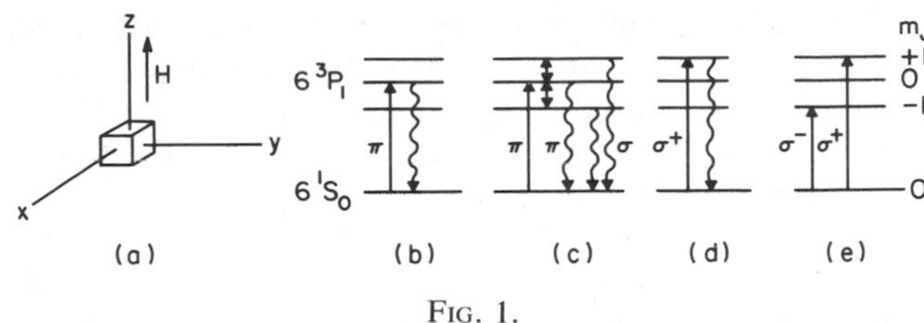

FIG. 1.

is a transition with $\Delta m_j = 0$, as shown in Fig. 1b for mercury isotopes with zero nuclear spin and, hence, no hyperfine interaction. If the excited mercury atoms in the sample emit radiation without experiencing a disorienting collision, the emitted fluorescence observed along the y axis will also be π type, polarized parallel to the magnetic field along z. This is one arrangement in which a specific sublevel of the excited state can be populated or optically pumped, in this case the $m_j = 0$ level of the $^3P_1$ state.

If transitions are induced to the other magnetic sublevels of the excited state, such as can be produced by a resonant radio-frequency field, the emitted radiation will contain some σ-type fluorescence with $\Delta m_j = \pm 1$ coming from the $^3P_1(m_j = \pm 1)$ levels, as depicted in Fig. 1c. Either the decrease in intensity of the π fluorescence or the increase in intensity of the σ fluorescence can be used to detect the magnetic resonance transitions between sublevels in the excited state of the mercury vapor. This experiment is one type of double resonance that can be performed—an optical resonance combined with a radio-frequency resonance, where the detection makes use of the high sensitivity with which the change in intensity of optical radiation can be detected.

Alternatively, as shown in Fig. 1d, circularly polarized or σ-type radiation propagating along z can produce an excitation to either the $m_j = +1$ or $m_j = -1$ level, depending on the direction of the circular polarization with respect to the direction of the applied magnetic field. This would correspond to an orientation in space of the components of angular momenta and of the magnetic moments of the mercury atoms. Still another distribution is achieved if both σ^+ and σ^- radiation is used for the excitation, as shown in Fig. 1e. Such will be the case if the mercury vapor is irradiated with the electric vector of the exciting light polarized perpendicular to the applied magnetic field. In this case there is no net magnetic moment in the excited vapor because of the equal populations of $m_j = +1$ and $m_j = -1$, and the situation is described as an "alignment" of the components of angular momentum of the excited state. In all of the foregoing cases it is possible to detect an excited-state magnetic resonance transition between the sublevels by observing the change in polarization of the atomic resonance fluorescence as the magnetic resonance condition is satisfied for the Zeeman splitting.

If a static electric field is applied to the mercury vapor, the Stark effect can be studied. For example, if the electric field is superimposed parallel to the magnetic field in Fig. 1a the intervals between $m = 0$ and $+1$ and between $m = 0$ and -1 are no longer equal, as they are in the pure Zeeman case. The double-resonance experiment will then yield two transitions split by the quadratic Stark effect.

In the presence of a nuclear spin magnetic moment the

magnetic sublevels are further split by the nuclear hyperfine interaction. In this situation not only Zeeman resonance transitions ($\Delta F=0$, $\Delta m_j=\pm 1$) can be detected by a change in intensity and polarization of the fluorescence, but hyperfine resonance transitions ($\Delta F=\pm 1$) can also be detected in a similar fashion. For nuclei with spin $I>\frac{1}{2}$ the perturbation produced by a nuclear electric-quadrupole interaction will further shift the hyperfine level structure.

The dynamic properties of the excited states of optically pumped atoms can be found from time-dependence studies of the emitted radiation. Alternatively, they can be investigated by probing with a second optical radiation source.

ATOMS IN GROUND STATES

Optical pumping is also used to produce orientation or alignment of atoms in their ground electronic states, either in atomic beams, where there are no collisions to affect the state populations, or in bulk vapor, for systems where the atomic ground states have small cross sections for disorientation when colliding with added buffer gases. The process is illustrated for alkali metal atomic vapors in which the nuclear hyperfine interaction is disregarded. Consider the experimental arrangement in Fig. 2a, in which D_1 radiation is circularly polarized and used to illuminate a cell containing the alkali metal vapor. A magnetic field H is applied to the vapor parallel to the direction of the pumping radiation, and the intensity of light transmitted through the cell is monitored. The transitions induced will have a selection rule of either $\Delta m_j=+1$ or -1, but not both, depending on the sense of circular polarization and direction of the applied magnetic field as illustrated in the energy-level diagram in Fig. 2b. Only one excitation transition can take place, removing atoms from only one sublevel of the ground state, whereas spontaneous emission can return the excited alkali metal atom to both ground-state sublevels. For an assembly of atoms the net effect in time will be a transfer of population from the $^2S_{1/2}(m_j=-\frac{1}{2})$ sublevel to the $^2S_{1/2}(m_j=+\frac{1}{2})$ sublevel. If the direction of circular polarization is reversed, there will be a transfer of population in the opposite direction, from $^2S_{1/2}(m_j=+\frac{1}{2})$ to $^2S_{1/2}(m_j=-\frac{1}{2})$, with a corresponding change in direction of magnetization of the vapor. The extent to which the pumping process can produce an atomic orientation depends not only on the rate at which the atoms can be excited but also on the relaxation processes that drives the $^2S_{1/2}(m_j=\frac{1}{2}$ and $-\frac{1}{2})$ levels back to thermal equilibrium. When the pumping radiation is first turned on, the vapor cell is opaque. The transmitted radiation increases in intensity as the atoms are pumped away from the absorbing level until equilibrium between the pumping and relaxation processes is established.

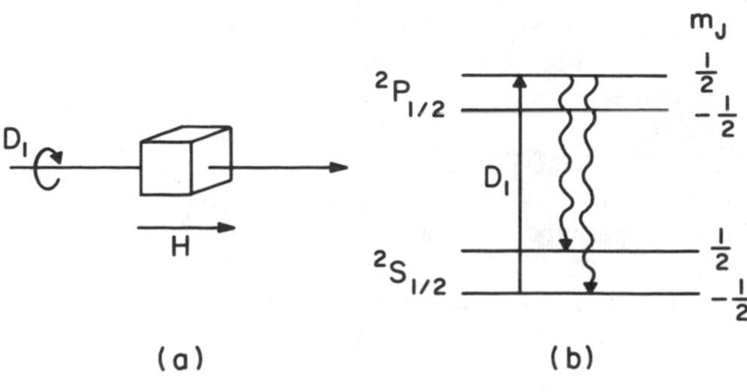

FIG. 2.

Radio-frequency or microwave resonances can be detected in optically pumped ground states since part of the pumped species is returned to the absorbing level. The magnetic resonance is detected as an increase in absorption by the vapor cell.

MOLECULES

All of the approaches used for atoms can be applied to molecules. Using tunable lasers and multiphoton techniques to produce stepwise excitation, or excitation followed by stimulated emission, both excited and ground electronic state levels can be pumped. However, the relaxation dynamics are quite different. For molecules in the gas phase it is very easy to reorient the various types of angular momenta. Rotational relaxation, for example, generally occurs with a single collision. Intramolecular energy transfer between the various internal degrees of freedom takes place with various time constants. These can be studied using molecular optical-pumping techniques.

Optically pumped molecules are the gain media for an important group of lasers. Most lasers that operate in the far-infrared wavelength region are molecular systems that are optically pumped by a CO_2 laser. The FIR laser action occurs between two molecular levels that are high above the ground state. The CO_2 laser radiation pumps the uppermost one of these. At the present time these lasers consist mainly of molecular systems where there is an accidental coincidence between a CO_2 laser line and a molecular transition that can produce the requirements for laser action.

The development of laser technology over the past quarter-century has made it possible to use optical pumping to study the dynamics as well as the spectroscopic structure of molecular systems. Using "pump" and "probe" techniques, state-to-state chemistry as well as the state structure of atoms and molecules can be studied. Systems in condensed phases can also be optically pumped, although the relaxation constraints are usually more severe than in the gas phase.

See also FINE AND HYPERFINE SPECTRA AND INTERACTIONS; LASER SPECTROSCOPY; RELAXATION PHENOMENA; RESONANCE PHENOMENA; ZEEMAN AND STARK EFFECTS.

BIBLIOGRAPHY

R. A. Bernheim, *Optical Pumping: An Introduction*. Benjamin, New York, 1965.

W. Happer, "Optical Pumping," *Rev. Mod. Phys.* **44**, 169–249 (1972).

W. Demtröder, *Laser Spectroscopy*. Springer-Verlag, Berlin, 1987.

Optical Spectroscopy *see* Visible and Ultraviolet Spectroscopy

Optics, Geometrical

R. Kingslake

Geometrical optics deals with those aspects of optics and optical systems that are not dependent in any way on the wave or quantum nature of light. In geometrical optics light is assumed to travel along rays, which are straight lines in a homogeneous medium, but they may be curved if the medium has a variable refractive index. The velocity of light in any transparent medium is less that its velocity in air, the ratio of the velocity in air to the velocity in a medium being known as the refractive index of the medium. Air is generally taken as the reference because all refractive index measurements are made in air and almost all lenses are used in air; on this basis the refractive index of vacuum is about 0.9998.

At any point along a fan of rays emitted by an object point there is a surface that is everywhere perpendicular to the rays; this surface is called a *wave front*, but the name is misleading because it implies the existence of light waves. Actually a wave front is the locus of points reached by the light after a given time along all possible ray paths. The optical path along a ray is the sum of the products of the geometrical path lengths multiplied by the refractive index of the medium in which each path lies, and a wave front is the locus of all points having the same optical path from a given point source. If the wave front emerging from a lens or other optical system is a true sphere, a perfect image will be formed. Any departure from a true sphere represents the presence of aberrations.

According to geometrical optics, each ray from an object point after passing through an optical system strikes a specified image plane at a single point. The assembly of all such points for all possible rays passing through the system constitutes the geometrical image of that object as formed by the system. In practice, of course, only a finite number of rays are actually traced, the assembly of their image points being called a *spot diagram*. It represents an outline picture of the image that will actually be formed by that lens, but it lacks the fine structure caused by diffraction and the interference effects which occur whenever two or more rays intersect each other. In spite of this limitation, however, the true image is always enclosed within the confines of the geometrical image, and hence a spot diagram, even when it is determined by tracing only a limited number of rays, is a fair representation of the actual image formed by the lens under discussion. Of course it is necessary to trace rays in several colors and to superpose the colored spot diagrams properly if it is desired to represent the image that would be formed in white light.

An optical system consists of an assembly of mirrors, lenses, prisms, and apertures, often with some additional components such as filters, beam splitters, polarizers, gratings, diffusers, fiber bundles, or lenticular plates or screens, together with light sources and radiation detectors. Lenses and image-forming mirrors generally have spherical surfaces because a sphere can be generated by a random motion between the tool and the work piece in surface contact during the grinding and polishing operations. However, aspheric surfaces of revolution about the lens axis are gradually coming into use, and some manufacturers have constructed machines for their regular production. The properties of conic-section mirrors have long been known and used in reflecting telescopes. Precise cylindrical surfaces can also be polished by surface contact if the polishing tool is constrained from rotating; these have found extensive application in anamorphic motion pictures. In the ophthalmic lens field, toric surfaces are commonly used, but here high precision is not required. Multifocal spectacle lenses are made either by grinding two or more separate surfaces on a single piece of glass, or by fusing a small segment of higher-index glass on a low-index blank and then polishing a single surface over the whole.

Some special-purpose lenses are composed of an assembly of prismatic segments having either flat or curved surfaces. Such assemblies are used in automobile headlight lenses to give a desired light distribution on the highway. When made with a rotational axis of symmetry, these assemblies become stepped or Fresnel lenses, which have found extensive application in lighthouses, in traffic signal lights, in Vuegraph projectors, and to brighten the corners of a ground-glass screen in a camera or projector.

A simple lens with spherical surfaces suffers from a number of well-recognized aberrations. The most familiar of these are

1. Spherical—different zones of the lens form images at different distances along the lens axis.
2. Coma—the different zones of the lens form images of different sizes.
3. Chromatic—the images in different colors are formed at different distances from the lens.
4. Lateral color—the images in different colors are of different sizes.

(Thus spherical aberration and coma bear the same relation as chromatic aberration and lateral color; in the first case the zone is varied while in the latter it is the color of the light.)

5. Distortion—the image magnification varies across the field so that straight tangential lines in the object appear as curved lines in the image.
6. Astigmatism—radial lines and tangential lines in the object are imaged at different distances from the lens. If the lens is well centered, astigmatism exists only in the outer parts of the field; but with cylindrical or toric surfaces, astigmatism appears also on axis.
7. Field curvature—this aberration is intimately mixed with astigmatism, since it implies that either the radial image locus or the tangential image locus (or both) is not on a flat image plane perpendicular to the axis but is on a curved surface that may be concave (inward) or convex (backward) relative to the lens itself.

In the presence of aberrations, the actual light distribution within an image caused by light waves having a finite wavelength is very complex, and it requires long and difficult calculations to determine it accurately. On the other hand, the tracing of a few rays is a trivial matter on even a small computer, and lens design is invariably carried out by purely geometrical procedures. However, if a lens is so good that all the geometrical aberrations are negligible, then it is necessary to take the wave nature of light into account. Such a lens is often called *diffraction limited*. It is extremely difficult to design and make a truly diffraction-limited lens, and very few practical optical systems deserve this title.

The theory of lenses is often broken down into orders of complexity. By first-order optics we refer to such matters as object and image distances, focal length, and relative aperture, assuming that no aberrations exist. The first-order properties of a system are found by tracing so-called paraxial rays lying throughout their length very close to the lens axis. Indeed, simplification is often carried still further by regarding the lens elements as being infinitely thin. Much can be learned about the properties of an optical system by first-order studies; for instance, the laws controlling the motion of the components of a zoom lens, or the layout of a submarine periscope.

The next stage in lens theory includes rays lying just outside the paraxial region, so that the first signs of aberration are beginning to appear. In this realm all aberrations when expessed as a transverse ray displacement are of the third order in terms of lens aperture and angular field; they are therefore called third-order or primary aberrations. Mathematical expressions for the magnitude of these aberrations are fairly simple, and they can be readily computed. Many lens systems are designed first by third-order methods, but it is necessary to leave small residuals of the various aberrations to offset the higher-order aberrations that will arise when the aperture and field are allowed to increase. Some workers have extended the theory to the next, or fifth, order, but now the number of different aberrations becomes so great and the mathematical expressions so complex that it is generally more convenient to go to direct ray tracing rather than to use algebraic theory to evaluate the performance of a lens, and to modify the system to improve it.

In modern computer-aided lens design procedures, the old classical aberrations are seldom used. Instead, any departure of a ray from its ideal location in the image plane is regarded as an aberration, and the sum of the squares of all these aberrations is used as a merit function to indicate the quality of the lens. A computer optimization program is then used to make whatever changes are required in the lens construction to minimize the merit function, staying always within the boundary conditions of the problem. These include such things as the focal length, the back focus, the physical size of the lens, and the permissible degree of vignetting or cutting of the oblique light beams by the apertures of the lens. Computer programs have completely revolutionized lens design, and thanks to the computer we now have zoom lenses, microscope objectives, and other optical systems of a far higher quality than ever before. Indeed, a zoom lens would be almost unthinkable were it not for the computer.

BIBLIOGRAPHY

A. C. Hardy and F. H. Perrin, *The Principles of Optics*. McGraw-Hill, New York, 1932. (E)

L. C. Martin, *Technical Optics*. Pitman, London, 1950. (I)

MIL-HDBK 141—Optical Design. U.S. Defense Supply Agency, Washington, D.C., 1962. (A)

W. J. Smith, *Modern Optical Engineering*. McGraw-Hill, New York, 1966. (I)

J. P. C. Southall, *Mirrors, Prisms and Lenses*. Macmillan, New York, 1933. (E)

Optics, Nonlinear

J. A. Giordmaine

The field of nonlinear optics deals with light-wave interactions and other optical phenomena arising from the nonlinear response of a medium to optical fields. Nonlinear optics includes, for example, processes in which energy is transferred between optical fields of different frequency or direction, as well as intensity-dependent scattering and absorption processes. Nonlinear optical effects become observable at light intensities high enough that the optical electric field is significant in comparison with the Coulomb field ($\approx 10^8$ V/cm) in atoms, molecules, or condensed matter. Monochromatic light of the required intensity, typically greater than 10^2 W/cm^2 ($\approx 10^2$ V/cm), first became available with the introduction of the laser in 1961. Nonlinear optics plays a major role in optical instrumentation and research and may in future offer a technology for all-optical switching and information processing.

The most important nonlinear optical effects can be described by a nonlinear optical polarization of the form

$$P=\chi^{(1)}E+\chi^{(2)}E^2+\chi^{(3)}E^3. \quad (1)$$

The dielectric polarization P is related to the displacement vector D and the electric field vector E through the constitutive equation $D=E+4\pi P$. The linear susceptibility $\chi^{(1)}$ is related to the dielectric constant ϵ through $\epsilon=1+4\pi\chi^{(1)}$. The material susceptibilities $\chi^{(2)}$ and $\chi^{(3)}$ are tensors with coefficients dependent on the frequencies of E and P.

Nonlinear processes differ dramatically from those of classical optics in their strong dependence on the intensity of light and in the coherent generation of light at combination frequencies. Conversion of light-wave energy with frequencies ω_1 and ω_2 to sum ω_+ and difference ω_- frequencies through the $\chi^{(2)}E^2$ polarization term corresponds to the appearance of combination frequencies $\omega_1+\omega_2\rightarrow\omega_+$ and $\omega_1-\omega_2\rightarrow\omega_-$ in the product E^2 of sinusoidal field amplitudes at ω_1 and ω_2. The power radiated at the combination frequencies from the polarization distribution P is proportional to the product of the incident powers at ω_1 and ω_2.

Nonlinear optical phenomena described by the $\chi^{(2)}E^2$ term include optical second harmonic generation ($\omega+\omega\rightarrow 2\omega$), parametric oscillation ($\omega_+\rightarrow\omega_1+\omega_2$), and optical rectification ($\omega-\omega\rightarrow 0$). The $\chi^{(3)}E^3$ term gives rise to optical phase conjugation, stimulated Raman and Brillouin scattering, two-photon absorption, self-focusing, and optical third harmonic generation.

The well-known electric field dependence of refractive index is a special case of Eq. (1). The $\chi^{(2)}E^2$ term describes the linear electro-optic (Pockels) effect when the E^2 term is the product of an optical field and a low-frequency or dc field. Similarly the $\chi^{(3)}E^3$ term includes the quadratic electro-optic (Kerr) effect as the product of an optical field and the square of a low-frequency or dc field. Because the electro-optic effects may contain important contributions from ionic motion or molecular reorientation, while optical harmonic generation involves primarily electronic response, the values of $\chi^{(2)}$ and $\chi^{(3)}$ describing optical nonlinearities are usually much smaller than the $\chi^{(2)}$ and $\chi^{(3)}$ for electro-optic effects.

The first observations of optical nonlinearities were made in 1961 by P. Franken and his collaborators (optical frequency doubling in quartz) and by W. Kaiser and C. G. B. Garrett (two-photon absorption in calcium fluoride). The subsequent period has seen the discovery of a wide variety of nonlinear optical phenomena. Numerous applications of nonlinear optics were made in laser spectroscopy, tunable and high-power light sources, generation of ultraviolet light in the 300 Å region, generation and measurement of optical pulses of 10^{-14}s duration, infrared difference-frequency generation, and optical devices. A detailed understanding of the electromagnetic aspects of the interaction of light waves was obtained, and considerable progress was made in explaining the nonlinear susceptibilities. Some of the principal nonlinear optical effects and their applications are described in greater detail below.

OPTICAL SECOND HARMONIC GENERATION

Second harmonic generation allows efficient conversion of visible or infrared laser light to higher frequencies where suitable monochromatic sources may be unavailable. Pulsed or continuous light with power greater than 1 W can be frequency doubled with 50–100% conversion efficiency in well-engineered optical systems with highly nonlinear crystals. Multiple-stage optical frequency doubling is a useful approach for upconverting visible laser light into the deep ultraviolet.

Important features of second harmonic generation are illustrated by the propagation in a nonlinear material of a plane wave optical field of the form $E_1 \cos(\omega_1 t - k_1 x)$. The propagation constant k_1 is related to the frequency ω_1 and phase velocity c/n_1 by $k_1 = n_1\omega_1/c$. According to Eq. (1) a polarization wave proportional to $\chi^{(2)}E_1{}^2 \cos(2\omega_1 t - 2k_1 x)$ is generated in the material, with the same phase velocity as the incident wave. This polarization is at the doubled frequency $2\omega_1$ and radiates a plane wave optical field of the form $E_2 \cos(2\omega_1 t - k_2 x)$, travelling at the new phase velocity c/n_2 of free waves at $2\omega_1$. Efficient conversion requires phase matching: $k_2 = 2k_1$ or $n_2 = n_1$. In frequency doublers optical birefringence is used to provide phase matching by exact compensation of the normally large dispersive difference between n_1 and n_2.

The frequency-doubled power W_2 from a crystal of length L, beam area A, and incident power W_1 is proportional to $(\chi^{(2)})^2 W_1{}^2 L^2/A$. High conversion efficiency requires optimum focusing, high $\chi^{(2)}$, and a long crystal, typically several millimeters. Important doubler crystals, discussed more fully below, include asymmetric, highly polarizable oxides in the perovskite, tungsten-bronze, iodate, and boron-oxide groups, such as lithium niobate, and conjugated organic molecules with donor-acceptor radicals, such as methylnitroaniline. These materials have $\chi^{(2)}$ coefficients in the range 10^{-9} to 10^{-7} esu.

OPTICAL PARAMETRIC OSCILLATORS

In a material with nonzero $\chi^{(2)}$, light of frequency ω_3 can exchange energy with pairs of light waves having frequencies $\omega_1 + \omega_2 = \omega_3$. Gain at the signal and idler frequencies ω_1 and ω_2 due to the pump at ω_3 leads to oscillation when the nonlinear medium is placed within a suitable optical resonator. The frequency of optical parametric oscillation is the frequency at which exact phase matching occurs, and is determined by dispersion and birefringence. Rotation of the oscillator crystal can provide extremely broadband tuning since the gain is not associated with discrete energy levels as in a laser. A lithium niobate oscillator pumped by a 1.06-μm neodymium laser tunes between 1.5 and 4.0 μm.

A wide variety of pulsed and cw optical parametric oscillators have been demonstrated in the visible and near infrared. Threshold cw pump powers in the mW range have been obtained; energy conversion efficiency as high as 50% is achieved in pulsed oscillators.

The mechanism of parametric gain involves the $\chi^{(2)}$ term in Eq. (1) in several stages. Pump and signal optical fields at ω_3 and ω_1 combine to generate polarization P_2 at the idler frequency $\omega_2 = \omega_3 - \omega_1$. The optical electric field radiated by P_2 at frequency ω_2, together with the pump field at ω_3, in turn generate polarization at the signal frequency $\omega_1 = \omega_3 - \omega_2$ in the correct phase to amplify the original signal field.

Degenerate parametric oscillators ($\omega_1 = \omega_2 = \omega_3/2$) produce light separable into two components called quadrature-phase amplitudes. The variance of the amplitude of the total field obeys the uncertainty principle. However the fluctuations in one of the components, called a squeezed state, can be greatly reduced. Squeezed light can be detected with signal-to-noise ratio in excess of the usual shot-noise limit. Parametric oscillators have become a useful tool in the study of the quantum statistical properties of light.

NONLINEAR OPTICAL SUSCEPTIBILITIES

In centrosymmetric crystals inversion symmetry causes the contributions to $\chi^{(2)}$ from adjacent microscopic regions to average exactly to zero. It follows that second harmonic generation and other $\chi^{(2)}$ effects require the absence of an inversion center, the same requirement as for piezoelectricity.

In principle the tensor $\chi^{(2)}$ relating the vectors $P(\omega_3)$, $E(\omega_1)$, and $E(\omega_2)$ contains 27 independent coefficients. For optical mixing in common transparent crystals, $\chi^{(2)}$ in fact is greatly simplified by point symmetry and energy considerations. In potassium dihydrogen phosphate (KDP), for example, the tensor $\chi^{(2)}$ is left with only one independent element d and Eq. (1) takes the form $P_x = 2dE_yE_z$, $P_y = 2dE_zE_x$, $P_z = 2dE_xE_y$.

The values of $\chi^{(2)}$ vary much more widely from material to material than do the values of $\chi^{(1)}$ [e.g. χ(Te)/ $\chi^{(2)}$(KDP)$\approx 10^4$]. Predictions of the magnitude of $\chi^{(2)}$ have been based on classical oscillator models, quantum-mechanical perturbation calculations, and models of bond polarizabilities.

Analysis of the nonlinear optical response based on classical anharmonic oscillator models suggests that relevant coefficients of the linear and nonlinear susceptibilities $\chi^{(1)}$ and $\chi^{(2)}$ should be related by $\chi^{(2)}=\delta(\chi^{(1)})^3$. Indeed the values of δ vary only little from one crystal to another and are in reasonable agreement with the value of $\approx 10^{-6}$ esu estimated from the oscillator models.

Complete formal expressions are available for $\chi^{(2)}$ based on quantum-mechanical perturbation theory. Approximations to these expressions give $\chi^{(2)}$ in terms of the moments of the electronic ground state charge distribution. Considerable progress has been made in calculating $\chi^{(2)}$ from the ground-state electronic wave functions using this approach.

Good agreement between predicted and experimental values of $\chi^{(2)}$ has been obtained from bond polarization models. One of these, the bond charge model, attributes the nonlinearity to redistribution of the bond charges and calculates $\chi^{(2)}$ explicitly in terms of measurable spectroscopic quantities such as the heteropolar, homopolar, and average band gaps. This approach has yielded particularly useful predictions of nonlinear coefficients for binary compounds.

STIMULATED RAMAN SCATTERING

Intense laser light may be efficiently converted to coherent light downshifted by frequencies corresponding to molecular or crystal lattice vibrations (stimulated Raman scattering). Stimulated vibrational Raman processes provide useful intense sources of pulsed radiation in the visible and near-IR spectra.

Stimulated Raman scattering is related to spontaneous Raman scattering in the same way that laser stimulated emission is related to fluorescence. The elementary downshifting (Stokes) scattering process, from a photon number point of view, consists of transfer of a photon from a mode at ω_1 of the incident radiation field to a mode at ω_2 of the scattered field, and promotion of a molecule to a vibrationally excited state of energy $\hbar(\omega_1-\omega_2)$. The probability of the process is proportional to $n_1(n_2+1)$, where n is the number of photons per mode. Threshold for the stimulated process is reached as n_2 exceeds one, when the scattering rate dn_2/dt becomes proportional to n_2 and n_2 grows exponentially.

The $\chi^{(3)}E^3$ term in Eq. (1) provides a complete description of stimulated Raman scattering through the process of $\omega_1-(\omega_1-\omega_2)\rightarrow\omega_2$. The gain of the vibrational process is typically about 1 cm^{-1} for laser light of 10^8 W/cm^2. Other processes such as self-focusing, self-phase modulation, and laser-induced breakdown often accompany stimulated Raman scattering.

Backward stimulated Raman scattering generates compressed optical pulses with peak intensities much higher than the incident laser power. Spin-flip Raman lasers have important spectroscopic applications. These sources, pumped by carbon dioxide or carbon monoxide lasers, are tunable in the 5–14 μm region.

Other nonlinear stimulated processes described by the $\chi^{(3)}$ term include stimulated Brillouin scattering (from acoustic phonons), stimulated Rayleigh wing scattering (from molecular reorientation), stimulated thermal Rayleigh scattering (from entropy fluctuations in liquids), and stimulated polariton scattering (coupled parametric and Raman processes).

OPTICAL PHASE CONJUGATION

Optical phase conjugation is a four-wave mixing process arising from $\chi^{(3)}E^3$, in which an input wave with phase $(\omega t-kx)$ generates a reflected phase conjugate wave with phase $(\omega t+kx)$. The phase conjugate wave can be understood as a product component of three fields with phases $(\omega t-ky)$, $(\omega t+ky)$ and $(\omega t-kx)$, the last being the input field and the first two being oppositely directed pump fields. The direction y is not critical. The phase conjugate wave can be considered as a time reversed form of the incident wave and has remarkable characteristics. For example, the phase conjugate field of light transmitted through a distorting medium is reconstructed to its original state by propagation back through the distorting medium. Phase conjugation has important applications in holography, and signal and image processing.

TWO-PHOTON ABSORPTION, SELF-FOCUSING, AND BISTABILITY

Two-photon absorption corresponds to the excitation by intense laser light of a material transition at twice the laser frequency. This process is described by a $\chi^{(3)}E^3$ term in Eq. (1) of the form $\omega+\omega-\omega\rightarrow\omega$, when the nonlinear polarization term is at the fundamental frequency and in phase for absorption. The absorption is proportional to the light intensity squared. Two-photon absorption provides a powerful new high-resolution spectroscopic tool because Doppler broadening is completely eliminated in the absorption of oppositely directed beams of the same frequency.

Self-focusing of light, the contraction and collapse of an intense light beam into filaments a few microns in diameter, is widely encountered in pulsed laser beams with power above a few megawatts. Power in the filaments may be of the order of 10^{10} W/cm^2 and the spectrum may be greatly broadened due to refractive-index transients. Self-focusing and self-trapping arise from an intensity-dependent refractive index through contributions from the $\chi^{(3)}$ term of the form $\omega+\omega-\omega\rightarrow\omega$. Self-focusing is related to the Kerr effect in liquids, and to the electrostrictive effect in solids. Self-focusing causes irreversible optical damage in solids and must be considered in the design of high-power lasers, and in the interpretation of all nonlinear optical experiments.

Optical bistability arises in saturable optical systems in which the output intensity determines the transmission. An example is a medium containing saturable absorption or dispersion located between parallel reflectors. At low intensities, reflection of incident light may be controlled primarily by the entrance mirror. At high intensities, bleaching of the absorber transfers control of reflection to the reflector combination. Such devices may exhibit fast switching and hysteresis, and provide a possible means for all-optical logic.

Nonlinear optics includes many other phenomena such as

coherent transient optical effects, nonlinear scattering in plasmas, nonlinear effects in optical waveguides, and multiphoton ionization and gas breakdown.

See also KERR EFFECT, ELECTRO-OPTICAL; LASER SPECTROSCOPY; LASERS; QUANTUM ELECTRODYNAMICS.

BIBLIOGRAPHY

Y. R. Shen, *The Principles of Nonlinear Optics*. Wiley, New York, 1984.

M. Schubert and B. Wilhelmi, *Nonlinear Optics and Quantum Electronics*. Wiley, New York, 1986.

M. D. Levenson and S. S. Kano, *Introduction to Nonlinear Laser Spectroscopy*, rev. ed. Academic Press, Boston, 1988.

A. Yariv, *Quantum Electronics*, 3rd ed. Wiley, New York, 1988.

Optics, Physical

Olof Bryngdahl

The classic definition of physical optics is that portion of optics which deals with those phenomena caused by electromagnetic radiation and manifested through its wave nature. The types of radiation involved have one particular property in common: the same speed of propagation in vacuum (2.997925×10^8 m/s). The spectrum comprises a range from short to long wavelengths including γ and x rays; ultraviolet, visible, and infrared light; and micro- and radio waves.

Lately, the term optical physics has been used to describe all the physical aspects of optical properties and phenomena. The fundamental question in physical optics, however, concerns the nature of the light itself. Whether the light has inherent particle or wave properties has lost its distinction in modern physics. However, the distinction has both a practical and historical importance in treatments of emission, propagation, interaction, and absorption of radiative energy. Thus, what is implied in physical optics is not only certain phenomena but also the means and methods by which these phenomena are described and treated. Three particular optical processes—polarization, interference, and diffraction—form the nucleus of physical optics. They are all variations of the same phenomenon, i.e., superposition of waves.

In physical optics, the radiation is considered to be an electromagnetic wave described by an **E** (electric) and **B** (magnetic) field. Maxwell's equations predict a transverse wave in free space with a propagation direction according to $\mathbf{E}\times\mathbf{B}$, where **E** and **B** are mutually perpendicular. The relationships between the space and time variations of these quantities are described by the wave equation which can be expressed in either vector or scalar form. For example, in Cartesian coordinates, all the components E_x, E_y, E_z, B_x, B_y, B_z, obey the scalar differential wave equation:

$$\frac{\partial^2\psi}{\partial x^2} + \frac{\partial^2\psi}{\partial y^2} + \frac{\partial^2\psi}{\partial z^2} = v^{-2}\frac{\partial^2\psi}{\partial t^2};$$

v is the speed of the wave. The electromagnetic wave transports energy, and its time-averaged value is what is generally observed. This flow of energy per unit area and unit time is the irradiance

$$I = \tfrac{1}{2}v^2\epsilon|\mathbf{E}\times\mathbf{B}| = \tfrac{1}{2}v\epsilon E^2 = v\epsilon\langle E^2\rangle,$$

where ϵ is the electric permittivity. The irradiance is proportional to the square of the electric field, E, which is commonly referred to as the optical field.

SUPERPOSITION OF WAVES

The wave equation is linear, i.e., containing only terms of the first power of ψ and its derivatives. Thus, if $\psi_1, \psi_2, \ldots$, are solutions, any linear combination of these is also a solution, i.e., $\psi = \Sigma c_i\psi_i$, where c_i represents arbitrary constants. Because of this principle of superposition, the resultant wave at any specific point is simply the algebraic sum of the individual waves. The superposition of sinusoids is usually treated by the algebraic method or by the addition of phasors (complex representation).

Harmonic waves, $E_n = E_{0n}\sin(\omega t - kz + \phi_n)$, propagating along the positive z axis are characterized by their amplitude, E_{0n}, their angular frequency, $\omega = 2\pi v/\lambda = kv$, and their state of polarization. Phase constants are represented by ϕ_n. Superposition of two waves with the same frequency results in

$$E = E_1 + E_2 = E_{01}\sin(\omega t - kz_1 + \phi_1) + E_{02}\sin(\omega t - kz_2 + \phi_2).$$

z_1 and z_2 are the distances from the sources of the waves to the point of observation. The resultant can be written in the form

$$E = E_0\sin(\omega t - \delta),$$

where

$$E^2_0 = E^2_{01} + E^2_{02} - 2E_{01}E_{02}\cos\Delta$$

in which $\Delta = k(z_1 - z_2) + \phi_2 - \phi_1$ is the phase difference. When the two waves are emitted by different sources, or even from different points in the same source, ϕ_1 and ϕ_2 change frequently by arbitrary amounts resulting in noncoherent waves. Then, only $E^2_{01} + E^2_{02}$ can be observed, i.e., the resulting irradiance is just the sum of the individual irradiances. If, on the other hand, the cross product in the expression for $E_0{}^2$ is not averaged out, the waves preserve a constant phase difference and are said to be coherent. Then, the irradiance is not simply the sum of the separate irradiances; an additional contribution occurs, the so-called interference term, $2E_{01}E_{02}\cos\Delta$. The waves will interfere constructively (be in phase) for $\Delta = 2p\pi$; they will interfere destructively (be out of phase) for $\Delta = (2p+1)\pi$, where p is an integer.

The case when two waves of the same frequency propagate in opposite directions is worth noting. Assume that a wave is reflected by a mirror (perfect conductor) at $z = 0$ and interferes with itself. Then,

$$E = E_{01}\sin(\omega t + kz) + E_{01}\sin(-\omega t + kz) = 2E_{01}\sin kz\cos\omega t,$$

because the boundary condition requires a zero **E** field component parallel to the surface. A standing wave is achieved. Its amplitude at any point is constant. $E = 0$ for $z = p\pi/k = p\lambda/2$ constitute the nodal points.

In a realistic situation, the superposing radiations are not strictly monochromatic but quasimonochromatic. The frequencies of the waves fall within a narrow range. Addition of the waves

$$E_1 = E_{01}\cos[(k+\Delta k)z - (\omega+\Delta\omega)t]$$

$$\text{and}\quad E_2 = E_{01}\cos[(k-\Delta k)z - (\omega-\Delta\omega)t]$$

results in

$$E = 2E_{01}\cos(kz-\omega t)\cos(\Delta kz - \Delta\omega t).$$

This can be interpreted as a single wave, $2E_{01}\cos(kz-\omega t)$, which has a modulation envelope, $\cos(\Delta kz-\Delta\omega t)$. $\Delta\omega$ is known as the beat frequency. The situation is similar for the case with many waves. The overall wave envelope does not propagate with the same velocity as the individual waves (phase velocity $=\omega/k$), but at a rate, $\Delta\omega/\Delta k$, called the group velocity. Furthermore, the wave trains in reality have finite lengths. They are nonperiodic functions in the sense that they do not repeat themselves to infinity. Application of Fourier analysis indicates that they can be represented as the summation of a continuous distribution of simple harmonic terms. The dominant contribution is centered around ω, and, as the length of the wave train increases, its frequency spectrum shrinks, and, finally, it closes down to a single tall spike for an infinitely long wave.

POLARIZATION

The electromagnetic wave can transport not only energy but also momentum. A left-circularly-polarized wave corresponds to photons with their spins aligned in the direction of propagation and a right-circularly-polarized wave to photons with reversed spin orientation.

A transverse electromagnetic wave is called linearly or plane polarized when the orientation of the **E** field is confined to one particular plane which contains both the electric field vector and the vector of propagation. Suppose two harmonic linearly polarized waves of the same frequency, but with their electric field directions orthogonal, propagate in the z direction:

$$\mathbf{E}_x = \hat{\imath}E_{0x}\cos(kz-\omega t)\quad\text{and}\quad \mathbf{E}_y = \hat{\jmath}E_{0y}\cos(kz-\omega t+\phi),$$

where $\hat{\imath}$ and $\hat{\jmath}$ are the unit vectors in Cartesian coordinates and ϕ is the relative phase difference between the waves. The resultant field is then

$$\mathbf{E} = \mathbf{E}_x + \mathbf{E}_y.$$

Depending on the value of ϕ, different states of polarization occur. For example, the two waves are in phase when $\phi = 2p\pi$, and the resultant wave, $(\hat{\imath}E_{0x}+\hat{\jmath}E_{0y})\cos(kz-\omega t)$, is linearly polarized. $\phi=(2p+1)\pi$ also results in a linearly polarized wave, $(\hat{\imath}E_{0x}-\hat{\jmath}E_{0y})\cos(kz-\omega t)$. Of course, the procedure can also be applied in reverse, i.e., any linearly polarized wave can be decomposed into two orthogonal components. In general, the summation of $\mathbf{E}_x$ and $\mathbf{E}_y$ will result in an elliptically polarized wave where **E** changes both magnitude and orientation with time. The linearly polarized wave is the special case with no rotation of **E**. Other special cases occur for $\phi=(2p-\frac{1}{2})\pi$ and $\phi=(2p+\frac{1}{2})\pi$ when $E_{0x}=E_{0y}$. In the former case, the wave is right-circularly-polarized and in the latter left-circularly-polarized. For example, in the right-circularly-polarized case, the wave is given by

$$E_{0x}\{\hat{\imath}\cos(kz-\omega t)+\hat{\jmath}\sin(kz-\omega t)\}.$$

The amplitude of this wave is constant. However, its direction changes with time in a rotating fashion. Just as a circularly-polarized wave can be synthesized from two orthogonal, linearly polarized waves, a linearly polarized wave can be synthesized from two opposing circularly-polarized waves. Natural light, which is also called unpolarized, can be represented by two orthogonal linearly or opposing circularly-polarized components which are incoherent (their relative phase difference varies randomly and rapidly).

Optical devices can influence the state of polarization of electromagnetic radiation. One class of device, called polarizers, is used to convert unpolarized waves into polarized ones. They are based on physical phenomena such as dichroism, reflection, scattering, or birefringence. A commonly used linear polarizer is the Polaroid which contains a sheet of polyvinyl alcohol impregnated with iodine and stretched in a certain direction. The **E** component of the incident field that is parallel to the stress direction is strongly absorbed. For radiation of longer wavelength, a grid of parallel conducting wires can be used; it functions in a similar fashion. There are also some materials which are inherently dichroic, e.g., tourmaline. Birefringent crystals like calcite and quartz are used in birefringent polarizers. Their optical anisotropy causes double refraction, i.e., the two orthogonal linearly polarized components of the incident wave undergo angular or lateral separation. Among the more known configurations are the Nicol, Wollaston, Glan–Foucault, and Glan–Thompson prisms.

Retarders are another class of optical elements which can be applied to change the polarization of an incident wave. They introduce a certain phase difference between two polarization states. Thus, they can be used to convert any given polarization state into any other. Certain wave plates like the quarter-, half-, and full-wave plates introduce phase differences of $\pi/2$, π, and 2π. With a quarter-wave plate it is possible to convert a linearly polarized wave into a circularly polarized one and vice versa. A half-wave plate can be used to rotate the plane of a linearly polarized wave. Devices that are capable of introducing a controllable retardance are called compensators, e.g., Babinet and Soleil compensators.

INTERFERENCE

Superposition of two waves with the same frequency (cf. above) results in an irradiance which is the sum of the individual irradiances of the individual waves, I_1+I_2, plus an interference term, $2\sqrt{I_1I_2}\cos\Delta$, where $\Delta = k(z_1-z_2)+\phi_2-\phi_1$. To form an interference pattern, the phase difference, $\phi_2-\phi_1$, between the two sources must remain constant in time, i.e., the waves have to be coherent. The common way to achieve this is to use two coherent secondary sources originating from the same source. According to the vector nature of the electromagnetic radiation, the two waves have to possess the same polarization state in order to interfere. To obtain contrasting interference patterns, the amplitudes of the superposing waves cannot be too unequal.

For $I_1 = I_2 = I_0$ the irradiance of the resulting field is $4I_0 \cos^2(\Delta/2)$ with a minimum of zero and a maximum $4I_0$.

Many arrangements have been suggested for producing and studying interference phenomena. These can be divided in several ways. The radiation from a source can be divided by wavefront splitting or amplitude splitting. In wavefront splitting, different portions of the original wavefront constitute the interfering waves. In amplitude splitting, some kind of beamsplitter is used to make the interfering wave occupy the same spatial portion of the original wavefront. An example of wavefront splitting is Young's experiment. The cylindrical wave from a line source falls on two narrow parallel slits spaced a distance d apart. These slits constitute two coherent secondary sources. If a screen is placed a distance D from the plane containing the slits, then constructive interference occurs at the locations on the screen where the difference in distance to the two slits is an integer number of wavelengths. Parallel fringes appear on the screen with a spacing $D\lambda/d$. The two slits can be replaced by a Fresnel double mirror or a Fresnel biprism. Examples of amplitude-splitting interferometers are the Michelson and Mach–Zehnder arrangements. Here the wavefront entering the interferometer is split into two waves by a beamsplitter. These waves then travel through different arms through the interferometer and are directed toward the same or a second beamsplitter with mirror arrangements so the waves still superpose. Two plane waves forming an angle ϕ relative to each other produce parallel interference fringes of a period $\lambda/\sin(\phi/2)$.

Some typical interference phenomena are observable in thin dielectric films. Different cases occur depending on illumination and observation conditions. When the interference pattern is dependent on angular parameters, fringes of equal inclination are formed and these fringes are localized fringes, i.e., localized to certain regions in space (Haidinger fringes). On the other hand, nonlocalized fringes of equal thickness (Fizeau fringes) are formed when the pattern is dependent only on the optical thickness of the film.

Interference phenomena can be arranged so that either two-beam or multiple-beam interference is obtained. In the former case, the fringes of the interference pattern have a sinusoidal profile; in the latter, they are sharper and more peaked.

Interferometers are used to test and examine properties of the radiation itself, shapes of surfaces, and qualities of optically transparent media. Another important application of these phenomena is the field of thin dielectric film coatings. For example, thin coatings can be used to eliminate unwanted reflections, and in the manufacture of dichroic mirrors, filters, heat reflectors, and solar cells.

DIFFRACTION

Diffraction deals with deviation of radiation from rectilinear propagation. It is a general characteristic of wave phenomena which occurs when a portion of a wavefront is obstructed in some way. If the amplitude and phase of the wavefront undergo spatial alterations, then contributions from these locations will interfere and form diffraction patterns as the radiation propagates. Thus, there is no real physical distinction between interference and diffraction. The difference is more historical and due to the traditional ways the phenomena are described and treated. If only a portion of an incident wavefront will pass through an optical instrument, it is important to consider the influence of diffraction when evaluating the results.

A common approach to treating diffraction problems is to use the Huygens–Fresnel principle: Each point of a wavefront acts as the source of secondary wavelets, and the amplitude of the radiation field is the superposition of these wavelets. Kirchhoff mathematically formulated this principle in the Fresnel–Kirchhoff diffraction theory, where the wavefield is decomposed into spherical waves. Sommerfeld modified it to a composition of plane waves in the Rayleigh–Sommerfeld diffraction theory. These approaches share some approximations and are adequate under most practical circumstances. Exact solutions of diffraction problems can be obtained utilizing electromagnetic radiation theory. These belong, however, to the optical problems that are most difficult to handle.

Diffraction phenomena are usually treated according to the region of interest and corresponding approximations; the entire space behind the diffractive object is called the Rayleigh–Sommerfeld region, excluding the space closest to the object results in the Fresnel region, and excluding even more the Fraunhofer region or far field. When both the source and the point of observation are far from the diffractive object, i.e., both incident and diffractive waves are essentially plane, then Fraunhofer diffraction occurs. Fresnel diffraction, on the other hand, occurs when the source and/or the point of observation are close to the object, i.e., the curvature of the waves cannot be neglected.

The slit and circular aperture are frequently used in optics. The distribution of radiation in the Fraunhofer case according to the Fresnel-Kirchhoff formulation is obtained by integrating the phase factor, e^{ikr}, over the aperture, where r is the distance from the point of observation to a point in the aperture. Illuminating a narrow slit of width d with coherent collimated radiation (plane wavefront) I_0 results in a diffraction pattern, $I = I_0(\alpha^{-1} \sin\alpha)^2$, where $\alpha = \frac{1}{2}kb \sin\phi$. ϕ is the angle from which the aperture is observed. This diffraction pattern has a maximum for $\phi = 0$ and goes to zero for $\phi = p\pi$. The diffraction pattern for a circular aperture of a radius R is

$$I = I_0[2J_1(\rho)/\rho]^2,$$

where $\rho = kR \sin\phi$. $J_1(\rho)$ is the Bessel function of the first kind. This diffraction pattern is common in optics; its central area is known as the Airy disk.

When the diffractive aperture consists of two parallel slits of width d separated by a distance D

$$I = I_0(\alpha^{-1} \sin\alpha)^2 \cos^2\beta,$$

where $\beta = \frac{1}{2}kD \sin\phi$. The factor $(\alpha^{-1} \sin\alpha)^2$ is the distribution function corresponding to a single slit and constitutes the envelope for the interference fringes, $\cos^2\beta$. The multiple-slit aperture (diffraction grating) is an important special case. A large number N of identical parallel slits results in

$$I = I_0(\alpha^{-1} \sin\alpha)^2(N^{-1} \sin^{-1}\beta \sin N\beta)^2.$$

Principal maxima occur within the envelope $(\beta^{-1}\sin\beta)^2$ at $p\lambda = D\sin\phi$, which is the so-called grating equation. The integer p is here called the order of diffraction.

An imaging technique called holography utilizes interference as well as diffraction phenomena. Both the amplitude and phase information of a radiation pattern are recorded as an interferogram (hologram), i.e., the interference pattern between the radiation field and a known reference wavefront. It is then possible to reconstruct the original radiation pattern by illuminating the recording with the known reference wavefront. The diffraction pattern caused by the hologram will then be identical to the original radiation field.

SCATTERING

The phenomenon that occurs when electromagnetic waves encounter obstacles much smaller then the wavelength is called scattering. Isolated scatterers absorb some energy from the incident radiation and reemit electromagnetic waves. The scattered irradiance is proportional to the fourth power of the frequency (Rayleigh scattering). An example of this type is scattering by the molecules of the atmosphere which causes the clear sky to appear blue to the human eye. On the other hand, clouds appear white, because their droplets are larger than the wavelengths of light so that ordinary reflection and refraction processes occur.

In physical optics the phenomena are treated on a macroscopic scale. Furthermore, little of what happens on the microscopic scale is understood. The types of waves generally encountered in optics are free propagating waves. Other types like guided and inhomogeneous (surface) waves also occur. The corresponding phenomena that those waves give rise to are described and treated in similar ways using terms similar to those mentioned here.

See also DIFFRACTION; ELECTROMAGMETIC RADIATION; INTERFEROMETERS AND INTERFEROMETRY; LIGHT SCATTERING; POLARIZED LIGHT; WAVES.

BIBLIOGRAPHY

M. Born and E. Wolf, *Principles of Optics,* 6th ed. Pergamon, Oxford, 1980.

E. Hecht, *Optics,* 2nd ed. Addison-Wesley, Reading, MA, 1987.

M. W. Klein and T. E. Furtak, *Optics.* Wiley, New York, 1986.

F. L. Pedrotti and L. S. Pedrotti, *Introduction to Optics.* Prentice-Hall, Englewood Cliffs, NJ, 1987.

K. Iizuka, *Engineering Optics.* Springer-Verlag, Berlin, 1983.

Optics, Statistical

Lorenzo M. Narducci

Amplitude and phase fluctuations are universal features of electromagnetic fields, especially in the visible region of the spectrum. The vast majority of natural sources radiate through a chaotic sequence of elementary emission processes, so that in this respect they resemble random signal generators much more than well-designed electrical oscillators. By contrast, the output of a stabilized cw laser can be made to approximate rather closely the ideal electromagnetic disturbance (i.e., a polarized monochromatic plane wave). Still, even the best laser sources are not immune from measurable fluctuations as a result of the effects of spontaneous emission, mechanical vibrations of the resonant cavity, and other unavoidable perturbations.

The task of statistical optics is to characterize the nature of these fluctuations and to indicate experimental procedures for their detection.

Traditionally this area of optics has been closely related to, if not completely identified with, coherence theory [1–3]. The notion of electromagnetic coherence brings to mind the formation of high-contrast fringes in interference and diffraction experiments. In fact, the early investigations of coherence from the mid-nineteenth century to the 1950s focused on the correlation between field variables at two space-time points, providing in the process a measure of electromagnetic coherence based on the visibility of interference fringes. As it turns out, the old criteria are not sufficiently general for a complete statistical classification of arbitrary fields.

It has been remarked that if a conventional thermal source and a well-stabilized laser were placed inside identical boxes and equipped with appropriate optical components (attenuators, collimators, polarizers, filters, etc.), so that their output intensities, wave fronts, and spectral distributions were perfectly matched, no classical interference measurement would enable us to discriminate between the two sources of radiation. Still, it is intuitively obvious that even under the idealized conditions just specified, we should be able to distinguish between an ordinary thermal lamp and a laser. The answer to this apparent paradox lies in the study of the higher-order correlation properties of the radiation or, to put it in simpler terms, in the analysis of the correlations between field powers of degree higher than first.

Historically, the existence of higher-order correlation effects in narrow-band thermal radiation was first demonstrated by Hanbury Brown and Twiss [4]. The subsequent rapid advances brought the following general conclusion into focus [2,3]: a complete statistical description of the electromagnetic field can be provided by an infinite hierarchy of nth-order coherence functions where, loosely speaking, the nth-order coherence function represents the ensemble average of a product of field variables that is proportional to the nth power of the intensity.

The central role played by the nth-order coherence function stems from the analysis of photoelectron counting, which is the basic detection process available for the observation of high-frequency electromagnetic fields. When a sensitive surface is illuminated by a light beam, photoelectrons are ejected with a probability per unit time that is proportional to the instantaneous light intensity at the photocathode. With excellent approximation, the photoelectron counting rate is given by

$$\begin{aligned} w^{(1)} &= sG^{(1)}(x,x) \qquad (1)\\ &= s\,\mathrm{tr}\{\rho E^{-}(x)E^{+}(x)\}, \quad x=\mathbf{r},t, \end{aligned}$$

where $E^{+}(x)$ is the positive frequency part of the electric field operator, $E^{-}(x)$ is its hermitian adjoint, and ρ is the density operator for the field. The parameter s measures the quantum efficiency of the detector. $G^{(1)}(x,x)$ is the first-order field correlation function evaluated at a single point and at a single time. It represents a special case of the two-point correlation function $G^{(1)}(x,x')$ that appears in the description of classical interference experiments.

Higher-order correlations and fluctuations can be detected by n-fold delayed coincidence experiments, i.e., by the measurement of the coincidences of the photon absorption processes taking place at different points in space and time. The n-fold delayed coincidence rate is given by

$$w^{(n)}(x_1, \ldots, x_n) = s^n G^{(n)}(x_1, \ldots, x_n, x_n, \ldots, x_1) \tag{2}$$

where

$$G^{(n)}(x_1, \ldots, x_n, x_n, \ldots, x_1) = \mathrm{tr}\{\rho E^-(x_1)\cdots E^-(x_n)E^+(x_n)\cdots E^+(x_1)\} \tag{3}$$

is the nth-order field correlation function.

In the classic experiments performed by Hanbury Brown and Twiss [4] the radiation coming from a discharge tube was split by a semisilvered mirror and brought to the surface of two photomultipliers. The average coincidence rate $w^{(2)}$ of the ejected photoelectrons from the two detectors was measured as a function of the optical path difference. It was found that, in addition to accidental background coincidences (which are independent of the optical delay), a small number of additional coincidences were detectable for small optical delays. The small size of the effect was due mostly to the finite response time of the instrumentation. The experiment was repeated by Martienssen and Spiller [5] using an interesting version of thermal source produced by laser light scattered by a rotating ground glass (Fig. 1). Since, in this case, the characteristic field fluctuation time could be controlled by varying the speed of the ground glass and, in particular, could be made as long as needed to ensure a satisfactory response of the instrumentation, the excess coincidence number for small delays was readily detected above the background contribution (Fig. 2).

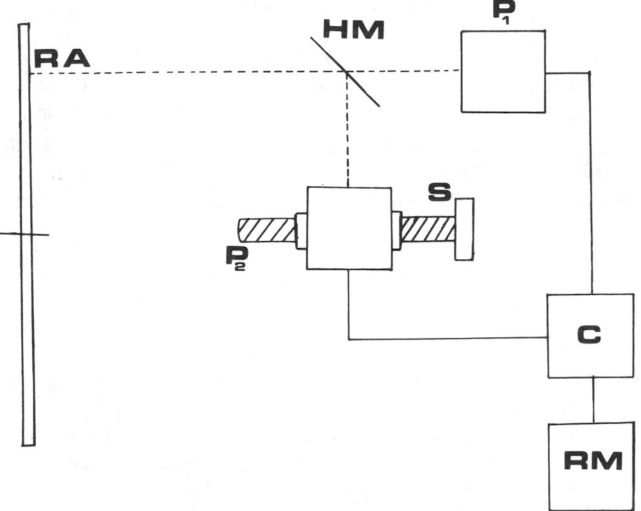

FIG. 1. Simplified diagram of the apparatus used in Ref. 5 to duplicate the Hanbury Brown and Twiss experiment. (Reproduced with permission from Ref. 5.)

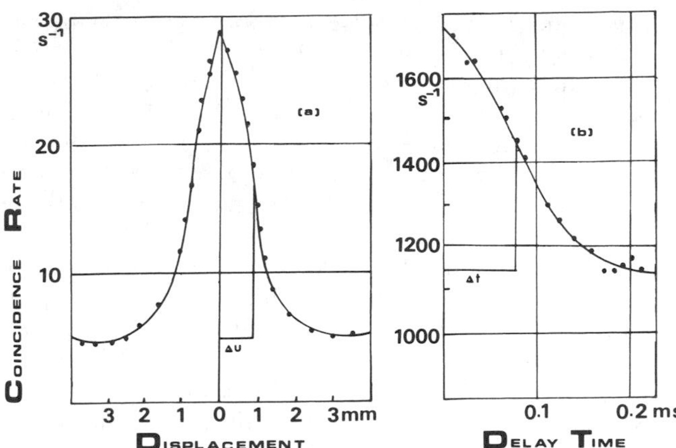

FIG. 2. (a) Coincidence rate as a function of position of the photomultiplier P_2. The field coherence time is 4 ms. (b) Coincidence rate between direct and delayed photomultiplier outputs. (Reproduced with permission from Ref. 5.)

Glauber proposed a description of the statistical properties of an arbitrary field in terms of the infinite hierarchy of correlation functions of the type given by Eq. (3). Since in the classic interferometric measurements the sharpest possible fringes are recorded if the condition

$$G^{(1)}(x_1,x_2) = \{G^{(1)}(x_1,x_1)G^{(1)}(x_2,x_2)\}^{1/2} \tag{4}$$

is satisfied (first-order coherence), the condition for Mth-order coherence was proposed to be

$$G^{(M)}(x_1, \ldots, x_M, x_M, \ldots, x_1) = \prod_{j=1}^{M} G^{(1)}(x_j, x_j). \tag{5}$$

In terms of delayed coincidence experiments, a field possesses Mth-order coherence if the M-fold coincidence rate is just the product of the counting rates that would be measured by the individual counters in the absence of all the others. We must conclude that in the classic two-fold delayed coincidence experiments by Hanbury Brown and Twiss the tendency toward statistical correlation for short delays was a reflection of lack of second-order coherence of the thermal source. By contrast, the same experiment performed with a single-mode laser well above threshold revealed no excess coincidences at all, thus ensuring second-order coherence for the source to within the experimental accuracy.

The development of low-noise quantum detectors and fast electronic instrumentation has been responsible for significant advances in statistical optics over the last decade. Extensive investigations have been carried out on the statistical properties of laser light scattered by random index of refraction fluctuations in transparent media. The dynamics of laser action has been studied experimentally for both single-mode and multimode operation over the entire range of conditions: below, at, and above threshold. Other types of radiation fields have recently become the center of attention. In particular, the fluorescence of atomic beams excited by laser light in resonance with a pair of atomic levels has been shown to give rise to a reduction of photoelectron coincidences for short delay times [6]. This effect, called photon antibunching, stands in sharp contrast with the opposite trend of the delayed coincidences observed with thermal radiation by Hanbury Brown and Twiss.

Since all the modern experimental techniques rely on the measurement of photoelectron statistics, it will be useful to survey the basic features of a typical detection system and the procedure by which information about a field can be extracted. The fluctuation of the photocurrent reflects the statistical behavior of the incident light if the recording process minimizes the amount of spatial and temporal integration effects. Under the ideal experimental conditions, the photosensitive area should be illuminated by a small uniform portion of the incident wave front so as to avoid spatial averaging, while the elementary counting interval should be much shorter than the characteristic fluctuation time of the light.

Under these conditions, usually referred to as detection within a coherence volume of the field, the probability of detection of n photoelectrons over a single counting time is given by

$$p(n)=\int_0^{\infty}\frac{(\alpha W)^n}{n!}e^{-\alpha W}P(W)\,dW \tag{6}$$

where W is the so-called integrated intensity and $P(W)\,dW$

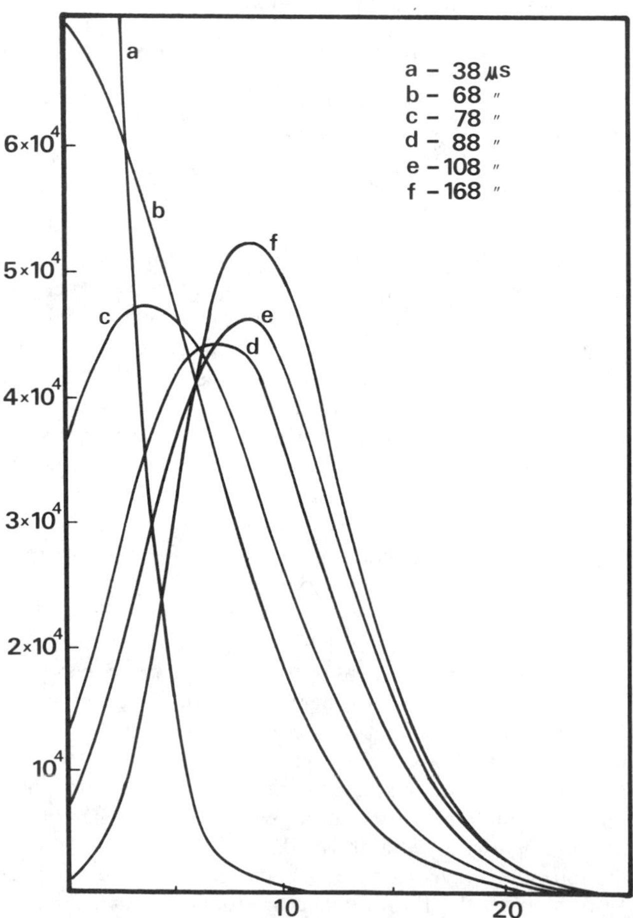

FIG. 3. Experimental photoelectron counting distribution obtained with a He-Ne laser under different operating conditions; curve (a) corresponds to the laser below threshold, curve (f) to the laser above threshold. (Reproduced with permission from Ref. 7.)

is the probability that W will take on a value in the range between W and $W + dW$. Thus the measured photoelectron probability distribution is directly related to the probability distribution of the light intensity by a Poisson transform, with the Poisson kernel playing the role of the photodetector transfer function.

If the source generates a thermal field and if the counting measurements are performed within a coherence volume, the predicted photoelectron distribution takes the form (Bose–Einstein distribution)

$$p(n)=\frac{1}{1+\langle n\rangle}\left(\frac{\langle n\rangle}{1+\langle n\rangle}\right)^n . \tag{7}$$

In the case of a single-mode laser operating well above threshold the expected distribution is (Poisson distribution)

$$p(n)=e^{-\langle n\rangle}\frac{\langle n\rangle^n}{n!} . \tag{8}$$

In Eqs. (7) and (8) the parameter $\langle n\rangle$ denotes the average number of photoelectrons detected over a single counting time. As the laser operating conditions are brought from below to above threshold, the initial thermal radiation, mostly due to spontaneous emission inside the cavity, turns into a coherent field [7]. The corresponding change of photoelectron distribution is beautifully displayed in Fig. 3.

A measurement of $p(n)$, as described above, provides information on what might be called the static equilibrium distribution of the field intensity, but no information on its correlation properties. These can be investigated by measuring the joint photoelectron counting distribution $p(n_1,n_2,\tau)$, i.e., the probability of recording n_1 photoelectrons over the first counting window and n_2 photoelectrons after a delay τ. By direct averaging over the experimental distribution $p(n_1,n_2,\tau)$, arbitrary two-time correlation functions of the type $\langle n_1^k n_2^h(\tau)\rangle$ can be constructed. The case corresponding to $k = h = 1$ is the one of most general interest because it provides a direct measure of the field intensity correlation function through the relation

$$\langle n_1 n_2(\tau)\rangle=\text{const.}\times\langle I(0)I(\tau)\rangle . \tag{9}$$

In practice, the computation of Eq. (9) requires an extensive amount of storage and data handling. Fast digital correlators and clipped correlation techniques have been developed for a more efficient processing of the data. A highly readable survey of the theoretical background and of the experimental techniques of photoelectron correlation spectroscopy can be found in Ref. 8.

See also FLUCTUATION PHENOMENA; LASERS.

REFERENCES

1. See, e.g., the resource letter QSL-1 on quantum and statistical aspects of light compiled by P. Carruthers in *Am. J. Phys.* **31**, 5 (1963). This resource letter, together with reprints from selected papers, has been published in the *Selected Reprints Series* of the American Association of Physics Teachers.
2. R. J. Glauber, "Optical Coherence and Photon Statistics," in *Quantum Optics and Electronics* (Les Houches Lectures 1964; C. de Witt, A. Blandin, and C. Cohen-Tannoudji, eds.), p. 63. Gordon & Breach, New York, 1965.

L. Mandel and E. Wolf, "Coherence Properties of Optical Fields," *Rev. Mod. Phys.* **37**, 231 (1965). (A)

3. An extensive collection of reprints and a detailed bibliography from 1850 and 1966 can be found in L. Mandel and E. Wolf, *Selected Papers on Coherence and Fluctuations of Light,* Vols. 1 and 2. Dover, New York, 1970.
4. R. Hanbury Brown and R. Q. Twiss, "Correlation between Photons in Two Coherent Beams of Light," *Nature* **177,** 27 (1956).
5. W. Martienssen and E. Spiller, "Coherence and Fluctuations in Light Beams," *Am. J. Phys.* **32,** 919 (1964).
6. H. J. Kimble, M. Dagenais, and L. Mandel, *Phys. Rev. Lett.* **39,** 691 (1977).
7. F. T. Arecchi and V. Degiorgio, "Statistical Properties of Laser Radiation during a Transient Buildup,"*Phys. Rev.* **A3,** 1108 (1971).
8. B. Chu, *Laser Light Scattering.* Academic Press, New York, 1974.

Order–Disorder Phenomena

Sandra C. Greer

The phrase "order–disorder phenomena" is used in the literature (1) to refer solely to the superlattice transition in binary alloys; (2) to refer to second-order phase transitions between states of different degrees of order; and (3) to refer to all phase transitions between states of different degrees of order.

In its original and most restrictive usage, order–disorder phenomenon refers specifically to the *superlattice* transition in a binary alloy. Such a transition in a binary solid solution is between (a) a less ordered structure for which each of the components of the alloy (e.g., iron and aluminum) is equally likely to be on a given lattice site and for which there is only short-range order, and (b) an ordered "superlattice" structure in which the components alternate on the lattice and there is long-range order. The less ordered state prevails at high temperatures and the ordered state appears upon cooling.

The extent of long-range order can be designated by the *long-range order parameter*, which is the fraction of the atoms of a component "in place" (i.e., on the alternate lattice sites) in excess of one half (which would be the number "in place" for a random arrangement in an equimolar mixture). Let P be the order parameter, a the number of atoms of component A "in place," and N the total number of atoms of component A. Then

$$a = \tfrac{1}{2}(1+P)N$$

and

$$P = \frac{2a}{N} - 1.$$

Thus for the highest degree of order $P=1$, and for complete disorder $P=0$. Even in the disordered state there is still some short-range or local order in that, on the average, an atom of one component is surrounded by atoms of the other component. A *short-range order parameter* σ can be defined as the fraction of nearest-neighbor pairs that consists of unlike atoms, in excess of one half (the fraction for a random arrangement). If g is the number of unlike pairs and r the total number of pairs, then

$$\sigma = \frac{2g}{r} - 1.$$

Thus $\sigma=1$ for greatest local order and $\sigma=0$ for local disorder.

Second-order transitions are those at which the second derivatives of the Gibbs free energy G are discontinuous or infinite (e.g., the heat capacity), while the first derivatives of G are continuous (e.g., the volume and the entropy). At a first-order transition the first derivatives of G are discontinuous. If the superlattice transition is considered as a function of the composition of the alloy, the compositions and transition temperatures at which the transition is second order are examples of *critical points*. At a critical point, the two phases are indistinguishable and can approach one another continuously and there exist anomalies in various thermodynamic properties. Similar critical points occur in many other kinds of physical systems and the phase transitions in them are quite analogous to the superlattice phase transition. Thus, it has become common to apply the designation order–disorder phenomena to all second-order phase transitions that involve a change in the degree of ordering rather than a displacement of molecules (as can occur in some solid–solid crystal transformations). The superlattice transition is a member of this more general class. A number of kinds of phase transitions can be so classified. For example, in a magnet, the electron spins can be aligned (ordered) or randomly oriented (disordered); the magnetization indicates the degree of alignment and is an order parameter, and the Curie point is a critical point. For the liquid–vapor transition, there also exists a critical point at a particular temperature, pressure, and density; the difference between the density of either the liquid phase or the coexistent vapor phase and the density at the critical point is a measure of the amount of (short-range) order and is also an order parameter.

The thermodynamic properties have the same general behavior in all these second-order transitions. Experiments show, for example, that at the critical point the specific heat at constant order parameter diverges, and that below the critical temperature the order parameter approaches zero approximately as the cube root of the difference in temperature from the critical temperature. This isomorphism of critical phenomena is most attractive: it means that the same theory can be used to understand many different kinds of phase transitions. The theories of critical phenomena are highly developed and are broadly applicable in the field of order–disorder phenomena.

In 1934 W. L. Bragg and E. J. Williams proposed a theory for the superlattice transition in which the solid is represented by a lattice, at each site of which there is an atom of one component or the other. It is assumed that the attraction between unlike molecules is greater than that between like molecules and that the ordered molecules are subject to a field that is an average of the interactions with all the other molecules. J. D. van der Waals had proposed an analogous theory for the liquid–gas transition in 1873; van der Waals accounted for the molecular interactions by assuming that each molecule is subject to a field that is an average of the fields due to all other molecules and is proportional to the density. Another such theory had also been proposed in 1907 by Pierre Weiss for the paramagnetic–ferromagnetic phase transition; this theory assumes that there is a lattice, at each site of which there is an electron spin that has one of two possible orientations, and that the energy is lower if adjacent spins are aligned. Again, a "mean field" due to the other molecules is assumed. Thus these theories all make the approximation of a "mean field" at each individual molecule due to all other molecules. At low temperature this mean field is sufficient to maintain order, but as the temperature is increased, the increased kinetic energy of the molecules causes more and more of them to get out of order. As the system becomes disordered, the average ordering field is diminished, so that it is then even easier for the system to become more disordered. (This "avalanche" effect has caused these phase transitions to be called *cooperative* phenomena.)

The methods of statistical mechanics can be applied to determine the thermodynamic behavior of these models. These *mean-field theories* (now called the "classical"—as opposed to the "modern"—theories of phase transitions) give rather good qualitative predictions of the phase transitions, but they fail in some essential ways. They all predict a simple jump discontinuity in the specific heat at constant order parameter at the critical point, whereas experiments on most such phase transitions show a divergence in the specific heat at the critical point. The classical theories predict that the order parameter should approach zero as $|T_c - T|^{\beta}$ where T_c is the critical temperature and β is 0.5, whereas experiments yield β near 0.3. The β is an example of a *critical exponent*. In other ways, too, we find that the classical theories, although useful, are not entirely adequate.

All classical theories have one assumption in common: they assume that the intermolecular forces in the system are sufficiently long ranged that the field acting on a given molecule is, indeed, an average of the forces of all the other molecules. Long-range forces are more amenable to "averaging." In our discussion of the "cooperative" effect near a critical point, we said that the "disordering" of one molecule makes it easier for the next one to be "disordered." This would be especially true for nearby molecules. Thus short-range interactions are especially important near the critical point. Since the two phases become more and more alike as the critical point is approached, it takes less and less energy to change a molecule or group of molecules from one phase to another, and the transition of any given molecule from one phase to another makes it easier for the next one to follow. Thus, fluctuations in the average order parameter occur more and more easily and themselves cause more fluctuations (i.e., the fluctuations are correlated). We can say that the classical mean-field theories ignore the correlations in fluctuations of the order parameter near the critical point. The distance over which the fluctuations correlate is called the *correlation length*. We expect the mean-field theories to work when the intermolecular forces act over a distance much greater than the correlation length, but to fail when the correlation length is comparable to the range of the intermolecular forces. At superconductor and ferroelectric phase transitions, the forces are sufficiently long ranged that mean-field theories do work. For binary alloys, ferromagnets, and the liquid–gas transition, the mean-field theories are inadequate.

A treatment that is more successful than the mean-field models is the *Ising model*. In the Ising model, as in the Bragg–Williams and Weiss models, the phase under consideration is represented by a lattice, at each site of which there is a molecule. The same model can be applied to different phase transitions by modifying the way the two site states are designated. For a binary alloy, the sites would be occupied by one or the other of the two components. For a magnet, the sites would be occupied by states "spin up" or "spin down." The Ising model has even been applied to the liquid–gas transition by designating the two site states as either "occupied" or "vacant"; this is the *lattice–gas* model. For a superlattice model, the interaction energy is taken to be lower between *unlike* molecules than between like molecules. If the energy is lower between *like* molecules, a model of separation into two phases of different compositions results; such "unmixing" occurs both in solid alloys and in liquid mixtures. In the Ising model, as opposed to mean-field models, the nearest-neighbor (short-range) interactions are specifically included. The methods of statistical mechanics can be applied (exactly for two dimensions and approximately for three dimensions) to obtain the thermodynamic properties of the Ising model. The resulting properties agree remarkably well with experimental results in fluids, some magnets, and alloys very near the critical point. For example, the specific heat of the three-dimensional Ising model diverages and the order parameter goes to zero as $|T_c - T|^{\beta}$ with $\beta = 0.32$.

The expression order–disorder phenomenon is often used in a still broader sense to refer to any transition between a state of matter that has some degree of order and a state that has a somewhat lesser degree of order. Such a general definition would certainly include all the second-order transitions mentioned earlier, and even more; it would also include many first-order transitions. For example, the transition between a crystalline solid and its melt is a transition between long-range order and short-range order. Still other transitions that can be called order–disorder transitions according to this broader definition are the conformational changes in biopolymers, the hydrocarbon disordering in biomembranes, the molecular reorientation in plastic crystals, and the molecular alignment in liquid crystals. Lattice models similar to the Ising model have been applied with profit to many of these cases, even for first-order transitions. On the other hand, first-order transitions lack the symmetry of second-order transitions and thus the development of theoretical descriptions of first-order transitions has lagged behind that of descriptions of second-order transitions.

See also CRITICAL POINTS; ISING MODEL; PHASE TRANSITIONS.

BIBLIOGRAPHY

C. Domb and M. S. Green, *Phase Transitions and Critical Phenomena*. Academic Press, New York, 1972, Vols. 1–8; C. Domb and J. L. Lebowitz, *Phase Transitions and Critical Phenomena*, Vols. 9–12. Detailed articles by experts. (A)

E. W. Elcock, *Order–Disorder Phenomena*. Wiley, New York, 1956. General discussion, followed by extensive treatment of binary alloys. (E–I)

H. S. Green and C. A. Hurst, *Order–Disorder Phenomena*. Wiley, New York, 1964. Extensive treatment of lattice models. (A)

A. I. Kitaigorodskiy, *Order and Disorder in the World of Atoms*. Springer, New York, 1967. Elementary discussions of a great variety of order–disorder phenomena. (E)

S.-K. Ma, *Modern Theory of Critical Phenomena*. Benjamin/Cummings, Reading, MA, 1976. (A)

H. E. Stanley, *Introduction to Phase Transitions and Critical Phenomena*. Oxford, London and New York, 1971. General discussion of gas–liquid and magnetic critical points with presentation of modern scaling theory. (I–A)

Organic Conductors and Superconductors

P. M. Chaikin

The materials categorized as organic superconductors form the most interesting and complex compounds yet discovered. While other classes of materials exhibit competitions between magnetism and superconductivity, or between metallic and insulating phases, the organic charge transfer salts exhibit a truly extraordinary variety of properties and transitions, often in a single compound [1]. Probably the most dramatic example is the Bechgaard salt $(TMTSF)_2ClO_4$, where TMTSF stands for tetramethyltetraselenafulvalene, which is variously metal, semimetal, insulator, antiferromagnet, spin-density-wave conductor, and superconductor, and also exhibits nonlinear transport, the integer, and possibly the fractional quantum Hall effect [2], and the effects of a quasiperiodicity, depending on temperature, pressure, and magnetic field. It also shows effects heretofore unseen in any other material: a cascade of magnetic-field-induced transitions, and a new type of quantum oscillation. The fascinating behavior is related to the anisotropic band structure which is quasi-one- or quasi-two-dimensional. The bandwidths in different directions are comparable to Coulomb interactions, magnetic field strengths, and temperature, hence small changes can make different interactions dominant.

The modern era in organic conductors dates to the early seventies and the discoveries on TTF-TCNQ [3] (tetrathiafulvalene tetracyanoquinedimethane). The search for organic superconductors was driven by the suggestion that they might be the ideal materials for predesigning a new type of pairing interaction (excitonic rather than the conventional phonon mechanism) which would yield high T_c's [4]. Previous organic charge transfer salts were plagued by problems of intrinsic disorder related to unsymmetric cations or anions. The disorder limited the mean free path and the conductivity, broadened any thermodynamic phase transitions, and led to complications in interpretation. TTF-TCNQ was sufficiently clean that it exhibited metallic conductivity down to 59 K, where a clear and abrupt metal-insulator transition occurred. This was the first (and remains the prototype) organic example of the instability of a one-dimensional metal to the formation of a density-wave insulator—the Peierls transition. The TTF molecule served as the basic building-block cation for all subsequent organic metals and superconductors (see Fig. 1).

The instability of a one-dimensional metal is simply related to the large number of degenerate states that are connected by the same wave vector. Degenerate states (those which have the same energy) are particularly susceptible to any perturbation which couples them. The energy ϵ vs. momentum P and occupation of P states for the electron in a one-dimensional metal is shown in Fig. 2. The Pauli exclusion principle prevents two electrons from being in the same state. The low-energy states are filled successively until all of the electrons are accounted for. The last energy level filled is called the Fermi energy; the surface in P space at this energy is called the Fermi surface (Fig. 2b). All P states lower than P_F are occupied, and all above are empty at zero temperature. What is peculiar to one dimension is that all of the degenerate states at the Fermi energy are connected by the same wave vector $q = 2P_F/\hbar$. Displacing the Fermi surface at P_F by this wave vector brings the two sheets into exact coincidence—they "nest" perfectly. The degree of mixing of degenerate states is independent of the amplitude of the mixing potential. Thus any size perturbation with period $2\pi/q$ will turn all of the travelling-wave states at $\pm P_F$ into standing-wave states which can take advantage of the periodic potential. The result is the opening of a gap at the Fermi energy. This yields an insulating state. All occupied states have their energy lowered while the unoccupied states are raised as shown in Fig. 2c. It costs some elastic energy to make the distortion of wave vector $2P_F$ but the lowering of the electronic energy always wins in one dimension. Thus at low temperature a one-dimensional metal is always unstable against a distortion forming an insulator. For many years the main research thrust was directed toward finding ways to suppress this metal-insulator instability so that the

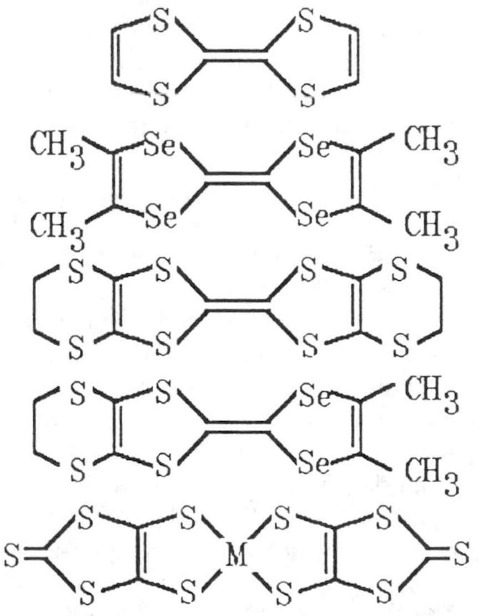

TTF

TMTSF

BEDT–TTF

DMET

dmit

FIG. 1. Cations used in organic superconductors and other conducting charge transfer salts. The abbreviations not already defined in text are as follows: BEDT, bis(ethylenedithio); DMET, dimethyl bis(ethylenedithio); and dmit, 1,3-dithia-2-thione-4,5-dithiolato. In the last case, the abbreviation properly refers only to one of the two entities attached to the metal atom, designated by M, in the center; so the compound shown is actually bis-dmit-M.

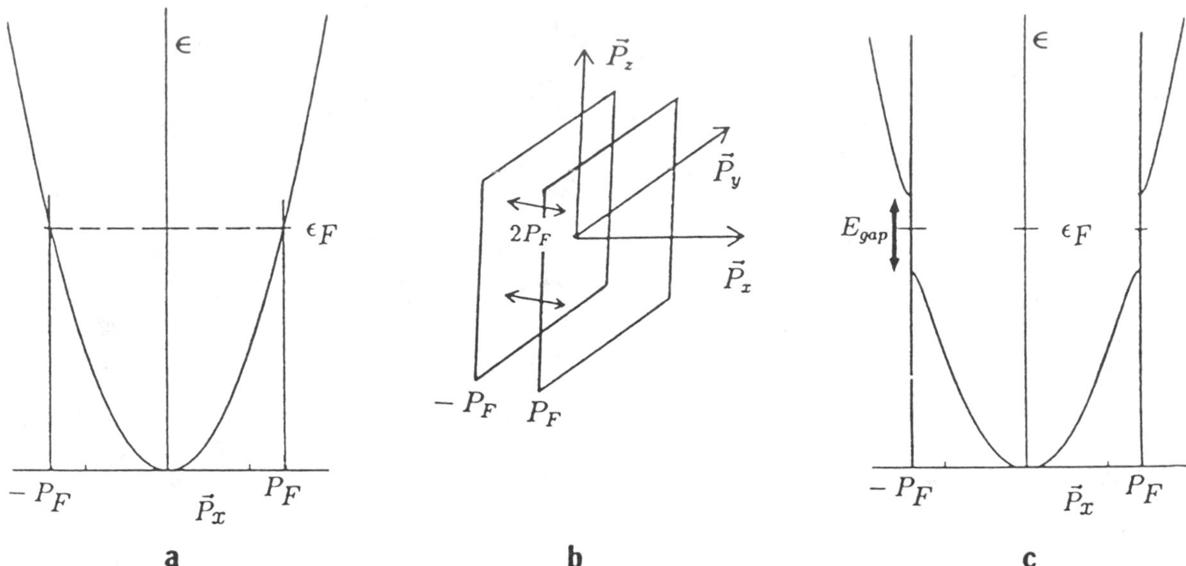

FIG. 2. a) Energy as a function of momentum for an electron gas at low temperature. All states below ϵ_F are occupied with electrons. b) The Fermi surface for the one dimensional electron gas. All states between the sheets are filled. c) A distortion of wavevector $2P_F/\hbar$ produces a gap at the Fermi energy and lowers the total energy of the system.

superconducting state might be obtained. One of the clearest directions was to make the materials less one dimensional.

The Bechgaard salts $(TMTSF)_2X$ ($X = PF_6$, ReO_4, NO_3, etc.) provided the next big breakthrough [5]. They were metallic at room temperature and remained metallic to lower temperature than any previous organic compounds, finally going insulating at ~12 K for the best metal. In the PF_6 salt, a moderate pressure suppressed the insulator and the compound went superconducting at ~1.2 K [6]. Soon after this discovery, most of the other members of this family were found to be superconducting under pressure with about the same T_c. The ClO_4 salt was found superconducting at ~1.4 K at ambient pressure. In all cases, further increase of pressure resulted in a decrease of the superconducting transition temperature.

The idea of suppressing the metal-insulator instability and finding superconductivity seemed to work. But the Bechgaard salts had lots of surprises. A series of experiments led to the suspicion that the insulating phase of the PF_6 salt was magnetic. This idea was confirmed by magnetization and antiferromagnetic resonance experiments [7]. The close proximity of the superconducting and spin-density-wave (SDW) phases in parameter space led to the speculation that the superconducting state might be exotic. This question remains for the newer materials as well.

Because the Bechgaard salts were the first organic superconductors, they were the focus of numerous studies to determine their band structure and other physical properties and parameters. One of the outcomes of these studies was the fact that although they were not one dimensional in the sense of a perfectly flat Fermi surface. The actual Fermi surface is similar to that shown in Fig. 3a consisting of two warped separate sheets at $\sim \pm P_F$ with no intersections. The warping is sufficiently strong in some compounds or under pressure that, as expected, the one-dimensional instability is suppressed and we have a metal which becomes superconducting at lower temperatures. However, there is another interesting feature of this Fermi surface; there is no possibility for an electron to move in a closed loop as it might in Figs. 3b or 3c.

In the presence of a magnetic field a charged particle moves on an orbit with constant energy (since the Lorentz force is always perpendicular to the velocity). In the more traditional cases the particle undergoes circular motion in the plane perpendicular to the field. The so-called cyclotron orbit involves periodic return to the same position and momentum with a well defined cylotron frequency ω_c. These orbits are sketched in Figs. 3b and 3c. We know from quantum mechanics that if a particle is localized it can have a discrete energy spectrum and that periodic motion is quantized. The electrons moving on the Fermi-surface orbits

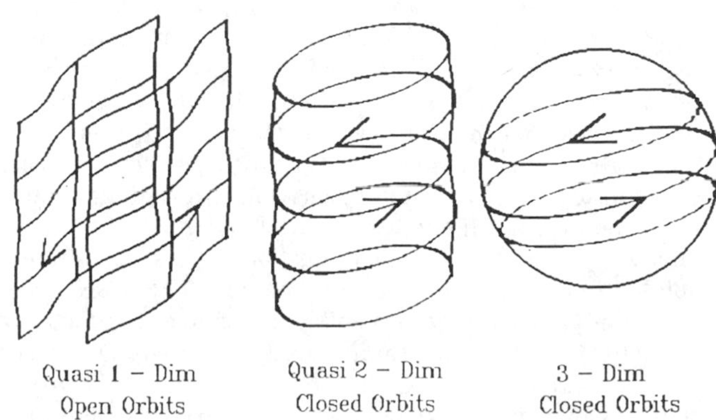

FIG. 3. The Fermi surfaces for quasi-one, quasi-two, and three dimensions, and the electron trajectories in a vertical magnetic field.

sketched in Figs. 3b and 3c will have energies which fall only in Landau levels $E_n = (n + \frac{1}{2})\hbar\omega_c$ where n is an integer. This Landau quantization has dramatic effects on the transport and the thermodynamic properties of conductors in high magnetic field. There are periodic oscillations in resistance and magnetization resulting from these discrete Landau levels crossing the Fermi energy. In degenerate semiconductors the consequences are even more exciting. If we can somehow arrange for the Fermi level to lie in the gap between two Landau levels (usually by having impurity states) we find the highly accurate Hall plateaus at integer multiples of e^2/h and vanishing resistance associated with the quantum Hall effect (QHE) [2].

Electrons in a magnetic field in the Bechgaard salts (Fig. 3a) cannot form closed orbits between the two sheets. Rather an electron traverses an open orbit which drifts off without ever returning to the same place. The delocalized particle must have a continuous spectrum and hence will not show Landau quantization, magneto-oscillations, or the QHE.

The Fermi surface in the metallic Bechgaard salts is warped enough to prevent nesting and the Peierls transition, and contains no closed orbits. There now comes an interesting interplay with the magnetic field. With no distortion from a density wave there are only open orbits, and a magnetic field will not give Landau levels. With a distortion and no magnetic field there is incomplete nesting, no gap, and not enough energy gain to stabilize the distortion. However, the distortion brings the two pieces of the Fermi surface together and creates closed orbits. If we have both a distortion and the magnetic field, then there are closed orbits and discrete energy levels spaced apart by $\hbar\omega_c$. We can have a gap over the entire Fermi surface and the energy gained by putting the Fermi level in the gap is sufficient to stabilize the distortion. Thus turning on a magnetic field makes the metal unstable and the SDW stable. We get both the SDW and the Landau quantization together; we cannot have them separately in this system. Moreover, as the magnetic field changes the system must continually adjust the distortion wave vector to keep the Fermi level in the gap between Landau levels. We get the QHE automatically since the energy stabilization for the SDW comes from having the Fermi level in the gap [8]. This is the only system that exhibits the QHE in a bulk crystal.

The second family of organic superconductors are the BEDT-TTF salts (see Fig. 1). The initial discovery of superconductivity in any sulfur-based salt was in 1983 in the $(BEDT\text{-}TTF)_2ReO_4$ salt which has a T_c of ≈ 2 K under a pressure of ≈ 4 kbar [9]. Unlike the Bechgaard salts which have remained with essentially the same T_c as originally discovered, the T_c of the ET salts has increased steadily and impressively with time. The first step was the I_3 salt with a T_c of 3.6 K in one phase and then 8 K in another phase. More recently the $Cu(SCN)_2$ salt has reported T_c's up to 11 K at ambient pressure [10]. T_c in these salts decreases rapidly with increasing pressure (~ 1 K/kbar). Because of their comparatively high T_c's the BEDT-TTF salts are presently the most widely studied organic charge transfer salts. As compared to the Bechgaard salts, these materials tend to be more varied in the number of stable compounds which can be formed with the same anion. For example the $(BEDT\text{-}TTF)_2I_3$ system has at least five structural different phases three of which are superconducting.

The BEDT-TTF salts are more two-dimensional than the Bechgaard salts. They are almost isotropic in the conducting plane but have a large anisotropy between the interplane and intraplane conductivities (and superconducting critical fields). The large mean free paths and the two-dimensionality of the Fermi surface in these salts has led to the ability to perform detailed Fermi surface studies for the first time in organic conductors. Precise and often spectacular magnetoresistance and magnetization oscillations (Shubnikov–de Haas and de Haas–van Alphen effects) have been performed on a number of these salts [11]. Comparisons with Fermi surface computations have allowed the first real test of our ability to calculate the band structure of complex organic molecular crystals.

There are also new and unusual magnetic field effects in the BEDT-TTF salts which have been seen in no other materials. Among these are the angle-dependent resonances seen in magnetotransport [12]. At particular "magic angles" there are sharp increases in the resistance in a field so that the resistance oscillates as a function of angle. The magnitude of the oscillations depends on the strength of the applied field and the temperature, but the "magic angles" are fixed for a particular material. The present understanding is that these resonances come from the tubular Fermi surface which is very similar to that schematized in Fig. 3b. The energy of a Landau level depends on its area. If the magnetic field is parallel to the tube axis in Fig. 3b then the cross section changes as one moves up and down the tube. However, if the magnetic field is tilted it is clear that the large-area orbits get smaller and the small-area orbits get larger. At some angle the spread in cross-sectional areas is minimum (depending on the exact band structure all areas may be the same at these angles) and this results in a large enhancement in the density of states. This is another effect which results from the organic conductors being not quite two-dimensional.

One of the original and still remaining questions concerning the superconductivity in the organics is whether it is conventional BCS type or exotic, involving new mechanisms or new symmetries, and *p*- or *d*-wave rather than *s*-wave pairing. The materials are highly anisotropic. One might expect and indeed one finds anisotropic superconducting properties. Critical fields parallel and perpendicular to the conducting planes differ by a factor of ~ 20. Specific heat jumps are generally in agreement with BCS predictions and fluctuations do not seem to significantly broaden the superconducting transitions. However, there have been several tantalizing experiments over the past several years in both the Bechgaard and BEDT salts which suggest that there may be places in momentum space where the superconducting energy gap disappears. This gives rise to power-law rather than exponential temperature dependences in several properties and these power-laws have been reported [1,10].

From the discovery of organic superconductivity in 1979 to the present there have been more than 35 organic compounds which have been discovered. The newest families involve the DMET and dmit (see Fig. 1) cations which have interesting new competitions with disorder and other metal-insulator transitions. We can expect the field of organic conductors and superconductors to yield new materials and exciting science for some time to come.

REFERENCES

1. Some good review articles are in NATO ASI Series, *Low Dimensional Conductors and Superconductors* (D. Jerome and L. G. Caron, eds.), Plenum, New York, 1987. D. Jerome and H. J. Schultz, *Adv. in Phys.* **31,** 299 (1982), and more recent results are in the proceedings of the International Conferences on Synthetic Metals, *ICSM Synth. Metals* **27** (1989).
2. See articles in *The Quantum Hall Effect* (R. E. Prange and S. M. Girvin, eds.), Springer-Verlag, New York, 1987.
3. L. B. Coleman et al., *Solid State Commun.* **12,** 1125 (1973) and F. Donoyer et al., *Phys. Rev. Lett.* **35,** 445 (1975).
4. W. A. Little, *Phys. Rev.* **134A,** 1416 (1964).
5. K. Bechgaard, C. S. Jacobsen, K. Mortensen, H. J. Pedersen, and N. Thorup, *Solid State Commun.* **33,** 1119 (1980).
6. D. Jerome, A. Mazaud, M. Ribault, and K. Bechgaard, *J. Phys. Lett.* **41,** L95 (1980).
7. W. M. Walsh et al., *Phys. Rev. Lett.* **45,** 829 (1980), K. Mortensen et al., *ibid.* **46,** 12338 (1981), and J. B. Torrance, et al., *ibid.* **49,** 881 (1982).
8. S. T. Hannahs, et al., *Phys. Rev. Lett.* **63,** 1988 (1989) and J. R. Cooper, et al. *ibid.* **63,** 1984 (1989).
9. S. S. P. Parkin, et al., *Phys. Rev. Lett.* **50,** 270 (1983).
10. J. M. Williams et al., *Prog. Inorg. Chem.* **35,** 51 (1987) and G. Saito, *Mat. Res. Soc. Symp. Proc.* **173,** 95, (1990), and M. Tokumoto et al., *ibid.*, p. 107.
11. K. Oshima et al., *Syn. Met.* **27,** A165, (1988).
12. K. Yamaji, *J. Phys. Soc. Japan* **58,** 1520 (1989).

Oscilloscopes

Oliver Dalton

INTRODUCTION

The cathode-ray oscilloscope is an electronic instrument that displays changing electrical events (signals) as a function of time on the screen of a cathode-ray tube (crt). The display is a graph, usually presented in Cartesian coordinates, with the horizontal (x axis) representing time and the vertical (y axis) input voltage. Electrical signals may be generated by transducers used to sense nonelectrical events, or they may occur directly in electrical circuits. For this article it will be assumed that the oscilloscope is connected to a time-varying electrical signal regardless of its origin.

Currently available oscilloscopes are capable of a very wide range of voltage amplitude and time measurements. No one oscilloscope covers the entire range. Table I summarizes these capabilities.

SELECTING AN OSCILLOSCOPE

Because of the wide range of oscilloscopes available it is not a simple matter to choose the best one for your needs. The first step is to determine, at least approximately, the characteristics of the signals that you will wish to measure. The five signal characteristics critical to making a correct choice are outlined below. The first three characteristics listed will determine the performance needed and include a discussion to guide the user. The last two determine the kind of oscilloscope required to achieve this performance; the user is referred to later sections describing oscilloscope systems.

1. *Signal Amplitude Range.* The specified deflection factors (in microvolts, millivolts, or volts per division) available in an oscilloscope can be used to determine its suitability. The lowest signal amplitude should provide over two divisions of deflection at the oscilloscope's minimum deflection factor. The highest signal amplitude should not cause more than full-screen deflection (typically eight divisions) at maximum deflection factor.
2. *Shortest Risetime of Signal.* The time for the shortest signal transition determines the needed oscilloscope's risetime. The specified oscilloscope risetime should be shorter than the signal risetime; how much shorter depends on the acceptable measurement error. If n is the maximum percentage error acceptable when viewing a signal having a risetime of t_s, then the oscilloscope's risetime must be no more than

   $$[(1+n/100)^2-1]^{1/2}\times t_s;$$

   e.g., if $n = 5\%$ and $t_s = 10$ ns, then the oscilloscope's risetime must be no more than 3.2 ns.
   If an oscilloscope's risetime is t_0 and the risetime measured on the oscilloscope is t_m, then the true signal risetime is $(t_m{}^2-t_0{}^2)^{1/2}$.
3. *Highest Frequency Component of Interest.* The bandwidth of an oscilloscope is the frequency at which its amplitude response is 3 dB (about 30%) below its low-frequency value. For accurate measurement of complex wave forms the oscilloscope's bandwidth should be much greater than that of the signal. The risetime and bandwidth of an oscilloscope are approximately related by

   $$\text{bandwidth (in MHz)} = \frac{350}{\text{risetime (in ns)}}.$$

Table I Approximate Range of Signal Characteristics Measurable with Available Oscilloscopes

	Signal	Minimum	Maximum	Corresponding range
General-purpose Oscilloscopes	Amplitudes	10 μV (10^{-5} V)	100 V	10^7:1
	Risetime	0.5 ns (5×10^{-10} s)	50 s	10^{11}):1
Sampling Oscilloscopes	Amplitudes	2 mV (2×10^{-3} V)	5 V	(2.5×10^3):1
	Risetime	50 ps (5×10^{-11} s)	50 ms	10^9:1

4. *Signal Repetition Rate*. The signal repetition rate is defined as the number of times per second that a signal repeats. A single-shot signal is one that occurs only once. If low-repetition-rate or single-shot signals are expected, read the sections on writing speed, storage oscilloscopes, dual-beam oscilloscopes, and sampling oscilloscopes.
5. *Number of Signals*. When it is necessary to view two or more signals simultaneously, then a multiple-trace or dual-beam oscilloscope is needed. Refer to later descriptions of these types.

MEASUREMENT ACCURACY

Most oscilloscopes can be used to measure time and voltage with an error of ±2–5%, except when they are being used near their limits of minimum deflection factor and risetime. There are special plug-ins available for some oscilloscopes that allow voltage measurements to be made with errors of less than ±1% and time errors of less than ±1 part in 10^6, the latter using digital techniques.

WRITING SPEED

To view the detail of a single-shot signal using a nonstorage oscilloscope it is necessary to photograph the wave form appearing on the crt. The writing speed of the oscilloscope, usually expressed in centimeters per microsecond, is the maximum speed of the crt spot that can be recorded using a specified camera and film. Most oscilloscopes, when equipped with an appropriate camera and high-speed film, are capable of recording the fastest signals to which they can respond.

The first five following sections briefly describe those parts of an oscilloscope of major concern to the user. The final four sections describe different kinds of oscilloscopes, including their capabilities and limitations.

Vertical Amplifiers

Vertical amplifiers provide the user with a range of deflection factors that allow him to select a convenient displayed wave-form amplitude. A wide range of amplifiers with differing performance and features is available (see the bibliography).

Multiple-Trace Vertical System

It is often desirable to view two or more signals simultaneously for comparison purposes. Many oscilloscopes have dual-trace capability and some can display up to eight traces. Multiple-trace displays are achieved by oscilloscope circuitry that switches rapidly between the various input signals.

Time Bases

An oscilloscope time base provides the capability of varying the scale of the horizontal (time) axis over a very wide range. A typical slowest speed is 50 s for a full scan. The highest speeds are arranged to give sufficient resolution for signals within the oscilloscope's risetime capability; they range up to 0.5 ns/cm for a 500-MHz (0.7-ns risetime) oscilloscope.

Delayed Time Base

Use of a second, delayed, time base enables a wave form to be examined in much increased detail on the horizontal axis. With it, any selected portion of the trace can be displayed magnified well over 1000 times.

Triggering

Adjustable trigger controls allow the user to select the point on his signal at which the start (left-hand edge) of the display in the horizontal axis occurs. Modern trigger circuits respond to all signal frequencies lying within the oscilloscope's vertical system bandwidth.

Storage Oscilloscopes

Storage oscilloscopes are capable of storing a single-shot signal trace on the crt screen for a considerable time after the event has ended. This feature allows visual study of such events without the user's having to take a photograph. If a permanent record is required, then a photograph may be easily made with just a simple camera and film of moderate speed. A stored trace may be rapidly erased at will in preparation for recording a new event. Storage oscilloscopes may be used in a nonstored mode when required.

Storage oscilloscopes are available that are capable of storage speeds approaching the recording ability of the highest-speed nonstorage oscilloscopes, cameras, and films.

Dual-Beam Oscilloscopes

Due to technical limitations, no single-beam, multiple-trace oscilloscope is capable of recording more than one single-shot event at the highest speed for which it is designed. A dual-beam oscilloscope is capable of recording two independent single-shot events simultaneously. Dual-beam oscilloscopes are generally larger and more expensive than single-beam equivalents.

Sampling Oscilloscopes

Sampling oscilloscopes are special-purpose instruments designed to overcome the bandwidth limitations of "conventional" oscilloscopes. Sampling oscilloscopes require repetitive signals in order to build up a complete wave form for display. They make use of the fact that it is possible to generate very short pulses, and that these pulses may be used to interrogate the signal wave form successively every time it occurs. Thus, with sufficient occurrences, a complete wave form can be reconstructed and displayed. Sampling oscilloscopes can extend the effective bandwidth to well over ten times that of "conventional" oscilloscopes.

Plug-in Oscilloscopes

Oscilloscopes primarily designed for research and engineering often make use of interchangeable (plug-in) vertical amplifiers and time bases. This feature gives the following advantages:

1. The user can configure his oscilloscope to optimally suit his measurement needs.
2. As new plug-ins become available the user can update his oscilloscope, within certain limitations.
3. In practice, plug-in oscilloscopes are the only oscilloscopes capable of making many types of measurements.

Plug-in oscilloscopes do have disadvantages; they generally are larger and more expensive than non-plug-in instruments of similar performance.

BIBLIOGRAPHY

General Descriptions of All Types of Oscilloscopes

Bernard M. Oliver and John M. Cage, *Electronics Measurements and Instrumentation,* Chapter 11. McGaw-Hill, New York, 1971. (I)

Clyde F. Coombs, *Basic Electronic Instrument Handbook,* Chapter 24. McGraw-Hill, New York, 1972. (I)

General Discussion of Selection of a Suitable Oscilloscope

Frank Elardo and Oliver Dalton, "Consider All the Alternatives When Choosing an Oscilloscope," *EDN* **20** (6), 22 (1975). (I)

Dual-beam and Storage Oscilloscopes

Vince Lutheran and Bernard Floersch, "Dual Beam: An Often Misunderstood Type of Oscilloscope," *EDN* **19** (15), 45 (1974). (I)

Steve Rosenthal, "Choose the Right Storage Oscilloscope," *Electronic Design* **22** (24), 150 (1974). (I)

Paramagnetism

J. E. Enderby

Substances with a positive magnetic susceptibility χ_m are said to be paramagnetic; that is, the magnetic moment induced by an applied magnetic field is in the direction of the field. Atomic paramagnetism (as opposed to nuclear effects, to which we return briefly at the end of this article) is electronic in origin and, where it exists, usually overcomes the more fundamental diamagnetism. Electronic paramagnetism is found in free atoms and molecules possessing an odd number of electrons or an unfilled inner shell. (There are in addition a few "even-electron" substances which are paramagnetic, for example, molecular oxygen, but these are somewhat exceptional.) The essential ingredient is that the atoms and molecules of the substance must possess a permanent magnetic dipole moment. A further important contribution to the magnetism of metallic conductors is the fact that the gas of free electrons, which is responsible for the characteristic properties of metals, is itself weakly paramagnetic. This paramagnetism is essentially independent of temperature T, which contrasts with atomic and ionic paramagnetism where $\chi_m \propto T^{-1}$ (Curie's law).

The classical microscopic theory of paramagnetism is due to Langevin. He noted that the tendency for the magnetic field B to align the elementary dipoles is countered by the random orientation induced by thermal agitation. This leads to a result for χ_m which can be written in the form

$$\chi_m = \frac{\mu_0 N m^2}{3kT}, \qquad (1)$$

where N is the number of dipoles per unit volume, each of which has moment m, k is Boltzmann's constant, and μ_0 is the permeability of free space. At very low temperatures and high magnetic fields the relationship between magnetization and applied field implied in Eq. (1) becomes more complicated and in fact may cease to be linear.

Paramagnetism can be investigated experimentally by measuring the weak force experienced by a paramagnetic material when placed in an inhomogeneous field. In the Curie method a small volume of material is placed in a region where the magnetic field is varying strongly with distance. The other method frequently used, the Gouy method, employs a long specimen, one end of which is in a field-free region, while the other is placed between the poles of a powerful electromagnet. In each case the force on the sample can be related to χ_m. Experimentally it is found that $\chi_m \propto T^{-1}$ for many substances, although for ferromagnetic materials above their Curie temperature θ, a modified law, originally due to Curie and Weiss and of the form $\chi_m \propto (T-\theta)^{-1}$, is often obeyed.

Quantum-mechanical corrections modify the Langevin formula in several significant ways. Not all directions are accessible to the dipole and the "quantization of space" introduces the notion of an effective dipole moment which depends on the total quantum number. At very low temperatures paramagnetic rare-earth salts show considerable departures from Curie's law. This happens because the effect of the crystalline electric field is to split certain energy levels even when there is no applied magnetic field. Furthermore, the values of χ_m vary with direction for much the same reasons.

The nuclei of many types of atoms possess angular momentum and this in turn gives rise to magnetic moments. Thus an assembly of such nuclear magnets obey laws similar to those for their atomic counterparts; in particular, nuclear paramagnetism obeys Curie's law. However, the contribution from the nuclei to the total susceptibility is only about 10^{-6} of the electronic part and therefore can, for many practical purposes, be neglected.

See also DIAMAGNETISM AND SUPERCONDUCTIVITY; FERRIMAGNETISM; FERROMAGNETISM; MAGNETIC MOMENTS; MAGNETS (PERMANENT) AND MAGNETOSTATICS.

BIBLIOGRAPHY

A. Abragam and B. Bleaney, *Electron Paramagnetic Resonance of Transition Ions*. Oxford Univ. Press, London, 1970.

J. A. McMillan, *Electron Paramagnetism*. Reinhold, New York, 1968.

R. S. Tebble and D. J. Craik, *Magnetic Materials*. Wiley, New York, 1969.

R. M. White, *Quantum Theory of Magnetism*. McGraw-Hill, New York, 1970.

D. E. G. Williams, *The Magnetic Properties of Matter*. Longmans, London, 1966.

B. D. Cullity, *Introduction to Magnetic Materials*, Addison-Wesley, Reading, MA, 1972.

Parity

Ernest M. Henley

INTRODUCTION

Physical invariance laws, which follow from symmetries of the physical world, lead to conservation principles. It is clear from visual examinations that not all symmetries in nature are exact; approximate symmetries lead to partial conservation principles. Rotational invariance follows from the isotropy of space and leads to the conservation of angular momentum. It is exact; that is, no breakdown has ever been observed. Mirror symmetry, together with rotational invariance, leads to *parity* conservation, which is approximate.

The parity operation is defined as a reflection of the coordinates through the origin (Fig. 1),

$$\mathbf{r} \to -\mathbf{r}. \tag{1}$$

Invariance under the parity operation means that no differences can be observed between a system that is reflected in the origin and the nonreflected one. Such a symmetry is often referred to as mirror symmetry, since the operation can be carried out by reflecting the system in a plane followed by a rotation of 180° about an axis perpendicular to the reflecting plane. The parity transformation is a discontinuous one, as opposed to one that can be carried out by a series of infinitesimal steps.

Historically, parity was introduced in atomic physics in 1924 when Laporte [1] discovered that atomic levels can be classified into two distinct types. Wigner [2] was able to show that the observations followed from invariance of the quantum-mechanical laws under space reflection. Subsequently parity invariance was assumed to hold for all atomic and subatomic phenomena; the symmetry simplifies observations that are required to characterize a system. Indeed, no measurements gave any definite contrary evidence. It was not until 1956 that Lee and Yang [3] suggested that parity might not be conserved in weak interactions of atomic and subatomic particles. Indeed, they found that the invariance had not been tested for the "weak" forces between particles and suggested experiments to test the symmetry. Their suggestion was made as a result of many puzzling observations made in the weak decays of what is now called the neutral kaon. It came as a great shock to most physicists that the suggestion was proven to be correct by definitive studies made in beta decay and in the decay of the pion. This was the first breakdown of a space-time symmetry, and it has taught physicists to be more wary of accepting what appear to be "beautiful theoretical" symmetries. The breakdown is, however, only a partial one and the parity symmetry remains an extremely useful concept. It is found to hold for the strong or hadronic (nuclear) interactions and to be broken maximally only for the weak interactions. For instance, atomic forces between electrons and the nucleus are invariant under the parity transformation, and only the weak forces allow us to distinguish an atomic system from its mirror image. Similarly, if we consider the forces between two protons, then the strong or hadronic (Yukawa) and the Coulomb forces respect mirror symmetry; however, the much weaker "weak force" between these two particles is not invariant under reflection.

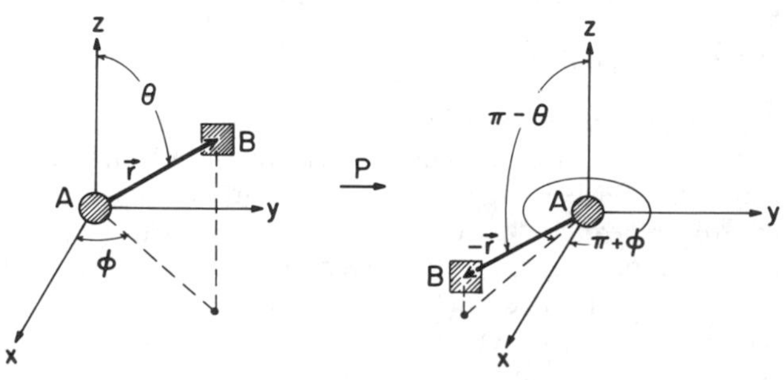

FIG. 1. The parity transformation.

THE PARITY OPERATION AND QUANTUM NUMBER

The operator that carries out the operation (1) is called the parity operator, P. It has the property that it transforms any true vector into its negative, and an axial vector, such as the angular momentum, into itself; e.g.,

$$P\mathbf{r}P^{-1} = -\mathbf{r}, \quad P\mathbf{p}P^{-1} = -\mathbf{p}, \quad P\mathbf{A}P^{-1} = -\mathbf{A}, \tag{2}$$
$$P\mathbf{E}P^{-1} = -\mathbf{E}, \quad P\mathbf{B}P^{-1} = \mathbf{B}, \quad P\mathbf{J}P^{-1} = \mathbf{J}$$

where $\mathbf{p}$ is the momentum, $\mathbf{A}$ the electromagnetic vector potential, $\mathbf{E}$ the electric field vector, $\mathbf{B}$ the magnetic induction vector, and $\mathbf{J}$ the angular momentum. Similarly, it transforms a scalar, e.g., $\mathbf{r}^2$, into itself, but a pseudoscalar, such as $\mathbf{J}\cdot\mathbf{p}$, into its negative.

Although parity is a quantum-mechanical operator, it should be noted that the underlying classical equations of motion of mechanics and electricity and magnetism are invariant under the transformation. That is,

$$m\frac{d^2\mathbf{r}}{dt^2} = \mathbf{F}; \quad \nabla\cdot\mathbf{E} = \rho, \quad \nabla\cdot\mathbf{B} = 0,$$
$$\nabla\times\mathbf{E} = -\frac{\partial\mathbf{B}}{\partial t}, \quad \nabla\times\mathbf{B} = \left(\mathbf{j} + \frac{\partial\mathbf{E}}{\partial t}\right)$$

are unchanged by the parity operation.

The operation P changes a spatial wave function $\psi(\mathbf{r})$ into one at the position $\mathbf{r}' = -\mathbf{r}$:

$$P\psi(\mathbf{r}) = \psi(-\mathbf{r}) \equiv \psi'(\mathbf{r}). \tag{3}$$

Since the wave function remains normalized, P is a unitary operator, $PP^{+} = P^{+}P = 1$. For nonrelativistic quantum-mechanical wave functions, which satisfy the Schrödinger equation, a second application of the operator P leads to the original state (if no degeneracy is present),

$$P^2\psi(\mathbf{r}) = P\psi(-\mathbf{r}) = \psi(\mathbf{r}). \tag{4}$$

If parity is a constant of the motion, i.e., does not change in time,

$$i\hbar\, dP/dt = [P,H] \equiv (PH - HP) = 0, \tag{5}$$

it follows that $[H,P] = 0$. Since the application of P^2 to a wave function leads back to the same wave function, we have $P^2 = PP^{+} = P^{+}P = 1$. The operator is therefore not only unitary but also Hermitian, $P = P^{+}$, and if it is conserved, i.e., $[H,P] = 0$, its eigenvalues are observables. Again since $P^2 = 1$, we have

$$P\psi(\mathbf{r}) = \psi(-\mathbf{r}) = \eta\psi(\mathbf{r}), \tag{6}$$

with eigenvalues η of $P = \pm 1$. The quantum number η is a multiplicatively conserved one; thus, if the initial state is a product of two wave functions, $\psi_a(\mathbf{r}_a)\psi_b(\mathbf{r}_b)$, we have

$$P\psi_a(\mathbf{r}_a)\psi_b(\mathbf{r}_b) = [P\psi_a(\mathbf{r}_a)][P\psi_b(\mathbf{r}_b)] = \psi_a(-\mathbf{r}_a)\psi_b(-\mathbf{r}_b)$$
$$= \eta_a\eta_b\psi_a(\mathbf{r}_a)\psi_b(\mathbf{r}_b). \tag{7}$$

The ideas above must be generalized for particles with spin. Since the spin is an internal angular momentum, it is not affected by the parity transformation. However, for a particle of half-integral spin, a rotation of 360° does not return the wave function to its original one and Eq. (4) needs to be generalized. A rotation of 720° or a fourfold reflection is required. The eigenvalues of P may thus be not only ± 1, but also $\pm i$.

In addition, we know that hadrons, particles that have strong interactions, possess structure. It is thus not surprising that their internal wave functions are also eigenstates of the parity operator to the extent that their structure is determined by a Hamiltonian that is invariant under the parity transformation,

$$PHP^{-1}=H \quad \text{or} \quad [P,H]=0.$$

This parity of particles or nuclei is called "intrinsic" parity. Since particles, e.g., mesons, can be produced and absorbed in a reaction, this parity is an important quantum number. However, the absolute intrinsic parity of a particle, defined to be that relative to the vacuum, can be determined only if its additive quantum numbers, such as charge, baryon number, and strangeness, are equal to zero. The reason is that conservation of additive quantum numbers forbids transitions between states that differ in these numbers. A particle whose intrinsic parity can be determined absolutely is, for instance, the neutral pion, which decays to two photons through an electromagnetic transition that conserves parity. Since there are two photons, their intrinsic parities play no role and that of the pion is found experimentally to be -1 [4]. On the other hand, it is not possible to determine the parity of the proton since it has both charge and baryon numbers that are nonzero. Protons cannot be produced or absorbed singly. We arbitrarily assign the proton a parity quantum number; by convention it is $+1$. The neutron, which differs from the proton in charge, should also be assigned a parity quantum number. Since it differs from the proton primarily through its charge and belongs to the same isospin multiplet, i.e., it is also a "nucleon," its parity quantum number is taken to be identical to that of the proton.

If particles A and B are bound or interact, then the parity of the system (AB) is the product of their intrinsic parities multiplied by that of their relative wave function, $\psi(\mathbf{r})$:

$$P\psi(AB)=\eta_A\eta_B P\psi(\mathbf{r})=\eta_A\eta_B\eta_\psi. \tag{8}$$

The parity of the relative motion is determined by the orbital angular momentum. If the relative position is $\mathbf{r}$ (see Fig. 1), then the parity transformation takes $\mathbf{r}$ into $-\mathbf{r}$, or in spherical coordinates $r\rightarrow r$, $\theta\rightarrow\pi-\theta$, $\phi\rightarrow\phi+\pi$. Since we have

$$Y_l^m(\pi-\theta,\pi+\phi)=(-1)^l Y_l^m(\theta,\phi), \tag{9}$$

the parity of the relative motion is determined solely by l and is given by $(-1)^l$. This knowledge can be used, for instance, to assign a parity to the deuteron, which consists of a proton and a neutron bound in a state of spin $J=1$ and of even orbital angular momentum. By Eqs. (8) and (9), we deduce that its parity is positive. This is confirmed experimentally.

We can make use of parity conservation to determine parity quantum numbers of particles and nuclei other than the proton and neutron. For strange particles, a further assignment is necessary. Any operator that is even under the parity transformation,

$$P\mathcal{O}_e P^{-1}=\mathcal{O}_e, \tag{10a}$$

such as the Hamiltonian, can only connect states of the same parity. Similarly, an odd operator,

$$P\mathcal{O}_o P^{-1}=-\mathcal{O}_o, \tag{10b}$$

such as the position operator, can only connect states of the opposite parity.

As an example, consider the parity of the π^- [4]. Since this particle differs from the π^0 only by charge, it also is expected to have negative parity. Because of its negative charge, however, its absolute parity cannot be determined, but that relative to nucleons can be found. The absorption of negative pions at rest by deuterium in the reaction

$$\pi^- + d\rightarrow n+n \tag{11}$$

allows us to deduce the parity of the pion relative to the deuteron. If the absorption occurs from an S $(l=0)$ Bohr orbit about the deuteron, the parity in the initial state is $\eta_\pi\eta_d(-1)^0=\eta_\pi$. Parity conservation requires that the final parity be equal to η_π. But angular momentum conservation and the Pauli principle require that the final two neutrons have $J=1$ and form a $^3P_1$ state of negative parity. Since the reaction (11) is observed to proceed at a rate consistent with a strong (parity-conserving) interaction, the parity of the π^- relative to the deuteron must be -1.

The parities of hadronically or electromagnetically unstable states can often be learned by studies of the decay. For electromagnetic decays with the emission of a photon, such as occur for many excited atomic and nuclear states, selection rules follow from parity conservation. For the emission (radiation) of a multipole L, the parities of the operators E_L and M_L are

$$\begin{aligned} &\text{Electric } 2^L \text{ multipole } E_L: && \eta_L=(-1)^L, \\ &\text{Magnetic } 2^L \text{ multipole } M_L: && \eta_L=(-1)^{L+1}. \end{aligned} \tag{12}$$

These transformation properties under the parity operation follow directly from the nature of the multipole operators; for instance, for an electric-dipole transition the operator is proportional to

$$E_1=\sum_i e_i\mathbf{r}_i, \tag{13a}$$

whereas for a magnetic dipole one it is proportional to

$$M_1=\sum_i (g_{il}\mathbf{L}_i+g_{is}\mathbf{S}_i) \tag{13b}$$

where e_i is the charge of the ith particle, $\mathbf{r}_i$ is its position, $\mathbf{L}_i$ $(\mathbf{S}_i)$ is its orbital (spin) angular momentum, and g is the corresponding gyromagnetic ratio. Measurements of the polarization of the emitted radiation in a transition differentiate between electric and magnetic multipoles and thus fix the parity of the decaying state if that of the final state is known. For instance, an E_1 decay to a ground state $J^\eta=0^+$ fixes the angular momentum and parity of the excited state to be 1^-.

Particle emission by a decaying state can also be used to assign parity. As a particular example of interest, consider some excited states of ^{16}O. As shown in Fig. 2, the state at 9.84 MeV has angular momentum 2 and is permitted by angular momentum and other selection rules to decay to the ground state of ^{12}C with the emission of an alpha particle,

$$^{16}\text{O}^*\rightarrow{}^{12}\text{C}+{}^4\text{He}. \tag{14}$$

However, the parities of both ^{12}C and ^{4}He in their ground states are $+1$. Since the orbital angular momentum l of the alpha particle relative to ^{12}C must be $2\hbar$ in order to conserve angular momentum, the decay can only proceed if the parity of the initial state is $(-1)^l=(-1)^2=+1$. The decay is observed with a width Γ of 0.8 keV, or a lifetime of $\hbar/\Gamma \simeq 8\times10^{-19}$ s, which is typical of that expected for a hadronically allowed transition. The parity of the 9.84-MeV state is therefore $+1$, as noted by the superscript in Fig. 2. By contrast, early sensitive searches for the decay of the state at 8.88 MeV to $^{12}C+^{4}He$ failed to reveal it. It was concluded that the parity of this state was negative, as noted in Fig. 2. This assignment was corroborated by γ-ray decays. A much more deliberate and exceedingly careful search for the decay found it to occur with a width of approximately 10^{-13} keV or a lifetime $\hbar/\Gamma \simeq 7\times10^{-6}$ s [5]. This very long lifetime gives further support to the negative parity assignment, but also demonstrates that parity conservation is not absolute but only approximate. The degree to which it fails at a rate approximately 1 part in 10^{13} is consistent with the order of magnitude expected from a breakdown in the weak interactions of nuclear particles.

TESTS OF PARITY CONSERVATION

A test for parity conservation is suggested by the foregoing method of assigning parity to given states of particles or nuclei. Selection rules that follow from parity conservation can be examined. An example is the alpha decay of the 8.88-MeV state in ^{16}O discussed above.

Another selection rule that follows from parity conservation is that the expectation values of operators that are odd under the parity transformation, Eq. (10b), must vanish. An example is the operator E_1, Eq. (13a),

$$\begin{aligned}\langle\psi(\mathbf{r})|E_1|\psi(\mathbf{r})\rangle &= \langle\psi|P^{-1}PE_1P^{-1}P|\psi\rangle \\ &= \eta_\psi{}^2\langle\psi|PE_1P^{-1}|\psi\rangle = \langle\psi|-E_1|\psi\rangle = 0. \qquad (15)\end{aligned}$$

Thus, parity conservation forbids a nondegenerate system from having a permanent electric-dipole moment (but not a magnetic one). The reverse statement does not follow; i.e., it is not true that the absence of an electric-dipole moment proves parity conservation. Indeed, time-reversal invariance also forbids such an electric-dipole moment [4]. Exceedingly careful and precise measurements of the neutron, which has no electric-monopole moment (charge), have failed to reveal an electric-dipole moment to a level of $\lesssim 4\times10^{-25}$ $e\cdot$cm [6]. Since the hadronic "size" of the nucleon is roughly 10^{-13} cm, this is impressive evidence for the conservation of P and of time-reversal invariance in the strong and electromagnetic interactions.

In general, however, a nonvanishing expectation value for an operator $\mathcal{O}_o$, which is odd under the parity transformation, shows that parity is not conserved. This is a generalization of the case of the static electric-dipole moment, Eq. (15), and can be proven in the same manner. A particularly useful example is that the expectation value of the helicity, $\mathbf{J}\cdot\hat{\mathbf{p}}$, namely, of the projection of the angular momentum on the linear momentum axis, must vanish. Thus, photons cannot be polarized longitudinally (circularly) through a parity-conserving interaction, nor can unpolarized particles scattered in a reaction acquire a polarization parallel or antiparallel to their momenta. Similarly, the cross section for incident particles that are polarized longitudinally, e.g., by an external magnetic field, cannot depend on the direction of the helicity (i.e., whether + or −) if the reaction proceeds via a parity-conserving interaction.

Indeed, it was a nonvanishing expectation value of a pseudoscalar observable akin to the foregoing kind that, following the suggestion of Lee and Yang, was used to demonstrate convincingly that parity conservation was violated maximally in a weak interaction. In this case, it was the expectation value of the spin $\mathbf{J}$ of an unstable nucleus along the momentum $\mathbf{p}$ of the electrons (beta rays) emitted in the decay, viz., $\langle\mathbf{J}\cdot\mathbf{p}_e\rangle$, that was found not to vanish [7]. The parent nucleus was polarized at low temperatures by a magnetic field, and the emitted electrons were detected. It was found that the electrons were emitted preferentially in a direction antiparallel to that of the polarization (i.e., direction of spin), whereas parity conservation demands that the rate be the same for electrons emitted parallel and antiparallel to the spin. This result clearly shows that weak interactions allow us to detect a difference between an object and its mirror image by means of the weak interactions, as is illus-

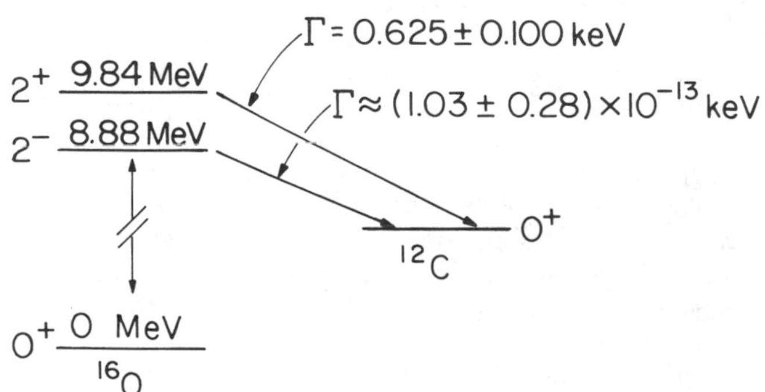

FIG. 2. Decay of two excited states of ^{16}O to $^{12}C+^{4}He$.

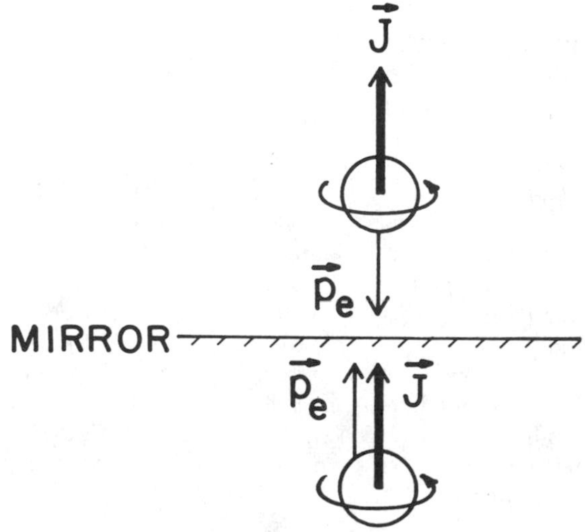

FIG. 3. Schematic view of nuclear polarization and preferential electron emission together with its mirror image.

trated in Fig. 3. The conclusion was corroborated by studies which demonstrated that muons emitted in the decay of charged pions were 100% longitudinally polarized and that electrons emitted in beta decays were similarly polarized, the degree and direction of polarization being $-v/c$, where **v** is their velocity [4]. Indeed, these experiments left no doubt whatsoever that parity is not conserved in the weak interactions. However, subsequent careful reexaminations demonstrated that parity was conserved in the strong and electromagnetic interactions.

What has been learned in the years since the downfall of parity conservation in 1956? In the intervening years many experiments have been carried out to elucidate the theory behind the breakdown of parity conservation in the purely leptonic sector, in semileptonic interactions, and in the nonleptonic sector. The latter is the most difficult one to relate to a basic theory because of the presence of strong interactions.

How can one test parity nonconservation? We have already shown that if the interaction that gives rise to a transition does not conserve parity, then it can connect states of the opposite parity. To wit, we have, with Eqs. (10),

$$\begin{aligned} \langle\psi_b{}^+|H_e+H_o|\psi_a{}^+\rangle &= \langle\psi_b{}^+|H_e|\psi_a{}^+\rangle, \\ \langle\psi_c{}^-|H_e+H_o|\psi_a{}^+\rangle &= \langle\psi_c{}^-|H_o|\psi_a{}^+\rangle \end{aligned} \tag{16}$$

where the superscript refers to the parity. However, if parity is not a constant of the motion, the initial and final states may no longer be exact eigenstates of the parity operator. Because of the weak interactions, for instance, the state labeled $\psi_a{}^+$ in Eq. (16) may acquire a small admixture F of the opposite parity:

$$\psi_a{}^+ \rightarrow \psi_\alpha = (1 - |F|^2)^{1/2}\psi_a{}^+ + F\psi_a{}^-. \tag{17}$$

The state has been relabeled α to indicate that certain quantum numbers, e.g., the energy, may change. It follows that for such a state, the expectation value of a parity-odd operator will not vanish. For the helicity operator, e.g., we have

$$\begin{aligned} \langle\psi_\alpha|\mathbf{J}\cdot\hat{\mathbf{p}}|\psi_\alpha\rangle = (1-|F|^2)^{1/2}(F\langle\psi_\alpha{}^+|\mathbf{J}\cdot\hat{\mathbf{p}}|\psi_\alpha{}^-\rangle \\ + F^*\langle\psi_\alpha{}^-|\mathbf{J}\cdot\hat{\mathbf{p}}|\psi_\alpha{}^+\rangle) \neq 0. \end{aligned} \tag{18}$$

Another effect is that both electric *and* magnetic multipoles of the same order L are allowed between two states ψ_α and ψ_β. Such an admixing gives rise to a circular polarization (nonvanishing helicity expectation value) of photons emitted in the transition. Indeed, this method of detecting parity nonconservation is a sensitive one and has been used for observing small effects in the presence of the strong and electromagnetic forces [4]. For small values of F, Eq. (18) shows that the effects are directly proportional to the parity-nonconserving (pnc) admixture F. High-energy electron scattering from H and ^{2}H [8] and atomic experiments [9] have been used to measure the strength of the weak interaction amplitude which interferes with the electromagnetic one. For momentum transfers of 1 GeV/c^2, the pnc asymmetry in *e-p* scattering is of the order of 10^{-4}. Atomic effects would be smaller, but experiments have been carried out in heavy atoms where there are enhancements due to nuclear coherence, relativity (the electrons must be at the nucleus), and other effects [9].

Although the weak force is only about 10^{-6} as strong as the hadronic one, the effects of the small resultant parity admixing in nuclear wave functions have been observed in a number of heavy and light nuclei. Heavy nuclei were the first to yield parity-nonconserving effects for hadrons because they can be enhanced in selective nuclei. However, most progress in understanding the weak interactions of hadrons has occurred from studies of parity nonconservation (PNC) in light nuclei, including proton–proton scattering at various energies. It is also parity admixing in wave functions that is responsible for the α decay of the 8.88-MeV level in ^{16}O. Here the decay proceeds through a parity-conserving interaction, but admixture of a small part, F, of 2^+ to the dominant 2^- state leads to a rate proportional to $|F|^2 \sim 10^{-13}$, consistent with experimental results [5].

BASIC THEORY OF PARITY NONCONSERVATION AND RECENT DEVELOPMENTS

The main advance that has occurred is that there is now a theory of the weak interactions [10], namely, the electroweak sector of the so-called "standard" model. The theory [10] is a gauge theory, for which spontaneous symmetry breaking gives rise to the masses of the weak gauge bosons, $W^\pm$ (82 GeV/c^2) and Z^0 (93 GeV/c^2). It is the large masses of these bosons that are reponsible for the weakness of the force at energies $E \ll m_W c^2$, and not the strength of the interaction which is determined by the fine structure constant $\alpha = e^2/\hbar c$. The electroweak interaction is a current–current one, but the weak sector has both vector and axial-vector currents, whereas the electromagnetic one has pure vector currents. At energies below several GeV, the weak interaction can be described by a point interaction

$$H_W = \frac{G_F}{\sqrt{2}c^2} \sum_{\mu=0}^{3} \int d^3x\, J^\mu(\mathbf{r}) J_\mu{}^\dagger(\mathbf{r}), \tag{19}$$

where J^μ is a four-vector $(c\rho, \mathbf{J})$ and G_F is the dimensional weak interaction strength, $G_F \approx 0.896 \times 10^{-4}$ MeV fm.^{3} In the framework of the electroweak theory, G_F is related to the electromagnetic interaction strength α and to the range of the weak force ($\hbar/m_W c \equiv R_W$) by

$$G = \frac{\sqrt{2}}{8} \frac{\alpha\hbar c R_W{}^2}{\sin^2\theta_W}, \tag{20}$$

where θ_W is the Weinberg angle, with $m_z = m_W/\cos\theta_W$, $\sin^2\theta_W \approx 0.22$. The weak current has both vector (V) and axial vector (A) parts,

$$J_\mu = V_\mu + A_\mu. \tag{21}$$

It is the term $(V^\mu A_\mu{}^\dagger + A^\mu V_\mu{}^\dagger)$ in the weak interaction, H_W, Eq (19), that basically is responsible for PNC. For a weak reaction, the interference of this part of H_W with the term $V^\mu V_\mu{}^\dagger$ or $A^\mu A_\mu{}^\dagger$ gives rise to pseudoscalar observables which herald PNC. In experiments with charged leptons the interference of $(V^\mu A_\mu{}^\dagger + A^\mu V_\mu{}^\dagger)$ with the electromagnetic current–current interaction, of the form $V^\mu_{\rm em} V^\dagger_{\mu,\rm em}$, gives rise to pnc observables.

The standard electroweak theory has both charged and neutral currents. In the hadronic sector, the current J_μ is more complicated. It has vector and axial-vector parts; it is composed of both charged and neutral currents. The charged currents not only have a part that conserves strangeness, charm, etc., but also have a strangeness-changing part. We will not write them down in detail [10].

For purely leptonic processes, such as $\nu_\mu e \rightarrow \nu_\mu e$, the PNC can be understood directly in terms of the basic theory [10], which has been corroborated by all experimental findings. Because a new feature of the standard model is the presence of both vector and axial-vector neutral currents, the vector part of which depends on the Weinberg angle θ_W, most recent experiments have focused on effects due to these currents. In the Weinberg–Salam theory [10], radiative corrections and second-order weak interaction effects are finite and calculable [11] and can be tested in accurate weak interaction and pnc experiments. Experiments such as $\nu_\mu e \rightarrow \nu_\mu e$ (Fig. 4a) can only proceed through neutral currents, whereas $\nu_e e \rightarrow \nu_e e$ (Figs. 4a and 4b), has contributions from both neutral and charged currents. These experiments, which include pnc effects, agree with the standard model and give a unique value of the Weinberg angle, $\sin^2 \theta_W = 0.22$.

In the semileptonic sector of the weak interaction, measurements of PNC observables have been carried out in electron scattering and atomic experiments. These experiments depend on the presence of neutral currents. High-energy, longitudinally polarized electrons were scattered off hydrogen and deuterium [8]. The difference in the cross section for right circularly polarized (parallel to the momentum) electrons and left circularly polarized electrons (antiparallel to the momentum) was measured [8] in the ratio

$$a = \frac{\sigma_R - \sigma_L}{\sigma_R + \sigma_L} . \tag{22}$$

The asymmetry was determined as a function of momentum transfer and was found to be $\sim 10^{-4}$; both the magnitude and the momentum transfer dependence of a, Eq. (22), agreed with the theory. At lower energies, $E \approx 0.5$–1 GeV, PNC experiments are under way at several laboratories, but the required accuracy has not yet been achieved. On the other hand, at extremely low energies, in the eV range, a number of atomic experiments [9] have found pnc effects that agree with the Weinberg–Salam theory. The experiments were carried out, for instance, by searching for the difference of the index of refraction for right and left circularly polarized laser light, i.e., for the expectation value of $\mathbf{J} \cdot \hat{p}$, at and near a resonant transition. By using heavy atoms, nuclear coherence and relativistic effects serve to enhance the small PNC and make the effect measurable. Calculations [9] show that the experimental results are in accord with the standard model.

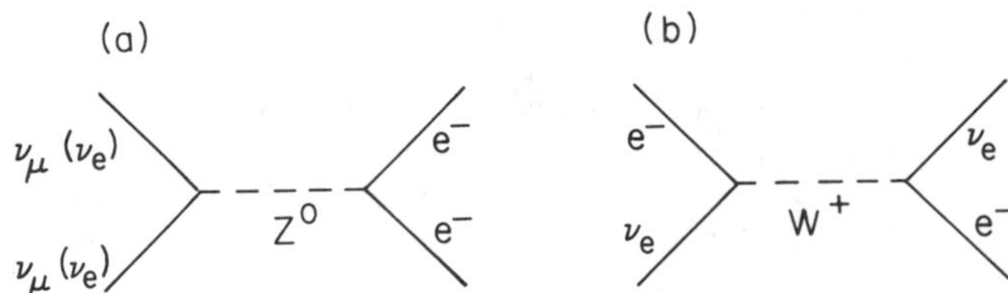

FIG. 4. Neutral and charged current contributions to νe^- scattering.

Experiments in the nonleptonic sector, due to the weak force between hadrons, such as those carried out in strange particle decays, are more difficult to interpret. The basic theory applies to the weak interactions of quarks. This interaction must be translated to hadrons and requires model-dependent assumptions. Strange particle decays occur only through the charged current part of the interaction. PNC due to neutral currents can be observed in nuclear experiments. These experiments have been carried out primarily at low energies ($\lesssim 1$ GeV) and are analyzed in terms of a weak pnc nucleon–nucleon potential. This weak interaction is calculated from the basic quark–quark weak interaction by way of models for the structure of the nucleon and meson exchanges [12]. The virtual mesons are coupled weakly through a pnc coupling to one nucleon and strongly to the other one. The weak coupling is model dependent. The state-of-the-art calculation gives values to the weak couplings that are listed in Table I. Nuclear pnc data are analyzed in terms of the pnc nucleon–nucleon potential, nuclear structure, and Eq. (18). The most interesting coupling is the weak pnc coupling of the pion to the nucleon because (a) of its sensitivity to neutral currents and (b) pion exchange is the longest-range component of the nuclear force. The data from light nuclei are summarized in Table II. The table shows light nuclear pnc asymmetries, which can be analyzed because the nuclear structure is known. In fact, in the case of fluorine, a relation between the measured pnc asymmetry and beta decay rates can be found [4], which decreases considerably the model dependence of the weak pion coupling constant deduced from the experiments. When interpreted in terms of weak coupling constants, the data are somewhat contradictory [4]. Also, since the major source of the weak pion–nucleon coupling is neutral currents, and this constant presently is consistent with zero, there still is no evidence for the presence of these currents in the hadronic sector of the weak interaction, i.e., in the weak interaction of hadrons. This is undoubtedly a problem of the strong interactions rather than a fault of the weak interaction, but the final word is not yet in hand.

Thus, although parity is not an exact symmetry of nature, it is conserved in hadronic and electromagnetic interactions, and therefore in the basic forces that determine the structures of nuclei, of atoms, and of matter. Moreover, the nonconservation of parity in the weak forces has opened a win-

Table I Best values and ranges of PNC coupling constants deduced by DDH (1980). The superscripts refer to isospin. All values are in units of 10^7.

Coupling Constant	Best Value	Range
$f^{(1)}_{wk}(NN\pi)$	4.6	0 to 12
$f^{(0)}_{wk}(NN\rho)$	−12	−32 to 12
$f^{(1)}_{wk}(NN\rho)$	−0.2	−0.4 to 0
$f^{(2)}_{wk}(NN\rho)$	−10	−12 to −8
$f^{(0)}_{wk}(NN\omega)$	−2	−10 to 6
$f^{(1)}_{wk}(NN\omega)$	−1.2	−2 to −0.8

Table II Parity asymmetries from selected experiments [4]

Experiment	Energy (MeV)	Asymmetry a Polarization P in Units of 10^{-6}
$\mathbf{p}p\ \sigma_{\text{total}}$	15	$a=-(0.17\pm0.08)$
	45	$a=-(0.15\pm0.022)$
	800	$a=0.24\pm0.11$
	5300	$a=2.65\pm0.6$
$np\rightarrow d\vec{\gamma}$	Thermal	$P_\gamma=0.18\pm0.18$
$p\alpha\ \sigma_{\text{elastic}}$	46	$a=-(0.334\pm0.093)$
$^{16}O(2^-)\rightarrow{}^{12}C(0^+)\alpha$		$\Gamma_\alpha=10\pm1$ eV
$^{18}F(0^-\rightarrow1^+)$		$\langle P_\gamma\rangle=80\pm390$
$^{19}F(\frac{1}{2}^-\rightarrow\frac{1}{2}^+)$		$\langle a\rangle=-(74\pm19)$
$^{21}Ne(\frac{1}{2}^-\rightarrow\frac{3}{2}^+)$		$P_\gamma=-(8\pm14)$

The recent precision 45-MeV **p**p scattering is by Kistryn *et al.* [*Phys. Rev. Lett.* **58,** 1616 (1987)].

dow that has been, and continues to be, exploited to disclose more about the basic theory of the weak interactions.

See also Conservation Laws; CPT Theorem; Currents in Particle Theory; Elementary Particles in Physics; Invariance Principles; Quantum Mechanics; Space-Time; Weak Interactions; Weak Neutral Currents.

BIBLIOGRAPHY

Basic Introduction

H. Frauenfelder and E. M. Henley, *Subatomic Physics.* Prentice-Hall, Englewood Cliffs, NJ, 1974 (revision in press).

More Advanced Studies

A survey of the literature up to 1970 is given by L. M. Lederman, "Resource Letter Neu-1, History of the Neutrino," *Am. J. Phys.* **38,** 129 (1970).

J. J. Sakurai, *Invariance Principles and Elementary Particles.* Princeton University Press, Princeton, NJ, 1964.

R. J. Blin-Stoyle, *Fundamental Interactions and the Nucleus.* Elsevier, North-Holland, New York, 1973.

H. Frauenfelder and E. M. Henley, *Nuclear and Particle Physics.* Benjamin (Advanced Book Program), Reading, MA, 1975.

E. D. Commins and P. H. Bucksbaum, *Weak Interactions of Leptons and Quarks.* Cambridge University Press, Cambridge, 1983.

V. Gottfried and V. F. Weisskopf, *Concepts of Particle Physics, Vol. I.* Oxford University Press, New York, 1984.

D. H. Perkins, *Introduction to High Energy Physics,* 3rd ed., Chap. 3. Addison-Wesley, Menlo Park, CA, 1987.

R. Novick (ed.), *Thirty Years Since Parity Nonconservation. A Symposium for T. D. Lee.* Birkhäuser, Boston, 1988.

Reviews

T. D. Lee and C. S. Wu, *Annu. Rev. Nucl. Sci.* **15,** 381 (1965); **16,** 471, 511 (1966).

E. M. Henley, *Annu. Rev. Nucl. Sci.* **19,** 367 (1969).

S. Weinberg, *Rev. Mod. Phys.* **46,** 225 (1974).

M. A. B. Bég and A. Sirlin, *Annu. Rev. Nucl. Sci.* **24,** 379 (1974).

E. N. Fortson and L. Wilets, *Adv. At. Mol. Phys.* **16,** 319 (1980).

E. N. Fortson and L. L. Lewis, *Phys. Rep.* **113,** 289 (1984).

E. G. Adelberger and W. Haxton, *Annu. Rev. Nucl. Part. Sci.* **35,** 501 (1985).

E. M. Henley, *Prog. Part. Nucl. Phys.,* **13,** 403 (1984); **20,** 387 (1987).

REFERENCES

1. O. Laporte, *Z. Phys.* **23,** 135 (1924).
2. E. Wigner, *Z. Phys.* **43,** 624 (1927).
3. T. D. Lee and C. N. Yang, *Phys. Rev.* **104,** 254 (1956).
4. E. G. Adelberger and W. Haxton, *Annu. Rev. Nucl. Part. Sci.* **35,** 501 (1985).
5. H. Hättig, K. Hünchen, and H. Wäffler, *Phys. Rev. Lett.* **25,** 941 (1971); K. Neubeck, H. Schober, and H. Wäffler, *Phys. Rev.* **C10,** 320 (1974).
6. S. Alterev *et al., Phys. Lett.* **102B,** 13 (1981).
7. See, e.g., H. Frauenfelder and E. M. Henley, *Subatomic Physics,* pp. 203–209. Prentice-Hall, Englewood Cliffs, NJ, 1974.
8. C. Y. Prescott *et al., Phys. Lett.* **77B,** 347 (1979); **82B,** 524 (1979).
9. L. L. Lewis, J. H. Hollister, D. C. Soreide, E. G. Lindahl, and E. N. Fortson, *Phys. Rev. Lett.* **39,** 795 (1977); P. E. G. Baird, M. W. S. M. Brimicombe, R. H. Hunt, G. J. Roberts, P. G. H. Sandars, and D. N. Stacey, *Phys. Rev. Lett.* **39,** 795 (1977); L. M. Barkov and M. S. Zolotorev, *Pis'ma Zh. Eksp. Teor. Fiz.* **27,** 375 (1978 [*JETP Lett.* **27,** 379 (1978)]; R. Conti *et al., Phys. Rev. Lett.* **42,** 343 (1979); E. N. Fortson and L. L. Lewis, *Phys. Rep.* **113,** 289 (1984).
10. S. Weinberg, *Phys. Rev. Lett.* **19,** 1264 (1967); A. Salam, in *Elementary Particle Physics: Relativistic Groups and Analyticity* (N. Svartholm, ed.), p. 367. Almqvist and Wiksells, Stockholm, 1968.
11. M. A. B. Bég and A. Sirlin, *Annu. Rev. Nucl. Sci.* **24,** 379 (1974).
12. B. Desplanques, J. F. Donoghue, and B. Holstein, *Ann. Phys. (NY)* **124,** 449 (1980); G. F. de Teramond and B. Gabioud, *Phys. Rev.* **C36,** 691 (1987).

Partial Waves

Paul L. Csonka

Any (separable) solution of the Schrödinger equation for a spinless particle in vacuum with kinetic energy E can be written as $\chi_E(t,r,\theta,\varphi)=e^{-iEt}\psi_k(r,\theta,\varphi)$, where

$$\begin{aligned}\psi_k(r,\theta,\varphi) &= \sum_{l,m=0}^{\infty} A_{l,m}(k)N_{l,m}Y_{l,m}(\theta,\varphi)\\ &\quad\times[j_l(kr)+B_{l,m}(k)n_l(kr)]\\ &\underset{r\to\infty}{\rightarrow}\frac{1}{kr}\sum_{l,m=0}^{\infty} a_{l,m}(k)N_{l,m}Y_{l,m}(\theta,\varphi)\\ &\quad\times\sin\left[kr-\frac{l\pi}{2}+\delta_{l,m}(k)\right].\end{aligned} \tag{1}$$

The notation is as follows: $\hbar=1$; the particle mass is M; $k=(2ME)^{1/2}$; t means time; r,θ, and φ are spherical coordinates (the radius vector, polar angle measured from the z axis, and azimuthal angle, respectively); $Y_{l,m}$ are spherical harmonics {normalized so that $Y_{l,0}(\theta,\varphi)=[(2l+1)/4\pi]^{1/2}\times P_l(\cos\theta)$ where P_l is the Legendre polynomial of order l}; $j_l(x)$ and $n_l(x)$ are the spherical Bessel and Neumann functions, respectively [normalized so that for $r\rightarrow\infty$, $j_l(x)\rightarrow x^{-1}\times\sin(x-l\pi/2)$, $n_l(x)\rightarrow -x^{-1}\cos(x-l\pi/2)$]. Any particular ψ can be specified by giving either all constants $A_{l,m}(k)$ and $B_{l,m}(k)$ or all constants $a_{l,m}(k)$ and $\delta_{l,m}(k)$, where by definition

$a_{l,m}(k) \equiv A_{l,m}(k)[\cos\delta_{l,m}(k)]^{-1}$, and $\tan\delta_{l,m}(k) \equiv -B_{l,m}(k)$. The normalization constant,

$$N_{l,m} \equiv [4\pi(l+|m|)!]^{1/2}[(2l+1)(l-|m|)!]^{-1/2},$$

is factored out for later convenience.

Each $(kr)^{-1}N_{l,m}Y_{l,m}(\theta,\varphi)\sin(kr - l\pi/2 + \delta_{l,m,k})$ is called a *partial wave*, and is an eigenfunction of kinetic energy, total angular momentum, and angular momentum projection along the z axis. Each $a_{l,m}(k)$ is referred to as a *partial wave amplitude*. The last form in Eq. (1) is the expansion of ψ into partial waves. Each $\delta_{l,m}(k)$ is called a *phase shift*, for reasons explained below. {Occasionally the $N_{l,m}Y_{l,m}(\theta,\varphi)[j_l(kr) + B_{l,m}(k)n_l(kr)]$ are referred to as "partial waves" and the $A_{l,m}(k)$ as "partial wave amplitudes." Sometimes the $N_{l,m}$ is chosen differently, which changes $A_{l,m}(k)$}.

When the particle has spin s, then ψ has $(2s+1)$ components. Each component can be expanded into partial waves—a straightforward generalization of the spinless case. Here we will discuss only spinless particles.

When ψ is cylindrically symmetric around the z axis, then each partial wave must be independent of φ; thus $a_{l,m}(k)$ must vanish whenever $m \neq 0$. Then $N_{l,0}Y_{l,0}(\theta) = P_l(\cos\theta)$; all arguments φ and subscripts m can be dropped in Eq. (1):

$$\psi_k(r,\theta) \underset{r\to\infty}{\longrightarrow} \frac{1}{kr}\sum_{l=0}^{\infty} a_l(k)P_l(\cos\theta)\sin\left[kr - \frac{l\pi}{2} + \delta_l(k)\right]. \tag{2}$$

An example is $\psi = \exp(ikz)$, a plane wave moving along the z axis. It can be written as $\exp(ikr\cos\theta)$, and with the well-known formula due to Bauer:

$$\begin{aligned}\exp(ikr\cos\theta) &= \sum_{l=0}^{\infty} i^l(2l+1)P_l(\cos\theta)j_l(kr)\\ &\underset{r\to\infty}{\longrightarrow} \frac{1}{kr}\sum_{l=0}^{\infty} i^l(2l+1)P_l(\cos\theta)\sin\left(kr - \frac{l\pi}{2}\right).\end{aligned} \tag{3}$$

Thus, for this function $a_l(k) = i^l(2l+1)$, and all $\delta_l(k) = 0$.

Partial waves are frequently used to describe the scattering of a cylindrically symmetric incoming wave, ψ_{in}, on a real (not complex) central potential, $V(r)$, centered on the axis of symmetry (which we choose to be the z axis). In the trivial case when $V(r) = 0$ everywhere, the solution ψ of the Schrödinger equation is equal to ψ_{in}. But when $V(r) \neq 0$ for some range of r, then $\psi \neq \psi_{\text{in}}$. The function $\psi - \psi_{\text{in}} \equiv \Delta\psi$ is called the *scattered wave*, because it is due to the presence of the potential. Now, because the whole geometry is cylindrically symmetric, all wave functions in the problem must have such symmetry, and thus Eq. (2) is valid for each. Usually we choose ψ_{in} to be a plane wave; then it can be expanded as in Eq. (3). When the potential has finite range, i.e., $V(r) = 0$ for $r > r_0$, then for $r > r_0$ (i.e., in a vacuum) the ψ must be of form given in Eq. (2). Therefore, the partial-wave expansion of $\Delta\psi$ is the right-hand side of Eq. (2) minus that of Eq. (3):

$$\begin{aligned}\Delta\psi(r>r_0,\theta) \underset{r\to\infty}{\longrightarrow} &\frac{1}{kr}\sum_{l=0}^{\infty} P_l(\cos\theta)\\ &\times \tfrac{1}{2}\{(-i)^{l+1}[a_l(k)\exp(i\delta_l(k)) - i^l(2l+1)]\exp(ikr)\\ &+ i^{l+1}[a_l(k)\exp(-i\delta_l(k)) - i^l(2l+1)]\exp(-ikr)\}.\end{aligned} \tag{4}$$

{Here we used $\sin x = (-i/2)[\exp(ix) - \exp(-ix)]$ and $\exp(il\pi/2) = i^l$.} On physical grounds we now impose the boundary condition that the scattered wave contains only outgoing waves but no ingoing waves; i.e., that the coefficient of $\exp(-ikr)$ in Eq. (4) vanish: $\exp[-i\delta_l(k)] = a_l^{-1}(k)i^l(2l+1)$. Using this equation, we rewrite the coefficient of $\exp(ikr)$:

$$\begin{aligned}\Delta\psi(r>r_0,\theta) \underset{r\to\infty}{\longrightarrow} &\frac{1}{kr}\exp(ikr)\sum_{l=0}^{\infty} P_l(\cos\theta)(2l+1)\\ &\times [\exp(i\delta_l(k))]\sin\delta_l(k).\end{aligned} \tag{5}$$

The scattering amplitude is commonly defined as $f = \lim_{r\to\infty} \Delta\psi\, r\exp(-ikr)$. With Eq. (5), we have

$$f(\theta) = k^{-1}\sum_{l=0}^{\infty} P_l(\cos\theta)(2l+1)[\exp(i\delta_l(k))]\sin\delta_l(k). \tag{6}$$

The differential cross section in the direction θ is $d\sigma(\theta)/d\Omega = kk_{\text{in}}^{-1}|f(\theta)|^2$, where k_{in} is the incoming momentum and k is the scattered momentum. For our case $k = k_{\text{in}}$. The total scattering cross section $\sigma \equiv \int[d\sigma(\theta)/d\Omega]\,d\Omega$, integrated over all solid angles Ω. We write $f(\theta)$ as in Eq. (6), square, then use $\int_0^\pi d\theta(\sin\theta)P_l(\cos\theta)P_{l'}(\cos\theta) = \delta_{l,l'}\,2(2l+1)^{-1}$, note that cylindrical symmetry gives $\int d\varphi = 2\pi$, and find that

$$\sigma = 4\pi k^{-2}\sum_{l=0}^{\infty}(2l+1)\sin^2\delta_l(k). \tag{7}$$

Observe that the lth partial wave contributes at most $4\pi k^{-2}(2l+1)$ to σ. Comparison of Eqs. (6) and (7) shows that $\sigma = 4\pi k^{-1}\,\mathrm{Im}[f(\theta=0)]$, a manifestation of the optical theorem.

The $\delta_l(k)$ can be calculated when $V(r)$ is known. We can show that when $V(r)$ has no bound states, $\delta_l(k) > 0$ for attractive $V(r)$, and $\delta_{l,k} < 0$ for repulsive $V(r)$—an intuitively appealing result. [The same $V(r)$ may be attractive for some l and repulsive for others.] For low k, usually (but not for resonances) only terms with $l \lesssim kr_0$ contribute significantly to scattering [otherwise $\sin\delta_l(k)$ is small], as also suggested by classical physics. Thus, the series in Eqs. (6) and (7) converge fast for low k, a great virtue of partial-wave analysis.

The $\delta_l(k)$ can be defined (because they then do not depend on r) also for an infinite-range potential, provided that as $r\to\infty$, $V(r)\to 0$ faster than r^{-1}. (The unshielded Coulomb potential requires special treatment because it violates this condition.)

Writing ψ in terms of exponentials [as $\Delta\psi$ is written in Eq. (4)], we observe that the coefficient of the outgoing wave $\exp(ikr)$ in the lth partial wave is $(-1)^{l+1}\exp(2i\delta_l(k))$ times the coefficient of the ingoing wave $\exp(-ikr)$. Since the outgoing amplitude is the S matrix times the ingoing amplitude, the lth diagonal element of the S matrix can be defined (by a suitable choice of phases) as

$$S_l(k) = \exp(2i\delta_l(k)).$$

The off-diagonal elements of S all vanish, because l (the angular momentum) is conserved. Thus for $r\to\infty$ the scattering has the effect of shifting the phase of partial-wave amplitudes by multiplying them by a phase factor $\exp(2i\delta_l(k))$. This is why the $\delta(k)$ are referred to as phase shifts.

When $V(r)$ is complex (not real), then the foregoing equations [except Eq. (7)] still hold provided that in them we substitute for $\delta_l(k)$ a complex phase shift $\epsilon_l(k) = \delta_l(k) + i\gamma_l(k)$. Defining $\eta_l(k) \equiv \exp[-2\gamma_l(k)]$, we now have $a_l(k) = i^l(2l+1)\eta_l^{1/2}(k)\exp(i\delta_l(k))$. The scattering is elastic when $\gamma_l = 0$ (i.e., $\eta_l = 1$); inelastic when $\gamma_l > 0$ (i.e., $\eta_l < 1$). The fraction of particles absorbed by the potential from each partial wave is $1-\eta^2$. The $\Delta\psi$ is now the wave function of the *elastically* scattered particles, $f(\theta)$ is the *elastic* scattering amplitude, and $kk_{in}^{-1}|f(\theta)|^2$ is the *elastic* differential scattering cross section. Deriving Eq. (7) we used complex conjugation, and thus we cannot generalize it by simply substituting ϵ_l for δ_l. Instead, $\int|f(\theta)|^2\, d\Omega$ now gives the total *elastic* scattering cross section

$$\sigma_{el} = 4\pi k^{-2}\sum_{l=0}^{\infty}\tfrac{1}{4}(2l+1)|\exp[2i\epsilon_l(k)]-1|^2.$$

The total *inelastic* cross section is

$$\sigma_{inel} = 4\pi k^{-2}\sum_{l=0}^{\infty}\tfrac{1}{4}(2l+1)[1-\eta_l^2(k)].$$

Clearly, whenever $\sigma_{inel} \neq 0$, the σ_{el} cannot vanish either. The *total* cross section $\sigma_{el} + \sigma_{inel}$ is

$$\sigma_T = 4\pi k^{-2}\sum_{l=0}^{\infty}\tfrac{1}{2}(2l+1)[1-\eta_l(k)\cos 2\delta_l(k)].$$

Now the optical theorem is $\sigma_T = 4\pi k^{-1}\,\mathrm{Im}[f(\theta=0)]$.

When V is noncentral (depends on angle), then an incoming partial wave with some l value may scatter into an outgoing partial wave with a different l. Now the scattered wave can be obtained by multiplying ψ_{in} by a matrix in a space labeled by l and m. Even if the interaction is described by a matrix in a space labeled also by the other quantum numbers of the particles (spin, particle type, etc.) but provided that the interaction falls off fast enough for $r\to\infty$, it can still be parametrized in terms of phase shifts. Thus if any interaction is negligible outside some "black box," we can summarize all its effects that are observable at $r\to\infty$ by giving the phase shifts.

Several related but distinct definitions of partial-wave resonance have been proposed, all in analogy with resonance in classical mechanics. According to one we say that the l_0th partial wave has a resonance at energy $E_0 = (2M)^{-1}k_0^2$ if $\delta_{l0}(k)$, as a function of k, increases and passes through 90° at $k = k_0$. If, furthermore, in such a case $\eta_{l0}(k)$ is a constant as a function of k when k is near k_0, then the term $\tfrac{1}{4}|\exp[2i\epsilon_{l0}(k)]-1|^2$ in σ_{el} assumes the locally maximum value of $\tfrac{1}{4}[\eta_{l0}^2(k_0) + 2\eta_{l0}(k_0) + 1]$ at k_0. But if $\eta_{l0}(k)$ varies with k near $k = k_0$, then this variation may mask the local maximum, making it difficult to detect resonances solely by looking at the behavior of σ_{el}. According to an alternative definition, a partial wave is said to have a resonance at k if near k_0 the Breit–Wigner formula holds: $(2i)^{-1}\{\exp[2i\epsilon_{l0}(k)]-1\} = \Gamma_{el}[2(E_0-E)-i\Gamma]^{-1}$ where Γ_{el} and Γ are constants called, respectively, the elastic and total width. We can show that in this case $\delta_{l0}(k)$ passes through 90° at k_0 when $\Gamma_{el}\Gamma^{-1} > \tfrac{1}{2}$, but passes through 0 when $\Gamma_{el}\Gamma^{-1} < \tfrac{1}{2}$. A third and more profound definition connects a resonance with the existence of a pole on the second sheet of the analytically continued partial-wave scattering amplitude.

We can conveniently visualize phase shifts by drawing a so-called Argand plot for each l value of interest. The diagram shows, in the complex plane, the value of $(2i)^{-1} \times \{\exp[2i\epsilon_l(k)]-1\} = (i/2) + \tfrac{1}{2}\eta\exp[-i(\pi/2) + 2i\delta_l(k)]$ for suitably chosen E values. Thus, each point in the diagram has a radius vector of length $\tfrac{1}{2}\eta_l$ as measured from the point $\tfrac{1}{2}i$ on the imaginary axis. The radius vector makes an angle $2\delta_l$ measured clockwise from any axis that is antiparallel to the imaginary axis. The corresponding value of E (or k) is written next to every point in the diagram. All points for which $\eta \leq 1$ must lie within or on the circle with radius $\tfrac{1}{2}$ drawn around the point $\tfrac{1}{2}i$. When for a certain E_0 the $\delta_l(k)$ goes through 90°, the corresponding point in the Argand plot lies on the imaginary axis, above the point $\tfrac{1}{2}i$. For $\delta_l = 0$, the point lies on the imaginary axis in the interval $[0, \tfrac{1}{2}i]$. It may happen that the interaction under study consists of two parts: one is a "background" producing a phase shift $\epsilon_l^{(1)}$ varying slowly with k in the vicinity of k_0, the other part produces a rapidly varying $\epsilon_l^{(2)}$ going through 90° at k_0. The total phase shift will then be $\epsilon_l^{(1)} + \epsilon_l^{(2)}$ and the corresponding curve in the Argand plot makes a "loop" near k_0.

See also SCATTERING THEORY.

BIBLIOGRAPHY

D. Bohm, *Quantum Theory*, pp. 557–581. Prentice-Hall, Englewood Cliffs, NJ, 1951. (E)

D. D. Brayshaw, *Phys. Rev. Lett.* **37**, 1329–1333 (1976). (Treats the definition of resonance.) (A)

R. H. Dalitz and R. G. Moorhouse, *Proc. Roy. Soc.* **A318**, 270–298 (1970). (A)

L. Fonda, *Fortschr. Phys.* **20**, 135–144 (1972). (A)

K. Gottfried, *Quantum Mechanics*, Vol. I, pp. 117–160. Benjamin, New York, 1966. (E)

P. Roman, *Advanced Quantum Mechanics*, pp. 143–200. Addison-Wesley, Reading, Mass., 1965. (E)

T. G. Trippe, *Rev. Mod. Phys., Suppl.* **48**, S1–S245 (1976). (Includes several Argand plots.) (A)

Partons

James D. Bjorken

Parton is a word invented by Richard P. Feynman to describe conjectured pointlike constituents within hadrons. Suppose it is possible to describe the strong interactions of hadrons by equations of motion stemming from the existence of a local conserved energy-momentum tensor $\theta_{\mu\nu}$; suppose also that $\theta_{\mu\nu}$ can be decomposed into two parts, one of which, $\theta_{\mu\nu}^{(0)}$, is that of a free-field theory and the other of which contains interaction terms. Then partons may be considered quanta of $\theta_{\mu\nu}^{(0)}$.

The parton concept is difficult to develop in a precise way. While nonrelativistic systems such as molecules, atoms, or nuclei have been successfully described in terms of constituents, this is harder to do for hadrons, because any such description must treat the partons relativistically. Motions of such constituents are difficult to follow and to describe simply when they themselves are moving at speeds close to the velocity of light.

The parton concept only flourished when it was fully realized that this difficulty could be largely (but not fully) circumvented by describing the internal constituents of a hadron, not in its rest frame, but in a frame of reference in which it is in extreme relativistic motion, i.e., its kinetic energy is very large compared to its rest energy. For most applications to high-energy collision processes, this is in fact an appropriate reference frame. In such an "infinite-momentum" frame, the internal motions of the partons and the rate at which they interact with each other are slowed down (frozen) because of the relativistic time-dilatation effect. Hence during a high-energy collision process the partons may be regarded as free quanta, and a description remarkably analogous to familiar nonrelativistic descriptions becomes possible.

Feynman originally developed the parton concept to describe the properties of high-energy collisions of hadrons. This description has been a very fruitful stimulus for the evolution of the present description of such collisions; however, it has remained a qualitative or at best a semiquantitative tool. For lepton–hadron collisions, the parton concept has proved a somewhat more quantitative tool. Here the incident lepton (electron, muon, or neutrino) may be regarded (like a parton) as pointlike and structureless. Furthermore, for the case of electron or muon interactions, the interaction between the lepton and the parton is the known electromagnetic force as given essentially by Maxwell's equations.

Thus an analogy may be drawn between an electron–hadron collision and the classical α-particle–atom collision studied by Rutherford. For Rutherford the α particle was the pointlike probe and the atom was the structured object. The description for, say, an electron–proton collision turns out to be very similar, as do the experimental results. A relatively large number of electrons are scattered into large angles, in a way compatible with the idea of a proton containing pointlike objects. This is to be compared with Rutherford's finding a large number of α particles scattered into large angles, indicating a "pointlike" nucleus within the atom.

Consequently we find that it is possible to describe the inelastic electron–proton collisions as follows: In an "infinite-momentum" frame of reference (in which the proton motion is extreme-relativistic) the proton is considered as a collection of pointlike noninteracting partons with a distribution of momenta to be experimentally determined. Experiment determines that:

1. The mean momentum of a charged parton is about 25% of that of the proton.
2. About 50% of the proton momentum is carried by charged partons.
3. The partons have no "size" or extension that is larger than 5–10% of the proton radius.
4. The spin angular momentum of the charged partons is most likely $\frac{1}{2}$.
5. The charge of the partons is fractional.

This last conclusion only follows when electron–nucleon scattering data are compared with neutrino–nucleon data. It is now generally accepted that charged partons within hadrons do exist and have the properties of the fractionally charged quarks proposed by Gell-Mann and Zweig.

This quark–parton concept provides such a simple and quantitative description of lepton–nucleon interactions that nowadays it is widely used as a tool in probing other less understood lepton–quark interactions, such as the neutral-current neutrino–nucleon collisions. It may also be used in describing electron–positron annihilation into hadrons; the process is viewed as annihilation first into a pair of partons, with subsequent evolution into hadrons. It has also been useful in the description of production of massive lepton pairs in hadron–hadron collisions. There the process is regarded as annihilation of a parton in one incident hadron with an antiparton in the other. And in high-energy hadron collider experiments clear evidence for strong parton–parton interactions also exist.

A basic distinction between the parton and other pointlike constituents of the past is that the parton cannot be fully isolated, but is found only within hadrons. This is related to its being a quantum associated with the "free" energy-momentum tensor $\theta^{(0)}_{\mu\nu}$, not the full energy-momentum tensor $\theta_{\mu\nu}$. Only in limited circumstances—namely, that the process is controlled by the dynamics at short distances—can the difference be ignored. If only because of the apparent fractional parton charge as compared with the integer charge of known quanta of $\theta_{\mu\nu}$, it is unreasonable to expect the difference between $\theta_{\mu\nu}$ and $\theta^{(0)}_{\mu\nu}$ to be negligible for isolated single partons. This leads to a major problem in understanding the relationship between the partons used in describing phenomena at short distances and the hadrons that evolve from such partons (at a much larger distance scale). There is a considerable descriptive understanding of this relationship, much of which is based on the generally accepted theory of the strong interactions, quantum chromodynamics. But much remains to be done in sharpening the concepts and giving them quantitative predictive power.

The generally accepted quantum field theory incorporating the parton concept is quantum chromodynamics. It has by now passed many crucial tests. The parton model has been incorporated into the fabric of that theory to the extent that some will argue that it is implied by it. However, it is this author's opinion that there is still much to be understood in the relationship of the parton model to quantum chromodynamics.

Another problem with the parton concept is to find a precise mathematical framework for its description. Although the "infinite-momentum" description is of advantage in admitting the parton concept, along with allowing a simple description of high-energy collisions, there are disadvantages. For example, there are symmetries that become obscured by the mathematical formalism. An elementary particle at rest is essentially round; when in relativistic motion, it is pancake-shaped. In the latter description, the symmetry that ensures the roundness of the particle in its rest frame is obscure, and at best complicated. As yet, the mathematical difficulties associated with this feature have not been fully overcome. The most promising candidate for a precisely formulated quantum field theory incorporating the parton concept is known as quantum chromodynamics; however, it too is not yet in a state of fully satisfactory development. Thus, although the parton concept has been a powerful intuitive tool and a useful phenomenological aid, it still must be considered as a provisional and speculative description of hadron structure.

See also HADRONS; QUARKS.

BIBLIOGRAPHY

R. P. Feynman, *Photon–Hadron Interactions*. Benjamin, Advanced Book Program, Reading, Mass. 1972.

Phase Transitions

Melville S. Green and J. M. H. Levelt Sengers

EXAMPLES: CLASSIFICATION

Perhaps the most far-reaching principles of physics are its conservation laws, which tell us that the sum of certain quantities—mass, energy, momentum, numbers of atoms of various species—remains constant in the transformations and vicissitudes of matter. Within the limitations imposed by the conservation laws, however, matter may undergo profound changes. Among the most striking of these changes are those called phase transitions. We need not go far from ordinary experience to encounter phase transitions. The precipitation of the water vapor of the atmosphere as rain and the subsequent evaporation of the precipitated water are examples of phase transitions. The characteristic feature is the existence of a substance in two very different forms—in this instance, a palpable liquid form and an impalpable vapor form—and the transformation of one into the other under suitable circumstances. Two other meteorological examples—the precipitation of water vapor in cold weather as snow—and its subsequent evaporation or sublimation, are instances of a solid-to-vapor transformation, whereas the freezing and melting of natural bodies of water is an example of a liquid-to-solid phase transition. An example from mineralogy is carbon, which may exist in the familiar black amorphous form or in the beautiful crystalline form of diamond. From our human point of view amorphous carbon and diamond are materials of very different economic and esthetic values, but from the dispassionate point of view of physics, they are forms of the same substance. The transformation of amorphous carbon into diamond occurs at temperatures and pressures that until recently could only be produced by geological forces, but the necessary conditions are now produced in the laboratory and in industrial processes.

In all of the examples we have just given, the phase transformation takes place portion by portion. Part of the water vapor in a cooling cloud condenses to raindrops. As the temperature drops below 0°C, ice starts forming on a lake, the sheet growing in thickness as the winter progresses. Diamonds are found in a matrix of amorphous carbon, suggesting that the high pressure in the earth transformed only a portion of the carbon. If two dissimilar phases can coexist, and transform into one another gradually by the change of a "field" variable such as pressure, temperature, chemical potential, a magnetic or electric field, we call the phase transition *first order*, because first-order derivatives of an appropriate thermodynamic potential, such as number, entropy, and energy densities, change discontinuously at such a transition.

There is another class of phase transitions, in which the two phases are only infinitesimally different at the transition, and the transformation takes place all at once. They are called *second order*; this name was chosen because it was (erroneously) believed that, although first-order derivatives are continuous, second-order derivatives, such as heat capacities, would jump. If, for instance, a strong container is filled about one-third with liquid, sealed, and heated, one finds that the properties of the coexisting liquid and vapor approach each other first gradually, then precipitously, so that the difference between the two phases disappears at the critical point, above which there is only one phase. At the critical point, the limiting values of the number, energy, and entropy densities of the two subcritical and the one supercritical phase are all the same. The constant-volume specific heat, however, does not experience a jump discontinuity, as suggested by equations of state of the van der Waals type, but was found experimentally to show a weak divergence, slightly stronger than logarithmic (Fig. 1). The spontaneous magnetization of a permanent magnet will, likewise, diminish as the temperature is raised, and become equal to zero at what is called the Curie temperature. Above this temperature, there is no spontaneous magnetization, and the material can acquire a magnetic moment only by being subjected to a magnetic field. Again the specific heat has been found to have a weak divergence at the Curie point, contrary to the jump discontinuity predicted by the Weiss or mean-field theory. Other examples of second-order phase transitions are superfluidity and superconductivity. Above a well-defined temperature, the lambda temperature, the flow properties of liquid helium-4 are qualitatively no different from those of any other liquid. Immediately below this temperature, however, liquid helium acquires frictionless superflow properties and its difference in behavior compared to normal helium increases sharply as the temperature drops. The specific heat has a near-logarithmic singularity at the superfluid phase transition. Superconducting materials are capable of carrying resistanceless current once they are cooled below a certain critical temperature T_c. They *do* show a jump discontinuity in the heat capacity.

Although first- and second-order phase transitions appear to be very different in character, they are, in fact, aspects of the same phenomenon viewed from different perspectives, as will be explained below.

PHASE RULE AND PHASE DIAGRAMS

The thermodynamics of phase transitions, as formulated by J. Willard Gibbs, reduces a bewildering variety of physical phenomena to a few deceptively simple concepts. One of these concepts is Gibbs' phase rule, which states that if a system of r different components exists in p phases, then p can be at most equal to $r+2$. In the case that p equals $r+2$, a change in any of the field variables will make at least one phase disappear. A state point for which $p-(r+2) = 0$ is

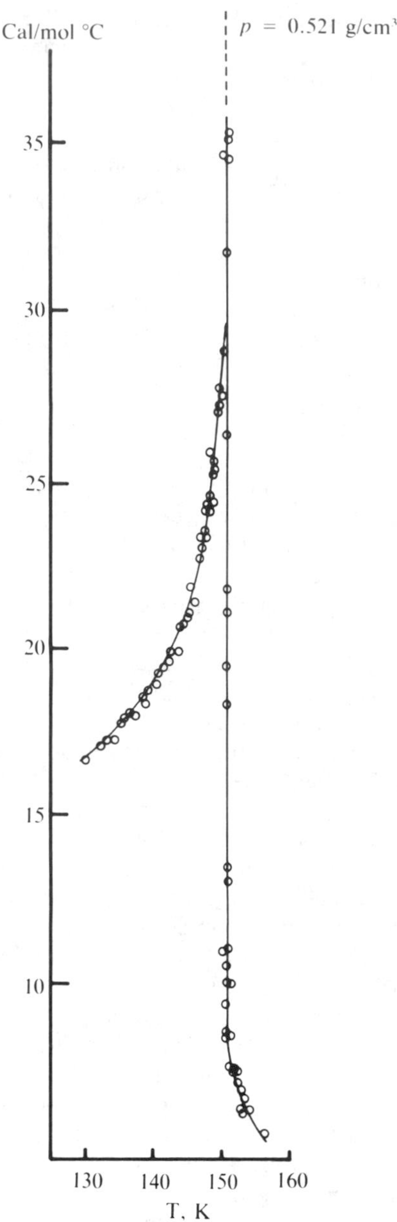

FIG. 1. Variation of C_v of argon with temperature at the critical density [after M. Bagatskii, A. Voronel, and V. Gusak, *Sov. Phys.-JETP* **25,** 72 (1963)]. An example of a near-logarithmic divergence of the specific heat at a second-order phase transition.

nonvariant. In a nonvariant system, one can still vary the relative amounts of each of the phases, but one cannot change the field variables without losing at least one phase. If $(r+2)-p=f>0$, then f, called the number of degrees of freedom, indicates how many field variables can be independently varied (within limits), without losing a phase. A system with $f=1$ is called univariant, and so on.

A phase diagram portrays the regions in the space of independent thermodynamic variables where the various phases exist, and the boundaries between these regions. Figure 2 shows a schematic phase diagram of water, water vapor, and several of the ices in *PT* space. Since water is a one-component fluid ($r=1$), it can coexist in at most three different phases in nonvariant points called triple points, indicated by triangles in Fig. 2. The most well-known is the temperature-scale fixed point at 0.01°C, representing equilibrium of ice I, liquid water, and water vapor, but there are many others, indicating equilibrium between two different ices and a fluid phase. Two phases can coexist along a so-called phase-equilibrium curve which is univariant. Many such curves are indicated in Fig. 2. They include the sublimation curve, the melting curves of the various ices, the equilibrium curves between pairs of ice phases, and the vapor-pressure curve. These univariant curves, separating different bivariant regions of single-phase behavior, are themselves loci of first-order phase transitions. When the system is made to cross such a curve, its number density and other physical properties undergo a discontinuous change. Thermodynamics relates the jumps in entropy, ΔS, in enthalpy, L (the latent heat), and in volume, ΔV, to the slope dP/dT of the phase-equilibrium curve by means of the Clausius–Clapeyron equation

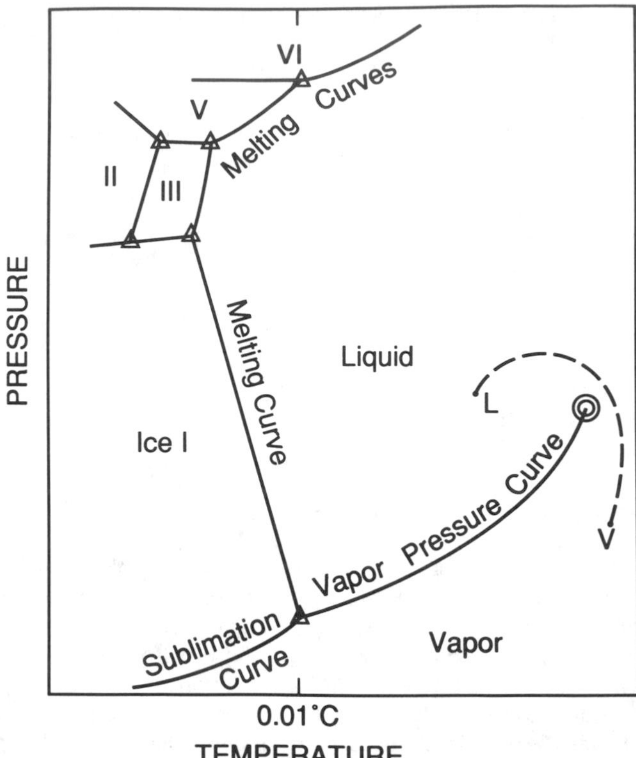

FIG. 2. Schematic phase diagram for water and ice, not to scale. The lines indicate two-phase coexistence, where any quantity of one phase may be in equilibrium with any quantity of the other. Triple points, where three phases coexist, are indicated by triangles. The circled point is the critical point where liquid and vapor become identical.

$$\frac{dP}{dT} = \frac{\Delta S}{\Delta V} = \frac{L}{T\Delta V}. \tag{1}$$

The transitions from solid to liquid, from solid to vapor, and from liquid to vapor take place from a less to a more disordered phase. Therefore, ΔS and L are both positive, and the sign of the associated volume change can be deduced from Eq. (1) and the slope of the appropriate phase boundary curve in Fig. 2. Water is one of the very few examples for which the melting curve has a negative slope, so that the specific volume of solid ice I is larger than that of the coexisting liquid, and ice floats on water.

In the phase diagram, Fig. 2, we find another nonvariant point, namely, the critical point at the end of the vapor–liquid coexistence curve. At this point, the coexisting liquid and vapor phases become identical to each other, thus eliminating the one degree of freedom on the vapor-pressure curve. The evanescence of the distinction between liquid and vapor is the cause of the very striking phenomenon of *critical opalescence*. The natural thermal fluctuations produce temporary liquid-like or vapor-like regions throughout the fluid that are large enough to scatter light and cause the otherwise transparent fluid to become cloudy or opalescent.

The connection between first- and second-order phase transitions is clear from Fig. 2. On any path that intersects the phase boundary curve, except at the critical point, the properties of the system change discontinuously, and the transition is first order. A second-order transition, in which the two coexisting phases become equal to each other and to the supercritical phase, is experienced if a path that coincides with the phase equilibrium curve (the vapor-pressure curve in Fig. 2) is followed to the critical point. In a fluid this path is one in which the overall density of the two-phase system equals the critical density. In many instances, such as the superfluid and the superconductor, the path intersecting the phase boundary is not physically accessible, and the second-order phase transition may be the only one seen in the laboratory.

Another very important insight can be gleaned from Fig. 2: although second-order phase transitions occur in conjunction with first-order ones, not all first-order transitions have associated second-order transitons. Only the vapor–liquid (or fluid–fluid) phase transition may end in a critical point; none of the fluid–solid phase boundaries in any system known today has ever shown a critical point. A solid and a fluid have an essential difference in symmetry that cannot gradually diminish and disappear, as required for a critical point.

Finally, from Fig. 2 it can also be seen that a liquid state can be transformed into a vapor gradually, on a path such as L–V, without a phase transition occurring on the way. Van der Waals called this the *continuity of states*. A fluid below its critical temperature can be liquefied by compression; for a fluid above its critical temperature this is not possible. Thus, fluids with low critical temperatures are hard to liquefy. Helium-4, with a critical temperature of 5.2 K, was liquefied by Kamerlingh Onnes in 1908; the availability of liquid helium as a coolant then led to the discoveries of superconductivity and of superfluidity. A superfluid phase transition has recently been found in liquid helium-3, at a few mK above the absolute zero of temperature.

MODELS OF PHASE TRANSITIONS; ORDER PARAMETER

Ferromagnet

The so-called Lenz–Ising model for the phase transition of the uniaxial ferromagnet has proven to be a conceptually simple but mathematically very challenging tool for understanding second-order phase transitions and for relating phase behavior in physically very different systems. Magnetic spins, capable only of pointing upward (+) or downward (−) are placed on a lattice of dimensionality d. For $d=2$, the model is shown in Fig. 3. The spins are assumed to interact by nearest-neighbor forces only, favoring parallel alignment. Below the critical temperature, the spins will tend to line up and one of two ordered states results, with a net spontaneous magnetization upwards (O1) or downwards (O2). The two low-temperature states shown are almost perfectly ordered: only one spin is out of order in the 16-spin domain. The spontaneous magnetization equals the excess number of spins pointing in one direction. The ratio of the spontaneous magnetization to that in the perfectly ordered state is called the *order parameter*. In a first-order phase transition, state O1 is transformed into state O2 by flipping the spins by means of a magnetic field. In a second-order phase transition, ordered states O1 and O2 are transformed into the disordered state D by increasing the temperature, which increases the thermal motion and induces individual spins to flip. The order parameter diminishes, disappearing at the Curie point, at which point the distinction between the phases O1, O2, and D disappears. Above the critical temperature, the tendency for nearest-neighbor spins to align in parallel is not completely lost, but the long-range order or spontaneous magnetization no longer exists; on the average, as many spins point up as down (Fig. 3, D).

The significance of the Lenz–Ising model is twofold. First of all, it has been solved exactly (by L. Onsager) in two dimensions and in zero magnetic field, while very good approximations exist for the three-dimensional model and for nonzero field. The exact solution displays the logarithmic specific-heat divergence not present in mean-field models. Second, it can be simply transformed to model phase transitions other than that of the uniaxial ferromagnet. For these reasons, it has played a central role in the understanding of second-order phase transitions.

Vapor–Liquid

The Lenz–Ising model is made into a crude model for the vapor–liquid phase transition by means of the transformation shown in Fig. 4. Each lattice site of Fig. 3 is replaced by a cell. A molecule is placed in cells containing upward-

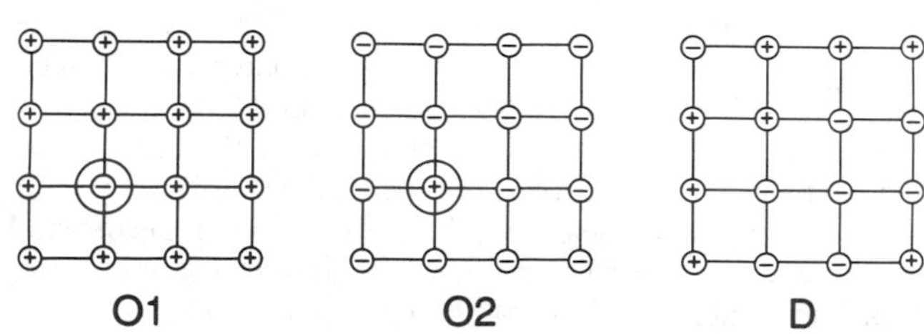

FIG. 3. The two subcritical (partially) ordered states, O1 and O2, and the supercritical disordered state of the Ising model for the ferromagnet. Neighboring spins favor parallel alignment. (+) upward-pointing magnetic spin; (−) downward-pointing magnetic spin. The circle indicates a site where a spin is out of order.

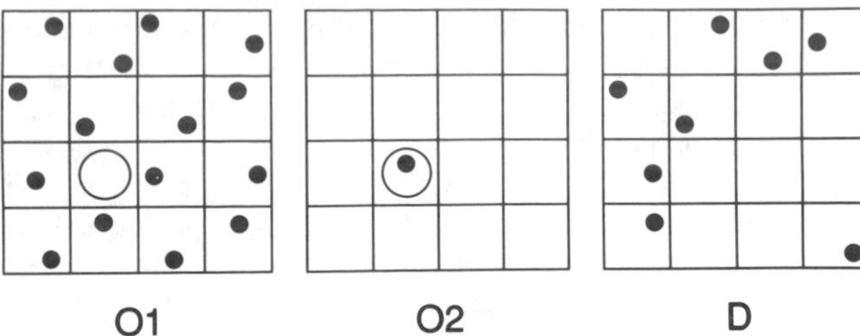

FIG. 4. Transcription of the Ising model to the lattice gas. Spins pointing up are replaced by filled cells, spins pointing down by empty cells. States O1 and O2 represent coexisting vapor and liquid. State D represents the supercritical homogeneous fluid at the critical density.

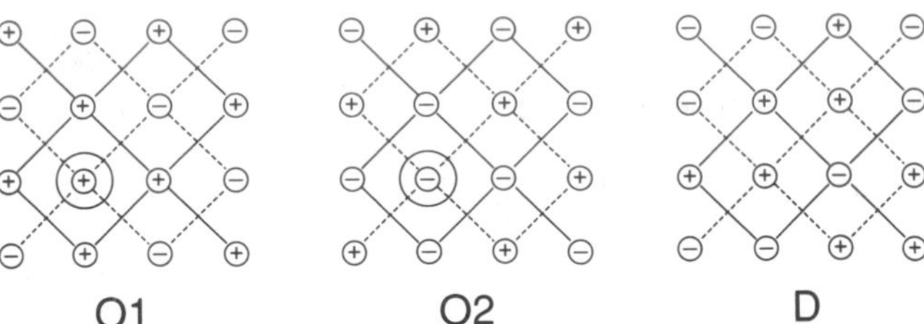

FIG. 5. Transcription of the Ising model to a model for an antiferromagnet, where spins favor opposite alignment. In the ordered states, two sublattices, α (full lines) and β (dashed lines), can be distinguished, on each of which spins are aligned preponderantly parallel. In the disordered state, spins are randomly oriented on each of the sublattices.

pointing spins, whereas cells containing downward-pointing spins are left empty. Below the critical point, two ordered phases, one (O1) with almost all filled cells, a liquid, and one (O2) with almost all empty cells, a vapor, coexist. In this case, the order parameter is defined as the difference between the density and the critical density, divided by the critical density. A slight change in the chemical potential can drive the system, through a first-order phase transition, from one state to the other. As the temperature is increased, and the coexisting phases approach the critical point, the liquid density decreases and the vapor density increases. Infinitesimally close to and above the critical point, half the cells are filled and half are empty.

Liquid–Liquid

If, in addition to the molecules put in the cells in Fig. 4, a different type of molecule is put into the empty cells, a crude model of a liquid–liquid phase separation is obtained, in which two liquids of different composition show partial miscibility below, and full miscibility above the critical point. The order parameter is the difference in composition from the critical composition.

Antiferromagnet

More subtle types of phase transitions, where no physical separation of the phases is apparent, can be obtained as indicated in Fig. 5. In the ordered phase of an antiferromagnet, the magnetic spins align antiparallel to their nearest neighbors, resulting in an orderly alternation of up and down spins. Again, the order is not yet perfect: one spin points in the wrong direction. Two fictitious sublattices are drawn in Fig. 5. In O1, the upward pointing spins have been placed on the α lattice and the downward ones on the β lattice. The other ordered phase, O2, has the opposite arrangement of the spins on the sublattices. The ordered phases have no net magnetization, and the order parameter in this case is defined not in terms of the total magnetization, but in terms of the sublattice magnetization. As the temperature increases, the sublattice magnetizations decrease, due to increased flipping of individual spins, until the preferential arrangement ceases. Above T_c, the disordered state, D, has equal numbers of up and down spins on each sublattice. X-ray diffraction demonstrates the reality of the phase transition: the disordered state has only half the lattice spacing of the ordered states, and therefore the disordered state has a higher translational symmetry. A near-logarithmic singularity appears in the specific heat.

Binary Alloy

By replacing the up spins by one type of atom, and the down spins by another, a model is obtained of the *order–disorder* phase transition in a binary alloy. Figure 6 illustrates the segregation of the two types of atoms on different sublattices below T_c, O1 and O2, and the equal distribution of both types on each sublattice above T_c, D. The order parameter is defined as the excess fraction of atoms of one type on each sublattice. In antiferromagnets and binary alloys, no physical field can transform state O1 to state O2; only the second-order transition can be observed in the laboratory.

In the Lenz–Ising model for uniaxial ferromagnets and its variants, the order parameter is a scalar. In the case of an *isotropic* ferromagnet or antiferromagnet, the order parameter is a three-dimensional vector quantity. A *planar* ferromagnet has a two-dimensional spin or order parameter, and so does the superfluid. In the case of superfluidity and superconductivity, the order parameter is a complex quantity with a magnitude and a phase angle, which may be in-

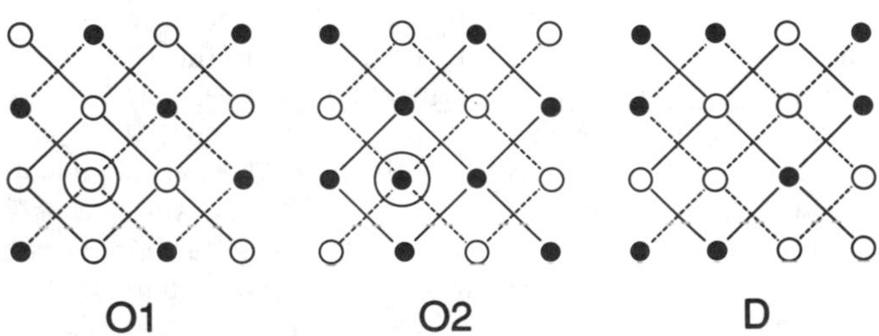

FIG. 6. Transcription of the Ising model for the antiferromagnet to a model for the order–disorder transition in a binary alloy. Upward-pointing spins have been replaced by atoms of one kind, downward ones by atoms of another kind. The atoms are segregated on two sublattices below the critical temperature, but are randomly distributed above the critical point.

terpreted as the wave function of the finite fraction of the particles that have zero momentum. The superflow properties may be considered a manifestation of the fact that a wave function with zero momentum cannot be localized.

The order parameter also determines the *universality class* of the transition. Systems that have the same space dimensions and the same tensorial character of the order parameter are said to belong to the same universality class, i.e., they have the same critical singularity in the thermodynamic potential. Ordinary critical points in fluids and fluid mixtures belong, with binary alloys, uniaxial ferromagnets, and antiferromagnets, in the Ising universality class. Phenomenological scaling laws, followed by the renormalization-group theory, characterize this universal behavior accurately. Experiment is in full agreement with the theoretical results.

PHASE TRANSITIONS INVOLVING FLUID PHASES

Fluid Mixtures

From the example of water, we have already noted that even a one-component system can have considerable complexity in its phase behavior if more than one solid phase is formed. As soon as a second component is added, however, the degree of complexity is increased in a major way, even if the solid phases are not considered. The additional degree of freedom transforms triple points and critical points into triple lines and critical lines. In addition to liquid–vapor, one may have liquid–liquid and gas–gas phase separation. It is possible for *three* fluid phases to coexist along a three-phase line, which may end in one or two critical end points, where two out of the three phases become identical to each other. Nonvariant points are quadruple points, where a total of four fluid and solid phases coexist. Adding a third component gives rise to a new type of critical point, namely, one where three phases become identical simultaneously. Such a point is called a *tricritical point,* and the critical anomalies differ from those near an ordinary critical point.

Complete classifications of the topology of phase behavior in binary mixtures do not exist for systems including solid phases. For three-component systems, classifying even the fluid-phase equilibria appears to be beyond present-day reach.

The superfluid transition of liquid helium-4 is shifted to lower temperatures by the admixture of helium-3. Finally, another type of tricritical point is reached, below which the mixture separates into two liquid phases, the one richer in helium-4 being superfluid. Thus, a second-order line goes over into a first-order one at this tricritical point.

Liquid Crystals

Fluid phases are characterized by the absence of positional order of the molecules. There are, however, other types of ordering that can occur. When the molecules are very asymmetric, with a long apolar tail and a polar head, they can order directionally but not positionally in various ways. The directional order is temperature dependent, and both first- and second-order phase transitions can occur. Such fluids are called *liquid crystals*; they are widely used in display screens for watches and portable TVs.

Adsorbed Layers

Fluid–solid phase transitions can also occur in two-dimensional systems. One example is that of an adsorbed layer on a substrate, where both ordered and disordered phases can be formed, and both first- and second-order phase transitions have been found. Another example is that of a surfactant layer on water. Depending on the surface coverage, the surfactant can exist in a gas-like, a liquid-like, and several ordered phases, in which the surfactant molecules raise their apolar tails out of the water and line them up. Only recently have several of these transitions been proven to be first order.

Surfactants

Systems containing surfactants, water, and oily components have many other possibilities of forming partially ordered phases; the surfactant can organize itself into aggregates of various shapes, called micelles, with a first-order phase transition between a dense and a dilute micellar phase as one of the possibilities. The surfactant can also collect at the interface between interpenetrating oil and water phases; such a system is called a microemulsion.

Numerous other fascinating physical phenomena have features in common with thermodynamic phase transitions: the formation of glassy states; the complex foldings that determine the biological functions of macromolecules; the transformation from a laminar to a turbulent flow regime. The study of these phenomena makes heavy use of the methods and concepts developed in the study of the simpler phase transitions discussed here.

See also CRITICAL POINTS; FERROMAGNETISM; HELIUM, LIQUID; LIQUID CRYSTALS; ORDER–DISORDER PHENOMENA; RENORMALIZATION; SUPERCONDUCTIVITY THEORY; THERMODYNAMICS, EQUILIBRIUM.

BIBLIOGRAPHY

C. Domb and M. S. Green (eds.), *Phase Transitions and Critical Phenomena*. Vols. 1–6; C. Domb and J. L. Lebowitz (eds.), Vols. 7–12. Academic Press, London, New York, 1972–1988.

J. W. Gibbs, "Equilibrium of Heterogeneous Substances," in *Collected Works*. Longmans, Green, New York, 1931.

M. S. Green and J. V. Sengers (eds.), *Critical Phenomena, Proceedings of a Conference*. NBS Miscellaneous Publication 273. U.S. GPO, Washington, DC, 1965.

D. L. Goodwin, *States of Matter*. Prentice-Hall, Englewood Cliffs, NJ, 1975.

M. Levy, J.-C. Le Guillou, and J. Zinn-Justin (eds.), *Phase Transitions*. Plenum Press, New York and London, 1980.

H. E. Stanley, *Introduction to Phase Transitions and Critical Phenomena*. Oxford, London and New York, 1971.

H. N. V. Temperley, *Changes of State*. Cleaver-Hune, London, 1956.

Philosophy of Physics

Peter Caws

PHYSICS AND PHILOSOPHY

The Greek word *phusis,* from which "physics" is derived, is generally translated as "nature," although its original meaning is perhaps closer to "growth"; it comes from a root verb *phuo,* meaning "engender." "Growth" nowadays would be construed as a term of biology rather than physics, but for the Greeks it had a much wider meaning, including all processes of change in the natural world. Change, or "coming to be," always involves motion, and the agents of change have to expend energy or perform work; all these are Greek notions. The *Physics* of Aristotle is a treatise about causality, motion, space, and time, all of them still problematic categories in what we would now call the philosophy of physics.

Philosophy itself began with speculation about the physical world (Aristotle calls the Milesian nature-philosophers the *phusikoi*). But its name hints at a question about the certainty of our knowledge of the world. The *Sophoi* were the wise men of antiquity, but it is said of Pythagoras that when somebody called him *sophos* he declined the compliment, saying that he preferred the title *philosophos,* one who cares about wisdom. Philosophers ever since have taken it as their task to challenge claims to knowledge or wisdom; if a scientist (etymologically "one who knows") makes a pronouncement about nature, for example, the philosopher's job is to ask what is really being claimed, how it is known, whether it is true or only probable, and so on. Philosophy proper follows these questions with others: what knowledge is and how it is possible, what sort of thing truth and probability are. The philosophy of physics, however, limits itself to raising questions about the propositions of physics: their status, their meaning, their confirmation, their logical organization, and their consequences for our knowledge of the world.

Physics and philosophy have maintained close links for nearly all of their joint history, with the exception of the thousand years of the Dark and Early Middle Ages. In early modern times Galileo, who can reasonably be thought of as the first major classical physicist, played a prominent part in the philosophical revolt against Aristotelianism, a scholastic movement that in spite of its name did little credit to Aristotle. Descartes and Leibniz engaged in speculation about physical and mathematical problems (Leibniz rivaled Newton as author of the differential calculus); Kant made a serious contribution to cosmology (the Kant–Laplace hypothesis). In our own century, relativity and quantum theory have posed deep philosophical conundrums, and it is safe to say that no physicist who has made really fundamental contributions to theory has been able to remain totally indifferent to the philosophical implications of his work. Einstein and Schrödinger are particularly good examples.

PHILOSOPHY OF SCIENCE AND PHILOSOPHY OF PHYSICS

Many philosophical problems that arise in connection with physics are relevant also to chemistry, biology, psychology, and so on. Some of the problems that fall into this category are, for example, the status of scientific theory—whether it is a description of an objective world, or a summary of empirical observations, or a conventional body of propositions in which scientists have a vested interest, or merely a device for prediction; the origin of scientific hypotheses—whether they are painstakingly drawn from experience by induction, or miraculously hit upon by genius, or refined from a natural human capacity for invention; the nature of causal determination; the structure of explanation and its relation to prediction; the forms of inference and the probability of its conclusions. Some of these questions would have been philosophical questions whether science as we know it had existed or not and have long histories in their own right (many of the major problems of the philosophy of science as inquiry and as reasoning are anticipated in Aristotle's *Posterior Analytics*).

The philosophy of science also has some specific things to say about the relation of physics to the other sciences, however, and about physics as such; the latter constitute philosophy of physics proper, although insofar as physics behaves like a typical science a great deal of the philosophy of science in the broader sense is *also* philosophy of physics. Physics is the most basic of the empirical sciences and thus has an asymmetrical relation to the others, because every object in the universe has physical properties and is acted upon by physical forces, whereas some objects—subatomic particles, for example—have no chemical properties and many undergo no chemical changes, whereas most objects in the universe have no biological, sociological, etc. properties. To put the same point in another way (and assuming no supernatural explanations), every event describable in chemical, biological, etc. terms is also in principle describable in physical terms, but the reverse is not the case. The description might of course be forbiddingly complex, so that for practical purposes the other sciences are necessary; but the philosophical thesis known as *physicalism* maintains that the basic assumptions of physics are sufficient in principle for the derivation of all the laws of all the sciences. This viewpoint reflects the conviction that everything in the universe has in fact evolved out of material or energy once having had only physical properties, and that this evolution has taken place in accordance with physical law.

We have to distinguish between the *ontological* priority of physical objects (their fundamental status among existing things) and an assumed *epistemological* priority (their being the basis of the rest of our knowledge). Strictly speaking, of course, we do not know physical things, only their appearances. The attempt to provide a purely phenomenological foundation for science, however, has been unsuccessful, and most philosophical accounts accept an initial hypothesis of matter in some form or other.

CLASSICAL AND MODERN PHYSICS

The rise of classical physics in the seventeenth century and its development in the eighteenth and nineteenth were as dramatically successful as they were because they were governed by a very simple set of philosophical assumptions and had very modest philosophical ambitions. The physical world was taken to be a large machine whose material parts interacted causally by exerting forces on one another; it occupied Euclidean space and endured through ''equably flowing'' time, to use an expression of Newton, whose ''Rules of Reasoning in Philosophy'' (by which he meant ''science'') sum up neatly the methodological presuppositions of his epoch; economy of explanation, assignment as far as possible of like causes to like effects, extrapolation of local results to the universe as a whole, and reliance on empirical evidence. Physical quantities were taken to be continuously variable and consequently to permit of infinitestimal changes. The modest tone was set by Galileo, who wished only to discover a geometrical representation that would actually fit some given motion, for example, the descent of a ball along an inclined plane. Newton disclaimed all hypotheses as to what gravitation really was; it was enough for him to describe how bodies moved in gravitational fields, and his successful marrriage of Kepler's astronomy and Galileo's mechanics stands as the greatest single intellectual achievement of all time. But the assumptions of classical physics, simple as they were, were so powerful that they came to dominate the whole world outlook of the Age of Enlightenment, much to the discomfort of some philosophers who believed that physical explanation was incompatible with the reality of human value.

The revolution in physics that occurred at the beginning of this century with Planck's quantum hypothesis and Einstein's theory of relativity overthrew the metaphysical assumptions of classical physics while leaving its methodological assumptions more or less intact. As is well known, the revolution occurred because classical physics could not explain certain perplexing phenomena, such as the energy distribution in blackbody radiation and the photoelectric effect, while some of its theoretical formulations led to paradoxes when pushed to their limit (e.g., the notion of a standing electromagnetic wave). These difficulties made it clear that, while there was still no reason to doubt that results obtained locally would hold, under similar conditions, anywhere else in the universe, local events themselves were not amenable to classical explanation for very small displacements, very high velocities, or very large distances, and that basic assumptions such as strict causal determinism, fixed objective values of measurable quantities, and so on could no longer be held uncritically.

It is to be noted that the observations on which these theoretical conclusions were based were not observations of very small displacements or very high velocities but observations of macroscopic magnitude. Even experimental particle physics, which began in the late nineteenth centuty with the investigation of cathode rays, observed only the macroscopic consequences of the microevents it postulated, and this of course is still true. However, these macroscopic events are very carefully contrived to force the rejection of various theoretical alternatives. Scientific progress consists in the *elimination* of hypotheses as much as in their devising. Modern physics tells us that the world does not behave classically on the submicroscopic or cosmic scales, although it does behave classically in the middle region that we inhibit and in which we have learned our descriptive language. The result is that we have no way of describing, in familiar qualitative terms, the events whose quantitative features form the subject matter of quantum physics, cosmology, and so on. This used to be a matter of some concern, for example, in connection with wave-particle dualism, but it should not be surprising. Our perceptions and naive thoughts are adapted to the scale of our bodies, our days, and our lives; relativistic and quantum phenomena have no direct bearing on them, and being equipped to envisage such phenomena would have been of no evolutionary advantage. But we could not have come to understand physics at all without the complex material equipment of the human organism, so that Schrödinger could answer the question ''Why is the atom so small?'' by pointing out that we are necessarily large. At the same time modern physics must not be incompatible with classical physics when the latter is restricted to its tested domain of applicability; this is the force of Bohr's correspondence principle.

PHILOSOPHICAL TASKS AND IMPLICATIONS

Philosophy can contribute nothing to the data of experimental physics, but it may make serious contributions to the organization of theoretical physics, to the interpretation of the relations between experiment and theory, and to the understanding of the enterprise as a whole. With respect to the first of these points, work in axiomatization and in what has been called ''foundations research'' aims to reduce the number of fundamental propositions of physics to a minimum and to clarify the meanings of its basic terms. With respect to the second, a long-standing debate, triggered by the crisis in turn-of-the-century physics that led to the emergence of quantum theory and relativity, pitted ''operationism''—a term due to P. W. Bridgman, signifying the positivist principle of refusing all physical concepts not constructible in terms of concrete operations, and therefore making no commitment to the objective reality referred to by those concepts—against an almost Platonic view of the ultimate reality and rational harmony of the physical world (consider Einstein's faith: ''God does not play dice with the universe,'' ''God may be subtle, but he isn't mean''). At stake is our conception of the objective status of truth, as well as the related question of whether the nature of things may be really indeterminate, so that probability is an ontological and not merely an epistemological category, as suggested by Heisenberg's uncertainty principle according to the interpretation of the Copenhagen school, or whether inquiry will always in principle reveal, under the statistical surface, a determinate microstructure, as Bohm and the Paris group have held. The success of thermodynamics in providing for itself, through statistical mechanics, just such a determinate basis encouraged the hope that the process might repeat itself on a smaller scale of phenomena, although further inquiry seems to lead to an infinitely regressing alternation of deterministic and stochastic levels. Following this line of development has produced theories of catastrophe and of chaos, with applications to very large and complex systems. Yet there really may be an end to the hierarchy of magnitudes; the quark might, for example, be the ultimate particle. That this remark may seem quaint to readers of this article in the not too distant future nevertheless attests to the vigor of physics as an active and changing discipline.

There remain some genuinely metaphysical puzzles when we come to the third point. Physicists are after all human organisms, and some physicists (notably Dicke and Wheeler) find the existence of human knowers far from accidental in the scheme of the physical world. The EPR paradox (due to Einstein, Podolsky, and Rosen) and the collapse of state functions under observation have been interpreted by some theorists (although certainly not by Einstein!) as suggesting that the observer, instead of being an outsider who at worst interferes with what he is observing, may have something to do with its constitution (the *anthropic principle*). As long as, like Mach, we take the physical universe as given, merely busying ourselves with its most economical description, we can avoid philosophical perplexity, but if we confront the question of its extent and origin, for example, unanswered questions abound. If the big bang obeyed physical laws, did those laws antedate the physical universe? If causality breaks down at the quantum level, why are all electrons, for example, alike? If the universe is a "timeless" space manifold, how do we account for our experience of time, for the reality of change?

The development of grand unifying theories (GUTs) promises new light on these questions, though there are conceptual difficulties in the very notion of a system that contains a total theory of itself. What does seem clear, especially in the light of recent work on the implications of the EPR paradox (e.g., Bell's theorem and the experimental results of Aspect and his group) is that local realism and the strong independence of distant events have to be given up at the quantum level. The physics of what I like to call the "flat region" cannot be made to yield a consistent world much more refined than the coarse-grained and approximate one from which it was derived, and an imagination trained there (as all our imaginations are) cannot encompass the "real" world as it must be if current cosmological and quantum conjectures are warranted. Physics can master the formal syntax of that world by means of mathematical conventions but cannot translate its material vocabulary, being driven to the metaphorical use of ordinary terms such as "color," "charm," "strangeness," and so on, in order to speak of it at all.

This does not mean, as has sometimes been suggested, that the world is at bottom "mathematical," only that physics has to be mathematical in order to describe it. This inaccessibility of the physically real to the material imagination has been taken by some writers as a ground for comparing it to other forms of transcendence, especially religious ones, but there is as far as I can see absolutely no justification for this or for the mysticism it encourages, whose claims are diametrically opposed to the modesty of science. The philosophical stance appropriate to physics refrains from offering hypotheses as conclusions. As Bertrand Russell once pointed out, philosophy cares less about the answers to its questions, "since no definite answers can, as a rule, be known to be true," than about the questions themselves; if some fundamental questions of physics prove, as they may, to be unanswerable, the ancient alliance of philosophy and physics will have a long future.

See also HISTORY OF PHYSICS.

BIBLIOGRAPHY

For an excellent review of work in the philosophy of science in general, with bibliographies, see Frederick Suppe (ed.), *The Structure of Scientific Theories,* 2nd ed. Urbana, IL, 1977.

David Bohm, *Causality and Chance in Modern Physics*. Van Nostrand-Reinhold, Princeton, NJ, 1957. (I)

Niels Bohr, *Atomic Physics and Human Knowledge*. Wiley, New York, 1958. (E)

H. Bondi, *Assumption and Myth in Physical Theory*. Cambridge, London and New York, 1967. (I)

Max Born, *Natural Philosophy of Cause and Chance*. Dover, New York, 1964. (A)

P. W. Bridgman, *The Logic of Modern Physics*. Macmillan, New York, 1927. (I)

Mario Bunge, *Foundations of Physics*. Springer-Verlag, Berlin and New York, 1967. (A)

Paul Davies, *Superforce; The Search for a Grand Unified Theory of Nature*. Heinemann, London, 1984. (I)

Sir Arthur Eddington, *The Nature of the Physical World*. Dent, London, 1935. (E)

Albert Einstein, *The World as I See It*. Covici-Friede, New York, 1934. (E)

Werner Heisenberg, *Physics and Philosophy*. Harper, New York, 1958. (E)

Henry Margenau, *The Nature of Physical Reality*. McGraw-Hill, New York, 1950. (A)

Ilya Prigogine, *Order Out of Chaos: Man's New Dialogue With Nature*. Bantam, New York, 1984. (I)

Michael Redhead, *Incompleteness, Nonlocality, and Realism: A Prolegomenon to the Philosophy of Quantum Mechanics*. Clarendon Press, Oxford, 1987. (A)

Erwin Schrödinger, *Science, Theory, and Man*. Dover, New York, 1957. (I)

Phonons

P. G. Klemens

Phonons are quanta of energy of the normal modes of vibration of a crystal lattice or of an elastic continuum. The concept of the phonon is closely analogous to that of the photon in electrodynamics. In each case we consider the quantization of the energy of a single wave. According to classical theory, the possible values of the energy content are continuously variable. According to quantum theory, the energy content of a wave, or of any other harmonic oscillator, of frequency ν can have only a discrete set of values $(N+\frac{1}{2})h\nu$, where N is an integer and h the Planck constant. The zero-point energy $h\nu/2$ is always present and can be regarded as part of the formation energy of the oscillator. The additional energy consists of an integral number N of quanta or phonons. Energy can be added to or removed from the oscillator only in integral numbers of phonons.

The concept of phonons refers to the quantization of the energy content of each oscillator, not to the nature of the oscillator itself. The oscillator can be a progressive displacement wave in a discrete lattice of atoms (lattice wave), or a displacement wave in an elastic continuum (sound wave), or even a localized vibration. It is not necessary to consider a discrete lattice to have the energy of each oscillator quantized. Phonons can exist in an elastic continuum just as photons exist in a continuous medium or in free space.

The quantization of the energy of each vibrational mode or oscillator has important thermodynamic consequences. While the number of phonons in each oscillator must be, at any instant, an integer N, the average number of phonons in an ensemble of identical oscillators, each of frequency ν, in thermal equilibrium at temperature T is given by the Planck distribution

$$N^0 = (e^{h\nu/kT} - 1)^{-1} \quad (1)$$

where k is the Boltzmann constant. The average energy content of each oscillator (without zero-point energy) is $N^0 h\nu$, and each oscillator contributes to the specific heat of the system an amount $h\nu(dN^0/dT)$. Unless $kT \gg h\nu$, the specific heat of the oscillator falls below the classical value of k. Hence, the specific heat of a solid containing G atoms and having the vibrational energy of $3G$ harmonic oscillators is well below the classical value of $3Gk$ at low temperatures. It is given by

$$C(T) = \int d\nu\, S(\nu) h\nu (dN^0/dT) \quad (2)$$

where $S(\nu)\, d\nu$ is the number of normal modes in the frequency interval $(\nu, \nu + d\nu)$. At low frequencies these modes or lattice waves are similar in character to plane elastic waves in a continuum; at higher frequencies the character and spectral distribution of the lattice wave are increasingly influenced by the crystal structure. At lowest frequencies, $S(\nu) \propto \nu^2$ and $C(T) \propto T^3$ (Debye specific heat). It is one of the objectives of lattice dynamics to relate $S(\nu)$, and hence the thermodynamic properties, to the crystal structure and the nature of the interatomic forces.

The concept of phonons is also important because the phonon can be looked upon, for many purposes, as a particle of fixed energy and momentum moving freely within the solid. Let the displacement field due to a lattice wave be of the form

$$\mathbf{u}(\mathbf{r}) = B(\mathbf{q})\mathbf{E} \exp[i(\mathbf{q}\cdot\mathbf{r} - 2\pi\nu t)] \quad (3)$$

where $\mathbf{u}$ is the displacement of a particle at position $\mathbf{r}$ and time t; $\mathbf{q}$ and ν are the wave vector and frequency, respectively; $\mathbf{E}$ is a unit vector denoting polarization direction (or a set of vectors if there is more than one atom per unit cell); and $B(\mathbf{q})$ is the amplitude. Each phonon contained in that wave acts like a particle of energy $h\nu$ and of "momentum" $h\mathbf{q}/2\pi$, moving with a velocity equal to the group velocity of the wave, i.e., $\mathbf{v} = 2\pi\, \partial\nu/\partial\mathbf{q}$, since in wave motion the group velocity governs the energy flow, and since the phonon is a quantum of energy. We can thus describe the transport of vibrational energy within a solid in terms of a flow of phonons and use the concepts of the kinetic theory of gases. The phonon can also be partly localized by quantizing not a single wave (3), but a wave packet.

The thermal conductivity of an insulator, or the lattice component of the thermal conductivity of solids that contain free electrons, is given by

$$\kappa = \tfrac{1}{3} C v l \quad (4)$$

where C is the lattice specific heat, v the average phonon velocity, and l the average phonon mean free path, or wave attenuation length. In a structurally perfect solid with truly harmonic interatomic forces, there would be no energy exchange between the lattice waves, and the phonon mean free path would be infinite. The interchange of energy among lattice waves can be pictured as the creation of some phonons and the destruction of others. The mean free path is limited by lattice defects, which scatter phonons from one wave into another, and by the anharmonic component of the lattice forces, which, to lowest order, leads to three-phonon interactions. In such an interaction process, a phonon of mode $\mathbf{q}$ and one of mode $\mathbf{q}'$ are destroyed and a phonon $\mathbf{q}''$ is created, or vice versa. Energy must be conserved in the process, so that

$$\nu + \nu' = \nu'', \quad (5)$$

while the uniformity of the medium requires that

$$\mathbf{q} + \mathbf{q}' = \mathbf{q}''. \quad (6a)$$

Equation (6a) expresses the conservation of momentum of the interacting phonons, just as momentum is conserved when free particles collide. These momentum-conserving processes (also called normal processes) are the only ones allowed in an elastic continuum. In a discrete lattice, however, additional processes are possible. Let $\mathbf{b}$ be an inverse lattice vector of the crystal, i.e., all properties of the static lattice can be expressed as Fourier series of the form $\Sigma_{\mathbf{b}} F(\mathbf{b}) \exp(i\mathbf{b}\cdot\mathbf{r})$. The additional three-phonon interaction processes satisfy a wave-vector selection rule

$$\mathbf{q} + \mathbf{q}' = \mathbf{q}'' \pm \mathbf{b} \quad (6b)$$

and are called Umklapp processes (because the net momentum of the participating phonons is "flipped over"). Each such process is a three-phonon interaction and a Bragg reflection combined. Thus the phonon momentum is not always conserved and is not a true momentum. Although momentum can be ascribed to a sound wave in a continuum, it is not possible in a crystal to apportion its overall momentum uniquely between the lattice and the lattice waves.

A theory of lattice thermal conductivity can be set up in terms of the phonon mean free path, limited by phonon–phonon interactions and by phonon–defect scattering, and using Eq. (4) or its appropriate spectral generalization.

Phonons can also interact with other excitations in a crystal, particularly electrons. If we treat the electron as a wave of wave vector $\mathbf{k}$, where $h\mathbf{k}/2\pi$ is its momentum, it can be scattered from state $\mathbf{k}$ into $\mathbf{k}'$ with the emission of a phonon $\mathbf{q}$, or vice versa. Again there are normal and Umklapp processes, satisfying

$$\mathbf{k} = \mathbf{k}' + \mathbf{q}(\pm\mathbf{b}). \quad (7)$$

These electron–phonon processes limit the electron mean free path; the latter enters into the description of electronic conduction processes. When the interaction involves mainly normal processes and the sum of the momentum of the electron and phonon gas is conserved, electrons and phonons exert a mutual dragging force on each other that becomes evident in the phonon-drag component of the thermoelectric power. In metals the electron gas interacts directly with all phonons; in semiconductors it can directly interact only with low-frequency phonons.

In principle, all mobile excitations can interact with phonons. Neutrons passing through the solid can be scattered by phonons analogously to (7). Their change in energy and momentum can be measured and attributed to $h\nu$ and $h\mathbf{q}/2\pi$ of the phonon. Inelastic or phonon-assisted neutron scattering is thus a powerful tool to determine the phonon dispersion curves, i.e., the dependence of ν on $\mathbf{q}$; at the same time these processes strikingly confirm the phonon concept.

Spin–lattice relaxation processes involve the emission or absorption of individual phonons at special centers in the solid. These can be studied either by monitoring the magnetic state of impurities or defects, or by measuring their effect on phonon flow (ultrasonic attenuation, thermal resistance).

At very low temperatures in single crystals of nonmetals, phonons can travel in straight lines from one external surface to another. This makes the apparent thermal conductivity depend on size and surface condition. At somewhat higher temperatures in perfect crystals, phonons can interact but conserve their momentum, so that phonon flow in a temperature gradient is analogous to Poiseuille flow in fluids.

Ballistic phonons can be generated by hot spots, heated films, or superconducting junctions. They travel in straight lines to other surfaces, to be detected by fast bolometers or superconductors. This allows studies of anisotropic group velocities, diffuse and specular reflection, and weak attenuation for phonons of frequencies above thermal. Special techniques allow visualization of anisotropic propagation patterns.

See also ACOUSTOELECTRIC EFFECT; BRILLOUIN SCATTERING; LATTICE DYNAMICS; NEUTRON DIFFRACTION AND SCATTERING; NEUTRON SPECTROSCOPY; QUASIPARTICLES.

BIBLIOGRAPHY

A. C. Anderson and J. P. Wolfe (eds.), *Phonon Scattering in Condensed Matter V*. Springer-Verlag, Berlin, 1986.

R. Berman, *Thermal Conduction in Solids*. Oxford, London and New York, 1976. (I)

J. S. Blakemore, *Solid State Physics*. Saunders, Philadelphia, PA, 1969. (I)

C. Kittel, *Introduction to Solid State Physics*, 4th ed. Wiley, New York, 1971. (E)

P. G. Klemens, "Thermal Conductivity and Lattice Vibrational Modes," in *Solid State Physics*, Vol. 7, pp. 1–98. Academic Press, New York, 1958. (A)

A. A. Maradudin, "Theoretical and Experimental Aspects of the Effects of Point Defects and Disorder on the Vibration of Crystals," in *Solid State Physics*, Vol. 18, pp. 274–420; Vol. 19, pp. 1–134. Academic, New York, 1966. (A)

J. A. Reissland, *The Physics of Phonons*. Wiley, New York, 1973. (I)

J. M. Ziman, *Electrons and Phonons*. Oxford, London and New York, 1960. (A)

Photoconductivity

Clarence R. Crowell

Photoconductivity is bulk electrical conductivity induced by visible light or infrared radiation. Photoconductivity has frequently been confused with photo(electric) emission, which is light-induced injection of carriers across a boundary that may be a solid–solid or solid–vacuum interface, and the photovoltaic effect, which involves the conversion of electromagnetic energy to electrical energy (e.g., a solar cell) and where the motion of light-generated minority carriers toward a solid–solid rectifying junction develops an electrochemical potential difference. The adjective photoelectric applies to all three phenomena [1]. A pure photoconductor should have nonrectifying (Ohmic) contacts that are not illuminated.

Photoconductivity was first reported by Smith [2] for selenium and has since been observed in most solids. In semiconductors and insulators, photoconductivity is described as intrinsic or extrinsic if it is excited, respectively, by photons of energy greater than or less than the energy band gap. Appreciable photoconductivity is, however, always associated with defect- or dopant-induced energy levels within the band gap.

Photoconductive gain G is the number of carriers that transit a structure per absorbed photon; G equals the ratio of the system response time τ_0 to the carrier transit time T_r. T_r varies inversely as the applied voltage until, in the absence of shallow traps or sensitizing centers, it equals the dielectric relaxation time τ_d. T_r determines the bandwidth ΔB of the detector response. When shallow traps are added, G is increased, but the maximum of $G\Delta B$ is invariant and a significant figure of merit. If sensitizing centers, a fraction f of which are charged with the carrier species, are added in numbers greater than the shallow traps, then [3]

$$G\Delta B = (2\pi\tau_d f)^{-1}.$$

Sensitization for visible light may be quenched by infrared radiation. Another figure of merit is the detectivity,

$$D^* \equiv \frac{S/N}{P_D}\left(\frac{\Delta B}{A}\right)^{1/2}$$

where S/N is the signal to available noise power ratio for a radiant power density P_D on a detector area A.[4]

See also ELECTRON ENERGY STATES IN SOLIDS AND LIQUIDS; PHOTOELECTRON SPECTROSCOPY; PHOTOVOLTAIC EFFECT.

REFERENCES

1. *IEEE Standard Dictionary of Electrical and Electronic Terms*. Wiley, New York, 1972.
2. W. Smith, *Nature* **7**, 303 (1873).
3. A. Rose, *Concepts in Photoconductivity and Allied Problems*. Wiley-Interscience, New York, 1963.
4. H. Levinstein, *Semicond. Semimetals* **5**, 3 (1970).

Photoelastic Effect

Donald F. Nelson

The *photoelastic effect* is the alteration of the optical propagation constants of a medium caused by mechanical stress or the resultant deformation. It is also called the *elastooptic effect* or the *piezooptic effect*. The *direct* photoelastic effect occurs in any material medium; in a piezoelectric medium an additional distinguishable *indirect* photoelastic effect occurs which consists of a succession of the piezoelectric and electrooptic effects. In general, the photoelastic effect is described by a fourth-rank tensor; in an isotropic medium such as a glass, plastic, or liquid it reduces to only two independent constants. Mathematically the photoelastic effect is characterized by a term in the electric field wave equation proportional to the electric field of the light wave and to a measure of the deformation of the medium. Since the latter may result from an acoustic wave, the photoelastic effect represents the lowest-order interaction of light and sound. Because of this dual dependence and its discovery date of 1815, the photoelastic effect is often called the oldest nonlinear optical effect.

The photoelastic effect is studied by three main techniques: piezobirefringence, Brillouin scattering, and acoustooptic diffraction. In the oldest technique, piezobirefringence, the birefringence induced in the medium by a static, uniform stress is found from the phase changes of the light wave. In Brillouin scattering, a light wave, typically in the form of a laser beam, is scattered in direction and shifted in frequency from the dynamic deformation associated with an acoustic excitation which arises from the thermal content of the medium. Acoustooptic diffraction is similar to Brillouin scattering except that the acoustic excitation is coherently generated by a transducer. In both Brillouin scattering and acoustooptic diffraction the scattered light wave direction satisfies a Bragg condition, or alternatively stated, a phase-matching or wave-vector conservation condition. While piezobirefringence measurements are directly proportional to the relevant photoelastic tensor component, both Brillouin scattering and acoustooptic diffraction are proportional to the square of the tensor component.

For most of this century the photoelastic effect has been used to study stress distributions in transparent objects that simulate machine parts or structural components. In the past two decades it has found use as a laser-beam deflector and modulator.

HISTORY

The photoelastic effect was discovered in 1815 by Brewster first in jellies and then in glasses and cubic crystals. Pockels gave a phenomenological formulation applicable to crystals of any symmetry in 1889 that was believed correct until 1970. In 1901 Coker initiated the use of the photoelastic effect for studying inhomogeneous stress distributions in transparent solids. Brillouin in 1922 made the very important prediction that light would be scattered from thermal fluctuations and be frequency shifted in the process. Though Gross claimed the first observations of Brillouin scattering in 1930, it is clear from the perspective of today that Meyer and Ramon were the first to do so in 1932. In that same year acoustooptic diffraction in liquids was first observed both by Debye and Sears and by Lucas and Biquard. In 1964 Chiao, Townes, and Stoicheff first observed stimulated Brillouin scattering. In the interaction geometry that they studied, an intense laser beam produced a coherent forward-propagating hypersonic wave and a coherent backward-propagating frequency-downshifted light wave.

In spite of the long history of the photoelastic effect Nelson and Lax found in 1970 that its formulation had a fundamental flaw. Whereas the strain had been taken as the independent variable characterizing the elastic deformation up to that time, they showed that in anisotropic media, e.g., crystals, the displacement gradient, which includes both strain and rotation effects, is the correct independent variable. The coupling to rotation reduces the symmetry of the Pockels photoelastic tensor. The numerical prediction of this coupling was verified by Nelson and Lazay by Brillouin scattering in rutile in 1970 and later in calcite. Interestingly, two situations that should have given equal scattering intensities according to the Pockels formulation were found to differ by almost 1000:1 in rutile.

MATHEMATICAL FORMULATION

Because of Pockels' influence the photoelastic effect is generally formulated in terms of a perturbation of the index ellipsoid,

$$\beta_{ij} d_i d_j = 1, \tag{1}$$

where β is the optical-frequency dielectric impermitivity tensor (inverse dielectric tensor) and $\mathbf{d}$ is a normalized electric displacement vector. The rectangular crystallographic coordinate system is used here. The lengths of the major and minor axes of an ellipse formed by the intersection of a plane with the index ellipsoid are the inverse squares of the refractive indices for the two light waves propagating in a direction normal to the plane. A perturbation of β by an elastic deformation leads to the replacement of $\beta \to \beta + \Delta\beta$ with

$$\Delta\beta_{ij} = p_{ijkl} u_{k,l}, \tag{2}$$

where p_{ijkl} is a Pockels-type tensor, but with lowered symmetry, and $u_{k,l} \equiv \partial u_k / \partial x_l$ is the displacement gradient. The p_{ijkl} tensor has interchange symmetry of its first two indices i, j exactly for a static interaction and to a high level of accuracy for dynamic interactions. Equation (2) can be reexpressed as

$$\Delta\beta_{ij} = p_{ij(kl)} s_{kl} + p_{ij[kl]} r_{kl}, \tag{3}$$

where s_{kl} is the infinitesimal strain,

$$s_{kl} \equiv u_{(k,l)} \equiv (u_{k,l} + u_{l,k})/2, \tag{4}$$

and r_{kl} represents the infinitesimal rotation,

$$r_{kl} \equiv u_{[k,l]} \equiv (u_{k,l} - u_{l,k})/2. \tag{5}$$

Note that parentheses (brackets) surrounding two tensor

subscripts indicate symmetry (antisymmetry) upon interchange.

The tensor $p_{ij[kl]}$ that couples to rotation is called the rotooptic tensor. Nelson and Lax showed that it could be expressed as

$$p_{ij[kl]} = \delta_{i[k}\beta_{l]j} + \delta_{j[k}\beta_{l]i}, \tag{6}$$

where δ_{ij} is the Kronecker delta. This expression shows that the rotooptic effect can be described as the rotation of the linear optical anisotropy. Since this is zero for cubic crystals, Eq. (6) vanishes in this case. For a uniaxial crystal, for example, the only nonzero components of Eq. (6) are

$$p_{(23)[23]} = p_{(13)[13]} = (n_e^{-2} - n_o^{-2})/2, \tag{7}$$

where n_e and n_o are the principal refractive indices.

In a piezoelectric crystal the effective photoelastic tensor is

$$p_{ijkl}^{\text{eff}} = p_{ijkl} - \frac{r_{ijm}a_m a_n e_{nkl}}{\epsilon_0 \mathbf{a}\cdot\boldsymbol{\kappa}\cdot\mathbf{a}} \tag{8}$$

where the second term is due to the indirect elastooptic effect expressed in terms of the electrooptic tensor r_{ijm}, the piezoelectric stress tensor e_{nkl}, the dielectric tensor κ_{ij}, the unit wave vector **a** of the acoustic wave, and the permittivity ϵ_0 of free space.

In nonlinear optics the photoelastic effect is often expressed by a nonlinear susceptibility

$$\chi_{mnkl} = \tfrac{1}{2}\kappa_{mi}\kappa_{nj}p_{ijkl}, \tag{9}$$

where the factor of $\frac{1}{2}$ arises from p_{ijkl} being defined for a static deformation and χ_{mnkl} for one of two frequency components, $\omega_3 = \omega_1 \pm \omega_2$, ω_1 being the frequency of the light wave and ω_2 that of the acoustic wave. The nonlinear polarization that drives the wave equation is then given by

$$P_i^{\omega 3} = \epsilon_0 D_{\chi ijkl} E_j^{\omega 1} u_{k,l}^{\omega 2}, \tag{10}$$

where $D = 1$ for $\omega_2 \neq 0$ and $D = 2$ for $\omega_2 = 0$.

BIBLIOGRAPHY

T. S. Narasimhamurty, *Photoelastic and Electro-Optic Properties of Crystals*. Plenum, New York, 1981.

D. F. Nelson, *Electric, Optic, and Acoustic Interactions in Dielectrics*, Chap. 13. Wiley, New York, 1979.

Photoelectron Spectroscopy

Dean E. Eastman and Franz J. Himpsel

The external photoeffect—in which electrons are ejected from matter which is irradiated—was discovered by Hertz in 1887 and led Einstein, in 1905, to postulate his quantum theory of light. However, it was not until after the mid-1940s that the utility of photoelectron spectroscopy—which is based on the external photoeffect—was recognized as an important technique for studying the electronic structure of all phases of matter [1–3]. Since then, tremendous progress both in understanding fundamental photoemission processes and in technological development has been occurring. Traditionally, several subfields of photoelectron spectroscopy have developed using different radiation sources, spectrometers, etc. These include x-ray photoelectron spectroscopy (XPS or ESCA) using x-ray sources [1], and ultraviolet photoelectron spectroscopy (UPS), which has been applied to molecules, solids, and surfaces [2–5]. Currently, with the advent of new sources such as synchrotron radiation, etc., the distinctions between these subfields are vanishing.

For all types of photoelectron spectroscopy, photons of a given energy are used to excite electrons above the vacuum energy level ϕ of a sample, and some of these excited electrons are emitted into vacuum, where their energy distribution is measured with an electron energy analyzer (Fig. 1). For a solid sample, there are several parameters which can be varied, namely, the photon energy $h\nu$, polarization $\hat{\epsilon}_p$, and direction $\hat{\Omega}_i$ of the incident radiation, and the energy E and direction $\hat{\Omega}_f$ of the emitted electron. Thus the photoelectron emission intensity per absorbed photon, N, depends on all these parameters and we can write $N = N(E,$

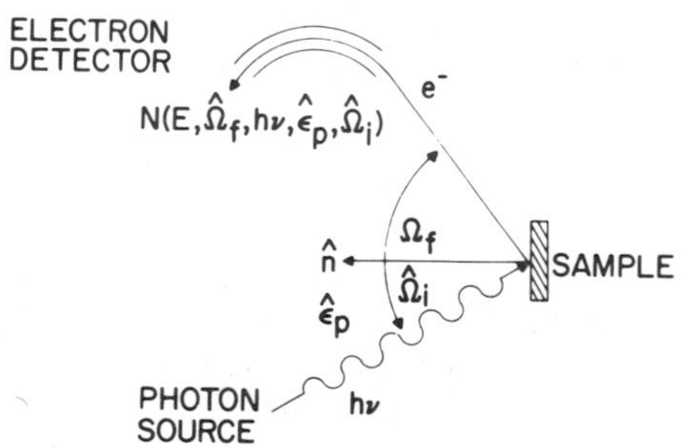

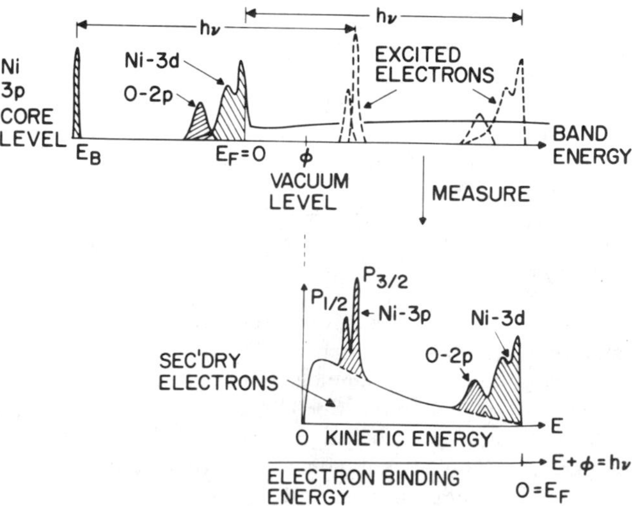

FIG. 1. Schematic experimental photoemission spectroscopy arrangement showing parameters that can be varied (upper), and schematic energy level picture of the photoemission process (lower).

$h\nu$, $\hat{\Omega}_f$, $\hat{\epsilon}_p$, $\hat{\Omega}_i$). We show the energetics of the photoemission process in the energy level diagram in Fig. 1, where several valence and core electron levels are schematically shown for a nickel surface with an adsorbed layer of oxygen. A photon of energy $h\nu$ is absorbed, exciting one of the electrons from a filled valence or core level, and thereby giving rise to a distribution of excited electrons (dashed lines). A small fraction of these electrons are excited sufficiently close to the surface (typically 5–20 Å) that they can escape the solid without loss of energy. The kinetic energy distribution of these "primary" electrons is measured (cross-hatched in Fig. 1) together with a smooth background of scattered secondary electrons.

Such an energy distribution directly reflects the energy level spectrum of the sample; namely one has the simple relation $E_i = h\nu - IP_i$, where IP_i is the ionization potential energy of the ith ion state of the system and E_i is the corresponding measured kinetic energy. For free atoms and molecules, vibrational and rotational excitations accompany each ion state, and one has the relation $IP_i = I_i + \Delta E_{\text{vib}} + \Delta E_{\text{rot}}$, where I_i is the adiabatic electronic ionization potential and ΔE_{vib} and ΔE_{rot} are vibrational and rotational excitation energies. For XPS and UPS studies of solids, the Fermi energy E_F, rather than the vacuum level, is the customary energy reference level, and one writes $IP_i = E_B^{(i)} + \phi$ where ϕ is the work function (vacuum level minus E_F) and $E_B^{(i)}$ is the electronic binding energy of the ith level. In UPS studies of valence electrons, binding energies are often called initial energies, i.e., $E_i = -E_B^{(i)}$, where E_i is usually used in a one-electron energy-band sense ($E_i < 0$ for occupied valence states). We next describe several examples of XPS and UPS applied to solids and molecules so as to illustrate a select few of the many types of applications of photoelectron spectroscopy.

High resolution is easily attained at low energies using He and Ne resonance lamps, as is seen in Fig. 2 for the photoelectron spectrum for the outer three valence orbitals $[(2\sigma_u 2s)^2(1\pi_u 2p)^4(3\sigma_g 2p)^2]$ of CO molecules [2]. Here the adiabatic ionization potentials are indicated, and numerous vibrational bands are seen on the high-binding-energy side of each ionization potential. The $(3\sigma_g 2p)^2$ level ($IP = 14.01$ eV) exhibits weak vibrational structure while the $(1\pi_u 2p)^4$ level exhibits strong vibrational structure; this occurs because the 3σ orbital is a nonbonding orbital while the 1π orbitals are bonding orbitals and thus couple electronic and vibrational excitations. In general, the valence molecular-orbital ionization potentials are sensitive to the electronic wave functions, and thus yield direct information concerning the electronic structure which determines the chemical and physical nature of the molecule.

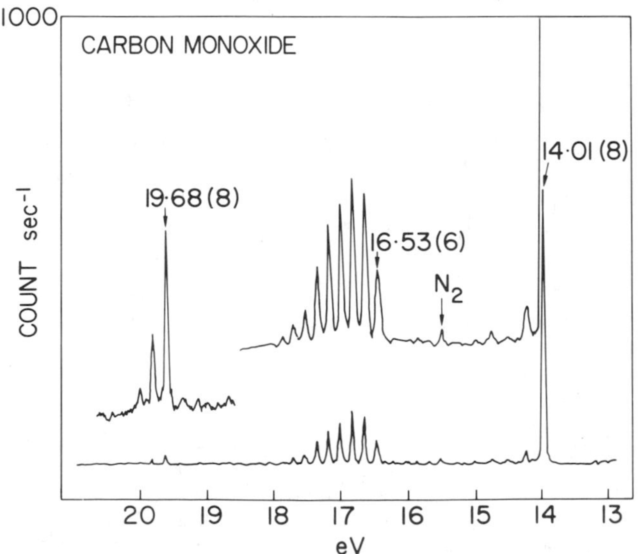

FIG. 2. Photoelectron spectrum for the outer three molecular orbitals of CO using a He I (21.2-eV) resonance lamp. (Reprinted with permission from Ref. 2.)

An important general application of x-ray photoelectron spectroscopy is the study of chemical shifts in binding energies for a core level of an atom in different chemical environments. Indeed, K. Siegbahn and his collaborators, who first developed x-ray photoelectron spectroscopy, coined the widely used acronym ESCA, which is an abbreviation for electron spectroscopy for chemical analysis [1]. An ESCA example showing chemical shifts in the carbon $1s$ levels in ethyl trifluoroacetate ($CF_3COOCH_2CH_3$) is shown in Fig. 3 [6]. All four carbon atoms have quite distinct $1s$ binding energies, with the C$1s$ level of the CF_3 fluoro group being shifted by about 8 eV compared with the CH_3 methyl group. This occurs because of the much larger negative charge transfer onto fluorine relative to hydrogen, and reflects the much larger electronegativity of fluorine. The measurement of such chemical shifts in molecules, solids, and surfaces gives direct information concerning the oxidation states, charge transfer for various chemical bonds, etc. [1].

For solids, valence electrons interact and form energy bands. An important area of application for XPS and UPS is the study of the electronic structure of such valence bands, which determine the physical and chemical properties of solids. In XPS one sees, to a good approximation, the density of states, weighted by atomic-like cross sections [7] for the different shells. In ultraviolet photoelectron spectroscopy, if one uses photon energies less than about 40–60 eV, spectra become more complicated than shown in XPS. At such energies the measured spectrum resembles a quantity usually called the energy distribution of the joint density of states, rather than the total density of states. This behavior occurs because the electron states of the solid have a well-defined crystal momentum, and both energy and crystal momentum are conserved in the optical excitation process. In this case, since the photon has negligible momentum (i.e., a wavelength long compared with atomic dimensions), the only electron–hole pairs that can be excited are those which have the same crystal momentum and conserve energy. Consequently, the UPS photoemission spectra change with photon energy and contain both electron energy and momentum information. A set of $h\nu$-dependent, angle-integrated spectra for polycrystalline copper is shown in Fig. 4, where it is compared with theoretical spectra [5]. As shown in Fig. 4, the photoemission spectra exhibit d-band emission features between -2 and -5 eV which change markedly with photon energy. These spectral features, which are due to direct interband transitions that conserve energy and crystal mo-

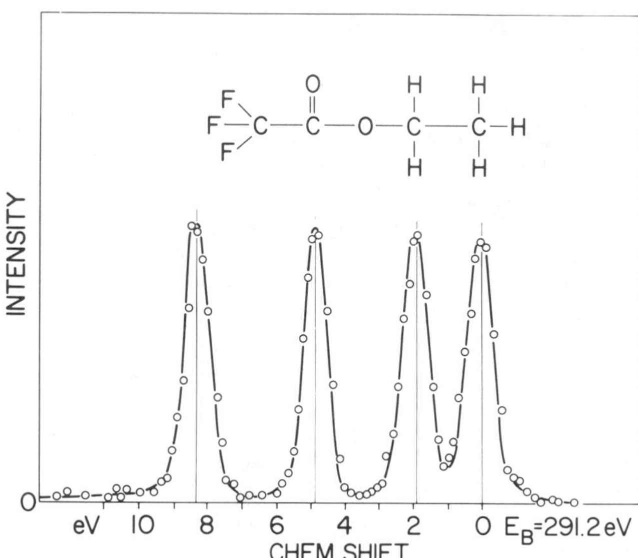

FIG. 3. X-ray photoelectron spectrum (ESCA) for ethyl trifluoroacetate showing the chemical shifts of the C 1*s* electron levels for carbon atoms having different chemical environments. (Reprinted with permission from Ref. 6.)

mentum, are seen to be described quite well by the set of theoretical spectra, which were calculated using an optimized Korringa–Kohn–Rostoker band model.

Even more detailed features than shown in Fig. 4 can be obtained by studying angle-resolved photoemission from single-crystal samples (Figs. 5 and 6 and Ref. [8]). In this case, the crystal momentum parallel to the surface ($\mathbf{k}_{\parallel}$) is conserved in the photoemission process [the final-state wave function is identical to a low-energy electron-diffraction (LEED) wave function] and thus both the energy and $\mathbf{k}_{\parallel}$ momentum of various electron states can be directly determined. The third momentum component $\mathbf{k}_{\perp}$ (perpendicular to the surface) is not conserved due to refraction of the photoelectron wave at the surface, but it can be varied by tuning the photon energy $h\nu$ (Fig. 5). Therefore, one is able to determine $E(\mathbf{k})$ energy-band dispersions [8,9], which characterize electron states in a crystalline solid. Such band structure data are being compiled for a large variety of materials [10]. A complete measurement of all the parameters involved in photoemission yields a complete description of the electronic states, which includes angular (point group) symmetry and electron spin (Fig. 6). The former is obtained via the polarization $\hat{\mathbf{e}}_p$ of the photon, the latter via the spin polarization σ of the photoelectron by using dipole-selection rules for optical transitions.

Photoelectron spectroscopy, especially UPS, has become a powerful tool for determining the electronic structure of surfaces, e.g., intrinsic surface states on semiconductors and metals and sorbed atoms and molecules on various surfaces. The sensitivity to observe surface states and monolayer or submonolayer quantities of adsorbates results from the very short (typically ~5–15 Å) escape depths of the photoejected electrons that are measured.

An example for a surface state can be seen in Fig. 5. In this set of spectra, there is a surface state at 0.4 eV below E_F. Two-dimensional surface states are distinguished from three-dimensional bulk states by their absence of an $E(\mathbf{k}_{\perp})$ band dispersion, which implies that the binding energy of the surface state is independent of the photon energy as $\mathbf{k}_{\parallel}$ is fixed.

An application to chemisorption is depicted in Fig. 7, where photoelectron spectra for clean single-crystal Ni(111) and for Ni(111) with a layer of chemisorbed benzene (C_6H_6) are shown, together with spectra for several layers of condensed benzene and for gas-phase benzene [12]. Figures 7(b) and 7(c) show the difference spectra for chemisorbed C_6H_6 + Ni(111) minus clean Ni(111) and condensed C_6H_6 + Ni(111) minus clean Ni(111), respectively. For both spectra, the five lower-lying peaks are associated with benzene molecular-orbital energy levels. Several messages concerning the application of photoelectron spectroscopy to chemisorption are illustrated in Fig. 7, namely, the use of photoemission as a "fingerprint" of the adsorbed molecular species, and the large changes in ionization energies of a nonbonding nature (relaxation shifts) that occur for adsorbed species. Concerning the "fingerprinting" aspect, the simi-

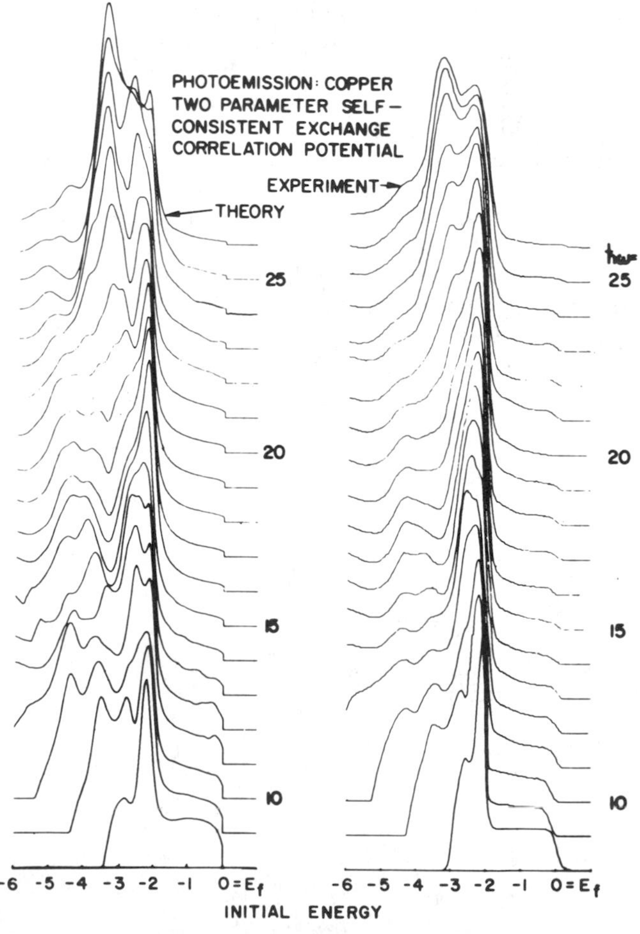

FIG. 4. A set of angle-integrated ultraviolet photoelectron spectra for copper with $8 \leq h\nu \leq 26$ eV together with calculated spectra [5]. Strong $h\nu$-dependent trends due to direct interband transitions are seen (see text).

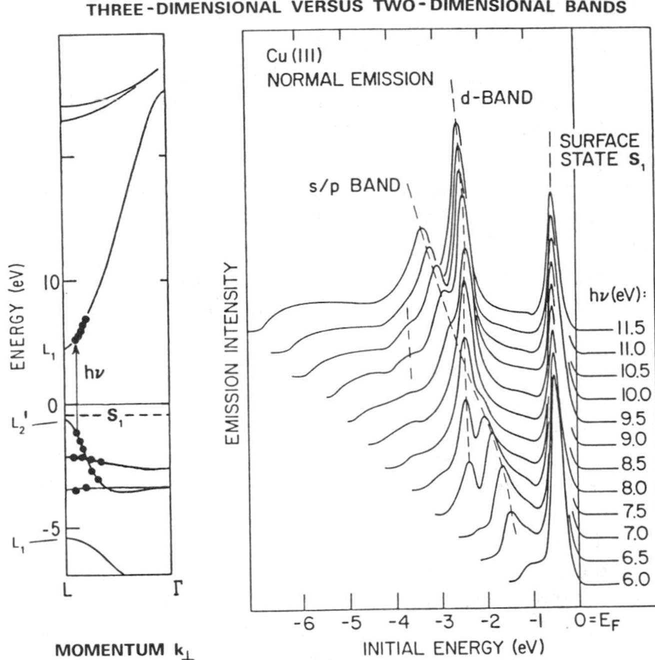

FIG. 5. Surface and bulk states at a Cu(111) surface, seen in angle-resolved photoemission. The structure at 0.4 eV below E_F is a surface state; the other peaks are bulk states. They change energy due to dispersion along the $k_\perp$ axis as shown on the left-hand side. Such measurements make it possible to map out energy band dispersions, which completely characterize the electronic structure of a solid (Ref. [9]).

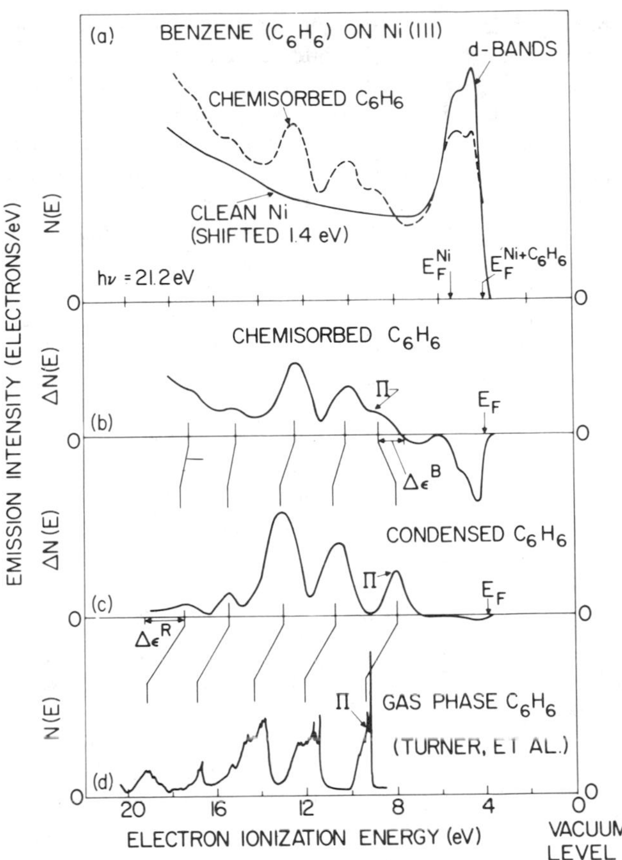

FIG. 7. Photoelectron spectra ($h\nu = 21.2$ eV) for clean Ni(111), chemisorbed benzene on Ni(111), condensed benzene, and gas-phase benzene [12].

larity of the spectrum for chemisorbed benzene to those for condensed and gas-phase benzene—in particular the lowest four energy levels which involve nonbonding σ orbitals—implies that the adsorbed benzene molecule remains structurally intact. At elevated temperatures (above ~470 K) benzene is observed to dehydrogenate and leave a carbonaceous surface residue. Concerning relaxation shifts, all the nonbonding lower-lying σ orbitals for chemisorbed benzene have ionization energies relative to the vacuum level which are reduced from those of molecular benzene by about 2 eV. This relaxation shift is due to increased screening of the excited hole state of the molecule by electrons from the metal substrate.

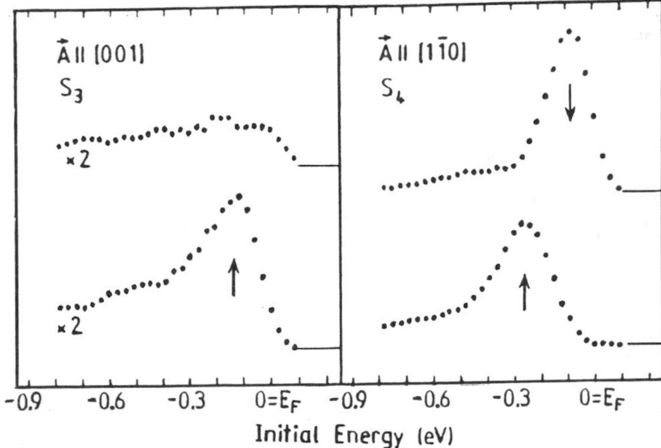

FIG. 6. Polarization-dependent, angle- and spin-resolved photoelectron spectra for ferromagnetic Ni(110). Electronic states of different symmetry (S_3 and S_4) are selected by varying the direction of the electric field vector **A**. Majority- and minority-spin bands are separated by the ferromagnetic exchange splitting. The minority-spin S_3 band is unoccupied at this k point (From Ref. [11]).

Photoelectron spectroscopy is mainly a technique for probing occupied electronic states. There exists a complementary technique for probing unoccupied states, i.e., inverse photoemission [13]. Thereby an electron impinges onto a solid and the emitted bremsstrahlung photons are detected. Thus, photon and electron interchange their role as incoming and outgoing particles. The energy distribution of the emitted photons reflects the density of unoccupied states. By using low-energy electrons (in the order of 10 eV) one can make the technique surface sensitive and capable of momentum-resolved measurements, like angle-resolved UPS. The counterpart of XPS is often called BIS (bremsstrahlung isochromat spectroscopy).

Apart from mapping electronic states, photoelectron spectroscopy can also be used for structure determination of surfaces by exploiting the effect of photoelectron diffraction [14]. Photoelectrons from a core level (with energies of a few hundred eV) are scattered by neighboring atoms and interfere with the directly emitted photoelectron wave, as in low-

low-energy electron diffraction (LEED). The resulting interference pattern contains information about interatomic distances and bond directions, which can be modeled by methods similar to a LEED intensity analysis. By choosing various core levels as source of photoelectrons one is able to probe the local structure around particular atoms. There exist various implementations of this technique, e.g., measuring the angular electron distribution at fixed photon energy or the change in emission intensity with photon energy at a fixed angle.

In summary, photoelectron spectroscopy has developed over the past decades into a powerful tool for probing the electronic structure of gases, liquids, solids, and their interfaces. In this brief article, only a few of the many applications and aspects of photoelectron spectroscopy have been touched upon. The rapid development and increase in the range of applications of photoelectron spectroscopy is continuing at present, especially with the advance in intense continuum synchrotron radiation sources.

See also ADSORPTION; CHEMICAL BONDING; ELECTRON ENERGY STATES IN SOLIDS AND LIQUIDS; LOW-ENERGY ELECTRON DIFFRACTION (LEED); PHOTON; SURFACES AND INTERFACES.

REFERENCES

1. K. Siegbahn *et al., ESCA: Atomic, Molecular, and Solid State Structure Studied by Means of Electron Spectroscopy*. Almqvist and Wiksells Boktryckeri Ab, Stockholm, 1967; K. Siegbahn, *J. Electron Spectrosc.* **5,** 3 (1974).
2. D. W. Turner *et al., Molecular Photoelectron Spectroscopy*. Wiley-Interscience, London, 1970.
3. B. Feuerbacher, B. Fitton, and R. F. Willis (eds.), *Photoemission and the Electronic Properties of Surfaces*. Wiley, New York, 1978; M. Cardona and L. Ley (eds.), *Photoemission in Solids I and II*. Topics in Applied Physics Vols 26 and 27. Springer-Verlag, Berlin, 1978, 1979.
4. D. E. Eastman, in *Techniques of Metals Research VI*, pp. 413–479 (E. Passaglia, ed.). Interscience, New York, 1972.
5. D. E. Eastman, in *Proceedings of the IV International Conference on Vacuum Ultraviolet Radiation Physics, Hamburg, 1974*, pp. 417–449 (E. E. Koch *et al.*, eds.). Pergamon, Vieweg, 1975.
6. U. Gelius *et al., A High Resolution ESCA Instrument with X-ray Monochromator for Gases and Solids*. Uppsala University Report UUIP-817, 1973.
7. J. J. Yeh and I. Lindau, *At. Data Nucl. Data Tables* **32,** 1 (1985).
8. E. W. Plummer and W. Eberhardt, *Adv. Chem. Phys.* **49,** 533 (1982).
9. F. J. Himpsel, *Adv. Phys.* **32,** 1 (1983).
10. *Landolt-Börnstein; Numerical Data and Functional Relationships in Science and Technology*. Group III, Vol. 23. Springer-Verlag, Berlin, 1989.
11. R. Raul, H. Hopster, and R. Clauberg, *Phys. Rev. Lett.* **50,** 1623 (1983).
12. J. E. Demuth and D. E. Eastman, *Phys. Rev. Lett.* **32,** 1123 (1974).
13. V. Dose, *Surf. Sci. Rep.* **5,** 337 (1985); F. J. Himpsel, *Comments Cond. Mat. Phys.* **12,** 199 (1986); N. V. Smith, *Rep. Prog. Phys.* **51,** 1227 (1988).
14. Y. Margoninski, *Contemp. Phys.* **27,** 203 (1986).

Photoionization

David L. Ederer

Photoionization is a process where a photon of frequency ν and energy $h\nu$ interacts with an atom or molecule to produce an ion and one or more electrons. The quantity h is Planck's constant and has a magnitude of 6.26×10^{-27} ergs. The interaction can be represented by the equation $h\nu+M\rightarrow M^{m+}+me$, where $m\geq1$. Since ionization energies (i.e., the minimum energy necessary to remove an outer electron from the system) lie in the range 8×10^{-12}–4×10^{-11} ergs, the corresponding minimum frequency of the photon to produce ionization is 1.3×10^{15}–6×10^{15} s^{-1}. This frequency corresponds to wavelengths* between 2000 and 500 Å (1 Å$=10^{-8}$ cm). Not all photon absorption processes produce ionization. In molecules, for example, dissociation occurs with high probability at low photon energies. In this process neutral or excited fragments are produced. As the photon energy increases above threshold it eventually exceeds the binding energy of the electrons in inner shells of the atom or molecule. (These shells are labeled *K, L, M*, etc., counting from the most tightly bound electrons.) Thus a very energetic photon, e.g., an x ray, can remove any electron whose binding energy is less than the photon energy.

The probability that an energetic photon will produce ionization is measured by the photoionization cross section $\sigma(\nu)$, which has the dimension of an area. The cross section is the quantity that reflects the electronic properties of an atom. The variation of the cross section with photon frequency is a window through which one can study the electronic nature of an atom subshell by subshell. This connection is established by expressing the cross section in terms of the absolute square of a matrix element, $\mathbf{D}_{\epsilon,0}(\nu)$, such that

$$\sigma(\nu)=\frac{e^2h^2}{2\pi m^2c\nu}\mid\mathbf{D}_{\epsilon,0}(\nu)\mid^2. \tag{1}$$

In this equation e, m, and c are the electronic charge, mass of the electron, and the velocity of light, respectively. The quantities 0 and ϵ define the initial state and final continuum state of the system. The matrix element $\mathbf{D}_{\epsilon,0}(\nu)$ is usually expressed in the dipole approximation where

$$\mathbf{D}_{\epsilon,0}(\nu)=\frac{4\pi^2m\nu}{h}\int u_\epsilon\sum_i x_iu_0\,d\tau. \tag{2}$$

The initial-state wave function, u_0, of the system and the final-state wave function of the ion, u_ϵ, are determined by solving the Schrödinger equation for the system. In fact, as these functions cannot be obtained exactly except for hydrogen, a great deal of insight is gained about the electronic properties of the atom by studying the degree of approximation required to give best agreement with experiment [1].

The number of ions N_x produced by N photons in a thin layer of atoms or molecules x cm thick is given by

$$N_x=Nn\sigma(\nu)x, \tag{3}$$

* The photon wavelength is related to the frequency by the formula $\lambda=c/\nu$, where c is the velocity of light.

where n is the number density of atoms or molecules. Thus the number of ions produced in a gas column of length L is obtained by integrating Eq. (3) to obtain

$$N_L = N\{1 - \exp[-n\sigma(\nu)L]\}. \quad (4)$$

From Eq. (4) the maximum number of ions that can be obtained by the primary photoionization process is just equal to the photon number N. However, in the case of molecules some of the photons may produce dissociation, reducing proportionally the number of ions produced. The magnitude of the photoionization cross section $\sigma(\nu)$ is a function of the photon energy and at the ionization threshold typically is of the order of the cross-sectional areas of the atom or about 10^{-17} cm^2. To illustrate the functional dependence of the photoionization cross section on the photon energy $h\nu$, the photoionization cross sections [2] of two atoms, helium and xenon, are shown in Fig. 1. Note that in the case of helium the cross section decreases monotonically, whereas for xenon there are abrupt changes in the magnitude of the cross section at photon energies corresponding to the binding energies of the L- and M-shell electrons. These discontinuities are called x-ray absorption edges. At photon energies corresponding to the binding energy of some of the N-shell electrons there is no abrupt change in the cross section; rather it rises to a prominent maximum as the photon energy increases. At other photon energies there are minima and secondary maxima as well as narrow features (width <0.1 eV)

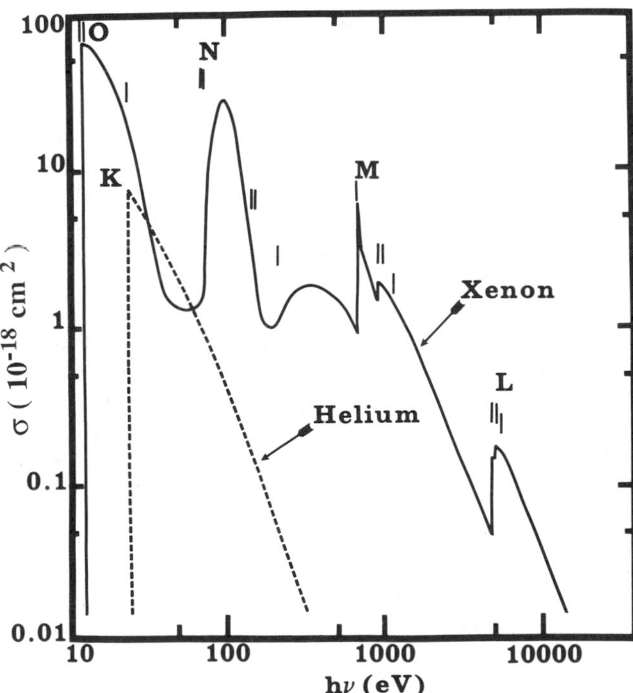

FIG. 1. The photoionization cross sections of helium and xenon. The photon energy is expressed in units of electron volts (1 eV $= 1.602 \times 10^{-12}$ erg). The binding energy of the subshell electrons (K, L, M, etc.) is denoted by the tick marks. The K binding energy of xenon (not shown) is about 34 keV and the maximum cross section at the K absorption edge is 7×10^{-21} cm^2.

due to a photoionization process called autoionization. The detailed study of these features in xenon and other atoms and molecules yields insight about the fundamental properties [3] of the electron cloud surrounding the nuclei.

The total photoionization cross section described in the preceding paragraphs includes the direct production of a multiply charged ion and several electrons. However, this cross section is typically one or two orders of magnitude smaller than the cross section for single ionization. In the x-ray region Compton scattering has an appreciable cross section. This process produces ionization but the photon is inelastically scattered rather than absorbed. Other related topics that may be of interest to the readers are predissociation, preionization, the Auger effect, shake-up, and shake-off.

See also AUGER EFFECT; COMPTON EFFECT; IONIZATION.

REFERENCES

1. A good review of the theoretical techniques used to calculate atomic photoionization cross sections can be found in A. F. Starace, Theory of Atomic Photoionization, in *Encyclopedia of Physics*, Vol XXXI (W. Mehlhorn, ed.), pp. 1–121. Springer-Verlag, Berlin, 1982.
2. Photoabsorption cross-section compilations can be found in the following articles: R. D. Hudson and L. J. Kieffer, *At. Data* **2**, 205 (1971); E. B. Saloman, J. H. Hubbell, and J. H. Scofield, *At. Data Nucl. Data Tables* **38**, 1 (1988). Multiphoton bibliographies 1977–1988 are contained in the following publications: J. H. Eberly, J. W. Gallagher, E. C. Beaty, N. Piltch, S. J. Smith, NBS Publication LP-92 and Suppls. 1–6. These publications are available from the JILA Atomic Physics data center, JILA, CB 404, University of Colorado, Boulder, CO 80309-0440.
3. Geoffrey V. Marr, *Photoionization Processes in Gases*. Academic Press, New York, 1967.

Photon

A. J. Fennelly

In 1900 light was considered a continuous electromagnetic wave. Three centuries of geometrical and physical optics had discredited all notions that light is corpuscular in nature. The continuous wave view of light uniquely determines its interaction with matter, i.e., the nature of its absorption and emission by atoms. As long as light is regarded as a continuous wave its interaction with matter is a continuous process. Then electromagnetic energy is absorbed or emitted as radiation of infinitesimal amounts. Hence, an arbitrary amount of energy can be exchanged between a radiation field and atoms at equilibrium.

Under the influence of the classical wave theory of light physics was unable to explain (1) the spectrum of radiation emitted by an object held at temperature T, (2) the emission of electrons from a metal's surface when light of a frequency ν, greater than a critical frequency ν_0, is incident on it (photoelectric effect), and (3) the discrete spectra of atoms.

Using classical radiation theory and oscillators with a continuous spectrum of energy states, Lord Rayleigh and Sir James Jeans obtained a description of the spectrum of thermal radiation which diverges catastrophically in the ultraviolet. Their result was that the energy density of radiation increases without limit as the frequency increases. Wien obtained an emprical formula which also failed to fit experiment, at the lower frequencies. Worse, the ultimate conclusion of the Rayleigh–Jeans law is that at equilibrium most of the universe's energy is at the highest frequencies. This would result, if true, in our rapid incineration.

In 1900 Max Planck formulated the correct description of thermal radiation. His equation for the energy density of radiation $R(\lambda)$ at wavelength λ from an object at temperature T is $R(\lambda)=C_1\lambda^{-5}[e^{C_2/\lambda T}-1]^{-1}$, where $C_1=2c^2h$ and $C_2=hc/k$ with c and k the speed of light and Boltzmann's constant, respectively. The constant h is called Planck's constant and is, in MKS units, equal to 6.626×10^{-34} J s. Planck derived this by assuming that a mode at frequency ν can change its energy only by an amount $Nh\nu$ where N is an integer. This first constrains the oscillator absorbing or emitting the radiation to exist only in states of energy $E\simeq Nh\nu+K$, where K is a constant for a given oscillator. More important, it constrains the radiation field to change its energy and hence frequency only by the discrete amounts $\Delta E=\Delta nh\nu$. The minimum element of energy, called by Planck the *quantum*, which may be absorbed or emitted is $\Delta E_{\min}=h\nu$. For visible light this is only $\sim10^{-19}$ J.

Once it was realized that oscillators have discrete states, this implied the same for light. It cannot interact with matter as a continuous wave. Rather, it comes and goes as a burst or lump of energy, determined by its frequency. Thus it must have some particle-like properties. In 1905, in order to describe the photoelectric effect, Einstein named these lumps *photons*. A photon has assigned to it frequency ν, an energy $h\nu$, a momentum $h\nu/c$, and an angular momentum $h/2\pi$ (thus independent of its energy). When light of frequency ν greater than a critical frequency ν_0 strikes a clean metal surface, *photoelectrons* are emitted with kinetic energy $W=h(\nu-\nu_0)$. The energy $W_0=h\nu_0$ is the *work function* for emission of a photoelectron. Therefore, we see no photoelectrons for frequencies $\nu<\nu_0$. The number of electrons released is proportional only to the number of photons absorbed and so is dependent on the energy density of the beam. But the electron energies are not dependent on the beam energy density, just on its frequency.

In 1912 Niels Bohr extended the concept of quantum states or quantization to the atom, reasoning that the electrons orbited at discrete radii, determined by the discrete amounts of energy absorbed from photons and by their initial or ground states. The frequency of light ν_{nm} emitted in passing from a state of energy E_n to a state of energy E_m is given by $\nu_{nm}=(E_n-E_m)/h$. This *Bohr frequency condition* is now accepted as basic to our understanding of discrete atomic spectra as the emission and absorption of photons by atom. As an atom absorbs a photon it goes to a state of higher energy and as it emits one, to a state of lower energy.

If the photon concept is true, it must lead to other consequences which have all been confirmed by experiment. Einstein in 1919 attempted a reconciliation of Planck's spectral distribution with Boltzmann's distribution of states of the atoms at equilibrium responsible for the thermal radiation. For a system with a number of atoms N_m in state m and N_n in state n there are three processes involving photons: (1) radiation absorption at rate $B_{mn}N_mR(\lambda)$, (2) spontaneous emission at rate $A_{nm}N_n$, and (3) stimulated emission at rate $B_{mn}N_nR(\lambda)$, where A and B are called Einstein coefficients (constants). If the absorption and total emission rates are equal (equilibrium), Einstein recovered Planck's law if $B_{nm}=B_{mn}=c^3A_{nm}/8\pi h\nu^3$. Sixty years later we now have the widespread application of this in the laser and maser (light/microwave amplification by stimulated emission of radiation): A specially prepared beam of photons stimulates the emission of more photons from a sample overpopulated in one state.

The *Compton effect* established the photon concept beyond a doubt in 1923. Compton observed a dependence of wavelength on scattering angle for x rays scattered from electrons in crystals. Conservation of total momentum $h\nu/c$ and energy $h\nu$ of the photon, before and after the collision with an electron at rest of mass m, require that the photon frequency change since its speed c does not. This leads to a variation of frequency with scattering angle which was precisely that measured by Compton.

Photons themselves behave like a packet of waves spreading through space after emission by an atom. We know this because a diffraction–interference experiment, performed by G. I. Taylor in 1908, showed that photons are *individually diffracted* and *individually interfere* with themselves. The experiment produced a diffraction pattern of a needle with the intensity so low that individual photons were encountering the needle *singly*. Yet the diffraction pattern remained detectable.

Extension of the concept of the photon to other areas has always been fruitful and verified by experiment. The study of real photons and their interaction with matter in the *quantum mechanics* developed by Dirac has led to our understanding of a large number of phenomena. These include the statistics of assemblies of large numbers of particles and photons; photon coincidences and coherence of electromagnetic waves; radiation effects in molecules, solids, and fluids; x-ray scattering; Rayleigh scattering and dispersion; the Raman effect in molecular spectroscopy; and so on.

When the possibility of annihilation and creation of fundamental particles is taken account of, *quantum electrodynamics* can explain the interaction of photons with matter at the nuclear and subnuclear level. Then the forces between particles are viewed as due to the exhange of *virtual photons* as well as real ones. The concept's accuracy is borne out by the experimental verification of quantum electrodynamics to extreme degrees of accuracy. A concept of such basic utility as the photon is rare. It seems more trustworthy with every further extension and application. It likely will remain useful for many decades to come.

With further progress in physics through and after the decade of the seventies, the concept of the photon has proven even more useful and durable. It has become both central to the interpretation of quantum mechanics, and the indispensible element of experimental tests of the reality of the currently accepted understanding of quantum phenomena. In the standard model of quantum electrodynamics, the electromagnetic field is viewed as composed of fluxes of virtual photons; and the electromagnetic interactions of particles, with charges and/or electromagnetic moments, are modeled as mediated by the exchange of virtual photons. This theory has been an extraordinary success in all its experimental tests, to extremely high accuracies. The concept has proven so useful and successful that it has been adapted to provide the basic model for the action of the other forces on particles and each other in the weak interactions (W and Z bosons), strong interactions (pions at low energies and gluons at high energies), gravitation (gravitons), and even the interaction among particles in solids (phonons, magnons, etc.). In all circumstances the models are both computationally efficient, and easily interpreted toward development of experimental tests. Those have all been developed as the last of theories called gauge theories.

In tests of our notions of quantum theory, counts of photons simultaneously emitted in pairs, from electron–positron annihilation, or from certain atomic transitions, and measurements of their polarization states, the concept of the photon has become central to our understanding of the quantum world. Furthermore, recent experiments in photon bunching and photon antibunching have further confirmed the usefulness and success of the concept. The fact that those experiments have no classical descriptions seems to confirm the absolute existence of the photon beyond its mere usefulness as a concept. The experiments involving the creation of squeezed states in lasers and quantum optical systems, which squeeze the fundamental Heisenberg uncertainty circles into ellipses in phase space so as to measure one of a pair of a canonical conjugate dynamical variables to an arbitrarily high degree of accuracy, also confirm the usefulness of the photon concept, and even its absolute existence. Indeed, one remarkable series of experiments has used ultrafast electronics to demonstrate that whereas individual photons transit a Michelson interferometer and can be counted individually, an interference fringe pattern is visible when they are not counted, and hence that single photons interfere with themselves, simultaneously traversing both paths of the interferometer. The photon thus continues to be an extremely useful concept, providing the major conceptual basis for many successful approaches to understanding the problems of physics. Indeed, the evidence in favor of the independent, absolute existence or reality of the photon continues to grow as the concept is applied in experiment, application, and theory, and is extended to ever greater tests of its consequences for physics.

See also BLACKBODY RADIATION; BOHR THEORY OF ATOMIC STRUCTURE; COMPTON EFFECT; LASERS; LIGHT; MASERS; QUANTUM ELECTRODYNAMICS; QUANTUM MECHANICS.

BIBLIOGRAPHY

A. A very elementary reference for the nonscientist which nevertheless contains much information concisely:

Clarence Rainwater, *Light and Color*. Golden Press, New York (Western Publishing Company, Inc., Racine, WI), 1971; Golden Science Guide. (EEE)

B. A comprehensive elementary reference for the nonscientist:

D. Falk, D. Brill, and D. Stork, *Seeing the Light*. Harper & Row, New York, 1986. (EE)

C. Standard optics texts with discussion of the photon:

F. A. Jenkins and H. E. White, *Fundamentals of Optics*, 4th ed. McGraw-Hill, New York, 1976. (E)

E. Hecht and A. Zajac, *Optics*. Addison-Wesley, Reading, MA, 1974. (E)

D. Popular books on the role of the photon and the quantum realms of physics:

J. Gribbin, *In Search of Schrödinger's Cat*. Bantam Books, New York, 1985. (E)

N. Herbert, *Quantum Reality*. Doubleday Anchor Press, New York, 1985. (E)

A. Rae, *Quantum Physics: Illusion or Reality?* Cambridge University Press, New York, 1986. (E)

J. C. Polkinghorne, *The Quantum World*. Princeton University Press, Princeton, NJ, 1984. (E)

T. Hey and P. Walters, *The Quantum Universe*. Cambridge University Press, New York, 1987. (E)

E. Introductory texts on quantum mechanics which discuss the photon:

R. P. Feynman, R. A. Leighton, and Matthew Sands, *The Feynman Lectures on Physics, Volume III Quantum Mechanics*. Addison-Wesley, Reading, MA, 1965. (I)

H. Haken and H. C. Wolf, *Atomic and Quantum Physics*. Springer-Verlag, New York, 1984. (I)

F. An advanced discussion on quantum notions, including photons:

B. d'Espagnat, *Conceptual Foundations of Quantum Mechanics*. 2nd ed. Addison-Wesley (Benjamin Books), Reading, MA, 1976. (A)

G. A few texts which address the photon and the quantum aspects of light:

M. Sargent III, M. O. Sully, and W. E. Lamb, Jr., *Laser Physics*. Addison-Wesley, Reading, MA, 1974. (A)

L. Allen and J. H. Eberly, *Optical Resonance and Two-Level Atoms*. Dover, New York, 1987. (I)

P. L. Knight and L. Allen, *Concepts of Quantum Optics*. Pergamon, Oxford, 1983. (A)

R. Loudon, *The Quantum Theory of Light*, 2nd ed. Oxford University Press, New York, 1983. (A)

H. Solid-state physics texts which illustrate the usefulness of the adaptations of the photon concept to phonons in solids:

J. Ziman, *Theory of Solids*. Cambridge University Press, Cambridge, 1965. (A)

C. Kittel, *Quantum Theory of Solids*, 2nd rev. ed. Wiley, New York, 1987. (A)

I. Quantum field-theory texts which illuminate the usefulness of the photon and boson concepts:

G. Kane, *Modern Elementary Particle Physics*. Addison-Wesley, Reading, MA, 1987. (I)

R. P. Feynman, *Quantum Electrodynamics*. Addison-Wesley, Reading, MA (W. A. Benjamin), 1961. (A)

R. P. Feynman, *The Theory of Fundamental Processes*. Addison-Wesley, Reading, MA, 1961. (A)

R. P. Feynman, *Photon-Hadron Interactions*. Addison-Wesley (W. A. Benjamin), Reading, MA, 1972. (A)

J. M. Ziman, *Elements of Advanced Quantum Theory*. Cambridge University Press, Cambridge, 1969. (A)

J. D. Bjorken and S. Drell, *Relativistic Quantum Mechanics*. McGraw-Hill, New York, 1964. (AA)
J. D. Bjorken and S. Drell, *Relativistic Quantum Fields*. McGraw-Hill, New York, 1964. (AA)
C. Itzykson and J. B. Zuber, *Quantum Field Theory*. McGraw-Hill, New York, 1980. (AA)
S. S. Schweber, *An Introduction to Relativistic Quantum Field Theory*. Harper & Row, New York, 1961. (AA)

Photonuclear Reactions

Evans Hayward

Photonuclear reactions are by definition those interactions in which a quantum of electromagnetic radiation disturbs the internal coordinates of a nucleus. The lowest energy at which such a reaction can occur is that of the first excited state of the nucleus, perhaps a few hundred keV in a heavy nucleus or a few MeV in a light one. Once excited, this state can only decay to the ground state. This simple process of absorption and reemission of a photon by a discrete, well-defined energy level is known as resonance fluorescence. Discrete energy levels at higher excitations can decay either by resonance fluorescence or through some sequence of lower-lying levels. The photon absorption cross section at low energies then consists of a number of sharp, well-defined peaks.

At some higher energy, perhaps 8 MeV in a heavy nucleus or 15 MeV in a light one, nucleon emission becomes possible. The energy levels broaden as the excitation energy increases until, finally, the cross section becomes smooth and continuous. The most outstanding feature of this continuum is the giant resonance, a peak in the photon absorption cross section that is so large that it must result from electric dipole absorption. The cross section integrated to 30 MeV is often measured in units of the dipole sum:

$$\int \sigma \, dE = \frac{2\pi^2 e^2 \hbar}{Mc} \frac{NZ}{A} = 0.06 \frac{NZ}{A} \quad \text{MeV b}.$$

Absorption cross sections integrated to 30 MeV for elements with $A \geq 100$ are approximately 1 dipole sum; for elements with $A \leq 40$ they can be as small as 0.6 of a dipole sum. The magnitude of the cross sections for $40 < A < 100$ are less well established, but presumably there is some kind of a smooth transition between the two regions.

Other features of the giant resonance, such as its energy and width, are also best discussed with respect to these three regions of A. For example, for $A \geq 100$ the energy E_0 at which the giant resonance peaks is near 15 MeV and varies inversely as the nuclear radius, as $A^{-1/3}$. In fact, for these nuclei the prediction of the hydrodynamic model,

$$E_0 = 80A^{-1/3} \quad \text{MeV},$$

is well supported by experiment. For the light nuclei, $A \leq 40$, there is no systematic dependence of the resonance energy on A; most have the giant resonances near 20 MeV. Again, the location of the resonance energy makes a smooth transition between $A = 40$ and $A = 100$.

The giant resonance width is apparently associated with the nuclear shape. Closed-shell nuclei such as ^{16}O, ^{40}C, ^{90}Zr, or ^{208}Pb are sometines viewed as rigid spheres; their giant resonances are the narrowest, 4 or 5 MeV wide. Nuclei having large intrinsic deformations, such as the rare earth nuclei or the actinides, have characteristically much wider giant resonances, about 8 MeV. Vibrational nuclei such as Cd or Ag, whose shape is changing, are said to be dynamically deformed. They too can have widened giant resonances because the nuclear surface is oscillating so as to make the nucleus appear deformed.

The photonuclear giant resonances display various structures, the best understood being the giant-resonance splitting for the heavy deformed nuclei. In the hydrodynamic model these two resonances are ascribed to charge oscillations along the one long and two short axes of the nuclear ellipsoid, their resonance energies corresponding to the two characteristic lengths. In fact, this energy difference yields one of the best values for the intrinsic quadrupole moment of the nuclear ground state; of course, this phenomenon is also responsible for the large giant-resonance widths for deformed nuclei. The giant resonances of some vibrational nuclei display some less pronounced undulations, which result from the oscillations of the nuclear surface and have been described in the so-called dynamic collective model, an outgrowth of the simple hydrodynamic model.

The giant resonances of the light nuclei display much more structure. This presumably results from the excitations of individual nucleons and their interaction with more complicated configurations in the continuum. In the nuclear shell model, the electric dipole giant resonance results from the sum of all the transitions of neutrons and protons to the next higher shell of opposite parity. Unfortunately, this scheme places the giant resonance at about half of its experimentally observed energy. This difficulty has been overcome in the particle–hole calculations, in which the energies of the individual transitions have been taken from experiment and their amplitudes added coherently, taking into account an appropriate force. The general features of these giant resonances have thus been reproduced, but the detailed complexity observed experimentally presents an almost hopeless problem to the theorist.

Several interesting phenomena occur at energies above the giant resonance. For example, in heavy nuclei there is apparently electric quadrupole strength mixed in with the tail of the dipole resonance. The high-energy photoprotons emitted following the absorption of photons near 25 MeV have angular distributions that are peaked forward of 90°. This well-established effect is usually attributed to $E1$–$E2$ mixing.

At still higher energies, above 50 MeV, the predominant absorption mechanism is through the quasideuteron effect. Here, a photon interacts with a neutron–proton pair while they are undergoing a collision inside the nucleus. If the neutron and proton do not undergo further interactions in the nucleus, they emerge with the energy and angular correlation appropriate to the deuteron photodisintegration. This has been observed in light nuclei. Of course, in heavy nuclei this correlation is usually destroyed, and the quasideuteron absorption can result in the emission of three, four, or even five nucleons. These high-energy phenomena are so important that their cross section integrated from 30 to 140 MeV is approximately 75% of a dipole sum.

There are now many data that extend the measurements of the total photonuclear cross sections up to above 400 MeV, well beyond the peak of the delta resonance. It has been established that for $A>100$ the total cross section is the cross section for the emission of at least two neutrons; for the actinides the fission cross section represents the total; and for $A<40$ the direct attenuation method leads to the total cross section. The result of this extensive series of measurements is that when the total cross section is divided by A we obtain a universal curve for nuclei extending from Li to U. When compared with the total cross section for the proton, this universal curve for complex nuclei is slightly wider but with a reduced peak cross section, the integrated cross section being about six dipole sums per nucleon.

See also NUCLEAR REACTIONS; NUCLEAR STRUCTURE; RESONANCES, GIANT.

BIBLIOGRAPHY

J. S. Levinger, *Nuclear Photodisintegration*. Oxford, London and New York, 1960. (I)

M. Danos and E. G. Fuller, *Annu. Rev. Nucl. Sci.* **15,** 29 (1965). (I)

M. G. Huber, *Am. J. Phys.* **35,** 685 (1967). (I)

B. M. Spicer, in *Adv. Nucl. Phys.* **2,** 1 (1969). (I)

F. W. K. Firk, in *Annu. Rev. Nucl. Sci.* **20,** 39 (1970). (I)

E. Hayward, *Photonuclear Reactions* (NBS Monograph 118). U.S. Govt. Printing Office, Washington, D.C. 20402 (1970). (I)

Proc. Int. Conf. Nucl. Structure Studies Using Electron Scattering and Photoreaction, Sendai, 1972 (K. Shoda and H. Ui, eds.) (Res. Rep. Lab. Nucl. Sci., Tohoku Univ., Tomizawa, Sendai, Japan) **5,** 1972. (I)

Proc. Int. Conf. Photonuclear Reactions and Applications, Asilomar Conf. Grounds, Pacific Grove, CA, 1973 (B. L. Berman, ed.). Lawrence Livermore Lab., Univ. of California, Livermore, CA, 1973. (I)

B. L. Berman and S. C. Fultz, *Rev. Mod. Phys.* **47,** 713 (1975). (I)

J. Ahrens, *Nucl. Phys.* **A446,** 229c (1985).

Photosensitive Materials *see* Light-Sensitive Materials

Photosphere

A. M. Title

The photosphere is the region of the sun from which the majority of solar radiation escapes. It is bounded below by a convection zone (6800 K) and above by the temperature minimum (4400 K). Horizontally, the photosphere is structured by thermal plumes which overshoot from the convection zone. The temperature difference between the center and edges of the plumes is about the photospheric temperature.

The solar photosphere is especially interesting because it is an observable interface between convection and radiation. Both the vertical density gradient and the photon mean free path are just below what can be resolved with ground-based telescopes, but are within the resolution of current technology space optical systems. Hence, at present the very best observations allow detailed measurements of convection and three-dimensional radiative transfer processes. Modern high-speed computers are just sufficiently fast to begin to make progress in detailed simulations of the top of the convection zone and the photosphere with spatial scales comparable with measurements.

When photographed in white light the photosphere appears to be a uniformly granulated surface with a typical spatial scale of 1500 km. However, accurate measurements of the size distribution of the surface intensity fluctuations shows that the number of structures increases with decreasing size to the limits of telescope resolution. The solar granulation was first reported by Herschel in 1801, but it has been only in the last decade that observed temporal and spatial scales have been well described by detailed theoretical models and numerical simulations because of the difficulties of constructing models which combine convection and radiation.

For thousands of years observers have noted dark spots on the solar surface. These sunspots, which wax and wane with a period of approximately 11 years, are a result of eruption of magnetic flux from below the surface. At present there do not exist good theoretical models of spots as stable magnetic structures embedded in the solar surface. This suggests that spots are essentially dynamic structures. It may also be that important geometrical properties of spots both above and below the surface have not yet been recognized. The magnetic fields associated with spots do not disappear when the spot vanishes, but they are distributed over the solar surface by a combination of surface flows and diffusion generated by the random convective flows.

Although when the magnetic field is confined in a spot, the surface darkens; when the spot disappears and the field is distributed, the surface brightens at the locations of the field. Because of the nonlinear relationship between confined and distributed fields the total solar flux drops with the appearance of a new sunspot group, but on average the surface brightness is proportional to the total flux measured on the solar surface. Thus, the brightness of the sun rises and falls in phase with the solar cycle.

Ground-based observations of the sun are complicated by the atmosphere which introduces both distortions and blurring, "seeing," which make time sequences of images very difficult to interpret. In 1985 a space shuttle-borne telescope obtained movies of the photosphere completely free of atmospheric effects. Based on the knowledge obtained from the seeing-free data it has been possible to gather new ground-based data of very high quality and then to properly separate solar and atmospheric effects.

One of the major difficulties in the interpretation of the granules, which are just the tops of the convective plumes, is that the convection is chaotic and a variety of processes are occurring simultaneously. Thus, the surface patterns vary locally and globally depending on the amount of surface magnetic field and larger-scale horizontal flow patterns. In addition a significant fraction of the granules expand radially, "explode," at the end of their lives, significantly affecting the granules in their neighborhood.

Observationally the evolution of the granules is masked to a large extent by the oscillations in the solar atmosphere. Convection produces surface displacements and acoustic noise. These displacements give rise to free-surface gravity waves which are analogous to ocean surface waves. The sharp drop in density at the photosphere and the increase in temperature toward the center of the sun form a resonant cavity for acoustic waves. The combination of a resonant structure, a large source of acoustic noise, and a medium with very little dissipation allows resonant wave patterns to develop in the sun. About 10 000 000 such acoustic modes coexist in the surface at any instant in time and cause the atmosphere to rise and fall with a speed of about 2 km s^{-1} in a period of about 5 min.

Because the dispersion relations of the seismic waves have now been well measured, it is possible to separate convective and wave motions. The study of the photospheric waves has created a new science—helioseismology. Just as earthquakes can be used to understand the center of the earth, the photospheric surface wave can be used to measure the temperature and constitution of the solar interior. In addition, the surface wave patterns allow the measurement of interior rotation as a function of radius, longitude, and latitude.

Although the top of the photosphere is about 4400 K, the atmosphere above it is much hotter. Immediately above the photosphere is the chromosphere with a temperature of about 20 000 K. Even higher in the atmosphere is the corona which has a temperature of several million degrees. Because all of the solar energy is generated internally, the high temperatures of the outer layers must be generated by nonthermal processes. Until the mid-1970s the heating was thought to be caused by photospheric acoustic waves which propagated outward and turned into shock waves which dissipated in the outer layers. Spacecraft instruments have been able to measure the amount of wave heating and have shown that it is not sufficient to explain the observed temperatures.

Currently the best explanation of the heating of the outer atmosphere is the interaction of the convective plumes with small-scale magnetic structure in the surface. However, the basic mechanisms of the heating are not yet understood and this is an area of very active research.

See also STELLAR ENERGY SOURCES AND EVOLUTION; SUN.

BIBLIOGRAPHY

J. O. Stenflo (ed.), *Solar Photosphere: Structure, Convection and Magnetic Fields,* IAU Symposium No. 138. Kluwer Academic Publishers, Dordrecht, 1990.

H. Spruit, A. Nordland, and A. Title, "Solar Convection," *Annu. Rev. Astron. Astrophys.* **28** (1990).

R. J. Bray, R. E. Loughhead, and C. J. Durrant, *The Solar Granulation.* Cambridge University Press, Cambridge, 1985.

H. Zirin, *Astrophysics of the Sun.* Cambridge University Press, Cambridge, 1988.

W. C. Livingston and A. N. Cox (eds.), *The Solar Interior and Atmosphere.* University of Arizona Press, Tucson, 1990.

Photovoltaic Effect

John J. Hall

The *photovoltaic effect* is the voltage produced between two dissimilar materials when their common junction is illuminated with radiation. It was first reported in 1839 by Becquerel, who observed that a potential difference was produced between two identical electrodes immersed in an electrolyte when but one of them was illuminated. Today, interest centers on the photovoltaic effect produced in totally solid systems, for which several important applications are found; among these are the photocell, which utilizes the photovoltage produced across a metal–semiconductor junction as a measure of the intensity of the incident light, and the solar cell, which uses the photovoltage produced by an irradiated *p-n* semiconductor junction as a source of power.

When light or other radiation is absorbed in matter, the energy of the absorbed photons or particles may be transferred to the electron system of the material, resulting in the creation of charge carriers. These may be electron–ion pairs in an electrolyte, for example, or electron–hole pairs in a semiconductor. If the carriers are created in a region of varying electrical potential, they will be accelerated by the resultant electric field and a current, the photocurrent, will flow. In photovoltaic systems, the electric field occurs at the interface of two dissimilar materials. The origin of this field is the difference in the chemical potential, or Fermi level, of the electrons in the two isolated materials. When they are joined, the system approaches a new thermodynamic equilibrium state which may only be achieved when the Fermi level is equal in the two materials. This occurs by the flow of electrons from one material to the other until a voltage difference is established between the two whose potential is just equal to the initial difference of the Fermi levels. It is this "contact potential" at the material interface which is the source of the electric field driving the photocurrent.

The photovoltaic effects produced at metal–semiconductor or semiconductor *p-n* junctions provide important and instructive examples. If a voltage V is applied across such a junction, in the absence of illumination, the resultant current may be expressed by the diode equation:

$$I = I_s[\exp(eV/kT) - 1],$$

where I_s is the reverse saturation current, and kT/e is the absolute temperature expressed in volts (300 K = 0.026 V). If the junction region is now illuminated to produce a constant photocurrent I_l, the total current through the junction becomes

$$I = I_s[\exp(eV/kT) - 1] - I_l.$$

The maximum photovoltage due to I_l is produced when the circuit is open and the total junction current is zero, under which condition we may easily solve for the open circuit photovoltage

$$V_{oc} = (kT/e)\ \ln(1 + I_l/I_s).$$

In real devices, the photocurrent is often several orders of magnitude greater than the reverse saturation current, so that photovoltages many times kT/e may be obtained. However, under conditions of constant illumination, I_l/I_s is a sufficiently strong function of the temperature that photocells and solar cells ordinarily show negative temperature coefficients of the photovoltage.

See also ELECTRON ENERGY STATES IN SOLIDS AND LIQUIDS; PHOTOCONDUCTIVITY; PHOTOELECTRON SPECTROSCOPY; SEMICONDUCTORS.

BIBLIOGRAPHY

M. Altman, *Elements of Solid-State Energy Conversion*, Chap. 4. Van Nostrand-Reinhold, New York, 1969. (E)

H. J. Hovel, "Solar Cells," in *Semiconductors and Semimetals*, Vol. 11, R. K. Willardson and A. C. Beer (eds.). Academic Press, New York, 1975. (A)

S. M. Sze, *Physics of Semiconductor Devices*, 2nd ed., Chap. 12. Wiley-Interscience, New York, 1981. (I)

Piezoelectric Effect

Donald F. Nelson

The *piezoelectric effect* is a linear, reversible electromechanical interaction occurring in materials possessing certain symmetry properties. The *direct piezoelectric effect* is the production of an electric polarization by a strain; the *converse piezoelectric efffect* is the production of a stress by an electric field. The direct effect is linear in the strain; the converse effect is linear in the electric field. Twenty of the thirty-two crystal classes are piezoelectric. All 20 piezoelectric classes lack a center of symmetry. The one remaining noncentrosymmetric crystal class, denoted O (Schönflies) or 432 (Hermann–Mauguin), possesses other symmetries which combine to exclude piezoelectricity. Important piezoelectric crystals include quartz, Rochelle salt, $(NH_4)H_2PO_4$ (ADP), $LiTaO_3$, and $LiNbO_3$. Other materials, e.g., electrically poled ferroelectric ceramics composed of $PbZrO_3$ and $PbTiO_3$, that possess less structural regularity than crystals may exhibit piezoelectricity. The common battery-operated "quartz watch" uses a piezoelectric resonator made of quartz to regulate the electronic clock mechanism.

HISTORY

The direct effect was discovered by Pierre and Jacques Curie in 1880. On the basis of thermodynamic reasoning G. Lippmann predicted the existence of the converse effect and that its interaction strength would be identical to that of the direct effect. This prediction was later verified by the Curie brothers.

The first practical use of piezoelectricity was made during World War I when P. Langevin produced ultrasonic waves in water surrounding a quartz plate (now called an *electromechanical transducer*) forced to vibrate by the application of an alternating voltage. Shortly after the war W. G. Cady invented the *piezoelectric resonator* which exploits the interaction of the piezoelectrically driven mechanical resonance of a crystal upon the electrical driving circuit. One application of this is in the frequency control of an electrical oscillator, another is in making efficient band-pass electrical filters.

MATHEMATICAL CHARACTERIZATION

The direct piezoelectric effect produces a contribution $e_{ijk}s_{jk}$ to the linear polarization P_i,

$$P_i = \epsilon_0 \chi_{ij} E_j + e_{ijk} s_{jk}, \tag{1}$$

where e_{ijk} is the *piezoelectric stress tensor*, s_{jk} is the strain tensor, χ_{ij} is the linear electric susceptibility tensor, E_j is the electric field, and ϵ_0 is the permittivity of free space. Summation over repeated tensor subscripts is implied. The converse effect produces a contribution $-e_{kij}E_k$ to the linear stress t_{ij},

$$t_{ij} = c_{ijkl} s_{kl} - e_{kij} E_k, \tag{2}$$

where c_{ijkl} is the stiffness tensor. The latter equation may be solved for the strain s_{mn},

$$s_{mn} = s_{mnij} t_{ij} + d_{kmn} E_k, \tag{3}$$

where $d_{kmn} \equiv s_{mnij} e_{kij}$ is the *piezoelectric strain tensor* and s_{mnij} is the compliance tensor.

Both piezoelectric tensors possess interchange symmetry on their last pair of indices,

$$e_{kij} = e_{kji}, \quad d_{kij} = d_{kji}. \tag{4}$$

This allows them to possess at most 18 independent components and to be recorded as 3×6 matrices. To do the latter, the last two tensor indices are contracted to one index having values 1, 2, 3, 4, 5, 6 corresponding to 11, 22, 33, 23 or 32, 13 or 31, 12 or 21, respectively. To make matrix multiplication equivalent to the scalar product of tensors, the value of d_{kl} is one-half of d_{kij} when $l = 4, 5, 6$; no such factors are needed for e_{kl}.

Recently a loss of the symmetry (4) was predicted (Nelson, 1988) when an optic mode of vibration of a piezoelectric crystal has a frequency somewhat near the acoustic wave frequency used for study. This is expected to occur near a second-order phase transition where a Brillouin-zone-center soft optic mode approaches zero frequency. In this situation the infinitesimal strain is no longer an adequate independent variable. Instead, the displacement gradient, which is the sum of infinitesimal strain plus infinitesimal rotation, must be used. The rotation couples to the angular momentum of the soft optic mode to cause the symmetry reduction.

SECONDARY PIEZOELECTRIC EFFECTS

A succession of the direct and converse piezoelectric effects produces a piezoelectric stiffening of the stiffness tensor. For an acoustic wave propagating in the direction **a** the effective stiffness tensor becomes

$$c_{ijkl}^{\mathrm{eff}} = c_{ijkl} + \frac{a_m e_{mij} a_n e_{nkl}}{\epsilon_0 a_p \kappa_{pq} a_q}, \tag{5}$$

where κ_{pq} is the dielectric tensor. Piezoelectricity also alters the dielectric tensor to be

$$\kappa_{pq}^{\text{low}} = \kappa_{pq} + (1/\epsilon_0) e_{pmn} d_{qmn} \tag{6}$$

for frequencies well below the mechanical resonances of the crystal. Piezoelectricity also leads to indirect contributions to the electro-optic effect, the elasto-optic effect, etc.

See also CRYSTAL SYMMETRY; ELASTICITY; TRANSDUCERS.

BIBLIOGRAPHY

W. G. Cady, *Piezoelectricity*. Dover, New York, 1964 (two volumes). (I) This has been the standard reference source since its first publication in 1946.

D. F. Nelson, *Electric, Optic, and Acoustic Interactions in Dielectrics*. Wiley, New York, 1979. (A) Chapters 10, 11, and 17 show that Eqs. (1) and (2) hold for nonpyroelectric piezoelectrics when components are referred to either the spatial (laboratory) or material coordinate systems; for pyroelectrics they are true only in the material coordinate system and only when the spontaneous electric field has been cancelled by extrinsic surface charge.

D. F. Nelson, *Phys. Rev. Lett.* **60,** 608 (1988).

J. F. Nye, *Physical Properties of Crystals,* Chap. VII. Clarendon, Oxford, 1957. (E)

Plasma Confinement Devices

H. P. Furth

1. GENERAL PRINCIPLES

A plasma consists of electrically charged particles: electrons of charge $q_e = -e$ and ions of charge $q_i = Ze$. In a magnetic field **B**, these charged particles are constrained to execute gyro-orbits of frequency $\Omega = qB/mc$ and radius $r_L = v_\perp/\Omega$, where $v_\perp$ is the particle velocity transverse to **B**; they move freely in the direction along **B**, at velocity $v_\parallel$.

An assembly of electrons and ions can be confined quasistatically by a magnetic field configuration provided that the field lines are closed within the confinement volume (as in a torus) or that the particle velocity $v_\parallel$ can be made to vanish at the ends of an open confinement volume (as in a mirror machine). Magnetic field configurations of these two types are used for the long-time containment of hot plasmas in controlled fusion research, and are sometimes referred to as "magnetic bottles." Other hot-plasma confinement possibilities include the utilization of finite particle inertia, radiation pressure, etc. In the following discussion, principal attention is given to quasistatic magnetic confinement.

Particle Drifts

Only in the case of a uniform magnetic field, and in the absence of other fields, do the orbits of plasma particles correspond to simple gyromotion. In a weakly nonuniform magnetic fiels, changing over some scale height s, the center of the gyromotion (the guiding center) moves perpendicular to **B** at a drift velocity reduced from the particle velocity v by a factor of order r_L/s. Specifically, there are drifts related to the gradient of the field strength,

$$\bar{v}_g = W_\perp \frac{c}{q} \frac{\mathbf{B} \times \nabla B}{B^3}, \tag{1}$$

and to the curvature of the field lines,

$$\bar{v}_c = 2W_\parallel \frac{c}{q} \frac{\mathbf{B} \times (\mathbf{B} \cdot \nabla)\mathbf{B}}{B^4}, \tag{2}$$

where $W_\perp = \frac{1}{2}mv_\perp^2$, $W_\parallel = \frac{1}{2}mv_\parallel^2$. In the presence of a weak electrostatic field, or a weak electric field induced by the time variation of **B**, there results the additional drift

$$\bar{v}_E = c\frac{\mathbf{E} \times \mathbf{B}}{B^2}. \tag{3}$$

When the magnetic field is nearly constant, in the sense that the gyromotion is only weakly perturbed, there is an adiabatic invariant of the motion

$$\mu = W_\perp/B \tag{4}$$

which corresponds to the action integral over the gyro-orbit and represents the (diamagnetic) dipole moment of the gyromotion. For fixed total kinetic energy $W = W_\perp + W_\parallel$, as in a static magnetic field, the conservation of μ implies that particles can be confined in a mirror machine (Fig. 1), provided that the mirror ratio $R_M = B_{\text{ends}}/B_{\text{middle}}$ exceeds the magnitude of $W/W_\perp$ taken at the middle of the device. The localization of particles by mirror trapping plays an important role in closed configurations, as well as in open ones.

When the periodicity of the bounce motion between mirrors is only weakly perturbed by the drift motion, there is a second adiabatic invariant

$$J = \oint v_\parallel \, ds \tag{5}$$

which is the action integral over the periodic motion along field lines. The particle drifts then lie on surfaces of constant J, which can be calculated from (5) using $v_\parallel = (2/m)^{1/2}[W - \mu B(s)]^{1/2}$. For axisymmetric systems, the constant-J surfaces are, of course, axisymmetric. In this special case, the orbit constraint actually follows from an *exact* invariant, the canonical angular momentum in the direction of symmetry.

Particle Collisions

The basic assumption of plasma physics is that the number of particles N in a Debye sphere [cf. Eq. (8)] is very large, so that collective interactions dominate scattering. In that case, small-angle scattering is also dominant over large-angle scattering by a factor of order ln$N$. The self-collision time of a particle population of temperature T, density n, mass m, and charge Ze is given by

$$\tau_c = \frac{m^{1/2}(3kT)^{3/2}}{8\pi n Z^4 e^4 \ln\Lambda} \tag{6}$$

where lnΛ is usually the order ln$N$ (i.e., about 10–20 for typical laboratory plasma). This is the time for like particle species to exchange energy and momentum. The time for ions of charge Ze to equilibrate their temperature with, or lose their momentum to, the neutralizing electrons is greater

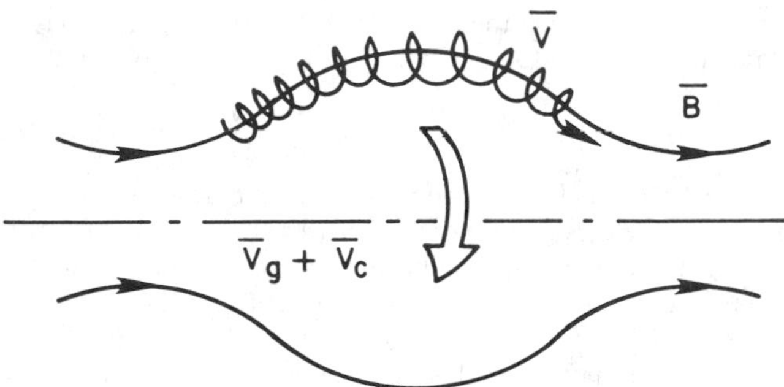

FIG. 1. Particle gyromotion and average drift in a mirror machine.

than the electron τ_{ce} by the factor $m_i/m_e z^2$, if it is assumed that T_i/T_e does not exceed m_i/m_e.

The effect of scattering on particle confinement is strongest for open-ended systems. The characteristic time for altering the magnetic moment μ is τ_c; since small-angle scattering predominates, τ_c is also the characteristic time scale for particle loss from mirror machines, even in the presence of very high mirror ratios. In closed confinement systems, scattering allows the particles to diffuse across the magnetic field by stepping from one orbit to another. If the gyroradius is small compared with the plasma size, the confinement time can be much greater than τ_c.

Plasma Diamagnetism

A plasma may be regarded as an aggregation of dipole moments, producing a diamagnetic effect

$$\frac{\delta B}{B} = -\frac{4\pi n\mu}{B} = -\frac{4\pi n W_\perp}{B^2}. \tag{7}$$

The associated fractional reduction of the magnetic field pressure $B^2/8\pi$ inside the plasma is measured by $\beta = 8\pi n W_\perp / B^2$, which is also the ratio of plasma pressure $nW_\perp$ to magnetic pressure. The gradient in the dipole population at the plasma edge gives rise to a circulating plasma current density, which is referred to as the plasma diamagnetic current J_d.

Charge Neutralization

The pressure required to separate plasma electrons and ions by a distance of one Debye length

$$\lambda_D = (kT/4\pi e^2 n)^{1/2} \tag{8}$$

is of the order of the thermal pressure $P = nkT$. The Debye length sets the maximum size for confinement of unneutralized plasmas; most confinement experiments use approximately charge-neutral plasmas that are much larger than λ_D.

Equilibrium and Gross Stability

The equilibrium and gross stability properties of confined plasmas can be treated by a fluid description, typically the magnetohydrodynamic (MHD) model. Here it is assumed that r_L and λ_D are small, with μ conservation and the drift equations [(1)–(3)] being valid. In closed confinement systems, the fluid pressure is normally allowed to be isotropic ($P = P_\perp \equiv n\langle W_\perp \rangle = P_\parallel \equiv 2n\langle W_\parallel \rangle$); in mirror confinement, the pressure is necessarily anisotropic ($P_\perp \neq P_\parallel$). The equilibrium equations balance these fluid pressures against the electromagnetic stress tensor.

Since plasma is diamagnetic, displacements into regions of reduced B are energetically favored; this is the principal source of difficulty in finding stable equilibria. In the presence of an inhomogeneous magnetic field, the associated ∇B and curvature drifts can generate charge separations at the plasma surface that increase with time. The resultant electric field transverse to B then acts to displace the entire plasma according to Eq. (3) in the direction of weaker magnetic field.

To provide equilibrium, it is sufficient for the particle orbits to close within the plasma, as, for example, in the case of the axisymmetric mirror machine shown in Fig. 1. *Stable* equilibria constitute a more restricted class; in the mirror machine of Fig. 1, a local outward plasma displacement conforming with the field lines (i.e., a flute perturbation) will tend to grow, since the particle drifts in the outwardly decreasing magnetic field produce an azimuthal electric field that further displaces the plasma along $-\nabla B$. Similar instabilities exist in toroidal configurations. When they are localized in the region of unfavorable curvature (i.e., on the large-R side of the torus), they are called "ballooning modes."

Energy Loss Processes

The thermal energy content of a plasma can be lost simply by escape of the hot particles. In that case, the energy confinement time τ_E is identical with the particle confinement time τ_p. More commonly, heat conduction also plays a role. If the transport of particles and of heat is due to single-particle collisions, it is called classical; if it is due to collective effects (i.e., MHD turbulence or plasma microturbulence), it is called anomalous.

Classical particle diffusion $D_\perp$ across the magnetic field is based on electron-ion collisions (since like-particle collisions do not serve to shift the mean guiding-center location). Classical ion heat conduction $\chi_{\perp i}$ across the magnetic field depends on ion-ion collisions; for $T_e \sim T_i$, $\chi_{\perp i}$ is much greater than the particle diffusion or the electron heat conduction $\chi_{\perp e}$, since $r_{Li} \gg r_{Le}$. The energy loss from hot, magnetically confined plasmas is usually dominated by nonclassical mechanisms of particle and heat transport, which are based on collective plasma fluctuations rather than on single-particle collisions (cf. Sec. 2).

The heat conduction $\chi_\parallel$ along field lines is the same as for a magnetically unconfined plasma, unless mirror confinement is effective (cf. Section 3). In hot plasmas, $\chi_\parallel$ is typically much larger than $\chi_\perp$ and tends to be dominated by the electrons, since their thermal velocity is greatest.

Energy can also be lost by processes other than plasma transport. Bremsstrahlung constitutes an unavoidable form of radiation loss. In nonhydrogenic or impure hydrogenic plasmas, there can be much larger additional losses due to line radiation. Synchrotron radiation in the confining magnetic field tends to become more important than bremsstrahlung at high electron temperatures. In laboratory plasmas, the reabsorption of radiation is not a significant favorable factor, except in the case of synchrotron radiation. In the presence of neutral atoms, ion energy can be lost by charge exchange, but hot, dense plasmas tend to "burn out" the neutral population.

Fusion Plasmas

Deuterium and tritium fuse more readily than other combinations of nuclei, and near-term fusion reactor designs are therefore generally based on this process. The reaction $D + T \rightarrow {}^4He(3.5\ MeV) + n(14.1\ MeV)$ has a peak cross section of order 5 barns at 100-keV deuteron energy. The Coulomb scattering cross section is always somewhat larger; nonetheless, a modest energy multiplication $Q = 17.6$ MeV/(input energy per fusion)$\gtrsim 1$ can be achieved even when the ions are lost after one scattering time. If the plasma is well confined against scattering processes, large Q values can be achieved at lower temperatures ($\gtrsim 10$ keV). If the alpha particles of the D-T reaction are confined and allowed to thermalize with the plasma, the Q value becomes infinite for $n\tau_E > 3 \times 10^{14}\ cm^{-3}\ s$ ($T \sim 10$ keV) and the plasma is then said to ignite. Bremsstrahlung cooling sets a minimum ignition temperature of 4.0 keV even for perfectly confined D-T plasmas.

2. *CLOSED CONFINEMENT*

Toroidal confinement geometries have the feature that the outward gradient of B cannot be everywhere positive at the plasma surface. As a result, the fundamental problem in toroidal confinement design is to provide plasma equilibrium

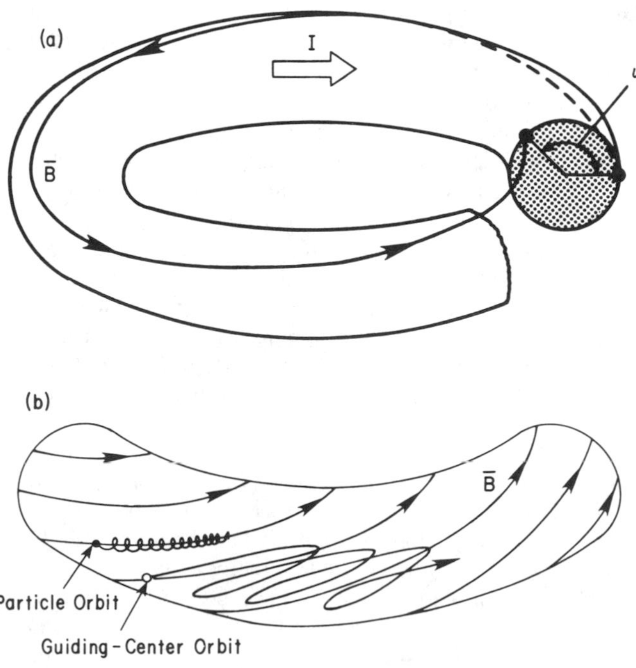

FIG. 2. (a) The tokamak configuration. (b) Motion of a mirror-trapped particle in the tokamak.

and stability against the tendency of the diamagnetic plasma to displace itself along $-\nabla B$, e.g., along its major radius R.

Tokamaks

The simplest closed configuration would be a pure toroidal magnetic field B_t; in this case, however, the particle drifts are incompatible with equilibrium. In the tokamak, equilibrium is provided by adding a poloidal field component B_p as in Fig. 2a, so that each field line is taken through a finite rotational transform angle ι on each transit around the torus. Conservation of the toroidal canonical angular momentum then ties the particle motion to the poloidal magnetic flux, producing the characteristic closed orbits shown in Fig. 2b.

The poloidal field of the tokamak is generated by a toroidally directed plasma current I, which can be induced by air-core transformer primary windings and/or by an iron core. The plasma pressure and poloidal magnetic pressure of the tokamak discharge create an outward force along R, which is balanced by an inward Lorentz force arising from an externally applied vertical magnetic field B_v.

Since the tokamak confining field depends critically on currents flowing in the plasma, it can be unstable against a variety of gross disturbances, in particular, helical plasma perturbations that conform approximately with the magnetic field lines (MHD kink modes). When $\iota = 2\pi$, a fundamental kink mode can arise, with poloidal periodicity $m = 1$ and toroidal periodicity $n = 1$. Even if the Kruskal–Shafranov condition $\iota < 2\pi$ is satisfied, there can arise weaker, higher-harmonic kink modes ($m/n = 2,3,4$ or $\frac{3}{2},\frac{4}{3}, \ldots$). These considerations impose strict limitations on the magnitude and radial profile of the plasma current and pressure.

As the β value of the tokamak plasma is raised, B_v must be raised as well; the result is a noncircular distortion of the poloidal field pattern. The permissible β is limited by this consideration, but special equilibrium solutions have been found with β values as large as 20–30%. The β value is also constrained by the requirement of stability against local plasma displacements into regions of weaker magnetic field (MHD kink, flute, and ballooning modes). MHD stability calculations give typical upper limits for the volume-averaged β in terms of the "Troyon formula" $\langle\beta\rangle = 3I/aB_t$, where I is the plasma current in megamperes, a is the minor radius in meters, B_t is the toroidal field in tesla, and $\langle\beta\rangle$ is given in percent. Present tokamak experiments agree fairly well with this formula and have reached values as high as 8% in the case of a vertically elongated D-shaped plasma cross section with low aspect ratio $R/a = 2.5$ (cf. Fig. 3 and Table I). For the fusion reactor application, where somewhat higher aspect ratios ($R/a \sim 3$–6) will be desired, the tokamak β value may be limited to about 5%—unless a successful transition can be made to the theoretically indicated "second stability regime," where the geometric distortions associated with the β-driven equilibrium shift are helpful in restoring MHD stability.

Strong geometric shaping of the tokamak minor plasma cross section incidentally lends itself to the formation of a "divertor," as illustrated in Fig. 3. The inner region of closed toroidal magnetic surfaces is bounded by a magnetic separatrix, beyond which the plasma outflow can be guided along

Table I Illustrative parameters of Tokamak experiments

	Minor cross section	R (cm)	a (cm)	B_t[a] (T)	I (MA)	n[a] (10^{13} cm^{-3})	T_e[a] (keV)	T_i[a] (keV)	τ_E (ms)	$\langle\beta\rangle$ (%)
T-3 (1969)	Round	100	15	3.5	0.1	7	1.2	0.5	7	0.1
PLT[b] (1978)	Round	132	40	3.5	0.45	4.5	4.0	6.5	20	0.3
Alcator C (1983)	Round	64	16	12	0.8	150	1.6	1.6	50	0.4
TFTR[b] (1988)	Round	250	85	5	1.4	8.7	8.5	27	180	1.2
JET[b] (1988)	Dee	296	125	3.4	6	6	6	18	330	0.8
DIII–D[b] (1988)	Dee	167	62	0.8	1.3	10	1.4	1.4	40	8

[a] Central values.
[b] With neutral-beam injection.

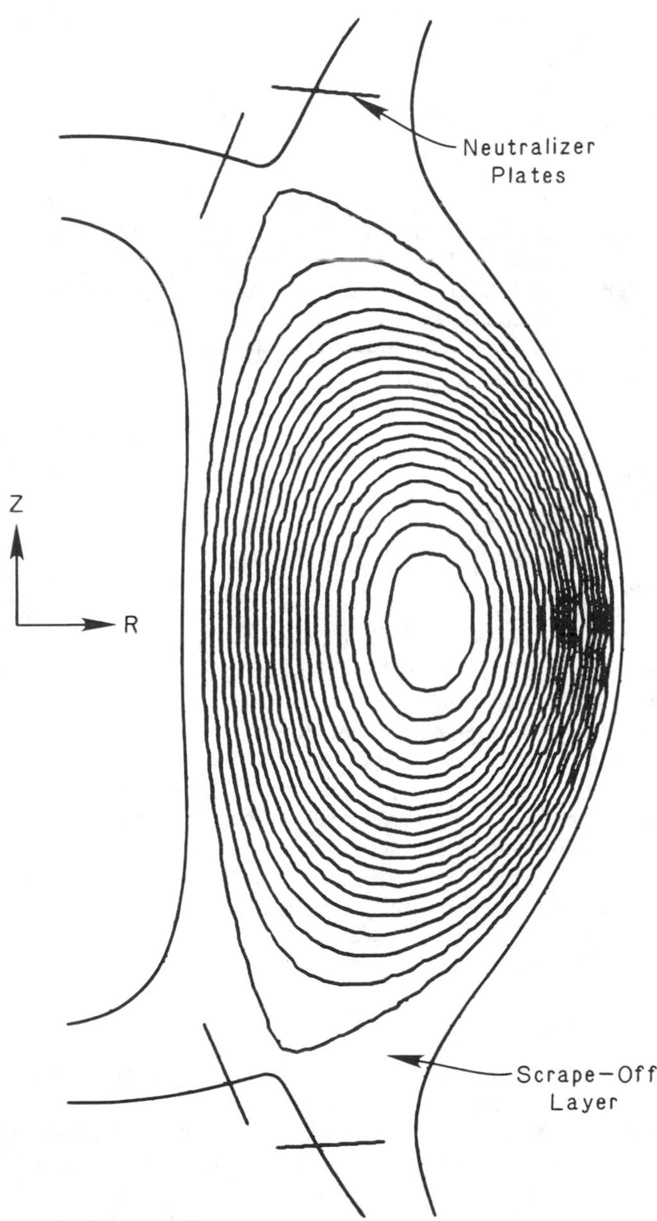

FIG. 3. D-shaped minor cross section of a tokamak plasma, bounded by a divertor separatrix.

open field lines (the "scrape-off layer") to a set of neutralizer plates. Bounding the tokamak plasma by a magnetic divertor, rather than by a simple mechanical "limiter" plate, has the advantage of reducing the uncontrolled recycling of neutral gas and impurity atoms at the plasma edge.

Parameters achieved by the T-3 experiment in 1969, and some illustrative present-day tokamak parameters, are shown in Table I. The tokamak plasma electrons can be heated Ohmically by the plasma current (Fig. 1) to electron temperatures in the multi-keV range; the ions are heated to somewhat lower temperatures by equilibration with the electrons. To progress from Ohmic-heating parameters to the range of fusion reactor interest, it appears that some form of auxiliary plasma heating is required. The principal techniques being developed for this purpose are the injection of energetic neutral-atom beams, and the damping of radio-frequency waves. Neutral-beam heating experiments on the TFTR device have reached ion temperatures up to 30 keV.

The energy loss of present-day tokamaks is due principally to plasma transport processes, with impurity-radiation and charge-exchange cooling playing secondary (but important) roles. Classical collisional transport in the tokamak parameter range, where the mean free path λ_e exceeds the plasma size, depends on details of the particle orbits, and can be much larger than simple classical transport in plane geometry. The predicted classical ion heat flow accounts fairly well for the measured $\chi_{\perp i}$ value in moderate-powered tokamak-heating experiments. The classical predictions for the electron heat flow and the particle flow are much smaller—but the experimentally measured values of $\chi_{\perp e}$ and $D_\perp$ usually turn out to be comparable to $\chi_{\perp i}$, thus creating a pronounced anomaly.

The theory of anomalous cross-field transport in toruses is well developed. The central feature is the excitation of electrostatic drift waves, propagating at the diamagnetic drift velocities of the plasma particles, and driven by various departures of the plasma from thermodynamic equilibrium: finite density and temperature gradients, etc. When $\lambda_c \gg R$, the mirror-trapping of particle orbits becomes of critical importance for anomalous transport, and gives rise to the so-called trapped-particle instabilities, which are dangerous potential loss mechanisms in the toroidal fusion reactor regime. Another important type of instability is the so-called η_i mode, which results from large values of $\eta_i = d \log T_i / d \log n$ and can therefore be avoided by increasing the peakedness of the plasma density profile $n(r)$. The enhancements in $\chi_{\perp i}$ that are seen in high-powered ion-heating experiments appear to be associated with this type of mode.

More detailed experimental documentation of the dominant anomalous transport processes and their scaling remains to be carried out, but the present empirical scaling laws for the tokamaks with reactor confinement requirements appear to be compatible.

Specifically, the "L-mode" scaling ($\tau_E \propto IR^{1.75} a^{-0.37} P^{-0.5}$, where P is the total heat outflow) proposed by Goldston in 1982 has provided quite accurate lower-limit predictions for τ_E in tokamak experiments, and would require approximately $IR/a \sim 80$ MA for an ignited tokamak D-T reactor. According to present economic projections, a full-scale tokamak fusion reactor (~1000 MW electric) could be allowed to have a current exceeding 20 MA, so as to meet the L-mode requirement for $R/a \sim 4$—but a current level in the 10-MA range would be more convenient, especially if the tokamak reactor is to operate in steady state. In the large tokamaks of the present generation, special experimental techniques, involving the use of a divertor separatrix (Fig. 3) and/or of centrally peaked density profiles, have typically succeeded in improving τ_E by factors of about two relative to the L-mode, thus lowering the reactor-current requirement by the same factor.

At the outset of tokamak research, the only known method for driving the plasma current around the torus was transformer-type magnetic induction, achieved by pulsing the poloidal-field coils. Since this approach tends to limit the tokamak plasma duration to tens of seconds in present-generation experiments and to a number of hours in projected reactor operation, there has been a strong motivation to develop techniques for noninductive steady-state current drive. Beams of tangentially injected energetic ions, as well as microwave beams in the lower-hybrid frequency range, have been used successfully for this purpose. Tokamak currents reaching 2 MA have been driven for several seconds in the JT-60, and smaller currents have been driven for several minutes in the TRIAM—both in Japan. While the energetic efficiency of noninductive current drive remains to be improved, the tokamak is currently regarded as a potential steady-state plasma-confinement device.

Pinches

The tokamak is the most successful member of a class of closed confinement systems known as pinches. The Z pinch uses a current-carrying plasma column (Fig. 4a), confined entirely by its own poloidal magnetic field. The simple configuration was popular in early fusion research but suffers from severe MHD instabilities of the flute and kink types. The θ pinch (Fig. 4b) is compressed by a longitudinal magnetic field, thus avoiding the most severe MHD modes. The basic heating method in Z and θ pinches is initial shock heating, followed by adiabatic compression.

In open-ended geometry the θ-pinch has proved effective in generating high-density (10^{16}–10^{17} cm^{-3}), high-temperature ($T_i = 1$–5 keV) plasmas, with β values approaching unity, but with confinement times limited by the end loss of these collision-dominated plasmas, which escape at thermal velocity. The end loss can be stopped by heaping the lon-

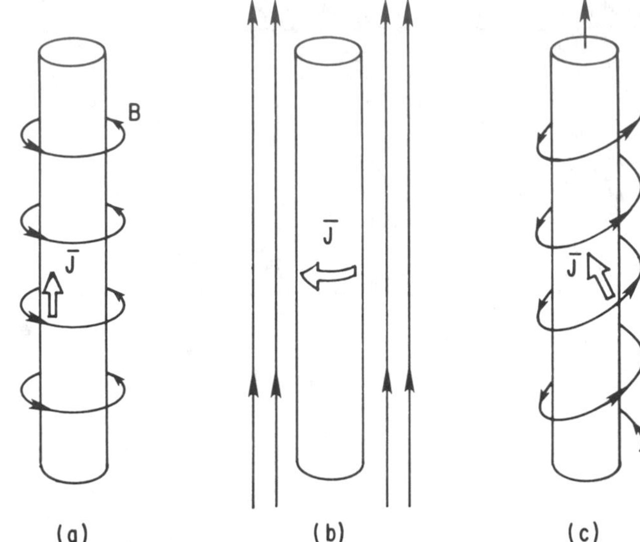

FIG. 4. Three types of pinch: (a) the Z pinch; (b) the θ pinch; (c) the "stabilized pinch."

gitudinal field lines back through the plasma interior—giving rise to an axially elongated closed magnetic toroid that is called the field-reversed configuration (FRC). New problems of MHD instability are introduced by the unfavorable curvature at the ends of the toroid, but special stabilizing techniques may prove effective.

Another adaptation of the θ-pinch uses a static longitudinal magnetic field and forms a high-β plasma column ($n \sim 10^{18}$ cm^{-3}) by means of a CO_2 laser beam. Recently, the simple Z-pinch geometry has been used to achieve similarly dense, anomalously stable plasma filaments of 10–100 μm diameter, by applying voltages in the megavolt range to drive 100-KA-range current pulses of submicrosecond duration through frozen hydrogen fibers.

The Z and θ pinches can be combined in the form of the so-called "stabilized pinch" (Fig. 4c), which resembles the tokamak in that it uses helical field lines and can achieve substantially MHD-stable configurations. The difference is that the stabilized pinch uses toroidal and poloidal fields of comparable strengths, i.e., rotational transform values well above 2π. As a result, higher β values can be accommodated, but stability is more precarious than for tokamaks. The best results are obtained for toroidal field profiles that reverse near the plasma edge; this approach is called the reverse-field pinch (RFP). Some representative parameters for recent experiments are given in Table II.

Stellarators

Toroidal plasma equilibrium can be created in a vacuum magnetic field by departing from axisymmetry. To generate a rotational transform in this case, a set of helical multipole windings can be used (Fig. 5a), or an equivalent magnetic field structure can be generated by means of a set of noncircular, nonplanar toroidal-field coils; such confinement schemes are called stellarators. In terms of relative experimental effort, the stellarator is the second most intensively

Table II Illustrative parameters of RFP and stellarator experiments

	Torus types	R (cm)	a (cm)	B_t^a (T)	I (MA)	n^a (10^{13} cm^{-3})	T_e^b (keV)	T_i^b (keV)	τ_E (ms)	$\langle\beta\rangle$ (%)
ETA-BETA (1982)	RFP	65	12	—	0.12	5	0.1	0.1	0.1	10
ZT-40 (1986)	RFP	114	20	—	0.34	8	0.3	0.3	0.7	18
WVII-A (1982)	Stellarator[c]	200	10	3.2	0	12	0.7	1	20	0.5
Heliotron E (1986)	Stellarator[c]	220	20	1.9	0	2.6	0.7	1.6	10	0.3

[a] Average values.
[b] Central values.
[c] With neutral-beam injection.

studied plasma-confinement configuration, after the tokamak. Some representative parameters are given in Table II. Plasma equilbrium can also be obtained in the absence of rotational transform (i.e., for field lines that close after a single toroidal circuit), provided that the integral $\oint dl/B$ is the same on all field lines comprising a constant-pressure contour of the plasma. (If this condition is met, the diamagnetic current J_d is divergence-free and does not give rise to growing charge separation.) A configuration that permits such equilibria is the bumpy torus of Fig. 5b.

In these nonaxisymmetric configurations, the constraint of canonical angular momentum conservation is lost, thus introducing a class of particle orbits making finite excursions even in the limit $r_L \rightarrow 0$. For nearly collisionless plasmas, this feature can give rise to large enhancements of the classical heat conduction $\chi_{\perp i}$. On the other hand, the predominant use of vacuum magnetic fields to provide confinement helps to eliminate MHD instabilities at low β values, and enhances the possibilities for high-β toroidal confinement. Some approaches to the toroidalization of the θ pinch have followed along this line.

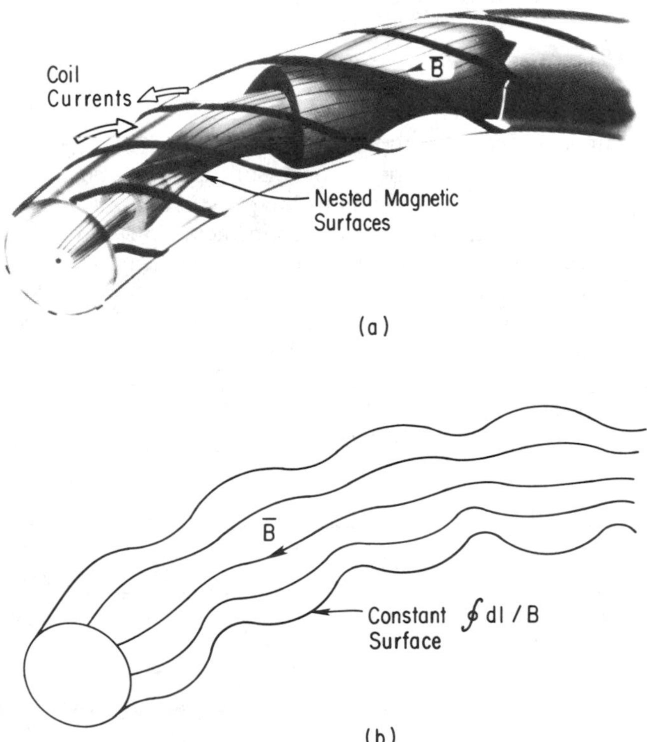

FIG. 5. (a) Stellarator with hexapole windings. (b) Bumpy torus.

3. *OPEN-ENDED CONFINEMENT*

Removal of the constraint of field-line closure greatly facilitates the achievement of MHD-stable equilibria. The central problem of open-ended configurations is the minimization of end loss; a number of different approaches have evolved.

Mirror Machines

As illustrated in Fig. 1, particles that conserve the adiabatic invariant μ can be confined by magnetic mirrors. The adiabaticity condition will be met if r_L is sufficiently small compared with $|B/\nabla B|$, and will hold for times of the order of the scattering time τ_c of Eq. (6). For comparable temperatures, the electron scattering time is shorter by the factor $(m_e/m_i)^{1/2}$ than the ion scattering time. The electrons thus tend to be lost first, but are restrained by the resultant positive electrostatic well ($e\phi_{max} \sim 4.7kT_e$). The case of principal interest is $T_e \ll T_i$; the ions can then be held by the magnetic mirrors against the electrostatic pull of the electrons, and the whole plasma is lost on the time scale τ_{ci}. If the plasma energy is introduced through the ions, the condition $T_e \neq T_i$ is admissible, since the electron–ion equilibration time typically exceeds τ_{ci}. The presence of the electrostatic well, however, has an appreciable adverse impact on ion confinement. For D-T fusion reactor parameters, energy multiplication factors of order $Q \sim 8$ would be possible in the range $T_i \gtrsim 100$ keV, but the presence of the electrons depresses the Q value to the neighborhood of unity. The electron effect on ion confinement can be turned to advantage in a triple-mirror scheme, the so-called "tandem mirror machine," where the positive electrostatic potentials of two dense end-mirror plasmas serve to provide confinement for an isotropic ion population in a larger central mirror region. This special approach creates the prospect of mirror systems with classical energy confinement times amounting to appreciable multiples of τ_{ci}, and with fusion reactor Q values well in excess of unity.

Initial experiments with mirror machines resembling Fig. 1 gave rise to severe nonclassical, or anomalous, particle losses due to MHD flute instability. The tendency toward fluting, or other types of gross fluid instability, can be suppressed in mirror-machine geometries designed to provide $\nabla B>0$ at the plasma surface (minimum-B configurations). The Ioffe machine of Fig. 6 is a particularly convenient embodiment of this idea; it has nested ellipsoidal constant-B surfaces, and the particles drift on axis-encircling constant-J surfaces. In mirror machines of this type, beginning in 1962 with the P-7 device, and continuing with the PR-5 device of Table II, the problem of anomalous loss was reduced to high-frequency microinstability arising from the inherently non-Maxwellian ion distribution.

Since $W/W_\perp$ must be less than the mirror ratio, the ion distribution function has a "loss cone" in velocity space around the $v_\parallel$ axis. In addition, there is a depletion of low-energy ions, since these have the shortest scattering times and also tend to be expelled electrostatically. A large variety of microinstabilities can be driven by such departures from thermal equilibrium. For the fusion reactor application, the most important modes are: the flutelike drift-cyclotron loss-cone mode, which arises in the range of the ion-cyclotron frequency from a coupling of electron and ion waves; and the high-frequency, convective-loss-cone mode, which propagates in the direction along **B** and amplifies in a laserlike manner. The drift mode can be stabilized by providing large values of a/r_L, where a is the plasma minor radius; stabilization of the convective mode requires that L/r_L should not be too large, where L is the axial plasma length.

Simultaneous freedom from all microinstabilities is difficult to provide, and mirror confinement experiments generally suffer from high-frequency activity and enhanced end-loss rates. A powerful stabilizing technique is to fill in the lowest-energy portion of the loss cone with unconfined "warm" ions in transit through the confined plasma. By this technique, the results given in Table III were obtained on the 2*XIIB* device, in a plasma resembling Fig. 6. Mirror-confined plasmas are generally built up by injection and trapping of accelerated ions or neutral atom beams; the former technique was used in the P-7 and PR-5 experiments, and the latter in 2*XIIB*. "Warm" transit ions can be injected by plasma guns. The tandem mirror machine, which has the best prospects of achieving large Q values in the face of classical scattering, is also especially well suited to combat microinstabilities, since the isotropic central plasma provides stabilizing "warm" particles for the anisotropic mirror-confined end plasmas.

The neutral-beam–injected tandem mirror experiments TMX and TMX-U (cf. Table III) consisted of long uniform-B central cells, bounded by two sets of appropriately shaped higher-B end cells, to create the desired ion-trapping potential. Central-cell ions were successfully confined for times longer than τ_{ci}. Anomalous cross-field losses were then found to become stronger than the normally dominant mirror end-losses, thus imposing a new type of limit on the energy confinement time. Experiments have not yet established the prospects for reactor-compatible cross-field confinement in tandem mirrors relative to tori.

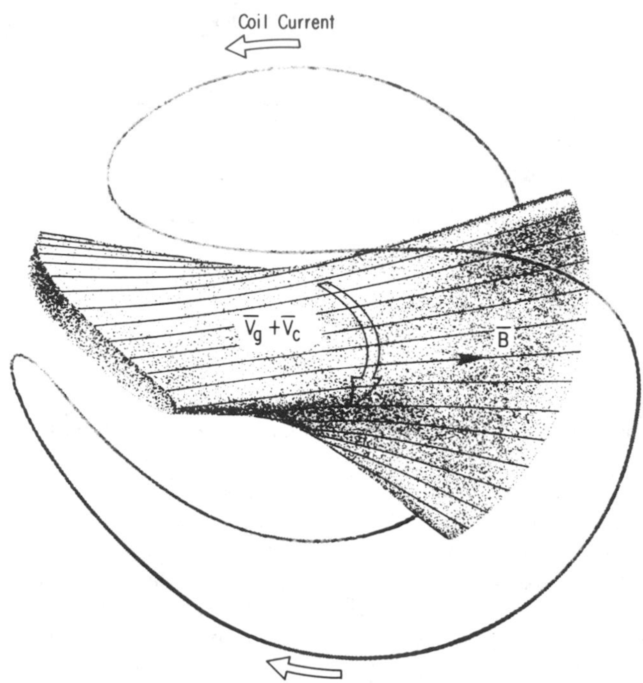

FIG. 6. Ioffe-type minimum-B mirror configuration, produced by a "baseball coil."

Multiple Mirrors, Free Flow

In dense, collisional plasmas, end loss can be reduced by using a large number of mirror regions connected in series. If the length of each region is comparable to the scattering mean free path, the plasma confinement time can be extended substantially beyond τ_c through the recapture of particles that have entered the loss cone.

Open-ended systems, such as θ pinches, which do not rely on any substantial end-stoppering effects, can still benefit significantly from magnetic confinement. Since the magnetic field reduces the free flow to a single dimension, it becomes possible to consider systems of sufficient length so that the

Table III Illustrative parameters of mirror experiments

	Mirror types	L^a (cm)	a (cm)	B^b (T)	R_M	n (10^{13} cm^{-3})	T_e (keV)	$\langle w_i\rangle^c$ (keV)	τ_E (ms)	β^b (%)
PR-5 (1965)	Minimum-B	120	7	0.5	1.7	0.01	0.05	1	0.1	0.1
2XIIB (1975)	Minimum-B	150	7	0.7	2.0	10	0.1	13	0.5	~100
TMX (1980)	Tandem	530	25	0.1	20	2	0.1	0.2	2	20
TMX-U (1986)	Tandem	800	30	0.3	7	0.3	0.3	2.5	5	5

[a] Central cell length.
[b] Central value.
[c] Average ion energy.

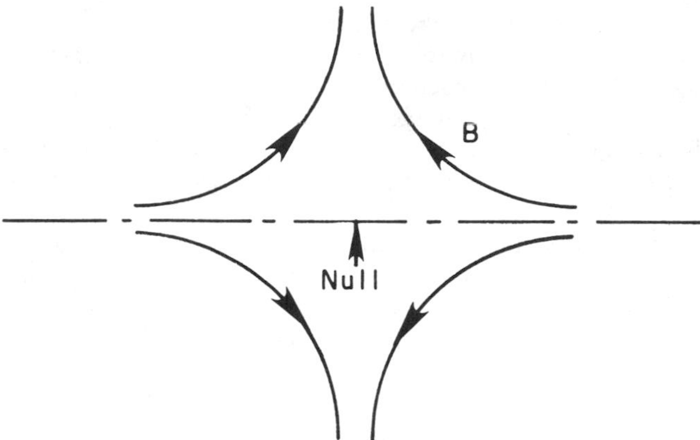

FIG. 7. The spindle cusp.

end loss in the open direction is tolerable. For the fusion reactor application, the minimum length envisaged for θ-pinch systems using conventional magnets is on the order of kilometers. By means of imploding, flux-compressing liquid-metal liners, this length could be reduced substantially.

Cusps

Open-ended configurations with very strongly favorable ∇B properties can be produced by joining two mirrors with oppositely directed field lines as in Fig. 7; in this case, an additional line mirror results around the midplane, and there is a central null-field point. "Cusp" confinement of this type is impaired by the nonadiabaticity of orbits passing near the central null. Since μ is not conserved, the central plasma could escape by free flow; good confinement can be obtained only if a $\beta \approx 1$ plasma is bounded by geometrically very narrow escape regions, as in Fig. 7.

New Approaches

There are many other confinement possibilities, some of them involving energetic-particle rings or levitated current-carrying mechanical rings inside the plasma. The illustrations given here serve to convey the basic physical problems and design configurations for magnetic confinement of hot plasmas.

See also MAGNETOHYDRODYNAMICS; PLASMAS; PLASMA WAVES.

BIBLIOGRAPHY

L. A. Artsimovich, *Controlled Thermonuclear Reactions* (A. C. Kolb and R. S. Pease, eds.). Gordon and Breach, New York, 1964.

D. Baldwin, *Rev. Mod. Phys.* **49,** 317 (1977).

F. L. Hinton and R. D. Hazeltine, *Rev. Mod. Phys.* **48,** 239 (1976).

N. A. Krall and A. W. Trivelpiece, *Principles of Plasma Physics.* McGraw-Hill, New York, 1973.

Proceedings of Tenth International Conference on Plasma Physics and Controlled Nuclear Fusion Research, London, 1984. IAEA, Vienna, 1985.

Proceedings of Eleventh International Conference on Plasma Physics and Controlled Nuclear Fusion Research, Kyoto, 1986. IAEA, Vienna, 1987.

Proceedings of Twelfth International Conference on Plasma Physics and Controlled Nuclear Fusion Research, Nice, 1988. IAEA, Vienna, 1989.

E. Teller *et al.* (eds.), *Fusion.* Academic Press, New York, 1981.

Plasmas

R. L. Morse

I. DEFINITION OF PLASMA PHYSICS AND MAJOR APPLICATIONS

Plasma is a state of matter. The term "a plasma" is used to refer to a quantity of matter in the plasma state. In the plasma state a significant number, if not all, of the electrons in matter are free, i.e., not bound to an atom or molecule. Rigorous definitions of plasma require that the average kinetic energy of the free electrons exceed some value which increases with their density. In practice we often say that matter is in the plasma state when there are enough free electrons to provide a significant electrical conductivity. Usually, only a small fraction of the electrons in matter need to be free to meet this criterion.

The most common reason for a large number of electrons being free is the collision-induced ionization caused by energetic thermal motions of atoms at high temperatures. Familiar examples are the material conditions in lightning and other intense electrical discharges. Large densities of free electrons are also found in metals at solid densities and in all materials at sufficiently high densities, regardless of temperature. This "pressure ionization" is responsible for the electrical conductivity of metals at room temperature. However, most terrestrial plasmas, with the exception of metals, which are not usually thought of as plasmas, are very hot and not very dense by terrestrial standards. With limited reservations it can be said that plasma is the highest temperature state of matter and, therefore, occurs at higher temperatures than does the gaseous state. Plasmas are, therefore, usually thought of as hot, ionized gases. For example, the hot, partially ionized, conducting gases in various electric discharge lamps have densities ranging from atmospheric density to much less. On the other hand, a census of matter in the universe shows that most of it is in the plasma state and much of that is in stellar interiors where the density of matter is so high that we would call it plasma almost independent of its temperature. Figure 1 shows the temperatures and densities of various plasmas. The measure of plasma density used in Fig. 1 is the total electron density, N_e, rather than just the free electron density, n_e. Curves are included in Fig. 1 which show the degree of ionization of hydrogen, i.e., the fraction of the electrons which are free, as a function of density and temperature. Degree of ionization is not shown for the higher densities because there the usual ideas of ionization do not apply. It is a fact, as shown in Fig. 1, that a given degree of ionization is achieved at a lower temperature when the density is lower. Also indicated is the electron Fermi energy (familiar from statistical and solid state physics) which depends on the density and is a rough measure of the temperature below which our usual ideas about ionization do not apply. Below this line ionization may be thought of as being caused by high density alone. Note for instance that inertial confinement plasmas are above this line, in spite of their high densities, because of their high temperatures.

The study of the physics of plasmas is interesting and important because of the desire to understand various, mostly extraterrestrial, natural phenomena, and because of certain technological applications of matter in the plasma state. As Fig. 1 indicates, these natural phenomena include the Earth's ionosphere, stars, and the solar wind. The technological applications of plasmas include the possibility of clean, economical electric power generation by the controlled thermonuclear fusion reaction of light elements (usually the heavy isotopes of hydrogen, deuterium and tritium), and it is this application which has been responsible for much of the considerable effort in plasma physics since about 1953. The temperatures and densities at which economically useful thermonuclear fusion reaction rates occur are in the ranges indicated in Fig. 1 for the different fusion methods, which requires that the fuel of prospective fusion power reactors be in the plasma state. This fact has caused plasmas to be studied intensely because various characteristic plasma properties, such as high thermal conductivities as well as high electrical conductivities, would strongly influence fusion reactor behavior. In the case of the magnetic confinement approach, Fig. 2a, a low density plasma of fusion fuel, which would otherwise cool much too fast by thermal conduction to the reactor walls, is insulated from these walls by interposing a magnetic field. The high electrical conductivity of the fuel plasma then prevents it from moving across the magnetic field and touching the walls. In the inertial confinement approach, Fig. 2b, small solid pellets containing fusion fuel are heated by intense beams from lasers or charged particle accelerators and the nuclear reactions occur during the brief time that the heated pellets are constrained from disassembly by their own inertia. In this approach, high thermal conductivity plays an important role in using ablation from the surface of a pellet to compress its core densities far above solid density where reaction rates are large enough to be economical.

A more quantitative discussion of the plasma state requires that we consider the collective motion of the free electrons. For this purpose we ignore at first the relatively sluggish motion of the much heavier ions. If the density of the negative electron charges in a region exceeds (or is less than) the density of positive ion charges, the region has a net negative (or positive) charge. Consequently, electric fields occur, according to Maxwell's equations, which tend to restore charge neutrality, i.e., zero net charge, in the re-

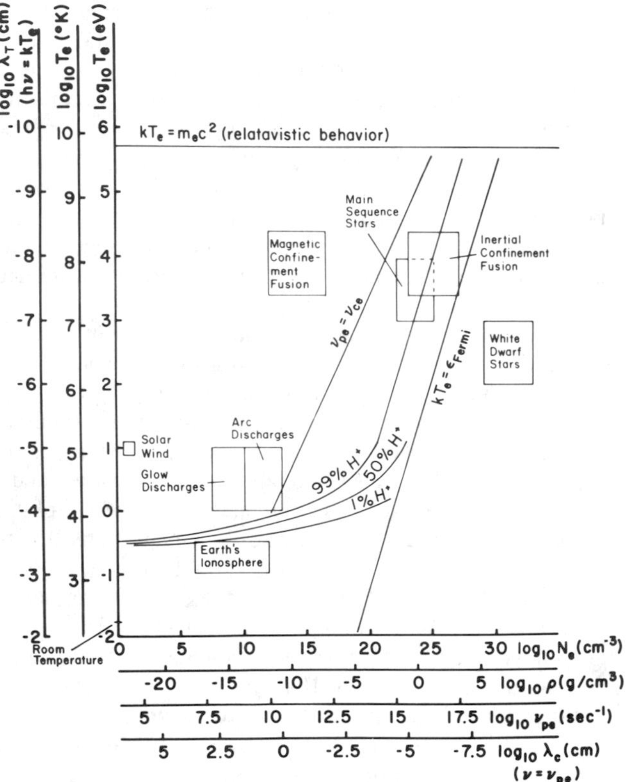

FIG. 1. The densities and temperatures of various natural and man-made plasmas are shown here. The densities are given in units of N_e, the total (bound and free) electron density; ρ, the mass density of hydrogen with this electron density; ν_{pe}, the electron plasma frequency at this electron density; and λ_c, the wavelength of light with frequency ν equal to ν_{pe}. The temperatures of the electrons, T_e, are given in electron volts (eV), in degrees Kelvin (°K), and in units of the wavelength of light λ_T at which the quantum energy, $h\nu$, equals the mean electron thermal energy kT_e. In many, but not all, plasmas the electron and ion temperatures are equal. The multiple scales also serve as conversion tables, i.e., nomographs, for the different variables with which a given axis is labeled. The line $\nu_{pe} = \nu_{ce}$ separates collisionless plasmas (upper left) from collision-dominated plasmas (lower right). The line $kT_e = E_{\text{Fermi}}$ separates density-ionized plasmas (lower right) from lower-density gases which become plasmas only when thermal ionization is sufficient. The % H^+ curves indicate the degree of ionization of hydrogen in the latter region. Near and above the line $kT_e = m_e c^2$ electron behavior is relativistic. The symbol k denotes the Boltzmann constant which converts temperature into an average energy per particle, and which is of course different for the different temperature scales.

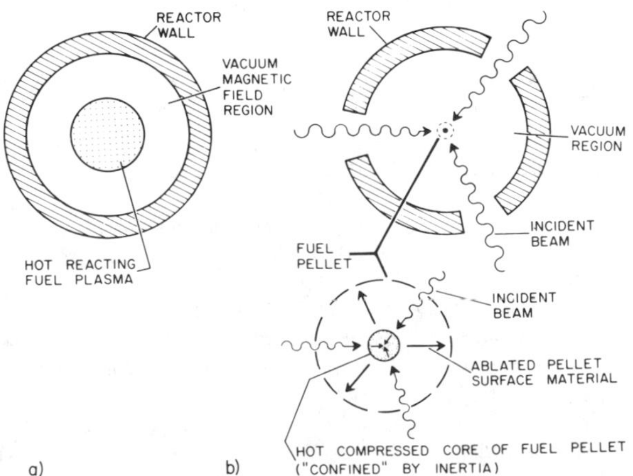

FIG. 2. Schematic diagrams of the two principal approaches to controlled thermonuclear fusion: (a) magnetic confinement and (b) inertial confinement.

gion. However, the inertia of the electrons causes this charge imbalance correcting procedure to be oscillatory with a frequency at or very close to what is called the plasma frequency,

$$\nu_{pe} = (n_e e^2/\pi m_e)^{1/2} \ (\text{s}^{-1}),$$

where e, m_e, and n_e are the electron charge and mass, and the free electron density, respectively, in cgs units. Note that ν_{pe} is independent of the dimensions of the region of charge imbalance. Figure 1 has the abscissa given in ν_{pe} and in the free space wavelength of light at this frequency, λ_Φ, as well as in units of N_e.

Here for simplicity, the total electron density, N_e, is used in place of n_e, to calculate ν_{pe} and λ_c. Strictly speaking therefore, these two abscissa scales only apply to the fully ionized region in the upper left of Fig. 1.

When an electromagnetic wave frequency is less than the local plasma frequency the wave cannot propagate. For this reason λ_c is sometimes called the cutoff wavelength. Waves incident on a region in which they cannot propagate for this reason tend to be reflected. This effect accounts for reflection of radio signals from the ionosphere (see the parameters of the ionosphere in Fig. 1).

In many conditions plasma oscillations are only weakly damped and are easily observed. The initial observation of these oscillations by Langmuir (1929) in a laboratory plasma led him to call ionized gases which supported such oscillations "plasmas" by analogy with the gelatinlike oscillatory behavior of biological substances of the same name.

If plasma oscillations occur with a maximum oscillatory velocity equal to the mean electron thermal velocity, V_e, then from harmonic oscillator relations the spatial displacement of these oscillation electrons is

$$\lambda_D = V_e/2\pi\nu_{pe}.$$

(This length is called the Debye length because essentially the same quantity was first used by Debye in the related context of the theory of electrolytes before its use in plasma physics.) Clearly, if λ_D is greater than the dimensions of the plasma then oscillations with a velocity V_e will move the electrons out away from the ions before the oscillations complete a full period. This is equivalent to saying that conservation of energy together with Maxwell's equations for electric fields do not force the plasma to remain charge neutral. Moreover, if the plasma is much smaller than λ_D, then the charged particles of which it is composed act independently rather than collectively. In this case plasma oscillations, which are a collective phenomenon, as well as most of the properties which we associate with plasmas below, disappear and what remains we do not call a plasma. A more complete definition of a plasma is, therefore, a quantity of material whose free electron density n_e is large enough to affect its electrical and thermodynamic properties significantly and whose dimensions are larger than λ_D. It is then a matter of definition to say that plasmas are, on the average, nearly charge neutral, and in fact as plasmas become much larger than λ_D nature requires plasmas to be very nearly charge neutral. These basic concepts are discussed further in Ref. 1. The physics of ionization and electrical discharges is discussed in Ref. 2.

II. FURTHER CHARACTERISTICS AND APPLICATIONS OF PLASMAS

A. Luminosity

One of the most striking characteristics of plasmas is their luminosity. Of the various light sources we know, most are plasmas. The free electrons in a plasma emit photons of electromagnetic radiation at and below their energy when they collide with ions, by a process called bremsstrahlung (German for "braking radiation"). The ordinate of Fig. 1, T_e, is also plotted in units of the wavelenth, λ_T, of light with photon energy, $h\nu$, equal to the average thermal energy, kT_e. Since the temperature in energy units is the average thermal energy of an electron in a plasma, we see that bremsstrahlung radiation from the various plasmas we know spans the spectrum from the far infrared ($\lambda \gtrsim 10^{-3}$ cm) through the visible ($\lambda \simeq 0.5 \times 10^{-4}$ cm) and ultraviolet ($\lambda \simeq 10^{-5}$ cm) and into the x-ray regime ($\lambda \lesssim 10^{-6}$ cm). In addition, free electrons radiate by recombining with, i.e., again becoming bound to, ions. Also, because of their high temperatures, many atoms in plasmas, whether neutral or partially ionized, are in energetically excited states from which they can drop to lower energy states (including the lowest or ground state) by emitting radiation energy. This is called bound–bound or line radiation because the radiation, which occurs only at those energies corresponding quantum mechanically to the differences between atomic energy levels, forms narrow lines on photographs of the emitted light spectrum. At the highest temperatures, bremsstrahlung dominates. In cooler, less fully ionized plasmas recombination becomes most important, and in the coldest plasmas line radiation dominates.

B. Collisionless Behavior

Another important characteristic of plasmas is their ability to exhibit what is called collisionless behavior. As plasmas become much hotter the ordinary binary collisions between electrons and ions become less important because of the long range nature of the $1/r^2$ force between charged particles. This is the opposite of what might be expected from thinking about the increase in the collision frequency of hard sphere-like particles as their relative thermal velocities increase. At a given density, as the temperature increases the binary collision frequency, which is approximately proportional to $n_e/T_e^{3/2}$, becomes smaller than various other characteristic frequencies of this plasma, such as the collective plasma oscillation frequency described above. When this happens we say that the plasma has become collisionless because the phenomena associated with the other frequencies have become more important than the collisions. Since there are various binary collision frequencies (e.g., electron–electron, electron–ion, ion–ion) and various characteristic frequencies associated with a plasma in addition to ν_{pe}, there are different definitions of "collisionless." However, perhaps the most fundamental definition is that the plasma frequency ν_{pe} exceed the electron–electron collision frequency ν_{ce}. The $\nu_{ce} = \nu_{pe}$ line for fully ionized hydrogen is plotted in Fig. 1. By this definition we would then say that plasmas above this line are collisionless. The basic physics of collisionless plasmas, including magnetic field effects and instability (below), is discussed in Ref. 3 and, at a more advanced level, in Ref. 4.

C. Interaction With Magnetic Fields

The interaction of plasmas with magnetic fields arises in significant ways in connection with extraterrestrial plasmas and inertial confinement fusion, and is especially important in magnetic confinement fusion. This interaction may be approached by thinking of a plasma as an electrically conducting fluid (this approach is called magnetohydrodynamics or MHD) or by thinking first of the motion of individual charged particles in a magnetic field and later considering the collective plasma character of the charged particles. From the individual charged particle point of view the important fact is that a charged particle moving in a magnetic field feels a force

$$\mathbf{F} = q\frac{\mathbf{V}}{c} \times \mathbf{B}$$

where c is the speed of light, q and $\mathbf{V}$ are the charge and vector velocity of the particle, $\mathbf{B}$ is the magnetic field vector (all in cgs Gaussian units), and the "×" between the two vectors indicates the vector product. That is, $\mathbf{F}$ is at right angles to $\mathbf{V}$ and $\mathbf{B}$ and only that component of $\mathbf{V}$ which is at right angles to $\mathbf{B}$ enters the multiplication. The magnitude of the force, F, is $F = qV_{\perp}B/c$. Here $V_{\perp}$ is the component of $\mathbf{V}$ perpendicular to $\mathbf{B}$, and $V_{\|}$, which is used below, is the parallel component. Consequently, a charged particle moving through a magnetic field in the absence of other force fields gyrates around magnetic field lines without changing the magnitude of its velocity, $\mathbf{V}$, i.e., with constant kinetic energy. If the magnetic field $\mathbf{B}$ is constant in time and space, then both $V_{\perp}$ and $V_{\|}$ are constant, i.e., the particle trajectory is a simple constant-pitch and -radius spiral parallel to $\mathbf{B}$ as shown in Fig. 3. The gyration frequency of the particle is given by

$$\Omega_g = qB/mc,$$

which is sometimes called the cyclotron or synchrotron frequency, and the radius of gyration is, consequently,

$$r_g = \frac{V_{\perp}}{\Omega_g} = \frac{V_{\perp}mc}{qB}.$$

Note that Ω_g is independent of velocity and, therefore, resonance between this natural frequency and any temporal oscillations of the $\mathbf{E}$ or $\mathbf{B}$ fields in the plasma occurs at this same frequency for all particles of the same charge-to-mass ratio, q/m, in a given region. This fact is the basis for various resonant plasma-heating schemes. Since Ω_{ge}, the gyro frequency for electrons, is a characteristic frequency of the plasma, it should also be noted that $\Omega_{ge} > \nu_{ce}$ is another definition of collisionless (see above). This criterion in fact indicates approximately when binary collisions are sufficiently weak to allow the magnetic field to confine the plasma. When this collisionless criterion is not met, the gyration of the electrons (and ions) about magnetic fields is so disrupted by collisions that the confining effect of the magnetic field is lost.

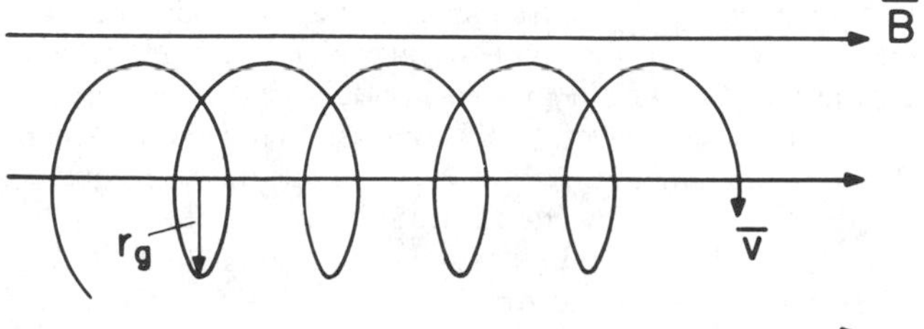

FIG. 3. Spiral, or gyrating, motion of a charged particle in a magnetic field **B**.

If the magnitude of $\mathbf{B}$, $|\mathbf{B}|$, changes as a function of distance along a magnetic field line while still remaining constant in time, then $V_{\perp}$ and $V_{\|}$ will in general change in time but still in such a way as to keep $V = (V_{\|}^2 + V_{\perp}^2)^{1/2}$ constant, and in general to increase $V_{\perp}$ in regions of increasing $|\mathbf{B}|$. Hence, to maintain the constancy of V the particle tends to be excluded from regions of sufficiently high $|\mathbf{B}|$. This effect is called "mirroring." If, on the other hand, $|\mathbf{B}|$ is constant along field lines but the lines have a constant curvature, like the field of a current carrying wire, then $V_{\perp}$ and $V_{\|}$ remain constant but the instantaneous center about which the particle is gyrating moves perpendicular to $\mathbf{B}$ and to the radius of curvature of $\mathbf{B}$. This motion of the center of gyration, or "guiding center," is called a "drift." The role of magnetic fields in guiding the motion of charged particles, and the additional flexibility and complexities introduced by "mirroring" and "drift" motions, are central to the use of magnetic fields to confine thermonuclear fusion plasmas. The confinement of these reactor plasmas from contact with cold reactor walls is discussed in more detail in the article on plasma confinement devices.

D. Instability

Perhaps the most important and complex characteristic of plasmas is instability. There are several types of plasma instabilities and many instabilities within each type. Since most plasmas are similar in some respects to fluids they are susceptible to instabilities similar to those which cause fluids at rest in unstable equilibrium or in laminar flow to become turbulent and to seek a different equilibrium or flow pattern. For instance, small ripples of the interface between a light fluid and a superposed heavier fluid grow, at first exponentially, in time. The heavier fluid will then fall through the lighter, usually in a seemingly disordered, i.e., in a turbulent, way until the system comes to rest with the heavier fluid on the bottom. This exponential growth of the ripples of the interface is called the Rayleigh–Taylor instability. A similar type of surface instability occurs when the interface between a confined plasma and its confining magnetic field is curved toward the plasma, as it must be in many reactor arrangements. The curvature-induced drifts of charged particles in magnetic fields discussed above play a role in the development of this instability. In the analogy to the Rayleigh–Taylor instability the plasma plays the role of the heavier fluid and the confining vacuum magnetic field that of the lighter. As a consequence of the instability the plasma passes through the field and to the reactor wall. The solution to this problem has been found in using more complex magnetic field shapes. In addition to such macroscopic fluidlike instabilities, in which the ripples that grow are usually much larger than r_g and λ_D for the plasma particles, collisionless plasmas are susceptible to microinstabilities which develop on the scale of these microscopic lengths. These microinstabilities, which are driven by a variety of phenomena such as large electric currents or heat flow that cause the plasma to depart from local thermodynamic equilibrium, have the effect of making collective oscillations grow and cause strongly enhanced density fluctuations. The density fluctuations in turn have the same effect on the plasma as an anomalously high binary collision frequency and, therefore, give rise to anomalously large electrical resistivity or anomalously reduced thermal conductivity. When these anomalous collision frequencies are driven by the electric currents that must flow in the interface between a confined plasma and its confining magnetic field, they can cause a breakdown of confinement just as if $\nu_{ce}>\Omega_{ge}$ (above). On the other hand, anomalous collision frequencies and the resulting enhanced resistivity can be beneficial if they are driven by currents passed through the bulk of a plasma to heat it because the heating rate is increased. This is called "turbulent heating."

This same mixed role of microinstabilities occurs in inertial confinement fusion where absorption of laser light incident on target pellets is enhanced by microinstabilities. On the other hand, microinstabilities also deter subsequent thermal conduction of the absorbed energy to the core of the pellet where it is needed to produce compression and burning of the thermonuclear fuel. Rayleigh–Taylor instability also occurs in inertial confinement systems (Fig. 2b). When compression of the core of a pellet occurs, it is necessarily accompanied by some inward acceleration. This acceleration is locally equivalent to a gravity (with the direction of the gravity vector opposite to that of the acceleration vector). Something very similar to the classic Rayleigh–Taylor unstable situation occurs because the effective gravity of the inward acceleration is directed outward from the heavy plasma of the pellet core toward the lighter plasma which has been ablated from the pellet surface. This unstable situation, which threatens to cause turbulent disruption of otherwise smooth, spherically symmetric compressions, could be troublesome because it might prevent reaching the high fuel densities necessary for the economical release of fusion energy. Pellets are currently being devised that consist of various outer layers of different materials chosen in such a way as to minimize or eliminate the problem.

The role of plasma instabilities in extraterrestrial plasmas is less clear at this time but is thought to be responsible for some otherwise inexplicable features of solar flares and the solar wind.

III. CURRENT RESEARCH IN PLASMA PHYSICS

The major objectives of current plasma research are to understand plasmas which occur in nature, for both intellectual satisfaction and application to space exploration, and to achieve the economical release of nuclear fusion energy. The primary tools employed for the former are visible and radio astronomical observation and, more recently, more direct measurements of plasma properties near the earth by satellites. Laboratory research, which is mainly directed toward controlled fusion, is conducted with large magnetic devices, or with lasers or high current charged–particle accelerators, depending on the approach. In both cases the size and complexity of these machines is approaching that of the largest accelerators for elementary particle research. A noteworthy laboratory problem is that the methods, called "diagnostics," for measuring the properties of the plasmas which are produced with these machines must be very fast. While there are some steady state, low-density and -temperature, magnetically confined plasmas, most have time durations ranging from 10^{-2} s down to 10^{-5} s. Inertial confinement plasma research is even more challenging in this respect because target pellet plasma remains in its compressed state for only of the order of 10^{-9} s.

Theoretical research into the behavior of plasmas relied at first on analytical mathematical methods and only limited use of electronic computers. These methods were very successful in calculating the "equilibrium" magnetic field configurations required to confine plasmas, in predicting the occurrence of instabilities of these configurations, and in the related area of predicting the myriad complexities of small amplitude waves in plasmas. For a discussion of this latter area see the article on plasma waves. However, for predicting the consequences of instabilities when the resulting disturbances grow to large amplitude and become turbulent, extensive use of numerical simulation techniques with large computers has been very useful (Ref. 5). Computer simulation has also been used extensively in calculating the compression and reaction processes in inertial confinement fusion pellets (Ref. 6).

A large portion of the plasma physics research currently directed toward the economical use of nuclear fusion energy is concerned with understanding and controlling the occurrence and effects of instabilities.

IV. ACCOMPLISHMENTS OF PLASMA PHYSICS

The following is a list of what the author believes are the major accomplishments of plasma physics since its beginnings in the 1920s.

1. The understanding of binary collisional and radiative processes in plasmas and the transition from collisional to collisionless behavior.
2. The understanding of collective modes of oscillation of collisionless plasmas.
3. Success in devising a variety of equilibrium magnetic confinement configurations.
4. Calculation of pellet heating and compression schemes to yield economical fusion energy release through inertial confinement.
5. The prediction and identification of the occurrence of many, if not all possible, instabilities of plasmas.
6. A qualitative understanding of the anomalies in plasma behavior caused by the occurrence of instabilities and the resulting plasma turbulence.

It is in this last area where the most complex problems lie and the most progress is still needed.

See also BREMSSTRAHLUNG; INERTIAL FUSION; IONOSPHERE; NUCLEAR FUSION; PLASMA CONFINEMENT DE-

VICES; PLASMA WAVES; PLASMONS; SOLAR WIND; STELLAR ENERGY SOURCES AND EVOLUTION.

REFERENCES

1. L. Spitzer, *Physics of Fully Ionized Gases,* 2nd rev. ed. Interscience, New York, 1962.
2. S. C. Brown, *Basic Data of Plasma Physics.* Technology Press of the Massachusetts Institute of Technology, Cambridge, 1959.
3. F. F. Chen, *Introduction to Plasma Physics.* Plenum Press, New York, 1974.
4. N. A. Krall and A. W. Trivelpiece, *Principles of Plasma Physics.* McGraw-Hill, New York, 1973.
5. *Methods In Computational Physics, Vol. 9, Plasma Physics.* Academic Press, New York, 1963.
6. T. P. Hughes, *Plasmas and Laser Light.* Wiley, New York, 1975.

Plasma Waves

Thomas H. Stix

Plasma is the state of matter at temperatures so high that an appreciable fraction of the molecules and atoms are dissociated into ions and electrons. Stellar and interstellar matter is mostly in the plasma state, as is matter in the magnetosphere, the ionosphere, flames, chemical and nuclear explosions, and electrical discharges. Matter in a controlled thermonuclear reactor would also be in the plasma state, and it has been largely the study of this abundant source of energy and electrical power that has led to our extensive current knowledge of the plasma state. In terms of gross properties, plasma differs from un-ionized gas by its high electrical and thermal conductivity, and by the emission of electromagnetic radiation such as light or bremsstrahlung. A plasma may also display a variety of unusual dielectric or refractive properties for intermediate-frequency radiation. The study of these properties falls within the discipline termed plasma waves.

Generally speaking, a wave or an instability describes the response of a medium to some kind of perturbation. A wave describes a nongrowing traveling response, an absolute instability describes a response that grows in time even at the location of the original disturbance, and a convective instability describes a response that, when viewed in a moving frame, grows in time. Knowledge of plasma waves is useful in a number of applications: astrophysics, auroral phenomena, ionospheric radio communication, microwave amplifiers, particle accelerators, and controlled fusion are a few examples. In controlled fusion experiments, for instance, low-intensity plasma waves are used as diagnostic tools, probing hot, magnetically confined plasma in a noninterfering manner to determine density, temperature, and the presence of nonthermal fluctuations. High-intensity waves are used to heat and to drive current through these plasmas, and much thought goes into apparatus design and the choice of operating parameters in order to avoid the plasma instabilities that might otherwise impair the needed excellent magnetic insulation.

On a microscopic scale, an important difference between plasma and un-ionized gas lies in the nature of the particle interactions—gas collisions are approximately hard-sphere collisions between neutral molecules or atoms, slightly modified at long range by very weak van der Waals forces, whereas plasma interactions stem entirely from the long-range Coulomb forces. The dominant collisions in a plasma, then, turn out to be the many small-angle scatters whose cumulative effect is greater than that of the relatively infrequent large-angle events, and the appropriate mathematical treatment for these small-deflection collisions invokes collision terms in the kinetic (Boltzmann) equation of the Fokker–Planck form. But another difference between plasmas and neutral gases is perhaps more important here: The Coulomb cross section decreases as the inverse square of the energy of the particles, and in high-temperature plasmas the collision frequencies are often low compared to other frequencies of interest and the mean free paths may be long compared to typical wavelengths or even plasma dimensions. A collision-free model then frequently offers an excellent first approximation for the hot plasma. Unlike what occurs in a neutral gas, however, particle interactions do not disappear in the collision-free plasma, because the charged plasma particles both induce and respond to the existing electric and magnetic fields. The mathematical analysis of the plasma then requires the self-consistent or simultaneous solution of the plasma equations plus the Maxwell equations. It turns out, however, that despite the vast difference in interparticle coupling, there may still be strong similarities in the gross behavior of a collision-free plasma and that of a neutral gas. For example, the ion acoustic wave in an unmagnetized plasma propagates at sound speed, the pressure fluctuations being transmitted by electric fields due to slightly unneutralized ion and electron space charge fluctuations, rather than by direct particle collisions. Also, it may be shown that the number of normal modes for the plasma is the same as that for a neutral gas with the same number of particles. On the other hand, new wave phenomena appear in plasmas that have no counterpart in neutral gases. These include plasma oscillations, collision-free absorption mechanisms, cyclotron resonance effects, finite-Larmor-radius effects, particle trapping, wave echoes, backward waves, mode conversion, and many other effects.

Quantitative study of plasma waves and instabilities generally starts with the set of S kinetic equations that describe the evolution of the single-particle distribution functions $f_s(\mathbf{r},\mathbf{v},t)$ for the various ion species ($s=1,2,\ \ldots,\ S-1$) in the plasma and for the electrons ($s=S$) with charge q_s and mass m_s,

$$\frac{\partial f_s}{\partial t} + \mathbf{v}\cdot\frac{\partial f_s}{\partial \mathbf{r}} + \frac{q_s}{m_s}\left(\mathbf{E} + \frac{\mathbf{v}}{c}\times\mathbf{B}\right)\cdot\frac{\partial f_s}{\partial \mathbf{v}} = \left(\frac{\partial f_s}{\partial t}\right)_{\text{coll}}.$$

Charge and current density in the plasma are given by

$$\sigma(\mathbf{r},t) = \sum_s \int q_s f_s(\mathbf{r},\mathbf{v},t)\, d^3\mathbf{v}$$

and

$$\mathbf{j}(\mathbf{r},t) = \sum_s \int q_s \mathbf{v} f_s(\mathbf{r},\mathbf{v},t)\, d^3\mathbf{v},$$

and the Maxwell equations complete the set.

In the collision-free limit, the right-hand side of the kinetic equation is equated to zero, giving an oft-used form called the Vlasov equation. For many purposes it suffices to work with the moments of the kinetic equation. The first two are

$$\frac{\partial n_s}{\partial t}+\nabla\cdot(n_s\mathbf{v}_s)=0,$$

$$n_s m_s\frac{\partial \mathbf{v}_s}{\partial t}+n_s m_s \mathbf{v}_s\cdot\nabla\mathbf{v}_s=$$

$$n_s q_s\left(\mathbf{E}+\frac{\mathbf{v}_s}{c}\times\mathbf{B}\right)-\nabla\cdot\mathbf{P}_s+\sum_r \mathbf{F}_{sr},$$

in which $\mathbf{F}_{sr}$ is the force per unit volume on particles of type s due to collisions with particles of type r, and

$$n_s(\mathbf{r},t)\equiv\int f_s(\mathbf{r},\mathbf{v},t)\,d^3v,$$

$$\mathbf{v}_s(\mathbf{r},t)\equiv n_s^{-1}\int \mathbf{v}f_s(\mathbf{r},\mathbf{v},t)\,d^3v,$$

$$\mathbf{P}_s(\mathbf{r},t)=m_s\int(\mathbf{v}-\mathbf{v}_s)(\mathbf{v}-\mathbf{v}_s)f_s(\mathbf{r},\mathbf{v},t)\,d^3v,$$

$$\sigma=\sum_s q_s n_s, \quad \text{and} \quad \mathbf{j}=\sum_s q_s n_s \mathbf{v}_s.$$

$\mathbf{P}_s$ is the pressure tensor for the "fluid" of s-type particles. When the pressure is isotropic, $\mathbf{P}_s$ is a diagonal tensor with three nonzero components, each equal to $p_s=n_s\kappa T_s$, and $\nabla\cdot\mathbf{P}_s\rightarrow\nabla p_s$. On the other hand, pressure anisotropies can persist for long times in a hot plasma, and the isotropic form is frequently inapplicable. In the presence of a strong magnetic field $\mathbf{B}(\mathbf{r})$, the characteristic motion of particles is a combination of streaming along $\mathbf{B}$, with velocity $v_\parallel$, and gyration around the magnetic lines of force, with velocity $v_\perp$, at the angular gyrofrequency $\omega_c=qB(\mathbf{r})/mc$. With slight inhomogeneity of $\mathbf{B}(\mathbf{r})$ these two motions will quickly symmetrize the pressure tensor in the directions normal to the local $\mathbf{B}$, whereas the relaxation between $T_\parallel$ and $T_\perp$ will, in a hot plasma, proceed at a much slower rate. A frequently appropriate pressure tensor is then $\mathbf{P}=\mathbf{1}p_\perp+\hat{B}\hat{B}(p_\parallel-p_\perp)$ in which $\mathbf{1}$ is the unit dyadic, $\hat{B}$ is the unit vector along $\mathbf{B}(\mathbf{r})$, $p_\parallel=n\kappa T_\parallel$, $p_\perp=n\kappa T_\perp$, and $T_\parallel$ and $T_\perp$ are the temperatures for motions parallel and perpendicular, respectively, to $\mathbf{B}$.

The third moment equation introduces heat flow. For many applications, however, the details of heat transfer are not important to wave or instability dynamics and the moment equations can be closed by an isothermal law (when the electron thermal velocity much exceeds the wave phase velocity along $\mathbf{B}$), or by an adiabatic law derivable from the condition of zero-divergent heat flow. With low collision frequencies, relaxation of velocity-distribution anisotropies during a single oscillation period may be very incomplete, and separate adiabatic laws can apply to $p_\parallel$ and to $p_\perp$. The appropriate fluid relations then are the double-adiabatic Chew–Goldberger–Low equations.

The most elementary considerations of waves usually pertain to small-amplitude steady-state oscillations in homogeneous media, in which case the media equations may readily be linearized and then Fourier-transformed in both time and space. There results a set of equations in the Fourier amplitudes of the field variables, and the condition for nontrivial solutions is that the determinant of this set of linear equations be zero. This condition necessarily introduces a dependence between angular frequency ω and wave number $\mathbf{k}$ such that $\omega=\omega(\mathbf{k})$. A well-known example of this dependency, called the dispersion relation, is the equation for light waves *in vacuo*, $\omega=|\mathbf{k}|c$. The dispersion relations for the two Alfvén modes are $\omega=|\mathbf{k}|c_A$ (compressional mode) and $\omega=(\mathbf{k}\cdot\mathbf{B}/|\mathbf{B}|)c_A$ (shear mode), $c_A{}^2\equiv c^2/(1+\Sigma_s 4\pi n_s m_s c^2/B^2)$, summed over the S species of charged particles in the plasma. For another well-known mode, Langmuir–Tonks plasma oscillations, the dispersion relation is simply $\omega^2=\omega_{pe}{}^2$, in which the plasma frequency ω_{pe} is given by $\omega_{pe}{}^2\equiv 4\pi n_e e^2/m_e$.

From the dispersion relation, which generally has more than one root, we can determine the relative phase velocities of the different modes (roots), and can then go back to the linearized wave equations to fix further properties of these modes. Of interest are the ratios of the various field amplitudes and their polarization, and these properties may be used to understand and to classify the modes. Based on this information, for instance, waves are sometimes labeled fast or slow, or ordinary or extraordinary, with linear, elliptical, or circular (right- or left-handed) polarization.

For a large class of plasma waves, finite-temperature effects, including pressure as well as heat flow, are unimportant. These are the waves derivable from "cold-plasma" theory, based on the first two moment equations above with $\mathbf{P}_s=0$ and $\mathbf{F}_{sr}=0$, plus the equations for charge and current density and the Maxwell equations. Included in this category are the Alfvén or hydromagnetic waves, the ordinary and extraordinary electromagnetic waves, Langmuir–Tonks plasma oscillations, ion cyclotron and electron cyclotron waves, lower and upper hybrid-frequency oscillations, and whistlers. The cold-plasma waves display an important pair of characteristic plasma phenomena, namely, plasma oscillations and cyclotron resonance effects. The former are high-frequency modes in which the ions are virtually motionless, but the electrons move coherently in simple compressional oscillations at $\omega=\omega_{pe}$ (mentioned above) with the electron kinetic energy flowing into the electrostatic energy of the space-charge field and back. Cyclotron resonance effects for both ions and electrons also play an important role for waves in a magnetized plasma. However, coherence effects—principally the electric fields induced by the local differences in space-charge density for the ion and electron "fluids"—modify the wave dynamics so that, for example, the mode frequencies of the so-called ion and electron cyclotron waves are actually somewhat shifted from the single-particle cyclotron frequencies. The shift might be greater, but the plasma can display great "ingenuity" to satisfy its basic craving for space-charge neutrality: in the ion cyclotron wave, for instance, relatively large space-charge fluctuations due to compressional motions of ions moving coherently *across* $\mathbf{B}$ are almost entirely neutralized by the mobile electrons streaming *along* $\mathbf{B}$ from half a wavelength away.

The mathematical analysis of cold-plasma waves in a uniform plasma is simple and straightforward, and is included in most elementary texts on plasma physics. The dispersion relation for these waves turns out to be a biquadratic equation in the refractive index, $n\equiv kc/\omega$, with coefficients depending on plasma density, on background magnetic field strength $|\mathbf{B}_0|$, on frequency, and on the angle between $\mathbf{k}$ and

$\mathbf{B}_0$. This general dispersion relation neatly lends itself to a comprehensive analysis via a certain representation in "parameter space" (dimensions of parameter space here are plasma density and background magnetic field strength) called the Clemmow–Mullaly–Allis (CMA) diagram, and with the aid of this CMA diagram the relationships between all of the cold-plasma waves are easily understood. As an example, CMA analysis shows that the shear Alfvén wave is never faster (at a given propagation angle) than the compressional Alfvén wave; that the wave electric field of the shear wave is left-hand polarized around the $\mathbf{B}_0$ vector; that the refractive index of the shear wave goes to infinity for certain directions of $\mathbf{k}$; that the shear Alfvén wave belongs to the same branch of the dispersion relation as the ion cyclotron wave; and that this branch becomes nonpropagating at frequencies greater than the ion cyclotron frequency.

Finite plasma temperature introduces three new phenomena into the simple cold-plasma model for plasma waves. The most obvious modification is the appearance of electron and ion pressure, which modifies the dispersion relation for each of the cold-plasma waves and which also introduces new branches into the dispersion relation of which the simplest is the ion acoustic wave, mentioned earlier. A second modification is the emergence of so-called finite-Larmor-radius effects as the radius of gyration of ions and electrons in the background magnetic field $\mathbf{B}_0$ becomes nonnegligible in comparison to the wavelength across $\mathbf{B}_0$. Third, and most important, is the explicit appearance of collision-free absorption in the linear theory. A careful examination based on causality will reveal absorptive components for the refractive index associated with the cold-plasma resonances, but hot-plasma theory makes the mechanism clear: Energy can be transferred between the wave and the particle distribution on a microscopic scale, the transfer taking place only to and from the "resonant" particles, i.e., to those ions and electrons that, in their own rest frame, "see" the wave electric field as a static field. For example, Landau damping is the collisionless process for transfer of energy to those ions or electrons that are moving parallel to $\mathbf{B}_0$ with $\mathbf{v}\cdot\hat{B}_0=\omega/(\mathbf{k}\cdot\hat{B}_0)$, and when the wave electric field "seen" by magnetized ions and electrons is properly analyzed, it is found that collisionless absorption also takes place at the particle cyclotron frequencies and at every integral harmonic!

Actually, if collisions are too infrequent, second-order wave effects will cause even the resonant particles to see a changing wave phase, and the efficiency of the energy transfer may be reduced or nullified. The process is also inefficient if collisions are too frequent and the collisional mean free path becomes comparable to a wavelength. But in the broad range of intermediate collisionality, the macroscopic effects of the so-called collision-free processes are, in fact, independent of the actual degree of collisionality.

It is worth remarking that collision-free absorption is a *nonlocal* plasma wave property. Particles in a hot, nearly collision-free plasma move approximately along free-streaming unperturbed (zero-order) trajectories that describe helical paths around the lines of force of the background magnetic field. Collision-free absorption is an *integrated* effect in which the charged particle "sees" a static first-order $\mathbf{E}$ field over a reasonably extended portion of its zero-order trajectory. This strange ability of particles in a hot plasma to "remember" their recent histories shows up in other plasma processes such as microinstabilities, trapped-particle modes, and plasma-wave echoes.

Inhomogeneity of the plasma or of the magnetic field can also introduce new plasma-wave phenomena. In the simplest case, waves can be followed through a slowly changing medium by using a local approximation for the refractive properties. The dispersion relation is then a slowly varying function of position, i.e., $\omega=\omega(\mathbf{x},\mathbf{k})$, and a pair of equations of Hamiltonian form will trace out the ray or wave-packet trajectories,

$$\frac{d\mathbf{x}}{dt}=\frac{\partial\omega}{\partial\mathbf{k}},\qquad \frac{d\mathbf{k}}{\partial\mathbf{k}}=-\frac{\partial\omega}{\partial\mathbf{x}}.$$

Near certain plasma-wave resonances, however, the refractive index may be rapidly changing even though the plasma parameters (e.g., n_e, T_e, or B_0) are slowly changing, and the adiabatic approximation breaks down. At such "critical layers," it is possible to get reflection, partial or complete transmission, partial or complete absorption, or conversion from a short-wavelength mode to a long-wavelength mode or vice versa.

In a different effect, inhomogeneity introduces into the linearized first moment equation a new term, $\mathbf{v}^{(1)}\cdot\nabla n^{(0)}$, which sometimes brings about major modifications of the wave dispersion relation. Ion acoustic waves in a uniform magnetized plasma, for instance, take on quite a new character in a nonuniform plasma and become "drift waves." Still another modification can appear: If the region of propagation is limited in spatial extent by the inhomogeneity, the dispersion relation typically changes from a continuous spectrum to a discrete spectrum. The same effect occurs, of course, with neutral-gas sound waves, and an organ pipe constitutes a familiar example.

With this awareness of plasma-wave processes, three broad classes of plasma instabilities may be understood. It may first be pointed out that in linear theory, an unstable equilibrium shows up as a root of the dispersion relation in which ω has an imaginary component corresponding to growth in time. The first type of plasma instability is the linear fluid instability, derived using the linearized moment equations. The driving force for fluid instabilities comes from a nonthermal arrangement of the plasma and the magnetic field in configuration space. For example, the acceleration of a sparse plasma by a dense plasma is subject to the well-known Rayleigh–Taylor instability, and an important variant is the well-known interchange instability, which appears in the case of plasma confinement by a magnetic field with lines of force locally concave with respect to the plasma. The picturesque name of this instability describes the behavior of unstable adjacent magnetic lines of force, which pass through each other like extended fingers on two hands. The Kelvin–Helmholtz or velocity-shear instability also occurs in plasmas, as do a number of modes without neutral-fluid counterparts. One of these is the magnetohydrodynamic (MHD) kink mode for a current-carrying plasma column, in which any bending of the current column is aggravated by increased magnetic pressure of lines of force squeezed together on the concave side of the bend. It is curious that dissipative processes such as resistivity, generally considered mechanisms for wave damping, may actually enhance MHD instability by eliminating perfect-conductivity constraints on the motion. Resistive MHD instabilities play an important role in phenomena of the earth's magnetosphere, such as aurora, and of the sun's corona.

The second major class of plasma instability is that involving collisionless energy transfer processes. For these so-called linear microinstabilities the driving force comes from a nonthermal particle velocity distribution. The simplest example is the two-stream instability, a long-wavelength purely growing mode for a plasma that contains counterstreaming beams of electrons; but much more delicate distortions of the velocity distribution can produce a growing envelope of otherwise stable plasma waves. Important sources of free energy for driving plasma microinstabilities include temperature anisotropies ($T_\perp \neq T_\parallel$), current flow, trapped-particle distributions, and gradients of density, velocity, or temperature. Given such a source, the actual occurrence of instability also requires that there exist an intermediary—generally some kind of plasma wave—through which the free energy is made accessible. For example, to release the energy associated with a velocity distribution, $f(E)$, which in some range $E \simeq E_0$ increases with particle energy $E = mv^2/2$, it could be necessary to have available an otherwise stable wave with phase velocity $\omega/k \simeq (2E_0/m)^{1/2}$. Because they can release free energy stored in quite subtle deviations from thermal equilibrium, microinstabilities are under intense study in connection with plasma confinement for controlled fusion. Beyond establishing threshold conditions for the unstable modes, it is also of interest to determine saturation mechanisms (e.g., mode–mode coupling, quasilinear diffusion, parametric decay) and saturation amplitudes, as well as to understand the effects on plasma transport rates.

The third major class of plasma instabilities is that of parametric phenomena; the driving force for these processes comes from the presence of a finite-amplitude wave or oscillation in the plasma, and the scattering of its photons in wave–wave or wave–particle interactions. Both macroscopic and microscopic plasma processes can be involved in this class of instabilities, which has important applications to the heating, confinement, and compression of plasmas by radio-frequency and laser means.

Another major area of plasma-wave research focuses on the use of high-intensity rf electromagnetic fields to heat plasmas and/or to drive unidirectional currents through them. In both cases, the essential mechanism invoked is collision-free absorption, described above. Landau damping and its first cousin, transit-time damping, as well as cyclotron and cyclotron-harmonic damping have all been used for these purposes. In tokamak plasmas, temperatures of many tens of millions degrees Kelvin have been achieved by these methods, as well as plasma currents of many hundred thousand amperes. A theoretical model that has been remarkably accurate in its reproduction of these experimental observations is based on the kinetic equation, above. The equation is averaged over a few rapid oscillations, in time or in space, of the rf field. The first-order terms disappear; in second order, one component of the term containing the rf **E** and **B** fields takes the form of a Fokker–Planck velocity-space diffusion term—this is the so-called quasilinear diffusion term. The evolution of the s-species distribution function may then be seen to result from a balance between enhanced velocity-space diffusion induced by the intense rf fields, and ordinary collisional relaxation from Coulomb interactions between particles.

Summarizing, matter at high temperatures can be appreciably ionized and is then said to be in the plasma state. Plasma waves are the response of the plasma medium to some kind of perturbation, and a plasma instability corresponds to a response or oscillation that grows in time or in space. In contrast to the mechanism for sound waves in a neutral gas, momentum transfer in plasma waves occurs via the coupling forces between the charged particles of the plasma and the self-consistent electrostatic and electromagnetic fields that they collectively produce; two-particle collisions play an interesting but only nondominant role. Nevertheless, despite the difference in microscopic transfer mechanisms, plasma waves exist that are very similar to neutral-gas sound waves. Many new properties show up, however, such as plasma oscillations, microinstabilities, collision-free absorption, particle trapping, wave echoes, and mode conversion. Moreover, in a strong background magnetic field, plasma-wave propagation becomes highly anisotropic and cyclotron resonances as well as finite-Larmor-radius effects can appear. Incorporating phenomena drawn from neutral-gas physics, from fluid mechanics, and from electromagnetic theory, bringing in single-particle and nonlocal effects such as cyclotron resonance, finite gyroradius, and particle "memory," introducing collective effects such as electrostatic plasma oscillations and hydromagnetic waves, and with special flourishes due to strongly anisotropic dielectric properties in a magnetic field, plasma inhomogeneity, dissipative and nonlinear processes, the study of plasma waves and instabilities offers the physicist an incredibly fertile area for experimental and theoretical research.

See also CYCLOTRON RESONANCE; INERTIAL FUSION; MAGNETOHYDRODYNAMICS; PLASMAS; PLASMA CONFINEMENT DEVICES.

BIBLIOGRAPHY

F. F. Chen, *Introduction to Plasma Physics and Controlled Fusion*, 2nd ed. Plenum, New York, 1984.

J. P. Freidberg, *Ideal Magnetohydrodynamics*. Plenum, New York, 1987.

N. A. Krall and A. W. Trivelpiece, *Principles of Plasma Physics*. McGraw-Hill, New York, 1973.

D. R. Nicholson, *Introduction to Plasma Theory*. Wiley, New York, 1983.

G. Schmidt, *Physics of High Temperature Plasmas*, 2nd ed. Academic, New York, 1979.

L. Spitzer, Jr., *Physics of Fully Ionized Gases*, 2nd ed. Wiley (Interscience), New York, 1962.

T. H. Stix, *Waves in Plasmas*. American Institute of Physics, New York, 1990.

R. B. White, *Theory of Tokamak Plasmas*. North-Holland, Amsterdam and New York, 1989.

Plasmons

David B. Chang and Betsy Ancker-Johnson

A plasmon is a quantum of the charge-density oscillations that occur in a plasma, i.e., in an electrically neutral collection of charges in which some of the charges are free to move in response to their Coulomb interactions. Plasmons involving charges in solids occur in metals, semimetals, and semiconductors, and they have been studied in large atoms and molecules. The plasmons can be classified according to the types of charges involved (electrons or holes, valence or conduction); the number of mobile charged components; the degree of departure from thermal equilibrium; the degree of degeneracy; and the dimensional constraints on the collective oscillations.

Some plasma oscillations in solids involve all the valence-band electrons. The corresponding plasmon energy is of the order of 10 eV for most solids and is given approximately by $E=\hbar(4\pi ne^2/m)^{1/2}$, where $\hbar$ is Planck's constant divided by 2π, n is the density of valence-band electrons, e is the electronic charge, and m is the free-electron mass. The expression is valid when E is larger than valence-band–unoccupied-band energy differences, e.g., for Al, Mg, Be, Si, and the alkali metals. For semiconductors and doped semimetals, in addition to these high-energy plasmons, there are low-energy plasmons associated with conduction-band electrons. These plasmons have energies ranging from hundredths to several tenths of an electron volt, depending on the temperature and the amount of doping. In metals, the plasma oscillation frequencies are much higher than collision frequencies, whereas in semiconductors and doped semimetals, the collision and conduction-band plasma oscillation frequencies can be comparable.

The high-energy plasmons have been observed in metals, semimetals, and semiconductors through the energy-loss spectra of high-energy (typically 1–10's of keV) electrons scattered by the plasmons and through the inelastic scattering of high-frequency electromagnetic radiation (for instance, Cu K_α x rays). In heavy atoms, photoabsorption cross sections in the UV to soft-x-ray region can be reasonably well explained in terms of the excitation of high-energy atomic plasmons. The low-energy plasmons in semiconductors and doped semimetals have been extensively studied by Raman scattering [typically of photons from neodymium-doped yttrium aluminum garnet (1.2 eV) and CO_2 (0.12 eV) lasers]. These studies give information on the plasma densities and plasmon dispersion properties and lifetimes resulting, e.g., from Landau damping, collisions, and interactions of the plasmons with each other and with acoustical and optical phonons. In addition, radiation from plasmons has been detected from surfaces, and plasmons have been indirectly observed through their effects on electron tunneling and on photoemission spectra.

A relatively incompressible plasma occurs when only one charge component is free to move, with the other charges tied to lattice positions: This one-component plasma occurs in simple metals, in extrinsic semiconductors, and in ionic crystals at temperatures where only one ion species is mobile. A relatively compressible plasma occurs when charge carriers of both signs are free to move: These multicomponent plasmas occur in pure semimetals, in intrinsic semiconductors, and in semiconductors and insulators into which equal densities of oppositely charged carriers have been injected or created by impact ionization. When a magnetic field B is applied to the plasma, the types of electromagnetic waves which the plasma can support at frequencies lower than the cyclotron frequency $\omega_c=eB/mc$—with c denoting the speed of light—depends markedly on whether the plasma is compressible or incompressible. In the former case, Alfvén waves are replaced by helicon waves with frequency inversely proportional to the square of the wavelength.

In semiconductors and insulators, the energy required to produce one-component plasmas is of the order of 10^{-3}–10^{-1} eV, and that required for the production of multicomponent plasmas is of the order of 0.1–5 eV. In metals and semimetals, free carriers exist at $T=0$ K. Consequently, solid-state plasmas in thermal equilibrium are attainable at room temperatures (0.025 eV) and below. Depending on the material and temperature, these equilibrium plasmas can be either degenerate (e.g., in metals, in semimetals, and in some heavily doped semiconductors) or Maxwellian (e.g., in some semiconductors). Nonequilibrium solid-state plasmas can be produced by a variety of means: by electron–hole injection, by electric-field-induced impact ionization (e.g., in InSb at 2×10^4 V/m), and by single-photon or multiphoton processes (e.g., plasma in PbTe produced by CO_2 laser irradiation).

In a Maxwellian plasma, the electric potential from a charge is effectively screened out at distances greater than the Debye length $\lambda_D=(\epsilon kT/4\pi ne^2)^{1/2}$, where ϵ is the dielectric constant, k is Boltzmann's constant, T is the absolute temperature, n is the electron density, and e is the electronic charge. In a degenerate plasma, the electric potential from the charge is effectively screened out at distances greater than a Fermi–Thomas length $\lambda_{FT}=(\epsilon E_F/6\pi ne^2)^{1/2}$, where E_F is the Fermi energy, although a small oscillatory tail dropping off as $1/r^3$ does exist at large distances $r\gg\lambda_{FT}$. These screening lengths in solid-state plasma range from the order of an angstrom to a micron. Consequently, solid-state plasma properties can often be calculated by assuming the plasma to be infinite in extent. However, some interesting effects occur when the boundedness of the plasma is taken into account. For instance, at the surface between a plasma and material of dielectric constant ϵ_0, a surface plasmon of energy $E_s=h\omega_p/(1+\epsilon_0)^{1/2}$ exists, which is confined to within a screening length of the surface. Surface plasmon detection in the energy-loss spectra of fast electrons passing through thin solid films, in optical reflection, in LEED and Auger electron spectroscopy, and in tunneling of electrons through thin insulating layers has proved useful for surface diagnostics.

Plasmons are responsible for a variety of physical effects. In metals, for instance, the electron plasma determines not only the electrical transport properties of the crystal but also the cohesive energy, crystal structure, and phonon frequencies. The van der Waals force between two blocks of metal at small distances of separation L varies as L^{-3} and is due to surface plasmons. The van der Waals force between two conducting chains varies as $L^{-3}(\ln L)^{-3/2}$, having a much longer range than the L^{-6} force between nonconducting chains. This long range is a consequence of the spatial dispersion of the conduction-electron plasmon modes in a chain. Pi electrons in cyclic polyene molecules have similar sound-wave-like plasmon dispersion relations, so these carcinogen-like molecules may also exhibit very long-range van der Waals interactions. Solid-state plasmas are involved in several practical applications, e.g., lasers based on the recombination of electron–hole plasmas, plasma-filled waveguides, microwave sources, low-voltage and low-noise plasma-coupled devices, and *p-n* junctions in the microwave breakdown region.

Recently a great deal of attention has been paid to the two-dimensional plasmas which occur in multiple quantum well heterostructures. These have been studied with a variety of techniques, including picosecond and subpicosecond luminescence spectroscopy, time-resolved Raman scattering, time-resolved absorption bleaching of laser probes, and rf spectroscopy. Several interesting effects have surfaced. Large differences in the momentum relaxation times of electrons and holes in GaAs quantum-well structures have been attributed to the degeneracy of the two-dimensional electron plasma at high densities. Plasma-enhanced magnetooptical polarization rotation and magnetoplasma oscillations have been observed, the latter even under quantum Hall effect conditions. Negative mobilities of electrons have been found in nonequilibrium electron–hole plasmas in GaAs heterostructures, due to carrier drag on the electrons by a high mobility hole plasma. Phase changes have been found to occur, e.g., in the two-dimensional electron plasma at a GaAs/GaAlAs heterojunction when a magnetic field induces a liquid to solid phase transition. Phase changes have also been observed in pulsed laser-, electron beam-, or ion beam-annealing of semiconductor surfaces, where high optical reflectivity with persistence times of the order of 100 ns has been attributed to a boson condensation to a superfluid phase.

See also ELECTRON ENERGY STATES IN SOLIDS AND LIQUIDS; PLASMAS; QUASIPARTICLES.

BIBLIOGRAPHY

J. A. Ball, J. A. Wheeler, and E. L. Fireman, "Photoabsorption and Charge Oscillation of the Thomas-Fermi Atom," *Rev. Mod. Phys.* **45**, 333–352 (1973).

Maurice Glicksman, "Plasmas in Solids," in *Solid State Physics* (H. Ehrenreich, F. Seitz, and D. Turnbull, eds.), Vol. 26, pp. 275–427. Academic Press, New York, 1971.

P. M. Platzman and P. A. Wolff, "Waves and Interactions in Solid State Plasmas," in *Solid State Physics* (H. Ehrenreich, F. Seitz, and D. Turnbull, eds.), Suppl. 13. Academic Press, New York, 1973.

Polarizability

Charles W. Drake

The polarizability is a microscopic electric property of dielectric media which relates the macroscopically observable electric susceptibility to the properties of the molecules or atoms comprising the media. In macroscopic terms, we may consider an electrically neutral dielectric to have electric dipole density **P** induced in the medium when an external electric field is applied (because of the negative and positive charge separation). As a function of the electric field **E** a component of **P** can be expanded in general as

$$P_i = \sum_j a_{ij}E_j + \sum_{j,k} b_{ijk}E_jE_k + \cdots.$$

It is found in practice that at normal temperatures and traditional electric field strengths only the leading term is necessary. For isotropic substances, **P** is parallel to **E** or $a_{ij} = \chi_e\delta_{ij}$, where $\delta_{ij} = 1$ or 0 as $i = j$ or $i \neq j$. χ_e is the electric susceptibility of the dielectric and is a constant with respect to geometric properties. Therefore

$$\mathbf{P} = \chi_e\mathbf{E}. \tag{1}$$

For anisotropic substances, such as many crystalline solids, there are at most six independent a_{ij}'s. We will confine ourselves to the isotropic case. If the medium is considered to be made up of N molecules per unit volume of a single kind, we may write $\mathbf{P} = N\langle\mathbf{p}\rangle_m$, where $\langle\mathbf{p}\rangle_m$ is the average effective dipole moment of a molecule. The polarizability α, or more specifically the molecular polarizability, can be defined as the dependence of the electric field on the molecular dipole moment or

$$\langle\mathbf{p}\rangle_m = \alpha\mathbf{E}_l. \tag{2}$$

$\mathbf{E}_l$ is the electric field at the position of the molecule. This local field will not in general be identical to the macroscopic field **E** which appears in Eq. (2). The local field $\mathbf{E}_l$ may be set equal to **E** for gases at low pressures, with the criterion that the distance between molecules be large compared to molecular dimensions. This criterion holds for gases near normal pressure and temperature. In more dense media the method due to Lorentz (Ref. 1, pp. 94–96; Ref. 3, sec. 4.5; Ref. 4, pp. 47–49) is usually employed; this approximation replaces $\mathbf{E}_l$ by $\mathbf{E} + 4\pi/3\mathbf{P}$ so that we may relate χ_e and α by Eqs. (1) and (2):

$$\mathbf{P} = N\alpha E_l = N\alpha(\mathbf{E} + \tfrac{4}{3}\pi\mathbf{P}) = \chi_e\mathbf{E},$$

giving

$$\chi_e = \frac{N\alpha}{1 - \frac{4}{3}\pi N\alpha}. \tag{3}$$

The susceptibility χ_e may be related in turn to the dielectric permittivity ϵ by relations among the electric displacement **D**, the polarization **P**, and the electric field **E**, i.e., $\mathbf{D} = \epsilon\mathbf{E} = \mathbf{E} + 4\pi\mathbf{P} = \mathbf{E} + 4\pi\chi_e\mathbf{E}$ so that $\epsilon = 1 + 4\pi\chi_e$. This may be combined with Eq. (3) and solved for α to give

$$\alpha = \frac{3}{4\pi N}\left(\frac{\epsilon - 1}{\epsilon + 2}\right), \tag{4}$$

which is known as the Clausius–Mosotti equation. For substances where $\mathbf{E}_l \approx \mathbf{E}$ (or $\epsilon \approx 1$) the equation simplifies to $\alpha = (\epsilon - 1)/4\pi N$. The relationship provides a method of measuring α, valid for gases, from the macroscopic measurable quantity ϵ.

We have neglected any time dependence in these electrical

quantities. We should however write for $\mathbf{D}(\mathbf{r},t)$:

$$\mathbf{D}(\mathbf{r},t)=\mathbf{E}(\mathbf{r},t)+\int_0^\infty G(\tau)\mathbf{E}(\mathbf{r},t-\tau)d\tau. \quad (5)$$

This shows that **D** depends not only on the present value of **E** but also on all values of **E** at prior times. We may write the Fourier transformation of Eq. (5) and get

$$\mathbf{D}(\mathbf{r},\omega)=\mathbf{E}(\mathbf{r},\omega)+4\pi\chi_e(\omega)\mathbf{E}(\mathbf{r},\omega),$$

where

$$4\pi\chi_e=\frac{1}{\sqrt{2\pi}}\int_{-\infty}^{\infty} G(\tau)e^{i\omega\tau}d\tau,$$

the Fourier transform of $G(\tau)$. We therefore are dealing with Fourier amplitudes of **E**, **D**, χ_e, ϵ, and α. The discussion here is limited to frequencies much less than any characteristic frequencies in the molecules.

In order to discuss an appropriate model for a molecule, we must consider two polarizability mechanisms. If the molecule has an electric dipole moment in the absence of an electric field in the reference frame in which the molecule is at rest, the molecule is classified as a polar molecule. There will not be a macroscopic polarization, since the dipole axes of the molecules will be randomly oriented, i.e., $\langle\mathbf{p}\rangle_m$ is zero. This is consistent with a general quantum-mechanical derivation that any isolated molecule (or any other quantum-mechanical system which is bound by the electromagnetic interaction) having a definite value of total angular momentum cannot have odd electric multipole moments if there is no degeneracy of states of different parity. A permanent electric dipole moment is also forbidden by time-reversal invariance. An electric field will, however, tend to align the molecules, producing a net polarization and an effective $\langle\mathbf{p}\rangle_m$ parallel to the field direction. This corresponds quantum-mechanically to mixing molecular states of opposite parity by the electric field.

The effective molecular dipole moment arising from the orientation of the permanent dipole moment of polar molecules should become completely aligned or saturated in sufficiently high electric fields. At ordinary fields and temperatures, saturation is prevented by the thermal motion of the molecules, which tends to produce random dipole orientation. The polarization may be calculated by finding the average angle between the applied electric field and the molecular dipole axis using the Maxwell–Boltzmann temperature-dependent distribution of directions. The result is

$$\langle\mathbf{p}\rangle_m=p_0\left[\coth\left(\frac{p_0E}{kT}\right)-\frac{kT}{p_0E}\right], \quad (6)$$

where p_0 is the permanent dipole moment, k is the Boltzmann constant, and T is the absolute temperature. Equation (6) is known as the Langevin formula. For $p_0E/kT\ll 1$ the formula reduces on the right to $p_0^2E/3kT$.

In nonpolar molecules, which do not have permanent electric dipole moments, induced moments can be produced in an electric field by distortion of the electronic charge distribution. This mechanism also applies to monatomic molecules or atoms and of course will also be present in a polar molecule. This induced polarizability is an additional mechanism producing a molecular polarizability, which can be calculated from various models, from a simple Lorentz atom to elegant quantum-mechanical schemes. The induced polarizability is not temperature dependent.

The polarizability from both effects, electronic deformation α_e and molecular orientation, is

$$\alpha=\alpha_e+p_0^2/3kT \qquad (p_0E/kT\ll 1). \quad (7)$$

From (7) it can be seen that if the polarizability of a gas is plotted as a function of $1/T$, a straight line should result. If the slope is not zero, the molecules must be polar, and from the slope, the permanent dipole moment, p_0, can be found. The intercept with the $1/T$ axis gives α_e.

The polarizability of nonpolar molecules and the permanent dipole moment of polar molecules can be measured on isolated, individual molecules using molecular-beam techniques. Experiments are done using electric and magnetic deflection of the molecules in inhomogeneous fields, and also resonance experiments are performed where the quantum-mechanical energy levels are measured. The tensor nature of α must be reconsidered when studying an isolated molecule. For example, with a diatomic molecule an $\alpha_\perp$ must be defined when the electric field is perpendicular to the intermolecular axis and an $a_\parallel$ when **E** is parallel to the intermolecular axis. Such experiments and appropriate techniques are described in Refs. 5–7.

The permanent-dipole-moment determinations are of prime importance in molecular-structure studies. The electronic polarizability measurements serve as excellent tests for molecular electronic wave function calculations.

ADDENDUM

The development of coherent optical sources and advances in associated optical techniques have made common the ability to irradiate materials with very intense optical and near-optical electric fields. These experimental advances have led to greatly increased understanding of a multitude of quantum-optical effects in materials which in turn has led to further applications. Many of these higher-order nonlinear processes are equivalent classically to the use of the higher-order terms in the expansion of the polarization, **P**. Some of these effects require terms up to coefficients of order E^5. Applications which have made use of high-order terms are in emission and absorption spectroscopy including various types of Raman spectroscopy and in wave-front modification producing phase-reversal reflections leading to optical correction schemes.

See also DIELECTRIC PROPERTIES; ELECTRIC MOMENTS; ELECTRODYNAMICS, CLASSICAL.

REFERENCES

1. J. R. Reitz and F. J. Milford, *Foundation of Electromagnetic Theory*. Addison-Wesley, Reading, Mass., 1960. (I)
2. D. J. Griffiths, *Introduction to Electrodynamics*, 2nd ed. Prentice-Hall, Englewood Cliffs, NJ, 1989. (I)

3. J. D. Jackson, *Classical Electrodynamics*, 2nd ed. Wiley, New York, 1975. (A)
4. W. F. Brown, "Dielectrics," in *Handbuch der Physik*, Vol. XVII. Springer-Verlag, Berlin, 1956. (A)
5. N. J. Ramsey, *Molecular Beams*. Oxford University Press, Oxford, 1955. (A)
6. P. Kusch and V. W. Hughes, "Atomic and Molecular Beam Spectroscopy," in *Handbuch der Physik*, Vol. XXXVII/1. Springer-Verlag, Berlin, 1959. (A)
7. T. C. English and J. C. Zorn, "Molecular Beam Spectroscopy," in *Methods of Experimental Physics*, Vol. 3, Pt. B (D. Williams, ed.). Academic Press, New York, 1974. (A)
8. T. L. Weatherly and Q. Williams, "Electric Properties of Molecules," in *Methods of Experimental Physics*, Vol. 3 (D. Williams, ed.). Academic Press, New York, 1962. (A)
9. J. H. Van Vleck, *The Theory of Electric and Magnetic Susceptibilities*. Oxford University Press, Oxford, 1932. (A)

Polarization

Homer E. Conzett

A spin-polarized beam or target of particles is one in which the particle spins are preferentially oriented, as contrasted with the random distribution of spin directions in an unpolarized ensemble. Although most polarized beams and targets are ensembles of spin-$\frac{1}{2}$ or spin-1 nuclei, systems with higher spin values are playing an increasingly important role in current nuclear physics research.

The ensembles of interest are characterized by spin orientations which possess cylindrical symmetry with respect to the orientation (quantization) axis. Then for spin I and fractional magnetic substate populations n_m, with $-I \leq m \leq I$ and $\sum_m n_m = 1$, the degree of orientation of rank λ is defined for $\lambda = 1,2$:

$$f_1(I) \equiv \frac{\langle I_z \rangle}{I} = \frac{1}{I} \sum_m m \cdot n_m,$$
$$f_2(I) \equiv \frac{\langle I_z^2 \rangle}{I^2} = \frac{1}{I^2} \left(\sum_m m^2 n_m - \tfrac{1}{3} I(I+1) \right). \tag{1}$$

$\langle I_z \rangle$ is the value, averaged over the ensemble, of the spin component along the quantization axis, z. For the more common spin-$\frac{1}{2}$ and spin-1 systems we then have the following:

spin-$\frac{1}{2}$:

$$\text{Polarization} = P_z(\tfrac{1}{2}) = f_1(\tfrac{1}{2}) = 2 \sum_m m \cdot n_m;$$

spin-1:

$$\text{Vector-polarization} = P_z(1) = f_1(1) = \sum_m m \cdot n_m,$$

Tensor-polarization, alignment

$$= P_{zz}(1) = 3 f_2(1) = 3 \left(\sum_m m^2 n_m - \tfrac{2}{3} \right).$$

Also, from

$$\sum_m n_m = 1,$$

$$P_{zz}(1) = 1 - 3 n_0.$$

So it is seen that the less-familiar tensor component is a measure of the difference of n_0 from the unpolarized value $\frac{1}{3}$. The photon carries a spin of 1, but since the electromagnetic field is transverse to the direction (z) of propagation, the photon spin substate (m) populations are limited to n_+ and n_-, corresponding to right and left circular polarizations, respectively. Since $n_0 = 0$, photon beams are, in a strict sense, always tensor polarized.

Ion sources of polarized spin-$\frac{1}{2}$, spin-1, and spin-$\frac{3}{2}$ nuclear particles have been developed, as have nuclear targets of even higher spin values. This makes it possible to provide energetic polarized beams at various research facilities, e.g., Van de Graaff accelerators, cyclotrons, and synchrotrons. These polarized beams and targets are crucial for experiments which seek to identify the various spin-dependent components of the nuclear interactions.

BIBLIOGRAPHY

S. E. Darden, "Description of Polarization," in *Polarization Phenomena in Nuclear Reactions* (H. H. Barschall and W. Haeberli, eds.), p. 39. University of Wisconsin Press, Madison, 1971. (I-A)

W. Haeberli, "Polarized Beams," in *Nuclear Spectroscopy and Reactions*, Part A (J. Cerny, ed.), p. 151. Academic Press, New York, 1974. (E-I)

G. G. Ohlsen, "Polarization Transfer and Spin Correlation Experiments in Nuclear Physics," *Rep. Prog. Phys.* **35**, 717 (1972). (I-A)

M. Simonius, *Theory of Polarization Measurements, Observables, Amplitudes and Symmetries* (Lecture Notes in Physics **30**; D. Fick, ed.), p. 38. Springer-Verlag, New York, 1974. (A)

Polarized Light

P. F. Liao

Sir William Bragg in his lectures in 1931 described the polarization of light as a "quality which is not indeed perceived by the eye but can be detected in other ways and must somehow be accounted for," and indeed light polarization had played an important role in the development of our present understanding of electromagnetic radiation. Efforts to understand double refraction brought Thomas Young in 1817 to first propose that light is a "transverse vibration" and that a constant direction of the vibration "is polarization." By 1864 James Maxwell had developed a theory that required the vibrations to be strictly transverse and that gave the connection between electricity and light. The results of this theory, which are known as Maxwell's equations, describe light as an electromagnetic wave consisting of electric and magnetic fields that are perpendicular to each other and to the direction of propagation. The direction of the electric field is called the polarization direction.

Light coming from ordinary light sources is normally unpolarized. The electric vector changes direction rapidly and randomly although it remains perpendicular to the direction of propagation. Linearly polarized light has a constant polarization direction. For monochromatic radiation, the wave solution to Maxwell's equations that corresponds to a wave propagating in the z direction can be written as

$$\mathbf{E}=\hat{x}E_x \cos(kz-\omega t+\phi_1)+\hat{y}E_y \cos(kz-\omega t+\phi_2) \quad (1)$$

where $k=\omega/c$ and c is the speed of light. If $\phi_1=\phi_2$, Eq. (1) represents a linearly polarized wave. If $|E_x|=|E_y|$ and $\phi_1=\phi_2+\pi/2$, the light is circularly polarized, and the electric field vector rotates with angular frequency ω. In the general case ($\phi_1\neq\phi_2$ and $|E_x|\neq|E_y|$) the wave is elliptically polarized.

Linearly polarized light can be produced by passing an unpolarized beam through a polarizer. The most common type is the dichroic polarizer (such as the Polaroid filter), which selectively absorbs one polarization while passing the other. The Nicol prism is a polarizing device based on the same phenomenon of double refraction that originally led Young to the concept of transverse polarization. The prism consists of two pieces of calcite cemented together as shown in Fig. 1. The anisotropic character of calcite crystal causes an unpolarized beam to break up into two beams, the ordinary (O) and extraordinary (E) beams, which propagate in different directions and have orthogonal polarizations. At the cemented interface the angle of incidence of the ordinary beam is such that it is totally reflected, while the extraordinary beam is transmitted.

The light reflected from a smooth interface between transparent media of different refractive indices n_1 and n_2 is also polarized because the reflection coefficient for light polarized in the plane of incidence (p polarized) differs from that for light polarized perpendicular (s polarized) to the plane of incidence. In fact it can be shown that at an angle of incidence (Brewster's angle) θ_B given by

$$\tan\theta_B=n_2/n_1 \quad (2)$$

the reflected light is completely s polarized, since at this angle p-polarized light is completely transmitted. The polarization of reflected light allows the use of polarizing sunglasses to reduce glare from horizontal surfaces. Circularly polarized light can be obtained with a birefringent plate called a quarter-wave plate. The refractive index in this material for light polarized along $\hat{x}$ differs by Δn from that polarized along $\hat{y}$. Hence if the length l of the plate is chosen such that $\Delta nl=\pi/2$, a linearly polarized beam ($\phi_1=\phi_2$, $|E_x|=|E_y|$) will be converted to a circularly polarized beam ($|\phi_1-\phi_2|=\pi/2$).

The polarization of light can be analyzed with a polarizer. If a beam of light is polarized in the y direction ($E_x=0$), the electric field that passes through an analyzing polarizer whose axis is an angle θ to the y direction is $E_y \cos\theta$. Hence the transmitted intensity, which is proportional to the square of the field, is given by $\cos^2\theta$.

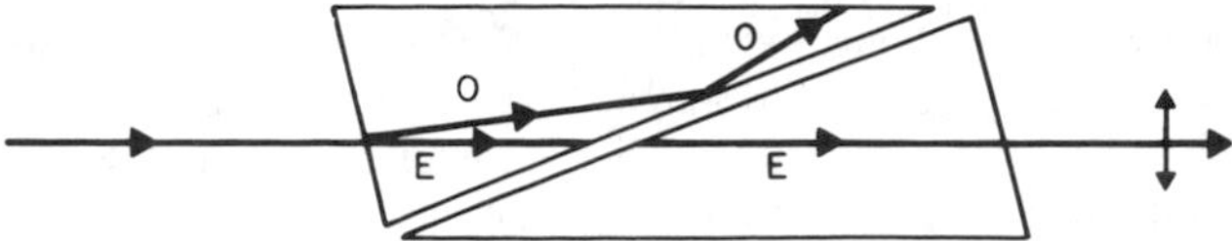

FIG. 1. The Nicol polarizing prism.

There are several important uses of polarized light. Such optical properties as birefringence and double refraction, which can be examined with polarized light, often provide important information about the crystal structure of a material. Strains that produce birefringence are easily observed by placing the strained material between a pair of crossed polarizers. Solutions of some organic compounds, such as sugar, are optically active, and cause the polarization direction of a linearly polarized beam to rotate as the light passes through the solution. A measurement of the rotation can be used as a measure of the solution concentration. Light polarization also plays an important role in determining the selection rules for absorption and emission of light by atoms and molecules and is an important tool in spectroscopy. Finally, the usefulness of polarized light is recognized by bees, who apparently use the polarization of skylight to assist them in navigation.

See also ELECTRODYNAMICS, CLASSICAL; ELECTROMAGNETIC RADIATION; KERR EFFECT, ELECTRO-OPTICAL; KERR EFFECT, MAGNETO-OPTICAL; LIGHT; MAGNETIC CIRCULAR DICHROISM; MAXWELL'S EQUATIONS; OPTICAL ACTIVITY; OPTICS, PHYSICAL.

BIBLIOGRAPHY

M. Born and E. Wolf, *Principles of Optics*. Pergamon, New York, 1965. (A)
William Bragg, *The Universe of Light*. Dover, New York, 1959. (E)
E. B. Brown, *Modern Optics*. Reinhold, New York, 1965. (I)
F. J. H. Dibbin, *Essentials of Light*. Cleaver-Hume Press Ltd., London, 1961. (E)
F. A. Jenkins and H. E. White, *Fundamentals of Optics*, 4th ed. McGraw-Hill, New York, 1976. (A)
M. V. Klein, *Optics*. Wiley, New York, 1970. (I)
B. Rossi, *Optics*. Addison-Wesley, Reading, Mass., 1957. (I)
W. A. Shurcliff, *Polarized Light*. Harvard Univ. Press, Cambridge, Mass., 1966. (A)
W. T. Welford, *Optics*. Oxford, London and New York, 1976. (E)

Polaron

Jozef T. Devreese

A charge placed in a polarizable medium will be screened. Dielectric theory describes the phenomenon by the induction of a polarization around the charge carrier. The induced polarization will follow the charge carrier when it is moving through the medium. The carrier together with the induced polarization is considered as one entity and was called a polaron by L. D. Landau.

A conduction electron in an ionic crystal or a polar semiconductor is the prototype of a polaron. Fröhlich proposed a simple model Hamiltonian for this entity through which the dynamics of the electron is treated quantum mechanically. The polarization, carried by the longitudinal optical phonons, is represented by a set of quantum oscillators with frequency ω and the interaction between the charge and the polarization field is linear in the field. This model (which up to now has not been solved exactly) has undergone extensive investigations [1–3]. The strength of the electron–phonon interaction is related to a dimensionless coupling constant α. When the coupling is weak (α small), the energy (ΔE) involved in the formation of a polaron, i.e., the self-energy, is proportional to α ($\Delta E = -\alpha h\omega$). The polaron mass m^*, which can be measured by cyclotron resonance experiments, is larger than the mass m of the charge carrier without self-induced polarization:

$$m^* = m(1 + \alpha/6).$$

When the coupling is strong (α large), a variational approach due to Landau and Pekar indicates that the self-energy is proportional to α^2 and the polaron mass is proportional to α^4.

Feynman used an exactly solvable model for the polaron that made possible variational calculations. He simulated the interaction between the electron and the polarization modes by harmonic interaction between a hypothetical particle and the electron. Using the solutions of this model, he was able to handle the Fröhlich Hamiltonian variationally. The resulting expressions for the self-energy and polaron mass are valid for arbitrary coupling and are quite accurate. Physically more directly accessible properties of the polaron, such as its mobility and optical absorption, have been investigated. In the framework of the Feynman approach, this information can be obtained from a frequency-dependent dissipation function. For relatively low temperatures T the mobility takes the form

$$u = \left(\frac{v}{w}\right)^3 + \frac{3}{4\alpha\beta}\, e^{\beta} \exp\{(v^2 - w^2)/w^2 v\}, \tag{1}$$

where $\beta = 1/kT$, and v and w are functions of α deriving from the Feynman model. Experimental work of F. C. Brown and co-workers on alkali halides and silver halides indicates that the mobility obtained from Eq. (1) describes the experimental results quite accurately (Fig. 1).

The magnetooptical absorption $\Gamma(\Omega)$ at the frequency Ω takes the form

$$\Gamma(\Omega) = \frac{\text{Im}\ \Sigma(\Omega)}{[\Omega - \omega_c - \text{Re}\ \Sigma(\Omega)]^2 + [\text{Im}\ \Sigma(\Omega)]^2}. \tag{2}$$

ω_c is the cyclotron frequency for a rigid-band electron, and

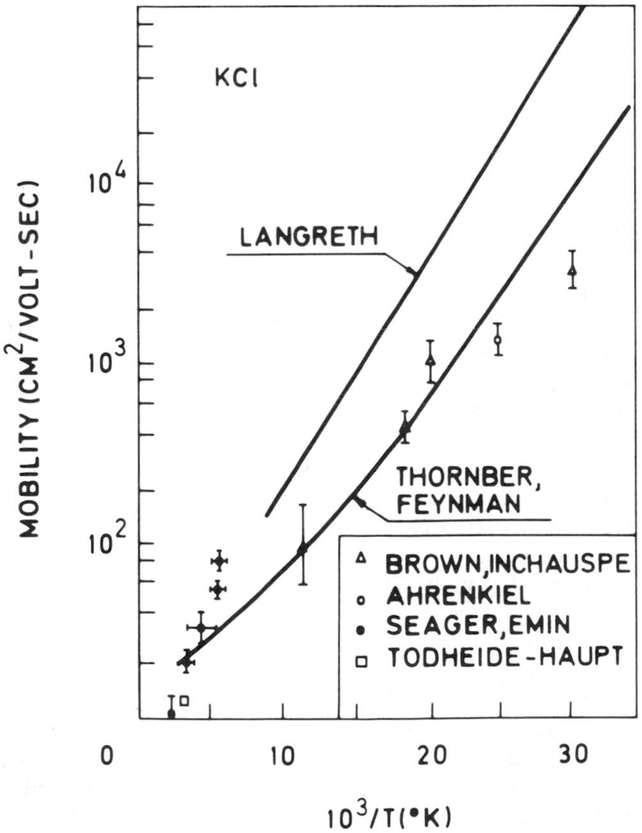

FIG. 1. Comparison of experimental and theoretical mobilities for KCl. The points shown by triangles and by open and closed circles are all photo-Hall data. The squares are drift mobility data. The two theoretical curves were drawn for $m/m_c = 0.412$ and $\theta = 300$ K. [From: F. C. Brown, *Point Defects in Solids*, Vol. 1 (J. Crawford and I. Slipkin, eds.), Plenum, New York, 1972].

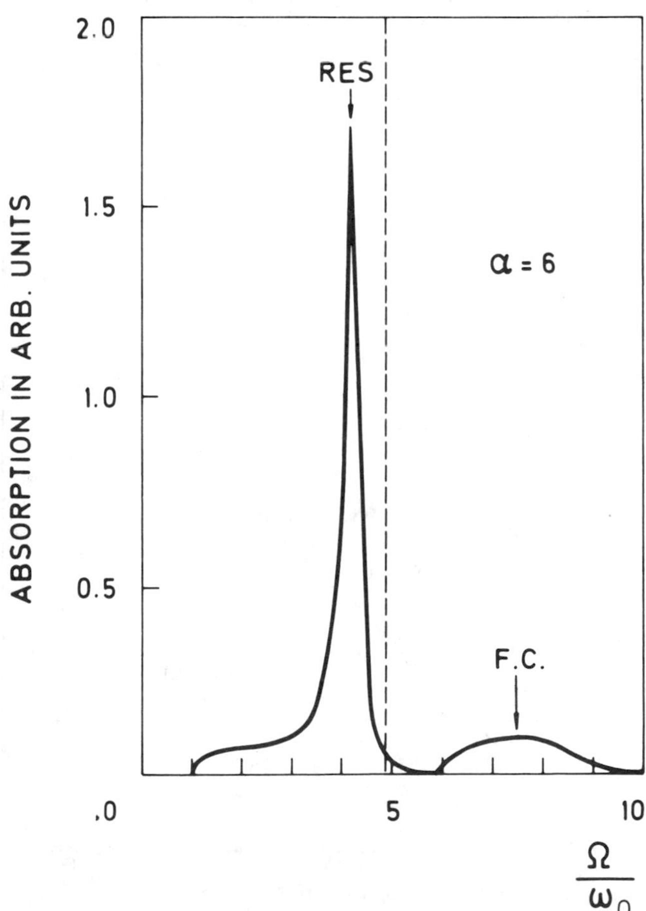

FIG. 2. Optical absorption of polarons at $\alpha = 6$. The RES peak is very intense compared with the Franck–Condon peak. The frequency $\Gamma = \nu$ is indicated. (See Ref. [2], p. 119.)

$\Sigma(\Omega)$ is the so-called "memory function" which contains the dynamics of the polaron. $\Sigma(\Omega)$ depends also on α and ω_c.

In the absence of an external magnetic field ($\omega_c = 0$) the spectrum (2) of the polaron at weak coupling is determined by the absorption of radiation energy which is reemitted in the form of longitudinal optical phonons. At larger coupling, $\alpha \geqq 6$, the polaron can make transitions toward a relatively stable internal excited state called the "relaxed excited state" (RES) (see Fig. 2). The RES peak in the spectrum also has a sideband which is connected with the possibility of a Franck–Condon-type transition.

Unfortunately the coupling constant α of the known ionic crystals is too small ($\alpha < 5$) to allow for the experimental detection of the relaxed excited state, i.e., the resonance condition $\Omega = \text{Re } \Sigma(\Omega)$ cannot be satisfied for $\alpha \leqq 6$.

The application of a sufficiently strong external magnetic field, however, allows one to satisfy the condition $\Omega = \omega_c + \text{Re } \Sigma(\Omega)$, which determines the polaron cyclotron resonance frequency. From this condition the polaron cyclotron mass can be derived. Using the most accurate polaron models to evaluate $\Sigma(\Omega)$, one obtains good agreement between theory and experiment [4] in determining the polaron cyclotron mass as shown in the case of AgBr ($\alpha = 1.53$) (Fig. 3).

It is very helpful to consider the polaron cyclotron resonance peak as due to a transition toward the internal relaxed excited state of the polaron, the latter being stabilized by the application of the external magnetic field. This concept is fruitful, especially if the polaron coupling constant α is sufficiently large.

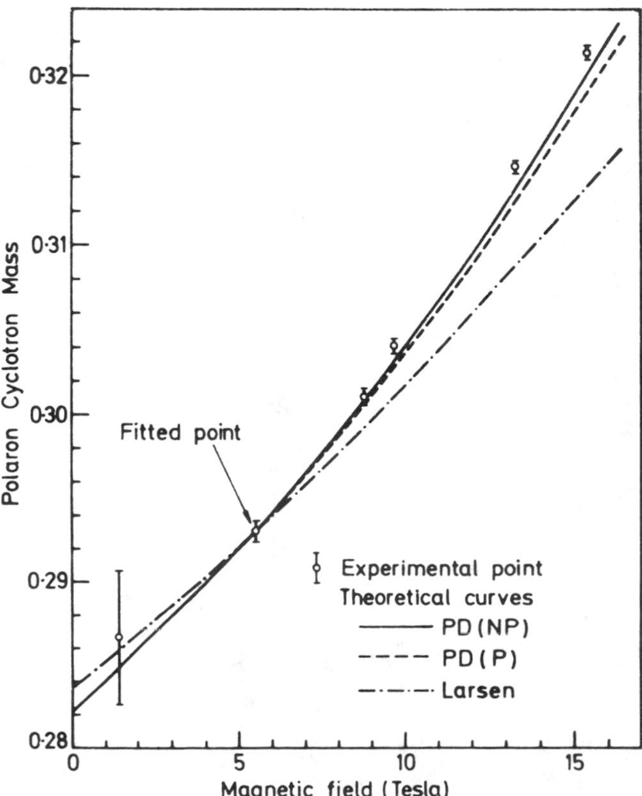

FIG. 3. Cyclotron resonance of electrons in AgBr: comparison of experiment and theory. PD (NP): Peeters and Devreese with two-band Kane corrections. PD (P): Peeters and Devreese with parabolic band. Larsen: Larsen with parabolic band. (See Ref. [4].)

POLARONS IN TWO DIMENSIONS (2D)

The great interest over the last decade in the study of the two-dimensional electron gas (2DEG) has also resulted in many investigations on the properties of polarons in two dimensions [5]. The Fröhlich-type polaron coupling does play a role in several polar heterojunctions (like GaAs–GaAlAs) and in MOSFETs where the electrons couple to the longitudinal optical phonons of the SiO_2 which has a large Fröhlich α.

A simple model for the 2D polaron system consists of an electron confined to a plane, interacting via the Fröhlich interaction with the LO phonons of a 3D surrounding medium. The mass of such a 2D polaron is no longer described by the expression valid in 3D; it is given by

$$\frac{m^*}{m} = 1 + \frac{\pi}{8}\alpha.$$

It has been shown that simple scaling relations exist, connecting the physical properties of polarons in 2D with those in 3D. An example of such a scaling relation is (Ref. [5], pp. 135–136)

$$\frac{m^*_{2D}(\alpha)}{m_{2D}} = \frac{m^*_{3D}(\frac{3}{4}\alpha)}{m_{3D}}, \tag{3}$$

where m_{2D} and m_{3D} are the electron-band masses in 2D and 3D, respectively.

The effect of the confinement of a Fröhlich polaron is to enhance the effective polaron coupling. However, many-particle effects tend to counterbalance this effect because of screening [5, 6].

Also, in 2D systems cyclotron resonance is an excellent tool to study polaron effects. Although several other effects have to be taken into account (nonparabolicity of the electron bands, many-body effects, form of the confining potential, etc.), the polaron effect is clearly revealed at the condition where the cyclotron frequency equals the LO phonon frequency. In Fig. 4 theory and experiment for the polaron cyclotron mass in a GaAs–GaAlAs heterojunction are compared.

For most of the solid-state systems in which, so far, two-dimensional electron (or hole) gases have been realized, the Fröhlich coupling is weak ($\alpha < 1$) so that polaron effects only occur in a perturbative sense.

An interesting 2D polaron system consists of electrons on liquid He. In this system the electrons couple to the ripplons of the liquid He. The effective coupling can be relatively high and, for some values of the parameters, even might lead to self-trapping.

Many other physical properties of polarons have been studied, including the possibility of localization, polaron transport, magnetophonon resonance, optical properties, etc.

Significant are also the generalizations of the polaron concept (see Table I). They have, for example, been successfully

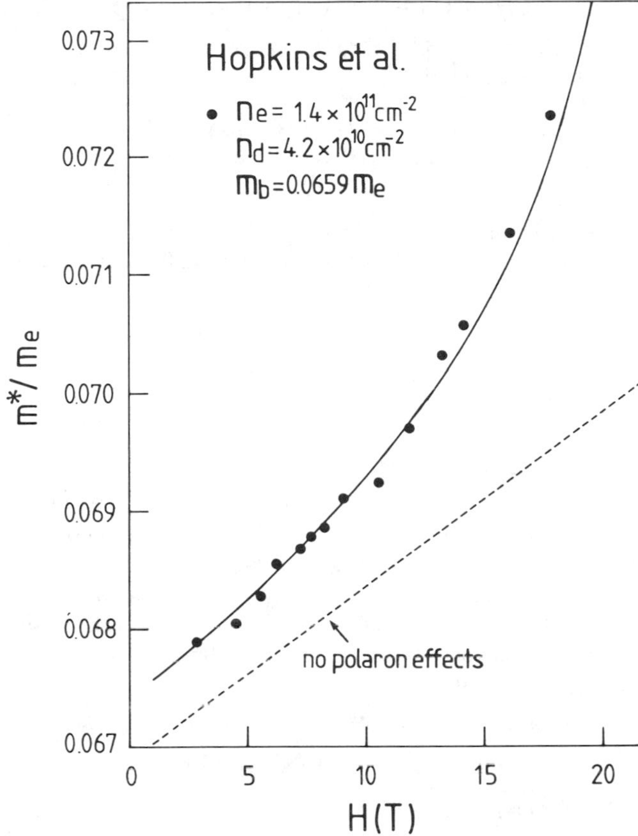

FIG. 4. Cyclotron mass as function of the magnetic field. The experimental results (solid dots) are from Hopkins *et al.* [*Superlattices Microstruct.* **2,** 319 (1986)]. The full curve is the theoretical result [see *Surf. Sci.* **196,** 437 (1988)] and the dashed curve corresponds to the theoretical result without polaron effects.

invoked to study the properties of polyacetylenes. Furthermore, bipolarons and their possible role in the high-T_c superconductors have been the subject of recent investigations.

In Table II the Fröhlich coupling constant is given for a few solids.

Table I Extensions of the polaron concept. Only the Fröhlich polaron is treated here

Concept	"Candidates"
Acoustic polaron	Hole in AgCl
Piezoelectric polaron	ZnO, ZnS, CdS
Electronic polaron	Band structure of all semiconductors
Spin polaron	Magnetic semiconductors
Bipolarons	
One center	Amorphous chalcogenides
Two center	Ti_4O_7, vanadium-bronzes
Small polarons and hopping	Transition-metal oxydes Polyacetylenes
Superconductivity and (bi)-polarons ("Localized electron pairs")	$BaPb_{1-x}$ Bi_xO_3 $LiTi_2O_4$ "high-T_c superconductivity"

Table II Fröhlich coupling constants

Material	α	Material	α
CdTe	0.31	KI	2.5
CdS	0.52	RbCl	3.81
ZnSe	0.43	RbI	3.16
AgBr	1.6	CsI	3.67
AgCl	1.8	TlBr	2.55
CdF_2	3.2	GaAs	0.068
InSb	0.02	GaP	0.201
KCl	3.5	InAs	0.052
KBr	3.05	$SrTiO_3$	4.5

See also CYCLOTRON RESONANCE; DIELECTRIC PROPERTIES; PHONONS; QUASIPARTICLES.

REFERENCES

1. G. C. Kuper and G. D. Whitfield (eds.), *Polarons and Excitons*. Oliver and Boyd, Edinburgh, 1963.
2. J. T. Devreese (ed.), *Polarons in Ionic Crystals and Polar Semiconductors*. North-Holland, Amsterdam, 1972.
3. T. K. Mitra, A. Chatterjee, and S. Mukhopadhyay, *Phys. Rep.* **153,** 2–3 (1987).
4. J. W. Hodby, G. P. Russell, F. Peeters, J. T. Devreese, and D. M. Larsen, *Phys. Rev. Lett.* **58,** 1471 (1987).
5. J. T. Devreese and F. M. Peeters (eds.), *The Physics of the Two-Dimensional Electron Gas,* Volume B157, ASI Series. Plenum, New York, 1987.
6. S. Das Sarma and B. A. Mason, *Ann. Phys.* (*NY*) **163,** 78 (1985).
7. S. A. Jackson and P. M. Platzman, *Phys. Rev.* **B24,** 499 (1981); see also F. M. Peeters in Ref. [5], p. 393.

Polymers

H. Mark

Polymers (high polymers, macromolecular substances) are of great fundamental importance for our existence and our culture. The human body, all animal and plant tissues, and most building substances in organic nature—such as proteins, wood, and chitin—consist of polymeric or macromolecular materials. Many minerals, such as silica and feldspar, are inorganic polymers, and numerous products of ancient and modern industry, such as porcelain, glass, textiles, paper, rubbers, and plastics, are either entirely or substantially polymeric. It was, however, only recently clearly recognized that all these substances possess one essential common feature, namely, that they consist of very large molecules.

DEFINITIONS AND CLASSIFICATION[*11*]

A polymer (Greek *polys*, many; *meros*, part or unit) is a substance consisting of molecules which are, at least approximately, multiples of low-molecular-weight units. The low-molecular-weight unit is the *monomer*. As long as the polymer is strictly uniform in molecular weight and molecular structure, its degree of polymerization is indicated by the Greek word for the number of monomers which it contains; thus we speak of a *dimer, trimer, tetramer, pentamer,* and so on. The term *polymer* designates a combination of an unspecified number of units. For example, trioxymethylene is the (cyclic) trimer of formaldehyde:

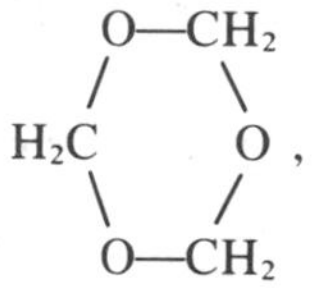

```
     O—CH2
    /     \
H2C        O ,
    \     /
     O—CH2
```

and polystyrene is the polymer of styrene:

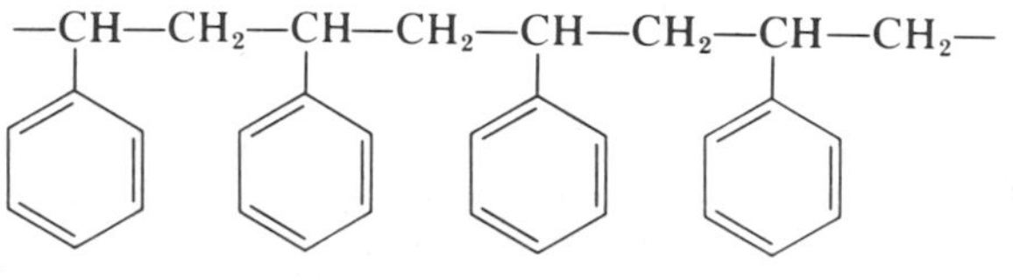

$$-CH(C_6H_5)-CH_2-CH(C_6H_5)-CH_2-CH(C_6H_5)-CH_2-CH(C_6H_5)-CH_2-$$

If the number of units becomes very large, one also uses the term *high polymer.* According to present-day usage, a polymer or high polymer need not consist of individual molecules which all have the same molecular weight, nor is it necessary that they all have the same chemical composition and molecular structure as each other or as the monomer unit. Natural polymers may exist in their native state, such as certain globular proteins or polycarbohydrates, in which the individual molecules all have the same molecular weight and molecular structure, but most synthetic and natural high polymers are obtained and investigated in a state where significant differences occur in the molecular weight of the individual macromolecules so that the material must be considered as a mixture of *homologous polymeric constituents.* The existence of a lesser or wider molecular-weight distribution is caused by our present inability to prepare polymers of exactly uniform character and by the lack of methods of resolving a homologous polymeric mixture into completely homogeneous fractions. The slight variability in chemical composition and molecular structure results from the presence of end groups, occasional branches, variations in the orientation of the monomeric units, and irregularity in the sequence of different types of these units in copolymers. A more rigorous definition and nomenclature, conforming to the practice in the domain of ordinary organic molecules, would be impracticable because the above variations always occur and their elimination, or even the quantitative determination of their nature or amount, is not possible by methods now available. The present usage of the words polymer and high polymer may further be justified on the basis that the above-mentioned variations often do not significantly affect the physical and chemical properties of the substance.

Isomeric polymers are polymers which have essentially the same percentage composition, but differ with regard to the arrangement of the individual atoms or atom groups in the molecules. Isomeric vinyl-type polymers may differ in the relative orientations (head to tail, head to head, and tail to tail or random mixtures of the two) of consecutive mers (monomeric units):

head to tail:

$$-CH_2-CHX-CH_2-CHX-CH_2-CHX-CH_2-CHX-,$$

head to head and tail to tail:

$$-CH_2-CHX-CHX-CH_2-CH_2-CHX-CHX-CH_2-,$$

or in the orientation of substituents or side chains with respect to the plane of the hypothetically extended backbone chain:

```
 H H X H H H X H
-C-C-C-C-C-C-C-C-.
 X H H H X H H H
```

Cis-trans isomerism may and probably does occur for any polymer containing double bonds other than those in pendent vinyl groups (those attached to the main chain):

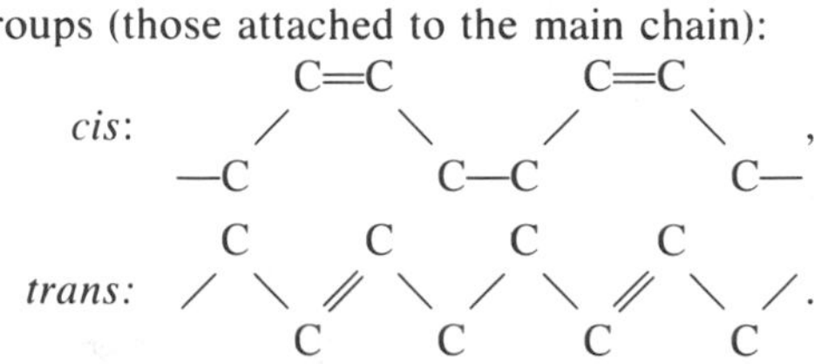

```
          C=C         C=C
cis:     /   \       /   \     ,
      —C      C—C         C—

        C    C    C    C
trans: / \ // \ / \ // \ /.
      C    C    C    C
```

Isomeric linear polymers from dienes, such as polybutadiene, polyisoprene, and polychloroprene, may and do result from different amount of 1,2- and 1,4-addition.

Isomeric copolymers may and probably do differ with respect to the way in which the different monomers are distributed along the chain even though their overall composition is the same:

regular alternation: –*A*–*B*–*A*–*B*–*A*–*B*–*A*–*B*–*A*–*B*–*A*–*B*–,

random alternation: –*A*–*B*–*B*–*A*–*A*–*A*–*B*–*A*–*B*–*B*–*A*–*B*–.

A *homopolymer* consists of macromolecules which are formed either by a single type of unit or by two (or more) chemically different types in regular sequence. Homopolymers may and do contain irregularities in minor amounts at the chain ends and at branch junctions.

Copolymers are macromolecules containing two or more chemically different monomeric units in a more or less irregular sequence. Normally, the relative numbers of the different types of units are not the same in different individual macromolecules nor even at different points in a single macromolecule. A mixture of macromolecules containing essentially only one type of monomeric unit with other macromolecules containing essentially only another type of monomeric unit is called a polyblend or a mixture of homopolymers. A *block copolymer* is a copolymer which contains longer stretches of two or more monomeric units linked together by chemical valences in one single chain:

—A-A-A-A-A-B-B-B-B-A-A-A-A—.

A graft copolymer contains branches of varying length made up of different monomeric units, on a common backbone or trunk chain:

```
    B-B-B-B-B-B-B-B-B-B—
    |
—A-A-A-A-A-A-A-A-A-A-A-A-A-A-A—.
```

Graft and block copolymers represent in their structure and in their properties a transition between normal copolymers and polyblends.

The detailed structure of block and graft copolymers depends essentially on the relative reactivity ratios of the various building units, which in turn are determined by the polarity and donor–acceptor characteristics of the individual monomers.

A tactic polymer consists of macromolecules whose monomeric units follow one another along the chain with steric configurations ordered according to some rule. The rule, or *tacticity*, may be simple or composed by few simple rules, but it must not have in any case random or statistic character. More complex tacticities can be originated not only by steric configuration of a few units, but also by programmatic arrangements of monomeric units having different configuration along the macromolecular chain.

A *eutactic polymer* is a tactic polymer whose monomeric units are *completely* ordered, so that it is devoid of any element of structural disorder. By way of example, a butadiene polymer is eutactic when all the monomeric units have the same *trans*-1,4 configuration or all have the *cis*-1,4 configuration.

An *atactic polymer* is a polymer whose macromolecules, although they have positional or structural arrangements, do not possess a steric order of the units. By way of example a vinyl polymer, obtained by conventional radical processes, is generally atactic; it may possess a head-to-tail arrangement, but it is disordered with regard to the steric configurations of the tertiary carbon atoms which follow each other along the main chains.

An isotactic polymer is one whose monomeric units are asymmetric units, ordered in such a way that when passing along a single chain from one monomeric unit to the next one finds repetition of the unit configuration. Isotactic polymers are usually eutactic. A *syndiotactic polymer* consists of macromolecules whose monomeric units are asymmetric and ordered in such a way that, when passing along the chain from one unit to the next, one finds a repetition of position and of structure, but *inversion* of the steric configuration. For example, a butadiene polymer formed through 1-2, head-to-tail enchainment, the successive monomeric units of which have alternately right and left configuration along the chain in respect of a selected direction, is a syndiotactic polymer.

Stereocopolymers are polymers with macromolecules that are formed by monomeric units which are chemically identical and which are ordered through positional and structural arrangements, but which do not have a unitary steric structure. *Stereoblock polymers* are stereocopolymers with molecules which are formed either by tactic blocks of at least two types, or by tactic and atactic blocks.

Polyelectrolytes are substances which, on dissolving in water or other ionizing solvents, dissociate to give polyions together with an equivalent amount of counterions (ions of small charge and opposite sign). Polyelectrolytes can be polyacids, polybases, polysalts, or polyampholytes.

Under certain conditions there exists the possibility for the establishment of chemical bonds between individual growing polymer chains. Such *cross links* can be formed during polycondensation reactions whenever tri- or tetrafunctional monomers are used, and during addition polymerization whenever dienes or trienes are involved. The initial result of a *cross-linking reaction* is the formation of small domains, within which the macromolecules consist of a random three-dimensional network of chemically linked chains; such domains are called *microgels*. They can be discovered by dynamic viscosity measurements, by turbidity, and by microfiltration. If a cross-linking reaction extends over macroscopic dimensions and if the polymerizing system is imbibed by a solvent or by the monomer, there results a gel that, after drying, is converted into an insoluble and infusible three-dimensional network, such as thermosetting resins like phenol- or urea-formaldehyde or copolymers of styrene and divinylbenzene. Similar insoluble and infusible systems are obtained if a vinyl derivative is polymerized in the presence of a small amount of a divinyl compound in such a manner that a local excess of free radicals prevails. The materials thus obtained are called *popcorn* polymers; their exact structure and properties are not yet completely elucidated.

POLYMERIZATION PROCESSES

There are many ways in which small molecules can be united with each other to form big ones. Their classification is usually based either on the *mechanisms* by which the monomeric units are joined together or on the *experimental conditions* which prevail during the reaction.

Mechanisms of Polymerization

The first principle distinguishes between polymerization effected through addition and polymerization through condensation mechanisms.

Typical cases of *addition polymerization* are the following.

(1) The addition of small molecules of one type to each other due to the opening of a multiple bond without elimination of any part of the molecule (vinyl-type polymerization):

$$H_2C{=}CHX + H_2C{=}CHX \rightarrow H_2C{=}CX{-}CH_2{-}CH_2X.$$

(2) The addition of small molecules of one type to each other by the opening of a ring without elimination of any part of the molecule (epoxide-type polymerization):

$$CH_3OH + (x+1)\underset{\diagdown\ O\ \diagup}{CH_2{-}CH_2} \rightarrow CH_3O[CH_2{-}CH_2{-}O{-}]_xCH_2{-}CH_2{-}OH.$$

(3) The addition of small molecules of one type to each other by opening of a multiple bond with elimination of part of the molecule (aliphatic diazo-type polymerization):

$$n\mathrm{RCH{=}N_2} \rightarrow (\mathrm{{-}CHR{-}})_n + n\mathrm{N_2}.$$

(4) The addition of small molecules of one type to each other by opening a ring with elimination of part of the molecule (α-aminocarboxy-anhydride-type polymerization):

$$\begin{array}{l} CH_2{-}CO \\ |\qquad\quad \diagdown \\ |\qquad\qquad O \rightarrow {-}CH_2CONH{-} + CO_2. \\ |\qquad\quad \diagup \\ NH{-}CO \end{array}$$

(5) The addition of small molecules of one type to another type by opening of a multiple bond (polyurethane, polyureide, and polyamide formation):

$HOCH_2CH_2OH + OCN(CH_2)_6NCO \rightarrow$

$$—OCH_2CH_2OCONH(CH_2)_6NHCO—.$$

(6) The addition to each other of biradicals formed by dehydrogenation (poly-*p*-xylylene polymerization):

$\cdot H_2C—C_6H_4—CH_2\cdot \rightarrow$

$$—CH_2—C_6H_4—CH_2—CH_2—C_6H_4—CH_2—.$$

Typical cases of *condensation polymerization* are the following:

(1) The formation of polyesters, polyamides, polyethers, polyanhydrides, etc., by elimination of water or alcohols from bifunctional molecules such as ω-hydroxy- or ω-aminocarboxylic acids or glycols, diamines, diesters, and dicarboxylic acids (polyester and polyamide-type polycondensation):

$HO(CH_2)_xOH + HOOC(CH_2)_yCOOH \rightarrow$

$$HO(CH_2)_xOOC(CH_2)_yCOOH + H_2O.$$

(2) The formation of polyhydrocarbons by the elimination of halogen or hydrogen halides with the aid of metal and metal halide catalysts (Friedel–Crafts and Ullmann-type polycondensation):

$$ClH_2C(CH_2)_xCH_2Cl \rightarrow H_2C(CH_2)_xCH_2CH_2(CH_2)_xCH_2—.$$

(3) The formation of polysulfides or poly-polysulfides by the elimination of sodium chloride from bifunctional alkyl or aryl halides with alkali sulfides or alkali polysulfides, or by the oxidation of dimercaptans (Thiokol-type polycondensation):

$ClH_2C(CH_2)_xCH_2Cl$

$$+ Na_2S_4 \rightarrow ClH_2C(CH_2)_xCH_2S_4CH_2(CH_2)_xCH_2Cl.$$

Experimental Conditions of Polymerization

From a more practical point of view, one frequently uses the *experimental conditions* as the principle for classifying polymerization processes.

(1) Polymerization in the *gaseous phase* at normal, reduced, or elevated pressure. The most important process of this type is the polymerization of ethylene, which is usually carried out at pressures above 300 atm in the temperature range between 150 and 250°C (see polyethylene). Poly-*p*-xylylene is prepared at reduced pressure by cooling the pyrolyzed products from 900°C to room temperature.

(2) Polymerization of one or more monomers in their *pure liquid phase* (bulk polymerization or block polymerization). Many vinyl-type polymerizations are carried out in this manner, particularly if it is desired to obtain large, transparent pieces (plates, rods, lenses, and so on) of the final product. Polyester and polyamide formation is almost exclusively carried out in the bulk phase, as is the polycondensation of alkylsilanediols or phenol-urea and melamine-formaldehyde addition products.

(3) Polymerization of one or more monomers by *dispersion* in the form of droplets of various size in a nondissolving liquid (suspension, bead, or pearl polymerization). Styrene, methyl methacrylate, and other monomers are polymerized in this manner to give beads of very uniform size and quality for injection and compression molding.

(4) Polymerization of one or more monomers in *solution*. If the polymer is soluble in its own monomer and in the solvent selected for the reaction, the system thickens increasingly as the reaction proceeds. If the polymer is insoluble in the liquid constituents, a slurry forms as the reaction proceeds and the polymer settles down as a more or less swollen powdery mass. Vinyl chloride, acrylonitrile, vinyl acetate, and other monomers are polymerized and copolymerized by this technique.

(5) Polymerization of one or more monomers in *emulsion*. It has been found that polymerization takes place in a particularly favorable manner if the monomer or monomers are brought together in the form of an aqueous emulsion with the aid of soaps, detergents, or emulsifying agents. The polymer or copolymer is then obtained in the form of a latex which can be either used as such or coagulated by the addition of acids or salts, or by centrifuging, to form lumps of the final product. Extensive research on the mechanism of emulsion polymerization indicated that the initiation and part of the propagation reaction occur essentially in the micelles formed by the emulsifying agent. The final growth of the macromolecules takes place in the polymer particles which are swollen by monomer transferred by diffusion from the monomer droplets. Most rubbery polymers and copolymers and some plastics are produced on a very large scale with the aid of emulsion polymerization.

There are a few other polymerization techniques, such as polymerization in the solid state, in a fine powder, or in a foam of small bubble size, but as yet they have not been worked out on a technical scale.

All techniques mentioned above can be carried out in batches or in a continuous manner. Batch processing still prevails in most of the large-scale operations, but there is a definite tendency to develop continuous schemes for all polymerization and polycondensation processes. Examples are the continuous polymerization of vinyl acetate, acrylonitrile, caprolactam, and GR-S (Buna-S, butadiene–styrene copolymer).

CATALYSIS

Catalysts play an important role in most polymerization reactions. Addition polymerization can be accelerated essentially by three different types of catalysis: free radical and two kinds of ionic catalysis (carbonium ion and carbanion).

Free-Radical Catalysis

Initiation of the reactions chains is by a free radical. The initiating free radicals can be either produced by decomposing a labile molecule by heat, light, or fast elementary particles or formed in the course of a chemical reaction. Typical cases of free-radical-forming catalysts are peroxides, such as acetyl peroxide or benzoyl peroxide; hydroperoxides, such as cumene hydroperoxide; and azo compounds, such as 2,2′-bisazoisobutyronitrile. (Substances of this type should strictly not be called "catalysts" because they take part permanently in the reaction and appear chemically combined with the resulting macromolecules. The word "initiator" expresses better the exact role which these molecules play in an addition polymerization.) The mechanism of their action can be expressed by the following equations:

$$C_6H_5{-}\overset{O}{\overset{\|}{C}}{-}O{-}O{-}\overset{O}{\overset{\|}{C}}{-}C_6H_5 \rightarrow 2C_6H_5{-}\overset{O}{\overset{\|}{C}}{-}O\cdot; \quad (1)$$

$$C_6H_5{-}\overset{O}{\overset{\|}{C}}{-}O{-}O{-}\overset{O}{\overset{\|}{C}}{-}C_6H_5 \rightarrow 2C_6H_5\cdot + 2CO_2; \quad (2)$$

$$C_6H_5{-}CH{=}CH_2 + h\nu \rightarrow C_6H_5{-}\dot{C}H{-}CH_2\cdot; \quad (3)$$

$$C_6H_5{-}CH{=}CH_2 + \text{heat} \rightarrow C_6H_5{-}\dot{C}H{-}CH_2\cdot; \quad (4)$$

$$(H_3C)(NC)(H_3C)C{-}N{=}N{-}C(CH_3)(CN){-}CH_3 \rightarrow N_2 + 2H_3C{-}C(H_3C)(NC)\cdot. \quad (5)$$

The initiator decomposes in the course of a monomolecular reaction, either by thermal action at a sufficiently high temperature or at low temperatures as a photosensitizer under the influence of a photon or a high-energy elementary particle, such as an electron, proton, alpha particle, or neutron. The activation energy to open the critical bond in the molecule of the initiator, which is usually a —O—O—, —C—N, or —C—O bond, amounts to about 15–30 kg cal. Consequently, there are initiators which act at relatively low temperatures, such as certain hydroperoxides and aliphatic azodinitriles (40–50°C), whereas others, such as peroxides or aliphatic azodiesters, may require temperatures as high as 150–175°C to decompose into free radicals. As in other monomolecular reactions, the frequency factor is of the order of 10^{12}–10^{14}.

$$C_6H_5{-}\overset{O}{\overset{\|}{C}}{-}O\cdot + H_2C{=}CHX \rightarrow C_6H_5{-}\overset{O}{\overset{\|}{C}}{-}O{-}CH_2{-}CHX\cdot \quad (6)$$

The free radical as produced in reaction (1) attacks the double bond of the monomer, attaches itself to it, and reproduces the unpaired electron at the other end of the adduct [Eq. (6)]. This process is called the *growth or propagation reaction* and consumes the molecules of the monomer to form macromolecules [Eq. (7)]; its activation energy and frequency constants have been found to be 5–10 kg cal. and 10^6–10^8, respectively.

$$C_6H_5{-}\overset{O}{\overset{\|}{C}}{-}O{-}CH_2\cdots CHX\cdot + C_6H_5{-}\overset{O}{\overset{\|}{C}}{-}O{-}CH_2\cdots CHX\cdot \quad (7)$$

$$\rightarrow C_6H_5{-}\overset{O}{\overset{\|}{C}}{-}O{-}CH_2{-}CHX\cdots CHX{-}CH_2{-}O{-}\overset{O}{\overset{\|}{C}}{-}C_6H_5.$$

Another interaction of the free radical at the end of a growing chain with a monomer molecule is to abstract a hydrogen atom from the latter and to transfer the free-radical character from the growing chain to the monomer [Eq. (8)]:

$$R{-}CH_2{-}CHX\cdot + CH_2{=}CHX \rightarrow R{-}CH_2{-}CH_2X + CH_2{=}CX\cdot. \quad (8)$$

Such processes are called chain transfer reactions; they occur between growing chains and molecules of the monomer, the already formed polymer, the catalyst, and the solvent. If two free-radical chain ends react with each other there results a pairing up of the lone electrons either by recombination or by disproportionation. This leads to the termination or cessation of the propagation step and finishes the polymerization reaction by the formation of stable macromolecules. The activation energies and frequency factors of typical termination reactions have been found to be 1–5 kg cal. and 10^9–10^{11}, respectively. The cooperation of the three reactions—(1), (6), and (7)—leads to average free lifetimes for the free-radical-type chain ends between 10^{-1} and 10 s and to steady-state free-radical concentrations of 10^{-6}–10^{-8} mol/liter.

If it is required to produce free radicals at lower temperatures, such as room temperature or below, the interaction of electron donors, such as Fe^{2+} or other reducing agents, with electron acceptors, such as H_2O_2 or other oxidizing agents, is utilized, setting up a *redox system* as known in the *Fenton reaction*,

$$H_2O_2 + Fe^{2+} \rightarrow Fe^{3+} + HO^- + HO\cdot,$$

that yields a small but relatively permanent source of free radicals, which initiate polymerization according to reaction (1) at temperatures as low as −20°C. This *reduction activation* is of great practical importance in emulsion polymerization, where many undesirable complications can be avoided or at least suppressed by working at low temperatures. Redox systems of practical importance are sodium bisulfite–potassium persulfate, cumene hydroperoxide–ferrous gluconate, potassium ferricyanide–sodium thiosulfate, and others.

Free-radical initiation is used in the bulk polymerization of styrene and methyl methacrylate, in the solution copolymerization of vinyl chloride–vinyl acetate, in the suspension polymerization of acrylonitrile and vinyl acetate, in the emulsion copolymerization of butadiene–styrene and butadiene–acrylonitrile, and in many other vinyl-type addition polymerizations.

Ionic Catalysis

Carbonium Ion Type. Certain vinyl type monomers, such as propylene, isobutylene, and most vinyl ethers, do not respond to free-radical initiation but can be initiated by Friedel–Crafts-type catalysts which produce protons by interaction with a cocatalyst. The protons attack the monomer to form a carbonium ion, which acts here as the carrier of the propagation. A typical case of such an ionic polymerization is the formation of polysiobutylene from isobutylene under the influence of boron trifluoride as catalyst and water as cocatalyst; it can be represented by the equations:

$$BF_3 + H_2O \rightleftharpoons H_2OBF_3 \rightleftharpoons H^+ + [HOBF_3]^-, \quad (9)$$

$$H^+ + CH_2{=}C(CH_3)_2 \rightarrow CH_3{-}C(CH_3)_2^+, \quad (10)$$

$$CH_3{-}C(CH_3)_2{-}CH_2\cdots C(CH_3)_2^+ + [HOBF_3]^- \rightarrow CH_3\cdots CH{=}C(CH_3)_2 + H_2OBF_3. \quad (11)$$

The interaction of catalyst and cocatalyst produces protons [Eq. (9)], the concentration of which is controlled by an equilibrium which depends on the concentration of the reactants and on temperature. Catalysts of this type are boron trifluoride, aluminum chloride, stannic chloride, and other acidic compounds; cocatalysts are water, ethers, and aminos. The proton adds to the monomer to form a carbonium ion [Eq. (10)], which propagates the chain by subsequent addition of other individual monomers. No formation of free radicals by splitting of an electron pair is necessary in this mechanism for initiation or propagation, but the displacement of an electron pair within the electronic cloud of the interacting particles suffices here to bring about the chemical changes which are necessary to combine the monomeric units to macromolecules. As a consequence, the activation energies for initiation and propagation are much lower here than in the free-radical mechanism, and it has been found that carbonium-type polymerization leads to a rapid formation of large molecules even at temperatures as low as —80°C.

The *termination* or *cessation* takes place by the reaction of the negatively charged complex ion with the carbonium-ion chain end, and can result in the formation of a terminal hydroxyl group or a terminal double bond [Eq. (11)]; it essentially restores the molecule of the catalyst and eventually even the catalyst–cocatalyst complex, so that it can attack another monomer and start another reaction cycle.

Carbanion Type. It has been found that negative ions, such as OH^-, NH_2^-, and R^-, can also initiate vinyl-type addition polymerization under certain conditions by forming a carbanion. This mechanism is of great importance and it has been established that most polymerizations which are initiated by alkali metals or alkali alkyls are of this type. In certain cases the lifetime of the carbanions is very large so that there is essentially no termination reaction. Polymers which form under such conditions are called "living polymers."

Most addition polymerization involving ring opening is subject to catalytic influence by positively or negatively charged ions. Thus the polymerization of ethylene oxide and other epoxides is strongly accelerated by bases, and the polymerization of α-amino acid–carboxylic anhydrides by acids. Both types are initiated by water, alcohols, or amines.

Coordination Complex Catalysts (Ziegler–Natta catalysts). A very important group of ionic initiators is formed if salts of transition metals—Ti, Zr, V, etc.—are reacted with metal alkyls—butyl-Li, $ZnEt_2$, $AlEt_3$, etc. In such cases one obtains a solid precipitate which polymerizes ethylenic compounds by a suspension-type process in which the initiation is caused by an *ion pair*. Polymerizations of this type take place very rapidly under mild conditions and lead to many products of great importance, e.g., linear polyethylene, isotactic polypropylene, *cis*-1,4-polybutadiene, and syndiotactic polybutylene.

Most polycondensation reactions are also catalytically influenced by acids and bases. In these cases one has to distinguish between equilibrium and rate conditions. Water-binding agents, such as phosphoric anhydride, sulfuric acid, or trifluoroacetic anhydride, very markedly increase the overall rate of polymer formation but do so by shifting the equilibrium in the direction of polycondensation rather than by any strictly catalytic influence on the elementary process of chain propagation. As most polycondensation reactions are carried out at rather elevated temperatures (200°C or higher), it is necessary to employ acids and bases of very low volatility, such as Sb_2O_3, BaO, PbO_2, or NaH.

MEASUREMENT OF HIGH MOLECULAR WEIGHTS

The outstanding property of all polymers is that they have molecules of high molecular weights; as a consequence, methods of measuring high molecular weights are of special interest in polymer research. Of the four methods to be mentioned, three permit the absolute determination of the various molecular weight averages of dissolved macromolecules; the fourth permits only relative molecular weight determinations but is frequently applied because of its experimental simplicity. The existing procedures are described below.

Osmotic Pressure

Up to molecular weights of 4000–5000, conventional cryoscopic, ebullioscopic, and vapor-pressure measurements are applicable. The results can be evaluated by equations based on Raoult's law and van't Hoff's law; low-molecular-weight impurities have a falsifying effect on the measurements and their interpretation.

No general experimental techniques exist for measuring molecular weights between 5000 and 15,000. Above 15,000 the direct measurement of the osmotic pressure with the aid of semipermeable membranes is applicable. Special types of osmometers have been developed for the purpose of working in the temperature range from −20 to 120 °C, using different types of solvents. The most useful membranes are denitrated cellulose nitrate, gel cellophane, and cross-linked polyvinyl alcohol. The method is limited at the low-molecular-weight end by noticeable diffusion of the solute through the membrane and at the high-molecular-weight end by the small magnitude of the osmotic pressure. No really reliable determinations can be made above molecular weights of about 1.5×10^6.

Solutions containing macromolecules with molecular weights above 15,000 do not obey van't Hoff's law and must be evaluated by the Huggins–Flory equation:

$$P/c = RT/M_n + Bc \quad (12)$$

where P is the osmotic pressure, c is the solute concentration in g/100 ml solution, R is the gas constant, T is the absolute temperature, M_n is the *number average molecular weight* of the solute, and B is an empirical constant characterizing the solubility of the particular polymer in the particular solvent.

In order to arrive at numerical values for M_n and B, it is necessary to carry out individual osmotic-pressure measurements at four or five different concentrations within the range of 0.1–0.5% and to plot P/c vs c. If the measurements are properly made and no disturbing influences interfere, the points are arranged on a straight line, whose slope is B and whose intercept is RT/M_n. A great deal of important information has been accumulated by this method.

Light Scattering

According to Einstein, Raman, and Debye, it is possible to determine the osmotic pressure of a solution from its capacity to scatter light. The total amount of laterally scattered intensity is measured by the turbidity τ of a system, which is defined by

$$\tau = (1/l) \ln (I/I_0), \tag{13}$$

where l is the length of the scattering system, and I_0 and I are the intensity of the beam as it enters and leaves the scattering system, respectively.

Turbidity and molecular weight of a polymer solution are related by the equation

$$Hc/\tau - RT/M_w + 2Bc, \tag{14}$$

where τ is the turbidity as defined in Eq. (13), c, R, T, and B have the same meaning as in Eq. (12), M_w is the *weight average molecular weight* of the solute, and H is a constant which depends on the wavelength of the scattered light and on the refractive indexes of the solvent and the solution; it has to be determined by a measurements independent of these quantities. To use Eq. (14) it is necessary to carry out individual turbidity measurements at four or five different concentrations within the range of 0.02–0.10%, and to plot Hc/τ vs c. If the measurements are properly made and no disturbing influences interfere, the points are arranged on a straight line, whose slope is $2B$ and whose intercept is RT/M_w. With sufficiently sensitive instruments it is possible to measure molecular weights as small as 3000–4000; in the domain of molecular weights above 300,000–400,000, the application of Eq. (14) is not permissible any more because the dimensions of the scattering particles become comparable with the wavelength of the scattered light. As soon as this is the case, a more complicated analysis is necessary which involves measurements of the scattered intensity at different concentrations and different angles. Debye, Doty, and Zimm have expanded the theory and the experimental techniques so that molecular weights up to several million can be successfully measured by the turbidity method.

Ultracentrifuge

Svedberg has originated the use of this instrument for the determination of the molecular weight of macromolecules. Two approaches can be used.

(1) Sedimentation and Diffusion Velocities. The molecular weight of a solute can be expressed as follows:

$$M_w = \frac{RTs}{D(1 - V\rho)}, \tag{15}$$

where M_w is the weight average molecular weight of solute, R and T have the same meaning as in Eq. (12), V is the partial specific volume of solute, ρ is the specific gravity of solvent, D is the diffusion rate constant, and s is the sedimentation rate constant.

In the case of globular macromolecules, such as certain proteins, D and s do not depend noticeably on concentration and can be obtained from a single measurement of the rates of diffusion and sedimentation at an arbitrary concentration in the range between 0.05 and 0.25%. For linear polymers with long flexible chains, D and s depend noticeably on solute concentration; four or five measurements of the rates of diffusion and sedimentation have to be carried out and the intrinsic values of D_0 and s_0 must be determined by extrapolation to infinite dilution. These values are then used in Eq. (15). The weight average molecular weights thus obtained, in general, agree well with those computed by Eq. (14).

(2) Sedimentation Equilibrium. The rate of concentration during sedimentation and back diffusion in a high-speed centrifuge establishes a final stationary state of solute concentration which permits determination of the molecular weight of the solute according to the following equation:

$$M_z = \frac{2RT \ln c/c_0}{(1 - V\rho)\omega^2(x^2 - x_0^2)}, \tag{16}$$

where R, T, V, and ρ have the same meaning as in Eq. (12) and (15), c_0 and c are the equilibrium solute concentrations at the distances x_0 and x from the axis of rotation, respectively, and M_z is the z-average molecular weight of the solute. This method has recently been greatly improved and expanded by Williams and Wales; it permits the determination of all three averages of the molecular weight of the solute and establishes valuable knowledge concerning the molecular-weight distribution of a given material.

Viscosity

Berl, Biltz, and Ostwald have pointed out that a relation exists between the molecular weight of a polymer and its capacity to produce solutions of high viscosity, and Staudinger has formulated a quantitative relation between the relative thickening power of a polymer in solution and its molecular weight. Kramer has expressed this relation in the form

$$[\eta] = K_m M_w, \tag{17}$$

where $[\eta]$ is the *limiting viscosity number* (or *intrinsic viscosity*), K_m is the Staudinger constant, and M_w is the weight average molecular weight of the solute; $[\eta]$ is obtained by extrapolating to zero concentration the Huggins equation for the *viscosity number* (or *specific viscosity*):

$$\frac{\eta_c - \eta_0}{\eta_0}\frac{1}{c} = [\eta] + k'[\eta]^2 c \tag{18}$$

where η_c is the viscosity of a solution of concentration c (in g/100 ml solvent), η_0 is the viscosity of the pure solvent, and k' is the empirical constant describing the influence of concentration on the thickening power of the polymer.

Having first determined $[\eta]$ from four or five individual viscosity measurements according to Eq. (18), and having calibrated the empirical constant K_m by any of the absolute methods previously described, the Staudinger equation (17) purportedly permits the determination of the molecular weight of a polymer by viscosity measurements. It has been shown, however, by subsequent, more detailed studies that Eq. (17) only holds for a few solute–solvent systems and has to be replaced in general by the relation:

$$[\eta]=KM^a, \tag{19}$$

where two constants, K and a, are necessary to convert limiting viscosity numbers into molecular weights. Equation (19) holds for all previously known polymer solutions over a considerable molecular-weight range and is therefore widely used to obtain molecular weights of macromolecules from viscosity measurements. The characteristic constants K and a have been determined for many solute–solvent combinations. A few particularly important cases are listed in Table I. To apply them, it is necessary to measure the solute concentration in grams of solute per 100 ml solvent, and to carry out the viscosity measurements in a concentration range below 2 g per 100 ml.

MOLECULAR HETEROGENEITY OF POLYMERS

Most polymers, natural or synthetic, are nonhomogeneous as far as their molecular weight is concerned; they represent a mixture of macromolecules with different degrees of polymerization. Several methods exist for determining the molecular-weight distribution in a given sample; some are of only analytical character, whereas others actually permit separation of the various molecular-weight fractions from each other in a preparative manner.

The relatively simplest way to get a first impression of the molecular heterogeneity of a polymer sample is to determine the different modes or averages of the molecular-weight distribution function. The first mode or number average molecular weight is defined by

$$\bar{M}_n=\frac{\int f(M)M\,dM}{\int f(M)\,dM} \tag{20}$$

where $f(M)$ represents the fraction of the material having a molecular weight M as a function of M. Experimentally, $\bar{M}_n$ can be obtained by all those methods which count the individual molecules in a solution of the sample, such as the analytical determination of end groups, the direct measurement of the osmotic pressure, the vapor-pressure lowering, the boiling-point elevation, or the melting-point lowering.

The second mode or weight average molecular weight is defined as

$$\bar{M}_w=\frac{\int f(M)M^2\,dM}{\int f(M)M\,dM}; \tag{21}$$

it can be obtained experimentally by measurement of the turbidity of a macromolecular solution, or by the combination of sedimentation and diffusion.

The third mode or z-average molecular weight is represented by

$$\bar{M}_z=\frac{\int f(M)^3M\,dM}{\int f(M)M^2\,dM}. \tag{22}$$

In order to illustrate the significance of the three averages $\bar{M}_n$, $\bar{M}_w$, and $\bar{M}_z$, let us assume that we have a mixture of four molecules having a molecular weight of 2 and four molecules having a molecular weight of 4. Then the three averages are given by

$$\bar{M}_n=\frac{(4\times 2)+(4\times 4)}{4+4}=\frac{24}{8}=3,$$

$$\bar{M}_w=\frac{(4\times 2^2)+(4\times 4^2)}{(4\times 2)+(4\times 4)}=\frac{80}{24}=3.33,$$

$$\bar{M}_z=\frac{(4\times 2^3)+(4\times 4^3)}{(4\times 2^2)+(4\times 4^2)}=\frac{288}{80}=3.6.$$

From this it can be seen that for any nonuniform polymer the z average is largest and the n average is smallest. The z average is experimentally accessible by sedimentation equilibrium measurements in the ultracentrifuge.

For a completely homogeneous material the three average molecular weights are identical; growing nonuniformity causes the ratios $\bar{M}_z$:$\bar{M}_w$:$\bar{M}_n$ to increase. If the molecular-weight distribution function is caused by complete randomness in the building up or degradation of linear macromolecules, the "normal" ratios are simply

$$\bar{M}_z:\bar{M}_w:\bar{M}_n=3:2:1.$$

By experimental determination of the values for $\bar{M}_z$, $\bar{M}_w$, and $\bar{M}_n$ one can therefore arrive at a conclusion whether a

Table I. Viscosity Average Molecular Weights

Polymer	Solvent	Temperature (°C)	K	a
Cellulose acetate	Acetone	25	1.9×10^{-5}	1.03
Cellulose nitrate	Acetone	25	3.8×10^{-5}	1.0
66 Nylon	90% Formic acid	25	11.0×10^{-4}	0.72
Polyvinyl acetate	Acetone	20	2.76×10^{-4}	0.66
Polystyrene	Benzene	30	1.7×10^{-4}	0.72
Polyisobutylene	Benzene	20	3.6×10^{-4}	0.64
Natural rubber	Toluene	25	5.02×10^{-4}	0.67
GR-S	Toluene	25	5.3×10^{-4}	0.67
Neoprene	Toluene	25	5.0×10^{-4}	0.62

given material is "normal" in its distribution or whether it has a narrower or wider distribution.

Figure 1 gives a picture of the three M_w averages as they exist in many addition and condensation polymers, e.g., in nylon, polyester and polystyrene. A very efficient method to arrive rapidly at such a result for any soluble polymer is the gel permeation chromatography (GPC) which presents automatically M_w distribution curves of the type shown in Fig. 1.

More quantitative statements as to the shape of $f(M)$ can be obtained by an analysis of the concentration gradient in the meniscus of a sedimentation-equilibrium run or in a series of sedimentation-velocity or -diffusion experiments. If the macromolecules carry electric charges, electrophoresis experiments can be used for the same purpose.

Another analytical method is the turbidity titration, which involves measurement of the turbidity increase of a dilute polymer solution on the gradual addition of a precipitant and fractional precipitation of the different molecular-weight species. The most accurate results are obtained when the individual fractions are separated by filtration or centrifugation. In this way these fractions are obtained in a pure state and their weight and molecular weight can be determined by independent measurements. Plotting the percentage weights of the individual fractions against their molecular weights, a differential molecular-weight distribution curve of the investigated material is obtained, which gives a fairly complete insight into the molecular heterogeneity of the sample.

GENERAL PROPERTIES OF HIGH POLYMERS IN BULK

The most important property of organic polymers as compared with ordinary organic substances is that they exhibit considerable mechanical strength in their bulk state. Hence they can be used to make fibers, films, plastics, rubbers, or coatings of commercially valuable properties. The degree to which mechanical properties are exhibited depends on several factors, the cooperation of which determines the most appropriate utilization of a given polymer and characterizes it as a typical material to form a fiber, a plastic, or a rubber.

One decisive factor for mechanical resistance is the *degree of polymerization* (or molecular weight) of the material under consideration. Extensive experiments with many polymers have shown that for each polymer there exists a certain critical molecular weight below which the material shows no mechanical strength at all. Above this critical value, mechanical strength is rapidly developed with increasing molecular weight, as shown by the relatively steep rise of the curve in Fig. 3. In the domain of higher degrees of polymerization, the increase of mechanical resistance with molecular weight becomes less pronounced, as represented by the flatter part of the curve in Fig. 2. An industrially useful material must, therefore, have a molecular weight somewhat above the inflection point of this curve. In principle, and as a matter of safety, it is advisable to choose a molecular weight well above this inflection point, but it has been found that materials having too high a degree of polymerization exhibit exorbitant viscosities in solution or in the molten state, which render such important processes as filtration, spinning, casting, and molding unduly difficult. A proper compromise between easy processibility and desirable ultimate properties must therefore be established by choosing an optimum molecular-weight range in each individual case. Another important factor for the development of characteristic mechanical properties is the *magnitude of the intermolecular forces between the individual macromolecules* of a given material. If the intermolecular attraction is small and the chains are difficult to fit laterally into a latticelike structure, then at a given temperature and stress the material will show a distinct tendency to return to its curled-up, relaxed state. Such conditions are typical for *rubbers*. On the other hand, if the molar cohesion is strong and the fine structure of the chains provides for easy lateral fitting into a latticelike arrangement, a state of high external crystallinity is favored and the material is a typical *fiber*. In intermediate cases, in which the forces are moderate and the geometry of the chains is moderately favorable for crystallization, the behavior of the material will depend greatly on external conditions such

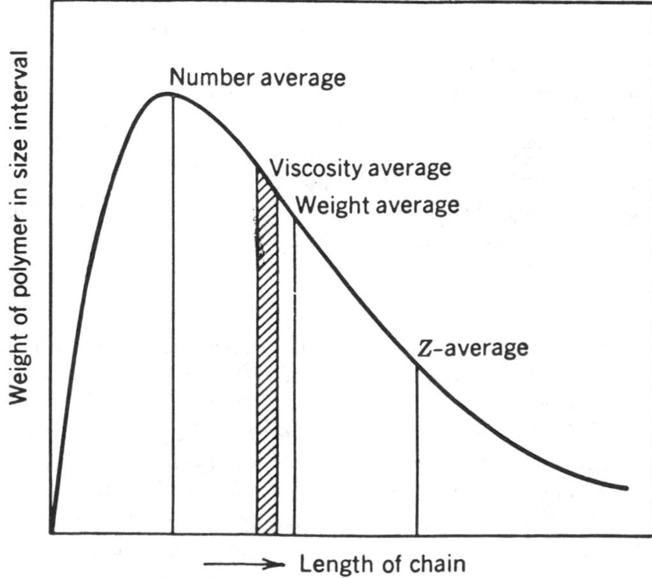

FIG. 1. M_w distribution in a typical polymer (from Van Krevelen and Hoftyzer).

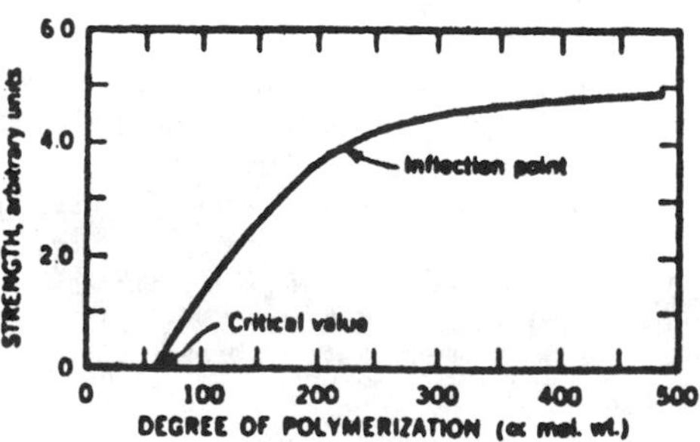

FIG. 2. Dependence of mechanical strength upon degree of polymerization (molecular weight).

as temperature and external mechanical forces. Such systems are typical *plastics*.

These qualitative considerations show that rubbers, plastics, and fibers are not intrinsically different materials. Their difference is rather a matter of degree and is produced by the way in which the intermolecular forces between the long chains and their general tendency to curl and fold cooperate in forming a certain mixture of crystallized and disordered portions in a given sample of the material. In general, it can be said that under given conditions, such as temperature and external forces applied to the sample, the tendency to crystallize is brought about by two different factors, which to a certain extent conflict with each other:

(a) The forces between the individual macromolecules. This influence corresponds to the ΔH term in the expression for the free-energy change during crystallization. If the forces are strong (above 5000 cal per mole of the group involved), then at ordinary temperatures they will preponderantly determine the behavior and the material can be expected to behave more or less as a typical fiber. This seems to be true for cellulose, proteins, nylon, Terylene (Dacron, Mylar), orlon acrylic fiber, etc.

(b) The geometrical bulkiness of the chains. This influence corresponds to the $T\Delta S$ term in the expression for the free-energy change during crystallization. Chains that fit easily into a crystal lattice (polyethylene, *trans*-polydiolefins) crystallize under the influence of comparatively weak forces and hence, in general, have a tendency to appear more fiberlike than would be expected if the forces alone were considered. On the other hand, materials which have bulky and irregular chains do not crystallize even if the intermolecular forces are quite strong. Hence they will be more rubberlike than would be expected on the basis of only the intermolecular forces (vinyl copolymers, cellulose mixed esters, etc.).

See also CATALYSIS; INTERATOMIC AND INTERMOLECULAR FORCES; LIGHT SCATTERING; RHEOLOGY; SEDIMENTATION AND CENTRIFUGATION; VISCOSITY.

BIBLIOGRAPHY

C. E. H. Bawn, *The Chemistry of High Polymers*. Butterworth, London, and Interscience, New York, 1948.

F. W. Billmeyer, *Polymer Chemistry*. New York, 1957

W. H. Carothers *Collected Papers*. New York, 1940

P. J. Flory, *Principles of Polymer Chemistry*. Cornell University Press, Ithaca, N.Y., 1953.

R. W. Lenz, *Organic Chemistry of Polymers*. Wiley, New York, 1967.

H. Mark, *Physical Chemistry*. New York, 1940

H. Mark and R. Raff, *High Polymeric Reactions* (Vol. III of *High Polymers*). Interscience, New York, 1941.

C. S. Marvel, *Organic Chemistry of Polymers*. Wiley, New York, 1959

K. H. Meyer, *Natural and Synthetic High Polymers* (Vol. IV of *High Polymers*). Interscience. New York, 1942.

K. H. Meyer and H. Mark, *Der Aufbau der hochpolymeren organischen Naturstoffe*. Akadem. Verlagsges., Leipzig, 1930.

H. Staudinger, *Die hochmolekularen organischen Verbindungen*. Springer, Berlin, 1932.

See also, particularly, *Encyclopedia of Polymer Science and Technology*. Wiley, New York, 1964–1972.

J. K. Stille, *Introduction to Polymer Chemistry*. Wiley, New York, 1962

Polymorphism *see* Allotropy and Polymorphism

Positron

A. Rich

Quantum mechanics, as conceived by Schrödinger, Heisenberg, et al. in the period 1925–1927, was essentially a nonrelativistic (NR) theory. Particles moving at relativistic (R) speeds (particle velocity v comparable to the speed of light c or particle kinetic energy $T \geqslant m_0c^2$, where m_0 is the particle rest mass) could not be accommodated within the framework of the theory. Even at the speeds characteristic of atomic electrons in hydrogen ($10^{-2}c$) serious corrections to the NR theory were necessary to account for observed atomic spectra. The transition from a nonrelativistic to a fully relativistic theory was accomplished by P. A. M. Dirac in 1928. Since Dirac's work led directly to the prediction of the positron (e^+), we discuss it briefly at this point.

Consider a particle of rest mass m_0 and charge q moving at velocity **v** (NR momentum $p_x = m_0v_x$, $p_y = m_0v_y$, $p_z = m_0v_z$; R momentum $p_x = \gamma m_0 v_x$, $\gamma = 1/(1 - v^2/c^2)^{1/2}$, etc., $p^2 = p_x^2 + p_y^2 + p_z^2$) in an electric potential $V(r)$ and in the absence of magnetic fields. Its NR Hamiltonian is $H = (p^2/2m_0) + qV(r)$. In the period 1925–1927 Schrödinger and others showed that atomic systems could be fairly successfully described by an equation obtained by converting p and H into operators and letting them operate on a wave function ψ. The solution of the resulting equation gives both $\psi(r,t)$ (the state of the system) and the energy W associated with a particular state ψ. Once ψ has been obtained, average values of other quantities of interest (momentum, angular momentum, etc.) may be computed according to the prescription: Average value of operator $= \int[\psi^* \text{ (operator) } \psi]\, dV$ where dV is the volume element.

The years 1926–1928 saw attempts by Schrödinger, Klein, and Gordon to arrive at a relativistic wave equation based on the classical covariant relativistic Hamiltonian, which in the absence of magnetic fields is

$$(H - qV)^2 = p^2c^2 + m_0^2c^4. \quad (1)$$

The solution to the wave equation (called the Klein–Gordon equation) resulting from the NR operator substitutions used by Schrödinger had two specific features not shared by the Schrödinger ψ. The nonlinear nature of H leads to the problem that it is impossible to determine meaningfully the average values of most dynamical variables according to the NR rule indicated earlier. A second difficulty arises from the fact that a mathematically valid solution of the Klein–Gordon equation exists for $W \leq -m_0c^2$. Such a solution cannot be arbitrarily excluded (as is possible classically), since induced transitions from positive- to negative-energy states should be possible.

In order to solve these problems, Dirac introduced in 1928 a suitably linearized Hamiltonian that yielded a wave equation (the Dirac equation) that solved the problem involving the average values of operators. The Dirac equation also led naturally to the concept of electron spin ($S = h/4\pi$, h is Planck's constant) and magnetic moment [$\boldsymbol{\mu} = (g/2)(q/m)\mathbf{S}$]. The magnetic moment may be thought of classically as due to the effective current loops created by the rotating charge. The so-called gyromagnetic ratio g expresses the distribution of charge within the spinning object and the structure of the object. These properties of the electron were introduced ad hoc into the NR theory by Uhlenbeck and Goudsmit in 1925 in order to explain the fine-structure splitting in alkali atoms (see ELECTRON).

Solutions of Dirac's equation for $W \leq -m_0c^2$ were, however, still mathematically admissible. The interpretation of these states of negative energy was put forth by Dirac, Weyl, Oppenheimer, and others in the years 1929–1931. The essence of the interpretation is that there do indeed exist states of negative energy and that normally all of these states are occupied by electrons. The Pauli exclusion principle forbids more than two electrons (one with spin up, the other with spin down) from occupying a given energy state. Thus, with all states occupied it is impossible for a positive-energy ("normal") electron to make a transition to a negative-energy state. If, however, a negative-energy electron is given enough energy (at least $2m_0c^2$) by, for example, a cosmic ray, it can make a transition to a positive-energy state. The vacancy or "hole" left in the sea of negative-energy electrons will behave like an antielectron, i.e., a particle of charge $+e$ (Dirac, 1929) and mass m_0 (H. Weyl, 1931). It is interesting to note that Dirac initially thought that the negative-energy solution might be a proton.

Dirac's hole theory of the positron was now essentially complete. Dirac even speculated on the existence of antiprotons, etc. The test of the theory was whether or not one could discover or produce positrons. Confirmation of Dirac's hypothesis came in 1932–1933 when C. D. Anderson and independently P. M. S. Blackett and G. P. S. Occhialini discovered positrons in cosmic-ray-created cloud chamber tracks. Positron tracks bent oppositely to those of electrons when a magnetic field was applied. The charge-to-mass ratio of the positron $(e/m)_+$ was measured to $\pm 15\%$ within 2 years of the positron's discovery and agreed with e/m for the electron. The latest electron–positron comparisons are detailed in Table I.

Many ramifications of the Dirac hole theory were explored in the years immediately subsequent to 1932 and in particular the treatment of the interaction of particles with the sea of negative-energy electrons and photons led directly to the development of the new field of quantum electrodynamics. One of the most important steps forward in our understanding of the positron occurred in 1949 when Feynman showed that a mathematical formalism could be set up in which the positron behaved like an electron traveling backward in time. The great advantage of Feynman's theory was calculational: the complicated integrations involving the sea of negative-energy electrons that Dirac's theory always necessitated did not have to be performed.

Table I Comparison of fundamental electron–positron properties

Charge-to-mass ratio[a]	$\dfrac{(e^-/m^-)-(e^+/m^+)}{(e^-/m^-)} = (1 \pm 1.3) \times 10^{-7}$
Mass[a]	$\dfrac{(m^- - m^+)}{m^-} = (1 \pm 1.3) \times 10^{-7}$
Magnetic moment	$\dfrac{(\mu^- - \mu^+)}{\mu^-} = (+0.5 \pm 2.1) \times 10^{-12}$

[a] The charge-to-mass ratio is obtained from a comparison of electron to positron cyclotron frequencies for electrons and positrons trapped in the same magnetic field. The mass comparison is obtained from the charge-to-mass ratio comparison by assuming that the electron and positron charges are identical.

A formal capstone to the theory was the work of J. Schwinger, B. Zumino, and G. Lüders, who showed (1951–1954) that a wide class of field theories have a symmetry (*CPT* invariance) that guarantees that any particle and its antiparticle have equal masses, equal lifetimes if the particle is unstable, and equal magnetic moments. The most fundamental experimental work connected with antimatter involves extremely precise experimental comparisons of the foregoing quantities with their values for ordinary matter. Any difference, no matter how minute, between the mass or magnetic moment for electron and positron would have enormous ramifications for physical theory. A list of the comparisons is given in Table I. As can be seen, up to the rather considerable accuracy achieved so far, positron and electron are indeed identical except for the sign of the charge.

See also ANTIMATTER; *CPT* THEOREM; ELECTRON; HAMILTONIAN FUNCTION; QUANTUM ELECTRODYNAMICS; QUANTUM FIELD THEORY; RELATIVITY, SPECIAL THEORY; SCHRÖDINGER EQUATION; SPIN.

BIBLIOGRAPHY

R. P. Feynman, *Quantum Electrodynamics*. Benjamin, Advanced Book Program, Reading, Mass., 1962.

N. R. Hanson, *The Concept of the Positron*. Cambridge, London and New York, 1963. (E)

J. J. Sakurai, *Invariance Principles and Elementary Particles*. Princeton Univ. Press, Princeton, NJ, 1964. (A)

L. I. Schiff, *Quantum Mechanics*. McGraw-Hill, New York, 1968. (I)

Positron Annihilation in Condensed Matter

Stephan Berko and Karl F. Canter

The positron is the antiparticle of the electron. When injected into ordinary matter it interacts with its surroundings prior to annihilating with an electron into gamma (γ) rays.

The behavior of positrons in condensed matter has been the subject of numerous experimental and theoretical investigations that have led to the development of the positron annihilation technique (PAT) for the study of various problems in condensed matter physics and chemistry. By now the applications of PAT divide into two main categories: (a) bulk studies, using mainly fast positrons from radioactive sources, and (b) surface and near surface studies, using the more recently developed slow, variable energy positron beams. By measuring the positron lifetime, the energy distribution, and the angular correlation of the annihilation γ's, one obtains valuable information about the behavior of the positron-electron pair prior to annihilation, and, by inference, about the electronic properties of the substance under investigation.

The positron (e^+) lifetime depends on the density of electrons at the site of the positron, and is of the order of 10^{-10}–10^{-9} s in liquids and solids. It can be accurately measured electronically by using e^+ sources such as ^{22}Na that emit an essentially prompt nuclear γ ray accompanying the positron. The time-delayed coincidence measurement between this nuclear "$t=0$ signal" and one of the annihilation γ's yields directly the positron lifetime. Time-bunched slow e^+ beams have also been developed recently for various time-dependent measurements. The angular correlation between the annihilation gamma rays (ACAR) in the 2γ annihilation mode reflects the momentum distribution of the positron-electron pair in the laboratory system. Typical deviations from collinearity between the two γ's are of a few milliradians. Using multiple γ detectors or position sensitive γ cameras in coincidence, placed at a large distance from the source-sample assembly, one obtains the 2D ACAR spectrum, i.e., the projection of the e^+–e^- momentum distribution onto a plane. The energy distribution of one of the gammas from the 2γ decay mode also depends, through the Doppler effect, on the momentum density of the annihilating pair.

In bulk measurements fast positrons, injected into the sample from a long-lived radioactive source such as ^{22}Na, ^{58}Co or ^{64}Cu, lose their energy by ionizing collisions and phonon scattering, and come to near thermal equilibrium with the solid prior to annihilation, after penetrating to mean depths of 10–100 μm. One notes that there is no Pauli exclusion between the electron and the positron. In well-annealed crystals of metals and semiconductors, the thermalized positron takes the form of a delocalized state, a Bloch wave, with maxima at interstitial regions due to core repulsion. In defected materials on the other hand, the positron can become localized (trapped) prior to annihilation in or around low density regions such as vacancies, dislocations and microvoids. Core repulsion also leads to the possibility of slow positrons being ejected from a surface due to their negative work function—an effect used in the production of slow e^+ beams.

The annihilation process, well understood theoretically (quantum electrodynamics), results in two quanta (2γ) from e^+–e^- spin-singlet overlap, or in three quanta (3γ) from spin-triplet overlap. In most materials 2γ decays predominate, since the 3γ decay is $\sim 10^3$ times less probable. In some liquids and insulating solids, the positron can capture an electron prior to annihilation, forming positronium (Ps), the hydrogen-like positron-electron atom (6.8 eV binding energy in vacuum). Positronium annihilates subsequently with a characteristic signature, and is thus detectable by studying the annihilation γ's.

Perhaps the most important applications of PAT to date have been in metal and semiconductor physics. High precision 2D ACAR experiments on well-annealed oriented single crystals have been performed to obtain the anisotropic momentum distribution of the conduction and valence electrons sampled by the delocalized positron. In metals, discontinuities in these ACAR spectra reflect directly the position of the Fermi surface. The shape and size of the Fermi surface can thus be obtained even in substances where the more standard solid state techniques such as magnetoresistance, the de Haas–van Alphen effect, etc., fail because of short electronic mean free paths. Results of detailed experiments in substitutionally disordered alloys have been compared with predictions of modern alloy theories, such as in models based on the coherent potential approximation. In simple metals like the alkali metals or aluminum, where the conduction electrons are nearly free, experimental positron lifetime and angular correlation results have served as a testing ground for the various theoretical many-body physics techniques used to study the importance of the many-body correlations in the interacting "one-positron, many-electron" system. By now the behavior of the positron is well enough understood, and the e^+–e^- interactions are sufficiently well accounted for in many metal systems, to permit the detailed comparison of the ACAR data with band theoretical computations, thus testing the nature of the electronic wavefunctions.

Spin-aligned positrons, as obtained from parity-nonconserving radioactive e^+ decay, can be shown to retain much of their spin orientation on thermalizing, and have been used to map out the spin dependent Fermi surfaces in ferromagnetic materials. 2D ACAR measurements have also been performed to obtain the complex Fermi surfaces of several superconducting materials, such as the technologically important A-15 compounds. Once high-quality single crystals of the new CuO-based high T_c superconducting compounds become available, 2D ACAR experiments will no doubt be helpful in elucidating the nature of their electronic structure.

The positron annihilation technique has been successfully applied to various defect studies and has been accepted as a standard tool in metallurgy. Once a positron is trapped in a low-density defect such as a vacancy, it sees a lower charge density and thus annihilates with a somewhat longer lifetime than from the delocalized state. The Doppler profile and the ACAR spectrum also show marked changes in the momentum density on positron trapping. These effects have been developed into sensitive tools for metallurgical problems. For example, by observing the change in the positron lifetime spectrum as vacancies are thermally activated upon heating the sample, the value of the vacancy formation energy can be obtained. Quenching and annealing experiments have been performed in various metals and alloy systems. Vacancy clustering and microvoid formation due to radiation damage by electrons and neutrons can also be observed by positron annihilation measurements; several research groups are trying to develop these effects for routine nondestructive testing of materials. The same techniques have been used more recently to study the important problem of vacancies in semiconductors such as Si, Ge, and GaAs. Detailed positron studies are in progress to obtain information about the technologically important questions of defect structures at various interfaces.

When positronium is formed in some condensed medium, the spin-triplet Ps can have its long-lived 3γ vacuum lifetime ($\sim 1.4 \times 10^{-7}$ s) shortened considerably by 2γ annihilations of its positron with an ''external'' electron, or by other triplet-singlet conversion mechanisms. Thus a complex 2γ decay lifetime spectrum is observed. The 2γ angular correlation also reflects a complex behavior: A broad distribution is characteristic of e^+ annihilation with the bound atomic electrons of the medium, while Ps, if formed, contributes a narrow peak corresponding to the center-of-mass momentum distribution of its spin-singlet state. The presence of Ps in liquids and solids can also be observed by the application of an external magnetic field known to admix the spin states via a second-order Zeeman effect.

The ability to distinguish experimentally between these modes of decay has led to important experiments designed to study the intrinsic properties of Ps, as well as the interaction of positrons and Ps with atoms and molecules. In gases the study of slow e^+ scattering, of Ps formation, and of Ps scattering is the focus of interest. The study of e^+ and Ps interactions with complex molecules, particularly in liquids, has led to the development of a new branch of chemistry—positronium chemistry.

In simple liquids, particularly in liquid He, positron annihilation exhibits a surprisingly rich variety of phenomena. There is experimental evidence that in liquid He Ps forms a self-consistent ''quantum bubble'' prior to annihilation, similar to the bubbles formed by free electrons in He. The angular correlation of the 2γ annihilation can be used to measure the zero-point energy of Ps in such a ''self-trapped'' state, yielding a value of the bubble radius to be compared with theory.

In organic solids, positron annihilation experiments reveal substantial Ps formation probabilities through the appearance of complex lifetimes and narrow angular correlation components. The sensitivity of the various lifetime components to temperature, pressure, and crystallinity resulted in the ''free-volume'' theory of Ps behavior in these solids. The affinity of the positron to low–atomic-density regions was used more recently to study phase transitions in liquid crystals and in polymers. In ionic crystals, besides annihilation from delocalized states, some positrons can be trapped in various defect centers; there is also indication that positronium can be formed, as well as bound states between a positron and negative ions. In oxides the situation is also complex; experiments in oxide powders indicate a large Ps-formation cross section, with Ps diffusing out of the powder grains. In some single crystals, like quartz, a delocalized Ps-like state similar to a delocalized exciton was observed and studied.

During the last decade the development of slow positron beams led to a study of a wealth of interesting surface physics phenomena. The production of slow, monoenergetic positrons relies on the use of ''fast-slow moderators'': when fast positrons are implanted into a well-annealed sample, those within ≈ 1000 Å of the surface can diffuse back and are ejected nearly monoenergetically by their negative work function. Typical moderator efficiencies are $\sim 10^{-4}$–10^{-3}. The e^+ beam energy is varied electrostatically and the beam is guided by magnetic or electrostatic optical systems onto the surface to be studied. The following picture emerges at a metal surface: positrons either (a) are ejected, or (b) are trapped into an ''image-potential induced'' surface state, or (c) pick up an electron to form a Ps of a few eV energy leaving the surface. In addition, upon heating the sample, the surface-trapped positrons can thermally ''desorb'' by electron capture, leading to a thermalized Ps emission. The study of these phenomena has been performed mainly with ''table-top'' e^+ beams yielding $\sim 10^5$–10^6 e^+/s. Slow e^+'s have been successfully used to study the e^+ surface states, near-surface and interface defects, ferromagnetic surfaces, etc.

Positron beam technology has been further improved by a unique ''brightness enhancement'' technique. An accelerated e^+ beam is focused down onto a secondary negative–work-function target and is reemitted at low energies, but from an area much smaller than that of the original moderator. Slow e^+ beams of ~ 1 μm diameter can be produced by multiple remoderation. Such high-brightness e^+ beams have been used for low-energy positron diffraction—LEPD—studies from surfaces. The differences between LEPD and its electron counterpart, LEED, are nontrivial. The Coulomb repulsion of the slow positrons by the ion cores and the lack of exchange leads to a significant simplification in the theoretical treatment of LEPD over that of LEED, as was recently demonstrated in the LEPD study of the surface structure of CdSe, a compound semiconductor. Possible applications of a novel ''positron reemission'' microscope based on brightness-enhanced e^+ beams are being presently explored at ~ 1 nm resolution.

The extension of the ACAR technique to the study of surfaces had to wait for the development of high-intensity slow e^+ beams. Beam facilities have now been built using as primary e^+ sources pair production from LINACs or high-intensity ^{64}Cu sources, reaching beam intensities of 10^8 e^+/s. The preliminary 2D ACAR experiments using these beams have indicated that angle-resolved positronium emission spectroscopy from clean single-crystal surfaces can become a sensitive technique for surface electronic structure studies. The feasibility of producing variable-energy positronium beams for future surface studies has also been recently demonstrated.

At present ultrahigh-intensity LINAC-based ''positron factories'' are being planned; a further 10–100 fold increase in e^+ beam intensity is technically feasible. Such future beams, incorporating brightness-enhanced microbeam technology, will provide more detailed information about the rich physics of e^+ and Ps surface interactions and lead to further surface study applications of the positron annihilation technique.

See also ANTIMATTER; ELECTRON ENERGY STATES IN SOLIDS AND LIQUIDS; FERMI SURFACE; HELIUM, LIQUID; POSITRONIUM; RADIATION DAMAGE IN SOLIDS.

BIBLIOGRAPHY

H. J. Ache (ed.), "Positronium and Muonium Chemistry," Advances in Chemistry Series, No. 175, American Chemical Society, Washington, 1978.

S. Berko, *Momentum Distributions,* R. N. Silver and P. E. Sokol (eds.). Plenum, New York, 1989.

W. Brandt and A. Dupasquier (eds.), *Positron Solid-State Physics, Proc. Int. School E. Fermi.* North-Holland, Amsterdam, 1983.

M. Dorikens, L. Dorikens-van Praet, and D. Segers (eds.), *Positron Annihilation, ICPA-8.* World Scientific, Singapore, 1989.

P. Hautojarvi (ed.), *Positrons in Solids.* Springer-Verlag, Berlin, 1979.

P. J. Schultz and K. G. Lynn, "Interaction of Positron Beams with Surfaces, Thin Films, and Interfaces," *Rev. Mod. Phys.* **60,** 701 (1988).

S. C. Sharma (ed.), *Positron Annihilation Studies of Fluids.* World Scientific, Singapore, 1988.

Positron–Electron Colliding Beams

Burton Richter

Positron–electron colliding-beam machines are devices to produce a large collision rate between antimatter (e^+) and matter (e^-). When antiparticles and particles collide, they can annihilate to produce a state of pure energy that can rematerialize into many different particles with very few restrictions on the kinds of particles that might be produced. When the antiparticle and particle are positron and electron, the intermediate pure energy state is particularly simple and well understood, its formation being described by the well-tested theory of quantum electrodynamics.

Most of the work with $e^\pm$ colliding beams has been the study of the properties and kinematic distributions of the particles produced when the intermediate pure-energy state rematerializes. The principal aim is to determine the interaction between these particles and to search for new particles that might exist with large rest masses. In the past years, experiments with $e^\pm$ colliding beams have uncovered many new particles, and taught us much about the interactions of these particles. New machines of higher energies are coming into operation and this branch of the accelerator art has become one of the most active parts of elementary-particle physics.

The basic design of the colliding-beam machine is dictated by the necessity of reaching a large enough interaction energy to produce the particles one wants to study. This interaction energy, or center-of-mass energy (E^*), for a particle of energy E_1 colliding with a particle of energy E_2, is given by

$$E^* = (4E_1E_2)^{1/2}. \qquad (1)$$

For electrons or positrons, the rest-mass energies are the same and are about equal to $\frac{1}{2}$ MeV (million electron volts), whereas the interaction energies of interest are many GeV (billion electron volts). For example, to reach an interaction energy of 100 GeV, where some of the most interesting new phenomena are expected, would require a positron beam of about 5×10^6 GeV incident on an electron at rest. The world's largest electron accelerator is the 3-km-long linear accelerator at the Stanford Linear Accelerator Center (SLAC), in California, that can produce a beam of up to 50 GeV energy. Scaling up this accelerator to 5×10^6 GeV would require a machine about 3×10^5 km long. The obviously more practical solution to the problem of achieving large interaction energies is to make two equal-energy beams collide with each other; for our example, this would require beam energies of only 50 GeV each.

Two techniques are currently in use to produce large interaction rates in $e^\pm$ collisions—the *storage ring* and the *linear collider*. The storage ring is the older technology. Construction of the first such device, the Princeton–Stanford storage ring (500 MeV beam energy, 12 m circumference) began in 1958, whereas the largest such machine, the LEP ring at CERN (60 GeV energy, 27 km circumference) was completed in 1989.

The linear collider was conceived as a less costly alternate to the storage ring for very high-energy machines. The first of these machines, the SLC at Stanford (50 GeV per beam, 3 km in length) was completed in 1987.

STORAGE RINGS

In an electron storage ring, beams of positrons and electrons circulate in opposite directions for periods of the order of several hours between refills, colliding only at specially designed interaction regions. The basic processes that come into play in building up the necessary beam current are as follows.

A short bunch of either electrons or positrons is injected into the machine by a pulsed injection system of some kind. The injected particles are guided around a roughly circular closed orbit by a collection of bending and focusing magnets. Particles that are close to but not exactly on the closed orbit execute stable *betatron oscillations* around that orbit. The particles must travel in a very good vacuum so that collisions of the electrons or positrons with molecules of the residual gas in the vacuum system do not cause the particles to scatter or lose large amounts of energy and so be lost to the beam. Pressures of 10^{-8} to 10^{-9} Torr are typical.

As the particles are bent around the design orbit by the magnetic field, they lose energy by *synchrotron radiation*. This energy loss can be very large for high-energy machines and is one of the problems that must be overcome in them. This energy loss is proportional to the fourth power of the beam energy divided by the radius of curvature in the bending magnets. The energy loss is made up by a radio-frequency (rf) accelerating system. The periodic rf accelerating field collects the particles into short bunches.

The interplay of the synchrotron-radiation processes, the bending and focusing fields, and the rf system results in a decay of all transverse and longitudinal components of a particle's momentum toward a final value with zero transverse momentum and a unique longitudinal momentum given by that corresponding to the ideal central orbit in the guide field. This time constant of this decay is the *radiation damping time*.

The radiation damping means that phase space is not conserved in the beam, and that particles with large amplitudes of oscillation with respect to the design orbit will move toward that orbit. In particular, particles put into the ring by the injection system with large amplitudes will have those amplitudes shrink and thus will move away from the injection system. When they have moved far enough, another pulse of particles can be injected into the ring, and so forth. In this way, large circulating currents can be built up.

The synchrotron radiation is not a continuous process but a quantum-mechanical process, and the fluctuations in the synchrotron radiation tend to make the beam size increase. The interplay between radiation damping and the quantum fluctuations results eventually in a statistically stationary distribution of particle amplitudes. The distributions in longitudinal and transverse amplitudes are Gaussian, and the standard deviations of these distributions give the *natural beam size*.

The interaction rate of e^+ and e^- beams is measured by the *luminosity* ($\mathcal{L}$), the interaction rate per unit cross section, given by

$$\mathcal{L} = (N_B{}^2/A) f_B, \tag{2}$$

where N_B is the number of particles in a bunch (assumed to be the same for the e^+ and e^- beams), A is the effective area of the bunch transverse to the direction of motion, and f_B is the collision frequency of the bunches in the machine. All of the present generation of $e^\pm$ colliding-beam machines use special focusing elements near the collision points to reduce A to as small a value as possible, and also use high-power rf systems to allow large values of N_B.

There is, however, a limit to the particle density (N_B/A) beyond which the colliding beams become unstable and particles are lost from the ring. This limit comes from the very strong electromagnetic fields a particle in one beam feels as it passes through the other beam. These forces are equivalent to a nonlinear focusing lens whose strength is largest for a test particle near the center of the other beam and smallest for one far from the center. It has been found experimentally that the upper limit to the particle density corresponds to that which changes the number of betatron oscillations per turn by about 0.05 per collision region.

In addition, N_B itself is limited for a given machine by the available rf power to make up for synchrotron-radiation losses and by the very large peak currents in the beams which produce electromagnetic fields in the vacuum chamber that can interact back on the beam in a destructive fashion. The average circulating currents in machines now operating range from 50 mA to 0.5 A, and peak currents are in the range of hundreds of amperes. The LEP machine requires 16 MW of rf power to operate at its maximum energy.

Typical luminosities of present-day storage rings are 10^{31}–10^{32} cm^{-2} s^{-1}, whereas the physics cross sections of interest are 10^{-32}–10^{-35} giving event rates of 1.0 to 0.001 s^{-1}. Special purpose machines are under study that might achieve luminosities as large as 10^{34}.

The highest-energy machine now operating (1989) is the LEP machine at CERN (45 GeV). Others are TRISTAN (28 GeV) at KEK in Japan; PEP (15 GeV) at Stanford and CESR (8 GeV) at Cornell in the United States; DORIS (5 GeV) at DESY in Germany; VEPP III (5 GeV) at Novosibirsk in Russia, and BEPC (2.5 GeV) at Beijing in the PRC.

Synchrotron radiation from the existing high-energy $e^\pm$ storage rings is the most intense source of x rays available in the region 0.1 to 50 keV. These x rays are being used for studies in solid-state physics, chemistry, biology, and medicine. The studies have been so successful that special-purpose electron storage rings are now under construction or being designed solely as sources of synchrotron radiation.

LINEAR COLLIDER

Because of the rapid increase of the energy loss through synchrotron radiation (proportional to E^4/R), power demands of storage rings increase very rapidly with energy. It can be shown on quite general grounds that the minimum-cost storage ring where circumference-dependent costs and rf-power-dependent costs are properly balanced has a scaling law that makes the size (and the cost) increase as the square of the beam energy. Thus, an electron storage ring 10 times the energy of LEP would have a hundred times the circumference (2700 km). Thus, storage rings much beyond the energy of LEP are not financially feasible.

The linear collider in its pure form consists of two linear accelerators firing beams of electrons and positrons at each other. Since there is no bending during the acceleration, there is no synchrotron radiation produced and the cost of these machines scales like the first power of the energy. It is generally agreed in the accelerator physics community that for energy significantly beyond that of LEP, linear colliders will clearly be the less costly alternative.

The same equation [Eq. (2)] governs the luminosity of linear colliders and storage rings. The number of particles per bunch (N_B) tends to be about the same in the two kinds of machines, the collision frequency (f_B) is typically larger in storage rings, but because of the absence of quantum effects and synchrotron radiation in the acceleration process the effective area (A) can be made much smaller in the linear collider. For example, the area of the beams at the collision point in the LEP storage ring is about a thousand times the area in the SLC.

There are three main components to a linear collider—the source, the main accelerator, and the final focus system. The source must produce beams that are capable of being focused to a very small size at the high-energy end. The figure of merit for the source is the invariant emittance (product of rms radius and transverse momentum of particles within the beam). The smaller the invariant emittance of the source, the smaller the beam that can be produced by the final focus system at the collison point. The sources for high-energy linear colliders all incorporate storage rings of 1–2 GeV energy that are designed to produce the required emittance through radiation damping (described above). The SLC damping rings produced beams with invariant emittance of about 1×10^{-5} m MeV/c. The higher-energy machines now under study will use sources with emittances of a factor of 10 smaller.

The main accelerator must boost the beam to the required final energy without diluting the small emittance produced by the source. The SLC uses a conventional linear accelerator. After many studies, it has been concluded that such exotic acceleration methods as lasers, plasmas, etc., are not practical for at least the foreseeable future and that the next generation of linear colliders will also use conventional linacs. The intensity of the bunches of particles required for the linear collider are very large (on the order of 10^{10} per bunch) and this brings in new problems. The most important of these are *wakefield* effects on the accelerated bunch itself. These wakefields are produced by interaction of the high-intensity bunch with the accelerating structure which, in turn, interacts back on the bunch that produced the field and can severely affect the beam quality. The transverse wakefield can produce the worst effects. Since its intensity is proportional to the distance the beam lies off the axis of the accelerator, the beam must be held close to the symmetry axis of the linac, within 200 μ in the SLC and within 20–50 μm in future machines that use beams of smaller invariant emittance.

The final focus system produces the tiny beam spots at the collision point that are required for high-luminosity operation. This is complicated because magnetic lenses used to focus beams have aberrations like those of glass lenses required to focus light beams. As with high-quality camera lenses, linear-collider final-focus systems combine many lenses of different properties to cancel aberrations and produce spots at the emittance limit, analogous to the diffraction limit in optical systems. The SLC produces a beam spot of about 2 μm radius. Future machines will require much smaller beam spots.

The SLC began an operation for physics research in 1989 and the results from the first experients were published in August of 1989. Already, experience with the SLC has taught the accelerator physics community much about the design of the next design of linear collider. The SLC will continue for many years as both a test bed for the development of the linear collider and as a physics research tool.

Groups in the United States, Russia, Japan, and Europe are working on the design of future machines of 500–2000-GeV center-of-mass energy. These high-energy machines require the development of new technology primarily in the main accelerator area to produce an energy-efficient and cost-effective acceleration system. The R&D program is international in scope with excellent information exchange and international collaboration on specific technology development projects. At the rate progress is being made, construction of a high-energy linear collider could begin before the middle of the 1990s.

See also ANTIMATTER; BETATRON; ELECTRON; ELEMENTARY PARTICLES IN PHYSICS; POSITRON; SYNCHROTRON RADIATION.

BIBLIOGRAPHY

Proceedings of the 1989 International Lepton and Photon Conference. SLAC, Stanford, California, 1989. Summary articles on the most recent results in the $e^{\pm}$ colliding beams field.

Proceedings of the IEEE 1989 Particle Accelerator Conference. March 20–23, 1989, Chicago, Illinois (to be published). The most recent of a regular series of conference reports. This and earlier volumes are where one can find descriptions of the latest colliding-beams projects and discussions of the most advanced level on accelerator problems.

M. Sands, in *Proceedings of the International School "Enrico Fermi" Course XLVI*. Academic Press, New York and London, 1971. The best introduction to storage rings.

J. R. Rees, in *Techniques and Concepts of High Energy Physics IV* (T. Ferbel, ed.). Plenum Press, New York, 1988. A good general introduction to linear colliders.

Positronium

Stephan Berko, Karl F. Canter, and A. P. Mills, Jr.

Positronium (Ps) is the hydrogen-like bound state of an electron and its antiparticle, the positron. Since its discovery in 1951 by Martin Deutsch, Ps has been the subject of numerous theoretical and experimental investigations aimed at testing quantum electrodynamics (QED), the most precise theory in physics to date. Together with muonium (the bound state between an electron and a positive muon), Ps serves as an ideal test of QED, since it is a purely leptonic system and the theoretical interpretation of its spectrum is unhindered by the presence of heavy particles. Positronium can be formed when a positron emitted by a radioactive source picks up an electron from an atom while slowing down in matter. The ultimate fate of Ps is self-annihilation into quanta of high-energy electromagnetic radiation (γ rays). The annihilation lifetime (10^{-10}–10^{-7} s) is sufficiently long to not affect substantially the energy levels of Ps, but does contribute to the width of the levels and hence limits the ultimate precision of our measurements.

Experimental information about Ps may be obtained by measuring the various properties of the annihilation γs. Since the positron and the electron have to be in close proximity to annihilate, annihilation occurs mainly from S states (angular momentum $l=0$). States with $l>0$ will decay to S states by the emission of light. Selection rules based on the conservation laws of angular momentum, parity, and charge conjugation lead to spin-dependent annihilation properties. States with the electron and positron spins "parallel" (orthopositronium, *o*-Ps) decay into an odd number of γs; one-photon annihilation is rare since it requires the presence of a third body, such as a nucleus. States with "antiparallel" spins (parapositronium, *p*-Ps) decay into an even number of γs. The annihilation cross section decreases rapidly with increasing number of γs; thus in its ground state *p*-Ps ($1\,^1S_0$, singlet) decays in 1.25×10^{-10} s into two γs, while the 3γ lifetime of the *o*-Ps ($1\,^3S_1$, triplet) state is 1.42×10^{-7} s. It was the appearance of a long-lifetime component in the time spectrum of positrons annihilating in a gas that led Deutsch to the discovery of Ps; the sensitivity of the long-lived $1\,^3S_1$ Ps state to the presence of molecules such as NO confirmed the existence of Ps and signaled the beginning of Ps physics and chemistry.

The main energy spectrum of Ps is hydrogen-like, scaled down by a factor of 2 due to the equality of the positron and electron masses (m_e); $E_n=-\frac{1}{2}hcR_\infty(1/n^2)$, where $hcR_\infty=\frac{1}{2}m_ec^2\alpha^2=13.606\ldots$ eV, α being the Sommerfield fine-structure constant ($\alpha = e^2/\hbar c \approx 1/137.036\ldots$). Thus the Ps binding energy is ~6.8 eV compared to ~13.6 eV for hydrogen (H). The fine structure of the Ps levels is, however, considerably different from that of H. The large difference between "fine structure" and "hyperfine structure" in hydrogen and in all other ordinary atoms disappears in Ps, since the magnitude of the magnetic moment of the positron is equal to that of the electron; the spin–spin interaction is of the same order of magnitude as the spin–orbit interaction. Ps being the bound state of a particle-antiparticle pair, the energy levels are further split, even in the lowest order, by the "virtual annihilation" interaction. The energy splitting of the $n=1$ levels, $\Delta E(1^3S_1 - 1^1S_0)$, is, in lowest order, $\frac{7}{6}\alpha^2 hcR_\infty$, where $\frac{4}{6}\alpha^2 hcR_\infty$ is due to the spin–spin (magnetic moment) interaction, whereas "virtual annihilation" gives rise to $\frac{3}{6}\alpha^2 hcR_\infty$.

The first spectroscopic measurement on Ps was that of $\Delta E(1\,^3S_1 - 1\,^1S_0)$. In a magnetic field B, the triplet and singlet quantum states corresponding to the magnetic quantum number $m=0$ are mixed, leading to an enhanced 2γ annihilation rate. A radio-frequency (rf) field induces transitions between the unmixed $m=\pm 1$ states and the mixed $m=0$ state, producing a resonant increase in the 2γ yield (decrease in the 3γ yield) at the proper rf frequency, which depends on B and $\Delta E(1^3S_1 - 1^1S_0)$. Thus, $\Delta E(1^3S_1 - 1^1S_0)$ can be measured to high precision. During the last four decades this technique has led to experimental $\Delta\nu = \Delta E/h$ values of ever increasing precision. At present (1990) two independent high-precision experiments yield $\Delta\nu(1^3S_1 - 1^1S_0) = (203\ 389.1 \pm 0.7)$ MHz and $\Delta\nu(1^3S_1 - 1^1S_0) = (203\ 387.5 \pm 1.6)$ MHz. The most recent theoretical computations yield

$$\Delta E(1\,^3S_1 - 1\,^1S_0) - \alpha^2 hcR_\infty \left[\frac{7}{6} - \frac{\alpha}{\pi}\left(\frac{16}{9} + \ln 2\right) - \frac{5}{12}\alpha^2 \ln\alpha + O(\alpha^2)\cdots\right],$$

corresponding to $\Delta E(1^3S_1 - 1^1S_0) = 203\ 400.3$ MHz plus terms of order $\alpha^4 hcR_\infty \approx 9$ MHz and higher. The latter have not been completely evaluated and can yield corrections of several MHz. Thus, theory and experiment are in reasonably good agreement.

In addition to the energy levels, the values of the annihilation rates themselves serve as sensitive tests of QED predictions. The value of the 2γ decay rate of the short-lived $1\,^1S_0$ state can be obtained indirectly from the width of the resonance line in the $\Delta\nu(1^3S_1 - 1^1S_0)$ measurement, and more precisely from a measurement of the magnetically quenched $m=0$ state lifetime. The longer 3γ lifetime of the unperturbed $1\,^3S_1$ Ps state can be measured directly and has been the center of attention since the first accurate measurements disagreed with the best QED calculations by several percent! A reevaluation of the theory brought it into better agreement with experiment; the theoretical decay rate (inverse lifetime) is presently given by

$$\Gamma_{3\gamma} = \Gamma^0_{3\gamma}\left[1 - \frac{\alpha}{\pi}(10.282 \pm 0.003) + \frac{1}{3}\alpha^2 \ln\alpha + \cdots\right] = 7.03830 \pm 0.00005\ \mu\text{s}^{-1},$$

where the lowest order rate is

$$\Gamma^0_{3\gamma} = \frac{\alpha^6 m_e c^2}{\hbar}\,\frac{2(\pi^2 - 9)}{9\pi} = 7.2112\ \mu\text{s}^{-1}.$$

At present, the annihilation rates are measured most precisely in a gas by extrapolation to zero pressure; the latest measurement yields $\Gamma_{3\gamma} = 7.0514 \pm 0.0014\ \mu\text{s}^{-1}$. There appears to be a disagreement with theory, but the final verdict must await the calculation of the $O(\alpha^2)$ terms in the theoretical expression. All that can be said at this time is that the coefficient multiplying the α^2 would have to be unusually large to remove the discrepancy.

Another property of Ps that serves as a test of QED is the fine-structure splitting in the excited states. The production and observation of the excited states of Ps eluded experimentalists for over 20 years, in spite of numerous experimental searches. Since Ps is formed by positrons with a few eV kinetic energy, the usual Ps gas experiments use the gas molecules as an energy "moderator" of fast positrons emitted by the radioactive source, as well as a source of electrons for Ps formation. At the required gas densities, Ps excited states, if formed, are easily broken up in subsequent collisions. Since the fast positrons are slowed down by ionization of the moderating gas, an intense radiation background is produced which prohibits observation of the characteristic decay light from the excited states.

The disadvantges of forming Ps with fast positrons have been overcome with recent advances in producing high-intensity, low-energy monochromatic positron beams. In particular, it has been found that if slow positrons (i.e., a few eV in energy) are directed at various solid surfaces, the positrons have a high probability of reemerging from the surface with a captured electron, i.e., as Ps. Thus, Ps can be formed in a vacuum external to a solid target bombarded with slow positrons. Further experiments indicated that a small fraction of the Ps atoms are formed in the first excited state. These excited Ps atoms were observed by detecting directly the ultraviolet (Lyman-α, 243 nm) photons emitted in the decay of the $2P$ states to the $1S$ states, in coincidence with the annihilation γs of the ground state. The detection of the excited state led to the measurement of the fine structure of the $n=2$ states of Ps, the equivalent of a "Lamb-shift" measurement in atomic hydrogen. The Ps atoms were produced in a rf cavity. Time-delayed coincidences between a photomultiplier detecting the Lyman-α photons and one of the annihilation γs of the ground state were monitored as a function of rf frequency. At the proper frequency the rf field induced conversion between the $2\,^3S_1$ and $2\,^3P_2$ states leading to an ~10% increase in the delayed coincidence signal. Thus, the value of the $2\,^3S_1$–$2\,^3P_2$ fine-structure splitting was measured and found to agree within 10 MHz with the QED theoretical prediction of 8625 MHz. More recently, all three $2\,^3S_1$–$2\,^3P_n$ fine-structure intervals were measured and found to be in agreement with theory also.

The measurements of the hyperfine interval and the fine-structure intervals are limited in accuracy by the 1280-MHz 2γ annihilation width of the 1^1S_0 state and the 50-MHz Lyman-α radiative width of the $2P$ states. On the other hand, the largest QED energy shifts are combined with the narrowest level widths in the triplet S states. Rather than using a neighboring singlet state or P state as a reference, one may measure the difference between two $n\ ^3S_1$ states directly, and have a natural linewidth of only 1 MHz. Such a measurement is not trivial because the most accessible interval, $\Delta\nu(1\ ^3S_1-2\ ^3S_1)$, is an optical transition forbidden in first order. However, optical measurements are now possible thanks to the development of high-power lasers, the perfection of first-order Doppler-free two-photon spectroscopy, the improvement of positron moderation techniques, and the availability of thermal-energy positronium in vacuum.

The first successful measurement of $\Delta\nu(1\ ^3S_1-2\ ^3S_1)$ used a high-power pulsed laser to induce the forbidden transition. To match the pulse structure of the light, a small laboratory positron beam was bunched to achieve a high instantaneous intensity, producing a burst of about 50 Ps atoms at a temperature of 600 K in a 1-cm^3 volume. The Ps was excited from the $1\ ^3S_1$ state to the $2\ ^3S_1$ state by two pulsed, counterpropagating laser beams. The laser wavelength, 486 nm, is half the positronium Lyman-α wavelength, and thus two photons are able to induce the $n=1$ to $n=2$ transition. The light was narrowed in frequency to about 25 MHz and scanned past the $1S$–$2S$ resonance frequency. Atoms that were excited to the $2S$ state were quickly ionized by the high-intensity light, and the liberated positrons were detected one at a time by an electron multiplier. A few-minute scan yields a resonance curve, the center of which may be determined to a few MHz relative to a calibration line from a Te_2 absorption spectrum. A series of scans allowed correction for the ac Stark shift introduced by the intense laser beams and the second-order Doppler shift (time dilation) due to the Ps motion relative to the laser frame of reference. The measurements were reduced to an absolute value of the $1\ ^3S_1$–$2\ ^3S_1$ interval by a series of measurements of the Te_2 reference line. The latest result is $\Delta\nu(1\ ^3S_1-2\ ^3S_1)=1\ 233\ 607\ 218.9\pm 10.7$ MHz. The agreement with theory is acceptable (within 20 MHz) considering that the order α^4R_∞ terms have not been calculated.

The introduction of accelerator-based high-intensity pulsed positron beams, the production of cryogenic positronium in vacuum, and the availability of new optical techniques for trapping and cooling of atoms will soon make possible a vast improvement in the Ps $1\ ^3S_1$–$2\ ^3S_1$ measurements. A precision of order 1 kHz is not an unthinkable goal. The most difficult challenge will be to match the experimental precision with theory. At present, theory and experiment are about equally precise, but no Ps calculation has ever succeeded in getting over the $\alpha^4R_\infty c\approx 9$-MHz barrier. Improvement of our knowledge of the positronium energy levels by four orders of magnitude is a challenge for the next decade and will no doubt lead to improved tests of QED and open up new areas of Ps research.

See also ELECTRON: FINE AND HYPERFINE SPECTRA AND INTERACTIONS; LEPTONS; MUONIUM; POSITRON; QUANTUM ELECTRODYNAMICS.

BIBLIOGRAPHY

V. W. Hughes, *Atomic Physics,* Vol. 3. Plenum, New York, 1972.

"Discovery of Positronium," in *Adventures in Experimental Physics,* Vol. 4, pp. 64–127 (B. Maglic, ed.). World Science Education, Princeton, NJ, 1975.

M. A. Stroscio, *Phys. Lett. C* **22,** 215–277 (1975).

T. C. Griffith and G. R. Heyland, *Nature* **269,** 109–112 (1977).

S. Berko, K. F. Canter, and A. P. Mills, Jr., "Positron Experiments," in *Progress in Atomic Spectroscopy,* Part B, pp. 1427–1452 (W. Hanle and H. Kleinpoppen, eds.). Plenum, New York, 1979.

S. Berko and H. N. Pendleton, *Ann. Rev. Nucl. Part. Sci.* **30,** 543 (1980).

A. Rich, *Rev. Mod. Phys.* **53,** 127 (1981).

A. P. Mills, Jr., in *Positron Solid State Physics* (W. Brandt and A. Dupasquier, eds.), p. 432. North-Holland, Amsterdam, 1983.

P. J. Schultz and K. G. Lynn, *Rev. Mod. Phys.* **60,** 701 (1988).

The Spectrum of Atomic Hydrogen: Advances (G. W. Series, ed.). World Scientific, Singapore, 1988.

Precession

Norman F. Ramsey

When a top gyroscope is rapidly spinning about its axis of symmetry and is subjected to a torque, such as the gravitational torque in Fig. 1, the plane containing the top axis and the vertical will rotate about the vertical axis. This motion of the top is called *precession.* If the initial conditions are suitably chosen, the precession angular velocity will be constant. If the initial conditions are not so chosen, the axis of the top will bob up, down, and sidewise, sweeping out a path such as the trace in Fig. 2. These periodic motions, superposed on the steady precession, are called *nutation.*

The physical reason for the precession can most easily be seen when the foregoing conditions for a constant precessional angular velocity are met. Gravity exerts a torque **N**, whose magnitude is given by

$$N=Mgl\sin\theta \tag{1}$$

where M is the mass of the top and l the distance from the

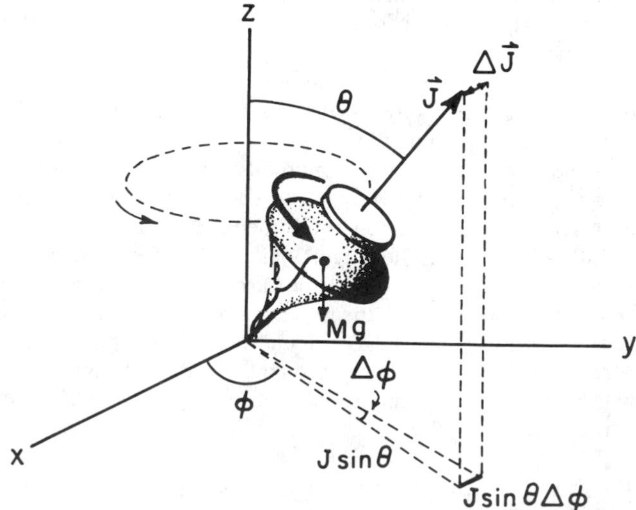

FIG. 1. Torque, angular momentum, and precession of a top in a gravitational field.

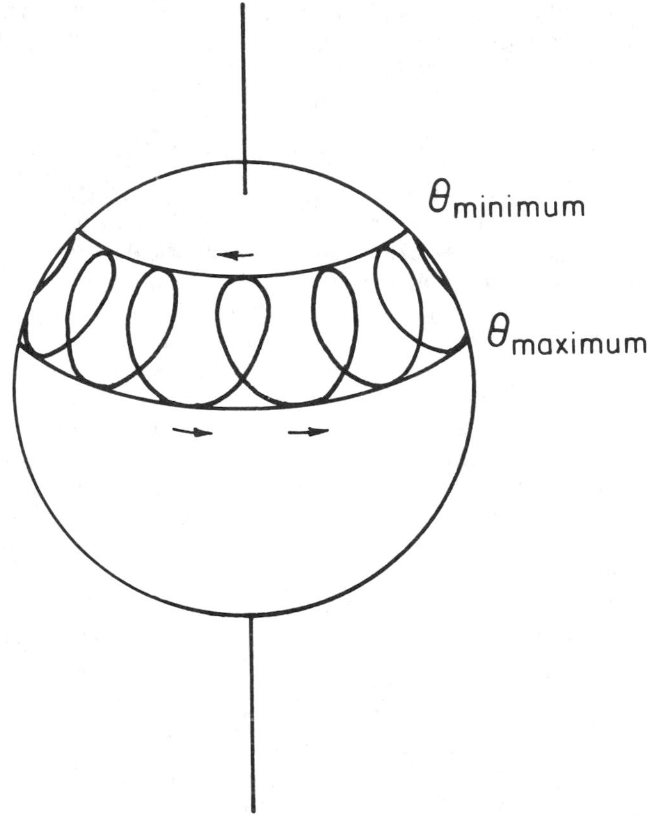

FIG. 2. Motion of axis of spinning top showing nutation.

point of support to the center of mass. The direction of the torque is perpendicular to the plane containing the vertical and the top axis. If the plane containing the top axis and the vertical precesses through an angle $\Delta\phi$ in time Δt and if **J** is the spin angular momentum of the top, the magnitude of the change in angular momentum in time Δt is $J \sin\theta\ \Delta\phi$. By the equation of motion for total angular momentum, the torque equals the rate of change of angular momentum, so that

$$N = J \sin\phi \frac{\Delta\phi}{\Delta t} = Mgl \sin\phi. \tag{2}$$

Consequently, the precessional angular velocity ω is given by

$$\omega = \frac{\Delta\phi}{\Delta t} = \frac{Mgl}{J}. \tag{3}$$

If the initial conditions are not such as to yield simple uniform precession, the quantitative treatment is more complicated and is discussed in standard texts on classical mechanics [1].

The term *precession* is also applied to analogous motions of systems other than tops in gravitational fields. Thus, if the angular momentum J is associated with a magnetic moment μ, the application of a magnetic field H will give rise to a precession, called the *Larmor precession*. The Larmor precession frequency ω_L can be obtained by noting that in this case the torque is $\mu H \sin\theta$, so μH corresponds to Mgl in Eq. (2). Therefore, from Eq. (3)

$$\omega_L = \frac{\mu H}{J}. \tag{4}$$

Likewise, in astronomy the *precession of the equinoxes* is a slow westward motion of the equinoctial points for the earth along the ecliptic. The largest part of this precession, called the *lunisolar* precession, is caused by the action of the sun and moon on the ellipsoid-shaped earth. The total precession is diminished by the smaller *planetary precession,* which is in the opposite direction and is caused by the gravitational attraction of the planets, which slightly reorients the plane of the earth's orbit.

Although the most familiar case of precession is that of a top primarily spinning about its axis of symmetry and with a torque supplied externally by, say, gravity, it should be noted in the general motion of a rigid body that even with no externally applied torques there is also a force-free precession of the angular velocity about the principal axes of the initial ellipsoid of the body when the initial conditions are more complicated than a simple rotation about a principal axis [1].

See also DYNAMICS, ANALYTICAL; ROTATION AND ANGULAR MOMENTUM.

REFERENCE

1. H. Goldstein, *Classical Mechanics* 2nd ed. Addison-Wesley. Reading, Mass., 1980.

Pressure *see* Ultrahigh-Pressure Techniques

Probability

J. M. Hammersley

Scientific method consists of a cycle of three steps: (i) the provisional consideration of a scientific hypothesis, namely, a set of assumed postulates about the nature of reality and how it works; (ii) a deductive theory, often mathematical, that derives the logical consequences of these assumed postulates; and (iii) an analysis of experimental data to discover if those data are in discord with the provisionally assumed postulates. We can never hope to prove a hypothesis true, merely to weed out and refine the untrue by iterating the foregoing cycle. In the part of science that has to deal with experimental errors or uncertain events, probability theory is the branch of mathematics that handles the deductive step (ii), while statistics is the methodology for the inferential step (iii).

LAWS OF PROBABILITY

A *random event* is an event that has a chance of happening, and *probability* is a numerical measure of that chance. We write $P(A)$ for the probability that the event A occurs; $P(A+B+\cdots)$ for the probability that at least one of the events $A, B\cdots$ occurs; $P(AB\cdots)$ for the probability that all the events $A, B, \ldots$ occur; and $P(A|B)$ for the probability that the event A occurs when it is known that B occurs. Every probability

P is a number lying between 0 and 1, both inclusive; and higher values of P indicate greater chances. The laws of probability state that

$$P(A+B+\cdots) \leq P(A)+P(B)+\cdots \tag{1}$$

and

$$P(AB)=P(A|B)P(B). \tag{2}$$

If only one of the events A, B, ... can occur, they are called *exclusive*, and equality holds in (1). If at least one of the events A, B, ... must occur, they are called *exhaustive*, and the left-hand side of (1) equals 1. $P(A|B)$ is called the *conditional probability of A given B*. If $P(A|B)=P(A)$, we say that A and B are independent events. The laws of probability are consistent with, but do not entail, the frequentist interpretation, wherein $P(A)$ is (thought of as) the proportion of occasions on which A occurs.

RANDOM VARIABLES

A *random variable* is a number Y that characterizes each of a set of exhaustive and exclusive events. Its (cumulative) distribution function $F(y)=P(Y \leq y)$ is the probability that Y does not exceed a prescribed y. The expectation of a function $g(Y)$ is $Eg(Y)=\int_{-\infty}^{\infty} g(y)\, dF(y)$. The characteristic function $\phi(t)=Ee^{itY}$, where $i^2=-1$, determines and is determined by the distribution F. The cumulant function is $K(t)= \log \phi(t)$. (These definitions extend to random vectors provided that $\mathbf{Y} \leq \mathbf{y}$ has coordinatewise interpretation and $\mathbf{tY}$ is an inner product.) The moments μ_r' and cumulants κ_r of Y are coefficients generated by $\phi(t)=\Sigma_r \mu_r'(it)^r/r!$ and $K(t)=\Sigma_r \kappa_r(it)^r/r!$ In particular, $\mu_1'=EY$ is the *mean* of Y; var $Y=\mu_2=\mu_2'-\mu_1'^2$ is the *variance* of Y; and $\sigma=\sqrt{\mu_2}$ is the *standard deviation* of Y. The *covariance* of two random variables X and Y is $\mathrm{cov}(X,Y)=E(XY)-E(X)E(Y)$. Their correlation coefficient is $\mathrm{cov}(X,Y)/(\mathrm{var}\ X\ \mathrm{var}\ Y)^{1/2}$. The variance–covariance matrix of a random vector $\mathbf{Y}$ has (in its ith row and jth column) the covariance of the ith and jth coordinates of $\mathbf{Y}$. Addition of independent random variables is isomorphic to multiplication of their characteristic functions. Standard distributions include (i) the *normal* (Gaussian or Laplacian) distribution

$$F(y)=(2\pi\sigma^2)^{-1/2}\int_{-\infty}^{y} \exp[-(x-\mu)^2/2\sigma^2]\, dx;$$

(ii) the *multinormal distribution*

$$|2\pi\mathbf{V}|^{-1/2}\int_{-\infty}^{\mathbf{y}} \exp[-\tfrac{1}{2}(\mathbf{x}-\boldsymbol{\mu})^{\prime}\mathbf{V}^{-1}(\mathbf{x}-\boldsymbol{\mu})]\, d\mathbf{x},$$

where $\mathbf{V}$ is the variance–covariance matrix of the vector $\mathbf{Y}$ and $|\cdots|$ denotes the determinant; (iii) the *binomial* distribution

$$\sum_{x \leq y} \binom{n}{x} p^x(1-p)^{n-x};$$

(iv) the *Poisson* distribution

$$\sum_{x \leq y} \frac{e^{-\lambda}\lambda^x}{x!};$$

and (v) the *uniform* distribution $(y-a)/(b-a)$ for $a \leq y \leq b$.

Distribution functions often contain constants, called *parameters* (e.g., μ and σ in (i). $\boldsymbol{\mu}$ and $\mathbf{V}$ in (ii), n and p in (iii), λ in (iv), and a and b in (v)).

CONVERGENCE OF SEQUENCES OF RANDOM VARIABLES

Let Y_r be random variables with respective distribution functions $F_r(y)$ $(r=1,2,\ldots)$; and let Y be a random variable with distribution function $F(y)$. The most important forms of convergence (for $Y_r \to Y$ as $r \to \infty$) are (i) convergence in distribution, namely, $F_r(y) \to F(y)$ for all y at which $F(y)$ is continuous; (ii) convergence in quadratic mean, namely, $E(Y_r-Y)^2 \to 0$; (iii) convergence in probability, namely, $P(|Y_r-Y|>\epsilon) \to 0$ for any prescribed $\epsilon>0$; and (iv) almost-sure convergence, namely, $P(|Y_r-Y| \to 0)=1$. There are various theorems that relate these forms of convergence: thus (iv) implies (iii); but (iii) only implies the existence of an almost-surely convergent subsequence. Suppose (for simplicity) that the Y_r are independent and identically distributed with common mean μ and variance σ^2; and let $S_n=Y_1+\cdots+Y_n-n\mu$. Then the *weak law of large numbers* asserts that $S_n/n \to 0$ in probability; the *strong law of large numbers* asserts that $S_n/n \to 0$ almost surely; the *law of the iterated logarithm* asserts that

$$P\left(\limsup_{n\to\infty} \frac{|S_n|}{(2n\sigma^2 \log\log n)^{1/2}}=1\right)=1;$$

and the *central limit theorem* asserts that $S_n/(n\sigma^2)^{1/2}$ converges in distribution to the normal distribution. These theorems are also true under a wide variety of less restrictive but more elaborate assumptions.

MARKOV PROCESSES AND FIELDS

A Markov process is a set of random variables, indexed by a scalar parameter t (usually signifying time), such that the conditional probability of future events given past events is equal to the conditional probability of these future events given only the most recent of these past events. A Markov field is an extension to the case where the indexing parameter ranges over a multidimensional space, and where the conditional probabilities are (sufficiently) determined by configurations on the spatial boundaries of the events in question. Every Gibbs ensemble is a Markov field, and (under suitable restrictions, notably the absence of zero probabilities) the converse also holds.

BAYES'S THEOREM

Suppose that A_1, A_2, ... are exclusive and exhaustive events, and that B is some other event. Then Bayes's theorem asserts that $P(A_r|B)=P(A_r)P(B|A_r)/\Sigma_r P(A_r)P(B|A_r)$. This is a simple consequence of the laws of probability. In a typical application, A_1, A_2, ... are alternative scientific hypotheses, while B denotes some observed experimental data. In such a context, $P(A_r)$ is the prior probability of the hypothesis A_r before the experiment is done; $P(A_r|B)$ is the corresponding posterior probability of the hypothesis in the light of the observations; and $P(B|A_r)$ is the likelihood of the observations assuming the hypothesis A_r. Usually the hypothesis specifies the distribution of the random variables that constitute the observed data; and then it is straightforward to write the likelihood. The difficulty of the theorem lies in interpreting what is meant by the prior and posterior "probabilities" of a hypothesis, though if these can be somehow surmised, the theorem tells us (or should tell us) how experiment ought to influence our attitude to a hypothesis. Presumably (though this presumption has been a battleground among scientists and philosophers for over 200 years) we should prefer the hypothesis with the greatest posterior probability. This, however, requires that we should be able to assess the relative magnitudes of the prior probabilities $P(A_r)$. In the "absence" of any preference among these priors, *Bayes's postulate* tells us to take the $P(A_r)$ equal to each other (though it should be added that Bayes himself was less convinced of this than many of his followers). In fact, it is very easy to get oneself into a logical and philosophical morass either by accepting Bayes's postulate or (what amounts to the same thing) by trying to define "relative scientific ignorance." Nevertheless *some* sort of hazy conative appreciation of Bayes's postulate and its consequences is worth possessing.

ESTIMATION

A typical scientific hypothesis specifies that the observations in an experiment are random variables with a given *parent* distribution containing some unknown parameter. For example, a hypothesis might specify that the number of particles recorded by a Geiger counter in one minute is a random variable having a Poisson distribution with unknown parameter λ. The experiment consists of taking independent observations $Y_1, Y_2, \ldots, Y_n$, each of the number of particles recorded in one minute. We want to use these data to estimate λ. The parameter is called an *estimand*, or the quantity to be estimated. To this end, we process the data to obtain some function $f=f(Y_1, Y_2, \ldots, Y_n)$ that will serve as an estimate of the estimand. To get a clear grasp of what is going on, it is important to distinguish between four different words: estimand (already defined), estimation (the processing of the data and the rationale behind such procedures), estimate, and estimator (both to be defined later). The function $f(Y_1, \ldots, Y_n)$ can be considered in two lights. In the first place, we can regard it as a function $f(\ldots,\ldots,\ldots)$, i.e., a recipe of *how* to process the data when once we have got them: for example, f might be the recipe "Take the arithmetic mean of the observations." In this sense, we refer to f as an *estimator*. Second, we can regard $f(Y_1, \ldots, Y_n)$ as the number obtained by applying the estimator to the data: such a number is called an *estimate*. Since the data $Y_1, \ldots, Y_n$ are random, the estimate f is also a random variable, and the distribution of the estimate is called the *sampling distribution* to distinguish it from the parent distribution. If the estimate is to be a good shot at the estimand, the sampling distribution should be clustered closely about the value of that estimand. The sampling distribution depends on the estimator and the parent distribution: its derivation is simply a question of integration over a suitable space (which may present mathematical difficulties, but no logical difficulties). The problem of estimation then is to choose, among all possible estimators, one that yields a "good" estimate according to some criterion. For example, if the mean of the sampling distribution equals the estimand, we say that the estimator is unbiased. Again, the standard deviation of the sampling distribution is called the standard error of the estimator; and we might seek an estimator with minimum standard error, in the expectation that it will provide estimates close to the estimand. An estimator with minimum standard error is called efficient (or rather, this is one of the principal meanings of the adjective). Using an inefficient estimator amounts to throwing away part of the information contained in an experiment, which may be an expensive sacrifice when observations are costly.

The determination of unbiased efficient estimators is sometimes a difficult exercise in the calculus of variations. A simpler method of estimation is the *method of maximum likelihood*: this amounts to choosing the value of the estimand that maximizes the likelihood of the data (and is thus a straightforward exercise in differentiation). The method of maximum likelihood is equivalent to accepting Bayes's postulate and maximizing the posterior probability. Under certain conditions it can be proved that a maximum-likelihood estimator is asymptotically efficient (i.e., very nearly efficient in a large enough sample of observations). In the example cited earlier (Poisson distribution of counts from a Geiger counter) it turns out that the arithmetic mean of the sample, $f(Y_1, Y_2, \ldots, Y_n)=(Y_1+Y_2+\cdots+Y_n)/n$, is both an efficient unbiased estimator and the maximum-likelihood estimator. But situations can also arise when different criteria of estimation lead to different estimates, and then we have to arbitrate between criteria.

REGRESSION

A linear hypothesis postulates a linear relationship between the coordinates of an experimentally observed random vector, the coefficients of this linear relationship being unknown estimands. Regression is the estimation of these estimands. The linearity is more apparent than real, since subsets of the coordinates of the random vector can be permitted to be nonlinear functions of each other. A common (because simple) procedure is to use linear estimators, i.e., to confine attention to estimators that are linear functions of the (possibly nonlinearly related) coordinates of the vector. In these circumstances, unbiased efficient estimation amounts to finding a *least-squares fit*, which can be expressed in terms of the variance–covariance matrix of the observed vector. An important special case, known as the *analysis of variance*, arises when all except one of the coordinates can take only the values 0 and 1: the analysis then investigates how the quantity represented by this exceptional coordinate depends on the presence or absence of factors (i.e., concomitant experimental conditions) represented by the remaining coordinates. The analysis of covariance is a halfway house between general regression and the analysis of variance, arising when a selected subset of coordinates is restricted to 0 and 1.

SIGNIFICANCE TESTS

A significance test is a procedure for allegedly judging a hypothesis. The problem of devising such a procedure remains unsolved in the present state of knowledge. There are three main contending theories, all three controversial and unsatisfactory. In *confidence theory*, we calculate the respective probabilities that the procedure will reject a true hypothesis or accept a false hypothesis; this fails because it gives only the right answer to the wrong (because irrelevant) question. In *fiducial theory*, we try to attach some sort of (ill-defined) "probability" directly to a hypothesis; this fails because it is known to be mathematically unsound and so gives a definitely wrong answer to the (possibly right?) question. In *decision theory*, we minimize a risk (usually financial) consequent upon accepting a false hypothesis; this fails when difficulties arise in allotting financial risks to scientific hypotheses.

SAMPLING AND EXPERIMENTAL DESIGN

Correct inferences can only be drawn from data if the data are representative and collected in a manner that will allow the proper comparisons to be made. Sampling theory and experimental design are the branches of statistics addressed to these issues, and they have an extensive literature.

See also ERROR ANALYSIS; MONTE CARLO TECHNIQUES; STATISTICS; STOCHASTIC PROCESSES.

BIBLIOGRAPHY

K. L. Chung, *A Course in Probability Theory,* 2nd ed. Academic Press, New York, 1974.

W. G. Cochran, *Sampling Techniques,* 2nd ed., Wiley, New York, 1963.

W. G. Cochran and G. M. Cox, *Experimental Designs,* 2nd ed. Wiley, New York, 1957.

W. Feller, *Introduction to Probability Theory and its Applications,* Vol. 1, 3d ed. 1968; Vol. 2 (2nd ed.), 1971. Wiley, New York.

M. G. Kendall and A. Stuart, *The Advanced Theory of Statistics,* Vol. 1, 4th ed., 1977; Vol. 2 (3d ed.), 1973; Vol. 3 (3rd ed.), 1976. Griffin and Co., London.

R. L. Plackett, *Principles of Regression Analysis.* Oxford, London and New York, 1960.

Proton

Henry Kasha

The proton (from Greek for "the first one") is the nucleus of the hydrogen atom and a constitutent of all other atomic nuclei. The atomic number of an element is equal to the number of protons in its nucleus. The proton was identified by Wien (1898) and Thomson (1919) and named by Rutherford (1920).

The mass of the proton is equal to 938.279 6(27) MeV/c^2 ($\approx 1.6726\times 10^{-27}$ kg) which is approximately 1836.15 electron masses. Its electric charge is equal in its absolute value to that of the electron but is of the opposite sign (positive by convention). In conjunction with an intrinsic angular momentum (spin) of $\frac{1}{2}h/2\pi$, where h is Planck's constant, this charge should give rise, according to Dirac's theory for point particles, to a magnetic moment of one proton magneton, equal to $eh/4\pi mc$, where m is the proton mass and c the velocity of light. The actual magnetic moment of the proton is, however, 2.792 845 6(11) times greater, indicating an extended structure (see below).

To discuss further the properties of the proton it is necessary to consider its interactions with fields other than the electromagnetic and gravitational fields, and its relationship to other elementary particles. The proton is a strongly interacting particle (a hadron); the strong interaction is responsible for binding protons and neutrons into atomic nuclei. As far as this interaction is concerned, the proton is indistinguishable from the neutron; the neutron and the proton would, presumably, have the same mass in the absence of other interactions, especially electromagnetism. The proton and the neutron are therefore collectively referred to as nucleons. In addition, the proton is related to a large group of particles heavier than itself, all having half-integral spin. These particles are characterized by a conserved quantum number, and are called baryons. The stability of the proton (the present experimental lower limit on its mean life is of the order of 10^{31} years) is a result of this conservation law and of the fact that it has the lowest mass of all the baryons.

The lighter baryons and the hadrons with integral spin ("mesons") have been successfully classified using symmetry schemes which include the symmetry group $SU(3)$. This classification has led to the hypothesis that all hadrons, and in particular the proton, are composites of a set of fundamental point-like particles ("quarks") bound by the strong force. Subsequent scattering experiments on protons of point-like probe particles (electron, muons, and neutrinos) have confirmed this hypothesis and have led to further refinement of the theory of quarks, hadrons, and the weak and strong interactions.

The baryons are now considered as bound states of three quarks; the proton is a bound state of two u ("up") quarks with fractional electric charge $Q = +\frac{2}{3}$ and baryon quantum number $B = \frac{1}{3}$ and one d ("down") quark with $Q = -\frac{1}{3}$ and $B = \frac{1}{3}$, giving the proton the required values of $Q = +1$ and $B = 1$.

As do other baryons and leptons, the proton has an antiparticle, called the antiproton (discovered in 1955), which differs from it by the sign of its electric charge and of its baryon number, being the bound state of three antiquarks, $\bar{u}\bar{u}\bar{d}$.

Protons constitute a substantial fraction of the primary cosmic radiation. Protons accelerated to high energies (up to thousands of GeV) are used in elementary-particle research.

See also ANTIMATTER; BARYONS; COSMIC RAYS—ASTROPHYSICAL ASPECTS; ELEMENTARY PARTICLES IN PHYSICS; FINE AND HYPERFINE SPECTRA AND INTERACTIONS; HADRONS; NUCLEON; QUARKS; SPIN; STRONG INTERACTIONS; SU(3) AND HIGHER SYMMETRIES; WEAK INTERACTIONS.

Pulsars

Hong Yee Chiu

Pulsars were discovered by Hewish *et al.* [1] in 1968 during a sky survey of scintillation phenomena due to interplanetary plasma in the radio-frequency range ~100 MHz. Among the expected random noises emerged "signals" timed at regularly spaced intervals, of periods ~1 s. These periods were soon established to an accuracy of six or seven digits, making them one of the best determined astronomical constants outside the solar system.

Since the discovery of the pulsars many theoretical papers have been written on this and related subjects. To date, the radiation mechanism remains an unresolved subject. In this paper, those interpretations of the properties of pulsars that will most likely survive the passage of time (in the author's opinion) will be discussed. The readers should beware of more recent developments in literature as this is still a very active subject of study.

1. EMISSION AND ORIGIN OF PULSARS

One may safely state that all pulsars emit in the radio-frequency range 20–10^4 MHz. Two pulsars, the Crab pulsar NP0532 and the Vela Pulsar, are known to emit in the optical, x-ray, and γ-ray regions. Other x-ray or γ-ray emitting pulsars are actually binary systems which will not be discussed here. Over several hundred pulsars have been discovered to date; the periods of these pulsars range between 1.5 m and 3.5 s. The pulse shape, structure, polarization, intensity, and spectrum show great variation among different pulsars. For the same pulsar the pulse intensity also exhibits temporal variations, sometimes even from pulse to pulse.

The intrinsic radiation rate of pulsars in the radio regime ranges from 10^{29} to 10^{33} ergs/s, but the total emission rate (from radio to γ ray) may be as high as 10^{38} ergs/s (Crab pulsar) and in the case where accretion disks are suspected, even greater total radiation rates have been reported. Because of the high brightness temperature attached to pulsars, it is believed that some coherent radiation mechanism must be operating. However, it may be said that there is a general idea of how this coherent radiation takes place but no detailed mechanism exists. [6]

In general, pulsars emit with a duty cycle less than 10%. A simple geometrical model of a pulsar is a highly collimated beam, in the shape of a searchlight beam or a fan, co-rotating with the object.

The pulsing nature of the emission can be detected only when the receiver bandwidth is small. This is caused by the effect of interstellar plasma on the propagation of electromagnetic waves. In a plasma the velocity of propagation v_p is frequency dependent:

$$v_p \approx (1-\omega_p{}^2/2\omega^2)c \qquad (\omega_p << \omega), \tag{1}$$

where $\omega_p=(4\pi Ne^2/m)^{1/2}$ is the plasma frequency in a medium of electron density N, ω is the angular frequency of radiation. The propagation delay Δt in terms of the differences in velocities of propagation, and the distance l, is

$$\Delta t \simeq l\Delta v_p/c^2. \tag{2}$$

In interstellar medium N varies between 0.01 and 1. As an example, over a distance of 10^3 light years for $\nu=2\pi\omega$ at 10^8 and 10^9 Hz, Δt may amount to 60 s if $N=0.01$. Equation (2) enables one to determine the distance l from the measured time delay, assuming that N is known or can be estimated. For all pulsars l is less than 10^5 light years, and hence all pulsars are located within our galaxy. Furthermore, Eq. (2) applies to all pulsar emissions studied so far, including the optical and x-ray emissions of the NP0532 pulsar. It can thus be said that all radiations from pulsars originate from one common spatial region. Conversely, Eq. (2) has been used to obtain an upper limit of the photon mass ($<<10^{-44}$ g).

Detailed time-structure studies of the pulsar NP0532 shows that rapid variations with time scales $\sim 10^{-4}$ s exist. Since the size of the emission region must be smaller than the time scale of the fastest variation times the velocity of light, the region of emission is then less than 30 km.

Although pulsars exhibit great regularity in their periods, the periods of some of them have been found to be increasing at a rate $\Delta P/P \sim 10^{-12}$–10^{-16}. The only rational interpretation to this increase in period is that the time-keeping mechanism of the pulsars is rotation and as rotational energy is dissipated the period increases. In order for a star to be stable against a fast rotation as in NP0532, the radius of the star must be less than 100 km. This, coupled with the requirement that the emission region must be less than 30 km, and the association of the pulsar with a supernova remnant, implies that the pulsar must be a neutron star.

During a collapse process which transforms a star into a neutron star, conservation of angular momentum requires that the rotational angular velocity $\Omega \propto R^{-2}$. Therefore, even for a star like our sun with a moderate rotation ($P \sim 26$ days), after a collapse which reduces its radius by a factor of 10^5 (from 10^{11} to 10^6 cm), the period will decrease by a factor of 10^{10}. The fastest rotational period of a neutron star is around 10^{-3} s. During this collapse a number of the physical processes (e.g., ejection of matter) can decrease the angular momentum of the resulting neutron star so that this minimum rotational period is not violated.

The rotational energy of a neutron star of mass M is fairly accurately estimated by the Newtonian expression:

$$E_r \sim \tfrac{1}{2}MR^2(2\pi/P)^2 = 4\times10^{46}(M/M_\odot)(R/10\text{ km})^2P^{-2}\text{ ergs}, \tag{3}$$

where $M_\odot$ is the mass of the sun and P is the period in seconds. A increase in period dP requires the dissipation of rotational energy by an amount $2(dP/P)E_r$. In the case of NP0532, $dP/P=5\times10^{-13}$, $P=0.033$ s. Hence

$$\frac{dE_r}{dt} \cong 6\times10^{38}(M/M_\odot)(R/10\text{ km})^2\text{ ergs/s}. \tag{4}$$

Since 1982 a new class of pulsars was discovered, the millisecond pulsars (the prototype is PSR 1937+21) [7]. According to Eq. (4), pulsars with periods in the millisecond range will be slowed down in a matter of months or at most years if the magnetic field is of the order of 10^{12} G, as for other pulsars. However, these pulsars show an extremely small slowdown rate, with $-(1/P)(dP/dt)$ in the range of 10^{-21} s^{-1}. It may thus be concluded that the magnetic field of this class of pulsars is small, around 10^9 G. A theory has been proposed to explain the origin of the millisecond pulsars

[8]. Briefly speaking, this new class of pulsars is long extinct pulsars (the extinction resulted from a combination of increase in rotation period and a decrease in magnetic field), but revivified through accretion of matter from a companion star. Angular momentum transferred from the companion star rapidly decreases the rotation period, making it possible for the pulsar to revive its radio emission even though the field strength is only 10^9 G [9]. It is probably the most accurate clock available in the universe, surpassing that of atomic clocks. Millisecond pulsars have been observed in binary systems. Extremely accurate timing measurements of a millisecond pulsar in a binary system (PSR 1913+16) have yielded precise orbit parameters, from which the decay of the binary orbit has been accurately explained in terms of gravitational radiation [10].

2. NEUTRON STARS AND MAGNETIC FIELDS

Neutron stars were discussed by Landau as early as 1932, soon after the discovery of the neutron. In a dense electron gas the Fermi momentum p_F is related to the electron density N by the relationship

$$N=(1/3\pi^2)\lambda_c^{-3}(p_F/mc)^2. \qquad (5)$$

At a sufficiently high density ($N \gtrsim 10^{36}$ cm^{-3}, corresponding to a matter density of 10^{12} g/cm^3 if $Z/A \sim \frac{1}{2}$) the total Fermi energy of the electrons will exceed the total binding energy of the nuclei. Inverse beta process will eliminate most electrons and protons, and the predominant composition of matter is the neutron. A small fraction of matter exists in the form of electrons and protons and some nuclei (1–12% depending on the density) to prevent the decay of the neutrons.

Stable neutron stars exist in the density range 10^{12}–10^{16} g/cm^3. The physics of neutron matter becomes unclear above a density of $\sim 10^{14}$ g/cm^3, from lack of knowledge of nuclear interactions and high-order particle interactions. The general relativity parameter GM/RC^2 of neutron stars is of the order of 0.1 or greater. General relativity theory has a strong effect on the stability of neutron stars, in that the stress energy of the neutron matter contributes to the gravitational field thus causing a runaway-type instability if the mass of the neutron star is too high. An instability occurs at $0.7M_\odot$ if interaction is neglected in the neutron gas. Uncertainties in the particle interactions may cause a change of the value of the instability mass but cannot eliminate this instability. Currently it is believed that no stable neutron stars exist beyond a mass of $3M_\odot$ or $4M_\odot$.

Magnetic fields of a neutron star may be as high as 10^{14} G. During compression of a plasma, magnetic fluxes are conserved according to the scaling relation $H \propto R^{-2}$. Thus, for a star with a nominal field strength ($\sim 10^2$ G) the field may be as high as 10^{12} G after it is compressed to the size of a neutron star. The presence of such a strong magnetic field is supported by several observational facts associated with the Crab Nebula and other pulsars.

First, in order to couple the rotational energy of a spinning neutron star strongly to relativistic charged particles, as in the case of the Crab Nebula, the star must possess an intense magnetic field. A rotating magnetic dipole radiates electromagnetic waves of the same frequency as rotation. The dissipation rate of rotational energy of a neutron star is

$$\frac{dE_r}{dt}=-\frac{B^2\Omega^4R^6}{c^3}f, \qquad (6)$$

where f is a scaling factor ~ 1, $\Omega=2\pi/P$. Translating (6) into the measured quantities $d\Omega/dt$ and Ω, one obtains

$$I\Omega\dot{\Omega} \sim B^2\Omega^4R^6c^{-3}, \qquad (7)$$

where I is the moment of inertia of the neutron star ($I \sim MR^2$). From (7) the magnetic fields of most pulsars are placed between 10^{12} and 5×10^{13} G. This dipole radiation, having a very long wavelength, can accelerate charged particles to relativistic energies, thus accounting for the origin of cosmic rays.

A spinning magnetized conductor in a vacuum will also have an external electric field for which $\mathbf{E}\cdot\mathbf{B} \neq 0$. At the surface of the conductor the normal components of **E** will pull electrons or ions off the surface, thus generating a magnetosphere. This magnetosphere may account for the supply of particles in the origin of cosmic rays.

Below the surface the structure of a neutron star is fairly complicated. At a density of 10^{14} g/cm^3, the remaining charged particles can condense into solidlike structures (as a result of minimization of the Coulomb interaction). The lattice structure corresponding to the minimum Coulomb interaction is the body centered cubic. The "melting temperature," or the temperature at which $kT \sim \hbar\omega_0$ where ω_0 is the vibration frequency of the lattice, is around 10^{10} K. The outer surface of a neutron star is thus composed of a solid crust. This conclusion is verified in the observation of the so-called star quakes. It has been observed for at least two pulsars, NP0532 and the Vela pulsar, that the period decreases abruptly by an amount $\sim 10^{-6}P$ (Vela) and $10^{-9}P$ (Crab). This is interpreted as the resettling of the outer crust of the star, resulting from the readjustment of the balance between centrifugal force of rotation, the hydrostatic pressure gradient, and the gravitational field as the star slows down. A liquid layer can adjust continually but in the case of a solid crust the adjustment takes place abruptly in the fashion of "quakes."

The identification of pulsars as magnetized neutron stars is one of the most exciting discoveries in astrophysics in recent years. Much of the physics of neutron stars is still unknown; the properties of neutrons at densities beyond 10^{14} g/cm^3 are in the frontier land of particle physics. However, as long as the radiation mechanism is not known, little can be correlated between the rich data and the properties of neutron matter.

See also ASTRONOMY, X-RAY; COSMIC RAYS—ASTROPHYSICAL EFFECTS; MILKY WAY; NEUTRON STARS.

REFERENCES

1. *Pulsating Stars* (*Nature* reprints of communications on the subject of pulsars), Vol. I, MacMillan, New York, 1968; Vol. II, Plenum, New York, 1969. Many of the original papers are included here; references to other papers in the discovery era of the pulsars can be found there.

2. M. Ruderman, "Pulsars," *Annual Reviews of Astronomy and Astrophysics,* Vol. 10. Annual Reviews Inc., Palo Alto, Calif., 1972.
3. V. Canuto, "Equations of State at Ultra High Densities," Parts I and II, in *Annual Reviews of Astronomy and Astrophysics,* Vols. 12 and 13. Annual Reviews Inc., Palo Alto, Calif., 1974, 1975.
4. V. Canuto and H. Y. Chiu, "Intense Magnetic Fields," *Space Science Rev.* **12,** 3 (1971).
5. V. L. Ginzburg and V. V. Zheleznyakov, "On the Pulsar Emission Mechanisms," in *Annual Reviews of Astronomy and Astrophysics,* Vol. 13. Annual Reviews, Palo Alto, Calif., 1975.
6. *Pulsars, 13 Years of Research on Neutron Stars* (W. Sieber and R. Wielebinski, eds.), IAU Symposium No. 95. Reidel, Dordrecht, 1981. H. Ögelman and E. P. J. van den Heuvel (eds.), *Timing Neutron Stars,* NATO ASI series. Kluwaer Academic Publishers, Boston, 1988.
7. D. C. Backer, S. R. Kulkarni, C. Heiles, M. M. Davis, and W. M. Gross, *Nature* **300,** 615 (1982).
8. For a recent summary, see D. Becker and S. R. Kulkarni, *Phys. Today* **43**(3), 26 (1990).
9. J. H. Taylor, D. R. Stinebring, *Annu. Rev. Astron. Astrophys.* **24,** 285 (1986).

Pyroelectricity

S. C. Abrahams

Pyroelectricity is a property of crystals with a net electric moment per unit cell (the polarization vector **P**) in which a temperature change ΔT produces a proportional polarization change ΔP, and conversely, as given by

$$\Delta P_i = p_i \, \Delta T$$

where p_i is one, two, or three independent coefficients that together specify the pyroelectric vector **p**. The vector **P** necessarily conforms to the crystal symmetry, and is nonzero only in the ten noncentrosymmetric point groups that possess a unique direction, namely, 1, 2, *m*, 2*mm*, 3, 3*m*, 4, 4*mm*, 6, and 6*mm*. Detection of pyroelectricity is a definitive test for noncentrosymmetry. The total pyroelectric coefficient measured at constant stress is the sum of the coefficient at constant strain (the primary pyroelectric coefficient) and the piezoelectric contribution due to thermal expansion (the secondary pyroelectric coefficient). The ratio of primary to secondary pyroelectric coefficient varies widely with material but is generally in a range on the order of 10^2 to 10^{-2}. The magnitudes of the total pyroelectric vector **p** and polarization vector **P** are also a function of the material and range from 10^{-2} to 10^{-8} C m^{-2} K^{-1} for **p** and from 1 to 10^{-3} C m^{-2} for **P**. Nonuniform heating causes nonuniform stresses and strains and can give rise to a polarization by means of piezoelectric effects; this is sometimes referred to as tertiary pyroelectricity.

The temperature dependence of the total pyroelectric coefficient is determined by the combination of primary and secondary coefficients; an example of a resulting dependence is given in Fig. 1 in which the relative sense of p_2 may be seen to change sign at 106 K although the absolute sense is indeterminate in the macroscopic measurement. The magnitude of the thermal dependence $(1/p)(dp/dT)$ in nonferro-

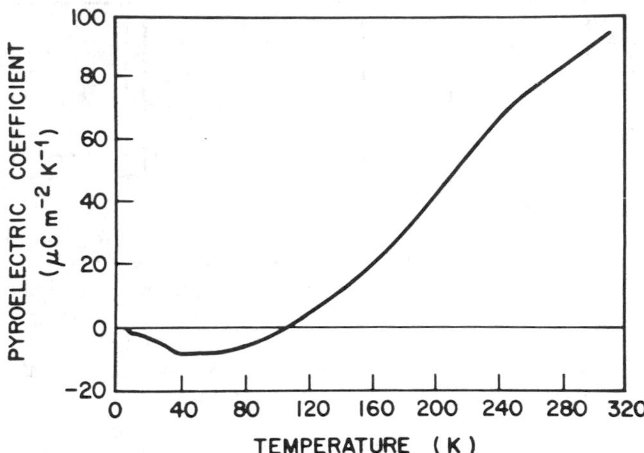

FIG. 1. Temperature dependence of the total pyroelectric coefficient p_2 of lithium sulfate monohydrate measured at constant stress [after S. B. Lang, *Phys. Rev.* **B4**, 3603 (1971)].

electric pyroelectric crystals typically ranges from 1 to 50 mK^{-1}. The absolute sense of the thermal dependence of **P** can be determined in a macroscopic experiment only if **P** becomes zero at a phase transition or if the sense of **P** can be reversed by application of an electric field. In such cases the crystal is ferroelectric and the spontaneous polarization $\mathbf{P}_s$ generally behaves as in Fig. 2, although it is possible for $\mathbf{P}_s$ to increase with increasing temperature as in Rochelle salt or ammonium sulfate.

The combined use of x ray or neutron anomalous scattering with measurement of the pyroelectric polarization allows the vector **p** to be determined in terms of the atomic arrangement, thereby giving the absolute sense of the pyroelectric coefficient. In ferroelectric crystals, the absolute sense of $\mathbf{P}_s$ can be similarly determined. In the case of $Li_2SO_4 \cdot H_2O$, for example, the absolute sense of p_2 has recently been shown to be negative, i.e., the crystal face (010) develops a negative polarity on heating. The vector **P** calculated from the atomic positions at 298 K is +32 mC m^{-2}

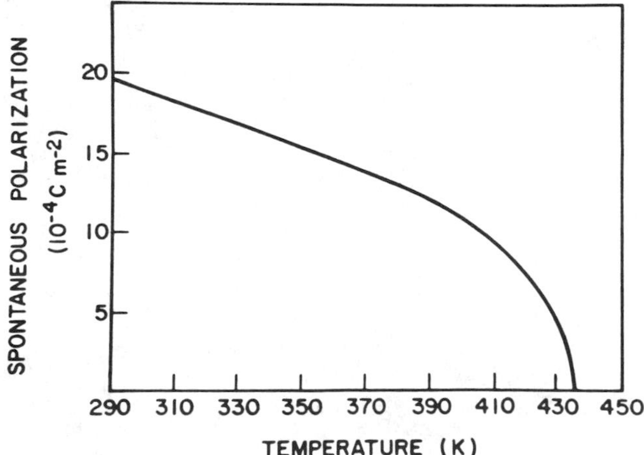

FIG. 2. Temperature dependence of spontaneous electric polarization $\mathbf{P}_s$ in terbium molybdate [after E. T. Keve, S. C. Abrahams, K. Nassau, and A. M. Glass, *Solid State Commun.* **8**, 1517 (1970)].

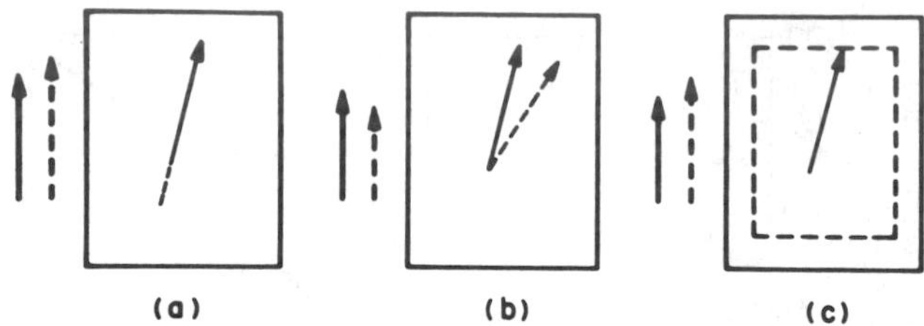

FIG. 3. Models for the origin of the pyroelectric effect; solid and broken lines represent states at different temperatures. (a) Change in *magnitude* of elementary dipole causing a primary pyroelectric effect; (b) change in *orientation* of elementary dipole causing a primary pyroelectric effect; (c) change in *volume* of unit cell causing a secondary pyroelectric effect. The polarization change ΔP is the difference between the polarization vector **P** at the two temperatures, represented by the heavy solid and broken arrows, along the (vertical) polar direction.

and at 80 K is $+44$ mC m^{-2}. The corresponding value of $\Delta \mathbf{P}$ on heating a crystal from 80 to 298 K is hence calculated to be -12 mC m^{-2} as compared with the experimental value taken from Fig. 1 of -10.4 mC m^{-2}. Similar agreement between macroscopic determinations of p and values of $\Delta \mathbf{P}$ calculated from the relative changes in atomic position and electronic charge distribution with temperature has been reported for an increasing number of inorganic pyroelectrics.

Typical relative displacements of ionic sublattices in nonferroelectric but pyroelectric crystals are on the order of 10^{-4} to 10^{-5} Å K^{-1}; in ferroelectrics, the displacements may be two or three orders of magnitude greater. Dipolar rotations are on the order of 10 to 10^{-1} arc min K^{-1}. A representation of the origin of the primary and secondary pyroelectric effects is illustrated in Fig. 3.

Numerous devices using the pyroelectric effect have been designed, including infrared and millimeter wave detectors, temperature sensors, calorimeters, thermal imagers, and television camera tubes. Many pyroelectric materials have been considered for such device applications. Among the more widely used are triglycine sulfate and lithium tantalate single crystals, lead zirconate titanate ceramics, and polyvinylidene fluoride (which can be manufactured in wide sheets with enhanced preferred orientation and polarity). Several figures of merit for thermal detection are commonly used in comparing materials, including p_i/C_p in the case of maximum current responsivity, $p_i/C_p\epsilon'$ for maximum voltage responsivity, and $p_i/C_p(\epsilon'')^{1/2}$ for highest detectivity, where p_i is the pyroelectric coefficient, C_p is the heat capacity, and ϵ' and ϵ'' are, respectively, the real and imaginary parts of the dielectric permittivity.

See also DIELECTRIC PROPERTIES; ELECTRIC MOMENTS; FERROELECTRICITY; POLARIZABILITY.

BIBLIOGRAPHY

Landolt-Börnstein, *Elastic, Piezoelectric, Pyroelectric, Piezooptic, Electrooptic Constants, and Nonlinear Dielectric Susceptibilities of Crystals*, III/16. Springer-Verlag, Berlin, Heidelberg and New York, 1984.

S. B. Lang, *Sourcebook of Pyroelectricity*. Gordon & Breach, New York, 1974.

M. E. Lines and A. M. Glass, *Principles and Applications of Ferroelectrics and Related Materials*. Oxford (Clarendon Press), London and New York, 1977.

J. F. Nye, *Physical Properties of Crystals*. Oxford (Clarendon Press), London and New York, 1957.

S. Ramaseshan and S. C. Abrahams, ed., *Anomalous Scattering*. Munksgaard, Copenhagen, 1975.

K. Toyoda, *Ferroelectrics*. Bibliography in each issue.

Quantum Electrodynamics

F. Rohrlich and A. Rich

DEFINITIONS

The theory of all electromagnetic interactions in which the electromagnetic field is treated according to the laws of quantum mechanics is called quantum electrodynamics (QED). The radiation field can in this theory be described in terms of quanta, called *photons,* thus having both a particle and a wave description just like the electron in ordinary quantum mechanics. The charged particles in QED may be treated in various approximations: on the classical or the quantum level, nonrelativistically or relativistically. For example, in quantum optics the electrons are usually described by nonrelativistic quantum mechanics.

In a narrower sense QED denotes the fundamental theory of electromagnetic interaction, parallel to the theories of strong, weak, and gravitational interactions. Among these, QED is by far the best-developed theory. Its predictions are accurate and, with only one possible exception to be discussed later, no clear discrepancy with experiment has been found even to the extremely high accuracy of present measurements (see below). In the classical limit QED becomes the electrodynamics of point charges in a vacuum. When averaged over the microscopic structure of matter it leads to the macroscopic Maxwell equations.

The interactions of electrons, muons, and photons are of special interest for testing QED because the electromagnetic interactions of protons, pions, and other hadronic particles are often partly masked by strong interactions, which are at present not as accurately describable.

In order to describe *pair creation* and *annihilation* (e.g., of electron–positron pairs) the charged particles must be treated by means of special relativity (q.v.) as well as quantum mechanics (q.v.). QED then becomes a relativistic quantum field theory (q.v.). This theory is beset by mathematical difficulties summarily denoted by "ultraviolet" divergences. In addition, the vanishing rest mass of the photon is responsible for further (less serious) difficulties, known as "infrared" divergences. The circumvention of these problems made the great success of QED possible.

HISTORY

The concept of quantization of the electromagnetic radiation field (i.e., the transition from its classical to its quantum description) had its origin in Planck's light quantum hypothesis (1900), which was advanced to explain blackbody radiation, and in the elevation of these quanta to particle status (photons) by Einstein's explanation of the photoelectric effect (1905) and Compton's experiment on scattering (1922). The statistical properties of the photon (which has spin $h/2\pi \equiv \hbar$ and no rest mass) were also known (Bose, 1924). But the quantum mechanics of the radiation field had to wait until the quantum mechanics of particles *with* rest mass was understood (1924–1926). The quantization of radiation was first developed by Dirac (1927) in a paper on the emission and absorption of radiation in atomic systems; but an earlier paper by Born, Heisenberg, and Jordan (1926) already contained the idea of treating photons similarly to harmonic oscillators. The first relativistic QED (involving spinless relativistic particles) was given by Heisenberg and Pauli (1929).

The development of QED involving charged relativistic particles of spin $\hbar/2$ required the knowledge of the Dirac relativistic wave equation (1928), which predicted the existence of the positron. The success of this equation in accounting for the fine structure of hydrogen, and the later discovery of the positron by Anderson (1932), established Dirac's equation despite the serious initial difficulties (negative-energy states, hole theory). In the same year (1928) a formalism was given by Jordan and Wigner to account for the statistical properties of spin-$\hbar/2$ particles. A set of anticommuting operators (creation and annihilation operators) would ensure the validity of the Pauli exclusion principle by providing antisymmetric wave functions. These symmetry properties could now be imposed on the *Dirac* wave function by requiring it to satisfy anticommutation relations. The electron wave function thus became the electron operator field ("second quantization"). Dirac and Heisenberg (1934) then developed the first QED of relativistic quantum-mechanical electrons.

A detailed study of QED in the 1930s led to the discovery of various mathematical difficulties such as the infinite electromagnetic self-energy of the electron (Weisskopf, 1939) and infinite vacuum expectation values (vacuum fluctuations). But it also led to very successful applications. Predictions of various atomic processes, such as the scattering of photons by bound and free electrons (Compton scattering, Klein and Nishina, 1929), bremsstrahlung, and pair production in the field of a nucleus (Bethe and Heitler, 1934), were all found to be in excellent agreement with experiments.

After World War II the discovery of Lamb and Retherford (1947) of a shift of the $2s_{1/2}$ level of hydrogen above the $2p_{1/2}$ level initiated rapid and amazing progress in QED. This discovery contradicted the prediction of the degeneracy for these two levels by the Dirac equation. The explanation of the shift was sought in the effects of the self-field of the electron (radiative corrections). These effects had so far not been computed because of the divergence difficulties encountered. That this was indeed the correct explanation was shown by Bethe (1947), who treated the electron nonrelativistically but produced the shift correctly to an accuracy of about 2%.

The radiative corrections could be found only by controlling the divergent terms. For this purpose a manifestly relativistically covariant formulation of QED was sought. Independently of each other, such formulations were developed by Tomonaga (1946–1948), Schwinger (1948–1949), and Feynman (1948–1949). These different formulations were shown to be equivalent by Dyson (1949). The relativistic calculations of the Lamb shift (Kroll and Lamb, French and Weisskopf, Schwinger, and Feynman) showed complete agreement with experiments. In addition, a correction to the Bohr magneton value of the electron's magnetic moment (anomalous magnetic moment) was found as a by-product by Schwinger in 1948 (see below).

The development of relativistic QED since that time has proceeded on two fronts. There was a steady improvement in the techniques of computing higher-order radiative corrections as well as more complicated processes; high-speed computing machines played an increasingly important role in these calculations. But there was also progress on the level of fundamental theory: The process of renormalization (see below) was developed to such an extent that the divergent terms can all be proven to be separable from the physical predictions in a consistent way. They can be lumped together in such a way that in the end only the mass and the charge of the electron have to be redefined. These "renormalized" quantities are then identified with the observed mass m and the observed charge e. Since no fundamental length would appear in the theory unless m is specified, the theory is in principle incapable of predicting m, which must therefore be taken from experiment. The charge could in principle be predicted in terms of $\hbar$ and c since the fine-structure constant $\alpha = e^2/(4\pi\hbar c)$ is dimensionless. But so far this has not been accomplished.

BASIC EQUATIONS

In the following the relativistic QED of electrons will be sketched. The theory of other charged particles is similar. Under consideration is a physical system of electrons (of negative and/or positive charge) represented by the field operator $\Psi(x)$, a quantized electromagnetic field (describing photons) represented by the vector potential operator $A^\mu(x)$, and an external (classical) electromagnetic field represented by the function $A^\mu_{\text{ext}}(x)$ in interaction. The field $A^\mu(x)$ describes the physical photons entering and leaving the system as well as the self-field of the electrons (virtual photons).

With the notation $x^\mu \equiv (x^0, \mathbf{x})$, $g^{\mu\nu} = \text{diag}(-1, +1, +1, +1)$, $\partial_\mu \equiv (\partial/\partial x^0, \nabla)$, $A_\mu B^\mu \equiv \mathbf{A}\cdot\mathbf{B} - A^0 B^0$, the basic equations of the theory are as follows. The (quantized) Maxwell equations (in rationalized units, $\hbar = 1$, $c = 1$)

$$\partial_\alpha F^{\alpha\mu}(x) = -J^\mu(x) \tag{1}$$

determine the field strength tensor $F^{\mu\nu} \equiv \partial^\mu A^\nu - \partial^\nu A^\mu$, which is produced by the electron current density

$$J^\mu(x) \equiv -\frac{ie_0}{2}[\overline{\Psi}(x), \gamma^\mu \Psi(x)]. \tag{2}$$

In this definition e_0 is the unphysical (unrenormalized) fundamental unit of charge; the γ^μ are the four 4×4 Dirac matrices satisfying

$$\{\gamma^\mu, \gamma^\nu\} = 2g^{\mu\nu}; \tag{3}$$

the adjoint $\overline{\Psi}$ is defined by $\overline{\Psi} = \Psi^\dagger A$ where A is the 4×4 Hermitian matrix that produces the Hermitian conjugate,

$$\gamma^\dagger_\mu = -A\gamma_\mu A^{-1}, \qquad A^\dagger = A. \tag{4}$$

The spinor indices and the γ-matrix element indices have been suppressed. The symbols [,] and {,} denote commutator and anticommutator; the commutator in (2) ensures charge-conjugation covariance and thus also the vanishing of the vacuum expectation value of J^μ. By using the matrix A we make the equations independent of the particular numerical representation of the defining algebra (3) for the Dirac matrices. In the representations in which $\gamma_\mu^\dagger = \gamma^\mu$ we have $A = i\gamma_0$.

The (quantized) Dirac equation

$$[\gamma^\mu \partial_\mu + ie_0\gamma^\mu(A_\mu(x) + A_\mu^{\text{ext}}(x)) + m_0]\Psi(x) = 0 \tag{5}$$

describes the action of both kinds of electromagnetic fields on the electron field. The constant m_0 is the unrenormalized mass (bare mass) of the electron. The signs are so chosen that the particle is the electron with negative charge, $-e_0$, while the antiparticle is the positron, with charge $+e_0$.

The external field may be expressed in terms of its generating external current,

$$\partial_\alpha F^{\alpha\mu}_{\text{ext}} = -J^\mu_{\text{ext}}. \tag{6}$$

The field equations (1) and (5) must be supplemented by both commutation and anticommutation relations. But there are formulations where these follow from the action integral (and certain other assumptions) together with the field equations. The (anti)commutation relations are given at equal times of the two operators involved and can be considered as initial conditions. For the Ψ field the *anti*commutation relations must be imposed, viz.

$$\begin{aligned} &\delta(x^0 - x'^0)\{\Psi(x), \Psi(x')\} = 0, \\ &\delta(x^0 - x'^0)\{\Psi(x), \overline{\Psi}(x')\} = i\gamma^0 \delta_4(x - x'). \end{aligned} \tag{7}$$

The commutation relations for the field A^μ depend on the gauge. There are covariant and noncovariant gauges. In the former, A^μ is a four-vector and all four components are operators; in the latter, the potential A^0 is not an operator and A^μ is not a four-vector. The most commonly used covariant gauge is the Feynman gauge, in which the commutation relations are

$$\begin{aligned} &\delta(x^0 - x'^0)[A^\mu(x), A^\nu(x')] = 0, \\ &\delta(x^0 - x'^0)[A^\mu(x), \partial_0' A^\nu(x')] = ig^{\mu\nu}\delta_4(x - x'). \end{aligned} \tag{8}$$

The usual noncovariant gauge is the Coulomb gauge (also called radiation gauge). In this gauge there are commutation relations only for the space part of the vector potential,

$$\begin{aligned} &\delta(x^0 - x'^0)[A^k(x), A^l(x')] = 0, \\ &\delta(x^0 - x'^0)[A^k(x), \partial_0' A^l(x')] = i(\delta^{kl} - \partial^k\partial^l\nabla^{-2})\delta_4(x - x'). \end{aligned} \tag{9}$$

Therefore $\mathbf{A}(x)$ is a nonlocal field: the operator ∇^{-2}, defined by

$$\nabla^{-2}f(\mathbf{x}) = -\frac{1}{4\pi}\int \frac{d^3x'}{|\mathbf{x}-\mathbf{x}'|}f(\mathbf{x}'),$$

is a nonlocal operator.

There are other important differences between covariant and noncovariant gauges. In a covariant gauge the Hilbert space $\mathcal{H}$ of state vectors on which the field operators act must have an indefinite metric, i.e., state vectors of negative and zero norm occur. In that case the Maxwell equations (1) cannot be satisfied as operator equations but only weakly, i.e., as matrix elements in a physical subspace of $\mathcal{H}$ that does have positive metric. The reason for this situation lies in the supplementary condition (Lorentz condition)

$$\partial_\mu A^\mu(x) = 0, \tag{10}$$

which in this gauge can only be satisfied weakly (Gupta–Bleuler condition); since the operator equation $\partial_\alpha\partial^\alpha A^\mu = -J^\mu$ holds, the weak equation (10) leads to a weak equation (1).

A positive-metric Hilbert space $\mathcal{H}$ (all of which is "physical") can be used only at the cost of describing the electromagnetic field by nonlocal operators such as is the case in the Coulomb gauge. The Maxwell equations (1) then hold as operator equations and so does the Coulomb gauge condition,

$$\nabla\cdot\mathbf{A} = 0. \tag{11}$$

It must be remembered that the gauge-dependent potentials A^μ are not observable, while the observable expectation values of the field strengths $F^{\mu\nu}$ and of their products are independent of the choice of gauge.

PERTURBATION SOLUTION

Except for very special cases no exact solutions of the basic equations of QED are known. We therefore resort to a power series expansion in e_0 whose convergence is, however, at best asymptotic. The starting solutions are then the free-field equations which are obtained from (1) and (5) by putting $e_0=0$. These can be solved easily in closed form (plane-wave solution). The commutation relations (7) and (8) can then be extended to all points x and x',

$$\{\psi(x),\psi(x')\}=0, \qquad \{\psi(x),\bar{\psi}(x')\}=iS(x-x'), \tag{12}$$

$$[\psi(x),a^\mu(x')]=0, \qquad [a^\mu(x),a^\nu(x')]=ig^{\mu\nu}D(x-x'). \tag{13}$$

The symbols ψ and a^μ denote the free fields; the invariant functions D and S are defined by

$$D(x) \equiv \frac{i}{(2\pi)^3}\int e^{ik\cdot x}\delta(k^2)\epsilon(k^0)\,d^4k = \frac{1}{2\pi}\epsilon(x^0)\delta(x^2) \tag{14}$$

with $\epsilon(t) = \pm 1$ for $t \gtrless 0$, and

$$S(x) = (\gamma^\mu\partial_\mu - m)\frac{i}{(2\pi)^3}\int e^{ip\cdot x}\delta(p^2+m^2)\epsilon(p^0)\,d^4p; \tag{15}$$

the latter can be expressed in terms of the cylinder functions J_1, N_1, and K_1.

Instead of expanding the fields in powers of e_0 it is more convenient to make a (formally) unitary transformation. In the Heisenberg picture that was used above, the time dependence was entirely in the fields, the vectors in $\mathcal{H}$ being independent of time. A time-dependent transformation can be made so that the new picture (interaction or Dirac picture) contains a divided time dependence: The time dependence of the fields is that of free fields, the time dependence of the vectors of $\mathcal{H}$ is due to the interaction. The fields are therefore linear functions of the creation and annihilation operators. In this picture the probability amplitudes for transitions from a state at $x^0 \to -\infty$ to a state at $x^0 \to +\infty$ are given by the matrix elements of the *scattering operator*

$$S = \frac{T_+\left(\exp[-i\int \mathcal{H}(x)\,d^4x]\right)}{\langle 0|T_+\left(\exp[-i\int \mathcal{H}(x)\,d^4x]\right)|0\rangle}, \tag{16}$$

which is known as the Dyson S operator. The operator $\mathcal{H}(x)$ is the interaction Hamiltonian in the Dirac picture,

$$\mathcal{H}(x) = -j_\mu(x)a^\mu(x) \tag{17}$$

with

$$j_\mu(x) \equiv -\frac{ie}{2}[\bar{\psi}(x),\gamma_\mu\psi(x)] = ie{:}\bar{\psi}(x)\gamma_\mu\psi(x){:}. \tag{18}$$

The symbol : : indicates an ordering of all creation operators to the left of all annihilation operators (normal or Wick ordering). The denominator in (16) is the vacuum expectation value of the numerator and is not equal to 1 because of the vacuum fluctuations. It ensures the correct normalization of the (formally) unitary operator S. The symbol T_+ indicates ordering of the operators $\mathcal{H}(x)$ according to their time argument; the larger x^0, the farther to the left the operator must stand when the exponential is written as a formal power series. This series is exactly equivalent to the power series in e_0 obtained for S by expanding the fields.

Each term in the expansion of (16) can be expressed as a finite sum of normal ordered products (Wick expansion). This permits us to select those terms that contribute to a given physical process of interest. The matrix element of such a normal ordered product between specified initial and final states is an integral that can be conveniently expressed in graphic form. These *Feynman diagrams* (q.v.) permit a one-to-one correspondence between a diagram and an integral and are a tremendous technical convenience.

In these diagrams each internal electron line of momentum p^μ represents the free *electron propagator* (the causal invariant function for the electron)

$$S_c = \frac{i\gamma_\mu p^\mu - m}{p^2+m^2-i\epsilon}, \tag{19}$$

each internal photon line of momentum k^μ represents the free *photon propagator* (in the Feynman gauge)

$$g^{\mu\nu}D_c(k) = \frac{g^{\mu\nu}}{k^2-i\epsilon}, \tag{20}$$

and each vertex represents a Dirac matrix γ_μ. The $-i\epsilon$ indicates how the poles are defined (causal or Feynman contour of integration). Given any Feynman diagram, the corresponding matrix element in momentum space is uniquely determined by certain rules involving these propagators.

When the self-field is taken into account to all orders, a given diagram is modified. The propagators S_c and D_c are replaced by radiatively corrected propagators S_c' and D_c'; at each vertex the matrix γ_μ is replaced by a matrix function Γ_μ of the electron momenta entering and leaving it. In each of the corresponding matrix elements the divergent terms (renormalization terms) must then be separated by an unambiguous carefully prescribed process. Only then can the physical predictions of the theory be obtained.

RENORMALIZATION

The perturbation solution requires renormalization. Just as a charged object moving through a medium will in general have an effective mass and an effective charge different from its mass and charge outside that medium, so will an electron with its self-field have a different mass and charge than without it. This holds whether or not divergent terms occur. QED is called a renormalizable theory because the (necessary) renormalization of mass and charge also suffices to remove all (ultraviolet) divergent terms. The rather difficult proof of renormalizability to all orders of the perturbation expansion culminates in the following result: The electromagnetic self-energy renormalization terms add to the unrenormalized mass m_0 (bare mass) a term involving m_{em}, the electromagnetic part of the mass, so that the effective mass $m = m_0 + m_{em}$, which is the physical (renormalized) mass, replaces the mass m_0 everywhere in the theory; after this mass renormalization the unrenormalized propagator parts and vertex parts of each diagram differ from their renormalized counterparts only by multiplicative constants:

$$S_c'(e_0) = Z_2 S_c^{ren}(e), \tag{21}$$

$$D_c'(e_0) = Z_3 D_c^{ren}(e), \tag{22}$$

$$\Gamma_\mu(e_0) = \frac{1}{Z_1}\Gamma_\mu^{ren}(e). \tag{23}$$

The effective charge e, which is the physical (renormalized) charge, is defined to be

$$e = (Z_3)^{1/2}\frac{Z_1}{Z_2}e_0. \tag{24}$$

Gauge invariance of the theory implies a relation between S_c' and Γ_μ (the Ward–Takahashi equation) which requires that two of the renormalization constants be equal,

$$Z_1 = Z_2. \tag{25}$$

By means of (21)–(24) it can then be shown that the S-matrix elements are independent of the renormalization constants Z_i. All ultraviolet divergences are thus lumped into m and e, which are now identified with their observed values.

INFRARED DIVERGENCES

The perturbation solution is physically unsatisfactory because it admits the self-field only in a "photon-by-photon" approximation. This approximation is unsuitable for a description of the self-field ("photon cloud") surrounding the physical electron; that cloud would contain a large number of very "soft" photons (photons of long wavelength) if the photon rest mass λ were very small. In the limit $\lambda \to 0$ that number becomes infinite. This is the origin of the infrared divergences in the scattering matrix. When the soft-photon part is treated nonperturbatively (e.g., by means of coherent states) these divergences disappear. In bound-state problems infrared divergences do not play a role in general.

PREDICTIONS AND EXPERIMENTAL CONFIRMATION

The early successes of QED, involving the Bethe–Heitler formulas for bremsstrahlung and pair production in the field of a nucleus, have been reexamined. Considerable extension of the old measurements to much larger energies and much higher accuracy continued to give excellent agreement. But only the calculation of radiative corrections (since 1947) opened the way for a comparison of QED with high-precision experiments. The following may serve as examples.

The magnetic moment of the electron (in units of the Bohr magneton μ_0) is in perturbation expansion predicted (Kinoshita, 1988) to be

$$\left(\frac{\mu}{\mu_0}\right)_{\text{Theory}} = 1 + 0.5\frac{\alpha}{\pi} - 0.328\,478\,965\ldots\left(\frac{\alpha}{\pi}\right)^2 + 1.175\,62(56)\left(\frac{\alpha}{\pi}\right)^3 - 1.47(15)\left(\frac{\alpha}{\pi}\right)^4 + O(\alpha^5). \tag{26}$$

Here α is the fine-structure constant ($e^2/4\pi\hbar c$) and the number in parentheses indicates the error in the last figures of the preceding number. This error in theoretical predictions is due to the numerical integrations involved. The best experimental value for α is at present based on the quantum Hall effect and is

$$\alpha^{-1} = 137.035\,997\,9(33). \tag{27}$$

The prediction (26) is then available to 11 significant figures as,

$$\left(\frac{\mu}{\mu_0}\right) = 1.001\,159\,652\,133(29), \tag{28}$$

whereas the present best experimental value (Van Dyck, Schwinberg, and Dehmelt, 1987) is

$$\left(\frac{\mu}{\mu_0}\right) = 1.001\,159\,652\,188(4) \tag{29}$$

in good agreement with (28).

Radiative corrections to the Dirac-equation prediction of the energy levels of hydrogen cause a raising of the $2\,^2s_{1/2}$ level and a slight lowering of the $2\,^2p_{1/2}$ level. The predicted $2\,^2s_{1/2}$–$2\,^2p_{1/2}$ difference (Lamb shift) is, according to the latest calculations (Mohr, 1983 and Johnson and Soff, 1985), 1057.87(1) MHz, whereas the most recent measurements yield 1057.845 (9) MHz (Lundeen and Pipkin, 1981) and 1057.851 (2) MHz (Pal'chikov, Sokolov, and Yakolev, 1983).

POSITRONIUM

Positronium is the hydrogen-like atom consisting of a positron and an electron. It is unstable, decaying either into three photons (for example, with a mean life of 1.4×10^{-7} s from its $1\ ^3S_1$ state) or into two photons (for example, with a mean life of 1.25×10^{-10} s from its $1\ ^1S_0$ state). Only recently was it possible to measure its Lyman-α line and to determine some of the ($n=1$)- and ($n=2$)-level structure and to measure the mean life of the $1\ ^1S_0$ and $1\ ^3S_1$ states with significant accuracy to test radiative corrections. The $1\ ^3S_1$–$1\ ^1S_0$ (hyperfine) separation is found to be 203.389 1 (0.7) GHz (Ritter et al., 1984) in good agreement with the best present QED prediction of 203.400 (10) GHz where the theoretical error is an estimate of the effect of the yet uncalculated $O(\alpha^2)$ radiative corrections. Similarly, recent (Conti et al., 1987) microwave spectroscopy experiments in the $n=2$ levels has yielded the results: $(2\ ^3S_1–2\ ^3P_2)=8619.6 \pm 2.7$ MHz, $(2\ ^3S_1–2\ ^3P_1)=13001 \pm 3.9$ MHz, and $(2\ ^3S_1–2\ ^3P_0)$—18 504 ± 10 MHz. The respective theoretical values for the above intervals are 8625 MHz, 13 011 MHz, and 18 496 MHz with typical error of ±6 MHZ again due to uncalculated $O(\alpha^2)$ radiative corrections.

A development of some interest in the positronium system is a nine standard deviation difference between the experimental result for the $1\ ^3S_1$ mean life which is $(141.82 \pm 0.028) \times 10^{-7}$ s (Westbrook et al., 1987) and the theoretical result (142.08×10^{-7} s) as calculated (Caswell and Lepage, 1979) to $O(\alpha)$. The difference could be due to uncalculated terms of $O(\alpha^2)$ in which case a coefficient of $+340(\alpha/\pi)^2$ would be needed in order to reduce the theoretical value to the experimental result. If this is the correct resolution of the experimental–theoretical difference, it would be of some interest because of the large size of the coefficient. This discrepancy is currently the most outstanding problem needing resolution in QED. In fact, all other tests of QED show good agreement between experiment and theory.

VACUUM POLARIZATION

Vacuum polarization is a typical QED effect. When a photon acts as an intermediary between two charges it provides for their electromagnetic interaction. The photon in this role is thereby in general not a real, physical particle; it is rather unobservable and does not satisfy the energy-momentum relation of a physical particle. Such "off-mass shell" particles are called "virtual" particles. Since such a photon spends part of its time as a virtual electron–positron pair, this interaction between the charges is thereby modified. Looked at from the point of view of the field (i.e., of the virtual photons) produced by a positive charge Q, the virtual pairs will be "polarized," the electrons being attracted by Q, the positrons repelled. This "polarization of the vacuum" thus shows similarity to the polarization in a dielectric medium.

The effects of vacuum polarization are observable. They are seen very close to the nucleus: A muonic atom has its energy levels shifted as a result of the modification of the nuclear Coulomb field by vacuum polarization effects. These effects also contribute to the energy-level shifts in hydrogen (27 MHz of the Lamb shift), and to radiative corrections in scattering.

A related prediction of QED that is qualitatively different from classical electrodynamics is the interaction of electromagnetic fields with one another, an effect that is absent in the classical theory because of its linear superposition of fields. The *nonlinear* QED effects arise again from virtual pair production and annihilation. Thus two photons have a nonvanishing scattering cross section (too small to have been measured so far), and a photon can scatter from the Coulomb field of a nucleus (Delbrück scattering). The latter effect has been observed as predicted.

QED is, of course, not restricted to electrons. It has been applied to the electromagnetic interactions of muons with the result that within present experimental error this particle differs in no way from a heavy electron. The application of QED to spinless charges (pions, for example) does not in general provide a good test for QED since these particles are all hadrons and are dominated by strong interactions.

The interplay of electromagnetic and strong interactions leads to "form factors" for hadrons, which are phenomenological descriptions of charge and magnetic moment distributions within the extended particle.

Present research is directed toward a unification of gravitational, electromagnetic, weak, and strong interactions. A future theory would then contain QED as a special case of a more general interaction dynamics that encompasses all interactions. It is hoped that such a unification might also help resolve the divergence difficulties of this otherwise successful theory.

See also ELECTRODYNAMICS, CLASSICAL; ELECTROMAGNETIC INTERACTION; ELECTRON; FEYNMAN DIAGRAMS; PHOTON; POSITRON; POSITRONIUM; QUANTUM FIELD THEORY; QUANTUM MECHANICS; QUANTUM OPTICS; RELATIVITY, SPECIAL THEORY; RENORMALIZATION; STRONG INTERACTIONS.

BIBLIOGRAPHY

Various books on quantum field theory also contain sections devoted to QED. The following references deal exclusively with the theoretical background of QED and with selected experimental tests of QED.

A. I. Akhiezer and V. B. Berestetskii, *Quantum Electrodynamics*. Wiley (Interscience), New York, 1965. (A)

J. M. Jauch and F. Rohrlich, *The Theory of Photons and Electrons*, 2nd expanded ed. Springer-Verlag, Berlin and New York, 1976. (A)

T. Kinoshita and J. Saperstein, "New Developments in QED," in *Proceedings of the 9th International Conference on Atomic Physics* (R. S. Van Dyck, Jr. and E. Norval Fortson, eds.). World Scientific Publishing, Singapore, 1984. (I)

B. E. Lautrup, A. Petermann, and E. de Rafael, "Recent Developments in the Comparison between Theory and Experiments in Quantum Electrodynamics," *Phys. Rep.* **C3**(4) (1972). (I)

A. Rich, "Recent Experimental Advances in Positronium Research," *Rev. Mod. Phys.* **53,** 1 (1981). (I)

J. Schwinger (ed.), *Quantum Electrodynamics*. Dover, New York, 1958. (A collection of original papers since 1927.) (A)

M. A. Stroscio. "Positronium: A Review of the Theory," *Phys. Rep.* **C22**(5) (1975). (I)

Quantum Field Theory

Michael E. Peskin

I. INTRODUCTION

Quantum field theory is the natural extension of quantum mechanics to dynamical fields, systems governed by partial differential equations for continuously varying functions of space and time. The original example of a quantum field theory, and still the prototype, is quantum electrodynamics (QED), the quantum theory of the electromagnetic field in interaction with a field obeying the Dirac equation. The quantized excited states of this system are photons and electrons. As a general framework, quantum field theory is the natural description of any system with many interacting quantum-mechanical particles, and it has had important applications to atoms, solids, and nuclei. However, the richest and most profound of its applications have come in systems in which the dynamical properties of the quantum field are illuminated by the symmetry of relativistic invariance, and this review will concentrate on the domain of relativistic quantum field theory [1].

Quantum field theory arose and differentiated itself from the general formalism of quantum mechanics in three distinct stages. The first of these occurred in the late 1920s, when the founders of quantum mechanics extended their formalism to the electromagnetic field by replacing the canonical field degrees of freedom—two per point in space—by operators [2]. They discovered that this was a profound generalization of the original quantum theory, introducing new questions, such as the interpretation of the zero-point motion of fluctuating electromagnetic fields, but also remarkable new results—stimulated emission, the existence of antiparticles, and the connection between spin and quantum statistics.

The next stage of development came in the late 1940s, in the work of Feynman, Schwinger, and Tomonaga. These authors set out to understand which problems of QED were truly fundamental questions and which were merely consequences of an inadequate formalism. They answered this question in a profound way, by reinventing quantum mechanics to remove the preferred role of time and developing in this way a formulation of QED which was manifestly relativistically invariant. The new formalism led to efficient methods for calculating complex quantum processes and thus to precision tests of the theory. In the past decade, the continuation of this program has produced two observables of QED, the anomalous magnetic moment of the electron and the hyperfine splitting of muonium, for which theory and experiment agree to parts per billion.

The success of quantum field theory in describing the electromagnetic interactions led to attempts to build quantum field theories for the strong and weak interactions. These attempts had early successes in Fermi's use of the neutrino field to construct a theory of β-decay and in Yukawa's postulate of a new scalar particle, later identified with the pion, to form a theory of the nuclear force. To bring these ideas to fruition, however, required profound advances in the theory. These advances, which culminated in the early 1970s with the contributions of 't Hooft and Wilson, among others, led to the identification of quantum field theories quite similar in structure to QED, yet capable of explaining the obvious qualitative differences among the strong, weak, and electromagnetic interactions [3]. Today, most particle physicists regard the quantum field theory models of the strong, weak, and electromagnetic interactions as experimentally established.

II. FORMULATION

The basic equations of quantum field theory can be written in three equivalent formulations. Let us describe these in turn:

1. States and Operators

The simplest way to make the transition from classical to quantum field theory is to follow the path of ordinary particle quantum mechanics in setting up a Hamiltonian description of the problem and then converting degrees of freedom to operators and Poisson brackets to commutators. The field itself provides the canonical coordinates

$$\{q_i\} \rightarrow \phi(\mathbf{x}):$$

the spatial position, $\mathbf{x}$, plays the role of a label indexing these coordinates. Corresponding to each coordinate, there is a conjugate momentum $\pi(\mathbf{x})$; for a simple scalar field, $\pi(\mathbf{x}) = \partial\phi(\mathbf{x})/\partial t$. The canonical commutation relations $[q_i, p_j] = i\hbar\delta ij$ take the natural form

$$[\phi(\mathbf{x}), \Pi(\mathbf{y})] = i\hbar\delta(\mathbf{x}-\mathbf{y}).$$

The Heisenberg equations of motion for the field are generated in the usual way as

$$i\hbar \frac{\partial}{\partial t} \phi(\mathbf{x},t) = [\phi(\mathbf{x},t), H],$$

where H is the Hamiltonian. If H takes the form of an integral of a local function,

$$H = \int d^3x \mathcal{H}(\mathbf{x}),$$

and the commutators of fields at equal time are delta functions as indicated above, then the Heisenberg equation of motion is a local partial differential equation for the field operator $\phi(x)$.

Following the conventional methods of quantum mechanics, one would next attempt to diagonalize the Hamiltonian and construct the eigenstates of lowest energy. Since the total momentum $\mathbf{P}$ is conserved, one might construct states which are simultaneously eigenstates of H and $\mathbf{P}$. Lorentz invariance, which has been obscured by the very different roles played by space and time, can be recognized if the states form classes with definite mass $m = (E^2 - p^2)^{1/2}$, such that unitary transformations within each class implement all possible boosts and rotations. The state of lowest energy is called the *vacuum*; in a realistic model, this state would literally describe empty space. The first excited states appear as localized excitations with definite energy, momentum, and mass—relativistic particles.

In quantum field theory, the vacuum is by no means simple. It is found as the solution to an eigenvalue problem involving an infinite number of degrees of freedom; thus, constructing the vacuum state explicitly is a problem at the same level of difficulty as that of constructing the exact thermodynamic state of a gas or liquid. In fact, the vacuum plays a role analogous to the equilibrium state in statistical mechanics, in that the essential physical predictions of a quantum field theory can be obtained from expectation values in this state.

The field operator $\phi(x)$ may be interpreted as the agent which creates and destroys excitations of the vacuum. For example, in the theory of a noninteracting scalar field, the field operator takes the form

$$\phi(x)=\int \frac{d^3p}{(2\pi)^3}\frac{1}{\sqrt{2E_p}}\,[a_p e^{-ip\cdot x/\hbar\phi} + a_p{}^{\dagger} e^{+ip\cdot x/\hbar}],$$

where $E_p=[p^2+m^2]^{1/2}$, the exponentials contain 4-vector products, and the operators $a_p{}^{\dagger}$ and a_p respectively create and destroy particles of definite momentum. The association of an operator a_p which destroys a particle with the wave function $e^{-ip\cdot x/\hbar}$ from which it is destroyed is the expression of quantum-mechanical wave-particle duality.

2. *Green's Functions*

It is often useful to convert the differential equations for the field operator into equations for matrix elements of products of field operators in the vacuum state. Since field operators at distinct points generally do not commute, an ordering must be specified; it is usual, for reasons explained below, to put the field in decreasing order of their time arguments. Define, then

$$G(x_1,\ldots,x_n)=\langle 0|T\{\phi(x)\cdots\phi(x_n)\}|0\rangle,$$

where the symbol T denotes time ordering and $|0\rangle$ is the vacuum state. $G(x_1,\ldots,x_n)$ obeys the partial differential equation satisfied by $\phi(x)$, with additional delta-function terms where two of its arguments coincide. For a free scalar field, for example, $G(x_1,x_2)$ satisfies

$$(\partial^2+m^2)G(x_1,x_2)=-i\hbar\delta^{(4)}(x_1-x_2).$$

Generalizing from this equation, we call $G(x_1,\ldots,x_n)$ a Green's function of the quantum field theory. In a relativistic quantum field theory, these Green's functions display manifest Lorentz covariance.

The differential equations for the Green's functions which follow from the operator equations of motion are called the *Schwinger–Dyson equations*. If the theory contains conserved charges and currents, the operator equation of current conservation $\partial_\mu j^\mu(x)=0$ leads to an especially simple set of equations for Green's functions known as the *Ward–Takahashi* identities.

In quantum electrodynamics and in other applications, a situation involving interacting quantum fields may usefully be considered as a problem of noninteracting fields, obeying linear field equations, coupled together by a weak nonlinear perturbation. In QED, for example, the strength of the nonlinear coupling is controlled by the dimensionless number $\alpha=e^2/\hbar c\approx 1/137$. Feynman, Schwinger, and Tomonaga showed that the Green's functions of QED could be calculated systematically in a perturbation expansion in α. Each term of this perturbation expansion is Lorentz covariant. Even more remarkably, each term has the interpretation of a dynamical process, in which the quantized field excitations—photons and electrons—propagate between points in space-time where, according to the nonlinear coupling, electrons absorb or emit photons. Feynman represented each of these processes by a *Feynman diagram,* thus providing a technique for visualizing the entire perturbation series, to arbitrarily high order. The technique of constructing a perturbation series from Feynman diagrams generalizes to all types of relativistic and nonrelativistic quantum field theories and provides the subject with a unifying methodology.

3. *Functional Integrals*

As a part of his electrodynamical researches, Feynman discovered that quantum mechanics could be reformulated by considering quantum-mechanical transition amplitudes to sample all possible intermediate states. Specifically, he found the following relation: Let $|\{q_i\}\rangle$, $|\{q_i'\}\rangle$ be two eigenstates of the coordinate operators, and let H be the Hamiltonian. Then

$$\langle\{q_i\}|e^{-iHT/\hbar}|\{q_i'\}\rangle=(\text{const})\int \mathcal{D}q(t)\,\exp\frac{i}{\hbar}\int dt\,L(q_i(t)),$$

where $L(q_i)$ is the corresponding Lagrangian and the integral is taken over all paths $\{q_i(t)\}$ which begin at $\{q_i'\}$ at $t=0$ and end at $\{q_i\}$ at $t=T$. The formula obviously has the correct classical limit: As $\hbar\to 0$ we may evaluate the integral by the method of stationary phase, and the stationarity condition is just

$$\delta\int dt\,L=0.$$

If $\hbar$ is not small, many different paths contribute, giving the familiar phenomena of quantum-mechanical interference.

This *functional integral* expression of a quantum-mechanical amplitude may be generalized to quantum field theory. The integral should run over all possible field configurations $\phi(\mathbf{x},t)$. In a local field theory, the Lagrangian takes the form $L=\int d^3x\,\mathcal{L}(x)$, and so the exponent of the integrand is $(1/\hbar)\int d^4x\,\mathcal{L}$. The question of the proper boundary conditions is more subtle. It may be shown, though, that (in general) the integral over all space and time is insensitive to the exact boundary conditions and yields the formula

$$G(x_1,\ldots,x_n)=\frac{1}{Z}\int \mathcal{D}\phi(x)\,\exp\left(\frac{i}{\hbar}\int d^4\,\mathcal{L}(\phi(x))\right)\phi(x_1)\cdots\phi(x_n),$$

where G is the Green's function defined above (with precisely the time-ordering prescription) and Z is the functional integral without the factors of $\phi(x)$.

We have now surveyed three equivalent formulations of quantum field theory. There has been considerable debate in the literature about which is the most fundamental description. Though the other two approaches often have calculational advantages, the functional integral formula has the advantage of making all symmetries of the Lagrangian manifest in the full quantum field theory. For those theories whose symmetry structure is quite subtle, such as gauge theories (defined below) or general relativity, it is most straightforward to begin from the functional formalism in which the presence of all relevant symmetries can be guaranteed.

The formula given above for quantum field theory Green's functions is remarkably similar to the standard formula for correlation functions in statistical mechanics. It is possible to make these formulas identical by the artifice of analytically continuing the time t to imaginary values: $t \rightarrow -it_E$. The Lorentz metric $|\mathbf{x}|^2 - t^2$ becomes a Euclidean metric $|\mathbf{x}|^2 + t_E{}^2$. To complete the identification, we need only exchange $\hbar \rightarrow kT$ to find the partition function of classical statistical mechanics in four space dimensions:

$$\int \mathcal{D}\phi(x) \exp\left(-\frac{1}{kT}\int d^4x\, \mathcal{H}(x)\right).$$

This identification, by analytic continuation, between quantum field theory in d space-time dimensions and classical statistical mechanics in d space dimensions has created an important interplay of ideas between the two fields. Some applications of this correspondence will be discussed below.

III. CHARACTERISTICS

From the general formalism of quantum field theory, one can derive the properties of systems of interacting fields. Many of these properties are general consequences of the principles of quantum field theory, rather than being specific to particular models. Let us discuss the most important of these:

1. Particle Production

Since particles appear in quantum field theory as the quantized excitations of fields, the theory is capable from the beginning of containing states with an arbitrary or even indefinite number of particles. The nonlinear interaction of fields can create and destroy particles. Any time-dependent external perturbation of a quantum field theory will produce additional particle production.

The ability of the field operators to create and destroy particles is necessary, in a relativistic quantum field theory, to allow the theory to have a well-defined ground state. The relativistic formula for the energy of a free particle is $E^2 = p^2 + m^2$; this has positive and negative solutions for E. Solutions of the relativistic wave equation which arbitrarily omit the negative solutions have instantaneous, and thus acausal, signal propagation. Thus, the quantum field must involve both positive- and negative-energy Fourier components. But particles must always have positive energy. The resolution of this paradox is shown in the formula above for the free scalar field operator: positive energy is associated with particle annihilation, negative energy with particle production. To make this assignment consistently, any field which creates particles must also be able to destroy particles. This implies that, for each type of particle which exists in nature, there is another particle of the same mass and the same spin, having opposite values of all conserved quantum numbers—an antiparticle. From more subtle arguments, one can derive two further, exceedingly powerful conclusions. The first of these is the *CPT theorem:* The dynamics of any quantum field theory is unchanged after the combined operations of space inversion (P), time reversal (T), and the replacement of particles by antiparticles (C). The second expresses a complication in the assignment of positive energy states to annihilation operators for fields which transform under Lorentz boosts and rotations: For fields with half-integer spin, this connection can be made consistently only if the canonical commutation relation above is replaced by an anticommutation relation:

$$\{\psi(\mathbf{x}), \Pi(\mathbf{y})\} = i\hbar\delta(\mathbf{x}-\mathbf{y}).$$

This implies the *connection between spin and statistics:* Particles of half-integer spin obey Fermi–Dirac statistics, whereas particles of integer spin obey Bose–Einstein statistics.

Because the number of particles can change in quantum field theory interactions, it can also vary in intermediate states in a perturbation-theory calculation. The space-time processes represented by Feynman diagrams usually involve additional particles which are created into intermediate states and then annihilated. These *virtual particles* may carry energy and momentum in a way unconstrained by the usual relation $E^2 = p^2 + m^2$; for example, a virtual photon may carry only 3-momentum and so induce the elastic electromagnetic scattering of charges. Characteristically, exchanges of virtual particles dictate the form of scattering amplitudes by producing poles or singularities in the kinematic variables at points where the virtual states can become real emitted particles satisfying the usual energy-momentum relation.

2. Infrared Divergences

When a pair of virtual particles is produced, these particles may have, individually, any possible 4-momentum. The complete effect of the pair creation is found by integrating over this momentum, and that integral diverges in the limit of large or small momentum. The two situations are physically distinct. A divergence at small momentum indicates an unexpected dependence on the minimum measurable energy of an emitted particle. A divergence at large momentum indicates a dependence on physics at very short distances.

In quantum electrodynamics, and in other theories of massless particles, infrared divergences appear because of the matching of quantum-mechanical results to the classical limit. In classical electrodynamics, when a charge is accelerated suddenly, it emits a finite energy per wave number at long wavelengths. This requires that an infinite number of soft photons be emitted. However, a Feynman diagram of order α^n can display the emission of at most n photons. The diagram then predicts the probability of this process to be infinite; similarly, diagrams with n virtual photons formally contribute negative infinite probabilities. Bloch and Nordsieck resolved this paradox by showing, first, that summing the perturbation series to all orders in α reproduced the classical spectrum of radiation, with sensible probabilities, and, second, that in each order of perturbation theory, quantities which are insensitive to the presence of soft photons, such as the total probability of scattering, are finite.

The classical correspondence generates further, and more observable, effects at high energy. The classical field of a relativistic charged particle may be interpreted as a sharing of the energy of this particle with a set of collinearly moving virtual photons. This sharing of energy, proportional to α multilog(E/m), has been observed in high-energy reactions of electrons (and forms the basis for studying two-photon collisions through electron–positron scattering). An analogous effect, discussed below, gives evidence for a quantum field theory description of the strong interactions.

3. *Ultraviolet Divergences*

Like classical field theories based in the continuum, quantum field theory contains quantities which are formally infinite because of the divergence of integrals at large momentum. In classical electrodynamics, as is well known, the electromagnetic self-energy is divergent. In quantum electrodynamics, both the mass and charge of the electron are divergent.

The analogy between quantum field theory and statistical mechanics, presented above, indicates that there is nothing wrong in principle with the presence of ultraviolet sensitivity. A continuum model of water would need to refer to the atomic scale for the values of the density and viscosity. In QED, one apparently has a similar situation, in which the mass and charge of the electron must be determined by new physics at very small distances or, alternatively, be fixed from experiment. It is crucial that the only ultraviolet divergences in the theory appear in these two quantities, so that once these quantities are fixed, the remaining predictions of the theory are finite and unambiguous. The proof of this statement is quite subtle: General scattering amplitudes contain ultraviolet-divergent contributions, and one must show that all of these contributions can be traced to the dependence of the scattering amplitude on the electron mass and charge. When indeed all ultraviolet divergences of a quantum field theory are removed by fixing a finite number of physical quantities, the theory is said to be *renormalizable*. Quantum electrodynamics, scalar field theories with interactions at most quartic in the field, and the generalizations of these theories discussed below have been proved to be renormalizable to all orders in perturbation theory. More general theories, including the quantum theory of general relativity, are not renormalizable and thus possess no predictive perturbation expansion.

4. *Coupling Constant Evolution*

The ultraviolet divergence of the electric charge in QED has an enlightening physical interpretation. Quantum field theory predicts that intermediate states of a quantum process contain virtual electron–positron pairs. These pairs can be polarized by an external charge, and so they make the vacuum of QED into a dielectric medium. QED predicts that the electric charge is weaker at large distances (though essentially constant for $r > \hbar/mc \sim 10^{-10}$ cm), and appears stronger as one penetrates the *vacuum polarization* of virtual charged pairs. At distances much smaller than the electron Compton wavelength, QED is approximately scale invariant, and so each unit of log r gives an equivalent amount of screening. This behavior is summarized by a *renormalization-group equation*:

$$-\frac{d}{d\log r}\,e(r)=\beta(e).$$

The right-hand side of this equation, the *Callan-Symanzik beta function*, is the contribution to the vacuum polarization from a given scale of distances.

The logarithmic dependence of coupling constants on distance scale is a crucial property of quantum field theories which is responsible for much of their qualitative behavior. In quantum electrodynamics, $\beta(e)>0$, the coupling constant e increases as $r\to 0$ and so must be determined by new physics at very short distances. However, the *non-Abelian gauge theories* discussed below have $\beta(e)<0$. In this case, integration of the renormalization-group equations gives $e(r)\to 0$ as $r\to 0$, a remarkable property called *asymptotic freedom*. In asymptotically free theories, replacing the fixed charge e by the solution $e(r)$ of the renormalization group equation (or, equivalently, summing the perturbation theory to all orders of α) removes all ultraviolet divergences.

IV. *INTERNAL SYMMETRIES*

To discuss the applications of quantum field theory to physics, it is necessary to discuss the symmetries of quantum field theories and the manner in which they might be realized. Because quantum field theories are abstract constructions, symmetry groups are embedded in them in particularly simple ways and often powerfully determine their physical properties. Relativistic field theories are always invariant under the Lorentz group. However, let us now consider theories which contain several fields, interrelated by symmetry operations acting independently of space-time. Such operations are called *internal symmetries*. In quantum field theory, internal symmetries can appear in four different ways:

1. *Manifest Global Symmetries*

This is the simplest situation; the various fields of the model form multiplets under a symmetry group, which also interrelates their couplings. *Noether's theorem* states that, to each global symmetry, there corresponds a conserved charge. In a local field theory, this charge will be always have the form

$$Q=\int d^3x\, j^0(\mathbf{x}),$$

where j^0 is the time component of a conserved current: $\partial_\mu j^\mu = 0$.

2. *Spontaneously Broken Symmetry*

It may happen that, even though a field transforms under a symmetry, it nevertheless acquires a nonzero expectation value in the vacuum state. For example, consider a set scalar field transforming as a vector $\boldsymbol{\phi}(x)$ under an internal symmetry of the Hamiltonian. One may construct a field theory of this type in which $\langle\phi^1(x)\rangle\neq 0$. Then the vacuum of the theory contains a condensate of ϕ particles or field which is asymmetric, and so the resulting quantum field theory will be qualitatively asymmetric. This phenomenon is similar to spontaneous magnetization, which picks a favored bulk spin direction in an otherwise symmetric material. (Through the relation between field theory and statistical mechanics discussed above, this analogy can be made precise.) When a continuous symmetry is spontaneously broken, a remnant of the original symmetry remains: *Goldstone's theorem* states that if any continuous symmetry is spontaneously broken, the spectrum of the field theory will contain a massless particle with the correct quantum numbers to be created by the conserved current associated with the symmetry generator.

3. *Local Gauge Symmetries*

Quantum electrodynamics possesses a higher type of internal symmetry, in which a different symmetry rotation can be applied at each point of space-time. This is the *gauge transformation*:

$$\psi(x)\rightarrow e^{i\alpha(x)}\psi(x),\ A_\mu(x)\rightarrow A_\mu(x)+\frac{\hbar c}{e}\,\partial_\mu\alpha(x).$$

Yang and Mills showed that any continuous group can be promoted to a local symmetry, as long as one adds to the theory one vector field for each symmetry generator and assigns this field an inhomogeneous transformation law similar to that of the photon field in QED. The vector fields are necessarily massless and acquire nonlinear self-interactions if the symmetry group is non-Abelian; the nonlinear terms are uniquely determined by invariance under the transformation law. These *non-Abelian gauge theories* have remarkable properties as quantum field theories. They are the only renormalizable theories of vector fields. The pure theories of vector fields alone are asymptotically free, and this property is often, though not always, preserved when the vector fields are coupled to additional fermions or scalars.

4. *Spontaneously Broken Local Symmetry*

If the symmetry of a Yang–Mills theory is spontaneously broken, the theory undergoes a further transformation, known as the Higgs mechanism. The massless vector boson associated with the spontaneously broken symmetry mixes with the Goldstone boson, and the two fuse into a massive vector boson. 't Hooft and Veltman proved that these theories are renormalizable; they provide the only renormalizable models of massive vector fields.

V. *APPLICATIONS OF QUANTUM FIELD THEORY*

The various applications of quantum field theory are discussed elsewhere in this volume. It should be noted here, however, that quantum field theory is a crucial element of our understanding of all aspects of the interaction of elementary particles. The earliest applications were made in quantum electrodynamics, and this theory has achieved stunning success in accounting for the detailed properties of electrons and simple atoms. More recently, quantum field theory descriptions have been found for the strong and weak interactions.

It is now accepted that a non-Abelian gauge theory known as quantum chromodynamics (QCD) gives the fundamental description of the strong interactions. The vector particles are known as gluons and act on the quarks which constitute hadrons. Because this theory is asymptotically free, it predicts that strong-interaction collisions simplify at large momentum transfer to elementary fermion–fermion and fermion–vector cross sections similar to those of lowest-order quantum electrodynamics. At the next level, these processes should display collinear radiation of gluons which degrades the energy spectrum of quarks progressively as the momentum transfer increases. Both effects are confirmed in high-energy lepton–nucleon and nucleon–nucleon scattering, and in the hadron distributions which result from e^+e^- annihilation.

The low-momentum behavior of the strong interactions is still not well understood, but the gauge theories give a new method for attacking this problem: Convert the gauge theory to an equivalent statistical mechanical system, approximate the statistical system by replacing space-time by a crystal lattice, and simulate this system on a large computer. Such a *lattice gauge theory* approach to the strong interactions is known to give a qualitatively correct spectrum of hadrons. In the next decade, it may give quantitative predictions for masses and matrix elements.

The weak interactions may be described by a non-Abelian gauge theory with spontaneously broken local symmetry, constructed originally by Glashow, Weinberg, and Salam. The vector particles of this theory are identified with the photon and the massive vector bosons $W^\pm$ and Z^0. The theory correctly predicted the masses of these bosons; at the time of this writing it is true, but only marginally, that the prediction requires accounting for the quantum field–theoretic coupling constant evolution. In the next few years, experiments on the decays of the Z^0 should give the first precision tests of quantum corrections in this theory.

Relativistic quantum field theory has also been tested on another front: Through the correspondence between quantum field theory and statistical mechanics discussed above, calculations in quantum field theory in three space-time dimensions yield predictions for idealized three-dimensional statistical mechanical systems. Wilson has argued that, when the correlations in a magnet or a fluid become much longer-ranged than the atomic spacing, as occurs at a critical point, this correspondence becomes exact; the atomic-scale physics, like any other ultraviolet regulator in a renormalizable field theory, can be eliminated from the description of the large-scale statistical fluctuations. Using this idea, one may compute the power laws, called *critical exponents,* with which thermodynamic quantities diverge at a critical point by Feynman-diagram computations in a quantum theory of scalar fields. The results agree with experimentally obtained exponents to a few percent accuracy.

VI. PROSPECTS

Quantum field theory is, then, a subject rich both in its formal development and in its application to physics. Over the past two decades, it has come to encompass the strong and weak interactions in addition to electromagnetism. However, this should not imply that quantum field theory is a closed subject. Of its many open problems, some suggest paths leading toward fundamental exensions of our understanding of physics. One such path, opened by Georgi and Glashow and Pati and Salam, is the attempt to unify the known interactions of elementary particles, all of which are described by gauge theories, into a common group structure. A second path approaches the long-standing problem of formulating a renormalizable quantum theory of gravity. A third path has been opened more recently by Atiyah, Witten, and others, who have uncovered an intimate connection between quantum field theory and higher geometry in which abstract spaces are classified by the properties of the quantum field theories constructed upon them. These threads seem to point onward to a rich synthesis which will extend, and may possibly transcend, quantum field theory as we know it today.

See also CPT THEOREM; CRITICAL POINTS; DISPERSION THEORY; FEYNMAN DIAGRAMS; FIELD THEORY, AXIOMATIC; GAUGE THEORIES; LATTICE GAUGE THEORY; QUANTUM ELECTRODYNAMICS; QUARKS; RELATIVITY, SPECIAL THEORY; RENORMALIZATION; S-MATRIX THEORY; STRONG INTERACTIONS; SYMMETRY BREAKING, SPONTANEOUS; WEAK INTERACTIONS.

REFERENCES

1. Two useful textbooks on quantum field theory are the set *Relativistic Quantum Mechanics and Relativistic Quantum Fields,* by J. D. Bjorken and S. D. Drell (McGraw-Hill, New York, 1964, 1965) and *Quantum Field Theory* by C. Itzykson and J.-B. Zuber (McGraw-Hill, New York, 1980). The last of these contains a comprehensive bibliography on the subjects reviewed in this article.
2. The most important papers in the early development of quantum field theory are collected in *Quantum Electrodynamics* (J. Schwinger, ed.). Dover, New York, 1958.
3. The most important papers of this period are collected in *Selected Papers on Gauge Theory of Weak and Electromagnetic Interactions* (C. H. Lai, ed.). World Scientific, Singapore, 1981.

Quantum Fluids

Alexander L. Fetter

Quantum fluids exhibit the remarkable property of remaining liquid at absolute zero temperature and zero pressure. This effect arises from their large zero-point energy and the small interatomic forces, both of which prevent the formation of a solid phase. More generally, a substance is a quantum fluid whenever the zero-point energy associated with the localization of a constituent particle by its neighbors is at least comparable with the mean interparticle potential energy, which renders it *fluid,* and with the thermal energy, which renders it *quantal*. To make this criterion quantitative, imagine each particle in an impenetrable cage of volume equal to the mean volume n^{-1} per particle, where n is the number density. Since the de Broglie wavelength λ will be approximately twice the typical linear dimension $n^{-1/3}$, the corresponding momentum $p_0 = h/\lambda$ is of the order of $\frac{1}{2}hn^{1/3}$, where h is Planck's constant, implying a zero-point energy $E_{zp} \approx p_0{}^2/2m = h^2n^{2/3}/8m$ per particle. A quantum fluid is one in which E_{zp} not only is a significant fraction of the mean potential energy per particle but also exceeds the mean thermal energy k_BT, where k_B is Boltzmann's constant and T is the temperature. Equivalently, if a substance remains liquid at a characteristic degeneracy temperature $T^* = E_{zp}/k_B = h^2n^{2/3}/8mk_B$ that depends on the mass and number density, it will be a quantum fluid in the domain $T \lesssim T^*$.

To predict the possible occurrence of quantum fluids, consider first a dilute classical gas at high temperatures. Cooling the system generally condenses it into a relatively dense liquid phase at a temperature far above the characteristic temperature T^* evaluated for the liquid density. In this case, the substance forms a *classical* liquid. On further cooling, the density and T^* remain essentially constant, but the material typically solidifies long before reaching T^*, precluding the transition to a *quantum* fluid. Evidently, quantum fluids are exceptional; for example, water has $T^* \approx 1.4$ K, whereas it actually solidifies at about 273 K.

How can this sequence (classical gas $\rightarrow$ classical liquid $\rightarrow$ solid) be evaded? Not only must the material have a high T^*; it must also fail to solidify. The first condition requires small mass and high number density. On the other hand, the second concerns solidification and melting, which occur primarily through the balance between thermal energy and the (classical) interaction energy between particles. In particular, strong interparticle potentials favor the solid phase; weak, the liquid. Other factors being equal, weak potentials maintain the liquid phase and therefore aid the formation of quantum fluids. In fact, several such quantum fluids exist, and Table I lists their typical parameters. As is evident from the values of T^*, they occur in three separate classes.

1. *^{3}He and ^{4}He*. The small mass of the helium isotopes raises the zero-point energy and T^* above the values typical of bulk matter. In addition, the closed 1*s* atomic shells weaken the interatomic potentials. Consequently these substances remain liquid even at zero temperature, unless an external pressure of 25 atm (^{4}He) or 35 atm (^{3}He) is applied to increase the density and interaction energy. In contrast, other noble gases (with smaller zero-point energy) solidify at much higher temperatures than the relevant T^*.

2. *Degenerate electrons*. The situation is quite different for electrons in normal metals, whose high electrical and thermal conductivity already suggest the existence of an electronic fluid. Although the number density n is comparable with that for He, the greatly reduced (electronic) mass raises T^* far above typical terrestrial temperatures. Con-

Table I Typical Quantum Fluids

	m (g)	n (cm^{-3})	T^* (K)	Statistics
^{3}He	5.01×10^{-24}	1.63×10^{22}	5.10	Fermion
^{4}He	6.65×10^{-24}	2.18×10^{22}	4.66	Boson
Electrons in Na	9.11×10^{-28}	2.65×10^{22}	3.88×10^4	Fermion
Nuclear matter	1.67×10^{-24}	1.95×10^{38}	8.00×10^{11}	Fermion
Neutron matter	1.67×10^{-24}	1.95×10^{38}	8.00×10^{11}	Fermion

sequently, such conduction electrons form a quantum fluid in both solid (e.g., Na) and liquid (e.g., Hg or molten Na) metals at all temperatures below the boiling point. It may be noted that electrons in white-dwarf stars form another, more exotic, quantum fluid, but the presence of gravity complicates the situation, and the system will not be considered here.

3. *Degenerate nucleons*. Our final example of a quantum fluid will be condensed nucleons, either protons and neutrons in large nuclei (nuclear matter) or neutron matter that is thought to occur in the interior of neutron stars (pulsars). In these cases, each particle's mass is again comparable with that of a helium atom, but the enormously high density yields $T^* \approx 10^{12}$ K. Moreover, the strong but short-range internuclear forces cannot solidify the material at the observed densities, giving rise to yet another quantum fluid (note that nuclei occur at room temperature and pulsars are thought to have $T \approx 10^8$ K.)

The preceding discussion is incomplete in one very important respect, for it omits the role of quantum statistics. This effect is seen most clearly in the behavior of an ideal monatomic gas.

(a) If the constituents are fermions, they obey the Pauli exclusion principle, which prohibits two particles from occupying the same quantum state or spatial location. This quantum-mechanical restriction is irrelevant in the classical limit (high temperature and low density) when $k_B T \gg E_{zp}$, but it becomes essential as T falls below T^*. In the extreme limit of zero temperature, an ideal classical gas (recall the equation of state $p = nk_B T$) would exert no pressure on the walls of its container, whereas in fact the Fermi statistics produce an effective interparticle repulsion and associated (degeneracy) pressure of the order of $nE_{zp} \approx h^2 n^{5/3}/8m$. For example, electrons in sodium exert a pressure of the order of 10^5 atm, which is balanced by the cohesive force of the ionic lattice.

(b) A very different situation arises in an ideal Bose gas, where the quantum statistics instead favor multiple occupation of the same quantum state or spatial location. If T falls below a critical temperature of the order of T^*, such quantum effects condense a finite fraction of all the particles into the lowest-lying (ground) state, which reduces the pressure *below* the classical value. Thus an ideal Bose gas at low temperature appears to have two distinct constituents—the excited particles that exert normal pressures and the condensed ones that exert no pressure.

Actual physical systems have relatively strong interparticle potentials. Nevertheless, the quantum fluids ^{3}He and ^{4}He indeed behave very differently below 1 K. In fact, liquid ^{3}He has many of the properties of a degenerate Fermi gas, like a linear specific heat and a small Pauli paramagnetism associated with the nuclear magnetic moment. It also displays normal flow properties with a well-defined viscosity. In contrast, liquid ^{4}He undergoes a dramatic phase transition at $T_\lambda \approx 2.17$ K, below which it acts like a mixture of two interpenetrating fluids—one normal and viscous, the other superfluid and inviscid. A conventional viscometer measures only the normal viscosity, for the superfluid component exerts no drag force. In flow through fine channels, however, the superfluid can pass through unhindered, leading to a vanishing effective viscosity. The evident contrast between the two isotopes provides strong evidence for the importance of quantum statistics in determining the properties of a quantum fluid. Moreover, the similarity with the corresponding ideal gases indicates that interactions merely modify the experimental behavior, leaving the effect of statistics essentially unchanged.

The helium isotopes form yet another quantum fluid, for up to 6% of ^{3}He is miscible in ^{4}He even as $T \to 0$ K. For $T \lesssim 1$ K, the ^{4}He is virtually pure superfluid and in effect forms an inert "vacuum" for the ^{3}He atoms. As a result, a given ^{3}He impurity experiences only the weak interactions from its relatively distant ^{3}He neighbors, and the system therefore may be considered a dilute weakly interacting gas, with an effective T^* given by $h^2 n^{2/3} x^{2/3}/8mk_B$ where x is the fractional concentration. By varying the temperature and concentration, it is possible to study the transition from a classical fluid to a Fermi quantum fluid in great detail; as expected from the discussion above, the transition closely resembles that in an ideal Fermi gas.

The contrast between pure ^{3}He and pure ^{4}He at $T \lesssim 1$ K might suggest that superfluidity is restricted to Bose quantum fluids. On the other hand, the electronic Fermi quantum fluid in many metals is known to exhibit frictionless flow associated with superconductivity (i.e., superfluidity in a charged system). Such behavior naturally raises the question of the relation between superfluidity and quantum statistics. To clarify this point, recall that a noninteracting Fermi system remains normal even at $T = 0$ K, with all the particles occupying the lowest available single-particle levels. Inclusion of interparticle potentials alters the situation markedly, for even a weak net attraction between the degenerate fermions leads to the formation of Cooper pairs. In certain ways, these pairs resemble diatomic molecules; for example, they consist of two identical fermions and therefore act like bosons. Each pair has a binding energy Δ, which implies a corresponding critical temperature $T_c \approx \Delta/k_B$ for their formation. (In practice, T_c is always far below T^*.) At low temperatures ($T \ll T_c$), thermal energies are too small to disrupt the pairs, which therefore constitute a passive background very like the condensate in a Bose quantum fluid. Bardeen, Cooper, and Schrieffer (1957) have shown how this very special property of fermions with attractive interactions can account for the electronic superconductivity in metals; in that case, the attraction arises from the electronic interaction with the background lattice.

The explanation of electronic superconductivity stimulated a search for similar superfluid behavior in other Fermi quantum fluids:

1. In 1972, liquid ^{3}He was finally observed to undergo a transition to a superfluid state at 2.7 mK, with the weak attractive van der Waals forces between the atoms responsible for the formation of Cooper pairs.
2. The situation in nuclear matter is much less clear, because there is no direct way to test for superfluidity in large nuclei. Nevertheless, the study of neutron transfer reactions and the comparison of energy levels between certain adjacent isotopes with even and odd numbers of neutrons provide some evidence for pairing states in nuclei.
3. The spin and charge dependence of the nuclear interactions indicates that pure neutron matter should become superfluid at densities typical of neutron stars. Since the binding energy of such Cooper pairs is of the order of k_B times 10^9–10^{10} K, the actual temperature (10^8 K) is too low to prevent the appearance of superfluidity. The observed angular deceleration of the Crab and Vela pulsars (especially following a sudden change) provides some evidence for this exotic superfluid phase.

Two other systems deserve mention as potential quantum fluids:

1. Hydrogen usually assumes the molecular form H_2, with the electrons tightly bound (their spins aligned in a singlet state). Thus, atomic hydrogen is found only under unusual circumstances, such as in a gas discharge. Recently, however, several laboratories have studied atomic hydrogen in a spin-polarized state (induced by a large magnetic field that favors the electronic triplet state). The corresponding triplet interatomic potential is much weaker than the singlet, so that the triplet atoms are unbound at most accessible temperatures. Thus, such polarized hydrogen atoms form a gas of weakly interacting particles, much like the original model of a weakly interacting Bose gas. Despite heroic experimental efforts, the critical density and temperature for the onset of Bose condensation have not yet been reached; nevertheless, several laboratories continue the search.
2. The new high-temperature superconductors differ in many ways from the usual BCS-type materials. Most importantly, they are insulators, or at best, poor conductors, in the normal state. In addition, they typically are antiferromagnets in the normal state. At present, no theory can provide a satisfactory account of these unexpected systems. It remains an open question whether the familiar BCS description can characterize them; indeed, several theorists have even suggested abandoning the usual Fermi-liquid description of the normal state, in favor of a wholly new electronic ground state. The situation remains uncertain, and much work remains to be done.

In summary, a system constitutes a quantum fluid if it remains liquid below a characteristic temperature T^*, where quantum degeneracy becomes significant. A Bose system, like ^{4}He, will be a superfluid. On the other hand, a Fermi system may be either normal or superfluid, depending on the sign and strength of the interparticle potential. Net attraction favors the appearance of Cooper pairs, which then form a superfluid condensate. ^{3}He and electrons in many metals exemplify this transition from normal to superfluid behavior, which generally occurs at a much lower temperature than T^*.

See also BOSE–EINSTEIN STATISTICS; FERMI–DIRAC STATISTICS; HELIUM, LIQUID; NEUTRON STARS; SUPERCONDUCTIVITY THEORY.

BIBLIOGRAPHY

General

C. Kittel, *Thermal Physics,* Chaps. 9–17. Wiley, New York, 1969. (E)

D. Pines and P. Nozières, *Quantum Liquids,* Benjamin, New York, 1966. (A)

F. Reif, *Fundamentals of Statistical and Thermal Physics,* Chap. 9. McGraw-Hill, New York, 1965. (I)

Liquid Helium

E. Feenberg, "Microscopic Quantum Theory of the Helium Liquids," *Am. J. Phys.* **38,** 684 (1970). (A)

C. T. Lane, "Resource Letter LH-1 on Liquid Helium," *Am. J. Phys.* **35,** 367 (1967). (E)

A. J. Leggett, "A Theoretical Description of the New Phases of Liquid ^{3}He." *Rev. Mod. Phys.* **47,** 331 (1975). (A)

F. London, *Superfluids,* Vol. II: *Macroscopic Theory of Superfluid Helium.* Dover, New York, 1964. (I)

J. C. Wheatley, "Dilute Solutions of ^{3}He in ^{4}He at Low Temperatures," *Am. J. Phys.* **36,** 181 (1968). (I)

J. C. Wheatley, "Experimental Properties of Superfluid ^{3}He," *Rev. Mod. Phys.* **47,** 415 (1975). (A)

J. Wilks, *An Introduction to Liquid Helium.* Oxford (Clarendon Press), London and New York, 1970. (I)

Electrons in Metals

D. M. Ginsberg, "Resource Letters Scy-1 and Scy-2 on Superconductivity," *Am. J. Phys.* **32,** 85 (1964); **38,** 949 (1970). (E)

D. N. Langenberg, "Resource Letter OEPM-1 on the Ordinary Electronic Properties of Metals," *Am. J. Phys.* **36,** 777 (1968). (E)

F. London, *Superfluids,* Vol. I: *Macroscopic Theory of Superconductivity.* Dover, New York, 1961. (I)

M. Tinkham, *Introduction to Superconductivity.* McGraw-Hill, New York, 1975. (A)

Nuclear Matter and Neutron Matter

H. A. Bethe, "Theory of Nuclear Matter," *Ann. Rev. Nucl. Sci.* **21,** 93 (1971). (A)

G. E. Brown, *Unified Theory of Nuclear Models and Forces,* 2nd ed., Chapters X–XII. North-Holland, Amsterdam, 1967. (I)

A. M. Lane, *Nuclear Theory,* Part I. Benjamin, New York, 1964. (A)

M. Ruderman, "Pulsars: Structure and Dynamics," *Ann. Rev. Astron. Astrophys.* **10,** 427 (1972). (I)

Spin-Polarized Atomic Hydrogen

I. F. Silvera and J. T. M. Walraven, in *Progress in Low Temperature Physics,* Vol X, p. 139 (D. F. Brewer, ed.). North-Holland, Amsterdam, 1986. (A)

REFERENCES

J. Bardeen, L. N. Cooper, and J. R. Schrieffer, "Theory of Superconductivity," *Phys. Rev.* **108,** 1175 (1957). (A)

Quantum Hall Effect *see* Hall Effect, Quantum

Quantum Mechanics

Ernest A. Seglie and Susan E. Fox

INTRODUCTION

Quantum mechanics is the theory of atomic and subatomic systems. Approximately the first 30 years of the 20th century represent the time of its conception and evolution.

Perhaps the complexity of the initial problem, blackbody radiation, was responsible for the long time it took to develop the early nonrelativistic quantum postulates. Rather than beginning with a chronological history, the route chosen in part I is the delineation of the ideas that preceded the notion of "quantized" energies. Contained within this first section are the developmental observations which lead into the theory and experimental evidence supporting it.

The second section describes the mathematical formalism: the fundamental equations of the Schrödinger wave theory. This section parallels the first by providing the symbolic language which best defines quantum theory.

The third section contains two of the most important extensions of the theory: first, to systems containing more than one particle, and then to systems for which relativistic considerations are necessary.

I.1. MATTER AND LIGHT: THE NATURE OF PARTICLES AND WAVES

The rise of quantum mechanics coincided with the failure of the classical concepts of waves or particles individually to describe events at the atomic and subatomic level. By 1900, one or the other of these concepts was the basis for the explanation of all known phenomena. Those occurrences regarding the behavior of matter (e.g., the motion of the planets, projectiles on the earth, etc.) are described by Newton's laws. The underlying assumptions are that matter consists of particles having mass and charge, and that the instantaneous position and momenta of these particles can be experimentally determined with arbitrary accuracy.

Those phenomena regarding the behavior of light are described by the wave theory of C. Huygens (further developed by G. Kirchhoff). The belief held by Newton was that light is composed of a flux of particles. The double-slit experiments of T. Young (1802) forced the abandonment of Newton's corpuscular theory since there was no way to explain the phenomenon of light interference. Huygens postulated that the propagation of light is analogous to the disturbance caused by a pebble thrown into a still pond. The resulting pattern is a series of concentric ripples (the wave front) whose crests are evenly separated by a distance called the wavelength. Huygens's principle specifies that each point on the wave front acts as a source for spherical waves. The new wave front is the sum of the contributions from all the previous points.

In the double-slit experiments, there are two possible paths that the rays of light can travel through to get to the same point on the detection screen. The wave amplitudes in each ray superpose (add) to produce a magnification or cancellation at particular points on the screen. When the interference is constructive, the waves add to yield a larger amplitude than either slit acting independently. When the interference is destructive, the waves act together to cancel the other's effect. Thus, each of the waves has a nonzero contribution; the combined effect is zero disturbance. The interference Young observed could not occur in the Newtonian corpuscular theory. In the early twentieth century, Young's observations were the foundation for crystal structure analysis based on x-ray diffraction. In diffraction from a crystal, it is the regular spacing of the atoms in the material which acts in a manner similar to the slits causing scattering in certain directions.

Experiments in two separate laboratories (C. J. Davisson and L. H. Germer, in the United States; G. P. Thomson, in Scotland) in 1927, showed that electrons when scattered off a crystal exhibit a diffraction pattern similar to the x-ray diffraction pattern just described. The equation for the wavelength (λ) associated with this scattering is

$$\lambda = h/p \tag{1}$$

(L. de Broglie, 1924), where h is the Planck constant with the value 6.6×10^{-34} J s, and p is the momentum (mv) of the particle. These experiments were the first proof that electrons, previously thought of as particles, also had wave aspects. The de Broglie hypothesis was expanded by E. Schrödinger in 1925 into an equation which completely defines the wavelike behavior of particles (Sec. II.4).

Einstein's discussion of the photoelectric effect (1905) recognized the complementary phenomenon of light waves behaving like particles. Specifically, the photoelectric effect describes the process by which light liberates electrons from the surface of a metal. It is necessary to assume that light energy is absorbed in discrete packets (light quanta, photons) by the electron. This energy is related to the wavelength of light by

$$E = hc/\lambda \tag{2}$$

(M. Planck, 1900), where c is the velocity of light, 2.99×10^{8} m/s, and h is the Planck constant. In this model, the intensity of light is proportional to the density of photons in the beam.

A. H. Compton (1923) experimentally observed the collision of light quanta with electrons. The successful analysis of this photon scattering depended on the concept of discrete packets. The photoelectric effect and the Compton effect support Newton's corpuscular theory of the nature of light. Yet, Young's experiments demonstrated the phenomena of diffraction and interference, which were explained only by wave theory.

There are several questions to be answered. Is the photon a wave or a particle? The answer is neither. The photon propagates like a wave but interacts with matter like a particle. The wave aspects and the particle aspects can only be reconciled by assuming (1) that the presence of wave phenomena in an experiment precludes the appearance of particlelike behavior, and (2) that experiments revealing the particle behavior of the photons exclude the observation of wavelike phenomena.

Are wave theory and corpuscular (particle) theory mutually exclusive? The answer to this is complex. The inability of a single, preexisting model to explain the results of all the experiments prompted N. Bohr to postulate it as a fundamental condition of nature. This is known as the principle of complementarity. Thus, the wave–particle duality observed came to be acknowledged. The complete specification of detail assumed in classical physics is not possible in atomic physics. Nowhere is this more evident than in W. Heisenberg's uncertainty principles.

I.2. THE UNCERTAINTY RELATIONS

The most famous of the uncertainty relations places a limit on the simultaneous specification of the position and momentum of a particle. The first class of uncertainty relations follows directly from the wave nature of the particles.

Consider a particle in one dimension known to be within a region Δx. The wave representing that particle must be confined within that region. The resulting wave cannot be described by a single wavelength, as is known from looking at the Fourier transform of the wave. Put another way, precisely where the wave begins and precisely where it terminates is hard to define. So a first approximation of the number of wavelengths (N) would be

$$N=\Delta x/\lambda,$$

which is better described by assigning an uncertainty of ± 1 to N. The estimation of the wavelength can now include the uncertainty in the number of wavelengths:

$$\lambda=\frac{\Delta x}{N}=\frac{\Delta x}{\Delta x/\lambda \pm 1}=\frac{\Delta x}{(\Delta x/\lambda)(1+\lambda/\Delta x)},$$

$$\lambda=\lambda \pm \lambda^2/\Delta x.$$

The uncertainty in the wavelength is introduced because of the finite extent of the wave:

$$\Delta\lambda=\lambda^2/\Delta x$$

or

$$(\Delta\lambda/\lambda^2)\Delta x=1. \tag{3}$$

From the de Broglie relation,

$$p_x=h/\lambda$$

differentiating p with respect to λ gives

$$\left|\frac{\Delta\lambda}{\lambda^2}\right|=\left|\frac{\Delta p}{h}\right|$$

and therefore,

$$\Delta p_x \Delta x \sim h. \tag{4}$$

There are position–momentum uncertainty relations for each of the three Cartesian coordinates.

Another class of uncertainty relations places a limit on the simultaneous specification of the lifetime of an excited state and the energy of the state. This energy is measured by the wavelength of the photon emitted in the deexcitation process. Here, the same analysis of the uncertainty in the photon wavelength leads to the equation,

$$(\Delta\lambda/\lambda^2)\Delta x=1,$$

where Δx measures the maximum photon length which cannot be longer than the product $c\Delta t$, where Δt is the time from excitation to deexcitation. Multiplying and dividing by c, the velocity of light, yields

$$c\,\frac{\Delta\lambda}{\lambda^2}\,\frac{\Delta x}{c}=1;$$

$\Delta x/c$ is the time it takes for the decay, since $\Delta x/\Delta t=c$. Differentiating E with respect to λ from Eq. (2) gives

$$\Delta E=(hc/\lambda^2)\Delta\lambda,$$

yielding

$$\Delta E\Delta t \sim h. \tag{5}$$

All these estimates represent a relation between the limits on the accuracy of physical measurements (recall that we took N with an uncertainty of only ± 1). So, in the literature, these relations appear with inequality signs: $\Delta x\Delta p_x \geq h$, $\Delta y\Delta p_y \geq h$, $\Delta z\Delta p_z \geq h$, $\Delta E\Delta t \geq h$.

I.3. QUANTIZATION OF ANGULAR MOMENTUM AND ENERGY LEVELS OF ATOMS

Having considered only the quantization of the energy in the photon, we will now consider two other properties which are quantized in nature. In particular, this includes the energy levels of bound systems (e.g., atoms or nuclei), also called stationary states. The states are confined by the attractive forces between them and are always a finite distance apart. In classical physics, any energy can exist for a system. In quantum theory, only certain energies are actually observed. These energies are referred to as the discrete spectrum. Angular momentum and those properties which relate to it (e.g., magnetic moments) are also quantized. Note that the Planck constant is specified in units of angular momentum. Quantization of the orbital angular momentum or of the energy levels of the bound states arises from the conditions imposed to prevent the wave from destructively interfering with itself.

The evidence that the z component of the magnetic dipole moment (and therefore the angular momentum of the atom) is quantized proceeds from the research of Stern and Gerlach in 1922. This is referred to as space quantization. Extensions of these experiments proved that in addition to the expected angular momentum associated with the orbital motion of the electron, the electron has intrinsic angular momentum. This intrinsic spin has possible projections on the axis of measurement of $\pm\frac{1}{2}\hbar$.

It is now known that all elementary particles have this intrinsic angular momentum (spin). If a particle has an integral spin, $(0,1,2,\ldots)\hbar$, it is called a boson. A particle with a half-integral spin, $(\frac{1}{2},1\frac{1}{2},2\frac{1}{2},\ldots)\hbar$, is known as a fermion. This is discussed in Section III.1 in greater detail.

N. Bohr first proposed the quantization of energy levels as a solution to interpreting the discrete line spectra of the hydrogen atom. From the work of E. Rutherford, it is known that the atom is composed of a positive, heavy center, the nucleus. The region of space around the nucleus is occupied by the electrons. These accelerate under the influence of the Coulomb attractive field. According to classical electromagnetic theory, an accelerating charge radiates and its orbit would spiral into the nucleus as its energy radiates away. This would take less than 10^{-9} s. To avoid this theoretical catastrophe, Bohr proposed that only certain orbits be allowed. This way, the continuous emission predicted by classical theory cannot occur.*

The allowed orbits are the ones in which the angular momentum is an integral multiple of $\hbar$. Bohr used classical physics to calculate the energy of the electron of the hydrogen atom in terms of the electron's angular momentum l, assuming the particle of mass m moving in a circle of radius r at a velocity v. Then $l = mvr$, and

$$E = -\frac{mZ^2e^4}{(4\pi\epsilon_0)^2}\frac{1}{2l^2}.$$

Bohr continued by assuming the angular momentum quantized as in

$$l = n\hbar. \tag{6}$$

This yields the possible energy levels of the hydrogen atom:

$$E_n = -\frac{mZ^2e^4}{(4\pi\epsilon_0)^2 2\hbar^2}\frac{1}{n^2} = (-13.6\text{ eV})\frac{1}{n^2}. \tag{7}$$

Transitions between the allowed energy levels are accompanied by the emission or absorption of photons with an energy

$$E_{\text{photon}} = E_{\text{final}} - E_{\text{initial}}$$
$$= (13.6\text{ eV})\left(\frac{1}{n_f^2} - \frac{1}{n_i^2}\right). \tag{8}$$

E_f and n_f are the energy and quantum number of the final state, respectively. E_i and n_i represent the same quantities for the initial state.

All the transitions with $n_f = 1$ are called the Lyman series. The radiation wavelength is in the ultraviolet region of the spectrum. The transitions with $n_f = 2$ are the Balmer series. These are in the visible range.

* The first direct experimental evidence showing that the internal states of the electron in the atom are quantized comes from an experiment performed by J. Franck and G. Hertz in 1914. These experiments involved absorption of energy in electron scattering.

The theory offered no way to calculate how quickly a given state decayed, as measured by the intensity of the spectral line, and it predicted too many lines. Addressing these theoretical omissions Bohr proposed a bipartite correspondence principle (1923). The first part asserts that in the limit of the large quantum numbers the predictions of quantum physics must correspond to those of classical physics. So the intensities are calculable for some transitions. In order to eliminate the prediction of too many lines, the second part asserts that if classical physics assign a zero intensity to a line, the quantum system will not radiate at that frequency. This principle somewhat extended the usefulness of Bohr's approach, but the theory still needed a better mathematical foundation in order to be applied to atoms with more than one electron. That formulation is the topic of the next part.

II.1. INTRODUCTION TO THE MATHEMATICAL FORMALISM

To combine the known wave and particle aspects of matter into a single theory demands that the defining equation (the Schrödinger equation) include the first derivative of time, not the usual second derivative. Since this equation also contains the imaginary number i (where $i = \sqrt{-1}$), the resulting solution is a complex-valued function not directly observable. The solution (called the wave function) is interpretable only as a probability amplitude for possible results of experiments. The expected result of a series of identical experiments is the average of the various possible outcomes each weighted by the square of the corresponding amplitude. In general, which of these outcomes will result from a single experiment is completely unpredictable. This differs from the situation in which probabilities normally apply.

The usual case for the introduction of probability is to account for some ignorance on the part of the observer. We use probabilities in the flip of a coin because the exact conditions of the coin toss are unknown. Of course, there are times when the stakes are great and the observer's ignorance can be replaced by determinations of the initial conditions. When this occurs, the outcome of the classical theory is not merely statistical prophecy but a complete determination. This is the difference between predicting the gross national product and landing a man on the moon! In quantum mechanics, the probabilities are irreducible.

II.2. THE STATE OF A SYSTEM IN CLASSICAL AND QUANTUM MECHANICS

In nonrelativistic classical mechanics, the state of a system of n particles is completely specified by the three components of the position and momentum of each of the particles. This is represented by a point in a $6n$-dimensional space, called the phase space of the system. Each component of position and momentum of each particle labels an axis: $3n$(position) + $3n$(momentum) = $6n$ dimensional. In classical mechanics, the uncertainty in each coordinate can be made arbitrarily small (limited only by the accuracy of the instruments used) so that the system is represented by a point. The time development of the state (i.e., its trajectory) in phase space can be calculated from Newton's laws or from any of several equivalent theoretical schemes (e.g., the Lagrangian or Hamiltonian approaches). Often, quantities like energy or angular momentum remain constant in time as the system evolves. These are called conserved quantities or constants of motion.

In quantum mechanics, the uncertainty principle modifies the above description of the state considerably. It allows only a probability description that the system lies within a $6n$-dimensional hypercube in phase space of volume h^{3n}. There is a fundamental requirement that the states can interfere with each other. This requires that the information about the states be statistical in nature and obey the superposition principle. Conserved quantities also exist in the quantum case but because of the uncertainty principle, they are, in general, fewer in number than in the corresponding classical case. The maximum set of conserved observables which are able to be specified with no uncertainty completely describes the quantum state. No other observables have definite values. Such a set of observables is called a complete set of commuting observables. The values the observables have are called the quantum numbers of the state.

II.3. THE WAVE FUNCTION AND EXPECTATION VALUES

For a particle with no internal degrees of freedom (e.g., no intrinsic spin) the function (of the coordinates) which contains all the information that can be known about the state of the system is called the wave function. By itself, it is not observable. For the bound states, the wave function can be fixed so that it is a real-valued function. In general, it is a complex function. As a matter of convenience, we can write it as

$$\Psi = \Psi_{\text{real}} + i\Psi_{\text{imaginary}}$$

or as

$$\Psi = Ae^{i\varphi}.$$

A is the modulus, and φ is the phase. Both are real functions of the coordinates. The most important property of the wave function is

$$P = \Psi^*(x,y,z)\Psi(x,y,z)\,dx\,dy\,dz$$

(M. Born, 1926), where P is the probability that the particle can be found within an infinitesimal volume element (dx by dy by dz) about the position (x,y,z). The asterisk indicates complex conjugation. Since only the modulus squared (A^2) enters the observable, the absolute phase of the total wave function cannot be determined; also, since the particle must be somewhere, the integral of P over all space, must be 1:

$$\int_{\text{all space}} \Psi^*(x,y,z)\Psi(x,y,z)\,dx\,dy\,dz = 1. \tag{9}$$

(normalization condition). The superposition principle is equivalent to the following: if Ψ_1 and Ψ_2 are the wave functions for two possible states of the system, then $N(\alpha\Psi_1 + \beta\Psi_2)$ is also a possible state of the system, where α and β are arbitrary constants and N is a number which guarantees that the wave function is normalized.† The basis set is defined as the set of wave functions

$$\Psi_1, \Psi_2, \Psi_3, \ldots, \Psi_n \ldots,$$

where each Ψ corresponds to a state in which each of the complete set of commuting observables has a definite value. An arbitrary wave function Ψ can be written as a unique sum of terms:

$$\Psi = \alpha_1\Psi_1 + \alpha_2\Psi_2 + \alpha_3\Psi_3 + \cdots + \alpha_n\Psi_n \cdots$$

(completeness assumption), where $\alpha_1, \alpha_2, \alpha_3, \ldots, \alpha_n, \ldots$ are called the expansion coefficients.

Once the basis set is specified, the expansion coefficients by themselves represent the state. Now, $\alpha_1, \alpha_2, \alpha_3, \ldots, \alpha_n, \ldots$ are the coordinates of a state vector in an abstract mathematical space, the Hilbert space, with axes labeled by the basis states Ψ_i. We define a scalar product in this space of the state Ψ with the state φ as

$$\langle\Psi,\varphi\rangle = \int \Psi^*(x,y,z)\varphi(x,y,z)\,dx\,dy\,dz.$$

If it is zero, the states are said to be orthogonal. The norm (N) of a state is the scalar product of a vector with itself. Since the normalization condition [Eq. (9)] requires the norm of all wave functions to equal 1, only the direction in Hilbert space is of consequence. Finally, the process of choosing a different basis set (a different, complete set of observables) is equivalent to choosing a different set of axes with which to describe the given direction. This change of coordinates resulting from such a choice can be represented as a unitary transformation (i.e., one which preserves lengths), and is similar to the change of coordinates of vector resulting from a rotation of axes in ordinary space.

The probability interpretation of the wave function provides a method to determine the most probable result (i.e., the expectation value) of a physical measurement. Thus, for example, the most probable result of a measurement x is calculated by multiplying x by the probability of finding the particle at x [i.e., $\Psi^*(x)\Psi(x)\,dx$] and summing (integrating) over all possible values of x. This expectation value of x is written

$$\langle x\rangle = \int xP(x)\,dx = \int \Psi^*(x)x\Psi(x)\,dx.$$

Any function of x can be similarly evaluated by appropriately replacing the central x by $f(x)$:

$$\langle f(x)\rangle = \int f(x)P(x)\,dx = \int \Psi^*(x)f(x)\Psi(x)\,dx.$$

More generally, the expectation value of an observable for a particle in a given state Ψ can be evaluated if the expansion of that state is known in terms of the complete set of states including the observable. Thus, if

$$\Psi = \alpha_1\Psi_1 + \alpha_2\Psi_2 + \cdots + \alpha_n\Psi_n + \cdots,$$

where each Ψ_n has a definite value of the observable f called f_n, the expectation value of f is

$$\langle f\rangle = \sum_n f_n|\alpha_n|^2, \tag{10}$$

where $|\alpha_n|^2$ clearly represents the probability that the system is in state n.

† For the principle of superposition to hold, all equations satisfied by the wave function must be linear in Ψ.

In this vector space we define an operator as a mathematical function which establishes a correspondence between each vector in the space and another vector in the same space. The necessity for the results of real experiments to yield real numbers ultimately requires the mathematical entities which represent the observables to be Hermitian operators within the Hilbert space. Since the expectation values are real, they are equal to their complex conjugate,

$$\int \Psi^* f_{op} \Psi \, dx\, dy\, dz = \int \Psi (f^*_{op} \Psi^*) \, dx\, dy\, dz,$$

where f^*_{op} denotes the operator which is the complex conjugate of f_{op}. This defines the operator as Hermitian.

Once the bases are chosen, Hermitian operators can be represented as matrices acting on the vectors in the space by matrix multiplication. The matrix has element f_{nm} in the nth row and mth column defined by

$$f_{nm} = \int \Psi_n^* f_{op} \Psi_m \, dx\, dy\, dz,$$

where Ψ_n and Ψ_m are the nth and mth states of the bases. The eigenfunctions (φ_k) of the operator are defined by the eigenvalue equation

$$f_{op}\varphi_k = f_k \varphi_k. \tag{11}$$

For the operations which represent observables, the set of all solutions to Eq. (11) form a complete set. Any function in the space can be expanded in terms of them. Each φ is orthogonal to all the others. If f_{op} is Hermitian, f_k is the eigenvalue, a real number. The Hermitian operators in the theory are linear:

$$f_{op}(\alpha\Psi_1 + \beta\Psi_2) = \alpha f_{op}\Psi_1 + \beta f_{op}\Psi_2.$$

It is assumed that the result of each individual experiment is one of the eigenvalues, f_k, of the observable f. Which of the eigenvalues (f_k) it is, can only be determined in the probabilistic sense mentioned earlier. Namely, an individual experiment will produce the result f_k with a probability equal to $|\alpha_k|^2$. The expectation value of an observable, as discussed previously, is the average value of identical repeated experiments on the same systems.

An important operator in the theory is the one representing the momentum of a particle. Its component in the x direction is given by

$$\hat{p}_x \equiv (p_x)_{op} = -i\hbar \frac{\partial}{\partial x};$$

similar equations hold for each other component. It is Hermitian since for any functions φ and ψ which vanish at infinity,

$$\int \varphi^* \hat{p}_x \psi \, dx = -i\hbar \int \varphi^* \frac{\partial \psi}{\partial x} dx = i\hbar \int \psi \frac{\partial \varphi^*}{\partial x} dx$$

$$= \int \psi \hat{p}_x{}^{\dagger} \varphi^* \, dx.$$

The eigenfunctions and eigenvalues of this operator‡ are determined from the eigenvalue equation

$$-i\hbar \frac{\partial}{\partial x} \psi = p_x \psi$$

‡ $\hat{f}$ represents the operator, f represents the eigenvalue.

yielding

$$\psi = N e^{i(p_x x/\hbar)}.$$

This has real and imaginary components:

$$\psi = N[\cos(2\pi p_x x/h) + i \sin(2\pi p_x x/h)].$$

Note that the wave repeats itself after a distance h/p (i.e., its wavelength). This is precisely the requirement of the de Broglie relation [cf. Eq. (1)].

The necessity of representing the momentum as a differential operator then demands that care be taken when two observables are considered simultaneously, for the order of the operations of multiplication by x and differentiation by x are not interchangeable,

$$\frac{\partial}{\partial x} x\psi \neq x \frac{\partial}{\partial x} \psi.$$

That is, the expectation value of $p_x x - x p_x$ is not zero but is, in fact, $-i\hbar$.

If F and G are two operators, then the quantity $FG - GF$ is called the commutator of F and G and is written $[F,G]$:

$$[F,G] = FG - GF. \tag{12}$$

If two operators have the same set of eigenfunctions, then the commutator is zero. The observables corresponding to these operators are then said to commute. Only commuting observables can be part of the set of observables used to specify the quantum state.

The commutator $[p_x, x] = -i\hbar$ is the operator which embodies the uncertainty principle. In terms of vector space, the eigenfunctions of position cannot also be eigenfunctions of momentum. Position and momentum are not part of the same complete set. If A is any function of p_x and x, then the commutators of p_x with A and x with A are

$$[p_x, A] = \frac{\hbar}{i} \frac{\partial A}{\partial x}, \tag{13a}$$

$$[x, A] = \frac{\hbar}{i} \frac{\partial A}{\partial p_x}, \tag{13b}$$

and similarly for the other components y and z.

II.4. THE SCHRÖDINGER EQUATION

The operator representing the kinetic energy

$$\frac{p^2}{2m} = \frac{1}{2m}(p_x p_x + p_y p_y + p_z p_z) \tag{14}$$

can now be written, considering the above discussion, as

$$\frac{p^2}{2m} = -\frac{\hbar}{2m}\left(\frac{\partial^2}{\partial x^2} + \frac{\partial^2}{\partial y^2} + \frac{\partial^2}{\partial z^2}\right) \equiv -\frac{\hbar}{2m}\nabla^2, \tag{15}$$

where ∇^2 is the Laplacian operator. The sum of the kinetic and potential energies is the total energy operator. When the potential is a function of position only, no momentum or time dependence, the total energy is

$$\hat{p}^2/2m + V(x,y,z) = E.$$

The eigenvalues of this operator are the energies of the system:

$$[(\hat{p}^2/2m)+V]\psi=E\psi. \tag{16}$$

This is the time-independent Schrödinger equation. The operator, the sum of the kinetic and potential energy, is the Hamiltonian H for the system. The time-dependent Schrödinger equation is given by

$$-\frac{\hbar}{2m}\left(\frac{\partial^2\psi(x,y,z,t)}{\partial x^2}+\frac{\partial^2\psi(x,y,z,t)}{\partial y^2}+\frac{\partial^2\psi(x,y,z,t)}{\partial z^2}\right)$$
$$+V(x,y,z,t)\psi(x,y,z,t)=i\hbar\frac{\partial}{\partial t}\psi(x,y,z,t), \tag{17}$$

or, an equivalent form,

$$H\psi=i\hbar\frac{\partial}{\partial t}\psi. \tag{18}$$

If the potential is time independent, then a time-independent equation can be obtained from the full equation by separating the variables using the substitution

$$\psi(x,y,z,t)=\Psi(x,y,z)e^{-iEt/\hbar}. \tag{19}$$

II.5. THE SCHRÖDINGER EQUATION AND CLASSICAL MECHANICS

The connection of the Schrödinger equation to classical mechanics depends on the consideration of the wave function written as $\psi=Ae^{iS/\hbar}$, where both A and S are real. Substitution into Eq. (18) yields

$$A\frac{\partial S}{\partial t}-i\hbar\frac{\partial A}{\partial t}+\frac{A}{2m}(\nabla S)^2-\frac{i\hbar}{2m}A\nabla^2S$$
$$-\frac{i\hbar}{m}(\nabla S)\cdot(\nabla A)-\frac{\hbar^2}{2m}\nabla^2A+VA=0.$$

Each of the real and imaginary terms of this equation must be equal to zero. The two resulting equations are

$$\frac{\partial S}{\partial t}+\frac{1}{2m}(\nabla S)^2+V-\frac{\hbar^2}{2mA}\nabla^2A=0, \tag{20}$$

$$\frac{\partial A}{\partial t}+\frac{A}{2m}\nabla^2S+\frac{1}{m}(\nabla S)\cdot(\nabla A)=0. \tag{21}$$

Neglecting the term containing $\hbar$ in the first equation, we obtain the familiar Hamilton–Jacobi equation for the action of the particle. This establishes the first part of the correspondence principle. The second equation, multiplied by $2A$, becomes

$$\frac{\partial(A^2)}{\partial t}+\mathrm{div}(A^2\nabla S/m)=0 \tag{22}$$

Since $\nabla S/m$ is the classical velocity (**V**) of the particle, this equation represents an equation of continuity which shows that the probability density ($\psi^*\psi=A^2$) moves according to the laws of classical mechanics with a classical velocity **V** at every point.

II.6. THE HEISENBERG EQUATION OF MOTION AND CLASSICAL MECHANICS

The mathematical formulation so far described is called the "Schrödinger picture." The basis vectors are fixed in time and the wave function for undisturbed motion is a function of time. We obtain the formal solution to the time-dependence problem by integrating Eq. (18):

$$\psi(x,t)=e^{iHt/\hbar}\psi(x,t=0),$$

whence we see that the Hamiltonian generates the time development of the wave function. The expectation of an operator F is written

$$\int(\psi e^{iHt/\hbar})^*F(e^{iHt/\hbar}\psi)\,dx\,dy\,dz.$$

Consider now the quantity $F_H=e^{-iHt/\hbar}Fe^{iHt/\hbar}$. The operator corresponds to the observable F at time t. Differentiating F_H generates its time dependence:

$$i\hbar\frac{dF_H}{dt}=[F_H,H_H]+i\hbar\frac{\partial F_H}{\partial t}. \tag{23}$$

This is the Heisenberg equation. The Heisenberg representation of an operator is related to the Schrödinger representation of the same operator by the equation

$$F_H=e^{-iHt/\hbar}Fe^{iHt/\hbar}.$$

In this "Heisenberg picture," the operators have all the time dependence, while the states are time independent. From Eq. (23), it is clear that an operator, not explicitly a function of time, which commutes with the Hamiltonian is a constant of the motion. That is, its derivative with respect to time is equal to zero. Actually, this result is true for both the Heisenberg and Schrödinger "pictures." Finally, using the commutator relation [Eq. (13)], x and p_x are written

$$\frac{dx}{dt}=\frac{1}{i\hbar}[q_i,H]=\frac{\partial H}{\partial p_x}, \tag{24}$$

$$\frac{dp_x}{dt}=\frac{1}{i\hbar}[p_x,H]=-\frac{\partial H}{\partial x}. \tag{25}$$

These are simply the Hamiltonian equations of motion found in classical mechanics. Equations (24) and (25) establish the formal correspondence between the classical Poisson bracket given by

$$\{A,H\}=\frac{\partial A}{\partial x}\frac{\partial H}{\partial p_x}-\frac{\partial A}{\partial p_x}\frac{\partial H}{\partial x},$$

and the commutator (viz.)

$$\{A,H\}\to[A_H,H_H]/i\hbar.$$

II.7. USING THE SCHRÖDINGER EQUATION

The usual task one encounters in applying quantum mechanics to a new physical situation is solving the Schrödinger equation for the given potential of that situation. Very few potentials have exact analytic solutions, and so numerical solutions are obtained, e.g., using computers to solve the differential equations. It is possible, however, that the solution of a problem with a slightly different potential is known. Then the problem's solution depends on the consideration of the small deviation (H') of the Hamiltonian as a perturbation to the Hamiltonian H_0 of the known problem. If the perturbation is small, the wave functions of the known problem are modified slightly by the addition of H' to H_0.

When we utilize the full Hamiltonian, the problem restatement is

$$H=H_0+H', \tag{26}$$

where the spectrum of H_0 is $(E_1,E_2,E_3,\ldots,E_n,\ldots)$ and its wave functions are $(\psi_1,\psi_2,\psi_3,\ldots,\psi_n,\ldots)$. To calculate the new spectrum and determine how the wave functions change, in general, the matrix of H is evaluated with the wave functions of H_0. The eigenvalues of H are the new spectrum. All the elements of H' are necessary:

$$H'_{mn}=\int \psi_m{}^*H'\psi_n\,dx\,dy\,dz. \tag{27}$$

If the original Hamiltonian does not have two identical eigenvalues, the elements H'_{mn} are interpretable in the following manner. The diagonal matrix elements H'_{mm} tell (to a first approximation) how much the energy of the state E_m moves. The new energy is equal to $E_m+H'_{mm}$. With the assumption that the system is in state n initially, the off-diagonal matrix elements, $m\neq n$, tell the rate of transitions to state m. Furthermore, if the density of states is expressed as $\rho(m)$ such that $\rho(m)dE_m$ is the number of states in the energy interval dE_m, then ω, the transition probability per unit time to one of the states in that region, is given by

$$\omega=\frac{2\pi}{\hbar}\rho(m)|H'_{mn}|^2. \tag{28}$$

This is called the golden rule.

III.1. EXTENSIONS TO THE MANY-BODY CASE

Most problems of interest in physics (including the hydrogen atom) involve more than one particle. Because of this, it is important to discuss the application of quantum mechanics in those situations. The Hamiltonian for more than one particle is the sum of the kinetic energy operators of each particle plus the total potential energy (from the external potentials and from the mutual interaction of the particles). This method applies as well to the Schrödinger equation for more than one particle. For instance, in the two-particle case, the resulting equation is

$$\left(-\frac{\hbar^2}{2m_1}\nabla^2_{(1)}+V_1-\frac{\hbar^2}{2m_2}\nabla^2_{(2)}+V_2+V_{12}\right)\Psi =i\hbar\frac{\partial}{\partial t}\Psi, \tag{29}$$

where Ψ is a function of the coordinates of particle 1 (x_1,y_1,z_1) and of particle 2 (x_2,y_2,z_2), and of the time. We obtain the time-independent equation by separating the variables using

$$\Psi=\psi(x_1,y_1,z_1,x_2,y_2,z_2)e^{-iEt/\hbar} \tag{30}$$

in Eq. (29), which yields

$$\left(-\frac{\hbar^2}{2m_2}\nabla^2_{(1)}+V_1+-\frac{\hbar^2}{2m_2}\nabla^2_{(2)}+V_2+V_{12}\right)\psi=E\psi. \tag{31}$$

A special case occurs when the particles are noninteracting.§ Then $V_{12}=0$. The equation separates into the sum of two pieces, each of which refers to a single particle. The product of the wave functions of each of the separated Hamiltonians yields the full wave equation:

$$\psi(x_1,y_1,z_1,x_2,y_2,z_2)=\psi(x_1,y_1,z_1)\psi(x_2,y_2,z_2). \tag{32}$$

Another special case occurs when the two or more particles are identical. The exchange of particles in the physical sense is equivalent to an exchange of coordinates in the wave function. An interchange of identical particles (whose coordinates¶ are ξ_1, ξ_2) produces

$$\psi(\xi_1,\xi_2)=\pm\psi(\xi_2,\xi_1). \tag{33}$$

This guarantees that the exchange of identical particles will not affect any physical properties because only the phases of the wave function are modified. The plus sign applies in all cases where the particles are bosons (possess integral spin). In all cases where the particles are fermions (possess half-integral spin), the minus sign applies. If two noninteracting fermions occupy single-particle states ψ_1 and ψ_2, the equation above implies that the full solution is

$$\Psi(\xi_1,\xi_2)=\psi_1(\xi_1)\psi_2(\xi_2)-\psi_1(\xi_2)\psi_2(\xi_1).$$

The minus sign between the terms gives the desired property when we interchange ξ_1 and ξ_2. This property has fundamental significance when we try to put two fermions in the same state. The resulting wave function is zero (Pauli exclusion principle); two fermions cannot exist in the same state. Restating this, two electrons cannot have the same quantum numbers. The development of atomic structure throughout the periodic table is a clear example of the Pauli exclusion principle. In general, for the case of two or more fermions, when ξ_1 interchanges with ξ_2, the wave function changes sign. Writing the wave function in the form of a determinant guarantees this:

$$\begin{aligned}\Psi&=N\begin{vmatrix}\psi_1(\xi_1) & \psi_2(\xi_1) & \psi_3(\xi_1)\\ \psi_1(\xi_2) & \psi_2(\xi_2) & \psi_3(\xi_2)\\ \psi_1(\xi_3) & \psi_2(\xi_3) & \psi_3(\xi_3)\end{vmatrix}\\ &=N\{\psi_1(\xi_1)\psi_2(\xi_2)\psi_3(\xi_3)+\psi_2(\xi_1)\psi_3(\xi_2)\psi_1(\xi_3)\\ &\quad+\psi_3(\xi_1)\psi_1(\xi_2)\psi_2(\xi_3)-\psi_1(\xi_3)\psi_2(\xi_2)\psi_3(\xi_1)\\ &\quad-\psi_2(\xi_3)\psi_3(\xi_2)\psi_1(\xi_1)-\psi_3(\xi_3)\psi_1(\xi_2)\psi_2(\xi_1)\}.\end{aligned} \tag{34}$$

If the particles are bosons, the wave function is symmetric under the interchange of ξ_i and ξ_j. The wave function for

§ The noninteracting problem is so much easier to solve that in most cases some approximation is used to reduce the full Hamiltonian to a more manageable sum composed of single-particle terms. Once we choose the best potential possible for each particle, it represents the "average interaction" of the given particle with all the other particles of the system. The criterion for choosing a single-particle V is that it minimize the "residual interaction" of the particle. The remainder of the Hamiltonian (i.e., the "residual interaction") is a perturbation. This is the basis of the Hartree equations.

¶ Note that the wave functions and coordinates can include the spin of the state.

the case of three particles in three states then looks like

$$\Psi(\xi_1,\xi_2,\xi_3)=\psi_1(\xi_1)\psi_2(\xi_2)\psi_3(\xi_3) + \psi_1(\xi_2)\psi_2(\xi_3)\psi_3(\xi_1)+\psi_1(\xi_3)\psi_2(\xi_1)\psi_3(\xi_2). \quad (35)$$

The modifications of the wave functions also influence the transition rates between states. The probability of a fermion transition from state i to state j diminishes by a factor $(1-n_j)$:

$$P_{j\leftarrow i}^{\text{fermion}}=(1-n_j)P, \quad (36)$$

where n_j is the occupation number of the state j. The probability of a transition into a state (here, state j) already occupied is zero by the Pauli principle. In an atom, this prevents transitions of all the electrons to the lowest Bohr orbit. There are already n_j bosons in the state j. The final state [the left-hand side of Eq. (37), see below] contains n_j+1 terms. Each term contributes a single-particle transition probability P per particle. The factor n_j+1 enhances the boson transition from state i to state j:

$$P_{j\leftarrow i}^{\text{boson}}=(n_j+1)P^{\text{single particle}}. \quad (37)$$

This property of the bosons tends to move them together. This phenomenon is responsible for superfluidity and laser action.

III.2. EXTENSION TO RELATIVITY THEORY

Until now the discussion of quantum mechanics has been limited to the nonrelativistic case where the system contains particles with velocities far below the speed of light, or equivalently, where the energies of those particles are much less than their rest mass energies, m_0c^2. The first application of relativity to quantum mechanics occurred in 1916 when A. Sommerfeld used the known relativistic variation of the mass of the electron to calculate a correction, known as the fine structure, to the Bohr theory energy levels [see Eq. (7)].

Relativity theory establishes a relationship between the electron's energy and its momentum. The Newtonian equation

$$p^2/2m=E \quad (38)$$

is replaced by the relativistic expression

$$c^2p^2+m_0{}^2c^4=E^2. \quad (39)$$

This latter equation treats the four quantities (p_x,p_y,p_z,E) as components of a single, but more abstract, quantity called a four-vector. The theory requires equal treatment of each four-vector component. As in Einstein's discussion of the space-time continuum, the three space dimensions and the time dimension are in equal apposition: The four coordinates of a particle in the space-time continuum are also a four-vector. Examining the similarities of the two sets of four-vectors suggests how to get from the time-independent Schrödinger equation (16),

$$H\Psi=E\Psi, \quad (40)$$

to the time-dependent equation,

$$H\Psi=i\hbar\frac{\partial}{\partial t}\Psi \quad (41)$$

by establishing a correlation between the four-vector analogs (x,y,z,ict) and $(p_x,p_y,p_z,iE/c)$. In this way (Secs. II.3, II.4),

$$p_x \text{ becomes } -i\hbar\frac{\partial}{\partial x},$$

$$p_y \text{ becomes } -i\hbar\frac{\partial}{\partial y},$$

$$p_z \text{ becomes } -i\hbar\frac{\partial}{\partial z},$$

$$iE/c \text{ becomes } i\hbar\frac{\partial}{\partial t}.$$

While relativity theory partially motivates the correlation between the different four-vectors, these relations are based on the Newtonian equation (38). In order to be relativistically correct, we should apply the four-vector analog to the appropriate equation (39) which yields

$$\nabla^2\psi-\frac{1}{c^2}\frac{\partial^2\psi}{\partial t^2}=\frac{m_0{}^2e^4}{\hbar^2}\psi. \quad (42)$$

This equation was first found by Schrödinger and is called the Klein–Gordon equation. The relativistically correct equation (42) has two serious deficiencies. The first is the objection to the existence of the negative-energy solutions first found in Eq. (39):

$$E=\pm[(cp)^2-m^2c^4]^{1/2}.$$

There is no simple interpretation to this. The second objection is the impossibility of using the wave function solution to Eq. (42) with the usual probability meanings. This is because the probabilities obtained are sometimes negative. By reinterpretation of Eq. (42), we eliminate these problems, but in so doing, it is no longer possible to interpret the Klein–Gordon equation as an equation for a single particle.

The origin of the above difficulties is that Eq. (42) contains a second derivative in time. In 1928, Dirac introduced an equation, linear in time, which gave only positive probabilities but did not eliminate the negative-energy solutions. The Dirac equation has a positive-energy spectrum commencing at the mass of the electron, m_0c^2, and going to positive infinity, $+\infty$. There exists, as well, a negative-energy spectrum. It begins at $-m_0c^2$, and goes to negative infinity, $-\infty$. If through a perturbation the electron makes a transition from the positive-energy to the negative-energy continuum, it would cascade downward, constantly emitting radiation. Since these transitions are not seen, Dirac reasoned that all the negative-energy states are filled. Then, by Eq. (36) no such transitions are possible. This filled negative-energy spectrum is called the Fermi sea. It pervades all of space. Excitation from the negative to the positive spectrum occurs through the addition of energy (e.g., from an x ray) to the electron in the amount of at least $2m_0c^2$. Then the excitation from a negative-energy state leaves a hole in the Fermi sea which behaves exactly as a particle having the mass of an electron, but with a charge of e. Such a particle is called a positron.

Experimental verification of the Dirac prediction (1928) came four years later from C. D. Anderson. Dirac theory also predicts the electron spin and magnetic moment, and the separation of states which are degenerate in the nonrelativistic approximation. Furthermore, the theory describes high-energy scattering and the creation of the "positron–electron pairs" by x-ray bombardment of matter. In the creation of the "pair," the total number of particles is not conserved, whereas the total charge is constant. This actually modifies the concept of the pure electromagnetic field, for the presence of an electric field alters the negative-energy solutions causing the states to readjust. Thus it is possible for the vacuum to polarize. Such an effect causes changes in the hydrogen spectrum (Lamb shift) and implies that photon–photon scattering is possible. Such scattering violates the superposition principle and represents a fundamental limitation to the classical theory of radiation.

The two relativistic equations are similar in that they start from the consideration of a single-particle equation but quickly lead to the inclusion of many-body aspects. The reformulation of these problems to deal better with their difficulties is the subject of quantum field theory.

See also ATOMIC STRUCTURE CALCULATIONS; BOHR THEORY OF ATOMIC STRUCTURE; COMPLEMENTARITY; COMPTON EFFECT; HAMILTONIAN FUNCTION; LIGHT; MOLECULAR STRUCTURE CALCULATIONS; OPERATORS; PHOTON; POSITRON; QUANTUM ELECTRODYNAMICS; ROTATION AND ANGULAR MOMENTUM; SPIN; UNCERTAINTY PRINCIPLE; WAVES.

BIBLIOGRAPHY

Elementary

Louis de Broglie, *Matter and Light,* 1st ed. (translated by W. H. Johnston). W. W. Norton and Co., New York, 1939).

Daniel T. Gillespie, *A Quantum Mechanics Primer.* International Textbook Co., Scranton, Pa., 1970.

Melvin W. Hanna, *Quantum Mechanics in Chemistry.* Benjamin, Menlo Park, Ca., 1969.

Banesh Hoffman, *The Strange Story of the Quantum,* 2nd ed. Dover Publications, Inc., New York, 1959.

Intermediate

Robert Eisberg and Robert Resnick, *Quantum Physics of Atoms, Molecules, Solids, Nuclei and Particles.* Wiley, New York, 1974.

L. D. Landau and E. M. Lifshitz, *Quantum Mechanics, Non-Relativistic Theory,* 2nd ed. (translated by J. B. Sykes and J. S. Bell), Vol. 3 in the *Course of Theoretical Physics.* Pergamon Press, Oxford, 1965.

M. Leon, *Particle Physics, An Introduction.* Academic Press, New York, 1973.

Albert Messiah, *Quantum Mechanics* Vols. I and II (translated by G. M. Temmer). Wiley, New York, 1965.

Leonard I. Schiff, *Quantum Mechanics,* 3rd ed. McGraw-Hill, New York, 1968.

George L. Trigg, *Quantum Mechanics.* Van Nostrand, Princeton, N. J., 1964.

Advanced

V. B. Berestetskii, E. M. Lifshitz, and L. P. Pitaevskii, *Relativistic Quantum Theory* (translated by J. B. Sykes and J. S. Bell), Vol. 4, parts 1 and 2, in the *Course of Theoretical Physics.* Pergamon Press, Oxford, 1971.

James D. Bjorken and Sidney D. Drell, *Relativistic Quantum Mechanics.* McGraw-Hill, New York, 1964.

P. A. M. Dirac, *The Principles of Quantum Mechanics,* 4th ed. Oxford University Press, Oxford, 1958.

Marvin L. Goldberger and Kenneth M. Watson, *Collision Theory.* Wiley, New York, 1967.

John C. Slater, *Quantum Theory of Matter.* McGraw-Hill, New York, 1951.

Philosophical Inquiries

Bryce S. DeWitt and Neill Graham, *The Many-Worlds Interpretation of Quantum Mechanics.* Princeton University Press, Princeton, 1973.

Robert G. Colodny (ed.), *Paradigms and Paradoxes, the Philosophical Challenge of the Quantum Domain,* Vol. 5, *University of Pittsburgh Series in the Philosophy of Science.* University of Pittsburgh Press, Pittsburgh, 1972.

Max Jammer, *The Philosophy of Quantum Mechanics: The Interpretation of Quantum Mechanics in Historical Perspective.* Wiley, New York, 1974.

Quantum Optics

John R. Klauder

Quantum optics deals with the quantum theory of the radiation field at or near optical frequencies, especially the interaction with matter for generation, propagation, and detection of optical fields. The quantum formulation is of necessity an extension of the classical formulation; the former is correct and more accurate than the latter for small numbers of photons per mode, but both approaches must give similar predictions for macroscopic states. The most convenient description of the classical theory is by means of *analytic signals,* which are a representation of the real field, depending on space ($\mathbf{r}$), time (t), and generally also on polarization (λ), by a complex function $V(x)\equiv V(\mathbf{r},t,\lambda)$ that contains only positive frequencies. Normalization is such that the instantaneous intensity $I(x)=|V(x)|^2$, an expression sensitive to the signal envelope and in which optical frequencies have effectively been averaged out. Just as hot bodies emit noise in the radio-frequency range, it can be expected that they also emit noise in the optical-frequency range. Consequently, it is appropriate to describe the output of an optical source not by a single wave form, but by an ensemble of wave forms each element of which fulfills the equation of propagation. For free space this equation is simply the wave equation. As with other stochastic processes, the descriptors of the system are the various correlation functions such as

$$\Gamma^{(m,n)}(x_1,x_2,\ldots,x_m;y_1,y_2,\ldots,y_n) \equiv\langle V^*(x_1)V^*(x_2)\cdots V^*(x_m)V(y_1)V(y_2)\cdots V(y_n)\rangle \quad (1)$$

for all $m,n\geq 0$, where $\langle\ \rangle$ denotes an ensemble average. For optical waves (with frequency $\nu\simeq 10^{15}$ Hz) there is no present-day detector that will follow the optical-frequency structure of $\Gamma^{(m,n)}$, and it is declared that even in principle the only measurable correlation functions are those with $m=n$. It is noteworthy for radio frequencies ($\nu\simeq 10^5$ Hz) that general m, n correlations are in principle observable.

The simplest detectors, such as photographic films and photodetectors, give responses proportional to the light intensity, at least within a large range of practical interest. By judicious use of slits and mirrors such detectors give information about the basic *mutual coherence function*

$$\Gamma^{(1,1)}(x,y) = \langle V^*(x)V(y)\rangle. \tag{2}$$

Application of the Schwarz inequality shows that the *complex degree of coherence*

$$\gamma(x,y) \equiv \frac{\Gamma^{(1,1)}(x,y)}{[\Gamma^{(1,1)}(x,x)\Gamma^{(1,1)}(y,y)]^{1/2}} \tag{3}$$

fulfills the bound $0 \leq |\gamma(x,y)| \leq 1$. The value of $|\gamma|$ represents a measure of coherence: $|\gamma| = 1$ indicates a form of full coherence; $|\gamma| = 0$ signifies a form of complete incoherence. For years the mutual coherence function was the only quantity regarding the optical radiation field that was measured, and it was therefore natural that coherence was decided on the basis of the only quantity measured.

All that changed in 1956 with the introduction of intensity interferometry by Hanbury Brown and Twiss. In their experiments measurements of

$$\Gamma^{(2,2)}(x,y;x,y) = \langle |V(x)|^2 |V(y)|^2\rangle \tag{4}$$

were made by correlating the output of two photodetectors each of which was sensitive to the signal intensity. Information about general four-field correlations follows from judicious use of slits and mirrors. Evidently utilization of a number of photodetectors, slits, and mirrors enables measurement of general correlations $\Gamma^{(n,n)}$ defined in (1), at least in principle. In practice, we generally settle for something less.

An important test of the statistics of a field is the photon-counting experiment, which measures the probability $P(m,T)$ for a counter to register m counts in a time interval T. A counter idealized to large bandwidth and zero recovery time records a count with a rate (probability per unit time) given by $\alpha|V(x)|^2$ where α denotes a counter efficiency factor. In combination with conventional statistical arguments it follows that

$$\begin{aligned} P(m,T) &= (m!)^{-1}\langle(\alpha U)^m e^{-\alpha U}\rangle \\ &\equiv (m!)^{-1}\int_0^\infty (\alpha U')^m e^{-\alpha U'}\rho_U(U')\,dU' \end{aligned} \tag{5}$$

where

$$U \equiv \int_t^{t+T} |V(x)|^2\,dt. \tag{6}$$

In the second relation in (5) use has been made of the fact that only the single positive random variable U enters, which must have some distribution function, denoted by ρ_U. Such an expression characterizes a compound Poisson process, since if ρ_U is a δ function for some value U_0, the counting distribution obeys the Poisson law. In many cases, radiation from a laser operating above threshold gives a Poisson counting distribution.

The concept of coherence must be reexamined when the measurability of higher-order correlation functions is admitted. A state of the electromagnetic field as represented by the ensemble of field functions is *fully coherent* if under the allowed measurements the results are identical to those that would arise for a *single* nonrandom field. Anything else represents *partial coherence*. Thus, for a fully coherent field, there is some nonrandom analytic signal V such that

$$\Gamma^{(1,1)}(x;y) = V^*(x)V(y) \tag{7}$$

and, for all n, that

$$\Gamma^{(n,n)}(x_1,\ldots,x_n;y_1,\ldots,y_n) = \prod_{j=1}^{n}\Gamma^{(1,1)}(x_j;y_j). \tag{8}$$

Note that the counting distribution for a fully coherent field is invariably a Poisson distribution. It should be emphasized that (7) and (8) do not imply that the ensemble really consists of a single field, since the results are invariant to overall phase transformations such as $V \rightarrow e^{-i\theta}V$. In fact it is common to assume complete overall phase uncertainty, namely, that θ is uniformly distributed. This assumption leads to the vanishing of $\Gamma^{(m,n)}$, $m \neq n$, consistent with the insensitivity of any detectors at optical frequencies.

Before the days of intensity interferometry when $\Gamma^{(1,1)}$ was the only quantity measured, Eq. (7) was the then-valid prescription for full coherence. At present a state that fulfills (7) is termed first-order coherent. More generally, a state that satisfies (7) and (8) for $n \leq N$ is said to be Nth-order coherent. The restrictions imposed by Nth-order coherence, $N < \infty$, are strong but they do not imply that the field is fully coherent. Hereafter a field that satisfies (7) and (8) for all n is simply called coherent.

Finally, before discussing the quantum theory, we consider the quantity

$$\Delta(x,y) \equiv \langle |V(x)|^2 |V(y)|^2\rangle - \langle |V(x)|^2\rangle\langle |V(y)|^2\rangle \tag{9}$$

relevant to two photodetectors. Intuitively, Δ vanishes for sufficiently large spatial or temporal separations, at least for any reasonable source. For coherent fields Δ vanishes identically, whereas for partially coherent fields Δ will generally be positive for sufficiently small spatial and temporal separations. An important class of partially coherent fields arises from thermal sources in which the field V is composed of an enormous number of independent contributions of comparable magnitude. Consequently, thermally generated fields have a normal distribution for which there is an identity such that

$$\Delta(x,y) = |\langle V^*(x)V(y)\rangle|^2. \tag{10}$$

Therefore measurement of the intensity correlation in (9) provides an indirect measurement of the modulus of the mutual coherence function. For stellar observations a direct measurement of the mutual coherence function may not be possible if the spatial separation exceeds a few meters, because of independent phase fluctuations introduced by the atmosphere. The indirect measurement provided by (9) enabled Hanbury Brown and Twiss to determine certain stellar diameters that could not be determined by more conventional means.

QUANTUM FORMULATION

The first task in a quantum analysis is the determination of an expression for the rate of multiple photon counting. Following Glauber, observe that photon detectors almost invariably operate by photon absorption, a process described quantum mechanically by means of the photon destruction operator

$$A(x)=\frac{1}{(2\pi)^{3/2}}\int \exp[i(\mathbf{k}\cdot\mathbf{r}-|\mathbf{k}|t)]a_{\lambda}(\mathbf{k})\frac{d^3k}{(2|\mathbf{k}|)^{1/2}}. \quad (11)$$

In this expression, units have been chosen so that $\hbar=c=1$, λ denotes a polarization index, and $a_{\lambda}(\mathbf{k})$ is a conventional destruction operator. In particular, if $a_{\lambda}^{\dagger}(\mathbf{k})$ denotes the adjoint operator, then the only nonvanishing commutator is

$$[a_{\lambda}(\mathbf{k}),a_{\lambda'}^{\dagger}(\mathbf{k}')]=\delta_{\lambda\lambda'}\delta(\mathbf{k}-\mathbf{k}'), \quad (12)$$

and if $|0\rangle$ denotes the vacuum or no-particle state, $a_{\lambda}(\mathbf{k})|0\rangle=0$. The space of vector states is spanned by repeated action of the creation operators $a_{\lambda}^{\dagger}(\mathbf{k})$ on $|0\rangle$.

The dominance of photon absorption in the measurement process means that the rate for single-photon absorption is essentially given by $\Sigma\alpha|\langle f|A(x)|i\rangle|^2$ where $|i\rangle$ and $|f\rangle$ denote initial and final states of the radiation field and α accounts for other factors. A summation over final states and an average over initial states is included. If α is essentially a constant, the expression for the single-photon counting rate is

$$\alpha\ \mathrm{Tr}[\rho A^{\dagger}(x)A(x)]\equiv\alpha\langle A^{\dagger}(x)A(x)\rangle. \quad (13)$$

Here $A^{\dagger}(x)$ is the adjoint operator, and while A (and thus $A^{\dagger}$) commutes with itself, $A^{\dagger}$ and A do not commute, in view of (12).

An analogous derivation of the n-photon counting rate, assuming that the detectors act independently, leads to

$$\alpha^n\ \mathrm{Tr}[\rho A^{\dagger}(x_1)\cdots A^{\dagger}(x_n)A(x_1)\cdots A(x_n)]$$
$$\equiv\alpha^n\langle A^{\dagger}(x_1)\cdots A^{\dagger}(x_n)A(x_1)\cdots A(x_n)\rangle. \quad (14)$$

This expression is a symmetric function of its arguments, but like (13) the operators are normal ordered (all $A^{\dagger}$ to the left of all A).

As in the classical case, slits, mirrors, and photodetectors provide measurement of general normal-order correlation functions of the form

$$G^{(n,n)}(x_1,\ldots,x_n;y_1,\ldots,y_n)$$
$$=\mathrm{Tr}[\rho A^{\dagger}(x_1)\cdots A^{\dagger}(x_n)A(y_1)\cdots A(y_n)]$$
$$\equiv\langle A^{\dagger}(x_1)\cdots A^{\dagger}(x_n)A(y_1)\cdots A(y_n)\rangle. \quad (15)$$

These functions provide certain information about the state of the system as represented by the density operator ρ. As before, a criterion for coherence can be adopted and its consequences analyzed. Since (15) represents an ensemble average just as much as (1), a coherent state ρ is defined as one for which there exists a function $V(x)$ such that for all n,

$$\langle A^{\dagger}(x_1)\cdots A^{\dagger}(x_n)A(y_1)\cdots A(y_n)\rangle$$
$$=V^*(x_1)\cdots V^*(x_n)V(y_1)\cdots V(y_n). \quad (16)$$

One example of a coherent state is provided by the pure "coherent state"

$$|V\rangle\equiv N_V\exp[-2i\int V(x)\dot{A}^{\dagger}(x)\,d^3x]|0\rangle \quad (17)$$

where the dot signifies time derivative and N_V is a factor ensuring normalization. For this vector

$$A(x)|V\rangle=V(x)|V\rangle,\quad \langle V|A^{\dagger}(x)=\langle V|V^*(x), \quad (18)$$

and with $\rho\equiv|V\rangle\langle V|$, (16) holds. The phase-uncertain state

$$\rho\equiv(2\pi)^{-1}\int_0^{2\pi}|e^{-i\theta}V\rangle\langle e^{-i\theta}V|\,d\theta \quad (19)$$

is also a coherent state since (16) holds.

In order to develop expressions for other than coherent states it is useful, following Sudarshan, to consider rather general linear combinations of the outer product $|V\rangle\langle V|$ that lead to density matrices in the form

$$\rho\equiv\int\phi(V)|V\rangle\langle V|\,\delta V. \quad (20)$$

The integral is to be understood as the limit as $N\rightarrow\infty$ of integrals over the first N expansion coefficients of a general V in a convenient but fixed orthonormal basis of functions. In terms of such a "diagonal representation" for ρ, the n-photon counting rate (14) is given by

$$\alpha^n\int\prod_{j=1}^{n}V^*(x_j)V(x_j)\phi(V)\,\delta V. \quad (21)$$

A similar computation for the counting distribution yields

$$P(m,T)=(m!)^{-1}\int(\alpha U)^m e^{-\alpha U}\phi(V)\,\delta V \quad (22)$$

where

$$U\equiv\int_t^{t+T}|V(x)|^2\,dt. \quad (23)$$

An alternative expression of the form

$$P(m,T)=(m!)^{-1}\int_0^{\infty}(\alpha U')^m e^{-\alpha U'}\hat{\rho}_U(U')\,dU' \quad (24)$$

holds with

$$\hat{\rho}_U(U')=\int\delta(U'-U)\phi(V)\,\delta V. \quad (25)$$

Equations (21) and (22) have the form of classical expectation values although the formulation is fully quantum mechanical. For a density matrix ρ for which $\phi(V)\geq0$ there exists a useful classical analog to the quantum theory. But expressions (21) and (22) remain valid for a density matrix ρ for which $\phi(V)$ is not strictly nonnegative. It is a remarkable fact due to the overcompleteness of the coherent states that a set of density matrices dense among all density matrices can be represented in the form (20) for smooth functions $\phi(V)$. As a consequence any density matrix ρ is expressible as the limit as $M\rightarrow\infty$ of a sequence $\rho_{(M)}$ each of which admits a diagonal representation with smooth functions $\phi_{(M)}(V)$. Convergence of such sequences can be shown to hold in the sense that

$$\mathrm{Tr}(\rho B)=\lim_{M\rightarrow\infty}\mathrm{Tr}(\rho_{(M)}B)$$
$$=\lim_{M\rightarrow\infty}\int\phi_{(M)}(V)\langle V|B|V\rangle\,\delta V \quad (26)$$

where B denotes an arbitrary bounded operator or one of a certain set of unbounded operators. The functions $\phi_{(M)}(V)$ generally converge to a kind of distribution $\phi(V)$, and with a certain but standard abuse of notation, (26) becomes

$$\mathrm{Tr}(\rho B)=\int\phi(V)\langle V|B|V\rangle\,\delta V. \quad (27)$$

Suppose further that by convention the operators $B=B(A^{\dagger},A)$ are assumed to be normal ordered. In that case

$$\mathrm{Tr}[\rho B(A^{\dagger},A)]=\int \phi(V)B(V^{*},V)\,\delta V. \tag{28}$$

Equation (28) representing the expected value of a normal-ordered operator $B(A^{\dagger},A)$ may be compared with the expected value of the corresponding classical expression $B(V^{*},V)$ in the relevant classical ensemble. Such an expectation is given by

$$\langle B(V^{*},V)\rangle=\int \rho(V)B(V^{*},V)\,\delta V \tag{29}$$

where $\rho(V)$ denotes the distribution function of the classical ensemble. The similarity of (28) and (29) is evident, and in this sense the quantum formulation admits an entirely classical-like description providing we are willing to admit field distributions that are more general than conventional distributions. For macroscopic states $\phi(V)$ is for all intents and purposes nonnegative and equivalent to a classical distribution. For states of a few photons the nonclassical behavior of $\phi(V)$ is important. The equivalence in formulation of the classical and quantum descriptions has been useful in analyzing a variety of problems both conceptually and pragmatically. This usefulness has been especially true for problems concerned with deriving the state of the radiation field from coupled dynamical equations involving source terms.

Recent developments in quantum optics have included theoretical and experimental studies of "squeezed states" of the radiation field. The principal distinction between coherent states and squeezed states relates to properties of the operator

$$X(\psi)\equiv\int [A(x)e^{-i\psi}+A^{\dagger}(x)e^{i\psi}]h(x)\,dx \tag{30}$$

for some mode function h. For coherent states the variance of X is constant in ψ, thus describing a circle as ψ varies; for squeezed states the variance of X varies with ψ and traces out an ellipse. The areas of the circle and of the (minimal) ellipse are the same in agreement with the Heisenberg uncertainty principle. As a consequence, the quantum fluctuations of a squeezed state for that ψ which minimizes the variance are less than the fluctuations of a coherent state, the output of an idealized laser. Such reduced quantum noise allows for improved precision in certain experiments, e.g., by tying the measurement time to when ψ achieves its optimal value. In addition the use of squeezed states in interferometers offers the potential of greatly enhanced sensitivity.

Squeezed states of the radiation field may be created from coherent states by evolution in a medium that induces a dilatation of the mode coordinate. For a single mode $a=a_{\lambda}(\mathbf{k})$ with a natural angular frequency ω, such a medium may be described phenomenologically by a Hamiltonian of the form $H=a^{\dagger}a^{\dagger}e^{-2i\omega t}+aae^{2i\omega t}$ which in a Schrödinger representation acts to change the variance of an initial Gaussian state. Such Hamiltonians are idealizations of strong pump, three- and four-wave mixing devices, and successful generation of squeezed states has occurred with both devices.

See also OPTICS, PHYSICAL; OPTICS, STATISTICAL; QUANTUM ELECTRODYNAMICS; STOCHASTIC PROCESSES.

BIBLIOGRAPHY

M. Born and E. Wolf, *Principles of Optics*, 5th ed. Pergamon, New York, 1975. (I)

R. J. Glauber, *Quantum Optics and Electronics* (C. DeWitt, A. Blandin, and C. Cohen-Tannoudji, eds.). Gordon & Breach, New York, 1964. (A)

J. R. Klauder and E. C. G. Sudarshan, *Fundamentals of Quantum Optics*. Benjamin, New York, 1968. (I)

H. M. Nussenzveig, *Introduction to Quantum Optics*. Gordon & Breach, New York, 1973. (I)

Quantum Statistical Mechanics

V. J. Emery

It is rare to have enough information to specify the state of a physical system completely. Even when the dynamical equations of motion can be solved, the task of making enough measurements to pick out a particular solution becomes prohibitive as the complexity of the system increases. It is then necessary to resort to a statistical method in order to calculate unbiased averages and correlations of physical quantities from whatever information can be obtained. This approach may be used for any system and it has been applied, for example, to atomic nuclei, but it is most successful when the number of degrees of freedom is very large and statistical fluctuations are correspondingly small. The most widely studied examples are macroscopic bodies such as gases, liquids, or solids at or close to thermal equilibrium, for which there are general thermodynamic relationships between physical quantities such as temperature and pressure. These relationships are not complete, and ultimately it is necessary to carry out a microscopic calculation to determine the statistical properties. This is the task of quantum statistical mechanics.

In practice, quantum effects are most important at low temperatures, where the energy of a particle due to its thermal motion is so low that its de Broglie wavelength becomes comparable in magnitude to other characteristic lengths such as the distance between the particles. They manifest themselves especially in the behavior of the "quantum fluids," liquid ^{3}He and ^{4}He, and of electrons and lattice vibrations in solids. In particular, the properties of superfluidity and superconductivity are spectacular examples of quantum mechanics on a macroscopic scale, and they can be understood by the methods of quantum statistical mechanics. At higher temperatures, it frequently is possible to use a classical approximation to describe the motion of part or all of a physical system, but quantum mechanics is truly the heart of statistical physics, for there are paradoxes and thermodynamic laws that cannot be understood on the basis of classical ideas alone.

In quantum mechanics, physical quantities are represented by operators $\hat{F}$ that have eigenvalues f_n and corresponding eigenstates $|n\rangle$. The expectation value of $\hat{F}$ in a repeated set of experiments is a weighted average of its eigenvalues:

$$\langle \hat{F} \rangle = \sum_n \rho_{nn} f_n$$

where, for simplicity of notation, it is assumed that $\hat{F}$ has a discrete spectrum. The weights ρ_{nn} are the diagonal elements $\langle n|\rho|n\rangle$ of the density operator $\hat{\rho}$ that characterizes the state of the system, and they satisfy $\Sigma_n\, \rho_{nn}=1$, for normalization. The density operator contains both dynamical and statistical information and, in general, it embodies an uncertainty in the value of physical quantities that is partly intrinsic to quantum mechanics and partly a consequence of having less than the maximum information that, in principle, could be obtained about the system in question.

The dynamical condition for $\hat{\rho}$ is known as Liouville's equation, which states that in the Heisenberg picture where operators evolve in time, $\hat{\rho}$ is a constant of the motion:

$$\frac{d\hat{\rho}}{dt} = 0.$$

This ensures that $d\langle\hat{F}\rangle/dt = \langle d\hat{F}/dt\rangle$ for every operator $\hat{F}$, so that the time evolution of the expectation value of $\hat{F}$ reflects the time dependence of $\hat{F}$ itself.

There are a number of special solutions of Liouville's equation that are particularly important in practice. The most common in quantum mechanics is a *pure state* for which $\hat{\rho}$ may be written $|\psi\rangle\langle\psi|$ or, if $|\psi\rangle = \Sigma a_n|n\rangle$, $\rho_{mn} = a_m a_n^*$. This form of $\hat{\rho}$ is a constant of the motion because in the Heisenberg picture, states do not evolve in time. The expectation value of $\hat{F}$ is $\Sigma_n|a_n|^2 f_n$, which will be recognized as the expression usually used in quantum mechanics. In this case, the information about the system is complete and any uncertainty in the value of $\hat{F}$ is intrinsic to nature. But it is unusual to have systems in pure states. In order to know that they are, it is necessary to construct a complete set of commuting observables and to measure their eigenvalues—a difficult if not impossible task that is seldom accomplished, even in the simplest situations. The density operator enables us to deal with this lack of information and may be used to describe a mixed (i.e., not pure) state of a relatively simple system or the thermodynamics of a macroscopic body.

The central concept in dealing with incomplete information is the microscopic definition of entropy or uncertainty introduced by von Neumann and by Shannon:

$$S = -k \sum_\alpha \rho_\alpha \ln \rho_\alpha$$

where k is known as Boltzmann's constant and the ρ_α are the eigenvalues of $\hat{\rho}$. By the methods of information theory it has been shown that, given a few quite reasonable assumptions about continuity and the composition of probabilities, this expression is an essentially unique specification of uncertainty. For a pure state with $\rho_{mn} = a_m a_n^*$, the ρ_α are 1 and 0 and the uncertainty vanishes, as it should. In other cases the procedure is to find the operator $\hat{\rho}$ that maximizes S subject to whatever information is available about the system. For a macroscopic body, the readily available information is usually related to general constants of the motion such as energy, number of particles, total momentum, and total angular momentum. Then the density matrix may be expressed as a function of these variables and it automatically satisfies Liouville's equation. For example, if the energy levels are E_α and the mean energy is known to be E, then by the method of Lagrange multipliers, $\hat{\rho}$ may be obtained by maximizing $s + \lambda E$ to find that

$$\rho_\alpha = \frac{e^{\lambda E_\alpha/k}}{Z}$$

and $\hat{\rho}$ has the same eigenstates as the Hamiltonian. The constant Z is known as the *partition function* and is chosen to make $\Sigma_\alpha\, \rho_\alpha = 1$, whereas λ is determined by the condition $\Sigma_\alpha\, \rho_\alpha E_\alpha = E$. Notice that in all of these summations over α, every member of a set of degenerate levels is to be included. From the minimization and the thermodynamic relation $T^{-1} = dS/dE$, which introduces the absolute temperature T, it follows that $\lambda = -T^{-1}$ and, substituting the expression for ρ_α into the definition of S using the thermodynamic definition of the free energy ($F = E - TS$), it can be seen that $Z = e^{-F/kT}$. This form of ρ_α is known as the *canonical distribution*. In practice, of course, we usually know the Lagrange multiplier T rather than the mean energy, but this interchange of independent variables is quite common in thermodynamics and does not affect the distribution.

The method is readily generalized to incorporate information related to other constants of motion and, for a set of commuting observables $\hat{A}_i$, $\hat{\rho}$ takes the form $\exp(\Sigma_i\, \mu_i \hat{A}_i)/Z$ where the μ_i are the Lagrange multipliers and Z is the corresponding partition function. A particularly important example is the *grand canonical distribution*, in which only the average number of particles is known. Then, following the method described above, we find

$$\rho_\alpha(N') = \frac{\exp[-(E_\alpha - \mu N')/kT]}{Z_G}$$

where the normalization constant Z_G is called the *grand partition function*. Here, the summation is to be carried out over states α and number of particles N'. The Lagrange multiplier μ is called the chemical potential and either μ or the mean number of particles N is specified. In practice, the choice between the canonical and grand canonical distributions is a matter of convenience because both give the same results in the thermodynamic limit $N \to \infty$, in which fluctuations of N are negligible.

From the preceding discussion it can be seen that as $T \to 0$, the states of lowest energy E_0 acquire a greater and greater weight. If E_0 is p-fold degenerate, then in the canonical distribution, $\rho_0 \to p^{-1}$ and $\rho_\alpha \to 0$ for $\alpha \neq 0$, so that $S \to k \ln p$ as $T \to 0$ and $S/N \to 0$ as $N \to \infty$ unless p is exceptionally large. The vanishing of the entropy at $T = 0$ is the content of Nernst's heat theorem, which therefore finds its microscopic justification in quantum statistical mechanics, although it is not a consequence of classical statistics.

In order to illustrate these ideas, it is simplest to evaluate the grand partition function of a noninteracting gas of particles obeying Bose or Fermi statistics. Suppose each single-particle state, labeled by subscript i, has energy ϵ_i and is occupied by n_i particles. Then the energy of this configuration is $\Sigma_i\, \epsilon_i n_i$ and the total number of particles is $\Sigma_i\, n_i$. Following the prescription described above, we obtain for the grand partition function

$$Z_G = \sum_{\{n_i\}} \exp\left[-\sum_i \frac{(\epsilon_i - \mu)n_i}{kT} \right]$$

where the summation is to be carried out over all allowed sets of integers n_i. For fermions without spin, each n_i can be 0 or 1, since Pauli's exclusion principle forbids two particles to occupy the same state. On the other hand, bosons can have n_i equal to any positive integer. The sum over n_i can then be performed for each i separately to give

$$\Omega \equiv -kT \ln Z_G$$

$$= \pm kT \sum_i \ln \left\{ 1 \pm \exp\left[-\frac{(\epsilon_i - \mu)}{kT} \right] \right\}$$

where the upper sign refers to fermions and the lower sign to bosons. The equation of state may now be obtained by using the thermodynamic relation $\Omega = -pV$ where p is the pressure and V the volume. The expression for the total number of particles is found by differentiating Ω with respect to μ:

$$N = \sum_i \frac{1}{\exp[(\epsilon_i - \mu)/kT] \pm 1}$$

and the summand in this equation is the Bose (minus sign) or Fermi (plus sign) distribution of an ideal gas.

The many important applications of these formulas, Planck's radiation law, Bose–Einstein condensation, Fermi degeneracy, and the thermodynamics of electrons and lattice vibrations in solids, demonstrate many of the peculiar features of quantum-mechanical systems. They are described in various articles in this encyclopedia. To go further, it is necessary to consider the effects of interactions and the approach to equilibrium. These topics are discussed in the sections on many-body theory and kinetic theory.

See also KINETIC THEORY; MANY-BODY THEORY; QUANTUM FLUIDS; QUANTUM MECHANICS; STATISTICAL MECHANICS; THERMODYNAMICS, EQUILIBRIUM.

BIBLIOGRAPHY

K. Huang, *Statistical Mechanics*. Wiley, New York, 1963.

L. D. Landau and E. M. Lifschitz, *Statistical Physics*. Addison-Wesley, Reading, Mass., 1969.

A. Katz, *Principles of Statistical Mechanics*. W. H. Freeman, San Francisco, 1967.

In particular, the book by Katz describes the information theory approach to statistical mechanics.

Quantum Theory of Measurement

Philip Pearle

Our knowledge of the small is obtained by observation of the large. Interaction between a microscopic system and a macroscopic apparatus is the subject of the quantum theory of measurement.

There have been two main attitudes in this field since the inception of quantum theory. One attitude *accepts the standard quantum formalism*. Various measurement processes are studied, either to elucidate their interesting features or to establish the consistency of the theory (Sections I–IV). Explanations are given of the relationship between nature and the formalism (Section V). Practitioners of this line include Bohr, Heisenberg, Von Neumann, Wigner, and Aharonov.

The other attitude is that the above program is unsuccessful in producing a satisfactory description of an individual system. Accordingly, attempts are made to *go beyond standard quantum theory* (Sections VI and VII). A modified formalism enables one to view quantum phenomena from a different perspective. Representatives of this line include Einstein, Schrödinger, deBroglie, Bohm, and Bell.

I. MEASUREMENT DESCRIPTION

The end product of a quantum theory calculation is the prediction of probabilities of possible outcomes of a measurement. One can usually do this without ever including the measurement apparatus in the calculation. (Perhaps for this reason many physicists think that quantum measurement theory is unnecessary.) But an apparatus is made out of particles, and particles are described by quantum theory. Von Neumann [1] pointed out that if quantum theory is to be *consistent,* it must be applicable to a description of the *whole* measurement process.

Thus, we are led to consider how the state $|s\rangle|A\rangle$ should evolve during a measurement; $|s\rangle$ and $|A\rangle$ are, respectively, the initial states of a system and of an apparatus. These states become *correlated* (or *entangled* to use Schrödinger's [1] more graphic term) by the measurement evolution:

$$|s\rangle|A\rangle = \Sigma\alpha_n|s_n\rangle|A\rangle \tag{1a}$$

evolves into

$$\Sigma\alpha_n|s_n\rangle|A_n\rangle \tag{1b}$$

In Eqs. (1), the physical property of the system (e.g., the spin of a particle), which the apparatus is designed to measure, can take one of the possible values $s_1 \ldots s_n$: these are the eigenvalues of an operator S with corresponding eigenvectors $|s_n\rangle$. $|A_n\rangle$ is the state of the apparatus which registers the value s_n.

The *probability rule* of quantum theory, applied to a state vector, gives the probabilities of outcomes of a future measurement, were one to take place. Applied to Eq. (1a), it gives $|\alpha_n|^2$ as the probability of the result s_n. Applied to Eq. (1b), it gives $|\alpha_n|^2$ as the probability both of the result s_n and of the apparatus registering the result s_n. These are obviously consistent, demonstrating the consistency of the probability interpretation.

Following the evolution (1), the *state-vector reduction rule* must be applied. According to standard quantum theory, the state vector evolves in two different ways. One is continuous evolution according to the Schrödinger equation, as in Eq. (1a)→Eq. (1b). The other is the reduction of the state vector. This is a discontinuous jump, following the measurement

(1b), into the state $|s_n\rangle|A_n\rangle$ consistent with the outcome of the measurement. Thus, the entangled state (1b), which does not allow a description of the physical state of the system independently of the apparatus, becomes unentangled, again allowing such a description.

More will be said later about the reduction of the state vector, which is held by some to be the Achilles heel of standard quantum theory.

It is often said that a measurement disturbs the system. This simply means that Eq. (1b) is different from Eq. (1a). If a property q (operator Q, with $[Q,S]=0$) of the system is measured before the evolution (1), the probability of the result q_n is $|\langle q_n|s\rangle|^2$; after the reduction following the evolution (1) it is $\Sigma|\alpha_n|^2|\langle q_n|s_n\rangle|^2$. These are usually different. (A notable exception is if $|s\rangle=|s_n\rangle$.)

II. MEASUREMENT MODEL

If the evolution (1) is necessary for the consistency of the rules of quantum theory, one had better be sure that such evolutions can actually take place. This leads to the construction and analysis of measurement models.

In the simplest model, the complexity of the apparatus is dispensed with in order to expose the mechanism whereby the correlation (1b) takes place. The apparatus is represented by a single particle of mass m, in an initial wave-packet state that is well localized in position and momentum (about mean values $\bar{x}$ and $\bar{p}$): call the state $|A;\bar{x},\bar{p}\rangle$. The initial state $|\psi,0\rangle=|s\rangle|A;\bar{x},\bar{p}\rangle$ evolves according to the Schrödinger equation

$$\frac{d|\psi,t\rangle}{dt}=-i\left[H_s+\frac{p^2}{2m}-g(t)Sx\right]|\psi,t\rangle, \tag{2}$$

where H_s is the system Hamiltonian. The coupling $g(t)$ between the system operator S and the particle's position x is turned on and off so rapidly that the free evolutions of the system and apparatus can be neglected during the measurement; the coupling can then be approximated by $g\delta(t)$ ($g=$constant).

The solution of Eq. (2) immediately after the measurement is $|\psi,0+\rangle=\exp(igSx)|\psi,0\rangle$ or, since x is the generator of momentum translation,

$$|\psi,0+\rangle=\Sigma\alpha_n|s_n\rangle|A;\bar{x},\bar{p}+gs_n\rangle. \tag{3}$$

Thus, the interaction exerts an impulsive force on the particle, changing its mean momentum. The initial wave packet splits into a number of different wave packets, one for each s_n. The different mean momenta $\bar{p}+gs_n$, and eventually different mean positions $\bar{x}+gs_nt/m$, act like the different positions of an apparatus "pointer." Equation (3) has the form of Eq. (1b), showing that the desired evolution has been achieved.

This model, or a variant of it, is an often employed useful tool for exploring measurement situations.

III. WIGNER–ARAKI–YANASE THEOREM

The possibility of a barrier to achieving the evolution (1) is raised by a theorem of Wigner [8]. It says that S must commute with every additive conserved quantity, or the evolution (1) is impossible. (An example of an additive conserved quantity is the x component of angular momentum $J_x=S_x+L_x$: it is additive because it can be written as the sum of the separate components S_x of system and L_x of apparatus.)

The theorem is best illustrated by an example. Suppose that $S=S_z$ is being measured: S does *not* commute with $J_x=S_x+L_x$. If the evolution (1) takes place, then $\exp(-iHt)|s_n\rangle|A\rangle=|s_n\rangle|A_n\rangle$ is a special case. The matrix element of J_x taken between two such expressions is

$$\langle A|\langle s_1|(S_x+L_x)|s_2\rangle|A\rangle=\langle A_1|\langle s_2|(S_x+L_x)|s_2\rangle|A_2\rangle, \tag{4}$$

where the left-hand side of Eq. (4) has been simplified by utilizing the conserved nature of J_x $[\exp(iHt)J_x \times\exp(-iHt)=J_x]$. When use is made of $\langle s_1|s_2\rangle=0$, $\langle A|A\rangle=1$, and $\langle A_1|A_2\rangle=0$, Eq. (4) becomes $\langle s_1|S_x|s_2\rangle=0$. But S_x's off-diagonal matrix elements *do not* vanish (else J_x would commute with $S=S_z$), so we have a contradiction.

Thus, the evolution (1) cannot take place. Does this mean that the quantum theory is consistent? There are at least three ways to avoid this conclusion.

It can be shown [9], for a large apparatus, that Eq. (1b) need only be slightly modified (the larger the apparatus, the smaller the modification): the amplitudes are slightly different from α_n, and a small term may be added that describes "erroneous" measurements (e.g., $\sim|s_1\rangle|A_2\rangle$). Or, the modification can instead be written so that the states $|s_n\rangle$ are replaced by slightly different states [10]: the measurement then has no error, but it is a "disturbing" measurement.

These two arguments imply an inconsistency in quantum theory, either of the probability rule or of the state-vector reduction rule, that is negligible FAPP ("For All Practical Purposes," a useful acronym coined by Bell [5]) only because apparatus are large.

The third argument, in the context of the above example, is that one cannot really measure S_z: what one measures is $\mathbf{S}\cdot\mathbf{N}$, where $\mathbf{N}$ is a dynamical variable that is part of the apparatus. Since $\mathbf{S}\cdot\mathbf{N}$ commutes with J_z, the theorem does not apply, and the evolution (1) occurs. This point of view, due to Aharonov (unpublished), is most satisfactory, in that there is no need for a large apparatus to make quantum theory consistent. Instead, a large apparatus is needed to make $\mathbf{N}$ point with high accuracy in the z direction throughout the measurement, i.e., to make $\mathbf{S}\cdot\mathbf{N}\approx S_z$.

This discussion illustrates how the consistency of quantum theory can be called into doubt and cleared of doubt by measurement theory considerations.

IV. TOPICS IN DISTURBANCE

Here are five examples of the use of quantum measurement theory. Examples (i), (ii), and (v) involve consistency considerations; (iii) and (iv) describe interesting unusual measurement situations.

(i) One of the first measurement analyses, by Heisenberg [1], was of a particle's position by a γ-ray microscope, in order to establish that "If there existed experiments which allowed a 'sharper' determination of p and q than $\Delta p \Delta q \approx h$ permits, then quantum mechanics would be impossible."

(ii) According to Rosenfeld [11], "...Landau and Peierls questioned the logical consistency of quantum electrodynamics by contending that the very concept of electromagnetic field is not susceptible, in quantum theory, to any physical determination by measurements." This led Bohr and Rosenfeld to examine the measurement of the electromagnetic field (system) by charged test bodies (apparatus). They contended that only the average of fields, over small space-time regions, is meaningful in quantum theory, and they had to be careful that the fields of their extended test bodies did not contaminate the results. Rosenfeld concludes that they found "...no limitation whatsoever to the definition of any single field component."

(iii) It can be argued that a continuously observed system never evolves [12]. *Continuous observation,* over the interval t, might be idealized as the $N \to \infty$ limit of a sequence of N free evolutions of $|s\rangle$ [$|s\rangle \to \exp(-iHt/N)|s\rangle$], with each evolution followed by an instantaneous measurement (and reduction) to determine if the system remains in state $|s\rangle$. The probability that the system is unchanged after one cycle is $P = |\langle s|\exp(-iHt/N)|s\rangle|^2 \approx 1 - \langle s|(H-\bar{H})^2|s\rangle(t/N)^2$, where $\bar{H} \equiv \langle s|H|s\rangle$. Because P is quadratic to lowest order in t/N, the probability that the system remains in the state $|s\rangle$ at time t as -$N \to \infty$ is $P^N \to 1$. This has been called watched-pot behavior ("the watched pot never boils") or the quantum Zeno's paradox. In more realistic models of the continuous interaction of a system with an apparatus, the evolution out of the state $|s\rangle$ is slowed down but not stopped.

(iv) However, it is possible to measure continuously a time-dependent quantity [corresponding to a Schrödinger operator $S(t)$, and Heisenberg operator $S_H(t) = \exp(iHt)S(t)\ \exp(-iHt)$] without measurements at times t and t' interfering, provided $[S_H(t), S_H(t')] = 0$. An example, for a harmonic oscillator, is $S(t) = x \cos \omega t - (p/m\omega) \sin \omega t$. This is called a *quantum nondemolition* measurement [13].

(v) When considering application of the reduction rule, the nature of the apparatus must be taken into account. If the apparatus is "too simple" (e.g., made from one particle as in Section II), premature reduction of (1b) could result in wrong predictions. Suppose (1b) evolves further and a subsequent measurement is made. This measurement outcome can depend upon *interference* between the states into which the states $|\alpha_n\rangle|A_n\rangle$ evolve. It is plausible [14] that with a "sufficiently complex" apparatus, whether man-made or a natural environment, such interference is unmeasurable FAPP so it does not matter when one chooses to reduce (1b). But consistency FAPP is not fundamental consistency. It appears to some that standard quantum theory has a flaw: the circumstances under which the state-vector reduction rule is to be applied are not precisely defined.

V. *INTERPRETATIONS*

Understanding state-vector reduction is often regarded as the central problem of quantum measurement theory. Various approaches to this problem are reflected in the various interpretations given to the state vector.

In Born's [15] interpretation, the state vector corresponds to human knowledge of the state of the system. In Bohr's [1] interpretation the state vector represents the possible responses of the system to any experiment that might be performed. State-vector reduction is naturally suited to these interpretations: it represents the updating of knowledge or the termination of a measurement. But some find these interpretations unsatisfactory. Why should it not be possible to describe nature itself, without people? Why should there be no description of nature in between measurements?

Interpretations have been given to quantum theory with no reduction rule, so that only Schrödinger evolution takes place. In Einstein's [16] interpretation, the state vector corresponds to an ensemble of identical objects. In Everett's [17] interpretation, the state vector describes an ensemble of real universes, only one of which we occupy. Einstein's own criticism may be applied to both: why cannot we have a theory in which the state vector describes the *individual reality* we see around us?

Quantum theory, without the reduction rule, does not allow this latter interpretation. This was made vivid by Schrödinger's "cat paradox" [1]. It is a hypothetical experiment in which a decaying atom with a half-life of 1 h is coupled by a lethal gadget to a cat, resulting after 1 h in a state vector

$$|\psi\rangle = (1/\sqrt{2})|\text{atom undecayed, cat alive}\rangle + (1/\sqrt{2})|\text{atom decayed, cat dead}\rangle \quad (5)$$

This cannot represent the state of an individual cat.

In what follows, attempts to impose the description of individual reality on quantum theory will be discussed.

VI. *HIDDEN VARIABLES THEORIES*

One can imagine that there is a classical point particle dancing around in space inside a wave packet. That this picture, or more generally, the classical picture of a point representing many particles dancing around in configuration space, is consistent with quantum theory is shown by theories of deBroglie [6], Bohm [1,2], and Nelson [18].

These theories employ classical interpretations of the imaginary and real parts of the Schrödinger equation. One is simply conservation of probability, $\partial\rho/\partial t + \nabla \cdot \mathbf{J} = 0$. The other is equivalent to Newton's second law with a strange kind of force. A deBroglie–Bohm particle undergoes a completely causal evolution under this "quantum force." An ensemble of such particles at any time has the quantum position probability density ρ, with particle velocity at a point given by $\mathbf{J}/\rho$, provided this is so initially. The Nelson theory enjoys a similar success in producing ρ, although a particle is subjected to the force of a randomly fluctuating "background field," so ρ satisfies a diffusion equation.

These theories may be criticized in that the source of the quantum force or background field is not given, and neither is the mechanism whereby the correct initial conditions are established. Such criticisms miss the point, which is that these theories explicitly demonstrate the falsity of Bohr's assertion [19], "...in quantum mechanics we are not dealing with an arbitrary renunciation of a more detailed analysis of atomic phenomena, but with a recognition that such an analysis is *in principle* excluded."

However, the picture of nature that emerges from "a more detailed analysis of atomic phenomena" may present aspects dramatically different from classical physics. Following the trail of an argument by Einstein, Podolsky, and Rosen [1], Bell [20] considered two widely separated particles in an entangled state. He showed that certain predictions of quantum theory cannot be reproduced by a hidden variables theory unless the particles interact nonlocally, a conclusion, he remarked, "Einstein would have liked least."

Bell's result stimulated experimental tests [21]. Does nature agree with the predictions of quantum theory or the predictions of a local hidden variables theory? The result, perhaps not too surprisingly, is that the quantum theory predictions are vindicated. But this does not validate Bohr's assertion. Rather, it shows that "a more detailed analysis" requires models that have the remarkable, but not unimaginable, property of nonlocal interaction.

VII. STATE-VECTOR REDUCTION AS A PHYSICAL PROCESS

Schrödinger [1] objected to the postulate of state-vector reduction following a measurement because "...observation is a natural process like any other and cannot *per se* bring about an interruption of the orderly flow of events." This suggests modifying Schrödinger's own state-vector evolution so that state-vector reduction is an "orderly" physical *process,* not a *postulate*. In this approach, individual reality is simply represented by the state vector; there is no attempt to introduce point particles as in Section VI.

In continuous state-vector reduction (CSR) [22] theories, a randomly fluctuating term is added to the Schrödinger equation so that the state vector in Eq. (1b) continues to evolve to one of the states $|s_n\rangle|A_n\rangle$ (the state depends upon the fluctuations). The predictions of quantum theory are obtained from the simple requirement that the modified Schrödinger equation must ensure the Martingale property $d\langle|\alpha_n(t)|^2\rangle_{\mathrm{av}}/dt=0$ ($\langle\ \rangle_{\mathrm{av}}$ denotes the average over all fluctuations). Since reduction takes place (i.e., $|\alpha_n(t)|^2\to 1$ or 0 as $t\to\infty$), the Martingale property implies that

$$|\alpha_n(0)|^2=\langle|\alpha_n(\infty)|^2\rangle_{\mathrm{av}}=1\cdot\mathrm{Prob}(|\alpha_n(\infty)|^2=1)+0\cdot\mathrm{Prob}(|\alpha_n(\infty)|^2=0). \quad (6)$$

Thus, $\mathrm{Prob}(|\alpha_n(\infty)|^2=1)=|\alpha_n(0)|^2$, which is the quantum theory prediction.

The process described by a CSR theory should display the following necessary behavior. Applied to a system in a superposition of *microscopically* distinguishable states such as Eq. (1a), the process should have negligible effect. Applied to a superposition of *macroscopically* distinguishable states such as Eq. (1b) [or the cat of Eq. (5)], the process should rapidly effect a reduction to one of the states $|s_n\rangle|A_n\rangle$, with the quantum probability $|\alpha_n|^2$.

A proposal, different from CSR, which however satisfactorily displays this behavior, was made by Ghirardi, Rimini, and Weber [23]. This spontaneous localization (SL) theory operates in the following way. Consider a particle in a "canonical two-packet state," a superposition of two wave packets separated by a distance much greater than 10^{-5} cm with each packet width much less than 10^{-5} cm. Randomly, of the order of once in 10^{16} s per particle, the wave function is suddenly "hit," multiplied by a Gaussian of width$\approx 10^{-5}$ cm. The nonlocal hit leaves essentially only one packet remaining, if the center of the Gaussian lies near that one packet. Moreover, this is most likely because the probability of a hit is proportional to the wave function squared norm after the hit, which is large only if the center of the Gaussian lies near a packet.

Now, this process scarcely affects an individual particle because it is so infrequent. However, the hit of any *one* particle of the apparatus in the state (1b) reduces the *whole* wave function. This can occur rapidly; for example, if the apparatus contains 10^{23} particles, a hit will occur on average in 10^{-7} s.

It has been possible to modify the Schrödinger equation so that the above reduction mechanism works in continuous fashion [22,23]. The modified Schrödinger equation for a single particle in this continuous spontaneous localization (CSL) theory is

$$d\psi(\mathbf{x},t)/dt=-iH\psi(\mathbf{x},t)+(w(\mathbf{x},t)-\lambda)\psi(\mathbf{x},t). \quad (7)$$

$w(\mathbf{x},t)$ is a randomly fluctuating white-noise function whose time average is 0, but it is likely to have the same value at nearby points, as it has a Gaussian correlation function of characteristic length 10^{-5} cm; $\lambda\approx 10^{-16}\ \mathrm{s}^{-1}$, and H is the usual Hamiltonian.

As the wave function $\psi(\mathbf{x},t)$ evolves, its norm changes. In addition to Eq. (7), the theory is defined by a probability rule: the probability that $\psi(\mathbf{x},t)$ occurs is equal to its squared norm $\int dx|\psi(\mathbf{x},t)|^2$ multiplied by the probability of occurrence of the function $w(\mathbf{x},t)$ that generates $\psi(\mathbf{x},t)$. [It is a consequence of Eq. (7) that the total probability $\langle\int dx|\psi(\mathbf{x},t)|^2\rangle_{\mathrm{av}}$ equals 1.] The wave function can be normalized at any time, as Eq. (7) is linear.

According to Eq. (7), during each dt the wave function is hit by infinitesimal Gaussians, and the result is added to $\psi(t)$ to give the wave function $\psi(t+dt)$. With overwhelming likelihood, the net effect on a canonical two-packet wave function is that one packet grows large while the other packet diminishes in size. Thus, the CSL theory reduces the wave function continuously. Its many-particle extension possesses the same mechanism of rapid reduction for a macroscopic object enjoyed by the SL theory.

There are experiments which, in principle, allow determination of whether the physical process of reduction actually takes place [24].

These are examples of how quantum theory of measurement considerations may give rise to new physical ideas.

See also ERROR ANALYSIS; QUANTUM MECHANICS; UNCERTAINTY PRINCIPLE.

REFERENCES

Good elementary books

P. C. W. Davies and J. R. Brown (eds.), *The Ghost in the Atom*. Cambridge University Press, Cambridge, 1986.

N. Herbert, *Quantum Reality: Beyond the New Physics*. Doubleday, New York, 1985.

A. I. M. Rae, *Quantum Physics: Illusion or Reality*. Cambridge University Press, Cambridge, 1986.

E. Squires, *The Mystery of the Quantum World*. Hilger, Bristol, 1986.

Collections of Papers

1. J. A. Wheeler and W. H. Zurek (eds.), *Quantum Theory and Measurement*. Princeton University Press, Princeton, 1983.
2. L. E. Ballentine, "Resource Letter IQM-2: Foundations of quantum mechanics since the Bell inequalities," *Am. J. Phys.* **55,** 785 (1987).
3. D. M. Greenberger (ed.), *New Techniques and Ideas in Quantum Measurement Theory," Annals N.Y. Acad Sci.* **480** (1986).
4. R. Penrose and C. J. Isham (eds.), *Quantum Concepts in Space and Time*. Clarendon, Oxford, 1986.
5. A. Miller (ed.), *Sixty-two Years of Uncertainty: Historical, Philosophical and Physics Inquiries into the Foundations of Quantum Mechanics*. Plenum, New York, 1990.
6. B. d'Espagnat (ed.), *Foundations of Quantum Mechanics*. Academic, New York, 1971.
7. J. S. Bell, *Speakable and Unspeakable in Quantum Mechanics*. Cambridge University Press, Cambridge, 1987.

Articles and Books

8. E. Wigner [1], p. 298.
9. H. Araki and M. Yanase [1], M. Yanase [1,6]; G. C. Ghirardi, F. Miglietta, A. Rimini, and T. Weber, *Phys. Rev.* **D24,** 347–353 (1981).
10. A. Shimony and H. Stein [6]; T. Ohira and P. Pearle, *Am. J. Phys.* **56,** 692 (1988).
11. L. Rosenfeld [1], p. 477; N. Bohr and L. Rosenfeld [1].
12. T. Sudbery [4], and references therein.
13. C. M. Caves, in *Quantum Optics, Experimental Gravitation and Measurement Theory* (P. Meystre and M. O. Scully, eds.). Plenum, New York, 1981.
14. A. Daneri, A. Loinger, and G. M. Prosper [1]; H. D. Zeh [1,6], Joos [3]; Zurek [3] and references therein.
15. M. Jammer [2], p. 38.
16. P. Pearle [2]; L. E. Ballentine [2].
17. H. Everett [1]; B. S. DeWitt [6]; B. S. DeWitt and N. Graham [2].
18. E. Nelson [3]; see references therein.
19. N. Bohr [1], p. 43.
20. J. S. Bell [1–3,5–7]; M. Redhead, *Incompleteness, Nonlocality and Realism*. Clarendon, Oxford, 1987; M. Horne, "Hidden Variables" in this encyclopedia.
21. A. Aspect and coworkers [1–4]; references in M. Horne [20].
22. P. Pearle [3,4] (for CSR), Ref. 5 for CSL.
23. G. C. Ghirardi, A. Rimini, and T. Weber (Ref. 5).
24. A. Zellinger [4]; P. Pearl [22].

Quarkonium

Elliott D. Bloom

Quarkonium is a simple atom made up of a heavy quark and its antiparticle (or antiquark) held together by the strong color force. Such a simple system of two objects bound together by a fundamental force provides a powerful experimental tool to investigate the characteristics of the objects and the force that holds them together. Numerous examples of such systems have been used in the past. The earth–moon system has been used to observe the gravitational force. The hydrogen atom, consisting of an electron and a proton bound together by the electromagnetic force, has been central in the development of quantum mechanics.

Quarks are elementary spin-$\frac{1}{2}$ particles (fermions) that are the fundamental building blocks of matter. Five flavors (different kinds) are now known: u, d, s, c, and b. The last two, c and b, are by far the most massive, with b being about three times more massive than c. A bound system of a quark and an antiquark constitutes a meson. However, experiments using "ordinary" mesonic states made up of the lighter u, d, and s quarks can yield only indirect information about the quarks and the interquark forces, because the mass of these quarks is comparable to the binding energy that holds the quarks together in the meson. As a result, the quarks in an ordinary meson move with a speed close to light, and calculations of their properties must be done quite exactly to be reliable. Such calculations for ordinary mesons are currently too difficult to be practical.

Mesons consisting entirely of the heavy quarks, c and b, are ideal in this respect because they can be quite accurately modeled, and hence the special name **quarkonium**. The c anti-c (denote by $\bar{c}$) and $b\bar{b}$ mesons can be produced in large numbers by existing accelerators. The $c\bar{c}$ mesons are called charmonium (c stands for charm), and the $b\bar{b}$ mesons are called beautonium (b stands for beauty). In these quarkonium systems, the binding energy of the quarks is small compared to the quark mass, and so the quarks move at a small fraction of the speed of light. Approximate methods of analysis are quite accurate in such cases, particularly for beautonium.

The use of charmonium as a test bed for the color force was initiated with the discovery of the J/ψ meson in 1974, the first of a spectrum of $c\bar{c}$ bound states, with a mass (in energy units) of 3095 MeV. The mass of the proton is about 30% the mass of the J/ψ. The meson was found almost simultaneously by two groups of experimenters. One group, led by Samuel Ting at the Brookhaven National Laboratory, named the new particle J. The other group, led by Burton Richter at the Stanford Linear Accelerator Center (SLAC), called it ψ; hence the name J/ψ. Richter and Ting jointly received the 1976 Nobel prize in physics for the discovery. The first of an even larger spectrum of beautonium states, a meson named Υ, was discovered in 1977 by a team, led by Leon Lederman, at the Fermi National Accelerator Laboratory (Fermilab). The mass of the Υ is 9460 MeV, or about 10 times the mass of the proton.

The general features of the quarkonium spectrum are illustrated in Fig. 1. The figure shows photon transitions be-

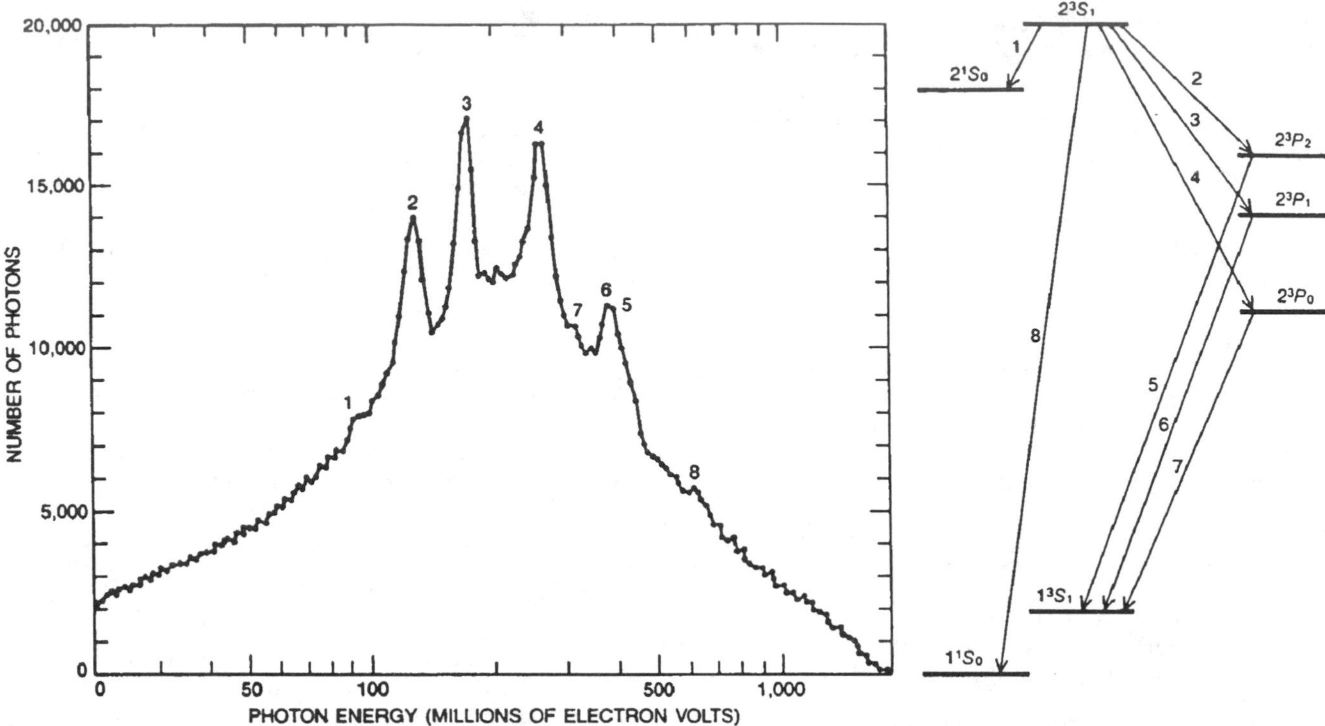

FIG. 1. Photon transitions between states of the charmonium system have been mapped with the Crystal Ball detector. Number of photons vs. photon energy, in MeV, is plotted. Only the $2\,^3S_1$ state is made directly in this experiment; the other states are formed when the $2\,^3S_1$ state decays by emitting a photon, which is detected by the Crystal Ball. The peaks in the spectrum of photons are keyed by the number to the transitions in the diagram of the charmonium spectrum (also included in the figure). The strength of the photon lines as well as their positions in photon energy are currently modeled with high accuracy.

tween the states of the charmonium system as mapped with the Crystal Ball detector, and a diagram of the charmonium spectrum. This detector is a NaI(Tl) device optimized to study the photon transitions in quarkonium, and the data shown were obtained in SLAC in 1978–1980. Analogous to the case of the hydrogen atom, the quantum-mechanical motions of the quark–antiquark in quarkonium result in a series of discrete stationary states (each a meson). These states have a narrow energy spread, both absolutely as compared to ordinary mesons and relatively as compared to their energy spacing (difference in masses), leading to a relatively large photon transition rate among many of the states. These features are due to the nature of the strong color force that binds quarkonium, as well as to the nature of the quarks themselves.

The photon line spectrum observed in Fig. 1 delineates the stationary states. Note that the photon energies involved are about 10^8 greater than in the case of hydrogen, but the underlying descriptions of the two systems have a great deal in common. The states of quarkonium are labeled by the spectroscopic notation developed for atomic physics. For example the $1\,^1S_0$ state consists of a c and $\bar{c}$ with zero relative angular momentum (S), antialigned spins (singlet), and principal quantum number equal to 1. (For hydrogen, the binding energy of the electron in a given stationary state is propor-

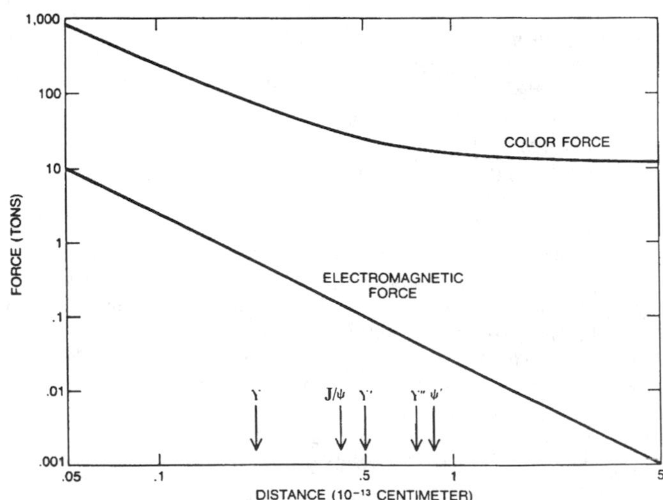

FIG. 2. The force between quarks has been mapped by studies of quarkonium. Distance between the quark and antiquark, in units of 10^{-13} centimeters, vs. force, in tons, is plotted. In addition to the strong color force is plotted the form of the electromagnetic force between the quarks that is also acting. Shown for reference is the size of a few of the quarkonium states used in the mapping. The states shown are $n\,^3S_1$ states and the primes denote n value (principal quantum number) with unprimed being $n = 1$.

tional to $1/n^2$ to good approximation, where n is called the principal quantum number.) The $1\,^3S_1$ state has aligned spins (triplet), and the $2\,^3S_1$ has principal quantum number equal 2. The P states have one unit of angular momentum; states with higher angular momentum are denoted by $D, F, \ldots,$ as in the case of the atomic spectroscopic notation.

The spectroscopy of beautonium has also been extensively studied by a number of experiments (including the Crystal Ball). The results of more than a decade of work have delineated details of quark properties and the strong color force. Figure 2 summarizes our present knowledge about the force between quarks. This force seems to obey a law significantly different from that describing the electromagnetic force, both in strength and in the dependence on the distance between quarks. At exceedingly short distances, as shown in the figure, the color force also seems to follow an inverse-square law as does electromagnetism. However, beyond 10^{-13} centimeters the force has a constant value (of about 16 tons) independent of the distance. The latter feature results in permanent confinement of the quarks in mesons, i.e., free quarks do not exist.

Particle physicists expect another and much heavier quark to exist, the so-called t quark (t stands for truth). It has yet to be discovered and experiments indicate that the t mass must be greater than 70 000 MeV. The discovery of truthonium may allow experiments to probe to even smaller distances in exploring the fundamental nature of quarks and the color force that binds them.

BIBLIOGRAPHY

Elliott D. Bloom and Gary J. Feldman, "Quarkonium," *Sci. Am.* **246**(5), 66–77 (1982).

Robert N. Cahn (ed.), e^+e^- *Annihilation: New Quarks and Leptons*. The Benjamin/Cumming Publishing Co., Redwood, CA, 1985.

For the most recent information on the charmonium (J/ψ, ψ', . . .) and beautonium (Υ, Υ', Υ'', . . .) properties, see "Review of Particle Properties," *Phys. Lett.* **B239** (1990).

Quarks

R. K. Adair

Before 1950, there was definitive evidence for the existence of only two sets of elementary particles which took part in the strong or nuclear force. There were two baryons, the neutron and proton, which seemed to differ only in their electric charges and electromagnetic properties, and there were three mesons, the positive, neutral, and negative pions (the π^+, π^0, and π^- mesons), which were also thought to differ only in their electric charges. According to the Yukawa model, the nuclear force between the baryons was analogous in important ways to the electromagnetic force between two charged objects. The strong force between the baryons was considered to result from the exchange of mesons between the baryons even as the electromagnetic force between two charged particles was known to be mediated by the exchange of photons between the two charges. Just as the strength of the electromagnetic forces between two objects is proportional to the product of the charges held by the objects, the strength of the force between two baryons was considered to be proportional to the product of the strong interaction charges held by the two baryons. Electric charge was observed to be conserved in all processes; the number of baryons was also seen to be conserved. However, photons, the quanta of light, were observed to be emitted and absorbed by electric charges; analogously, mesons, considered to be the quanta of the strong interactions, were observed to be emitted and absorbed by baryons. The nuclear or strong interactions thus seemed to be similar to the electromagnetic forces but stronger at small distances while weaker at large distances. The greater strength at small distances was considered to follow from the larger size of the strong-interaction unit charge, whereas the weaker strength at large distances followed from the short range of the force that would follow naturally—as shown by Yukawa—from the large mass (about 270 times that of the electron) of the π mesons.

However, in spite of the many similarities between electromagnetism and the strong interactions as described in this manner, there were also profound differences. The electric field at any point must be described by three numbers which transform as the components of a vector; the electromagnetic field was a vector field. However, the strong interaction field as described in this picture was determined by but one number (like a pressure field); the strong field was a (pseudo)scalar field. The equivalent statement in quantum field theory was that the spin of the photon—the quantum of the electromagnetic field—was $1\hbar$, whereas the spin of the pion, the quantum of the strong field was $0\hbar$.

However, by 1960, this simple attractive picture of the strong interacting particles, collectively called hadrons, was in disarray. Many more baryons and many more mesons had been discovered, some holding a new quality conserved in strong interactions which is now labeled *hypercharge*. Could so many particles be fundamental? Certainly some simpler unifying description of hadrons was necessary.

Such a simpler description of hadrons as elements of a group, with the technical label $SU(3)$, was found in 1961 by Murray Gell-Mann and by Yuval Ne'eman. Previously, quite obvious patterns of light nuclei that followed from the approximate equivalence of the *two* particles—the neutron and proton, differing only in charge—which made up these nuclei were described by the group $SU(2)$. Eugene Wigner showed that somewhat broader and less transparent patterns of these nuclei, conforming to the group $SU(4)$, followed from the equivalence of *four* nearly equivalent "particles" that made up those nuclei—protons with spin up and with spin down, and neutrons with spin up and with spin down.

Although a few physicists explored the possibility that existence of the $SU(3)$ patterns (or *supermultiplets*) of hadrons and mesons indicated that a "fundamental" set of *three* particles must exist, the specific schemes that were proposed using known particles as the triplet did not work and the most attractive base triplet from a mathematical view would result in the assignment of fractional electric charges of $-\frac{1}{3}e$ and $+\frac{2}{3}e$, where e is the unit electric charge. To the few physicists that had reached that point, the fractional charges seemed to rule out such a basic set. Then, quite independently, Gell-Mann and George Zweig made the bold suggestion—buttressed by detailed arguments—that such a triplet of fractionally charged particle was quite probably real. Gell-Mann then named the particles *quarks* (from a phrase in *Finnegans Wake,* by James Joyce). Zweig called them *aces*, but quarks they became.

According to the description of Gell-Mann and of Zweig, there were three quarks, now labeled *u* (for up), *d* (for down), and *s* (for strange), where the *d* and *s* quarks have electric charges of $-\frac{1}{3}$ and the *u* quark has a charge of $+\frac{2}{3}$. All quarks are spin-$\frac{1}{2}$ fermions. The three different quarks are equivalent in their strong interactions differing only in their masses; the *u* quark is a few MeV/c^2 heavier than the *d* quark, whereas the *s* quark is a few hundred MeV/c^2 heavier than the *u* and *d*. The names of the *u* and *d* quarks followed from certain mathematical similarities of the set of two with the spin-up and spin-down states of spin-$\frac{1}{2}$ fermions such as the electron or proton; "strange" followed a historic nomenclature now abandoned.

In this quark description, the hadrons are combinations of three quarks, whereas the mesons are combinations of a quark and an antiquark. Here, the proton is basically the combination *uud*, whereas the neutron is *ddu*; the π^+ is $u\bar{d}$, the π^- is $d\bar{u}$, and the π^0 is made up equally of the pairs $u\bar{u}$ and $d\bar{d}$ where the overbar signifies antiparticle. Hypercharge was then understood as essentially the property of containing an *s* quark and hypercharge was conserved by the strong interactions even as the different quark varieties (called *flavors*) are conserved by the strong interactions.

In the following decade, numerous experiments failed to discover free quarks in bulk matter or to detect their production in the interactions of high-energy particles from accelerators and in cosmic rays. However, an array of experimentally observed properties of elementary particles was explained in terms of quark properties. N. Cabbibo demonstrated that the weak (beta-decay-type) decays of elementary particles could be explained, rather precisely, as the decays of their quark constituents. The beta decay of the neutron (*udd*) to the proton (*uud*) with the emission of an electron and a neutrino is then the consequence of the decay of a *d* quark, found in the neutron, to a *u* quark with the emission of an electron and a neutrino.

There was also important evidence of another kind that made the "particle" identification of quarks more secure. Measurements of the scattering of very high-energy electrons, accelerated by the Stanford Linear Accelerator (SLAC), from neutrons and protons demonstrated that the nucleons contained, or were made of, point-like charged particles which scattered the electrons strongly. Through analyses of the magnitude of the scattering, which depends upon the size of the charges held by the partons, the partons were identified as the fractionally charged quarks.

Many less striking results were also fitted by the quark hypothesis and helped to define the character of quarks and their interactions. However, with all of the positive results, several puzzling problems remained. The quark model of the baryons required as many as three identical particles to lie in the same quantum state (or orbit), violating the Pauli exclusion principle. The weak interaction of the *s* quark seemed most singular and the absence of the weak decays to neutral combinations of leptons (the absence of weak neutral currents) seemed arbitrary and anesthetic. And despite considerable experimental effort, free quarks were not found.

Later work, which enlarged the set of known particles, clearly demonstrated the existence of two more quarks. The work of S. Ting and his colleagues at Brookhaven National Laboratory, and of B. Richter and coworkers at SLAC, revealed in 1974 the existence of a fourth quark, labeled *c*, with an electric charge of $+\frac{2}{3}$. This quark, with a mass about 1.5 times that of the proton, is a kind of partner of the *s* quark even as the *u* and *d* quarks are partners. With the introduction of this partner to the *s*, the absence of certain observed neutral weak interactions was understood (as a cancellation of *u–d* and *s–c* contributions) and the unnaturally lonely position of the *s* quark was removed. The symbol *c* stands for "charm"—for no very good reason.

Then in 1947, L. Lederman and colleagues led a search at Fermilab (the Fermi National Accelerator Laboratory) that uncovered a fifth quark, *b* for "bottom" in parallel with "down," with a charge of $-\frac{1}{3}$ and a mass about equal to 5 times the proton mass. Even as the *d* with a charge of $-\frac{1}{3}$ has a (slightly) heavier partner *u*, with a charge of $+\frac{2}{3}$, and the *s* and the heavier *c* form a similar partnership, we are confident that the *b* has a heavier partner *t* (for "top" in parallel with "up"), with a charge of $+\frac{2}{3}$ which has not yet been found (as of 1989). The mass of the *t* must be as large as 100 times the mass of the proton (i.e., a silver atom) and may be twice as heavy.

About this time a series of experiments—first at the CEA (Cambridge Electron Accelerator) then, most definitively, at SLAC—measured the ratio *R* of the production of hadrons to the production of muon pairs in the collisions of electrons and positrons. In a useful approximation, the probability of a pair of particles—quark and antiquark, or negative muon and positive (anti)muon, is proportional to the square of the charge of the particles. The production of hadrons was found to be three times as great as suggested by the simple model of *u*, *d*, *c*, and *s* quarks (at the collision energies achieved, the heavier *b* and *t* quarks would not contribute). This augmented other evidence to the effect that there were three kinds of each of the six flavors of quark. There were three, almost identical, varieties of the *d* quark, of the *u* quark, etc. Even as the kind of quark (*u*, *d*, *s*, etc.) was described as the "flavor," the three new varieties of each quark were labeled as "colors." Hence, instead of 6 quarks, there were 18; *u-red, u-yellow, u-blue, d-red,* etc., each holding one of six flavors and one of three colors. There are also 18 antiquarks with 6 antiflavors and 3 anticolors. (Obviously, quark flavor has nothing to do with gustatory flavor nor are quark colors related to visual colors.)

With three colors, the occupation of the same state by three quarks of the same flavor, which had been observed, no longer violated the Pauli exclusion principle; the three quarks had different flavors and, hence, were not identical.

There are now three known generations of quark families—with quite different mass scales. How many generations are there in all? We have only explored the mass region under 100 GeV and have not yet even found the second member, the t quark, of our third-generation family. Even as we know of three generations of quarks, we have seen three families, three generations, of leptons. Leptons are spin-$\frac{1}{2}$ fermions that do not take part in the strong interactions (that is, they do not hold any color charge) but play a role in the weak interactions similar to that of the quarks. There are six leptons: the electron and electron–neutrino, the muon and muon–neutrino, and the tau and tau-neutrino. The electron, muon, and tau have a charge of -1, the neutrinos all have 0 charge.

Even as the mass of the quarks making up the different quark families increases strongly as one goes from the u–d family to the c–s family to the t–b family, the mass of the charged lepton member of the lepton pairs varies similarly. The mass of the electron is 0.53 MeV, the muon mass is 105 MeV, and the tau mass is 1785 MeV; all of the neutrino masses are small and possibly zero. We now believe that these sets of quarks and sets of leptons are probably conjoined in three generations of quark–lepton families; (u,d,e^-,ν_e), (c,s,μ^-,ν_μ), and (t,b,τ^-,ν_τ). We have ways to determine the number of neutrino generations and if these families are joint quark–lepton sets as we believe, this would also serve to set the number of quark generations.

There are (at least) two important measures of the number of neutrino generations. Astrophysicists have determined this number through application of their understanding of cosmology and the early universe and the knowledge we have of the relative abundances of primordial hydrogen, deuterium, and helium. In that early universe (after the first few minutes), the energy of three generations of neutrinos and antineutrinos would account for about 49% of the energy of the universe with the photons taking 51% (the mass of the relatively few nucleons can be neglected). If there were four generations, the neutrino energy be 56% of the total energy leaving the photons 44%, etc. The abundances of deuterium and helium are sensitive to the energy partition, and the measured values of those relative abundances then establish limits on the number of generations of neutrinos—and through the quark-lepton family conjecture, the number of generations of quarks. The best fit to the data seems to suggest that the three generations that we know is all that nature has provided, and there are certainly no more than five generations.

A second method requires no confidence in models of the early universe. The Z^0 boson, which mediates the weak interactions, decays equally to the different varieties of quarks and antiquarks and leptons and antileptons that have masses small compared to the 91-MeV mass of the Z. If there are more generations of neutrinos, there will be more ways for the Z to decay, the decay rate will be larger, and the complementary energy width of the state will be larger. Present measurements of that width at SLAC appear to be consistent with three generations of neutrinos but not with any number greater than five.

The description of the quark–quark forces that led to stable baryon systems of three quarks and of stable meson systems of quark and antiquark (but not of other combinations) now was less similar to electromagnetism than the nucleon–pion model on the surface, but was much closer in fundamentals since both descriptions now took the form of vector fields. According to this model, the quarks—with their color charge—emit and absorb massless spin-one *gluons* in analogy to the emission and absorption of massless spin-1 photons by particles with an electric charge. Moreover, just as charge is conserved in the charge-photon electromagnetic interaction, both charge and flavor are conserved in the quark–gluon strong interaction.

Moreover, even as two electric charges attract or repel according to the charge combinations (like repel, unlike attract), different color combinations attract and repel. Of the nine color combinations possible for a quark and antiquark, eight related combinations (a color octet) repel and one combination (a color singlet) attracts. Indeed, only color singlets lead to attractive forces. Since color singlets can be formed from quark–antiquark combinations and from three-quark combinations—and not from other primitive ensembles—we find only three-quark baryons and quark–antiquark mesons in nature.

With all of the similarities, there is an important difference between the gluon-mediated strong interaction between quarks holding color charge and the photon-mediated electromagnetic interaction between particles holding electric charge: The gluons carry color charge while photons are electrically neutral. The force between two electrical charges falls off as the inverse square power of the distance between the charges; there are no electrical forces exerted on the photons which can then be considered—in a somewhat simplistic picture—to be emitted isotropically by each charge, hence the inverse square dependence with distance of the electric field strength and the force between two charges. For quarks in an attractive color singlet state, the gluons are attracted by the quark charges and by the color charges held by the other gluons. As a consequence, the gluon field between quarks forms a kind of flux tube of relatively constant cross section where the field strength is independent of distance. Hence, the force between the two quarks is independent of the distance between the two particles and it takes an infinite energy to separate the two particles to an infinite distance. Hence quarks cannot be separated; quarks are *contained* and we can never expect to see a free quark or, for similar reasons, a free gluon.

Since the strong force between quarks is almost independent of the distance between the quarks, the force is much larger than the electromagnetic force at moderate distances though the value of the color charge seems to be about the same as the value of the electric charge. At very small distances, vacuum polarization effects of the kind that increase the effective electric charge at small distances are countered by forces between the emitted gluons in such a manner that the total force between quarks is sharply decreased. At very small distances, reached by very high interactions, the effective force is almost zero, an effect called *asymptotic freedom*.

In the early 1970s work by Steven Weinberg, Sheldon Glashow, Abdus Salam, and others, led to an understanding of the weak and electromagnetic forces as two closely related parts of a unified electroweak whole. Almost immediately it became clear that the quark–gluon strong color interactions could also be joined to the electroweak interaction in a *Standard Model* of the three forces. In this view, held by almost all physicists today, quarks and leptons are closely related source or matter particles and the photon, the W^+, W^-, and Z^0 (quanta of the weak interactions), and the gluon are closely related field or force particles. Although we are not so certain that an equally complete understanding of the place of gravity in the anticipated unified description of matter and force is yet in hand, we feel that the goal of Einstein, to understand all of the forces and particles of nature as aspects of just one unified force, may be in sight.

See also FIELD THEORY; HADRONS; PARTONS; QUANTUM ELECTRODYNAMICS; STRONG INTERACTION; $SU(3)$ AND HIGHER SYMMETRIES; WEAK INTERACTIONS.

Quasars

E. Margaret Burbidge

Quasars (quasistellar objects, QSOs) are astronomical objects characterized by the following properties: they appear star-like and are usually rather faint on sky-survey photographic plates; they are extragalactic, i.e., they lie outside our galaxy or Milky Way; they emit nonthermal continuum radiation which can be at all frequencies (radio, infrared, optical or visual, ultraviolet, x ray). Their UV, optical, and near-IR spectra show thermally produced broad emission lines of the common elements. These lines are shifted from their laboratory wavelengths toward longer wavelengths by large amounts. Many quasars also show thermally produced absorption lines in their spectra, shifted from their laboratory wavelengths toward longer wavelengths by amounts that are usually less than the shifts exhibited by the emission lines. The quasars are commonly variable on short time scales and by large amounts. Variability has also been detected in the intensities of the emission lines in some quasars, and in the intensities of the absorption lines in a few cases. The continuum radiation may be polarized and, if so, the angle and amount of polarization is usually variable. The nonthermal character of the optical radiation causes quasars in general to have excess blue and ultraviolet radiation and sometimes excess infrared radiation as compared with stars. The radiation is thought to be due to high-energy electrons interacting with magnetic fields and is called synchrotron radiation.

DISCOVERY

Accurate positions of small-diameter radio sources tabulated earlier in the 3C (Third Cambridge) catalogue led during 1960–1963 to the tentative identification of some sources with blue stellar objects (BSOs) having properties unlike normal stars. The spectrum of the relatively bright (12.8 mag) BSO identified with strong radio source 3C 273 revealed broad emission lines of hydrogen, and showed that 3C 273 is extragalactic and resembles the nuclei of active galaxies, e.g., some radio galaxies and Seyfert galaxies (*see* GALAXIES). The nuclei of such active galaxies contain diffuse hot gas excited by a central nonthermal energy source that generates high-energy particles responsible for the radio emission and nonthermal continuun. Following the identification as quasars of many radio sources catalogued by various radio observatories worldwide, searches for BSOs and unusual infrared stellar objects led to the discovery of a large number of radio-quiet quasars. Photographic survey plates taken through objective prisms or grating/prisms, which provide low-resolution spectra of all stars, etc., in fields many square degrees in size, have been used to identify many more radio-quiet quasars, which are more numerous than radio-emitting quasars by a large (still uncertain) factor. Some x-ray-emitting quasars have been discovered serendipitously by the Einstein High Energy Astronomical Observatory (HEAO) in fields around known quasars which were being studied by the HEAO. The total number of known quasars for which optical spectroscopy has been obtained is about 5000 (as of 1989). Quasars are usually catalogued by their coordinate designations, although many also have names deriving from radio-source catalogues or from other means of discovery.

REDSHIFTS

The shifts of the spectral lines from their laboratory values λ_0 to the observed values λ are denoted by $z=(\lambda-\lambda_0)/\lambda_0$. They are usually attributed to the Doppler effect (*see* DOPPLER EFFECT) due to motion of the quasar away from the observer. For small $\lambda-\lambda_0$, we have $z=v/c$, where v is the line-of-sight velocity and c is the velocity of light. For large z, however, the special relativistic relation must be used:

$$\frac{v}{c}=\frac{(1+z)^2-1}{(1+z)^2+1}.$$

For galaxies, there is a good correlation between redshifts and apparent brightness, i.e., between redshifts and distance. Following Hubble, this is interpreted as due to uniform expansion of the universe and enables redshift to be used as a measure of distance (*see* HUBBLE EFFECT). Quasars, however, do not show such a clear-cut correlation, and it is not universally accepted that their redshifts in all cases have the same cause as those of galaxies, although they are usually interpreted thus. The distribution of redshifts, $N(z)$, peaks at about $z=2.2$, and declines steeply for larger z, although (as of 1989) there are about a dozen quasars with z greater than 4. The quasar with the largest known redshift (as of 1989) is 2203+292 (coordinate designation), which is radio emitting and has $z=4.399$. If due to a Doppler shift, this corresponds by the formula above to a recession velocity $v=0.93c$.

ABSORPTION LINES

In 1966 a quasar with $z\approx2$ was found to have narrow absorption lines in its spectrum, in addition to the broad emission lines. Gas that is relatively cooler than the source of the continuum radiation must be present in the line of sight from observer to quasar to produce absorption lines in the spectrum of a quasar. Since the first discovery in 1966, almost all large-redshift quasars have been observed to have absorption lines; in some cases there are many, and a few low-redshift quasars also have them. The absorption lines, when identified, have been found to be due to resonance lines of the same common elements as produce the emission lines, but at many redshifts usually with $z_{abs}<z_{em}$. Some quasars have dozens of such sets of absorption lines. Two further absorption-line phenomena have been observed. Some quasars (possibly 10% of all) have broad absorption lines in their spectra adjacent on the short-wavelength side to the corresponding emission lines, and these extend to decreasing wavelengths that, on the Doppler wavelength-shift interpretation, imply that the absorbing gas is flowing out from the quasar at velocities $\sim c/10$ or larger. This phenomenon resembles on a larger scale that seen in the spectra of supernovae (explosions of stars in final stages of stellar evolution with the massive ejection of gas). The second observed spectrographic phenomenon is the presence, on the short-wavelength side of the resonance Lyman-α emission line of hydrogen, of large numbers of (mostly) narrow absorption lines, named "the Lyman-α forest." These are due to gaseous clouds in the line of sight to the quasars. Some of these hydrogen lines are accompanied by absorption lines of the common heavier elements, probably in lower abundances relative to hydrogen than the solar abundance ratios. The broad absorption lines or absorption troughs are currently interpreted as due to outflowing gas from the quasar and in some cases have been shown to have time-variable intensities. The narrow absorption lines at $z_{abs} \ll z_{em}$ have usually been interpreted as due to galaxies, galactic haloes, or intergalactic clouds lying in the line of sight between the quasar and observer. The origin of the "Lyman-α forest" is controversial (as of 1989) and the subject of current research, but these hydrogen lines, or at least many of them, may be primordial gas left over from the formation of the quasar.

PHYSICAL MODELS

No completely satisfactory physical model for the quasars has yet been constructed, although the amount of observational data gathered, especially during the past 10 years, is large. The source of the enormous energy output ($\sim 10^{47}$ erg/s if the redshifts are cosmological and indicate distance) is thought to be the gravitational collapse of a large mass, perhaps into a black hole (*see* BLACK HOLES). The means by which such collapse leads to generation of high-energy particles has not been satisfactorily worked out. Good models exist, however, for the ionized gas which produces the broad emission lines. The central energy source is believed to be surrounded by numerous clouds which produce the broad emission lines (the broad emission line region or BELR). The spread in velocity of these clouds is ~ 1000 km/s and may be due to rotation, random motions, or radial (in and out) motions, or any combination thereof. The gas composing these clouds is believed to be photoionized by the central source; temperatures are $\sim 30\,000$ K and electron densities $\sim 10^9$ cm^{-3} or possibly higher. The chemical composition is similar to the Sun's. The time-variability of emission lines observed in several cases has set rather stringent limits to the size of the BELRs, ~ 1 light-year or possibly even less. Quasar spectra show, in addition to the broad emission lines, narrow emission lines due to "forbidden" transitions in the common elements, i.e., transitions for which dipole radiation is forbidden by the quantum-mechanical selection rules. Such radiation can only occur in low-density regions, where the electron density is too low for collisional depopulation of the excited levels to occur. Forbidden lines in quasar spectra have been observed to remain constant over many years, and are believed to be produced in a large low-density region enveloping the BELR. The broad absorption lines (absorption "troughs" or BALs) that have been observed adjacent to the broad emission lines and extending toward shorter wavelengths are interpreted as due to outflowing gas, with velocities ranging up to $0.1c$ relative to the BELR or even larger in some cases. The mechanism of acceleration of gas up to these velocities is not understood. The outflowing gas does not completely cover the source of continuum radiation and may be in a cone or beam, but the geometry is not understood. Some BALs show much structure, and variability on a time scale of several years has been detected. The outflowing gas at such highly supersonic velocities is likely to be unstable, and the structure in BALs and the changing structure observed in some cases may indicate breakup of the flow due to such instabilities.

Narrow absorption lines can occur either at redshifts close to the redshifts of the corresponding emission lines, or at $z_{abs} \ll z_{em}$. The former are thought to be related to the BAL phenomenon and are produced in gas associated with the quasar, with abundances of heavier elements relative to hydrogen probably somewhat smaller than the Sun's. Those at $z_{abs} \ll z_{em}$ are described in the section entitled "Absorption Lines."

Those quasars which show very rapid continuum variability and large, variable polarization probably emit beamed radiation. Various hypotheses have been put forward for producing collimated beams of charged particles and radiation. Observations at radio frequencies using the Very Long Baseline Interferometer technique (VLBI) have detected expanding structure which, if the quasars are assumed to be at the cosmological distances implied by the expanding universe interpretation of their redshifts, requires expansion at velocities apparently exceeding the velocity of light. However, satisfactory theoretical explanations have been formulated to avoid "superluminal" velocities, although these do require highly relativistic, directed velocities at near the velocity of light.

GRAVITATIONAL LENSING

Some close pairs of quasars with identical spectra have been discovered. These have been interpreted as images of a single quasar, undergoing gravitational lensing by a massive galaxy in or very nearly in the line of sight from quasar to observer. The bending of a light ray passing close to a massive object was predicted by Einstein and verified by the very small displacement of stellar images when their light passes close to the Sun during solar eclipse. The theory of gravitational lensing can, depending on the orientation of quasar, lensing object, and observer, produce multiple images. The lensing object is difficult to observe in the presence of relatively bright quasar images, but this has been achieved and it is generally accepted that gravitational lensing of quasar images does occur.

ASSOCIATION WITH GALAXIES

The relationship of quasars with galaxies is the most controversial aspect of the study of quasars. If all redshifts of quasars are Doppler shifts due to the expansion of the Universe, then the apparent cutoff in redshift between $z=4$ and $z=5$ suggests that this corresponds to the principal epoch of quasar formation some 10–15 billion years ago, depending on the cosmological model. Whether quasar formation precedes, follows, or occurs as an alternative path to galaxy formation is in the realm of cosmogony rather than astrophysics. Among the low-redshift quasars, some have been found near small groups of galaxies having about the same redshift as the quasar, indicating that the standard interpretation of the quasar redshift is correct and the quasars are physically associated with the galaxies. Also some quasars are surrounded by faint diffuse luminosity corresponding to a galaxy in which the quasar is embedded. Quasars have not, however, been found to be associated with rich clusters of galaxies. Some statistical studies have indicated association of high-redshift quasars with galaxies of much smaller redshift, and some cases of bridges that appear to connect such low-redshift galaxies with high-redshift quasars are known. If evidence along these lines persists and grows, then another cause for the redshifts is needed. It has long been known that gravitational redshifts are insufficient, but there is currently no generally accepted theory of noncosmological redshifts for quasars.

See also BLACK HOLES; DOPPLER EFFECT; GALAXIES; HUBBLE EFFECT; SYNCHROTRON RADIATION.

BIBLIOGRAPHY

Monograph (Discovery, Early Observations, etc.):

G. Burbidge and M. Burbidge, *Quasi-Stellar Objects*. W. H. Freeman, San Francisco, CA, 1967.

Books (General):

D. W. Weedman, *Quasar Astronomy*. Cambridge University Press, Cambridge, 1986.

H. Arp, *Quasars, Redshifts and Controveries*. Interstellar Media, Berkeley, CA, 1987.

Catalogs:

A. Hewitt and G. Burbidge, "A New Optical Catalog of Quasi-Stellar Objects." *Astrophys. J.* (*Suppl.*) **63,** 1 (1987).

A. Hewitt and G. Burbidge, "The First Addition to the New Optical Catalog of Quasi-Stellar Objects." *Astrophys. J.* (Suppl.) **69,** 1 (1989).

Conference Proceedings:

J. S. Miller (ed.), *Astrophysics of Active Galaxies and Quasi-Stellar Objects*. University Science Books, Mill Valley, CA, 1985.

G. Swarup and V. K. Kapahi (eds.), *Quasars,* International Astronomical Union Symposium No. 119. Reidel, Dordrecht, 1986.

J. A. Zensus and T. J. Pearson (eds.), *Superluminal Radio Sources,* Proc. Workshop at Big Bear Solar Observatory, California. Cambridge University Press, Cambridge, 1987.

P. Osmer, A. Porter, R. Green, and C. Foltz (eds.), *Proceedings of a Workshop on Optical Surveys for Quasars*. Astronomical Society Pacific Conference Series, 1988.

Popular Articles:

R. Preston, "Beacons in Time: Maarten Schmidt and the Discovery of Quasars," *Mercury* **17,** 2 (1988).

G. Burbidge, "Quasars in the Balance," *Mercury* **17,** 136 (1988).

Technical Reviews with Full Bibliographies:

R. J. Weymann, R. F. Carswell, and M. G. Smith, "Absorption Lines in the Spectra of Qausi-Stellar Objects," *Annu. Rev. Astron. Astrophys.* **19,** 41 (1981).

D. E. Osterbrock and W. G. Matthews, "Emission-Line Regions of Active Galaxies and QSOs," *Annu. Rev. Astron. Astrophys.* **24,** 171 (1986).

Quasiparticles

P. W. Anderson

The concept of quasiparticles in solid-state and many-body physics is closely related to high-energy physicists' idea of a "physical" or "dressed" *renormalized* particle, as opposed to the '"bare" or "undressed" particles that appear in the unrenormalized theory. The idea is that very often a many-body system will have excitations that behave like free elementary particles but are actually very complicated in their internal structure, which hides possibly large effects of interparticle interaction. The clearest exposition of the idea is Landau's in his "Fermi liquid theory" [1,2], in which he proves that at sufficiently low temperatures an interacting sea of Fermi particles such as electrons or He_3 atoms behaves in many essential respects as though it were composed of weakly interacting free fermions, with a well-defined, sharp Fermi surface and a constant mass (not equal to the bare one) and spin susceptibility (also not equal to the bare one). He introduces a quasiparticle distribution function $n(p,\sigma)$ for quasiparticles of momentum p and spin σ, and assumes that overall $n(p,\sigma)$ differs very little from the value for a perfectly free Fermi gas at $T\to 0$,

$$\begin{aligned} n_0(p,\sigma) &= 1, \quad p<p_F, \\ &= 0, \quad p>p_F, \end{aligned} \tag{1}$$

where p_F is the Fermi momentum. We can then write the energy of the total system as a power series in $\delta n = n - n_0$:

$$\begin{aligned} E(\delta n) = E_0 &+ \sum_{p,\sigma} E_p \delta n(p,\sigma) \\ &+ \sum_{pp'\sigma\sigma'} F(p,p',\sigma,\sigma')\delta n(p,\sigma)\delta n(p',\sigma') + \dots, \end{aligned} \tag{2}$$

and Landau shows that only these terms play a role as $T\to 0$. The quantities F depend only on the directions of p and p', not on their distance from the Fermi surface, and can be expressed in terms of a relatively few "Landau parameters," which determine the renormalizations of mass, susceptibility, etc. The important idea is that (2) is valid and useful even though the distribution of actual real, "bare" Fermi particles does not resemble (1) at all, as is true in liquid He_3, for instance.

The concept has been generalized to many other situations, the most important being the BCS theory of superconductivity, where Schrieffer's book [3] gives a beautiful exposition of the quasiparticle concept. In the BCS theory the quasiparticle states are elementary equivalent fermions that are neither pure electrons nor pure holes but a linear combination of both. Nucleons in the shell model of the nucleus can also be understood as occupying quasiparticle states.

In general, the quasiparticle is a fermion; it is a special case of the concept, also due to Landau, of an "elementary excitation" of a many-body system, which may be either a boson (phonon, roton, plasmon, spin wave, etc.) or a fermion. By very general arguments [4] it can be shown that the very lowest-energy excited states of any such system will usually take on the very simple form of an apparently noninteracting gas of free objects, which can be bosons, fermions, or both.

See also Excitons; Phonons; Plasmons; Polaron; Quantum Electrodynamics; Renormalization; Superconductivity Theory.

REFERENCES

1. L. D. Landau, *Sov. Phys. JETP* **3**, 920 (1956); **5**, 101 (1957).
2. For an elegant, simple and complete review, see P. Nozières, *Interacting Fermi Systems*. Benjamin, New York, 1963.
3. J. R. Schrieffer, *Theory of Superconductivity*. Benjamin, New York, 1964.
4. P. W. Anderson, *Concepts in Solids*. Benjamin, New York, 1962.

Radar

Maurice W. Long

The word *radar* is an acronym for "radio detecting and ranging." Most radars emit radio waves in short bursts (pulses), with a relatively long interval between pulses. Then, they detect the pulses which have been reradiated from objects (targets). The time delay between the transmitted and received pulses is a measure of the distance to the target. Target bearing is usually determined from knowing the pointing direction of a narrow radio-wave beam produced by the radar antenna.

The civilian and military applications of radar for surveillance, navigation, air-traffic control, weather tracking, early warning, and missile guidance are well known. Radar is also a major tool for scientific investigations of the earth's atmosphere, the ionosphere, and the planets. Applications include mapping; assessment of crops, forests, and flood disasters; topographic studies for water resources and minerals; and iceberg and ocean-surface surveillance.

The first radars used wavelengths of several meters, but by 1940 microwaves were being used. Now there are radars also operating at millimeter and optical wavelengths. The advantage of the shorter wavelengths is narrower beam widths for a given overall antenna size. Disadvantages include greater loss of signal strength because of increased atmospheric absorption and scattering plus various equipment limitations. Some side-looking airborne radars use the echo Doppler frequency caused by the aircraft motion to reduce the along-track, i.e., azimuthal, beam width. These radars are said to have "synthetic apertures" because the apparent beam widths are narrower than the diffraction-limited beam widths.

The maximum detection range for microwave radar is limited by the earth's curvature, and for normal atmospheric conditions the radar horizon exceeds the optical horizon by 7%. There is a renewed interest in decimeter waves because they propagate over the horizon via ionospheric refraction. With these ionospheric "bounces," ground-based radars have sensed ocean-surface conditions over huge areas, at distances of several thousand kilometers. In fact, such over-the-horizon (OTH) radars are important elements of early-warning defense systems for a number of nations.

Radar imagery is complementary to photographs obtained with infrared, visible, or ultraviolet waves. Since the atmosphere is essentially transparent for microwaves, radar permits acquisition of information under "all weather," day or night conditions that are impossible with aerial photography. This has permitted several million square kilometers of South America, usually concealed by a heavy cloud cover, to be mapped. Although in many respects similar to a photograph, radar imagery highlights different features. For example, radar is sensitive to vertical dimensions and, therefore, it emphasizes topographic features. Airborne radar has proved useful in studying oceans and their currents and for applications in hydrology, mineral exploration, topographic mapping, land-use monitoring, agriculture, and forestry.

The U.S. National Aeronautics and Space Administration (NASA) has supported the development of a variety of radars for airborne and space-based applications. A space-based radar mapped the Moon during the Apollo 17 flight in December 1972. The first space radars used to sense the earth were altimeters on the satellites Skylab and GEOS-3, launched in 1973 and 1975, respectively. A significant step toward providing large-scale features of the earth began in June 1978 with the launch of the NASA SEASAT satellite. It housed three radars specifically designed for ocean surveillance (a precision altimeter for topography and wave heights, a "scatterometer" for measuring radar-echo strength and thereby estimating wind speeds at sea, and a synthetic aperture radar for depicting sea surface contours).

Spaceborne synthetic aperture radars (SARs) with fine range and cross-range resolution have provided useful images of the earth. As already noted, the first spaceborne SAR was aboard SEASAT, a free-flying satellite. The later SARs—SIR-A (1981) and SIR-B (1984)—were aboard the Space Shuttle and are called Shuttle Imaging Radars. The next step in the series is SIR-C, with its first flight scheduled for August 1991. This will be a joint United States/Germany/Italy experiment with three radars: two U.S. radars operating at L (1.2 GHz) and C (5.3 GHz) bands, and a German/Italian radar, called X-SAR, operating at X band (9.6 GHz). These U.S. radars can use various combinations of horizontal and vertical polarizations for transmit and receive. For X-SAR vertical polarization is used both for transmit and receive.

Radar images have proven to be particularly useful for surface mapping, e.g., geological exploration and terrain classification, of features dependent on surface roughness and/or slope variation. Additionally, useful imagery has been obtained for ocean-surface waves and their patterns. Other significant developments include imagery that show the feasibility of precision, large-scale polar ice mapping and tracking. Also, effects of elevated soil moisture due to rainfall have been seen on images over large regions of the midwestern U.S. farm belt, where the terrain is flat. At the opposite extreme of water content, SIR-A penetrated sands in hyperarid regions of the eastern Sahara Desert and the images revealed radar echoes from previously unknown subsurface features such as buried valleys, geologic structures, and possibly Stone Age occupation sites. The SIR-B radar provided for the first time stereo images from space that have been used to generate topographic contours and perspective views of terrain.

Electromagnetic waves are highly attenuated (attenuation increases with an increase in frequency) when propagating through the earth, due to moisture and dense objects. Even so, ground probing at short ranges has proved useful for a number of civilian applications. Typically a ground-probing radar is portable or mobile, uses an antenna mounted at or near the ground, and has high-range resolution through use of an extremely wide bandwidth (e.g., a few hundred megahertz). Example applications include:

- Detection of buried pipes (metallic and nonmetallic) and utility cables
- Archeological mapping and geophysical prospecting
- Location of voids and determination of other subsurface conditions under highways and bridges
- Detection of hidden objects and voids in walls and tunnels
- Measurement of ice thickness and location of permafrost

Well-known to the television viewer are the real-time weather-radar reflectivity distributions of rain, which have been available to research meteorologists since the mid-1940s. Since the late 1960s pulse Doppler radars have provided additional research data in the form of velocity maps. These radars provide, as a function of spatial coordinates, the "weather" velocity relative to the radar (radial velocity), and the spread in this radial velocity. Mean velocity can highlight a potentially dangerous condition, and the spread in velocity is a measure of turbulence. Thus, the microwave beam of Doppler radar can be used, day or night, to penetrate clouds, rain showers, and thunderstorms to reveal the dynamical structure within an otherwise unobservable event. In fact, an attribute of Doppler radar is the provision of advance warning on the development of tornados via the detection of thunderstorm cyclones. A large-scale network of Doppler weather radars, called Nexrad, is planned for installation in the early 1990s with sites in the continental United States, Alaska, Hawaii, Puerto Rico, and certain overseas military bases.

See also DOPPLER EFFECT; MICROWAVES AND MICROWAVE CIRCUITRY.

BIBLIOGRAPHY

D. K. Barton, *Modern Radar System Analysis.* Artech House, Norwood, MA, 1988. (I,A)

J. B. Cimino, B. Holt, and A. H. Richardson, "The Shuttle Imaging Radar B (SIR-B) Experiment Report," Jet Propulsion Laboratory Publication 88-2, March 15, 1988. (I)

D. J. Doviak, D. S. Zrnic, and D. S. Sirmans, "Doppler Weather Radar," *Proc. IEEE* **67**, 1522 (1979). (I)

C. Elachi, "Radar Images of the Earth from Space," *Sci. Am.* **54** (December 1982). (E)

K. Iizuka and A. Freundorfer, "Detection of Nometallic Buried Objects by a Step Frequency Radar," *Proc. IEEE* **71**, 276 (1983). (I)

Jet Propulsion Laboratory, "The Second Spaceborne Imaging Radar Symposium (April 28–30, 1986)," JPL Publication 86-26, December 1, 1986. (A)

M. W. Long, *Radar Reflectivity of Land and Sea.* Artech House, Norwood, MA, 1983. (I,A)

S. P. Parker (ed.), *Encyclopedia of Science and Technology,* 6th ed. McGraw-Hill, New York, 1987. (E)

M. I. Skolnik, "Fifty Years of Radar," *Proc. IEEE* **73,** No. 2, (1985). (I)

R. Watson-Watt, *The Pulse of Radar.* The Dial Press, New York, 1959. (E)

Radiation Belts

J. A. Van Allen

A *magnetosphere* is that region surrounding a planet within which its magnetic field has an important role in physical phenomena involving electrically charged particles. A *radiation belt* is a more limited, interior feature of a magnetosphere comprising a population of energetic, electrically charged particles (electrons, protons, and heavier atomic ions) durably trapped in the external magnetic field of the planet. The term energetic in this context conventionally means kinetic energy $E \geq 30$ keV. A radiation belt is of toroidal shape, encircling the planet, with its axis of rotational symmetry coincident with the magnetic dipolar axis of the planet. To a first approximation, each particle therein moves with constant energy and independently of all other particles along a helical path encircling a magnetic line of force, subject only to the Lorentz force of a static magnetic field on a moving electrically charged particle. The pitch of the helix is the steepest at the magnetic equator and becomes 90° at its mirror or reflection points at the northern and southern ends of its latitudinal excursion. The helix drifts slowly in longitude so as to generate the overall toroidal shape of the trapping region. An isolated, nonradiating particle has an infinite residence time in the simplified, idealized case of motion in a vacuum in a dipolar magnetic field, as shown theoretically by Stoermer in 1905. Departures from this case in real magnetospheres are attributed to the presence of thermal and nonthermal ionized gas (plasma) which causes a large variety of cooperative physical phenomena.

The magnetospheric properties of a planet are an essential part of its gross, phenomenological character. Also, the energetic particle population may place important constraints on the practicality of *in situ* measurements and on the survival of electronic and optical equipment, human flight crews, animals, and other living materials flown therein.

Natural radiation belts are *not* comprised of radioactive nuclei, nor does the population of energetic particles shield the planet from external radiations.

The diverse particle phenomena in the Earth's magnetic field have been studied intensively, both observationally and theoretically, since the discovery of their existence in 1958 by James Van Allen of the University of Iowa. In addition, a series of artificial radiation belts were produced by the United States and Russia in 1958 and in 1962. The energetic particles (principally electrons) in these artificial belts were the decay products of radioactive fission nuclei injected into the magnetic field by nuclear bomb bursts at high altitudes.

The entire body of knowledge on both natural and artificial radiation belts and the associated body of plasma physical phenomena at Earth define a prototypical planetary magnetosphere.

The physical mechanisms for the creation of magnetospheric phenomena are of an electromagnetic nature. The minimum condition for the existence of a planetary radiation belt is that the planet's dipole magnetic moment be sufficiently great that the flow of the solar wind (tenuous ionized gas from the solar corona, flowing outward through the solar system) is arrested before it reaches the top of the appreciable atmosphere or surface of the planet. Durable trapping of charged particles is possible only if this condition is met. But even when the condition is not met, important plasma physical phenomena occur—as have been observed near the Moon, Mercury, Venus, Mars, and the comets Giacobini–Zinner and Halley.

Beyond the elementary consideration mentioned above, there are very complex physical processes of thermalization and capture of the solar wind (plasma), convective transport of it by the combination of the planetary magnetic field and induced electric fields, and the acceleration and diffusion of charged particles by fluctuating electric and magnetic fields—all of which contribute to the development of the total magnetospheric system. A reasonable level of understanding of all of these processes at Earth has been achieved and certain scaling principles can be used as guidelines for conjecture on the nature of the magnetospheres of other planetary bodies. Of the seven planets thus far investigated, only Jupiter, Saturn, Uranus, and Earth have well-developed radiation belts.

The radiation belts and other features of Earth's magnetosphere are shown to approximate scale in the noon–midnight meridian plane cross section of Fig. 1. The inner and outer radiation belts are two distinct features, defined by the intensity of particles capable of penetrating a specific shield ($\approx$1 g cm^{-2} of aluminum). In a generalized sense, there are as many different radiation belts as there are different species of particles and energy ranges that one wishes to distinguish. The principal sources of particles for the outer belt are the solar wind and the ionosphere; and for the inner belt, electrons and protons from the in-flight radioactive decay of neutrons from nuclear reactions produced by galactic cosmic rays and solar energetic particles in the tenuous gas of the upper atmosphere. The eventual fate of magnetospheric particles is either to become part of the atmosphere or to escape into space. The first two sources of particles are responsible for most of the gross geophysical manifestations of the magnetosphere (aurorae, geomagnetic storms, and heating of the upper atmosphere). The third is responsible for the relatively stable population of very energetic protons and some of the energetic electrons in the inner radiation belt.

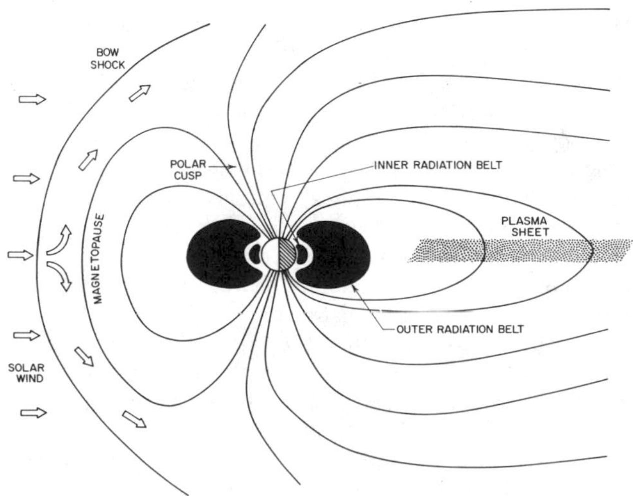

FIG. 1. Principal features of Earth's magnetosphere to approximate scale in the noon–midnight meridian plane cross section.

The residence times of individual particles in the radiation belts of Earth are controlled by ionization losses in the atmosphere near the earth (altitudes of $\leq$400 km), increase rapidly to the order of years at a radial distance of about 8000 km, then decline in a complex and time-variable way to values of the order of weeks, days, minutes, and seconds in the outer fringes. There are quite low intensities of radiation-belt particles within a spherical shell of about 400 km thickness around Earth. This is the region of space flight that is relatively safe from the radiation point of view. The inner radiation belt extends from this lower boundary to an equatorial radial distance of about 12 000 km and the outer radiation belt from this point outward to $\approx$60 000 km. There is considerable overlap of the two principal belts and a complex and time-variable structure in the outer one. Some sample omnidirectional intensities are $J=2\times10^4$ (cm^2 s)$^{-1}$ of protons $E_p>30$ MeV in the most intense region of the inner belt; and $J=3\times10^8$ (cm^2 s)$^{-1}$ of protons $E_p>0.1$ MeV, $J=2\times10^8$ (cm^2 s)$^{-1}$ of electrons $E_e>0.04$ MeV, and $J=1\times10^4$ (cm^2 s)$^{-1}$ of electrons $E_e>1.6$ MeV in the most intense region of the outer belt.

Great advances in knowledge of planetary radiation belts and magnetospheres have been achieved by appropriately instrumented spacecraft in their close encounters with Jupiter in 1973 (Pioneer 10), 1974 (Pioneer 11), and 1979 (Voyagers 1 and 2); with Saturn in 1979 (Pioneer 11), 1980 (Voyager 1), and 1981 (Voyager 2); and with Uranus in 1986 (Voyager 2). A fly-by of Neptune by Voyager 2 will occur in August 1989 to complete exploratory surveys of all of the major planets except Pluto. The four now known planetary magnetospheres have certain basic features in common but each exhibits distinctive features.

Magnetospheres are thought to be ubiquitous in remote astrophysical systems of stars and planets.

See also AURORA; IONOSPHERE; MAGNETOSPHERE; SOLAR WIND.

BIBLIOGRAPHY

W. N. Hess, *The Radiation Belt and Magnetosphere*. Blaisdell, Waltham, MA, 1968.

J. A. Van Allen, *Origins of Magnetospheric Physics*. Smithsonian Institution Press, Washington, DC, 1983.

Ed. by T. Gehrels (ed.), *Jupiter*. University of Arizona Press, Tucson, 1976.

T. Gehrels and M. S. Matthews (eds.), *Saturn*. University of Arizona Press, Tucson, 1984.

Radiation Chemistry

Harold A. Schwarz

Radiation chemistry is the study of chemical effects produced by absorption of the energy of ionizing radiations in matter. Note that this definition does not include the chemical fate of radioactive nuclei recoiling from formation processes, which is part of the study of hot-atom chemistry.

Radiation chemical effects are produced almost entirely by fast charged particles, regardless of the original nature of the radiation impinging on the system. X rays and γ rays interact with matter by the Compton effect, the photoelectric effect or, at higher energies, pair production. In each case energetic electrons are produced with about half or more of the energy of the x rays or γ rays. Fast neutrons lose energy by hard-sphere collisions with nuclei in the substrate and thus produce energetic positive ions. If the substrate contains hydrogen atoms then most of the energy goes into proton formation, as the masses of the neutron and proton are nearly the same.

Fast charged particles lose energy by knock-on collisions with electrons, which can be expressed in terms of Rutherford cross sections, and by glancing collisions with more distant electrons. As long as the velocity of the charged particle is larger than the orbital velocity of the electrons, the energy loss is indiscriminate. This means that the amount of energy originally absorbed by each component in a mixture depends largely on the fraction of all electrons in the mixture which belong to that component.

The initial products of the energy absorption are low-energy electrons (eventually thermalized), positive ions, and excited states of the substrate, many of which dissociate into free radicals. Radiation chemical effects are strongly phase dependent. For example, many ions produced in the gas phase fragment as a result of excess energy left behind in their formation. This fragmentation requires some time, however, and cannot usually compete with deactivation by collisions when the ions are produced in the liquid phase. Thus, about half of the *n*-butane ions, $C_2H_{10}^+$, formed in the gas phase will decompose to a propyl ion, $C_3H_7^+$, and a methyl radical, $CH_3^{\cdot}$, but only 2% fragment in the liquid phase. On the other hand, exothermic ion–molecule reactions which are rapid in the gas phase, such as

$$H_2O^+ + H_2O \rightarrow H_3O^+ + OH^{\cdot},$$

can be safely assumed to be rapid in the liquid phase as well. Most studies have been made in liquids and this fact will be reflected in the rest of the discussion.

RADIATION SOURCES

The majority of radiation chemical studies utilize either γ-ray sources or electron accelerators. The γ-ray sources are mainly used to irradiate samples for final product analysis. The electron accelerators are used for pulse radiolysis studies of free radical and ionic intermediates. Short bursts of electrons are delivered to the sample in this technique. Pulse lengths as short as 5 ps are available, though 10 ns is more common. The most generally used technique for observing the intermediates is to follow light absorption, usually in the visible or near-UV portion of the spectrum. The time resolution of this technique can be as short as 5 ps, when the analyzing light is Cherenkov light generated in xenon by the electron pulse itself before irradiating the sample. Otherwise, 1 to 10 ns is more common. Conduction techniques, with time resolutions of a few nanoseconds or longer, are also used, particularly for studying the properties of electrons. Of course, many intermediates can be studied at leisure by irradiating the sample in a glassy state at low temperature, thus eliminating the need for short pulses and fast equipment.

ELECTRONS AND IONS

Much of the interest in radiation chemistry centers around the electrons and ions produced in liquids. Electrons last until they react with the solvent or with impurities, and lifetimes of microseconds or even much longer can be obtained in many liquids, thus making these studies possible. It is found that there is a wide variation in the mobility and other properties of electrons in different liquids. These effects can be rationalized by a two-state model for the electron: a conduction band, similar to electrons in semiconductors, and a trapped state in which the electron is localized by polarization of the solvent. In polar liquids such as water the electron is strongly trapped and so the mobility is about 10^{-3} $cm^2\ V^{-1}\ s^{-1}$, similar to that of ordinary ions. In nonpolar liquids, such as hydrocarbons, the energy level of the trapped state is only slightly lower than that of the conduction band, so that the electrons spend a significant fraction of time in the conducting state. This results in high mobilities for the electron, such as $0.09\ cm^2\ V^{-1}\ s^{-1}$ in *n*-hexane and $55\ cm^2\ V^{-1}\ s^{-1}$ in neopentane, $C(CH_3)_4$. In liquids with mobilities greater than $10\ cm^2\ V^{-1}\ s^{-1}$ the electrons are mainly in the conducting or "quasi-free" state in which mobility is limited by scattering of the electrons by potentials arising from fluctuations in the density of the liquid. At mobilities below about $1\ cm^2\ V^{-1}\ s^{-1}$, the electron is mostly localized in the trapped state. High-pressure effects on the mobilities of such electrons give the partial molar volume of trapped electrons to be $-20\ cm^3$, which indicates the solvent is becoming more dense around the electron because of electrostriction. Such a change would be expected if the electron is localized in a volume of about 5-Å radius.

The conduction-band energy (V_0) of the electron relative to the gas phase is strongly correlated with the mobilities, i.e., mobility increases as V_0 decreases. This is in accord with the two-state model: As V_0 decreases, the conduction-band energy approaches or even goes below the energy level of the trapped state.

The conduction-band energy is generally determined by measuring photoelectric thresholds in the liquid and in vacuum, though other techniques give similar results. In hydrocarbon liquids, V_0 generally lies between $+0.1$ and -0.5 eV, and correlates with the symmetry of the molecule. Tetrahedral molecules such as neopentane and tetramethylsilane, TMS, have low V_0. Other highly branched hydrocarbons have intermediate V_0, and straight-chain hydrocarbons have the highest V_0.

High positive ion mobilities are less commonly found in liquids than are high hole mobilities in semiconductors. The reason is that only positive ions with the same structure as the parent molecule can move by hole conduction, and in liquids the positive ions commonly react to form other species. Hole mobility has been seen in a few liquids such as decalin.

Electrons and positive ions must be formed in pairs, of course, and so considerable recombination might be expected because of the Coulombic attraction of the pair. Actually, in water and other very polar liquids the range of attraction is small, being reduced to about thermal energy in 7 Å. Consequently little recombination of geminate pairs is expected or observed. In hydrocarbons, however, the energy of attraction is not reduced to thermal level until the pair has separated by about 300 Å and so recombination is dominant. In most hydrocarbons 97% of the ions recombine. This recombination has been observed directly on a picosecond time scale by several techniques: light absorption, microwave conductivity, and delayed ejection by intense infrared lasers. There is also a correlation in hydrocarbons between mobility and free-ion yields, with about 25% of the ions separating in high-mobility liquids such as neopentane and tetramethylsilane.

TRACK EFFECTS

The principal feature which distinguishes radiation chemistry from all other methods of activation which might produce the same species is that energy deposition by ionizing radiation is very nonuniform. For instance, an α-particle will produce about five ion pairs and radical pairs per nanometer of track length, and half of these are formed within 1 nm of the center line of the track. The resulting concentrations of species are so large that the overall chemistry is dominated by reactions between species as they diffuse away from the track. These reactions can occur between species produced in different pairs, as well as the same pair, and so many products can be made, not just recombination to the original substrate. Even with sparsely ionizing radiation, such as 1-MeV electrons for which energy loss events are separated by several hundred nanometers, many of the energy losses result in secondary electrons with sufficient energy to produce more intermediates within a few nanometers of the original event. This produces a "spur" of radicals and ions and these species can also react while diffusing into the bulk of the solution. As a result of spur reactions, about half the radicals initially produced by 1-MeV electrons disappear in a few nanoseconds and track reactions cause about 95% of the radicals generated by α-particle tracks to recombine. In water, there is a yield of hydrogen, H_2, of 0.45 molecule per 100 eV of energy absorbed, most of which is due to spur reactions. The yield of this product increases to 1.3 in α-particle tracks. Spur reactions have been observed directly in optical absorption studies made in the time region from 5 ps to 10 ns, and they also leave their signature in inverted spin population seen in magnetic resonance studies.

APPLICATIONS OF RADIATION CHEMISTRY

Radiation chemistry has several applications to other areas of science. Perhaps the most extensive use is made in studying reaction mechanisms involving oxidation and reduction. These studies are particularly successful in aqueous solution because the species formed by radiolysis of water, hydrated electrons (e^-_{aq}), hydroxyl radicals (OH), and hydrogen atoms, are very well characterized.

The high mobility of electrons in some hydrocarbons is finding uses in large-scale detectors for high-energy physics. For this application it is fortunate that high mobility, which allows fast time response, is associated with high yield, which increases sensitivity.

See also ELECTRON ENERGY STATES IN SOLIDS AND LIQUIDS; ELECTRON SPIN RESONANCE; HOT-ATOM CHEMISTRY; KINETICS, CHEMICAL; RADIATION INTERACTION WITH MATTER.

BIBLIOGRAPHY

R. A. Holroyd and W. F. Schmidt, "Transport of Electrons in Nonpolar Fluids," *Ann. Rev. Phys. Chem.* 40, 439–468 (1989).

Farhataziz and M. A. J. Rodgers (eds.), *Radiation Chemistry, Principles and Applications*. VCH Publications, New York, 1987.

Radiation Damage in Solids

David O. Welch

The physical properties of solids can be altered considerably by exposure to electromagnetic and particle radiation, in general by the disruption of the solid's crystal structure and the alteration of its electronic structure, hence the term radiation "damage." The extensive study of this phenomenon since its prediction in 1943 by Eugene Wigner has been motivated principally by its consequences for materials used in the construction of nuclear reactors and the storage of radioactive waste. Also, the study of radiation effects in semiconductors and insulators has been stimulated by the deleterious effects of radiation from the Van Allen belts, the solar wind, and the explosion of nuclear weapons upon the properties of solid-state electronic devices in artificial satellites and spacecraft. More recently, radiation damage has emerged as an important consideration in the design of ion-implantation processes in the fabrication of microelectronics. However, radiation damage in solids can also be utilized beneficially: some examples include the production of microporous filters and of new metastable materials by ion-beam mixing, applications in geochronology, the study of meteorites, paleontology, in particle detectors and dosimeters in nuclear physics, and in information-storage devices.

Even though the study of radiation damage in solids has largely paralleled the development of nuclear energy and space technology, the first recognition that particle irradiation can disrupt the orderly arrangement of atoms in crystals, and thus strongly affect their physical properties, seems to have been made in 1914 by Hamburg, who suggested that properties of so-called metamict minerals could be explained by the disruptive action of α-particles emitted by the decay of radioactive isotopes incorporated into the crystal structure of the minerals. At the time of its proposal, Hamburg's hypothesis was not generally accepted. Today, a wide variety of experimental and theoretical evidence supports the view that the primary cause of the alteration of the properties of crystalline material by exposure to radiation is the displacement of atoms from their normal locations in the crystal structure to sites that are normally empty in unirradiated crystals, i.e., the creation of lattice defects.

A complete description of the process of radiation damage and its consequences requires an understanding of the mechanisms by which energy is transferred from the radiation to the atoms of a solid, and of the mechanisms by which the transferred energy results in the creation of lattice defects. In addition, it is necessary to know the type, distribution, and stability of the defects created and how defects, in turn, affect the macroscopic properties of materials, such as mechanical strength, electrical conductivity, and optical properties. Ideally, this understanding can be incorporated into a quantitative model that can be used, for example, to predict the evolution of the mechanical properties of a structural member in a nuclear reactor over the lifetime of the reactor. Unhappily such a comprehensive model does not yet exist, even for such ideal, well-characterized materials as pure, single crystals of copper, much less for more complex materials such as structural steel, although useful empirical models have been developed to aid in the development of nuclear reactors and in studies of their safety. However, even though a detailed, comprehensive understanding of radiation damage is not yet complete, some aspects of the problem are quite well understood for specific classes of materials: for example, electron damage to pure, face-centered-cubic metals at low temperatures. Furthermore, a significant amount of the difficulty arises because of the lack of a quantitative understanding of the behavior of unirradiated materials in some areas, such as mechanical properties.

THE INTERACTION OF RADIATION WITH SOLIDS

The first step in the radiation damage process is the transfer of energy from the radiation to the solid. An atom of a crystal that receives sufficient energy, typically a few tens of electron volts, will be permanently displaced from its lattice site. If the displaced atom (a "primary knock-on") has sufficient energy, it can, in turn, displace other atoms ("secondaries"), and so on, creating a cascade of displaced atoms. The mechanisms by which atoms receive energy from the radiation depend on the nature of the radiation.

A beam of energetic, charged particles, such as electrons, protons, or fission fragments, loses energy primarily by exciting the electrons of the solid to higher-energy levels and by collisions with the nuclei of the atoms of the irradiated solid. At sufficiently high energies, energy loss by electronic excitations dominates, but as the particle slows down, energy transfer to nuclei becomes increasingly important until this mode of energy transfer dominates near the end of the particle's range. In metals, only the energy transferred to the nuclei by the irradiating particles is effective in creating lattice defects. In some insulators, however, such as the alkali halides, lattice defects can be created by the energy released upon the deexcitation of electrons from excited states. Furthermore, in polymers, molecular crystals, and other covalently bonded materials the state of chemical bonding can be altered by electronic excitations.

Neutrons interact essentially only with the nuclei of the irradiated solid. In addition to elastic collisions in which there is a transfer of kinetic energy, nuclear reactions can also be important in the radiation damage process. Of particular importance are (n,p) and (n,α) processes, since the hydrogen and helium formed by these reactions are effective in stabilizing the atomic displacement damage that results from elastic collisions. The transmutation products which result from such nuclear reactions alter the chemistry of the material on a microscopic scale and can thereby cause deleterious effects such as embrittlement caused by the precipitation of small particles of new chemical phases. As a consequence of the short range of nuclear forces compared with that of the Coulomb interaction, the mean spacing between collisions in which an atom is displaced from its lattice site is substantially larger for a neutron than for a proton of the same energy. This difference is about three orders of magnitude for 1-MeV particles in a typical metal, in which case the mean spacing between displacements for neutron bombardment is about 5 cm.

Energetic photons displace atoms indirectly through the action of electrons excited by Compton or photoelectric processes, while at high photon energy (e.g., greater than about 5 MeV) the creation of electron–positron pairs begins to make a significant contribution to the creation of displacement damage. Direct interaction of the photons and nuclei makes a negligible contribution if the photons are less energetic than about 100 MeV. In contrast to metals, for which photons of γ-ray energy are required to cause atomic displacements, some insulators, notably the alkali halides, suffer displacement damage from photons in the ultraviolet range, in which case the necessary kinetic energy for displacement is supplied to the atoms by radiationless deexcitation of electrons in excited states. In metals the energy must be supplied by Coulomb scattering of fast electrons excited by the irradiating photons.

THE NATURE OF RADIATION DAMAGE IN SOLIDS

The elementary units of radiation damage in simple crystalline solids are normally occupied crystal lattice sites from which atoms are missing (vacancies) and atoms located in normally unoccupied positions in the lattice (interstitials). In insulators and semiconductors, vacancies and interstitials can exist in a variety of charge states. In addition to such displacement damage, chemical impurities resulting from nuclear transmutations may be present in appreciable quantities. The physical properties of irradiated crystals are determined by the density, distribution, state of aggregation, and electronic state of vacancies and interstitials. These quantities depend on the conditions of irradiation, such as temperature, irradiation intensity, and length of exposure, and on the nature of the radiation. For example, a beam of 1-MeV electrons irradiating a metal produces isolated pairs of vacancies and correlated interstitials, whereas a beam of

1-MeV neutrons produces isolated cascades of radiation damage, about 1000 atomic volumes in size, in which the defect distribution is highly correlated. This difference in defect distribution is attributable primarily to the difference in mean energy of atoms recoiling from their lattice sites following collisions with the irradiating particles. In the case of the 1-MeV electrons, the mean recoil energy is of the order of 50 eV; hence primary recoil atoms usually have insufficient energy to produce further displacements, and each displacement event generally consists of the creation of one interstitial–vacancy pair. However, in the case of 1-MeV neutrons, the mean primary recoil energy is of the order of 100 keV; thus primary knock-ons have ample energy to displace highly energetic secondary knock-ons, thereby generating a cascade of displaced atoms. The size and structure of such cascades, and hence their effect on material properties, are quite dependent on the spectrum of energies of the incident neutrons; thus the nature of radiation damage in thermonuclear fusion reactor materials cannot be inferred, without further experimentation, from data on the radiation damage which occurs in fission reactions because the spectrum of energies of the radiation in the two types of reactors is quite different.

The temperature dependence of the state of radiation damage can be illustrated by the behavior of a crystal of platinum irradiated with fast electrons at 4 K and heated slowly to a temperature of several hundred degrees following the irradiation. At the low temperature of irradiation there is insufficient thermal energy for atomic motion; therefore until the crystal is warmed, the state of damage remains as it was created by the radiation: isolated interstitial atoms separated by a small distance from their original lattice sites, now vacant. When the temperature is raised to about 10 K, there is sufficient thermal energy for atomic motion such that the most closely separated interstitials and vacancies begin to recombine and thus be annihilated. By 30 K, interstitials can migrate freely through the lattice by diffusion until they either find vacant sites and are annihilated or become trapped at impurity atoms or by other interstitial atoms to form clusters.

As the temperature increases, the interstitial clusters grow in size, reaching about 10 or so atoms per cluster by 500 K. At this temperature, thermal fluctuations can supply enough energy for the atoms on normal lattice sites in the crystal structure to exchange places with adjacent vacancies. The vacancies thus become mobile and migrate through the crystal, occasionally combining with, and thereby gradually annihilating, interstitial clusters. By 700 K, the radiation damage has been removed through the annihilation of vacancies and interstitial atoms, and the platinum crystal is returned to an undamaged state. In other materials, the details of the recovery from the damaged state created by fast electrons at low temperature can differ somewhat from the events just described for platinum, but the broad details of the recovery process are the same for a wide variety of materials. Studies of recovery characteristics following irradiation with fast electrons at low temperatures have provided much of the existing knowledge of the properties of elementary lattice defects, such as vacancies and interstitials, in metals and elemental semiconductors.

In general, the state of damage as a function of temperature and mode of irradiation is determined by a balance between the rate of production of elementary defects by the radiation and the competing processes of defect aggregation and annihilation. The latter processes depend critically on the migratory mobility of defects, which, in turn, is a sensitive function of temperature, and on the concentration and spatial distribution of defects. Such a competition between creation, annihilation, and aggregation is an example of a dissipative system, such as a system of coupled chemical reactions far from thermodynamic equilibrium. Such systems often exhibit the phenomenon of self-organization, in which spatial and temporal structures emerge. In the case of radiation damage, such self-organized patterns also exist, e.g., the void lattice discussed below.

An excellent illustration of the complex relationship between the various factors involved in the creation and evolution of radiation damage is provided by the phenomenon of the swelling in volume of metals subjected to a high exposure of fast neutrons at temperatures of the order of half the melting point. Since interstitials and vacancies are quite mobile at these temperatures, it would be expected naively that annihilation processes would readily occur as the interstitials and vacancies wander through the material and destroy each other and thus that radiation effects would be minor; on the basis of such overly simple arguments, swelling would not occur. Indeed, despite decades of research on the effects of neutrons on the properties of metals (albeit under different conditions), the phenomenon of swelling was completely unexpected when it was discovered in 1966. The resulting volume changes can exceed 10% and pose serious technological problems for the development of fast breeder and fusion reactors. The swelling occurs because under the conditions of the irradiation, vacancies aggregate to form empty cavities or "voids." This occurs largely because, compared to vacancies, interstitial atoms are preferentially attracted to "sinks," i.e., locations of other lattice defects such as dislocation where they may be absorbed from the crystal lattice, thus leading to an excess of vacancies, because of the absence of interstitials to annihilate them, and their subsequent aggregation into voids. This preferential attraction is thought to occur because interstitial atoms distort the geometry of the crystal lattice more severely than do vacancies, and it is these distortions that cause the defects to be attracted by internal stress fields surrounding the sinks. In addition, it is thought that helium atoms produced by (n,α) reactions play a critical role in the nucleation and stabilization of voids. Interestingly, all of the properties of vacancies and interstitials that give rise to the formation of voids were known prior to 1966, but the behavior of the system as a whole under the appropriate conditions had not been considered. In some cases the voids which form are organized into well-defined, periodic lattices, similar to the crystal lattices formed by the atoms of the metal itself; various explanations have beem proposed for the appearance of such surprisingly perfect structures, called void lattices, ranging from explanations based on the existence of a long-range interaction between the voids as a consequence of the elastic distortion of the metal to explanations based on dynamical effects which are a consequence of the details of the mechanism of the thermally activated diffusive motion of interstitials.

The lattice distortions caused by the radiation-induced defects described above, together with the disruption to the periodicity of the lattice by missing or improperly located atoms, result in the scattering of the conduction electrons of metals, thus causing an increased electrical resistivity. Furthermore, the scattering of electrons by lattice distortions provides the basis for the study of radiation-produced defects by transmission-electron microscopy. Both of these effects have been the principal tools for the study of radiation effects in metals. Recently studies of the diffuse scattering of x rays by the defect-produced lattice distortions have helped to resolve a long-standing controversy over the structure of interstitials and the interpretation of recovery during annealing following irradiation of metals.

Lattice defects in ionic crystals and semiconductors can trap electrons and holes in the valence band, consequently giving rise to well-defined optical absorption and emission and to magnetic resonance effects. Because of these characteristics, defects, and thus the radiation damage process, can be studied in these materials by a wide variety of experimental techniques. As a consequence, probably more is known about the structure and properties of lattice defects and their production by radiation for the alkali halide family than for any other class of materials. As in the case of metals, the elementary units of damage are vacancies and interstitials, although in insulating materials these can exist in a variety of charge states. For example, in NaCl, anion vacancies can be empty or can trap either one or two electrons.

The state of radiation damage in nonconductors, as in metals, under given conditions reflects a dynamic balance between the creation of vacancies and interstitials and their aggregation or annihilation. In contrast to pure metals, however, many nonconducting crystals (Al_2O_3, MgO, SiC, $SrTiO_3$, and SiO_2, among others) can be converted from the crystalline state to an amorphous glass-like state by sufficiently long exposure to radiation. Covalently bonded materials appear to be particularly susceptible to amorphization by ion implantation, whereas very high doses are required for materials whose bonding is basically ionic. It is not required that the material be nonconducting to be susceptible to amorphization by radiation, since some intermetallic compounds with a narrow compositional range, such as Zr_3Al, are susceptible to this phenomenon as well. A phenomenon peculiar to insulating materials is the production of linear trails of atomic disorder or "tracks" in the wake of massive charged particles. Etching solutions preferentially attack the damaged region, producing large holes that are observable by optical microscopy. The track phenomenon has been exploited for a variety of applications, ranging from solid-state detectors and the production of small holes of controlled size in membranes to geochronology.

Finally, a wide variety of chemical effects can be produced by electronic excitation by ionizing radiation in solid organic and polymeric materials and in pseudostable inorganic materials such as ammonium perchlorate. Some of these effects are decomposition to form gaseous products, polymerization, creation of cross links, and breaking of primary links in high polymers.

Acknowledgment

This research was performed under the auspices of the U.S. Department of Energy, Division of Materials Sciences, Office of Basic Energy Science under Contract No. DE-AC02-76CH00016.

See also COLOR CENTERS; CRYSTAL DEFECTS; RADIATION CHEMISTRY; RADIATION INTERACTION WITH MATTER.

BIBLIOGRAPHY

D. S. Billington and J. H. Crawford, Jr., *Radiation Damage in Solids*. Princeton University Press, Princeton, NJ, 1961.

R. O. Bolt and J. G. Carroll (eds.), *Radiation Effects on Organic Materials*. Academic, New York, 1963.

J. W. Corbett, *Electronic Radiation Damage in Semiconductors and Metals*. Academic, New York, 1966.

G. J. Dienes and G. H. Vineyard, *Radiation Effects in Solids*. Wiley (Interscience), New York, 1957.

A. E. Hughes and D. Pooley, *Real Solids and Radiation*. Wykeham Publications (London) Ltd.; Springer-Verlag, New York, 1975.

G. C. Messenger and M. S. Ash, *The Effects of Radiation on Electronic Systems*. Van Nostrand Reinhold, New York, 1986.

N. L. Peterson and R. W. Siegel (eds.), *Properties of Atomic Defects in Metals*. North-Holland, Amsterdam-New York–Oxford, 1978. Reprinted from *Journal of Nuclear Materials*, Vols. 69–70.

C. W. White, C. J. McHargue, P. S. Sklad, and L. A. Boatner, "Ion Implantation and Annealing of Crystalline Oxides." *Mater. Sci. Rep.* **4**, 41–146 (1989).

"Radiation Effects on Materials" (report to the American Physical Society by the study group on physics problems relating to energy technologies), *Rev. Mod. Phys.* **47**, Suppl. No. 3 (1975).

E. Sonder and W. A. Sibley, "Defect Creation by Radiation in Polar Crystals," in *Point Defects in Solids*, Vol. 1 (J. H. Crawford, Jr. and L. M. Slifkin, eds.). Plenum, New York, 1972.

M. W. Thompson, *Defects and Radiation Damage in Metals*. Cambridge, London and New York, 1969.

Radiation Detection

G. Davis O'Kelley and Andrew P. Hull

Much of the present success in the sensitive, accurate, and convenient measurement of nuclear radiation is due to continuing advances in detector technology, coupled with progress in the development of improved semiconductor electronics and digital computers. Because the recording and analysis of data is so specialized, the discussion that follows will be restricted to the characteristics of radiation detectors and will not treat the ancillary equipment required for a complete system.

GAS-FILLED DETECTORS

Although gas-filled counters are among the oldest of radiation detectors, they continue to be used extensively in nuclear science. These devices make use of ionization, which may be produced directly by charged particles such as electrons, protons, or very heavy ions. Uncharged species such as gamma rays or neutrons must first interact with the detector to produce charged particles by secondary ionization.

The *ionization chamber* is an instrument designed to measure the ionization produced by nuclear radiation in an insulating substance such as a gas. The two principal arrangements of ionization chambers are shown in Fig. 1. If an electric field is maintained between two electrodes, then the electrons and positive ions that appear along the path of an energetic charged particle or photon will move toward the two electrodes, causing a current to flow in the external circuit. If ionization is produced at a steady rate within the chamber, it will be observed that the current flowing through resistance R in Fig. 1 from collection of charge will rise as the applied potential V_P is increased, until it reaches a limiting value, the *saturation current,* which is proportional to the source intensity. The currents of interest typically lie in the range of about 10^{-8}–10^{-14}A.

When the rate of arrival of ionizing radiation is too low to permit current measurements, or when it is necessary to determine the energy distribution of particles stopped in the gas, an ionization chamber may be operated in the pulse mode. The resulting pulse is relatively complex, since the electrons are collected rapidly; but the positive ions, which move about 10^3 times slower, may continue to influence the charge at the collecting electrode by electrostatic induction for hundreds of microseconds. Thus, the external pulse circuits are usually designed to respond only to the electron collection portion of the pulse; however, the amplitude and shape of the electron pulse depends on the location of the

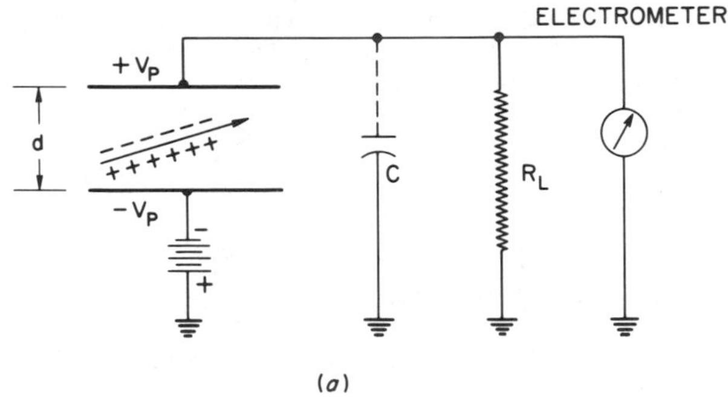

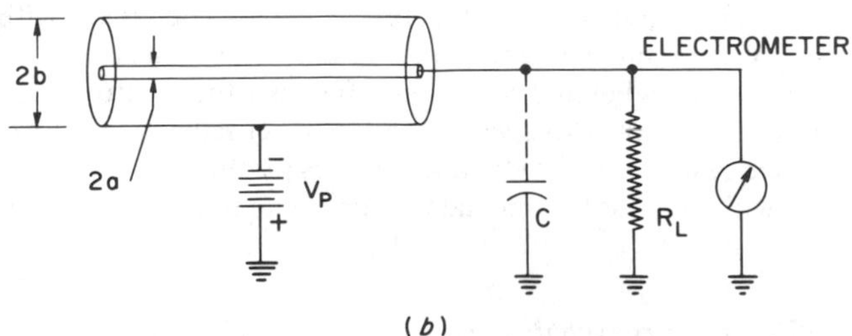

FIG. 1. Schematic representations of ionization chamber designs: (a) parallel-plate configuration; (b) cylindrical geometry. The electric field at a radial distance r from the anode is given by $\mathscr{E} = (V_P/r)\ln(b/a)$.

ion pairs when they were formed. In parallel-plate ionization chambers some improvement can be achieved by making the collecting electrode very small and collimating the incident radiation. The most satisfactory technique for removing the effect of positive-ion induction was proposed by Frisch, who added a grid between the high-voltage electrode of a parallel-plate chamber and the collecting electrode. The collecting electrode is shielded by the grid from the influence of the positive ions, so the charge at the collecting electrode is equal to the total ionization.

Parallel-plate chambers are widely used as fission counters. Chambers containing ^{235}U or other standard fissionable material are employed frequently as neutron monitors for reactor control and for personnel protection. Some neutron-sensitive chambers are filled with BF_3 gas or lined with boron or lithium; the ionization is produced by alpha particles and recoil nuclei from the (n,α) reaction on ^{10}B or ^{6}Li. Cylindrical current chambers may be used for assaying sources of low-energy beta particles by introducing the radioactive materials as gases. Gamma-ray measurements can be made on chambers filled with a gas to high pressure; such chambers are extensively used in nuclear medicine, where they are known as "dose calibrators," high-pressure large ion chambers with a volume of several liters. Ionization chambers operated in the pulse mode are frequently used as windowless alpha counters. Although the use of Frisch grid chambers has largely been eclipsed by the modern semiconductor counter, the large active volumes possible make such pulse chambers useful in nuclear physics as high-resolution, heavy-ion spectrometers. In health physics a modification of the parallel-plate chamber known as an "air-wall" chamber is used for primary standardization of radiation exposure rate. Air-filled cyclindrical current chambers are employed in portable survey instruments, and pocket dosimeters in the form of an ionization chamber with a self-contained quartz-fiber electroscope and transparent scale have been used for more than 35 years.

If the diameter of the anode in Fig. 1b is very small compared to that of the cathode, it is possible to achieve very high electric-field strengths with modest values of the applied voltage V_P. As the applied voltage is advanced beyond the ionization-chamber saturation region, electrons produced in ionization processes acquire sufficient energy near the anode to produce additional ionization of the counter gas by collision; this process is termed *gas multiplication*. A detector that makes use of gas limits multiplication is called a *proportional counter,* because the height of a signal pulse is proportional to the initial ionization. This suggests that a proportional counter can be operated to count alpha particles in the presence of electrons, which usually have lower energy.

Proportional counters are relatively easy to design and are stable in performance; thus, they still find considerable application in nuclear science. Proportional counters with filling gas continuously flowing are used in several geometries. The 4π beta counter is used for absolute measurement of the disintegration rates of beta-ray sources; this counter is, in effect, two proportional counters whose outputs are summed and that view opposite sides of a thin source on a low-mass backing. A gas-flow proportional counter in a 2π geometry makes a very convenient windowless counter for routine assay of both alpha and beta radioactive sources. Flow-type proportional counters with end windows are also used for routine measurements of radioactivity and in some alpha-sensitive health-physics survey instruments. Tissue-equivalent gas-filled proportional counters are employed for microdosimetry. Proportional detectors filled with krypton or xenon are used for measurement of energy spectra and intensities of low-energy x rays and gamma rays in the presence of high-energy electrons and electromagnetic radiation. Such detectors have recently found application in satellite and rocket measurements of x rays in space and in studies of Mössbauer phenomena. Two of the principal neutron detectors in use are proportional counters filled to several atmospheres of pressure with ^3He or BF_3. Neutrons produce ionization by nuclear reactions with the filling as: $^3He(n,p)^3H$ and $^{10}B(n,\alpha)^7Li$. Finally, an important new application to nuclear structure studies is in the construction of position-sensitive detectors, which, for example, may be used to replace photographic plates at the focal plane of a magnetic spectrometer. The simplest design for such a detector uses an anode of very high resistance and electrical connections to either end. By means of appropriate circuitry the position of the incident radiation can be determined to better than 1 mm in a counter that is 300–400 mm long.

If the voltage applied to a proportional counter is increased, then the secondary ionization produced by gas multiplication spreads outward from the anode until at some critical value of the electric field a single electron in the counter will initiate a continuous discharge unless a quenching agent is present in the counter gas or some external circuit is provided to reduce the applied voltage momentarily below the threshold value. The height of the electrical pulse from such a counter is very large, often several volts, but is independent of the initial ionization. Detectors operating in this mode are called *Geiger counters*.

It is apparent that Geiger counters cannot be used to measure directly the energy of an incident particle, so their use in nuclear-structure physics is limited. Their large signal amplitudes make Geiger counters of interest in some arrays containing large numbers of detectors, since auxiliary circuitry can be simplified. For similar reasons, they are extensively employed as detectors in portable health-physics survey instruments and in noncritical systems for assay of radioactivity. By the selection of appropriate materials, the sensitivity of a Geiger tube to photons may be made proportional to their energy.

SEMICONDUCTOR DETECTORS

A semiconductor detector is the solid-state equivalent of the gas-filled ionization chamber. The development of semiconductor detectors has become the most important innovation in radiation detector technology of recent years. These devices are discussed as a separate topic in this encyclopedia.

SCINTILLATION DETECTORS

A functional diagram of a modern scintillation counter is shown in Fig. 2. Incident radiation deposits energy into the

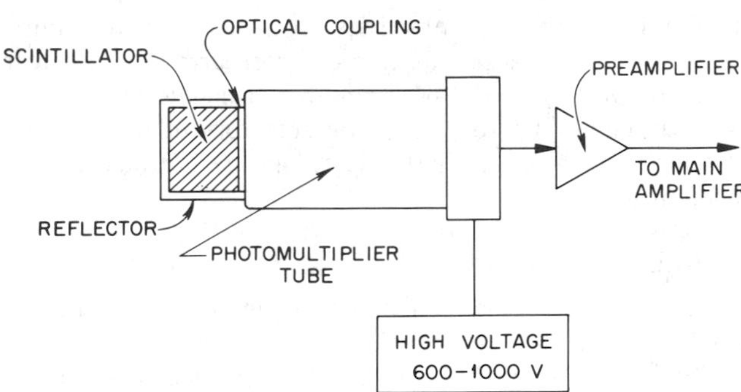

FIG. 2. Schematic diagram of the essential components of a scintillation detector system.

scintillator. A percentage of the energy dissipated is converted to light photons, which are transmitted to the photocathode of a photomultiplier tube, aided by an efficient optical reflector and coupling combination. The light energy is converted to a burst of photoelectrons at the cathode, and the number of electrons is increased within the multiplier structure of the photomultiplier tube by a factor of about 10^6. The negative output signal at the anode is of sufficient amplitude to be amplified and processed by standard electronic data-recording systems.

The choice of a scintillator depends on the requirements of the experiment. Because of their low effective atomic number and consequently low backscattering, *organic scintillators* are preferred for detection and spectrometry of electrons. Organic scintillators are available either as pure single crystals, as liquid solutions, or as solid solutions (plastics). The decay of the fluorescent light from organic scintillators is very rapid, and in combination with high-gain photomultiplier tubes with small transit-time spread, detector pulses can be generated with fast rise times suitable for recording high counting rates or for coincidence measurements at time resolutions of less than 10^{-9} s. Because of their high speed and hydrogenous composition, organic scintillators are used as proton-recoil detectors in fast-neutron time-of-flight studies. Organic scintillators also find extensive application in liquid scintillation counting, in which the radioactive sample to be measured is dissolved in an organic liquid scintillator, which overcomes the effects of scattering and self-absorption inherent in other geometries.

The decay of the fluorescent light emitted from *inorganic scintillators* such as NaI(Tl) is much slower than for most organic scintillators; however, the photopeak-to-Compton ratio is much higher, so the energy resolution is greatly improved over that of the organic scintillators. Additionally, NaI(Tl) is available in large, clear single crystals, has the highest light output of the room-temperature inorganic scintillators, and has a high effective atomic number. All of these characteristics make crystals of NaI(Tl) very suitable as high-efficiency gamma-ray detectors. Scintillation detectors using a 7.6-cm-diameter, 7.6-cm-long NaI(Tl) crystal are still widely used for gamma-ray assay measurements because of their low cost and moderate energy resolution, although they are rapidly being supplanted by Ge(Li) semiconductor detectors of much higher resolution. Again because of their low cost and high efficiency, NaI(Tl) gamma-ray detectors are employed in high-sensitivity, low-background gamma counting and in large multiple-detector arrays in nuclear structure physics. High-Z bismuth germanate, known as BGO, is employed for arrays of detectors in tomography systems, and in several large detectors used at high-energy accelerators.

OTHER DETECTORS

A few more specialized detection methods will be described briefly.

When a charged particle passes through a transparent dielectric material with a velocity greater than the velocity of light in that material, a cone of electromagnetic radiation is emitted (Čerenkov radiation). The Čerenkov light is usually detected by photomultiplier tubes, as in scintillation counters, except that in a *Čerenkov counter* the optical collection systems often are arranged to accommodate only one angle of emission.

The *spark chamber* is primarily an instrument for high-energy physics experiments. It differs from the detectors described above in that it possesses high spatial resolution. The passage of a charged particle through the chamber is usually announced by an array of external detectors, often plastic scintillation detectors, and a simultaneous high-voltage pulse sets up a high electric field within the chamber. A spark forms along the track of ionization left by the particle, accompanied by the emission of intense visible and ultraviolet light, by the production of a shock wave, and by intense electric and magnetic fields. These properties are the basis for several data-recording systems.

Photons interacting with crystals such as LiF and $CaSO_4$ leave electrons in traps, from which they may subsequently be released by heating and in this manner give off light. Thermoluminescent dosimeters have virtually supplanted photographic film for personnel dosimetry.

Charged particles impinging on solids such as mica, glass, or a film of high polymer produce trails of radiation damage that can be "developed," or chemically etched, to reveal the track of the particle. The tracks in such *dielectric track detectors* can be measured individually or counted by means of a number of microscope and electronic techniques. Much of the utility of this method lies in the ability of dielectric track detectors to respond to very heavily ionizing particles and yet remain insensitive to lightly ionizing particles. They have been very useful in studies of fission, in studies of heavy ions produced in nuclear reactions, in measurements of the mass and energy distributions of the galactic cosmic rays, in determinations of cosmic-ray exposure ages, in radioactive dating in neutron dosimetry, and in alpha track detectors for the measurement of the concentration of radon and its daughters indoors.

See also CLOUD AND BUBBLE CHAMBERS; COUNTING TUBES; SCINTILLATION AND ČERENKOV COUNTERS; SEMICONDUCTOR RADIATION DETECTORS; THERMOLUMINESCENCE.

BIBLIOGRAPHY

J. B. A. England, *Techniques in Nuclear Structure Physics*, Part 1. Wiley, New York, 1975. (A)

J. Krugers, ed., *Instrumentation in Applied Nuclear Chemistry*. Plenum, New York, 1973. (I)

P. W. Nicholson, *Nuclear Electronics*. Wiley, New York, 1974. (A)

P. J. Ouseph, *Introduction to Nuclear Radiation Detectors*. Plenum, New York, 1975. (E)

W. J. Price, *Nuclear Radiation Detection*, 2nd ed. McGraw-Hill, New York, 1964. (I)

Radiation Detectors *see* Semiconductor Radiation Detectors

Radiation Interaction with Matter

J. H. Crawford, Jr., and B. T. Kelly

The term "radiation," as used here, includes both electromagnetic and particulate forms. Matter, in general, encompasses nuclei, atoms, molecules, etc., but in this article principally refers to solids and the consequences of their exposure to radiation. The excitation and ionization of atomic and molecular systems are only briefly considered.

I. ELECTROMAGNETIC RADIATION (PHOTONS)

The interaction processes characteristic of photons depend upon their energy (E_p). It is convenient to separate the effects into those associated with two ranges: (a) high energies for which $E_p > 10^4$ eV, and (b) low energies for which $E_p < 10$ eV. The energy range between these two values is associated with the far-ultraviolet region and can be studied using monochromatic radiation from synchrotron sources; it is not considered further in this brief review.

A. High-Energy Photons (X Rays and γ Rays) [1–3]

High-energy photons undergo three dominant interaction processes, each of which produces energetic electrons: the photoelectric (PE) effect, the Compton effect (CE), and pair production (PP). The PE effect is dominant at the low-energy end of the range, where photons are absorbed to eject deep-shell electrons with kinetic energies $E_e = E_p - W$, where W is the electron binding energy. This leads to secondary processes with the emission of x rays and Auger electrons as the deep-shell vacancies in the ionized atoms are filled. The cross section for the PE process (probability of event/unit time/unit photon flux) increases approximately with the fifth power of the atomic number Z and decreases from the K-, L-, and M-shell threshold energies with increasing E_p as $E_p^{-7/2}$. The PE cross section can be significant for large values of Z, even for $E_p \sim 1$ MeV.

In the range 0.1 MeV$\leqq E_p \leqq$10 MeV the interactions are generally dominated by CE, where the photons can be treated as particles. The probability of a Compton-scattering event scales with the Z of the atomic system. The CE electrons cover a continuous spectrum from zero to a maximum:

$$E_e(\max) = 2E_p^2(mc^2 + 2E_p)^{-1}, \tag{1}$$

where m is the electron mass and c the velocity of light. The creation of electron–positron pairs (PP) is possible when E_p exceeds 1.02 MeV, that is twice the energy of the electron rest mass mc^2. The cross section for pair production rises slowly from this threshold and becomes comparable with CE at about 10 MeV, while the reaction is dependent upon Z^2. The kinetic energy of each member of the pair produced is $\frac{1}{2}(E_p-2mc^2)$.

Other less prevalent reaction mechanisms include direct Compton scattering of photons by the nucleus and photo-disintegration of the nucleus. The first of these requires very large values of E_p and the second is sensitive to the details of the nuclear structure.

B. Low-Energy Photons (Infrared, Visible, and Near Ultraviolet) [3,4]

Low-energy photons interact principally by ionization and excitation of the outer orbitals or valence-band electrons in a solid. In the vacuum ultraviolet region ($E_p>10$ eV) photons are efficiently absorbed by all but the lightest atoms and most molecular and condensed systems. The details of the absorption spectra (as in PE) are sensitive to the electronic structure of the absorber, and absorption spectral measurements constitute one of the primary tools for electronic structure investigation. Insulator crystals, the so-called wide-band-gap materials such as alkali and alkaline earth halides and many oxides, are transparent in the visible range. The absorption edge or threshold corresponds to the minimum E_p required to excite and/or ionize the constituent atoms of the solid. The first absorption processes result in the creation of excitons, i.e., excitations which are analogous to excited states of isolated atoms but which in all but certain halides are not localized but belong to the entire crystal. Photoproduction of quasifree electrons and holes occurs just beyond the narrow excitation interval and photoconductivity results. The recombination of electron–hole pairs and decay of excitons may generate either photons (luminescence) or phonons (crystal lattice vibration). Crystals become opaque beyond the absorption edge. Localized absorption processes can occur at imperfections in semiconductor crystals such as impurity atoms, ions, or lattice vacancies which either are ionized or can be excited by photon energies less than the absorption edge. The photon energy E_p diminishes as the energy band gap decreases, passing through the visible into the infrared. Most commonly used semiconductors exhibit an edge lying in the near infrared (0.8 eV for Ge) and which is indicative of the relatively weak binding of electrons in semiconductors. In metals the band gap is zero and photons of all energies can be absorbed to increased the kinetic energy of conduction electrons or holes; this kinetic energy is rapidly transferred to the crystal lattice as heat (phonon generation) by scattering processes.

Photons of suitable energy are capable of exciting vibrations directly in both molecules and crystal lattices. The excitation may occur by either photon absorption or scattering (Raman process). The phonon energies lie in the range 10–100 meV, and thus photons well into the infrared are required for the absorption process. The propagating wave must couple to the vibration via its oscillating electric field; hence only those vibrations which generate an alternating current electric field can be excited. In polar or ionic solids these are the so-called longitudinal optical (LO) or transverse optical (TO) phonons and are responsible for the long-wavelength cutoff of infrared transmission in such crystals which is called the "Restrahl" frequency. In homopolar materials, there are no electric-dipole vibrations and photon-absorption excitation of polar modes does not occur. Acoustic or nonpolar vibrations may be excited in both polar and nonpolar materials by the scattering of visible photons, a process known as Raman scattering in which the photon energy is changed by plus or minus the amount of the vibrational energy. Scattering by the Mie or Rayleigh processes does not change the photon energy and is not further discussed.

C. Radiolysis [5]

An important consequence of exposing certain solids and gases to ionizing radiation such as x rays or γ rays is radiochemical decomposition or radiolysis. The decomposition may be transitory in the absence of any other materials, for gases, but it is possible to obtain significant chemical reaction rates at temperatures well below those expected from thermal reaction rates. The deposition of 100 eV of energy in the gas produces ~1–10 ionized species which can react chemically. Important examples from the nuclear industry are the radiolytic reaction between graphite and carbon dioxide, and the decomposition of water.

In solid materials such as the alkali and alkaline earth halides the irradiation produces lattice vacancies and interstitials which are manifested by creation of color centers. In essence these defects are created during the recombination of electrons with halogen molecular ions and the decay of excited halogen pairs (excitons).

II. CORPUSCULAR RADIATION [1,6,7]

A. Fast Neutrons

Neutrons created in the fission or fusion processes have energies in the range 1–20 MeV. These neutrons up to a few MeV interact with atoms by hard-sphere collisions, whereas at higher energies anisotropic and inelastic scattering become important. The collision of fission neutrons with nuclei is the process by which neutrons are moderated in a thermal nuclear reactor. The average energy transferred in a hard-sphere collision is given by

$$\bar{T}=\tfrac{1}{2}T(\max)=\frac{2E_nMn}{(M+n)^2}, \tag{2}$$

where E_n is the neutron energy and M and n are the masses of the target atom and the neutron, respectively. $T(\max)$ is the energy transferred in a head-on collision. A typical neutron–atom collision with $E_n=2$ MeV and $M=100$ atomic mass units gives a nucleus with a recoil energy of $\sim 10^4$ eV. The energy required to eject an atom irreversibly from its crystal lattice site E_d is in the range 10–50 eV for most solids, and thus not only is the primary recoil atom displaced but it becomes a particle capable of producing secondary and tertiary displacements which initiates a displacement cascade characterized by a relatively high density of lattice defects (interstitial atoms and lattice vacancies) and of dimensions of ~100 lattice spacings. The crystal lattice damage results in stored (Wigner) energy, and changes in the physical properties (electronic, optical, magnetic, and mechanical) which are structure sensitive. Inelastic scattering, where the excited nucleus emits a γ ray of energy E_γ, leads to a reduced mean recoil energy:

$$\bar{T}=\frac{2E_nMn}{(M+n)^2}-\frac{E_\gamma}{M+n}. \quad (3)$$

Anisotropic scattering must be considered also at the higher neutron energies, nucleus by nucleus. The number of atoms displaced by a knock-on primary atom of energy E_p is roughly $E_p/2E_d$ for energies below that at which the moving atom loses its energy principally by electronic excitation L_c. A crude estimate of L_c is A keV, where A is the atomic weight.

B. *Charged Particles*

When incident upon a solid, energetic charged particles (protons, deuterons, and heavier ions with energies in the MeV range) lose most of their energy by ionization of lattice atoms, which results in trapped charge and possibly radiolysis (see above) in insulators but which has as the major effect the heating of the solid. A significant fraction of this energy, however, is dissipated via elastic collisions. In contrast to neutrons which undergo hard-sphere collisions, energy transfer results from charge interaction between the moving ion and the nucleus of the target atom, i.e., Rutherford scattering. The primary characteristic of this process is that small energy transfers are strongly favored. Hence the average kinetic energy of an atom recoiling from a proton collision is much less than for a collision involving a neutron of the same energy. Therefore, on the average, the lattice damage will consist of small clusters of lattice defects rather than the extensive disordered regions characteristic of fast neutrons.

C. *High-Energy Electrons*

Because of the rapid increase in mass with energy in the relativistic energy range, MeV electrons are also capable of displacing atoms in solids. This process is a convenient means of both producing isolated interstitial–vacancy pairs and determining the magnitude of E_d. The relativistic expression for the energy transferred in a head-on collision between an electron of energy E_e and an atom of mass M is

$$T(\max)=\frac{2(E_e+2mc^2)}{Mc^2}E_e. \quad (4)$$

The value of E_d for a particular crystalline solid can thus be determined from the lowest-energy electron capable of inducing a change in some physical property sensitive to displaced atoms. Since absorption of high-energy photons (γ rays) by solids produces eleectrons with energies in the MeV range (see Section IA), the effect is that of internal electron bombardment; hence, interstitial–vacancy pairs can be created by γ irradiation via the displacement process.

The physical sputtering from solid surfaces due to ion bombardment [8] has become a subject of some interest with regard to fusion systems for power generation, as has the implantation of atoms at controlled depths from surfaces to modify materials properties. The first of these is principally concerned with energetic light ions from a high-temperature plasma. The number of sputtered atoms for each incident ion increases with ion energy to values ~0.02. Analytic formulas have been proposed to describe this sputtering process. Compilations now exist of the staffing and range of ions in solids [9].

D. *Channeling [10,11]*

When energetic protons or heavy ions are incident upon a single crystal along some relatively open crystallographic direction, the loss of energy to the lattice is minimized and anomalously large penetration depths are observed. This process is called channeling and may involve an open low-index lattice channel or the space between two lattice planes. Channels can be blocked by interstitials and oversized impurities, and the number and identity of these may be investigated by measuring the intensity and the energy of back-scattered ions which would be "channelled" in the absence of these blocking imperfections.

See also CHANNELING; COMPTON EFFECT; EXCITONS; LUMINESCENCE; PHONONS; PHOTOCONDUCTIVITY; PHOTONS; RADIATION DAMAGE IN SOLIDS; RAMAN SPECTROSCOPY; SYNCHROTRON RADIATION.

REFERENCES

1. D. S. Billington and J. H. Crawford, Jr., *Radiation Damage in Solids,* Chap. 2. Princeton University Press, Princeton, NJ, 1961. (I)
2. O. S. Oen and D. K. Holmes, *J. Appl. Phys.* **30,** 1289 (1959). (A)
3. C. Kittel, *Introduction to Solid State Physics,* 4th ed, Chap. 18. Wiley, New York, 1971. (E)
4. B. DiBartolo, *Optical Interactions in Solids.* Wiley, New York, 1968. (A)
5. E. Sonder and W. A. Sibley, *Point Defects in Solids.* Vol. 1, Chap. 4 (J. H. Crawford and L. M. Slifkin, eds.). Plenum Press, New York, 1972. (E)
6. B. T. Kelly, *Irradiation Damage to Solids.* Pergamon Press, Oxford, 1966. (E)
7. M. W. Thompson, *Defects and Radiation Damage in Metals.* Cambridge University Press, London, 1969. (I)
8. J. Roth, *Physics of Plasma-Wall Interactions in Controlled Fusion,* p. 351. Plenum, New York, 1986.
9. J. F. Ziegler, J. P. Biersack, and U. Littmark, *The Stopping and Range of Ions in Solids,* Vol. 1. Pergamon, New York, 1985.
10. D. V. Morgan, *Channeling.* Wiley, New York, 1973. (I)
11. S. T. Picraux, E. P. Er Nisse, and F. L. Vook, *Applications of Ion Beams to Metals.* Plenum, New York, 1974. (A)

Radioactivity

Bernard G. Harvey

INTRODUCTION

Radioactivity is a property of the atomic nucleus. Many unstable nuclei spontaneously disintegrate, forming different nuclei by emitting some form of matter and energy. When they are sufficiently energy excited, all nuclei can disintegrate, but the term radioactivity is usually reserved for the decay of nuclei with little or no excitation energy. For example, an unexcited nucleus of ^{226}Ra (nuclear charge +88) disintegrates into ^{222}Rn (charge +86) and ^{4}He (charge +2). The ^{4}He nucleus (the α-particle) carries off as kinetic energy nearly all the 4.86 MeV of energy liberated in the decay. There are many other ways, described in what follows, by which unstable nuclei can decay.

Radioactive decay is *spontaneous*. It can occur even when the nucleus is totally isolated from external influences, although the presence of atomic electrons is sometimes required. Unlike most chemical reactions, the decay is not triggered by the absorption of energy from external sources. An unstable nucleus may live for billions of years before it suddenly and spontaneously disintegrates.

Radioactivity was discovered accidentally by Henri Becquerel in 1896. At that time it was known that x rays cause glass to fluoresce and that they also blacken photographic plates. Becquerel therefore investigated a possible connection between these two effects by placing a fluorescent uranium compound on a sensitive plate. The plate was blackened, but he soon found that even nonfluorescent compounds of uranium produced the same effect but by the emission of radiations capable of penetrating black paper and pieces of silver. In the following four decades, studies of radioactivity led to the discovery of several new chemical elements (Po, Ra, Rn, Ac, Fr), the discovery and investigation of the atomic nucleus (Rutherford, 1911), the production and identification of the neutron, and the discovery of artificially induced nuclear transformations and of nuclear fission.

RADIOACTIVE DECAY LAW

For a given unstable nucleus and a given type of decay to a specific final product, there is a definite probability that the nucleus will disintegrate within a time interval dt. Since the decay is a property of the *isolated* nucleus (or atom), the number dN of nuclei disintegrating in a time interval dt is proportional to the number N of nuclei present in the sample:

$$dN/dt = \lambda N,$$

where λ is the proportionality constant (decay constant). Integration of this equation gives the number N_t of nuclei remaining at time t out of a number N_0 present at $t=0$:

$$N_t = N_0 e^{-\lambda t}.$$

By setting the remaining number equal to one-half of the initial number, the time $t_{1/2}$ for one-half of the nuclei to decay is obtained:

$$t_{1/2} = (\ln 2)/\lambda = 0.693/\lambda.$$

The quantity $t_{1/2}$ is called the *half-life*. Values for known unstable nuclei range from about 10^{-12} s to 10^{20} years, an enormous range. The exponential decay law has been tested out to 45 half-lives, at which point only $2.8 \times 10^{-14} N_0$ nuclei remain in the sample.

Because of the statistical nature of the decay process, the number of decays dN in time interval dt is an *average* quantity subject to fluctuations resembling those of death rates in a biological population. When sufficiently large numbers N of decay events are observed in successive measurements, the fluctuations in N obey a Gaussian distribution with standard deviation equal to $\sqrt{N}$.

TYPES OF RADIOACTIVE DECAY

In addition to the emission of ^{4}He nuclei (α-particles), unstable nuclei may decay in a variety of other ways that can be classified according to the fundamental force that is responsible. The strong *nuclear force* between the nuclear constituents (neutrons and protons) causes the emission of α-particles. The same force can also bring about other decay modes. For example, nuclei that contain an unusually high ratio of protons to neutrons can, in a few cases, decay by proton emission (e.g., ^{19}Na which contains 11 protons and only 8 neutrons). Nuclei with an unusually high ratio of neutrons to protons sometimes decay by the emission of a neutron (e.g., ^{10}Li which has 7 neutrons and only 3 protons). In rare cases, the emission of heavier fragments is observed. For example, the nucleus ^{222}Rn occasionally decays by emitting a nucleus of ^{14}C rather than the much more probable α-particle.

The much weaker *electromagnetic force* can cause an unstable nucleus to emit a photon (γ ray). This can occur, though, only when the nucleus is excited to a energy above its lowest (ground) state. Frequently, a nucleus will decay by particle emission to an excited state of the product nucleus, which in turn will emit a γ ray. Alternatively, the electromagnetic force can cause a nucleus to interact with, and eject, an electron from an atomic orbital as an alternative to the emission of a γ ray. This process is known as *internal conversion* (IC). The Coulomb (electric) part of the electromagnetic force may become so strong in nuclei with many protons that it overcomes the strong force that tends to hold nuclei together. In this case, the heavy nucleus may break into two smaller nuclei of roughly equal mass and charge by a process known as *spontaneous fission* (SF). About 200 MeV of energy is released, mainly as kinetic energy of the two fission fragments.

An extremely weak force (known simply as the *weak force*) causes the emission of either negatively charged electrons (β^- particles) or positively charged electrons (β^+ particles, positrons). In many cases, an unstable nucleus captures an atomic electron as an alternative to emitting a positron. This process is called *electron capture* (EC). EC, β^+ or β^- emission, collectively known as β decay, are accompanied by the emission of a neutrino or an antineutrino, (ν or $\bar{\nu}$), particles with zero electric charge and very small, perhaps zero, mass. The nucleus ^{82}Se has been observed to decay by the simultaneous emission of two β^--particles and two antineutrinos. The half-life for this decay is enormously long—1.1×10^{20} years. It is theoretically possible for certain nuclei to decay by emitting two β-particles with no neutrinos but this process has not yet been observed.

Compared with the time scale of most nuclear events, radioactive decay is an extremely slow process. For the emis-

Table I. Summary of the modes of radioactive decay of unstable nuclei with charge Z and mass A

Decay mode	Emitted particle	New Z	New A	Force responsible
α decay	α, (^{4}He)	$Z-2$	$A-4$	Strong
Proton decay	^{1}H, (proton)	$Z-1$	$A-1$	Strong
Neutron decay	neutron	Z	$A-1$	Strong
Fragment Z_1, A_1	e.g., ^{14}C	$Z-Z_1$	$A-A_1$	Strong
β decay	β^- (electron), $\bar{\nu}$	$Z+1$	A	Weak
	β^+ (positron), ν	$Z-1$	A	Weak
	ν only (EC)	$Z-1$	A	Weak
	$2\beta^-$, $2\bar{\nu}$	$Z+2$	A	Weak
γ decay	γ (photon)	Z	A	Electromagnetic
	electron (IC)	Z	A	Electromagnetic
Spontaneous fission	Nucleus breaks into two parts of various sizes	$Z/2$	$A/2$	Coulomb (electromagnetic)

sion of charged nuclear particles such as protons, α-particles, and heavier fragments, and for SF, the slowness is caused by a Coulomb potential energy barrier that inhibits the separation of the unstable nucleus into two charged parts. In fact, all nuclei with charge greater than about 66 are unstable to α decay, but unless the energy release is greater than 4 MeV, the Coulomb barrier makes the half-life so long that the decay is unobservable.

The processes of β decay and γ decay are slow because the fundamental forces that are responsible are so weak. In no case does the gravitational force play a detectable role. Each type of decay leaves a residual nucleus that is related in a definite way to the atomic number (nuclear charge, Z) and mass number A of the initial unstable nucleus. Table I summarizes the various decay modes.

Many radioactive nuclei are unstable with respect to decay by more than one mode. A ^{238}U nucleus, for example, can emit an α particle and produce a nucleus of ^{234}Th. Alternatively, it is also capable of spontaneously fissioning into two roughly equal parts as a result of the Coulomb forces arising from its high nuclear charge. Each mode of decay has its own characteristic half-life (its partial half-life). For ^{238}U, they are 4.51×10^9 years (α decay) and 8.0×10^{15} years (SF). The α-decay process is thus much more probable in this case. Very often, a nucleus produced by the decay of another is itself unstable. The ^{234}Th produced by the α decay of ^{238}U undergoes β^- decay ($t_{1/2} = 24.1$ days) to form ^{234}Pa, itself unstable. The chain of decays continues until finally the stable nucleus ^{206}Pb is reached. The products of spontaneous fission (as well of those of neutron-induced fission in nuclear power plants) are nearly always unstable. Most of them decay by the β^- process.

Very frequently, the decay of an unstable nucleus leads to the formation of a second nucleus in an energy-excited state. For example, the nucleus ^{56}Co (half-life 77 days) decays by positron emission (β^+ decay) or by electron capture (EC) to one of several excited states of ^{56}Fe, which in turn decays by γ-ray emission or by internal conversion to states of lower energy, and finally to the completely stable lowest-energy state (the ground state) of ^{56}Fe (the main component of ordinary iron). One of the excited states of ^{56}Fe (0.847 MeV above the ground state) decays to the ground state with a half-life of 6×10^{-12} s. Excited states with measurable half-lives are known as isomers (or isomeric states). The γ rays from ^{56}Co decay were observed from the 1987 supernova explosion SN 1987a. Radioactive decays to excited states have been an important source of information about nuclear spectroscopy and structure.

See also Alpha Decay; Beta Decay; Gamma Decay; Isomeric Nuclei; Nuclear Properties.

BIBLIOGRAPHY

R. D. Evans, *The Atomic Nucleus*. McGraw-Hill, New York, 1955. Particularly useful for statistics of radioactive decay.

J. K. Tuli, Nuclear Wallet Cards (1985). Available from National Nuclear Data Center, Brookhaven National Laboratory, Upton, NY 11973: a pocket-sized compilation of nuclear and radioactivity data.

E. Brown and R. B. Firestone, *Table of Radioactive Isotopes*. Wiley, New York, 1986.

W. M. Cottingham and D. A. Greenwood, *An Introduction to Nuclear Physics*. Cambridge University Press, New York, 1986.

K. S. Krane, *Introductory Nuclear Physics*. Wiley, New York, 1987.

S. B. Patel, *Nuclear Physics: An Introduction*. Wiley, New York, 1988.

Radioastronomy *see* Astronomy, Radio

Radiochemistry

Walter D. Loveland

Radiochemistry is the chemical manipulation of radioactivity. It is done as part of an attempt to solve problems in diverse fields such as physics, chemistry, biology, medicine, etc. Radiochemistry is often confused with *nuclear chemistry,* the study of nuclear phenomena by chemists, although there is no necessary connection between the subjects. (Chemists may study nuclei without utilizing chemical techniques or radioactivity, whereas the aim of a radiochemical procedure may have nothing to do with gaining insight into nuclear behavior.)

The chemical manipulation of radioactivity is similar to ordinary chemistry except the quantities of radioactive material are very small, usually unweighable. To avoid the complications of dealing with such small numbers of atoms, one dilutes the radioactive atoms with stable atoms of the same chemical species (a *carrier*). The assumption is that the stable and radioactive atoms will behave the same chemically. The inactive species will "carry" the radioactive species along during chemical procedures. Since most chemical phenomena depend only upon the electron configurations of the atoms and molecules, this assumption is usually justified. (Exceptions can occur in light atoms, such as hydrogen, where the difference in mass between different isotopes can affect the vibrational frequencies of chemical bonds, etc., leading to different chemical behavior.) For quantitative chemical work, the carrier must mix completely with the radionuclide. This mixing is often accomplished by making a series of oxidation/reduction reactions with the carrier/tracer system to ensure the carrier and tracer are in the same chemical oxidation state.

Frequently one will use radiochemical procedures to separate a desired radioactive species from a mixture of other radionuclides and inactive material. One evaluates these radiochemical separations by their: (a) chemical yield (fraction of the desired product recovered), (b) separation factor (which is a measure of the chemical purity of the product), and (c) decontamination factor (which is a measure of the radioactive purity of the product). The chemical yield of a separation is the fraction of the carrier recovered at the end of the separation. Chemical yields are measured using gravimetric methods, instrumental methods, tracer techniques, or activation analysis. Chemical yields need not be 100%, but they must be known. The separation factor, SF, is the ratio of the concentrations of the desired [D] and undesired species [UD], before and after the separation

$$SF = ([UD]_i/[D]_i)([UD]_f/[D]_f),$$

where i and f denote initial and final concentrations. Quite often, however, one is more interested in the radiopurity rather than the chemical purity of the sample. One will use the *decontamination factor,* DF, which is the ratio of the radioactivities of the desired and undesired species before and after the separation:

$$DF = (A_{UD}{}^i/A_D{}^i)(A_{UD}{}^f/A_D{}^f).$$

Radiochemical separations typically involve decontamination factors of 10^3 or more. These large decontamination factors result from repeated applications of a separation method. The overall decontamination factor is the product of the decontamination factors of the individual steps.

In the study of environmental radioactivities and in the search for unusual phenomena, such as the detection of solar neutrinos, *low-level radiochemistry* is used. Low-level radiochemistry involves the chemical manipulations of radionuclides whose counting rates are of the order of or less than background rates. In such procedures, one must pay special attention to the radioactivity levels present in all chemical reagents used, counter materials, etc.

In all radiochemical procedures, high- or low-level, a final step is the preparation of a carrier sample in a form suitable for radiation measurement. The sample size, weight, and thickness and its effect on the radiation being detected are important considerations in this step.

Radiochemistry involves the use of many different separation methods. Examples of these include precipitation, ion-exchange chromatography, gas chromatography, extraction chromatography, solvent extraction, distillation, and electrodeposition. Quite often these classical chemical techniques are combined with small laboratory computers to produce highly automated, rapid radiochemical separations. For example, the use of a supersonic jet of helium gas to transport radionuclides along with a microcomputer-controlled automated high-performance liquid chromatography system has allowed the separation of the actinide elements from each other within a few minutes. Similarly solvent extraction procedures using high-speed centrifuges for phase separation can effect chemical separations in fractions of a second.

The following examples are presented to illustrate the usefulness of radiochemistry in diverse applications.

(1) *The study of the actinide elements.* All the actinide elements are radioactive. Study of their chemical or nuclear properties or their utility as materials must involve radiochemistry to some extent. For many of the transuranium elements, a chemical identification of the atomic number of the element was a crucial part of its discovery. Continuing radiochemical studies of the chemical properties of the transuranium elements are part of the forefront of chemical research to understand relativistic effects upon atomic structure. Even studies directed toward understanding the structure of actinide nuclei quite often involve radiochemical procedures to prepare accelerator targets, etc.

(2) *The nuclear fuel cycle.* The nuclear fuel cycle consists of (a) the processing of all material necessary for nuclear power production, including uranium exploration, mining, refinement, and fuel element fabrication and (b) the reprocessing of spent fuel and radioactive waste handling.

During uranium exploration, mining, and refining, the uranium ore is crushed and concentrated. The uranium is leached from the ore using sulfuric acid to extract $UO_2(SO_4)_2{}^=$. This uranyl sulfate complex is then removed from aqueous solution by solvent extraction with a tertiary amine such as trioctyl amine. Removal of uranium from the organic solvent through a series of steps leads to the intermediate compounds, $Na_2U_2O_7$, "yellow cake;" UO_3, "orange oxide;" and UF_4, "green salt." The final products are UO_2, UC, UF_6, and uranium metal which are fabricated into fuel elements.

Following their use as reactor fuels, the spent fuel elements contain a radioactive mixture of actinide elements and fission products. These fuel elements are reprocessed to extract valuable products and to lessen the hazards associated with nuclear waste. [Reprocessing of fuel from military applications occurs in many countries (including the United States) as does commercial reprocessing (with the exception of the United States)]. In fuel reprocessing, the fuel elements are chopped up and dissolved in HNO_3. The separation of U, Pu, and the fission products is most commonly done by a solvent extraction process known as the Purex process (*Pu-U*ranium-*Ex*traction). The extracting agent is tributyl phosphate (TBP) dissolved in kerosene. The separation is based upon the differing chemical properties of the oxidation states of Pu, U, and the fission products. In a first step, Pu(IV) and U(IV) in ~6*M* HNO_3 are extracted into TBP-kerosene, leaving 99% of the fission products behind in the aqueous solution. Then, the plutonium is reduced to the +3 state. In a second extraction step, the uranium remains in the organic phase while the plutonium is extracted into the aqueous phase. Subsequent processes purify the plutonium and uranium, treat the high-level waste, and recover process chemicals.

(3) *Isotope dilution.* The measurement of the volume of liquid or gases in large or irregularly shaped objects can often be conveniently determined by the isotope-dilution method. A small amount of a solution containing a known amount of a radioactive tracer is added to the unknown object. After sufficient time has elapsed to ensure complete mixing of the tracer solution in the object, a small, measured volume is withdrawn. The volume of the object is then the ratio of the activity of the initial tracer solution added to the activity in the volume withdrawn multiplied by the volume of the withdrawn solution. Water, which has been labeled with tritium, ^{3}H, is frequently used for isotope-dilution measurements such as blood volumes in animals, or volumetric determination in large bodies of water.

(4) *Tracer studies of reaction mechanisms.* A frequent question which arises in studies of the reaction between chemical compounds concerns the fate of particular atoms. Consider the decarbonylation of benzoylformic acid to yield benzoic acid and carbon monoxide:

$$C_6H_5\text{-}C(=O)\text{-}{}^{14}C(=O)OH \xrightarrow{H_2SO_4} C_6H_5\text{-}C(=O)OH + {}^{14}CO$$

Benzoylformic acid — Benzoic acid — Carbon monoxide

Without the use of a ^{14}C tracer, it is not possible to decide whether the carbon monoxide comes from the carbonyl group or the carboxyl group. When the reaction is carried out with a ^{14}C label on the carboxyl group, the label appears on the carbon monoxide, indicating that the carboxyl group was the origin of the ^{14}CO. Alternatively, a label on the carbonyl group results in retention of the ^{14}C radioactivity in the benzoic acid. Tracer studies of this type have had wide application in determining the detailed fate of atoms and molecular fragments in chemical systems.

(5) *Activation analysis.* The transmutation or ''activation'' of stable isotopes into radioactive ones permits quantitative determination of the amounts of elements present in matcrials by the methods of activation analysis. In most cases this activation is achieved by irradiation of the material in a reactor, since several important physical phenomena combine to advantage in this technique. Almost all elements undergo activation by thermal neutron capture. Coupled with this is the fact that modern nuclear reactors are capable of producing intense fluxes of neutrons ($\Phi \sim 10^{15}$ neutrons/cm^2 s) that easily penetrate into matter. Thus, simultaneous determination of many elemental concentrations with high sensitivity is possible. In certain favorable cases for elements with large neutron capture cross sections, sensitivities as low as 10^{-12} g/g of sample can be attained. In addition, in most cases the analysis can be performed by a comparison method in which the activation of an element in the sample to be analyzed is compared with the extent of activation of a known amount of the same element in a standard sample. As long as the sample and standard are irradiated for the same time in an identical environment, detailed knowledge of the neutron flux and the neutron capture cross section is unnecessary since the extent of activation will be in proportion to the weights of the element undergoing activation in the two materials. The extent of activation is determined quantitatively by measuring the activity of some characteristic radiation (usually γ rays). The uniquely high sensitivity of activation analysis has been applied to the determination of trace-element abundances in numerous materials, e.g., semiconductors, lunar and meteoritic materials, high-purity metals, and environmental samples. Forensic activation analysis seeks to use the pattern of trace-element abundances, much as one does with a fingerprint, to ascertain the source of materials or to authenticate, in the case of artifacts and art objects, the provenance of these objects.

The most important limitations to the technique include the poor activation properties of some of the light elements, e.g., H, He, Be, C, N, and O, the inability of activation analysis to distinguish chemical states, and the difficulties of the analysis of complicated γ-ray spectra when samples contain many different elements.

(6) *Radioactive dating.* The characteristic half-lives of radioactive nuclei can, in favorable cases, be used to date objects. The ^{14}C dating technique has had numerous applications in archaeology. Interaction of cosmic rays with the earth's atmosphere produces small amounts of ^{14}C. Living organisms continuously exchange this carbon activity with their environment during their lifetimes. Once the organism dies, however, the exchange ceases and the activity level in the dead organism decreases with the half-life of ^{14}C ($T_{1/2} = 5730$ years). Thus the specific activity of ^{14}C can be used as a measure of the time which has elapsed since the organism's death. Bones and other organic material from prehistoric sites can frequently be dated by this technique, provided some ^{14}C still remains in the material.

On a longer time scale, the naturally occurring radioactivity of uranium and thorium can be used to ascertain the geological age of minerals. The end products of the ^{238}U and ^{232}Th decay chains are ^{206}Pb and ^{208}Pb, respectively. For those minerals which do not contain lead from other sources, the relative amounts of these stable lead decay products to their radioactive precursors are a measure of the minerals' geological age. Similar radioactivity clocks can be developed from measurements of the ratio of other long-lived and stable isotopic species, e.g., $^{40}K/^{40}Ar$ or $^{87}Rb/^{87}Sr$.

BIBLIOGRAPHY

C. H. Wang, D. L. Willis, and W. D. Loveland, *Radiotracer Methodology in the Biological, Environmental, and Physical Sciences.* Prentice-Hall, Englewood Cliffs, N.J., 1975. (E). An introduction to the use of radiotracers, nuclear science and radiation detection.

G. Friedlander, J. W. Kennedy, E. S. Macias, and J. M. Miller,

Nuclear and Radiochemistry, 3rd ed. Wiley, New York, 1981. (I). The classic textbook of radiochemistry.

G. R. Choppin and J. Rydberg, *Nuclear Chemistry, Theory and Applications*. Pergamon, Oxford, 1980. (I). Especially recommended for its discussions of the applications of radiochemistry.

P. Kruger, *Principles of Activation Analysis*. Wiley-Interscience, New York, 1971. (I). Contains an excellent discussion of radiochemistry.

Radiological Physics

Colin G. Orton and Gail D. Adams

INTRODUCTION

Medical physics is the application of physics and physical principles to the diagnosis and treatment of disease. Radiological physics is that branch of medical physics which involves the use of radiation, both ionizing and nonionizing. Such radiation includes the whole spectrum of electromagnetic radiation from microwaves through x and γ rays together with other radiation such as ultrasound.

Radiological physics as a science started in earnest soon after the discovery of x rays and radioactivity at the end of the 19th century when these new modalities began to be used for diagnostic imaging and radiotherapy. By the 1930s, with the advent of particle accelerators, the 1940s with the birth of nuclear medicine, and the 1950s with the widespread employment of ^{60}Co teletherapy units, physicists, and especially atomic and nuclear physicists, began to enter the radiological physics profession in large numbers. New applications of radiation in medicine were developed and by the 1960s departments and divisions of medical or radiological physics began to be created, initially primarily in Europe where several educational programs were started, and later in North America. There are now numerous educational programs available worldwide ranging from nondegree training opportunities through postdoctoral fellowships. Active areas of application and research include diagnostic radiology, radiation therapy, nuclear medicine, nuclear magnetic resonance, lasers, hyperthermia, thermography, radiation dosimetry, and neuromagnetometry, and the use of computers in all of the above.

Most radiological physicists work directly with patients and physicians. They are often directly responsible for the quality of care delivered to individual patients, yet they are not legally allowed to practice medicine. However, in many countries radiological physicist certification and licensure (or equivalent) is becoming prevalent. For example, in Canada you can become board certified by examinations conducted by the Canadian College of Physicists in Medicine, and in the United States physicists are certified by the American Board of Medical Physics, the American Board of Radiology, the American Board of Health Physics, or the American Board of Scientists in Nuclear Medicine, and state licensure is just beginning.

RADIATION THERAPY

Roentgen discovered x rays in 1895. By 1897, x rays were being used to treat cancer. Although some benign conditions occasionally benefit from ionizing radiation, about 99% of a department's workload remains in the treatment of cancer.

Cancer treatments are based on two observations: (a) the deposition of energy from ionizing radiation may inhibit cell division and (b) cancerous and normal tissues have different capacities for recovery from ionizing radiation exposures. There is a relationship between the amount of energy deposited and the degree of induced inhibition of cell division. Further, differing cell types vary in sensitivity and in capacity to repair radiation damage. The success of radiation therapy depends on dosimetry, but it also depends on accurate localization of the cancer volume and on the differential recovery pattern between cancerous and normal tissues when a total radiation dose is spread over several weeks.

Dosimetry, the determination of energy absorbed in a volume of interest, is a central consideration; absolute units are required. Ionization chambers and solid-state detectors are used for measurements. The typical situation involves x rays in an energy range where Compton attenuation predominates. Field sizes are neither so small as to have negligible scatter contamination nor so large as to have maximum scatter.

The objective of radiation therapy is to deliver a high dose of radiation to the tumor without exceeding the tolerance dose to intervening and surrounding normal tissues. For superficial tumors this is easily achieved by the use of low-energy x rays or electron beams which are poorly penetrating. For deeper tumors, a cross-firing (or multiple-beam) technique is employed such that the cancer is always irradiated by all beams but the normal tissues are exposed to only some of the radiation. Higher-energy beams are employed for these treatments, such as γ rays from ^{60}Co units and x rays from linear accelerators and microtrons, with maximum energies as high as 50 or even 70 MeV (Fig. 1).

One of the principal tasks of the radiotherapy physicist is to determine what radiation to use for individual patients, and how best to arrange the radiation fields in order to avoid overexposure of sensitive organs and tissues that are "in the way." This is called "treatment planning." Due to the complexity of such work, this is normally done with the aid of dedicated computers and the development of software for these treatment-planning computers is a continuing source of employment for radiological physicists in radiotherapy facilities, where new treatment techniques are always under development (see below).

Dynamic Radiotherapy

Dynamic (or conformation) radiotherapy involves the continuous manipulation of the spatial orientation of the beam and the patient in order to obtain the required shape of the high-dose region in the patient. Typically this is done using a rotational technique in which the therapy beam is rotated around the patient while at the same time various other parameters are changed under computer control. These param-

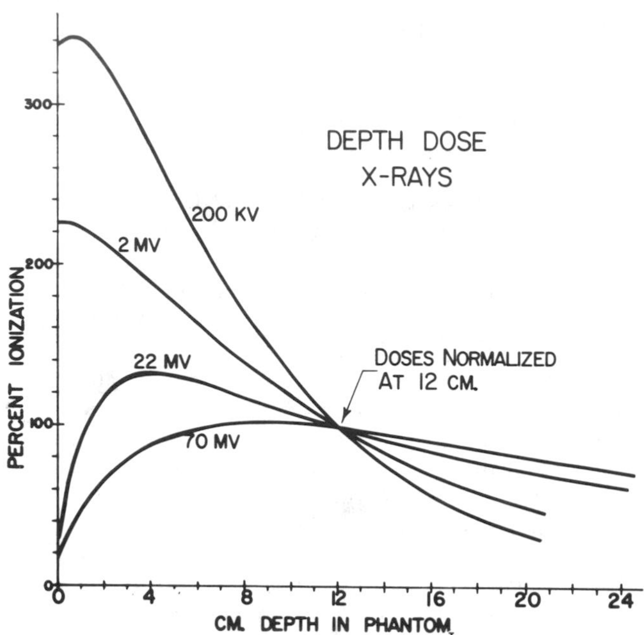

FIG. 1. Comparison of relative doses delivered in water along the central axis for x-ray beams generated by electrons of energies from 0.2–70 MeV incident upon W or Pt targets. In radiation therapy it is frequently necessary to deliver a required dose to a deep-seated tumor (in this case at a depth of 12 cm), hence the rationale for dose normalization.

eters include some or all of the following: beam size, beam output (for an accelerator) or rotational speed (for a ^{60}Co unit), beam orientation, and patient orientation (*X*-, *Y*-, and *Z*-axis linear movements).

Planning of dynamic treatments is extremely tedious as are the associated quality-assurance procedures. Consequently, dynamic therapy has not become commonplace, despite the fact that the dose distributions obtained appear to be superior to those normally used for patients. However, linear accelerator manufacturers are beginning to build such computer-controlled motion capabilities into their machines so dynamic therapy should soon develop into a more clinically acceptable procedure.

Intraoperative Radiotherapy

An alternative procedure for putting the radiation where the cancer is without excessive irradiation of surrounding healthy tissues is to perform intraoperative radiotherapy (IORT) whenever possible. With IORT, a beam of radiation, usually electrons, is applied directly to the tumor while it is exposed during an operative procedure. This is normally done as a boost therapy to supplement a normal course of external beam treatments. Ideally, this procedure should be performed in the operating room suite of the hospital. However, to equip and shield such rooms is expensive so most IORT is performed in specially designed rooms in the radiotherapy department. When not used for IORT, these rooms and machines are used for conventional treatments.

Stereotactic Radiosurgery

Another beam-shaping technique which is gaining in popularity is stereotactic radiosurgery. This involves the use of multiple small-pencil beams of radiation fired from many different directions and all aimed at the tumor. Presently this is restricted to the treatment of brain lesions such as small cancers and A-V malformations. Machines used include the "gamma-knife," with several hundred small, high-activity ^{60}Co sources and, more commonly, conventional linear accelerators equipped with specially designed stereotactic hardware. Radiological physicists have done considerable developmental work to make this into a technique achievable and affordable in any modern radiotherapy facility.

Brachytherapy

For some treatments, it is possible to place the sources of radiation directly within the tumor. This is called brachytherapy and employs radioactive plaques, needles, tubes, wires, or small "seeds" made of such radionuclides as ^{137}Cs, ^{60}Co, ^{192}Ir, ^{125}I, and ^{90}Sr. These sources are either placed over the surface of a tumor (surface plaques or moulds), implanted within the tumor (interstitial implants), or placed within a body cavity surrounded by the tumor (intracavitary therapy). Normally these treatments utilize photon- or β-particle-emitting isotopes, although some work has demonstrated the utility of other radionuclides such as ^{252}Cf which emits neutrons from spontaneous fission.

Heavy-particle Therapy

The reason for using neutrons in radiotherapy is mainly radiobiological. Neutrons and other heavy particles are found to be more effective than conventional radiations at damaging cells in poorly oxygenated regions of tumors. It is these hypoxic cells which are thought to make some tumors resistant to normal radiotherapy treatments. This has led to significant research into the use of neutrons, π^- mesons, and stripped nuclei for cancer therapy. Since neutrons provide the least expensive form of these alternative radiations, several high-energy neutron therapy facilities have been established with 50-MeV beams capable of penetration similar to that of 4–8-MeV x rays (Figs. 1 and 2). The π^- mesons and stripped nuclei have an additional advantage of a Bragg peak in their tissue absorption curve (Fig. 2). The depth of this peak depends on the beam energy, which is readily adjustable to correspond to the depth of the tumor. Unfortunately, the cost of the accelerators required to produce such beams is excessive, so they are not widely employed and, second, the Bragg peaks are too narrow to treat most tumors so they have to be widened by range-shifting filters which reduce both the physical and biological advantages of these beams. However, the Bragg peak phenomenon *is* used with beams of high-energy protons (100–300 MeV) although presently only for the treatment of small tumors such as ocular melanomas. Only a few machines capable of large-field proton therapy exist and these are very expensive.

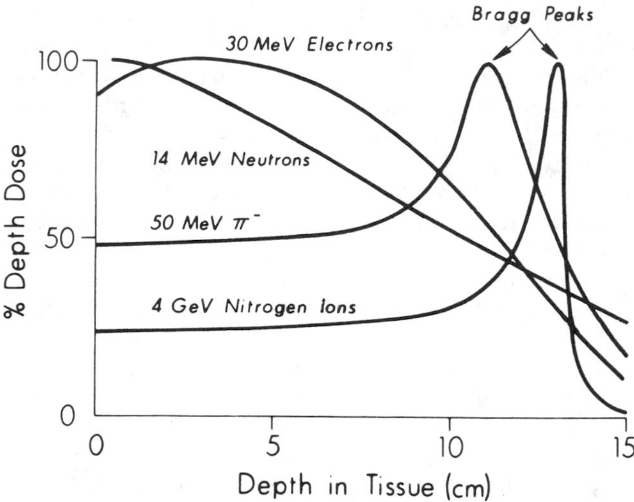

FIG. 2. Depth–dose curves for beams of high-energy particles.

Monoclonal Antibody Therapy

An exciting new development in radiotherapy is the application of monoclonal antibodies tagged with high activities of suitable radionuclides. These tumor-specific antibodies are derived from the patient's own cancer and hence they selectively target this tumor when injected into the patient. The selection of the appropriate radionuclides (usually emitters to restrict the range of irradiation) and the associated dosimetry when they are deposited in cells in the tumor is an active area of research in radiological physics. When tagged with emitters, these monoclonal antibodies are also useful for defining the position and extent of tumors using nuclear medicine procedures.

Hyperthermia

Hyperthermia is the application of heat to destroy cancer cells and is often used in conjunction with radiotherapy. Temperatures of 43°C and above sustained for up to 1 h will destroy cells, both cancerous and normal. When combined with radiation, these high temperatures are believed to enhance the relative sensitivity of otherwise resistant cells, and this is the rationale for such hyperthermia cancer therapy. Physicists are developing various heating mechanisms utilizing microwaves, radio-frequency radiations, and ultrasound and are researching methods to accurately monitor the temperature of tissues.

Photodynamic Therapy

Photodynamic therapy (PDT) is the application of light to destroy cancer cells which have been made sensitive to such irradiation by the use of certain drugs called hematoporphyrins. Radiological physicists are researching ways to deliver such light treatments in the most effective manner, often by the use of lasers.

DIAGNOSTIC RADIOLOGY

Radiation therapy relates to the application of radiation to treat disease situations that have been diagnosed; in contrast, diagnostic radiology relates to the application of radiation to humans to assist in determining the existence or extent of disease. In this case, the radiation is only a means to an end. X rays are used; maximum energy of the bremsstrahlung ranges from 20 keV to 2 MeV, but most often is 40–120 keV. Differential penetration of these x rays through the body produces a pattern attributable to variations in density and atomic number of body tissues. This pattern may be recorded by a detector and later studied by the physician to elicit possible disease status.

The physician requires an image of the subject that has "diagnostic value." Some of the uneasy questions posed to the investigator in this area are: What is required to achieve diagnostic value? What range of diagnoses is possible in varying circumstances? How should conflicts be resolved between the various possible ways to record images, including assessment of differing diagnostic values? How much radiation dose should a patient receive to achieve an acceptable level of diagnostic value? Obviously, solid answers are needed.

The overall problem has national population significance. There are over 500 million x-ray films taken on the U.S. population each year, with varying contributions to the dose to the population. Thus, we should strive to secure the greatest diagnostic value with minimum radiation exposure.

Attacks on the problem may be sorted into four groups: theory of images (discussed later), development of imaging detectors, identification and adoption of quality assurance programs, and introduction of new techniques.

Radiological physicists have developed a wide variety of imaging devices. The simplest, and still the workhorse of radiology, is radiographic film with or without intensifying fluorescent screens. Ultrafast screens have been developed which have significantly reduced patient radiation doses. Other devices, using electrostatic imaging such as xerography and electron radiography, have proven useful for special applications. Recently, so-called *filmless radiography* has been developed using phosphor-coated reusable materials as imaging devices.

Electronic systems based upon image amplifiers have replaced conventional fluoroscopy for real-time imaging and, combining this technology with computers, the field of *digital imaging* has emerged. With digital imaging systems, contrast and intensity can be readily manipulated and, most importantly, images can be transmitted electronically to remote terminals. Many radiology departments are converting from conventional to digital imaging for most radiological procedures. Also with digital techniques it is possible to electronically overlay images before and after contrast media have been injected. This is the principle of digital subtraction angiography.

Computerized tomography (CT) remains the most significant development in radiology since the discovery of x rays. In modern CT scanners, narrow "pencil" beams of x rays pass through the patient and the transmitted intensity of each beam is measured by a series of opposing small detectors.

This entire array of pencil beams and detectors is rotated rapidly around the patient and transmission data is collected and computer analyzed for each 1° of rotation. Mathematical transformation of all this information results in a grey scale or color image representing the attenuation characteristics of all tissues in the body section (or slice) studied. These images have considerably higher contrast than conventional radiographs and are produced in seconds. An example is shown in Fig. 3. CT scanners have revolutionized the practice of radiology and have completely replaced many previous procedures.

Quantitative CT scanning, in which the electron density of tissues is determined, is an active area of radiological physics research and application. It is applied in radiation therapy to account for the differential attenuation of radiations as they pass through the body and also in *bone densitometry* in the study of osteoporosis. Often dual-energy scanning is required in order to achieve adequate sensitivity to small changes in bone density. Dual-energy radionuclide scanning is also used for bone densitometry.

NUCLEAR MEDICINE

The term nuclear medicine includes all uses other than brachytherapy of incorporated radioactive material (RM) in the diagnosis and treatment of disease. Most radionuclides used for radiopharmaceuticals are products of fission processes; a few are cyclotron produced. Notable at present is the availability of ^{99}Mo contained on an alumina frit, from which may be eluted ^{99m}Tc. The 140-keV γ from ^{99m}Tc is especially useful for detection; the short half-life minimizes the radiation dose to the patient.

Procedures involve the localization of RM as a result of some physiological activity, for example, as the iodide ion in the thyroid gland or as colloidal particles in reticuloendothelial cells.

Diagnostic procedures employing RM may be segregated according to the kind of information sought: quantization of the amount of activity or recording of activity distribution.

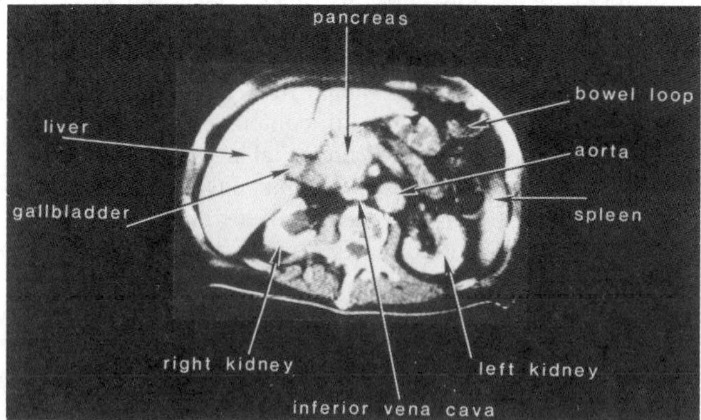

FIG. 3. Image of upper abdomen of a patient, obtained using a computerized tomography (CAT) scanner. Several of the major internal structures are noted. Some individual muscles and sections of ribs may also be seen. (Courtesy of Dr. E. C. McCullough and the Mayo Clinic.)

Detectors are mainly NaI(Tl) crystals in appropriate geometries, with solid-state crystals (e.g., intrinsic germanium) achieving some importance. The associated electronics normally contains single-channel or multichannel analyzers and includes dedicated minicomputers in many instances.

The recording of activity distribution is the production of images. A moving-probe scanner with output related to counting rate was used for many years to give images of reasonably static RM distribution patterns. It was a slow, time-consuming procedure—one that demanded no small cooperation by the patient. The development of the gamma camera considerably reduced the time required for static imaging, such as in Fig. 4; it has made available the capability of recording a sequence of images (at, say, 2-s intervals). This introduces dynamic function studies. But, of course, with such sophistication there coexists the need for quality control of the overall system to protect both the patient and data.

Dosimetry in nuclear medicine is of consequence. In fact, the details, such as the role of Auger showers, are more highly developed for RM than for the photon beams of radiation therapy. As in the case of diagnostic radiology, the effort is to achieve maximum information of diagnostic value for minimum radiation dose to the patient (and to other nearby persons as well). The need for assessment of activity in absolute units with sustained accuracy over periods of years has required the inception of quality-assurance programs relating to measurements.

Many significant developments in nuclear medicine have been made by radiological physicists, probably the most notable being the *radioimmunoassay* (RIA) technique by Rosalyn Yalow and Solomon Berson. This method is the basis of diagnostic and physiological assays used throughout the world and earned the developers the Nobel Prize.

Tomographic imaging in nuclear medicine can be achieved by either *single photon emission computerized tomography* (SPECT) or *positron emission tomography* (PET) scanning. SPECT scanning involves the rotation of detectors around the patient and is analogous to CT imaging with x rays. PET scanning, however, achieves its tomographic capabilities by using an array of stationary detectors around the patient and using the spatial 180° opposing properties of the 0.511-MeV annihilation radiation from positron-emitting radiopharmaceuticals deposited in the organ or region of interest. A variety of biologically important compounds labeled with positron-emitting nuclides of carbon, nitrogen, oxyen, and fluorine are used to study metabolic processes and organ function in patients. The positron emitters are very short-lived and are usually produced by an in-house cyclotron, often the same cyclotron as used for neutron therapy.

NUCLEAR MAGNETIC RESONANCE

The most recent major breakthrough in radiological physics was the development of nuclear magnetic resonance techniques for the diagnosis of disease. In *magnetic resonance imaging* (MRI) the patient is placed in a strong magnetic field of 0.5 T up to 2 T and sometimes even higher in superconducting machines. This magnetic field induces a net magnetization of the nuclei in the patient. A short rf pulse in

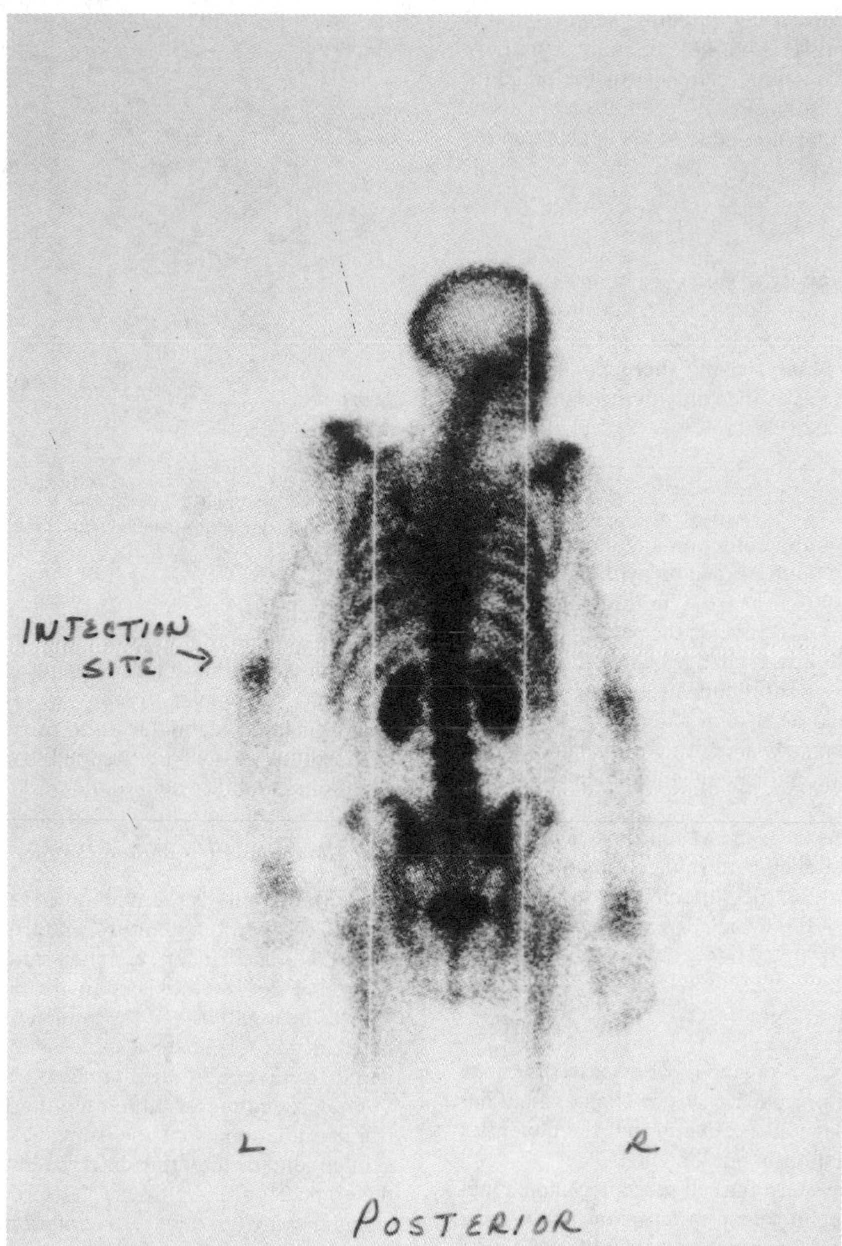

FIG. 4. Image of a "total-body" scan, in this case searching for metastatic spread of a cancer to bone. ^{99m}Tc-labeled diphosphonate had been administered; the bony uptake is in relation to metabolic activity in the bone. Two vertebra and three ribs are clearly involved. The kidneys and the urinary bladder are also prominent, being the elimination route for the radiopharmaceutical. (Courtesy of Dr. C. W. Smith.)

applied at the Larmor frequency of the precessing nuclei which, in turn, emit an rf signal of the same frequency. This is detected and gives a "fingerprint" of the chemical environment of the nucleus being studied. This information is one dimensional but is converted into a two-dimensional anatomical image by adding a gradient to the applied magnetic field which results in a frequency modulation of the emitted rf signal. A series of such measurements is analyzed by computer to generate the image. MRI is somewhat more versatile than CT in that the contrast between anatomical structures can be manipulated by selection of different timing sequences of the excitation rf pulses resulting in so-called T_1 and T_2 images, which can be quite different showing different features of the anatomy. Also, metabolic processes can be studied by the use of ^{31}P imaging. The basic difference between CT and MRI is that CT provides anatomical structure data only, whereas MRI also provides information on physiological, functional, and metabolic properties.

Magnetic resonance spectroscopy (MRS) is just beginning to be used to study biochemical and metabolic characteristics of diseased tissue in humans *in vivo*. Chemical shift spectra are generated by the use of surface coils and very high field (up to 4 T) superconducting magnets. It is possible to assay levels of various metabolites and to determine tissue acidity noninvasively in just a few minutes. With appropriately designed surface coils it is possible to study separately the spectra from specific volume elements of the patient's tissues and organs. The study of tumors, their metabolism, and their reaction to treatment is an obvious application receiving great interest.

LASERS

Lasers are used for a variety of purposes in medical treatment. They are used in high intensities to destroy tissues blocking blood vessels and bronchial passages. In lower intensities they are used in photodynamic therapy. One branch of radiological physics involves the study of the transmission of laser light through tissues.

NEUROMAGNETOMETRY

Neuromagnetrometry is the detection and mapping of the extremely weak magnetic fields associated with the electrical currents flowing within active neurons in the human brain. Neuromagnetic imaging (NMI) reveals the pattern of activity in the brain and how this pattern changes with external stimuli. The NMI has been used to delineate the regions of the brain associated with disorders such as epilepsy and is expected to become increasingly useful as a diagnostic tool, especially as a means of monitoring the effectiveness of therapeutic procedures.

Because of the weakness of these neuromagnetic fields, physicists have had to develop ultrahigh sensitivity measurement techniques such as the superconducting quantum interference device (SQUID) which is extensively used in NMI equipment.

ULTRASONOGRAPHY

Ultrasound is used both in diagnosis of disease situations and in the evaluation of normal variations such as fetal development. It is particularly attractive in that the biological risk is minimal; the radiation is nonionizing.

There are three basic modes of use at present. Simple ranging (recording times of echo return in terms of the distance of an echo-generating surface from the transmitter–receiver probe) is called A-mode. If the signal, as on an oscilloscope, is presented only as a bright dot when the echo is received, the system is B-mode. Moving the probe in any scanning motion will enable a body-section image to be presented from a two-dimensional sequence of echoes as illustrated in Fig. 5. If the structure producing one or more of the echoes changes location with time, particularly if quasiperiodic, it is useful to record B-mode dots on a moving photosensitive strip, thus yielding M-mode.

Overlaid with the basic presentation modes is the application of the Doppler effect. An output signal may relate to the existence of frequency change, to the amount of frequency change, or to a time-dependent pattern of frequency changes. Doppler ultrasound finds particular use in the detection and characterization of fetal life and in studies of the cardiovascular system.

Images in ultrasonography are of tissue interfaces; the

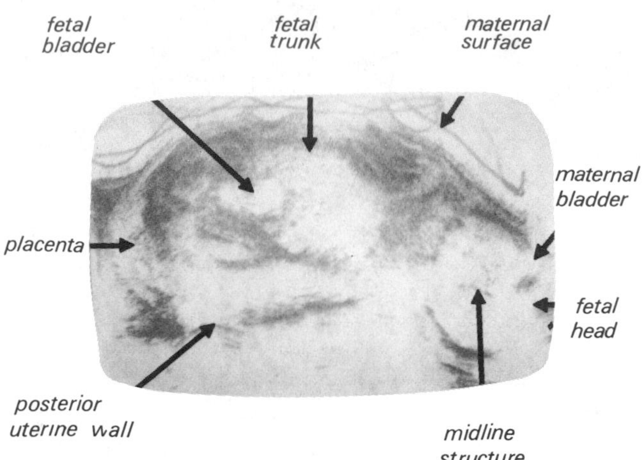

FIG. 5. Ultrasonograph in sagittal section of a fetus in the mother's abdomen. Prominent structures are labeled. (Courtesy of Dr. R. E. Brown.)

available clarity of details is generally less than in x-ray applications. However, recent developments incorporating digital imaging techniques and improvements in the design of transducer arrays have tended to improve the overall quality of ultrasound images.

THERMOGRAPHY

Thermography is the science and practice of recording images of surface temperature patterns in a subject. The military have used it for reconnaissance, engineers have used if for heat-flow studies, and medicine is using it for assessing surface heat patterns of the human body. Abnormal vascular patterns may indicate a cancerous growth; inflammations and arthritic conditions may also be recognized.

The production of a thermogram requires detection of infrared radiation from the subject. A variety of solid-state semiconductor materials have been and are continuing to be developed for this purpose. To produce a two-dimensional signal display (an image), a camera must either scan the subject, using a small-signal detector, or focus the two-dimensional emission pattern on a detector array. Infrared camera development and evaluation is an active area of research.

The importance of the technique can be seen in its demonstrated value in mass screening for the early detection of breast cancer. It carries no possible radiation hazard. Figure 6 shows a thermogram from which a suspected cancer diagnosis was made, a diagnosis proven to be correct at surgery.

IMAGES

As seen in the foregoing, the formation of images having diagnostic value is a central issue in diagnostic radiology, nuclear medicine, ultrasonography, and thermography. A substantial research effort is proceeding toward understanding the principles of image formation and the properties of the various transducer elements interposed between the ra-

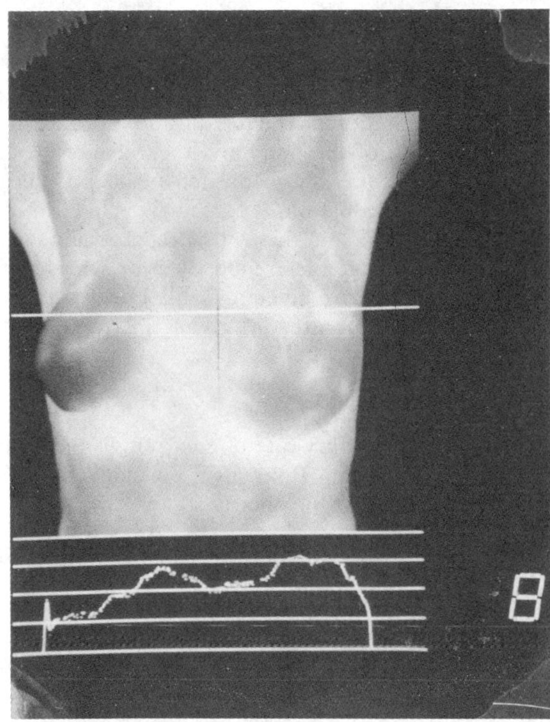

FIG. 6. Thermogram (infrared image) of a woman with an area on the left breast (on the viewer's right) of elevated temperature, hence suspect for cancer. Higher temperatures radiate more, thus appear lighter. The graph at the bottom presents a temperature scan along the line shown. (Courtesy of Dr. T. J. Love.)

diant energy from the subject and the formation of the output presentation.

A major portion of the present attack is adapted from optics. The dimension-space characteristics of the subject are transformed to frequency space and the capability of the transducer elements (or overall system) is determined in frequency space. The image differs from the subject because of the inability of the transducer to transmit all frequencies present without attenuation.

PROTECTION

The willful application of ionizing radiation to people carries the concomitant responsibility of exposure evaluation to the patient, to radiation workers, and to any member of the general population who may be nearby. Particularly in these last two categories there is an area in common with health physics.

The particular activities include design and testing of protective enclosures for radiation generators and emitters, measurement of radiation quality and quantity emitted from a generator or source, evaluation of operational characteristics of radiation generators (especially of diagnostic x-ray generators), evaluation of special-situation exposures, and maintenance of records relevant to radiation safety activities.

SPECIAL INTERESTS

Applications of physics to medicine continue to spread into new areas. Some years ago, lithium fluoride and other thermoluminescent materials were introduced into the armamentarium of dosimeters. More recently, other radiation-activated properties of solids have been investigated, such as radiation-induced conductivity.

In fact, *radiation dosimetry* has been an active area of research and development throughout the history of radiological physics. The quest for accurate, reproducible methods to measure the dose to patients, especially in radiotherapy, has led to the study of cavity ionization theory and the development of several national and international protocols for the measurement of photon, electron, and neutron radiations.

Electron-spin resonance is of interest, at least in that some cancerous tissue is recognizably different from normal tissue of the same organ.

The detection and quantitation of small quantities of abnormal materials is receiving attention. Variously, flame atomic absorption, proton-induced x-ray excitation, and x-ray fluorescence are under investigation.

In addition to some of the intimations in the preceding sections, mathematical developments are brought to physiological processes, to decision making, and to observer evaluation.

COMPUTERS

No review of radiological physics would be complete without mention of the role of computers, especially dedicated minicomputers.

As second-generation computers were becoming available (in the 1950s) they were being used in medicine, as elsewhere, as fast arithmetic processors. In this respect, they have made the most contribution in therapy where they are used to produce two-dimensional isodose displays for treatment planning. Almost all departments now have minicomputers dedicated to this purpose.

As succeeding generations of computers are developed, their utilizations become widespread: image manipulation and analysis; scheduling patients for procedures and recording their performance; patient histories, with sophisticated retrieval techniques; simultaneous acquisition of data from a random-event source while processing other data; and process control. It was necessary in several descriptions in this article to note the involvement of dedicated minicomputers in routine work. They are, of course, addressable also for investigative activities.

See also RADIATION INTERACTIONS WITH MATTER; RADIOCHEMISTRY; ULTRASONICS; X-RAY SPECTRA AND X-RAY SPECTROSCOPY.

BIBLIOGRAPHY

General (free from distributors)

Bibliography of Books in Print in Medical Physics and Related Fields. Medical Physics Publishing Corporation, Madison, WI.

The Medical Physicist and *The Roles, Responsibilities, and Status of the Clinical Medical Physicist*. American Association of Physicists in Medicine, New York.

Specialized

F. H. Attix, *Introduction to Radiological Physics and Radiation Dosimetry*. Wiley, New York, 1986.

T. S. Curry *et al.*, *Christensen's Introduction to the Physics of Diagnostic Radiology*, 3rd ed. Lea and Febiger, Philadelphia, 1984.

H. E. Johns and J. R. Cunningham, *The Physics of Radiology*, 4th ed. Thomas, Springfield, IL, 1983.

F. Khan, *The Physics of Radiation Therapy*. Williams and Wilkins, Baltimore, 1984.

F. Mettler, *Magnetic Resonance Imaging and Spectroscopy*. Churchill, Livingstone, New York, 1986.

J. A. Sorenson and M. E. Phelps, *Physics of Nuclear Medicine*, 2nd ed. Saunders, Philadelphia, 1986.

P. N. T. Wells, *Scientific Basis of Medical Imaging*. Churchill, Livingstone, London, 1982.

Radiometry

William L. Wolfe

INTRODUCTION

Radiometry is the science of the measurement of radiation. The main applications are in the visible and near-visible portions of the electromagnetic spectrum, from about 0.1 to about 1000 μm. The portion of radiometry that deals with visual responses and illumination is called photometry. The main subjects are the generation and transfer of radiation, interaction with matter, standards, calibration, and measurement.

GENERATION AND SOURCES

The fundamental source is the theoretical construct of a blackbody, one that absorbs all the radiation incident on it and one whose radiation characteristics are dependent only on the temperature of the body and the wavelength of the radiation. The flux per unit area of such a blackbody is given in terms of Planck's equation. It can be given in terms of the flux rate of photons or the flux rate of energy and in terms of the independent-variable frequency ν or wavelength λ. The different versions are given in Table I. The total power or the total photon rate is given in terms of the Stefan–Boltzmann law:

$$\Phi=\sigma T^4 A, \qquad \Phi_q=\sigma' T^3 A$$

where Φ is the power (flux rate of energy), Φ_q the flux rate of photons, σ the Stefan–Boltzmann constant, σ' the Stefan–Boltzmann constant for photons, and A the area of the source.

The maximum of the blackbody curve is given by the Wien displacement law. For power this is

$$\lambda_m T=2898\ \mu\text{m K}.$$

For photons it is

$$\lambda_m T=3670\ \mu\text{m K}.$$

The maxima depend on the different forms of the Planck function.

No material and no source is actually a perfect blackbody; its variation is described by its emissivity or emittance. The accepted definition of (hemispherical) emissivity is the ratio of power per unit area emitted into a hemisphere by the body in question to the same quantity emitted by a blackbody at the same temperature

$$\epsilon(\lambda\ T)=\Phi/\Phi^{BB}.$$

In general, emissivity is a function of the temperature of the

Table I Density of states function D in the Planck function. The independent variable is given as y. The functions are number density n_y, energy density w_y, photon exitance M_{qy}, radiant exitance M_y, photon radiance, and radiance. Each entry is D, the multiplier of $(e^x-1)^{-1}$ in the Planck expression.

y	k	$\tilde{\nu}$	ν	$x=h/kT$	λ	ω
$n_y=N_y/v$	k^2/π^2	$8\pi\tilde{\nu}^2/c^3$	$8\pi\nu^2/c^3$	$8\pi\left(\frac{kT}{ch}\right)^3 x^2$	$8\pi\lambda^{-4}$	$\omega^2/c^3\pi^2$
$w_y=h\nu n_y$	$chk^3/2\pi^3$	$8\pi ch\tilde{\nu}^3$	$8\pi h\nu^3/c^3$	$8\pi ch\left(\frac{kT}{ch}\right)^4 x^3$	$8\pi ch\lambda^{-5}$	$h\omega^3/2\pi^3c^3$
$M_{qy}=\dot{n}c/4$	$ck^2/4\pi^2$	$2\pi c\tilde{\nu}^2$	$2\pi\nu^2/c^2$	$2\pi c\left(\frac{kT}{ch}\right)^3 x^2$	$2\pi c\lambda^{-4}$	$\omega^2/4\pi^2c^2$
$M_y=u_yc/4$	$c^2hk^3/8\pi^3$	$2\pi c^2h\tilde{\nu}^3$	$2\pi h\nu^3/c^2$	$2\pi c^2h\left(\frac{kT}{ch}\right)^4 x^3$	$2\pi c^2h\lambda^{-5}=c_1\lambda^{-5}$	$h\omega^3/8\pi^3c^2$
$L_{ny}=M_{ny}/\pi$	$ck^2/4\pi^3$	$2c\tilde{\nu}^2$	$2\nu^2/c^2$	$2\pi c\left(\frac{kT}{ch}\right)^3 x^2$	$2c\lambda^{-4}$	$\omega^2/4\pi^3c^2$
$L_y=M_y/\pi$	$c^2hk^3/8\pi^4$	$2c^2h\tilde{\nu}^3$	$2h\nu^3/c^2$	$2ch\left(\frac{kT}{ch}\right)^4 x^3$	$2c^2h\lambda^{-5}$	$h\omega^3/8\pi^4c^2$

body, the wavelength of the radiation, the surface condition, and other properties of the material. Some materials have an emissivity that is constant (less than 1) over considerable spectral intervals; these are called "gray" bodies. Others (notably gases) have considerable spectral structure and are called "colored" or "spectral" emitters.

Attempts to realize a standard radiator, one that closely approaches a blackbody, are usually based on a geometrical approach. A hollow chamber that is spherical, conical, cylindrical, or some combination of these is heated as uniformly as possible. It is provided with an aperture that has an area that is quite small compared to the area of the enclosed volume. The "blackness" of such a cavity source is based on obtaining many reflections of the radiation inside the cavity before it can exit from the aperture. An emissivity of 0.999+ can be obtained with sources of reasonable size, depending on the requirements for power, temperature, and aperture area. The next most common sources are tungsten and metals covered with paints that have been measured to be black over the spectral region of interest.

INTERACTION WITH MATTER

Radiation, when incident on a sample of material, is reflected, transmitted, and absorbed. The ratio of the reflected portion to the incident portion (in equivalent units) is the reflectance; similarly, there are transmittance and absorptance ratios. Kirchhoff's law states that emissivity and absorptance (or absorptivity) are equal for any sample under identical conditions of temperature and at the same wavelength or over the same wavelength interval. This must of course apply to the flux per unit area into a hemisphere. The absorbed radiation is found from the Beer–Bouguer law that equal fractions of radiation are absorbed by a given thickness of material:

$$-\frac{d\Phi}{\Phi} = \alpha x, \qquad \Phi = \Phi_0 e^{-\alpha x}$$

where $-d\Phi/\Phi$ is the fractional absorbed flux, x the sample thickness, α the absorption coefficient, and Φ_0 the initial flux.

GEOMETRICAL CONCEPTS IN RADIATION TRANSFER

The flux Φ varies in general over area and in direction. A defined quantity called radiance represents the flux per unit area per solid angle in a given direction. Radiance L can be defined in terms of a double differential as

$$L = \frac{\partial^2\Phi}{\partial\Omega\, \partial A \cos\theta}$$

where $\partial\Omega$ is the differential solid angle and $\partial A \cos\theta$ is the projected differential area.

The exchange of flux from an area A_1 to A_2 is calculated as follows.

$$\Phi(A_2) = \int_{A_1} \int L_1(\theta, \phi)\, dA_1 \cos\theta_1\, d\Omega$$

where $\Phi(A_2)$ is the flux on all of surface A_2, L_1 the radiance of surface A_1, and θ_1 the angle between the surface normal of A_1 and the line of interaction. This formula can also be written

$$\Phi(A_2) = \int_{A_1} \int_{A_2} \frac{L_1(\theta, \phi)\, dA_1 \cos\theta_1\, dA_2 \cos\theta_2}{d^2}$$

where A_2 and θ_2 are the quantities equivalent to A_1 and θ_1, and d is the interaction distance. In some cases the integration can be simplified. In others, the essential features can be found in tables of configuration factors.

The equations show the radiance as a function of angle, and in general it is. In the special case that the radiance from a surface is independent of angle that surface is called Lambertian. A blackbody is Lambertian.

If the dimensions of a source are small compared to the distance of measurement or the field of view of the instrument, the area integral of radiance is the important radiometric quantity; it is called radiant intensity and given the symbol I. Sometimes the solid-angle integral of radiance is important; it is called flux density and usually emitted flux density and received flux density are distinguished by the names radiant emittance (or exitance) and radiant incidance or irradiance.

Transmittance, reflectance, and emissivity all have directional properties. Directional emissivity is the ratio of the radiance of a material to that of a blackbody. Directional reflectivity is a function of the angles of incidence and of reflection and is defined as the reflected radiance divided by the incident irradiance. Scattering samples transmit in the beam and out of the beam, and the degree to which they keep radiation in the beam is a measure of their image-spoiling properties.

MEASUREMENTS AND CALIBRATION

A calibration is usually a measurement made under known and controlled conditions. The first rule of measurement is to make it under conditions as much like those of the calibration as possible. The next is similar: Account for the differences. A radiometer used for a radiometric measurement in general is affected by the position of focus on the detector; the angle from which the radiation comes; the temporal, spectral, spatial, and polarization properties of the radiation; the nature of the background; the temperature of the instrument; and in some cases the phase of the moon. Certainly atmospheric transmission can have a profound effect. As many of the parameters and variables as possible should be eliminated by calibration and the effects of the rest should be considered very carefully.

STANDARDS

Two types of radiometric standards have been used by the standardizing agencies of the world: the standard source and the standard detector. The first of these is the cavity simulator discussed earlier under sources; the second is a cavity-type detector that is as black as possible and has an electric heating coil. First, radiation excites the detector. Then a monitored amount of electric current is used to obtain the same detector output. These self-calibrated detectors seem to be the more successful standards.

See also BLACKBODY RADIATION; SPECTROPHOTOMETRY.

BIBLIOGRAPHY

M. A. Bramson, *Infrared: A Handbook for Applications*. Plenum, New York, 1968.

Fred E. Nicodemus, *Radiometry*. Selected Reprints, Am. Assoc. Physics Teachers, New York, 1970.

A. Drummond, *Advances in Geophysics*, Vol. 14. Academic Press, New York, 1970.

Clair L. Wyatt, *Radiometric Systems Design*. Macmillan, New York, 1987.

William L. Wolfe, "Radiometry," in *Applied Optics* and *Optical Engineering*, Vol. VIII, Chap. 5 (R. R. Shannon and I. C. Wyart, eds).

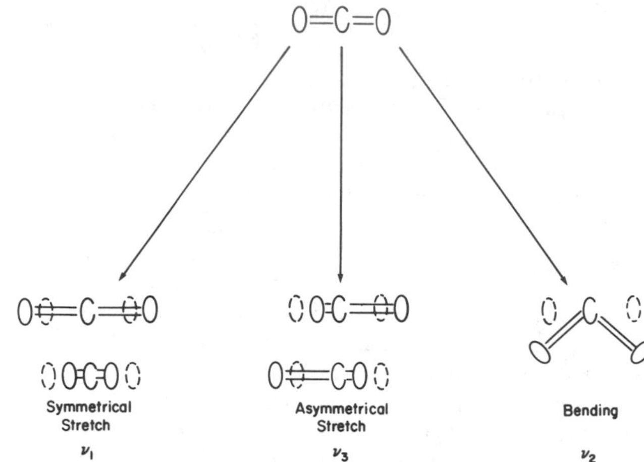

FIG. 1. The three nuclear vibration of CO_2. C = carbon atom, O = oxygen atom. The dotted oxygen symbol indicates the equilibrium position.

Raman Spectroscopy

W. G. Fateley and F. F. Bentley

Imagine ourselves crossing an old bridge and feeling the normal slow oscillation of the bridge under our steps; then, there arrives a horse to share our bridge route. We now feel the vibration from our steps and the intense oscillation generated by the racing horse. In the Raman effect, molecules are undergoing a similar experience. Here the oscillation of the molecule during nuclear vibration is quite slow compared to the very rapid oscillation of a light wave passing by the molecules. However, the molecule, like ourselves on the bridge, is not only experiencing motion from the slow nuclear vibration but it is also affected by the very rapid oscillating electric field of light. Since the electronic cloud follows the nuclei, we find fluctuations in shape of the electronic cloud that surrounds the molecule from both the nuclear vibrations and the light wave. But first let us describe the vibration of the molecule.

If we choose carbon dioxide as an example to demonstrate the three normal vibrations, these vibrations are illustrated in Fig. 1. Although the molecule is continually undergoing all three of these vibrations, for simplicity let us consider only the symmetrical stretching oscillation. We note that in Fig. 1 we have identified the position of the nuclei of the carbon, C, and oxygen, O, atoms during their vibration; also, from our knowledge of this molecule, we know that there are a total of 22 electrons surrounding CO_2. With some care and imagination, we can draw an ellipsoid of this electron cloud about the molecule, i.e., electronic cloud (O=C=O). This cloud represents the most likely areas containing the electrons belonging to the CO_2 molecule. We can imagine that the cloud of electrons tends to follow the vibrating nuclei as shown in Fig. 2. Here the ellipsoid cloud is elongated and compressed during the symmetrical vibration. Therefore, our best picture of the electron cloud is a pulsating ellipsoid deformed in the direction of the moving nuclei during the vibration oscillation. As previously discussed we realize that the pulsations of the electronic cloud will be influenced by a light wave passing by the molecules. The addition of two different perturbing effects can be represented as the summation of waves; see presented in Fig. 3.

We realize that the electric component of the light (or electromagnetic radiation) will induce a change in the electron cloud of the molecule. The light beam would perturb the electronic cloud and induce an instantaneous dipole moment. The oscillating frequency of this induced dipole will be exactly the frequency of the perturbing light radiation, ν_0. We now have the complication of the slow pulsations of the electronic cloud (due to nuclear vibrations) interacting with the rapid oscillation of the light waves (electrical component). Figure 3a illustrates the very slow oscillation of the electronic cloud from nuclei motion while the rapid oscillation of the electronic cloud from the electrical component of light is shown in Fig. 3b. The combination of these waves to illustrate the complexity of perturbing forces influencing the electronic cloud surrounding the molecule CO_2 is given in Fig. 3c.

Applying a more classical or analytical approach, how does light radiation react with this pulsating electron cloud? Smekal (1925) considered the theoretical problem of a light wave passing by this vibrating molecule. He concluded that the conservation of energy must hold for such an inelastic

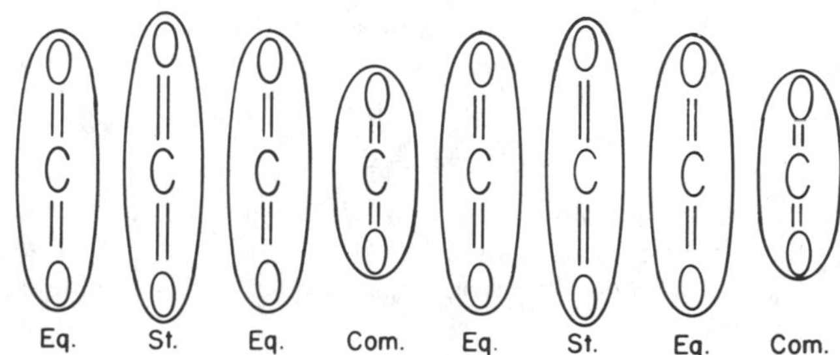

FIG. 2. The stretching and compression during the υ_1 vibrational frequency: Eq = equilibrium position, St = stretching from equilibrium, Com = compression of the molecule.

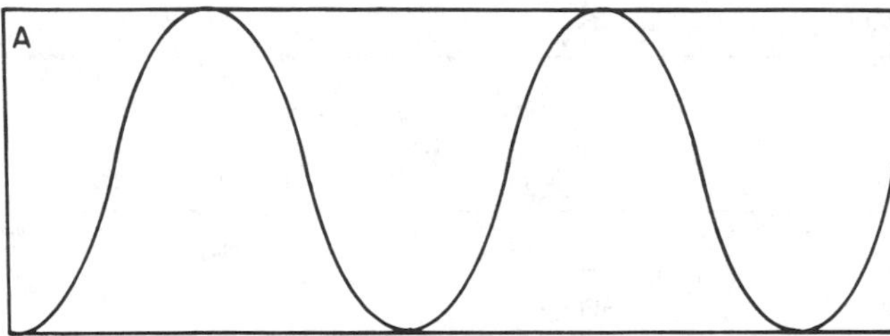

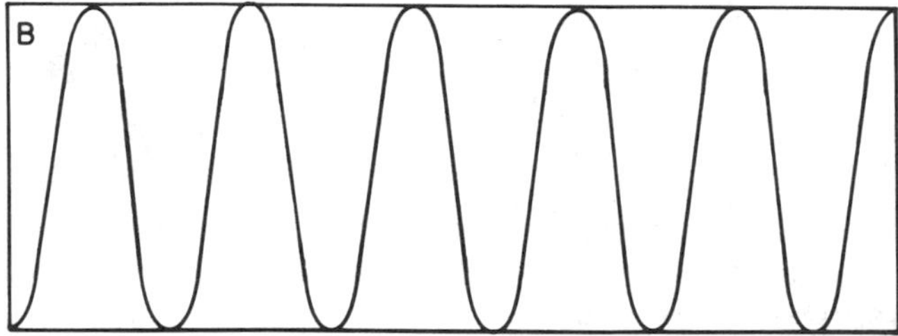

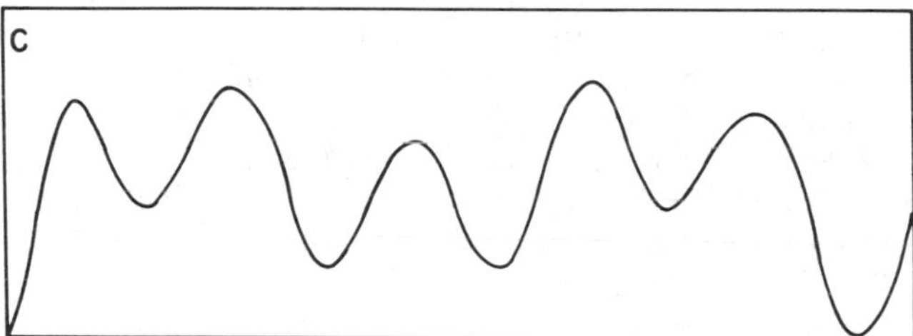

FIG. 3. Part A is the normal oscillation of a stretching vibration; part B is the frequency of oscillations indicating the light wave; part C is the summation of these two perturbing efforts.

collision.[1] Equation (1) reflects the differences in energies before and after a collision with a light wave:

$$\begin{bmatrix} h\nu_0 \text{ (photon energy)} \\ +\frac{1}{2}mv_0^2 \text{ (molecular translation energy)} \\ +E_0 \text{ (vibrational energy)} \end{bmatrix}_{\text{before collision}}$$

$$= \text{constant} = \begin{bmatrix} h\nu_1 \\ +\frac{1}{2}mv_1^2 \\ +E_1 \end{bmatrix}_{\text{after collision}} \qquad (1)$$

(h = Planck's constant, m = mass of the molecule, v_1^2 = average velocity of the molecule squared, ν_1 = frequency in waves/sec). Of course, this is simply a restatement of the conservation of energy laws; and, in addition, we would expect the translational energies to be unaffected by the light wave, i.e., $\frac{1}{2}mv_0^2 = \frac{1}{2}mv_1^2$. Then Eq. (1) reduces to

$$h\nu_0 + E_0 = h\nu_1 + E_1 \qquad (2)$$

or, rearranging

$$(E_0 - E_1) = h(\nu_1 - \nu_0).$$

Here, the change in internal energy $E_0 - E_1$ given in Eq. (2) is the result of a change of vibrational energy, ν_1 of the symmetrical mode, Fig. 1.

To understand the electromagnetic radiation's effect on the molecule let us consider the light from a laser whose frequency is ν_0 waves/s (where the energy and frequency are related by $E = h\nu_0$) passing through a sample of CO_2 gas. For simplicity we will again consider only the symmetrical stretch vibration ν_1 (note the vibrational frequency $\nu_1 = 4.02 \times 10^{13}$ waves/s while a light wave from a 488.0-nm laser line is $\nu_0 = 6.1484 \times 10^{16}$ waves/s). The easiest way to illustrate the symmetrical stretching molecular vibration is to describe this motion as a function of R, the interatomic distance between the oxygen atoms, see Fig. 4. The nuclear vibrational frequency ν_1 represents the number of times we have both a complete compression and a stretching of the C–O bonds in 1 s. In Fig. 4 we see that the energy necessary to compress the bonds and reduce R increases drastically the closer the nuclei approach one another. Likewise additional energy is necessary to stretch or lengthen R from the molecular equilibrium shape to a point where complete dissociation of CO_2 begins to occur. This is represented by the right side of the curve in Fig. 4. A simpler picture of this vibration is given in Fig. 5. We have drawn horizontal lines on this curve in Fig. 5 to note the extent of vibration (or change in R) in the simplest or ground vibrational energy state, $v = 0$. Also higher vibrational energy levels occur and they are marked $v = 1$, $v = 2$, ... As quantum mechanics is applied to the vibrating molecules, the difference in energy of a vibration, $\Delta E = h\nu_1$, is characterized by the energy level difference between the levels $v = 0$ and $v = 1$. This energy difference, $h\nu_1$, is what is directly observed in the infrared spectrum (see section on infrared spectroscopy), and by the Raman shift in frequency from the Rayleigh line (Fig. 6).

The interaction of the light wave does not change the oscillation frequency or distance R, but excites the electronic cloud of the molecules to a much higher energy.

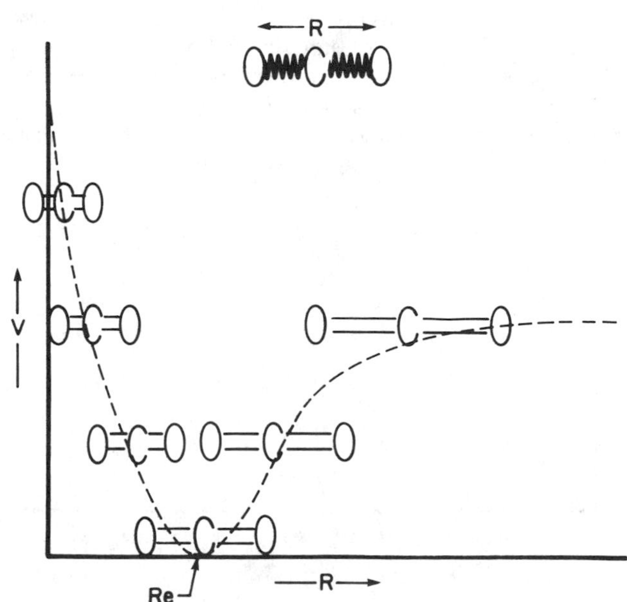

FIG. 4. The potential V for the symmetrical stretching vibration ν_1.

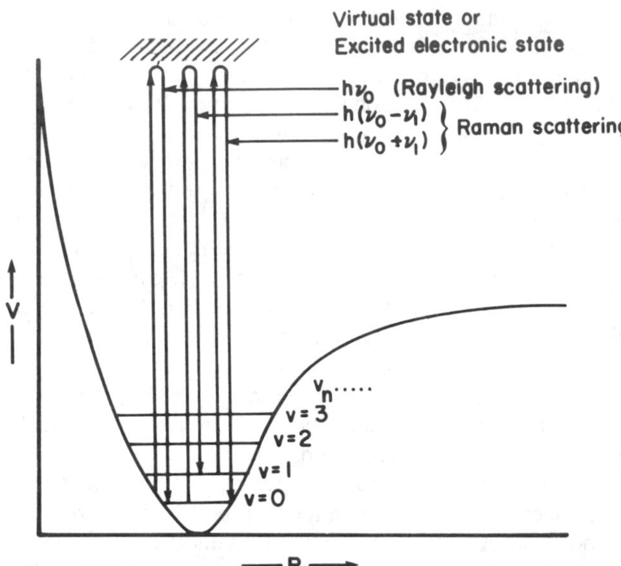

FIG. 5. The excitation and emission of photons.

The light interacting with the molecule can be observed in the following ways.

1. The molecule can be excited to a very high energy state, indicated by the area designated as the excited state in Fig. 5. The amount of energy necessary to reach this excited state is $h\nu_0$. Therefore, the return or relaxation of the molecule to the ground-state vibrational energy level $v=0$ results in the emission of a photon of energy $h\nu_0$. This emission is usually observed in the visible spectral region and is called Rayleigh scattering.

2. There will be times when the molecule is relaxing from the excited state and does not completely relax to the ground state, $v=0$, but stops at $v=1$, or even a higher energy level. This is illustrated in Fig. 5, and we can see analytically that the energy emitted in this process is decreased by $h\nu_1$; therefore, the photon energy of emission is $h(\nu_0-\nu_1)$.

3. Yet another situation can occur when the molecule is excited from a higher vibrational energy level, say $v=1$. The energy absorbed in this process is still $h\nu_0$. The molecule can relax to the original $v=1$ vibration energy level and emits a photon $h\nu_0$; however, the relaxation can be to the ground vibrational state $v=0$. The return to the ground state $v=0$ results in the emission of a photon which is $h\nu_1$ greater than the exciting laser energy $h\nu_0$. This can be described analytically in Fig. 5 as the photon energy emitted $h(\nu_0+\nu_1)$. This energy difference is also represented in Fig. 6.

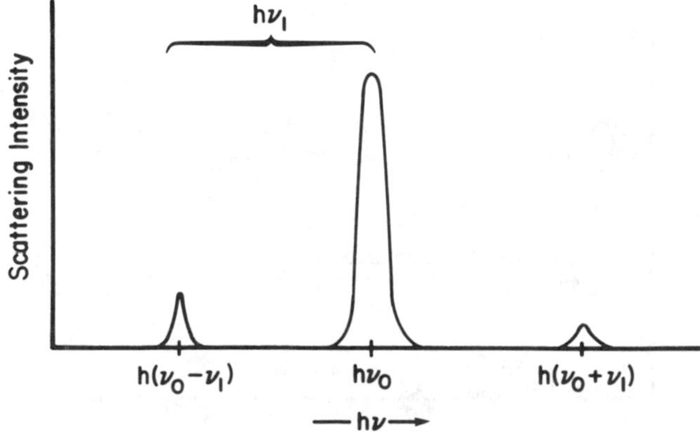

FIG. 6. The detection of scattering intensity as a function of $h\nu$. The displacement between the laser scattering $h\nu_0$, $h(\nu_0+\nu_1)$, and $h(\nu_0-\nu_1)$ is the Raman shift $h\nu_1$.

Table I

			Vibrations with spectral activity		
			Infrared	Raman	Coincidences
(a)	BF_3: F above B, two F below	planar	3	3	2
(b)	BF_3: B above three F	pyrimidal	4	4	4
(c)	F—B—F with F below B	90° model	6	6	6

We can summarize these effects in Fig. 6 by the emission photon $h\nu_0$, $h(\nu_0-\nu_1)$ and $h(\nu_0+\nu_1)$. The intensity of this radiation follows the order $h\nu_0 >>> h(\nu_0-\nu_1) > h(\nu_0+\nu_1)$. The emission of the photons $h(\nu_0-\nu_1)$ and $h(\nu_0+\nu_1)$ is the so-called Raman effect, although we must acknowledge that Smekal first predicted the possibility of this observation.

The instrumentation necessary to observe the Raman effect is relatively simple. The emission and scattering from the sample is usually collected by a lens and focused into a monochromator. The laser line is easily detected. At displacements from the laser line of $\pm h\nu_i$, where ν_i is the vibrational frequency of spectral active vibration, one observes the Raman effect. Gases, liquids, and solids can be observed by this technique, although a fluorescence background can present some difficulties.

The vibrational spectrum can be observed by both infrared and Raman spectroscopy. The advantage of both techniques is realized in the elucidation of the structure of matter. The spectral selection rules, i.e., what vibrational modes are infrared and Raman active or inactive, are very important information in determining the structure of a molecule.[2] For example, BF_3 is known through our knowledge of bonding theory to have at the center boron atom with three fluorine atoms attached to this centered atom. But what about the molecular structure? The selection rules for various possible structures predicted the spectral activity shown in Table I. Coincidences, in Table I, describe the number of vibrational motions which can be observed, i.e., spectrally active, by both the infrared spectroscopy and by the Raman effect. In the case of planar BF_3, two of the three vibrations will yield identical frequencies in the infrared and Raman spectrum. Note that none of the other possible structures have this identical property. For the vibrational spectrum we find BF_3 to be planar.

We note that the infrared and Raman effect is an excellent

way for "fingerprinting" molecules.[3] With rare exceptions no two molecules have identical infrared and Raman spectra. Also, great importance is given to the frequency of various vibrations for this is a great aid in the elucidation of the structure of matter.

See also INFRARED SPECTROSCOPY; MOLECULAR SPECTROSCOPY.

REFERENCES

1. General: Marvin C. Tobin, *Laser Raman Spectroscopy,* 35 of *Chemical Analysis.* Wiley-Interscience, New York, 1971.
2. What vibrations are spectral active: W. G. Fateley, F. R. Dollish, N. T. McDevitt, and F. F. Bentley, "Infrared and Raman Selection Rules for Molecular and Lattice Vibration: The Correlation Method." Wiley-Interscience, New York, 1972.
3. To identify a Raman spectrum: F. R. Dollish, W. G. Fateley, and Freeman F. Bentley, *Characteristic Raman Frequencies of Organic Compounds.* Wiley-Interscience, New York, 1974.

Rare Earths

K. A. Gschneidner, Jr.

The rare earths are a group of 17 metallic elements which are located in the third column (elements 21, 39, and 57) and in the extended sixth row (elements 58 and 71) of the periodic table. The elements 57 through 71 are called the lanthanides. All of the rare earths form trivalent compounds with three electrons in their valency shell, and as metals most of them have three electrons in their conduction bands. Normally they exist as trivalent positive ions in aqueous solutions. For the properties of the rare earths, see Table I.

The rare earths are not rare. Cerium and yttrium are more abundant in the earth's crust than lead and tin, and even the rarer elements, thulium, europium, and lutetium, are much more abundant than the platinum group elements. Without exception, the rare earths have never been found in nature as individual elements, but only as mixtures in the form of a chemical compound. The relative abundance of the individual rare-earth elements varies considerably in these mixtures, depending on their geological environment. The most important minerals are monazite, xenotime, and bastnasite, which were precipitated from either the molten rocks or from superheated brines under pressure.

The rare earths were discovered by Lt. Karl Axel Arrhenius in a quarry in the village of Ytterby near Stockholm in 1787. The first individual element (yttrium) was isolated in 1794 as an impure oxide. Since yttrium oxide resembled the "common earths" (oxides of magnesium, aluminum, and calcium), it was called a "rare" earth, and since then this name has been applied to yttrium and its closely related cogeners.

In the lanthanide series the electrons are added to the inner $4f$ shell of the atom, leaving the number of valence electrons undisturbed. Since this inner shell is shielded by the completed $5s^2$–$5p^6$ subshells, the $4f$ electrons play a minor role in determining the chemical properties of the elements. This inner shell holds up to 14 electrons, giving rise to the 14 elements of atomic number 58 through 71. But since the $4f$ electrons are screened by the inner filled electron subshells, the extra positive charge on the nucleus as the atomic number increases is not completely balanced by the negative charge of the added $4f$ electron resulting in an effective increasing nuclear charge which causes the electron cloud to contract. This contraction is known as the lanthanide con-

Table I. Properties of the rare earths

Rare earth	Symbol	Atomic number	Atomic weight[a]	No. 4f Electrons	Metallic radius (Å)	Atomic volume $\left(\frac{cm^3}{mol}\right)$	Density (g/cm)	Melting point (°C)	Boiling point (°C)
Scandium	Sc	21	44.95591	0	1.6406	15.039	2.989	1541	2836
Yttrium	Y	39	88.90585	0	1.8012	19.893	4.469	1522	3338
Lanthanum	La	57	138.9055	0	1.8791	22.602	6.146	918	3464
Cerium	Ce	58	140.115	1	1.8247	20.696	6.770	798	3443
Praseodymium	Pr	59	140.90765	2	1.8279	20.803	6.773	931	3520
Neodymium	Nd	60	144.24	3	1.8214	20.583	7.008	1021	3074
Promethium	Pm	61	(145)	4	1.811	20.24	7.264	1042	3000[b]
Samarium	Sm	62	150.36	5	1.8041	20.000	7.520	1074	1794
Europium	Eu	63	151.965	6	2.0418	28.979	5.244	822	1527
Gadolinium	Gd	64	157.25	7	1.8013	19.903	7.901	1313	3273
Terbium	Tb	65	158.92534	8	1.7833	19.310	8.230	1356	3230
Dysprosium	Dy	66	162.50	9	1.7740	19.004	8.551	1412	2567
Holmium	Ho	67	164.93032	10	1.7661	18.752	8.795	1474	2700
Erbium	Er	68	167.26	11	1.7566	18.449	9.066	1529	2868
Thulium	Tm	69	168.93421	12	1.7462	18.124	9.321	1545	1950
Ytterbium	Yb	70	173.04	13	1.9392	24.841	6.966	819	1196
Lutetium	Lu	71	174.967	14	1.7349	17.779	9.841	1663	3402

[a] 1987 standard atomic weights.
[b] Estimated.

traction, which is observed not only in the pure elements, but also in all of their compounds.

The rare earths usually have three electrons in their valence shell, and differ only in their $4f$ electron count, ranging from 0 to 14. There are, however, several lanthanides in which there is not much difference in the binding energy of a $4f$ electron and a $5d$ valence electron, and a $4f$ electron may be promoted to the valence shell making a tetravalent ion, as in the case of cerium, praseodymium, and terbium; or the $5d$ valence electron may be demoted to the $4f$ shell, thereby making the ion divalent, as for samarium, europium, and ytterbium. Because of this valence change it is possible to separate cerium, europium, and ytterbium from the other lanthanides by first putting them in their tetravalent or divalent state and then removing them with a simple chemical operation. This valency change has been used by the rare-earth industry to effect a simple, reliable, and inexpensive method for separating these elements from a naturally occurring mixture of the rare-earth elements. Furthermore, the dual valency of these latter three elements and also that of samarium and thulium has been the subject of a large number of scientific studies over the last 15 years. Of particular interest has been the nonintegral valences between 3 and 4 for cerium, and between 2 and 3 for the other elements. Of these elements, cerium has been studied the most and many unusual physical behaviors and new phenomena have been observed, especially at low temperatures (<20 K).

Because of the nearly identical atomic sizes and the same chemical valences, one rare-earth ion can readily be substituted in a crystal lattice of another rare earth with very little strain. As a result of this similarity there is only a little difference in the aqueous solubilities of two rare-earth elements, especially adjacent ones, for a given salt. In order to separate adjacent elements, scientists developed fractional crystallization or precipitation methods. For most elements it took thousands of fractionations to isolate a relatively impure individual element—indeed scientists spent a lifetime to obtain such a precious product.

During World War II while working on the separation of fission products as part of the atomic bomb project scientists developed the elution chromatographic ion exchange method for separating the rare-earth elements, which was a great improvement. A major breakthrough occurred in the 1950s when Frank H. Spedding and coworkers developed the band-displacement ion-exchange process, which was capable of producing macro quantities of extremely pure individual elements. Industry adopted this method and produced them in large quantities at a reasonable price, but within 10 years liquid–liquid extraction methods were developed which provided even lower priced individual-rare-earth elements. With the availability of pure and reasonably priced materials a worldwide research effort was launched to determine the properties of rare-earth materials and to understand their fundamental nature. From this fountainhead of knowledge many new and exciting high-technological applications flowed forth.

Basic studies of the optical properties of the rare-earth compounds led to new phosphors and lasers: (1) europium as the red phosphor and yttrium oxide as the host material in color television picture tubes; (2) neodymium as the lasing ion in YAG (yttrium aluminum garnet) and glass lasers; and (3) rare-earth phosphors in fluorescent tubes. A recent development is the use of rare-earth phosphors to reduce the exposure to x rays of medical patients by a factor of 100. An important fact, not fully appreciated, is that rare-earth materials, whether they are used as phosphors or as hosts, must be extremely pure—of the order of 99.999% pure with respect to all impurities—to maintain optical (spectral) purity and their high efficiencies.

Investigations of the magnetic properties of rare-earth materials led to the discovery of: (1) the superior permanent magnet properties of $SmCo_5/Sm_2Co_{17}$ and Nd–Fe–B permanent magnet materials; (2) garnet materials, such as GGG (gadolinium gallium garnet), for magnetic storage bubble devices; (3) grant magnetostrictions in RFe_2—100 to 1000 times larger than had been observed previously; (4) the electrooptic behaviors of amorphous R–Co (or Fe) alloys for information storage; and (5) the high-temperature superconducting ceramic oxides, such as $YBa_2Cu_3O_{7-x}$ with a transition temperature >90 K. Other studies on the chemical, metallurgical, ceramic, thermodynamic, and electrical properties of rare-earth materials resulted in the use of (1) yttria-stabilized zirconia as oxygen sensors for controlling automotive exhaust emissions and in improving gasoline consumption; (2) mixed rare earths in zeolite cracking catalysts, which improved the efficiency of the manufacture of gasoline by a factor of 3; and (3) the addition of small amounts of lanthanum or yttrium to superalloys to improve their oxidation and/or corrosion resistance for high-temperature applications such as aircraft jet engines and turbine engines.

See also Atoms; Elements; Heavy-Materials Fermion; Luminescence; Mixed Valence Materials; Permanent Magnets; Superconductivity; Transition Elements.

BIBLIOGRAPHY

K. A. Gschneidner, Jr., and L. Eyring (eds.), *Handbook on the Physics and Chemistry of Rare Earths,* Vols. 1–12. North-Holland, Amsterdam, 1978–1989. (A)

K. A. Gschneidner, Jr. (ed.), *Industrial Applications of Rare Earth Elements,* ACS Symposium Series No. 164. American Chemical Society, Washington, DC, 1981. (I)

K. A. Gschneidner, Jr., and J. Capellen (eds.), "1787–1987: Two Hundred Years of Rare Earths", IS-RIC-10, a pamphlet from the Rare-earth Information Center, Iowa State University, Ames, IA, 1988. (E,I)

Rare Gases and Rare-Gas Compounds

Neil Bartlett

THE GASES

The monatomic gases helium (He, 2), neon (Ne, 10), argon (Ar, 18), krypton (Kr, 36), xenon (Xe, 54), and radon (Rn, 86) (chemical symbols and atomic numbers in parentheses) are often called the noble gases or inert gases because of their low chemical activity. Indeed He and Ar are not rare, being relatively abundant terrestrially. All of the gases occur naturally, although all isotopes of Rn are unstable. He is the most abundant cosmically and is estimated to constitute 23 wt.% of all matter. The abundances of the other gases relative to one atom of He are: Ne, 2.8×10^{-3}; Ar, 4.9×10^{-5}; Kr, 1.7×10^{-8}; Xe, 1.3×10^{-9}. In air, the abundances (in parts per million by volume) are: ^{3}He, 7×10^{-6}; ^{4}He, 5.24; Ne, 18.2; Ar, 9320; Kr, 1.14; Xe, 0.086; Rn, 6×10^{-14}. The terrestrial abundance of Ar is high because of the beta decay of ^{40}K. The known abundances in terrestrial igneous rocks (in wt.%) are: He, 3×10^{-7}; Ne, 7×10^{-9}; Ar, 4×10^{-6}; Rn, 1.7×10^{-14}. In certain minerals the He concentration is high, one sample of thorianite having yielded 10.5 cm^3 of gas/g. The gases occur in meteorites, the concentration depending on the history of the meteorite. The isotopic ratios have cosmological significance and can provide an age for the material of the meteorite. He often occurs in natural gas and often constitutes 1% or more by volume, with concentrations as high as 8.9% having been reported from a well in New Mexico.

The high atmospheric abundance of Ar led Lord Rayleigh, a British physicist, to observe that "nitrogen" obtained from air was denser than nitrogen obtained by chemical generation. His collaboration with Professor W. Ramsay led to the identification of Ar as a new element and this in turn led Ramsay to the discovery of He, which was obtained in 1895 by heating the mineral cleveite. In 1898 Ramsay and Travers discovered Ne, Kr, and Xe by fractionating liquid air and thus established the existence of a new group of elements in the periodic table.

In the periodic table each rare gas stands between a highly electropositive element, an alkali metal, and (with the exception of helium) a highly electronegative element, a halogen. The near inability of these elements to enter into chemical combination with other elements, or even with themselves, is compatible with this periodic table placement and indicates that the ground-state electron configuration is a highly favorable one. These electron configurations are: He, $1s^2$; Ne, $1s^22s^22p^6$; Ar, $1s^22s^22p^63s^23p^6$; Kr, $1s^22s^22p^63s^23p^6$-$3d^{10}4s^24p^6$; Xe, $1s^22s^22p^63s^23p^63d^{10}4s^24p^64d^{10}5s^25p^6$; Rn, $1s^22s^22p^63s^23p^63d^{10}4s^24p^64d^{10}4f^{14}5s^25p^65d^{10}6s^26p^6$. The ns^2np^6 outermost electron shell of all of the gases, except He, constitutes the electron "octet," which has been a key concept in the electronic theory of valence. The stability of this outermost electron shell is not constant, however, as the first ionization potentials of the gaseous ground-state atoms, given in Table I, indicate. The ionization potential decreases as the effective diameter of the rare-gas atom increases. Radon has the lowest ionization potential (10.75 eV) and is undoubtedly the biggest atom. The increase in polarizability with atomic number, like the decrease in ionization potential, can be associated with the lower effective nuclear charge experienced by the ns^2np^6 electrons in the heavier atoms. The greater the polarizability of a rare-gas atom, the greater its heat of adsorption on charcoal. Greater polarizability is also associated with greater solubility in liquids and greater ease of clathrate (or cage-compound) formation. Thus the clathrates of radon are, of the rare gases, the most readily formed and most stable, whereas clathrates of helium are unknown. High polarizability, greater solubility, and perhaps even clathrate formation may account for the greater narcotic effects to be found for the heavier gases; thus helium (because it is not very soluble in body fluids) is a valuable diluting gas for oxygen in deep sea diving, whereas xenon (which is rather soluble in many fluids) is an excellent anesthetic.

Because of their chemical inertness He, Ne, and Ar are useful as shrouding or diluent gases (particularly the abundant He and Ar) whenever an inert atmosphere is required, e.g., in arc welding and in the machining and molding of reactive metals. Moreover, the small size and high atomic velocity of helium make it an excellent leak-detector gas for vacuum or pressure systems and also an excellent heat-transfer agent. The low chemical reactivity of the rare gases combined with their molecular simplicity causes their gas-discharge spectra to be relatively simple and usually characteristically colored, and this has led to their widespread use in colored lighting, e.g., neon signs. The gases are also increasingly important in gas lasers and promise to be important in high-power lasers. Liquid He is an important cryogenic liquid. Being much less dense than air and chemically inert, gaseous helium is ideal for lighter-than-air transport vehicles.

Compounds

The first true rare-gas chemical compound was prepared in 1962 when Xe gas was oxidized by PtF_6 vapor. Shortly thereafter the xenon fluorides were made. The known compounds involve, so far (1989), Kr, Xe, and Rn, and the atoms or groups (ligands) bonded to the rare-gas atom in these com-

Table I First ionization potential, atomic diameter, static polarizability and heat of adsorption on activated charcoal for rare gases

	First ionization potential (eV)	Diameter in crystal (Å)	Static polarization (Å^3)	Heat of adsorption (kcal/per g atom) at temperature (deg absolute in parentheses)
He	24.586	—	0.204	0.54 (few deg)
Ne	21.563	3.2	0.392	1.13 (91)
Ar	15.759	3.84	1.63	3.93 (168)
Kr	13.999	3.96	2.465	5.32 (223)
Xe	12.129	4.36	4.01	8.74 (248)

pounds are always highly electronegative. Although unligated oxygen atom is itself sufficiently electronegative to bond to xenon, krypton oxides are so far unknown. For a nitrogen atom to bond effectively to xenon it is necessary to ligate the nitrogen with two highly electronegative groups such as $-SO_2F$. The only established carbon to noble-gas-atom bond so far is that in the pentafluorophenylxenon(II)fluoroborate, prepared in 1988. The $[C_6F_5Xe]^+$ of this salt is isoelectronic with C_6F_5I and the Xe–C bond is therefore a classical electron-pair bond. The established ligands for each of the gases are given in parentheses: Kr (–F); Xe [–F, –O, –Cl, –Br, $-OSO_2F$, $-OClO_3$, $-OTeF_5$, $-OSeF_5$, $-OC(O)CF_3$, $-N(SO_2F)_2$, $-C_6F_5$ (to Xe^+)]; Rn (–F). Xe forms three binary fluorides, XeF_2, XeF_4, and XeF_6, all being colorless solids at room temperature. Two oxides, XeO_3 and XeO_4, have also been established. Because of the small size and high electronegativity of the F ligand and the low bond energy of the F_2 molecule, the fluorides of Kr, Xe, and Rn are, for each element, the most thermodynamically favorable compounds. The enthalpies of dissociation of the difluorides KrF_2, XeF_2, and RnF_2 into atoms are given as 23, 65, and (estimated) 95 kcal mol^{-1}, respectively. The enhancement in stability with increase in mass of the rare gas correlates with the fall in first ionization energy of the rare-gas atom. Thus the ionization potential of Xe is 43 kcal mol^{-1} lower than that of Kr and the total bond energy in XeF_2 is 42 kcal higher than that in KrF_2. This correlates with Mössbauer and x-ray photoelectron spectroscopic evidence which indicate that the noble-gas atom has lost appreciable electron density to the F ligands in these compounds. Thus formulation of the difluorides as single-electron-bonded molecules, in which each atom has an approximate octet, has been suggested as a simple bonding model, viz., $\mathrm{F} \cdot \ddot{\mathrm{G}} \cdot \mathrm{F}$, or alternatively as a resonance hybrid of the canonical forms $(F{-}G)^+F^-$ and $F^-(G{-}F)^+$. Physical evidence also indicates that the bond polarity and bond strength are higher in the xenon oxides (which, because of the strong bond in O_2, are detonators) than in the xenon fluorides and suggests that, if the Xe–F bond is a single-electron bond; then the Xe–O bond must be a two-electron bond. Thus for electron bookkeeping purposes XeO_3 may be represented as $\ddot{\mathrm{X}}\mathrm{e}(:\mathrm{O})_3$—again each atom maintaining an octet valence-electron configuration.

The shapes of many of the rare-gas compound molecules have been defined. All compounds of formula GL_2 or GLL' involve a linear disposition of a central rare-gas atom and its two linked atoms, the molecules being dumbbell in shape. XeF_4 is square planar (D_{4h}), XeO_3 is a triangular pyramid (C_{3v}) with Xe at the apex, and XeO_4 is tetrahedral. The $XeOF_4$ molecule is a square pyramid (C_{4v}) with an O atom at the apex and resembles a XeF_4 molecule to which an O atom has been attached from one side on the fourfold axis. XeF_6 is a "molecular jelly-fish" in which several geometries, C_{3v}, C_{2v}, and C_s, appear to be energetically very similar. In the solid phase XeF_6 is apparently a cluster of XeF_5^+ and F^- ions.

In cations such as $(XeF)^+$ and $(KrF)^+$ the positive charge appears to be centered largely on the noble-gas atom and *trans* to the F ligand. Such cations form complexes with Lewis bases exemplified by the N atoms of HCN and CH_3CN, this donor atom being in linear array with the $(GF)^+$ as in $(HCN{:}KrF)^+$. Such nitrogen bases have little impact on the bonding of G to F but anionic bases such as F^-, OSO_2F^- or $OTeF_5^-$ themselves bond more strongly to G, and simultaneously weaken the G to F bonding of the cation.

As befits their relatively weak chemical bonds and low thermodynamic stability, the rare-gas compounds are, potentially, powerful oxidizers. Thus salts derived from XeF_2 and containing the XeF^+ ion are able to oxidize radon and fix it. Similarly, salts derived from krypton difluoride oxidize molecular O_2 to form O_2^+ salts, Xe to form XeF^+ salts, and Rn to form RnF^+ Salts. Xenon difluoride is kinetically stable in neutral aqueous solution but it was used to prepare the first samples of perbromates (BrO_4^-) by oxidizing bromate in aqueous media. XeF_2 has also been used to fluorinate aromatic molecules and large fused-ring systems. The carbon–xenon bonded species $[C_6F_5Xe]^+$ is an effective source of $C_6F_5^+$ and, e.g., gives $[(C_6F_5^+)_2I]^+$ salts on interaction with C_6F_5I.

Because of their lower thermodynamic stability, krypton compounds are more potent oxidizers than their xenon relatives and salts of KrF^+ are effectively sources of F^+ in the synthesis of salts of ClF_6^+, BrF_6^+, and ReF_6^+. Solutions of KrF_2 in liquid HF (in the presence of a fluorobase such as an alkali fluoride) in most cases generate the highest attainable oxidation state of an element at 20°C or lower, examples being $[AgF_4]^-$, $[AuF_6]^-$, and $[NiF_6]^{2-}$ salts.

See also ATOMS; CHEMICAL BONDING; ELEMENTS; MOLECULES; POLARIZABILITY.

BIBLIOGRAPHY

For a comprehensive treatment of the properties of the rare gases up to 1961 the reader is referred to the two volume work *Argon, Helium and the Rare Gases* (G. A. Cooke ed.). Interscience Publishers, New York and London, 1961, Vol. I, "History Occurrence and Properties," and Vol. II, "Production, Analytical Determination and Uses."

For a short account of the properties of the gases see A. H. Crockett and K. C. Smith, "The Monatomic Gases" in *Comprehensive Inorganic Chemistry* (J. C. Bailor *et al.*, eds.). Pergamon Press, Oxford, 1973, Vol. 1, pp. 139–211.

For information up to 1974 the Abstracts (Library of Congress Card No. 75-27055) of the Symposium entitled *Noble Gases* (R. E. Stanley and A. A. Moghissi, eds.), sponsored by the U.S. Environmental Protection Agency's National Environmental Research Center at Las Vegas, should be consulted.

For the history of the discovery of the rare gases see M. W. Travers, *Life of Sir William Ramsay*. Edward Arnold, London, 1956.

The following may be consulted for further information on the rare-gas compounds: (a) *Noble-Gas Compounds* (H. H. Hyman, ed.), The University of Chicago Press, Chicago and London, 1963 (good for the reports of the theoretical and experimental work of the first few months of rare-gas chemistry). (E, I, and A) (b) N. Bartlett and F. O. Sladky, "The Chemistry of Krypton, Xenon and Radon" in *Comprehensive Inorganic Chemistry* (J. C. Bailor *et al.*, eds.). Pergamon Press, Oxford, 1973, Vol. 1, pp. 213–330. (I) (c) N. Bartlett "Noble-Gas Compounds," *Endeavour* **XXXI,** 107 (1972). (I) (d) *Noble-Gas Chemistry and Its Significance*. Göttingen Akademie-Jahrbuch, 1961, p. 22. (E)

Rayleigh Scattering

D. W. Berreman

Rayleigh scattering is the name applied to the incoherent scattering of light by particles all of whose dimensions are much smaller than the wavelength of the light. Such scattering was first explained quantitatively by Lord Rayleigh [1] in 1871. The phenomenon accounts for the blue color of clear sky in daylight. Rayleigh showed that scattering by submicroscopic particles is much greater for light of short wavelength than long. The scattering intensity is inversely proportional to the fourth power of the wavelength of the light, so that blue of light of 4500-Å wavelength is scattered more intensely than red light of 6500-Å wavelength by a ratio of $(6500/4500)^4$, or about 5 to 1.

The intensity of Rayleigh-scattered light polarized with its electric field parallel to the plane of the scattering angle is also proportional to the square of the cosine of that angle, while the other polarized component has an intensity independent of the scattering angle. Consequently, light scattered at right angles from the source, where the cosine is zero, is completely plane-polarized. The polarization of forward- and backward-scattered light is unaltered.

Not only foreign particles in the air, such as dust and very small droplets of fluid, contribute to Rayleigh scattering. There is also a contribution from the gas molecules of the air itself when, in the course of their random motion, a larger- or smaller-than-average number occupy any particular very small volume of space. On clear days the contribution of such fluctuations in air density to Rayleigh scattering may be greater than that from solid and liquid particles.

When particles in the air are not so small compared to the wavelength, the scattered intensity depends in a complicated way on wavelength and particle shape. Polarization is also less pronounced. The first systematic study of scattering by larger particles was done by Mie in 1908 [2]. Such scattering is often called Mie scattering. Although Mie's study was directed specifically toward scattering by spherical, colloidal metal particles, it applies also to liquid aerosols and other spherical particles. Mie's work gives a qualitative understanding of scattering by nonspherical particles, and it reduces to Rayleigh's theory when the spheres are very small.

Particles with dimensions near the wavelength of light tend to scatter different wavelengths in more nearly equal proportion. Consequently, the sky is more grey than blue when it has larger dust, smoke, or liquid droplets in it. The very small size of the random fluctuations in air density make the sky appear extraordinarily blue on days when larger foreign particles are few and Rayleigh scattering predominates.

Various colorful, polarization-dependent light-scattering phenomena that are related to Rayleigh and Mie scattering in the atmosphere and in water are called Tyndall effects [3], after John Tyndall, whose experimental investigations preceeded Rayleigh's theoretical work.

See also Atmospheric Physics; Light Scattering; Polarized Light; Scattering Theory.

REFERENCES

1. J. W. S. Rayleigh, *Philos. Mag.* **XLI,** 107–120 (1871) [Reprinted in *Scientific Papers by Lord Rayleigh (John William Strutt),* Vol. 1, pp. 87–99 (Paper 8). Dover, New York, 1964.]
2. G. Mie, Ann. Phys. **25**(4), 377–445 (1908) [Described in M. Born and E. Wolf, *Principles of Optics,* 2nd ed., pp. 633ff. Macmillan, New York, 1964.]
3. John Tyndall, *Philos. Trans. R. Soc. (London)* **160,** 333–366 (1870).

Reflection

Elsa Garmire

Reflection occurs whenever electromagnetic radiation interacts with an interface consisting of a change in refractive index which takes place in a distance less than the electromagnetic wavelength. Specular reflection occurs from a flat abrupt surface, giving rise to reflected radiation at equal angles, θ_1, as shown in Fig. 1. Some portion of the incident light is transmitted at angle θ_2. Both incident and transmitted rays lie in the same plane as the incident ray, called the plane of incidence. The laws of specular reflection hold for all wavelengths and are determined by the refractive indices of the two media. These laws are derived from Maxwell's equations for electromagnetic radiation in matter, by applying the appropriate boundary conditions.

Consider first cases in which both media are transparent. The angle of the transmitted ray is given by Snell's law:

$$n_1 \sin \theta_1 = n_2 \sin \theta_2. \tag{1}$$

The reflectivity and transmission of the interface are given by the ratios of the electric field of the reflected ray E_1 and transmitted ray E_2 to the incident field E_0. The reflectivity is different for light polarized in the plane of incidence and normal to the plane of incidence. Incident light polarized normal to that plane has an electric field component transverse to the plane of incidence annd is called TE (transverse electric). Light polarized in the plane of the drawing has a magnetic field component transverse to the plane of incidence and is called TM (transverse magnetic).

The electric field ratios are given by Fresnel's reflection formulas:

$$\left.\begin{aligned} \frac{E_1}{E_0} &= \frac{\cos\theta - \sqrt{n^2 - \sin^2\theta}}{\cos\theta + \sqrt{n^2 - \sin^2\theta}} \\ \frac{E_2}{E_0} &= \frac{2\cos\theta}{\cos\theta + \sqrt{n^2 - \sin^2\theta}} \end{aligned}\right\} \text{TE}$$

$$\left.\begin{aligned} \frac{E_1}{E_0} &= \frac{n^2\cos\theta - \sqrt{n^2 - \sin^2\theta}}{n^2\cos\theta + \sqrt{n^2 - \sin^2\theta}} \\ \frac{E_2}{E_0} &= \frac{2n\cos\theta}{n^2\cos\theta + \sqrt{n^2 - \sin^2\theta}} \end{aligned}\right\} \text{TM} \tag{2}$$

In these equations the quantity $n \equiv n_2/n_1$, $\theta \equiv \theta_1$.

The reflectivity is defined as the ratio of the reflected light intensity to the incident light intensity: $R = |E_1{}^2/E_0{}^2|$. The

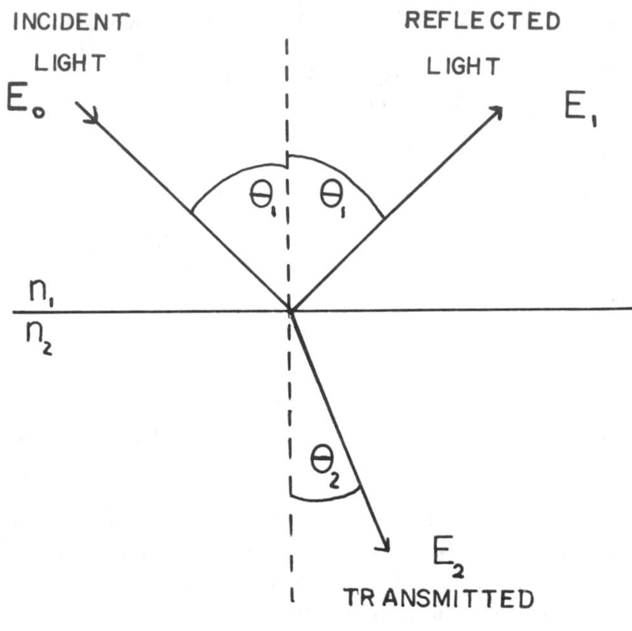

FIG. 1. Incident, reflected, and transmitted rays.

transmission is given by the ratio of the transmitted light intensity to the incident light intensity: $T=|E_2^2/E_0^2|$. If there are no losses, $T+R=1$. One particularly simple case is the reflectivity at normal incidence ($\theta=0$). In this case there is no distinction between TE and TM polarizations and

$$R=\left(\frac{n_1-n_2}{n_1+n_2}\right)^2. \tag{3}$$

When light is incident from less dense to more dense matter, this is called external reflection and $n>1$ ($n_2>n_1$). Figure 2 shows a plot of external reflectivity for visible light reflecting from glass. It can be seen that the reflectivity is higher for TE polarization than for TM, particularly at large angles, which means that light becomes partially polarized upon reflection. Furthermore, for the TM polarization, there is one angle, called the polarizing angle or Brewster's angle, at which there is no reflection. This angle is given by

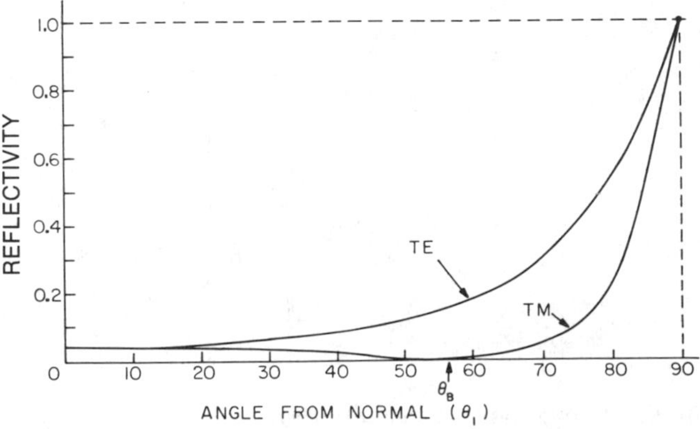

FIG. 2. External reflectivity as a function of angle for visible light reflecting from glass ($n=1.5$).

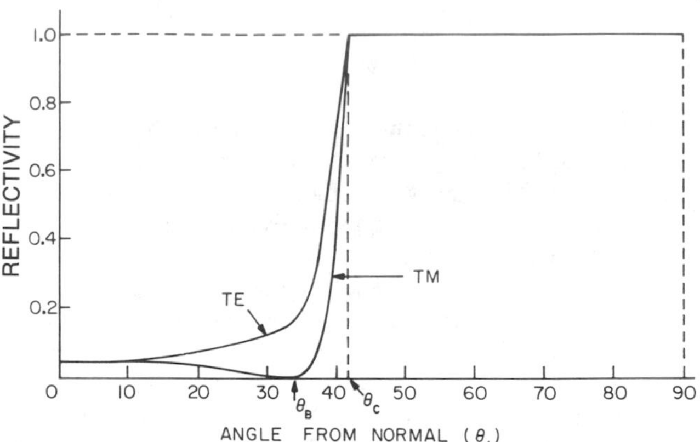

FIG. 3. Internal reflectivity as a function of angle for visible light in glass ($n=0.67$).

$$\tan\theta_B=n_2/n_1. \tag{4}$$

At this angle the reflection is completely TE polarized and there is total transmission of TM polarization.

When light is incident from a more dense to a less dense medium, $n<1$ and we have internal reflection. Figure 3 shows a plot of internal reflectivity for visible light in glass ($n=0.67$). In this case we note that beyond a certain critical angle, defined by

$$\sin\theta_c=n_2/n_1. \tag{5}$$

we have total internal reflection for both polarizations, and $R=1$. We note that, as in external reflection, the TM polarization experiences no reflection at the polarizing angle.

In the regime of total internal reflection, light experiences a phase change upon reflection given by

$$E_1/E_0=\exp(i\delta),$$

where

$$\delta_{\mathrm{TE}}=-\tan^{-1}\left\{\frac{\sqrt{\sin^2\theta-n^2}}{\cos\theta}\right\},$$

$$\delta_{\mathrm{TM}}=-\tan^{-1}\left\{\frac{\sqrt{\sin^2\theta-n^2}}{n^2\cos\theta}\right\}.$$

The sign shown here assumes a time dependence given by $\exp(-i\omega t)$. This phase change is important in understanding dielectric waveguides.

When a medium is not transparent, its refractive index can be considered complex: $n=n'+in''$. This is a convenient way of including both conductivity and dielectric constant in the properties of the material. Solution of Maxwell's equations for complex refractive index (including complex dielectric constant or complex conductivity), introducing appropriate complex boundary conditions, yields exactly the same Fresnel's formulas as Eq. (2), but with a complex refractive index n.

The reflectivity as a function of angle from an absorbing medium is shown in Fig. 4 for the two polarizations. It can be seen that the general shape is the same as for transparent media, but that the reflectivity of absorbent media is generally higher than for transparent media. The limiting case

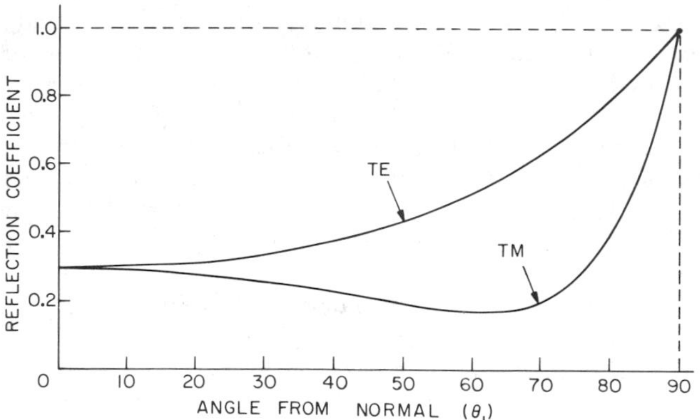

FIG. 4. External reflectivity as a function of angle of an absorbing medium.

of an absorbing medium is metallic reflection, in which the reflectivity is large at all angles.

The particularly simple case of reflectivity at normal incidence is given by

$$R = \frac{(2-n')^2+n''^2}{(1+n')^2+n''^2}.$$

We see that for $n' \gg 1$, $R \rightarrow 1$, as is the case in metals.

See also ELECTROMAGNETIC RADIATION; MAXWELL'S EQUATIONS; POLARIZED LIGHT; REFRACTION.

BIBLIOGRAPHY

M. Born and E. Wolf, *Principles of Optics*, pp. 36–51, 611–624. (Pergamon Press, New York, 1970).

Grant R. Fowles, *Introduction to Modern Optics*, pp. 49–54, 164–168. (Holt, Rinehart and Winston, New York, 1968).

J. D. Jackson, *Classical Electrodynamics*, pp. 269–284 (Wiley, New York, 1975).

F. A. Jenkins and H. E. White, *Fundamentals of Optics*, pp. 509–534 (McGraw Hill, New York, 1957).

Reflection High-Energy Electron Diffraction (RHEED)

Richard W. Vook

RHEED (reflection high-energy electron diffraction) is a special case of HEED in which the incident electron beam strikes the surface of a specimen at grazing incidence. Diffraction patterns from monocrystalline, polycrystalline, and amorphous structures can be obtained. The RHEED pattern consists of approximately one half of a transmission electron-diffraction (TED) pattern bounded by a "shadow edge" parallel to the surface. There are, however, important differences, which depend on the angle of incidence, surface topography, and refractive index μ of the diffracting material, that strongly influence the character of the RHEED pattern.

The closest resemblance to half of a TED pattern is obtained in the case of a rough surface, where the pattern is produced by electrons that penetrate only the extremities of the surface protrusions. As the surface becomes flatter and smoother, refraction and surface effects become important. The angle of incidence then determines the depth in the crystal from which coherent diffraction takes place, and the Bragg law becomes

$$(\lambda/2d)^2 = (\mu^2 - 1) + \sin^2\phi \tag{1}$$

where $\mu(=\lambda/\lambda') \simeq 1 + U/2V > 1$; λ and λ' are the electron wavelengths in vacuum and in the crystal, respectively; d is the interplanar spacing; ϕ is the grazing angle of incidence and reflection; U is the mean inner potential; and V is the accelerating voltage. Refraction shifts the reflections toward the shadow edge by an amount

$$Z - Z' = L^2(\mu^2 - 1)/(Z + Z') \tag{2}$$

where L is the distance from the specimen to the photographic plate, and Z' and Z are the positions of the diffracted beam with and without refraction. The method may be used to determine inner potentials.

With very flat surfaces, the penetration of the electron beam in a direction perpendicular to the surface is very small. The surface layer of the crystal that contributes to producing coherently diffracted electrons with detectable intensity may be as little as a few monatomic layers, resulting in symmetric line broadening along a line perpendicular to the surface of the crystal. For monocrystalline materials, diffraction spots are converted to parallel streaks. The flat surfaces needed to see such effects have been obtained by vacuum-deposited thin-film epitaxial growth techniques using as substrates mica or films grown on mica. Mechanically polished faces that have been cleaned and annealed have also been used. A similar effect occurs for faceted monocrystalline surfaces. The streaks are located perpendicular to the facet planes. Their orientation can be used to determine the Miller indices of the crystal plane of the facet.

The sensitivity of RHEED to adsorbed monocrystalline surface layers is comparable to that of LEED, but flat smooth surfaces are required in the former case. Then a few tenths of an average monolayer (around 10^{-8} g/cm^2) can be detected using photographic recording techniques.

RHEED instruments are usually constructed so that the specimen surface can be rotated about its normal by up to 360°. The angle of grazing incidence can be varied by several degrees, and the specimen itself can be translated by at least 0.5–1 cm perpendicular to its surface. The diffraction pattern is usually recorded photographically, either by direct electron impingement on the plate or by photographing the pattern formed on a fluorescent screen. Faraday cages or other electron detectors have also been used for more quantitative work. The camera constant ($L\lambda$) is usually determined from standard specimens. Because of the small angle of incidence, the area of the sample surface from which the diffracted beams arise is much larger than in TED. Consequently, interplanar spacing measurements tend to be less accurate. However, all of reciprocal space can be sampled by rotating the crystal about an axis perpendicular to its surface. Also, the beam intensity at a point on the surface of a sample is much less than in TED, leading to lower heating and radiation damage effects.

A recent important application of RHEED is in molecular-beam epitaxy studies where it is used in monitoring monatomic layer growth from changes in surface atomic roughness. As the film thickens during growth, the intensities of the specular and diffracted electron beams oscillate with a period equal to the atomic or molecular layer thickness. High scattered intensities are associated with flat surfaces, which occur when an atomic layer is complete. Low intensities correspond to atomically rough surfaces characteristic of a surface with a high density of monolayer-thick islands, in the ideal case. This technique is especially useful in monitoring the growth of superlattice films.

See also ELECTRON DIFFRACTION; LOW-ENERGY ELECTRON DIFFRACTION (LEED); SURFACES AND INTERFACES.

BIBLIOGRAPHY

Richard Beeching, *Electron Diffraction*. Methuen, London, 1950. (E)

John M. Cowley, *Diffraction Physics*. North-Holland, Amsterdam, 1981. (A)

P. J. Dobson, N. G. Norton, J. H. Neave, and B. A. Joyce, Vacuum **33,** 593 (1983). (E,I)

Z. G. Pinsker, *Electron Diffraction* (transl. by J. A. Spink and E. Feigl). Butterworths, London, 1953. (I)

H. Raether, "Elektroneninterferenzen," *Hand. Phys.* **32,** 433 (1957). (I–A)

T. B. Rymer, *Electron Diffraction*. Methuen, London, 1970. (E–I)

R. W. Vook, "Microstructural Characterization of Thin Films," in *Treatise on Materials Science and Technology* (H. Herman, ed.), Vol. 4, Academic Press, New York 1 (1974). (E–I)

Refraction

John D. Reichert

Since the very earliest times, the everyday experience of mankind has prejudiced a belief that light travels in straight line paths. This is apparent as one locates his home campfire or notices the sharp edges of shadows. Indeed, any case in which light appears to move in other than straight lines is curious enough to merit a special name for the effect: refraction. Mirages, lens effects, highway shimmer, the mysterious mislocation of fish, and the bending of starlight by gravity all evoke this curiosity and are included in the category of refraction. Indeed, our most sophisticated views of nature stipulate that light does, in a sense, always travel along a shortest path between two points, and this path can be defined to be the "real" straight line. This viewpoint, of course, leads to the uncomfortable conclusion that it is man and the space around him that are curved.

The curvature of the "real" straight lines in the mirage-type effects seems to be a rather arbitrary description because it is easily understood that the light is moving through a medium, such as air, with a nonuniform, spatially varying index of refraction. The effect is actually explained more simply by accounting for the index n as a function of position. On the other hand, the bending of starlight by gravity occurs in vacuum and may be described more fundamentally as an effect of the curvature of space as described by the theory of relativity. In either case, the effect is a refraction of light because the light appears to bend and follow a curved path.

It can be rather difficult to distinguish refraction cleanly from reflection or, even, from diffraction and interference effects. Close examination of the edges of shadows, for example, reveals that they are not completely sharp. Some of the light must have been bent by the object casting the shadow. Traditionally, one considers this to be a diffraction effect and removes it from the refraction category. Similarly, some of the light incident on a glass sphere will be directed back toward the source of the light. This portion of the redirected light can be abitrarily classified either as reflected or as refracted. Other aspects of the pattern of the light redirected by the sphere might be classified as diffraction or interference effects. In this case, however, tradition would tend to refer to this event as refraction by light by a sphere.

A reasonable point of view is simply to recognize certain broad but overlapping categories called refraction, reflection, diffraction, and interference and to realize that the tradition that introduced these categories has been useful, but is older than the electromagnetic description of light. The behavior of light is completely specified by Maxwell's equations, and the solution of those equations includes all the subtlety of the phenomena without artificial division into fuzzy categories. Said another way, solution of Maxwell's equations illustrates the inherent fuzziness of the categories.

Certain effects, however, are easily recognized as traditional examples of refraction. The distinction is of some importance, calculationally. A simple method for obtaining approximate solutions of Maxwell's equations is known as ray tracing and has proved, under proper conditions, to be

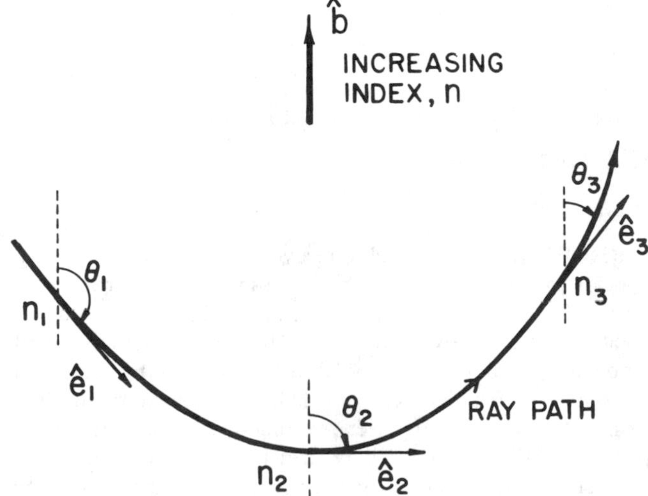

FIG. 1. Ray in stratified medium. At all points on the path, $n \sin\theta$ is constant, and the path curves toward the direction $\hat{b}$ of increasing n.

highly accurate. Thus, a traditional refraction effect can be and is most simply analyzed with ray-tracing techniques. Because ray tracing accounts only for lowest-order effects produced by a spatially varying index of refraction, interference and diffraction effects are automatically eliminated from the analysis and, correspondingly, the analysis fails in situations in which such effects are prominent. One earmark of difficulty in a ray-tracing calculation is the crossing of rays. Great caution and some experience are required to deal accurately with caustics, surfaces where rays cross. In such regions, the ray-tracing procedure can fail to produce accurate approximations to solutions of Maxwell's equations.

When Maxwell's equations are solved in the "slowly varying amplitude (SVA, sometimes called *eikonal*) approximation," the energy flux (Poynting vector) can be described by a "ray path vector" $\mathbf{r}(s)$. The parameter s of the curve $\mathbf{r}(s)$ is understood to be the arc length along the curve. The resulting differential equation for $\mathbf{r}$ is (in the high-frequency limit, $\lambda \to 0$, of the approximation)

$$\frac{d}{ds}\left(n\,\frac{d\mathbf{r}}{ds}\right) = \nabla n, \tag{1}$$

where n is the index of refraction of the medium at location $\mathbf{r}$. Since s is the arc length, $d\mathbf{r}/ds$ is a unit vector tangent to the ray path:

$$\frac{d\mathbf{r}}{ds} \equiv \hat{e}(s) \quad \text{and} \quad \frac{d\mathbf{r}}{dt} = v\hat{e}(s),$$

where $v \equiv c/n(\mathbf{r})$ is the speed of light at point r in the medium. The vector $\hat{e}(s)$ gives the direction of the ray at point $\mathbf{r}$ on the path. In any region in which n is constant, Eq. (1) is trivially integrated, and the path is a straight line. In general, the paths determined by (1) are the optical geodesics in the media and, as mentioned above, it is possible to define these curves to be the "straight lines" in the media.

For the special case of a stratified medium in which n varies in only one direction, specified by a unit vector $\hat{b}$,

$$\nabla n \equiv g(\mathbf{r})\ \hat{b} \equiv g(s)\hat{b},$$

one can integrate (1) along a ray path to obtain

$$n_2\hat{e}_2 - n_1\hat{e}_1 = \hat{b}\int_{s1}^{s2} g(s)ds. \tag{2}$$

It is clear from (2) that $\hat{e}_2, \hat{e}_1$, and $\hat{b}$ are coplanar so that the ray moves in a single plane specified by $\hat{e}_1$ and $\hat{b}$. Furthermore, from (2) one obtains

$$n_2(\hat{b}\times\hat{e}_2) = n_1(\hat{b}\times\hat{e}_1),$$

from which

$$n_2 \sin\theta_2 = n_1\sin\theta_1, \tag{3a}$$

where θ_2 and θ_1 are, respectively, the angles of inclination of $\hat{e}_2$ *and* $\hat{e}_1$ with respect to $\hat{b}$.

Equation (2) may be rewritten

$$\hat{e}_2 = (n_1/n_2)\hat{e}_1 + [\cos\theta_2 - (n_1/n_2)\cos\theta_1]\hat{b}. \tag{3b}$$

The path of a ray in a stratified medium is illustrated in Fig. 1. Mirage-type effects arise in this way when air is less dense near a hot surface. A similar effect occurs when light from a star enters the earth's atmosphere. This astronomical refraction causes celestial objects to appear slightly higher above the horizon than their true elevation. Highway shimmer and the twinkling of stars result in the same way due to motions of the air.

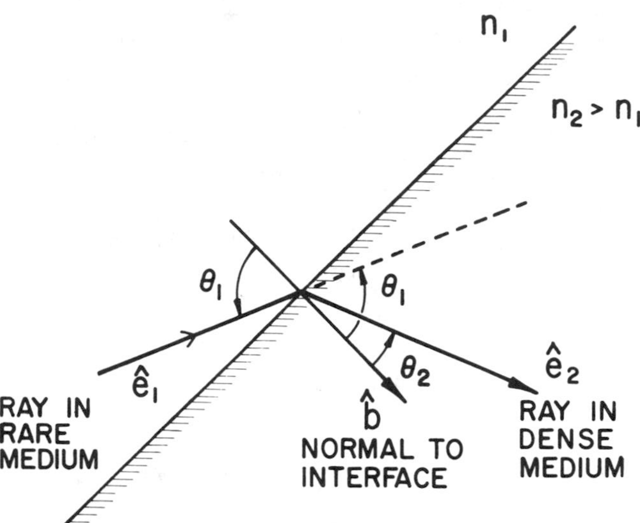

FIG. 2. Refraction at an interface. Snell's law of refraction states that, at an interface of two media, the angles of incidence and refraction are related by $n_1\sin\theta_1 = n_2\sin\theta_2$.

Snell's law of refraction for the sharp bending of a ray path at the interface of two media, each with constant index of refraction, is a special case of (3) above, where $\hat{b}$ is a unit vector normal to the interface. This situation is illustrated in Fig. 2. The momentum carried by the electromagnetic field is proportional to $n\hat{e}$ at the interface, and so Snell's law simply states that the transverse component of the field momentum (the component parallel to the interface) is continuous across the interface.

If the index of refraction depends on the wavelength of the light, then the amount of bending is different for rays corresponding to different wavelengths. This aspect of the refraction phenomenon is known as dispersion, and media characterized by a wavelength-dependent index of refraction are termed dispersive media. In gases and glasses, liquids and everyday materials, violet light has a slightly larger index and bends more than the longer-wavelength red light.

Eyeglasses are, for many of us, the most important application of refraction, but movie projectors, cameras, telescopes, and microscopes are all refraction devices. Surely, the most beautiful example of refraction is the rainbow. In addition to its reputation in folklore, the rainbow has an important place in the history of physics. In *The Day the Universe Changed,* James Burke makes the case that when, sometime during the first decade of the 14th century, Theodoric of Freiburg discovered what causes the rainbow, he was conducting the first properly scientific experiment in Western European history! Such reputation justifies a brief analysis—to savor the romance and to illustrate refraction computations.

The optics of the primary rainbow is illustrated in Fig. 3. For sun elevation, a, above the horizontal, the elevation of a feature of the rainbow is $(\alpha - a)$, where α is the angular

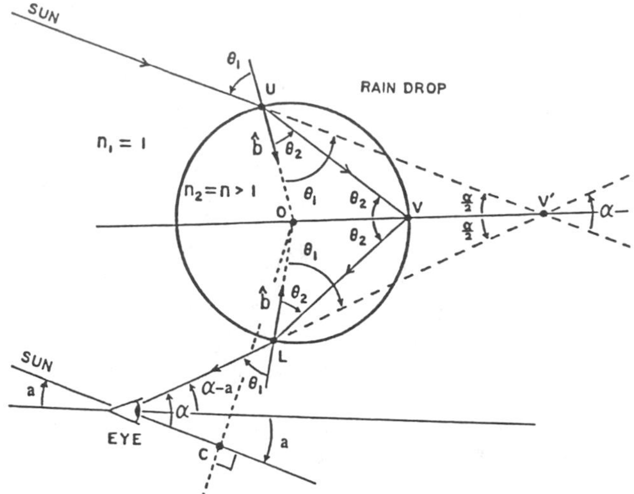

FIG. 3. Geometry of a rainbow. A ray from the sun enters the eye after refractions at U and L and internal reflection at V in a rain drop. The net effect is an apparent reflection from V', behind the water droplet.

radius of the feature measured from the center, C, of the rainbow. The surface normals, $\hat{b}$, point to the center, O, of the spherical droplet. Since ΔUOV and $\Delta UOV'$ have a common angle, it is easy to see that

$$2\theta_2 = \theta_1 + \frac{\alpha}{2}. \tag{4a}$$

Snell's law, (3a), is also obeyed:

$$n \sin \theta_2 = \sin \theta_1. \tag{4b}$$

From (4), zeroing the slope of α with respect to θ_1, one finds that α is largest when

$$\tan \theta_1 = \sqrt{\frac{4-n^2}{n^2-1}} \equiv Q, \tag{5a}$$

where

$$n \equiv \sqrt{\frac{Q^2+4}{Q^2+1}}, \tag{5b}$$

and the auxiliary quantity Q^2 is approximately equal to 3 for water. The largest value is

$$\alpha_M = 2 \tan^{-1} \left\{ \frac{Q^3}{3Q^2+4} \right\}. \tag{6a}$$

If α_M is measured, then the refractive index may be obtained by solving the cubic equation

$$Q^3 - \left(\tan \frac{\alpha_M}{2} \right) (3Q^2+4) = 0 \tag{6b}$$

for Q, and then using (5b).

If, for example, one finds that the tops of the red and violet bands are at

$$\alpha_M(\text{red}) = 42.5° \quad \text{and} \quad \alpha_M(\text{violet}) = 40.5°$$

then

$$Q^2(\text{red}) = 2.90 \quad \text{and} \quad Q^2(\text{violet}) = 2.72,$$

and

$$n(\text{red}) = 1.330 \quad \text{and} \quad n(\text{violet}) = 1.344.$$

Thus, the rainbow flaunts a 1% effect in refractive index!

NOTE: One often sees Fig. 3 drawn with a ray entering at L, emerging at U, and crossing the input ray on the way to the eye. Such drawings have propagated for nearly 700 years, even though this would reverse the colors and cannot happen for $n < \sqrt{2}$. Rays do cross in the secondary rainbow, which has two internal reflections in the droplet.

See also DIFFRACTION; DISPERSION THEORY; ELECTROMAGNETTIC RADIATION; LIGHT; MAXWELL'S EQUATIONS; REFLECTION; RELATIVITY, GENERAL THEORY.

BIBLIOGRAPHY

Max Born and Emil Wolf, *Principles of Optics*, 5th ed. Pergamon Press, New York, 1975. (A)

James Burke, *The Day the Universe Changed*. Little, Brown and Co., Boston, 1985. (E)

John David Jackson, *Classical Electrodynamics*. Wiley, New York, 1962. (A)

Francis A. Jenkins and Harvey E. White, *Fundamentals of Optics*. McGraw-Hill, New York, 1957. (E)

Demetrius T. Paris and F. Kenneth Hurd, *Basic Electromagnetic Theory*. McGraw-Hill, New York, 1969.

Julius Adams Stratton, *Electromagnetic Theory*. McGraw-Hill, New York, 1941. (A)

Refractories *see* Ceramics and Refractories

Regge Poles

R. Blankenbecler

The theory of Regge poles is an attempt to relate the high-energy behavior of scattering cross sections to the spin properties of low-mass particles (or resonances). This unexpected relation between high and low energy is based on the fundamental idea that the force between two strongly interacting particles is due in turn to the exchange of strongly interacting particles. This picture, that the force arises from the emission and absorption of particles, leads to many experimental predictions. The exchanges are the mechanism by which momentum is transferred between the projectile and the target in a scattering experiment. Other quantities, such as charge and strangeness, can also be exchanged. It should be stressed that the exchanged particles are virtual (not real) in the sense that their creation violates energy–momentum conservation, but because of the Heisenberg uncertainty principle, they can be produced for a sufficiently short interval. This theory is named after the physicist T. Regge, who first gave a general discussion of quantum-mechanical scattering using these terms and concepts.

It is convenient and customary to consider the cross section for the scattering of two particles as a Lorentz-invariant function of s, the square of the initial energy in the center-of-mass system, and t, the negative of the square of the momentum transfer in the same system. The differential cross section $d\sigma/dt$ is proportional to the square of the invariant scattering amplitude divided by s^2; furthermore, the optical theorem states that the imaginary part of the forward ($t = 0$) scattering amplitude divided by s is proportional to the total cross section. Thus if the scattering amplitude behaved as a power of s, say $s^{\alpha(t)}$, the differential cross section would behave as $s^{2\alpha(t)-2}$ and the total cross section as $s^{\alpha(0)-1}$. Note that t is negative in a physical scattering process.

The connection between the spin of low-mass resonances and the large-energy behavior arises in a simple manner. If an exchanged particle has spin zero, then it can only transmit information about the momentum transfer that is its momentum; hence $\alpha = 0$, since the amplitude must be independent of the energy. If the exchanged particle has spin, however, it can carry additional directional information, and we find that if it has a spin of J, then $\alpha = J$. The basic assumption in Regge theory is that there is an interpolation function $\alpha(t)$ that is equal to an integer J when t is equal to the square of the mass of a particle whose spin is J. The physical particles with differing J's that are described by the same $\alpha(t)$ form a Regge family and the function $\alpha(t)$ is called the Regge trajectory. The coherent exchange of such a family of particles produces a scattering amplitude of the form $\beta(t)s^{\alpha(t)}$, where $\beta(t)$ is termed the Regge residue. Both the residue and trajectory depend on the quantum numbers of the exchanged particles. Since several particle types may contribute, the scattering amplitude will be a sum of terms of the foregoing form with differing functions $\alpha(t)$ and $\beta(t)$.

In addition to the force arising from the exchange of particles, there can also be scattering arising from resonances. In this case the projectile and target fuse into a single unstable state that has a mass equal to the incident center-of-mass energy. After a short time, this state then decays into the final particles of interest. Since the known resonances occur in a rather limited mass range, this contribution can be important only at the lower energies. These resonances should form Regge families and should be described by trajectory functions. Therefore, as discussed earlier, they may be exchanged as well and their resulting Regge contributions at high energy must be consistent with low-energy information.

One further important property, called factorization, follows from the fact that since the emission and absorption processes are independent, the residue function $\beta(t)$ must factor into a function depending only on the projectile properties times a separate function of the target. This feature can be experimentally checked and seems to be well satisfied (at least to the 10% level of accuracy). One also sees that since $\alpha(t)$ is in general not a constant (it is expected to be an increasing function of t for t near zero), the detailed shape of the t distribution of $d\sigma/dt$ is expected to depend on the energy; its width should decrease logarithmically with increasing energy. This effect is also seen experimentally and the dependence of α on t for negative t is consistent with the masses and mass differences observed at positive t, as are the values of $\alpha(0)$ needed to fit the energy dependence of total cross sections.

In this picture of the origin of interparticle forces, the incident particles can exchange any particle allowed by the specific reaction of interest. Since this includes themselves, it is possible that the final particles are simply the incident particles interchanged. This "exchange force" then treats even and odd angular momentum states quite differently and they must be described by quite independent force laws. Therefore the Regge families actually separate into two independent groups, each described by its own trajectory function that corresponds to a physical particle at every other integer J value. These two groups are said to have opposite signature.

If the incident and/or final particles themselves have spin, then additional structure must in general appear in the Regge residues. The residues must contain explicit polynomials in the angular momentum, so that the contributions vanish from those states that are not allowed by the laws of the addition of angular momentum. Furthermore, factorization conditions must also hold between different initial and final spin states of the projectile and target. These conditions can considerably restrict the possible behavior of amplitudes, especially for small α values.

One of the most striking features of high-energy scattering processes is that total cross sections are remarkably constant over a large energy range. This is usually ascribed to the existence of an even-signatured Regge trajectory with the value $\alpha(0) = 1$ that carries no other quantum number. This is called the vacuum or Pomeron (after the Russian physicist, S. Pomeranchuk) trajectory. No simple trajectory can be larger than this value at $t = 0$ because it has been proven from very general principles that the magnitude of the forward amplitude as s goes to infinity must be less than $s \ln^2 s$, the Froissart bound. However, the theory becomes quite singular if any trajectory becomes equal to 1 for zero momentum transfer, and a precise description of the predictions or the theory in this situation is not yet available. In processes involving strongly interacting particles, if the exchange of one Regge family is important, the simultaneous exchange of another (or several more) should *a priori* be important. It can be shown, however, that the double-Regge-exchange contribution to the forward amplitude varies essentially as s to the power $\alpha_1(0)+\alpha_2(0)-1$. Therefore, if $\alpha_1(0)$ and $\alpha_2(0)$ are smaller than 1, then this contribution will become negligible compared to the single Regge exchanges. However, if one of the α's is unity, then this double-exchange contribution has the same power behavior as a single exchange. The effect of the exchange of multiple Pomerons on a given single Regge contribution is termed shadowing, or absorption, and is the subject of intense study.

Regge theory can also be applied to selected types of inelastic reactions, especially the so-called inclusive reactions in which we detect a selected type of particle in the final state but do not specify what else may have been produced. Regge theory predicts not only the dependence on the incident energy but also the dependence on the momentum of the detected particle (which is simply related to the total mass of the unobserved particle system). Regge theory can also be utilized as a theoretical model to develop quite general descriptions of production processes, such as the multi-Regge or multiperipheral models, and used for further theoretical study.

The invariant amplitude that describes the process $A+B \rightarrow A+B$ in which s is the square of the energy also describes the process $A+\bar{A} \rightarrow \bar{B}+B$ in which t is the square of the energy. The Regge exchanges in the former process produce direct-channel resonances in the latter. If the latter amplitude is expanded in partial waves, then there is a pole when the angular momentum J equals $\alpha(t)$. This is the origin of the term Regge pole, and its contribution to the full amplitude for large s takes the form

$$R(t)(s)^{\alpha(t)}[1 \pm e^{-i\pi\alpha(t)}]/\sin\pi\alpha(t),$$

where the upper (lower) sign corresponds to positive (negative) signature. This formula exhibits the expected energy dependence and the poles at every other J value. Recall that t is negative for physical processes in which s is the square of the energy, and $\alpha(t)$ cannot be larger than 1 here. For the physical processes in which t is positive, $\alpha(t)$ can be arbitrarily large, reflecting the possible angular momenta of particles. However, when t is in this physical region, the trajectory function $\alpha(t)$ can be shown to be complex, so that the scattering amplitude cannot become infinite here. When the real part of $\alpha(t)$ is close to the integer J, the foregoing formula reduces to the familiar Breit–Wigner resonance form, and the imaginary part of the $\alpha(t)$ is proportional to the total width of the resonance.

See also PARTIAL WAVES; RELATIVITY, SPECIAL; ROTATION AND ANGULAR MOMENTUM; SCATTERING THEORY; UNCERTAINTY PRINCIPLE.

BIBLIOGRAPHY

L. Bertocchi, "Theoretical Aspects of High Energy Phenomenology," *Proc. Heidelberg Int. Conf. Elementary Particles* (H. Filthuth, ed.). North-Holland, Amsterdam, 1968. (A)

S. C. Frautschi, *Regge Poles and S-Matrix Theory*. Benjamin, New York, 1963. (I)

Relativity, General

James L. Anderson

Of the various theories of gravity that have been proposed to date, the general theory of relativity is almost universally acknowledged to be the most satisfactory. It, among all of these theories, requires the smallest number of basic assumptions and contains the fewest number of adjustable parameters. It is also by far the richest in the kind and number of its predictions. And finally, unlike other theories, it agrees with observation today in almost all cases to within a tenth of a percent. In what follows we shall try to justify these assertions by setting forth the principal features of the general theory as we understand them today.

SPECIAL RELATIVISTIC THEORIES OF GRAVITY

Until the beginning of this century the most successful theory of gravity was the one set forth by Newton in the *Principia*. With only one exception, a small anomalous precession of the perihelion (closest approach) of the planet Mercury, it described accurately all gravitational phenomena known at this time. But with the development of special relativity in 1905 it became clear that some modification of the Newtonian theory was necessary; the Poisson equation

$$\nabla^2\phi = 4\pi G\rho \tag{1}$$

(G is the gravitational constant), which relates the gravitational potential ϕ to the mass density ρ of matter, is not Lorentz invariant, as required by the special theory.

It is a relatively simple matter to modify Eq. (1) so that it is Lorentz invariant; we replace the Laplacian operator ∇^2 in Eq. (1) by the d'Alembertian operator

$$\Box \equiv \frac{1}{c^2}\frac{\partial^2}{\partial t^2} - \nabla^2,$$

where c is the velocity of light, and we take ρ to be the energy density of matter as measured in the local rest frame of the matter. For matter that is moving slowly compared to c, this modified equation reduces to Eq. (1). However, it does not agree with observation; among other things, it predicts the wrong value and direction for the perihelion motion of Mercury. It also does not lead to a bending of light as it passes near the edge of the sun.

Although it is possible to construct a special relativistic theory of gravity that is consistent with the observed values of the perihelion advance and the bending of light, and indeed several such theories have been proposed, they all contain features that make them less than satisfactory. Any theory of gravity must, of course, agree with the known observational facts and reduce to the Newtonian theory in the appropriate limit. In particular it must take account of the fact that, unlike electromagnetism, gravity is a purely attractive force—all gravitational "charges" are positive. Even more important, though, is the fact that the ratio of the inertial mass of a body to its gravitational charge appears to be a universal constant; in particular, and again unlike the electromagnetic case, there are no gravitationally uncharged bodies. In all special relativistic theories these facts are not unavoidable consequences, as they are in general relativity, but must be imposed from the outside. In addition, these theories all contain adjustable parameters that do not appear in general relativity. They also all appear to suffer from one or another form of the so-called negative-energy disease: They predict that a gravitationally bound system such as a double star will *gain* energy as time goes on rather than lose it, as is the case for electromagnetically bound systems.

There is a final feature of gravity that should be discussed in regard to special relativistic theories. In the Newtonian theory of gravity the gravitational "charge" of a body is taken to be proportional to its inertial mass, so that the ρ appearing in Eq. (1) is just the mass density of whatever matter is present. However, special relativity requires that all forms of energy possess an equivalent mass, according to the famous formula $E=mc^2$. Thus, all forms of binding energy in matter should contribute to its inertial mass and hence to its gravitational charge. Indeed, the most accurate experimental checks by Dicke and others have confirmed that with one possible exception, this is the case. The exception is the gravitational binding energy itself, which in ordinary matter is some 40 orders of magnitude smaller than the electrical binding energy and hence is much too small to be detected by currently available techniques. Thus, observation does not compel us to include gravitational energy in the computation of the gravitational charge of a body nor does any special relativistic theory of gravity do so. Nevertheless, it would be strange indeed if, of all forms of energy, gravitational energy alone did not contribute to the gravitational charge of a body.

We can include gravitational energy as a source of gravitational charge in a special relativistic theory even though we are not forced to do so. When we do, however, the resulting equations for the gravitational field are no longer linear because the gravitational binding energy is at least quadratic in the gravitational field itself. But what is most peculiar is that when the gravitational energy is consistently included in the gravitational charge, the resulting equations for the gravitational field in addition to becoming nonlinear have essentially the same form as they do in general relativity.

THE PRINCIPLE OF GENERAL RELATIVITY

In the years following his development of the special theory of relativity Einstein set himself two tasks: to construct a theory of gravity that was consistent with the principles of special relativity, and to find a generalization of the special theory that did away with the special role played by the inertial frames of reference in that theory. Even though he was unaware of many of the problems associated with the construction of a relativistic theory of gravity discussed earlier, Einstein soon realized that there was more involved than merely finding an appropriate Lorentz-invariant set of equations for the gravitational field and that the two tasks were in fact intimately related to one another. In order to understand how this realization came about and had as its consequence the general theory of relativity, it will prove helpful to review briefly the space-time structures of Newtonian physics and special relativity.

The world pictures of both Newtonian and special relativistic space-time were endowed with certain absolute geometrical elements. In the Newtonian picture these consisted of planes of absolute simultaneity and straight lines. A plane of absolute simultaneity consists of all those space-time events that were simultaneous with each other. Since these planes can not intersect, each one can be assigned a unique time coordinate t and together they constitute the absolute time of Newton. It was implicitly assumed that there was an objective, observer-independent way of deciding whether or not two events were simultaneous. The straight lines were also assumed to be physically indentifiable, being the trajectories of free bodies, that is, bodies on which no forces acted. Newton also assumed that a subset of these trajectories corresponded to states of absolute rest. This last assumption, however, proved to be completely unnecessary for the description of any known physical system, so that there proved to be no physically objective means of identifying the states of absolute rest. A formal assertion of this fact is known as the principle of *Galilean relativity*.

The transition from the Newtonian world picture to that of special relativity came with the realization by Einstein that in addition to the impossibility of objectively identifying states of absolute rest it was also impossible to identify objectively the Newtonian planes of absolute simultaneity. In place of these planes the special theory substituted light cones: two space-time events lie on the same light cone if a light signal from one can reach the other, and the light cone through a given space-time point consists of the collection of all other points that can be connected to it by a light ray. If the point in question is taken to be the origin of a Cartesian system of coordinates (t,x,t,z), then the equation of the light cone through this point is given by

$$t^2-\frac{1}{c^2}(x^2+y^2+z^2)=0. \tag{2}$$

Note that in the limit in which c, the velocity of light, is allowed to approach infinity, Eq. (2) reduces to

$$t=0, \tag{3}$$

which is the equation of the plane of simultaneity through the origin. And indeed, in this limit, the light-cone structure of special relativity is transformed into the Newtonian planes of absolute simultaneity.

While special relativity replaced the planes of absolute simultaneity of the Newtonian world picture by light cones, it retained the straight lines of that picture; and of course it insisted upon the impossibility of physically distinguishing any subset as corresponding to states of absolute rest. Thus, the transition from the Newtonian world picture to that of special relativity consisted, in the final analysis, of replacing one set of absolute objects, the plans of absolute simultaneity, by another, the light cones. Of course, the form of the laws of physics that can be formulated with the latter are quite different from those formulated with the former and it was the experimental verification of these differences that led to the replacement of Newtonian physics by special relativity. But it must be emphasized that in one respect the two theories are quite similar: They both contain what we have called absolute objects, in one case the planes of absolute simultaneity and the straight lines, and in the other the light cones, plus again the straight lines.

The nature of absolute objects is quite different from that of the other kinds of objects used in constructing physical theories, which we call dynamical: The former are given once and for all and are unaffected by the presence of other physical systems or the particular states these systems find themselves in. The latter, on the other hand, depend on the former and vary with the state of the physical system with which they are associated. Furthermore, the uniqueness of a set of absolute objects such as the light cones and straight lines allows us to single out from the totality of all space-time observers a special subclass by the requirement that they all measure the same values for the absolute objects in question. In the case of the absolute objects of Newtonian physics and special relativity, one such subclass constitutes the so-called inertial observers. Because the absolute objects have the same values for all such observers there is no objective physical way to distinguish among them. Since the inertial observers move with respect to each other with uniform velocity we are led to the notion of the relativity of uniform motion, referred to elsewhere in this article as the principle of Galilean relativity.

It can be argued that the existence of absolute objects violates a kind of generalized law of action and reaction—if A can affect B, then B should be able to affect A—and that consequently absolute objects are only approximately so; that in fact all physical objects are actually dynamical. However appealing this argument may appear, it does not itself constitute a proof of the nonexistence of absolute objects; only a direct appeal to observation can do that. Einstein's rejection of the absolute objects of special relativity and hence the special status of inertia observers was based on the by then well-established universality of the ratio of inertial mass to gravitational charge. Although the fact that all bodies fall with the same acceleration in the earth's gravitational field had been known since the time of Galileo, no special significance was attached to it. It was left to Einstein to recognize its profound implications for the nature of physical space-time. It meant in fact that there were no free particles in nature; all particles were acted on by gravity in the same way and consequently could not be used to identify objectively the straight lines of either the Newtonian or the special-relativistic world picture.

Although the nonexistence of free particles argued strongly for the unobservability of the absolute straight lines of special relativity, it did not rule out the possibility that some other physical system or combination of systems might be used for the purpose. Nor did it rule out the absolute nature of the light cones of that world picture. However, Einstein realized in effect that if these objects were dynamical, then there would no longer be an objective means of uniquely identifying inertial observers. In that case all observers would be equivalent and hence all motion would be relative. This was just the generalization that he had been searching for and that he called the principle of *general relativity*.

GEOMETRY AND GRAVITY

If the absolute objects of special relativity are in principle unobservable, then either they must not appear in any dynamical law or else they themselves must become dynamical elements, to be determined by dynamical laws like all other such objects. Since the dynamical laws of special relativity require light cones and straight lines for their formulation and since these laws must hold in some appropriate approximation, it was the second possibility that Einstein chose. The problem then was to find a satisfactory set of dynamical laws for these quantities. It was this problem that occupied him for most of the 10 years between 1905 and 1915 when he published the general theory. Since it would take us too far afield to review his many attempts to find these dynamical laws and since we are not strictly speaking following the historical development of the subject here, we shall be content to try to describe the final results of his investigations.

The light cones and straight lines of special relativity can be characterized in two different but completely equivalent ways: We can specify directly the points constituting these objects by means of a set of algebraic expressions. For example, for an inertial observer the light cone through the origin would be given by Eq. (2). Or we can introduce onto the space-time manifold a flat, Riemannian metric. Given this metric, we can construct uniquely the light cones and straight lines of special relativity and vice versa. It is this characterization that can most easily be used in the construction of the dynamical laws for these objects that are required in general relativity.

The notion of a Riemannian metric arose in the study of non-Euclidean geometries and is used to generalize the Pythagorean expression for the distance between two neighboring points in a Euclidean space. If we label the points of the four-dimensional space-time manifold by four coordinate values and if x^μ and $x^\mu + dx^\mu$ ($\mu = 0,1,2,3$) are the coordinates of two neighboring points of the space-time manifold, then the distance ds between these points is given in terms of a Riemannian metric tensor $g_{\mu\nu}$ at x^μ by the expression

$$ds^2 = \sum_{\mu=0}^{3} \sum_{\nu=0}^{3} g_{\mu\nu}(x)\, dx^\mu\, dx^\nu. \tag{4}$$

A space-time metric is said to be flat if it is possible to find a coordinate system in which its off-diagonal components vanish and its diagonal components have the values $g_{00} = -g_{11} = -g_{22} = -g_{33} = 1$. A flat metric with these values for its components is said to be Minkowskian and an observer who measures these values for $g_{\mu\nu}$ is defined to be an inertial observer. We can also charactrerize a flat metric in a coordinate-independent manner by requiring that it satisfy a set of differential equations that have the same form in all coordinate systems. A set of equations with this property is said to be *generally covariant*. It can then be shown that these two characterizations of a flat metric are completely equivalent.

We can also rewrite the special relativistic equations of motion of a system in generally convariant form with the help of $g_{\mu\nu}$. These equations, together with the equations for $g_{\mu\nu}$ that imply its flatness, then constitute a generally covariant description in terms of an arbitrary system of coordinates. In such a covariant description the metric is an absolute object that can be used to distinguish the class of inertial observers. Thus the requirement of general covariance by itself does not imply the principle of general relativity—as long as the metric is flat and hence absolute, it can be used to define a family of inertial observers.

In order to satisfy the principle of general relativity. Einstein realized that not only must the equations of physics be generally covariant but also that the space-time metric must be a dynamic object on the same footing as all other dynamical objects in physics. Once the metric becomes dynamical it will in general no longer be flat and hence cannot be used to single out a family of inertial observers. Having come to this realization, Einstein was then confronted with the task of constructing a set of generally covariant equations of motion for $g_{\mu\nu}$. Although he spent considerable effort to obtain such equations, he found that if he restricted himself to equations of second differential order, they were unique. They contained only two adjustable constants, the so-called cosmological constant and a single coupling constant κ that determines the strength of the interaction between the metric and the various kinds of matter and radiation that exist in nature. It is this uniqueness of the equations of motion for the metric that above all else distinguishes general relativity from other theories of gravity.

At the beginning of this article we said that general relativity was primarily a theory of gravity. It remains therefore to describe how the gravitational field is characterized in this theory. To do so in fact requires the introduction of no new objects; the gravitational field is described by the same symmetric tensor field $g_{\mu\nu}$ that characterizes the metric of space-time and that is needed for the covariant formulation of the dynamical laws of all other physical systems.

Einstein arrived at this identification of gravity and the geometry of space-time as a consequence of his principle of *equivalence*. In effect, this principle postulated the impossibility of distingishing inertial (geometrical) effects such as the Coriolis and centripetal forces from the effects of suitably chosen gravitational fields. If the former cannot be distinguished from the latter, it follows that one and the same object suffices to describe both. Indeed, in the limit of weak gravitational fields and slowly (compared to the velocity of light) moving bodies the predictions of general relativity agree completely with Newtonian gravitational theory if the coupling constant κ referred to earlier is taken equal to $8\pi G/c$. In particular, in this limit

$$g_{00} = 1 + \frac{2}{c^2}\,\phi, \tag{5}$$

where ϕ is the gravitational potential and satisfies Eq. (1).

To sum up, we see that general relativity is based on three principles: The principle of general covariance, the principle of general relativity, and the principle of equivalence. The first of these principles imposes strong restrictions on how the geometry of space-time affects the behavior of all other physical systems. In particular it requires that all such systems must couple at least minimally to this geometry. The second principle requires that this geometry must be dynamically determined like all other physical objects, and the third principle identifies this geometry with gravity.

TESTS AND PREDICTIONS OF GENERAL RELATIVITY

Unlike other theories, general relativity is a very restrictive theory. It allows for very few possibilities for the coupling of the gravitational field metric to matter and radiation and for the form of its equations of motion. Aside from the value of the coupling constant κ, which is fixed by the requirement that in the appropriate limit the predictions of general relativity should agree with those of Newtonian theory, these equations contain only the cosmological constant, which must have such a small (if not zero) value that it has no observable consequences on any but a cosmological scale. Thus, general relativity, in contrast to alternative theories of gravity, yields essentially unique predictions for the effects of gravity on other physical systems.

There are two rather different kinds of predictions that can be derived from general relativity. One kind concerns very small deviations from the predictions of Newtonian theory. These deviations are very difficult to observe since they usually amount to at most a few percent of the Newtonian predications. Nevertheless there are several such predictions that can be tested in our solar system and as well as on Earth. The theory predicts a small precession of the perihelion of planetary orbits. For Mercury it amounts to some 43 seconds of arc per century. It also predicts a small bending of light in the gravitational field of the sun of 1.75 seconds of arc for light that just grazes its edge. In addition it predicts a small delay in the time it takes light signals to pass through the Sun's gravitational field. Finally, general relativity predicts that clocks in gravitational fields run at a somewhat slower rate than similar clocks in the absence of such fields. All of these effects have been observed, the first three in the solar system and the latter on the earth, and their values agree with the predictions of general relativity to within the experimental accuracy of what is today less than 1%. This accuracy, coupled with other observations described below, is sufficient to enable us at the present time to rule out all other competing theories of gravity.

Unlike Newtonian theory, general relativity also predicts the existence of gravitational radiation. As might be expected, these waves are all very weak and therefore extremely difficult to detect. For example, the Earth–Sun system radiates only about 200 W of gravitational energy. However, there is every expectation that improved technology based on the pioneering work of Joseph Weber will lead to the unambiguous detection of gravitational waves incident on the earth.

In addition to their direct detection on earth, there is now strong indirect evidence for the existence of such waves. Any system that radiates gravitational energy must perforce make up for this loss through a decrease in its own energy, kinetic plus potential. The effect is called gravitational radiation damping and it too is, for most systems, far too small to be observed. However, the discovery of the binary pulsar PS1319+16 in 1974 by Joseph Taylor and his coworkers provides us with a system in which the rate of energy loss due to gravitational radiation is of the same order of magnitude as the energy loss due to electromagnetic radiation of the Sun, that is, approximately 10^{27} W. The net effect of such an energy loss is to produce a gradual decrease in the orbital period of the system of about 4×10^{-4} s/year. An increase of this amount has in fact been measured and agrees with the prediction of general relativity to within the present experimental error of about $\frac{1}{2}$%. Whereas other theories of gravity predict the existence of gravitational radiation, most of them give results that are not in agreement with these observations and hence can be ruled out on physical grounds.

The other kind of prediction made by general relativity is qualitatively different from those of the Newtonian or special relativistic theories of gravity. It is these latter predictions that give general relativity its special interest and make it such an exciting field of study. Chief among such predictions is the phenomenon of gravitational collapse leading to the creation of black holes. Because of the nonlinearity of the field equations for the gravitational field, gravity can act in effect as its own source. Once the gravitational potential ϕ becomes of the order of c^2 [see Eq. (5)], these nonlinear effects become important and cause the field to grow without bound. The investigation of the properties of the resulting black holes has become one of the most active areas of research in physics today.

The role played by general relativity in the evolution of the universe has, in recent years, taken on a renewed interest due to the symbiotic relation that has developed between cosmology and elementary particle theory. As one extends backwards in time our knowledge of the universe a point is finally reached where the size of the present visible universe is less than a Planck length of 10^{-33} cm. Beyond that point it is believed that gravity can no longer be treated as a classical field but must in some way be "quantized." The quantization of the gravitational field and its unification with the other elementary forces in nature today stands as perhaps the greatest challenge facing physics.

See also BLACK HOLES; GRAVITATION; GRAVITATIONAL WAVES; LORENTZ TRANSFORMATIONS; RELATIVITY, SPECIAL THEORY; SPACE-TIME.

BIBLIOGRAPHY

A popularized account of general relativity can be found in Peter G. Bergmann's *The Riddle of Gravitation* (Scribner's, New York, 1968).

An excellent popular account of recent advances in general relativity is given by William J. Kaufmann III in *The Cosmic Frontiers of General Relativity* (Little, Brown, Boston, 1977).

An introductory account of general relativity is contained in Wolfgang Rindler's *Essential Relativity* (Van Nostrand-Reinhold, Princeton, N.J., 1969).

Advanced Texts

R. Adler, M. Bazin, and M. Schiffer, *Introduction to General Relativity*, 2nd ed. McGraw-Hill, New York, 1975.

James L. Anderson, *Principles of Relativity Physics*. Academic, New York, 1967.

L. D. Landau and E. M. Lifshitz, *The Classical Theory of Fields*, 3d rev. English ed. Addison-Wesley, Reading, Mass., 1971.

C. W. Misner, K. S. Thorne, and J. A. Wheeler, *Gravitation*. W. H. Freeman, San Francisco, 1973.

S. Weinberg, *Gravitation and Cosmology*. Wiley, New York, 1972.

Relativity, Special Theory

Claude Kacser

The special theory of relativity is a formulation of the relationship between physical observations and the concepts of space and time proposed by Albert Einstein (1879–1955) in 1905. The theory with all its implications has been thoroughly tested and is accepted by nearly all physicists. It is one of the two cornerstones of modern physics, the other being wave (or quantum) mechanics.

OVERVIEW

The theory arose out of contradictions between electromagnetism and Newtonian mechanics and has great impact on both those areas. The original historical issue was whether it was meaningful to discuss the electromagnetic wave-carrying "ether" and motion relative to it. Also whether one could detect such motion, as unsuccessfully attempted in the Michelson–Morley experiment (q.v.). Einstein "annihilated" these questions and the ether concept, in his Special Relativity.

However, his basic formulation does not involve detailed E.M. theory. It arises out of the question, "What is time?" Newton, in the *Principia* (1686), had given an unambiguous answer: "Absolute, true, and mathematical time, of itself, and from its own nature, flows equably without relation to anything external, and by another name is called duration."

This definition is basic to all of classical physics, Einstein had the genius to question it, and found that it was not correct. Instead, each "observer" necessarily makes use of his (or her) own scale of time. Further, for two observers in relative motion, their time scales will differ. This induces a related effect on distance.

Both space and time become *relative* concepts, fundamentally dependent on the observer. Each observer generates his own space-time framework or coordinate system. All observers have equal validity, there being no absolute frame of reference. Motion is relative, but only relative to other observers. What is absolute is stated in Einstein's *first relativity postulate*: The basic laws of physics are identical for two observers who have a constant relative velocity with respect to each other. (In 1916 Einstein was able to generalize this further, to deal with all states of relative motion including acceleration. This became the general theory of relativity (q.v.), which conjointly is a theory of gravitation, and supersedes Newton's universal theory of gravitation.)

Recap of Newtonian Relativity; also on Observers

Newtonian mechanics itself contains a relativity principle, Galilean relativity, in which time and space are absolute. In that system, consider two observers A and B, with B having a *constant* velocity **V** with respect to (hereafter abbreviated w.r.t.) A. In the following, we always make use of these same two observers and their associated frames of reference. *Throughout this article,* we consider only observers whose relative velocity **V** is constant—for a changing **V** one must refer to Einstein's general relativity (q.v.). Always, by "an observer" (call him or her A) is meant a (possibly infinite) *set* of observers {A} spread throughout space, all relatively at rest, all with identical clocks that are synchronized w.r.t. each other. This situation is equivalent to a complete space-time coordinate grid, and the existence of such a grid is already implied when we discuss motion of an object w.r.t. an observer. Most emphatically, we are not concerned with the visual appearance that is actually seen by *one* observer. Rather initially, we do correct out for well-understood physical phenomena like parallax and perspective; *and* for the "understood" temporal delays in receiving the light signals from distant events that must clearly have taken place earlier in time. But how much earlier?—that will turn out to be the crucial question. That is why we need an infinite set of observers.

For ease (but not necessity) we assume that observers A and B both use the same units of space and of time, and use parallel sets of Cartesian coordinate axes, and that each observer has chosen his origin of coordinates so that the two space–time origins coincide at time $t=0$. Then, if any event E has space-time coordinates $(\mathbf{r},t)\equiv(x,y,z,t)$ according to observer A, and coordinates $(\mathbf{r}',t')\equiv(x',y',z',t')$ according to observer B, Galilean (or Newtonian) relativity asserts that

$$\begin{array}{ll} \mathbf{r}'=\mathbf{r}-\mathbf{V}t, & \mathbf{r}=\mathbf{r}'+\mathbf{V}t' \\ t'=t, & t=t'; \end{array} \quad \text{Galilean transformation.} \tag{1}$$

Here and always, "primed" measurements will refer to B, and unprimed ones to A. Notice in particular that both observers have the *same* $t=t'$.

From this follows directly that if a "particle" P is located instantaneously at space–time event E and is moving with velocity $\mathbf{v}$ according to A, then according to B it has velocity $\mathbf{v}'=\mathbf{v}-\mathbf{V}$. That is,

$$\mathbf{v}'=\mathbf{v}-\mathbf{V}, \qquad \mathbf{v}=\mathbf{v}'+\mathbf{V} \quad \text{(Galilean velocity addition).} \tag{2}$$

These results agree with everyday common experience of relative velocity for everyday speeds, but that does *not* give us the right to assume they hold for *all* speeds.

ELECTROMAGNETIC VIEWPOINT

Einstein, in his 1905 paper, was attempting to reconcile Eqs. (1) and (2) with the laws of electromagnetism, in particular, Maxwell's equations for the propagation of light in a vacuum. The latter predict a speed of light c that seems to make no reference to the motion of the observer; i.e., the equations should apply equally for observer A *and* for observer B, and predict the *same* speed of light c for *both* A *and* B. That is, they predict the *same* speed c for the *same* light pulse w.r.t. A, *and* w.r.t. B. This contradicts Eq. (2); so previous workers had assumed that Maxwell's equations only applied directly to the frame of reference that was stationary w.r.t. the ether—a mysterious ethereal substance that could not be observed directly, but that carried the electromagnetic waves.

Various experiments had been performed with the attempt of detecting motion w.r.t. the ether. The most famous of these was the Michelson–Morley experiment (q.v.), first performed in 1887 and often repeated, which had the paradoxical result that, however the earth moved around the sun and spun on its own axis, the earth was apparently always at rest w.r.t. the ether—*or* that the speed of light has always *exactly* the same value w.r.t. the earth, regardless of the earth's motion.

Einstein's attempt to reconcile electromagnetism with the Galilean transformation addressed the same issue. Einstein cut through all the obfuscation by ignoring it, and instead proposed the *second relativity postulate*: the speed of light in vacuum is an absolute constant for *all* observers, independent of the velocity of the light source, or the velocity of the observer.

This postulate contradicts both common sense and the Galilean transformation equations (1) and (2). Instead, it is fully in accord with Maxwell's equations, and with the null result of the Michelson–Morley experiment. Einstein realized that time could *not* be absolute. He was thus led to develop a replacement for the Galilean transformation, which replacement is traditionally (though erroneously) called the Lorentz transformation. For ease of presentation we now assume that $\mathbf{V}$ is directed along the $+x$ (and $+x'$) axes. For that case, Einstein derived from his *two* postulates the *Lorentz transformation*:

$$\begin{array}{ll} x'=\gamma(x-Vt), & x=\gamma(x'+Vt'), \\ y'=y, & y=y', \\ z'=z, & z=z', \\ t'=\gamma\left(t-\dfrac{Vx}{c^2}\right), \text{ and } & t=\gamma\left(t'+\dfrac{Vx'}{c^2}\right), \end{array} \tag{3a}$$

where

$$\gamma=\frac{1}{\sqrt{1-(V^2/c^2)}}\geq 1. \tag{3b}$$

These new transformation equations have the property that when *all* speeds are much less than the speed of light, they effectively reduce to the Galilean transformation (1). Thus, they do not contradict classical physics; rather they show how to *extend it* correctly when speeds comparable to that of light are involved. (Special relativity is an evolution, not a revolution.)

All the kinematic consequences of special relativity are already contained in the Lorentz transformation equations. However, true physical insight into the various phenomena and consequences is only obtained by a detailed, physically oriented discussion of the derivation of the Lorentz transformation, and of their consequences. Lack of space prevents such a full discussion here; instead, we briefly present the principal consequences of the theory.

VELOCITY ADDITION

In place of Eqs. (2) we find

$$\mathbf{v}=\frac{1}{1+[(\mathbf{V}\cdot\mathbf{v}')/c^2]}\left\{\mathbf{V}\left[1-\frac{\mathbf{v}'\cdot\mathbf{V}}{V^2}\left(\frac{1}{\gamma}-1\right)\right]+\frac{1}{\gamma}\mathbf{v}'\right\} \tag{4}$$

(very unwieldy) or, in component form, the relativistic velocity "addition" (better termed "compounding")

$$v_x=\frac{v_x'+V}{1+(v_x'V/c^2)},$$

$$v_y=\frac{v_y'}{\gamma[1+(v_x'V/c^2)]},$$

$$v_z=\frac{v_z'}{\gamma[1+(v_x'V/c^2)]}, \tag{5}$$

with $\gamma=[1-(V^2/c^2)]^{-1/2}$.

For the case when $\mathbf{v}'$ is in the same direction as $\mathbf{V}$, this simplifies to "*addition*" *of parallel velocities*:

$$v=\frac{v'+V}{1+(v'V/c^2)}. \tag{6}$$

In *all* these cases, if both v' and V are much smaller than c, we regain the Galilean result (2). However, when the "particle" is really light, with $v'=c$, then (surprisingly?) we find $v=c$. This must be, since it was put into the Lorentz transformation by the second relativity postulate itself—light always has velocity c for *all* observers. (There is v, v' symmetry in the above.)

Further, if a particle has a speed less than c for one observer, it also has a speed less than c for *any* other possible observer whose relative velocity V is less than c. Thus, the velocity of light seems to be a maximum limiting speed. (We discuss this again later under dynamic aspects.)

While the theory can contain speeds greater than c [particles with such speeds have been called "tachyons" (q.v.)], they are not a natural part of it. None have yet been observed, and attempts to incorporate such tachyons into full quantum field theory have so far failed. Regardless, to avoid causality paradoxes, relativity *must postulate* that no *signal* can travel faster than c.

LENGTH CONTRACTION

Consider a rod is at rest w.r.t. observer B with length L_0 according to B. Then according to observer A, for whom the rod has velocity $\mathbf{V}$, the length of the moving rod is given by

$$L=\gamma^{-1}L_0=\left(1-\frac{V^2}{c^2}\right)^{1/2}L_0 \quad \text{(length contraction).} \qquad (7)$$

This moving length is contracted. This effect is known as the Fitzgerald contraction (or the Lorentz–Fitzgerald contraction). It arises due to measurement/synchronization issues.

It is important to notice that this effect is symmetrical. A rod that is stationary w.r.t. A has a shorter contracted length according to B, w.r.t. whom it is moving. *All* special relativity effects have this symmetry. *Neither* observer is preferred. All that matters is the state of motion of the observed object relative to the observer.

TIME DILATION

Suppose, according to observer B, that two events occur at the *same* location and are separated by B's time interval τ. Then observer A will find that the time interval between the events is longer than τ, and this time dilation is given by

$$t=\gamma\tau=[1-(V^2/c^2)]^{-1/2}\tau. \qquad (8)$$

Again, there is complete symmetry between A and B. Always it is the moving clock that runs slow, in the following sense. A has his complete set of observers and clocks. He sees a clock that is moving and associated with B, and that counts τ clock-tick *intervals* between events E_1 and E_2. (An event is a "happening" at a definite location and a definite time.) According to B these events take place at the same location, and separated by τ seconds. According to A, however, they occur at different spatial locations and are separated by a longer time $\gamma\tau$ seconds. A concludes that B's moving clock is in fact running slowly, at the rate of 1 tick every γ seconds.

Yet, if instead B observed a clock associated with and at rest w.r.t. A, B would assert that A's now moving clock was running slowly, again by the factor γ. *All* relativity effects are symmetrical in this manner A makes the same assertion about B's moving clocks running slowly, as B makes about A's moving clocks.

Further understanding of time dilation can be obtained by discussion of the "light clock." This basically is a device in which light goes back and forth along a straight rod between two mirrors. Each to-and-fro constitutes one equal time interval, and so the device is a clock. It is discussed in several of the references.

THE TWIN PARADOX

A particularly paradoxical consequence of time dilation is known as the twin paradox (or clock paradox). See separate article.

SIMULTANEITY, CAUSALITY, AND THE LIGHT CONE

It is clear that the Lorentz transformation equations (3) interrelate space and time. Some people refer to a four-dimensional space-time continuum, but this is misleading since space and time are very different. Nonetheless, we can readily see that if two events E_1 and E_2 have times t_1' and t_2', which are equal, so that they are simultaneous w.r.t. B, they need *not* be simultaneous w.r.t. A. In fact, by suitable choice of the sign of $\mathbf{V}$, we can make either event E_1 or event E_2 be the earlier, when viewed by A.

Hence the concept of simultaneity becomes only relative, and we must ask, "Simultaneous w.r.t. which observer?" However, this ambiguity does not violate causality, as we will discuss below.

It is useful to define various *light-cone* regions w.r.t. an event E (see Fig. 1). Suppose event E has coordinates $(\mathbf{r}_E,t_E)$ according to observer A. Consider another event P, with coordinates $(\mathbf{r}_P,t_P)$; we can define $\Delta\mathbf{r}=\mathbf{r}_P-\mathbf{r}_E$, and $\Delta t=t_P-t_E$. Then four space-time regions (shown in Fig. 1) can be constructed as follows:

$|\Delta\mathbf{r}|^2<c^2(\Delta t)^2$ *and* $\Delta t>0$, P lies within future light cone of E; or P is in E's "future";

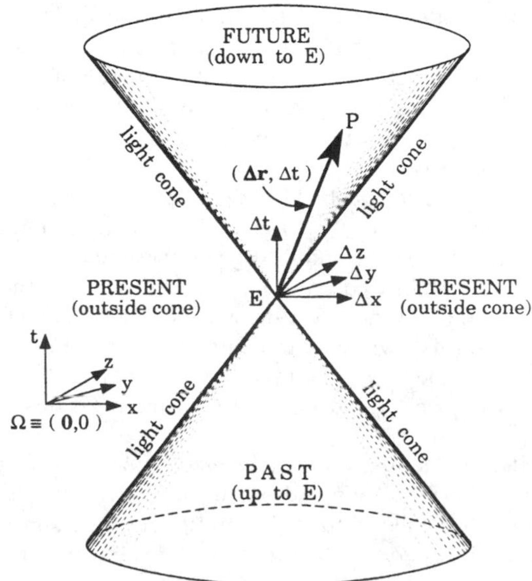

FIG. 1. The light cone, past, present, and future, shown relative to the event E. P is a general other event. The observer's reference space-time origin is $\Omega=(\mathbf{0},0)$. The cone surface is given by $(\Delta x)^2+(\Delta y)^2+(\Delta z)^2=c^2(\Delta t)^2$. The figure is a two-dimensional representation of a four-dimensional space.

$\|\Delta\mathbf{r}\|^2<c^2(\Delta t)^2$ *and* $\Delta t<0$,	P lies within past light cone of E; or P is in E's "past";
$\|\Delta\mathbf{r}\|^2=c^2(\Delta t)^2$,	P lies on E's light cone;
$\|\Delta\mathbf{r}\|^2>c^2(\Delta t)^2$,	P lies outside light cone of E; or lies in E's "present."

The names "future," "past," and "present" will be explained immediately.

A direct result of the Lorentz transformation equations (3) is that if an event P lies in a particular one of these four regions w.r.t. the event E as seen by one observer A, then it lies in the *same* region for *any* other observer B, whatever B's velocity $\mathbf{V}$ provided $V<c$. (Naturally, values of $\Delta\mathbf{r}$ and Δt do change.)

Back to causality. If event R is the result of a causative event C, then R must be located in space-time so that a causing *signal* can travel from the *earlier* event C. For self-consistency, relativity theory must assert that no *signal* can travel faster than the speed of light c. Hence with $\Delta\mathbf{r}=\mathbf{r}_R-\mathbf{r}_C$ and $\Delta t=t_R-t_C$, it must be the case that $\Delta t>0$, and that $|\Delta\mathbf{r}|\leq c\ \Delta t$. That is, the result R must lie on or *within* the future (or future light cone) of the cause C. Hence (previous paragraph) event R lies in the future of event C according to *all* observers. Thus, the possibility of R's being caused by C is an invariant statement, and all observers will agree on it.

Similarly, if C is itself caused by some other prior event P, *all* observers will agree that P lies in the past of C.

Finally, all events L outside C's light cone may properly be stated to be in "limbo" w.r.t. C. No such event L can communicate with the event C, *nor* vice versa. (Traditionally limbo is called the "present" or "now," but these are poor names.) Consider an event L in limbo w.r.t. C; then some observers will find that L occurred before C, whereas others will find that C occurred before L. The time sequence is reversible, but it is immaterial—no signal can go *either* way.

OTHER KINEMATIC EFFECTS

There are many other important kinematic consequences of the special theory, all derivable from the Lorentz transformation equations (3). Among these are stellar (and other angular) aberration the relativistic Doppler effect (q.v.), the transformation of acceleration, and the visual appearance of rapidly moving objects.

EXPERIMENTAL TESTS OF KINEMATIC EFFECTS

There are numerous verifications of the foregoing predictions. Since the magnitude of any effect depends on how close v^2/c^2 is to unity, we usually need extremely high speeds ($c=3\times10^8$ m/s$=186{,}000$ miles/s). Thus, most tests have used short-lived elementary particles. Both length contraction and time dilation (which are really two aspects of the same phenomenon) are "tested" daily by high-energy experimentalists who use high-speed beams of unstable particles. As a result of the exponential unstable particle-decay law, any such beam has a particle count that decreases exponentially along the beam length, but the effective laboratory lifetime is always found to be dilated; or equivalently, the beam length as seen by the decaying particle is always contracted compared to its laboratory length.

Since atomic clocks (q.v.) are accurate to parts in 10^{10}, several successful tests at "Newtonian" speeds have been performed. Thus, a direct test of the twin-paradox prediction has been carried out to about 10% accuracy (1976) by flying a very accurate atomic clock in an airplane in a long racetrack path, and comparing the clock rate with that of an identical one stationary on ground level. A similar more accurate but less direct experiment (1960) made use of the Mössbauer effect. That Mössbauer experiment compared the frequency of atomic transition radiation in ^{57}Fe for iron on the end of a spinning centrifuge arm, with that for iron at rest at the center of the centrifuge. Many other less startling predictions have also been verified. For instance, one form of velocity addition is tested in the Fizeau experiment. No educated physicist can doubt the truth of the kinematic predictions of the special theory.

We now turn to the dynamic predictions. If anything, they are even more startling.

DYNAMICAL EFFECTS: RELATIVISTIC MOMENTUM AND RELATIVISTIC KINETIC AND TOTAL ENERGY

If special relativity changes the relationship between space and time, and changes the relationships between velocities as seen by different observers, it is almost obvious that it must have a profound effect on Newtonian dynamics, and particularly on Newton's second law $\mathbf{F}=m\mathbf{a}$. Very significant changes occur.

Rest frame. This important concept refers to the frame in which a particle is (instantaneously) at rest—and to all quantities measured by observers in this frame.

Mass

A notational issue—yet profoundly important. In many relativity presentations (but generally not in Einstein's own works), a misleading set of mass definitions was created—rest mass, relativistic mass (an abomination), transverse mass, etc. It has been strongly and correctly argued by Okun that these confusions should *not* be propagated. So here I will use m as the one-and-only mass of a particle, being what is often called the rest mass and written m_0. (Most equations in the literature can be obtained from mine by the substitution of m_0 for my m.) This mass m (by others often called m_0 or the rest mass) is the same as the Newtonian mass at low velocities (*if* this case is actualizable). Most important, m is a scalar or invariant—it has the same value for *all* observers of the particle, and is a constant parameter for the particle. It is to be determined by experiment, and by use of relativistic dynamics as outlined below.

A particle may have either finite or zero mass. As we will soon see, finite-mass particles necessarily travel at speeds less than c. Zero-mass particles necessarily travel at speed c w.r.t. all observers (and light itself is one important case). Thus, some equations differ for these two cases.

Relativistic Momentum

A particle of finite mass m, when traveling at velocity $\mathbf{v}$, has momentum

$$\mathbf{p}=m\gamma\mathbf{v}=\frac{m\mathbf{v}}{[1-(v^2/c^2)]^{1/2}}. \tag{9}$$

Note. We always use v (lower case) for a particle velocity, and V (upper case) for the relative velocity of two frames or observers. Confusingly but traditionally, in dynamics we use γ as the above function of the particle velocity v, when in cinematics it was a function of V (see above).

Relativistic Kinetic Energy

The same finite mass particle has kinetic energy (symbol K.E. or T) given by

$$\text{K.E.} = mc^2(\gamma - 1) = mc^2\{[1-(v^2/c^2)]^{-1/2} - 1\}. \tag{10}$$

However, this concept is *not* useful in relativity. Instead, it is useful to define two new quantities.

RELATIVISTIC REST ENERGY

For a finite-mass particle (whatever its velocity), we define this most important *rest frame concept* as

$$E_0 = mc^2. \tag{11}$$

This is the rest energy of the particle, in its rest frame. (m is an invariant scalar, but E_0 is actually a particular value of a variable E, in a particular frame, the rest frame)

RELATIVISTIC TOTAL ENERGY (OR SIMPLY ENERGY)

This, for a finite mass particle, is

$$E = \text{K.E.} + E_0 = m\gamma c^2 = \frac{mc^2}{[1-(v^2/c^2)]^{1/2}}. \tag{12}$$

This total energy, from now on simply called energy, turns out to be a key concept.

MOMENTUM–ENERGY RELATIONS

One readily obtains

$$E^2 - p^2c^2 = m^2c^4 \quad \text{and} \quad p = v\frac{E}{c^2}. \tag{13}$$

These are most important and useful relationships; in particular they hold for *both* finite and zero mass particles (see later). They hold for all observers [see Eq. (17)].

NEWTONIAN DYNAMICS ISSUES

Notice that the expressions for **p** and for K.E. are such that for $v \ll c$, they reduce to the nonrelativistic (Newtonian) definitions. Thus the relativistic definitions do not so much supersede the Newtonian ones as extend them in a new way into thc high-speed domain. (Again evolution, not revolution.)

It then turns out that Newton's second law is best *extended* as

$$\mathbf{F} = \frac{d\mathbf{p}}{dt} = \frac{d}{dt}\left\{\frac{m\mathrm{v}}{[1-v^2/c^2]^{1/2}}\right\}, \tag{14}$$

and that the work–energy theorem for one particle is best written:

$$\text{work done} = W = \int \mathbf{F}\cdot d\mathbf{r} = \Delta\text{K.E.} = \Delta E$$

$$= \Delta(m\gamma c^2) = \Delta\left\{\frac{mc^2}{[1-(v^2/c^2)]^{1/2}}\right\}, \tag{15}$$

where Δ means "change in ..., after minus before."

In (14) and (15) **F** is now the relativistic force, and must be suitably determined in any given case. For the motion of a particle of charge q in electric and magnetic fields **E** and **B**, **F** is the Lorentz force $\mathbf{F} = q(\mathbf{E} + \mathbf{v}\times\mathbf{B})$.

We notice that Eq. (15) states that it requires an infinite amount of work to accelerate a particle with finite mass m, from rest to a speed c. Note also in Eqs. (9) and (10) that p, and K.E. go to infinity as v approaches c for finite m. Put differently, we assert that no particle with nonzero rest (mass) can be accelerated to a speed as great as c; that is, the velocity of light is a limiting upper velocity. (Recall prior discussion.)

Equations (10)–(16) are appropriate to relativistic dynamics. They lead, for instance, to relativistic momentum conservation in two-body collisions, and other straightforward *extensions* of Newtonian ideas. Yet it is Eqs. (11) and (12) that lead to a whole new startling development. But first a digression.

RELATIVISTIC TRANSFORMATIONS OF MOMENTUM AND ENERGY

Before we leave **p** and E, we simply record the transformation equations for (**p**, E) as seen by our two different observers A and B. We find the *transformation of momentum and energy*:

$$p_x' = \gamma\left(p_x - \frac{VE}{c^2}\right), \qquad p_y' = p_y, \qquad p_z' = p_z,$$

$$\frac{E'}{c^2} = \gamma\left(\frac{E}{c^2} - \frac{Vp_x}{c^2}\right). \quad \text{with} \quad \gamma = \left(1 - \frac{V^2}{c^2}\right)^{-1/2}. \tag{16}$$

These equations are very simple, yet tell us a most remarkable fact: the four quantities (p_x, p_y, p_z, E/c^2) transform exactly like the basic four coordinates (x, y, z, t). There is a simple reason for this, which, however, takes us too far afield.

For all observers, we have the invariant

$$m^2c^4 = E^2 - p^2c^2 = E'^2 - p'^2c'^2. \tag{17}$$

ZERO-MASS PARTICLES

Newtonian dynamics cannot support particles of zero mass. Remarkably, relativistic dynamics does. Let us set $m = 0$, *and* also $v = c$; then we find

$$p = E/c, \quad \text{with } p \text{ along the velocity } \mathbf{v} = c,\ m = 0. \tag{18}$$

As a consequence of the second postulate (implemented via the Lorentz transformation), different observers will find different directions of the particle velocity **v**, yet all find the *same* magnitude $\mathrm{v} = c$. All observers will use Eqs. (13) and (17) consistently; Eqs. (9), (10), and (12) are now meaningless, since $m = 0$ and $\gamma = \infty$.

Indeed such postulated particles have zero (rest) mass; *they also have no rest frame*. Such zero-mass, speed c particles do seem to exist—the neutrinos (q.v.), of which two have been clearly detected, and there is strong evidence also for the existence of a third. There remains experimental uncertainty, yet they do have zero mass within error.

But perhaps more important, light is also such a zero-mass *particle*—the photon (q.v.)—as Einstein helped to demonstrate in another context, the photoelectric effect. He also discussed it within special relativity (see below). Light always travels with the same speed c w.r.t. *all* observers, but its direction will differ. Its energy and momentum correspond to the classical E.M. energy and momentum flux (see below).

MASS-ENERGY EQUIVALENCE

We return to mass and energy. Further analysis (see bibliography) leads to the remarkable conclusion that mass and energy are intimately related. This has been stated loosely in the famous equation $E=mc^2$. But this powerful symbolic equation is misleading. It was not used by Einstein, who wrote

$$E_0=mc^2. \tag{19}$$

Here m is our mass (others' rest mass) and E_0 is the energy in the *rest* frame of the *system* (if it has one).

The incredible consequences and underlying principles are that:

1. In the rest frame of any system, any form of "classical" energy, e.g., thermal/heat, kinetic, chemical, or nuclear, has a mass associated with it. Such a mass persists and is the same (is invariant) for all observers w.r.t. whom the system has velocity (Recall we use mass as the invariant value best measured in the *rest* frame.) For low or zero system velocities, this mass has inertia and gravitational interactions exactly as given by Newton; for large system velocities the dynamics and gravitational issues are as determined by special and general relativity for such a mass. In all respects this mass-of-energy is a contribution to the total mass.
2. The sum of the masses of the constituents of a system is *not* necessarily a constant. Mass is *not conserved*; nor is the sum of the "classical" energies. Classical energy is *not* conserved. What is constant (i.e., conserved) is the system *rest*-frame *total* energy:

   $$E_0=\{\sum \text{(rest mass of each constituent)}c^2 + \sum \text{(all ``classical'' energies)}\}=\text{const.} \tag{20}$$

   From Eq. (19) this, of course, is just mc^2 for the total system.
3. It is possible to destroy some constituent mass of a system and convert it to corresponding constituent kinetic energy, e.g., in a nuclear reactor (q.v.) or "atomic" bomb.
4. It is also possible to do the converse: to create actual constituent mass (constituent *matter*) by converting constituent (kinetic) energy to mass. Such created matter will have all usual material properties (beginning alchemy).

This last process is experimentally one of the primary tools of high-energy physicists. For instance, a proton (p) traveling at high speed can be made to hit another proton at rest, and *create* a new pion (or π^0 meson) out of some of the incident kinetic energy. That is,

$$p+p+\text{energy}\rightarrow p+p+\pi^0+\text{less energy}.$$

All such predictions have been amply verified, from the atomic bomb (q.v.) to particle creation. It is clear that energy and mass are equivalent and interchangeable.

ELECTROMAGNETIC ASPECTS OF SPECIAL RELATIVITY; THE PHOTON

Since relativity grew out of the conflict beteen Newtonian mechanics and Maxwell's relations for the electromagnetic field, there naturally are many implications of the theory for electromagnetism.

Maxwell's theory predicts the existence of E.M. waves. *In vacuum* these have speed c (the speed of light), carry energy density $U=(\frac{1}{2})(\epsilon_0E^2 + B^2/\mu_0)$, and carry momentum density $P=U/c$. (The last is related to the Poynting vector $\mathbf{S}=\mathbf{E}\times\mathbf{H}$). Relativity then *must* view this combination of energy density U and momentum density P as together related to a physical *system* of *zero* mass, since Eq. (13) gives $U^2-P^2c^2=m^2c^4=0$. Necessarily the system is traveling at speed c, from Eq. (18). This system corresponds to a superposition of photons (see above).

The photon is a particle with *no* rest frame, and it travels at c according to *all* observers. However, its energy and frequency and wavelength vary from observer to observer. (See "Zero-mass Particles" above. See also PHOTON and DOPPLER EFFECT.)

Other electromagnetic aspects are less basic, but still striking. One is the idea that magnetism is only a "derived" effect, caused by the motion of an observer through an electric field.

HAS THE SPECIAL THEORY OF RELATIVITY BEEN PROVED?

Yes. While macroscopic tests of kinematic effects are scarce or nonexistent the totality of success in experimental high-energy physics (q.v.) and in high-energy accelerator design (q.v.) overwhelmingly relies on the predictions and calculations of this theory. The theory is true.

See also ACCELERATORS; DOPPLER EFFECT; ELECTROMAGNETIC INTERACTIONS; ELEMENTARY PARTICLES; LORENTZ TRANSFORMATIONS; MAXWELL'S EQUATIONS; MICHELSON–MORLEY EXPERIMENT; NEUTRINO; NEWTON'S LAWS; NUCLEAR REACTOR; PHOTON; RELATIVITY, GENERAL; SPACE-TIME; TACHYONS; TWIN PARADOX.

BIBLIOGRAPHY

There are far too many books and other sources available, many satisfactory. Many physics textbooks have chapters on the special theory.

Primarily Historical, Biographical, and/or Philosophical

Albert Einstein and L. Infeld, *The Evolution of Physics*, 2nd ed. Simon and Schuster, New York, 1966.

Gerald Holton, "Resource Letter SRT-1 on Special Relativity

Theory," *Am. J. Phys.* **30,** 462 (1962). (A comprehensive list of historical references.)

Cornelius Lanczos, *Albert Einstein and the Cosmic World Order.* Wiley, New York, 1965. (A somewhat panegyric but very interesting nonmathematical account of the significance of Einstein.)

Abraham Pais, *Subtle is the Lord ..., The Science and Life of Albert Einstein.* Oxford University Press, Oxford and New York, 1982. (This complete personal and scientific biography presents much detailed context and development.)

Hans Reichenbach, *The Philosophy of Space and Time* (transl. by Maria Reichenbach and John Freund). Dover, New York, 1958. (A very complete philosophical discussion in paperback, of many aspects of special and general relativity.)

Paul Arthur Schilpp (ed.), *Albert Einstein: Philosopher-Scientist,* 2 vols. Tudor, New York, 1951; Harper Torchbooks, New York, 1959.

Explicit, Not Very Mathematically Demanding Treatments

Hermann Bondi, *Relativity and Common Sense,* 2nd ed. Doubleday, Garden City, New York, 1980. Very incomplete, yet very beautiful treatment of many important points. Enphasizes a modern viewpoint based on *one* observer with one clock, who uses radar methods. Develops the related mathematical tool known as the **k**-calculus; paperback.

Max Born, *Einstein's Theory of Relativity* (rev. ed. prepared with the collaboration of Gunther Liebfried and Walter Biem). Dover, New York, 1965. (A very good account of the prior Newtonian world view, of the early "ether" theories, and of Einstein's special and general theories of relativity, placed in the overall scheme of physics; highly recommended.)

Lev D. Landau and G. B. Rumer, *What Is Relativity?* (transl. by N. Kemmer). Basic Books, New York, 1961; Fawcett, Greenwich, Conn., 1966. (A lighthearted Russian presentation with witty illustrations; paperback.)

N. David Mermin, *Space and Time in Relativity.* McGraw-Hill, New York, 1968. (Very concerned with clarifying possible misconceptions, errors of thought, and other paradoxical aspects; paperback.)

Introductory (Freshman Physics Major) Textbooks

David Bohm, *The Special Theory of Relativity.* Benjamin, New York, 1965. [Very good background philosophical discussion, together with explicit algebraic treatment; includes the **k**-calculus (cf. comment after Bondi, above); paperback.]

A. P. French, *Special Relativity.* Norton, New York, 1968. (One of the MIT series. The treatment is complete but somewhat brief, and misleadingly straightforward.)

Claude Kacser, *Introduction to the Special Theory of Relativity.* Prentice-Hall, Englewood Cliffs, NJ, 1967.

Robert Resnick, *Introduction to Special Relativity.* Wiley, New York, 1968. (Competent but somewhat unexciting.)

Robert Resnick, *Basic Concepts in Relativity and Early Quantum Theory,* 2nd ed. Wiley, New York, 1985.

Ray Skinner, *Relativity.* Blaisdell, Waltham, MA, 1969. [Some very nice features, particularly world-line discussion of particle interactions. However, often uses the less intuitively understandable four-vector formulation (four-vectors are very nice in a second treatment).]

James H. Smith, *Introduction to Special Relativity.* Benjamin, New York, 1965. (Rather appealing introductory treatment, but often uses four-vector formulation; paperback.)

Edwin F. Taylor and John Archibald Wheeler, *Spacetime Physics.* W. H. Freeman, San Francisco, 1966. [A very stimulating and novel presentation. Wheeler is a world expert on special and general relativity. The book is very demanding, although at an introductory level. It is full of thought-provoking questions and paradoxes. Well worth the effort of study, but it helps also to have also a mundane approach to refer to in difficulty. (The solution manual, included in the second printing, is an essential part of the book.) Beware: their definition of inertial frame is very unusual, but valid from a general relativity viewpoint.] (A revised second edition is in process.)

More Advanced

Peter G. Bergmann, *Introduction to the Theory of Relativity.* Prentice-Hall, Englewood Cliffs, NJ, 1942; Dover, New York, 1976.

Albert Einstein, *Relativity, the Special and General Theory* (transl. by R. W. Lawson). Methuen, London, 1920; reprinted as a University Paperback, 1960, and by Crown, New York, 1961. (Not really a textbook; quite difficult in places.)

Albert Einstein, *The Meaning of Relativity,* 5th ed. Princeton University Press, Princeton, NJ, 1956. (Includes general relativity.)

H. A. Lorentz, A. Einstein, H. Minkowski, and H. Weyl, *The Principle of Relativity: A Collection of Original Memoirs* (transl. by W. Perrett and G. B. Jeffrey). Methuen, London, 1923; paperback reprint, Dover, New York, 1968.

C. Møller, *The Theory of Relativity.* Oxford University Press, Oxford, London and New York, 1952. (Includes general relativity.)

Wolfgang Pauli, *Theory of Relativity* (transl. by G. Field), Pergamon, New York, 1958. (Includes general relativity.)

W. Rindler, *Special Relativity,* 2nd ed. Wiley (Interscience), New York, 1966. (A good intermediate-level text.)

W. G. V. Rosser, *An Introduction to the Theory of Relativity.* Butterworths, London, 1964. (A detailed intermediate-level text.)

W. G. V. Rosser, *Introductory Relativity.* Butterworths, London, 1967. (A shortened simplified version of the foregoing.)

Other

Special Relativity, selected reprints published for the A.A.P.T. by the American Institute of Physics, New York, 1962. (Particularly thorough on the twin paradox.)

Charles Kittel, Walter D. Knight, and Malvin A. Ruderman, *Mechanics* (Berkeley Physics Course), Vol. 1, 2nd ed., rev. by A. Carl Helmholz and Burton J. Moyer. McGraw-Hill, New York, 1973. (Some nice chapters on special relativity at an introductory level.)

Lev B. Okun, "The Concept of Mass." *Phys. Today,* pp. 31–36 (June 1989).

A novel diagrammatic way of presenting the content of the Lorentz transformation has been developed by R. W. Brehme, *Am. J. Phys.,* **30,** 489 (1962). It has been described and used in the textbook, *University Physics,* by Francis W. Sears and Mark W. Zemansky (Addison-Wesley, Reading, MA., 1963), but *not* in recent editions. See also Albert Shadowitz, *Special Relativity* (Saunders, Philadelphia, PA, 1968), for a related diagrammatic approach due to E. Loedel.

George Gamow, *Mr. Tompkins in Paperback.* Cambridge University Press, Cambridge, London, and New York, 1965. [A lighthearted story of a wanderer in Relativity (and Quantum) Land; beware, some of the discussion on the visual appearance of moving objects is technically erroneous.]

Relaxation Phenomena

R. Orbach

A physical system disturbed from an equilibrium configuration may relax back to equilibrium by means of contact with a thermal reservoir. The rate of approach to equilibrium, most often characterized by a relaxation time τ, can yield information concerning the coupling between the system and the reservoir, and concerning the microscopic properties of the system itself. Intermediate steps may be involved, and the equilibrium distribution may be only metastable, a position of only relative minimum free energy.

It is convenient in many cases to cast the description of the system and reservoir into the form of a density matrix ρ (see the book by Tolman[1] for a detailed description of the density matrix and its uses). The equilibrium configuration is defined as that configuration for which the off-diagonal density matrix elements, ρ_{nm}, $n \neq m$, are zero, while the diagonal elements are given by the Boltzmann relation,

$$\rho_{mm} = \frac{\exp(-E_m/k_BT)}{\sum_n \exp(-E_n/k_BT)}. \tag{1}$$

Here, E_m is the energy of the mth state, and T is the thermal equilibrium temperature. It proves useful in some cases to define a system temperature T_s (e.g., the spin temperature for nuclear spins in a solid or liquid; see the book by Abragam[2] for details), where T_s may differ from the reservoir temperature, but the system has come into internal equilibrium. In such a case, ρ^s_{nm}, $n \neq m$, vanishes, and

$$\rho^s_{mm} = \frac{\exp(-E_m/k_BT_s)}{\sum_n \exp(-E_n/k_BT_s)}. \tag{2}$$

For systems with finite numbers of energy levels, T_s can even be negative, referring to an inversion of level occupancy. Relaxation to equilibrium with the reservoir takes the route

$$T_s<0 \text{ to } T_s \sim \infty \text{ to } T_s = T.$$

The actual contact between the system and reservoir may not be direct. Conduction-electron spins in a metal typically relax much more rapidly to the kinetic energy of the electrons than to the external environment (the heat bath with which the metal is in contact). This leads to a picture of connected reservoirs of finite heat capacity, ultimately coupled to a very large reservoir of unlimited heat capacity as the relaxation route. Each of the intervening reservoirs may be described by its own local temperature. Thus, nuclear spins interacting with a large external magnetic field may be described in terms of their Zeeman level occupancies (the Zeeman reservoir). In turn, the nuclei may interact with one another via dipolar spin forces. This interaction can relax the Zeeman level occupancies to equilibrium with the "dipolar" reservoir. Finally, the latter may be in contact with a large thermal bath. The specific heats of the first two reservoirs may be comparable, leading to an exchange of energy or relaxation from the Zeeman reservoir to the dipolar reservoir to the thermal bath. The Zeeman relaxation path is thereby indirect, and can even lead to important phase-memory effects when the dipolar reservoir is only weakly coupled to the thermal bath. (In such a case, the use of a dipolar temperature is even wrong. The off-diagonal density matrix elements are most certainly not zero. See Abragam[2] for a fuller discussion of this important case.)

Many times, the intermediate reservoir cannot accommodate the energy being dumped into it if its coupling to the thermal bath is weak. This leads to a "bottleneck," and a heating of the intermediate reservoir. An example is the normal electrons (the quasiparticles) in a superconducting metal at finite temperatures. These electrons are in intimate contact with the lattice vibrations of energy $\gtrsim 2\Delta$, where Δ is the energy gap. Excitation of the quasiparticles can occur by any number of methods, one such being the breaking of "Cooper pairs" of the superconducting condensate (an excellent physical description of these processes can be found in the book by de Gennes[3]). The excited quasiparticles relax by generation of lattice vibrations of energy $\gtrsim 2\Delta$, which in turn can break other Cooper pairs. This results in a trapping of these participating lattice vibrations, not unlike the trapping of resonance radiation (see the articles by Holstein[4] and the book by Mitchell and Zemansky[5]). The resultant heating of the intermediate reservoir (in this case the lattice vibrations of energy $\gtrsim 2\Delta$) is relieved only by the relatively slow process of scattering into other vibrations of differing energy (anharmonic scattering), or by direct flow into the thermal bath.

The approach to equilibrium in general depends on the microscopic nature of the system and the character of the equilibrating interactions. For example, the collision term in the Boltzmann equation for the six-dimensional distribution function f (see the article on transport theory in this volume[6]) is often written in the relaxation-time approximation,

$$\left.\frac{\partial f}{\partial t}\right|_{\text{collisions}} = -\frac{f-\bar{f}}{\tau(E)}, \tag{3}$$

where $\bar{f}$ is the equilibrium distribution function and $\tau(E)$ is a relaxation time dependent on the energy of the relaxing particle (gas molecule, electron, etc.). The time $\tau(E)$ can only be defined under special circumstances. For most systems, a relaxation time exists only for quasielastic scattering. Thus, for electrons in metals, $\partial f/\partial t|_{\text{collisions}}$ can be written in the form (3) only for impurity scattering, or upon scattering off of lattice vibrations at temperatures $T > \Theta_D$, the Debye temperature.

Formal expressions for the collision term can be separated into processes which scatter out of the state defined by the position vector $\mathbf{r}_i$ and velocity vector $\mathbf{c}_i$; and for scattering into that state. The subscript i designates the ith particle. The latter processes interfere with the ability to define a $\tau(E)$, and in many cases cause the collision term to generate an integral equation for the distribution function f.

The simple form of (2) must be abandoned when considering relaxation of higher-order multipoles. In magnetic systems, one must deal with the relaxation of the magnetic moment. The characteristics of both magnitude and direction of the magnetization vector require a physical condition to be placed upon the relaxation process. For example, in ferromagnetic resonance[7] the length of the magnetization vector $\mathbf{M}$ may be preserved in the relaxation process, and relaxation may occur to the instantaneous direction of the field experienced by $\mathbf{M}$. The relaxation term in the torque equation for $\mathbf{M}$ is then of the Landau–Lifshitz type:

$$\left.\frac{d\mathbf{M}}{dt}\right|_{\text{relaxation}} = -\frac{\lambda}{M^2}[\mathbf{M}\times(\mathbf{M}\times\mathbf{H})], \tag{4}$$

where λ is a constant and $\mathbf{H}$ is the field experienced by $\mathbf{M}$. The double cross product ensures constancy of $\mathbf{M}$ and relaxation toward $\mathbf{H}$. In the absence of driving terms, the magnetization vector experiences damped precession as exhibited in Fig. 1. An alternate form of the damping, proportional to $d\mathbf{M}/dt$, has been proposed by Gilbert:

$$\left.\frac{d\mathbf{M}}{dt}\right|_{\text{relaxation}} = \frac{\alpha_0}{M}\left(\mathbf{M}\times\frac{d\mathbf{M}}{dt}\right). \tag{5}$$

It is a simple matter to express (4) and (5) in the same form if one is willing to alter the free precession constant. For large damping (the constant α_0 large) in, for example, pulsed remagnetization of thin films, Eq. (5) seems to give better agreement with experiment. The time for total magnetization reorientation is proportional to the damping parameter α_0.

Relaxation for a variety of both paramagnetic and ferromagnetic systems can be described in terms of the Bloch conditions:

$$\left.\frac{dM_z}{dt}\right|_{\text{relaxation}} = -\frac{M_z-\bar{M}_z}{T_1}, \tag{6a}$$

$$\left.\frac{dM_{x,y}}{dt}\right|_{\text{relaxation}} = -\frac{M_{x,y}}{T_2}, \tag{6b}$$

where $T_2 \leqq 2T_1$ for sensible response. These famous terms rely for their applicability on the special character of the z direction, as would be the case for a strong steady magnetic field pointing along the z direction, with weak perturbing alternating fields along the perpendicular x and y directions. The quantity $\bar{M}_z$ refers to the equilibrium value of the z component of magnetization. Clearly, Eq. (6) does not conserve the length of the magnetization vector M. Furthermore, it assumes weak alternating driving fields compared to intermoment coupling terms (e.g., dipolar forces). If the converse is true, relaxation may occur to the instantaneous direction of the field experienced by the magnetic system in a plane rotating with the angular frequency of the alternating fields. (This limit is discussed in detail by Redfield.[8]) In more concentrated nuclear-magnetic-moment systems, the longitudinal relaxation time T_1 is caused by fluctuating transverse dipolar interactions, the frequency of fluctuation being equal to the Zeeman precessional rate (for a full discussion of dipolar relaxation, see Abragam[2] and Slichter[9]). The transverse time T_2 is the consequence both of transverse dipolar fluctuations at the Zeeman precessional rate and of longitudinal dipolar fluctuations at zero frequency. The latter are termed the frequency-modulation effect; the former are termed the spin-flip effect.

In more dilute nuclear magnetic and in electronic magnetic systems, longitudinal relaxation is primarily through interaction with the lattice vibrational modes. These give rise to very rapid temperature dependences for the relaxation times. (Experimental examples can be found in the collection of Manenkov and Orbach.[10])

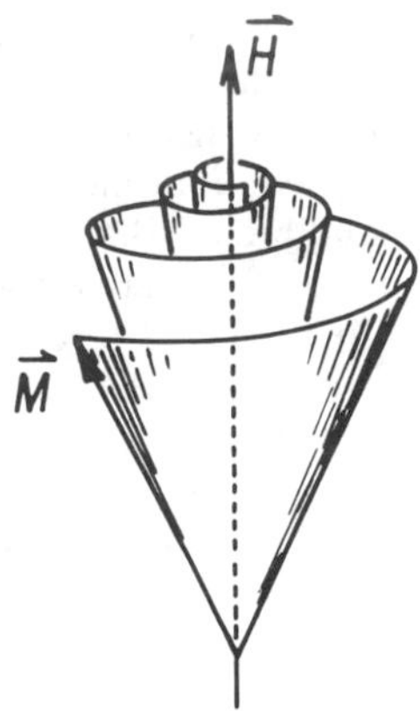

FIG. 1. Damped precession of the magnetization vector **M** according to the Landau–Lifshitz relaxation term in the presence of a steady magnetic field, in the absence of alternating fields.

The relaxation times caused by interaction with the lattice vibrations turn out to be very important for determination of the operating characteristics of a variety of physical devices, in particular masers and lasers. The same can be said for atomic or molecular collisions for gaseous devices. The study of relaxation processes is, therefore, important not only for understanding fundamental interactions in gases, liquids, and solids, but also of great practical importance.

See also ELECTRON SPIN RESONANCE; LASERS; LATTICE DYNAMICS; NUCLEAR MAGNETIC RESONANCE; RESONANCE PHENOMENA; SUPERCONDUCTIVITY THEORY.

REFERENCES

1. R. C. Tolman, *The Principles of Statistical Mechanics*. Oxford University Press, New York, 1938. (A)
2. A. Abragam, *The Principles of Nuclear Magnetism*. Clarendon Press, Oxford, 1961. (A)
3. P. G. de Gennes, *Superconductivity of Metals and Alloys*. Benjamin, New York, 1966. (I)
4. T. Holstein, *Phys. Rev.* **72,** 1212 (1947). (A)
5. Allan C. G. Mitchell and Mark W. Zemansky, *Resonance Radiation and Excited Atoms*. Cambridge University Press, Cambridge, England, 1961. (I)
6. Sidarshan K. Loyalka, *Transport Theory* [this volume]. (E)
7. S. V. Vonsovskii (ed.), *Ferromagnetic Resonance*. Pergamon Press, Oxford, 1966; M. Sparks, *Ferromagnetic Relaxation Theory*. McGraw-Hill, New York, 1964. (I)
8. A. G. Redfield, *IBM J. Res. Devel.* **1,** 19 (1957). (A)
9. C. P. Slichter, *Principles of Magnetic Resonance*. Harper and Row, New York, 1963. (I)
10. A. A. Manenkov and R. Orbach (eds.), *Spin-Lattice Relaxation in Ionic Solids*. Harper and Row, New York, 1966. (I,A)

Renormalization

David R. Nelson

The concept of renormalization in physics has its origins in 19th century hydrodynamics. It was discovered that large objects moving slowly through a viscous fluid behave in some ways as if they possess an enhanced mass due to the fluid particles they drag along. We would now say the mass of such objects is renormalized away from the "bare" value it has in isolation by interactions with the medium. This idea finds modern expression in Landau's quasiparticle picture of Fermi liquids and in the band theory of metallic conduction. Landau argued that liquid ^{3}He could be described at low temperatures by a dilute gas of quasiparticle excitations near a spherical Fermi surface. The quasiparticles were to be viewed as "dressed" versions of the constituent ^{3}He fermions, with altered masses, interactions, etc. The dynamics of conduction electrons in metals is often parametrized in a similar fashion by an effective electron mass describing interactions with the periodic lattice potential, and with other electrons.

In all these examples, the basic idea is to replace a complicated many-body problem by a simpler system in which interactions are absent or negligible. Complicated many-body effects are absorbed into redefinitions of masses and coupling constants. For objects moving slowly through a viscous medium, the mass renormalization can be calculated perturbatively as a power series in the Reynolds number. Even if interactions are strong enough to make perturbation theory intractable, as is the case for liquid ^{3}He, phenomenological renormalized parameters taken from experiments provide a very useful description of the physics.

Problems developed when these ideas were applied to quantum electrodynamics. Calculations of the renormalized electron mass due to its interaction with electromagnetic radiation were carried out perturbatively in powers of the fine structure constant. At every order in perturbation theory, infinities due to interactions at very short distances were discovered. The renormalized mass and coupling constant, for example, diverged as the cutoff Λ (the largest allowed inverse length scale) was taken to infinity. Such infinities can be viewed as arising from equivalent contributions to the renormalized mass and coupling constant from many different length scales. Although Feynman, Schwinger, and Tomanaga showed how to extract meaningful results from such poorly behaved series, their renormalization procedure does not give a great deal of physical insight.

A somewhat different way of approaching such problems was subsequently explored by Gell-Mann and Low, who reformulated renormalized perturbation theory in terms of a cutoff-dependent coupling constant, $g(\Lambda)$. This enabled them to study the gradual evolution of bare, unrenormalized quantities into dressed, renormalized ones. The set of transformations which gradually change g from a bare to a dressed quantity is called a renormalization group. These ideas were developed further in the 1960s by Wilson, Callan, Symanzik, and others. An equation, now called the Callan–Symanzik equation, was derived for the n-point functions of renormalizable field theories by demanding that the physical values be asymptotically independent of the cutoff. One is led quite naturally to consider the logarithmic derivative of $g(\Lambda)$, often called the "β function,"

$$\beta(g)=\frac{\partial g(\Lambda)}{\partial \ln\Lambda}. \tag{1}$$

A spectacular application of these ideas came with the discovery by Politzer and by Gross and Wilczek of asymptotic freedom. It was found that the functions $\beta(g)$ for a select set of candidates for strong interaction field theories (describing "quantum chromodynamics" or QCD) were such as to drive the renormalized couplings to zero at short distances, or, equivalently, at very high energies. The idea of asymptotically free theories of the strong interactions led to an understanding of experiments on electron–proton scattering which showed that protons behave at high energies as if they were composed of weakly interacting constituents called quarks or partons.

Modern renormalization ideas were generalized and used to construct a theory of critical point singularities in 1971 by K. G. Wilson. Building on work by Kadanoff, Wilson was able to explain the singularities near critical points, and show how to calculate them. In doing so, he showed renormalization group ideas had applicability far beyond high-energy physics, and placed a powerful mathematical tool at the disposal of the rest of the scientific community. Critical point phenomena provide a particularly simple context in which to explain this new perspective on renormalization theory.

Phase changes of matter in thermodynamic equilibrium are often characterized by sharp discontinuities in quantities such as the magnetization of a ferromagnet or the density of a liquid–vapor system. Usually, these phase transitions can be made continuous by adjusting an external parameter like the temperature or pressure. The point at which the discontinuity vanishes is called a critical point.

The behavior of a liquid–gas system near such a singular point is particularly striking. The fluid suddenly becomes milky and opaque, and scatters laser light intensely in the forward direction. It seems clear that some sort of cooperative phenomenon is taking place, and that the particles of fluid are strongly correlated over distances comparable to the wavelength of the laser light. The range of correlation $\xi(T)$ is a measure of this cooperativity, and is expected to diverge strongly at T_c,

$$\xi(T)\sim|T-T_c|^{-\nu}. \tag{2}$$

This divergence introduces a range of length scales into the problem not unlike that encountered when the cutoff is sent to infinity in quantum field theory.

The striking thing about this, and other anomalies near critical points, is that it appears to be universal. A wide variety of liquid–vapor systems and ferromagnets exhibit correlation lengths which appear to diverge in precisely the same way. A second curious fact is that universal numbers like ν are not simply one-half or unity, although most simple theories lead inevitably to these values. In fact, it is now believed that ν lies in the range

$$0.64\lesssim\nu\lesssim0.70 \tag{3}$$

in most three-dimensional systems. Modern renormalization theory focuses on the calculation of such numbers. The exponent ν turns out be closely analogous to a number which determines the way masses scale with cutoff in quantum field theories.

To illustrate the renormalization-group method, we consider a simple model of magnetism, a two-dimensional

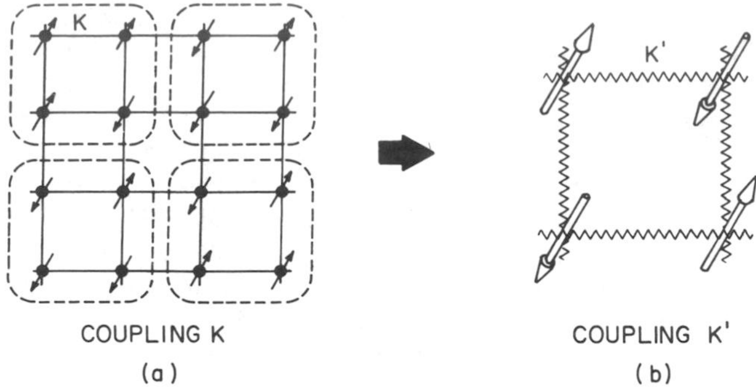

FIG. 1. Kadanoff block scheme. Dashed lines enclose groups of spins on lattice (a), which is characterized by nearest-neighbor spin coupling K. Out of each block, we construct a collective spin variable by integrating over the internal degrees of freedom. The new spin degrees of freedom populate a lattice (b) with twice the old lattice spacing, and interact as if they had coupling K'.

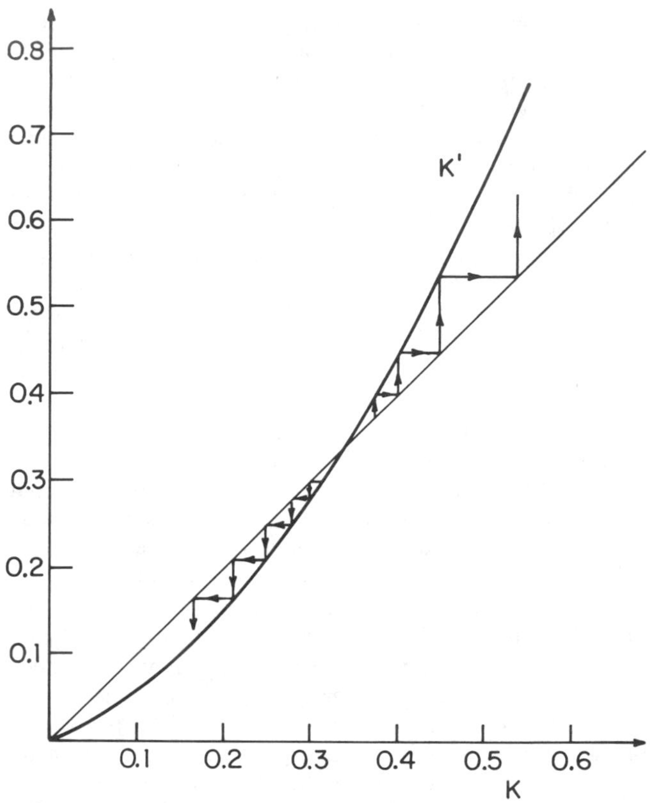

FIG. 2. Approximate recursion relation for Ising spins, with a fixed point at $K^*=K_c=0.34$. The "ladder" construction shows successive temperatures produced by repeated iterations of the transformation. The slope of the function $K'(K)$ through its fixed point is related to the critical exponent ν.

square lattice of Ising spins $\sigma_i=\pm 1$ (see Fig. 1). To each configuration $\{\sigma_i\}$ of spin we assign an energy

$$H[\{\sigma_i\}]=-J\sum_{\langle i,j\rangle}\sigma_i\sigma_j, \tag{4}$$

where i and j label lattice sites, and the sum is restricted to nearest-neighbor pairs of spins. The calculation of the partition sum associated with Eq. (4) as a function of the dimensionless inverse temperature $K=J/k_BT$ is a very difficult task. It was, in fact, computed exactly for this two-dimensional problem by Onsager in 1944. Onsager's work, which remains today a true mathematical *tour de force*, eventually led to the exponent prediction

$$\nu=1. \tag{5}$$

An analytical evaluation of such partition sums in three dimensions or in much more general situations appears hopeless, however.

Rather than attempting to partition sums exactly, we can, in fact, make progress by merely thinning out the spin degrees of freedom slightly. To do this, we follow the original suggestion of Kadanoff and group the spins on the original lattice (which is at coupling K) into blocks as shown in Fig. 1. With some ingenuity, it turns out to be possible to integrate approximately over the internal degrees of freedom within each block. We are then left with a new statistical-mechanical problem with twice the lattice spacing. Provided the interactions between spins in the new lattice are of the same form as in the old, couplings in the new lattice can be characterized simply by a new temperature K'.

This sort of program was first carried successfully for Ising spins on a triangular lattice by Niemeyer and Van Leeuwen. A very primitive version of their transformation gives a relationship between the new and old temperatures, namely

$$K'=2K(e^{3K}+e^{-K})^2/(e^{3K}+3e^{-K})^2. \tag{6}$$

Equation (6) is depicted graphically in Fig. 2, together with a "ladder" construction which shows how the temperature changes with repeated iterations of the transformation. The recursion formula (6) has a fixed point at $K^*\approx 0.34$, and it seems natural to identify this isolated point with the critical point coupling K_c of the spin system.

We can see the utility of this procedure of "thinning out" spins as follows: even an approximate calculation of the partition function is difficult near T_c because of the large correlation length $\xi(T)$ in this region. However, note from Fig. 2 that the renormalization transformation *increases* temperatures which, initially, are slightly above T_c. By repeatedly iterating the transformation (6), we can relate a difficult calculation near T_c to a more tractable high-temperature problem. A very accurate calculation of the partition sum is possible at high temperatures by expanding in powers of K. In a similar fashion, we see that the transformation (6) *decreases* temperatures initially slightly below T_c, and eventually produces a more manageable (low-temperature) problem.

The physics behind a renormalization transformation is in the reduction of the correlation length. If the lattice spacing is changed by a factor b by the thinning out process, it follows the correlation length transforms according to

$$\xi(K')=\xi(K)/b. \tag{7}$$

For the square lattice spin system shown in Fig. 2, $b=2$, while for the triangular lattice of Niemeyer and Van Leeuwen it turns out that $b=\sqrt{3}$. Equation (6) can be used in conjunction with (7) to produce a critical exponent. Near the fixed point $K^* = K_c = 0.34$, Eq. (6) can be written in the approximate form,

$$K' - K_c = b^{0.89}(K - K_c), \tag{8}$$

which when combined with (7) gives a functional equation for $\xi(K - K_c)$,

$$\xi(K - K_c) = b\xi(b^{0.89}(K - K_c)). \tag{9}$$

It is easy to check that the solution of this functional equation is $\xi(T) \sim |K - K_c|^{-\nu}$ with $\nu \approx 1.12$, which is not too different from the exact result $\nu = 1.0$.

In more sophisticated calculations, it is not possible to describe a renormalization transformation in terms of a single coupling constant. More complicated couplings such as next-nearest-neighbor and multispin interactions are generated after one iteration. When these extra couplings are taken into account, they not only produce more accurate estimates of critical exponents, but also provide a qualitative explanation of universality: If several couplings are included in the calculation, the renormalization transformation can be viewed as a mapping of one Hamiltonian onto another in a multidimensional coupling-constant space. As K is adjusted toward K_c, it turns out that any initial set of couplings "flows" toward a unique fixed point under repeated iterations of the transformation. Just as in the one-interaction-constant example, critical exponents are related to the eigenvalues of the transformation about this fixed point.

From its origins in 19th century hydrodynamics, renormalization theory has developed into a powerful and sophisticated tool for understanding complicated problems in physics. Much of the recent progress in this area has resulted from a fertile interaction of statistical mechanics with quantum field theory. In its modern form, renormalization theory is especially useful for any problem involving multiple length scales. Anderson applied renormalization group methods to the Kondo problem in condensed-matter physics about the time Wilson applied them to critical phenomena. Such methods have also proved useful in solving certain hydrodynamic problems, in studies of electron localization in metals, and in problems of insect-population biology. There is some hope that they can be used to understand the multiple length scales encountered in turbulent fluids at high Reynolds numbers.

BIBLIOGRAPHY

N. N. Bogoliubov and D. V. Shirkov, *Introduction to the Theory of Quantized Fields*. Interscience, New York, 1959.

S. Coleman, in *Properties of the Fundamental Interactions, 1971* (A. Zichichi, ed.). Editrice Compositori, Bologna, Italy, 1972.

M. E. Fisher, *Rev. Mod. Phys.* **42,** 597 (1974).

S.-K. Ma, *Modern Theory of Critical Phenomena*. Benjamin, Advanced Book Program, Reading, Mass., 1976.

D. R. Nelson, *Nature* **269,** 379 (1977).

P. Pfeuty and G. Toulouse, *Introduction to the Renormalization Group and Critical Phenomena*. Wiley, New York, 1977.

H. D. Politzer, *Phys. Rev.* **C14,** 131 (1974).

K. G. Wilson and J. Kogut, *Phys. Rep.* **C12,** 77 (1974).

Resistance

Arthur F. Kip

The simplest electric circuit can be described as a source of voltage, V (as in an electric battery), with its two terminals connected to a current-carrying element, through which a current, i (measured in amperes) flows. The current which flows is given by

$$i\ (\text{amperes}) = \frac{V\ (\text{volts})}{R\ (\text{ohms})}, \tag{1}$$

where R is the *resistance,* measured in *ohms*. This simple relationship, known as Ohm's law, is based on the idea that materials that can carry currents contain mobile electric charges (positively or negatively charged, or both). In the presence of the electric field produced by the voltage source, the mobile charges drift in the direction of the electric force on them. Although in many cases the actual current-carrying mobile charges are electrons (as in the usual case of metallic conductors), positive current is arbitrarily defined in the direction that *positive* charges would move. Thus the current flows out of the positive terminal of a battery connected to a conducting circuit and back into the negative terminal.

The essential feature of Ohm's law is that the current in a circuit is proportional to the applied voltage. This relationship holds for metals and for many other materials. The value of R depends on the circuit material involved and on its shape and size, and varies with the temperature of the material.

Theory and experiment show that with uniform conducting material the resistance of a long cylinder or wire is given by

$$R = \rho \frac{L}{A}, \tag{2}$$

where L is the length, A the cross-section area, and ρ the *resistivity* of the material (measured in ohm-meters). The resistivity measures the intrinsic behavior of a given material, independent of its shape and size.

The resistivity of metals at room temperature ranges from 10^{-4} to 10^{-2} $\Omega\cdot$m. Very pure metals may have their room-temperature resistivities reduced by many orders of magnitude when cooled to very low temperatures.

Dielectric materials (also called insulators) may have resistivities as high as 10^{20} $\Omega\cdot$m. The resistivity of semiconductors is intermediate between that of metals and that of dielectric materials, and is drastically affected by small admixtures of impurities.

The behavior of materials that obey Ohm's law can be understood on the basis of a simple model. Full understanding of all metallic properties depends on more detailed consideration of quantum ideas. In the simple model we image conduction electrons continuously moving at high velocities in a random fashion between collisions with impurity atoms or other lattice imperfections. An applied potential difference produces an electric field, which accelerates the electrons while in flight and gives a net mean drift velocity along the field direction. The extra kinetic energy picked up by the electrons while in flight is lost to the lattice on each collision. This model is in agreement with Ohm's law and is also consistent with Eq. (2).

The latter equation provides the basis for understanding the effect of combining resistances in series or parallel. A group of resistances $R_1, R_2, \ldots, R_i$ connected in series has a total resistance of

$$R = \sum_i R_i. \tag{3}$$

When connected in parallel, their total resistance is given by

$$1/R = \sum_i 1/R_i. \tag{4}$$

For resistances of uniform cross section A, the current is given by

$$i = nevA, \tag{5}$$

where n is the density of mobile charge carriers, e is their charge, and v is their drift velocity, as caused by the applied electric field.

In dc circuits, the voltage/current ratio is completely fixed by the total resistance in the circuit. In ac circuits two new quantities also affect the voltage/current ratio. These are *capacity* and *inductance*. The net effect of all three elements in called the *impedance* of the circuit, and from this the voltage/current ratio can be found.

See also SEMICONDUCTORS; SUPERCONDUCTORS.

Resonance Phenomena

W. Happer

The term resonance originated in acoustics where the audible resonances of musical instruments have been known since antiquity. A physical system is said to exhibit a resonance if it can exist in a well-defined state for a time much longer than the period for internal motion in that state. A resonant system can also be excited efficiently by a relatively weak periodic driving force in a narrow band of frequencies close to the resonance frequency. More generally, a resonant system can be efficiently formed by a monoenergetic beam of particles impinging on a target when the center-of-mass energy of the colliding pair is equal to the energy of a quasistationary state of the compound system of the target and the particle.

A simple example of a resonant system is shown in Fig. 1. A mass m is attached to a spring of force constant k. Frictional dissipation of energy is represented by a viscous damping coefficient η in the equation of motion

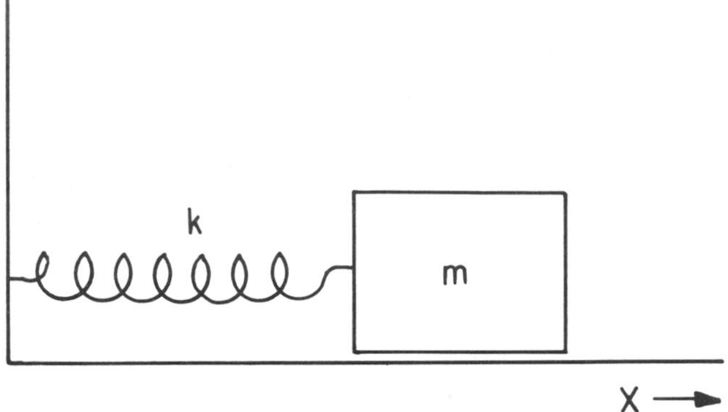

FIG. 1. A mechanical resonator.

$$m\ddot{x} = -kx - \eta\dot{x}. \tag{1}$$

The solution of Eq. (1) is

$$x = A\cos(\omega_0 t - \phi)\exp(-\gamma t/2), \tag{2}$$

where the resonant frequency is

$$\omega_0 = \left[\frac{k}{m} - \left(\frac{\eta}{4m}\right)^2\right]^{1/2} \tag{3}$$

and the damping rate is

$$\gamma = \eta/m \tag{4}$$

The quality factor Q of the system is

$$Q = \omega_0/\gamma. \tag{5}$$

The amplitude A and phase ϕ are constants of integration.

The energy of the oscillator is the sum of kinetic energy and potential energy which, for large Q, are very nearly

$$E_k = \tfrac{1}{2}m\omega_0{}^2A^2\sin^2\omega_0 t\exp(-\gamma t), \tag{6}$$

$$U = \tfrac{1}{2}kA^2\cos^2\omega_0 t\exp(-\gamma t). \tag{7}$$

Since $m\omega_0{}^2 \simeq k$, the peak kinetic energy is equal to the peak potential energy and the oscillator energy is transformed periodically from kinetic to potential energy. The total energy, the sum of Eqs. (6) and (7), decays exponentially at a rate γ:

$$E = \tfrac{1}{2}kA^2\exp(-\gamma t). \tag{8}$$

The lifetime τ of the excited system is

$$\tau = 1/\gamma. \tag{9}$$

In subsequent discussion we shall assume that Q is much greater than unity; then the system can oscillate for many cycles before appreciable damping occurs.

Resonant systems can be efficiently excited by a periodic driving force. For example, if the mass of Fig. 1 is subject to an external force

$$F = f\cos\omega t, \tag{10}$$

the system will settle down after a time on the order of γ^{-1} to a steady-state oscillating motion described by

$$x = A\cos(\omega t - \phi), \tag{11}$$

where the amplitude A is

$$A = \frac{f}{[(k - m\omega^2)^2 + \omega^2\eta^2]^{1/2}} \tag{12}$$

and the phase shift is

$$\phi=\tan^{-1}\frac{\omega\eta}{k-m\omega^2}. \tag{13}$$

At low frequencies, $\omega^2 \ll k/m$, the phase shift is negligible and the mass is driven in phase with the force (10) with a peak displacement of f/k. For $\omega^2=k/m \simeq \omega_0^2$, the phase shift is $\pi/2$ and the peak displacement of the mass is Qf/k, a factor of Q times larger than for a slowly oscillating or time-independent force. For very large frequencies the phase shift is π, the amplitude decreases as ω^{-2}, and the mass responds as if it were not attached to the spring. The resistance to the driving force comes mainly from the spring at low frequencies and mainly from the mass at high frequencies. For high-Q systems and nearly resonant driving frequencies Eqs. (11) and (12) can be well represented by

$$A^2=\left(\frac{f}{2m\omega_0}\right)^2\frac{1}{(\omega-\omega_0)^2+\gamma^2/4}, \tag{14}$$

$$\phi=\tan^{-1}\frac{\gamma}{2(\omega_0-\omega)}. \tag{15}$$

The oscillator can be driven to large amplitudes over a band of frequencies of width

$$\Delta\omega=\gamma; \tag{16}$$

consequently, the damping rate γ is often called the width of the resonance. The phase ϕ varies by $\pi/2$ over the resonance width. From Eqs. (16) and (5) one can see that Q is a measure of the sharpness of the response curve for a sinusoidal driving force. The amplitude and phase of the driven oscillator are sketched in Fig. 2.

Another simple resonant system is the series electrical circuit sketched in Fig. 3. The behavior of this circuit is completely analogous to that of the mechanical oscillator of Fig. 1. If the driving emf is removed from the circuit, the circuit equation is

$$\frac{q}{C}=L\frac{di}{dt}+Ri. \tag{17}$$

The charge q on the capacitor is related to the current i in the circuit by

$$i=-\frac{dq}{dt}. \tag{18}$$

The differential equation for the series RLC of Fig. 3 is thus equivalent to the differential equation for the mechanical oscillator of Fig. 1 if one makes the replacements $q\to x$, $C\to 1/k$, $L\to m$, $R\to\eta$. Making these substitutions one finds that the resonant frequency is

$$\omega_0=\left[\frac{1}{LC}-\left(\frac{R}{4L}\right)^2\right]^{1/2}, \tag{19}$$

$$\gamma=R/L. \tag{20}$$

For a high-Q circuit the resistance R must be much less than $\omega_0 L$.

In the electrical circuit of Fig. 3 the energy is transferred periodically from the electrical energy stored in the capacitor to magnetic energy stored in the inductor. If the circuit is driven by a sinusoidal voltage source, the peak charge on the capacitor and the phase shift between the peak charge and the driving voltage are described by curves similar to those in Fig. 2 for A and ϕ.

The resonators illustrated in Figs. 1 and 3 are called lumped-parameter resonators since the two forms of energy (kinetic and potential or magnetic and electrical) are stored in well-defined system components (e.g., a spring or a capacitor). Many important resonators do not allow for a clear separation of different forms of energy, and such systems are called distributed-parameter resonators. A violin string is an example of a distributed-parameter resonator since the kinetic energy of the vibrating string and the potential energy

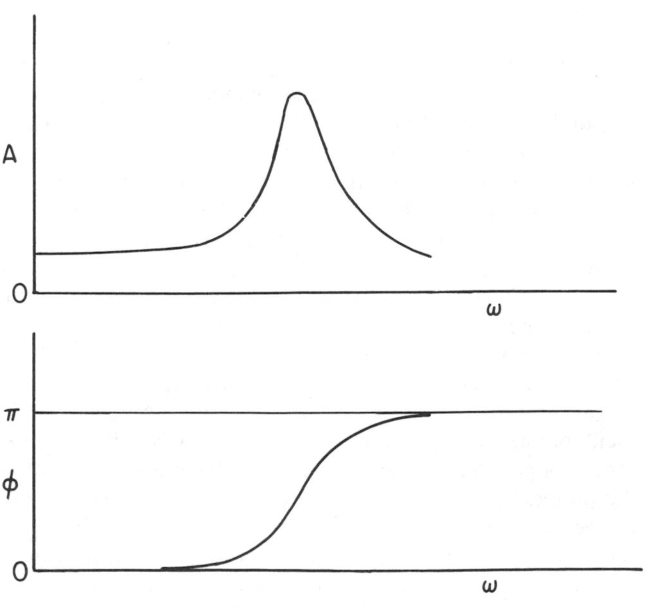

FIG. 2. Resonance response curves.

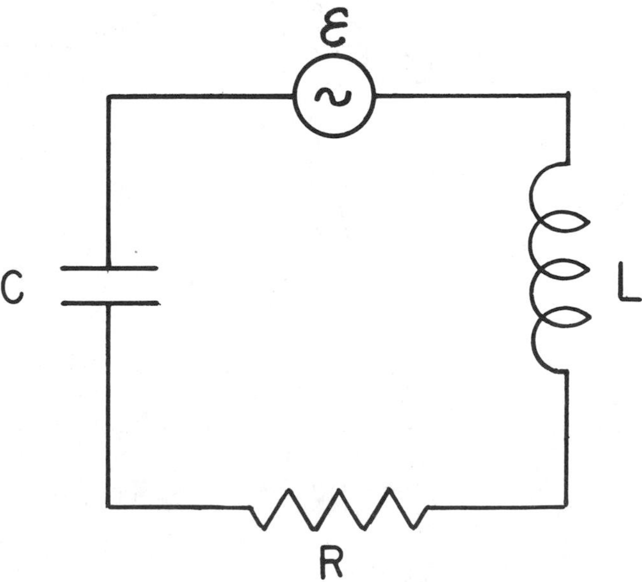

FIG. 3. A resonant electrical circuit.

of the stretched string are both distributed along the entire length of the string. An electrical transmission line can support distributed-parameter resonances very similar to those of a violin string.

Almost all physical systems exhibit multiple resonances. For example, a violin string can vibrate at the fundamental frequency or at one of the overtone frequencies. Each resonance has a characteristic frequency and width. Often certain resonance modes have much higher Q values than others in the same system, and some modes may be particularly insensitive to external influences like ambient temperature. These mode properties may be used to advantage in many cases. For example, a quartz crystal oscillator may be cut in such a way that certain modes are nearly insensitive to the temperature of the crystal.

Multiple resonances are particularly important in the microscopic realm of molecules, nuclei, and elementary particles. In microscopic physics the concept of resonance is tied to the idea of stationary or quasistationary states of a quantum-mechanical system. For example, consider a sodium atom in the $4P$ excited state as sketched in Fig. 4. We may characterize the $4P$ state by its energy E and by its lifetime τ. The $4P$ state can decay radiatively by electric dipole transitions to the lower-lying $3S$, $4S$, or $3D$ states. Each decay mode is called a channel and each contributes a partial rate to the total decay rate of the $4P$ state, i.e.,

$$\gamma = \gamma(4S) + \gamma(3S) + \gamma(3D). \tag{21}$$

One decay mode $4P \rightarrow 3P$ is so strongly forbidden compared to the others that it can be neglected. The total lifetime τ of the excited state is related to the decay rate γ by (9).

Many different decay modes can contribute to the lifetime of a level. For example, a heavy nucleus in an excited state may emit a photon, it may emit a neutron or an alpha particle,

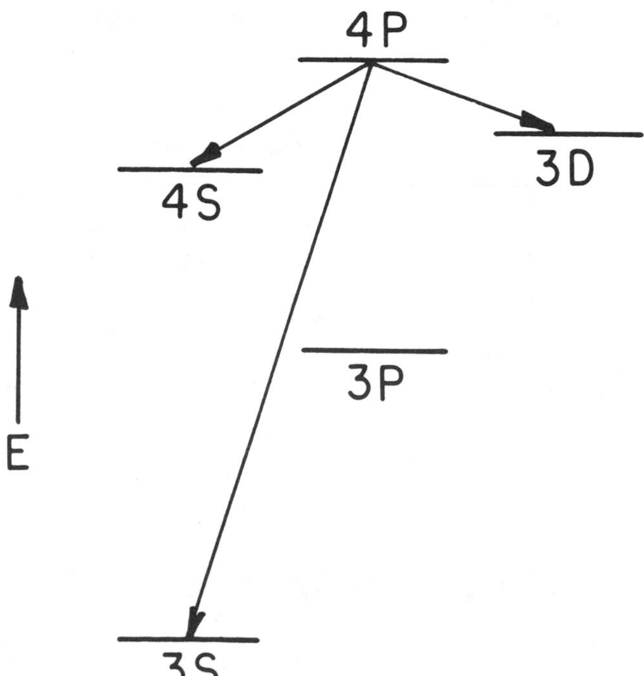

FIG. 4. The major decay channels of the $4P$ state of sodium.

or it may fission, and other decay channels are also possible. Each channel shortens the lifetime of the excited state. The relative number of decays in each channel is called the branching ratio. In some cases most decay channels are closed by selection rules and the excited state then has an unusually long lifetime.

When an atom decays radiatively from an upper to a lower energy state it emits a photon in a narrow band of frequencies centered about the Bohr resonance frequency ω_0

$$\hbar\omega_0 = E_u - E_l, \tag{22}$$

where E_u and E_l are the energies of the upper and lower states. The bandwidth of the radiation is

$$\Delta\omega = \gamma_u + \gamma_l. \tag{23}$$

Note that both the upper- and lower-state widths contribute to the observed bandwidth $\Delta\omega$. Thus, the photons emitted in different branches from the same excited state can have different bandwidths because they connect the excited state to final states of different lifetimes. Completely analogous considerations hold if the excited state decays by emitting an electron, a neutron, or some other particle rather than a photon.

Expression (23) is a consequence of the uncertainty relation

$$\Delta E\tau \simeq \hbar \tag{24}$$

between the lifetime of τ of a state and the energy uncertainty ΔE of the state. The energies E_u and E_l are uncertain by $\hbar/\tau_u = \hbar\gamma_u$ and $\hbar/\tau_l = \hbar\gamma_l$, respectively, and the Bohr frequency condition (22) can be satisfied for a band of frequencies ω_0.

Resonance behavior is often observed in the cross section for various scattering events in atomic, nuclear, or high-energy physics. For example, the radiative capture cross section for a slow neutron by a nucleus is often described very well by the Breit–Wigner formula

$$\sigma = \frac{\lambda^2}{4\pi} \frac{\Gamma_x \Gamma_y}{(E - E_0)^2 + (\Gamma/2)^2}, \tag{25}$$

where γ is the de Broglie wavelength of the incident neutron and E is the neutron energy. The resonance energy E_0 corresponds to a quasistationary state of the target nucleus plus the incident neutron, the total width (in energy units) of the state is Γ, Γ_x is the partial width of the input channel, and Γ_y is the partial width for radiative capture of the neutron. Note the close similarity of the Breit–Wigner resonance shape (25) to the classical response function (14) of a driven oscillator. The phase shift ϕ of the wave function of the scattered neutron also varies with energy like the phase shift (15) of a classical oscillator near resonance.

The resonances in a scattering experiment are interpreted as the formation of a compound system composed of the incident particle and the target. This compound system lives much longer than the transit time of the projectile through the target. In the case of high-energy scattering experiments the resonances are often regarded as particles in their own right. For example, a resonant increase in the scattering of π^+ mesons by protons at a center-of-mass pion energy of 159 MeV is due to the formation of the $N^*_{3/2}(1236)$ particle, an excited state of the nucleon with spin $\frac{1}{2}$, isospin $\frac{3}{2}$, and even parity. The rest mass of the particle is 1236 MeV and the width is 125 MeV. Since no selection rules forbid the decay of this particle, it rapidly decomposes into a proton and pion again.

The response of a resonant system to a sinusoidal driving force is characterized by two real functions of the driving frequency ω, for example, the phase ϕ and the ratio A/f of the displacement magnitude A to the sinusoidal force magnitude f. These are frequently combined to form a complex number $\alpha(\omega)=Ae^{i\phi}/f$ with real and imaginary parts $\alpha'(\omega)$ and $\alpha''(\omega)$. For example, $\alpha(\omega)$ might represent the polarizability of an atom in an electromagnetic wave of frequency ω. The polarizability α is analogous to the quantity iZ^*/ω, formed from the complex impedance Z of the electrical circuit of Fig. 3. The complex conjugate Z^* appears rather than Z since conventional circuit theory deals with quantities proportional to $e^{i\omega t}$ whereas physical calculations usually consider quantities proportional to $e^{-i\omega t}$. For the idealized mechanical oscillator of Fig. 1, the polarizability can be calculated from Eqs. (12) and (13), but for real mechanical oscillators, or other resonant systems like atoms or nuclei, the system may be so complicated that it is necessary to resort to experimental measurements to determine the response. Fortunately, it is not necessary to measure both the real and imaginary parts of α since they are related by the Kramers–Kronig formulas

$$\alpha''(\omega) = -\frac{\mathcal{P}}{\pi}\int_{-\infty}^{+\infty}\frac{\alpha'(x)\,dx}{x-\omega}, \tag{26a}$$

$$\alpha'(\omega) = \frac{\mathcal{P}}{\pi}\int_{-\infty}^{+\infty}\frac{\alpha''(x)\,dx}{x-\omega}. \tag{26b}$$

Here $\mathcal{P}$ denotes the principal value of the integrals, which extend along the entire real ω axis. Thus, if either the real or imaginary part of the polarizability function can be measured, the other part can be calculated with the aid of Eqs. (26). The Kramers–Kronig formulas are valid if the resonator is a passive linear system which satisfies causality, that is the response $x(t)$ at time t is a linear combination of the forces $F(t')$ at all prior times $t'<t$. This is equivalent to the statement that the complex function $\alpha(\omega)$ of the complex variable ω has no poles in the upper half-plane where $\omega''>0$. Dispersion relations similar to Eqs. (26) are widely used in science and engineering to analyze resonant systems, for example, to infer the optical absorption coefficient of a semitransparent medium from measurements of the index of refraction of the medium. Electrical circuits which satisfy Eqs. (26) are often called minimum-phase filters. Systems which violate dispersion relations like Eqs. (26), for example, lasers or active electrical filters, have internal sources of energy.

The collection of all resonances of a physical system is called the spectrum. The many varieties of spectroscopy, ranging from studies of the slow periodic motions of celestial bodies to the extremely rapid internal motions of elementary particles, are designed to reveal the laws which govern a physical system by inference from the spectrum of resonances. Some of the most important spectroscopic techniques are radiofrequency and microwave spectroscopy, which deal with resonances of nuclear and electronic angular momenta, infrared spectroscopy which deals with vibrational resonances of molecules or solids, optical spectroscopy which deals with the resonances of the valence electrons of matter, ultraviolet and x-ray spectroscopy which deal with resonances of inner-shell electrons in matter, and gamma-ray spectroscopy which deals with nuclear resonances. Electrons, protons, neutrons, and pions are but a few of the many particle beams which are used to investigate resonance phenomena in condensed matter, gases, plasmas, atoms, nuclei, and elementary particles.

See also ACOUSTICS; DISPERSION THEORY; ELECTRON SPIN RESONANCE; MUSICAL INSTRUMENTS; NETWORK THEORY, ANALYSIS AND SYNTHESIS; NUCLEAR MAGNETIC RESONANCE; NUCLEAR SCATTERING; QUANTUM STATISTICAL MECHANICS; VIBRATIONS, MECHANICAL.

Resonances, Giant

E. G. Fuller

A giant resonance is said to exist in a nucleus or nuclear model when a major fraction of the total transition strength for an electromagnetic interaction operator is found to be located within an isolated and relatively narrow range of excitation energy. While the meanings of the words "major" and "narrow" are by no means well-defined, a giant resonance is usually considered to exist if more than 30% of the appropriate energy-weighted sum-rule strength for an operator is found to exist within an energy interval that is less than 30% of the mean excitation energy for the transitions induced by the operator. The most prominent and well-studied giant resonance is that associated with electric-dipole transitions and photonuclear reactions. This "resonance," while not so well defined in the lightest nuclei (the criteria just mentioned are not completely satisfied), is well established by the end of the p shell (^{12}C and ^{16}O). It is the most prominent feature of the nuclear absorption cross section for photons as a function of energy. Recently, both inelastic electron and hadron (proton, deuteron, alpha-particle, etc.) scattering experiments have proved to be very fruitful sources of information about other angular momentum modes of nuclear excitation.

The energy-weighted sum rule used to define the strength of a "resonance" is defined by

$$S(\sigma\lambda) = \sum_f (E_f - E_0)B(\sigma\lambda,\omega;0\rightarrow f) \tag{1}$$

where σ is either E or M, depending on whether the transition is electric or magnetic of order λ. The excitation energy is $E_f-E_0=\hbar\omega$ and the sum is taken over the appropriate set of final states that can be reached by transitions of order λ. For electric-multipole transitions,

$$B(E\lambda,\omega;0\rightarrow f)$$
$$= \sum_{M_f m} |\langle I_f M_f;T_f T_3| \sum_i \frac{e}{2} r_i^{\lambda} Y_{\lambda m}(\hat{r}_i)(1-\tau_3^{i})|I_0 M_0;T_0 T_3\rangle|^2, \tag{2}$$

where I, T, M, and T_3 represent, respectively, the nuclear spin, isospin, and their projections. The quantity τ_3 is equal to -1 for protons and $+1$ for neutrons. The sum i is over all nucleons in the nucleus. The electric multipole operator $r^{\lambda}Y_{\lambda m}$ is valid in the long-wavelength limit. It consists of two terms; the first is a scalar and the second, $\tau_3{}^i$, is one component of a vector in isospin space. In a nucleus with a ground-state isospin of $T_0=T_z=(N-Z)/2$, the selection rules for transitions induced by these two parts of the operator are, for isoscalar transitions, $\Delta T=0$; and for isovector transitions, $\Delta T=0, \pm 1$ with no $\Delta T=0$ for $T_0=T_z=0$ nuclei. In general, when $T_z\neq 0$, considerable mixing is expected between the isoscalar and isovector $\Delta T=0$ transitions. Only for self-conjugate nuclei, i.e., $T_z=0$, or for very restrictive nuclear models is it possible to make a meaningful separation of the total sum rule into isoscalar and isovector components. The sum rule resulting from this matrix element is

$$S(E\lambda)=\frac{\hbar^2}{2M}\frac{\lambda(2\lambda+1)^2}{4\pi}e^2\langle r_p{}^{2\lambda-2}\rangle_0 Q(N,Z), \tag{3}$$

where

$$\langle r_p{}^{2\lambda-2}\rangle_0 \equiv \frac{1}{Z}\langle 0|\sum_i r_i{}^{2\lambda-2}\left(\frac{1-\tau_3{}^i}{2}\right)^2|0\rangle$$

is the mean $(2\lambda-2)$ moment of the charge radius of the nucleus. For a uniformly charged sphere of radius R, $\langle r_p{}^{2\lambda-2}\rangle_0 = 3R^{2\lambda-2}/(2\lambda+1)$. In addition to equivalent expressions for $E0$ and $M1$ multipoles, Table I gives values of $Q(N,Z)$ for various electric multipoles as well as brief notes on their validity and the models used to obtain these factors. The factors $Q(N,Z)$ are all obtained by assuming that only the kinetic energy term in the nuclear Hamiltonian is quadratic in the nucleon momenta and that the Hamiltonian commutes with the charge operator $(1-\tau_3{}^i)$. The result is obtained by making use of closure and the commutation properties of the Hamiltonian with the electric-multipole operator. An equivalent expression in terms of an energy-weighted integral over energy of the photon absorption cross section associated with a given multipole is

Table I Energy-weighted sum rules and giant resonance energies[a]

σλ	Name	Energy [MeV($A^{-1/3}$)]	Expressions and comments
$E0$	Isoscalar (Suzuki; Ferrell)	60	$\sum_f (E_f-E_i)\lvert M_{if}\rvert^2=(\hbar^2/M)A\langle r_m{}^2\rangle_0$. For self-conjugate nuclei, $\langle r_m{}^2\rangle_0$ is mean-square matter radius.
$E0$	Isovector (Suzuki)	178	
$E0$	Hydrodynamic	168	An isovector mode. Energy 2.16 × giant dipole energy.
$M1$	Isoscalar (Gell-Mann and Telegdi)		$\int \frac{\sigma(M1)}{E_\gamma}dE_\gamma=\pi^2\frac{e^2}{\hbar c}\frac{1}{4}\frac{\langle(\mathbf{L}+(\mu_p+\mu_n)2\mathbf{S})^2\rangle_0}{(Mc)^2}$. Self-conjugate nuclei. Model needed to calculate ground-state expectation values.
$M1$	Isovector (Kurath)		$\sum_f (E_j-E_i)B(M1,i-f)\simeq -a\left(\mu_n-\mu_p+\frac{1}{2}\right)^2\left\langle\sum_i^A \mathbf{l}_i\cdot\mathbf{s}_i\right\rangle_0$ where a is the spin–orbit splitting parameter. Self-conjugate nuclei. Model needed to calculate ground-state expectation values.
$E1$	Classical (Levinger and Bethe)		$Q(N,Z)=NZ/A$. Nuclear recoil included.
$E1$	Hydrodynamic	78	$Q(N,Z)=0.86NZ/A$. Fundamental dipole mode. Nuclear surface fixed. Internal isovector oscillation.
$E2$	Hydrodynamic	125	An isovector mode. Energy 1.6 × giant dipole energy.
$E2$	Total		$Q(N,Z)=Z$. Small nuclear recoil effects neglected. No model dependence.
$E2$	Pure Isoscalar (Gell-Mann and Telegdi)		$Q(N,Z)=A/4$. Self-conjugate nuclei. For $T=T_z\neq 0$, isoscalar–isovector mixing is probable.[b]
$E2$	"Isoscalar" (Nathan and Nilsson)	58	$Q(N,Z)=Z^2/A$. Model dependent. No isoscalar–isovector mixing. Mode is surface oscillation, neutron and proton fluids move together in phase.[b]
$E2$	"Isovector"	135	$Q(N,Z)=NZ/A$. Model dependent. No isoscalar–isovector mixing. Neutron and proton liquids each maintain constant density, move bodily with respect to each other. Strength given by difference between total and "isoscalar."[b]

[a] For electric multipoles, $\lambda\geq 1$, see Eqs. (3) and (4) for complete expression.
[b] Assumes that nuclear charge and matter distributions are the same.

$$\int \frac{\sigma(E\lambda)\, dE_\gamma}{E_\gamma^{2\lambda-2}} = \pi^2 \frac{e^2}{\hbar c} \frac{(\hbar c)^{4-2\lambda}}{Mc^2} \frac{(\lambda+1)}{[(2\lambda-1)!!]^2} \langle r_p^{\,2\lambda-2} \rangle_0 Q(N,Z). \tag{4}$$

Unfortunately, there has been some confusion in the literature as to which of the sum rules indicated in Table I should be used to analyze giant resonance data. In a number of instances, results have been presented with only the simple statement that they are normalized to "the energy-weighted *E*2 sum rule," with no accompanying expression or reference to indicate which sum rule was used. As a result, comparisons between different early measurements can be uncertain by factors of the order of 2.

Except for electric-monopole transitions, which are forbidden for photons (but not for inelastic electron scattering), the most unambiguous determination of the strength of a giant resonance is obtained when it is possible to determine the photon absorption cross section for a particular multipole. The strength can then be determined by carrying out the integral indicated in Eq. (4). Unfortunately, this is experimentally practical only for the electric-dipole case (the photonuclear giant resonance) and for a few magnetic-dipole transitions in light nuclei where the strength is concentrated in a few strong levels lying below the neutron and proton separation energies.

Data on *E*2 giant resonances come almost exclusively from the analysis of inelastic-scattering experiments, which lead directly to values for $B(E\lambda,\omega)$. Since in inelastic hadron scattering the excitation of a nuclear multipole is via the strong nuclear interaction, the analysis of these data requires the use of a specific nuclear model. The relationship of the nuclear-multipole mass modes excited by these reactions to the electromagnetic modes is not always obvious. Inelastic electron scattering, on the other hand, directly excites the electromagnetic modes. With good enough experimental data there is no inherent need to invoke a nuclear model to extract a value of $B(E\lambda,\omega)$ for a particular resonance. For technical reasons, however, it has always been necessary to use models to carry out this analysis. Unfortunately, the values of $B(E\lambda,\omega)$ derived from an electron-scattering experiment can differ by a factor of 2 when different models for the photonuclear giant resonance are used in the analysis.

There are sufficient experimental data available on electric-dipole and -quadrupole giant resonances to permit the extraction of trends with mass number *A*. The peak, neglecting fine structure of the photonuclear giant resonance, is at about 24 MeV in carbon and oxygen. It falls slowly to 19.9 MeV for $A=90$ and is then given approximately by the expression $E_0=77.5A^{-1/3}$ MeV. The parameter 77.5 MeV varies by less than 2% between $A=90$ and $A=209$. The integrated absorption cross section over the giant resonance (integral to 30 MeV) increases from about 0.6 times the classical dipole sum-rule result for carbon and oxygen to about 1.25 times this value for $A \geq 120$. Inelastic electron and hadron scattering experiments have shown that for $A \geq 20$ there are two electric-quadrupole giant resonances, one (presumably an isoscalar resonance) located at $63A^{-1/3}$ MeV, the other (presumably isovector) located at $120A^{-1/3}$ MeV. The lower peak is observed in both hadron and electron scattering experiments. Essentially, all the information for the higher-energy resonance comes from inelastic electron scattering experiments. That this peak is not seen with hadrons may be associated with the fact that it is a quadrupole mode associated with internal degrees of freedom of the nucleus rather than one associated with nuclear surface modes. Due to the strong absorption of hadrons by nuclei, their scattering is presumably the result of surface interactions only. While the data on the strengths of these resonances are not very consistent, there is no question that they exhaust a large fraction of the total *E*2 sum rule.

See also PHOTONUCLEAR REACTIONS; SUM RULES.

BIBLIOGRAPHY

Resonance Energies

*E*0 M. Danos, *Nucl. Phys.* **5,** 23 (1958); T. Suzuki, *Nucl. Phys.* **A217,** 182 (1972).

*E*1 H. Steinwedel and J. H. D. Jensen, *Z. Naturforsch.* **5a,** 413 (1950).

*E*2 M. Danos, *Ann. Phys.* (Leipzig) **10,** 265 (1952); A. Bohr and B. Mottelson, *Nuclear Structure* Vol. II, Chap. 6. Benjamin (Advanced Book Program), Reading, Mass., 1975.

Sum Rules, General

E. K. Warburton and J. Weneser, *Isospin in Nuclear Physics,* Chap. 5.4, p. 195. North-Holland, Amsterdam, 1969.

Amos deShalit and Herman Feshbach, *Theoretical Nuclear Physics, Vol. I: Nuclear Structure,* Chap. VII, Sections 7 and 8, pp. 706–715. Wiley, New York, 1974.

Sum Rules, Specific

*E*0 R. A. Ferrell, *Phys. Rev.* **107,** 1631 (1957).

*E*1 J. S. Levinger and H. A. Bethe, *Phys. Rev.* **78,** 115 (1950).

*M*1 D. Kurath, *Phys. Rev.* **130,** 1525 (1963).

*M*I and *E*2 M. Gell-Mann and V. L. Telegdi, *Phys. Rev.* **91,** 169 (1953).

Electron Scattering

W. C. Barber, *Ann. Rev. Nucl. Sci.* **12,** 1 (1962).

Herbert Überall, *Electron Scattering from Complex Nuclei (Part B).* Academic, New York, 1971.

M. Sasao and Y. Torizuka, *Phys. Rev.* **C15,** 217 (1977).

Hadron Scattering

E. C. Halbert, J. B. McGrory, and G. R. Satchler, *Nucl. Phys.* **A245,** 189 (1975).

G. R. Satchler, *Phys. Rev.* **14C,** 97 (1974).

F. E. Bertrand, *Ann. Rev. Nucl. Sci.* **26,** 457 (1976).

Data

*M*1 L. W. Fagg, *Rev. Mod. Phys.* **47,** 683 (1975).

*E*2 F. E. Bertrand, *Ann. Rev. Nucl. Sci.* **26,** 457 (1976).

Deuteron M. P. de Pascale *et al., Phys. Lett.* **119B,** 30 (1982).

10–120 MeV Critical Review

*E*1 J. R. Calarco, B. L. Berman, and T. W. Donnelly, *Phys. Rev. C* **27,** 1866 (1983).

Evaluated Data for He[4]

*E*1 E. G. Fuller, *Phys. Rep.* **127**(3) (1985).

Evaluated Data for C^{12}, N^{14}, *and* O^{16}

*E*1 S. S. Dietrich and B. L. Berman, *At. Data Nucl. Data Tables* **38,** 199 (1988).
A compilation of data from all sources measured with monoenergetic photons.

Rheology

F. R. Eirich

Rheology (*rheo*—flow; *logos*—knowledge), the interdisciplinary science of irreversible processes underlying the phenomena of inelasticity and flow of condensed matter, was so named by E. C. Bingham, who founded the Society as well as the *Journal of Rheology* in 1929. Overlapping with areas of physical chemistry, solid state, polymer, colloid, surface, and tribophysics, with biophysics, fluid and applied mechanics, dislocation and plasticity theory, and fracture mechanics, rheology deals with frictional (viscous or diffusional) transport phenomena and deformations of shape, dimensions, or positions relative to external and internal coordinates, and with the conversion of imparted mechanical energy into heat (internal friction and irreversible thermodynamics). In contrast to fluid dynamics, including hydrodynamics, which treats the masses in motion as continua, rheology considers the mechanics of flow (below the onset of turbulence) as the result of reversible and irreversible molecular processes. Thus, it has common aspects with the mechanics of highly compressible fluids (gases), fluid dynamics, and the study of elasticity (the molecular and/or continuum mechanics of full recovery of the original shape of given bodies, and of the invested energy). Since rheology is the principal science concerned with the stress–strain and strain-rate-dependent molecular behavior, it follows that the molecular features of mechanical hysteresis and relaxation, of flow as a nonconservative process, of all mechanical testing up to and including failure and fatigue, as well as the engineering of industrial processes, lie within its concerns. The same is finally true for the movements of fluids within living organisms (biorheology), for much of geophysics, and for some metallurgy.

Irreversible processes, as they occur during macroscopic flow, mechanical aging, or as dislocations during the transient of permanent deformation of solids, are the responses of matter being sheared. "Simple shearing" proceeds along equidistant, rectilinear flow lines (laminar flow); other forms of flow occur in tension ("pure shearing," see below), compression, radial, torsional, conical, and biaxial flows, and various forms of extrusion, often affected by the rate (or frequency, if oscillatory) of the enforced motions and their duration. Since all atomic molecular motions and their space requirements (volume) are affected by temperature and pressure, all forms of flow must be studied under isothermal and isobaric conditions. Where measurable temperature increases due to flows occur, their effects have to be studied as an integral part of the rheological process.

Macroscopic flow is described by constitutive equations and parameters which relate observed stresses and rates of straining to material properties. In principle, knowing the relevant molecular properties (weight, size, shape, polarity, polarizability, friction factors, etc.), the parameters of the continuum's mechanical equations of motion should be obtainable. In fact, most constitutive equations are derived on the basis of mass and momentum balances and considerations of rate effects. The first and simplest constitutive equation of flow, for readily mobile liquids of constant volume in simple shear, was formulated empirically by Isaac Newton, who stipulated that a constant force acting on the *y-z* end of an imaginary stack of horizontal *x-z* fluid lamellae of area *A* causes a steady-state velocity difference along *y*, i.e., a rate of shear gradient that causes steady unlimited deformation (flow) in the *x-y* planes:

$$\frac{F_{xy}}{A_{xy}} = \sigma_{xy} = \frac{\eta d(dx/dy)}{dt} = y\left(\frac{dv_x}{dt}\right),$$

where $F_{xy}/A_{xy} = \sigma_{xy}$ is the stress, acting along x and causing imaginary *x-z* planes of the fluid to move differentially. η is a material parameter (constant), the coefficient of internal friction or viscosity, often just called the "viscosity." Its reciprocal, ϕ, is the fludity. The gradient of the velocity vector, dv_x/dy, is often given the symbol D_{xy}, or simply D. Since $dx/dy = \gamma$ is the shear strain, $\dot{\gamma} = d(dx/dy)/dt$ is the rate of shear identical with D_{xy}. The unit of the viscosity is the poise (dyn s/cm^2); in international units it is one pascal-second, equal to 10 poises.

Such "simple" shearing flows, symbolized by a uniform shifting rate of a deck of cards, is in reality complicated because it entails an internal rotation of angular velocity, $w = \dot{\gamma}/2$, easily shown by the rotation of the axes of a square that were originally parallel to the *x-y* coordinates. In "pure" shearing motion, on the other hand, the x dimensions of an *x-y* plane increase in length, the y dimensions shrink, while the axes stay perpendicular so that now

$$\dot{\gamma} = \frac{dv_x}{dy} + \frac{dv_y}{dx}.$$

For flow in more than one plane, the stresses and gradients become sets of three-dimensional vectors, i.e., tremors. If one stipulates a velocity, v, which is nonsteady but limited to small Reynolds numbers, $R = (v\rho r/y)$, l, with v, the velocity of mass flow, ρ. the constant density of the fluid, and r, the radius of an object, or of the radius of the curvature of the conduit (no turbulence), the equation of continuity is $\bar{V}v = 0$, and the Navier–Stokes equations have to be applied (ρ = density, F = body force, P = pressure):

$$\rho \frac{dv}{dt} = \rho F + \bar{V}P + \eta \bar{\nabla} v^2$$

of which Newton's equation is a special case. To solve for a particular flow, one has to apply the pertinent equation of motion and to introduce the boundary conditions. Exact solutions exist for the viscometric flows, where the stresses are defined by shear rates only, and vice versa. For example, for the steady-state flow of a Newtonian liquid under pressure P through a tube, i.e., the coaxial frictional flow of thin cylindrical lamellae in telescopic motion with a boundary layer of the fluid adhering to the wall, one obtains Hagen–Poiseuille's law:

$$Q/t = \pi PR^4/8\eta l.$$

Q/t is the rate of volume flow, R is the diameter of the tube, and l is its length. The flow profile (velocity distribution) is parabolic, and the stress at the wall is $\sigma_R = PR/2l$. If two liquids flow through the same capillary, the relative viscosity corrected for the density (the "kinematic" viscosity $=\eta/\rho$) is $\eta_{rel}=\eta/\eta_0=t/t_0$, where t_0 and η_0 are the flow time and viscosity of the reference liquid, and t and η the time and viscosity of the unknown liquid. Other viscometric flows are those due to shearing between parallel plates, those around spheres as described by Stokes' law:

$$v=F/6\pi\eta r$$

(where v and r are the relative velocity and the radius of the sphere, F is the force acting on it, and $6\pi\eta r=f$ is the friction factor), and extensional flows during uniaxial pulling by a force, F:

$$F_x=\frac{\lambda d(d\epsilon/dx)}{dt}=\lambda\left(\frac{dv}{dx}\right),$$

where ϵ is the longitudinal strain defined by the relative extension in length, dl/l, in the tensile direction x, and $\lambda=3\eta$ is Trouton's coefficient of extensional viscosity.

Nonviscometric flows are those in which the flow of a volume element is determined by the history of its flow and of its surroundings. This includes most industrial forming operations. For some liquid solutions, the solute may become sufficiently adsorbed at the air–liquid interface so that a very thin surface layer may exhibit its own "surface viscosity," usually measured by the torsional forces on a suspended disc.

Complying with the Navier–Stokes equations are the rather simple, low-molecular-weight liquids whose viscosity is independent of the shear rate. The same equations are also obeyed by practically all irreversible deformations which are slow enough for directional molecular diffusion to accommodate the enforced motion, i.e., to exclude inertial forces. A large body of theoretical rheology deals with more complicated types, the so-called non-Newtonian fluids and flows, employing the fundamental principles of stress (or energy) balances, coordinate invariance, etc., for their description. For memory fluids, e.g., the stresses are no longer linear functions of the gradients, the viscosities become functions of the stresses or gradients and their derivatives, stresses normal to the flow lines appear, and the gradient history becomes important. In simple shearing flow, one may write for the viscosity function of such fluids

$$\eta(\dot\gamma)=\sigma(\dot\gamma)/\gamma,$$

and for the concurrent first and second normal stress differences

$$t_{11}-t_{22}=t_{11}-t_{33}=\sigma_1(\gamma)$$

and

$$t_{22}-t_{33}=\sigma_2(\gamma).$$

These stress differences, perpendicular to the flow direction, are due to tensions along the stream lines and are often approximately proportional to γ^2. The general viscosity function for tube flow is obtained by a log–log plot of σ_R vs. $4(Q/t)/\pi R^3$, which gives $f(\eta)=(\sigma_R/D)[\frac{3}{4}+\frac{1}{4}(d\ \ln D/d\ \ln \sigma_R)]$.

Other ways to describe non-Newtonian flow include the use of the limits of the viscosities at zero and infinite shear rates, η_0 and η_{00}, the differential viscosity, $\eta_{diff}=d\sigma/d\gamma$, and "power laws" such as

$$\sigma=\eta^0(D/D^0)^n;\qquad \eta=\eta^0(D/D^0)^{n-1}.$$

In this, η^0 and D^0 are chosen for a standard shear rate and n is a measure of the deviation from Newtonian flow.

Liquids for which η decreases with D, usually because of a breaking up of particles, are called shear-thinning. When η increases, usually by structure formation by elongated particles, they are called shear-thickening. Stable liquid threads are formed in extensional flow, i.e., fibers can be spun when Trouton's viscosity increases with the flow rate, or the viscosity of the fluid increases by cooling or evaporation of solvents. Shear-thickening accompanied by volume expansion is called dilatancy. It occurs in suspensions of such high concentration of solid particles that the latter must follow curved flow lines when passing one another, and thus increase their distances. In the important field of electrorheology, structures are induced, and the viscosity thus varied, by applying electrical fields to suspensions of responsive particles.

True liquids show no lower stress limit for the initiation of flow. Solids in shear or extension above a yield stress σ_0 undergo a variety of extensive, irreversible, deformations collectively called "plastic" flow (tertiary creep). Plastic flow is not homogeneous and not viscometric. In contrast to the viscous case, plastic flow stresses have a low dependence on shear rate and external pressure. Materials, often two-phase or multiphase solids which liquify above σ_0, may exhibit true flow. If the latter is viscometric, they are called Bingham plastics:

$$\sigma-\sigma_0=\eta\dot\gamma.$$

If solids that are capable of yielding under the prevailing stresses are forced through a tube, liquefaction occurs first at the wall where σ_{xy} is highest, and thus give rise to "plug" flow. On increasing the pressure, the yield stress σ_0 will eventually reach the axis of the tube and all the material will then participate in the shear flow. Time-dependent liquefaction under shear and isothermal return from the fluid to the solid state after stopping the force constitutes thixotropy.

When modeling the fluid flow on the molecular level, one must take into account that the atoms or molecules in the fluid rest state move by irregular thermal movements to and fro, randomly changing their positions. In liquids, i.e., the dense fluids considered here, the free volume, V_f, between the molecules is only a fraction, usually 3–7%, of the total volume. The resulting packing, though random over greater distances or time intervals, shows locally quasiordered domains which fluctuate in dimension and density on a characteristic time scale (see below).

Simple or pure shearing causes the flow of materials by the superposition of a bias onto the random rest movements, i.e., causing excess directional motion by overbalancing the randomizing molecular collisions. The same thermal motion, incidentally, does not only randomize speeds and directions, but also uneven concentrations of solutes, whereby the coefficient that characterizes the motion of a given particle, the coefficient of diffusion, $\mathcal{D}$, is inversely proportional to the viscosity of the environment. According to Einstein, $\mathcal{D}=kT/6\pi\eta r$, where kT is the thermal energy of a diffusing sphere.

Because of the randomizing effects of thermal motion and diffusion, flow continuously dissipates mechanical energy. This model of flow affords a second definition of viscosity as the ratio of the work dissipated per unit volume and time to the square of the shear gradient:

$$\eta=\dot{W}/D^2.$$

The process of viscous, internal (or fluid) friction in bulk flow thus consists of a momentum transfer and consequent drag or acceleration between molecular layers of different velocities by cross-difussion of molecules. For the momentum exchange of the freely moving molecules of gaseous (compressible) fluids, one obtains from the kinetic theory of gases

$$\eta=m\bar{c}/3\sqrt{2}\pi d^2,$$

where m and d are molecular mass and diameter and $\bar{c}$ is the average speed. In dense fluids (liquids) the molecules are coupled by attractive potentials and move by diffusive steps between positions of the quasilattices. From the perturbation of the distribution function in the rest state by the velocity gradient created by the pressure tensor, Born and Green found for the viscosity of liquids

$$\eta=f(r/V_m)[m\varphi(r)^{1/2}]e^{-\varphi r/\mathrm{kT}},$$

where r is a measure of the intermolecular distances, V_m and m are the molecular volume and mass, and φ, is the intermolecular potential. On the other hand, employing the theory of chemical rate processes, Eyring *et al.* based their derivation of η on the existence of typical molecular jump frequencies with activation energies, E_{act}, over distances of the free volume to the $\frac{1}{3}$ power:

$$\eta=(V_f{}^{1/3}/V_m)(2mkT)^{1/2}e^{-E_{\mathrm{act}}/kT}.$$

The dependence on r, or $V_f{}^{1/3}$, gives the dependence of η on the external pressure. Both derivations are prototypes of many molecular theories, some of which are quite successful in predicting the viscosity of simple liquids, and even of mixtures and solutions. Empirically, one often finds the fluidity, ϕ, of mutually miscible liquids to be additive:

$$\phi=\sum w_i\phi_i$$

where the w_i are the species weight fractions and ϕ_i the individual fluidities. All derivations must reproduce the empirical temperature dependence of the viscosities, approximately represented by the Arrhenius equation:

$$\eta=Ae^{E/RT}$$

where E is an activation energy related to the heat of evaporation (mole cohesion) and A is a weakly temperature-dependent preexponential factor. A and E are more nearly constant for viscosities measured at constant volume than at constant pressure.

Molecular and, as will be seen below, also mechanical models permit rationalization of all types of deformation, utilizing the concepts of a time-dependent elasticity and of spectra of characteristic relaxation times, τ. In the ideal elastic deformation cycle, a solid system is forced to go through a continuous series of reversible changes of increasingly higher free energy (nonrelaxing states of strain, $\tau\rightarrow\infty$) and, on removal of force, returns to its former state by giving up the imparted energy as heat or work. A strain may become large enough to loosen the local order and to allow the movement of molecules and domains in plastic flow, or the order may break down by melting. The forced state of higher free energy in such deforming or flowing bodies is, as stated, one of decreased randomness of the (rest) molecular motion and/or positions, maintained against diffusion by the applied force, but instantly ($\tau\rightarrow 0$) beginning to return to randomness or new positions of lower potential energy when the force stops. The difference between the quasiordered domains or motions in the strained state and that at rest will decline, and the energy dissipate, as fast the randomizing back-diffusion (relaxation) proceeds. Depending on the extent of the imposed order and on the times of observation as compared with the rates of strain relaxation, one may see ideally fluid, plastic, or partially elastic, rate and/or time dependent, i.e., non-Newtonian or inelastic behavior, the more so the more complicated the prevailing molecular structure is. The relaxation time τ, the time after the cessation of the force required to allow the number n_0 of biased molecules to decay by a first-order process (self-diffusion) to n_0/e, is defined by

$$\phi=\sum w_i\phi_i$$

The corresponding time-dependent modulus, $G(t)$, may be established by observing the stress relaxation with time at shear strain γ, or tensile strain ϵ,

$$E(t)=\sigma_t/\epsilon_t=(\sigma_0/\epsilon_0)e^{-t/\tau},$$

$$\eta=\frac{1}{3}\int_0^\infty E(t)\,dt.$$

At constant volume, $G(t)=\frac{1}{3}E(t)$. At constant stress one may observe continuous flow (primary creep); if a constant weight is applied, the requirement of constant volume of an unconstrained, i.e., continuously extending, sample of constant modulus or viscosity reduces the cross section eventually to the point of instability and rupture.

Experimentally, a wide spread of τs, ranging from milliseconds to thousands of hours in liquids, plastics, and viscoelastic bodies, may be observed. The time-dependent shear moduli are dimensionally related to τ and η as

$$G\doteq\eta/\tau,$$

or, in tension, as $\eta=\frac{1}{3}\int E(t)\,dt$. Bodies which relax too fast within the time span of the applied force do not sustain shear or tension. However, when flow is prevented and premature cavitation protected against, tensile forces can be measured even in mobile liquids and are of the order of 10^{11} dyn/cm^2. For viscosities of, e.g., 10^{11} poses, τ (a measure of domain lifetime) is of the order of 10^{10} s, about 1000 times the period of thermal oscillations. Similarly, $\tau_B=\eta_B/K$ has been formulated for the occurrence and relaxation of volume strains (K, bulk modulus; η_B, the viscosity of volume changes). In liquids of high viscosity (small V_f or V_m, and/or large E_{act}), strained positions may decay so slowly after stress removal that "frozen strains" persist for long periods as a record of previous thermal, or strain, history. Where temporary strained states are perturbed by new forces, new strains will be superimposed. If this is additive, the material obeys Boltzmann's superposition principle; if the duration of the new force was shorter than the period for the first strain relaxation, the previous strain decay curve will be returned to, a phenomenon often called mechanical memory.

Real materials span the whole range of relaxation behavior as a function of their chemistry, morphology, or of temperature and pressure. "Solids" owe their time-independent elasticity and moduli to extensive internal order which hampers self-diffusion and extends relaxation beyond the time of observation. However, if the applied stresses exceed the yield value, σ_0, the dislocations and other imperfections move fast enough to cause "plastic" flow. "Fluids" exhibit complete relaxation and rate-dependent energy dissipation as the basic response to shearing stresses. Combinations of both modes of response, exhibiting elements of time-dependent strained states (including normal stresses), are typical for "viscoelasticity," often exhibited by two-phase, or multiphase, materials, or by homogeneous systems of molecules which are capable of time- and stress-dependent states of conformation, or of association by entanglement. Prominent among the latter are materials of crystallizable and noncrystallizable macromolecules with flexible long chains whose low symmetry and extensive entanglements require cooperative local motions to respond to external force, or to fall into a crystalline pattern during cooling and freezing. Even though their decline of V_f with falling temperatures is smaller than that of easily crystallizable materials, the need for cooperative motion causes the relaxation times to remain long. On further cooling, the mobility of the molecular segments in a macromolecular mass will become blocked below a relatively narrow volume and temperature range, the glass transition temperature, T_G, at which the free volume has typically dropped to 2.5% of the total volume. Even crystallizable molecules may interlock randomly but rigidly when cooled rapidly enough, and thus also form a "glass." Any glass transition is accompanied by substantial changes in viscosity and elastic moduli. Blends of material that do not mix on the molecular level show a multiplicity of transitions. Studies and interpretation of the infrastructure of materials in terms of their mechanical responses are the field of rheological spectroscopy.

The constitutive equations of simultaneously viscous and linear-elastic bodies in the area of validity of Boltzmann's superposition principle are expressed by

$$\sigma_{xy}(\tau) = \int_{-\infty}^{t} G(t-t')\dot{\gamma}(t')\, dt',$$

with integration from past times up to the present time, t. Alternatively, this behavior can be expressed by Maxwell's equations and modeled by elastic springs and viscous dashpots coupled in series or parallel. An alternative representation uses electrical capacitances plus resistances. The simplest model for elastico-viscous liquids with instant elastic response and subsequent flow (primary creep) is the Maxwell model described by

$$\dot{\gamma}_M = \left(\frac{1}{G}\right)\frac{d\sigma}{dt} + \sigma/\eta,$$

whereas the Voigt element represents the simplest case of viscoelastics solids

$$\sigma_{xy} = G\gamma + y\dot{\gamma}.$$

This equation shows that at stress–strain equilibrium, when $\dot{\gamma}$ goes to 0, $\sigma = G\gamma$, and γ arrives at a finite value by a time-dependent retarded elasticity ("secondary creep"):

$$\gamma_v = (\sigma/G)(1 + e^{-t/\tau}),$$

where τ is now the retardation time. This is followed by a retarded strain recovery after sudden stopping of a tensile stress application at a strain ϵ_t:

$$\epsilon = \epsilon_0 e^{-t/\tau}.$$

Most materials, even if consisting of only one component, exhibit a spread of τs attributable to various modes of motion of large molecules and to accidents of morphology (history). Similarly, inhomogeneous multiphase, multicomponent materials exhibit very broad relaxation spectra and, thus, require a combination, or set, of equations with corresponding distributions of moduli, viscosity coefficients, and τs for the description of their mechanical behavior. There are graphical and mathematical methods available to analyze creep and relaxation data in terms of creep or relaxation functions, or as spectra of retardation or relaxation times (see below). Bodies with relaxation times shorter than, or comparable to, the duration of the experiment will dissipate part of the strain energy input per volume, of $E_{v_\epsilon} = \int_0^\epsilon (\sigma/\epsilon)\, d\epsilon$, during a cycle from zero stress–strain to the final maximal strain and back to zero. To the extent that the return curve of such a cycle has a lower modulus and lies below the extension curve, forming a loop, the area of the loop measures the hysteresis, the loss of the invested work per volume as a function of temperature or strain rate. Hysteresis shows a maximum when a material is in transition between two states, especially at the glass transition, where viscosities, strain rates, and strains (amplitudes during oscillations) are intermediate.

Depending on whether solid or fluid domains form the continuous phase, dispersions may possess overall properties ranging from solidity to Newtonian flow. They include polymers, soaps, latexes, pastes, paints, filled elastomers, concrete, muds, biological systems, etc. Their flow behavior may be quite spectacular, as the climbing-up of a polymeric solution on the stem of a stirrer, or the tubeless syphon, when a fluid continues to be drawn into a pipette long after the latter was raised above its level. Such non-Newtonian behavior calls for experiments which determine more than one component of the stress tensor, methods that in contrast to the "viscometry" of Newtonian liquids belong to "rheometry."

In sufficient dilution, the contributions of dispersed particles to the energy dissipation of a flowing dispersion are additive. Einstein based his pioneer calculations on the excess dissipation of energy by the perturbation of the flow of Newtonian continua by the presence of discrete particles. He derived for the viscosity of dilute suspensions of solid spheres

$$\eta = \eta_0(1 + 2.5\varphi),$$

where η_0 is the viscosity of the pure liquid and $\varphi = v/V$ is the volume concentration of the disperse phase. This approach has been expanded to cover higher powers of φ, as well as the contribution of fluid and of nonspherical particles. For the latter, the coefficients of φ become functions of the particle axis ratios. The intrinsic viscosity is defined as the relative viscosity minus one for $\varphi \to 0$, i.e. as

$$[\eta] = (\eta/\eta_0 - 1), \ \varphi \to 0$$

and has a characteristic value for any solution. However, the coils of long random chains pervade a large volume, proportional to $N^{1/2}$, with N as the number of statistically active chain links, so that the range of viscosities linear with φ is very limited and terms of φ^2 and higher powers must be considered. Many equations have been developed for concentrated dispersions, most of which can be reduced to the form

$$\eta/\eta_0 = 1/(1 - \varphi/A)^n,$$

where A is a constant to be evaluated for each system, usually of the order of 1. For $n = 2.5$ and $A = 1$, the first term of the expansion agrees with Einstein's equation. For high values of φ (e.g., upper limit 0.74 for spheres), η goes to infinity, i.e., the dispersions will turn into pasty, usually thixotropic solids.

Macromolecular chains in solution have been modeled as strings of beads (monomeric units) of various degrees of stiffness, including random coiling. For the flow around individual beads, friction factors were adopted as, e.g., Stokes' f for spheres. The flow resistances of strung-up beads, i.e., of a coherent swarm, are summed up over their number $N = M/m$ (m = the monomer unit and M = the macromolecular weight) and for their hydrodynamic interaction to obtain a theoretical equation for the intrinsic viscosity. Interestingly, macromolecular coils may become impenetrable for flow through their interior, and then behave like fluid droplets. The most frequently used theoretical derivation for the intrinsic viscosity of such solutions is that by Flory, valid for molecular weights larger than about 50 000:

$$[\eta] = \Phi(R_0{}^2/M)^{3/2} M\alpha^3 = KM^{1/2}\alpha^3,$$

where Φ and K are constants, R_0 is the end-to-end distance of the average statistical chain multiplied by factors which account for structural features, and α is a factor depending on the molecular (coil) expansion in a given solvent.

More sophisticated theories of macromolecular behavior in solution introduce spring forces between the beads or chain sections to account for the extra energy dissipation when the coils in the shear field contract and expand with the frequency $\dot{\gamma}/2$ from their equilibrium dimensions. The corresponding average coil anisotropy, or the aperiodic rotatory orientation of rod-like molecules in competition with Brownian motion, are further cause for $[\eta]$ to vary with $\dot{\gamma}$ and for the appearance of flow birefringence, offering an opportunity to study macromolecular shapes and internal mobility. A particular model for the motion of long-chain molecules in solution, or melts, is that of a segment-by-segment proceeding snake-like diffusion in the direction of the most extended chain axis, called reptation, which has become the basis for a number of theories of chain behavior.

Long-chain molecules often help to reduce the drag of macroscopic turbulent fluid friction in pipes, or around submerged bodies, possibly because they extend over longer regions than the dimensions of the incipient fluid regions in which inertia begins to dominate. Frequently, the rheology of macromolecular dispersions and melts also deals with the effects of transient chain entanglements: for shorter, largely slipping, chains, $\eta \alpha M$, while for the entangled chains of length greater than a critical molecular weight, $\eta \infty M^{3.4}$. Here, and in the presence of additional permanent (chemically cross-linked) networks embedded in the liquid, applied forces cause elastic strain, and viscous flow becomes encumbered by friction against the net, causing a viscoelastic response. Most technological solutions, plastisols, adhesives, filled elastomers, gels, and biological systems contain some types of transient or permanent networks, often in the form of microscopic gel particles.

Much insight into the nature of viscoelastic behavior has been derived from research observing the effects of short-time, small-strain sinusoidal vibrations in materials at rest, or superimposed on transient or steady states, and studying phase relations, wave propagation, and damping. For perfectly elastic materials, the stress is in phase with the strain, there are no lattice dislocations, the rate of straining has no effect, and $\eta = \infty$. For inelastic materials (viscoelastic, or elastico-viscous), the stress is not in phase with the strain, $\gamma_{xy}(t) = \gamma^0_{xy} \sin \omega t$, where γ^0 is the maximum amplitude of the strain, ω is the frequency, and t is the time. The accompanying rate of straining is zero at the peaks and valleys of the strain-time periodic motion, and largest at $\gamma = 0$, i.e., is 90° out-of-phase with its strain:

$$\dot{\gamma}_{xy} = \omega\gamma^0_{xy} \cos \omega t.$$

Strain and strain rate evoke simultaneous elastic and viscous stresses which add up to

$$\sigma_{xy} = \gamma^0_{xy}(G' \cos \omega t + G'' \sin \omega t),$$

with G' as the elastic (energy storing) and G'' as the inelastic (energy dissipating) loss modulus. Since the stresses are shifted from the strains (the phase of the enforced oscillation) by some phase angle, θ,

$$\sigma_{xy} = \sigma^0_{xy} \sin(\omega t + \theta).$$

It follows that $G' = G^0 \cos \theta$, $G'' = G^0 \sin \theta$, and $G''/G' = \tan \theta$, the "loss tangent." $G^0 = \sigma^0_{xy}/\gamma^0_{xy}$ is the absolute modulus. G' and G'' are frequency dependent and interdependent and, since they act 90° out-of-phase, can be related by complex notation,

$$G^* = G' + iG'',$$

with G^* as the "complex" modulus. The ratio of imparted

energy converted into heat to stored energy is $\dot{W}/W = (\pi/2)$ tan θ, and changes with tan θ from 0 to 100%, corresponding to the mechanical hysteresis loop of a transient stress–strain cycle. The greater tan θ, the greater the self-heating of a material during continuous cycling. If the time for sample relaxation is longer then the stress cycle, the material will remain at some strain level while the cycling goes on, a condition which aggravates fatigue (aging) in crystalline materials, but reduces the danger of it in crystallizable elastomers.

The course of G'' and tan θ with temperature or frequency reveals much about the molecular response mechanisms of viscoelastic and thus, to some extent, of all materials. At low temperature, when V_f is small and G' large, and at high frequencies when the strain amplitudes are small because of the shortness of time for a cycle, tan θ and G'' will be small and the response largely elastic. For high temperatures, or very low frequencies, the molecular responses are largely viscous, but the energy losses, $G'' = \omega\mu$, remain small because of the low stresses and shear rates, and G' will approach 0. In transition ranges such as T_G, i.e., during the most rapid change in the character of the molecular motions, tan θ and the losses will peak. For a material with a single relaxation time, this will occur at the midpoint of the rise in G' with frequency or falling temperature, which is also the point where the imposed frequency becomes equal to $1/\tau$. Multicomponent and multiphase materials may show a characteristic sequence of multiple loss peaks whose determination by rheological spectroscopy becomes a preferred method of structural "finger printing." If inertial effects are included, the peaks of tan θ are also the areas of resonance.

The course of the storage modulus G' with temperature T and ω describes the course of behavior of viscoelastic materials. At temperatures much below T_G, the storage modulus for crystalline materials is always high and declines only slightly with temperature up to the liquefaction at the melting point T_m. Glasses show a marked drop of G' at T_G (up to a factor of 10^{-5}) depending on the chemistry and the degree of amorphous packing. Between T_G and T_m, amorphous long-chain materials exhibit a second plateau, or shoulder, which depends on the density of the real or transient network which survived T_G; if the relaxation times of these network links are sufficiently short, there will be a rapid further drop of G' at T_m to the levels of mobile elastic liquids. For long relaxation times, e.g., chemical cross-links, or in the presence of fillers, the plateau of the second modulus will be horizontal or even rising (constituting the "rubber range") up to and beyond T_m, followed by a final drop when the network breaks down at still higher temperatures. In some elastomers and plastics, the cross-links break under stress and reform in relaxed positions, whereas the individual molecular chains remain intact, thus showing truly plastic flow of a nature which has been called chemorheological.

The reciprocity of the effects of frequency (or rate of deformation) and temperature on the mechanical behavior of viscoelastic materials above the glass transition T_G (or a related reference temperature) means that changes in the molecular response can be made by equivalent rises in temperature or lowering of frequency (rate of straining). Furthermore, since major internal mobility begins at and above T_G, it should be possible to correlate various viscoelastic states by their position on a temperature or frequency scale relative to T_G or T_m. Thus, using a system of reduced variables, the behavior of all linear viscoelastic materials becomes comparable at corresponding states. According to this time–temperature superposition principle, sections of log–modulus curves of a material at constant temperature versus log frequency can be shifted by a factor a_T, derived by Williams, Landel, and Ferry, to merge into a single modulus–frequency curve over a wide range of frequencies, creating "master" curves of moduli. Such master curves permit prediction of the linear (for small amplitudes) rheological responses under most practical conditions and have become the premise of much viscoelastic engineering.

Whether during cycling or during transient force application, all viscoelastic responses become nonlinear for large strain amplitudes. Normal stresses will also appear at large strains (contributing to lubrication, see the elastic–hydrodynamic theories) and the mathematical and physical interpretations of observed behavior become much more difficult. Typical rubbers (elastomers) possess molecular networks above T_G of randomly coiled long chains which are cross-linked and do not flow, but are capable of very high (up to 1000%) reversible extensions ("high elasticity") proportional to $N^{1/2}$, the number of statistical units in the links of the network. They exhibit restoring forces that are largely proportional to the entropy decrease of the extending network: $F = -T(\partial s/\partial l)_{T,p}$. Energy dissipation in rubbers is considerable around T_G, but subsides as the usage temperature rises to 50°C, or more, above T_G. Very large strains cause appreciable volume changes through chain alignment (Poisson ratios smaller than $\frac{1}{2}$) even in shear, requiring constitutive equations which contain not only nonlinear strain or stress-rate-dependent terms, but include further the effects of the bulk modulus. Large strains tend also to render internal and/or domain organization more anisotropic, approaching the order of nematic liquid crystals.

The most frequent consequences of subjecting viscoelastic or rubbery materials to large strains are, first, a modulus or viscosity drop by the breaking of internal structures, or by a slip that causes molecular orientation, often followed by work-hardening if mobile dislocations and entanglements become locked, or when denser packing, or crystallization, are induced. Finally, large strains cause crazes, around which stress concentrations facilitate rupture. It is the microrheology of the domains in front of growing cracks, domain work-hardening, and the extent of energy dissipation with their effects on brittle versus ductile behavior, which determines the strength of high-molecular-weight materials. For elastomers, even the ultimate properties (critical stress and elongation) are amenable to the time–temperature superposition, and hence show, on a plot of stress-at-break versus strain-at-break, a characteristic failure locus called the failure envelope. A striking response of elastic liquids is found when their dwell time in a force field, say the time of passage of a batch through a piece of processing equipment, is short compared to the relaxation time (small Deborah number). The strain recovery then continues beyond the end of the operation and leads to distortion, or even fragmentation, of extended shapes (melt "swell," or "fracture"). Unrecovered strains during cold-forming will recover on heating or annealing. The rheology of pastes, doughs, of highly and multiply filled matrices, and of block polymers and composites is further complicated by their inhomogeneous nature. Beside localized energy storage and dissipation, highly nonlinear flow and elasticity of the components and interfaces must be considered and finite element analysis be employed.

See also ANELASTICITY; ELASTICITY; INTERNAL FRICTION IN CRYSTALS; RELAXATION PHENOMENA; TRANSPORT THEORY; TRIBOLOGY; VISCOSITY.

BIBLIOGRAPHY

R. B. Bird, R. C. Armstrong, O. Hassager, and C. F. Curtiss, *Dynamics of Polymer Liquids*. Wiley, New York, 1977.
R. B. Bird, W. E. Stewart, and E. N. Lightfoot, *Transport Phenomena*. Wiley, New York, 1960.
F. R. Eirich (ed.), *Rheology, Theory and Applications*. Vols. I–V. Academic Press, New York, 1956–1969.
J. D. Ferry, *Viscoelastic Properties of Polymers*, 2nd ed. Wiley, New York, 1970.
C. D. Han, *Rheology in Polymer Processing*. Academic Press, New York, 1976.
R. Tanner, *Engineering Rheology*. Clarendon Press, Oxford, 1985.
A. B. Metzner, "Rheology of Suspensions in Polymeric Liquids," *J. Rheology* **29**(6), 739–775 (1985).

Rotation and Angular Momentum

H. van Dam, J. D. Louck, and *L. C. Biedenharn*

I. IN CLASSICAL MECHANICS

A. Introduction to Rotations

Classical mechanics is based on Newton's laws of motion; a reference frame for describing motions for which Newton's laws are valid is, by definition, an *inertial reference frame*. Inertial reference frames are related to each other by a group of transformations. This group consists of *spatial translations* (displacing the reference frame parallel to itself), *time translations, rotations* (changes in orientation of the frame), and *boosts* (giving a constant velocity to the frame). This group of invariance transformations of Newtonian physics is called the *Galilei* group. Rotations form a subgroup of the Galilei group, and we begin by describing this concept geometrically.

Consider Fig. 1. From the viewpoint of inertial frames, it is the relation between two arbitrary frames (right-handed triads of perpendicular unit vectors), shown as $(\hat{e}_1,\hat{e}_2,\hat{e}_3)$ and $(\hat{f}_1,\hat{f}_2,\hat{f}_3)$, that is important. This relation is expressed by $\hat{f}_i=\Sigma_j S_{ij}\hat{e}_j$ $(j=1,2,3)$, where the direction cosines $S_{ij}=\hat{f}_i\cdot\hat{e}_j$ are the elements of a real, proper orthogonal matrix $S=(S_{ij})$. This *transformation between frames* is called a (frame) rotation. Using the geometry of the figure, we can then express each S_{ij} in terms of the *Euler angles* α,β,γ; we denote the corresponding matrix by $S(\alpha\beta\gamma)$. Each set of Euler angles thus specifies a relation between two inertial frames, and conversely. It is this viewpoint of the rotation subgroup that has been used above in the description of the Galilean group.

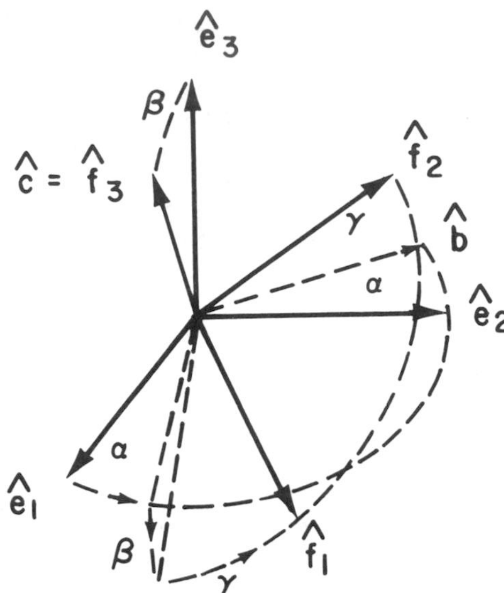

FIG. 1. The Euler-angle relation between two inertial frames $(\hat{e}_1,\hat{e}_2,\hat{e}_3)$ and $(\hat{f}_1,\hat{f}_2,\hat{f}_3)$.

In a somewhat different but related sense, a rotation $\mathcal{R}$ may be described by specifying its action on an arbitrary vector **r** as shown in Fig. 2. Here each rotation is characterized by a direction in space $\hat{n}$ (unit vector) and an angle ϕ. $\mathcal{R}$ is called a rotation by ϕ about $\hat{n}$. The relation between **r** and $\mathcal{R}$**r** is

$$\mathcal{R}\mathbf{r}=\mathbf{r}\cos\phi+(\hat{n}\times\mathbf{r})\sin\phi+\hat{n}(\hat{n}\cdot\mathbf{r})(1-\cos\phi). \quad (1)$$

Each unit vector $\hat{n}$ and each angle ϕ defines a rotation $\mathcal{R}$ (not necessarily distinct), and we often write $\mathcal{R}(\phi,\hat{n})$ for $\mathcal{R}$.

In this second description of a rotation, the frame $(\hat{f}_1,\hat{f}_2,\hat{f}_3)$ of Fig. 1 may be obtained from the frame $(\hat{e}_1,\hat{e}_2,\hat{e}_3)$ by the rotation $\hat{f}_i=\mathcal{R}(\alpha\beta\gamma)\hat{e}_i$ $(i=1,2,3)$ where $\mathcal{R}(\alpha\beta\gamma)$ is a sequence of three rotations (read from right to left) given by

$$\mathcal{R}(\alpha\beta\gamma)=\mathcal{R}(\gamma,\hat{c})\mathcal{R}(\beta,\hat{b})\mathcal{R}(\alpha,\hat{e}_3)=\mathcal{R}(\alpha,\hat{e}_3)\mathcal{R}(\beta,\hat{e}_2)\mathcal{R}(\gamma,\hat{e}_3) \quad (2)$$

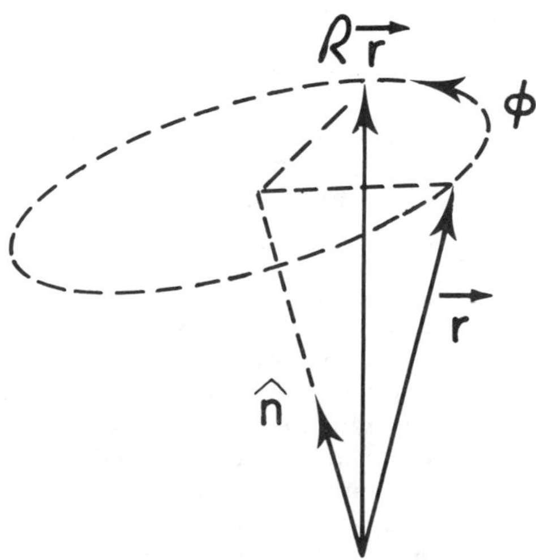

FIG. 2. Geometrical interpretation of the action of a rotation $\mathcal{R}(\phi,\hat{n})$ on an arbitrary vector **r**.

where $\hat{b}=\sin\alpha\hat{e}_1+\cos\alpha\hat{e}_2$, $\hat{c}=\cos\alpha\ \sin\beta\hat{e}_1+\sin\alpha\ \sin\beta\hat{e}_2+\cos\beta\hat{e}_3$. The matrix $S(\alpha\beta\gamma)$ is then related to the rotation $\mathcal{R}(\alpha\beta\gamma)$ by $S_{ij}(\alpha\beta\gamma)=[\mathcal{R}(\alpha\beta\gamma)\hat{e}_i]\cdot\hat{e}_j$.

A′. Digression on "turns"

For linear momentum the essential object is the displacement vector. Is there an elementary object for rotations which corresponds to the displacement vector? Such an object is the "turn" represented by a directed arc (length $\phi/2$) of a great circle on a unit sphere. The normal to the plane of the great circle is the axis of the rotation generated by the turn and the rotation angle is ϕ, with the sense of the rotation being given by the direction of the arc. Note the factor $\frac{1}{2}$ which is necessary to get the commutation relations (see below) uniformly right! Two turns are considered to be the same if by translation along their common great circle they may be brought into coincidence. Note also that a given rotation may be represented by two distinct turns, as a result of the peculiar fact that there are *two* distinct turns (0 and π) equivalent to *no* rotation (the double covering of the rotation group).

To find the commutation rules, one examines the analog of the parellelogram law. Turns are composed by placing them head to tail where their great circles intersect; it is clear that $AB\neq BA$. As the separate turns A and B get smaller, however, the turns AB and BA approach parallel vectors, which, on a sphere, implies that the turns AB and BA intersect as far away as possible ($\pi/2$). A diagram will readily convince one that this leads to the law $\mathbf{J}\times\mathbf{J}=i\mathbf{J}$ for the generators of infinitesimal rotations.

This simple and intuitive "explanation" of the commutation relations for angular momentum is rather more fruitful than it might appear. For example, it shows rather clearly that for a sphere of very large radius the infinitesimal turns approach infinitesimal displacement vectors on the appropriate tangent plane to the sphere. Also, using the calculus of geodesic triangles on the surface of a sphere, one has an explicit way of calculating the parameters (turn) which corresponds to the product of two successive rotations. This method is superior to the usual approach.

B. Connection between Invariance for Rotations and Conservation of Angular Momentum

If we limit ourselves to those Newtonian systems that can be obtained from Lagrange's equations (this restricts the forces to be derivable from potentials), then there is a close connection between invariance transformations and conservation laws. This connection was first noted by C. G. J. Jacobi, but is now commonly called the Noether theorem. To be precise, energy is conserved if Lagrange's equations are invariant for time translations, momentum is conserved if Lagrange's equations are invariant for space translations, and finally angular momentum is conserved if Lagrange's equations are invariant for rotations. Moreover, the quantities that are to be conserved are themselves defined by these same invariances. This beautiful connection between conservation laws and invariances is not valid for Newton's equations themselves. [This point (due to E. P. Wigner) is not emphasized in the literature; further discussion, and references, may be found in the article by Houtappel *et al.* cited in the bibliography.] Even for rotationally invariant equations, angular momentum is not necessarily conserved unless we assume that the forces can be derived from potentials that involve only the position coordinates, or unless we assume "central forces," as we shall illustrate in Sec. D.

In introducing the special theory of relativity, Einstein noted (from Maxwell's equations) that the group of invariance transformations had to be changed from the Galilei group to what is now called the Poincaré group (inhomogeneous Lorentz group). If we assume an action principle (i.e., the validity of Lagrange's equations), then once again we have the validity of Noether's theorem, giving both the conservation laws and the conserved quantities: linear momentum, energy, angular momentum. Since a different invariance group is involved, the conserved quantities differ from the Newtonian (or Galilean) ones.

C. Noninertial Frames, Centrifugal Force, Coriolis Force, Mach's Objections

If $(\hat{e}_1,\hat{e}_2,\hat{e}_3)$ is an inertial frame and $\mathcal{R}(t)$ is a rotation depending on time, then the frame $\hat{f}_i=\mathcal{R}(t)\hat{e}_i$ $(i=1,2,3)$ is a *noninertial frame*. The laws in such a frame must be found by transforming Newton's laws from the original (inertial) frame. For instance, if we assume that the Euler angles α and β are constant in time, but that γ is proportional to time, $\gamma=\omega t$, then in the primed frame Newton's laws are modified by the appearance of centrifugal and Coriolis forces. The forces $\mathbf{F}_{\text{rot}}$ on a particle m at position $\mathbf{r}$ in the rotating frame are given by adding the Coriolis and centrifugal forces to the forces $\mathbf{F}$ in the inertial frame: $\mathbf{F}_{\text{rot}}=\mathbf{F}-2m(\mathbf{r}\times\mathbf{v})-m\boldsymbol{\omega}\times(\boldsymbol{\omega}\times\mathbf{r})$, where $\boldsymbol{\omega}=\mathcal{R}(\alpha,\beta,\omega t)\hat{e}_3$ in the case described.

These additional forces appear because the f frame is rotating with respect to absolute Euclidean space. Many, including Leibniz and Mach, objected to the presence of this "absolute space" in Newtonian mechanics. They argued that the concept "rotation with respect to absolute space" must be replaced by, say, "rotation with respect to the distant stars." The special theory of relativity does not meet this objection and even the general theory does not meet it. In the latter theory, space-time has become an object that is now to be influenced itself by the objects (masses) in it. However, space-time acts back on the bodies in it, and it is still meaningful to talk about a (local) reference frame rotating with respect to this space-time. Thus, space-time is still absolute in general relativity, and does not meet the objection stated above.

D. Conservation of Angular Momentum for Central Forces

Let us consider Newton's laws for the special case in which the primitive elements are point masses and in which the forces between these point masses are pairwise equal and opposite, and directed along the line connecting the points of each pair ("central forces"). The linear momentum of particle i is given by the vector $\mathbf{p}_i = m_i\mathbf{V}_i = m_i d\mathbf{X}_i/dt$. For an isolated system of N such particles, the force on particle i is then $d(m_i\mathbf{V}_i)/dt = \Sigma_{j(j\neq i)}\mathbf{F}_{ij}$ with $\mathbf{F}_{ij} = -\mathbf{F}_{ji}$; $\mathbf{F}_{ij} = f_{ij}(\mathbf{X}_i - \mathbf{X}_j)$, $i = 1,2,\ldots,N$. This force law can be put in words in the following way: During every infinitesimal increment in time dt, an infinitesimal amount of linear momentum is exchanged between each pair of particles i and j; i receives $dt\ \mathbf{F}_{ij}$ and j receives $dt\ \mathbf{F}_{ji} = -dt\ \mathbf{F}_{ij}$. In this way it is obvious that the total linear momentum is conserved, since all that happens is an exchange of linear momentum between particles making up the system:

$$\frac{d}{dt}\left(\sum_{i=1}^{N} m_i\mathbf{V}_i\right) = \frac{d}{dt}(\mathbf{P}) = \mathbf{0}. \tag{3}$$

The angular momentum ℓ_i of particle i with respect to the origin of the inertial frame is defined by $\ell_i = \mathbf{X}_i \times m_i\mathbf{V}_i$. We conclude that the total angular momentum is also conserved for an isolated system of N particles: during dt an amount of linear momentum $dt\ \mathbf{F}_{ij}$ leaves i and arrives on j, and during dt an amount of angular momentum $\mathbf{X}_i \times dt\ \mathbf{F}_{ij}$ leaves i and an amount $\mathbf{X}_j \times dt\ \mathbf{F}_{ij}$ arrives on j. Since $\mathbf{F}_{ij}$ is parallel to $(\mathbf{X}_i - \mathbf{X}_j)$, we have that $\mathbf{X}_i \times dt\ f_{ij}(\mathbf{X}_i - \mathbf{X}_j)$ leaves i and $\mathbf{X}_j \times f_{ij}(\mathbf{X}_i - \mathbf{X}_j)$ arrives on j. The difference between these quantities is $(\mathbf{X}_i - \mathbf{X}_j) \times dt\ f_{ij}(\mathbf{X}_i - \mathbf{X}_j) = \mathbf{0}$. In other words, nothing is lost, and we have a conservation law for total angular momentum:

$$\frac{d\mathbf{L}}{dt} = \mathbf{0}, \quad \mathbf{L} = \sum_i \ell_i. \tag{4}$$

E. Rotation of Solid Bodies

The rotational motion of a solid body having a fixed point O may be described with the aid of a rotating frame $\hat{f}_i = \mathcal{R}(\phi,\hat{n})\hat{e}_i$ $(i = 1,2,3)$, where $(\hat{e}_i,\hat{e}_2,\hat{e}_3)$ is an inertial frame and ϕ and $\hat{n}$ are functions of time. In this case we fix the frame in the body. The angular velocity $\boldsymbol{\omega} = \Sigma_i\ \omega_i\hat{f}_i$ is then related to the motion of the frame by $d\hat{f}_i/dt = \boldsymbol{\omega} \times \hat{f}_i$ so that $(d\hat{f}_i/dt)\cdot\hat{f}_j = \epsilon_{ijk}\omega_k$. In terms of the angle ϕ and the components $n_i = \hat{e}_i\cdot\hat{n}$, we find $\omega_i = n_i(d\phi/dt) + (d\hat{n}/dt)\cdot[\hat{e}_i \sin\phi + (\hat{n}\times\hat{e}_i)(1-\cos\phi)]$. The angular momentum $\mathbf{M}$ and the kinetic energy T may then be written out with the aid of the inertial tensor (dyadic) $\boldsymbol{I} = \Sigma_{ij}\ \boldsymbol{I}_{ij}\hat{f}_i\hat{f}_j$, where $\boldsymbol{I}_{ii}$ is the moment of inertia about axis f_i and $-I_{ij}$ is the product of inertia for axes $\hat{f}_i$ and $\hat{f}_j$: $\mathbf{M} = \boldsymbol{I}\cdot\boldsymbol{\omega}$ and $T = (\boldsymbol{\omega}\cdot\boldsymbol{I}\cdot\boldsymbol{\omega})/2$.

II. IN QUANTUM THEORY

A. Increased Importance of Invariances in Quantum Theory

In quantum mechanics rotational invariance and angular momentum conservation are far more important concepts than in classical (Newtonian) mechanics. There are two reasons for this: First, Bohr's quantization (1913) was itself a quantization of orbital angular momentum (Planck's constant h is dimensionally an angular momentum); this structure is subsumed under quantum mechanics. The second, and much deeper, reason is that for quantum mechanics, because of the superposition postulate, the set of allowed states has the linear structure of Hilbert space. In classical theory rotational invariance leads to the fact that for every state of the system there exists a rotated state that has similar properties. Unless the state is invariant for rotations, every rotation leads to a new state, and in this way there is an infinite number of states, all of which have similar properties. Similarly, in quantum theory every rotation leads to a new state, and again there is an infinite number of such new states. However, these states form a linear manifold, and this infinity of allowed states can be expressed linearly in terms of basis states, a much smaller subset of states, often finite in number. It is this structure which leads to the far-reaching applicability of the theory of linear representations of groups to quantum mechanics.

B. Unitary Representations of SU(2) Appear

In quantum mechanics only the absolute values of the inner products between pairs of states in Hilbert space are observable. Accordingly, a state vector ψ is physically equivalent to the state vector $e^{i\phi}\psi$; the physical states are therefore *equivalence classes* or rays in a *projective space*. We are thereby necessarily led to projective representations of the rotation group. Wigner has shown that these projective representations of the rotation group are unitary representations of the group SU(2) of 2×2 unitary unimodular matrices. For this reason, the group SU(2) must be admitted into the study of rotationally invariant quantum mechanics. It is through this mathematical structure that the quantum concept of *spin* of a point particle makes its appearance. In quantum theory a particle can carry orbital angular momentum (the analog of the classical quantity $\mathbf{r}\times\mathbf{p}$) and an intrinsic spin angular momentum.

The relation between the group of rotations and SU(2) may be made by associating a 2×2 Hermitian matrix $X = x_1\sigma_1 + x_2\sigma_2 + x_3\sigma_3 = \mathbf{x}\cdot\boldsymbol{\sigma}$ with each point (x_1,x_2,x_3) of Euclidean three-space. The σ_i denote the Pauli matrices, which are a basis for all 2×2 traceless, Hermitian matrices:

$$\sigma_1 = \begin{pmatrix} 0 & 1 \\ 1 & 0 \end{pmatrix}, \quad \sigma_2 = \begin{pmatrix} 0 & -i \\ i & 0 \end{pmatrix}, \quad \sigma_3 = \begin{pmatrix} 1 & 0 \\ 0 & -1 \end{pmatrix}. \tag{5}$$

Each 2×2 unitary matrix similarity transformation transforms a traceless Hermitian matrix into a new traceless Hermitian matrix; that is, the transformation $\mathbf{x}'\cdot\boldsymbol{\sigma} = U(\mathbf{x}\cdot\boldsymbol{\sigma})U^{\dagger}$ defines a mapping $(x_1,x_2,x_3)\rightarrow(x_1',x_2',x_3')$ of Euclidean space onto itself.

We readily work out that this mapping is $x\rightarrow x' = Rx$ (column matrix notation), where R is the real, proper orthogonal matrix with elements $R_{ij} = \frac{1}{2}\mathrm{tr}(\sigma_i U\sigma_j U^{\dagger})$. This mapping of elements of SU(2) onto elements of SO(3) is two-to-one: both U and $-U$ map to the same R. A very useful result coming from this relation is the two-to-one mapping $\{e^{-i\phi\hat{n}\cdot\sigma}, -e^{-i\phi\hat{n}\cdot\sigma}\}\rightarrow R(\phi,\hat{n})$, where $R(\phi,\hat{n})$ may be interpreted either as the matrix representing $\mathcal{R}(\phi,\hat{n})$ on the basis $(\hat{e}_1,\hat{e}_2,\hat{e}_3)$ or as the matrix expressing the relation between two frames $\hat{f}_i = \Sigma_j R_{ij}(\phi,\hat{n})\hat{e}_j$.

While the foregoing relations between SU(2) and SO(3) may be used to unify the treatment of angular momentum, it is still useful to consider the quantum theory of orbital angular momentum on its own.

Let us consider the Schrödinger equation $H\psi(\mathbf{r},t)=i\hbar\,\partial\psi(\mathbf{r},t)/\partial t$, where H is the Hamiltonian operator, obtained from the classical Hamiltonian by replacing $\mathbf{p}$ by $-i\hbar\nabla$. Making the same replacement in $\mathbf{L}=\mathbf{r}\times\mathbf{p}$, we obtain the orbital angular momentum operator:

$$\mathbf{L}=-i\hbar\mathbf{r}\times\nabla \tag{6}$$

Apart from the extra factor $\hbar$, the operator $\hat{n}\cdot\mathbf{L}$ is just the generator of a rotation about $\hat{n}$ if applied to functions $\psi(\mathbf{r})$ (see below). As such, the components $L_i=\hat{e}_i\cdot\mathbf{L}$ of $\mathbf{L}$ relative to an inertial frame have the commutation relations summarized by the vector equation

$$\mathbf{L}/\hbar\times\mathbf{L}/\hbar=i\mathbf{L}/\hbar. \tag{7}$$

These commutation relations are characteristic of rotation generators.

The relation of the operators (6) to rotations is most easily comprehended by applying the linear operator

$$T(\phi,\hat{n})=\exp(-i\phi\hat{n}\cdot\mathbf{L}) \tag{8}$$

to a function $\psi(x)=\psi(x_1,x_2,x_3)$ ($\mathbf{r}=\Sigma_i\, x_i\hat{e}_i$). The result is (for a suitable class of functions)

$$\psi'(x)=T(\phi,\hat{n})\psi(x)=\psi(x') \tag{9}$$

where $x'=R^{-1}(\phi,\hat{n})x$.

The Euler-angle rotation $\mathcal{R}(\alpha\beta\gamma)$ given by Eq. (2) is represented on the space of wave functions by a sequence of three operators of the type (8) corresponding to the rotations appearing in Eq. (2):

$$\begin{aligned}T(\alpha\beta\gamma)&=T(\alpha,\hat{e}_3)T(\beta,\hat{e}_2)T(\gamma,\hat{e}_3)\\&=e^{-i\alpha(L_3/\hbar)}e^{-i\beta(L_2/\hbar)}e^{-i\gamma(L_3/\hbar)}\end{aligned} \tag{10}$$

where the action of this operator on $\psi(x)$ is determined by Eq. (9).

The operator $(\hat{n}\cdot\mathbf{L})/\hbar$ is called the generator of a rotation about $\hat{n}$. Three independent generators are $L_i/\hbar=(\hat{e}_i\cdot\mathbf{L})/\hbar$ ($i=1,2,3$). If the Schrödinger equation has spherical symmetry around some origin, then for every solution ψ of the Schrödinger equation, ψ' [obtained by (9)] must also be a solution for all $(\phi,\hat{n})$. This implies that $\mathbf{L}$ commutes with the Hamiltonian. Hence, since $\mathbf{L}$ does not involve the time explicitly, it is a conserved quantity. Here we see the close connection between rotational invariance and the conservation laws of angular momentum. From the invariance of the Schrödinger equation for rotations, we derive that the independent generators of these rotations are conserved quantities; they are (apart from a constant $\hbar$) just the angular momenta. Noting that the invariance itself defines the conserved quantities, we see that the proper expressions for the angular momenta could be derived without the necessity of using $\mathbf{p}\rightarrow-i\hbar\nabla$.

The transformation given by Eq. (9) actually defines a unitary representation of the rotation group, but it is a very complicated representation composed in general of many irreducible parts (which play the role of elementary systems for angular momentum). Although deep theorems (due to Weyl) guarantee complete reducibility into denumerably many parts, it is customary in physics to proceed differently and (following Heisenberg, Born, and Jordan) construct directly the irreducible systems for angular momentum.

This method consists of postulating the commutation relations (setting $\hbar=1$ for convenience)

$$\mathbf{J}\times\mathbf{J}=i\mathbf{J} \tag{11}$$

as the fundamental definition of angular momentum [$\mathbf{L}=-i\mathbf{r}\times\nabla$ then appears as a special realization of operators obeying (11)]. The idea is to find all possible realizations of the commutation relations (11) subject only to the conditions that the J_i ($i=1,2,3$) be linear Hermitian operators on a Hilbert space.

To solve this problem we notice that $J^2=J_1^2+J_2^2+J_3^2$ commutes with each J_i. This signifies that J^2 together with, say, J_3 may be simultaneously diagonalized. With this observation, we now state the unique (up to equivalence) solution of the problem posed above: On a Hilbert space on which the J_i are Hermitian operators, the only eigenvalues of J^2 that can occur are of the form $j(j+1)$, where j belongs to the set $\{0,\frac{1}{2},1,\frac{3}{2},2,\ldots\}$; for each j that occurs there exist $2j+1$ orthonormal eigenkets of J_3 having eigenvalues $m=j,j-1,\ldots,-j$. Furthermore, if we denote the orthonormal eigenkets by $|jm\rangle$, then the orthonormality property is expressed by $\langle j'm'|jm\rangle=\delta_{j'j}\delta_{m'm}$, and the action of the J_i on this basis is expressed in terms of the operators $J_\pm=J_1\pm iJ_2$ by

$$J_\pm|jm\rangle=[(j\mp m)(j\pm m+1)]^{1/2}|jm\pm1\rangle, \tag{12a}$$

$$J_3|jm\rangle=m|jm\rangle. \tag{12b}$$

In Sec. B above, we noted that rotationally invariant quantum mechanics leads to unitary representations of SU(2). This means that we must consider the representations of the J_i for the half-integer values of j. For example, to give the electron spin $\frac{1}{2}$, we need to modify the Schrödinger equation as was first done by Pauli in a straightforward way, and in a much deeper, relativistic way by Dirac. The total angular momentum for the general case consists of an orbital angular momentum $\mathbf{L}$ and a spin $\mathbf{S}$, the total angular momentum $\mathbf{J}$ being given by $\mathbf{J}=\mathbf{L}+\mathbf{S}$. The generator of rotations for this structure is $\mathbf{J}$, which satisfies the commutation relations (20). Note that $[\mathbf{L},\mathbf{S}]=0$ and $[\mathbf{P},\mathbf{S}]=0$ so that the spin $\mathbf{S}$ may be characterized as intrinsic.

C. Explicit Irreducible Representations of SU(2)

We now present the inequivalent irreducible unitary representations of SU(2) in several useful forms by giving explicitly the elements of the unitary matrices representing $U\in$ SU(2).

We have previously noted that each unitary unimodular matrix may be written in the form $U(\theta,\hat{n})=\exp(-i\theta\hat{n}\cdot\boldsymbol{\sigma}/2)=\sigma_0\cos(\beta/2)-i\hat{n}\cdot\boldsymbol{\sigma}\sin(\beta/2)$ (σ_0 = identity), $0\leqslant\theta\leqslant2\pi$, $\hat{n}\cdot\hat{n}=1$. The elements of the unitary matrix representing $U(\theta,\hat{n})$ are

$$D^j_{m'm}(\theta,\hat{n})=\langle jm'|e^{-i\theta\hat{n}\cdot J}|jm\rangle, \tag{13}$$

although this form is not yet very explicit, but will be made so below.

In terms of the Euler-angle parametrization of SU(2), we have $U(\alpha\beta\gamma) = \exp(-i\alpha\sigma_3/2)\exp(-i\beta\sigma_2/2)\exp(-i\gamma\sigma_3/2)$, where the domain of definition of the Euler angles is $0\leqslant\alpha\leqslant 2\pi$, $0\leqslant\beta\leqslant\pi$, $2\pi\leqslant\beta\leqslant 3\pi$, $0\leqslant\gamma\leqslant 2\pi$. The elements of the unitary matrix representing $U(\alpha\beta\gamma)$ are

$$D^j_{m'm}(\alpha\beta\gamma) = e^{-im'\alpha} D^j_{m'm}(\beta,\hat{e}_2) e^{-im\gamma}. \tag{14}$$

The functions (13) are given explicitly by

$$D^j_{m'm}(U) = [(j+m')!(j-m')!(j+m)!(j-m)!]^{1/2}$$
$$\times \sum_s \frac{(u_{11})^s (u_{12})^{j+m'-s} (u_{21})^{j+m-s} (u_{22})^{s-m'-m}}{s!(j+m'-s)!(j+m-s)!(s-m'-m)!}, \tag{15}$$

where the u_{ij} are the elements of the matrix $U(\theta,\hat{n})$. In particular, the middle factor in Eq. (14) is obtained by putting $u_{11}=u_{22}=\cos(\beta/2)$, $u_{21}=-u_{12}=\sin(\beta/2)$ in Eq. (15).

D. Wigner Coefficients

The discussion above concerns the properties of a single angular momentum operator **J**. If the physical system is composed of two or more particles, or if we consider the angular momentum of a single particle as composed of two separate parts, for example, orbital and intrinsic spin, then the problem of compounding the different angular momenta arises.

Two angular momenta $\mathbf{J}_1,\mathbf{J}_2$ that can be described independently at any instant of time are called *kinematically independent* momenta. Quantum mechanically, kinematic independence is expressed by the statement

$$[\mathbf{J}_1,\mathbf{J}_2]=0 \qquad \text{(meaning componentwise commutation)}$$

where the expression $[\mathbf{J}_1,\mathbf{J}_2]$ is the commutator of the operators $\mathbf{J}_1$ and $\mathbf{J}_2$. The angular momentum $\mathbf{J}\equiv\mathbf{J}_1+\mathbf{J}_2$ of a system composed of two kinematically independent angular momenta $\mathbf{J}_1$ and $\mathbf{J}_2$ will have eigenfunctions $\psi_{j_1 j_2 : jm}$ corresponding to the operators J_1^2, J_2^2, J^2, J_z (we now use x,y,z to denote components), that are expressible in terms of the eigenfunctions $\psi_{j_1m_1}$ and $\psi_{j_2m_2}$, corresponding to the operators J_1^2, J_{1z} and J_2^2, J_{2z}, respectively. The $\psi_{j_1 j_2 : jm}$ are linear combinations of $\psi_{j_1m_1}$ and $\psi_{j_2m_2}$ of the form

$$\psi_{j_1 j_2 : jm} = \sum_{m_1m_2} C^{j_1 j_2 j}_{m_1 m_2 m} \psi_{j_1m_1}\psi_{j_2m_2} \tag{16a}$$

where the coefficients are zero unless the quantum numbers j_1, j_2, j, m_1, m_2, m satisfy

$$m=m_1+m_2 \quad\text{and}\quad j\in\{j_1+j_2, j_1+j_2-1,\ldots,|j_1-j_2|\}. \tag{16b}$$

The coefficients $C^{j_1 j_2 j}_{m_1 m_2 m}$ that characterize the addition, or coupling, of two angular momenta are called (inaccurately) Clebsch–Gordan coefficients or (more properly) Wigner coefficients.

In addition to being coupling coefficients, the coefficients $C^{j_1 j_2 j}_{m_1 m_2 m}$ play a second fundamental role in the quantum theory of angular momentum. An irreducible tensor operator **T** under SU(2) is defined as a family of operators $\{T(JM)\}$, with $-J\leqslant M\leqslant J$, that satisfy the commutation rules

$$[J_\pm, T(JM)] = [(J\mp M)(J\pm M+1)]^{1/2} T(JM\pm 1),$$
$$[J_z, T(JM)] = MT(JM). \tag{17}$$

Each of the operators $T(JM)$ has a matrix representation whose matrix components are denoted by $\langle j'm'|T(JM)|jm\rangle$. The statement of the fundamental Wigner–Eckart theorem is then given by

$$\langle j'm'|T(JM)|jm\rangle = C^{jJj'}_{mMm'}\langle j'\|T(J)\|j\rangle. \tag{18}$$

Here $\langle j'\|T(J)\|j\rangle$ is a number called the *reduced matrix element* of the operator $T(JM)$. [In Eq. (18) all quantum numbers that refer to characteristics of a physical system other than angular momenta have been suppressed.]

According to this theorem, the matrices of a tensor operator **T** can be factored into two parts, the first of which is the Wigner coefficient (which can be calculated entirely from symmetry considerations), and the second of which is a matrix that contains specific information about the physical properties of the particular operator, and that is completely independent of the quantum numbers m, M, m' (hence, the second term is orientation independent).

The conditions stated by Eq. (16b) contain the conservation laws of angular momentum; accordingly, the Wigner coefficients contain the selection rules that govern atomic spectra. That almost all the rules of spectroscopy follow from the symmetry of the problem is perhaps the most remarkable result of the quantum theory of angular momentum.

E. Generalizations

If, instead of two angular momenta, three angular momenta $\mathbf{J}_1$, $\mathbf{J}_2$, $\mathbf{J}_3$ are added to form the momentum **J**, then the coupling of the three is not unique. For example, we might first add $\mathbf{J}_1$ and $\mathbf{J}_2$ to form an intermediate momentum $\mathbf{J}_{12}$ and then form $\mathbf{J}=\mathbf{J}_{12}+\mathbf{J}_3$. Alternatively, we might form $\mathbf{J}_{23}=\mathbf{J}_2+\mathbf{J}_3$ and then $\mathbf{J}=\mathbf{J}_1+\mathbf{J}_{23}$. The eigenfunctions of the first coupling scheme will be different from the eigenfunctions of the second, but the first may be transformed into the second by means of a unitary transformation. The matrices of this transformation are denoted by

$$[(2j_{12}+1)(2j_{23}+1)]^{1/2} W(j_1 j_2 j j_3; j_{12} j_{23})$$

where $W(j_1 j_2 j j_3; j_{12} j_{23})$ are the *Racah coefficients*. The 6-j symbols are related to the Racah coefficients by

$$\begin{Bmatrix} j_1 & j_2 & j_3 \\ l_1 & l_2 & l_3 \end{Bmatrix} = (-1)^{j_1+j_2+l_1+l_2} W(j_1 j_2 l_2 l_1; j_3 l_3).$$

Closely related to these coefficients are the $\bar{Z}$ coefficients, which are defined by

$$\bar{Z}(a,b,c,d;e,f)$$
$$= [(2a+1)(2b+1)(2c+1)(2d+1)]^{1/2} W(abcd;ef) C^{acf}_{000}.$$

These coefficients are important in applications of physical interest.

Coupling schemes involving arbitrarily many momenta may be constructed. An important case involves four momenta, and the transformation coefficients are called the 9-j symbols.

CONCLUDING REMARK

Recently there has been considerable discussion of particles with *fractional* angular momenta. Such particles cannot exist in three-dimensional space where the commutation relations, Eq. (11), rigorously imply that angular momenta are restricted to integers or half-integers only (the double-covering of the rotation group (see Section A′) is responsible for the half-integers). In two spatial dimensions (where there is but a single rotation operator), however, one *can* have an arbitrary (real) eigenvalue for the angular momentum. (This reflects the fact that the covering group of rotations in two-space is infinitely cyclic.) Some experimental situations (high-temperature superconductivity, fractional quantized Hall effect) are believed to be effectively two dimensional, hence allowing fractional angular momenta. This can only be approximate and not rigorously true.

See also ANGULAR CORRELATION OF NUCLEAR RADIATION; ATOMIC SPECTROSCOPY; BOHR THEORY; CONSERVATION LAWS; CORIOLIS ACCELERATION; DYNAMICS, ANALYTICAL; GROUP THEORY IN PHYSICS; INVARIANCE PRINCIPLES; MOMENT OF INERTIA; MOMENTUM; QUANTUM MECHANICS.

BIBLIOGRAPHY

L. C. Biedenharn and H. van Dam (eds.), *Selected Papers on the Quantum Theory of Angular Momentum,* Academic, New York, 1965.

L. C. Biedenharn and J. D. Louck, *Angular Momentum in Quantum Physics, Encyclopedia of Mathematics and its Applications,* Vol. 8, Addison-Wesley, Reading, MA, 1981.

V. Bargmann, "On the Representation of the Rotation Group," *Rev. Mod. Phys.* **34,** 829 (1962).

L. C. Biedenharn, J. M. Blatt, and M. E. Rose, "Some Properties of the Racah and Associated Coefficients," *Rev. Mod. Phys.* **24,** 249 (1952).

M. Born and P. Jordan, *Elementare Quartenmechanik,* Vol. 1x, Springer, Berlin, 1930.

E. Cartan, *The Theory of Spinors,* MIT Press, Cambridge, 1966.

A. R. Edmonds, *Angular Momentum in Quantum Mechanics,* Princeton University Press, Princeton, NJ, 1957.

U. Fano and G. Racah, *Irreducible Tensorial Sets,* Academic, New York, 1959.

R. M. F. Houtappel, H. van Dam, and E. P. Wigner, "The Conceptual Basis and Use of the Geometric Invariance Principles," *Rev. Mod. Phys.* **37,** 595 (1965).

A. F. Nikiforov, V. B. Uvarov, and Y. L. Levitan (eds.), *Tables of Racah Coefficients,* Macmillan, New York, 1965.

G. Racah, "Theory of Complex Spectra II," *Phys. Rev.* **62,** 438 (1942).

T. Regge, "Symmetry Properties of Clebsch-Gordan Coefficients," *Nuovo Cimento* **10,** 544 (1958).

M. E. Rose, *Elementary Theory of Angular Momentum,* Wiley, New York, 1957.

M. Rotenberg *et al., The 3-j and 6-j Symbols,* Technology Press, Cambridge, 1959.

T. Ischidzu *et al., Tables of Racah Coefficients,* Pan-Pacific Press, Tokyo, 1960.

E. P. Wigner, *Group Theory and Its Applications to Atomic Spectra,* Academic, New York, 1959.

S-Matrix Theory

Geoffrey F. Chew

The most firmly established aspects of quantum theory are expressible through the concept of the scattering matrix [1]—*S* matrix for short—identified by J. Wheeler in 1937 and first illuminated in depth by W. Heisenberg in 1943. Recognizing all physical observations to be on the macroscopic scale, and eschewing reference to unobservable microscopic variables, *S*-matrix theory steers clear of mathematical and conceptual difficulties that have plagued quantum equations of motion for microscopic degrees of freedom. The degrees-of-freedom concept—traditional fabric for models of nature—is absent from *S*-matrix theory and, lacking dynamical variables, there need be no equation of motion. The power of *S*-matrix theory stems from the commonly perceived causal relationship of (large-scale) events in relativistic space-time, in combination with the quantum principle of superposition. The full extent of this power is not understood—there having arisen a variety of conjectures that will be described below.

Common to all forms of quantum theory is the feature that experimental information, no matter how precise and exhaustive, does not allow prediction of the result of a subsequent experiment but only a statement of the *probability* that a certain result will be found [1]. Each "complete" experiment* yields a result that may be associated with a "wave function" and the probability P_{BA} of obtaining a result associated with wave function B, given that a "prior" experiment has yielded a result associated with wave function A,† may be written in Dirac (bra-ket) notation [2] as

$$P_{BA} = |\langle B|S|A\rangle|^2,$$

where S is a linear operator in the Hilbert space of all those (macroscopic) wave functions that correspond to results from conceivable experiments. The matrix elements of this operator—collectively encompassing all possible experimental knowledge—constitute the S matrix. Physical measurements being necessarily macroscopic, the S matrix is defined only in the "large," in contrast to many of the operators that arise in less general forms of quantum theory [3].

Any linear superposition of macroscopic wave functions is supposed to correspond to a conceivable experimental result, while the probabilities of different "final" results following any arbitrary "initial" result must sum up to unity. Taken together these two considerations imply that the S matrix is *unitary*, a nonlinear constraint.

An especially useful set of macroscopic wave functions is associated with the concept of a freely moving particle, which corresponds in "standard" quantum theory to a nondegenerate stationary state [4]. Each particle is characterized by mass and spin, i.e., by energy and angular momentum in its rest frame of reference. No distinction is made between "elementary" and "composite" particles, electrons, nuclei, atoms, and molecules all sharing the same role, although it should be remembered that the *classical* notion of particle (e.g., a baseball) corresponds to a coherent superposition of many nearly degenerate stationary states. An experimental result may be expressed in terms of the number and types of particles detected, as well as by their positions and momenta. All this information resides in the associated wave function whose space-time content is connected to its energy–momentum content by Fourier transformation.‡ It is in this transformation that the Planck constant $\hbar$ makes its appearance in *S*-matrix theory—as a factor needed for dimensional consistency.

The homogeneity of space-time implies that a common displacement of both initial and final experiments, either in space or in time, should lead to no change in P_{BA}. When expressed in the particle momentum basis this requirement translates into the vanishing of S_{BA} unless $p_A = p_B$ and $E_A = E_B$, where $p_{A,B}$ and $E_{A,B}$ are the total momenta and energies of initial and final states. That is, energy and momentum are conserved. According to the principle of Lorentz invariance, furthermore, the value of S_{BA} is unchanged if a common shift of inertial frame of reference is applied both to experiment A and to experiment B. These two properties together reflect the *Poincaré invariance* of the S matrix [5]. Other symmetries of nature, whether exact or approximate, are reflected by corresponding *S*-matrix invariance properties.

* In practice experiments are inevitably incomplete, yielding less than the maximum possible information, so that one deals not with "pure" wave functions but with the density matrix, a concept described in most standard treatments of quantum mechanics. See, for example, K. Gottfried, *Quantum Mechanics*, Sec. 20. Benjamin, New York, 1966. To simplify the discussion here we consider only the ideal case of maximally informative experiments.

† Strictly speaking, the existence of long-range electromagnetic and gravitational effects precludes a clean distinction between "initial" and "final" measurements, but such effects are often sufficiently weak to allow usefulness for the *S*-matrix concept. It should be noted that the measurement concept itself depends on the existence in nature of weak, long-range forces.

‡ See any text on quantum theory—such as Gottfried, *op. cit*. The Heisenberg uncertainty relation is thereby automatically fulfilled. An important aspect of physical measurements is that, although arbitrarily sharp spatial localization of a particle is unachievable, arbitrarily precise momentum definition may be approached as the size of the measuring apparatus becomes larger and larger. This asymmetry between position and momentum measurement is a relativistic phenomenon.

If a result B is not causally compatible with a result A and with physical space-time particle trajectories (e.g., if the measurement corresponding to B is made at a time earlier than that needed for particles to reach apparatus B), then P_{BA} must vanish. Analysis of this causality requirement [5], using the Fourier-transform connection between space-time and momentum–energy, has shown that in the momentum basis S-matrix elements are *analytic* functions of particle momenta with singularities that correspond to mechanisms for transmitting signals over macroscopic distances. Transmission by a single intermediate particle, for example, corresponds to a simple pole—the position in the complex momentum space being determined by the particle mass. The pole residue has a *factorizable*§ dependence on initial and final measurements, corresponding to the physical characteristic of any particle that, once created, its future does not depend on the details of its past. The only information that can be transmitted by a particle resides in its momentum and its spin orientation plus quantum numbers, such as electric charge. Macroscopic signal-transmitting mechanisms that involve several particles (multiple collisions) lead to branch-point singularities rather than poles, but each branch point is determined in position and character by classical space-time causality considerations. Collectively, causality-related poles and branch points are known as "Landau singularities." A widely employed although still imprecise conjecture known as "maximal analyticity" asserts that *all* S-matrix singularities in the complex momentum space, including those at nonreal values of momenta, are deductible through analytic continuation from a combination of causality and unitarity considerations once the particle poles are known [6]. Another widely used assumption—proved in some cases from unitarity—is that all poles have factorizable residues even if the pole position does not correspond to a real value for a particle mass [7]. A pole corresponding to a complex mass may then be associated with the concept of *unstable particle*; when sufficiently small, the imaginary part of the mass can be interpreted in suitable units as the inverse of the particle "lifetime."

The possibility that unrelated particle reactions may occur "simultaneously" in different regions of space leads to the decomposability of the S matrix into clusters [5]: The S matrix may be expressed as a sum of products of analytic "connected parts" each associated with a physically distinct reaction. Poincaré invariance together with analyticity leads to a property of connected parts called *crossing* [5], which requires the existence of antiparticles. According to the crossing principle, after the energy of a particle entering a reaction ("ingoing" particle) has been analytically continued within the connected part from a positive value to a negative value, the connected part describes a *different* reaction—where the originally ingoing particle has become an outgoing antiparticle.

Cauchy formulas expressing an analytic connected part in terms of its poles and branch points are known as *dispersion relations* [8] and can be used to crosscheck and enlarge experimental information about the S matrix. When combined with unitarity, dispersion relations possess physical predictive power analogous to that of the Schrödinger or Dirac equation, certain S-matrix singularities through the crossing principle playing the role of interparticle "force" [9]. It has been conjectured that equations of motion may be superfluous, all successful predictions that flow from such equations being deducible from S-matrix analyticity and unitarity.

The use of Cauchy formulas as substitutes for equations of motion invokes the assumption that connected parts in certain large-momentum limits exhibit a type of asymptotic behavior identified in 1958 by T. Regge. It is not known whether *Regge behavior* [8] is a necessary consequence of unitarity and anayticity, but this behavior seems to be exhibited by all hadronic connected parts. A corollary of Regge behavior is the unique analytic continuability of particle poles to nonphysical values of *angular* momentum—where they are called "Regge poles." Landau branch points also may be continued in angular momentum. It is the singularities in the complex angular-momentum plane that determine the precise form of Regge asymptotic behavior. Conversely, measurements of reaction probabilities at very high energy give information about singularities in angular momentum [10].

Because of the nonlinear character of the unitarity condition no explicit mathematical construction has yet been achieved of an analytic S matrix that is simultaneously unitary and Poincaré invariant. Investigations have revealed that the poles (particles) may not be arbitrarily specified but are mutually constraining. It remains unknown how much arbitrariness is allowed by the general principles. According to the "bootstrap" conjecture, the S matrix that governs massive particles is *uniquely* determined [6]. A more conservative conjecture is that although numerical parameters (e.g., pole positions and residues) are determined, the specification of certain internal quantum numbers (e.g., electric charge and baryon number) must be added to Poincaré invariance, causality, and superposition. The traditional position goes beyond S-matrix theory and demands addition of an equation of motion for degrees of freedom such as fields.

Massless particles such as photons are often more effectively handled not individually but in large aggregates that build *coherent states*. Weinberg [11] has shown masslessness of particles with spin 1 and spin 2 to require a special property called "S-matrix gauge invariance" when such particles are "soft"—i.e., with extremely small energy. Stapp [12] has deduced one consequence of such "gauge invariance" to be that soft photons collectively build coherent states equivalent to classical electromagnetic fields coupled to charged particles. These coherent states superpose *differing* numbers of photons; average photon number, when large, relates to classical-field strength. It is plausible, although not yet demonstrated, that through coherent states S-matrix theory can accommodate both classical electromagnetism and gravity.

§ A functional dependence of the form $F(A,B)=f_A(A)f_B(B)$ implies an absence of correlation between A and B.

See also Dispersion Theory; Fourier Transforms; Invariance Principles; Lorentz Transformation; Quantum Mechanics; Quantum Theory of Measurement; Regge Theory.

REFERENCES

1. H. P. Stapp, *Phys. Rev.* **D4,** 1303 (1971).
2. P. A. M. Dirac, *Quantum Mechanics,* 3rd ed., p. 18. Oxford University Press, London and New York, 1947.
3. P. A. M. Dirac, in Ref. 2, p. 34.
4. P. A. M. Dirac, in Ref. 2, p. 116.
5. For a careful statement of general *S*-matrix principles, see D. Iagolnitzer, *The S Matrix,* North Holland, Amsterdam (1978).
6. G. F. Chew, *Physics Today* **23,** 23 (October, 1970).
7. R. J. Eden, P. V. Landshoff, D. I. Olive, and J. C. Polkinghorne, *The Analytic S Matrix,* p. 247. Cambridge University Press, Cambridge, 1966.
8. W. S. C. Williams, *An Introduction to Elementary Particles,* Chap. 12. Academic Press, New York and London, 1971.
9. G. F. Chew, *The Analytic S Matrix,* Chap. 9. Benjamin, New York, 1966.
10. R. J. Eden, High Energy Collisions of Elementary Particles, Chap. 9. Cambridge University Press, Cambridge, 1967.
11. S. Weinberg, *Phys. Rev.* **135,** B1049 (1964).
12. H. P. Stapp, *Phys. Rev.* **D28,** 1386 (1983).

Scanning Tunneling Microscopy

Robert J. Hamers

The scanning tunneling microscope (STM) is a device for studying the properties of the surfaces of materials on an atomic distance scale. The STM is inherently a surface-sensitive microscope and is capable of imaging individual atoms at the surfaces of a wide variety of electrically conductive samples. It was invented in the early 1980s by Gerhard Binnig and Heinrich Rohrer, of the IBM Corporation's Research Division, who were awarded the 1986 Nobel Prize in Physics for their invention.

The physical basis of the STM is the quantum-mechanical tunneling of electrons between an extremely sharp tip and the surface of the sample of interest. If an extremely sharp tip (so sharp that it essentially ends in a single atom) is brought to within 10 angstroms of the sample, the wave functions of the sample and tip will overlap slightly even though the sample and tip are not actually touching. Electrons will then be able to pass between the sample and tip by tunneling. When a small bias (typically 10 mV on metallic samples and up to 3 V for semiconductors) is applied between the sample and tip, a net flow of current results.

The magnitude of this tunneling current depends strongly on the amount of overlap between the sample and tip wave functions, which decrease exponentially into the ≈10 Å-wide barrier between them. For a fixed bias voltage the tunneling current I is proportional to

$$I \propto \exp(-A\sqrt{\Phi}Z),$$

where Φ is the average work function of sample and tip and the constant A is given by

$$A = 2\sqrt{2m}/\hbar = 1.025 \text{ eV}^{-1/2} \text{ Å}^{-1}.$$

For typical work functions of 4–5 eV, this corresponds to roughly an order-of-magnitude change in the tunneling current for each angstrom change in the sample–tip separation Z. Because of this strong dependence of the tunneling current on the sample–tip separation, the magnitude of the tunneling current is a very sensitive probe of the separation between the sample and tip, typically allowing separation changes as small as 0.005 Å to be observed.

To use this tunneling process in a scanning tunneling microscope, a sharp tip is mounted on a three-dimensional piezoelectric scanner assembly, which allows the tip position to be controlled in all three dimensions with sub-angstrom precision by applying appropriate control voltages to the scanners. As the scanners move the tip parallel to the surface (X–Y) plane, fluctuations in the tunneling current can be observed as the single atom at the end of the tip scans over the individual atoms of the sample. To prevent the tip from touching the surface, the STM is conventionally operated in a "constant current" mode, utilizing a feedback system. As the tip position is scanned in the X, Y plane, the feedback system (using the Z scanner) moves the tip toward or away from the surface in order to keep the current constant. A small computer records the surface height changes as the tip position is scanned and presents the resulting image on a computer graphics screen.

The lateral resolution of the STM is sample dependent, but atoms separated by less than 2 Å have been resolved on graphite. Because of the need to keep the sample and tip less than 10 Å apart (without touching, which would damage both sample and tip), isolation of the STM from external vibrations and acoustic noise is important.

Tips used in the STM are usually made by electrochemical etching of tungsten wire, although Pt and Pt–Ir alloys are also used. The configuration of the atoms at the end of the tip is generally not well defined. Instead, one relies on the statistical probability that one atom or cluster of atoms will protrude slightly farther than the rest to constitute the "single-atom" tip required for atomic resolution imaging of surfaces. The inability to prepare tips of known, stable atomic configurations results in variable resolution and some irreproducibility in scanning tunneling microscopy. Samples to be studied with the STM must usually be quite flat; otherwise, the image obtained will be a convolution of the topography of the sample with the three-dimensional shape of the tip.

The "bumps" observed in STM images do not always correspond to the positions of the atoms themselves. Rather, the bumps are locations where the wave functions of the sample and tip overlap most strongly. Particularly on semiconductor surfaces, the electronic states at the surface may have a complicated dependence both on spatial position and on the energy of the state. This sensitivity to local electronic structure sometimes produces changes in the STM images as the sample–tip bias is changed. While this makes it difficult to accurately determine the exact atomic positions, it also enables one to distinguish among chemically inequivalent surface atoms.

More detailed information on the energies and positions of the surface electronic states is obtained from measurements of the tunneling current I as a function of the applied voltage V. Analysis of the resulting I–V curves reveals structure which reflects the energy-dependent density of states. Such "tunneling spectroscopy" measurements can be used

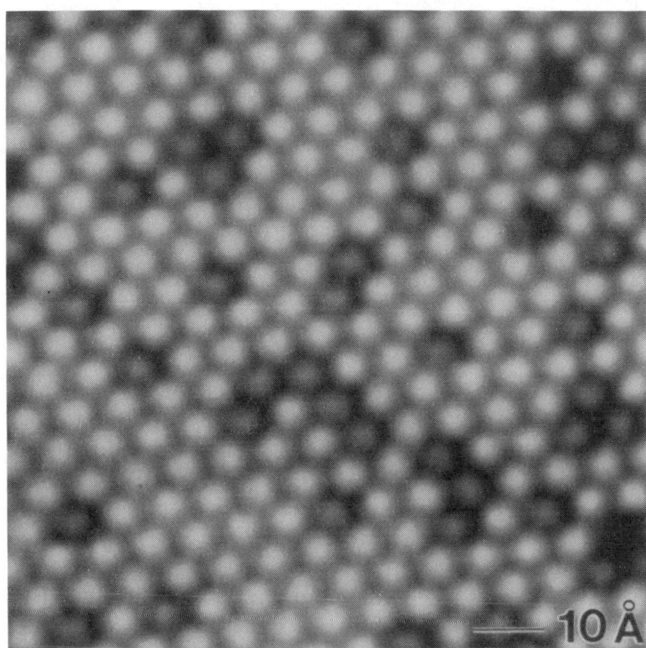

FIG. 1. Scanning tunneling microscope image of ordered aluminum atoms atop a silicon surface. Individual substitutional defects can also be observed, in which silicon atoms replace aluminum atoms in the outermost atomic layer of the surface.

to determine the electronic structure at atomic-sized defects and other surface inhomogeneities.

Although most STM studies are performed in vacuum, atomic resolution can be achieved under a variety of other conditions, including normal atmospheric environments and in liquids. Large biological molecules have also been imaged by depositing them onto flat substrates, such as single-crystal graphite. Although still a relatively young technique, STM has already enjoyed a wide variety of applications in physics, chemistry, materials science, and biology. Undoubtedly, many more new aand exciting applications of this technique will yet be found.

BIBLIOGRAPHY

G. Binnig and H. Rohrer, *Sci. Am.* **253,** 40 (1985).
R. J. Hamers, *Adv. Phys. Chem.* **40,** 231 (1989).
P. K. Hansma and J. Tersoff, *J. Appl. Phys.* **61,** R21–R23 (1987).

Scattering Theory

Kenneth M. Watson

DESCRIPTION OF SCATTERING EVENTS

The fundamental structure and interactions of matter are investigated primarily by scattering experiments. Scattering theory provides the framework for systematic analysis of such experiments. Several references are available that provide detailed descriptions of scattering theory [1–4] as applicable to elementary particle phenomena. Other applications of scattering include the use of radar, acoustic imaging of the seafloor, etc. In this review we shall first discuss the quantum-mechanical formulation of particle scattering.

Such scattering typically involves the collision of two kinds of particles, say P_1 and P_2, with respective initial momenta $\mathbf{p}_1$ and $\mathbf{p}_2$. (Collisions of three or more particles are of interest for studying chemical reactions and will not be considered here; see, e.g., [3, Appendix B].) We shall refer to P_1 as beam, P_2 as target particles. In the laboratory frame of reference $\mathbf{p}_2=0$. Scattering events are classified on the basis of the final states observed after the collision has occurred. In *elastic* scattering, particles P_1 and P_2 are merely deflected, being observed to have final moments $\mathbf{q}_1$ and $\mathbf{q}_2$ (indicated as $P_1+P_2 \rightarrow P_1+P_2$; an example is $e+\mathrm{H} \rightarrow e+\mathrm{H}^*$).

In a scattering reaction new particles emerge from the collision (indicated as $P_1+P_2 \rightarrow Q_1+Q_2$, $P_1+P_2 \rightarrow Q_1+\cdots+Q_N$; examples are $\mathrm{H}+\mathrm{H} \rightarrow p+e+\mathrm{H}$ and $p+p \rightarrow p+n+\pi^+$). Since a given collision, involving P_1 and P_2 as the initial particles, can in general result in several alternative kinds of final states, it has been convenient to introduce the notion of "reaction channels" to classify the final state following the scattering. A particular channel is specified by the set of particles $Q_1, \ldots, Q_N$ in the state.

The results of observations of scattering events are usually presented as scattering cross sections (when spin polarization is observed, other quantities are needed). To describe the scattering cross section, we suppose that a uniform flux F_B of beam particles P_1 (F_B represents the number of particles per unit area) has bombarded a "thin target" of particles P_2 ($\mathbf{p}_2=0$). In the course of the experiment ΔN_{sc} scattering events have been detected. The scattering cross section $\Delta\sigma$ is defined then as

$$\Delta\sigma = \Delta N_{\mathrm{sc}}/F_B. \tag{1}$$

The detectors used in the experiment will ordinarily detect only certain specified kinds of particles and certain ranges of momenta for these, which accounts for the Δ in $\Delta\sigma$. If we sum over all ranges of final moments and all possible kinds of final particles (channels), we obtain from Eq. (1) the total cross section

$$\sigma = \sum_{\text{momenta}} \sum_{\text{channels}} \Delta\sigma. \tag{2}$$

The number of events ΔN_{sc} is evidently a Lorentz invariant, being the same number to all observers. The beam flux F_B is also unchanged if we transform to another Lorentz frame moving parallel to $\mathbf{p}_1$. The *center-of-mass, or barycentric,* frame of reference is that in which $\mathbf{p}_1=-\mathbf{p}_2\equiv\mathbf{k}$. Evidently, $\Delta\sigma$ is the same in both the laboratory ($\mathbf{p}_2=0$) and the barycentric frames of reference. [It is customary to define $\Delta\sigma$ to be a Lorentz invariant with respect to an arbitrary Lorentz transformation. This is done by replacing F_B in Eq. (1) by an invariant quantity F_1 that is equal to F_B in the laboratory frame of reference: see, e.g., Eq. (3.142) of Ref. 3.]

There are some useful general properties of $\Delta\sigma$, which we now describe. First, the total energy E and the total momentum $\mathbf{P}$ must be conserved in a collision. Thus, $\Delta\sigma$ must contain a factor

$$\delta(\mathbf{P}_f-\mathbf{P}_i)\delta(E_f-E_i), \tag{3}$$

where we have represented "initial" and "final" by the respective subscripts i and f. We have noted that the detectors of final-state particles will ordinarily record events for a restricted domain only of the momenta $\mathbf{q}_1 \cdots \mathbf{q}_N$. We therefore expect $\Delta\sigma$ to involve an integration.

$$\int d^3q_1 \cdots d^3q_N \cdots,$$

over this domain set by the detectors. Now Eqs. (1) and (3) are Lorentz invariants, and so is the ratio $d^3\mathbf{q}_1/\epsilon_1$, where $\epsilon_{1_{q1}}$ and M_1 are the energy and rest mass of Q_1]. We can therefore write $\Delta\sigma$ in an evident invariant form as

$$\Delta\sigma = \int \frac{d^3q_1}{\epsilon_{1_{q1}}} \cdots \frac{d^3q_N}{\epsilon_{N_{qN}}} \, \delta(\mathbf{P}_f - \mathbf{P}_i)\, \delta(E_f - E_i) I. \tag{4}$$

The quantity I is an invariant function of $\mathbf{p}_1 \cdots \mathbf{q}_N$. It is dependent on the detailed dynamics of the collision.

For elastic scattering in the center-of-mass frame we have $\mathbf{p}_1 = \mathbf{k} = -\mathbf{p}_2$; $\mathbf{q}_1 = \mathbf{k}' = -\mathbf{q}_2$, $\mathbf{k}' = \mathbf{k}$, and $\mathbf{k}\cdot\mathbf{k}' = k^2 \cos\theta$. The invariant quantity I may then be taken as a function of the two variables k and θ, or $I = I(k,\theta)$. Alternatively, we can express I as a function of the two Mandelstam [5] variables

$$\begin{aligned} s &= (p_1 + p_2)^2 = (q_1 + q_2)^2, \\ t &= (p_1 - q_1)^2 = (p_2 - q_2)^2. \end{aligned} \tag{5}$$

[Here we have written the four-vector $(\mathbf{p}_1, \epsilon_{1_{p1}})$ simply as p_1, etc.]

For elastic scattering in the barycentric frame (4) has the form

$$\begin{aligned} \Delta\sigma &= \int \frac{d^3q_1 d^3q_2}{\epsilon_{1_{q1}}\epsilon_{2_{q2}}} \delta(\mathbf{q}_1 + \mathbf{q}_2)\, \delta(\epsilon_{1_{q1}} + \epsilon_{2_{q2}} - \epsilon_{1k} - \epsilon_{2k}) I \\ &= \int \frac{d\Omega_1 q_1^2\, dq_1}{\epsilon_{1_{q1}}\epsilon_{2_{q1}}} \delta(\epsilon_{1_{q1}} + \epsilon_{2_{q1}} - \epsilon_{1k} - \epsilon_{2k}) I. \end{aligned}$$

If the observed solid angle is sufficiently small that I can be evaluated at a single angle θ, we can remove $d\Omega_1$ from the integral and evaluate the q_1 integral to give the differential center-of-mass cross section as

$$\left(\frac{d\sigma}{d\Omega_1}\right)_{\text{c.m.}} = \left[\frac{(k/c^2)}{(\epsilon_{1k} + \epsilon_{2k})}\right] I(k,\theta). \tag{6}$$

The corresponding differential cross section in the laboratory frame may also be easily evaluated from (4). The laboratory frame moves parallel to k with a velocity

$$c\beta = c^2 k/\epsilon_{2k}. \tag{7}$$

The total energies (kinetic plus rest energy) in the two frames are related by the useful expression

$$E^2_{\text{c.m.}} = c^2[c^2(M_1{}^2 - M_2{}^2) + 2M_2 E_{\text{lab}}], \tag{8}$$

In the laboratory frame (4) is

$$\Delta\sigma = \int \frac{d^3q_1\, d^3q_2}{\epsilon_{1_{q1}}\epsilon_{2_{q2}}} \delta(\mathbf{q}_1 + \mathbf{q}_2 - \mathbf{p}_1) \times \delta(\epsilon_{1_{q1}} + \epsilon_{2_{q2}} - \epsilon_{1_{p1}} - M_2 c^2) I.$$

Integration over q_2 gives the result that $q_2 = p_1 - q_1$. Division by the solid angle $d\Omega_1$ of particle P_1 and integration over q_2 gives us the differential cross section in the laboratory frame as

$$\left(\frac{d\sigma}{d\Omega_1}\right)_{\text{L}} = \frac{(q_1/c^2) I(k,\theta)}{\epsilon_{1_{q1}}\epsilon_{2|\mathbf{p}_1 - \mathbf{q}_1|}} \left[\frac{\mathbf{q}_1}{\epsilon_{1_{q1}}} - \frac{\mathbf{q}_1\cdot(\mathbf{p}_1 - \mathbf{q}_1)}{q_1 \epsilon_{2|\mathbf{p}_1 - \mathbf{q}_1|}}\right]^{-1}. \tag{9}$$

Here $\epsilon_{1_{q1}} + \epsilon_{2|p_1 - q_1|} = \epsilon_{1_{p1}}$.

Since $I(k,\theta)$ is the same quantity in Eqs. (6) and (9), it may be eliminated to give the transformation equation between the differential cross section in the laboratory and barycentric frames.

The kinematics of more general reactions may be discussed in a similar manner using Eq. (4). (This is done, e.g., in Chapter 2 of Ref. 1 or Chapters 3 and 6 of Ref. 3.)

A powerful formal description, applicable to relativistic and many-particle collisions, has developed following work of Heisenberg [6], Moller [7], Lippmann and Schwinger [8], Gell-Mann and Goldberger [9], and many others.

POTENTIAL SCATTERING

Nonrelativistic elastic scattering by a local, central potential $V(r)$ provides the simplest example of a collision of two particles. In the barycentric coordinate system the wave function $\psi_{\mathbf{K}}{}^+$ may be expressed as a function of the relative particle coordinate r. The Schrödinger equation describing this interaction is the notation of [3].

$$[\nabla_r{}^2 + K^2 - v(r)]\psi_{\mathbf{K}}{}^+(\mathbf{r}) = 0. \tag{10}$$

Here $v(r) = 2M_r V(r)/\hbar^2$, with M_r the reduced mass of P_1 and P_2, and $\hbar\mathbf{K}$ is the momentum of P_1 prior to the collision. Although scattering can be studied under more general conditions, the properties of (10) are simplified when

$$\begin{aligned} &\int_0^\infty dr\, r|v(r)| < \infty, \\ &\int_0^\infty dr\, r^2|v(r)| < \infty. \end{aligned} \tag{11}$$

The continuum wave functions $\psi_{\mathbf{K}}{}^+$ are conventionally normalized so that

$$(\psi_{\mathbf{K}}{}^+, \psi_{\mathbf{P}}{}^+) = \delta(\mathbf{K} - \mathbf{P}). \tag{12}$$

The $\psi_{\mathbf{K}}{}^+$ and the bound-state wave functions (if any) form a complete set of states in the reduced space from which the center-of-mass coordinates have been eliminated. The full two-body wave functions are of the form

$$\psi = (2\pi h)^{-3/2} \exp(i\mathbf{P}'\mathbf{R})\psi_{\mathbf{K}}{}^+(\mathbf{r}) \tag{13}$$

where $\mathbf{P}$ is the total momentum and $\mathbf{R}$ is the center-of-mass coordinate of P_1 and P_2.

The asymptotic form of $\psi_{\mathbf{K}}{}^+$, as Kr becomes very large, is

$$\psi_{\mathbf{K}}{}^{+} \rightarrow (2\pi h)^{-3/2}[e^{i\mathbf{K}\cdot\mathbf{r}} + (e^{iKr}/r)f(\hat{\mathbf{K}}\cdot\hat{\mathbf{r}})]. \tag{14}$$

Here $f(\hat{\mathbf{K}}\cdot\hat{\mathbf{r}})$ is the scattering amplitude and describes the deflection of P_1 from the direction $\hat{\mathbf{K}}$ to the direction $\hat{\mathbf{r}}$. The barycentric scattering cross section (6) in this case has the explicit form

$$\left(\frac{d\sigma}{d\Omega_1}\right)_{\text{c.m.}} = |f(\hat{\mathbf{K}}\cdot\hat{\mathbf{r}})|^2. \tag{15}$$

The quantity I in (6) may be expressed in terms of $|f|^2$ using (15) and the cross section in the laboratory frame obtained using (9).

The partial-wave-expansion of $\psi_k{}^+$ is of the form

$$\psi_{\mathbf{K}}{}^{+}(\mathbf{r}) = \sum_{l=0}^{\infty} \left[\frac{(2l+1)}{(4\pi Kr}\right] i^l \exp[i\delta_l(K)] \times P_l(\hat{\mathbf{K}}\cdot\hat{\mathbf{r}}) w_l(K;r). \tag{16}$$

Here P_l is the Legendre polynomial of order l and $\delta_l(K)$ is the scattering phase shift. The radial wave function w_l satisfies the ordinary differential equation

$$\left[\frac{d^2}{dr^2} + K^2 - \frac{l(l+1)}{r^2} - v(r)\right] w_l = 0 \tag{17}$$

and is regular at the origin. (For a more complete discussion see, e.g., Eq. (6.120) of Ref. 3.) For Kr large, w_l has the asymptotic form

$$w_l(K;r) = (2/\pi)^{1/2} \sin[Kr - \pi l/2 + \delta_l(K)]. \tag{18}$$

The quantity

$$S_l(K) = \exp[2i\delta_l(K)] \tag{19}$$

is an eigenvalue of Heisenberg's S matrix (See, e.g., Chapter 7 of Ref. 2 for a thorough discussion of the S-matrix.)

Another class of useful solutions of Eq. (15) are the Jost functions $f_l(\pm K;r)$. These satisfy the boundary conditions that as $Kr \rightarrow \infty$,

$$f_l(\pm K;r) \rightarrow i^l e^{\mp iKr}. \tag{20}$$

When conditions (11) are satisfied the function $f_l(K;r)$ is analytic in K for $\mathrm{Im}(K) \leftarrow 0$. Study of the Jost functions has given insight into many general properties of collision processes and has stimulated extensions to relativistic quantum theory. (See, e.g., [2, Chapter 12] or [3, Chapter 6]).

SCATTERING BY NONCENTRAL FORCES

We now suppose that P_1 and P_2 have respective spins S_1 and S_2, with z components v_1 and v_2, and that the forces responsible for scattering also act on the spin orientation. If we write $u(v_1,v_2)$ for the spin wave function of P_1 and P_2, the asymptotic wave function (12) is modified to read

$$\psi^{+}{}_{\mathbf{K}v_1v_2} \rightarrow (2\pi h)^{-3/2}[e^{i\mathbf{K}\cdot\mathbf{r}} u(v_1,v_2) + (e^{iKr}/r)\langle v_1{}',v_2{}'|\mathbf{f}(\hat{\mathbf{K}}\cdot\hat{\mathbf{r}})|v_1,v_2\rangle u(v_1{}',v_2{}')]. \tag{21}$$

The scattering amplitude $\mathbf{f}$ is now a $(2S_1+1)\times(2S_2+1)$ square matrix. The cross section in the barycentric frame, with initial spin orientation (v_1,v_2) and final spin orientation $(v_1{}',v_2{}')$, is

$$\left(\frac{d\sigma}{d\Omega_1}\right)_{\text{c.m.}} = |\langle v_1{}',v_2{}'|\mathbf{f}(\hat{\mathbf{K}}\cdot\hat{\mathbf{r}})|v_1,v_2\rangle|^2. \tag{22}$$

For a scattering experiment in which $v_1{}'$ and $v_2{}'$ are not observed (22) should be *summed* over the $(2S_1+1)\times(2S_2+1)$ values of $v_1{}'$ and $v_2{}'$. For an experiment with an unpolarized beam (22) should be averaged over v_1 and v_2.

When P_1 and/or P_2 are polarized prior to collision, it is often convenient to introduce a density matrix ρ_i for the initial state:

$$\boldsymbol{\rho}_i = \overline{u(v_1,v_2)u^{\dagger}(v_1,v_2)}, \tag{23}$$

where the bar indicates a suitably weighted average over v_1 and v_2. Evidently $Tr(\rho_i) = 1$. For unpolarized beam and target particles, $\boldsymbol{\rho}_i$ is just the identity matrix divided by $(2S_1+1)\times(2S_2+1)$.

The final-state density matrix is

$$\boldsymbol{\rho}_f = \mathbf{f}\boldsymbol{\rho}_i\mathbf{f}^{\dagger}. \tag{24}$$

The scattering cross section, summed over all final spin orientations, is

$$\left(\frac{d\sigma}{d\Omega_1}\right)_{\text{c.m.}} = \mathrm{Tr}(\boldsymbol{\rho}_f). \tag{25}$$

The average value of the spin vector $\mathbf{S}_1$, for example, is

$$\tilde{\mathbf{S}}_1 = \mathrm{Tr}(\mathbf{S}_1\boldsymbol{\rho}_f)/\ \mathrm{Tr}(\boldsymbol{\rho}_f). \tag{26}$$

Following two sequential scatterings, the density matrix is

$$\boldsymbol{\rho}_f = \mathbf{f}_2\mathbf{f}_1\boldsymbol{\rho}_i\mathbf{f}_1{}^{\dagger}\mathbf{f}_2{}^{\dagger}. \tag{27}$$

Here $\mathbf{f}_1$ is the scattering amplitude for the first and $\mathbf{f}_2$ that for the second scattering. (Chapter 18 of [1] and Chapter 7 of [3] provide more detailed descriptions of scattering by noncentral forces.)

SCATTERING FROM COMPLEX PARTICLES

Scattering of two particles when one or both are complex systems (such as atoms or molecules) involves a multiparticle interaction. Collisions involving complex particles may be elastic (in which case P_1 and P_2 may formally be treated as if simple, noncomplex particles) or inelastic, or lead to reactions in which different particles Q_1, Q_2, ..., Q_N appear. Exact analytic solutions cannot be given when three or more particles interact. Thus, the scattering of complex systems must be treated by approximate and/or numerical methods.

When the kinetic energy of the colliding particles is high compared with their internal binding energies, the *impulse approximation* ([11]; see also, e.g., [3, Chapter 11] is often useful.

The *optical model* (see, e.g., [1, Chapter 20] or [3, Chapter 11]) provides a technique that is frequently convenient for describing elastic scattering from complex particles.

An elegant technique for studying three-particle interactions (e.g., $e+\mathrm{H}$ or $n+\mathrm{D}$ scattering) has been proposed by Faddeev (extensive references and an excellent review are given in [1, Chapter 19]; see also [13]). Numerical techniques have been applied to Faddeev's equations with some success (see, e.g., [14]).

A very extensive literature exists relating to the scattering of electrons, protons, ions, and atoms by atoms or molecules [14,15]. Relatively good approximations (usually relying on numerical computation) are available for studying scattering of electrons by the lighter atoms [14,13,14]. Semiclassical techniques may be applicable for studying slow collisions of protons and ions with light atoms [14,16]. The more complex phenomena that occur when heavy atoms or molecules scatter, or when the complex structure of both P_1 and P_2 must be taken into account, are often studied by heuristic methods, crude approximations, and/or experiments.

RADAR

The development of radar during World War II led to very practical applications of electromagnetic wave scattering. Radars mounted on satellites have proven practical for observing the surface of the oceans and land [17,18].

Radar scattering from the ocean surface has received considerable attention (see, for example, the review by Valenzuela [19]). An important mechanism is Bragg scattering from ocean waves. The Bragg scattering condition is that

$$\lambda_R = 2\lambda \sin\theta \tag{28}$$

where λ_R is the electromagnetic wavelength, λ that of the water waves, and θ the angle of incidence [20]. The radar cross section for specular backscatter (defined by convention as 4π times the cross section [15]) from a facet with principal radii of curvature R_1 and R_2 is [21].

$$\omega_f = \pi R_1 R_2$$

A "two scale" scattering model has been proposed by Bass et al [22].

ACOUSTIC SCATTERING

The detection and imaging of scattered acoustic waves has had important applications in such diverse fields as medicine and geophysics. Acoustic scattering has been of special significance in oceanography, since optical signals are attenuated by 10 to 20 meters of sea water.

Numerical solution of the scalar wave equation is usually considered too complex, except for simple scattering geometries. A general discussion of scattering from various structures is given by Junger and Feit [23]. (See also, Wilton [24].)

The Helmholtz integral equation is often used as a basis for formulating approximation. An analysis of approximation methods using this has been given by Thorsos [25]. Complications associated with eigenfunctions of the homogeneous integral equation are discussed by Schenk [26].

An application to sea-floor scattering is illustrated in the paper by Stanis et al. [27].

See also CENTER-OF-MASS SYSTEM; COLLISIONS, ATOMIC AND MOLECULAR; ELECTRON BOMBARDMENT OF ATOMS AND MOLECULES; LORENTZ TRANSFORMATIONS; NUCLEAR REACTIONS; NUCLEAR SCATTERING; PARTIAL WAVES; *S*-MATRIX THEORY.

REFERENCES

1. C. J. Joachain, *Quantum Collision Theory*. North-Holland, Amsterdam, 1975.
2. R. G. Newton, *Scattering Theory of Waves and Particles*. McGraw-Hill, New York, 1986.
3. M. L. Goldberger and K. M. Watson, *Collision Theory*. Krieger, Huntington, N.Y. 1964.
4. N. F. Mott and H. S. W. Massey, *The Theory of Atomic Collisions*. Oxford, London and New York, 1965.
5. S. Mandelstam, *Phys. Rev.* **112,** 1344 (1958).
6. W. Heisenberg, *Z. Naturforsch.* **1,** 608 (1946).
7. C. Moller, K. *Dan Vidensk. Selsk. Mat-Fys. Medd.* **23**(1), (1945).
8. B. Lippmann and J. Schwinger, *Phys. Rev.* **79,** 469 (1950).
9. M. Gell-Mann and M. Goldberger, *Phys. Rev.* **91,** 398 (1953).
10. R. Omnes and M. Froissart, *Mandelstam Theory and Regge Poles*. Benjamin, New York, 1963. V. De Alfano, S. Fubini, G. Furlan, and C. Rossetti, *Currents in Hadron Physics,* North-Holland, Amsterdam, 1973.
11. G. F. Chew, *Phys. Rev.* **80,** 196 (1950); G. F. Chew and M. Goldberger, *Phys. Rev.* **87,** 778 (1952); J. Askin and G. C. Wick, *Phys. Rev.* **85,** 686 (1952).
12. K. Watson and J. Nuttall, *Topics in Several Particle Dynamics*. Holday-Day, San Francisco, 1967.
13. J. C. Y. Chen, "Faddeev Equations for Three-Body Coulomb Dynamics," in *Case Studies in Atomic Physics,* Vol. 3, p. 307. North-Holland, Amsterdam, 1973.
14. B. H. Bransden, *Atomic Collision Theory*. Benjamin, New York, 1970.
15. W. H. Miller, *Acc. Chem. Res.,* **4,** 161 (1971).
16. J. C. Y. Chen and K. M. Watson, *Phys. Rev.,* **188,** 236 (1969).
17. R. Stewart. *Methods of Satellite Oceanography*. University of California Press, Berkeley, 1985.
18. R. Gasparovic, J. Apel, and E. Kaiseschke, *J. Geophys. Res.* **93**(12), 304 (1988).
19. G. R. Valenzuela, *Boundary-Layer Meterology,* **13,** 61 (1978).
20. J. W. Wright, *IEEE Trans.,* **AP-16,** 217 (1968).
21. K. M. Siegel, *Appl. Sci. Res. Bull.,* **7,** 293 (1958).
22. F. Bass, I. Kalmykiv, A. Ostrovsky, and A. Rosenberg, *IEEE Trans.,* **AP-16,** 554 (1968).
23. M. Junger and D. Feit, *Sound Structures, and Their Interaction*. MIT Press, Cambridge, Mass., 1986.
24. D. T. Wilton, *Int. Jour. for Numerical Methods in Eng.,* **13,** 123 (1978).
25. E. I. Thorsos, *J. Acoust. Soc. Am.,* **83,** 78 (1988).
26. H. A. Schenck, *J. Acoust. Soc. Am.,* **44,** 41 (1967).
27. S. Stanis, K. Briggs, P. Fleische, R. Ray, and W. Sawyer, *J. Acoust. Soc. Am.,* **83,** 2134 (1988).

Schrödinger Equation

Edwin F. Taylor

The fundamental "wave equation" of nonrelativistic quantum mechanics was developed by Erwin Schrödinger at the end of 1925. For a single particle of mass m in a potential $V(\mathbf{r})$ it takes the time-dependent form

$$-\frac{\hbar^2}{2m}\nabla^2\Psi(\mathbf{r},t) + V(\mathbf{r})\Psi(\mathbf{r},t) = i\hbar\,\frac{\partial\Psi(\mathbf{r},t)}{\partial t},$$

where $\hbar = h/2\pi = 1.05459 \times 10^{-27}$ erg s and h is Planck's constant.

This equation and its various extensions may be used to describe the whole of nonrelativistic dynamics, including systems that are free [$V(\mathbf{r}) = 0$ everywhere] as well as bound.

For a bound particle in an energy eigenstate of energy E_n the wave function may be written

$$\Psi(\mathbf{r},t) = \psi_n(\mathbf{r})e^{-E_n t/\hbar}.$$

In this case the time-independent Schrödinger equation is, by substitution,

$$-\frac{\hbar^2}{2m}\nabla^2\psi_n(\mathbf{r}) + V(\mathbf{r})\psi_n(\mathbf{r}) = E_n\psi_n(\mathbf{r}).$$

The latter is an operator equation analogous to the classical energy equation, with the operator $i\hbar\nabla$ replacing p in the classical expression for kinetic energy, $p^2/2m$. This analogy may be used to generalize the equation to describe more than one particle, including cases in which particles interact with one another. For example, two charged particles of equal mass m and charges q_1 and q_2, respectively, that interact with one another electrically as well as interacting individually with a binding charge Q fixed at the origin are described by the energy eigenvalue equation

$$-\frac{\hbar^2}{2m}\nabla_1{}^2\psi_n(\mathbf{r}_1,\mathbf{r}_2) - \frac{\hbar^2}{2m}\nabla_2{}^2\psi_n(\mathbf{r}_1,\mathbf{r}_2) + \left(\frac{q_1q_2}{r_{12}} + \frac{q_1Q}{r_1} + \frac{q_2Q}{r_2}\right)\psi_n(\mathbf{r}_1,\mathbf{r}_2) = E_n\psi_n(\mathbf{r}_1,\mathbf{r}_2).$$

Here ∇_1 implies differentiation with respect to the coordinates of particle 1 only, r_{12} is the distance between the two particles, r_1 is the distance of particle 1 from the charge Q at the origin, and so forth. The electrical units are Gaussian.

The meaning of the various wave functions, denoted by Ψ above, has been the subject of some controversy. Max Born originated the interpretation according to which the squared magnitude of the wave function represents a spatial probability density. Thus, for a single particle, the expression

$$|\Psi(\mathbf{r},t)|^2 d\tau$$

represents the probability of finding a particle in the volume element $d\tau$ at position $\mathbf{r}$ at time t. For a many-particle wave function, the corresponding expression

$$|\Psi(\mathbf{r}_1,\mathbf{r}_2,\cdots,t)|^2 d\tau_1 d\tau_2\cdots$$

gives the joint probability at time t of finding particle 1 in $d\tau_1$ at $\mathbf{r}_1$ *and* particle 2 in $d\tau_2$ at $\mathbf{r}_2$ and so forth.

It is clear that predictions based on such probabilities must apply only statistically, that is to a large number of identically prepared systems. The centrality of probability at this fundamental level of physics, with the consequent lack of determinism in predicting the result of a single observation, has troubled many, including Albert Einstein. Nevertheless, the routine use of this interpretation over the years has led to no experimental inconsistencies, and it remains the consensus of practicing physicists.

At the time of the development of the Schrödinger equation, the analytic solution of differential equations was already a highly developed art. The exact solution for the single electron in the electrical potential of the nucleus in the hydrogen atom was carried out by Schrödinger himself. The helium atom, with its two interacting electrons (electrical interactions described by the two-particle equation above under the assumption of a fixed nucleus), proved to be intractable to analytic solution; but approximate solutions of high accuracy were obtained by numerical methods. More recently the computer has been harnessed to provide approximate solutions for the electronic configurations of many of the atoms in the periodic table and the properties of chemical bonds.

See also ATOMIC STRUCTURE CALCULATIONS; EIGENFUNCTION; HAMILTONIAN FUNCTION; MOLECULAR STRUCTURE CALCULATIONS; QUANTUM MECHANICS.

BIBLIOGRAPHY

Erwin Schrödinger's original papers may be found in *Collected Papers on Wave Mechanics*. Blackie and Son, London, 1929, reprinted by Chelsea Pub. Co., New York, 1978.

Some insights into the personalities of the original workers in the field are given in K. Przibram (ed.), *Letters on Wave Mechanics*. Philosophical Library, New York, 1967.

Scintillation and Čerenkov Counters

Kenneth J. Foley

SCINTILLATION COUNTERS

Early experiments with radioactivity showed that a single α particle striking an activated zinc sulfide screen caused a visible flash of light to be emitted. Experiments using this principle to measure the relative probability of the scattering of α particles through different angles led to the discovery of the nucleus of the atom. In spite of the introduction of newer detectors, modern developments have kept the technique of scintillation counters at the forefront of nuclear and elementary-particle physics.

The basic process in scintillation counters is the loss of kinetic energy by charged particles when passing through matter. Most of the energy loss is due to collisions with electrons in the medium, leading to excitation and ionization of some molecules. The subsequent deexcitation and recombination lead necessarily to the emission of light quanta with a characteristic fluorescent spectral distribution. The distinguishing features of useful scintillation-counter material are the emission of a significant fraction of the light in the visible region and the transparency of the material to the emitted light. The details of the light emission process will not be discussed here; interested readers will find detailed descriptions in the literature [1].

Visual observation of scintillation light is rarely employed nowadays. Instead, a photomultiplier tube is used to detect the light pulse. The principle of the photomultiplier tube is illustrated in Fig. 1. Light quanta incident on the photo-

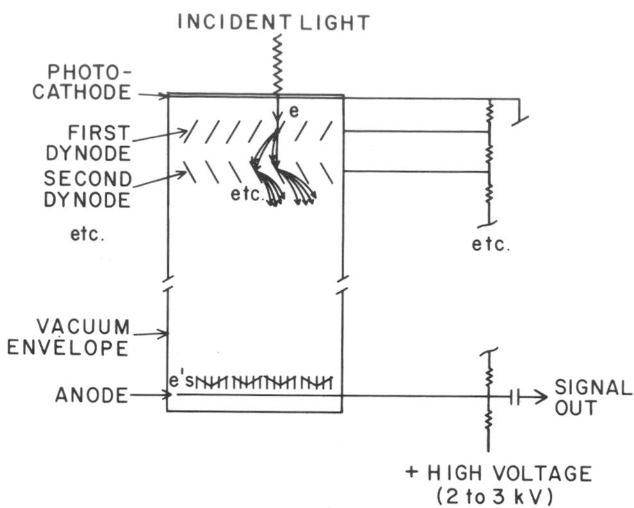

FIG. 1. A schematic illustration of the principle of the photomultiplier tube.

cathode produce secondary electrons via the photoelectric effect. These electrons are then accelerated by the electric field and strike the first "dynode," the surface of which is coated with a material having a large secondary emission coefficient; typically, four secondary electrons are emitted for each incident electron. These electrons are then accelerated to the next dynode, where further amplification occurs, etc. After amplification, the electrons are collected at the anode. A photomultiplier tube with 12 amplification stages gives an overall gain of about $4^{12} \approx 10^7$. Since the duration of the pulse of electrons collected at the anode is typically 10^{-8} sec, a single electron from the photocathode produces a current of about 150 μA. This level is quite suitable for modern electronic circuits, which can then be used to determine the amplitude of the electrical pulse and the time of arrival of this signal. These are directly related to the intensity of the light pulse emitted by the scintillator and the time at which the charged particle struck the scintillator.

Let us consider some practical applications of the technique.

If a charged particle is not sufficiently energetic to pass through a block of scintillator, then the energy deposited in the scintillator is the total kinetic energy carried by that particle. Thus, in principle, the intensity of the emitted light will give a direct measurement of the kinetic energy (E) of the particle. If, on the other hand, the scintillator is very thin, such that the particle passes through losing only a small fraction of its energy, then we measure the differential energy loss (dE/dx). This depends on the energy of the particle, its charge, and its mass. In cases where the kinetic energy is small compared to the rest mass of the particle, if we measure dE/dx (in a thin scintillator) and then E (in a thick scintillator), we can also identify the particle. There are practical difficulties in that most scintillators exhibit saturation, whereby the light output is not proportional to the energy deposited when the rate of deposition is high; nevertheless, this technique has been very valuable in nuclear physics experiments.

If a scintillator has a short fluorescent decay time, the arrival time of a particle at the scintillator can be determined very precisely; modern plastic scintillators have decay times of about 10^{-9} sec. Hence, a pair of scintillation counters permits the measurement of the velocity of a particle that passes through both scintillators. If this is coupled with a measurement of the momentum of the particle, its mass can be determined. Straightforward experiments using this technique were used to study the production of unstable particles (mesons, etc.) when accelerators were able to produce protons with kinetic energies comparable to their rest mass. In order to set a scale for such apparatus, it should be noted that the velocity of light in vacuum (the limiting velocity of particles) is about 30 cm in 10^{-9} sec; typical flight paths of several meters are employed in such experiments.

When the kinetic energy of the particle passing through the scintillator is comparable to, or larger than, the rest mass of the particle, dE/dx is almost independent of E, as is the velocity of the particle. This is the situation in high-energy physics where the principal use of scintillation counters is to use the extremely good time resolution to select events of specific topology that occur very infrequently. Usually "simultaneous" pulses in sets of scintillators are used to select relevant events. Other detectors with better spatial resolution are then used to measure the parameters of the interaction.

The foregoing applications refer to measurements of the properties of charged particles. An important use of scintillation counters is in the measurement of the energy of γ rays. In this case a counter of high atomic number and large physical size is desirable: the former to enhance the conversion of γ rays to a shower of electrons and the latter to ensure the deposition in the counter of a large fraction of the electron energy. The most popular material chosen is NaI (with a small "impurity" content of thallium, which increases the light output and reduces the fluorescent decay time); for efficient detectors single crystals must be used. Modern technology permits the production of NaI crystals of dimensions of 30 cm or more.

It is also possible to detect fast neutrons in scintillation counters. Here we measure ionization from recoil charged particles produced in interactions of the neutrons. Since the interaction probability of fast neutrons is small, even a thick scintillator produces little information on the neutron energy. In general we must rely on the good time resolution of fast scintillators and use time of flight to determine the energy of the neutron.

Table I lists some of the scintillators used in nuclear and elementary-particle physics, along with the fluorescent decay times and brief comments on their particular merits.

ČERENKOV COUNTERS

An electromagnetic phenomenon of great significance for experiments in elementary-particle physics occurs when the velocity of a charged particle (v) approaches that of light in vacuum (c). In a medium of refractive index n, the velocity of light is c/n. If the velocity of a particle passing through this medium exceeds c/n, electromagnetic radiation ("light") is emitted at a characteristic angle θ given by the equation

$$\cos\theta = 1/\beta n \qquad (1)$$

Table I

Type	Decay time (sec)	Comments
Plastic scintillator	$\sim 2 \times 10^{-9}$	Available in large sizes, easily machined
Liquid scintillator	$\sim 10^{-9}$	Inexpensive; used in large counter arrays, particularly for neutron counters
NaI	$\sim 0.25 \times 10^{-6}$	Excellent γ-ray detector; needs careful handling since these detectors are hygroscopic single crystals.

This is Čerenkov radiation. A physically meaningful value of θ is obtained only if $\beta \geq 1/n$ (note that $n \geq 1$ and $\beta \leq 1$). The theory of Čerenkov radiation, including polarization effects, has been studied extensively [2], but we shall concern ourselves here with practical applications.

In order to set a scale, we refer to Table II, where the refractive index (n) and the minimum relative velocity at which Čerenkov radiation is emitted (β_t) are listed for various transparent substances. We also list the momentum (p) for electrons, pions, and protons for that value of β_t ("thresholds").

Before discussing applications it is necessary to consider several other properties of Čerenkov radiation. The intensity of light (I) is given by the relation

$$I \propto \sin^2\theta. \tag{2}$$

In addition, as might be expected, the intensity of the light is proportional to the length of the medium traversed by the charged particle.

Following the customary procedure, we separate Čerenkov counters into two categories, threshold and differential.

Threshold Čerenkov Counters

Let us suppose that we have a "beam" of charged particles produced by a high-energy accelerator. The momentum and charge of the beam have been selected by bending magnets and collimators, but the composition of the beam is not unique: The positively charged beam will contain pions, kaons, and protons, while the negatively charged beam will contain pions, kaons, and antiprotons (I shall not discuss the identification of the small beam contamination of electrons and muons). Because of their different masses, the different types of particles have different values of β (for a given momentum). Threshold counters can be used with electronic circuits to pick out the particles desired for a particular experiment. A simple threshold counter is shown in Fig. 2. Čerenkov light emitted by the particle is reflected by the mirror onto the face of the photomultiplier tube, which produces a current pulse at its output. As an example of how threshold counters are used, consider the identification of kaons in a positively charged beam. For this, two threshold Čerenkov counters are needed: one with a medium in which pions give Čerenkov radiation while kaons and protons do not, the other using a medium in which kaons and pions give Čerenkov radiation while protons do not: the medium is usually gas, the refractive index being selected by choosing the operating pressure. By demanding a pulse from the second counter and the absence of a pulse in the first counter, kaons are selected. Such systems are common in elementary-particle physics since their construction is relatively simple. The major drawback is that at high momentum, pions passing through a threshold Čerenkov counter set below the kaon threshold will produce light at a very small angle. Equation (2) indicates that the intensity of the light will be relatively low; the requirement of high efficiency then leads to counters that exceed 50 m in length at momenta of ~200 GeV/c. Careful beam design is needed to accommodate such counters.

Table II

Substance	n	β_t	Thresholds p_e (GeV/c)	p_π (GeV/c)	p_p (GeV/c)
Lucite	1.50	0.67	0.00045	0.12	0.84
Water	1.33	0.75	0.00057	0.16	1.07
Freon-114[a]	1.0014	0.9986	0.0094	2.6	17.7
Air[a]	1.00029	0.99971	0.021	5.7	39
Helium[a]	1.000035	0.999965	0.060	16	112

[a] At standard temperature and pressure.

Differential Čerenkov Counters

In differential counters the refractive index is such that several types of particles produce light, but at different angles according to Eq. (1). In this case a slit and optical system is used to collect light emitted at a specific angle, chosen to correspond to the desired particle (in a focusing optical system, the Čerenkov light produced at an angle θ forms a ring image of radius $f \tan\theta$, where f is the focal length of the system). Such counters are shorter than threshold counters, but suffer from difficulties due to divergence of the beam, optical aberrations, and optical dispersion. However, sophisticated counters have been built to operate at momenta up to 400 GeV/c.

The preceding examples emphasized the use of Čerenkov counters to identify beam particles. They are also used extensively to identify particles produced in collisions of high-energy particles. Due to the broad angular spread of such particles, threshold counters have usually been employed. However, recent developments should change the situation dramatically. Large differential counters ("ring imaging Čerenkov counters") have been built in which the position and radius of the Čerenkov radiation ring are used to identify particles even in interactions with many charged particles

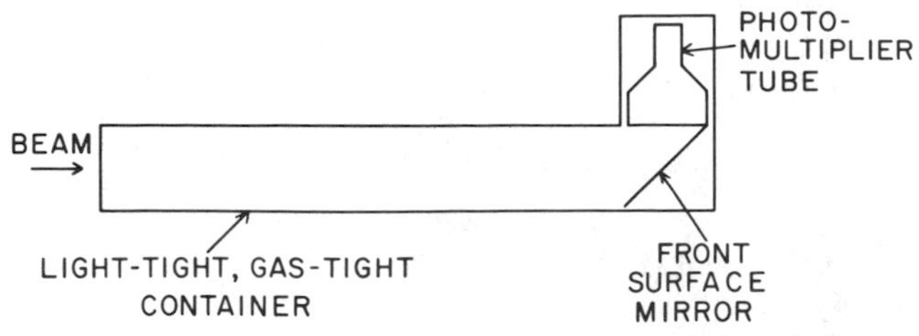

FIG. 2. A simple threshold Čerenkov counter.

hitting the detector. The devices are rather exotic, employing ultraviolet light in the range 1500–2000 Å. Light of this wavelength can produce free electrons from certain gases via the photoelectric effect; the electrons are then detected in multiwire proportional chambers, effectively measuring the location of the Čerenkov photons. With this technique, large areas can be covered with relatively inexpensive detectors. These counters will play a large part in experiments at high-energy accelerators.

See also ČERENKOV RADIATION; GAMMA-RAY SPECTROMETERS; LUMINESCENCE; NEUTRON SPECTROSCOPY; RADIATION INTERACTION WITH MATTER; SECONDARY ELECTRON EMISSION.

REFERENCES

1. J. B. Birks, *The Theory and Practice of Scintillation Counters.* Pergamon, New York, 1964.
2. V. P. Zrelov, "Čerekov Radiation in High Energy Physics," AEC TR-7099, published for USAEC and NSF by Israel Program for Scientific Translation, 1970.

Second Sound

Henry A. Fairbank

Second sound, the name given by Landau to a type of wave propagation in superfluid ^{4}He, is easily generated by a heat pulse or by periodic fluctuations of the temperature. It is detected by any temperature-sensitive receiver. Temperature fluctuations propagate obeying a wave equation instead of the usual dissipative equation of heat conduction. The pressure and density fluctuations that accompany these waves are extremely small. Thus second-sound waves are more accurately described as thermal or temperature waves.

The behavior of second sound in superfluid helium is described well by the two-fluid model and the Landau phonon–roton excitation spectrum of the liquid. In this description the liquid is composed of two interpenetrating components. The normal component, of density ρ_n, carries the entropy and thermal excitations of the liquid. The superfluid frictionless component, of density ρ_s, is the zero-entropy background component of the liquid. ρ_n increases from zero to ρ, the total density, as the temperature goes from zero to the λ point. In second sound the velocity fields of these two components are out of phase with each other by π and a temperature change is propagated without dissipation and with virtually no fluctuation in the density and pressure of the liquid. The velocity of second sound, as given by this model, is

$$u_2 = \left(\frac{\rho_s}{\rho_n} \frac{T}{C} S^2 \right)^{1/2}$$

(where T, C, and S are the temperature, specific heat, and entropy), in good agreement with experiment. Second-sound measurements have provided one of the best precision probes of this remarkable fluid.

The presence of ^{3}He in superfluid ^{4}He dramatically affects the second-sound properties. At temperatures below about 0.5 K the ^{3}He excitations provide the major contribution to the entropy and energy content of the fluid and the second-sound velocity is determined by the velocity of these excitations. Thus at low temperatures the velocity is approximately proportional to the square root of the temperature.

Is second sound a phenomenon of other superfluids? The answer is "yes" for pure He^3 which (due to a pairing interaction of the nuclear spins) becomes a superfluid below 2.7 mK. Second sound has been generated and detected in this remarkable anistropic liquid in both A_1 and the B phases.

Following theoretical suggestions, second sound has been looked for in substances other than superfluids. In a dielectric solid, heat pulses are carried exclusively by phonons. If resistive phonon processes, such as Umklapp, impurity, boundary, and imperfection scattering processes dominate, the phonons will propagate diffusively. However, if the phonon mean free path for these processes is large and the mean free path for normal (nondissipative) processes is small, heat should be propagated as second sound obeying the wave equation. These conditions are realized in pure ^{4}He and ^{3}He single crystals in the temperature range 0.4–1.0 K where second sound indeed is observed, with properties predicted by this theory. At higher temperatures, heat pulses are propagated diffusively, as in other solids. At lower temperatures, where the normal-process mean free path becomes long, the heat pulses propagate ballistically with the speed of sound.

See also HELIUM, LIQUID; QUANTUM FLUIDS; PHONONS.

BIBLIOGRAPHY

J. Wilks, *An Introduction to Liquid Helium.* Clarendon Press, Oxford, 1970.

S. J. Putterman, *Superfluid Hydrodynamics.* North-Holland, Amsterdam, 1974.

H. A. Fairbank, in *Near Zero, New Frontiers of Physics,* p. 141. W. H. Freeman and Company, New York, 1988.

L. R. Corruccini and D. D. Osheroff, *Phys. Rev. Lett.* **45,** 2029 (1980).

S. T. Lu and H. Kojima, "Proc. 18th Int. Conf. on Low Temperature Physics, Kyoto, 1987," *Jpn. J. Appl. Phys.* **26,** 199 (1987).

C. T. Lane, *Superfluid Physics.* McGraw-Hill, New York, 1962. (I)

F. London, *Superfluids,* Vol. 2. Wiley, New York, 1954. (A)

J. Wilks, *The Properties of Liquid and Solid Helium.* Oxford (Clarendon Press), London and New York, 1967. (A)

Secondary Electron Emission

P. E. Best

Electrons which emerge from a solid as the result of bombardment by energetic particles are known as secondary electrons. Here the term is used for situations in which the incident or primary particles are electrons. The secondary yield δ is the number of electrons emerging from the solid for each incident electron that enters it. One characterization of the secondary-emission properties of a sample is the plot of δ vs. K, the kinetic energy of the incident electrons at the surface. If δ_m is the maximum value of δ, occurring at energy K_m, then plots of δ/δ_m vs. K/K_m have a similar form for all solids (Fig. 1). Values of δ_m range from 0.4 for specially prepared porous surfaces, through 0.5 to 1.8 for metals, and about 1 to 20 for insulators prevented from charging, up to 1000 for samples with negative electron-affinity surfaces. Corresponding values of K_m range from 0.2 to greater than 20 keV.

For solids with $\delta_m>1$ the lower and upper values of K for which $\delta=1$ are called K_{I} and K_{II}, respectively. These values have particular significance for solids which are electrically isolated, including insulator surfaces. At equilibrium in an electron beam, the net current entering such a sample is zero. To achieve this condition in a chamber at ground potential, the sample becomes charged to a potential which causes K to be zero or K_{II}, depending on whether the initial value of K is less than K_{I}, or greater than K_{II}, respectively. For an initial K between K_{I} and K_{II} the value of δ is reduced to unity by a relatively slight positive change of sample potential which prevents the slowest secondary electrons from escaping.

The secondary emission from a solid depends on the bulk and surface electronic structures, the energy and angle of incidence of the primary beam, and the surface topography. A quantitative understanding of secondary electron emission requires solution of the transport equations which describe the cascade processes involved. Qualitatively, the main features in secondary-emission measurements are understood by means of a three-step model in which electrons are excited into high-energy conduction states directly or indirectly by the incident beam; they then move to the surface, finally emerging across the surface barrier into the vacuum. The role that one step plays in the emission is probed by experiments in which variation due to the other steps is a minimum. For the data in Fig. 1 the transport and emergence steps can be represented by an average escape depth which varies little with change of K. The escape depth is a measure of the maximum distance from the surface from which an excited electron can reach the surface and escape into the vacuum. The general form of Fig. 1 can then be understood by use of the concept of penetration depth, which is the maximum distance from the surface at which appreciable excitation of valence electrons can be caused by the incident beam. The penetration depth increases with K. For small K the escape depth is greater than the penetration depth, and a constant fraction of electrons excited by the incident beam can emerge into the vacuum. The total number of valence electrons excited by the incident beam is proportional to K, and so the yield rises linearly with K for $K<K_m$ (Fig. 1). For K greater than K_m the penetration depth is greater than the escape depth. Furthermore, the average distance traveled by incident electrons between inelastic scattering events, the mean free path, increases with increase in K for $K>200$ eV. Therefore, δ decreases as K increases for $K>K_m$ (Fig. 1).

The wide variation of values of δ_m for different samples is mainly a result of the wide variation of escape depths. A large escape depth implies a large value of δ_m. For a metal, the strong electron–electron interaction causes electrons originally excited to states above the vacuum level to lose energy, and the ability to escape into the vacuum, in a short distance (Fig. 2a). The escape depth is less than 10 nm; δ_m is small. In the insulator (Fig. 2b), electrons in the solid with energy just above the vacuum level have insufficient energy above the bottom of the conduction band to excite a valence electron across the band gap. These electrons lose energy by phonon excitation, a process with relatively long mean free path and with small energy loss per collision. This insulator will have an escape depth and δ_m larger than those for metals. Figure 2c represents the situation for a p-type single-crystal semiconductor covered with cesium to produce a negative electron-affinity surface. For such samples the escape depth can be of the order of microns, the diffusion length for electrons in the bulk semiconductor: δ_m is large. Most studies and applications of secondary emission have been made in the reflection mode. The large escape depth for negative-affinity emitters permits these materials to be used in the transmission mode, in which the secondaries emerge from the face opposite that which the incident beam enters.

A majority of secondary electrons have kinetic energies of less than 50 eV. By far the most prominent peak in the energy spectrum of the secondary electrons is the "slow"

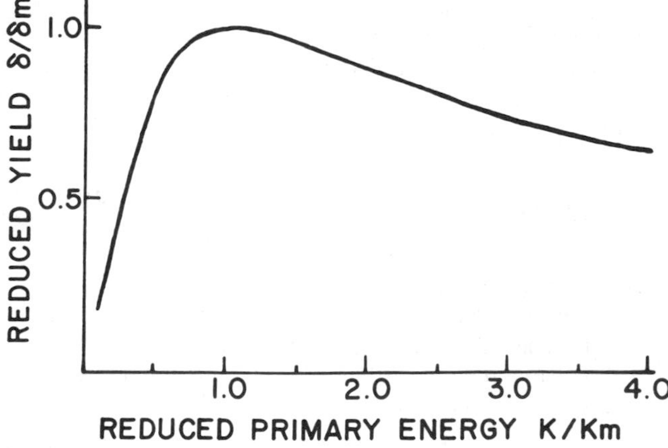

FIG. 1. Plot of reduced secondary yield, δ/δ_m, versus reduced primary energy, K/K_m. This curve has similar form for all solids.

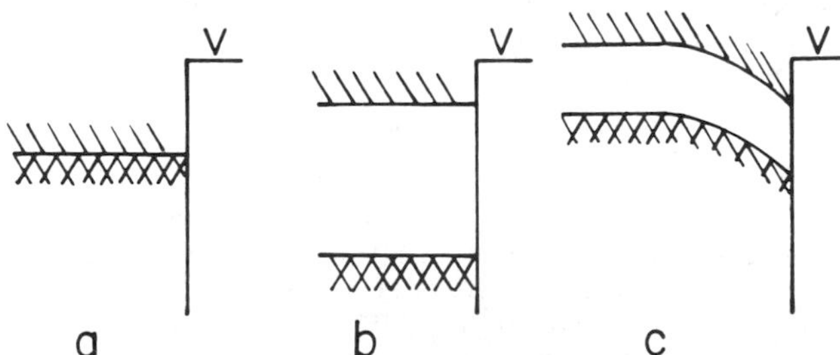

FIG. 2. Simple energy band models for (a) a metal, (b) an insulator, and (c) a semiconductor with a negative electron-affinity surface. The disposition of the energy bands with respect to the vacuum level is the main factor in determining the average escape depth, and hence δ_m. In this figure V is the vacuum level, the hatch marks are at the bottom of the unoccupied band, and the cross-hatching is at the top of the occupied band.

peak, which occurs at very low energies, less than 6 eV. The shape of the leading edge of the slow peak is strongly affected by surface dipole layers.

Very low-intensity peaks in the energy spectrum can be due to the Auger process, in which electrons are emitted in the radiationless decay of inner atomic vacancies. Auger peaks are used to identify atomic species present at surfaces and to estimate their relative abundance.

Measurements with angular resolution have been made of secondary emission from single crystals. The variation of δ with angle of incidence, at fixed K, is interpreted in terms of a penetration depth that varies with the direction of the incident beam in the crystal. Measurements of angle-resolved energy spectra at fixed input-beam conditions support a theory of excitation in which structure in the density of states is prominent in the energy distribution of the excited electrons. For lower incident-beam energies such measurements have also provided evidence for an emission process in which electrons are excited into surface resonances before emerging into the vacuum.

Currently, the most rapidly growing area of secondary electron research is concerned with the emission of spin polarized electrons from magnetic materials. The dependence of the spin polarization of emitted electrons has been measured as a function of the energy of both the secondary and the primary electrons for a number of magnetic crystals, with and without adsorbates. The results of such experiments are compared with results of theoretical calculations to better understand the transport and emission processes. An original aim of such studies was to develop a method of magnetic depth profiling.

The degree of spin polarization of secondary electrons has been used as a signal in scanning electron microscopes. Such instruments have probed the magnetic domain structure of samples, as well as elucidating the magnetization distribution of 180° domain walls at single-crystal surfaces.

Secondary emission is essential to the operation of electron multipliers and photomultipliers. The emission is used to distinguish qualitatively between different surfaces in scanning electron microscopy and is important in the operation of a number of vacuum tube devices.

See also Auger Effect; Electron and Ion Impact Phenomena; Electron Energy States in Solids and Liquids; Electron Microscopy.

BIBLIOGRAPHY

R. U. Martinelli and D. G. Fisher, "The application of semiconductors with negative electron affinity surfaces to electron emission devices," *Proc. IEEE* **62,** 1339–1360 (1974).

J. Schäfer, R. Schoppe, J. Hölzl, and R. Feder, "Experimental and theoretical study of the angular resolved secondary electron spectroscopy (ARSES) for W(100) in the energy range $0 \leq E \leq 20$ eV." *Surf. Sci.* **107,** 290–304 (1981).

R. J. Celotta and D. T. Pierce, "Polarized electron probes of magnetic surfaces," *Science* **234,** 333–340 (1986).

H. P. Oepen and J. Kirschner, "Magnetization distribution of 180° domain walls at Fe(100) single-crystal surfaces," *Phys. Rev. Lett* **62,** 819–922 (1989).

Sedimentation and Centrifugation

Ronald J. Gibbs

The process of sedimentation of one material from some other substance has been used since antiquity for the separation of materials. The separation can be of solid particles from a liquid or a gas, separation of two liquids that are immiscible, or the removal of gas bubbles from liquid. The physical process is the same for all of these examples.

When a substance settles in a medium, it is acted on by the force of gravity ($\frac{4}{3}\pi r^3 d_p g$) and at the same time by the buoyant force of the medium ($\frac{4}{3}\pi r^3 d_m g$) given by Archimedes' principle, which results in a net force $\frac{4}{3}\pi r^3 (d_p - d_m)g$, where r is the radius of particle; d_p is the density of particle; d_m is the density of medium; and g is the acceleration of gravity. As the particle moves there is a resistance to motion for all particles. The character of this resistance varies depending on the velocity and the physical properties of the medium. In the case of small particles (<60 μm in diameter) settling in water, Stokes's law applies, for the particles are settling through the fluid with laminar flow lines around the particle. The resistance which a fluid offers to the movement of a particle suspended in the fluid is

$$R = 6\pi r\eta v, \tag{1}$$

where R is the resistance in g cm/s^2; η is the viscosity of the fluid; and v is the velocity of the particle. When the resistance R equals the net force, the velocity of the particle becomes constant and remains so. Therefore the equation is

$$6\pi r\eta v = \tfrac{4}{3}\pi r^3 (d_p - d_m)g. \tag{2}$$

By solving this equation for v we obtain

$$v = \frac{2(d_p - d_m)gr^2}{9\eta}, \tag{3}$$

which is considered Stokes's law.

As the settling velocity increases because of larger particles in the fluid, higher density, or different properties of the medium, the settling is no longer laminar in nature and the resistance term changes. The change in settling velocity in water for various size spheres (Fig. 1) illustrates the change in the resistance term (Gibbs *et al.*, 1971). In this case note the transition from the laminar flow to the turbulent flow around the particle over a wide region. An equation to fit the data was obtained which for low velocities is simply Stokes's law and for higher velocities takes into account the turbulent nature of the process (Gibbs *et al.*, 1971). This equation is

$$V = \frac{-3\eta + [9\eta^2 + gr^2 d_m(d_p - d_m)(0.015476 + 0.19841r)]^{1/2}}{d_m(0.011607 + 0.14881r)}. \tag{4}$$

Another approach is often used in working with particles that settle in a nonlaminar manner, although it is not as accurate for special cases as the spheres in water given above. The drag force for the settling of these particles is Newton's law, often given as

$$F_d = C_d A d_m (v^2/2), \tag{5}$$

where F_d is the drag force; d_m is the density of the medium;

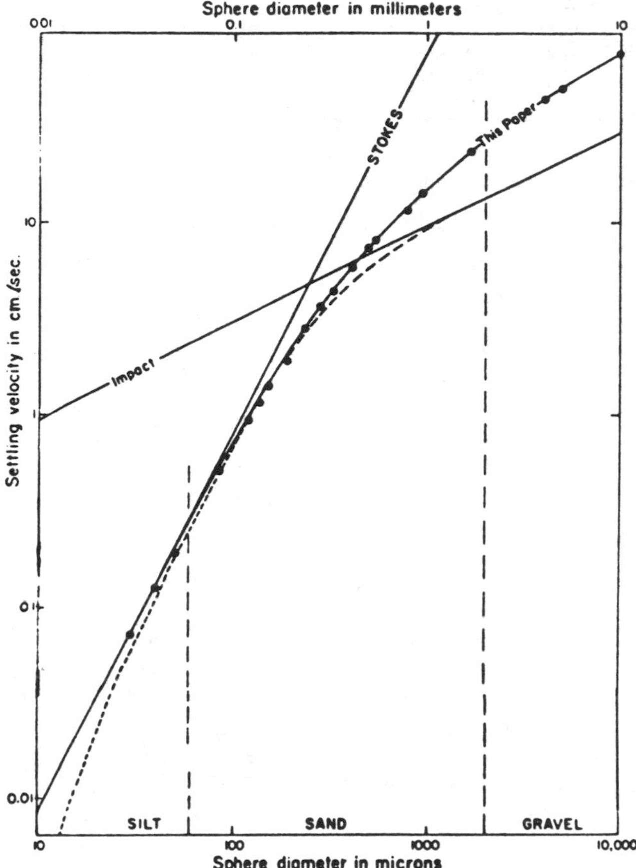

FIG. 1. Settling velocity in water for various size spheres. [Used with permission of the Society of Economic Paleontologists and Mineralogists, from R. J. Gibbs *et al.*, *J. Sedimentary Petrology* **41**, 7–18 (1971).]

A is the projected area of the particle perpendicular to the direction of motion; v is the particle velocity; and C_d is the drag coefficient. The drag is equal to the net motion force on the particle when the terminal velocity is attained. The net motion force is the combination of gravitational force and Archimedes buoyancy:

$$(\pi/6)D^3(d_p - d_m)g, \tag{6}$$

where D is the particle diameter. The terminal velocity is attained when the drag force [eq. (5)] balances the net motion force [Eq. (6)] and is expressed as

$$v^2 = \frac{4(d_p - d_m)gD}{3d_m C_d}. \tag{7}$$

The Reynolds' number (Re) is the ratio of inertial forces to viscous forces in a system and is defined as

$$Re = d_m vD/\eta, \tag{8}$$

where η is the coefficient of viscosity of the liquid. Schillar and Nauman (1933) proposed the following relationship between Re and C_d for use in determining the velocity:

$$C_d = (Re/24)(1 + 0.15Re^{0.687}). \tag{9}$$

One technique proposed by Heywood (1962) to obtain the velocity for particles that settle in a nonlaminar manner utilizes tables he prepared, along with corrections for various nonspherical particles.

For aggregates of particles settling in water, Gibbs found $V = 1.73D_f^{0.78}$, where V is the aggregate settling velocity in cm s^{-1} and D is the aggregate diameter in cm. These aggregates have irregular shape and high porosity.

The process of sedimentation for small particles has a very low settling velocity and therefore any usage entails a long time interval. A commonly used method of speeding up the sedimentation process is the application of centrifugation techniques to increase the gravitational term in the equations given. The centrifugal force is given by

$$\text{centrifugal force (dynes)} = (2\pi N)^2 R, \tag{10}$$

where N is the number of revolutions per second and R is the radius of rotation in centimeters. A more useful form is to express the centrifugal force as the number of times that gravity has been increased, expressed as follows:

$$\text{RCF (or } Xg) = (2\pi N)^2 R/980, \tag{11}$$

where RCF is relative centrifugal force, Xg is the number of times gravity has been increased, 980 is the acceleration of gravity expressed in dynes per gram. As the particle moves in the centrifuge tube the radius of rotation will change during the experiment so that, in actual use, this changing radius must be corrected for by integrating the various radii. The resulting "Xg" term is then simply incorporated into any of the settling equations given above to obtain the velocity needed. This procedure must be done for each centrifuge for all have different radii of rotation.

See also RHEOLOGY; TURBULENCE; VISCOSITY.

BIBLIOGRAPHY

R. J. Gibbs, *J. Geophys. Res.* **90,** 3249–3251 (1985).

R. J. Gibbs, M. D. Matthews, and D. A. Link, *J. Sedimentary Petrology* **41,** 7–18 (1971).

H. Heywood, *Symposium on the Interaction between Fluids and Particles,* p. 1. Institute of Chemical Engineers, 1962.

Sir I. Newton, *Principia,* Lib. 11, loc. cit. 21.

Sir G. G. Stokes, *Mathematical and Physical Papers,* 11.

Seismology

W. M. Adams

DEFINITION AND SUBDIVISIONS

Seismology is the study of vibrations. Vibrations can be classified according to the location of occurrence, for example, in a planet or in a machine. Vibrations in planets, planetquakes, can be classified according to the planet. On the Moon, there are moonquakes; on Earth, there are earthquakes. (However, an earthquake felt at sea is called a seaquake.) Earthquakes are classified according to the causative agent, man-made or natural. Natural earthquakes are further subdivided by the cause as: volcanic, due to subterranean magma movement, or tectonic, due to abrupt relative movement of two land masses. This ordering of vibrations is summarized in Table I.

ATTRIBUTES OF TECTONIC EARTHQUAKES

Coordinates

A tectonic earthquake is characterized by eight parameters. The location of the initial abrupt relative movement is called the focus of hypocenter and is fixed by three space coordinates and one time coordinate. The space coordinates are the latitude, longitude, and the depth. The position on the surface of the Earth directly above the focus is called the epicenter and is specified geographically by only the latitude and longitude. The epicenter of an earthquake may lie anywhere, but the focus will have a depth less than 750 km. The time coordinate is the origin time, usually given in Universal Time (formerly called Greenwich Mean Time). Earthquakes may occur at any time.

Size

The size of a tectonic earthquake is characterized by the amount of energy released as vibrations. This energy is stated in ergs or joules (10^7 ergs equal one joule). The seismic energy of a tectonic earthquake may range from less than 10^{15} ergs to more than 10^{22} ergs. An estimation of the energy is quite difficult; hence an operational definition of size and magnitude has evolved and will be defined later.

Strike and Dip

The abrupt, relative movement of two earth masses occurring on a planar surface is called a fault. Specification of the orientation of the fault plane requires two parameters: the azimuth of the fault trace of the strike, and the dip of the fault. Faults are classified according to the orientation of the relative motion of the two land masses: strike-slip or dip-slip. Strike-slip faulting does relatively more damage to buildings; dip-slip faulting causes relatively greater tsunamis, often called, erroneously, tidal waves.

Plunge

The other parameter characterizing a tectonic earthquake is the direction of the motion vector in the fault plane. This can be parametrically specified by giving the dip of the motion vector, called the plunge.

To summarize, the eight parameters describing a tectonic earthquake arc:

1. latitude } epicenter } focus
2. longitude } epicenter } focus
3. depth } focus
4. origin time
5. size (energy or magnitude)
6. strike (or azimuth)
7. dip
8. plunge

Table I Partial Ordering of Vibrations

Vibrations	Planetquake	Earthquake	Natural	Tectonic
	Machinery	Moonquake	—	—
	Other	Other	Man-made	Volcanic

Estimation of Attributes of a Tectonic Earthquake

The manner in which each of these properties is determined or estimated will now be described. The interrelationships among the parameters will then be considered.

Seismology actually involves determining both the parameters of earthquakes and the parameters characterizing the velocity structure of the medium in which the seismic energy has been propagated. For simplicity, the velocity structure of the Earth is assumed known and the earthquake parameters determined. The earthquake parameters could be assumed known and the velocity structure estimated. Indeed, seismology may be considered as an alternating iteration between these two approaches.

The study of earthquakes, assuming a known velocity structure, involves estimating the earthquake parameters, determining the interrelationships of the parameters, and investigating the possible physical meaning of these findings. Such study may use the sense of man aided by instruments; this may be called sensory seismology or observational seismology (Adams, 1964). If special instruments are used to augment the sense and provide a geographic historical record, then instrumental seismology is involved. (The study of the instruments *per se* is seismometry [Sohon, 1942].) Sensory seismology is still important since tectonic earthquakes are uncontrollable events not producible on demand, hence of notable rarity—fortunately for society, unfortunately for seismologists. This rarity makes even vague historical information of importance.

SENSORY SEISMOLOGY

An earthquake is the vibration of the earth *per se*. Phenomena directly related to this ground vibration are to be *primary effects*. All other phenomena are said to be *secondary effects*. In particular, the response of anything to the ground motion is a secondary effect, e.g., a building collapsing.

Any particular fact about an earthquake can be classified as sensory and primary, sensory and secondary, instrumental and primary, or instrumental and secondary. This classification will be covered in order.

Sensory Study of Primary Effects of Earthquakes

The sensory study of earthquake effects may be arranged according to the bodily sense involved.

Sound. There is often a sound like thunder or a gunshot accompanying the tremor. The sound may be a slow rolling sound or a short quick clap. (Sounds of buildings are secondary effects). Similar sounds, called brontides, have been heard without accompanying ground motion.

Sight. Some earthquakes have been accompanied by lights. These have ranged from a soft sky glow like the aurora to bright streaks across the heavens. Evaluation of sensory perceptions is often difficult; for example, the following "sight":

GREAT EARTHQUAKE AT KAIMU, 1823

We also examined the effects of an earthquake experience in this place about two months before. We were informed that it took place about ten o'clock in the evening. The ground, after being agitated some minutes with a violent tremulous motion, suddenly burst open, for several miles in extent, in a direction from north to east, from south by west, and emitted, in various places at the same instant, a considerable quantity of smoke and luminous vapour, but none of the people were injured by it. A stone wall, four feet thick, and six feet high, enclosing a garden at the north end of the village, was thrown down.

A chasm about a foot wide marked distinctively its course; this was generally open, though at some places it seemed as if the earth had closed up again.

AN EARTHQUAKE CHASM

We entered a house, sixteen feet by twelve on the inside, through which it has passed. Ten persons, viz. one man, six women, and three children, were asleep here at the time it occurred. They were lying on both sides of the house, with their heads towards the centre; some of them very near the place where the ground was rent open. The trembling of the ground, they said, awoke them, but before they could think what it was that had disturbed them the earth opened with a violent percussion; a quantity of sand and dust were thrown up with violence, and smoke and steam were at the same time emitted.

After a short interval, a second percussion was felt, vapour again rose, and at the opposite end of the house to that in which they were lying, they saw a light blue flame, which almost instantly disappeared.

We asked them if they were not alarmed? They said they were at first, but after remaining awake some time, and finding the shock was not repeated, they lay down and slept till morning, when they filled up the fissure with grass and earth.

(From the *Journal of William Ellis* [1827] Advertiser Publishing Co., Ltd., Honolulu, Hawaii 1963.)

No similar observation is known.

The passage of waves on the surface of the ground, looking like the ocean swell at sea, has been reported for very large earthquakes. The opening and closing of crevasses has been reported. Both types of reports seem reliable.

Smell. All reports of odors related to earthquakes, e.g., gases released from water ejected in sand blows, appear to be of secondary effects.

Taste. All reports of tastes related to earthquakes appear to be of secondary effects, e.g., new crevasses allowing sea water to contaminate a coastal water well.

Touch (or Kinesiologic). During very large earthquakes, it may be impossible to stand, walk, or run.

Sensory Study of Secondary Effects of Earthquakes

For most civilized persons, the response of man-made objects to ground motion is the most noticeable feature of an earthquake. Indeed, a scale primarily based on such effects is used to measure the intensity of the earthquakes at a given location. This is called the Modified-Mercalli Scale given in Table II in an abridged form. The intensity level is deter-

Table II Modified Mercalli Intensity Scale of 1931 (Abridged)

I. Not felt. Marginal and long-period effects of large earthquakes.

II. Felt by persons at rest, on upper floors, or favorably placed.

III. Felt indoors. Hanging objects swing. Vibration like passing of trucks. Duration estimated. May not be recognized as an earthquake.

IV. Hanging objects swing. Vibration like passing of heavy trucks; or sensation of a jolt like a heavy ball striking the walls. Standing motor cars rock. Windows, dishes, doors rattle. Glasses clink. Crockery clashes. In the upper range of IV wooden walls and frame creak.

V. Felt outdoors; direction estimated. Sleepers wakened. Liquids disturbed, some spilled. Small unstable objects displaced or upset. Doors swing, close, open. Shutters, pictures move. Pendulum clocks stop, start, change rate.

VI. Felt by all. Many frightened and run outdoors. Persons walk unsteadily. Windows, dishes, glassware broken. Knickknacks, books, etc., off shelves. Pictures off walls. Furniture moved or overturned. Weak plaster and masonry D cracked. Small bells ring (church, school). Trees, bushes shaken.

VII. Difficult to stand. Noticed by drivers of motor cars. Hanging objects quiver. Furniture broken. Damage to masonry D, including cracks. Weak chimneys broken at the roof line. Fall of plaster, loose bricks, stones, tiles, cornices. Some cracks in masonry C. Waves on ponds; water turbid with mud, Small slides and caving-in along sand or gravel banks. Large bells ring. Concrete irrigation ditches damaged.

VIII. Steering of motor cars affected. Damage to masonry C; partial collapse. Some damage to masonry B; none to masonry A. Fall of stucco and some masonry walls. Twisting, fall of chimneys, factory stacks, monuments, towers, elevated tanks. Frame houses moved on foundations if not bolted down; loose panel walls thrown out. Decayed piling broken off. Branches broken from trees. Changes in flow or temperature of springs and wells. Cracks in wet ground and on steep slopes.

IX. General panic. Masonry D destroyed; masonry C heavily damaged, sometimes with complete collapse; masonry B seriously damaged. Frame structures, if not bolted, shifted off foundations. Frames racked. Serious damage to reservoirs. Underground pipes broken. Conspicuous cracks in ground. In alluviated areas sand and mud ejected, earthquake fountains, and craters.

X. Most masonry and frame structures destroyed with their foundations. Some well-built wooden structures and bridges destroyed. Serious damage to dams, dikes and embankments. Large landslides. Water thrown on banks of canals, rivers, lakes, etc. Sand and mud shifted horizontally on beaches and flat land. Rails bend slightly.

XI. Rails bend greatly. Underground pipelines completely out of service.

XII. Damage nearly total. Large rock masses displaced. Lines of sight and level distorted. Objects thrown into the air.

Masonry A. Good workmanship, mortar, and design; reinforced, especially laterally, and bound together by using steel, concrete, etc.; designed to resist lateral forces.

Masonry B. Good workmanship and mortar; reinforced, but not designed in detail to resist lateral forces.

Masonry C. Ordinary workmanship and mortar; no extreme weaknesses like failing to tie in at corner, but neither reinforced nor designed against horizontal forces.

Masonry D. Weak materials, such as adobe; poor mortar; low standards of workmanship; weak horizontally.

mined by the phenomena observed during the earthquake at a location; the intensity is not indicative of the size of the earthquake. For example, at a given place a small earthquake nearby may produce the same intensity as a large earthquake far away.

Intensity may be mapped by collecting observations over an area about the epicenter.

The values of intensity are plotted on a map of the epicentral area and equal values of intensity, called isoseismals, contoured. If isoseismal maps for many earthquakes are available for a given region, then a seismic risk map may be constructed. The highest intensity ever experienced at a given location is plotted at that location. Such a map of maximum experienced intensity may be indicative of relative seismic risk over the region. The approximate number of years between observation of that maximum intensity is indicated by subnumerals.

Sensory Study of Earthquake Parameters

The location of the epicenter can be estimated as the center of the concentric, usually elliptical, isoseismals. The depth of the focus can be estimated from the spacing of the isoseismals: the wider the spacing the deeper the focus. The energy release may be estimated from values for depth and maximum intensity. The strike of the fault-plane is given by the major axis of the isoseismals. An estimate of the dip can even be obtained, assuming equal movement on the entire fault plane, by comparing the gradient of the isoseismals on opposite sides of the epicenter: the fault dips toward the side with higher gradient, i.e., more area at higher intensity. The plunge has not been determined from meaningful sensory data. For details, see Davison (1921).

INSTRUMENTAL STUDY OF PRIMARY EFFECTS OF EARTHQUAKES

The measurement of ground motion with seismographs is the instrumental study of primary earthquake effects. Infrasounds have been measured from some large earthquakes. No lights, smells, or odors have been measured.

INSTRUMENTAL STUDY OF THE ATTRIBUTES OF A TECTONIC EARTHQUAKE

The focus and origin time are estimated by a least-squares procedure. The size is characterized by the magnitude which is defined in terms of the amplitude recorded at a specific distance on a specified instrument having specified characteristics and on specified geological structures. A Wood-Anderson torsion seismometer, having a dynamic magnification of 2800, a period of 0.8 sec., and 0.8 damping is assumed to be located on basement rock at a distance of 100 km from the epicenter. An actual observation is reduced to this standard by correction for:

1. geologic structure underlying the seismological observation,
2. characteristics of the seismograph used,
3. the phase (or the component of the phase) measured,
4. the depth of the focus, and
5. the epicentral distance.

Let the reduced amplitude be A; then the magnitude is defined as

$$M = \log_{10}(A/A_0)$$

where A_0 is the reference record amplitude of one micron (10^{-6} m). The reference amplitude, corresponding to a "zero magnitude" earthquake, was originally chosen to be smaller than the smallest measurable recording. Since then, high-gain seismographs permit amplitude measurements corresponding to negative magnitudes.

In practice, values of the magnitude range from less than 0 to over 8. However, as can be seen from the definition, theoretically there is no limit, either upper or lower. Increasing the amplitude by a factor of ten increases the magnitude one unit.

An empirical relationship between magnitude and energy has been estimated (Richter, 1949):

$$\log_{10}E = 11.8 + 1.5M$$

Theoretically the energy should be proportional to the square of the amplitude (Bullen, 1953, p. 126). The discrepancies may be due to the effects of earthquake mechanism on magnitude being ignored.

The strike, dip, and plunge may be estimated by reduction of seismograph observations back to the motion of a sphere of unit radius about the focus (Adams, 1958). The recordings are corrected for orientation of the seismographs, relative magnification and phase shifts of the seismographs, surface reflecting effects, and any propagation effects. For certain reasonable source mechanisms the distribution of motions on the unit focal sphere is known, and conversely. The determination of strike, dip, and plunge are not yet routine practice; special study is still required.

INSTRUMENTAL STUDY OF SECONDARY EFFECTS OF EARTHQUAKES

There is only one known case of the Earth "swallowing" a person during an earthquake. The greatest loss of life is usually due to the collapse of buildings upon the occupants and bystanders. Modern construction does not necessarily mean earthquake-resistant construction. Properly constructed buildings, even on material which liquifies due to the vibration (said to be thixotropic material), may have surviving occupants. The buildings are practically monolithic "ships." Very few records of buildings' response to earthquakes have been obtained. To alleviate this lack of information, all buildings constructed in Los Angeles higher than a given level must contain a three component strong-motion seismograph. The Uniform Building Code contains a Seismic Risk Map of the United States, which guides engineering design efforts. The higher the number, the greater the risk, so the better the required construction. The Russian practice has been outlined by Medvedev (1962).

INTERRELATIONSHIPS OF EARTHQUAKE ATTRIBUTES

Size versus Location

The number of earthquakes of size larger than magnitude 7 has been determined for the period 1897–1964. For an equal-area division of the earth (area of each division 7×10^5 km^2), for the period 1897–1964, we have determined that the number of earthquakes of magnitude greater than 7 and, using the following equation, an energy index:

$$K = 10 \log_{10} \left[\sum_{i=1}^{h} 10\{11.8 + 1.5M(i)\} \right]$$

From this and other seismicity maps, a general rule has evolved:

The greater the surface gradient, the greater the seismicity.

Furthermore, studies relating secular vertical velocity to seismicity provide another rule-of-thumb:

The greater the secular vertical velocity, the greater the seismicity.

Number versus Size

For a given area and depth range, the general finding is that the greater the number (i.e., frequency of occurrence) of a given size, then the greater the extreme size for the given time interval. The form of the general equation for annual frequency of occurrence is:

$$\log_{10} N = a + b(8 - M)$$

where the constants a and b relate to the level of seismicity and to physical properties and state-of-stress at the focus. The value of b is inversely related to the state-of-stress (Berg, 1968). Tremors, shortly before a major earthquake, called foreshocks, have b values of 0.6 or less; tremors after a major earthquake, called aftershocks, have b values from 0.7 to 1.1. As a world average for earthquakes with foci less than 60 km deep,

$$\log_{10} N = 0.5 + 0.9(8 - M)$$

Hence, it is found that a decrease of one in magnitude means an eightfold increase in number.

Number versus Depth

The number of earthquakes in the entire world between 1904 and 1949 has been plotted versus depth. There are three distinct maxima, near the surface, at 400 km, and at 600 km. Detailed examination of specific tectonic areas indicate that the number and depth of maxima vary. A maximum at depth has been correlated with rate of sea-floor spreading (Isacks, et al., 1968).

Energy Release versus Time

The energy released in a given area is primarily in the large earthquakes, despite the greater number of small earthquakes. Whereas a decrease of 1 in the magnitude is an eightfold increase to the number, it corresponds to a decrease of about 40 in the energy released. The energy may be accumulated over a period of time. An apparent change in worldwide seismicity about 1949 is possible. No reasonable explanation is available.

Fault–Plane Orientation versus Position

The merger results of earthquake mechanism studies have been interpreted by some to confirm the "sea-floor spreading" hypothesis; then the principal stress axis in the elastic stress field of the earth can be mapped. Such interpretations, however, should still be considered tentative. For example, the sea-floor spreading hypothesis necessitates one of two stress fields. In either case, the axes of principal stresses for intermediate and deep earthquakes must be the same. However, there are significant active areas, Japan, for example, which have the pressure axis horizontal for shallow earthquakes, dipping from continent to ocean for intermediate earthquakes, and parallel to the tectonic axis for deep earthquakes!

Clearly, many exciting discoveries are yet to be made in seismology.

Energy Release versus Space and Time. Intensive study of the seismicity over a tectonic region at several adjacent regions often indicates the possibility that the zone of high energy is migrating (Blot, 1963). More work is appropriate.

Periodicity of Large Earthquakes. Several intensive studies show no short-term periodicity of earthquakes. However, there may be a very long periodicity (hundreds of years) of extremely large earthquakes in a given region. For example, in the western Iberian Peninsula there have been earthquakes of magnitude greater than 8 in 1146, 1344, 1531, 1755, and 1968.

See also GEOPHYSICS.

BIBLIOGRAPHY

K. E. Bullen, *An Introduction to the Theory of Seismology*, 2nd ed. Cambridge, London and New York, 1953.

L. Cagniard, *Reflection and Refraction of Progressive Seismic Waves* (transl. and rev. by E. A. Flinn and C. Hewitt Dix). McGraw-Hill, New York, 1962.

W. C. Elmore and M. A. Heald, *The Physics of Waves*. McGraw-Hill, New York, 1969.

B. Gutenberg and C. F. Richter, *Seismicity of the Earth and Associated Phenomena*. Princeton Univ. Press, Princeton, NJ, 1954.

H. Kolsky, *Stress Waves in Solids*. Dover, New York, 1963.

L. D. Leet, *Practical Seismology and Seismic Prospecting* (The Century Earth Science Series, K. F. Mather, ed.). Appleton, New York, 1938.

J. B. Macelwane and F. W. Sohon, *Introduction to Theoretical Seismology,* Pt. I, *Geodynamics;* Pt. II. *Seismometry.* St. Louis Univ., Saint Louis, MO, 1932.
A. W. Musgrave, ed., *Seismic Refraction Prospecting.* Society of Exploration Geophysicists, George Banta, Menasha, WI, 1967.
C. B. Officer, *Introduction to the Theory of Sound Transmission with Application to the Ocean.* McGraw-Hill, New York, 1958.
C. F. Richter, *Elementary Seismology.* W. H. Freeman, San Francisco, 1958.
J. S. Rinehart, *Stress Transients in Solids.* Hyper Dynamics, Santa Fe, NM, 1975.
R. Van Nostrand, ed., *Seismic Filtering* (N. Rothenburg, transl.). Society of Exploration Geophysicists, Tulsa, OK, 1971.
J. E. White, *Seismic Waves: Radiation, Transmission, and Attenuation.* McGraw-Hill, New York, 1965.

Semiconductor Radiation Detectors

H. W. Kraner

INTRODUCTION

Semiconductors have been actively used as a solid medium for radiation detection for about 20 years. Silicon and germanium have been most extensively (but not exclusively) used not only because of their intrinsic solid-state properties but also because of the high state of development of the arts of producing pure materials. Semiconductors provide a complete solid analog to a gas-filled ionization chamber in which not only can radiation be registered as an event but other properties such as its energy deposition and time of arrival can be directly determined from the ionization produced in a specific sensitive volume. As in a gas-filled chamber, an electric field is used to sweep out and collect the ionization (both positive and negative charges: holes and electrons) caused by impinging nuclear radiation. The great advantage of a solid over a gas is that it provides a radiation stopping power enhanced by at least the density ratio of $\sim 10^3$. The volume in which the electric field exists is the sensitive region of the detector and is usually configured in either "parallel-plate" geometry with planar electrodes opposing each other at a given spacing or in coaxial cylindrical geometry with the electrodes being inner and outer cylindrical shells. An internal field of these shapes can be produced in a semiconductor as the reverse-bias condition in a diode structure. A material doped such that positive holes are the majority charge carrier is called *p* type and a material having negative electrons as the majority carrier is called *n* type. A diode structure is formed by the union of *p*- and *n*-type materials which permits current (positive charge) to flow only if the potential of the *p*-type side is raised (biased) relative to the *n*-type side. If the device is biased with the *p*-type side negative with respect to the *n*-type side, conduction current cannot flow and the bias potential appears across the region of the physical *p*-to-*n* transition. The collection field is of the order of the applied bias (V_A, volts) divided by the "depletion" depth W or extent of material depleted of free carriers by the field. In an abrupt *p-n* junction, W extends from the junction into the material according to

$$W = (2\epsilon\rho\mu V_A)^{1/2}, \qquad (1)$$

where ϵ (F/cm) is the dielectric constant of the material, ρ is the resistivity (Ω cm), and μ is the mobility of the majority carrier (cm^2/V s). Usually the impurity concentration of one side is much greater than the other; for example if $\rho_p \ll \rho_n$, the active region extends primarily into the *n*-type side of the device and it is referred to as a p^+n structure. To further the example, in a p^+n silicon device where μ is μ_e^-, equal to about 1400 cm^2/V s at 300 K, and taking $\rho = \rho_n \approx 500\ \Omega$ cm with $\epsilon = 12 \times \epsilon_0 = 12 \times 8.85 \times 10^{-14}$ F/cm, we find that a 100-V reverse bias will produce a depletion depth or active region of about 120 μm. This device would be appropriate to detect natural alpha radioactivity.

In addition to the ability to support internal fields, the second critical quality of the semiconductor is that it offers charge-carrier lifetimes (τ) and mobilities sufficient to permit traversals of relatively large sensitive regions (~1 cm). The $\mu\tau E$ product, where E is the electric field, is a length which if greater than the detector dimensions assures full collection of all ionization produced regardless of position in the active region. This is generally the case in Ge and Si. As charge is collected from the active region, an induced charge appears on the terminals of the device. The output rise time is of the order of the active width divided by a carrier saturation velocity of $\sim 10^7$ cm/s.

The amount of charge collected from the ionizing event is related to the energy loss of the event (for example, a γ ray of energy E, if the γ ray is totally absorbed in the sensitive volume) by the *average* energy required to create an electron–hole pair, ϵ. Thus, $Q = qE/\epsilon$, where $\epsilon_{Si} = 3.64$ and $\epsilon_{Ge} = 2.96$ eV/ion pair and q is the electronic charge. These values are not greatly changed over a wide range of temperature or for various ionizing projectiles, electrons, protons, α-particles, etc., and are about one-tenth that for a gas. The average energy loss per ion pair, ϵ, is often taken to be ~3 times the semiconductor band gap (or ionization potential) with some theoretical justification.

The energy resolution of a charge collecting device is described by the observed spread of energy observations about an average energy loss, E_0, in which the average number of charge carriers collected per event is E_0/ϵ. The distribution about E_0 is Gaussian with a full width at half-maximum (FWHM) of $2.35\sqrt{\epsilon E_0}$. To correct for the fact that the individual ionizations in each event are not independent (but must add up to exactly the total energy loss), a factor F, called the Fano factor, is introduced into the variance, so that the expected full width at half-maximum of a distribution is $2.35\sqrt{F\epsilon E_0}$ where for small volumes of good material $F \approx 0.1$ but may range from 0.07 to 0.12.

The statistical fluctuations just described must be added in quadrature to the "noise" of the electronic system attached to record the collected charge per event. Figure 1 shows the expected energy resolutions for silicon and germanium devices of two important cases: (a) a small silicon detector configured to minimize the electronic or system resolution, taken to be 100 eV, and (b) a much larger germanium detector and system with an electronic resolution of 500 eV. A Fano factor of 0.1 is used and these values of energy resolution are realized in practice. In the first case it is important

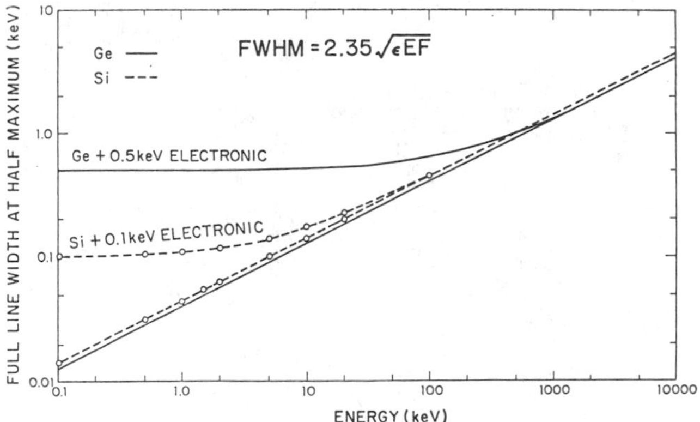

FIG. 1. Theoretical energy resolutions (full width at half-maximum) of silicon and germanium semiconductor detectors as a function of energy loss combined with typical electronic or system contributions.

to note that the silicon detector is capable of detecting radiations down to ≤ 1 keV with an energy resolution of ~120 eV, which permits the distinction of the characteristic x rays of elements over the entire periodic table down to the element sodium and is useful for elemental analysis. It is also interesting to note from Fig. 1 that it is the carrier statistics which limit detector resolution above ~300 keV for germanium detectors where they are most used for γ-ray spectroscopy.

CONFIGURATIONS

Before describing the several practical configurations of semiconductor devices, two more general concepts must be introduced: the practical requirements of sensitive volume and leakage current. In a junction detector, the sensitive volume is limited by the depletion depth which, in turn, is dependent upon the applied reverse bias and the resistivity or chemical purity of the available material. Very large sensitive volumes or depletion depths are often desirable, especially for γ-ray detection, which are not permitted by the practical limits of these parameters. This problem was first surrmounted in both Si and Ge by the technique of lithium drifting in which the *n*-type dopant, Li, is diffused into the crystal as the *n* contact on *p*-type material. Because Li is an interstitial dopant it can be easily "pulled" through the crystal with a biasing field where it selectively pairs with the *p*-type impurities to achieve a compensated high-resistivity material. Thus, compensated regions between 5 and 15 mm can be obtained, creating large active volumes of semiconductor; in a coaxial or cylindrical geometry, volumes of over 100 cm^3 in germanium can be obtained which provide substantial counting efficiency for γ rays over the energy range of interest for nuclear physics, 50 keV to 10 MeV. Germanium detectors of large volumes revolutionized nuclear spectroscopy during the late 1960s and 1970s. Lithium-drifted silicon detectors in a planar geometry have been long used as thick charged-particle detectors and high-resolution x-ray detectors. As x-ray detectors, about 1 mm of silicon is adequate for detection efficiency; however, to achieve low-noise performance, as indicated in Fig. 1, the detector capacitance must be kept low restricting the active area to <1 cm^2 and the thickness to >3 mm, requiring lithium drifting. This planar detector has found wide usage as an x-ray spectrometer with the capabilities given by Fig. 1 and shown in Fig. 2C.

Although lithium drifting was used extensively to produce large-volume germanium detectors, it is a difficult technology and considerable effort was put into producing very high purity which was shown to be theoretically possible in the range of 10^{10} impurities/cm^3. Today, by careful attention to the sources of impurities and physical defects, large germanium crystals are produced with such low impurity content that depletion depths of over 2 cm can be achieved with applied potentials of several thousand volts [Eq. (1)]. Crystal diameters approach 8 cm which permits large high-resolution γ-ray spectrometers having full energy peak efficiencies equal to a 3×3-in. NaI(T1) detector for the 1.17-and 1.33-MeV γ rays of ^{60}Co. The charge collection properties (i.e., absence of charge trapping) of high-purity germanium is also generally excellent so that the energy resolution predicted in Fig. 1 is realized in practice. Some high-purity silicon is available; however, lithium drifting is still (only) used for thick silicon detectors.

The first and still common detector configuration used is that of the planar junction detector having an active volume simply defined by the applied bias or by the physical thickness of the material itself, if the depletion depth extends from the rectifying contact to the opposite edge of the device. In such a totally depleted device the opposite or back contact must be Ohmic and not inject carriers into the active volume.

Silicon detectors formed as *p-n* junctions with bias- and material-dependent depletion depths ranging from 10 to 200 μm are useful for a variety of radiations ranging from fission fragments to high-energy heavy ions and electrons. They are planar structures with the junction formed by either opposite impurity diffusion, ion implantation, or the formation of a Schottky surface barrier. A Schottky surface-barrier junction structure is formed by the combined effects of the work function of a given metallization and the surface states of the semiconductor to yield a given, fixed surface charge that is opposed to that of the bulk material forming a junction at the surface. Gold and palladium form *p*-type surfaces on silicon and are used on *n*-type material; aluminum forms an *n*-type surface and is used as an "Ohmic" or nonrectifying contact to *n*-type material or as a weak barrier contact on *p*-type material. Both surface-barrier diodes and ion-implanted junction detectors can yield devices with very thin (<100 Å) entrance windows or inert layers. Evaporated palladium Schottky barriers are also used as rectifying contacts on high-purity germanium detectors.

Semiconductor devices are operated as reverse-based diodes and draw leakage current with typical diode reverse direction characteristics. The fluctuation of the leakage current is the noise over which a charge pulse must be distinguished and is proportional to the leakage current itself. Near the ambient temperature range the leakage current in silicon and germanium is due to carrier generation by thermal excitation over the semiconductor band gap with a $e^{-E_g/2kT}$ dependence, where E_g, the band gap, in silicon is 1.14 eV and 0.68 eV in germanium.

Due to the lower band gap of germanium, germanium spectrometers must be cooled to reduce leakage currents to below ~1 nA which gives comparitively negligible current-generated noise. Li-drifted silicon detectors for x-ray spectroscopy with the capability of 100 eV system resolution must also be cooled to yield subpicoampere leakage currents. Convenience has been persuasive in the use of liquid nitrogen as a cryogenic coolant (77 K) which provides an acceptably low temperature for both silicon and germanium devices. Part of the Li-drifting technique for germanium involves a careful final cooling to the liquid-nitrogen operating temperature; the device must then remain at this temperature over its operating lifetime. However, now that most germanium spectrometers are fabricated from high-purity germanium, using more stable contact dopants as well, the device need only be cooled in the cryostat during use and may be stored indefinitely at room temperature. Cooled devices must be contained in an evacuated container called a cryostat to prevent heat input from thermal conduction of a gas ambient. The cryostat vacuum is generally maintained by cryopumping of an internal organic molecular sieve held at the coolant temperature. The clean vacuum condition of the cryostat is also helpful to maintain the very high surface cleanliness of the semiconductor required to keep surface leakage currents low.

Another configuration of junction detectors for charged particles has recently found application to register not only the event itself, but its spatial position on the detector. Several position-sensitive detectors sensitive to a particle position in the plane transverse to the incident direction can be placed in a row or stack along the trajectory to measure the track of a penetrating particle. Silicon-junction detectors of large area comprising the entire area of a 2- or 3-in. diam crystal cut into a thin wafer, which are position sensitive, have found large-scale recent application as tracking devices in high-energy physics experiments which must sort out the details of events which produce many minimum ionizing particles. The methodology of integrated circuit manufacture has been used to produce many minute and separate contact areas on a silicon wafer using the electrical isolation provided by the oxide passivation and the oxide technology of integrated circuit fabrication. A common one-dimensional device is known as a "strip detector" and consists of narrow active junction strips formed by photolithography which will collect charge from a particle traversal and register a "hit" along that strip (analogous to a wire in a multiwire gaseous detector). Figure 2 shows several such detectors with part A being a typical strip detector having 2400 individual strips, 12 μm wide on a 25-μm pitch across most of the 3-in.-diam wafer. The individual strips are aluminum wire bonded to the copper lines of the printed circuit mother board, which in turn must then be connected to an external electronics chain of preamplifier, amplifier, etc. Each strip is, of course, not visible; however, the "fan-out" of the lines from each strip gives some idea of the complexity of this device; it should be noted that only every sixth strip is actually read out in this device, giving 400 output lines, and the hits on the included five strips are registered by charge division between the read out strips. Detector C is another strip detector built on a 2-in. wafer with 50-μm strip pitch in which every strip is read out with the fan-out divided between top and bottom; an intermediate mother board is shown.

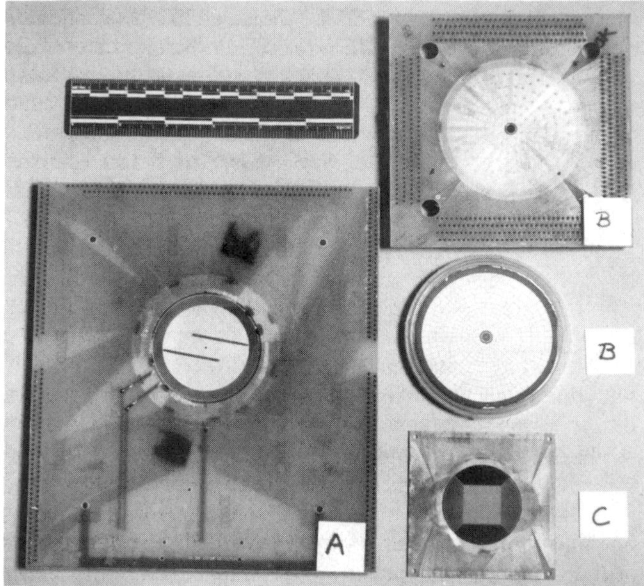

FIG. 2. Silicon position-sensitive detectors: **A** is a strip detector built on a 3-in.-diam wafer with 2400 vertical strips of 25-μm pitch. The wafer is mounted on a printed circuit board with the traces wire bonded to the silicon traces. An evaporated pair of horizontal germanium stripes are visible which connects unconnected strips to keep them at a uniform potential. **C** is a smaller strip detector with 50-μm pitch built on a 2-in. silicon wafer. The strips are horizontal and each strip is read out alternatively left and right. **B** is a "pad" detector with 512 separate devices on a 3-in.-diam wafer (lower) mounted on a printed circuit board (upper) with connections to each pad being made by wire bonding from traces on the board down through holes in the board to each active pad.

Detectors labeled B in Fig. 2 represent another configuration of position-sensitive detectors which is divided into small areas called "pads." There are 512 individual active detectors which will register not only the position of a particle, but its energy loss in the silicon which contributes to the information provided by other detectors to determine quickly if a high-energy event is of interest to be fully recorded (a trigger detector). Contact to each pad is made by wire bonding from lines on a printed circuit mother board through holes which overlay the corners between four pads. Again, the connectors and cabling out to the electronics from this board is extensive and not shown.

The requirements for high spatial resolution of a few microns tracking detectors in high-energy physics has fostered considerable communication with the semiconductor device industry such that sensors developed for other applications have found direct uses in experiments. One such sensor is the charge-coupled device (CCD) which is widely used as an imaging element in TV cameras. The impact of this and other sensors is to reduce the readout complexity that accompanies conventional detectors such as strip or pad detectors. CCDs do provide resolution of 20 by 20 μm, the active cell size, and can retain the information of the particle hit, but they are relatively slow to read out and have application for only specific experiments. Another recent development for position sensing is the semiconductor drift detector which is a silicon-technology-based device with electrodes on opposing surfaces of, for example, a 300-μm-thick wafer that both totally deplete the wafer and also produce a field gradient along the area that conducts electrons from the position of particle traversal to a collecting anode. The field conducting the carriers along the area is low enough to permit the electrons to "drift" at constant velocity which gives their drift length or the hit position through the time interval between an externally derived start and the collection stop at the anode. One-dimensional position measurement to 10 μm has been demonstrated by this technique and cylindrical geometries and multiple anode detectors have been produced that provide two-dimensional information. An important feature of this detector is the fact that carriers may be collected on an anode separated from the active region and of almost any shape; thus very small, low-capacitance collecting anodes are possible. A low-capacitance point-of-charge collection translates into low-noise opportunities for the processing electronics and advantage will be taken of this fact in developing on-detector integrated signal-processing electronics with extraordinary noise performance.

Communication with the integrated circuit industry has also provided electronics processing chains miniaturized as integrated circuits having preamplifier, shaping amplifier, a sample and hold stage, and a multiplexed serial readout of digitized pulse height information. Other electronic readout functions are being developed within the application specific integrated circuit (ASIC) industry. Readout "chips" with 128 channels per chip have been produced and bonded directly to strip detectors to reduce the cabling complexity. Efforts in the future will no doubt see the electronic package fabricated directly on the detector wafer; however, some material incompatibility between the detector substrate and the electronic substrate will have to be addressed.

RESULTS

It is impossible in a short article to fully express the impact that high-resolution semiconductor detectors have had on nuclear spectroscopy and related technologies. The energy spectra of several radiations shown in Fig. 3 may give some appreciation in three areas which have been mentioned. These results are typical in that they are not those of the "best possible" system but of systems that are readily attainable in most laboratories.

At the top, the spectra of α-particles from the α decay of ^{241}Am to the several low-lying levels in ^{237}Np is shown as observed with the surface barrier detector of Fig. 2. The system has an electronic resolution of ~14 keV and alpha groups separated by only ~40 keV are easily observed. The thin ^{241}Am source and the detector were in vacuum to prevent undue α-particle energy loss to an air ambient.

The spectra of several γ-ray-emitting nuclides are shown in the middle view measured with a large high-purity germanium coaxial detector. The energy resolution of the 1332.48-keV ^{60}Co peak is 1.85 keV which is not far from the statistical limit of ~1.5 keV given in Fig. 1. The residual events along the abscissa are due to Compton γ-ray-scat-

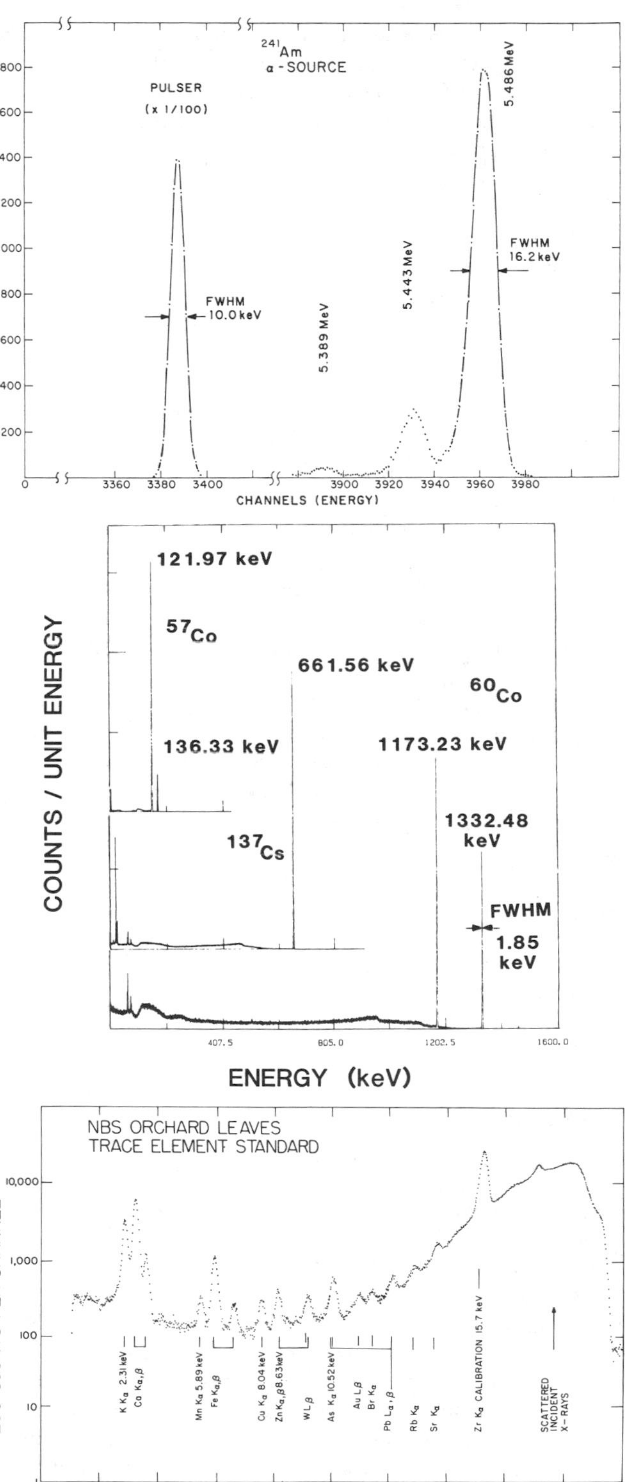

FIG. 3. Typical energy spectra from semiconductor detectors. *Top:* ^{241}Am α-particles observed with a silicon surface-barrier detector. *Middle:* Gamma-ray spectra of ^{57}Co, ^{137} Cs, and ^{60}Co observed with a large high-purity coaxial germanium detector having 20% efficiency (relative to a 3×3-in. NaI(Tl) detector for the ^{60}Co γ rays). These spectra were provided by the courtesy of Ortec Inc., Oak Ridge, TN, whose help is gratefully acknowledged. *Bottom:* X-ray fluorescence of trace element standard taken with a high-resolution lithium-drifted silicon x-ray spectrometer.

tering interactions in the detector which do not result in the full γ-ray energy loss; the kinematic Compton "edges" are visible in the ^{137}Cs and ^{60}Co spectra.

The bottom spectrum was observed with a cooled, lithium-drifted silicon x-ray spectrometer and associated low-noise electronic system. This spectrometer provides energy resolutions (FWHM) as low as 120 eV at low photon energies in the x-ray region. Trace amounts of several elements present in a slurry of ground-up grasses ("orchard leaves") have been detected by their characteristic x-ray emissions when fluoresced by a filtered and "hardened" x-ray source of about 25 keV. The abscissa spans the energy range from 0 to 25 keV and we clearly see the elements from potassium to lead present in the grass slurry. This target has been standardized and the amounts of Mn, Fe, Cu, and Pb present and observed are known to be 90, 300, 12, and 44 parts per million, respectively. The energy resolution of this system allows the distinction of characteristic x rays down to about sodium ($Z = 12$) in atomic number. This analytical technique is a relatively quick and efficient procedure to measure trace amounts (1–100 ppm) of a broad range of elements nondestructively and without elaborate sample preparation.

In conclusion, it should be mentioned that other semiconductor materials have received considerable attention but less than widespread usage. This group includes the compounds GaAs, CdTe, and HgI_2 which have produced working detectors but spectrometers of poorer energy resolutions than with silicon or germanium. Efforts are generally in the direction of larger band-gap semiconductors of high atomic number to achieve good γ-ray detection efficiency with a room-temperature device.

See also COUNTING TUBES; CRYSTAL DEFECTS; SEMICONDUCTORS; TRANSISTORS.

Semiconductors, Amorphous

G. Lucovsky and P. C. Taylor

INTRODUCTION

In the last 10–15 years there has been a significant upsurge of interest in the tetrahedrally bonded amorphous semiconductors, primarily hydrogenated amorphous silicon, a-Si:H, and alloys based on materials such as *a*-Si,Ge:H, and *a*-Si,C:H. This situation derives primarily from the realization that the electronic properties of *a*-Si:H and amorphous-silicon-based alloys could be changed in a reproducible manner by doping, and that the incorporation of hydrogen in these materials played a key role in reducing defects and making the *n*- and *p*-type doping possible. The evolution of this material from 1976 to the present represents an interesting interplay between refinements in material preparation, characterization, and new experimental techniques for studying amorphous materials on the one hand, and the evolution of theory and microscopic models for the electronic states, including the related optical and electronic properties, on the other. This combination has also played a determinant role in the development of emerging device technologies that utilize the unique properties of this class of amorphous materials. During this same period, an earlier focus on chalcogenides and pnictides, that was also driven by expectations for thin-film electronic devices, has waned considerably. Limitations in device performance due to material properties, coupled with the rapid advances in the amorphous-silicon-based materials, have permanently shifted the focus in this field. Nonetheless, there have been important advances in the understanding of the chalcogenide and pnictide materials, mostly in terms of their atomic structure, and in associated phenomena that are photoinduced, including photodarkening and photoassisted diffusion.

The historical evolution of this field can be traced through the proceedings of the International Conference on Amorphous and Liquid Semiconductor or ICALS meetings, held at Edinburgh, 1977; Cambridge, MA, 1979; Grenoble, 1981; Tokyo, 1983; Rome, 1985; Prague, 1987; and Asheville, 1989. The 1979 and 1983–1989 meetings were published as issues of the *Journal of Non-Crystalline Solids* [1], whereas the 1977 and 1981 meetings were published as conference proceedings by G. G. Stevenson Ltd., Dundee [2] and the *Journal de Physique* [3], respectively. During this same period, there were two major reviews devoted solely devoted to *a*-Si:H: one published by Springer-Verlag and edited by J. D. Joannopolous and G. Lucovsky [4], and the second published by Academic Press and edited by J. I. Pankove [5]. A more general review on noncrystalline solids was edited by M. Pollack, and published by the CRC Press [6]. In addition, there has been a continuing series of symposia on amorphous silicon sponsored by the Materials Research Society; the proceedings of these symposia have been published in the MRS Symposium Proceedings series [7]. Research on chalcogenides, pnictides, and other amorphous materials, including the tetrahedrally bonded semiconductors, is also described in the textbooks by N. F. Mott and E. A. Davis [8], R. Zallen [9], and S. R. Elliott [10]. Finally, a recent, and continuing series of papers on *Advances in Amorphous Semiconductors,* edited by H. Fritzsche, contains information on a wide range of amorphous semiconductors [11].

AMORPHOUS SILICON

Progress in the understanding of the fundamental properties of amorphous silicon, and its potential impact on device technologies, has been based to a large extent on the development and understanding of approaches to thin-film deposition and their relationship to bonded hydrogen incorporation in the deposited thin films. For example, it was established early on that there was a close relationship between hydrogen incorporation and the reduction of native bonding defects, mostly dangling bonds. The early work in this field was done by W. Paul and collaborators, with the emphasis on hydrogenated amorphous Ge, rather than Si. However, the two milestones in the subsequent focus on hydrogenated amorphous Si, (i) the development of solar cell structures, first disclosed in a U.S. patent issued to D. E. Carlson [12] and (ii) the first literature discussion of doping of amorphous silicon by W. E. Spear and P. G. LeComber [13], both occurred at a time when the important role of hydrogen incorporation in reducing native defects was not completely understood.

Hydrogenated amorphous silicon has been prepared by a variety of different techniques including the glow discharge decomposition (GD) of silane (SiH_4), and various silane derivatives (including SiF_4), reactive sputtering, thermal chemical vapor deposition (CVD) with postdeposition hydrogenation, and most recently by remote plasma enhanced CVD (remote PECVD). The amount of hydrogen incorporated into the film, and the local bonding configurations of the silicon–hydrogen groups, were found to vary with deposition conditions. The amount of hydrogen was first accurately determined by nuclear reaction analysis, and more recently by secondary ion mass spectrometry (SIMS) and ^1H nuclear magnetic resonance (NMR). These techniques were then used as primary standards for the calibration of infrared (IR) absorption measurements, which could additionally distinguish between different local bonding arrangements, monohydride or SiH groups, dihydride or SiH_2 groups, and polyhydride or $\{SiH_2\}_n$ configurations [see Figs. 1a and 1b)]. The IR measurements could also be used as a secondary standard for determination of the total bonded hydrogen concentration. Hydrogen concentrations have also been studied by thermal evolution techniques. One important result that has been derived from these studies of the deposition of *a*-Si:H is the identification of "processing windows" or conditions for obtaining electronic grade material. These include: (i) limiting the bonded hydrogen concentration to about 6–15 at.%; (ii) minimizing or eliminating bonded hydrogen in polyhydride configurations; and (iii) using a deposition process in which the substrate temperature was in the range of ~200 to ~300°C. In addition, it has also been established that concentrations of oxygen, nitrogen, and carbon in excess of about $10^{19}/cm^3$ can degrade the electronic properties of the deposited thin films.

Deposited films of hydrogenated amorphous silicon have been studied extensively by transmission electron microscopy (TEM). These studies have indicated that films can (i) be "homogeneous" on a scale of dimensions greater than about 10 Å; (ii) display a microstructure; or (iii) be diphasic with a columnar structure. The films which display a homogeneous structure via the TEM characterization are generally those films which contain bonded hydrogen predominantly in the monohydride bonding arrangements. These films also have electronic properties which can be characterized as being either photovoltaic (PV) quality or, more generally, device quality. The level of structural inhomogeneity generally increases with increasing hydrogen concentration; the diphasic materials typically display islands of homogeneous material, dominated by monohydride bonding, connected by "tissue" material, which contains higher concentrations of bonded hydrogen mostly in polyhydride bonding arrangements. Nuclear magnetic resonance (NMR) has been used to probe the homogeneity of *a*-Si:H films, by establishing correlations between the local bonding arrangements of incorporated hydrogen (or deuterium) atoms and the microstructures as defined by TEM. In this way NMR has played a complementary role to TEM in characterizing the different types of microstructural inhomogeneities that can be present in the deposited thin films (see Fig. 2).

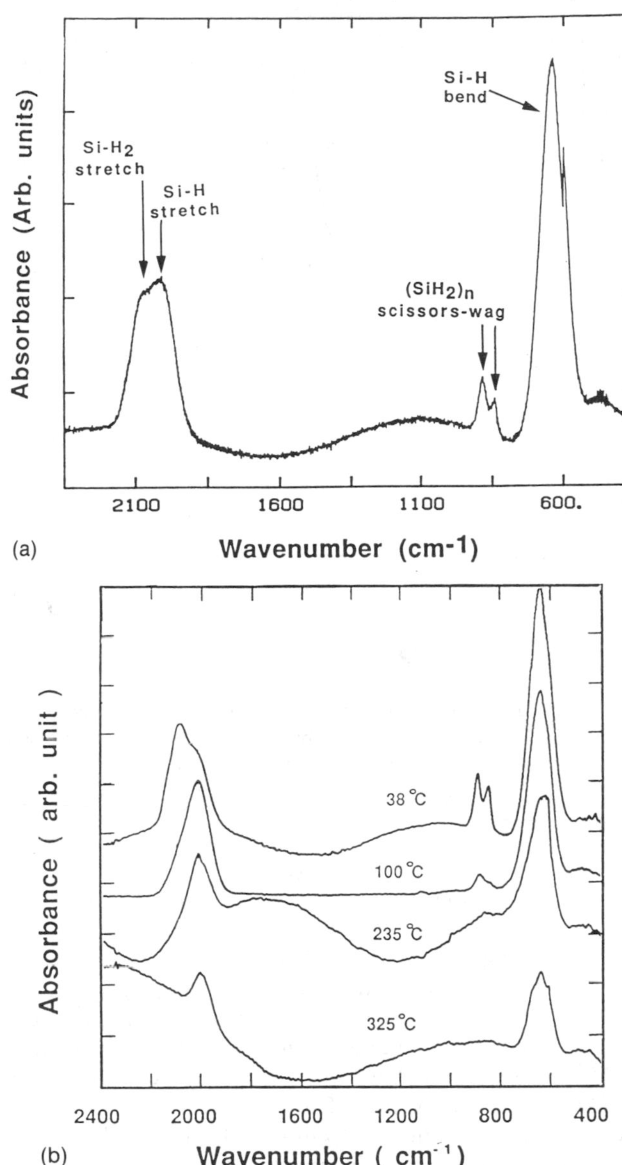

Fig. 1. (a) Infrared absorption spectrum of an *a*-Si:H film deposited by remote PECVD at a substrate temperature of 70°C. The vibration mode assignments are indicated. (b) Infrared transmission spectra of four remote PECVD *a*-Si:H films deposited at different substrate temperatures. As the temperature decreases from 325 to 38°C, the bonding configurations giving rise to the features in the spectra change from those associated with monohydride bonding at 2000 and 630 cm^{-1}, to those associated with polyhydride bonding at 2090, 890/845, and 630 cm^{-1}.

Substitutional doping of *a*-Si:H films has been accomplished by several different approaches. For example, *n* doping and *p* doping, using phosphorus and boron, respectively, as the substitutional dopant atoms, has been achieved in the GD and remote PECVD processes by adding phosphine and diborane to the silane or the silane derivative in the processing gas mixtures. Figures 3a and 3b illustrate the conductivity and conductivity activation energy changes that take place upon adding phosphine and diborane to the silane gas in the remote PECVD process. These changes are almost identical to those that take place for similar doping gas/silane

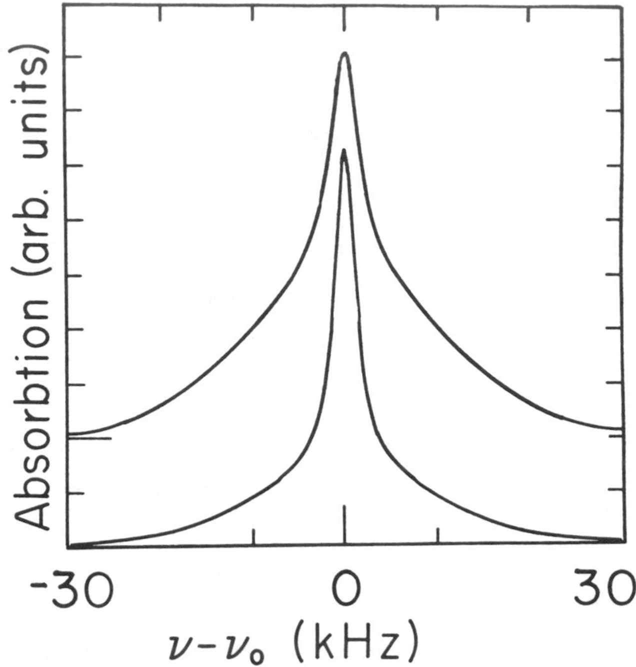

FIG. 2. ^{1}H NMR line shapes for two representative samples of *a*-Si:H. The line shapes contain a broad and a narrow component which are due to clustered and dilute bonded hydrogen, respectively. The broad feature is more dominant in the top trace.

ratios in the GD process. Doping has also been accomplished using doped sputtering targets. Finally, postdeposition doping has been accomplished using ion implantation. At the highest doping levels achieved, the doping efficiency, defined as the ratio of the number of active doping species to the total number of "impurity" atoms in the thin film, has been determined to be approximately 30–40%. The highest levels of doping have resulted in very large changes in the resistivity, between values of 10^{10} Ω cm for undoped, or so-called "intrinsic" material, to 10–100 Ω cm for both *n*-doped and *p*-doped materials. The doping process also adds defects to the *a*-Si:H host material. This association of doping and defect generaton, has been described in terms of a valence alternation pair model, where each active dopant atom incorporated in a fourfold-coordinated bonding environment also induces a threefold-coordinated Si native bonding defect. The combined effect of both of these bonding arrangements in *n*- and *p*-doped material is to promote a movement of the Fermi level toward the respective conduction or valence band edges, and hence an increase of the respective dark conductivities.

The electronic states of hydrogenated amorphous silicon have received considerable attention. These states can be separated into three groups: (i) extended states, lying well above the valence and conduction band edges; (ii) localized band-tail states, defining the respective band edges; and (iii) localized states within the forbidden gap. The extended states have been studied by optical absorption and reflection, and by direct and inverse photoemission; the band-tail states by absorption and photoconductivity; and the states within the band gap by a variety of optical and electronic techniques including photothermal deflection spectroscopy (PDS), the constant photocurrent method (CPM), deep-level transient spectroscopy (DLTS), and electron-spin resonance (ESR). The totality of these experiments has demonstrated that all of these electronic states are sensitive to the deposition conditions, specifically (i) the hydrogen concentration and bonding arrangements of the incorporated hydrogen and (ii) any thermally driven relaxation effects that can take place either during or after film deposition. Typical densities of states are (i) $\sim 10^{22}/cm^3$ eV for the extended states; (ii) 10^{19}–$10^{21}/cm^3$ eV for the band-tail states; and (iii) 10^{15}–$10^{18}/cm^3$ eV for the states within the gap. As noted above, native bonding defects are also introduced in the doping process, and these have been identified by their effects on the distribution, and the number of electronic states both within the band tails and the forbidden gap.

The electronic states of amorphous silicon have been studied using, among other techniques, steady-state and transient photoconductivity, and photoluminescence. Transient photoconductivity studies have yielded detailed information about the hole and electron mobilities and the lifetimes for deep trapping. Both the hole and electron mobilities are controlled by shallow traps in the vicinity of room temperature,

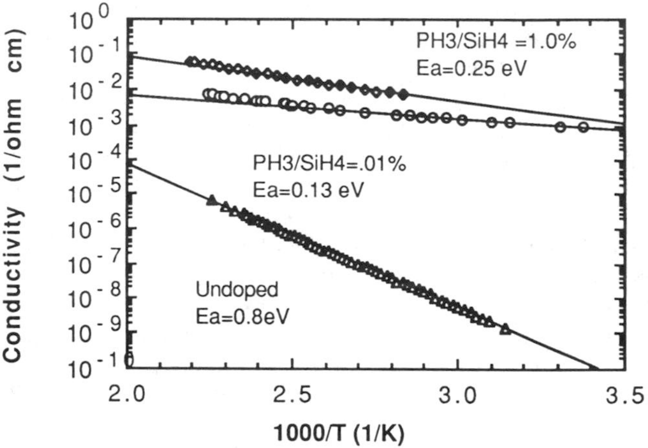

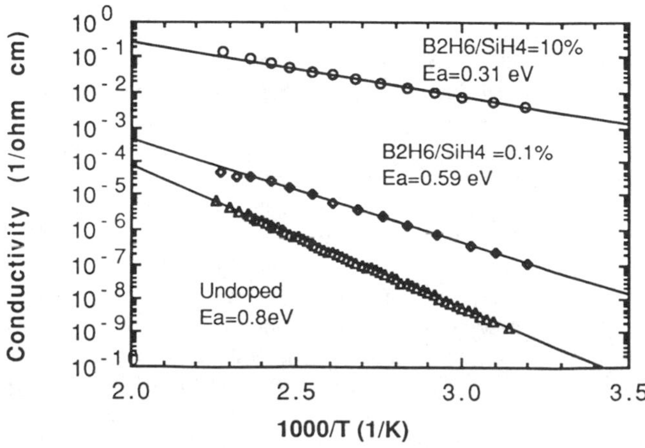

FIG. 3. Dark conductivity versus $1/T$ for (a) the formation of *n*-type *a*-Si:H by using phosphine/silane mixtures and (b) the formation of *p*-type *a*-Si:H by using diborane/silane mixture.

but show transitions to dispersive transport as the temperature is reduced. The room-temperature electron mobility is about 0.5 cm^2/V s, with an activation energy of about 0.2 eV and a prefactor, the extended-state mobility, of about 5–10 cm^2/V s. The hole mobility at room temperature is about two orders of magnitude less and has a higher activation energy, approximately 0.35 eV. Trapping times vary with hydrogen content, doping, and residual O, N, and C concentrations. There are good correlations between trapping times, as determined from photoconductivity, the efficiency of photoluminescence, and the transient light-induced ESR signal, which provides a measure of the defects with unpaired spins.

The density of defect states in *a*-Si:H can also be affected by exposure to high-intensity illumination, the so-called Stabler–Wronski effect, and by the application of stress bias, as in thin-film transistor structures (TFTs). The effects of light soaking and bias soaking can be reversed by thermal annealing. Studies of the defects induced in light-soaked materials, using ESR (see Fig. 4), PDS, CPM, etc., to probe the defect states, indicate that these defects are essentially the same as the native defects found in materials produced under nonoptimized deposition conditions, i.e., they are primarily dangling bond defects. There has recently been a proposal that overcoordinated Si atoms, called floating bonds, might also be responsible for some of the characteristic intrinsic defect behavior. However, the existence of these floating bond defects has not to this date been confirmed; in fact the most recent studies by ESR appear to rule out any significant contribution of this type of bonding defect to the observed ESR signals.

The optical properties of *a*-Si:H have been studied in great detail. Like all amorphous semiconductors the optical absorption edge exhibits three general regions that involve transitions between specific pairs of the three types of electronic states discussed above (see Fig. 5). The region at highest energies, which can be used to provide an estimate of an optical energy gap, is slowly varying with energy via a power

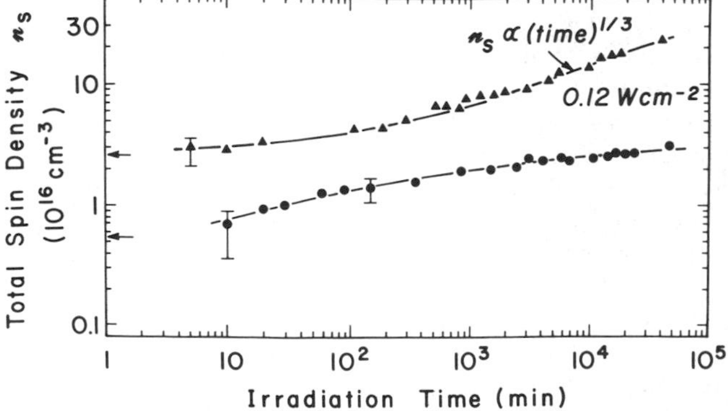

Fig. 4. Total spin densities, by ESR, for two thin films of *a*-Si:H on quartz substrates. The triangles represent data for a film with an initial spin density of $\sim 2.7\times10^{16}$ cm^{-3} (as indicated by the top arrow). The circles represent data for a film with an initial spin density of $<6\times10^{15}$ cm^{-3} (as indicated by the bottom arrow). The two samples were each irradiated with 0.12 W cm^{-2} of light from a tungsten source. This increase in the defect density after optical excitation is known as the Stabler–Wronski effect.

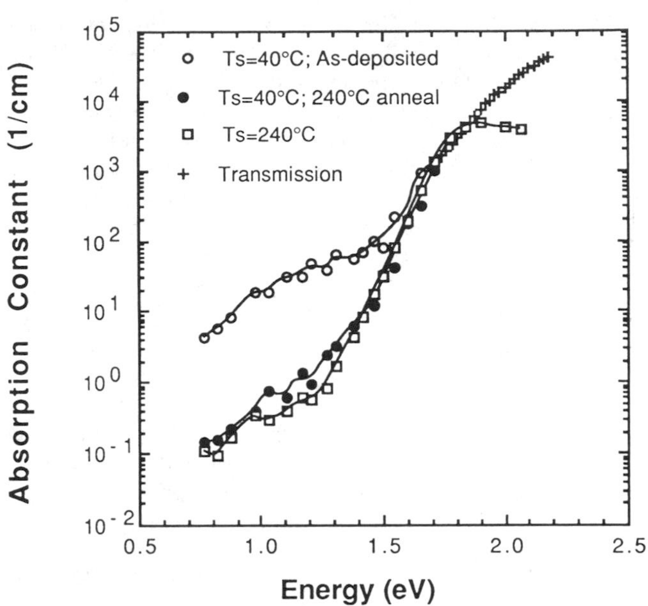

Fig. 5. Optical absorption spectra, absorption constant α versus photon energy $h\nu$, for *a*-Si:H indicating the three regions of absorption: (i) band-to-band transitions, $h\nu>1.9$ eV; (ii) band-tail absorption, $1.9>h\nu>1.5$ eV; and (iii) sub-band-gap absorption (by CPM), $h\nu<1.5$ eV.

law which depends roughly on the energy squared. Absorption in this region is between extended states in the valence and conduction bands. The second region is the so-called Urbach edge region which is named by analogy to a similar behavior which can occur in crystalline solids with high densities of defect states. In this region the optical absorption coefficient depends exponentially on the energy. In device quality *a*-Si:H, the slope of this edge is always about 50 meV. The transitions giving rise to this behavior involve the band-tail states, and the value of approximately 50 meV then reflects a density of states average of the slopes of the valence and conduction band edges. The third region of the optical absorption spectrum consists of a slowly varying absorption which extends well below the Urbach edge down to energies which are approximately half of the optical energy gap. This absorption is due to transitions between localized states and states in the respective band tails. In *a*-Si:H and related alloys, this absorption is thought to involve native bonding defects which consist primarily of silicon atoms with dangling bonds. The position of the energy gap, sometimes defined by the photon energy E_{04} at which the optical absorption constant is 10^4 cm^{-1}, the slope of the Urbach edge, and the magnitude of the below band-gap absorption all depend strongly on the specific parameters which are employed in depositing the films. However, in device-quality *a*-Si:H, which can now be made by several different techniques, including GD, remote PECVD, and reactive sputtering, the optical properties associated with each of these three regions of absorption are all essentially identical among the different preparation techniques.

The absorption in the Urbach tail, and particularly in the below-band-gap region, is not easily measured using standard optical transmission techniques because *a*-Si:H can only be made in thin-film form, and the absorption in this region is not strong enough to be measured directly. For this reason, several alternative techniques have been developed which have greater sensitivity, including such techniques as PDS and CPM. In using any of these techniques, care has to be taken to distinguish between bulk and surface contributions to the extrapolated optical absorption constants.

Perhaps the most technologically important aspect of the optical properties of *a*-Si:H and related alloys is the existence of metastable increases in the below-band-gap absorption under the influence of light, electrical bias, or any other external influence which produces carrier recombination. This recombination-induced increase in the below-band-gap absorption, which was first discovered by Stabler and Wronski as a metastable decrease in the photoconductivity, has a deleterious effect on electronic devices because the increased absorption is associated with an increase in the density of Si dangling bonds which reduce the lifetimes of photogenerated (as in PV devices) or electronically generated (as in TFTs) free carriers. The Stabler-Wronski effect remains one of the most significant unsolved puzzles which has hindered the widespread utilization of *a*-Si:H in large-scale photovoltaic applications.

The Stabler-Wronski effect also shows up as an increase in the ESR signal after optical excitation (see Fig. 4). These increases in the ESR signals scale accurately with observed increases in the below-band-gap absorption. Although the saturation of this effect is still a matter of some controversy, ESR and absorption experiments on device-quality films indicate that saturated densities of about 10^{17} cm^{-3} are to be expected.

Some recent studies of the time-resolved changes in the optical properties of *a*-Si:H and related alloys have shed light on the conduction mechanisms in these amorphous solids. On a picosecond time scale, the electron mobilities are still as low as about 10 cm^2/V s. This value indicates that there is rapid thermalization of any photoexcited free carriers into the band tails immediately following the optical excitation process.

Before we begin a discussion of various heterojunctions and device structures based on *a*-Si:H and its alloys, we discuss some of the more important Si-based alloy materials, and also another form of Si that has played an important role in the development of the device technologies, namely, microcrystalline Si, or simply μc-Si.

AMORPHOUS-SILICON-BASED ALLOYS AND MICROCRYSTALLINITY

Much of the work on amorphous-silicon-based alloys such as *a*-Si,Ge:H and *a*-Si,C:H was stimulated by the need for materials with band gaps other than that of *a*-Si:H (~1.7–1.8 eV) to optimize the performance of photovoltaic devices. This need has stimulated work on wide-band-gap alloys (E_g>1.9 eV), such as *a*-Si,C:H, *a*-Si,N:H, and *a*-Si,O:H, and narrow-band-gap alloys (E_g<1.5 eV), such as *a*-Si,Ge:H and *a*-Si,Sn:H. Alloying has had the general effect of degrading many of the electronic properties with respect to their values in *a*-Si:H. These properties include the carrier mobilities and lifetimes. In spite of these decreases in electronic quality, a number of these alloy materials have been used in device structures. The most important alloys include (i) intrinsic *a*-Si,Ge:H, which has found applications in the *i* regions of *p-i-n* photovoltaic structures; (ii) *n*-doped and *p*-doped *a*-Si,C:H which has found applications in the *n* and *p* regions of the same *p-i-n* structures; and (iii) *p*-doped *a*-Si,O:H which has been used as a photoreceptor for electrophotography. The deposition of alloy films with optimized electronic properties has been primarily by the GD method, with fluorides of silicon being used in addition to the hydrides that have been employed for the intrinsic and doped *a*-Si:H alloys.

Other complementary studies have also focused on microcrystalline ternary alloys, primarily for the *n* and *p* regions of photovoltaic device structures. By using significant hydrogen dilution, coupled with relatively high power, films produced by the GD and remote PECVD processes display a transition from an amorphous to a microcrystalline structure. This transition from an amorphous to a microcrystalline structure is easily detected by x-ray diffraction, electron diffraction, TEM imaging, or Raman scattering. In addition, increases in the electrical conductivity by factors greater than about 100 have been achieved in both *n*- and *p*-doped material by promoting this transition to a microcrystalline phase.

HETEROSTRUCTURES AND DEVICE STRUCTURES

As noted above, the stringent requirements which must be met before photovoltaic devices based on *a*-Si:H will be attractive for large-scale terrestrial power applications have necessitated the consideration of heterostructures in which *a*-Si:H is combined with other alloys, such as *a*-Si,Ge:H. In addition, other novel-device structures have suggested the use of "superlattices" based on *a*-Si:H and *a*-Si,Ge:H alloys. It is not at all obvious that with the very short scattering lengths for carriers in amorphous solids, on the order of interatomic spacings, quantum confinement in superlattice structures can in fact occur in these materials. Nonetheless, recent modulated optical experiments have strongly suggested that quantum size effects can be observed in thin amorphous semiconducting layers, and recent photovoltaic device structures have shown that superlattices based on amorphous semiconductors may be useful in improving efficiencies. For these reasons studies of amorphous layered structures will continue to be important in the future.

As both photovoltaic and thin-film transistor devices improve, interfaces are becoming increasingly more technologically important. Not just the initial properties of these interfaces are significant but also the metastable changes which often can occur under operational conditions are of real importance. There is mounting evidence that changes like those which yield the Stabler–Wronski effect in the bulk of these amorphous layers also occur at important interfaces, such as those between intrinsic *a*-Si:H and dielectric materials in TFT structures. In fact, in the high-quality TFT devices, any bias or light-induced defect generation at the interfaces may be more important than similar defect production in the bulk in degrading the electrical performance.

The initial emphasis on *a*-Si:H device applications was focused on photovoltaics. The driving force for this interest has been and continues to be the increased optical absorption in the visible in *a*-Si:H, relative to crystalline Si. The first device structures were small, on the order of square millimeters to about one square centimeter. These devices displayed conversion efficiencies of about 5%. At the present state of the art, which has been extended to double and triple tandem-cell structures, efficiencies as high as 13% have been achieved in cells with square centimeter dimensions, and 6–8% in solar panels that are as large as one square foot.

The second important application of *a*-Si:H has been in thin-film transistors (TFTs), where one of the more important applications is for the driving elements in large-area liquid-crystal displays. TFTs have also found numerous other applications in device structures that require direct coupling between light sensors and electronics. Other applications of amorphous silicon that have received attention are as photoreceptors for electrophotography, as charged-particle detectors, and as photosensors for facsimile. It is likely that this list will grow as the technology for improving amorphous-silicon-based alloys improves and as multichamber integrated processing continues to develop as a fabrication technique for the formation of heterostructure devices comprised of semiconducting and dielectric materials.

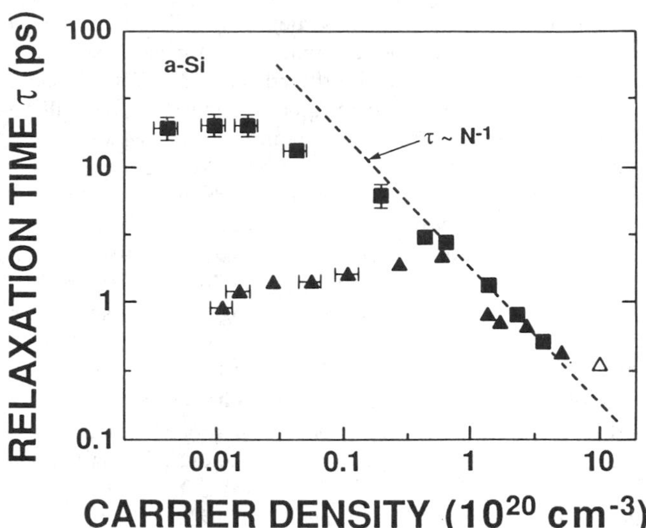

FIG. 6. Initial recovery time, τ, of photoinduced reflectivity changes versus the photogenerated free-carrier density, N. The dashed line indicates the bimolecular process that dominates at the highest optical pumping levels. The triangles are for nonhydrogenated *a*-Si and the squares for *a*-Si:H. The differences between the low carrier density saturation values for the squares and triangles reflect differences in the shallow trap concentrations between *a*-Si and *a*-Si:H.

RECENT ADVANCES IN AMORPHOUS SILICON

The most recent ICALS and ICAST (International Conference on Amorphous Semiconductor Technology) conferences have included several new and interesting areas of research in *a*-Si:H. One of these areas deals with the applications of femtosecond spectroscopy for studying free-carrier recombination mechanisms. These studies have indicated a transition between a bimolecular process at the highest levels of free-carrier generation ($>10^{20}$ cm^3) to a band-tail trapping mechanism at lower free-carrier generation levels (see Fig. 6). The issues yet to be resolved are the specific nature of the bimolecular process, and whether the properties of the optically generated free carriers can indeed be described in terms of a plasmon model.

Other studies have focused on relaxation processes, possibly involving hydrogen motion, and motion of defects, that can take place during or after film deposition. These relaxation phenomena clearly impact on the electronic properties of the materials, but they are not understood in terms of a microscopic picture, even though many models have been advanced.

Finally, studies of carrier transport in *a*-Si:H have been extended to very low temperatures, <10 K, and these have revealed a surprising increase in the mobility of electrons. The experimental results have been confirmed in a number of laboratories, and at least one model has been proposed. We see this as an active area of research, one which will receive considerable attention at the 1991 and 1993 International Conference on Amorphous Semiconductor (ICAS) meetings at Garmisch–Partenkirchen, FRG, and Cambridge, UK, respectively.

OTHER TETRAHEDRALLY BONDED MATERIALS

There have been studies of other tetrahedrally bonded materials, e.g., *a*-Ge, *a*-C, and the amorphous III–Vs. The driving force for the research on *a*-Si:H has been the device applications, and none of the other tetrahedrally bonded materials has shown promise for the electronic quality required for these applications. Amorphous Ge has been doped, but the values of the carrier mobilities and lifetimes are sufficiently lower than those of *a*-Si:H so as to rule out any significant use in PV or TFT applications. Amorphous C has received interest, mostly as a hard coating, and also in the context of an interest in producing thin films of diamond. There have been continuing studies of III–V materials, such as GaAs, GaP, etc., and also of ternary and binary alloys in the system Cd,Ge,As; however, none of these studies has revealed electronic properties that are in the range required for device applications. In general, research in these non-*a*-Si-based tetrahedrally bonded materials has decreased over the last 5–10 years, with a larger fraction of the community focusing on basic and applied studies of *a*-Si, μ*c*-Si, and closely related amorphous and microcrystalline alloys.

CHALCOGENIDES AND PNICTIDES

In the last decade amorphous semiconductors based on the Group-VI (chalcogenide) elements (S, Se, Te) and the Group-V (pnictide) elements (P, As, Bi) have been studied less intensively than amorphous semiconductors based on Group-IV elements (C, Si, Ge), primarily because the practical applications for the amorphous chalcogenides and pnictides are presently less important than for the Group-IV amorphous semiconductors, particularly *a*-Si and *a*-Si-based alloys. The chalcogenide and pnictide amorphous semiconductors generally cannot be doped, and therefore are less ideally suited for many electronic applications. As will be discussed below, there are some applications which do not require doping, and these applications have been pursued.

The chalcogenide glasses are perhaps best known because of the use of glassy or vitreous selenium, generally alloyed with small amounts of arsenic, as a photoreceptor in photocopying applications. The first detailed studies of the electronic and optical properties of chalcogenide glasses began in the 1950s, and the first potential electronic application for these materials occurred in the late 1960s with the development of threshold and memory switches by S. R. Ovshinsky and his coworkers. Although these electronic applications have not proved to be commercially viable, the device studies, and the possibilities of thin-film integrated electronic structures, encouraged a significant amount of research in the chalcogenides throughout the early 1970s. Recently, research on the chalcogenide and also the pnictide amorphous semiconductors has been driven by their potential applications in other fields, such as photoreceptors, photoresists, nonlinear optical materials, vidicons, and infrared-transmitting optical windows and fibers. Because these applications do not have nearly as wide a range of potential commercialization as those of *a*-Si:H, the research on the chalcogenide and pnictide amorphous semiconductors has not been nearly as extensive into the 1980s.

There are two standard preparation methods for making chalcogenide and pnictide amorphous semiconductors. These are (i) quenching from liquid melts and (ii) thin-film deposition, generally using physical deposition techniques such as evaporation or sputtering. The local structural order of the resulting amorphous materials is known to vary significantly with the method and the particular conditions of preparation. For example, even with the same method of preparation, such as quenching from the melt, the local structural order is often dependent on specific parameters of the preparation, such as the quenching rate.

Recently, the structural properties of these classes of amorphous semiconductors have been characterized using several experimental techniques such as Raman scattering, infrared absorption, x-ray and neutron scattering, as well as several more exotic techniques such as x-ray-absorption fine-structure (EXAFS). The short-range order in the materials has been well understood for some time. The local nearest-neighbor coordination numbers, Z, for glasses containing elements from Group IV through Group VII of the periodic table are given by $Z=8-N$, where N is the column number of the element in the periodic table. This so-called "$8-N$" rule was first suggested by N. F. Mott in the 1960s.

Structural studies of these amorphous semiconductors have recently concentrated on established order which can sometimes occur on a scale of second-, third-, or higher-nearest neighbor distances. Most chalcogenide or pnictide amorphous semiconductors exhibit order at this so-called intermediate or medium range. For example, there is an eight-membered ring structure and a helical chain structure which coexist in amorphous selenium, and there may exist analogous ring and chain-like or layer structures in such well-studied glasses as As_2S_3 and As_2Se_3. Intermediate range order, in the form of four-membered ring structures, has also been shown to exist in other well-studied chalcogenide glasses such as GeS_2 and $GeSe_2$.

The optical properties of the chalcogenide and pnictide amorphous semiconductors have also been studied in some detail in last decade. As is the case for tetrahedrally coordinated amorphous semiconductors, many of the optical properties of the chalcogenide and pnictide amorphous semiconductors are known to change under the influence of optical excitation. However, unlike the tetrahedrally coordinated amorphous semiconductors, any optically induced metastabilities are in this case greatly influenced by a very strong electron–lattice interaction which dramatically affects the resulting defect structure. The most general consequence of this strong electron–lattice interaction is that in most chalcogenides and in many pnictide amorphous semiconductors the primary defects consist of undercoordinated or overcoordinated atoms and the lowest-energy states of these defects consist of electronic states which are either doubly occupied or empty. This type of occupation results from the fact that the lattice relaxation which occurs on occupying a given defect with a second electron is so strong that the energy gained, via a structural relaxation, is actually greater than the additional Coulomb energy which is required to place the second electron on the defect. This situation can be described in a tight-binding formalism as a negative effective electron–electron correlation energy, or simply a negative-U bonding configuration. The inability to dope these amorphous solids has been shown to be directly related to the negative-U nature of these intrinsic bonding defects. The native bonding defects in *a*-Si:H have been shown to have a positive U, and this is one of the primary differences in the electronic behavior of these two different classes of amorphous semiconductors.

Because of the dominance of the negative U, the defects in the chalcogenide glasses are diamagnetic in the ground state. However, some of these defects can be rendered paramagnetic by optical excitation, and these paramagnetic states have been shown to be metastable at low temperatures. Irradiation with light of band-gap energies produces both an ESR signal and a below-band-gap optical absorption. There is also often a decrease in the intensity of a characteristic photoluminescence band after optical excitation. These metastable changes can be reversed both by thermal cycling and by excitation with light of energies below the optical gap, so-called optical bleaching. In some of the pnictide amorphous semiconductors, such as amorphous arsenic and amorphous phosphorus, these reversible, metastable effects also occur, but in these systems there exist both below-band-gap optical absorption and an ESR signal even before excitation with band-gap light. This experimental observation has been interpreted as being due to the presence in these elemental pnictide amorphous semiconductors of both positive-U and negative-U defects.

One important optical effect which has been studied in some detail is the so-called photodarkening (PD) effect. The PD effect is a metastable shift of the optical absorption edge

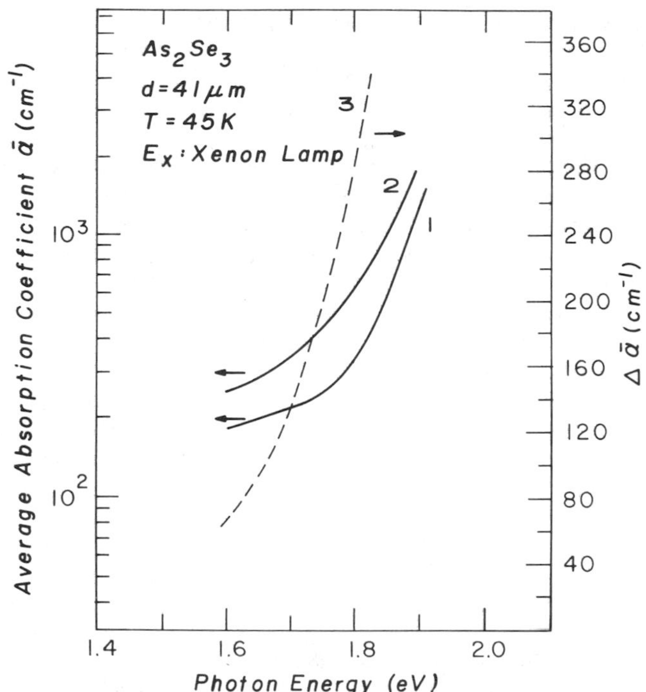

FIG. 7. Optical absorption edge in glassy As_2Se_3 (curve 1, left-hand scale). After optical excitation at 45 K with light from a xenon lamp the absorption changes to curve 2 (left-hand scale). This effect, which is metastable, is called photodarkening. The difference between curves 2 and 1 is shown by the dashed curve (curve 3, right-hand scale).

to lower energies after irradiation with light of energy greater than or approximately equal to the band-gap (see Fig. 7). The PD effect is a subtle one tied to changes in the nonbonding valence band which occurs in those chalcogenide glasses where all the chalcogen atoms are twofold coordinated. X-ray scattering and EXAFS studies of the subtle bonding changes which occur during PD suggest that the changes involve at least second-nearest neighbors. When some of the chalcogen atoms become threefold or fourfold coordinated, a bonding situation that can be brought about by alloying with metal atoms, the PD effect then disappears.

In addition, and particularly in deposited thin-film chalcogenide materials, there is also an irreversible shift of the optical absorption edge to lower energies which cannot be annealed. This irreversible effect is a photostructural change as evidenced by changes observed in the local structural order using different probes such as infrared spectroscopy, nuclear quadrupole resonance (NQR), x-ray scattering, and EXAFS. Well-annealed bulk samples of chalcogenide glasses do not exhibit similar irreversible photostructural effects.

Another interesting effect which has received some attention recently is the so-called photodoping process where metal atoms such as silver or copper can be made to diffuse into chalcogenide glasses by first evaporating a metal film onto the glass and then applying light to initiate the diffusion of the metal atoms through the glass. The details of this process are not well understood, but the term "photodoping" is clearly a misnomer. Although the metal atoms change the structure of the glass, they do not substitutionally dope these materials in the same way that doping is achieved in a tetrahedrally bonded amorphous or crystalline material.

Essentially all chalcogenide glasses exhibit *p*-type conductivity; however, there is one exception to this general rule. In germanium selenide glasses containing bismuth the dominant conductivity is *n* type. This change of conductivity type, from a *p*-type conductivity before alloying with Bi to an *n*-type conductivity upon alloying with Bi has sometimes been called doping, but once again the term is a misnomer. The Fermi level is still very close to midgap in the *n*-type material so the change in the conductivity type has more to do with the position of the dominant deep defect levels with respect to the middle of the energy gap than it does with substitutional doping. Nonetheless, this system is interesting and some rudimentary *p-n* junction structures have been constructed using the bismuth germanium selenide glasses as the *n* layer.

More important device applications of the chalcogenide glasses include their traditional use: (i) as photoreceptors for such applications as electrophotography; (ii) as optical fibers for infrared transmitting applications; (iii) as vidicons; and (iv) as inorganic photoresists for masking in submicron device structures. Although selenium alloyed with arsenic has been replaced in many photocopying applications by other chalcogenide and nonchalcogenide photoreceptors, including amorphous silicon and organic photoconductors, there are some instances where amorphous-selenium-based chalcogenide alloys are still used. In addition, arsenic trisulfide, As_2S_3, and alloys based on this composition are being used as infrared transmitting fibers and infrared window materials. Finally, the chalcogenide glasses are not yet widely used as inorganic photoresists mainly because the contrast ratios between exposed and unexposed regions are not very great. However, because the photoinduced changes which drive changes in the etch rates are of a local molecular nature, these materials exhibit excellent resolution for small-scale-device applications, and may therefore become interesting as device dimensions continue to push deeper into the submicron regime.

CONCLUDING COMMENTS

The evolution of the field of amorphous semiconductors has been very closely coupled with the anticipated and actual development of device technologies. Prior to the late 1960s, there was relatively little interest in chalcogenide amorphous semiconductors, even though *a*-Se and *a*-Se alloys were studied extensively at the Xerox Corporation and employed as the photoreceptor element in plain paper copiers since the early 1960s. This particular and specialized application was not sufficient to stimulate a great deal of general interest in the field. The announcements in the late 1960s by Ovshinsky and his coworkers of thin-film electronic switching and memory devices based on amorphous semiconductors, mostly chalcogenides, provided the stimulus for a rapid growth of this field. The scientific interest extended to the Group-IV

Group-IV materials, as well as the chalcogenides. However, when thin-film device technologies, based on the chalcogenides, did not meet the expectations of device technologists, interest in the field of amorphous semiconductors appeared to be slowing down. The field was in effect "reborn" by the almost simultaneous announcements of the Carlson patent for *a*-Si solar cells [12] and the successful *n* and *p* doping of amorphous Si by Spear and LeComber [13]. The changes of focus that have evolved within this field can be tracked by the Tables of Contents in the ICALS (and ICAST) Conference Proceedings [1–3]. For example in the early and mid-1980s, it had become apparent that interest in devices using *a*-Si had spread well beyond the solar cell technology, with the most important applications being based on amorphous-silicon TFTs. This, in fact, prompted the call for a new conference, run in parallel with ICALS 13, ICAST 1, the First International Conference on Amorphous Semiconductor Technology. The amorphous-semiconductor community has very recently combined the ICALS and ICAST conferences back into a single conference, but with a new name and subtitle: the International Conference on Amorphous Semiconductors (ICAS): Science and Technology. This merger and renaming of the original ICALS series establishes that just as in the field of crystalline semiconductors, the science and technology of the amorphous and microcrystalline semiconductors cannot be treated separately, and that progress in each of these areas will benefit from close coupling with one another.

See also GLASSY METALS; SEMICONDUCTORS, CRYSTALLINE; THIN FILMS.

REFERENCES

1. *J. Non-Crys. Solids* **35, 36** (1980); **59, 60** (1983); **77, 78** (1985); **97, 98** (1987); **114, 115** (1989).
2. *Amorphous and Liquid Semiconductors*, W. E. Spear (ed.), G. G. Stevenson Ltd., Dundee, 1977.
3. *J. Phys. (Paris)* **42** (1981).
4. J. D. Joannopoulos and G. Lucovsky (eds.), *The Physics of Hydrogenated Amorphous Silicon, Parts I and II*. Spring-Verlag, Berlin, 1984.
5. J. I. Pankove (ed.), *Semiconductors and Semimetals, Volume 21, Hydrogenated Amorphous Silicon,* Parts A–D. Academic Press, Orlando, FL, 1984.
6. M. Pollak (ed.), *Noncrystalline Semiconductors*. CRC Press, Boca Raton, FL, 1987.
7. *Materials Research Society Symposium Proceedings* **49** (1985); **70** (1986); **95** (1987); **118** (1988); **149** (1989).
8. N. F. Mott and E. A. Davis, *Electronic Processes in Non-Crystalline Materials,* 2nd ed. Clarendon Press, Oxford, 1979.
9. R. Zallen, *The Physics of Amorphous Solids*. John Wiley, New York, 1983.
10. S. R. Elliott, *Physics of Amorphous Materials*. Longman, London, 1984.
11. H. Fritzsche (ed.), *Advances in Amorphous Semiconductors,* Vol. 1. World Scientific, Singapore, 1989; and Vol. 2 (to be published).
12. D. E. Carlson, U.S. Patent 4,064,521 (1977).
13. W. E. Spear and P. G. LeComber, *Solid State Commun.* **17,** 1193 (1975).

Semiconductors, Crystalline

Heinrich J. Welker and *Walter Kellner*

1. *INTRODUCTION*

Since semiconducting phenomena can be detected in a great variety of matter, including glassy and amorphous solids and even liquids, any definition of semiconductors which includes all matter showing semiconducting phenomena can be only very general and imprecise. In this article, we have preferred to present a rather narrow definition more applicable to the "typical" semiconductors and exclude all other materials from our consideration.

Definition: Semiconductors are crystalline solids with predominantly covalent bonding. They differ from metals in that they have an energy gap $E_g>0$ between the valence and conduction bands. A sufficiently pure semiconductor will show no electrical conduction at $T=0$, while at $T>0$ electronic (electron or hole) conduction occurs due to impurities, thermal activation, or external influence (radiation).

In this definition, the term crystalline excludes any material with an unclearly defined structure. "Predominantly covalent bonding" means that the semiconductor properties disappear gradually as other bonding mechanisms take over. On transition from a group-IV crystal (e.g., Si, Ge) to a III-V (e.g., GaAs), II-VI (e.g., ZnS), and finally I-VII (e.g., NaCl) compound, the bonding changes from covalent to ionic and the crystal structure from a diamond to a NaCl type, the latter being an ionic conductor. Another example that illustrates the importance of crystal structure for the conduction mechanism rather than the position of an element in the periodic table is the radically different behavior of the two modifications of carbon: diamond and graphite. While diamond is a semiconductor with a crystal structure as described below, in the "sheet lattice" of graphite only three electrons per atom form covalent bonds toward neighboring atoms. One electron per atom can move freely within the lattice, resulting in a quasimetallic conductivity.

Although the criterion $E_g>0$ clearly separates metals from semiconductors, it is not a simple matter to distinguish semiconductors from insulators. Indeed, semiconductors may be used as insulating materials (e.g., semi-insulating GaAs or diamond) and many insulators show semiconductor properties (e.g., in photoconductivity).

2. *CRYSTAL STRUCTURE*

Because of the nature of the covalent bond, which is formed by two electrons with opposite spins from two neighboring atoms, the important semiconductor crystals are based on tetrahedral elements: Each atom is surrounded by four equidistant nearest neighbors lying at the corners of a tetrahedron. Figure 1 shows the diamond lattice (IV elements), the zinc-blende lattice (typical for III-V compounds), and the wurtzite lattice. Some of the III-V compounds, e.g., GaN, crystallize in the wurtzite lattice, while many of the II-VI compounds crystallize in both modifications, the zinc-blende and the wurtzite lattices.

In a crystal it is possible to construct a primitive unit cell which has all the symmetry properties of the crystal: the

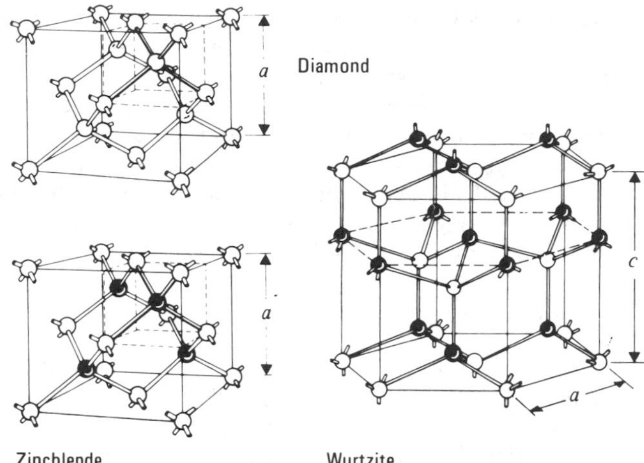

FIG. 1. Lattice structures of important semiconductors.

Wigner-Seitz cell. The Wigner-Seitz cell of the reciprocal lattice, lying in the so-called **k** space, is called the first Brillouin zone. For a given set of basis vectors $\mathbf{a}_i$ of the real lattice, the basis vectors of the reciprocal lattice $\mathbf{b}_j$ are defined by

$$\mathbf{a}_i \cdot \mathbf{b}_j = 2\pi\delta_{ij}, \qquad \delta_{ij} = \begin{cases} 0 & \text{for } i \neq j, \\ 1 & \text{for } i = j. \end{cases} \tag{1}$$

Figure 2 shows the first Brillouin zone for diamond and zincblende lattices.

3. *ENERGY BAND STRUCTURE*

To determine the energy band structure of a semiconductor, Schrödinger's equation

$$\Delta\phi_{\mathbf{k}}(\mathbf{r}) + (2m/\hbar^2)[E_{\mathbf{k}} - V(\mathbf{r})]\phi_{\mathbf{k}}(\mathbf{r}) = 0 \tag{2}$$

has to be solved. Bloch's theorem states that for a periodically varying potential $V(\mathbf{r})$, the solutions $\phi_{\mathbf{k}}(\mathbf{r})$ are of the form

$$\phi_{\mathbf{k}}(\mathbf{r}) = e^{j\mathbf{k}\cdot\mathbf{r}}U_{\mathbf{k}}(\mathbf{r}), \tag{3}$$

where $U_{\mathbf{k}}(\mathbf{r})$ is periodic in **r** with the periodicity of the crystal lattice. It can be shown that the energy $E_{\mathbf{k}}$ is periodic in the **k** space. Thus it is sufficient to determine $E_{\mathbf{k}}$ within the first Brillouin zone.

For calculating the band structure of semiconductors the *orthogonalized-plane-wave* method and the *pseudopotential* method are used most frequently.[2] Some results are shown in Fig. 3.[14]

The lower bands, the valence bands, are filled with the valence electrons of the semiconductor, one *s* and three *p* electrons. The upper bands, the conduction bands, are empty at $T=0$. The separation between the energy of the lowest conduction band and the highest valence band is called the band gap E_g. No electron states exist in this forbidden energy region.

The two top valence bands can be approximated at Γ ($\mathbf{k}=0$) (for definitions see Fig. 2) by two parabolas of different curvature. This corresponds to two types (*light* and *heavy*) of holes with different effective masses m, given by

$$m = \hbar^2/(\partial^2 E/\partial k^2) \tag{4}$$

The most significant difference between the three semiconductors of Fig. 3 is the position of the lowest minimum in the conduction band, lying at L for Ge, on the Δ axis close to X for Si, and at Γ for GaAs. (For definitions see Fig. 2.)

Because of the shift of the conduction-band minimum from $\mathbf{k}=0$, Si and Ge are called indirect semiconductors, while GaAs is called a direct semiconductor. In direct semiconductors band-to-band recombination of an electron and a hole with simultaneous emission of a photon is possible. In indirect semiconductors such a transition is possible only when a phonon is involved, which takes up the difference in the momenta of electron and hole.

Another important difference between direct and indirect semiconductors is the anisotropy of effective mass in indirect semiconductors. While in direct semiconductors the constant-energy surfaces are spheres at the zone center, in indirect semiconductors these surfaces are ellipsoids with their centers lying at $\mathbf{k}\neq 0$. In accordance with (4) this leads to a longitudinal (in the direction of the symmetry axis of the ellipsoid) and a transversal effective mass.

With the aid of an isoelectronic scheme (Table I) a systematic decrease in separation of the valence and conduction bands at corresponding points with the average number of electrons per lattice site can be observed for group-IV elements and for III-V compounds.

Since the subbands of the conduction band meet or overlap at certain points in **k** space, no forbidden energy region exists within the band and the band may be considered as continuous in real space. The same is true for the valence band. This forms the basis of a simplified band diagram, where electron energy is plotted versus distance (Fig. 4).

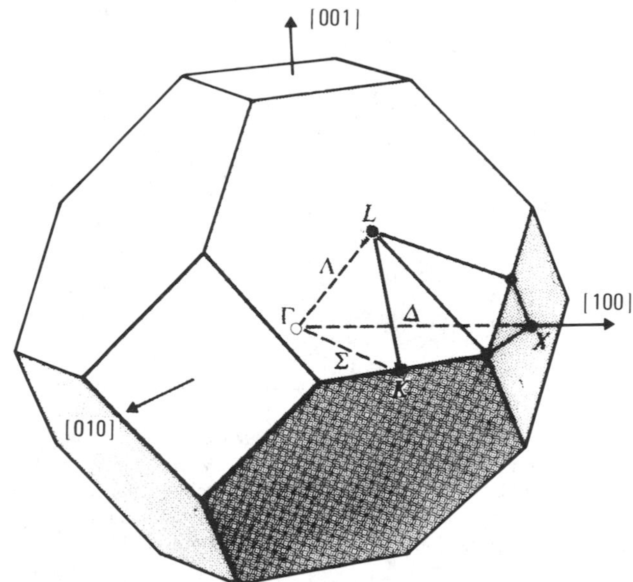

FIG. 2. First Brillouin zone for diamond and zinc-blende lattice showing important symmetry points and lines.

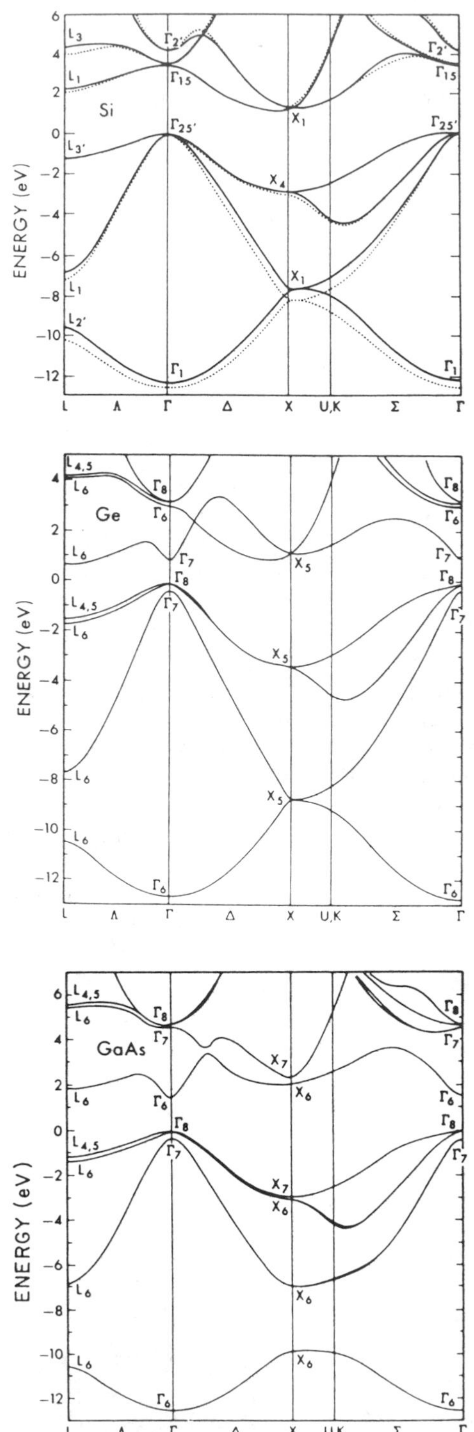

FIG. 3. Band structures of Ge, Si, and GaAs at $T=0$. The band gap E_g is the energy difference $X_1-\Gamma_{25'}=1.17$ eV (Si), $L_6-\Gamma_8=0.76$ eV (Ge), and $\Gamma_6-\Gamma_8=1.51$ eV (GaAs). The satellite valley for GaAs at L_6 is 0.31 eV above the conduction-band minimum Γ_6. The dashed lines represent earlier calculations for Si. (From Ref. 14.)

This representation is very useful in semiconductor device physics.[10]

Table II lists some properties of important semiconductors. Note that many semiconductors show room-temperature mobilities between 10^3 and 10^4 cm^2/V s, two orders of magnitude above the mobilities of metals. These high mobilities correspond to effective masses which are smaller than the electron rest mass.

The band structure is of fundamental importance for the device application of a material. With the development of modern crystal growth techniques such as molecular beam epitaxy (MBE) it has become possible to modify the band structure for a special purpose (band-structure engineering[13]). Some examples will be mentioned in Section 8.

4. *REAL CRYSTALS, DEFECTS, IMPURITIES*

Real crystals differ from the ideal ones discussed so far in many respects. At the surface, the energy band structure is altered by "dangling bonds" which can produce surface states within the energy gap. At $T>0$ thermal vibrations of the lattice atoms, called phonons, occur. When a photon with an energy comparable to the band gap is absorbed, excitons can be created. These are electron-hole pairs which attract each other by Coulomb interaction forming a state within the band gap.

Real crystals also show irregularities in their lattice structure. Some of these *crystal defects* are the following:

Point defects: (a) Intrinsic defects: vacancy, interstitial atom, pair of vacancy and interstitial atoms (Frenkel defect); (b) chemical defects (impurities): shallow impurity, deep impurity, isovalent impurity, pairs of impurities.

Line defects: Edge and screw dislocations.

Area defects: Grain boundaries, stacking faults.

The technically most important group among these defects are the substitutional impurities, i.e., impurities on lattice sites. Controlled incorporation of impurities is of fundamental importance in semiconductor device technology. The purest silicon crystals available today contain approximately 10^{12} impurities/cm^3, which is lower by a factor of 5×10^{10} than the density of the Si atoms. The purest semiconductor crystals produced so far have been Ge crystals with less than 10^9 impurities/cm^3. Within the bulk crystal, impurities can be incorporated either from the melt during crystal growth or by nuclear reactions [e.g., Si(n,γ)P[15]].

For device applications, monocrystalline layers with the required impurity concentration (*doping*) are grown by

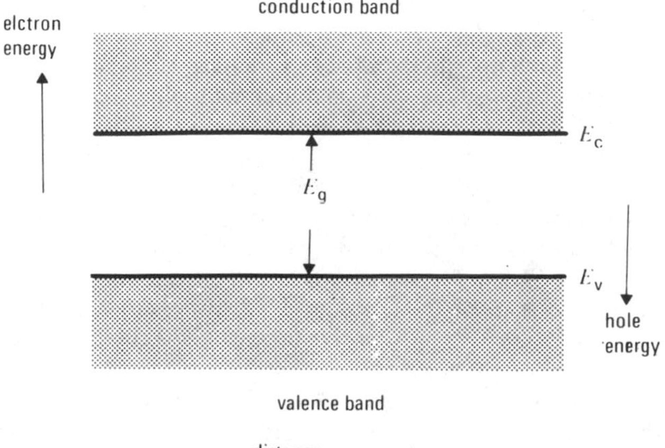

FIG. 4. Simplified band diagram.

Table I Isoelectronic scheme for IV elements and III–V compounds[a]

N_e	IV Elements	III–V Compounds				
14	Si (3.43)			AlP (~5)		
23			AlAs (3.1)		GaP (2.8)	
32	Ge (0.90)	AlSb (2.3)		GaAs (1.51)		InP (1.42)
41			GaSb (0.8)		InAs (0.42)	
50	Sn (−0.42)			InSb (0.25)		

[a] Numbers in parentheses give the energy gap (in eV) at $T = 0\ K$ for the zone center Γ of the Brillouin zone. Underlined values represent the band gap. N_e is the average number of electrons per lattice site.

epitaxy[16] from liquid phase, vapor phase, or molecular beam. Doped layers with certain doping profiles are also achieved by alloying, diffusion,[17] or ion implantation.[18]

Impurities introduce energy levels within the forbidden band. Donor levels are neutral when filled by an electron and positively charged when empty. Acceptor levels are neutral when empty and negatively charged when filled by an electron. A group-V element (P, As, Sb) on a lattice site in Si tends to "donate" an electron to the conduction band. A group-III element (B, Al, Ga) in Si tends to "accept" an electron to form four covalent bonds, thus leaving a "hole" in the valence band. Donors create n-type conduction and acceptors p-type conduction.

In III-V compounds group-II elements on III lattice sites act as acceptors and group-VI elements on V lattice sites act as donors. Group-IV elements can act either as donors (on III lattice sites) or as acceptors (on V lattice sites). The latter impurities are called amphoteric.

Shallow impurities are either donors with an energy level close to the conduction band or acceptors with an energy level close to the valence band. The order of magnitude for the ionization energy of a shallow impurity can be calculated from the ionization energy of a hydrogen atom when the free-space permittivity ϵ_0 is replaced by the permittivity ϵ of the semiconductor and the electron mass m_0 by the effective mass m:

$$E_i = \left(\frac{\epsilon_0}{\epsilon}\right)^2 \frac{m}{m_0} E_H. \tag{5}$$

E_H is the ionization energy of the hydrogen atom (13.56 eV). For donors these ionization energies are about 0.03 eV for Si and about 0.01 eV for Ge and GaAs. For acceptors the values are 0.015 eV for Ge and 0.05 eV for Si and GaAs.

Impurities with higher ionization energies than those of shallow impurities are called deep impurities. These play an important role as recombination centers or as electron or hole traps; e.g., in bipolar silicon transistors a gold diffusion process is used to reduce the minority-carrier lifetime, thereby yielding shorter switching times. In semi-insulating GaAs deep levels are responsible for reducing the number of free carriers and thereby increasing the resistivity to $\geq 10^7$

Table II Properties of important semiconductors (at 300 K) (from Ref. 19).

	Semiconductor[a]		Crystal structure[b]	Lattice constant (nm)	Bandgap[c] (eV)		Mobility ($cm^2\ V^{-1}\ s^{-1}$) Electrons	Holes	Effective mass, m/m_0[d] Electrons	Holes	Density ($g\ cm^{-3}$)	Dielectric constant (static)
Element	Si	14	D	0.543	1.12	i	1500	475	0.98/0.19	0.16/0.49	2.33	11.9
	Ge	32	D	0.565	0.66	i	3900	1900	1.6/0.08	0.04/0.3	5.32	16.0
III–V	GaP	23	Z	0.545	2.24	i	110	75	1.5/0.18	0.14/0.86	4.13	11.0
	AlSb	32	Z	0.614	1.63	i	200	420	1.5/0.21	0.11/0.5	4.26	14.4
	GaAs	32	Z	0.565	1.43	d	8800	400	0.066	0.4	5.31	13.2
	InP	32	Z	0.587	1.34	d	4600	150	0.080	0.8	4.79	12.4
	GaSb	41	Z	0.610	0.67	d	4000	1400	0.045	0.06/0.33	5.61	15.7
	InAs	41	Z	0.606	0.33	d	33000	460	0.027	0.41	5.67	14.5
	InSb	50	Z	0.648	0.16	d	78000	750	0.014	0.4	5.78	17.7
II–VI	ZnS	23	Z	0.541	3.66	d	165		0.28	0.59	4.09	8.6
	CdS	32	W	a, 0.416 c, 0.676	2.42	d	300	50	0.20	0.7	4.82	8.6
	CdSe	41	Z	0.606	1.73	d	800		0.11	0.45	5.74	9.4
IV–VI	PbTe	67	C	0.645	0.19	i	6000	4000	0.24		8.16	

[a] The figure following the chemical formula gives the average number of electrons per lattice site.
[b] D: diamond, Z: zincblende, W: wurtzite, C: cubic NaCl type.
[c] i: indirect; d: direct semiconductor.
[d] Two numbers separated by a slash indicate longitudinal/transversal electron effective mass or light/heavy hole effective mass, respectively.

Ω cm. These GaAs crystals are suitable substrates for high-frequency devices such as metal semiconductor field effect transistors (MESFETs). In many other applications, however, the effect of traps is detrimental to device performance (e.g., decrease of current gain in bipolar transistors, low efficiency of light-emitting diodes due to nonradiative recombination).

Isovalent impurities have the same number of valence electrons as the replaced lattice atom. A nitrogen atom on a phosphorus site in GaP forms an isovalent acceptor. Other, more complicated types of impurities are polyvalent impurities, pairs or clusters of impurities, or of impurities plus defects (vacancies, dislocations, etc.).

5. *CARRIER CONCENTRATIONS*

In thermal equilibrium the probability of a state of energy E to be occupied by an electron is given by the "Fermi-Dirac distribution function,"

$$W(E)=\left[1+\exp\left(\frac{E-E_F}{kT}\right)\right]^{-1} \tag{6}$$

where E_F is the *Fermi energy* also called the *chemical potential.* For $E=E_F$, $W(E)$ equals 0.5. For $|E-E_F| \geq 4kT$, $W(E)$ can be approximated by the *Maxwell-Boltzmann distribution function* with an error <2%:

$$W(E) \cong \exp\left(-\frac{E-E_F}{kT}\right) \quad \text{for} \quad E-E_F \geq 4kT \tag{7}$$

and

$$1-W(E) \cong \exp\left(-\frac{E_F-E}{kT}\right) \quad \text{for} \quad -(E-E_F) \geq 4kT; \tag{8}$$

$1-W(E)$ can be regarded as the distribution function for holes.

If the Fermi level is within the energy gap and separated by more than $4kT$ from either band edge, the semiconductor is called *nondegenerate* and the Maxwell-Boltzmann distribution may be used. In this case the electron density n is given by

$$n=\int_{E_c}^{\infty} N(E)W(E)dE \tag{9}$$

where $N(E) \sim (E-E_c)^{1/2}$ is the density of states in the conduction band. n is calculated to be

$$n=N_c \exp(E_F-E_c)/kT, \tag{10}$$

where N_c is the effective density of states in the conduction band:

$$N_c=2(2\pi mkT/h^2)^{3/2}. \tag{11}$$

In a similar way one obtains the hole concentration

$$p=N_v \exp[(E_v-F_F)/kT], \tag{12}$$

where N_v is the effective density of states in the valence band given by a formula similar to (11).

From (10) and (12) the mass-action law is obtained:

$$np=N_cN_v \exp(-E_g/kT). \tag{13}$$

In an intrinsic semiconductor conduction takes place by thermal generation of electron-hole pairs, hence $n_i=p_i$. Since (13) is valid also in this case, we have

$$np=n_i^2. \tag{14}$$

The extrinsic case is characterized by $|N_D-N_A| \gg n_i$ and

$$n=N_D-N_A, \quad p=n_i^2/(N_D-N_A) \tag{15}$$

for n-type conduction and

$$p=N_A-N_D, \quad n=n_i^2/(N_A-N_D) \tag{16}$$

for p-type conduction. A schematic representation of the band diagram, density of states, Fermi-Dirac distribution, and carrier concentrations is given in Fig. 5.

In Fig. 6 the electron concentration for an n-type Si sample is plotted versus $1/T$. The intrinsic case is characterized by the temperature dependence $\exp(-E_g/2kT)$ and the saturation range by Eq. (15). At lower temperatures the Fermi level approaches the impurity level and the probability that an electron occupies the impurity level increases: The impurities are frozen out. The slope in this range is of the order of magnitude of E_i from Eq. (5).

6. *CARRIER TRANSPORT*

For an extrinsic n-type semiconductor the current density $\mathbf{j}$ due to an external electric field E is given by

$$\mathbf{j}=-nq\mathbf{v}. \tag{17}$$

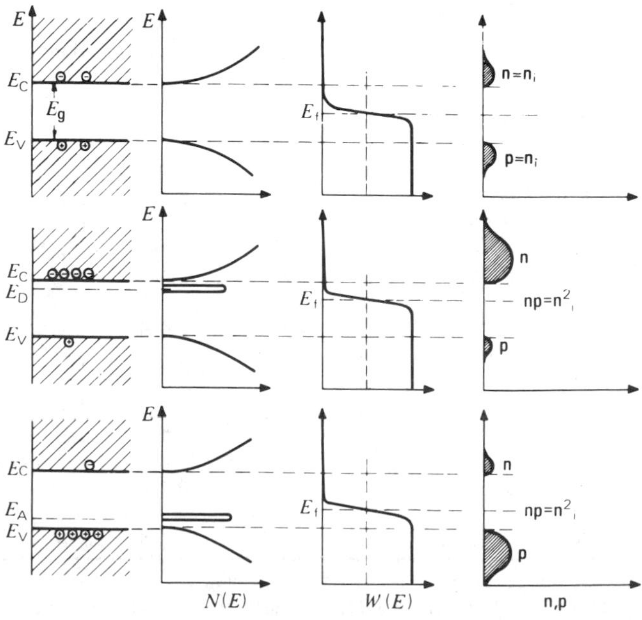

FIG. 5. Schematic band diagram, density of states, Fermi-Dirac distribution, and the carrier concentrations for V intrinsic, n-type and p-type semiconductors at thermal equilibrium. After Ref. 11.

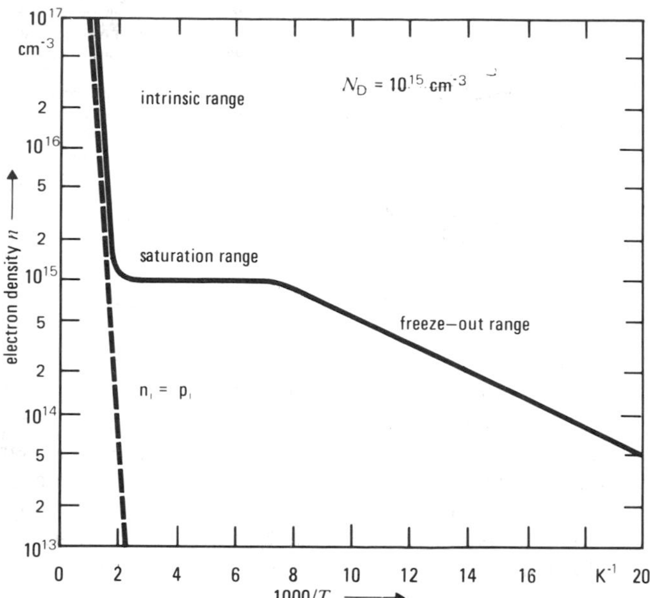

FIG. 6. Electron density as a function of temperature for a Si sample with a donor impurity concentration of 10^{15} cm^{-3}. After Ref. 11.

The drift velocity **v** is opposite to **j** in the case of negative-charge transport.

The electric field causes an accelerated motion of the electrons which is interrupted every time an electron suffers a collision with a lattice atom or any other scattering event. Integration of the equation of motion yields

$$\mathbf{v} = -(q\tau_m/m)\mathbf{E} \tag{18}$$

or

$$\mathbf{v} = -\mu_n \mathbf{E}, \tag{19}$$

with

$$\mu_n = q\tau_m/m. \tag{20}$$

μ_n is called the electron *mobility* and τ_m is introduced as an average *momentum relaxation time*. Relaxation means return to equilibrium. τ_m depends on the scattering mechanism involved. From (17) and (18) it follows that

$$\mathbf{j} = \sigma \mathbf{E}, \tag{21}$$

where the conductivity σ is given by

$$\sigma = nq\mu_n = nq^2\tau_m/m. \tag{22}$$

In a semiconductor there are always both electrons and holes present. Thus, in the more general case of *mixed* conduction (22) has to be replaced by

$$\sigma = q(n\mu_n + p\mu_p), \tag{23}$$

where the *hole mobility* μ_p is given by an equation analogous to (20).

Some of the important scattering mechanisms affecting the mobility are:

Acoustical phonon scattering with $\mu \sim m^{-5/2}T^{-3/2}$;

Ionized impurity scattering with $\mu \sim m^{-1/2}N_I^{-1}T^{3/2}$ (N_I is the ionized impurity density);

Optical phonon scattering with $\mu \sim m^{-5/2}f(T)$, where $f(T)$ decreases monotonically with temperature;

Piezoelectric scattering with $\mu \sim m^{-3/2}T^{-1/2}$, important in compound semiconductors at low temperatures;

Neutral impurity scattering with μ independent of temperature;

Carrier-carrier scattering;

Intervalley scattering;

Intravalley scattering.

When two or more scattering mechanisms act independently from each other the resulting mobility is given by

$$\frac{1}{\mu} = \frac{1}{\mu_1} + \frac{1}{\mu_2} + \cdots; \tag{24}$$

(24) is known as Mathiessen's rule.

Above a certain field strength the charge carriers acquire on average more energy from the electric field than they lose in scattering events, e.g., with optical phonons, which means that the effective electron temperature is higher than the lattice temperature (*warm* and *hot* carriers). In this range Ohm's law is no longer valid and the mobility decreases with increasing electric field. Some experimental curves of velocity versus electric field are shown in Fig. 7. The carrier velocity decreases at low fields with increasing doping concentration. At high field, however, the drift velocity is essentially independent of doping concentration. Intervalley scattering from the Γ valley into valleys of higher energy is responsible for the peak and the negative differential mobility of the GaAs and InP curve.

The electron velocity as shown in Fig. 7 is observed in samples with a large drift region. When the high-field region in a semiconductor is made very small (typically ≤ 0.5 μm) higher electron velocities may be observed due to nonstationary transport: Whenever the time of flight of an electron through the high-field region is shorter than the time constant for intervalley scattering, the electron will not be scattered

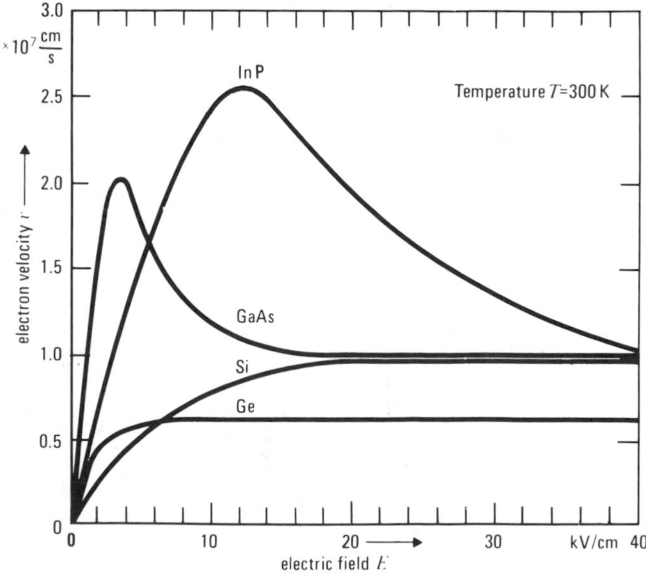

FIG. 7. Electron velocity versus electric field for high purity Ge, Si, GaAs, InP.[20–22]

into a higher-energy valley. This effect is clearly seen in III–V devices where the electrons staying in the central valley remain "light" electrons with high mobility and velocity. The increase in drift velocity leads to higher current, transconductance, and cut-off frequency of a device.

7. *EXPERIMENTAL METHODS*

A brief list of the methods most frequently used is given in this section.

Conductivity σ or resistivity $\rho = 1/\sigma$ can be obtained from a *four-point probe* measurement (Fig. 8). For a thin sample with probe spacing $s \gg w$ the resistivity is given by

$$\rho = \frac{\pi w}{\ln 2} \frac{V}{I}. \tag{25}$$

In *van der Pauw's method* (Fig. 9) a plane parallel disk with four point contacts at the circumference is employed. For measuring resistivity there is a current flow I_{AB} between two neighboring contacts and a voltage drop V_{CD} between the two opposite contacts. Then, by definition,

$$R_{AB,CD} = | V_{CD}/I_{AB} |. \tag{26}$$

In a second experiment

$$R_{BC,DA} = | V_{DA}/I_{BC} | \tag{27}$$

is measured. Van der Pauw proved that

$$\exp(-\pi w R_{AB,CD}/\rho) + \exp(-\pi w R_{BC,DA}/\rho) = 1. \tag{28}$$

From (28) we obtain

$$\rho = \frac{\pi w}{2 \ln 2} (R_{AB,CD} + R_{BC,DA}) f, \tag{29}$$

where the *van der Pauw factor* f depends on the ratio $R_{AB,CD}/R_{BC,DA}$. With the same sample the *Hall effect* can be measured when the current flows from A to C and the voltage is measured between B and D. The resistance

$$R_{AC,BD} = V_{BD}/I_{AC} \tag{30}$$

is changed by an amount $\Delta R_{AC,BD}$ when a homogeneous magnetic field is applied perpendicular to the sample surface. The Hall coefficient R_H is then given by

$$R_H = w \Delta R_{AC,DB}/| \mathbf{B} |. \tag{31}$$

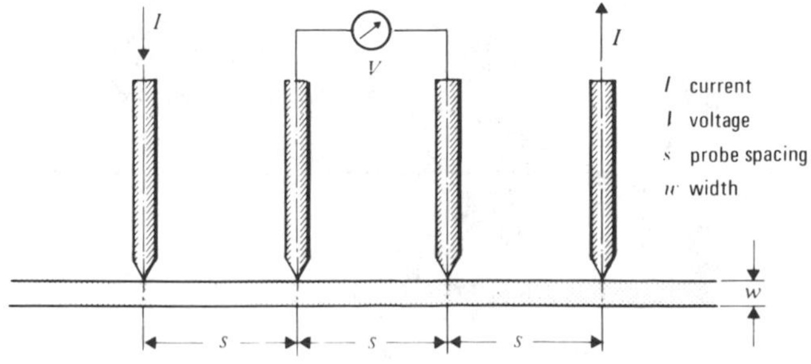

Fig. 8. Setup of four-point probe resistivity measurement.

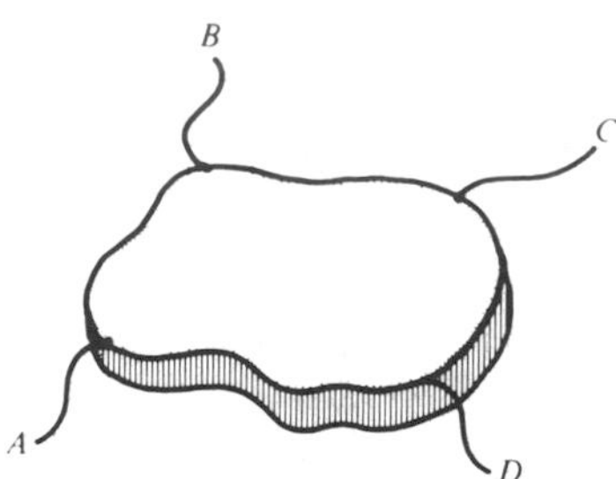

Fig. 9. Plane parallel disk for *van der Pauw* measurements.

The sign of R_H depends on the conductivity type: $R_H < 0$ for electron conduction and $R_H > 0$ for hole conduction.

Hall-effect and conductivity measurements are used to determine the carrier concentration and mobility. From their temperature dependence the band gap and the impurity concentration may be obtained. In Hall experiments, the magnetic field B is chosen so small that there is no change in conductivity. Under strong magnetic fields, however, a significant increase in resistivity ($\rho \sim | \mathbf{B} |^2$) is observed, the so called *magnetoresistance effect*, which is important in determining the band structure.

Carrier concentration can be obtained from capacitance versus voltage measurements of reverse-biased *Schottky contacts* (rectifying metal–semiconductor contacts):

$$n = \frac{2}{q\epsilon} \frac{dV}{d(1/C^2)}. \tag{32}$$

From the polarity of reverse bias the conductivity type is determined.

The *Seebeck effect* is also used to determine the conductivity type. *Minority carrier lifetime* (e.g., the lifetime of electrons in a p-type sample) is an important parameter in many device applications. It is generally measured using *photoconductivity* experiments, the *photovoltaic effect*, the photoelectromagnetic (PEM) effect, or the Haynes-Shockley experiment. The latter is mainly used to measure *drift velocity*.

The *effective mass* is obtained from *cyclotron resonance*, the *Faraday effect* or the Shubnikov–de Haas effect (oscillations of magnetoresistance of high magnetic fields and low temperatures).

Measurements of *optical absorption and reflection* provide information on band structure and energy levels in semiconductors: *Fundamental absorption* is based on the fact that a semiconductor, while opaque for light with an energy greater than the band gap, becomes transparent for light of lower energy. The absorption spectrum is modified when impurities or excitons have to be taken into account.

In the presence of a strong magnetic field *Landau levels* are introduced in the conduction and valence band. These levels cause a shift of the optical absorption edge. A strong electric field also causes a shift of fundamental absorption (Franz-Keldysh effect).

8. *SEMICONDUCTOR DEVICES*

Among semiconductor devices, bulk-effect devices, diodes, bipolar transistors, and unipolar transistors can be distinguished. Table III gives an overview of the most important devices for electronic, optoelectronic, and other applications.

Bulk-effect devices use the basic semiconductor effect as listed above. For example, a current limiter uses the effect of velocity saturation at high electric field (Fig. 7). In a III–V semiconductor like GaAs the electrons are scattered from a low-energy, high-mobility valley to a high-energy, low-mobility valley. This leads to the formation of accumulation layers of electrons which generate microwave oscillations; such a device is called a transferred electron device. In a photoconductor, incident light generates carriers and increases the conductivity. Magnetoresistors and Hall effect sensors are being used to detect changes in the magnetic field: These devices can serve as position sensors in machines or cars.

Diodes are the building blocks for the more complex transistors as well as devices for special applications. A diode can be formed by changing the doping in the semiconductor from *p* type to *n* type (*p-n* junction diode) or by contacting a semiconductor with metal [metal–semiconductor diode (MES diode)] or by building a metal–insulator semiconductor (MIS) structure. *p-n* junction diodes are used as rectifiers, varistors, or varactors. Transit time diodes such as IMPATT (impact-ionization avalanche transit time) or BARITT (barrier injection transit time) produce microwave radiation.

Conversion of electrical energy into light is performed by light emitting diodes (LEDs). LEDs typically show linewidths from 10 to 50 nm. With special device structures the linewidth can be narrowed down to 0.01–0.1 nm in LASER diodes. For detection of photons various types of photodiodes are available. A photodiode has a depleted semiconductor region with a high field in order to separate electron–hole pairs generated by light. The intrinsic (*i*) region in a *p-i-n* photodiode is that depleted region. MES or MSM (metal–semiconductor–metal) and avalanche photodiodes have been developed for detecting light signals modulated at microwave frequencies.

Table III Overview on semiconductor devices

Device	Application: Electronic	Application: Optoelectronic + other
Bulk effect devices	Resistor	Optical filter
	Current limiter	Photoconductor
	Transferred electron device	Thermistor
		Magnetoresistor
		Hall effect sensor
Diodes	*p-n* junction diode	Light emitting diode
	MES diode	Laser
	MIS diode	Photodiode
	Rectifier	Avalanche photodiode
	Varistor	Solar cell
	Varactor	
	Tunnel diode	
	Transit time diode	
Bipolar transistors	Bipolar junction transistor	Phototransistor
	Thyristor	
	Heterojunction bipolar trans.	
	Resonant tunneling bipolar transistor	
Unipolar transistors	MOSFET	Chem FET
	MESFET, JFET	
	HFET	

Solar cells convert sunlight directly into electric energy. At present solar cells are important as power supplies in satellites, space vehicles, and in special small-scale terrestrial applications. Research and development of low-cost flat panel arrays, of thin-film devices, and of many new concepts has made significant progress which will enhance the use of solar cells in terrestrial applications.

Bipolar transistors are *pnp* or *npn* structures. The output current can be modulated with a small input current, i.e., the device shows current gain. This can be used to build, e.g., amplifiers for low-noise or high-power applications, phototransistors, or digital circuits. The first monolithically integrated circuits were built with silicon bipolar transistors. This technology is still attractive for high-speed digital and analog integrated circuits. Thyristors are basically four-layer devices (*p-n-p-n*) that can be switched between a low-current "off" state and a high-current "on" state. They are able to switch very high currents (up to 5 kA) and to withstand high voltages (up to 10 kV). Recent improvements of bipolar transistor performance at microwave frequencies has been demonstrated with the use of heterojunctions such as AlGaAs/GaAs, GaInAs/InP, or SiGe/Si. The incorporation of a resonant tunneling diode into the emitter of a bioplar transistor yields a resonant-tunneling bipolar transistor (RBT). It is a multifunctional device, i.e., a simple device can perform a function of a circuit. The RBT and the heterojunction bipolar transistor are presently investigated in many research laboratories.

In unipolar transistors or field-effect transistors (FETs) the output current flows through a semiconductor with only one type of conductivity (*p* type or *n* type). The current flow between source and drain contacts is modulated by a transverse electric field originating from the gate electrode. As the gate current is usually small, the drain current is basically controlled by gate voltage. Current control can be achieved either by a MIS diode, by a MES diode, or by a *p-n* junction diode. The silicon FET with a MIS diode is called MOSFET (O = oxide as insulator). This device is the most important one for achieving the highest levels of integration as required, e.g., in memories or microprocessors. In 1990, 4-megabit memories are commercially available. With the present trend 128 megabits will be available by the year 2000. MESFET and JFET on GaAs are attractive devices for very high-speed analog and digital circuits. The millimeter-wave frequency range ($>$ 30 GHz) will be the domain of heterostructure FETs based on III–V materials such as AlGaAs/InGaAs/GaAs or AlInAs/InGaAs/InP.

Field-effect transistors are being developed for use as chemical sensors to detect gases. In such devices the gate region is covered with a material which reacts with the gas to be detected. Electric charge from the chemical reaction produces the electric field for changing the output current.

The reader is referred to Refs. 10 and 11 for an introduction to the physics of semiconductor devices. State of the art results are published regularly in Ref. 23.

See also CHEMICAL BONDING; CRYSTAL BINDING; CRYSTAL DEFECTS; CRYSTALLOGRAPHY, X-RAY; CYCLOTRON RESONANCE; DE HAAS–VAN ALPHEN EFFECT; ELECTRON ENERGY STATES IN SOLIDS AND LIQUIDS; EXCITONS; GALVANOMAGNETIC EFFECTS; HALL EFFECT; INSULATORS; LASERS; LIQUID CRYSTALS; MAGNETORESISTANCE; PHONONS; PHOTOCONDUCTIVITY; PHOTOVOLTAIC EFFECT; SEMICONDUCTORS, AMORPHOUS; THERMOELECTRIC EFFECTS.

REFERENCES

1. O. Madelung, "Halbleiter," in *Handbuch der Physik*. (S. Flügge, ed.), Vol. XX, pp. 1–245, Springer, Berlin, 1957. (I)
2. J. C. Slater, "The Electronic Structure of Solids," in *Handbuch der Physik* (S. Flügge, ed.), Vol. XIX, pp. 1–136. Springer, Berlin, 1956. (A)
3. C. Kittel, *Introduction to Solid State Physics,* 4th ed. Wiley, New York, 1971. (I)
4. C. A. Wert and R. M. Thomson, *Physics of Solids*. McGraw-Hill, New York, 1964. (I)
5. O. Madelung, *Physics of III–V Compounds*. Wiley, New York, 1964. (I/A)
6. J. C. Phillips, *Bonds and Band in Semiconductors*. Academic Press, New York, 1973. (A)
7. E. Spenke, *Elektronische Halbleiter*. Springer, Berlin/Heidelberg/New York, 1965. (I)
8. J. Tauc (ed.), *Amorphous and Liquid Semiconductors*. Plenum Press, London/New York, 1974. (A)
9. K. Seeger, *Semiconductor Physics*. Springer, Wien/New York, 1973. (I/A)
10. A. S. Grove, *Physics and Technology of Semiconductor Devices*. Wiley, New York, 1967. (I)
11. S. M. Sze, *Physics of Semiconductor Devices*, 2nd ed. Wiley, New York, 1981. (I/A)
12. E. H. Putley, *The Hall Effect and Semiconductor Physics*. Dover Publications, New York, 1968. (A)
13. C. Hilsum, "Band Structure Engineering," *American Institute of Physics Conference Series No. 12,* pp. 77–86. AIP, New York, 1971. (Review)
14. J. R. Chelikowski and M. L. Cohen, "Nonlocal Pseudopotential Calculation for the Electronic Structure of Eleven Diamond and Zinc-Blende Semiconductors," *Phys. Rev. B* **14,** 556–582 (1976).
15. W. E. Haas and M. S. Schnöller, "Silicon Doping by Nuclear Transmutation," *J. Electr. Mat.* **5,** 57–68 (1976).
16. L. Hollan, J. P. Hallais, and J. C. Brie, *The Preparation of Gallium Arsenide,* Chap. 1, pp. 1–218 (E. Kalids, ed.), Current Topics in Materials Science, Vol. 5. North-Holland Publishing Company, Amsterdam, 1980. (I/A)
17. B. I. Boltaks, *Diffusion in Semiconductors*. Academic Press, New York, 1963. (I)
18. (a) G. Dearnealy, J. H. Freeman, R. S. Nelson, and J. Stephen, *Ion Implantation*. North-Holland Publishing Company, Amsterdam, 1973. (I/A) (b) H. Ryssel and I. Ruge, *Ion Implantation*. Wiley, Chichester, 1986. (I/A)
19. (a) *Handbook of Electronic Materials, Vol. 2: III–V Semiconducting Compounds* (M. Neuberger, ed.), *Vol. 5: Group IV Semiconducting Materials* (M. Neuberger, ed.). IFI/Plenum, New York, 1971. (b) H. Wolf, *Semiconductors*. Wiley, New York, 1971. (c) H. Hartman, R. Mach, and B. Selle: *Wide Gap II–VI Compounds as Electronic Materials* (E. Kaldis, ed.), Current Topics in Materials Science, Vol. 9. North-Holland Publishing Company, Amsterdam, 1982.
20. C. B. Norris and J. F. Gibbons, "Measurement of High-Field Carrier Drift Velocities in Silicon by a Time-of-Flight Technique," *IEEE Trans. Electr. Dev.* **ED-14,** 38–46 (1967).
21. A. Neukermans and G. S. Kino, "Measurement of the Electron Velocity-Field Characteristics in Germanium Using a New Technique," *Phys. Rev.* **B7,** 2693–2703 (1973).
22. W. Fawcett and D. C. Herbert, "High-Field Transport in GaAs and InP," *J. Phys. C: Solid State Phys.* **7,** 1641–1654 (1974).
23. International Electron Device Meeting organized by IEEE and held annually in December in Washington DC or San Francisco, CA.

Servomechanism

G. J. Thaler

A servomechanism is a feedback control system in which the controlled variable is a mechanical position. Thus a satellite tracking station uses a servomechanism to keep an antenna pointed at the satellite; a magnetic disk memory uses a servomechanism to position the read/write head over a specified data track on the disk; an automatic milling machine uses servomechanisms to position the material and the milling cutter; etc.

The basic configuration of a servomechanism is shown by the solid lines in Fig. 1. A position control system may be required to drive the output, C, to some commanded position R. It may also be required to act as a regulator and keep the output C at the desired R despite the application of some load disturbance D (a typical load disturbance is a wind load on a radar antenna). A basic requirement is that the output C be measured at least as accurately as required by the positioning specifications. This measurement of C is then compared with the desired position R by subtraction. If there is an error, $E=R-C$, the error signal is amplified and applied to the actuator, which drives the output C in the proper direction to reduce the error E.

The requirement for position accuracy is normally a *steady-state* requirement; e.g., the cutting tool on a milling machine must be accurately positioned with respect to the work before a cut is made. However, steady state may be interpreted as a steady dynamic condition; e.g., a radar antenna must be pointed at a moving target within specified position accuracy, which implies that the antenna itself is moving. In order to achieve desired accuracy, the most commonly used technique is to increase the amplifier gain (by an amount that can be calculated). In some cases this tech-

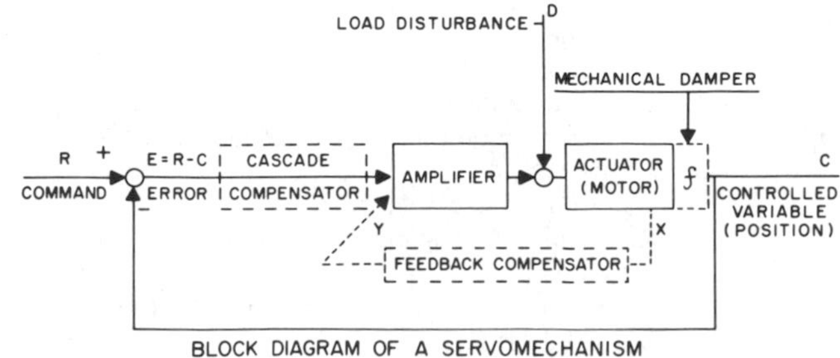

FIG. 1. Block diagram of a servomechanism.

nique cannot work, and desired steady-state accuracy is then achieved by introducing pure integrators in cascade with the error signal.

Both gain adjustment and introduction of integrators tend to destabilize the system; i.e., the closed loop becomes an oscillator and as such is not a useful positioning system. The engineer is then faced with a design problem; he must stabilize the system and provide an acceptable transient response while retaining those features (gain and/or integrators) required to meet the accuracy specification. To achieve the desired results, additions (shown by the dotted blocks in the diagram) must be made to the system. Either a cascade compensator or a feedback compensator may be used, or some combination of the two if the situation warrants it. Occasionally it is advantageous to use a mechanical damper (dashpot, eddy current device, etc.) or a tuned damper (mass, spring, friction unit).

For small positioning systems the compensators may be thought of as electric circuit filters. In practice they are often implemented with resistors, capacitors, and operational amplifiers. Various design techniques are available but are beyond the scope of this discussion. It may be pointed out, however, that the cascade filter design is usually accomplished using some transfer function method, and the designer is required to choose the number of zeros and poles to be used in the compensator and to determine their locations such that the system is stabilized and acceptable transient behavior is achieved. For the feedback compensator design transfer function methods may be used, but state variable methods are also possible. In either case appropriate signals must be measured and processed before summing them into the amplifier.

When the servomechanism is part of a large and complex system the information it requires may well be processed by a digital computer. Thus R may be in digital form, C and other signals may be converted to digital form, and the compensators used may be digital filters (either subroutines in the control computer or small special-purpose computers).

The accuracy required of a servomechanism depends on the nature of the application, and the accuracy achievable is limited in part by nonlinear characteristics in the components and in part by the accuracy of available measuring devices. Theoretically a steady-state error of *zero* is possible for cases in which R (or D) is a constant, a step, a ramp, a sine wave, or other deterministic input. Practically, the condition of zero steady-state error is seldom realistic. Certainly position errors of only a few microinches have been achieved in mass-produced position controls, and even greater accuracy is certainly achievable, but undoubtedly very expensive.

BIBLIOGRAPHY

J. L. Bower and P. M. Schultheiss, *Introduction to the Design of Servomechanisms*. Wiley, New York, 1958.

V. W. Eveleigh, *Introduction to Control Systems Design*. McGraw-Hill, New York, 1972.

H. L. Hazen, "Theory of Servomechanisms," *J. Franklin Inst.* **218**, (1934).

G. J. Thaler, *Design of Feedback Systems*. Dowden, Hutchinson & Ross, Stroudsburg, PA, 1973.

Shock Waves and Detonations*

Marvin Ross

A shock wave is a disturbance propagating at supersonic speed in a material, accompanied by an extremely rapid rise in pressure, density, and temperature. Such waves may be generated by the sudden release of a large amount of energy in a limited space, as in the detonation of high explosives (chemical energy), the passage of supersonic aircraft (mechanical energy), or the discharge of lightning bolts in a narrow channel (electrical energy). A shock wave not sustained loses energy through viscous dissipation and reduces to a sound wave (e.g., thunder).

A detonation is a wave in which an exothermal reaction takes place and that moves with supersonic velocity into the undetonated material. In a detonation, a shock wave compresses, heats, and ignites an explosive mixture that gives off sufficient energy and expanding reaction products to sustain the shock. The detonation process requires a shock wave. A flame, or deflagration, differs from a detonation because it propagates at subsonic speeds and is not accompanied by a shock wave.

SHOCK WAVES

Figure 1 illustrates the formation of a one-dimensional, planar shock wave generated by accelerating a piston in small increments from zero velocity to some final constant velocity. The first infinitesimal compression at the piston face results in the propagation of a sound wave. However, subsequent compressions at the piston face take place with the material at higher densities and result in higher local sound speeds. This produces a train of waves in which the first is at the speed of sound in the undisturbed material and the last, closest to the piston face, is supersonic. Because the last wave can catch but not pass the first, all the waves eventually coalesce into a single, steep wave front across which exists a sharp discontinuity in pressure, density, and temperature. The width of the discontinuity is generally a few molecular mean-free-path lengths. Behind the piston, the reverse process of gas expansion creates a rarefaction wave that moves in a direction opposite to the shock wave and piston motion. A rarefaction wave cannot form a shock front.

The change in physical properties across the shock front is described by the Rankine–Hugoniot (R–H) equations: conservation of mass,

$$\frac{V_0-V}{V_0}=\frac{U-U_0}{S}; \tag{1}$$

conservation of momentum,

$$P-P_0=S(U-U_0)/V_0; \tag{2}$$

and conservation of energy,

$$E-E_0=\frac{(P+P_0)(V_0-V)}{2}, \tag{3}$$

* Work performed under the auspices of the U.S. Department of Energy by Lawrence Livermore Laboratory under contract #W-7405-Eng-48.

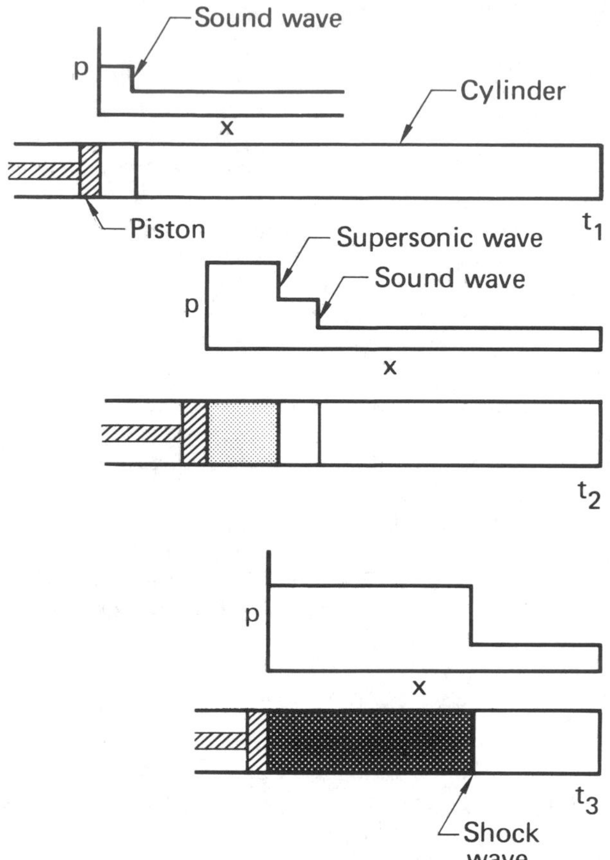

FIG. 1. Formation of a one-dimensional planar shock wave. (Shown by pressure–distance p–x plots at three successive times, t_1 to t_3.)

where S is the shock velocity, and E, V, P, and U are, respectively, the energy, specific volume, pressure, and material or piston velocity in the shocked states. The subscript zero indicates the initial states. The R–H equations represent the locus of all final states that can be reached by shock-compressing a material from the same initial state; the resultant curve of pressure against volume is known as a Hugoniot (See Fig. 4). If the initial state is known, then the final-state properties may be determined from a measurement of any two of these five properties. Of these, the shock velocity is the most commonly measured. Similar equations can be derived to describe conical shocks generated by a sharp-nosed projectile or spherical shocks that travel symmetrically outward in three dimensions. In these cases, the shock wave expands radially and decreases in intensity as the front propagates from its point of origin.

As do other wave phenomena, shock waves reflect, refract, and diffract when colliding with rigid walls, when moving from one medium to another, and when passing geometrical objects. A shock wave traveling in material A will always cross an interface and transmit a shock to an adjacent material B. In addition, a rarefraction wave will be reflected back into A if the impedance of A (initial density × shock velocity) is greater than the impedance of B; or a shock wave will be reflected back into A if the impedance of B is greater. No reflection occurs at equal impedance. For example, a shock wave traveling from a solid to a gas will be reflected from the interface as a rarefaction wave and will be transmitted ahead as a shock wave. In traveling from a gas to a solid, both the reflected and transmitted waves will be shocks.

Shock waves set up in supersonic flows that change direction (Fig. 2) are referred to as oblique. The pressure increases across the shock front because of the extra compression at the compression corner (Fig. 2a) and decreases over a broadened front as the flow expands at the expansion corner (Fig. 2b). Oblique shocks occur on the surfaces of transonic and supersonic aircraft, and because their pressure gradients influence overall structural integrity, they have been studied extensively.

The R–H equations are widely used to study plane shock waves propagating in an ideal gas. For this case, the pressure, volume, and temperature ratios across the shock front are as follows:

$$\frac{P}{P_0} = \frac{2\gamma M^2 - (\gamma - 1)}{(\gamma + 1)}, \tag{4}$$

$$\frac{V}{V_0} = \frac{(\gamma - 1)M^2 + 2}{(\gamma + 1)M^2}, \tag{5}$$

and

$$\frac{T}{T_0} = \frac{PV}{P_0 V_0} \tag{6}$$

where the Mach number M is the shock velocity divided by the speed of sound in the undisturbed medium; γ is the specific heat ratio, C_p/C_v, for the gas. For real gases, the equations must be modified to account for vibration, rotation, electron excitation, and dissociation. A unique feature of shock waves in gases is the high final temperature, up to 15 000 K, that can be generated. This feature is largely responsible for the widespread application of shock tubes to

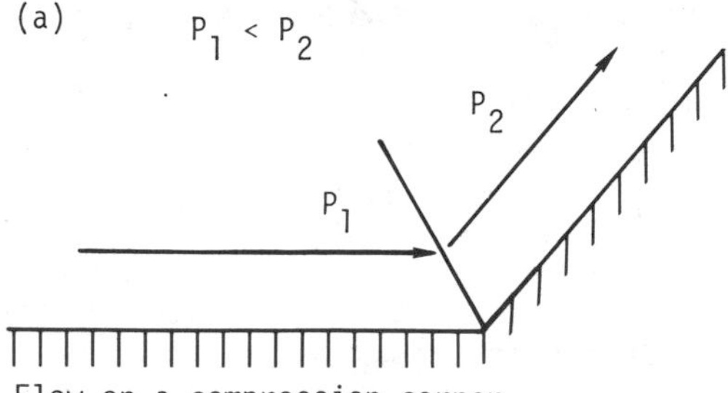

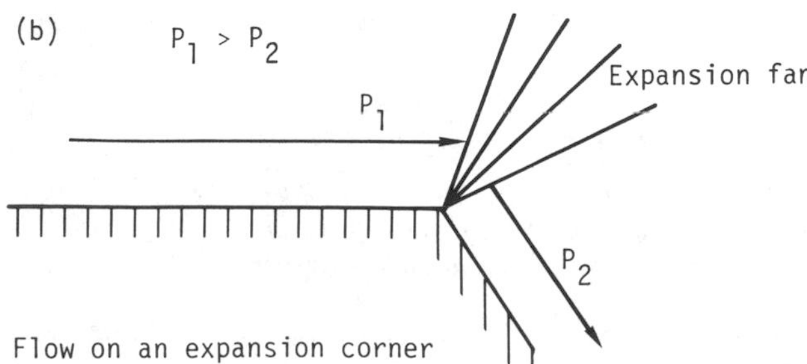

FIG. 2. Formation of oblique shock waves in gases on nonplanar boundaries.

the study of fundamental problems in science and aeronautics. Measurements have been made of chemical reaction rates, vibrational relaxation rates, and the rates and energies of dissociation and ionization.

In aerodynamics, modified shock tubes are used as short-duration wind tunnels to produce the high-Mach-number ($M>16$), high-temperature (6000 K) environments needed to study the gas dynamics encountered by a missile or reentry vehicle—environments that conventional wind tunnels ($M<7$) cannot achieve.

In the shock tube shown in Fig. 3, a one-dimensional shock wave is generated by the sudden bursting of the diaphragm separating a high-pressure driver gas from a low-pressure gas. After bursting, the driver gas acts as a piston and forms a compression wave that rapidly steepens to form a shock front in the low-pressure gas. Simultaneously, a rarefaction forms and travels back into the high-pressure end of the tube. On reaching the end of the tube the shock wave is reflected and moves back into the already shock-heated gas, further raising its temperature, pressure, and density. For the low-density gases used in shock-tube experiments, only the shock velocity need be measured to determine the final P, V, and T.

The shock velocity can be measured by a number of methods that detect the characteristic large, sharply defined changes in material properties. Because the available experimental time is short (10^{-6}s), all instrumentation is designed for fast response. Piezoelectric pressure transducers or heat transfer gauges placed along the tube indicate shock arrival times, and hence velocity. Optical methods are widely used because of their fast response times and noninteractive nature. The large density gradient at the shock front leads to a refractive index gradient that deflects a passing light beam. This deflected light is used by the schlieren method to photograph shock structure and by interferometry to provide quantitative density profiles. If the shock is luminous, a streaking camera can record a continuous measurement of the shock position with time. Chemical composition can be studied with absorption, emission, and Raman spectroscopy.

The generation of shock waves in solids and liquids, because of their higher density, requires more energy than for generation in gases. This energy can be supplied by explosives or high-velocity guns. The most common method of measuring shock-wave velocities in solids or liquids is the use of electrical probes that short-circuit at the passing of the shock. The high surface luminosity of shock-compressed solids and liquids permits the use of streak cameras. Some experimental Hugoniots for metals are shown in Fig. 4.

The relationship between pressure and density is well known for gases but not for liquids and solids. Consequently, most shock-wave experiments on condensed materials have centered on pressure–volume determinations by measuring shock and piston velocity and using the R–H equations. With this method, pressures up to about 1 TPa (10 Mbar) are measured. The entire range of pressures existing within the Earth's interior, including the core (about 3.5 Mbar), is accessible to laboratory shock-wave studies.

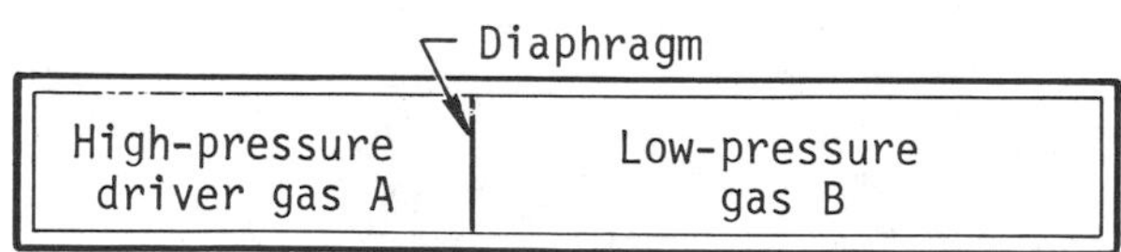

(a) Shock tube and gases at initial state, before rupture of diaphragm;

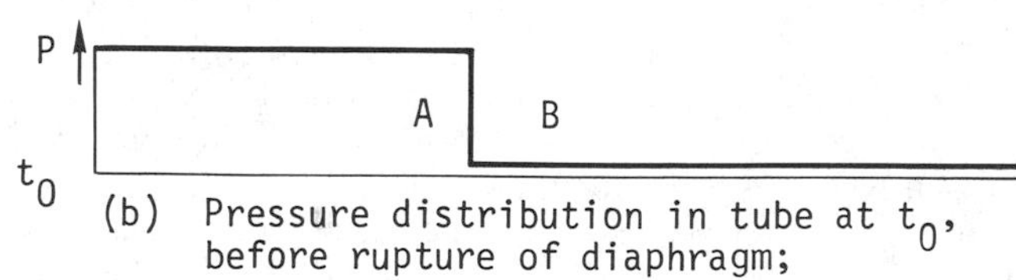

(b) Pressure distribution in tube at t_0, before rupture of diaphragm;

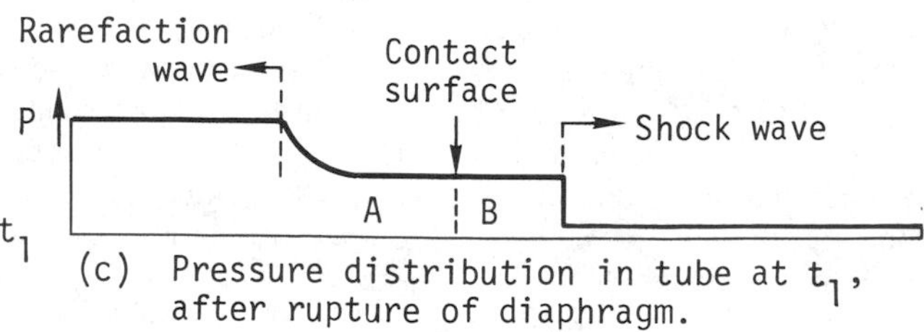

(c) Pressure distribution in tube at t_1, after rupture of diaphragm.

FIG. 3. Generation of a plane shock wave in a simple shock tube.

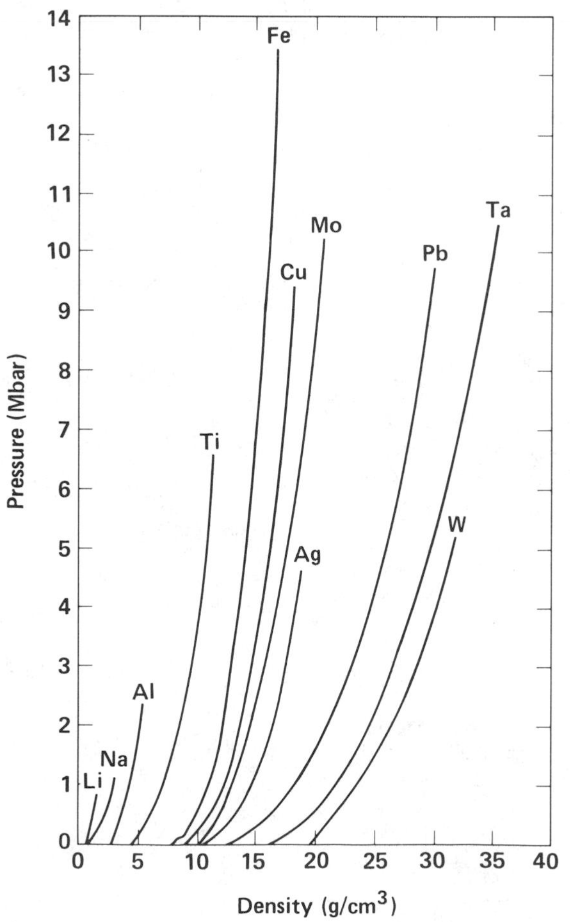

FIG. 4. Some experimental Hugoniots for metals.

DETONATIONS

Gases were first detonated in the laboratory in 1881, when an explosive mixture in a long uniform tube was ignited at one end and the initial combustion wave was observed to accelerate rapidly to a high constant speed (detonation velocity) over the remaining length. The detonation velocity is independent of the method of ignition, tube material, or tube diameter (beyond a certain minimum), but is dependent on the chemical composition of the gas mixture. A stable detonation cannot be generated outside certain ranges of composition (called limits of detonability). Typical detonation velocities, temperatures, and pressures in gas mixtures are, respectively, 2000 m/s, 3000 K, and 2 Pa (20 bars), and are relatively insensitive to the initial conditions. For comparison, burning rates are 0.1 m/s. Detonation velocities in gases can be accurately computed with the Chapman–Jouquet (C–J) theory.

For a shock wave without a chemical reaction (Fig. 5), the Hugoniot curve passes smoothly from its initial state (P_0, V_0). This is not so for a detonation, which can be thought of as a two-step process in which a chemical reaction releases energy in a constant-volume explosion and the reaction products are then shock-compressed to some final state. The velocity of this final state is proportional to the slope of the line through the initial and final states (Rayleigh line) and is given by Eq. (2). Because the R–H equations cannot by themselves predict which of the Rayleigh lines (*OA* or *OB*) corresponds to the unique detonation velocity, an additional assumption is necessary. This is provided by C–J theory, which assumes that the detonation wave proceeds at the minimum attainable velocity. This is given by the tangent to the reacted Hugoniot curve drawn from the initial state, the line *OA*. The point of tangency is called the C–J point. Whereas C–J theory is successfully applied to gases, difficulties arise when the detonation properties of solid or liquid explosives are computed. Here, the products form a very dense gas for which the *P–V–E* relationship is not well known and, hence, the computed properties are less accurately predicted. Typical velocities and pressures in solid explosives are 8000 m/s and 30 kPa (300 kilobars).

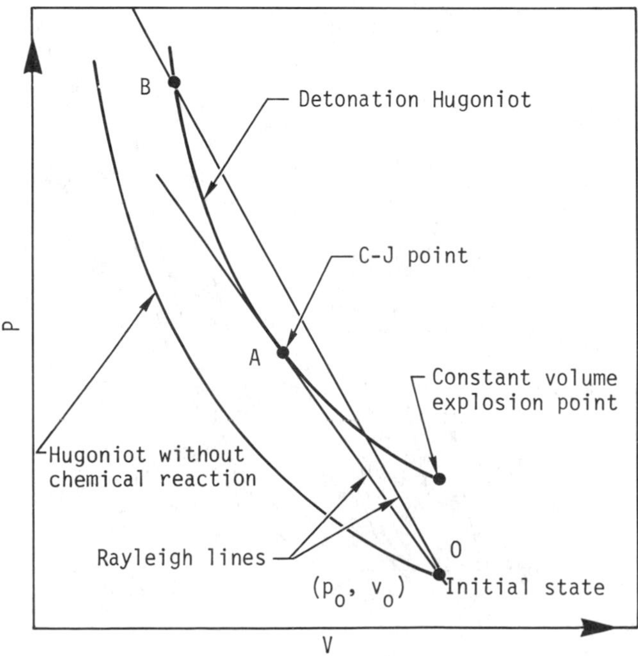

FIG. 5. Hugoniot curve for a detonation.

The detailed structure of a detonation wave is visualized as a shock front followed by a chemical reaction zone. Pressure and density peak at the shock front and fall as the reaction proceeds. In gases at atmospheric pressure, this reaction zone is about 1 mm thick. Beyond this zone, the product gas expands isentropically and acts as a continuously moving piston that propagates the shock at constant velocity. Detonation is initiated by the introduction of a shock wave or a local heat source. The methods used to study detonations are generally the same used for shock waves. Research is directed to determining C–J points, product composition, wave structure, and studying the initiation or onset.

See also ACOUSTICS; EQUATIONS OF STATE; WAVES.

BIBLIOGRAPHY

M. A. Cook, *The Science of High Explosives*. Krieger, Huntington, N.Y., 1971. (I)

I. I. Glass, *Shock Waves and Man*. Univ. of Toronto Institute for Aerospace Studies, Toronto, 1974. (E)

J. K. Wright, *Shock Tubes*. Methuen, London, 1961. (I)

Ya. B. Zeldovich and Yu. P. Raizer, *Physics of Shockwaves and High Temperatures Hydrodynamic Phenomena*, Vols. 1 and 2. Academic Press, New York, 1967. (A)

Soil Physics

Don Kirkham

Soil physics, as discussed here, deals with physical properties and conditions of soil related to plant growth. Soil mechanics, a discipline of soil engineering in which soil is considered a building material, will not be discussed.

The physical nature of a soil depends on its composition. An average crop soil at an average time during a growth season will consist of about 50% solids, 25% air, and 25% water, all volume percentages. The air and water space will fluctuate as the plant roots take up water and as rainfall or irrigation supplies additional water. The bulk density of such an average soil will be about 1.3 g/cm^3 of air dry soil referred to a volume of 1 cm^3 of the bulk soil. Soils when cultivated may have bulk densities of 1–1.2 g/cm^3 and compacted soils 1.4–1.6 g/cm^3 or larger. In compacted soils, roots cannot proliferate because of high root impedance. Also, compacted soils do not imbibe rainfall easily because it runs off, and do not supply enough air to plant roots. If the temperature is too low or too high, plants will not grow. From these remarks it is evident that soil impedance, soil water, soil air, and soil temperature are the principal physical factors that govern crop growth. Soil physics deals largely with these four factors.

Soil impedance depends on the soil moisture content and may be measured by penetrometers. Needle penetrometers are used to simulate the forces encountered by a growing root. As a root pushes forward and outward to increase its volume, the work done will be given by $W = PV$, where P is the pressure of the soil the root is working against and V is the volume increase. Farm tractor tires compress the soil with pressures of about 3 kg/cm^2. When tractor wheels, as in plowing, ride in furrow bottoms, plow soles with high plant root impedance may result.

Soil water retention and movement are major areas in soil water physics. After a rain or irrigation, water tends to be pulled into the deeper soil by gravity. Capillary forces react against gravity forces to retain the water. If too much water is retained, there is too little air space in the soil for proper root aeration. If too little water is retained, the soil will be droughty. Tensiometers are used to measure the tension of water held in the soil at various moisture contents. Tensiometers work up to about 0.8 atm (80 kPa) of soil water suction or even 100 atm (10 MPa) through a properly supported plastic membrane. Plants can take water from the soil up to a tension of about 15 atm (1.5 MPa). Plants wilt when water is held more tightly in the soil.

Soil water movement is governed by Darcy's law, which says that the volume of water flowing across a unit area of bulk soil per unit time is directly proportional to the potential energy gradient that drives the water. The factor of proportionality is called the hydraulic conductivity. The potential energy gradient consists, in most soil-water flow problems, of a pressure and a gravity component; the two components add up to the hydraulic gradient, which is then the potential energy gradient. In a soil the hydraulic gradient may be measured by standpipes called piezometers if the soil is water saturated and by tensiometers if the soil is unsaturated. For saturated soil the factor of proportionality in Darcy's law, the hydraulic conductivity, is a constant, and the law, when applied to the equation of continuity for the soil water, gives Laplace's equation of potential theory, so that the methods of potential theory become available in soil physics. Soil physics problems of potential theory are complicated by unusually shaped boundaries of the water-saturated flow medium that exist due to water tables in the soil that are often curved and moving. The water tables move slowly so that sets of quasi-steady-state solutions are obtained. In unsaturated soil the coefficient of proportionality in Darcy's law is not a constant but varies rapidly with soil moisture content. Nonlinear mathematics results. Progress is being made on solving unsaturated flow problems by use of digital computers. Practical applications of soil-water movement theory are in the drainage and irrigation of land. Artificial drainage is required in both humid and arid lands. In humid lands, excess water must be removed, especially in the early growing season, by soil drain tubes or ditches; and in arid lands, the excess irrigation water that must be applied to keep the soil from salinizing must also be removed by soil drain tubes or drainage ditches. For reclaiming salted soil it is found that the salt is removed most economically by applying water to move largely in an unsaturated state.

Soil aeration is tied up with soil water. Oxygen from the air must move through the soil profile to the plant roots. This process is largely by diffusion, for which Fick's law and the associated mathematics are applicable. The oxygen passes through water-emptied soil pores and then through a layer of water about the root to the root. It has been found that about 10% by volume of air space is required in crop soils to assure continuous aeration paths to the plant roots. When this much air space is not available, plants must get oxygen from that dissolved in soil water.

Soil temperature is the fourth soil physical factor. In the spring, a warm soil temperature is desired. Corn (maize) seedlings at 15°C may have their growth rate more than doubled for a 10°C increase in temperature; temperatures above 30°C decrease the seedling growth rate. Soil temperature can be physically controlled by straw mulches and by drainage. In the spring, soil that is drained does not require the sun's energy to evaporate this drained water; the sun's energy is then used to warm up the soil. Fourier's heat flow law may be used to predict soil temperatures. Use of the law is complicated by the movement of water and vapor that occurs when there is a soil thermal gradient.

Isotopes and atomic radiation have been used since about 1950 in studying soil impedance, water, aeration, and temperature. Radioactive gold can evaluate root impedance; neutron probes measure soil moisture; gamma rays, soil density; beta rays, leaf turgor as influenced by soil moisture. Oxygen-18, deuterium, tritium, and carbon-14 and other isotopes act as tracers for water, air, and plant nutrients, in particular, when temperature changes are involve. The International Atomic Energy Agency, Vienna, is a source of publications.

In recent years soil physicists have paid much attention to environmental problems. Agricultural fertilizers, pesticides, and herbicides cause non-point-source pollution of groundwater. Point-source pollution, as from feedlots, landfills, or industrial plants, is also under research. The location and containment of nuclear wastes is a soil physics concern. Infiltration relates to all of these pollution problems. Water, once in the soil and in the saturated state, is subject to potential flow theory. Soil physicists have developed a formula and instrumentation for determining and controlling the extent of this pollution or potential pollution. More sophisticated mathematical techniques, including modeling, require increased interdisciplinary cooperation. Soil physics today is related to chemistry, geophysics, geology, hydrology, agronomy, and civil engineering.

See also CAPILLARY FLOWS; FLUID PHYSICS.

BIBLIOGRAPHY

Jacob Bear and Arnold Verrijt, *Modeling Groundwater Flow and Pollution* (with computer programs for sample cases). Dordecht, Boston, and Reidel, Norwell, MA, 1987. (Sold and distributed by Kluwer Academic Publishers, New York.) (A)

Daniel Hillel, *Introduction to Soil Physics*. Academic Press, New York, 1982. (E)

Arnold Klute (ed.), *Methods of Soil Analysis. Part 1, Physical and Mineralogical Methods,* 2nd ed., American Society of Agronomy, Madison, WI, 1986. (I)

Don Kirkham and W. L. Powers, *Advanced Soil Physics*. Wiley-Interscience, New York, 1972. Reprint edition (revised), Krieger, Malabar, FL., 1984. (A)

Theo John Marshall and John Winspere Holmes, *Soil Physics*. Cambridge University Press, Cambridge 1979; second edition 1988. (I)

Solar Energy

Lloyd O. Herwig

INTRODUCTION

The radiant energy entering the earth's upper atmosphere from the sun is a huge and almost constant solar energy source having essentially unlimited lifetime with respect to mankind's foreseeable energy needs. It is, has been, and will continue indefinitely to be the earth's primary source of energy that dwarfs all other forms in magnitude and versatility.

In passing through the earth's atmosphere to the earth's surface, this incident radiant energy is subject to absorption and scattering processes that selectively reduce and diffuse it. The solar energy density reaching the vicinity of the earth's surface is about 75% of the incoming density at the top of the atmosphere. This remaining energy provides the warmth we feel, the light we see, the natural energy giving rise to atmospheric circulation, weather, and seasons, and, most importantly, the energy required by photosynthetic processes to sustain all the earth's life forms including the production of food and fibrous materials for mankind's use. The sun's radiant energy is the basic physical ingredient sustaining the natural environment and all life on earth.

At any given location or time on the earth's surface, the incident solar energy is variable in both quantity and composition. This variability is largely due to changes in lower atmospheric conditions and to the rotation and orbit of the earth relative to the sun, which gives rise to diurnal (day/night) and seasonal variations, respectively. Some of the atmospheric absorption and scattering processes are relatively constant in time and depend largely on the density, composition, and path length through the air mass (including water vapor, ozone, and other gases). Other processes are geographically and temporally highly variable because of the presence of dust, clouds, and turbulence in the lower regions of the atmosphere, giving rise to large variations in total incident energy as well as in its composition in terms of direct (unscattered) and diffuse radiation components.

The incidence of the sun's radiant energy at the earth's surface and the complex interactions of this radiation with the earth's environment give rise to two general classes of solar energy resources that can be categorized as either direct or indirect. The category of direct (primary) solar energy resources is associated exclusively with the incident solar radiation (insolation) coming to any given location where it is directly absorbed and converted to produce thermal or electrical energy. Solar technologies based upon the direct conversion of the radiant energy to produce useful thermal energy include water heating, heating and cooling of buildings, crop drying, industrial process heat, and production of electricity using heat engine cycles (solar thermal electric systems). Those based upon the direct conversion of the radiant energy to useful electrical energy include the production of electricity using semiconductor materials and devices (photovoltaic systems).

The indirect solar energy resources are associated with the absorption and storage of the incident radiant energy by the earth's environment. These resources include energy collected by the atmosphere (winds), oceans, and biomass. There are solar technologies associated with these resources to provide useful thermal energy, electric power, or fuels for society. One of the currently best known of the solar technologies based upon indirect solar resources is the production of electricity by water power, which involves solar energy evaporation of ocean water, wind transportation of water vapor, condensation and rain over land, water run-off and storage behind dams, and use of flowing water to generate hydropower.

THE EARTH'S PRIMARY SOLAR ENERGY RESOURCE

Space measurements of the rate of solar radiant energy entering the earth's upper atmosphere give a value of about 1350 W/m^2 (135 mW/cm^2). Figure 1 shows schematically some of the absorption and scattering processes and their quantitative effects for average cloudless days within about plus or minus 50 degrees latitude worldwide. The direct solar radiation component (direct rays from the sun giving rise to shadows when an opaque object is interposed) at the earth's surface can vary from more than 80% (frequent U.S. southwest conditions) to less than 20% (occasional U.S. northeast conditions) whereas the diffuse component (scattered light coming from all directions) arriving at the earth's surface can vary from about 15% to 80%. In the presence of clouds, dust, and other atmospheric pollutants, the direct solar radiation component at a given location can be reduced from the cloudless day values to zero and the diffuse component can

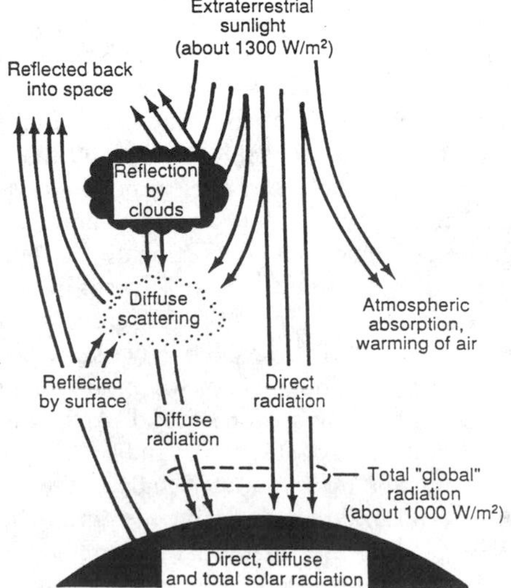

Fig. 1. A schematic diagram showing the major factors affecting sunlight (solar radiation) as it passes through the earth's atmosphere to the earth's surface.

vary over a wide range depending upon the scattering and absorption of the intervening atmospheric conditions.

Solar resource measurements made at many locations over decades show that the average daily and seasonal solar energy availability is reasonably predictable statistically from year to year and even from month to month. Thus, solar power systems can be practically designed to meet projected needs.

Radiant energy from the sun is continuously incident on the earth's atmosphere (the one-half facing the sun at any given time) at a power level about 10,000 times the continuing total world thermal energy use from all other energy sources. Its rate of energy delivery to the surface of the continental United States is equivalent to more than 800 times the current rate of U.S. thermal energy usage from all fossil (coal, oil, and gas), nuclear, hydro, and other energy resources.

It is impressive to realize that the average daily solar radiation (1,480 Btu/square foot/day or 16,800 kJ/m^2/day) falling upon 1 acre (0.4047 hectare) of the continental United States is equivalent in total energy content to about 11 barrels of oil. In many southern and western locations, the daily average incident solar radiation per acre is equivalent to 15 to 20 barrels of oil whereas in most northern areas it averages 5 to 10 barrels. At current prices of crude oil ($15 to 20 per barrel), the average worth of solar energy per acre of U.S. surface area could be considered to be more than $165 per day at 100% collection and utilization efficiency.

This huge radiant energy resource is an unusually available, versatile, and high-quality form of energy. It is virtually inexhaustible with time; transportable at high speeds through air or transparent matter; easily and efficiently convertible to thermal energy by absorption on dark surfaces or in semi-transparent matter; uniquely vital for energizing photosynthesis processes in plants and other living cells; capable of producing high-energy photochemical reactions in matter; and capable of initiating electronic processes in semiconductor materials (devices) to produce electricity directly. In addition, the direct (or unscattered) radiance from the sun can be redirected by means of mirrors (reflection) and lenses (focusing) in new directions or to increased radiant energy density. Concentration (focusing) of the direct radiance from the sun by mirrors and lenses can readily produce high radiance densities, which can be converted to high thermal temperatures at an absorbing surface.

THE EARTH'S INDIRECT SOLAR ENERGY RESOURCES

The sun's radiant incident energy is continually converted to several important indirect solar energy resources as it interacts with the earth's atmosphere and surface features. The most important indirect, earth-based solar energy resources include plant and animal matter produced by photosynthesis (biomass energy), movements of the earth's air mass (wind energy), movement and temperature differences in the oceans (ocean energy), and water and hydroelectric power. The uneven absorption of incident radiant energy in the atmosphere and at the earth's surface produces thermal gradients giving rise to circulation in the atmosphere and the oceans, and the earth's weather generally. In turn, these atmospheric and oceanic circulations along with the myriad of thermal processes and energy transfers on the earth's surface and in the atmosphere give rise to rainfall over land areas and to water (hydro) power, and even to the ocean waves produced by the moving atmosphere over water bodies.

All of the indirect solar energy resources involve various degrees of energy storage in the conversion of incident radiant energy. The storage includes relatively short-term (days to weeks) wind storage in atmospheric thermal gradients, longer-term (years to decades) storage of biomass in its many forms (e.g., wood, grass, living organisms, and organic wastes), and very long-term (decades to centuries) thermal gradients and currents in the huge masses of ocean water. Hydroelectric power systems as a solar energy-based technology incorporates storage through confinement of run-off water behind dams for use as required for electricity production. This more or less natural energy storage in the indirect resources allows more flexibility in the design of power systems based on them since the system may not have to incorporate as much energy storage to overcome the rigid diurnal cycle inherent in the radiant resource. In the case of biomass and ocean-based power systems, baseload (up to 24 hours per day) power generation is quite feasible.

Renewable energy resources has become a popular terminology for a broad range of energy resources including incident solar radiation, indirect solar resources (e.g., wind, ocean, biomass, and hydro resources), and geothermal resources. The discussion in this article does not include the geothermal resource.

In the main, solar power systems based upon the conversion of incident radiant energy are designed to collect and absorb the radiation and directly produce either thermal or electrical power. Absorption of radiation using flat-plate and concentrating solar collectors can produce a wide range of temperatures for a broad range of power needs (e.g., heating and cooling of buildings, industrial process heat, drying of products, production of electricity through thermodynamic cycles, and domestic water heating). These collectors can also produce electricity directly if the dark absorbing surface is replaced by special semiconductor devices that directly produce electric power in photovoltaic power systems. In these solar thermal or electric power systems, there is no natural storage available in the daily resource and the system design must include provisions for matching the load with the daily resource, or of providing storage either internal to the systems (thermal or electric storage) or externally through backup systems.

SOLAR ENERGY POWER SYSTEMS AND ENVIRONMENTAL IMPACTS

Minimal environmental impacts are present in the solar power systems used to convert the incident solar energy resources to useable power for the thermal and electrical needs of society. The operation of solar power systems does not add overall heat to the atmosphere; release noxious combustion gases (e.g., acid rain precursors) or particulate matter to the atmosphere; produce carbon dioxode, other greenhouse gases, radioactive gases, or fuel wastes (e.g., ashes or radioactive solid waste); or require extensive mining facilities (with land disruptions) or fuel transportation support systems. These solar power systems do require construction materials and activities, which can produce some small initial overall environmental impacts. Overall land use (area-wise) is not greatly different from that required for fossil or nuclear power plants when plant and mining areas required for fuel production are included in an analysis.

In the conversion of the indirect solar energy resources to useful power, there are again minimal environmental impacts. Solar power systems based upon wind, ocean, and hydro energy resources have much the same low environmental impacts as those based upon the incident solar energy, as described above. Overall environmental impacts for biomass-based power systems vary somewhat from other solar- (or renewable-) energy-based power systems since burning or chemical treatment of biomass products is often used to obtain thermal energy directly or to produce gaseous, liquid, or solid fuels for a wide range of energy needs. Since biomass growth requires carbon dioxide from the environment, the return of the carbon dioxide in power system operations results in no net addition over the relatively short fuel cycle time.

Of all the energy resources, solar energy is the most familiar to people, the most environmentally acceptable to society, and at the same time the most underutilized basic energy resource available today to a large majority of people all over the world. For society, solar energy has had a profound influence in the shaping of social structures and life on earth as we know it. In the future, its increased utilization by society is expected to have ever more far-reaching positive effects on mankind and the earth's environment.

Some of the approaches to the utilization of solar energy for terrestrial needs will be described in the next sections.

SOLAR ENERGY RESOURCES AND CONVERSION TECHNOLOGY

In the design of solar power systems based upon absorption of incident radiation, it is necessary to realize this radiation consists of two components—direct and diffuse. The direct radiation is the component coming directly to the earth without being scattered and the one producing sharp shadows when it is interrupted by an opaque object. On a very clear day, the direct to scattered (diffuse) components have a ratio up to about 7; whereas, on an overcast day, the ratio is very low.

In general, flat-plate collectors absorb both direct and diffuse incident radiation and perform effectively under a range of atmospheric and insolation conditions. On the other hand, concentrating collectors based upon either mirror or lens configurations (commonly point- or line-focus systems) generally function only on the direct incident radiation. For that reason, concentrator systems are more limited in their effectiveness and are most economic in those geographic regions where direct radiation is available for a high percentage of a year; e.g., the southwestern United States.

The concentration of direct solar radiation by focusing mirror or lens systems can be utilized to produce relatively high thermal temperatures in an absorber placed at the optical focus (up to several thousand degrees centigrade for point-focus collectors, if desired). These higher temperatures from concentrator collectors can be applied, in conjunction with a working fluid and the boiler and condenser components for a heat engine cycle, to produce mechanical or electrical power. This technology of converting direct incident radiation to thermal energy using concentrating collectors is commonly referred to as solar thermal conversion.

Solar Heat Conversion

The incident radiant energy coming to a given location on the earth's surface during the day can be captured and converted to thermal energy or electrical energy by a wide range of collectors. The simplest thermal energy collector is a dark surface that faces the sun, absorbs much of the radiation incident on its surface, and converts it to heat energy, e.g., a black asphalt pavement on a sunny day.

A relatively efficient, low-temperature flat-plate collector widely used in solar heating systems consists of a dark, flat surface enclosed in the sun's direction by a transparent thermal insulator, such as alternating layers of glass and air (transparent glazings), and enclosed on all other sides by insulating materials. The incoming radiation is absorbed on the dark surface and converted to heat energy in the enclosure, a form of the greenhouse effect. A pumped or naturally circulated fluid, such as air or water, in close thermal contact with the absorber element can then be used to transport the heat energy to a nearby location for use. (A car exposed to sunshine through its closed windows has the basic aspects of a solar collector, except for heat removal and utilization.) Flat-plate collectors with glazings can readily provide thermal energy at temperatures well above 50°C with conversion efficiencies greater than 50% to provide hot water, heat for building space, and many other uses.

There are many other types of collectors that concentrate (focus) the solar radiation by means of cylindrical or spherical mirrors (or lenses) to line or point absorber surfaces, respectively, which become exceedingly hot. As water, air, or special fluids are moved through the area of concentrated energy, the acquired thermal energy can be transported for use at a high temperature as industrial process heat or for producing electricity using various types of heat engine cycles, depending upon a system's design.

Solar Electric Conversion

The conversion of incident solar radiation, wind energy resources, and ocean energy resources to electricity includes a wide range of resources, technologies, and systems, including photovoltaic, solar thermal, wind, and ocean thermal systems. Photovoltaic systems convert incident solar radiation directly to electricity by means of special semiconductor materials and devices involving no moving parts in the conversion processes.

Wind energy resources arise due to the uneven absorption and conversion of solar radiation to thermal energy at the earth's surface and in the atmosphere. The heating processes and transfers affect air density and water content giving rise to atmospheric pressure differences (forces) and atmospheric movements (winds). Wind energy systems convert the kinetic energy of the atmosphere to useful mechanical or electrical energy by means of wind turbines. Well-known wind conversion systems include sailing vessels, water-pumping farm windmills, and wind-generated ocean waves.

The ocean surface waters serve as a giant solar energy collector and storage unit. Absorption of solar radiation on tropical ocean areas results in warm surface waters whereas the low intensity of solar radiation at the earth's poles gives rise to huge quantities of cold polar water. As this polar water sinks and flows in deep ocean currents toward the equator, warm surface and cold deep-water sources are formed in low-latitude deep ocean areas. An energy conversion system operating from these vertically adjacent warm and cold sources applied in a thermodynamic cycle (heat engine cycle) can produce mechanical or electrical power. This solar application area is designated as ocean thermal conversion.

Biomass Conversion

Solar energy fixed in biomass resources by photosynthesis processes provides a wide-ranging and large array of indirect solar energy resources that can be converted to useful power directly (e.g., burning of wood) or can be converted to gaseous, liquid, or solid fuels for storage or immediate use. Biomass energy conversion systems utilize organic waste materials (urban, agricultural, industrial, and animal) and organic feedstock (crops) grown on terrestrial or marine energy farms. The conversion of grains to ethyl alcohol (ethanol) by fermentation for use as a motor vehicle fuel additive (to produce gasohol) or as alcohol fuel has been widely implemented in recent years. Production of electricity from biomass combustion in conjunction with the Rankine (water/steam) heat engine cycle has also been widely applied in recent years.

SOLAR ENERGY AND WORLD USE OF ENERGY

The world use of primary energy resources for all human-directed activities is estimated to be about 335 quads annually, of which approximately 80 quads are used annually in the United States. One quad of energy is defined as 10^{15} Btu (British thermal units) and is equivalent to about 170 million barrels of crude oil or about 40 million tons of coal. The 80 quads of primary energy used in the United States includes fossil fuels about 86% (oil—41%, gas—22%, coal—23%), nuclear fuels about 7%, and solar-based energy resources about 7% (hydropower—3.5%, biomass—3.5%). In the United States use of primary energy, about 31 quads (39%) is used to meet thermal energy needs, 28 quads (35%) for electricity generation needs, and 21 quads (26%) for transportation needs.

Solar-based energy resources also play a significant role in the economies of countries around the world. In particular, hydropower is used extensively wherever flowing water is relatively available, and substantial resources still remain to be harnessed in some areas, many of which are quite remote from existing population centers. In most of the developed countries, and especially the United States, most of the large hydroelectric resources have already been harnessed. The only other significant use of solar energy resources is present in the many arid and remotely located countries of the developing world where cooking of meals and heating of living quarters is largely done by burning wood and other biomass products. Unfortunately, the methods of utilization and the nonrenewal of wood resources has led to regional (and even global) problems of deforestation and desertification because of population pressures and the basic societal needs for cooking and heating energy.

A recent study of accessible energy reserves of the United States determined that the inclusion of renewable energy-based resources (mainly solar-based resources) raised the total United States energy reserves to more than 600,000 quads, of which 90% is represented by renewable energy resources. This estimate of accessible resources (those that are economically extractable by current and developing technologies) contrasts with the long-standing estimates of U.S. energy reserves of 6,400 quads (including more than 80% attributed to coal reserves and only 11% attributed to renewable energy), based mainly on economic extraction using current technologies. Thus, solar-based energy resources constitute a huge acccessible resource for the United States, in particular, and for most of the world, in general.

REVIEW OF SOLAR-BASED TECHNOLOGIES

The solar technologies to be briefly described in the next sections of this article are capable of meeting a large fraction of the future U.S. (and world) needs for energy resources. They include solar heat, solar electric, and biofuel technologies.

Most solar energy systems based upon these technologies have important and relatively unique attributes in their capabilities for meeting societal energy needs. In the United States and many locations around the world, solar energy resources are locally available for conversion to local energy needs. Generally, the solar power systems can be engineered to provide energy forms that fit the immediate needs of a user, whether it is a family, community, or business. That is, systems can be designed to provide thermal, electricity, or other types of energy requirements, including gaseous, liquid, or solid fuels. In most of the solar technologies, the size of the system can be adapted well to the size of the need because the system components are modular and can be designed and assembled in either large or small systems. This also means that smaller systems need not cost a great deal more per unit of energy delivery than larger systems. As pointed out previously, the environmental impacts of solar energy systems are minimal for either large or small systems.

Solar Heat Technologies

Solar heat for buildings includes the following applications: water heating, active heating, active cooling, passive heating, passive cooling, and daylighting. Figure 2 shows some of the applications for a building.

Solar water heating in residential structures is best known to the public, along with the heating of swimming pools and of hot water for schools, hospitals, and commercial buildings. Many types of solar collectors are used in these ap-

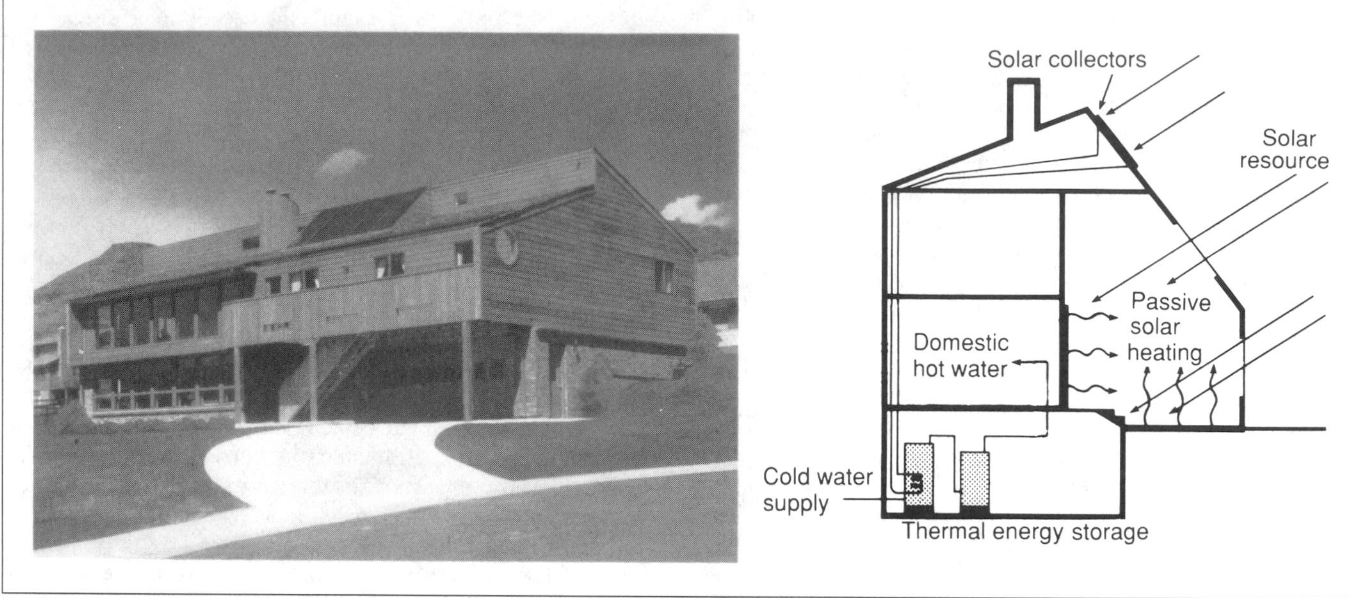

FIG. 2. A photograph of a solarized residential building, and a schematic diagram of its functions.

plications, but the most common one is the flat-plate, glazed collector, described earlier. Potable water is heated directly in many systems as it is moved through the collector by electric pumps or by natural circulation. In most locations subject to freezing weather, a nonfreezing working fluid is used in the collector in conjunction with a heat exchanger to heat the potable water in a storage tank. The batch heating of water in a collector incorporating a large water storage container has become a popular alternative for many residential installations. Frequently, two-tank hot water systems are used where the solar collector preheats the water in the first tank, which then feeds warm or hot water into the conventional home water heater to assure relatively uniform water temperature and availability.

There are estimated to be more than one million solar water heating installations in U.S. homes today, which is only a few percent of the total that might benefit from this technology. Reliable and economical solar water heating systems supplying the order of 50% of water heating needs for residential and commercial buildings can be purchased in many regions of the United States. Also, there are many types of solar collectors available for solar water heating and other applications beside flat-plate collectors. They include evacuated glass tube (high-temperature, high-performance), plastic absorber sheets without glazings (low-temperature for swimming pool heating), concentrator, and many other collector variations. A discussion of the many collector types, systems, and specific applications is beyond the scope of this article.

Active heating and cooling systems utilize the sun's incident energy to perform the functions needed to maintain thermal comfort in buildings by actively adding or removing heat. In carrying out these functions, the systems use solar collectors to collect and store energy in much the same way as for heating water. In general, active solar heating systems are much simpler in operation, better developed, and accepted as commercial products than are active solar cooling systems, which require higher-temperature and higher-performance collectors to operate the relatively complex cooling equipment, e.g., absorption, desiccant, and heat engine cooling. The number of active solar heating systems currently operating in the United States is estimated to be less than 100 000 buildings, and the number of active solar cooling systems to be only a few thousand buildings. In 1990, the U.S. commercial market for new solar heating or cooling systems for buildings is small.

Passive heating and cooling systems utilize and integrate various architectural features and building components into a building structure to stabilize interior temperature and provide thermal comfort for occupants. In the case of passive heating, the system designer utilizes energy controls and building structures such as south-facing glazings (windows), thermal insulation, and storage materials. The heating design relies on the natural heat transfer processes of radiation, convection, and conduction to collect, store, distribute, and control the sun's energy. There is usually no auxiliary power required for regular system operation. In the case of passive cooling design, heat avoidance, natural lighting, and natural cooling methods are used to minimize a building's energy cconsumption while improving the indoor comfort level. Natural cooling methods include natural ventilation, night cooling, and radiative cooling, and are designed to use conventional building materials and minimal mechanical operational assistance.

Passive heating and cooling technology has advanced rapidly over the past 10 years and has been achieving growing acceptance in the architectural and building communities. By 1990 in the United States there are several hundred thousand buildings that incorporate one or more passive features by design and plan to reduce energy requirements, stabilize interior temperatures, and increase the occupant's comfort. Many, if not most, of the prominent passive features are cost-effective now, and there is a great expectation among solar energy experts that most buildings within about 10 years will incorporate passive heating and cooling principles and components in new buildings. It is estimated that building energy requirements can be reduced by a minimum of about 40% with careful passive system design and with little extra cost.

Daylighting involves the conscious and skillful use of natural light to provide a building's illumination during its daytime use. Increasing the daylight availability at the perimeter of buildings is most readily accomplished and involves effective use of wall penetrations, glazings, and interior light redirection to control and distribute incoming daylight. Daylighting the core of buildings is a larger challenge involving collection and ducting of daylight to building interiors for distribution, use of roof penetrations, and incorporation of atrium structures.

Other uses of solar energy collected at low (less than 100°C) to medium (in the range of several hundred degrees) temperatures include agricultural applications (e.g., heating of animal shelters and drying of crops) and process heat applications (e.g., leaching of minerals from raw materials and processing of food).

Solar Thermal Technologies

Solar thermal systems generate medium (several hundred degrees) to high (up to several thousand degrees) temperatures for production of electricity by various heat engine cycles or for process heat applications. The achievement of medium to high temperatures in solar thermal collectors requires the use of concentration by lens or reflector configurations to converge the incident solar radiation to a point or line focus. An absorbing element at the focus position is raised to very high temperatures where a heat transfer fluid is pumped to remove the heat energy for a wide variety of uses. Concentration ratios for solar thermal applications range from a few suns to thousands of suns. In some advanced solar thermal applications, it is the highly intense concentration of radiation itself that enters into unique chemical reactions and provides desirable results, as opposed to the more normal mode of generating high temperatures upon absorption at a surface.

Solar thermal energy systems concentrate the sun's radiation to generate electricity, produce process heat, or provide high radiant flux for use in catalyzing chemical reactions or producing transportable fuels. Large solar thermal power systems are usually characterized as either distributed receiver systems or central receiver systems. Distributed receiver systems are characterized by large numbers of individual collector/receiver (conversion) units combined to work in parallel with each other in large power installations to produce electricity, process heat, or some other energy product. A central receiver system consists of a large field of ground-mounted, tracking mirrors (heliostats) that reflect (focus) the image of the sun to a central receiver (usually elevated well above the ground). At the central receiver, the concentrated radiant energy is absorbed and converted to an energy product, usually in concert with a thermodynamic working fluid in a heat engine cycle to convert the aggregated heat energy to electricity, but in some cases to utilize the strongly concentrated radiant energy for chemical processing.

Distributed receiver systems are based upon a variety of concentrator collectors including parabolic troughs, parabolic dishes, line- and point-focus Fresnel lens, and other optical approaches (e.g., compound parabolic reflectors and holographic methods). Generally, the individual collector/receiver units are engineered to produce power from a few kilowatts up to tens of kilowatts, which can be replicated and combined in large distributed fields to produce hundreds of kilowatts (or hundreds of megawatts). Figure 3 shows a distributed receiver installation of more than 50 MW (peak electrical power capacity) produced by a large number of parabolic reflecting troughs, a working fluid heated in its passage through a pipe receiver element at the trough-focus line, and the thermal energy (hundreds of degrees Centigrade) applied to a central Rankine cycle engine. In other operating installations, the collected thermal energy is used for process heat needs.

Another common form of distributed receiver system is based on the use of a reflecting, tracking parabolic dish that incorporates a small heat engine (e.g., Stirling cycle) at the focus to produce electricity. Since the temperature of a receiver and working fluid at the dish focus can be operated at temperatures up to the order of 1000°C, a dish collector unit can produce power at efficiencies of 30% or higher. Dish collector units ranging in diameter from 1 m to about 20 m have been demonstrated and can produce peak electric power from a few hundred watts to more than 50 kW, respectively. A large field of dish collector units can be designed to produce a great deal of electricity, thermal energy, or specialized products.

FIG. 3. A photograph of a small section of the largest (200 MW of peak electric output in 1990) commercial solar thermal power plant in the world. This power plant is located in central California and is a distributed, parabolic, reflecting-trough collector system feeding power to a utility grid.

Solar central receiver power systems have received a great deal of attention over the past 15 years and are in essence a large-scale spherical mirror configuration for achieving very high concentrations of incident solar radiation along with very high collector and system operating temperatures. The working configuration uses a field of large computer-guided mirrors, or heliostats, to redirect the direct component of the sun's energy from each mirror to an elevated central receiver. Each mirror in the field delivers a large fraction of the incident solar energy to the receiver so that thousands of mirrors in the field surrounding the central receiver result in high concentration of solar energy at the central receiver's surface. Most of the concentrated redirected energy is absorbed on the receiver where it is converted into sensible heat and transported by a working fluid to machinery for utilization. Power levels of 50–1000 MW (thermal) are feasible at working temperature up to 600°C for water-cooled receivers and steam Rankine heat engines, and above 1000°C for gas-cooled receivers and Brayton cycle heat engines.

Numerous central receiver power system experiments have been conducted since 1977 in the United States and around the world, but commercial systems for economic use in electric utility grids have not materialized. The largest experiment involved a prototype 10-MW (peak electric) installation (Solar One) funded by the U.S. Department of Energy and constructed at Barstow, California. The power system was based upon a water/steam receiver and a steam-driven electric turbine. Figure 4 is a photograph of the installed system, which was operated successfully for more than 5 years.

Other solar thermal technologies include solar ponds and total energy systems. The solar pond system consists of a shallow basin of water that collects the sun's radiant energy at temperatures well below 100°C for use as industrial process heat, electricity production, or other low-temperature applications. Its most common and developed form is called the salt gradient pond, in which a gradient of dissolved salt (higher densities at the bottom) is used to suppress the normal convection currents present in heated fresh water. As the solar radiation is transmitted through the shallow water, absorbed on the dark bottom, and converted to thermal energy in the bottom layer of water, the heat is stored and insulated from the atmosphere by the salt gradient. This heated bottom layer can be pumped from large pond structures and used for many applications.

Total energy systems based upon solar thermal technologies involve the use of thermal energy in conjunction with the production of electricity to obtain greatly increased efficiencies in the use of solar energy. These systems become cost-effective when the user has a balanced need for the two forms of energy generated.

The commercial impact of solar thermal systems through 1990 has been largely in the production and sale of electricity in the California utility market using parabolic trough distributed collector systems. Commercial installations in the Southern California area have reached more than 300 MW (peak electric) and plans are underway to install another several 100 MW in the next 3 years. Although research progress in central receiver technology and experiments has been steady through the years, there is little current interest among industrial or utility organizations to undertake commercial activities. Likewise, the extensive commercialization of solar pond and total energy systems has not materialized.

FIG. 4. A photograph of the largest experimental solar thermal central receiver plant built to date (1990) but deactivated in the late 1980s. This power plant produced about 10 MW of peak electric power for a utility grid based upon a water–steam (Rankine) conversion cycle and the radiation received from more than 1800 heliostats (about 50 MW of radiant energy received at the surface of the central receiver).

Solar Electric Technologies

Photovoltaic energy systems provide a nonthermal, nonmechanical means of converting sunlight directly to electrical energy. These systems depend upon solid-state devices, called photovoltaic (PV) cells, precisely fabricated from various combinations of semiconductor materials to absorb and transform incident light into direct current (dc) electric power. A cell can be made by forming a close electrical contact (junction) between two thin layers of different semiconductor materials (e.g., two types of silicon differing only in the kind of impurity atoms present), one exhibiting negative electronic properties and one positive, to form an electrically active junction. This junction introduces strong internal electric fields that, in the presence of light to eject electrons from semiconductor atoms, give rise to a built-in potential difference (voltage) and electric current, provided electrical contacts have been attached to the two sides of the junction and an external circuit is closed. As long as there is a source of light of appropriate wavelength incident on the collecting surface, the device behaves as a battery or generator, delivering dc electric power.

There are two general types of photovoltaic arrays: flat plate and concentrator. In flat-plate arrays, the sunlight falls directly upon the photovoltaic cells, whereas concentrator arrays involve optical systems to concentrate sunlight on the cells. Research progress has been substantial in both general types of arrays although various economic and technical tradeoffs are associated with the commercial development of each type of array. In 1990, the industrial interest and market development of flat-plate arrays has far exceeded that of concentrator arrays.

The single most important component in the continuing success of both flat-plate and concentrator systems is PV cell technology, which must be advanced to obtain improving cell performance while reducing manufacturing costs substantially. Cell technology for flat-plate systems is substantially different from that for concentrator systems, and the development paths and cell designs are also substantially different.

There have been rapid advances in the conversion efficiency and overall performance of PV cells designed for use in flat-plate systems. Flat-plate PV cells of major current interest fall in four broad approaches differing in cell materials and basic structures. The crystalline silicon wafer cell has a long history of development beginning with its use in single-crystal wafer form (thickness about 300 μm) for powering space satellites beginning in the late 1950s. In 1990, it is still the leading commercial PV product for photovoltaic power system applications. The other three approaches are characterized as thin-film cells and involve silicon, copper indium diselenide, and cadmium telluride as the dominant photovoltaic materials.

Amorphous silicon thin-film PV cells have emerged from the research stage to become an important commercial product in the 1980s, particularly for powering consumer products (e.g., battery chargers, garden lights, calculators, and watches). Amorphous silicon cells differ substantially from crystalline silicon cells in being a radically different form of the material and in being a so-called thin-film PV cell, only a few microns in cell thickness. The deposition of amorphous silicon on a glass or plastic substrate and its fabrication into large-area modules is radically different from the processing of crystalline silicon cells and modules. A PV module is a large-area (up to 2 m^2 in some commercial products) packaged assembly of PV cells to facilitate construction of large PV power arrays and reduction of production and assembly costs.

Two other polycrystalline thin-film PV cells are also entering the commercial marketing stage in the late 1980s—copper indium diselenide and cadmium telluride modules. The long-term advantages of developing thin-film PV cells are considered to include reduced material costs and continuing, automated low-cost fabrication of large-area modules.

Progress in advancing the flat-plate (one sun = 1X) conversion efficiencies of both silicon wafer and all of the thin-film cells has been particularly rapid during the past 5 years. Laboratory silicon wafer cell efficiencies have been measured as high as 22% for cell areas of a few square centimeters, whereas commercially available cells with areas of about 100 cm^2 have top efficiencies of about 16%. The efficiency for the best commercial silicon wafer module approaches 15%, whereas average modules measure 11–12%. For the best amorphous silicon and polycrystalline thin-film laboratory cells, the small-area (1 cm^2) efficiencies have been measured at about 12%. A laboratory-fabricated thin-film polycrystalline silicon on a ceramic substrate has been measured to have an efficiency greater than 15%. Thin-film laboratory modules of about 1000 cm^2 have efficiencies of about 10%. Commercially available thin-film modules of 1000 cm^2 or larger have stable efficiencies of greater than 5%. The goal of producing low-cost, thin-film commercial modules with stable efficiencies of about 10% is considered to be within reach by the mid-1990s.

Whereas flat-plate collector technology requires large areas of PV cells to produce 1 kW (peak electric) power (e.g., about 10 m^2 for a module efficiency of 10%), the same amount of power with a concentrator array can be accomplished with a PV cell area at least 100 times smaller. However, the concentrator modules can only utilize the direct component of the incident sunlight and require optical concentrating systems in conjunction with the small PV cells and mechanical tracking of the array to follow the sun accurately through its constant movements. Thus, concentrator cell designs and materials are very different from those for flat-plate modules.

Concentrator PV cell designs are currently (1990) based upon single-crystal silicon or various III–V (e.g., gallium arsenide) wafer materials. Laboratory cell conversion efficiencies at sunlight concentrations of 100 to 200X have been measured as high as 30% for single-junction cells and approaching 35% for multiple-junction cells. Commercially available concentrator cells are currently greater than 20% and 25%, respectively, for the same concentration ratios.

The worldwide production and sale of PV modules in 1989 has been estimated to be more than 40 MW (peak electric) with U.S.-based manufacturers leading with about 35%, a growth over 1988 of about 20% worldwide and 33% in the United States. Since the mid-1970s, about 200 MW have been manufactured and installed around the world in tens of thousands of small power systems (capacities of a few tens of watts to tens of kilowatts) and in tens of millions of consumer products (capacities of tens of milliwatts for calculators and watches to tens of watts for battery chargers). These PV systems and applications based upon flat-plate cells are obviously economic.

Only about four experimental megawatt-scale PV systems have been installed and operated in electric utility grid systems, all in the United States. Figure 5 shows a 1-MW experimental power system. It is estimated that PV systems for utility applications can provide huge new markets though the 1990s as PV electricity production costs can be sharply reduced with scale-up of manufacturing volume, widening of economic applications, and continuing improvements in PV module performances.

Wind energy conversion sytems convert the kinetic energy in winds (the motion of moving air masses) into a number of useful forms of energy including lifting of water by small windmills and production of electric power (kilowatts to multiple megawatts) by wind turbines. There are two basic approaches for converting wind to useful energy: horizontal-axis wind turbines in which the axis of rotation is parallel to the wind stream and the ground (like farm windmills); and, vertical-axis wind turbines, in which the axis of the rotor's rotation is perpendicular to the wind stream and ground (Darrieus, or eggbeater configuration).

For both types of turbines, the rotor is the most critical element in determining economic impact at a good wind location. The rotors consist of one or more blades (usually two or three for modern high-performance rotors) with airfoil (wing-like) cross sections designed to extract the most en-

FIG. 5. A photograph of the first major experimental photovoltaic (PV) power plant feeding power to a utility grid. This power plant operates automatically (no routine operator personnel), consists of about 100 heliostats (sun trackers) mounting about 10 000 m^2 of flat-plate, crystalline-silicon PV modules.

ergy from a site's average wind conditions. For steady wind conditions, the rotor's airfoil shape is designed to create net pressure differential (lower pressure on the curved top side and higher pressure on the bottom) and effective aerodynamic lift so that the primary force acting on the rotor causes it to turn. The rotating blades drive a gearbox (and generator) to produce mechanical energy (or electricity).

The amount of power available to a rotor in a free-flowing air stream is proportional to the swept area of the rotor and the cube of the wind speed. As a result, the wind power density increases from about 50 W m^{-2} meter for wind speeds of 4.5 m s^{-1} (about 10 miles h^{-1}) to about 1400 W m^{-2} for speeds of 13 m s^{-1} (about 30 miles h^{-1}). Interestingly, the wind energy density for a 30-mile h^{-1} wind is about the same magnitude as the sun's energy density incident at the top of the earth's atmosphere. The theoretical maximum fraction of the power density that could be extracted from the windstream by a 100% efficient rotor is 0.593, known as the Betz coefficient. A limit of this magnitude can be rationalized because the wind speed behind the rotor blades could not be zero or the air mass passing through the swept area would pile up and impede further air mass flow.

Farm and residential-scale systems with two-bladed rotors typically have rotor diameters under 9 m and output power ratings up to 15 kW (electric) for wind speeds of about 7 m s^{-1} (about 15 miles h^{-1}). Utility-scale wind systems with two-bladed rotors can have rotor diameters up to about 90 m and have output ratings up to about 2.5 MW (electric) for wind speeds of about 12 m s^{-1} (about 27 miles h^{-1}). Figure 6 shows the current largest operating wind turbine.

Wind resources in the United States and around the world vary greatly from region to region, season to season, and diurnally and generally have the highest annual values along the shorelines of large bodies of water and in and around mountainous regions. Thus, the economic use of wind turbines is restricted to these areas with high average annual wind speeds or to those remote areas where wind turbines might be the technology of choice. Generally, locations having average annual wind speeds of more than 4.5 m s^{-1} (10 miles h^{-1}) are needed to produce economic electric power in most areas of the United States.

During the past 10 years, the installation of wind turbines in so-called wind farms reached a zenith in the early to mid-1980s in California, as a result of temporary tax incentives. As of the beginning of 1990, there are about 17 000 wind turbines installed there having a total capacity of more than 1600 MW (electric), which represents about 90% of all the utility-connected wind turbines in the world. Denmark is the second ranking country with an estimated 6%. In the late 1980s, in particular, the design and operating results of wind turbine farms has demonstrated that on-line availability can be greater than 90%, turbines in the capacity range of several hundred kilowatts can be installed for the order of $1,000 per kilowatt, and electric power can be delivered to a utility grid for much less than 10 cents per kilowatt hour in good wind regimes, which makes them potentially good investments.

Ocean thermal energy systems are based upon the use of ocean temperature differences between warm surface waters in the tropics and cold water from the underlying depths to operate a heat cycle that generates electricity. These systems utilize the huge, stored ocean resources of warm and cold water that make it possible to produce baseload (24-hour per day production, if needed) electricity from a solar-based indirect resource.

The two principal power cycles applying this concept are the closed cycle, utilizing a working fluid with a low boiling

FIG. 6. A photograph of the largest operating, grid-connected wind turbine generator system. This wind turbine system produces peak electric power of about 3.2 MW in winds of about 30 miles per hour and is located in Hawaii.

point (e.g., ammonia), and the open cycle, using sea water itself as the working fluid. In both cycles, the warm water pumped from the surface is applied to vaporize the working fluid on the high-pressure side of a large turbine, whereas the cold water pumped up from the ocean depths is applied to condense the working fluid on the low-pressure side of the turbine. For the closed-cycle application, it is necessary to have large heat exchangers for both the vaporizing and condensing sides of the system; whereas for the open cycle, the warm sea water vaporizes to create the steam vapor pressure (under modest vacuum conditions supplied by a large vacuum pump) that after pushing through the turbine blades is condensed by a cold water spray or by a cold water heat exchanger surface. In the latter case, the condensed vapor can be collected from the heat exchanger surface as potable water, which can have high value in tropic areas and islands.

Since the surface of the earth consists mainly of ocean water, most of the solar energy that falls on earth is absorbed by the oceans of the world. In some areas, particularly those within about 30 degrees latitude relative to the equator, the surface water is as much as 25°C warmer than the waters 1000 m below. A closed-cycle theoretical efficiency (Carnot) for production of power in a Rankine cycle is about 9% for a 25°C temperature difference. In practice, a practical cycle efficiency of 3–5% can be achieved by using very large heat exchangers and a very large low-pressure turbine. Though huge volumes of warm and cold seawater are required per kilowatt of plant capacity, the water is readily available.

Based upon the vast ocean areas and quantities of warm and cold waters, this energy resource is huge in magnitude but is not conveniently located to serve local needs. Thus, the long-range concept of an ocean thermal system is of an array of huge floating structures grazing on the tropic ocean surface (with large long pipes suspended under the structures to the deep cold water below), producing electricity, and electrolyzing seawater to produce hydrogen for transport to the mainlands for a myriad of power and chemical uses. Or, the electricity generated on ocean sites could be combined with electricity-intensive floating industrial plants (e.g., aluminum refineries) or be used to power a floating industrial complex or large city.

In the near-term practical application of ocean thermal technology, the plants would be constructed on land-based, near-shore platforms close to steep falloffs and nearby cold deep water (e.g., Pacific or Caribbean Islands) where cold water piping could be anchored to the terrain. Such an electric power plant could supply not only electricity but also potable water from the condenser, cold discharge water for cooling buildings, and nutrient-rich water discharge into a lagoon or an aquatic seawater hatchery to produce prodigious quantities of seafood. This type of on-land, multiple-use island power plant is under development and early testing in the Hawaiian Islands.

Despite the huge ocean thermal resources available in the tropics and their potential for island power plants for many strategic purposes, there has been only modest U.S. government support for developing the practical technology base and surprisingly little U.S. commercial interest or development. Except for the modest efforts in Hawaii to undertake the construction and testing of a 165-kW (electric) experimental, on-shore plant, the technology is now being developed most strongly by Japanese and French engineers and companies.

Other ocean energy technologies have been developed to produce power from ocean waves, currents, and tides. In particular, power systems utilizing the ocean waves and the ocean tides have been relatively successful, and a few power systems based upon them have been operated for extended periods in Europe. Numerous other tidal power plants, which, of course, derive their energy from the motion of the moon and sun around the earth, are being considered in Europe, Canada, the United States, and other locations.

Biomass energy conversion systems can produce thermal energy, biofuels, or chemicals from crops and vegetation grown on land or water and from organic waste materials and residues arising from agricultural, animal, crop, urban, and industrial sources. In this discussion, biomass is taken to include any organic material that has originated from photosynthesis. Figure 7 shows the complex interactions involved in the utilization of biomass sources to serve societal energy needs based upon the broad range of biomass conversion processes and energy products.

The biomass conversion processes include combustion (burning in air) to product heat for many uses, biochemical processes to produce fuels (e.g., fermentation of corn to produce alcohol and anaerobic digestion of animal manure to produce methane gas), and thermochemical processes (gasification, pyrolysis, and liquefaction) to produce fuel gases, methanol, crude oils, chemical compounds, and charcoal. The combustion of forest and agricultural residues already in 1990 contributes about 4% of the primary energy used in the United States as process heat or as steam to produce electricity. This contribution is expected to be extended substantially as more residues are utilized and as huge quantities of burnable feedstocks from municipal solid waste collection and later from managed biomass farms become available.

The broad range of energy needs and products from biomass energy conversion systems include industrial process heat, electricity, heat for homes and agricultural needs, biofuels (gaseous, liquid or solid fuels) for transportation and other fuel needs, and chemical feedstocks. Since field stands of biomass (living or dead) have nominal lifetimes ranging from a few days to one hundred years or longer (e.g., trees), the biomass resource is often available in various stored forms to fuel a baseload power plant, if desired. Many sources of organic waste material are available from storage or are continuously collected so that a power system based on them can be operated as a baseload power facility also.

Fermentation is an enzymatically controlled anaerobic breakdown of energy-rich organic compounds to produce ethyl alcohol (ethanol), which is a versatile type of biofuel produced by biochemical processes. Sugar, starch, and many forms of biomass (e.g., crops, trees, and organic residues and wastes) can all be fermented to produce ethanol, which can be used directly as an automobile fuel or a blend, or further reacted to form ethyltertiarybutyl ether, a high-octane blending agent.

Gasification, pyrolysis, and liquefaction technologies all involve heating carbonaceous material (e.g. biomass) in an oxygen-limited or oxygen-free environment to produce mix-

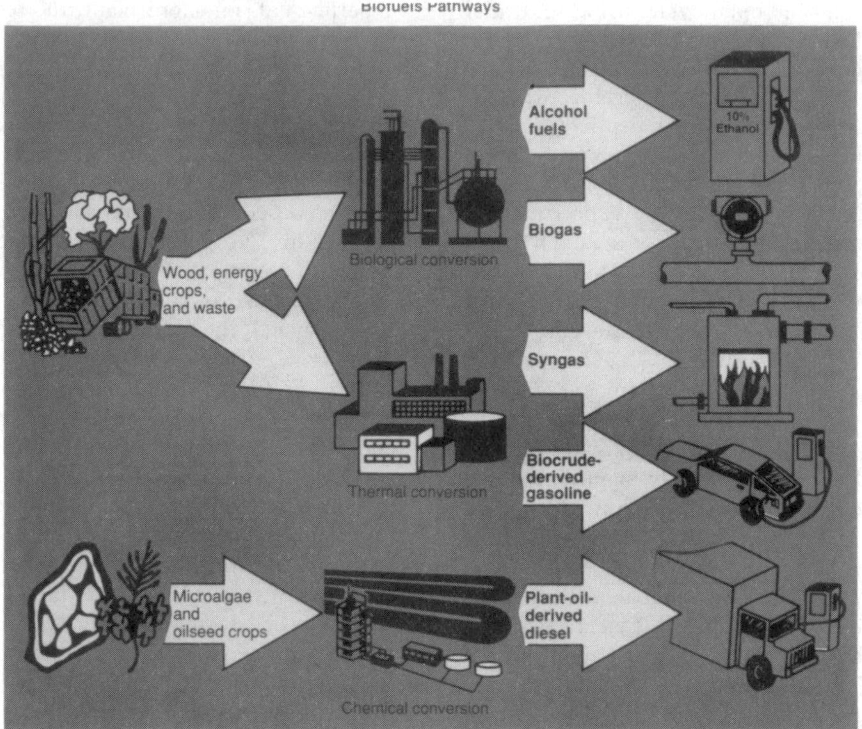

FIG. 7. A schematic diagram of the biofuels pathways. It show the many paths for obtaining biomass resources, converting them to useful energy forms, and utilizing the energy products.

tures of fuel gas, tars, oils, and char. The process equipment and conditions used with each technology are quite different depending upon the major products desired. Air, oxygen, and indirectly heated gasification processes are used to produce fuel gases. Pyrolysis technology has evolved from the production of charcoal by burning (smoldering) biomass in restricted-oxygen enclosures. Pyrolysis of biomass in reactors results in the carbonization and collection of pyroligneous acids, which then can be distilled to obtain methanol, acetic acid, ethyl acetate, and other chemicals. Liquefaction covers a wide range of approaches to converting biomass to various liquids under high-pressure and high-temperature conditions in reactors with the use of reduction gases (e.g., hydrogen and carbon monoxide gases).

Biofuels represent a class of high-value gaseous, liquid, and solid fuels that derive their energy indirectly from the sun through photosynthesis and biomass conversion. Both waste and virgin organic matter can be converted to fuel forms much like conventional fossil fuels by means of chemical or biological processes, and thus serve as renewable alternatives to the conventional fossil fuels. These conversion processes can accomplish transformations in a relatively short period of time (often in a few hours) compared to the millions of years for nature to accomplish the same thing for fossil fuels.

Biomass energy technology also includes the growth of biomass in large quantities (biomass farms) for the specific purpose of converting it to needed energy or fuels. A great deal of research on rapid growth of biomass has been conducted over the past 15 years, and experimental and prototype biomass farms have been undertaken. Among the biomass energy crops that have been studied relatively extensively are short-rotation (4 to 10 years) woody trees and shrubs, high-yielding herbaceous plants (grasses and legumes) that can be harvested one to three times per year, managed forests (silvaculture), and numerous aquatic plants (including algae, kelp, and various water weeds) grown in fresh, coastal, brackish, or ocean waters. In the case of tree growth and tree farm research, the usual natural forest production of wood biomass, which averages 1 to 2 dry tons per acre year, has been raised to as high as 15 dry tons per acre year for clonal plantings of Populus hybrid trees under managed farm conditions.

The production of ethanol fuel from surplus corn and other grain food crops has become an important industrial commodity since its serious consideration started in the mid-1970s. It is being produced in 1990 at a rate of about 900 million gallons annually of which a large fraction is used in gasoline. The ethanol is used for octane enhancement and for producing a popular fuel called gasohol (90% gasolinc and 10% ethanol), which has penetrated about 10% of the gasoline fuel market in the United States, largely because of a reduction of federal (and some state) taxes on the use of gasoline. Ethical issues concerning the diversion of food crops for the production of motor fuels have been raised, but the use of spoiled or inferior potential food products has been largely noncontroversial. Brazil has been the world leader in adopting ethanol fuel, which is produced largely from surplus sugar crops, as a 100% fuel for automobile engines especially adapted for it.

CONCLUSION

Solar energy is a huge, accessible, versatile, and worldwide energy resource that is largely untapped, virtually unlimited in lifetime, highly useable for a wide variety of local societal energy needs, and essentially free of environmental pollution effects. The resource is widely available as direct radiation and after absorption by the earth's environment as a number of indirect forms. It is converted into useful energy through absorption of the direct radiation in solar thermal and photovoltaic systems to produce thermal and electrical power, respectively. It is also converted from its indirect forms (running water, wind, ocean temperature difference, and biomass resources) into valuable power. The energy impact of solar-based technologies on the total primary energy use in the United States in 1990 is approaching 9%, or about 7 quads, largely due to its indirect use in hydroelectric power plants and wood waste burning in the forest products industry, but also significant use in the production of alcohol fuels and burning of municipal and agricultrual wastes. On the other hand, the potential future energy impact of solar-based technologies in the United States and around the world is relatively unbounded and very much a function of economic competitiveness (and long-term availability) among the fuel resources and their related power systems and of a growing number of environmental considerations associated with them.

SOURCES FOR FURTHER INFORMATION

Solar Energy Research Institute, Public Information Division, 1617 Cole Blvd., Golden, CO, 80401. Source of summary and research publications in all of the solar energy technologies.

National Technical Information Service (NTIS), Department of Commerce, 5285 Port Royal Road, Springfield, Virginia 22161. Source of hundreds of research and general reports on solar technologies.

Superintendent of Cocuments, U.S. Government Printing Office, Washington, D.C., 20402. Source of selected reports on solar technologies.

See also Atmospheric Physics; Heat Engines.

The Solar System

Eugene H. Levy

The solar system consists of the sun and the planetary system. The sun is an ordinary star, one of at least several $\times 10^{11}$ stars in the Milky Way Galaxy, located in the galactic disk, some 10 000 parsecs from the center (1 parsec $\sim 3\times 10^{18}$ cm). The surrounding planetary system comprises nine known major planets and myriad smaller objects, including planetary satellites, comets, and asteroids. The major solar system objects and their properties are listed in Table I.

The elemental compositions of the sun, the planets, and other cosmical objects are inferred from analyses of the spectra of light emitted from and absorbed by the outermost layers, supported by other means including direct measurements when possible and theoretical interior models that are constrained by notions about cosmic elemental abundances, meteoritic elemental abundances, and by measurements of average density and moments of inertia, where those are available. By mass, the sun consists of approximately 74% hydrogen, 24% helium, and 2% of the remaining heavier elements, among the most abundant of which are carbon, nitrogen, oxygen, neon, magnesium, silicon, sulfur, and iron.

The solar system exhibits a number of striking structural regularities. The planets all orbit the sun in the same direction, on nearly circular, concentric trajectories, and in approximately the same plane. The plane of the planetary orbits nearly coincides with the plane of the sun's equator, and the direction of planetary revolution is the same as the direction of the sun's rotation. Planets also *tend* to rotate with the same sense as their orbital motion and the spin of the sun. Generally speaking, the planets divide into two broad classes: the terrestrial planets (Mercury, Venus, Earth, and Mars), occupying the inner solar system within about 1.5 astronomical units from the sun (1 AU$\sim 1.5\times 10^{13}$ cm), and the jovian planets (Jupiter, Saturn, Uranus, and Neptune), occupying the outer solar system and ranging from about 5 AU to about 30 AU in distance from the sun. The remaining planet—Pluto—is a small anomalous system, averaging some 40 AU distance from the sun. The asteroids comprise a numerous class of objects, which orbit the sun, mainly in a broad band between the orbits of Mars and Jupiter. The largest asteroid, Ceres, is nearly 1000 km in diameter; and the asteroids scale to smaller and smaller sizes in larger and larger numbers. At the present time, comets (perhaps 10^{12} in number) spend most of their time in the so-called Oort cloud, at very large distances from the sun—ranging to as far as 50 000 AU. Other small- to medium-sized objects also exist in the solar system, moving on a variety of orbits, including many, undoubtedly, that have not yet been discovered.

The terrestrial planets (so called because of their generic similarity to Earth) are composed largely of silicate rock and metal. The jovian planets (labeled for their generic similarity to Jupiter) have several Earth-masses of rock and metal in their interiors, as well as abundant quantities of more volatile substances, such as water, and more or less thick, massive atmospheres rich in hydrogen and helium. The asteroids occur in a variety of types and compositions and are dominantly composed of rock and metal, with admixtures of water and organic carbon compounds. Comets are accumulations of ice, rock, and dust—including organic compounds—ranging, typically, from a few kilometers to perhaps several tens of kilometers in diameter. When a comet approaches close to the sun, the solar heat vaporizes its volatile ice; the resulting emissions of gas and dust reflect sunlight and produce the spectacular illuminated displays, which may stretch an appreciable fraction of an astronomical unit across the sky. Meteorites, which occasionally fall to earth, are thought to be mostly fragments of asteroids scattered into Earth-orbit crossing trajectories by gravitational resonance interactions, and pieces of comets and other debris moving through the vicinity of Earth. Analyses of meteorites have

Table I Principal bodies of the solar system and approximate characteristics

Body	Mass (g)	Mean density (g cm^{-3})	Rotational period (days)	Radius (cm)	Distance from Sun (AU)	Orbital period (years)
Sun	2.0×10^{33}	1.4	25.36	6.9×10^{10}	—	—
Mercury	3.3×10^{26}	5.4	58.66	2.4×10^{8}	0.39	0.24
Venus	4.9×10^{27}	5.2	242.98	6.1×10^{8}	0.72	0.61
Earth	6.0×10^{27}	5.5	1.00	6.4×10^{8}	1.00	1.00
Mars	6.5×10^{26}	3.9	1.03	3.4×10^{8}	1.52	1.87
Jupiter	1.9×10^{30}	1.3	0.41	7.1×10^{9}	5.20	11.86
Saturn	5.7×10^{29}	0.7	0.44	6.0×10^{9}	9.54	29.47
Uranus	8.8×10^{28}	1.2	0.7	2.6×10^{9}	19.19	84.06
Neptune	1.0×10^{29}	1.6	0.7	2.5×10^{9}	30.06	164.81
Pluto	1.3×10^{25}	2.1	6.39	1.2×10^{8}	39.53	248.54

Note: Some entries in this table should be regarded as tentative or very approximate. Diameters and rotation periods for some outer planets are, in various instances, difficult to measure or define. For example, the sun and the jovian planets—with no solid surfaces—have rotation rates that depend sensitively on the definition and measurement approach.

provided important clues about the formation and early history of the solar system.

Many of the planets are centers of their own satellite systems. Planetary satellites occur in several varieties: large planet-like satellites, small asteroid-like satellites, and diffuse rings consisting of many small bodies. The largest satellites are Earth's Moon, the so-called Galilean satellites of Jupiter—Io, Europa, Ganymede, and Callisto—Saturn's moon Titan, and Neptune's moon Triton. Each of these major planetary satellites can be considered a small planet in its own right, with internal evolutionary processes, some with substantial atmospheres. Io has recently been discovered to be heated internally as a result of tidal interaction; this internal heating is manifested by abundant volcanism seen on Io's surface. Many of the planetary satellites apparently accumulated in place around their parent planets. The most complete manifestation of this is seen in the jovian system, where the Galilean satellites show evidence of having accumulated in a localized jovian subnebula; these satellites show a compositional trend with distance from Jupiter that mimics aspects of the compositional trend of the planets. On the other hand, some satellites are relatively small (tens of kilometers in diameter), inert rocks, which are captured remnants of early solar system accretion. At least four planets have satellite rings of diffuse matter: Jupiter, Saturn, Uranus, and Neptune. In relative terms, the largest planetary satellite, when measured as a fraction of its parent planet, is Charon, the recently discovered satellite of Pluto—approximately half Pluto's size; the Pluto–Charon system could as easily be characterized as a small double planet.

The sun's gravitational attraction is the central organizing influence in the solar system, keeping the planets, asteroids, and comets in their places by balancing the inertial forces of their orbital motions. Most of the sun's emitted energy, approximately 4×10^{33} ergs s^{-1}, escapes as radiation, and most of that occurs at visible wavelengths between 0.4 and 0.7 μm, althugh significant amounts of radiation are emitted at both longer, infrared, and shorter, ultraviolet, wavelengths. At the orbit of Earth, solar radiation flux is about 1.4×10^6 ergs cm^{-2} s^{-1}, or 1.4 kW m^2. The sun also emits a continual, radial, supersonic wind of gas because the solar atmosphere is not bound to the sun and expands into space. This "solar wind" of ionized gas streams through space at speeds normally of 300–400 km s^{-1}. At the orbit of Earth, the solar wind density is typically 5–10 hydrogen atoms per cm^3, corresponding to a rate of mass loss from the sun of some few $\times 10^{12}$ g s^{-1}, about the same as the rate at which the sun loses mass from radiation alone. The sun's magnetic field is entrained by the solar wind and stretched out, away from the rotating sun, through interplanetary space in a grand spiraling shape. At Earth's orbit, this interplanetary magnetic field has an intensity of about 5γ (5×10^{-5} G). The solar wind and the interplanetary magnetic field undergo large fluctuations resulting ultimately from variable conditions near the sun's surface, including solar storms or flares.

The sun interacts with the planets both through its radiation and through the solar wind. The longer wavelengths of radiation provide energy that affects the heat balance and temperatures of planets; shorter-wavelength radiation is also important because photon-induced chemical reactions influence the ionization and chemistry of planetary atmospheres, particularly at high altitudes. The solar wind and interplanetary magnetic field interact with planetary atmospheres and planetary magnetic fields (magnetospheres). On Earth, this interaction controls a number of upper atmospheric manifestations, including the north and south polar aurorae, as well as geomagnetic storms. There is evidence that terrestrial climate is correlated in some way with 11- and 22-year solar magnetic sunspot cycles, but the mechanisms underlying such an effect are not understood.

Most of the major bodies of the solar system possess internally generated magnetic fields. The sun, Mercury, Earth, Jupiter, Saturn, and Uranus are known to have such magnetic fields. Of the planets that have been scrutinized closely, only Venus and Mars do not have strong internally produced magnetic fields at the present time. Earth's Moon shows no indication of contemporary magnetic field generation; however, the abundance of magnetized rocks on

Moon's surface hints at the possible presence of a lunar magnetic field early in Moon's history. Primitive meteorites carry remanent magnetization that suggests the presence of a magnetic field in the nebula from which the solar system formed. Such cosmical magnetic fields are thought to be generated through a self-excited hydromagnetic dynamo process: convection of an electrically conducting fluid, combined with the effects of rotation, interacts with a magnetic field to regenerate electrical currents that balance the dissipation of the magnetic field caused by electrical resistance.

Measurements of radioactive isotope abundances in solar system matter indicate that the solar system formed approximately 4.55×10^9 years ago. Insofar as is known, the planetary orbits have been generally stable over that period of time, and major reshuffling of the planetary positions has not occurred. It is believed that the sun and planetary system formed at about the same time from a single precursor object called the protosolar nebula. The protosolar nebula is thought to have resulted from the gravitationally induced collapse of a rotating interstellar cloud. Because of the rotation and energy dissipation, the collapsing nebula settled into a flattened disk. Evolutionary processes in this rotating, disk-shaped nebula are thought to have resulted in the formation of the sun and planetary system with the structural features described above. The composition of the sun is representative of the overall composition of the gas and dust from which the system formed. The rock and metal-dominated compositions of the terrestrial planets resulted because these planets accumulated from the solid dust component in the precursor material; at the high temperatures that prevailed in the inner solar system, only the least volatile of mineral species could persist in the solid form and be accumulated during planetary accretion. The formation of the jovian planets is also believed to have initiated through the accumulation of solid matter; however, because of the lower temperatures that prevailed in the outer solar system, far from the sun, even such relatively volatile chemicals as water and ammonia existed in the condensed, solid-ice form. This accounts for the high abundance of "ices" in outer solar system objects. Because of their large masses and resulting strong gravities, the jovian planets captured and retained substantial hydrogen and helium-rich gaseous atmospheres directly from the surrounding nebula. Jupiter and Saturn captured the most massive atmospheres, whereas Uranus and Neptune have far less substantial hydrogen and helium-rich envelopes. It is possible that the gaseous nebula had already been largely dispersed by the time Uranus and Neptune had accumulated; planetary accretion is likely to have been prolonged in the distant outer solar system, where orbital angular velocities were slower and dynamical time scales thus longer.

Current ideas suggest that the final stages of planet accumulation may have involved violent collisions between a relatively small number of large protoplanetary objects. On this basis, a number of the solar system's otherwise difficult to explain deviant features may become explicable; among these are the variable tilt angles of planetary spins, including the extreme 98° obliquity of Uranus, and the formation of Earth's moon. Impacts have played a continuing role in the evolution of planets. On Earth, impacts of comet-like objects may have played a role in bringing water and organic material to the planet, thus creating an environment for the origin and evolution of life. Even in relatively recent geological times, impacts may have caused significant disturbances of the terrestrial environment and influenced the continuing evolution of life.

The physical conditions that produced the solar system are thought to be characteristic of star formation more generally throughout the universe. Observations also show that the physical processes inferred to have been involved in the formation of our own planetary system prevail in numerous instances of star formation where sufficient detail can be observed. This suggests that planetary sysems may be prevalent in the universe and that many systems may have characteristics grossly similar to those of our own solar system. As this article is written, no planetary systems are known other than the solar system. However, during the next few decades, advances in measurement techniques promise the capability to discover and study what other planetary systems exist around several hundred stars within about 10 parsecs of the sun.

See also SOLAR WIND; SUN.

BIBLIOGRAPHY

The solar system is described at an introductory level in numerous general textbooks and references on astronomy and space science, e.g., G. Abell, D. Morrison, and S. Wolff, *Exploration of the Universe*, 5th ed. Saunders College Publishing, Philadelphia, 1987; or D. Morrison and T. Owen, *The Planetary System*. Addison-Wesley, Reading, MA, 1988.

A somewhat more advanced treatment of several solar system subjects and related questions can be found in F. Shu, *The Physical Universe*. University Science Books, Mill Valley, CA, 1982.

Specified subjects are covered in detail in a number of specialized books; among them H. J. Melosh, *Impact Cratering: A Geologic Process*. Oxford University Press, New York, 1989; D. Turcotte and G. Schubert, *Applications of Continuum Physics to Geological Problems*. Wiley, New York, 1982; W. B. Hubbard, *Planetary Interiors*. Van Nostrand Reinhold, New York, 1984; J. Lewis and R. Prinn, *Planets and their Atmospheres*. Academic Press, New York, 1984; E. N. Parker, *Cosmical Magnetic Fields*. Oxford University Press, New York, 1979; L. J. Lanzerotti, C. F. Kennel, and E. N. Parker, *Solar System Plasma Physics*. North-Holland, Amsterdam, 1980.

Topical reviews of solar system subjects are regularly published in *Annual Review of Earth and Planetary Sciences* and *Annual Review of Astronomy and Astrophysics*, Annual Reviews, Inc., Palo Alto, CA.

A continuing series of comprehenssive treatments of selcted subjects pertinent to solar system science is published as the Space Science Series, University of Arizona Press, Tucson. Readers interested in the formation of the solar system will find *Protostars and Planets* (T. Gehrels, ed., 1978), *Protostars and Planets II* (D. Black and M. Matthews, eds., 1985), and *Meteorites and the Early Solar System* (J. Kerridge and M. Matthews, eds., 1988) to be especially useful. Other volumes in the series cover a variety of additional subjects.

Solar Wind

J. R. Jokipii

The main solar wind is a continuous, radial, supersonic outflow of plasma from the top of the solar atmosphere. This flow penetrates beyond Earth to the farthest reaches of the solar system. At the orbit of Earth it is observed to be composed chiefly of protons and electrons with a mean kinetic temperature of approximately 100 000 K, a mean density of approximately 10 particles cm^{-3}, and a mean radial flow velocity of several hundred kilometers per second. Deviations from radial flow are small and irregular, with transverse velocities being less than about 10 km $\sec^{-1}$. The plasma is regarded as "collisionless" since collision mean free paths are larger than relevant macroscopic length scales. The wind is quite inhomogeneous and turbulent, with its parameters exhibiting large fluctuations over a variety of length and time scales.

The source of energy for this supersonic flow is the high ($\sim 10^6$ K) temperature in the solar corona. The solar gravitational field is insufficient to hold the corona in static equilibrium and, as first demonstrated by E. N. Parker in 1957, the appropriate fluid equations show that the corona must expand supersonically to form the solar wind. The equations are nonlinear and any realistic solutions must be obtained by computer. A number of detailed solutions have been obtained, and it is now clear that the gross average physical characteristics can be understood in terms of spherically symmetric fluid flow with reasonable parameters for the solar corona. In these solutions the flow accelerates from subsonic flow near the sun, passes through the sound speed at some 10–20 solar radii, and then flows at a nearly constant speed of some 350–450 km $\sec^{-1}$ out to large distances from the sun. At a distance of the order of 50–100 AU (1 AU $= 1.495 \times 10^8$ km), the flow must decelerate because of the resistance of the interstellar medium. The characteristics of this transition are still a matter of considerable speculation; current ideas favor a standing shock wave where the flow makes a transition to subsonic velocity, with a consequent heating of the plasma. Beyond this point the solar material slowly merges into the interstellar gas.

As the gas flows out from the sun it drags with it solar magnetic field lines (the flow satisfies the conditions under which magnetic field lines are frozen into the fluid). At distances greater than about 10 solar radii, the kinetic energy density of the flow, $\frac{1}{2}\rho v^2$, exceeds the magnetic-field energy density, $B^2/8\pi$, so that the magnetic-field stresses do not appreciably affect the flow. The structure of the field for a uniform, time-independent solar wind with constant velocity V_0 is readily worked out. The field may be regarded as being combed out to be nearly radial at some radius r_0 near the sun, in which case the magnetic field as a function of heliocentric spherical coordinates is given by

$$B_r(r,\theta,\phi) = B_r\left(r_0,\theta,\phi - \frac{(r-r_0)\Omega t}{V_0}\right)\left(\frac{r_0}{r}\right)^2,$$

$$B_\theta(r,\theta,\phi) = 0,$$

$$B_\phi(r,\theta,\phi) = -B_r\left(r_0,\theta,\phi - \frac{(r-r_0)\Omega t}{V_0}\right)\left(\frac{r_0}{r}\right)^2 \frac{r\Omega \sin\theta}{V_0},$$

where Ω is the angular velocity of solar rotation (period $= 27$ days). The field lines form Archimedean spirals on cones of constant heliographic latitude θ. The twist of the spirals is a direct consequence of solar rotation. Observations in the vicinity of Earth's orbit, in the ecliptic plane, show that the average field is reasonably well represented by the above equations, with the field magnitude being about 5×10^{-5} G, and the spiral angle $\sim 45°$ relative to the radial direction. The sense of the field is observed to remain constant (either directed out from the sun or in) for several days at a given point, with the transition between outward and inward magnetic field being very abrupt. The picture in the solar equatorial plane then is one of ~ 4 inward or outward magnetic sectors distributed around the sun, with relatively thin boundary regions separating them, as indicated schematically in Fig. 1. The whole structure rotates with the sun.

Observations taken from spacecraft up to some 25° away from the solar equatorial plane reveal that the three-dimensional structure of the magnetic field is quite simple, except when the sunspots are near their maximum (every 11 years). The sector structure disappears above some latitude, which is about 10° near sunspot minimum and which increases toward sunspot maximum. Above this latitude the field becomes of one sign (directed inward or outward along the Archimedean spiral), with the fields in opposite hemispheres being of opposite sign. The field in the northern hemisphere was outward from the sun prior to the last sunspot maximum and is now directed inward, oscillating with a 22-year period. Separating the two regions of oppositely directed magnetic field is a thin electrical current sheet which oscillates above and below the equatorial plane up to the limiting latitude discussed above. The sector structure observed near the equatorial plane is caused by this thin current sheet crossing the point of observation, as it rotates with the sun. The solar-wind speed is a minimum at the current sheet and increases toward the poles.

The above paragraphs have summarized the steady-state, laminar solar wind, which corresponds reasonably well to the average properties of the observed solar wind. However, it must be emphasized that the solar wind is observed to be quite turbulent and all of the fluid variables fluctuate irregularly over a wide variety of time and length scales. Recently it has been increasingly recognized that the fluctuations play a major role and affect the average properties of the wind by, for example, changing the transport coefficients and influencing the energy balance.

The fluctuations in the solar wind may be conveniently grouped into three major categories. At the largest time and length scales the whole character of the solar wind may be regarded as changing with time. Included in this category would be the 11-year, solar-cycle-related variation. Also, on a slightly shorter time scale, the solar wind tends to be emitted in fast and slow streams which interact with each other as a result, for example, of solar radiation. The interaction between streams is indicated in Fig. 1.

The second category of fluctuations may be termed transient disturbances which originate at the sun and propagate

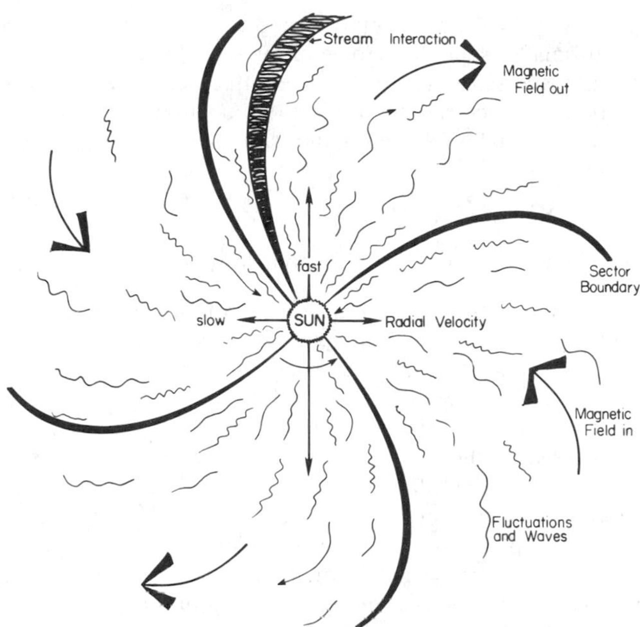

FIG. 1. Schematic illustration of the solar wind in the solar equatorial plane. The sun rotates counterclockwise and twists the interplanctary magnetic field into a spiral. Four magnetic sectors are shown, with the average field in each pointing alternately in or out as indicated by large arrows. The wind flows radially, being emitted in faster or slower streams as a function of angle around the sun. As the sun rotates, fast streams overtake slow streams to produce regions of stream interaction. The flow is quite turbulent or irregular on a small scale.

out in the fluid. The prime example of such a phenomenon is the solar-flare-caused blast wave which propagates outward from the sun and causes a variety of transient efforts throughout the solar system. The total energy involved in such a flare-associated shock can be as large as $\sim 10^{32}$ ergs.

The third kind of fluctuations occur on shorter time and length scales (on the order of hours or less) and can be associated with waves and "turbulence" in the fluid, although the precise wave modes present are not yet clear. It has been demonstrated that Alfvén waves propagating outward from the sun contribute substantially to the observed fluctuations. The observed fluctuation power spectrum as a function of wave number in this range is a power law $Ak^{-\alpha}$ with index $\alpha \sim 1.6 \pm 0.3$, which is surprisingly close to the $Ak^{-5/3}$ expected for fluid mechanical turbulence (the Kolomogorov spectrum).

The solar wind provides a nearby laboratory where many complex astrophysical phenomena can be studied *in situ*, through the use of space probes. It also is the medium through which many effects of the sun get transported to Earth, and hence is of considerable importance to our environment. Thus, flare-associated shock waves propagate to Earth and cause magnetic storms, short-wave fade-outs, and aurorae. Recent evidence suggests a connection between magnetic sectors and terrestrial weather. The interaction between the solar wind and other bodies in the solar system is a whole new discipline of planetary physics, one spectacular manifestation of which is the tail of a comet, which can be considered a solar wind shock.

In short, the solar wind is exceedingly rich in physics which touches on many aspects of our environment, from terrestrial weather to the distant stars.

See also INTERSTELLAR MEDIUM; PLASMAS; SUN; TURBULENCE.

BIBLIOGRAPHY

John C. Brandt, *Introduction to the Solar Wind*. W. H. Freeman, San Francisco, 1970. (I)

A. J. Dessler, "Solar Wind and Interplanetary Magnetic Field," *Rev. Geophys.* **5**(1), 1–41 (1967). (A)

A. J. Hundhausen, *Coronal Expansion and Solar Wind*. Springer-Verlag, New York, 1972. (A)

J. R. Jokipii, "Turbulence and Scintillations in the Interplanetary Plasma," in *Annual Review of Astronomy and Astrophysics*, Vol. 11, pp. 1–28. Annual Reviews, Palo Alto, CA, 1973. (A)

E. N. Parker, "The Solar Wind," *Sci. Am.* **226**, 66–76 (April 1964); reprinted in *Frontiers of Astronomy*. W. H. Freeman, San Francisco, 1971. (E)

V. J. Pizzo, T. E. Holzer, and D. G. Sime (eds.), *Proceedings of the Sixth International Solar Wind Conference*. NCAR Technical Note NCAR/TN-306+Proc, May 1988. (A)

James A. Van Allen, "Interplanetary Particles and Fields," *Sci. Am.* **235**, 161–173 (September 1975). (E)

Solid-State Physics

Conyers Herring

DEFINITION

Solid-state physics is the study of what may be called the material properties of solid bodies.

The term material properties refers to those properties that characterize any large quantity of a particular material, whatever its shape (e.g., any large volume, or a large area of an interface between two substances). Thus, for example, solid-state physics includes such things as the measurement and explanation of the dielectric constant of a crystal, but it does not include the study of the electrostatic field distribution when a specimen of a particular shape is acted on by a particular arrangement of external charges.

The dividing line between "solids" and other substances is vague, and its accepted position has varied a little over the years. To the layman and to early scientists, a solid has simply been something relatively unyielding, or in a more refined concept, something that returns to its original configuration when weak deforming forces are removed. But it has been found that nearly all bodies undergo at least a slight plastic deformation when strained, and we also know that glassy materials, which are very hard and elastic when cold, soften continuously as they are heated. In the nineteenth century these facts, combined with natural philosophers' systematization of the many properties of crystals, led to a greater and greater identification of solids with crystalline materials. In 1912 and 1913 von Laue and the Braggs verified, by x-ray diffraction, the concept that crystals consist of atoms stacked in regular arrays, a concept that had been developed in some depth as a theoretical speculation, out of roots going back to Newton and earlier, during the whole nineteenth century. Following this verification, and especially after the advent of quantum mechanics in 1925, concepts based on crystalline order contributed so much to the understanding of crystals that it became fashionable simply to define solids as crystalline or polycrystalline materials. In recent decades, an increasing interest in amorphous materials and in the role of defects in crystals has led the solid-state physics community to consider as its domain all materials in which atoms spend most of their time vibrating about fixed sites, whether regularly or irregularly arranged. Indeed, since there is an increasing interaction between the literature of solid-state physics and that devoted to the atomic understanding of liquids, the enlarged concept of a field called the *physics of condensed matter* is increasingly used.

DISTRIBUTION AND EXTENT OF RESEARCH ACTIVITY

For the last several decades solid-state physics as we have defined it has been accounting for over a third of all research publication in physics. For most of the first half of the twentieth century the corresponding fraction was fairly constant at a somewhat smaller value, of the order of a fifth, and in the nineteenth century the fraction seems to have been still smaller. The proportion of current research papers that are purely theoretical is less than a third, lower than for any of the other major subfields of physics.

Of the types of materials studied, metals and inorganic semiconductors currently get the most attention; the literature of inorganic insulators is rather smaller, but still larger than that of noncrystalline solids and organic insulators combined. The kinds of properties of any of these materials that can be studied are so diverse that only a few gross groupings of them can be indicated here. One convenient distinction is that between the properties of surfaces, films, and interfaces, to which about a quarter of current solid-state research papers are devoted, and bulk properties. Another useful distinction separates properties that can be adequately described in terms of positions and motions of entire atoms from those that depend in an essential way on the quantum-mechanical behavior of electrons. Although the distinction between these "atom-unit" and "electronic" properties is often rather fuzzy (e.g., electron motions determine interatomic forces), it is real enough to have divided the solid-state research community into groups that often feel quite remote from each other. The most prolific areas of atom-unit study are, in rough order, crystal structure (often considered to belong to a separate science, crystallography), structure and kinetics of surfaces and interfaces, the production and motion of bulk defects and impurities, and plastic deformation (which merges into the fields of metallurgy and materials science). In the domain of electronic properties, the most prolific areas are, again in rough order, optical properties, magnetism and magnetic resonance, superconductivity, normal electronic conduction, and electronic properties of surfaces. The areas named account for nearly four-fifths of current research publication; the remainder falls into many smaller subfields. Some of the latter, like neutron diffraction and scattering, Mössbauer effect, calorimetry, elasticity, and dielectric and ferroelectric polarization, have important ties to both the atom-unit and the electronic groups.

KEY CONCEPTS

Order and Defects in Spatial Atomistic Structure

Fundamental to much of solid-state physics is the understanding of the geometrical nature of crystalline order. This understanding, already developed in the nineteenth century, led at once to a working out of the constraints crystalline symmetry imposes on the tensors describing various material properties. In the last half-century it has come to be recognized that crystalline order is most appropriately described in terms of correlations between the situation at one point $\mathbf{r}$ of space and that at another point $\mathbf{r}'$: when $\mathbf{r}'$ and $\mathbf{r}$ are near each other such correlations are called *local order*, whereas if the correlations persist for arbitrarily large distances of $\mathbf{r}'$ from $\mathbf{r}$ they are described as *long-range order*. A crystal has a periodic long-range order, e.g., in its mass density or its electron density, whereas an amorphous substance has only local order. In the early decades of the present century, the translational symmetry of crystals, and the rotational symmetry that often accompanies it, played very important roles in simplifying the mathematics of classical and quantum-mechanical theories of crystals, and thus in helping solid-state physics get started as a quantitative science.

In recent decades it has been found that some crystals can have periodic long-range order with two incommensurate periods in the same direction; e.g., the electron spin density in certain antiferromagnetic crystals can have an approximately sinusoidal spatial variation with a wavelength incommensurate with the average interatomic spacing. More recently still, some metallic alloys have been found to have mutually incommensurate periodicities in several directions at once. Unlike classical crystals, which have nearly perfect long-range periodicity, these substances, called *quasicrystals,* have only partial correlations in their atomic positions, but these partial correlations persist to long distances. These properties enable quasicrystals to manifest symmetries (icosahedral, decagonal, etc.) that are not possible for classical crystals with perfectly periodic order.

Having come to understand ideal crystalline order, we are led next to examine the simplest types of departure from such order, i.e., *localized imperfections*. Special interest naturally attaches to those imperfections that, because of the geometrical nature of the crystal lattice, cannot be eliminated by any purely local shifts of atomic positions, etc. Such "irreducible imperfections" can be classified by dimensionality as point, line, or planar defects. The simplest point defects correspond to having too many or too few atoms of one type or another in a given number of cells of the crystal: a missing atom is called a *vacancy*; an extra atom is called an *interstitial* if it is squeezed in between the normal atomic sites, or an *interstitialcy* if two or more atoms share symmetrically the space that one would normally occupy; an atom of a species not normally present in the crystal is an *impurity,* and may be further classified as interstitial (if it is simply additional to the atoms of the perfect crystal) or substitutional (if it replaces one of the latter). Any of these types of point imperfections can be moved to a neighboring location by displacements of atoms in its vicinity (hence it can diffuse by the action of thermal agitation), but it cannot be created or annihilated except at a boundary.

The most important line imperfections are *dislocations.* Suppose we were to travel around a closed loop in a crystal lattice, avoiding any imperfections, and count how many net lattice spacings we have traversed in each direction of the compass. If the crystal is perfect everywhere, the net (integer) number of lattice translations traversed in each direction must return to zero after we have returned to our starting point. If after such return the net number of lattice translations is not zero, there must be an imperfection in the crystal structure somewhere in the region enclosed by the circuit, and since this imperfection must be encountered in any attempt to shrink the circuit to a point, via any route, the imperfection must be localized in the neighborhood of a line that cannot end inside the crystal, i.e., the line must either extend from boundary to boundary, or form a closed loop. Imperfections of this type are called dislocations. They play an essential role in plastic deformation, since the boundary between a region of a crystal in which a slip by one lattice spacing has occurred and a region in which it has not must be a dislocation. It turns out, too, that dislocations often play an important role in many other phenomena besides plastic deformation.

Planar imperfections include *grain boundaries,* which separate crystalline regions of different orientation, and *stacking faults,* where the registry of adjacent parallel planes, though still such as to make a good fit locally, is in some respect different from that in the perfect crystal. The atomic arrangement at a grain boundary is usually somewhat disordered, but for what is called a *twin boundary* there is no disorder, the boundary being simply a plane of atoms that is shared in common by the crystal lattices of the two grains, which thereby have a special orientational relationship.

Further Aspects of Order and Defects

After introducing the concept of long-range order at the start of the preceding section, we concentrated on the atomic-scale periodicities that constitute crystalline order. But this concept is also important on the macroscopic scale, where a solid medium can be treated as a continuum. Such a property as the local magnetization vector in a ferromagnet, or the complex scalar amplitude describing superconducting order (see below), forms a continuous classical field. This field, called an *order parameter,* can vary from point to point, and it may or may not show long-range order. But even if it does not have long-range order, in the previously defined sense of having nonzero correlations at arbitrarily large ranges, it can sometimes show *topological order* in the sense of varying smoothly everywhere, with no singularities. Changes of temperature or other parameters can cause such topological order to appear or disappear. The concept is particularly important for two-dimensional systems, where sometimes any nonzero temperature suffices to destroy long-range order in the correlational sense. Clearly the concepts of order-parameter fields and topological order are not limited to solid-state physics, but often apply to liquids, especially to the orientationally ordered phases known as *liquid crystals.*

Glasses and other systems with randomness in their structure often have occasional regions where the structure, instead of setting into a thermally stable configuration at low temperatures, fluctuates thermally between (usually) two nearly equally stable configurations. Such regions, called *two-level systems,* can be found at arbitrarily low temperatures, where they can dominate the specific heat, thermal resistance, and other properties of glassy materials.

Elementary Excitations and Defect-Associated States

The dynamical properties of crystals, whether treated by classical mechanics (often a good approximation for nuclear motions) or quantum mechanics (necessary for the electrons) can become very complicated to describe because many particles interact simultaneously. But considerable understanding can be achieved by use of approximations that bypass these complications: if the vibrations of the atoms are treated as harmonic, they can be resolved into normal modes; if the electrons are assumed to move independently of each other except for the constraint of the exclusion principle, their behavior can be described in terms of wave functions of single electrons. In both cases the symmetry and regularity of the crystal lattice provide a very important framework for the description of the normal modes or the electronic eigenstates. That is, these can be taken in the form of traveling waves, with a modulation having the periodicity of the crystal lattice and with a fundamental wave vector lying within a polyhedron that for historical reasons is called the *first Brillouin zone*; waves with wave vectors outside this polyhedron can always be treated as waves with wave vectors inside it, with a superposed modulation having the periodicity of the lattice. In the vibrational case, a quantum of occupation of such a normal mode is called a *phonon*: a quantum state of this type for an electron is called a *Bloch wave.*

The energy for a Bloch wave or the frequency of a phonon is a continuous though multivalued function of its wave vector inside the first Brillouin zone, and at the boundary of this polyhedron joins continuously to its value at the opposite boundary. Thus, in general, the spectrum of electron energies or phonon frequencies consists of an infinite or a finite number, respectively, of continuous bands, which may be separated by forbidden gaps. This fact is especially important in the electronic case: the ground state of a crystal, in the independent-electron approximation, is one in which all Bloch-wave states that have energies below some limiting value, the *Fermi energy,* are occupied while all higher ones are empty. If the Fermi energy lies in the forbidden gap between two bands, i.e., if there are just enough electrons to exactly fill an integral number of allowed energy bands, the crystal will be an insulator. If the electron count is sufficient to fill only part of its highest occupied band, or if the Fermi energy is in a range where two or more bands overlap, the crystal will be a metal. In the latter case the Bloch-wave states having energies equal to the Fermi energy will have wave vectors lying on a surface in the first Brillouin zone known as the *Fermi surface*. It turns out that the size, shape, and topology of this surface have an important influence on many properties of metals.

Another important property of Bloch-wave states is their response to uniform electric and magnetic fields. It turns out that wave packets made of these states are accelerated by the electrostatic and Lorentz forces just as free particles would be, except that the ratio of force to acceleration is an *effective mass,* dependent on the crystal potential, in general anisotropic and often negative. The states at the bottom of any band have positive effective masses, those at the top negative; from this it follows that the "hole" left when an electron is removed from near the top of an otherwise full band behaves like a positive charge with a positive mass.

Since the electrostatic interactions of the different electrons in a solid are in fact very strong, we might well question the utility of concepts like those just described, which treat different electrons one at a time. For some problems these concepts do indeed break down, but it can be shown that with relatively minor modifications these concepts have a validity for interacting crystal electrons that is exact in a certain asymptotic limit. That is, the low-lying excited quantum states of a real metal—the states that will be appreciably occupied at low temperatures—can be put into a one-to-one correspondence with those of the independent-electron model described earlier, hence can be described in terms of equivalent one-electron states, or *quasiparticles,* the excitation consisting in occupation of a few such states just outside the Fermi surface (which turns out to have a precise conceptual meaning even in the presence of strong interactions) and the vacating of a corresponding number just inside. This is the basis of Landau's *Fermi-liquid theory* of metals. For an insulator, such a description—a representation of excited states in terms of electrons near the bottom of the normally empty bands and holes near the top of the normally filled ones—can be shown to be even more accurate.

The quasiparticles just described, and the phonons mentioned previously, are special cases of a more general concept, that of *elementary excitations*. These are approximate excited quantum states of a many-body system that form a continuous family describable in relatively simple physical terms, and whose difference from true eigenstates is small in the sense that their level widths are much smaller than their energies of excitation; some authors prefer to restrict the term to cases where the ratio of these two quantities approaches zero as the excitation energy decreases. The most important type of elementary excitation, after those named, is the *magnon* or *spin wave*, an excitation in which the direction of magnetization in a ferromagnetic material, or that of a sublattice moment in an antiferromagnetic, is spatially nonuniform and propagates as a wave. Another important elementary excitation, though not asympotically exact in the sense mentioned above because its minimum excitation energy is rather large, is the *plasmon*, a wave of charge accumulation. Other more exotic types of excitations are sometime discussed in the literature. For example, there are sometimes irregularities in the distribution of electronic charge that correspond to a local excess charge accumulation of $\pm e/2$ (or some other simple fraction) and which can propagate freely in an otherwise periodic crystal. In three-dimensional systems all elementary excitations obey either Fermi or Bose statistics (electrons and holes are Fermi; phonons, magnons, and plasmons are Bose).

Many phenomena that involve small but finite energies of excitation of a solid can be well described in terms of these various kinds of elementary excitations and interactions between them. A systematic though quasiphenomenological description of these interactions, for electron and hole quasiparticles, is a part of the Landau *Fermi-liquid theory* referred to above. Examples of effects due to the interactions of these and other kinds of elementary excitations are the following: their scattering by each other; renormalization of their energies; binding of an electron to a hole in an insulator to form what is called an *exciton*; and hybridization of different kinds of excitations (phonons with light waves, magnons with phonons, etc.).

The imperfections in crystal structure that were discussed earlier have characteristic consequences for the vibrational motion and electronic levels of solids. By destroying the perfect periodicity of the crystal, they can scatter elementary excitations that travel as waves (phonons, electrons, magnons, etc.) and they can bind such excitations into localized states. Thus we can have localized vibrational modes, localized magnon modes, and localized electronic levels; all of these are possible not only at point defects or impurities, but also at dislocations, planar defects, or surfaces. Two fields in which most phenomena are dominated by electronic impurity or defect centers are luminescence and semiconductor physics, the former because of the wealth of possibilities presented by optical transitions at the many possible types of impurities in many possible hosts, the latter because impurities in a nonmetal that bind an electron or hole only weakly can easily be thermally excited to produce free electrons or holes and thus to make the crystal a conductor.

Statistics, Fluctuations, and Response Theory

Besides elementary excitations, there are a number of other even more general theoretical tools that are very valuable for an understanding of the properties of solids.

Thermodynamics and equilibrium statistical mechanics are of course fundamental, but often need to be formulated in more general terms than is customary in textbooks and courses in order to take account of effects of crystal anisotropy, anisotropic stresses, and electric and magnetic fields. A related discipline, which can again be formulated in either macroscopic (thermodynamic) or statistical-mechanical language, is the theory of irreversible processes, and especially its branch known as *linear-response theory*. This discipline describes the time-dependent behavior of any measurable property of a system after it has been stimulated out of an equilibrium state by an infinitesimal time-varying perturbation of some parameter. Particularly important is the fact that if either the in-phase or the out-of-phase part of the response is known at all frequencies, the other can be calculated. This result, called the *Kramers–Kronig* theorem, can be shown to follow from the mere fact that cause must precede effect. It is invaluable, for example, in the analysis of optical absorption and dispersion.

Another useful aspect of linear-response theory is the relating of the dissipative response in a process like electrical conduction to the magnitude of certain fluctuations in thermal equilibrium (the *fluctuation–dissipation* theorem). But besides thermal averaging and its associated thermal fluctuations, which usually have a short time scale, there is another type of averaging, with associated fluctuations, that is often important in solid-state theory. This is averaging over the atomic descriptors of a "random" system, e.g., the exact positions of impurity atoms in a crystal. These may be constant over the duration of an experiment if the temperature is low enough; yet if successive experiments on different specimens seem reproducible, it is tempting to assume that the measured result represents the ensemble average of results that would be obtained for all the various sets of descriptors, weighted by the likelihood of their occurrence. It turns out that this is true in many cases, but not in all, depending on such factors as dimensionality. And often the fluctuations in the measured quantity from one set of descriptors to another are measurable and important, e.g., the low-temperature resistance of a wire can change perceptibly and by a universal order of magnitude when the position of a single impurity atom is changed.

Phases and Phase Transitions

In solid-state physics, as indeed in condensed-matter physics generally, the occurrence of sharp phase transitions has fascinated theorists for many decades. These are clearly cooperative phenomena, involving correlations between properties measured at points remote from each other, i.e., long-range order as we defined it earlier. Such order can occur, for example, in atomic positions, in the alternation of different chemical constituents, or in the magnetization of atomic spins. When variation of some physical parameter like temperature or presssure causes one phase to change into another, the change in physical properties is sometimes completely discontinuous (first-order phase change, with a latent heat) and sometimes continuous, though with discontinuous or singular derivatives (higher-order phase change). In the latter case the thermal fluctuations in the local state of the material become of longer range as the transition is approached and, indeed, at the transition point any quasi-macroscopic property (i.e., a quantity like magnetization density averaged over a region containing many atoms) shows a pattern of fluctuations that looks the same irrespective of spatial scale, i.e., when examined with a "microscope" of arbitrary magnification. This remarkable fact makes possible an application of renormalization-group concepts borrowed from elementary-particle theory. In the last two decades these have been used to elucidate details of the variation of various physical properties as the parameters of a system (temperature, magnetic field, etc.) approach values at which a higher-order phase transition will take place. Many properties, such as specific heat or magnetic susceptibility, become infinite (or have singular contributions that become zero) as the phase transition is approached, and this behavior is to leading order proportional to some power, called a *critical exponent,* of the distance from the transition, e.g., of the difference $|T-T_c|$ between the temperature and that of the transition. Scaling behavior and critical exponents can be defined for response functions, like diffusivity and conductivity, as well as for thermodynamic descriptors.

True singularities occur only in what is called the *thermodynamic limit,* i.e., in the limiting behavior of some sort of average quantity (e.g., the energy density) as the size of the system goes to infinity. However, there is an important adjunct of the theory of phase transitions, dealing with the leading-order *finite-size* corrections required when one is dealing with large but finite systems.

Spatial ordering is however not the only kind of discontinuous phenomenon that can take place in the thermodynamic limit: a freezing in time is also possible, the most studied example of which is found in the systems known as *spin glasses*. These are typically dilute populations of magnetic atoms randomly positioned in a nonmagnetic host metal and interacting weakly with each other because of their large separations. In such situations the interactions turn out to be of a random nature, favoring sometimes a parallel orientation of a pair of atomc moments, sometimes an antiparallel. At high temperatures the moments of all the magnetic atoms fluctuate randomly in direction in the course of time. But below a certain critical temperature T_g and in the limit of an infinitely large system, the biasing effect of the interactions with distant atoms can so outweigh thermal fluctuations that the moment of a particular atom can maintain a statistical preference for a particular orientation for an infinitely long time. What this preferred direction is will in general depend on the detailed history of the system prior to its cooling below T_g and can be different in repetitions of the same experiment. This phenomenon of a freezing in time below a certain finite temperature seems to be encountered in other types of random or glassy systems.

Propagation in Random Media versus Localization

We have already noted consequences of the fact that wave-like disturbances (e.g., elastic waves or electron waves) propagate uniformly in a uniform or perfectly periodic medium, but are scattered by imperfections. When the medium contains many imperfections, randomly distributed, a wave initiated in a given region might naively be expected, in the absence of dissipation, to propagate diffusively to ever larger distances, the range being asymptotically proportional to the square root of the time. This is indeed what happens in three-dimensional systems if the imperfections scatter sufficiently weakly. However, for sufficiently strong scattering a phenomenon called *Anderson localization* can take place, which can keep most of the integrated wave intensity localized within a distance from the starting point that remains finite at all times. The criterion for such localization depends on the energy or frequency of the wave and on the dimensionality of the system, and for one-dimensional systems a sufficiently random distribution of even arbitrarily weak scatterers can ultimately lead to localization of all waves.

This phenomenon has obvious implications for the electrical conductivity of glasses and solids with random impurities, especially at low temperatures where elastic scattering of electronic carriers by impurities or inhomogeneities predominates over dissipation due to coupling of the carriers to lattice vibrations or other such degrees of freedom. In situations where the carriers can be treated as moving independently of each other, a simple random medium can have critical energies, called *mobility edges,* for electrons or holes, below which they are localized and above which they can migrate. In such cases the limiting behavior of the material at very low temperatures will be metallic or insulating according to whether the Fermi level lies to one side or the other of the mobility edge, and changing parameters such as doping or pressure can cause a *metal–insulator transition.* In many cases, however, one must also take account of the mutual interactions of the carriers, which can sometimes greatly modify the conditions under which a metal–insulator transition occurs. (For example, such transitions can occur in nearly perfect crystals, where there is no localization phenomenon.)

Even when localization does not occur, there are important effects associated with the statistical fluctuations in the propagation of randomly scattered waves. These effects, sometimes called *weak localization,* cause the previously mentioned sensitivity of the low-temperature resistance of a conductor with random scatterers to tiny changes in the position or nature of even a single scatterer and cause both random fluctuations and periodic oscillations in resistance when an external magnetic field is varied.

Superconductivity

The most spectacular of the effects produced by the interaction of elementary excitations is the phenomenon of *superconductivity,* the complete disappearance of electrical resistance that occurs for most metals at extremely low temperatures. Discovered in 1911, this phenomenon came to be recognized as characteristic of an equilibrium thermodynamic phase, thanks to the discovery in 1933 of the *Meissner effect,* the expulsion of a previously existing weak external magnetic field from the interior of a macroscopic sample when the latter becomes superconducting. Superconductivity was considered the outstanding mystery of solid-state physics until 1957 when Bardeen, Cooper, and Schrieffer showed that it could be understood as the result of an association of electrons into pairs, mediated usually by their coupling to phonons, and a preferential condensation of these pairs into a single pair state. More explicitly stated, a superconducting phase of a metal is one in which very many pairs of electrons—a macroscopically large number—occupy exactly the same two-particle quantum state. (Such macroscopic occupation would be impossible for a single-electron state because of the exclusion principle, but it is possible for the helium atoms in superfluid ^{4}He, which obey Einstein–Bose statistics, and it is possible for *pairs* of electrons.) In many situations the atomic-scale descriptors of the state of the metal can be assumed to equilibrate rapidly with the local average values of the amplitude and phase of the pair wave function, so that the latter can serve as an order parameter in terms of which such macroscopic variables as free-energy density, electric current density, etc., can be expressed. (Multicomponent pair wave functions, if they should be found to occur, could be used similarly.) Further details of the behavior of superconductors can be described in terms of quasiparticle-type excitations which play a role like that of the electron and hole quasiparticles of normal metals.

Much use is made of the *Ginzburg–Landau* formulation, which postulates an analytically simple form for the dependence of the free energy on the quasimacroscopic superconducting order parameter and on the magnetic vector potential, a form which should become valid as the transition temperature is approached. An applied weak external magnetic field is excluded from the interior of a superconductor by currents localized within a thin *penetration depth* of the surface. With increasing field, either of two things can happen, depending on the parameters of the metal. In *Type-I* superconductors, the superconductivity can suddenly disappear in all or part of the specimen, wherever the field exceeds a temperature-dependent limit H_{c1}. In Type-II superconductors, the field can penetrate the superconductor along a family of closely spaced lines (often curved) called *vortices,* between which supercurrents can still flow and circulate around the vortices; only above a second critical field, H_{c2}, does superconductivity disappear. Lateral motion of vortices causes energy dissipation, hence resistance, but is often largely suppressed by a pinning effect of impurities or imperfections.

When there is a weak electrical connection between two bulk superconductors, a current can flow from the one to the other either by the transfer of single electrons or by the transfer of electron pairs. In the former case, or in the case where one of the metals is nonsuperconducting, the transfer changes the occupation of electronic quasiparticles (many-body excitations above the ground state, present due to thermal agitation or other stimulation) in the superconductor. This type of transfer is dissipative and gives information on the quasiparticle spectrum and on the energy gap, the (usually nonzero) amount of energy required to break up an electron pair into two quasiparticles. Charge transfer by pairs is nondissipative at low static currents, but at higher currents manifests *Josephson effects,* described in terms of the equations of motion for the number of pairs transferred and the relative phase of the pair wave function in the two bulk superconductors, these two quantities being canonically conjugate variables in quantum mechanics. In the simplest such effect, the *dc Josephson effect,* maintenance of a finite voltage V between the metals results in a current oscillation at a frequency $\nu = 2eV/h$, h being Planck's constant.

The discovery of high-temperature superconductivity (often above liquid-air temperature) in 1987 has led to a tremendous amount of speculation regarding the forces responsible for the pairing interaction in the new materials. Since all these materials are crystals containing fairly well-separated planes of copper atoms connected by bridging oxygens, attention has been focused on the electronic structure of two-dimensional systems, with strong interactions favoring an antiferromagnetic alignment of Cu^{++} spins. A number of mechanisms have been proposed that might cause superconductive pairing due to electron–electron interactions alone, without the need for electron–phonon coupling. An additional challenge to theory is posed by the unusual properties of the new materials at temperatures above the superconductivity range, e.g., a large resistivity nearly linear in temperature. Many intriguing new concepts have been introduced, e.g., elementary excitations with spin $\frac{1}{2}$ but no charge (*spinons*), or with charge but no spin (*holons*), excitations with statistics intermediate between Fermi and Bose (*anyons*), and other departures from the Fermi-liquid picture of metals—but there is as yet no consensus as to their roles in the understanding of the new materials.

Other Special Regimes

Besides the superconducting state, there are several other regimes, encountered in solid systems, whose understanding has required noteworthy extensions of conventional physical concepts. One such is seen in the low-temperature conduction of two-dimensional systems of highly mobile electrons, such as those in the thin space-charge region adjoining a planar semiconductor surface or interface, when a magnetic field H is applied normally to the conducting plane (x-y plane). When a current I is made to flow in the x direction through a region of width Δy, a Hall voltage V_H is measured in the y direction across this region. For classical macroscopic conduction in a uniform plate V_H would be proportional to IH, at least in the low-field region. But for systems of the sort described above, it is often found that the graph of V_H against H shows plateaus where V_H/I is independent of H and equal, with remarkable accuracy, to $h/\nu e^2$, where h is Planck's constant, e is the electronic charge, and ν is a small integer. This plateau phenomenon is called the *quantum Hall effect*. In the same range of H where the plateau occurs, the resistance in the x direction shows a broad minimum and approaches zero as temperature is decreased.

These phenomena represent general properties of two-dimensional conduction that are, within limits, unaffected both by impurities or imperfections and by electron–electron interactions. For a two-dimensional electron gas in a magnetic field and without such imperfections or interactions, the electronic states could be described in terms of one-electron *Landau levels,* with energies $(n+\frac{1}{2})h\omega_c$, where n is an integer and ω_c the cyclotron frequency, or, in the presence of an electric field E, by drifting "cycloidal" states with mean velocity cE/H. Scattering by imperfections will blur this picture and cause an Anderson-type localization of most of the states; electron–electron interactions will complicate the picture further. However, it can be shown that irrespective of such complications, a Hall voltage plateau will occur as an exact consequence of the way in which the vector potential of the electromagnetic field enters into the Schrödinger equation whenever the Fermi level of a two-dimensional electronic system lies in an energy range where all electronic quasiparticle states are localized, i.e., a range roughly midway between two of the Landau levels of the idealized system. In terms of the latter system, the integer ν in $V_H/I=h/\nu e^2$ is identifiable with the number of completely filled Landau levels (of either spin), and hence is called the *filling factor*. The low longitudinal resistance arises naturally because in this regime dissipation (migration along the Hall field) is possible only via an inefficient trickle through the localized states at the Fermi level, or by excitation into such states from the current-carrying states, a process requiring a sizable activation energy.

Even more remarkable is the *fractional quantum Hall effect,* the occurrence of plateaus in the magnetic-field dependence of the Hall voltage and of corresponding minima in the longitudinal resistance, with simple fractional values for the filling factor ν. These plateaus arise when electron interaction effects are stronger than the level broadening due to scattering. They are pictured as due to a mechanism, analogous to that described above, for the integral quantum Hall effect, except that the role played there by ordinary electronic quasiparticles in states derivable from ordinary Landau levels is now played by many-body quasiparticles of the strongly interacting systems, for which a fractional electric charge has been proposed.

There are some metals in which the normal crystallographic periodicity in the atomic positions is overlain by an, in general, incommensurate periodicity in the charge density of the conduction electrons with accompanying slight shifts in the atomic positions. This periodic fluctuation occurs at wave vectors $\mathbf{Q}$ capable of translating a sizable portion of the Fermi surface of the undistorted crystal to a position in wave-vector space very close to another such sizable portion; when such a $\mathbf{Q}$ exists, creation of holes just inside the one Fermi surface portion and of new electrons just outside the other will cost very little in quasiparticle excitation energy, and this may be outweighed by a lowering of potential energy, especially if the lattice is allowed to distort. The combined distortion is called a charge-density wave (CDW); an analogous effect involving a periodic spin polarization of the electrons is called a spin-density wave (SDW). Both have been seen for metals in equilibrium. In the CDW case, a further possibility occurs: the CDW can be set in motion by an electric field if this exceeds a threshold value E_T determined by the pinning of the CDW due to crystal imperfections. This causes the metal to show non-Ohmic conduction at quite low fields and generates noise at a frequency proportional to the CDW velocity.

Finally, there are a number of categories of phenomena of especial physical interest which, though by no means unique to solid-state physics, often occur in solid systems, where the stability and reproducibility of solid media provides a good environment for studying them. For example, the realm of *macroscopic quantum phenomena* can be studied in Josephson junctions, where under suitable conditions the difference of superconducting phase, a macroscopic quantity, can be observed to "tunnel" through a range of values that would be energetically inaccessible in a quasi-classical description. It has been suggested that the distribution of magnetization in ferromagnets can sometimes behave similarly. Many other interesting phenomena occur that involve quantum and statistical-fluctuation effects in *mesoscopic systems,* systems that are of much larger than atomic size but not yet large enough to be treated as macroscopic continua. Another realm with its own characteristic features is that of *aggregation,* the growth of a structure, such as a dendritically branched crystal, by the successive addition of atoms or larger particles to an entity growing under nonequilibrium conditions. Related is the phenomenon of *percolation,* the manner in which small units or "building blocks" placed at random in a space can form large clusters of blocks each in contact with one or more others, which for a sufficiently high concentration of blocks can extend to infinity.

EXPERIMENTAL TOOLS

All the traditional types of physical measurements—mechanical, acoustic, calorimetric, electrical, magnetic, optical, etc.—are used and indeed are essential in the study of solids. In addition, a number of specifically modern tools have come to be especially useful. Some examples of these are as follows: lasers, important both for their high resolution and for their great intensity; synchrotron radiation, which fills the gap between optical and x-ray spectroscopy and provides great intensity, tunability, and time resolution; radio frequency and microwave fields, which can, among other things, probe the constituents of solids in exquisite detail by electron-spin resonance, nuclear magnetic resonance, double resonances, cyclotron resonance, and dimensional resonances associated with the orbits of the electrons of metallic slabs in a magnetic field; diffraction and inelastic scattering of neutrons, which yield information about crystal structure, magnetic order, and dispersion relations of phonons and magnons; ultralow temperatures; ultrahigh-vacuum technology, a necessary prerequisite to the controlled study of surfaces; ultrahigh magnetic fields, now widely available thanks to superconducting magnets; SQUIDs, superconducting devices that can measure magnetic fields with great sensitivity by utilizing the Josephson effect; ultrahigh pressures, which now can be especially effectively exploited thanks to the development of diamond-anvil presses which allow light and x rays to pass in and out; *positron annihilation,* which yields information both on electron momenta and on the properties of defects that can trap positrons; spin precession of positive muons implanted into solids, and angular correlation of gamma rays emitted by excited nuclei in solids, both of which give information on hyperfine fields; scattering of fast ions from atoms of a solid, and their relatively free passage along channels between atomic planes in a crystal; tunneling of electrons through thin insulating layers between conductors or superconductors, a rich source of information about impurities, elementary excitations, etc.; the Mössbauer effect, which gives information about magnetic fields and electron densities at nuclear sites and about lattice vibrations.

Especially noteworthy has been the steady development over the years of capabilities for measuring structures and phenomena on finer and finer scales of space and time. *Electron microscopy,* which has for some decades been able to map the configurations of dislocations and some other defects in crystals, can now clearly see individual atom rows from an end-on perspective. Individual atoms on surfaces can be detected by *field ion microscopy*, and great detail about their positions and electronic structure is revealed by *scanning tunneling microscopy*. This is a technique in which electrons tunnel through a few angstroms of vacuum (or insulating fluid) between a metal tip of atomic dimensions and a conducting substrate. As this current is tremendously sensitive to the distance of the tip from the substrate and to its proximity to asperities on the latter, the variation of current when the tip is moved controllably by a piezoelectric drive yields a wealth of information. *Atomic-force microscopy* undertakes to measure forces exerted on a similarly controlled tip by interaction with substrate atoms. Time resolution of changes in solid and molecular systems is now possible on scales much smaller than typical atomic vibrational frequencies, thanks to the development of ultrashort laser pulses.

Modern developments in thin-film technology have helped in several ways to widen the scope of basic research on solids. New phases, or phases outside the composition range of thermodynamic stability, can be synthesized by low-temperature deposition. Known phases can be prepared not only as thin films, but as quasi-one-dimensional "wires" hundreds or perhaps even tens of atom spacings across. "Superlattices" composed of very many periodically alternating layers of two or more different materials can be tailor-made to any of a range of specifications and used to study the dependence of various properties of a synthetic "crystal" on different parameters.

RELATION TO OTHER AREAS OF PHYSICS AND OTHER SCIENCES

Among the other areas of physics, the ones whose subject matters most often overlap solid-state physics (as measured, e.g., by the frequency of research papers that could legitimately be assigned to either field, or by the frequency with which workers switch from the one field to the other) are optics and atomic, molecular, and electron physics. As for areas outside physics, the strongest overlaps are with materials science (metallurgy, ceramics), chemistry (physical chemistry or chemical physics, polymer science), and electrical engineering (solid-state devices). Though overlap with the earth and planetary sciences is much smaller in total volume, the common area of properties of solids at very high pressures is an important one. In many of the areas of overlap, about the only distinction between solid-state physics on the one hand and chemistry or engineering sciences on the other is that the former tends to emphasize simple and understandable cases, whereas the latter venture farther afield into complex problems and systems of technological importance.

Less quantifiable but sometimes more spectacular are the influences different fields have on each other through instrumentation on the one hand and major novel concepts on the other. Solid-state physics has been both donor and beneficiary of many such influences. For example, it has received very powerful experimental tools from nuclear physics in the Mössbauer effect and pile neutrons; it has made various contributions by way of repayment, perhaps most notably by making possible the development of solid-state counters, but also by various other contributions to techniques of measuring properties of nuclei, especially their magnetic-dipole and electric-quadrupole moments. Elementary-particle physics has enriched the outlook of solid-state theorists through the notions of the Green's functions and renormalization groups; experimental particle physics has benefited, as have many other areas of science and technology, from the invention of superconducting magnets. The concept of superfluidity due to fermion pairing, first developed by solid-state physicists to account for superconductivity, has found application in the theory of heavy nuclei and in the theory of neutron stars. Moreover, from origins in solid-state physics, semiconductor electronics, with its progeny microelectronics and optical electronics, has become a major branch of engineering, and through the collaboration of electrical engineers, materials scientists, and mathematicians has brought us into the computer age, which has revolutionized all fields of science and technology, including solid-state physics itself.

See also CRITICAL POINTS; CRYSTAL DEFECTS; CRYSTAL GROWTH; CRYSTALLOGRAPHY, X-RAY; CRYSTAL SYMMETRY; EXCITONS; FERMI SURFACE; GLASSY METALS; HALL EFFECT, QUANTUM; INSULATORS; LATTICE DYNAMICS; METALLURGY; METALS; PHASE TRANSITIONS; PHONONS; PLASMONS; QUASIPARTICLES; SEMICONDUCTORS, AMORPHOUS; SEMICONDUCTORS, CRYSTALLINE; SUPERCONDUCTING MATERIALS; SUPERCONDUCTIVITY THEORY; SURFACES AND INTERFACES.

Solid-State Switching

Melvin P. Shaw

Switching in an electronic device or system is defined as the induced transition from one steady state on the operating characteristic of a device to another. Figure 1 shows the terminology that is required to understand the process for the simple two-terminal case where the device current, I, is controlled by a bias voltage, V_B. The load resistor is R. The equation of the dc load line (dcll) is, from Kirchoff's Voltage Law, $V_B = IR + V$, where V is the voltage across the device. The steady-state dc operating point is shown in Fig. 1. For a voltage V across the device, it carries a current corresponding to that at the operating point. If we change V_B to some other value, the device current and voltage will switch to a new value, and there will be a new operating point.

Switching can occur by making, breaking, or changing the connections in a circuit, or, as shown above, by changing the dc bias level between two values. It can also occur if an imposed ac signal level is changed between two different values. Switching is of fundamental importance because it is the principal operation in computing, information processing, and automatic control. Since solid-state switches are small, fast, reliable, and inexpensive, they are incorporated in most modern systems.

Many solid-state devices used as switches have characteristics that also make them desirable for use as amplifiers, oscillators, detectors, or modulators. Most solid-state switches are made of the elemental semiconductor silicon, although compound semiconductors such as gallium arsenide and superconducting materials also are being used.

A useful switch must be able to remain in and then rapidly change between at least two distinct operating points in response to different input signals. The different operating points must differ in some way, such as resistance, dynamic resistance, current, or voltage. In Fig. 1, where we showed a nonlinear device by way of example, all four of these parameters generally differ at different points on the characteristics. Perhaps the most useful transition is the one that can readily represent OFF and ON states, or the digits 0 and 1 of the Boolean algebra used in computer logic. In order to change and read a logic gate, the least amount of energy needed is about $\Delta E = kT$, where k is the Boltzman constant and T is the absolute temperature. Employing the Heisenberg uncertainty principle along with this energy consideration tells us that the fastest expected time of any switching transition is about 10 ps, neglecting any inherent circuit limitations.

It is often the case that the two distinct states through which useful semiconductor devices switch correspond to different spatial distributions and local densities of electrons and/or holes. These charge distributions must be able to be changed very rapidly in order to make the switching transition. The time it takes to neutralize injected charges depends on the mechanism by which they dissipate: dielectric relaxation, recombination, drift, or cooling. All four of these mechanisms in general determine the ultimate switching speed of a particular device.

As discussed above, many switching devices can be classified simply as systems that switch from one resistive state to another under an appropriately large bias voltage or similar excitation. In making these transitions, some devices might switch from one quadrant of their current–voltage, $I(V)$, curve to another, such as the switch from saturation to cut-off in the three-terminal bipolar junction transistor (BJT). However, many other devices switch because of a region of negative differential conductivity (NDC) in their constitutive current density–electric field, $j(E)$, characteristic, such as a *pnpn* diode. It turns out that the representation of an arbitrary switching device as having a region of NDC is quite useful for studying switching in general. Hence, in what we shall discuss here, we shall proceed along this path, keeping in mind that the circuit treatment we will

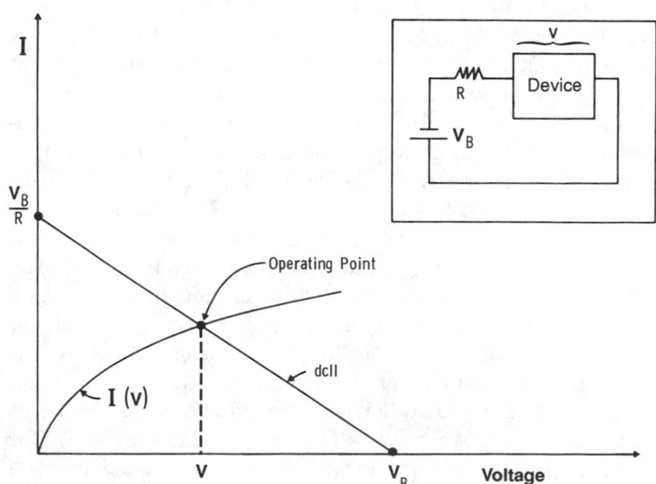

FIG. 1. Load line (dcll) and current–voltage, $I(V)$, curve for a device (nonlinear) placed in the circuit shown in the inset.

present is general and holds whether or not the switch device contains a region of NDC.

NDC curves have the two general forms shown in Fig. 2: N shaped (NNDC) or S shaped (SNDC). If we imagine a dcll as in Fig. 1 on this figure, and a specific operating point on either the NNDC or SNDC curve, we see that by increasing V_B from zero we can reach operating points that with a further increase in V_B will "jump" along the dcll to a value substantially different from the original operating point. It is clear that the nature of this jump or switching transition depends upon the load resistor R. For example, consider the NNDC curve. For a sufficiently light load (large slope of the dcll) we might be able to have the operating point trace out the entire $j(E)$ curve. For a heavy load (small slope of the dcll), we would reach the peak point (zero slope) of the $j(E)$ curve and then switch along the dcll to a new operating point at substantially higher E and somewhat smaller j.

Although we shall discuss the standard switching devices shortly, it is worth pointing out at this time specific examples of NDC devices. Some NNDC devices are the tunnel diode, Gunn diode, and resonant tunneling device. Some SNDC devices are the *p-i-n* diode, *pnpn* diode, the amorphous chalcogenide thin-film switch, and the heterojunction hot electron device.

The existence of a region of NDC in a constitutive relationship admits of a large number of instabilities, including switching, circuit oscillations, amplification, the propagation of internal space-charge modes, and chaos. In fact, an NDC region can be shown by Maxwell's equations to be unstable against both the formation of high-electric-field domains and high-current-density filaments. The manifestation of the instability is determined primarily by the shape of the entire curve (S or N) and the transverse and longitudinal boundary conditions. From the standpoint of switching, however, the major concern is the transient response of the system, and this is governed in general by the local circuit environment. Indeed, to understand the problem in detail, we need to focus on the bulk properties, the contact region, and the circuit.

In order to understand the transient response, it is important to identify the major local circuit components. To do this, we must appreciate that we will be doing so in a lumped element approximation, so effects at frequencies of tens of gigahertz or more could be substantially different from what we will predict.

Let us define the basic switching device as an NDC element. In order to apply a bias, it must be contacted by a metal. This, in general, produces a nonlinear voltage-dependent contact resistance, R_c. Next, the NDC element must be held in a package of some sort, and this introduces a package inductance, L_p, and a package capacitance C_p. The package is then wired to the battery circuit; this produces a lead inductance L_l, lead plus load resistance R_L, and power source internal resistance R_B. The driving voltage is V_B. Maxwell's equations then tell us that the NDC element itself can be represented by an intrinsic inductance L_i in series with a constitutive curve $i_c(V)$ (for NNDC) or $V_c(i)$ (for SNDC), the pair in parallel with an intrinsic capacitance C_i. The $V_c(i)$ curve represents the lossy response of the charge carriers in the NDC element to all driving forces or fields: potential gradients, carrier concentration gradients, temperature gradients; it can exhibit hysteresis because of relaxation effects. Each specific device has one or more mechanisms of charge dynamics that determine the $V_c(i)$ curve; we will describe them in our discussion of each particular device.

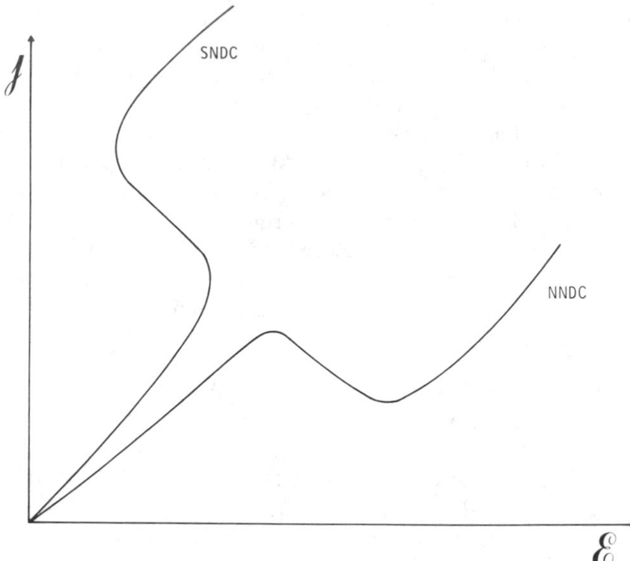

FIG. 2. Example of SNDC (current controlled) and NNDC (voltage controlled) constitutive $j(E)$ curves.

The local primary circuit environment contains five reactive components: L_p, L_i, L_l; C_i, C_p, plus a generally nonlinear $V_c(i)$ curve. Hence, the differential equation governing the temporal response of the system is of fifth order; general solutions must therefore be obtained with the aid of a computer. However, in a large number of NDC elements, reasonably good solutions can be obtained under the following approximations.

In a large number of NNDC elements the switching transitions involve relatively large changes in voltage and relatively small changes in current, i, through the NNDC element. In these situations L_p and L_i can be neglected, and C_i and C_p can be put together in parallel to provide a new effective capacitance C. If we write $L=L_p$ and $R=R_L+R_B$, it is easy to show that the differential equation governing the transient response of the system is

$$LC\frac{d^2V}{dt^2}+\left[L\frac{di_c(V)}{dV}+RC\right]\frac{dV}{dt}+\left(1+\frac{Ri_c(V)}{V}\right)V=V_B, \quad (1)$$

where V is the voltage across C and t is the time. Equation (1) closely resembles the equation of a nonlinearly damped anharmonic oscillator.

For many SNDC elements the changes in current, i, are relatively large and the changes in voltage are relatively small. Here we can neglect L_p and C_i, and define $L=L_i+L_p+M$, where M is any mutual inductance that might exist between L_i and L_p. The differential equation governing the system is

$$LC\frac{d^2i}{dt^2}+\left[C\frac{dV_c(i)}{di}+\frac{L}{R}\right]\frac{di}{dt}+\left(1+\frac{V_c(i)}{iR}\right)i=\frac{V_B}{R}, \qquad (2)$$

where i is the current through the $V_c(i)$ curve.

Note that Eq. (1) is stable if $RC>L|(di_c(V)/dV)|$, and Eq. (2) is stable if $L/R>C|(dV_c(i)/di)|$. (However, if saddle-point instabilities are also included, it is additionally required for the stability of the NNDC case that $1+R\ di_c/dV>0$ and for the SNDC case that $R+dV_c/di>0$.) That is, under these conditions, operating points on the NDC region are circuit stable. Circuit oscillations will not occur. However, if these points are bulk controlled, they will be unstable against domain and filament formation.

The circuits from which Eqs. (1) and (2) are derived are the duals of each other; the voltage response of an NNDC element in its primary circuit transforms to the current response of an SNDC element in its primary circuit, and vice versa.

The switching time is determined by the solution of Eqs. (1) or (2), given that the $V_c(i)$ curve has a known limiting temporal response governed by change-transfer effects, thermal considerations, etc. The switching time is the time it takes for the NDC element to make a transition. However, in order to transfer information, the switched signal must propagate to other points in the circuit. The fastest speed by which this can occur is the speed of light in vacuum. In practice, however, the transmission line is limited by the resistance of the metallic propagation line and its distributed capacitance, and the dielectric properties of the line.

A convenient way of representing the switching transition is to solve Eqs. (1) and (2) in each region that the device passes through from its initial to its final state. This can be accomplished analytically by linearization of the $V_c(i)$ curve. Once the equations are solved in each region, the regional solutions are joined together, the other variable (current or voltage) is found, and the time eliminated to obtain the current–voltage Lissajous-type figure that represents the switching transition. Figure 3 shows this for an SNDC element that exhibits current filamentation in the ON state. Here the ON and OFF states are noted, along with the dcll. V_t, I_t is the threshold point. The Lissajous-type figure (thin solid line) represents the circuit response, seen here as a damped oscillation as the energy is exchanged during the switching event between the local reactive components L and C. V_h, I_h is the holding point, which is circuit dependent. There is a minimum holding current, I_{hm}, below which the device always switches back to the OFF state. A curve of arrows is also shown, which represent the $V_c(i)$ curve for uniform current conditions, were they to occur in this device that is subject to current filamentation. Thus, the arrows represent the trajectory that is followed by the transport component of the current during the switching transition and while the current filament is forming.

The Lissajous-type figure is shown in the plane of conduction current, i, versus voltage, whereas the dcll line corresponds to the plane of total current, I, which includes the displacement current. If the first-cycle circuit response would have fallen below I_h, rather than undergoing a switching transition to the operating point shown as the heavy dot on the ON state, the circuit response would have returned the current to the OFF state; a relaxation oscillation then results.

The following describes the operation of a wide variety of switching devices and the mechanisms that produce specific $I_c(V)$ or $V_c(i)$ curves.

The Schottky diode is a rectifying metal–semiconductor (MS) contact. Interface states and to a small degree the difference in work functions between the metal (M) and the semiconductor (S) give rise to a Schottky barrier of height ϕ, and depletion of majority carriers in the barrier region. The observed current I is the difference between majority carrier flow in opposite directions across the barrier. The flow from M to S is impeded by $\exp(-\phi/kT)$, the flow from S to M by $\exp[-(\phi-qV)/kT]$ because of the bias potential qV in S. This leads to the rectifier characteristic

$$I=I_r\left[\exp\left(\frac{qV}{nkT}\right)-1\right], \qquad (3)$$

where n is an ideality factor equal to unity for a perfect diode, and I_r is the reverse leakage current. Switching times less than 0.1 ns are achieved for small (5-μm-diam) devices.

The p-n junction diode is formed at the interface of an n- and p-type semiconductor. Electrons diffuse from the n to the p material and holes from the p to the n material until an internal electric field is built up which causes the resultant drift current to cancel the diffusion current. This prevents further increase of the equal and opposite space charges. The energy barrier in the high-resistance intrinsic junction is about equal to the energy gap and nearly twice as big as the ϕ of Schottky diodes. Equation (3) also holds for p-n junc-

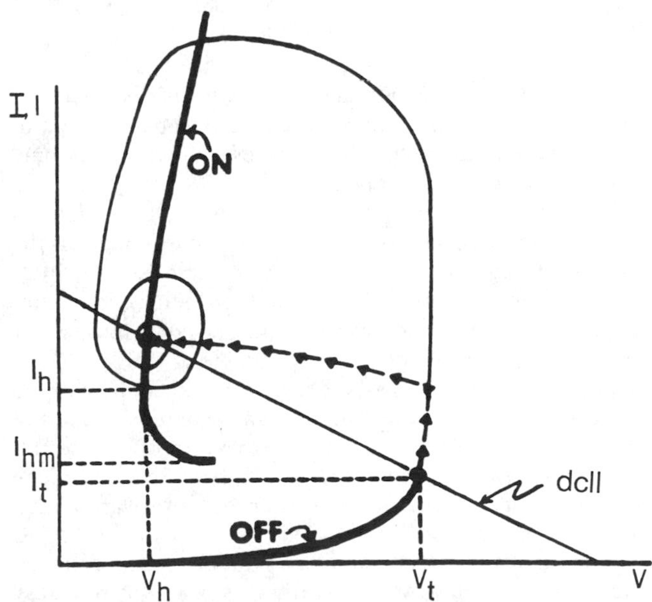

FIG. 3. Estimated $V(I)$ and $V(i)$ curves for an SNDC switching event. (All parameters are defined in the text.) The circuit oscillations damp, a filament forms, and switching occurs to the ON state, shown as the heavy dot. As V_B is varied, the dot traces out a filament characteristic (ON-state curve). The switch occurs in a damped oscillatory fashion; spiraling into the ON state occurs. In the current–time profile this appears as a damped circuit ringing.

tions, but with much smaller values of I_r. This leads to a larger forward voltage drop V_F. For a 10-μA current, $V_F \approx 0.25$ V for a Schottky diode and $V_F \approx 0.55$ V for a silicon *p-n* junction. For $V>0$, majority carriers flow across the junction. For $V<0$, they are pulled away from the junction and only the small minority-carrier density contributes to I_r until the breakdown voltage of the diode is reached, where I increases rapidly due to Zener breakdown or avalanche multiplication of carriers in the high field of the reverse-biased junction. The turn-on time of a *p-n* diode is only a few nanoseconds. A turn-off time of microseconds is needed to sweep out the excess minority carriers which are stored in the junction during conduction.

Step-recovery diodes are *p-n* junctions whose graded doping profile minimizes the stored-minority-carrier transit time. They are used for fast and microwave switching.

The Zener diode is a *p-n* junction with a sharp and well-controlled reverse-bias avalanche breakdown voltage (0.2 V to several hundred volts). Its fast switching, voltage sensitivity, and low resistance in the ON state make it useful as an overload and transient suppressor and in control circuits.

The p-i-n diode is a current-controlled microwave switch. It has a thin layer of undoped (intrinsic) region between strongly doped *p* and *n* layers. At microwave frequencies the intrinsic region acts like a capacitor. Forward bias (dc or low frequency) produces a carrier avalanche and makes the diode conductive. It sometimes shows SNDC characteristics.

The tunnel diode is a *p-n* junction in which both sides are so heavily doped that the Fermi level lies in the valence band on the *p* side and in the conduction band on the *n* side. At forward bias this overlap of the band edges allows electrons from the *n* side to tunnel quantum mechanically through the junction barrier into empty valence-band states without changing their energy. At higher bias the electron energies are too high for energy-conserving tunneling and the current drops through an NNDC range. The current rise at larger biases (>0.4 V) is the conventional forward-bias current of a *p-n* junction. The switching time is less than 1 ns since only majority carriers are involved.

The junction transistor consists of two *p-n* junctions in close proximity. Under normal conditions one junction is the forward-biased emitter, the other the reverse-biased collector. The transistor is OFF when both collector and emitter are reverse biased (base current zero). The collector–emitter leakage current is about 10 nA for Si. In its ON state both junctions are forward biased ($V_{CE}<1$ V), injecting carriers into the base region which reaches saturation as a result of space-charge and Ohmic effects. For fast turn-off speeds saturation must be avoided by, for example, a Schottky diode between collector and base which becomes forward biased before saturation is reached. This device exhibits SNDC characteristics under breakdown conditions in the common emitter model. However, it is used as a switch primarily between cut-off (OFF) and saturation (ON).

The avalanche transistor is a transistor structure with alloyed step junctions and a base-layer resistivity that is larger than that of both the collector and emitter. With increasing V_{CE} avalanche breakdown of the collector occurs ($V_{CE} \approx 60$–80 V). The emitter current is first low, but with increasing current through the base the base potential increases and the emitter becomes forward biased, which causes a large emitter–collector current to flow at reduced $V_{CE} \approx 30$ V. An SNDC region is traversed as the emitter becomes forward biased.

The pnpn diode has a reversed-biased collector junction, which yields a high OFF-state resistance (>10^8 Ω) between two forward-biased junctions. After onset of avalanche in the middle junction at the breakdown or switch-ON voltage, the two emitters saturate the two base regions and make the middle junction appear forward biased. A low-voltage ON state results. The device returns to its OFF state when the current falls below a holding current value. It is an SNDC device for one polarity of voltage.

Silicon-controlled rectifier (SCR). Since switching of the *pnpn* diode is initiated by the condition that the sum of the common base current gains α_{npn} and α_{pnp} becomes unity, rather than by the onset of avalanching, this condition can be achieved and the magnitude of the switch-ON voltage controlled by changing the current of either one of the two forward-biased junctions. This is done in the SCR by an additional terminal to one of the base regions. This SNDC device is a three-terminal *pnpn* structure.

The silicon-controlled switch (SCS) has terminals to each of the two base regions of a *pnpn* transistor. One permits control of ON switching, the other controls the holding current and thus OFF switching. This SNDC device is a four-terminal *pnpn* structure.

The triac is an SCR device built such that operation with either voltage polarity yields a nearly identical SCR characteristic.

The unijunction transistor (UJT) is an *n*-type semiconductor base with two Ohmic contacts, between which is placed a *p-n* junction (anode). Forward biasing of the anode injects minority carriers (holes) into the base, which decreases the resistance between the anode and the negative base contact. The resultant shift of the base potential below that of the anode increases its forward bias and thus the minority-carrier injection. This leads to an SNDC characteristic. The switching speed (ON to OFF) is orders of magnitude lower than that of bipolar transistors because of large charge-storage effects in the high-resistivity base region.

Thyristors. The SCR, SCS, and triac are a few of several solid-state switches having thyratron-like SNDC characteristics. Some can also be light activated. Their low power consumption, trigger sensitivity, and reliability make them useful in many control circuits.

The field-effect transistor (FET) is a majority-carrier (unipolar) device, in which the conductance between the source and drain electrodes along a thin semiconductor channel made of *n* or *p* material is controlled by the voltage difference between this channel and an adjacent gate electrode. Insulation between the channel and the gate is provided by a *p-n* junction (junction FET) or by a thin oxide layer yielding a metal–oxide–semiconductor (MOS)FET, also called an insulated gate (IG)FET. The gate voltage either enhances or depletes the carrier concentration in the channel and thus alters its conductance between low (OFF) and high (ON) values. The power consumption is very low, as it is essentially the change in energy stored in the device capacity. Because of device capacitances, the switching speed (particularly OFF to ON) is slower than that of fast bipolar devices. However, the small size of MOS devices, their self-isolation, and ease of manufacture permit large-scale integration of complex digital and memory circuits on one small silicon chip. The work-function difference and hence the switching voltage (about 1.5 V) is decreased to 0.6 V by using p-doped polycrystalline Si instead of Al as a gate electrode. Such silicon-gate IGFETs have higher speed and use even less power.

The MNOS memory switch is a MOS device with a silicon nitride dielectric layer (N) between the thin oxide and the gate electrode. Charges trapped in the silicon nitride layer hold the conductance state in the channel and thus provide the memory action: 5–10 V "write" and "erase" pulses of opposite polarity change the charge state of the traps. The memory access time is under 50 ns: the access time is at least twice as great in ferrite core memories. However, the MNOS memory is volatile, which means that it requires circuit voltage to keep its memory.

The amorphous thin-film threshold switch. These SNDC devices contain as their active part a noncrystalline (amorphous) semiconductor. In an Ovonic threshold switch a layer of chalcogenide (Se, Te, or S) alloy glass, for example, $Te_{50}As_{30}Si_{10}Ge_{10}$, is sputtered or evaporated between two nonreacting electrodes which are about 1 μm apart. The normal (OFF) resistance is of over 1 MΩ. If an applied voltage of either polarity exceeds the threshold voltage (2–200 V), the device switches in times as fast as 0.1 ns from the threshold point to its ON state. In this state the voltage drop is about 1 V, independent of current. As soon as the current falls below a holding current of about 1 mA, the device returns to its OFF state (see Fig. 3). Under pulse conditions the ON state is reached after a delay time of between 10 and 1000 ns, depending on the pulse height above threshold. The OFF state is the normal resistance of the semiconducting glass. In the ON state injection from the contacts, trap filling, and the resultant increase in mobility produce a filamentary conducting region typically less than 10 μm in diameter. This device is useful as a surge suppressor.

Amorphous thin-film memory switch. This switch is similar to the Ovonic threshold switch except for the use of a bistable chalcogenide glass, for example, $Ge_{14}Te_{83}Sb_3$, which can be reversibly altered by electrical pulses between a resistive noncrystalline state and a conductive microcrystalline state. Amorphous hydrogenated silicon thin-film devices are also useful, but they cannot be reset conveniently and are therefore primarily used as PROM.

The Josephson superconducting junction. When two superconductors are separated by a sufficiently thin nonsuperconducting region (less than about 50 Å), the tunneling of electron pairs produces a magnetic-field-dependent current at zero voltage, called the Josephson current. Switching can be achieved from this state to a voltage state by either current overdrive or a magnetic field. The switching time approaches several picoseconds and the switching event requires extremely low energy; it is our fastest known electronic switch, with power-delay products (less than about 1 fJ) well below those of either Si or GaAs devices. Because of the presence of the zero-voltage current, the Josephson element can be classified as an NNDC device. The recent breakthrough in high T_c superconductors may resuscitate this device as a possible competitor with conventional semiconductor devices.

The Gunn diode. This NNDC device is used mainly for microwave oscillation and amplification. However it is also used as a switch when the contact/doping profile/length interrelated parameters are designed to produce a high electric field at the anode contact for sufficiently high values of bias. Under these conditions the device switches from a low-voltage–high-current state to a high-voltage–reduced current state. The mechanism for producing NNDC is the field-induced intervalley transfer of electrons from a high-mobility to low-mobility valley in the conduction band, and the manifestation of the current instability, e.g., oscillations or switching, is determined primarily by the cathode contact boundary condition.

Resonant tunneling devices. These NNDC devices are generally fabricated in the form of semiconductor heterostructures having two or more barriers, although single-barrier heterostructures also produce NNDC effects. GaAs and $Ga_xAl_{1-x}As$ are generally employed. The basic principle of operation of these devices involves the creation of a resonant tunneling channel through a multiple quantum-well structure, and the coupling of the channel to an emitter and collector. The resonance involves the tunneling transmission probability as a function of the energy of the incident electrons, which is related to the width and height of the quantum wells. The structures can often be thought of as multiwell, variably (in general) spaced superlattice energy filters.

Heterojunction hot electron diode. This SNDC device is fabricated from $GaAs/Ga_xAl_{1-x}As$ interfaces. Carriers are at first retarded from a heterojunction barrier. The barrier is subsequently thinned and reduced by the applied bias, and the field is increased in the narrow-gap (GaAs) region. Eventually, the field in the region becomes large enough to heat carriers to sufficiently high temperatures so as to emit them thermionically over the barrier, thereby increasing the current, which in turn causes an even larger field in the narrow-gap region. This produces the feedback mechanism required for stabilization of the SNDC curve.

Metal-Oxide Varistor. Here small metal particles (Sn, Ni, Nb, etc.) are surface oxidized and subsequently sintered. The transport mechanism involves an increase in current as the carriers tunnel through the oxide regions separating the particles. The current–voltage characteristics are generally rather soft and vary gradually, as opposed to structures that exhibit sharp discontinuous-type changes.

Optical mass memories. Generally made from chalcogenide materials, these thin-film structures offer great promise for extremely high-density reversible memories. A focused laser beam is used to change the optical properties of the material in a microscopic region; the region is then read optically. Amorphous films are usually fabricated in disc form; they can be used as is, with the laser either changing the local region to crystalline, or vaporizing the region and creating a bubble in the material. The films are often heated after deposition to form crystallites and then induced to change regionally into the amorphous phase by elevating the local temperature optically and then rapidly removing the excitation (quenching). Several megabytes of information can be reversibly stored in a region of a few square inches.

Optoelectronic devices use light-emitting and -detecting diodes in close proximity. Light coupling achieves high electrical isolation between input and output.

Integrated logic circuits. The switching elements in digital electronics are no longer discrete devices but instead integrated circuits: gates, clocks, counters, and memory arrays, each containing numerous devices. These integrated elements are used together with drivers, encoders, and other circuitry to prepare the integrated circuits which comprise the random-access memories (RAM), electrically programmable memories (PROM), shift registers, and other peripherals needed to build, for example, a microprocessor or microcomputer. Even complex computing and data-handling systems use only two basic logic operations, counting and gating. The basic logic elements of gating are the operations of coincidence starting "AND"; of anticoincidence or sorting out, "OR"; and of prohibition, "NOT."

The Real-Space-Transfer Device is the real-space analogue of the Gunn diode. It is based on the real-space transfer of electrons in a modulation-doped heterostructure. The conduction-band edge of a $GaAs/Al_xGa_{1-x}As$ heterolayer, for instance, has a discontinuity ΔE_c due to the larger bandgap of Al_xGa_{1-x} as compared to GaAs, e.g., $\Delta E_c = 250$ meV for $x=0.3$. The AlGaAs layer is heavily doped, while the GaAs layer is undoped. Therefore, the ionized-impurity scattering leads to a much smaller mobility in AlGaAs than in GaAs. Since the mobile carriers fall into the GaAs well, they are separated in real space from their parent donors which are responsible for scattering. This results in very high current densities in the GaAs layer if an electric field is applied parallel to the layers. As the field is increased, the carriers are heated up strongly, resulting in a thermionic emission current from GaAs into AlGaAs. This real-space transfer reduces the average mobility and thus can lead to NNDC. The extremely high mobility in the GaAs (>5000 cm^2/Vs at 300 K) singles out this effect for applications in fast electronic switches and oscillators. A particularly important application is the High Electron Mobility Transistor (HEMT), where the channel between source and drain contains a modulation-doped GaAs/AlGaAs heterojunction.

The incredibly rapid pace of the development of integrated digital technology strives toward increasing speed and reducing cost, achieved by the higher packing density of very large-scale integration (VLSI). The fastest gating (less than 1 ns) is presently possible with bipolar junction transistors in the emitter-coupled logic (ECL) mode [less than 3 ns in the transistor–transistor logic (TTL) mode]. However, the lowest cost is achievable with the slower MOS technology with its breakthroughs into LSI and VLSI during the past 20 years. MOS technology has limitations in that its gate propagation delay is greater than 10 ns. However, although ECL is the fastest logic switch, it also has the highest static power dissipation (about 40 mW per gate). The power dissipation for TTL is about 2 mW, and here is where the use of CMOS becomes very desirable, since its power dissipation is about 10^{-3} mW. Small power dissipation is very important for future ULSI (ultra) applications involving millions of components and GSI (giga), with billions of drivers on the same chip. GSI will require feature sizes of about 0.1 μm. TTL gates require a chip area per gate of about 3×10^4 μm^2, compared to 15×10^2 μm^2 for CMOS. CMOS gates also have the lowest power-delay product of any semiconductor device, nearly 0.6 pJ.

Recent research indicates that vacuum microelectronic devices, which are based on micromachined on-chip miniature "vacuum tubes" with cold electron emitters, will have gate delays in the femtosecond range. At high temperatures, for high-speed applications it is expected that diamond semiconducting films will play a major role. For low-temperature, high-speed electronics, we expect that high-T_c superconducting devices, both as leads and as Josephson junctions, could contribute substantially to the solid state switching technology.

See also CIRCUITS, INTEGRATED; DIELECTRIC PROPERTIES; MAGNETIC DOMAINS AND BUBBLES; RELAXATION PHENOMENA; SEMICONDUCTORS, AMORPHOUS; SEMICONDUCTORS, CRYSTALLINE; SUPERCONDUCTORS; TRANSISTORS.

BIBLIOGRAPHY

A. Blicher, *Thyristor Physics*. Springer-Verlag, New York, 1976. (A)

J. D. Greenfield, *Practical Digital Design Using IC's*, 2nd ed. Wiley, New York, 1983. (I)

E. L. Johnson and M. A. Karim, *Digital Design. A Pragmatic Approach*. PWS Engineering, Boston, 1987. (I)

A. Madan and M. P. Shaw, *The Physics and Applications of Amorphous Semiconductors*. Academic, New York, 1988. (A)

M. Morris Mano, *Digital Logic and Computer Design*. Prentice-Hall, New York, 1979. (I)

E. Schöll, *Nonequilibrium Phase Transitions in Semiconductors*. Springer-Verlag, Berlin, 1987. (A)

M. P. Shaw, H. L. Grubin, and P. R. Solomon, *The Gunn–Hilsum Effect*. Academic, New York, 1979. (A)

M. P. Shaw, V. V. Miten, E. Schöll, and H. L. Grubin, *The Physics of Instabilities in Solid State Electron Devices*. Plenum, New York, 1991.

S. Sze, *Physics of Semiconductor Devices*, 2nd ed. Wiley, New York, 1981. (A)

P. D. Taylor, *Thyristor Design and Realization*. Wiley, New York, 1987. (I)

E. S. Yang, *Microelectronic Devices*. McGraw-Hill, New York, 1988. (I)

Solitons

Harvey Segur

Given a partial differential equation involving both time and space (or a system of such equations), a *solitary wave* is a special solution that: (i) is localized in space; (ii) has a fixed velocity; and (iii) has permanent form (or at least exhibits some kind of coherence). Except for degenerate cases, equations admitting solitary waves are nonlinear, so typical solitary waves are inherently nonlinear phenomena. As examples, the solitary wave for the *Korteweg–de Vries* (*KdV*) *equation*,

$$\partial_t u + 6u\, \partial_x u + \partial_x^3 u = 0, \quad (1)$$

is $u(x, t) = 2\kappa^2 \operatorname{sech}^2\{\kappa(x - 4\kappa^2 t + x_0)\}$. For the *nonlinear Schrödinger equation,*

$$i\, \partial_t \phi = \partial_x^2 \phi + 2|\phi|^2 \phi, \quad (2)$$

the solitary wave is a localized wave packet:

$$\phi(x, t) = A \exp\{iBx + i(B^2 - A^2)t + it_0\} \operatorname{sech}\{A(x + 2Bt + x_0)\}.$$

The word "soliton" is intended to suggest a solitary wave that acts like a particle; i.e., it maintains a localized, coherent identity even in the course of interaction with other localized objects. Two definitions of "soliton" are commonly used.

STRICT DEFINITION

A soliton is a localized nonlinear wave that regains asymptotically (as $t \to +\infty$) its original ($t \to -\infty$) shape and velocity after interacting with any other localized disturbance; the only long-term effect on the soliton from the interaction is a phase shift.

For most equations, such interactions are impossible and there are no solitons. Two remarkable facts are (i) some equations, including Eqs. (1) and (2), *do* admit solitons and (ii) some of these equations describe physical phenomena.

Mathematical Stucture

Equations that admit solitons are said to be *integrable* or *completely integrable*. Every integrable equation has a long list of miraculous properties, including the following. Associated with each integrable equation is a pair of linear differential equations containing a free parameter. The integrable equation is equivalent to the requirement that these two linear equations be compatible for all permissible values of the parameter. For example, the linear equations associated with Eq. (1) are

$$-\partial_x^2 \psi - u\psi = \lambda\psi, \quad (3)$$

$$\partial_t \psi = (\partial_x u)\psi + (4\lambda - 2u)\, \partial_x \psi. \quad (4)$$

One verifies directly that for fixed values of λ, $\partial_x^2 \partial_t \psi = \partial_t \partial_x^2 \psi$ if and only if $u(x, t)$ satisfies Eq. (1).

A consequence of this equivalence is that every integrable equation can be *integrated* (i.e., solved) exactly, as an initial-value problem. The method of solution, sometimes called the *inverse scattering transform,* is a generalization of the method of Fourier transforms.

How do the solutions look? For most integrable equations, smooth initial data on $-\infty < x < \infty$ that vanish rapidly as $x \to \pm\infty$ evolve as $t \to \infty$ into a finite number (N) of solitons, typically traveling at N different speeds, plus an oscillatory wave train that decays in amplitude as it spreads out in space. The oscillatory wave train is qualitatively similar to the solution of the linearized problem, obtained by simply omitting the nonlinear terms. The solitons have no counterpart in the linearized problem.

More than a century before solitons were discovered, Liouville had proved that certain finite-dimensional Hamiltonian systems could be integrated completely if there exists a canonical transformation from the original coordinates to what are called *action-angle variables*. Neglecting a few subtleties, we may assert that equations that admit solitons are Hamiltonian, and that the inverse-scattering transform is a canonical transformation to the action-angle variables of the problem. Equations admitting solitons are nontrivial examples of infinite-dimensional Hamiltonian systems that are completely integrable in the sense of Liouville.

The initial-value problem also can be solved with smooth, spatially periodic initial data. In this case, the solution is periodic in space and almost periodic in time; there is no limiting behavior as $t \to \infty$.

Complete integrability is not restricted to partial differential equations. Parallel theories exist for differential-difference equations, integro-differential equations, etc.

Physical Applications

Some integrable equations describe certain physical phenomena in which dissipation is negligible. The KdV equation, (1), describes approximately the evolution of long waves of moderate amplitude, propagating in one dimension in any one of several media, including: ion-acoustic waves in a plasma; long, gravity-induced waves in shallow water; and longitudinal elastic waves in a long rod. The nonlinear Schrödinger equation, (2), describes approximately the evolution in one dimension of a nearly monochromatic wave train of moderate amplitude, with applications to optical pulses in a long fiber, high-frequency vibrations on a crystal lattice, Langmuir waves in a plasma, and surface waves in deep water. Both equations involve 1(space) + 1(time) dimensions and some external constraint must be imposed so that the physical waves propagate in only one spatial dimension. Other applications of integrable equations in $(1+1)$ dimensions include the dynamics of certain one-dimensional crystal lattices, and stationary, axisymmetric solutions of Einstein's equations of relativity in a vacuum.

Most of the known integrable equations involve $(1+1)$ dimensions. In $(2+1)$ dimensions, there are a few known integrable equations of physical interest; in higher dimensions, the self-dual Yang–Mills equations are essentially the only known example.

When an integrable equation in $(1+1)$ dimensions is embedded in a physical problem in $(3+1)$ dimensions, the one-dimensional solitons may or may not be stable to transverse perturbations. Even if they are stable, they are localized in only one direction. Practical applications have been most successful where the physical problem permits wave propagation in essentially only one dimension.

LOOSE DEFINITION

A soliton is a spatially localized, coherent object (i.e., a solitary wave) that is dynamically and structurally stable. This definition has been particularly useful for applications in condensed matter, where the spatial and statistical properties of the solitons are more important than their dynamical interactions. In these applications a soliton represents a *domain wall*: a thin region separating two different equilibria that are energetically equivalent. These solitons may move, or they may be "pinned" by impurities. Individual solitons account for many of the phenomena observed at moderate excitation energies. At higher energy levels, some thermodynamic properties of the material can be deduced from the statistical mechanics of a large number of solitons.

BIBLIOGRAPHY

M. J. Ablowitz & H. Segur, *Solitons and the Inverse Scattering Transform*. SIAM, Philadelphia, 1981.

G. L. Lamb, Jr., *Elements of Soliton Theory*. Wiley-Interscience, New York, 1980.

S. P. Novikov, S. V. Manakov, L. P. Pitaevskii, and V. E. Zakharov, *Theory of Solitons* (translated from Russian). Plenum, New York, 1984.

S. E. Trullinger, V. E. Zakharov, and V. L. Pokrovsky, *Solitons*. North-Holland, Amsterdam, 1986.

Note:

(i) Lamb's book is more elementary than any of the other three.
(ii) In terms of the two definitions of solitons given in the accompanying article, the first three books listed here use the strict definition, while the last book uses the loose definition.
(iii) The last book is a collection of 16 articles by different authors on physical applications of solitons.

Sound, Underwater

R. J. Urick

Underwater sound is the science of sound in the sea. Sound travels through the sea as well as, or better than, it does through the air; in both media it is used for communication and for the detection and location of objects of many kinds.

Of all the forms of energy that exist, sound travels through the sea with the least attenuation. For example, the attenuation of sound waves is less, by several orders of magnitude, than electromagnetic waves of the same wavelength. Because of being able to travel so readily through the sea, underwater sound has a wide variety of applications, ranging from underwater acoustic beacons that mark a particular spot in the sea to the detection of enemy submarines hundreds of miles away.

Underwater sound systems are called *sonars,* an acronym that stands for *so*und *na*vigation and *r*anging. Sonars are of two basic types: active and passive. *Active* or *echo-ranging* sonars are essentially underwater radars using sound instead of electromagnetic waves. They send out a pulse of sound by means of a *transducer* (Latin, "to take across") that converts electricity into sound. This pulse of sound travels through the sea to the target, where it is reflected and travels as an echo back to the transducer. Here it is converted back again to an electrical signal. The target echo is then amplified and altered by various techniques known as *signal processing,* and presented to the observer on a *display* where it may be seen or heard. Passive sonars, on the other hand, utilize the underwater noise made by their target, such as a distant ship or a school of fish; this noise is received on an *array* of transducers, and is again processed and displayed to the observer.

An important kind of passive sonar uses transducer arrays laid on the deep ocean floor. This system, called *sosus,* is located in many ocean areas, where it serves to detect and track noisy submarines out to long distances by means of sound traveling over deep, favorable transmission paths.

Unfortunately, the sea is full of noises of various kinds that obscure, and interfere with, the echo or the wanted noise of the target. These unwanted sounds have many origins: far-distant ship traffic, biological noisemakers such as fish and porpoises, and the noise made by the rough sea surface. The surface of the sea makes noise by some yet not entirely understood process; it pours down, like rain, upon a receiving transducer beneath the sea. Distant unidentifiable ship traffic is an ever-present form of man-made noise pollution of the underwater environment in which active and passive sonars must operate.

Another kind of unwanted background is *reverberation*—the slowly-decaying quivering tonal blast that follows a sonar pulse and often obscures the echo from the target. In active sonars reverberation is caused, like the reverberation in rooms, by the back-scattering of sound from the boundaries of the medium. In addition, there exist scatterers in the body of the sea itself, in the form of fish and other kinds of marine life, that add to the reverberation caused by the sea surface and bottom. Fortunately, any motion of the target causes the echo to be slightly different in frequency from the background of reverberation in which it occurs, so that the echo can often by separated from the reverberation by filtering.

Underwater sound signals and echoes are made evident to a human observer by a variety of audio and visual displays. Earphones and loudspeakers are still useful accessories to more complex presentations. Oscilloscope displays are commonly used in many sonar systems, as are paper-tape recordings of such characteristics as the bearing-time history of the target.

The military applications of sonar include the acoustic mine, passive and active systems for submarine detection, torpedoes that "home" on the echo or the radiated noise of the target, and mine-hunting sonars, which seek mines resting on the ocean floor. These applications cover about six decades of frequency, from subsonic frequencies to frequencies of several hundred kilohertz. Military sonars range in size from compact *sonobuoys,* which detect underwater sounds and transmit them to an aircraft by means of a radio link, to large powerful echo-ranging sonars on surface ships. A photograph of the bulbous bow containing an echo-ranging sonar transducer on a destroyer in dry dock is shown in Fig. 1.

A wide variety of peaceful, or nonmilitary, applications of underwater sound exist at the present time. The oldest of these are *depth sounders,* in which the depth of water beneath a ship is determined by timing the echo from the bottom. Acoustic beacons, transponders, fish-finders, and speedometers, which measure the components of ship speed over the bottom, are all commercially available. Side-scan sonars and subbottom profilers are used to probe the ocean floor to the side of the vessel and to explore what lies beneath the ocean floor. Such uses of underwater sound may be ex-

FIG. 1. The bulbous bow-dome of the AN/SQS-26 sonar on a ship in drydock. The sonar transducer is located inside the dome. Its size is evident when compared to the workman nearby. Photograph courtesy of the B. F. Goodrich Company.

pected to increase in keeping with man's ever increasing exploitation of the seas.

See also ACOUSTICS; TRANSDUCERS.

BIBLIOGRAPHY

R. J. Urick, *Principles of Underwater Sound,* 2nd ed. McGraw-Hill, New York, 1975.

Space-Time

Joshua N. Goldberg

Our conception of space as three dimensional comes from our awareness of a spatial translation group in our local vicinity. That is, we learn at an early age that rigid displacements can take place; that these displacements can be made continuously, that an arbitrary displacement can be reached by motions in three standard directions relative to our bodies (right–left, front–back, and up–down); that these results are independent of where we happen to be; and that except for an external force, which we call gravity, there is no preferred direction of making a displacement. From these observations, we conclude that space is a homogeneous and isotropic three-dimensional continuum.

Time impinges on our senses in the ordering of events into earlier and later. It assigns a precedence to occurrences. In our daily experience this assignment is absolute. That is, the decision as to which of two events is earlier can be made independent of who makes the decision.

Space and time, as described above, are independent of one another and form the Newtonian understanding of an absolute space and an absolute time. In this picture the set of physically equivalent observers share the same space and the same time. They are related to each other by the Galilean group of transformations: rigid spatial translations and rotations, rigid translations in time, and uniform changes in velocity given by

$$x' = x + vt, \tag{1}$$

which maps space points at a given time into space-points *at the same time*. Furthermore, space is endowed with a Euclidean metric (measure of length)

$$\Delta l^2 = \Delta x^2 + \Delta y^2 + \Delta z^2 \tag{2}$$

and intervals of time are measured by a periodic system which is assumed to remain uniform. These measurements, which are made by rigid rods and by clocks, reflect the homogeneous and isotropic properties of space and time. The existence of inertial observers related by the Galilean group, Eq. (1), indicates the lack of a state of absolute rest in the universe, but does not alter the absolute nature of space and time sparately.

The Michelson–Morley experiment, which showed that the speed of light is independent of the observer's velocity, found its explanation in Einstein's two postulates for special relativity: (1) The laws of nature are the same for all uniformly moving observers; (2) the speed of light is isotropic and a universal constant independent of the motion of its source. These postulates implied that *simultaneity* (when two events occur *at the same time*) depends on the state of motion of the observer. Thus the ordering of events is not absolute and, in fact, the separation of space and time breaks down. This result follows because the events which one observer would identify as "at the same time" are shifted in time for those events which a second observer moving relatively to the first would so identify.

In the Einsteinian picture there is *no invariant separation of space and time* so the set of physically equivalent observers share a four-dimensional continuum, *space-time*. They are related to each other by the group of Poincaré transformations: rigid translations in space and time, and the Lorentz transformations, which consist of the rigid spatial rotations and the uniform changes in velocity ($\beta = v/c$, c is the speed of light):

$$\begin{aligned} x' &= \frac{x - \beta ct}{(1-\beta^2)^{1/2}}, \qquad y' - y, \\ ct' &= \frac{ct - \beta x}{(1-\beta^2)^{1/2}}, \qquad z' = z, \end{aligned} \tag{3}$$

where for simplicity we have taken the relative velocity, v, to be along the x axis. Note that the set of points defined by t' = constant are not identical to any set with t = constant—there is no invariant separation of space and time. However, there is an invariant geometrical structure, the set of points defined by

$$c^2t^2 - x^2 - y^2 - z^2 = 0, \tag{4}$$

which is mapped into itself by a Lorentz transformation. These points define the set of light rays which impinge on the origin from the past and go out from the origin into the future, hence, the *light cone*. The light cone separates space-time into its interior, which consists of the *past* and *future* of the origin, and the exterior, which is sometimes referred to as *elsewhere*. For any point in the interior of the light cone, there exists a Lorentz transformation such that the point has a purely time-like separation from the origin ($x' = y' = z' = 0$, $t' \neq 0$). Since the past is mapped into the past and the future into the future, for such points, precedence is preserved. For any point in the exterior, there exists a Lorentz transformation, such that the point has a pure space-like separation from the origin ($x'^2 + y'^2 + z'^2 \neq 0$, $t' = 0$). For such points, precedence is not preserved by a Lorentz transformation. These notions allow us to introduce an invariant metric (measure of length) on the four-dimensional space-time:

$$\Delta s^2 = c^2\Delta t^2 - \Delta x^2 - \Delta y^2 - \Delta z^2. \tag{5}$$

For two points with a time-like, light-like (null), or space-like separation, Δs^2 is, respectively, greater than zero, equal to zero, or less than zero. Space-time with the above metric is called *Minkowski space* after H. Minkowski who first elaborated its geometry.

Minkowski space is an *absolute space-time*. Every observer related by a Poincaré transformation has the same spatial geometry and it is a global geometry—it extends to all of space-time. Physical processes take place within the framework of the Minkowski geometry, but have no effect on its structure.

General relativity was constructed to allow matter to interact with and to affect the geometry of space-time—in particular, to make gravity a manifestation of a geometry with curvature. Each freely falling observer, to first order, can introduce in his vicinity a Minkowski frame. However, because of the curvature of space-time the frame cannot be extended globally. Frames associated with different freely falling observers at a point are related by Lorentz transformations but, in general, there is no global mapping of such freely falling observers at different points into one another.

The geometry is described by 10 functions $g_{\mu\nu}(x)=g_{\nu\mu}(x)$, the metric tensor,

$$ds^2=g_{\mu\nu}dx^\mu\, dx^\nu, \tag{6}$$

where the repeated indices μ and ν are summed over 0, 1, 2, 3 [the x^μ are to be identified with (ct, x, y, z), respectively]. In accordance with the previous discussion, at any one point the metric may be given the Minkowski form

$$ds^2=c^2\, dt^2-dx^2-dy^2-dz^2. \tag{7}$$

Metrics with this property are called *Lorentzian with signature* -2 and a space-time with a Lorentzian metric is said to be *hyperbolic*. This guarantees a light-cone structure with the local properties found in Minkowski space. However, the geodesics, the straight lines of a space with curvature, will no longer be linear and may even be closed or nearly closed curves. The great circles on a sphere are geodesics.

The Einstein theory is the generally accepted theory of gravitation today. It gives a system of differential equations for the metric tensor by relating 10 components of the Riemann curvature tensor to the distribution of matter. Symbolically we write these equations

$$G_\mu{}^\nu=-\;8\pi k T_\mu{}^\nu, \tag{8}$$

where $T_\mu{}^\nu$ is the energy-momentum tensor for the physical matter. By virtue of differential identities, the Bianchi identities, satisfied by the left-hand side, we have the requirement

$$T_\mu{}^\nu{}_{;\nu}=0, \tag{9}$$

where the semicolon denotes the covariant derivative with respect to the Riemannian connection defined by the metric tensor. This equation is important because it represents conservation of energy, momentum, and angular momentum.

With the introduction of general relativity, space-time becomes more abstract and at the same time more physical. It is more abstract because it is related to a system of nonlinear differential equations which do not have a unique solution. It is more physical because the geometry of space-time participates in the dynamic physical processes of the universe. Physicists and astrophysicists try to find phenomena in nature to determine the geometrical properties which define the physical space-time. At the same time one studies the solutions of the Einstein equations in the hope that they will inform us about the kind of phenomena we should be looking for.

Thus, the notion of a space-time is no longer that of the unique environment in which we live. The solutions of the Einstein equations give us a variety of models of possible *physical* geometries. Therefore, we must define *a space-time* rather than *the space-time*. Before doing so, we shall discuss some of the properties we need to include.

The fact that the physical space-time has curvature means that the metric components $g_{\mu\nu}$ cannot be reduced to constants everywhere. That, in turn, means that one cannot expect a single coordinate system to cover the entire space-time. In Minkowski space (and in our everyday three-dimensional Euclidean space), we can imagine that a system of four (or three) perpendicular directions can define a rectangular coordinate system that covers the entire space-time without intersections or other singularities. On the surface of a sphere, however, that cannot be done. There will be a singularity at the north or at the south pole. Therefore, one requires a minimum of two coordinate patches to cover the sphere without singularity. One patch covers the north pole and can reach down below the equator. The second patch covers the south pole and can extend above the equator. In the overlap region, one compares the two patches by means of coordinate transformation functions.

In a general curved space-time, one finds the same problem. Coordinate patches are needed to cover the space-time to avoid the occurrence of singularities in the coordinate system. Where the patches overlap, differentiable transformation functions relate the respective coordinates. And because of the curvature, in general there are no symmetries to help us pick out a preferred set of coordinate systems like the rectangular system of Minkowski space or our Euclidean 3-space. This means that the coordinates do not have an intrinsic geometrical or physical meaning. As a result, physically meaningful quantities have to be defined in a way which is coordinate independent. In particular, the concept of a space-time itself is to be understood without reference to any system of coordinates. In its simplest form a space-time is a four-dimensional manifold M with a Lorentzian metric $g_{\mu\nu}$ assigned to it. We want the manifold to be continuous so that every neighborhood has an infinite number of points and we want these points to be distinguishable. An abstract space-time may be unrelated to our observable universe. To be physically reasonable, we believe that the manifold and its metric should be a solution of the Einstein equations. But, not all solutions are physically reasonable. While there are generalizations and alternatives to the Einstein theory which are under consideration, recent observations by I. I. Shapiro and his collaborators have placed the classical tests of general relativity within 1% of the Einstein prediction. Furthermore, observations of the binary pulsar by J. H. Taylor give remarkable agreement with the perihelion precession predicted by the Einstein theory and even show evidence of a period change due to gravitational radiation in good agreement with the rate expected from the theory.

Therefore, we may define a space-time (M, g) to be a four-dimensional manifold which is endowed with a Lorentzian metric tensor $g_{\mu\nu}$, a solution of the Einstein equations, whose first derivatives are continuous, but whose second derivatives may have discontinuities. The discontinuities can occur at the boundary of matter or across the front of a gravitational wave.

Given a space-time (M, g), what kind of singular situations might one find? First of all, gravity is universally attractive. Therefore, matter tends to collapse. The Einstein equations tell us that as long as the local energy density remains positive, the collapse will continue indefinitely, eventually resulting in a singularity. Even before forming a singularity, the matter becomes so compact that light is unable to escape from its surface. This is the formation of a black hole. The singularity itself is revealed by the existence of null or time-like geodesics which have a finite length. This implies the abrupt ending of a physical process. Study of the Einstein equations suggests that singularities form under quite general conditions.

In a reasonable space-time, one would not expect problems with causality. However, there exist solutions which contain time-like and null curves which either are closed or enter the neighborhood of some point more than once. The Gödel metric is an example of the first situation and that of Taub–Newman–Unti–Tamburino is an example of the second. Such space-times are nonphysical in the sense that they require a periodicity where events are so ordered that the future prepares the present which is to evolve to the future. We do not conceive of the universe as being that well ordered, particularly at the quantum-mechanical level where fluctuations occur. Therefore, we expect the physical space-time to satisfy a *stable causality* condition. That is, small variations of the metric should not lead to closed time-like or null geodesics.

We have been discussing the properties of a space-time at the macroscopic level. At the microscopic level quantum effects become important. This is expected to occur at the Planck length, $l_p \approx 10^{-33}$ cm. This is the realm of quantum gravity. Here even the manifold structure may break down and the metric tensor may lose its meaning although the concept of curvature may remain. Much work is under way to explore these possibilities in a theory of quantum gravity. At this level one also expects gravitational interactions to be important in understanding the structure of the fundamental particles. To explore this relationship, higher-dimensional theories are being constructed to allow for the additional degrees of freedom. While these theories (string theory, supergravity, and Kaluza–Klein theories) may have the formal appearance of defining a space-time, they go beyond the above definition by encoding the properties of the fundamental particles into the extra dimensions of an internal space. This is both beyond the scope of this article and also at a very early stage of development.

See also Black Holes; CPT Theorem; Dynamics, Analytical; Invariance Principles; Lorentz Transformations; Michelson–Morley Experiment; Relativity, General; Relativity, Special Theory; Time; Twin Paradox.

Spectrophotometry

David L. MacAdam

The procedure used to measure transmittance or reflectance as a function of wavelength is called spectrophotometry.

Etymologically, spectrophotometry is a misnomer, because (a) photometry, literally and historically, means measurement of light, whereas spectrophotometry is not confined to the visible (light) region of the spectrum (~0.4 to ~0.7 μm); and (b) spectrophotometry measures not even radiant power but the ratio of transmitted or reflected power to incident power. Nevertheless, commercially available instruments that measure spectral transmittance and/or reflectance for wavelengths from 0.25 to 10 μm (or longer) are called spectrophotometers. They should not be confused with spectroradiometers, which measure radiant power (not ratios) as functions of wavelength.

Specialized research instruments, both spectrophotometers and spectroradiometers, have been built for use at shorter and longer wavelengths. Research instruments of these kinds are, more commonly than commercial instruments, called more appropriate names—spectroradiometers, spectral transmissometers, or spectroreflectometers.

Spectrophotometry is widely used in chemical analysis, metallurgy, mineralogy, biology, and medicine. The most sensitive, and most common, techniques of spectrophotometry for chemical analysis and medical applications involve measurement of spectral transmittance.

An increasingly frequent application of spectrophotometry is measurement of color and formulation of mixtures of dyes, paints, or pigments for production of colored materials. Most commercially available spectrophotometers for the visible range (~0.4 to ~0.7 μm) were designed for such applications. Most usually, color measurement requires measurement of spectral reflectance.

The great sensitiveness of people for detection of very slight differences of color, and their tendency to associate colors with quality, freshness, and desirability (or undesirability) of substances or things, impose severe requirements on the accuracy of spectrophotometry. For color measurement, the average wavelength of each measured spectral band should be accurate to ±0.5 nm, and reflectance should be accurate to ±0.2%.

Every spectrophotometer contains a source of radiant power more or less continuously distributed over the spectral range for which the instrument is designed. Because the instrument measures only the (inverse) ratio of the incident power and the power transmitted (or reflected) by the sample being measured, the spectral distribution of the built-in source is not important, so long as there is measurable power in each spectral band selected by the spectrophotometer for incidence on the sample.

The spectral band is selected by a dispersing component (the monochromator) of the spectrophotometer. In some spectrophotometers, the dispersing device is a prism (or two prisms in succession, with an intermediate slit to increase spectral purity).

Diffraction gratings, including replicas and holographs, are coming into more common use for dispersion. Linear-array detectors placed in the focal plane of the spectrum of the reflected (or transmitted) light eliminate moving parts, otherwise needed to scan the wavelength range. In combination with computers, such linear arrays can evaluate reflectances (or transmittances) of many wavelengths practically simultaneously. Thus the colors of rapidly changing substances can be recorded at relevant moments. Alternatively, the spectral reflectances, at many wavelengths, of unchanging substances can be remeasured many times in a short time and averaged, to increase accuracy. The number of wavelengths at which such reflectance data can be determined simultaneously is fixed by the number of detector elements in the diode array (e.g., 16, . . . , 1024 or more).

By use of suitable programs in the computer associated with the linear array, reflectances at different groups of wavelengths can be combined according to prescribed algorithms. Thus, color tristimulus values can be determined practically instantaneously from spectral reflectances, by use of the weighted-ordinate method, or the selected-ordinate method. The latter would make full use of a 1024 (or more) element array, by merely adding together, simultaneously, the outputs of three different groups of 100 or more different array elements. Determination of sets of tristimulus values for different illuminants (of the sample, not the source built into the spectrophotometer) and for a different standard observer would require only selection of other programs, which can run simultaneously (or nearly simultaneously). Of course, as is customary with earlier spectrophotometers, spectral reflectance data at predetermined wavelengths (e.g., at equal 20-nm, 10-nm, 5-nm, 2-nm, or 1-nm intervals, or any other fixed wavelengths, such as the weighted-average wavelengths of responses of the separate elements of the linear array) can be stored in the memory of the computer, for use later for computation of tristimulus (color) values for any subsequently chosen illuminant and observer, or even for nonvisual reactions to the light, such as erythemal or agricultural.

For color measurement, the bandwidth of the monochromator should not be greater than 10 nm. Some recent spectrophotometers, especially some designed for measurement of spectral transmittance, have bandwidths as small as 1 nm.

Spectral reflectance is measured by comparison of the energy reflected from the sample to that reflected from a standard (white) reflecting material. Opaque layers of MgO or $BaSO_4$ or Halon (a polyfluorocarbon) powder are commonly used for such working standards. By international agreement, reflectances are expressed as percent of the reflectance of a theoretical perfectly diffuse complete reflector, on the basis of which the spectral reflectance of the working standard can and should be calibrated.

For many materials, especially those with glossy or semiglossy surfaces, reflectance depends on the geometrical distributions of incident and collected reflected energy. Standard conditions of incidence and collection have been recommended by the International Commission on Illumination (CIE).

Instruments designed for measurement of spectral reflectance or spectral transmittance with any arbitrary angle of incidence of collimated light on the sample, and arbitrary (same or different) angle of reflection (whether in or out of the plane of incidence), are called goniospectrophotometers.

Some spectrophotometers record transmittance or reflectance continuously, as curves, versus wavelength of the incidence power. In some cases, as mentioned for linear arrays, reflectances at discrete wavelengths are transmitted to computers for on-line numerical analysis.

Spectrophotometers for industry and commercial color measurements usually incorporate computers that perform the calculations recommended by the CIE. They can be programmed to report any desired selection of derived results, such as the color difference from a previously measured standard, in internationally agreed-upon terms and units.

The stability of the reflectance-measuring component of the spectrophotometer should be sufficient to avoid any spurious values that might arise from the necessary time difference between reception of power reflected from the sample and that from the working standard. Most spectrophotometers used for color measurement alternate the incidence of power on the sample and on the standard at rates at least as high as 60 Hz, which greatly reduces errors due to slow drifts of light-source intensity or detector or amplifier efficiencies.

Fluorescent substances greatly complicate spectrophotometry. In principle, monochromatic energy should be incident on such samples; for unit radiant power of each incident spectral band (from 0.3 to 0.7 μm for color measurement), the complete spectral distribution of the reflected and fluorescently emitted power should be measured. This requires spectroradiometry, not spectrophotometry, of the power received from the sample. The results of such a measurement cannot be represented by a single curve; they form a family of curves of emitted power versus emitted wavelengths, the average wavelength of the incident spectral band being the parameter of each separate curve. Alternatively, excitation curves of emitted power versus average wavelength of incident power might be plotted, with average wavelength of the emitted power as the parameter. Because of the short duration of emission of many so-called fluorescent materials, accuracy of measurement cannot be assured by alternation ("flicker") of the radiant power incident on the working standard and sample.

The CIE has suggested a simplification for practical measurement of fluorescent materials by prescribing the spectral distribution of undispersed incident radiant power. Designated D65, this distribution is intended to represent average daylight. If a source that has an adequate approximation of D65 is used, a spectroradiometer (not a spectrophotometer) can measure the total of the reflected and emitted power.

Because of the commercial importance of fluorescent materials, which are increasingly used for outdoor signs, safety and rescue devices, and paper, fabric, and laundry brighteners, and the urgent need for measurements, many attempts have been made to simplify further the methods of measurement. Such simplifications, however, require assumptions and approximations, the validities of which are limited and applications of which are frequently doubtful.

See also VISIBLE AND ULTRAVIOLET SPECTROSCOPY; RADIOMETRY; REFLECTION.

BIBLIOGRAPHY

Deane B. Judd and Gunter Wyszecki, *Color in Business and Industry*, 3d ed. Wiley (Interscience), New York, 1975.

David L. MacAdam, "The Influence of Professor A. C. Hardy and His Spectrophotometer on Colorimetry," *CIE J.* **1**, 8 (1982).

Spin

Richard H. Sands

Spin refers to the intrinsic angular momentum of an object in the frame in which the center of mass is at rest. Classically one thinks of it as caused by a rotation of a macroscopic body about an axis through the center of mass; however, this classical view meets with difficulty when dealing with the elementary particles.

It is useful in understanding the extension of this concept of spin to the elementary particles to consider first a system of these particles, such as an atomic nucleus, which executes some motion as a whole. The internal state of the nucleus is not in general determined solely by the energy because the "internal" angular momentum I of the nucleus due to the angular momentum of the particles in their motion about the center of mass of the nucleus may still have various directions in space. We have no reason to suppose *a priori* that this variable is absent when the system is an elementary particle itself. Hence we see that in quantum mechanics the possibility of some "intrinsic" angular momentum S must be ascribed to an elementary particle and experiment must decide its value. It is unnecessary and, in fact, meaningless to attribute this angular momentum of an elementary particle to the rotation about an axis through its own center of mass. In accord with the Bohr correspondence principle, one would expect the latter interpretation to be true only in the limit of large quantum number but not for $S=\frac{1}{2}\hbar$.

For an elementary particle with mass M, charge Q, and intrinsic angular momentum (spin) S, symmetry requires that any associated magnetic moment $\boldsymbol{\mu}$ be parallel or antiparallel to $\mathbf{S}$ (Wigner–Eckart theorem). Dimensional analysis show that $\boldsymbol{\mu}$ will be proportional to $(Q/Mc)\mathbf{S}$. This may be expressed as an equality,

$$\boldsymbol{\mu} = g(Q/2Mc)\mathbf{S},$$

where the constant of proportionality, g, is dimensionless and is referred to as the g factor.

It is possible, in principle, to calculate the magnetic moment of an elementary particle from a knowledge of its internal structure and the interactions of the particle with its own fields; therefore, an experimental measurement of g provides a means of testing the validity of the theory. Such calculations and tests based on current models for the so-called "elementary" particles will be discussed later.

SPIN, ELECTRON

The hypothesis that an electron has intrinsic spin and magnetic moment was independently proposed by several scientists based on indirect evidence; for example, A. H. Compton in 1921 postulated that the phenomenon of ferromagnetism appeared to be associated with an intrinsic magnetic moment of the electron rather than with the orbital moment of the electrons and that the magnitude of the ferromagnetism required an effective charge-to-mass ratio substantially greater than e/m (i.e., an electron g factor considerably greater than unity). Although he was reasoning classically and hence erroneously, he noted that an electron spin of approximately $\hbar$ would account for the necessary magnetic moment.

The concept first attracted wide attention, however, in 1925 when G. Uhlenbeck and S. Goudsmit, by postulating that the electron had an intrinsic spin of $S=\frac{1}{2}\hbar$ and $g=2$, were able to use the Bohr–Sommerfeld quantum theory to explain the fine structure of atomic spectral lines and their behavior in applied magnetic fields.

Orbital angular momentum has the very interesting feature discovered by Bohr in 1913 that it is quantized in the sense that any changes in the component of angular momentum along one axis can occur only in integer units of a smallest value, $\hbar$. This also applies to spin. If we denote spin by S expressed in units of $\hbar$, then $S=\frac{1}{2}$ for each electron and the component of S along a magnetic field direction (z axis) has only two possible values: $m_z=+\frac{1}{2}$, $m_z=-\frac{1}{2}$. Note that to go from one value to another requires a change of one unit of $\hbar$.

The experimental fact that angular momentum is quantized was demonstrated in a simple and straightforward manner in 1922 by O. Stern and W. Gerlach, although the proper interpretation of that experiment in terms of electron spin did not occur for several years afterward. Stern and Gerlach designed and performed an experiment to measure the magnetic moment of individual silver atoms. They produced a beam of silver atoms by evaporating silver in a hot oven and letting some of the atoms escape through a line of small holes. This beam was directed (x direction) between the pole tips of a magnet designed to produce a large gradient of the magnetic field (z direction) perpendicular to the beam direction (see Fig. 1). In the classical theory, the magnetic energy of a given silver atom in the magnetic field is given by $U=-\boldsymbol{\mu}\cdot\mathbf{B}=-\mu_zB_z$. If B_z changes as a function of the vertical position z, then there will be a vertical force on each silver atom given by $F_z=-\partial U/\partial z=\mu\cos\theta(\partial B_z/\partial z)$, where θ is the angle between $\boldsymbol{\mu}$ and $\mathbf{B}$. Thus classically one would have predicted that because the silver atoms would emerge from the oven with their magnetic moments in random directions, all values of θ would occur and the beam would be spread vertically in a continuous fashion. What Stern and Gerlach observed, however, was that the silver atoms were separated vertically into two discrete groups, indicating that the z component of the magnetic moment was *quantized*. We

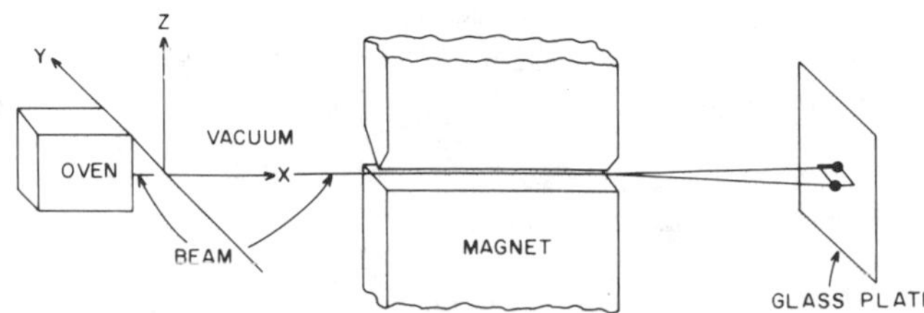

FIG. 1. Schematic drawing of apparatus for the Stern–Gerlach experiment.

now know that these two states correspond to the two possible values of the projection of the $S=\frac{1}{2}$ along the z direction; namely, $m_z=+\frac{1}{2}$ and $-\frac{1}{2}$. At the time, it was thought that these corresponded to the projections of orbital angular momentum, $L=1$, where $L_z=+1$ and -1 were the only allowed states. (The old quantum theory did not allow $L_z=0$.) This was a most amazing result! That the spin projections are quantized is an experimental fact which the physicist now records as a law of nature. Such recording does not detract from the amazement.

Spin angular momentum also serves as the distinguishing attribute for the division of all elementary particles into two basic families: All particles having half-odd integer angular momentum $\frac{1}{2}\hbar$, $\frac{3}{2}\hbar$,..., obey Fermi–Dirac statistics and are called *fermions*, and those having integer spin angular momentum $0,\hbar,2\hbar,\ldots$, obey Bose–Einstein statistics and are called *bosons*. This fundamental division of the elementary particles is based on the quantum theory wherein the wave function describing a collection of particles must be either symmetric or antisymmetric under the interchange of any two identical particles. If the wave function is antisymmetric (changes sign) under the interchange, then the particles are said to obey Fermi–Dirac statistics. The connection between the symmetry properties of the wave function and the statistics is most easily seen by noting that from the antisymmetry of the wave function for fermions, the wave function must vanish if the coordinates of the two identical particles assume the same value. This means physically that two such identical particles can never be found exactly at the same place. One can show that they are found mostly at distances larger than a characteristic distance d given by the expression $d{\sim}\hbar/p$, where p is the *relative* momentum of the particles. In other words, if one makes a six-dimensional phase space of three ordinary space coordinates and the three components of the momentum and considers cells of dimension d and momentum spread $\Delta p{\sim}h/d$, no two fermions may be found within one cell. This delimits the statistical distribution of such particles, whereas one may place as many bosons in one cell as desired. In determining whether two particles are identical one must specify the projection of the spin along some axis; thus the rule that no two fermions are to be found at a relative distance much less than $\hbar/p$ applies only if their spin projections are the same.

This connection of spin with particle statistics has many important consequences. One area where these consequences are readily apparent is that to which we have already alluded, namely, that of atomic structure and spectroscopy. Because of the importance of spectroscopy to the historical development of the concept of spin and of quantum mechanics, it is useful to summarize the important features connecting spin with atomic structure. In an atom containing several electrons the spin vectors s and orbital angular momenta l combine to give the total angular momentum J of the electronic structure; therefore, J will have integral values in units of $\hbar$ when the number of electrons is even and half-integral values when the number of electrons is odd. The relation between spin and statistics determines the chemical properties of the elements; i.e., the identity of two electrons is determined by the quantum numbers n, l, s, m_l, and m_s and because electrons are fermions, no two may be identical. This forces a shell structure on the atom, thus resulting in the periodic table of the elements.

Because the electron has a magnetic moment accompanying its spin, for a single electron the energy will be lower when l and s are in opposite directions ($j=l-s$) than when they are parallel ($j=l+s$); this follows from the fact that when the electron is moving around the positively charged (Ze) nucleus with a speed v at a distance r, it experiences a magnetic field $Ze\mathbf{r}\times\mathbf{v}/r^3c$ G, and the magnetic moment of the spinning electron has a different energy in the two positions it can assume with respect to this field. The resulting energy difference (doublet fine structure) is

$$\Delta E=h\Delta\nu=\frac{hcR\alpha^2Z^4}{n^3l(l+1)}\ \text{erg},$$

where R is the Rydberg constant and α is Sommerfeld's fine-structure constant:

$$R=2\pi^2me^4/\hbar^3c=109{,}737.31534(13)\ \text{cm}^{-1}$$

[strictly speaking, this value of R is only correct for a heavy nucleus surrounded by a single electron; for a nucleus of mass M one must replace m in the Rydberg constant by the reduced mass $\mu=mM/(M+m)$],

$$\alpha=e^2/\hbar c=1/137.0359979(33).$$

The energy E of an electron in a state with quantum numbers n, l, and j ($j=l\pm\frac{1}{2}$) including relativistic and spin effects is

$$E=-hc\frac{RZ^2}{n^2}+hc\frac{R\alpha^2Z^4}{n^3}\times\left[\frac{3}{4n}-\frac{1}{l+\frac{1}{2}}+\frac{j(j+1)-l(l+1)-s(s+1)}{2l(l+\frac{1}{2})(l+1)}\right]\ \text{erg}.$$

From this it can be seen that the total spin–orbit energy is usually quite small compared to the energy of the electron in the central Coulomb field but increases relatively with increasing nuclear charge. This expression for the spin–orbit energy includes an important relativistic correction (the Thomas factor of 2), which contributes an equal amount to the precession and the energy, as in the simple picture of a magnetic moment precessing in the magnetic field caused by its motion.

The type of coupling between s and l to give J depends on the relative size of the spin–orbit interaction energy of the individual electrons and the interaction between different electrons. In some atoms the interaction between the electrons is of such a nature that the total angular momentum J can be considered as the vector sum of a spin vector S, which is the resultant of all the spins s, and an orbital vector L, the resultant of all the individual orbital angular momenta l (so-called Russell–Saunders couplings). In other cases J is the resultant of the individual vectors $j=l+s$ of each electron (j,j coupling), and a variety of intermediate cases exist.

The angular momentum of the electronic structure of an atom is caused only by electrons in unfilled shells, because for closed electron shells the angular momentum vectors add up to zero. The spin accounts in detail quantitatively for the fine structure of atomic energy levels; e.g., for a single electron the spin causes all energy states (except those which have $l=0$) to be split into two levels (doublets); for two electrons outside closed shells S can have the values 0 and 1 and the levels are unsplit (singlets) or split into three (except for $l=0$) (triplets); and for higher values of S one obtains higher multiplets.

The g factor for an energy state of resultant spin S and orbital angular momentum L with resultant J is given by

$$g_J = 1 + \frac{J(J+1) - L(L+1) + S(S+1)}{2J(J+1)}.$$

The magnetic moment is $g_J J$ Bohr magnetons.

In 1927, P. A. M. Dirac showed that $S=\hbar/2$ and $g_S=2$ followed from a relativistically covariant formulation of Schrödinger's quantum theory. Following these developments, calculations by other workers using these properties of the electron led to a satisfactory understanding of many previously unexplained phenomena.

By 1947 experiments of increasing accuracy revealed small but significant discrepancies from the predictions of the Dirac theory. Willis Lamb and R. C. Retherford showed that the $2^2S_{1/2}$ and $2^2P_{1/2}$ states of hydrogen were separated by about 1000 MHz in contrast to their being degenerate as predicted by the Dirac theory. Again in 1947, precision measurements of the hyperfine structure intervals in hydrogen and deuterium by J. E. Nafe, E. B. Nelson, and I. I. Rabi showed a 0.2% discrepancy from predictions based on $g=2$. These experiments set the stage for the introduction of the present theory of quantum electrodynamics, which takes into account the interaction of the elementary particle with the surrounding electromagnetic field.

Shortly after these experiments G. Breit suggested that there should be a correction to the electron g factor of order α (the fine-structure constant) due to the above effect. By November of 1947, P. Kusch and H. M. Foley reported the first direct measurement of the electron g factor, $g=2.00238(10)$. In December 1947, J. Schwinger applied the concepts of mass and charge renormalization first suggested by H. A. Kramers to eliminate the divergences which led to earlier failures of quantum electrodynamics and obtained the lowest-order radiative correction to the electron g factor, $g=2(1+\alpha/2\pi)=2.00232$. Since that time, theory and experiment have continued to improve with the fantastic agreement to nearly $1:10^{11}$ and calculations underway to sixth order in α/π. The present measured value is

$$g = 2.002\,319\,304\,376(8).$$

SPIN, NUCLEAR

The spin of the proton was postulated by several scientists in the mid-1920s, but it was first demonstrated by D. M. Dennison in 1927 when he found that he could account for the curve of specific heat versus temperature for molecular hydrogen gas if he assumed that the proton had spin $\frac{1}{2}$. This eventually led to the discovery by Bonhoeffer and Harteck in 1929 of the two kinds of hydrogen molecule, *ortho-* and *para*-hydrogen, corresponding to the two proton spins being parallel or antiparallel, as postulated by Dennison.

Somewhat previously, W. Pauli had suggested that the hyperfine structure of atomic spectral lines might be caused by an angular momentum I of the nucleus and an accompanying magnetic moment. In 1927 Back and Goudsmit measured and analyzed the hyperfine structure of bismuth and its Zeeman effect and thus concluded that the bismuth nucleus had spin $I=\frac{9}{2}$ and demonstrated that Pauli's assumption was correct.

SPIN, ELEMENTARY PARTICLES

It is interesting to note that the presently accepted theory of the so-called elementary particles (proton, neutron, mesons, etc.) indicates that, except for the leptons and photons, they are made up of other particles called quarks. Thus the use of the word spin to denote the intrinsic angular momentum of the elementary particles is on the same footing as the use of spin to denote the total angular momentum I of the nucleus, which is the resultant of the spin of the protons and neutrons and their orbital angular momenta inside the nucleus. Efforts to date to estimate the magnetic moments (g factors) of the elementary particles other than the electron and positron have met with varied success (~1 to 20% even with optimum parameter fitting). This merely indicates that calculations within the framework of the new theory, quantum chromodynamics, are much more difficult than for quantum electrodynamics.

The spins and magnetic moments of the elementary particles are shown in Table I.

SPIN–SPIN INTERACTION

The connection of spin with particle statistics, mentioned earlier in this article, permits a portion of the electrostatic Coulomb interaction between two fermions to be expressed in terms of their relative spin orientations, namely $\mathcal{H}=2\mathcal{J}\mathbf{S}_1\cdot\mathbf{S}_2$. This electrostatic interaction, written in this form, is known as the *spin–spin* or *exchange interaction*. One may choose to include magnetic interactions as well, in which case $\mathcal{J}$ becomes a tensor and the interaction may be written $\mathcal{H}=-2\mathbf{S}_1:\mathcal{J}:\mathbf{S}_2$. Such interactions are important in ferromagnetism, but space does not permit further discussion here.

Spin effects have been observed in high-momentum-transfer collisions between spin-polarized protons at 10–30-GeV energies available at the Argonne and Brookhaven National Laboratories. These spin effects have been tentatively as-

Table I Table of spins

	S	μ	
e	$\frac{1}{2}$	1.00115965241(20)	$e/2m_ec$
μ	$\frac{1}{2}$	1.001165922(9)	$e/2m_\mu c$
π	0		
p	$\frac{1}{2}$	2.7928456(11)	$e/2m_pc$
n	$\frac{1}{2}$	−1.91304211(8)	$e/2m_pc$
Λ^0	$\frac{1}{2}$	−0.6138(47)	$e/2m_pc$
Σ^+	$\frac{1}{2}$	2.33(13)	$e/2m_pc$
Σ^-	$\frac{1}{2}$	−1.46(25)	$e/2m_pc$
Σ^0	$\frac{1}{2}$	No measurement	
Ξ^0	$\frac{1}{2}$	−1.22(2)	$e/2m_pc$
Ξ^-	$\frac{1}{2}$	−1.85(75)	$e/2m_pc$
Ω^-	$\frac{3}{2}$	No measurement	
Neutrino	$\frac{1}{2}$	No measurement	
Photon	1		

signed by some to spin–spin interactions between constituent particles within the proton (e.g., quarks, which are believed to be fermions according to the theory of quantum chromodynamics).

See also Bose–Einstein Statistics; *CPT* Theorem; Electron Spin Resonance; Elementary Particles; Fermi–Dirac Statistics; Magnetic Moments; Nuclear Magnetic Resonance; Quantum Electrodynamics; Rotation and Angular Momentum; Quantum Chromodynamics.

Stark Effect *see* Zeeman and Stark Effects

Statics

Norman Feather

The subject of dynamics (or mechanics), which deals generally with the motion of bodies under the influence of forces, is traditionally subdivided into *statics* and *kinetics*. Statics deals with all those cases in which the motion is unaccelerated; kinetics treats essentially of accelerated motion.

The modern concept of force originated with Newton; the subject kinetics, therefore, effectively took shape with the publication of the *Principia* (1687). The subject of statics, however, insofar as it deals with bodies at rest, has a much longer history. The Greek philosophers entertained intuitive notions of force, and Archimedes (287–212 B.C.) in particular, in his consideration of the lever, enunciated the "law of moments," which finds a place in the modern formulation. Similarly valid contributions were made at a later date by Leonardo da Vinci (1452–1519) and Simon Stevin (Stevinus, 1548–1620).

After Newton, there was no logical reason why the two subjects, statics and kinetics, should remain distinct—the former was seen as no more than a special case of the latter—but tradition and convenience have maintained the distinction between them. Moreover, the civil engineer is very intimately concerned with the effects of forces in structures at rest; applied statics, therefore, provides a substantial component of his professional training.

In modern statics the validity of Newton's third law is accepted at the outset: It is accepted that, as between two bodies, the only forces that can operate are paired forces, each such pair consisting of equal forces, acting in opposite directions along the same straight line, between one of the "ultimate particles" of one body and one of the ultimate particles of the other. In any situation, therefore, any body, according to this analysis, is acted on by a multitude of forces. In order to make progress, then, it is first necessary to simplify this description. To do so consistently it is necessary to introduce the fiction of the ideal rigid body. It is assumed that any solid body under consideration is completely immutable as to size and shape under all conditions. (As occasion arises, it is also assumed that all strings and ropes are completely flexible and inextensible.)

The simplest situation occurs when the lines of action of the "component" forces acting from outside on the ultimate particles of a body are parallel. Then it may be shown that the overall effect is the same as would be produced by a single "resultant" force of magnitude equal to the algebraic sum of the magnitudes of the actual forces, and with a line of action (parallel to the actual lines of action) that is completely determinate. A special case occurs when the component forces, acting on all particles of the body in the same sense along parallel lines, are in magnitude proportional to the masses of the particles. This is the situation in relation to the gravitational forces that constitute the weight of a (not too large) body near the surface of the earth. Then the line of action of the resultant passes through a unique point in the body, whatever the body's orientation. In this connection this point is referred to as the "center of gravity" of the body. Essentially this result was deduced by John Wallis in 1671, although the concept of mass had not at the time been fully developed by Newton. The position of the center of gravity is uniquely determined by the distribution of mass in the body; for that reason, and with wider relevance, the point is designated the center of mass of the body.

Suppose now that the external forces acting on a body reduce to two equal forces acting in opposite senses along parallel straight lines. This situation is not intelligibly covered by the previous result, which would point to a zero-magnitude resultant acting along a line at infinity. Obviously, however, such a pair of forces must tend to turn the body about an axis perpendicular to the plane in which they act. This turning effort is referred to as the "moment of the couple" constituted of the two forces, and its magnitude is given by the product of either force and the distance separating their lines of action. It is a property of the two forces constituting a couple that the sum of the individual moments of the two forces about any axis in the body perpendicular to their plane of action is constant.

In the most general case of a rigid body acted on by external forces of arbitrary specification, those forces may be shown to reduce to a single force (say, F) acting along a determinate line in the body (Poinsot's central axis) and a single couple (say, of moment G) having its axis parallel to that line. Such a force–couple pair is referred to as a "wrench." Frequently, specification of a wrench is carried out in terms of $\mathbf{F}$ (a localized vector quantity) and p ($\equiv G/F$), a linear parameter referred to as the "pitch" of the wrench. In the two special cases already described, the pitch of the wrench-resultant is zero and infinite, respectively.

Statics treats of bodies under the influence of external forces so disposed that they have no accelerating effect, in respect either of linear or of rotational motion. In such circumstances both the force and the couple of the wrench-resultant acting on the body must be zero. In the practical development of the subject, this condition is usually expressed in terms of force components. Rectangular axes are taken, arbitrarily oriented through an arbitrary point in the body, and it is laid down (a) that the algebraic sum of the force components shall be zero when taken in the direction of each axis in turn, and (b) that the algebraic sum of the moments of the force components shall be zero about each axis.

There is no logical reason why the subject of statics should be concerned with solid bodies exclusively: Isolated volumes of liquid or gaseous matter are bodies in the physical sense. Traditionally, the subject of hydrostatics treats of the equilibrium behavior of such systems. Their characteristics, therefore, have not been referred to in this article.

See also HYDRODYNAMICS; KINETICS AND KINEMATICS; NEWTON'S LAWS.

BIBLIOGRAPHY

H. Lamb, *Statics*, 3d ed. Cambridge Univ. Press, London and New York, 1928. (I)

F. Miller, Jr., *A Student's Introduction to Physics*, Chap. 5. Harcourt, New York, 1959. (E)

E. J. Routh, *A Treatise on Analytical Statics*, 2 vols. Cambridge Univ. Press, London and New York, 1892.

Statistical Mechanics

J. R. Dorfman

The aim of statistical mechanics is to predict the properties of systems composed of large numbers ($\sim 10^{23}$) of particles in terms of the mechanical properties of the individual particles and of the forces between them. Statistical mechanics can be divided into two main areas, each of which attempts to explain a body of experimental results for macroscopic systems. These areas are (1) *equilibrium statistical mechanics,* often called *statistical thermodynamics,* which deals with derivation of the laws of thermodynamics and calculation of the values of the thermodynamic functions and other equilibrium properties of systems in terms of their microscopic properties; and (2) *nonequilibrium statistical mechanics,* which deals with the approach of systems to thermodynamic equilibrium, derivations of the macroscopic transport equations, such as the equations of fluid dynamics, and calculations of the values of the transport coefficients and other nonequilibrium properties of systems in terms of their microscopic properties.

Statistical mechanics was developed in the nineteenth century, largely by the work of Maxwell, Boltzmann, and Gibbs. It is based on the following central idea: Suppose a person is interested in computing some property of a system consisting of a large number of particles that depends in some way on the dynamical processes in the system, e.g., the force in some direction exerted on a piston placed in a fluid system. It is not possible to solve the dynamical equations of motion for this system in order to calculate this property exactly, since the equations are in general much too complex, and the initial conditions required to determine the exact solution are usually not known. Instead, the property of interest is computed by determining its *average* value, taken over a suitable ensemble of similarly prepared systems. Statistical mechanics is successful in predicting properties of large systems because (1) the values of most of the quantities of experimental interest are not sensitive to the particular microscopic state of the system, and (2) the ensemble average value is usually much easier to compute than the exact value. Fluctuations about the average value are usually small provided the system is sufficiently large and the ensemble has been properly constructed. The fact that the system is large, while leading to enormous complexity on a microscopic scale, leads to regular, simply describable behavior on a macroscopic scale.

To construct an ensemble for a particular calculation, we consider an arbitrarily large number of independent replicas of the system, which satisfy the same macroscopic constraints as the system but may differ in their microscopic states. For example, the system might be a gas that is isolated from its surroundings and whose macroscopic state at a given instant is defined by the condition that the gas particles are all to be found in one corner of the container. The ensemble would include each of the possible microscopic states of the system that satisfy these macroscopic conditions.

The microscopic states available to the system are described by specifying all the possible quantum states available to the system at the initial instant, with the given constraints, by the set of wave functions $\{\psi_i(\mathbf{q})\}$ corresponding to these states. Here $\mathbf{q}$ represents all the variables—positions or momenta, spins, etc.—that appear in the wave function. Then to form an ensemble average of some quantity, a statistical weight W_i must be assigned to each of these quantum states. This set of wave functions is determined by the conditions imposed at the initial instant, but the wave functions will vary with time according to Schrödinger's equation so that after a time t, $\psi_i(\mathbf{q})$ will have been transformed to $\psi_i(\mathbf{q},t)$. The ensemble average of any dynamical quantity $\mathbf{F}$ will, in general, be time dependent, and we denote it by $\langle F(t)\rangle$. All such ensemble averages can be expressed in terms of a time-dependent density matrix $\rho_{kl}(t)$ as

$$\langle F(t)\rangle = \sum_i W_i \int d^N q\, \psi_i^*(\mathbf{q},t)\mathbf{F}\psi_i(\mathbf{q},t) = \mathrm{Tr}(\boldsymbol{\rho}\mathbf{F}) = \sum_{k,l} \rho_{kl}(t)F_{lk}$$

where F_{lk} is independent of time and is given by

$$F_{lk} = \int d^N q\, \phi_l^*(\mathbf{q})\mathbf{F}\phi_k(\mathbf{q}),$$

$\phi_k(\mathbf{q})$ form a complete set of wave functions for the system, which are summed over in forming the trace, and $\rho_{kl}(t)$ is given by

$$\rho_{kl}(t) = \sum_i W_i(a_l^{(i)}(t))^* a_k^{(i)}(t).$$

Here $a_k^{(i)}(t)$ are the expansion coefficients for the wave functions $\psi_i(\mathbf{q},t)$ in terms of $\phi_k(\mathbf{q})$; i.e.,

$$\psi_i(\mathbf{q},t) = \sum_k a_k^{(i)}(t)\phi_k(\mathbf{q}).$$

Using the fact that $\psi_i(\mathbf{q},t)$ satisfies Schrödinger's equation, we can show that $\rho_{kl}(t)$ satisfies the equation

$$\frac{\partial \rho_{kl}(t)}{\partial t} = \frac{i}{\hbar}\sum_n (\rho_{kn}H_{nl} - H_{kn}\rho_{nl})$$

where

$$H_{nl} = \int d^N q\, \phi_n^*(\mathbf{q})\mathbf{H}\phi_l(\mathbf{q})$$

and $\mathbf{H}$ is the Hamiltonian operator of the system. This equation for ρ_{kl} is called the quantum-mechanical Liouville equation.

For many purposes the full quantum treatment of the microscopic states of the system is not required. For example, we can use classical mechanics to treat the translational motion of the particles in gases and liquids at room temperature, and quantum mechanics to treat their internal states. Where there is a very slow exchange of translational and internal energies, a treatment of many of the bulk properties of the fluid can be based on classical statistical mechanics, either by considering the classical limit of the quantum ensemble described above, or by using classical ensembles from the beginning.

To set up a classical ensemble, we consider a collection of N particles, each with f position and f momentum coordinates, and set up a coordinate space, called a phase space or Γ *space*, with $2Nf$ dimensions. One point in Γ space then specifies the position and momentum for every particle of the system at one instant of time. The time development of the system is represented by a trajectory in Γ space, which is determined by the solutions of the classical equations of motion for the system. An ensemble is constructed by considering all possible replicas of the system that satisfy the same macroscopic conditions at the same initial instant of time, and assigning a probability distribution $\rho(\Gamma)$ such that $\rho(\Gamma)\, d\Gamma$ is the probability that one of the replicas, chosen at random, will be in the $2Nf$-dimensional volume element $d\Gamma$ about the point Γ in Γ space at the initial instant. The fact that each of the systems in the ensemble develops in the course of time implies that the probability distribution function for finding a system in a given region of phase space will in general be time dependent, having the value $\rho(\Gamma)$ only at the initial instant. It can easily be shown from the classical equations of motion that this time-dependent distribution function $\rho(\Gamma)$ satisfies the equation

$$\frac{\partial\rho(\Gamma,t)}{\partial t}+\{\rho(\Gamma,t),H\}=0$$

where H is the classical Hamiltonian for the N-particle system and $\{\rho(\Gamma,t),H\}$ is the Poisson bracket of ρ and H. This equation is called the classical Liouville equation.

It follows from both the classical and quantum versions of Liouville's equation that $\rho(\Gamma)$ or ρ_{kl} will not depend on time if it is a function only of the conserved quantities of the motion, or in the quantum case, of the matrices of the conserved quantities. Such distribution functions or density matrices describe *stationary ensembles*.

The methods of equilibrium statistical mechanics are based on the following stationary ensembles. We consider an isolated system whose energy lies in the range E to $E+\delta E$. Then we assign equal weights to all quantum states of the system whose energies lie in the desired range, i.e.,

$$W_i=\frac{1}{\Omega}\quad \text{if } E\le E_i\le E+\delta E$$
$$=0\quad \text{otherwise}$$

where Ω is the total number of quantum states with energies in the given range. For a classical system, we choose the phase-space distribution function $\rho(\Gamma)$ to have the form

$$\rho(\Gamma)=\begin{cases}\text{constant} & \text{for}\quad E\le H(\Gamma)\le E+\delta E\\ 0 & \text{otherwise}\end{cases}$$

where $H(\Gamma)$ is the total energy of the system at the phase point Γ. The ensembles just described are called the quantum and classical *microcanonical ensembles*, respectively. The principal justification for the use of these ensembles is that the macroscopic properties of systems calculated by using them are entirely in accord with experimental results. However, it should be possible to base their use on the fundamental principles of mechanics. One approach, *ergodic theory*, tries to show that the average of any dynamical quantity of an isolated system taken over a sufficiently long time will be equal to the value predicted on the basis of the microcanonical ensemble; recently Ya. G. Sinai announced a proof of this for a simplified classical system of hard spheres, but a general proof for more realistic systems or for quantum systems is not available. Another approach forms the subject of nonequilibrium statistical mechanics and will be discussed further on. At any rate, the microcanonical ensembles provide a simple statistical description of isolated systems, and they predict that the probability of observing a particular macroscopic state of the system is proportional to the number of different microscopic states in which the macroscopic state can be realized, which is a reasonable starting point for a statistical theory of equilibrium.

The microcanonical distribution is used to give a statistical description of the thermodynamics of systems when the total energy E, volume V, and number of particles of each species α, N_α, are specified. In addition, it may be necessary to specify certain intensive quanties, denoted by $z_1,\ldots,z_j$, such as the values of external electric, magnetic, and gravitational fields, to determine completely the thermodynamic state of the system. The connection between thermodynamics and statistical mechanics is given in the microcanonical ensemble by the relation

$$S(E,V,\{N_\alpha\},z_1,\ldots,z_j)=k_B\ln\Omega(E,V\{N_\alpha\},z_1,\ldots,z_j).$$

Here S is the thermodynamic entropy as a function of E, V, $\{N_\alpha\},z_1,\ldots,z_j$; k_B is a numerical constant, called the Boltzmann constant,

$$k_B=1.38044\times10^{-16}\text{ erg deg}^{-1};$$

and $\Omega(E,V,\{N_\alpha\},z_1,\ldots,z_j)$, the *microcanonical partition function*, is equal to the total number of quantum states of the system with energies from E to $E+\delta E$. If the system is large enough, the entropy defined this way is not sensitive to the width δE of the energy shell. The statistical expressions for the other thermodynamic quantities can be obtained by using ordinary thermodynamic relations. For example, the thermodynamic temperature T is given by

$$T^{-1}=(\partial S/\partial E)_{V,\{N_\alpha\},z_1,\ldots,z_j}.$$

The prescription for taking the classical limit of the relation between the entropy and the number of quantum states is to use the fact that in the classical limit there are

$$\prod_\alpha (N_\alpha!)^{-1}\frac{d\Gamma}{h^{Nf}}$$

quantum states in a region of $d\Gamma$ in phase space, where h is Planck's constant. The factors $(N_\alpha!)^{-1}$ take into account that for identical particles, a permutation of particle indices does not lead to a new quantum state.

The fundamental relation between the macroscopic and statistical formulations of thermodynamics, $S=k_B\ln\Omega$, provides a microscopic interpretation for the fact, embodied in the second law of thermodynamics, that the entropy of an isolated system increases whenever any of the constraints imposed on the system are relaxed without changing the energy, and the system spontaneously proceeds to a new equilibrium state. Very simply, the relaxing of the constraint has the effect of increasing the number of quantum states available.

The microcanonical ensemble can be used to obtain a statistical-mechanical description of a system in contact with a heat bath, with which it can exchange energy or energy and particles. The procedure is to regard the system plus the heat bath as a large isolated system and use the microcanonical distribution for the whole system. If the system exchanges only energy with the heat bath, the probability P_i that the system is in a particular quantum state i with energy E_i is given by

$$P_i = Z^{-1} e^{-\beta E_i}$$

with $Z = \Sigma_i \, e^{-\beta E_i}$, and the summation is carried out over all quantum states of the system. This distribution is called the canonical distribution, and Z is the *canonical partition function*. Here it is assumed that the system interacts only weakly with the heat bath; i.e., the perturbations on the E_i due to the system's interactions with the heat bath can be neglected. The E_i are taken to depend on $\{N_\alpha\}, V$, and $z_1, \ldots, z_j$. Here β is a quantity that is characteristic of the heat bath, and the heat bath is supposed to be so large that it can absorb any amount of energy from the system without any essential changes in its properties. The connection between the thermodynamic properties of systems in contact with heat reservoirs and the canonical partition function is given by

$$F(T, V, \{N_\alpha\}, z_1, \ldots, z_j) = -k_B T \ln Z$$

where $\beta = (k_B T)^{-1}$, F is the Helmholtz free energy of the system, and T is the thermodynamic temperature. In the classical limit

$$Z \to \prod_\alpha (N_\alpha!)^{-1} \frac{1}{h^{Nf}} \int d\Gamma \, e^{-\beta H(\Gamma)}$$

and the relation between Z and F remains the same.

In computing the canonical partition function, the summation is over all the quantum states of a system, not over all the energy levels; thus each energy level is weighted according to its degree of degeneracy in carrying out the sum. Further, we must take into account the quantum statistics obeyed by the particles that make up the system when enumerating the allowed quantum states. If nuclear or chemical reactions are taking place in the system, the set $\{N_\alpha\}$ is defined to be the numbers of the constituent particles that are unaffected by the reactions (e.g., protons, neutrons, electrons, or atoms of each element), and the numbers of the reacting chemical or nuclear species are regarded as thermodynamic parameters fixed by the equilibrium condition.

If the system has a fixed volume V but can exchange energy and particles with the heat bath, and the heat bath is sufficiently large, we can use the microcanonical distribution for the system plus the heat bath to show that the probability of finding the system in a state where it has N_α particles of each species α and is in quantum state i with energy $E_i(V, \{N_\alpha\}, z_1, \ldots, z_j)$ is given by

$$P(E_i, \{N_\alpha\}) = \Xi^{-1} \exp(\beta \sum_\alpha N_\alpha \mu_\alpha) \exp(-\beta E_i)$$

where

$$\begin{aligned} \Xi = & \sum_{N_1=0}^{\infty} \cdots \sum_{N_k=0}^{\infty} \exp(\beta \sum_\alpha N_\alpha \mu_\alpha) \sum_i \\ & \qquad \exp[-\beta E_i(V, \{N_\alpha\}, z_1, \ldots, z_j)] \\ = & \sum_{N_1=0}^{\infty} \cdots \sum_{N_k=0}^{\infty} \exp(\beta \sum_\alpha N_\alpha \mu_\alpha) Z(T, V, \{N_\alpha\}, z_1, \ldots, z_j). \end{aligned}$$

Here there are k distinct types of particles. This distribution is the *grand canonical distribution*, and Ξ, the corresponding partition function. Again, β and $\{\mu_\alpha\}$ are parameters characteristic of the reservoir. The grand canonical distribution is used to derive the thermodynamic properties of open systems, and the relation between the statistical and thermodynamic description of such systems is provided by

$$P(T, V, \{\mu_\alpha\}, z_1, \ldots, z_j) V = kT \ln \Xi$$

where P is the pressure, $T = (k_B \beta)^{-1}$ is the temperature, and the μ_α are the chemical potentials of the species α. It is generally believed that if the system is large enough, its properties should not depend sensitively on whether it is exchanging particles and/or energy with a heat bath. Consequently, the calculated values of the thermodynamic properties of a large system should be the same regardless of the ensemble used in their computation. This is borne out by explicit calculations, but there is no general proof of this *equivalence of ensembles* for all systems of interest.

Finally, we are interested in computing values of the thermodynamic functions for systems that are large enough that the effects of the boundaries can be neglected. Then we consider the *thermodynamic limit* where for the microcanonical ensemble $E \to \infty$ and $V \to \infty$ in such a way that the energy per unit volume E/V remains fixed at the value of the finite system; for the canonical ensemble $V \to \infty$ and the numbers $N_\alpha \to \infty$ in such a way that the particle densities N_α/V remain fixed; and for the grand canonical ensemble, $V \to \infty$, and the mean numbers $\langle N_\alpha \rangle \to \infty$ such that the mean densities $\langle N_\alpha \rangle / V$ are fixed. The quantities of interest in the limiting situation are the specific values of the thermodynamic functions, such as the free energy per particle. There is a considerable literature detailing the microscopic properties a system must have for its specific thermodynamic properties to exist in the thermodynamic limit, and establishing the equivalence of various ensembles for calculating the specific quantities. The conditions on the microscopic properties, such as the interaction potentials between particles, are sufficiently general that the existence of the thermodynamic limit is assured for a wide variety of systems of physical interest.

The simplest applications of equilibrium statistical mechanics are calculations of thermodynamic properties of systems of noninteracting particles, e.g., dilute gases of monatomic or polyatomic molecules, blackbody radiation (where the particles are photons), and harmonic crystals (where the particles are phonons, i.e., the quantized lattice vibrations). It is more difficult to compute the properties of systems of interacting particles, but there have been some notable successes. For example, to derive the thermodynamic properties of dense gases, we must take into account the interactions of the gas molecules. It is possible to expand the pressure of the gas or other thermodynamic quantities as a power series in the density, the *virial expansion*. The first term gives the ideal, noninteracting gas result, the second term takes into account interactions between two particles, and the successively higher-order terms take into account interactions between three, four, ... particles.

One of the major areas of research in statistical mechanics is to extend the theory beyond dense gases so as to include the gas–liquid phase transition, the liquid state, the liquid–solid phase transition, and the solid state beyond the harmonic crystal approximation. For the gas–liquid phase transition it is possible to make quantitative predictions about the behavior of the thermodynamic functions near the critical point. Moreover, it has been shown that the gas–liquid phase transition has many properties in common with other phase-transition phenomena, such as the paramagnetic–ferromagnetic phase transition in metals and phase transitions in binary liquid mixtures. The theories for these phase transitions indicate that near their critical point, there are correlations between particles extending over distances considerably larger than the average interparticle spacing or the range of the interparticle potential. These correlations are responsible for the collective behavior of the system necessary for a change of phase. Although these correlations are propagated by the interparticle potential, they are not very sensitive to the details of the potential, so that many qualitatively different systems have similar phase-transition properties. Although considerable progress has been made in computing equilibrium properties of fluids, a great deal remains to be done; the liquid–solid phase transition is still only poorly understood, and our understanding of liquid structure, even for simple nonpolar liquids, is far from complete.

The theories for simple systems, and especially for the liquid and solid states, have been considerably aided by the use of computers. In addition to the fact that many difficult calculations are now a routine matter for computers, the use of computer simulations of molecular dynamics has led to a deeper understanding of the properties of fluids and solids.

Some of the most fruitful applications of statistical mechanics have been in explaining the electronic properties of solids. Many properties of metallic conductors and semiconductors can be understood on the basis of the statistical mechanical theory of Fermi–Dirac particles, the electrons and holes, in a solid. The theories for superconductors and superfluids are applications of statistical mechanics to various quantum systems. Among the many other problems that have been successfully treated within equilibrium statistical-mechanical theory are the conformations of polymer molecules in solution, the denaturation of protein molecules, the helix–coil transition in DNA-type molecules, and the properties of magnetic systems of various types, of nuclear matter, and of stellar matter.

We now turn our attention briefly to the statistical mechanics of nonequilibrium processes. Nonequilibrium statistical mechanics deals with the time development of an ensemble of similar systems described at some initial time by a nonequilibrium distribution function or density matrix, and attempts to describe the average behavior of this ensemble. The basic equation for describing the time dependence of the ensemble distribution function $\rho(\Gamma,t)$ or of the density matrix is the Liouville equation. The basic problem in the nonequilibrium theory is to describe the approach of a system to equilibrium, as well as the behavior of time-dependent fluctuations about equilibrium. Since it is well known from experiments that in a wide variety of circumstances, the approach of a system to equilibrium is governed by the hydrodynamic equations, or in general by the relations of nonequilibrium thermodynamics, a major goal of nonequilibrium statistical mechanics is to provide a microscopic counterpart to these macroscopic equations.

In the main, two approaches have been developed for the treatment of this problem. One approach applies to dilute or dense gases, and is part of the subject matter of kinetic theory of gases. In this approach we try to develop an equation for the single-particle distribution function $f(\mathbf{r},\mathbf{v},t)$, defined so that $f(\mathbf{r},\mathbf{v},t)\, d^3r\, d^3v$ is the number of particles of the gas in the region d^3r about $\mathbf{r}$ and d^3v about $\mathbf{v}$ at time t. The major contribution to the theory of this function was made by Boltzmann, who used plausible arguments about the number of collisions taking place in the gas in a small time interval to derive an equation for $f(\mathbf{r},\mathbf{v},t)$ in the case that the gas is dilute. This equation, the *Boltzmann transport equation*, predicts that for an isolated system $f(\mathbf{r},\mathbf{v},t)$ decays in the course of time to its equilibrium value, and allows the construction of a nonequilibrium analog for the thermodynamic entropy, the negative of the so-called H function, which increases monotonically in time as the system approaches equilibrium. Moreover, we can derive the hydrodynamic equations for a dilute gas together with expressions for the associated transport coefficients, such as the coefficients of thermal conductivity and viscosity, in terms of the intermolecular potential. These expressions describe the experimental results for dilute gases so well that they are used now to test various intermolecular potential models. In spite of the considerable success of the Boltzmann equation, a complete theory of nonequilibrium processes in gases requires that we be able to derive the Boltzmann equation from the Liouville equation directly, i.e., to replace the plausible argument about the number of collisions by a plausible construction of an initial ensemble, and to extend the Boltzmann equation to higher orders in the density so as to treat dense gases. Such a derivation is necessary in order to understand how we can reconcile the irreversible behavior of macroscopic systems with the reversible microscopic equations of motion.

There have been satisfactory derivations of the Boltzmann equation from the Liouville equation. However, the extension of the Boltzmann equation to higher orders in the density is quite involved and not yet completely clear; e.g., the transport coefficients of a dense gas cannot be expanded in a power series in the density; instead, logarithmic terms appear. Computer simulation of molecular dynamics has considerably aided the development of the kinetic theory of dense gases by allowing a detailed examination of the dynamical processes taking place in the gas and by providing quantitative information as to the effects of these processes.

The other approach to the theory of nonequilibrium processes that has been especially successful is called the time correlation function theory. This approach seeks special solutions of the Liouville equation for a system that is close to a total equilibrium state from which it has been perturbed by an imposed external field or a gradient in the velocity or temperature of the system. One then studies the behavior of this perturbed equilibrium system with time; shows that, on the average, the hydrodynamic equations are obeyed; and obtains explicit expressions for the transport coefficients. The expression for a typical transport coefficient τ for a classical system has the form

$$\tau = \int_0^\infty dt \, \langle J(0)J(t) \rangle .$$

Here J represents an N-particle current that depends on the positions and velocities of all N particles; $J(0)$ is the value of this current at some initial time $t=0$; $J(t)$ is its value a time t later, obtained by solving the equations of motion for the N-particle system; and the angular brackets denote an average over an equilibrium canonical ensemble, taken in the thermodynamic limit. A similar expression can be given for quantum systems. These expressions are very similar to the Einstein formula for the diffusion coefficient of a particle undergoing Brownian motion. Since the derivations of the time correlation function formulas are quite general, they apply to a wide variety of systems. However, because we must use the dynamics of an N-particle system to evaluate them, explicit calculation of these formulas presents a formidable problem, similar to but more difficult than the evaluation of the partition function. Nevertheless, approximation methods have been developed to evaluate these functions for a number of systems, particularly gases and liquids. Moreover, these functions are particularly suited for computer studies, where the computer follows the dynamics of the system and calculates $\langle J(0)J(t) \rangle$ directly. In addition, the functions can be studied experimentally by means of light-scattering or neutron-scattering methods. Each of these various approaches—theory, computer studies, and experiment—stimulates progress in the other two areas.

To conclude this brief description of nonequilibrium statistical mechanics, we mention that a very basic understanding of the foundations of statistical mechanics is likely to develop from modern research on ergodic theory as well as on nonlinear dynamics and chaos. Some important areas in which new and interesting applications of nonequilibrium statistical mechanics have been made include such topics as crystal growth and pattern formation, the theory of electric conduction in solids, flow properties of superconductors and superfluids, the dynamics of polymers, and the theory of coagulation and polymerization. See the appropriate entries in this encyclopedia for further details.

See also CRITICAL POINTS; EQUATIONS OF STATE; ERGODIC THEORY; *H* THEOREM; KINETIC THEORY; PHASE TRANSITIONS; QUANTUM STATISTICAL MECHANICS; THERMODYNAMICS, EQUILIBRIUM; THERMODYNAMICS, NONEQUILIBRIUM.

BIBLIOGRAPHY

David Chandler, *Introduction to Modern Statistical Mechanics*. Oxford University Press, New York, 1987.

J. Kestin and J. R. Dorfman, *A Course in Statistical Thermodynamics*. Academic Press, New York, 1971.

R. Kubo, *Statistical Mechanics*. North-Holland, Amsterdam, 1967.

L. D. Landau and E. M. Lifshitz, *Statistical Physics*, 2nd ed. Addison-Wesley, Reading, Mass., 1969.

R. Kubo *et al.*, *Statistical Physics*. Springer-Verlag, Berlin, 1985 (2 volumes).

F. Reif, *Fundamentals of Statistical and Thermal Physics*. McGraw-Hill, New York, 1965.

R. C. Tolman, *The Principles of Statistical Mechanics*. Oxford, London and New York, 1939.

Statistics

Morris H. DeGroot and Paul Shaman

Statistics is the science concerned with making inferences and decisions under conditions of uncertainty or partial ignorance. Broadly speaking, the subject of statistics includes the design of effective experiments as well as the measurement, collection, and analysis of experimental data.

Statistical theory is founded on the modern theory of probability and deals with the development and justification of methods for use in experimentation, inference, and decision making. In actual practice, the precise dictates of statistical theory are often followed somewhat loosely, and statistics is aptly termed an art as well as a science.

Statistics has become increasingly important and relevant in scientific work, especially with the advent of new areas of experimentation and the widespread use of high-speed computers. In the past, experimental physical data of high precision could often be obtained inexpensively, and simple statistical methods sufficed. Today, in areas such as high-energy physics, experimental data are often very costly and may be in short supply, may lack desired precision, and may be distorted by extraneous noise. As a result, the use of sophisticated methods of statistical analysis can be helpful.

An experiment that leads to the collection of data is central to our discussion, and the initial concern must be the plan of the experiment. A good experimental design should be simple enough to be understandable to others and capable of execution within time and budgetary limits. It should permit the application of well-understood statistical methods in the analysis of the data and, of primary importance, it should allow the user to draw inferences that directly relate to the goals of the experiment. The design will usually specify what types of observations to record and how many to take.

Statistics comes into play when experimental outcomes are regarded as being governed by probability distributions over sets of possible values. A common description of a simple experiment supposes that n independent, identical trials are performed, with outcomes $x_1,\ldots,x_n$. Before the experiment is performed, these outcomes are *random variables*, each of which is assumed to be governed by the same probability distribution. Under these conditions, the random variables $x_1,\ldots,x_n$ are said to form a random sample from that distribution. The distribution is often specified up to the values of certain unknown *parameters*. The basic problem of statistics is to make inferences about these unknown parameters from the observed values of $x_1,\ldots,x_n$.

A quantity that can be calculated directly from the sample $x_1,\ldots,x_n$ is called a *statistic*. A technique often useful in statistical analysis is to reduce the data in a sample to a few statistics that can serve as summary values. The choice of appropriate statistics is an important problem and depends on the assumptions that are made about the distribution from which the sample is drawn.

Several distributions are widely used as probability models in statistical applications. In many problems a discrete distribution, for which all the probability is concentrated on a finite number or an infinite sequence of possible outcomes, is used and is described by a probability mass function. The binomial and Poisson distributions are examples of such discrete distributions. For these two particular distributions, the sample mean

$$\bar{x} = \frac{1}{n}\sum_{i=1}^{n} x_i$$

is a sufficient statistic in the sense that it contains all the information about the unknown parameters that is present in the sample values.

In other problems, a continuous distribution is used and is described by a probability density function (pdf). The most famous and important distribution of this type is the normal, or Gaussian, distribution, for which the pdf is

$$f(x) = \frac{1}{\sqrt{2\pi}\sigma}\exp\left[-\frac{1}{2\sigma^2}(x-\mu)^2\right], \quad -\infty < x < \infty.$$

The parameters μ and σ^2 are, respectively, the mean and variance of the distribution ($-\infty < \mu < \infty$, $\sigma > 0$). For the normal distribution, the sample mean x and the sample variance

$$s^2 = \frac{1}{n-1}\sum_{i=1}^{n}(x_i - \bar{x})^2$$

are jointly sufficient statistics. For a random sample from a uniform distribution over an interval with unknown end points, the largest and smallest values in the sample are jointly sufficient statistics. For other distributions, the sample median and other sample percentiles are appropriate statistics, even though they may not be sufficient.

The normal distribution for which $\mu = 0$ and $\sigma^2 = 1$ is called the standard normal distribution. Various distributions derived from the standard normal, such as the chi squared (χ^2) and the Student t distributions, play an important part in statistical theory. Tables of probabilities for these distributions and the standard normal are given in many statistics texts and books of mathematical tables, and are accessible from hand-held calculators.

In the context of statistical inference the notation $f(x;\theta)$ is used to designate a probability density function or probability mass function with unknown parameter or set of parameters θ. A common problem is the estimation of θ, given a random sample of observations from a distribution characterized by $f(x;\theta)$. A statistic must be chosen to serve as the estimate of θ. Before the experiment is performed and data values are collected, this statistic is itself a random variable governed by a probability distribution. The term *estimator* is used for this random variable. Estimators $\hat{\theta}$ are described and compared according to various criteria. A *Bayes estimator* of a parameter minimizes the expected value of some suitably defined loss function. In particular, a Bayes estimator with respect to squared error loss minimizes the expected value of the mean squared error calculated with respect to a specified probability distribution over the possible values of θ. An estimator $\hat{\theta}$ (or, more precisely, a sequence of estimators of θ) is called *consistent* if it converges in probability to θ as the sample size increases, and it is *mean square consistent* if its mean squared error tends to 0.

For large sample sizes, the properties of an estimator $g(\hat{\theta})$ of $g(\theta)$ when $\hat{\theta}$ is an estimator of θ may often be studied using the delta method. For suitable functions g and estimators $\hat{\theta}$, expansion of $g(\hat{\theta})$ in a Taylor series about θ with the first-order derivative and a remainder permits approximate evaluation of the mean, variance, and distribution of $g(\hat{\theta})$. This technique is especially useful when θ is a vector of parameters. The variance formula obtained by the delta method is called the *law of propagation of errors*. Quite generally jackknife and bootstrap methods may be employed to assess the statistical error of an estimator. These techniques are especially useful when the estimator is a complicated or nontractable function of the observations. Their application involves resampling from the observed data. Although the jackknife and bootstrap may require considerable computational effort, the availability of inexpensive and fast computation has made their use practical.

The method of maximum likelihood is commonly used to derive estimators of parameters. Given a random sample $x_1,\ldots,x_n$, the likelihood function is defined to be

$$L(\theta) = \prod_{i=1}^{n} f(x_i;\theta),$$

and a maximum likelihood estimator is any value of θ for which $L(\theta)$ is maximum. Maximum likelihood estimators may not be unique, may be difficult or impossible to determine analytically, or may even fail to exist. In some cases maximum likelihood estimators are clearly inferior to other estimators according to certain comparison criteria. Nonetheless, under widely applicable regularity conditions, maximum likelihood estimators are consistent estimators of parameters and asymptotically possess a normal distribution with minimum mean squared error uniformly minimum in θ. In problems that are nontractable analytically, it is common to locate a maximum by using the Newton–Raphson method or other similar iterative techniques to solve the likelihood equation $L'(\theta) = 0$.

The methods discussed above yield point estimates of parameters. Sometimes it is desirable instead to estimate an interval or set in which θ is likely to lie. A set $I(x_1,\ldots,x_n)$ of possible values of θ is said to be a confidence set estimator for θ with confidence coefficient $1-\alpha$ if the probability of the set of sample values for which $I(x_1,\ldots,x_n)$ contains θ is $1-\alpha$ for all possible θ. The most common example is a confidence interval for a single parameter. The proper interpretation of a set estimator I with confidence coefficient $1-\alpha$ is that prior to collection of the sample values, the probability is $1-\alpha$ that I will cover the actual value of θ. Thus, in the long run $100(1-\alpha)\%$ of the sets so constructed can be expected to cover the true value of θ. Decreasing α increases the probability of coverage, but at the same time enlarges I, thus giving a less precise estimator. After the sample values $x_1,\ldots,x_n$ have been obtained and the set I has been calculated, it is not possible to specify the probability that θ actually lies in I without recourse to the Bayesian methods described later.

Another context for estimation arises when measurements are taken simultaneously on several variables, $y,x^{(1)},\ldots,x^{(m)}$, and it is desired to estimate or fit a functional relationship between y and the x variables. In a linear model, y is treated as a linear combination of nonrandom, known functions of the x variables with unknown coefficients to be estimated, plus a random error term. The linear combination of functions of the x variables may be viewed as a signal component and the error term as an additive noise component. A simple model involves a polynomial in a single variable x plus the error term e, $y=\beta_0+\beta_1x+\cdots+\beta_kx^k+e$. Given n pairs of measurements (x_i,y_i), $i=1,\ldots,n$, the error variables $e_1,\ldots,e_n$ are often assumed to be uncorrelated with mean 0 and constant variance. The parameters $\beta_0,\beta_1,\ldots,\beta_k$ may then be estimated by the least squares method, which chooses values minimizing

$$\sum_{i=1}^{n}(y_i-\beta_0-\beta_1x_i-\cdots-\beta_kx_i^k)^2.$$

If, moreover, the errors e_i are assumed to be normally distributed, then the least squares estimators are maximum likelihood estimators, and we can construct confidence sets for the parameters and test hypotheses about their values.

More general versions of the linear model allow the errors e_i to be correlated. A nonlinear model postulates a relationship $y=h(x;\beta)+e$ where the parameter β is involved nonlinearly. In another specification, the relationship is nonparametric, such as $y=h(x)+e$, and the curve $h(x)$ is assumed to be smooth. Models of this general type are usually called *regression* models, and the parameters in the linear models are called regression coefficients. In practice, the x variables may be random variables whose values are observed along with the values of y, or they may be control variables whose values are chosen by the experimenter. Linear models are also treated in the analysis of variance, where observations are collected in groups and the parameters are the group means or linear combinations of them.

Another topic widely studied and used in statistical applications is the classical theory of hypothesis testing. The methodology is also widely misunderstood and misused. A statistical hypothesis is usually some assertion about the values of certain parameters in a probability model within the context of a scientific experiment. Statisticians speak of testing a null hypothesis H_0 versus an alternative hypothesis H_1, the latter often including all possibilities not encompassed by H_0. A test is a rule based on the experimental data and used to accept or reject H_0. A *significance level* α, usually a number such as 0.10, 0.05, or 0.01, is chosen in advance, and H_0 is rejected if the observed sample values fall in a region chosen to have probability α when H_0 is true and to consist of values that are most unlikely under H_0. Choice of an appropriate value of α should be based on the costs of making incorrect decisions and the available sample size.

If H_0 is rejected, the result is said to be "statistically significant at level α." Statistical significance is not the same as practical significance. A test may reject the hypothesis that $\theta=\theta_0$ when in fact θ is $\theta_0+\delta$ and δ is too small to be of any practical significance. This may occur when the sample size is large and the test therefore has high power, i.e., ability to discriminate between a null hypothesis and close alternative values of the parameter.

A common misinterpretation of a statistical test is to regard the smallest value of α for which H_0 can be rejected, given the data values, as the probability that H_0 is true. It is not this probability, but simply the probability of obtaining a certain set of sample values when H_0 is true. After the sample has been obtained, it is not possible to specify the probability that H_0 is true without using Bayesian methods. Another confusing matter concerns which hypothesis to label the null, as opposed to the alternative. The conclusion reached by a test procedure may depend on this arbitrary assignment. Also it should be noted that the null hypothesis does not necessarily specify that the data were generated merely by "pure chance." Examples of popular tests are the chi square *goodness of fit* test, which is used to test whether the sample came from a particular type of distribution, and the *t-test*, which is used to test hypotheses about the unknown value of a mean or a regression coefficient when the sample is normally distributed.

Bayesian statistical methods are based on the notion that probability is subjective and that any uncertainty about the unknown value of a parameter or other variable can be represented by a subjective probability distribution. The Bayesian approach provides a unified theory of statistical inference. Before an experiment is performed, uncertainty about the value of a parameter θ is represented by a *prior* probability distribution. The likelihood function $L(\theta)$ determined by the sample outcomes is then combined with this prior distribution to obtain a *posterior* distribution for θ, in accordance with Bayes's theorem. In turn, this posterior distribution serves as a prior distribution for any subsequent experimentation involving θ. Thus, knowledge and belief about θ are summarized at any stage by the current posterior distribution of θ. The probability that the unknown value of θ lies in any specified region can be calculated, and inferences and decisions about θ can be based on these probabilities.

The model of an experiment that yields a random sample generated by a fixed number of independent and identical trials can be generalized in several ways. If each outcome x is a vector, *multivariate* analysis is appropriate. If the total number of trials is not fixed in advance, but is determined by inspection of the sample values as they are collected, then *sequential* analysis is appropriate. If the observations $x_1,\ldots,x_n$ are obtained sequentially in time and are not independent, then *time series* analysis is appropriate.

In time series analysis a probability model usually involves a stationary stochastic process, and statistical problems deal with inference about parameters characterizing the correlation structure. A common procedure is to take the discrete Fourier transform of the random sample values, since the Fourier coefficients are approximately uncorrelated variables. Statistical methodology based on these Fourier coefficients is called inference in the frequency domain and is used to derive estimates of the spectral density and related parameters. Models with parameters occurring in the time domain, such as autoregressive processes, are also widely utilized in time series analysis.

See also PROBABILITY; STOCHASTIC PROCESSES.

BIBLIOGRAPHY

M. H. DeGroot, *Optimal Statistical Decisions*. McGraw-Hill, New York, 1970. (I; Bayesian methods and sequential analysis.)

M. H. DeGroot, *Probability and Statistics*, 2nd ed. Addison-Wesley, Reading, Mass., 1986. (E; a general treatment.)

W. J. Dixon and F. J. Massey, Jr., *Introduction to Statistical Analysis*, 3rd ed. McGraw-Hill, New York, 1969. (E; emphasizes use of statistical methodology.)

N. R. Draper and H. Smith, *Applied Regression Analysis*, 2nd ed. Wiley, New York, 1981. (E)

B. Efron, *The Jackknife, the Bootstrap, and Other Resampling Methods*. SIAM, Philadelphia, 1982. (A)

B. Efron and G. Gong, "A Leisurely Look at the Bootstrap, the Jackknife, and Cross-Validation," *Am. Statist.* **37**, 36 (1983). (I)

T. S. Ferguson, *Mathematical Statistics: A Decision Theoretic Approach*. Academic Press, New York, 1967. (A)

G. M. Jenkins and D. G. Watts, *Spectral Analysis and Its Applications*. Holden-Day, San Francisco, 1968. (I; a frequency domain treatment of time series analysis.)

P. W. M. John, *Statistical Design and Analysis of Experiments*. Macmillan, New York, 1971. (I; analysis of variance models.)

S. J. Press, *Applied Multivariate Analysis*. Holt, Rinehart & Winston, New York, 1972. (I)

C. R. Rao, *Linear Statistical Inference and Its Applications*, 2nd ed. Wiley, New York, 1973. (A; a comprehensive survey of statistical theory.)

Stellar Energy Sources and Evolution

R. L. Sears

The simplest observational fact about stars is that they shine. The sources of this energy and the changes in stellar structure as the sources are used up are the subjects of this article. At the present time the theory of stellar evolution is sufficiently developed so that each of the stars we can see can be identified with a particular evolutionary stage.

The three principal energy sources, in order of occurrence during a typical star's life, are (1) gravitational potential energy, (2) thermonuclear reactions, and (3) thermal energy (kinetic energies of particles). Each of these can be estimated for the sun, an average star: its mass $M_\odot$ is 1.99×10^{33} g, its radius $R_\odot$ is 6.96×10^{10} cm, its luminosity $L_\odot$ (total electromagnetic energy being radiated) is 3.86×10^{33} erg s^{-1}, and its chemical composition is about 91% hydrogen, 9% helium, and 0.1% of all the heavier elements, by numbers of atoms. The gravitational potential energy of the sun may be taken approximately as $-GM_\odot^2/R_\odot$, where G is Newton's gravitation constant; this figure, -3.8×10^{48} erg, represents the gravitational energy released in contraction from infinity to the present radius. Alternatively, the available gravitational energy between the present state and the final state, which for the sun is a white dwarf, can be estimated from the difference in radii: it is $+4.7\times10^{50}$ erg. The thermonuclear energy of the sun may be taken approximately as $fXMc^2$, from the Einstein relation, where $f = 0.71\%$ is the fraction of mass converted to energy when hydrogen is burned (by nuclear reactions) into helium and $X = 71\%$ is the fraction of the solar mass that is hydrogen; this gives 9.0×10^{51} erg. (The additional energy available from burning of helium to heavier elements is only about one-tenth of this.) The thermal energy content of the sun may be taken as (minus) half the gravitational potential energy released, from the virial theorem; this gives 2×10^{48} erg.

The relative importance of each of these energy sources can be illustrated by dividing the amounts by the present rate of energy outflow. For the sun, gravitational contraction can have supplied energy to keep it shining at its present rate for only the past 3×10^7 years, a time much shorter than the age of Earth. The thermal energy content corresponds to a similarly brief period. The nuclear source (even if only about one-tenth of it is used before the sun changes its structure markedly, as the theory of stellar evolution described below indicates) can provide sufficient energy for the sun to shine over some 7×10^9 years; this is a comfortable time scale, as pointed out long ago by Eddington. Although gravitation and heat are important factors in certain stages of a star's life, nuclear reactions provide the principal energy source. The most important series of these are hydrogen burning and helium burning.

Hydrogen burning takes place through two sets of processes, first described by H. A. Bethe and others in the late 1930s. The principal reactions in the *proton–proton chain* are as follows:

$$^1\mathrm{H}+{}^1\mathrm{H}\rightarrow{}^2\mathrm{D}+e^++\nu_e,$$

$$^2\mathrm{D}+{}^1\mathrm{H}\rightarrow{}^3\mathrm{He}+\gamma,$$

$$^3\mathrm{He}+{}^3\mathrm{He}\rightarrow{}^4\mathrm{He}+2p.$$

The net effect is to convert four protons into one α-particle, with the release of energy in the form of γ rays (including those from the annihilation of the positron with an electron) and kinetic energies of product particles (including neutrinos). The total energy released is 26.73 MeV per chain (6.40×10^{18} erg g^{-1} gram of hydrogen consumed), of which all but about 2% from neutrinos contributes to the luminosity of a star. The proton–proton chain has two side branches. Instead of completion through the last step listed above, the following processes provide about 15% of the terminations in models of the sun:

$$^3\mathrm{He}+{}^4\mathrm{He}\rightarrow{}^7\mathrm{Be}+\gamma,$$

$$^7\mathrm{Be}+e^-\rightarrow{}^7\mathrm{Li}+\nu_e,$$

$$^7\mathrm{Li}+{}^1\mathrm{H}\rightarrow 2\,{}^4\mathrm{He}.$$

The third branch occurs in place of the ^{7}Be electron capture:

$$^7\mathrm{Be} + {}^1\mathrm{H} \rightarrow {}^8\mathrm{B} + \gamma,$$

$$^8\mathrm{B} \rightarrow {}^8\mathrm{Be}^* + e^+ + \nu_e,$$

$$^8\mathrm{Be}^* \rightarrow 2\ {}^4\mathrm{He}.$$

This provides only about 0.02% of the terminations in solar models, but the energetic neutrinos from the ^{8}B β decay are the primary ones in the Brookhaven and Kamiokande II experiments which have detected solar neutrinos, in an amount distinctly lower than predicted.

The proton–proton chain operates at temperatures from about 8×10^6 to 15×10^6 K, and thus in main-sequence stars from $0.08 M_\odot$ to $1 M_\odot$. At higher temperatures the protons can overcome higher Coulomb barriers, and thus in more massive hydrogen-burning stars the *carbon–nitrogen cycle* provides most of the energy. The principal reactions are as follows:

$$^{12}\mathrm{C} + {}^1\mathrm{H} \rightarrow {}^{13}\mathrm{N} + \gamma,$$

$$^{13}\mathrm{N} \rightarrow {}^{13}\mathrm{C} + e^+ + \nu_e,$$

$$^{13}\mathrm{C} + {}^1\mathrm{H} \rightarrow {}^{14}\mathrm{N} + \gamma,$$

$$^{14}\mathrm{N} + {}^1\mathrm{H} \rightarrow {}^{15}\mathrm{O} + \gamma,$$

$$^{15}\mathrm{O} \rightarrow {}^{15}\mathrm{N} + e^+ + \nu_e,$$

$$^{15}\mathrm{N} + {}^1\mathrm{H} \rightarrow {}^{12}\mathrm{C} + {}^4\mathrm{He}.$$

The net effect is the same as that of the proton–proton chain, with, however, a 6% loss of energy in neutrinos. The cycle has a side branch: about one per thousand $^{15}\mathrm{N} + {}^1\mathrm{H}$ reactions goes to $^{16}\mathrm{O} + \gamma$ instead of completing the cycle directly; but it is subsequently completed through $^{17}\mathrm{O} + {}^1\mathrm{H} \rightarrow {}^{14}\mathrm{N} + {}^4\mathrm{He}$ or $^{18}\mathrm{O} + {}^1\mathrm{H} \rightarrow {}^{15}\mathrm{N} + {}^4\mathrm{He}$, and so is sometimes referred to overall as the "CNO tri-cycle."

Helium burning takes place at temperatures exceeding 10^8 K, in the hydrogen-exhausted cores of red giant stars. The *triple-alpha reaction* is $3\ {}^4\mathrm{He} \rightarrow {}^{12}\mathrm{C} + \gamma$; this proceeds through two stages involving the formation of a small amount of ^{8}Be which, before it breaks up, can capture a third α-particle, at sufficiently high densities, 10^3 g cm^{-3} or more. This produces 7.27 MeV per reaction or 5.84×10^{17} erg g^{-1} of helium consumed. When the helium abundance is low and the carbon abundance is high, another reaction, $^{12}\mathrm{C} + {}^4\mathrm{He} \rightarrow {}^{16}\mathrm{O} + \gamma$, becomes competitive, with the consequence that helium burning leads to a core containing part carbon and part oxygen. Subsequent energy-production processes involve burning of carbon, neon, oxygen, and silicon, during short-lived phases of advanced evolution described below. These have important roles in nucleosynthesis.

In order to explain quantitatively the luminosity of a star, one incorporates the rates of the energy-production processes in a mathematical model of a stellar interior. This is based on the hypotheses of hydrostatic equilibrium (conservation of momentum), conservation of mass, an energy transport mechanism—radiation, convection, or conduction—and conservation of energy. These can be stated as four first-order differential equations which can be solved by techniques described in the texts listed in the Bibliography. The solution depends basically on the total mass of the star and the distribution of chemical composition throughout the mass, which may be set by the previous evolutionary development of the star. The solution provides values of the luminosity L and the effective surface temperature T_e, which can be compared with observations. Another important application of the solution comes from the values of the energy-generation rate, which, in the case of thermonuclear energy production, represents the rate of change of chemical composition at each point. One may choose a time step and compute a new distribution of composition; this determines the solution for the next model which can then be computed. In this way one can build an evolutionary sequence of stellar interior models. The changes with time of variables such as L and T_e can be plotted as evolutionary tracks, as in Fig. 1.

Before describing the results and current situation of the computations, it is appropriate to consider an aspect of the theory of great influence on late stages of stellar evolution. The equation of state at low densities is that for an ionized ideal gas plus radiation; at high densities, above 10^4 or 10^5 g cm^{-3}, the effect of electron-degenerate pressure becomes important. This arises from the Pauli exclusion principle which limits the number of electrons per unit volume of phase space (coordinates and momenta) to $2/h^3$, where h is Planck's constant. In a region of limited space volume, high gas densities force electrons to high momenta, thus providing much more of the pressure than the nuclei, which follow the ideal gas law. Completely degenerate stellar models including only the electron pressure were found long ago by S. Chandrasekhar to fit the white dwarf stars. The radius decreases with increasing mass and, when relativistic electron momenta predominate, the radius vanishes at a finite mass. This Chandrasekhar limiting mass is about $1.4 M_\odot$. A dense star or core of higher mass cannot be maintained in equilibrium by electron-degenerate pressure but must contract to higher densities. A similar situation exists for a neutron gas at a density appreciably greater than nuclear, about 2×10^{14} g cm^{-3}; present uncertainty in the baryon–baryon interaction, however, makes uncertain the corresponding limiting mass for neutron stars, which is estimated at around $3 M_\odot$. A configuration of mass higher than that for a neutron star, bereft of nuclear energy sources, is thought to continue gravitational contraction indefinitely as a black hole.

The computed evolutionary development of most single stars, including the sun, may be described in four stages, which correspond to four different sources of energy. A star is formed from a density compression of interstellar matter (gas and grains); it contracts and radiates away energy provided by gravitation. This gravitational-contraction stage ends when the star's central temperature becomes high enough for hydrogen fusion to begin. The main-sequence core-hydrogen-burning stage represents the largest fraction of a star's life. The third stage, advanced evolution, is characterized by helium burning and reactions of heavier elements; such stars are generally high-luminosity, low-T_e red giants. The final stage differs among stars depending on their masses: white dwarf, neutron star, black hole, or perhaps no remnant core at all.

These four stages of stellar evolution are successively passed through by most single stars within a wide range of masses. A lower limit occurs at $0.08M_{\odot}$; stars of mass less than this can be formed from interstellar matter, but before gravitational contraction raises the temperature sufficiently to ignite hydrogen the increasing density leads to electron degeneracy: contraction ceases and the star cools and radiates the thermal energy of its nuclei for an indefinite time. Such stars—"brown dwarfs"—are currently being searched for in the infrared and in perturbations of companions. The upper limit of stellar masses is about $100M_{\odot}$; stars of larger mass evidently are not found because in the gravitational-contraction stage the denser core forms first and provides sufficient heat and radiation pressure to disperse the remaining matter. Between these two limits, stars of different masses proceed through the stages outlined above.

Results are illustrated in Figs. 1 and 2, based on computations by B. Paczyński. Figure 1 has the coordinates of a Hertzsprung–Russell diagram; it shows evolutionary tracks for seven different masses of solar-type composition, beginning with the hydrogen-burning stage at the lower left. The hooks shortly after the dots represent the phase where hydrogen burning ceases in the core, containing about 10% of the stellar mass; a brief overall contraction of the star occurs and heats the hydrogen in a thin shell outside the core to

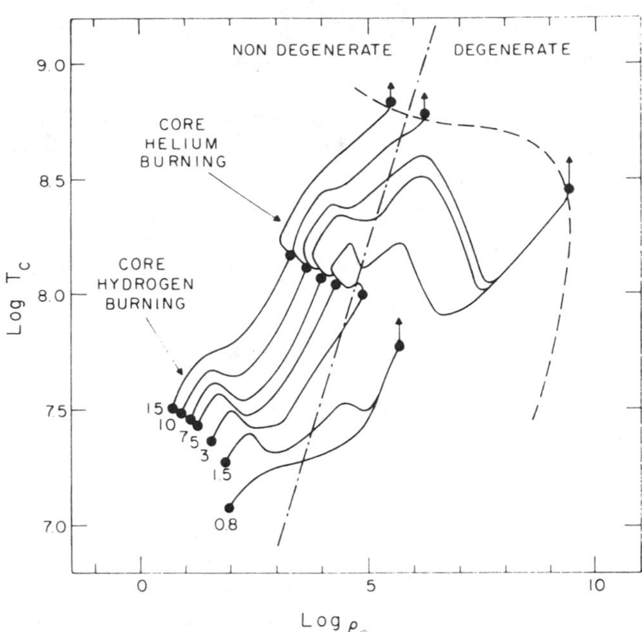

FIG. 2. Evolutionary tracks of the central temperatures and densities of the stars in Fig. 1. At the dashed line, carbon-burning energy production overtakes the neutrino energy loss. [Reproduced with permission from Sears (1974).]

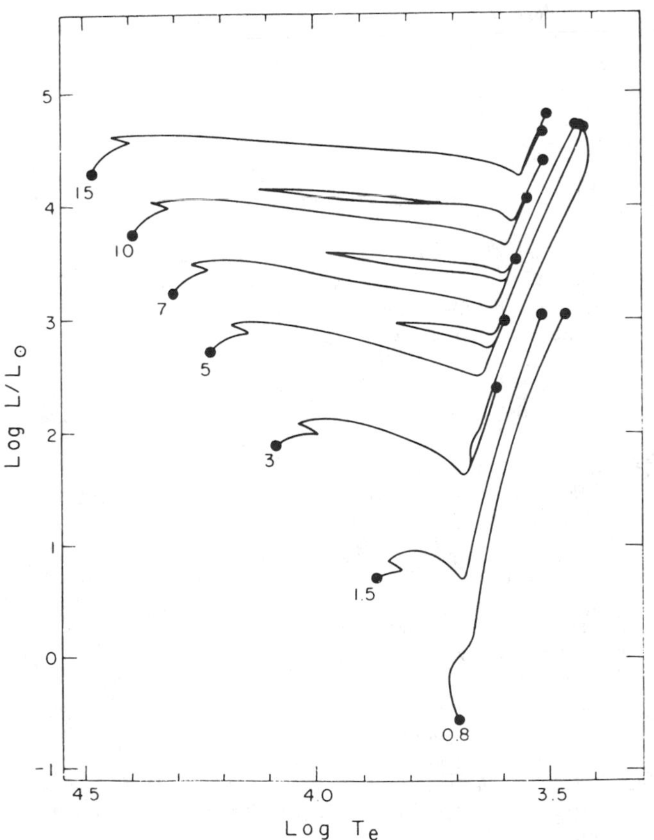

FIG. 1. Evolutionary tracks of stars with originally chemically homogeneous solar composition, computed by B. Paczyński. Numbers indicate the mass of each star in units of the solar mass. Large dots on each track indicate successively ignition of hydrogen, helium, and carbon. [Reproduced with permission from Sears (1974).]

ignition temperature. (The lowest-mass star is partially degenerate at the center and does not contract noticeably.) The evolution to this phase represents most of the lifetime of the star; correspondingly, in observed Hertzsprung–Russell diagrams of various samples most stars are found in the main-sequence band. Subsequent hydrogen shell burning carries the stars to the red-giant region. Figure 2 shows the evolutionary changes of central temperature and central density for the stars in Fig. 1. T is in degrees Kelvin and ρ is in g cm^{-3}; both of these generally increase with time as a nuclear energy source is used up and gravitation compresses the core. Figure 2 partially illustrates the differences in advanced and final stages for two main mass ranges of stars.

Lower-mass stars ($0.08M_{\odot}$ to about $8M_{\odot}$, including the sun) go through hydrogen- and helium-burning stages. Central helium burning begins in the red-giant region in Fig. 1, building up a carbon–oxygen core. In the two lowest-mass stars, helium ignition occurs in a degenerate region, leading to a thermal flash; the subsequent evolution, however, follows that of the more massive stars. In the H–R diagram, the stars go through various excursions ("blue loops" or "horizontal branches"), and then return to the red-giant region on the "asymptotic giant branch" as the C–O core builds up. Energy is produced in an outer hydrogen-burning shell and an inner helium-burning shell, alternatively in time. The latter undergoes thermal pulses which produce convection that can circulate material and bring neutron-producing reactions, such as $^{13}C + {}^{4}He \rightarrow {}^{16}O + {}^{1}n$, into play. Neutron captures can lead to noticeable nucleosynthesis of heavier elements at this stage; the unstable element technetium has long been observed in some red-giant spectra. At some point, the radius and luminosity of the star are so large that a good fraction of the outer mass is weakly bound and can be ejected by various mechanisms, producing a planetary nebula. (In the figures, the models for $7M_\odot$ and below have been computed ignoring the mass loss, leading to carbon ignition in a degenerate region; it now appears that mass loss will circumvent this event.) The remaining core contracts rapidly at high luminosity until electron degeneracy obtains throughout and inhibits the contraction. The final state is a slowly cooling white dwarf with the thermal motions of the C and O ideal-gas nuclei providing the energy source for the diminishing luminosity.

Higher-mass stars (about $8M_\odot$ to around $100M_\odot$) follow the stages of the lighter stars through helium burning but their mass is sufficient to retain the outer envelope. Carbon burning begins and builds up a core of neon, sodium, and magnesium, via, for example, $^{12}C + {}^{12}C \rightarrow {}^{20}Ne + \alpha$. This and subsequent processes are accompanied by thermal neutrino emission, resulting from e^+e^- pair production by hot photons. The channel $e^+ + e^- \rightarrow \nu_e + \bar{\nu}_e$ is rare, but the neutrinos can escape directly from the star; in fact, its neutrino luminosity exceeds its surface photon luminosity in carbon burning and later stages, consequently accelerating the evolution. The next principal nuclear fuel is oxygen which burns to silicon and neighboring elements; and the last is silicon which proceeds through a series of reactions to iron-peak elements. [These processes involve several subsidiary reactions such as (γ, α) and (α, γ).] For iron or heavier nuclei, thermonuclear processes are endoergic. The iron core contracts, raising its temperature and density. Unlike the previous history of the star, no further exoergic nuclear reactions are available to halt the contraction; instead, two effects can occur which precipitate a collapse. If the density reaches a sufficiently degenerate situation where the electrons have enough energy to be captured by the protons in the iron nuclei, then the electron-degenerate pressure support of the core is reduced. If the temperature reaches a sufficiently high value, about 5×10^9 K, for the iron nuclei to be photodisintegrated into helium and neutrons, then the radiation-pressure and gas-pressure support is reduced in this endothermic process. Both of these effects occur more or less simultaneously in current computations; the result is a collapse of the core. The electron-capture effect, $e^- + p \rightarrow n + \nu_e$, leads to "neutronization" of the core material: the increasingly neutron-rich nuclei eventually drip off free neutrons, which may form the basis of a neutron star. Outside the core, the envelope of the star is in a precarious situation. There is no general consensus on the details at present but it is likely that the envelope will be blown off by core implosion, bounce, and explosion, perhaps driven by a high-temperature shock wave, augmented by nuclear energy input from the heated envelope material and by thermal neutrino emission. Current investigations of the phenomenon are complicated, involving high-temperature hydrodynamics and a large network of nuclear reactions; but the goal is a description of a supernova explosion and its consequences, such as synthesis of heavier elements, dispersal of synthesized elements into the interstellar medium, and acceleration of cosmic rays. The remnant core of these massive stars, depending on its mass, may be a neutron star or a black hole.

A spectacular confirmation of the theory of stellar evolution occurred with the detection of some 20 neutrinos, by the Kamiokande II and Irvine–Michigan–Brookhaven experiments, from Supernova 1987A in a nearby galaxy. These thermal neutrinos ($\bar{\nu}_e$ were detected) arrived over a time of some 10 s, indicating that they arose from cooling off of the hot proto-neutron star formed just after the collapse. Their number and energy, in the range from 10 to 40 MeV, are consistent with the gravitational energy released in the formation of a neutron star.

The preceding outline of the various stages of stellar evolution represents more of a framework for current research rather than a canonical text of conventional wisdom. Some active areas have not been described, such as the evolutionary interactions of close binary stars, which lead to many bright x-ray sources in the sky. The historical application to astronomical objects of the concepts and methods of physics has proceeded on no broader a front than at the present time.

See also Black Holes; Cosmic Rays, Astrophysical Effects; Interstellar Medium; Neutrino; Neutron Stars; Nucleosynthesis; Pulsars; Sun.

BIBLIOGRAPHY

G. O. Abell, D. Morrison, and S. C. Wolff, *Exploration of the Universe*, 5th ed. Saunders College Publishing, Philadelphia, 1987. A college-level general astronomy textbook, with good chapters on stellar structure and evolution. (E)

C. W. Allen, *Astrophysical Quanitites*, 3rd ed. Athlone Press, London, 1973. A standard compilation of constants and data used by astronomers. (I)

W. D. Arnett, J. N. Bahcall, R. P. Kirshner, and S. E. Woosley, "Supernova 1987A," *Annu. Rev. Astron. Astrophys.* **27,** 629 (1989). Review of observations at all wavelengths and of theory. (A)

J. N. Bahcall, *Neutrino Astrophysics*. Cambridge University Press, New York, 1989. Thorough review of the solar neutrino problem, including solar models and planned experiments. (A)

H. A. Bethe and G. Brown, "How a Supernova Explodes," *Sci. Am.* **252,** 60 (1985). Authoritative summary. (E)

A. Burrows, "The Birth of Neutron Stars and Black Holes," *Phys. Today* **40,** 28 (September, 1987). Events leading to SN 1987A. (E)

D. D. Clayton, *Principles of Stellar Evolution and Nucleosynthesis*. McGraw-Hill, New York, 1968; reprinted by University of Chicago Press, 1983. A clearly written textbook. (I)

J. P. Cox and R. T. Giuli, *Principles of Stellar Sturcture*. Gordon and Breach, New York, 1968. An extensive, detailed treatise. (I)

M. J. Harris, W. A. Fowler, G. R. Caughlan, and B. A. Zimmerman, "Thermonuclear Reaction Rates, III," *Annu. Rev. Astron. Astrophys.* **21,** 165 (1983). Standard compilation and reference. (A)

I. Iben, Jr., and A. Renzini, "Asymptotic Giant Branch Evolution and Beyond," *Annu. Rev. Astron. Astrophys.* **21,** 271 (1983). Review of detailed calculations. (A)

C. E. Rolfs and W. S. Rodney, *Cauldrons in the Cosmos*. University of Chicago Press, 1988. Reference and text on nuclear astrophysics. (I)

M. Schwarzschild, *Structure and Evolution of the Stars*. Princeton University Press, Princeton, NJ, 1958; reprinted by Dover Publications, New York, 1965. The standard work–clear, enthusiastically written, and authoritative. (I)

R. L. Sears, "Origin of the Chemical Elements and Stellar Evolution," *J. R. Astron. Soc. Canada* **68,** 1 (1974). A general review. (I)

S. L. Shapiro and S. A. Teukolsky, *Black Holes, White Dwarfs, and Neutron Stars*. Wiley-Interscience, New York, 1983. A textbook on the physics of compact objects. (I)

Stochastic Processes

Irwin Oppenheim, Kurt E. Shuler, and George H. Weiss

The theory of stochastic processes deals with random processes that develop in time. If $x(t)$ is such a process, then without further assumptions, a complete description of all statistical properties of $x(t)$ requires specification of an infinite ensemble of probability densities $p_n(x_1,t_1;x_2,t_2;\ldots;x_n,t_n)$ where $p_n(x,t)\ dx_1\ dx_2\cdots dx_n$ is the joint probability that $x_1<x(t_1)\leq x_1+dx_1$, $x_2<x(t_2)\leq x_2+dx_2$, and so forth. In practice no process requiring this amount of information in its description can be handled. Simplifying assumptions that restrict the class of stochastic process must be introduced to develop a theory leading to usable results. The simplest possible assumption is that of independence. Independent processes are those for which $p_n(x,t)$ can be factored as

$$p_n(x_1,t_1;x_2,t_2;\ldots;x_n,t_n)=p_1(x_1,t_1)p_1(x_2,t_2)\cdots\ p_1(x_n,t_n).$$

For these processes (e.g., coin tossing) knowledge of $x(t)$ at one time t does not imply knowledge about $x(t')$ at any other time t'.

The simplest assumption that introduces the notion that the value of $x(t)$ at any given time can influence the properties of $x(t')$ at a later time is the Markov assumption. If a Markov process is observed at a sequence of times $t_1,t_2,\ldots,t_n$, then a prediction of statistical properties of $x(t)$ for $t>t_n$ requires only knowledge of $x(t_n)$, knowledge of the earlier values of $x(t)$ being irrelevant. A simple example of a Markov process is a coin-tossing game in which a player earns a dollar on the appearance of a head and loses a dollar when a tail appears. The current state of the player's fortune is $x(t)$. If the tosses are made at 1,2,3,... seconds, then knowing $x(n)$ we can say that $x(n+1)=x(n)+1$ or $x(n)-1$, either outcome occurring with probability $\frac{1}{2}$. In this process the knowledge of $x(j)$ for any $j<n$ is irrelevant for the prediction of statistical properties of $x(n+1)$.

To define the Markov process more precisely let $w_2(x,t|y,s)dx$ be the probability that $x<x(t)\leq x+dx$ given that $x(s)=y$, where $t>s$. Let $t_1<t_2<t_3<\cdots<t_n$ be an ordered sequence of times. The Markov assumption is equivalent to the relation

$$p_n(x_1,t_1;x_2,t_2;\ldots;x_n,t_n)=p_1(x_2,t_1)w_2(x_2,t_2|x_1,t_1)$$
$$\times\ w_2(x_3,t_3|x_2,t_2)\cdots x_2(x_n,t_n|x_{n-1},t_{n-1}).$$

Thus, the single-step transition probabilities (the $w_2(x,t|y,s)$) are the important functions in the characterization of a Markov model. The simplest Markov processes are Markov chains. These are defined on a set of discrete states in discrete time. An example of a Markov chain is the coin-tossing game mentioned earlier. In a Markov chain, the transition probabilities are the elements of a transiton matrix $\mathbf{P}=(p_{ij})$, where p_{ij} is the probability that $x(t+1)$ will equal j when $x(t)=i$. The p_{ij} satisfy $p_{ij}\geq 0$ and $\sum_j p_{ij}=1$ where the sum is over all possible final states. The two-step transition probabilities, i.e., the probability that $x(t+2)=J$ given that $x(t)=i$, are $\sum_k p_{ik}p_{kj}$, which is the ijth element of $\mathbf{P}^2$. Similarly, n-step transition probabilities are the elements of $\mathbf{P}^n$. The n transition proabilities therefore satisfy $p_{ij}^{(n)}=p_{ik}^{(s)}p_{kj}^{(n-s)}$ where $0<s<n$. This is a consequence of the matrix identity $\mathbf{P}^n=\mathbf{P}^s\mathbf{P}^{n-s}$. More generally, Markov processes satisfy the *Chapman–Kolmogorov* identity,

$$w_2(x,t|y,s)=\int w_2(x,t|z,u)w_2(z,u|y,s)\ dz, \tag{1}$$

where the integration is over all possible intermediate states z, and where u is any arbitrary time between s and t. This equation is a starting point for the derivation of the master equation and the Fokker–Planck equation, both of which are discussed below.

A class of models widely applied in many areas of physics is that of random walks. These may either be Markovian or non-Markovian. In the random walk, a particle makes random displacements $\mathbf{r}_1\mathbf{r}_2,\ldots$ at times $t_1,t_2,\ldots$. Various properties of the sum

$$\mathbf{R}_n=\sum_{j=1}^{n}\mathbf{r}_j \tag{2}$$

are usually sought, but other parameters of interest may be suggested by particular applications. When the $\mathbf{r}$'s are independent, identically distributed random variables with finite first and second moments, one can develop an asymptotic theory for the distribution of $\mathbf{R}_n$ for large n. More specifically, suppose that $p(\mathbf{r})$ is the probability density for a single displacement, with $p(\mathbf{r})d^m\mathbf{r}$ being the joint probability that the components of a single step of the random walk in m-dimensional space fall in the intervals (x_1,x_1+dx_1), $(x_2,x_2+dx_2),\ldots,(x_m,x_m+dx_m)$. The first and second mixed moments of $p(\mathbf{r})$ are defined by

$$\langle x_i\rangle=\int x_ip(\mathbf{r})d^m\mathbf{r}, \tag{3a}$$

$$\langle x_i,x_j\rangle=\int x_ix_jp(\mathbf{r})d^m\mathbf{r}. \tag{3b}$$

When these moments are finite, the probability density of the end-to-end vector of an n-step random walk, $g_n(\mathbf{R})$, approaches the Gaussian form

$$g_n(\mathbf{R})=(2\pi n\ \det\mathbf{M})^{-m/2}\exp\left[-\frac{1}{2n}(\mathbf{R}-n\langle\mathbf{r}\rangle)\times\mathbf{M}^{-1}(\mathbf{R}-n\langle\mathbf{r}\rangle)\right], \tag{4}$$

where the ijth element of $\mathbf{M}$ is $m_{ij}=\langle x_ix_j\rangle-\langle x_i\rangle\langle x_j\rangle$, det $\mathbf{M}$ is the determinant of $\mathbf{M}$, and $\langle\mathbf{r}\rangle$ is a vector whose jth element is $\langle x_j\rangle$. Many generalizations of this result can be proved.

The expression in Eq. (4) implies that

$$\langle \mathbf{R} \rangle = n\langle \mathbf{r} \rangle \qquad \sigma^2(\mathbf{R}) = n \det \mathbf{M}, \tag{5}$$

i.e., both the average displacement in n steps and the variance of this displacement are proportional to n. This property is satisfied, at least for large n, by all Markovian random walk models.

An application in which random walks play a central role is found in the theory of polymer chain configurations. In the random walk model of polymer configurations $\mathbf{r}_j$ is the vector separating the jth and $(j+1)$st monomer, and the end-to-end vector, $\mathbf{R}$, is the vector separating the monomers on either end of the chain. The theory is useful for the derivation of molecular parameters that can be measured by physical techniques such as light scattering.

One of the most intensely studied generalizations of the random walk problem is the excluded volume random walk or self-avoiding walk (often abbreviated SAW). This model constrains the random walk to never occupy the same position twice. The study of such a model is suggested by the physical consideration that no two atoms can occupy the same position. The SAW is obviously a non-Markovian process, since a knowledge of all prior sites visited by the random walker is required to calculate the transition probability on a given step. For this reason the dependence of the σ^2 on n differs from that given in Eq. (5). An exact theory is not yet available for the excluded-volume random walk but many studies suggest that in two dimensions $\sigma^2 \sim n^{3/2}$ and in three dimensions $\sigma^2 \sim n^{1.18}$. Monte Carlo studies indicate that the asymptotic form for the probability density of the end-to-end vector differs from the Gaussian. The general area of excluded-volume problems is still a fruitful one for research.

There are two principal formulations of Markovian problems in continuous time that frequently arise in physical applications. The first leads to the *master equation* and the second to the *Fokker–Planck equation*. We outline these briefly for one-dimensional processes. Both formulations can be derived from the Chapman–Kolmogorov equation in Eq. (1), which is, by itself, too general to handle specific problems. A derivation of the master equation starts from the observation that the transition probabilities $w_2(x,t|y,s)$ must satisfy $w_2(x,t|y,t) = \delta(x-y)$ in a continuous one-dimensional state space. To derive the master equation we assume the validity of the expansion.

$$w_2(x,t+\Delta t|y,t) = \delta(x-y) + A(x,y;t)\,\Delta t + o(\Delta t), \tag{6}$$

where $o(\Delta t)$ is a term having the property that $o(\Delta t)/\Delta t \to 0$ as $\Delta t \to 0$. The important assumption here is the existence of an infinitesimal rate $A(x,y;t)$. When the *Ansatz* of Eq. (6) is combined with Eq. (1) we find that

$$w_2(x,t+\Delta t|y,s) = \int [\delta(x-z) + A(x,z;t)\Delta t + o(\Delta t)]\, w_2(z,t|y,s)\, dz \tag{7}$$

$$= w_2(x,t|y,s) + \Delta t \int A(x,z;t|y,s)\, dz + o(\Delta t)$$

or in the limit $\Delta t \to 0$

$$\frac{\partial w_2(x,t|y,s)}{\partial t} = \int A(x,z;t) w_2(z,t|y,s)\, dz \tag{8}$$

The integrodifferential equation (8) is the master equation in continuous one-dimensional state space. Although it might seem like circular reasoning to use Eq. (6), because A, the rate function, is derived from w_2, in applications one often calculates the rate A, in terms of physical properties of the system being studied. It is this fact that makes the master equation useful and the Chapman–Kolmogorov equation essentially useless in specific applications.

As a simple example formulated in terms of the master equation, consider a stochastic theory of the isomerization reaction $X \xrightarrow{k} Y$. The deterministic theory of chemical kinetics implies that the number of X molecules, $n_x(t)$, decreases exponentially as $n_x(t) = N \exp(-kt)$ where N is the initial number of X molecules. To formulate this in stochastic terms allowing us to calculate a measure of the fluctuations in the number of X molecules one allows there to be any number of X molecules present at an arbitrary time. Define $p_n(t)$ to be the probability that there are n molecules of X at time t. In a Markovian description of the kinetics of the process, since $p_n(t)$ is to be found subject to the initial condition $p_n(0) = \delta_{n,N}$ one sets the conditional probability $p_n(t) = w_2(n,t|N,0)$. The rate of conversion of X molecules at time t is taken, by analogy with the mass action law, to be kn. The master equation for this system is readily shown to be

$$\dot{p}_n = k(n+1)p_{n+1} - knp_n \tag{9}$$

which must be solved subject to the initial condition stated earlier. The form of this equation implies that $\langle n(t) \rangle \equiv \sum_n np_n$ satisfies the equation $\langle \dot{n} \rangle = -k\langle n \rangle$, which has the same form as the deterministic rate equation. The solution to Eq. (9) is found to be

$$p_n(t) = \binom{N}{n} (1 - e^{-kt})^n e^{-(N-n)kt}. \tag{10}$$

Moments of the number of molecules of X present at time t are easily calculated from this expression. We find that the mean number of X molecules, $\langle n(t) \rangle$, is

$$\langle n \rangle = N \exp(-kt) \tag{11}$$

and the variance is

$$\sigma^2(t) = \langle n^2(t) \rangle - \langle n(t) \rangle^2 = Ne^{-kt}(1 - e^{-kt}). \tag{12}$$

Thus, as noted earlier, the expression for $\langle n(t) \rangle$ agrees with the result derived from the deterministic rate equation. The variance, $\sigma^2(t) \equiv \langle n^2(t) \rangle - \langle n(t) \rangle^2$, is a measure of the size of fluctuations. The standard deviation normalized with respect to the mean, $\sigma/\langle n \rangle$, is of order $N^{-1/2}$, as can readily be seen from Eqs. (11) and (12). For values of N of the order of 10^{23}, the fluctuations about the mean are not experimentally observable. This type of result can also be shown to be valid in the case of more complicated kinetic schemes.

The Fokker–Planck (FP) equation is one of the transport or evolution equations for stochastic processes in continuous space and time. Sufficient conditions that imply the form of the FP equation (in one dimension) require the existence and finiteness of functions $a(x,t)$ and $b(x,t)$ defined by the infinitesimal moments

$$\begin{aligned} a(x,t) &= \lim_{\Delta t\to 0} \frac{1}{\Delta t} \int (y-x)w_2(y,t+\Delta t|xt)\, dy, \\ b(x,t) &= \lim_{\Delta t\to 0} \frac{1}{\Delta t} \int (y-x)^2 w_2(y,t+\Delta t|x,t)\, dy. \end{aligned} \tag{13}$$

In addition, the higher moments defined in the same way must satisfy

$$\lim_{\Delta t\to 0} \frac{1}{\Delta t} \int (y-x)^n w_2(y,t+\Delta t|x,t)\, dy = 0 \qquad n=3,4,\ldots . \tag{14}$$

In these circumstances the FP equation for $p(x,t) = w_2(x,t|x_1,0)$ is

$$\frac{\partial p}{\partial t} = \frac{1}{2}\frac{\partial^2}{\partial x^2}(bp) - \frac{\partial}{\partial x}(ap). \tag{15}$$

Similarly, a multidimensional parabolic equation can be found for processes in higher dimensions provided that conditions similar to Eqs. (13) and (14) are satisfied. Supplementary boundary conditions are required when one or more surfaces or points are absorbing or reflecting. A surface is said to be absorbing when a particle impinging on it adheres to it thereafter, and it is reflecting when the particle is reflected from it. The master equation, Eq. (8), reduces exactly to the FP equation when the conditional probability $w_2(x,t+\Delta t|y,t)$ of Eq. (6) is of such a form that Eqs. (13) and (14) hold. Under a number of physical conditions in which $w_2(x,t+\Delta t|y,t)$ is sharply peaked around $y=x$, the FP equation may be a good approximation to the master equation.

As in the master equation technique, the coefficients a and b are generally found in terms of models that incorporate physical considerations. A common technique for deriving such coefficients is the use of a *Langevin equation*, which combines a stochastic element with the dynamics of a system. A typical Langevin equation for the motion of a point particle in a viscous medium has the form

$$m\dot{u} + \gamma u = F(t) \tag{16}$$

where m is mass, u the velocity, γu a damping force whose form is suggested by Stokes's law, and $F(t)$ is a rapidly fluctuating random force that can be introduced in several ways. All one needs to do to derive the FP equation is to specify properties of the first two moments of $F(t)$ appropriately. We choose $F(t)$ to be a Gaussian process whose moments satisfy

$$\langle F(t)\rangle = 0, \qquad \langle F(t)F(t')\rangle = 2D\delta(t-t'), \tag{17}$$

where D is a constant. That is to say, at any given time the average value of $F(t)$ is zero (a suitable correction can be made when $\langle F(t)\rangle =$ constant) and is uncorrelated at different times. When Eq. (17) is valid, one can solve Eq. (16) explicitly for $u(t)$ in terms of $F(t)$ and the infinitesmal moments required for Eq. (13) can be calculated in terms of suitable averages of the solution. We find for the Langevin equation in Eq. (16) that the functions a and b are given by $a=\alpha u$ where $\alpha = \gamma/m$ and $b=2D$,

$$\frac{\partial p}{\partial t} = D\frac{\partial^2 p}{\partial u^2} + \alpha \frac{\partial}{\partial u}(up). \tag{18}$$

This is known as the Ornstein–Uhlenbeck equation which can be used to discuss Brownian motion in a one-dimensional velocity space. Furthermore, in this case the parameter γ is given, in terms of $\beta = (k_B T)^{-1}$, by

$$\gamma = \beta \int_0^\infty \langle F(t)F(0)\rangle\, dt = \beta D. \tag{19}$$

The Langevin equation, Eq. (16), can be derived from molecular theory for a Brownian particle.

There have been a number of attempts to provide a phenomenological generalization of the Langevin equation. If one assumes that the random function $F(t)$ is a Gaussian with the moment properties

$$\langle F(t)\rangle = 0, \qquad \langle F(t)^* F(t')\rangle = g(|\, t-t'\,|) \tag{20}$$

where $g(t)$ is an arbitrary function of time that goes to 0 as $t\to\infty$, then Eq. (16) is replaced by

$$m\dot{u}(t) + \beta \int_0^t u(t-\tau)g(\tau)\, d\tau = F(t) \tag{21}$$

in which $u(t)$ is the velocity. A particular form that has been used in a number of investigations is

$$g(t) = D\theta \exp\{-\theta|\, t\,|\}, \tag{22}$$

where D and θ are constants. There are some special cases in which equations similar to Eq. (21) can be justified from molecular theory, e.g., harmonic oscillator systems or systems in which there is a weak coupling between the bath and the system. However, there are many applications in which the validity of Eq. (21) is assumed without physical justication. In some instances this can lead to incorrect physical conclusions being drawn from its use.

Other generalizations of the Langevin approach to the calculation of functions appearing in the FP equation have been extensively studied both by mathematticians and physicists under the heading of stochastic differential equations. In such a formulation one starts, for example, by writing a differential equation of the form (in one dimension)

$$\dot{u} = f(y,t) + g(y,t)\xi(t) \tag{23}$$

in which $f(y,t)$ and $g(y,t)$ are deterministic functions and $\xi(t)$ is a random function with specified properties, which may be more general than those shown in Eq. (17). A variety of partial differential equations or integrodifferential equations describing the distribution of $u(t)$ can then be found, among which is the FP equation. The particular evolution equation that is found depends on the assumed properties properties of $\xi(t)$. Again, in most cases these equations are purely phenomenological and lack justification from a more detailed physical model.

There are several generalizations of Eq. (15) which may be used to study multidimensional diffusion processes. These may, for example, describe the evolution of the probability density of a point in phase space. As a second example of the use of the FP equation, the simplest version of the stochastic theory of irreversible thermodynamics can be based on a multivariate generalization of Eq. (18). The usual assumption made is that the vector of fluctuations from equilibrium, α, satisfies an equation of the form $\dot{\alpha} = L\alpha$, where L is a constant matrix. The stochastic theory is derived by assuming that α satisfies the Langevin equation $\dot{\alpha} = L\alpha + \epsilon(t)$, where $\epsilon(t)$ is a zero-mean random vector whose components satisfy $\langle\epsilon_i(t)\epsilon_j(t')\rangle = 2D_{ij}\delta(t-t')$, the D_{ij} being constants. One can show that the joint probability density of the $\alpha_i(t)$ satisfies a multivariate FP equation that generalizes the Ornstein–Uhlenbeck equation in Eq. (18).

A common class of problems finding wide application in the physical sciences is that of *first passage times*. Assume that a surface is absorbing. In a situation in which the particle moves subject to random forces one is often interested in finding statistical properties of the time at which the particle first reaches the absorbing surface. For example, the stochastic model for the dissociation of diatomic molecules treats the bonding energy as a random variable subject to change by thermal fluctuations. Dissociation occurs when the energy reaches a critical value. The calculation of the mean time to dissociation, which is the inverse of the rate constant, can be carried out in the framework of first-passage-time methodology and allows one to relate a macroscopic parameter to properties of a microscopic model.

One of the most useful results in the theory of first passage times is that when the probability density for the position of a randomly moving particle at time t whose initial position was $\mathbf{r}_0$, $p(\mathbf{r},t|\mathbf{r}_0,0)$, satisfies a generalized diffusion equation of the form

$$\frac{\partial p}{\partial t} = \mathcal{L}(\mathbf{r},t)p, \tag{24}$$

where $\mathcal{L}(\mathbf{r},t)$ is a linear operator, then the probability density for the first passage time starting from some initial position $\mathbf{r}_0$ satisfies the equation

$$\frac{\partial p}{\partial t} = \mathcal{L}^{\dagger}(\mathbf{r}_0,t)p, \tag{25}$$

where $\mathcal{L}^{\dagger}(\mathbf{r}_0,t)$ is the operator which is the mathematical adjoint to $\mathcal{L}(\mathbf{r},t)$. This relation allows one to simplify the calculation of rates using either a Markovian or non-Markovian description of the kinetics of the process.

There are many problem areas in physics that are invariably studied within the framework of the theory of stochastic processes. These include the phenomena of electrical noise, the theory of turbulence, the theory of chemical reaction rates, and transport properties of random media. For example, a common formulation of the theory of turbulence tries to find the statistical characteristics of a fluid moving in accordance with the Navier–Stokes equation, but with a random initial condition. Other areas of research where the stochastic approach has proven to be of great value include polymer physics, spin relaxation, and energy relaxation in general. In the sense that stochostic equations such as the master equation, the FP equation, and the Langevin equation can be found using appropriate restrictions on the Liouville equation, all dynamic processes of physics described by a Liouville equation can also be formulated in stochastic terms. The applications listed here and the methodology and utility of the stochastic approach to these problems are discussed in detail in the works in the Bibliography.

See also FLUCTUATION PHENOMENA; MONTE CARLO TECHNIQUES; PROBABILITY; STATISTICS.

BIBLIOGRAPHY

L. Arnold, *Stochastic Differential Equations, Theory and Applications.* Wiley, New York, 1974. (A)

G. K. Batchelor, *The Theory of Homogeneous Turbulence.* Cambridge University Press, London and New York, 1956. (A)

M. J. Beran, *Statistical Continuum Theories.* Wiley (Interscience), New York, 1968. (A)

D. R. Cox and H. D. Miller, *The Theory of Stochastic Processes.* Wiley, New York, 1965. (I)

W. Feller, *An Introduction to Probability Theory and Its Applications.* Wiley, New York, 1953 (Vol. 1, 3d ed.); 1971 (Vol. 2). (I,A)

C. W. Gardiner, *Handbook of Stochastic Methods for Physics, Chemistry, and the Natural Sciences.* Springer-Verlag, New York, 1985. (I,A)

M. Kac, *Probability and Related Topics in Physical Sciences.* Wiley (Interscience), New York, 1959. (A)

N. G. van Kampen, *Stochastic Processes in Physics and Chemistry.* North-Holland, Amsterdam, 1981. (I,A)

L. Mandel and E. Wolf (eds.), *Selected Papers on Coherence and Fluctuations of Light.* Dover, New York, 1970. (A)

E. W. Montroll and B. J. West, in *Fluctuation Phenomena*, p. 134. North-Holland, Amsterdam, 1979.

I. Oppenheim, K. E. Shuler, and G. H. Weiss, *The Master Equation in Chemical Physics.* MIT Press, Cambridge, MA., 1977. (I,A)

A. Papoulis, *Probability: Random Variables and Stochastic Processes.* McGraw-Hill, New York, 1965. (I)

S. A. Rice, *Chemical Kinetics.* Elsevier, Amsterdam, 1985. (I,A)

Z. Schuss, *Theory and Applications of Stochastic Differential Equations.* Wiley, New York, 1980. (I,A)

N. Wax (ed.), *Selected Papers on Noise and Stochastic Processes.* Dover, New York, 1954. (A)

G. H. Weiss and R. J. Rubin, *Adv. Chem. Phys.* **52,** 363 (1983). (I)

H. Yamakawa, *Modern Theory of Polymer Solutions.* Harper and Row, San Francisco, 1971. (I,A)

String Theory

John H. Schwarz

Traditional studies of the relativistic quantum physics of elementary particles assume that the particles can be described as mathematical points without any spatial extension whatsoever. This approach has had impressive success, but appears to break down at extremely high energies or short distances where gravitational forces become comparable in strength to the nuclear and electromagnetic forces that act between particles.

In 1974 the late Joël Scherk and I proposed overcoming this limitation by basing a unified description of all elementary particles and the forces that act among them on fundamental one-dimensional curves, called "strings," rather than on point particles. String theories appear to be free from the inconsistencies that have plagued all previous attempts to construct a "unified field theory" that describes gravity together with the other forces. "Superstring" theories, which contain a special kind of symmetry called supersymmetry, show the most promise for giving realistic results.

THREE STRING THEORIES

There are many consistent quantum field theories for point particles, although none contains gravity. It is extremely difficult, on the other hand, to formulate a theory of elementary extended objects that is consistent with the usual requirements of quantum theory. In the case of strings (one-dimensional elementary objects), there are a few schemes that appear to be consistent. It is not known whether there are any at all for objects of more than one dimension (such as two-dimensional membranes), but this appears to be very unlikely. The existence of string theories depends on special features that do not generalize to higher-dimensional objects.

Remarkably, every classical solution of each of the known string theories gives rise to a particle spectrum that contains exactly one massless spin-2 graviton. Moreover, this graviton interacts in accord with the dictates of general covariance, which implies that general relativity gives a correct description at low energies. The characteristic length scale of the strings is the Planck length.* This is determined by requiring the gravitational coupling to have the usual Newtonian value.

Strings can occur in two distinct topologies called open and closed. Open strings are line segments with two free ends, whereas closed strings are loops with no free ends. Depending on the theory, strings may have an intrinsic orientation or not. For each solution of a particular string theory, there is a corresponding spectrum of elementary particles given by the various quantum-mechanical excitations (normal modes) of the string. These include rotational and vibrational excitations as well as excitations of various "internal" degrees of freedom that can reside along the string. The internal degrees of freedom can describe Lie group symmetries, supersymmetry, and so forth. In string theory, one has a unified view of the rich world of elementary particles as different modes of a single fundamental string. String states that have mass much below the Planck mass are finite in number and should correspond to observable particles. There are also an infinite number of modes with mass comparable to or larger than the Planck mass that are probably not observable. In general, they are unstable and decay into the light modes, although there could be some with magnetic charge or fractional electric charge or some other exotic property that are stable. Since we are unlikely to be able to make such superheavy particles, it would only be possible to observe them if they already exist in sufficient number as remnants of the big bang.

At the present time there are three known consistent string theories (listed in Table I). The type I superstring theory consists of both open and closed strings that are unoriented. The type II superstring theory and the heterotic string theory consist of oriented closed strings only. (They are distinguished by different internal degrees of freedom.) Each of the three theories is completely free of adjustable dimensionless parameters or any other arbitrariness. Thus, aside from this threefold choice, one has a completely unique theory that consistently incorporates quantum gravity.

To know the right theory is not enough. After all, nature is described by *solutions* of the equations. Normally, what is relevant is the quantum state of lowest energy (called the "vacuum") and low-lying excited states (which can be interpreted in terms of particle creation). It can happen that a theory has many different possible vacuum configurations. In this case one must make an arbitrary (phenomenological) choice to describe the experimental data—perhaps even adjusting a number of parameters—despite the fact that the underlying theory is itself unique. Precisely this problem seems to be facing us in the case of string theory. Despite the near uniqueness of the theory, there seems to be a very large number of possible solutions, any one of which is theoretically acceptable. If the choice can only be made phenomenologically, that would be disappointing. Thus, many string theorists speculate that when "nonperturbative" effects are properly understood, all but one (or a few) of the solutions will turn out to be inconsistent or unstable.

In a theory of gravity, the dynamics determines the geometry of space-time as part of the characterization of the vacuum configuration. We would like to derive the geometry of four-dimensional Minkowski space or a realistic cosmology. In fact, there are classical solutions to each of the three theories with any space-time dimension $D \leq 10$. Thus, the dimension of space-time is properly regarded as a property of the solution and not of the theory itself. Many of the solutions with $D<10$ can be interpreted as having a ten-dimensional space-time manifold in which $10-D$ spatial dimensions form a compact space K, so that altogether the space-time is $M_D \times K$—a direct product of D-dimensional Minkowski space and K. However, there are other classes of solutions with $D<10$ that do not seem to admit such an interpretation.

The case $D=10$ is special in that it is the largest value possible. This is progress, but it would be much more satisfying to understand why $D=4$ is necessary, which has not yet been achieved. Indeed, each of the theories admits solutions with $D = 10$, listed in Table II, that are consistent

Table I Consistent string theories

Name	String topology	String orientation
Type I	Open and closed	Unoriented
Type II	Closed	Oriented
Heterotic	Closed	Oriented

* The Planck length is formed from the most fundamental constants—Planck's constant ($\hbar$), the speed of light (c), and Newton's constant (G). Specifically,

$$L_{\rm Pl} = \left(\frac{\hbar G}{c^3}\right)^{1/2} \sim 1.6 \times 10^{-33} \text{ cm}.$$

Table II $D = 10$ solutions

Theory	Symmetry group	Space-time supersymmetry
Type I	$SO(32)$	$N = 1$
Type II	—	$N = 2A$
Type II	—	$N = 2B$
Heterotic	$E_8 \times E_8$	$N = 1$
Heterotic	$SO(32)$	$N = 1$
Heterotic	$SO(16) \times SO(16)$	$N = 0$

so far as we can tell. These solutions are certainly not realistic, but they do seem to be of fundamental importance from a theoretical point of view. It is a real challenge to find a good reason to exclude them as potential vacuum configurations. For $D<10$ the number of vacuum configurations is much larger. It is a big industry to try to construct a complete list and identify the ones that could be realistic. At the moment the heterotic theory seems to offer the best prospects for realistic solutions, but it is not out of the question that the type I or type II superstring theories could also yield phenomenologically viable solutions.

Let us turn to the question of whether string theory can be tested. It seems to me that there are several promising possibilities. The first is to use it to calculate the properties of elementary particles at ordinary energies. After all, if the theory is unique and the solutions to the theory do not have too much freedom, then a great deal of particle physics data should be calculable. There is no reason that "low-energy" phenomena should be especially difficult to extract. A second possibility is that some Planck scale particles (such as magnetic monopoles) were formed early in the Big Bang and survive to the present epoch as observable stable entities. A related possibility is that characteristic features of superstring theory are required for a successful understanding of the cosmology of the very early universe. Our present understanding of string theory is not yet sufficient to make definitive testable predictions in any of these areas. But with all the brainpower that is being brought to bear, the rate of progress is very impressive. Pessimism about eventual testability is probably unwarranted. (As Witten has noted, general relativity gave rise to various predictions that seemed quite hopeless to verify when they were made. These included neutron stars, black holes, gravitational radiation, and gravitational lenses. There is now substantial observational evidence for all of these.)

THE STRUCTURE OF INTERACTION

In the perturbation expansion treatment of quantum field theory, point–particle interactions are represented by Feynman diagrams. The history of the motion of a particle is a trajectory in space-time called the world line of the particle. Interactions are represented by joining or bifurcating world lines. The complete interaction amplitude for a given set of incoming particles and a given set of outgoing particles is given by a sum of contributions associated with all allowed diagrams with the chosen initial and final states. In particular, they must include all possible interactions appropriate to the theory in question. The diagrams can be classified by their topology and the contributions from diagrams of any particular topology is given by a finite-dimensional integral. The integrals usually diverge, but in renormalizable theories there is a well-defined prescription for extracting finite results unambigously.

String interactions can be formulated in an analogous manner. The space-time trajectory of a string is a two-dimensional surface called the *world sheet*. Feynman diagrams are two-dimensional surfaces with specific incoming and outgoing strings, once again classified by their topology. The possible world-sheet topologies are more limited in the case of the type II and heterotic theories, which only contain closed oriented strings, than in the case of the type I theory. Therefore, the discussion that follows will be restricted to these theories. (The basic ideas are essentially the same in the type I theory. There are just some additional allowed topologies.)

The type II and heterotic string theories have a single fundamental interaction. It can be described by a portion of world sheet, called the "pants diagram," depicted in Fig. 1. When the diagram is intersected by a plane representing a time slice T_1, one sees two closed strings. Intersecting the surface by a time slice at time T_2 reveals just one closed string. Clearly, at intermediate times the two closed strings approached one another, touched, and joined. The reverse process in which two closed strings join to give one is also allowed.

The interaction structure described by the pants diagram differs in fundamental respects from interactions in point–particle theories. To explain the distinction, consider Fig. 2, where a point–particle vertex and the pants diagram are drawn. In each case, we can ask at what space-time point the interaction that turns two particles into one takes place. We can also represent the time slices corresponding to two observers in distinct Lorentz frames by the lines $t=$const. and $t'=$const. drawn in each case. In the point–particle theory there is a definite space-time point at which the interaction occurs that is unambiguously identified by all observers. In the string case, on the other hand, the interaction point corresponds to the point at which the time slice is tangent to the surface, and this differs from one observer to another.

Clearly there is a fundamental difference between the interactions depicted in Figs. 2a and 2b. In the point–particle

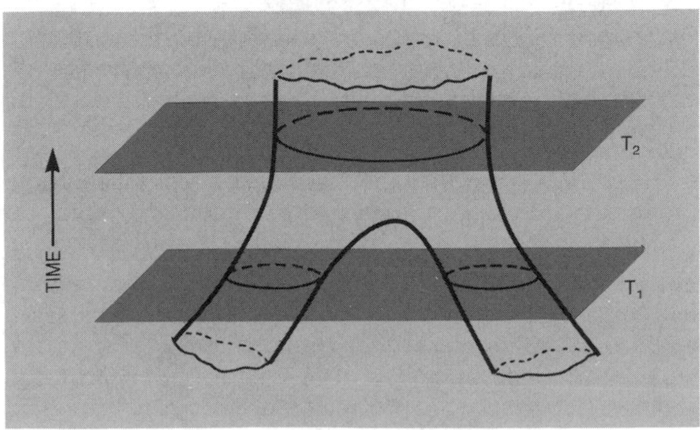

FIG. 1. Pants diagram.

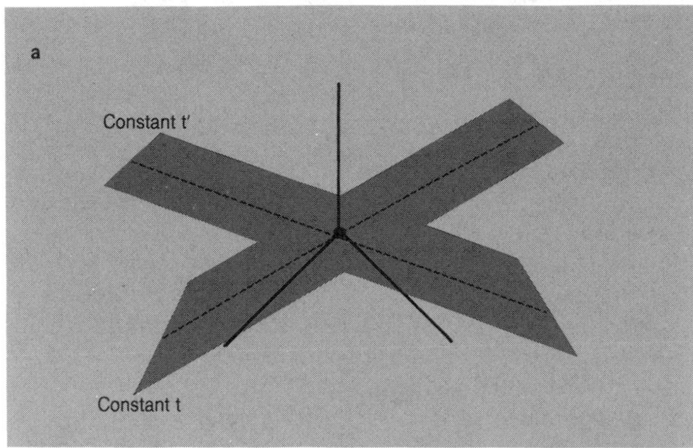

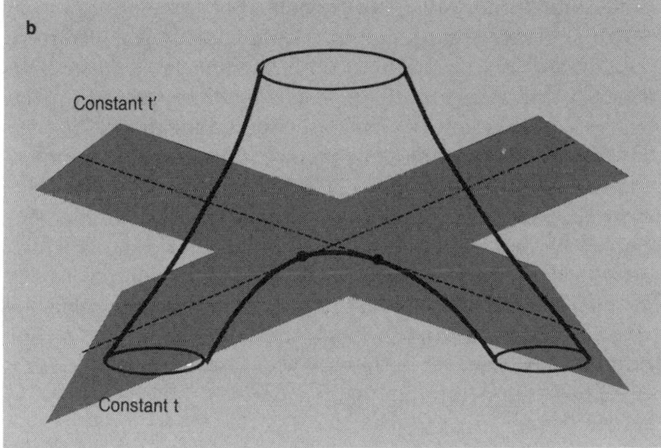

FIG. 2. Point–particle vertex (a) and pants diagram (b).

case (Fig. 2a) the "manifold" of lines is singular at the junction, which is a special point. Arbitrary choices are possible in the association of interactions with such vertices. This is part of the reason why ordinary quantum field theory has so much freedom in its construction. The string world sheet (Fig. 2b) is a smooth manifold with no preferred points. The fact that it describes interaction is purely a consequence of the topology of the surface. The nature of the interaction is completely determined by the structure of the free theory with none of the arbitrariness that exists in the point–particle case.

String world sheets are two-dimensional surfaces that can be described as Riemann surfaces using techniques of complex analysis. This means that (at least locally) one can use complex coordinates z and $\bar{z}$. A fundamental feature of string theory is that world sheets that can be related by a conformal mapping $z \rightarrow f(z)$ are regarded as equivalent. Thus, in performing the sum over distinct geometries only surfaces that are conformally inequivalent should be included. Fortunately, for each topology, the conformally inequivalent geometries can be characterized by a finite number of parameters, and thus the Feynman integrals are finite dimensional.

The topological classification of the Feynman diagrams is especially simple in the case of the type II and heterotic theories. The world sheets are characterized by a single integer, the genus, which is the number of holes in the surface. The external strings can be represented as points on the surface, since this is conformally equivalent to tubes extending off to infinity. The genus corresponds to the number of loops, i.e., the power of $\hbar$ in the perturbation expansion. It is remarkable that there is just one diagram at each order of the perturbation expansion, especially as the number of them in ordinary quantum field theory is very large indeed.

Not only is the number of diagrams much less than in ordinary quantum field theory, but the convergence properties of the associated integrals are much better. The properties of multiloop (genus $g>1$) amplitudes are not fully understood. The analysis involves various sophisticated issues at the frontiers of the theory of Riemann surfaces, algebraic geometry, and maybe even number theory. However, it appears extremely likely that the following is true: The only divergences that occur are ones of very well understood and inevitable origin. The types of divergences that result in parameters becoming arbitrary in renormalized quantum field theories, or amplitudes becoming completely undefined in nonrenormalizable field theories, have no counterparts in string theory.

How can it be that general relativity, interpreted as a quantum theory, has nonrenormalizable divergences, whereas string theory, which agrees with it at low energies, is nonsingular? The essential reason can be traced to effects at the Planck scale that are present in string theory but not in general relativity. In particular, there is an infinite spectrum of string modes corresponding to particles whose mass is of the order of, or greater than, the Planck mass. These states contribute as virtual particles in scattering processes to produce subtle patterns of cancellations that soften the high-momentum ("ultraviolet") behavior of the Feynman integrals.

REMAINING CHALLENGES

A great deal of effort is being expended on the development of fundamental principles and a more geometric formulation. The history of string theory can be contrasted with that of general relativity. In that case, Einstein began by formulating certain far-reaching principles—the equivalence principle and general covariance—then finding their proper mathematical embodiment in the language of Riemannian geometry. This led to dynamical equations and experimental predictions, many of which have been tested and verified. In string theory, we have not yet identified the fundamental principles that generalize the equivalence principle and general coordinate invariance. These must surely exist, however, since general relativity is a low-energy (long distance) approximation to string theory. These principles, whatever they are, are likely to require a new kind of geometry, perhaps an infinite-dimensional generalization of Riemannian geometry, for their implementation. Some specific suggestions along these lines can be found in the recent literature, but our understanding is still far from complete.

Once the correct geometric formulation incorporating the fundamental principles of string theory in a comprehensible form is achieved, we should be in a good position to answer many profound questions. It should be possible to study nonperturbative effects and possibly understand why a particular solution with four-dimensional space-time and the phenomenologically required symmetries and particles is selected. It will also be interesting to study how string theory modifies classical general relativity at short distances. In particular, it would be interesting to know whether the singularity theorems are still valid. We could then also investigate how some of the profound issues of quantum gravity are resolved.

In a theory without adjustable parameters, any dimensionless number in nature should be calculable. Some of them are extremely small. For example, the cosmological constant, expressed in Planck units, is observed to be smaller than 10^{-120}. We might hope to identify a symmetry principle that forces it to be exactly zero, but none is known. Some theorists consider this the single most challenging problem in physics. Prior to string theory the cosmological constant was not calculable, and therefore the problem could not even be studied.

I find it remarkable that there now seems to be a reasonable chance that we will find a unique fundamental theory of nature. Certainly, it would have been considered pure folly to express such a hope 10 years ago. I think it is unrealistic to expect too much too soon, however. It will probably take a few decades of hard work to obtain a satisfactory understanding of what string theory is really all about. This will require substantial advances in mathematics. Also, the experimental results that can be expected during the next 10–20 years are likely to play an important role in shaping our ideas.

See also SUPERSYMMETRY AND SUPERGRAVITY.

BIBLIOGRAPHY

1. P. C. W. Davies and J. Brown (eds.), *Superstrings: A Theory of Everything?* Cambridge University Press, Cambridge, 1988. (E)
2. M. Kaku and J. Trainer, *Beyond Einstein*. Bantam Books, New York, 1987. (E)
3. M. B. Green, "Superstrings," *Sci. Am.* **255,** 48–60 (September 1986). (I)
4. J. H. Schwarz, *Superstrings—The First 15 Years of Superstring Theory*. World Scientific, Singapore, 1985. Reprints and commentary in 2 vols. (A)
5. M. B. Green, J. H. Schwarz, and E. Witten, *Superstring Theory*, 2 vols. Cambridge University Press, Cambridge, 1987. (A)
6. P. Ramond, "Dual theory for free fermions," *Phys. Rev.* **D3,** 2415 (1971); A. Neveu and J. H. Schwarz, "Factorizable dual model of pions," *Nucl Phys.* **B31,** 86 (1971). (A)
7. J. Scherk and J. H. Schwarz, "Dual models for non-hadrons," *Nucl. Phys.* **B81,** 118 (1974). (A)
8. M. B. Green and J. H. Schwarz, "Anomaly cancellations in supersymmetric $D = 10$ gauge theory and superstring theory," *Phys. Lett.* **149B,** 117 (1984); "Infinity cancellations in $SO(32)$ superstring theory," *Phys. Lett.* **151B,** 21 (1985). (A)
9. D. J. Gross, J. A. Harvey, E. Martinec, and R. Rohm, "Heterotic string theory (I). The free heterotic string," *Nucl Phys.* **B256,** 253 (1985); "Heterotic string theory (II). The interacting heterotic string," *Nucl. Phys.* **B267,** 75 (1986). (A)
10. P. Candelas, G. Horowitz, A. Strominger, and E. Witten, "Vacuum configurations for superstrings," *Nucl. Phys.* **B258,** 46 (1985); E. Witten, "Symmetry breaking patterns in superstring models," *Nucl. Phys.* **B258,** 75 (1985). (A)

Strong Interactions

Kameshwar C. Wali

INTRODUCTION

Long-range forces of gravity and electromagnetism account for large-scale macroscopic phenomena. On the small scale of atomic nuclei, the two other known types of interactions, "strong" and "weak," assume a primary role. In recent years, weak and electromagnetic interactions have found a unified description within the framework of a non-Abelian gauge theory with spontaneous symmetry breaking. It is based on the gauge group $SU(2) \times U(1)$ and uses the Higgs mechanism of spontaneous symmetry breaking to separate the weak and the electromagnetic interactions. Strong interactions were originally conceived as forces responsible for the binding of nucleons to form a nucleus and also for a wealth of phenomena observed in high-energy collisions of strongly interacting particles (hadrons). Since the 1930s, the theoretical description of strong interactions has undergone several phases. The modern theory of strong interactions is the theory called quantum chromodynamics. Like the electroweak interactions, it is a non-Abelian, quantum field theory which describes the interactions between quarks which are the constituents of nucleons. The conventional strong interactions (nuclear forces) between the observed hadrons are to be derived from QCD as residual interactions just as one derives Van der Waals–type electromagnetic interactions between neutral atoms.

HISTORICAL DEVELOPMENT

The classic α-particle experiments of Rutherford established that the bulk of atomic mass was concentrated in a very small nucleus of charge equal to the atomic number. The electrostatic repulsive forces in such a small system had to be overcome by stronger attractive forces to prevent its spontaneous disintegration. Moreover, the scattering experiments also showed that, given sufficient energies, the α-particles could overcome the electrical forces, penetrate the atom, and interact with the nucleus. One could infer from the experiments that the new nuclear forces and interactions were of short range ($\sim 10^{-13}$ cm), and approximately one thousand times stronger than the long-range electromagnetic forces. Hence the name "strong interactions." Further study of nuclear static and dynamic properties led to rapid developments in both theory and experiments. The discovery of the neutron in 1932 completed the picture of the nucleus as a composite of strongly bound neutrons and protons and solved the difficulties of the earlier model of the nucleus as a bound state of protons and electrons. A study of nuclear energy levels and scattering cross sections showed the equality of strong forces between protons and neutrons and hence their charge independence. Heisenberg introduced the concept of isotopic spin to characterize this property. Viewed in an abstract space, proton and neutron were up and down components of one and the same particle, the nucleon. The abstract space expressed internal degrees of freedom; isotopic spin was a new kind of symmetry in that it was not linked with the conventional space-time symmetries such as rotations and translations. Thus, the concept of an internal symmetry of an elementary particle was born. Isotopic spin symmetry postulated that the strong interactions were invariant under rotations in an abstract three-dimensional space provided much weaker electromagnetic and still much weaker weak interactions were neglected. In 1938, Yukawa provided a theoretical basis by postulating the existence of a new particle of finite mass whose exchange between the strongly interacting particles produced the short-range nuclear force. Its existence was subsequently confirmed in 1948 in cosmic ray experiments. It existed in three charge states, $+1, 0, -1$, in conformity with isospin symmetry. Known as π-mesons, they provided the foundation for meson theory of strong interactions.

However, meson field theory constructed along the lines of quantum electrodynamics (QED) proved inadequate to deal with the strong interactions, whose coupling strength measured in terms of a dimensionless parameter (known as Yukawa coupling constant) was of an order of magnitude of 10. Hence perturbation theory so successful in the case of QED was not of much help. In the meantime, it became possible to manufacture meson beams by using nuclear collisions in laboratory accelerators and do scattering experiments with π-mesons on protons. In the early 1950s, Fermi and his collaborators discovered by such experiments a short-lived resonant state (N^*), which could be considered as a new species of nucleon. This marked the beginning of a proliferation of so-called elementary particles. Experiments outpaced theory and the rapid discoveries of new particles and resonant states by 1960–1961 suggested larger families and higher internal symmetries of their interactions. A new additive quantum number called the "strangeness" was introduced by Gell-Mann and Nishijima to explain the strong production but weak decays of a new species of particles (K, Λ, Σ, and Ξ particles). Strict conservation of strangeness became a property of strong interactions not shared by weak interactions. In the collisions of ordinary particles with strangeness quantum number equal to zero, the strange particles had to be produced in association so that the total strangeness of the product particles was zero. The experimental verification of this phenomenon of associated production established strangeness as a new internal symmetry feature of strong interactions. Gell-Mann and Ne'eman combined strangeness and isotopic spin into a higher symmetry based on the group $SU(3)$, which became quite successful in classifying the hadonic multiplets. The dramatic prediction of the Ω^- particle to complete a baryonic decuplet and its subsequent discovery, and likewise the prediction of the η-meson and its subsequent discovery, launched $SU(3)$ symmetry as a very good approximate symmetry of strong interactions. Subsequently, even a higher symmetry, $SU(6)$, came into play when ordinary spin was combined with $SU(3)$ symmetry and the baryonic octet (spin $\frac{1}{2}$) and decouplet multiplets (spin $\frac{3}{2}$) found themselves together in the **56** multiplet of $SU(6)$. Although a less precise symmetry than $SU(3)$, $SU(6)$ classification of N and N^* into a single multiplet led to some remarkable correlation of their properties. However, the relativistic extension of the theory ran into conceptual as well as experimental difficulties and had to be abandoned. In its failure, however, it left behind a new concept, the concept of color. The **56** multiplet of $SU(6)$ was totally symmetric in its spin and $SU(3)$degrees of freedom.In order to reconcile this fact with the generalized Pauli's exclusion principle for the fermions, a new degree of freedom was necessary and that was provided by $SU(3)$ color. More about this later.

Two alternative theoretical viewpoints on strong interactions emerged in the 1970s. One was the quark model and the other was the S-matrix theoretic approach. The quark model with fractionally charged triplets and antitriplets proved eminently useful for symmetry considerations and explaining the fact that the observed hadrons appeared as only singlet, octet, nonet, or decuplet multiplets of $SU(3)$. This was a simple consequence of the quark model if mesons were bound states of quarks and antiquarks, and baryons were bound states of three quarks. With the color degree of freedom, each quark came in three colors. Stated more precisely, quarks (antiquarks) belonged to the triplet (antitriplet) representations of $SU(3)$ color ($SU(3)_c$). The extension of the quark model to dynamics, that is, calculations of decay rates, selection rules, cross sections in various reactions, etc., gave encouraging results in agreement with a wide variety of data. However, the main successes of the quark model were in the low-energy regime. It did not shed much light on relativistic high-energy processes. Moreover, experimental searches for isolated fractionally charged quarks yielded only negative results.

The S-matrix approach, popularly known as the bootstrap approach, was an outgrowth of the dispersion-theoretic investigations of Chew, Goldberger, Low, and many others. From the proliferated families of hadrons, it became meaningless to categorize a subset of them as elementary and the others as composites. Chew hypothesized that all hadrons were equally elementary or equally composite and that all hadrons were composites of other hadrons bound together by attractive forces originating from the exchanges of hadrons. With dispersion relations, analyticity in energy and momentum variables, and unitarity as tools, attempts were made to set up self-consistent bootstrap models to explain the observed hadrons as poles in the analytically continued S-matrix amplitudes. However, the approximation schemes necessary to carry out calculations in practice turned out to be totally inadequate to explain the observed complexity in hadron spectroscopy, and the bootstrap approach slowly faded away. In the meantime, the concept of Regge poles, which may be regarded as an extension of particle-exchange concepts to high-energy collisions, entered in hadron physics. A Regge pole exchange in the t channel was equivalent to a sum of resonances in the direct or s channel. This concept of duality culminated in the so-called Veneziano model which was a highly constrained nonunitary first approximation, but one which embodied the bootstrap idea and related the high-energy scattering in one channel to properties of low-energy resonances in the crossed channel. In the hands of Nambu, the dual model of Veneziano was transformed into a Lagrangian formalism based on the hadronic string model and a mathematical framework to derive N-particle, factorized S-matrix elements came into existence. Certain consistency requirements led to the necessity of introducing more space-time dimensions (26 and 10 in bosonic and fermionic strings, respectively). The whole scheme, however, faced severe difficulties owing to the presence of tachyons and the theory remained dormant until its reinterpretation in the 1980s as the superstring theory, a theory not of the observed hadrons, but the theory of all the underlying elementary constituents, in fact a theory of everything. A discussion of this development is beyond the scope of this review.

MODERN THEORY

The current or modern theory of strong interactions reverts back to the old-fashioned relativistic quantum field theory with the underpinning of gauge invariance, locality, causality, and renormalizability. The theory is called quantum chromodynamics (QCD). The local gauging of the previously mentioned $SU(3)$-color degree of freedom provides the strong field just as the local gauging of $U(1)$ electric charge provides the electromagnetic field. Quarks are the sources and sinks of the color field interacting with each other through the emission and absorption of eight massless quanta called gluons just as charged particles interact with each other by emission and absorption of massless photon—the quantum of the electromagnetic field. QCD in some respects is similar to QED. But a profound difference arises between the two because unlike electric charge, color is not a scalar. Quarks can change color by emitting and absorbing gluons which carry color. Such theories are called non-Abelian gauge theories and their unexpected properties seem to fit what is needed for the description of strong interactions.

COMPARISON BETWEEN THEORY AND EXPERIMENT

Quark Confinement

Since isolated quarks are not observed, QCD must give rise to forces that eternally confine the quarks within the hadrons. The non-Abelian nature of the QCD interactions do provide a hope for such confinement. It is expected that the interactions can give rise to a linearly rising potential requiring infinite energy to separate colored quarks from each other. However, when the quarks or quarks and antiquarks form color singlets, the interactions disappear and the color singlets can escape the confinement. These are the observed hadrons. The conventional strong forces between the hadrons are to be interpreted as Van der Waals–type forces between neutral atoms. Nonperturbative QCD formulated on a lattice has confirmed at least qualitatively some of the above-stated expectations regarding confinement.

Deep Inelastic Scattering; Quark–Parton Structure

Deep inelastic scattering and hadron–hadron collisions (high energy, small momentum transfer) suggested point-like constituents inside the hadrons leading to Feynman's parton parton model. The partons were soon identified as quarks and the quark–parton model was strikingly successful in a phenomenological explanation of high-energy electron–hadron, neutrino–hadron, and some features of hadron–hadron scattering. This intuitive quark–parton model is given a microscopic theoretical basis by QCD because of two of its remarkable properties. One, as already stated, is the property of color confinement. The other is the so-called asymptotic freedom. Unlike QED, coupling and mass parameters in QCD are not constants. They scale with energy. Coupling strength in particular becomes weaker and in the extreme high-energy limit it vanishes. Quarks and partons, therefore, behave as free constituents when explored at high energies or short distances. Because of this property, QCD allows systematic perturbative methods by means of which one can calculate the scaling behavior in various reactions. It also allows one to derive parton model sum rules as a first approximation and compute corrections.

e^+e^- Annihilation

The energy distribution of hadron products in e^+e^- annihilation can be predicted from scaling and asymptotic freedom properties of QCD. To a first approximation, e^+e^- annihilation proceeds by "two-jet" formation; the jets are produced by a quark and an antiquark. Three-jet formation processes attributable to quark, antiquark, and a gluon have also been observed. Detailed analyses of the characteristics of jets confirm the hypothesis that they originate in spin-$\frac{1}{2}$ quark or spin-1 gluon.

Quarkonium

Bound states of heavy quark–antiquark pairs annihilate when these particles approach within a Compton wavelength of each other. Perturbative QCD can be applied to predict the ratio of hadronic annihilation (via the predominant three-gluon intermediary) to the electromagnetic annihilation into $\mu^+\mu^-$. Theoretical predictions are consistent with the experimental findings in the case of ψ, ψ', and γ decays.

SUMMARY

A vast amount of experimental data (e^+e^- anninilation jets, deep inelastic scattering, quarkonium decay) has been analyzed using perturbative QCD. It is fair to say that the results are consistent with QCD. However, as a fundamental theory of strong interactions, QCD has yet to pass severe tests. Confinement of colored states is yet to be established rigorously either through further developments of the theory and/or more extended and refined calculations on a lattice. QCD has nothing whatsoever to say about the quark mass spectrum or the details of internal structure of the hadrons. To relate experimental information obtained from hadronic reactions to quarks and compare the results with QCD, one needs to make assumptions that go outside the framework of QCD. Further, within the QCD framework, there is no natural explanation for isospin conservation, approximate $SU(3)$ flavor symmetry, etc., which play such an important role in hadronic phenomena.

See also DISPERSION THEORY; ELECTROMAGNETIC INTERACTION; GAUGE THEORIES; GRAVITATION; HADRONS; ISOSPIN; LATTICE GAUGE THEORY; MESONS;

QUARKS; REGGE THEORY; S-MATRIX THEORY; STRING THEORY; $SU(3)$ AND HIGHER SYMMETRIES; WEAK INTERACTIONS.

BIBLIOGRAPHY

General

Kurt Gottfried and Victor F. Weisskopf, *Concepts of Particle Physics,* Vols. 1 and 2. Clarendon Press, Oxford, and Oxford University Press, New York, 1984.

S Matrix and Regge Poles

S. C. Frautschi, *Regge Poles and S Matrix Theory*. W. A. Benjamin, San Francisco, 1963.

P. H. Frampton, *Dual Resonance Models*. W. A. Benjamin, San Francisco, 1974.

Quark Model and Higher Symmetries

J. J. J. Kokkedee, *The Quark Model*. W. A. Benjamin, San Francisco, 1969.

F. J. Dyson, *Symmetry Groups in Nuclear and Particle Physics*. W. A. Benjamin, San Francisco, 1966.

Quantum Chromodynamics

W. Marciano and H. Pagels, *Phys. Rep.* **36,** 137 (1978).

F. Wilczek, *Annu. Rev. Nucl. Sci.* **32,** 177 (1982).

Sum Rules

J. B. French

STRENGTH DISTRIBUTIONS

The most important use of sum rules is in studying the excitation or decay modes of a system. Our purpose may be to learn more about the excitation itself (how, for example, is the electromagnetic-quadrupole excitation related to that induced by inelastic scattering of particles?); or we may wish to study some of the states of a system, or its Hamiltonian, by using sum rules to measure occupancies or energies of single-particle states; or we may want to determine either the adequacy of a particular model space used to describe the system (is a restricted set of orbits adequate for the low-energy states of an atom? Must we consider mesonic degrees of freedom in a nuclear process?); or the existence of relevant symmetries or other algebraic structures.

The essential idea is that if $|W\rangle$ is a stationary state and O an excitation operator, then $O|W\rangle$, not in general stationary, may be expanded in the stationary states:

$$O|W\rangle = \sum_{W'} \langle W'|O|W\rangle \cdot |W'\rangle. \tag{1}$$

We can define $|\langle W'|O|W\rangle|^2 = R(W',W)$, the square of the expansion coefficient (Fig. 1), as the *strength* for the process $|W\rangle \xrightarrow{O} |W'\rangle$, or the relative probability that under the action of O the system will make the transition $|W\rangle \rightarrow |W'\rangle$. Then the *nonenergy-weighted sum rule*

$$\mathcal{M}_0(W) = \sum_{W'} R(W',W) = \sum_{W'} |\langle W'|O|W\rangle|^2 = \langle W|O^\dagger O|W\rangle \tag{2}$$

simply expresses the square of the norm of $O|W\rangle$ in terms of the total strength starting with $|W\rangle$ and also via an ex-

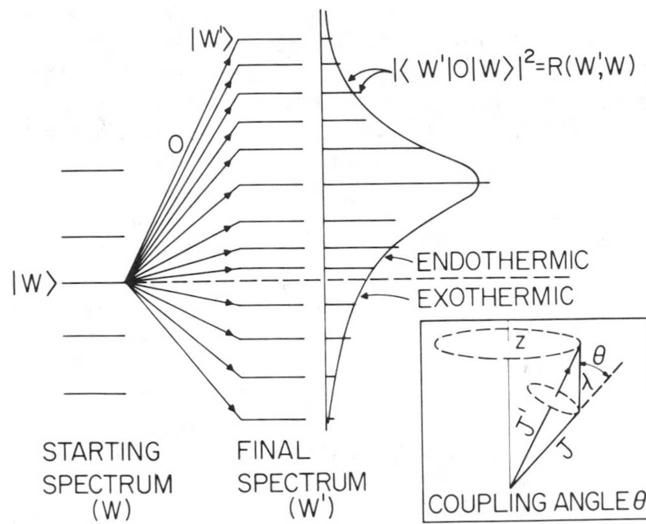

FIG. 1. For an operator O acting on the state $|W\rangle$ the strength is indicated by the length of the horizontal line drawn at each final state $|W'\rangle$; the curve gives the smoothed strength. It is not essential that O correspond to an observable excitation. The final spectrum is divided according as $(W'-W)\gtrless 0$. If the states carry definite angular momenta (J,J'), the strength distribution breaks into a set of subdistributions, one for each J'. If O also carries angular momentum λ, the decomposition gives the "angular distribution" of the strength according to the coupling angle θ shown in the inset $(2[J(J+1)\lambda(\lambda+1)]^{1/2}\cos\theta = J'(J'+1) - J(J+1) - \lambda(\lambda+1))$.

pectation value in $|W\rangle$. In a wide range of circumstances, usually by invoking some version of perturbation theory, the strengths can be deduced from experimental cross sections, lifetimes, etc., and then (2) gives us information about the starting state $|W\rangle$. Formally we regard $|W\rangle$ as a member of a model space in which O and the Hamiltonian H operate. Then if, for example, the total measured strength $\mathcal{M}_0(W)$ (or even, as sometimes happens, the strength for a single transition) is larger than the maximum eigenvalue of $O^\dagger O$, the space must be inadequate. As another example, for excitation operators $O^{(+)} = a_i^\dagger$, the fermion creation operator, and $O^{(-)} = a_i$, $\mathcal{M}_0^{(-)}(W)$ is the occupancy of single-particle state No. i in the (many-fermion) state $|W\rangle$. Moreover, $\mathcal{M}_0^{(-)}(W) + \mathcal{M}_0^{(+)}(W) = 1$, which (a) can be used to test the method of determining the strengths (in nuclei this is done via nucleon-transfer experiments); (b) demonstrates the fundamental commutation law $[a_i, a_i^\dagger]_+ = 1$.

The sum rule (2) takes account of the total strength originating with $|W\rangle$ but not of its energy distribution. We can do better by considering the *energy-weighted sum rules*

$$\mathcal{M}_p(W) = \sum R(W',W)(W')^p = \langle W|O^\dagger H^p O|W\rangle, \tag{3}$$

in which we recognize the $\mathcal{M}_p(W)$ $(p \geq 0)$ as the moments of the strength that starts with $|W\rangle$.

In practice we are restricted to low-order sum rules, say $p \leq 4$. Experimentally the high-order energy weighting emphasizes small parts of the strength, which are not measurable because of contributions from other processes. Theoretically $O^\dagger H^p O$, for high p, involves many-body interactions, and model states that might be available will be inadequate to deal with them (model states always promise more than they can deliver in the way of high-order correlations). If, however, a known mechanism essentially fixes the shape, the distribution is determined by the $p=0-2$ sum rules (the centroid is $\mathcal{M}_1/\mathcal{M}_0$ and the width $[\mathcal{M}_2/\mathcal{M}_0-(\mathcal{M}_1/\mathcal{M}_0)^2]^{1/2}$). This happens, for example, if the strength is large and confined to a narrow energy band, thereby forming a "giant resonance." Then, for a resonance built on the ground state, $[H,O]|W_g\rangle \equiv -i\hbar\dot{O}|W_g\rangle \simeq \hbar\omega|(W_g+\hbar\omega)\rangle$ with $\hbar\omega$ the resonance energy. If also $O^\dagger|W_g\rangle=0$, we have an analog of the harmonic oscillator, $H=\hbar\omega(a^\dagger a+\frac{1}{2})$ with $O=a^\dagger$ a boson creation operator. In complex systems we may seek an O satisfying these equations in order to produce a theory of harmonic vibrations.

For another example consider the Thomas–Reiche–Kuhn sum rule: $p=1$; $O=O^\dagger=$ electric-dipole operator $=\Sigma e_i \mathbf{r}_i$. R is measurable for an atom via the integrated photoelectric cross section. $2OHO=[O,[H,O]_-]_- + [OO,H]_+$ and if $H|W\rangle=0$ and $H=T+V$ where $[V,O]=0$, then $O^\dagger HO \equiv (\hbar^2/2M_e)z=\mathcal{M}_1(W)$. In nuclei the sum rule is disturbed by the space-exchange part of V (which does not commute with O and therefore gives corrections to $\mathcal{M}_1$) and modified by the presence of the neutrons. The $E1$ strength forms a high-lying giant resonance describable classically as a vibration of protons with respect to neutrons.

In finite-dimensional model spaces the strength distribution is determined by the proper moments ($p\geq 0$) but it is often useful to consider inverse energy weighting ($p<0$); this is true *a fortiori* for more general spaces. Many atomic quantities (refractive indices, stopping power, etc.) are simply expressible in terms of moments, including $p<0$, of the electric-dipole oscillator strength $(W'-W)R(W',W)$, and similarly for parameters that describe collective motions in many-particle systems. Often, however, we can measure only the exothermic (or only the endothermic) strengths (Fig. 1), and indeed some of the $|W\rangle$ states may not even be physically realizable in the process.

SUM RULES AND ALGEBRAIC STRUCTURES

If the excitation carries a definite angular momentum λ we can refine Eqs. (1)–(3) by regarding the (Clebsch–Gordan) angular momentum coupling as a tensor multiplication, $O|W\rangle$ being replaced by $(O^\lambda \times \psi^J(W))^{J'}$ with J' the resultant. Ignoring statistical and phase factors, we find (writing $\bar{\psi}$, $\bar{O}$ for the adjoints, and $\langle\ \rangle$ as the matrix element) that

$$(O^\lambda \times \psi^J(W))^{J'} \approx \sum_{W'} \langle (\bar{\psi}^{J'}(W') \times O^\lambda \times \psi(W))^0 \rangle \psi^{J'}(W'), \quad (4)$$

$$\mathcal{M}_p(W,J\rightarrow J') \approx \sum_\nu \langle (\bar{\psi}^J(W) \times (\bar{O}^\lambda \times H^p \times O^\lambda)^\nu \times \psi^J(W))^0 \rangle U(J\lambda J\lambda;\, J'\nu). \quad (5)$$

The Racah coefficient in (5) corresponds to P_ν (cosθ) (Fig. 1), so that (5) gives the "angular dependence" of the strength moment via a P_ν expansion. A measurement of the total strengths for each J' determines, by a Racah–Legendre inversion of (5) for $p=0$, the coupled matrix element for each ν value. We see that $\bar{\psi}^J(W)$ and $\psi^J(W)$ must also couple to angular momentum ν so that the fixed-J' measurements sense also the nonscalar many-particle density, in contrast to the scalar sum rules (3). With single-particle transfer we determine not just the occupancies but also certain contributions to the electromagnetic moments, while electric moment and transition-rate (especially $E2$) measurements determine much about the nuclear shapes.

The J_μ operators (rotation-group generators) are closed under commutation (defining therefore a Lie algebra and associated symmetry) and, for most systems, they commute with H so that the symmetry is good. For other good symmetries we can proceed as above, though for the simple representations found at low energies it is often easier to select the final-state symmetry via a projection operator $\mathcal{P}$ (whence $\langle W|O^\dagger H^p O|W\rangle \rightarrow \langle W|O^\dagger H^p \mathcal{P} O|W\rangle$ in (3)).

We have a broken symmetry if the generators of a Lie algebra do not all commute with H. One such is nuclear spin–isospin $SU(4)$, whose algebra includes the allowed Fermi and Gamow–Teller β-decay operators. The Fermi operators generate an almost good isospin $SU(2)$ subgroup that gives a very sharp β-decay resonance whose $p=0$ sum rule is almost exactly calculable. In general the most interesting sum rules arise when the symmetry breaking is small enough to treat by perturbation theory; a major example is the Adler–Weisberger sum rule relating the axial-vector β-decay coupling constant to an energy integral involving $\pi^\pm + p$ cross sections.

STRENGTHS, SUM RULES, AND STATISTICS

In finite-dimensional many-particle model spaces (important, e.g., for nuclei) there are remarkable connections between strengths and sum rules, symmetries, and statistical behavior. These follow from central limit theorems connected with the group $U(N)$ of unitary transformations in the N-dimensional single-particle space and with its various subgroups. To within fluctuations, they determine $R(W',W)$ itself (and therefore resolve the endothermic–exothermic problem and generate the inverse-energy-weighted sum rules) and all without the explicit construction of the many-particle states, which is only too often not feasible.

Energy-level fluctuations in a quantum system are describable by n-point correlation functions, the simplest and often the only one giving rise to measurable quantities being the $n=2$ function $S(x,y)=\langle\rho(x)\rho(y)\rangle-\langle\rho(x)\rangle\langle\rho(y)\rangle$; here the microscopic density function $\rho(x)$ defines the spectrum and $\langle\ \rangle$ indicates ensemble or spectral averaging (often related by ergodicity). The comparative study of experimental and theoretical "fluctuation measures," which are sum-rule quantities defined as weighted integrals of $S(x,y)$, represents one of the few ways presently available for defining and studying quantum chaos and its relationship to classical chaos. The same formalism is encountered, for example, in the dynamical theory of liquids for which $\rho(x,t)$ is the microscopic particle density at time t. One often deals there with the "dynamic structure factor," $S(k,\omega)$, the double (space and time) Fourier transform of the $n=2$ time-dependent S function. This is well adapted, for example, to the analysis of inelastic neutron scattering by liquids (k, ω then defining momentum and energy transfer) and gives rise to a family of directly applicable sum rules. It is obvious that sum rules arise in many domains of physics; the significant ones are those which can be experimentally measured or theoretically evaluated or both.

See also BETA DECAY; NUCLEAR REACTIONS; PHOTONUCLEAR REACTIONS; ROTATION AND ANGULAR MOMENTUM.

BIBLIOGRAPHY

Various aspects of sum rules for simpler systems are discussed by H. A. Bethe and E. E. Salpeter, *Handb. Phys.* **35,** 88; U. Fano and J. W. Cooper, *Rev. Mod. Phys.* **40,** 441 (1968), give a very complete discussion of atomic oscillator-strength distributions; a detailed application is by J. L. Dehmer, M. Inokuti, and R. P. Saxon, *Phys. Rev.* **A12,** 102 (1975). An extended account of multipole sum rules is given by J. B. French, in *Proc. Fermi Int. Summer School,* Course 36 (Academic Press, New York, 1966). A beautifully clear account of broken symmetries and sum rules is by N. Cabibbo in *Recent Developments in Particle Symmetries,* A. Zichichi, ed. (Academic Press, New York, 1966). Determination of nuclear shapes by sum-rule analyses of electromagnetic moments and strengths is treated by D. Cline and C. Flaum, in *Proc. Int. Conf. on Nuclear Structure Studies (Sendai 1972),* K. Shoda and H. Ui, eds. (Tohoku University, 1972).

For analysis of energy-level fluctuations see R. U. Haq, A. Pandey, and O. Bohigas, *Phys. Rev. Lett.* **48,** 1086 (1982). For liquids an elegant account is given in *Atomic Dynamics in Liquids* by N. H. March and M. P. Tosi (Wiley, New York, 1976). A very detailed treatment of excitation sum rules in nuclei is given by E. Lipparini and S. Stringari, *Phys. Rep.* **175,** 103 (1989).

Sun

Thomas P. Caudell and *Henry A. Hill*

The sun is, from a galactic perspective, an ordinary star, being distinguished in its brightness only among the many faint red dwarf stars in the "local" stellar neighborhood. Albeit ordinary, the sun is *our* star and is responsible for the existence and maintenance of all known living organisms on this planet. This fact alone makes the observation and understanding of our star one of the most relevant goals in modern astrophysics today.

The close proximity of the sun allows detailed investigation of the intricate solar surface, as well as global structures and properties. Much has been learned about the sun through the observation of its surface and the light emitted there, tracking the motion of solar system objects around the sun and the measurements of the solar diameter. These observations tell us the mass [5] of the sun,

$$M_{\odot} = 1.989 \times 10^{33} \text{ g};$$

the radius [5],

$$R_{\odot} = 6.9598 \times 10^{10} \text{ cm};$$

and luminosity [5],

$$L_{\odot} = 3.90 \times 10^{33} \text{ ergs/s}.$$

Through spectroscopic observation of the visual surface of the sun, called the photosphere, the composition and run of temperature through the solar atmosphere may be inferred. The composition [3] is by mass approximately hydrogen, 78%; helium, 20%; others, 2%. The temperature varies from nearly 7000 K at the bottom of the photosphere, a thin layer some 500 km thick, to a minimum of 4300 K near the top. Higher up, in the chromosphere and corona, where the atmosphere is *very* thin and subject to the whims of solar magnetic fields, the kinetic temperature rockets up to a maximum of over 10^6 degrees. The chromosphere and corona are easily visible during total solar eclipses but are otherwise too faint to see.

Sunspots are magnetically concentrated cool regions located within the photosphere. Their dark shade is due to their relatively low temperature [3] in relation to that of the surrounding gas, some 4000 K opposed to 7000 K. Observing sunspots over a period of days will plainly show solar rotation. Recording the rotational period of sunspots as a function of solar latitude reveals the interesting result that the equator rotates faster than high latitude regions; the period of rotation [3] varies from 26 days at the equator to almost 32 days at the poles. The origin of this differential rotation is as yet unexplained.

The structure of the solar interior must be inferred from theory. Accurate computer calculations "evolve" the sun from zero age to its present state using the current radius, mass, luminosity, and composition as boundary conditions. By placing the sun within the larger scheme of observed stars, a somewhat coherent picture of the birth of the sun has been pieced together. Starting within a large interstellar cloud of hydrogen gas and dust in a spiral arm of our galaxy, as a denser than average knot, perhaps similar to the Bok globules observed in the Milky Way [2], the infant sun began to contract under the force of its own gravity. Although these early stages of collapse are not well understood, it is suspected that such an object will emit infrared light as gravitational potential energy is converted into thermal kinetic energy. As the central regions contract faster than the surface layers, the central temperature quickly rises to the ignition temperatures for thermonuclear reactions [3], nearly 10^7 degrees. Converting hydrogen into helium through nuclear fusion in its core, with the release of radiant and thermal energy, the body establishes a pressure gradient which slows and eventually stops the contraction leading to a state of quasiequilibrium. The internal structure is delicately balanced such that the release of energy in the core just equals the radiant energy escaping into space. At this point the sun is said to have reached its main sequence phase at zero age.

Mathematically evolving this zero-age main-sequence model of the sun to the present age, believed to be 4.7×10^9 years, one finds the interior stratifying into four major shells, the core, mentioned previously, the intermediate interior, the convection zone, and the photosphere. The average properties [3] of each are given in Table I. The mechanism through which the thermonuclear energy is transported down the temperature gradient from the core to the surface partially defines the distinction between layers. Energy is transported by radiation, absorption and reemission, through the core and intermediate zones. In the convection zone, the gradients are such as to lead to an instability which produces

Table I Average properties of Sun's major shells

	Thickness	Average temperature (K)	Average density (g/cm³)
Photosphere	$7\times10^{-4}R_\odot$	5.4×10^{3}	2×10^{-7}
Convection zone	$0.15R_\odot$	0.25×10^{6}	5×10^{-3}
Intermediate zone	$0.60R_\odot$	4×10^{6}	10
Core	$0.25R_\odot$	11×10^{6}	89

turbulent flow of the gas and transporting of the energy through hot convective gas cells. This energy is deposited in the layers immediately below the photosphere where radiative processes transport the energy into the photosphere. The photosphere is defined as the layer from which practically all light which escapes to space is finally reemitted or scattered. The light we detect at earth from the photosphere [3], if assumed to be emitted in thermal equilibrium, has a blackbody temperature of 5780 K. Although the light is at best formed in local thermal equilibrium, the spectral characteristics of the sun and the blackbody radiator at 5780 K are remarkably similar.

Theories of the processes occurring in the photosphere and the upper atmosphere may be directly detected in detail through observations. Theories of the interior, on the other hand, have not been testable in detail in the past. That condition has been rapidly changing in the last few years. A subatomic particle which is a by-product of the nuclear fusion process in the core, known as the neutrino [1], should be detectable on earth at a certain flux level according to calculations for a standard solar model. To date these neutrinos have not been observed at fluxes consistent with the calculated flux, but much lower levels. This result implies that perhaps the standard model is not correct in the central regions where the neutrinos are produced and that the temperature there should be less. Much theoretical effort has recently gone into constructing models of the sun with cooler cores while maintaining the correct radius, mass, and luminosity, a nontrivial task.

The sun has been discovered to oscillate in small-amplitude global normal modes with periods ranging from a few minutes to several hours (see Ref. 4 for review of the subject). Unlike the neutrino measurement, observations of global oscillations have the potential of giving information and therefore model constraints on *all* layers in the sun. The periods as well as the spatial surface properties of oscillations depend on the details of the model and as such will eventually allow detailed tests of models of the solar interior.

Not more than a half a century has passed since Sir Arthur Eddington speculated that the energy source for the sun and stars was of nuclear origin. Until a decade ago the sun was considered understood. The theory of stellar evolution had yielded the gross properties of our star and no contradictory observations existed. Within the last 10 years many new observations and experiments have been performed; the unobservable is being brought into the open and we find that our knowledge of the solar interior is not adequate. The implications of the inadequacy go well beyond the sun and may have important consequences in other areas of astrophysics.

See also MILKY WAY; NUCLEOSYNTHESIS; PHOTOSPHERE; STELLAR ENERGY SOURCES AND EVOLUTION.

REFERENCES

1. J. Weneser and G. Friedlander, *Science* **235,** 755 (1987).
2. B. J. Bok and P. F. Bok, *The Milky Way*. Harvard University Press, Cambridge, Mass., 1976.
3. E. G. Gibson, *The Quiet Sun*. NASA SP-303, 1974.
4. Arthur N. Cox, William C. Livingston, and Mildred S. Matthews (eds.), *The Solar Interior and Atmosphere,* University of Arizona State Science Series. University of Arizona Press, Tucson, 1990.
5. S. Weinberg, *Gravitation and Cosmology, Principles and Applications of the General Theory of Relativity*. Wiley, New York, 1972.

Superconducting Materials

D. K. Finnemore

With the discovery of superconductivity in the copper oxide perovskite-like materials, a new era of superconducting materials research was opened. This class of materials not only shows high-temperature superconductivity but they also show a whole range of interesting new normal-state properties. The chemical bonding, the mechanical metallurgy, the methods for processing, and the physical properties of these compounds are sufficiently different from the the classical superconduction that it is necessary to delve deeper into our understanding of both the basic phenomena that cause superconductivity and the preparation procedures needed for these materials. One of the many ways in which these materials are unusual is that a small change in the oxygen content or a small change in cation doping level can transform the material from a magnetic insulator, first to a spin glass, then to low-carrier-density metal, and finally to a superconductor. All of these occur with a few percent change in doping. These copper oxide compounds would be very interesting new materials even if they were not high-temperature superconductors. This article describes some of the basic ideas that are important in the development of new materials and compares the classical superconductors such as Nb or Nb_3Sn with the more intricate copper oxide materials.

RAISING THE TRANSITION TEMPERATURE

After the initial discovery of superconductivity by Kamerlingh Onnes in 1911, there was a 75-year period in which there was a rather steady rate of discovery of new superconducting materials. About half the elements of the periodic table were determined to become superconducting under the proper conditions. For cases like tungsten metal, it may require very low temperatures; for cases like silicon, it may take very high pressures; or for cases like bismuth, it may require an amorphous state. In one form or another, however, many elements show the effect. In the early days, the search for new superconductors concentrated on the study of elements but a vast array of alloys also soon were found. In 1932, the discovery of superconductivity in NbC at 11 K ushered in a new era in which many compounds with the NaCl or B1 crystal structure were found to be superconducting. In 1941, NbN at 15 K was discovered and in 1953 $NbN_{0.7}C_{0.3}$ at 17 K was discovered. That same year V_3Si at 17 K ushered in another era of discoveries centered on materials having a beta-tungsten or A-15 structure. These include Nb_3Sn at 18 K in 1954 and Nb_3Ge at 23 K in 1973. Interspersed along the way, of course, was the discovery of superconductivity in many other materials at lower temperature, most notably the organic superconductors which have T_c's in the range of 10 K. For the first 75 years there was a relatively steady increase in the highest T_c of about 3 K per decade. We call these the classical superconductors.

Then came the discovery of superconductivity in a new class of copper oxide metals by Bednorz and Müller [1]. In 1973, Johnston and coworkers [2] had discovered the first oxide superconductor with $LiTi_2O_4$ and in 1975, Sleight and coworkers [3] had opened the field of perovskite superconductors with $(Pb\text{-}Bi)BaO_3$, but it was the discovery of $(LaBa)_2CuO_4$ near 36 K that started the revolution. Within 2 years, the yttrium-based compounds were discovered by Chu and coworkers [4] at 92 K, the Bi-based materials by Maeda and coworkers [5] at 115 K, and the Tl-based materials by Sheng and Hermann [6] at 125 K. The copper oxide compounds seemed to be a special new class that might involve a new interaction causing the high degree of correlation among the electrons. Along with the rise in T_c came corresponding progress raising some of the other performance parameters as well. The upper critical field, H_{c2}, and the superconducting energy gap, Δ, scale with the transition temperature, T_c, so they also are extremely high. In practical applications this means, for example, a new regime of magnetic fields for magnets and a new regime of frequency for infrared detectors. The primary difficulty with these materials has been the very brittle mechanical behavior and the rather low critical current density which the materials will carry at high temperature and high magnetic field.

FUNDAMENTAL MATERIAL PROPERTIES

A nearly free-electron or band theory picture describes the normal state of the classical superconductors rather well and the BCS [1] theory describes the superconducting state. Within this picture, the superconducting ground state is composed of a coherent mixture of normal state wave functions in which states are occupied in pairs with opposite momentum and spin ($k\uparrow, -k\downarrow$). In classical superconductors, it is the electron–phonon interaction that causes the electron waves to add coherently at one place in space to create a Cooper pair wave packet. It should be noted, however, that this coherent wave packet could be caused by an attractive interaction. In the high-T_c materials, for example, it has been suggested that a magnetic interaction involving the spin on the copper site could also lead to a coherent superconducting state; all the features of the BCS theory go through for any attractive interaction.

To understand how superconducting electrons lock into a phase coherent unit, an analogy may be useful. In some ways, a superconductor is like a piece of rope. In a rope, the strand-to-strand coupling is rather weak but there are thousands of strands overlapping any one strand so the cooperative strength is large. In a superconductor, the pairing forces are rather weak but there are millions of pairs overlapping any one pair so the cooperative coupling force is strong. It is helpful therefore to have a long coherence distance and lots of overlap of the Cooper pairs to get strong pair–pair correlations.

The two-fluid model provides a very convenient way to think about the electronic properties of superconductors. In this model the electron gas is composed of a normal fluid with density n_n, in which electrons behave just as they do in the normal state. There also is a superfluid with density n_s, which shows all the superconducting properties. In the normal state $n_n = 1$ and all the electrons are normal. As a superconductor is cooled through T_c, n_s begins to grow and n_n decrease such that $n_n + n_s = 1$. The superfluid density rises quickly at T_c and $n_s \sim 0.9$ by the time the reduced temperature $t = T/T_c = 0.6$. For most superconducting applications it is necessary that most of the electrons be in the superconducting ground state or the superfluid state. In addition, the excitations out of the ground state or the normal fluid should be a small fraction of the total. For most practical applications, T/T_c should be on the order of 0.6 which is 75 K for the $Tl_2Ba_2Ca_2Cu_3O_{10}$ material with $T_c = 125$ K. For $Y_1Ba_2Cu_3O_7$ with $T_c = 92$ K the operating temperature should be about 55 K.

A key parameter determining the performance of high-performance materials is the size of the Cooper pair wave packet (which is just the coherence distance). It is given by the distance an electron traveling at the Fermi velocity, v_F, will travel in the lifetime of the virtual particle exchanged to give the correlated state $\xi = v_F \Delta t = v_F \hbar/\Delta E \sim \hbar v_F/kT_c$. This means that high values of T_c lead to short coherence distances and small correlation volumes. The only way to alleviate this short ξ problem in high-temperature superconductors is to have materials with large v_F. In the classical superconductors, ξ ranges from about 4 nm in a *d*-band compound like Nb_3Sn to about 1000 nm in an *s-p* band metal like Al. In the high-temperature superconductors, $\xi \sim 0.5$ to 2 nm so the correlation volumes, which go as ξ^3, are very small. If a room-temperature superconductor were found with the same v_F as for $Tl_2Ba_2Ca_2Cu_3O_{10}$, ξ would be less than one unit cell and the pair–pair overlap would be very small. Hence, long-range phase locking among the electrons might be difficult to maintain.

A second key materials property is the superconducting penetration depth, λ. It is the characteristic distance in which an applied magnetic field will change in a superconductor and it goes inversely as the square root of the density of carriers, $\lambda^2 \sim n_s^{-1}$. For the classical materials this is typically 50 nm and for the high-T_c materials it is 120 nm. In this regard the high-T_c materials are similar to classical materials. For a penetration depth of a few hundred nm, one needs about 10^{21} carriers per cm^3. Both ξ and λ can be changed by varying the electron mean free path, l, in that $\xi = (\xi_0 l/3)^{1/2}$ *and* $\lambda = (\lambda_0/l)^{1/2}$, where ξ_0 and λ_0 are the pure metal values.

A third important material variable is the Ginsburg–Landau variable, κ. It is just the ratio of the above two distances,

$\kappa \approx \lambda/\xi$, and determines the response of the material to a magnetic field. If $\kappa < 2^{-1/2}$, the material is a type I superconductor and it excludes all flux up to the thermodynamic critical field H_c, where the material goes totally normal. If $\kappa > 2^{-1/2}$, then the material excludes all flux up to a lower critical field, H_{c1}, where flux enters the material in the form of a triangular array of vortices in which each vortex contains one quantum of flux, $\phi_0 = hc/2e = 2.07 \times 10^{-7}$ G cm^2. As the field increases, vortices nucleate and move into the superconductor until the upper critical field, H_{c2}, is reached where superconductivity disappears. A useful relation is $H_c^2 \approx H_{c1}H_{c2}$. For Nb, $\lambda \approx 0.8$; for NbTi, $\kappa \approx 10$; for the high-T_c materials, $\kappa \approx 100$. Hence, the vortex state covers almost the entire area of the superconducting part of the H–T plane for high-T_c superconductors.

ANISOTROPIC PROPERTIES

The crystal structures of these new high-T_c compounds are highly anisotropic, a feature which has important implications for both the physical and mechanical properties. The unit cell is a multilayer modified perovskite structure in which divalent or trivalent ions occupy the corners of the perovskite cube, oxygen occupies the cube face centers, and Cu occupies the cube center. The primary modification from the perovskite structure in these materials arises because many oxygen atoms are missing from very specific layers. This gives a very two-dimensional character to the structure. Taking the 125-K superconducting phase of $Tl_2Ba_2Ca_2Cu_3O_{10}$ as a typical example, there is a five-layer sandwich in the middle of the unit cell composed of the sequence CuO_2–Ca–CuO_2–Ca–CuO_2 in which the copper-containing sheets have the full complement of oxygen. The two Ca sheets, however, are totally devoid of oxygen. Hence, the two outside sheets of Cu atoms are coordinated with five oxygens and the central Cu sheet is coordinated with four oxygens. On both sides of this sandwich are BaO and TlO layers to complete the nine-layer unit cell. The layer character is very clear here.

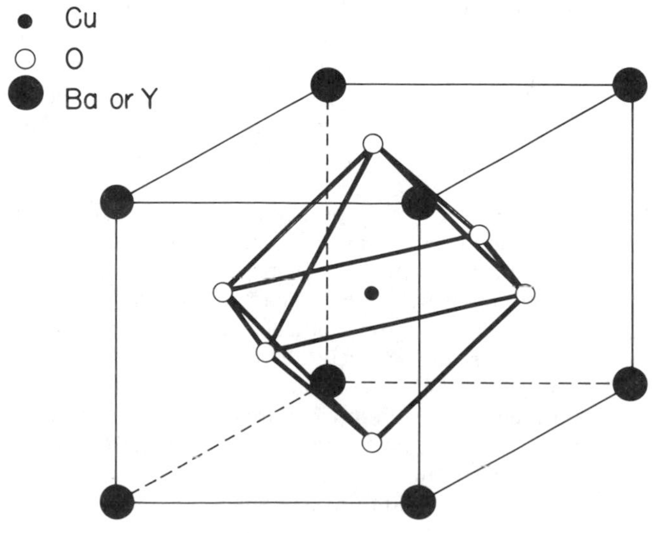

FIG. 1. Perovskite structure.

The bonding in these materials has a highly directional and rather covalent character. This, in turn, leads to behavior rather different from the *s-p*-band metals such as Pb or intermetallic compounds such as Nb_3Sn where there are metal–metal bonds. In high-T_c metals, Hall effect and electron energy loss spectroscopy (EELS) have established that electrical current is carried by holes induced in the oxygen sites of the CuO_2 sheets. The conduction is highly anisotropic with much higher conductivity parallel to the CuO_2 planes than perpendicular to the planes. In addition the two fundamental electrodynamic superconducting parameters, ξ and λ, also are highly anisotropic. The coherence distance, $\xi = 0.18\hbar v_F/kT_c$, parallel to the a–b plane is typically 2 to 3 nm and parallel to the c axis it is typically 0.5 to 0.7 nm. In addition to being anisotropic, the values are very small because the Fermi velocity, v_F, is rather low and T_c is high. This also means that the size of the Cooper pair is comparable to the unit cell and there are relatively few pairs overlapping one another. The phase locking among the electrons which arises from the pair–pair correlations is rather weak. This is to be contrasted with the classical superconductors where the coherence distance is dozens of nm and there are millions of pairs overlapping the space occupied by any given pair.

The mechanical properties of the high-T_c materials also are very anisotropic. For the $Y_1Ba_2Cu_3O_7$ material in particular, the lattice contracts far more along the a–b planes than along the c axis in cooling from the crystal formation temperature to T_c. This leads to large stresses in granular material and if the grain size is greater than 2 to 5 μm, microcracks are pervasive. This microcracking is a central issue to be solved in any scheme to be used for processing these materials.

FLUX PINNING

The lateral motion of vortices in a superconductor gives a time-varying magnetic field which, by Faraday's law, produces a voltage and power dissipation if a current is flowing. In practical devices, it is absolutely essential to control the motion of vortices and it is here that many of the most difficult materials preparation problems arise. The problem is to create a metallurgical microstructure which permits large supercurrents to flow where you want them to flow and yet pin the vortex lattice so it will not move and create an unwanted voltage. The microstructure must be compatible with pinning the flux line lattice and still have a favorable topology for strong phase locking over long distances and massive supercurrent transport.

Roughly speaking, a vortex is composed of a core of radius ξ where n_s goes to zero, and a surrounding circulating supercurrent region where the magnetic field and supercurrents fall off in a distance λ. In a large-κ material like the high-temperature superconductors, the vortices repel one another and stack into a triangular lattice where the lattice spacing, a, is given by $\phi_0 = B(3/4)^{1/2}a^2$. At 5 T, where a typical magnet might operate, the vortex spacing is about 20 nm. This spacing is larger than the core radius of 5 nm for NbTi and much larger than the 1.5 nm of the high-T_c materials. At each core there is a small island where n_n increases to 1. The vortex spacing, however, is smaller than λ so the magnetic field and supercurrent density are rather smoothly varying. In classical superconductors a major source of pinning is the free energy needed to move a normal vortex core from one place to another. Typically the free energy per unit length, $\delta G/l$, is the free energy per unit volume, $H_c^2/8\pi$, times the area of the core, $\pi\xi^2$. For a classical superconductor with $H_c = 0.1$ T and $\xi = 10$ nm, the pinning potential, U_0, is on the order of 1 eV compared to a thermal energy of 0.003 eV at 4.2 K. This gives $U_0/kT_c \sim 30$. For the high-T_c materials the corresponding number is $U_0/kT_c \sim 2$. In the high-T_c materials, the flux lattice is much more mobile and there is a large amount of flux creep over large portions of the H–T plane.

WEAK LINKS

In classical superconductors, weak links are not a major problem because the chemical bonding is predominantly metal–metal bonding. The electrons are very delocalized and flow from atom to atom rather easily; the muffin tin potential used in band theory works well. In the high-T_c materials, the bonding is much more covalent and directional and the electronic wave functions are much more localized. Small distortions in the bonding thus lead to a larger disruption in the flow of electrons in the high-T_c materials. In classical materials, the electrons cross grain boundaries with ease. In high-T_c materials, the grain boundaries are a major impediment to supercurrent flow. Even a tilt boundary with a few degrees mismatch reduces supercurrent flow by orders of magnitude. Ceramic materials often seem to behave as though there were an insulating barrier around every grain. They are much like granular Al, a material prepared by evaporating Al in an oxygen atmosphere to give a thin Al_2O_3 barrier between grains. For granular Al, the interior of the grains goes superconducting at T_c of Al but long-range phase locking does not set in until a lower temperature where the Josephson coupling energy across the Al_2O_3 barriers becomes comparable to kT. Granular Al, in fact, is the material people use when they want a weak pinning material. A major research goal in the high-T_c materials is to eliminate these weak links.

There is a remarkable contrast here. In classical superconductors, defects such as grain boundaries are very beneficial to J_c because they increase the pinning. In the high-T_c materials, defects are detrimental because they provide weak links that reduce J_c. How can this be? In classical superconductors, there are millions of pairs overlapping any given pair and the phase locking is very tight. Critical current densities would be 10^8 A/cm^2 due to the basic pair breaking caused by the drift kinetic energy of the electrons reaching a value equal to the energy gap. In classical superconductors, J_c is limited by depinning of the vortex lattice which occurs two orders of magnitude below the depairing limit. In this environment, a defect such as a grain boundary may cause a little weakening of the capacity to carry supercurrents but there is plenty to spare on the scale of 10^6 A/cm^2. Therefore these weak spots can pin the flux line lattice without suppressing J_c.

In the high-temperature superconductors, however, the phase locking is much more fragile because the coherence length is so short. Here ξ is only a few unit cells (1.5 nm) long in the a–b plane and less than a unit cell (0.7 nm) along the c axis. There are only a dozen Cooper pairs overlapping the volume of any given pair. In this environment, a defect such a grain boundary has a size comparable to ξ and can cause a major disruption of the superfluid density.

In the high-temperature superconductors, defects such as grain boundaries and twin planes serve as weak links rather than pinning centers. For example, decoration experiments have shown that vortices easily slide along the twin or grain boundaries. Regions of suppressed superfluid density which might have been good pinning centers in a superconductor with a long coherence distance actually are detrimental to J_c if the coherence distance is sufficiently short compared to the size of the defect.

FLUX CREEP

Because $U_0 \sim 10^{-1}$ eV, there are large region of the H–T plane where thermally activated flux flow is an important feature of the electromagnetic response of the superconductor. As the magnetic field and temperature rise, the superfluid density and the binding energy per pair are reduced so it is easier for vortices to be thermally activated out of pinning sites and there are large portions of the H–T plane where the flux moves reversibly. The irreversibility line in $Y_1Ba_2Cu_3O_7$ goes approximately as $H_{irr} \sim (1 - T/T_c)^{3/2}$ and passes through 3 T at about 70 K. Mechanical oscillator experiments, in fact, have shown a closely related phenomenon of flux line lattice melting. For $Bi_2Sr_2Ca_1Cu_2O_8$ single crystals the flux lattice melts at about 40 K. Some new mechanism such as surface pinning must be devised to inhibit flux creep. A major challenge in materials processing is to find a way to stabilize the flux line lattice without creating unacceptable weak links in materials with a coherence comparable to the unit cell size.

See also ELECTRON ENERGY STATES IN SOLIDS AND LIQUIDS; GLASSY METALS; JOSEPHSON EFFECTS; SUPERCONDUCTIVITY THEORY; TUNNELING.

REFERENCES

1. J. G. Bednorz and K. A. Müller, *Z. Phys.* **B64,** 189 (1986).
2. D. C. Johnston, H. Prakash, W. H. Zachariasen, and R. Viswanathan, *Mat. Res. Bull.* **8,** 777 (1973).
3. A. W. Sleight, J. L. Gillson, and P. E. Bierstedt, *Solid State Comm.* **17,** 27 (1975).
4. M. K. Wu, J. R. Ashburn, C. J. Torng, P. H. Hor, R. L. Meng, L. Gao, Y. Q. Wang, and C. W. Chu, *Phys. Rev. Lett.* **58,** 908 (1987).
5. H. Maeda, Y. Tanaka, M. Fukutomi, and T. Asano, *Jpn. J. Appl. Phys.* **27,** 209 (1988).
6. Z. Z. Sheng and A. M. Hermann, *Nature* **332,** 138 (1988).
7. J. Bardeen, L. N. Cooper, and J. R. Schrieffer, *Phys. Rev.* **108,** 1175 (1957).
8. Y. Yeshurun and A. P. Malozemoff, *Phys. Rev. Lett.* **60,** 2853 (1988).
9. P. L. Gammel, L. F. Schneemeyer, J. V. Waszczak, and D. J. Bishop, *Phys. Rev. Lett.* **61,** 1666 (1989).

Superconductive Devices

John Clarke

Superconducting electronics, stimulated largely by the discovery of the Josephson effects, began in the early 1960s. Since then a small number of devices have emerged that, operating at or below the boiling point of liquid ^{4}He, 4.2 K, are superior to competing technologies and are used very successfully in a wide range of applications—from the detection of weak far-infrared signals from outer space to the measurement of tiny magnetic fields emanating from the human brain. Needless to say, the advent of oxide superconductors with transition temperatures (T_c) well above the boiling point of liquid N_2, 77 K, has added enormous impetus to the field. Although it is too soon for these materials to have had much impact on practical applications, over the next decade one might expect to see a substantial growth in both the variety of devices used and their range of applications. This article briefly describes the present state of the art and gives some assessment of the possible impact of high-T_c superconductors.

SUPERCONDUCTING QUANTUM INTERFERENCE DEVICES

The most widely used superconducting devices are superconducting quantum interference devices (SQUIDs). The two types of SQUID, the dc SQUID and the rf SQUID, are the most sensitive instruments available for the measurement of a variety of physical quantities, including magnetic field, magnetic field gradient, magnetic susceptibility, voltage, and displacement.

The dc SQUID consists of two Josephson junctions mounted in parallel on a superconducting loop (Fig. 1a). The device shown in Fig. 2, which is typical of devices presently in use, is constructed from thin films of Nb, with Al_2O_3 barriers for the junctions. A constant current I_0, greater than the maximum zero-voltage current of the two junctions, biases the junctions in the voltage state. This voltage is periodic in the magnetic flux, Φ, applied to the loop with a period of one flux quantum Φ_0. However, one can detect a change $\delta\Phi$ very much smaller than Φ_0 using the flux-locked loop shown in Fig. 1a. The feedback circuit generates a current in a coil coupled to the SQUID so as to generate a flux, $-\delta\Phi$, thereby maintaining the total flux in the SQUID at a constant value. The output voltage V_0 is proportional to $\delta\Phi$. Typically, the dynamic range is as high as 10^7 $Hz^{1/2}$ in a 1-Hz bandwidth, and the flux resolution approaches 10^{-6} Φ_0 $Hz^{-1/2}$ at frequencies above 1 Hz.

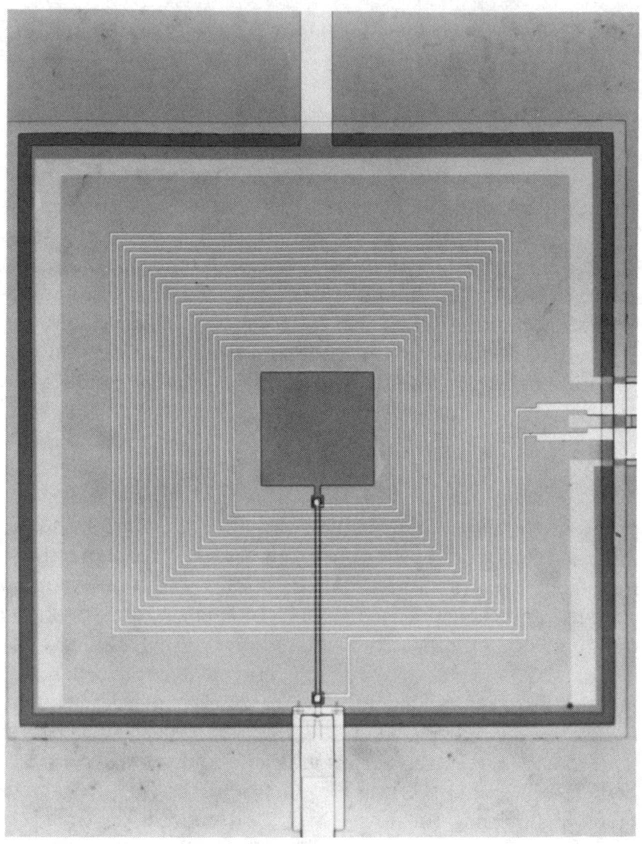

FIG. 2. Thin-film dc SQUID. The two Josephson junctions are fabricated one on each side of the slit in the Nb square washer, near the lower edge. The 20-turn spiral input coil, also of Nb, is electrically insulated from the washer, but tightly coupled to it magnetically. The washer is about 1 mm across. (Courtesy Non Fan, University of California, Berkeley.)

(a)

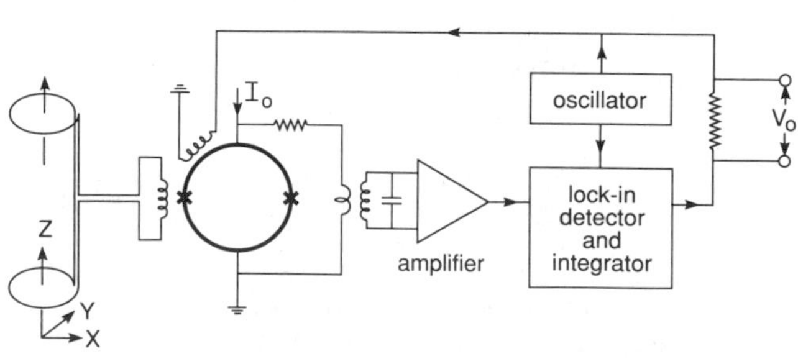

(b)

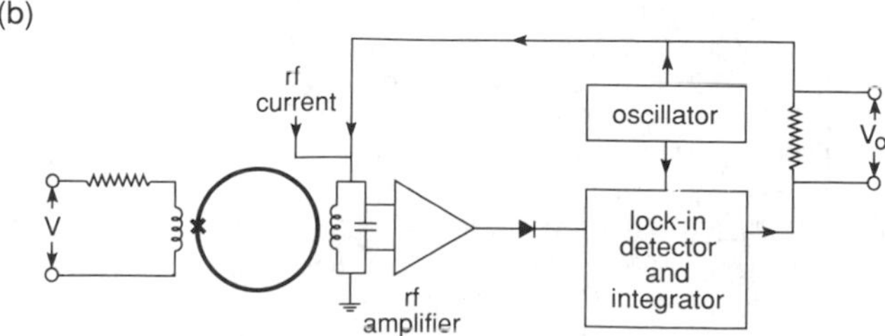

FIG. 1. (a) Simplified schematic of flux-locked dc SQUID used as gradiometer; (b) flux-locked rf SQUID used as voltmeter. Symbol x represents a Josephson junction.

The rf SQUID (Fig. 1b) consists of a single Josephson junction incorporated into a superconducting loop. The loop is coupled to the coil of a cooled LC resonant circuit excited at its resonant frequency, typically 30 MHz. The amplitude of the rf voltage across the resonant circuit is periodic in Φ with period Φ_0. After suitable demodulation of the rf signal, the output is used to flux-lock the SQUID in the same way as the dc SQUID. The rf SQUID is significantly less sensitive than the dc SQUID, but has been the mainstay of the field until quite recently. A commercially available version is shown in Fig. 3.

SQUIDs are generally used in conjunction with a superconducting flux transformer, for example, the gradiometer shown in Fig. 1a. A change in the magnetic field gradient $\partial H_z/\partial z$ produces a supercurrent in the transformer and hence a voltage at the output of the flux-locked SQUID, whereas a change in the uniform magnetic field produces no response. A typical sensitivity is 10^{-13} T m^{-1} $Hz^{-1/2}$. Other configurations can be used to measure $\partial H_z/\partial x$ or $\partial^2 H_z/\partial z^2$. A flux transformer with a single pickup loop is a magnetometer with a sensitivity that can be as high as 10^{-15} T $Hz^{-1/2}$. One can use a configuration similar to that in Fig. 1a to measure magnetic susceptibility by placing the sample in one of the pickup loops. The application of a magnetic field H_z produces an output from the flux-locked SQUID proportional to the susceptibility of the sample. The configuration in Fig. 1b is a voltmeter, with a resolution usually limited by Johnson noise in the resistor: for example, with a resistor of 10^{-8} Ω at 4 K, the resolution is about 10^{-15} V $Hz^{-1/2}$. The dc SQUID, without feedback, has been used as an rf amplifier for frequencies up to 100 MHz or higher, with a noise level lower than any competing device.

The applications of SQUIDs are far ranging. The largest single application is in biomagnetism, where gradiometers are used to map the sources of tiny spontaneous or evoked magnetic fields generated by the brain. Other applications include geophysical surveying, nondestructive testing, nuclear magnetic resonance, and displacement sensors in gravity wave antennas and monopole detectors.

The first devices of any kind made from high-T_c superconductors were dc and rf SQUIDs. The junctions are formed by the boundaries between two grains of superconducting material. These devices can be operated at 77 K, but their low frequency noise is at present substantially higher than in the low-T_c devices. However, there has been considerable progress in understanding and reducing this noise, and it is very likely that practical devices with sensitivities high enough for many applications will emerge in the near future. The principal advantage of higher-temperature operation lies in the fact that liquid N_2 boils much more slowly than liquid He. As a result, it is reasonable to consider operating relatively small cryostats unattended for periods of a year or more, thereby offering much greater flexibility in operating SQUIDs in remote areas.

DETECTORS OF MICROWAVE AND SUBMILLIMETER RADIATION

A very successful device is the superconductor–insulator–superconductor (SIS) quasiparticle tunnel junction used as a mixer of high-frequency electromagnetic radiation (Fig. 4). Such detectors can be used as the front end of highly sensitive receivers for frequencies ranging from tens to hundreds of gigahertz. The incoming signal at frequency f_S is mixed with the output of a local oscillator at frequency f_{LO} to produce a mixed-down signal at an intermediate frequency $f_{IF} = |f_S - f_{LO}|$ that is subsequently amplified. The mixing process involves a nonlinearity in the current–voltage (I–V) characteristic of the mixing element. In the SIS

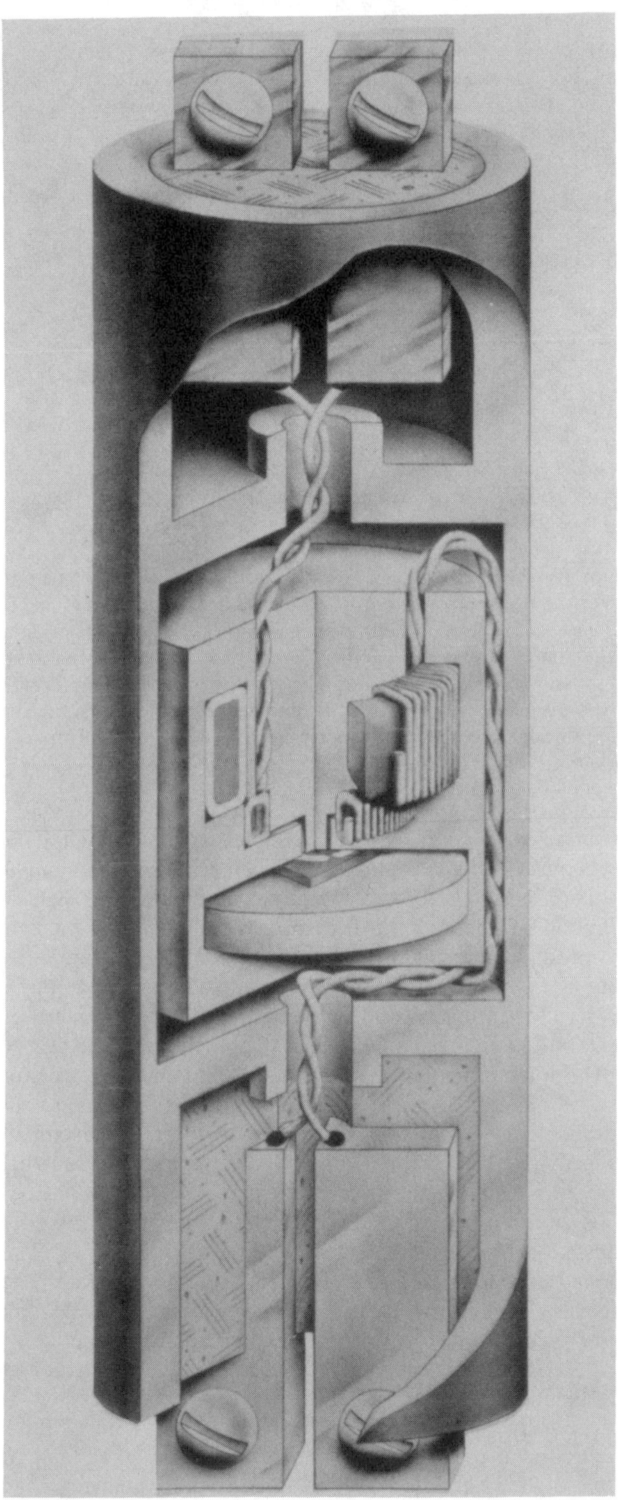

FIG. 3. rf SQUID, machined in toroidal configuration from solid rod of Nb. The single, thin-film junction is on a separate chip near the middle. Terminals at the upper end supply feedbck current and rf bias to one toroidal coil; these at the lower end are connected to the input coil, which is also toroidal. (Courtesy BTi, Inc. San Diego, California.)

junction, this nonlinearity occurs at a voltage corresponding to the sum of the energy gaps where the current rises sharply with voltage. The local oscillator produces current steps at

FIG. 4. Two halves of microwave circuit for SIS mixer operating over the frequency range from 85 to 110 GHz. Signal and local oscillator enter through the wwaveguide formed by the two blocks when they are screwed together. The junction and a thin-film radiofrequency filter are located on a suspended strip line transverse to the waveguide. The IF output is obtained from the end of the strip line through a connector. (Courtesy P. L. Richards, University of California, Berkeley.)

voltages nhf_{LO}/e (n is an integer) above and below the current onset, corresponding to emission and absorption of photons of energy hf_{LO}/e by the tunneling electrons. When one applies a signal as well, it is mixed with the local oscillator to produce an oscillatory voltage across the junction at frequency f_{IF}; this voltage is coupled into a low-noise amplifier. With careful design and implementation, these mixers can exhibit useful levels of gain, up to 10 dB, and at 36 and 100 GHz noise temperatures have been obtained that are close to the quantum limit, $hf_{\mathrm{S}}/k_{\mathrm{B}}$. In some applications it is desirable to use series arrays of junctions, thereby substantially improving the dynamic range of the receiver without degradation of the gain or noise.

At frequencies up to several hundred gigahertz mixers of this kind are the most sensitive detectors available. They have been installed as receivers on a number of radio telescopes, working mostly at frequencies around 100 GHz, and are typically used to observe molecular lines. Operated at 4.2 K, SIS mixers have an assured future. Prospects for high-T_c devices, however, are rather limited, partly because of the considerable difficulties in making any kind of tunnel junction, and partly because operation at higher temperatures will inevitably degrade the noise performance.

A quite different device is the Josephson parametric amplifier, which consists of a single junction shunted with a superconducting inductor (as in the rf SQUID). In the degenerate mode in which the amplifier has been most successfully used, the device is pumped at a frequency f_{P} equal to $2f_{\mathrm{S}}$, where f_{S} is the signal frequency. The pump current in the loop modulates the nonlinear inductance of the Josephson junction, producing a time-varying inductance that amplifies the signal. The output signal is thus at the same frequency, f_{S}, as the input. A device of this kind operated at about 19 GHz has been shown to be at the quantum limit and has been used to demonstrate "squeezing" of thermal noise.

The two devices described above belong to the class of coherent detectors that are generally preferred at frequencies up to a few hundred gigahertz. At higher frequencies, however, one uses incoherent detectors, usually preceded by optical filters to define the bandwidth. A good example of this class of detector is the bolometer, which consists of an absorbing element and a thermometer with weak thermal coupling to a heat bath. When electromagnetic radiation falls on the bolometer, its temperature rises by an amount proportional to the power absorbed. In the mid 1970s, a successful bolometer was developed that used the sharp resistive transition of an Al film as the thermometer, but this was superseded by semiconducting thermometers which have a less restricted range of working temperature. However, the advent of high-T_c superconductivity has renewed interest in the superconducting bolometer, operating at liquid N_2 temperatures. At these temperatures and at wavelengths greater than about 20 μm where photodetectors are no longer practicable, this device may well prove to be substantially better than any competing detector.

JOSEPHSON VOLTAGE STANDARD

One of the earliest applications of Josephson tunneling was to the voltage standard. One irradiates a Josephson tunnel junction with microwaves of frequency f, typically about 100 GHz, in the absence of any current bias. The microwaves generate a series of supercurrent steps, crossing the zero current axis, at voltages $nhf/2e$. The steps arise from the absorption or emission of microwave photons by the tunneling Cooper pairs. Since frequency can be determined to extremely high accuracy, the voltages at which the steps appear are also precisely defined. Thus, Josephson junctions can be used as convenient and very stable voltage standards, although it should be emphasized that they are used to maintain rather than define the volt. Many national laboratories around the world rely on this standard.

However, the voltages at which the steps appear on a single junction are relatively small: at 100 GHz the interval is about 200 μV, and one is limited to values of n corresponding to voltages below the sum of the energy gaps, a few millivolts. For this reason there has been considerable effort to make very large series of arrays of junctions, containing typically 2000 junctions, but in some cases as many as 20 000. The voltage across these arrays can be between 1 and 10 V, greatly facilitating comparison with other voltage standards. Figure 5 shows a typical array.

To use high-T_c junctions at 77 K in this application it will first be necessary to develop a technology for high-quality tunnel junctions.

DIGITAL DEVICES

Early in the history of Josephson devices it was recognized that the time required for a junction to switch from the zero voltage state to the voltage state could be exceedingly short, a few picoseconds. Furthermore, the dissipation in such a

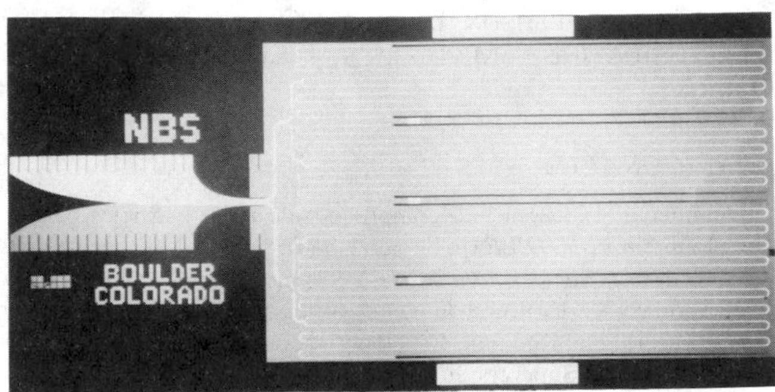

FIG. 5. Integrated circuit layout for an 18 992-junction array of Josephson junctions designed for use as a 10-V standard. The chip is 19 mm long (Courtesy C. A. Hamilton, NIST, Boulder.)

junction, typically a few microwatts, is very low compared with that in typical semiconductor devices, enabling one to achieve high packing density of junctions on a chip and reducing the signal transit times between devices. Another important advantage offered by low-temperature operation is the ability to use superconducting transmission lines to connect junctions; such connections are virtually dispersion-free. Consequently, Josephson devices are very attractive elements for fast computers and other digital circuits, and substantial development efforts were mounted in the United States in the 1970s and early 1980s. However, this program was largely abandoned in 1983, and today there is substantial work in this field only in Japan.

Materials have been a major factor in the development of the technology. Earlier devices were fabricated from Pb alloys and tended to deteriorate with storage at room temperature or on cycling between room temperature and liqiud ^{4}He temperatures. Furthermore, the reproducibility of the critical current from junction to junction and from chip to chip was inadequate. However, the introduction of Nb-Al_2O_3-Nb junctions in which the barrier is made by oxidizing Al deposited on the Nb base electrode has overcome both problems, and processing no longer appears to be a major difficulty.

Broadly speaking, there are two types of circuit element required for a computer: logic and memory. Josephson logic gates differ from semiconducting gates in that they are latching, that is, they do not return to their original state after the input signal is removed. As a result, the Josephson logic gate is biased by pulses rather than steady voltage, and the pulse frequency determines the clock frequency. Logic gates in turn, also fall into two classes: those with magnetic coupling, in which the gate is controlled by the magnetic field generated by the input current, and those with current injection, in which the input current is injected directly into the gate. A wide variety of logic devices has been developed. A simple magnetically coupled device consists, in essence, of a dc SQUID in which the two Josephson junctions have hysteretic current–voltage characteristics. The junctions are initially biased with a current below their critical current so that they are in the zero voltage state. In an OR gate, there is a single input of magnetic flux to the loop that reduces the critical current to a value below the bias current, causing the junctions to switch to the voltage state. In an AND gate, two simultaneous inputs are required to produce switching. To make an AND gate with current injection, one passes a current pulse through the junctions and, simultaneously, a current into one arm of the SQUID loop. The fastest gate delay reported so far, 1.5 ps, was achieved with a hybrid device involving both magnetic coupling and current injection.

There are two types of memory. The cache memory, which is coupled directly to the central processing unit, is relatively small but very fast. The main memory, which is fed by the cache memory, is large, but has a slower access. Each memory cell consists of either a one- or two-junction SQUID in which a persistent current may be established around the loop. A "0" or "1" is stored either as a clockwise or anticlockwise current or, alternatively, as a zero or nonzero current. One writes a digit in the cell by momentarily switching the junction of the one-junction SQUID or one of the junctions of the two-junction SQUIDs into the resistive state and resetting it to the superconducting state with an appropriate bit stored. The contents of the cell are read either by sensing the magnetic flux produced by the SQUID or by measuring the persistent current directly. To date, although it was not completely functional, the most sophisticated circuit produced was a 1-K-bit random access memory with a minimum access time of about 0.5 ns. There are good prospects for cache memories of ths kind, but the outlook for main memory is much less certain. It seems likely that in a large computer with superconducting logic and cache memory one would use semiconducting main memory.

A number of high-speed superconducting circuits involving various logic and memory devices have been demon-

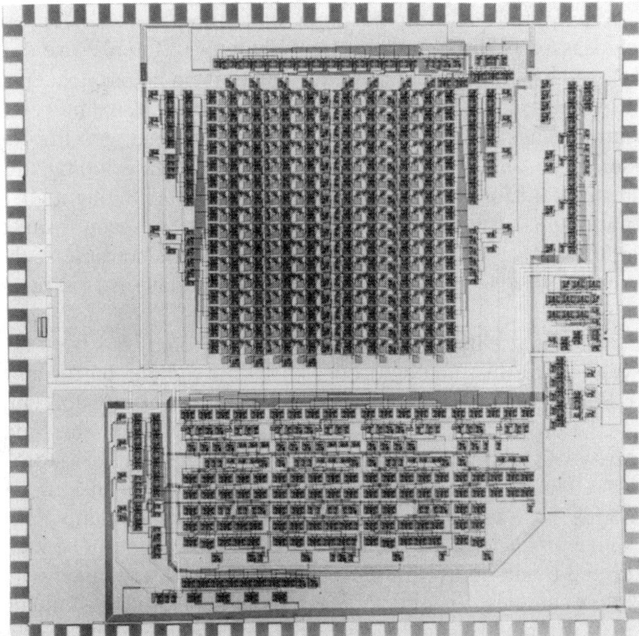

FIG. 6. 4-bit microprocessor fabricated with a 2.5-μm linewidth niobium technology. The chip is 5 mm × 5 mm and includes 5011 Nb-Al_2O_3-Nb Josephson tunnel junctions. The microprocessor operates with a clock frequency of 770 MHz and dissipates 5 mW. (Courtesy S. Hasuo, Fujitsu Laboratories, Atsugi, Japan.)

ing various logic and memory devices have been demonstrated, including multipliers, arithmetic logic units, and a 4-bit microprocessor that operated at 770 MHz (shown in Fig. 6). This is an order of magnitude faster than its semiconducting counterpart.

Other potentially important devices include shift registers and analog-to-digital (A-to-D) converters. These circuits involve Josephson junctions, often in the configuration of a type of A-to-D converter is based on the dc SQUID (Fig. 1a), which converts input magnetic flux into pulses across one junction or the other, depending on whether the signal is increasing or decreasing. Systems have been operated that are capable of digitizing, for example, 6 bits at 100 MHz or 4 bits at 280 MHz. Improvements of factors of 3 or 4 in the frequency are expected, and if achieved would yield performances comparable with or slightly better than conventional A-to-D converters. Superconducting A-to-D converters have also been proposed that offer very high accuracy, perhaps as high as 30 bits, at low frequencies.

At present, some of the digital circuits hold considerable promise for small-scale applications, for example, microprocessors. However, the development of a mainframe computer still appears to be far away. High-T_c superconductivity will not have an impact on this field until a very reliable and reproducible junction technology becomes available.

ANALOG SIGNAL PROCESSING

Several devices have been developed for processing high-speed analog signals. One, available commercially, is the Josephson sampler, which relies on a very fast dc SQUID as a sampling gate. Sampling techniques are commonly used to determine the waveforms of high-speed repetitive signals: the amplitude is sampled for a short time interval stepped successively through the pulse. When these samples are displayed as a function of time, for example on an oscilloscope, a picture of the waveform is reconstituted. Sampling resolution times as low as 2 ps have been achieved with the Josephson sampler. It would be worthwhile developing this instrument for operation at 77 K, thereby facilitating its use with signal sources at room temperature. A second, quite different, sampling method combines optoelectronic techniques with a Nb transmission line to achieve a 1-ps resolution.

Recently, a new superconducting technology has been developed for fast analog processing, for example, performing a Fourier transform of a signal to analyze its spectral content, or the cross-correlation of two pulses to determine the similarity of their shape. In this technique, the signal is stored on a Nb stripline, wound in a spiral typically 1 m long. A series of "taps" along the line enables one to sample the signal at successive time intervals. Each sample can be multiplied by an appropriate factor before being combined with other samples, as appropriate for the application. Although in its infancy, this technique is predicted to have bandwidths as large as 10 GHz and computational rates equivalent to 10^{12} arithmetic operations per second. The development of this technique at 77 K is quite attractive, since one could combine superconducting strip lines with fast semiconductor processing devices.

See also JOSEPHSON EFFECTS; QUASIPARTICLES; SUPERCONDUCTING MATERIALS; SUPERCONDUCTIVITY THEORY.

BIBLIOGRAPHY

A number of excellent and comprehensive review articles can be found in *Proc. IEEE* (to be published); NATO Advanced Study Institute: *Superconducting Electronics* (to be published by Springer-Verlag); and in *Superconducting Devices* (edited by Ruggerio and Rudman, to be published). Recent proceedings of the Applied Superconductivity Conferences give excellent coverage: *IEEE Trans. Magn.* **MAG-17** (1981), **MAG-19** (1983), **MAG-21** (1985), **MAG-23** (1987), and **MAG-25** (1989).

Superconductivity Theory

Philip B. Allen

Superconductivity is a property shared by many metals at low temperature T. As T is reduced the resistivity ρ of a normal metal decreases. For a nonsuperconducting metal like Cu, ρ settles to a constant value ρ_0 at low T; ρ_0 can be made smaller by purifying the metal, but can never completely disappear. In a superconductor, by contrast, a critical temperature T_c is reached below which ρ abruptly becomes zero, as shown in Fig. 1. The value of T_c varies widely: for Al, $T_c=1.2$ K; for Pb, $T_c=7.2$ K; and for the intermetallic compound Nb_3Sn, $T_c=18.0$ K. Until 1986, the highest known value was 23 K. The search for materials with higher T_c has been an obsession for many years, and was dramatically vindicated by the discovery of a new class of superconductors containing metallic layers of copper oxide with T_c up to and exceeding 100 K. Selected values of T_c are shown in Fig. 2.

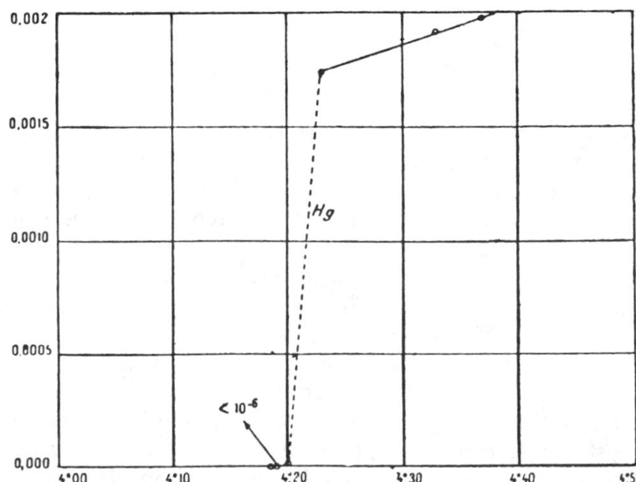

FIG. 1. Resistance of mercury from 4.0 to 4.5 K as measured by H. Kamerlingh Onnes in 1911. This was the first observation of superconductivity. (Reprinted from *Superconductivity*, AIP Selected Reprints.)

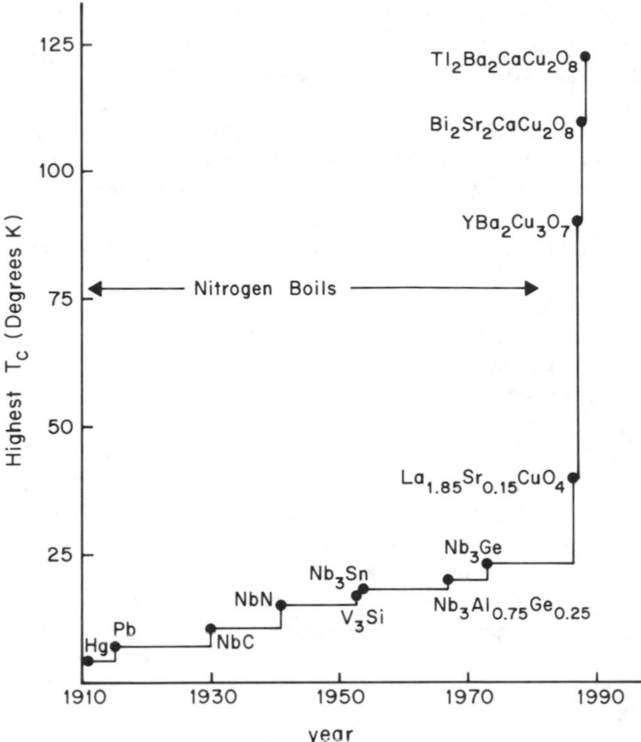

FIG. 2. Historical map of highest known superconducting transition temperature.

Besides the absence of electrical resistivity, superconductors have another unexpected property: an aversion to magnetic fields, expressed in one of two different ways, called type I (seen in most pure elemental superconductors) or type II (seen in many intermetallic compounds and alloys). In the case of type I materials (and also type II in weak fields) an external magnetic field B_{ext} is completely cancelled (or "expelled") from the interior of the superconductor by the opposite field of a solenoidal supercurrent at the surface of the sample. This is known as the "Meissner effect" (see Fig. 3) and costs the superconductor an energy $B_{\text{ext}}^2/8\pi$. For fields

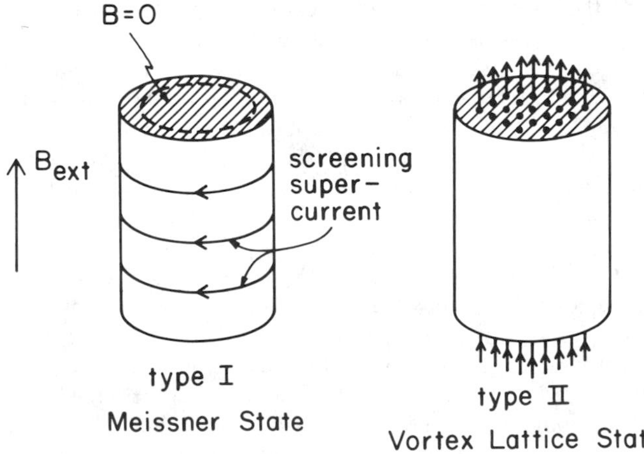

FIG. 3. Meissner state and vortex lattice state illustrated for the case of cylindrical samples in an axial applied magnetic field.

greater than a critical field H_c, there is insufficient energy available and the metal goes back into its "normal" resistive state. Thus $H_c^2/8\pi$ measures the free-energy difference between the normal and superconducting states. As T increases to T_c, this energy and the "thermodynamic" critical field H_c both go to zero. At $T=0$, H_c is usually fairly small (e.g., ~0.1 T for Pb) so type I superconductors are not usable for high-field applications such as magnet windings.

If a field B is applied to a superconductor which was previously superconducting in a field-free region, Faraday's law says that the time rate of change of flux through the superconductor, $\dot{\Phi}$, should remain zero because vanishing resistivity ρ implies vanishing induced E field. Thus, the Meissner state seems to be a consequence of $\rho=0$. However, if the B field is first applied to a metal in its normal state (i.e., $T>T_c$) and then T is reduced below T_c, $\dot{\Phi}=0$ would imply that the B field would be trapped, whereas expulsion actually occurs. Thus, the Meissner effect requires additional explanation beyond $\rho=0$. London showed that the effect follows from a phenomenological equation $\mathbf{j}\propto\mathbf{A}$ where $\mathbf{j}$ and $\mathbf{A}$ are current density and vector potential, respectively, with gauge freedom exploited to choose $\nabla\cdot\mathbf{A}=0$. London observed that $\mathbf{j}\propto\mathbf{A}$ could be understood from the quantum mechanics of electrons provided the wave function is postulated to have a "rigidity" ($\nabla\psi$ should be unaffected by $\mathbf{B}$).

London also noticed that these ideas require that if a superconductor with a hole is in the Meissner state, the magnetic flux Φ through the hole should be quantized in integral multiples of the "quantum of flux," $\Phi_0=h/e^*$. The appropriate charge is now known to be $e^*=2e$ because the relevant wave function describes pairs of electrons. Quantization of flux in units of $\Phi_0=h/2e=2.07\times10^{-15}$ T m^2 has been seen experimentally.

A type I superconductor has a positive energy to form an interface between normal and superconducting regions, and the Meissner state avoids the creation of such an interface. A type II superconductor has a negative energy for interface formation, and this leads to the vortex lattice state (Fig. 3) where the field penetrates the material in the form of "flux tubes" containing normal metal and a single quantum Φ_0 of flux. Outside the flux tubes the material is superconducting and field free. This behavior occurs only for external fields in an interval $H_{c1}\leq B_{\text{ext}}\leq H_{c2}$ (Fig. 4), where the "lower critical field" H_{c1} is less than the "thermodynamic" critical field H_c, and the "upper critical field" H_{c2} is greater. As the field B_{ext} increases toward H_{c2}, the size of the superconducting region (between the flux tubes) shrinks to zero, and the sample makes a continuous transition back to the normal state. Figure 4 shows how alloying converts Pb from type I to type II. In some cases H_{c2} can be very large, >20 T. Thus, type II materials are more useful than type I for high-field applications as in magnets, motors, and generators.

The abrupt disappearance of superconductivity at temperatures T above T_c is not completely mysterious—there is an exact analogy with many other phase transitions, such as the abrupt disappearance of spontaneous magnetism in a ferromagnet at the Curie temperature. However, the absence of electrical resistance is a much more difficult problem. It is *not* explainable either by the classical (Drude–Lorentz) theory, which was available when Onnes discovered super-

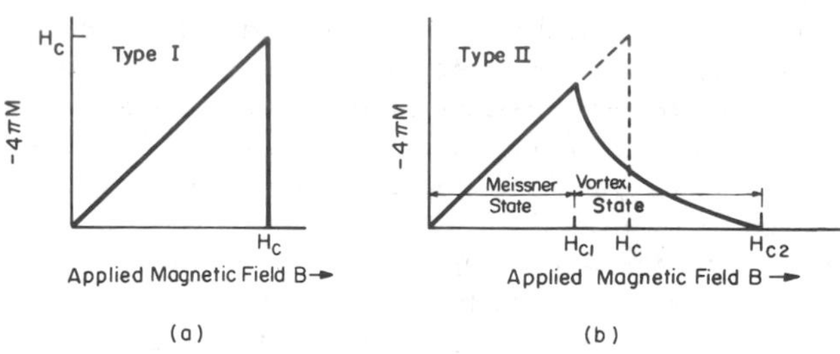

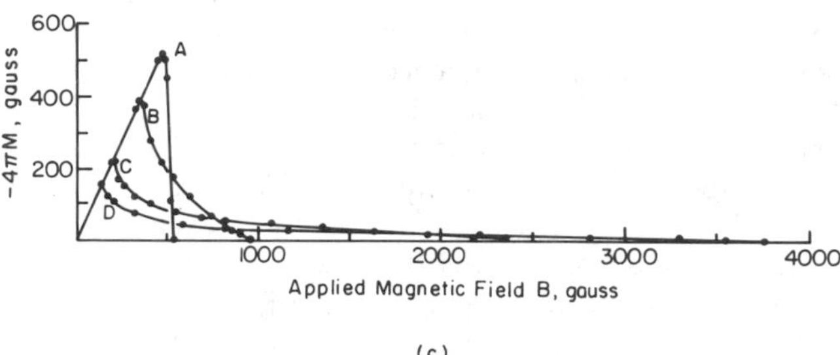

FIG. 4. (a) Magnetization M versus applied field B for a bulk (type-I) superconductor. The slope of 1 corresponds to a susceptibility $\chi = -1/4\pi$. (b) Magnetization curve for a type-II superconductor. The "vortex" state exists for applied fields between H_{c1} and H_{c2}. Below H_{c1} there is complete flux expulsion (Meissner state). (c) Experimental magnetization curves of Pb (A) and Pb-In alloys of increasing In concentration (B,C,D) showing the onset of type-II behavior in the alloys. [These figures are adapted (with permission) from C. Kittel, *Introduction to Solid State Physics*, 5th ed. Chap. 12. Wiley, New York, 1976. The data come originally from J. D. Livingston, *Phys. Rev.* **129,** 1943 (1963).]

conductivity in 1911, or by the quantum theories which Bloch and Grüneisen developed around 1930 to explain conductivity in normal metals. The mystery was not solved until the BCS (Bardeen–Cooper–Schrieffer) theory in 1957. The BCS theory and its subsequent extension have provided a very nearly complete description of a remarkable range of physical properties and effects. Prominent among these are the effects predicted by Josephson. On a deeper level the BCS theory has introduced new ideas into the structure of quantum theory which have resulted in a deeper understanding of broken symmetry, and important developments in nuclear theory and field theory. Furthermore, the BCS theory strongly suggests that the instability of the normal electron system (which leads to superconductivity in metals) might well occur in other systems of Fermi particles, such as nuclear matter in neutron stars and liquid ^{3}He. The latter conjecture has been verified by the discovery by Osheroff *et al.* in 1972 of several new phases of ^{3}He at temperatures below 3 mK. Thus, the BCS theory of superconductivity can be regarded as one of the most important and far-reaching achievements of modern physics.

The essence of BCS theory is the appearance of a subtle new field, the "pair field." This field is the "order parameter" of the superconducting state, just as the magnetization is the "order parameter" of a ferromagnet. This field is both a superconducting wave function and a macroscopic observable field, although less accessible to direct observation than magnetization because it does not couple to an easily tuned external field in the way that magnetization couples to B fields. Normally a quantum wave function is not a macroscopic observable. The overall phase, for example, is completely hidden from observation. In superconductors, the relative phase $\phi_1 - \phi_2$ of two superconductors is related to the amount of supercurrent which can tunnel through a thin barrier between them, according to the Josephson relation $I = I_0 \sin(\phi_1 - \phi_2)$. Thus, the phase ϕ_1 is not equivalent to $\phi_1 + \Delta\phi$, or in more obscure language, a superconductor exhibits "broken gauge symmetry."

The possibility of a macroscopic quantum wave function for electrons was unprecedented before BCS theory. However, for particles of integer spin obeying Bose statistics (like photons) the idea is more familiar. A classical **E** field is just such a macroscopic manifestation of a quantum photon wave function. Superfluid ^{4}He, which consists of atoms which are "composite bosons" having total spin 0, can also be described by a macroscopic wave function. But electrons have half-integer spin and obey Fermi statistics, which prevents more than one electron occupying the same state. In order to generate a macroscopic wave function, it is necessary to "fool" the Fermi statistics by organizing the electrons into singlet pairs of spin 0 (triplet pairs of spin 1 are also possible in principle, and occur in superfluid ^{3}He and possibly also in some exotic metals with low T_c). The pairs (called "Cooper pairs" for reasons to be explained shortly) are not "good" composite bosons like ^{4}He atoms. They are spatially strongly intertwined with each other, and their pairing disappears in the normal state. Pairing in superconductors is a collective phenomenon, like magnetization in a magnet. Only the presence of other pairs (with the phases of the pair wave functions aligned) encourages the formation of new pairs.

An important clue about how pairing occurs was provided by Cooper. He considered the problem of an idealized quantum gas of noninteracting electrons at $T = 0$. Two extra electrons are added which can interact with each other via a potential energy V. The extra electrons are affected by the noninteracting gas because of the Pauli principle which forbids the extra electrons to occupy any of the filled states with energy ϵ less than the "Fermi energy" ϵ_F. Cooper showed that this restriction has the unexpected effect that the added electrons can form a quantum bound state (with energy less than $2\epsilon_F$ by an amount Δ_0) for an arbitrarily weak attractive potential. The same behavior occurs for an isolated pair in two dimensions but an isolated pair in three spatial dimensions has no bound state until the attractive potential exceeds a critical value V_c. The density of states $D(\epsilon)$ of electrons goes like $\epsilon^{(d/2)-1}$ in a space of dimension d. Because of the blocked states with $\epsilon < \epsilon_F$, the density of available states jumps discontinously from 0 to $D(\epsilon_F)$, an effect which mimics what happens at $\epsilon = 0$ in $d = 2$ even with no blocked states. Cooper's result shows that the filled Fermi sea of noninteracting electrons will become unstable if even a weak attractive interaction is turned on. However, at $T > 0$, the sharp discontinuity of $D(\epsilon)$ for unoccupied states is smoothed out, which causes the binding to vanish if $k_B T \gtrsim \Delta$.

BCS theory shows how this "pairing instability" leads to the formation of a superconducting state.

The BCS state is characterized by a gap $\Delta(T)$ in the spectrum of electronic excitations. As temperature increases to T_c, Δ goes to zero. The size of the gap is proportional to $\exp[-1/D(\epsilon_F)|V|]$ which shows that there is no possible Taylor expansion or perturbation theory in powers of V starting from the noninteracting case $V=0$.

The actual construction of BCS theory requires still an additional trick. A variational ground state of the form

$$\Psi(r_1, \ldots, r_N)=\hat{A}\{\chi(r_1-r_2)\chi(r_3-r_4)\ldots\chi(r_{N-1}-r_N)\} \quad (1)$$

is needed to describe pairing at $T=0$. Here $\chi(r_1-r_2)$ is a 2-body wave function chosen to minimize the energy, and $\hat{A}$ is an antisymmetrizing operator. Unfortunately, this type of wave function is extremely awkward to use. To make a more usable wave function, first write Eq. (1) in second-quantized form

$$|\Psi\rangle=(\sum_k \chi_k c_{k\uparrow}^+ c_{-k\downarrow}^+)^{N/2}|\text{vac}\rangle, \quad (2)$$

where $|\text{vac}\rangle$ is the vacuum state. The operator in parentheses creates a pair of electrons with net spin 0 in a state with pair wave function $\chi=\sum_k \chi_k e^{ik(r_1-r_2)}$. Now replace Eq. (2) by the choice

$$|\Psi_{\text{BCS}}\rangle=\exp(\sum_k \chi_k c_{k\uparrow}^+ c_{-k\downarrow}^+)|\text{vac}\rangle. \quad (3)$$

The wave function $|\Psi\rangle$ is the projection of $|\Psi_{\text{BCS}}\rangle$ onto the space with N electrons. Equation (3) can now be written as

$$\begin{aligned}|\Psi_{\text{BCS}}\rangle&=\prod_k \exp(\chi_k c_{k\uparrow}^+ c_{-k\downarrow}^+)|\text{vac}\rangle\\ &=\prod_k (1+\chi_k c_{k\uparrow}^+ c_{-k\downarrow}^+)|\text{vac}\rangle.\end{aligned} \quad (4)$$

The last step follows because of the Fermionic nature of the c^+ operators, i.e., $(c^+)^2=0$. Finally, a normalized version of Eq. (4) is

$$|\Psi_{\text{BCS}}\rangle=\prod_k (u_k+v_k c_{k\uparrow}^+ c_{-k\downarrow}^+)|\text{vac}\rangle. \quad (5)$$

This is the BCS wave function, with u_k and v_k variational parameters chosen to minimize energy subject to the constraint $|u_k|^2+|v_k|^2=1$. It can be viewed as an algebraically manageable mutation of Eq. (1). This wave function has the property of mixing states with different numbers of electrons. This is not only a computational convenience, but in fact a necessity for giving a state with well-defined phase. The relative phase of u_k and v_k, $\phi=\text{Im}[\ln(v_k/u_k)]$, must be independent of the single-particle eigenlabel k in order to minimize energy. The BCS wave function then can be written

$$|\Psi_{\text{BCS}}\rangle=\sum_{n=0}^{\infty} e^{in\phi}|n\rangle,$$

where $|n\rangle$ is the part of $|\Psi_{\text{BCS}}\rangle$ [Eq. (5)] which contains exactly n pairs of $2n$ electrons. The fluctuation $\overline{n^2}-\overline{n}^2=\sum_k |u_k v_k|^2$ is comparable to the corresponding fluctuation $\sum_k f_k(1-f_k)$ {where $f_k=[1+\exp(\epsilon_k/k_B T)]^{-1}$ is the Fermi–Dirac occupation function} found for a noninteracting gas of electrons in the canonical ensemble. The fluctuations are negligible for a macroscopic sample.

The wave function (5) is a very good choice for a particular class of interacting electron problems. Specifically, it gives the exact ground state (in the thermodynamic limit) for Hamiltonians of the form

$$\mathcal{H}_{\text{red}}=\sum_k \epsilon_k(c_{k\uparrow}^+ c_{k\uparrow}+c_{-k\downarrow}^+ c_{-k\downarrow}) + \sum_{kk'} V_{kk'} c_{k\uparrow}^+ c_{-k\downarrow}^+ c_{-k'\downarrow} c_{k'\uparrow}, \quad (6)$$

where ϵ_k is the single-particle energy measured relative to the Fermi energy. Equation (6) is known as the BCS "reduced" Hamiltonian. A complete Hamiltonian would contain many additional interaction terms $V_{kk'}(q)c_{k+q}^+ c_{-k}^+ \times c_{-k'} c_{k'+q}$ for all q. The omitted terms break pairs, whereas the terms kept in H_{red} merely transfer pairs from $(k'\uparrow, -k'\downarrow)$ to $(k\uparrow, -k\downarrow)$. The complete thermodynamics of Eq. (6) are worked out in BCS theory and provide an excellent model for superconductivity. Thus, the microscopic theory rests on an important (and apparently correct) assumption, namely, that the interactions omitted from Eq. (6) do not qualitatively alter the behavior.

In a mean field model of a magnet, spin–spin interactions like $J\sigma_i\sigma_j$ are replaced by $J[\sigma_i\langle\sigma_j\rangle+\langle\sigma_i\rangle\sigma_j-\langle\sigma_i\rangle\langle\sigma_j\rangle]$. This neglects terms quadratic in the fluctuations $\delta\sigma_i=\sigma_i-\langle\sigma_i\rangle$ and turns the interacting problem into a noninteracting problem. Each spin feels only a "mean field" rather than the actual interaction with its neighbor. The mean field must be calculated self-consistently, leading to an equation like

$$\langle\sigma\rangle=\tanh(zJ\langle\sigma\rangle/k_B T) \quad (7)$$

with z equal to the number of nearest neighbors. This has a nontrivial solution $\langle\sigma\rangle\neq 0$ only for $T<T_c=zJ/k_B$. The solution of Eq. (6) for $\langle\sigma\rangle$ versus T is shown in Fig. 5. In a magnet the number of neighbors z is fairly small (e.g., ~8) which allows the fluctuations $\delta\sigma_i$ to have large effects, especially in lower-dimensional problems. Mean field theory then makes serious quantitative and even qualitative errors.

The BCS problem [Eq. (6)] is much more amenable to a mean field solution than a typical magnet because the number of pair states k' which interacts with a given k is large. The pair creation operator $c_{k\uparrow}^+ c_{-k\downarrow}^+$ can be replaced by its mean value $F_k^*=\langle c_{k\uparrow}^+ c_{-k\downarrow}^+\rangle$ plus a fluctuation $c_{k\uparrow}^+ c_{-k\downarrow}^+ - F_k$, and terms quadratic in the fluctuations are usually ignorable. Then the BCS Hamiltonian becomes

$$\mathcal{H}_{\text{red}}=\sum_k [\epsilon_k(c_{k\uparrow}^+ c_{k\uparrow}+c_{-k\downarrow}^+ c_{-k\downarrow}) + \Delta_k c_{k\uparrow}^+ c_{-k\downarrow}^+ + \Delta^*_k c_{-k\downarrow} c_{k\uparrow}]+\text{const}, \quad (8)$$

where Δ_k is related to the "pair field" F_k by

$$\Delta_k=\sum_{k'} V_{kk'} F_{k'}$$

and must be calculated self-consistently. The quadratic form (8) can be diagonalized by making a unitary transformation

$$\gamma_{k0}=\cos\theta_k\, c_{k\uparrow}^+ + \sin\theta_k\, c_{-k\downarrow}, \quad (9)$$

$$\tan 2\theta_k = \Delta_k/\epsilon_k, \tag{10}$$

$$\mathcal{H}_{\rm red} = \sum_k E_k(\gamma_{k0}^+\gamma_{k0} + \gamma_{k1}^+\gamma_{k1}), \tag{11}$$

$$E_k = (\epsilon_k{}^2 + \Delta_k{}^2)^{1/2}, \tag{12}$$

with a similar equation for γ_{k1}. E_k and $\gamma_k{}^+$ are quasiparticle energy and creation operators, respectively, and Δ_k is the minimum excitation energy, or gap. Using Eqs. (9)–(12), the mean F_k can be calculated, and a self-consistency equation results:

$$\Delta_k = -\sum_{k'} V_{kk'} \frac{\Delta_{k'}}{2E_{k'}} \tanh\left(\frac{E_{k'}}{2k_BT}\right). \tag{13}$$

This is the famous BCS gap equation. It is the superconducting analog of Eq. (8) for a magnet. Nontrivial solutions exist only for T less than a critical temperature T_c. An analytic solution for T_c can be obtained for the "BCS model interaction"

$$(V_{kk'})_{\rm BCS} = \begin{cases} -V & \text{if } |\epsilon_k|<k_B\theta \text{ and } |\epsilon_{k'}|<k_B\theta \\ 0 & \text{otherwise.} \end{cases} \tag{14}$$

The answer is [provided $D(\epsilon_F)V \ll 1$]

$$T_c = 1.13\theta \exp[-1/D(\epsilon_F)V]. \tag{15}$$

Another result is that the gap $\Delta(T)$ at $T=0$ is given by $2\Delta(0) = 3.52k_BT_c$. The full solution of Eqs. (13) and (14) for $\Delta(T)$ is a universal curve (plotted in Fig. 5) when $\Delta(T)$ is normalized to $\Delta(0)$ and T is normalized to T_c. The universal behavior of $\Delta(T)$ in BCS theory provides a microscopic explanation for the "law of corresponding states," a phrase denoting the experimental observation of universality in many physical properties of superconductors [$H_c(T)$, for example].

We have not yet considered the origin of the interaction V or the meaning of the cutoff θ. These questions are independent of the mathematical theory described above. An important experimental clue was the discovery in 1950 by Maxwell and Serin and others of the "isotope effect," the dependence of T_c on isotopic mass. This suggested that the electron–phonon interaction might play an important role. Following theoretical work of Fröhlich, Bardeen, and Pines, BCS identified V as the interaction between electrons from lattice polarization, a second-order effect of the electron–phonon interaction. This leads to the identification of the cutoff θ with the Debye temperature Θ_D, and the formula (15) helps explain why the observed values of T_c are usually so low.

The subsequent development of BCS theory has been a rich field of contemporary physics. For example, the understanding of the nature of the coupling V has improved. The need for refinements of BCS theory was evident almost immediately because of the behavior of the so-called "bad actors" Pb and Hg, which show deviations from the law of corresponding states. Also Giaever discovered "bumps" corresponding to phonon energies in the differential conductance of superconducting tunnel junctions. These bumps did not have a very natural explanation in the BCS equations. It was soon recognized that the difficulty lay in the BCS

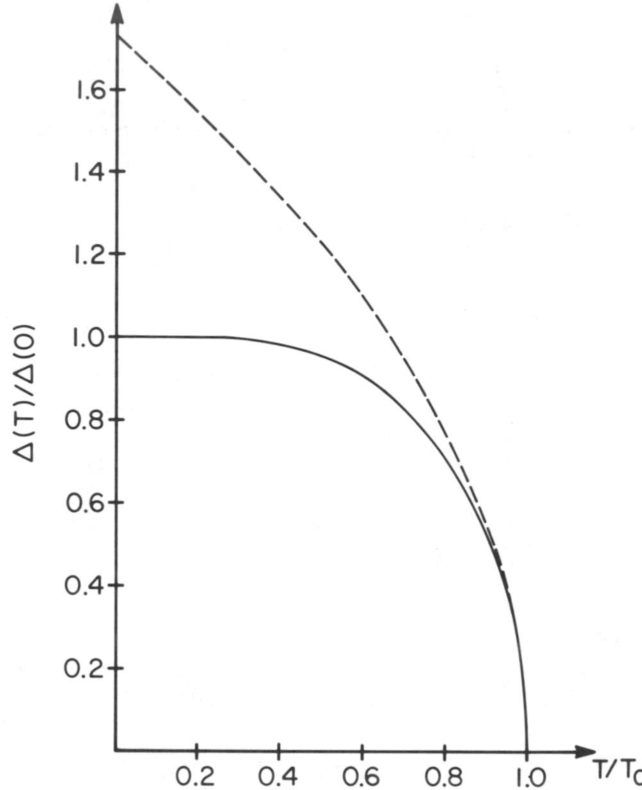

FIG. 5. Reduced value of the BCS energy gap $\Delta(T)/\Delta(0)$ versus reduced temperature T/T_c (solid curve), and approximate form $\Delta(T)/\Delta(0) \approx 1.74\,(1-T/T_c)^{1/2}$ (dashed curve). To better than 1% accuracy (i.e., the thickness of the line), these curves also represent the magnetization of a spin-$\frac{1}{2}$ ferromagnet in mean-field approximation [Eq. (7)].

assumption of an instantaneous interaction [proportional to $\delta(t-t')$]. Relativistic theory shows that all interactions are retarded at least by $\delta(t-t'-|\mathbf{x}-\mathbf{x}'|/c)$, but the phonon-induced interaction is much more severely retarded than this. Physically what happens is that the first electron at $(\mathbf{x}t)$ interacts with the lattice by a coupling constant M, creating a disturbance which can be Fourier analyzed in terms of phonons. This disturbance develops and propagates in space and time at a rate which is two or three orders of magnitude slower than electron Fermi velocities ($\sim 10^8$ cm/s). In order to interact efficiently with this disturbance, the second electron at $(x't')$ must wait for the disturbance to develop. This allows the first electron time to move away. Thus two electrons are able to interact by "leaving messages for each other" in the crystal lattice, never meeting directly (which would be disadvantageous because the direct interaction is dominated by a strong Coulomb repulsion). The form of the actual interaction is

$$\begin{aligned} V(xt, x't') &= -\sum_Q M_Q(x)M_Q{}^*(x') \\ &\quad \times \sin \omega_Q(t-t')e^{-\gamma_Q(t-t')} \qquad \text{if } t'>t, \\ &= 0 \qquad \text{otherwise.} \end{aligned}$$

The interaction oscillates in time at frequencies determined by the phonon frequency ω_Q. The interaction responsible for superconductivity occurs primarily during the first half-cycle while V is attractive and before the weakening effects of phonon damping γ_Q and dephasing (caused by dispersion in γ_Q) occur. A generalized version of BCS theory powerful enough to include retarded interactions was first written down by Eliashberg. This theory has a time-dependent pair field. The gap $\Delta(\omega)$, which is measured by an external ac probe at frequency ω, is essentially the Fourier transform of the pair field. The gap is complex, the real and imaginary parts corresponding, respectively, to the parts of the pair field which are in phase and out of phase with the interaction. The gap $\Delta(\omega)$ has bumps when the applied frequency is in resonance with the phonon frequencies. These correspond to Giaever's bumps in tunneling conductance, because the dc voltage across a tunnel junction behaves like an ac probe at frequency $\hbar\omega = eV$. The careful measurement of these bumps has been developed into a powerful spectroscopic tool by Rowell and McMillan and has provided detailed confirmation of the Eliashberg theory for many superconductors.

Other effects besides the electron–phonon interaction may in principle contribute to the attractive interaction needed in BCS theory. The "high T_c" copper-oxide-based superconductors do not show a normal isotope effect, and may require either supplementary attractive interactions or else an alternative route altogether.

The BCS theory has also been extended to cases where the superconducting order parameter $\Psi(x) = \langle\psi_\uparrow(x)\psi_\downarrow(x)\rangle$ (the Fourier transform of F_k) varies in space because of external perturbations or boundary conditions. The equations are extremely complicated, except near T_c where Ψ is small and a series expansion can be used. The resulting "Landau–Ginzburg" equations essentially duplicate a theory which was correctly guessed by L. D. Landau and V. L. Ginzburg in 1950:

$$\alpha\Psi + \beta|\Psi|^2\Psi + \tfrac{1}{2}m^*\left(\frac{\hbar}{i}\nabla - \frac{2e}{c}\mathbf{A}\right)^2\Psi = 0,$$

where α is proportional to $T - T_c$. The first two terms are derivable from the theory of Eqs. (7)–(15), whereas the last term requires adding a magnetic field term to the Hamiltonian. Using this theory, Abrikosov predicted the existence of type II superconductors and the flux lattice state. This theory contributed to a revolutionary improvement in understanding of magnetic properties of superconductors and the development of the first superconducting magnets around 1960.

The understanding of superconductivity does not appear to be complete. The high-T_c copper oxide superconductors, for example, were not anticipated prior to 1986. What can be anticipated is continued discovery of new materials, and further developments both in theory and applications.

See also Helium, Liquid; Josephson Effects; Phase Transitions; Superconducting Materials; Symmetry Breaking, Spontaneous.

BIBLIOGRAPHY

Some of the most important original papers are reprinted in *Superconductivity, Selected Reprints* (American Institute of Physics, New York, 1964). Other collections of reprints are: D. Pines, *The Many Body Problem* (Benjamin, New York, 1962); N. N. Bogoliubov, *The Theory of Superconductivity* (Gordon and Breach, New York, 1968).

An English translation of the paper by Landau and Ginzburg is available in *Men of Physics: Landau*, Vol. II (D. ter Haar, ed.) (Pergamon, New York, 1965). The Nobel addresses of Bardeen, Cooper, Schrieffer, Giaever, Josephson, and Bednorz and Müller are highly recommended, and can be found in *Phys. Today* 23–46 (July 1973); *Science* **183,** 1253 (1974); **184,** 527 (1974); *Rev. Mod. Phys.* **60,** 585 (1988).

Elementary (the algebra of BCS theory is not given)

H. M. Rosenberg, *Low Temperature Solid State Physics*, Chap. 6. Oxford, London, 1963.

A. W. B. Taylor, *Superconductivity*. Wykeham Publications, London, 1970.

D. R. Tilley and J. Tilley, *Superfluidity and Superconductivity*, 2nd ed. Adam Hilger, Bristol, 1986.

J. E. C. Williams, *Superconductivity and its Application*. Pion Ltd., London, 1970.

Intermediate

C. Kittel, *Quantum Theory of Solids*. Chap. 8. Wiley, New York, 1963.

C. G. Kuper, *An Introduction to the Theory of Superconductivity*. Oxford, London, 1968.

E. A. Lynton, *Superconductivity*, 3rd ed. Methuen, London, 1969.

M. Tinkham, *Introduction to Superconductivity*. Krieger, Malabar, 1980.

Advanced

P. G. de Gennes, *Superconductivity of Metals and Alloys*. Addison-Wesley, Redwood City, CA, 1989.

R. D. Parks (ed.), *Superconductivity*. Marcel Dekker, New York, 1969.

G. Rickayzen, *Theory of Superconductivity*. Wiley-Interscience, New York, 1965.

J. R. Schrieffer, *Theory of Superconductivity*. Benjamin, New York, 1964.

Specific Classes of Superconductors

J. C. Wheatley, "Experimental Properties of Superfluid ^{3}He," *Rev. Mod. Phys.* **47,** 415–470 (1975).

A. J. Leggett, "A Theoretical Description of the New Phases of ^{3}He," *Rev. Mod. Phys.* **47,** 331–414 (1975).

G. R. Stewart, "Heavy Fermion Systems," *Rev. Mod. Phys.* **56,** 755–787 (1984).

D. M. Ginsberg (ed.). *Physical Properties of High Temperature Superconductors I*. World Scientific, Singapore, 1989.

Superheavy Elements

Walter Greiner

At present, an extensive search for superheavy elements (SHE) is going on. While the search during the early 1970s concentrated on their discovery in nature, it now focuses on the production of SHE by complex nucleus–nucleus collisions. These efforts were provoked by theoretical studies on the stability of very heavy and superheavy elements, which started in 1966 [1–4]. The underlying theoretical idea for the existence of SHE is the following: The nuclear liquid drop model (LDM), which views the nucleus as a fluid consisting of neutrons and protons held together by a surface tension, is based on the empirical Bethe–Weizsäcker mass formula. The most improved versions of the latter are due to Seeger [5] and Myers and Swiatecki [6]. They predict that nuclei with $Z \geq 100$ become unstable against fission because the Coulomb repulsion of the Z protons compensates—and exceeds—the stabilizing effect of the surface tension. This would terminate the periodic system of stable elements at $Z \approx 100$, i.e., at the element fermium. There occurs, however, a second stabilizing effect due to nuclear shell structure, which markedly modifies this result, since protons and neutrons are especially strongly bound together if their Z or N numbers (or even better, both) are magic, namely, $Z, N = 2$, 8, 20, 50, 82. These numbers characterize the nucleonic "rare-gas configurations"—in analogy to the saturated electron shells of atoms—and led Mayer and Jensen [7] to the nuclear shell model. Hence, even though a nucleus is unstable against fission according to the LDM, it can still be stabilized beyond $Z \approx 100$ if its shell structure is magic or doubly magic. This appears for the following superheavy, doubly magic nuclei:

$Z = 114$, $N = 184$ or 196 (center of first island of stability),
$Z = 164$, $N = 272$ (center of second island of stability).

Nuclei with slightly different combinations of N and Z in the vicinity of these doubly magic configurations are also stable against fission, but to a lesser extent. Thus $Z = 114$ and 164 form the centers of the so-called islands of stability. The situation is summarized in Fig. 1, which shows the "geography of nuclei" with proton number Z as ordinate and neutron number N as abcissa. The magic neutron and proton numbers are indicated by horizontal and vertical lines, respectively. The doubly magic nuclei appear at the intersections.

The "reef" of the stable nuclei within this "sea of instability" is clearly visible. The neutron-rich nuclei form a kind of "shallow water" on the lower side of the reef, because they decay by β^- decay $n \rightarrow p + e^- + \nu$) and neutron emission [characterized by the condition that the neutron binding energy becomes zero, i.e., $E_B(n) = 0$]. The proton-rich nuclei also form a shallow-water area on the upper side of the reef. They decay by β^+ decay $p \rightarrow n + e^+ + \nu$) and proton emission [$E_B(p) = 0$]. The heaviest stable nucleus on the reef is uranium with a half-life of $\tau = 4.51 \times 10^9$ yr. Beyond it, nuclei decay predominantly because of spontaneous nuclear fission. Beyond this natural end of the stable periodic elements (i.e., the reef) appear the two islands of stability. They are separated from the reef by shallow ($Z = 114$) and deep ($Z = 164$) waters, respectively, just as Sicily and Sardinia are separated from the Italian peninsula. It is obvious that these islands of stability can be reached from the main land only by bridging the "dividing waters" with bigger nuclei, i.e., by bombarding heavy stable nuclei with other stable nuclei. The stability relations of the first island are quantitatively summarized, according to calculations of Ref [2], in Fig. 2. Obviously various decay channels exist, like fission, α decay, and electron capture (β^+ decay). According to these calculations total half-lifes of about 1 yr are expected for nuclei around $Z = 112$. Odd nuclei may live longer because of additional stability due to the odd nucleon. Nevertheless, it should be kept in mind that these calculations contain great uncertainty (of the order $\pm 10^3$–10^4) resulting from the large extrapolations on which they are based and from theoretical difficulties. Indeed, various researchers have obtained quantitatively different results (longer lifetimes) from those quoted here. The main practical problem lies nowadays in the experimental way of fusing superheavy nuclei. There have been various proposals for achieving this:

1. The Berkeley group (Ghiorso *et al.* [8]) bombards heavy nuclei with light ones in order to keep Coulomb excitation energy and transferred angular momentum small.

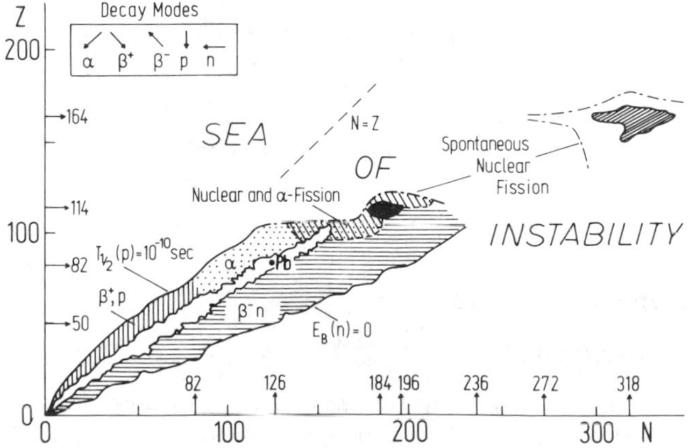

FIG. 1. Geography of nuclei: The sea of instability, the reef of stability, and the superheavy island of stability (densely shaded) at the intersections $Z = 114$, $N = 184$ and $Z = 164$, $N = 318$, respectively.

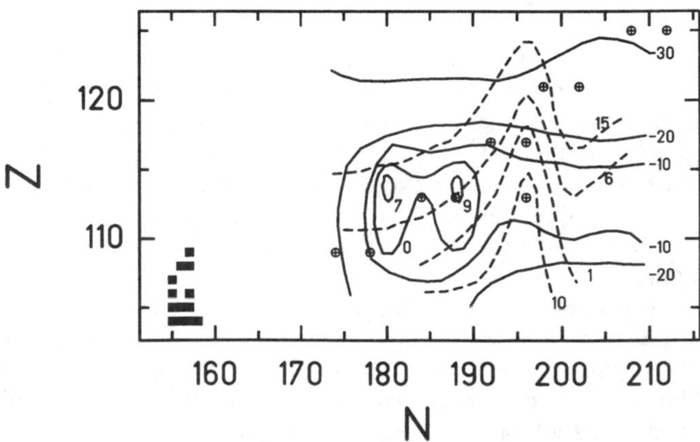

FIG. 2. The lifetimes (in $\log_{10}T$) of the nuclei of the first island of stability: solid lines, against fission; dashed lines, against α decay. The solid circle signs indicate stability against β^+ decay (electron capture). The squares represent the heaviest experimentally synthesized trans-actinide nuclei.

2. The Dubna group (Flerov *et al.* [9]) tries to amalgamate larger spherical nuclei with other large spherical fragments. The idea is that the compound nucleus of two touching spherical nuclei will have little deformation energy and hence a good chance for survival (i.e., fusion).
3. The difficulty in fusing superheavy elements lies in the fact that a large amount of internal energy and angular momentum are introduced into the compound system during the reaction. Consequently the main mechanism used for the synthesis of the heaviest known elements was the "cold fusion" of complex nuclei. In this type of reaction a large spherical nucleus like ^{208}Pb or ^{209}Bi merges with accelerated ions heavier than argon to form slightly excited compound nuclei, which emit few neutrons and γ rays. This projectile–target combinations and a carefully chosen bombarding energy lead to the minimal possible excitation of the created heavy nucleus. This mechanism was proposed by Gupta, Parvulescu, Sandulescu and W. Greiner [10], who calculated the potential energy (PES) for all possible projectile-target combinations leading to a given SHE. They found local minima in this PES, where at least one of the reaction partners is spherical. During a heavy-ion collision every deviation from these positions causes a driving force toward this local minimum and consequently a dynamical change of the shape of the compound nucleus. A large amount of energy is then transferred into the excitation of surface vibrations and a drastical reduction of lifetimes follows. Only if the target and the projectile are chosen with respect to the minima in the PES would the excitation be minimum.

The newest trans-actinide elements ($Z=107$–109) have been produced using exactly the best projectile–target combination predicted by Gupta *et al.* [10]. In this procedure, they undergo mainly α decay and show enhanced stability against spontaneous fission.

The heaviest of these short-lived elements now known is $Z=109$, which was discovered at GSI (Darmstadt, Germany) from one single correlated decay sequence following the fusion of ^{209}Bi with ^{58}Fe in 1982 by Münzenberg *et al.* [13]. Two more decay chains were observed in 1988. The observed lifetime against α decay is 3.4 ms. This feature may be interpreted as an indication for the onset of the magic stabilizing effect of the $Z=114$ shell. The identification chain of the newly produced presuperheavy elements 107–109 is illustrated in Fig. 3. We might also consider overshooting the island of stability. The decay products of such a reaction should also contain superheavy elements. Reactions of this type (Pb–Pb, U–Pb) using the new generation of heavy-ion colliders seem promising. The UNILAC at GSI, south of Frankfurt-am-Main in Germany [11], and SuperHILAC at Berkeley in the United States [12] are the first to accelerate uranium nuclei to such high energies that the Coulomb barrier to any other nucleus can be overcome (7–10 MeV/nucleon). There have also been extensive searches for superheavy elements in nature. See Ref. [13] for a full discussion of these attempts; that they were unsuccessful is, in view of the expected lifetimes (see Fig. 2), not surprising.

The importance of superheavy elements is twofold: First, their existence would prove the basic nuclear shell model substantially beyond its present range of validity. For example, it would be a great surprise with significant implications for our understanding of nuclear shell structure if the next magic proton number beyond lead should not be 114, but perhaps 120 or 126. Second, the SHE would extend the periodic system significantly and contribute to chemistry by continuing the exciting and successful discovery of the transplutonic elements started by Seaborg.

Related to, though quite distinct from, SHE are the intermediate superheavy molecules (ISM), which are formed during the collision of, e.g., Pb with Pb, Pb with U, or U

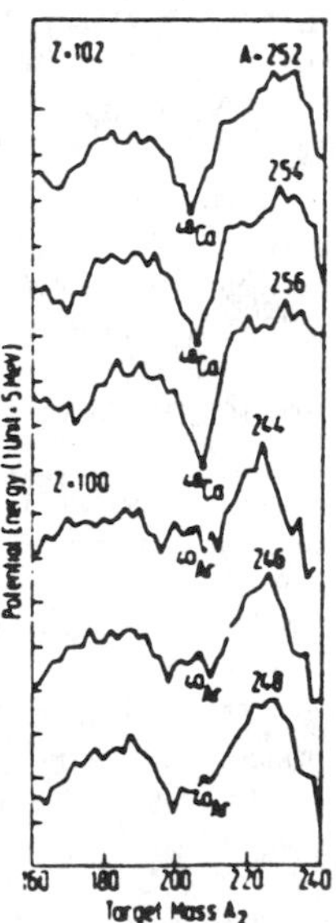

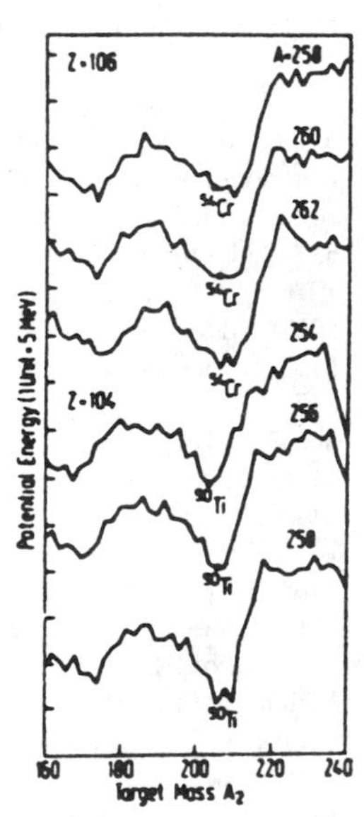

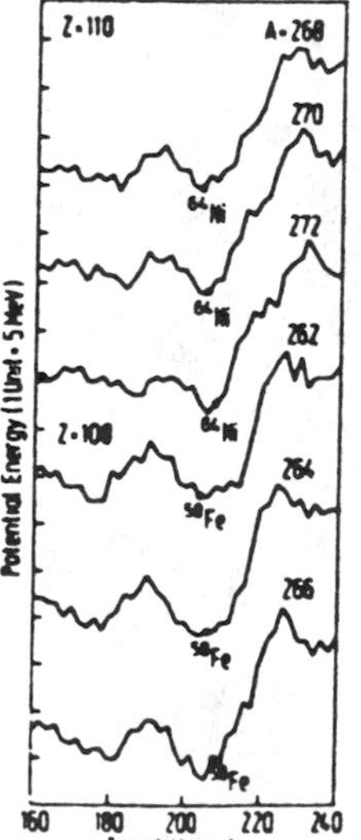

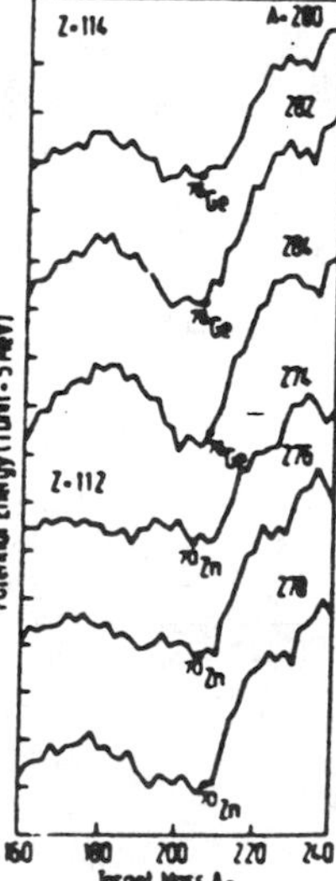

FIG. 3. Potential energy (PES) as a function of the target mass A_2 and the relative separation R for the compound system (A) with (a) $Z=100$–102, (b) $Z=104$–106, (c) $Z=110$, (d) $Z=112$–114. The projectile nuclei relevant to the Pb and Bi targets are shown.

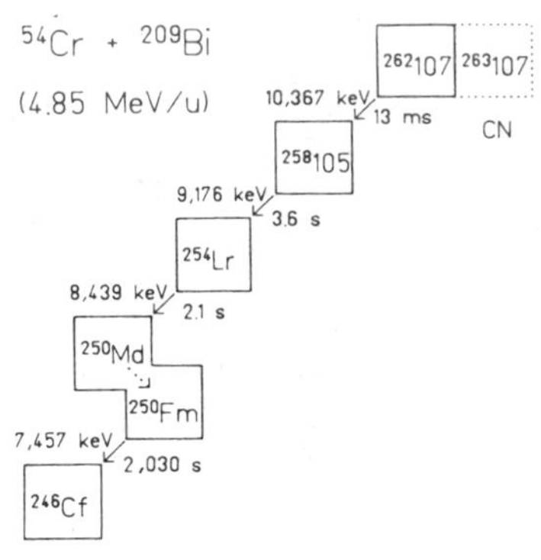

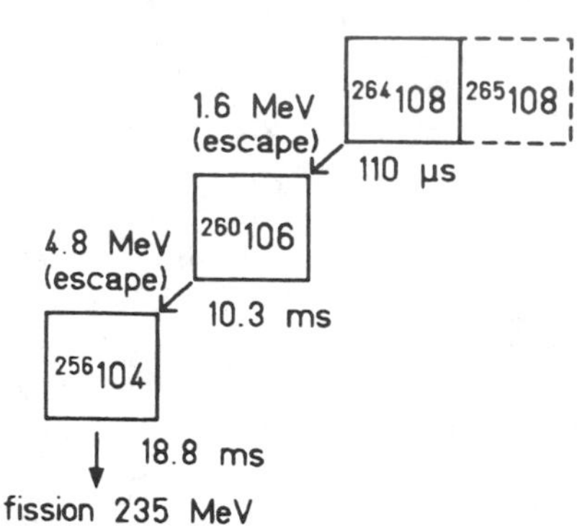

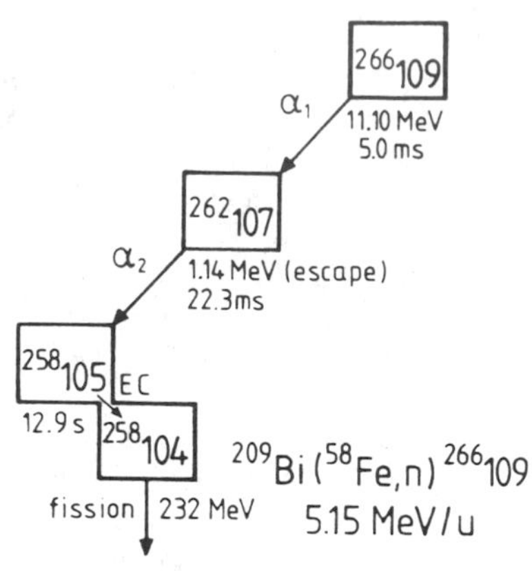

FIG. 4. Decay series for elements 107, 108, and 109

with U. While the velocity of the colliding nuclei below their Coulomb barrier is $v_{\text{ion}} \sim c/20$, the velocity of the surrounding inner shell electrons is $v_{\text{el}} \sim c$. Hence, during the collision the electrons can form molecular (two center) orbitals. As the distance between the nuclei becomes smaller, the ISM becomes an intermediate superheavy atom (ISA). For a short while the electrons "see" the united charge of the two colliding nuclei. Thus, by bombarding heavy ions with heavy ions, superheavy atoms up to charges $Z = 184$ (U–U) can be simulated. For lighter systems (Br–Br, Ni–Ni, I–Au) these intermediate molecules have been experimentally verified via the so-called quasimolecular x radiation. The ISM are of basic importance for testing and verifying the predictions for quantum electrodynamics of strong fields. That is, because of the very strong electric fields around superheavy atoms, the vacuum (more precisely the electron–positron vacuum) becomes unstable for superheavy nuclei with $Z > Z_{\text{crit}} \approx 170$ [14]. While the undercritical vacuum is neutral, the overcritical vacuum is charged. This is illustrated in Fig. 5, which shows the electronic energy levels for superheavy atoms. The filled negative energy levels (Dirac sea) represent the

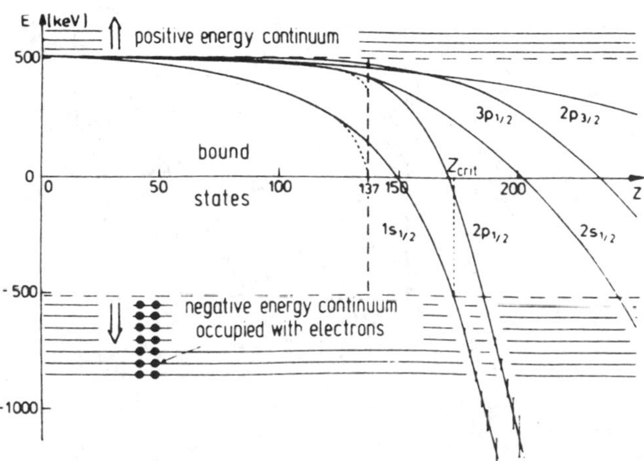

FIG. 5. The electron levels in superheavy atoms. Up to $Z_{\text{crit}} \approx 170$ the electron-filled negative energy continuum represents the neutral vacuum. Above Z_{crit} the vacuum becomes charged, because of the "diving" of *K*-, *L*-, . . . shell electrons.

neutral vacuum. For $Z > Z_{\text{crit}}$ the unfilled $1s\sigma$ state joins the negative-energy continuum, which leads to energyless e^+–e^- pair creation: One electron moves into the empty 1*s* orbit while its hole is repelled as a positron from the nucleus. The positron emission manifests the recharge of the vacuum, i.e., the phase transition from the neutral to the charged vacuum. This very fundamental process appears analogously also for the pseudoscalar pion field, the scalar σ-meson field, and the Schwarzschild-type gravitational field. The first two of these phase transitions are known as pion condensation and the Lee–Wick phenomenon, respectively [15].

See also NUCLEAR PROPERTIES; NUCLEAR STRUCTURE.

REFERENCES

1. S. G. Nilsson, I. R. Nix, A. Sobiczewski, Z. Szymanski, S. Wycech, C. Gustavson, and P. Möller, *Nucl. Phys.* **A115,** 545 (1968).
2. U. Mosel and W. Greiner, *Z. Phys.* **217,** 256 (1968); **222,** 261 (1969).
3. Yu. A. Muzychka, V. Pashkevich, and V. Strutinski, *Sov. J. Nucl. Phys.* **8,** 417 (1969).
4. For reviews see G. Hermann, *Phys. Scripta* **10A,** 71 (1974); T. Johanssen, S. G. Nilsson, and Z. Szymanski, *Ann. Phys. (Paris)* **5,** 377 (1970); J. R. Nix, Proc. Leysin Conf. 1970 (CERN 70-30, 1970), p. 605; *Annu. Rev. Nucl. Sci.* **22,** 65 (1972); *Phys. Today* 30 (April 1972).
5. P. A. Seeger, Proc. Leysin. Conf. 1970 (CERN 70-30, 1970), p. 217.
6. W. D. Myers and W. J. Swiatecki, *Ark. Fys.* **36,** 343 (1967).
7. M. G. Mayer and J. H. D. Jensen, *Elemetary Theory of Nuclear Shell Structure.* Wiley, New York, 1956.
8. A. Ghiorso, M. Nurmia, J. Harris, K. Eskola, and P. Escola, *Phys. Rev. Lett.* **22,** 1317 (1969).
9. Yu. Ts. Oganessian, Yu. V. Lobanov, M. Hussonnois, Yu. P. Kharitonov, B. Gorski, O. Constantinescu, A. G. Popeko, H. Bruchertseifer, R. N. Sagaidak, S. P. Tretyakova, G. V. Buklanov, A. V. Rykhlyuk, G. G. Gulbekyan, A. A. Pleve, G. N. Ivanov, and V. M. Plotko, *International School on Physics "Enrico Fermi",* Varenna, Italy, 23 June–3 July, 1987.
10. R. K. Gupta, C. Parvulescu, A. Sandulescu and W. Greiner, *Z. Phys.* **A283,** 217 (1977).

11. Ch. Schmelzer, *Proceedings of the International Conference on Reactions Between Complex Nuclei, Nashville, 1974* (J. Hamilton, ed.). North-Holland, Amsterdam, 1974.
12. High Intensity Uranium Beams from the SuperHILAC and the Bevalac, Proposal-32, Lawrence Berkeley Laboratory, Berkeley. CA, May 1975.
13. G. Münzenberg, W. Reisdorf, S. Hofmann, Y. K. Agarwal, F. P. Heßberger, K. Poppensieker, J. R. H. Schneider, W. F. W. Schneider, K.-H. Schmidt, H.-J. Schött, and P. Armbruster, *Z. Phys.* **A315**, 145 (1984).
14. B. Müller, H. Peitz, J. Rafelski, and W. Greiner, *Phys. Rev. Lett.* **28**, 1235 (1972); for a review see W. Greiner and W. Scheid, in *Heavy Ion Collisions* (R. Bock, ed.), Vol. 3. North-Holland, Amsterdam, 1980.
15. See, e.g., the review by J. M. Irvine, "Pion Condensation and Other Abnormal States of Matter," *Rep. Prog. Phys.* **38**, 1385 (1975).

Supersymmetry and Supergravity

C. P. Burgess

In three spatial dimensions all particles must have the quantum-mechanical statistics of either fermions or bosons, depending on whether or not their wave function changes sign when any two particles are interchanged. In a relativistic theory this statistical property is related to the particle's spin by the spin-statistics theorem [1]. This theorem states that all bosons have integer spins and all fermions have half-odd-integer spin. Bosons and fermions therefore necessarily have different spins.

Supersymmetry is a remarkable symmetry that relates the properties of particles having differing statistics and differing spins. It is the only quantum-mechanical symmetry known which relates particles with different spins in a way that is consistent with special relativity and involves a finite number of particle types. It is widely believed to be the only possible such symmetry.

At the time of this writing (January 1990) it remains an open question whether supersymmetry plays a role in the formation of the laws of nature. There is at present no experimental evidence in its favor. However its rich and tightly constrained structure holds the promise of relating otherwise disparate features of particle physics, restricting the divergences that commonly arise in field theories, and possibly providing a framework for unifying gravity with everything else. These remarkable properties have sparked an immense amount of theoretical activity over the years since its discovery more than 15 years ago.

HISTORICAL BACKGROUND

The idea of a symmetry that can relate particles with different spins has had many precursors over the years. One of its early guises arose within the context of the nonrelativistic quark model [2] during the 1960s. At that time much effort was being spent trying to understand the great variety of strongly interacting particles, or hadrons, that had been produced by cosmic rays or in accelerators over the preceding decades. Eventually it was found that many of the properties of these particles could be accounted for by the quark model, in which the observed hadrons were considered as bound states of more fundamental spin-$\frac{1}{2}$ constituents called quarks.

Within the quark model each of the three types of quarks known at the time, u, d and s, can have its spin along any fixed direction in space pointed up ($\uparrow$) or down ($\downarrow$). Furthermore, if the two possible spin states of the three quarks types are written as a six-component column, as follows:

$$q = \begin{pmatrix} u\uparrow \\ d\uparrow \\ s\uparrow \\ u\downarrow \\ d\downarrow \\ s\downarrow \end{pmatrix}, \qquad (1)$$

then many of the features of the strong interactions could be understood if it were assumed that the underlying physics were approximately invariant under the multiplication of this column, q, by an arbitrary 6×6 unitary matrix, U, with unit determinant. The collection of all such transformations forms a mathematical group known as $SU(6)$. One of the most interesting properties of this type of symmetry is that it relates particles of differing spins, such as the $u\uparrow$ and $d\downarrow$ quarks for example.

In its original formulation, the $SU(6)$ quark model was a purely nonrelativistic theory. Given its phenomenological successes, many attempts were made [3] to find its relativistic generalization. The difficulty in doing so led to the suspicion that such a generalization might not be possible. This conclusion was ultimately borne out with the proof of various "no-go" theorems, culminating with that of Coleman and Mandula [4] in 1967.

A symmetry may be defined as a set of transformations that (i) take one-particle states to one-particle states, (ii) acts on many-particle states as if they are products of one-particle states, and (iii) leaves scattering amplitudes unchanged (and so commutes with the scattering matrix, S). The Coleman–Mandula theorem asserts on very general grounds that, for an interacting theory, it is impossible to find a connected symmetry group, G, that contains the Poincaré transformations of special relativity and has transformations (other than the usual Lorentz transformations) that can change the overall spin of any particle state.

This seemed to close the book on the matter until the advent of supersymmetry several years later. Supersymmetry evades the conclusions of the Coleman–Mandula theorem by slipping through a technical loophole. The loophole is that Coleman, Mandula, and their contemporaries assumed that any symmetry would have the mathematical structure of a Lie group, or Lie algebra. This assumption turns out to fail for supersymmetries, which instead lie within the class of *graded* Lie algebras. The graded nature of the algebra is intimately connected to the fact that supersymmetry always relates particles whose spins differ by one-half and so necessarily have opposite statistics. Some features of Lie algebras and graded Lie algebras are described in more detail below. A generalization of the Coleman–Mandula analysis to graded Lie algebras was done by Haag, Łopuszanski, and Sohnius [5] and leads to the conclusion that only a very restricted form for supersymmetry transformations is possible.

Supersymmetry in its present form has its roots [6] within the dual models that were actively investigated during the late 1960s and early 1970s. Although there were some earlier anticipations [7], serious investigations began with the first presentation of a renormalizable, supersymmetric four-dimensional field theory [8] by Wess and Zumino in 1974. This theory enjoyed a "global" supersymmetry in which the same supersymmetry transformation is performed everywhere throughout space-time. The first theory with a "local" supersymmetry, that is with supersymmetry transformations that vary with position, was constructed shortly thereafter [9]. A remarkable feature of locally supersymmetric theories is that they necessarily must involve gravity, hence being known as supergravity theories.

As their features became better understood, the phenomenology of supersymmetric theories became more and more actively investigated, starting with early work in the 1970s and continuing throughout the 1980s.

This work has recently come full circle with the emergence of superstring theory as an extremely promising candidate for a truly unified theory of all known interactions. Superstring theory is the modern offspring of the original dual models within which supersymmetry was first discovered. As alluded to in the name, superstring theories naturally include supersymmetry and indeed rely on it for their very consistency.

Graded Lie algebras have found applications other than as symmetry transformations that commute with the time evolution of a physical system. They have also been used as *spectrum-generating* algebras within nuclear physics [10]. Since the supersymmetry algebra is not required in these applications to be relativistic symmetries, the restrictions coming from the Haag–Łopuszanski–Sohnius result need not apply.

It is likely that supersymmetric theories will continue as an active field of research for the forseeable future.

MATHEMATICAL FRAMEWORK: Z_2-GRADED ALGEBRAS

A Lie *group* is a mathematical group whose elements, $g(\alpha)$, are labeled by a set of parameters, α^i, that take continuous values. The structure of the group is largely determined by the properties of the first derivatives of the group elements with respect to these parameters evaluated at the identity transformation: $T_i \equiv -i\,\partial g/\partial\alpha^i\,|_{g=1}$. The group multiplication law, $g(\alpha)g(\beta) = g(\gamma)$, may be expressed in terms of these *generators*, T_i, as *commutation relations* of the form

$$T_iT_j - T_jT_i = ic^k_{ij}T_k, \tag{2}$$

where the coefficients (*structure functions*) c^k_{ij} are a set of given numbers which define the group. The indices i, j, and k all take integer values between 1 and N, where N is the group's dimension defined as the number of independent parameters needed to label the elements of the group. Einstein's summation convention, in which there is an implied sum over any repeated indices, is used in Eq. (2) and throughout what follows. The collection of all of the generators, T_i, that satisfy Eq. (2), together with all of their possible sums and products forms the *Lie algebra* associated with the Lie group in question.

Lie groups are ubiquitous within physics because of the important role played by symmetries. The set of all symmetry transformations of a particular physical system has all of the properties of a group. Because many common symmetry transformations, such as rotations or translations, are labeled by continuous parameters, like angles or distances, the relevant group is often a Lie group.

Supersymmetry is an exception to this rule, since it is does not form a Lie algebra defined by conditions of the form of Eq. (2). Instead, supersymmetry transformations form what is known as a *Z_2-graded algebra*. This means that although some symmetry transformations within the supersymmetry algebra, the so-called "even" elements, satisfy relations of the form of Eq. (2), others, the "odd" elements, satisfy instead *anticommutation* relations among themselves:

$$T_iT_j + T_jT_i = d^k_{ij}T_k. \tag{3}$$

The d^k_{ij} are again known constants. The odd generators satisfy relations like Eq. (2) with the even generators. The difference in sign between Eqs. (2) and (3) is required if the odd generators are to transform bosons into fermions and vice versa.

The most general form for a symmetry algebra that is a Lie algebra, i.e., satisfies Eq. (2), is given, subject to some technical assumptions, by the Coleman–Mandula theorem. Its generators must be a linear combination of the four-momentum, P^μ, that generates space-time translations, the angular momentum, $J^{\mu\nu}$, that generates the Lorentz transformations of special relativity, and any number of "internal symmetries," Z^a, that must necessarily commute with P^μ and $J^{\mu\nu}$. (If all particle states are precisely massless, this conclusion must be slightly modified to include conformal transformations.) Since the Z^a's must commute with the angular-momentum generator, $J^{\mu\nu}$, they cannot relate particles with differing spins. More precisely, any two states, $|\psi\rangle$ and $|\phi\rangle$ say, related by the action of one of the Z^a's, $|\psi\rangle = Z\,|\phi\rangle$, must have the same spin.

The generalization of this result to include graded algebras that involve relations like Eq. (3) implies similar restrictions on the algebras that are physically acceptable. In this case the Coleman–Mandula result applies directly to the "even" part of the algebra and so implies that the most general "even" generators are given P^μ, $J^{\mu\nu}$, and Z^a. The new feature is that it is possible to have nontrivial "odd" generators, Q^i_α. These *supercharges* or *supersymmetry* generators must commute with P^μ and must transform under Lorentz transformations as spinors. The label α denotes the four independent components of such a (Majorana) spinor. In general, there may be more than one supersymmetry generator, and this is indicated by the index $i = 1, \ldots, N$. The simplest case of only one supersymmetry is often known as simple, or $N = 1$, supersymmetry. *Extended* supersymmetry, on the other hand, indicates a graded symmetry involving more than one superchange: $N > 1$. The most general form that Eq. (3) may take turns out to be

$$Q^i_\alpha \overline{Q}^j_\beta + \overline{Q}^j_\beta Q^i_\alpha = -2i(\gamma_\mu)_{\alpha\beta}P^\mu\delta^{ij} + [i(A_a)^{ij}\delta_{\alpha\beta} + (B_a)^{ij}(\gamma_5)_{\alpha\beta}]Z^a. \quad (4)$$

In this expression γ_μ and γ_5 denote 4×4 Dirac *matrices* while $\delta_{\alpha\beta}$ and δ^{ij} represent the components of the unit matrix of appropriate dimension. $\overline{Q}$ is the spinor that transforms conjugately to Q under Lorentz transformations. The real coefficients $(A_a)^{ij} = -(A_a)^{ji}$ and $(B_a)^{ij} = -(B_a)^{ji}$ are arbitrary apart from one restriction: they must vanish unless the corresponding internal-symmetry generator, Z^a, commutes with all of the other Z's. A generator with this property is known as a "central charge."

Notice that if there is only one supersymmetry generator ($N = 1$ supersymmetry), then the antisymmetry of $(A_a)^{ij}$ and $(B_a)^{ij}$ under the interchange of i and j implies that they must vanish. In this case Eq. (4) reduces to the algebra

$$Q_\alpha \overline{Q}_\beta + \overline{Q}_\beta Q_\alpha = -2i(\gamma_\mu)_{\alpha\beta}P^\mu. \quad (5)$$

As may be seen from either Eq. (4) or (5), a supersymmetry transformation Q^i_α is in some sense the "square root" of a space-time translation, P^μ. This relationship is responsible for many of the unique features that arise within supersymmetric theories.

These conditions determine, among other things, the types of particles that can be related by supersymmetry transformations. Because the supercharges must transform as a spinor under the Lorentz group, each supersymmetry transformation must change a particle's spin by one-half. In extended supersymmetries this is true separately for each supercharge, Q^i. While $N = 1$ supersymmetry can only relate particles whose spins differ by one-half, extended supersymmetries with N supercharges have representations that can contain particles whose spins run from s to $s + N/2$ in steps of one-half. If the theory is required to involve only particles whose spins lie between -2 and 2, then (in the absence of central charges) it would follow that there can be no more than $N = 8$ supersymmetries.

PHENOMENOLOGICAL IMPLICATIONS

Although the tightly constrained form enjoyed by supersymmetric theories has immense theoretical appeal, in physics the bottom line must always be found in contact with experiment. Much effort has been therefore devoted to attempts to explore the phenomenological prospects of supersymmetric theories. The results so far are disappointing, but not entirely discouraging.

Supersymmetric theories all share the same most fundamental prediction: For every particle known, there must be (at least) one other particle, its superpartner, which has opposite statistics, and is related to the first by a supersymmetry transformation. Supersymmetry implies that the physical properties of particles and their superpartners are directly related.

Virtually all theoretical work on the phenomenological aspects of supersymmetry to date has involved simple, or $N = 1$, supersymmetry. This is because it is generally true that extended supersymmetries require the left- and right-handed spin-$\frac{1}{2}$ particles have the same couplings to spin-1 particles. According to our present understanding of high-energy phenomena as embodied within the very successful "standard model" [1] of high-energy physics, however, left- and right-handed spin-$\frac{1}{2}$ particles have spin-1 couplings that are fundamentally different. This argues that extended supersymmetries are unlikely to be relevant at energies that are of experimental interest in the near future.

If only a single supersymmetry exists, then every particle must have precisely one superpartner. The particle and its superpartner must couple in the same way to other particles. Their electric charges would, for example, be required to be equal. If the ground state of the system is invariant under supersymmetry transformations, then every particle and its superpartner must also have exactly the same mass. Any such superpartner for a known particle, the electron for instance, would have been easily observed by now in many experiments and so can be ruled out. It follows that if supersymmetry is to be a fundamental symmetry of nature, the supersymmetry must be broken by the ground state. The supersymmetry is then said to be *spontaneously broken*. The present failure to observe any supersymmetric counterparts to any of the known particles implies a lower bound on superparticle masses roughly in the range of 40 to several hundred times the mass of a proton. The precise limit depends on the quantum numbers and couplings of the particles of interest.

The signatures from supersymmetric particles to be expected at high-energy accelerators are now fairly well understood [12]. In a wide class of models [13] all superpartners of the presently known particles turn out to be odd under a conserved multiplicative quantum number known as "R-parity." All of the "ordinary" particles that have been (or, like the top quark and Higgs bosons, are expected to be) found are even under this quantum number. Conservation of this quantum number implies that superpartners must be pair-produced, and that the lightest superpartner must be stable. Since this lightest superpartner is, in most models, electrically neutral and only interacts weakly with ordinary matter, it is expected to escape unseen from any particle detector. The telltale footprint of supersymmetry would then be an event with unseen missing energy and momentum.

The bad news is that, despite extensive searches, no such signal has yet been seen. This indicates that the mass of any superpartner must be higher than is presently experimentally accessible.

The good news, on the other hand, is that supersymmetric theories seem to solve some perceived theoretical difficulties with the standard model of high-energy particle interactions. The principal difficulty with which supersymmetry seems to help is known as the "hierarchy problem." This is the puzzle of why gravity is so very weak compared to all of the other known interactions. More quantitatively, the problem is to explain why it should be that the ratio of the coupling constant (G_F) for the second-weakest force, the weak interactions, is some 10^{33} times larger than the coupling constant for gravity, Newton's constant (G_N) assuming that both are measured in fundamental units with Planck's constant ($\hbar$) and the speed of light (c) taken equal to 1.

At present the smallness of this ratio is understood as being due to the vast difference between the fundamental distance scales appropriate to each of these forces. The strength of the weak interactions is related to a length scale, l_W, associated with the masses of the intermediate weak bosons, $W^{\pm}$ and Z^0: $l_W \approx 10^{-16}$ cm. The length scale at which gravitational interactions become strong is the Planck length, $l_P \approx 10^{-33}$ cm. It is this incredible difference in length scales that is responsible for the feeble strength of gravity relative to the other interactions. This manifests itself physically as the enormous difference between the sizes of typical astrophysical objects and those of common experience since astrophysical scales are fixed by a competition between gravity and the other interactions.

What is not understood is how such a small ratio of fundamental distance scales can arise within any reasonable theory. Supersymmetry seems to solve at least part of this problem, since the extra particles that would be implied in a supersymmetric theory tend to allow such an extreme hierarchy of scales. This success is only possible, however, if the scale of any spontaneous supersymmetry breaking is not much higher than the present experimental limits. Either supersymmetry is found in the reasonably near future or it cannot help understand the hierarchy problem.

SUPERGRAVITY

Another interesting consequence of the symmetry algebra, Eq. (4), is that it *must* be realized locally rather than globally. This is because we know that any candidate theory of nature must include gravity, and so must be, according to general relativity, invariant under arbitrary changes of space-time coordinates. Since an infinitesimal general coordinate transformation may be thought of as a translation through a distance that varies with position in space-time, and since Eq. (4) implies that the repetition of two supersymmetry transformations is a translation, consistency of the supersymmetry algebra with general covariance requires invariance with respect to local, or space-time-dependent, supersymmetry transformations.

As is always the case, invariance with respect to space-time-dependent transformations involves the introduction of a new "gauge" particle that couples to the conserved current associated with the symmetry. The properties of the gauge particle may be understood in terms of the properties of the conserved current to which it couples.

Consider therefore the conserved current to which this gauge particles must couple in supersymmetric theories. It transforms as a spinor-vector $U_\alpha^\mu(x)$ under Lorentz transformations, and satisfies the conservation condition

$$\partial_\mu U_\alpha^\mu = 0. \qquad (6)$$

In order to couple to such a current, the gauge particle associated with local supersymmetry invariance must have spin-$\frac{3}{2}$ and so must be a fermion. Its partner under supersymmetry is the spin-2 graviton that arises from general covariance. This spin-$\frac{3}{2}$ particle is usuallly known as the *gravitino*. It is required by the supersymmetry algebra to couple universally to all matter with gravitational strength. It is therefore very feebly coupled indeed. The gravitino mass is proportional to the size of the order parameter that is responsible for spontaneous supersymmetry breaking. In most models in which supersymmetry survives down to accelerator energies, this mass is no heavier than several hundred times the proton mass.

Early in their development it was hoped that supergravity theories might solve some of the divergence problems that notoriously plague general relativity at high energies. It was subsequently found that although the extra supersymmetries ameleorate these difficulties somewhat, they are not completely eliminated.

There does appear, however, to be a single candidate theory that does not suffer from these problems. This is the theory of superstrings. Superstring theory postulates that the basic building blocks of matter are not point particles, as has been otherwise assumed, but are rather one-dimensional "strings" that sweep out two-dimensional surfaces through space-time. An important feature of these string theories is that they cannot be formulated without automatically incorporating supergravity. It also appears that calculations within these theories are free from divergences. It is remarkable that supersymmetry appears as a necessary ingredient at a very fundamental level. Indeed, superstring theories predict that very many supersymmetries are present at the Planck scale, although likely only one of these can survive down to presently accessible energies.

Up to the present, supersymmetry remains a fascinating theoretical tool whose ultimate role, if any, within our understanding of physical laws eludes us. Its beautiful features suggest that this role is there to be found, if not now then perhaps sometime in the future.

See also GAUGE THEORIES; GRAVITATION; LIE GROUPS; QUARKS; SPIN; STRING THEORY; SYMMETRY BREAKING, SPONTANEOUS.

REFERENCES

1. R. F. Streater and A. S. Wightman, *PCT, Spin, Statistics, and All That*. Benjamin/Cummings, Reading MA, 1964.
2. F. E. Close, *An Introduction to Quarks and Partons*. Academic Press, London, 1979.
3. F. J. Dyson, *Symmetry Groups in Nuclear and Particle Physics*. Benjamin, New York, 1966.
4. S. Coleman and J. Mandula, *Phys. Rev.* **159,** 1251 (1967).
5. R. Haag, J. Łopuszanski, and M. Sohnius, *Nucl. Phys.* **B88,** 257 (1975).
6. J.-L. Gervais and B. Sakita, *Nucl. Phys.* **B34,** 632 (1971).
7. Yu. A. Gol'fand and E. P. Likhtman, *JETP Lett.* **13,** 323 (1971); D. V. Volkov and V. P. Akulov, *Phys. Lett.* **46B,** 109 (1973).
8. J. Wess and B. Zumino, *Nucl. Phys.* **B70,** 39 (1974); *Phys. Lett.* **49B,** 52 (1974).
9. D. Z. Freedman, S. Ferrara, and P. van Nieuwenhuizen, *Phys. Rev.* **D13,** 3214 (1976); S. Deser and B. Zumino, *Phys. Lett.* **62B,** 335 (1976).
10. F. Iachello, *Phys. Rev. Lett.* **44,** 772 (1980).
11. S. Weinberg, *Phys. Rev. Lett.* **19,** 1264 (1967); *Phys. Rev.* **D5,** 1412 (1972); A. Salam, in *Elementary Particle Theory: Relativisitic Groups and Analyticity* (*Nobel Symposium No. 8*) (N. Svartholm, ed.). Almqvist & Wiksell, Stockholm, 1968; S. Glashow, J. Iliopoulos, and L. Maiani, *Phys. Rev.* **D2,** 1285 (1970).
12. H. P. Nilles, *Phys. Rep.* **C110,** 1 (1984); H. E. Haber and G. L. Kane, *Phys. Rep.* **C117,** 75 (1985).
13. L. Hall, "Alternative Low-Energy Supersymmetry" *Mod. Phys. Lett. A* (to be published)

SUPPLEMENTARY READINGS

M. Green, "Superstrings" *Sci. Am.*, 48 (September 1986). (E)
M. Green, J. Schwarz, and E. Witten, *Superstring Theory*, Vols. 1 and 2. Cambridge University Press, Cambridge, 1987. (A)
Howard E. Haber and Gordon L. Kane, "Is Nature Supersymmetric?" *Sci. Am.*, 52 (June 1986). (E)
P. van Nieuwenhuizen, *Phys. Rep.* **C68,** 189 (1981). (A)
J. Wess and J. Bagger, *Supersymmetry and Supergravity*. Princeton University Press, Princeton, NJ, 1983. (A)

Surface Tension

F. M. Fowkes

At surfaces and interfaces, discontinuities in kinds of materials or in densities give rise to interfacial forces which result in surface tension or interfacial tension. At surfaces, discontinuities in density produce gradients in cohesive forces which tend to diminish the density in the surface region and produce the two-dimensional "surface tension." At interfaces between immiscible materials, discontinuities in cohesive forces tend to produce tension in the interfacial region of both materials, but in addition there are adhesional forces between the dissimilar materials which tend to decrease the interfacial tension.

Surface and interfacial tensions are readily observed with liquids, but are more difficult to observe with solids. Little stretching of the surface region can occur on rigid solids even though gradients in cohesive forces are present. Even an enhanced surface concentration of lattice vacancies may not result in the kind of surface tension observed with liquids or softer solids. However, on heating rigid solids to higher temperatures sufficient to provide atomic mobility, surface tensions are indeed observable.

The magnitude of surface tension γ is proportional to the magnitude of cohesive forces and is a useful measure of cohesive forces. Interfacial tension (γ_{12}) is a measure of the tension in the surface region of the two materials ($\gamma_1+\gamma_2$) minus the adhesion force of attraction or "work of adhesion," W_{12}:

$$\gamma_{12}=\gamma_1+\gamma_2-W_{12} \qquad \text{(Dupre, 1869)}$$

Surface or interfacial tension in liquids is measurable by the pressure difference across a curved interface:

$$\Delta F=\gamma(1/R_1+1/R_2) \qquad \text{(Young, 1805: Laplace, 1806)},$$

where R_1 and R_2 are the two radii of curvature. Application of this principle lies behind the various methods of boundary-tension measurement (capillary height, drop or bubble shape methods, ring method, grain boundary widening, maximum bubble pressure, oscillating jet, etc.).

Surface or interfacial tensions (together with the surface area α) contribute a work term $\gamma\ d\alpha$ in thermodynamics which increases both the internal energy U and Helmholtz free energy A:

$$\left(\frac{\partial U}{\partial \alpha}\right)_{S,V,ni} = \left(\frac{\partial A}{\partial \alpha}\right)_{T,V,ni} = \gamma.$$

On the other hand, the enthalpy H and Gibbs free energy G are area independent but dependent on boundary tension:

$$\left(\frac{\partial H}{\partial \gamma}\right)_{S,P,ni} = \left(\frac{\partial G}{\partial \gamma}\right)_{T,P,ni} = -\alpha.$$

Surface or interfacial tensions and contact angles (θ) can be used to predict the difference in "surface free energy" (F_S) of a solid when immersed in various liquids:

$$\gamma_L \cos\theta=F_{SV}-F_{SL} \qquad \text{(Young, 1805)},$$

where the subscripts refer to liquids L, vapor V, and solid S. The work of adhesion to the solid is then

$$W_{SL}=\gamma_L(1+\cos\theta)+(F_S-F_{SV}) \qquad \text{(Harkins and Livingston, 1942)}.$$

Surface tension and the work of adhesion W_{12} or work of cohesion ($W_{11}=2\gamma_1$) result from the attractive forces between molecules, which include the ever-present van der Waals attractions (mainly the London dispersion forces) and often certain specific interactions such as hydrogen-bonding or other electron donor–acceptor (Lewis acid–base) interactions. In mercury the surface tension is the sum of the contributions from the metallic bond γ^m and from dispersion-force interactions γ^d. At 20°C, $\gamma_1^m=284$ mJ/m^2 and $\gamma_1^d=200$ mJ/m^2. For water at 20°C, $\gamma_2^d=22.0$ mJ/m^2, and the acid–base contribution $\gamma_2^{ab}=50.8$ mJ/m^2. The work of adhesion between mercury and water involves only van der Waals interactions, and so it is calculable from

$$W_{12}^d=(W_{11}^d W_{22}^d)^{1/2}=1(22.0\times 200)^{1/2}=132 \text{ mJ/m}^2 \qquad \text{(Fowkes, 1964)}$$

which is borne out by experiment, for the above prediction is within less than 1% of the experimental value.

BIBLIOGRAPHY

Arthur W. Adamson, *Physical Chemistry of Surfaces*, 4th ed., Chaps. 1 and 2. Wiley, New York, 1982
John F. Padday, "Theory of Surface Tension," in *Surface and Colloid Science*, Vol 1, pp. 39–251 (E. Matijevic, ed.). Wiley-Interscience, New York, 1969.
Frederick M. Fowkes, *Ind. Eng. Chem.* **12,** 40 (1964).

Surface Waves on Fluids

W. K. Melville

A surface separating two fluids may support wave modes in which the velocity field is confined to the neighborhood of the surface. Such waves result from the balance between the fluid inertia and the restoring forces of gravity and surface tension, both of which tend to make the interface flat. The most common example of these waves occurs at the interface between air and water. Due to the large density difference between the fluids, the dynamics of the air may often be neglected entirely and the free surface waves in the water studied separately. With the approximations that the fluid is incompressible, inviscid, and its motion began from a state of rest, the kinematics of the fluid are described by the potential flow (Laplace's) equation. However, this linear equation is subject to nonlinear kinematic and dynamic boundary conditions at the surface which render intractable the general solution to the problem of the evolution of arbitrary initial disturbances.

With the assumptions that the slope of the surface is everywhere small (i.e., $\ll 1$) the boundary conditions may be linearized. In linear sinusoidal waves the fluid particles describe elliptical orbits; the eccentricity of the orbits decreases as the fluid depth increases, becoming circular when the depth is very much greater than the wavelength, λ. In this case the velocity field decreases exponentially away from the surface with a length scale $k^{-1}=\lambda/2\pi$. Free surface waves propagate with a phase speed c which is a function of the wave number k, with a parametric dependence on the gravitational acceleration g, the surface tension divided by the fluid density γ, and the undisturbed depth of the fluid, h:

$$c^2=\{g/k+\gamma k\}\tanh(kh).$$

If $k^2 \ll g/\gamma$, the effects of surface tension are negligible and the waves are commonly called gravity waves. If $k^2 \gg g/\gamma$, the effects of gravity are negligible and the waves are called capillary waves.

In general, surface waves are dispersive, with the phase speed varying as a function of the wave number, k. For capillary waves the phase speed decreases as the wavelength increases, for gravity waves the phase speed increases. In deep water (i.e., $kh \gg 1$) there is a minimum phase speed when $k=\sqrt{g/\gamma}$, corresponding to a wavelength of 17 mm and a phase speed of 23 cm/s. To a good approximation, long gravity waves (i.e., $kh \ll 1$) are nondispersive and $c^2 \approx gh$.

While the phase (or crest) of a sinusoidal component travels at the speed c, the amplitude and therefore the energy associated with the linear wave travels at the group velocity

$$c_g = \frac{\partial(c/k)}{\partial k}.$$

For gravity waves $c_g<c$, whereas for capillary waves $c_g>c$. While the evolution of linear dispersive water waves may be predicted in principle by integral transform methods, detailed analytical results are usually approximate and employ asymptotic methods based on long-time spatial separation of Fourier components due to dispersion. Approximate analytical solutions may be found for initial- and boundary-value problems, including evolution of arbitrary initial disturbances, propagation over and past spatially varying topography, and scattering by obstacles. The availability of supercomputers has led to a relaxation of the assumption of linearity, and "exact" numerical solutions may be obtained for problems involving strongly nonlinear free waves.

If the dynamics of the fluid (usually air) above the surface is significant, as in the case of waves directly generated by the wind, the dynamic boundary condition becomes more complicated, and the assumptions of linearity and potential flow may be less tenable. Interest in surface waves in the last three decades has centered largely on the problems associated with wave generation by wind and nonlinear aspects of free waves. Interest has been stimulated by microwave remote sensing of the oceans and other natural water bodies from space. Considerable progress has been made in understanding the stability of weakly nonlinear waves, and in exploiting the properties of solitary waves in both deep and shallow water. While rational theories of wave generation by wind have been developed, there is no complete theory, and difficulties associated with the nonlinear interaction between waves of different scales remain. The strongly nonlinear process of deep-water wave breaking is poorly understood.

See also Fluid Physics; Hydrodynamics; Nonlinear Wave Propagation; Waves.

BIBLIOGRAPHY

Horace Lamb, *Hydrodynamics*. Cambridge University Press, Cambridge, 1932. (Reprinted in Dover Publications, New York.) A classic in the fluid mechanics literature; dated in style but still a rewarding reference.

P. H. LeBlond and L. A. Mysak, *Waves in the Ocean*. Elsevier, Amsterdam, 1978. Oceanographic applications of surface wave theory, including the influence of Earth's rotation.

James Lighthill, *Waves in Fluids*, Chaps 3 and 4, and Epilogue. Cambridge University Press, Cambridge, 1978. A novel combination of physical insight and analysis marks this treatment of acoustic, surface, and internal waves in fluids.

C. C. Mei, *The Applied Dynamics of Ocean Surface Waves*. Wiley, New York, 1982. A sophisticated mathematical treatment for applications of surface wave theory in engineering and oceanography.

O. M. Phillips, *The Dynamics of the Upper Ocean*, 2nd ed., Chaps 3 and 4. Cambridge University Press, Cambridge, 1977. An application of linear and weakly nonlinear wave theory to the description of oceanic surface waves.

G. B. Whitham, *Linear and Nonlinear Waves*. Wiley, New York, 1974. A general treatise on wave motion, with surface wave applications dispersed throughout.

Surfaces and Interfaces

F. Cyrot-Lackmann and M. C. Desjonquères

The last 20 years have been a time of increasing interest in the fundamental studies of surfaces and interfaces. This is clearly due to the development of new techniques, using experiments performed in ultravacuum conditions, which have made possible the characterization of the surface at a microscopic level, and to the very great importance of many phenomena that occur on solid surfaces. Indeed, it is essential to acquire a deeper knowledge of basic surface phenomena in order to understand in a fundamental way heterogeneous catalysis, epitaxial growth, and corrosion, as well as many solid-state electronic devices or biological systems.

The surface of a solid may be defined as the outermost atomic layer(s) including, possibly, foreign atoms. It can be characterized in the same manner as any solid or molecule. One must know the chemical identity of the atoms present and the structural arrangement of these atoms, their vibration states, and the energetic and spatial distribution of electrons.

The *composition and structure of the surface* may differ considerably from the perfect termination of the bulk structure (Fig. 1). For a surface free of impurities, the outermost atomic plane(s) may be displaced from the bulk position (relaxation) or may have a new periodicity (reconstruction). Foreign atoms may be bound to the surface via van der Waals forces (physisorption) or chemical interactions (chemisorption) in a finite number or form an ordered or disordered overlayer. At the surface of an alloy, the concentration in one of the components can be larger than in the bulk (segregation).

To identify the chemical nature of surface atoms, one principally uses Auger electron spectroscopy. In this method, one uses the fact that electrons ejected from surface atoms as they relax after ionization of a core electron have characteristic energies depending on core levels and, consequently, on chemical identity. This is the usual technique for checking the cleanliness of a surface.

The surface crystallography can be investigated by various techniques observing directly either the real or the reciprocal lattice or by indirect methods. Field ionization micrographs give, under some conditions, an image of the surface lattice or the adsorption site of adsorbates. Scanning tunneling microscopy, based on the tunneling of electrons between a metallic tip and the sample surface, is a probe of the local surface density of electronic states at the Fermi level which contains information on the atomic structure. The surface reciprocal lattice is directly obtained in diffraction experiments [with low-energy electrons (LEED), neutral atoms, or grazing x rays] from the pattern of the diffracted beams, the determination of actual surface atom position and distance relative to inner planes being derived from a careful comparison of the oberved and calculated variations of the intensity of the diffracted beams with the energy of the incident particle. The most commonly used indirect methods are the study of surface-extended x-ray absorption fine structure (SEXAFS) or of photoelectron diffraction (PhD) and apply mainly to the determination of adsorption geometry. In SEXAFS the oscillations of the absorption cross section as a function of the energy of the incident photon, above a given threshold characteristic of adsorbed atoms, are analyzed and, since they arise from the backscattering of the outgoing electron wave by the neighbors of the excited atom, this analysis gives the bond length and coordination number of the adsorbate. In PhD a core electron is excited by a photon and the variations of the emitted current with the analyzer direction, due to the scattering of the outgoing electron wave by the neighbors of the excited atom, is recorded and compared with calculations performed for various atomic geometrics. Using these techniques, one has thus found that reconstruction occurs frequently for covalent materials, for example, the well-known 7×7 silicon (111), and more rarely on metals. For adsorbed layers, a large number of well-ordered periodic arrangements have been observed (Fig. 2).

Another problem related to the surface atomic structure is the surface stability and the equilibrium shape of small aggregates. It obviously depends on the surface free energy γ, i.e., the work, necessary to create a unit area of free surface. The dependence of this quantity on the crystallographic orientation is given by the γ plot, i.e., the surface $\gamma=\dot{\gamma}(\phi,\theta)$ which, at least at low temperatures, presents cusps, due to the free energy for step formation, in directions corresponding to rather close-packed surfaces when the temperature increases; due to entropy effects, this step energy may vanish above a given temperature. In this case the flat surface is no more at equilibrium and becomes rough at the atomic scale (roughening transition). In addition some facets disappear in the equilibrium shape (constructed from Gibbs Wulff theorem) of small aggregates which becomes rounder. This roughening transition has been observed in particular on ^{4}He crystals.

The existence of the surface modifies the *vibration spectrum of atoms (phonons)*. Theoretical studies are usually performed neglecting relaxation and reconstruction and with central forces (sometimes corrected by angular interactions) between nearest neighbors. One finds:

(i) bulk modes, with a modified amplitude and density, which are coupled by the surface;
(ii) surface modes (including Rayleigh waves) with an amplitude exponentially decreasing into the bulk;
(iii) virtual or resonant modes with an enhanced amplitude at the surface but which are nevertheless extended states because of their coupling with bulk modes.

Recently the Rayleigh mode dispersion curves have been measured, using inelastic interaction of electrons (electron energy loss spectroscopy) or neutral atoms (He scattering) with the surface in which surface phonons are created or annihilated, showing a rather good agreement with calculations. The mean square displacements, specific heat, and entropy have also been calculated with a reasonable agree-

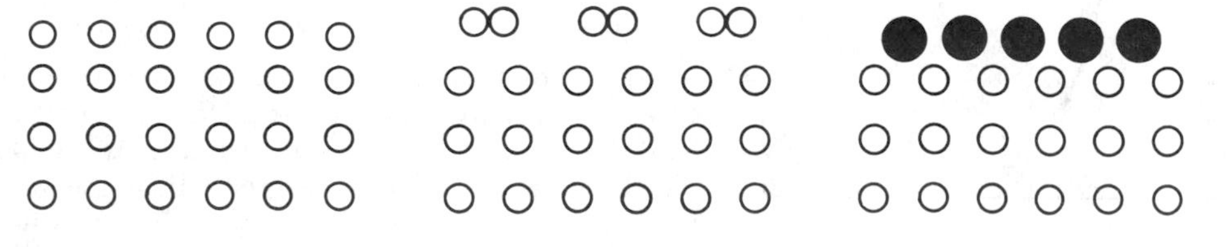
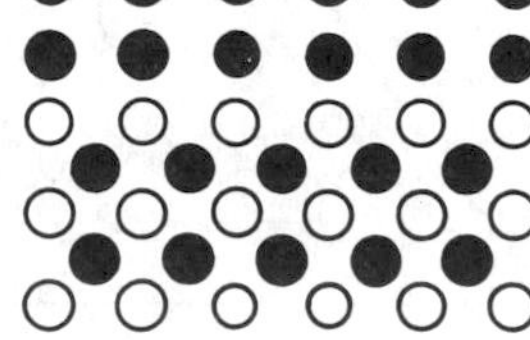

FIG. 1. Some possible structures of a crystal surface.

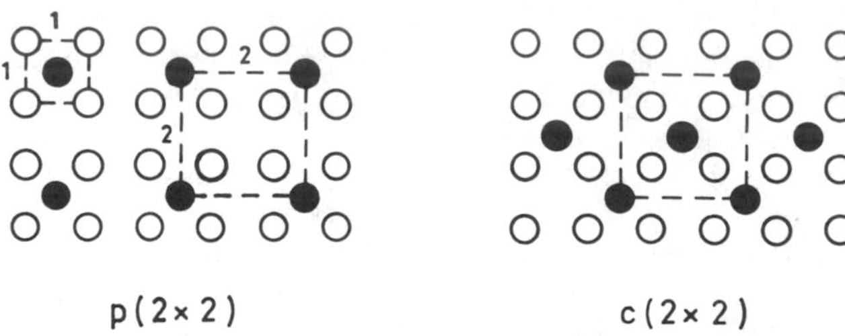

FIG. 2. (100) surface of a cubic crystal showing some frequently observed overlayer structures. The pattern of foreign atoms has a unit mesh twice the size of the (1 × 1) mesh of substrate atoms. They are centered, *c* (2 × 2), if the mesh has a foreign atom at its center, or primitive, *p* (2 × 2), if it does not.

ment with experiments (LEED, measurement of specific heat of small particles, and thermal variation of surface energy, respectively).

The knowledge of the *electronic structure at the surface* is of very great interest since it should have an influence on chemical reaction at surfaces but it is much more difficult to attain than for the bulk because of the loss of symmetry. Qualitatively, one expects that the valence energy level spectrum will have, at least, the same extension as in the bulk since the surface region is only a small fraction of the whole solid. But, similar to the case of vibration states, the modification of boundary conditions can lead, in addition to the perturbed Bloch bulk waves, to the existence of localized or surface states decaying exponentially into the bulk or of virtual surface states (resonances) which are extended states with a large amplitude near the surface and are then somewhat localized. These extra states may show up as a peak in the local density of states at the surface, i.e., the density of electronic levels weighted by the magnitude of the wave functions at the surface (Fig. 3). Both types of surface states have been observed on Al, Si, Ni, W, Mo, . . .) The main techniques are x-ray and electron spectroscopies (ESCA, ultraviolet photoemission, field emission, ion neutralization) which, though they are mainly sensitive to the surface layer,

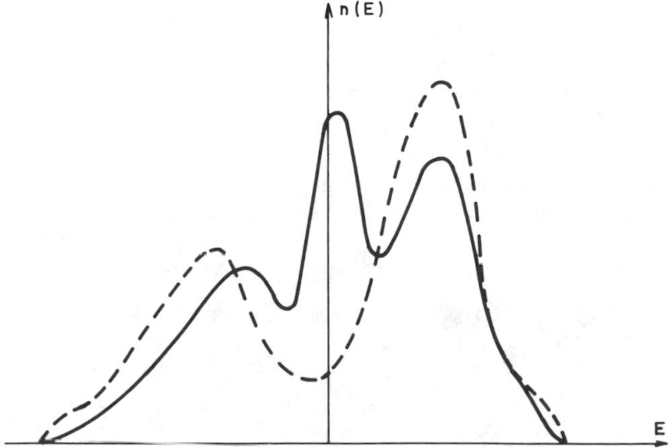

FIG. 3. Schematic local density of states on a (100) surface (solid line) of a bcc transition metal compared to the bulk one (dashed line); the central peak present in the surface case is due to surface states and resonances.

involve bulk contributions which are not always easy to disentangle from surface ones. An additional effect is slight displacement of the center of gravity of the local density of states at the surface relative to the bulk one resulting from the modified environment of surface atoms (broken bonds). This displacement is followed almost rigidly by core levels. It has been confirmed experimentally by core level photoemission spectroscopy.

All these properties (atomic structure, vibrational and electronic states) are obviously modified when adsorbates are present on the surface; for instance, some (vibrational or electronic) state may be localized in an adsorbed overlayer.

In the foregoing, we were concerned with the free surface of solids. Another interesting and related topic is that of *interfaces*. This is a very broad field with numerous applications in metallurgy, crystal growth, wetting of surfaces, etc. There is much work currently in progress in this domain, especially to understand the formation of metal–semiconductor or semiconductor–semiconductor junctions or heterojunctions. They involve a difference in potential, known as the Schottky barrier, and whose properties are widely used in the microelectronic devices. Moreover, another field of large interest is the formation of quantum wells between two semiconductors, i.e., the possibility of "graving" circuits in one dimension. The study of this geometry with very high magnetic field, the "quantum Hall effect," has led to the Nobel Prize for Prof. Von Klitzing (1985).

Let us remark on interfaces between crystals. There are, however, other contacts between solids of interest. Thus, the amorphous-crystal interface covers the problem of the surface treatment of semiconductors, the use of the tunnel effects through the insulating layers in MOS (metal–oxide–semiconductor) devices, and the adhesion of glues of plastics on metals, but it is not well understood. It is the same for the solid–liquid contact, which is the key to crystal growth or wetting of surfaces.

When the contact is between two crystals of the same phase, it is called a grain boundary. The problem is then defined by the relative position of the two crystals. Grain boundaries can be classified into

(i) large-angle incoherent grain boundary where there is no special relationship between the crystals (such a configuration appears clearly in field ion microscopy);
(ii) low-angle grain boundary where the two crystals have nearly the same orientation;
(iii) stacking faults for structures which can be described as a stacking of close-packed planes;
(iv) twins which are contacts of two crystals which can be deduced from each other by a symmetry operation around a plane or an axis.

The existence of these various boundaries plays an important role in plasticity. Large-angle grain boundaries and twins harden the material at low temperature by stopping dislocations. At higher temperature, self-diffusion is faster along the boundary than in the volume and, on the whole, soften the material.

Very little is known about their electronic structure and the possible existence of states localized near the interface.

Only some evaluation of the grain boundary energy has been made which can be compared to some measurements deducted from the width of splitting of a dislocation in partials or from the study of stable triple points.

A second type of interface is the epitaxial interface. Two crystals in exact epitaxy are in contact along a plane where two of their (small) periods coincide. Cases of perfect epitaxy are known in Guinier–Preston zones as, for example, in Al alloys. The energy of this interface is important in adhesional properties of metals and its probably governed by the sharing of electrons between the two metals. This epitaxial interface governs the formation of metallic multilayers, which lead to "artificial" structure for the metals, i.e., metastable structures with very peculiar properties (superconducting, magnetic, etc.). It is an overall field which now opens due to the modern deposition techniques, such as molecular beam epitaxy or coevaporation ones.

See also ADSORPTION; AUGER EFFECT; CATALYSIS; CRYSTAL DEFECTS; CRYSTAL GROWTH; ELECTRON ENERGY STATES IN LIQUIDS AND SOLIDS; FIELD-ION MICROSCOPY; LOW ENERGY ELECTRON DIFFRACTION; PHOTOELECTRON SPECTROSCOPY.

BIBLIOGRAPHY

Atomic Structure and Mechanical Properties of Metals, Proceedings, International School of Physics "Enrico Fermi", Course LXI, Varenna, July 8–20, 1974 (G. Caglioti, ed.). North-Holland, Amsterdam, 1976. (I, A)

CRC Crit. Rev. Solid State Mat. Sci., **4**(3), August (1976). (A)

Dynamic Aspects of Surface Physics, Proceedings, International School of Physics "Enrico Fermi", Course LVIII, Varenna, 1973 (F.O. Goodman, ed.). Editrice Compositori, Bologna, 1975. (I, A)

Homer D. Hagstrum, "Electronic Characterization of Solid Surfaces," *Science* **178**, 275–282 (1972). (E)

Special Issue on Surface Physics, *Phys. Today* **28**(4) (April 1975).

A. Zangwill, *Physics at Surfaces*. Cambridge University Press, Cambridge, 1988.

G. Höhler, *Solid Surface Physics*. Springer-Verlag, New York, 1979.

L. Fedman and J. W. Mayer, *Fundamentals of Surface and Thin Film Analysis*. North-Holland, Amsterdam, 1986.

SU(3) and Symmetry Groups

S. L. Glashow and J. F. Donoghue

A priori, one would expect little relation between the mathematics of group theory and the physics of the fundamental forces and elementary particles. However, in reality there is a surprising connection: Nature appears to be organized using symmetry groups. There are several manifestations of groups, such as in the gauge theories of the fundamental forces and in the organization of the spectrum of elementary particles. The same mathematical group structures, $U(1)$, $SU(2)$, and $SU(3)$, appear more than once in different functions, and it is important to keep the usages distinct. For example, the group $SU(3)$ appears both as a classification scheme for hadrons [often called $SU(3)$ of "flavor" in modern usage], and as the gauge theory of the strong interactions [quantum chromodynamics, referred to as $SU(3)$ of "color"]. Historically, it was as a classification scheme that the group $SU(3)$ first entered particle physics.

It was known in the 1950s that all strongly interacting elementary particles (i.e., hadrons) appear as multiplets with definite isospin T and hypercharge Y. Nucleons are a $T=\frac{1}{2}$, $Y=1$ doublet with $T_3=\frac{1}{2}$ for a proton and $T_3=-\frac{1}{2}$ for a neutron; the hyperons Σ^-, Σ^0, Σ^+ are a $T=1$, $Y=0$ triplet, etc. In other words, strong interactions seem to be invariant, or almost invariant, under the isospin-hypercharge group $SU(2)\times U(1)$, and observed multiplets correspond to irreducible representations of this group.

Many physicists believed in the existence of a higher symmetry group G, which would contain isospin-hypercharge as a subgroup and would lead to the grouping together of several isotopic multiplets into larger supermultiplets of hadrons. These supermultiplets, corresponding to irreducible representations of G, would include all or several of the observed spin-$\frac{1}{2}$ hadrons: nucleons, Λ and Δ hyperons, and Ξ's. Because these particles are only crudely degenerate in mass, G was not regarded as an exact symmetry group of strong interactions.

Many possibilities for the group G were examined: $SO(7)$, $SU(2)\times SU(2)\times SU(2)$, and exceptional group G_2 are examples that led nowhere. In retrospect, Sakata's suggestion [1] of 1956 was seminal. He believed that the nucleons and the Λ hyperon are truly fundamental, and that all other hadrons are composites of just these three. Later, the idea of charge independence of nucleon–nucleon forces was generalized [2] to apply to interactions among the allegedly fundamental p, n, and Λ. Thus, $SU(3)$ as a higher symmetry group was a natural extension of isotopic $SU(2)$. These ideas were the logical precursors to what Gell-Mann termed the eightfold way and to the notion of quarks. There do indeed seem to be fundamental constituents to hadrons, but they are the quarks, not p, n, and Λ. The original Sakata idea is simply wrong. If p, n, and Λ constitute an $SU(3)$ supermultiplet, then to what supermultiplet do Σ and Ξ belong? There is no adequate answer.

$SU(3)$ is the group of unitary modular 3×3 matrices, and its defining representation is three dimensional. Now consider a set of eight traceless and linearly independent 3×3 matrices λ_a $(a=1,\ldots,8)$. Let U be any unitary unimodular 3×3 matrix. Observe that the eight matrices $\lambda'_a=U\lambda_a U^{-1}$ are traceless, and are thus linear combinations of the λ_a:

$$\lambda'_a = \sum_b \lambda_b R_{ba}(U).$$

The matrices $R(U)$ define the irreducible eight-dimensional representation of $SU(3)$, the adjoint representation.

What distinguishes the eightfold way from the Sakata model is the assignment of the eight spin-$\frac{1}{2}$ baryons $(n,p,\Lambda,\Sigma^-,\Sigma^0,\Sigma^+,\Xi^-,\Xi^0)$ to the eight-dimensional representation of $SU(3)$ as first suggested by Gell-Mann [3] and Ne'eman [4]. Rather than treating p, n, and Λ separately, the eightfold way regards all eight baryons as a single $SU(3)$ supermultiplet.

The isospin-hypercharge subgroup of $SU(3)$ may be identified with the unitary matrices $\exp[i(\theta\cdot T+\phi Y)]$ with

$$T_a = \frac{1}{2}\begin{bmatrix} \sigma_a & 0 \\ 0 & 0 \end{bmatrix}, \qquad Y = \frac{1}{3}\begin{bmatrix} 1 & & \\ & 1 & \\ & & -2 \end{bmatrix},$$

where σ_a are the three Pauli matrices and θ_a, ϕ are parameters. T_a and Y are said to be the generators of the isospin and hypercharge subgroups. If the basis elements of the eight-dimensional representation are chosen to be eigenstates of T_3 and Y, they may be neatly displayed on a graph:

Y
n p
Σ^- Σ^0 Λ Σ^+ T_3
Ξ^- Ξ^0

where we have indicated the identification of the eight baryons with these elements. (Both Λ and Σ_0 lie at the origin.)

Should the eight baryons be grouped together to form the 3×3 traceless matrix

$$B = \begin{bmatrix} \left(\frac{1}{\sqrt{2}}\Sigma^0 + \frac{1}{\sqrt{6}}\right) & \Sigma^+ & p \\ \Sigma & \left(-\frac{1}{\sqrt{2}}\Sigma^0 + \frac{1}{\sqrt{6}}\Lambda\right) & n \\ \Xi & -\Xi^0 & -\sqrt{\frac{2}{3}}\Lambda \end{bmatrix},$$

the $SU(3)$ transformations under which strong interactions are to be approximately invariant are of the form $B \to UBU^{-1}$.

Like the spin-$\frac{1}{2}$ baryons, pseudoscalar mesons also transform like an octet,

Y
K^0 K^+
π^- π^0 η π^+ T_3
K^- $\bar{K}^0$

and

$$M = \begin{bmatrix} \left(\frac{1}{\sqrt{2}}\pi^0 + \frac{1}{\sqrt{6}}\eta\right) & \pi^+ & K^+ \\ \pi^- & \left(-\frac{1}{\sqrt{2}}\pi^0 + \frac{1}{\sqrt{6}}\eta\right) & K^0 \\ K^- & \overline{K}_0 & -\sqrt{\frac{2}{3}}\eta \end{bmatrix},$$

The most general $SU(3)$-invariant Yukawa coupling between mesons and baryons is simply

$$a\ \mathrm{Tr}(\overline{B}_{\gamma_5} BM) + b\ \mathrm{Tr}(\overline{B}_{\gamma_5} MB).$$

Thus, the 12 distinct meson–baryon couplings that are compatible with isospin-hypercharge symmetry are reduced to just two in the limit of exact $SU(3)$ symmetry.

In the $SU(3)$ classification scheme, all hadrons must appear as $SU(3)$ supermultiplets. Experimentally, this indeed seems to be the case. Mesons of given spin and parity always come in singlets and octets. There is a pseudoscalar singlet η' in addition to the octet. The $K^*, \overline{K}^*$, and ρ are seven members of a vector meson octet, while ω and ϕ are mixtures of an $SU(3)$ singlet and the $T = Y = 0$ member of the octet. Similarly, there is a mixed singlet and octet of spin-2 mesons.

Baryons show somewhat more variety. There are several known octets, like the spin-$\frac{1}{2}$ baryons. But the even-parity spin-$\frac{3}{2}$ baryons are something quite different. They transform according to a 10-dimensional representation of $SU(3)$. Here is the T_3–Y plot for this decuplet:

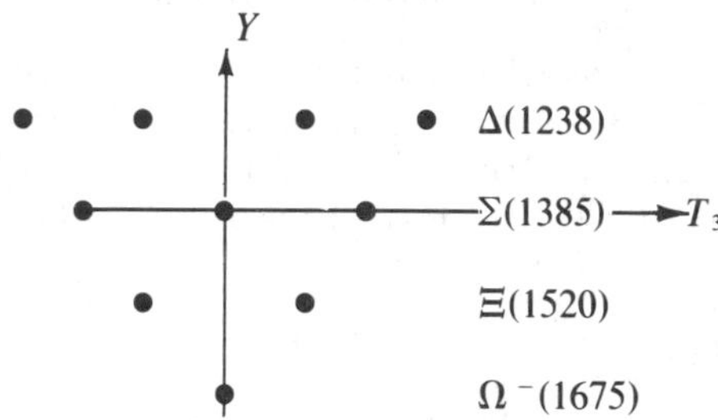

Indeed, it was the discovery [5] of the Ω^-, once the lone missing but predicted member of the decuplet, that convinced everyone of the correctness of the $SU(3)$ scheme.

Group-theoretic methods can also be used to analyze the physics that breaks the $SU(3)$ symmetry. Suppose the mechanism that breaks $SU(3)$ symmetry is both small and simple: small enough for first-order perturbation theory in the symmetry-breaking Hamiltonian $\mathcal{H}_{sb}$ to be useful, with $\mathcal{H}_{sb}$ transforming as simply as possible under $SU(3)$, like the $T = Y = 0$ member of an octet. A consequence of this assumption is that the mass term in the effective Lagrangian describing meson and baryon octets must be approximately of the form

$$a\ \mathrm{Tr}(M^2) + b\ \mathrm{Tr}(M^2 Y) + A\ \mathrm{Tr}(\overline{B}B) + B\ \mathrm{Tr}(\overline{B}BY) + C\ \mathrm{Tr}(\overline{B}YB).$$

From this we may deduce [7] mass formulas relating squares of pseudoscalar-meson masses:

$$M_K^2 = \tfrac{3}{4}M_\eta^2 + \tfrac{1}{4}M_\pi^2$$

and spin-$\frac{1}{2}$ baryon masses:

$$\tfrac{1}{2}M_\Xi + \tfrac{1}{2}M_N = \tfrac{3}{4}M_\Lambda + \tfrac{1}{4}M_\Sigma.$$

For the baryon decimet, stronger mass formulas are obtained:

$$M_{\Sigma^*} - M_\Delta = M_{\Xi^*} - M_{\Sigma^*} = M_\Omega - M_{\Xi^*}.$$

All the formulas are in rather good agreement with observed masses.

What is the origin of this $SU(3)$ symmetry and why is it (only) approximately correct? We now know that it is the result of the existence of three light quarks with masses that are not too different from each other. The quarks themselves form a triplet of $SU(3)$,

$$\Psi = \begin{pmatrix} u \\ d \\ s \end{pmatrix}$$

The underlying strong interaction treats each of these the same, such that their only difference comes from their different masses. If the masses were the same, there would be an exact $SU(3)$ symmetry $\Psi \rightarrow U_\Psi$ with U being a 3×3 $SU(3)$ matrix transformation. Because this acts on the "flavors" or types of quark, it is referred to as "flavor $SU(3)$". The quarks bind in $Q\overline{Q}$ and QQQ combinations. Group theoretically, this produces

$$3\otimes\overline{3} = 1\oplus8,$$

$$3\otimes3\otimes3 = 1\oplus8\oplus8\oplus10,$$

as the representations describing the bound state. The $SU(3)$ quantum numbers of the elementary particles have been determined by their quark substructure. In addition, the approximate nature of the $SU(3)$ symmetry is due to the fact that the masses of the quarks are not all the same. The strange quark mass is heavier than those of the u,d quarks, so that baryons and mesons which contain one or more strange quarks are heavier than other members of an $SU(3)$ multiplet. This mechanism works quantitatively, and by now the dynamical quark model has become a well-developed scheme.

The weak and electromagnetic properties of baryons and mesons also show well-defined properties under the $SU(3)$ classification scheme. As with the masses, these can also be understood as the remnant of the $SU(3)$ properties of the weak and electromagnetic properties of quarks. The group theoretic language is extremely useful in summarizing and relating the many matrix elements of hadronic transitions.

A very different use of the group $SU(3)$ is found in the gauge theory of the strong interactions—quantum chromodynamics. There *each* quark is seen to carry an internal quantum number, called color, which comes in three varieties. The properties of each color of quark are identical, and hence the world is invariant under a unitary transformation of the three colors of quarks, i.e., under $SU(3)$ of color. This group is an exact symmetry of nature. Within quantum chromodynamics there is also a set of particles in the octet representations. These are the eight gluons, which are spin-1 particles with interactions similar in character to that of the photon. It is thought that only color singlet representations of color $SU(3)$ occur as free particles in nature. This is the hypothesis of color confinement, which requires that quarks and gluons form bound states, rather than being seen as free particles.

This pattern of dual uses of symmetry groups occurs in other cases also. $SU(2)$ appears as a classification group called "isospin," which describes the similarity of particles composed of u and d quarks. It also appears as a gauge symmetry group for the weak interactions, with the gauge bosons being basically the W^+,W^-, and Z^0 intermediate vector bosons. The group $U(1)$ describes mathematically the symmetries that lead to various conserved quantum numbers, such as baryon number or lepton number. It is also the gauge group underlying electromagnetism. More exotic groups are often used in theories which attempt to go beyond the present standard model of particle physics.

See also GAUGE THEORY; HADRONS; ISOSPIN; QUANTUM CHROMODYNAMICS; QUARK.

REFERENCES

1, S. Sakata, *Prog. Theor. Phys.* **16,** 686L (1956).
2. M. Ikeda, S. Ogawa, and Y. Ohnuki, *Prog. Theor. Phys.* **22,** 715 (1959); Y. Yamaguchi, *Prog. Theor. Phys.* Suppl. 11, 1 (1960).
3. M. Gell-Mann, California Institute of Technology Synchroton Laboratory Rep. CTSL-20 (1961), unpublished. Reprinted in *The Eightfold Way* by M. Gell-Mann and Y. Ne'eman (Benjamin, New York, 1964).
4. Y. Ne'eman, *Nucl. Phys.* **26,** 222 (1961).
5. V. E. Barnes *et al., Phys. Rev. Lett.* **12,** 204 (1964).
6. M. Gell-Mann, *Phys. Lett.* **8,** 214 (1964); G. Zweig, unpublished (1964).
7. M. Gell-Mann, *Phys. Rev.* **125,** 1067 (1962); S. Okubo, *Prog. Theor. Phys.* **27,** 949 (1962).

Symbols, Units, and Nomenclature

E. Richard Cohen

INTRODUCTION

Even though material published throughout the world may be written in different languages, the use of an internationally accepted set of units and internationally recognized symbols for physical quantities increases their ability to be more generally understood. This uniformity has been achieved by the introduction of the International System of Units (*Système Internationale d'Unités,* or SI) and by general adherence to the recommendations of the International Organization for Standardization (ISO) and the International Union of Pure and Applied Physics (IUPAP). Although their recommendations are not universally followed, the symbols given in the tables in this article are those recommended by IUPAP.

GENERAL RECOMMENDATIONS

Numbers

The usual decimal indicator in both British and American English is a dot on the line, while in most European languages (including Russian and other languages using the Cyrillic alphabet) the decimal indicator is a comma. Therefore, to avoid ambiguity, neither symbol should be used to group digits in long numbers; groups of three digits may be separated by a thin space, but no comma or point should be used other than the decimal sign. Instead of a single final digit, the last four digits may be grouped (e.g., 1.234 567; 3.141 5927). (The ISO recommendation is that the comma be used as the decimal sign even for texts written in English, but this is not standard in the United States.)

Physical Quantities

A physical quantity is represented by a symbol and is expressed as the product of a numerical value (i.e., a pure number) and a unit:

$$\text{physical quantity} = \text{numerical value} \times \text{unit}.$$

Although the numerical value assigned to a given physical magnitude will in general depend on the system of units being used, neither the name nor the symbol for a physical quantity should imply any particular choice of unit.

Examples: $E=200$ J $F=27$ N

$n=1.55$ (refractive index).

Physical quantities can be combined by addition or subtraction only if they have the same unit. When physical quantities combine by multiplication or division, the usual rules of arithmetic apply both to the numerical values and to the units. A quantity that arises from dividing one physical quantity by another with the same unit has a unit that may be symbolized by the number 1. Such a unit often has no special name or symbol. The physical quantity is equivalent to a pure number and is invariant to a change in the system of units, but might be changed by a change in the system of base quantities.

Symbols for Physical Quantities

The symbols for physical quantities are single letters of the Latin or Greek alphabet. Either or both forms of a Greek letter (e.g., θ,ϑ; ϕ,φ) may be used. The form ϖ of the letter π may be used as if it were a distinct letter.

An exception to the single-letter rule is the set of two-letter symbols used to represent dimensionless combinations of physical quantities (see Table VI).

The symbols may be modified by subscripts or superscripts that should be printed in italic (*sloping*) type, if they are symbols for physical quantities or for numerical indices; otherwise they are printed in roman (upright) type. Thus one writes E_k if k represents a numerical index but E_{k} if k indicates "kinetic."

It is convenient to use symbols with distinctive typefaces in order to distinguish between the components of a vector (or a tensor) and the vector (or tensor) as an entity in itself, or to avoid the use of subscripts. The following standard conventions should be adhered to when the appropriate typefaces are available:

(a) Vectors should be printed in bold italic type, e.g., ***a***, ***A***.

(b) Tensors should be printed in slanted bold sans serif type, e.g., ***S***, ***T***.

The mathematical operations of addition, subtraction, multiplication, and division of physical quantities are indicated by the usual symbols:

$$H+h;\quad v-v_0;\quad ma,\ m\cdot a,\ m\times a;\quad \frac{F}{m},\ F/m,\ Fm^{-1},\ F(1/m).$$

This can be extended to cases where one of the quantities or both are themselves products, quotients, sums, or differences of other quantities. If brackets are necessary, they should be used in accordance with the rules of mathematics. When a solidus is used to separate numerator and denominator, brackets should be inserted if there is any doubt where the numerator starts or where the denominator ends:

$$\frac{L}{d_1+d_2},\ L/(d_1+d_2),\ \sin[2\pi(v-v_0)t+\phi]$$

Products of vectors and tensors have their own conventions:

Scalar product of vectors ***A*** and ***B***	$\boldsymbol{A}\cdot\boldsymbol{B}$
Vector product of vectors ***A*** and ***B***	$\boldsymbol{A}\times\boldsymbol{B}$
Dyadic product of vectors ***A*** and ***B***	$\boldsymbol{AB}$
Product of tensor ***R*** and vector ***A***	$\boldsymbol{R}\cdot\boldsymbol{A}$
Scalar product of tensors ***S*** and ***T***	$\boldsymbol{S}:\boldsymbol{T}$
Tensor product of tensors ***S*** and ***T***	$\boldsymbol{S}\cdot\boldsymbol{T}$

Symbols for Atoms and Nuclear Particles

Symbols for chemical elements should be written in roman (upright) type. The first letter is always capitalized; the second letter, if any, is always in lowercase type even in titles where all words are in capital letters (e.g., H, He, Li, Be, B). The symbol is not followed by a full stop. A nuclide is specified by the chemical symbol of the element with the mass number placed as a left superscript (e.g., ^{12}C). The right subscript position is used to indicate the number of atoms of a nuclide or element in a molecule (e.g., $H_2{}^{16}O$). The right superscript position may be used to indicate a state of ionization (e.g., $Ca_2{}^+$, $PO_4{}^{3-}$) or an excited *atomic* state (e.g., He*). A metastable *nuclear* state, however, is often treated as a distinct nuclide: e.g., either $^{118}Ag^m$ or ^{118m}Ag.

Roman numerals are used in two different ways:

1. The spectrum of an ionized atom is indicated by the small capital roman numeral one larger than the state of ionization, written on the line with a thin space following the chemical symbol (e.g., Ca II for once-ionized calcium)
2. Roman numerals in right superscript position are used to indicate the oxidation number (e.g., $Pb_2{}^{II}Pb^{IV}O_4$).

Particles and quanta appearing as projectiles or products in nuclear reactions are designated by the following symbols:

photon	γ	nucleon	N
neutrino	$\nu,\nu_e,\nu_\mu,\nu_\tau$	neutron	n
electron	e,β	proton ($^1H^+$)	p
muon	μ	deuteron ($^2H^+$)	d
tauon	τ	triton ($^2H^+$)	t
pion	π	helion ($^3He^{2+}$)	h
		alpha particle ($^4He^{2+}$)	α

The charge of a particle may be indicated by adding a superscript $^+,^0,^-$ to the symbol for the particle.

Examples: π^+,π^0,π^- e^+,e^- β^+,β^-

If no charge is indicated in connection with the symbols p and e, these symbols refer to the positive proton and the negative electron respectively. The bar ¯ or the tilde ˜ above the symbol for a particle is used to indicate the corresponding anti-particle; the notation $\bar{p}$ is preferable to p^- for the antiproton, but both $\bar{e}$ and e^+ (or $\bar{\beta}$ and β^+) are commonly used for the positron.

A letter symbol indicating a quantum number of a *single particle* should be printed in lower case upright type. A letter symbol indicating a quantum number of a *system* should be printed in capital upright type.

In atomic spectroscopy, the letter symbols indicating the orbital angular momentum quantum number are

$l,L=$	0	1	2	3	4	5	6	7	8	9	10	11	...
symbol for l	s	p	d	f	g	h	i	k	l	m	n	o	...
symbol for L	S	P	D	F	G	H	I	K	L	M	N	O	...

A right subscript attached to the angular momentum symbol indicates the total angular momentum quantum number j or J; A left superscript indicates the spin multiplicity $2S+1$, e.g., $p_{3/2}$, $^2P_{3/2}$.

An atomic electron configuration is indicated symbolically by:

$$(nl)^k(n'l')^{k'}\ldots$$

in which $k,k',\ldots$ are the numbers of electrons with principal quantum numbers $n,n',\ldots$ and orbital angular momentum quantum numbers $l,l',\ldots,$ respectively. Instead of $l=0,1,2,3,\ldots$ one uses the quantum number symbols s, p, d, f, ... and the parentheses are usually omitted (e.g., the atomic electron configuration: $1s^22s^22p^3$).

Nomenclature Conventions in Nuclear Physics

A species of atoms identical as regards atomic number (proton number) and mass number (nucleon number) should be indicated by the word "nuclide," not by the word "isotope." Different nuclides having the same mass number are called *isobaric nuclides* or *isobars*. Different nuclides having the same atomic number are called *isotopic nuclides* or *isotopes*. (Since nuclides with the same number of protons are *isotopes*, nuclides with the same number of neutrons have sometimes been designated as *isotones*.)

The symbolic expression representing a nuclear reaction should follow the pattern:

$$\text{initial nuclide}\left(\begin{matrix}\text{incoming particle}\\ \text{or quantum}\end{matrix},\ \begin{matrix}\text{outgoing particle(s)}\\ \text{or quanta}\end{matrix}\right)\text{final nuclide}$$

Examples: $^{14}N(\alpha,p)^{17}O$ $^{23}Na(\gamma,3n)^{20}Na$

DEFINITION OF UNITS AND SYSTEMS OF UNITS

In the system consisting of a set of physical quantities and the relational equations connecting them, a certain number of quantities are considered to form a set of *base quantities* for the whole system and are regarded by convention as dimensionally independent. All other physical quantities are *derived quantities,* defined in terms of the base quantities and expressed algebraically as products of powers of the base quantities.

In the same way, a *system of units* is based on a set of defined samples of the base quantities chosen by convention to be the independent base units, and all units for derived quantities are expressed as products of powers of the base units. The system and its units are said to be *coherent* if the derived units are expressed in terms of the base units by relations with numerical factors equal to unity. Derived units and their symbols are expressed algebraically in terms of base units by means of the mathematical signs for multiplication and division.

The product of two units may be written as N m or as N·m. The quotient may be written m/s, $\frac{m}{s}$ or the product of m and s^{-1}. Not more than one solidus should be used in an expression: e.g., not m/s/s but m/s^2, not $W/sr/m^2$, but $W/(sr\cdot m^2)$.

The International System of Units (SI)

The General Conference of Weights and Measures, to which the United States adheres by treaty, established the Système International d'Unités (International System of Units) with the international abbreviation SI in 1960. It is a coherent system based on seven base units:

length	metre	m
mass	kilogram	kg
time	second	s
electric current	ampere	A
thermodynamic temperature	kelvin	K
amount of substance	mole	mol
luminous intensity	candela	cd

The full name of a unit is always printed in lower case roman (upright) type and is not capitalized. If the name is derived from a proper name, its abbreviation is a one- or two-letter symbol whose first letter is capitalized. Except for the symbol L for liter (as an alternative to l, in order to avoid confusion with the number 1), the symbol for a unit whose name is not derived from a proper name is always printed in lower case roman type.

Prefixes that may be used with a unit to indicate decimal multiples and submultiples are given in Table I. Because the base unit of mass contains the prefix "kilo-," multiples and submultiples are formed by attaching the prefix to "gram": e.g., use "gram" (g) not "millikilogram" (mkg), "milligram" (mg) not "microkilogram" (μkg). Compound prefixes formed by the juxtaposition of two or more prefixes should not be used (e.g., use pF not μμF, GW not kMW). When a prefix symbol is used with a unit symbol, the combination is considered as a single symbol that can be raised to a positive or negative power without using brackets.

Examples: cm^3 means $(0.01\ m)^3 = 10^{-6}\ m^3$, not $0.01\ m^3$

μs^{-1} means $(10^{-6}\ s)^{-1} = 10^6\ s^{-1}$, not $10^{-6}\ s^{-1}$

In addition to the names and symbols for base units, specific names and symbols have been given to several coherent derived SI units; these derived units are listed in Table II.

It is a general rule of SI that the use of non-SI units should be discontinued. However there are some important instances where this is either impractical or inadvisable. Certain important and widely used units do not properly fall

Table I Standard prefixes

Factor by which the unit is multiplied	Name of factor	Symbol
10^{18}	exa	E
10^{15}	peta	P
10^{12}	tera	T
10^{9}	giga	G
10^{6}	mega	M
10^{3}	kilo	k
10^{2}	hecto	h
10	deka	da
10^{-1}	deci	d
10^{-2}	centi	c
10^{-3}	milli	m
10^{-6}	micro	μ
10^{-9}	nano	n
10^{-12}	pico	p
10^{-15}	femto	f
10^{-18}	atto	a

within the SI. The special names and symbols of units that have been accepted for continuing use and the corresponding units of the SI are listed in Table III. Although the use of these units is acceptable, their combination with SI units to form incoherent compound units should be avoided.

Decimal multiples or submultiples of the time units listed in Table III should not be formed by using the prefixes given in Table II. Forming symbols for decimal multiples or submultiples of units by using the symbols of those prefixes is also not possible with superscript symbols, such as °, ′, and ″ for angle units.

In addition to the units given in Table III there are other non-SI units that should logically be abandoned, but whose continued use in special fields is accepted for a limited transitional period: angstrom (Å $=10^{-10}$ m), barn (b $=10^{-28}$ m^2), bar (bar = 100 kPa), atmosphere (atm = 101.325 kPa), calorie (cal = 4.184 J), curie (Ci $=3.7\times10^{10}$ s^{-1}), roentgen (R $=2.58\times10^{-4}$ C/kg).

Other Systems of Units

SI has to a great extent replaced the older MKS metric system, but the CGS system (based on the centimeter, gram, and second as units in mechanics) is still widely used in some branches of physics. In everyday life in the United States, the customary units are still based on the foot, pound-force, and the second (although the foot and the pound are exactly defined, respectively, in terms of the metre and the kilogram). These units are almost never used in physics except for the description of equipment (e.g., a ¼-in drill).

TABLES OF SYMBOLS AND UNITS

The most common symbols for expressing the mathematical relationships in physics are presented in Table IV. An extensive (but of course, incomplete) list of the symbols and SI units of physical quantities is given in Table V. Many of the symbols listed are general; they may be made more specific by adding superscripts or subscripts or by using both lower and upper case forms. Where more than one symbol is given there is no implied preference in the ordering. Table VI presents the most commonly used standard symbols for dimensionless parameters in physics and engineering. It is recommended by ISO that these two-letter parameters be printed in *sloping* type in the same way as single-letter quantities; when such a symbol is a factor in a product it should be separated from other symbols by a thin space, a multiplication sign or brackets. This disagrees with some journals that set two-letter symbols in roman type to distinguish them from ordinary products. In this table *sloping roman* is used to distinguish a two-letter symbol from the product of two *italic* single-letter symbols.

Table II Derived units

Quantity	Name	Symbol	Expression in terms of SI base or derived units
Plane angle	radian	rad	m/m
Solid angle	steradian	sr	m^2/m^2
Frequency	hertz	Hz	s^{-1}
Force	newton	N	m kg s^{-2} = J/m
Pressure	pascal	Pa	m^{-1} kg s^{-2} = N/m^2 = J/m^3
Energy, work, quantity of heat	joule	J	m^2 kg s^{-2} = N m
Power, radiant flux	watt	W	m^2 kg s^{-3} = J/s
Quantity of electricity, electric charge	coulomb	C	A s
Electric potential, potential difference, electromotive force	volt	V	m^2 kg s^{-3} A^{-1} = W/A = J/C
Capacitance	farad	F	m^{-2} kg^{-1} s^4 A^2 = C/V
Electric resistance	ohm	Ω	m^2 kg s^{-3} A^{-2} = V/A
Conductance	siemens	S	m^{-2} kg^{-1} s^3 A^2 = A/V = Ω^{-1}
Magnetic flux	weber	Wb	m^2 kg s^{-2} A^{-1} = V s
Magnetic flux density	tesla	T	kg s^{-2} A^{-1} = Wb/m^2
Inductance	henry	H	m^2 kg s^{-2} A^{-2} = Wb/A
Celsius temperature	degree (Celsius)	°C	K
Luminous flux	lumen	lm	cd sr[a]
Illuminance	lux	lx	m^{-2} cd sr = lm/m^2 [a]
Activity	becquerel	Bq	s^{-1}
Absorbed dose[b]	gray	Gy	m^2 s^{-2} = J/kg
Dose equivalent[b]	sievert	Sv	m^2 s^{-2}

[a] The symbol sr must be included here to distinguish luminous flux (lumen) from luminous intensity (candela).

[b] The dose equivalent is equal to the absorbed dose multiplied by dimensionless factors defining the relative biological effectiveness of the radiation. Although the gray and the sievert have the same expression in terms of base units, they measure conceptually distinct quantities.

Table III Non-SI units

Quantity	Unit	Symbol	Magnitude in SI units
Plane angle	degree	°	(π/180) rad
	minute (of angle)	′	(π/10 800) rad
	second (of angle)	″	(π/648 000) rad
Time	minute	min	60 s
	hour	h	3600 s
	day	d	86 400 s
	year	y, a	31 556 952 s[a]
Length	astronomical unit	AU	$1.495\ 978\ 70 \times 10^{11}$ m
	parsec	pc	$3.085\ 677\ 57 \times 10^{16}$ m
Volume	litre	l, L	$1\ \mathrm{dm}^3 = 10^{-3}\ \mathrm{m}^3$
Mass	tonne	t	1 Mg = 1000 kg
	atomic mass unit	u	$m(^{12}\mathrm{C})/12 \approx 1.660\ 5402 \times 10^{-27}$ kg
energy	electronvolt	eV	$(e/\mathrm{C})\ \mathrm{J} \approx 1.602\ 177\ 33 \times 10^{-19}$ J
	watt hour	Wh	3.6 kJ

[a] Mean Gregorian year, 365.2425 days.

Table IV Recommended mathematical symbols

a. General symbols

ratio of the circumference of a circle to its diameter	π
base of natural logarithms	e
infinity	∞
equal to	$=$
not equal to	$\neq$
identically equal to	$\equiv$
by definition equal to	$\overset{\mathrm{def}}{=}$, $:=$
corresponds to	$\triangleq$
approximately equal to	$\approx$
asymptotically equal to	$\simeq$
proportional to	$\propto$
approaches	$\rightarrow$
greater than	$>$
greater than or equal to	$\geq$
much greater than	$\gg$
less than	$<$
less than or equal to	$\leq$
much less than	$\ll$
plus	$+$
minus	$-$
plus or minus	$\pm$
a multiplied by b	ab, $a \cdot b$, $a \times b$
a divided by b	a/b, $\frac{a}{b}$, ab^{-1}
a raised to the power n	a^n
magnitude of a	$\lvert a \rvert$
square root of a	$\sqrt{a}$, $\surd a$, $a^{1/2}$
mean value of a	$\bar{a}$, $\langle a \rangle$
factorial p	$p!$
binomial coefficient: $n!/[p!(n-p)!]$	$\binom{n}{p}$

Table IV (*continued*)

b. Letter symbols

Although the symbols for mathematical *variables* are usually set in sloping or italic type, the symbols for the common mathematical *functions* are always set in roman (upright) type.

exponential of x	$\exp x$, e^x
logarithm to the base a of x	$\log_a x$
natural logarithm of x	$\ln x$, $\log_\mathrm{e} x$
common logarithm of x	$\lg x$, $\log_{10} x$
binary logarithm of x	$\mathrm{lb}\ x$, $\log_2 x$
sine of x	$\sin x$
cosine of x	$\cos x$
tangent of x	$\tan x$, $\mathrm{tg}\ x$
cotangent of x	$\cot x$, $\mathrm{ctg}\ x$
secant of x	$\sec x$
cosecant of x	$\mathrm{cosec}\ x$, $\csc x$

For the *hyperbolic functions* the symbolic expressions for the corresponding circular functions are followed by the letter h.

Examples: $\sinh x$, $\cosh x$, $\tanh x$, etc.

The shortened forms $\mathrm{sh}\ x$, $\mathrm{ch}\ x$, and $\mathrm{th}\ x$ are also used.

For the *inverse circular functions* the symbolic expressions for the corresponding circular functions are preceded by the letters arc.

Examples: $\arcsin x$, $\arccos x$, $\arctan x$, etc.

For the *inverse hyperbolic functions* the symbolic expression for the corresponding hyperbolic function are preceded by the letters ar.

Examples: $\mathrm{arsinh}\ x$, $\mathrm{arcosh}\ x$, etc. (or $\mathrm{arsh}\ x$, $\mathrm{arch}\ x$, etc.)

summation	$\sum$
product	$\prod$
finite increase of x	Δ Greek capital δ, not a triangle.
variation of x	δx
total differential of x	$\mathrm{d}x$
function of x	$f(x)$
composite function of f and g: $g(f(x))$	$g \circ f$
convolution of f and g:	$f * g$

$$f*g = g*f = (f*g)(x) = (g*f)(x) = \int_{-\infty}^{\infty} f(x-t)g(t)\,\mathrm{d}t$$

limit of $f(x)$	$\lim_{x\rightarrow a} f(x)$, $\lim\limits_{x\rightarrow a} f(x)$
derivative of $f(x)$	$\frac{\mathrm{d}f}{\mathrm{d}x}$, $\mathrm{d}f/\mathrm{d}x$, f'
time derivative of $f(t)$	$\dot{f}$
partial derivative of f	$\frac{\partial f}{\partial x}$, $\partial f/\partial x$, $\partial_x f$, f_x
total differential of f: $\mathrm{d}f(x,y) = (\partial f/\partial x)_y\,\mathrm{d}x + (\partial f/\partial y)_x\,\mathrm{d}y$	$\mathrm{d}f$
variation of f	δf
Kronecker delta symbol	δ_{ij}
Dirac delta function: $\delta(\mathbf{r}) = \delta(x)\delta(y)\delta(z)$	$\delta(x)$, $\delta(\mathbf{r})$
signum a: $\begin{cases} a/\lvert a\rvert & \text{for } a \neq 0, \\ 0 & \text{for } a = 0 \end{cases}$	$\mathrm{sgn}\ a$
greatest integer $\leq a$	$\mathrm{ent}\ a$, $[a]$

For $a \neq$ integer, $[-a] = -([a]+1)$; e.g., $[-3.14] = -4$

c. Symbols for special values of periodic quantities.

A quantity whose time dependence is such that $x(t+T) = x(t)$, where T is the smallest strictly positive constant value for which this relation holds for all t, is said to vary periodically with period T.

instantaneous value	x, $x(t)$
maximum value	$\hat{x}$, $x_{\max}$
minimum value	$\check{x}$, $x_{\min}$

Table IV (*continued*)

mean value: $\frac{1}{T}\int_0^T x(t)\,dt$	$\bar{x}$, $\langle x\rangle$
root mean square (rms) value: $\left[\frac{1}{T}\int_0^T [x(t)]^2\,dt\right]^{1/2}$	X, $\tilde{x}$, x_{rms}, (x_{eff})
d. Complex quantities	
imaginary unit: $(i^2=-1)$	i, i, j, j
real part of z	Re z, z'
imaginary part of z	Im z, z''
(The notation z', z'' is used primarily for physical quantities, e.g., the complex representation of the dielectric constant: $\epsilon=\epsilon'+i\epsilon''$.)	
modulus of z	$\|z\|$
phase, argument of z: $z=\|z\|e^{i\phi}$	ϕ, arg z
complex conjugate of z, conjugate of z	z^*, $\bar{z}$
e. Vector calculus	
vector	$\boldsymbol{A}$, $\boldsymbol{a}$
absolute value	$\|\boldsymbol{A}\|$, A
unit vector: $\boldsymbol{a}/\|\boldsymbol{a}\|$	$\boldsymbol{e}_a$, $\hat{\boldsymbol{a}}$
unit coordinate vectors	$(\boldsymbol{e}_x, \boldsymbol{e}_y, \boldsymbol{e}_z)$, $(\boldsymbol{i}, \boldsymbol{j}, \boldsymbol{k})$
scalar product of $\boldsymbol{a}$ and $\boldsymbol{b}$	$\boldsymbol{a}\cdot\boldsymbol{b}$
vector product of $\boldsymbol{a}$ and $\boldsymbol{b}$	$\boldsymbol{a}\times\boldsymbol{b}$, $\boldsymbol{a}\wedge\boldsymbol{b}$
dyadic product of $\boldsymbol{a}$ and $\boldsymbol{b}$	$\boldsymbol{ab}$
differential vector operator, nabla	$\partial/\partial\boldsymbol{r}$, ∇
gradient	grad ϕ, $\nabla\phi$
divergence	div $\boldsymbol{A}$, $\nabla\cdot\boldsymbol{A}$
curl	curl $\boldsymbol{A}$, rot $\boldsymbol{A}$, $\nabla\times\boldsymbol{A}$
Laplacian	$\Delta\phi$, $\nabla^2\phi$
Dalembertian: $\nabla^2\phi-c^{-2}\partial^2\phi/\partial t^2$	$\Box\phi$
second order tensor	A
scalar product of tensors S and T: $(\sum_{i,k}S_{ik}T_{ki})$	S:T
tensor product of tensors S and T: $(\sum_k S_{ik}T_{kl})$	S·T
product of tensor S and vector A: $(\sum_k S_{ik}A_k)$	S·A
f. Matrix calculus	
matrix: $\begin{pmatrix} a_{11} & \cdots & a_{1n} \\ \vdots & & \vdots \\ a_{m1} & \cdots & a_{mn}\end{pmatrix}$	A, $\{a_{ij}\}$

Table IV (*continued*)

product of A and B	AB
inverse of A	A^{-1}
unit matrix	E, I
transpose of matrix A: $(A^{\mathrm{T}})_{ij}=A_{ji}$	A^{T}, $\tilde{A}$
complex conjugate of A: $(A^*)_{ij}=A^*_{ij}$	A^*
Hermitian conjugate of A: $(A^\dagger)_{ij}=A^*_{ji}$	$A^\dagger$
determinant of A	det A
trace of A: $\sum_i A_{ii}$	Tr A
g. Symbolic logic	
conjunction: $p\wedge q$ means "p and q"	$\wedge$
disjunction: $p\vee q$ means "p or q or both"	
negation	$\neg$
implication	$\Rightarrow$
equivalence, bi-implication	$\Leftrightarrow$
universal quantifier	$\forall$
existential quantifier	$\exists$
h. Theory of sets	
is an element of: $x\in A$	$\in$
is not an element of	$\notin$
contains as element: $A\ni x$	$\ni$
set of elements	$\{a_1, a_2, \cdots\}$
empty set	$\emptyset$, $\varnothing$
the set of positive integers and zero	N
the set of all integers (positive and negative and zero)	Z
the set of rational numbers	Q
the set of real numbers	R
the set of complex numbers	C
set of elements of A for which $p(x)$ is true	$\{x\in A\|p(x)\}$
is included in: $B\subseteq A$	$\subseteq$, $\subset$
contains: $A\supseteq B$	$\supseteq$, $\supset$
is properly contained in	$\subset$, $\subsetneq$
contains properly	$\supset$, $\supsetneq$
union: $A\cup B=\{x\|(x\in A)\vee(x\in B)\}$	$\cup$
intersection: $A\cap B=\{x\|(x\in A)\wedge(x\in B)\}$	$\cap$
difference: $A\backslash B=\{x\|(x\in A)\wedge(x\notin B)\}$	$\backslash$

Table V Physical quantities and symbols[a,b,c]

Quantity	Symbol	Unit
a. Space and time		
space coordinates	(x, y, z)	m
relativistic coordinates: $x_0=ct$, $x_1=x$, $x_2=y$, $x_3=z$, $x_4=ict$	(x_0, x_1, x_2, x_3)	m
	(x_1, x_2, x_3, x_4)	m
position vector	$\boldsymbol{r}$	m
length	l	m
breadth	b	m
height	h	m
radius	r	m
thickness	d, δ	m
diameter: $d=2r$	d	m
element of path	ds, dl	m
area	A, S	m^2
volume	V, v	m^3
plane angle	α, β, γ, θ, ϕ	rad
solid angle	Ω, ω	sr
wavelength	λ	m
wave number: $\sigma=1/\lambda$	σ[d]	m^{-1}
wave vector	σ	m^{-1}

Table V (*continued*)

Quantity	Symbol	Unit
angular wave number: $k=2\pi/\lambda$	k	m^{-1}
angular wave vector, propagation vector	$\boldsymbol{k}$	m^{-1}
time	t	s
period, periodic time	T	s
frequency: $f=1/T$	f, ν	s^{-1}
angular frequency: $\omega=2\pi f$	ω	s^{-1}
relaxation time: $F(t)=\exp(-t/\tau)$	τ	s
damping coefficient: $F(t)=\exp(-\delta t)\sin\omega t$	δ, λ	s^{-1}
growth rate: $F(t)=\exp(\gamma t)\sin\omega t$	γ	s^{-1}
logarithmic decrement: $T\delta=T/\tau$	Λ	
speed: ds/dt, v/c, (c = speed of light in vacuum)	v, β	m/s
velocity and its components: $d\boldsymbol{s}/dt$	$\boldsymbol{u}$, $\boldsymbol{v}$, $\boldsymbol{w}$, $\boldsymbol{c}$, (u, v, w)	m/s
angular velocity: $d\phi/dt$	ω	s^{-1}
acceleration: $d\boldsymbol{v}/dt$	$\boldsymbol{a}$	m/s^2
acceleration of free fall	g	m/s^2
angular acceleration: $d\omega/dt$	α	s^{-2}
b. Mechanics		
mass	m	kg
(mass) density: m/V	ρ	kg/m^3
relative density: $\rho/\rho(H_2O)$	d	
specific volume: $V/m=1/\rho$	v	m^3/kg
reduced mass: $m_1m_2/(m_1+m_2)$	μ, m_r	kg
momentum: $m\boldsymbol{v}$	$\boldsymbol{p}$	kg m/s
angular momentum: $\boldsymbol{r}\times\boldsymbol{p}$	$\boldsymbol{L}$, $\boldsymbol{J}$	m^2 kg/s
moment of inertia: $\int(x^2+y^2)\,dm$	I, J	kg m^2
force	$\boldsymbol{F}$	N
weight	W, G, P	N
moment of force	$\boldsymbol{M}$	N m
torque, moment of a couple	$\boldsymbol{T}$	N m
pressure	p, P	Pa
normal stress	σ	Pa
shear stress	τ	Pa
linear strain, relative elongation: $\Delta l/l_0$	ϵ, e	
modulus of elasticity, Young's modulus: σ/ϵ	E, Y	Pa
shear strain	γ	
shear modulus: τ/γ	G, μ	Pa
stress tensor	τ_{ij}	Pa
strain tensor	ϵ_{ij}	
elasticity tensor: $\tau_{ij}=c_{ijkl}\epsilon_{lk}$	c_{ijkl}	Pa
compliance tensor: $\epsilon_{kl}=s_{klji}\tau_{ij}$	s_{klji}	Pa^{-1}
volume strain, bulk strain: $\Delta V/V_0$	θ	
bulk modulus: $p=-K\theta$	K, κ	Pa
Poisson ratio	μ, ν, σ	
viscosity	η, μ	N s/m^2
kinematic viscosity: $\nu=\eta/\rho$	v	m^2/s
friction coefficient	μ, f	
surface tension	γ, σ	J/m^2
energy	E, W	J
potential energy	E_p, V, Φ, U	J
kinetic energy	E_k, T, K	J
work: $\int\boldsymbol{F}\cdot d\boldsymbol{s}$	W, A	J
power: dE/dt	P	W
generalized coordinate	$\boldsymbol{q}$, q_i	—
generalized momentum: $\partial L/\partial q_i$	$\boldsymbol{p}$, p_i	—
action integral: $\oint p\,dq$	J, S	J s
Lagrangian function, Lagrangian: $T(q_i,\dot{q}_i)-V(q_i,\dot{q}_i)$	L	J
Hamiltonian function, Hamiltonian: $\sum_i p_i\dot{q}_i-L$	H	J
Hamilton's principal function: $\int L\,dt$	W, S_p	J s
Hamilton's characteristic function: $2\int T\,dt$	S	J s
c. Statistical physics		
number of particles	N	

Table V (*continued*)

Quantity	Symbol	Unit
number density of particles: N/V	n	m^{-3}
particle position vector (position components)	$\boldsymbol{r}$, (x, y, z)	m
particle velocity vector (velocity components)	$\boldsymbol{c}$, $\boldsymbol{v}$, $\boldsymbol{u}$; (c_x, c_y, c_z), (v_x, v_y, v_z), (u_x, u_y, u_z)	m/s
particle momentum vector (components)	$\boldsymbol{p}$, (p_x, p_y, p_z)	kg m/s
average velocity	$\boldsymbol{c}_0$, $\boldsymbol{v}_0$, $\langle \boldsymbol{c} \rangle$, $\langle \boldsymbol{v} \rangle$	m/s
average speed	$\bar{c}$, $\bar{v}$, $\langle c \rangle$, $\langle v \rangle$, u	m/s
most probable speed	$\hat{c}$, $\hat{v}$	m/s
mean free path	l, λ	m
interaction energy between particles i and j	ϕ_{ij}, V_{ij}	J
velocity distribution function: $n = \int f \, dc_x \, dc_y \, dc_z$	$f(\boldsymbol{c})$	s^3/m^6
Boltzmann function	H	—
volume in γ phase space	Ω	—
canonical partition function	Z	—
microcanonical partition function	Ω	—
grand canonical partition function	Ξ	—
symmetry number	s	
diffusion coefficient	D	m^2/s
thermal diffusion coefficient	D_{td}	m^2/s
thermal diffusion ratio	k_T	
thermal diffusion factor	α_T	
characteristic temperature	Θ	K
rotational characteristic temperature: $h^2/8\pi^2 kI$	Θ_{rot}	K
vibrational characteristic temperature: $h\nu_z/k$	Θ_v	K
Debye temperature: $h\nu_D/k$	Θ_D	K
Einstein temperature: $h\nu_E/k$	Θ_E	K
d. Thermodynamics [e]		
quantity of heat	Q	J
work	W	J
thermodynamic temperature	T	K
Celsius temperature	t, θ [f]	°C
entropy	S	J/K
internal energy	U	J
Helmholtz function: $U - TS$	A, F	J
enthalpy: $U + pV$	H	J
Gibbs function: $H - TS$	G	J
Massieu function: $-A/T$	J	J
Planck function: $-G/T$	Y	J
pressure coefficient: $(\partial p/\partial T)_V$	β	Pa/K
relative pressure coefficient: $(1/p)(\partial p/\partial T)_V$	α_p, α	K^{-1}
compressibility: $-(1/V)(\partial V/\partial p)_T$	κ_T, κ	Pa^{-1}
linear expansion coefficient	α_l	K^{-1}
cubic expansion coefficient: $(1/V)(\partial V/\partial T)_p$	α_V, γ	K^{-1}
heat capacity	C_p, C_V	J/K
specific heat capacity: C/m	c_p, c_V	J/(kg K)
Joule-Thomson coefficient	μ	K/Pa
isentropic exponent: $-(V/p)(\partial p/\partial V)_S$ [g]	κ	
ratio of specific heat capacities: $c_p/c_V = (\partial V/\partial p)_T(\partial p/\partial V)_S$ [g]	γ, κ	
heat flow rate	Φ, q	W
density of heat flow rate	$\boldsymbol{q}$, ϕ	W/m^2
thermal conductivity	κ, k, K, λ	W/(m K)
thermal diffusivity: $\lambda/\rho c_p$	a, D	m^2/s
e. Electricity and magnetism[h]		
quantity of electricity, electric charge	Q, q	C
charge density	ρ	C/m^3
surface charge density	σ	C/m^2
electric current	I, i	A
electric current density	j, J	A/m^2
electric potential	V, ϕ	V
potential difference	U, V	V

Table V (*continued*)

Quantity	Symbol	Unit
electromotive force	E, $\mathcal{E}$	V
electric field (strength)	$\boldsymbol{E}$	V/m
eleectric flux	Ψ	C
magnetic potential difference; magnetomotive force: $\oint H_s\, ds$	U_m, F_m	A
magnetic field (strength)	$\boldsymbol{H}$	A/m
electric dipole moment	$\boldsymbol{p}$	C m
dielectric polarization	$\boldsymbol{P}$	C/m^2
permittivity: $\boldsymbol{D}=\epsilon\boldsymbol{E}$ [i]	ϵ	F/m
electric susceptibility[i]	χ_e	
polarizability[i]	α, γ	
electric displacement: $\epsilon_0\boldsymbol{E}+\boldsymbol{P}$	$\boldsymbol{D}$	C/m^2
relative permittivity: ϵ/ϵ_0	ϵ_r, K	
magnetic vector potential	$\boldsymbol{A}$	Wb/m
magnetic induction, magnetic flux density	$\boldsymbol{B}$	T
magnetic flux	Φ	Wb
permeability: $\boldsymbol{B}=\mu\boldsymbol{H}$ [i]	μ	H/m, N/A^2
relative permeability: μ/μ_0	μ_r	
magnetization: $\boldsymbol{B}/\mu_0-\boldsymbol{H}$	$\boldsymbol{M}$	A/m
magnetic susceptibility[i]	χ, χ_m	
magnetic dipole moment	$\boldsymbol{m}$, $\boldsymbol{\mu}$	A m^2
capacitance	C	F
resistance	R	Ω
reactance	X	Ω
impedance: $R+jX$	Z	Ω
loss angle: $\arctan(X/R)$	δ	
conductance	G	S$=\Omega^{-1}$
susceptance	B	S
admittance: $1/Z=G=+jB$	Y	S
resistivity	ρ	Ω m
conductivity: $1/\rho$	γ, σ	S/m
self-inductance	L	H
mutual inductance	M, L_{12}	H
coupling coefficient: $L_{12}/(L_1L_2)^{1/2}$	k	
electromagnetic energy density	w, u	J/m^3
Poynting vector	$\boldsymbol{S}$	J/m^2
f. Radiation and light[j]		
refractive index: c/c_n	n	
radiant energy	Q, Q_e, W	J
radiant energy density	w	J/m^3
spectral concentration of radiant energy density (in terms of wavelength): $\int w_\lambda\, d\lambda$	w_λ	J/m^4
radiant (energy) flux, radiant power: $\int \Phi_\lambda\, d\lambda$	Φ, Φ_e, P	W
radiant flux density: $\Phi=\int \phi\, dS$	ϕ	W/m^2
radiant intensity: $\Phi=\int I\, d\Omega$	I, I_e	W/sr
spectral concentration of radiant intensity (in terms of frequency): $I=\int I_\nu\, d\nu$	I_ν, $I_{e,\nu}$	W s/sr
irradiance: $\Phi=\int E\, dS$	E, E_e	W/m^2
radiance: $I=\int L\cos\vartheta\, dS$	L, L_e	W/(sr m^2)
radiant exitance: $\Phi=\int M\, dS$	M, M_e	W/m^2
emissivity: M/M_B (M_B: radiant exitance of a blackbody radiator)	ϵ	
luminous efficacy: Φ_v/Φ_e	K	
spectral luminous efficacy: $\Phi_{v,\lambda}/\Phi_{e,\lambda}$	$K(\lambda)$	lm/W
maximum spectral luminous efficacy	K_m	lm/W
luminous efficiency: K/K_m	V	
spectral luminous efficiency: $K(\lambda)/K_m$	$V(\lambda)$	
quantity of light	Q, Q_v	lm s
luminous flux	Φ, Φ_v	lm
luminous intensity: $\Phi=\int I\, d\Omega$	I, I_v	cd
spectral concentration of luminous intensity (in terms of wave number): $I=\int I_\sigma\, d\sigma$	I_σ, $I_{v,\sigma}$	cd m
illuminance, illumination: $\Phi=\int E\, dS$	E, E_v	lx

Table V (*continued*)

Quantity	Symbol	Unit
luminance: $I = \int L \cos\vartheta \, dS$	L, L_v	cd/m^2
luminous exitance: $\Phi = \int M \, dS$	M, M_v	lm/m^2
linear attenuation coefficient	μ	m^{-1}
linear absorption coefficient	a	m^{-1}
absorptance: Φ_a/Φ_o	α [k]	
reflectance: Φ_r/Φ_o	ρ [k]	
transmittance: Φ_{tr}/Φ_o	τ [k]	
g. Acoustics		
sound particle velocity	$\boldsymbol{\mu}$	m/s
velocity of sound	c	m/s
velocity of longitudinal waves	c_l	m/s
velocity of transverse waves	c_t	m/s
group velocity	c_g	m/s
acoustic pressure	p	Pa
sound energy flux, acoustic power	W	W
reflection coefficient: W_r/W_0	ρ	
acoustic absorption coefficient: $1-\rho$	α_a, α	
transmission coefficient: W_{tr}/W_0	τ	
dissipation factor: $\alpha_a - \tau$	ψ, δ	
loudness level	L_N	dB
sound power level	L_W	dB
sound pressure level	L_p	dB
h. Quantum mechanics		
wave function	Ψ	m$^{-2/3}$
momentum operator in coordinate representation	$(\hbar/i)\boldsymbol{\nabla}$	kg m/s
probability density: $\Psi^*\Psi$	P	m^{-3}
probability current density: $(\hbar/2im)(\Psi^*\boldsymbol{\nabla}\Psi - \Psi\boldsymbol{\nabla}\Psi^*)$	S	m^{-2} s^{-1}
charge density of electrons: $-eP$	ρ	C/m^3
current density of electrons: $-eS$	j	A/m^2
Dirac bra vector	$\langle \ldots \|$	
Dirac ket vector	$\| \ldots \rangle$	
commutator of A and B: $AB - BA$	$[A, B]$, $[A, B]_-$	—
anticommutator of A and B: $AB + BA$	$[A, B]_+$	—
matrix element: $\int \phi_i^*(A\phi_j) \, d\tau$	A_{ij}	—
expectation value of A: Tr(A)	$\langle A \rangle$	—
Hermitian conjugate of operator A: $(A^\dagger)_{ij} = A_{ji}^*$	$A^\dagger$	—
annihilation operators	a, b, α, β	—
creation operators	$a^\dagger$, $b^\dagger$, $\alpha^\dagger$, $\beta^\dagger$	
i. Atomic and nuclear physics		
nucleon number, mass number	A	
proton number, atomic number	Z	
neutron number: $A - Z$	N	
nuclear mass (of nucleus ^{A}X)	m_N, $m_N(^A\mathrm{X})$	kg
atomic mass (of nuclide ^{A}X)	m_a, $m_a(^A\mathrm{X})$	kg
(unified) atomic mass constant: $\frac{1}{12}m_a(^{12}\mathrm{C})$	m_u	kg
relative atomic mass: m_a/m_u	A_r, M_r	
molar mass: $N_A m_a$	M	kg/mol
mass excess: $m_a - Am_u$	Δ	
quantum numbers:		
principal quantum number	n, n_i	
orbital angular momentum	L, l_i	
electron spin	S, s_i	
total angular momentum	J, j_i	
magnetic	M, m_i	
nuclear spin	I, J [l]	
hyperfine	F	
quadrupole moment	Q [m]	C m^2
magnetic moment	μ	A m^2
g-factor: $\mu/I\mu_N$	g	
gyromagnetic ratio, gyromagnetic coefficient: ω/B	γ	C/kg
Larmor circular frequency	ω_L	rad/s
level width	Γ	J

Table V (*continued*)

Quantity	Symbol	Unit
reaction energy, disintegration energy	Q	J
cross section	σ	m^2
macroscopic cross section: $n\sigma$	Σ	m^{-1}
impact parameter	b	m
scattering angle	ϑ, θ	rad
internal conversion coefficient	α	
mean life	τ, τ_m	s
half life	$T_{1/2}$, $\tau_{1/2}$	s
decay constant, disintegration constant	λ	s^{-1}
activity	A	Bq, s^{-1}
Compton wavelength: $\lambda_C = h/mc$	λ_C	m
linear attenuation coefficient	μ, μ_l	m^{-1}
atomic attenuation coefficient	μ_a	m^2
mass attenuation coefficient	μ_m	m^2/kg
linear stopping power	S, S_l	J/m
atomic stopping power	S_a	J m^2
linear range	R, R_l	m
recombination coefficient	α	m^3/s
j. Molecular spectroscopy		
quantum numbers:		
electronic spin	S	
nuclear spin	I	
vibrational mode	v	
total angular momentum,[n]		
(LM and STM; excluding electron and nuclear spin)	N	
(excluding nuclear spin): $\boldsymbol{N}+\boldsymbol{S}$ [o]	J	
(including nuclear spin): $\boldsymbol{J}+\boldsymbol{I}$	F	
vibrational angular momentum (LM)[n]	l	
component of $\boldsymbol{J}$ in the direction of an external field	M, M_J	
component along the symmetry axis,		
of electronic orbital angular momentum	Λ, λ_i	
of electronic spin	Σ, σ_i	
of total electronic angular momentum: $\lvert\Lambda+\Sigma\rvert$	Ω, ω_i	
of angular momentum,		
(excluding electron and nuclear spin); for LM: $K=\lvert\Lambda+l\rvert$	K	
(excluding nuclear spin, tightly coupled electron spin); for LM: $P=\lvert\Lambda\pm l\rvert$; for STM: $P=\lvert K+\Sigma\rvert$	P	
degeneracy of vibrational mode	d	
electronic term: E_e/hc [p]	T_e	m^{-1}
vibrational term: E_{vibr}/hc [p]	G	m^{-1}
rotational term: E_{rot}/hc [p]	F	m^{-1}
total term: T_e+G+F [p]	T	m^{-1}
principal moments of inertia: $I_A \le I_B \le I_C$ [q]	I_A, I_B, I_C	kg m^2
rotational constants: $A = h/8\pi^2 c I_A$, etc.[p]	A, B, C	m^{-1}
k. Solid state physics		
lattice vector: a translation vector which maps the crystral lattice onto itself	$\boldsymbol{R}$, $\boldsymbol{R}_0$	m
fundamental translation vectors of a crystal lattice: $\boldsymbol{R}=n_1\boldsymbol{a}_1+n_2\boldsymbol{a}_2+n_3\boldsymbol{a}_3$, ($n_1$, n_2, n_3, integers)	$\boldsymbol{a}_1$, $\boldsymbol{a}_2$, $\boldsymbol{a}_3$, $\boldsymbol{a}$, $\boldsymbol{b}$, $\boldsymbol{c}$	m
(circular) reciprocal lattice vector: $\boldsymbol{G}\cdot\boldsymbol{R}2\pi m$, where m is an integer	$\boldsymbol{G}$	m^{-1}
(circular) fundamental translation vectors for the reciprocal lattice: $\boldsymbol{a}_i\cdot\boldsymbol{b}_k=2\pi\delta_{ik}$, where δ_{ik} is the Kronecker delta symbol[r]	($\boldsymbol{b}_1$, $\boldsymbol{b}_2$, $\boldsymbol{b}_3$), ($\boldsymbol{a}^*$, $\boldsymbol{b}^*$, $\boldsymbol{c}^*$)	m^{-1}
lattice plane spacing	d	m
Miller indices	h_1, h_2, h_3; h, k, l	
single plane or set of parallel planes in a lattice[s]	(h_1, h_2, h_3), (h, k, l)	
full set of planes in a lattice equivalent by symmetry[s]	{h_1, h_2, h_3}, {h, k, l}	

Table V (*continued*)

Quantity	Symbol	Unit
direction in a lattice[s]	$[u, v, w]$	
full set of directions in a lattice equivalent by symmetry[s]	$\langle u, v, w\rangle$	
Bragg angle	ϑ	rad
order of reflection	n	
short range order parameter	σ	
long range order parameter	s	
Burgers vector	$\boldsymbol{b}$	m
particle position vector[t]	$\boldsymbol{r}, \boldsymbol{R}$	m
equilibrium position vector of an ion	$\boldsymbol{R}_0$	m
displacement vector of an ion	$\boldsymbol{u}$	m
normal coordinates	Q_i, q_i	—
polarization vector	e	
Debye–Waller factor	D	
Debye angular wave number	q_{D}	m^{-1}
Debye angular frequency	ω_{D}	rad/s
Grüneisen parameter: $\alpha/\kappa\rho c_V$ (α: cubic expansion coefficient; κ: compressibility)	γ, Γ	
Madelung constant	α	
mean free path of electrons	l, l_{e}	m
mean free path of phonons	Λ, l_{ph}	m
drift velocity	v_{dr}	m/s
mobility	μ	$\mathrm{m^2/(V\ s)}$
one-electron wave function	$\psi(\boldsymbol{r})$	$\mathrm{m}^{-3/2}$
Bloch wave function: $\psi_{\boldsymbol{k}}(\boldsymbol{r}) = u_{\boldsymbol{k}}(\boldsymbol{r}) \exp(\mathrm{i}\boldsymbol{k}\cdot\boldsymbol{r})$	$u_{\boldsymbol{k}}(\boldsymbol{r})$	$\mathrm{m}^{-3/2}$
density of states: $\mathrm{d}N(E)/\mathrm{d}E$	N_E, ρ	J^{-1}
(spectral) density of vibrational modes	g, N_ω	s
exchange integral	J	—
resistivity tensor	ρ_{ik}	Ω m
electric conductivity tensor	σ_{ik}	S/m
thermal conductivity tensor	λ_{ik}	W/(m K)
residual resistivity	ρ_{R}	Ω m
relaxation time	τ	s
Lorentz coefficient: $\lambda/\sigma T$	L	$\mathrm{V^2/K^2}$
Hall coefficient	A_{H}	$\mathrm{m^3/C}$
Ettinghausen coefficient	A_{E}	$\mathrm{K\ m^3/J}$
Nernst coefficient	A_{N}	$\mathrm{m^3/J}$
Righi-Leduc coefficient	A_{RL}	$\mathrm{K\ m^3/J\ V}$
thermoelectromotive force between substances a and b	$E_{\mathrm{ab}}, \Theta_{\mathrm{ab}}$	V
Seebeck coefficient for substances a and b: $\mathrm{d}E_{\mathrm{ab}}/\mathrm{d}T$	$S_{\mathrm{ab}}, \epsilon_{\mathrm{ab}}$	V/K
Peltier coefficient for substances a and b	Π_{ab}	V
Thomson coefficient	μ, τ	V/K
work function: $\Phi = e\phi$ [u]	ϕ, Φ	V
Richardson constant: $j = AT^2 \exp(-\Phi/kT)$	A	$\mathrm{A/(m^2\ K^2)}$
electron number density[v]	n, n_{n}, n_-	m^{-3}
hole number density	p, n_{p}, n_+	m^{-3}
donor number density	n_{d}	m^{-3}
acceptor number density	n_{a}	m^{-3}
instrinsic number density: $(np)^{1/2}$	n_{i}	m^{-3}
energy gap	E_{g}	J
donor ionization energy	E_{d}	J
acceptor ionization energy	E_{a}	J
Fermi energy	$E_{\mathrm{F}}, \epsilon_{\mathrm{F}}$	J
particle angular wave vector, propagation vector	$\boldsymbol{k}$	m^{-1}
phonon angular wave vector, propagation vector	$\boldsymbol{q}$	m^{-1}
Fermi angular wave vector	$\boldsymbol{k}_{\mathrm{F}}$	m^{-1}
electron annihilation operator	a	
electron creation operator	$a^\dagger$	
phonon annihilation operator	b	
phonon creation operator	$b^\dagger$	
effective mass	$m_{\mathrm{n}}^*, m_{\mathrm{p}}^*$	
mobility ratio: $\mu_{\mathrm{n}}/\mu_{\mathrm{p}}$	b	
carrier life time	$\tau_{\mathrm{n}}, \tau_{\mathrm{p}}$	s

Table V (*continued*)

Quantity	Symbol	Unit
diffusion length: $\sqrt{D\tau}$	L_n, L_p	m
characteristic (Weiss) temperature	Θ, Θ_W	K
Néel temperature	T_N	K
Curie temperature	T_C	K
superconductor critical transition temperature	T_c	K
superconductor (thermodynamic) critical field strength	H_c	A/m
superconductor critical field strength (type II)[w]	H_{c1}, H_{c2}, H_{c3}	A/m
superconductor energy gap	Δ	J
London penetration depth	λ_L	m
coherence length	ξ	m
Landau-Ginzburg parameter: $\lambda_L/\sqrt{2}\xi$	κ	
l. Chemical physics[x]		
relative atomic mass, ("molecular weight")	A_r, M_r	
amount of substance	n, ν [y]	mol
molar mass	M	kg/mol
concentration: $c = n/V$	c	mol/m^3
molar fraction	x	
mass fraction	w	
volume fraction	ϕ	
molar ratio of solution	r	
molality of solution	m	mol/kg
chemical potential (referred to one particle)	μ	J
absolute activity: $\exp(\mu/kT)$	λ	
relative activity	a	
reduced activity: $(2\pi mkT/h^2)^{3/2}\lambda$	z	m^{-3}
osmotic pressure	Π	Pa
stoichiometric number of substance B	ν_B	
extent of reaction: $d\xi_B = dn_B/\nu_B$	ξ	mol
(thermodynamic) equilibrium constant: $\prod_B \lambda_B{}^{\nu_B}$	K	
charge number of an ion	z	–
ionic strength: $\frac{1}{2}\sum_i m_i z_i{}^2$	I	mol/kg
m. Plasma physics		
charge of a particle	q	C
energy of a particle	ϵ	J
Debye length	λ_D	m
dissociation energy (of molecule X)	E_d, $E_d(X)$	J
electron affinity	E_{ea}	J
ionization energy	E_i	J
charge number of ion (positive or negative)	z	
number density of ions of charge number z	n_z [z]	m^{-3}
degree of ionization	x	
degree of ionization, charge number $z \geq 1$: $n_z/(n_z + n_{z-1})$	x_z	
neutral particle temperature	T_n	K
ion temperature	T_i	K
electron temperature	T_e	K
electron number density	n_e	m^{-3}
electron plasma angular frequency: $n_e e^2/\epsilon_0 m_e$	ω_{pe}	rad/s
electron cyclotron angular frequency: eB/m_e	ω_{ce}	rad/s
ion cyclotron angular frequency: zeB/m_i	ω_{ci}	rad/s
collision frequency	ν_{coll}, ν_c	s^{-1}
mean time interval between collisions: $1/\nu_{coll}$	τ_{coll}, τ_c	s
cross section: $1/nl$	σ	m^2
(electron) ionization efficiency: $(\rho_0/\rho)\, dN/dx$ (dN: number of ion pairs formed by an ionizing electron traveling through dx in the plasma at gas density ρ; ρ_0: gas density at $p_o = 133.322$ Pa, $T_o = 273.15$ K)	s_e	m^{-1}
one-body rate coefficient: $-dn_A/dt = k_m n_A$	k_m	s^{-1}
binary rate coefficient, two-body rate coefficient (e.g., $X + Y \rightarrow XY + h\nu$): $dn_{XY}/dt = k_b n_X n_Y$	k_b	m^3/s
ternary rate coefficient, three-body rate coefficient (e.g., $X + Y + M \rightarrow XY + M^*$): $dn_{XY}/dt = k_t n_M n_X n_Y$	k_t	m^6/s
relaxation time: $\tau = 1/k_m$	τ	s

Table V (*continued*)

Quantity	Symbol	Unit
positive, negative ion diffusion coefficient	D_+, D_-	m^2/s
electron diffusion coefficient	D_e	m^2/s
ambipolar (ion–electron) diffusion coefficient: $(D_+\mu_e+D_e\mu_+)/(\mu_++\mu_e)$	D_a, D_{amb}	m^2/s
characteristic diffusion length	L_D, Λ	m
ionization frequency	ν_i	s^{-1}
ion–ion recombination coefficient: $dn_-/dt=-\alpha_i n_- n_+$	α_i	m^3/s
electron–ion recombination coefficient: $dn_e/dt=-\alpha_e n_e n_+$	α_e	m^3/s
plasma pressure	p	Pa
magnetic pressure: $\boldsymbol{B}\cdot\boldsymbol{H}$	p_m	Pa
magnetic pressure ratio: p/p_m	β	
magnetic diffusivity: $1/\mu\sigma$	ν_m, η_m	m^2/s
Alfvén speed: $B/(\mu\rho)^{1/2}$ (ρ: mass density; μ: permeability)	v_A	m/s

[a] Symbols may be made more specific by adding superscripts or subscripts or by using both lower and upper case forms.
[b] Where more than one symbol is given there is no implied preference in the ordering.
[c] Either or both forms of a Greek letter (e.g., θ, ϑ; ϕ, φ) may be used.
[d] In molecular spectroscopy $\bar{\nu}$ is also used.
[e] The index m is added to a symbol to denote a molar quantity; uppercase letters may be used for extensive quantities and lowercase letters for specific quantities.
[f] When symbols for both time and Celsius temperature are required, use t for time and θ for temperature.
[g] For an ideal gas the isentropic exponent κ is equal to the ratio of heat capacities γ.
[h] The relationships are given here in the rationalized 4-dimensional Système International.
[i] In anisotropic media quantities such as permittivity, susceptibility and polarizability are second-rank tensors; component notation should be used if the tensor character of these quantities is significant, e.g., ϵ_{ij}.
[j] The word 'light' refers both to the electromagnetic spectrum of all wavelengths and to that portion of it that produces a response in the human eye. In describing light, the same symbols are often used for the corresponding radiant, luminous and photonic quantities. Although the symbols are the same, the units and dimensions of these three quantities are different; subscripts e (energetic), v (visible) and p (photon) should be added when it is necessary to distinguish among them.
[k] $\alpha(\lambda)$, $\rho(\lambda)$, and $\tau(\lambda)$ designate spectral absorptance $\Phi_a(\lambda)/\Phi_0(\lambda)$, spectral reflectance $\Phi_r(\lambda)/\Phi_o(\lambda)$, and spectral transmittance $\Phi_{tr}(\lambda)/\Phi_o(\lambda)$, respectively.
[l] I is used in atomic physics, J in nuclear physics.
[m] A quadrupole moment is actually a second-rank tensor; if the tensor character is significant the symbol should be $\boldsymbol{Q}$ or Q_{ij}.
[n] LM = linear molecule. STM = symmetric top molecule. DM = diatomic molecule. PM = polyatomic molecule. For further details see: *Report on Notation for the Spectra of Polyatomic Molecules* (Joint Commission for Spectroscopy of IUPAP and IAU, 1954), J. Chem. Phys. **23,** 1997 (1955).
[o] Case of loosely coupled electron spin.
[p] All energies are taken with respect to the ground state as the reference level.
[q] For diatomic molecules, use I and $B=h/8\pi^2cI$.
[r] In crystallography, $\boldsymbol{a}_i\cdot\boldsymbol{b}_k=\delta_{ik}$.
[s] When the letter symbols in the bracketed expressions are replaced by numbers, the commas are usually omitted. A negative numerical value is commonly indicated by a bar above the number; e.g., $(\bar{1}10)$.
[t] Lowercase and uppercase letters are used, respectively, to distinguish between electron and ion position vectors.
[u] The symbol W is used for the quantity $\Phi+\mu$, where μ is the electron chemical potential which, at $T=0$ K, is equal to the Fermi energy E_F.
[v] In general, the subscripts n and p or − and + may be used to denote electrons and holes, respectively.
[w] H_{c1}: for magnetic flux entering the superconductor; H_{c2}: for disappearance of bulk superconductivity, H_{c3}: for disappearance of surface superconductivity.
[x] In general, the attribute X of chemical species B is denoted by the symbol X_B, but in specific instances it is more convenient to use the notation X(B); e.g., $X(CaCO_3)$ or X (H_2O: 250°C).
[y] ν may be used as an alternative symbol for amount of substance when n is used for number density of particles.
[z] If only singly charged ions need be considered, n_{-1} and n_{+1} may be represented by n_- and n_+.

Table VI Dimensionless parameters[a]

The symbols used in these definitions have the following meanings:

- a, thermal diffusivity ($K/\rho c_p$)
- c, velocity of sound
- c_p, specific heat capacity at constant pressure
- f, a characteristic frequency
- g, acceleration of free fall
- h, heat transfer coefficient: heat/(time × cross sectional area × temperature difference)
- k, mass transfer coefficient: mass/(time × cross sectional area × mole fraction difference)
- l, a characteristic length
- v, a characteristic speed
- x, mole fraction
- B, magnetic flux density
- D, diffusion coefficient
- K, thermal conductivity
- $\beta' = -\rho^{-1}(\partial\rho/\partial x)_{T,p}$
- γ, cubic expansion coefficient: $-\rho^{-1}(\partial\rho/\partial T)_p$
- η, viscosity
- λ, mean free path
- μ, magnetic permeability
- ν, kinematic viscosity: η/ρ
- ρ, mass density
- σ, surface tension; electric conductivity
- Δp, pressure difference
- Δt, a characteristic time interval
- Δx, a characteristic difference of mole fraction
- ΔT, a characteristic temperature difference

a. Dimensionless constants of matter		
Prandtl number	$\nu/a = \eta c_p/K$	*Pr*
Schmidt number	ν/D	*Sc*
Lewis number	$a/D = Sc/Pr$	*Le*
b. Momentum transport		
Reynolds number	vl/ν	*Re*
Euler number	$\Delta p/\rho v^2$	*Eu*
Froude number	$v(lg)^{-1/2}$	*Fr*
Grashof number	$l^3 g\gamma\,\Delta T/\nu^2$	*Gr*
Weber number	$\rho v^2 l/\sigma$	*We*
Mach number	v/c	*Ma*
Knudsen number	λ/l	*Kn*
Stouhal number	lf/v	*Sr*
c. Transport of heat		
Fourier number	$a\,\Delta t/l^2$	*Fo*
Péclet number	$vl/a = Re \cdot Pr$	*Pe*
Rayleigh number	$l^3 g\gamma\,\Delta T/\nu a = Gr \cdot Pr$	*Ra*
Nusselt number	hl/K	*Nu*
Stanton number	$h/\rho v c_p = Nu/Pe$	*St*
d. Transport of matter in binary mixtures		
Fourier number for mass transfer	$D\,\Delta t/l^2 = Fo/Le$	*Fo**
Péclet number for mass transfer	$vl/D = Pe \cdot Le$	*Pe**
Grashof number for mass transfer	$l^3 g\beta'\,\Delta x/\nu^2$	*Gr**
Nusselt number for mass transfer	$kl/\rho D$	*Nu**
Stanton number for mass transfer	$k/\rho v = Nu^*/Pe^*$	*St**
e. Magnetohydrodynamics		
Magnetic Reynolds number	$v\mu\sigma l$	*Rm*
Alfvén number	$v(\rho\mu)^{1/2}/B$	*Al*
Hartmann number	$Bl(\sigma/\rho\nu)^{1/2}$	*Ha*
Cowling number (second Cowling number)	$B^2/\mu\rho v^2 = Al^{-2}$	Co, Co_2
first Cowling number	$B^2 l\sigma/\rho v = Rm \cdot Co_2 = Ha^2/Re$	Co_1

[a] Symbols recommended in International Standard ISO31/12-1981, ISO Standards Handbook 2, (second edition, 1982)

BIBLIOGRAPHY

ISO Standards Handbook 2, Units of Measurement, 2nd ed. International Organization for Standardization, Geneva, 1982.

IUPAP, *Symbols, Units, Nomenclature and Fundamental Constants in Physics,* 1987 rev. *PHYSICA,* **164A,** (1987). Used with permission.

Symmetry Breaking, Spontaneous

Ling-Fong Li

I. INTRODUCTION

The term spontaneous symmetry breaking (SSB) refers to a class of relativistic quantum field theories where the ground state of the system does not have the symmetries that are present in the Lagrangian. Consequently, even though the basic interaction is invariant under the symmetry transformation, the symmetry is no longer manifest in the physical states. This mechanism and the principle of gauge invariance are the two most important ingredients in the modern theories of fundamental interactions, including electromagnetic, weak, and strong interactions [1].

The spontaneous symmetry breaking has its origin in the nonrelativistic many-body problems. The simplest example is the infinite ferromagnets [2] in which the interaction between spin vectors associated with each lattice point is invariant under rotation. Below the Curie temperature T_C, the ground state of the system exhibits a spontaneous magnetization where all spin vectors line up in some direction, and is not rotationally invariant. The dynamical reason for this spontaneous breaking of the rotational symmetry arises from the effective interaction between neighboring spin vectors which favors this parallel alignment of the spins.

One of the important features of the SSB is the Goldstone theorem which states that the spontaneous breaking of a continuous symmetry implies the existence of massless scalar particles, known as Goldstone bosons. The study of this connection was initiated by Nambu [3] and subsequently the proofs with various degree of rigor and generality were provided by Goldstone [4] and others [5,6]. The fact that no such massless particles are known experimentally has been viewed as a handicap for the application of SSB to particle physics. However, the idea that the pions whose masses are much smaller than all the other hadrons can be considered as almost Goldstone bosons has been quite successful in the understanding of the chiral symmetries of the strong interaction [7].

When SSB takes place in theories with local symmetries, something rather dramatic happens: the zero-mass Goldstone bosons can combine with the massless gauge bosons to form massive vector bosons. This removes the undesirable feature of massless particles in both theories with SSB and theories with local symmetries. This idea also originates from the study of the nonrelativistic many-body problems. It turns out that in the theory of superconductivity the SSB does not give rise to Goldstone excitations. This feature is directly traceable to the presence of the long-range Coulomb interaction between electrons [8]. The generalization of this phenomenon to the relativistic field theory [9–11], called the Higgs mechanism, has become an important tool in the application of theories with local symmetries.

Without question the most important application of SSB has been the construction of a unified theory of weak and electromagnetic interactions [12,13]. This theory has been very successful in explaining low-energy weak-interaction phenomenology. By now, the Higgs mechanism is a standard tool in the description of the elementary-particle interactions.

II. GOLDSTONE THEOREM

To illustrate the main features of SSB, we will outline the proof of the Goldstone theorem and ignore some of the subtleties. Any continuous symmetry of the Lagrangian density, by Noether's theorem, implies the existence of a conserved current,

$$\partial^{\mu} J_{\mu}(x) = 0$$

where

$$\partial^{\mu} = \left(\frac{\partial}{\partial t}, -\nabla\right) \quad \text{and} \quad J^{\mu} = (J^0, \mathbf{J}). \tag{1}$$

The symmetry charge is defined by

$$Q(t) = \int d^3x \; J_0(x). \tag{2}$$

Using the current conservation in (1), we can see that the charge $Q(t)$ is conserved,

$$\frac{d}{dt}Q(t) = \int d^3x \; \partial_0 J_0(x) = -\int d^3x \; \nabla\cdot\mathbf{J}$$
$$= -\int d\mathbf{S}\cdot\mathbf{J}(\mathbf{x},t) = 0. \tag{3}$$

where we have imposed appropriate boundary conditions so that the surface integral vanishes. For any local operation $\phi(x)$ which is not a singlet under the symmetry transformation we can write the transformation law as

$$\phi(y) \mapsto \phi'(y) = e^{i\alpha Q(t)}\phi(y)e^{-i\alpha Q(t)} = \phi(x)$$
$$+ i\alpha[Q(t),\phi(y)] + \cdots, \tag{4}$$

where α is an arbitrary constant. For convenience, set $y = 0$ and write

$$[Q(t),\phi(0)] = \eta. \tag{5}$$

From (3), we see that the operator η is independent of t.

The condition for the spontaneous symmetry breaking is the existence of some local operator $\phi(0)$ such that the vacuum expectation value (VEV) of this commutator is nonzero,

$$\langle 0 \mid [Q(t),\phi(0)] \mid 0\rangle = \langle 0 \mid \eta \mid 0\rangle \neq 0. \tag{6}$$

Note that this condition implies that

$$Q(t) \mid 0\rangle \neq 0 \tag{7}$$

which means that the vacuum (ground state), $|0\rangle$, is not invariant under the symmetry transformation, $e^{i\alpha Q}|0\rangle \neq |0\rangle$. Inserting a complete set of intermediate states, we can write relation (6) as

$$\sum_n (2\pi)^3\delta^3(\mathbf{P}_n)\langle 0|J_0(0)|n\rangle\langle n|\phi(0)|0\rangle e^{-iE_nt} - \langle 0|\phi(0)|n\rangle\langle n|J_0(0)|0\rangle e^{iE_nt} = \langle 0|\eta|0\rangle \quad (8)$$

where we have used the translational invariance to write

$$J_0(x) = e^{iP\cdot x}J_0(0)e^{-iP\cdot x}$$

Here P^μ are the energy-momentum operators. Since the right-hand side of (8) is independent of t and is nonzero, this relation can be satisfied only if there exists an intermediate state with the property

$$E_n = 0 \qquad \text{for } \mathbf{P}_n = 0. \quad (9)$$

This is the massless Goldstone boson. This rather general result is valid independently of the perturbation theory.

Note that the operator η, which has nonzero VEV, is required to be a Lorentz scalar (or pseudoscalar) so that the Lorentz invariance is not spontaneously broken. The local operator $\phi(x)$ can be either an elementary field of the theory or a composite operator made out of the elementary fields. In the former case the Goldstone boson is an elementary particle, whereas in the latter case it is a bound state of the elementary particles.

We will now elucidate the Goldstone theorem in a simple example. In particular, we will see how the symmetry breaking condition arises. Consider a Lagrangian density of scalar fields with $SO(2)$ symmetry given by

$$\mathcal{L} = \tfrac{1}{2}[(\partial_\mu\phi_1)^2 + (\partial_\mu\phi_2)^2] - V(\phi_1,\phi_2) \quad (10)$$

with

$$V(\phi_1,\phi_2) = -\frac{\mu^2}{2}(\phi_1^2+\phi_2^2) + \frac{\lambda}{4}(\phi_1^2+\phi_2^2)^2. \quad (11)$$

It is easy to see that $\mathcal{L}$ is invariant under the $SO(2)$ transformation given by

$$\begin{pmatrix}\phi_1\\ \phi_2\end{pmatrix} \rightarrow \begin{pmatrix}\phi_1'\\ \phi_2'\end{pmatrix} = \begin{pmatrix}\cos\alpha & \sin\alpha\\ -\sin\alpha & \cos\alpha\end{pmatrix}\begin{pmatrix}\phi_1\\ \phi_2\end{pmatrix}, \quad \alpha\text{: constant}, \quad (12)$$

which is just a rotation in the (ϕ_1,ϕ_2) plane. The corresponding Hamiltonian is of the form

$$H = \int d^3x\{\tfrac{1}{2}[(\partial_0\phi_1)^2 + (\partial_0\phi_2)^2 + (\nabla\phi_1)^2 + (\nabla\phi_2)^2] + V(\phi_1,\phi_2)\}. \quad (13)$$

The classical effective potential can be identified as

$$U(\phi_1,\phi_2) = \tfrac{1}{2}[(\nabla\phi_1)^2 + (\nabla\phi_2)^2] + V(\phi_1,\phi_2). \quad (14)$$

Note that we have to take λ in (11) to be positive, so that $U(\phi_1,\phi_2)$ is bounded below. The classical ground state corresponds to the field configurations which minimize the potential energy given in (14). Since the $(\nabla\phi_i)^2$ terms in (14) are non-negative, the minimum of $U(\phi_1,\phi_2)$ will have the property that $\nabla\phi_1 = \nabla\phi_2 = 0$ with constant values of ϕ_1 and ϕ_2 given by those minimizing $V(\phi_1,\phi_2)$. It is easy to see that for $\mu^2>0$, the absolute minima of $V(\phi_1,\phi_2)$ are those values of ϕ_1 and ϕ_2 satisfying

$$\phi_1^2 + \phi_2^2 = v^2 \qquad \text{with } v = \sqrt{\frac{\mu^2}{\lambda}}, \quad (15)$$

i.e., the minimum consists of points on a circle with radius v in the (ϕ_1,ϕ_2) plane and they are related to each other through $SO(2)$ rotations. Thus, the classical ground state is not unique and each point on this circle can be used as a classical ground state to construct a complete quantum theory. For example, if we choose the classical ground state to be

$$\phi_1 = v, \qquad \phi_2 = 0,$$

then in the quantum theory where ϕ_i' become operators, the vacuum (ground state) is required to have the property that

$$\langle 0|\phi_1|0\rangle = v, \qquad \langle 0|\phi_2|0\rangle = 0. \quad (16)$$

The nonzero VEV for ϕ_1 is the symmetry breaking condition needed for the Goldstone theorem. Note that different choices of classical ground state are all equivalent through $SO(2)$ rotation; the corresponding quantum theories are also equivalent.

To find the particle spectra in perturbation theory, we consider small oscillations around the true minimum by defining the new fields

$$\phi_1' = \phi_1 - v, \qquad \phi_2' = \phi_2 \quad (17)$$

and write the Lagrangian density in terms of ϕ_1' and ϕ_2',

$$\mathcal{L} = \frac{1}{2}[(\partial_\mu\phi_1')^2 + (\partial_\mu\phi_2')^2] - \mu^2\phi_1'^2 - \lambda v\phi_1'[(\phi_1')^2 + (\phi_2')^2] - \frac{\lambda}{4}[(\phi_1')^2 + (\phi_2')^2]^2. \quad (18)$$

Since there is no quadratic term in the ϕ_2' field, ϕ_2' corresponds to massless Goldstone boson, whereas ϕ_1' is a particle with mass $\sqrt{2}\mu$.

Note that if we replace μ^2 by $-\mu^2$ in $V(\phi_1,\phi_2)$ given in (11), then the classical minimum is at $\phi_1=\phi_2=0$ which is unique. In this case the corresponding quantum theory will have $\langle 0|\phi_1|0\rangle = \langle 0|\phi_2|0\rangle = 0$. Then there is no spontaneous symmetry breaking and ϕ_1,ϕ_2 will be degenerate (have the same mass μ) as a consequence of the $SO(2)$ symmetry.

In our simple example, the $SO(2)$ symmetry is broken completely. In the more general case, depending on the representation content of the scalar fields, it is possible to have some residual symmetry left over after the spontaneous symmetry breaking. For example, if we have a set of scalar fields $\boldsymbol{\phi} = (\phi_1,\phi_2,\phi_3)$ transforming as a vector under the rotation group $SO(3)$, the scalar potential is of the form

$$V(\boldsymbol{\phi}) = -\frac{\mu^2}{2}\boldsymbol{\phi}\cdot\boldsymbol{\phi} + \frac{\lambda}{4}(\boldsymbol{\phi}\cdot\boldsymbol{\phi})^2. \quad (19)$$

It is easy to see that the minimum is at

$$\phi_1^2+\phi_2^2+\phi_3^2=v^2 \qquad \text{with } v=\sqrt{\frac{\mu^2}{\lambda}}, \tag{20}$$

i.e., points on the surface of a sphere with radius v in (ϕ_1,ϕ_2,ϕ_3) space. A convenient choice is

$$\phi_1=v, \qquad \phi_2=\phi_3=0. \tag{21}$$

It is clear that this choice is invariant under the rotation in the (2,3) plane. Thus, there is a $SO(2)$ symmetry left over after SSB.

Here we see that the pattern of symmetry breaking depends on the transformation properties of the scalar fields. One of the general features is that the number of Goldstone bosons is the same as that of the broken symmetry generators. For more discussions on the group theory aspects of SSB see Ref. [14].

III. HIGGS PHENOMENON

When global symmetries are extended to local symmetries by introducing the gauge fields, the implication of SSB is quite different. The massless Goldstone boson will combine with a massless gauge boson to form a massive vector boson. This is known as the Higgs mechanism, which solves the problem of unobserved massless gauge bosons. This makes the gauge theories (in particular the non-Abelian gauge theories), which have the property of universal couplings, very useful frameworks for describing the elementary particle interactions.

We will illustrate the Higgs mechanism in the simple example used in Section II. Define the complex field by $\phi=(1/\sqrt{2})(\phi_1+i\phi_2)$. Then the symmetry transformation in (12) becomes a $U(1)$ phase transformation,

$$\phi \rightarrow \phi' = e^{i\alpha}\phi. \tag{22}$$

The gauge field A_μ can be introduced in the same way as the photon field. The resulting Lagrangian is

$$\mathcal{L}=(D_\mu\phi)^+(D^\mu\phi)+\mu^2\phi^+\phi-\lambda(\phi^+\phi)-\tfrac{1}{4}F_{\mu\nu}F^{\mu\nu}, \tag{23}$$

where

$$D_\mu\phi=(\partial_\mu-igA_\mu)\phi, \qquad F_{\mu\nu}=\partial_\mu A_\nu-\partial_\nu A_\mu. \tag{24}$$

This Lagrangian is invariant under the local symmetry transformation (or gauge transformation)

$$\phi(x)\rightarrow\phi'(x)=e^{i\alpha(x)}\phi(x), \tag{25}$$

$$A_\mu(x)\rightarrow A_\mu(x)=A_\mu(x)+\frac{1}{g}\partial_\mu\alpha(x).$$

Here $\alpha(x)$ is an arbitrary function of space-time in contrast to the global symmetry transformation where α is a constant independent of x. Note that the gauge boson mass term of the form $(A_\mu)^2$ is not invariant under the transformation in (25). So the gauge boson is massless. As before the SSB is generated by the nonzero VEV of the scalar field

$$|\langle 0|\phi|0\rangle| = \frac{v}{\sqrt{2}}. \tag{26}$$

The important new feature here is that the $|D^\mu\phi|^2$ term in $\mathcal{L}$ will give rise to a quadratic term $(A_\mu)^2$ through SSB,

$$|D_\mu\phi|^2=|(\partial_\mu-igA_\mu)\phi|^2=\tfrac{1}{2}v^2g^2A_\mu^2+\cdots. \tag{27}$$

Thus the gauge boson acquires a mass $M=gv$

To see the physics more clearly, we will parameterize the complex field $\phi(x)$ in the polar variables,

$$\phi(x)=\frac{1}{\sqrt{2}}[v+\eta(x)]e^{i\xi(x)/v}, \tag{28}$$

where $\eta(x)$ and $\xi(x)$ are quantum fields with zero VEV. We can define new fields by gauge transformation with $\xi(x)/v$ as the gauge function:

$$\phi''(x)=e^{-i\xi(x)/v}\phi(x)=\frac{1}{\sqrt{2}}[v+\eta(x)],$$

$$B_\mu(x)=A_\mu(x)-\frac{1}{gv}\partial_\mu\xi(x). \tag{29}$$

In terms of these new fields the Lagrangian density is given by

$$\mathcal{L}=\frac{1}{2}(\partial_\mu\eta)^2-\mu^2\eta^2-\frac{1}{4}(\partial_\mu B_\nu-\partial_\nu B_\mu)^2+\frac{1}{2}(gv)^2B_\mu B^\mu$$

$$+\frac{1}{2}g^2B_\mu B^\mu\eta(2v+\eta)-\lambda v^2\eta^3-\frac{\lambda}{4}\eta^4. \tag{30}$$

Note that the ξ field has disappeared as a consequence of the local symmetry. In (30) we see that there is a massive vector boson with mass $M=gv$ and a real scalar particle with mass $\sqrt{2}\mu$. Observe that we start with two real scalar fields, ϕ, and ϕ_2, and one massless gauge boson, A_μ (with two polarization states). After SSB we have only one scalar particle η and one massive vector boson B_μ (with three polarization states). What happens is that, as seen in (29), the scalar field $\xi(x)$ has combined with massless gauge field $A_\mu(x)$ to form a massive vector field B_μ and there are no massless particles at the end. This is the Higgs phenomenon.

IV. APPLICATION

We will discuss two important applications of the Higgs phenomenon.

1. Weak Interactions

The first successful application of the Higgs phenomenon is the area of weak interactions. Here the non-Abelian $SU(2)\times U(1)$ gauge group provides a unified description of weak interactions and electromagnetism. The gauge bosons are W_μ^i, $i=1,2,3$ for the $SU(2)$ group with gauge coupling g, and B_μ for the $U(1)$ group with gauge coupling g'. For simplicity, we will discuss the simple case of one generation of quarks and leptons, consisting of a u quark, a d quark, an electron, and its neutrino (u,d,e,ν_e). The left-handed fermions transform as doublets under $SU(2)$, whereas the right-handed fermions are $SU(2)$ singlets:

$$\begin{pmatrix} u \\ d \end{pmatrix}_L \begin{pmatrix} \nu_e \\ e \end{pmatrix}_L \quad u_R,d_R,e_R. \tag{30}$$

The $U(1)$ quantum number is given by $Y=2(Q-T_3)$, where Q and T_3 are the electric charge and the third component of $SU(2)$ charges, respectively.

The spontaneous symmetry breaking is generated by a complex scalar doublet

$$\phi=\begin{pmatrix}\phi^+\\ \phi^0\end{pmatrix} \tag{31}$$

which develops a vacuum expectation value given by

$$\langle 0 \mid \phi \mid 0\rangle=\frac{1}{\sqrt{2}}\begin{pmatrix}0\\ v\end{pmatrix} \tag{32}$$

with $v\simeq 250$ GeV. This breaks the local symmetry from $SU(2)\times U(1)$ down to electromagnetic $U(1)$ symmetry. Of four scalar fields (ϕ and its complex conjugate), three are combined with the original massless gauge fields to form massive vector bosons, $W^{\pm}$ and Z^0, given by

$$W_\mu^{\pm}=\frac{1}{\sqrt{2}}(W_\mu^1\pm iW_\mu^2),\quad Z_\mu=-\sin\theta_w B_\mu+\cos\theta_w W_\mu^3 \tag{33}$$

with masses

$$M_w=\tfrac{1}{2}gv,\qquad M_z=\frac{gv}{2\cos\theta_w}, \tag{34}$$

where $\theta_w=\tan^{-1}(g'/g)$ is called the Weinberg angle. The massless photon A_μ corresponds to the combination

$$A_\mu=\cos\theta_w B_\mu+\sin\theta_m W_\mu^3. \tag{35}$$

The extension to include other known leptons and quarks is straightforward.

The experimental discoveries of a whole class of neutral current phenomena [15] (those processes mediated by the neutral gauge boson Z_μ) and the existence of W and Z vector bosons [16,17] give very strong support for this model. This model has been so successful in explaining all the low-energy weak-interaction phenomenology that it is now referred to as the "standard model of electroweak interaction."

One of the remnants of the SSB in the standard model is the presence of the leftover scalar particle, called the Higgs particle. This particle is quite elusive experimentally and its confirmation will be of great importance.

2. *Grand Unified Theories*

The idea that electromagnetic, weak, and strong interactions can be unified into one framework comes from the fact that they are all described by gauge theories; electromagnetic and weak interactions by $SU(2)\times U(1)$ gauge theory with SSB, the strong interaction by $SU(3)$ gauge theory of the color (QCD). In this unification scheme (called grand unified theories), these three interactions all evolve from a unified interaction through spontaneous symmetry breaking. The simplest model [18] along this line is based on the gauge group $SU(5)$ with a single gauge coupling. The SSB comes in two stages. At energies about 10^{15} GeV, the local $SU(5)$ symmetry is broken spontaneously down to $SU(2)\times U(1)$ symmetry of electroweak theory and $SU(3)$ of strong interaction. Then at energies around 250 GeV, the $SU(2)\times U(1)$ symmetry is broken down to $U(1)$ symmetry of electromagnetic interactions.

Since the energy at which unification occurs is more than 10 orders of magnitudes higher than the presently available energies, in general it is very difficult to test these theories in the laboratory. However, one interesting consequence of the grand unification is that the baryon number is no longer conserved. This implies that the proton is unstable but with rather long lifetime. The prediction of the proton lifetime in the simplest model is only slightly larger than the previous experimental limit. This has caused great excitement in the experimental search for the proton decay. Unfortunately more recent results all give negative results, which has ruled out the simplest model. One can certainly modify the simplest model to accommodate this new limit but the hope of finding positive experimental evidence for the unification is fading away. On the other hand, the grand unification theories have many interesting applications in cosmology where the universe at very early time can have energy as high as the grand unification energy.

See also FIELD THEORY, UNIFIED; GAUGE THEORIES; NONLINEAR WAVE PROPAGATION; QUANTUM FIELD THEORY; SUPERCONDUCTIVITY THEORY.

REFERENCES

1. For a survey, see, e.g., T. P. Cheng and L. F. Li, *Gauge Theory of Elementary Particle Physics*. Clarendon, Oxford, 1984.
2. W. Heisenberg, *Z. Phys.* **49,** 619 (1928).
3. Y. Nambu, *Phys. Rev. Lett.* **4,** 380 (1960).
4. J. Goldstone, *Nuovo Cimento* **19,** 154 (1961).
5. J. Goldstone, A. Salam, and S. Weinberg, *Phys. Rev.* **127,** 965 (1962).
6. S. Bludman and A. Klein, *Phys. Rev.* **131,** 2363 (1963).
7. For a survey, see, e.g., S. Adler and R. Dashen, *Current Algebra and Application to Particle Physics*. Benjamin, New York, 1968.
8. P. W. Anderson, *Phys. Rev.* **112,** 1960 (1958).
9. P. W. Higgs, *Phys. Rev. Lett.* **13,** 508 (1964).
10. F. Englert and R. Brout, *Phys. Rev. Lett.* **13,** 321 (1964).
11. G. S. Guralnik, C. R. Hagen, and T. W. B. Kibble, *Phys. Rev. Lett.* **13,** 585 (1964).
12. S. Weinberg, *Phys. Rev. Lett,* **19,** 1264 (1967).
13. A. Salam, in *Elementary Particle Physics* (Nobel Symp. No. 8) (N. Svartholm, ed.). Almquist and Wilsell, Stockholm, 1968.
14. L. F. Li, *Phys. Rev.* **D9,** 1723 (1974).
15. For a review see U. Amaldi *et al., Phys. Rev.* **D36,** 1385 (1987).
16. G. Arnison *et al., Phys. Lett.* **122B**, 103 (1983).
17. G. Arnison *et al., Phys. Lett.* **126B**, 398 (1983).
18. H. Georgi and S. L. Glashow, *Phys. Rev. Lett.* **32,** 438 (1974).

Synchrotron

E. M. McMillan and J. M. Peterson

GENERAL OVERVIEW

Synchrotrons comprise a large class of ring-shaped particle accelerators used principally for particle-physics research, and more recently, for the production of synchrotron radiation. The particle beams also have important applications in medical treatment and in medical and biological research. The name "synchrotron" refers to the synchronous nature of the acceleration process in which the beam particles are automatically kept in step (or "in phase") with the oscillating radio-frequency accelerating voltage as they circulate in the accelerator ring. This synchronous property, called "phase stability," will be discussed more fully later. It is essential in maintaining a particle beam, which in a typical accelerating synchrotron must travel millions of kilometers in an evacuated pipe only a few centimeters in diameter (and even many billions of kilometers in a storage-ring synchrotron).

Phase stability was discovered independently by E. M. McMillan (who proposed the name synchrotron) in the United States and V. I. Veksler in the Soviet Union in 1945. McMillan immediately started the design of an electron synchrotron with a beam-particle energy of 300 MeV (million electron volts) at the Lawrence Berkeley Laboratory (present name) in California. Similar machines were soon started also at Cornell University, Purdue University, and the Massachusetts Institute of Technology. The Berkeley machine came into operation in 1949 (Fig. 1). Meanwhile, in 1946 the synchrotron principle was confirmed experimentally by F. K. Goward and D. E. Barnes in England; they modified an existing 4-MeV betatron into an 8-MeV synchrotron. Soon after, a betatron being built at the General Electric Company in Schenectady, New York, was successfully converted into a 70-MeV synchrotron, from which synchrotron radiation was first observed in 1947.

During the next 40 years several generations of synchrotrons of increasing size and sophistication have evolved. As of 1989 some 17 electron and 20 proton synchrotrons having particle energies of 1 GeV [giga (10^9) electron volts] or greater have been built around the world for research in nuclear and particle physics. Of these 17 electron synchrotrons, 9 are now extensively used for the production of synchrotron radiation. In addition, about 12 electron (or positron) synchrotrons have been built for exclusive use as sources of synchrotron radiation and 18 more are under construction. This field of synchrotron radiation has had almost an explosive growth over the past decade. Several proton synchrotrons have been modified to accelerate several species of heavy ions. Partially ionized atoms as heavy as uranium have been accelerated in the Bevatron synchrotron at the Law-

FIG. 1. Vacuum chamber of the Berkeley electron synchrotron, 2 m in diameter. The accelerating cavity was formed by copper plating on one of the fused-silica sections. (LBL Photo.)

rence Berkeley Laboratory, which was originally designed to accelerate protons (Fig. 2).

The largest synchrotron now in operation is the 1-TeV [tera (10^{12}) electron volts] proton synchrotron at the Fermi National Accelerator Laboratory in Illinois (Fig. 3); it is 6.3 km in circumference. Under construction in the USSR is a 3-TeV proton synchrotron that is 21 km in circumference, and in Switzerland and France at the CERN Laboratory a 100-GeV electron–positron synchrotron 27 km in circumference is expected to be in operation in 1989. The largest now conceived is a 20-TeV proton synchrotron 87 km in circumference (The Superconducting Supercollider), which is scheduled to be built near Dallas, Texas, although not yet authorized by the United States Congress.

ACCELERATOR BASICS

The characteristics of all particle accelerators using magnetic guide fields are governed by a small number of equations that show how the principal parameters are interrelated. The size is determined by the radius of curvature R of a beam particle of final momentum p in a magnetic field of strength B, the relationship being (in Gaussian units)

$$R = pc/qB, \tag{1}$$

which c is the velocity of light and q is the net electric charge of the beam particle. The particle kinetic energy T is usually a more convenient parameter than the momentum, the exact relationship between them being

$$(pc)^2 = T^2 + 2TM_0c^2, \tag{2a}$$

where M_0 is the rest mass of the beam particle (in g). The quantity M_0c^2 is the particle rest mass in energy units. For the proton, M_0c^2 is 938 MeV, and for the electron it is 0.511 MeV. [In laboratory practice it is convenient to express particle energies in units of electron volts (eV). One eV is the energy gained by a particle with the charge of one electron from the electric field between two electrodes whose difference in electric potential is 1 V. To obtain the energy in ergs, which is the Gaussian unit of energy, multiply the number of eV by 1.60×10^{-12} ergs eV^{-1}.]

FIG. 2. The Bevatron before installation of the concrete radiation shield, showing the original injection system, which has since been greatly modified. Protons were brought to 460 keV by the Cockcroft–Waltor accelerator at the right, then to 10 MeV by the linear accelerator in the center, and put into orbits in the ring at the left (38 m in diameter) by a pulsed electrostatic inflector. (LBL Photo.)

FIG. 3. Aerial view of the FNAL proton synchrotron. The underground main ring, 2 k in diameter, is roughly outlined by an access road and cooling water pond on the surface. The smaller ring in the center foreground is the booster synchrotron, which raises protons to 8 GeV before injection into the main ring; a circular auditorium and a 16-story building are at the left. (FNAL Photo.)

In the nonrelativistic limit, where the particle kinetic energy is much smaller than its rest mass, the relationship (2a) simplifies to

$$(pc)^2 \approx 2TM_0c^2, \tag{2b}$$

and in the relativistic limit, where the kinetic energy is much larger than the rest mass, it simplifies to

$$pc \approx T. \tag{2c}$$

Thus, from Eqs. (1) and (2b) one sees that for a given maximum strength of the magnetic guide field, the characteristic size of an accelerator must increase as the square root of the required final beam energy in the nonrelativistic regime, whereas in the relativistic regime [Eq. (2c)] it increases linearly with the required beam energy. In a 20-MeV cyclotron with a 15-kg iron magnet, for example, the final orbit radius is only 43 cm; in the 20-TeV Supercollider with a 66-kg superconducting magnet, the magnetic radius of curvature is 10 km, and the machine circumference is 87 km! If superconducting magnets were not available, the Supercollider dimensions would have to be about three or four times larger.

To accomplish acceleration as the beam particles circulate in the magnetic guide field, one can arrange that they pass through a voltage gap in an electromagnetic resonator that is tuned to a multiple of the revolution frequency of the beam. E. O. Lawrence was inspired to invent the cyclotron when he realized that the revolution frequency f_0 in a constant, uniform magnetic field was approximately independent of the particle energy in the nonrelativistic regime:

$$f_0 = v/2\pi R = pqB/2\pi Mpc = qB/2\pi Mc, \tag{3}$$

where v, the particle velocity, has been expressed as the ratio p/M, and the radius R as pc/qB from Eq. (1). Equation (3) is exact if the particle mass M is taken as the *total* relativistic mass, which includes both the rest mass M_0 and the mass due to its kinetic energy T:

$$M = M_0 + T/c^2. \tag{4}$$

In the nonrelativistic regime the kinetic energy is much less than the rest energy, so that in this regime the total mass is *almost* independent of energy.

This small variation of the mass of a particle as it is accelerated limits the cyclotron as originally conceived. As the particle mass increases, its revolution frequency decreases correspondingly, and so it slowly gets out of step (out of phase) with the constant accelerating frequency. The small decrease in magnetic field strength with increasing radius, which is needed for focusing the beam (discussed below), also contributes to this effect. In order to minimize this limitation, the acceleration must be carried out in as few revolutions as possible, which requires relatively high accelerating voltages. It was realized that this phase slippage due to the relativistic mass increase and to the decrease in magnetic field strength with increasing radius could be avoided by using not a *constant* accelerating frequency but rather a *varying* frequency—one that decreased during the accelerating cycle according to Eq. (3). In this way the beam particles could, in principle, be kept in step (in phase) with the radio-frequency accelerating voltage throughout the accelerating cycle. However, it required the later introduction of the synchrotron principle, or phase stability, by McMillan and Veksler to make this method of frequency variation a practical, workable method of acceleration.

ORBIT STABILITY

Phase stability is one aspect of the general problem of orbit stability in accelerators. In order to accelerate efficiently a beam of particles, it is essential that the particles with small deviations from the ideal orbit not be lost before the acceleration cycle is complete. Orbit stability is the property of automatic compensation for small errors in position or direction (transverse stability) or in phase or energy (phase stability).

Transverse Stability

Automatic compensation for beam deviations in the transverse directions is referred to as "focusing," which is completely analagous to optical focusing. The simplest form of magnetic field arrangement that will properly focus a particle beam is a field that is uniform in the azimuthal direction and decreases slowly with increasing radial position R, as

$$B=B_0(R_0/R)^n$$

with the "field index" n greater than zero but less than 1.0. B_0 is the field strength at the average beam radius R_0. For small deviations in either the radial or the axial directions, the beam particles will oscillate stably about the average beam orbit at radius R_0. The wavelength of the axial oscillations is $2\pi R_0/\sqrt{n}$, and that of the radial oscillations is $2\pi R_0/\sqrt{1-n}$. These transverse oscillations about a central, ideal orbit are called "betatron oscillations," after the betatron accelerator where they were first analyzed, but they are common to all types of accelerators having transverse stability. In a magnetic field with the index n greater than 1.0, radial stability is lost; a particle with a small deviation in the radial direction will be defocused—i.e., it will diverge from the central orbit at R_0 and be lost. Similarly, in a field in which the index n is less than zero (a field that increases in strength with radius), a particle beam will be defocused in the vertical direction. This focusing property of magnetic fields that are azimuthally uniform but slowly decreasing in strength with radius is called "weak focusing."

Weak focusing was used in all of the early cyclotrons and betatrons and also the first synchrotrons. It has the disadvantage of requiring relatively large magnetic apertures to accommodate the beam size and angular divergence available from practical ion sources. The advent of "strong focusing" had a large impact on accelerator design; it produced particle beams of much smaller cross section so that the magnet apertures could be correspondingly smaller, thus making large accelerators more practical to build.

The invention of "strong focusing" is credited both to N. Christofilos, who in 1950 applied for a patent on this method of focusing, and to E. D. Courant, M. S. Livingston, and H. S. Snyder, who, being unaware of Christofilos' work, published their discovery in 1952. Instead of using a magnetic field that is azimuthally uniform and slowly decreasing in the radial direction, which is typical of a weak focusing system, a strong focusing system uses an arrangement of magnetic fields whose radial variations along the particle orbit alternate between regions of large positive and large negative values. A short length L of magnet with a large *decreasing* radial variation ($n \gg 1$) *focuses* the *axial* (usually vertical) particle motion with a focal length equal to $BR/(L\ dB/dR)$ but *defocuses* the *radial* (usually horizontal) motion with a negative focal length of about the same value. (dB/dR is the magnitude of the radial variation, or gradient, of the magnetic field strength.) Conversely, a similar length of magnet with a large *increasing* radial variation ($n \ll -1$) *defocuses* the *axial* motion but *focuses* the *radial* motion with the same focal lengths. However, as is well known in optics, a combination of two equal and opposite lenses, somewhat separated, has, overall, a *focusing* action. Thus, a succession of magnets with alternating gradients has the required property of transverse orbit stability, both horizontally and vertically. Strong focusing, or "alternating-gradient" focusing, has resulted in great reduction in required magnet apertures and now is the normal method of focusing in accelerators and in beam-transport lines.

Strong focusing has several forms. Radial ridges in the iron pole faces of a cyclotron produce fields of alternating gradient as traversed by the beam and thus can produce the required transverse stability, even with an average field strength that increases with radius so as to compensate the relativistic increase in particle mass with energy [and thereby maintain a constant frequency of revolution, Eq. (3)]. The use of ridges that spiral as they go out from the center of the magnet make this type of focusing even more effective. The first strong-focusing synchrotrons used magnets which combined the functions of bending and focusing—i.e., the magnets had both strong average field strengths and strong radial gradients, the gradients alternating between positive and negative values along the beam path. A somewhat more flexible method is to separate the functions by using uniform dipole magnets to provide the bending and quadrupole magnets (of alternating sign) to provide the focusing.

Phase Stability

Phase stability is complementary to transverse stability; it indicates automatic compensation of deviations in the longitudinal direction—i.e., deviations in phase (timing with respect to the accelerating voltage waveform) or in energy from that of an ideal, central particle that is in perfect synchronism with the sinusoidal accelerating waveform. Such deviations in time and energy are considered "longitudinal" effects because they affect the relative longitudinal position of a particle.

Consider a particle that has, say, a small error in phase. It crosses the radio-frequency acclerating electrode at a slightly incorrect time and so gains a slightly incorrect amount of energy, either too much or too little, depending on the particular arrangement. A particle with too much energy (and thus a positive velocity error) will travel at a slightly larger-than-ideal average radius. Thus, such a particle travels along a longer orbit (but at a higher velocity) and consequently will, in general, arrive back at the accelerating structure after one turn with a slightly different phase and so get another, this time slightly different, gain in energy.

Thus, an error in phase leads to an error in energy, which in turn leads to a different error in phase, and so on. For any given magnetic guide-field arrangement and particle energy, the phase of the ideal, synchronous particle can be chosen so that any particles with small deviations in energy and/or phase will simply oscillate in energy and phase about the ideal, synchronous values. Since such phase, or "synchrotron," oscillations result in a stable, long-lasting situations, they represent the property of phase stability.

ACCELERATORS WITH PHASE STABILITY

The realization of the principle of phase stability spawned a variety of "magnetic resonance" accelerators—i.e., machines using magnetic guide fields and radio-frequency accelerating fields operating at the revolution frequency of the particle beam, or a multiple of that frequency. The original cyclotron operated with a fixed, weak-focusing magnetic field and fixed accelerating frequency. As was noted earlier, its operation was limited because the particle revolution frequency gradually decreased as the particles were accelerated from small to large radii, both because of the relativistic increase in the mass of the accelerated particles and because of the radial decrease in magnetic field strength. This change in frequency can cause the phase of the particle to slip relative to the fixed-frequency accelerating voltage to the point where acceleration is no longer possible. In fact, if the particle beam is not intercepted by a target before its maximum energy, the continuing phase slippage would eventually produce *deceleration*, causing the beam to spiral back to the center of the cyclotron. To minimize this problem, the early cyclotrons were designed with relatively high accelerating voltages, so as to limit the number of revolutions of the beam during acceleration, and, thereby, the total amount of phase slippage.

Even before the synchrotron principle was discovered, J. R. Richardson had pointed out that the phase slippage problem could be avoided if the accelerating frequency were smoothly changed so as always to match the particle revolution frequency, as given in Eq. (3). However, it was still thought that the required frequency match was so critical that a small number of revolutions and relatively high accelerating voltages were still required. With the advent of phase stability came the realization that high voltages were not necessary. As long as the frequency of the accelerating voltage was continuously changed so as to match the revolution frequency of the central, or average, beam particle as it was accelerated, then the other beam particles with slightly different energy or phase would oscillate stably about the average values and not be lost no matter what rate of acceleration was used.

Phase Stability through Frequency Modulation—Synchrocyclotrons

There are two basic methods for maintaining the match between the particle revolution frequency and the frequency of the accelerating voltage. One is to use a fixed magnetic field strength and to vary the accelerating frequency throughout the cycle according to Eq. (3). This method of providing phase stability through modulation of the accelerating frequency is used in large relativistic cyclotrons, which are referred to as synchrocyclotrons, or FM (frequency-modulated) cyclotrons. The great advantages of the synchrocyclotron relative to the original fixed-frequency cyclotron are that much higher beam energies are attainable at much lower accelerating voltages and that accelerator operation is simpler and easier. The main disadvantage is that lower average beam currents are produced in the synchrocyclotron because the acceleration process is cyclic, rather than continuous as in the fixed-frequency cyclotron. For example, in the 184-in. (magnet pole diameter) synchrocyclotron at the Lawrence Berkeley Laboratory the accelerated beam is produced only in bursts at the modulation rate of $60\ s^{-1}$.

The same principle of frequency modulation is used in microtrons (sometimes called electron cyclotrons). In these accelerators the orbit length changes discontinuously at each turn by an integral number of radio-frequency wavelengths because of the substantial energy gain per turn. These changes in orbit length change the accelerating frequency as seen by the beam, even though the electrical frequency is held constant.

Phase Stability through Magnet Modulation—Synchrotrons

The other basic method of maintaining the match between the accelerating frequency and the particle revolution frequency is through variation of the strength of the magnetic guide field so as to maintain a constant (or almost constant) radius of curvature. To achieve a constant radius (R) independent of momentum (p), Eq. (1) indicates that the average magnetic field strength must increase in direct proportion to the momentum. The attractive feature of a constant average radius throughout the acceleration cycle is that the magnets can be relatively small—that they can be in the form of a narrow ring of small magnets at the average radius R, rather than a very large magnet system extending from a small injection radius to a large final radius, as is required in synchrocyclotrons and other cyclotrons. This class of constant-radius, ring-shaped, phase-stable accelerators is now generally referred to as synchrotrons. There are several categories of synchrotrons according to the type of beam particle (electron, proton, or heavy-ion synchrotrons) or according to the type of focusing (weak-focusing or strong-focusing synchrotrons) but all are characteristically ring shaped. Because of the radiation hazard produced by stray beam losses around the ring or by the beam-striking experimental targets, synchrotrons are usually housed in underground tunnels or in structures formed from concrete shielding blocks.

In principle, the accelerating frequency in a synchrotron must be varied as well as the magnetic field strength, as indicated in Eq. (3). However, since the average radius is constant, the required frequency variation is simply proportional to the variation in the beam velocity. If the beam energy is in the relativistic regime throughout the acceleration cycle, the beam velocity is always approximately equal to the velocity of light, so that little or no frequency variation is required. Thus, in most electron synchrotrons, whose injection energies are usually quite relativistic, no variation in the accelerating frequency is required.

The great advantage of the synchrotron relative to other forms of magnetic resonance accelerators is the relatively small size of the required magnet system and, consequently, the lower cost for a given beam energy. For this reason most high-energy accelerators in the world are synchrotrons. The principal disadvantage is the relatively low level of beam current that can be produced in the synchrotron, which, as in the synchrocyclotron, is due to the intermittent, cyclic nature of the process of acceleration. For special applications requiring larger beam currents, special forms of cyclotrons or linear accelerators have been built that have large duty cycles—i.e., ouptut beams that are continuous, or nearly so.

STORAGE AND COLLIDER RINGS

Synchrotrons have recently been extensively used as storage rings and collider rings, and further developments are actively underway. A storage ring serves to store a particle beam for relatively long periods, typically several hours or days. To achieve such long storage times requires a very good vacuum in the acceleration chamber and a very low level of orbit-perturbing electromagnetic fields, including fields produced by the beam itself. Storage rings are used principally as sources of synchrotron light and as collider rings for studying nuclear reactions.

A collider synchrotron ring is an arrangement in which two counterrotating beams are brought into collision at one or more points in the ring. If the two types of particle beams are antiparticles of each other—say, protons and antiprotons, or electrons and positrons—and are to have the same energy, then one synchrotron ring can accelerate and store both beams in the same vacuum chamber. The 400-GeV Super Proton Synchrotron at the CERN Laboratory near Geneva, Switzerland, is being used as a proton/antiproton collider, as is the Tevatron, a 1-TeV proton synchrotron at Fermilab near Chicago, Illinois (Fig. 4). The 30-GeV Intersecting Storage Ring at CERN consisted of two synchrotron rings built to study proton–proton reactions at several points where the two beam orbits crossed. Beam lifetimes on the order of months were observed in this machine. The Superconducting Super Collider will be a two-intersecting-ring arrangement for providing proton–proton collisions at a beam energy of 20 TeV. The large electron/positron collider rings PEP (at the Stanford Linear Accelerator Center near Palo Alto, California), PETRA (at the DESY Laboratory in Hamburg, Germany), and LEP (at CERN) are single-ring synchrotrons that accelerate, store, and collide counterrotating electron and positron beams.

A colliding-beam system has the advantage with respect to a system using a beam of the same energy striking a fixed target that much more energy is available in the collision for producing nuclear reactions. At high energy the beam particles become very massive, as implied by Einstein's equivalence of mass and energy. A 1-GeV proton, for example, is about 1000 times as heavy as a proton at rest, and a 1-TeV proton is about 1 000 000 times as heavy. Thus, when

FIG. 4. The tunnel at Fermilab now houses both the original (upper) ring of iron magnets that form the 400-GeV "Main Ring" and the added (lower) ring of superconducting magnets that comprise the 1-TeV Tevatron. In collider operation the main ring receives sequentially several batches of proton and antiproton beams from a chain of smaller accelerators. It then accelerates these beams to 150-GeV energy and transfers them to the Tevatron, which in turn accelerates them to about 0.9 TeV and then stores and collides the two counterrotating beams for period of several hours.

an energetic proton hits a stationary proton in a fixed-target system, it is analagous to a heavy bowling ball hitting, say, a light ping-pong ball. For fundamental kinematic reasons very little energy can be transferred in such an unbalanced collision. On the other hand, in a collider arrangement in which the two colliding beam particles have equal masses, the entire energy of both particles is in principle available for producing nuclear reactions. This relative advantage increases in proportion to the square root of the beam energy, being a factor of about 46 in available energy for a 1-GeV proton collider and about 1460 at 1 TeV.

See also BETATRON; CYCLOTRON; SYNCHROTRON RADIATION.

Synchrotron Radiation

Robert L. Johnson and Manuel Cardona

Synchrotron radiation (SR) derives its name from having been first observed in a type of electron accelerator called a synchrotron. It is the electromagnetic radiation which is emitted when charged particles moving at close to the velocity of light (i.e., relativistic velocities) are accelerated. For example, relativistic electrons passing through a uniform magnetic field move on a circular trajectory and emit SR due to the acceleration caused by the Lorentz force. Such radiation was first generated on earth as a by-product in high-energy accelerator rings; however, there are many astronomical sources of SR, e.g., the pulsar in the Crab nebula and SR emission from relativistic electrons in the galactic magnetic field.

The power radiated by a relativistic charged particle of mass m, with energy E, moving in a circle of radius R, is proportional to $(E/m)^4/R^2$. Thus, significant radiated power is only obtained for light particles (i.e., electrons or positrons) accelerated to high energies, typically greater than 500 MeV. In accelerator, or storage, rings the radiation is a result of the centripetal acceleration of the electrons or positrons in the powerful magnets used to guide them on a closed orbit. Viewed from alongside the ring, the charges oscillate to and fro in much the same way as those on a radio antenna. Since the particles whiz around the ring several million times a second, one might expect radiation at the frequency of the particle orbit, called the cyclotron frequency ω_0. An observer moving with the particle would, in fact, observe the classical dipole radiation pattern with this frequency as indicated in Fig. 1a. Since the charged particles in the synchrotron and storage rings have velocities close to that of light, an observer at rest in the laboratory frame of reference observes synchrotron radiation which is strikingly different from conventional dipole radiation. The Lorentz transformation of the radiation pattern results in a very narrow pencil of light, shown schematically in Fig. 1b, with a maximum in the forward direction tangential to the orbit and two side lobes which contain the backward radiation of Fig. 1a bent into the forward direction. The angular width of the forward beam is given approximately by $\Theta = \gamma^{-1}$, where $\gamma = E/mc^2$, the electron rest mass energy $mc^2 = 0.511$ MeV, so the beam is collimated vertically to typically 0.1–1 mrad. Since the emission takes place along a circular arc inside each of the bending magnets, a flat fan of radiation is emitted. The natural collimation of synchrotron radiation is one of its most useful characteristics which results in its high brightness when compared to conventional light sources which emit uniformly in all directions.

Even more remarkable than the transformation of the radiation pattern is the transformation of the frequency spectrum from the single cyclotron frequency into a quasicontinuum composed of a high number of harmonics of ω_0, extending from radio frequencies (ω_0) into the ultraviolet, x-ray, and even γ-ray regions. The transformation can be understood by considering the duration of the pulse of radiation seen by a stationary observer as a single electron flashes past. A short pulse contains harmonics extending to very high frequencies. The duration of the pulse is the difference between the time for the particle to move along the arc of the orbit viewed by the observer and the time for light to travel along the chord of the arc. For relativistic particles it is clear that the time interval will be very short and is, in fact, proportional to γ^{-3}. Thus, the spectrum extends up to a nominal "cut-off" frequency ω_c where $\omega_c \cong \omega_0\gamma^2 = c\gamma^3/R$.

Synchrotron radiation is produced by an electron interacting with a uniform magnetic field. Since this represents the simplest process of light emission by the interaction of matter with radiation, it is possible to calculate the energy distribution of the emitted radiation exactly. Thus, SR sources can be used as radiometric standards over a wide spectral range. The SR spectrum shown in Fig. 2 is completely characterized by the critical energy $\epsilon_c = 3\hbar c\gamma^3/2R$ or, equivalently, the critical wavelength $\lambda_c = 4\pi R/3\gamma^3$. Half the total power is radiated above the critical energy and half below. The maximum photon flux is obtained at about half the critical wavelength with a rapid fall-off on the high-energy side of the spectrum, but only a slow variation on the low-energy side. Synchrotron light extends from the far-infrared into the x-ray region in a smooth continuum.

The radiation emitted in the plane of orbit is linearly polarized in the plane of the ring. At angles above and below the plane of orbit there are contributions from the side lobes which are polarized perpendicular to the plane of the ring and phase shifted by 90° with respect to the central maximum. Thus, in general, the light is elliptically polarized with one of the axes of the polarization ellipse always in the orbital plane. The high degree of linear polarization makes SR invaluable for polarization-dependent studies in the vacuum ultraviolet and x-ray regions.

The radiation has a well-defined pulsed time structure which depends on the length of the electron bunches, the number of bunches, and the circumference of the ring. Since bunch lengths of a few centimeters can be achieved, the corresponding pulses last $\cong 10^{-10}$ s, and for a single orbiting bunch the repetition rate is the cyclotron frequency (MHz).

In circular high-energy accelerators the energy loss caused

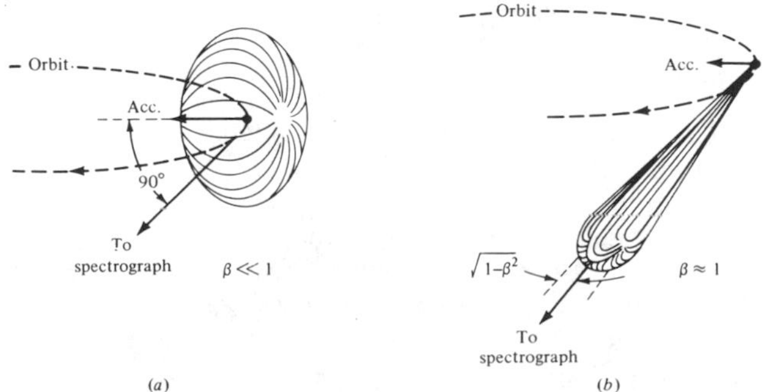

FIG. 1. Radiation pattern of an electron moving in a uniform magnetic field. The dipole pattern from a slow electron on a circular orbit (a) is distorted into a narrow cone in the instantaneous direction of motion for a relativistic electron moving with a velocity v close to c—the velocity of light [$\beta = v/c \cong 1$ and $\gamma = (1-\beta^2)^{-1/2}$].

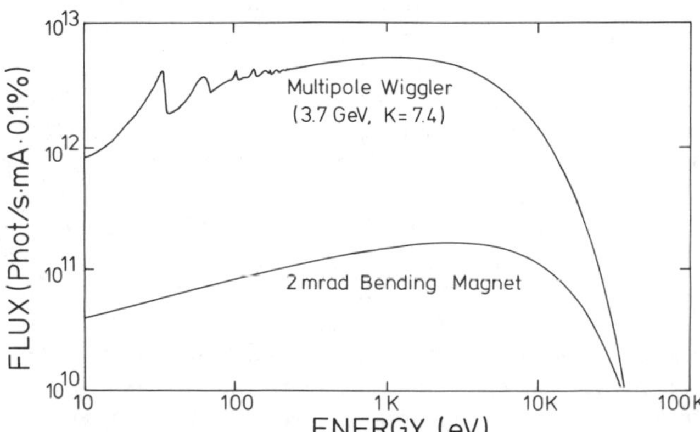

FIG. 2. Characteristic shapes of the synchrotron radiation spectra produced by a bending magnet and a multipole wiggler. The intensity from the wiggler is markedly higher than that from the bending magnet. At low photon energies the wiggler acts as an undulator producing even higher intensities.

by synchrotron radiation has to be compensated by the rf power introduced in accelerating cavities. In the higher-energy machines several megawatts are dissipated in the form of SR around the ring. These losses are reduced if the bending radius is large, so particle physicists have built larger and larger machines to minimize the SR losses, e.g., LEP at CERN in Geneva has a circumference of 27 km.

The excellent characteristics of synchrotron radiation for a wide range of studies became apparent in the early experiments performed parasitically on the elementary particle physics accelerators. The continuous nature of the SR spectrum coupled with a suitable monochromator with a narrow bandpass make SR unique for experiments requiring a continuously tunable source in the vacuum ultraviolet and x-ray regions. The natural collimation, polarization, and time structure are all useful characteristics for experiments in diverse fields. Early work was performed by atomic physicists and spectroscopists; as the instrumentation improved solid-state and surface scientists were able to perform absorption, reflection, luminescence, photoemission, and photodesorption experiments in the vacuum ultraviolet (VUV) region (2000–6000 Å). It is in the x-ray region where the unique properties of SR really stand out and there exists a large interdisciplinary community of biologists, physicists, and chemists which exploits the high brightness and tunable x rays for a wide range of structure determination techniques, e.g., protein crystallography and extended x-ray absorption fine-structure (EXAFS) measurements. Medical applications have been proposed and the technique of noninvasive coronary angiography, which exploits the contrast enhancement at the iodine *K*-absorption edge, is currently being tested. Biologists use small-angle-scattering techniques to study muscle fibers and the same method is exploited by polymer scientists. The techniques of soft x-ray microscopy are being developed which will enable cell biologists to study living cells in an aqueous environment. A lateral resolution of about 400 Å has already been achieved which is sufficient to glean valuable information. The technique of x-ray lithography has great potential for microfabrication and for the production of ultralarge-scale integrated circuits with linewidths of 0.25 μm and below. Prototype memory chips have already been made using x-ray lithography so industrial applications of the technique for electronic chip production is no longer just a remote possibility. Currently, synchrotron sources and associated instrumentation are being developed for use in the production environment.

The many and varied demands for synchrotron light could not be satisfied by parasitic use of the early machines, so rings were built specially designed as sources of synchrotron radiation. The earliest dedicated rings were small VUV machines with energies <500 MeV. In the mid 1970s it became clear that there was an overwhelming need for higher-energy dedicated rings for x-ray experiments. Such machines were proposed and funded in Britain, Japan, and the United States and became operational in the early 1980s. These machines were so superior to previous sources that the demand for SR continued to grow and the user community expanded enormously. The experiments became increasingly sophisticated and demanding with more stringent requirements on photon flux, spectral resolution, spatial resolution, and timing. At the present time, third-generation storage rings are being constructed to satisfy these demands.

Many of the operational synchrotron sources listed in Table I are second-generation storage rings which were optimized to provide high-brightness radiation from bending magnets. The third-generation machines are designed to produce significantly higher brightness from so-called insertion devices. The insertion devices are periodic arrays of magnets, as shown in Fig. 3, which produce no overall deviation of the charged particles, so they are inserted in straight sec-

Table I Storage Ring Synchrotron Radiation Sources (August 1989)

Location	Ring (Lab)	Electron Energy (GeV)	Notes
Brazil			
Campinas	LNLS-1	1.0	Dedicated*
Campinas	LNLS-2	2.0	Proposed
China (PRC)			
Beijing	BEPC (IHEP)	2.2–2.8	Partly dedicated
Hefei	HESYRL (USTC)	0.8	Dedicated
China (ROC-Taiwan)			
Hsinchu	SRRC (SYNC. RAD. RES. CTR.)	1.3	Dedicated*
Denmark			
Aarhus	ISA	0.6	Partly Dedicated*
England			
Daresbury	SRS (DARESBURY)	2.0	Dedicated
France			
Grenoble	ESRF	6.0	Dedicated*
Orsay	DCI (LURE)	1.8	Dedicated
	SUPERACO (LURE)	0.8	Dedicated
Germany			
Bonn	ELSA	3.5	Partly dedicated
Dortmund	DELTA	1.5	Design/FEL use
Hamburg	DORIS II (HASYLAB)	3.5–5.5	Partly dedicated
West Berlin	BESSY	0.8	Dedicated
	BESSY II	1.5–2.01	Design/dedicated

Table I (*continued*)

Location	Ring (Lab)	Electron Energy (GeV)	Notes
India			
Indore	INDUS-I (CAT)	0.45	Design/dedicated*
	INDUS-II (CAT)	2.0	Design/dedicated*
Italy			
Frascati	ADONE (LNF)	1.5	Partly dedicated
Trieste	ELETTRA	1.5–2.0	Dedicated*
Japan			
Kansai Area	STA RING	8.0	Design/dedicated
Okasaki	UVSOR (IMS)	0.75	Dedicated
Tokyo	SOR-RING (ISSP)	0.4	Dedicated
Tsukuba	TERAS (ETL)	0.6	Dedicated
	PHOTON FACTORY (KEK)	2.5	Dedicated
	ACCUMULATOR RING (KEK)	6.0–8.0	Partly dedicated
	TRISTAN MAIN RING (KEK)	8.0–30.0	Planned use
Korea			
Pohang	POHANG LIGHT SOURCE	2.0	Design/dedicated*
Netherlands			
Eindhoven	EUTERPE	0.4	Partly dedicated*
Sweden			
Lund	MAX (LTH)	0.55	Dedicated
USA			
Argonne, IL	APS (ANL)	7.0	Design/dedicated
Baton Rouge, LA	LSU	1.4	Design/dedicated*
Berkeley, CA	ALS (LBL)	1.5	Dedicated*
Gaithersburg, MD	SURF II (NIST)	0.28	Dedicated
Ithaca, NY	CESR (CHESS)	5.5–8.0	Partly dedicated
Stanford, CA	SPEAR (SSRL)	3.0–3.5	Partly dedicated
	PEP (SSRL)	5.0–15.0	Partly dedicated
Stoughton, WI	ALADDIN (SRC)	0.8–1.0	Dedicated
Upton, NY	NSLS I (BNL)	0.75	Dedicated
	NSLS II (BNL)	2.5	Dedicated
USSR			
Karkhov	N-100 (KPI)	0.10	Dedicated
Moscow	SIBERIA I (Kurchatov)	0.45	Dedicated
	SIBERIA II (Kurchatov)	2.5	Dedicated*
Novosibirsk	VEPP-2M (INP)	0.7	Partly dedicated
	VEPP-3 (INP)	2.2	Partly dedicated
	VEPP-4 (INP)	5.0–7.0	Partly dedicated

* In construction or approved for construction as of August 1989.
Compiled by Herman Winick, SSRL.

tions in the machine. Although the idea of using such periodic magnet arrays to make the electron beam snake to and fro was suggested by Ginzburg in 1947, it is only recently that suitable magnetic materials (e.g., $SmCo_5$, Sm_2Co_{17}, or $Nd_2Fe_{14}B$) for short-period devices have become available. The advantage of the insertion devices is the higher intensity produced on axis due to the superposition of the radiation from all of the poles. If the magnetic field is strong the deviation angle of the electrons will be large compared to the emission cone of the radiation and there will be no coherence between the radiation emitted from different poles. This is the case for wiggler devices, where the gain in intensity is proportional to the number of periods N and the spectrum retains the characteristic shape shown in Fig. 2. In undulators, the magnetic field is weak, so the deviation of the electron trajectory is small compared to the emission cone of SR and the amplitudes from the different poles add coherently so that interference effects dominate the emitted spectrum.

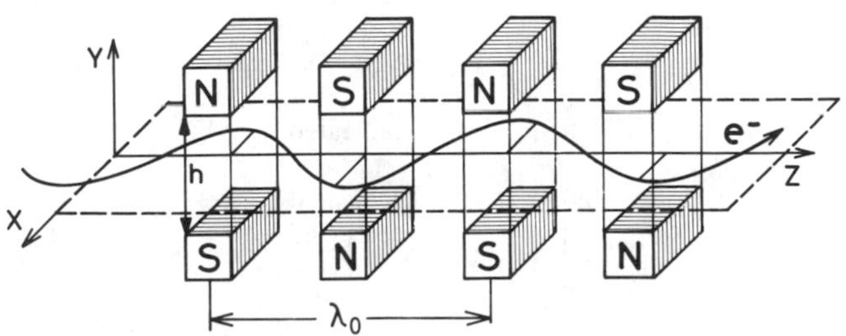

FIG. 3. Schematic of the planar magnet structure used in wigglers and undulators consisting of N periods of length λ_0. Helical and crossed-field magnetic structures can be used to produce circularly polarized light.

Insertion devices are characterized by the magnetic period length λ_0 and the magnetic field strength which can be conveniently described by the dimensionless wiggler parameter K, where $K = e\lambda_0 B_0/2\pi mc$ (e and m are the electron charge and mass and B_0 is the amplitude of the periodic magnetic field on axis). The periodicity of the magnet structure leads to a corresponding periodicity in the emitted electromagnetic radiation. In a frame of reference moving along the z axis with the relativistic electron, the magnet structure is Lorentz contracted and changed into a very strong electromagnetic wave which causes the electrons to move in a figure-eight trajectory. For small K a dipole radiation pattern is obtained which when transformed to the lab frame of reference produces the first undulator maximum. As the K value is increased, the radiation patterns become more complex and significant intensity is generated in higher-order harmonics. For a symmetric device, only odd harmonics n are obtained on axis at wavelengths given by

$$\lambda_n = \frac{1}{n}\frac{\lambda_0}{2\gamma^2}\left(1 + \frac{K^2}{2}\right), \qquad n = 1, 3, 5, \ldots$$

or corresponding photon energies

$$E_n = n\frac{hc}{\lambda_0}\frac{2\gamma^2}{1 + K^2/2}.$$

The width of the undulator maximum is proportional to $1/nN$ and the position of the maximum can be changed by altering K through variation of the gap between the poles of the permanent magnets. For a well-collimated electron beam, the increase in spectral brilliance is proportional to N^2.

The maximum deflection angle of the electron trajectory is given by K/γ and the angular opening of the radiation cone is $\cong \gamma^{-1}$, so one can classify undulators as devices with $K \leq 1$ and wigglers with $K \gg 1$. The total radiated power is proportional to K^2, so wigglers are high-power broad-band devices, whereas undulators provide a quasimonochromatic beam in a very small solid angle. The high brightness and coherence of undulator radiation makes it particularly suitable for x-ray microscopy and x-ray holography as well as a wide range of other applications. It is important to note that the amplitude of the excursions of the electron beam in an undulator is typically only 10 μm which means that optimum performance of the undulator can only be obtained if the electron beam has extremely small diameter and divergence, and is very stable. The third-generation machines are now being designed and constructed to meet these requirements at Argonne (APS), Berkeley (ALS), Berlin (BESSY II), Grenoble (ESRF), Trieste (ELETTRA), and in Japan (Kansai Area).

See also BREMSSTRAHLUNG; POLARIZED LIGHT; SYNCHROTRON.

BIBLIOGRAPHY

C. Kunz (ed.), *Synchrotron Radiation—Techniques and Applications*, Topics in Current Physics, Vol. 10. Springer-Verlag, Berlin, Heidelberg, New York, 1979.

H. Winick and S. Doniach (eds.), *Synchrotron Radiation Research*. Plenum Press, New York, 1980.

Handbook on Synchrotron Radiation

Vol. 1, E.-E. Koch (ed.), North-Holland, Amsterdam, 1983;
Vol. 2, G. V. Marr (ed.), North-Holland, Amsterdam, 1987;
Vol. 3, D. Moncton (ed.), to be published.

Tachyons

G. Feinberg

Tachyons are hypothetical objects whose speed is always greater than that of light, $c=3\times10^5$ km/s. One way to see how the possibility of tachyons arises is to consider the expressions for energy and momentum given by special relativity, viz.,

$$E=\frac{m_0c^2}{(1-V^2/c^2)^{1/2}},$$

$$\mathbf{p}=\frac{m_0\mathbf{V}}{(1-V^2/c^2)^{1/2}}.$$

Upon taking $V>c$ and m_0 to be an imaginary number, say $i\mu$, it is found that E and $\mathbf{p}$ are real, as required for measurable quantities. Hence tachyons may be thought of as objects with imaginary rest mass m_0. Since V/c always remains greater than 1, the rest mass is not directly measurable, and an imaginary value for it creates no difficulty.

Several properties of tachyons follow directly from these equations. For example, the energy of a tachyon decreases as its speed increases, approaching zero as the speed becomes infinite. On the other hand, the momentum of a tachyon cannot vanish, and its minimum value is given by μc, again at infinite speed. The quantity $E^2-p^2c^2$ is a constant and invariant quantity, for tachyons as for ordinary particles, but for tachyons its value is negative $(=-\mu^2c^4)$.

Another novel property of tachyons concerns the manner in which some of their attributes vary when viewed by different observers in relative motion. Consider a tachyon produced at some point in a space A at some time, carrying a positive energy, and absorbed at another point B, at a later time. As viewed by an observer in motion (with $u<c$!), this process can instead appear as the production of an antitachyon at B, followed at a later time by the absorption of the antitachyon at A. This interchange of emission and absorbtion can only be carried through in a quantum theory of tachyons, where it is implemented by allowing the exchange of some creation operators with annihilation operators under Lorentz transformations. This possibility, with the resultant change of the tachyon number densities from observer to observer, allows for the elimination of various paradoxes that would otherwise beset a description of tachyons.

In some of the recent string theories of subatomic particles, tachyon states occur naturally. However, little effort has been made to interpret these states physically. Instead, attention has focused on string theories in which tachyon states are absent.

Various experimental searches have made for tachyons, in which some of their novel properties are the focus of the search. Electrically charged tachyons could be produced in pairs by gamma rays, and would emit Čerenkov radiation in a vacuum, because they satisfy the condition ($V>c$/index of refraction) there. A search for particles with this property gave no indication that they exist. A search for neutral tachyons, produced in hadron collisions, used the fact that $E^2-p^2c^2$ would be negative for tachyons. In a sample of about 5000 neutral particles produced in K-meson–proton scattering, and in proton–antiproton annihilations, no particles with this property were found, indicating a low or vanishing rate for tachyon production. Other searches have also not detected tachyons, and various indirect arguments have been given that imply that if they exist, their interactions with ordinary particles are very weak. Therefore, at present, tachyons remain a hypothetical possibility, with no indication that nature has utilized that possibility.

See also RELATIVITY, SPECIAL THEORY; ELEMENTARY PARTICLES IN PHYSICS.

Temperature

Robert Weinstock

Temperature as a scientific concept is a quantitative refinement and generalization of the universal human experience of hotness and coldness. Assignment of numerical values to such subjective conditions as "cold," "cool," "lukewarm," etc., is possible because objects undergo measurable quantitative changes as they pass from one of these subjectively perceived conditions to another. A metallic wire, for example, normally exhibits an increased electrical resistance as it becomes "hotter"; a gas confined at constant pressure expands as it gets hotter, contracts as it grows cooler. Based on the latter phenomenon, the first temperature-measuring devices we know of were fashioned in the early seventeenth century, with air the confined gas: To different volumes assumed by the body of air in its different conditions of hotness and coldness the instrument maker would assign, almost arbitrarily, different "degrees of temperature." Such devices were called thermometers as early as 1626.

The liquid-in-sealed-glass-tube thermometer common today was probably invented by Ferdinand II, Grand Duke of Tuscany, around 1654. Temperature measurements by it are based on the propensity of liquids, like gases and solids, to expand/contract, normally, as they become hotter/colder.

Scientific use of temperature is subject to two obvious requirements: (1) universal agreement on a temperature scale, with well-defined means for translating readings from one to another of several such scales; and (2) technology sufficiently refined to produce thermometers that give identical readings when subjected to identical conditions. The early eighteenth century saw significant progress toward achievement of both requirements: Wide adoption of the Fahrenheit scale (°F) was promoted by reliable mercury-in-glass thermometers manufactured in Amsterdam by Danzig-born craftsman Daniel Gabriel Fahrenheit from about 1717 until his death in 1736, and by his direct successors for some years thereafter. Having obtained his first serious thermometric instruction and early inspiration directly from Ole Roemer—professor of astronomy at Copenhagen, first to establish the finiteness of the speed of light—Fahrenheit evidently based his scale on two "fixed points": the melting of ice ("32 degrees") and "the healthy human body" ("96 degrees"). Soon after his death the higher fixed point was replaced by the boiling point of water ("212 degrees"). Intermediate Fahrenheit temperatures were defined by the length of the mercury column through linear interpolation; linear extrapolation defined temperatures as low as zero, as high as 600 degrees.

By late 1741 Anders Celsius, professor of astronomy at Uppsala, had introduced a temperature scale on which the "degree" was one hundredth of the interval between the freezing of water and its boiling. Celsius designated boiling temperature as 0, freezing as 100; the reversal that led to the Celsius (formerly centigrade) scale of today (°C) was accomplished soon after Celsius's death in 1744.

With the significant exception of Robert Hooke's generally ignored ideas on the subject, early speculation on the intrinsic meaning of temperature was wildly off; yet it served to sharpen focus on the quest for an "absolute" temperature scale—one based, that is, on principles and phenomena independent of special properties of particular substances used in thermometric instruments. The quest met success in 1848, finally, when William Thomson (later Lord Kelvin) proposed a universal temperature scale based on the ideal reversible heat engine introduced and analyzed in 1824 by Sadi Carnot. The Carnot engine—a theoretical entity, one should keep in mind—operates cyclically between two fixed temperatures, absorbing in each cycle amount Q_1, say, of energy (heat) from a reservoir maintained at hotter temperature t_1, and yielding an amount Q_2 of energy (heat) to a reservoir maintained at cooler temperature t_2; during each cycle the engine "converts" the difference $(Q_1 - Q_2)$ into an equal amount of mechanical work that it performs externally. Using his remarkable intuitive understanding of thermodynamic phenomena that anticipated by a quarter-century formal enunciation of the first and second laws of thermodynamics, Carnot demonstrated in 1824 that the ratio Q_1/Q_2 depends only on the respective reservoir temperatures t_1 and t_2—independently of the engine's particular working substance. More precisely, $Q_1/Q_2 = f(t_1)/f(t_2)$, where f is a function whose form depends only on the temperature scale on which t_1 and t_2 are measured. The final form of Thomson's idea was to choose as the "absolute temperature" scale one for which $f(T) = T$ for all temperatures T measured on it. Thus, if the ideal Carnot engine operates between absolute temperatures T and T_0, where T_0 is arbitrarily assigned to a particular reproducible condition (e.g., the melting point of ice at 1 atmosphere of pressure), then $T = (Q/Q_0)T_0$, where Q/Q_0 is the amount of heat the engine must absorb from source at T in order to yield one unit of heat to reservoir at T_0. Thomson's proposal was promptly accepted; by 1887 it was the subject of international agreement. Since 1954, the universally accepted absolute scale—Kelvin temperature (K)—has been Thomson's with $T_0 = 273.16$ K assigned to the "triple point" of water—i.e., to the condition in which liquid water, its solid phase (ice), and its vapor can remain isolated together without macroscopic changes in the three amounts present. (An equivalent definition of the Kelvin scale is based directly on the second law of thermodynamics and the concomitant entropy concept.)

Since the Kelvin scale is based on a theoretical entity, its utility depends on the existence of actual temperature scales that coincide with it, within measurement precision, over appreciable ranges. One such is related to the Kelvin scale through mediation of another theoretical entity, the ideal gas: At constant volume an ideal gas exerts pressure proportional to its Kelvin temperature; fundamental understanding of the material constituency of real gases, well buttressed by empirical evidence, enforces conviction that ideal-gas behavior is approximated arbitrarily closely by that of a real gas at sufficiently low density and pressure. Indeed, the 1887 international agreement designated the constant-volume hydrogen-gas thermometer (appropriately extrapolated to zero density) as *defining* measurements on the Kelvin scale—a convention that persists today in all its essentials.

Another theoretical entity involved in scientific use of the Kelvin scale is a hypothetical body that absorbs all electromagnetic radiation impinging upon it: a blackbody. A blackbody maintained at any temperature radiates electromagnetic energy distributed among wavelengths as described by a universal formula—first deduced by Max Planck in 1900—in terms of wavelength, Kelvin temperature, and certain physical constants. There is sufficient empirical evidence to indicate that, in rather broad circumstances, the electromagnetic spectrum radiated by an actual hot body is close enough to the blackbody ideal to enable one to infer the body's temperature from intensity-versus-wavelength measurements on its radiation. At temperatures above 1337.58 K, this fact served to define, via Planck's formula, the Kelvin temperature of a very hot body, by international agreement since 1948. Extension of the use of Planck's law down to 773.15 K dates from 1 January 1990.

In order to define the temperature of a material system too "cold" for direct gas-thermometer use, one must understand the submicroscopic nature of the system sufficiently well to establish, on firm theoretical grounds, a mathematical relationship between measurable parameters and the system's Kelvin temperature. Temperatures have been measured in this way below 0.001 K.

The actual measurement of temperature (with which we are not particularly concerned here) is accomplished, at all but the very lowest temperatures—down to 13.81 K by international agreement since 1968 and to 0.65 K after 1 January 1990—by means of precisely specified instruments such as resistance thermometers, thermocouples, and optical pyrometers. Each of these has been in principle calibrated against a defining scale with the aid of a number of fixed points: the triple point of hydrogen (13.81 K), boiling point of neon (27.102 K), triple point of water (273.16 K = 0.01°C), freezing point of gold (1337.58 K), and several others. In the lowest range use is made of vapor–pressure measurements on helium (both ^{4}He and ^{3}He).

While one might infer from the foregoing that "temperature" is merely "that which the thermometer measures," a more intuitive realization of the concept has developed along with our understanding of submicroscopic phenomena and knowledge of fundamental laws of nature during the past century: What the thermometer measures, roughly speaking, is an average mechanical energy of the particles constituting the system under study—which is essentially the view, expressed even more roughly, published by Robert Hooke in his *Micrographia* in 1665. For a particular system this rough formulation can be refined with a quantitative precision that is the greater the simpler the system and the more completely we understand its submicroscopic constitution. The Kelvin temperature of an ideal monatomic gas (thus of a real monatomic gas at sufficiently low density and pressure), for example, is given by $T = (2/3k)\overline{E}$, where $\overline{E}$ is the mean kinetic energy per atom and k the universal Boltzmann's constant. More complicated relations hold for more complicated systems; they all support the generalization that higher temperature is a macroscopic manifestation of more energetic molecular and submolecular motion. This generalization is for many scientists the essence of the temperature concept.

See also BLACKBODY RADIATION; ENTROPY; EQUATIONS OF STATE; HEAT ENGINES; KINETIC THEORY; MAGNETIC COOLING; THERMODYNAMICS, EQUILIBRIUM; THERMOMETRY.

BIBLIOGRAPHY

Herbert Arthur Klein, *The Science of Measurement*. Dover, New York, 1988. (E)

Leeds and Northrup Tech. J., Spring (No. 6, 1969). Leeds and Northrup Co., North Wales, PA. (A)

W. E. Knowles Middleton, *A History of the Thermometer and Its Use in Meteorology*. Johns Hopkins, Baltimore, 1966. (E)

T. J. Quinn, *Temperature*. Academic Press, London, 1982. (I)

Robert Resnick and David Halliday, *Physics*, 2nd ed., Chap. 21. Wiley, New York, 1966. (E)

Mark W. Zemansky, *Heat and Thermodynamics* 5th ed., Chap. 1. McGraw-Hill, New York, 1957. (I)

Tensors *see* Vector and Tensor Analysis

Thermal Analysis

Paul D. Garn

Thermal analysis is a collective term describing techniques in which measurement of some property of the sample is performed continuously as the sample is being heated or cooled in a regular manner. The principal thermoanalytical methods are differential thermal analysis (DTA), differential scanning calorimetry (DSC), thermogravimetry (TG), evolved-gas detection (EGD), and evolved-gas analysis (EGA); but other techniques such as monitoring x-ray diffraction intensity, electrical resistivity, or thermal expansion are also practiced. An older use of the term referred exclusively to cooling curves (of temperature versus time).

Differential thermal analysis comprises the measurement of the difference in temperature between the sample and a reference material while they are being heated simultaneously either in separate wells of a block or in contiguous sample holders. So long as no physical or chemical change is taking place, the temperature difference (ΔT) is small and nearly constant (the base line). However, when the sample reaches a temperature at which an endothermic process such as melting, decomposition, or a solid–solid transition takes place, the thermal energy must not only heat the sample but also carry out the change in state. As a result, the sample will not be heated as rapidly as the reference and a temperature difference appears. When the process is completed, the temperature difference will approach zero again because the sample and reference specimens are in the same thermal environment. Similarly, if a process gives up heat, as in crystallization or oxidation, the sample will gain in temperature—compared to the reference—during the process. The temperature difference will again approach zero after the process is complete. (See Fig. 1.)

The return of the signal to a stable base line makes possible the observation of a second, third, etc., process with high sensitivity. For some materials, quantities as small as a few micrograms can be detected. DTA, like the other thermoanalytical methods, is specific for *compounds* as compared to the elements in the compound. By selection of furnace and sample-holder types, the technique can be made quantitative, in most cases to <5%, in some cases to about 1% for moderate sample sizes (5–100 mg). Other sample-holder types may be used to improve resolution of close-lying peaks.

The exact form of the observed peak and the relation of the measured temperature to the equilibrium temperatures of reversible processes are somewhat apparatus dependent but nevertheless highly reproducible. Reversible decompositions will be atmosphere dependent, and so information

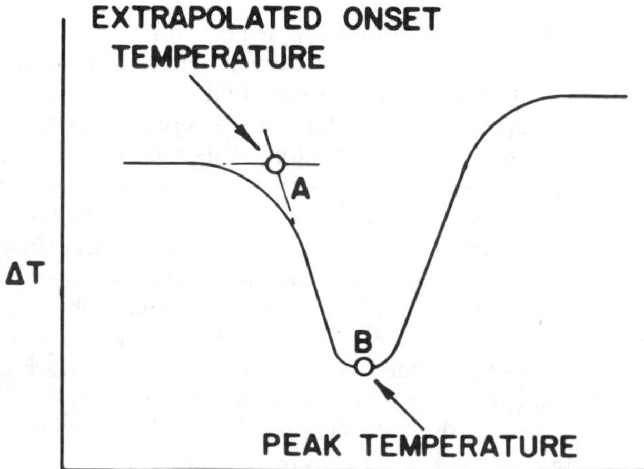

FIG. 1. DTA curve with single peak, showing the common measured points. For some materials as many as seven peaks may be observed.

can be gained from changes in atmosphere and/or pressure. Also atmosphere control can be used to simulate larger-scale processes, so that thermal reactions can be studied in some detail before scale-up.

An abrupt change in heat content or enthalpy is most easily detected and measured. Nonetheless, the change in heat capacity at a second-order or a glass transition can be detected as a change in slope of the base line or as a shift of the base line to a slightly different value.

A closely related technique is differential scanning calorimetry, in which the same phenomena are detected but by a different measurement. A steadily increasing flow of heat is supplied to the sample and reference specimens identically and the temperature-difference signal between points close to each is fed into an amplifier, which supplies extra energy (from a separate heater) to maintain the difference near zero. This amplifier output is the signal displayed as the heat input in millicalories per second—as compared to the ΔT in DTA. DSC instruments *are* designed to be accurately quantitative; DTA instruments *may* be designed to be accurately quantitative.

Thermogravimetry is the measurement of weight during the heating (and sometimes cooling) of the specimen. The specimen is generally hung from or supported by a balance, but springs are also used.

The TG measurement is typically weight versus time with a simultaneous recording of temperature, but it may also be plotted directly as weight versus temperature. Through electronic differentiation of the weight with respect to time, a rate of weight change can also be plotted.

TG can provide a better measure of the quantity of a material than can DTA or DSC, but only if the weight change from its vaporization, decomposition, or reaction is completely resolved from any other weight change. It can be used to measure very slow reactions and to ascertain the products of a reaction by the relative amount of weight gained or lost. (See Fig. 2.)

These techniques may be carried out in ambient air, in vacuum, or under controlled pressures of a known atmosphere. TG and DTA may be performed simultaneously on the same specimen. Each of the techniques can be combined with EGA or EGD. There is no demonstrable superiority in combined techniques but there may be practical advantages if the amount of sample is limited.

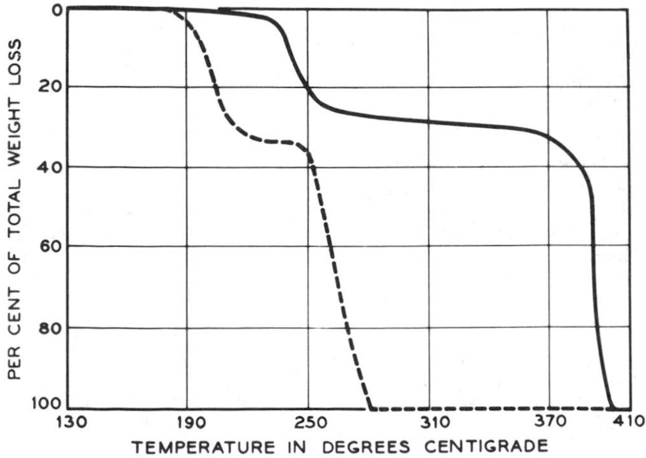

FIG. 2. Thermogravimetric curves of cobalt(II) oxalate dihydrate in an open - - - and a closed but not sealed ——, vessel. The exposure to the atmosphere in the open vessel allows easy escape of gaseous products and entry of oxygen. The reactions proceed differently.

Evolved-gas detection comprises the measurement of changes of composition of a stream of gas that flows through, over, or around the specimen. Typical gas-chromatographic detectors are used—for example, the thermal conductivity, gas density, or flame ionization detectors. The gas is swept through the sample chamber (or DTA, DSC, or TG furnace assembly) into the detector, whose response is plotted against time (with a simultaneous temperature plot) or directly against temperature. If there is a single gaseous product in the detector at any one time and the response of the detector to that product is known, a time-based plot can be used for quantitative measurement (EGA).

Evolved-gas analysis provides identification and/or quantitative measurement of the gaseous products. When two or more gaseous products are being evolved, either separation (by gas chromatography, e.g.) or selective measurement (by mass spectrometry, e.g.) can be used.

The International Confederation for Thermal Analysis has published recommended nomenclature [1–3] and reporting practices [4–6] and has certified reference materials (distributed by the U.S. National Bureau of Standards) for use with dynamic techniques.

Much of the recent commercial instrumentation makes use of computer-based data acquisition, treatment, and presentation. Digital data acquisition enables modification of the presentation mode (printout) of the stored data so that inspection of particular temperature regions can be done with enhanced sensitivity when needed. Comparison of records is easy. Mathematical operation can be used for smoothing or base-line correction.

The ease of mathematical manipulation of data has enabled almost indiscriminate application of arbitrary kinetic equations to thermoanalytical data. Their applicability to crystalline solids is questionable, but many noncrystalline samples may be describable. There are simple tests [7,8] that can be used to guard against misapplications of a kinetic equation.

Another benefit of advancing technology is the ability to use infrared spectroscopy with thermal methods. The development of very sensitive detectors, along with the small high-speed computer, enables Fourier transform spectroscopy in short enough times to produce data that can be correlated well with the thermal data. The spectra of the specimen may be obtained as the temperature is varied or the spectra of gaseous products (generally from another technique such as DTA or TG) obtained.

See also CALORIMETRY; PHASE TRANSITIONS; THERMAL EXPANSION.

REFERENCES

1. R. F. Schwenker, Jr., and P. D. Garn, eds., *Thermal Analysis: Proceedings of the Second International Conference* pp. 683–691. Academic Press, New York, 1969.

2. *Talanta* **16,** 1227 (1969).
3. *J. Thermal Anal.* **4,** 343–347 (1972).
4. *Analyt. Chem.* **39,** 543 (1967).
5. *Analyt. Chem.* **44,** 640–641 (1972).
6. *Analyt. Chem.* **46,** 1146–1147 (1974).
7. *J. Thermal Anal.* **13,** 581–593 (1978).
8. *Thermochimica Acta* **135,** 71–77 (1988).

BIBLIOGRAPHY

Techniques and Instrumentation

Paul D. Garn, *Thermoanalytical Methods of Investigation,* Academic Press, New York, 1965.

Current Development and Applications

Analytical Chemistry, Reviews Issue. Thermal analysis is reviewed in alternate years.

Thermal Expansion

Richard K. Kirby

When the temperature of a substance is increased, the volume that it occupies is usually increased. A measure of this change is the coefficient of thermal expansion,

$$\beta = \frac{1}{V}\left(\frac{\partial V}{\partial T}\right)_p$$

where V is the volume at temperature T and constant pressure P. In the case of ideal gases the volume at constant pressure is proportional to the average translation kinetic energy per particle, $V \propto \frac{1}{2}mv^2 = \frac{3}{2}kT$, so that the value of β is equal to T^{-1}. At room temperature and atmospheric pressure $\beta \sim 3400 \times 10^{-6}\ \mathrm{K}^{-1}$ for nearly all gases. Because of the short-range structural ordering that takes place in liquids their thermophysical behavior is more complex. As a result the temperature dependence of β usually takes the empirical form $\beta = a + bT + cT^2$. Three examples are given in Table I.

In solids practically all of the thermal energy goes into the motion of atoms around their equilibrium positions in the structure and the thermal expansion of that structure is the result of these vibrations. To account for this expansion, however, the thermal vibrations of the atoms about their equilibrium positions must be assumed to be anharmonic. Qualitatively, the time-average atomic position in a harmonic potential is independent of the amplitude of the oscillation, which is related to the temperature; an asymmetric potential is required to cause a shift in the average position of an atom as the temperature changes. The structure can expand differently in different directions depending on the type of crystalline structure. The coefficient of linear thermal expansion in the ith direction is defined as

$$\alpha_i = \frac{1}{L_i}\left(\frac{\partial L_i}{\partial T}\right)_p .$$

The sum of three orthogonal linear coefficients of a crystal is equal to its volume coefficient: $\beta = \alpha_1 + \alpha_2 + \alpha_3$. Some examples are given in Table II.

The temperature dependence of β for solids is very complex but can be approximated by the Grüneisen equation, $\beta = \gamma C_p / VB_s$, where γ is the Grüneisen parameter, C_p is the heat capacity at constant pressure, and B_s is the adiabatic bulk modulus. Since the product γ / VB_s is nearly constant over a wide temperature range, the value of β has about the same temperature dependence as C_p; that is, at low temperatures $\beta \to 0$ as $T \to 0$ and at high temperatures β is slowly increasing. It has been found that most metals expand about 7% on heating from 0 K to their melting temperature. Therefore, a metal with a low melting point T_M has a large coefficient of thermal expansion, and vice versa; see Table III.

In the case of metals at $T < 20$ K the heat capacity can be expressed as $C_p = a_{el}T + a_{lat}T^3$, where the two terms indicate the contributions of the conduction electrons and the lattice, respectively. Through Grüneisen's equation the thermal expansion can be expressed in a similar manner:

$$\beta = \frac{1}{VB_s}(\gamma_{el}\alpha_{el}T + \gamma_{lat}\alpha_{lat}T^3).$$

High-sensitivity methods with which β can be determined with a precision of $10^{-10}\ \mathrm{K}^{-1}$ can be used to measure these

Table I Coefficients of thermal expansion of liquids

Temperature (K)	Water ($\times 10^{-6}\ \mathrm{K}^{-1}$)	Mercury ($\times 10^{-6}\ \mathrm{K}^{-1}$)	Pentane ($\times 10^{-6}\ \mathrm{K}^{-1}$)
273	−64	181.7	1465
283	+85	181.8	1530
293	194	181.8	1610
303	263	181.9	1695

Table II Coefficients of linear thermal expansion of several solids

		200 K	293 K	400 K
NaCl (fcc)	$\alpha_1 = \alpha_2 = \alpha_3$	35.3	39.7	43.0
Fe (bcc)	$\alpha_1 = \alpha_2 = \alpha_3$	10.1	11.8	13.2
SiO_2 (hex)	$\alpha_1 = \alpha_2$	4.9	7.4	8.8
	α_3	10.3	13.6	15.6
In (tetr)	$\alpha_1 = \alpha_2$	39.5	52.9	77.3
	α_3	5.6	−9.6	−34.8
U (orth)	α_1	22	23	24
	α_2	1	0	−1
	α_3	17	19	22
SiO_2 (glass)	$\alpha_{isotropic}$	−0.13	+0.41	0.55
Teflon (polymer)	$\alpha_{isotropic}$	85	>500[a]	150

[a] Phase transition at 293 K causes the coefficient to be very large over about 10 degrees. All values should be multiplied by $10^{-6}\ \mathrm{K}^{-1}$.

Table III Relationship between thermal expansion and melting points

Metal	T_M (K)	β at 293 K ($\times 10^{-6}\ \mathrm{K}^{-1}$)
Lead	601	86.1
Copper	1357	50.1
Platinum	2042	26.7
Tungsten	3653	13.5

contributions to thermal expansion as well as those caused by other physical interactions. These include magnetic interactions, the splitting of molecular energy levels, and the resonant tunneling between equilibrium positions in the structure.

The combination of optical and x-ray diffraction techniques to measure thermal expansion has been used to determine the number of thermal vacancies that are generated at temperatures approaching T_M. The number of these vacancies is proportional to the difference between the linear coefficients of the bulk material and the crystal lattice at the same temperature. In the case of copper at the melting point this difference is 1.5×10^{-6} K^{-1}.

A variety of high-speed and noncontact measurement techniques have been employed to investigate the thermal expansion of molten materials at high temperatures. Rapid resistive self-heating of specimens and a variety of diagnostics, such as high-speed optical pyrometry, laser interferometry, and laser photography, have been used to determine the expansion in both solid and liquid phases as the material is heated through melting. Many metals exhibit $\beta \sim 10^{-4}$ K^{-1} in the liquid phase just above the melting point.

See also CERAMICS; METALS.

BIBLIOGRAPHY

Theory

J. F. Nye, *Physical Properties of Crystals,* p. 322. Oxford, London and New York, 1960.

J. G. Collins and G. K. White, "Thermal Expansion of Solids," in *Progress in Low Temperature Physics,* Vol. IV, p. 450. 1964.

T. H. K. Barron and R. W. Munn, "Analysis of the Thermal Expansion of Anisotropic Solids, Application to Zinc," *Phil. Mag.* **15,** 85 (1967).

F. W. Sheard, "Calculation of the Thermal Expansion of Solids from the Third-Order Elastic Constants," *Phil. Mag.* **3,** 1381 (1958).

G. K. White, "Thermal Expansion at Low Temperatures," *Nature* **187,** 927 (1960).

G. K. White, "Thermal Expansion of Vitreous Silica at Low Temperatures," *Phys. Rev. Lett.* **34,** 204 (1975).

W. E. Schoknecht and R. O. Simmons, "Thermal Vacancies and Thermal Expansion," *Proc. 3d AIP Conf. on Thermal Expansion,* p. 169 (1972).

R. S. Krishnan, R. Srinivasan, and S. Devanarayanan, *Thermal Expansion of Crystals.* Pergamon Press, New York, 1979.

Measurement Techniques

R. K. Kirby, "Thermal Expansion of Ceramics, Mechanical and Thermal Properties of Ceramics," NBS Spec. Publ. 303, p. 41 (1969).

T. A. Hahn and R. K. Kirby, "Thermal Expansion of a Borosilicate Glass from 80 to 680 K—Standard Reference Material 731," *Proc. AIP Conf. on Thermal Expansion* **17,** 93 (1974).

G. R. Gathers, J. W. Shaner, and R. L. Brier, "Improved Apparatus for Thermophysical Measurements on Liquid Metals up to 8000 K," *Rev. Sci. Instrum.* **47,** 471 (1976).

W. D. Drotning, "Thermal Expansion of Iron, Cobalt, Nickel, and Copper at Temperatures up to 600 K above melting," *High Temp.-High Pres.* **13,** 441 (1981).

A. P. Miller and A. Cezairliyan, "Transient Interferometric Technique for Measuring Thermal Expansion at High Temperatures: Thermal Expansion of Tantalum in the Range 1500–3200 K," *Int. J. Thermophys.* **3,** 259 (1982).

Data and General Techniques

Thermophysical Properties of Matter (The TPRC Data Series), Vol. 12, *Thermal Expansion of Metallic Elements and Alloys;* Vol. 13, *Thermal Expansion of Nonmetallic Solids.* Plenum, New York, 1976.

R. S. Krishnan, R. Srinivasan, and S. Devanarayanan, *Thermal Expansion of Crystals.* Pergamon Press, New York, 1979.

Journal

International Journal of Thermophysics (A. Cezairliyan, ed.) Plenum Press, New York.

Symposia Proceedings

Proceedings of the International Thermal Expansion Symposium, 1970–1988.

Thermionic Emission

I. J. D'Haenens

Thermionic emission is a term used to describe the emission of electrons (and/or ions) from a solid when it is heated in vacuum. Thermionic emission can be viewed as an evaporation process, the temperature dependence of which is accurately described by the Richardson equation:

$$j_R = A(1-r)T^2 \exp(-e\phi/kT), \tag{1}$$

where j_R is the current density of emitted electrons, A is the emission constant, $1-r$ is the transmission coefficient of the surface barrier for electrons, T is the temperature, and $e\phi$ is the work function, which can be interpreted as an electron latent heat of vaporization.

The Richardson equation is the fundamental equation of thermionic emission, and can be derived from thermodynamic [1] or quantum statistical [2] considerations of the electrons in the metal. The quantum statistical derivation of the Richardson equation proceeds from a consideration of the thermal population of the electronic states in the metal. The metal is represented in the Sommerfeld approximation, which assumes a boxlike potential well in which the barrier is formed by joining the average lattice potential energy $(-W_a)$ to the image motive* energy $(-e^2/16\pi\epsilon_0 x)$, as represented schematically in Fig. 1. Furthermore, the electrons are assumed to obey Fermi–Dirac statistics, which gives for the energy distribution of electrons in the metal

* The motion of electrons in the space just outside the surface of metal is greatly influenced by the attraction of each electron to its image charge with a force equal to $-e^2/16\pi\epsilon_0 x^2$. Thus, the electron behaves as if it were moving in an electrostatic field whose potential is the actual potential plus $-e/16\pi\epsilon_0 x$. This fictitious potential is called the *image motive*. The motive is defined as that quantity, the gradient of which at any point gives $1/e$ times the force on an electron. The image motive is not a true electrostatic potential since the definition of true potentials involves a time average of force per unit charge in the limit $q \to 0$, where q is the electric charge at point x. For the image force, $F = -q^2/16\pi\epsilon_0 x^2$, and $\lim_{q\to 0} dF/dq = 0$. In spite of this conceptual difficulty, it is legitimate to use the image motive to calculate the wave function of an electron outside of a metal surface [e.g., J. Bardeen *Phys. Rev.* **58,** 727 (1940)].

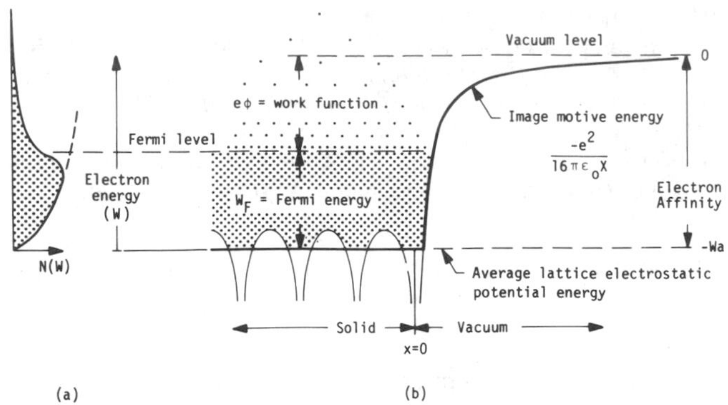

FIG. 1. Sommerfeld model of a metal. (a) Energy distribution of electrons as given by Eq. (2). (b) Boxlike potential well defined by the average lattice electrostatic potential and image motive energies indicated by heavy outline.

$$N(W)=4\pi\left(\frac{2m}{h^2}\right)^{3/2}\frac{W^{1/2}}{[e^{(W-W_F)/kT}+1]}. \tag{2}$$

The current density j will now be given by

$$j=e\int N(W)v_z dW, \tag{3}$$

where v_z is the component of electron velocity normal to the surface, and the integration is performed over those electrons which can escape from the metal, i.e., whose normal kinetic energy, $\frac{1}{2}mv_z^2$, is greater than the barrier height, $W_F+e\phi$. To effect the integration, we use the following identities:

$$W=\tfrac{1}{2}mv^2=\tfrac{1}{2}m(v_x^2+v_y^2+v_z^2); \tag{4a}$$

$$W^{1/2}dW=2(\tfrac{1}{2}m)^{3/2}v^2dv\,\frac{1}{2\pi}=(2\pi)^{-1}(\tfrac{1}{2}m)^{3/2}dv_x dv_y dv_z; \tag{4b}$$

and the fact that $e^{(W-W_F)/kT}\gg 1$. The current density is thus given by

$$j=2e\left(\frac{m}{h}\right)^3\int_{-\infty}^{\infty}e^{-(m/2kT)v_x^2}dv_x \times\int_{-\infty}^{\infty}e^{-(m/2kT)v_y^2}dv_y \times\int_{[2(W_F+e\phi)/m]^{1/2}}^{\infty}e^{W_F-(m/2kT)v_z^2}v_z dv_z. \tag{5}$$

The first two integrals are of the form

$$\int_{-\infty}^{\infty}e^{-ax^2}dx=\left(\frac{\pi}{a}\right)^{1/2}\rightarrow\left(\frac{2\pi kT}{m}\right)^{1/2}$$

so that

$$j_R=(4\pi mek^2/h^3)(1-r)T^2e^{-e\phi/kT}. \tag{6}$$

Comparison of this result with Eq. (1) indicates that the emission constant A is given by $4\pi mek^2/h^3=120.4$ A/(cm K)^2.

The application of an electric field to the surface of a metal results in an increase in thermionic current density. This is referred to as the Schottky effect, and is illustrated in Fig.

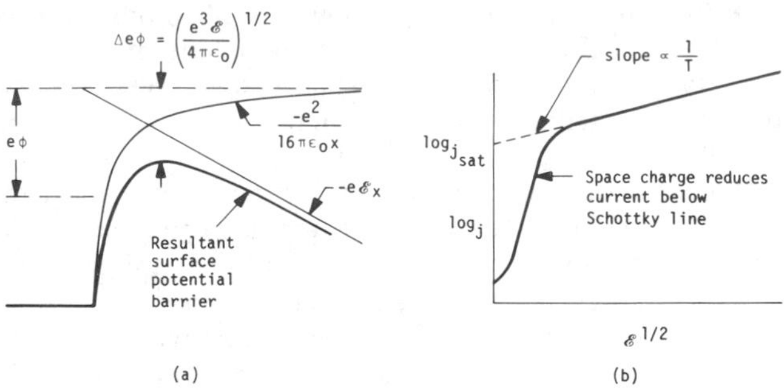

FIG. 2. Schottky effect. (a) Lowering of the potential barrier by image motive and applied fields. (b) Illustrative Schottky plot showing the effect of space charge at low fields, and the definition of saturation current density, j_{sat}.

2, which shows how the image motive and the applied electrostatic field combine to lower the surface potential barrier. The barrier is lowered an amount

$$\Delta e\phi=\left(\frac{e^3\mathscr{E}}{4\pi\epsilon_0}\right)^{1/2}=3.8\times10^{-5}\mathscr{E}^{1/2}\text{ eV}. \tag{7}$$

This results in a modification of the equation for thermionic current density to

$$j_{RS}=\frac{4\pi mek^2}{h^3}[1-\bar{r}(\mathscr{E})]T^2\exp\left[\frac{-e\phi+(e^3\mathscr{E}/4\pi\epsilon_0)^{1/2}}{kT}\right] \tag{8}$$

$$=j_R\exp\left(\frac{e^3\mathscr{E}}{4\pi\epsilon_0}\right)^{1/2}\left(\frac{1-\bar{r}(\mathscr{E})}{1-\bar{r}(0)}\right), \tag{9}$$

where $1-\bar{r}(\mathscr{E})$ is the transmission coefficient of the modified surface barrier, averaged over the thermionic electron energy. This is known as the Richardson–Schottky equation and predicts the log of thermionic emission current density versus the square root of the applied electrostatic field to be a straight line with a slope that varies inversely with temperature:

$$\log\left(\frac{j_{RS}}{j_R}\right)=\frac{0.1912}{T}\mathscr{E}^{1/2}+\log\left(\frac{1-\bar{r}(\mathscr{E})}{1-\bar{r}(0)}\right). \tag{10}$$

Experimentally, at low values of applied electrostatic field $\mathscr{E}$, the actual current density falls below the Schottky line because of the effect of space charge. The space-charge-limited current in a plane-parallel vacuum diode is given by the Child–Langmuir law:

$$j_{CL}=\frac{4\epsilon_0}{9}\left(\frac{2}{mx}\right)^{1/2}\mathscr{E}^{3/2}, \tag{11}$$

where x is the cathode–anode separation.

Experimental values of the work function ($e\phi$) may be determined from a Richardson plot: a graph of the values of $\log(j_{sat}/T^2)$ vs $1/T$, where the saturation current density j_{sat} is obtained by linear extrapolation of the Schottky line to $\mathscr{E}=0$. In general, for metals a straight Richardson line is obtained. The slope of this line is $-e\phi/2.3k=-5040\phi$, and the intercept extrapolated to the $\log(j_{sat}/T^2)$ axis is defined as $\log A$. Thermionic emission parameters thus obtained are labeled $e\phi_R$ and A_R; A_R contains factors depending on the nonuniformity of real surfaces and the temperature dependence of the work function [2].

See also METALS; SURFACES AND INTERFACES; WORK FUNCTION.

REFERENCES

1. P. W. Bridgman, *The Thermodynamics of Electrical Phenomena in Metals*. MacMillan, New York, 1934. (I)
2. C. Herring and M. H. Nichols, *Rev. Mod. Phys.* **21**, 185–270 (1949). (A)

Thermodynamic Data

Robert D. Freeman

The power of thermodynamic analysis can be exploited only if adequate data—i.e., numerical values for the various pertinent thermodynamic functions—are available. Fortunately, there are many compilations of critically evaluated thermodynamic data and many data centers continually producing additional compilations. Equally fortunately, there is a directory for these data centers and compilations (Freeman, 1984).

A "typical" compilation provides for each listed chemical substance a numerical value for these thermodynamic functions: heat capacity C_p, entropy S, Planck function $Y(T;0,0) = -[G(T)-H(0)]/T - S(0)$, enthalpy $H(T)-H(0)$, and the change in enthalpy ΔH and in the Gibbs energy (or Gibbs function) ΔG for the reaction in which the given substance is formed from its constituent chemical elements in their defined standard reference state. Values are typically provided at 100-K intervals from 0 or 300 K to 1500, 3000, or 6000 K. These data enable us to calculate the feasibility of proposed reactions/processes (reactions with positive ΔG do not occur), to calculate the equilibrium state for known reactions (to what extent does liquid sodium dissolve carbon from steel?; what is the maximum specific impulse available from a given rocket-fuel-oxidizer combination?), and, of course, to calculate enthalpy (ΔH) or energy losses/gains resulting from chemical reactions.

For the first four functions listed, the data in such compilations are, for solids and liquids, derived from experimental, usually calorimetric, measurements of the heat capacity C_p or of the enthalpy $[H(T)-H(298.15\text{ K})]$ over the range from 0 to 400–600 K for C_p, 500–1500 K for enthalpy, with extrapolation to higher temperatures. Recently developed high-speed pulse techniques extend experimental measurements to the melting point for electrical conductors (e.g., 3700 K for tungsten). For gaseous substances statistical thermodynamic calculations provide more precise data than does direct experimental measurement—*if* good spectroscopic data are available for molecular parameters: moment(s) of inertia, vibrational frequencies, electronic energy levels. Values for ΔH for formation ($\Delta_f H$) are obtained from direct calorimetric measurement of heat of combustion (in O_2 or F_2) or of the heat of other appropriate chemical reaction; another source of data for $\Delta_f H$ is the analysis of the variation with temperature of the equilibrium constant for a reaction. Values for $\Delta_f G$ are usually obtained from combination of data for $\Delta_f H$ and entropies ($\Delta_f G = \Delta_f H - T\ \Delta_f S$), although essentially direct determination is possible by measurement of the emf E of a reversible galvanic cell in which the desired reaction occurs ($\Delta G = -nFE$); the difficulty lies in creating the appropriate reversible cell. Another major source of ΔG data is the experimental determination of the equilibrium constant (K) for a reaction/process ($\Delta G = -RT\ \ln K$).

There are a variety of other types of compilations. Some deal with PVT data for fluids (including the critical region), with $PVTy_i$ data for mixtures, and with PTx_iy_i data for vapor–liquid equilibria ($x_i[y_i]$ = mole fraction of component i in the liquid [vapor] phase). These data are particularly important in the design of plants/processes operating at high pressures, and/or involving separation of components, as in petroleum processing. There are compilations that provide comprehensive data for specific substances: steam, methane, ethylene, liquified natural gas, minerals, molten salts (density, electrical conductivity, viscosity, surface tension, reversible electrode potentials, vapor pressure), gaseous ions (energy change for formation), substances derived from coal (coal gasification/liquefaction). Aqueous solutions play a major role in chemistry, geochemistry, biochemistry, and the environment; many compilations devoted to their properties are available. Phase diagrams are available for metallurgical and ceramics systems; of particular current interest are phase diagrams for various metal–hydrogen alloys that may be useful in hydrogen storage systems. Some compilations are concerned with specific ranges of temperature/pressure (cryogenic properties; high-pressure properties). Other compilations provide data for melting and boiling points, vapor pressure as a function of T, the solubility of a wide variety of substances in various solvents, thermal conductivity, etc.

Execution of a thermodynamic analysis often requires a variety of data *other* than the typical "thermochemical" data tabulated in the sources described above. Obvious examples of needed data are molecular structural parameters, bond dissociation energies, and atomic energy levels. At the September 1979 Grenoble Meeting of the CODATA Task Group on Internationalization and Systemization of Thermodynamic Tables (later renamed the TG on Chemical Thermodynamic Tables), the Task Group expressed a desire to have available in one place a bibliography of sources of these auxiliary data, with particular emphasis on those sources which have been found especially useful by experienced data evaluators/compilers. The *first and second approximations* to the desired bibliography have been published (Freeman, 1979, 1983).

Both of the guides/bibliographies described above refer to, and *do not repeat* the contents of a Bibliography that is available, free, from the Office of Standard Reference Data/NIST (formerly NBS). This Bibliography is a 140-page book and a 35-page supplement which lists all publications issued under the U.S. National Standard Reference Data System from its inception in 1964 through 1986; NSRDS-NBS publications and *Journal of Physical and Chemical Reference*

Data reprints are included. This bibliography, which is extensively indexed and cross-indexed, contains *many* articles relevant to thermodynamic data/analyses, including many critically evaluated compilations of data.

BIBLIOGRAPHY

R. D. Freeman, "Chemical Thermodynamics," in *CODATA Directory of Data Sources for Science and Technology* (E. F. Westrum, Jr., ed.). Pergamon, New York, 1984. Available as CODATA Bulletin No. 55, April 1984.

R. D. Freeman, "Sources of Adjuvant Data for Thermochemical Analyses," *Bull. Chem. Thermodyn.* **22,** 499–505 (1979); **26,** 569–576 (1983).

J. C. Sauerwein and G. B. Dalton, "Standard Reference Data Publications," NBS Special Publication 708 (1964–1984) and SP 708 Supplement 1 (1985–86). Office of Standard Reference Data, National Institute of Standards and Technology, Gaithersburg, MD 20899.

Thermodynamics, Equilibrium

Robert B. Griffiths

I. INTRODUCTION

Thermodynamics deals with certain macroscopic properties of bulk matter, such as pressure and density, in circumstances where temperature is a significant variable. It provides a complete description of these properties under conditions of equilibrium, and provides a starting point for the investigation of nonequilibrium phenomena such as hydrodynamics, transport properties of solids, and chemical reactions. Thus it has very broad applications both in pure science and in engineering. Indeed, wherever "temperature" is part of a scientific description, one or more thermodynamic principles is involved.

The bridge between thermodynamics, a purely macroscopic discipline, and atomic physics is provided by statistical mechanics. The basic concepts of thermodynamics were invented and confirmed experimentally well before the advent of a satisfactory atomic physics, and do not depend on the details of atomic interactions. Nonetheless, statistical mechanics has provided many insights into thermodynamic relations as well as formulas for calculating (in principle!) the thermodynamic functions of a system in terms of properties of its constituent atoms. The relation between thermodynamics and statistical mechanics is discussed in Refs. 3–6.

II. EQUILIBRIUM AND FUNDAMENTAL RELATIONS

For simplicity, we shall limit our discussion to a pure fluid in the absence of electric, magnetic, and gravitational fields. For extensions to other systems, see Sec. VII. A fluid left undisturbed in a closed container eventually reaches a state of *thermodynamic equilibrium* in which macroscopic motion ceases and the temperature and pressure are uniform. Under these conditions its macroscopic properties are determined by a small number of independent variables, in particular the volume V, energy U, and quantity of matter N (in moles), assuming a pure fluid.

The science of thermodynamics is based on the existence of an additional macroscopic variable for a system in equilibrium, the entropy S, which for a pure fluid is a function of U, V, and N:

$$S = S(U, V, N) \tag{1}$$

This function is an example of a *fundamental relation* that contains *all* the thermodynamic information about a system in equilibrium. For example, the absolute temperature T is given by

$$1/T = (\partial S/\partial U)_{N,V}, \tag{2}$$

and other quantities such as pressure and heat capacity can also be expressed in terms of S and its partial derivatives.

Because T is always positive for a fluid, (1) can be inverted to yield a fundamental relation with U a function of S, V, and N:

$$U = U(S, V, N). \tag{3}$$

Its partial derivatives are

$$T = \left(\frac{\partial U}{\partial S}\right)_{N,V}, \quad p = -\left(\frac{\partial U}{\partial V}\right)_{S,N}, \quad \mu = \left(\frac{\partial U}{\partial N}\right)_{S,V} \tag{4}$$

where p is the pressure and μ the chemical potential. If S, V, and N are smooth functions of a parameter τ, the chain rule applied to (3) yields

$$\frac{dU}{d\tau} = T\frac{dS}{d\tau} - p\frac{dV}{d\tau} + \mu\frac{dN}{d\tau}. \tag{5}$$

It is customary to omit the $d\tau$ and write this equation as a relationship among differentials:

$$dU = T\,dS - p\,dV + \mu\,dN. \tag{6}$$

The result (6), which may be thought of as an abbreviation for (5), provides a concise summary of (4) and is easy to manipulate.

III. VARIABLES: EXTENSIVE, INTENSIVE, DENSITIES, FIELDS

The variables S, U, V, and N are called *extensive* because their values for a thermodynamic system are sums of contributions from the different (macroscopic) parts of which it is composed. This implies, in particular, that the fundamental relation (3) for a pure fluid in a container must satisfy the functional equation

$$\lambda U(S, V, N) = U(\lambda S, \lambda V, \lambda N) \tag{7}$$

for any positive number λ. Upon differentiating (7) with respect to λ and setting $\lambda = 1$, we obtain the Euler equation:

$$U = TS - pV + \mu N \tag{8}$$

The differential of (8), combined with (6), yields the *Gibbs–Duhem* relation

$$S\,dT - V\,dp + N\,d\mu = 0, \tag{9}$$

If in (7) we let λ be $1/N$, we obtain

$$U(S, V, N) = NU(S/N, V/N, 1) = Nu(s, v) \tag{10}$$

where $u = U/N$, $s = S/N$, and $v = V/N$ are examples of *densities*, quantities that are ratios of extensive variables. The differential of u is

$$du = T\,ds - p\,dv. \tag{11}$$

Since $u(s, v)$ contains the same information as $U(S, V, N)$, apart from the total size of the system, it is often (but not always) advantageous to replace extensive variables by densities in thermodynamic calculations.

The quantities T, p, and μ in (4) are examples of *field* variables and should be distinguished from densities, because fields and densities play different roles in thermodynamics. Both fields and densities are examples of *intensive* thermodynamic variables. (Some authors use "intensive" in a sense equivalent to our use of "field," but others employ it for both densities and fields.)

IV. CONDITIONS OF EQUILIBRIUM

Consider two containers of fluid, α and β, isolated from their environment and initially from each other. Let extensive variables for the two containers be indicated by superscripts α and β. The combined energy U_t and entropy S_t of both containers are given by

$$U_t = U^\alpha + U^\beta, \tag{12}$$

$$S_t = S^\alpha(U^\alpha, V^\alpha, N^\alpha) + S^\beta(U^\beta, V^\beta, N^\beta), \tag{13}$$

if we assume that the fluid in each container is in equilibrium. Suppose that the containers are now connected by a thermal link that permits the flow of energy (heat), but not matter, from one to the other, and allowed to reach equilibrium. The new values of U^α and U^β must be such that U_t in (12) remains unchanged, by the conservation of energy (the *first law* of thermodynamics), and such that S_t is a maximum (a consequence of the *second law* of thermodynamics). Under the assumption that the functions S^α and S^β are differentiable, this maximum occurs when

$$\partial S^\alpha/\partial U^\alpha = \partial S^\beta/\partial U^\beta, \tag{14}$$

which means, by (2), that the *temperatures* of the two containers are *equal*.

Analogous considerations (given in detail in Ref. 2) show that if a linkage permits matter as well as energy to flow between the containers, both μ and T will be equal in equilibrium, and a mechanical connection that permits a decrease in V^α with an identical increase in V^β leads to equal pressures. Thus conditions of equilibrium give rise to equalities among field variables. In particular, T, μ, and p will be uniform inside a single container for a system in equilibrium. This holds true even if phase separation into liquid and vapor occurs, in which case the density variables will not be uniform.

V. THERMODYNAMIC STABILITY

Equation (14) results from noting that, in equilibrium, S_t must be an extremum. In fact, the second law imposes a stronger condition, that S_t be a maximum, and this can be used to derive (as in Ref. 2) the *stability conditions* on the fundamental relations in (1) and (3). Stated concisely, the stability requirement is that S be a concave function of U, V, and N or, equivalently (for $T \geq 0$), that U be a convex function of S, V, and N. Convexity of U is, in turn, equivalent to convexity of $u(s, v)$. If u is twice continuously differentiable, convexity of u is the same as saying that the matrix of its second derivatives (the Hessian),

$$\begin{pmatrix} \partial^2 u/\partial s^2 & \partial^2 u/\partial s\partial v \\ \partial^2 u/\partial v\partial s & \partial^2 u/\partial v^2 \end{pmatrix}, \tag{15}$$

be a nonnegative quadratic form. That is, the eigenvalues of (15), regarded as a matrix of real numbers, must be nonnegative. In particular, the following inequalities must hold:

$$\partial^2 u/\partial s^2 \geq 0, \quad \partial^2 u/\partial v^2 \geq 0, \tag{16}$$

$$(\partial^2 u/\partial s^2)(\partial^2 u/\partial v^2) \geq (\partial^2 u/\partial s\partial v)^2. \tag{17}$$

Note that the conditions (16) on the diagonal elements of (15) are necessary, but not sufficient, to ensure convexity.

In physical terms, the stability requirements imply that the heat capacities at constant volume and pressure, and the adiabatic and isothermal compressibilities, are everywhere positive, or at least nonnegative. There are additional implications (which we shall not discuss) in the case of fluid mixtures and other more complicated systems.

VI. LEGENDRE TRANSFORMATIONS AND THERMODYNAMIC POTENTIALS

Additional fundamental relations can be obtained from (1) or (3) using Legendre transformations. Those obtained from (1) are called Massieu functions, and those from (3) are known as thermodynamic potentials. As an example of the latter, consider the Helmholtz free energy A defined by

$$A = A(T, V, N) = U(S, V, N) - TS \tag{18}$$

where S on the right-hand side is to be expressed as a function of T, V, and N by inverting the first equation in (4). The relation (18) expressing A in terms of T, V, and N (*not* in terms of S, V, N) is a fundamental relation containing the same information as (3). All other thermodynamic quantities may be expressed in terms of A and its partial derivatives. The differential of (18) combined with (6) yields

$$dA = -S\,dT - p\,dV + \mu\,dN. \tag{19}$$

Other Legendre transformations yield the enthalpy $H(S, p, N)$ and Gibbs free energy $G(T, p, N)$. The basic strategy is to replace an extensive independent variable by its conjugate field [i.e., the one paired with it in (6) or (8)]. The stability conditions for these potentials can be summarized by the statement that each is a *convex* function of its extensive arguments if its field arguments are held fixed, but a *concave* function of its field arguments if its extensive arguments are held fixed.

Whereas Legendre transformations yield no new thermodynamic principles or information, they do permit us to use sets of independent variables that may in practice be much more convenient than S, V, and N. It is also worth noting that the logarithms of the partition functions for the different equilibrium ensembles in statistical mechanics are related to each other by Legendre transformations (they are actually Massieu functions) in the thermodynamic limit.

In discussions of systems undergoing phase transitions, (18) should be replaced by

$$A(T, V, N) = \inf_{S}[U(S, V, N) - TS] \tag{20}$$

Further comments on transformations of this type can be found in Ref. 6.

VII. COMPLEX SYSTEMS

The basic thermodynamic principles discussed above can be extended, with greater or lesser ease, to a wide variety of systems that are more complex than a pure fluid, at the cost of introducing additional extensive thermodynamic variables (or densities) along with corresponding fields. Provided the system is homogeneous and in equilibrium, all thermodynamic properties are determined by expressing the entropy or energy in terms of the other extensive variables, as in (1) or (3), or by use of a fundamental relation obtained through a Legendre transformation. Specific examples are:

1. *Fluid mixtures.* For a mixture containing N_1 moles of the first chemical component, N_2 of the second, etc., the variable N in (1) and (3) must be replaced by the set of variables $N_1, N_2, \ldots$, and Eq. (6) becomes

$$dU = T\,dS - p\,dV + \sum_j \mu_j\,dN_j \tag{21}$$

where μ_j is the chemical potential for the jth component.

2. *Solids.* As long as only isotropic stresses are considered, the thermodynamic description of a solid is the same as that of a fluid. In the presence of anisotropic stress (shearing forces), however, it is necessary to introduce the components of the strain tensor as additional density variables, and the components of the stress tensor as additional fields. For further details, see Ref. 7.

3. *Systems in electric and magnetic fields.* The electric and/or magnetic polarizations provide new extensive thermodynamic variables, while the electric and magnetic fields are also fields in the thermodynamic sense. However, the long-range character of electric and magnetic forces leads to complications in that the energy of a sample may depend on its shape, and the field inside a sample can be different from that which would be produced by the same external charges and currents in the absence of the sample. There are several equivalent ways of formulating the thermodynamic properties of bodies in static electric and magnetic fields but, unfortunately, no uniform notation. See Refs. 2, 7.

VIII. ADDITIONAL TOPICS

Some additional topics that belong to the subject of the thermodynamics of equilibrium, but have not been discussed here, are treated in the References:

1. Equilibrium in chemical reactions [2];
2. Third "law" of thermodynamics [10(a), 11];
3. Negative temperatures [10(b)];
4. Surfaces and interfaces [4, 8].

See also ENTROPY; FREE ENERGY; STATISTICAL MECHANICS; THERMODYNAMICS, NONEQUILIBRIUM.

REFERENCES

A. *General Thermodynamics*

1. E. Fermi, *Thermodynamics*. Dover, New York, 1956. (E)
2. H. B. Callen, *Thermodynamics*. Wiley, New York, 1963. (I) (See also Ref. 5.)

B. *Thermodynamics and Statistical Mechanics*

3. C. Kittel and H. Kroemer, *Thermal Physics,* W. H. Freeman, San Francisco, 1980. (E)
4. L. D. Landau and E. M. Lifschitz, *Statistical Physics*. Pergamon, New York, 1969. (I)
5. H. B. Callen, *Thermodynamics and an Introduction to Thermostatistics,* Wiley, New York, 1985. (I)
6. R. B. Griffiths, in *Phase Transitions and Critical Phenomena* (C. Domb and M. S. Green, eds.), Vol. 1, p. 7. Academic, New York, 1972. (A)

C. *Special Topics*

7. J. F. Nye, *Physical Properties of Crystals*. Oxford, London and New York, 1957. (A)
8. R. Defay, I. Prigogine, A. Bellemans, and D. H. Everett, *Surface Tension and Adsorption*. Wiley, New York, 1966. (A)
9. J. S. Rowlinson and B. Widom, *Molecular Theory of Capillarity*. Oxford University Press, Oxford, 1982. (A)
10. (a) R. B. Griffiths, in *A Critical Review of Thermodynamics* (E. B. Stuart, B. Gal-Or, and A. J. Brainard, eds.), p. 101. Mono, Baltimore, MD, 1970. (I) (b) N. F. Ramsey, in *A Critical Review of Thermodynamics* (E. B. Stuart, B. Gal-Or, and A. J. Brainard, eds.), p. 217. Mono, Baltimore, MD, 1970. (I)
11. J. Wilks, *The Third Law of Thermodynamics*. Oxford, London and New York, 1961. (I)

Thermodynamics, Nonequilibrium

P. Glansdorff and I. Prigogine

The physical media on our scale, such as liquids or solids, are made up of a considerable number of interacting particles. Under this microscopic aspect they can be looked upon as complex mechanical systems which involve a great number of degrees of freedom and whose processes of evolution ultimately have to be consistent with the fundamental laws of dynamics.

But following the macroscopic point of view the same physical media can be thought of as continua whose properties of evolution are given by phenomenological laws between directly measurable quantities on our scale, such as, for example, the pressure, the temperature, or the concentrations of the different components of the media. The macroscopic perspective is of interest because of its greater simplicity of formalism and because it is often the only view practicable.

Thermodynamics is a part of this macroscopic physics. It is based on two principles originating during the 19th century from technical considerations concerning the work and the efficiency of thermal engines. The efficiency is not directly related to the fundamental principles of dynamics, but has since developed parallel and somewhat separately from the great simultaneous achievements of analytical dynamics.

As a consequence, this abnormal situation, which was bound inevitably to divide scientists into partisans and detractors, has been prolonged until recently because the detractors were actually uncovering a much more profound difficulty with the theory, which was that the principles of thermodynamics explicitly make apparent the concept of irreversibility and along with it the concept of dissipation and temporal orientation which were ignored by classical (or quantum) dynamics, where the time appears as a simple parameter and the trajectories are entirely reversible.

Is it possible to unite the two disciplines? If not, how can we explain the existence of two distinct disciplines for the same object? The essential contribution of the Austrian physicist L. Boltzmann (1844–1906) concerns the mechanical interpretation of the theory of heat. In his second paper "Mechanische Bedeutung der Zweiten Hauptzatses" in 1866 (*Wien. Ber.* **53,** 199) this concern is manifest.

Boltzmann's work was to become a considerable success, notably because of his contribution to the development of the theory of gases and for his microscopical interpretation of the entropy concept introduced by R. Clausius (1822–1888); nevertheless, it was not able to clarify completely the whole question, essentially because of the additional hypothesis concerning the concept of molecular chaos and the notion of state probability, both of which were alien to the postulates of dynamics.

Some recent contributions, specially devoted to this question, are given in the references. They show the channels through which these difficulties were at last overcome. They permit us today to interpret the thermodynamics of irreversible processes as a consequence of an appropriate dynamics concerning the study of complex media. This is far more than a mere probabilistic information theory [1–3].

The present statement is based on accepting such an interpretation. We would point particularly to the recent progress in the nonlinear domain, to the stability conditions concerning the processes in this region under the influence of fluctuations, to the developments related to symmetry breaking, and, more especially, to the new concept of *dissipative structure,* which in turn introduces the notion of order in the theories of evolution.

On the other hand, the classical properties arising from the thermodynamics of equilibrium states are supposed to be known.

FUNDAMENTAL PRINCIPLES

The irreversibility concept first appeared in 1824 in the primary formulation of the second principle by Sadi Carnot (1796–1832). In 1850, Clausius derived from it a mathematical formulation using the entropy concept. In the first place, along a reversible process, i.e., one made up of a succession of equilibrium states, the entropy S of the system is defined by its differential increment:

$$dS = dQ/T; \tag{1}$$

while, in the nonequilibrium domain, i.e., for irreversible processes, the second principle is formulated by the Carnot–Clausius relation

$$dS > dQ/T$$

or (2)

$$dS = dQ/T + dQ'/T \quad (dQ' > 0).$$

The quantities dQ, dQ', and T represent, respectively, the heat furnished by the outside world during the infinitesimal time interval dt; the corresponding noncompensated heat, which characterizes the intrinsic irreversibility of the system; and the absolute temperature introduced by Lord Kelvin (W. Thomson, 1824–1907).

It is notably due to relation (2) that one deduces the property according to which an isolated system ($dQ = 0$) evolves with an increasing entropy toward a maximum at the equilibrium state.

Moreover, this relation shows that two contributions are responsible for the entropy change; one is the external heat flow, and the other is generated by the internal transformations associated with the noncompensated heat. It can then be written quite generally as a balance equation [4]:

$$dS = d_eS + d_iS, \tag{3}$$

where the external contribution, the *entropy flow,* is given by

$$\Phi \equiv \frac{d_eS}{dt} = T^{-1}\frac{dQ}{dt}, \tag{4}$$

and the *entropy production,* a measure of irreversibility, by

$$P \equiv \frac{d_iS}{dt} = T^{-1}\frac{dQ'}{dt} \geqslant 0. \tag{5}$$

One then concludes that the second principle of thermodynamics expresses not only the fact that the entropy is maximum for an isolated system in equilibrium, but more generally the property of *minimum entropy production* valid for all equilibrium states whether isolated or not ($P = 0$). However, such an extension is still incomplete because the fundamental Carnot–Clausius inequality (2) concerns only the so-called *closed* systems, i.e., those without exchange of matter with the surroundings.

For *open* systems, which undergo energy as well as mass exchanges with the outside world, the general balance equation (3) remains valid, but the entropy flow then has to contain additional contributions such as due to mass exchange through convection and diffusion. This extension is of particular interest for the study of biological processes because these are generally *open systems.*

But because of the qualitative feature of the second principle as well as the absence of any relation between the inequality (5) and the characteristic quantitative laws of evolution of the media, the development of thermodynamics has been long limited to the equilibrium domain alone, while the Carnot–Clausius inequality survived as a simple curiosity.

The thermodynamics of irreversible processes took a few steps forward only when new advances were made following isolated attempts. One can mention in this respect the work of Lord Kelvin on the thermoelectric effect; that of Lord Rayleigh (1842–1919) on the dissipation of energy in a viscous fluid; that of P. Duhem (1861–1917) on chemical friction; and the microscopic definition of entropy,

$$S = k \log W, \tag{6}$$

established by Boltzmann in his theory of gases. The quantity denoted by W represents the thermodynamic probability of the state of the system, i.e., a quantity associated with the number of possibilities which, on the molecular scale, permit the realization of a given state; k is the universal gas constant, also called Boltzmann's constant. The criterion of increasing entropy in isolated systems then acquires the physical significance of an order → disorder evolution and, consequently, the equilibrium state corresponds to complete chaos (Boltzmann H theorem) [5].

Practically speaking, the most important contribution of the second principle to the study of the physicochemical processes is the relation established by Th. De Donder (1872–1957), the founder of the Brussels school of thermodynamics, between the sign of the noncompensated heat, or the entropy production defined in (5), and the old qualitative concept of chemical affinity (Boerhaave, Leiden, 1733). The fundamental phenomenological relation introduced by De Donder can be written, using the entropy production,

$$P = T^{-1}\mathcal{A}w \geq 0, \tag{7}$$

where $\mathcal{A}$ is the affinity and w is the chemical reaction rate. One observes that the sign of the rate w, which characterizes the "direction" of this reaction, is effectively prescribed by the sign of the affinity $\mathcal{A}$. In the thermodynamical sense of the term, the latter is defined as a *state function*.

The bilinear form thus obtained was later extended to more general situations, including a variety of irreversible effects, such as, for example, several simultaneous chemical reactions, or transport phenomena like thermal conduction, diffusion, and viscous flow.

For homogeneous media, the extension of the form (7) may be written as [6]

$$P = \sum_{\alpha} X_\alpha J_\alpha \geq 0, \tag{8}$$

where X_α represents a generalized force [e.g., $T^{-1}\mathcal{A}$ in (7)] and J_α represents a flow [e.g., w in (7)]. In heterogeneous media, the same relation takes the form of a volume integral,

$$P = \int_V \sum_{\alpha} x_\alpha j_\alpha dV \geq 0, \tag{9}$$

where the local generalized forces x_α are, for example, the gradients associated with the various transport phenomena (heat, diffusion, convection), as well as local chemical affinities. On the other hand, by definition, the equilibrium state corresponds to cancellation of all the flows j_α.

However, this important improvement in the expression of the second law implies the introduction of a restrictive hypothesis usually called the *local equilibrium* assumption. Indeed, in order to obtain the above bilinear expressions, one has to suppose that the local state of a nonequilibrium system can be described by the same quantities used to define the equilibrium state, and that it is governed by the same equation of state [7].

This assumption is justified when the local distribution functions differ only slightly from their equilibrium value as a result of the multitude of molecular interactions. However, as a small deviation from statistical equilibrium can give birth to important macroscopic flows, the local equilibrium assumption covers practically a large domain of thermodynamics of irreversible processes, especially in physical chemistry and biophysics.

Practically, only the following applications are to be excluded:

(1) high vacuum or very low temperatures, because of the average insufficiency of interactions;
(2) nonequilibrium physical systems whose behavior implies the introduction of supplementary variables or gradients into the state equation.

At the present level of development, the thermodynamics of irreversible processes is not yet able to eliminate completely all empiricism from complex problems such as one encounters in rheology [8,9].

In fact, the local equilibrium assumption, as it is used below, means the validity at each point of the Gibbs law of systems in equilibrium. The latter introduces an extended definition of the local entropy, valid for a large class of nonequilibrium states independent of the Boltzmann definition recalled in (6). It can be written

$$dS = T^{-1}dE + pT^{-1}dV - \sum_{\gamma} \mu_\gamma T^{-1} dm_\gamma \tag{10}$$

for a state characterized by its energy E. its volume V, and the mass m_γ of each of its constituents γ, at the local pressure p, temperature T, chemical potentials μ_γ.

LINEAR THERMODYNAMICS

The flows J_α are related to the generalized forces X_α by phenomenological or kinetic laws. In the immediate neighborhood of equilibrium, they take the linear form of an analytical development limited to the first-order terms, the zero-order terms vanishing identically. One then obtains relations of the form

$$J_\alpha = \sum_{\beta} L_{\alpha\beta} X_\beta \quad (J_{\alpha\,\mathrm{eq}} = X_{\alpha\,\mathrm{eq}} = 0), \tag{11}$$

where the constants $L_{\alpha\beta}$ are phenomenological coefficients characterizing the media. In particular, the well-known Fourier law of heat conduction, Fick's law of diffusion, and Newton's law of viscous fluids are typical examples of (11). As an offset, the domain of application of such linear relations between rates of reaction and chemical affinities is much more narrow and without great practical interest [4, 6, 7].

In the framework of this linear theory, the entropy production defined in (8) becomes a positive definite quadratic form which can be written as

$$P = \sum_{\alpha\beta} L_{\alpha\beta} X_\alpha X_\beta \geq 0 \tag{12}$$

and similarly for the functional (9), concerning nonhomogeneous media. As the positive sign has to be satisfied for all values of the X variables, a well-known theorem of algebra on quadratic forms implies that sign conditions have to be imposed on phenomenological coefficients $L_{\alpha\beta}$. For example, in the case of two flows ($\alpha,\beta=1,2$), the prescribed conditions are

$$L_{11}>0; \quad L_{22}>0; \quad (L_{12}+L_{21})^2<4L_{11}L_{22}. \tag{13}$$

Hereby, a first link is established between the requirements of the second law of thermodynamics and the properties of the system. For instance, the thermal conductivity and the diffusion coefficients, which are of the $L_{\alpha\alpha}$ type, have to be positive, while the thermodiffusion coefficient similar to $L_{\alpha\beta}$ has no definite sign.

Actually, the linear theory has practically reached its full development, thanks to the two following important supplementary contributions.

First of all, the so-called reciprocity relations were established by Onsager for the coupling coefficients in 1931 [5, 10]:

$$L_{\alpha\beta}=L_{\beta\alpha} \quad (\beta\neq\alpha) \tag{14}$$

in the absence of magnetic field. This symmetrization of the matrix of phenomenological coefficients leads to the exclusion of any possibility of rotation around the equilibrium state in the linear region of the space spanned by the generalized forces. On the other hand, it was these relations which led to the correct interpretation of Lord Kelvin's work on the thermoelectric effect. Verified experimentally, notably through the symmetry of the thermal conduction tensor in anisotropic media [9], the Onsager relations have contributed greatly to the interpretation of multiple coupling effects (thermodiffusion, active transport, electro-osmosis, . . .).

The second typical contribution of the linear domain concerns the theorem of minimum entropy production at the stationary state for given external constraints (Prigogine, 1954) [4]. This property, obvious according to (5) for the equilibrium state, subsists thus for the stationary states, but in this case the minimum value is no longer zero; it depends essentially on the imposed constraint which prevents the system from reaching equilibrium.

This theorem results directly from the condition (14). It is interpreted as a variational formulation of the second principle in the linear domain, and is then applicable to the stationary states under weak constraints. As an example, for a two-component fluid initially distributed uniformly between two reservoirs interconnected by a capillary, each of which is maintained at a uniform but different temperature, a stationary state sets in when the mass flow vanishes, and then yields

$$L_{12}X_1+L_{22}X_2=0. \tag{15}$$

The prescribed constraint here is the "thermal force X_1." The minimum condition then requires that the entropy production derivative with respect to X_2, i.e., $(\partial P/\partial X_2)_{X_1}=2J_2$, vanish. The latter condition is indeed verified.

More concretely, the minimum entropy production expresses a property common to all stationary states belonging to the linear domain. Accordingly, the behavior of all the above states tends to approach as near as possible the equilibrium state, that is the complete degradation. It is only the imposed constraint which prevents the system from reaching equilibrium. As a matter of fact, such behavior results directly from a dynamical property of complex systems leading to the concept of the so-called *punctual attractors* to characterize an equilibrium state as a fixed point in the space of states, near which the whole set of the dynamical trajectories seems to be attracted, independently of the initial conditions. Moreover, as we shall see in the following paragraph, these stationary states are stable and no fluctuation can lead to instability beyond which one could expect a different and better organized behavior. In other words, all random fluctuations decay and disappear.

Thus one observes that the whole set of stationary states of the linear region belongs to a stable branch, called the *thermodynamic branch* in opposition to quite different situations of the nonlinear domain, where this branch can become unstable starting from some marginal state and may be replaced by one or several other stable branches, characterized by an entirely different internal and sometimes higher organization such are respectively the so-called *strange attractors* and *dissipative structures* quoted below. This possibility then permits a close connection with the concept of biological order, while the domain of weak constraints of linear thermodynamics can in no way contain such organizations.

The other implications of the linear thermodynamics, as, for example, the study of the coupling effects, are developed today in detail in several handbooks, such as those of De Groot and Mazur (1962) or Prigogine (1968) [4, 6].

NONLINEAR THERMODYNAMICS

This domain, situated farther away from the equilibrium state than the preceding one, has been explored only recently and only for the case of systems for which the local equilibrium assumption is valid.

The first attempt concerns the establishment of an evolution criterion for systems upon which strong and permanent constraints are imposed. Its mathematical expression is deduced from the separation into two terms of the differential dP of the entropy production (8). The first term is d_XP (or JdX), by contraction to be compact, and the second d_JP (or XdJ). One then deduces from the mass and energy balance equations that a dissipative system with stationary constraints evolves according to the general criterion (Glansdorff and Prigogine, 1954) [7].

$$d_XP\leqslant 0. \tag{16}$$

The equality sign corresponds to the stationary state if it exists. In contrast, the d_JP term has no definite sign. The same property has been later extended to dissipative–convective systems using in addition the momentum balance equation (1964). One sees immediately that this criterion is a generalization of the minimum entropy production theorem, to which it reduces, by the way, in the linear domain. Indeed, one just has to observe that because of the phenomenological laws (11) and the reciprocity law (14), the above criterion directly restores this theorem under the form

$$dP=2d_XP=2d_JP\leqslant 0 \tag{17}$$

and, by the same token, limits its validity to the linear domain, a limitation often lost sight of.

But the essential feature of the evolution criterion (16), as far as its interpretation in the nonlinear domain is concerned, arises from its nonvariational formulation in the general case. Indeed, in contrast to dP the quantity d_XP is not necessarily the exact differential of some scalar uniform potential as was the case of the entropy production in linear thermodynamics, or that of free energy in the equilibrium theory. More precisely, two distinct situations may occur.

In the first case, the mathematical expression of the kinetic laws between flows and generalized forces do permit the reduction of d_XP to an exact differential of a kinetic potential Φ, eventually using a positive integrating factor λ. This gives rise to a variational law of evolution as

$$\lambda d_XP = d\Phi \leq 0, \tag{18}$$

which is a direct generalization of the minimum entropy production theorem in the stationary state. As a result any evolution with permanent constraints is still oriented in the direction of the latter state. Moreover, the Le Châtelier–Braun principle concerning the stability of equilibrium governs as well the stationary states of this class.

In the second case, d_XP is by no procedure reducible to an exact differential expression, which excludes the existence of any uniform scalar potential, but permits, however, rotations around the stationary state, such as, for example, the predator–prey cycles of the Lotka–Volterra type [7, 11, 12].

Nevertheless, in spite of the absence of a uniform potential, the evolution criterion (16) enables us to obtain a formulation of the stability conditions for a stationary state, owing to a method already used in the classical case of equilibrium. The latter method expresses the stability of some reference state through the impossibility of any process starting from such a state due to the requirements of the evolution criterion. Therefore, in the present case, one deduces after using (16) the condition $\delta_XP>0$, in the neighborhood of the reference state.

Besides, it is easy to establish the equality $\delta_XP=0$ for the stationary state itself, selected as the starting point for the perturbation. Therefore, the stability criterion (16) for infinitesimal disturbances around a steady state may be expressed by means of the second-order bilinear forms as

$$\sum_\alpha \delta J_\alpha \cdot \delta X_\alpha > 0 \tag{19}$$

for the homogeneous dissipative media considered in (8). This quantity, often useful in nonlinear stability theory, has been called the *excess entropy production*. The generalization to nonhomogeneous and continuous media is straightforward due to (9) [7].

As a first consequence, the criterion (19) proves the stability of the thermodynamic branch and confirms the absence of any critical marginal state in the linear domain. Indeed, the relation (11) inserted into (19) gives rise to a definite quadratic expression of the form $\Sigma L_{\alpha\beta}\delta X_\alpha \delta X_\beta$, whose positive sign ensuring stability is identically satisfied as a result of the second law (12), dealing with the entropy production.

On the other hand, inequality (19) can be written, after simple manipulations, equivalently as

$$d_t(\delta^2 S) > 0, \tag{20}$$

which merely expresses the infinitesimal stability condition of Liapounov, since $\delta^2 S$ expressed in terms of the independent variables E, V, m_γ, by means of the Gibbs law (10) is a negative definite quadratic form in consequence of the well-known stability conditions for the equilibrium state ($c_v>0$, . . .) assumed from the outset.

The relation (20) concerns essentially dissipative media under stationary constraints. However, the extension of the Liapounov criterion to convective effects is rather straightforward [7].

Application to specific situations shows that in the nonlinear domain, i.e., far from equilibrium, the stability criterion is no longer identically satisfied. This may arise, for example, from convective effects, such as the onset of free convection in a horizontal layer of a fluid at rest and heated from below (Bénard problem, critical Rayleigh number), or from the laminar–turbulent flow transition (critical Reynolds number). Also autocatalytic effects in a sequence of coupled chemical reactions occurring inside a system may generate a similar chemical instability.

This last observation suggests a comparison between the manifestations of order by fluctuations in macroscopic physics and those of biological processes in which autocatalytic effects are often present. This has clearly been demonstrated, for example, for the glycolytic cycle for which the stability criterion (19) written here in the form $\Sigma_\rho \delta w_\rho \delta \mathcal{A}_\rho > 0$, is no longer satisfied. The demonstration that such evolutions actually occur in a region where the thermodynamic branch has become unstable has to be considered as a major success, which favors the latter comparison [13].

Briefly, the stability criterion is no longer fulfilled at the critical marginal state for which the relation (19) reduces to an equality. This possibility results from the fact that in the nonlinear region, the left-hand side presents itself often as a difference between two groups of terms of opposite signs, indicating the existence of a competitive effect between the stabilizing action due to the dissipation and the destabilizing action arising either from convection or from catalytic processes, i.e., symbolically,

$$\{\text{dissipation}\} - K\{\text{convection or catalysis}\} \geq 0. \tag{21}$$

The constant K represents the constraint associated with the nonequilibrium state; it is usually a nondimensional ratio. Its critical value corresponds to the marginal state, i.e., to the equality sign.

DISSIPATIVE STRUCTURES AND FLUCTUATIONS

Beyond the instability, that is, for stronger constraints, the system evolves according to the second principle, but it generally no longer complies with the minimum entropy production theorem. On the contrary, as happens frequently, an internal organization generated by *symmetry breaking* appears inside the system, which thus becomes more orderly than the states belonging to the thermodynamic branch. Such organized situations have been named *dissipative structures*. In actuality they are either temporal (chemical clocks, limit cycles), spatial (Bénard convection cells), or spatiotemporal (chemical waves and quasi-waves; see Fig. 1). One actually knows several examples of each case [14]. In mathematical analysis, the study of the characteristic branches of the nonlinear domain belongs to bifurcation theory [15].

One observes an apparent similarity between these new symmetry breakings and the classical phase transitions at equilibrium (such as crystals). However, the distinction is essential as the first survive uniquely due to the permanent constraints and the suitable irreversible exchanges of matter and energy with the outside world.

In addition, successive critical states, corresponding for example to a sequence of bifurcations, may also occur as a result of a progressive increasing of the external constraints, which thus permit the system to attain, after several bifurcations, a highly organized state. This possibility draws the concept of dissipative structure still nearer to that of biological order and confers on it, at the same time, a certain historical background. Then, the deviation from the equilibrium state can be interpreted as a measure of order arising from the nonlinearity of the evolution laws and from the existence of the bifurcation points which result from it.

The mechanism responsible for the instabilities, and accordingly for the existence of bifurcations with creation of dissipative structures, arises from the existence of fluctuations. Every physical macroscopic system necessarily undergoes local fluctuations. This is a direct consequence of any theory based on mean values deduced in turn from microscopic random quantities. The fluctuations are governed by the laws of chance, but the system reacts following its own macroscopic laws of evolution. This reaction can entail the regression and ultimately the disappearance of the fluctuations. The evolution law is therefore stable and the very existence of the fluctuations can be practically ignored. This is actually the case of most of classical statements dealing with linear macroscopic physics.

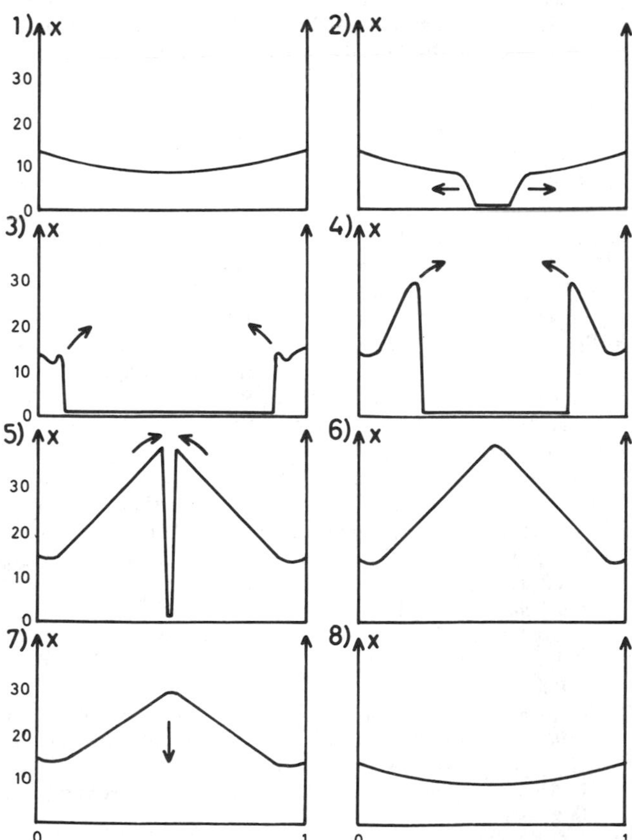

FIG. 1. Plane chemical wave for prescribed values of the concentrations on the boundaries 0 and 1. The diagrams (1)–(8) are successive representations of the concentration X, related to an intermediate component. Comparison between diagrams (1) and (8) establishes the periodic character of this spatiotemporal dissipative structure.

As an offset, instabilities correspond to an amplification of fluctuations which permits them to attain macroscopic values. Through this mechanism, a dissipative structure may be interpreted as a giant fluctuation stabilized by the energy and mass flows as imposed by the boundary conditions.

The consideration of the nonlinear domain leads us then to confer on fluctuations an essential role within a discipline where, previously, their role was always secondary. Consequently, the strictly dynamic characteristic laws of the thermodynamics governing the behavior of the dissipative structure is more or less tempered by the stochastic aspect of the fluctuations.

A formula due to Einstein, resulting from the study of Brownian motion, tells us that the probability of a fluctuation around the equilibrium state of an isolated system is given by the relation

$$\mathrm{Pr} \propto \exp(\Delta S/k), \tag{22}$$

ΔS measuring the deviation from the maximum of S at equilibrium ($\Delta S<0$). For small fluctuations, one can replace the finite increment by an expansion limited to the second-order term and then write (22) in the form [16]

$$\mathrm{Pr} \propto \exp(\delta^2 S/2k), \tag{23}$$

since the first-order term vanishes identically, i.e., $(\delta S)_{\max}=0$.

The validity of (23), but not that of (22), has been extended to nonequilibrium stationary states. This leads to an interpretation of the instability mechanism. Indeed, with $\delta^2 S<0$ and $d_t \delta^2 S>0$, the reference state is stable and the fluctuation decays ($\delta^2 S \rightarrow 0$), and this unperturbed state is also the most probable among all the neighboring states. On the other hand, with $\delta^2 S<0$ and $d_t \delta^2 S<0$, the state of reference is unstable, the fluctuation amplifies, and the perturbed state drifts away toward a new higher maximum, which is, consequently, more probable than the initial state of reference but situated at a finite distance from the first [7].

A more detailed study concerning the classification of the different types of fluctuations has been established in the case of dissipative media, starting from a stochastic analysis of their random behavior [17]. This study enables us to associate, to each value of the constraint, the type of fluctuations generating instability. The latter are classified by order of magnitude using a coherence length, a characteristic of their range inside the system.

For the weak constraints of the thermodynamic branch, fluctuations decay for all values of this length, because stability is herein ensured. But beyond this threshold, in the

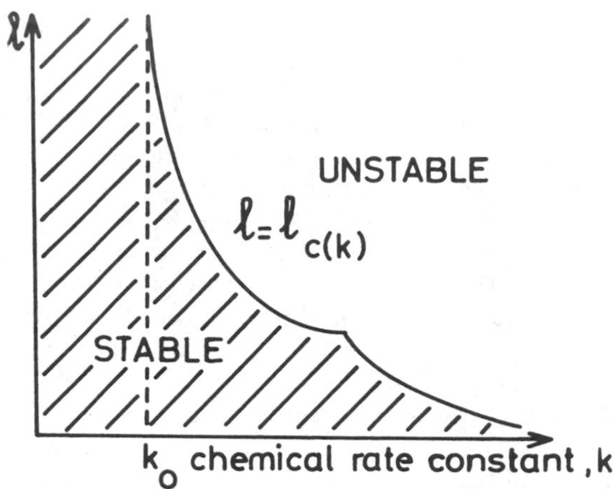

FIG. 2. Coherence length l of a fluctuation versus a characteristic chemical constraint k at the neutral stability. The macroscopic process becomes unstable for $k>k_0$.

domain of dissipative structures, one observes that the most important fluctuations are the most destablilizing.

Therefore, for each type of constraint, one observes the existence of a threshold or a *nucleation length,* determining the boundary between sizable destabilizing fluctuations and the others. When the distance from the equilibrium state increases, the nucleation length decreases gradually, until finally only the thermal fluctuations remain stabilizing (see Fig. 2).

STRANGE ATTRACTORS AND CHAOTIC MOTION

According to the above commentaries on linear thermodynamics a state of equilibrium may be interpreted in the phase space as a *punctual attractor* with respect to the set of trajectories (sink). Similarly the single periodic orbit observed in the previous section occurs as a *linear attractor* (limit cycle). Both alike correspond to simple motions. For a two-dimensional flow, only these two simple attractors may exist, which is known for a long time.

However, in at least three-dimensional phase space, a very different type of attractor named *strange* or *chaotic attractor* may occur after successive bifurcations due to unstable dynamical processes. The definition of a such strange attractor is linked to that of a chaotic behavior, that is in the present case, a complicated geometrical motion where nearby dynamical trajectories, though bounded, diverge exponentially according to the Liapounov relaxation time.

Only a few properties of the strange attractors have been actually explored. Among these one may be quoted a *self-similar* character which means that partial structures are repeated again and again on a finer and finer scale. Another property concerns its dimension which usually turns out to be fractal, i.e., not an integer [18, 19].

Nevertheless strange attractors hold a leading part in the present time to interpret various nonlinear physical as well as chemical behaviors. Such are, for example, the case related to the development of the turbulent motion in fluid flow, and on the other hand, autocatalytic effects involved in a system of chemical reactions [13, 20]. But probably the most important conclusion to be drawn from the appearance of these strange attractors involves a reconciliation between the foundations of dynamics and those of thermodynamics. Indeed, the impossibility of tracing back to the initial conditions after unstable dynamical processes introduces surprisingly into dynamics the concept of arrow of time for the slightest manifestation of dissipation.

It is well known that the lack of the notion of past and future in dynamics, either classical, relativistic, or quantum, constituted until now a fundamental, and apparently irreducible opposition to thermodynamics of irreversible processes [21].

See also ENTROPY; EQUATIONS OF STATE; FLUCTUATION PHENOMENA; PHASE TRANSITIONS; STASTISTICAL MECHANICS; THERMODYNAMICS, EQUILIBRIUM; TRANSPORT THEORY.

REFERENCES

1. I. Prigogine, C. George, F. Henin, and L. Rosenfeld, "A unified formulation of dynamics and thermodynamics," *Chem. Scripta* **4,** 5–32 (1973). (A)
2. P. Glansdorff and I. Prigogine, *Entropie, Structure et Dynamique,* Table ronde à l'Ecole Polytechnique: Sadi-Carnot et l'Essor de la Thermodynamique, Paris, 11–13, June 1974. (I).
3. J. Meixner, *Entropy and Entropy Production,* International Symposium on the Foundations of Continuum Thermodynamics, Bussaco, Portugal, 1973. The MacMillan Press, New York, 1974. (A).
4. I. Prigogine, *Introduction to Thermodynamics of Irreversible processes.* Wiley, New York, 1967. (I).
5. M. W. Zemansky, *Heat and Thermodynamics.* McGraw-Hill, New York, 1957. (I).
6. S. R. De Groot and P. Mazur, *Non-Equilibrium Thermodynamics.* North Holland, Amsterdam, 1962. (A).
7. P. Glansdorff and I. Prigogine, *Thermodynamic Theory of Structure, Stability and Fluctuations.* Wiley, London, 1971. (A)
8. S. Nemat-Nasser, *On Nonequilibrium Thermodynamics of Viscoelasticity and Viscoplasticity.* International Symposium on the Foundations of Continuum Thermodynamics, Bussasco, Portugal, 1973. The MacMillan Press, New York, 1974. (A)
9. R. S. Schechter, *The Variational Method in Engineering. McGraw-Hill, New York, 1967. (I)*
10. L. Onsager, "Reciprocal relations in irreversible processes." *Phys. Rev.* **37,** 405 (1931); **38,** 2265 (1931). (A)
11. A. J. Lotka, *J. Am. Chem. Soc.* **42,** 1595 (1920). (A)
12. V. Volterra, *Théorie mathématique de la lutte pour la vie.* Gauthier-Villars, Paris, 1931. (A)
13. A. Goldbeter, "Mechanism for oscillatory synthesis of cyclic AMP in *Dictyostelium discoideum.*" *Nature* **253,** 540–542 (1975). (A)
14. G. Nicolis, *Spatio-temporal Dissipative Structures in Non-equilibrium Systems.* Physikalische Blätter, 31, Jahrgang, Dezember, Heft 12, 1975. (A)
15. N. Minorsky, *Nonlinear Oscillations.* Van Nostrand, New York, 1962. (A)
16. H. Callen, *Non-Equilibrium Thermodynamics Variational Techniques and Stability.* University of Chicago Press, Chicago and London, 1965. (A)
17. G. Nicolis, "Dissipative structures with applications to chemical reactions," in *Cooperative Phenomena* (H. Haken, ed.). North-Holland, Amsterdam, 1974. (A)
18. A. J. Lichtenberg and M. A. Lieberman, *Regular and Stochastic*

Motion (Applied Mathematical Sciences 38). Springer-Verlag, New York, 1983.
19. B. B. Mandelbrot, *The Fractal Geometry of Nature*. Freeman, New York, 1983.
20. J. P. Eckmann and D. Ruelle, "Ergodic Theory of Chaos and Strange Attractors," *Rev. Mod. Phys.* **57**, 617–656 (1985).
21. I. Prigogine and I. Stengers, *Between Time and Eternity*. Shambala, Boston, 1989.

Thermoelectric Effects

C. L. Foiles

Thermoelectric effects occur when mobile charge carriers in liquids and solids are subjected to the influence of temperature gradients and/or particular electrical potential gradients. In the absence of a magnetic field there are three thermoelectric effects; the Seebeck (discovered in 1822 by T. J. Seebeck), the Peltier (discovered in 1834 by J. C. A. Peltier), and the Thomson (predicted in 1854 and experimentally documented a few years later by Wm. Thomson) effects.

The Seebeck effect is the basis for thermocouples. Consider the circuit shown in Fig. 1. Two different materials, A and B, are joined with their junctions maintained at temperatures T_1 and T_2 while a break in the circuit is maintained at T_0. An open-circuit voltage ΔV is generated if $T_1 \neq T_2$. For homogeneous materials this voltage is independent of the precise details of the temperature gradients and of T_0. Only the difference $T_2 - T_1$ is important. ΔV is linearly proportional to small temperature differences,

$$\Delta V_{AB} = \alpha_{AB}(T_2 - T_1), \tag{1}$$

but it is more generally described as

$$\Delta V_{AB} = \int_{T_1}^{T_2} [S_A(T) - S_B(T)]\, dt \equiv \int_{T_1}^{T_2} S_{AB}(T)\, dT. \tag{2}$$

α_{AB} is the Seebeck coefficient for the AB couple and $S_m(T)$ is the thermoelectric power, or thermopower, of material m. (Note: Thermopower is a commonly accepted misnomer.) Contrary to an oft stated misconception, $S_m(T)$ is a unique, well-defined property of a single homogeneous material and does not depend on a second material for its existence. This misconception arises from the practical constraint of measuring $S_{AB}(T)$, which necessitates prior knowledge of either $S_A(T)$ or $S_B(T)$ before a separation of terms is possible. Confusion about the sign of thermopower is also common; if $T_2 = T_1 + \Delta T$ causes the polarity shown in Fig. 1, then $S_A > S_B$.

The Peltier and Thomson effects refer to reversible heat processes. Passage of an electrical current density J across an isothermal boundary formed by two different materials causes the generation or absorption of a Peltier heat P_p at the junction. Specifically, if current flow from material A into material B generates heat, then reversing the current direction causes heat to be absorbed. Similarly, passage of an electrical current through a homogeneous material subjected to a temperature difference ΔT produces a Thomson

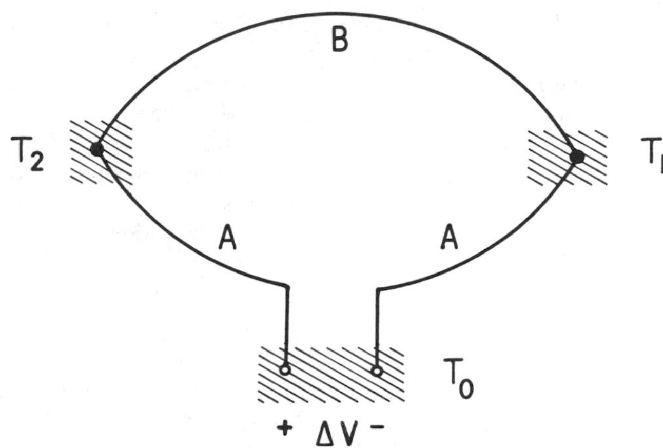

FIG. 1. Thermoelectric circuit from materials A and B. The shaded regions are isothermal regions having the indicated temperatures.

heat P_t. Whether P_t is generated or absorbed depends on the direction of J relative to ΔT. The dependence of both P_p and P_t on current direction permits the unique separation of these heats from irreversible Joule heating, which depends upon J^2 and is always a generated heat. The Peltier coefficient π_{AB} and the Thomson coefficient σ_t are proportionality constants in the expressions

$$P_p = \pi_{AB} J, \tag{3}$$

$$P_t = -\sigma_t J \Delta T. \tag{4}$$

The preceding three coefficients are not independent but are, instead, related through the Kelvin relations

$$\pi_{AB} = T(S_a - S_B), \tag{5a}$$

$$\sigma_{tA} = T \frac{dS_A}{dT}, \tag{5b}$$

which were developed by Thomson (later Lord Kelvin) on the basis of thermodynamic arguments. The original thermodynamic arguments are not rigorous, but the same results emerge from a rigorous analysis using irreversible thermodynamics. Equation (5b) can be integrated to obtain

$$S_A(T) = \int_0^T \left(\frac{\sigma_{tA}(T')}{T'} \right) dT' \tag{6}$$

where $S_A(0)$ is set equal to zero to comply with the third law of thermodynamics. Thus, a careful measurement of the Thomson heat as a function of temperature can be used to determine the absolute thermopower of any material. A measurement of this type provides the basis for absolute values of thermopower and serves as direct refutation of the misconception discussed below Eq. (2).

Thus far the discussion incorrectly implies that thermoelectric effects are scalar quantities. A general analysis relates electrical and heat current densities, **J** and **U**, respectively, to electrical potential and temperature gradients, ∇V and ∇T, respectively, and produces thermoelectric effects which are tensors. For isotropic and cubic materials these tensors reduce to scalars. For anisotropic materials and for materials in the presence of a magnetic field (i.e., thermomagnetic effects) the tensor form is necessary. The study of thermomagnetic effects is becoming increasingly important, but the complexity of the topic precludes any mention or discussion of specific effects.

Turning from the preceding macroscopic viewpoint to a microscopic viewpoint results in thermoelectric effects being associated with deviations from equilibrium distributions of charge carriers. These deviations are known to occur in two ways. Thermal and potential gradients directly affect the distributions and produce "diffusion contributions" to thermoelectric effects. In solids the presence of thermal gradients also causes the phonon distribution to differ from its equilibrium value. This difference means that the phonons contribute to the heat current and that interactions between nonequilibrium phonons and charge carriers alter the distributions of the latter. This alteration produces additional thermoelectric contributions, called "phonon-drag contributions." Both types of contributions have been well documented in semiconductors, semimetals, and metals.

The temperature dependence of thermoelectric effects makes a concise summary of magnitudes impossible. For many materials even the sign of an effect changes with temperature. However, the general pattern of significant, low-temperature phonon-drag contributions which decrease at higher temperatures permits a crude comparison of thermopowers near room temperature. Metals have values of either sign ranging from ~ 0.1 to nearly 40 $\mu V/K$. Semiconductors produce values of either sign from ~ 1 to 10^3 $\mu V/K$. The essential difference in the physics of these materials is the thermal statistics: Semiconductors require Maxwell–Boltzmann statistics while metals require Fermi–Dirac statistics.

It is commonly stated (and generally believed) that the thermopower is identically zero in superconductors but these materials require very careful consideration. This statement is an excellent practical guide but it is not true in general. Thermal gradients in superconductors modify the distributions of charge carriers. If the superconductor is anisotropic, or if boundary conditions disturb the surface sheath that would produce zero magnetic flux within an isotropic superconductor, or if magnetic flux is present in the superconductor, then the charge carrier modifications produce thermoelectric effects. Our present understanding of these effects is limited but it is likely to be expanded in the next few years. The new high-T_c superconductors (with their strong anisotropy, short coherence lengths, and greater diversity in flux effects) and the heavy-fermion superconductors (with their large mass enhancements) provide a rich but as yet untapped opportunity to explore thermoelectric effects in superconductors.

The utility of thermoelectric effects can be divided into two standard categories, fundamental and applied. A free-electron model predicts that $S(T)$ is related to the sign of the dominant charge carrier, but $S(T)$ is also sensitive to the details of scattering mechanisms. The latter dependence is more important and causes a number of errors in sign prediction. Thus, from a fundamental perspective, thermoelectric and thermomagnetic effects provide complicated but sensitive probes for studying scattering mechanisms, band structure, and Fermi-surface effects. From an applied viewpoint, thermoelectric effects provide a form of thermometry (thermocouples), a mechanism for converting heat energy into electrical energy, and a reversible method for heating or cooling. Metals are generally used for the thermocouple application and semiconductors for the other two applications.

See also ELECTRON ENERGY STATES IN SOLIDS AND LIQUIDS; PHONONS.

BIBLIOGRAPHY

F. J. Blatt, P. A. Schroeder, C. L. Foiles, and D. Greig, *Thermoelectric Power of Metals and Alloys*. Plenum, New York, 1976. (Intermediate level review.)

C. A. Domenicali, *Rev. Mod. Phys.* **26,** 237 (1954). (Excellent treatment from perspective of irreversible thermodynamics, advanced.)

T. C. Harman and J. M. Honig, *Thermoelectric and Thermomagnetic Effects and Applications*. McGraw-Hill, New York, 1967. (Advanced and detailed treatment stressing energy conversion processes.)

R. R. Heikes and R. W. Ure, Jr., *Thermoelectricity: Science and Engineering*. Interscience Publishers, New York, 1961. (Intermediate to advanced treatment emphasizing semiconductors and applications.)

D. K. C. MacDonald, *Thermoelectricity: An Introduction to the Principles*. Wiley, New York, 1962. (Elementary.)

A. H. Wilson, *The Theory of Metals,* 2nd ed. Cambridge University Press, Cambridge, 1953 (Immediate treatment for metals, semimetals, and semiconductors.)

P. J. Hirschfeld, *Phys. Rev.* **B38,** 9331 (1989); R. P. Huebener, *Magnetic Flux Structures in Superconductors*. Springer-Verlag, Berlin, Heidelberg, New York, 1979. (Discussions of thermoelectric effects in superconductors.)

Thermoluminescence

Paul W. Levy

The basic thermoluminescence phenomenology is well illustrated by the following example. It applies to a large fraction of the transparent crystals which can be prepared with one or more different kinds of impurities and to a much smaller fraction of the so-called "pure" crystals. If these crystals are heated soon after they are grown, they will not emit light. However, if they are subsequently exposed to a few rad of ionizing radiation and then reheated, light will be emitted. Thus, thermoluminescence is the light which is emitted by a substance when it is heated and which can be attributed to previous exposure to ionizing radiation. The usual thermoluminescence measurement is described by a curve of light intensity vs. temperature, called a *glow curve,* containing one or more "peaks," called *glow peaks*. Natural quartz, calcite, i.e., $CaCO_3$:Mn, KCl:Tl, Al_2O_3:Cr, and ZnS:Cu are examples of materials in which the thermoluminescence is related to the presence of impurities. Many of the alkali halides exhibit thermoluminescence which is associated with the presence of defects and color centers.

With one important exception, most of the basic physical processes involved in thermoluminescence are reasonably well understood and are described in well-known books [1–

5]. However, these books do not adequately describe the physical processes responsible for glow curves containing more than one glow peak. This is an important reservation, since glow curves containing a single glow peak are rarely observed, if at all. However, for ease of explanation, consider first glow curves containing a single glow peak. Such curves are the subject of almost all thermoluminescence "theory" in the literature. When exhibiting a single glow peak the thermoluminescent solid must contain two different types of atomic or molecular defects or impurities. One type functions as an electron trap or (electronic) hole trap. The second essential defect or impurity is the center responsible for the emission of light. Consider first the role of the charge-trapping centers. When the crystal is exposed to ionizing radiation, electron-hole ionization pairs are formed. Usually a very large fraction of the ionization-induced charges recombine in milliseconds or less at a variety of defects and impurities termed recombination centers. Often the recombination process is accompanied by light emission, which is classified as fluorescence, phosphorescence, or collectively, as radioluminescence. Important for the thermoluminescence process are the remaining electrons and holes, not undergoing immediate recombination, which are captured (trapped) by defects or impurities, usually referred to as electron, hole, or charge traps. Consider next the centers that emit light or thermoluminescence when the crystal is subsequently heated. The most easily visualized luminescent center, and one of the most commonly occurring, is also a hole trap. When previously irradiated crystals are heated, electrons released thermally from traps "wander about" until they interact with a hole trapped on an emission center. As part of the electron-hole-recombination, or charge-neutralization, process the electron makes one or more electronic transitions that are accompanied by photon emission. Such centers can be called recombination luminescent centers. A second type of luminescent center is an electron trap which contains a ground state and at least one or more excited states. When a thermally released electron encounters such a center, it is initially trapped in one of the upper levels. Usually the lifetime of the excited state is very short and the trapped electron quickly makes a transition to the ground state with the emission of light. When the crystal reaches a higher temperature, the trapped electron may be thermally released, perhaps to enter into a second light emission process. However, at higher temperatures it is more likely that the trapped electron will recombine with a thermally released hole, with or without the emission of light. When functioning as an emission center, this type of defect may be called a transition luminescent center. Thus, a single type of defect or impurity may be an electron trap, a luminescent center, a recombination center, or some combination of these.

To an appreciable extent, there is a duality or reciprocal character associated with the sign of the charge which interacts with the various centers. In the preceding paragraphs, the words electron and hole can usually be interchanged without violating physical principles. However, a few of the processes occasioned by this charge interchange procedure would be encountered only rarely.

The defects and impurities responsible for luminescence and thermoluminescence introduce localized electronic states into the crystal lattice [4–6]. Qualitatively such centers are easily understood. If a normal lattice atom is replaced substitutionally by an atom whose valence is either more or less than the replaced atom, a local charge unbalance is created. Substitutional impurities lacking an electron would tend to trap ionization electrons and impurities containing an excess electron (lacking a hole) would tend to lose an electron (capture a hole). Equally important are lattice defects that create local charge unbalance. They also preferentially trap whatever ionization charge restores local charge neutrality. For example, the defect formed by a missing negative ion in an alkali halide, such as NaCl or KBr, tend to trap an electron. The well-known F center is formed when an ionization electron is trapped on a Cl^- ion vacancy in an alkali halide, or on similar negative-ion vacancies in other types of crystals. F centers may be, in the proper circumstances, electron traps (to form dielectron or F' centers), recombination centers, and both transition- and recombination-type luminescent centers.

Another type of center important for thermoluminescence processes is the substitutional impurity having the same valence as the atom, or molecular ion, it replaces. In these cases, the atom can function as an electron trap, as a hole trap, or in both ways. The properties of such centers depend on the detailed electronic interactions between the impurity atoms and the neighboring atoms [6].

THERMOLUMINESCENCE KINETICS

The theory describing thermoluminescence glow curves, the shape of the individual glow peaks, the dependence on heating-rate, pre-heating irradiation dose, and physical parameters, such as trapped charge concentration, activation energy, etc., is traditionally described as thermoluminescence kinetics, not as thermoluminescence theory. This is a consequence of the fact that almost all thermoluminescence theory describes kinetic processes.

Kinetics of Glow Curves Containing a Single Glow Peak

Until recently almost all thermoluminescence theory described glow curves that contained only a single glow peak. It cannot be emphasized too strongly that only one or two materials appear to exhibit single peak glow curves (unless single-peak curves are created by controlled heating) and these do not appear to have the characteristics, particularly the shape, expected from the single glow peak kinetics (theory) described below. More than a thousand materials exhibit glow curves containing more than one peak; a rough average is 3 to 5 peaks. Since single glow peak theory is used in nearly all published thermoluminescence papers, it is included here. Moreover, it is the basis for kinetics describing glow curves containing more than one peak.

The existing thermoluminescence theory (kinetics), for both one glow peak and multiple glow peak glow curves, employs one or the other of two quite distinct approaches. One approach can be called the "rate equation" method. This approach is based on equations that describe the flow of charges from the various traps to the conduction and valence bands, from the bands to traps, i.e., retrapping, and to the light emitting recombination luminescence centers, as the temperature is increased (usually linearly). This approach can be used to describe almost every conceivable thermoluminescence material and/or trap configuration. The resulting equations require a rate constant for each process. This method suffers from at least two major, and several minor, difficulties. First, except in the simplest cases it is usually difficult, and occasionally almost impossible, to meaningfully relate the solutions in terms of rate constants to physical parameters, such as the probability that a given type of trap will capture free charges. Second, it is often difficult to obtain solutions of the rate equations by computerized numerical techniques. Except for all but very simple cases, this is attributable to the occurrence of controlling terms describing conduction band concentration(s), and related processes, that are numerically many orders of magnitude smaller than other terms. (In principle these difficulties can always be overcome by known programming techniques and the use of sufficient computer time.) The rate equation approach will not be described additionally in this article.

The second approach to thermoluminescence kinetics for single and multiple glow peak curves, the one detailed here, does not have a generally accepted name but can be called the "traditional" or "competition equation" approach. It is based on the original equation that is the basis for the well-known first- and second-order thermoluminescence kinetics for one glow peak, introduced by Randall and Wilkins [7] and Garlick and Gibson [8]. This approach is readily generalized to situations at least as general as those described by the rate equation approach.

Consider, as described in references 7 and 8, the thermoluminescence originating in a solid containing a single type of electron (or hole) trap and a single type of hole (or electron) trap that is also a recombination center that emits light as a result of electron-hole recombination, i.e., a luminescent center. Let N_0 be the concentration of electron traps in the solid and n_0 the concentration of electrons, introduced by ionizing radiation, that are trapped on the N_0 traps. This approach requires a number of conditions that are usually not stated. For example: 1) The hole trap concentration is at least n_0. 2) The trapped electron and hole concentrations are equal; as would be required by charge neutrality. 3) The electron and hole trap concentrations must be constant, i.e., not changed by irradiation; this is required for the resulting equation(s) to apply to different values of n_0. 4) n_0 does not change between irradiation and measurement. Other implied conditions could be mentioned. For example, traps are usually populated at temperatures at which all thermal untrapping is negligible. Consider next the processes occurring when the sample is heated. Also, assume that the kinetic parameters, described below, are such that only electrons escape into the conduction band. The flow of electrons to the conduction band, i.e., number per unit time, is given by the well known term $ns \exp(-E/kT)$, where n is the trapped charge concentration at temperature T, s the preexponential or "attempt-to-escape" frequency, E the activation energy for thermal untrapping, k the Boltzmann constant, and T the temperature in degrees Kelvin. These definitions require that certain other conditions also apply. For example, both s and E are independent of temperature, an assumption that appears to be quite good near and above room temperature. Electrons escaping into the conduction band are subject to only two competing processes. 1) They may be retrapped in empty traps, or 2) they may recombine with trapped holes to produce luminescence. ("Double charge" trapping on traps containing charges is also assumed negligible.) The probability that an electron in the conduction band is retrapped is proportional to $\sigma_t(N_0-n)$ where σ_t is the cross section for any conduction band electron to be trapped in the (N_0-n) empty electron traps. Likewise the probability that a conduction band electron enters into light producing recombination is proportional to $\sigma_r n$ where σ_r is the recombination cross section and n is the concentration of trapped holes. For this simple case the trapped hole concentration is equal to the concentration of trapped (and the negligible conduction band) electrons. Then the probability that a conduction band electron will enter into either processes, i.e., retrapping or recombination, is proportional to $\sigma_t(N_0-n)+\sigma_r n$.

The equation describing the flow of electrons to recombination centers during heating, which gives the thermoluminescence light intensity, is

$$I(t)=-\frac{dn}{dt}=nse^{-E/kT}\frac{\sigma_r n}{\sigma_r n+\sigma_t(N_0-n)} \tag{1}$$

The first factor in the right-hand expression is the number of electrons reaching the conduction band in time increment dt and the remaining part gives the fraction of these entering into thermoluminescence, or light producing recombination.

In the usual thermoluminescence measurement the sample is irradiated at a temperature low enough to ensure that the term $\exp(-E/kT)$ is negligible and light emission (phosphorescence) is undetectable. Subsequently, during a measurement the sample temperature is increased in a controlled manner as the light emission is recorded. Usually the temperature is increased linearly. There are two reasons for this. First, a precise and reproducible linear temperature increase is the easiest to achieve experimentally. Second, a linear temperature dependence leads to simple solutions of Eq. (1). In this case temperature and time are related by the simple expression $T=T_0+\beta t$ or $dT=\beta\, dt$, where β is the heating rate in degrees per unit time.

Using a linear temperature increase Eq. (1) becomes

$$I=-\frac{dn}{dT}=n\frac{s}{\beta}e^{-E/kT}\frac{\sigma_r n}{\sigma_r n+\sigma_t(N_0-n)}. \tag{2}$$

To avoid confusion (see below), Eq. (2) will be called the General One Trap, or GOT, equation. It was originated by Randall and Wilkins [6], who emphasized only the case that is well known as first-order thermoluminescence kinetics. This is obtained by assuming retrapping is negligible, or $\sigma_r n \gg \sigma_t(N_0-n)$. For negligible retrapping, Eq. (2) becomes

$$I=-\frac{dn}{dT}=n\frac{s}{\beta}\exp(-E/kT), \tag{3}$$

which has the solution

$$I=n_0 s\exp(-E/kT)\exp\left(-\int_{T_0}^{T}\frac{s}{\beta}\exp(-E/kT)\,dT\right) \tag{4}$$

Equation (4), or (3) and (4), are usually referred to as the first-order thermoluminescence equation(s) or the Randall–Wilkins equation(s). It usually applies only when recombination is very much more likely than retrapping. If a constant fraction of the recombination events produce luminescence, the shape of an observed glow curve described by (4) would not change, a fact not determinable from glow curves.

When retrapping is nonnegligible, Eq. (1) must be replaced by an approximation that includes this process. The very widely used approximation that includes recombination is obtained by assuming $\sigma_t = \sigma_r$, i.e., that the recombination and retrapping cross sections are equal. Currently, this assumption does not appear to be widely applicable. Using $\sigma_t = \sigma_r$, Eq. (1) yields

$$I = -\frac{dn}{dt} = \frac{n^2}{N_0} s \exp(-E/kT). \qquad (5)$$

This equation differs from (3) in that the n term is replaced by an n^2 term. In this case, the dependence of thermoluminescence intensity on temperature obtained with the usual linear temperature rise is given by

$$I = n_0{}^2 s \exp(-E/kT) \times N_0 \left[1 + \frac{n_0}{N_0} \int_{T_0}^{T} \frac{s}{\beta} \exp(-E/kT)\, dT\right]^{-2}. \qquad (6)$$

All of the parameters in Eq. (4) have have been defined above, including N_0, which is the concentration of electron traps.

Equation (6), or (5) and (6), are usually referred to as the second-order thermoluminescence equation(s) or the Garlick–Gibson equation(s). Equations (3)–(6) are often referred to as first- and second-order equations since the concentration, n, appears with power 1 in Eq. (3) and with power 2 in Eq. (5), in agreement with common "kinetic" usage. The principal features associated with first- and second-order glow peaks are illustrated by Fig. 1. The total area under both the first-order and second-order kinetic curves is proportional to the initial trapped charge concentration n_0. Most importantly, the shape of the first-order curve is independent of n_0, i.e., the shape and the temperature of the peak maximum do not depend on the ionizing radiation exposure used to create the initial trapped charge concentration. In contrast, the second-order curve shapes depend on the ratio n_0/N_0, which in turn depends on the exposure used to establish the initial trapped-charge population. Also, the temperature of the maximum depends on the ratio n_0/N_0. This dependence on n_0/N_0 can often be used to differentiate between the two types of kinetics.

The properties of the General One Trap (GOT) equation, which is the general case from which the first- and second-order kinetics are derived, have only recently received the attention they deserve [9]. Since it is unlikely that the approximations used to obtain Eqs. (4) and (6) will apply to more than a small fraction of any materials that exhibit single glow peaks, it is likely the GOT kinetics will have wide applicability to any glow curves that truly contain a single glow peak. Also, the GOT case is intimately related to the first- and second-order cases. Numerical solutions of the GOT equation, i.e., Eq. (2), are shown in Fig. 2 for "typical"

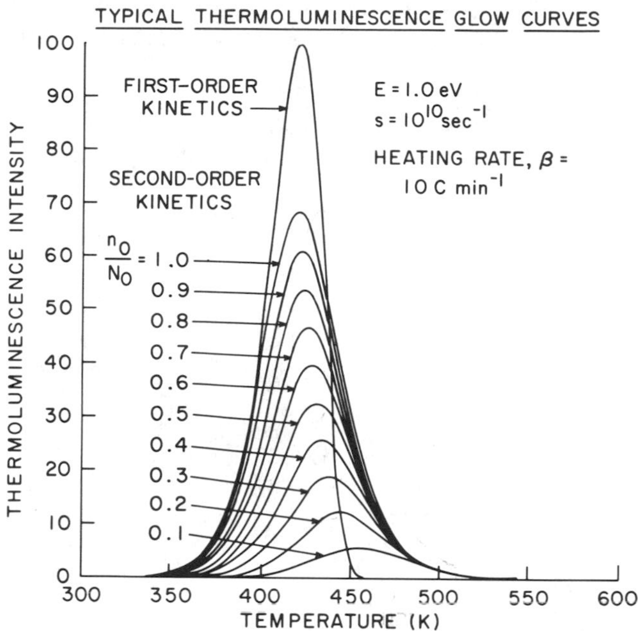

Fig. 1 First- and second-order glow curves containing a single glow peak computed with "typical" parameters. The first-order curve shape and peak temperature are independent of the initial trap charge concentration. The second-order curve shape and peak temperature depend on the fraction of traps initially populated.

parameters. The analytic solution(s) of the GOT equation(s) are involved and transcendental. Numerical examples obtained from analytical solutions are at least as difficult to obtain, and apply to data, as direct numerical solutions of (the differential) Eq. (2). This figure shows that the glow curves computed from the GOT equation for high initial electron trap occupancy, $n \sim N_0$, (but $n < N_0$), closely resemble first-order curves over the range where most of the trapped charge is released, and corresponds to the temperature range that provides sufficient intensity to measure experimentally. At appreciably lower, i.e., intermediate, n_0 values the easily measurable part of the glow curves do not resemble either the first- or second-order curves. At very low n_0 values, the easily measurable part of the glow peaks closely resemble the second-order curves. Both of these cases are obtained analytically from Eq. (2) by considering the case $n_0 \simeq N_0$ and $n_0 \ll N$, or equivalently $\sigma(N-n) \ll N_r$ and $\sigma(N-n) \gg n_r$, where $\sigma = \sigma_t/\sigma_r$ [9]. These relations between GOT and first- and second-order thermoluminescence kinetics emphasize the fact that experimentally recorded glow curves can be readily compared only in the region where the intensity is "out of the-noise." The high temperature "tails" differ but are "in the-noise" and are very difficult to study experimentally.

Kinetics of Glow Curves Containing Two or More Glow Peaks

As emphasized above, until recently the existing thermoluminescence theory, i.e., kinetics, has been concerned almost exclusively with single glow peak cases. This has persisted even though very few, or possibly not one, material has been found that exhibits only a single glow peak when

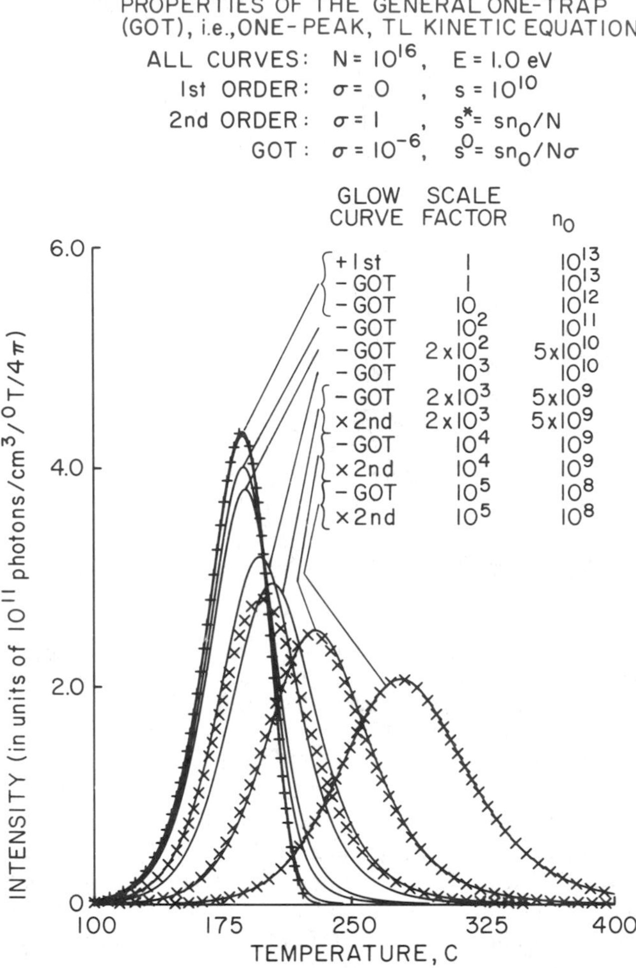

FIG. 2. Properties of the General One Trap (GOT) thermoluminescence equation—for a single peak glow curve—as demonstrated by comparison with first- and second-order glow peaks for the same initial trapped charge concentrations, i.e., doses, and other parameters. At high (low) initial trapped charge the GOT curves resemble first-order (second-order) curves in the *regions where the intensity is large enough to readily measure*. Thus the measured single peak glow curves may appear to be first- or second-order, even though the retrapping cross sections requirements for first- and second-order *are not fulfilled*.

irradiated at room temperature. Also, very few studies have been made on single peak systems, even though it is relatively simple to create single peak data by removing the lower temperature peaks by controlled heating at temperatures below those seriously depleting the highest temperature peak. The kinetics detailed in this article that describe glow curves with more than one glow peak and describe individual peaks that are not completely independent of each other, have been termed interactive kinetics. The interactive kinetic equations are obtained by extending the "traditional" or "competition equation" approach used to derive the GOT equation, Eq. (2), and the first- and second-order equations, to systems with more than one type of electron and/or hole trap. In interactive kinetics each type of electron trap (and hole trap) is characterized by its own set of parameters. Trap conc (concentration is abbreviated conc) N_{ei}; trapped electron conc n_{ei}; initial trapped electron conc before heating n_{0ei}; activation energy for thermal untrapping E_{ei}; preexponential factor s_{ei}; cross section for retrapping with conduction band electrons σ_{tei}; and cross section for radiative or nonradiative recombination of trapped holes with valence band holes σ_{rei}. The corresponding parameters for each type of hole trap are obtained by exchanging e for h and conduction band for hole band. It must be emphasized that electrons (or holes) thermally excited to the conduction (or valence) band may enter into a number of different interactions. In particular, both retrapping and recombination may occur and charges released from one type of trap may be retrapped in deeper traps that contribute to the glow only at high temperatures, or momentarily trapped in previously "emptied" traps. Also, and importantly, if more than one recombination process occurs, it is possible that one or more process may be nonradiative and not contribute light to the observed emission. However, such recombination must be included in the kinetic equations to maintain charge balance.

The equations given below do not include all possible processes, and a number of unlikely interactions are excluded. For example: 1) Conduction to valence band recombination is not included since it occurs very rarely in thermoluminescence, if at all. 2) It is assumed that the trap concentrations remain constant after premeasurement irradiation. This may not occur if one of more types of traps are "annealed out" by heating. 3) It is assumed that the premeasurement irradiation does *not* introduce traps. This assumption is, almost certainly, not valid for many thermoluminescence applications in geology. It does not apply to many materials subject to doses greater than 10^5 to 10^7 rad and to materials such as the alkali halides that are susceptible to the ionization damage process. (See COLOR CENTERS and RADIATION DAMAGE IN SOLIDS articles.) 4) Processes that cause two or more charges to be trapped on the same trap (e.g., electron trapping on F centers to create F' centers) are excluded.

The general interactive kinetic equations, given below, apply to almost all thermoluminescence materials. The restrictions given above are included to illustrate their limitations.

$$\frac{dn_{ei}}{dt} = -n_{ei}s_{ei}e^{-E_{ei}/kT} + E\,\frac{\sigma_{tei}(N_{ei}-n_{ei})}{R_e+U_e} - H\,\frac{\sigma_{rei}n_{ei}}{R_h+U_h} \tag{7}$$

$$\frac{dn_{hi}}{dt} = -n_{hi}s_{hi}e^{-E_{hi}/kT} + H\,\frac{\sigma_{thi}(N_{hi}-n_{hi})}{R_h+U_h} - E\,\frac{\sigma_{rhi}n_{hi}}{R_e+U_e}, \tag{8}$$

where

$$E = \sum_{i=1}^{u} n_{ei}s_{ei}e^{-E_{ei}/kT} \tag{9}$$

$$H = \sum_{i=1}^{v} n_{hi}s_{hi}e^{-E_{hi}/kT} \tag{10}$$

$$R_e = \sum_{i=1}^{v} \sigma_{rhi} n_{hi} \tag{11}$$

$$R_h = \sum_{i=1}^{u} \sigma_{rei} n_{ei} \tag{12}$$

$$U_e = \sum_{i=1}^{u} \sigma_{tei}(N_{ei} - n_{ei}) \tag{13}$$

$$U_h = \sum_{i=1}^{v} \sigma_{thi}(N_{hi} - n_{hi}). \tag{14}$$

The sums $i = 1...u$ and $i = 1...v$ refer to the u different types of electron traps and the v different types of hole traps. Charge neutrality requires

$$\sum_{i=1}^{u} \frac{dn_{ei}}{dt} = \sum_{i=1}^{v} \frac{dn_{hi}}{dt}. \tag{15}$$

Equations (7) through (15) describe charge conservation during the interactive thermoluminescence process. The corresponding glow curves can be calculated in two ways. First, by superimposing the net untrapping from each type of electron and/or hole trap that results in recombination, and differentiating between radiative and nonradiative recombination processes, the equations for a single peak glow curve, described above, result from this approach and assume the recombination is radiative. Second, the recombination at each type of trap can be computed and the glow curve obtained by superimposing only the radiative recombination contributions. In other words the thermoluminescence glow curve is given by

$$I(t) = \sum_{i=1}^{u'} \delta_{ei} H \frac{\sigma r_{ei} n_{ei}}{R_h + U_h} + \sum_{i=1}^{v'} \delta_{hi} E \frac{\sigma_{rhi} n_{hi}}{R_e + U_e}, \tag{16}$$

where δ_{ei}, $\delta_{hi} = 1$ for radiative recombination and 0 for nonradiative recombination, and u and v have been replaced by u' and v' to emphasize the fact that the sums include only the radiative recombination contributions to the glow curve. This procedure is illustrated in the example given below. In applying these equations more than half of the terms can usually be neglected, e.g., hole untrapping can be neglected for deep traps in which all of the trapped holes are "used up" by recombination with electrons released from shallow traps at temperatures lower than those causing the holes to untrap.

If more than one recombination process contributes to the thermoluminescence, it is highly probable that more than one emission band will be observed. To analyze data, it may be necessary, or informative, to compute separate $I(t)$ curves for each band.

Applications of Interactive Kinetics to Observed Glow Curves

The interactive kinetics approach explains numerous features of measured glow curves that were regarded as anomalous. This is illustrated by a study of a system containing three types of electron traps and one type of hole trap, in which the hole trap is also a radiative recombination center. Also, in this example it is assumed that holes are not thermally released from the hole trap, which would occur if the appropriate activation energy were sufficiently large. The pertinent equations are obtained by putting $u = 3$ into (7) through (15) $\delta_{ei} = 0$ and $\delta_{h1} = 1$ in (16) and not including equations describing hole untrapping. In this case (as in almost all cases), it is convenient to express all cross sections as ratios, i.e., $\sigma_i = \sigma_{tei}/\sigma_{rh1}$. Figure 3 shows numerically computed glow curves for the given initial trap charge concentrations and other numerical values given in the caption. Also shown are first- and second-order glow curves computed with the appropriate corresponding numerical values. Figure 3 demonstrates numerous usually unexplained features of measured glow curves found in the literature; only the more important are mentioned here: 1) The glow peak shapes, peak temperatures, and relative heights depend

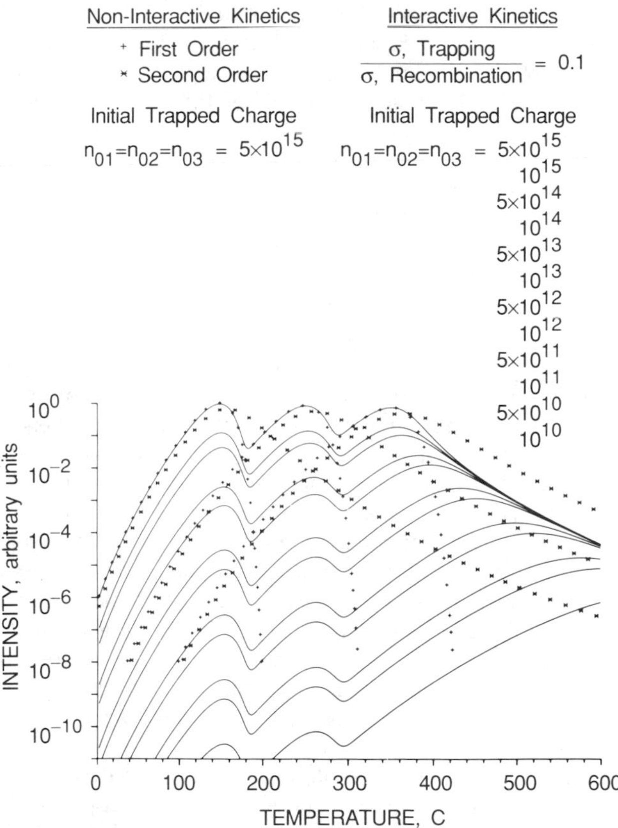

FIG. 3. Properties of the Interactive Thermoluminescence Equation—as demonstrated by a system containing three types of shallow electron traps and one type of deep hole trap recombination center—for the indicated initial trapped charge concentrations. The other parameters are: $N_{01} = N_{02} = N_{03} = 10^{16}$; all $\sigma_{tei}/\sigma_{rei} = 0.1$; $s_i = 10^{10}$; $E_1 = 1.0$, $E_2 = 1.25$, and $E_3 = 1.50$ eV. Shown for comparison are first- and second-order curves for comparable parameters. Note the large differences between curves that would not be detected with conventional thermoluminescence apparatus. Interactive kinetics "explains" a large fraction of the anomalous features in published glow curves.

strongly on the initial trapped charge conc. 2) With increasing trapped charge conc, the curves become increasingly like first- and second-order curves. 3) At low initial trapped charge conc, the height of the low-temperature peaks is so much lower than the high-temperature peak that the low-temperature peaks may not be observed with conventional thermoluminescence apparatus. 4) Peak height vs. "dose" curves may not be linear; in particular they may be supralinear.

A very large fraction of glow curves with more than one peak can be "explained" by interactive kinetics. This can be demonstrated by regarding glow curves, such as those contained in Fig. 3, as "data," and analyzing each glow peak using procedures based on the first- and second-order kinetic expressions [10]. Usually this is best done by computerized best-fit procedures. For example, this type of analysis of the curves in Fig. 3 shows: 1) While the curves shown were computed with activation energies of 1.00, 1.25, and 1.50 eV, the best-fit first-order analysis gave corresponding E values that ranged from 0.978 to 1.002 for the 1.00 eV peak, from 0.789 to 1.841 for the 1.25 eV peak, and from 1.025 to 1.31 for the 1.50 peak. The corresponding best-fit second-order analysis gave E values of 1.25 to 1.43 for the 1.00 eV peak, 1.89 to 1.25 for the 1.25 peak, and 1.49 to 2.125 for the 1.50 peak. 2) While "s" values of 10^{10} were used for all computed "data," the "s" values obtained from the first-order analysis ranged from 4.7 to 6.9×10^{9}, 5.4×10^{9} to 3×10^{11}, and 4.0×10^{4} to 2.4×10^{8} for the three peaks. Similarly, while the "s^*" ($s^*=sn_0/N$) values for second-order fits should be $<10^{10}$, the fitting procedure gaves values ranging from 2.0×10^{7} to 2.7×10^{15}.

Thus, interactive thermoluminescence kinetics, and in particular the study of glow curves computed from interactive kinetics and analysed—to simulate the analyses of measured glow curves described in the literature—by techniques that assume they are described by first- and/or second-order kinetics, provides explanations for most of the anomalous thermoluminescence results in the literature. For example: 1) Interactive kinetics explains why the glow peaks in glow curves containing more than one peak (practically all glow curves) always overlap. 2) It explains why the temperature of the glow peak maxima, the peak shapes, and the relative intensities usually depend, often strongly, on the premeasurement dose, i.e., on the trapped charge concentration when the measurement is started. 3) It accounts for curves of peak height, or area, vs. dose that exhibit supralinearity. 4) Interactive kinetics explains why the kinetic parameters, E, s, s^*, s', etc., obtained by using analyses based on first- and second-order kinetics are often not physically realistic and, in addition, why they often depend strongly on the premeasurement trapped charge concentration or dose. In particular, it explains why many published s, s^*, or s' values are often orders of magnitude outside of the physically realistic range. In other words, first- and second-order kinetics do not account for retrapping by more than one type of trap, cases where $\sigma_r\neq\sigma_t$, etc., and this changes, often appreciably, the shape of the glow curves. This shape change occurs even though the untrapping of individual charges is described by the usual $s\exp(-E/kT)$ term. 5) The interactive kinetic equation that applies to single glow peak systems, the GOT equation, is the equation from which first- and second-order kinetics are obtained as special cases. The use of the GOT equation to analyze experimental data is difficult. Mathematically, GOT glow curves are always different from the first- and second-order curves. However, the difference(s) is in the low intensity "wings," where it is usually obscured by measurement "noise." Furthermore, in the high-dose limit (large initial trapped charge), the GOT equation approaches the first-order equation and in the low-dose limit (low initial trapped charge) it approaches the second-order equation. 6) Lastly, although not described in this article, the interactive kinetics concept provides explanations for many observations on thermoluminescence materials resulting from optical bleaching, premeasurement thermal treatment, "aging," heating rate effects, etc.

EXPERIMENTAL KINETIC PARAMETER DETERMINATIONS

As explained, the published thermoluminescence research often includes attempts to determine if particular glow peaks are best described by first- or second-order kinetics, or occasionally some variant such as partial-order kinetics, and to determine the kinetic parameters E, s, s^*, s', etc. It must be strongly reiterated that all such kinetic determinations are based on the assumption (usually not mentioned) that the physical process producing each peak is independent of the processes producing the other peaks. Also, as emphasized above, many of the physically unrealistic parameters reported in the literature can be explained by interactive kinetics. Nevertheless it is quite likely that some peaks can be closely described by first- and second-order kinetics in the temperature range where the measurable intensity is sufficiently "out-of-the-noise." For example: 1) The easily measurable part of some glow peaks may be well described by first-order kinetics if the recombination cross section is so large retrapping is negligible; an example is given in reference 11. 2) In systems with more than one peak the highest temperature peak may occur at temperatures where all retrapping (or the effects of) is negligible and the peak is described by GOT, first- or second-order kinetics.

Currently, techniques for analyzing single glow peaks can be divided into two groups. 1) Procedures using computerized "best-fit" methods. Or, 2) procedures used for many years for determining glow peak kinetics and parameters from measured peak properties, such as peak temperature, full width at half maximum, variation in heating rate, etc. Procedures in the second category are described, in detail, in the book by Chen and Kirsh [5]. Clearly, there are advantages in analyzing glow curves containing one glow peak, or more than one glow peak, by computerized best-fit procedures. Examples of kinetic parameters obtained by best-fit methods are given in references 9, 10, and 11. The use of computerized best-fit procedures for glow curve analysis usually does not require specially written software, such as that employed for reference 11. The kinetic equations can be inserted into the best-fit procedures available for use on large computers. To maximize the usefulness of computer analysis the measured glow curves should be recorded digitally at closely spaced temperature intervals.

THERMOLUMINESCENCE MEASUREMENTS

Until recently almost all thermoluminescence measurements were made with apparatus in which a fraction of the light emitted by the sample impinged directly on the photodetector. The spectral-distribution information contained in the emitted light was ignored. Occasionally, emission spectra were obtained from efficient emitters, such as LiF TLD-100 dosimeter material, over the entire range of glow curve temperatures, or, in a few very favorable cases, the spectra were recorded over a single glow peak. With very few exceptions, the usual thermoluminescence data consisted solely of a curve of photodetector current vs. sample temperature or heating time. When the photodetector current is not proportional to the number of light-emitting processes occurring per unit time, such data are not suitable for kinetic determinations. This is in accord with the well-known principle. To determine the kinetics describing any process, the signal analyzed must be strictly proportional to the number of "reactions" occurring per unit time. For example, if the emission spectrum changes with temperature, especially in a region where the photodetector sensitivity changes rapidly with wavelength, curves of phototube current vs. temperature cannot be expected to provide reliable kinetic analyses. Actually, much published thermoluminescence data is unsuitable for reliable kinetic analysis. However, for many applications, including thermoluminescence dosimetry, dating of archaeological samples, and the authentication of fine art objects, photodetector current curves are quite adequate.

It is now possible to determine the spectral distribution of the thermoluminescent emission from a large number of materials at closely spaced temperature intervals. There are two reasons for this. In many samples the thermoluminescence intensity can be greatly enhanced by sufficiently large exposure to x rays, gamma rays, and other types of radiation. The second, and more important, is the development of highly sensitive apparatus for making spectral-distribution measurements as the samples are heated. The light gathering power of the apparatus can be enhanced by using large aperture optics. An approximate spectral distribution may be obtained with diode detector arrays. However, the most precise spectral information is obtained by using scanning spectrometers, whose wavelength dependent "thruput" and detector sensitivity have been determined. For maximum sensitivity photon counting can be used. And, most importantly, the measurements are controlled and the data recorded—at closely spaced intervals—with a small computer. For example, a 1024-point spectrum can be recorded at approximately 0.2°C temperature intervals. A typical spectrum obtained from such equipment (at Brookhaven National Laboratory) is shown in Fig. 4. In addition, inasmuch as the spectral sensitivity of this equipment can be determined, the recorded data can be converted to curves of absolute emission intensity vs. wavelength or photon energy. From such data glow curves can be obtained, for each observed emission band, that are eminently suitable for reliable kinetic analysis.

A large fraction of the individual features that are observed in thermoluminescence measurements are illustrated in Fig. 4. They include (1) The emission spectra contain both broad continuous bands which are characteristic of defects and certain impurities, e.g., in the 3.3-eV region, and sharp line spectra which are produced by rare-earth elements, e.g., at approximately 2.2, 2.3, and 2.6 eV. Also, there are regions where the sharp line spectra are superimposed on continuous spectra. (2) Some emission bands occur in certain temperature ranges, but not in others. (3) The glow curves corresponding to some of the emission bands differ from those obtained from other bands. (4) When different emission bands produce glow curves in the same temperature range, the individual glow peaks often have precisely the same shape. In other words, the glow curves reflect the charge untrapping process and are independent of the type of emission center. (5) An examination of Fig. 4 illustrates that many thermoluminescence emission features are obscured when only detector current is recorded, as is the case in most thermoluminescence measurements.

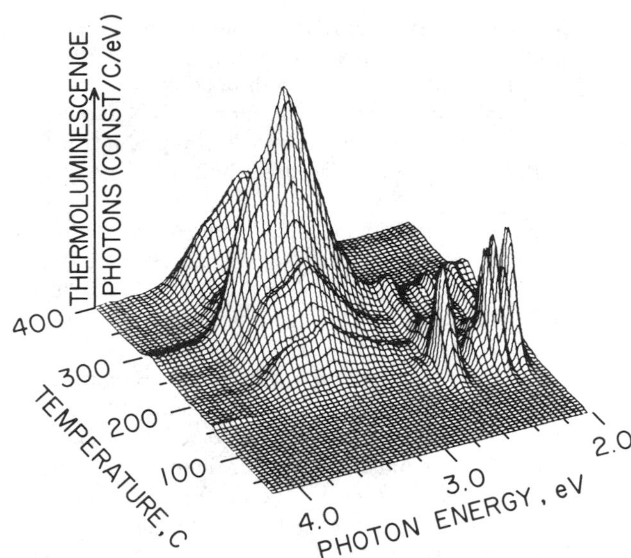

FIG. 4. A "three-dimensional" plot showing thermoluminescence intensity vs. photon energy (wavelength) and sample temperature. The measurements were made on a piece of lavender-colored fluorite, from Korea, after a 10°R-^{60}Co gamma-ray irradiation. This plot shows many of the various combinations of emission bands, line spectra, and glow peaks at various temperatures that can be observed during thermoluminescence measurements.

APPLICATIONS OF THERMOLUMINESCENCE

Currently, the most widely used application of thermoluminescence is in radiation dosimetry [2]. Ideally, for dosimetry the thermoluminescence signal, i.e., the glow-peak height or area, should increase linearly with the total exposure. In principle, this can be achieved with a material exhibiting first-order kinetics, as described above, in which the trapped-charged population increases in direct proportion to the total ionization. However, many of the available thermoluminescence dosimeters appear to be subject to a number of difficulties. The response depends on the nature of the radiation, it may be altered by exposure to light or may decay with time, etc. The signal vs. dose curve may be nonlinear and display "supralinearity." In the latter case, the response curve may increase linearly from "zero" dose to a given dose, curve upward with dose over a restricted dose range and, for larger doses, increase linearly with a slope larger than that observed initially. Supralinearity can be attributed to the retrapping described by interactive kinetics.

A second well-known application is the dating and/or authentication of fine art and archaeological objects. Pottery shards are the best example in this category. When the pottery was originally fired, the natural thermoluminescence was removed from any thermoluminescent materials present; usually this includes appreciable quantities of quartz crystallites. After firing, the quartz and other mineral components are continuously subjected to radiations from the potassium, uranium, and thorium impurities and their daughters contained in the object and, for best results, the surrounding material such as soil. To determine the age of the object, e.g., a shard, the thermoluminescence of a small sample e.g., 1–50 mg, is measured with high-sensitivity apparatus. Next, the sample is subjected to a series of laboratory irradiations and the intensity vs. dose, or sensitivity, characteristics are determined. From the sensitivity curve, it is a simple matter to determine the dose received by the object since it was last heated. Last, the natural radioactive impurity content of the object and, if possible, the immediate surroundings are determined by chemical or activation analysis, counting, or other means. Alternatively, in some instances the exposure rate at the point where the object was found may be determined with dosimeters. Once the radioactive impurity content and the radiation from the surroundings are known the annual dose may be computed. The age is the total exposure acquired by the object since it was fired divided by the annual exposure. At present, accurate age determinations are beset by numerous practical difficulties. However, in many instances the age may be determined accurately enough to determine if an object is authentic. For example, if an object appears to be Mayan, and is purported to have been excavated recently, but exhibits thermoluminescence appropriate for an age of 30 years it is clearly not authentic.

During the past decade, geological applications of thermoluminescence have increased appreciably. For example, thermoluminescence can be used to explore for lead-zinc deposits in carbonate rock. The laboratory radiation induced, or artificial, thermoluminescence intensity of host carbonate rock depends on the distance from the lead-zinc mineralization. It can be 3–5 times "background" 100 meters from the deposit, and increase to 10–100 times "background" near the deposit. Similar changes occur in the natural thermoluminescence of the quartz component of rock containing uranium ore. Also, a number of different kinds of rock deposits may be dated using techniques similar to those used to date archaeological artifacts. Lastly, thermoluminescence is now widely used, especially in Europe, to date air-borne (loess) and water-borne deposits. However, in this application the charge traps in the minerals are not "emptied" by heating, as in pottery dating, but by the optical bleaching caused by exposure to sunlight as the particles are being transported to the deposit site. Unfortunately, however, often the traps are not completely emptied. The trapped charge concentration reaches a (comparatively low) level that is a consequence of retrapping during exposure to sunlight. This retrapping is entirely analogous to the retrapping controlling all thermoluminescence kinetics except first-order kinetics. It appears that the use of thermoluminescence, and the closely related electron spin resonance (ESR) based methods, will increase appreciably in the near future.

See also Color Centers; Crystal Defects; Electron Energy States in Solids and Liquids; Luminescence; Radiation Damage in Solids.

REFERENCES

1. G. F. J. Garlick, *Luminescent Materials*. Oxford University Press, Oxford, 1949.
2. J. R. Cameron, N. Suntharalingam, and G. N. Kenney, *Thermoluminescent Dosimetry*. University of Wisconsin Press, Madison, 1968.
3. D. Curie, *Luminescence in Crystals*, translated by G. F. J. Garlick. Methuen and Co., London, 1963.
4. P. Braunlich, ed. *Thermally Stimulated Relaxation in Solids*. Springer-Verlag, Berlin, 1979.
5. R. Chen and Y. Kirsh, *Analysis of Thermally Stimulated Processes*. Pergamon, Oxford, 1981.
6. A. M. Stoneham, *Theory of Defects in Solids: Electronic Structure of Defects in Insulators and Semiconductors*. Oxford University Press, Oxford, 1975.
7. J. T. Randall and M. H. F. Wilkins, *Proc. Roy. Soc. (London)*, **A184,** 336 (1945).
8. G. F. J. Garlick and A. F. Gibson, *Proc. Roy. Soc.* **A60,** 574 (1948).
9. P. W. Levy, *J. Lumin.*, **31/32,** 133 (1984).
10. P. W. Levy, *Phys. Res.* **B1,** 436 (1984).
11. P. L. Mattern, K. Lengweiler, and P. W. Levy, *Rad. Eff.* **26,** 237 (1975).

Thermometry

R. P. Hudson

Thermometry, the measurement of temperature, dates from the early years of the seventeenth century when quantitative observations were first applied to the thermoscope of the ancient Greeks [1]. Thermometers for which the theory connecting temperature and the observed (temperature-indicating) physical quantity is complete (e.g., constant-volume gas pressure, velocity of sound in a gas, total radiation, "noise" power in a resistor)—termed *primary thermometers*—furnish the bases for the international standards but are generally too cumbersome for everyday use. They are chiefly used to establish the temperatures of occurrence of selected sharply defined physical phenomena (changes of state such as solid–liquid, liquid–vapor, and, very recently, normal-metal–superconductor transitions, three-phase equilibrium, or triple points), which are then used as "fixed points" for reference tie points of the International Temperature Scale [2], or ITS. Definition of this scale is completed through recommended *secondary thermometers* and procedures, for the purpose of interpolation between these fixed points. From time to time the International Committee of Weights and Measures, through its Advisory Committee on Thermometry, evaluates new data and revises the ITS: This was done most recently in 1989 and the years immediately preceding, leading to a scale designated ITS-90 which came into force on 1 January 1990 [3]. By "realizing" the latter scale and directly comparing it with the indications of any useful practical thermometer, or *sensor,* the latter may be calibrated, after which its temperature indications, T_{90}, are approximations to thermodynamic temperatures and of a quality determined by a wide variety of degrading factors. These include thermodynamic shortcomings in the ITS itself, errors in realization thereof, errors in the comparison (calibration) procedure, and instability in the sensor. Desirable qualities in a sensor include high sensitivity, stability (reproducibility), small size, fast response, ease of calibration and utilization, and low cost. Numerous physical properties may be utilized to indicate, and hence to measure, temperature; but these properties will vary in the ease with which they are measured, and hence it is always a further practical advantage when any necessary ancillary equipment is relatively simple and inexpensive.

PRIMARY THERMOMETERS

Gas Thermometer

Classic—This makes use of the pressure–volume–temperature relationship of a "permanent" gas, namely, PV/T=constant in the limit of vanishingly small pressure. In the commonest mode of employment (constant volume), pressure P is proportional to absolute temperature T. For any real gas, the proportionality deviates somewhat with the pressure, and hence this effect has to be corrected for. For the most exacting standards work, other corrections are also of great importance, viz., thermal expansion of the bulb material, adsorption effects, behavior of the gas in connecting tubes running to the pressure-measuring instrument, and many others [2]. To a large extent, ITS-90 has been established on temperatures of fixed points as measured by the gas thermometer; these range from 13.8 K, or the triple point of hydrogen, to 1235 K, the freezing point of silver. [At the scale's low-temperature end the ^{3}He vapor pressure thermometer (q.v.) defines the scale between 0.65 and 5 K, and furnishes (an infinity of) fixed points near 3 K for use with an interpolating gas thermometer to span the range 3 K to the neon triple point at 24.6 K. Above the silver point, ITS-90 is defined in terms of the Planck radiation law and realized using an optical pyrometer (q.v.).] The gas thermometer is normally used to measure pressure *ratios,* which are then equated to temperature ratios; the reference temperature is that of the triple point of water, 273.16 K by definition, and the pressure in the bulb when at that temperature is the reference pressure.

The thermometer may also be used in a true "absolute" mode if the amount of gas introduced into and extracted from the bulb can be determined with adequate accuracy. Then, at a stable (but unknown) temperature T, an isotherm is traced out as the pressure is decreased by decreasing the content of the bulb; extrapolation to zero pressure permits a calculation of T [4]. The method automatically yields information on the departure of the gas from "ideal" behavior as the pressure is increased, information that is of crucial importance to the highest-accuracy utilization of the constant-volume gas thermometer (above).

Acoustic—The adiabatic compressibility K of a gas determines the speed of sound transmission S within it and thus S may also be used to determine temperature. It is given by the relation $S^2=K/\rho=\gamma\mathfrak{R}T/M$ (where ρ is the density, γ is the ratio of specific heats, C_p/C_v, $\mathfrak{R}$ is the gas constant, and M is the molecular weight). Again, as with the gas thermometer, refinements to the simple equation are needed to take into account nonideal effects, which only vanish at zero pressure. An actual device is usually constructed in the form of an acoustic resonator. Here, the wavelength λ of sound of known frequency ν is determined by detecting the structure of a standing-wave pattern; then $S=\nu\lambda$. There are many subtleties affecting the accuracy of this method [5], some of which are not yet fully resolved. The device has been chiefly employed as a standard, to complement the gas thermometer, in the 2- to 30-K region [6]; its sensitivity is greatest at low temperatures. It measures temperatures per se, not ratios, and the ultimate accuracy depends on the accuracy with which $\mathfrak{R}$ (or, rather, $\mathfrak{R}/M$) is known. One temperature (viz., 273.16 K) is, however, set *by definition*. Hence the method may be "turned around" to yield a value for $\mathfrak{R}$ if S is determined at the triple point of water [6]. Most recently, the *spherical* acoustic resonator has been developed to the point where it may be employed to make measurements of thermodynamic temperature in the region of 30°C and with an accuracy at least as good as that of the constant-volume gas thermometer [7].

Total Radiation Thermometry

The total radiation $E(T)$ emitted by a blackbody at a temperature T is given by

$$E(T)=\sigma T^4, \tag{1}$$

where σ is the Stefan–Boltzmann constant.

As, in a situation analogous to that encountered with the gas thermometer, there are difficulties in applying Eq. (1) directly, one may, instead, measure the ratio $E(T)/E(T_0)$, where T_0=273.16 K. This method has been employed [8] to measure temperature in the range −130 to +100°C and so to achieve (with comparable accuracy) a very important check upon the results of constant-volume gas thermometry in that region.

Noise Thermometer

Thermal fluctuations in an electrical resistance R give rise to a fluctuating "noise voltage," the mean square value of which, for a particular frequency interval dv, is given by the Nyquist relation, $\langle V^2_{f}\rangle=4RkT\ dv$, where k is Boltzmann's constant. The *maximum* noise power available is given by $kT\,d\upsilon$, or $\sim 4\times 10^{-21}$ W per unit bandwidth at room temperature, and is seen to be independent of the actual value of R. Measurement complications arise because all electrical measuring instruments add their own internal noise to the primary signal, which is, as we have just seen, extremely small. The intrinsic signal is largest for high temperatures and advances in modern technology offer some hope that the method will evolve into a primary standard for high temperatures in the not-too-distant future [9]. Rather surprisingly, the greatest advance to date has come at extremely low temperatures (0.01–1 K). This rests upon the development of the "superconducting weak link" or "Josephson junction." Such a junction, when biased by a dc voltage V, emits monochromatic electromagnetic radiation according to the relation $h\upsilon = 2eV$, hence at a frequency of 456 THz per volt. The noise voltage in the biasing resistor is converted into a noise frequency modulation of the pure signal and the measurement is reduced to a (sophisticated) frequency–technology process [10,11]. A temperature so derived is thermodynamic and, again, ultimately dependent only on the value used for $\mathcal{R}$ (entering here as k, i.e., scaled by Avogadro's number, N).

Gamma-Ray Anisotropy Thermometer

This highly specialized sensor is useful only at very low temperatures (to date, well below 1 K), and is dependent on the phenomenon of nuclear orientation. At high temperatures, nuclear magnetic moments are randomly oriented and the pattern of emitted gamma rays is spatially isotropic. In suitable solid environments, at very low temperatures, it is possible to line up the small nuclear magnets, and then the emission pattern becomes anisotropic. The "anisotropy," which is readily measured, depends only on details of the decay scheme and the atomic environment—all of which can be measured—and on the absolute temperature (actually kT) [11,12].

SECONDARY THERMOMETERS

Secondary thermometers must be calibrated. If this is done in accord with the procedures specified in the text of the ITS, the thermometer readings should be a reasonably accurate representation of thermodynamic temperature (see above). Each stage of separation from such a primary calibration may be taken to increase the measurement uncertainty by a factor of 3, as a rough estimate.

The first two sensors described next are essentially primary thermometers except that the physical magnitude observed depends on geometrical details of the instrument that cannot be measured with great accuracy. This "scaling factor" is eliminated by measuring only temperature ratios but because, practically, it is impossible to use the instruments at 273.16 K, the thermometers are calibrated at other, more convenient, temperatures and therefore "piggyback" on some other (primary) instrument(s). (The gas thermometer itself is used in this way for work at low temperatures.)

Optical Pyrometer

Above the silver point (1234 K, 961°C), the temperature T_{90} is defined in terms of Planck's law and the "spectral concentration" of the radiance of a blackbody (see TEMPERATURE). In practice, standard lamps—which are easier to operate than blackbody furnaces—are used for maintaining radiance standards. Early pyrometers used the eye as detector to match the brightness of the subject body to that of a standard lamp incorporated in the instrument. More satisfactory models were later developed which employed a photoelectric detector in place of the eye [13]. Sectored discs were used in the case of a very bright source to reduce the signal by a known factor and so maintain the detector in its optimum working range. In the most modern designs, the detector is a silicon cell which is linear over a wide range of received signals. The standard lamp is external to the pyrometer and is used intermittently for checking the instrument. [The temperature of the silver point itself was established, for use in ITS-90, chiefly by measurements with an optical pyrometer calibrated at the zinc point ($\approx$420°C).]

If the body under study is not a blackbody, its emissivity ϵ_λ will be less than unity and its true temperature T will be higher than the indicated, or brightness, temperature $T_{b\lambda}$; their relationship is given by

$$(T^{-1} - T_{b\lambda}^{-1}) = (\lambda \ln \epsilon_\lambda)c_2^{-1}, \tag{2}$$

where c_2 is the second radiation constant, hc/k.

Nuclear Resonance Thermometer

Many atomic nuclei possess magnetic dipole moments which interact very weakly with each other and hence, in diamagnetic metals or insulators, give rise to a very accurately Curie-law susceptibility, $\chi = \text{const.} \times T^{-1}$. Because of the very small Curie constant, practical thermometers based on this principle are limited to very low temperatures; they require a single-point calibration. Advantage is taken of the high sensitivity of the nuclear magnetic resonance technique to measure this nuclear paramagnetism. The continuous-wave approach requires rather strong magnetic fields, which are frequently undesirable, and a low-field pulsed method has been developed specifically for thermometry [11,14]. (The advent of very high-sensitivity magnetometers has made it feasible to make nonresonant susceptibility determinations up into the liquid helium region; the nuclear resonance approach is suited to the region below 0.1 K.)

Magnetic Thermometer

This title traditionally applies to the measurement of electronic (i.e., nonnuclear) paramagnetism and most commonly entails, as sensor material, a few grams of inorganic "salt." Such salts are compounds with many "waters of crystallization"; hence they tend to be magnetically dilute, i.e., the magnetic ions are far (6–8 Å) apart. Many ions, and in a variety of compounds, can be used for magnetic thermometry—which is usually practiced in the range 0.001–1 K, but has been pursued in a few cases to 50 K and higher—but selection criteria developed over the years have narrowed the field of common usage to relatively few, viz., cerous magnesium nitrate, CMN (Ce^{3+}), potassium and methylammonium chromic alums (Cr^{3+}), and lanthanum-diluted CMN.

The most satisfactory way of measuring paramagnetic susceptibility in this thermometric application is to have the sensor salt form the core of a mutual inductance whose coefficient is then determined by means of an ac bridge or a SQUID magnetometer. The thermometer must be calibrated at two well-separated temperatures (usually in the 1- to 4-K region) in order to determine constants A and B in the equation $T = A(N-B)^{-1}$ where N is the bridge reading at balance for temperature T. Outside the calibration range, a temperature determined by extrapolating (usually downward) this formula is termed the magnetic temperature, T^*. T^* will tend to diverge progressively from T as the latter falls farther below 1 K; this deviation has been accurately determined for the first three of the salts mentioned earlier [11,15].

^{3}He Melting Curve Thermometer

The melting pressure of solid ^{3}He is a sensitive thermometric parameter and its application has been growing steadily [11,16]. It is, in fact, likely that it will be used, eventually, to extend the practical temperature scale below 0.65 K down to the region of 0.001 K. In several aspects it is similar to the helium vapor-pressure thermometer (q.v.), manifesting a $P(T)$ relation in accord with the Clausius–Clapeyron equation and showing high resolution, insensitivity to magnetic fields (below 0.5 T) and rf interference, and having essentially zero power dissipation. It requires ^{3}He at pressures ranging from 29 to 40 atm and a suitable transducer, which is calibrated *in situ*. Although (again, analogously to the vapor-pressure thermometer) it is primary in principle, the melting-curve thermometer seems to deliver the highest accuracy when it is used as a secondary instrument, making use of a standard P-vs.-T relation.

Vapor-Pressure Thermometer

The saturation vapor pressure (s.v.p.) of a liquid is a sensitive function of temperature and can be very conveniently used for measurable pressure up to the fluid's critical point. The s.v.p. is given approximately by the relation $\log p = A - BT^{-1}$ and standard tables enable one to read off T, given p; the tables are normally prepared by using a different thermometer to indicate T. For the lowest portion of T_{90}, however, it was necessary to produce p-vs.-T relations (for ^{3}He and ^{4}He) that were as thermodynamically correct as possible. This was done [17] by calculating $p(T)$ from first principles and ascertaining that it could be reconciled very closely with the indications of a magnetic thermometer (q.v.) that had been calibrated against a rhodium–iron thermometer (see *Resistance Thermometers*) that was itself the repository of a gas-thermometer-based scale.

Resistance Thermometers

1. *Metal.* The most important thermometer in this class is the standard platinum resistance thermometer [18], which is used (in conjunction with thermometric fixed points) to define the ITS between 13.81 K and 961°C. Its outstanding qualities of sensitivity and reproducibility can only be maintained, however, by very careful handling; more rugged, lower-quality platinum resistance thermometers are coming into wide industrial use. Other metals (e.g., copper, nickel, and indium) and a few alloys are also used.

Fow the low-temperature region (and particularly for 1 to 30 K) an important advance was achieved with the development of the rhodium–iron thermometer which was wire formed from Fe containing 0.5 at .% of Rh [19]. This sensor is adequately sensitive and is very stable. It has proved particularly useful for "storing" a gas-thermometer-based scale, and has been used in connection with setting up vapor-pressure (q.v.) scales for He that form the 0.65–5-K portion of ITS-90.

2. *Semiconductors.* These have a negative (approximately exponential) temperature characteristic and are best suited to low temperatures, below 20 K. At very low temperatures, the resistance becomes prohibitively large and the "working range" of a particular resistor is set by its composition (parent material and "doping"). The most commonly encountered thermometers are germanium and carbon, the latter being slightly the less reproducible but having the practical advantages of ready availability and low cost.

Various (proprietary) oxide mixtures are formed into "thermistors" for wide-ranging temperature measurement and control applications. Their main advantages are small size (sometimes pinhead or even smaller), high sensitivity, and low cost; generally speaking, their reproducibility is poor. Silicon p-n junction diodes are in quite wide use in commercial "electronic" thermometers—chiefly for control purposes—in the range 1 K to 300 K.

Thermocouple

Two dissimilar metals joined together and heated at the junction develop a thermoelectric potential difference between the unconnected ends (Seebeck effect) that is widely employed in thermometry [20]. Typical outputs are 10–40 $\mu V \cdot K^{-1}$. A thermocouple formed from platinum and from platinum–10%-rhodium alloy was used in all international temperature scales through 1989 to define the scale between 630.74°C and the gold point. It was limited, however, to an accuracy of ± 0.2°C and the successful development (after many years of effort) of the high-temperature platinum resistance thermometer permitted building the latter into ITS-90, if only up to the silver point. For very high temperature, above the melting point of platinum (about 1769°C), tungsten–rhenium alloys have been shown to be very useful. New alloys of nickel, chromium, and silicon (Nisil and Nicrosil) have been shown to offer long-term high-temperature (~1100°C) stability greatly superior to that of the commonly used Chromel–Alumel thermocouples. Copper–Constantan thermocouples are frequently employed in the middle-temperature range and gold–cobalt alloys (which tend to suffer from brittleness) have been developed for low temperatures (helium–hydrogen region).

Liquid Expansion Thermometer

Commonly termed "liquid-in-glass" or, in one of the most accurate forms, "clinical", these thermometers are among the most venerable and familiar. The volume expansion of a fluid such as mercury, or alcohol, is transformed into the movement of a thread of the fluid along a uniform-bore capillary tube. The dimensions of the device determine the working range and after calibration has permitted temperatures to be "marked" at each extremity of the capillary section, the interval between is marked off in appropriate divisions, assuming perfect linearity. The validity of this assumption can only be ascertained by a much more detailed calibration. The instrument should be calibrated in the same way in which it is intended to be employed thereafter, i.e., with bulb only immersed, or with all the fluid including the capillary section immersed [21].

Other Thermometers

Any physical property that varies sensitively and reproducibly with temperature and that can be readily measured is suitable for thermometric application. For example, thermal expansion is much utilized—that of metals in bimetallic strips, and that of (specially cut) quartz crystals for which the resonant frequency, falling in the megahertz region, changes approximately linearly with temperature in the middle-temperature range. A fairly recent invention is an entirely optical thermometer, developed for medical application, that makes use of the temperature-varying rotation of the plane of polarization of light in certain crystals.

See also BLACKBODY RADIATION; CONDUCTION; CRITICAL POINTS; EQUATIONS OF STATE; HIGH TEMPERATURE; JOSEPHSON EFFECTS; MAGNETIC COOLING; NOISE, ACOUSTICAL; NUCLEAR MAGNETIC RESONANCE; PHASE TRANSITIONS; TEMPERATURE; THERMOELECTRIC EFFECTS.

REFERENCES

1. W. E. Knowles Middleton, *A History of the Thermometer*. Johns Hopkins Press, Baltimore, Md., 1966.
2. J. A. Beattie, "Gas Thermometry" in *Temperature: Its Measurement and Control in Science and Industry* (H. C. Wolfe, ed.), Vol. 2, Chapter 5. Reinhold, New York, 1955.
3. "International Temperature Scale of 1990," *Metrologia*.
4. See, e.g., K. H. Berry, *Metrologia* **15,** 89 (1979).
5. A. R. Colclough, *Metrologia* **9,** 75 (1973).
6. T. J. Quinn, A. R. Colclough, and T. R. D. Chandler, *Phil. Trans. Roy. Soc.* **A283,** 367 (1976).
7. M. R. Moldover and J. P. M. Trusler, *Metrologia* **25,** 165 (1988).
8. J. E. Martin, T. J. Quinn, and B. Chu, *Metrologia* **25,** 107 (1988).
9. C. J. Borkowski and T. V. Blalock, *Rev. Sci. Instr.* **2,** 151 (1974); L. Crovini and A. Actis, *Temperature Measurement, 1975* (Conf. Ser. No. 26, Inst. of Phys., London, 1975), p. 398.
10. R. A. Kamper, in *Proc. Symp. Phys. Superconducting Devices* (Univ. of Virginia, Charlottesville, Va., 1967), pg. M-1; available as document No. AD661848 from National Technical Information Service, Springfield, Va.
11. See, e.g., R. P. Hudson, H. Marshak, R. J. Soulen, Jr., and D. B. Utton, "Recent Advances in Thermometry below 300 mK," *J. Low Temp. Phys.* **20,** 46 *et seq.* (1975).
12. H. Marshak, *J. Res. Natl. Bur. Stand.* **88,** 175 (1983).
13. See, e.g., T. P. Jones and J. Tapping, *Metrologia* **18,** 23 (1982).
14. R. E. Walstedt, E. L. Hahn, C. Froidevaux, and E. Geissler, *Proc. Roy. Soc. London* **A284,** 499 (1965).
15. For a general discussion see, e.g., R. P. Hudson, *Principles and Application of Magnetic Cooling*. North-Holland, Amsterdam, 1972.
16. R. A. Scribner and E. D. Adams, *Temperature: Its Measurement and Control in Science and Industry, Vol. 4* (H. H. Plumb, ed.), p. 37. Instrument Society of America, Pittsburgh, 1972; see also D. S. Greywall, *Phys. Rev. B* **33,** 7520 (1986) and earlier articles dicited therein.
17. M. Durieux and R. L. Rusby, *Metrologia* **19,** 67 (1983).
18. See, e.g., J. L. Riddle, G. T. Furukawa, and H. H. Plumb. *Platinum Resistance Thermometry,* NBS Monograph 126, U.S. Dept. of Commerce, Washington, D.C., 1973.
19. R. L. Rusby, *Temperature Measurement, 1975* (Conf. Ser. No. 26), p. 125. Institute of Physics, London, 1975; *Temperature: Its Measurement and Control in Science and Industry, Vol. 5* (J. F. Schooley, ed.), p. 829. American Institute of Physics, New York, 1982.
20. R. L. Powell, W. J. Hall, C. H. Hyink, Jr., L. J. Sparks, G. W. Burns, M. G. Scroger, and H. H. Plumb, NBS Monograph 125, U.S. Dept. of Commerce, Washington, D.C., 1974.
21. J. A. Wise, *Liquid-in-Glass Thermometry,* NBS Monograph 150, U.S. Dept. of Commerce, Washington, D.C., 1976.

Thin Films

Myron Strongin, D. L. Miller, R. C. Budhani, and M. W. Ruckman

INTRODUCTION

In the trivial case, thin films can be used to make very small amounts of a given substance with essentially the same properties as larger samples of the bulk material. The real fascination with the physics and metallurgy of films, however, involves physical phenomena and materials that are unique to films and the multilayer structures produced using thin films. In this very brief survey we intend to discuss some of the more interesting properties of films and some important technological applications.

PREPARATION

Films can be prepared in a variety of ways [1]. All thin film processing consists of three basic steps: synthesis or generation of the depositing species, transport from source to substrate, and nucleation and growth of the film on the substrate. A process in which these three steps can be varied independently offers greater flexibility, and a large variety of materials with tailored properties can be deposited. Some common methods of depositing films are electroplating, chemical vapor deposition (CVD), and vacuum evaporation. The need to deposit clean materials for semiconductor device applications [2] and the study of intrinsic materials properties led to great advances in ultrahigh-vacuum technology during the 1970s and 1980s. Vacuum evaporation by electron beams and resistively heated Knudsen cells at pressures as low as 10^{-10} Torr is routine in commercially available molecular-beam epitaxy systems. The generation of a depositing species is also realized by physical sputtering of a target of the material of interest. Both rf and dc glow discharges are used for the sputter-deposition purposes. The deposition rates in sputtering can be increased tremendously, if suitable magnetron designs or electron-coupled resonance (ECR) systems are used to increase the plasma density at the target. Ceramic and compound films can be deposited by introducing a gas of one of the constituent elements into the plasma [3]. The use of electron-impact dissociation in a glow discharge of the gases used in a conventional CVD process provides a versatile low-temperature deposition process for technologically important elemental and compound semiconducting and metallic films. In addition to the basic evaporation, CVD, and sputtering techniques, several hybrid processes such as ion-beam sputtering, activated reactive evaporation, ionized cluster beam deposition, laser CVD, and laser ablation are also being used to control specific properties of thin films [1–3].

CHARACTERIZATION OF FILMS

Films can be characterized by a number of metallurgical techniques, optical and electron microscopy, ellipsometry, and x-ray diffraction. Surface-sensitive techniques [4], such as Auger electron spectroscopy, secondary-ion mass spectroscopy, and the spectroscopy of photoemitted electrons, can be especially useful for very thin films. Also useful is the technique of Rutherford backscattering (RBS) [5]. In backscattering experiments, high-energy particles are scattered from the films and an analysis of the energy of the backscattered particles allows a determination of the elements in the sample. Apart from the elemental analysis, the RBS ion-channeling experiments provide useful information about the degree of epitaxial growth on single-crystal substrates. This technique, as well as the surface-sensitive spectroscopies, has been developed to give a profile of composition as a function of depth into a film. The structure and epitaxial growth of ultrathin films can be studied by techniques such as low-energy electron diffraction (LEED), reflection high-energy electron diffraction (RHEED), and high-energy electron diffraction (HEED) [6]. Moreover, recent developments in the area of high-power x-ray sources and synchrotrons have enabled scientists to apply x-ray diffraction [6] and x-ray absorption to study the structure of ultrathin films [7]. The recent development of atomic level microscopies (e.g., scanning tunneling microscopy and high-resolution transmission electron microscopy) now makes it possible to study atomic level processes at thin-film surfaces and interfaces directly [8].

STRUCTURAL PROPERTIES OF FILMS

The structural properties of films are intimately connected with the properties of the substrate, deposition rate, gaseous impurities, and substrate temperature. By varying these parameters, unique materials and structures have been stabilized in films [9].

Probably the most dramatic structural changes in films are those observed in amorphous or microcrystalline metals. These materials are made by depositing metals onto substrates that are cooled to the temperature of liquid helium, which is 4.2 K above absolute zero [10]. At these temperatures the metal atoms lack the thermal energy necessary for extensive diffusion, and extremely disordered structures can be obtained. In the case of Bi deposited at these temperatures it appears that the Bi actually is in an amorphous phase, since at some temperature near 30 K it rapidly crystallizes into its bulk structure [11]. With other materials, grains gradually grow as the temperature is raised, and there is no sharp transformation. These films are probably composed of small grains having the same structure as the bulk material. Finally, it is worth mentioning that many phases can be formed in films that are either difficult or impossible to make by bulk metallurgical techniques; this includes amorphous alloys of semiconductors and metals that at present can only be made in their films.

Another problem involved with film structure is the study of the interaction of metal atoms on various substrates and especially how this interaction affects the structural properties of the film. This constitutes the study of epitaxy [12] and the related problem of film nucleation [13]. In most cases the initial atoms hitting a substrate form small islands of metal about 20–30 Å in diameter. These islands then grow and at some stage they touch and begin to coalesce into larger islands. As the large islands grow together, channels are formed that are then covered with newly deposited metal until a uniform film is finally formed. These studies have traditionally been made by using an electron microscope [14]. In recent years low-energy electron diffraction (LEED) has been used to study monolayers of metals on various substrates. This technique has shown that it is possible in certain cases to achieve uniform crystalline monolayers or near monolayers of a metal on a particular substrate. That the diffracted electrons have low energy causes them to interact strongly with the specimen. In some ways this makes a cursory interpretation difficult; however, detailed analysis of the intensity of LEED patterns can give detailed information about atomic positions on surfaces in some cases. A LEED pattern of an epitaxial thin metal layer is shown in Fig. 1 [15]. Perhaps the most exciting development in the area of epitaxial growth is deposition of semiconducting [16] and metallic [17] superlattices which are periodic structures having dimensions which approach atomic spacing. The periodic modulation in the structure and chemical composition of a superlattice leads to a vast array of interesting physical properties in these structures. Superlattices have been grown for III–V compound semiconductors (GaAs–$Ga_xAl_{1-x}As$), II–VI semiconductors, elemental semiconductors (Si–Ge), semiconductor–insulator (Si–CaF_2), semiconductor–metal (Si–$NiSi_2$), metal–metal (Cu–Ni, Nb–Ta, Y–Gd, etc.), and ceramic–ceramic ($YBa_2Cu_3O_7$–$DyBa_2Cu_3O_7$).

In contrast to the cases just discussed which have atomically sharp interfaces, adiatom-induced disruption occurs. This is the most common behavior for metal–semiconductor bilayers [18] and metal–metal junctions where a high surface energy metal is deposited on a low surface energy substrate [e.g., Nb/Pd(111) or Pd/Al(100)] [19]. Interfacial disruption is temperature dependent and produces a thin disordered alloy or intermetallic surface compound. Control of the sub-

FIG. 1. The LEED pattern from a PbTe single-crystal substrate (top) and an epitaxial Pb film (bottom) grown on this substrate by evaporation of Pb. The Pb film is less than 15 Å thick. The energy of the diffracted electron beam is 94 eV.

strate temperature or deposition parameters leads to the growth of ordered surface compound layers that can be put to a number of applications [e.g., $NiSi_2$/Si(100)].

SOME PHYSICAL PHENOMENA IN THIN FILMS

Besides the convenience of using films in certain applications, there are phenomena that are best studied in, and are even unique to, ultrathin films. Films, of course, provide a way of studying phenomena in approximately two dimensions. We say approximately since the effect of the substrate can be nontrivial. Both magnetic phase transitions and superconducting phase transitions have been studied in this quasi-two-dimensional case. Problems of nucleation and the film–substrate interaction are crucial in the interpretation of these results. For example, results of Ni films and iron–nickel alloys show that the magnetization and Curie temperature stay constant in films down to about 50 Å thick. Below this value various investigators obtain different results; however, the films are granular in this regime and there is no longer an ideal film geometry. In a similar way, experiments where Ni is grown epitaxially on Cu show no magnetic behavior below 7 Å [20], although it is not clear what effect the copper substrate has. A similar situation is found in the investigation of superconducting materials. Below some critical temperature superconducting metals go into an ordered state, where they have zero electrical resistance. As superconducting films approach thicknesses below about 50 Å the transition temperature of some metals show a large decrease [21]. It is likely that the interaction with the substrate plays a role in the degradation of these superconducting properties. It should also be mentioned that phase transitions in two dimensions have been studied in rare gas films and in films of superfluid helium. The recent text by Dash [22] serves as an excellent introduction to this area.

Another class of problems that are of interest involves changes in the transport properties of metal films due to the film boundaries. If the films can be made thin enough so that the distance between internal electron scatterings becomes greater than the film thickness, it is clear that boundary scattering will greatly influence the transport properties [23], and this field has been studied experimentally and theoretically for many years. The problem of how electrons scatter from film boundaries, and the effect of this scattering on transport properties, has also been studied by using magnetic fields, which bend the electron orbits away from the surface, causing a decrease in resistance. This effect has been observed in wires and films.

Another class of problems, somewhat related to the foregoing, involves changes of the transport properties of films due to quantization of electron orbits in the direction perpendicular to the plane of the film. This effect results from the wave nature of the electrons in the film and the fact that the electron waves must fit within the small dimension of the film in discrete units of the electron wavelength perpendicular to the plane of the film. Clearly, if the film is much thicker than the electron wavelength this effect becomes negligible. In the case where the thickness is only a few wavelengths, however, this effect becomes significant. Since the characteristic wavelength for electrons in a metal is only on the order of a few angstroms, or an interatomic spacing, this effect is difficult to observe. In Bi the situation is somewhat different since the films grow on mica in such a way that the electron wavelength perpendicular to the film is on the order of a few hundred angstroms. Films can be made conveniently in this range and the foregoing effects of "size quantization" have been studied in this regime.

The physics behind electrical, optical, and semiconducting properties of thin films deposited in superlattice form is not well understood [24]. In the case of semiconducting superlattices, composition-related modulation in the electron affinity confines the electrons in layers with greater affinity. When the confined-layer thickness is small, electrons are distributed as size quantized waves in two dimensions. Tunneling across the low electron affinity layers becomes important for smaller barrier thicknesses. Effects of size quantization have been observed in MBE-grown GaAs superlattices by optical absorption, photoluminescence, and Raman scattering. The two-dimensional nature of electrons in high affinity layers of the superlattices can be enhanced by modulation doping. The mobility of electrons in these layers is extremely high (10^5 cm^2 V^{-1} s^{-1}) [24]. In the presence of a magnetic field perpendicular to the plane of this 2D system, carriers execute cyclotron motion with a frequency eB/m^*, where m^* is the effective mass. Landau quantization of this cyclotron motion splits the 2D continuum of states into a series of discrete Landau levels with a well-defined gap in the excitation spectrum (minimum disorder-induced broadening). This new state of the electrons leads to the fractional quantum Hall [24] effect and the dissipationless flow of the electron current.

True metallic coherent superlattices of Nb–Ti, Cu–Ni, Ag–Fe, and Y(Gd, Dy, Er, Ho) have been grown successfully. The (Gd, Dy, Ho, Er) superlattices which grow on bcc (110) Nb single crystals show interesting magnetic ordering effect as a function of the chemical modulation wavelength [17]. In the case of Gd/Y system, for fixed Gd layer thickness, the ferromagnetic coupling across the nonmagnetic layer changes from parallel to antiparallel with the increasing Y layer thickness. The Dy/Y system on the other hand shows helical ordering of the spin moments within the Dy sublattice and the magnetic phase coherence is maintained between the successive Dy layers. The transition-metal superlattices also show interesting hydrogen storage characteristics and supermodulus effects.

In this section we have tried to give some inkling of basic physical phenomena that can be studied with thin films. Of course, a discussion of many unique phenomena in films has been left out, such as the high-field behavior of superconducting films, the optical properties of films, the use of films for studying materials by the technique of electron tunneling, the Josephson effect in superconducting films, the photovoltaic effect in films, adsorption and catalysis on films, amorphous alloys, magnetic films, and many other fascinating phenomena. It is worth mentioning that monolayer structures of a transition metal, on a transition-metal substrate (such as Pd on Nb), cause a dramatic change in the electronic and catalytic properties of the overlayers, with drastic changes in chemical reactivity. This is a promising area for tailoring catalytic properties.

UNIQUE MATERIALS USING FILMS

Perhaps the most important modern materials that involve thin film technology are the new multilayer structures involved in integrated circuits [25], where hundreds of components are placed on a silicon chip whose size is of the order of 2.5 mm × 2.5 mm. In some sense this is a unique material that is made by using films, and in fact these devices have revolutionized modern electronic technology. Nevertheless, nearly all the properties of these devices can be understood in terms of the behavior of the bulk materials from which they are fabricated. The very clever thin-film techniques used to make integrated circuits are primarily concerned with reducing device size.

A further important technological use of thin films is found in photovoltaic devices. Again the phenomenon itself is not unique to films; it appears, however, that the economic use of photovoltaic devices could well depend on film technology. The present approaches to photovoltaic device manufacture involve silicon production technology and the technology of producing III–V and II–VI compounds [26]. For GaAs the absorption edge is much sharper than for silicon, and whereas a wafer thickness greater than 10^{-3} cm is necessary for reasonable absorption efficiency with silicon, only about 10^{-4} cm is necessary for GaAs. Hence even though silicon is an intrinsically cheaper material it appears that polycrystalline thin-film GaAs detectors can compete with silicon devices. These considerations are too detailed to be dealt with here, but they essentially involve the conditions on the grain size in these polycrystalline films in relation to their optical absorption lengths and charge-carrier diffusion lengths. Passivation of dangling bonds in amorphous silicon films by incorporation of hydrogen and resulting substitutional dopability has made hydrogenated amorphous silicon (*a*-SiH) an interesting photovoltaic material [27]. The understanding of defect passivation by hydrogen has also been an interesting topic in the physics of noncrystalline semiconductors.

Multilayered materials made from materials having large differences in their x-ray scattering powers (e.g., W and C) are being used in high angle of incidence x-ray reflectors and monochromators. They represent an important technological advance for x-ray optics in that the reflectivity of single-layer reflectors at high angles of incidence is $\sim 10^{-4}$ and, unlike naturally occurring crystals, the spacing of the reflecting surfaces can be tailored to diffract longer-wavelength x-ray or vacuum ultraviolet radiation. The use of figured multilayer structures in a telescope was recently demonstrated by Walker *et al.*, [28] who used a pseudo-Cassegrain telescope with spherical multilayer optics to obtain high-resolution soft-x-ray solar images.

In concluding this section we briefly discuss thin-film materials that remain superconducting to the highest temperatures, and that promise to be extremely important technologically. Prior to the discovery of metal-oxide higher-T_c materials, the highest transition temperature superconductors had the A-15 structure, and Nb_3Ge had the highest known superconducting transition temperature near 23 K and, in some reports, as high as 25 K. Nb_3Sn, which is the easiest niobium-based A-15 superconductor to make, has a transition temperature T_c of about 18 K. An interesting feature of these materials is that the higher T_c's appear in the less stable materials. While Nb_3Sn can be made by various bulk metallurgical techniques, at present Nb_3Ge has only been prepared as a thin film by deposition from the vapor phase onto substrates held at a high temperature [29].

Since the discovery of high-temperature superconductivity in $La_xBa_{1-x}CuO_4$ by Bednorz and Müller in 1986, a large number of perovskites featuring $Cu-O_2$ planes and T_c's ranging in some cases as high as 127 K have been discovered. The ceramic form of these materials has highly anisotropic properties and poor grain boundary coupling which limits technological application. In thin-film form it is possible to grow oriented single-crystal material which shows great technological promise [30]. Epitaxial thin films of high-T_c oxides with critical current $>10^6$ A/cm^2 at 77 K grown on lattice-matched substrates such as $SrTiO_3$ have been deposited by a number of deposition processes. These films provide the basis for the development of a variety of active and passive rf and microwave superconducting devices. In addition to these applications, the epitaxial films also provide clean systems to study the physics of superconductivity at unusually high temperatures.

SUMMARY

We have tried to give some idea of unique phenomena that occur in films from the point of view of a solid state physicist. In conclusion we emphasize that films form the basis for many established and developing technologies. In the simplest case, metal layers have always been important for optical coatings, and even as protective coatings for various materials. In the most sophisticated cases, films and their interaction on silicon surfaces form the basis of modern electronic technology. Furthermore, films of silicon, GaAs, and composites of these materials promise to lead to practical photovoltaic devices. Thin-film materials provide a basis for the practical application of high-temperature superconductors. It is fair to say that besides being a field of great interest in modern physics, the study of thin films also has contributed vital elements to the most sophisticated commercial technologies.

See also ADSORPTION; AUGER EFFECT; CATALYSIS; ELECTRON AND ION IMPACT PHENOMENA; JOSEPHSON EFFECTS; LOW-ENERGY ELECTRON DIFFRACTION; PHOTOELECTRON SPECTROSCOPY; SUPERCONDUCTING MATERIALS; SURFACES AND INTERFACES; TUNNELING.

REFERENCES

1. J. L. Vossen and W. Kern (eds.), *Thin Film Processes*. Academic, New York, 1978.
2. S. K. Ghandi. *VLSI Fabrication Principles*. Wiley, New York, 1983.
3. R. F. Bunshah (ed.), *Deposition Technologies for Films and Coatings*. Noyes Publishing, New Jersey, 1982.
4. K. N. Tu and R. Rosenberg (eds.), *Analytical Techniques for Thin Films,* Treatise on Materials Science and Technology, Vol. 27. Academic, New York, 1988.
5. J. F. Ziegler, "Materials Analysis by Nuclear Backscattering," in *New Uses of Ion Accelerators* (J. F. Ziegler, ed.), pp. 75–103. Plenum, New York, 1975.
6. R. Feidenhans'l, *Surf. Sci. Rep.* **10,** 105–188 (1989).
7. D. C. Koningsberger and R. Prins (eds.), *X-ray Absorption: Principles, Applications, and Techniques of EXAFS, SEXAFS, and XANES*. Wiley, New York, 1988.
8. R. M. Tromp, R. J. Hamers, and J. E. Demuth, *Science* **234,** 304–309 (1986).
9. K. L. Chopra. *Thin Film Phenomena*. McGraw-Hill, New York, 1969.
10. N. V. Zavaritsky, *Dokl. Akad. Nauk SSSR* **86,** 501 (1952); W. Buckel and R. Hilsch, *Z. Phys.* **132,** 420 (1952). Earlier work by Shalnikov on amorphous films dates to the late 1930s.
11. R. Hilsch, *Noncrystalline Solids*, V. D. Freschette (ed.). Wiley, New York (1958).
12. J. W. Matthews (ed.), *Epitaxial Growth*. Academic, New York, 1975.
13. J. P. Hirth and K. L. Moazed, in *Physics of Thin Films* (G. Hass and R. E. Thein, eds.), Vol. 4, pp. 97–136. Academic, New York, 1967.
14. A detailed review of this area is provided by the articles by C. A. Neugebauer and S. Mader in *Handbook of Thin Film Technology* (L. I. Maissel and R. Glang, eds.), pp. 8-3 to 8-44. McGraw-Hill, New York, 1979.
15. M. Sagurton, M. Strongin, F. Jona, and J. Colbert, *Phys. Rev. B* **28,** 4075–4078 (1983).
16. A. Y. Cho and J. R. Arthur, in *Progress in Solid State Chemistry* (G. Somorjai and J. McCaldin, eds.), Vol. 10. Pergamon, New York, 1975.
17. B. Y. Jin and J. B. Ketterson, *Adv. Phys.* **38,** 189–366 (1989).
18. J. H. Weaver. *Synchrotron Radiation Photoemission Studies of Interfaces* in ibid ref. 4.
19.
20. C. A. Neugebauer, *Z. Angew. Phys.* **14,** 182 (1962).
21. Myron Strongin, R. S. Thompson, O. F. Kammerer, and J. E. Crow. *Phys. Rev. B* **1,** 1078 (1970); D. L. Miller, Myron Strongin, O. F. Kammerer, and B. G. Streetman, *Phys. Rev. B* **13,** 4834 (1976).
22. J. G. Dash, *Films on Solid Surfaces*. Academic, New York, 1975.
23. K. Fuchs, *Proc. Cambridge Philos. Soc.* **34,** 100 (1938); E. H. Sondheimer, *Adv. Phys.* **1,** 1 (1952).
24. D. C. Tsui, in *Proceedings of the 17th International Conference on the Physics of Semiconductors* (J. D. Chadi and W. A. Harrison, eds.), pp. 247–254. Springer-Verlag, New York, 1985; A. C. Gossard, in *Preparation of Thin Films* (K. N. Tu and R. Rosenberg, eds.), *Treatise on Materials Science and Technology*, Vol. 24, pp. 13–65. Academic, New York, 1982.
25. For a review of the use of thin films in integrated circuits see I. Blech, Harry Sello, and L. V. Gregov, in *Handbook of Thin Film Technology* (L. I. Maissel and R. Glang, eds.), pp. 23-1 to 23-22. McGraw-Hill, New York, 1970.
26. A review of the recent outlook for compound semiconductor photovoltaics is given by J. M. Woodall and H. J. Hovel in *Proc. of the Symposium on Films for Solar Energy, J. Vac. Sci. Technol.* **12,** 1000 (1975).
27. For a review of amorphous semiconductors see David Adler, *Sci. Am.* **236** (5), 36 (1977).
28. A. B. C. Walker, Jr., T. W. Barbee, Jr., R. B. Hoover, and J. F. Lindblom, *Science* **241,** 1781 (1988).
29. J. R. Gavaler, *Appl. Phys. Lett.* **23,** 480 (1973); L. R. Testardi, *Proc. IEEE Trans.* Magn. **MAG-11,** 197 (1975).
30. P. Chaudhari, R. H. Koch, R. B. Laibowitz, T. R. McGuire, and R. J. Gambino, *Phys. Rev. Lett.* **58,** 2684 (1987).

Three-Body Problem: Gravitational

Victor Szebehely

In his classical masterpiece *Analytical Dynamics*, Sir Edmond Whittaker (1904) describes the problem of three bodies as the most celebrated of all dynamical problems and mentions Poincaré's proof (1890) that the problem is not integrable, i.e., it has no generally valid analytical solutions. The gravitational applications of this problem include the motion of space probes, of the Sun–Earth–Moon system, of triple stars, and of triple galaxies. Because of the combination of its unsolvability, its great scientific importance, and many practical applications, the problem received the attention of the most famous scientists in the past 300 years, beginning with Newton's *Principia* (1687). The above combination also shows how well nature hides its true face when one of the most important generating problems of dynamics and celestial mechanics is not readily solvable.

A comparison between the two-body and three-body problems clarifies the unsolvability of the latter. The problem of two bodies was for all practical purposes completely solved by Kepler when, in 1619, he announced his laws of motion of the planets around the Sun, neglecting the effects of the planets on each other. The motion of Earth around the Sun as influenced by a third body, say Jupiter, becomes an unsolvable problem since precise predictions for arbitrary long times are not possible. The motion of the Moon around Earth not influenced by other bodies is completely solved but this solution is only approximate since it neglects the important perturbations of the Sun. Our correct intuition regarding two-body motions is taken for granted, as any ballplayer knows. We do not possess such ability about the motion of three bodies. An important theoretical difference between the two- and three-body problems exists at close approaches. When two bodies pass each other at a small distance, the computation of the orbits presents no problems. On the other hand, when three bodies approach each other closely, the computation becomes extremely difficult in principle and in practice. In mathematics we refer to the collision as a singularity which can be eliminated by regularization for the two-body problem, whereas the triple collision is, in general, an essential singularity.

We distinguish between the restricted and the general problems of three bodies. In the general problem, three bodies with masses of the same order of magnitude attract each other according to the Newtonian gravitational law. The restricted problem of three bodies, on the other hand, requires the determination of the motion of only one body whose mass is much smaller than that of the other two bodies whose motions are known. Consequently, the general problem is described by an eighteenth-order system of ordinary differential equations, whereas the restricted problem is given by a sixth-order system. The restricted problem was originated by Euler (1772) in connection with his lunar theory. The Moon's mass is much smaller than the mass of the Earth and the Sun which influence its motion and whose motions are assumed to be known. On the other hand, the motion of three stars forming a triple system requires the determination of all three orbits simultaneously. The importance of the coupling of the equations in the general problem distinguishes it from the restricted problem of three bodies. In spite of the fact that the restricted three-body problem is considerably simpler, it shares the nonintegrability property with the general problem.

Since generally valid analytical solutions of our problem do not exist, approximations for finite duration of time must be considered. Such solutions can be obtained by analytical approaches, known as general perturbation methods, or by numerical techniques, known as special perturbations. There are a number of practically important problems which can be "solved" by either of these techniques, subject to the restriction that the validity of these solutions is limited. In some special cases qualitative predictions of long-time validity can be made.

Concerning the circular restricted problem of three bodies, in which the two bodies with large masses move on circles, we have a generally valid functional relation between the velocity and the position of the third body known as the Jacobian integral. This relation does not give precise predictions but it offers permitted and forbidden regions for the motion. In this way some very useful information can be obtained about the stability of the Moon's orbit as influenced by the Earth and the Sun. This approach might also be used to establish stable regions of planetary orbits in binary star systems with applicability to locate planets (and possibly life) in the galaxy. In space research this technique allows us to establish stable regions of artificial satellites, starting conditions for lunar probes, etc.

For the general problem of three bodies certain qualitative results are now available which allow the classification of the dynamical behavior of three bodies with masses of the same orders of magnitude. For positive values of the total energy, the system of three bodies will explode and all bodies depart to infinity. If the total energy of the system is negative, the possible outcomes are binary formation, revolution, or interplay. The formation of a binary with the third star escaping the system is the most frequent outcome, offering an explanation for the existence of a large number of binary stars in the galaxy. The configuration known as revolution occurs when two stars form a binary and the third star revolves, outside of their orbits. Interplay is the type of motion when two stars form a temporary binary, whereas the third star departs to a relatively short distance and then it returns to join the binary. This is followed by another departure performed either by the same star or another participant of the system. No permanent binary is formed and no permanent escape occurs.

The two types of special solutions of our problem are periodic orbits and equilibrium solutions. Thousands of periodic orbits were computed beginning in 1913 at the Copenhagen Observatory and the work is still going on, utilizing high-speed computers in many centers of celestial mechanics with applications to space research and stellar dynamics. Lagrange had predicted analytically the existence of equilibrium solutions of the problem of three bodies in 1772 and the observational establishment of the Trojan asteroids verified his solutions in 1906. Some of the periodic solutions and of the equilibrium solutions are stable but no general statements can be made concerning the longtime behavior of our system of three bodies at the present time.

BIBLIOGRAPHY

For the mathematical aspects see: G. D. Birkhoff, Dynamical Systems, Vol. IX. American Mathematical Society *Colloquium Publ,* Providence, RI, 1927 (A); C. L. Siegel and J. K. Moser, *Lectures on Celestial Mechanics*. Springer-Verlag Berlin, 1971 (A); A. Wintner, *The Analytical Foundations of Celestial Mechanics*. Princeton University Press, Princeton, NJ, 1941. (A)

For celestial mechanics see: D. Brouwer and G. M. Clemence, *Methods of Celestial Mechanics*. Academic Press, New York, 1961 (I); H. C. Plummer, *An Introductory Treatise on Dynamical Astronomy*. Cambridge University Press, London and New York, 1918 (I); H. Poincaré, *Les Méthodes Nouvelles de la Mécanique Céleste,* Vol. 1 (1892), Vol. 2 (1893), Vol. 3 (1899). Gauthier-Villars, Paris. Reprinted by Dover, New York, 1957 (A); F. Tisserand, *Traite de Mécanique Céleste,* Vol. 1 (1889); Vol. 2 (1891), Vol. 3 (1894), Vol. 4 (1896). Gauthier-Villars, Paris (I); V. Szebehely, "Mechanics, Celestial," in *Enclyclopedia Britannica*, 1974 (I); Y. Hagihara, *Celestial Mechanics*, Vols. 1 and 2, Massachusetts Institute of Technology, Cambridge, Mass., 1970–1972; Vols. 3–5, Japan Society for Promotion of Science, Tokyo, Japan, 1974–1976; J. Lagrange, *Oeuvres*, Gauthier-Villars, Paris, 1867–1892, and *Mécanique Analytique*, Veuve Desaint, Paris, 1788.

For the lunar theory see: G. W. Hill, *Collected Mathematical Works*, Vol. 1, 1905; Vol. 2, 1906, Vol. 3, 1906, Vol. 4, 1907. Carnegie Institute of Washington, Washington, D.C. (I); E. W. Brown, *An Introductory Treatise on Lunar Theory*. University Press, Cambridge, 1896. Reprinted by Dover, New York, 1960.

For introductory treatments see: Danby (1962, 1989), McCuskey (1963), Moulton (1960), Pollard (1966), Roy (1978), Stumpff (1959–1974) and Szebehely (1989).

For historical reviews see: A. Gautier, *Essai historique sur la problème des trois corps*. Courcier, Paris, 1817; G. W. Hill, *Bull. Am. Math. Soc.* **2**, 125 (1896); E. T. Whittaker, *Rept. 69th Meeting of the Brit. Assoc. Advan. Sci.* Murray Publ., London, 1900; E. O. Lovett, *Quart. J. Math.* **42**, 252 (1911); R. Marcolongo, *Il Problema dei Tre Corpi*. Hoepli, Milano, 1919.

For the recently established classification of possible motions see publications of the *Leningrad Observatory* by Agekyan, Anosova and Orlov and papers in the journal *Celestial Mechanics* by Harrington, Marchal, Standish, and Szebehely.

For the restricted problem see: C. V. L. Charlier, *Die Mechanik des Himmels*. Viet and Co., Leipzig, 1902–1907; E. Finlay-Freundlich, *Celestial Mechanics*. Pergamon, New York, 1958; V. Szebehely, *Theory of Orbits*, Academic, New York, 1967.

For the dynamics see: E. T. Whittaker, *A Treatise on the Analytical Dynamics of Particles and Rigid Bodies*. Original first edition, Cambridge University Press, London, and New York, 1904. First American Printing by Dover, New York, 1944. L. A. Pars, *Treatise on Analytical Dynamics*. Heinemann, London, 1965.

Three-Body Problem: Quantum Mechanical

B. F. Gibson

The quantum-mechanical three-body problem compounds the difficulty of the classical problem, that of overlapping interactions among the pairs of particles, with the complexities of quantum mechanics. Nevertheless, much progress has been made in understanding the formal structure of the nonrelativistic problem in both hadronic and atomic physics as well as in generating numerical solutions for particular problems in these areas. The formal work includes recasting the Schrödinger equation into a mathematically valid integral equation or differential equation plus boundary conditions for the scattering problem and extracting useful information about the structure of the solution. Numerical success has resulted from the development of sensible approximations that enable one to reduce the calculation to the capabilities of modern supercomputers. Approximation schemes are sometimes based more upon physical insight than upon mathematical justification and often work better than they ought, particularly in atomic physics where the difficulties associated with the long-range Coulomb interaction have stymied development of a general scattering theory. Finite-element (e.g., splines) and Padé-type numerical methods (including Lanczos and continued fraction algorithms) have proven quite useful in the effort to produce benchmark solutions.

The traditional approach to nuclear physics has been formulated in terms of nuclear Hamiltonians in which nonrelativistic nucleons interact via pairwise (two-body) forces, and subnuclear degrees of freedom are ignored. The simplifications inherent in this model are enormous, and this semiphenomenological description of nuclear structure and nuclear scattering and reactions has enjoyed considerable success, at least in the qualitative description of a large volume of data. There is little theoretical justification for these assumptions other than the rough consistency that is achieved between theoretical predictions and experimental data. The implication is that forces depending simultaneously upon the coordinates and quantum numbers of three nucleons (three-body forces) are relatively unimportant. The existence of a three-body force is required by any theory of strong or Coulomb interactions based upon the exchange of virtual quanta, as in a proper relativistic field theory. However, it is only recently that nontrivial tests of the traditional model have revealed the inadequacy of the pairwise, local potential assumption in the nonrelativistic Hamiltonian picture. The fault could lie in many places—three-body forces, relativistic corrections, subnuclear degrees of freedom, nonlocal interactions, etc. (These concepts are not mutually exclusive.) However, the fact that computational sophistication has improved to the point that one can quantitatively challenge experimentally the theoretical predictions of the traditional nuclear physics model is a noteworthy achievement.

Because the bound-state problem is formally much simpler than any three-body continuum problem (the wave function vanishes asymptotically in all directions), most numerical work has been in this area. Variational methods have been supplanted by direct solution of the three-body Schrödinger equation, usually following the decomposition proposed by Faddeev's seminal work on implementing the proper continuum boundary conditions which are known as the Faddeev equations. Benchmark calculations of the trinucleon bound-state properties have shown that contemporary, realistic two-body force models underbind the triton (BE = 8.48 MeV) by the order of 1 MeV, generate a theoretical root-mean-square charge radius ($\langle r^2(^3\text{He})\rangle^{1/2} = 1.93$ fm, $\langle r^2(^3\text{H})\rangle^{1/2} = 1.81$ fm) which is too large by some 10%, produce a ^{3}He–^{3}H Coulomb energy difference (760 keV) that falls short by about 100 keV, and lead to an asymptotic normalization constant (spectroscopic factor) ratio (*d* state/*s* state: $C_2/C_0 \simeq 0.044$) which is 10% too small. The correlation of these trinucleon asymptotic properties with the binding energy is such that the addition of a phenomenological three-body force that accounts for the missing binding energy will also account for the charge radii and asymptotic normalization constant ratio, but a deeper understanding of the meson exchange dynamics of such an interaction is needed before the problem is solved. The failure to predict the Coulomb energy difference is a strong indication that charge symmetry in nuclear forces does not hold at this level.

For the atomic and molecular bound-state problem, remarkably accurate variational calculations exist. In particular, the atom and H^- ion seem well modeled quantitatively, as an accuracy of better than 10^{-10} has been achieved with variational functions of more than 300 terms. Subtleties such as electromagnetic fine-structure splitting and quantum electrodynamic effects can be studied. Muonic molecular structure calculations ($p\mu^-p$, $d\mu^-t$) have also been pushed to a high degree of accuracy and are important ingredients in our attempt to understand the physics of muon-catalyzed fusion (e.g., $d+\mu^- + t \to \alpha + n + \mu^-$).

The three-body continuum problem requires somewhat more sophistication to solve than does the bound-state problem. For the two-body problem, one can transform the Schrödinger equation into an integral equation that properly incorporates the scattering boundary condition, the Lippmann–Schwinger equation. For three particles the corresponding integral equation is mathematically ill behaved, because the kernel in all its iterations contains parts that only describe the configuration in which a single pair interacts while the spectator particle is unaffected and can be indefinitely far removed. Mathematically, the solution to the bound-state equation is also a solution to this continuum equation implying an unphysical nonuniqueness for the solution. This can be resolved in a number of ways ranging from imposing supplemental boundary conditions upon the solution to rearranging the three-body scattering equation so that the uninterrupted interaction of pairs of particles is summed. The best known and mathematically most thoroughly studied scheme is of the latter type and results in the Faddeev equations. In solving the disconnectedness problem, the integral formulation of the Faddeev equation restates the two-body dynamics in terms of full off-shell two-body t matrices rather than the (equivalent) two-body potentials. In the hadronic three-body problem this can be an advantage because the phenomenological strong-interaction potentials are often somewhat singular due to strong short-range repulsion. However, the integro-differential form of the Faddeev equation which works directly with the potentials has now been solved numerically and shown to yield the same benchmark results as are obtained from the integral formulation. In hadronic problems, the pair interactions are often dominated by a few partial waves, and the principal feature of these is a weakly bound state (e.g., the deuteron in the n–p system) or a resonance (e.g., the Δ in the π–N system). In such cases the two-body t matrix can be well represented by one term or by a sum of separable terms. Such a separable form for the t matrix, when inserted into the Faddeev equations, reduces them after partial wave decomposition to a coupled set of one-variable integral equations that are easily solved using standard numerical methods. A wide range of three-hadron problems have been attacked in this manner with remarkable success. These include elastic scattering, polarization, and breakup reactions for n–d, d–α, π–d, and K–d scattering. In a related relativistic formalism, features of π–d, π–N, and π–π scattering have been investigated. The most detailed comparison between the nucleon–deuteron scattering data and model calculations utilizing realistic local potential representations of the N–N data have been performed using the full two-variable Faddeev equation with no separable expansion. Coulomb effects have been shown to be significant, especially at low energy as one would anticipate. In all cases, the major features of the physics emerge as consequences of the properties of the two-body interaction and the three-body dynamics.

In the atomic problem the long-range nature of the Coulomb force introduces singularities in the off-shell t-matrix, so that the Faddeev program fails, particularly at energies above the threshold to produce three "free" particles. However, approximation schemes (e.g., Glauber and first Born approximations) have enjoyed some success for e^-–H scattering above 100–200 eV. The close coupling scheme, in which the continuum function is expanded in a set of bound states (or pseudostates) of one electron and the motion of the other is solved for, appears to work well near threshold. The calculations reproduce many of the resonances in the e^-–H system and provide physical insight into their origins.

The molecular three-body system is really a many-body problem in which the basic forces are Coulombic. One reduces the calculational difficulty by employing the molecular Born–Oppenheimer approximation, where one solves the electron problem first to obtain a complicated effective potential energy surface on which the nuclei move. Because there are usually many excited states, breakup channels are neglected in full three-dimensional calculations. Three-body forces are very important because the three nuclei reside in an electron soup. Although the Born–Oppenheimer approximation is quite accurate, the electron part of the problem cannot be solved exactly, which implies that the resulting three-body problem for the nuclear motion can be solved only approximately. When model results fail to agree with experiment, one cannot be certain why. Nevertheless, one has recently achieved impressive results for the rearrangement scattering problem $A+BC \to AB+C$ or $AC+B$ for a number of different atoms. It is in this area of reaction scattering in which formal developments and supercomputer advances have led to real progress in understanding the resonances and rotational–vibrational states of molecular systems.

In spite of the existence of a formally correct starting point and many successful detailed three-body calculations, the general structure of quantum-mechanical three-body states and their implications of the structure for phenomenology are not well understood. Difficulties plague our understanding over the whole range of physics, from problems in analyzing nuclear reactions such as $\pi + d \to N + N$ to our inability to understand $\mu^- + p + d \to {}^3\text{He} + \mu^-$ and to predict the threshold ionization properties of $e^- + \text{H} \to p + e^- + e^-$. Still, some general features of three-body systems have been studied and yielded surprising outcomes. Nuclear forces were shown to have a nonzero range else the triton bound state would collapse to a point. Under special conditions, a three-body system interacting via short-range forces can have an infinite number of bound states even though the two-body interaction does not. In the low-energy ${}^3\text{H} + \gamma$ photonuclear reaction, it was found that the $d+n$ channel robs most of the strength from the $p+n+n$ channel with the same quantum numbers due to the dynamics in the off-shell rescattering. Such surprises provide a measure of the richness of the quantum-mechanical three-body problem.

BIBLIOGRAPHY

Nuclear and Particle

L. D. Faddeev, *Mathematical Aspects of the Three-Body Problem in the Quantum Scattering Theory*. Israel Program for Scientific Translations, Jerusalem, 1965.

R. D. Amado, "The Three-Body Problem," in *Elementary Particle Physics and Scattering Theory*. Gordon & Breach, New York, 1970.

E. W. Schmid and H. Ziegelmann, *The Quantum Mechanical Three-Body Problem*. Vieweg Tracts in Pure and Applied Physics, Vieweg, Braunschweig, 1974.

A. W. Thomas, *Modern Three-Body Physics*. Springer-Verlag, Berlin-Heidelberg-New York, 1977.

R. C. Newton, *Scattering Theory of Waves and Particles*. Springer-Verlag, New York-Heidelberg-Berlin, 1982.

W. Glöckle, *The Quantum Mechanical Few-Body Problem*. Springer-Verlag, Berlin-Heidelberg-New York-Tokyo, 1983.

J. L. Friar, B. F. Gibson, and G. L. Payne, "Recent Progress in Understanding Trinucleon Properties," *Annu. Rev. Nucl. Part. Sci.* **34**, 403–33 (1984).

Atomic and Molecular

E. Gerjuoy, "Three-Particle Problem in Atomic Physics," in *Atomic Physics 2*. Plenum, New York, 1971.

A. Kuppermann, "Accurate Quantum Calculations of Reactive Systems," in *Theoretical Chemistry Advances in Perspectives* (D. Henderson, ed.). Vol. 6A, 79–164 (1981).

D. C. Clary, *The Theory of Chemical Reaction Dynamics*. Reidel, Boston, 1986.

R. T. Pack and G. A. Parker, "Quantum Reactive Scattering in Three Dimensions Using Hyperspherical Coordinates Theory," *J. Chem. Phys.* **87**, 3888–3921 (1987).

D. J. Kouri *et al.*, "New Time-Dependent and Time-Independent Computational Methods for Molecular Collisions," in *Mathematical Frontiers in Computational Chemical Physics*. Springer-Verlag, New York-Heidelberg-Berlin, 1988.

Time

P. C. W. Davies

Time is the most fundamental aspect of our experience. Its properties have intrigued and baffled philosophers, theologians, and scientists for centuries. Questions about the nature of time, its global and microscopic structure, continue to fill books and provoke experimental research. The long-standing controversies have centered on the following questions: Was there a *beginning* of time (and all things)? Will there be an end? What distinguishes past from future? What is the "present" moment and why does it "move"? Does time even *exist* as a separate physical entity, or is it merely a construct of our world description? Remarkably, progress in physical science during the last century has gone a long way to answering some of these questions, though not all.

THE BEGINNING AND END OF TIME

Was there a first moment? Astronomers have discovered that the universe is engaged in a systematic pattern of expansion. The expansion rate suggests that about ten billion years ago all the matter in the universe was compressed into a highly dense, intensely hot fireball. Our present understanding of gravity indicates that this fireball could not have remained static, but must have been expanding exceedingly rapidly. There is apparently no point in the past at which this expansion could have been arrested. It follows that the fireball must have been in a state of infinite density a finite time ago. According to the theory of relativity, this event, called a singularity, represents an edge or boundary of time. It is not possible to continue known physics through this boundary. For this reason the singularity at the beginning of the fireball phase is usually referred to as the creation of the universe. This singular event represents the creation not only of all the matter in the universe, but of space and time as well. What preceded the creation was not just a featureless void, but literally nothing at all.

Great interest attaches to whether there will likewise be a singularity in the future that will bring about the end of the universe in a similar fashion. This could occur if the present phase of expansion were to be reversed and give way to collapse. The universe would then meet its demise in another fireball, shrinking catastrophically toward a disappearance at the final singularity. Present observational evidence is somewhat ambiguous about whether the expansion rate of the universe is really slowing down enough to bring about this eventual collapse.

Some cosmologists have suggested alternatively that the universe may be infinitely old, either by speculating that singularities would not actually form, or that not all of the universe would encounter them. Such model cosmologies raise difficult questions about their thermodynamics (see below).

TIME ASYMMETRY

Everyday experience with physical processes reveals a strong asymmetry between past and future time directions. Many phenomena in physics (e.g., the mixing of two gases) are often described as *irreversible*. This means that they are observed to proceed in one time direction only. We do not observe gases *spontaneously* unmixing.

Other examples of unidirectional changes are the emission of retarded radiation, gravitational collapse to a black hole, and the measurement of a quantum-mechanical observable.

There is a long-standing paradox that the laws of physics that describe all these asymmetric processes do not themselves show a time asymmetry. The reverse of every physical process (except an unimportant case that involves exotic elementary particles) is quite consistent with these laws. Thus, every molecular collision that causes two gases to mix together would, if the molecular trajectories were reversed, proceed in the opposite direction along exactly the same paths. This underlying time symmetry implies that a gas would, if left completely isolated for an enormous duration, unmix itself again as a result of stupendously rare statistical fluctuations. It follows that the time asymmetry that is observed when two gases mix is really a consequence of their initial conditions; the unmixed gases are much more *ordered* at the beginning than when they have mixed together and reached equilibrium. In short, the initial state is a very remarkable one, whereas the fully mixed state is much more typical. This tendency for improbable macroscopic order of a system to give way to disorder is one expression of the second law of thermodynamics.

How did the universe achieve its high state of order in the

first place, if this is such an improbable condition on average? Apparently, it was not simply made that way, because there is evidence that during the primeval fireball the universe was in a condition of thermodynamic equilibrium. The answer to the question is actually very contentious. There is general agreement that the expansion of the universe plays a decisive role, and upsets the initial equilibrium.

Two distinct types of process contribute to the growth of disorder in the universe. One is the emission of starlight. The thermal disequilibrium in the vicinity of the sun generates nearly all the time-asymmetric processes on earth (e.g., erosion of rocks, biological activity, melting snow, flowing rivers). The other process is gravitational collapse, which causes stars to implode. Both processes cause highly organized energy in stars to become spread across the universe in the form of disordered radiation.

Sometimes it is proposed that the progress from order to disorder will be reversed at some future epoch of the universe. This would have the effect of interchanging past and future. Such a situation would require the universe to have started in a very special state. According to the more conventional view, the universe will slowly "run down" and approach a condition of complete disorder (thermal equilibrium), after which nothing much of interest will happen. This is called the heat death. The heat death will not occur if the universe recontracts again.

MICROSCOPIC TIME

Most physicists assume that time is a continuum. That is, it may be infinitely divided into progressively smaller intervals. High-energy physics confirms that time is "smooth" down to about 10^{-25} s. However, at sufficiently short durations, around 10^{-43} s, it is expected that a profound disruption of the "smoothness" of time takes place. This is because the Heisenberg uncertainty principle of quantum mechanics allows the fleeting appearance of energy for small time intervals. The amount of this "borrowed" energy is inversely proportional to the time interval for which it exists. For durations of about 10^{-43} s these energy fluctuations are enormous—so large, in fact, that their gravity can cause severe distortion of space and time on a microscopic scale. It is conjectured that the fluctuations may be so violent that space and time actually start to break up completely. In that case, the smoothness of time is only a feature of intervals longer than 10^{-43} s.

RELATIONAL TIME

A deep and enduring dilemma about time is whether it is a real entity of some sort, or simply a way of describing an ordered sequence of events. Isaac Newton believed that time existed in its own right, quite independently of whether or not things happen in the universe. Others have argued that we only obtain information about time by examining the events that occur, such as the successive chimings of a clock. Time is then merely a way of expressing a relation between these events—a linguistic convenience.

According to this view, it is just as nonsensical to invent a mysterious physical entity called time to explain the relation between events as to invent a physical substance called "citizenship" to explain the relation between Englishmen.

Although a fascinating dispute, the outcome is unlikely to have much relevance for the physics of time as such.

PSYCHOLOGICAL TIME

One fundamental aspect of time has not appeared in the physical description already given. The human mind divides time into past, present, and future. Everyone is aware of a flow, or flux, of time toward the future. This flow appears to sweep the past out of existence and bring the future into being. Such temporal activity is not a feature of time in physics, at least as understood at present. Its appearance in the human mind is sometimes dismissed as purely psychological, and sometimes conjectured to indicate a great unsolved problem of physics. In either case it is a mystery.

See also COSMOLOGY; UNCERTAINTY PRINCIPLE; UNIVERSE.

BIBLIOGRAPHY

M. Capek, *The Concepts of Space and Time*. Reidel, Boston, 1976. (A)

P. C. W. Davies, *The Physics of Time Asymmetry*. Univ. of California Press, Berkeley, 1974. (A)

R. P. Feynman, "The Distinction Between Past and Future" in *The Character of Physical Law*. B. B. C. Publications, Cox and Wyman, London, 1965. (E)

J. T. Fraser and N. Lawrence, eds., *The Study of Time*. Springer, New York, Vol. 1, 1972; Vol. 2, 1975; Vol. 3, 1979. (A)

T. Gold, ed., *The Nature of Time*. Cornell Univ. Press, Ithaca, 1967. (I)

D. Layzer, "The Arrow of Time," in *Sci. Amer.*, **233** (6), 56 1975. (I)

H. Reichenbach, *The Direction of Time*. Univ. of California Press, Berkeley, 1956. (A)

———, *Philosophy of Space and Time*. Dover, New York, 1958. (A)

J. J. C. Smart, ed., *Problems of Space and Time*. Macmillan, New York, 1964. (I)

R. Morris, *Time's Arrows*. Simon and Schuster, New York, 1984.

R. Flood and M. Lockwood (eds.), *The Nature of Time*. Blackwell, Oxford, 1986.

Transducers

Eugene Mittelmann and Carol Z. Rosen

These are essentially translating devices, capable of translating, or transducing, the magnitude of one physical quantity into a similar magnitude of another physical quantity in such a manner that the relationships between the two quantities follow a definite mathematical expression, usually referred to as the input–output or transducer characteristic of the device. Accordingly, we speak of linear or square law transducers, or transducers defined by any other relationship described by a known mathematical function having the magnitude of the physical quantity to be measured as the independent and the translated magnitude as the dependent variable.

Practically all modern transducers attempt to translate the magnitude of the physical quantity to be measured, or controlled, into a corresponding electrical output signal, voltage or current. However, in a general sense, any device performing translation can be correctly designated as a transducer, even though in the current literature the term is restricted to devices with an electrical output. As an explanatory example, we might consider the well-known mercury column thermometer as a transducer in the broad, general sense. It translates equal amounts of temperature changes into equal amounts of changes of the length of a mercury column, thus being characterized as a linear transducer. The more specific counterpart, yielding an electrical signal as a linear output of a temperature-measuring transducer, is the thermocouple. It consists of two thin wires of dissimilar metals joined together. When exposed to the temperature of the medium surrounding it, or attached to a body whose temperature is to be measured, it will generate an electrical signal in the form of a direct-current voltage, directly proportional to the temperature to be measured.

Characteristically, the thermocouple's output voltage becomes available directly, without the need for any additional power source. Such transducers are referred to as self-generating devices. The majority of transducers, however, require an additional power source, either the direct- or alternating-current type, to make the translation from one physical quantity to an electrical signal possible. These transducer types are the so-called excited transducers.

Still remaining in the field of temperature measurement and control, we can cite the example of infrared temperature indicators and controllers as representative of transducer types with a nonlinear relationship between input and output, requiring not only additional power sources for excitation but also additional devices to "condition" the output signal, making it possible to derive a final output, in the form of either current or voltage, that has a linear or other arbitrarily chosen relationship, conforming to a "design equation" between input (temperature) and final output. Every body, raised to a certain temperature, will emit electromagnetic radiation in the infrared range (at wavelengths of the order of magnitude of micrometers) that is an exponential function of the temperature, determined by Planck's law. Letting this radiation impinge on the surface of a photoelectric cell responsive to the spectral range of the radiation emanating from the observed body results in an output voltage that is a modified exponential function of temperature, determined also by the response characteristic of the photocell. To obtain a linear relationship between temperature and the final output signal, a so-called linearizing amplifier is inserted between the photocell's output and the final output terminals, both amplifying and simultaneously extracting the logarithm of the output by means of special electronic circuitry, thus providing a linear input–output characteristic.

The principles allowing transduction of the measurand into an electrical signal are numerous, limited only by the ingenuity of the designer. A few typical ones, widely used in industrial applications, particularly in the chemical industries and in mechanical measurements, are mentioned. Flow, temperature, and pressure are of paramount interest in the control of chemical processes. Small turbines inserted into a flow pipe will respond with a linear speed variation to flow velocity variations, capable of being translated into electrical signals with the help of revolution counters or tachometer generators. Electromagnetic flowmeters permit us to measure flow rates without interfering with the flowing liquid's velocity distribution inside a circular pipe. They are based on the Faraday principle that a conductor moving in a magnetic field will have a voltage induced in it proportional to the velocity of movement and the intensity of the magnetic field. Disposing magnetic coils around the periphery of the flow pipe, we apply a current, which excites a magnetic field that penetrates the flow pipe perpendicularly to the flow direction; two electrodes contacting the moving liquid in the immediate vicinity of the pipe's inside wall are then the source of a signal voltage directly proportional to velocity of the flowing liquid.

Differential transformers are widely used for the measurement and control of small mechanical displacements. They are based on the principle that the voltage induced in the secondary coil of a magnetic structure depends on the coupling between the exciting primary and the secondary, determined by the relative geometry of a ferromagnetic structure coupling primary and secondary. If, for instance, a cylindrical tubing of insulating material has an exciting primary coil wound over its center part, with two equal secondary coils placed at a symmetrical distances on each side, and a cylindrical, ferromagnetic core shorter than the distance between the ends of the two outer coils being movable along the tube's axis inside the tube, an alternating voltage applied to the primary will produce different voltages in both coils, depending on the relative position of the ferromagnetic core with respect to them. The difference between the two voltages is a measure of the core's displacement from the symmetrical center position. Differential transformers are used to detect motions or measure displacements.

Various transducers make use of the principle of capacity variations. The capacity of two metallic surfaces (electrodes) facing one another, such as plates or one hollow cylinder inside an other, is determined by the total area of the surfaces, the spacing between them, and the intervening dielectric constant, partly or completely filling the space between the electrodes. Characteristic examples are the remaining fuel indicators in use on many aircraft, consisting of two coaxial cylinders immersed in the fuel tank. The height of the fuel along the cylinders' axis is directly proportional to the capacity of the cylinder and is measured either in a self-balancing capacity bridge or forms part of an electronic circuit generating oscillations with a frequency determined by the value of capacity of the two cylinders. Because the frequency of the oscillator is proportional to the inverse square root of capacity, special signal-conditioning circuits are needed to obtain a linear relationship between fuel quantity and the system's output signal.

Piezoelectricity is a linear coupling between electric and material variables. The direct effect is produced by applying a mechanical stress to crystals. In the converse effect a strain is produced. The converse effect, for example, transduces incoming atmospheric electromagnetic signals, with wavelengths hundreds of meters long, into acoustic signals with wavelengths on the order of 10^{-3} m and velocities reduced by 10^5. Such acoustic signals can be coded, filtered, time compressed, and delayed by 10^{-6} s in a path length of 10^{-2} m. The negligible acoustic attenuation of these bulk and Rayleigh surface elastic waves make piezoelectrics excellent high-Q vibrators. Piezoelectric crystals, ceramics, composites, and polymers find applications in radio, telecommunications, radar, and in seismology, sonar, and accelerometry. They are used as audio-transducers, resonators, filters, frequency standards, delay lines, and actuators and can be fabricated into transformers, ultrasonic motors, detectors for nondestructive testing, acoustic microscopes, and light-diffracting acoustic gratings. The anisotropic property of piezoelectricity permits a wide variation in device geometry and in selection and design of materials.

Many other transducer principles, too numerous to be dealt with here, are feasible and are in use.

BIBLIOGRAPHY

Harry N. Norton, *Handbook of Transducers for Electrical Measuring Systems,* Prentice-Hall, 1969.

Kurt Lion, *Instrumentation in Scientific Research,* McGraw-Hill Co., 1959.

Frank Oliver, *Practical Instrumentation Transducers,* Hayden Book Comp. 1971.

Instrument Society of America, *Transducer Handbook,* Plenum Press, 1957.

Werner Holzblock, *Instruments for measurement and control,* Reinhold Publ. 1957.

Donald Fink, *Electronic Engineering Handbook,* Chapter 10, McGraw-Hill, 1975.

C. Rosen, B. Hiremath, and R. E. Newnham (eds.), *Piezoelectricity.* American Institute of Physics, New York, 1990.

Transistors

R. M. Ryder

A transistor is an electronic amplifying device, using controlled electron streams inside semiconducting materials. The transistor and the electronics derived from it have truly revolutionary importance for human society, even greater than the Industrial Revolution of the 18th–19th century. Through improved communication, electronic control systems, and computers, the "Transistor Revolution" has multiplied the power of man's *mind* many times, just as the Industrial Revolution, based on the steam engine, multiplied the power of man's *muscles* many times. The explosive growth of computers, and computer-controlled machinery of various kinds, is a result of the development of suitable transistors and integrated circuits.

1. PRINCIPLES OF TRANSISTOR OPERATION

Transistors operate by means of novel physical principles, discovered through study of the quantum mechanics of solids. A controlled stream of electronic charges, either negative conduction electrons or positive holes, can travel within a solid semiconductor, usually silicon. "Bipolar" transistors are so called because both kinds of charge carriers take part in the action, but in the "unipolar" transistors only one sign of charge carrier is active. The holes are vacancies in the structure of valence electrons which holds the solid material together. Both holes and electrons respond to an applied electric field by moving with an average velocity characteristic of the material. For small fields, this velocity is proportional to the field, with the constant of proportionality being the mobility of the holes or electrons.

In a typical *npn* bipolar transistor, a thin region of "p" type material, conducting by (positive) holes, is constructed between two n regions which conduct by (negative) electrons. The electric potential on the "base" (the p region) controls the number of charges (electrons) which pass through from the more highly doped n region (the emitter) to the less highly doped n region (the collector). In a *pnp* transistor, all the polarities are reversed, that is, a stream of holes passes from emitter p region to collector p region in response to the potential on the thin n-type base region. In either case, under suitable conditions the electron stream at the collector can deliver much more power to an external load than is required to activate the base, so that the external load has impressed upon it an amplified replica of whatever signal waveform was applied to the base. Power amplification by a factor of 100 or more is typical for low-frequency transistors.

When the base region is sufficiently thin (a fraction of a micrometer), the speed of the transistor action can be very high, sufficient to amplify or control microwave frequencies up to several times 10^9 Hz. Still higher frequencies, up to more than 10^{11} Hz, can be generated and controlled by semiconductor diodes (two-electrode devices) of various types. The useful frequency spectrum for communications has been greatly extended to include infrared and optical, and even X-ray, sensors. Thus the entire presently useful radiofrequency spectrum can be advantageously instrumented with solid-state devices, which have completely superseded vacuum tubes for all low-power applications of receiving type, such as ordinary radio receivers. Power transistors of modest capability, up to a few hundred watts, can also be used for communications frequencies. But for very high-power applications of transmitting type, such as radio broadcasting, vacuum tubes are still necessary to produce the desired tens or hundreds of kilowatts of power.

Field-effect transistors (FETs) are so called because the stream of charge carriers is controlled by a different mechanism, that of induction by the electric field of the control electrode. In an FET, the flow of carriers from a "source" region is attracted or inhibited by the electric field from a nearby "gate" electrode; the third electrode, which terminates the stream of carriers in the triode (three-electrode) device, is called the "drain." For FETs also, amplification results when the amplified replica (output signal) at the drain contains more power than the input signal applied to the gate. FETs are called unipolar because only one of the two types of charge carrier (electron or hole) is involved in the action: the control electrode (gate) is an external metal electrode, not an internal region of opposite conductivity type, like the base of the bipolar unit. Note that in both kinds of transistors, the active stream of carriers actually passes through the interior of the semiconductor—much like the stream of carriers in a vacuum tube, but with much lower velocity and energy.

Depending on just how the gate electrode affects the carrier stream, FETs are further subdivided into two classes. In an insulated gate FET (IGFET), the gate metal is insulated by a thin film between the metal electrode and the semiconductor, so that the flowing charge stream attracted (or repelled) by the gate resides in a thin subsurface layer of the semiconductor, called a "channel." The IGFET is also called a MOSFET (metal-oxide-semiconductor-field-effect-transistor), when the insulating film is (as usual) an oxide. In a "junction FET" (JFET), on the other hand, the gate metal is a rectifying junction contact to the semiconductor surface, which must be reverse-biased to avoid drawing current, and which therefore *repels* carriers from a depletion region channel just below the gate, and thereby also controls the current flowing from source to drain through the channel.

For most communications applications involving amplification, such as radio and telephone amplifiers, bipolar transistors are used because they have greater gain and power capability. JFETs have have some use as low-noise, low-distortion preamplifiers for TV and FM sets; also JFETs in gallium arsenide material (GaAs FETs), which use Schottky-barrier rectifying electrodes as gates, are currently coming into use as microwave amplifiers at frequencies above 2 GHz, up to about 10 GHz (in 1976). IGFETs are not much used in transmission applications because of their high electrical noise, but they are very popular for integrated circuits, especially for logic and computer applications. The reason is that if (as usual) utmost speed is not required and noise is not critical, then the IGFET fabrication process often requires fewer and simpler steps than the bipolars. A typical application is to the handheld calculator, which can tolerate very much slower operation than is needed for large, fast computers, and which needs the simpler processing and high yield that leads to low-cost integrated circuits.

2. *POWER TRANSISTORS*

Most transistors are advantageously very small, economical of both space and power. Often, especially in slow computer applications, they operate at "micropower" levels, of the order of microwatts per transistor. Nevertheless, it is true that when output power is needed, transistors can be built big enough to drive locomotives. Power transistors fall into two classes: (1) linear amplifiers, whose output is a faithful replica of the input; and (2) nonlinear power units, which operate by being triggered on and off, the so-called "switching" mode of operation. It is the latter which can deliver very large output, but only at very low frequencies; they can be thought of as power *controllers* rather than amplifiers, and are very suitable for driving variable-speed electric motors in sizes up to about 1000 kilowatts.

The linear power transistors, intended for use as power output stages of amplification, are simply large-area units, almost always bipolar silicon. They can deliver up to a few hundred watts per unit at audio (audible sound) frequencies, in such typical applications as drivers for loud speakers in public-address systems. When designed for higher frequencies, up to a few times 10^9 Hz, they have to be progressively smaller as the frequency rises, but some amplifiers can still put out 10 W at 2 GHz. For the microwave frequencies especially, there is a complicated fabrication problem in that the emitter and base electrodes have to be very narrow, closely juxtaposed, and still comparatively large in area. Consequently, a large high-frequency power transistor can be thought of as a mosaic of a large number of small transistors, interconnected by metallic leads running over the surface in a complicated form which often looks something like the skeleton of a microscopic fish. Frequently, the emitter and base electrodes are "interdigitated" or interleaved, to keep the emitter–base distance very small; ordinarily ballast resistors are built into the leads or into the emitters to keep the current uniformly distributed, and avoid local high current concentrations which would burn out the transistor.

In other words, a power transistor of this kind is really a special form of integrated circuit, which contains, besides transistors proper, other elements, such as leads, resistors, and heat-distributing elements ("heat sinks"), all built together as a unitary structure. Units of this kind work very well as high-fidelity amplifiers, telephone amplifiers, and small radio transmitters, but are about two orders of magnitude too small for such high-power uses as broadcast radio transmitters.

The nonlinear power units are usually called semiconductor controlled rectifiers (SCRs) or "thyristors," because of the way they operate. They may be thought of as rectifying diodes, capable of producing large direct currents by rectification of ordinary alternating-current power, but which are held "off" for a controllable time by an auxiliary electrode (the "gate"). When the forward-biasing applied alternating voltage becomes sufficiently large, the SCR is triggered "on" and conducts thereafter in the same way as a usual diode power rectifier. In this way, by varying the gate control voltage the SCR can vary the current for a variable-speed electric motor, even in sizes large enough to drive locomotives. Such solid-state control is used today to drive anything needing either manual or automatic control of mechanical motion.

Since the simplest SCRs work under only one polarity of potential, a further development, known as a Triac, has become popular, since it will draw current in either polarity and respond to gating voltage of either polarity. Such units use two SCRs connected together in opposite polarity, and responding to a common gate electrode; the whole structure can be built into a single chip of silicon. Triacs are widely used for driving small motors from electronic control systems.

3. *ELECTRONIC DEVICES CLOSELY RELATED TO TRANSISTORS*

Amplification is usually obtained via a three-electrode (triode) device, the transistor. But different numbers of electrodes can be used for other purposes, and such solid-state devices find numerous uses in transistor electronics. For example, some two-electrode units (diodes) have extended the usable radiofrequency spectrum to over 100 GHz, while other units can be used to make a variety of electronic sensors for use in electronic control systems. The latter can respond to heat, light, sound, alpha radiation, pressure, or other external influences. Some diodes can emit light, and are widely used in displays or signaling; some diodes are lasers, optical generators or amplifiers.

The simple diode *rectifier* permits current flow in only one direction of polarity, thus converting alternating to direct current. Since alternating current is almost universally used for electric power transmission, while direct current is needed at the point of use for many applications (powering electronic systems, motor drives, electroplating, etc.), diode rectifiers find a myriad of uses. Small rectifiers, known as detectors, can detect the presence of radiofrequency waves up to the highest frequencies used in communications. Superheterodyne detection, in which such a detector is driven by a local oscillator at a slightly different frequency, can increase the sensitivity to radio waves by a large factor, of the order of 100,000. Such diodes, called "mixers" because they combine the weak signal with the local oscillator drive, are almost universally used in all radio receivers.

Somewhat similar structures, called "varactors" (variable reactance diodes), can give ultralow noise "parametric" amplification, and for that reason can be used in radio astronomy and for reception of the very weak signals from distant space travelers. For the utmost in low-noise reception, the whole apparatus can be cooled to liquid-helium temperatures. Ultrafast circuits also sometimes use diodes to avoid the complication of the third electrode of the transistor. Varactors are also useful in power mixers whose object is to produce very accurate control of high frequencies for superheterodyne reception. Still other varactors can produce fine-tuning control for heterodyne oscillators, and are widely used for channel selection in radio receivers.

Diodes for electronic pulse shaping and switching are widely used in logic circuits for computers as are transistors also. Specified combinations of input pulses are organized to produce specified outputs, resulting in addition, subtraction, multiplication, and division of numbers, as well as other computer logic operations such as switching and routing of information. Similar *p-i-n* "plasma" diodes can generate very high-speed pulses, or switch and control radiofrequency signals.

Microwave diode oscillators of various kinds are preferred local oscillators for superheterodyne reception above about 10 GHz. Among these are "transferred electron" diodes, which utilize the negative resistance consequent to transferring to different energy states of the solid in order to generate microwave oscillations; also "IMPATT" diodes, which utilize for the same purpose a combination of transit time of the charge through the device, together with impact ionization of valence electrons by the fast-moving carriers, to produce new carriers and therefore multiply the electrical conductivity. It was the IMPATT diodes that opened up the extended frequency range for radio to above 100 GHz.

Modern astronomy makes more and more extensive use of electronic techniques. Among these are use of sensors, rather than film or the human eye, to pick up an order of magnitude or even more in sensitivity for observation of very faint stars; use of television techniques in taking pictures; and computer processing of images, to bring out very faint details. Of course "remote control" techniques have to be used for operation of advanced mechanisms in space, such as the Hubble telescope (launched in May 1990) to observe x-ray and ultraviolet sources invisible from ground-level observation. Perhaps the best-known diode radiation sensor is the "solar cell," which has been widely used in space vehicles for the generation of electric power directly from sunlight. The light energizes valence electrons so that they become conduction electrons, at the same time leaving mobile holes in the semiconductor material; each solar cell thereby can develop about 1 V, depending on what material is employed. The solar cells are similar in principle to the photocells long used in photography and for such industrial uses as controlling the closing of elevator doors; but the solar cells use high-quality single-crystal material available as a result of the transistor development program. They are therefore substantially larger in size and efficiency, and can produce enough power to operate fairly large electronic systems for use in satellites and space vehicles. For use in earth satellites, the solar cells have to be "radiation hardened" to avoid substantial reduction of performance under bombardment by the "solar wind" of high-velocity particles from the sun. Terrestrial use is also possible, and may become large if development to reduce costs is sufficiently successful. The new photodiodes in small sizes are also used as detectors for light beams used in optical communications. Some other types of radiation, such as x rays and nuclear alpha particles, are also detectable by semiconductor diodes.

The light-emitting diodes (LEDs) depend on the fact that in some semiconductors the recombination of holes and electrons can emit light with high efficiency. For indicator lights featuring long useful life and reliability, the LEDs are useful; but for routine room-temperature use in portable equipment, they have been superseded by liquid-crystal displays (LCDs) which draw essentially no current from the long-life batteries.

Semiconductor lasers emit light of high spectral purity and high intensity, useful for optical signaling applications; they demand exceptionally high crystal quality, and therefore never could be made or even dreamed of until after adequate methods for making high-quality semiconductor material had been developed for transistor use. The LEDs and lasers are easily modulated at very fast rates by changing the applied voltage, and therefore are extensively used in optical-fiber communications which transmit information via modulated light beams.

In the examples cited, the semiconductor devices are widely useful in the inputs and outputs of electronic systems. One can say that they facilitate communication between electronic systems and the external world. One can also say that the sensors and amplifiers permit the world to transmit information from place to place via electronics. While en route, the information can also be processed by computers which "think" about it. When the information arrives at its destination, electronic actuators can carry out various kinds of appropriate action; or the information can be stored electronically in a memory or library, which is itself also often electronic. By such means, electronics has already greatly increased the power of the human mind.

4. SCIENTIFIC ASPECTS OF TRANSISTORS

New scientific knowledge of solid-state physics was necessary in order to make transistors possible; but this debt to pure science has been repaid many times over by the new scientific knowledge resulting from the transistor program. In addition, the transistor program has led to new scientific instrumentation, to the ubiquitous computers now used for analyzing the information gained, to new, highly pure, and well-controlled materials, and to "high technology" useful for fabricating all sorts of sophisticated syntheses. Only a sampling of a few highlights can be given here.

Viewed as a scientific achievement, the invention of the transistor showed for the first time how to control the interaction of two types of charge (negative electrons and positive holes) in order to achieve amplification via a solid. Much further knowledge of semiconductors has resulted, including the injection, collection, generation, and recombination of holes and electrons; the absorption and emission of light and other radiation; the interaction with vibrational acoustic energy; and the structure of the solid material itself (band theory), including the calculation of its properties from fundamental quantum mechanics.

Along with improved material knowledge has come more powerful use of the material properties in electronic-device applications. As a result of his analysis of the behavior of the first, point-contact transistors, Shockley was able to show the existence of the phenomenon of carrier injection from electrodes into the material, and from that could predict the existence of the "junction" transistors (*npn* and *pnp*, described above). This fine example of successful *prediction* in science was followed by experimental verification, and the very successful bipolar transistors were the result.

One of the most interesting, and for a long time most puzzling, features of the transistors was the behavior of the free surfaces of the semiconductor, and the interfaces with other materials, both metals and insulators. Transistors are very sensitive to surface effects, because the electrical conductivity of the semiconductor typically depends on very small and closely controlled amounts of impurities, called "dopants," often less than one part per million. A correspondingly small amount of charge present on the surface—maybe as little as 1/10,000 of a monolayer of atoms—can completely upset or even invert the conductivity, and produce corresponding drastic changes in the properties of the device. When such a change happens spontaneously at an uncontrolled surface, the device may be regarded as unreliable by the user. Considering that almost any common object not isolated in high vacuum quickly picks up one or several monolayers of adsorbed gas from the ambient atmosphere, one can see that the requirement that the surface be neutral, or at least stable, is often impossibly onerous when the surfaces are exposed to air.

In the early 1950s it was discovered that freshly exposed transistor surfaces were indeed very sensitive, and later it was found that water vapor and sodium ions, both very common contaminants, could produce instability on transistor surfaces. At that point of the development, the only way to keep the surfaces adequately clean was to enclose the transistors in glass–metal hermetically sealed encapsulations. This need meant a great additional expense, and also at the time thoroughly inhibited the development of integrated circuits.

In the late 1950s it was found that a passivating coating of silicon dioxide could give a more stable surface, but even yet, the transistors were sensitive to charged ions on the *outside* of the oxide; also, it was found that sodium and water vapor could actually penetrate through the thin oxide coating, and affect the silicon within. Because of these effects, very small and difficult to control, the manufacture of transistors required dust-free "clean" rooms and carefully filtered air; even so, there was and still is much difficulty in achieving adequate freedom from surface impurities. After further years of effort it was found that surfaces of oxide-coated silicon transistors could be successfully sealed either by metals or by a thin insulating coating of silicon nitride; even then, the outer surfaces had to be protected against electrolytic corrosion. All this took a long time to discover, because the effects are small, occur only slowly, and only under adverse circumstances, such as high ambient humidity or other contamination. But the overall result was eventually to develop, in the best transistors and integrated circuits, a totally unprecedented degree of reliability, very long life and stability in electronic equipment—once the "bugs" are out of it!

During the course of these investigations, it became possible for the first time to make successful field-effect transistors. The FET had been predicted and even patented many years before, but for completely mysterious reasons had never worked. From known electromagnetic theory it was certain that the gate electrode must induce charge in the material, but the induced charge did not show the mobility of the charge ordinarily present. An explanation was provided by Bardeen's theory of "surface states," which postulated that "traps" on the surface of the semiconductor held the induced charge immobile. While this language of "surface states" or "interface states" is still used to describe the effect, the true explanation is much simpler. The semiconductor is always covered with a thin layer of different material, usually an oxide; and charge on the outside of the oxide, or even in the interior of the oxide, is equally effective as charge on the electrodes in inducing charge in the semiconductor. This effect could not be circumvented until the invention of the MOSFET by Atalla and Kahng in 1960, in which the interfering charge was held to insignificant values—or, at least, stationary values!—by a surface coating of very clean "hard" silicon oxide.

Eventually the MOSFETS proved very successful in "integrated" circuits for computers, but only after prolonged and expensive experimentation found ways to produce a surface both clean and fully "passivated" after the very many necessary processing steps. Success here was signalled by the spectacular and precipitous fall in prices of electronic calculators in the early 1970s. In all, it took about 20 years and hundreds of millions of dollars to conquer the surface sensitivity of semiconductors, so that MOSFET integrated circuits finally became successful. Even now, it requires superlative care for successful fabrication of such units.

From the scientific point of view, the sensitivity of the surfaces, so troublesome technologically, was a great scientific opportunity to understand surface science. For the first time, a surface became available (the interface between silicon and its oxide) which was stable enough to be measured repeatedly for quantities like numbers and energies of surface charges and surface traps. These studies have helped to elucidate not only transistors but also scientific problems like chemical catalysis.

Another lesson to be drawn here is the complete failure of the Edisonian, or empirical, approach to invention of the FET. Only the scientific approach through detailed theory and experiment was able to discover the reasons for surface sensitivity of semiconductors, which so long impeded the successful use especially of field-effect transistors.

Another aspect of materials science was the preparation of the semiconductor materials, which had to be single crystals of uniform purity even when doped to less than one part per million with appropriate impurities to control the conductivity. Noteworthy here was the new process of zone refining, in which a liquid zone is caused to move slowly through the material by progressive melting and resolidification, sweeping along as it goes most of the impurities in the material. By this uniquely powerful method, materials as pure as one part in 10^{10} of electrically significant impurities have been obtained. Zone refining is now a standard way of producing ultrapure materials of many other kinds, not restricted to semiconductors.

Growth of single crystals of large size and high crystalline perfection has also been developed for use with transistors. Of particular importance is epitaxial growth, in which a new layer of controlled thickness and impurity content is laid down over a single-crystal substrate of the same material, and acquires the same crystal orientation. Such "epi-layers" are extensively used in silicon transistors and integrated circuits. They may be grown from materials in the vapor phase, from the liquid phase, or from a molecular beam of the semiconductor material.

5. *DIGITAL AMPLIFICATION*

When the transistor project began, long-distance telephony used "frequency multiplex" amplifiers, which amplified hundreds of telephone conversations at once, and which demanded high-quality, high-performance electron tubes for the purpose. The "Morton tube," which first made transcontinental television possible (in 1948), was such a tube. Its technology was so difficult that no manufacturer except Western Electric, advised by Bell Laboratories, ever attempted to make it.

"Time-division multiplex" amplifiers became more cost-effective in the 1950s. Here the idea was to *code* the telephone conversations into pulses, or "streams of bits," which could be decoded at the far end to reproduce the original conversation. Beside the cost saving for telephone equipment, the pulses could give orders of magnitude better accuracy in numerical representation, simply by using more and more digits to represent each signal. Consequently the digital method was immediately adopted for computers as well, and has been standard ever since.

The bit streams could be carried long distances over the same facilities as previously used. But with the use of transistors, vast simplifications became possible for localized equipment. Even the first transistors could operate with ten or a hundred times less power and voltage than any electron tube. After the surface problems of transistors were stabilized in the late 1960s, it became possible to do digital processing for even such examples as digital watches, which can operate for several years on a tiny "lithium" cell, smaller than a fingernail. Remote-controlled equipment, operating at the bottom of the sea or in distant outer space, became possible. Even the familiar telephone bell had to be eliminated, because it took a hundred times more power than was required to transmit the voice alone over the telephone.

Instead of having to control the analog *shape* of pulses very accurately, the amplifiers need only to count the digital *number* of pulses correctly, with *very accurate timing* to keep the bit streams separated. Philosophically, one can say that the telephone business has come back to the technique first used by Samuel F. B. Morse in 1844, when transmission methods were so crude that even speech had to be "Morse coded" in order to be transmitted from one railroad station to another, and telephone service was still forty years in the future.

6. *FABRICATION TECHNIQUES*

Many experimental techniques, originally used in advanced physics research, have been adapted for production of transistors and other modern products. Collectively, these techniques are known as the "high technology" in which the United States led the world in the 1950s and 1960s. Particularly prominent are high-vacuum processes and elaborately instrumented measurements. Some examples are: *vacuum evaporation* (to make thin films of metals or other materials); *sputtering* in gas at low vacuum, to add or remove thin films; *electron beam evaporation*, in which the heating is applied locally by a beam of electrons, thus dispensing with the need to find a crucible material which can hold the liquid metal without contaminating it; and ultraclean techniques, including dust-free rooms, to accomplish assembly operations without injury to minute electrodes or sensitive surfaces. Advanced techniques of analysis include the use of electron beams to produce x rays indicating the chemical composition of test surfaces, neutron activation to generate radioactive counts for analysis, and scanning electron microscopes which use television techniques and electron beams to reveal submicroscopic details of surfaces. Many other computer-aided instruments have achieved great improvements in sensitivity, resolving power, flexibility, capability, accuracy, and sheer quantity of scientific measurements. Such techniques permit fabrication and control of sophisticated structures by using knowledge which once was completely inaccessible to measurement.

The "beam-lead, sealed junction" technique

In 1963, the first highly reliable technique for making transistors in microscopic sizes was perfected by M. P. Lepselter at Bell Labs. The trick was to make a "window" by localized etching through a "hard" oxide, then to make a contact through the window to the semiconductor, and to cover the entire area with the contact metal, so that the transistor beneath the window could be contacted, yet completely protected against outside contamination.

By using two different protective coatings—usually "hard" silicon oxide (SiO_2) and what is somewhat inaccurately called silicon nitride (SiN)—one can make elaborate computer-controlled patterns over the semiconductor. The hard oxide is normally used next to the semiconductor, because the silicon–silicon oxide interface is uniquely free of unstable interface states right next to the semiconductor. Also, the combination is very refractory, and can be processed at high temperatures to anneal, or "diffuse," junctions into the material. Windows can be made in SiO_2 by a unique solvent, hydrofluoric acid (HF), while SiN can be windowed by phosphoric acid, which does not attack the oxide.

To make the complicated microscopic patterns required for transistors and integrated circuits, masking and photolithography are the usual methods. A thin layer of a photosensitive material is laid down over the semiconductor wafer, and then patterned by exposure to light, in a way similar to an ordinary photograph. X rays may be used instead of light, when the pattern is to be especially fine; or computer-controlled electron beams in high vacuum are used, when the pattern is especially complicated, or arbitrarily changeable. After development, the complicated photolithographic pattern becomes differentially soluble, so that after leaching with an appropriate chemical solvent the developed and fixed photosensitive material remains present only in the desired complicated pattern. The underlying material can then be subjected to chemical etching which reproduces the mask pattern in the semiconductor material, or more often in other masking layers over the material. Silicon oxide and silicon nitride masks patterned in this way are often used to produce desired patterns of high-temperature diffused areas or local variations in epi-layer thickness or composition; metal layers are used to produce contacts or interelectrode conductors. Multiple masks are often used, in which the successive patterns must be superposed so that they register with each other to very close tolerances, sometimes to a few micrometers or even a fraction of a micrometer; in integrated circuit use, sometimes as many as 8 or 10 masks must be matched to such tolerances. Considering the close tolerances, the high sensitivity to surface contamination, the elaborate nature of the patterns, and the variety of materials and processes, it is no wonder that such a process is called "high" technology!

7. CONCLUSIONS

Transistors and the resulting electronics are beginning a profound, even a revolutionary change in human society, by increasing the power of human communication and human thought. The way in which this has come about is also worthy of intense sociological study of the mechanisms by which science, technology, and society interact; in other words, the societal organization of scientific research and development. Meanwhile, the transistor revolution is beginning to play an indispensable part in improving human knowledge to the point where it may become capable of designing a better society. The wise use of knowledge is alone the unique hope for better understanding and better control of the political and economic processes which determine the fate of the world.

See also CIRCUITS, INTEGRATED; ELECTRON BEAM TECHNOLOGY; ELECTRON ENERGY STATES IN SOLIDS AND LIQUIDS; LASERS; PHOTOVOLTAIC EFFECT; SEMICONDUCTORS, CRYSTALLINE; SURFACES AND INTERFACES.

BIBLIOGRAPHY

A. S. Grove, *Physics and Technology of Semiconductor Devices*. Wiley, New York, 1967.

S. M. Sze, *Physics of Semiconductor Devices*, 2nd ed. Wiley, New York, 1981.

S. M. Sze, editor, about 50 books on details of the semiconductor technology.

Transition Elements

E. Bucher

1. BASIC CONCEPTS AND DEFINITION

The Pauli principle states that each electronic quantum state characterized by n, l, m, s (the principal, orbital, magnetic, and spin quantum numbers, respectively) can be occupied by one electron only. Since the orbital state $l=2$ contains $2l+1$ m substates ($m=-2, -1, 0, +1, +2$) each of which can be occupied by $s=+\frac{1}{2}, -\frac{1}{2}$, a transition series is expected to contain $2(2l+1)=10$ elements. Electrons in the $l=2$ state are called d electrons (regardless of all other quantum states); $n=3$ is the lowest principal quantum number for which a d state is occupied. Among all the known chemical elements (naturally occurring and synthetic elements) we can therefore distinguish four series of transition elements: the $3d$ ($n=3$, $l=2$) the $4d$ ($n=4$, $l=2$), the $5d$ ($n=5$, $l=2$), and the $6d$ ($n=6$, $l=2$) elements. The following definition of a transition element is commonly in use: A transition element is a chemical element in its ground state with a partially filled d shell, i.e., with a number of d electrons Z_d restricted by $0<Z_d<10$.

A first complication occurs in the $5d$ series where the $4f$ electron shell ($n=4$, $l=3$) is filled up first after the first $5d$ transition element (lanthanum). This leads to a subgroup of 14 elements, called rare earths or lanthanides, among the $5d$ transition elements. Only after the $4f$ shell is filled up are more $5d$ electrons added. The second complication appears in the $6d$ series in which after the initial buildup of the $6d$ shell (Ac, Th) the fourteen $5f$ electrons ($n=5$, $l=3$) appear. These elements form a second subgroup among the transition elements, called actinides. Most of them are synthetically formed transuranium elements.

2. CHEMICAL PROPERTIES

We list below the whole complex of transition elements, including 4*f* and 5*f* subgroups:

Group number:	IIIB	IVB	VB	VIB	VIIB	VIIIB		
3*d* transition series:	Sc	Ti	V	Cr	Mn	Fe	Co	Ni
4*d* transition series:	Y	Zr	Nb	Mo	Tc	Ru	Rh	Pd
5*d* transition series:	La–Lu	Hf	Ta	W	Re	Os	Ir	Pt
5d-4f series (lanthanides or rare earths):	La, Ce, Pr, Nd, Pm, Sm, Eu, Gd, Tb, Dy, Ho, Er, Tm, Yb, Lu							
6d transition series (actinides):	Ac	Th						
6*d*-5*f* series:	Th, Pa, U, Np, Pu, Am, Cm, Bk, Cf, Es, Fm, Md, No, Lr, Ku							

Except for the beginning of the *d* series, the transition elements are well known for their complex behavior in that they form compounds in many different valence states (e.g., in Ru and Os, all eight valence states have been observed[1]). Among the rare earths, it is sometimes energetically more favorable to convert a 5*d* electron into a 4*f* electron. These divalent rare earths (Sm, Eu, Tm, Yb in general) resemble more closely the preceding Ba in their chemical behavior and do not exhibit transition-element properties, since they are stripped of their only 5*d* electron.

All the elements listed above are metals with relatively high melting points (900–1000°C or higher) and crystallize in relatively simple structures[1] [fcc, bcc, hcp, or sequences thereof (La–Sm)], except Mn and most elements of the actinide series. They show in general a wide range of solid solutions and form numerous intermetallic compounds and complex phases[2,3] (except group III with groups IV, V, VI). In pure bulk form most transition elements are relatively stable in air against corrosion. The lower groups usually form a protective oxide layer (except Eu), whereas the higher group belongs to the noble metals except 3*d* elements Fe, Co, Ni.

3. PHYSICAL PROPERTIES

In the atomic state (e.g., in the vapor state) the *d* energies are relatively sharp. In the condensed state, however, the *d* levels broaden into an energy band[4] which has a width typically of 2.5–3 eV, 4–4.5 eV, 5–5.5 eV, and 6 eV for 3*d*, 4*d*, 5*d*, and 6*d* elements. Depending on crystal symmetry the *d* level undergoes a Stark effect, i.e., an energy splitting due to electric fields generated by the charge distribution of the nearest (or higher) neighbors.[5,6] The interaction of this electric potential with the outermost *d* electrons is so strong that the orbital quantum number is not preserved and the orbital moment is quenched.[4–6] Depending on crystal symmetry, several *d* subbands may form, which in general overlap.[5,6] Elements in which the highest occupied energy level of the electrons is below the maximum energy of a band represent metals. According to our definition all transition elements exhibit metallic behavior and, therefore, are usually called transition metals (*T* metals in general or 3*d*, 4*d*, 5*d*, 6*d* metals more specifically).

At low temperature the transition metals show a number of unique features:

(1) The interaction of the *d* electrons with the lattice vibrations in which the density of states, i.e., the number of allowed states per unit of energy, is involved leads in general to superconductivity, in the transition metals themselves as well as in their alloys and metallic compounds with other transition and nontransition elements.[4]

(2) The occurrence of superconductivity is limited by the strength of the electron–electron interaction. In the 3*d* elements, this leads to ferromagnetism in Fe, Co, Ni; to antiferromagnetism in Mn; and to a rather complex spin-density wave state in Cr below 338°K. All these elements are nonsuperconducting and Ti and V are the only superconductors among the 3*d* elements.

In contrast, all 4*d* elements are paramagnetic; however, the electron–electron interaction in the beginning (Y) and at the end (Pd) of the 4*d* series is still strong enough to prevent superconductivity. These elements show an enhanced paramagnetic susceptibility. In particular, Pd is considered as a model for a paramagnon material, i.e., a material in which electron–electron interactions are important but not strong enough to induce a ferromagnetic transition. All 5*d* elements are also superconducting except the nonmagnetic Yb, Lu, and Pt and the magnetic rare earths in which the magnetic moments arise from the 4*f* electrons. In the latter, the orbital moment is well preserved and the interaction with the crystal-field potential is relatively weak.[5,6] In the analogous 5*f* series the situation is much more complex. Depending on the individual compounds, the properties of 5*f* electrons in the lighter actinides can vary between the limits of complete delocalization into a relatively narrow 5*f* band to strongly localized 5*f* electrons with well-preserved quantum states *L*, *S*, *J*. The actinides are usually very complex because they show several different structures and the physical properties strongly depend on them. Superconductivity has been found in certain phases of Pa, U, and Np, whereas in compounds they may be magnetic. It is striking that in Sect. 2 the transition series contains only eight, rather than ten elements as one would expect from the definition in Sect. 1. This arises from the fact that certain configurations of *d* and *s* electrons do not occur in the ground state for energetic reasons which are too difficult to predict from first-principles consideration (minimization of energy for the whole system of electrons) (see Table I of transition elements with their electron configurations). In the metallic solid state we cannot in general attribute an integral number of *d* electrons and *s* electrons to each element.

In the band picture the character of *s* and *d* electrons is not preserved and their properties are admixed. Therefore, in the metallic state we consider as transition elements those elements with unfilled *d* bands, unlike in the atomic state where an integral number between 1 and 10 *d* electrons is required. This implies that other elements such as Cu and Ag may behave as transition elements in chemical compounds if encountered in a proper valence state, e.g. Cu^{2+} or Ag^{2+} will be in a $3d^9$ or $4d^9$ configuration, respectively, although these elements are not considered as transition elements in the metallic state.

4. APPLICATIONS

The technical applications of transition elements are nearly inexhaustible. Best known are alloys of 3*d* elements for steel production: Cr–Ni–Fe alloys with various additions

Table I. Chemical Symbols and Outer Electron Configuration Names of all Transition Elements

		Outer atomic configuration	Solid-state configuration in general
3d Series:	Sc = Scandium	$3d^1 4s^2$	$(3d4s)^3$
	Ti = Titanium	$3d^2 4s^2$	·
	V = Vanadium	$3d^3 4s^2$	·
	Cr = Chromium	$3d^5 4s^1$	·
	Mn = Manganese	$3d^5 4s^2$	·
	Fe = Iron	$3d^6 4s^2$	·
	Co = Cobalt	$3d^7 4s^2$	·
	Ni = Nickel	$3d^8 4s^2$	$(3d4s)^{10}$
4d Series:	Y = Yttrium	$4d^1 5s^2$	$(4d5s)^3$
	Zr = Zirconium	$4d^2 5s^2$	·
	Nb = Niobium	$4d^4 5s^1$	·
	Mo = Molybdenum	$4d^5 5s^1$	·
	Tc = Technetium	$4d^5 5s^2$	·
	Ru = Ruthenium	$4d^7 5s^1$	·
	Rh = Rhodium	$4d^8 5s^1$	·
	Pd = Palladium	$4d^{10} 5s^0$	$(4d5s)^{10}$
5d and *4f* Series:	La = Lanthanum	$4f^0 5d^1 6s^2$	$4f^0(5d6s)^3$
	Ce = Cerium	$4f^1 5d^1 6s^2$	$4f^1(5d6s)^3$ $4f^0(5d6s)^4$
	Pr = Praseodymium	$4f^2 5d^1 6s^2$	$4f^2(5d6s)^3$
	Nd = Neodymium	$4f^4 5d^0 6s^2$	$4f^3(5d6s)^3$
	Pm = Promethium	$4f^5 5d^0 6s^2$	$4f^4(5d6s)^3$
	Sm = Samarium	$4f^6 5d^0 6s^2$	$4f^5(5d6s)^3$ $4f^6(5d6s)^2$
	Eu = Europium	$4f^7 5d^0 6s^2$	$4f^7(5d6s)^2$ $4f^6(5d6s)^3$
	Gd = Gadolinium	$4f^7 5d^1 6s^2$	$4f^7(5d6s)^3$
	Tb = Terbium	$4f^8 5d^1 6s^2$	$4f^8(5d6s)^3$
	Dy = Dysprosium	$4f^9 5d^1 6s^2$	$4f^9(5d6s)^3$
	Ho = Holmium	$4f^{10} 5d^1 6s^2$	$4f^{10}(5d6s)^3$
	Er = Erbium	$4f^{11} 5d^1 6s^2$	$4f^{11}(5d6s)^3$
	Tm = Thulium	$4f^{13} 5d^0 6s^2$	$4f^{12}(5d6s)^3$ $4f^{13}(5d6s)^2$
	Yb = Ytterbium	$4f^{14} 5d^0 6s^2$	$4f^{13}(5d6s)^3$ $4f^{14}(5d6s)^2$
	Lu = Lutetium	$4f^{14} 5d^1 6s^2$	$4f^{14}(5d6s)^3$
	Hf = Hafnium	$4f^{14} 5d^2 6s^2$	$(5d6s)^4$
	Ta = Tantalum	$5d^3 6s^2$	·
	W = Tungsten	$5d^4 6s^2$	·
	Re = Rhenium	$5d^5 6s^2$	·
	Os = Osmium	$5d^6 6s^2$	·
	Ir = Iridium	$5d^7 6s^2$	·
	Pt = Platinum	$5d^9 6s^1$	$(5d6s)^{10}$
6d and *5f* Series:	Ac = Actinium	$6d^1 7s^2$	$(6d7s)^3$
	Th = Thorium	$6d^2 7s^2$	$(6d7s)^4$
	Pa = Protactinium	$5f^2 6d^1 7s^2$	$5f^0(6d7s)^5$
	U = Uranium	$5f^3 6d^1 7s^2$	$5f^0(6d7s)^6$ $5f^1(6d7s)^5$ $5f^2(6d7s)^4$
	Np = Neptunium	$5f^4 6d^1 7s^2$	$5f^0(6d7s)^7$ $5f^4(6d7s)^3$
	Pu = Plutonium	$5f^6 6d^1 7s^2$	$5f^1(6d7s)^7$ $5f^5(6d7s)^3$
	Am = Americium	$5f^7 6d^0 7s^2$	
	Cm = Curium	$5f^7 6d^1 7s^2$	
	Bk = Berkelium	$5f^8 6d^1 7s^2$	

Table I.—(Continued)

	Outer atomic configuration	Solid-state configuration in general
Cf = Californium	$5f^9 6d^1 7s^2$	?
Es = Einsteinium	$5f^{10} 6d^1 7s^2$	
Fm = Fermium	$5f^{11} 6d^1 7s^2$	
Md = Mendelevium	$5f^{12} 6d^1 7s^2$	
No = Nobelium	$5f^{14} 6d^0 7s^2$	?
Lr = Lawrencium	$5f^{14} 6d^1 7s^2$	?
Ku = Kurchatovium	$5f^{14} 6d^2 7s^2$	?

such as Co, Ti, V, Mn, Mo, Nb, and rare earths for improvement of strength and/or anticorrosion. Ferromagnetic alloys from Fe–Co–Ni with Si additions are used for transformer cores and Fe–Co alloys for electromagnets. Among permanent magnets, $SmCo_5$ and $Nd_2Fe_{14}B$ exhibit energy products as high as 25×10^6 and 49×10^6 Oe G, respectively. Superconducting solenoids are still manufactured from NbTi alloys (approximately 1:1 ratio) and Nb_3Sn intermetallics. Recently discovered high-T_c superconductors (T_c above 70K) $Ba_2YCu_3O_7$, $Bi_2Sr_2Ca_nCu_{n+1}O_{2n+6}$, $Tl_2Ba_2Ca_2Cu_3O_{10}$, Pb_2Sr_2–$Y_{0.75}Ca_{0.25}Cu_3O_8$ invariably contain Cu^{2+}–O systems. For high-temperature furnaces, Ni–Cr and Kantal are used up to 1100°C, and $Pt_{1-x}Rh_x$ and $MoSi_2$ up to 1700°C as heaters in air, and Nb, Ta, Mo, and W under vacuum or noble gas for temperatures above 2000°C. Many other well-known technical alloys are Invar (low thermal expansion), Manganin, Inconel, Kovar, Ti–Al–Be (for supersonic airplaines), Pt/Pt–Rh, and W/W–Re alloys for high-temperature thermocouples, Pt for exhaust catalysts, etc.[7]

See also ELECTRON ENERGY STATES IN SOLIDS AND LIQUIDS; ELEMENTS; FERROMAGNETISM; PARAMAGNETISM; RARE EARTHS; SUPERCONDUCTIVITY THEORY; TRANSURANIUM ELEMENTS.

REFERENCES

1. For description of other properties see e.g. the *CRC Handbook of Chemistry and Physics,* 56th ed. Chemical Rubber Publishing Co., Cleveland, Ohio, 1975. (I)
2. W. B. Pearson, *Handbook of Lattice Spacings and Structure of Metals and Alloys,* Vol. 2. Pergamon, New York, 1967. (I)
3. M. Hansen (ed.), *Constitution of Binary Alloys,* Vol. 1. Mc-Graw-Hill, New York, 1958; R. P. Elliot (ed.), *Constitution of Binary Alloys,* Suppl. 1. McGraw-Hill, New York, 1965; F. A. Shunk (ed.), *Constitution of Binary Alloys,* Suppl. 2. Mc-Graw-Hill, New York, 1970. (I)
4. C. Kittel, *Introduction to Solid State Physics,* 4th ed. Wiley, New York, 1971. (A)
5. M. T. Hutchings, in *Solid State Physics,* F. Seitz and D. Turnbull (eds.), Vol. 16. Academic Press, New York, 1964. (A)
6. A. H. Morrish, *The Physical Principles of Magnetism.* Wiley, New York, 1965. (A)
7. G. S. Brady, *Materials Handbook,* 10th ed. McGraw-Hill, New York, 1971.

Transmission Lines and Antennas

R. W. P. King

As illustrated in Fig. 1, communication by radio or television requires **(1)** a generator of currents suitably modulated with the information to be sent along **(2)** a *transmission line* to **(3)** a transmitting *antenna*. In this the oscillating currents generate **(4)** electromagnetic waves that travel outward in space to **(5)** a receiving *antenna* in which they induce currents that travel along **(6)** another *transmission line* to **(7)** a receiver that reconstructs the original information contained in the modulated current.

The simplest and most common transmission lines consist of two conductors separated by an insulator and arranged either as a two-wire line **(6)** or coaxial line **(2)** as shown in Fig. 1. The two conductors carry equal and opposite currents that obey the differential equations

$$-\frac{\partial I_x(x)}{\partial x}=zV(x), \quad -\frac{\partial V(x)}{\partial x}=yI_x(x), \tag{1}$$

where $V(x)$ is the potential difference between the two conductors at a distance x from the generator, $I_x(x)$ is the current in one of the conductors [that in the other is $-I_x(x)$], $z=r+j\omega l$ is the impedance, and $y=g+j\omega c$ is the admittance per unit length; r, l, c, and g are, respectively, the resistance, inductance, capacitance, and leakage conductance per unit length. The propagation constant $\gamma=\alpha+j\beta=\sqrt{zy}$ and the characteristic impedance $Z_c=\sqrt{z/y}$ characterize the electrical properties of the transmission line; α is the attenuation constant, β is the phase constant. The solution of these equations can be expressed in the following general form[1]:

$$V(x)=V_0^e\,\frac{\sinh\theta_0\cosh(\gamma w+\theta_s)}{\sinh(\gamma s+\theta_0+\theta_s)}, \tag{2}$$

$$I(x)=\frac{V_0^e}{Z_c}\,\frac{\sinh\theta_0\sinh(\gamma w+\theta_s)}{\sinh(\gamma s+\theta_0+\theta_s)}, \tag{3}$$

where V_0^e is the emf applied at the input terminals AA' of the transmission line and $I(0)$ is the current entering terminal A. The complex terminal function $\theta_0=\coth^{-1}(Z_0/Z_c)$ involves the impedance Z_0 of the generator connected across AA'. Similarly, $\theta_s=\coth^{-1}(Z_s/Z_c)$, where $Z_s=R_s+jX_s$ is the impedance terminating the transmission line at $x=s$. It is the input impedance of the antenna a in **(3)** of Fig. 1. θ_0 and θ_s are related to complex reflection coefficients by the formula $\Gamma=|\Gamma|e^{j\psi}=e^{-2\theta}$; $\theta=\rho+j\Phi$. In Eqs. (2) and (3) $w=s-x$ is the distance from the load. The transmission line is matched and losses due to dissipation in the line are minimized when the standing-wave ratio $SWR=I_{\max}/I_{\min}=\coth\rho_s$ is equal to 1. This occurs when $\rho_s=\infty$, $\Gamma_s=0$, and $Z_s=Z_c$. The efficiency of transmission is: $\text{Eff.}=\sinh 2\rho_s/\sinh 2(\alpha s+\rho_s)$.

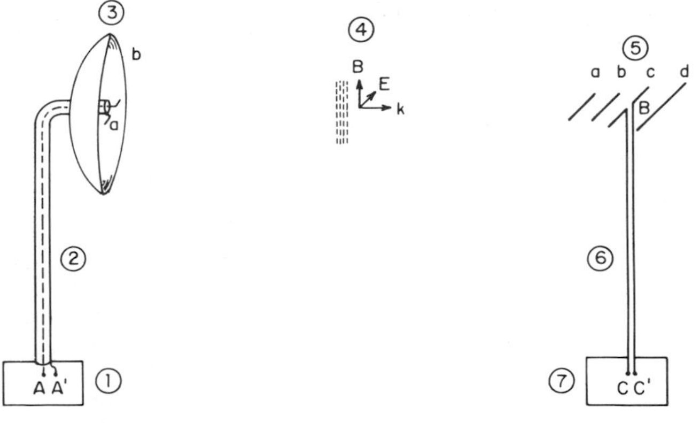

FIG. 1. Transmitting and receiving system.

For use in integrated circuits at frequencies of 1–100 GHz, a transmission line consisting of a thin metal strip on a dielectric-coated conducting plane is in common use. It is known as microstrip. Because of the presence of the air–substrate boundary, the longitudinal currents in the strip line are governed only approximately by Eq. (1) in a restricted range of frequency for which γ and Z_c are calculated with an empirically determined, frequency-dependent, effective dielectric constant for the substrate. The field associated with the longitudinal and transverse currents in the strip includes not only the transverse electric and magnetic components of the TEM mode, but also the longitudinal components of a surface (lateral) wave that propagates along the boundary in the air.

The transmitting antenna (a in **3** of Figure 1) is symmetrically connected to the two conductors of the feeding transmission line which maintains the voltage V across its input terminals. The current in a highly conducting, center-driven dipole with half-length h and radius a ($a \ll h$) (in the absence of the reflector b in **3** of Fig. 1) satisfies the following integral equation[2]:

$$\int_{-h}^{h} I(z')K(z,z')\,dz' = (-j4\pi/\zeta_0)[C_1\cos kz+(V/2)\sin|kz|], \tag{4}$$

where $K(z,z')=\exp(-jkR)/R$, $R=[(z-z')^2+a^2]^{1/2}$, $\zeta_0\doteq 120\pi\ \Omega$. C_1 is a constant to be determined from $I(h)=0$. Approximate solutions of this equation have been derived in different forms and by a variety of methods. A simple, reasonably accurate form for dipoles with $kh\le 5\pi/4$ is[3]

$$I(z)=(j4\pi V/\zeta_0\Psi\cos kh)[\sin k(h-|z|)+T_U(\cos kz-\cos kh)+T_D(\cos\tfrac{1}{2}kz-\cos\tfrac{1}{2}kh)], \tag{5}$$

where Ψ is a real constant, T_U and T_D are complex constants that involve the length and the radius of the antenna. The input impedance of the antenna is $Z_{\text{in}}=I(0)/V$. For many purposes the two-term form of (5) obtained with $T_D=0$ is sufficiently accurate. For calculating the distant electromagnetic field the first term (with $T_U=T_D=0$) is usually adequate. For long antennas an alternative formula is available.[4]

The outward-traveling electromagnetic waves generated by the oscillating currents in the antenna constitute the electromagnetic field. The angular distribution of the magnitude of this field at distant points is known as the far-field pattern. The electric far field due to the current in (5) is

$$E_\theta=\frac{j\zeta_0 I(0)}{2\pi}\,\frac{e^{-jkR}}{R}\,f_I(\theta,kh), \tag{6}$$

where

$$f_I(\theta,kh) = \frac{F_m(\theta,kh) + T_U G_m(\theta,kh) + T_D D_m(\theta,kh)}{\sin kh + T_U(1-\cos kh) + T_D(1-\cos \frac{1}{2}kh)}. \qquad (7)$$

The angle θ is measured from the axis of the dipole. In (7),

$$F_m(\theta,kh) = [\cos(kh \cos \theta) - \cos kh]/\sin \theta. \qquad (8)$$

The functions $G_m(\theta,kh)$ and $D_m(\theta,kh)$ are similar, longer expressions that involve kh and θ.[3] Approximate far-field patterns obtained with $T_U = T_D = 0$ are shown in Fig. 2 for dipoles with three different lengths. These patterns are rotationally symmetric about the axis of the dipole. In them the radial distance from the origin 0 at any angle θ to a point on one of the patterns is proportional to the intensity of the electric field in the direction of the specified angle θ. For example, with $kh = \pi$ the direction in which the outward-traveling waves have their greatest intensity is in the equatorial plane defined by $\theta = 90°$. As seen from (6), the intensity of the far field in all directions decreases as the reciprocal of the distance R from the center of the antenna.

When a center-loaded receiving antenna (like c in **5** of Fig. 1 but in the absence of a, b, and d) with $kh = \pi$ is illuminated by an electromagnetic wave from a distant transmitter, the voltage $V(0)$ across the load impedance Z_L is largest when the axis of the antenna is parallel to the incident electric field, E^{inc}. When this is true, the induced voltage across the load is

$$V(0) = (2h_e E^{inc} Z_L)/(Z_0 + Z_L), \qquad (9)$$

where Z_0 is the input impedance and $2h_e$ is the complex effective length of the antenna.[2]

In the circuits of Fig. 1, $V(0)$ is the voltage impressed across the input terminals B of the two-wire line (**6**). This can be used in (2) and (3) as $V_0^e = V(0)$ to determine the voltage across and the current in the receiver terminals CC' of **7** in Fig. 1.

In the discussion to this point the transmitting antenna and the receiving antenna have been assumed to consist only of the dipoles a in **3** and c in **5**. In order to concentrate the radiated field of the transmitting antenna into a narrow beam, a metal reflector like b in **3** of Fig. 1 may be used at high frequencies. At low frequencies effective metal reflectors are too large to be practical. However, arrays of dipoles or monopoles erected vertically on the earth may be used to obtain the desired directional properties. The pattern of a broadside array of half-wave dipoles with spacing b equal to a half-wavelength is shown in Fig. 3a. The directivity of a receiving antenna like **5** in Fig. 1 can also be increased, for example, with a Yagi array of metal rods—a single reflector d and from 1 to 10 or more directors like a and b in **5**. A directional pattern of the received voltage for a 10-element Yagi array is shown in Fig. 3b.

An insulated antenna[5] is a conductor (radius a, conductivity σ_c) in a tube of dielectric (radius b, wave number $k_d = \omega[\mu_0(\epsilon_d - j\sigma_d/\omega)]^{1/2}$, $\sigma_d \sim 0$) in an ambient medium like the earth or sea (wave number $k_1 = \omega[\mu_0(\epsilon_1 - j\sigma_1/\omega)]^{1/2}$). Often it is used at a distance d from the boundary with air or the rock of the sea floor (wave number $k_2 = \omega[\mu_0(\epsilon_2 - j\sigma_2/\omega)]^{1/2}$; for air $k_2 = \omega(\mu_0\epsilon_0)^{1/2}$). It has the properties of a transmission line but with the resistance per unit length consisting of two

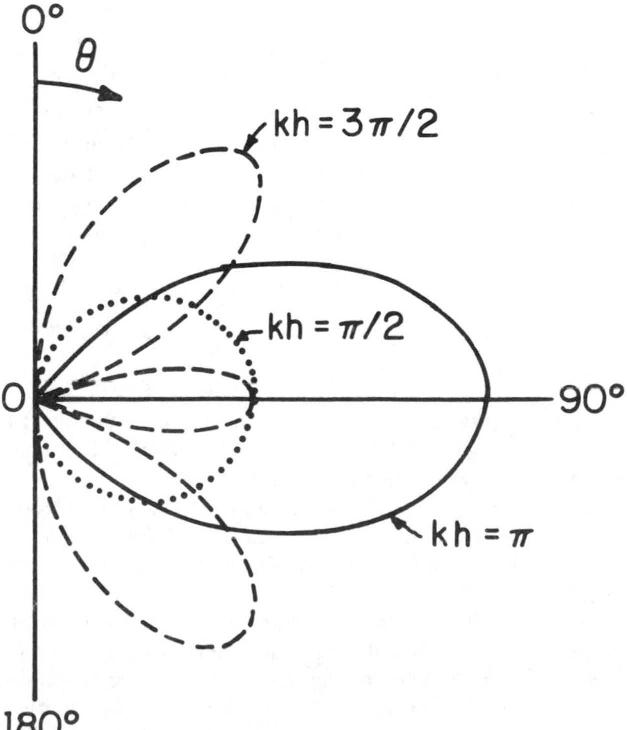

FIG. 2. Far-field patterns of dipole antenna. The radial distance from the origin 0 to the graphs in the direction θ is proportional to $|E_\theta|$ in that direction.

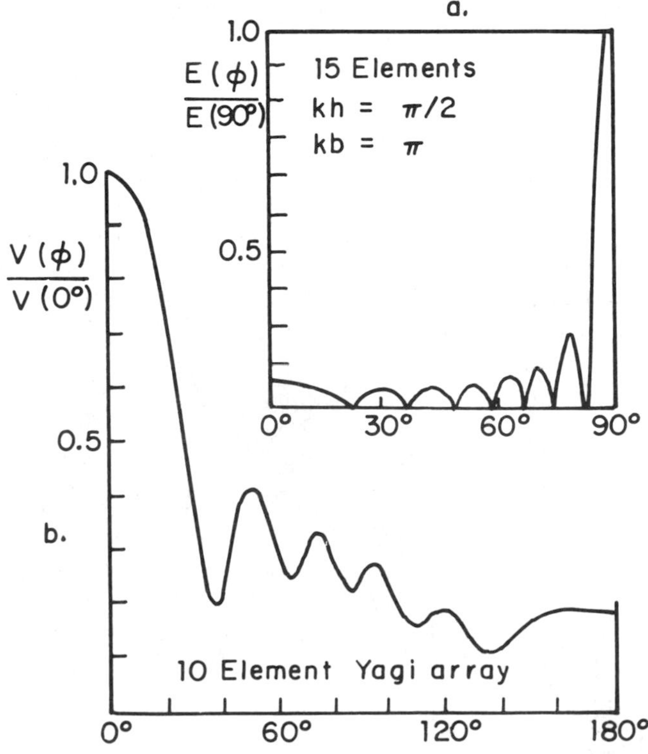

FIG. 3. Directional patterns of (a) broadside array; (b) Yagi–Uda array.

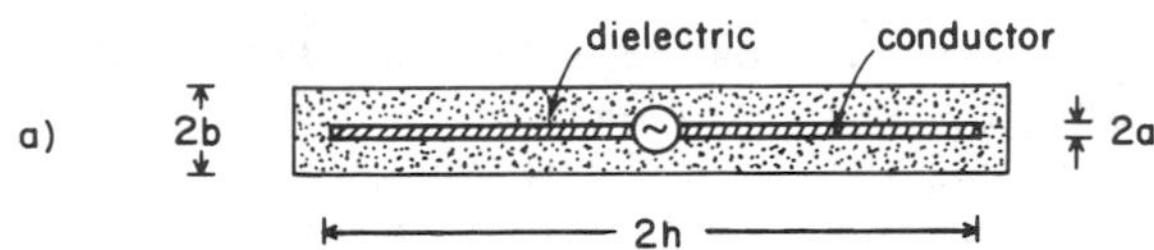

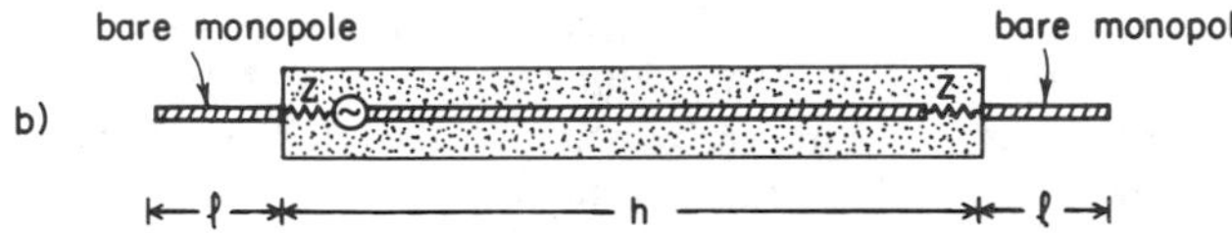

FIG. 4. Insulated antennas. (a) Completely insulated dipole. (b) Terminated insulated antenna. Ambient medium: earth, sea water.

parts: the usual internal (Ohmic) resistance r^i and an external (radiation) resistance r^e. When implanted in biological tissues for the hyperthermia treatment of tumors at very high frequencies, or when used in the earth for subsurface heating, communication, or remote sensing, or when used in the sea for communicating with submarines at frequencies in the kilohertz or megahertz range, radiation dominates and $r^e \gg r^i$. When used on the sea floor for the exploration of the oceanic crust at $f \sim 1$ Hz, radiation is negligible and $r^i \gg r^e$. The current in the conductor is given by Eq. (3) with $\gamma = jk_L$, $k_L = \sqrt{-z_L y_L}$, and $Z_{cL} = \sqrt{z_L/y_L}$. When center-driven as in Fig. 4a, $\theta_0 = j\pi/2$ and $\theta_s = 0$ so that (3) reduces to

$$I_x(x) = -\frac{jV_0^e}{2Z_{cL}} \frac{\sin k_L(h-|x|)}{\cos k_L h}. \tag{10}$$

When terminated as in Fig. 4b, $\theta_0 = j\pi/2$ and $\theta_s \sim \infty$ so that Eq. (3) becomes

$$I_x(x) = \frac{V_0^e}{Z_{cL}} e^{-jk_L x}. \tag{11}$$

When $|k_1 b| < 1$, the admittance and impedance per unit length are $y_L = g_L + j\omega c_L$, $z_L = z^i + z^e + z_{12}$, where $z^i \sim r_0 + (j\omega\mu_0/8\pi)$, $z^e \sim (\omega\mu_0/8) + (j\omega\mu_0/2\pi)[\ln(2/|k_1 b|) - 0.5772]$, $z_{12} \sim (\omega\mu_0/8) + (j\omega\mu_0/2\pi)[\ln(1/|k_1 d|) - 0.5772]$, $g_L = 2\pi\sigma_d/\ln(b/a)$, $c_L = 2\pi\epsilon_d/\ln(b/a)$, and $r_0 = (\pi a^2 \sigma_c)^{-1}$. The formulas for k_L and Z_{cL} are

$$k_L = k_d \left\{ 1 + \frac{1}{\ln(b/a)} \left[\ln\frac{2}{|k_1 b|} + \ln\frac{1}{|k_1 d|} - 0.90 - j\left(\frac{2\pi r_0}{\omega\mu_0} + \frac{\pi}{2}\right) \right] \right\}^{1/2}, \tag{12}$$

$$Z_{cL} = \frac{\omega\mu_0 k_L}{2\pi j_d^2} \text{lb}(b/a). \tag{13}$$

At very low frequencies, these reduce to $k_L = (1-j)(\omega c_L r_0/2)^{1/2}$ and $Z_{cL} = (1-j)(r_0/2\omega c_L)^{1/2}$. The properties of the insulated antenna can be varied greatly by changes in the thickness $(b-a)$ of the dielectric.

In general, antennas are conductors with many shapes and sizes depending on the particular application and the frequency. Their function is to provide conducting paths for currents that generate an electromagnetic field or are induced by such a field.

See also ELECTROMAGNETIC RADIATION; MICROWAVES AND MICROWAVE CIRCUITRY.

REFERENCES

1. R. W. P. King, *Transmission-Line Theory*. Dover Publications, New York, 1965.
2. R. W. P. King and C. W. Harrison, Jr., *Antennas and Waves: A Modern Approach*. M.I.T. Press, Cambridge, Mass., 1969.
3. R. W. P. King, R. B. Mack, and S. S. Sandler, *Arrays of Cylindrical Dipoles*. Cambridge University Press, New York, 1968.
4. C. L. Chen and T. T. Wu, in *Antenna Theory, Part I*, Chap. 10, R. E. Collin and F. J. Zucker (eds.). McGraw-Hill, New York, 1969.
5. R. W. P. King and G. S. Smith, *Antennas in Matter*, MIT Press, Cambridge, MA., 1981.

Transport Properties

W. H. Butler

When a material is subjected to an external field, currents of charge or energy may flow within the material in response to the field. These various responses to applied fields are important characteristics of the material and are referred to as its transport properties. The topic encompasses a wide variety of phenomena. The currents may consist of charge or heat flows. The current carriers may be electrons, ions, ionic vibrations (phonons), or light waves (photons). The applied fields may be electric, magnetic, or thermal gradients and they may be static or time dependent. A number of interesting effects occur when two fields are applied simultaneously as in the Hall effect or when the application of one kind of field elicits a field of different kind as in the thermoelectric effect.

METALS, INSULATORS, AND SEMICONDUCTORS

In regard to their response to a static electric field, solids are classified as metals, semiconductors, or insulators. When a steady electric field is applied to a metal, its electrons respond by producing a charge current. No such electron current is elicited in an insulator by a static electric field because its electrons are too tightly bound to their ions to move through the material. This difference in response to an electric field is due to differences in the electronic structures of metals and insulators. In an insulator the quantum state of lowest energy is separated by a finite energy gap from any higher-energy current-carrying state. In a metal, the energy gap between the lowest-energy quantum state and a current-carrying state is infinitesimal so an applied field can lead to its occupation and an associated flow of current.

The electronic structure of a pure semiconductor at zero temperature is like that of an insulator in that there is a finite energy gap which prevents a current from flowing in response to a static electric field. This energy gap is smaller than in an insulator, however, and thermal energies may be sufficient to raise some electrons into higher-energy states that can carry a current. Most of the practical applications of semiconductors arise from this small but nonzero energy gap that can be easily modified by the addition of small amounts of dopants or applied electric fields, thereby significantly altering their transport properties.

Some materials that conduct electricity resonably well are not metals. In these materials the electrical current is carried by the ions or by charged defects such as impurities or vacancies that are able to move through the material in response to an applied field.

All solids respond to a temperature gradient with a current of thermal energy. In an insulator this heat current is carried by the vibrations of the ions and, at sufficiently high temperatures, by photons (light waves). In a metal the heat current is carried by electrons and by ionic vibrations.

In addition to normal metallic conduction described above, there is another type of electron transport called superconductivity which is observed in certain metals at sufficiently low temperatures. In a normal metal the charge current carried by the electrons is proportional to the applied electric field and the heat current is proportional to the temperature gradient. In a superconductor a charge current can flow in the absence of an electric field and no heat current is associated with this "super" current. The superconducting state results from a fundamental change in the nature of the quantum states of the electrons. Transport properties of superconductors are not correctly described by the semiclassical theory described below.

THEORY OF TRANSPORT PROPERTIES

The modern theory of transport properties includes a wide variety of complicated phenomena. The foundations of transport theory lie in the difficult area of nonequilibrium statistical mechanics. An exact and rigorous theory of transport phenomena in solids would be particularly involved because the current carriers, electrons, phonons, and photons, have wave characteristics which should be treated using quantum mechanics.

Fortunately it is known from experiment that most transport properties can be understood in terms of semiclassical Boltzmann theory which is based on the assumption that the current carriers are particles that can be described by a distribution function, $f(\mathbf{k},\mathbf{r},t)$, which gives the number of particles per unit volume having momentum $\mathbf{p}(=\hbar\mathbf{k})$ at position $\mathbf{r}$ and time t. The wave nature of the current carriers is accounted for in this treatment by associating the momentum with the inverse of the wavelength, λ, through the de Broglie relation, $|\mathbf{p}|=h/\lambda$, where $h(=2\pi\hbar)$ is Planck's constant. The wave vector, $\mathbf{k}$, which appears in semiclassical Boltzmann theory is proportional to the inverse of the wavelength, $|\mathbf{k}|=2\pi/\lambda$.

The assumption that the wave nature of the current carriers can be fully accounted for by this procedure is certainly not self-evident, and is not always a good approximation to reality. Nevertheless, the relatively simple Boltzmann theory which will be outlined below is often quite accurate for such common transport properties as the electrical resistivity of metals and the thermal conductivity of both metals and insulators. After we give a brief summary of the more important aspects of the Boltzmann picture we will mention some of its limitations. The original Boltzmann transport equation was given by Ludwig Boltzmann in 1872. The semiclassical extension was published by Felix Bloch in 1928, shortly after the discovery of quantum mechanics.

The Boltzmann theory for classical particles is described in the following article (*see* Transport Theory). The semiclassical theory is quite similar. The particles in a small volume of the six-dimensional phase space which encompasses both position $\mathbf{r}$ and momentum $\hbar\mathbf{k}$ can leave this volume either by "flowing" out or by being scattered. Thus, the rate of change of the distribution function is

$$\frac{df(\mathbf{k},\mathbf{r},t)}{dt} = \nabla_{\mathbf{k}}f(\mathbf{k},\mathbf{r},t)\cdot\frac{d\mathbf{k}}{dt} + \nabla_{\mathbf{r}}f(\mathbf{k},\mathbf{r},t)\cdot\frac{d\mathbf{r}}{dt} + \left(\frac{\partial f(\mathbf{k},\mathbf{r},t)}{\partial t}\right)_{\mathrm{scatt}}. \quad (1)$$

In the absence of any applied fields the system of particles will, given sufficient time, reach equilibrium and the particle distribution function will be that characteristic of equilibrium, f_0. For a Fermi gas, e.g., electrons in a normal metal, the equilibrium distribution is given by the Fermi function

$$f_0^{\mathrm{F}}(\epsilon_{\mathbf{k}},T) = \{\exp[(\epsilon_{\mathbf{k}}-\mu)/k_{\mathrm{B}}T] + 1\}^{-1} \quad (2)$$

and for a system of bosons, e.g., phonons, it is given by the Bose–Einstein distribution

$$f_0^{\mathrm{B}}(\epsilon_{\mathbf{k}},T) = [\exp(\epsilon_{\mathbf{k}}/k_{\mathrm{B}}T)-1]^{-1}. \quad (3)$$

In addition to Boltzmann's constant, k_{B} $(=1.381\times10^{-23}$ J/K, and the absolute temperature, T, the equilibrium distribution functions depend on the function $\epsilon_{\mathbf{k}}$ which gives the relation between the energy of an electron or phonon and its momentum and is known as the dispersion relation. The particle velocity or, more accurately, the wave group velocity is defined in terms of the dispersion relation through $\hbar\mathbf{v}_{\mathbf{k}}=\nabla_{\mathbf{k}}\epsilon_{\mathbf{k}}$. For temperatures lower than the melting points of most metals, the Fermi function is approximately unity if the energy $\epsilon_{\mathbf{k}}$ is less than the chemical potential, μ, and approximately zero for energies greater than μ.

In the presence of an applied field, the system will eventually reach a steady-state condition in which the distribution function is independent of time. In this case, however, the distribution function will differ from the equilibrium one by an amount that depends on the applied field. Thus, the Boltzmann equation (1) can be simplified by setting the left-hand side equal to zero and substituting

$$f(\mathbf{k},\mathbf{r},t)=f_0(\epsilon_{\mathbf{k}})+g_{\mathbf{k}} \quad (4)$$

on the right-hand side. The function $g_{\mathbf{k}}$ which describes the deviation from equilibrium will be proportional to the applied fields unless they are extremely strong. If we retain only linear terms in the applied fields, the Boltzmann equation becomes

$$\left[q\mathbf{E} + \frac{q}{c}\mathbf{v}\times\mathbf{B} - \frac{\epsilon_{\mathbf{k}}-\mu}{T}\nabla_{\mathbf{r}}T\right]\cdot\mathbf{v}_{\mathbf{k}}\frac{\partial f_0}{\partial\epsilon} = -\left(\frac{\partial g_{\mathbf{k}}}{\partial t}\right)_{\mathrm{scatt}}. \quad (5)$$

If this equation can be solved for $g_{\mathbf{k}}$, the charge current can be calculated from

$$\mathbf{J}_q = \sum_{\mathbf{k}} q\mathbf{v}_{\mathbf{k}}g_{\mathbf{k}}, \quad (6)$$

and the heat current can be calculated from a similar expression in which the particle energy, $\epsilon_{\mathbf{k}}$, replaces the charge, q. The sum over $\mathbf{k}$ in Eq. (6) represents a sum over all of the electron states. Equation (5) can be simplified slightly since the term involving the magnetic field vanishes because the Lorentz force, $(q/c)\mathbf{v}\times\mathbf{B}$, is perpendicular to the carrier velocity. Thus, to treat transport properties arising from magnetic fields, it is necessary to retain terms that are higher order in the applied fields than those retained here.

A rigorous solution of Eq. (5) requires a detailed understanding not only of the dispersion relation $\epsilon_{\mathbf{k}}$, but also of the scattering mechanisms. In metals the primary current carriers are usually electrons which are scattered primarily by impurities and phonons. The probability for an electron to scatter off impurities is essentially independent of temperature, whereas the scattering rate for scattering off phonons is zero at 0 K, increases rapidly at temperatures below the maximum phonon frequencies and finally varies linearly with temperature at higher temperature. Phonons, which contribute to the heat current, scatter off impurities, and off electrons with a scattering rate that is essentially independent of temperature. The scattering rate for phonons to scatter off other phonons increases with temperature due to the increase in the number of phonons with temperature.

Let us consider the simple case of a metallic conductor in the presence of an electric field but without a thermal gradient. In this case the linearized Boltzmann equation for the electrons becomes

$$q\mathbf{E}\cdot\mathbf{v}\,\frac{\partial f_0}{\partial\epsilon} = -\left(\frac{\partial g_{\mathbf{k}}}{\partial t}\right)_{\text{scatt}}. \tag{7}$$

This is still a formidable equation to solve in terms of the microscopic scattering processes, but a simple and qualitatively correct approximate solution can be obtained by assuming that an electron with energy $\epsilon_{\mathbf{k}}$ can travel freely for a time $\tau_{\mathbf{k}}$ between scattering events. During this time it is acted upon by the applied field which causes a change in its energy given by classical mechanics as

$$\delta\epsilon_{\mathbf{k}} = q\mathbf{v}_{\mathbf{k}}\cdot\mathbf{E}\tau_{\mathbf{k}}. \tag{8}$$

This change in energy corresponds to a change in the distribution function

$$f_0(\epsilon_{\mathbf{k}}+\delta\epsilon_{\mathbf{k}}) = f_0(\epsilon_{\mathbf{k}}) + \frac{\partial f_0}{\partial\epsilon}\,q\mathbf{v}_{\mathbf{k}}\cdot\mathbf{E}\tau_{\mathbf{k}}. \tag{9}$$

By comparing Eq. (9) with Eq. (4) it can be seen that the second term on the right of Eq. (9) is an approximate expression for $g_{\mathbf{k}}$ expressed in terms of the still unknown electron "lifetime," $\tau_{\mathbf{k}}$. This same expression for the deviation from equilibrium could have been obtained by assuming that the deviation function $g_{\mathbf{k}}$ would relax to zero during a time $\tau_{\mathbf{k}}$ according to

$$g_{\mathbf{k}}(t) = (g_{\mathbf{k}})_E e^{-t/\tau} \tag{10}$$

if the field were turned off at time $t=0$. For this reason this approximate solution to the linearized Boltzmann equation is sometimes called the "relaxation time" approximation.

Once the change in the distribution function induced by the applied field $g_{\mathbf{k}}$ is known, the current can be calculated using Eq. (6):

$$\mathbf{J}_q = \sum_{\mathbf{k}} q^2 v_{\mathbf{k}} v_{\mathbf{k}}\cdot\mathbf{E}\,\frac{\partial f_0}{\partial\epsilon}\,\tau_{\mathbf{k}}. \tag{11}$$

Unless the temperature is extremely high, the energy derivative of the Fermi function, $\partial f_0/\partial\epsilon$, is sharply peaked at $\epsilon=\mu$, the physical consequence of which is that only electrons with an energy approximately equal to the Fermi energy μ contribute to the currents in a metal.

Since the current in direction i, J_{qi}, is related to the applied field in direction j, E_j, through Ohm's law,

$$J_{qi} = \sum_{j=1,3} \sigma_{ij}E_j, \tag{12}$$

the conductivity is given by

$$\sigma_{ij} = \sum_{\mathbf{k}} q^2 v_{\mathbf{k}i} v_{\mathbf{k}j}\,\frac{\partial f_0}{\partial\epsilon}\,\tau_{\mathbf{k}}. \tag{13}$$

In general the current need not be in the same direction as the applied field, hence σ in Eq. (13) is a tensor. For cubic systems, however, it can be shown that σ is diagonal.

The electron lifetime τ can be calculated in terms of the microscopic scattering probabilities. If, for example, the scattering is isotropic the lifetime can be approximated by

$$\frac{1}{\tau_{\mathbf{k}}} = \sum_{\mathbf{k}'} P_{\mathbf{k}\mathbf{k}'}, \tag{14}$$

where $P_{\mathbf{k}\mathbf{k}'}$ is the probability of an electron scattering from state $\mathbf{k}$ to state $\mathbf{k}'$. For a perfect crystal without impurities or other imperfections, at zero temperature, the scattering probability is zero and the electron lifetime is infinite which implies that the conductivity will be infinite. Surprisingly, this is true even though the ions may be displaced from their lattice sites because of their zero-point motion. Zero-point motion results from the fact that even when the ions are in their lowest energy state their positions are not precisely fixed because this would violate the Heisenberg uncertainty principle. For weak perturbations the scattering probability can be obtained from second-order perturbation theory:

$$P_{\mathbf{k}\mathbf{k}'} = \frac{2\pi}{\hbar}\left|\langle\mathbf{k}\,|V|\,\mathbf{k}'\rangle\right|^2\delta(\epsilon_{\mathbf{k}}-\epsilon_{\mathbf{k}'}), \tag{15}$$

where $\langle\mathbf{k}|V|\mathbf{k}'\rangle$ is the matrix element of the perturbing potential, V. Usually, however, the electron lifetime is treated as a parameter whose qualitative behavior is known or as an empirical parameter that can be determined from experiment.

Current areas of research on transport properties emphasize situations in which the semiclassical theory, outlined above, is inadequate. If the scattering rate is very strong so that the electron mean free path defined by $\mathbf{v}_{\mathbf{k}}\tau_{\mathbf{k}}$ is comparable to its wavelength $\lambda=2\pi/|\mathbf{k}|$, the semiclassical picture breaks down because the electron does not travel a full wavelength between scattering events. In this case, λ and hence k are not well defined.

A second limitation of the semiclassical theory is that it ignores quantum interference effects. These effects arise because electrons propagate between scattering events as waves rather than as particles. A wave may propagate between two points by many different paths which depend on the random locations of the scattering centers. Usually the interference terms that arise from the different paths will vanish when one averages over the positions of the scattering centers. There are, however, a few paths which involve the same scatterers taken in different orders which may yield nonvanishing interference terms. Usually these terms decrease the conductivity. Quantum interference effects are most important at low temperature and in one- and two-dimensional systems.

Another area of current research is concerned with the effect of strong fields on transport properties. For this problem the semiclassical picture may be valid, but it is necessary to avoid the linearization step which was used to obtain Eq. (5) from Eq. (1).

See also CONDUCTION; ELECTROCHEMISTRY; ELECTRON ENERGY STATES IN SOLIDS AND LIQUIDS; GALVANOMAGNETIC AND RELATED EFFECTS; HALL EFFECT; INSULATORS; LATTICE DYNAMICS; METALS; PHONONS; QUANTUM STATISTICAL MECHANICS; QUASIPARTICLES; SEMICONDUCTORS, CRYSTALLINE; STATISTICAL MECHANICS; SUPERCONDUCTING MATERIALS; SUPERCONDUCTIVITY THEORY; THERMOELECTRIC EFFECTS; TRANSPORT THEORY.

BIBLIOGRAPHY

B. L. Altshuler, A. G. Aronov, D. E. Khmelnitskii, and A. I. Larkin, "Coherent Effects in Disordered Conductors," in *Quantum Theory of Solids* (I. M. Lifshitz, ed.). MIR, Moscow, 1982. (A)

N. W. Ashcroft and N. D. Mermin, *Solid State Physics*. Holt, Rinehart & Winston, New York, 1976. (I)

F. J. Blatt, *Physics of Electronic Conduction in Solids*. McGraw-Hill, New York, 1968. (I)

C. Kittel, *Introduction to Solid State Physics*. Wiley, New York, 1971. (E)

G. D. Mahan, *Many Particle Physics*. Plenum, New York, 1981. (A)

J. M. Ziman, *Electrons and Phonons*. Oxford University Press, London, 1960. (A)

Transport Theory

Sudarshan K. Loyalka

Motion of a given species of particles, in a host medium, is a problem of considerable interest in several branches of physics. Studies of scattering and absorption of light (photons) by atmospheric nuclei, evolution of neutron distribution in a nuclear reactor, penetration of γ or x rays (photons) through a given material, diffusion of molecules of a given species of gas through a gas mixture, and motions of ions, atoms, and electrons in a plasma have several common features. Physical models, mathematical equations, and techniques that have been developed to understand the above phenomena form the subject of transport theory.

A description of the motion of each particle in an N-particle system leads to a set of $6N$ first-order differential equations (neglecting internal degrees of freedom), and it is futile to attempt solving such a system for realistic problems. Therefore, the motion of particles is usually described by a single-particle distribution function as this quantity is generally sufficient for evaluating macroscopic properties (heat, mass, momentum, charge transport) of particle motion. Historically, some of the first attempts in this direction were made by Maxwell and Boltzmann, and it was Boltzmann (1872) who introduced the equation

$$\frac{\partial f_i}{\partial t} + c_i \frac{\partial f_i}{\partial \mathbf{r}} + \mathbf{F} \cdot \frac{\partial f_i}{\partial \mathbf{c}_i} = \sum_{j=1}^{n} \iiint (f_i' f_j' - f_i f_j) g_{ij} \sigma_{ij}(g_{ij}, x)\, d\Omega\, d\mathbf{c}' \quad (1)$$

for a description of molecular motion in a "dilute" gas mixture. Here $f_i(\mathbf{r},\mathbf{c}_i,t)$ is the expected number of the molecules of species i at position $\mathbf{r}$ in $d\mathbf{r}$ with velocity $\mathbf{c}_i$ in $d\mathbf{c}_i$ at time t. The integral term on the right-hand side describes the gain and loss of particles of species i from the phase-space volume due to molecular collisions, and is dependent on intermolecular potentials through the collision cross section σ_{ij}. This equation and several variants of it are usually referred to as the transport equations.

It is clear that a complete evaluation of $f_i(\mathbf{r},\mathbf{c},t)$ would require specification of the initial distribution $f_i(\mathbf{r},\mathbf{c},0)$ and boundary conditions on the surface enclosing the gas. The latter constitute an interesting feature of the transport theory in that the reflected molecular distribution is specified in terms of the molecular distribution incident on the surface and gas–surface interaction properties. Thus, while in the bulk of the gas the molecular distribution is smooth and largely determined by intermolecular potentials, the molecular distribution is discontinuous (in the velocity space) near the surface and is determined by the gas–surface interactions.

An initial boundary-value problem of this type is quite difficult to solve, and in fact in many practical cases it is not even necessary to consider such problems in their full generality. It is convenient here to consider a parameter $\mathrm{Kn}=l/d$ (known as the Knudsen number), where l is a mean free path of molecules and d is a characteristic dimension of the flow system. Thus, if $\mathrm{Kn}\ll 1$ (the continuum regime), then in the bulk of the gas the effect of gas–surface interaction is relatively unimportant, and the molecular distribution is more or less completely determined by intermolecular potentials. Actually, by the use of proper asymptotic expansions in this limit, the Boltzmann equation can be shown to yield the phenomenological laws of heat transfer (Fourier), momentum transfer (Newton), and diffusion (Fick). These procedures also provide transport coefficients (thermal conductivity, viscosity, diffusion coefficients) in terms of intermolecular potentials and have been highly successful in computation of transport coefficients and the modeling of intermolecular potentials by comparison of theoretical results with experimental data for transport coefficients. Many interesting problems of particle transport, however, also occur in the regime Kn~1 (the transition regime) and Kn≫1 (the free molecular regime), and it becomes necessary to consider solutions of integrodifferential transport equations with associated initial-boundary conditions.

Although there had been numerous attempts at solving the integrodifferential transport equation with associated initial-boundary conditions, both in the kinetic theory of gases and in radiative transfer, the subject advanced rapidly with the advent of the nuclear, space, and computer eras. A neutron can undergo absorption or elastic or inelastic scattering and cause fission and a host of other reactions in a variety of materials. Usually, the neutron density is much lower than the density of the host nuclei, neutron–neutron interactions can be neglected, and the single-particle neutron distribution can be described by the neutron transport equation (or just the transport equation):

$$\frac{\partial f(\mathbf{r},\mathbf{c},t)}{\partial t} + \mathbf{c} \cdot \frac{\partial f(\mathbf{r},\mathbf{c},t)}{\partial \mathbf{r}} = -c\, \Sigma_t(\mathbf{r},\mathbf{c},t) f(\mathbf{r},\mathbf{c},t) + \int \Sigma_s(\mathbf{r},\mathbf{c}\rightarrow\mathbf{c},t) f(\mathbf{r},\mathbf{c},t) d\mathbf{c} + S(\mathbf{r},\mathbf{c},t), \quad (2)$$

where Σ_t is the cross section (cm^{-1}) for neutron interaction with host nuclei per unit length of neutron travel, Σ_s is the scattering kernel and includes contributions of fissions and other reactions. S represents neutron sources. Associated with the above are initial-boundary conditions of the type

$$f(\mathbf{r},\mathbf{c},0) = G(\mathbf{r},\mathbf{c}), \quad (3)$$

$$f(\mathbf{r},\mathbf{c},t) = B(\mathbf{r},\mathbf{c},t), \quad \text{for } \mathbf{r}\in\partial s,\ \mathbf{c}\cdot\mathbf{n}_r>0, \quad (4)$$

where ∂s indicates the surface of the host medium, and $\mathbf{n}_r$ (cm^{-1}) is the inward normal at ∂s. If there are no inwardly directed neutron sources at the surface of the host medium, a situation of common occurrence, then

$$f(\mathbf{r},\mathbf{c},t) = 0, \quad \text{for } \mathbf{r}\in\partial s,\ \mathbf{c}\cdot\mathbf{n}_r>0. \quad (5)$$

Equation (2), although linear, has quite a complicated mathematical structure. The cross sections are dependent on the energy and angle of scattering and must be determined by extensive experimental measurements. Also, boundary conditions (4) and (5) introduce discontinuities in the distribution (particularly near the surface) and preclude the use of several conventional techniques of mathematical physics.

It was realized quite early that while neutron distributions in large reactors could be crudely described by the diffusion equation, the weapons and the sophisticated reactor designs required accurate solutions of the transport equation. This need has led to a variety of models of the transport equation that can be solved analytically or "exactly" by various numerical methods. Several numerical schemes that provide approximate solutions of the transport equation have also been developed.

Generally the scattering kernel is very complicated, and most methods for solving transport equations use some approximate representation of this kernel (for example, a degenerate angular-spherical harmonics-energy-multigroup expansion). One of the simplest models of the transport equation corresponds to the one-speed, isotropic-scattering, slab-geometry case, which leads to a problem of the type

$$\mu \frac{\partial \Psi(x,\mu)}{\partial x} + \Psi(x,\mu) = \frac{c}{2} \int_{-1}^{1} \Psi(x,\mu')\, d\mu' + S(x,\mu'), \quad x\in(-a,a), \quad \mu\in(-1,1),$$
$$\Psi(\pm a,\mu) = 0, \quad \mu \lessgtr 0, \quad (6)$$

Equation (6) can be studied by use of either Wiener–Hopf or singular eigenfunction expansion techniques. The latter method and some variants of it have received extensive attention in the recent past, and these investigations have provided much insight into the nature of the solutions of the transport equation.

The analytical techniques are, however, not feasible for most transport problems of practical interest. Rather, it has been necessary to devise several approximate methods that include spherical harmonic expansions of $f_i(\mathbf{r},\mathbf{c}_i,t)$ and moments methods, finite-difference discrete-ordinate methods (S_N method), variational methods (finite-element techniques), conversion of integrodifferential equations to integral forms, invariant embedding, Monte Carlo methods, etc. The discrete-ordinate S_N methods and the Monte Carlo methods are particularly suited to high-speed computer calculations. Many standard computer programs are now available for routine solutions to a variety of neutron-transport problems. The S_N method is prone to give erroneous results in some cases but is, computationally, very efficient. The Monte Carlo method is very versatile, as one can simulate the single-particle distribution by using a large sample of neutrons and tracing the exact history of each neutron as it undergoes scattering and absorption, and causes fissions and other reactions. The method is adaptable to arbitrary geometries and, in principle, it is possible to use point-scattering cross sections. Both the S_N and the Monte Carlo methods are also suitable for parallel processors, and considerable progress is being made in development of fast algorithms.

The methods that have been developed to study neutron transport apply equally well to many problems of radiative transfer, kinetic theory of gases, and plasma transport, and vice versa. The subject of transport theory is still quite young, and one can anticipate continual improvements in the analytic and numerical techniques as well as in the characterization of new transport phenomena (such as thermal diffusion). The theory is rich in its mathematical content and applications, and we can expect deeper explorations of analytical and numerical questions associated with solving the transport equation.

See also DIFFUSION; HEAT TRANSFER; HYDRODYNAMICS; STATISTICAL MECHANICS; TRANSPORT PROPERTIES.

BIBLIOGRAPHY

G. I. Bell and S. Glasstone, *Nuclear Reactor Theory*. Van Nostrand Reinhold, New York, 1970. (E)

G. A. Bird, *Molecular Gas Dynamics*. Clarendon Press, Oxford, 1976. (I)

K. M. Case and P. F. Zweifel, *Linear Transport Theory*. Addison-Wesley, Reading, MA, 1967. (A)

C. Cercignani, *Theory and Applications of the Boltzmann Equation*. Elsevier, New York, 1975. (A)

S. Chandrasekhar, *Radiative Transfer*. Clarendon Press, Oxford, 1957. (A)

S. Chapman and T. G. Cowling, *The Mathematical Theory of Non-Uniform Gases*. Cambridge University Press, Cambridge, 1939, 1952, 1970. (I)

B. Davison, *Neutron Transport Theory*. Clarendon Press, Oxford, 1957. (A)

J. H. Ferziger and H. G. Kaper, *Mathematical Theory of Transport Processes in Gases*. North-Holland, Amsterdam, 1972. (I)

J. Spanier and E. M. Gelbard, *Monte Carlo Principles and Neutron Transport Problems*. Addison-Wesley, Reading, MA, 1969. (I)

M. M. R. Williams, *Mathematical Methods in the Particle Transport Theory*. Butterworth, London, 1970. (A)

Transuranium Elements

Richard L. Hahn

The elements with atomic numbers (Z) greater than 92 (which is the Z of uranium) are all radioactive. Because their half-lives in general are much shorter than the age of the universe (about 4 billion years), they no longer exist in nature, although traces of an isotope of plutonium ($Z=94$) with a mass of 244 and half-life of 80 million years have been found. Thus, all of these elements are what might be termed artificial; i.e., they have been produced by scientists in nuclear reactions. The currently known (or claimed) transuranium elements are neptunium ($Z=93$, chemical symbol Np), plutonium (94, Pu), americium (95, Am), curium (96, Cm), berkelium (97, Bk), californium (98, Cf), einsteinium (99, Es), fermium (100, Fm), mendelevium (101, Md), nobelium (102, No), lawrencium (103, Lr), and elements 104, 105, and 106, for which no names have been accepted as yet because of disputes over the priority of discovery. During the past several years, claims have also been put forth for the discovery of elements 107 through 109. A heroic attempt to synthesize element 110 was unsuccessful.

The longest-lived or most stable isotope that is known for each of these elements is ^{237}Np (half-life of 2 million years), ^{244}Pu (80 million years), ^{243}Am (7,000 years), ^{248}Cm (300,000 years), ^{247}Bk (1,400 years), ^{251}Cf (800 years), ^{254}Es (276 days), ^{257}Fm (80 days), ^{258}Md (54 days), ^{259}No (1 hr), ^{260}Lr (3 min), ^{261}104 (70 s), and ^{262}105 (40 s). For the higher atomic numbers, newly discovered nuclides have been claimed, with the following masses and half-lives: element 106, mass 259 (7 ms), 260 (20 ms) and 263 (0.9 s); 107, 261 (1 ms), 262 (5 ms), and 267 (1 s); 108, 265 (2 ms); and 109, 266 (4 ms). While many of these nuclide discoveries have not been confirmed, the trend seems clear: as the atomic numbers and masses of the transuranium isotopes increase, their stability, as measured by their relative half-lives, decreases. A clear trend is exhibited by these data, namely that the transuranium elements become less stable as the atomic number and mass increase. Thus, while the half-life of ^{238}U is so long, 4.5 billion years, that it is found in the earth's crust, where it is even more abundant than such elements as iodine, mercury, silver, or bismuth, the half-lives of isotopes of elements 102 and above are on the order of minutes or less.

The transuranium elements are the heaviest elements that exist, and, as a result, have some very special characteristics. In terms of nuclear behavior, these elements are unique in that they are unstable to radioactive decay by alpha-particle emission and by spontaneous fission, as well as to decay by beta-particle emission or orbital electron capture; in addition, all of these elements are readily fissionable when excited in a nuclear bombardment (the word "spontaneous" is used to describe the radioactive decay process because no external source of energy is required to cause the nucleus to fission). In terms of their chemistry, the 15 elements with atomic numbers from 89 (actinium, Ac) through 103 behave very similarly, one element to the next, and are called members of the actinide series. This behavior is analogous to what is observed for the rare-earth elements, or lanthanide series (Z from 57 through 71); it is explained in terms of the filling of an inner electron shell, the 5f shell, in going across the series, while the outer or valence electron configurations do not change appreciably.

For the lower actinides, from actinium through plutonium, oxidation states from +3 to +6 are known, but beginning with americium, the predominant oxidation state in solution is +3, with mendelevium and nobelium exhibiting +2 valence states as well. A report has also been published for the preparation of a +1 oxidation state of mendelevium, which would be the first incidence of such a low valence state in the actinides. However, other scientists have not been able to verify this claim. Among the many chemical compounds of the actinides that have been studied are the oxides, hydroxides, halides, pnictides (compounds formed with phosphorus, nitrogen, arsenic, antimony, or bismuth), chalcogenides (formed with sulfur, selenium, or tellurium), compounds formed with inorganic oxo-acids, such as carbonates, nitrates, halates, etc., and various organic complexes and chelates. Element 104 is the first of the transactinide elements. In the few chemical experiments that have been done with the transactinides, element 104 has been found to behave similarly to its homolog, hafnium (element 72), by having a stable +4 valence state (so that element 104 is sometimes called eka-hafnium), and element 105 has been found to be similar to tantalum (73). Elements 106 through 109 are thus expected to be chemically similar to elements 74 through 77.

Another well-known chemical characteristic of the actinide (or lanthanide) elements is also due to the filling of the inner 5f (or 4f) electron shell: as the atomic number of the element increases in going across the series, the ionic radius (for a given oxidation state) is observed to decrease. This effect is called the actinide (or lanthanide) contraction. It is explained by noting that as electrons are added to the f shell in going across the series, the number of protons in the nucleus increases correspondingly, causing an increase in the electrical attraction between the relatively unchanged outer electron shells and the atomic nucleus, and hence a decrease in the atomic or ionic radius.

The transuranium elements have been synthesized in a variety of ways. Table I lists details of the original discovery experiments for elements 93–107. At present, their production methods can be divided into two groupings: (1) The elements up to and including Fm can be produced in weighable amounts that vary from many grams down to about 10^{-12} g

Table I Discovery of the transuranium elements

Element, atomic number, and name	Country and year of discovery experiment	Production method	Method of identification
93, neptunium	U.S.A., 1940	Cyclotron bombardment. $^{238}U + n \rightarrow ^{239}U \xrightarrow{\beta-} ^{239}Np$	Beta-decay of known isotope ^{239}U.
94, plutonium	U.S.A., 1940	Cyclotron bombardment. $^{238}U + d \rightarrow ^{238}N \xrightarrow{\beta-} ^{238}Pu$	Beta-decay of known isotope ^{238}Np.
95, americium	U.S.A., 1944	Reactor irradiation. Successive neutron captures $^{239}Pu(n,\gamma)^{240}Pu(n,\gamma)^{241}Pu \xrightarrow{\beta-} ^{241}Am$	Chemical separation of Am from Pu. Beta-decay of purified Pu fraction.
96, curium	U.S.A., 1944	Cyclotron bombardment. $^{239}Pu + \alpha \rightarrow ^{242}Cm$	Chemical separations. Alpha-decay of ^{242}Cm to known isotope ^{238}Pu.
97, berkelium	U.S.A., 1949	Cyclotron bombardment. $^{241}Am + \alpha \rightarrow ^{243}Bk$	Chemical separations. Detection of decay by alpha-particle and x-ray emission.
98, californium	U.S.A., 1950	Cyclotron bombardment. $^{242}Cm + \alpha \rightarrow ^{245}Cf$	Chemical separations. Alpha decay of ^{245}Cf.
99, einsteinium	U.S.A., 1952	Thermonuclear explosion. Multiple neutron captures and β decays (beginning with uranium) $\rightarrow ^{253}Es$	Chemical separations, using debris from explosion. Alpha decay of ^{253}Es.
100, fermium	U.S.A., 1953	Thermonuclear explosion. (Same as for einsteinium) $\rightarrow ^{255}Fm$	Chemical separations. Alpha decay of ^{255}Fm.
101, mendelevium	U.S.A., 1955	Cyclotron bombardment. $^{253}Es + \alpha \rightarrow ^{256}Md$	Chemical separations. Decay (orbital electron capture) to isotope of fermium, ^{256}Fm.
102, nobelium	U.S.A., 1958 (earlier claim, 1957, from Sweden later was shown to be incorrect).	Linear accelerator (LINAC) bombardment $^{246}Cm + ^{12}C \rightarrow ^{254}No$	Alpha decay of ^{254}No to known isotope, ^{250}Fm.
103, lawrencium	U.S.A., 1961	LINAC bombardment. Mixture of ^{249}Cf, ^{250}Cf, ^{251}Cf, and ^{252}Cf bombarded with ^{10}B and $^{11}B \rightarrow ^{257}Lr$.	Variation of production cross sections with energy of bombarding ions (excitation functions).
104, kurchatovium[a]	U.S.S.R., 1964	Cyclotron bombardment. $^{242}Pu + ^{22}Ne \rightarrow ^{260}104$	Measured excitation function for observed spontaneous fission decay. Same activity was not produced in bombardments with other ion–target combinations (cross bombardments).
104, rutherfordium[a]	U.S.A., 1969	LINAC bombardment. $^{249}Cf + ^{12}C \rightarrow ^{257}104$	Observed alpha decay to known isotope, ^{253}No.
105, nielsbohrium[a]	U.S.S.R., 1970	Cyclotron bombardment. $^{243}Am + ^{22}Ne \rightarrow 105$	Excitation function of spontaneous fission decay. Cross bombardments. Variation of yield of product with change in angle of emission. Also observed alpha decay.
105, hahnium[a]	U.S.A., 1970	LINAC bombardment. $^{249}Cf + ^{15}N \rightarrow ^{260}105$	Alpha decay to known ^{256}Lr. Cross bombardments.
106[a]	U.S.A., 1974	LINAC bombardment. $^{249}Cf + ^{18}O \rightarrow ^{263}106$	Observed alpha decay sequence to known descendants: $^{263}106 \rightarrow ^{259}104 \rightarrow ^{255}No$.
106[a]	U.S.S.R., 1974	Cyclotron bombardment. ^{206}Pb, ^{207}Pb, and ^{208}Pb bombarded with $^{54}Cr \rightarrow 106$.	Observed spontaneous fission decay. Cross bombardments.
107[a]	Federal Republic of Germany (F.R.G.), 1981	LINAC bombardment. $^{209}Bi + ^{54}Cr \rightarrow ^{262}107$	Velocity filter separation of desired products. Observed correlated alpha-particle decays leading to known nuclei.
108[a]	F.R.G., 1984	LINAC bombardment. $^{208}Pb + ^{58}Fe \rightarrow ^{265}108$	Velocity filter separation of desired products. Observed correlated alpha-particle decays leading to known nuclei.
109[a]	F.R.G., 1982	LINAC bombardment. $^{209}Bi + ^{58}Fe \rightarrow ^{266}109$	Velocity filter separation of desired product. Observed correlated alpha-particle decays leading to known nuclei.

[a] Conflicting claims to priority of discovery of elements 104, 105, 106, and 107 have not yet been resolved. Discovery of elements 108 and 109 has only been claimed by one group.

(or 10^9 atoms). To produce these elements, targets such as uranium are irradiated in a high-flux reactor, where they undergo a series of multiple neutron captures and (negative) beta-particle radioactive decays to form elements with higher masses and atomic numbers than the targets. Chemical processing of the irradiated targets separates the different elements formed by this process. (2) Elements with Z of 101 and above can only be obtained in trace amounts, from accelerator bombardments of targets with charged particles. To form the heaviest elements, beams of nuclei, which are called heavy ions because they have atomic masses greater than that of helium-4, must be used to introduce several protons into the target nucleus at one time. Reactions such as $^{18}O + {}^{249}Bk \rightarrow {}^{262}105 + 5n$ (neutrons) or $^{12}C + {}^{249}Cf \rightarrow {}^{257}104 + 4n$ have often been used, in which the excess energy that is available in the reaction is released by the emission of several neutrons. Because the chance of forming a desired nuclear product decreases as the energy content of the nucleus increases, a new approach has been used to form the elements above 105. Called "cold nuclear fusion," the process involves bombarding especially stable target nuclei, such as ^{208}Pb or ^{209}Bi, with beams such as ^{54}Cr or ^{58}Fe to form products with low excess energy, as indicated by the fact that only one or two neutrons are emitted in the reaction.

The properties of the transuranium elements have several practical implications:

(1) Because they are radioactive, they must be handled with special care and techniques. In particular, alpha-particle-emitting nuclei can pose special health hazards. Because their chemical behavior is such that they tend to be deposited in bone tissue, the transuranium elements can cause long-term radiation damage in the human body. So, in addition to using the usual radiation shielding for nuclei that emit beta-particles and gamma rays, scientists studying the transuranium elements must work in special closed facilities, such as glove boxes, to prevent the inadvertent airborne spread of any fine particles that they might inhale or ingest.

(2) The energy released in the radioactive decay of these elements is often so large that decomposition, or radiation damage, of their chemical compounds occurs. Such elements and compounds clearly present peculiar problems to scientists trying to investigate their chemical properties.

(3) Added to these problems are the special difficulties encountered in trying to study the behavior of elements that exist only for seconds or minutes. For example, to study the chemical properties of elements 103 or 104, one has to work at an accelerator, produce a few hundred atoms (at most) in an irradiation, purify the desired element, perform a chemical measurement, and then begin the procedure over with a new irradiation prior to making another chemical measurement. Such methodology has been dubbed "one-atom-at-a-time" chemistry.

(4) Because some isotopes of the transuranium elements are easily fissionable, emitting enough neutrons to sustain a chain reaction, they have been used in nuclear weapons and, in a controlled way, as fuels in nuclear reactors. As is well known, immense amounts of energy are liberated in nuclear fission. Indeed nuclear reactors are used in many parts of the world today to produce electricity commercially. However, in recent years, public concern has grown about the long-term safety of the nuclear power industry, caused in part by accidents at two different reactors: the partial fuel melt-down at the Three Mile Island reactor in the United States and by the far more serious explosion at the Chernobyl reactor in the Soviet Union, which resulted not only in the wide spread of high levels of radioactive contamination, but in the loss of human lives. Serious questions have been raised in public forums about (a) the operating conditions at nuclear reactors, especially those that have been in service for many years, as well as about (b) the safe long-term storage of long-lived radioactive materials from nuclear reactors, such as nuclear fuel assemblies, and (c) the chemical and physical behavior of the transuranium elements and of their fission products, if they are accidentally introduced into the natural environment. Scientists and engineers today are studying these questions intensively, to determine viable solutions to these problems. Valuable information relevant to questions (b) and (c) has been obtained from scientific studies of a fascinating phenomenon, a naturally occurring nuclear reactor called Oklo, which achieved criticality (without any human intervention!) in a uranium ore deposit in Gabon, Africa, several million years ago.

The fact that very short half-lives are encountered in the heaviest elements has led many scientists to conclude that element 109 may be at or near the limit of nuclear stability, and that it may be very difficult, or even impossible, to produce elements with even higher atomic numbers. However, several years ago, an idea was developed about a possible new region of nuclear stability (called the "island of stability") that might exist far from the elements known today. Based on our knowledge of nuclear structure and stability, this new region of "superheavy elements" was predicted to occur around atomic numbers 114 or 126. The prospect of finding such new, relatively stable, elements was especially exciting to many scientists, because it would allow them to test and to extend their ideas, not only about nuclear properties, but about chemical behavior as well. Many ingenious experiments were carried out, to try to produce isotopes of these elements in nuclear bombardments, or to detect extremely long-lived varieties of the "SHE's" in nature (by taking advantage of the fact that elements that are members of a particular chemical "family" all have similar chemical properties, and tend to be found together in ore deposits, purified metals, etc.). Although several optimistic claims have been made for the discovery of superheavy elements, none of them unfortunately has survived the scrutiny of the scientific community. Because of this lack of success, interest in this research area has waned in recent years. Many scientists have concluded that either the SHE's are much less stable than originally thought, or that the conditions that have been available in nuclear bombardments were simply not optimum for reaching the sought-after island of stability.

See also ATOMS; NUCLEAR FISSION; RADIOACTIVITY; SUPERHEAVY ELEMENTS; TRANSITION ELEMENTS.

BIBLIOGRAPHY

K. W. Bagnall, *The Actinide Elements*. Elsevier, Amsterdam, 1972.
V. I. Goldanskii and S. M. Polikanov, *The Transuranium Elements*, translated by J. E. S. Bradley. Consultants Bureau, New York, 1973.
E. K. Hyde, I. Perlman, and G. T. Seaborg, *The Nuclear Properties of the Heavy Elements*, Vol. I–III. Prentice-Hall, Englewood Cliffs, N. J., 1964.
J. J. Katz and G. T. Seaborg, *The Chemistry of the Actinide Elements*. Methuen, London, 1957.
C. Keller, *The Chemistry of the Transuranium Elements*. Verlag Chemie, Weinheim, 1971.
J. R. Nix, "Predictions for Superheavy Nuclei," *Physics Today*, p. 30, April (1972).
G. T. Seaborg and J. L. Bloom, "The Synthetic Elements," *Scientific American*, **220** (4), 56 (April, 1969).

Tribology

R. G. Munro

Tribology is the science of interacting material interfaces in relative motion. Attention is often focused specifically on the friction and wear properties of the materials and the control of those processes through lubrication. Every object with moving parts, whether in a machine or a biological body, experiences the consequences of those interactions. While some of the resulting effects are desirable, e.g., traction and polishing, others cause detrimental energy losses and material degradation. Tribologically generated energy losses in the United States alone are estimated to occur at a rate exceeding 10^{18} J/year, which is approximately the annual energy output of 27 large electrical power plants.

Tribological interactions begin as two surfaces are pressed together and are dependent on the microscopic characteristics of the contacting surfaces. The microscopic topography of any surface has a definable roughness that may be similar to either gently rolling hills and valleys or to a rugged mountain range. In either case, the force between the two bodies is supported by microscopic surface irregularities, called asperities, having typical linear dimensions on the order of a few micrometers with the peaks being scattered irregularly at distances of tens of micrometers. Thus, when such surfaces have a relative speed of a few meters per second, the interactions at the interface of two asperities are of fleeting duration, about a millionth of a second. Yet, the interactions are rich in complexity.

Stress levels at the asperities may exceed 10 GPa, a pressure comparable to 1×10^5 earth atmospheres. The interacting asperities respond to this stress by deforming elastically until the material plastic or yield stress limit is reached. Then the material must flow or fracture, creating wear debris. In brittle, granular material such as structural ceramics, grains may be pulled apart or individually broken. Resulting microscopic, instantaneous, "hot-spot" temperatures have been estimated to be as high as 2000°C. Energy released during these processes, therefore, may promote chemical reactions among the surface materials, the gaseous environment, and any lubricating fluid. As a result, the surface may oxidize, material properties or phases may change, material may be transferred between the surfaces, or a reaction product may be deposited on one or both surfaces.

At the macroscopic level, the frictional force developed between two sliding surfaces has long been known to obey two empirical laws rather faithfully: friction is proportional to the applied load, and friction is independent of the apparent contact area. Based on the first law, the coefficient of friction, μ, is defined as the ratio, F/N, of the frictional force, F, tangential to the interface, and the force, N, applied normal to the interface. In practice, measured values of μ are average values determined over a period of time that is long compared to the duration of the interaction of any two asperities. The second law results from the fact that the actual contact between surfaces occurs only at scattered asperities. The true area of contact, A, is the sum of the interface areas of the contacting asperities. Each asperity can only support stress up to its yield strength, P, with the number of asperities being such that $PA = N$. In the adhesion theory of friction, the contacting asperities weld together at this limit. Hence, the two surfaces can continue in motion only if a shear stress equal to the shear strength, S, of the material acts at the asperity interfaces. The total frictional force then must be $F = SA$, from which it follows that $\mu = F/N = SA/PA = S/P$, independent of the area of contact.

Wear of a material occurs when any material transport occurs during a tribological contact. Metals wear by various mechanisms including adhesion, abrasion (third-body interactions), corrosion (chemically accelerated wear), and fatigue (cyclic loading). For metals, the total volume removed from a surface sliding against an unworn surface tends to increase proportionately with the normal force and the total sliding distance, and to decrease with the hardness of the material. For ceramics, the situation appears to be more complex and depends on the microstructure, the interface

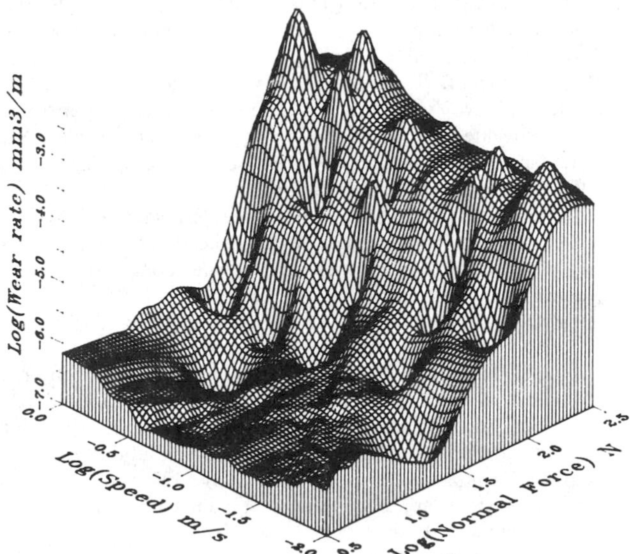

FIG. 1. Advanced ceramics may have complex wear behaviors involving dramatic transitions from low to high wear regions.

forces and grain boundary energies, the material properties such as flexural strength, hardness, and toughness, and the creep and crack propagation characteristics. Figure 1 shows the wear of an alumina material as a function of normal force and the sliding speed when the contact is lubricated with paraffin oil. The wear varies dramatically depending on the speed and load conditions.

Attempts to control the wear of a material may involve tailoring materials properties or, more commonly, using a lubricant. Lubricants are useful for reducing friction, decreasing surface temperatures, removing wear debris, inhibiting surface reactions, and supporting part of the applied load through elastohydrodynamic effects. Under sufficiently high loads as in bearings and gears, a chemical reaction may occur resulting in a protective boundary-layer film that is a product of the lubricant, the surface material, and gaseous molecules. The effectiveness of boundary lubricants depends critically on the tribochemical reactions and requires highly specialized lubricant formulations that may depend on the chemical composition and the microstructure of the material.

Studies in tribology, therefore, are highly interdisciplinary and are directed toward the synthesis of a broad range of physical information to understand the mechanisms by which the interactions at interfaces lead to energy consumption and material losses.

See also FATIGUE; FRICTON.

BIBLIOGRAPHY

A. Beerbower, "On Becoming a Tribologist," in *Mechanical Engineering,* Vol. 107, No. 2, pp. 62–69. American Society of Mechanical Engineers, New York, 1985. A personal perspective on the evolution of tribology. (E)

E. R. Booser (ed.), *Theory and Practice of Tribology*. CRC Press, 1983. Fundamentals and design guides. (I)

F. P. Bowden and D. Tabor, *The Friction and Lubrication of Solids,* Parts I and II. Clarendon Press, Oxford, 1954, 1964. A comprehensive treatment of experimental studies of metals and some nonmetals. (I)

F. P. Bowden and D. Tabor, *Friction: An Introduction to Tribology*. Anchor, New York, 1973. A descriptive treatment. (E)

E. R. Braithwaite (ed.), *Lubrication and Lubricants*. Elsevier, Amsterdam, 1967. A descriptive treatment of the mechanisms of lubrication and the classes of lubricants. (I)

B. J. Briscoe and M. J. Adams, *Tribology in Particulate Technology*. Adam Hilger, Philadelphia, 1987. A collection of papers on tribological aspects of powders and powder compacts, slurries, densification, and cohesion. (I)

D. H. Buckley and E. Rabinowicz, "Fundamentals of the Wear of Hard Materials," in *Science of Hard Materials,* pp. 825–849. Adam Hilger, Bristol, 1985. The relation of wear to the mechanical properties and surface chemistry of hard metals. (I)

A. Cameron and C. M. Ettles, *Basic Lubrication Theory*. Ellis Horwood, Chichester, 1981. Theory of fluid film lubrication. (I)

H. Czichos, *Tribology*. Elsevier, New York, 1978. An interdisciplinary systems approach to understanding and using tribology fundamentals. (I)

D. Dowson, *History of Tribology*. Longman, Inc., New York, 1979. A review of progress and developments into the twentieth century. (E)

J. H. Dumbleton, *Tribology and Natural and Artificial Joints*. Elsevier, New York, 1981. A survey of the role of tribology in the performance of prosthetic materials and devices. (I)

T. S. Eyre, "The Mechanisms of Wear," in *Tribology International*, Vol. 11, pp. 91–96. Butterworth, Guildford, 1978. A classification of the wear of metals in industrial situations. (I)

R. S. Fein, "A Perspective on Boundary Lubrication," in *Industrial Engineering and Chemical Fundamentals*, Vol. 25, No. 4, pp. 518–524. American Chemical Society, Washington, DC, 1986. A qualitative review of boundary lubrication phenomena. (I)

R. Gohar, *Elastohydrodynamics*. Ellis Horwood, Chichester, 1988. Theory and experiments in film lubrication with descriptive illustrations. (I)

S. C. Lim and M. F. Ashby, "Wear-Mechanism Maps," *Acta Metall.* **35,** 1–24 (1987). A review of the dominant mechanisms of wear for metals under various operating conditions. (I)

F. F. Ling and C. H. T. Pan (eds.), *Approaches to Modeling of Friction and Wear*. Springer-Verlag, New York, 1988. Assessments of models of sliding friction and wear. (I)

D. F. Moore, *Principles and Applications of Tribology*. Pergamon, New York, 1975. Unified treatment of tribology theories and practical applications. (I)

R. G. Munro and S. M. Hsu, *Advanced Ceramics: A Critical Assessment of Wear and Lubrication*. U. S. Department of Commerce NISTIR 88-3722, 1989. A comprehensive review of issues and state-of-the-art tribology data on ceramics. (I)

M. B. Peterson and W. O. Winer (eds.), *Wear Control Handbook*. American Society of Mechanical Engineers, New York, 1980. A handbook of theory and engineering applications. (I)

E. Rabinowicz, *Friction and Wear of Materials*. Wiley, New York, 1965. Theories relating tribological and mechanical properties.

A. D. Sarkar, *Friction and Wear*. Academic, New York, 1980. An introduction to principles with industrial examples. (I)

L. M. Sheppard, "Ceramics for the Ultimate in Wear Resistance," in *Advanced Materials and Processes* **1**(1), 44–51 (1985). A descriptive survey of ceramics in advanced tribological applications. (E)

F. A. Smidt and P. J. Blau (eds.), *Engineered Materials for Advanced Friction and Wear Applications*. ASM International, Metals Park, OH, 1988. A collection of state-of-the-art papers. (A)

N. P. Suh, *Tribophysics*. Prentice-Hall, Englewood Cliffs, NJ, 1986. Conceptual aspects and fundamental understanding. (I)

Tunneling

J. M. Rowell

An insulator can be distinguished from a conductor by its inability to carry an electric current, that is, to allow the movement of electrons (see *Electron Energy States in Solids and Liquids*). However, if an insulator is made very thin, this simple distinction begins to fade. For example, if two copper wires are brought close together in a vacuum no current should flow through the vacuum, which is an insulator, until the wires actually touch each other; but when this experiment is carried out, a small current is observed when the wires are still about 50 Å (50×10^{-8} cm) apart, and this current increases rapidly as the wires are moved closer. This phenomenon of current flow through thin insulators, where such a flow would not be expected, is known as electron tunneling.

The concept of tunneling is much more general than this example and applies to all objects meeting a barrier, such as a ball thrown against a wall. In this case, if the wall is thin enough, the ball has a chance of traveling directly through the wall. This alien notion has its roots in quantum mechanics, and was used in the late 1920s and early 1930s to explain a number of phenomena, but was most dramatically illustrated for electrons tunneling through semiconductors by Esaki in 1957 and for electrons penetrating insulators by Giaever in the early 1960s. Giaever also showed that if these electrons tunneled either out of (or into) a superconducting metal (see SUPERCONDUCTIVITY THEORY), then remarkable information about the superconductor could be obtained. In 1962 Josephson predicted that the *pairs* of electrons present in superconductors could similarly tunnel through insulators (see JOSEPHSON EFFECTS; SUPERCONDUCTIVE DEVICES). For their work on electron tunneling, Esaki, Giaever, and Josephson were awarded the Nobel Prize in 1973.

To obtain an idea of how small this effect generally is, consider an electron approaching an insulator which presents a barrier of height h (eV) to its motion and is d Å thick. This is represented in Fig. 1, where the top of the barrier is the energy where electrons can move easily through the insulator, just as a ball can be thrown over a wall if given enough energy. The electron approaching the barrier with energy h (eV) less than the top of the barrier will always bounce back, however many times the approach is made, if it is considered in classical terms. But in quantum-mechanical language one would say that the wave nature of the electron does not stop abruptly at the insulator but decays through the barrier and, if this barrier is thin enough, does not disappear entirely even on the other side. Hence the electron has a chance of appearing on the opposite side. This chance, known as the tunneling probability for the electron through the barrier, can be written approximately as

$$P = \exp(-dh^{1/2}),$$

with d in Å and h in eV. Typically h is 2 eV, hence P is 10^{-62} when d is 100 Å, 10^{-18} when d is 30 Å, and 10^{-6} when d is reduced to 10 Å, that is, still only one in a million when the insulator is a few layers of atoms thick (the spacing between atoms in a crystal is often about 3 Å). With such a small chance for one electron to penetrate the barrier, it is surprising that tunneling can ever be observed, but if the single electron is replaced by a metal with many electrons (about 10^{22} in a cubic centimeter of the metal) continually attempting to tunnel, then a detectable current can be measured. For example, in Fig. 2a the same barrier is shown, but now metals A and B have a certain density of electron states, which is plotted horizontally against the energy on the vertical axis. The electrons in A and B occupy the lowest states, and these states are therefore filled up to the "Fermi energy" (see *Electron Energy States in Solids and Liquids*). When no voltage is applied from A to B, the Fermi levels of A and B are aligned, as in Fig. 2a. If, by means of an external circuit, a battery is connected across from A to B, then the Fermi level of A is raised relative to that of B. The difference

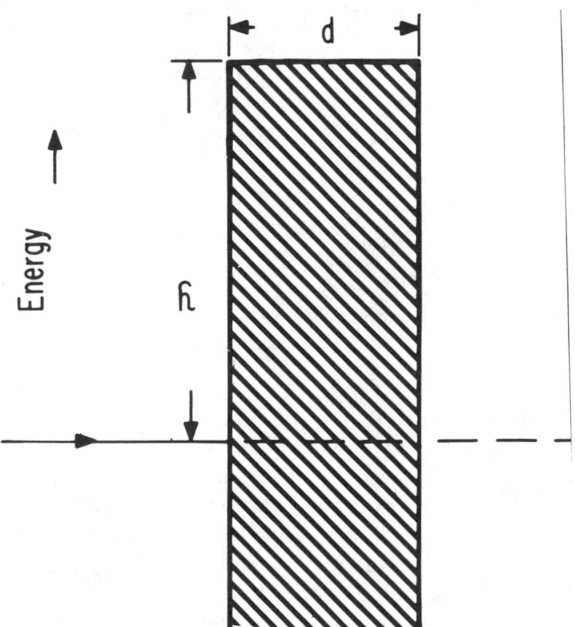

FIG. 1. Representation of an electron meeting a barrier of thickness d and height h above the energy of the electron.

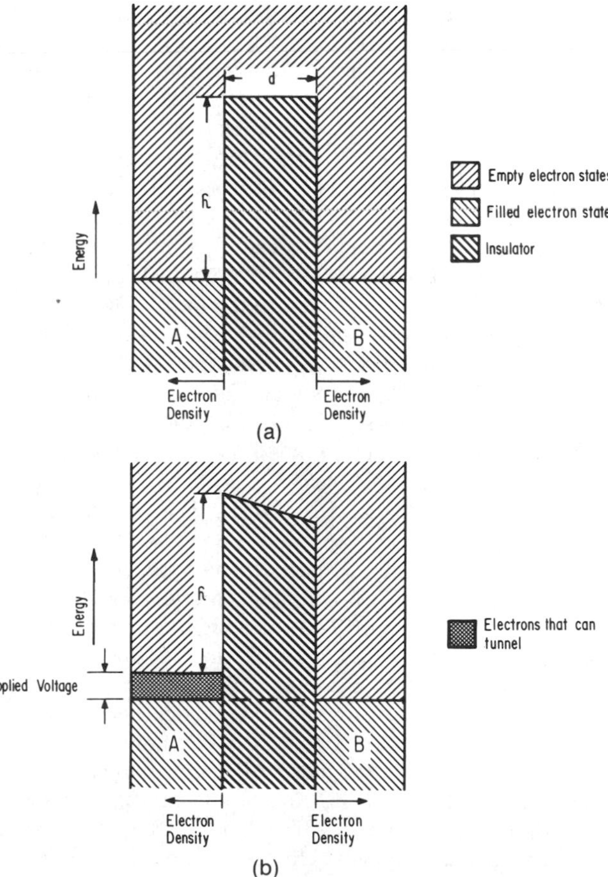

FIG. 2. Schematic of two metals A and B separated by an insulating barrier. The filled and empty electron states are shown. In (a) no potential difference has been applied between A and B, whereas in (b) an applied voltage has shifted the Fermi level of A above that of B. The electrons that can now tunnel are indicated.

in the two levels is equal to the applied potential difference from the battery, as shown in Fig. 2b. The result is that some electrons in *A* are now opposite to empty electron states in B. These electrons, indicated in Fig. 2b, each have a chance *P* of tunneling into *B*. A tunneling current therefore flows through the insulator, which is the net effect of very many electrons all trying to tunnel, but with any one electron having a minute chance of doing so. As long as the potential difference *eV* applied between *A* and *B* is a small fraction of the barrier height *h*, then the current density is

$$J = \left[3 \times 10^{10} \frac{h^{1/2}}{d} \exp(-dh^{1/2})\right] V \text{ A/cm}^2.$$

If *h* is again 2 eV and the voltage *V* is 0.01 V, *I* is 10^{-55} A/cm^2 when *d* is 100 Å, and increases to 5×10^{-12} A/cm^2 when $d = 30$ Å and to 30 A/cm^2 when *d* is 10 Å. This illustrates the point that very thin layers of insulator, approximately 10–30 Å thick, have to be made in order to see electron tunneling. If the potential difference *eV* is larger than the barrier height *h*, then the barrier is very distorted in shape, as shown in Fig. 3, and some electrons actually tunnel into allowed states in the insulator, rather than into *B*. As the thickness of the tunnel barrier is then changed by the applied voltage, the current increases exponentially as *V* is increased.

While it might appear impossible to prepare insulators as thin as 10–30 Å, this is not so as nature is, for once, cooperative. If a metal is heated in a vacuum, it first melts and then "boils," with metal atoms traveling in all directions through the vacuum. The atoms stick to whatever they meet, and if a piece of glass is placed in the vacuum it is soon covered with a film of metal. This process, known as vacuum evaporation, gives metal films with clean surfaces. If such a film, directly after being deposited, is exposed to the air, then it rapidly becomes covered with a layer of oxide which grows by reaction of oxygen with the metal. Oxides are generally good insulators, and Giaever showed that such oxide films were often of the correct thickness to allow electron tunneling. Conveniently, they are formed on a metal, which can then become metal *A* of Fig. 2 for example. For example, a freshly made aluminum film will grow a layer of aluminum oxide of the correct thickness if it is exposed to air for a few minutes. If a second metal film (*B* say) is then deposited over this oxide by vacuum evaporation, a sandwich of two metals separated by a thin oxide insulator is produced. This structure is generally called a tunnel junction and two typical examples are shown in Fig. 4. In Fig. 4a a cross geometry of two films is indicated, whereas in Fig. 4b the two films have been evaporated in a line. These are the two most common types of tunnel junctions using oxide insulators.

As he increased the voltage across aluminum–aluminum oxide–metal tunnel junctions of the kind shown in Fig. 4a, Giaever saw the two behaviors of current discussed above, the linear dependence at low voltages and the exponential increase at high voltages. However, it was still hard to believe that thin enough oxide layers had been made to allow electron tunneling to be observed. An even more convincing proof that tunneling through the oxide was actually taking place was obtained when one of the metal films was a superconductor. In such metals there is a gap (called the energy gap) in the electron states, as shown in Fig. 5. This gap occurs symmetrically about the Fermi level and has a total width 2Δ. The electrons which would occupy the region of states with Δ below the Fermi level are piled into a very strong narrow peak just below the gap, as shown. Similarly there is a peak of empty electron states just above the gap. In Fig. 5a there is no applied voltage between *A* and *B*. The electrons in the normal metal *A* cannot flow into the gap of superconducting metal *B*, as no electrons are allowed within this gap region. If a battery is now used to raise the potential

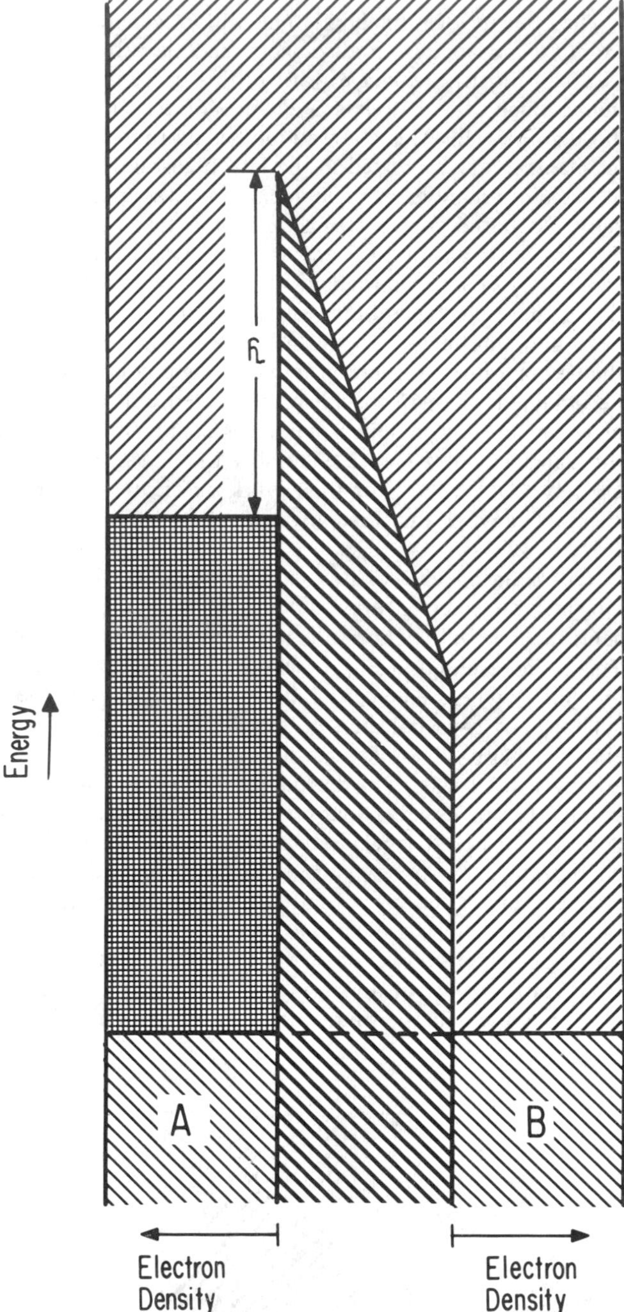

FIG. 3. The effect of applying a very large potential difference, greater than the barrier height, to a tunnel junction of the type shown in Fig. 2. The cross hatchings also have the same meanings as in Fig. 2.

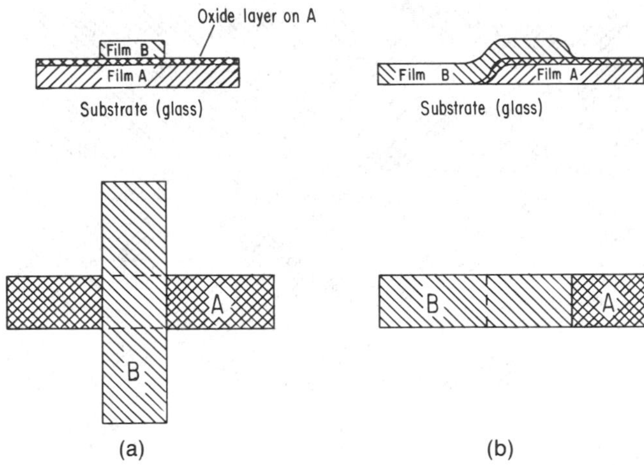

FIG. 4. Two common ways of making thin-film tunnel junctions are indicated, the cross and in-line geometries.

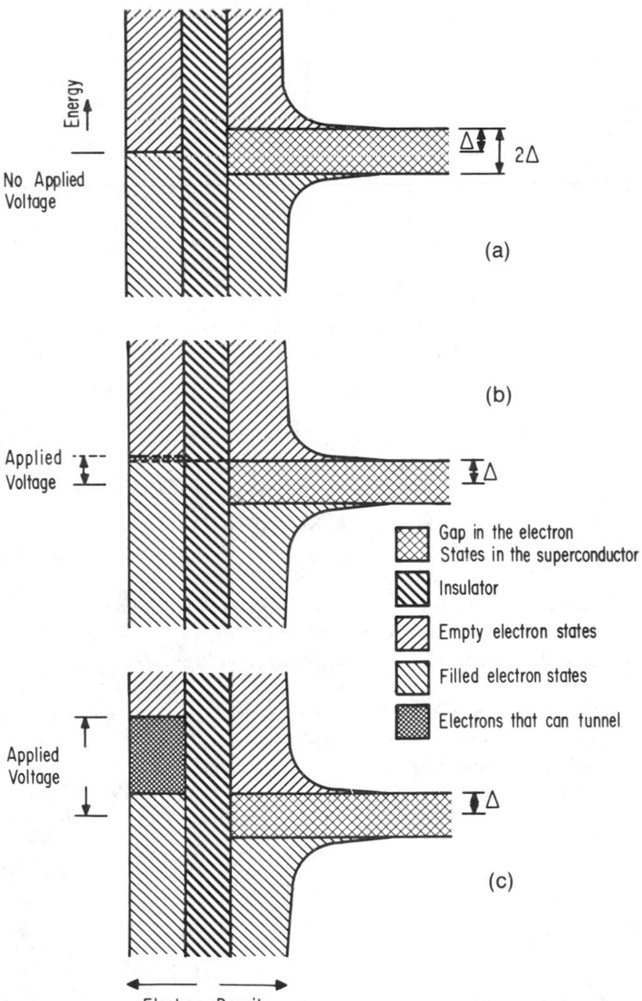

FIG. 5. Representation of a tunnel junction consisting of one normal metal and one superconducting metal, with an energy gap 2Δ. The effect of applied voltage, shown in (b) and (c), is to allow electrons to tunnel into states above the energy gap.

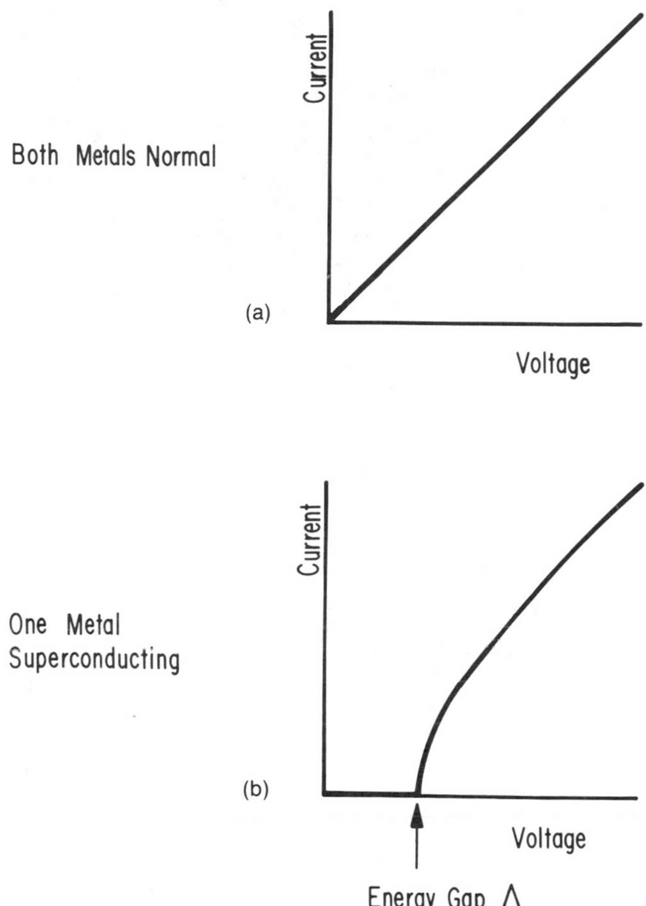

FIG. 6. The tunnel current resulting from a voltage applied to a junction, when both metals are normal (upper trace) and when one metal is superconducting (lower trace).

of A relative to B by an amount eV which is just greater than Δ, then a few electrons from A can tunnel into the large peak in the density of electron states in B. As the battery voltage is increased (Fig. 5c), electrons tunnel into states high above the gap and the effect of the gap on the total tunnel current is soon lost. In an experiment, the tunnel current is plotted against the voltage measured from A to B. If both metals are normal (i.e., not superconducting), the current rises linearly with voltage, as explained above and shown in Fig. 6a. But when metal B is cooled enough to be superconducting, then the energy gap opens. At very low temperatures the tunnel current is zero until the applied voltage is close to the energy gap and then the current rises rapidly as the situation of Fig. 5b is reached. Thus with a superconductor to tunnel into, the current depends on voltage as in Fig. 6b. In this way, Giaever observed the effect of the superconductor's energy gap in the tunneling current–voltage plot.

From the same arguments used for Figs. 2, 3, and 5, where the voltage from a battery can be regarded as having the effect of sliding the electron density of metal A up (or down) relative to B, it should now be possible to deduce that if *both* A and B are superconductors with equal energy gaps (Fig. 7a), then the current will depend on voltage in the way shown in Fig. 7b. However, an even more interesting effect in such

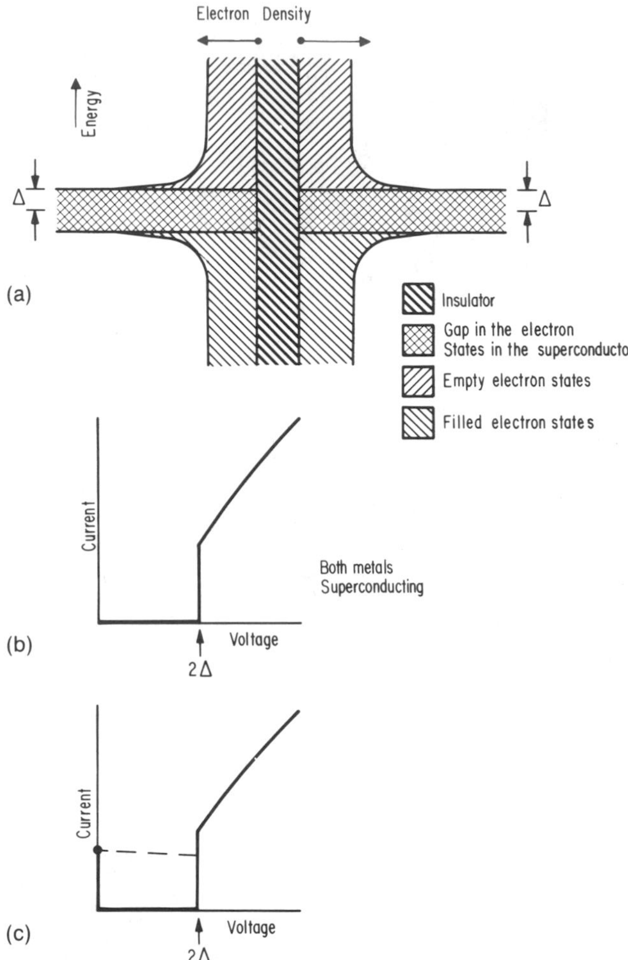

FIG. 7. (a) Representation of a tunnel junction with both metals superconducting. (b) The tunnel current against voltage plot for such a junction. (c) The tunneling of pairs of electrons results in a current flow at zero voltage, as shown.

a tunnel junction was predicted by B. D. Josephson in 1962. In superconductors there are pairs of electrons; and Josephson showed that, in addition to the tunneling of single electrons discussed above, these pairs of electrons should also be able to tunnel through the oxide without being broken apart. This tunneling of pairs, known as the Josephson effect, results in a number of phenomena, but the simplest is an extra current which flows, up to a certain limiting value, without any voltage appearing across the oxide insulator, as in Fig. 7c. The oxide can be said to have become a weak superconductor, in that it carries a "supercurrent."

The ideas of Figs. 2–5 serve as an introduction to tunneling in semiconductor diodes, which was actually the first system where electron tunneling was observed convincingly. Such a diode is shown in Fig. 8, where the electrons in regions of *p*- and *n*-type semiconductor (see SEMICONDUCTORS) are separated by the region where the material changes from *p* to *n* type. Again, by considering electrons from filled states in *A* tunneling to empty states in *B* when the applied voltage is in one direction, and electrons tunneling from *B* to *A* when the voltage is applied in the opposite direction, it will be seen that the current–voltage plot will have a region of "negative

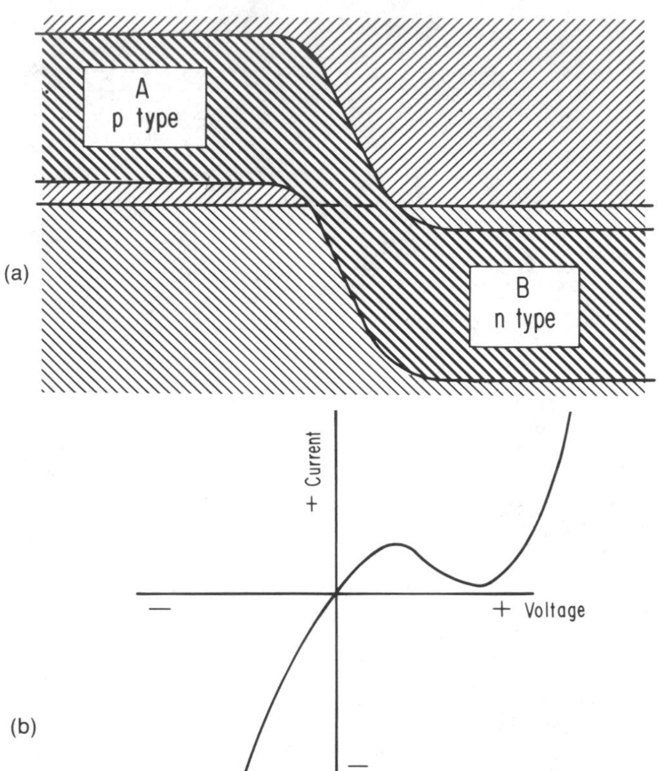

FIG. 8. A schematic of the filled and empty electron states in a junction of *p* and *n* semiconductor layers. The cross hatched regions have the same meanings as in Fig. 2. In the lower plot the dependence of the tunnel current on applied voltage is shown.

resistance" (where current decreases as voltage increases). This effect is indeed observed in tunnel diodes, also known as Esaki diodes, as illustrated in Fig. 8b.

See also ELECTRON ENERGY STATES IN SOLIDS AND LIQUIDS; INVERSION AND INTERNAL ROTATION; JOSEPHSON EFFECTS; SEMICONDUCTORS, CRYSTALLINE; SUPERCONDUCTIVE MATERIALS; SUPERCONDUCTIVITY THEORY.

BIBLIOGRAPHY

E. Burstein and S. Lundquist, (eds.), *Tunneling Phenomena in Solids*. Plenum Press, New York, 1969. (I,A)

C. B. Duke, *Tunneling in Solids*. Academic Press, New York, 1969. (A)

D. R. Lamb, *Electrical Conductive Mechanisms in Thin Insulating Films*. Methuen, London, 1967. (I)

L. Solymar, *Superconductive Tunneling and Applications*. Wiley, New York, 1972; Chapman and Hall, London, 1972. (I,A)

Paul K. Hausma (ed.), *Tunneling Spectroscopy*. Plenum Press, New York, 1982.

E. L. Wolf, *Principles of Electron Tunneling Spectroscopy*. Oxford University Press, New York, 1985.

Turbulence

C. H. Gibson

Turbulence is a random, rotational state of fluid motion that arises when shear instabilities from the nonlinear terms in the momentum equation overcome viscous damping at small scales to form eddy-like motions [1,9,12,14,15,17,18]. The eddies interact and grow to larger scales by entrainment until the vortex-inertial forces of the eddies are constrained by boundaries, or buoyancy, electromagnetic, or Coriolis forces, or by some other damping mechanism, and the kinetic energy of the turbulence is converted to internal waves, Coriolis-inertial waves, or other forms of nonturbulence. The standard state of natural fluids in motion such as the atmosphere, ocean, stars, and galaxies is a competition between turbulence and its constraints [16]. Without turbulence in a gravitational field, shears concentrate in the vicinity of those embedded surfaces most strongly buoyancy stabilized by vertical density gradients, forming vertical shear zones that approach nearly horizontal vortex sheets of ever-increasing shear instability. Finally, turbulence appears in patches that begin at the smallest (viscous) scales, entrain the kinetic energy of the shear zone, and mix away the stabilizing density gradients. The turbulence patches grow until the largest eddies, and eventually all eddies, are constrained by buoyancy and converted to internal waves. If horizontal shears exist, the entrainment process continues horizontally, forming anisotropic "2D" turbulence eddies that grow until the horizontal energy supply from the shear is exhausted or a boundary or Coriolis force constraint is encountered. Thereafter, eddy growth ceases and turbulence "collapse" begins as the eddy kinetic energy is converted to Coriolis-inertial waves (eddies), internal waves, and heat. Energy dissipation by friction continues within the turbulent patch throughout its evolution at a monotonically decreasing average rate per unit mass ϵ that tends to be homogenized by the stirring and thus characterizes the turbulence at every stage.

The most important practical property of turbulence is the strongly diffusive, stirring action of the turbulent eddies during its active phase. Turbulent diffusion is enormously more effective than molecular diffusion in most natural and engineered flows. Even though a small fraction of the ocean and atmosphere is actively turbulent at a given time, the diffusion and eventual mixing of fluid properties is completely dominated by turbulence through every layer except the earth–sea and sea–air interfaces for the ocean and at some unknown high altitude for the atmosphere. Much of the mixing and diffusion induced by turbulence may actually occur in patches of stirred and partially mixed (scrambled) scalar property microstructure that persist as fossil turbulence after the turbulence itself has been damped (see Fig. 1). Fossil turbulence is defined [7] as fluctuations of flow or physical fields produced by active turbulence which persist after the fluid is no longer actively turbulent at the scale of the fluctuations. Jet contrails above the inversion layer, refractive index fluctuations observed by radar scattering far downstream in the wake of mountains, and oceanic patches of temperature microstructure devoid of detectable velocity microstructure are examples of fossil turbulence. Active turbulence stirs and scrambles scalar fields like temperature in the ocean, but the irreversible mixing by molecular diffusion to produce entropy of mixing continues later in the fossil turbulence regime, and the diffusion continues by nonturbulent advection and intrusions.

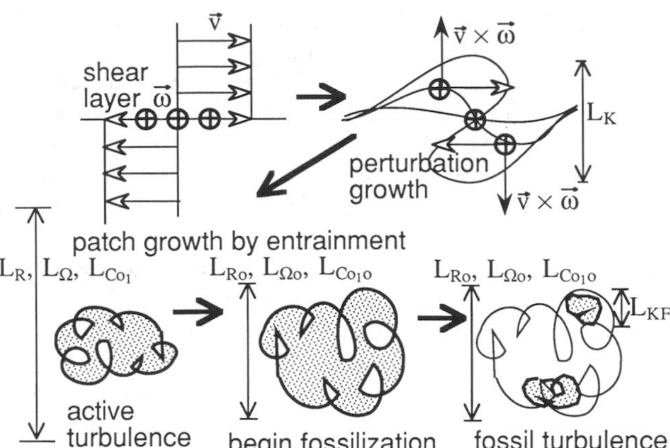

FIG. 1. Evolution of a turbulence patch. The shear layer in the upper left is unstable because inertial-vortex forces amplify perturbations to form eddies, as shown in the upper right. Eddies at the Kolmogorov scale form most rapidly, pair, and entrain nonturbulent fluid to form an actively turbulent patch, with L_O less than constraining scales as shown at lower left. When the largest eddy scale L_O equals a constraining scale fossilization begins, and the turbulence "collapses" leaving fossil turbulence remnants, or partially mixed microstructure in scalar fields like temperature, at larger scales.

A flow is actively turbulent only if the eddies are dominated by vortex-inertial forces of order $\rho U^2 L^2$, where ρ is the density of the fluid and U is a characteristic velocity on scale L. Eddies formed by viscous flow over wall cavities are nonturbulent because their viscous forces $\mu(U/L)L^2$ are larger than the inertial forces $\rho U^2 L^2$. Osborne Reynolds (1883) discovered the ratio of these forces $\rho UL/\mu = UL/\nu$ must be larger than about 2000 for turbulence to develop in pipe flow where μ is the dynamic viscosity and ν is the kinematic viscosity, so this dimensionless group is called the Reynolds number Re. Buoyancy forces of order $g\Delta\rho L^3 = gL^4\rho_{,z} = \rho N^2 L^4$ develop to resist the formation of turbulent eddies in a stable (downward) density gradient, where g is gravity, $N^2 \equiv g\rho_{,z}/\rho$ and $\rho_{,z}$ is the downward gradient of density. The Froude number $\mathrm{Fr} \equiv U/LN$ is the square root of the inertial force $\rho U^2 L^2$ divided by the buoyancy force $\rho N^2 L^4$. The Rossby number $\mathrm{Ro} \equiv U/Lf$ is $\rho U^2 L^2$ divided by the fictitious Coriolis force $\rho f UL^3$ for the flow in a rotating spherical coordinate system with Coriolis parameter $f = 2\Omega \sin\theta$, where Ω is the (constant) angular velocity and θ is the latitude. The first Cowling number $\mathrm{Co}_1 \equiv \sigma B^2 L/\rho U$ is the inverse of $\rho U^2 L^2$ divided by the electromagnetic force $\sigma UB^2 L^3$, where σ is the electrical conductivity and B is the magnetic field strength, and the magnetic Reynolds number $\mathrm{Re}_m \equiv UL/\eta$ is small as in laboratory flows of liquid metals, where η is the magnetic diffusivity. By the definition of turbulence given, in order for a rotating, stratified flow in a magnetic field to be considered turbulent, the Reynolds number, Froude number, Rossby number, and inverse first Cowling numbers of the flow must all be larger than critical values Re_{crit}, Fr_{crit}, Ro_{crit}, and Co_{1crit}^{-1}. For large Re_m, turbulence requires that a critical Alfvén number A_{crit} must be exceeded, where $A^2 \equiv \rho U^2/\sigma\eta B^2$ is the inverse second Cowling number Co_2^{-1}.

Many phenomena are identified as turbulence which are

not. Most "turbulence" motions referred to by aircraft crew members are actually internal waves, and most temperature and refractive-index fluctuations identified as turbulence by oceanographers and astrophysicists are probably fossil turbulence. Viscous eddies, Coriolis-inertial eddies, and geostrophic eddies are not turbulence. Just as all eddy motions are not turbulence, all random fluid motions are not turbulence. Internal waves and viscous sublayers with random forcing are random but are not turbulence. Because turbulence in natural fluids may be strongly constrained, and therefore patchy in space and intermittent in time, it is important to be able to distinguish turbulence from fossil turbulence to be sure that some of the scalar microstructure patches which dominate the average scalar dissipation rate, and therefore the average inducated scalar diffusivity, are in their original actively turbulent state, and thus avoid undersampling errors. The hydrodynamic state of buoyancy constrained 3D turbulence generated microstructure can be classified as active turbulence ($Fr/Fr_{crit}>1$, $Re/Re_{crit}>1$), active-fossil turbulence ($Fr/Fr_{crit}<1$, $Re/Re_{crit}>1$), or completely fossil turbulence ($Fr/Fr_{crit}<1$, $Re/Re_{crit}<1$) by means of hydrodynamic phase diagrams inferred from fossil turbulence theory which compare the Froude number of the microstructure flow to the critical value Fr_{crit} at the buoyant–inertial transition, and the Reynolds number to the buoyant–inertial–viscous transition value Re_{crit}. The flow of mesostructure fluctuations produced by 2D turbulence in rotating and magnetohydrodynamic flows may be similarly classified with hydrodynamic phase diagrams using Ro/Ro_{crit} and $Co_1-1/Co_{1crit}-1$ compared to Burger number ratios Bu/Bu_{crit}, where $Bu \equiv L_O/L_{crit}$ is a ratio of turbulent energy to turbulent transition-length scales.

Several transition-length scales arise in turbulence and turbulent mixing. The best known is the Kolmogorov scale $L_K \equiv (\nu^3/\epsilon)^{1/4}$, where $\epsilon \equiv 2\nu e_{ij}^2$ and e_{ij} are the components of the rate-of-strain tensor $(u_{ij}+u_{j,i})/2$ which marks the inertial-viscous transition at Re_{crit}. It may be derived by setting the inertial force ρU^2L^2 equal to the viscous force $\mu(U/L)L^2$ at small scales noting that the velocity scale $U \sim Le_{ij} \sim L(\epsilon/\nu)^{1/2}$ at viscous scales, and solving for the turbulence transition length L_K. The buoyancy–inertial transition length, or Ozmidov scale, $L_R \equiv (\epsilon/N^3)^{1/2}$ is derived by setting ρU^2L^2 equal to the buoyancy force ρN^2L^4 using the Kolmogorov–Oboukhov law $U \sim (\epsilon L)^{1/3}$ to find the turbulent velocity scale U. L_R is an upper bound for turbent energy, or Oboukhov, scales L_O in stratified fluids. The buoyancy–inertial-viscous scale $L_{KF} \equiv (\nu/N)^{1/2}$ is the size of the smallest eddy in a fossil turbulence patch at the point when fossilization is completed at all scales (L_{KF} is $(\nu/f)^{1/2}$ and $[\nu/(\sigma B^2/\rho)]^{1/2}$ for Coriolis or electromagnetic constraints at the Coriolis-inertial–viscous and electromagnetic-inertial–viscous points, respectively). L_{KF} is derived by setting all three forces equal, and occurs when $\epsilon=\epsilon_F \approx 30\nu N^2$ (or $\epsilon_F \sim \nu f^2$ or $\epsilon_F \sim \nu(\sigma B^2/\rho)^2$). The Coriolis-inertial scale $L_\Omega \equiv (\epsilon/f^3)^{1/2}$ arises by setting the inertial force ρU^2Lh of a 2D turbulent eddy of horizontal extent L and vertical extent h equal to the Coriolis constraining force ρfUL^2h with $U \sim (\epsilon L)^{1/3}$. L_Ω is an upper bound for horizontal turbulent energy scales in a rotating flow with vertical rotation vector. Similarly, for $Re_m \ll 1$ the electromotive-inertial scale $L_{Co} \equiv [\epsilon/(\sigma B^2/\rho)^3]^{1/2}$ is an upper bound for turbulent energy scales in the plane perpendicular to a constant magnetic field $\mathbf{B}$. For $Re_m \ll 1$, magnetic forces do not constrain large-scale turbulent motions but have complex effects at small scales as the magnetic field is scrambled and locally intensified. Turbulence-wave transitions occur when the energy scale L_O of a growing turbulence patch with decreasing ϵ equals one or the other of the constraining scales L_R, L_Ω, or L_{Co} (all $\sim\epsilon^{1/2}$) at which fossilization of the turbulence begins. The hydrodynamic state of fluid motions can also be classified by comparing turbulent energy, overturn, and transition scales; for example, a buoyancy constrained flow with constant N is actively turbulent if $L_R>L_O>L_K$, active fossil if $L_{Ro} \approx L_T > L_R \approx L_O > L_{KF}$, and fossil if $L_{Ro}>L_{KF}>L_R$, where L_{Ro} indicates the Ozmidov scale when fossilization begins with $\epsilon=\epsilon_0>\epsilon_F$ (ϵ_0 is inferred from the overturn scale L_T of the fossil turbulence patch since $L_T \approx L_{Ro}$ after fossilization and $L_T \approx L_O$ before fossilization). Froude number and Reynolds number ratios can be estimated from the length scales using $Fr/Fr_{crit} \approx (L_R/0.6L_T)^{3/2}$ and $Re/Re_{crit} \approx (L_R/5L_{KF})^2$ [7].

The momentum equation describing rotating, stably stratified, incompressible flows such as the ocean and atmosphere is

$$\frac{\partial \mathbf{v}}{\partial t} = -\nabla B + \mathbf{v}\times\boldsymbol{\omega} + \mathbf{g}' - 2\boldsymbol{\Omega}\times\mathbf{v} + \nabla\cdot\left(\frac{\boldsymbol{\tau}}{\rho}\right), \quad (1)$$

where $\mathbf{v}$ is the velocity, $B \equiv \nabla p/\rho + v^2/2 + gz$ is the Bernoulli group of mechanical energy terms, $\boldsymbol{\omega}$ is the vorticity, $\mathbf{v}\times\boldsymbol{\omega}$ is the inertial-vortex force, $\mathbf{g}' \equiv [(\rho-\rho_0)/\rho_0]\mathbf{g}$ is the force of gravity per unit mass, $\mathbf{g}$ is gravity, ρ_0 is the ambient density of depth z, $-2\boldsymbol{\Omega}\times\mathbf{v}$ is the Coriolis-inertial force, $\boldsymbol{\tau}$ is the viscous stress tensor, and $\nabla\cdot(\boldsymbol{\tau}/\rho)$ is the viscous force per unit mass. Turbulence occurs because shear layers are absolutely unstable, meaning that the direction of the vortex-inertial forces that arise when a shear layer is perturbed are in the direction of the perturbation and will thus be amplified. Figure 1 is a schematic representation of the evolution of a turbulence patch in a fluid flow with constraints from the original vortex sheet to complete fossilization, as described above.

The first papers on turbulence [4] displaying significant insight into the appropriate physical and mathematical formulation of the problem were by G. I. Taylor in 1921 ("Diffusion by Continuous Movements") and in 1935 ("Statistical Theory of Turbulence"). In the second paper, actually a series of four, Taylor pointed out the tendency of turbulence toward isotropy and the fact that the mean dissipation rate per unit mass ϵ depends on the characteristic velocity U and the length scale L as $\epsilon \sim U^3/L$, but is independent of viscosity. This equation is sometimes called the Kolmogorov–Oboukhov law [12]. On the basis of these ideas Kolmogorov (1941) [4] proposed universal similarity hypotheses for the probability functions describing velocity differences over various separation distances in high Reynolds number turbulence. The first Kolmogorov similarity hypothesis assumes a universal critical Reynolds number for the relative velocity at small separation differences in fully developed turbulence, so that the flow will be turbulent in regions larger than the Kolmogorov length scale L_K, where $L_K \equiv (\nu^3/\epsilon)^{1/4}$ and ν is the kinematic viscosity, and it will be laminar, or viscous, on smaller scales. Transformation of length and time scales to values normalized with the Kolmogorov length L_K and time $T_K \equiv (\nu/\epsilon)^{1/2}$ should therefore cause the form of the probability laws for small-scale velocity differences to collapse to universal forms for all turbulent flows with separation lengths smaller than the scale of the largest eddy L_O.

Kolmogorov's second similarity hypothesis assumes not only homogeneity and isotropy, but that the separation distances are large compared to L_K (but still smaller than L_O). In this case, the probability laws should be independent of viscosity, reproducing Taylor's result that the characteristic velocity difference $U \sim (\epsilon L)^{1/3}$. Kolmogorov's hypotheses are quite general. Since the probability laws are universally similar, all resulting statistical parameters such as power spectra, cross-spectra, correlation functions, or structure functions must also have universal forms when the coordinates are transformed by Kolmogorov length and time scales. A refinement, sometimes called Kolmogorov's third similarity hypothesis (1962) [10], was advanced 20 years later to take into account the tendency of the local dissipation rate to become increasingly intermittent (patchy) in space and time with increased Reynolds number. The effect of intermittency corrections is imperceptibly small for most predictions of the first two hypotheses, at least for the lower-order statistical parameters like spectra that have been most thoroughly tested by measurements.

Because of its four-dimensional stochastic nature, the general mathematical description of turbulence can be formidable. Manipulation of the unbounded set of tensor relations and their isotropic invariants is a challenging mathematical task [1,9,13,17]. Selecting a measurable subset of such parameters with physical significance is the essence of most experimental turbulence research [2,3,14,15,18]. The minicomputer has become an indispensable tool in experimental turbulence by permitting estimation of many previously inaccessible statistical parameters from measurements and by permitting the high speeds of complex data acquisition and processing necessary to adequately characterize turbulent flows. The supercomputer is rapidly developing the capacity to completely simulate turbulence and turbulent mixing [8,11] with sufficiently large Reynolds numbers to be relevant to more than simply transition processes. This is particularly valuable when applied to problems such as turbulent combustion, high Mach number turbulence, flows with peculiar boundary conditions, and mixing of low Prandtl number scalars which are difficult to study experimentally. Simulations using approximate forms of the nonlinear equations describing turbulence of ever-increasing sophistication have been made possible with supercomputers and give ever-improved insights into the complexities of turbulent flows.

For practical applications in engineering, meteorology, oceanography, combustion, and chemically reacting systems, the more complicated turbulent transport and diffusive processes are often dealt with in an empirical manner. Empirical data and theoretical models are used to identify critical processes and the relevant dimensional parameters, and these are organized by dimensional analysis. This results in various dimensionless functions and parameter curves of drag, heat, and mass transfer coefficients in terms of the remaining relevant dimensionless groups. In a more refined approach, the describing equations are "Reynolds averaged" about various means and trends peculiar to the process to give "Reynolds fluxes" of properties down mean gradients proportional to "eddy diffusivity" coefficients. Such attempts to model turbulent diffusion by analogy with molecular diffusion have met with limited success because of the great differences in physical mechanisms involved, particularly the fact that the "mixing length" for turbulence depends on the flow and tends to expand to fill the space available unless constrained. The analogy improves for turbulent boundary layers and turbulent diffusion over scales larger than a constraining scale, such as vertical diffusion in the ocean and atmosphere through layers much thicker than maximum values of L_R.

Turbulent mixing of dynamically passive scalar properties of a fluid such as temperature or species concentration can also be described by universal similarity hypotheses [5,6] using many of the same concepts introduced by Kolmogorov for the velocity field. Interaction of turbulence with dynamically active scalar and vector fields such as density, strong electric or magnetic fields, or chemically reacting species are obviously subjects of great importance, but are quite poorly understood at present.

See also FLUID PHYSICS; HYDRODYNAMICS; MAGNETOHYDRODYNAMICS; STOCHASTIC PROCESSES; VORTICES.

REFERENCES

1. G. K. Batchelor, *Theory of Homogeneous Turbulence*. Cambridge University Press, London, 1967.
2. P. Bradshaw, *An Introduction to Turbulence and its Measurement*. Pergamon Press, London, 1971.
3. S. Corrsin, "Turbulence: experimental methods," in *Handbuch der Physik*, Vol. 8, Pt. 2, pp. 524–590. Springer-Verlag, Berlin, 1963.
4. S. K. Friedlander and L. Topper (eds.), *Turbulence: Classic Papers on Statistical Theory*. Interscience, New York, 1961; G. I. Taylor, "Diffusion by Continuous Movements," *Proc. London Math. Soc. Ser. 2*, **20**, 196–211 (1921); G. I. Taylor, "Statistical Theory of Turbulence. Parts I–IV," *Proc. R. Soc.* **A151**, 421–478 (1935); A. N. Kolmogorov, "The Local Structure of Turbulence in Incompressible Viscous Fluid for Very Large Reynolds Numbers," Comptes Rendus (Doklady) de l'Academie des Sciences de l'U. R. S. S. **30**, 301–305 (1941).
5. C. H. Gibson, "Fine Structure of Scalar Fields Mixed by Turbulence: I. Zero-Gradient Points and Minimal Gradient Surfaces," *Phys. Fluids* **11**, 2305–2315 (1968).
6. C. H. Gibson, "Fine Structure of Scalar Fields Mixed by Turbulence: II. Spectral Theory," *Phys. Fluids* **11**, 2316–2327 (1968).
7. C. H. Gibson, "Internal Waves, Fossil Turbulence, and Composite Ocean Microstructure Spectra," *J. Fluid Mech.* **168**, 89–1171 (1986).
8. C. H. Gibson, W. T. Ashurst, and A. R. Kerstein, "Mixing of Strongly Diffusive Passive Scalars Like Temperature by Turbulence," *J. Fluid Mech.* **194**, 261–293 (1988).
9. J. O. Hinze, *Turbulence*, 2nd ed. McGraw-Hill, New York, 1975.
10. A. N. Kolmogorov, "A Refinement of Previous Hypotheses Concerning the Local Structure of Turbulent Boundary Layers," *J. Fluid Mech.* **30**, 741–776 (1962).
11. R. M. Kerr, "Higher-Order Derivative Correlations and the Alignment of Small-Scale Structures in Isotropic Numerical Turbulence," *J. Fluid Mech.* **153**, 31–58 (1985).

12. L. D. Landau and E. M. Lifshitz, *Fluid Mechanics*. Pergamon Press, London, 1959.
13. J. Lumley, *Stochastic Tools of Turbulence*. Academic Press, New York, 1970.
14. A. S. Monin and A. M. Yaglom, *Statistical Fluid Mechanics,* Vol. 1. The MIT Press, Cambridge, MA, 1971.
15. A. S. Monin and A. M. Yaglom, *Statistical Fluid Mechanics,* Vol. 2. The MIT Press, Cambridge, MA, 1975.
16. O. M. Phillips, *The Dynamics of the Upper Ocean*. Cambridge University Press, London, 1966.
17. H. Tennekes and J. L. Lumley. *A First Course in Turbulence*. The MIT Press, Cambridge, MA, 1972.
18. A. A. Townsend, *Structure of Turbulent Shear Flow*. Cambridge University Press, London, 1976.

Twin Paradox

Joseph Dreitlein

A direct and unambiguous consequence [1] of the theory of relativity is the relativity of the time interval between two different events. If two identically constructed clocks move between two events in space-time along different world lines, the time lapse measured by each of the two will in general be different. This conclusion, so counter to the intuitive Newtonian concept of space-time structure, is paradoxical.

If, to accentuate the paradox, the clocks are taken to be the biological clocks of two identical twins, one accelerating twin can travel into the other twin's distant future with an arbitrarily short lapse of his own *biological* time. It is implicitly assumed that living processes, including consciousness and cognition, proceed at a rate completely determined by physical processes such as chemical reaction rates. However, no direct experimental verification of this relativity phenomena for biological processes is currently available.

Two distinct cases of the twin paradox are distinguished, depending on whether a true gravitational field characterized by tidal forces (a nonzero Riemann tensor) is present. In the absence of a gravitational field, the (proper) time τ read by a clock that moves between the two events of departure (D) and rendezvous (R) is given by

$$\tau = \int_D^R \left(1 - \frac{v^2}{c^2}\right)^{1/2} dt. \tag{1}$$

The time t is the synchronized time read by clocks in a chosen inertial frame, and $\mathbf{v}(t)$ records the history of the velocity of the moving clock measured in this frame. The longest time recorded by a clock (A) is that of a clock at rest and experiencing no acceleration between D and R. Another clock (B) traveling between D and R along a different world line must accelerate somewhere along the trajectory. Special relativity fully and adequately treats accelerated motion with the experimentally tested hypothesis: the instantaneous rate of a shock-proof accelerating clock is the same as the rate of a nonaccelerating clock with respect to which it is instantaneously at rest. General relativity, the theory of gravity, contrary to opinions often voiced, has no additional light to cast upon accelerated motion in situations where true gravitational effects on the motion are negligible. Since, in the inertial reference system in which A is at rest, the acceleration of B implies that $\mathbf{v}(t)$ is somewhere nonzero, the clock B will necessarily measure a shorter lapse of time between D and R than A, whose velocity $\mathbf{v}(t)$ is always zero; see Eq. (1). To mitigate the force of the paradox, it might be remarked that each twin can absolutely detect with accelerometers a different history of acceleration. The relativity of duration [2], in effectively gravitationally free situations, has been repeatedly experimentally tested by observing both the dilation in the lifetime of swiftly moving unstable particles and the transverse Doppler effect, logically equivalent to time dilation.

In the presence of gravity, clock rates exhibit a behavior, even more counterintuitive. Clocks close to a gravitating body run more slowly than distant clocks, as evinced by comparisons carried out theoretically using coordinate time units measuring time lapse at coordinate infinity. The observed gravitational redshift confirms the predicted behavior [3]. Furthermore, two twins in a gravitational field can simultaneously depart (event D), execute different, *possibly free-fall* orbits, and meet again (event R) but with different aging [4]. Doppler frequency shifts of light from distant stars will convince the twins that they executed different orbits even in the case of zero accelerometer readings of each. The general expression for the proper time lapse between two events (D and R) is

$$\tau = \int_D^R \left(g_{\mu\nu} \frac{d}{d\lambda} x^\mu \frac{d}{d\lambda} x^\nu\right)^{1/2} d\lambda. \tag{2}$$

Here $x^\mu(\lambda)$ is the space-time coordinates of the orbit parameterized with the parameter λ. It is of significance to note that, in contrast to special relativistic situations, the value of τ need not be a maximum but only a stationary value for a free-fall orbit. There exist now several experimental tests concerned with these clock comparisons [5].

The most paradoxical time behavior occurs in certain geometries of space-time that are solutions of the Einstein field equations. In the Gödel universe [6], for example, no unambiguous concept of the past or future exists. There are world lines that take an observer backward in time, so that one twin could travel into another twin's past. The paradoxes associated with such possibilities are so counter to the commonly held concept of time that such models are frequently declared unphysical by hypothesis. Yet it remains to be seen whether future radical restructuring of the concepts of simultaneity, temporal duration, and temporal ordering will destroy the remnants of the intuitive grasp of time acquired from everyday experience.

See also RELATIVITY, GENERAL; RELATIVITY, SPECIAL THEORY; SPACE-TIME; TIME.

REFERENCES

1. A. Einstein, *The Principle of Relativity,* p. 35. Dover, New York.
2. L. Marder, *Time and the Space-Traveller*. University of Pennsylvania Press, Philadelphia, 1974. Over 300 references on the twin paradox are cited in this book.
3. Charles W. Misner, Kip S. Thorne, and John Archibald Wheeler, *Gravitation,* p. 1055. Freeman, San Francisco, 1971.
4. Joseph Dreitlein and John Frazzini, *Am. J. Phys.* **43,** 596 (1975).
5. A. Held (ed.), *General Relativity and Gravitation,* Volume 2, p. 472. Plenum Press, New York, 1980.
6. Kurt Gödel, *Rev. Mod. Phys.* **21,** 447 (1949). See also Kurt Gödel, *Albert Einstein: Philosopher Scientist,* Vol. 2, p. 555. Open Court, La Salle, IL, 1970.

Ultrahigh-Pressure Techniques

John Schroeder and George C. Kennedy†

Ultrahigh pressure is the pressure region from 500 kbar (50 GPa) up to about 2.5 Mbar (250 GPa). This discussion is limited to five fundamentally different types of apparatus that allows one to cover the pressure range from about 10 kbar (1 GPa) to 2.5 Mbar (250 GPa). A further division will be made by discussing optical access (including x-rays) or electrical access to the high-pressure cell. We do not cover the vast array of equipment that is available for the pressure range less than 10 kbar and consists mostly of some type of cell where pressure is generated by hydraulic means.

Several distinctly different types of systems are readily available for research in the pressure range from 500 kbar (50 GPa) down to 10 kbar (1 GPa). The 500 kbar to 10 kbar region may be explored by the use of an opposed beveled diamond anvil device or a diamond indenter–diamond anvil apparatus. In that pressure range, the single crystal diamond device has competition from the opposed anvil device made of diamond compact (100–500 kbar) and the compressible gasket apparatus (100–400 kbar), while the extruded gasket apparatus (60–100 kbar) and the piston–cylinder apparatus (10–60 kbar) are applicable at lower pressures. Above 500 kbar the diamond anvil device and diamond indenter–diamond anvil apparatus are unique, and up to now pressures of 2.5 Mbar (250 GPa) have been reached with such pressure cells. For completeness we note that shock velocity measurements enable pressure–volume measurements at pressures as high as 200 GPa, but this article will be limited to the discussion of static generation and measurement techniques.

SINGLE CRYSTAL BEVELED DIAMOND-OPPOSED ANVIL APPARATUS

The first use of single-crystal diamonds as opposed anvils for high-pressure cell construction was by Van Valkenburg [1], but the device did not reach wide scale use until Piermarini et al. [2] showed that the frequency shift of the ruby R_1 and R_2 fluorescence lines under pressure is usable as a secondary pressure calibration. The use of opposed diamond anvil cells above 1 Mbar in pressure was proposed and demonstrated by Mao, Bell, Shaner and Steinberg [3] around 1978. The opposed diamond anvil cell lends itself very nicely to many forms of optical probing, since the single-crystal diamonds are good optical windows through which the samples in the pressure cell can be monitored by various optical techniques, such as Brillouin scattering, Raman scattering, resonant Raman scattering, photoluminescence, transmission, and absorption measurements. Nuclear spectroscopy, such as Mössbauer measurements and positron annihilation studies, has been done up to high pressure using diamond anvil cells. True hydrostatic conditions (no pressure gradients) in this pressure cell can be maintained up to at least 600 kbar and a quasihydrostatic environment at least up to the Mbar range. The two major drawbacks of the diamond anvil pressure cell are the extremely small sample volume ($\sim 10^{-6}$–10^{-7} cm^3) and the difficulty of bringing electrical leads into the small sample space especially at the higher pressures. Either a nonmetallic gasket must be employed, limiting the pressure range to less than 300 kbars, or the lead-in wire must be electrically isolated from the gasket and diamond surfaces, again a high-pressure limiting process.

The principle and design of the diamond anvil apparatus is quite simple, but assembly and alignment for the highest pressures is often a highly taxing affair. Either a single-levered (Fig. 1a) or double-levered mechanical press (Fig. 1b) is used on the cylinder-piston device holding the opposed diamond anvils. The two single-crystal diamond anvils are of gem quality [4], brilliant cut, flawless, and without visible inclusions. For Brillouin scattering (or any process where luminescence is of no concern) diamonds with nitrogen platelets are desirable, since they are somewhat stronger than the ultrapure type IIa, which must be used for photoluminescence or high frequency Raman work. These diamonds are glued on cemented tungsten carbide (WC) rocker-backing blocks with epoxy or other adhesive materials. Between the diamond table and the WC rocker a thin sheet of zirconium metal may be placed to diminish any effects of surface irregularities in the cemented tungsten carbide rockers. These rockers are situated on the top of the piston and the bottom of the cylinder. The fit between cylinder and piston must be ground and polished to have a clearance that is around 8 to 10 μm. The diamond anvils can be made concentric by horizontal translational motion of the backing blocks in perpendicular directions to each other. The diamond anvil faces can be made parallel by rotating the tungsten carbide rockers about their specific azimuth angles. The piston and cylinder are fitted in the lever device where the fulcrum is at the extreme right of the main body, the pushing block for the piston near the fulcrum and the advancing screw to the left. The advancing screw is kept tight by a series of Belleville washers that are compressed as the screw is tightened, the force coupling block is pulled up, and the force is transmitted to the lever arm. A more mechanically balanced configuration is the version with two lever arms situated symmetrically on both sides of the piston-cylinder assembly. Both the single lever and double lever device have a lever ratio of about 1:5. Figure 1a portrays a typical single levered diamond anvil press device with the details of the piston-cylinder of the diamond anvil press shown in Fig. 2.

† Deceased

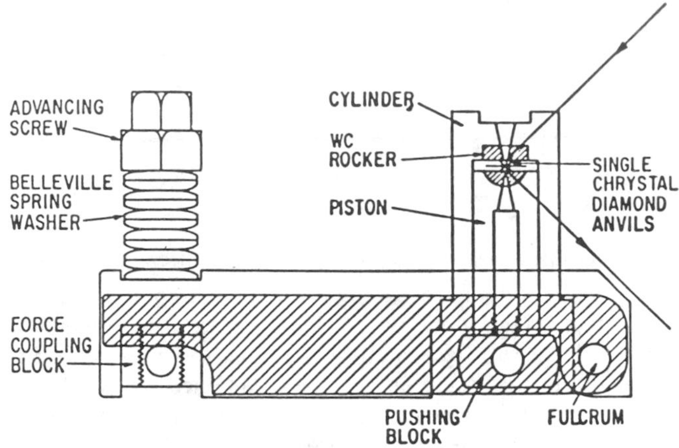

FIG. 1a. Single-levered diamond press.

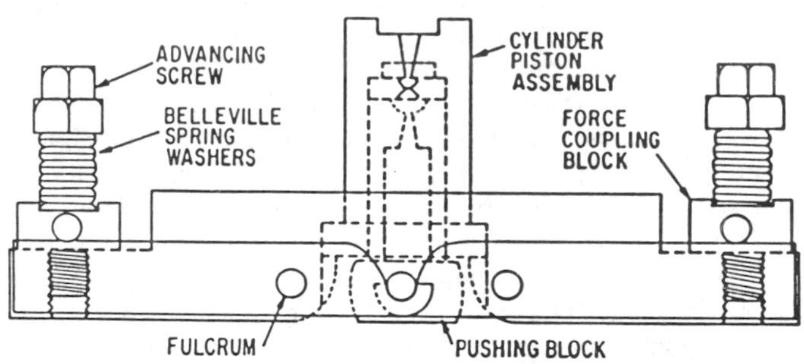

FIG. 1b. Double-levered diamond press.

Diamond anvil presses may be used for x-ray diffraction studies at high-pressure or for Brillouin or Raman spectroscopy (also photoluminescence). The typical configuration for x-ray diffraction is shown in Fig. 3. Here an 80° angled slot was made at the end of the cylinder with a corresponding slot in the upper WC rocker. The x-ray beam can be adjusted to pass the center of the diamond anvil faces by using a collimator made of a lead glass thermometer tube which is contained in the piston. For Brillouin scattering a typical pressure cell configuration with six optical access ports is shown in Fig. 2. Brillouin scattering allows the measurements of the sound velocity as a function of pressure in one configuration, and sound velocity and refractive index as a function of pressure with a different geometrical set up as given in Fig. 2. The empirical equation of state of the compressed material inside the diamond anvil cell may thus be calculated from the Brillouin scattering results. Raman scattering and photoluminescence measurements as a function of pressure may also be done with the configuration in Fig. 2.

Critical to achieving ultrahigh pressure in a beveled opposed diamond anvil cell is the alignment of the diamond pressure faces and the gasket material choice and preparation. The more accurately parallel the diamond anvil faces (culet), the higher the pressures that may be produced. Proper beveling of the anvil faces also influences the maximum pressure value, since from finite element analysis of the elasticity of diamonds, the diamond anvils with beveled anvil faces can take more face pressure with less octahedral shearing stress along the anvil faces than can anvils with flat faces [5]. Bruno and Dunn [5] find that an optimum bevel angle of 15° does give a minimization of the stresses at the outer corner of the diamond faces.

The parallelism of the diamond faces may be tested by simply bringing the diamond faces in contact with each other. If no interference fringes are evident, then the anvil faces (culet faces) are parallel to each other within roughly 1 μm. If interference fringes exist, then realignment of the anvil faces is necessary. Anvil alignment may also be tested by using a film of silver iodide (AgI) between the pressure faces. AgI has the zincblende structure at atmospheric pressure and room temperature, but it transforms to the rock salt structure (NaCl) at about 5 kbar with a large red shift of the absorption edge. The red shift continues with increasing pressure until 97 kbar, where a first order phase transition with a distinct blue shift occurs [6]. The 5 and 97 kbar phases are distinctly visible under a microscope with transmitted light illumination. Hence, the degree to which both phase boundaries are centrosymmetric may then be used to determine the parallelism of both pressure faces to each other. Rotation of both WC rockers allows precise adjustment of the parallelism of both pressure faces.

The choice and preparation of the proper gasket material is crucial to attaining the highest hydrostatic pressures with a diamond anvil cell. The gasket needs to be as strong and as tough as possible, yet capable of plastic deformation to provide an even pressure distribution on the anvil face and to also provide the side support for the tapered pavilion faces of the diamond anvils. Satisfactory candidates for gasket material are work-hardened stainless steels, Inconel, Udiment-

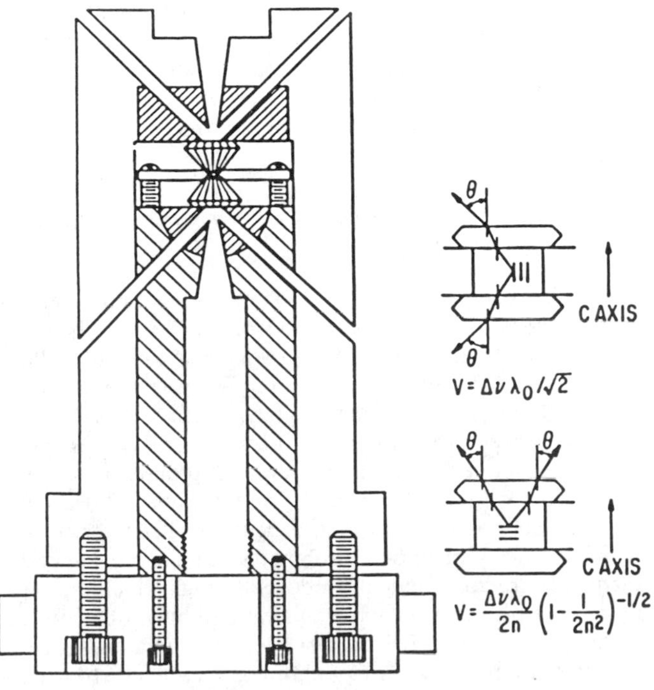

FIG. 2. Diamond press for Brillouin and Raman scattering studies with six optical access ports.

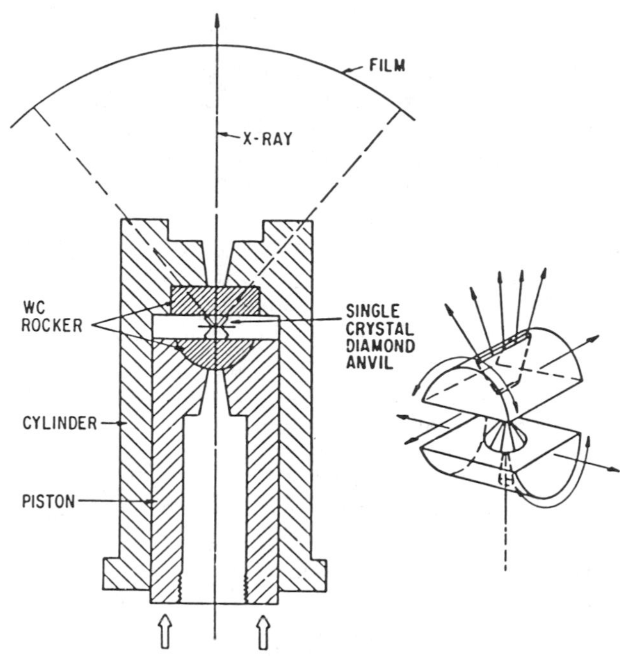

FIG. 3. Diamond press for x-ray diffraction studies.

700, René 41, Nimonic-90, 17-7 PH (precipitation hardened) stainless steel, etc. Figure 4 gives a sectional view of a typical deformed metal gasket without a bevel angle on the diamond anvil face. Several techniques are presently in use to make the sample chamber in the metal gasket. One method is to pre-indent a circular flat hole on the metal gasket with the diamond anvils and then drill a hole of about 127 μm diameter through the flat area. A second method would be to punch a 250 μm diameter hole in a 0.152 mm-thick metal sheet. The gasket is placed between the anvils and centered with the anvil faces and the hole is filled with a suitable liquid to prevent inward collapse. Pressure is applied so that the metal gasket intrudes inward symmetrically until the desired diameter is reached, while the gasket thickens around the outer edge of the diamond faces. To reach the highest pressures possible the gasket material must provide support on the tapered pavilion facet sides of the anvil. Hard materials cannot provide this support and gaskets made of nonmetallic materials usually are limited to around 300 kbar.

Since the diamond anvils are quite transparent, the pressure monitoring in a diamond anvil cell is done by spectroscopic techniques. A small chip of ruby (about 5 μm in diameter) is placed in the sample chamber and from the shift of the ruby-*R* fluorescence line, one may determine the pressure in the sample chamber. The accepted value for the ruby-*R* line shift is 0.365 Å $kbar^{-1}$ or 0.753 cm^{-1} $kbar^{-1}$ [7]. The ruby fluorescence shift is quite sensitive to temperature, showing a positive shift with temperature. Linearity of the ruby scale is found to be valid up to about 300 kbar, but it has been used up to 1 Mbar [8]. Mao et al. [8] have determined that a small positive nonlinearity exists in the pressure dependence of the wavelength shift of the *R* lines, hence the linear scale underestimates the pressure. They [8] proposed an empirical relationship for pressure versus the ruby line shift ($\Delta\lambda$), as $P\text{ (Mbar)} = 3.808[(\Delta\lambda/6942 + 1)^5 - 1]$. Below 300 kbar the deviation from linearity of the ruby fluorescence shift with pressure is negligible, while at about one Mbar the error in using the above equation is stated to be less than 3% [7].

To achieve hydrostatic conditions in a diamond anvil cell a number of organic liquids are employed as pressure media. Among these are methanol : ethanol : water in the ratio of 16 : 3 : 1, which gives hydrostatic conditions up to 145 kbar and nearly hydrostatic conditions up to ~200 kbar; various liquified rare gases such as He, Ne, Ar, Xe; and hydrogen and deuterium have also been proved useful for the attainment of hydrostatic conditions in the pressure cell. Both helium and hydrogen have been used above 600 kbar, and the resulting cell environment was nearly hydrostatic.

Pressure in excess of 2 Mbar has been generated by pressing spherically tipped diamonds with radii less than 1 μm against a flat diamond anvil surface [9,10,11]. Such a device is called a diamond indenter–diamond anvil apparatus. These large pressures are possible with the indenter–anvil technique due to the extremely small pressurized region which, because of its smallness, can be dislocation free [9]. The drawback of the diamond indenter device is its small size and lack of hydrostatic pressure conditions in the sample space. The pressure is calculated in the diamond indenter–diamond anvil cell from the Hertz relations (Hertz contact theory) where linear, isotropic elastic behavior is assumed for the diamonds used in this technique (the assumption that diamond is elastically isotropic is approximately correct). The diamond indenter–diamond anvil pressure cell was pioneered by A. L. Ruoff at Cornell University in the 1970s. The device lends itself to study the insulator to metal transition in a number of materials such as xenon [11], silicon, ZnS, and other semiconducting materials [9]. In the pressure

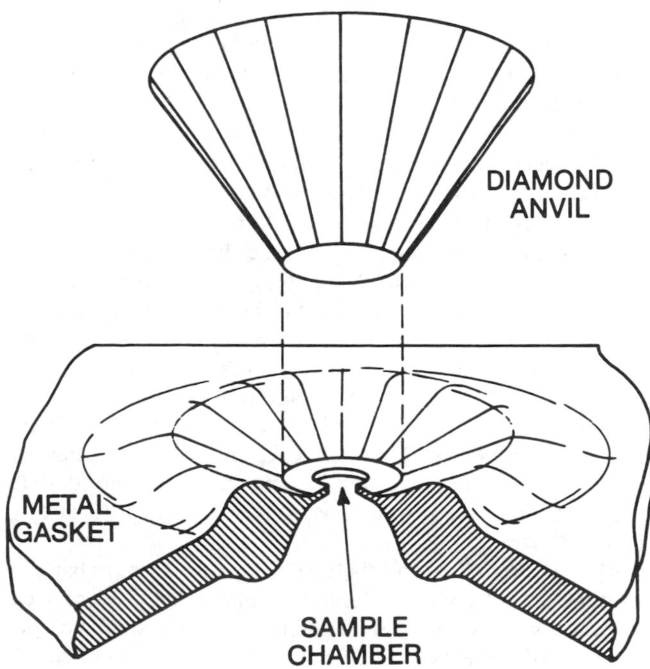

FIG. 4. A cross-sectional view of a typical deformed metal gasket without a bevel angle on the diamond anvil face.

cell the conductivity of the sample as a function of pressure is measured with an interdigitated-electrode technique. The actual electrode is produced either by photolithography or by electron beam lithography [11].

OPPOSED-ANVIL APPARATUS

In 1941 P. W. Bridgman [12] designed an opposed-anvil apparatus in which two conical pistons of tungsten carbide were forced together and the sample was trapped between the piston faces. Pressures as high as 100 kbar have been obtained, although high pressure gradients and pressure uncertainties exist. Permutations of this apparatus have since been developed. Griggs and Kennedy [13] did the obvious experiment of heating the anvils, and a number of reactions of geological interest were studied. Unfortunately, with this method pressure uncertainties and pressure gradients are severe. Shearing stresses may or may not displace equilibrium boundaries. Thus, most phase equilibrium work done with an opposed-piston apparatus has been discarded and replaced by better measurements, which have been made in the piston–cylinder apparatus.

Certain obvious advantages, however, exist in the use of opposed anvils. High pressure can be obtained in a small annulus of boron nitride trapped between opposed anvils with a sample present inside an annulus. X-ray diffraction patterns of the minerals can be taken at high compressions. Estimates of the compressibility and observations of phase changes at high pressures can be made.

COMPRESSIBLE GASKET APPARATUS

A properly designed compressible gasket apparatus can achieve very high pressures if the piston faces are very small and minute quantities of sample are confined between them, and if a very large fraction of the total thrust applied to the system is applied to the gasket area, thus furnishing graded support to the tungsten carbide anvils.

H. G. Drickamer [14] of the University of Illinois developed a very sophisticated compressible gasket apparatus that achieves pressures in the 300- to 400-kbar range. However, the piston faces are minute, no more than 1–2 mm on a side, and thus the sample that can be studied is also minute. X-ray properties, optical properties, and resistance of samples as a function of pressure can be observed in this apparatus.

An advanced compressible gasket apparatus has been developed by F. P. Bundy [15]. Bundy, with meticulous attention to detail of design, has been able to demonstrate pressures of about 400 kbar achieved between the tips of polycrystalline diamond anvils supported by compressible gaskets. Again the pistons are exceedingly small, normally less than 1 mm in diameter.

The big limitation on this type of high-pressure apparatus involving the trapping of minute samples between carbide anvils or polycrystalline diamond masses is the small volume of sample that can be studied. It is difficult to carry out chemical experiments in these environments, and the properties of matter that can be observed are somewhat limited.

EXTRUDABLE GASKET APPARATUS

A vast family of apparatus designs based on the principle of an extrudable gasket have been developed. The first apparatus of this type, designed by H. Tracy Hall [16], is illustrated in Fig. 5. Pressures are achieved by compressing and extruding a pyrophyllite gasket. The disadvantages of this apparatus are obvious. Pressure inside the capsule is unknown because part of the force is distributed on the pyrophyllite gasket and part of the force is distributed on the sample. Calibrations of pressure on such apparatus at high temperature are exceedingly difficult. Reactions that involve a volume change cannot be followed very readily because the piston anvils lock up tight against the die walls. Pressures achieved or claimed in this kind of apparatus can be taken in many instances with a grain of salt.

A number of permutations of the extrudable gasket apparatus have been built. The apparatus shown, with two anvils, compresses the sample in the central cylindrical region. However, devices employing four anvils meeting to form a tetrahedron, six anvils to form a cube, eight anvils to form an octahedron, etc., have all been devised. Each of these various versions of the extrudable gasket apparatus has its own proponents. The simple design shown by Tracy Hall is probably the best, and there is no advantage and some decided disadvantage associated with the proliferation of anvils and gaskets.

If volumes of 10 cm^3 or more are to be studied, the maximum pressure attainable with an extrudable gasket apparatus does not seem to differ very much from the maximum pressures obtainable in the piston–cylinder apparatus; and disadvantages, such as the inability to measure pressure and inability to determine volume changes associated with phase changes, militate against the choice of a large-belt apparatus for high-pressure research.

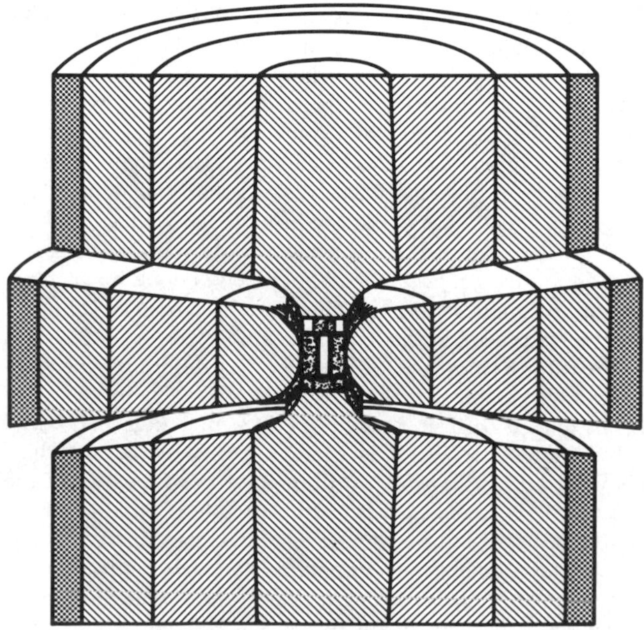

FIG. 5. A schematic of the extrudable gasket apparatus as designed by H. Tracy Hall [16].

PISTON–CYLINDER APPARATUS

The principle of a piston–cylinder apparatus is quite simple. Pressures are generated by forcing a piston into a cylinder. If proper care is taken with the design of this apparatus, pressures somewhat in excess of 60 kbar can be achieved simultaneously with temperatures in excess of 2000°C. The first apparatus of this kind was described by Boyd and England in 1960 [17]. The illustration from the original Boyd and England article is shown here as Fig. 6.

Great sophistication in design is required for a piston–cylinder apparatus to work well. The pressure vessel is normally made of tungsten carbide bonded together with 13–15% cobalt, supported by one or more hardened steel supporting rings. The supporting rings are best made of a maraging (tough and hard) steel. The supporting rings are normally tapered at an angle of 1–1.5° and are press-fitted together, giving the tungsten pressure vessel an external support of as much as 250 000 lb/in.^{2} of carbide surface. Certain designs in current operation, working at 60 kbar, will accommodate as much as 50 cm^3 of sample. Total weight of such a pressure vessel, including the steel supporting rings and the tungsten carbide core, may be as much as 1 ton. Special equipment for handling the pressure vessel, loading it, and placing it in the press is required. The stored energy in a vessel of this size may be in excess of a million ft/lb. Obviously, safety precautions are required.

The tungsten carbide pressure vessel requires not only radial support, as is provided by the hardened steel supporting rings, but also substantial end support, normal to the axis of the pressure vessel in order to keep the pressure vessel from splitting at right angles to its cylindrical axis. This end-load support is normally provided by a high-thrust ram as shown in Fig. 6.

Pressures are generated by forcing a tungsten carbide piston into the cylinder by use of an auxiliary ram, and pressures are measured by simple force/area calculations. The tungsten carbide piston used to generate the pressure is of a quite different grade of carbide from that used in the cylinder; about 3% cobalt bonder appears to be a preferred grade, since high compressive strength rather than ductility is required in the piston. Special attention must be paid to the exposed, unsupported length of the piston and the details of piston construction.

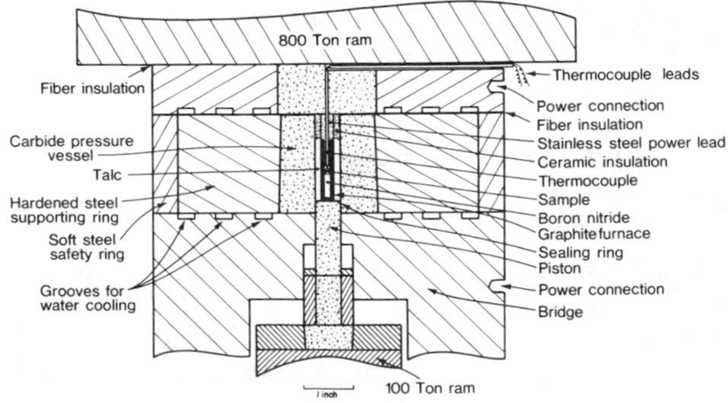

FIG. 6. Typical piston–cylinder apparatus taken from reference [17].

Cells for research with a piston–cylinder apparatus range widely in design and can become exceedingly complex. High-temperature work is normally done in solid-pressure media cells. Temperatures are obtained by passing a low-voltage, high-amperage current through a graphite tube that serves as a heater. Obviously, the tube must be insulated from the tungsten carbide pressure wall. The insulation should be designed to be mechanically weak at the temperatures involved, so that a disproportionate part of the piston force is not carried by the insulation bushing. Thus, mechanically strong insulating bushings such as those made of aluminum oxide, fired pyrophyllite, or glass are not acceptable. Insulating bushings made of various pressed salts seem to serve well. A typical and a rather elaborate one is illustrated in Fig. 7 [18].

Certain experiments can only be done under truly hydrostatic conditions. A mixture of 4 parts methanol to 1 part ethanol is commonly used as a pressure-transmitting medium.

Precautions should be observed in the use of the piston–cylinder apparatus if reasonable life at reasonable pressures is to be expected from the rather expensive tungsten carbide pressure vessel. Alkali metals, mercury, indium, gallium, and other softer metals should be kept from contact with the carbide. For inexplicable reasons these substances seem to penetrate the carbide and rapid deterioration takes place. Hydrocarbons also readily penetrate the pressure vessel at high pressures with equally disastrous results.

Tungsten carbide is apparently exceedingly sensitive to hydrogen embrittlement. A single drop of machine-shop oil left in the center of a run to be taken to high temperature will evolve enough hydrogen to cause almost instantaneous total destruction of the pressure vessel.

The limitation of the attainment of pressures in the tungsten carbide piston–cylinder apparatus where large volumes can be studied and the physical environments can be controlled is set by the crushing strength of tungsten carbide,

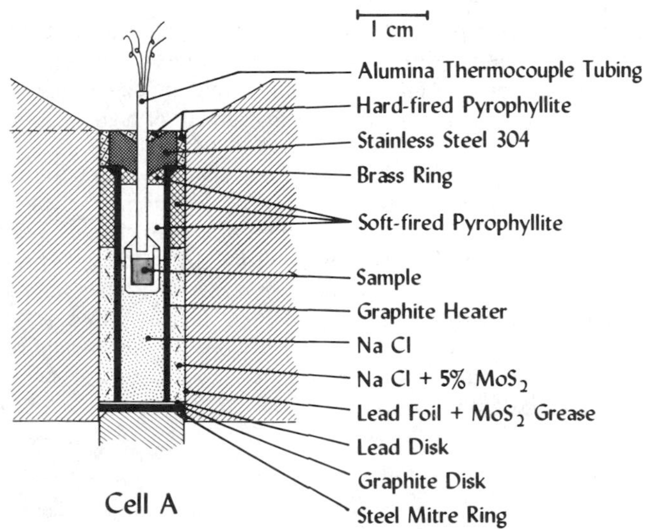

FIG. 7. Schematic of the typical research cell for the piston–cylinder apparatus.

about 65 kbar. The future of high-pressure research where large volumes at high temperatures can be studied lies in the development of better materials than tungsten carbide. Fortunately, they appear to be on the way. The development of pistons of bonded diamond and segmented pressure vessels made of bonded diamond is actively under way in two or three different laboratories around the world. With the advent of this new material, it seems possible that high-pressure studies of several hundred kilobars in substantial volume can be made.

SUMMARY AND CONCLUSIONS

This article would not be complete without highlighting some selected experimental results obtained with the described high-pressure techniques. Diamond anvil cells have stimulated the quest for metallic hydrogen [19,20]. Other substances that are favorites for metallization are deuterium, nitrogen, xenon, cesium iodide, etc. [11,28] and these have been extensively investigated with diamond anvil pressure devices. Resonant Raman scattering induced by pressure tuning applied to semiconductors in diamond anvil pressure cells has become a useful technique in studying the local environment of defects in these materials, both in the bulk, colloid, and nanocrystalline semiconductor-glass composites [21–23]. All of the above pressure cells have been used in measurements to establish melting curves and phase boundaries between solid phases. The high-pressure phenomena of the electronic transition in a variety of materials such as potassium, rubidium, cesium, rare earth metals, etc., have been studied with the various high-pressure cells [24]. Spectroscopic measurements at very high pressures have only become possible since the advent of the single-crystal opposed diamond anvil cell. Optical absorption and reflection studies now are possible with relative simplicity. Raman scattering and Brillouin scattering on glasses and crystalline solids at high pressure are carried out in many laboratories throughout the world. High-pressure x-ray diffraction to look at phase changes and compression of solids are practiced by many researchers. There exists not an area of condensed matter physics where high-pressure measurements are not being used to provide significant tests of theoretical predictions and within a short time frame the diamond anvil cell has become a most popular instrument of high-pressure research.

A number of books have appeared that are of general interest. The classic in the field of high-pressure investigation is Bridgman's 1931 book on high-pressure physics [25]. Other useful books have been published by Bradley [26] and by H. Li. D. Pugh [27]. Several good review articles on diamond anvil pressure cells have been published by Jayaraman [7,28]. The references contained in both the Jayaraman reviews are a valuable source pertaining to the state of the art of diamond anvil pressure devices.

REFERENCES

1. A. Van Valkenburg, *High Pressure Measurement* (Giardini and Lloyd, eds.), p. 87. Butterworth, Washington, DC, 1963.
2. G. J. Piermarini, S. Block, J. D. Barnett, and R. A. Forman, *J. Appl. Phys.* **46,** 2774 (1975).
3. H. K. Mao, P. M. Bell, J. W. Shaner, and D. J. Steinberg, *J. Appl. Phys.* **49,** 3276 (1978).
4. Supplied by Drukker Corp. Amsterdam (also Doubleday-Harris, N.J.); and also by Lazarus and Kaplan, of New York City.
5. M. S. Bruno and K. J. Dunn, *M. R. S. Symposia Proceedings (High Pressure in Science and Technology)* **22,** 239 (1984).
6. H. G. Drickamer, *Solid State Physics* (Seitz and Turnbull, eds.), p. 46. Academic Press, New York, NY (1965).
7. A. Jayaraman, *Rev. Mod. Phys.* **55,** 65 (1983).
8. H. K. Mao and P. M. Bell, in *Carnegie Institution of Washington Year Book* **75,** 824 (1978).
9. A. L. Ruoff, *High Pressure Science and Technology* (B. Vodar and Ph. Marteau, eds.), Vol. 1, p. 127. Pergamon Press, Oxford, 1980.
10. A. L. Ruoff, "High Pressure in Research and Industry," *Proceedings of the 8th AIRAPT Conference (1981)* (C. M. Backmann, T. Johannisson, and L. Tegner, eds.), p. 108. Uppsala, Sweden, 1982.
11. D. A. Nelson and A. L. Ruoff, *Phys. Rev. Lett.* **42,** 383 (1979).
12. P. W. Bridgman, *J. Appl. Phys.* **12,** 461 (1941).
13. D. T. Griggs and G. C. Kennedy, *Am. J. Sci.* **254,** 722 (1956).
14. H. G. Drickamer and A. S. Balchan, *Modern Very High Pressure Techniques* (R. H. Wentorf, Jr., ed.), p. 25. Butterworths, London, 1962.
15. F. P. Bundy, *Rev. Sci. Instrum.* **45,** 1318 (1975).
16. H. T. Hall, *J. Chem. Edu.* **38,** 484 (1961).
17. F. R. Boyd and J. L. England, *J. Geophys. Res.* **65,** 741 (1960).
18. Peter W. Mirwald, Ivan C. Getting, and George C. Kennedy, *J. Geophys. Res.* **80,** 1519 (1975).
19. H. K. Mao and R. J. Hemley, *Science* **244,** 1462 (1989).
20. I. F. Silvera, *Bull. Am. Phys. Soc.* **35,** 195 (1990).
21. X. S. Zhao, J. Schroeder, T. G. Bilodeau, and L. G. Hwa, *Phys. Rev. B* **40,** 1257 (1989).
22. X. S. Zhao, J. Schroeder, P. D. Persans and E. Lu, submitted to *Phys. Rev. B.* (1990).
23. X. S. Zhao, J. Schroeder, P. D. Persans, and T. G. Bilodeau, *Bull. Am. Phys. Soc.* **35,** 714 (1990).
24. H. G. Drickamer and C. W. Frank, *Electronic Transitions and High Pressure Chemistry and Physics of Solids*. Chapman and Hall, New York, Halstead Press, 1973.
25. P. W. Bridgman, *The Physics of High Pressure*. G. Bell & Sons, London, 1931.
26. C. C. Bradley, *High Pressure Methods in Solid State Research*. Plenum, New York, 1969.
27. H. L. Pugh, *The Mechanical Behavior of Materials Under Pressure*. Elsevier, Amsterdam, 1970.
28. A. Jayaraman, *Rev. Sci. Instrum.* **57,** 1013 (1986).

Ultrashort Optical Pulses

C. L. Tang

In the case of optical pulses, the meaning of *ultrashort* has been changing continuously in recent years as shorter and shorter pulses are generated using various laser techniques. [1–5] At the time of this writing near the end of 1989, it is finally close to the ultimate limit: a few optical cycles [3] in the visible. For the ultraviolet [4] and the infrared [5] spectral ranges, it is still in the tens of femtoseconds range. The availability of ultrashort optical pulses has led to the study of a wide variety of ultrafast physical and chemical processes in the femtosecond time domain. For example, molecular vibrations have been observed directly in the time domain using such sources.

Ultrashort optical pulses are typically generated using *mode-locked* dye lasers followed by pulse compression [6,7] to achieve the shortest possible pulse length. The basic idea of the femtosecond dye laser is shown schematically in Fig. 1. The conventional type of laser consists of a gain medium, which emits and amplifies light, and a pair of laser-cavity mirrors (M1 and M2) facing each other, which feed the light repeatedly back into the laser medium for amplification. The output is generally a continuous wave. In mode-locked dye lasers, there is in the cavity, in addition to the gain medium, a nonlinear *saturable absorber*, which has the unusual nonlinear property that its optical loss is less for more intense light than for weaker light. As the emitted light, which is initially in the form of random noise spikes, circulates in the laser cavity, the highest spike gains in energy at the expense of the others and eventually becomes the sole survivor. As this single high-intensity spike circulates in the cavity and impinges periodically on the output mirror (M2) of the laser, a periodic train of short pulses is emitted. This explains how the pulses are formed, but it does not explain why the pulses can be as short as, for example, 30 fs coming directly out of such a laser. This short-pulse duration is a result of a delicate balance of the gain, the nonlinear saturation, loss, and dispersion in all the various components in the laser cavity. To reach a few optical cycles, 6 fs [3] at 620 nm, for example, additional means to compress further the pulses external to the laser cavity is used. To compress the pulses, the pulses coming out of the mode-locked dye laser are first amplified and sent through a nonlinear medium, for example, a section of optical fiber, in which the index of the medium changes with the light intensity. This introduces a time-varying frequency chirp on the pulse. If such a frequency-chirped pulse is sent through a dispersive medium or structure such as a pair of gratings, [6,7] the back-end of the pulse can catch up with the front-end leading to compression of the pulse.

Since there is no detector that has a fast enough response time to measure such short optical pulses directly in the time domain, how are such pulses measured? The pulse length is usually determined by measuring the spatial extent of the pulse [8]. In such a measurement, the short pulse is first split into two through the use of a beam splitter and then recombined with a variable delay between the two pulses (see Fig. 2). These two pulses are then sent into a thin nonlinear optical crystal suitably oriented for optical second-harmonic generation. The second-harmonic output intensity as a function of the delay between the two pulses gives the autocorrelation of the pulse. From the autocorrelation of the pulse, the width of the original pulse can be calculated. The measurement is, therefore, basically a dc measurement and no fast detector is needed. The time resolution of the measurement is determined by how accurately the delay between the two pulses can be determined spatially, which can be down to 1 fs.

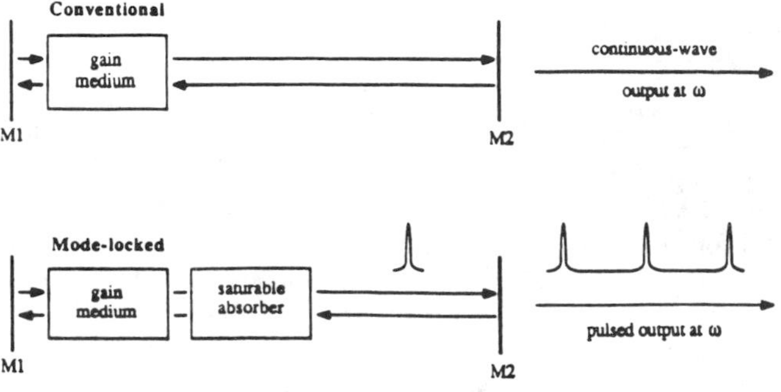

FIG. 1. Schematic of a mode-locked laser.

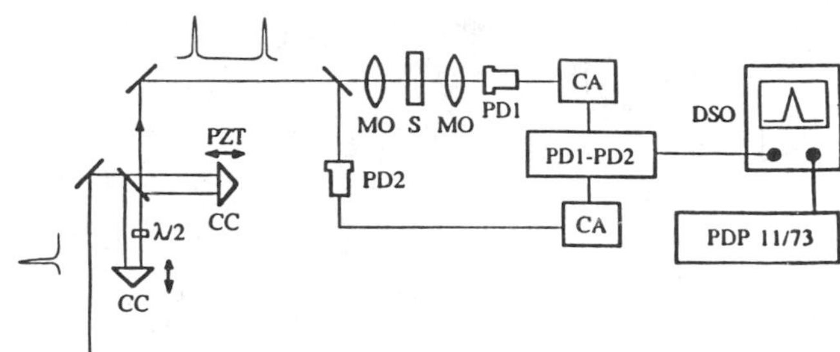

FIG. 2. Schematic of setup for measuring the pulse width and for studying the relaxation dynamics of the sample (S) in the femtosecond time domain. [10] For measuring the pulse width, the sample is replaced by a thin nonlinear optical crystal and only PD1 is needed. Key: CC = corner cube; $\lambda/2$ = half-wave plate; MO = microscope objective; S = thin sample or second-harmonic generation crystal; PD1, PD2 = photodiodes; PZT = piezoelectric translation; CA = current amplifier; DSO = digital oscilloscope.

A slightly modified version of such a setup is used in optical-correlation spectroscopy [9] for studying the ultrafast dynamics in a variety of materials. Suppose the second-harmonic crystal is replaced by a thin sample of a medium with an optical absorption in resonance with the ultrashort-pulse frequency. Because of the saturation effect, when the excitations of the two pulses see each other, the total transmission of the two pulses through the resonance-absorption medium will be larger than that when they do not see each other. Thus, the total transmission as a function of the delay between the pulses, or the transmission-correlation peak, gives information on the relaxation dynamics of the excited state. In fact, the transmission-correlation peak is the convolution of the autocorrelation of the pulse and the response function of the medium. [9]

An example of the transmission-correlation peak at 620 nm of a thin jet of dye molecules, ethyl violet, dissolved (10^{-3} *M*) in ethylene glycol is shown in Fig. 3A. The ~150-fs oscillations, or quantum beats, [10] correspond to molecular vibrations as seen directly in the time domain following coherent wave-packet excitation by femtosecond optical pulses. This can be explained qualitatively in terms of the schematic diagram shown in Fig. 3B. Physically, the first femtosecond pulse excites the molecules from the ground electronic level to an excited electronic level, leaving the individual molecular configuration intact but changing the

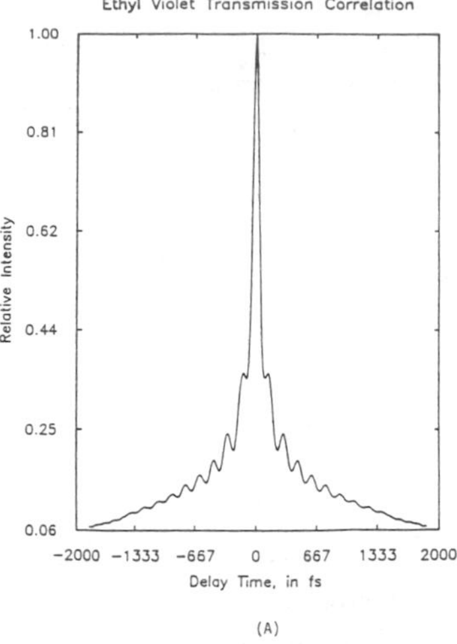

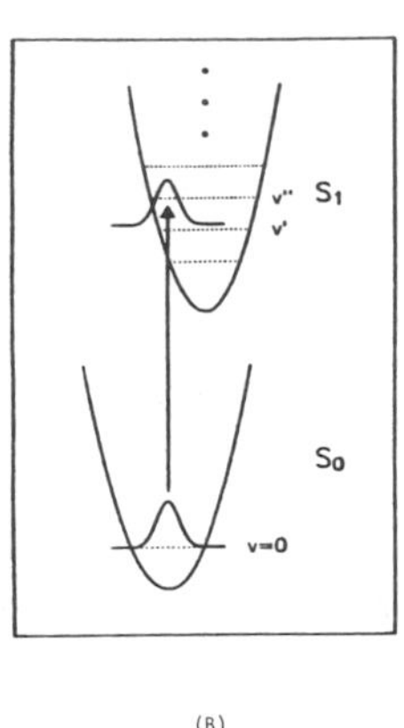

FIG. 3. (A) Transmission-correlation peak for ethyl violet showing quantum beats. (B) Schematic showing the excitation of coherent wave packet in a molecule. The molecule is initially in the ground vibrational state of the ground electronic level. A short pulse excites the molecule into a coherent-superposition vibrational state (or coherent vibrational wave packet) of the excited electronic level. This wave packet will oscillate at the vibrational frequency of the molecule, leading to quantum beats in the femtosecond time domain.

equilibrium configuration of the individual molecules. The second femtosecond pulse sees the subsequent coherent vibrations of the molecules around the new equilibrium configuration. This gives rise to the oscillations in the response or the transmission-correlation peak of the dye molecules. Similar oscillations have now been seen in a variety of molecules. [10,11]

REFERENCES

1. A. J. DeMaria, D. A. Stetser, and H. Heynau, *Appl. Phys. Lett.* **8,** 174 (1966).
2. R. L. Fork, B. I. Greene, and C. V. Shank, *Appl. Phys. Lett.* **38,** 671 (1981).
3. R. L. Fork, C. H. Brito Cruz, P. C. Becker, and C. V. Shank, *Opt. Lett.* **12,** 483 (1987).
4. D. C. Edelstein, E. S. Wachman, L. K. Cheng, W. R. Bosenberg, and C. L. Tang, *Appl. Phys. Lett.* **52,** 2211 (1988).
5. D. C. Edelstein, E. S. Wachman, and C. L. Tang, *Appl. Phys. Lett.* **54,** 1728 (1989).
6. S. H. Nakatsuka, D. Grischkowsky, and A. C. .Balant, *Phys. Rev. Lett.* **47,** 910 (1981).
7. C. V. Shank, R. L. Fork, R. Yen, R. H. Stolen, and W. J. Tomlinson, *Appl. Phys. Lett.* **40,** 761 (1982).
8. E. P. Ippen and C. V. Shank, *Ultrashort Light Pulses* (S. Shapiro, ed.). Springer-Verlag, New York, 1977.
9. A. J. Taylor, D. J. Erskine, and C. L. Tang, *J. Opt. Soc. Am. B* **2,** 663 (1985).
10. M. J. Rosker, F. W. Wise, and C. L. Tang, *Phys. Rev. Lett.* **57,** 321 (1986); F. W. Wise, M. J.
11. J. Chesnoy and A. Mokhtari, *Phys. Rev. A* **38,** 3566 (1988).

Ultrasonic Biophysics

Floyd Dunn

The field embraces the interaction of ultrasound and biological media, viz., the modes of interaction, the biological consequences of such interactions, and the applications of these interactions to biology and medicine. The five major topics concern (1) the ultrasonic propagation properties of living systems, (2) the physical mechanisms of interaction of ultrasound and biological structures, (3) the tissue-modifying effects (including toxicity) of exposure to ultrasound (largely with regard to humans in medical situations), (4) aspects of dosimetry, and (5) ultrasonic microscopy. Investigations herein contribute widely as, for example, in yielding details of reactions occurring in the time range 10^{-3}–10^{-10} s and in providing the basis for ultrasonic medical diagnosis and therapy.

The *propagation properties* of biological media are generally described in terms of the velocity or speed of the ultrasonic wave process and the attenuation and/or absorption of the wave energy as these are measurable. The soft vertebrate tissues, with the exception of pulmonary tissues, exhibit velocities very nearly those of water, viz., about 1500 m s^{-1} and with little dispersion. Tissues with high collagen content exhibit speed of sound values approximately 10–15% greater. The speed of sound in the lung is very dependent upon inflation and frequency and may be considerably less than that of the soft tissues, whereas mammalian bone yields velocities nearly twice the latter, all in the low megahertz frequency range. The frequency dependence of attenuation in these soft tissues, excepting the lung, has the approximate value of 0.1 Np cm^{-1} MHz^{-1}. Lung exhibits values some 50 times greater, in the low-megahertz frequency range, but with an exponential frequency dependence. Bone attenuation is approximately an order of magnitude greater than that of the soft tissues. The fraction of the attenuation ascribed to absorption processes (intrinsic energy dissipation) varies with tissues; more than 90% for most soft tissues, and infinitesimal for the lung. Architectural features such as cellular structure and cellular organization are believed to account for scattering (possibly due to collagenous structuring) and any other nondissipative loss processes to make up the remainder. Among the soft tissues, both attenuation and speed of sound increase as water content decreases and collagen content increases. The absorption mechanisms are intimately related to protein content, being directly proportional to its concentration and independent of its cellular relationship. Investigations of macromolecular solutions have implicated the polypeptide chain as largely responsible for the broad spectrum relaxational absorption, with such mechanisms as proton transfer, conformational changes, disturbance of solute–solvent equilibria, and molecular volumetric expansion invoked to account for the observed frequency dependence. (See Refs. [1, 2, 4, 5] for details.)

Three classes of *physical mechanisms of interaction* are believed to promote the reversible and irreversible changes produced in biological media by ultrasound of significant amplitude, viz., thermal, mechanical, and cavitation. The mechanisms appear related to the ultrasonic dose considered delivered to the site of the action and which is described in terms of the energy flux per unit time, viz., the acoustic intensity, and the duration of the exposure. Details are available only for a few tissues and, for example, for mammalian brain, the acoustic intensity threshold, I, for irreversible structural changes is a function of the time of irradiation, t, according to the relation $It^{1/2} = 200$ W cm^{-2} s$^{-1/2}$, in the approximate frequency range 1–10 MHz. The constant ranges from 50, for murine neonatal spinal cord, to 400, for mammalian liver. The thermal mechanisms are known to be responsible at intensities less than about 10^2 W cm^{-2} and for durations of exposure greater than seconds, under which conditions adequate opportunity exists for temperatures to increase, by virtue of molecular absorption processes, beyond damaging levels. This has led to hyperthermia as a cancer treatment in which the temperature of a tumor volume is increased to approximately 43°C and maintained for periods of the order of 1 h. At intensities greater than about 10^3 W cm^{-2}, where the exposure durations at threshold become of the order of milliseconds, transient cavitation is responsible for the irreversible damage. Here, the temporal and spatial sequence of histological events is completely different from those associated with the other mechanisms, viz., variations of tissue sensitivities are not observed and fluid-filled regions appear to provide cavitation nuclei. Transient cavitation may be the mechanism involved in lithotripsy wherein nephrotic calculi are disrupted into pieces sufficiently small to be excreted with urine, obviating the necessity for scalpel surgery. Mechanical mechanisms, largely unidentified but probably associated with nonlinear processes promoting unidirectional forces such as shear stresses, seem to be implicated in the threshold region intermediate between the thermal and cavitation regions. The extension of the threshold region to much reduced intensities and greatly increased exposure time, and to include other tissues, yields segments having different I–t relationships and possibly culminating with a time-independent intensity of approximately 10 mW cm^{-2}, whereas the heat produced by absorption processes is easily accommodated by homoiotherms. The initial anatomical site of action in the absence of cavitation appears to be the level of subcellular structures, viz., synapses and mitochondria. Cavitation is a necessary condition for degradation of molecular species of approximate molecular weight 10^5 and smaller (in solution). Little is known of stable cavitation in tissues and of associated nuclei (see Refs. [2, 3, 5] for details).

The *toxic effects* of ultrasound appear to be negligible for the exposures employed in clinical diagnosis, viz., time-average intensities of about 100 mW cm^{-2} for relatively short exposure durations at single anatomic sites. Whole-body radiation in humans can only occur during fetal exposure during early pregnancy. Epidemiological data are not available. Ultrasound is an extraordinarily inefficient agent for producing genetic effects as this has only been observed in a dosage-unrelated study with a primitive plant. Teratogenesis has been reported only with rare occasion in laboratory studies involving rodents, and then only at supraclinical dosages (see Refs. [2–5] for details).

The limited studies dealing with *ultrasonic dosimetry*, viz., the absorbed dose necessary to produce suprathreshold lesions of specified volume in mammalian brain, suggest a universality over some soft tissues and independence of frequency. Inability to (1) account for the fate of the wave energy propagating in living systems, (2) identify uniquely the biological response, (3) produce fields of desired distribution in the heterogeneous systems, and (4) specify completely, by a measuring process, the field present and responsible for the observed effects, partially define the difficulties inherent in this area. (See Refs. [2, 3, 5] for details.)

Ultrasonic microscopy is a means for observing spatial distributions of the elastic constants and, thereby, provides information on a very small scale that is unavailable by other means. (For example, a resolution of the order of 1 μm results with an operating frequency at 1.5 GHz in a medium, such as water, for which the speed of sound is 1.5×10^5 cm s^{-1}.)

See also ACOUSTICS; ULTRASONICS.

REFERENCES

1. F. Dunn, P. D. Edmonds, and W. J. Fry, "Absorption and Dispersion of Ultrasound in Biological Media," in *Biological Engineering* Chap. 3, pp. 205–332 (H. P. Schwan, ed.). McGraw-Hill, New York, 1969. (I)
2. F. Dunn and W. D. O'Brien, Jr., *Ultrasonic Biophysics*. Dowden, Hutchinson and Ross, Stroudsburg, PA, 1976. (A)
3. W. J. Fry and F. Dunn. "Ultrasound: Analysis and Experimental Methods in Biological Research," in *Physical Techniques in Biological Research*, Vol. 4, Chap. 6, pp. 261–394 (W. L. Nastuk, ed.). Academic Press, New York, 1962. (I)
4. S. A. Goss, R. L. Johnston, and F. Dunn, "Comprehensive Compilation of Empirical Ultrasonic Properties of Mammalian Tissues," *J. Acoust. Soc. Am.* **64,** 423–456 (1978); **68,** 93–108 (1980). (I)
5. C. R. Hill (ed.), *Physical Principles of Medical Ultrasound*. Wiley, New York, 1986. (I)

Ultrasonics

Warren P. Mason and R. N. Thurston

The science of producing and transmitting sound waves in materials has become a flourishing art with many practical applications. Among the low-amplitude uses are underwater sound transmission for locating submarines, measuring the depth and detail of ocean bottoms, flaw detection in materials, delay lines for storing information and for performing many processing calculations, and many medical applications such as locating cancers and other imperfections in the human body. Ultrasound waves have been used in many physical investigations.

When the amplitude of the sound wave becomes large, some nonlinear effects occur, such as the production of cavities in liquids and fatigue in solids. These effects give rise to another series of applications, such as ultrasonic cleaning, emulsification of liquids, machining of materials, and tests for fatigue in materials, particularly when a large number of cycles are required. Biological applications, such as the destruction of bacteria and the use of focused ultrasound as a surgical knife, are also possible.

ULTRASONICS IN UNDERWATER SOUND AND RADAR SYSTEMS

The first application, and for many years the only application, was the production of sound in seawater for the detection of submarines. Prof. Paul Langevin [1] produced the first underwater sound transducer by using the piezoelectric effect in quartz—the only type available at that time. He has been called the father of ultrasonics [1]. Much work has been done on underwater detection and destruction of submarines by using ultrasonic waves. The World War II work is available [2,3] but later work is highly classified.

During World War II ultrasonic waves were also used as an adjunct to radar, particularly in moving-target radar systems, which accentuate the appearance of moving targets by minimizing the appearance of reflections due to stationary targets. In this use one frame containing all the reflections during one pulse period is stored in an ultrasonic-wave delay line having a delay time equal to the period between successive pulses. The stored information is used to balance out radar reflections. Discriminations of 30–40 dB have been obtained.

Another use of ultrasonic waves is in high-power radars [4]. By use of the dispersive characteristic of waves in a wire or strip, the amplifier power output can be made to cover a good share of the repetition cycle rather than just the initial pulse. The original pulse can be recovered by sending the reflection from the surrounding terrain, collected by a second antenna, through a network having the inverse dispersion.

New transducer materials in the form of ferroelectric ceramics, such as barium titanate, lead titanate, or lead zirconate [5] (PZT), have increased the electromechanical coupling factor, which determines the transducer loss. These materials are used in underwater sound and radar systems as well as for other applications.

ULTRASONICS IN PHYSICAL INVESTIGATIONS

Ultrasonic waves have been applied as a tool for investigating a large number of physical phenomena. These cannot be discussed in detail here [6] but a partial list includes oscillatory magnetoacoustic phenomena in metals; giant quantum oscillations; effects of dislocations and impurities on sound wave propagation; relaxation processes in gases, liquids, and solids; paramagnetic spin–phonon and nuclear spin interactions in solids; Fermi surfaces of metals; propagation of sound in liquid and condensed helium, and sound propagation in liquid crystals. Two of the simplest interactions, namely, the interactions of acoustic waves with phonons, which carry the thermal energy of a solid, and with electrons, which carry the electrical energy in a metal, will be illustrated here.

Figure 1 shows one of the best measurements [7] of the attenuation of a dielectric crystal, in this case Al_2O_3. Between 0 and 30 K the attenuation is independent of the temperature and is thought to be due to scattering loss caused by imperfections in the crystal. Between 30 and 100 K, the attenuation rises rapidly and is assumed to be due to the direct interaction of the sound waves with thermal phonons. Above 100 K the attenuation levels off and is due to the Akhieser effect, which involves a strain-induced frequency shift of the thermal phonon normal modes. Because their occupation numbers cannot adjust instantaneously to the new equilibrium distribution required by the normal mode frequency shift, the process dissipates energy from the sound waves [10] to the thermal phonons.

Figure 2 [8] shows the effect of electrons on the attenuation of sound waves in a tin single crystal. The effect is small until the temperature is below 12 K. At 10 K the attenuation is proportional to the square of the frequency and can be accounted for by an electron viscosity effect. Just above the superconducting temperature, 3.73 K, the attenuation is proportional to the frequency. Below the superconducting temperature, the attenuation decreases, in agreement with the Bardeen–Cooper–Schrieffer theory of superconductivity.

HIGH-AMPLITUDE ULTRASONIC VIBRATIONS

As the amplitude of the motion increases, several new effects appear which have been used in practical devices. One is cavitation in liquids, which has been used in ultrasonic cleaning, in producing chemical changes, and in producing biological changes.

By use of the large motions possible with mechanical transformers driven by transducers, machining of brittle

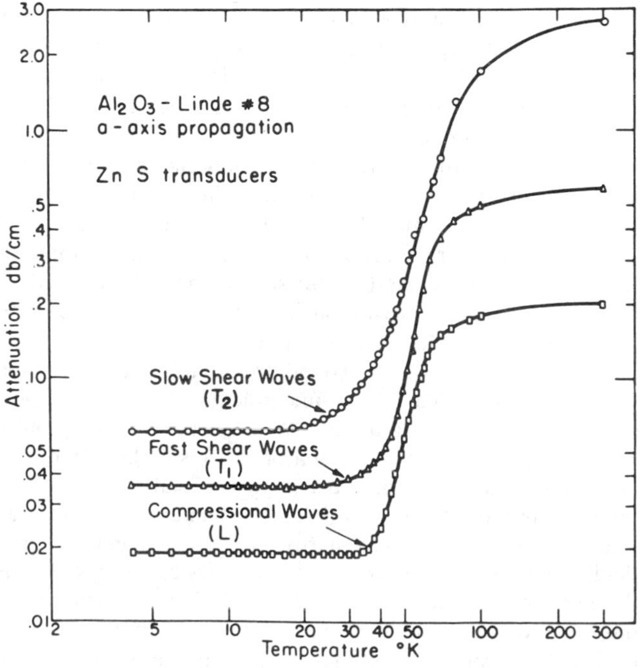

Fig. 1. Attenuation of shear and compression waves in Al_2O_3 at 1 GHz (after J. De Klerk [7]).

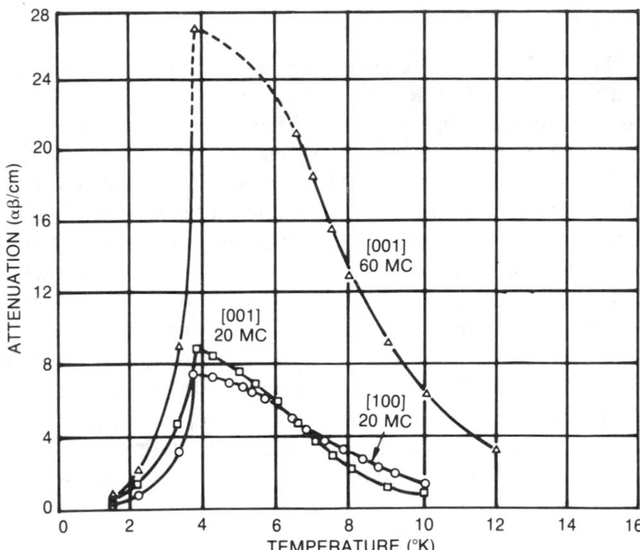

FIG. 2. Attenuation of shear waves for a single crystal of tin as a function of the temperature and frequency (after Bömmel [8]).

materials and welding of thin sheets of plastics and metals can all be accomplished. High-amplitude devices have also been used to study internal friction, acoustic emission, and fatigue in metals and rocks. A transducer, transformer, and specimen that have been used for this purpose [9] are shown in Fig. 3. This device has been used up to strains of 3×10^{-3}. The internal friction in this high-amplitude range is caused by the breakaway of dislocations from pinning points and by the generation of Frank–Read loops, which result in plastic strain. These cause slip bands, which eventually join up to produce fatigue cracks. Ultrasonic frequencies are useful for studying fatigue, since a large number of cycles can occur in a reasonable time.

Acoustic emission—the noise in the sample associated with dislocation motion or other stress-induced events—can be studied by putting transducers on the sample or by optical detection of the emitted vibrations [11].

WAVES IN CRYSTALS; ANISOTROPY

Elastic waves are either transverse or longitudinal in unbounded materials that are isotropic, and along simple directions in unbounded crystals. However, corresponding to a general propagation direction in an unbounded crystal are three orthogonal directions of the displacement, and none of these possible displacement directions need be parallel or perpendicular to the propagation direction [12].

GUIDED WAVES

Guided waves have their origin in reflection and refraction at boundary surfaces or interfaces, or refraction associated with continuously varying elastic properties or density. Examples of structures that support guided waves are rods, strips, plates, tubes, and bores [13]. Related phenomena occur in the atmosphere, in ocean acoustics, and seismology. In both optics and acoustics, the guidance phenomenon provided by a cladding of faster propagation velocity than the core is used to isolate a wave from the outer surface of a fiber or rod in order to prevent loss of energy from the wave into the surrounding medium or supporting structure. In optical fibers, this is necessary in order to achieve low-loss transmission over tens of kilometers. In ultrasonics, a possible application is to acoustic sensors that pass through unfriendly environments.

A guided elastic wave along a free surface is called a Rayleigh surface wave, or surface acoustic wave (SAW). In addition to confinement by the surface, lateral guidance, if desired, can be obtained by means of a ridge structure or by diffusion of a substance that changes the propagation velocity. SAWs are important in seismology and in modern ultrasonic devices used in wave filters and other signal processing applications [14]. SAWs have the advantage of being excited and detected by interdigital transducers made by the widely used planar photolithographic fabrication methods. The transducers may be fabricated on a piezoelectric crystal (such as quartz or $LiNbO_3$) or on a nonpiezoelectric substrate, such as glass with coupling provided by a piezoelectric film (e.g., ZnO), deposited between the transducer fingers. SAW bandpass filters have been used for channel selection, television intermediate frequency filtering, timing recovery in optical fiber digital transmission systems, and a number of other applications. The practical range of center frequency is from about 20 MHz where the devices are inconveniently large to a few GHz where the transducer features are inconveniently small and the material loss per wavelength (which is usually proportional to the frequency in this frequency range) is becoming large [15].

NONCONTACTING TRANSDUCERS; OPTICAL EXCITATION AND DETECTION

The advent of lasers and other advances in physics now make possible the controlled optical excitation and detection

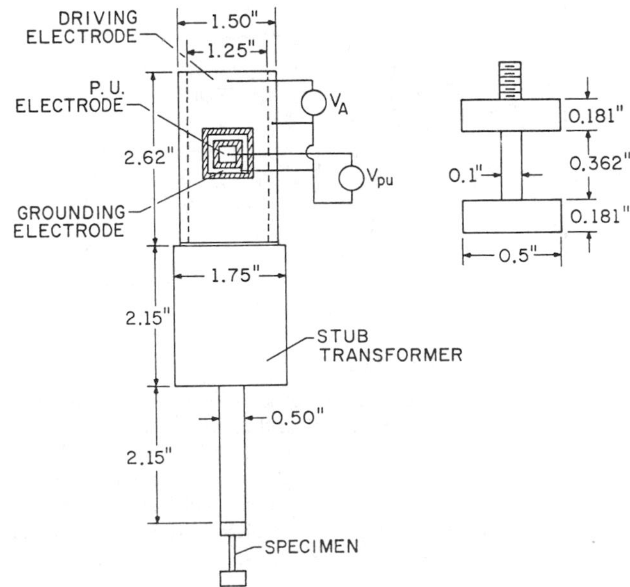

FIG. 3. High-amplitude device employing a ceramic transducer, a mechanical transformer, and a specimen shape for increasing the strain in the midsection (after Mason [9]).

of ultrasonic waves. Excitation involves the absorption of energy (which could be given up by a beam of electrons or other particles as well as a laser beam) and the subsequent propagation of waves arising from the mechanical response to this energy input. Such processes are the basis of photoacoustic spectroscopy [16]. Optical detection of ultrasound typically uses the surface of the vibrating solid as a mirror which either deflects the light as the mirror tilts, or is incorporated in an interferometer [11]. Other noncontacting methods use an electrostatic transducer or an electromagnetic-acoustic transducer (EMAT). An electrostatic transducer is simply a capacitor, while an EMAT is a coil and magnet arranged to generate a voltage when the wave passes by. An advantage shared by all noncontacting methods of detection is that the wave is probed without any appreciable influence of the measurement on the wave. For this reason, when detection sensitivity permits, noncontacting transducers are sometimes preferred in studies of acoustic emission and in nondestructive evaluation of materials and structures.

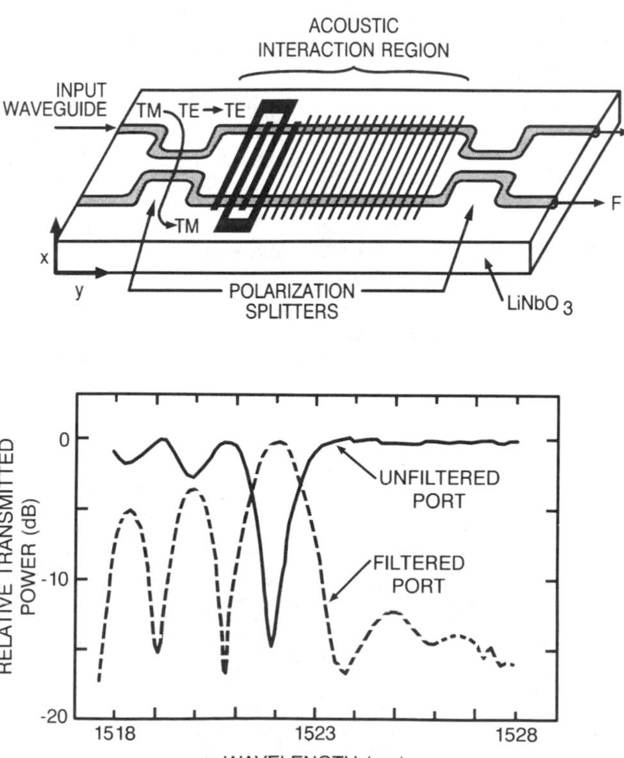

FIG. 4(a) Schematic of a polarization-independent acoustically tuned optical filter (after Smith et al. [19]). Not to scale! The optical waveguides are 8 microns wide, separated by 6.5 microns in the polarization splitters and 40 microns in the acoustic interaction region. At 1523 nm the beat length for pi phase difference of the TE and TM modes is about 20 microns, which corresponds to a half wavelength of the acoustic wave at 180 MHz. The overall length is about 5 cm. (b) Transmitted power (on a log scale) from filtered (F) and unfiltered (U) ports for an unpolarized broadband LED source. Switching with 15 dB extinction ratio indicates 97% efficiency of the polarization-independent filter.

INTERACTIONS

The mechanical stress and strain fields of an acoustic wave can in principle interact with (affect or be affected by) imperfections and other physical fields [17]. It is such interactions that allow the generation and detection of ultrasonic waves and the useful physical investigations mentioned above. The *acousto-optic* interaction arises because the index of refraction depends on the density changes or strain in the acoustic wave. The spatially varying refractive index distribution induced by the acoustic wave diffracts the light [18], and is used in such applications as the imaging of elastic waves by Bragg diffraction, acousto-optic light modulators, optical beam deflectors, switches, spectrum analyzers, and optical information processing. Acousto-optic beam deflectors are widely used in the modelocking of lasers.

An application that combines SAWs and acousto-optics is a polarization-independent acoustically tuned optical filter [19]. The device illustrated in Fig. 4 was fabricated on x-cut $LiNbO_3$, with both the acoustic waves and the guided optical waves propagating in the *y* direction. The acousto-optic interaction region functions as a TE-TM polarization mode converter, converting those frequencies which are selected to be phase matched by choice of the acoustic frequency from the TE (horizontal) optical waveguide mode to TM (vertical) in one of the optical waveguides and TM to TE in the other. Suppose that an optical fiber inputs both TE and TM components to the planar optical waveguide. The first part of the pictured device is a TE-TM mode splitter, which has the property that the TE mode passes through in the same guide, but the TM mode crosses to the other guide. In the acoustic interaction region, each of these modes is converted to the other, but only for the narrow band of optical frequencies selected by the SAW frequency through the phase matching condition. The filtered frequencies go to one output (F) of the second splitter, while the unselected frequencies go to the other output (U).

See also ACOUSTICS; FATIGUE; INTERNAL FRICTION IN CRYSTALS; MAGNETOACOUSTIC EFFECT; RADAR; RELAXATION PHENOMENA; SOUND, UNDERWATER; TRANSDUCERS; ULTRASONIC BIOPHYSICS.

REFERENCES

1. P. Langevin, *J. Phys. (Paris) Colloq.* **C6**, Suppl. Nos. 11–12 (1972).
2. A. C. Keller, "Submarine Detection by Sonar," *Trans. AIEE* **66,** 1217 (1947).
3. M. B. Gardner, "Mine Mark 24, World War II Acoustic Torpedo," *J. Audio Eng. Soc.* **22,** 614–626 (1974).
4. J. R. Klauder *et al.*, "Theory and Design of Chirp Radars," *Bell Syst. Tech. J.* **39,** 745 (1960).
5. Don Berlincourt, "Piezoelectric Crystals and Ceramics," Chapter 2, pp. 63–124 in *Ultrasonic Transducer Materials* (Oskar E. Mattiat, ed.). Plenum, New York, 1971.
6. A large number of these phenomena and their relation to ultrasonics are discussed in the series *Physical Acoustics* (W. P. Mason and R. N. Thurston, eds.), Vols. 1–19. Academic Press, New York, 1964–1990.
7. J. De Klerk, in *Physical Acoustics* (W. P. Mason, ed.), Vol. 4A, Chapter 5. Academic, New York, 1966.

8. H. Bömmel, *Phys. Rev.* **96,** 220 (1954).
9. W. P. Mason, in *Microplasticity* (Charles J. McMahon, Jr., ed.), Vol. 2, pp. 287–363. Wiley (Interscience, New York, 1968.
10. With *wave*, we in this article use the words *sound, acoustic, elastic, mechanical,* and *ultrasonic* without any essential distinction.
11. James W. Wagner, in *Physical Acoustics,* Vol. 19, Chapter 5, *Ultrasonic Measurement Methods* (R. N. Thurston and Allan D. Pierce, eds.). Academic Press, San Diego, 1990.
12. F. I. Federov, *Theory of Elastic Waves in Crystals.* Plenum Press, New York, 1968.
13. R. N. Thurston, *J. Acoust. Soc. Am.* **64,** 1 (1978).
14. B. A. Auld, *Acoustic Fields and Waves in Solids.* Wiley, New York, 1973.
15. Robert L. Rosenberg, in *Miniaturized and Integrated Filters* (S. K. Mitra and C. F. Kurth, eds.), Chapter 8. Wiley, New York, 1989.
16. *Physical Acoustics* (W. P. Mason and R. N. Thurston, eds.), Vol. 18. Academic Press, San Diego, 1988.
17. W. P. Mason, *Crystal Physics of Interaction Processes.* Academic Press, New York, 1966.
18. Adrian Korpel, *Acousto-optics.* Marcel Dekker, New York, 1988.
19. D. A. Smith, J. E. Baran, J. J. Johnson, and K. W. Cheung, *J. Lightwave Technology* **8** (1990).

Ultraviolet Spectroscopy *see* Visible and Ultraviolet Spectroscopy

Uncertainty Principle

Elliott H. Lieb

Proposed by Heisenberg [1], and later proved by Kennard [2] and Weyl [3] among others, the uncertainty principle states that the momentum and position of a quantum-mechanical particle cannot be simultaneously determined sharply. For any normalized wave function $\psi(\mathbf{x})$, the fluctuations of these quantities (in three dimensions) satisfy the inequality

$$\langle\psi,(\mathbf{p}-\bar{\mathbf{p}})^2\psi\rangle\langle\psi,(\mathbf{x}-\bar{\mathbf{x}})^2\psi\rangle\geq 9\hbar^2/4. \tag{1}$$

Here $\mathbf{p}=-i\hbar\nabla$ is the momentum operator, $\mathbf{x}$ is the position operator (in x space) and $\bar{\mathbf{p}}=\langle\psi,\mathbf{p}\psi\rangle$ and $\bar{\mathbf{x}}=\langle\psi,\mathbf{x}\psi\rangle$ are, respectively, the average momentum and position. For any operator (observable) A, the expected value of A is $\langle\psi,A\psi\rangle=\int\psi(\mathbf{x})^*(A\psi)(\mathbf{x})\,d^3x$. The normalization condition is $\langle\psi,\psi\rangle=\int|\psi(\mathbf{x})|^2\,d^3x=1$. Weyl's proof uses the commutation relation $\mathbf{p}\cdot\mathbf{x}-\mathbf{x}\cdot\mathbf{p}=-3i\hbar$ and then the Schwarz inequality. (1) becomes an equality for any Gaussian, $\psi(\mathbf{x})=(a/\pi)^{3/4}\exp(-a|\mathbf{x}|^2/2)$, and it also holds if the terms $\bar{\mathbf{p}}$ and $\bar{\mathbf{x}}$ are omitted. In fact, we can always reduce to the case $\bar{\mathbf{p}}=\bar{\mathbf{x}}=0$ by replacing $\psi(\mathbf{x})$ by $\phi(\mathbf{x})=\exp(-i\bar{\mathbf{p}}\cdot\mathbf{x}/\hbar)\psi(\mathbf{x}+\bar{\mathbf{x}})$.

Inequality (1) is also called the principle of *indeterminacy*. This terminology presents several logical pitfalls which should be avoided:

1. Although it is true that momentum and position cannot be simultaneously specified sharply (so that there are no fluctuations in either quantity), this fact should not be confused with the probabilistic interpretation of quantum mechanics, i.e., that the theory only predicts the probability, and not the certainty, of an experimental outcome.
2. Given $\psi(\mathbf{x})$ at any time, the Schrödinger equation gives a unique, *deterministic* time evolution for $\psi(\mathbf{x})$, and hence for any $\langle\psi,A\psi\rangle$. All (1) means is that the classical quantities $\mathbf{p}$ and $\mathbf{x}$ do not (nor should they be expected to) carry over sharply into quantum mechanics.
3. Classically, a particle has six degrees of freedom ($\mathbf{p}$ and $\mathbf{x}$). Inequality (1) does not imply that a particle has fewer degrees of freedom in quantum mechanics; on the contrary, it has infinitely many, since a state is described by a function whose value $\psi(\mathbf{x})$ must be given for each $\mathbf{x}$.

While (1) is simple, it does not accurately reflect the basic fact that the kinetic energy, $T=\langle\psi,\mathbf{p}^2\psi\rangle/2m$ where m is the mass, increases when a wave function is compressed *anywhere*. Indeed, (1) would permit ψ to consist of two highly compressed, but well-separated, wave packets, and yet have low kinetic energy because $\langle\psi,\mathbf{x}^2\psi\rangle$ is large. There exist better, and technically more useful, uncertainty principles. By this is meant a lower bound for $(2m/\hbar^2)T=\int|\nabla\psi(\mathbf{x})|^2\,d^3x$ in terms of an integral involving $\rho(\mathbf{x})$ and $\mathbf{x}$, where $\rho(\mathbf{x})=|\psi(\mathbf{x})|^2$ is the probability density for finding the particle at $\mathbf{x}$. We always assume that $\int\rho(\mathbf{x})\,d^3x=1$ and that space is three-dimensional.

The Sobolev inequality[4,5] is in some sense the strongest and most useful:

$$(2m/\hbar^2)T\geq K_3\left[\int\rho(\mathbf{x})^3\,d^3x\right]^{1/3}, \tag{2}$$
$$K_3=3(\pi^2/4)^{2/3}\approx 5.478.$$

This can be generalized[5] for $1<q\leq 3$ to

$$(2m/\hbar^2)T\geq K_q\left[\int\rho(\mathbf{x})^q\,d^3x\right]^r \tag{3}$$

where $r=\frac{2}{3}(q-1)^{-1}$. In particular, when $q=\frac{5}{3}$ and $r=1$, $K_{5/3}=9.578$. For $q<3$, K_q must be determined from the numerical solution of a nonlinear Schrödinger equation. The right-hand sides of (2) and (3) are not simply the expectation values of an operator, but they do have the desired feature of showing that the compression of a wave function anywhere increases T. There are other inequalities of this kind.[6]

Inequality (2) can be used[7-9] to find simple lower bounds for the ground-state energy E_0 of a Schrödinger Hamiltonian $H=\mathbf{p}^2/2m+V(x)$. For any normalized ψ,

$$\langle\psi,H\psi\rangle\geq(\hbar^2/2m)K_3\left[\int\rho(\mathbf{x})^3\,d^3x\right]^{1/3}+\int\rho(\mathbf{x})V(\mathbf{x})\,d^3x. \tag{4}$$

The right-hand side of (4) can easily be minimized subject to $\int\rho(\mathbf{x})\,d^3x=1$. As an example, when $V(x)=-Ze^2|x|^{-1}$ (hydrogen atom), the minimum is $E_0'=-\frac{2}{3}m\hbar^{-2}Z^2e^4$ while the correct value is $E_0 = -\frac{1}{2}m\hbar^{-2}Z^2e^4$. Inequality (1) is not capable of yielding a finite lower bound to E_0 for the hydrogen atom.

For N particles that are *fermions* we can obtain stronger estimates on T than that obtained by considering one particle at a time[5,7,8]. This is so because of the Pauli exclusion principle. If $\psi(\mathbf{x}_1,\ldots,\mathbf{x}_N;\ \sigma_1,\ldots,\sigma_N)$ is a normalized N-particle wave function that is antisymmetric in the space-spin variables $(\mathbf{x}_i,\sigma_i)$, then

$$T \equiv N\frac{\hbar^2}{2m}\sum_{\sigma_i=1}^{s}\int |\nabla_1\psi(\mathbf{x}_1,\ldots,\mathbf{x}_N;\ \sigma_1,\ldots,\sigma_N)|^2\, d^3x_1\cdots d^3x_N$$

is the kinetic energy; s is the number of spin states per particle (2 for electrons). The probability density function for finding a particle at x is

$$\rho(\mathbf{x}) = N\sum_{\sigma_i=1}^{s}\int |\psi(\mathbf{x},\mathbf{x}_2,\ldots,\mathbf{x}_N;\ \sigma_1,\ldots,\sigma_N)|^2\, d^3x_2\cdots d^3x_N$$

and

$$\int \rho(\mathbf{x})\, d^3x = N.$$

Then

$$T \geq (\hbar^2/2m)L\int \rho(\mathbf{x})^{5/3}\, d^3x$$

where $L = (1.50)(3\pi)^{2/3}(2s)^{-2/3}$.

See also HAMILTONIAN FUNCTION; QUANTUM MECHANICS; SCHRÖDINGER EQUATION.

REFERENCES

1. W. Heisenberg, "Über den anschaulichen Inhalt der quantentheoretischen Kinematik und Mechanik," *Z. Phys.* **43** 172–198 (1927). (I) See also W. Heisenberg, *The Physical Principles of the Quantum Theory*. Dover, New York, 1930. (E)
2. E. H. Kennard, "Zur Quantenmechanik einfacher Bewegungstypen," *Z. Phys.* **44** 326–352 (1927). (I)
3. H. Weyl, *Gruppentheorie und Quantenmechanik*, p. 272. Hirzel, Leipzig, 1928. English translation by H. P. Robertson (Dover, New York), p. 393. (I)
4. S. L. Sobolev, *Mat. Sb.* **46** 471–497 (1938). (A)
5. E. H. Lieb and W. E. Thirring, "Inequalities for the Moments of the Eigenvalues of the Schrödinger Hamiltonian and Their Relation to Sobolev Inequalities," in *Studies in Mathematical Physics* (E. H. Lieb, B. Simon, and A. S. Wightman, eds.), pp. 269–303. Princeton Univ. Press, Princeton, NJ, 1976. (I)
6. V. Bargmann, "Note on Some Integral Inequalities," *Helv. Phys. Acta* **45**, 249–257 (1972). (I)
7. E. H. Lieb, "The Stability of Matter," *Rev. Mod. Phys.* **48,** 553–569 (1976). (I)
8. E. H. Lieb and W. E. Thirring, "Bound for the Kinetic Energy of Fermions Which Proves the Stability of Matter," *Phys. Rev.* **35,** 687–689 (1975). (I)
9. W. G. Faris, "Inequalities and Uncertainty Principles," *J. Math. Phys.* **19,** 461–466 (1978). (A)

Underwater Sound *see* Sound, Underwater

Units *see* Symbols, Units, and Nomenclature

Universe

George O. Abell

According to Webster, the universe is the whole body of things and phenomena observed or postulated. To the Greeks of antiquity, the universe was the spherical earth and the crystalline *celestial sphere* that surrounded the earth and rotated daily on an axis through its center. Pythagoras envisioned the sun, moon, and planets as carried by separate transparent spheres just inside the celestial sphere, whose independent motions accounted for the gradual wandering of the planets among the stars in the sky.

Copernicus, in 1542, changed the antique view of the universe by placing the sun at its center and putting the earth, along with the other planets, in orbit about the sun. The stars, in the Copernican view, were enormously farther away than Saturn, the most distant planet then known. The observations of the motions of the planets by Tycho Brahe, and the analysis of those observations by Johannes Kepler in the early seventeenth century, led to the realization that the true orbits of the planets in space are ellipses. This was the clue that led Newton to the derivation of his law of gravitation more than half a century later.

Meanwhile, Galileo, a contemporary of Kepler, described telescopic observations of the sky for the first time. He found the Milky Way to be composed of stars too faint to be seen with the unaided eye. A century later it was generally accepted that the stars are, in fact, suns. Thomas Wright suggested, in 1750, that the sun was part of a great system of stars (our galaxy), and in 1755 Immanuel Kant proposed that there were other galaxies, or "island universes," like our own but far beyond it. Late in the eighteenth century, William Herschel made systematic star counts, and demonstrated that Wright was essentially correct. Herschel's model of the Galaxy was like a gigantic "grindstone" of stars, several thousand light-years across, with the sun near the center. (One *light-year*, the distance traveled by light in a year, is 9.46×10^{12} km; by comparison, the distance from the earth to the sun is approximately 1.5×10^8 km, or about 8 light-*minutes*.) The Milky Way is the light from stars in our line of sight as we look edge-on through our disk-shaped galaxy.

Many of the stars in the Galaxy are concentrated into clusters. Some of these, the *globular clusters*, contain tens of thousands of stars. Harlow Shapley, from his analysis of the distribution of globular clusters in space, showed (in 1920) that the Galaxy is far larger than had been supposed; modern estimates put its diameter at about 10^5 light-years. Moreover, Shapley showed that the solar system is *not* at the center of the Galaxy but some two-thirds of the distance from the center to one rim. The vast majority of stars in the Galaxy could not be seen by Herschel because the interstellar space contains matter in the form of very tenuous gas (on the average, about 1 atom per cubic centimeter) and even sparser solid microscopic grains called *interstellar dust*. The dust obscures the light of most stars in the Galaxy that are more than about 3000 light-years from the sun. Today it is known that the entire Galaxy contains more than 10^{11} stars and is rotating, with the sun taking about 2×10^8 years to revolve about the center.

In 1924, Edwin Hubble showed that Kant was correct—ours is only one of billions of galaxies. The nearest are a few hundred thousand light-years away, and the most distant yet observed have distances in excess of 5×10^9 light-years. Many galaxies are associated with strong sources of radio radiation, among them the quasars, now thought to be highly luminous explosive events occurring in the central parts of very remote galaxies.

Galaxies tend to occur in clusters, ranging from groups of a dozen or more (like the *Local Group*, of which our galaxy is a member) to great clusters of thousands of galaxies. These clusters are typically several million light-years in diameter. Recent studies show that there are even larger units of matter—great inhomogeneities averaging perhaps 3×10^8 light-years in diameter, containing many clusters and groups of galaxies, probably individual galaxies, and possibly some sparse intergalactic gas. On a scale much larger than these *superclusters*, however, space appears to be homogeneous.

Einstein's general relativity predicts gravitation on a cosmic scale. Unless there is an unknown cosmic repulsive force to balance gravitation, therefore, the universe should collapse—unless it is expanding at great speed. By 1931 Hubble and Milton Humason, from an analysis of the line-of-sight velocities of distant galaxies (derived from the Doppler shifts of lines in their spectra), showed that the universe is indeed expanding. Present-day observations can be understood if we posit that, at a time about 10–20 thousand million years ago, all of the matter in the universe was packed close together. The currently popular cosmological models, based on general relativity, hypothesize that a cosmic explosion began the expansion. At first the universe consisted only of certain subnuclear particles and radiation, but as it expanded the universe cooled and atomic nuclei, and later atoms, formed. The neutral atoms no longer interacted with the radiation, and so the models predict that the radiation should still be present, flowing past us from all directions in space. This prediction appears to be verified by the observation, since the late 1960s, of extremely isotropic microwave radio radiation. If our interpretation is correct, this radiation is that released from the primeval fireball that started the expansion of the universe.

It remains to be determined whether the universe will expand forever, or whether the mean density of matter in space is great enough to produce enough gravitation to eventually stop it, and produce a subsequent contraction. In mid-1980, the observations slightly favored a universe that will expand forever, but the measurements are too uncertain, and the theoretical assumptions too unverified, for us to regard this prediction as more than extremely tentative.

See also ASTRONOMY, RADIO; COSMOLOGY; GALAXIES; HUBBLE EFFECT; INTERSTELLAR MEDIUM; MILKY WAY; QUASARS.

BIBLIOGRAPHY

General Astronomical Texts

G. O. Abell, *Exploration of the Universe*, 3rd ed. Holt, Rinehart & Winston, New York, 1975. (I)

G. O. Abell, *Realm of the Universe*, 2nd ed. Saunders, Philadelphia, 1980. (E)

R. Jastrow, and M. G. Thompson, *Astronomy, Fundamentals and Frontiers*. McGraw-Hill, New York, 1974. (E)

Elske v. P. Smith, and K. C. Jacobs, *Introductory Astronomy and Astrophysics*. Saunders, Philadelphia, 1973. (I)

Solar System

W. M. Kaula, *An Introduction to Planetary Physics*. Wiley, New York, 1968. (A)

D. Menzel, *Our Sun*. Harvard Univ. Press, Cambridge, Mass., 1959. (E)

F. Whipple, *Earth, Moon, and Planets*. Harvard Univ. Press, Cambridge, Mass., 1968. (E)

Stellar Astronomy and Astrophysics

L. H. Aller, *Atoms, Stars, and Nebulae*. Harvard Univ. Press, Cambridge, Mass., 1971. (I)

B. J. Bok and P. E. Bok, *The Milky Way*, 4th ed. Harvard Univ. Press, Cambridge, Mass., 1973. (E)

J. C. Brandt, *The Sun and Stars*. McGraw-Hill, New York, 1966. (I)

W. K. Rose, *Astrophysics*. Holt, Rinehart & Winston, New York, 1973. (A)

T. L. Swihart, *Astrophysics and Stellar Astronomy*. Wiley, New York, 1968. (A)

Galaxies and Cosmology

E. Hubble, *The Realm of the Nebulae*. Dover, New York, 1958. (I)

P. J. E. Peebles, *Physical Cosmology*. Princeton Univ. Press, Princeton, NJ., 1971. (A)

D. W. Sciama, *Modern Cosmology*, Cambridge Univ. Press, London. (I)

S. Weinberg, *The First Three Minutes*, Basic Books, New York, (1977). (E)

Vacuums and Vacuum Technology

C. M. Van Atta and M. Hablanian

A vacuum can be produced in an isolated chamber by a few basic methods: positive displacement by a variety of mechanisms such as reciprocating or rotary pistons; momentum transfer, where a solid surface or a fluid jet moving at high velocities propels the pumped gas out of the chamber; cryogenic means, where the gases are simply frozen into a solid or into an absorbed state; chemical means, where the gas is removed from space as a result of a chemical reaction having a solid residue; and ionization means, where gas molecules are first ionized and then driven into a solid substrate by the action of high-voltage electric fields. All of these methods are used in practice singly or in combination to achieve pressures as low as 10^{-12} Torr or 10^{-10} Pa (133 Pa = 1 Torr).

A conventional high-vacuum system is illustrated in Fig. 1. The system is initially pumped down by a coarse vacuum pump to a pressure (0.005–0.200 Torr) at which a high-vacuum pump can be put into operation. The transfer pressure depends on the gas evolution and pump capacity.

The gas *throughput* Q, which is proportional to the mass flow, is defined as

$$Q = P_c S = P_c \frac{dV}{dt} \text{ Torr L s}^{-1}, \tag{1}$$

in which P_c is the pressure in the vacuum chamber and S is the *pumping speed* of the system,

$$S = \frac{dV}{dt} = \frac{Q}{P_c} \text{ L s}^{-1}, \tag{2}$$

the volume of gas per second removed from the chamber at the pressure P_c.

On the high-vacuum side of the high-vacuum pump, the pressure is generally less than 10^{-5} Torr, and the mean free path is long compared with the diameter of the connecting duct. In this pressure regime, gas flows by *molecular diffusion* from a region of higher to a region of lower molecular density. The impedance to gas flow due to constrictions or obstacles in the flow path can be particularly critical.

The *conductance* C between two points along the flow path, at which the pressures are P_1 and P_2, is defined as

$$C = Q/(P_1 - P_2) \text{ L s}^{-1}. \tag{3}$$

For a pipe of inside diameter D cm and length L cm, the conductance according to Knudsen is approximately

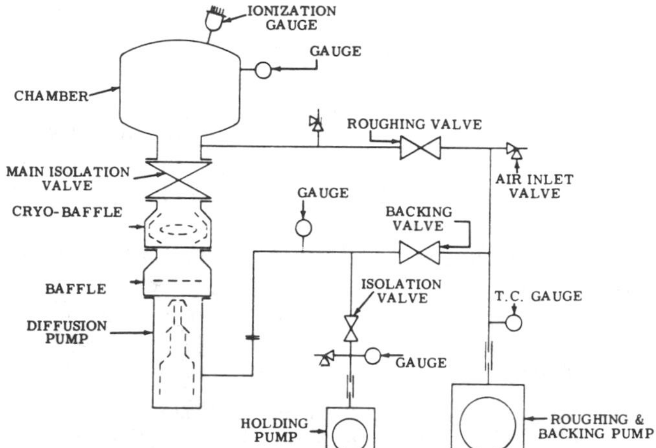

FIG. 1. Conventional high-vacuum system.

$$C = 3.81 \left(\frac{T}{M}\right)^{1/2} \frac{D^3}{L + 4D/3}, \qquad C(\text{air}) = 12.1 \frac{D^3}{L + 4D/3}, \tag{4}$$

in which T is the gas temperature in kelvins, M is the molecular weight of the gas, and $T = 20°\text{C} = 293$ K is assumed for air in the second expression. Several conductances C_1, C_2, C_3, etc., in series result in a combined conductance C given (for long pipes) by

$$\frac{1}{C} = \frac{1}{C_1} + \frac{1}{C_2} + \frac{1}{C_3} + \cdots, \tag{5}$$

and the pumping speed S resulting from a vacuum pump of pumping speed S_p connected through a conductance C is approximately given by

$$\frac{1}{S} = \frac{1}{S_p} + \frac{1}{C}. \tag{6}$$

(Corrections are required when the pipes are short.) Consider a 15-cm-diam high-vacuum pump of 100 L s^{-1} pumping speed connected to a chamber by a 15-cm length of 15-cm-diam pipe, for which the conductance is about 817 L s^{-1} for air [Eq. (4)]. The resulting pumping speed is approximately

$$\frac{1}{S} = \frac{1}{1000} + \frac{1}{817} \quad \text{or} \quad S = 450 \text{ L s}^{-1}, \tag{7}$$

which is a loss of over one-half the pump's pumping speed.

Gas entering a high-vacuum pump is propelled toward the pump outlet into the forevacuum region by action of three or four stages of annular jets of heavy molecules. In the case of a diffusion pump the working fluid is highly refined hydrocarbon oils or high-molecular-weight synthetic "oils" of low vapor pressure at room temperature. The vapor pressure in the jet stack is about 2 Torr at operation temperature of about 200–250°C. The vapor from the jets is condensed on the pump wall which is cooled by water. Some molecules of the working fluid are scattered upward and out the inlet of the pump. To prevent the *backstreaming* vapor from contaminating surfaces in the vacuum chamber, a baffle assembly is installed at the entrance to the diffusion pump. For many applications a single water-cooled or refrigerated baffle array is sufficient; but for processes sensitive to hydrocarbon contamination a baffle at a water-cooling temperature and a second baffle at liquid-nitrogen temperature (−186°C) may be required. At best it is possible to design an effective vapor baffle system with conductance about equal to the speed of the pump. The pumping speed is thus reduced to one-half that of the pump by the baffle system alone. The pumping speed versus pressure of a typical diffusion pump is shown in Fig. 2.

Oil-sealed, rotary vacuum pumps are available in single-stage units with displacements from about $\frac{1}{2}$ to 450 L s^{-1} and in double-stage units from about $\frac{1}{2}$ to 50 L s^{-1}. The *permanent gas* ultimate pressure of single-stage pumps is typically 5×10^{-3} Torr, and that of double-stage pumps about 1×10^{-4} Torr. The pumping speed of single-stage pumps decreases slowly with decreasing pressure from about 0.90 of the displacement at atmospheric pressure to about 0.70 at an inlet pressure of 0.25 Torr. The pumping speed of double-stage pumps is about 0.87 of the displacement at atmospheric pressure and 0.80 at 0.25 Torr.

A vacuum system as described up to this point can reach and maintain a pressure of about 10^{-8} Torr with no additional accessories. The vacuum chamber, preferably made of stainless steel or aluminum, should be as smooth and free of oil as possible. The flanges for assembling the components of the system may be sealed with O-ring gaskets placed in grooves of such depth that metal-to-metal contact between the flanges results in about a one-third compression of the O-rings. The most commonly used elastomer for O-rings is Buna-N because of its excellent mechanical properties and resistance to deterioration when exposed to oil. However, butyl, Neoprene, and Viton A are far better for vacuum service because of lower degassing rates and tolerance of high temperatures.

Materials condensed and adsorbed on the interior surfaces of a vacuum system are slow to leave the surfaces at room temperature and thus prolong the pumping time required to reach a desired ultimate pressure. The rate of outgassing of these surface contaminants is a very sensitive function of the temperature, so that systems designed for service below 10^{-8} Torr should be provided with means of heating the chamber and as much of the system as possible. A mild baking temperature of 100°C will speed up outgassing considerably. However, by using Viton A O-rings and baking the system at 200°C, pressures down to 10^{-9} Torr or lower can be attained. Significantly lower pressures can be attained by using metal gaskets and bakable valves with no elastomers at the seals. Such systems can be designed for baking at temperatures up to 400°C and for base pressures of 10^{-10} Torr or less.

Pumping means other than those described above are available. The Roots blower type of rotary mechanical booster pump illustrated in Fig. 3 is most useful in large systems to facilitate speedy pumpdown to operating pressure and pumping of large throughputs. This type of positive-displacement mechanical pump requires no sealing oil, but depends on the low conductance through small radial and axial clearances at pressures such that the molecular mean free path is greater than about one-twelfth the clearances. At pressures below this critical value the internal leakage back through the pump is a minimum, and the pump can sustain a compression ratio P_{out}/P_{in} of about 50 for air. Mechanical booster pumps are normally backed by single-stage, oil-sealed mechanical pumps with a displacement staging ratio of about 10:1. A mechanical booster pump of 580 L s^{-1}

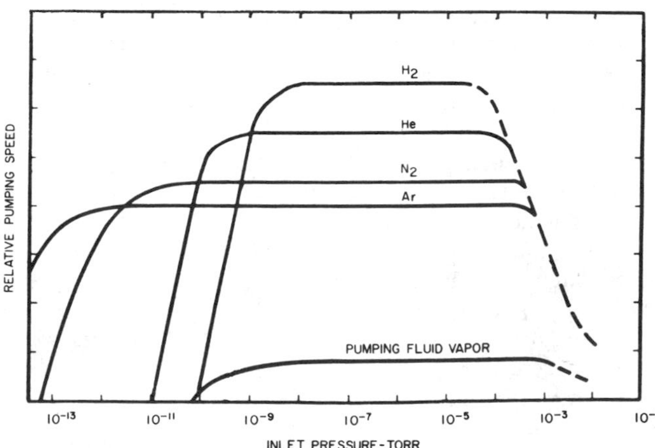

FIG. 2. Pumping speeds of a diffusion pump for various gases.

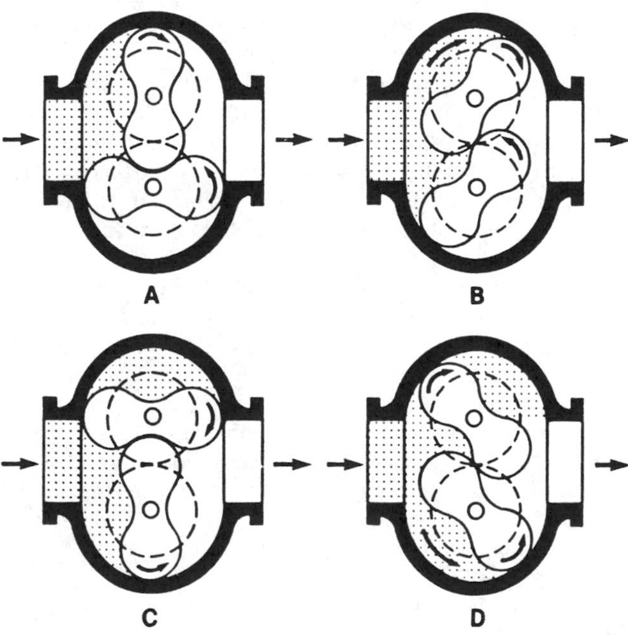

FIG. 3. Cross section of a Roots pump.

displacement backed by a single-stage oil-sealed pump of 61 L s^{-1} displacement has a pumping speed for air of about 470 L s^{-1} from an inlet pressure of about 1 Torr down to about 5×10^{-3}, below which the speed decreases to zero at about 1×10^{-4} Torr. For inlet pressures above 1 Torr, the pumping speed decreases with increasing pressure, reaching one-half its maximum value at about 25 Torr. Since a mechanical booster pump is not oil sealed, it can be comparatively free of contaminating oil. However, even the reduced amounts of oil reaching the inlet of the Roots pump can be disturbing in some vacuum process work. To eliminate the presence of liquid oil in the vacuum pumping train, multistaged Roots pumps are used. In four, five, or six stages such pumps can reach a vacuum of about 10 mTorr while discharging to atmosphere. Roots pumps contain lubricants in their bearings and synchronizing gear boxes. Vapors from these lubricants can seep through seals and even liquid oil can transfer through the seals as a result of a malfunction. To avoid oil entirely, completely oil-free compressors can be used as vacuum pumps, again producing 10 to 20 mTorr vacuum in four stages. Reciprocating piston pumps lend themselves better to a completely dry operation because both the piston and the cylinder can be easily cooled by the surrounding air (the piston internally and the cylinder externally). The pistons of such devices are lined by a composite elastomeric material having a low coefficient of friction.

The modern turbomolecular pumps were developed in the 1960s. The design consists of a cylindrical housing with bearings supporting an axial shaft carrying rotating disks which alternate with stationary disks mounted in the housing. The rotational speed of a small unit with disk diameter of 17 cm is 16 000 rpm. The inlet is at the top, leading into the arrays of disks on the shaft and housing. The stationary disks have radial slots cut at an angle to transmit predominantly those molecules moving within an angular range favorable to be caught by similar, oppositely oriented slots in the moving disks and bounced toward the next stationary disk. The general arrangement of the turbomolecular pump is illustrated in Fig. 4. In a small model of turbomolecular pump, there are 9 to 15 disk pairs or stages. The device is capable of maintaining a compression ratio of 10^4 for hydrogen, about 10^8 or more for air, and much higher for heavy molecules. When it is backed by a double-stage, oil-sealed mechanical pump, pressure of the order of 10^{-11} Torr can be attained after mild baking of the vacuum chamber and central portion of the turbomolecular pump housing. Backstreaming from the backing pump is nearly eliminated. The pumping speeds of turbomolecular pumps range from 50 to 9000 L s^{-1}.

Turbomolecular pumps developed in the 1980s often have a molecular drag stage at the exit which achieves exit pressures in the range of 20–40 Torr. The drag stage usually consists of a cylinder attached to the main rotor. In close proximity to the cylinder, there are spiral grooves in the stator in which the pumped gas molecules are propelled toward the exit. In order to free the turbopump from any possibility of backstreaming from its bearing lubricating oil or grease, some pumps are made with magnetic bearings. The magnetic bearings and the higher permissible exhaust pressure permit the use of completely oil-free backing pumps (for example, diaphragm pumps), i.e., an entirely oil-free pumping train.

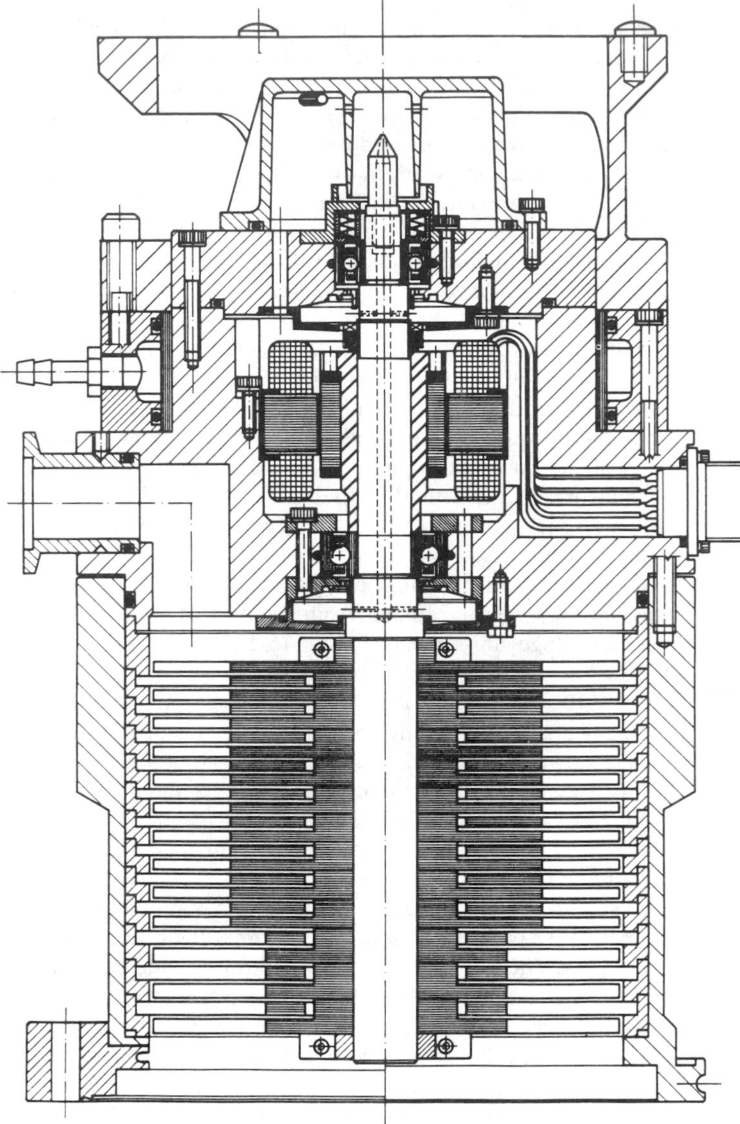

FIG. 4. General arrangement of a turbomolecular vacuum pump.

In recent years, there has been rapid development of getter pumping by evaporation of reactive metals such as titanium, zirconium, and molybdenum on vacuum-exposed surfaces. This procedure results in especially high pumping speeds for most gases—particularly if the coated surfaces are cooled to liquid-nitrogen temperature. Auxiliary pumping, such as a turbomolecular pump, is needed for the initial pumpdown of the system and for subsequent pumping of helium, neon, and methane.

Getter-ion pumps ionize gas molecules and accelerate them into titanium surfaces, partly gettering and partly burying them under sputtered metal. Helium, neon, and methane are problem gases for which auxiliary pumping means are required. Because of extreme sensitivity to hydrocarbon contamination, clean auxiliary pumping such as a sorption pumping device (see below) must be provided.

Artificial zeolite ''molecular sieve'' sorption pumps have been developed to provide a contamination-free means of pumping a system down from atmospheric pressure to about 0.1 Torr. The zeolite is thoroughly outgassed by heating to a temperature of 350°C for several hours, the exhaust valve is then closed and the unit cooled by liquid nitrogen, and finally the inlet valve to the vacuum chamber is opened. For large vacuum chambers, several sorption pumps are connected to a manifold and (after activation) are opened to the chamber, one at a time, until the chamber pressure is low enough for the getter-ion pump to be put into operation.

Cryogenic pumping consists of exposing surfaces maintained at low temperature (20 K or lower) to the gases in a vacuum chamber so that they condense with very low vapor pressure on the cold surfaces. At this temperature, all gases except neon, helium, and hydrogen are condensed with vapor pressures well into the high-vacuum range. In most cryogenic pumps the adsorbing surfaces are cooled by the use of a mechanical refrigeration principle using helium as the heat-transfer fluid. A dual-stage expander section is attached to the vacuum system and the compressor is contained in a separate unit which is connected to the expander by flexible hoses. There are three "freezer" sections. The first stage, producing temperatures near that of liquid nitrogen, is used primarily for pumping water vapor (and other gases of lower vapor pressure). The second stage, at a temperature of 15–20 K, is used to pump most of the gases contained in air. The third stage is also at 15–10 K but it contains molecular absorber bonded to the cooled metallic surface. Such absorbers (usually charcoal) have a very high internal surface area and pores approaching atomic dimensions. Gases such as helium and hydrogen are pumped by cryosorption at the third stage. A typical cryogenic pump is shown in Fig. 5. Very large throughput at low operating pressure is practicable with cryogenic pumping combined with gettering either by getter-ion pumps or by evaporation of reactive metals.

Various combinations of the procedures described above are possible with the result that large gas throughputs can be pumped in the ultrahigh-vacuum range below 10^{-9} Torr. Alternatively, base pressures of 10^{-11}–10^{-12} Torr can be attained in systems of low throughput.

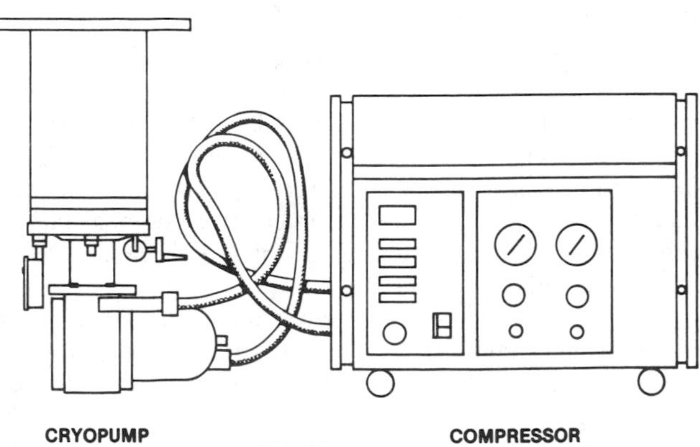

FIG. 5. Construction of a typical cryopump.

BIBLIOGRAPHY

J. F. O'Hanlon, *A User's Guide to Vacuum Technology*. Wiley, New York, 1989.

A. Roth, *Vacuum Technology*. North-Holland, New York, 1976.

G. L. Weissler and R. W. Carlson (eds.), *Vacuum Physics and Technology*. Academic, New York, 1979.

G. Lewin, *Fundamentals of Vacuum Science and Technology*. McGraw-Hill, New York, 1965.

A. Guthrie, *Vacuum Technology*. Wiley, New York, 1965.

G. F. Weston, *Ultrahigh Vacuum Practice*. Butterworths, London, 1985.

P. A. Redhead, J. P. Hobson, and E. V. Kornelsen, *The Physical Basis of Ultrahigh Vacuum*. Chapman & Hall, London, 1968.

A. Berman, *Total Pressure Measurements in Vacuum Technology*. Academic, Orlando, FL, 1985.

S. Dushman, *Scientific Foundations of Vacuum Technique*, 2nd ed., revised by J. M. Lafferty. Wiley, New York, 1962.

C. M. Van Atta, *Vacuum Science and Engineering*, pp. 185–205. McGraw-Hill, New York, 1965.

Vapor Pressure

D. Ambrose

All substances evaporate, and the pressure exerted by the vapor in equilibrium with the solid or liquid phase of a substance is its vapor pressure. It is affected by the curvature of the surface from which evaporation takes place, and the vapor pressure of microscopic droplets is higher than the normal value; this affects the formation of clouds and rain.

The vapor pressures of different substances vary widely. At 25°C, for example, the vapor pressures of many involatile substances are too low to be measurable, whereas that of a volatile substance such as carbon dioxide is about 6 MPa. Vapor pressure depends on temperature, and the temperature at which the vapor pressure of a substance is 1 atm (101 325 Pa) is defined as its normal boiling point; boiling points range from 4.2 K for helium up to, for example, 6000 K for tantalum. Vapor pressures may be measured by a static determination of the pressure exerted or by determination of the temperature of boiling at a known pressure. These techniques become progressively more difficult as the pressure is lowered, and since at pressures below 100 Pa the property of interest is often the concentration of the substance in the gas phase above it, measurement of this concentration is one method by which low vapor pressures are determined.

If vapor pressure is plotted against temperature, the curves for the solid and liquid intersect at the triple point with a discontinuity of slope. Supercooled liquid, which is metastable, has a higher vapor pressure than the stable solid. The exact relationship between vapor pressure p and temperature T is complex, but for many purposes the curves may be adequately described by the approximate equation

$$\log p = A - B/T \tag{1}$$

where A and B are constants for each substance and phase.

The vapor pressure of a substance continues to increase as the temperature is raised until at its critical temperature the properties of vapor and liquid become identical, and at this temperature the vapor pressure is known as the critical pressure. The critical pressures of the majority of substances do not exceed 5 MPa, although a few are much higher than this (e.g., water, 22.05 MPa). Critical properties of only a few hundred elements and compounds have been measured because for many, particularly the involatile elements, the temperatures are too high to be experimentally accessible, and most compounds decompose before the critical temperature is reached.

So far consideration has been restricted to substances that vaporize without decomposition. If the substance vaporizing decomposes reversibly, as do many inorganic compounds at high temperatures, there are different chemical species in the liquid and vapor phases. For example, bismuth telluride Bi_2Te_3 dissociates at 900 K into its elements; the vapor pressure is then the sum of the partial pressure of the species present in the vapor.

The vapor pressure of a mixture of substances is the sum of the partial pressures of its components. The partial pressure p_i of a component i is expressed by the equation

$$p_i = f_i p_i^0 x_i, \tag{2}$$

where x_i is the mole fraction of the component, p_i^0 its vapor pressure in the pure state at the prevailing temperature, and f_i its activity coefficient. The activity coefficient depends on the interactions between the components of the mixture and is a function of the composition. However, if one component is present in very small concentration (solute), its activity coefficient may be independent of the composition (Henry's law), whereas the activity coefficient of the major component (solvent) is unity (Raoult's law); if the components are sufficiently similar in character, Raoult's law may be applicable at all compositions. The concentrations of the components in the vapor phase usually differ from their concentrations in the liquid phase, and if this is so the components may be separated by distillation.

The vapor pressure p of a solution consisting of an involatile solute B in a volatile solvent A is less than that of the solvent p_A^0, and the relative lowering $(p_A^0 - p)/p_A^0$ is proportional to x_B, the mole fraction of the solute.

See also PHASE TRANSITIONS; CRITICAL POINTS; EQUATIONS OF STATE.

BIBLIOGRAPHY

P. W. Atkins, *Physical Chemistry*, 3rd ed. Oxford University Press, London, New York, 1986. (A)

C. N. Hinshelwood, *The Structure of Physical Chemistry*. Oxford University Press, London, New York, 1951. (A)

L. Pauling, *General Chemistry*, 2nd ed. Freeman, San Francisco, 1954. (E)

J. S. Rowlinson and F. L. Swinton, *Liquids and Liquid Mixtures*, 3rd ed. Butterworth, London, 1982. (A)

Variables, Hidden *see* Hidden Variables

Vector and Tensor Analysis

Domina Eberle Spencer

For the mathematical representation of the fundamental concepts of physics, it is desirable to employ *holors* [1]: that is, single entities that are defined by the specification of one or more *merates*. The simplest holor is specified by a single number, ϕ. Univalent holors may be written with a superscript as u^i or with a subscript as v_i, where the index $i = 1, 2, \ldots, n$. Bivalent holors may be written as s^{ij}, t_{ij}, or δ^i_j, where indices $i, j = 1, 2, \ldots, n$. More complicated holors can be constructed with an unlimited number of indices, such as Γ^i_{jk}, R^i_{jkl}, R_{ijkl}, and δ^{ijk}_{lmn}.

Tensors are holors that have a particularly simple transformation equation. A nilvalent tensor ϕ is called a *scalar*. It has the same numerical value in all coordinates, so its transformation equation is

$$\phi = \phi' \tag{1}$$

where ϕ is the value of the scalar in coordinate system x^i (say, spherical coordinates) and ϕ' is the value of the scalar in coordinate system $x^{i'}$ (say, rectangular coordinates). A quantity such as temperature or electric potential has the same value at point P no matter what coordinate system is used to identify the point P.

Another special case of a tensor is a *vector* [2]. A *vector* is a univalent holor that has a particularly simple transformation equation when the point coordinates x^i are changed to another system of point coordinates $x^{i'}$. The relation between the merates of a *contravariant* vector $u^{i'}$ in the primed coordinate system and the merates in the unprimed coordinate system u^i must be a linear homogeneous equation with coefficients the partial derivatives of the point coordinates $\partial x^{i'}/\partial x^i$,

$$u^{i'} = \frac{\partial x^{i'}}{\partial x^i} u^i. \tag{2}$$

Equally important is the *covariant* vector, for which the transformation coefficients are the inverse matrix of $\partial x^{i'}/\partial x^i$,

$$v_{i'} = \frac{\partial x^i}{\partial x^{i'}} v_i. \tag{3}$$

The inverse relationship of the two sets of coefficients is expressed by the equation

$$\frac{\partial x^{i'}}{\partial x^i}\frac{\partial x^i}{\partial x^{j'}} = \delta^{i'}_{j'}, \tag{4}$$

where $\delta^{i'}_{j'}$ is the Kronecker delta.

A tensor may have any number of indices. Its transformation equation is required to be linear homogeneous and its coefficients must contain one matrix coefficient $\partial x^{i'}/\partial x^i$ for each contravariant index and one inverse matrix coefficient $\partial x^i/\partial x^{i'}$ for each covariant index. For instance, the Riemann–Christoffel tensor R^i_{jkl} has the transformation equation

$$R^{i'}_{j'k'l'} = \frac{\partial x^{i'}}{\partial x^i}\frac{\partial x^j}{\partial x^{j'}}\frac{\partial x^k}{\partial x^{k'}}\frac{\partial x^l}{\partial x^{l'}} R^i_{jkl}. \tag{5}$$

Not all holors [3] that occur in the study of physics are tensors. Other important transformation equations that occur may contain a multiplicative term on the right (as with akinetors) or an additive term.

The algebra of holors is a fertile discipline with a wide variety of possible products. Of particular importance are the *scalar product* of two vectors, which transforms as a scalar, and the *vector product* of two vectors, which transforms as a vector. The scalar product of two contravariant vectors is defined in terms of the *metric coefficients*

$$g_{ij} = \sum_{i'=1'}^{3'} \frac{\partial x^{i'}}{\partial x^i}\frac{\partial x^{i'}}{\partial x^j}, \tag{6}$$

which are the coefficients in the expression for the element of distance ds,

$$(ds)^2 = g_{ij}\, dx^i\, dx^j. \tag{7}$$

The scalar product S of two vectors u^i and v^j is written

$$S = g_{ij} u^i v^j. \tag{8}$$

The vector product is defined in terms of the trivalent alternator

$$\delta^{ijk}_{123} = \begin{cases} 0 & \text{if any two indices are identical} \\ +1 & \text{if } ijk \text{ is a positive permutation of 123,} \\ -1 & \text{if } ijk \text{ is a negative permutation of 123.} \end{cases}$$

The vector product of two covariant vectors is written

$$w^i = \frac{1}{\sqrt{g}}\, \delta^{ijk}_{123}\, u_j v_k, \tag{9}$$

where $g = |g_{ij}|$ is the determinant of the metric tensor. Other products called *gamma products* can be defined to fit the physical application [4]. Other scalar products of contravariant vectors u^i and v^j can be written

$$\gamma_{ij} u^i v^j = S, \tag{10}$$

where the gamma coefficients γ_{ij} can be chosen to fit the application. Other vector products can be written for a covariant vector u_i and a contravariant vector v^j such as

$$w^k = \gamma^{ki}_j u_i v^j. \tag{11}$$

There are many other gamma products.

In a metric space, any vector has both contravariant merates v^i and covariant merates v_j which are related by the metric coefficients

$$v_i = g_{ij} v^j \quad \text{and} \quad v^i = g^{ij} v_j, \tag{12}$$

where g^{ij} is the inverse of g_{ij} and is defined by the equation

$$g^{ij} g_{jk} = \delta^i_k. \tag{13}$$

A vector can also be defined in terms of components $(v)_i$:

$$\mathbf{V} = \mathbf{a}_1 (v)_1 + \mathbf{a}_2 (v)_2 + \mathbf{a}_3 (v)_3, \tag{14}$$

where $\mathbf{a}_i$ are unit vectors in the three coordinate directions and the components $(v)_i$ are related to the contravariant and covariant merates by the equations

$$\begin{aligned} (v)_1 &= \sqrt{g_{11}}\, v^1 = \sqrt{g_{11}}(g^{11}v_1 + g^{12}v_2 + g^{13}v_3), \\ (v)_2 &= \sqrt{g_{22}}\, v^2 = \sqrt{g_{22}}(g^{21}v_1 + g^{22}v_2 + g^{23}v_3), \\ (v)_3 &= \sqrt{g_{33}}\, v^3 = \sqrt{g_{33}}(g^{31}v_1 + g^{32}v_2 + g^{33}v_3). \end{aligned} \tag{15}$$

Most derivatives of tensors do not transform as tensors. Vector calculus [2] deals with some particular combinations of derivatives that transform as scalars or vectors. The gradient of a scalar ϕ transforms as a covariant vector:

$$\text{grad}\ \phi = \frac{\partial \phi}{\partial x^i} = \frac{\partial x^{i'}}{\partial x^i}\frac{\partial \phi}{\partial x^{i'}}. \tag{16}$$

In terms of components the gradient of a scalar is written

$$\begin{aligned} \text{grad}\ \phi = {} & \mathbf{a}_1 \sqrt{g_{11}}\left(g^{11}\frac{\partial \phi}{\partial x^1} + g^{12}\frac{\partial \phi}{\partial x^2} + g^{13}\frac{\partial \phi}{\partial x^3}\right) \\ & + \mathbf{a}_2 \sqrt{g_{22}}\left(g^{21}\frac{\partial \phi}{\partial x^1} + g^{22}\frac{\partial \phi}{\partial x^2} + g^{23}\frac{\partial \phi}{\partial x^3}\right) \\ & + \mathbf{a}_3 \sqrt{g_{33}}\left(g^{31}\frac{\partial \phi}{\partial x^1} + g^{32}\frac{\partial \phi}{\partial x^2} + g^{33}\frac{\partial \phi}{\partial x^3}\right). \end{aligned} \tag{17}$$

The divergence of a vector is a combination of derivatives that transforms as a scalar,

$$\nabla_i v^i = \frac{1}{\sqrt{g}}\frac{\partial}{\partial x^i}(\sqrt{g}\, v^i). \tag{18}$$

In terms of components, the divergence is written

$$\operatorname{div}\mathbf{v}=\frac{1}{\sqrt{g}}\left\{\frac{\partial}{\partial x^1}\left[\left(\frac{g}{g_{11}}\right)^{1/2}(v)_1\right]+\frac{\partial}{\partial x^2}\left[\left(\frac{g}{g_{22}}\right)^{1/2}(v)_2\right]+\frac{\partial}{\partial x^3}\left[\left(\frac{g}{g_{33}}\right)^{1/2}(v)_3\right]\right\}. \tag{19}$$

The curl of a vector is a combination of derivatives that transforms as a vector,

$$w^i=\frac{1}{2}\frac{1}{\sqrt{g}}\delta^{ijk}_{123}\left(\frac{\partial v_k}{\partial x^j}-\frac{\partial v_j}{\partial x^k}\right). \tag{20}$$

In terms of components, the curl of a vector is written

$$\begin{aligned}\operatorname{curl}\mathbf{v}=\mathbf{a}_1\left(\frac{g_{11}}{g}\right)^{1/2}&\left[\frac{\partial}{\partial x^2}\left(g_{31}\frac{(v)_1}{g_{11}}+g_{32}\frac{(v)_2}{g_{22}}+g_{33}\frac{(v)_3}{g_{33}}\right)-\frac{\partial}{\partial x^3}\left(g_{21}\frac{(v)_1}{g_{11}}+g_{22}\frac{(v)_2}{g_{22}}+g_{23}\frac{(v)_3}{g_{33}}\right)\right]\\ +\mathbf{a}_2\left(\frac{g_{22}}{g}\right)^{1/2}&\left[\frac{\partial}{\partial x^3}\left(g_{11}\frac{(v)_1}{g_{11}}+g_{12}\frac{(v)_2}{g_{22}}+g_{13}\frac{(v)_3}{g_{33}}\right)-\frac{\partial}{\partial x^1}\left(g_{31}\frac{(v)_1}{g_{11}}+g_{32}\frac{(v)_2}{g_{22}}+g_{33}\frac{(v)_3}{g_{33}}\right)\right]\\ +\mathbf{a}_3\left(\frac{g_{33}}{g}\right)^{1/2}&\left[\frac{\partial}{\partial x^1}\left(g_{21}\frac{(v)_1}{g_{11}}+g_{22}\frac{(v)_2}{g_{22}}+g_{23}\frac{(v)_3}{g_{33}}\right)-\frac{\partial}{\partial x^2}\left(g_{11}\frac{(v)_1}{g_{11}}+g_{12}\frac{(v)_2}{g_{22}}+g_{13}\frac{(v)_3}{g_{33}}\right)\right].\end{aligned} \tag{21}$$

Essential in the formulation of the differential equations of mathematical physics are the scalar and vector Laplacians. The *scalar Laplacian* is defined as

$$\nabla^2\phi=\operatorname{div}\operatorname{grad}\phi. \tag{22}$$

and the *vector Laplacian* as

$$\text{✡}\ \mathbf{v}=\operatorname{grad}\operatorname{div}\mathbf{v}-\operatorname{curl}\operatorname{curl}\mathbf{v}. \tag{23}$$

Tensor derivatives of tensors are called *covariant derivatives*. For any type of tensor the covariant derivatives are defined in terms of the linear connection Γ^i_{jk}, which is defined only by its transformation equation:

$$\Gamma^i_{jk}=\frac{\partial x^i}{\partial x^{i'}}\frac{\partial x^{j'}}{\partial x^j}\frac{\partial x^{k'}}{\partial x^k}\Gamma^{i'}_{j'k'}+\frac{\partial^2 x^{i'}}{\partial x^j\partial x^k}\frac{\partial x^i}{\partial x^{i'}}. \tag{24}$$

In the general case in which $\Gamma^i_{jk}\neq\Gamma^i_{kj}$, there are two [1] covariant derivatives

$$\overset{(1)}{\nabla}_j v^i=\frac{\partial v^i}{\partial x^j}+v^l\Gamma^i_{jl}, \tag{25a}$$

and

$$\overset{(2)}{\nabla}_j v^i=\frac{\partial v^i}{\partial x^j}+v^l\Gamma^i_{lj}, \tag{25b}$$

which transform as mixed tensors. For a covariant vector u_i, there are two [1] covariant derivatives

$$\overset{(1)}{\nabla}_j u_i=\frac{\partial u_i}{\partial x^j}-u_l\Gamma^l_{ji}, \tag{26a}$$

and

$$\overset{(2)}{\nabla}_j u_i=\frac{\partial u_i}{\partial x^j}-u_l\Gamma^l_{ij} \tag{26b}$$

which transform as bivalent covariant tensors. The covariant derivatives are related by

$$\overset{(2)}{\nabla}_j v^i=\overset{(1)}{\nabla}_j v^i-2v^l S^i_{jl}, \tag{27a}$$

$$\overset{(2)}{\nabla}_j u_i=\overset{(1)}{\nabla}_j u_i+2u_l S^l_{ji}, \tag{27b}$$

where

$$S^i_{jk}=\tfrac{1}{2}(\Gamma^i_{jk}-\Gamma^i_{kj}). \tag{28}$$

The concept of the covariant derivative is readily extended to tensors of any valence. There are 2^m possible covariant derivatives of a tensor of valence m, each containing m terms involving the linear connection. For example, for a mixed tensor of valence 2 there are four [1] covariant derivatives:

$$\overset{(1)}{\nabla}_k w^i_j=\frac{\partial w^i_j}{\partial x^k}+w^l_j\Gamma^i_{kl}-w^i_l\Gamma^l_{kj}, \tag{29a}$$

$$\overset{(2)}{\nabla}_k w^i_j=\frac{\partial w^i_j}{\partial x^k}+w^l_j\Gamma^i_{kl}-w^i_l\Gamma^l_{jk}, \tag{29b}$$

$$\overset{(3)}{\nabla}_k w^i_j=\frac{\partial w^i_j}{\partial x^k}+w^l_j\Gamma^i_{lk}-w^i_l\Gamma^l_{kj}, \tag{29c}$$

$$\overset{(4)}{\nabla}_k w^i_j=\frac{\partial w^i_j}{\partial x^k}+w^l_j\Gamma^i_{lk}-w^i_l\Gamma^l_{jk}. \tag{29d}$$

Essential to the development of tensor calculus are the two Riemann–Christoffel tensors [1]

$$R^i_{jkl}=\frac{\partial\Gamma^k_{jl}}{\partial x^i}-\frac{\partial\Gamma^k_{il}}{\partial x^j}+\Gamma^m_{jl}\Gamma^k_{im}-\Gamma^m_{il}\Gamma^k_{jm}, \tag{30a}$$

$$T^k_{ijl}=\frac{\partial\Gamma^k_{lj}}{\partial x^i}-\frac{\partial\Gamma^k_{li}}{\partial x^j}+\Gamma^m_{lj}\Gamma^k_{mi}-\Gamma^m_{li}\Gamma^k_{mj}. \tag{30b}$$

The Riemann–Christoffel tensors can be shown to be the key to three important problems of tensor calculus:

1. commutivity of the covariant derivative,
2. integrability,
3. curvature of space.

Thus, the entirc subject of tensor calculus depends on an essential holor Γ^i_{jk} that does not transform as a tensor.

Contracted covariant derivatives are also of importance [5]. When applied to a contravariant vector v^i in a metric space where Γ^i_{jk} is symmetric in j and k, this process gives the divergence [2] of the vector v^i. However, there are eight distinct contracted covariant derivatives of a contravariant bivalent tensor T^{kj} and four distinct contracted covariant derivatives of a mixed tensor T^k_j. These are similar to the divergence only in rectangular coordinates in Euclidean space [1].

BIBLIOGRAPHY

1. Parry Moon and Domina Eberle Spencer, *Theory of Holors*. Cambridge University Press, Cambridge, 1986; Dirk Jan Struik, *Theory of Linear Connections*. Springer-Verlag, Berlin and New York, 1934; Jan Arnoldus Schouten, *Ricci-calculus*. Springer-Verlag, Berlin and New York, 1954.
2. Parry Moon and Domina Eberle Spencer, *Vectors*. Van Nostrand Reinhold, Princeton, NJ, 1965; University of Connecticut Paperback, Storrs, CN, 1980.
3. Shama Y. Uma, "The Transformation of the Electromagnetic Field Vectors in Accelerated Systems," Ph.D. Thesis, University of Connecticut, Storrs, CN, 1987.
4. Parry Moon and Domina Eberle Spencer, "A New Mathematical Representation of Alternating Currents," *Tensor* **14,** 110 (1963).
5. Philip J. Mann and Domina Eberle Spencer, "The meaning of the 'divergence' of a bivalent tensor," presented to the American Mathematical Society, Albany, NY, August 9, 1983.

Vibrations, Mechanical

S. I. Hayek

Vibration is a phenomenon that occurs when a mechanical system has a restoring force that is proportional to system displacement from a state of stable equilibrium. Mechanical vibration occurs in many elastic systems such as structures (buildings, bridges, foundations, and ships) and systems with moving parts such as engines, automobiles, conveyor belts, chains, circular saws, and computer disks. While there are some useful uses of vibration, it is generally considered to have destructive and harmful consequences. Excessive vibration could lead to fatigue in metal structures and the consequent failure after many years of use. Structures excited to resonance lead to sudden and fatal structural failure due to rupture of one or more components of that structure. Vibrating structures cause waves in air that result in annoying noise.

There are three distinct types of vibration:

1. *forced* vibration, where a disturbing external force causes the vibratory motion;
2. *free* vibration, where the structure vibrates without the influence of an external force;
3. *random* vibration, where the external forces are random.

Examples of *forced vibration* are (a) those caused by rotating machinery with a mass unbalance such as engines, motors, rotors, and fans; (b) torsional vibration caused by unbalanced torque driving shafts in machinery; (c) longitudinal vibration caused by reciprocating forces applied to linkages; (d) flexural vibration in structural beams and machinery foundations caused by transverse forces originating from mass unbalance in rotating machinery; (e) vibration of stretched guitar strings excited by strumming; and (f) base-excited vibration of simple oscillators such as a tire–shock absorber system excited by changes in road profile.

For a simple oscillator, consisting of a mass m, spring with spring constant k, and a dashpot with a viscous damping coefficient b, being excited to vibration by a harmonic external force of amplitude F_0 and frequency ω, the displacement of the mass $x(t)$ is governed by

$$m\frac{d^2x}{dt^2}+b\frac{dx}{dt}+kx=F_0 \sin \omega t. \tag{1}$$

The solution for the displacement of the simple oscillator due to the force F_0 is given by

$$x(t)=\left(\frac{F_0}{k}\right)\sin(\omega t-\phi)\left\{\left[1-\left(\frac{\omega}{\omega_0}\right)^2\right]^2+\left(\frac{2\zeta\omega}{\omega_0}\right)^2\right\}^{-1/2}=X(\omega)\sin(\omega t-\phi), \tag{2}$$

where ω_0 is the resonance (natural) frequency given by $\omega_0^2=k/m$, $\zeta=0.5b/\sqrt{km}$, and ϕ is the phase angle given by

$$\tan\phi=\frac{2\zeta\omega_0\omega}{\omega_0^2-\omega^2}. \tag{3}$$

It can be seen that when the frequency of the forced excitation ω equals the resonance frequency ω_0, the displacement amplitude $X(\omega)$ is amplified to levels that can cause excessive motion. At very low frequencies ($\omega \ll \omega_0$), the response is controlled by the stiffness k, whereas at high frequencies, the response is controlled by the mass m. The response frequency spectrum $X(\omega)$ is sharp and highly magnified if the damping coefficient b is low. Damping is measured by the quality factor Q, defined by $Q=1/2\zeta$ which also indicates the amplification factor of the response at resonance, i.e.,

$$X(\omega_0)\cong Q(F_0/k). \tag{4}$$

Lightly damped structures with $Q\cong 1000$ could fatigue under sustained excitation near resonance, due to the stresses caused by the excessive vibration. Some structures (wooden, fiber composite, or welded) have Q's of 50–100 due to inherent damping mechanisms. Designers of such structures strive to lower the Q's of a system to minimize the stresses induced by excessive vibration by the addition of damping treatments. These could take the form of sheets or devices made from elastomers, polymers, or plastic materials.

For forced vibration of an undamped continuous system, the displacement of a structure is governed by the wave equation

$$\nabla^2 u=\frac{(\partial^2 u/\partial t^2)}{c^2}-F, \tag{5}$$

where the displacement u is the displacement of the structure from equilibrium, c is the sound (phase) speed of propagation in the structure, F represents distributed mechanical forces, and ∇^2 is the Laplacian operator. For transverse vibration of a stretched string, longitudinal vibration of a bar, or torsional vibration of torsion bar, the motion obeys the one-dimensional equation with $u=u(x,t)$. The phase speed c depends on the inertial and elastic parameters of the system.

Flexural undamped vibration of a structure is governed by

$$\nabla^4 u + \frac{(\partial^2 u/\partial t^2)}{\alpha^2} = F, \tag{6}$$

where ∇^4 is $\nabla^2\nabla^2$, α^2 is a constant related to the structure's inertial and elastic parameters, and F represents the distributed external mechanical forces on the structure.

Vibration of structures involves the superposition of elastic waves propagating through the structure. These propagating waves establish *standing waves* which have a spatial distribution called *normal modes*. A simple example of this type of vibration is the response of a stretched string of length L fixed on both ends. Its normal modes $u_n(x)$ have the form

$$u_n(x) = \sin(n\pi x/L), \qquad n = 1,2,..., \tag{7}$$

which describes the natural response of the string at different resonance frequencies ω_n given by

$$\omega_n = n\pi c/L, \qquad n = 1,2,.... \tag{8}$$

Free vibration occurs when a system has been excited by an external mechanical force but the force has been removed or when a system is excited to vibration by imparting to it an initial displacement or velocity. The motion of a system is then free of external forces and tends to respond in a decaying sinusoidal pattern with a frequency equal to that of its resonance (natural) frequency. For a simple oscillator, the free vibration of the mass is given by

$$x(t) = X_0 e^{-\zeta\omega_0 t} \sin[\omega_0 t\sqrt{1-\zeta^2} + \phi], \tag{10}$$

where X_0 is amplitude of the motion and ϕ is the phase angle.

Random vibration is caused by nondeterministic force fields that result in nondeterministic (random) response of a system. Typical of these external forces are those generated by airborne pressure of jet engines, pressure fluctuation of turbulent flow of fluids, and the pressure of gusts of winds. This means that the time history of the random force contains a wide spectrum of frequencies. When a system is subjected to such forces, it will oscillate at one or more of its natural frequencies and thus extract energy from the force field at those frequencies.

See also FATIGUE; MUSICAL INSTRUMENTS; TURBULENCE.

BIBLIOGRAHY

William T. Thomson, *Theory of Vibrations with Applications*. Prentice-Hall, Englewood Cliffs, NJ, 1988. (I)

S. Timoshenko, *Vibration Problems in Engineering*. D. Van Nostrand, Princeton, NJ, 1937. (I)

Eugen Skudrzyk, *Simple and Complex Vibratory Systems*. Pennsylvania State University Press, University Park, PA, 1968. (A)

J. C. Snowdon, *Vibration and Shock in Damped Mechanical Systems*. Wiley, New York, 1968. (I)

M. C. Junger and D. Feit, *Sound-Structures and Their Interactions*. MIT Press, Cambridge, MA, 1986. (A)

Viscosity

Robert F. Berg

Viscosity is most often used for the description of *Newtonian* fluids, for which viscosity is independent of the fluid's shear stress and history. Figure 1 shows the simplest situation, Couette flow, where fluid is contained between two parallel plates of area A and separation Δz. The force F required to drag the upper plate with a velocity Δv relative to the lower plate defines the viscosity η according to

$$F = \eta A \frac{\Delta v}{\Delta z}. \tag{1}$$

The *Navier–Stokes* equation describes the motion of a viscous fluid of density ρ due to external forces $\mathbf{K}$ and gradients of the pressure P:

$$\rho\left[\frac{\partial}{\partial t}\mathbf{v} + (\mathbf{v}\cdot\nabla)\mathbf{v}\right] = \rho\mathbf{K} - \nabla\cdot P + \eta\nabla^2\mathbf{v} + \left(\kappa + \frac{\eta}{3}\right)\nabla(\nabla\cdot\mathbf{v}). \tag{2}$$

The *bulk viscosity* κ, responsible for dissipation upon dilatation of the fluid, is negligible in most contexts. For an incompressible fluid $\nabla\cdot\mathbf{v} = 0$ and the last term disappears, leaving the flow determined by the *kinematic viscosity*

$$\nu = \frac{\eta}{\rho}. \tag{3}$$

The accurate calculation of viscosity from first principles has proved intractable except for simple gases at low density where the Chapman–Enskog theory predicts

$$\eta = \frac{5}{16\pi^{1/2}} \frac{(k_B T)^{1/2}}{\sigma^2 \Omega^{(2,2)*}}. \tag{4}$$

where k_B, T, m, and σ are Boltzmann's constant, the temperature, and the molecular mass and diameter. The collision integral $\Omega^{(2,2)*}$, derived from the intermolecular potential, equals 1 for hard spheres but goes roughly as $T^{-1/2}$ for more realistic potentials. Equation (4) is famous for its correct prediction of the density independence of the viscosity of gases.

Liquids, unlike gases, generally show decreasing viscosity with increasing temperature. Many physically motivated empirical equations have been used, the most popular being the Andrade equation:

$$\eta = Ae^{B/T} \tag{5}$$

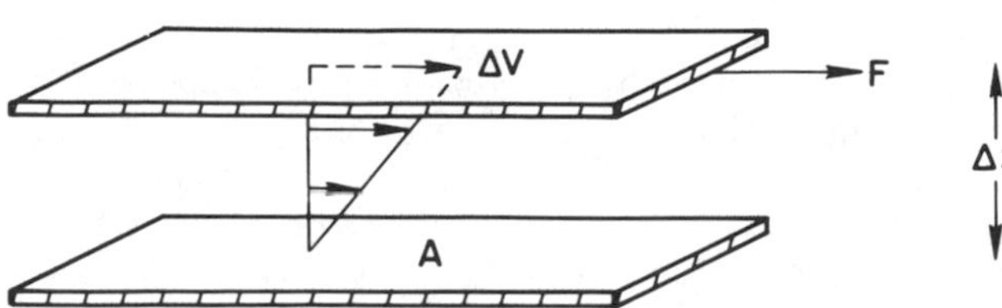

FIG. 1 Laminar shear flow between parallel plates.

Much of the temperature dependence of the viscosity of dense liquids occurs because the viscosity measurements are usually at constant pressure rather than constant volume; this allows the strong density dependence of viscosity to be felt through the liquid's thermal expansion. For many simple liquids, the viscosity can be correlated better through their volume V than their temperature, the simplest form being

$$\eta = \frac{A}{(V - V_0)}, \quad (6)$$

where V_0 is a minimum volume where vitrification is implied to occur. However, water is an interesting counterexample; its viscosity near room temperature actually decreases when the pressure is raised by a few tens of MPa.

The equation of motion, Eq. (2), which allows only temperature and density dependence of the viscosity, does not apply for *non-Newtonian* fluids which can show shear rate dependence and *viscoelasticity*. These common generalizations of viscosity can be required in polymer solutions and melts, colloidal suspensions, supercooled liquids, fluids near critical points, and biological fluids. Viscoelasticity, where the fluid retains a memory of earlier stresses, is often handled in the equation of motion by generalizing the viscosity to a complex number

$$\eta^* \equiv \eta_1 - i\eta_2. \quad (7)$$

In general, if a fluid has an internal relaxation time τ, effects such as shear thickening will be strongest for shear rates S such that $S\tau \approx 1$ and viscoelastic effects will be most pronounced near frequencies ω satisfying $\omega\tau \approx 1$. Generalization of viscosity to a tensor quantity is required for anisotropic fluids such as liquid crystals, magnetized plasmas, and superfluid ^{3}He.

Although a viscosity measurement is sometimes required solely as input to a theory describing, for example, fluid convection or crystal growth, viscosity can also yield information about phase changes, glass transitions, and microscopic effects such as polymer shape changes. Viscometers require controlled flow and can employ vibrating wires, cylinders, plates, or spheres as well a steady flow between cylindrical shells, through capillaries, or around falling spheres. Systematic errors are usually larger than the experimental precision and can be caused by geometrical inaccuracy (including bubbles or particles in liquids), secondary flow (unwanted inertial flows), and sample temperature and composition errors. The fluid velocity is usually assumed to match that of the solid boundaries, an assumption which fails for rarified gases, for example, when the molecular mean free path is comparable to experimental lengths.

The SI units of viscosity are kg m^{-1} s^{-1} abbreviated Pa s (0.1 Pa s = 1 Poise ≡ 1 g cm^{-1} s^{-1}). The dynamic and kinetic viscosities of some common fluids are given below.

Fluid	η (Pa s)	ν (m^2 s^{-1})
^{4}He (4.2 K, liquid)	3.33×10^{-6}	2.67×10^{-8}
Air (20°C)	1.83×10^{-5}	1.52×10^{-5}
Water (20°C)	1.00×10^{-3}	1.00×10^{-6}
Glycerol (20°C)	1.41	1.12

See also ANELASTICITY; FLUID PHYSICS; GLASS; HYDRODYNAMICS; KINETIC THEORY; POLYMERS; RHEOLOGY.

BIBLIOGRAPHY

R. B. Bird, W. E. Stewart, and E. N. Lightfoot, *Transport Phenomena*. Wiley, New York, 1960. Textbook with many examples of particular flow configurations.

S. G. Brush, "Theories of liquid viscosity," *Chem. Rev.* **63,** 513 (1963). Still an excellent introduction.

R. Darby, *Viscoelastic Fluids*. Dekker, New York, 1976. Textbook beginning with basic concepts and proceeding through to generalized rheological equations of state.

J. D. Ferry, *Viscoelastic Properties of Polymers*, 3rd ed. Wiley, New York, 1980. Concepts, models, and experimental techniques.

J. O. Hirschfelder, C. F. Curtiss, and R. B. Bird, *Molecular Theory of Gases and Liquids*. Wiley, New York, 1964. Comprehensive work including detailed expositions of kinetic theories of transport.

J. Kestin and R. DiPippo, "Viscosity of Gases," in *American Institute of Physics Handbook*, 3rd ed. (D. E. Gray, coordinating editor). McGraw-Hill, New York, 1972. Includes carefully assessed data on the absolute viscosities of gases.

L. D. Landau and E. M. Lifshitz, *Fluid Physics*. Pergamon, New York, 1959. Excellent and concise text on the fundamentals of hydrodynamics.

R. S. Marvin and D. L. Hogenboom, "Viscosity of Liquids" in *American Institute of Physics Handbook*, 3rd ed. (D. E. Gray, coordinating editor). McGraw-Hill, New York, 1972; R. S. Marvin, *J., Res. Natl. Bur. Stand.* **75A,** 535 (1971). Especially useful for their discussions on the expected accuracy of viscosities reported in the literature.

D. A. McQuarrie, *Statistical Mechanics*. Harper & Row, New York, 1976. Introductions to kinetic theory and continuum mechanics.

R. C. Reid, J. M. Prausnitz, and B. E. Poling, *The Properties of Gases and Liquids*, 4th ed. McGraw-Hill, New York, 1987. Contains useful data and equations for estimating viscosity when given limited or no experimental data.

Y. S. Touloukian, S. C. Saxena, and P. Hestermans, *Viscosity (Vol. 11 of Thermophysical Properties of Matter)*. IFI/Plenum, Washington, DC, 1975. Text and extensive bibliography on theory, estimation, and measurement plus data on fluids and fluid mixtures.

D. S. Viswanath and G. Natarajan, *Data Book on the Viscosity of Liquids*. Taylor and Francis (Hemisphere), New York, 1989. Summary correlations and plots for about 900 fluids.

Visible and Ultraviolet Spectroscopy

Stanley S. Ballard and Donald C. O'Shea

Spectroscopy is the study of the spectrum of the component colors or wavelengths into which a beam of light can be dispersed. Perhaps the most familiar spectrum is the rainbow, formed when sunlight is dispersed by water droplets in the atmosphere. A rainbow gives rather broad color bands that blend into each other, whereas the techniques of modern spectroscopy can sort out component wavelengths very exactly, often obtaining sharply defined "spectrum lines" whose wavelength can be measured with great accuracy.

WAVELENGTH RANGES

The visible region is a relatively small portion of the broad electromagnetic spectrum, but it is an important region because it includes the wavelengths to which the human eye is sensitive. The wavelength range for the eye is conventionally taken as extending from 400 nm at the violet end to 700 or perhaps 750 nm in the extreme red (1 nm $=10^{-9}$ m).

Beyond the red is the broad infrared region that extends to microwaves, with a nominal wavelength of separation between the two of 10^6 nm or 1 mm. Light of wavelength shorter than visible is termed ultraviolet, and this region overlaps at its short end with the x-ray region, at wavelengths of a few nanometers. The same general experimental techniques are used in the ultraviolet down to 200 nm as in the visible region, but this is not the case for the infrared. Going to even shorter wavelengths introduces difficulties: Air becomes opaque, so vacuum instruments must be used for these shorter wavelengths. Quartz, which largely replaces glass as a transmitting optical material in the ultraviolet, becomes opaque in the "vacuum region" as do most other materials, so reflection rather than refraction must be employed to focus the rays.

SOURCES

For an atom to emit light, which can then be analyzed spectroscopically, it must be excited: that is, it must be raised to an energy state above its most stable, or ground, state. Excitation can be by thermal or electrical means. Heating a solid to a few hundred degrees temperature causes it to glow, as in the tungsten filament of an ordinary lamp bulb. When the light from such an incandescent source is examined with a spectroscope, it disperses into a continuous spectrum, reminiscent of the rainbow. A heated gas or vapor, on the other hand, gives off the specific radiations that are characteristic of the excited atoms. An example is the bright yellow flame of a Bunsen burner when some ordinary salt is sprinkled on it: the yellow is due to the strong resonance lines of sodium. Other sources of radiation are gas discharges (such as a neon sign), fluorescent materials, and electric arcs and sparks. The sources for stellar spectra, of such great value in astrophysics, are the bright stars themselves.

In spectrochemical analysis the sample, which may be a few milligrams of a powdered substance, can be placed in the hollowed end of the lower electrode of a carbon arc or of a high-voltage spark between two metal electrodes. Of course the spectrum of the electrode material is also obtained.

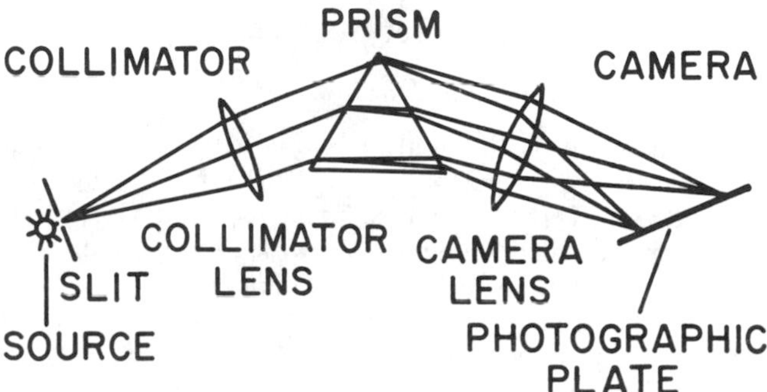

FIG. 1. Optical system of a simple prism spectrograph.

The above are sources of emission spectra. Information on atomic structure is also given by absorption spectra. These are obtained by passing light from an incandescent source through a length of gas or liquid, or sometimes a solid. Lines and bands characteristic of the sample are dark against the bright continuous spectrum of the source. Examples are the Fraunhofer lines in the sun's spectrum; they establish the fact that many familiar elements are present in the cooler outer atmosphere—the photosphere—of the sun.

DISPERSING ELEMENTS AND SPECTROGRAPHS

The optical system of a typical prism spectrograph is sketched in Fig. 1. The "source" may be any of the items mentioned above, or other, more esoteric light sources. Light strikes the narrow vertical slit, enters the collimator, and is rendered parallel by the collimator lens. It then strikes the prism, made of glass for the visible region, usually quartz for the ultraviolet, and perhaps rock salt or calcium or lithium fluoride for the near infrared. Following the prism is the camera lens, which brings light to a focus on the photographic plate, which records the spectrum. (The plate is tilted as shown because the camera lens is not corrected for chromatic aberration.) It should be noted that a *spectrum line* is hence a fairly monochromatic image of the spectrograph slit; in high-quality, high-resolution instruments it could better be referred to as a diffraction pattern of the slit image, formed in light of a narrow wavelength band.

Of practical interest is the dispersion of a spectrograph: this is $dl/d\lambda$, or the distance along the plate that corresponds to a given wavelength interval. Often the reciprocal is used, $d\lambda/dl$, and the units might be nm/mm. Of even greater importance is the resolving power, which is a measure of the sharpness of the spectrum lines and of the smallest wavelength difference that will register as two separate lines rather than one broader one.

A device with larger dispersion and resolving power, replacing the prism of Fig. 1, is the grating. It uses the phenomena of interference and diffraction, whereas the prism employs refraction and its variation with wavelength. The spectrograph of Fig. 1 could be simplified by replacing the collimator and camera lenses and the prism with a grating having a concave mirror surface that brings the diffracted light to a focus on the photographic plate. There are many different types of so-called grating mounts, or spectrographs; they are described in the books by Sawyer and by Harrison *et al.* Since light passes through no refractive elements, from slit to focal plane (photographic plate), a concave grating spectrograph is not limited by transmission considerations and can in principle cover the entire spectral range from soft x rays through the ultraviolet, visible, and infrared to the microwave region.

Prism spectroscopes are still used because of their compactness, simplicity, and stability. They give only a single spectrum, whereas gratings give several orders of spectra, as well as the zero order: the direct image of the slit; also, gratings may form false lines called *ghosts*. It is now possible to shape the grooves of a grating so that much of the incident light is diffracted into a single order, thus competing all the more favorably with prisms. Recently, such gratings can also be manufactured by photographing extremely fine interference fringes. The plates are developed and bleached, giving a very efficient *holographic* grating that does not possess ghosts.

A dispersing device with even higher resolving power is the interferometer, of which there are several different forms. However, an interferometer must always be used with a prism or grating of much lower dispersion, in order to get spectra that can be interpreted.

DETECTORS

Until the *photographic plate* was perfected, early spectroscopists had to rely on their eyes and drawing skills to record spectra. Photography brought with it not only permanent, objective recordings, but also detection in the ultraviolet and the near-infrared regions, where the eye is insensitive. Special dyes can be used in the manufacture of spectrographic plates to increase their sensitivity in certain spectral bands; different emulsions are available to give a range of graininess, which affects the resolution obtainable and, inversely, the photographic speed or length of exposure time. Additional advantages of the photographic plate are that it is cumulative, increasing the signal response with exposure time; it is also comprehensive, recording a large number of emission lines simultaneously. One disadvantage is its nonlinearity, which makes calibration of intensity measurements difficult. The necessary processing of the plate eliminates the possibility of real-time analysis.

The *photomultiplier tube* provides an increase in sensitivity over the photographic plate. Light incident upon a sensitive surface may cause electrons to be ejected from the surface. The resulting photocurrent can be amplified by accelerating the photoelectrons through an electric field onto a dynode, a surface which ejects secondary electrons. These electrons are further amplified by a chain of additional dynodes, so that a pulse of electrons is formed.

When there is enough light, the current from the multiplier can be used to measure the amount of radiation. When the light is faint, as with many weak emission lines, the individual pulses of electrons are counted using standard nuclear-counting electronics. By selecting a particular range of pulse heights, spurious pulses due to noise or cosmic-ray events can be rejected from the signal count.

The sensitivity of the photomultiplier tube depends on wavelength. Usually it is more sensitive in the shorter wavelength (blue and ultraviolet) regions of the spectrum although tubes with extended red sensitivity are available. However, this higher sensitivity at longer wavelengths is a mixed blessing, since electrons can also be released as a result of ambient thermal energy even when the tube is not illuminated. These *dark counts* or *dark currents,* which determine the lower limit of detectability of the photomultiplier, can be reduced by cooling the tube below the ambient temperature.

The advantages of photomultiplier tubes are almost the reverse of the photographic plate. The phototube response is linear over many orders of magnitude, in distinct contrast to the nonlinear response of the photographic plate. In addition, the photomultiplier tube offers real-time response over a narrow band of wavelengths isolated by a slit placed in the image plane of the spectrograph. An array of phototubes, detecting light at several points in the spectrum, can increase the coverage but cannot provide the comprehensive record of the photographic plate.

The most recent advance in detectors, the *electronic photoarray,* combines the comprehensiveness of a photographic plate and the linearity of a photomultiplier tube. An offshoot of low-light-level television, the detector consists of a photoemissive surface, a silicon target, and a readout beam, plus beam control and detection electronics. Photoelectrons from the emissive surface are accelerated and imaged onto the silicon target: a two-dimensional array of photodiodes. The energetic photoelectrons, hitting a diode, cause the surface charge at that point to be depleted. In regions where no photoelectrons strike the surface charge remains unaffected. At this stage the silicon target resembles an undeveloped photographic plate. The electron beam is scanned across the target, restoring the charge in the depleted region and completing an electrical circuit with the scanned diodes. The current needed to replace the surface charge is sensed, amplified, and converted to a digital signal and stored in a computer memory location corresponding to the beam position on the target. The electron beam, in effect, "develops" the silicon target "plate" and, by restoring the surface charge, provides a new "plate." Data are read out of the computer memory using standard numerical and graphical methods.

Such a target, when coupled with a photoemissive surface, is called a silicon-intensified target (SIT). If an image intensifier is placed before the SIT, providing a brighter image for the photosurface, the device is called an intensified silicon-intensified target (ISIT). Other solid-state detectors are under development, so that the ultimate sensitivity and utility of these electronic photoarrays are unknown. The major disadvantage of the electronic photoarray is its high cost.

APPLICATIONS

The two major applications of spectroscopy have been the determination of atomic and molecular structure, and the analysis of "unknown" samples for chemical content. These two are often called *spectrum analysis* and *spectrochemical analysis,* respectively. In the early decades of the 20th century, great advances were made in our understanding of the electron structure of atoms and molecules; for the latter, the infrared spectral region was also important.

Spectrum Analysis

When the spectral features have been detected and converted to some sort of graphical (photographic plate or strip-chart record) or digital readout, the spectroscopist can analyze the results. There are three pieces of information for each spectral feature: its central wavelength or frequency, it breadth or linewidth, and its relative intensity. (Under certain conditions the light may be polarized, an additional piece of information.) Most modern spectrometers are calibrated so that the wavelengths and linewidths are easily determined. If there is no calibration, a standard source with many lines of known wavelengths, such as an iron arc, can provide a calibration spectrum. Information on relative intensities is easily acquired if a linear detector is used; the comparisons are more difficult with a photographic plate. Absolute intensity measurements are considerably more difficult since all light losses and wavelength dependences must be taken into account and the detector must be calibrated with a standard source.

Within an atomic spectrum, the frequencies of emission can, in many cases, be related to one another by simple sums and differences. This is a result of the electronic structure of an atom. By analyzing the spectra of atoms and their ions of successively higher ionization states, experimental evidence is accumulated that enables theorists to compare it with their quantum-mechanical calculations, testing their assumptions and their understanding of the atomic systems.

Spectrochemical Analysis

Since each atomic species emits light with a unique set of wavelengths and relative intensities, a sample of mixed atomic species can be analyzed by vaporizing a small amount of the material and recording a spectrum of the emitted light. Using wavelength tables listing the more prominent lines in the spectra of various atoms, one can first identify the presence of a particular species and then, using intensity ratios and premixed samples of known composition, provide a quantitative measure of the amounts of the various species in the sample. Spectrochemical analysis has become a routine tool in chemical laboratories and manufacturing plants for characterizing many materials and products. Molecular species can be similarly identified and analyzed, but the problem is made more difficult by the complicated nature of molecular spectra.

See also BALMER FORMULA; ELECTROMAGNETIC RADIATION; FRAUNHOFER LINES; GRATINGS, DIFFRACTION; INTERFEROMETERS AND INTERFEROMETRY; MOLECULAR SPECTROSCOPY; PHOTOELECTRON SPECTROSCOPY; PHOTOIONIZATION; RADIOMETRY.

BIBLIOGRAPHY

G. R. Harrison, R. C. Lord, and J. R. Loofbourow, *Practical Spectroscopy*. Prentice-Hall, New York, 1948. (I)

Gerhard Herzberg, *Atomic Spectra and Atomic Structure*, 2nd ed. Dover, New York, 1944). (I)

Ralph A. Sawyer, *Experimental Spectroscopy*, 3rd ed. Dover, New York, 1963. (I)

Vision and Color

Edwin H. Land

A room illuminated to a level of 20 footcandles and observed through tightly fitting goggles with lenses of optical density 4.5 (transmitting 0.003 of 1% of visible light, thus reducing the effective level of illumination to 1/1500 of a footcandle) will become visible in about 30 minutes. In spite of its content of colored objects, the room and its contents will be colorless, with a range of lightnesses from white to black by way of intermediary lightnesses. The lightnesses will be related to the colors of the objects in the fully seen room (i.e., without the goggles) in approximately the way that they would be in a black and white panchromatic photograph taken through a green color-separation filter. In other words, the reds will appear very dark, the greens lighter, far blues dark, whites light, and blacks dark. However, whereas the fully illuminated room regarded through the green-separation filter shows residues of color, the room seen through the density 4.5 goggles after 30 minutes of wearing them is peculiarly and uniquely colorless. It is under these conditions that the observer can study the nature of image formation and the meaning of an image produced with the intermediacy of only one set of receptors. The beginning of the study of color in full images is best made in this situation, which is completely devoid of and completely uncomplicated by the experience of color.

In this colorless world it is found that the nature of the image is not determined by the flux at the observer's eye: the illumination can be easily arranged so there is more flux from a very dark region than from a very light region, whether these regions be three-dimensional objects in the room or artifacts contrived with a montage of dark and light papers. The paradox immediately arises that each of the objects, the papers for example, whether dark or light or in between, maintains its lightness without significant change as it is moved around the scene into regions of higher or lower flux. Light papers will be seen as light and dark papers as dark, simultaneously, even with the same flux coming from each of them to the eye. Strong gradients of flux across the field will be only weakly apparent, if at all.

Furthermore, in an intricate collage of areas of various lightnesses, sizes, and shapes, the lightness of a given member does not change visibly as it is relocated in any part of the collage and associated with a new arbitrary surround. When a small area is *completely* surrounded by a large area, the lightness of the small area will change somewhat depending on whether the large area is darker or lighter than the small area. The impressive fact is that, in general, lightness is not modified by the immediately surrounding areas nor is it modified by the areas surrounding them.

In the very special case of uniform illumination, it can be shown that the lightness of all areas except the lightest is determined by the ratio of flux across the two sides of the edge between the lightest area and any other area if the latter is hypothetically transported to contiguity with the lightest area. While this ratio will be a ratio of reflectances at the contiguous edges, in the much more interesting general case of oblique or nonuniform illumination, reflectance falls away and the lightness of any bounded area is shown as a first approximation to be quite simply the product of all the ratios of flux along a pathway originating in the lightest area and traveling to the area in question. A satisfactory computer program is readily designed to recommend the number of paths, their lengths, convolutions, the threshold value of ratios, and perhaps most important how to utilize all the pathways starting in all areas. These very simple programs provide entirely reliable models in which the lightness of any area will behave according to the rules in the preceding paragraphs. If the further assumption is made that the output of an individual receptor is proportional to the logarithm of the absorbed flux, then the ratio-taking program can become a series of subtractions and the multiplication can become a series of additions if, with the progress of retinal anatomy, subtraction and addition are seen to be more appropriate than division and multiplication. With either of these essentially equivalent mechanisms, however, it is possible to eliminate the role of flux at the very earliest stage after the initial absorption.

An important consequence of an evolutionary program for relating objects to images by subordinating flux and emphasizing lateral ratios is that the extensive and unpredictable variation in illumination, in geography, and in time does not camouflage the objects. The task of delineating the physical characteristics of the objects is handled with the purely physical techniques of lateral standardization (e.g., the product of ratios along the paths from all areas of highest lightness to the given area). All this occurs within the image itself in as yet undetermined parts of the retinal–cortical pathway so that there is a minimum of psychological or intuitive demand on the observer at this early and vital stage of total image formation. As indicated above, the lateral standardization does not utilize averaging of areas or averaging of flux. The process does, however, call for an arithmetic that extends over the whole visual field. Furthermore, since the relevant phenomena are seen in a pulse, all of the computations and conclusions about lightness must be carried out in a fraction of a second without dependence on eye movement. With a single pulse, eye movement, by definition, is not necessary: with continuous illumination, it would appear that the normal quick motions of the eye are required to maintain the process *fresh*.

Thus, what we call an image of an object is not a replication of it, not a metered duplication, rather, it is a shrewd *description* that permits a reliable operational relationship between an animal and the observed environment. One should note in passing that the notional idea that a photograph is a test of "reality" is merely a further demonstration of the competence for *description* possessed by the visual system: The photograph, at best, maintains some of the significant visual criteria in spite of further modulations and deformations of patterns of luminosity and reflectance.

For the observer wearing the density 4.5 goggles the principal phenomena in the vocabulary of the *description* are those in the lightness–darkness series, those in plasticity (e.g., sphericity, cylindricalness), and those associated with the endless array of *Gestalten* (e.g., circle, square, field-and-ground, etc.). It is the first of these categories, namely, the lightness–darkness series, that becomes the basis of color when the observations move to a higher effective level of illumination.

When a black and white photograph is taken with a combination of filter and emulsion sensitizers so chosen as to have a net response curve as a function of wavelength that is the same as that for the set of receptors used with the 1/1500 of a footcandle illumination, lightnesses in the photograph can be examined alongside the lightnesses of the same objects in the room and, after that, the photograph can be examined under normal room illumination without the goggles. Since the lightnesses in the photograph appear essentially the same under both illuminations, a photograph can be used as a convenient record of the initial experience at low light levels (Fig. 1d). If all four sets of receptors could, for experimental purposes, be used independently, then the photograph would be merely a convenience. In fact, however, the only set of receptors that can be used independently is the one employed in the low-light-level experiment because it is isolated from the others by its hypersensitivity: It can be engaged in the image-forming function at a level of illumination at which the other three sets of receptors are not yet operative (Fig. 2).

In recent times measurements have been made of the three other retinal pigments than the one involved with the first

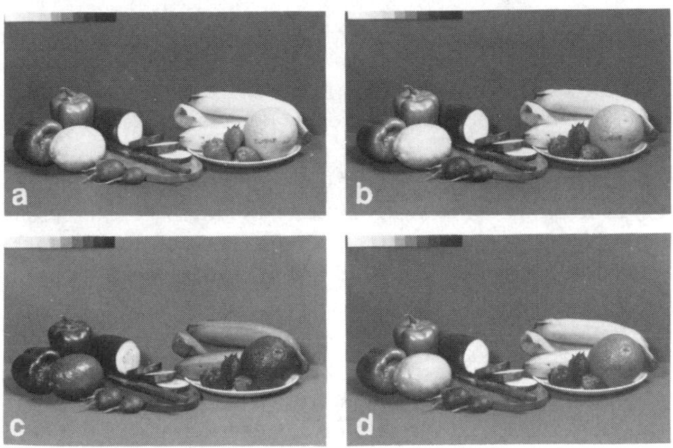

FIG. 1. Photographs of a still life taken with film-filter combinations shown in Fig. 3: (a) long-wave pigment record, (b) middle-wave pigment record, (c) short-wave pigment record, and (d) rod record.

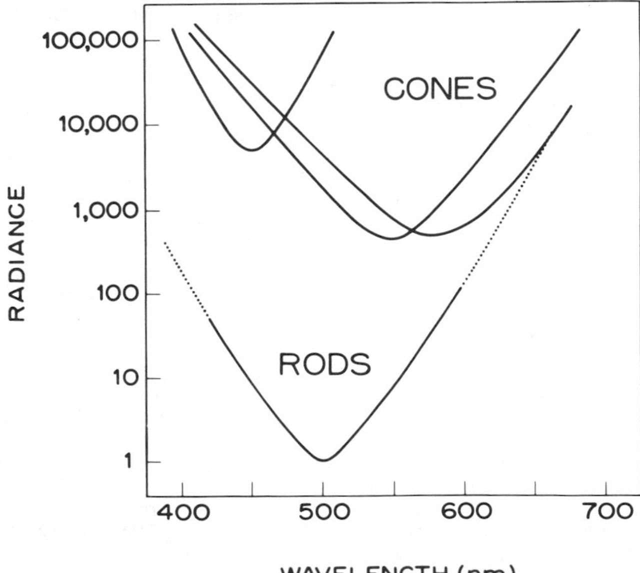

FIG. 2. Plot of wavelength versus the radiance necessary to excite a threshold response of the rods and the various types of cones in the retina. These curves are adapted from G. Wald, *Science* **101**, 653–658 (1945); **145**, 1007–1017 (1964).

set of receptors which we have been studying, that is, the three pigments presumably determining the responses of the three other sets of receptors. Figure 3 shows the response suggested by these measurements. Because of the extensive overlap of these curves, it is not possible to carry out the analogous simple experiment described in the beginning of this article: We cannot see directly the array of lightnesses that are generated to form each of the separate three images. However, since we have established, with the set of receptors that we *could* isolate, that a photograph gives a useful representation of whatever the stimuli are that determine image formation by a set of receptors, we can now make the analogous three photographs, with appropriate sensitizers and filters, that show each of the separate images in terms of the lightness formation by each set of receptors. Figures 1a, 1b, and 1c show this set of photographs.

While the photograph is useful, it is gravely limited by the fact that it is a flux-sensitive medium, the amount of silver deposited being roughly proportional to the amount of flux. Within its limitations the photograph can reproduce many of the gradients and many of the edges, but in the intellectually critical situation in which as much flux is coming to the eye from a jet black region as is coming from a pure white region, the photographic *print* fails. The whole area surrounding the jet black region will be completely overexposed—a phenomenon that does not exist for the essentially flux-insensitive eye. The photographic *transparency* to succeed must have a very great density range, perhaps as great as 4, and be projected with correspondingly bright illumination. All these difficulties are obviated by the visual receptor system in which *ratios* of flux are the first determinants in the lightness process. It is now apparent that color-separation photographs used in color photography as invented by Maxwell deal with the special case of reasonably uniform illumination and are therefore not challenged to perform in the manner of the eye. Conversely, however, this does not *at all* imply that the eye works in the manner of the photograph. With

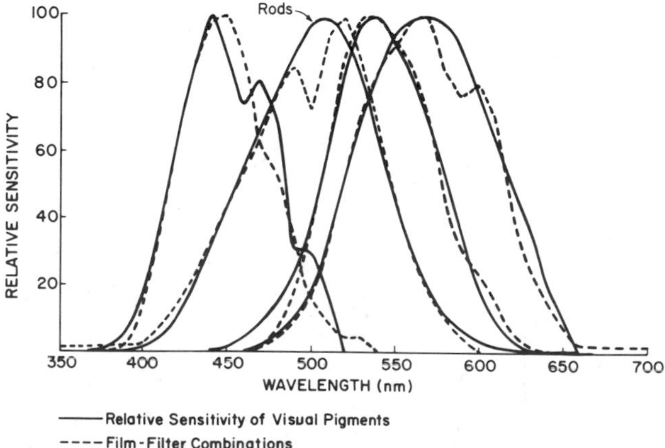

FIG. 3. The normalized spectral sensitivity of the four film–filter combinations used to match the spectral sensitivities of the four pigments in the retina. The film–filter combinations are shown with the dotted lines. The sensitivities of the pigments are shown with solid lines. The cone data are from P. Brown's normalized measurements multiplied by the transmission of the eye and macular pigment. The rod data are the CIE scotopic luminosity curve.

these precautions in mind, the four photographs in Fig. 1 can be examined.

After the three lightnesses have been determined for each set of receptors utilizing whatever interaction between the retina and cortex is required, no further information is necessary to characterize the color of any object in the field of view. Any specific color is a report on a trio of three specific lightnesses. There is a specific and unique color for each trio of lightnesses (Fig. 4).

We have noted that the first experiment at low light level shows that an isolated single system generates an image in terms of lightnesses that is free from color. It has been shown that if, guided by Fig. 2, a colored display is illuminated first by one waveband at 550 nm, the illumination being raised to just beyond the amount necessary to make the display visible after the observer has been in the dark (a condition under which only one receptor system will be operating) and, secondly, by a wave band at 656 nm, and then the illuminants at both wave bands are turned on together, the display will appear in color. The array of lightnesses which characterizes the image formed by the set of hypersensitive receptors is now displaced by an array of colors reporting on the relative lightnesses at any region in the image for the *two* sets of receptors. The significance of using as one member of a pair the hypersensitive set of receptors is that its image can be seen alone and fully described and studied without the complication of color so that when it is then associated as described above with an image on the long-wave set of receptors, the emergence of variegated color can be ascribed to the compared lightnesses for these two sets of receptors. This result makes it plausible that when three images comprising, respectively, the lightnesses of the short-wave set, medium-wave set, and long-wave set are associated to give a full-colored image, it is the comparison of the respective lightnesses, region by region, that determines the color of those regions. These colors, like the colors in the lightness color-space (Fig. 4), are essentially independent of ratios of flux because the lightnesses both in the images and in the lightness color-space are independent of flux.

A dramatic example of this independence is shown when an array of variegated colors and shapes is placed alongside a duplicate array and both arrays are then illuminated each with its own set of bandpass interference filters—long, medium, and short wave. Two areas, a green in one display and a yellow in the other display, are observed while the illuminators are so adjusted that both areas are sending to the eyes exactly the same composition and quantity of flux; the yellow continues to look yellow and the green continues to look green. If the yellow and green areas are examined successively with illuminators on one wave band at a time, each in the midst of its own array, the lightnesses of the yellow paper and the lightnesses of the green paper will fall in the places demanded by the positions of these lightnesses in the lightness color-space. *The specific experience of the color of any region is the consequence of the simultaneous presence of the three specific lightnesses or lightness precursors.*

When color-separation photographs (Figs. 1a, 1b, and 1c) are made using the emulsion-sensitizer cum filter which correspond to each cone-pigment's absorption in Fig. 3, there is at first glance a sense of concern about the smallness of

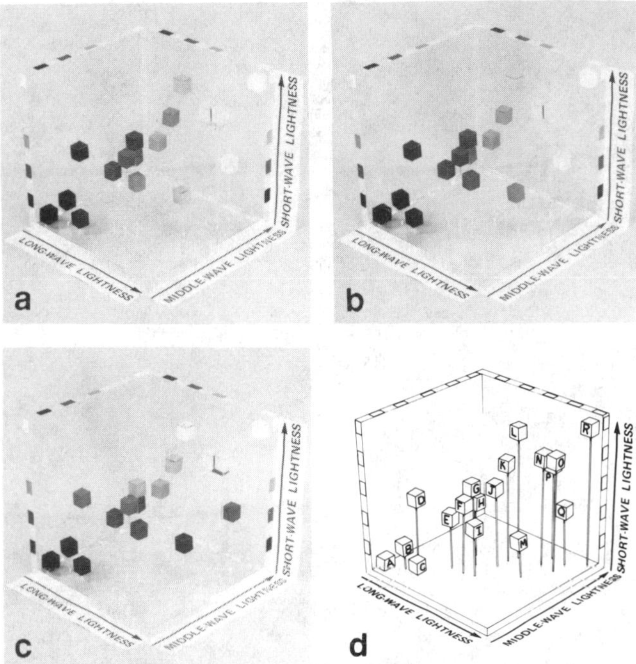

FIG. 4. Photographs of a model of the lightness color-space taken with the film–filter combinations shown in Fig. 3. (a) Long-wave pigment record, (b) middle-wave pigment record, (c) short-wave pigment record, and (d) key to photographs of lightness color-space.

Position	Lightness on a Scale 1–9			Color (Munsell Notation)
	Long	Middle	Short	
A	1	1	1	Black velvet
B	2	2	2	Dark gray
C	3	2	1	Dark brown (5 R 2.5/2)
D	2	3	5	Dark blue (7.5 PB 2.5/10)
E	4	4	4	Medium-dark gray
F	5	4	5	Magenta (2.5 RP 4/8)
G	5	5	6	Blue (5 PB 4/10)
H	4	6	4	Green (2.5 G 5/8)
I	6	4	4	Red (5 R 5/12)
J	6	6	6	Medium-light gray
K	6	7	7	Cyan (5 BG 6/6)
L	6	8	9	Sky blue (5 B 7/6)
M	8	6	3	Orange (2.5 YR 6/14)
N	8	8	8	Light gray
O	9	8	8	Pink (7.5 RP 8/6)
P	8	9	7	Spring green (7.5 GY 7/10)
Q	9	9	4	Yellow (5 Y 8.5/12)
R	9	9	9	White

the difference between the lightness in the long-wave-system photograph and the lightness in the middle-wave-system photograph for each of the objects. There is a dramatic difference for most objects between the lightness in the photograph representing the *short-wave* system and the lightness in either of the other two photographs. And yet it is the differences between the *long* and *middle* lightnesses that are largely responsible for the experience of vivid reds and greens. Although these differences seem small, it is reassuring to observe that they are reliably identifiable by inspection and that on the basis of inspection of the lightnesses in the photographs, the colors of the original objects can be reported by an observer who has not seen the original objects. Furthermore, when an observer is comparing lightnesses from the same object in two different photographs, the photographs must need be separated from each other, that is, side by side or else seen successively in time. Without presuming to postulate an anatomy, one can still define the desired structure as contiguous cortical strata in which the three sets of lightness data are distributed for simultaneous intercomparison. Such a structure has compelling intuitive appeal for providing strong color sensations from those differences in lightness that seem small when one inspects separate photographs.

Such reliable and sensitive responsiveness to small lightness difference becomes the basis for the color of a *spot* of light in a total surrounding area free of all light. When the flux from such a spot is measured at the eye and then changed in quantity, it is possible to estimate the change in lightness of the spot. Measurements show this estimated lightness to be a very weak function of flux, changing very slowly with enormous changes in flux (Fig. 5). If a spot of light is composed of one wavelength band, say a long wavelength, then we can expect that because of the shapes of the three cone-pigment curves and the locations of their three peaks at long, middle, and short wavelengths, long-wavelength flux will be absorbed significantly more by the long-wave pigment than by the other two. Thus, in Fig. 5, if L is

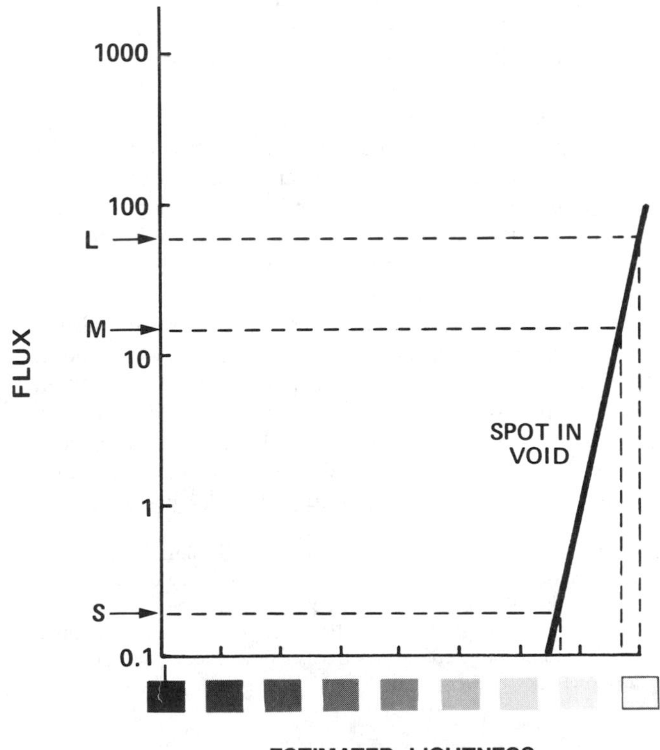

FIG. 5. Change of lightness of a spot of white light in a void with change in flux. (When the spot of light is a band of long wavelengths only, L, M, and S indicate the amount of that long-wave flux absorbed by the long-, middle-, and short-wave receptor systems.)

the amount of long-wave flux absorbed by the long-wave system, M the amount of long-wave flux absorbed by the middle-wave system, and S the amount of long-wave flux absorbed by the short-wave system, then the three lightnesses of the spot would be the three positions on the horizontal axis—9 for long, $8\frac{1}{2}$ for middle, and $7\frac{1}{2}$ for short. This combination of lightnesses is seen as a light reddish-orange (not ordinarily seen in objects under "room conditions" unless the surfaces are fluorescent).

The sensitivity curves in Fig. 3 have been normalized quite arbitrarily so that their *shapes* can be compared. It is usually assumed that there is continuous dynamic change in effective concentration when they are irradiated in use—the so-called bleaching process. This leads to unpredictability of relative net absorption which makes the color of a spot of light somewhat unpredictable because each of the three lightnesses *in a spot*, being a function of flux only, is therefore a function of absorbed flux, which in turn is a function of instantaneous concentration of absorber.

The spectrum, a strikingly anomalous display compared to the arrays which the eye evolved to see, can be regarded as a series of laterally contiguous spots each behaving like the spot in Fig. 5. However, each spot consists of radiation from a wave band displaced from its preceding neighbor so that L, M, and S will move to new positions for each spot.

Figure 6 shows the absorption of the three (cone) retinal pigments described in Fig. 3 and indicates at any wavelength the three relative absorptions that determine L, M, and S for that wave band (prior to bleaching). Thus, L, M, and S in Fig. 5 might describe the absorption of a wavelength 600 nm. At 490 nm, S would be the highest absorption and L the lowest, so that L would be the darkest of the three and S the lightest, leading to the blue experience.

Thus, for spots in the void, and for the spectrum regarded as a chain of spots in a void, color is determined by flux and wavelength which in turn determines the lightness as described in Fig. 5. This mechanism for generating lightness is entirely different from the mechanism which obtains in all ordinary vision as described in the first demonstration in this paper.

In both cases color is determined by comparison of lightnesses, but in the case of the spot, the lightness and the consequent color are determined almost inadvertently by differences in absorbed flux, whereas in the whole-field image, the role of flux vanishes with the taking of the edge ratios.

The first observation pointedly relevant to the mechanism of image and color was reported in 1672 by von Guericke.

> This is how it happens that in the early-morning twilight a clear blue shadow can be produced upon a white piece of paper: when a finger or other object is held between a lighted candle and the paper beneath in such a way that it projects a shadow upon the paper, under these conditions that shadow appears not black, but a perfect sky-blue.
>
> *Otto von Guericke, 1672*

This important experiment, a variation of which is shown in Fig. 7, we now know depicts the most elementary example of generating three different lightnesses on the three receptor systems and the consequential blue–green shadow which

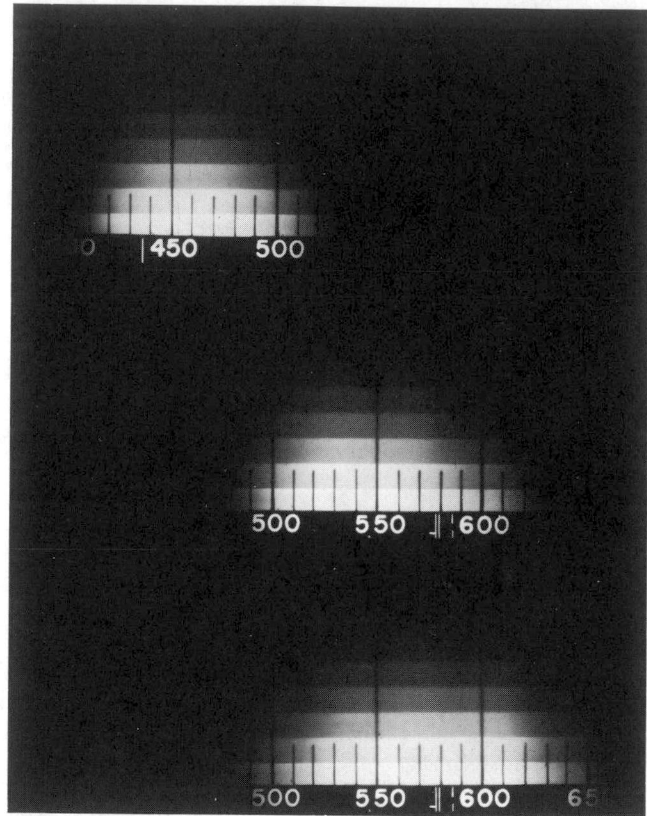

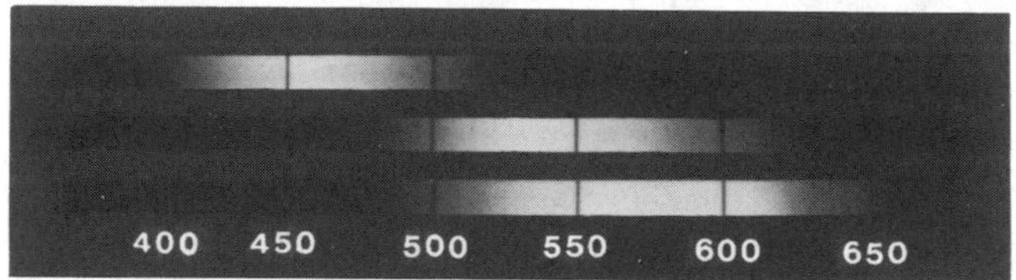

FIG. 6. Parts (a) and (b) show spectrograms made witih the film–filter combinations which match the sensitivity of the three cone pigments (see Fig. 3).

takes its correct place in the lightness color-space (Fig. 4). Similarly, the blue–green afterimage that follows the fixed regard of a red area on a white background when the red area is removed will be the correct outcome of the ratio on the long-wave system at the edge of the bleached and hence less absorbing area where the red square rested. Contrariwise and most important, exposing the total area of the retina to the long-wave radiation to produce moderate *overall* bleaching will have no noticeable effect because the ratios across the edges will not be changed significantly and no modification of the lightnesses in the long-wave image will result.

Contrary to one's naive intuition of "reality," it is the colored shadow that is the minimum scientific sample of the mechanism of color vision rather than a spot-in-a-void comprising flux compounded of various wavelengths. The next step in the analysis is the use of two black and white color-separation transparencies, one taken with a green filter in front of the camera lens and the other with a red filter. Superimposed projection of these two transparencies with white light and red light respectively carries the "color shadow" to a richly variegated family of colors no longer in shadows, but as images. The colors are readily predictable by extending the reasoning in Figs. 7a and 7b and selecting the colors in the lightness color-space. In this important transition from "green shadow" to colored image, the extended ratio taking and multiplication of ratios determines the lightness—not the ratio to the unique immediate neighborhood. Finally, all of these principles are applied in everyday ternary vision which creates a distinct lightness image for each of the three sensitivity systems and then compares them to generate color. A detailed embodiment of these principles is found in Retinex Theory which seeks to quantify color vision observations.

This article has not attempted to bridge the gap between the physics description of color vision and the rich field of discovery in cellular physiology and anatomy. David Hubel (1988) provides a trenchant introduction to this large field. A dramatic example of retinex principles appears in Semir Zeki's (*Nature*, 1980) experiments in which individual monkey brain cells were discovered which responded to an extended color display in close and simultaneous correspondence to a human's color perception. Other pioneering work in this search for the conjunction between physics and biology can be found in the Bibliography.

STAGE I: Projection Screen With Shadow-Image

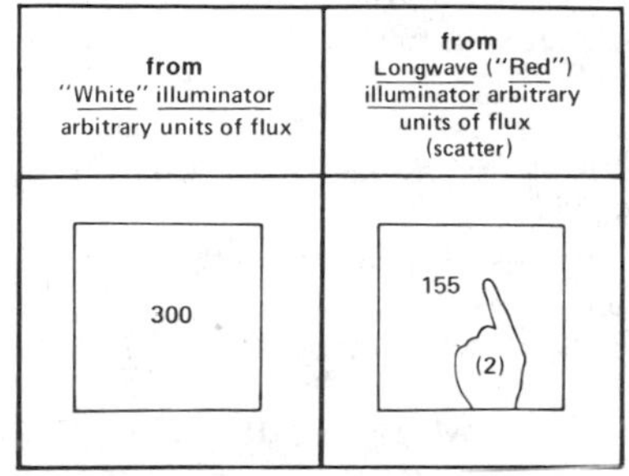

STAGE II: From Absorbed Flux On Three Receptor Systems To Three Lightnesses To Color.

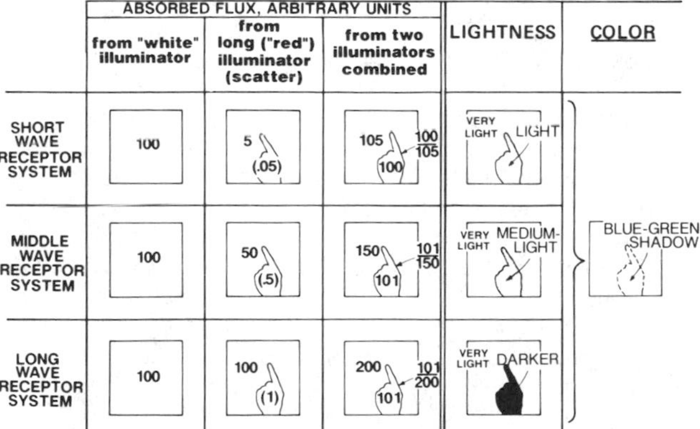

FIG. 7. (a) Stage I. In stage I a screen is illuminated by two illuminators with their beams superimposed. One illuminator causes the screen to send 300 units of "white" light to the eye (per unit area). The other illuminator, equipped with a filter passing a band of long wavelengths, causes the screen to send 155 units of long-wave light to the eye (per unit area). A pointing finger is placed in the beam of long-wave light causing a shadow image of the finger to be seen on the screen. The energies contributed by the two projectors for the background and for the image of the finger are depicted here. (b) Stage II. Stage II depicts what is happening on each of three separate receptor systems when the shadow image is viewed. The energy ratio at the edge between the shadow and the lightest place in the scene determines the lightness for the shadow on each separate receptor system. The shadow image will be light on the short-wave system, medium-light on the middle-wave system, and darker on the long-wave system. Therefore the shadow will be blue–green. (See lightness-color space, Fig. 4). The lightness of the lightest place in the scene for each receptor system will be near the top of the lightness scale being determined by flux in the same way that a spot has its lightness determined by flux.

BIBLIOGRAPHY

M. Aguilar and W. S. Stiles, "Saturation of the Rod Mechanism of the Retina at High Levels of Stimulation," *Opt. Acta* **1,** 59–65 (1954).

H. B. Barlow, R. Fitzhugh, and S. W. Kuffler, "Maintained Activity in the Cat's Retina in Light and Darkness," *J. Gen. Physiol.* **40,** 683–702 (1957).

P. K. Brown and G. Wald, "Visual Pigments in Single Rods and Cones in the Human Retina," *Science* **144,** 45–52 (1964).

F. W. Campbell and J. G. Robson, "Application of Fourier Analysis to the Visibility of Gratings," *J. Physiol.* **197,** 551–566 (1968).

M. E. Chevreul, *De la Loi du Contraste Simultané des Couleurs.* Pitois-Levrault, Paris, 1839.

N. W. Daw, "Neurophysiology of Color Vision," *Physiol. Rev.* **53,** 571–611 (1973).

N. W. Daw, "The Psychology and Physiology of Color Vision," *Trends Neurosci.* **7,** 330–335 (1984).

R. L. DeValois and P. L. Pease, "Contours and Contrast: Responses of Monkey Lateral Geniculate Nucleus Cells to Luminance and Color Figures," *Science* **171,** 694–696 (1971).

O. von Guericke, *Experimenta Nova (ut vocantur) Magdeburgica de Vacuo Spatio,* Chapter XII, pp. 141–142. Apud: Joannem Janssonium à Waesberge, Amstelodami, 1672.

W. A. Hagins, W. E. Robinson, and S. Yoshikami, "Ionic Aspects of Excitation in Rod Outer Segments," in *Energy Transformation in Biological Systems.* CIBA Foundation Symp. **31,** 169–189 (1975).

H. K. Hartline and F. Ratliff, *Studies on Excitation and Inhibition in the Retina.* Rockefeller University Press, New York, 1974.

S. Hecht, "Rods, Cones and the Chemical Basis of Vision," *Physiol. Rev.* **17,** 239–290 (1937).

S. Hecht, S. Shlaer, and M. H. Pirenne, "Energy, Quanta and Vision," *J. Gen. Physiol.* **25,** 819–840 (1942).

H. von Helmholtz, *Helmholtz's Treatise on Physiological Optics, Vol. II: The Sensations of Vision* (J. P. C. Southhall, ed.) Optical Society of America, New York, 1924.

E. Hering, *Outlines of a Theory of the Light Sense* (translated by L. M. Hurvich and D. Jameson). Harvard University Press, Cambridge, MA, 1964.

D. H. Hubel, *Eye, Brain and Vision.* Scientific American Library, New York, 1988.

D. H. Hubel and T. N. Wiesel, "Receptive Fields, Binocular Interaction and Functional Architecture in the Cat's Visual Cortex," *J. Physiol.* **160,** 106–154 (1962); "Receptive Fields and Functional Architecture of Monkey Striate Cortex," *J. Physiol.* **195,** 215–243 (1968); "Laminar and Columnar Distribution of Geniculo-Cortical Fibers in the Macaque Monkey," *J. Comp.*

Neurol. **146,** 421–450 (1972); "Sequence Regularity and Geometry of Orientation Columns in the Monkey Striate Cortex," *J. Comp. Neurol.* **158,** 267–294 (1974); "Functional Architecture of Macaque Monkey Visual Cortex," *Proc. R. Soc. London.* **198B,** 1–59 (1977).

A. C. Hurlbert, "Formal Connections Between Lightness Algorithms," *J. Opt. Soc. Am. A* **3,** 1684–1693 (1986).

A. C. Hurlbert and T. A. Poggio, "Synthesizing a Color Algorithm from Examples," *Science* **239,** 482–485 (1988).

D. Ingle, "The Goldfish as a Retinex Animal," *Science* **227,** 651–654 (1985).

B. Julesz, *Foundations of Cyclopean Perception.* University of Chicago Press, Chicago and London, 1971.

S. W. Kuffler and J. G. Nicholls, *From Neuron to Brain.* Sinauer Assoc., Inc., Sunderland, MA, 1976.

E. H. Land, "Color Vision and the Natural Image, Parts I and II," *Proc. Natl. Acad. Sci. USA* **45,** 115–129, 636–644 (1959); "Colour in the Natural Image." *Proc. R. Inst. Gr. Br.* **39,** 1–15 (1962); "The Retinex," *Am. Sci.* **52,** 247–264 (1964); "Lightness, Brightness and Reality," Albert A. Michelson Award Address (1966); "The Retinex Theory of Colour Vision," *Proc. R. Inst. Gr. Br.* **47,** 23–58 (1974); "Recent Advances in Retinex Theory and Some Implications for Cortical Computations: Color Vision and the Natural Image," *Proc. Natl. Acad. Sci. USA* **80,** 5163–5169 (1983); "An Alternative Technique fo the Computation of the Designator in the Retinex Theory of Color Vision," *Proc. Natl. Acad. Sci. USA* **83,** 3078–3080 (1986); "Recent Advances in Retinex Theory," *Vision Res.* **26,** 7–21 (1986).

E. H. Land and N. Daw, "Colors Seen in a Flash of Light," *Proc. Natl. Acad. Sci. USA* **48,** 1000–1008 (1962).

E. H. Land, D. H. Hubel, M. S. Livingstone, S. H. Perry, and M. M. Burns, "Colour-generating Interactions Across the Corpus Callosum," *Nature* **303,** 616–618 (1983).

E. H. Land and J. J. McCann, "Lightness and Retinex Theory," *J. Opt. Soc. Am.* **61,** 1–11 (1971).

C. J. Lueck, S. Zeki, K. J. Friston, M.-P. Deiber, P. Cope, V. J. Cunningham, A. A. Lammertsma, C. Kennard, and R. S. J. Frackowiak, "The Colour Centre in the Cerebral Cortex of Man," *Nature* **340,** 386–388 (1964).

W. B. Marks, W. H. Dobelle, and E. F. MacNichol, Jr., "Visual Pigments of Single Primate Cones," *Science* **143,** 1181–1183 (1964).

J. C. Maxwell, *Scientific Papers of James Clerk Maxwell* (W. D. Niven, ed.). Dover Publications, New York, 1965.

J. J. McCann, "Rod-Cone Interactions: Different Color Sensations from Identical Stimuli," *Science* **176,** 1255–1257 (1972); "Human Color Perception," in *Color: Theory and Imaging Systems*, pp. 1–23. Society of Photographic Scientists and Engineers, Washington, D.C., 1973.

J. J. McCann and J. L. Benton, "Interaction of the Long-Wave Cones and the Rods to Produce Color Sensations," *J. Opt. Soc. Am.* **59,** 103–107 (1969).

J. J. McCann and K. L. Houston, "Color Sensation, Color Perception and Mathematical Models of Color Vision," in *Colour Vision: Physiology and Psychophysics* (J. D. Mollon and L. T. Sharpe, eds.), pp. 535–544. Academic Press, London, 1983.

J. J. McCann, E. H. Land, and S. M. V. Tatnall, "A Technique for Comparing Human Visual Responses with a Mathematical Model for Lightness," *Am. J. Optom.* **47,** 845–855 (1970).

J. J. McCann, S. P. McKee, and T. H. Taylor, "Quantitative Studies in Retinex Theory," *Vision Res.* **15,** 445–458 (1976).

S. P. McKee, J. J. McCann, and J. L. Benton, "Color Vision from Rod and Long-Wave Cone Interactions: Conditions in Which Rods Contribute to Multicolored Images." *Vision Res.* **17,** 175–185 (1977).

S. P. McKee and G. Westheimer, "Specificity of Cone Mechanisms in Lateral Interaction," *J. Physiol.* **206,** 117–128 (1970).

H. R. Moskowitz, B. Scharf, and J. C. Stevens (eds.), *Sensation and Measurement, Papers in Honor of S. S. Stevens*, "Smitty Stevens' Test of Retinex Theory," pp. 363–368. D. Reidel, Dordrecht, 1974.

I. Newton, *Opticks*, 4th ed. (1730). Dover Publications, New York, 1952.

S. L. Polyak, *The Retina.* Chicago University Press, Chicago, 1941.

F. Ratliff, *Mach Bands: Quantitative Studies on Neural Networks in the Retina.* Holden-Day, San Francisco, London, and Amsterdam, 1965.

L. A. Riggs, F. Ratliff, J. C. Cornsweet, and T. N. Cornsweet, "The Disappearance of Steadily Fixated Visual Test Objects," *J. Opt. Soc. Am.* **43,** 495–501 (1953).

C. Rumford, *The Complete Works of Count Rumford*, Vol. IV, pp. 49–62. The American Academy of Arts and Sciences, Boston, MA, 1875.

W. A. H. Rushton, "The Difference Spectrum and the Photosensitivity of Rhodopsin in the Living Human Eye," *J. Physiol.* **134,** 11–29 (1956).

W. A. H. Rushton, W. A. Campbell, Hagins, and G. S. Brindley, "The Bleaching and Regeneration of Rhodopsin in the Living Eye of the Albino Rabbit and of Man," *Opt. Acta* **1,** 183–190 (1955).

M. Schultz, *Arch. Mikroskop. Anal.* **II,** 247–261 (1966).

W. S. Stiles, "Increment Thresholds and the Mechanisms of Color Vision," *Documenta Ophth.* **73,** 138–163 (1949).

G. Wald, "Human Vision and the Spectrum," *Science* **101,** 653–658 (1945); "The Receptors of Human Color Vision," *Science* **145,** 1007–1017 (1954).

F. S. Werblin and J. E. Dowling, "The Organization of the Retina of the Mud Puppy," *J. Neurophysiol.* **32,** 339–355 (1969).

G. Westheimer, "Modulation Thresholds for Sinusoidal Light Distributions on the Retina," *J. Physiol.* (*London*) **152,** 67–74 (1960); "Spatial Interaction in the Human Retina During Scotopic Vision," *J. Physiol.* (*London*) **181,** 881–894 (1965); "Spatial Interaction in Human Cone Vision," *J. Physiol.* (*London*) **190,** 139–154 (1967).

W. D. Wright, *The Measurement of Colour.* Hilger and Watts Ltd., London, 1964.

G. Wyszecki and W. S. Stiles, *Color Science: Concepts and Methods, Quantitative Data and Formulas.* Wiley. New York, 1967.

T. Young, "On the Theory of Light and Colours," *Philos. Trans.* **92,** 12–48 (1802).

S. M. Zeki, "Cortical Projections from Two Prestriate Areas in the Monkey," *Brain Res.* **34,** 19–35 (1971); "Color Coding in Rhesus Monkey Prestriate Cortex," *Brain Res.* **53,** 422–427 (1973); "Color Coding in the Superior Temporal Sulcus of Rhesus Monkey Visual Cortex," *Proc. R. Soc. London* **B197,** 195–223 (1977); "The Representation of Colours in the Cerebral Cortex," *Nature* **284,** 412–418 (1980); "The Response of Color Specific Cells in Monkey Visual Cortex to Colors Produced by Shadows," *J. Physiol.* (*London*) **306,** 29P–30P (1980).

Vortices

P. G. Saffman

The velocity of a fluid can be described by a vector function of space (**x**) and time (t), written as $\mathbf{u}(\mathbf{x},t)$. Its curl,

$$\boldsymbol{\omega}(\mathbf{x},\mathbf{t}) = \nabla \times \mathbf{u}(\mathbf{x},t), \tag{1}$$

is called the vorticity. In incompressible fluid, the relation between velocity and vorticity is analogous to that between magnetic field and current density. In this article, it will be supposed that the density ρ is constant unless the contrary is stated explicitly. Vorticity can be interpreted physically as an angular-momentum density; a spherical fluid particle, instantaneously frozen without loss of angular momentum, rotates with angular velocity $\frac{1}{2}\boldsymbol{\omega}$. The term vortex motion refers to flows in which the vorticity is confined to finite regions, called vortices, inside which the motion is said to be rotational. In a barotropic fluid (where ρ is uniquely determined by the pressure p) a fluid particle without vorticity can acquire it only by viscous diffusion or the action of nonconservative forces. Consequently, the study of vortices is primarily, but by no means exclusively, of interest for incompressible fluids of small viscosity moving under the action of conservative forces, since vorticity can then be associated with fluid particles.

Examples of vortex motion are vortex rings (smoke rings), which can be visualized by carefully dropping dye into a beaker of water, the bathtub vortex, the Kármán vortex street in the wake of a cylinder, the row of vortices formed by the Kelvin–Helmholtz instability at the surface between two streams of different velocity, the trailing vortices behind an aircraft (often visible as vapor trails) dust devils and tornadoes, and the quantized vortices of He II.

The mechanics of vortex formation is in most cases not well understood. Vortices cannot be created in a perfect fluid, and their existence is due to subtle effects of viscosity, or the application of nonconservative forces usually associated with baroclinic density variations, density gradients in supersonic flow with shock waves, or (for He II) quantum-mechanical effects. Experimental observations are difficult; the flows are often unsteady and nonintrusive measuring devices are needed. The laser Doppler anemometer has been a useful tool in recent years. Flow visualization techniques are commonly employed, but may suffer from uncertainties in interpretation. For example, the extent to which the vorticity is marked by dye can be questioned. Care needs to be taken in identifying a vortex with an extremum of the vorticity field. The streamlines at such a point in the plane perpendicular to the vorticity in a frame moving with the extremum will have a center and signify a vortex if the pressure is a minimum; a maximum of pressure would correspond to a saddle point for the streamlines.

Theoretical studies have tended to concentrate on the behavior of straight or circular vortex filaments, vortex patches, and vortex sheets. A vortex filament is a vortex in the shape of a tube of small cross section. The Helmholtz laws of vortex motion for fluids of negligible viscosity state that a filament is always composed of the same fluid particles and its strength is constant, this being the flux of vorticity through the cross section. An alternative form of the Helmholtz laws is that $\omega \propto dl$, where dl is the material line element of a vortex line. This equation can be generalized to $\omega/\rho \propto dl$ for barotropic fluid. The amplification of vorticity by the stretching of vortex lines is an important physical effect in turbulence and rotating fluids. The limit of a vortex filament of finite strength and zero cross section is called a line vortex. Rectilinear line vortices are studied extensively in two-dimensional flows, where the absence of stretching makes vorticity a conserved quantity. The Helmholtz laws imply that a rectilinear line vortex moves with the fluid unless external forces act on the core. The vortex is said to be bound if the external forces keep it at rest. Line vortices are a less useful concept in three dimensions because the velocity field produced by a curved line vortex is infinite on the vortex, giving rise to an infinite self-induced velocity. Vortex patches are finite areas of vorticity in two-dimensional flow. Most often, the vorticity is taken to be piecewise constant as then powerful analytical and numerical methods exist for the study of the structure and stability of the flow.

A vortex sheet is the limit in which the vorticity is compressed into a surface of zero thickness. The velocity normal to the sheet is continuous; the tangential component is discontinuous, the magnitude of the discontinuity being the strength of the sheet. The flow past a wing or lifting body can be described by bound vortices in the wing together with a trailing vortex sheet downstream, which rolls up into a pair of opposite rotating vortex tubes.

The kinematics and dynamics of vortex motion in uniform inviscid fluid follow in principle from Eq. (1) and

$$\frac{d\boldsymbol{\omega}}{dt} = \boldsymbol{\omega}\cdot\nabla\mathbf{u} + \nabla\times\mathbf{F} \tag{2}$$

($\mathbf{F}$ is the external body force density). Associated with a vortex are its impulse $\mathbf{I}$ and angular impulse $\mathbf{A}$, defined as

$$\mathbf{I} = \tfrac{1}{2}\rho\int_V \mathbf{r}\times\boldsymbol{\omega}\, dV, \qquad \mathbf{A} = -\tfrac{1}{2}\rho\int_V r^2\boldsymbol{\omega}\, dV. \tag{3}$$

These can be interpreted physically as the impulse and moment of impulse of the body forces that must be applied to generate it in fluid initially at rest. Confusion often occurs between the impulse and the momentum of the fluid. The latter is given by a conditionally convergent integral in an unbounded fluid, and is zero if the fluid is confined within a rigid container. Expressions for the kinetic energy are somewhat more complicated. For a three-dimensional bounded vorticity distribution in unbounded fluid, the kinetic energy is $\rho\int\mathbf{u}\cdot\mathbf{r}\times\boldsymbol{\omega}\, dV$. In two-dimensional flow, the impulse is given by Eq. (3) without the factor $\frac{1}{2}$, because of contributions over surfaces at infinity. The sums of $\mathbf{I}$ and $\mathbf{A}$ over all the vortices in an unbounded fluid are constant.

The velocity field produced by a vortex follows from the integration of Eq. (1) and is

$$\mathbf{u}(P) = \frac{1}{4\pi}\int_V \frac{\boldsymbol{\omega}(Q)\times \mathbf{r}_{PQ}}{r_{PQ}^3}\, dV_Q \tag{4}$$

($\mathbf{r}_{PQ} = \mathbf{r}_P - \mathbf{r}_Q$). The contribution from a line vortex of strength κ is given by the Biot–Savart law

$$\mathbf{u}(P) = \frac{\kappa}{4\pi}\oint\frac{d\mathbf{s}_Q\times\mathbf{r}_{PQ}}{r_{PQ}^3} = \frac{\kappa}{4\pi}\nabla\Omega, \tag{5}$$

where Ω is the solid angle subtended at P by the line vortex. This integral is divergent when P lies on the line vortex unless the radius of curvature R at the point is infinite. As $P\to Q$,

$$\mathbf{u}(P) \sim \frac{\kappa}{2\pi} \frac{\mathbf{t} \times \mathbf{r}_{PQ}}{r_{PQ}^2} + \frac{\kappa \mathbf{b}}{4\pi R} \log \frac{R}{r_{PQ}}, \tag{6}$$

where **t** and **b** are the tangent and binormal to the line at Q. The first term describes the circulatory motion about the vortex. The singularity in the second term is removed when the vortex is given a finite cross section and causes a self-induced translation of a curved filament in the direction of its binormal at a rate depending logarithmically on the core radius.

A stationary axisymmetric rectilinear vortex filament of strength κ in unbounded fluid has the velocity field

$$u_r = 0, \qquad u_\theta = V(r), \qquad u_z = W(r), \tag{7}$$

where V and W are arbitrary functions satisfying $rV \to \kappa/2\pi$ and $W \to 0$ as $r \to \infty$. The circulation about the vortex is $\lim_{r\to\infty} 2\pi r V = \kappa$. The streamlines are helices about the axis, circles when $W=0$. The cases most commonly considered are uniform core of radius a, with uniform vorticity and axial velocity,

$$\begin{aligned} V &= \frac{\kappa r}{2\pi a^2}, & W &= W_0, & r &< a, \\ V &= \frac{\kappa}{2\pi r}, & W &= 0, & r &> a; \end{aligned} \tag{8}$$

and hollow or stagnant core,

$$\begin{aligned} V &= 0, & W &= 0, & r &< a, \\ V &= \frac{\kappa}{2\pi r}, & W &= 0, & r &> a. \end{aligned} \tag{9}$$

Putting $a=0$ gives a rectilinear line vortex. A nonaxisymmetric filament is not steady. For uniform core, Kirchhoff gave an exact solution with elliptical cross section of semi-axes a and b, which rotates with constant angular velocity $\kappa/\pi(a+b)^2$.

The diffusion of vorticity by viscosity in a rectilinear filament is described by

$$\frac{\partial \omega}{\partial t} = \nu \nabla^2 \omega, \tag{10}$$

where ν is the kinematic viscosity. The solution,

$$\omega = \frac{\kappa}{4\pi\nu t} \exp\left(\frac{-r^2}{4\nu t}\right), \qquad u_\theta = \frac{\kappa}{2\pi r}\left[1 - \exp\left(\frac{-r^2}{4\nu t}\right)\right], \tag{11}$$

describes the viscous decay of a rectilinear line vortex. Little is known about the diffusion of a vortex filament with a turbulent core, except that the mean tangential velocity at some radius r will exceed $\kappa/2\pi r$ (i.e., there is a circulation overshoot) if the turbulence causes the vorticity to spread faster than it does as a result of molecular diffusion. An exact steady solution of the Navier–Stokes equations exists for an axisymmetric vortex in a uniform axisymmetric straining field $u = -\alpha x$, $v = -\alpha y$, $w = 2\alpha z$ $(\alpha > 0)$. The vorticity is $\omega = \Omega \exp[-\alpha(x^2+y^2)/2\nu]$ in the z direction, where Ω is arbitrary.

The velocity fields of rectilinear line vortices can be linearly superposed. Let (x_i, y_i) be the coordinates of the ith vortex of strength κ_i, whose core is subject to an external force per unit length with components (F_i, G_i). Then

$$\kappa_i \frac{dx_i}{dt} = \kappa_i U_i - G_i, \qquad \kappa_i \frac{dy_i}{dt} = \kappa_i V_i + F_i, \tag{12}$$

where (U_i, V_i) is the irrotational velocity at the vortex produced by the other vortices and boundaries. The force exerted by a stream on a bound vortex is $(\kappa_i V, -\kappa_i U)$ and is termed either the Magnus force or the Kutta lift. When the velocity field is produced entirely by the line vortices, the stream function at P, distance r_{Pi} from ith vortex, is

$$\psi(P) = \sum \frac{\kappa_i}{2\pi} \log r_{Pi},$$

and

$$\kappa_i U_i = -\frac{\partial W}{\partial y_i}, \qquad \kappa_i V_i = \frac{\partial W}{\partial x_i}, \tag{13}$$

where

$$W = \frac{1}{2\pi} \sum_{i \neq j} \kappa_i \kappa_j \log r_{ij} = \frac{1}{2} \sum \kappa_i \psi_i. \tag{14}$$

If the fluid is bounded, contributions from an image system are added to W. The motion of N vortices is given by the integration of $2N$ ordinary differential equations and lends itself to high-speed computing. Two-dimensional inviscid flows with continuous vorticity distributions can be modeled by large numbers of line vortices. In principle, viscosity can be incorporated by adding a random component to the velocity of each vortex, but accurate representation of the viscous modifications to the positions of the vortices requires that the number of vortices exceeds the Reynolds number. To overcome numerical difficulties, the velocity field of the line vortex is approximated by that of vortices of circular cross section. When the external forces are conservative or absent, Eq. (12) is Hamiltonian in form, and the methods of statistical mechanics can be applied for large N. A temperature can be defined which when positive favors a well-mixed state and when negative leads to states where vortices of like rotation clump together. Numerical calculations show this behavior.

Equilibrium configurations, with the vortices in uniform translation or rigid-body rotation, have the property, stated by Kelvin, that W is stationary subject to fixed **I** and **A**. The stability of rows and lattices of rectilinear line vortices to two-dimensional disturbances has been studied. Rows are generally unstable. The configuration of N equal vortices at the vertices of a regular polygon is stable for $N \leq 6$ and unstable for $N \geq 7$. Triangular lattices are stable but square and honeycomb lattices are unstable.

The stability to three-dimensional disturbances is more obscure because effects of finite core size and core structure are crucial. An isolated vortex filament with uniform core and no axial velocity is stable to infinitesimal disturbances (i.e., displacement much smaller than core size). The oscillation frequencies were found by Kelvin. It is unstable in an external flow to disturbances with axial wavelength comparable to core size. Hollow vortices do not have this instability. Snake-like disturbances, with displacement much less than core size, can be studied using the equations of motion of curved vortex filaments. Such analyses show instability of Kármán vortex sheets and vortex pairs of opposite strengths. The latter instability is believed to describe the often visible growing sinusoidal oscillations of vapor trails.

Rectilinear filaments of finite core size disintegrate if exposed to too large an external straining field. Two rectilinear vortices amalgamate if placed too close together. These properties have generally to be studied by numerical means, but analytical results exist for special cases. Hamiltonian methods show promise; they have been used to demonstrate that finite core size does not affect qualitatively the stability of the infinite Kármán vortex sheet.

Vortices of toroidal shape are called vortex rings. Kelvin's theory of vortex atoms stimulated nineteenth-century study. In terms of cylindrical polar coordinates (r,θ,z), the vorticity and swirl velocity in the core are such that ω_θ/r and ru_θ are constant on streamlines moving with the vortex in steady inviscid flow. For thin rings with core radius a small compared with the ring radius R, the following formulas give the velocity V, kinetic energy E, and impulse P of the ring:

$$V=\frac{\kappa}{4\pi R}\left(\log\frac{8R}{a}-\frac{1}{2}+\frac{2\pi}{\kappa^2}\int_C(u_r^2+u_z^2-2u_\theta^2)\,dS\right), \qquad (15)$$

$$E=\frac{\rho}{2}\kappa^2R\left(\log\frac{8R}{a}-1+\frac{2\pi}{\kappa^2}\int_C(u_r^2+u_z^2+u_\theta^2)\,dS\right), \qquad (16)$$

$$P=\rho\pi\kappa R^2; \qquad (17)$$

κ is the circulation about the core. The integral is over the cross section of the core. For uniform core without swirl, $\omega/r=\text{constant}$, $u_\theta=0$, the value of the integral is $\kappa^2/8\pi$, leading to Kelvin's expression for the velocity. According as $V>\kappa/2R$ or $<\kappa/2R$, the irrotational fluid carried along with the ring is ring shaped or oval. The Hamiltonian relation $V=\partial E/\partial P$ is satisfied for variations that leave κ, a^2R, ω_θ/r, and ru_θ constant. The relative error in these expressions is $O[(a^2/R^2)\log(R/a)]$.

For uniform core without swirl, numerical work has found the shapes of fat rings up to the limit of the Hill spherical vortex, which is an exact solution with the core a sphere of radius a. For this flow

$$V=\tfrac{2}{5}\kappa/a, \qquad E=\tfrac{8}{35}\pi\rho a\kappa^2, \qquad P=\tfrac{4}{5}\rho\pi a^2\kappa. \qquad (18)$$

It can be generalized to include swirl. Formulas for vortex rings correct to $O[(a^2/R^2)\log(R/a)]$ have been calculated analytically. Suitably normalized, they agree to 6% with the numerical calculations as far as the spherical vortex.

The stability and periods of vibration of thin vortex rings have been studied. Kelvin argued that the uniform core without swirl was stable to axisymmetric disturbances, but left open the question of stability to three-dimensional disturbances. There is evidence of instability to twisting motions, i.e., those in which the cross section remains circular but the circular axis of the core is deformed.

The interaction of vortex rings with one another or with rigid or free boundaries has applications to He II. Vortex rings produced in the laboratory tend to be fattish and subject to viscosity and turbulence. The effect of viscosity on thin rings has been studied. Definitive results for fattish rings are sparse. Vortex line breaking and reconnection occurs in real fluids and leads to fusion or fission of vortex rings. Vortex rings in stratified fluid are of interest for environmental engineering.

The steady motion of thin rings is a special case of the motion of curved vortex filaments. The velocity of a point P on the filament axis is given to first approximation by

$$\frac{d}{dt}\mathbf{r}_P=\mathbf{U}_{\text{ext}}+\frac{\kappa}{4\pi}\int_{[\delta]}\frac{d\mathbf{s}_Q\times\mathbf{r}_{PQ}}{r_{PQ}^3}+\frac{\mathbf{t}\times\mathbf{F}}{\rho\kappa}, \qquad (19)$$

where $\mathbf{U}_{\text{ext}}$ is the external irrotational velocity field produced by other vortices or bodies; $[\delta]$ denotes that a length 2δ centered on P is omitted from the Biot–Savart integral; and $\mathbf{F}$ is external force/unit length acting on the core. The cutoff length δ is given by

$$\delta=\frac{1}{2}a\exp\left(\frac{1}{2}-\frac{2\pi^2a^2\overline{v}^2}{\kappa^2}+\frac{4\pi^2a^2\overline{w}^2}{\kappa^2}\right), \qquad (20)$$

where a is the core radius and $\overline{v}^2$ and $\overline{w}^2$ are the average values in the core of the squares of the swirl velocity about the axis and axial flow parallel to the axis. A physical interpretation of the cutoff can be given in terms of a tension in the vortex filament. As the motion develops, the cutoff changes and has to be calculated from equations describing the changes in core structure. If $\overline{w}^2\gg\overline{v}^2$, the equation fails and the filament is better regarded as a weakly swirling jet. Filaments with significant axial velocity may undergo an instability or transition, characterized by large changes in core radius and reverse axial flow, known as vortex breakdown or bursting, whose explanation is controversial.

The dynamics of vortex sheets presents problems of extreme difficulty. Infinitesimal sinusoidal disturbances of wavelength λ on a plane sheet of strength U grow exponentially at a rate $\pi U/\lambda$. There is convincing evidence that the mathematical problem of the evolution of a vortex sheet is in general ill-posed, and that singularities develop in a finite time. Representation of the sheet by a row of point vortices produces a finite amplitude rollup into a row of vortices. The physical validity of this approximation is questionable, and the roles of viscosity and finite sheet thickness need elucidation. The same approximation is used to study the rollup of a finite vortex sheet and the structure of vortex sheets behind sharp-edged bodies in streams. A class of similarity solutions exists for which the sheet is a spiral. These help with the understanding of the singularities at the centers of the vortices formed as the sheet rolls up. The mathematical problem can be reduced to the solution of the nonlinear integrodifferential equation

$$\frac{\partial Z^*}{\partial t}(\Gamma,t)=-\frac{i}{2\pi}\mathrm{P}\int\frac{d\Gamma'}{Z(\Gamma,t)-Z(\Gamma',t)}, \qquad (21)$$

where $Z=X+iY$ is the parametric equation of the sheet and Γ is the integrated strength. Similar methods are used when there is conical symmetry, such as for the vortex sheets emanating from the leading edges of a delta wing. Three-dimensional sheets of general shape are presently intractable by methods in which a reasonable degree of confidence exists. Geophysical applications and investigations of the Rayleigh–Taylor instability of the interface between accelerated fluids lead to the study of sheets between fluids of different density.

See also FLUID PHYSICS; HYDRODYNAMICS; QUANTUM FLUIDS; TURBULENCE.

BIBLIOGRAPHY

G. K. Batchelor, *Fluid Dynamics*. Cambridge, London and New York, 1967. (E)

H. Helmholtz, "On Integrals of the Hydrodynamical Equations which Express Vortex Motion" (translated by O. G. Tait), *Philos. Mag.* **33,** 485–512 (1867). (H)

Lord Kelvin, *Mathematical and Physical Papers,* Vol. IV. Cambridge, London and New York, 1910. (H)

H. Lamb, *Hydrodynamics*, 6th ed. Cambridge, London and New York, 1932. (I)

L. Prandtl and O. C. Tietjens, *Fundaments of Hydro and Aeromechanics*. Dover, New York, 1934. (E)

Advanced work appears principally in *J. Fluid Mech., Proc. Roy. Soc., Phys. Fluids,* and *Phys. Rev.*

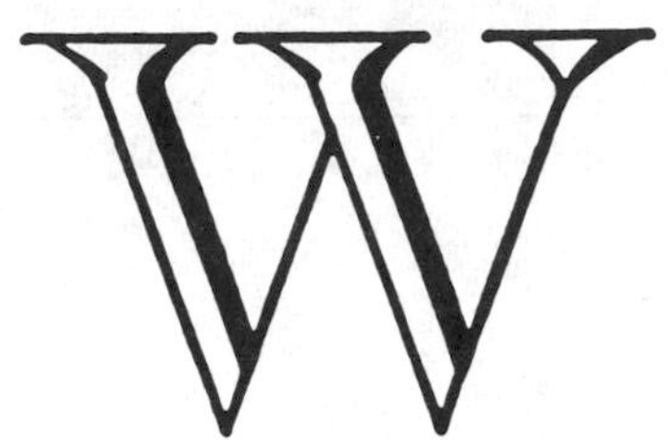

Water

Frank H. Stillinger

Because of the large amount of water present on our planet, especially in liquid form, this substance has become centrally important for many aspects of science and technology. This importance is connected partly to the peculiar behavior of pure water, partly to its qualifications as a liquid solvent, and partly to its role as a fluid medium for support of life.

The most prominent peculiarities exhibited by pure water are the reduction in volume upon melting at 0°C (by 8.3%), followed by further shrinkage to a density maximum as the liquid is heated to 4°C. These attributes are also shared by D_2O (m.p. 3.8°C, density max. at 11.2°C) and by T_2O (m.p. 4.5°C, density max. at 13.4°C). Although rare, these observations are not unique with water; the elements Si, Ge, and Bi also shrink upon melting, while In_2Te_3 appears to have a liquid-phase density maximum.

Additional water anomalies are (a) large number of ice polymorphs (including those that form at high pressure); (b) tendency toward reduced viscosity when liquid water below 30°C is compressed; and (c) minimum in isothermal compressibility $[-(\partial \ln V/\partial p)_T]$ for the liquid at 46°C.

Observable properties of water in pure form and as a solvent stem from the structure of the individual water molecules and from the way that intermolecular forces between those molecules cause aggregation into liquid and solid.

The isolated H_2O molecule is shaped like a wide-open V, with the oxygen nucleus at the central bend and hydrogen nuclei forming "arms" of length 0.96 Å. The HOH angle is 104.5°. These dimensions can vary slightly as the molecule vibrates and interacts with neighbors in a crystal or the liquid, but the overall shape remains.

The dominant effect in water molecule interactions is the formation of hydrogen bonds. When two water molecules form a hydrogen bond, one (the hydrogen donor) points one of its OH groups toward the back side of the oxygen atom of the second (the hydrogen acceptor). This arrangement is illustrated in Fig. 1. The oxygen–oxygen lengths of these bonds normally lie in the range 2.7–3.0 Å, so that the donated hydrogen resides only about one third of the way between oxygens, and so still "belongs" to the donor.

The maximum hydrogen bond strength is achieved when the molecules are arranged as shown in Fig. 1. This strength is about 4.2×10^{-13} ergs (6.0 kcal/mole of bonds), and exceeds thermal energy $k_B T$ by a factor of 10 at room temperature. The existence of these relatively strong hydrogen bonds between water molecules explains the relatively high melting and boiling temperatures for water, compared to other substances of comparable molecular weight (e.g., Ne, CH_4, NH_3, O_2, CO).

In a large aggregate of water molecules, optimum hydrogen bonding is achieved if each water molecule hydrogen-bonds to four others. Toward two of these four it donates its hydrogens, while it accepts hydrogens from the other two. This fourfold bonding is present in ordinary ice, causing formation of hydrogen bond hexagons. Without disturbing hydrogen bond strengths substantially, four-coordinated water networks can also form, which additionally incorporate squares, pentagons, and heptagons of hydrogen bonds. These patterns also exist in high-pressure ice polymorphs and in hydrate crystals.

Evidently the capacity for water molecules to form a diverse collection of three-dimensional networks of hydrogen bonds, while maintaining fourfold bonding at each molecule, has structural relevance for the liquid. Currently available evidence, both experimental and theoretical, indicates that liquid water consists of a structurally random network of hydrogen bonds uniformly filling the volume occupied by the liquid. That random network incorporates strained and broken hydrogen bonds, with greater frequency the higher the temperature. Furthermore, the network is labile, with bonds breaking in one place and reforming nearby, so that normal liquid flow and molecular diffusion are possible.

An isolated water molecule has dipole moment 1.86 debye (D), with hydrogens acting as though each bore one third of a protonic charge, and the oxygen as though it bore minus two thirds. Neighboring molecules in the liquid tend to have their dipoles somewhat aligned, so as to act in concert under the polarizing influence of an electrical field. The net result of this alignment, and of molecular polarizability, is a large

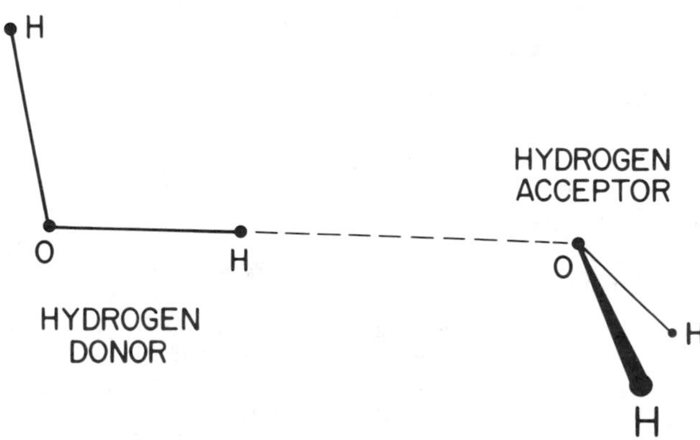

FIG. 1. Hydrogen bond (dashed line) linking two water molecules.

static dielectric constant (88.0 at 0°C, but declining to 55.3 at 100°C).

The ease with which water dissolves many ionic crystals, such as the alkali halides, stems partly from its high static dielectric constant. However, it is also connected with the relatively small size of the water molecules, which permits them to approach ions closely, solvate them strongly, and thus overcome the largely electrostatic binding of the ionic crystals.

The solvating power of water for ions facilitates the dissociation of water molecules themselves into H^+ and OH^- ions. In the liquid at room temperature, roughly one molecule in 55 million will have dissociated. The H^+ and OH^- formed in this way can readily be incorporated into the liquid's random hydrogen bond network, and they tend to form shortened hydrogen bonds in their vicinity. Both H^+ and OH^- have high apparent mobilities in water, due to the possibility of moving a succession of hydrogens along a chain of hydrogen bonds so as to cause a net transfer of ionic charge along that chain.

Nonionic substances with high solubility in water tend to have molecules with which water can hydrogen-bond. Usually this requires that molecules of those "hydrophilic" substances contain oxygen or nitrogen atoms.

Hydrocarbons (such as methane, hexane, acetylene, benzene) form an important group of "hydrophobic" molecular substances that are sparingly soluble in water. They cannot form hydrogen bonds with water strong enough to compete with those already present in that liquid itself. Consequently the random water network is obliged to restructure around the rare dissolved hydrocarbons so as to form a "cage" of hydrogen bonds of the required size. The corresponding geometric constraints on the water network cause entropies of solution for hydrocarbons in water to be negative.

Biologically important molecules (e.g. lipids, enzymes, RNA, hemes) often contain both hydrophilic and hydrophobic chemical groups. Consequently the biologically active conformations of these molecules, to the extent it is possible, place the hydrophilic groups on the outside to be in contact with water while the hydrophobic groups cluster within to avoid water contact. Since conformation is crucial to operation in most cases, it is obvious that specific solvation properties of water have profound effects in biology and doubtless have exerted a powerful influence on the course of chemical evolution from the first rudimentary "protolife" to present complex biochemistry.

See also HYDROGEN BOND; ICE.

BIBLIOGRAPHY

D. Eisenberg and W. Kauzmann, *The Structure and Properties of Water*. Oxford, London and New York, 1969. (I,A)

F. Franks, ed., *Water, A Comprehensive Treatise*, Vols. I–V. Plenum, New York, 1972–1975. (I,A)

R. A. Horne, ed., *Water and Aqueous Solutions*. Wiley (Interscience), New York, 1972. (I,A)

Waves

D. R. Tilley

1. INTRODUCTION

A simple wave is a single-frequency disturbance traveling at some speed v; the disturbance might be elastic strain, as in an acoustic wave; a combination of electric and magnetic fields, as in light; or some other quantity. All traveling waves transport energy. Single-frequency waves are of basic importance, but it is sometimes necessary to consider more general disturbances. Because of the pervasiveness and importance of the concepts, general undergraduate texts [1,2] contain major sections on waves, and a number of texts specifically on the topic have been written [3–6].

This review starts, in Sec. 2, with waves that are both *linear* and *nondispersive*. That is, amplitudes can be added (linearity) and velocity is independent of frequency (nondispersive propagation). Section 3 discusses the implications of linearity alone. Section 4 treats dispersive propagation in linear media, and Sec. 5 introduces nonlinearity.

2. NONDISPERSIVE MEDIA

The displacement u for a wave moving in the x direction in a nondispersive medium satisfies the wave equation

$$\frac{1}{v^2}\frac{\partial^2 u}{\partial t^2} = \frac{\partial^2 u}{\partial x^2}. \tag{1}$$

This applies to electromagnetic waves in vacuum [7,8], in which case $v=c$ and u is an electric or magnetic field vector in the y–z plane (transverse wave). It also applies to long-wavelength acoustic waves, which may be longitudinal (u along x) or transverse. For acoustic waves, $v^2=C/\rho$, where ρ is the density and C is an appropriate elastic modulus.

The general solution of (1) is

$$u(x,t)=f(x-vt)+g(x+vt) \tag{2}$$

where f and g are arbitrary (twice-differentiable) functions. The expression $f(x-vt)$ represents a disturbance $f(x)$ traveling to the right with speed v; we can see this by sketching $f(x-vt)$ at successive instants. The disturbance $g(x+vt)$ travels to the left with speed v. With single-frequency time dependence, the two terms take the form

$$u(x,t)=u_0\exp(\pm ikx-i\omega t). \tag{3}$$

Here the convention is that the real part of the right-hand side gives the physical displacement u, and the complex amplitude u_0 may include a constant phase factor. The parameters k and ω are the angular wave number and angular frequency, respectively; also used are wave number $K=k/2\pi$ and frequency $f=\omega/2\pi$. Equation (3) represents a traveling sinusoidal wave. Displacement $u(x,t)$ repeats after distance λ such that $k\lambda=2\pi$; thus wavelength $\lambda=2\pi/k=1/K$. Similarly, periodic time $T=2\pi/\omega=1/f$. From (2) and (3),

$$v=\omega/k=f\lambda. \tag{4}$$

It was mentioned that with the appropriate v, Eq. (1) describes either a wave in which u is the longitudinal displacement or a wave in which u is the transverse displacement. Not all waves are either simply longitudinal or simply transverse. In a surface wave on water and in the Rayleigh surface wave on an elastic solid, for instance, the displacement is part longitudinal, part transverse [9]. When light propagates through an anisotropic crystal, vector **D** is transverse, because $\nabla\cdot\mathbf{D}=0$ implies $\mathbf{k}\cdot\mathbf{D}=0$, but **E**, which is not parallel to **D**, is part longitudinal, part transverse [7]. In transverse, or partially transverse, waves, the displacement can take any direction in the plane normal to the propagation direction. One must then discuss the polarization of the wave, that is, the orientation of the displacement in that plane.

The time-averaged energy density $\langle E\rangle$ in an acoustic wave is a sum of kinetic and potential (strain) contributions, and the rate of transport of energy, or intensity $I=v\langle E\rangle$, of (3) is found to be [6] $I=\frac{1}{2}Z|u_0|^2\omega^2$ where $Z=\rho v=(\rho C)^{1/2}$ is the impedance. If a traveling wave (3) in a medium of impedance Z_1 meets an abrupt interface with a medium of impedance Z_2, the transmitted and reflected amplitudes are [6] Tu_0 and Ru_0, with $T=2Z_1/(Z_1+Z_2)$ and $R=(Z_1-Z_2)/(Z_1+Z_2)$. Similar expressions hold for electromagnetic waves. If $Z_1=Z_2$, there is impedance matching at the interface, and $T=1$, $R=0$. Frequently, as in optical systems, it is desirable to maximize transmission and minimize reflection at the interface between two different media. The simplest way to achieve this form of impedance matching is by the deposition of a thin *blooming layer* at the interface. In optics, a blooming layer on a lens of refractive index $(n_1n_2)^{1/2}$ and thickness $\lambda/4$, where n_1 and n_2 are the refractive indices of the lens and of the medium in contact and λ is the optical wavelength in the layer, gives zero reflection.

3. SUPERPOSITION

The wave equation (1) is linear in u, so that the sum of two solutions is itself a solution, as in (2); there is a *superposition principle*. Superposition holds in many media besides the nondispersive media of Sec. 1. We now discuss the consequences of superposition for waves of a single frequency ω, and the discussion includes dispersive media.

The simplest superposition is of two sinusoids of opposite velocities and equal amplitudes:

$$u(x,t)=A\sin(kx-\omega t)+A\sin(kx+\omega t)$$
$$=2A\sin kx\cos\omega t. \quad (5)$$

This is a *standing wave*, namely, the fixed spatial form $\sin kx$ oscillating with variable amplitude $2A\cos\omega t$. At points $kx_n=n\pi$, the *nodes*, u is always zero, while at $kx_a=(n+\frac{1}{2})\pi$, the *antinodes*, u oscillates over a maximum range. The energy density at a point on a standing wave oscillates between kinetic and potential, but there is no transport of energy, as is clear from the nature of the superposition in (5). Standing waves are excited in a medium of fixed length (more generally, volume), and boundary conditions determine where a node or antinode lies relative to the end points. Thus if transverse waves are excited on a string stretched between fixed points L apart, the end points must be nodes. The *modes* of vibration then have successively 1, 2, 3, ..., N, ... antinodes, wavelengths $2L/N$, and discrete frequencies $N\pi v/L$, as is seen from (4) in the form $\omega=2\pi v/\lambda$. Often, standing waves have low N, but in typical lasers $N\sim10^5$. This leads to some complications [6].

A second application of superposition is to interference, in which waves traveling by different paths produce interference fringes. Consider Young's slits experiment (Fig. 1), by which the wave picture of light was finally established. The waves from slits S_1 and S_2 are focused by lens L on screen F. If S_1 and S_2 radiate unending sine waves, with the same phase at each slit, the amplitude at F is

$$u=u_0 2^{1/2}(1+\cos\delta)^{1/2}\sin\omega t \quad (6)$$

where u_0 is the amplitude at S_1 and $\delta=2\pi d\sin\theta/\lambda$ [6]. Intensity I is proportional to $\langle u^2\rangle$, the time average, and so

$$I=2I_0(1+\cos\delta)=2I_0[1+\cos(2\pi d\sin\theta/\lambda)]. \quad (7)$$

What appear on F are fringes equally spaced in $\sin\theta$. In making the foregoing calculation we assume that the only effect of the different path lengths from S_1 and S_2 is to produce a phase difference, and that amplitude differences due to inverse-square-law diminution are negligible. This is called the *Fraunhofer condition*. Some beautiful examples of Young's fringes and other interference effects are found in Ref. [10].

We now return to the assumption that S_1 and S_2 radiate unending sine waves. They do not if G is an ordinary source, which emits short ($\sim10^{-9}$ s) wave trains of random relative phase. The purpose of S_0 is to produce spatial coherence between S_1 and S_2. Waves may travel through S_1 and S_2 from any part of S_0. If the final phase difference for paths from the top of S_0 is essentially the same as that for paths from the bottom, there is spatial coherence. Thus S_0 must be narrow. The simple calculation above gives the interference for a single wave train at S_0. The observed interference pattern is the sum of many single-train patterns, and is still given by (7).

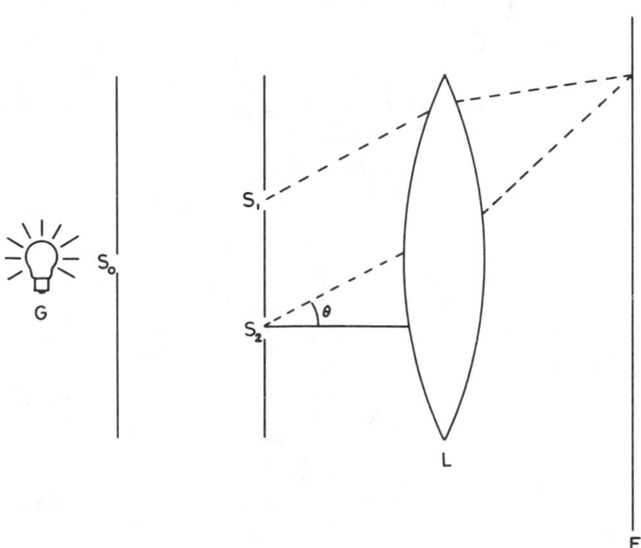

FIG. 1. Young's slits experiment.

Spatial coherence, as described, is a correspondence in phase between different points in space. For some experimenrts, such as the Michelson interferometer [7], the requirement is for *temporal coherence,* that is, a definite phase relation between values of displacement u at different time but at one particular point. Thorough treatments of coherence are given elsewhere [7,11].

Diffraction, for example by a single slit, a pinhole, or a grating, is handled similarly to Young's slits [6,7]. Thus if slits S_1 and S_2 in Fig. 1 are replaced by a central slit of width d, a calculation similar to that for Young's slits shows that the Fraunhofer diffraction pattern is $I \propto \sin^2 p/p^2$ with $p = \pi d \times \sin\theta/\lambda$. If we define an aperture function for the diffracting obstacle as equal to unity for transmitting regions and to zero for opaque regions, then for the single slit (as for other configurations) the intensity is the squared modulus of a Fourier transform of the aperture function [7,12]. Of particular interest is the *diffraction grating,* which consists of N parallel slits D apart. The diffraction pattern is a series of lines at positions $\sin\theta = m\lambda/D$ ($m=$ integer). Since the spacing depends on λ, the grating can be used to resolve lines of different wavelength, and is therefore a fundamental spectroscopic tool.

Diffraction sets a limit to the resolution of optical instruments. For instance, if a telescope of aperture d is used at wavelength λ, the angular spread of the image to the first minimum of the diffraction pattern is of the order of $\theta_d = \lambda/d$. Consequently, objects with an angular separation much less than θ_d are not resolved.

4. DISPERSIVE MEDIA: GROUP VELOCITY

As mentioned, the wave equation (1) has the special property that all sinusoids travel at the same speed v independent of ω. In general, v does depend on ω, and in this case of *dispersive propagation* u obeys some equation other than (1). Thus in the mass–spring system of Fig. 2, which is a model for longitudinal acoustic waves in a crystal, the displacement u_n of the nth mass satisfies

$$m\frac{d^2u_n}{dt^2} = C(u_{n-1} - 2u_n + u_{n+1}) \tag{8}$$

where C is the spring constant. Again, Schrödinger's equation for a free particle of mass m is

$$i\hbar\frac{\partial\psi}{\partial t} = -\frac{\hbar^2}{2m}\frac{\partial^2\psi}{\partial x^2} \tag{9}$$

where ψ is the probability amplitude. For light in a refracting medium, it is normally said that (1) holds with a frequency-dependent velocity, but this statement is only true for single-frequency light, in which case (1) may be replaced by the ordinary differential equation

$$-\frac{\omega^2}{v^2}u = \frac{d^2u}{dx^2} \tag{10}$$

and (3) is the general solution.

The basic result for dispersive media concerns a *wave packet*

$$u(x,t) = \int f(\omega)e^{i(kx-\omega t)}\,d\omega. \tag{11}$$

With an appropriate $f(\omega)$, this represents a fairly localized disturbance. Each sinusoid travels at its own phase velocity $v_p = \omega/k$; consequently the relative phases change with time and the packet changes shape as it travels. For short times the change of shape is negligible, and the peak travels at the group velocity

$$v_g = d\omega/dk. \tag{12}$$

The peak corresponds to in-phase superposition, so that the phase is stationary for variations of k and ω, $d(kx - \omega t) = 0$ or $x\,dk - t\,d\omega = 0$, which gives (12). This argument, based on Ref. [13], is due to Professor R. B. Dingle; the elementary proof uses the superposition of two sinusoids [6]. The distinction between group and phase velocity can be seen in an excellent film [14] of the motion of wave packets along a water trough.

A dispersive medium is characterized by the dispersion relation $\omega = \omega(k)$ from which v_p and v_g can be derived. For the mass and spring system of Fig. 2, substitution of (3) in (8) yields

$$\omega = 2\left(\frac{C}{ma}\right)^{1/2}\sin\left(\frac{ka}{2}\right). \tag{13}$$

Graphs of this dispersion relation, and of the dependence of v_g and v_p on k, are shown in Fig. 2. For the Schrödinger equation, (9), substitution of (3) yields $\omega = \hbar k^2/2m$. The group velocity is $\hbar k/m$, equal to the classical particle velocity once $\hbar k$ is identified as the momentum [6]. Other examples of dispersion relations can be found from many branches of physics [3–6].

In isotropic media it is normally stated that v_g is equal to the velocities v_E and v_I for transport of energy and information; the statement is justified by the observation that a localized disturbance moves with velocity v_g. A neat elementary proof that $v_E = v_g$ for waves in a bulk medium with small dissipation was given by Rayleigh in the Appendix to

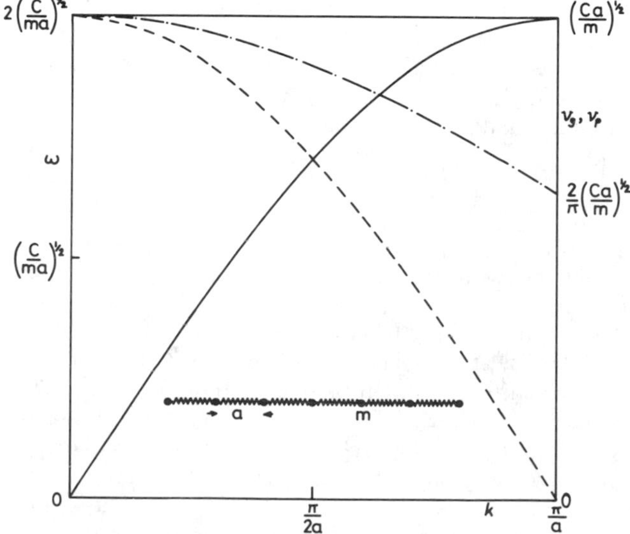

FIG. 2. Dispersion curve (———) of Eq. (13), with group (– – –) and phase (–·–·–·) velocities. Inset shows spring–mass system under consideration.

Vol. 1 of his famous book [15]. For optics, a careful discussion needs to take account of the fact that optical dispersion has its origin in the series of resonant absorptions that characterize the medium, and for frequencies very close to a strong absorption v_E cannot be clearly defined. General proofs that $v_E=v_g$ when v_E is defined in a bulk dispersive optical medium have been given [16,17]. It has been shown analytically [18] that $v_E=v_g$ for an optical wave (surface polariton) localized on the surface of a dispersive medium, and numerical investigation [19] gives the same result for waves traveling along an optically dispersive fiber. However, a proof applicable to all propagation geometries has not yet been given.

It is a famous principle of the theory of relativity that information cannot travel faster than the velocity of light, c. With v_I taken equal to v_g, this means that $v_g \leq c$, and it is important to realize that this places no restriction on the value of the phase velocity v_p. What the phase velocity describes is the passage of crests and troughs at the single frequency ω, and in order for information to be conveyed this passage must be modulated in some way; the modulation travels at the group velocity. An example is given by a simple waveguide, in which the dispersion is $\omega^2=\omega_c{}^2+c^2k^2$ with ω_c constant; the same relation holds propagation in an ionized plasma. It follows from the dispersion equation that $v_g v_p=c^2$ with $v_g<c$ and $v_p>c$.

5. NONLINEAR WAVE EQUATIONS: SOLITONS

The previous sections dealt with linear systems, and it was seen that in a dispersive linear medium, a general wave form changes shape as it travels. In a nonlinear system, by contrast, solutions can be found that maintain shape. For example, the Korteweg–de Vries (KdV) equation

$$\frac{\partial u}{\partial t}+\alpha u\frac{\partial u}{\partial x}+\frac{\partial^3 u}{\partial x^3}=0 \qquad (14)$$

(α is a constant), used to describe shallow-water waves, has the solitary wave solution

$$u(x,t)=3\alpha^{-1}v\,\mathrm{sech}^2[v^{1/2}(x-vt)/2] \qquad (15)$$

traveling at speed v with amplitude proportional to v. This can be verified by substitution or derived if we seek solutions that depend only on $x-vt$ [20–23]. A striking property of some solitary waves is that if two collide, each emerges from the collision unaltered in shape; solitary waves of this kind, such as (15), are called *solitons*. This property, hardly expected in a nonlinear system, emerges from computer experiments, and can be seen in some explicit "two-soliton" solutions that have been found.

Because of their stability, solitons have been used as models of elementary particles. More generally, they are expected to occur in a wide variety of nonlinear physical systems including optical communication fibers.

See also ACOUSTICS; DIFFRACTION; DISPERSION THEORY; ELECTROMAGNETIC RADIATION; GRATINGS, DIFFRACTION; INTERFEROMETERS AND INTERFEROMETRY; MAXWELL'S EQUATIONS; NONLINEAR WAVE PROPAGATION; OPTICS, GEOMETRICAL; OPTICS, NONLINEAR; POLARIZED LIGHT; REFLECTION; SCHRÖDINGER EQUATION; SURFACE WAVES ON LIQUIDS.

REFERENCES

1. H. C. Ohanian, *Physics*. Norton, New York and London, 1985. (E)
2. D. Halliday and R. Resnick, *Fundamentals of Physics*. Wiley, New York, 1986. (E)
3. F. S. Crawford, *Waves* (Berkeley Physics Course, Vol. 3). McGraw-Hill, New York, 1965. (E)
4. A. P. French, *Vibrations and Waves* (M.I.T. Introductory Physics Series). Van Nostrand Reinhold, London, 1982. (E)
5. I. G. Main, *Vibrations and Waves in Physics*. Cambridge University Press, London and New York, 1984. (E)
6. D. R. Tilley, *Waves*. Macmillan, London, 1974. (E)
7. S. G. Lipson and H. Lipson, *Optical Physics*. Cambridge University Press, London and New York, 1981. (A)
8. J. D. Jackson, *Classical Electrodynamics*. Wiley, New York, 1975.
9. M. G. Cottam and D. R. Tilley, *Introduction to Surface and Superlattice Excitations*. Cambridge University Press, London and New York, 1989. (A)
10. M. Cagnet, M. Françon, and J. C. Thrierr, *Atlas of Optical Phenomena*. Springer-Verlag, Berlin and New York, 1962. M. Cagnet, M. Françon, and S. Mallick, *Supplement to Atlas of Optical Phenomena*. Springer-Verlag, Berlin and New York, 1971.
11. R. Loudon, *The Quantum Theory of Light*. Oxford University Press, London and New York, 1983. (A)
12. D. C. Champeney, *Fourier Transforms and Their Physical Applications*. Academic Press, New York, 1973. (A)
13. H. and B. S. Jeffreys, *Mathematical Physics*. Cambridge University Press, London and New York, 1972.
14. E. David and G. Bekow, *Gruppen und Phasengeschwindigkeit*. Film C614 of the Institut für den Wissenschaftlichen Film, Göttingen.
15. J. W. S. Rayleigh, *The Theory of Sound*, 2 vols. Dover, New York, 1945.
16. L. Brillouin, *Wave Propagation and Group Velocity*. Academic Press, New York, 1960. (A)
17. R. Loudon, *J. Phys. A* **3**, 233 (1970).
18. J. Nkoma, R. Loudon, and D. R. Tilley, *J. Phys. C* **7**, 3647 (1974).
19. H. Khosravi, R. Loudon, and D. R. Tilley, *J. Opt. Soc. Am.* (to be published).
20. A. C. Scott, F. Y. F. Chu, and D. W. McLaughlin, *Proc. IEEE* **61**, 1443 (1973).
21. R. K. Bullough, "Solitons," in *Interaction of Radiation with Condensed Matter* (Proc. Winter College, Trieste, 1976), vol. 1, p. 382. IAEA, Vienna, 1977.
22. G. L. Lamb, *Elements of Soliton Theory*. Wiley, New York, 1980.
23. R. K. Dodd *et al., Solitons and Nonlinear Wave Equations*. Academic, New York, 1984.

Weak Interactions

Val L. Fitch

More than 80 years of painstaking experimental work and theoretical development culminated in the early 1970s with a highly successful theory, called the standard model, which shows that the weak and electromagnetic interactions are separate manifestations of one common electroweak force. The joining of the weak and electromagnetic phenomena into one interaction constitutes the first unification of different forces since Maxwell showed electricity and magnetism to be different manifestations of the same phenomenon.

The weak interactions refer to a class of forces which are 10^{14} times weaker than the strong forces which hold the nucleus together. The neutrino, a spin-$\frac{1}{2}$ particle without electrical charge, best exemplifies the weak interactions because it reacts with other matter only through the weak forces. Whereas a neutron, a strongly interacting particle, will travel on the average through 10 cm of iron before scattering, the neutrino will travel through 10^{16} cm. Recalling that the earth is largely iron with a diameter of 1.2×10^4 km, we see that the probability of the neutrino scattering in going through the earth is only one part in 10^6. The earth is essentiallly transparent to neutrinos! Indeed, at low energies the sun is nearly transparent. These are not irrelevant observations since the thermonuclear processes in stars lead to copious production of neutrinos. The mass of the neutrino has never been measured but the existing experimental limits indicate that it is tiny if not zero.

Despite the considerable weakness of the interactions we have been able to learn much about them because of two fortuitous circumstances. First, nuclear reactors and accelerators are such intense sources of neutrinos that the tiny probability of a single neutrino scattering in a detector of reasonable size is compensated by enormous fluxes. Second, the weak interactions manifest themselves in a large variety of rather common decay processes, many of which have been studied exhaustively. Due to these weak effects many radioactive nuclei decay with the emission of an electron and neutrino (the emitted electrons are called beta rays—the process is called nuclear beta decay). During the decay process, a constituent neutron transforms to a proton which remains as a part of the new nucleus, an isobar of the original. And because of the abundance of neutrons in nuclear reactors it has been possible to study the decay of the free neutron to electron, proton, and antineutrino (why antineutrino instead of neutrino will become apparent later)

$$n \rightarrow p + e^- + \bar{\nu}. \tag{1}$$

The mean life for this decay is 15 min. In addition, since a reactor core has the most intense concentration of neutrons, their decay provides a rich source of antineutrinos.

A well-known example of nuclear beta decay is ${}^{60}\mathrm{Co} \rightarrow {}^{60}\mathrm{Ni} + e^- + \nu$. The mean life is 144 days. The cobalt is produced as the result of fission of uranium nuclei. Another example is a naturally occurring beta emitter, ${}^{40}\mathrm{K}$. Its mean life is 4.5×10^9 years. This isotope of potassium constitutes about 0.01% of the total potassium in the earth and is a source of part of the natural radioactive background. The completely stable isotopes of potassium, ${}^{39}\mathrm{K}$ and ${}^{41}\mathrm{K}$, have relative abundances of 93 and 7%. When the earth was formed, one expects that ${}^{40}\mathrm{K}$ was formed with a relative abundance between that of its two neighbors. The fact that, because of its decay, it now constitutes only one part in ten thousand of the total potassium on the earth enables us to calculate the age of the earth, a measure of the time that has elapsed since the materials of the earth were formed.

In some nuclei the relative energies of the pertinent nuclear levels permit one of the orbiting electrons to be captured by one of the constituent protons which becomes a neutron and the electron becomes a neutrino:

$$e^- + p \rightarrow \nu + n. \tag{2}$$

This process is called *K* capture because it is the *K*-shell electrons which are usually involved. One should note the similarity between this reaction and neutron decay. The two processes can be related exactly when the energies of all the particles are taken into account.

Despite the complications associated with the decaying particles being submerged inside nuclei, the study of nuclear beta decay has been extraordinarily fruitful. The energy of the electrons and neutrinos is typically $\frac{1}{2}$ to 10 MeV. Their momenta are sufficiently small that the two leptons generally do not carry off any orbital *angular* momentum. Such decays are called "allowed" transitions. However, the combined spins of the two spin-$\frac{1}{2}$ leptons may total either 0 or 1 unit of angular momentum. Correspondingly, the nuclear spin may change by 0 or 1 unit in the decay. Those beta-decay processes where the leptons come off with their spins antiparallel, 0 angular momentum, are known as Fermi (F) transitions. When one unit of angular momentum is carried off by the leptons, the spin of the nucleus can change by 0 or 1 unit; the process is called a Gamow–Teller (GT) transition. The decay of ${}^{35}\mathrm{S}$ to ${}^{35}\mathrm{Cl}$ is an example of an F transition, while the decay of ${}^{60}\mathrm{Co}$ (spin 5) to ${}^{60}\mathrm{Ni}$ (spin 4) is a pure GT transition. Neutron decay is a mixed GT and F transition. In this case, the spin of the nucleon may or may not flip when changing from neutron to proton and the lepton spins may be parallel or antiparallel.

The energy spectrum of the electrons emitted in so-called allowed beta decays is dominated by statistical considerations (phase space with rather obvious corrections for Coulomb effects—the electron being attracted by the positive nucleus from which it is escaping). For those beta decays where the spin of the nucleus changes by 2 units, the leptons must also carry off *orbital* angular momentum and the process is suppressed. These are referred to as first forbidden transitions. The example of ${}^{40}\mathrm{K}$ mentioned above is a case of a third forbidden transition (spin change is 4). This accounts for the extraordinarily long mean life. In the simplest nuclei the study of beta decay has illuminated the weak interactions. In more complicated nuclei the beta-decay process has been used as a probe of nuclear structure.

A most important characteristic of the beta-decay process is that the leptons are emitted with the polarization equal to $-v/c$, where v/c is their speed relative to that of light. Since neutrinos have a very tiny or zero rest mass their velocity is equal to or nearly equal to c and they are correspondingly highly polarized along their line of flight. The spin of a lepton in relation to its flight direction is as a left-handed screw. Leptons are said to have left helicity. Correspondingly, antileptons are right-handed; they have right or positive helicity. The net polarization of the electron (or neutrino) emitted in beta decay is a manifestation of parity nonconservation in the weak interactions.

From reactions (1) and (2) we see that not only neutrinos but also electrons (or positrons), protons, and neutrons engage in weak interactions. A distinction is made between those weakly interacting particles which also interact strongly and those which do not. The electron and neutrino are the leptons. The neutron, proton, π, and K mesons are examples from a large class of strongly interacting particles called hadrons. In a sense the electron and its neutrino are two charge states of the same object as are the neutron and proton. Other kinds of charged leptons exist: the muon has a mass 207 times that of the electron and the tau, 3490 times, or 1.9 times the mass of the proton. Each variety of charged lepton is associated with a neutrino distinct from all the other neutrinos. The electron and mu neutrinos have been directly detected whereas the existence of the tau neutrino is still only inferred. Modern particle accelerators produce large numbers of π and K mesons both of which decay quickly to muons and neutrinos. Taus are made in electron–positron colliding-beam machines.

All evidence supports the view that the number of leptons minus the number of antileptons is a conserved quantity (conservation of leptons). Arbitrarily, the electron has been called a particle, the positron an antiparticle. The conservation of leptons in reaction (1) forces that neutrino to be an antiparticle.

Parity nonconservation is most dramatic in the decay of the muon to an electron plus two neutrinos,

$$\mu^- \rightarrow e^- + \bar{\nu}_e + \nu_\mu . \qquad (3)$$

Polarized muons with negative charge decaying at rest emit more electrons in the hemisphere opposite to the direction of the spin than in the reverse direction. The inverse is true for positive muons. In Fig. 1, the length of the vector indicates the number of electrons (positrons) from polarized μ^+ (μ^-) decay as a function of angle. The figure shows that the decay process is not invariant to a reflection about the line, a characteristic of parity nonconservation. It also shows that the process is not invariant under charge conjugation—the operation that transforms world to antiworld, in this case, μ^- to μ^+.

When it was first discovered that parity nonconservation occurred, leading to asymmetric decays, it was thought the process could be used to establish an absolute direction in space in violation of Mach's principle. However, it is noted from Fig. 1 that the combination of parity reversal, P, and charge conjugation, C, restores the fundamental symmetry. The fact that one cannot, with perfect CP symmetry, distinguish a distant galaxy from a distant antigalaxy would prevent one from establishing a preferred direction in space.

An important example of a weak interaction is the decay of the K meson (kaon) to two or three π mesons (pions). Indeed, the observation that the kaon, which has zero spin, could decay to three pions in a state of zero relative angular momenta and also decay to two pions was the first evidence that parity was not conserved in the weak interactions (the pion has odd intrinsic parity). Detailed studies show that not only is parity not conserved in this decay but also the com-

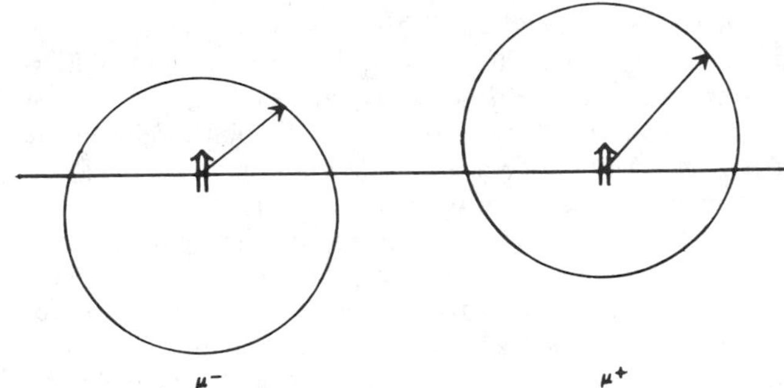

FIG. 1 The relative intensity of electrons (positrons) emitted from a source of polarized mu-minus (mu-plus) mesons as a function of the angle from the direction of polarization.

bined operations of charge conjugation and parity, CP, is not an exactly conserved quantity. The effect is small (the CP-nonconserving effect is about 2×10^{-3} of the CP-conserving process) but it has been exhaustively studied and the detailed parameters are extremely well known.

Modern field theory depends on the validity of a very general theorem which says that all interactions must be invariant under the combined operations of charge conjugation, parity, *and* time reversal, CPT. Accordingly, a violation of C and P invariance requires a compensating violation of time-reversal invariance. In fact, detailed experimental studies of the neutral kaon decay strongly support the CPT theorem. This means that, to a small but finite degree, fundamental interactions are different depending on the direction of time. The absence of complete CP symmetry also reopens the question of the complete validity of Mach's principle.

As noted above, nuclear reactors are sources of antineutrinos and, because of their abundance, the reaction

$$\bar{\nu}_e + p \rightarrow e^+ + n \qquad (4)$$

has been extensively studied. Indeed, the production of positrons using reaction (4) provided the first direct evidence for the existence of neutrinos.

It was thought for many years that the exchange of electrical charge in reaction (4) between the leptons was a unique characteristic of weak forces. However, it is now known that the reaction proceeds without charge exchange about 25% of the time. Similar reactions occur for the mu-like neutrinos, viz.,

$$\nu_\mu + n \rightarrow \mu^- + p \qquad (5)$$

or

$$\bar{\nu}_\mu + p \rightarrow \mu^+ + n \qquad (6)$$

and without charge exchange, e.g.,

$$\nu_\mu + p \rightarrow \nu_\mu + p . \qquad (7)$$

The theoretical understanding of the weak interactions was guided from the earliest days by the work of Fermi. The Fermi theory was drawn in close analogy with electrodynamics. As a charged particle interacts with the electromagnetic (photon) vector field, so a weakly interacting particle, e.g., a neutron, interacts with a lepton field. The theory enjoyed enormous success. The only change since the original work of Fermi was the addition of axial-vector interactions with a strength nearly equal to the original vector interaction assumed by Fermi. This addition produced the effects of parity nonconservation. The modified Fermi theory was very successful in accounting for low-energy weak interaction phenomena but predicted disturbing behavior at high energies—for example, a neutrino–nucleon cross section growing with energy without limit. To avoid these problems it was proposed that the weak interactions were mediated by massive bosons. These bosons could either be charged and account for the charge changing reactions through "charged currents" or neutral in analogy with the interaction of "neutral" currents in electrodynamics. The standard model evolved out of these observations.

Despite their similarity, these two forces are grossly different with respect to their range. The electromagnetic interaction is long range (the electrical potential depends on distance as I/r), whereas the weak interaction is short range, indeed, much shorter even than the strong interactions. The electrical force between charged particles is due to the exchange of photons which have zero rest mass. A shortened range can be arranged by requiring that the exchanged particle have a nonzero mass. From the simplest point of view, the exchange of a particle of rest mass m leads to an interaction potential energy

$$U(r)=(e^2/r)\exp(-rmc/\hbar), \tag{8}$$

where r is the range, $\hbar$ is Planck's constant, and c is the velocity of light. As $m\rightarrow 0$ it reduces to the Coulomb potential. By an increase in the mass of the exchanged particle the range of the interaction can be reduced to any arbitrary value. For example, if m equals the mass of the pion, the range is characteristic of nuclear forces, 1.4×10^{-13} cm. In applying these ideas to the weak interactions, to produce an effect with a given magnitude, one may vary the weak charge e_w or the range, i.e., the mass of the exchanged particle. For the same effect the "weak charge" could be very large if the characteristic range of the force is small. Therefore, unlike the electrical charge, it is necessary to specify not only the magnitude of the "weak" charge but also the volume in which it is effective. The weak charge squared is measured to be $e_w^2=0.88\times 10^{-37}$ eV cm (for comparison, the electromagnetic charge squared is $e^2=1.44\times 10^{-7}$ eV cm.) Over a typical nuclear volume this leads to a weak interaction potential of about 1 eV. However, experiments show the weak forces to be confined to very much smaller distances than nuclear dimensions and the question is—just how short is the range? Because the fundamental structure of the weak and electromagnetic interaction is similar, aside from the range of the force, a natural possibility is that the basic strength of the weak and electromagnetic charge is the same. This necessitates that the characteristic range of the weak force be set in the region of 3×10^{-16} cm corresponding to a mass of 70–80 GeV for the exchanged particle.

The standard model leads to the prediction of a charged vector boson, the $W^{\pm}$, which accounts for those weak interactions in which the lepton charge is changing, and the Z_0 which is involved in the neutral interactions. The theory predicts that the masses of these vector bosons are related through the Weinberg mixing angle, θ_W, viz.,

$$m_w=137.3 \text{ GeV}/\sin\theta_W \quad \text{and} \quad m_z=m_w/\cos\theta_W.$$

The angle, θ_W, has been measured in a wide variety of the neutral current processes and is found to be $\sin^2\theta_W=0.230\pm 0.005$.

With the advent of particle accelerators which have sufficient energy, both of these particles are observed with $m_w=81\pm 1.4$ GeV and $m_z=91.10\pm .05$ GeV. These measured values are in striking agreement with those predicted from the standard model.

Since hadrons are composed of quarks, the weak interactions involving hadrons are now described in terms of these more fundamental entities. A neutron is composed of two d quarks and one u quarks, whereas the proton is made of two u quarks and one d quark. The beta decay of the neutron then involves, in this picture, a transformation of a d- to a u-type quark, viz.,

$$d\rightarrow u+e^-+\nu. \tag{9}$$

Because they are composed of a limited number of primitive quarks, striking symmetries appear among the hadrons and since weak interactions occur between quarks as well as leptons, similar symmetries in the interactions appear. A table of quarks and leptons displays a remarkable correspondence.

Lepton	*Charge*	*Quark*	*Charge*
ν_e	0	u	$\frac{2}{3}$
e	-1	d	$-\frac{1}{3}$
ν_μ	0	c	$\frac{2}{3}$
μ^-	-1	s	$-\frac{1}{3}$
ν_τ	0	t	$\frac{2}{3}$
τ^-	-1	b	$-\frac{1}{3}$

As seen from the table, for each lepton there appears a corresponding quark with its electrical charge displaced by $+\frac{2}{3}$. This striking symmetry between the leptons and quarks suggests that the strong as well as the weak and electromagnetic interactions be included in the unification process. The symmetry also suggests that should new leptons be discovered the chances are high that corresponding quarks exist.

A question of most fundamental importance is whether the table above is complete or whether there exist still heavier charged leptons with their associated neutrinos. Studies of the Z_0 decay reveal that, most likely, there exist *no more than the three generations of leptons* as listed in the table. This conclusion comes from the following considerations. Since it couples to neutral currents, the Z_0 can decay to neutrino–antineutrino pairs. In fact, pure neutrino decay contributes significantly to the decay rate and therefore to the width of the Z_0 resonance which is observed in electron–positron colliders. Summarizing all the data currently available (1989) the conclusion is that there are 3.1 ± 0.2 generations of neutrinos. Such conclusions should hold except in the unlikely event that any new neutrino would be so massive that the Z_0 could not decay into them.

The old Fermi theory, modified with axial-vector contributions, required that the neutrino mass be strictly zero. The standard model imposes no such constraint and there is great interest in the question of neutrino masses. It is now known that the electron neutrino has a mass less than 10 eV, the mu neutrino less than 0.25 eV, and the tau neutrino less than 35 MeV. The neutrino masses are of special interest in cosmology since a nonzero rest mass would contribute significantly to the total mass of the universe and perhaps account for all of the missing dark matter. Pursuing the parallelism between quarks and leptons noted in the table, since the weak interactions allow transitions between the different flavored quarks could it not be that transitions also occur between the different generations of neutrinos especially since they can have finite and different masses? Such transitions would lead to oscillations between the different generations of neutrinos. Searches for such neutrino oscillations are now taking place, so far with null results.

The weak interaction between the quarks is quantified through a mass mixing matrix which arises because the mass eigenstates are not the same as the weak eigenstates. All of the mixing is expressed through a 3×3 unitary matrix operating on the charge $\frac{1}{3}$ quark states (d, s, and b), viz.,

$$\begin{pmatrix} d' \\ s' \\ b' \end{pmatrix} = \begin{pmatrix} V_{ud} & V_{us} & V_{ub} \\ V_{cd} & V_{cs} & V_{cb} \\ V_{td} & V_{ts} & V_{tb} \end{pmatrix} \begin{pmatrix} d \\ s \\ b \end{pmatrix}.$$

This matrix, the Kobayashi–Maskawa or K–M matrix, was remarkably prescient. When it was first suggested, only three flavors of quarks were known; the u, d, and s. It was proposed because it was noted that with at least six flavors of quarks there is room in the matrix for a CP-violating term. Now, five flavors of quarks have been discovered and it is fully expected that the sixth (the t or top) exists but at such a large mass it is beyond the reach of current accelerators. All of the weak interactions between the quarks are interpreted in terms of the K–M matrix.

Current experimental effort is directed toward evaluating the various elements in the matrix. These are determined from the weak decay rates of the relevant quarks, and in some instances, from neutrino scattering. Requiring the matrix to be unitary provides another constraint. The details are beyond the scope of this review but the details are contained in the references below. Of special significance is the decay of the neutral B meson (the meson containing the b quark). It is expected to show large violations of CP. To date, CP violation has been observed only in the neutral K-meson system.

See also Beta Decay; Currents in Particle Theory; Electron; Elementary Particles in Physics; Grand Unified Theories; Leptons; Mesons; Neutrino; Parity; Positron; Quarks; Weak Neutral Currents.

BIBLIOGRAPHY

E. D. Commins and P. H. Buchsbaum, *Weak Interactions of Leptons and Quarks*. Cambridge University Press, Cambridge, New York and Melbourne, 1983.

Particle Data Group, *Review of Particle Properties*. North-Holland, Amsterdam, 1988.

Weak Neutral Currents

Alfred K. Mann

I. INTRODUCTION

Weak neutral currents were first observed and their strength determined quantitatively in 1973 [1]. The interaction between any two weak neutral currents gives rise to much sought-after and initially elusive reactions characterized in present thinking by the exchange between the currents of a virtual, massive, electrically neutral vector boson, the Z^0. It had long been known that the exchange of a virtual, massless, electrically neutral vector boson, the photon, between electromagnetic currents—which are "neutral" in this terminology—described electromagnetic reactions. However, prior to 1973, only the reactions arising from the interactions of two weak charged currents, mediated by the exchange of virtual, massive, electrically charged vector bosons, the $W^{\pm}$, had been observed. Feynman diagrams schematically illustrating these processes are shown in Fig. 1.

Somewhat earlier, a unified theory of quantum electrodynamics and quantum weak dynamics had been formulated [2] in which an integral constituent was the weak neutral current. The success of this theory, now known as the electroweak theory (EWT), became apparent with the experimental discovery of weak neutral currents (WNC) and is exhibited in the ability of the EWT to describe in precise detail the wide variety of phenomena which have been intensively studied experimentally since 1973. The junction of the EWT and experiment culminated in the direct observation in 1983 of the massive vector bosons, the $W^{\pm}$ and Z^0, at the mass values predicted by the EWT [3], the predictions of which require as input the measured strength of the WNC found in WNC experiments.

More generally, the EWT provides a unified, renormalizable, gauge-invariant description of weak charged current (WCC) and electromagnetic neutral current (ENC) interactions, as well as WNC interactions [4]. It is, however, WNC phenomena and the $W^{\pm}$ and Z^0 masses which contribute the major quantitative tests of the predictions of the EWT. These extend over the wide range of squared momentum transfer from 10^{-6} $(\text{GeV}/c)^2$ to 10^4 $(\text{GeV}/c)^2$ and encompass such diverse processes as deep inelastic neutrino and longitudinally polarized electron scattering from isoscalar and nonisoscalar targets, ν-p scattering, ν-e scattering, parity violation in atoms via an induced transition electric dipole moment, asymmetries in high center-of-mass energy $e^+e^- \rightarrow \mu^+\mu^-$ reactions, and the masses of $W^{\pm}$ and Z^0.

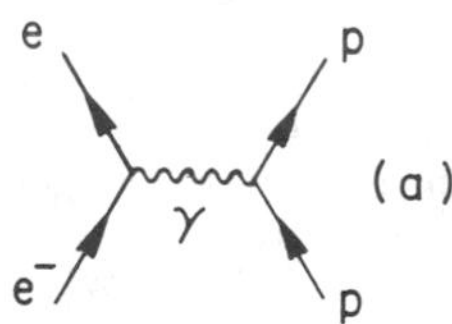

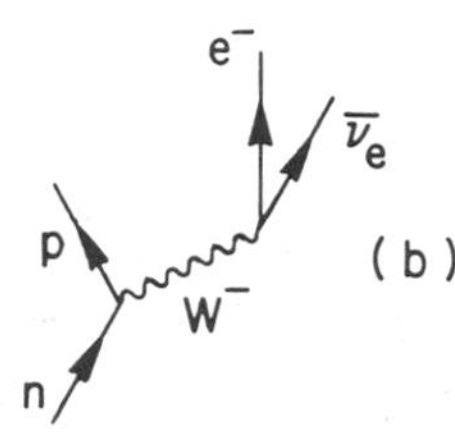

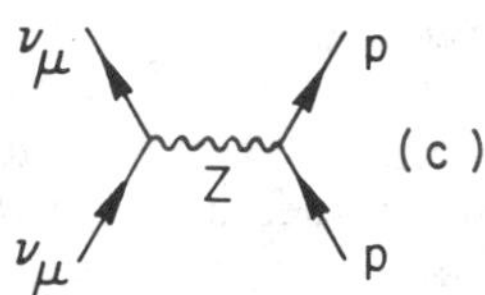

FIG. 1. Feynman diagram for photon exchange in quantum electrodynamics (a). In the unified electroweak theory, the weak interactions are mediated by massive vector bosons: the charged W for β-decay (b), and the neutral Z for elastic neutrino scattering (c).

In what follows we record in Section II the EWT expressions for various WNC interactions [5], primarily to illustrate their structure and emphasize the importance of radiative corrections in the theory. Section III summarizes the comparisons of EWT predictions with experimental data. Finally, in Section IV a short discussion is given of future exploration of still higher-precision WNC measurements as more stringent tests of the EWT and as sensitive probes of new physics beyond the EWT.

II. EWT EXPRESSIONS FOR WNC INTERACTIONS [5]

The effective Lagrangian for WNC interactions between massless neutrinos (ν) and quarks (q) is

$$-L^{\nu H}=\frac{G_{\rm F}}{\sqrt{2}}\,\bar{\nu}\gamma^{\mu}(1+\gamma_5)\nu J^{H}_{\mu}, \qquad (1)$$

where $G_{\rm F}$ is the Fermi coupling constant ($=10^{-5}/M_p^2$), γ^{μ} and γ_5 are Dirac matrices, and

$$J^{H}_{\mu}=\sum_i \bar{q}_i\gamma_{\mu}(g^i_V+g^i_A\gamma_5)q_i. \qquad (2)$$

Here, identical V, A couplings are assumed for all neutrino flavors, and, with $g^i_{V,A}\equiv\epsilon_L(i)\pm\epsilon_R(i)$, flavor independence of the quark couplings is also assumed, i.e., $\epsilon_{L,R}(u)=\epsilon_{L,R}(c)$, etc., where u and c are "up" and "charm" [5] quarks, respectively.

In the EWT the $\epsilon_L(i)$ and $\epsilon_R(i)$ are completely specified in terms of constants, calculable radiative-correction parameters, and the only undetermined parameter in the theory, the relative coupling strength of the WNC, usually written as $\sin^2\theta_W$.

The effective Lagrangian for the WNC interaction between muon-type neutrinos (and autineutrinos), ν_{μ} (and $\bar{\nu}_{\mu}$), and electrons is

$$-L^{\nu e}=\frac{G_{\rm F}}{\sqrt{2}}\,\bar{\nu}_{\mu}\gamma^{\mu}(1+\gamma_5)\nu_{\mu}J^{e}_{\mu}, \qquad (3)$$

where

$$J^{e}_{\mu}=\bar{e}\gamma_{\mu}(g^e_V+g^e_A\gamma_5)e \qquad (4)$$

and g^e_V and g^e_A are, again in the EWT, given in terms of $\sin^2\theta_W$ and radiative-correction parameters.

The parity violating (but CP conserving) interaction between electrons and quarks is given by

$$-L^{eH}=\frac{G_{\rm F}}{\sqrt{2}}\sum_i\,(C_{1i}\bar{e}\gamma_{\mu}\gamma_5 e\bar{q}_i\gamma^{\mu}q_i+C_{2i}\bar{e}\gamma_{\mu}e\bar{q}_i\gamma^{\mu}\gamma_5 q_i). \qquad (5)$$

In the EWT, the constants C_{1u}, C_{1d}, and C_{2u} depend on $\sin^2\theta_W$ and radiative-correction parameters.

Finally, in the EWT the $W^{\pm}$ and Z^0 masses are predicted to be

$$M_W=\frac{A_0}{\sin\theta_W(1-\Delta r)^{1/2}}, \qquad (6)$$

$$M_Z=\frac{M_W}{\cos\theta_W}, \qquad (7)$$

where

$$A_0=\left(\frac{\pi\alpha}{\sqrt{2}G_{\rm F}}\right)^{1/2}=37.281\ {\rm GeV}/c^2, \qquad (8)$$

and Δr is a calculable radiative-correction parameter.

It is customary to define the so-called renormalized weak angle in the EWT as

$$\sin^2\theta_W\equiv 1-M_W^2/M_Z^2. \qquad (9)$$

The effective Lagrangians in Eqs. (1) through (8) and the definition of $\sin^2\theta_W$ in Eq. (9) are sufficient to describe quantitatively all the WNC reactions mentioned in the Introduction. In principle, once a value of $\sin^2\theta_W$ is extracted from a given class of experiments and an appropriate radiative correction applied to the empirical value, all other classes of experiments then yield (when radiatively corrected) values of $\sin^2\theta_W$ which serve to test in detail the correctness and self-consistency of the EWT over a multiplicity of phenomena and a wide range of energies and momentum transfers. This is the strategy employed to compare the results of one class of experiments with another class and to facilitate and make clear comparison of experiment with the EWT.

III. WNC DATA AND THE EWT

In this section we briefly summarize WNC experimental results and compare them with the predictions of the EWT.

In Fig. 2 are shown values of $\sin^2\theta_W$ obtained from various classes of WNC experiments and measurements of the W and Z masses plotted against Q^2, the square of the four-

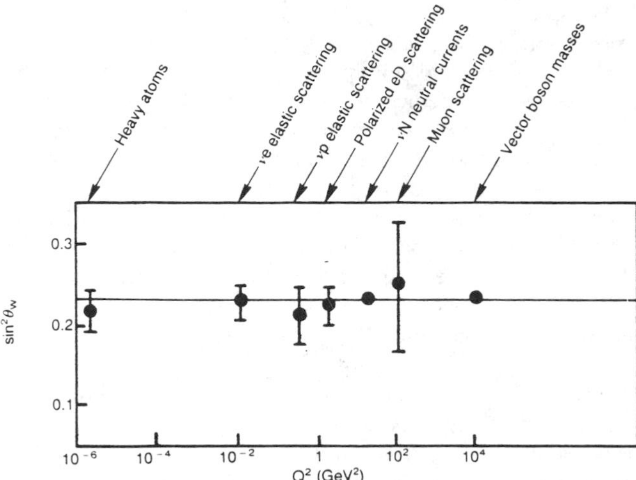

FIG. 2. Various determinations of $\sin^2 \theta_W$, the fundamental unspecified parameter of the electroweak theory, plotted against Q^2, the square of the four-momentum transfer typical of each class of experiments. Each determination is labeled by experiment type (see Table I). The horizontal line is at $\sin^2 \theta_W = 0.230 \pm 0.0048$. Reprinted from Ref. [7].

momentum transfer typical of each class of experiment [7]. Each point in the plot is labeled by the experiment type that produced it. Note the wide range of Q^2 and the convergence of the different determinations of $\sin^2 \theta_W$ on the single, universal value shown by the horizontal line in Fig. 2 at $\sin^2 \theta_W = 0.230 \pm 0.0048$.

The data in Fig. 2 are also summarized in Table I [7] which gives, in addition to the precise numerical values of $\sin^2 \theta_W$ from the different classes of experiments, values of $\sin^2 \theta^0$, i.e., the values obtained directly from experiment without the radiative corrections provided by the EWT. One sees the importance of the radiative corrections which are approximately 4% in the determination of $\sin^2 \theta_W$ from the parity violation in heavy atoms, approximately 7% in extracting $\sin^2 \theta_W$ from measurement of W and Z masses, and less than 1% in $\nu_\mu e$ scattering.

Table I. Determination of $\sin^2 \theta_W$ from Various Reactions [7]

Reaction	$\sin^2 \theta_W$ [a]	$\sin^2 \theta^0$ [b]
Deep inelastic $\nu_\mu N$ scattering	$0.233 \pm 0.003 \pm [0.005]$	0.242
$\nu_\mu p \rightarrow \nu_\mu p$	0.210 ± 0.033	0.208
$\nu_\mu e \rightarrow \nu_\mu e$	$0.223 \pm 0.018 \pm [0.002]$	0.221
W and Z masses	$0.229 \pm 0.007 \pm [0.002]$	0.214
Parity violation in heavy atoms	$0.220 \pm 0.007 \pm [0.018]$	0.212
Polarized eD scattering	$0.221 \pm 0.015 \pm [0.013]$	0.226
All data	0.230 ± 0.0048	

Data from Ref. [5]. See also Fig. 2.

[a] Where two uncertainties are shown, the first is experimental, and the second, in squared brackets, is theoretical. The latter includes the effect of letting the unknown masses of the top quark and Higgs boson range widely. The central values assume $M_{\mathrm{I}} = 45$ and $M_{\mathrm{II}} = 100$ GeV.

[b] Values that would be obtained from the data without radiative corrections.

Table II. Values of the Model-Independent Neutral-Current Parameters Compared With the Prediction for $\sin^2 \theta_W = 0.230$. After Ref. [7].

Quantity	Experimental Value	Prediction
$\epsilon_L(u)$	0.339 ± 0.017	0.345
$\epsilon_L(d)$	-0.429 ± 0.014	-0.427
$\epsilon_R(u)$	-0.172 ± 0.014	-0.152
$\epsilon_R(d)$	$-0.011 \pm^{0.081}_{0.057}$	0.076
g_A^e	-0.498 ± 0.027	-0.503
g_V^e	-0.044 ± 0.036	-0.045
C_{1u}	-0.249 ± 0.071	-0.191
C_{1d}	0.381 ± 0.064	0.340
$C_{2u} - \frac{1}{2}C_{2d}$	0.19 ± 0.37	-0.039

In addition to the wide range of Q^2 shown in Fig. 2, the wide diversity of the phenomena and the experimental techniques represented by the different classes of measurement in that figure should be emphasized. For example, parity violation in heavy atoms, at the extreme low end of momentum transfer, originates through a modified transition probability for the scattering of circularly polarized photons by a gas of heavy atoms. The modification is due to an induced transition electric dipole moment generated by the parity-violating WNC through exchange of a virtual Z^0 boson between electron and nucleus in addition to virtual photon exchange. The experiment consists in measuring very small rotations of the axis of polarization of the incident circularly polarized light following its passage through the atomic gas. Compare this with production of the massive Z^0 boson ($M_Z \simeq 91$ GeV/c^2) in $\bar{p}p$ collisions at center-of-mass energy of 560 GeV, and detection of the multi-GeV e^+ and e^- from the subsequent decay, $Z^0 \rightarrow e^+e^-$.

To compare experiment and theory in still greater detail, we show in Table II the experimental values of the quantities appearing in the effective Lagrangians in Section II and the values predicted by the EWT [7] when the value of $\sin^2 \theta_W$ is taken to be 0.230.

The content of Fig. 2 and Tables I and II may be summarized as demonstrating the "symbiotic" relationship between WNC phenomena and the EWT. The experimental tests provided by these phenomena and passed successfully by the EWT help to establish the theory as the quantitatively correct description of weak and electromagnetic interactions at the present level of experimental error.

IV. FUTURE PROGRESS

There exists the prospect that the precision of several of the classes of experiments discussed in the last section might be substantially improved in the coming decade. The precision of $\sin^2 \theta_W$ determined from a single class of experiments is of the order of 5–7% at present, but improvements to the level of 1% or better are likely to be made in the near future. This increased precision will in turn stimulate calculations of higher-order radiative corrections. Taken together, the result will be that WNC processes will serve not only as tests of the validity of the EWT, but also as probes of new physics beyond the EWT if any significant discrepancy between experiment and theory is found.

See also Currents in Particle Theory; Elementary Particles in Physics; Grand Unified Theories; Weak Interactions.

REFERENCES

1. F. J. Hasert *et al.* (Gargamelle Collaboration). *Phys. Lett.* **B46,** 121, 138 (1973); A. Benvenuti *et al.* (HPW Collaboration), *Phys. Rev. Lett.* **32,** 800, 1454, 1457 (1974).
2. S. L. Glashow, *Nucl. Phys.* **22,** 579 (1961). S. Weinberg, *Phys. Rev. Lett.* **19,** 1264 (1967). A. Salam, in *Elementary Particle Theory* (N. Svartholm, ed.), p. 367. Almqvist and Wiksells, Stockholm, 1969.
3. G. Arnison *et al.* (UA1 Collaboration), *Phys. Lett.* **B166,** 484 (1986); C. Albajar *et al., Z. Phys.* **C44,** 15 (1989); R. Ansari *et al.* (UA2 Collaboration), *Phys. Lett.* **B186,** 440 (1987).
4. P. W. Higgs, *Phys. Rev. Lett.* **12,** 132 (1964); **13,** 321 (1964); *Phys. Rev.* **145,** 1156 (1966); G.'t Hooft and M. Veltman, *Nucl. Phys.* **B50,** 318 (1972), and references therein.
5. U. Amaldi *et al., Phys. Rev.* **D36,** 1385 (1987); G. Costa *et al., Nucl. Phys.* **B297,** 244 (1988); P. Langacker, *Phys. Rev. Lett.* **63,** 1920 (1989).
6. S. L. Glashow, J. Iliopoulos, and L. Maiani, *Phys. Rev.* **D2,** 1285 (1970); J. J. Aubert *et al., Phys. Rev. Lett.* **33,** 1404 (1974); J. E. Augustin *et al., Phys. Rev. Lett.* **33,** 1406 (1974); A. Benvenuti *et al., Phys. Rev. Lett.* **34,** 419 (1975); E. J. Cazzoli *et al., Phys. Rev. Lett.* **34,** 1125 (1975).
7. P. Langacker and A. K. Mann, *Phys. Today* **42,** 22 (1989).

Whiskers

R. V. Coleman

The term *whisker* describes any fibrous growth of a solid, and such forms were studied extensively in relation to the development of microscopic theories of crystal growth, particularly those involving screw dislocations. If crystal growth proceeds by the motion of a dislocation step and if conditions are such that the rate of step generation dominates over the motion of steps away from the dislocation source, then a whisker profile results. This is a simple and elegant explanation for the growth of single-crystal whiskers, but many modifications and alternative theories involving oxidation, solid-state diffusion, stress recrystallization, or some combination of these have been developed. Impurities may also play a significant role.

Whisker crystals can be produced by vapor deposition, chemical reaction, electrolytic deposition, or oxidation of surfaces, and at vapor–liquid–solid interfaces. A famous form of whisker growth is that of Sn whiskers growing from tin-plated metal where applied stress can enhance the growth rate, and this has led to the term *squeeze whisker*. Low-melting-point metals such as Zn, Cd, Mg, Hg, and K have been grown from vapor, while higher-melting-point metals and semiconductors such as Fe, Co, Ni, Cu, Au, Ag, Pt, Si, and Ge are often grown by the hydrogen reduction of metallic salts. Whiskers of oxides, carbides, nitrides, metallic salts, graphite, polymers, organic materials, and metallic alloys have been reported.

Whisker-like crystals also result from the growth of very anisotropic solids such as the quasi-one-dimensional solids $NbSe_3$ and TaS_3 which contain linear chains of metal atoms. This extreme structural anisotropy is also reflected in the electronic structure which undergoes a phase transition to a charge-density-wave state at lower temperature characterized by a superlattice of charge modulation. Quasi-one-dimensional organic compounds such as $(TMTSF)_2PF_6$ also grow as whiskers and exhibit a phase transition to a spin-density-wave state characterized by a superlattice modulation of spin. Quasi-two-dimensional compounds such as the high-T_c superconductor $Bi_2Sr_2Ca_1Cu_2O_x$ can also be grown in single-crystal whisker form.

The high surface and volume perfection of whiskers originally made them very useful for the study of mechanical properties. Stress–strain curves were obtained with elastic strain regions extending to 4 or 5%. The crystal perfection of whiskers has also made them excellent samples for the study of a wide range of electric and magnetic properties in solids. Many of these experiments are discussed in the review articles listed in the Bibliography, which also contain extensive reference to the original work.

See also Crystal Defects; Crystal Growth.

BIBLIOGRAPHY

R. H. Doremus, B. W. Roberts, and D. Turnbull (eds.), *Growth and Perfection of Crystals*. Wiley, New York, 1958, and Chapman and Hall, London, 1958. (A)

J. J. Gilman (ed.), *The Art and Science of Growing Crystals*. Wiley, New York and London, 1963. (A)

E. M. Nadgornyi, "The Properties of Whiskers," *Sov. Phys.–Usp.* **5,** 462 (1962). (I)

R. V. Coleman, "The Growth and Properties of Whiskers," *Metall. Rev.* **9,** 261 (1964). (I)

C. C. Evans, *Whiskers*. Mills and Boon, London, 1972. (I)

Work Function

T. M. Donovan and A. D. Baer

The work function W is the minimum energy required to remove an electron at the Fermi energy from a solid. Figure 1 shows a simplified model of the surface of metal, with electronic states in the metal occupied up to the Fermi level. The work function is

$$W = E_v - E_f,$$

where E_v is the energy of an electron at rest outside the solid, and E_f is the energy of an electron at the Fermi energy inside the solid.

The work function is determined experimentally by measuring the minimum amount of excitation energy required to emit an electron from the solid. There is a high density of filled electronic states near the Fermi level of a metal, and clear-cut theoretical relationships exist which relate photoemission and thermionic emission of electrons from these

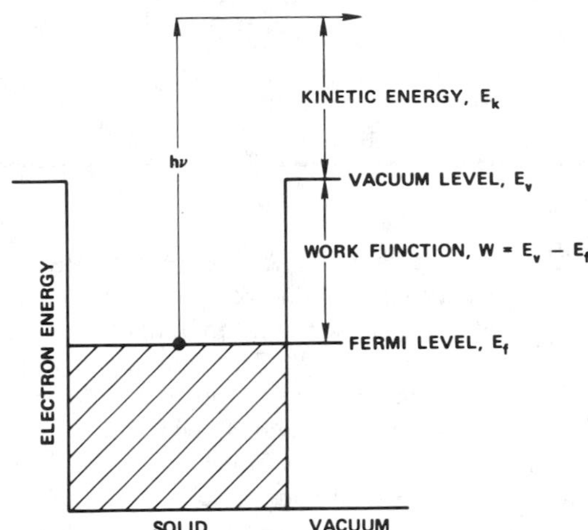

FIG. 1. Simplified model of electrons in a metal showing electron emission.

states to the work function [1–3]. For example, in Fig. 1, if an electron is excited from the Fermi level by a photon of energy $h\nu$, and the electron is emitted from the solid into vacuum without energy loss, then

$$W = h\nu - E_k,$$

where E_k is the kinetic energy of the electron outside the solid.

Since the Fermi level of a semiconductor or insulator generally lies in the band gap, the work function of a nonmetal sample is usually measured by first determining the work function of a metal surface and then measuring the difference between the work functions of the metal and the sample. The difference is equal to the contact potential difference which can be measured directly in a Kelvin probe [2,3] or retarding-potential experiment [4,5].

See also ELECTRON ENERGY STATES IN SOLIDS AND LIQUIDS; PHOTOELECTRON SPECTROSCOPY; THERMIONIC EMISSION.

REFERENCES

1. R. H. Fowler, *Phys. Rev.* **38,** 45 (1931). (I)
2. A. J. Decker, *Solid State Physics,* pp. 211, 230. Prentice-Hall, Englewood Cliffs, N. J., 1957. (E)
3. J. C. Rivière, *J. Solid State Surf. Sci.* **1,** 179 (1969). (A)
4. L. Apker, E. Taft, and J. Dickey, *Phys. Rev.* **74**, 1462–1474 (1948). (I)
5. C. N. Berglund and W. E. Spicer, *Phys. Rev.* **136,** A1030, (1964); **136,** A1044 (1964). (I)

X-Ray Astronomy *see* Astronomy, X-Ray

X-Ray Spectra and X-Ray Spectroscopy

Elisabeth Källne

Since Wilhelm Konrad Röntgen in 1895 found x rays with a discharge tube, this field of physics has been expanding with respect both to methods for generating x rays and to field of applications. This article will describe the physics behind the x-ray processes, the use of them in experimental investigations, and some of the fields of applications. The article is concluded with some useful references for further studies.

PRODUCTION OF X RAYS

The classical way to obtain x rays is to accelerate electrons from a hot filament in an x-ray tube onto a sample. In the collision with the sample atoms, the incident electrons experience continuous energy losses, which give rise to *bremsstrahlung*, and discrete energy losses caused by atomic excitations, which in turn give rise to *characteristic radiation*. The latter is peculiar for the sample used, since this x-ray radiation originates from filling of vacancies, usually in the innermost electron shells, where the initial vacancy is created by the electron bombardment. The simplest spectrum is the result of an atomic transition from an initial single vacancy in an inner shell to a final single vacancy in a less firmly bound shell and one speaks of transitions between single-hole states. An x-ray emission line spectrum originates from such simple transitions. The x-ray energy is the difference between the initial and final atomic states, $E_{h\nu} = E_{\text{initial}} - E_{\text{final}}$. Charged particles other than electrons may equally well be used for the excitation process. At the energies required, the bombarding electrons are always relativistic, while heavier particles can be of variable and lower energy, which, for instance, admits velocity reaction dependences to be studied. At low incoming velocities the projectile and the target atom may combine in various quasi-molecular systems, while at higher incoming velocities the process can be described as a fast knockout process (also see below). All charged particles perturb the atomic system in an undesirable way, which can be largely avoided if photons are used as projectiles. The photons can emanate from the emission of an electron-excited target, from relativistic electrons being accelerated in a magnetic field (synchrotron radiation), or, in the future perhaps, from an x-ray laser.

Besides the radiative x-ray emission process, the deexcitation of an excited atom can proceed through other channels like nonradiative Auger process, Coster–Kronig interband transition, or simultaneous emission of x ray and Auger electron, the so-called radiative Auger process. The relative probability for decay into the different channels depends on the atomic number (Z) of the atom; nonradiative processes, for instance, dominate for $Z<20$. The total rate at which the deexcitation occurs depends on the manifold of open exit channels and accessible final states, which in turn determines the lifetime of the atomic states involved. The transition from an initial hole state to a certain final hole state gives an x-ray line of shape given by a Lorentzian distribution and of a width determined by the lifetimes of the two involved states. Figure 1 shows an x-ray spectrum from copper, where the initial state has a vacancy in the deepest core shell, $1s$, and the final state has a vacancy in the $2p$ shell. The half-width of the line ($\Gamma_{K\alpha 1} = 2.7$ eV) gives an estimate of the lifetimes of the $1s$ and $2p$ hole states through Heisenberg's uncertainty principle, $\hbar < \Gamma t$, $t \sim 10^{-15}$ s. In the depicted spectrum there appears a certain background due to continuous bremsstrahlung and an asymmetry of the lines due to the fact that the initial state is not a pure single-hole state ($1s^{-1}$), but there is a small probability for multiple-vacancy admixture ($1s^{-1}2p^{-1}$).

The x-ray emission is governed by the transition operator, which is dominated by its dipole part. Given the initial state, the dipole operator will project out only certain of all available final states, i.e., x-ray emission is a selective process. Since x-ray spectra belonging to different initial hole states appear separate in energy regions, one gets the final state of different l symmetry nicely selected. Other probes of

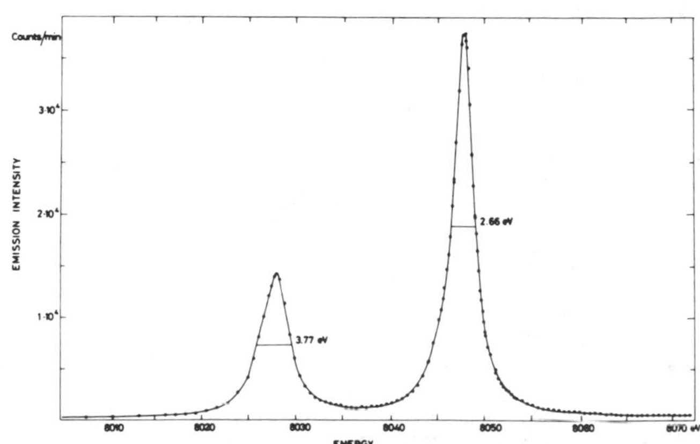

FIG. 1. High-resolution Cu $K\alpha_1\alpha_2$ ($1s^{-1} \rightarrow 2p_{3/2}{}^{-1}2p_{1/2}{}^{-1}$) x-ray emission spectrum from an electron-excited copper target. The spectrum was recorded with a 2-m curved quartz crystal spectrometer of Johann type. The x-ray tube was typically run at 20 kV and 40 mA. The vacuum in the spectrometer was approximately 10^{-5} Torr. The difference in half-width between the two lines is due to the possibility of filling the $2p_{1/2}$ hole by an interband transition (Coster–Kronig), thereby reducing the lifetime and increasing the half-width.

atomic structure are governed by other selection rules, so that other transitions stand out in those spectra. The combined use of spectra of complementary atomic probes is many times the best way to gain information on atomic structure.

The process with which the initial vacancy is created will necessarily cause a disturbance to the whole electronic system. On the one hand, the disturbance is rapid, and the process can be described as taking place in a frozen electronic system (the sudden limit). The orbits then do not have time to react on the removal of the inner-shell electron before the deexcitation takes place. However, some of the remaining $Z-1$ electrons may have changed states, i.e., been "shaken" to other states by the projectile or the fast outgoing electron (shake-up effect). This shows up as satellites in the x-ray spectrum. On the other hand, if the remaining $Z-1$ electrons have time to respond to the initial electron removal, the electrons will stay in the same orbits, although the orbits have adjusted themselves to the new potential (the adiabatic limit), and the spectrum will only show the main x-ray line without any satellites.

The classification of x-ray spectra was introduced by C. G. Barkla in 1908, viz., the $K, L, M, \ldots$, spectra, which is now in parallel use with the atomic orbit assignments, $1s$, $2s$, $2p$, etc. The x-ray spectrum belonging to an initial hole in the $1s$ orbit is called the K spectrum with the lines $K\alpha$: $1s^{-1} \rightarrow 2p^{-1}$, $K\beta$: $1s^{-1} \rightarrow 3p^{-1}$, etc., and similarly for the $n=2$ vacancy one obtaines the L spectrum. W. L. Bragg found these x-ray spectra to be groups of emission lines characteristic of the target element when studying the spectra with his crystal spectrometer. H.G.J. Moseley found regularities in the emission lines by plotting the square root of the frequency versus Z, the Moseley diagram. This provided the first definitive method for determining atomic numbers of elements. Conventionally, these x-ray lines are called diagram lines in contrast to nondiagram lines, i.e., satellites. Satellites supply a source of information on more subtle features of atomic structure. Besides the mentioned shake-up effects, they have a direct bearing upon effects such as (i) single or double electron transitions between multiply excited or ionized states, (ii) radiative Auger transitions, (iii) cross transitions between atoms in a molecule, or (iv) transitions due to the interaction between a hole in an inner shell and electrons in an outer partially filled shell. Some of these effects are present in the K spectrum of neon shown in Fig. 2.

RECORDING OF X-RAY SPECTRA

There are three essential parts in a setup for an x-ray experiment: a source, an analyzer, and a detector (see Table I).

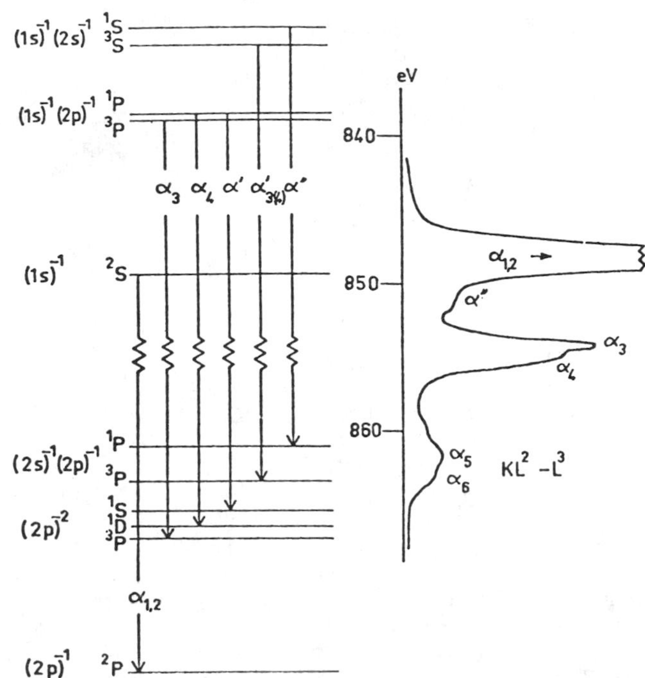

FIG. 2. The neon K x-ray emission spectrum recorded with a Bragg-crystal spectrometer. The level diagram to the left explains the initial and final hole states of the main line $K\alpha_{1,2}$ ($1s^{-1} \rightarrow 2p^{-1}$) and the satellites with initial multiple vacancies. (From T. Åberg, *X-Ray Spectra and Electronic Structure of Matter*, München, 1968.)

Figure 3 shows three types of common geometries for an x-ray spectrometer. The Cauchois and the Johann types of spectrometers use crystals as analyzers and operate for x rays in the wavelength region 1–100 Å. The effective wavelength regions for a certain crystal are determined by its lattice spacing d, as given through Bragg's diffraction law $n\lambda = 2d \sin\theta$. The extension to longer wavelengths, $\lambda > 20$ Å, has been possible by the advent of techniques for growing organic crystals with large lattice spacing. The grazing-incidence geometry uses a grating as analyzer and is effective in the 20–500-Å region. The development of blazed replica gratings, as well as of holographic transmission and reflection gratings, has provided for improved efficiency.

The classical photographic detection system is an effective time-integrated detector, although photoelectric methods have lately come into more frequent use. The latter methods have allowed for recording spectra during short time periods where either many separate detectors or a position-sensitive one is placed along the focal plane. Continuous progress in fields of applications and instrumental techniques have developed under mutual promotion. One recent example is in

Table I Essential parts for an x-ray experimental setup

Source	Analyzer	Detector
X-ray tube (electrons, photons)	Crystal (natural, organic)	Photographic plate
Ion beam	Grating (blazed replica, holographic)	Proportional counter
Synchrotron radiation		Geiger–Müller tube
Plasma		Scintillator, solid state

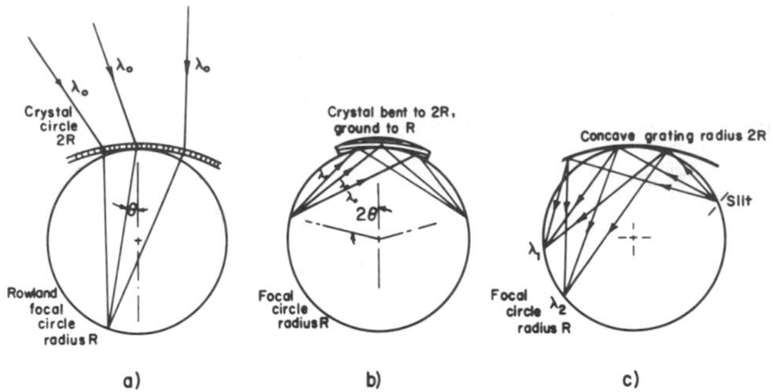

FIG. 3. Three types of focusing geometries for x-ray spectrometers. (a) Cauchois transmission type. A convergent beam is diffracted by the crystal and focused onto the focusing circle, the Rowland circle. This system is especially adapted for short wavelengths as angles of $\theta<45°$ should be chosen to minimize focusing defects. (b) Johann geometry utilizing a curved, ground crystal to give ideal focusing. The atom planes have been bent to a cylindrical surface of radius $2R$. A divergent beam from the point source on the Rowland circle is being dispersed and focused onto the Rowland circle. This system is effective for high resolution in the soft x-ray region (~1–20 Å). (c) Grazing-incidence geometry using a concave grating as dispersive element. The resolution of this instrument is determined by the radius R, as well as by the spacing of lines in the grating.

x-ray astronomy, where the limited instrumental space and the short recording times have resulted in the development of small, high-resolution spectrometers with rapid detection systems; these features are now being exploited in the growing field of x-ray spectroscopical studies of plasma physics.

X-RAY EMISSION

In order to understand the x-ray emission processes, it is suitable to study pure elements (in gaseous or solid state). This can be exemplified by some recent experiments using energetic ion beams of nickel or iron bombarding nickel or iron foils. In such experiments, the probability for initial multiple excitation is much higher than for electron or photon projectiles. Thus, there is a finite probability for exciting both of the two $1s$ electrons, which results in an x-ray emission spectrum with so-called hypersatellites. The deexcitation of the atom by a two-electron, one-photon emission gives an extra satellite which was observed for the first time in the mentioned experiment.

Assuming that the x-ray processes are known, the x-ray spectra give information on the electronic structure (core levels and the outer valence-electron distribution) of the element, compound, or alloy under study. Upon formation of a compound, there is a change in the valence-electron distribution, as well as in the core levels, as compared with the structure of the individual elements. In Fig. 4 the K emission spectra for aluminum and aluminum oxide are shown using electrons as exciting particles. The spectra are typical examples of the kind of information to be found in x-ray emission spectra. Some features should be pointed out starting from the Al $K\alpha_{1,2}$ emission lines representing the $1s^{-1}\rightarrow 2p^{-1}$ transitions. There is a change in energy of this structure upon formation of the oxide, which can be ascribed to a changed potential for the aluminum atom when the oxygen atoms are bound to the aluminum. The next structure, Al $K\alpha_3\alpha_4$, is produced by the $1s^{-1}\rightarrow 2p^{-1}$ transition with initial multiple vacancies ($1s^{-1}2p^{-1}$, $1s^{-1}2p^{-2}$). The relative intensity of this structure is significantly changed from the pure metal to the oxide. This is a phenomenon which can be utilized in quantitative analysis of samples. The last structure to be mentioned is the one representing the transition from the $1s^{-1}$ state to a final state with a vacancy in the valence band (band spectrum). This part of the spectrum is changed completely in shape as well as in energy position. The intensity distribution is related to the product of the transition probability and the density of states. From this spectrum it is possible to draw conclusions on the density of states in the pure metal and in the oxide. This part of the emission spectrum is the most useful one for studying gross features of the electronic structure of metals, compounds, and alloys.

The sample studied in x-ray spectroscopy can be in any of the four aggregation forms, i.e., solid, liquid, gas, or plasma. An x-ray emission spectrum from a tokamak plasma in the Fe $K\alpha$ region is shown in Fig. 5, where the single $1s^{-1}\rightarrow 2p^{-1}$ transition is followed by satellite structure on the high-energy side. This is due to initial multiple vacancies in the emitting atoms. With such a knowledge of the relation between the cause and appearance of an x-ray spectrum it is possible to use the x-ray spectrum as a diagnostic tool for determining the electron temperature in a plasma.

X-RAY ABSORPTION SPECTRA

Besides the x-ray emission spectra, there are methods pertaining to the absorption of x rays of which an example will be given here. Just as the valence-band emission spectra gives information on the filled part of the density of states, the unfilled part can be studied by exciting core electrons to previously unfilled states and studying the transmitted or emitted photon distribution. There are several similar methods, i.e., x-ray absorption, bremsstrahlung isochromat, characteristic isochromat, appearance potential spectra, which all give information on the unfilled part of the density of states. However, in the absorption process, there will also be effects due to the scattering of the outgoing electrons by neighboring atoms, and this can be studied in the extended x-ray-absorption fine structure (EXAFS). This has, during the last few years, proved to be a useful tool for studies of crystal structures. Particularly, very small samples as well as unstable and short-lived samples can be studied with

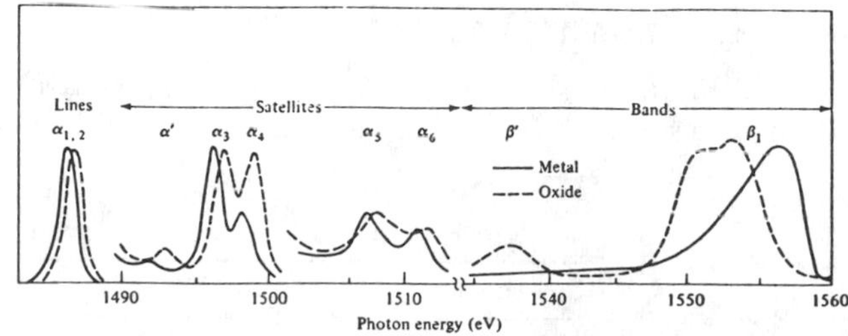

FIG. 4. Aluminum K emission spectrum emitted from a pure aluminum sample and from Al_2O_3. The spectra were measured with a Bragg-crystal spectrometer using an ammonium dihydrogen phosphate analyzing crystal. The atoms were excited by a 20-kV electron beam. [From G. L. Glen, *J. Appl. Phys.* **39**, 5377 (1968)].

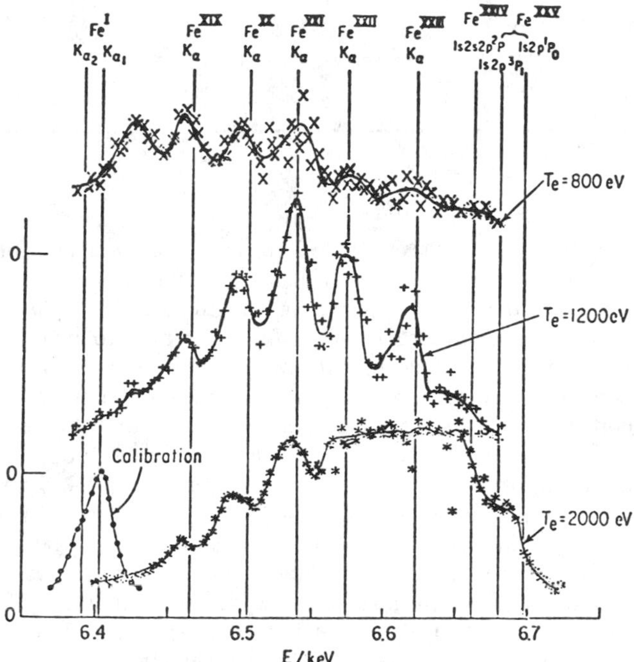

FIG. 5. X-ray emission spectrum from the Princeton tokamak plasma in the Fe $K\alpha$ region. The spectrum was recorded with a Bragg crystal spectrometer. The calibration curve shows the resolution of the spectrometer, as well as the position of the $1s^{-1} \rightarrow 2p^{-1}$ transition from the unstripped iron atom. Transitions from iron atoms with up to 24 electrons stripped have been assigned. (From F. Buchheit, courtesy of the Princeton University Plasma Physics Laboratory.)

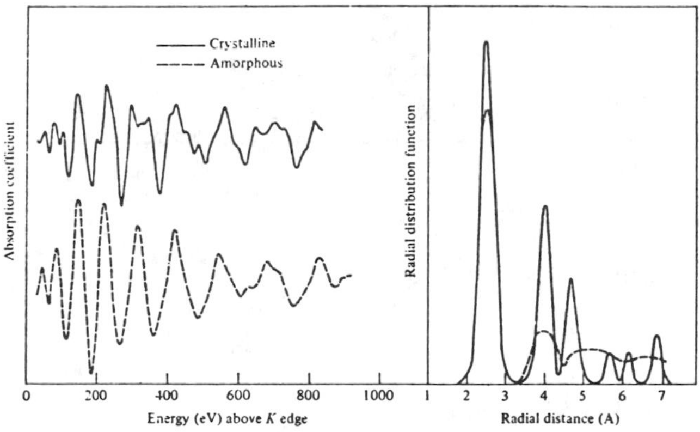

FIG. 6. Extended x-ray absorption fine structure for crystalline and amorphous germanium. The structure is measured by analyzing a photon beam transmitted through the sample with a Bragg-crystal spectrometer. The fine structure is analyzed with Fourier transformations to get the radial distribution functions shown in the figure. (From E. A. Stern, University of Washington, Seattle.)

EXAFS when the high-intensity photon beams available from electron storage rings are used. It has, for instance, been possible to study the structure of biological proteins as well as complicated crystal structures at high temperatures. In Fig. 6 is shown the x-ray-absorption fine structure for crystalline and amorphous germanium, together with the necessary Fourier analysis; this allows the determination of the nearest-neighbor distance and thereby the crystal structure. For studying the structures of regular and stable crystals, the conventional x-ray diffraction is still a versatile tool.

X-RAY SPECTRA FOR QUANTITATIVE ANALYSIS

The attenuation of x rays penetrating matter is used for several applications such as radiography, medical analysis, and material composition and homogeneity analysis. Electron microprobe instruments are now widely used for quantitative determination of sample compositions. The microprobe measures the intensity of the characteristic x-ray spectra emitted by each component in a sample, when bombarded by a well-focused electron beam. When protons are used instead of electrons, increased sensitivity is achieved in the quantitative analysis because of increased excitation cross section. This has made it possible to analyze very small samples (microns) and small concentrations of atoms in a sample (parts per million), i.e., trace analysis.

Finally, the use of intense photon beams from electron storage rings has opened up new possibilities for x-ray topographical and lithographical studies. Exposure times of a few seconds at the synchrotron produce comparable information to that acquired by conventional techniques in one day. This great increase in speed has been exploited to observe phase transitions in dielectric materials when thermally cycled.

See also ATOMIC SPECTROSCOPY; ATOMS; AUGER EFFECT; BREMSSTRAHLUNG; SYNCHROTRON RADIATION.

BIBLIOGRAPHY

L. V. Azaroff, *X-Ray Spectroscopy*. McGraw-Hill, New York, 1974.

C. Bonnelle and C. Mande, *Advances in X-Ray Spectroscopy*. Pergamon Press, Oxford, 1982.

B. Crasemann, *Atomic Inner-Shell Processes,* Vols. I, II. Academic Press, New York, 1975.

S. Flügge, "X-Rays," in *Encyclopedia of Physics,* Vol. XXX. Springer-Verlag, Berlin, 1957.

P. Lagarde, F.-J. Wuilleumier, and J.-P. Briand, "X-Ray and Inner-Shell Processes-X87," *J. Phys.* (Paris) Coll C9, **48** (1987).

R. Marrus, *Atomic Physics of Highly Ionised Atoms*. Plenum Press, New York, 1989.

G. Schmal and D. Rudolph, *X-Ray Microscopy*. Springer-Verlag, Berlin, 1984.

I. I. Sobelman, *X-Ray Plasma Spectroscopy and the Properties of Multiply-Charged Ions*. Nova Science Publishers, New York, 1988.

Z

Zeeman and Stark Effects

C. K. Jen

I. CHARACTERISTICS COMMON TO THE ZEEMAN AND STARK EFFECTS

There are many phenomenological features which are common to both the Zeeman and the Stark effects. We will first describe these common points and then deal with the individual characteristics later.

1. The two effects belong to the same general phenomenon of the "splitting" of a spectral line under the influence of an external field. The Zeeman effect represents the splitting of a spectral line by an external steady *magnetic* field. The Stark effect represents the splitting of a spectral line by an external steady *electric* field.

2. In each case, the effect originates from a perturbing potential energy resulting from the presence of an external magnetic (Zeeman) or an external electric (Stark) field. The perturbing potential energy in a moderately weak field is usually smaller than the *fine structure energy* in the unperturbed atomic or molecular system.

3. The eigenvalues of the split energy levels should logically be calculated by the exact solution of a secular equation in quantum mechanics, or, if that is not readily tractable, by the well-known methods of the perturbation theory directly for the nondegenerate case or by a generalized perturbation theory for the degenerate case. A radiative transition is possible (depending on the selection rules) between the split sublevels of the same originally degenerate energy level or between the split sublevels of different unperturbed energy levels. The transition frequency is equal to the difference in energy between the participating levels divided by Planck's constant (h). The intensity and polarization of the split component of a spectral line are calculable from the matrix elements of transition of the electric and/or magnetic moments.

4. The eigenvalue of a split energy level can be expressed as the sum of terms in the ascending order of the field strength. The coefficient of the field strength in the first-order term (if not zero) is the *permanent dipole moment* (magnetic for Zeeman and electric for Stark) of the atom or molecule. The second-order perturbing energy term is quadratic in field strength; hence, the coefficient of the linear field factor of the quadratic form is itself proportional to the field and is commonly known as the *field-induced* dipole moment. The higher-order terms beyond the second are usually of no practical significance.

5. In an atom or molecule, one can construct a vector diagram in which each vector may represent an electron orbital or spin angular momentum, molecular rotation or nuclear spin angular momentum, etc. Such vectors may be spatially added successively to form a *resultant* total angular momentum vector, which in quantum mechanics must be space quantized with respect to the external field axis. These *vector models* have been successfully employed to evaluate the *net* dipole moments associated with the resultant angular momentum vector and the space-quantized components along the field axis. In moderately weak fields, one can use the vector coupling schemes to calculate the contributions of the net dipole moment for either the Zeeman or the Stark effect. In sufficiently strong fields, however, the individual angular momentum vectors are no longer coupled together but are, instead, *separately* space quantized with respect to the external field. In the magnetic case, the phenomenon is called the Paschen–Back effect in contrast to the Zeeman effect. In the electric case, the physical phenomenon is analogous to the magnetic situation, but no specific name has been given to the strong electric field effect.

II. THE ZEEMAN EFFECT

P. Zeeman discovered in 1896 the phenomenon of the broadening of the yellow sodium D lines in a flame held between strong magnetic poles. The effect of spectral broadening in a magnetic field was later found to be well-resolved splitting of spectral lines, known ever since as the Zeeman effect.

The Zeeman effect was initially explained by H. A. Lorentz in his theory of electrons, which were believed to execute simple harmonic motions in a central field and which would emit radiation at the harmonic-motion frequency ν_0 for vibratory motions parallel with the magnetic field H_0 but would emit radiation at frequencies $\nu_0 \pm \nu_L$ for vibratory motions perpendicular to the field, where $\nu_L = eH_0/4\pi mc$ is known as the Larmor precession frequency. The spectral lines bearing these three frequencies were known as the Lorentz triplet. The Lorentz theory and the experimental results on the Zeeman effect were thought to be in general agreement at that time. (Zeeman actually had an axial hole through his magnet to observe the predicted sense of circular polarization, which was consistent with the notion of a negative charge for the electron according to the Lorentz theory.)

The earliest quantum-mechanical theory on the Zeeman effect gave essentially the same results as the classical Lorentz theory when only the *orbital* electronic moment was considered. But the theory of this *normal* Zeeman effect was woefully inadequate to explain various complicated Zeeman spectral patterns experimentally observed in many atoms. This great confusion began to fade away after the successful application of the theory of the *electron spin* by Uhlenbeck and Goudsmit in 1925.

The above new situation can be best explained by examining the magnetic moment of a spinning electron relative to that of an orbital electron. The magnetic moment of an orbital electron is $\boldsymbol{\mu}_L = -g_L(e/2mc)\mathbf{L}\hbar$, where g_L is the orbital "g factor" of an electron and is a dimensionless quantity having the value of unity, $e/2mc$ is the gyromagnetic ratio which is the quotient of the orbital magnetic moment to the angular momentum, and $\mathbf{L}$ is the orbital angular momentum with $\mathbf{L}$ representing an angular momentum vector in the units of $\hbar = h/2\pi$. One can simply write $\boldsymbol{\mu}_L = -\beta\mathbf{L}$, where β is the Bohr magneton. By a similar procedure, one can write the magnetic moment of a spinning electron as $\boldsymbol{\mu}_s = -g_s\beta\mathbf{S}$, where g_s is the g factor of the electron spin and is taken to be 2 for all usual computations but is close to 2.0023 for more refined calculations, and $\mathbf{S}$ is the electron spin angular momentum in the units of $\hbar$. The great significance of the electron spin theory (which is now universally accepted) lies in the fact that $g_s \simeq 2$ and $\mathbf{S}$ has components $M_s = \pm\frac{1}{2}$ along the magnetic field axis. It is this important concept and its corresponding consequences that led to the elucidation of the true interpretation of the Zeeman effect. However, it has unfortunately been labeled as the "anomalous" Zeeman effect.

1. The Zeeman Effect in Atoms

Take the simplest case of an atom with only one electron (i.e., the hydrogen atom). Let n be the principal quantum number, L be the orbital (angular momentum) quantum number, S be the electron spin (angular momentum) quantum number ($S=\frac{1}{2}$ for one electron), J be the quantum number representing the resultant angular momentum ($\mathbf{J}=\mathbf{L}+\mathbf{S}$), and M_J be the magnetic quantum number representing the projection of $\mathbf{J}$ along the field axis. A quantum state of the atom with a vector coupling is completely specified by the label of quantum numbers as $nLSJM_J$. The reader is referred to the usual vector model for the situation and the fact that, for a given n, L can take on all integral values from 0 up to $n-1$, J can take the values $L\pm\frac{1}{2}$ (for $S=\frac{1}{2}$), and M_J can take any of the values $-J, -J+1, \ldots, J-1, J$.

The perturbing Hamiltonian for the Zeeman effect is

$$\mathcal{H}_Z = \beta(\mathbf{L}+2\mathbf{S})\cdot\mathbf{H}_0 \tag{1}$$

where $\mathbf{H}_0$ is the external steady magnetic field. Either by finding the *average* value of the magnetic moment along the field with the aid of the vector model or by calculating quantum mechanically the expectation value of $\mathcal{H}_Z$ one can get the following:

$$W_Z = M_J g_J \beta H_0 \tag{2}$$

where

$$g_J = 1 + \frac{J(J+1)+S(S+1)-L(L+1)}{2J(J+1)}. \tag{3}$$

Equation (3) expresses the famous Landé g factor for the general case of the "LS" coupling. More specifically for the hydrogen or hydrogenic atom, where $S=\frac{1}{2}$, Eq. (3) is simplified to

$$g_J = 1 \pm \frac{1}{2L+1} \text{ for } J = L\pm\tfrac{1}{2}. \tag{4}$$

Figure 1 illustrates the three cases for the Zeeman effect on a spectral line. Case (a) shows the spectral line at frequency ν_0 in zero magnetic field. It is therefore the unshifted spectral line. Its intensity should be equal to the sum of the intensities of all the split components in a magnetic (or electric) field. Case (b) shows the normal Lorentz triplet in a magnetic field. The symbols π and σ denote polarizations of the radiation respectively parallel and perpendicular to the magnetic field. The π component corresponds to a quantum transition of $\Delta M_J = 0$, and the σ components correspond to transitions of $\Delta M_J = \pm 1$. In this case the π component is undisplaced from the original ν_0 position, and each σ component is displaced from ν_0 by the Larmor frequency ν_L, which is approximately 1.4 MHz per Oersted of magnetic field. When observed in the direction transverse to the field, the three components are all linearly polarized (the π polarization is naturally perpendicular to the σ polarization). But when observed in the direction longitudinal to the field, the π component is not visible while the σ_+ component ($\nu = \nu_0 + \nu_L$) corresponding to $\Delta M = -1$ for *emission* is right-handedly circularly polarized and the σ_- component ($\nu = \nu_0 - \nu_L$) corresponding to $\Delta M = +1$ for *emission* is left-handedly circularly polarized relative to the field. Historically, good examples for the normal Zeeman effect have been

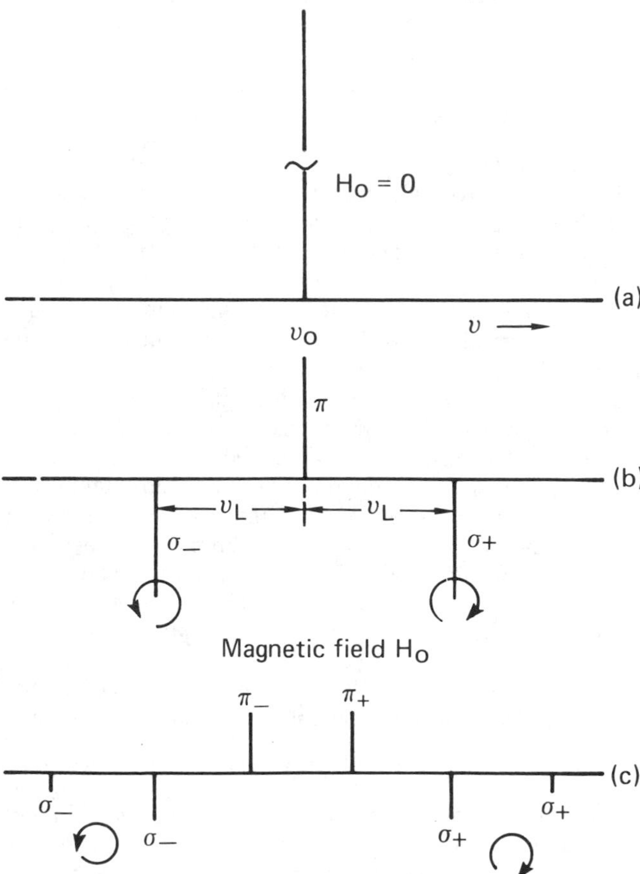

FIG. 1. Zeeman effect on a spectral line: (a) unshifted line at $H_0 = 0$ (drawn as a broken line to save space); (b) normal Zeeman effect H_0; (c) anomalous Zeeman effect at H_0 for a $^2P_{3/2} \rightarrow {}^2S_{1/2}$ transition. The magnetic field H_0 points perpendicularly into the paper.

found to be exceedingly rare; they are found only in cases like Cd for the $^1D \rightarrow {}^1P$ transition and similarly for Zn. Zeeman's experimental results on Na D lines should really belong to the anomalous case as we will soon see. Case (c) shows the anomalous Zeeman effect for the transition $^2P_{3/2} \rightarrow {}^2S_{1/2}$. Examples of this case are found in the D_2 line of sodium at 5890 Å or an equivalent type of transition in any of the atoms like hydrogen or the alkali atoms. In all these cases, the Landé g factor proves its validity in calculating the spectral splitting by the formula

$$\nu_{0z} = \nu_0 + (M' g'_{J'} - M g_J)\beta H_0/h \tag{5}$$

where ν_{0z} is the total frequency of one Zeeman component of a spectral line, $\nu_0 = (W' - W)/h$ is the transition frequency between the unperturbed levels of W' and W, and the last term is the difference between the Zeeman splitting frequencies at W' and W. The selection rule for the last (Zeeman) term is $\Delta J = J' - J = 0, \pm 1$. In spite of the more complicated spectral pattern for the anomalous Zeeman effect, the previous description on the polarization properties of the π and σ components for the normal Zeeman effect applies equally well to the anomalous effect.

Naturally, there are second order and even higher orders for the Zeeman effect, as can be seen from the perturbation theory. However, in most cases when the first-order (linear) Zeeman effect is present, the second order or higher is usually found to be not very significant.

2. *The Zeeman Effect on Molecules in Ground State*

Here, we have two major classes of molecules: those which are electronically paramagnetic and those which are not. In general, a molecule (or an atom) which has an odd number of electrons must necessarily be electronically paramagnetic (Kramer's rule). There is a small class of stable molecules like NO, NO_2, ClO_2, etc., in addition to a large class of generally unstable fragments of a molecule (commonly known as *free radicals* due to the existence in each case of an unpaired electron), which all belong to the odd-electron paramagnetic species. There are also exceedingly few cases of even-electron molecules that are found to be paramagnetic in their ground states, the most famous of which is O_2. In the other extreme, there is a very large class of molecules which do not have electronic paramagnetism in their ground state. We will later very briefly touch on the Zeeman effect of such molecules which are only weakly paramagnetic because of their molecular rotation and/or nuclear magnetism.

For the case of electronically paramagnetic molecules, one can generally write an admittedly over-simplified equation

$$W_Z = -M_J g_J \beta H_0, \tag{6}$$

where J represents the quantum number of $\mathbf{J}$, which is the resultant of several vectors consisting of the orbital angular momentum (mean value), the spin angular momentum of the unpaired electron(s), and the rotational angular momentum of the molecule. The factor g_J is the *net* g factor of the molecule along $\mathbf{J}$. The most often cited studies are those of diatomic or linear molecules (e.g., NO, OH, etc.) where there is an axial symmetry and various types of vector couplings (Hund's rule). The g_J factor may be much closer to 1 than 2, if the magnetic moment contribution of the orbital component weighs much more strongly than that of the spin component. For an asymmetric-top molecule having unpaired electron(s) but no particular symmetry, the g_J factor is principally contributed by the electron spin(s), since the mean value of the orbital angular momentum is most always zero.

Almost *all* molecules in their natural ground states are not electronically paramagnetic (they are known as the singlet Σ molecules). Such molecules have only the *rotational* magnetic moment due to the rotation of the positively charged nuclei and extranuclear electrons in the molecule. The rotational magnetic moment of a molecule can be either positive or negative and has a magnitude proportional to the rotational angular momentum of the molecule but its g factor is usually of the order of magnitude of a *nuclear magneton* (not Bohr) per unit angular momentum. (The smallness of the rotational magnetic moment is due to the near cancellation of the magnetic moments generated by the rotation of the positive and negative charges.)

It often happens that some nuclei in the molecule have magnetic moments of their own. Then there is a *net* magnetic moment for the whole molecule due to the rotational and nuclear magnetic contributions corresponding to the coupling of the molecular rotation and nuclear-spin angular momenta. The expression for the Zeeman energy is

$$\mathcal{H}_Z = -(g_J \mathbf{J} + g_I \mathbf{I}) \cdot \beta \mathbf{H}_0, \tag{7}$$

where $\mathbf{J}$ is the molecular *rotational* angular momentum, $\mathbf{I}$ is the nuclear-spin angular momentum, g_J is the *rotational* g factor, and g_I is the nuclear-spin g factor. Both g_J and g_I are approximately 2000 times lower than the g factor of an orbital electron. The type of the Zeeman effect as exemplified in Eq. (7) has been much studied by the molecular-beam magnetic resonance experiments and by microwave spectroscopy.

Finally, before we leave the scene of the Zeeman effect, it is well worth noting that the Zeeman splitting in paramagnetic atoms or molecules has been most actively studied in the *electron paramagnetic resonance* (EPR) research during the past three decades. This gives rise to a new branch of spectroscopy, which the pioneers of the Zeeman effect would hardly have anticipated.

III. THE STARK EFFECT

It seems somewhat surprising that the Stark effect was discovered as late as in 1913, almost 17 years later than the discovery of the Zeeman effect. This fact was principally due to the early experimental difficulty of observing the Stark effect while using a high electric field in the midst of flames. Fortunately, it happened that the Stark effect was discovered on the Balmer series of hydrogen in the same year that Bohr announced his theory of the hydrogen atom. This discovery gave an even greater boost to the Bohr theory than the Zeeman effect, because the Bohr theory was the only one that could satisfactorily explain the Stark effect, whereas the Lorentz theory could already largely (though crudely) explain the Zeeman effect at an earlier time.

1. The Stark Effect in Atoms

The perturbing Hamiltonian for the Stark splitting in a one-electron atom is

$$\mathcal{H}_S = e\mathbf{r}\cdot\mathbf{E}_0 = ezE_0, \tag{8}$$

where $-e\mathbf{r}$ is the electric dipole moment due to the electron and E_0 is the steady electric field along the z axis. We will treat the simplest case of a hydrogen atom with each of the eigenfunctions (ψ) in a nondegenerate or degenerate energy level designated by the symbols n,l,m which are, respectively, the principal, azimuthal, and magnetic ($m = m_l$) quantum numbers. It is understood that the electron-spin quantum number $S = \frac{1}{2}$ and its magnetic components $m_s = \pm\frac{1}{2}$ do not play an essential part in calculations on the Stark effect.

Let us treat the first-order Stark energy in the ground state, where $nlm = 100$. Then

$$W_S{}' = \int \psi^*_{100} ezE_0 \psi_{100} d\tau \tag{9}$$

Since $\psi^*_{100}\psi_{100} = \psi^2_{100}$ has even parity and ezE_0 has odd parity, the value for $W_S{}'$ in Eq. (9) is *zero*. Therefore, the ground state of the hydrogen atom has *no first-order* Stark effect.

One can obtain, however, a second-order Stark effect for the above situation by setting up a secular equation involving the matrix elements of the total energy of the hydrogen atom including the Stark perturbation with only three eigenfunctions ψ_{100}, ψ_{210}, and ψ_{310}. We can thereby obtain the solution for the sum of the unperturbed and perturbed (Stark) energies. The result for the second-order Stark energy is $W_S{}'' = -\frac{9}{4}a_0{}^3E_0{}^2$, where a_0 is the radius of the first Bohr orbit of hydrogen. If we equate $W_S{}'' = -\frac{1}{2}\alpha E_0{}^2$ as the energy of polarization by the field, then $\alpha = \frac{9}{2}a_0{}^3 = 0.677\times10^{-24}$ cm^3 is the well-known polarizability of the normal hydrogen atom.

So far, we have only dealt with the (orbital) nondegenerate case of the hydrogen atom. If, however, we should treat the case of $n=2$ of hydrogen, then there are four degenerate eigenfunctions $\psi(200)$, $\psi(210)$, $\psi(21\overset{+}{1})$, and $\psi(21\overset{-}{1})$. It so happens there is only one nonzero matrix element of ezE_0 between $\psi(200)$ and $\psi(210)$. As a consequence, the unperturbed $n=2$ of hydrogen is now split by the electric field into three sublevels, one of which remains in the unperturbed energy position with a twofold degeneracy [$\psi(21\overset{+}{1})$ and $\psi(21\overset{-}{1})$], which is a good example of the presence of some nonsplitting of the degenerate levels. The energy of another Stark sublevel is represented by a *linear* function of the field as $3ea_0E_0$. The energy of the third Stark sublevel follows another *linear* relation $-3ea_0E_0$ below the unperturbed position.

The second-order Stark effect of a nondegenerate or degenerate state is often handled by applying the usual or generalized perturbation theory. The second-order Stark effect on the helium atom has been very extensively studied by many people using various forms of the variational principle.

2. The Stark Effect on Molecules in Ground State

The Stark effect has been experimented in the optical range for molecules like H_2, HCl, etc., but the Stark splitting has generally been found to be small. On the other hand, the molecular-beam electric resonance method and microwave or submillimeter spectroscopy have successfully been used to study the Stark effect for a large variety of molecules like CsF, OCS, HCN, NH_3, H_2O, H_2O_2, CH_3F, etc.

The Stark effect in diatomic or linear molecules does not have a first-order or linear effect because the electric dipole moment $\boldsymbol{\mu}$ is perpendicular to the *rotational* angular momentum $\mathbf{J}$, hence the mean value of $\boldsymbol{\mu}$ along the field axis is zero. There is, however, a second-order or quadratic effect given by

$$W_S{}'' = \frac{[J(J+1) - 3M^2]}{2hB\,J(J+1)(2J-1)(2J+3)}\,\mu^2E_0{}^2, \tag{10}$$

where B is the rotational constant $h/8\pi^2I$ and I is the moment of inertia. The level is split into $J+1$ sublevels because of its dependence on M^2. The selection rules of transition are $\Delta J = -\pm1$, $\Delta M = 0, \pm1$.

The Stark effect in symmetric top molecules (e.g., NH_3) has a first-order, or *linear*, effect, because the molecule has enough symmetry to make the electric dipole moment have a constant component along the direction of the rotational angular momentum. One may thus have a linear Stark splitting energy of the form

$$W_S{}' = -\frac{MK\mu E_0}{J(J+1)}, \tag{11}$$

where K is the component of $\mathbf{J}$ along the symmetry axis and M would have $2J+1$ values from $-J$ to $+J$ along the field axis. The selection rules for transition are $\Delta J = 0$, ±1,

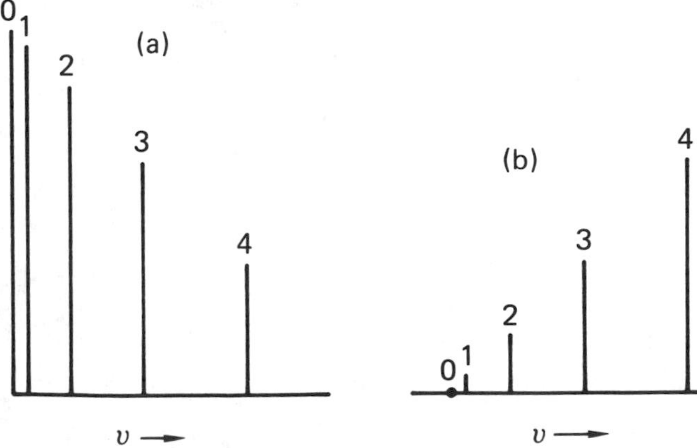

FIG. 2. Patterns of Stark components for $\Delta M = 0$ transitions in the rotation of an asymmetric top molecule: (a) $J = 5 \to 4$ transition, (b) $J = 4 \to 4$ transitions. Letters above the lines denote the values of M. The electric field E_0 is large enough for complete spectral resolution.

$\Delta K=0$, and $\Delta M=0, \pm 1$. The second-order Stark effect is usually much smaller than the first order and follows the same form as that of asymmetric top molecules.

The Stark effect of an asymmetric top molecule is always of the *second order* unless there is a near symmetry or accidental degeneracy. The general form of the Stark energy shift is

$$W_S{}'' = (A + BM^2)\mu^2 E_0{}^2, \tag{12}$$

where A and B are constants which may be complicated functions of the quantum labels characteristic of the particular level and the components of the electric dipole moments along certain molecular axes. The magnetic quantum number M is as always the component of the rotational angular momentum **J** along the field axis. There are thgerefore $J+1$ split levels since the Stark energy depends only upon $|M|$. The selection rules for transition are $\Delta J=0, \pm 1$ and $\Delta M=0, \pm 1$.

Figure 2 shows the Stark spectral patterns specifically for the special case of $\Delta M=0$, since that is the only case useful for a "Stark modulation spectrometer" (a spectrometer making use of Stark modulation field at a certain low frequency to increase the sensitivity of detection). It is noted that patterns of the Stark components are strikingly different for the $\Delta J = \pm 1$ transitions as compared with the $\Delta J=0$ transitions.

See also ATOMIC SPECTROSCOPY; MOLECULAR SPECTROSCOPY.

BIBLIOGRAPHY

H. A. Bethe and E. E. Saltpeter, *Quantum Mechanics of One- and Two-Electron Atoms*. Academic Press, New York, 1957.

G. W. Chantry, *Submillimetre Spectroscopy*. Academic Press, New York, 1971.

E. U. Condon and G. H. Shortley, *The Theory of Atomic Spectra*. Cambridge University Press, London, 1935.

R. M. Eisberg, *Fundamentals of Modern Physics*. Wiley, New York, 1964.

C. K. Jen, "The Zeeman Effect in Microwave Molecular Spectra," *Phys. Rev.* **74,** 1396–1406 (1948).

C. K. Jen, "Molecular and Nuclear Magnetic Moments in Microwave Zeeman Spectra," *Physica* **17,** 378–385 (1951).

G. V. Marr, *Plasma Spectroscopy*. Elsevier, New York, 1968.

E. Merzbacher, *Quantum Mechanics*. Wiley, New York, 1961.

L. I. Schiff, *Quantum Mechanics*. McGraw-Hill, New York, 1955.

T. M. Sugden and C. N. Kenney, *Microwave Spectroscopy of Gases*. Van Nostrand, New York, 1965.

CONTRIBUTORS

GEORGE O. ABELL
Late of Department of Astronomy, University of California, Los Angeles
Universe

MARK J. ABLOWITZ
School of Science, Clarkson University
Nonlinear Wave Propagation

S. C. ABRAHAMS
AT&T Bell Laboratories, Murray Hill, New Jersey
Ferroelasticity
Pyroelectricity

R. K. ADAIR
Physics Department, Yale University
Quarks

GAIL D. ADAMS
Pagosa Springs, Colorado
Radiological Physics

W. M. ADAMS
Department of Geology, Western Washington University
Seismology

IRSHAD AHMAD
Argonne National Laboratory, Argonne, Illinois
Alpha Decay

THOMAS J. AHRENS
Department of Geophysics, California Institute of Technology
Geophysics

D. E. ALBURGER
Brookhaven National Laboratory, Upton, New York
Beta Decay

B. ALDER
Lawrence Livermore National Laboratory, University of California, Livermore
Liquid Structure

GEORGE A. ALERS
Magnasonics, Inc., Albuquerque, New Mexico
Mechanical Properties of Matter

R. CASANOVA ALIG
David Sarnoff Research Center, Princeton, New Jersey
Electrostatics

PHILIP B. ALLEN
Department of Physics, State University of New York at Stony Brook
Superconductivity Theory

L. H. ALLER
Department of Astronomy, University of California, Los Angeles
Astrophysics

D. AMBROSE
Surbiton, Surrey, England
Vapor Pressure

BETSY ANCKER-JOHNSON
General Motors Technical Center, Warren, Michigan
Plasmons

A. C. ANDERSON
Department of Physics, University of Illinois
Cryogenics

JAMES L. ANDERSON
Department of Physics, Stevens Institute of Technology
Field Theory, Unified
Relativity, General

P. W. ANDERSON
Department of Physics, Princeton University
Quasiparticles

R. A. ANDERSON
Sandia Laboratories, Albuquerque, New Mexico
Insulators

J.-P. ANTOINE
Institut de Physique Théorique, Université Catholique de Louvain, Belgium
Group Theory in Physics

J. A. ARNAUD
Université des Sciences et Techniques du Languedoc, Montpellier, France
Klystrons and Traveling-Wave Tubes

A. S. ARROTT
Department of Physics, Simon Fraser University, Burnaby, British Columbia
Ferromagnetism

J. D. AXE
Brookhaven National Laboratory, Upton, New York
Ferroelectricity

A. D. BAER
U.S. Naval Weapons Center, China Lake, California
Work Function

JOHN N. BAHCALL
Department of Astrophysical Science, Princeton University
Astronomy, Neutrino

K. M. BAIRD
Applied Physics Division, National Research Laboratory, Ottawa, Canada
Interferometers and Interferometry

STANLEY S. BALLARD
Department of Physics, University of Florida
Visible and Ultraviolet Spectroscopy

C. J. BALLHAUSEN
Kemisk Laboratorium, University of Copenhagen, Denmark
Crystal and Ligand Fields

R. C. Barber
Department of Physics, University of Manitoba, Winnipeg
Isotopes

J. A. Barker
IBM Alamaden Research Center, San Jose, California
Interatomic and Intermolecular Forces

James A. Barnes
National Institute of Standards and Technology
Clocks, Atomic and Molecular

Richard G. Barnes
Department of Physics, Iowa State University
Nuclear Quadrupole Resonance

Stephen M. Barr
Bartol Research Institute, University of Delaware
Grand Unified Theories

Ernesto Barreto
Atmospheric Sciences Research Center, State University of New York at Albany
Corona Discharge

Neil Bartlett
Department of Chemistry, University of California, Berkeley
Rare Gases and Rare-Gas Compounds

Uwe H. Bauder
Lehrstühl für Technische Elektrophysik, Technische Universität München, Germany
Arcs and Sparks

James E. Bayfield
Department of Physics, University of Pittsburgh
High-Field Atomic States

Gordon Baym
Department of Physics, University of Illinois
Neutron Stars

Earl C. Beaty
Joint Institute for Laboratory Astrophysics, University of Colorado
Electron Bombardment of Atoms and Molecules

R. F. Bedford
National Research Council, Ottawa, Canada
Blackbody Radiation

A. H. Benade
Late of Department of Physics, Case Western Reserve University
Musical Instruments

James N. Benford
Beam and Plasma Research Group, Physics International Company, San Leandro, California
Electron and Ion Beams, Intense

Henry A. Bent
Chemistry Department, University of Pittsburgh
Chemical Bonding

F. F. Bentley
Formerly of Department of Chemistry, U.S. Wright–Patterson Air Force Base, Ohio
Raman Spectroscopy

Robert F. Berg
National Institute of Standards and Technology
Viscosity

Karl Berkelman
Newman Laboratory, Cornell University
Compton Effect

Stephan Berko
Department of Physics, Brandeis University
Positron Annihilation in Condensed Matter
Positronium

A. E. Berkowitz
Center for Magnetic Recording Research, University of California, San Diego
Magnets (Permanent) and Magnetostatics

Joan B. Berkowitz
Risk Science International, Washington, D.C.
High Temperature

Robert A. Bernheim
Department of Chemistry, Pennsylvania State University
Optical Pumping

D. W. Berreman
AT&T Bell Laboratories, Murray Hill, New Jersey
Rayleigh Scattering

Daniel Bershader
Department of Aeronautics and Astronautics, Stanford University
Fluid Physics

P. E. Best
Institute of Materials Science, University of Connecticut
Secondary Electron Emission

Robert T. Beyer
Department of Physics, Brown University
Acoustics

L. C. Biedenharn
Department of Physics, Duke University
Rotation and Angular Momentum

Yvon G. Biraud
Astronomie Infrarouge LAM, Observatoire de Meudon, France
Fourier Transforms

James D. Bjorken
Fermi National Accelerator Laboratory, Batavia, Illinois
Patrons

R. Blankenbecler
Stanford Linear Accelerator Center, Stanford University
Regge Poles

George E. Blomgren
Technology Laboratory, Eveready Battery Company Inc.,Westlake, Ohio
Electrochemical Conversion and Storage

Frank A. Blood, Jr.
Department of Physics, Rutgers University
Brownian Motion

Elliott D. Bloom
Stanford Linear Accelerator Center, Stanford University
Quarkonium

A. R. Bodner
Argonne National Laboratory, Argonne, Illinois
Hypernuclear Physics and Hypernuclear Interactions

Arno Bohm
Department of Physics, University of Texas, Austin
Momentum

Bart J. Bok
Late of Seward Observatory, University of Arizona
Milky Way

John J. Bollinger
National Institute of Standards and Technology
Clocks, Atomic and Molecular

David W. Bonnell
National Institute of Standards and Technology
High Temperature

Nicholas Bottka
U.S. Naval Research Laboratory
Dielectric Properties

Edward A. Boudreaux
Chemistry Department, University of New Orleans
Diamagnetism and Superconductivity

B. N. Brockhouse
Department of Physics, McMaster University, Hamilton, Ontario
Neutron Spectroscopy

W. E. Bron
Department of Physics, University of California, Irvine
Elasticity

Robert W. Brown
Department of Physics, Case Western Reserve University
Gauss's Law

Ronald Bryan
Department of Physics, Texas A&M University
Nuclear Forces

Olof Bryngdahl
University of Essen, Germany
Optics, Physical

Jeffrey Bub
Department of Philosophy, Princeton University
Complementarity

E. Bucher
Department of Physics, University of Constance, Germany
Transition Elements

R. C. Budhani
Brookhaven National Laboratory, Upton, New York
Thin Films

MARTIN J. BUERGER
Late of Department of Earth and Planetary Sciences, Massachusetts Institute of Technology
Crystal Symmetry
Crystallography, X-ray

R. BULLOUGH
United Kingdom Atomic Energy Authority, Oxfordshire, England
Deformation of Crystalline Materials

E. MARGARET BURBIDGE
Center for Astrophysics and Space Science, University of California, San Diego
Quasars

C. P. BURGESS
Department of Physics, McGill University, Montreal, Quebec
Supersymmetry and Supergravity

EDWARD A. BURKE
Late of Department of Physics and Astronomy, Adelphia University
Bohr Theory of Atomic Structure

W. H. BUTLER
Martin Marietta Systems, Inc., Oak Ridge, Tennessee
Transport Properties

DAVID O. CALDWELL
Department of Physics, University of California, Santa Barbara
Hyperons

JOSEPH CALLAWAY
Department of Physics and Astronomy, Louisiana State University
Electron Energy States in Solids and Liquids

KARL F. CANTER
Department of Physics, Brandeis University
Positron Annihilation in Condensed Matter
Positronium

MANUEL CARDONA
Max-Planck-Institut für Festkörperforschung, Germany
Synchrotron Radiation

PETER A. CARRUTHERS
Department of Physics, University of Arizona
Disperson Theory

THOMAS P. CAUDELL
Boeing Aircraft Corporation, Seattle, Washington
Sun

PETER CAWS
Department of Philosophy, George Washington University
Philosophy of Physics

P. M. CHAIKIN
Department of Physics, Princeton University
Organic Conductors and Superconductors

BRUCE CHALMERS
Division of Engineering and Applied Physics, Harvard University
Metallurgy

OWEN CHAMBERLAIN
Department of Physics, University of California, Berkeley
Nucleon

B. S. CHANDRASEKHAR
Walther-Meissner Institut für Tieftemperaturforschung der Bayerischen Akademie der Wissenschaften, Garching, Germany
Magnetostriction

DAVID B. CHANG
Hughes Aircraft Corporation, Long Beach, California
Plasmons

ALAN J. CHAPMAN
Department of Mechanical Engineering, Rice University
Heat Transfer

GEORGES CHARPAK
European Organization for Nuclear Research, Geneva, Switzerland
Counting Tubes

GEOFFREY F. CHEW
Lawrence Berkeley Laboratory, University of California, Berkeley
S-Matrix Theory

HONG YEE CHIU
Goddard Space Flight Center, Greenbelt, Maryland
Pulsars

ANDREAS CHRAMBACH
National Institute of Child Health and Human Development, Bethesda, Maryland
Electrophoresis

STEVEN CHU
Department of Physics, Stanford University
Laser Spectroscopy

JOHN CLARKE
Department of Physics, University of California, Berkeley
Superconducting Devices

E. RICHARD COHEN
Science Center, Rockwell International, Thousand Oaks, California
Symbols, Units, and Nomenclature

R. V. COLEMAN
Department of Physics, University of Virginia
Whiskers

STIRLING A. COLGATE
Los Alamos National Laboratory, Los Alamos, New Mexico
Cosmic Rays—Astrophysical Effects

ESTHER M. CONWELL
Xerox Webster Research Center, Webster, New York
Acoustoelectric Effect

HOMER E. CONZETT
Lawrence Berkeley Laboratory, University of California, Berkeley
Polarization

BERNARD R. COOPER
Department of Physics, West Virginia University
Fermi Surface

F. V. CORONITI
Department of Physics, University of California, Los Angeles
Magnetosphere

BERND CRASEMANN
Department of Physics, University of Oregon
Auger Effect

J. H. CRAWFORD, JR.
Oak Ridge National Laboratory, Oak Ridge, Tennessee
Radiation Interaction with Matter

GERARD M. CRAWLEY
Department of Physics and Astronomy, Michigan State University
Charged-Particle Spectroscopy

MICHAEL CREUTZ
Brookhaven National Laboratory, Upton, New York
Lattice Gauge Theory

CLARENCE R. CROWELL
Department of Materials Science, University of Southern California
Photoconductivity

PAUL L. CSONKA
Institute of Theoretical Science, University of Oregon
Partial Waves

N. E. CUSACK
School of Mathematics and Physics, University of East Anglia, Norwich, England
Liquid Metals

F. CYROT-LACKMANN
Laboratoire des Propriétés Electronique des Solides, Centre National de la Recherche Scientifique, Grenoble, France
Surfaces and Interfaces

OLIVER DALTON
Formerly of Tektronix, Inc., Beaverton, Oregon
Oscilloscopes

MICHAEL DANOS
National Institute of Standards and Technology
Čerenkov Radiation

J. G. DASH
Department of Physics, University of Washington
Adsorption

J. P. DAVIDSON
Department of Physics and Astronomy, University of Kansas
Nuclear Structure

P. C. W. DAVIES
Department of Theoretical Physics, The University, Newcastle-upon-Tyne, England
Time

J. M. DAVIS
Rockwell Hanford Operations, Richland, Washington
Hot Cells and Remote Handling Equipment

P. G. DE GENNES
Collége de France, Paris, France
Liquid Crystals

Morris H. DeGroot
Department of Physics, University of Nijmegen, The Netherlands
Statistics

E. der Mateosian
Brookhaven National Laboratory, Upton, New York
Gamma Decay

Malcolm Derrick
Argonne National Laboratory, Argonne, Illinois
Mesons

M. C. Desjonquées
Commissariat à l'Energie Atomique, Paris, France
Surfaces and Interfaces

R. W. Detenbeck
Department of Physics, University of Vermont
Light Scattering

Jozef T. Devreese
Universitaire Instelling Antwerpen, Belgium
Polaron

I. J. D'Haenens
Hughes Research Laboratories, Malibu, California
Thermionic Emission

J. F. Dillon, Jr.
AT&T Bell Laboratories, Murray Hill, New Jersey
Kerr Effect, Magneto-optical

F. J. DiSalvo
Department of Chemistry, Cornell University
Allotropy and Polymorphism

J. F. Donoghue
Department of Physics, University of Massachusetts
SU(3) and Symmetry Groups

T. M. Donovan
U.S. Naval Weapons Center, China Lake, California
Work Function

J. R. Dorfman
University of Maryland
Statistical Mechanics

David A. Dows
Department of Chemistry, University of Southern California
Infrared Spectroscopy

Charles W. Drake
Department of Physics, Oregon State University
Polarizability

G. W. F. Drake
Department of Physics, University of Windsor, Ontario
Fine and Hyperfine Spectra and Interactions

Joseph Dreitlein
Department of Physics, University of Colorado
Twin Paradox

G. Dresselhaus
Francis Bitter National Magnet Laboratory, Massachusetts Institute of Technology
de Haas-van Alphen Effect

M. S. Dresselhaus
Departments of Electrical Engineering and Physics, Massachusetts Institute of Technology
de Haas-van Alphen Effect

Alan Dressler
The Observatories, Carnegie Institution of Washington, Pasadena, California
Galaxies

William D. Drotning
Sandia National Laboratories, Albuquerque, New Mexico
Thermal Expansion

C. B. Duke
Pacific Northwest Laboratories, Richland, Washington
Low-Energy Electron Diffraction (LEED).

R. A. Dunlap
Department of Physics, Dalhousie University, Halifax, Nova Scotia
Hysteresis

Floyd Dunn
Department of Electrical and Computer Engineering, University of Illinois
Ultrasonic Biophysics

James R. Durig
Department of Chemistry, University of South Carolina
Microwave Spectroscopy

Dean E. Eastman
IBM Research Center, Yorktown Heights, New York
Photoelectron Spectroscopy

Philippe Eberhard
Lawrence Berkeley Laboratory, University of California, Berkeley
Center-of-Mass System

Alan S. Edelstein
U.S. Naval Research Laboratory
Kondo Effect

David L. Ederer
National Institute of Standards and Technology
Photoionization

T. Egami
Department of Materials Science and Engineering, University of Pennsylvania
Glassy Metals

F. R. Eirich
Department of Chemistry, Polytechnic Institute of New York
Rheology

Leonard Eisenbud
Department of Physics, State University of New York at Stony Brook
Eigenfunctions

T. Embleton
Division of Physics, National Research Council, Ottawa, Ontario
Noise, Acoustical

Guy T. Emery
Department of Physics, Bowdoin College
Gamma-Ray Spectrometers

V. J. Emery
Brookhaven National Laboratory, Upton, New York
Quantum Statistical Mechanics

P. R. Emtage
Westinghouse Electric Corporation, Research and Development Center, Pittsburgh, Pennsylvania
Hall Effect

J. E. Enderby
H. H. Wills Physics Laboratory, University of Bristol, England
Paramagnetism

Hermann Engelhardt
Department of Geology, California Institute of Technology
Ice

Stanley Engelsberg
Department of Physics and Astronomy, University of Massachusetts, Amherst
Metals

Gerhard Ertl
Fritz-Haber Institut der Max-Planck-Gesellschaft, Berlin, Germany
Catalysis

Allen E. Everett
Department of Physics and Astronomy, Tufts University
Lorentz Transformations

Henry A. Fairbank
Department of Physics, Duke University
Second Sound

W. G. Fateley
Department of Chemistry, Kansas State University
Raman Spectroscopy

Norman Feather
Late of University of Edinburgh, Scotland
Statics

G. Feinberg
Department of Physics, Columbia University
Tachyons

A. J. Fennelly
Teledyne Brown Engineering, Huntsville, Alabama
Photon

T. Ferbel
Universities Research Association, Berkeley, California
Inclusive Reactions

Fereydoon Family
Department of Physics, Emory University
Fractals

Herman Feshbach
Department of Physics, Massachusetts Institute of Technology
Nuclear Scattering

Alexander L. Fetter
Department of Physics, Stanford University
Quantum Fluids

D. K. Finnemore
Department of Physics, Iowa State University
Superconducting Materials

Val L. Fitch
Department of Physics, Princeton University
Weak Interactions

C. L. Foiles
Department of Physics, Michigan State University
Thermoelectric Effects

Leslie L. Foldy
Department of Physics, Case Western Reserve University
Faraday's Law of Electromagnetic Induction

Kenneth J. Foley
Brookhaven National Laboratory, Upton, New York
Scintillation and Čerenkov Counters

Simon Foner
National Magnet Laboratory, Massachusetts Institute of Technology
Magnetic Fields, High

G. W. Ford
Department of Physics, University of Michigan
H *Theorem*

Joseph Ford
School of Physics, Georgia Institute of Technology
Ergodic Theory

F. M. Fowkes
Department of Chemistry, Lehigh University
Surface Tension

Susan E. Fox
Formerly of Yale University
Quantum Mechanics

Donald R. Franceschetti
Department of Physics, Memphis State University
Electrochemistry

Jack H. Freed
Department of Chemistry, Cornell University
Gyromagnetic Ratio

Robert D. Freeman
Department of Chemistry, Oklahoma State University
Thermodynamic Data

A. P. French
Department of Physics, Massachusetts Institute of Technology
Atoms

J. B. French
Department of Physics and Astronomy, University of Rochester
Sum Rules

F. J. Friedlaender
School of Electrical Engineering, Purdue University
Electromagnets

J. P. Friedberg
Plasma Fusion Center, Massachusetts Institute of Technology
Nuclear Fusion

John Friedman
Department of Physics, University of Wisconsin
Gravitation

E. G. Fuller
National Institute of Standards and Technology
Resonances, Giant

Thomas Fulton
Department of Physics, The Johns Hopkins University
Feynman Diagrams

H. P. Furth
Plasma Physics Laboratory, Princeton University
Plasma Confinement Devices

F. L. Galeener
Department of Physics, Colorado State University
Absorption Coefficient

M. Garbuny
Westinghouse Research and Development Center, Pittsburgh, Pennsylvania
Heat Engines

Elsa Garmire
Center for Laser Studies, University of Southern California
Reflection

Paul D. Garn
Department of Chemistry, University of Akron
Thermal Analysis

S. Gasiorowicz
School of Physics and Astronomy, University of Minnesota
Elementary Particles in Physics

Peter P. Gaspar
Department of Chemistry, Washington University
Hot Atom Chemistry

B. D. Gaulin
Department of Physics, McMaster University, Hamilton, Ontario
Neutron Spectroscopy

J. D. Gavenda
Department of Physics, University of Texas, Austin
Magnetoacoustic Effect

Ronald J. Gibbs
College of Marine Studies, University of Delaware
Sedimentation and Centrifugation

B. F. Gibson
Los Alamos National Laboratory, Los Alamos, New Mexico
Three-Body Problem: Quantum Mechanical

C. H. Gibson
Scripps Institution of Oceanography, University of California, San Diego
Turbulence

J. A. Giordmaine
NEC Research Institute, Princeton, New Jersey
Optics, Nonlinear

P. Glansdorff
University of Brussels, Belgium
Thermodynamics, Nonequilibrium

S. L. Glashow
Lyman Laboratory, Harvard University
SU(3) and Symmetry Groups

Henry R. Glyde
Department of Physics, University of Alberta, Edmonton
Helium, Solid

Hubert F. Goenner
Institut für Theoretische Physik de Universität, Göttingen, Germany
Mach's Principle

Joshua N. Goldberg
Department of Physics, Syracuse University
Space-Time

Walter I. Goldburg
Department of Physics and Astronomy, University of Pittsburgh
Dynamic Critical Phenomena

Alfred S. Goldhaber
Institute of Theoretical Physics, State University of New York at Stony Brook
Magnetic Monopoles

B. Golding
AT&T Bell Laboratories, Murray Hill, New Jersey
Glass

Herbert Goldstein
Department of Mechanical and Nuclear Engineering, Columbia University
Dynamics, Analytical

J. A. Golovchenko
Department of Physics, Harvard University
Channeling

Myron L. Good
Department of Physics, State University of New York at Stony Brook
Diffraction

R. H. Good, Jr.
Department of Physics, Pennsylvania State University
Hamiltonian Function

James P. Gordon
AT&T Bell Laboratories, Holmdel, New Jersey
Masers

Paul Gorenstein
Smithsonian Astrophysical Observatory, Cambridge, Massachusetts
Astronomy, X-ray

N. B. Gove
Oak Ridge National Laboratory, Oak Ridge, Tennessee
Binding Energy

C. D. Graham, Jr.
Department of Materials, University of Pennsylvania
Magnetic Materials

Andrew V. Granato
Department of Physics, University of Illinois
Internal Friction in Crystals

Victor L. Granatstein
Department of Electrical Engineering, University of Maryland
Light

Melville S. Green
Late of Department of Physics, Temple University
Phase Transitions

O. W. Greenberg
Department of Physics, University of Maryland
Bose-Einstein Statistics

Joshua E. Greenspan
J. G. Engineering Research Associates, Baltimore, Maryland
Acoustics, Linear and Nonlinear

Sandra C. Greer
Department of Chemistry and Biochemistry, University of Maryland
Order-Disorder Phenomena

Walter Greiner
Institut für Theoretische Physik, University of Frankfurt, Germany
Superheavy Elements

Thomas J. Greytak
Department of Physics, Massachusetts Institute of Technology
Fluctuation Phenomena

Robert B. Griffiths
Department of Physics, Carnegie-Mellon University
Thermodynamics, Equilibrium

C. C. Grimes
AT&T Bell Laboratories, Murray Hill, New Jersey
Cyclotron Resonance

Franz Gross
On leave from Department of Physics, College of William and Mary
Bethe-Salpeter Equation

K. A. Gschneidner, Jr.
Ames Laboratory, Iowa State University
Rare Earths

Feza Gürsey
Physics Department, Yale University
Invariance Principles

R. A. Guyer
Department of Physics and Astronomy, University of Massachusetts, Amherst
Helium, Liquid

Steven W. Haan
Lawrence Livermore National Laboratory, University of California, Livermore
Inertial Fusion

M. Hablanian
Vacuum Products Division, Varian Associates, Inc., Lexington, Massachusetts
Vacuums and Vacuum Technology

W. Haeberli
Department of Physics, University of Wisconsin
Nuclear Polarization

Richard L. Hahn
Brookhaven National Laboratory, Upton, New York
Transuranium Elements

Yukap Hahn
Department of Physics, University of Connecticut
Multipole Fields

John J. Hall
Shearwater Company, Brooklyn, New York
Photovoltaic Effect

Arthur M. Halpern
Department of Chemistry, Northeastern University
Luminescence (Fluorescence and Phosphorescence)

M. Hamermesh
School of Physics and Astronomy, University of Minnesota
Lie Groups

Robert J. Hamers
IBM Research Center, Yorktown Heights, New York
Scanning Tunneling Microscopy

J. M. Hammersley
Mathematical Institute, University of Oxford, England
Probability

Richard S. Handley
Lumigen, Inc., Detroit, Michigan
Chemiluminescence

J. H. Hannay
H. H. Wills Physical Laboratory, University of Bristol, England
Geometric Quantum Phase

Theo W. Hänsch
Department of Physics, University of Munich, Germany
Doppler Effect

W. Happer
Department of Physics, Princeton University
Resonance Phenomena

P. Hariharan
CSIRO Division of Applied Physics, National Measurement Laboratory, Lindfield, Australia
Holography

Hanns L. Harney
Max-Planck-Institut für Kernphysik, Heidelberg, Germany
Isobaric Analog States

Frank E. Harris
Department of Physics, University of Utah
Molecules

Franklin S. Harris, Jr.
Rockville, Utah
Aerosols

J. C. Harrison
Geodynamics Corporations, Santa Barbara, California
Gravity, Earth's

Bernard G. Harvey
Lawrence Berkeley Laboratory, University of California, Berkeley
Radioactivity

John W. Hastie
National Institute of Standards and Technology
High Temperature

O. Häusser
TRIUMF, Vancouver, Canada
Nuclear Moments

Stephen W. Hawking
Department of Applied Mathematics and Theoretical Physics, University of Cambridge, England
Black Holes

S. I. Hayek
Pennsylvania State University
Vibrations, Mechanical

Evans Hayward
National Institute of Standards and Technology
Photonuclear Reactions

Raymond W. Hayward
National Institute of Standards and Technology
CPT *Theorem*

Dennis J. Hegyi
Department of Physics, University of Michigan
Interstellar Medium

Ernest M. Henley
Department of Physics, University of Washington
Parity

R. G. Herb
Department of Physics, University of Wisconsin
Accelerators, Potential-Drop Linear

Frank Herman
IBM Almaden Research Center, San Jose, California
Atomic Structure Calculations—One-Electron Models

J. C. Herrera
Brookhaven National Laboratory
Electromagnetic Radiation

Conyers Herring
Department of Applied Physics, Stanford University
Solid-State Physics

Lloyd O. Herwig
U.S. Department of Energy, Washington, D.C.
Solar Energy

Jacqueline N. Hewitt
Department of Physics, Massachusetts Institute of Technology
Gravitational Lenses

HENRY A. HILL
Department of Physics, University of Arizona
Sun

FRANZ J. HIMPSEL
IBM Research Center, Yorktown Heights, New York
Photoelectron Spectroscopy

DAVID G. HITLIN
California Institute of Technology
Baryons

JOSEPH V. HOLLWEG
Department of Physics, University of New Hampshire
Cosmic Rays: Solar System Effects

MICHAEL HORNE
Department of Physics, Stonehill College
Hidden Variables

JOHN N. HOWARD
Newton, Massachusetts
Atmospheric Physics

R. P. HUDSON
Bureau International de Poids et Mesures, Sèvres, France
Thermometry

VERNON W. HUGHES
Physics Department, Yale University
Muonium

ANDREW P. HULL
Brookhaven National Laboratory, Upton, New York
Radiation Detection

JAMES L. HUNT
Department of Physics, University of Guelph, Ontario
Brillouin Scattering

H. B. HUNTINGTON
Department of Physics and Astronomy, Rensselaer Polytechnic Institute
Diffusion

C. M. HURD
National Research Council, Ottawa, Ontario
Galvanomagnetic and Related Effects

TAKANOBU ISHIDA
Department of Chemistry, State University of New York at Stony Brook
Isotope Separation

WERNER ISRAEL
Department of Physics, University of Alberta, Edmonton
Black Holes

J. D. JACKSON
Lawrence Berkeley Laboratory, University of California, Berkeley
Electrodynamics, Classical

J. A. JACOBS
Department of Geology, University of Wales, Aberystwyth
Geomagnetism

GEORGE J. JANZ
Molten Salts Data Center, Rensselaer Polytechnic Institute
Molten Salts

CARSON JEFFRIES
Department of Physics, University of California, Berkeley
Excitons

C. K. JEN
Silver Spring, Maryland
Zeeman and Stark Effects

ROBERT L. JOHNSON
Deutsches Elektronen-Synchrotron, Hamburg, Germany
Synchrotron Radiation

J. R. JOKIPII
Committee on Theoretical Astrophysics, University of Arizona
Solar Wind

B. R. JUDD
Department of Physics and Astronomy, The Johns Hopkins University
Franck-Condon Principle
Jahn-Teller Effect

DAVID L. JUDD
Lawrence Berkeley Laboratory, University of California, Berkeley
Cyclotron

M. N. KABLER
U.S. Naval Research Laboratory, Washington, D.C.
Electroluminescence

MARK KAC
Late of Rockefeller University
H *Theorem*

CALUDE KACSER
Department of Physics and Astronomy, University of Maryland
Relativity, Special Theory

RAFAEL KALISH
Department of Physics, Technion University, Haifa, Israel
Nuclear Properties

ELISABETH KÄLLNE
Physics Department I, The Royal Institute of Technology, Stockholm, Sweden
X-ray Spectra and X-ray Spectroscopy

M. H. KALOS
Theory Center, Cornell University
Monte Carlo Techniques

ISABELLA KARLE
Materials Science and Component Technology Directorate, U.S. Naval Research Laboratory
Crystallography, X-ray

JEROME KARLE
Materials Science and Component Technology Directorate, U.S. Naval Research Laboratory
Crystallography, X-ray

HENRY KASH
Physics Department, Yale University
Proton

O. LEWIN KELLER, JR.
Oak Ridge National Laboratory, Oak Ridge, Tennessee
Elements

P. L. KELLEY
Lincoln Laboratory, Massachusetts Institute of Technology
Lasers

WALTER KELLNER
Siemens AG, Munich, Germany
Semiconductors, Crystalline

GARY L. KELLOGG
Sandia National Laboratories, Albuquerque, New Mexico
Field-Ion Microscopy

B. T. KELLY
United Kingdom Atomic Energy Authority, Salwick, England
Radiation Interaction with Matter

FRANCIS E. KENNEDY, JR.
Thayer School of Engineering, Dartmouth College
Friction

GEORGE C. KENNEDY
Late of Institute of Geophysics and Planetary Physics, University of California, Los Angeles
Ultrahigh-Pressure Techniques

DIETER P. KERN
IBM Research Center, Yorktown Heights, New York
Electron Beam Technology

EDWARD H. KERNER
Department of Physics, University of Delaware
Kepler's Laws

DONALD W. KERST
Department of Physics, University of Wisconsin
Betatron

R. W. P. KING
Division of Applied Sciences, Harvard University
Transmission Lines and Antennas

R. KINGSLAKE
Rochester, New York
Optics, Geometrical

GORDON S. KINO
Department of Electrical Engineering, Stanford University
Microscopy, Optical

TOICHIRO KINOSHITA
Laboratory for Nuclear Studies, Cornell University
Electromagnetic Interaction

ARTHUR F. KIP
University of California, Berkeley
Resistance

DON KIRKHAM
Department of Agronomy, Iowa State University
Soil Physics

S. KIRKPATRICK
IBM Research Center, Yorktown Heights, New York
Conduction

JOHN R. KLAUDER
AT&T Bell Laboratories, Murray Hill, New Jersey
Quantum Optics

Louis T. Klauder, Jr.
Louis T. Klauder and Associates, Philadelphia, Pennsylvania
Ampere's Law

P. G. Klemens
Department of Physics, University of Connecticut
Phonons

Daniel Kleppner
Department of Physics, Massachusetts Institute of Technology
Fine and Hyperfine Spectra and Interactions

G. M. Klody
National Electrostatics Corporation, Middleton, Wisconsin
Accelerators, Potential-Drop Linear

E. A. Knapp
Universities Research Association, Washington, D.C.
Accelerators, Linear

H. W. Kraner
Brookhaven National Laboratory, Upton, New York
Semiconductor Radiation Detectors

Edgar A. Kraut
Rockwell International Science Center, Thousand Oaks, California
Matrices

Lawrence C. Krisher
Institute for Physical Science and Technology, University of Maryland
Inversion and Internal Rotation

D. Kurath
Argonne National Laboratory, Argonne, Illinois
Nuclear States

Edwin H. Land
The Rowland Institute for Science, Cambridge, Massachusetts
Vision and Color

Kenneth R. Lang
Department of Physics and Astronomy, Tufts University
Hubble Effect

P. Langacker
Department of Physics, University of Pennsylvania
Elementary Particles in Physics

Marvin A. Lanphere
Geological Survey, United States Department of the Interior, Menlo Park, California
Geochronology

Jon M. Lawrence
Department of Physics, University of California, Irvine
Intermediate Valence Compounds

J. D. Lawson
Rutherford Appleton Laboratory, Chilton, England
Charged-Particle Optics

P. L. Leath
Serin Physics Laboratory, Rutgers University
Alloys

Alfred Leitner
Department of Physics, Rensselaer Polytechnic Institute
Fraunhofer Lines

Lionel M. Levinson
General Electric Research and Development Center, Schenectady, New York
Ceramics

Eugene H. Levy
Department of Planetary Sciences, University of Arizona
Solar System

Paul W. Levy
Brookhaven National Laboratory, Upton, New York
Color Centers
Thermoluminiscence

Ling-Fong Li
Department of Physics, Carnegie-Mellon University
Symmetry Breaking, Spontaneous

P. F. Liao
Bell Communications Research, Red Bank, New Jersey
Polarized Light

David R. Lide, Jr.
CRC Press, Inc., Gaithersburg, Maryland
Molecular Spectroscopy

Elliott H. Lieb
Department of Physics, Princeton University
Uncertainty Principle

M. H. Lietzke
Oak Ridge National Laboratory, Oak Ridge, Tennessee
Demineralization

M. E. Lines
AT&T Bell Laboratories, Murray Hill, New Jersey
Ferroelectricity

J. D. Litster
Department of Physics, Massachusetts Institute of Technology
Critical Points

P. B. Littlewood
AT&T Bell Laboratories, Murray Hill, New Jersey
Charge-Density Waves

C. J. Lobb
Department of Physics, Harvard University
Conduction

E. G. Loewen
Analytical Products Division, Milton Roy Company, Rochester, New York
Gratings, Diffraction

Maurice W. Long
Consultant, Atlanta, Georgia
Radar

A. E. Lord, Jr.
Department of Physics and Atmospheric Science, Drexel University
Levitation, Electromagnetic

J. D. Louck
Los Alamos National Laboratory, Los Alamos, New Mexico
Rotation and Angular Momentum

O. V. Lounasmaa
Low Temperature Laboratory, Helsinki University of Technology, Finland
Magnetic Cooling

Walter D. Loveland
Department of Chemistry, Oregon State University
Radiochemistry

Per-Olov Löwdin
Florida Quantum Theory Project, University of Florida
Molecular Structure Calculation

Sudarshan K. Loyalka
Department of Nuclear Engineering, University of Missouri, Columbia
Transport Theory

G. Lucovsky
Department of Physics, North Carolina State University
Semiconductors, Amorphous

Allen Lurio
Late of IBM Research Laboratory, Yorktown Heights, New York
Atomic Spectroscopy

H. Lüth
Direktor, Institut für Schicht- und Ionentechnik der Kernforschungsanlage Jülich, Germany
Ellipsometry

David L. MacAdam
Institute of Optics, University of Rochester
Spectrophotometry

J. Ross Macdonald
Department of Physics and Astronomy, University of North Carolina
Electrochemistry

Joseph Macek
Department of Physics and Astronomy, University of Tennessee
Electric Moments

Douglas E. MacLaughlin
Department of Physics, University of California, Riverside
Heavy-Fermion Materials

Alfred K. Mann
Department of Physics, University of Pennsylvania
Weak Neutral Currents

Robert H. March
Department of Physics, University of Wisconsin
Coriolis Acceleration

William J. Marciano
Brookhaven National Laboratory, Upton, New York
Gauge Theories

H. Mark
Polytechnic Institute of New York
Polymers

L. Marton
Late of *Advances in Electronics and Electron Physics,* Washington, D.C.
Electron Diffraction

Warren P. Mason
Late of Columbia University
Ultrasonics

J. R. Matthews
United Kingdom Atomic Energy Authority, Oxfordshire, England
Deformation of Crystalline Materials

D. C. Mattis
Department of Physics, University of Utah
Many-Body Theory

Leonard C. Maximon
National Institute of Standards and Technology
Bremsstrahlung

James A. McCray
Department of Physics and Atmospheric Science, Drexel University
Electronics

E. M. McMillan
Department of Physics, University of California, Berkeley
Synchrotron

John P. McTague
IBM Almaden Research Center, San Jose, California
Interatomic and Intermolecular Forces

R. L. Melcher
IBM Research Center, Yorktown Heights, New York
Magnetoelastic Phenomena

Heiner W. Meldner
Lawrence Livermore Laboratory, University of California, Livermore
Inertial Fusion

J. C. Melrose
Department of Petroleum Engineering, Stanford University
Capillary Flow

Robert J. Meltzer
The RJM Consultancy, Kirkland, Washington
Optical Activity

W. K. Melville
Department of Civil Engineering, Massachusetts Institute of Technology
Surface Waves on Fluids

N. D. Mermin
Department of Physics, Cornell University
Maxwell-Boltzmann Statistics

H. O. Meyer
Cyclotron Facility, Indiana University
Nuclear Polarization

D. L. Miller
University of Pennsylvania
Thin Films

A. P. Mills, Jr.
AT&T Bell Laboratories, Murray Hill, New Jersey
Positronium

Eugene Mittelmann
Consulting Engineer, Chicago, Illinois
Transducers

K. D. Moeller
Department of Physics, Fairleigh Dickinson University
Far-Infrared Spectra

J. W. Morris, Jr.
Department of Materials Science and Mineral Engineering, University of California, Berkeley
Heat

R. L. Morse
Department of Physics, University of Arizona
Plasmas

N. F Mott
Department of Physics, University of Cambridge, England
Metal-Insulator Transitions

Raymond D. Mountain
National Institute of Standards and Technology
Equations of State

R. G. Munro
National Institute of Standards and Technology
Tribology

Wolfgang Nadler
Department of Physics and Atmospheric Science, Drexel University
Electronics

Lorenzo M. Narducci
Department of Physics and Atmospheric Science, Drexel University
Optics, Statistical

Mark Nelkin
Department of Applied Physics, Cornell University
Hydrodynamics

David R. Nelson
Department of Physics, Harvard University
Renormalization

Donald F. Nelson
Department of Physics, Worcester Polytechnic Institute
Photoelastic Effect
Piezoelectric Effect

R. K. Nesbet
IBM Almaden Research Center, San Jose, California
Atomic Structure Calculations—Electronic Correlation
Magnetic Moments

Alfred O. Nier
School of Physics and Astronomy, University of Minnesota
Mass Spectroscopy

Thomas D. Northwood
Division of Building Research, National Research Council, Ottawa, Ontario
Acoustics, Architectural

A. S. Nowick
Henry Krumb School of Mines, Columbia University
Anelasticity

Richard M. Noyes
Department of Chemistry, University of Oregon
Kinetics, Chemical

R. J. Oakes
Department of Physics, Northwestern University
Currents in Particle Theory

G. David O'Kelley
Oak Ridge National Laboratory, Oak Ridge, Tennessee
Radiation Detection

Paul S. Olmstead
Winter Park, Florida
Error Analysis

Irwin Oppenheim
Department of Chemistry, Massachusetts Institute of Technology
Stochastic Processes

R. Orbach
Department of Physics, University of California, Los Angeles
Relaxation Phenomena

Colin G. Orton
Harper-Grace Hospital, Detroit, Michigan
Radiological Physics

Donald C. O'Shea
School of Physics, Georgia Institute of Technology
Visible and Ultraviolet Spectroscopy

Konrad Osterwalder
Department of Mathematics, Eidgenössische Technische Hochschule, Zurich, Switzerland
Operators

Edward Ott
Laboratory for Plasma Research, University of Maryland
Chaos

Thornton Page
Nassau Bay, Texas
Newton's Laws

A. Pais
Department of Physics, Rockefeller University
Electron

R. L. Parker
Consultant, Washington, D.C.
Crystal Growth

Philip Pearle
Hamilton College
Quantum Theory of Measurement

Norman Pearlman
Department of Physics, Purdue University
Heat Capacity

P. J. E. Peebles
Department of Physics, Princeton University
Cosmology

A. A. Penzias
AT&T Bell Laboratories, Murray Hill, New Jersey
Astronomy, Radio

Martin L. Perl
Stanford Linear Accelerator Center, Stanford University
Hadrons
Leptons

G. J. Perlow
Argonne National Laboratory, Argonne, Illinois
Mössbauer Effect

Michael E. Peskin
Standford Linear Accelerator Center, Stanford University
Quantum Field Theory

J. M. Peterson
Universities Research Association, Berkeley, California
Synchrotron

J. C. Phillips
AT&T Bell Laboratories, Murray Hill, New Jersey
Crystal Binding

Melba Phillips
New York, New York
Electric Charge

George C. Pimentel
Department of Chemistry, University of California, Berkeley
Hydrogen Bond

F. Plasil
Oak Ridge National Laboratory, Oak Ridge, Tennessee
Nuclear Fission

P. M. Platzman
AT&T Bell Laboratories, Murray Hill, New Jersey
Fermi-Dirac Statistics

John Preskill
Department of Physics, California Institute of Technology
Cosmic Strings

I. Prigogine
University of Brussels, Belgium
Thermodynamics, Nonequilibrium

Morris Pripstein
Lawrence Berkeley Laboratory, University of California, Berkeley
Center-of-Mass System

Jacques Prost
Groupe de Physico Chimie Théorique, Ecole de Physique et Chimie, Paris
Liquid Crystals

Lewis Pyenson
Department of History, University of Montréal, Quebec
History of Physics

Norman F. Ramsey
Department of Physics, Harvard University
Precession

A. Rassat
Chemistry Laboratory, École Normale Supérieure, Paris, France
Electron Spin Resonance

R. Ronald Rau
Deutsches Elektronen-Synchrotron, Hamburg, Germany
Hadron Colliders at High Energy

W. A. Reed
AT&T Bell Laboratories, Murray Hill, New Jersey
Magnetoresistance

John D. Reichert
Department of Electrical Engineering and Computer Engineering, University of New Mexico
Refraction

Frederick Reines
Department of Physics, University of California, Irvine
Neutrino

A. Rich
Department of Physics, University of Michigan
Positron
Quantum Electrodynamics

Burton Richter
Stanford Linear Accelerator Center, Stanford University
Positron-Electron Colliding Beams

P. H. Roberts
Department of Mathematics, University of California, Los Angeles
Magnetohydrodynamics

F. Rohrlich
Department of Physics, Syracuse University
Quantum Electrodynamics

Carol Z. Rosen
Plessey Electronic Systems Corporation, Totawa, New Jersey
Transducers

Fred J. Rosenbaum
Department of Electrical Engineering, Washington University
Microwaves and Microwave Circuitry

Marvin Ross
Lawrence Livermore Laboratory, University of California, Livermore
Shock Waves and Detonations

Laura M. Roth
Department of Physics, State University of New York at Albany
Faraday Effect

Walter G. Rothschild
Scientific Laboratories, Ford Motor Company, Dearborn, Michigan
Far-Infrared Spectra

J. M. Rowell
Conductors, Inc., Sunnyvale, California
Tunneling

M. W. Ruckman
Brookhaven National Laboratory, Upton, New York
Thin Films

M. E. Rudd
Department of Physics and Astronomy, University of Nebraska
Ionization

Arnold Russek
Department of Physics, University of Connecticut
Collisions, Atomic and Molecular

John D. Ryder
Ocala, Florida
Electron Tubes

R. M. Ryder
Summit, New Jersey
Transistors

P. G. Saffman
Department of Applied Mathematics, California Institute of Technology
Vortices

Richard H. Sands
University of Michigan
Spin

D. J. Scalapino
Department of Physics, University of California, Santa Barbara
Josephson Effects

A. Paul Schaap
Lumigen, Inc., Detroit, Michigan
Chemiluminescence

Gertrude Scharff-Goldhaber
Brookhaven National Laboratory
Isomeric Nuclei

L. B. Schein
IBM Almaden Research Center, San Jose, California
Electrophotography

John P. Schiffer
Argonne National Laboratory, Argonne, Illinois
Nuclear Reactions

K. E. Schmidt
Department of Physics, Arizona State University
Monte Carlo Techniques

H. W. Schmitt
Atomic Science, Inc., Oak Ridge, Tennessee
Nuclear Fission

John Schroeder
Department of Physics, Rensselaer Polytechnic Institute
Ultrahigh-Pressure Techniques

Harold A. Schwarz
Brookhaven National Laboratory, Upton, New York
Radiation Chemistry

John A. Schwarz
California Institute of Technology
String Theory

R. L. Sears
Department of Astronomy, University of Michigan
Stellar Energy Sources and Evolution

D. J. Seery
United Technologies Research Center, East Hartford, Connecticut
Combustion and Flames

B. G. Segal
Late of Barnard College
Balmer Formula

Ernest A. Seglie
Office of the Secretary of Defense, Washington, D.C.
Quantum Mechanics

Harvey Segur
Department of Applied Mathematics, University of Colorado
Solitons

D. J. Sellmyer
Department of Physics and Astronomy, University of Nebraska
Galvanomagnetic and Related Effects

M. H. Levelt Sengers
National Institute of Standards and Technology
Phase Transitions

R. B. Setlow
Brookhaven National Laboratory, Upton, New York
Biophysics

Jagdeep Shah
AT&T Bell Laboratories, Holmdel, New Jersey
Electron-Hole Droplets in Semiconductors

Paul Shaman
Department of Statistics, University of Pennsylvania
Statistics

S. M. Shapiro
Brookhaven National Laboratory, Upton, New York
Neutron Diffraction and Scattering

M. P. Shaw
Department of Electrical and Computer Engineering, Wayne State University
Solid-State Switching

D. Sherrington
Department of Physics, Imperial College of Science and Technology, London, England
Magnetic Ordering in Solids

G. Shirane
Brookhaven National Laboratory, Upton, New York
Neutron Diffraction and Scattering

Ferdinand J. Shore
Department of Physics, Queens College of the City University of New York
Nuclear Reactors

Howard A. Shugart
Department of Physics, University of California, Berkeley
Beams, Atomic and Molecular

Kurt E. Shuler
Department of Chemistry, University of California, San Diego
Stochastic Processes

R. P. Shutt
Brookhaven National Laboratory, Upton, New York
Cloud and Bubble Chambers

Walter P. Siegmund
Schott Fiber Optics, Inc., Southbridge, Massachusetts
Fiber Optics

R. H. Siemann
Department of Physics, Cornell University
Novel Particle Acceleration Methods

John A. Simpson
National Institute of Standards and Technology
Metrology

David J. Smith
Center for Solid State Science, Arizona State University
Electron Microscopy

Domina Eberle Spencer
Department of Mathematics, University of Connecticut
Vector and Tensor Analysis

H. E. Spencer
Department of Chemistry, Oberlin College
Light-Sensitive Materials

William Spindel
National Academy of Sciences
Isotope Separation

John L. Stanford
Department of Physics, Iowa State University
Meteorology

Anthony F. Starace
Department of Physics and Astronomy, University of Nebraska
Atomic Spectroscopy

Rolf M. Steffen
Formerly of Department of Physics, Purdue University
Angular Correlation of Nuclear Radiation

Philip Stehle
Department of Physics and Astronomy, University of Pittsburgh
Maxwell's Equation

Gary Steigman
Department of Physics, Ohio State University
Antimatter

Philip J. Stephens
Department of Chemistry, University of Southern California
Magnetic Circular Dichroism

Frank H. Stillinger
AT&T Bell Laboratories, Murray Hill, New Jersey
Water

Thomas H. Stix
Plasma Physics Laboratory, Princeton University
Plasma Waves

M. Strasberg
Acoustical Society of America, Washington, D.C.
Acoustical Measurements

Myron Strongin
Brookhaven National Laboratory, Upton, New York
Thin Films

E. C. G. Sudarshan
Department of Physics, University of Texas, Austin
Kinematics and Kinetics

L. S. Swanson
Department of Applied Physics, Oregon Graduate Center
Field Emission

L. S. Swenson, Jr.
Department of History, University of Houston
Michelson-Morley Experiment

Victor Szebehely
Department of Aerospace Engineering and Engineering Mechanics, University of Texas, Austin
Three-Body Problem, Gravitational

Michael Tabor
AT&T Bell Laboratories, Murray Hill, New Jersey
Dynamics, Analytical

C. L. Tang
School of Electrical Engineering, Cornell University
Ultrashort Optical Pulses

Frank R. Tangherlini
Department of Physics, College of the Holy Cross
Moment of Inertia

B. N. Taylor
National Institute of Standards and Technology
Constants, Fundamental

Cyrus Taylor
Department of Physics, Case Western Reserve University
Fields

Edwin F. Taylor
Department of Physics, Massachusetts Institute of Technology
Schrödinger Equation

P. C. Taylor
Department of Physics, University of Utah
Semiconductors, Amorphous

G. J. Thaler
United States Naval Postgraduate School
Servomechanism

Edward W. Thomas
School of Physics, Georgia Institute of Technology
Electron and Ion Impact Phenomena

R. N. Thurston
Bell Communications Research, Red Bank, New Jersey
Ultrasonics

D. R. Tilley
Department of Physics, University of Essex, Colchester, England
Waves

Laszo Tisza
Department of Physics, Massachusetts Institute of Technology
Entropy

A. M. Title
Lockheed Palo Alto Research Laboratory, Palo Alto, California
Photosphere

Tommaso Toffoli
Laboratory for Computer Science, Massachusetts Institute of Technology
Cellular Automata

Juergen Tonndorf
College of Physicians and Surgeons, Columbia University
Acoustics, Physiological

Gérard Toulouse
Ecole Normale Supérieure, Paris
Catastrophe Theory

Sol Triebwasser
IBM, Thornwood, New York
Circuits, Integrated

James W. Truran
Department of Astronomy, University of Illinois
Nucleosynthesis

WALLACE TUCKER
Smithsonian Astrophysical Observatory, Cambridge, Massachusetts
Astronomy, X-ray

B. F. TURNER
National Radio Astronomy Observatory, Charlottesville, Virginia
Astronomy, Radio

J. ANTHONY TYSON
AT&T Bell Laboratories, Murray Hill, New Jersey
Gravitational Waves

MARTIN A. UMAN
Department of Electrical Engineering, University of Florida
Lightning

R. J. URICK
Silver Spring, Maryland
Sound, Underwater

J. A. VAN ALLEN
Department of Physics and Astronomy, University of Iowa
Radiation Belts

C. W. VAN ATTA
Department of Applied Mechanics and Engineering Sciences, University of California, San Diego
Boundary Layers
Vacuums and Vacuum Technology

H. VAN DAM
Department of Physics, University of North Carolina
Rotation and Angular Momentum

JOHANNES A. VAN DEN AKKER
Appleton, Wisconsin
Mass

W. ALEXANDER VAN HOOK
Department of Chemistry, University of Tennessee
Isotope Effects

MICHEL VAN HOVE
Lawrence Berkeley Laboratory, University of California, Berkeley
Electron Diffraction

R. N. VARNEY
Palo Alto, California
Kinetic Theory

G. P. VELLA-COLEIRO
AT&T Bell Laboratories, Murray Hill, New Jersey
Magnetic Domains and Bubbles

KLAUS VON KLITZING
Max-Planck Institut für Festkörperforschung, Stuttgart, Germany
Hall Effect, Quantum

RICHARD W. VOOK
Department of Physics, Syracuse University
Reflection High-Energy Electron Diffraction (RHEED)

JAMES T. WABER
Michigan Technical University
Atomic Structure Calculations, Relativistic Atoms

KAMESHWAR C. WALI
Department of Physics, Syracuse University
Strong Interactions

JEARL D. WALKER
Department of Physics, Cleveland State University
Energy and Work

RICHARD F. WALLIS
Department of Physics, University of California, Irvine
Lattice Dynamics

R. E. WALSTEDT
AT&T Bell Laboratories, Murray Hill, New Jersey
Nuclear Magnetic Resonance

GEORGE D. WATKINS
Department of Physics, Lehigh University
Crystal Defects

KENNETH M. WATSON
Scripps Institution of Oceanography, University of California, San Diego
Scattering Theory

ALFONS WEBER
National Institute of Standards and Technology
Molecular Spectroscopy

JOHANNES WEERTMAN
Department of Materials Science, Northwestern University
Fatigue

LOUIS WEINBERG
Department of Electrical Engineering, The City College of the City University of New York
Network Theory: Analysis and Synthesis

DON WEINGARTEN
IBM Research Center, Yorktown Heights, New York
Field Theory, Classical

GABRIEL WEINREICH
Department of Physics, University of Michigan
Free Energy
Musical Instruments

J. WEINSTOCK
National Oceanic and Atmospheric Administration, Boulder, Colorado
Ionosphere

ROBERT WEINSTOCK
Department of Physics, Oberlin College
Temperature

GEORGE H. WEISS
National Institutes of Health, Bethesda, Maryland
Stochastic Processes

DAVID O. WELCH
Brookhaven National Laboratory, Upton, New York
Radiation Damage in Solids

HEINRICH J. WELKER
Siemens AG, Erlangen, Germany
Semiconductors, Crystalline

EDGAR F. WESTRUM, JR.
Department of Chemistry, University of Michigan
Calorimetry

CLYDE E. WIEGAND
Lawrence Berkeley Laboratory, University of California, Berkeley
Muonic, Mesonic, and Other Exotic Atoms

A. S. WIGHTMAN
Department of Physics, Princeton University
Field Theory, Axiomatic

SIR DENYS WILKINSON
University of Sussex, Falmer, Brighton, England
Isospin

W. P. WOLF
Section of Applied Physics, Yale University
Ferrimagnetism

WILLIAM L. WOLFE
Optical Sciences Center, University of Arizona
Radiometry

LINCOLN WOLFENSTEIN
Department of Physics, Carnegie-Mellon University
Conservation Laws

G. WOUCH
Lisle Technical Center, R. R. Connelly & Sons, Lisle, Illinois
Levitation, Electromagnetic

FRANCIS J. WRIGHT
Department of Applied Mathematics, Queen Mary College, London, England
Catastrophe Theory

FA YUEH WU
Department of Physics, Northeastern University
Ising Model

H. J. ZEIGER
Lincoln Laboratory, Massachusetts Institute of Technology
Lasers

MARK W. ZEMANSKY
Late of The City College of the City University of New York
Carnot Cycle

E. C. ZIPF
Department of Physics, University of Pittsburgh
Aurora

DAVID M. ZIPOY
Astronomy Program, University of Maryland
Astronomy, Optical

INDEX

Main entries are listed in CAPITAL LETTERS with the first page number in **bold face**. Other entries are listed with the first page number in which they appear in the article.